McGraw-Hill Concise Encyclopedia of Science & Technology

Fourth Edition

McGraw-Hill Concise Encyclopedia of Science & Technology

Fourth Edition

Sybil P. Parker
Editor in Chief

McGraw-Hill
New York San Francisco Washington, D.C. Auckland Bogotá Caracas
Lisbon London Madrid Mexico City Milan Montreal New Delhi San Juan
Singapore Sydney Tokyo Toronto

On the front cover: Streamer chamber photograph of 220 charged subatomic particles spilling out from the collision of a high-energy oxygen nucleus with a nucleus in a lead target. The experiment was performed at CERN, the European particle physics laboratory, located near Geneva, Switzerland. (*CERN/Science Photo Library/Photo Researchers*)

Cover design: pinpoint Design and Advertising

Library of Congress Cataloging in Publication Data

McGraw-Hill concise encyclopedia of science & technology / Sybil P. Parker, editor in chief.—4th ed.
p. cm.
Includes bibliographical references and index.
ISBN 0-07-052659-1 (hardcover)
1. Science—Encyclopedias. 2. Technology—Encyclopedias.
I. Parker, Sybil P.
Q121.M29 1998
503—dc21 98-3875

McGraw-Hill

A Division of The ***McGraw-Hill*** *Companies*

1 2 3 4 5 6 7 8 9 0 DOW/DOW 9 0 3 2 1 0 9 8

This book was printed on acid-free paper.

It was set in Helvetica Black and Souvenir by The PRD Group, Shippensburg, Pennsylvania.

The book was printed and bound by R. R. Donnelley & Sons Company, The Lakeside Press.

CONTENTS

STAFF

EDITORIAL STAFF

Mark D. Licker, Publisher
Jonathan Weil, Editor
Sally R. Felsenfeld, Editor
Thomas P. Martin, Editor
Patricia W. Albers, Editorial Administrator
Barbara C. Lai, Editorial Assistant

EDITING, DESIGN, & PRODUCTION STAFF

Roger Kasunic, Director of Editing, Design, and Production

Joe Faulk, Editing Manager
Frank Kotowski, Jr., Senior Editing Supervisor

Ron Lane, Art Director
Vincent Piazza, Assistant Art Director

Thomas G. Kowalczyk, Senior Production Manager
Suzanne W. B. Rapcavage, Production Manager

Consulting Editors

For the McGraw-Hill Encyclopedia of Science & Technology

Dr. S. C. Jong. *Senior Staff Scientist and Program Director, Mycology and Protistology Program, American Type Culture Collection, Rockville, Maryland.* MYCOLOGY.

Prof. Gary Judd. *Dean of the Faculty, Rensselaer Polytechnic Institute, Troy, New York.* METALLURGICAL ENGINEERING.

Prof. Cornelis Klein. *Department of Earth and Planetary Sciences, University of New Mexico, Albuquerque.* GEOLOGY (MINERALOGY AND PETROLOGY).

Dr. John A. Knauss. *Research Associate, Scripps Institution of Oceanography, La Jolla, California.* OCEANOGRAPHY.

Prof. Shin Liu. *Chairman, Thomas C. Jenkins Department of Biophysics, The Johns Hopkins University, Baltimore, Maryland.* BIOPHYSICS.

Prof. C. L. Liu. *Associate Provost and Professor of Computer Science, University of Illinois, Champaign.* COMPUTERS.

Prof. Karl E. Lonngren. *Department of Electrical & Computer Engineering, University of Iowa, Iowa City.* PHYSICAL ELECTRONICS.

Dr. Philip V. Lopresti. *Retired; formerly, Engineering Research Center, AT&T Bell Laboratories, Princeton, New Jersey.* ELECTRONIC CIRCUITS.

Dr. Michael L. McKinney. *Department of Geological Sciences, University of Tennessee, Knoxville.* INVERTEBRATE PALEONTOLOGY.

Prof. Marcia McNutt. *Department of Earth, Atmospheric, and Planetary Sciences, Massachusetts Institute of Technology, Cambridge.* GEOPHYSICS.

Dr. George L. Marchin. *Associate Professor of Microbiology and Immunology, Division of Biology, Kansas State University, Manhattan.* MICROBIOLOGY.

Prof. Melvin G. Marcus. *Deceased; formerly, Department of Geography, Arizona State University, Tempe.* PHYSICAL GEOGRAPHY.

Dr. Henry F. Mayland. *Soil Scientist, Northwest Irrigation and Soils Research Laboratory, USDA-ARS, Kimberly, Idaho.* SOILS.

Dr. James A. Moore. *Department of Chemistry, University of Delaware, Newark.* ORGANIC CHEMISTRY.

Prof. Conrad F. Newberry. *Department of Aerospace and Astronautics, Naval Postgraduate School, Monterey, California.* AERONAUTICAL ENGINEERING AND PROPULSION.

Dr. Gerald Palevsky. *Consulting Professional Engineer, Hastings-on-Hudson, New York.* CIVIL ENGINEERING.

Prof. Jay M. Pasachoff. *Director, Hopkins Observatory, Williams College, Williamstown, Massachusetts.* ASTRONOMY.

Prof. David J. Pegg. *Department of Physics and Astronomy, University of Tennessee, Knoxville.* ATOMIC, MOLECULAR, AND NUCLEAR PHYSICS.

Dr. William C. Peters. *Professor Emeritus, Mining and Geological Engineering, University of Arizona, Tucson.* MINING ENGINEERING.

Prof. W. D. Russell-Hunter. *Professor of Zoology, Department of Biology, Syracuse University, New York.* INVERTEBRATE ZOOLOGY.

Dr. Andrew P. Sage. *Founding Dean Emeritus and First American Bank Professor, University Professor, School of Information Technology and Engineering, George Mason University, Fairfax, Virginia.* CONTROL AND INFORMATION SYSTEMS.

Prof. Gerald Schatten. *Department of Zoology, University of Wisconsin, Madison.* CELL BIOLOGY.

Prof. Peter A. Scholle. *Department of Geological Sciences, Southern Methodist University, Dallas, Texas.* GEOLOGY (PHYSICAL, HISTORICAL, AND SEDIMENTARY).

Prof. Augusta Schurrer. *Department of Mathematics, University of Northern Iowa, Cedar Falls.* MATHEMATICS.

Mel Schwartz. *Materials Consultant, United Technologies Corporation, Stratford, Connecticut.* MATERIALS SCIENCE AND ENGINEERING.

Prof. Susan R. Singer. *Department of Biology, Carleton College, Northfield, Minnesota.* DEVELOPMENTAL BIOLOGY.

Prof. William A. Steel. *Retired; formerly, Department of Chemistry, Pennsylvania State University, University Park.* THERMODYNAMICS.

Prof. Thomas N. Taylor. *Department of Biology, University of Kansas, Lawrence.* PALEOBOTANY.

Prof. Marlin U. Thomas. *Head, School of Industrial Engineering, Purdue University, West Lafayette, Indiana.* INDUSTRIAL AND PRODUCTION ENGINEERING.

Prof. John F. Timoney. *Department of Veterinary Sciences, University of Kentucky, Lexington.* VETERINARY MEDICINE.

Prof. Romeo T. Toledo. *Department of Food Science and Technology, University of Georgia, Athens.* FOOD ENGINEERING.

Prof. Joan S. Valentine. *Department of Chemistry and Biochemistry, University of California, Los Angeles.* INORGANIC CHEMISTRY.

Dr. Blaire Van Valkenburgh. *Department of Biology, University of California, Los Angeles.* PALEONTOLOGY (VERTEBRATE).

Prof. David B. Wake. *Director, Museum of Vertebrate Zoology, University of California, Berkeley.* EVOLUTION.

Prof. Donald B. Wetlaufer. *Department of Chemistry and Biochemistry, University of Delaware, Newark.* BIOCHEMISTRY.

Prof. Frank M. White. *Department of Mechanical Engineering, University of Rhode Island, Kingston.* FLUID MECHANICS.

Prof. Richard G. Wiegert. *Institute of Ecology, University of Georgia, Athens.* ECOLOGY AND CONSERVATION.

Prof. Frank Wilczek. *Institute for Advanced Study, Princeton, New Jersey.* THEORETICAL PHYSICS.

Prof. W. A. Williams. *Department of Agronomy and Range Science, University of California, Davis.* AGRICULTURE.

Prof. George S. Wilson. *Department of Chemistry, University of Kansas, Lawrence.* ANALYTICAL CHEMISTRY.

Richard E. Wyman. *Wyman & Associates, Salt Lake City, Utah.* PETROLEUM ENGINEERING.

Dr. Terry L. Yates. *Chairman, Department of Biology, University of New Mexico, Albuquerque.* ANIMAL SYSTEMATICS.

PREFACE

For more than 35 years, in eight editions, the *McGraw-Hill Encyclopedia of Science & Technology* (20 volumes) has been recognized as the leading international scientific encyclopedia. Authoritative, comprehensive, up to date, and eminently readable, that encyclopedia has served the continuing reference needs of professionals, students, librarians, technical writers and editors, and the general public. However, the needs of many other readers are met by a concise encyclopedia still covering the full breadth of science and technology.

With this in mind, the editors conceived a shorter version of the multivolume work that would retain the authoritativeness, accuracy, clarity, recency, quality, and coverage in a convenient, concise format. The result is the *McGraw-Hill Concise Encyclopedia of Science & Technology*. To achieve this convenient single volume, the editors extracted the essential text from each article in the parent work while retaining the same proportionality between subjects. The length of each article suits the importance and complexity of the subject. The material has been condensed so that it is appropriate to the likely requirements of the reader seeking helpful knowledge without extensive detail. The articles retain the identity of the original authors, all recognized experts, whose affiliations are included in a complete alphabetical listing.

There are 7800 alphabetically arranged text entries. Virtually every entry includes cross references to other articles to guide the reader to related subjects. Many of the articles are illustrated with photographs, maps, graphs, diagrams, and drawings. A second color is used for most illustrations to enhance comprehension. Dual measurement units (U.S. Customary and International System) are used throughout the text to meet all users' needs.

The encyclopedia features an extensive Appendix with a variety of useful information. The Bibliographies section lists hundreds of current books and journals grouped according to scientific fields and disciplines. The Scientific Notation section compares the International, U.S. Customary, and metric systems of measurement; clarifies chemical nomenclature; and provides lists of symbols. The Appendix also includes mathematical notation, fundamental constants, and a geologic time chart. The Biographical listing identifies hundreds of scientists and their achievements. Finally, the 30,000-entry Index provides the reader with instant access to specific information in the text.

This reference was created to meet the need for current, accurate information on classical and newsworthy concepts in science and engineering in a practical, economical format. It is a starting point for anyone with a serious interest in the sciences, and even for scientists and other professionals looking for information on subjects outside their specialty. As such, the *McGraw-Hill Concise Encyclopedia of Science & Technology,* Fourth Edition, will be an invaluable addition to the reference shelf of any library.

Sybil P. Parker
Editor in Chief

ORGANIZATION OF THE ENCYCLOPEDIA

Alphabetization. The 7800 article titles are sequenced on a word-by-word basis, not letter by letter. Hyphenated words are treated as separate words. In occasional inverted article titles, the comma provides a full stop. The Index is alphabetized on the same principles. Readers can turn directly into the pages for much of their research. An example of sequencing is:

Air
Air brake
Air-cushion vehicle
Air pollution
Air pollution, indoor
Air pressure
Air-traffic control
Airborne radar
Aircraft
Aistopoda

Cross references. Virtually every article has cross references set in CAPITALS AND SMALL CAPITALS. These references offer the user the option of turning to other articles in the volume for related information.

Measurement units. Since some readers prefer the U.S. Customary system while others require the International System of Units (SI), measurements in the Encyclopedia are given in dual units. Insight into the measurement systems is provided by the discussion in the Appendix, which also has handy conversion tables.

Contributors. The authorship of each article is specified at its conclusion, in the form of the contributor's initials for brevity. The contributor's full name and affiliation may be found in the "List of Contributors" section at the front of the volume.

Appendix. Every user should explore the variety of succinct information supplied by the Appendix, which includes measurement tables, mathematical notation, fundamental constants, world telescope descriptions, and a biographical listing of scientists. Users wishing to go beyond the scope of this Encyclopedia will find recommended books and journals listed in the Bibliographies section; the titles are grouped by subject area.

Index. The 30,000-entry Index offers the reader the time-saving convenience of being able to quickly locate specific information in the text, rather than approaching the Encyclopedia via article titles only. This elaborate breakdown of the volume's contents assures both the general reader and the professional of efficient use of the *McGraw-Hill Concise Encyclopedia of Science & Technology.*

CONTRIBUTORS

Each article in the Encyclopedia is signed with the contributor's initials; coauthors' initials are separated by semicolons; initials of an original author are separated by a solidus from the initials of a validator who may have modified the text in a later printing. The section below, Contributors' Initials, gives all such initials, alphabetized according to the capital letters in each initials combination (lowercase letters have been added to differentiate contributors with like initials). The contributor's name is provided. The contributor's affiliation can then be found by turning to the following section, List of Contributors.

CONTRIBUTORS' INITIALS

A

A.A.B. **A. A. Bergh**
A.A.B.P. **A. Alan B. Pritsker**
A.A.G. **Albert A. Gunkler**
A.A.M. **Aly A. Mahmoud**
A.A.McK. **Alexander A. McKenzie**
A.A.W. **Arthur A. Welch**
A.B. **Alexander Baumgarten**
A.Ba. **Arthur Bank**
A.Bi. **Arthur Bienenstock**
A.Bre. **Alan Breier**
A.Bu. **Allison Butts**
A.B.C. **Ali B. Cambel**
A.B.G. **Alfred B. Garrett**
A.B.Gu. **Ashley B. Gurney**
A.B.M. **Alton B Moody**
A.B.McG. **Ann B. McGuire**
A.B.W. **A. B. Watts**
A.B.S. **Andrew B. Smith**
A.C. **Alice Chenualt**
A.Ch. **Alan Chodos**
A.Ci. **Alex Ciegler**
A.Cou. **Arnold Court**
A.Cr. **Arthur Cronquist**
A.C.B. **A. Clarke Beiler**
A.C.C. **Andrew C. Campbell**
A.C.Ca. **A. Charles Catania**
A.C.Cl. **Anthony C. Clement**
A.C.D. **A. C. Dickieson**
A.C.Do. **A. C. Downton**
A.C.G. **Arch C. Gerlach**
A.C.Gi. **Arthur C. Giese**
A.C.Go. **Arthur C. Gossard**
A.C.H. **Anthony C. Hearn**
A.C.N. **A. Conrad Neumann**
A.D. **Albert Damon**
A.Do. **Ali Dogramaci**
A.D.A. **Aaron D. Alexander**
A.D.F. **Alan D. Fiala**
A.D.H. **Arthur D. Hasler**
A.D.M. **Andrew D. Miall**
A.D.P. **Allan D. Pierce**
A.D.S. **A. Douglas Stone**
A.E. **Arne Engstrom**
A.Er. **A. Erdélyi**
A.Erd. **Arthur Erdman**
A.E.B. **Alex E. Brodhag, Jr.**
A.E.Ba. **A. Earle Bailey**
A.E.Br. **Arthur E. Bryson, Jr.**
A.E.C. **A. E. Cameron**
A.E.D. **A. E. Drake**
A.E.deM. **Augustus E. de Maggio**
A.E.F. **A. E. Fiore**
A.E.Fr. **A. E. Freeman**
A.E.G. **Arnold E. Greenberg**
A.E.J. **Amos E. Joel, Jr.**
A.E.J.E. **A. E. J. Engel**
A.E.L. **Albert E. Lintherland**
A.E.M. **A. E. Martell**
A.E.Mo. **Alice E. Moore**
A.E.S. **A. E. Siegman**
A.E.T. **Anthony E. Tancreto**
A.E.Ta. **Angus E. Taylor**
A.E.V.D. **Albert E. Von Doenhoff**
A.E.W. **Albert E. Wackerman**
A.E.Wo. **Albert E. Wood**
A.F. **A. Fiscarelli**
A.Fl. **Abraham Fleminger**
A.F.B. **A. F. Banfield**
A.F.H. **A. F. Hagner**
A.F.N. **Arthur F. Niemoeller**
A.F.P. **Albert F. Puchstein**
A.F.R. **Austen F. Riggs**
A.F.-T.C. **Albert F.-T. Chen**
A.G. **Anne Geller**
A.Gh. **Albert Ghiorso**
A.G.C. **Albert C. Conrad**
A.G.H. **Arthur G. Hansen**
A.G.He. **Angus G. Hepburn**
A.G.Ho. **Albert G. Holzman**
A.G.Hu. **Arthur G. Humes**
A.G.L. **Allan G. Lochhead**
A.G.McN. **Alvin G. McNish**
A.G.P. **Albert G. Petschek**
A.G.S. **Arthur G. Steinberg**
A.G.W.C. **A. G. W. Cameron**
A.H. **Arthur Haut**
A.He. **Alexander Hellemans**
A.Hy. **Adin Hyslop**
A.H.B. **Alan H. Brush**
A.H.G. **Arthur H. Graves**
A.H.J. **Aldon H. Jensen**
A.H.K. **Alan H. Karp**
A.H.M. **Anne H. Morgan**
A.H.R. **Austin H. Riesen**
A.H.Ro. **Arthur H. Robinson**
A.H.S. **Amelie H. Scheltema**
A.H.Sn. **Arthur H. Snell**
A.H.So. **Alfred H. Sommer**
A.H.St. **Alfred H. Sturtevant**
A.H.T. **Alan H. Tuttle**
A.H.W. **Alfred H. Wolferz**
A.I. **Alexander Illis**
A.I.S. **Anneila I. Sargent**
A.J.B. **Anthony J. Brazel**
A.J.Br. **Anthony J. Brammer**
A.J.C. **Alan J. Charig**
A.J.Co. **Almon J. Corson**
A.J.D. **Adrianus J. Dekker**
A.J.F. **A. John Ferguson**
A.J.G. **A. J. Grobecker**
A.J.H. **Armin J. Hill**
A.J.He. **Alan J. Heeger**
A.J.Ho. **Alan J. Hoffman**
A.J.Hu. **A. J. Hundhausen**
A.J.L. **Anthony J. Leggett**
A.J.R. **A. J. Rowell**
A.J.T. **Aaron J. Teller**
A.K. **Arthur Kelman**
A.K.M. **Alfred K. Mann**
A.L. **Anton Lang**
A.La. **Alan Lawley**

A.Lo. **Alan Longacre**
A.L.A. **Albert Louis Alfred**
A.L.B. **Aaron L. Brody**
A.L.DeV. **Arthur L. DeVries**
A.L.deW. **A. L. de Weck**
A.L.G. **Arnold L. Gordon**
A.L.H. **Allen L. Hanson**
A.L.K. **Arthur L. Kohl**
A.L.M. **A. L. Myers**
A.L.O. **Ada L. Olins**
A.L.P. **A. L. Ponikvar**
A.L.S. **Arthur L. Schawlow**
A.L.Sh. **Alex L. Shigo**
A.M. **Alfred Makulec**
A.Ma. **Albert Marden**
A.Mo. **Albert Moscowitz**
A.M.C. **Arthur M. Chickering**
A.M.D. **Andrew M. Davis**
A.M.D'A. **Amelio M. D'Arcangelo**
A.M.G. **Alan M. Gaines**
A.M.H. **A. M. Hartley**
A.M.I. **Aly M. Ibrahim**
A.McI. **Andrew McIntyre**
A.M.M. **Alexander M. Mood**
A.M.Ma. **Abe M. Macher**
A.M.P. **Arthur M. Perrin**
A.M.Pl. **Allison M. Platt**
A.M.W. **Andrew M. Weitzenhoffer**
A.N. **Allen Newell**
A.No. **Aaron Novick**
A.N.C. **Archie N. Carter**
A.N.Co. **Arthur N. Cox**
A.N.L. **Arnold N. Lowan**
A.N.S. **Albert N. Sayre**
A.O.N. **Alfred O. Nier**
A.P. **Allen Pearson**
A.Pe. **Astor Perry**
A.Pu. **Andy Purvis**
A.P.A. **Arnold P. Appleby**
A.P.M. **Alan P. Marchand**
A.P.S. **Andrew Paul Somlyo**
A.P.Sa. **Andrew P. Sage**
A.R. **Arthur Roberts**
A.Re. **Abram Recht**
Al.R. **Alexander Rich**
Ar.R. **Arthur Rich**
A.R.A. **Arthur R. Ayers**
A.R.B. **Anna Ruth Brummett**
A.R.C. **Anthony R. Cooper**
A.R.E. **Arthur R. Eckels**
A.R.J. **Arthur R. Jensen**
A.R.M. **Armand R. Maggenti**
A.R.P. **Allison R. Palmer**
A.R.P.R. **A. R. P. Rau**
A.R.R. **Augustus R. Rogowski**
A.S. **Abdus Salam**
A.Sa. **Allan Sandage**
A.Sc. **Anton Schindler**
A.Sch. **Arnold Schecter**
A.Sm. **Andrew Smith**
A.So. **Alfred Sommer**
A.St. **Alfred Steinschneider**
A.S.C. **Alden S. Crafts**
A.S.Co. **Alan S. Cohen**
A.S.G. **Alfred S. Goldhaber**
A.S.L. **A. Starker Leopold**
A.S.L.H. **A. S. L. Hu**
A.S.N. **Alberto S. Nieto**
A.S.P. **Albert S. Palmerlee**
A.S.R. **Alfred S. Romer**
A.S.Ru. **Allen S. Russell**
A.S.W. **Arthur S. Wightman**
A.Ste. **Askel Stenderup**
A.T. **Albert Tyler**
A.To. **Alfonso Torres**
A.T.A. **Alfred T. Anderson, Jr.**
A.T.W. **Andrew T. Weil**
A.T.Z. **Andrew T. Zander**
A.V.C. **A. V. Corlett**
A.W. **Abraham White**
A.W.A. **Arthur W. Adamson**
A.W.B. **Alan W. Bernheimer**
A.W.C. **Arthur W. Cooper**
A.W.Ca. **A. Welford Castleman, Jr.**
A.W.Co. **Anthony W. Confer**
A.W.C.V.G. **Alenxander W. C. von Graevenitz**
A.W.F. **Arthur W. Francis**
A.W.J. **A. Walter Jacobson**
A.W.K. **Anthony W. Knapp**
A.W.N. **Arch W. Naylor**
A.W.O. **Albert W. Overhauser**
A.W.W. **Andrew W. Wayne**

B

B.A.B. **B. Austin Barry**
B.A.G. **Bernard A. Galler**
B.A.S. **B. A. Summers**
B.B. **Benjamin Bederson**
B.Ba. **Brenda Bass**
B.B.B. **Blake B. Brantley, Jr.**
B.B.C. **Braxton B. Carr**
B.B.Cr. **Burton B. Crocker**
B.B.Cu. **Burris B. Cunningham**
B.B.M. **Bennett B. Murdock, Jr.**
B.C. **Benjamin Chu**
B.C.C. **Bryan C. Chakoumakos**
B.C.H. **Bruce C. Heezen**
B.C.S. **B. C. Sutton**
B.D. **Bruce Dan**
B.DeF. **Brian DeFacio**
B.Du. **Bruce Dunn**
B.D.S. **Blair D. Savage**
B.E. **Baltus Erasmus**
B.E.B. **Breck E. Byers**
B.E.Bo. **Bruce E. Bowler**
B.E.P. **Bernard E. Proctor**
B.E.R. **Bryant E. Rees**
B.E.S.G. **Brian E. S. Gunning**
B.E.W. **Bertram E. Warren**
B.F. **Bernard Frank**
B.Fr. **Benjamin Fruchter**
B.F.H. **Benjamin F. Howell, Jr.**
B.F.J. **B. Frank Jones, Jr.**
B.F.T. **Benjamin F. Trump**
B.F.W. **B. F. Wilson**
B.G. **Bernard Goodman**
B.Gi. **Barrie Gilbert**
B.G.B. **Brian G. Bagley**
B.G.D. **B. Gale Dick**
B.G.M. **Bruce G. Malloy**
B.H. **Bernard Haber**
B.He. **Bernd Heinrich**
B.Hoc. **Brian Hocking**
B.Hop. **Bruce Hopper**
B.H.B. **Benjamin H. Beard**
B.H.Bi. **Bruce H. Billings**
B.H.K. **Bostwick H. Ketchum**
B.H.McC. **Bayard H. McConnaughey**
B.H.W. **Byron H. Waksman**
B.I. **Brent Iverson**
B.I.S. **Bernard I. Spinrad**
B.J.B. **Bart J. Bok**
B.J.Ba. **B. Jayant Baliga**
B.J.D.M. **Bastiaan J. D. Meeuse**
B.J.F. **Barry J. Feldman**
B.J.K. **Brian J. Kilbey**
B.J.M. **Basil J. Mason**
B.J.Mu. **Brent J. Muus**
B.J.W. **Ben J. Wattenberg**
B.J.Wo. **B. J. Wood**
B.J.Z. **Bruno J. Zwolinski**
B.K. **Blair Kinsman**
B.Ku. **Bernhard Kummel**
B.L. **Bernard Lewis**
B.L.A. **Benjamin L. Averbach**
B.L.B. **Blaine L. Blad**
B.L.D. **Barry L. Doolan**
B.L.I. **B. Lynn Ingram**
B.L.Kl. **Benny L. Klock**
B.L.R. **Burtis L. Robertson**
B.L.S. **Bernard L. Segal**
B.M. **Bruce Margon**
B.Mi. **Barry Miller**
B.M.C. **Brian M. Chapman**
B.M.G. **Beverly M. Guirard**
B.M.L. **Bernard M. Leadon**
B.M.McC. **Barry M. McCoy**
B.M.S. **Bruce M. Sankey**
B.N.B. **Bertram N. Brockhouse**
B.O.B. **Berthold O. Bergh**
B.P. **Burton Paul**
B.Pi. **Benjamin Pinkel**
B.Po. **Bogdan Povh**
B.R. **U.S. Bureau of Reclamation**
B.Ro. **Barry Rosen**
B.Ru. **Brace Runnegar**
B.R.W. **Bennie R. Ware**
B.S. **Bobb Schaeffer**
B.Si. **Barry Simon**
B.St. **Brian Stevens**
B.S.D. **Bernard S. Dudock**
B.S.G. **Bernard S. Greensfelder**
B.S.M. **Bernard S. Meyer**
B.S.-N. **Bodil Schmidt-Nielsen**
B.S.O. **Bruce S. Old**
B.T.S. **Bradley T. Scheer**
B.U.H. **Bilal U. Haq**
B.V. **Bernard Vonnegut**
B.V.B. **Burton V. Barnes**
B.V.Bu. **Bohdan V. Burachinsky**
B.W. **Bonnie Webber**
R.W.But. **Robert W. Butler**
B.W.G. **Bruce W. Gonser**
B.W.N. **Benjamin W. Niebel**
B.-Y.T. **Bor-Yeu Tsaur**

C

C.A.	Chris Adams
C.A.A.	Chester A. Arnold
C.A.C.	Carleton A. Chapman
C.A.Co.	Charles A. Cohen
C.A.E.	C. A. Evans, Jr.
C.A.K.	C. Arthur Knight
C.A.M.	C. Ajmone Marsan
C.A.Me.	Charles A. Mebus
C.A.Mo.	Charles A. Moore
C.A.R.	Charles A. Ross
C.A.Re.	Charles A. Reed
C.A.S.	Charles A. Scarlott
C.A.Sch.	Charles A. Schroeder
C.A.T.	Cornelius A. Tobias
C.A.V.	Chris A. Veale
C.A.Z.	Carl A. Zapffe
C.B.	Charles Baltay
C.Ba.	Catherine Badgley
C.Be.	Clyde Berg
C.Bu.	Charles Butler
C.B.B.	Charles B. Beck
C.B.C.	Charles B. Curtin
C.B.Cl.	Craig B. Clemmens
C.B.D.	C. B. Duke
C.B.G.	C. B. Gillies
C.B.O.	Charles B. Ogburn
C.B.P.	Cornelius B. Philip
C.B.S.	C. B. Skotland
C.B.T.	Charles B. Tompkins
C.B.V.N.	Cornelis B. Van Niel
C.B.W.	Curtis B. Wilson
C.C.	Charles Collinson
C.C.B.	Clanton C. Black, Jr.
C.C.C.	C. C. Cutler
C.C.Du.	Charles C. Duncan
C.C.H.	Christos C. Halkias
C.C.Ho.	C. Clayton Hoff
C.C.K.	Clifford C. Klick
C.C.L.	Ching Chun Li
C.C.W.	Carl C. Wamser
C.D.	Carl Dover
C.D.B.	Carl D. Broadbent
C.DeB.	Carl de Boor
C.D.C.	C. Denise Caldwell
C.DeD.	Christian de Duve
C.D.G.	C. D. Gelatt, Jr.
C.D.Gu.	C. David Gutsche
C.D.M.	C. D. Moak
C.D.O.	Charles D. Oviatt
C.D.T.	C. Donnell Turner
C.DiT.	Cynthia Di Tallo
C.E.	U. S. Army Corps of Engineers
C.E.A.	Charles E. Applegate
C.E.B.	Clark E. Bricker
C.E.C.	Chester E. Cross
C.E.H.	Carl E. Howe
C.E.L.	Cecil E. Land
C.E.La.	Charles E. Lapple
C.E.LaM.	C. E. LaMotte
C.E.M.	Charles E. Moyer, Jr.
C.E.P.	Charles E. Packard
C.E.S.	Charles E. Slonecker
C.E.Sm.	Carl E. Smith
C.E.T.	Clair E. Terrill
C.E.Tu.	Charles E. Turner
C.E.V.	Cecil E. Vanderzee
C.E.W.	Clyde E. Wiegand
C.F.	Cyrus Feldman
C.Fr.	Clifford Frondel
C.F.A.	C. F. Andrus
C.F.C.	Charles F. Curtis
C.F.E.R.	Clyde F. E. Roper
C.F.F.	Carl F. Floe
C.F.G.	Clarence F. Goodheart
C.F.K.	Carl F. Kayan
C.F.N.	Charles F. Niven, Jr.
C.F.Q.	C. F. Quate
C.F.S.	Charles F. Squire
C.F.S.S.	C. F. Stewart Sharpe
C.F.T.	C. F. Tsang
C.F.W.	Carl F. Wellstead
C.G.C.	Charles G. Cook
C.G.G.	Chauncey G. Goodchild
C.G.S.	C. George Segeler
C.G.St.	Christos G. Stergis
C.H.	Cadet Hand
C.H.Ha.	C. Howard Hamilton
C.Ho.	Cecil Hougie
C.H.B.	Catherine H. Bailey
C.H.C.	Clarence H. Cleminshaw
C.H.Ch.	Charles H. Chestnut III
C.H.Co.	Carl H. Cotterill
C.H.D.	Charles H. Drake
C.H.H.	Clarence H. Hanson
C.H.He.	Charles H. Henry
C.H.L.	Choh Hao Li
C.H.M.	Carl H. Meyer
C.H.S.	Christopher H. Scholz
C.H.Sp.	Charles H. Sprankle
C.H.St.	Calvin H. Stevens
C.H.T.	Charles H. Townes
C.I.	C. Issel
C.I.K.	Clarence I. Kado
C.J.A.	Constantine J. Alexopoulos
C.J.B.	Charles J. Baer
C.J.G.	Charles J. Goebel
C.J.Ga.	Charles J. Gatchell
C.J.Go.	Clarence J. Goodnight
C.J.H.	C. Jack Hearn
C.J.He.	Carl J. Helmers
C.J.Her.	C. J. Herman
C.J.K.	Charles J. Kensler
C.J.L.	Carol J. Lonsdale
C.J.P.	Charles J. Papuchis
C.J.R.	Chalmer J. Roy
C.K.	Cornelis Klein
C.K.B.	Charles K. Bradsher
C.K.J.	C. K. Jen
C.K.N.P.	C. K. N. Patal
C.K.T.	Charles K. Taft
C.K.W.	Charles K. Weichert
C.L.	Carl Lamanna
C.L.A.	Charles L. Alley
C.L.B.	Charles L. Bauer
C.L.C.	Charles L. Camp
C.L.Ch.	Charles L. Christ
C.L.D.	Clive L. Dym
C.L.M.	Charles L. Mantell
C.L.Ma.	Clement L. Markert
C.L.Mas.	Colin L. Masters
C.L.N.	Curtis L. Newcombe
C.L.P.	C. Ladd Prosser
C.L.R.	Charles L. Rulfs
C.L.W.	C. Langdon White
C.M.	Charles Mangion
C.Ma.	C. Martinson
C.M.A.	Charles M. Antoni
C.M.H.	Cyril M. Harris
C.-M.S.	Cheng-Mei Shaw
C.M.Si.	Chester M. Sinnett
C.M.Y.	C. Maurice Yonge
C.N.	Claes Nylander
C.N.G.	Charles N. Gaylord
C.N.Sh.	Carl N. Shuster
C.N.Y.	C. N. Yang
C.O.	Cristian Orrego
C.O.D.	Carl O. Dunbar
C.O.hE.	Colm O. hEocha
C.O.Q.	Calvin O. Qualset
C.P.	Carl Pfaffmann
C.P.B	Charles P. Brewer
C.P.L.	Charles P. Lyman
C.P.P.	Charles P. Pfleeger
C.P.R.	Clark P. Read
C.P.S.	Charles P. Slichter
C.P.Sm.	C. Price Smith
C.P.Sw.	Carl P. Swanson
C.Q.	Chris Quigg
C.Qu.	Carol Quaife
C.R.	Claudio Rebbi
C.Ru.	Charles Rulfs
C.R.C.	Clark R. Chapman
C.R.F.	Charles R. Fourtner
C.R.L.	Chester R. Longwell
C.R.Lo.	Christopher R. Lowe
C.R.M.	Charles R. Marsland
C.R.Ma.	C. R. Martinson
C.R.S.	Charles R. Stephens
C.R.W.	Charles R. Wilke
C.S.	Christen Skaar
C.S.C.	Charles S. Cox
C.S.Cr.	Calvin S. Cronan
C.S.G.	Chester S. Gardner
C.S.H.	C. S. Hammen
C.S.He.	Carl S. Herz
C.S.Hu.	Cornelius S. Hurlbut, Jr.
C.S.P.	Carl S. Pederson
C.S.T.	Chen S. Tsai
C.S.V.	Carl S. Vestling
C.-S.Y.	Ching-Sung Yu
C.T.	Curt Teichert
C.T.C.	C. T. Chen
C.T.Cr.	Clayton T. Crowe
C.T.K.	Charles T. Kowal
C.T.S.	Chester T. Sims
C.T.T.	Charles T. Tart
C.V.C.	Charles V. Crittenden
C.V.S.	C. V. Shank
C.W.	Colin White
C.Wi.	Conrad Wickstrom
C.W.B.	Carl W. Boothroyd
C.W.Bu.	Charles W. Burnham
C.W.By.	Charles W. Byers
C.W.D.	Clifton W. Draper
C.W.E.	C. Wayne Ellett
C.W.H.	C. W. Hesseltine
C.W.K.	Chung W. Kim
C.W.N.	Chester W. Newton

C.W.Ni. **Charles W. Nixon**
C.W.O. **Christopher W. Olsen**
C.W.S. **C. W. Siegmund**
C.W.Si. **C. W. Sisler**
C.W.Sw. **Charles W. Swithinbank**
C.W.T. **Chales W. Turner**
C.W.V. **Charles W. Vokac**

D

D.A. **David Adler**
D.Al. **David Altman**
D.Ar. **Donald Armstrong**
D.Ay. **D. Aylmer**
D.A.B. **D. Allan Bromley**
D.A.Be. **David A. Bemis**
D.A.B.M. **David A. B. Miller**
D.A.C. **David A. Clark**
D.A.E. **David A. Ehrmann**
D.A.K. **David A. Keith**
D.A.L. **Donovan A. Ljung**
D.A.Lo. **D. A. Lowe**
D.A.M. **Douglas A. Marsland**
D.A.Me. **Duncan A. Mellichamp**
D.A.P. **David A. Palzkill**
D.A.Re. **David A. Reay**
D.A.Ro. **David A. Ross**
D.A.S. **David A. Shirley**
D.A.W. **Donald A. Wilbur, Sr.**
D.A.Y. **David A. Young**
D.B. **David Barr**
D.Be. **David Beaglehole**
D.Br. **Dirk Brouwer**
D.Bu. **Donald Burdick**
D.Bur. **Dave Bursky**
D.B.B. **Daniel B. Blake**
D.B.C. **David B. Cohen**
D.B.D. **Daniel B. Dallas**
D.B.H. **David B. Harrison**
D.B.J. **Deane B. Judd**
D.B.M. **David B. Meyer**
D.B.P. **David P. Pisoni**
D.B.S. **Donald B. Sinclair**
D.B.W. **Douglas B. Webster**
D.C. **David Cohen**
D.Ca. **David Casasent**
D.C.B. **D. C. Beaumariage**
D.C.C. **David C. Coleman**
D.C.G. **Detlef C. Gloge**
D.C.H. **David C. Hazen**
D.C.J. **Donald C. Jackson**
D.C.W. **David C. Webber**
D.C.Wh. **David C. White**
D.D.D. **D. Dwight Davis**
D.D.Do. **Donald D. Dodge**
D.D.F. **Dudley D. Fuller**
D.D.H. **David D. Houghton**
D.D.J. **Dennis D. Juranek**
D.D.MacC. **Donnell D. MacCarthy**
D.D.McC. **Dennis D. McCarthy**
D.D.P. **David D. Perkins**
D.D.R. **D. D. Robb**
D.D.Ra. **David D. Ragg**
D.D.Ru. **David D. Rubis**
D.D.S.T. **Donovan Des S. Thomas**
D.E. **Dorothy Eichorn**
D.E.A. **David E. Alburger**
D.E.B. **Donald E. Becker**
D.E.Br. **Don E. Brownlee, II**
D.E.E. **David E. Ellis**
D.E.F. **David E. Fogarty**
D.E.G. **David E. Giannasi**
D.E.Go. **David E. Goldberg**
D.E.H. **Dennis E. Hayas**
D.E.K. **Dale E. Kester**
D.E.Ki. **David E. Kissel**
D.E.Kl. **Douglas E. Kleinmann**
D.E.O. **Donald E. Olins**
D.E.P. **David E. Pritchard**
D.E.R. **Duane E. Roller**
D.E.Re. **D. Eugene Redmond, Jr.**
D.E.S. **Donald E. Savage**
D.E.T. **Douglas E. Duttle**
D.E.W. **Daniel E. Whitney**
D.F. **Denis Forster**
D.F.O. **Donald F. Othmer**
D.F.Ow. **Denis F. Owen**
D.F.P. **Donald F. Poulson**
D.F.S. **Dale F. Stein**
D.G. **D. Gauss**
D.G.F. **Donald G. Fink**
D.G.Fre. **David G. Frey**
D.G.H. **Dennis G. Hall**
D.G.K. **Dennis G. Kovar**
D.G.S. **Demetrios G. Samaras**
D.G.Sc. **Donald G. Schueler**
D.H. **Dorothy Hill**
D.H.A. **David H. Abrahams**
D.H.B. **Daryle H. Busch**
D.H.D. **David H. Douglass**
D.H.G. **David H. Goodstein**
D.H.K. **David H. Klein**
D.H.Ka. **David H. Katz**
D.H.L. **Derrick H. Lehmer**
D.H.S. **Daniel H. Sheingold**
D.H.T. **D. H. Tarling**
D.H.Th. **David Hurst Thomas**
D.H.W. **Denys H. Wilkinson**
D.J. **David Joravsky**
D.J.C. **Donald J. Cohen**
D.J.Ca. **D. J. Carlsson**
D.J.Dep. **Donald J. DePaulo**
D.J.E. **Don J. Easterbrook**
D.J.Ev. **Doyle J. Evans, Jr.**
D.J.F. **Dennis J. Frailey**
D.J.Fu. **Douglas J. Futuyma**
D.J.H. **Dennis J. Herrell**
D.J.Ho. **D. J. Horen**
D.J.Hor. **David J. Horn**
D.J.J. **Daniel J. Jones**
D.J.M. **D. J. Morré**
D.J.N. **David J. Nesbitt**
D.J.O. **Donald J. O'Connor**
D.J.P. **Donald J. Patton**
D.J.Pe. **David J. Pegg**
D.J.R. **Dewey J. Raski**
D.J.S. **David J. Sellmyer**
D.J.S.B. **Donald J. S. Barr**
D.J.T. **Douglas J. Theis**
D.J.W. **Daniel J. Weiler**
D.J.Wa. **Deborah Jean Warner**
D.J.Wi. **David J. Wineland**
D.K. **Daniel Kaizer**
D.K.S. **David K. Scott**
D.L. **David Lagunoff**
D.La. **David Layzer**
D.Lal. **D. Lal**
D.Laz. **David Lazarus**
D.Le. **D. Lewis**
D.Led. **Daniel Lednicer**
D.Lo. **Donald Louria**
D.Low. **David Lowenthal**
D.L.A. **Daniel L. Azarnoff**
D.L.An. **Donald L. Anglin**
D.L.D. **David L. Dalrymple**
D.L.G. **Dwight L. Glasscock**
D.L.H. **Donald L. Holt**
D.L.I. **Douglas L. Inman**
D.L.J. **Donald Lee Johnson**
D.L.P. **Donald L. Price**
D.L.T. **Donald L. Turrell**
D.L.W. **Donald L. Waidelich**
D.L.We. **David L. Weaver**
D.M. **Daniel Mazia**
D.Mo. **David Morrison**
D.M.D. **David M. Dawson**
D.M.DeL. **Dwight M. DeLong**
D.M.F. **David M. Farmer**
D.M.Fr. **David M. Fradkin**
D.M.G. **David M. Greenberg**
D.M.Ga. **Donald M. Gallant**
D.M.Gr. **Dieter M. Gruen**
D.M.H. **David M. Hodgin**
D.M.Hu. **Donald M. Hunten**
D.M.L. **David M. Larsen**
D.McN. **Donald McNellis**
D.M.P. **Daniel M. Popper**
D.M.Po. **David M. Pozar**
D.M.R. **David M. Rust**
D.M.S. **David M. Smith**
D.M.V. **Daniel M. Viccione**
D.M.W. **D. M. Wiles**
D.M.Y. **Don M. Yost**
D.N. **David Nachmansohn**
D.N.H. **Douglas N. Halbe**
D.N.K. **David N. Keast**
D.N.L. **D. N. Langenberg**
D.N.La. **Daniel N. Lapedes**
D.O.H. **David O. Haynes**
D.P. **David Park**
D.P.A. **Donald P. Abbott**
D.P.Ad. **Douglas P. Adams**
D.P.An. **D. Philip Ankeney**
D.P.B. **David P. Barash**
D.P.G. **Daniel P. Griffin**
D.Q.L. **D. Q. Lamb**
D.R. **Donald Raum**
D.Ru. **Dorothea Rudnick**
D.R.C. **Donald R. Calvert**
D.R.Co. **Dean R. Collins**
D.R.D. **Donald R. Dimmel**
D.R.F.H. **Donald R. F. Harleman**
D.R.J. **David R. Jones**
D.R.N. **David R. Nygren**
D.R.P. **Donald R. Peacor**
D.R.S. **Donald R. Stranks**
D.R.V. **David R. Veblen**
D.S. **Dave Scott**
De.S. **Dennis Stanford**
D.Sh. **Dan Shechtman**
D.Si. **David Siminovitch**
D.Sim. **Daniel Simberloff**
D.St. **Donald Stansbarger**

D.Ste.	Dewey Stewart
D.Sw.	Daniel Swern
D.S.B.	Douglas S. Billington
D.S.Br.	David S. Breslow
D.S.C.	Dennis S. Charney
D.S.G.	D. S. Grosch
D.S.Ge.	Donald S. Gemmel
D.S.H.	Donald S. Hayes
D.S.He.	David S. Hepler
D.S.Hi.	Daniel S. Hirschberg
D.S.K.	Douglas S. Kellogg
D.S.L.	Daniel S. Lehrman
D.S.P.	David S. Palermo
D.S.V.F.	D. S. Van Fleet
D.T.	David Turnbull
D.T.L.	David T. Lykken
D.T.R.	Donald T. Reay
D.T.S.	Donald T. Sawyer
D.V.M.	Dal V. Maddalon
D.V.Mo.	Donald V. Moore
D.V.W.	David V. Widder
D.W.	Daniel Wellner
D.Wi.	Dudley Williams
D.Wis.	Dennis Wisnooky
D.Wo.	David Wollesheim
D.W.B.	Donald W. Banner
D.W.C.	David W. Curkendall
D.W.E.S.	David W. E. Smith
D.W.F.	Don W. Fawcett
D.W.K.	Donald W. Kerst
D.W.Ki.	Donald W. King
D.W.Ku.	Donald W. Kupke
D.W.L.	David W. Latham
D.W.M.	David W. Mizell
D.W.S.	David W. Steadman
D.X.F.	Daniel X. Freedman

E

E.A.	Elihu Abrahams
E.Ad.	Elijah Adams
E.An.	Everett Anderson
E.At.	Elisha Atkins
E.A.A.	Edward A. Ackerman
E.A.Ad.	Edward A. Adelberg
E.A.B.	Edward A. Boyden
E.A.C.	Edsel A. Caine
E.A.F.	E. A. Freundt
E.A.G.	Earl A. Gulbransen
E.A.H.	Eugene A. Hollowell
E.A.K.	Elbert A. King
E.A.L.	Edmund A. Laport
E.A.S.	Edward A. Smuckler
E.B.C.	Edward B. Curdts
E.B.Cu.	Edward B. Cutler
E.B.F.	E. B. Ford
E.B.R.	Evans B. Reid
E.B.W.	E. Bright Wilson, Jr.
E.C.	Ellen Clarke
E.Co.	Ernest Coleal
E.C.A.	Ellsworth C. Alvord, Jr.
E.C.J.	Earl C. Joseph
E.C.Jo.	Edwin C. Jones, Jr.
E.C.L.	Edward C. Luckenbach
E.C.M.	Earl C. Mansfield
E.C.O.	Everett C. Olson
E.C.P.	E. C. Pielou
E.C.Pol.	Ernest C. Pollard
E.C.R.-R.	Edward C. Roosen-Runge
E.C.S.	Eugene C. Starr
E.C.St.	Edward C. Stevenson
E.C.Stu.	Edwin Charles Stumm
E.C.T.C.	Edward C. T. Chao
E.D.	Eric Delson
E.Do.	Erling Dorf
E.Dr.	Eric Drexler
E.D.A.	E. Dwight Adams
E.D.B.	Eugene D. Becken
E.D.C.	Eugene D. Commins
E.D.G.	Edward D. Goldberg
E.E.C.	Ernest E. Charlton
E.E.Co.	Earl E. Coulter
E.E.F.	Eberhard E. Fetz
E.E.S.	Esmond E. Snell
E.E.U.	Erwin E. Underwood
E.E.W.	E. Eugene Weaver
E.Fi.	Eric Firing
E.Fr.	Ernesto Freire
E.F.A.	Earl F. Arbuckle, III
E.F.B.	Edmour F. Blouin
E.F.H.	Edward F. Hammel
E.F.Ha.	Edward F. Harford
E.F.L.	Edward F. Leonard
E.F.McB.	Earle F. McBride
E.F.N.	Ernest F. Nippes
E.F.S.	Emil F. Steinert
E.F.T.	Edward F. Tedesco
E.F.W.	Elliott F. Wright
E.G.	Edward Gerjuoy
E.Gr.	Emil Grosswald
E.Gru.	Emanuel Grunberg
E.G.B.	Elmer G. Butler
E.G.C.	Elizabeth G. Cutter
E.G.H.	E. G. Heyne
E.G.N.	Elton G. Nelson
E.G.P.	Edward G. Partridge
E.G.Pr.	Ernest G. Pringsheim
E.G.R.	Edward G. Ramberg
E.G.Ro.	Eugene C. Rochow
E.G.S.	Emilio G. Segrè
E.G.St.	Edward G. Stuart
E.G.W.	Ernest G. Wever
E.H.	Edward Harrison
E.H.A.	Elbert H. Ahlstrom
E.H.Ap.	Evan H. Appelman
E.H.B.	Earl H. Beistline
E.H.Br.	Edgar H. Bristol
E.H.C.	Edward H. Cole
E.H.H.	Elbert H. Hadley
E.H.Ha.	Edwin H. Hammond
E.H.Q.	Edith H. Quimby
E.H.R.	Ernest H. Runyon
E.H.S.	Edgar H. Sibley
E.H.St.	Elmer H. Stotz
E.H.W.	Earl H. Wood
E.I.	Earl Ingerson
E.J.	Erik Jarvik
E.Jo.	Edward Johnson
E.Ju.	Evan Just
E.J.A.	E. James Archer
E.J.B.	Edward J. Beckhorn
E.J.Bo.	Edgar J. Boell
E.J.F.W.	Edward J. F. Wood
E.J.G.	Edward J. Groth
E.J.L.	Eugene J. Limpel
E.J.M.	Earl J. Merwin
E.J.Q.	Edward J. Quirin
E.J.T.	Edward J. Triebe
E.K.	Ernest Kent
E.Ke.	Edwin Kessler
E.Ki.	Eileen Kintsch
E.Kr.	Ernest Kramar
E.K.H.	Earl K. Hyde
E.K.K.	Edward K. Kaprelian
E.L.	E. Lazarides
E.Le.	Edward Leete
E.L.B.	Ernst L. Biberstein
E.L.C.	Earl L. Core
E.L.E.	Ernest L. Eliel
E.L.G.	Edward L. Ginzton
E.L.Ga.	Emil L. Gavlick
E.L.H.	Edwin L. Hamilton
E.L.Hi.	Edward L. Hill
E.L.L.	Elbert L. Little, Jr.
E.L.P.	E. Lowe Pierce
E.L.Po.	Ernest L. Pollitzer
E.L.S.	Everett L. Saul
E.L.T.	Edwin L. Turner
E.L.V.C.	Ernest L. Van Campenhout
E.L.W.	Erwin L. Weber
E.L.Wr.	Edward L. Wright
E.L.Y.	Ellis L. Yochelson
E.M.	Ernesto Marcus
Ev.M.	Eveline Marcus
E.Me.	Edward Meyers
E.M.B.	Elliott M. Blass
E.M.Br.	Edward M. Breinan
E.M.C.	Edward M. Cohart
E.M.Co.	Esther M. Conwell
E.M.E.	Eugene M. Ettenberg
E.M.Ey.	Edward M. Eyring
E.M.H.	Elaine M. Hendry
E.M.L.	Edwin M. Larsen
E.M.M.	Emil M. Mrak
E.M.R.	Eugene M. Rasmusson
E.M.S.	Edith M. Sheppard
E.M.We.	Elizabeth M. Wenzel
E.M.Wi.	Eva M. Winkler
E.M.Wis.	Edmund M. Wise
E.M.Y.	Edward M. Young
E.N.	Ernest Naylor
E.Ne.	Erwin Neter
E.No.	Eskel Nordell
E.N.G.	Edgar N. Gilbert
E.N.H.	E. Newton Harvey
E.N.S.	E. N. Skinner
E.O.D.	Edward O. Dodson
E.P.	Elliot Postow
E.Pa.	Eric Paterson
E.P.E.	Elbridge P. Eaton
E.P.G.	E. Patrick Groody
E.P.O.	Eugene P. Odum
E.P.P.	Edwin P. Plueddemann
E.P.S.	Ellis P. Steinberg
E.R.	Ernest Rabinowicz
E.Ra.	Erwin Raisz
E.Ro.	Elliott Robbins
E.R.W.	Ewald R. Weibel
E.R.B.	Elery R. Becker
E.R.C.	Earle R. Caley
E.R.Ca.	E. R. Carson
E.R.Co.	E. Richard Cohen
E.R.D.	Ellen R. Dirksen

E.R.G.	**Edwin R. Gilliland**
E.R.H.	**Egbert R. Hardesty**
E.R.J.	**E. Ruffin Jones**
E.R.R.	**Elmar R. Reiter**
E.S.	**Ernest Schweizer**
E.Se.	**Emilio Segrè**
E.S.B.	**Elso S. Barghoorn**
E.S.Ba.	**Edward S. Barnitz**
E.S.F.	**Edward S. Fry**
E.S.G.	**Emery St. George, Jr.**
E.S.R.	**Edward S. Ross**
E.S.S.	**Edward S. Sarachik**
E.S.Y.	**Edward S. Yang**
E.T.	**Erik Trinkaus**
E.T.C.	**Emil T. Chanlett**
E.U.C.	**E. U. Condon**
E.W.	**Ernest Wenkert**
E.Wa.	**Everett Waters**
E.Wi.	**Earle Williams**
E.W.A.	**Edward W. Allen, Jr.**
E.W.B.	**Edward W. Baker**
E.W.C.	**Edward W. Comings**
E.W.G.	**Edward W. Gore, Jr.**
E.W.Ge.	**Erwin W. Gelfand**
E.W.H.	**E. William Heinrich**
E.W.He.	**E. Wendell Hewson**
E.W.K.	**Edward W. Kimbark**
E.W.M.	**Erwin W. Müller**
E.W.Me.	**E. W. Merrill**
E.W.T.	**Elmer W. Travis**
E.Y.	**Eiju Yatsu**
E.Y.C.	**Edgar Y. Choueiri**

F

F.A.B.	**Frank A. Brown, Jr.**
F.A.C.	**Frank A. Carone**
F.A.Ch.	**Fenner A. Chace, Jr.**
F.A.G.	**Frank A. Geldard**
F.A.J.	**Francis A. Jenkins**
F.A.L.	**Franklin A. Long**
F.A.L.C.	**F. A. L. Clowes**
F.A.M.	**Frederick A. Murphy**
F.A.R.	**Francis A. Richards**
F.B.	**Fred Barker**
F.Ba.	**Fred Basolo**
F.B.B.	**France B. Berger**
F.B.G.	**F. B. Golden**
F.B.Go.	**Frank B. Golley**
F.B.P.	**Frederick B. Pogust**
F.B.S.	**Frank B. Salisbury**
F.B.Sh.	**Frederick B. Shuman**
F.B.T.	**Franz B. Tuteur**
F.B.V.H.	**Franklyn B. Van Houten**
F.B.W.	**Frank B. Wood**
F.C.	**Francis Clark**
F.Cr.	**Friedrich Cramer**
F.C.B.	**Frank C. Benner**
F.C.Br.	**Frederick C. Brown**
F.C.H.	**F. Clark Howell**
F.C.M.	**Frank C. Mock**
F.C.R.	**Fredrick C. Redlich**
F.C.S.	**Ferris C. Standiford**
F.C.St.	**Frank C. Sturges**
F.D.	**Farrington Daniels**
F.Du.	**Floyd Dunn**
F.D.B.	**Frank D. Brooks**
F.D.C.	**Franklin D. Cooper**
F.delaC.	**Felix de la Cruz**
F.D.DeV.	**Fred D. DeVaney**
F.D.K.	**Fred D. Kochendorfer**
F.D.L.	**Frank D. Lewis**
F.D.P.	**Frank D. Popp**
F.E.E.	**Frank E. Egler**
F.E.F.	**Floyd E. Ford**
F.E.Fo.	**F. E. Foss**
F.E.G.	**Frank E. Gardner**
F.E.J.F.	**Fredrick E. J. Fry**
F.E.N.	**Fred E. Nicodemus**
F.E.S.	**Frank E. Snodgrass**
F.F.	**Ferdinand Freudenstein**
F.F.E.	**Fredric F. Ehrich**
F.F.R.	**Frank F. Richards**
F.F.W.	**Felix F. Wu**
F.G.B	**Frederick G. Baily**
F.G.C.	**Forrest G. Chumley**
F.G.D.	**Frank G. Dollear**
F.G.E.	**F. G. Ely**
F.G.Et.	**Frank G. Ethridge**
F.G.Ev.	**F. Gaynor Evans**
F.G.H.	**Frank G. Hawksworth**
F.H.B.	**Frank H. Bower**
F.H.F.	**Frederick H. Fisher**
F.H.Fr.	**F. H. Froes**
F.H.L.	**Fritz H. Laves**
F.H.Lu.	**Frank H. Ludlam**
F.H.M.	**F. H. May**
F.H.R.	**Frank H. Rockett**
F.H.-Ro.	**Frederick Hayes-Roth**
F.H.S.	**Frank H. Shelton**
F.H.Sp.	**Frank H. Spedding**
F.H.T.	**Frank H. Taylor**
F.I.	**Franco Iachello**
F.J.A.	**Frederick J. Almgren, Jr.**
F.J.An.	**Francis J. Anscombe**
F.J.B.	**Frank J. Blatt**
F.J.D.	**Frank J. DiSalvo**
F.J.E.	**F. J. Edeskuty**
F.J.G.	**Fred J. Gruenberger**
F.J.J.	**Francis J. Johnston**
F.J.L.	**Frank J. Lockhart**
F.J.R.	**Fred J. Rosenbaum**
F.J.S.	**F. J. Simpson**
F.K.	**Fritz Kalhammer**
F.Ke.	**Frederic Keffer**
F.K.H.	**Forest K. Harris**
F.K.L.	**Frank K. Lawler**
F.K.O.	**Fredrick K. Orkin**
F.L.	**Fritz Laves**
F.Li.	**Folke Linde**
F.L.H.	**F. L. Hermach**
F.L.S.	**Frederic L. Schwab**
F.L.W.	**Frank L. Warner**
F.M.	**Freydoon Mansouri**
F.M.B.	**Frances M. Brodsky**
F.M.C.	**Frank M. Carpenter**
F.M.F.	**Frederick M. Fowkes**
F.McL.M.	**Frank McL. Mallet**
F.M.P.	**Francis M. Pipkin**
F.M.R.	**Francis M. Rogalloh**
F.M.W.	**Floyd M. Wahl**
F.M.We.	**Frances M. Weesner**
F.M.Wh.	**Frank M. White**
F.O'D.	**Franklin O'Donnell**
F.P.	**Francesco Paresce**
F.P.S.	**Francis P. Shepard**
F.R.	**Floyd Ratlifff**
F.Ri.	**Frank Riley**
F.Ro.	**F. Rohrlich**
F.Ru.	**Fritz Ruch**
F.R.B.	**F. R. Boyd**
F.R.F.	**F. R. Fosberg**
F.R.H.	**Frederick R. Hanson**
F.R.K.	**Fazlur R. Khan**
F.R.Ko.	**F. Ralph Kotter**
F.R.M.	**Frederick R. Miller**
F.R.McF.	**Forest R. McFarland**
F.R.V.	**Frank R. Voorhees**
F.R.Vo.	**Fred R. Volkmar**
F.S.	**Frederick Sanders**
F.St.	**Fred Sterzer**
F.Ste.	**Fred Stern**
F.S.G.	**Frederick S. Goulding**
F.S.J.	**Francis S. Johnson**
F.S.M.	**Frederick S. Merritt**
F.S.S.	**Frederick S. Szalay**
F.T.A.	**Fredrich T. Addicott**
F.T.L.	**F. Thomas Ledig**
F.T.T.	**Fredrich T. Thwaites**
F.V.	**Felix Viro**
F.Vo.	**F. Vögtle**
F.V.G.	**Fred V. Grau**
F.V.K.	**Frank V. Kosikowski**
F.V.L.	**F. V. Lenel**
F.W.	**Frank Wagner**
F.We.	**Fritz Weigel**
F.Wh.	**Frank Whiting**
F.Wi.	**Fred Wilt**
F.Wil.	**Franz Wilczek**
F.Wu.	**Fred Wudl**
F.W.B.	**Fred W. Billmeyer, Jr.**
F.W.H.	**Friedrich W. Hoffmann**
F.W.K.	**F. W. Karasek**
F.W.Ko.	**Frederick W. Koerker**
F.W.K.F.	**Frank W. K. Firk**
F.W.M.	**Franklin W. Martin**
F.W.S.	**Fred W. Smith**
F.W.Sw.	**F. W. Sweat**
F.W.T.	**Fred W. Tanner, Jr.**
F.W.Z.	**Frank W. Zink**

G

G.	**Govindjee**
G.A.	**Gordon Atkinson**
G.A.A.	**Gregory A. Ahern**
G.A.D.	**Gregory A. Davis**
G.A.E.	**George A. Economou**
G.A.H.	**George A. Hawkins**
G.A.Ho.	**G. A. Horton**
G.A.L.	**George A. Ledingham**
G.A.M.	**Gordon A. Macdonald**
G.A.Ma.	**George A. Maul**
G.A.Mo.	**Gary A. Moll**
G.A.R.	**G. Alan Robison**
G.A.S.	**G. Alan Solem**
G.A.W.	**Gary A. Williams**
G.B.	**Gunnar Böckstrom**

G.Bi.	Garrett Birkhoff
G.B.H.	George B. Hess
G.B.P.	Gayle B. Priester
G.C.G.	Gerard C. Gambs
G.C.K.	George C. Kennedy
G.C.Ke.	George C. Kent, Jr.
G.C.M.	George C. Messenger
G.C.N.	George C. Nelson
G.D.	Gerhard Derge
G.De.	Guy Deutscher
G.D.B.	Glenn D. Breuer
G.D.C.	Gene D. Cohen
G.D.F.	George D. Fulford
G.D.G.	George D. Garland
G.D.Go.	Gary D. Gordon
G.D.S.	Guy D. Smith
G.DeV.	Gerard De Vaucouleurs
G.E.E.	George E. Ewing
G.E.L.	George E. Liedholm
G.E.Li.	Gene E. Likens
G.E.Liv.	Gideon E. Livingston
G.E.P.	George E. Pake
G.E.Pe.	Gertrude E. Perlmann
G.E.S.	Garland E. Scott
G.E.St.	George E. Stapleton
G.E.Str.	Gary E. Striker
G.E.T.	George E. Templeton
G.E.W.	Gary E. Wnek
G.F.	Gioacchino Failla
G.Fa.	Gerhard Fankhauser
G.Fau.	Gunter Faure
G.Fr.	Geoffrey Fryer
G.F.B.	George F. Bobart
G.F.Be.	George F. Bertsch
G.F.Bi.	Giovanni F. Bignami
G.F.F.	Gordon F. Ferris
G.F.L.	George F. Ludvik
G.F.O.	Gilbert F. Otto
G.F.P.	George F. Pinder
G.F.R.	George F. Reddish
G.F.W.	George F. Wislicenus
G.G.	Gerson Goldhaber
G.Gr.	Gene Grey
G.Gu.	G. Guzinski
G.G.A.	George G. Adkins
G.G.B.	Gabriel G. Barna
G.G.K.	George G. Karady
G.G.L.	Gordon G. Lill
G.G.N.	Graham G. Nicholson
G.G.T.	Gerald G. Thorne
G.H.	George Harvey
G.He.	Gilbert Held
G.Ho.	Gerald Holton
G.Hot.	Grosvenor Hotchkiss
G.H.B.	Glenn H. Brown
G.H.Ba.	Grant H. Barlow
G.H.C.	Gilbert H. Cady
G.H.D.	George H. Davis
G.H.Di.	G. H. Dieke
G.H.F.	George H. Fried
G.H.M.	Glenn H. Miller
G.H.Mo.	George H. Morrison
G.H.My.	George H. Myer
G.H.R.	Gilbert H. Reiling
G.J.G.	Gerald J. Gastony
G.J.W.	George J. Wilder
G.J.Z.	George J. Zissis
G.K.	Georges Knaysi
G.Kr.	George Krauss
G.K.C.	George K. Celler
G.K.Cz.	Gerald K. Czamanske
G.K.S.	Gerald K. Sims
G.L.B.	Gordon L. Brownell
G.L.Be.	Gerald L. Benny
G.L.C.	Guy L. Childberg
G.L.Cl.	George L. Clark
G.L.E.	Gunther L. Eichhorn
G.L.G.	George L. Gaines
G.L.Gi.	Gerald L. Gilardi
G.L.K.	Gerald L. Klerman
G.L.S.	Gordon L. Shaw
G.L.St.	G. Ledyard Stebbins
G.L.T.	Gerald L. Thompson
G.L.V.	Gilbert L. Voss
G.M.	George Mandler
G.Mi.	Gardner Middlebrook
G.Mo.	Giuliana Moreno
G.Mu.	Glenn Murphy
G.M.B.	George M. Bouton
G.M.Bo.	Grayce M. Booth
G.M.C.	Gerald M. Clemence
G.M.G.	Glenn M. Glasford
G.M.Gu.	Gay M. Guzinski
G.M.H.	Gary M. Hieftje
G.M.K.	Glen M. Kohls
G.McP.	George McPherson, Jr.
G.M.R.	Gardner M. Reynolds
G.M.S.	George M. Savage
G.M.Se.	Gerhard M. Sessler
G.N.R.	G. N. Richards
G.O.	Gerald Oster
G.Ow.	Guillermo Owen
G.O.A.	George O. Abell
G.O.B.	Glenn O. Bressler
G.O.D.	Gunter O. Dietrich
G.O.R.	Gilbert O. Raasch
G.P.	George Papageorgiou
G.Po.	Guido Pontecovo
G.Pos.	German Posada
G.P.K.	Gerard P. Kuiper
G.P.M.	George P. Moore
G.P.S.	George P. Sutton
G.R.	George Rapp, Jr.
G.R.C.	Georgeanne R. Caughlan
G.R.G.	G. Ronald Gray
G.R.H.	George R. Harrison
G.R.N.	Gerald R. North
G.R.P.	G. R. Peirce
G.R.S.	G. R. Schoonmaker
G.R.St.	George R. Sturtevaut
G.R.St.P.	George R. St. Pierre
G.S.	George Sistare, Jr.
G.Sh.	G. Shanmugam
G.Sl.	George Slomp
G.Sm.	Gaylord Smith
G.Sp.	Garrison Sposito
G.St.	Gus Stavis
G.Sw.	George Switzer
G.S.C.	George S. Carter
G.S.G.	Gary S. Grest
G.S.H.	G. S. Hurst
G.S.M.	George S. Mill
G.S.T.	Gordon S. Taylor
G.T.	Geza Teleki
G.T.G.	Gerald T. Garvey
G.T.J.	Gideon T. James
G.T.S.	Glenn T. Seaborg
G.V.D.	George Van Dyne
G.V.M.	Gerard V. Middleton
G.W.	Gerald Weiss
G.W.B.	George W. Barlow
G.W.Br.	George W. Brindley
G.W.Bu.	G. W. Butler
G.W.DeV.	George W. DeVore
G.W.G.	Gordon W. Groves
G.W.H.	Gertrude W. Hinsch
G.W.K.	George W. Kessler
G.W.M.	George W. Mackey
G.W.O'D.	George W. O'Dom
G.W.P.	Gerald W. Prescott
G.W.R.	G. Wilse Robinson
G.W.S.	George W. Stroke
G.W.Si.	George W. Simon
G.Y.C.	Gilbert Y. Chin

H

H.A.B.	Horace A. Barker
H.A.Be.	Hans A. Bethe
H.A.F.	Henry A. Fairbank
H.A.L.	Herbert A. Laitinen
H.A.R.	Herman A. Rediess
H.A.W.	Harley A. Wilhelm
H.B.	Hans Brintzinger
H.Br.	Harold Brown
H.B.A.	H. Burnham Allport
H.B.F.	Howard B. Fell
H.B.G.	H. Bentley Glass
H.B.Gr.	Harry B. Gray
H.B.H.	H. B. Huntington
H.B.J.	Howard B. Jensen
H.B.M.	Harold B. Maynard
H.B.Mi.	Harlow B. Mills
H.B.S.	Howard B. Sprague
H.B.T.	Harold B. Tukey
H.B.W.	Harry B. Whittington
H.B.Wi.	Homer B. Willis
H.C.	Henry Charnock
H.Cl.	Hartwig Cleve
H.Co.	Helen Couclelis
H.Cr.	Howard Crum
H.Cu.	Harold Cummins
H.C.B.	Herbert C. Brown
H.C.C.	H. C. Casey, Jr.
H.C.Co.	Herbert C. Corben
H.C.N.	Harold C. Neu
H.C.R.	Harold C. Ries
H.C.Ri.	Harold C. Riggs
H.C.S.	H. C. Schutt
H.C.W.	Harold C. Weber
H.C.Wi.	Hurd C. Willett
H.C.Y.	Harry C. Yeatman
H.D.	H. Date
H.De.	Hans Dehmelt
H.Di.	Hugh Dingle
H.D.D.	H. D. Dooley
H.D.H.	Harold D. Hauf
H.D.He.	Hollis D. Hedberg
H.D.Ho.	Heinrich D. Holland
H.D.J.	Hugh D. Jones
H.D.O.	Harold D. Orville
H.D.P.	Harry D. Patton
H.D.T.	Horace D. Taft
H.E.	Henry Eyring
H.E.C.	Harold E. Campbell

H.E.Ca.	Herbert E. Carter
H.E.D.	Henry E. Duckworth
H.E.E.	Harold E. Edgerton
H.E.McG.	Harold E. McGannon
H.E.N.	Homer E. Newell
H.E.R.	Harold E. Rosenberger
H.E.S.	Hans E. Suess
H.E.T.	Helmet E. Thiess
H.E.V.G.	Henning E. Von Gierke
H.E.W.	Harvey E. Wegner
H.F.	Herman Feshbach
H.Fi.	Harold Fischer
H.Fo.	Herbert Fox
H.Fr.	Hans Frebold
H.Fre.	Herbert Frey
H.F.G.	Harvey F. George
H.F.K.	Harold F. Klock
H.F.Kr.	Henry F. Krous
H.F.O.	Harry F. Olson
H.F.R.	Harry F. Remde
H.F.S.	Henry F. Schaefer, III
H.F.Sh.	Hugh F. Shannon
H.F.V.	Herbert F. Volk
H.F.W.	Harold F. Walton
H.G.	Harold Goodglass
H.Gr.	H. Gränicher
H.G.B.	Harriet G. Barclay
H.G.Ba.	Herbert G. Baker
H.G.Bo.	Holbrook G. Botset
H.G.C.	Harold G. Corwin, Jr.
H.G.M.	H. G. Marshall
H.G.S.	Heinz G. Sell
H.G.Sh.	Harley G. Sheffield
H.G.St.	H. G. Stubbings
H.H.	Henry Hellmers
H.H.B.	Harrison H. Barrett
H.H.F.	H. Hugh Fudenberg
H.H.H.	Henry H. Hausner
H.H.He.	Hans H. Hecht
H.H.Hec.	Harry H. Heckman
H.H.Ho.	Herman H. Hobbs
H.H.J.N.	Herbert H. J. Nesbitt
H.H.K.	Herbert H. Kellogg
H.H.L.	Heinz H. Lettau
H.H.M.	Hilton H. Mollenhauer
H.H.Mu.	Haydn H. Murray
H.H.R.	Herbert H. Ross
H.H.S.	Harry H. Sisler
H.H.Sk.	Hugh H. Skilling
H.H.Sm.	Harold H. Smith
H.H.St.	Henry H. Storch
H.H.Su.	Hans H. Suter
H.H.T.	H. Holden Thorp
H.J.	Hans Jakobi
H.J.A.	Henry J. Albert
H.J.C.	H. John Carew
H.J.Cl.	Harry H. Clausen
H.J.E.	H. John Evans
H.J.G.	Harold J. Gibson
H.J.H.	Harvey J. Hoge
H.J.L.	Harry J. Lipner
H.J.McL.	Hugh J. McLellan
H.J.P.	Herman J. Phaff
H.J.R.	Herbert J. Reich
H.J.S.	Harlan J. Smith
H.J.W.	Herbert J. Welshimer
H.J.Wi.	Herold J. Wiens
H.J.Wir.	Henry J. Wirry
H.K.	Herwig Kogelnik
H.K.C.	H. Keith Chenault
H.K.W.	Helmut K. Weickmann
H.K.Wi.	H. K. Wickramasinghe
H.L.	Helge Larsen
H.La.	Harry Lass
H.Le.	Henry Leland
H.Lem.	Harvey Lemont
H.Li.	Helmut Lieth
H.Lip.	Harry Lipner
H.L.B.	H. L. Beggs
H.L.Ba.	Harry L. Baldwin, Jr.
H.L.Bo.	Harry L. Bowman
H.L.C.	Helen L. Cannon
H.L.D.	Herbert L. Dupont
H.L.F.	Harold L. Friedman
H.L.G.	Henry L. Gabelnick
H.L.G.S.	Henry L. G. Stroyan
H.L.H.	Howard L. Hamilton
H.L.He.	Harry L. Helms
H.L.Ho.	Harold L. Hoffman
H.L.J.	Hans L. Jensen
H.L.K.	Henry L. Kinnier
H.L.O.	H. L. Ostergaard
H.L.-Sc.	Hilary Lappin-Scott
H.L.Sh.	Harry L. Shipman
H.L.St.	Herbert L. Stahnke
H.M.	Herbert Maisel
H.Ma.	Henry Margenau
H.Mar.	Harvey Marshak
H.Mo.	Hugh Morrow, III
H.M.B.	Harris M. Burte
H.M.F.	Harry M. Freeman
H.M.G.	Hyatt M. Gibbs
H.M.H.	Howard M. Hickman
H.M.J.	H. Mark Johnston
H.M.M.	H. M. Munger
H.M.S.	Herman M. Schwartz
H.M.T.	Howard M. Temin
H.M.Ts.	Henry M. Tsuchiya
H.N.P.	Henry N. Pollack
H.N.W.	Henry N. Wagner, Jr.
H.O.H.	Harlyn O. Halverson
H.P.B.	Harland P. Banks
H.P.H.	Harry P. Hale
H.-P.S.	Hans-Peter Schultze
H.P.T.	Henry P. Treffers
H.P.W.	Hilda P. Willett
H.P.Yu.	Horace P. Yuen
H.R.	Hans Rademacher
H.Ra.	Hans Ramberg
H.Ras.	Howard Rasmussen
H.Ri.	Herbert Riehl
H.R.B.	Howard R. Batchelder
H.R.C.	Hollis R. Cooley
H.R.J.	Howard R. Jory
H.R.L.	Hans R. Lindner
H.S.	Hymin Shapiro
H.Sa.	Helmut Sauer
H.Si.	Herschel Sidransky
H.Sm.	Harry Smith
H.So.	Harry Sohon
H.S.A.	H. S. Aldwinckle
H.S.B.	Howard S. Bean
H.S.Bl.	Harold S. Black
H.S.Br.	Harry S. Brigham
H.S.H.	Helen S. Hogg
H.S.I.	Herbert S. Isbin
H.S.K.	Hugh S. Knowles
H.S.L.	Harry S. Ladd
H.S.La.	Hugh S. Landes
H.S.M.C.	H. S. M. Coxeter
H.S.P.	Harvey S. Price
H.S.S.	H. Subak-Sharpe
H.T.	Helen Tappan
H.Ta.	Henry Taube
H.T.H.	Henry T. Harrison
H.T.Ha.	Hudson T. Hartman
H.T.N.	H. Thomas Norris
H.V.	Howard Vollum
H.V.B.	Henry V. Borst
H.V.B.K.	Hibberd V. B. Kline, Jr.
H.V.C.	Homer V. Craig
H.W.	Herbert Wells
H.Wh.	Harry Wheeler
H.Wi.	Hugh Wilcox
H.Win.	Herman Winick
H.W.B.	Horace W. Babcock
H.W.Be.	H. Wayne Beaty
H.W.Bl.	H. Weston Blaser
H.W.Br.	Henry W. Brandli
H.W.Br.	Harold W. Brown
H.W.C.	Howard W. Campbell
H.W.F.	Henry W. Fischer
H.W.K.	Harold W. Keller
H.W.Kr.	Harold W. Kroto
H.W.M.	Hans W. Meissner
H.W.Me.	Henry W. Menard, Jr.
H.W.Mo.	Harland W. Mossman
H.W.O.	H. W. Oliver
H.W.P.	Hermann W. Pfefferkorn
H.W.Pi.	Hal W. Pilgrim
H.W.R.	H. W. Ritchey
H.W.Ru.	Howard W. Russell
H.W.S.	Horace W. Stunkard
H.W.W.	Howard W. Wainwright
H.Y.	Howard Yokelson
H.Y.F.	Hsu Y. Fan
H.Y.M.	Herbert Y. Meltzer
H.Z.	Hubert Ziegler
H.Zi.	Harold Zirin

I

I.A.B.	Irving A. Breger
I.A.L.	I. A. Lesk
I.A.S.	Ivan A. Sellin
I.B.	Ivan Barnes
I.Ba.	Itzhak Bars
I.Bl.	Ira Block
I.D.	Ingrid Daubechies
I.D.A.J.	Ivan D. A. Johnston
I.D.H.	Ian D. Hutcheon
I.E.L.	I. E. Levine
I.E.Li.	Irvin E. Liener
I.E.O.	I. Edgar Odom
I.F.	Irving Fox
I.Fr.	Irwin Fridovich
I.F.A.	Isabel F. Ahlgren
I.F.K.	Isaac F. Kinnard
I.G.	Isidore Gersh
I.Gi.	Ian Gibbons
I.Go.	Ivan Goodbody
I.Gor.	Isabella Gordon
I.H.G.	Irvin H. Gatzke
I.J.G.	Irving J. Good

I.K.F.	Irene K. Fischer
I.K.S.	Ivan K. Schuller
I.L.	Irving Lefkowitz
I.Le.	Ilse Lehiste
I.L.K.	Irving L. Kosow
I.L.T.	Ivan L. Tyler
I.M.B.	I. M. Bernstein
I.M.L.	I. Michael Lerner
I.M.T.	Ian M. Tattersall
I.N.L.	Ira N. Levine
I.P.	Igor Paul
I.P.K.	Ivan P. Kaminow
I.P.L.	I. Paul Lew
I.P.T.	Irwin P. Ting
I.R.	Irvin Rock
I.S.	Irving Sheft
I.St.	Isles Strachan
I.S.G.	Irving S. Goldstein
I.Z.	Israel Zelitch
I.Zl.	I. Zlotnik

J

J.A.	Jaime Amorocho
J.As.	James Aston
J.A.A.	John A. Allen
J.A.B.	J. A. Bolt
J.Ba.	John Bally
J.A.Br.	Joseph A. Brink, Jr.
J.A.C.	Joseph A. Cestone
J.A.Cl.	John A. Clements
J.A.Co.	Jonathan A. Coddington
J.A.D.	James A. Donaldson
J.A.G.	James A. Graham
J.A.H.	John A. Harvey
J.A.Hr.	John A. Hrones
J.A.Hy.	J. Allen Hynek
J.A.L.	James A. Lane
J.A.M.	John A. Manson
J.A.Mo.	James A. Moore
J.A.N.	James A. Noble
J.A.O.	John A. Osterman
J.A.P.	Joseph A. Pask
J.A.Ph.	James A. Phillips
J.A.S.	J. Allen Scott
J.A.Sh.	John A. Shellenberger
J.A.Si.	Jack A. Simon
J.A.Sm.	John A. Smith
J.A.Sw.	John A. Swets
J.A.W.	John A. Wheeler
J.A.Wr.	James A. Wright
J.B.	Jack Bass
J.Be.	Jesse Beams
J.Ben.	Jesús Benito
J.Bi.	Joseph Bicknell
J.Bid.	Jesse Bidanset
J.Big.	Jacob Bigeleisen
J.B.B.	J. Bruce Beckwith
J.B.C.	J. B. Cheatham, Jr.
J.B.E.	James B. Evans
J.B.F.	Jack B. Fisher
J.B.Fr.	J. Bruce French
J.B.H.	Jacques B. Hadler
J.B.Ha.	Jeffrey B. Harborne
J.B.J.	J. B. Jennings
J.B.K.	J. B. Ketterson
J.B.Ka.	James B. Kaler
J.B.L.	James B. Lackey
J.B.Lo.	Joseph B. Long
J.B.P.	Jon B. Pangborn
J.B.Po.	James B. Pollack
J.B.S.	John B. Scalzi
J.B.T.	Joyce B. Turner
J.B.V.S.	John B. Vander Sande
J.B.W.	John B. Woodward
J.C.	Joseph Callaway
J.Ca.	John Catchpole
J.Ch.	James Christensen
J.Cha.	Julia Chase
J.Cl.	John Clarke
J.Co.	Jerome Cohen
J.C.B.	John C. Brandt
J.C.Bu.	John C. Burgess
J.C.C.	John C. Couch
J.C.D.	John C. Drake
J.C.F.	John C. Ferm
J.C.G.	Jeffery C. Gibeling
J.C.L.	John C. Lane
J.C.M.	Julian C. Miller
J.C.Mo.	John C. Moore
J.C.N.	John C. Niedermair
J.C.O.	J. C. Overley
J.C.P.	John C. Polking
J.C.R.	John C. Reilly, Jr.
J.C.Ri.	J. C. Ritchie
J.C.S.	John C. Schelleng
J.C.Sc.	Jack C. Schultz
J.C.Sm.	Julian C. Smith
J.C.Sp.	John C. Spindler
J.C.St.	James C. Sternberg
J.C.W.	James C. Warf
J.C.Wy.	James C. Wyant
J.D.	Jack Davis
J.D.A.	John D. Anderson
J.D.B.	J. D. Bjorken
J.D.Br.	John D. Briggs
J.D.Bu.	James D. Burke
J.D.C.	J. Donald Capra
J.D.Co.	James D. Cobine
J.D.G.	J. D. Gannon
J.D.H.	J. Donald Harris
J.D.He.	Julius D. Heldman
J.D.J.	James D. Johnston
J.D.Jo.	Jon D. Johnson
J.D.L.	J. Dennis Lawrence
J.DeL.	Jules de Launay
J.D.M.	J. Derral Mulholland
J.D.Ma.	J. D. Maynard
J.D.P.	James D. Palmer
J.D.R.	James D. Reilly
J.D.Ro.	J. David Robertson
J.D.S.	James D. Schoeffler
J.D.Si.	John D. Singleton
J.E.	John Ehrlich
J.Ei.	John Einset
J.El.	Joseph Elgomayel
J.Ev.	John Everetts, Jr.
J.Ew.	John Ewing
J.E.A.	James E. Anders, Sr.
J.E.B.	James E. Bayfield
J.E.Bi.	John E. Biegel
J.E.Bl.	John E. Blair
J.E.Bla.	James E. Blankenship
J.E.G.	John E. Gibson
J.E.Gr.	Jonathan E. Grindlay
J.E.Gu.	James E. Gunckel
J.E.H.	John E. Haines
J.E.Ho.	J. Edgar Hoover
J.E.J.	J. Egif Juliussen
J.E.L.	John E. Lamar
J.E.Lo.	James E. Loughlin
J.E.M.	Joseph E. Marian
J.E.Ma.	James E. May
J.E.McC.	John E. McCarty
J.E.McD.	Joseph E. McDade
J.E.O'L.	John E. O'Leary
J.E.R.	Johannes E. Rijnsdorp
J.E.S.	James E. Schindler
J.E.Sc.	John E. Scott, Jr.
J.E.St.	John E. Stevens
J.E.U.	John E. Ullmann
J.E.We.	J. E. Webb
J.E.Wi.	Judith E. Winston
J.E.Wo.	James E. Woodall
J.F.	Jerry Feldman
J.Fo.	Joseph Ford
J.F.B.	Joseph F. Bunnett
J.F.C.	James F. Connors
J.F.Cr.	James F. Crow
J.F.Di.	J. F. Dillon, Jr.
J.F.E.	John F. Endicott
J.F.F.	James F. Ferry
J.F.H.	John F. Hahn
J.F.J.	Julian F. Johnson
J.F.Ja.	John F. Jarvis
J.F.L.	J. F. Loneragan
J.F.M.	José F. Maldonado-Moll
J.F.Ma.	John F. Marcelleti
J.F.Me.	John F. Meyer
J.F.Mo.	J. F. Moser, Jr.
J.F.McM.	John F. McMahon
J.F.McP.	Joseph F. McPartland
J.F.R.	John F. Reintjes
J.F.S.	James F. Scott
J.F.Se.	Joseph F. Sebald
J.F.T.	John F. Timoney
J.F.W.	John F. Wallace
J.F.We.	Jonathan F. Weil
J.F.Wo.	J. F. Worley
J.G.	John Garcia
J.Ge.	John Gergely
J.Go.	Joseph Golden
J.Gr.	Jerry Grey
J.G.B.	James B. Blencoe
J.G.C.	J. G. Carter
J.G.E.	J. Gordon Elder
J.G.F.	J. G. Futral
J.G.G.	Joseph G. Gall
J.G.H.	James G. Horsfall
J.G.J.	James G. Johnson
J.G.L.	John G. Liou
J.G.M.	John G. Malm
J.G.Ma.	John G. Maisey
J.G.P.	June G. Pattullo
J.G.R.	Jerome G. Rozen, Jr.
J.G.S.	James G. Speight
J.G.T.	John G. Trump
J.G.Tr.	John G. Truxal
J.G.W.	J. Guy Woodward
J.H.	John Handin
J.Ha.	John Hansen
J.Har.	J. Harris
J.Hare.	Jonathan Hare
J.Hart.	Jack Hartstein
J.Hu.	Jack Huebler

J.H.B.	**J. Harlen Bretz**
J.H.Bu.	**John H. Busser**
J.H.C.	**Joseph H. Camin**
J.H.Ca.	**James H. Calderwood**
J.H.Car.	**Jack H. Carlson**
J.H.Cl.	**J. Harold Clarke**
J.H.Cla.	**John H. Clarke**
J.H.Co.	**J. H. Collins**
J.H.Cr.	**John H. Crowe**
J.H.D.	**John H. Davis**
J.H.F.	**J. Homer Ferguson**
J.H.Fe.	**Janos H. Fendler**
J.H.Fi.	**Joseph H. Field**
J.H.H.	**Joseph H. Hamilton**
J.H.Ha.	**John H. Harley**
J.H.Hi.	**Joel H. Hildebrand**
J.H.L.	**Jean H. Langenheim**
J.H.M.	**John H. Miller**
J.H.N.	**John H. Nair**
J.H.Ne.	**James H. Nelson**
J.H.O.	**John H. Ostrom**
J.H.S.	**James H. Schulman**
J.H.Si.	**John H. Silliker**
J.H.S.B.	**J. H. S. Blaxter**
J.H.T.	**J. Herbert Taylor**
J.H.Ta.	**Joseph H. Taylor, Jr.**
J.H.W.	**J. H. Walker**
J.H.We.	**James H. Webber**
J.H.Z.	**John H. Zifcak**
J.I.K.	**Joseph I. Karash**
J.I.Y.	**John I. Yellott**
J.J.	**Jules Janick**
J.J.B.	**John J. Bleiweis**
J.J.C.	**Jeffrey J. Clarke**
J.J.Ca.	**James J. Carberry**
J.J.F.	**J. J. Fawcett**
J.J.G.	**Janos J. Gertler**
J.J.H.	**Joseph J. Horzepa**
J.H.Hr.	**J. J. Hromadik**
J.J.J.	**Julian J. Jaffe**
J.J.K.	**Joseph J. Katz**
J.J.L.	**J. J. Lagowski**
J.J.M.	**Joseph J. Moder**
J.J.Ma.	**John J. Marcholonis**
J.J.McD.	**John J. McDonald**
J.J.McK.	**John J. McKetta**
J.J.R.	**James J. Ryan**
J.J.S.	**John J. Schule, Jr.**
J.J.Sh.	**J. J. Sheehan**
J.J.Shi.	**John J. Shilling**
J.K.	**Jan Klein**
J.K.B.	**J. Kelly Beatty**
J.K.F.	**Jacob K. Frenkel**
J.K.G.	**John K. Galt**
J.K.H.	**Jack K. Holmes**
J.K.Ha.	**Janine K. Hasey**
J.K.L.	**James K. Lewis**
J.K.-M.	**June Kan-Mitchell**
J.K.M.P.	**John K. M. Pryke**
J.K.R.	**J. K. Roberts**
J.K.S.	**James K. Skilling**
J.K.T.	**J. Kerry Thomas**
J.L.	**Joseph Lach**
J.Le.	**Jonathan Leiby**
J.Lei.	**Joseph Lein**
J.Ly.	**John Lyman**
J.L.B.	**J. Laurens Barnard**
J.L.Ba.	**Jeffrey L. Bada**
J.L.Bl.	**J. Lewis Blackburn**
J.L.D.	**Joseph L. Doob**
J.L.G.	**John L. Gardiner**
J.L.Gr.	**Jesse L. Greenstein**
J.L.Gro.	**Jonathan L. Gross**
J.L.G.F.P.	**James L. G. FitzPatrick**
J.L.H.	**J. L. Harley**
J.L.J.	**Julian L. Jenkins**
J.L.K.	**J. Laurence Kulp**
J.L.Ka.	**Jack L. Katz**
J.L.L.	**Jack L. Lambert**
J.L.La.	**J. L. Larimer**
J.L.Lo.	**John L. Loeb**
J.L.M.	**John L. Margrave**
J.L.Me.	**Joseph L. Melnick**
J.L.McG.	**James L. McGaugh**
J.L.McH.	**J. L. McHugh**
J.L.R.	**James L. Rau**
J.L.Re.	**Joseph L. Reid**
J.L.St.	**Jacob L. Stokes**
J.L.Sta.	**J. L. Stanford**
J.L.Sto.	**Jeffrey L. Stott**
J.L.Str.	**Jack L. Strominger**
J.L.T.	**J. L. Turk**
J.L.T.W.	**John L. T. Waugh**
J.L.W.	**Joseph L. Walsh**
J.L.We.	**James L. Weaver**
J.L.Wi.	**James L. Wilson**
J.Ma.	**Jack Major**
J.Mar.	**John Markus**
J.Mau.	**John Mauchline**
J.Me.	**Jane Meinklejohn**
J.Mei.	**Jerome Meisel**
J.Mein.	**J. Meinwald**
J.Meix.	**Josef Meixner**
J.Men.	**Joshua Menkes**
J.Mes.	**John Meszar**
J.Mi.	**Jacob Millman**
J.M.A.	**John M. Armentrout**
J.M.A.D.	**J. M. A. Danby**
J.M.B.	**John M. Bird**
J.M.Bi.	**Jeannette M. Birmingham**
J.M.Bo.	**James M. Bobbitt**
J.M.C.	**John M. Carroll**
J.M.Ch.	**John M. Christie**
J.M.D.	**James M. Daniels**
J.M.Di.	**James M. Dille**
J.-M.F.	**Jean-Marie Franc**
J.Mi.F.	**J. Michael Fritsch**
J.M.H.	**John M. Hunt**
J.M.K.	**John M. Kays**
J.M.Ke.	**Joseph M. Keller**
J.M.Ky.	**John M. Kyle**
J.McK.	**James McKenzie**
J.M.L.	**Jack M. Linthicum**
J.M.M.	**James M. Manning**
J.M.MacD.	**J. M. MacDonald**
J.M.Mi.	**J. Murrary Mitchell, Jr.**
J.M.Mit.	**J. M. Mitchison**
J.M.McK.	**James M. McKenzie**
J.M.O.	**Jane M. Oppenheimer**
J.M.P.	**Jay M. Pasachoff**
J.M.S.	**Jay M. Savage**
J.M.T.	**John M. Teal**
J.M.To.	**Julian M. Tobias**
J.M.W.	**J. Marvin Weller**
J.M.Wi.	**Jack M. Williams**
J.M.Wo.	**John M. Woods**
J.M.Wor.	**John M. Woram**
J.N.	**Jerome Namias**
J.N.B.	**John N. Bahcall**
J.N.C.	**John N. Couch**
J.N.L.	**James N. Little**
J.N.Lu.	**James N. Luthin**
J.N.P.	**Joseph N. Payne**
J.O.B.	**J. Oscar Blew, Jr.**
J.O.C.	**John O. Corliss**
J.O.Cu.	**Joseph O. Culbertson**
J.O.E.	**John O. Edwards**
J.O.H.	**J. O. Hirschfelder**
J.O.K.	**John O. Kraehenbuehl**
J.O.S.	**J. O. Scanlan**
J.P.	**John Polking**
J.Pe.	**Jeff Perkins**
J.Po.	**John Postgate**
J.P.D.H.	**J. P. Den Hartog**
J.P.F.	**J. Paul Fitzsimmons**
J.P.Fr.	**Jeremiah P. Freeman**
J.P.G.	**J. P. Gerrity**
J.P.Go.	**J. P. Gordon**
J.P.H.	**John P. Hagen**
J.P.O'N.	**Jerome P. O'Neill, Jr.**
J.P.P.	**Joseph P. Porter**
J.P.S.	**John P. Scott**
J.P.T.	**John P. Tully**
J.P.V.E.	**James P. Van Etten**
J.R.	**John Rodgers**
J.Re.	**John Reeve**
J.R.A.	**John R. Anderson**
J.R.-An.	**Jaime Ramirez-Angulo**
J.R.Ap.	**John R. Apen**
J.R.B.	**James R. Bamburg**
J.R.Bo.	**James R. Bolton**
J.R.Bov.	**Joseph R. Bove**
J.R.Br.	**James R. Bright**
J.R.C.B.	**J. R. Casey Bralla**
J.R.D.	**James R. Daniel**
J.R.D.McC.	**J. R. D. McCormick**
J.R.F.	**J. R. Fulks**
J.R.H.	**James R. Holton**
J.R.Ha.	**Jeffrey R. Hazel**
J.R.Hu.	**John R. Huizenga**
J.R.M.	**Joan R. Marsden**
J.R.O'C.	**Joseph R. O'Connor**
J.R.P.R.	**June R. P. Ross**
J.R.R.	**James R. Rae**
J.R.Ra.	**John R. Raper**
J.R.S.	**James R. Sanford**
J.R.Se.	**John R. Sellars**
J.R.See.	**John Richard Seed**
J.R.Sm.	**John R. Smythies**
J.R.Su.	**J. R. Sutherland**
J.R.S.F.	**J. R. S. Fincham**
J.R.T.	**Jack R. Thompson**
J.R.Ta.	**Jane R. Taylor**
J.R.V.W.	**John R. Van Wazer**
J.R.Z.	**John R. Zimmerman**
J.S.	**Julius Schachter**
J.Sch.	**Jack Schubert**
J.Se.	**Josephine Semmes**
J.Sh.	**John Shortreed**
J.Si.	**Joseph Silk**
J.Sil.	**Joseph Silverman**
J.Sim.	**Joanne Simpson**
J.Simo.	**John Simonsen**
J.Sp.	**James Sprinkle**
J.Su.	**John Sullivan**
J.Sup.	**John Suppe**
J.S.B.	**John S. Belrose**
J.S.Br.	**Jerome S. Bruner**
J.S.F.	**James S. Frame**
J.S.Fe.	**Jay S. Fein**

J.S.Fr. **James S. Fritz**
J.S.G. **Joseph S. Gots**
J.S.Gr. **Jack S. Greenberg**
J.St.J. **Joseph St. Jean, Jr.**
J.S.L. **J. S. Lindsey**
J.S.McC. **Jack S. McCormick**
J.S.Mo. **Jeffrey S. Moore**
J.S.O. **Jerry S. Olson**
J.S.R. **John S. Ryland**
J.S.S. **Jeffry S. Shaw**
J.S.V. **James S. Vanick**
J.T. **Juergen Tonndorf**
J.T.G. **Joseph T. Gregory**
J.T.H. **Jens T. Holm**
J.T.K. **John T. Kuo**
J.T.L. **John T. Lucker**
J.T.O. **James T. Oris**
J.T.S. **James T. Staley**
J.T.St. **John T. Stock**
J.T.T. **Julius T. Tou**
J.T.W. **John T. Walkup**
J.V. **Joseph Veverka**
J.V.H. **Joseph V. Hollweg**
J.V.P. **Jesco Von Puttkamer**
J.W. **John Weaver**
J.Wi. **Jennifer Widom**
J.Wy. **John Wyke**
J.W.A. **James W. Angus**
J.W.B. **Joseph W. Barker**
J.W.Bl. **J. W. Blanton**
J.W.C. **J. W. Costerton**
J.W.E. **John W. Evans**
J.W.F. **J. W. Faller**
J.W.Fo. **Jackson W. Foster**
J.W.Fu. **James W. Fulton**
J.W.G. **John W. George**
J.W.Ge. **James W. Gewartowski**
J.W.Go. **John W. Gowen**
J.W.Goo. **Joseph W. Goodman**
J.W.H. **J. Woodland Hastings**
J.W.He. **Joel W. Hedgpeth**
J.W.J. **John W. James**
J.W.L. **John W. Lynn**
J.W.McC. **John W. McCain**
J.W.O. **John W. Olcott**
J.W.Ol. **J. W. Olness**
J.W.R. **James W. Rohlf**
J.W.S. **John Ward Smith**
J.W.St. **John W. Stewart**
J.W.W. **John W. Wentworth**

K

K.A. **Knut Aagaard**
K.Ar. **Kunitaka Arimura**
K.Arn. **Karl Arnstein**
K.At. **Kenji Atoda**
K.A.A. **Kenneth A. Arndt**
K.A.B. **Keith A. Browning**
K.A.E. **Kraft A. Ehricke**
K.A.Em. **Kerry A. Emanuel**
K.A.Er. **K. A. Erb**
K.A.McC. **Kathryn A. McCarthy**
K.A.P. **Keats A. Pullen, Jr.**
K.A.R. **Kenneth A. Roe**
K.A.S. **Kaj Aa. Strand**
K.B. **Klaus Brodersen**
K.B.C. **Kenneth B. Cumberland**
K.B.L. **Kenny B. Lipkowitz**
K.C.P. **Kenneth C. Parkes**
K.C.T. **Ker Clive Thomson**
K.D. **Keith Davies**
K.D.K. **Karl D. Kryter**
K.D.W. **K. D. Williamson**
K.E. **Katherine Esau**
K.E.L. **Kenneth E. Lassila**
K.E.M. **Kenneth E. Moyer**
K.E.S. **Karl E. Schaefer**
K.E.Sch. **K. E. Schoenherr**
K.F. **Kenneth Franzese**
K.F.K. **Karl F. Koopman**
K.F.M. **Karl F. Meyer**
K.G. **K. Grasshoff**
K.G.F. **Karen Grady Ford**
K.G.S. **Kenneth G. Stone**
K.H. **Kenneth Hunkins**
K.Hu. **Keith Hutchison**
K.H.M. **Kenneth H. Mann**
K.H.P. **Kenneth H. Purser**
K.J. **Kjell Johansen**
K.Jo. **Keith Johnston**
K.J.A. **Karl J. Astrom**
K.J.F. **Kenneth J. Frey**
K.J.H. **Katherine J. Hansen**
K.J.M. **Kenneth J. Morrison**
K.J.McN. **Kenneth J. McNamara**
K.J.S. **Kenneth J. Stuckas**
K.J.W. **Kimerly J. Wilcox**
K.Kr. **Karl Kristofferson**
K.Kri. **Kevin Krisciunas**
K.K.W. **K. K. Wang**
K.L. **Karl Lang**
K.Le. **Kornelius Lems**
K.Lo. **Kenneth Longmore**
K.L.D. **Kenneth L. Duke**
K.L.K. **K. L. Kliewer**
K.M.N. **Kenneth M. Noll**
K.N.O. **Kenneth N. Ogle**
K.N.-Z. **Klaus Napp-Zinn**
K.P.A. **Kenneth P. Able**
K.P.D. **Kenneth P. Davis**
K.P.S. **Klaus P. Schaal**
K.P.W. **K. Preston White, Jr.**
K.R. **Karl Rundman**
K.Ru. **Kathryn Ruoff**
K.R.D. **K. R. Dyer**
K.R.McC. **Ken R. McConnell**
K.R.P. **Keith R. Porter**
K.S. **Kai Siegbahn**
K.Sh. **Ken Shortman**
K.Si. **Kerry Sieh**
K.S.K. **Kaiser S. Kunz**
K.S.R. **Kenneth S. Rawson**
K.S.S. **Kenneth S. Suslick**
K.S.T. **Kip S. Thorne**
K.S.W. **Katherine S. Waldmann**
K.T. **Keith Thomson**
K.T.P. **Kevin T. Potts**
K.U.S. **Karl U. Smith**
K.V. **Kálman Vánky**
K.V.K. **Karl V. Krombein**
K.V.M. **Kenneth V. Manning**
K.W. **Klaus Wyrtki**
K.Wi. **Kevin Winn**
K.W.B. **Kenneth W. Bagnall**
K.W.F. **Karl W. Flessa**
K.W.H. **Keith W. Hipel**
K.W.J. **K. W. Johnson**
K.W.K. **Kerry W. Kemper**
K.W.L. **Keith W. Lawrence**
K.W.Li. **Kirk W. Lindstrom**
K.W.P. **Kenneth W. Perkins**
K.Y.T. **K. Y. Tang**
K.Z.M. **Karl Z. Morgan**

L

L.A. **Lester Andrews**
L.Ap. **L. Apker**
L.A.B. **Leslie A. Bryan**
L.A.C. **Louis A. Chiodo**
L.A.G. **Leslie A. Guildner**
L.A.K. **Lee A. Kilgore**
L.A.Ka. **Leonard A. Katz**
L.A.R. **Lorrin A. Riggs**
L.B. **Lane Barksdale**
L.Be. **Louis Berkofsky**
L.Ber. **Lewis Berner**
L.Bl. **Leo Blitz**
L.Br. **Louis Brand**
L.B.A. **Leslie B. Arey**
L.B.C. **Leallyn B. Clapp**
L.B.E. **Louis B. Early**
L.B.H. **Louis B. Holdeman**
L.B.Ho. **Lipke B. Holthuis**
L.B.L. **Luna B. Leopold**
L.B.Lo. **Lewis B. Lockwood**
L.B.Lu. **Lee B. Lusted**
L.B.S. **Louis B. Slichter**
L.B.T. **Louise B. Tyrer**
L.C. **Lionel Crawford**
L.C.A. **Leland C. Allen**
L.C.B. **L. C. Biedenharn**
L.C.G. **Louis C. Green**
L.C.Gi. **Larry C. Giunipero**
L.C.P. **Leanne C. Pitchford**
L.C.R. **Lemuel C. Robertson**
L.D. **Loyal Durand, III**
L.D.F. **Lloyd D. Fosdick**
L.D.H. **Leonor D. Haley**
L.D.K. **Lewis D. Kaplan**
L.D.M. **Lawrence D. Miles**
L.D.R. **Louis D. Roberts**
L.E. **Leo Esaki**
L.E.C. **Lee E. Cuckler**
L.E.K. **Lawrence E. Kinsler**
L.E.M. **Lawrence E. Marks**
L.E.Mo. **Leonard E. Mortenson**
L.E.R. **Leo E. Reichert, Jr.**
L.E.T. **Lawrence E. Tannas, Jr.**
L.E.Th. **L. E. Thelemaque**
L.F. **Louis Feit**
L.Fe. **Louise Ferguson**
L.Fi. **L. Finkelstein**
L.F.A. **Lyle F. Albright**
L.F.C. **L. F. Cleveland**
L.F.D. **Lawrence F. Dahl**
L.F.H. **L. F. Hough**
L.F.Hu. **Lester F. Hubert**

L.F.P.	**L. Fletcher Prouty**
L.F.S.	**Lee F. Schuchardt**
L.G.	**Leonard Goland**
L.Go.	**Louis Gordon**
L.Gr.	**Lawrence Grossman**
L.Gro.	**Lee Grodzins**
L.G.H.	**Llewellyn G. Hoxton**
L.H.	**Leonard Hayflick**
L.H.A.	**L. H. Allen, Jr.**
L.H.Al.	**Lawrence H. Allen**
L.H.L.	**Lawrence H. Lattman**
L.H.MacD.	**Laurence H. MacDaniels**
L.H.P.	**L.H. Princen**
L.H.R.	**Louis H. Roddis, Jr.**
L.H.V.V.	**Lawrence H. Van Vlack**
L.H.Z.	**Leroy H. Zimmerman**
L.J.A.	**Leslie J. Audus**
L.J.B.	**Louis J. Battan**
L.J.Ba.	**Leonard J. Banaszak**
L.J.C.	**Lawrence J. Campbell**
L.J.Ca.	**Linda J. Cassens**
L.G.DeG.	**Leslie J. DeGroot**
L.J.E.	**L. J. Ehrenberger**
L.J.K.	**Leonard J. Koch**
L.J.O.	**Leonard J. Obery**
L.J.P.	**Louis J. Pignataro**
L.J.P.M.	**L. J. Patrick Muffler**
L.K.	**Louis Kaplan**
L.K.L.	**Low K. Lee**
L.L.	**Louis Landweber**
L.L.B.	**Leo L. Beranek**
L.L.D.	**Louis L. Dienes**
L.L.L.	**Leroy L. Langley**
L.L.N.	**Lewis L. Nettleton**
L.L.S.	**L. L. Sloss**
L.M.	**L. Madestau**
L.M.B.	**Lloyd M. Beidler**
L.M.Bl.	**Leonard M. Blumenthal**
L.M.C.	**Lucy M. Cranwell**
L.M.Co.	**Lisa M. Cohen**
L.M.G.	**Lawrence M. Graves**
L.M.H.	**Lawrence M. Halpern**
L.M.J.	**Leonard M. Joseph**
L.M.K.	**Lyman M. Kells**
L.M.Kr.	**Lawrence M. Krauss**
L.M.L.	**Leon M. Lederman**
L.M.N.	**Leland M. Nicolai**
L.M.O.	**Linda Moretti Ojemann**
L.M.Ol.	**Leonard M. Olmsted**
L.M.P.	**Lawrence M. Palmer**
L.M.R.	**Louis M. Roth**
L.M.S.	**Louisa M. Slowiaczek**
L.N.	**Leo Nedelsky**
L.N.B.	**Lewis N. Blair**
L.N.K.	**Laveen N. Kanal**
L.N.Kr.	**Lloyd N. Krause**
L.-N.L.	**Lin-Nan Lee**
L.N.McC.	**Leslie N. McClellan**
L.O.B.	**Lawrence O. Brockway**
L.P.C.	**Leo P. Clements**
L.P.E.	**Lewis P. Emerson**
L.P.H.	**Lloyd P. Hunter**
L.P.M.	**Laurence P. Madin**
L.P.R.	**Louis P. Reitz**
L.R.	**L. Rozeanu**
L.R.M.	**Lee R. Mahoney**
L.R.S.	**L. R. Shannon**
L.R.W.	**Lawrence R. Walker**
L.R.Wi.	**Leonard R. Wison**
L.S.	**Louis Schmerling**
L.Sm.	**Levering Smith**
L.Smi.	**Lloyd Smith**
L.S.B.	**L. S. Birks**
L.S.C.	**Leonard S. Cutler**
L.S.L.	**L. Sigfred Linderoth, Jr.**
L.S.McC.	**Leland S. McClung**
L.S.N.	**Leon S. Nergaard**
L.S.O.	**Lindsay S. Olive**
L.S.R.	**Lewis S. Ramsdell**
L.S.S.	**Lee S. Schroeder**
L.T.R.	**Leon T. Rosenberg**
L.V.B.	**Loyal V. Bewley**
L.V.V.	**Leigh Van Valen**
L.W.C.	**Loren W. Crow**
L.W.J.	**Lawrence W. Jones**

M

M.A.B.	**Morton A. Bell**
M.A.Br.	**M. A. Breazeale**
M.A.C.	**M. A. Chappell**
M.A.D.	**Michel A. Duguay**
M.A.H.	**Mel A. Hagood**
McA.H.H.	**McAllister H. Hull, Jr.**
M.A.K.	**Michael A. Kinch**
M.A.L.	**Max A. Lauffer**
M.A.M.	**Milton A. Miller**
M.A.R.	**Martin A. Rizack**
M.A.Ri.	**Mark A. Riddle**
M.A.S.	**Michael A. Saad**
M.A.W.	**Michael A. Walsh**
M.B.	**Mark Barton**
M.Be.	**Manson Benedict**
M.Bo.	**Mark Bothwell**
M.Boh.	**M. Böhm**
M.Br.	**Mead Bradner**
M.B.B.	**Michael B. Bever**
M.B.M.	**Mark B. Moffett**
M.B.Ma.	**M. Brian Maple**
M.B.McC.	**Mary B. McCann**
M.B.S.	**Mary Betty Stevens**
M.C.	**Michel Copel**
M.Ca.	**Michael Callaham**
M.Ch.	**Martha Christensen**
M.C.C.	**M. C. Chang**
M.C.G.	**Martin C. Gutzwiller**
McC.G.	**McChesney Goodall**
M.C.H.	**Malcolm C. Holtje**
M.C.L.	**Myron C. Ledbetter**
M.C.McK.	**Malcolm C. McKenna**
M.C.W.	**Mark C. Willingham**
M.D.	**Michael Doudoroff**
M.Di.	**Michael Dine**
M.Dr.	**Max Dresden**
M.D.A.	**Mark D. Ardema**
M.D.B.	**Mark D. Barton**
M.D.C.	**M. Donald Cave**
M.D.H.	**M. Duffield Harsh**
M.D.Ha.	**M. D. Hassialis**
M.D.L.	**Michael D. Lebowitz**
M.D.P.	**M. David Potter**
M.D.Pa.	**Michael D. Papagiannis**
M.D.Po.	**Marc D. Porter**
M.D.R.	**Marvin D. Rausch**
M.D.S.	**M. D. Sturge**
M.D.W.	**M. D. Weiser**
M.E.	**Merrill Eisenbud**
M.Ek.	**Marit Ek**
M.Ew.	**Maurice Ewing**
M.E.B.	**Milton E. Bailey**
M.E.H.	**Mason E. Hale, Jr.**
M.E.Hi.	**M. E. Hines**
M.E.J.	**Murray E. Jarvik**
M.E.L.	**Malcolm E. Lines**
M.E.M.	**Margaret E. Meyer**
M.E.P.D.	**Micheal E. P. Davis**
M.E.R.	**Mary E. Rawles**
M.E.Re.	**M. E. Reichmann**
M.E.Ri.	**Mary E. Rice**
M.E.T.	**Michael E. Tauber**
M.E.V.V.	**M. E. Van Valkenburg**
M.E.W.	**Michael E. Wright**
M.F.	**Milton Fingerman**
M.Fi.	**Maxwell Finland**
M.Fo.	**Michael Fowler**
M.F.B.	**Mark F. Bocko**
M.F.D.	**M. F. Doherty**
M.G.	**Martin Gibbs**
M.Go.	**Marjorie Goldfarb**
M.Gr.	**Martin Greenspan**
M.G.A.-C.	**M. G. Audley-Charles**
M.G.F.	**Mark G. Foster**
M.G.G.	**Max G. Gergel**
M.G.M.	**M. G. Mellon**
M.G.P.	**Michael G. Parsons**
M.G.Po.	**Maurice G. Pollard**
M.G.S.	**Michael G. Safonov**
M.G.T.	**Monica G. Turner**
M.H.	**Michael Haratunian**
M.Ho.	**Michael Horan**
M.H.B.	**Michael H. Bruno**
M.H.C.	**Marvin H. Carruthers**
M.J.A.	**Malcolm J. Abzug**
M.J.C.	**Michael J. Camp**
M.J.Co.	**Michael J. Coyne**
M.J.Cr.	**Malcolm J. Crocker**
M.J.Cru.	**Michael J. Cruickshank**
M.J.F.	**Mitchell J. Feigenbaum**
M.J.G.A.	**M. J. G. Appel**
M.J.H.	**Max J. Herzberger**
M.J.K.	**M. J. Kopac**
M.J.Kr.	**Mary Jeanne Kreek**
M.J.Ku.	**Michael J. Kurylo**
M.J.K.H.	**Michael J. K. Harper**
M.J.L.	**M. J. Lewis**
M.J.M.	**Merritt J. Murray**
M.J.Po.	**Margaret J. Polley**
M.J.Se.	**M. J. Selby**
M.J.Si.	**Michell J. Sienko**
M.J.St.	**Martin J. Steindler**
M.J.W.	**Marvin J. Weinstein**
M.K.	**Matti Krusius**
M.K.S.	**Morton K. Schwartz**
M.K.W.	**Michael K. Wilkinson**
M.L.	**Martin Lotz**
M.Lo.	**Maichael Lopiano**
M.L.B.	**Milton L. Blanc**
M.L.Be.	**Myron L. Bender**
M.L.Bo.	**Marci L. Bortman**
M.L.G.	**Marie L. Goodnight**
M.L.K.	**M. L. Knotek**
M.L.M.	**M. L. Meade**
M.L.N.	**Mark L. Nichols**
M.L.P.	**Martin L. Perl**
M.L.S.	**Murrell L. Salutsky**
M.L.Sp.	**M. Leroy Spearman**
M.M.	**Manley Mandel**

M.M.B.	M. M. Brooke
M.M.Bu.	Maurice M. Bursey
M.M.C.	M. M. Coombs
M.M.G.	Madan M. Gupta
M.M.M.	Manfred M. Mayer
M.M.Mi.	Maynard M. Miller
M.M.Mid.	M. Mark Midland
M.M.R.	Mark M. Rochkind
M.M.S.	Mel M. Schwartz
M.M.Se.	Machelle M. Seibel
M.M.-Y.	Murray Moo-Young
M.M.W.	Mary M. Walczak
M.N.K.	Milton N. Kabler
M.O'B.	Michael O'Brien
M.O.McW.	Michael O. McWilliams
M.P.	Michael Perch
M.Ph.	Montgomery Phister, Jr.
M.Pl.	Mirek Plavec
M.Po.	Michael Pope
M.P.B.	Marland P. Billings
M.P.Ba.	M. Pauline Baker
M.R.	Mario Rabinowitz
M.R.G.	Melvin R. George
M.R.Gu.	Mark R. Guidry
M.R.H.	Milton R. Hales
M.R.M.	Michael R. Mulvihill
M.R.R.	Murray R. Robinovitch
M.S.	Max Samter
M.Sa.	Murray Sargent, III
M.Sh.	Maurice Shamma
M.Si.	Marshall Sittig
M.Sn.	Martin Snelgrove
M.So.	Myron Solberg
M.Sou.	Mott Souders
M.St.	Martin Stiles
M.-S.C.	Mo-Shing Chen
M.S.F.	Millard S. Firebaugh
M.S.G.	Michael S. Gazzaniga
M.S.K.	Marilyn S. Kerr
M.S.Kn.	Morris S. Knebelman
M.S.M.	Max S. Marshall
M.S.Mi.	Malcolm S. Mitchell
M.T.	Morris Tager
M.Ti.	M. Tigner
M.Ts.	Minoru Tsutsui
M.T.H.	Michel T. Halbouty
M.V.B.	M. V. Brian
M.V.L.B.	Michael V. L. Bennett
M.V.N.	Milos V. Novotny
M.W.	Martin Walt
M.Wu.	M. Wurm
M.W.A.	M. W. Adams
M.W.F.	Merrill W. Foster
M.W.M.	Margaret W. Mayall
M.W.Mi.	Martin W. Miller
M.W.R.	Marvin W. Rowe
M.W.S.	Malcolm W. Sinclair
M.Y.	M. Yamaguchi
M.Ya.	Masayuki Yamane
M.Ye.	Marvin Yelles
M.Yo.	Meir Yoeli
M.Z.	Milton Zaitlin
M.Ze.	Michael Zeilik

N

N.A.	Nelle Ammons
N.A.B.	Neil A. Benfer
N.A.Bl.	Neal A. Blake
N.A.C.	Noel A. Clark
N.A.M.	N. A. Mitchison
N.B.	Norman Balabanian
N.Be.	Neil Berglund
N.Bo.	Nestor Bohonos
N.B.A.	Norman B. Akesson
N.B.R.	Norman B. Reilly
N.C.P.	Niels C. Pedersen
N.C.R.	Norman C. Rasmussen
N.DeC.	Nicholas De Claris
N.D.G.	Neville D. Grace
N.D.L.	Norman D. Levine
N.D.Z.	Norton D. Zinder
N.E.	Niles Eldredge
N.E.S.	Norman E. Steenrod
N.-F.B.	Nichole-Fr. Bolender
N.F.Bu.	N. F. Buchwald
N.F.R.	Norman Ramsey
N.F.S.	Norman F. Schneidewind
N.G.D.	Nancy G. Dengler
N.G.H.	Narain G. Hingorani
N.G.L.	N. Gary Lane
N.H.	Neville Hogan
N.Ho.	Nigel Holder
N.H.B.	Norman H. Boke
N.H.N.	Norman H. Nachtrieb
N.I.	Nakao Ishida
N.J.	Nathan Jacobson
N.Je.	Neil Jesperson
N.J.G.	Nicholas J. Giarman
N.J.H.	Nicholas J. Hoff
N.J.R.	Norman J. Rosenberg
N.J.S.	Nathan J. Smith
N.J.T.	Nicholas J. Turro
N.J.Y.	N. J. Young
N.K.	Norman Kharasch
N.Ko.	Noémie Koller
N.K.M.	N. Karle Mottet
N.L.	Ningyi Luo
N.L.C.L.	Naomi L. C. Luban
N.L.K.	N. L. Kusters
N.L.N.	Neil L. Norcross
N.M.	N. Mandava
N.Mo.	Nicolas Mokhoff
N.M.B.	Norbert M. Bikales
N.MacC.	Neil MacCoull
N.M.F.	Nelson M. Fraiman
N.N.Li.	Norman N. Li
N.N.P.	Norman N. Potter
N.O.	Norman Olson
N.Ot.	Norman Otto
N.P.	Norbert Pfennig
N.Pe.	Nasser Peyghambarian
N.P.S.	Nicholas P. Samios
N.R.	Nathaniel Robbins, Jr.
N.Ru.	Norman Runham
N.R.B.	Norman R. Bell
N.R.G.	N. R. Greening
N.R.K.	Noel R. Krieg
N.S.	Nelson Stein
N.Sh.	Nathan Sharon
N.S.B.	Norbert S. Baer
N.S.F.	Nelson S. Fisk
N.S.G.	Newell S. Gingrich
N.S.J.	Norman S. Jones
N.S.S.	Norman S. Stoloff
N.T.	Neil Turok
N.To.	Nicholas Toth
N.Ty.	Nathan Tyler
N.T.S.	Nelson T. Spraft, Jr.
N.U.	Norbert Untersteiner
N.U.R.	Natalie U. Roy
N.W.P.	Norman W. Patrick
N.W.R.	Norman W. Radforth

O

O.A.	Otto Appenzeller
O.A.L.	Oscar A. Lorenz
O.B.R.S.	Oliver B. R. Strimpel
O.C.F.	Orval C. French
O.D.R.	Oscar D. Ratnoff
O.E.L.	Otto E. Lowenstein
O.E.N.	Olin E. Nelsen
O.H.	Olga Hartman
O.K.M.	Oliver K. Manuel
O.L.B.	Orus L. Bennett
O.L.G.	Oluf L. Gamborg
O.N.A.	Oscar N. Allen
O.P.S.	Otto P. Strausz
O.R.A.	O. Roger Anderson
O.S.	O. Siddiqi
O.S.T.	Olive S. Tattersall
O.W.B.	Orlan W. Boston
O.W.G.	O. W. Greenberg
O.W.R.	Oscar W. Richards

P

P.A.	Peter Andrews
P.Ar.	Peter Armbruster
P.A.F.	Paul A. Fleury
P.A.G.	Paul A. Giguère
P.A.H.	Per A. Holst
P.A.L.	P. A. Lemke
P.A.M.	Peter A. Marks
P.A.Ma.	Preston A. Marx
P.A.McL.	Patsy A. McLaughlin
P.A.O.	Paul A. Obrist
P.A.S.	Paul A. Souder
P.A.V.	Paul A. Vestal
P.A.Va.	Peter A. Vandenbergh
P.A.W.	Patrick A. Wegner
P.B.B.	Paul B. Beeson
P.B.C.	Philip B. Cowles
P.B.H.	Peter B. Hukill
P.B.K.	Peter B. Kaufman
P.B.M.	Paul B. Moore
P.B.P.	P. Buford Price, Jr.
P.C.	Praveen Chaudhari
P.Cl.	Preston Cloud
P.C.B.	Peter C. Badgley
P.C.H.	Perry C. Holt
P.C.J.	Philip C. Johnson
P.C.P.	Philip C. Peters
P.C.Pa.	Peter C. Patton
P.C.S.	Peter C. Sandretto
P.C.Si.	Paul C. Silva

P.De.	**Paul Delahay**
P.Du.	**Paul Duby**
P.DeB.	**Paul De Bach**
P.D.S.	**P. Dwight Sherman, Jr.**
P.D.St.	**Perry D. Strausbaugh**
P.DeT.A.	**Paulo De T. Alvim**
P.D.V.M.	**Paul D. V. Manning**
P.D.W.	**P. D. Walton**
P.E.	**Paul Evenson**
P.E.B.	**Patrick E. Barry**
P.E.Bi.	**Pierre E. Biscaye**
P.E.Bl.	**Philip E. Bloomfield**
P.E.D.	**Pol E. Duwez**
P.E.F.	**Paul E. Fanta**
P.E.Fe.	**Pierre E. Ferrier**
P.E.H.	**Philip E. Hicks**
P.E.L.	**Paul E. Lepard**
P.Fe.	**Peter Felker**
P.Fr.	**Peter Frumhoff**
P.F.C.	**Philip F. Clapp**
P.F.H.	**Paul F. Hoffman**
P.F.K.	**Paulden F. Knowles**
P.F.M.	**Paul F. Maderson**
P.F.S.	**Philip F. Spaulding**
P.F.W.	**Perry F. Williams**
P.G.	**Philip Garman**
P.Gr.	**Peter Gray**
P.G.B.	**Peter G. Bergmann**
P.G.K.	**Peter G. Kappus**
P.G.Kl.	**Paul G. Klemens**
P.G.M.	**Peter G. Martin**
P.G.R.	**Paul G. Ragland**
P.G.Re.	**Placidus G. Reischmann**
P.G.S.	**Paul G. Smith**
P.H.	**Philip Hodge**
P.H.B.	**Paul H. Black**
P.H.C.	**Philip H. Cook**
P.H.E.	**Philip H. Enslow, Jr.**
P.H.E.M.	**Paul H. E. Meijer**
P.H.G.	**P. H. Groggins**
P.H.H.	**P. H. Heckel**
P.H.Ha.	**Paul H. Harvey**
P.H.Ho.	**Peter H. Homann**
P.H.O.	**Philip H. Osberg**
P.H.R.	**Paul H. Ribbe**
P.H.S.	**Peter H. Schur**
P.H.Sm.	**Philip H. Smith**
P.H.St.	**Paul H. Stelson**
P.H.T.	**Paul H. Tracy**
P.H.W.	**Paul H. Wender**
P.I.S.	**Peter I. Somlo**
P.J.B.	**Paul J. Bender**
P.J.Br.	**Peter J. Brofman**
P.J.D.	**Peter J. Davies**
P.J.K.	**Philip J. Klass**
P.J.R.	**Paul J. Roper**
P.J.W.	**Peter J. Walsh**
P.J.Wy.	**Peter J. Wyllie**
P.J.Z.	**Paul J. Zwerman**
P.K.	**Peter Karlson**
P.Kl.	**Paul Klieger**
P.Ku.	**Polykarp Kusch**
P.K.J.	**Peter K. Johnson**
P.K.S.	**P. K. Seidelmann**
P.K.V.	**Pramode K. Verma**
P.L.A.	**Philip L. Alger**
P.L.B.	**Paul L. Boisvert**
P.L.M.	**Peter L. Marshall**
P.M.	**Peter Mohr**
P.M.A.	**Paul M. Anderson**
P.McG.	**Preston McGrain**
P.M.H.	**Patrick M. Hurley**
P.M.K.	**Peter M. Koch**
P.M.M.R.	**Peter M. M. Rae**
P.M.R.	**Patricia M. Rodier**
P.M.S.	**Philip M. Stehle**
P.M.VanP.	**Peter M. VanPeteghem**
P.O.E.	**Patrick O. Egan**
P.P.V.A.	**Paul P. Van Arsdel, Jr.**
P.R.	**Philip Richardson**
P.Ra.	**Pierre Ramond**
P.R.A.	**Peter R. Almond**
P.R.B.	**Philip R. Brooks**
P.R.E.	**Paul R. Ehrlich**
P.R.F.	**Paul R. Fields**
P.R.G.	**Peter R. Griffiths**
P.R.N.	**Philip R. Nachtsheim**
P.R.V.	**Peter R. Vogt**
P.S.	**Peter Satir**
P.Sh.	**Peter Shaw**
P.S.B.	**Peter S. Barracca**
P.S.DeC.	**P. S. De Carli**
P.-S.S.	**Pill-Soon Song**
P.T.	**Paul Tabor**
P.To.	**Pearl Toy**
P.T.J.	**Phyllis T. Johnson**
P.T.M.	**Paul T. Magee**
P.V.	**Peter Villiger**
P.V.deK.	**Peter van de Kamp**
P.V.R.	**P. Venkat Rangan**
P.W.	**Paul Weaver**
P.Wo.	**Peter Wolf**
P.W.A.	**P. W. Atkins**
P.W.B.	**Philip W. Brandt**
P.W.Br.	**Percy W. Bridgman**
P.W.P.	**Peter W. Price**
P.W.R.	**Paul W. Richards**
P.W.S.	**Paul W. Schmidt**
P.W.Si.	**Philip W. Signor**
P.W.W.	**Pedro W. Wygodzinsky**
P.Z.	**Petr Zuman**

Q R

Q.V.W.	**Quentin Van Winkle**

R.A.	**Richard Askey**
R.Ad.	**Robert Ader**
R.Au.	**Robert Austrian**
R.A.A.	**Rudolph A. Abramovitch**
R.A.An.	**Richard A. Anthes**
R.A.B.	**Richard A. Bambach**
R.A.Be.	**Ross A. Beaumont**
R.A.Bo.	**Rush A. Bowman**
R.A.Bu.	**Ralph A. Burton**
R.A.C.	**Richard A. Chapman**
R.A.Cl.	**Robert A. Clifton**
R.A.Co.	**Richard A. Colonna**
R.A.Cr.	**Roy A. Crowson**
R.A.D.	**Richard A. Davis, Jr.**
R.A.De.	**Richard A. Deno**
R.A.D.W.	**R. A. D. Wentworth**
R.A.F.	**Robert A. Fisher**
R.A.Fi.	**Russell A. Fisher**
R.A.G.	**Richard A. Gould**
R.A.Gr.	**Robert A. Grimm**
R.A.H.	**Robert A. Hodgson**
R.A.Ho.	**Richard A. Hoffman**
R.A.Hu.	**Roland A. Hultsch**
R.A.J.	**R. Alan Jones**
R.A.K.	**Robert A. Kaplan**
R.A.Kr.	**Richard A. Kretchmer**
R.A.L.	**Roy A. Larson**
R.A.Li.	**Roy A. Lindberg**
R.A.M.	**R. A. Miller**
R.A.Mo.	**Roger A. Morse**
R.A.Mu.	**Richard A. Müller**
R.A.P.	**Robert A. Paterson**
R.A.Pe.	**Robert A. Penneman**
R.A.Pi.	**Robert A. Pierotti**
R.A.R.	**Richard A. Rohde**
R.A.S.	**Robert A. Shaw**
R.A.Sp.	**Robert A. Spurgin**
R.A.St.	**Ruben A. Stiron**
R.A.T.	**Richard A. Tapia**
R.A.V.	**Romeo A. Vidone**
R.A.W.	**Robert A. Whitehouse**
R.A.We.	**Robert A. Weller**
R.A.Web.	**Richard A. Webb**
R.B.	**Robert Bakish**
R.Ba.	**Robert Barer**
R.Be.	**Richard Bellman**
R.Ber.	**Robert Beringer**
R.Bet.	**Russell Betts**
R.Br.	**Richard Bradfield**
R.Bro.	**Rollins Brook**
R.Bu.	**Ron Burghard**
R.B.C.	**Richard B. Couch**
R.B.H.	**Rolla B. Hill**
R.B.He.	**Ronald B. Herberman**
R.B.L.	**R. Bruce Lindsay**
R.B.N.	**Richard B. Nelson**
R.B.Ni.	**Richard B. Nichols**
R.B.P.	**Robert B. Plat**
R.B.S.	**Richard B. Sessions**
R.C.	**Richard Comeau**
R.Co.	**Roland Contreras**
R.C.D.	**Richard C. Dorf**
R.C.Du.	**Roger C. Duffield**
R.C.E.	**Robert C. Edlerfield**
R.C.F.	**Reynold C. Fuson**
R.C.G.	**Robert C. Gunning**
R.C.Go.	**R. C. Gonzalez**
R.C.K.	**R. C. Kellison**
R.C.L.	**Richard C. Lord**
R.C.M.	**Roy C. Mast**
R.C.MacD.	**Robert C. McDonald**
R.C.Mo.	**Raymond C. Moore**
R.C.N.	**R. C. Newton**
R.C.P.	**Roy C. Page**
R.C.Pu.	**Russell C. Putnam**
R.C.S.	**Randy C. Stauffer**
R.C.T.	**Richard C. Tilton**
R.C.Tr.	**Robert C. Truax**
R.D.A.	**Raymond D. Adams**
R.D.Al.	**Robert D. Allen**
R.D.An.	**Ronald D. Andreas**
R.D.Anw.	**Robert D. Anwyl**
R.D.B.	**Robert D. Barnes**
R.D.Bl.	**Roger D. Blandford**
R.D.C.	**Richard D. Cadle**
R.D.Ch.	**Robert D. Chapman**

R.D.Che. **Robert D. Chellis**
R.D.Co. **Robert D. Compton**
R.D.D. **Raymond D. Douglass**
R.D.E. **Robley D. Evans**
R.D'E.A. **Robert d'E. Atkinson**
R.DelF. **Raymond Del Fava**
R.D.H. **Rollin D. Hotchkiss**
R.D.K. **Roger D. Kempers**
R.D.M. **Reginald D. Manwell**
R.D.Mu. **R. D. Mushlitz**
R.D.McC. **R. D. McCright**
R.D.McI. **Richard D. McIver**
R.D.P. **Richard D. Purchon**
R.D.Pi. **Roger D. Pience**
R.D.R. **R. Dale Reed**
R.D.Ru. **Rogers D. Rusk**
R.D.T. **Ruth D. Turner**
R.D.W. **Robert D. Waldron**
R.E. **Richard Estes**
R.E.A. **Robert E. Apfel**
R.E.B. **Ralph E. Bennett**
R.E.Be. **Royce E. Beckett**
R.E.Bo. **Ray E. Bolz**
R.E.C. **Richard E. Chipkin**
R.E.Cr. **Ralph E. Crabill, Jr.**
R.E.DeM. **Robert E. DeMar**
R.E.F. **Richard E. Faw**
R.E.Fu. **Ralph E. Fuhrman**
R.E.G. **Robert E. Green, Jr.**
R.E.Gr. **Ralph E. Grim**
R.E.K. **R. E. Kallio**
R.E.Ke. **Richard E. Keating**
R.E.Ku. **R. E. Kuhn**
R.E.L. **Robert E. Lund**
R.E.Le. **Richard E. Leach**
R.E.M. **Roy Earl Morse**
R.E.McC. **Richard E. McCrosky**
R.E.P. **Raymond E. Peck**
R.E.R. **Ralph E. Ricksecker**
R.E.R.-H. **Robert E. Reed-Hill**
R.E.S. **R. E. Seaman**
R.E.Sh. **Robert E. Sheriff**
R.E.Sp. **Raymond E. Sprenkle**
R.E.T. **R. E. Taylor**
R.E.Tr. **Robert E. Treybal**
R.E.W. **R. E. Walstedt**
R.E.Wy. **Richard E. Wyman**
R.E.Z. **Rodger E. Ziemer**
R.F.B. **Robert F. Black**
R.F.Bu. **Rointan F. Bunshah**
R.F.Bun. **Robert F. Buntschuh**
R.F.Bus. **R. Frank Busby**
R.F.C. **Richard F. Chandler**
R.F.Cl. **Richard F. Clark**
R.F.D. **Rowland F. Davis**
R.F.Di. **Robert F. Dill**
R.F.F. **Richard F. Fenske**
R.F.Fl. **Richard F. Flint**
R.F.Fr. **Raymond F. Fremed**
R.F.G. **Robert F. Goddu**
R.F.H. **R. F. Horsnail**
R.F.J. **Robert F. Jones**
R.F.L. **Robert F. Legget**
R.F.N. **Ross F. Nigrelli**
R.F.P. **Richard F. Post**
R.F.R. **Reed F. Riley**
R.F.S.H. **R. F. S. Hearmon**
R.G. **Raymond Giese**
R.Ga. **Ronald Gandelman**
R.Gl. **Richard Glicksman**
R.Go. **Ronald Gold**
R.Gom. **Robert Gomer**
R.G.B. **Robert B. Benedict**
R.G.Bo. **Richard G. Bowman**
R.G.Bu. **Roger G. Burns**
R.G.C. **Robert G. Carroll, Jr.**
R.G.G. **R. G. Goodrich**
R.G.Gr. **R. G. Grogan**
R.G.H. **Raymond G. Herb**
R.G.J. **Roscoe J. Jackson, II**
R.G.Jo. **Robert G. Joppa**
R.G.K. **Richard G. Klein**
R.G.L. **Robert G. Larsen**
R.G.N. **Robert G. Neuhauser**
R.G.Ne. **Roger G. Newton**
R.G.P. **Robert G. Parr**
R.G.Pe. **Ralph G. Pearson**
R.G.Pr. **Ronald G. Prinn**
R.G.R. **R. G. Roble**
R.G.W. **Richard G. Wiegert**
R.G.Wi. **Rolf G. Winter**
R.G.Z. **Richard G. Zweifel**
R.H. **Richard Hazen**
Ri.H. **Richard Hong**
R.Ha. **Ronald Hazen**
R.Hac. **Ralph Hackl**
R.Hi. **Ron Hilliard**
R.Ho. **Robert Hofstadter**
R.Hog. **Robert Hogan**
R.Hol. **Robert Holt**
R.H.A. **Ross H. Arnett, Jr.**
R.H.B. **Robert H. Biggs**
R.H.Bi. **R. H. Bing**
R.H.Bo. **Richard H. Boyd**
R.H.Bu. **Richard H. Bube**
R.H.C. **Robert H. Cannon, Jr.**
R.H.Cl. **Richard H. Clarke**
R.H.D. **Richard H. Dalitz**
R.H.De. **Robert H. Denison**
R.H.Do. **Robert H. Dodds**
R.H.Dot. **Robert H. Dott, Jr.**
R.H.E. **Richard H. Eyde**
R.H.F.H. **R. H. F. Hunter**
R.H.G. **Roy H. Gigg**
R.H.Go. **Roland H. Good**
R.H.Goo. **Richard H. Goodwin**
R.H.H. **Reginald H. Haskins**
R.H.He. **Rolfe H. Herber**
R.H.J. **Richard H. Jahns**
R.H.K. **R. H. Kaufmann**
R.H.L. **Ralph H. Luebbers**
R.H.McC. **Richard H. McClelland**
R.H.N. **R. H. Noble**
R.H.P. **R. H. Pearce**
R.H.R. **Robert H. Rownd**
R.H.S. **Ralph H. Sharpe**
R.H.T. **Richard H. Tedford**
R.H.W. **Robert H. Wentorf, Jr.**
R.H.Wh. **Robert H. Whittaker**
R.I.B. **Robert I. Bernstein**
R.I.D. **Ralph I. Dorfman**
R.I.F. **Richard I. Felver**
R.I.H. **R. I. Hamilton**
R.I.Ha. **Robert I. Harker**
R.I.MacD. **Ruby I. MacDonald**
R.I.S. **Robert I. Stirton**
R.I.T. **Robert I. Tilling**
R.J.A. **Ronald J. Adler**
R.J.B. **Robert J. Bryan**
R.J.Ba. **R. J. Bandoni**
R.J.C. **Richard J. Campana**
R.J.Ch. **Roger J. Chaffin**
R.J.Co. **R. J. Collier**
R.J.Cu. **Roger J. Cuffey**
R.J.D. **Raymond J. Deland**
R.J.Do. **R. J. Doviak**
R.J.F. **Robert J. Foster**
R.J.G. **Raymond J. Gagne**
R.J.Gi. **Ronald J. Gillespie**
R.J.H. **Richard J. Higgins**
R.J.L. **R. J. Lowin**
R.J.M. **Roelof J. Meijer**
R.J.Me. **Robert J. Menzies**
R.J.Mo. **R. J. Motekaitis**
R.J.Mu. **R. J. Murgatroyd**
R.J.N. **Robert J. Naumann**
R.J.Ne. **Raymond J. Nelson**
R.J.R. **Rupert J. Riedel**
R.J.S. **R. J. Stephenson**
R.J.T. **Robert J. Thorn**
R.J.Ta. **Robert J. Taylor**
R.J.Tr. **Robert J. Tracy**
R.J.W. **Robert J. Weaver**
R.J.Wi. **Rodger J. Williams**
R.K. **Roger King**
R.Ko. **Richard Koral**
R.Kr. **Robert Kreilik**
R.Krz. **Roman Krzysztofowicz**
R.K.A. **Robert K. Adair**
R.K.Ba. **Robert K. Barnes**
R.K.C. **Richard K. Cook**
R.K.D. **R. K. Downey**
R.K.E. **Russell K. Edwards**
R.K.G. **Robert K. Godfrey**
R.K.H. **R. K. Horst**
R.K.L. **Ronald K. Linde**
R.K.Li. **Ray K. Linsley**
R.K.M. **Raman K. Mehra**
R.K.MacC. **Robert K. MacCrone**
R.K.Mo. **Richard K. Moore**
R.K.Mu. **Roger K. Murray, Jr.**
R.K.S. **Robert K. Soberman**
R.K.So. **R. K. Soost**
R.L. **Rolf Landshoff**
R.Le. **Roberto Lee**
R.Len. **Richard Lennert**
R.Lev. **Rachmiel Levine**
R.Lh. **Roger Lhermitte**
R.L.A. **Richard L. Armstrong**
R.L.An. **Robert L. Anderson**
R.L.At. **Ronald L. Atwood**
R.L.B. **Robert L. Bates**
R.L.Be. **Ronald L. Beatty**
R.L.Bi. **Raymond L. Bisplinghoff**
R.L.Br. **Robert L. Bradley, Jr.**
R.L.Bri. **Robert L. Brickley**
R.L.Bro. **Robert L. Brown**
R.L.Bu. **Robert L. Burwell**
R.L.C. **Robert L. Carroll**
R.L.Ch. **Roy L. Chenault**
R.L.D. **Robert L. Daugherty**
R.L.Du. **Raynor L. Duncombe**
R.L.F. **R. L. Fleischer**
R.L.Fi. **Robert L. Fisher**
R.L.Fr. **Ralph L. Freeman**
R.L.G. **Robert L. Gluckstern**
R.L.Ge. **Randall L. Geigel**

R.L.H.	**Roger L. Harned**
R.L.Ha.	**Roy L. Harrington**
R.L.Ho.	**Richard L. Hoffman**
R.L.Hu.	**Robert L. Hulbary**
R.L.J.	**Russell L. Jones**
R.L.K.	**Richard L. Koral**
R.LaM.	**Robert LaMotte**
R.L.Ma.	**Rolland L. Mays**
R.L.Mo.	**Richard L. Mower**
R.L.Moe	**Richard L. Moe**
R.L.N.	**Raymond L. Nace**
R.L.P.	**Robert L. Peaslee**
R.L.Pe.	**R. L. Peterson**
R.L.Pr.	**Robert L. Protell**
R.L.R.	**Robert L. Ramey**
R.L.S.	**Robert L. Scott**
R.L.Sw.	**R. L. Swanson**
R.L.T.	**Richard L. Tomasetti**
R.L.W.	**Ray L. Watterson**
R.L.Wh.	**Roy L. Whistler**
R.M.	**Roy Middleton**
R.M.A.	**Ronald M. Atlas**
R.Mi.	**Rodger Mitchell**
R.Mu.	**Richard Muther**
R.M.B.	**Reeve M. Bailey**
R.M.Ba.	**Richard M. Bateman**
R.M.Bo.	**Robert M. Boynton**
R.M.C.	**Raymond M. Cable**
R.M.F.	**Robert M. Finks**
R.M.Fo.	**Robert M. Fox**
R.M.H.	**Ralph M. Hardgrove**
R.M.K.	**Richard M. Klein**
R.M.Kl.	**Robert M. Kleinpell**
R.M.M.	**Richard M. Mitterer**
R.M.N.	**Robert M. Norris**
R.McN.A.	**R. McNeill Alexander**
R.M.No.	**Richard M. Noyes**
R.M.P.	**Robert M. Page**
R.M.Ph.	**Richard M. Phelan**
R.M.R.	**Roger M. Reeve**
R.M.S.	**Robert M. Sawyer**
R.M.Sch.	**Robert M. Schultz**
R.M.Sm.	**Raymond M. Smullyan**
R.M.So.	**Robert M. Sorensen**
R.M.St.	**Richard M. Stephenson**
R.M.T.	**Robb M. Thomson**
R.M.W.	**R. M. Walker**
R.M.Wa.	**Robert M. Warner**
R.M.Wi.	**Robert M. Wienski**
R.N.	**Ronald New**
R.Nu.	**Richard Nuccitelli**
R.N.B.	**Roy N. Barnett**
R.N.H.	**Rodney N. Hader**
R.N.M.	**Robert N. Mayall**
R.N.S.	**R. Norris Shreve**
R.O.	**Rufus Oldenburger**
R.O.R.	**Rodney O. Radke**
R.O'Ra.	**Ronan O'Rahilly**
R.O.Re.	**Robert O. Reid**
R.P.	**Robert Pecora**
R.Pa.	**Robert Paine**
R.Pr.	**Robert Priestley**
R.P.B.	**Richard P. Buck**
R.P.D.	**Robertson P. Dinsmore**
R.P.F.	**Roy P. Forster**
R.P.G.	**Robert P. Giblon**
R.P.H.	**Richard P. Hall**
R.P.Hi.	**Robert P. Higgins**
R.P.Hu.	**Ralph P. Hudson**
R.P.L.	**R. Paul Larsen**
R.P.Lu.	**Raymond P. Lutz**
R.P.S.J.	**R. P. S. Jeffries**
R.P.U.	**Ronald P. Uhlig**
R.P.V.	**Richard P. Voelker**
R.P.W.	**Robert P. Wagner**
R.P.Wh.	**Roy P. Whitney**
R.P.Wi.	**Ralph P. Winch**
R.R.	**R. Radebaugh**
R.Ro.	**Richard Roberts**
R.R.C.	**Rita R. Colwell**
R.R.H.	**Robert H. Hibbard**
R.R.McM.	**Robert R. McMath**
R.R.R.	**R. R. Rogers**
R.R.S.	**Robert R. Shannon**
R.R.Sh.	**R. R. Shively**
R.S.	**Rudolf Schmid**
R.Se.	**Robert Searles**
R.Sh.	**Robert Shprintzen**
R.Si.	**Raymond Siever**
R.Sim.	**Robert Simpson**
R.St.	**Robert Stratton**
R.Str.	**Reuben Straus**
R.Su.	**Richard Sudds**
R.S.B.	**Robert S. Boynton**
R.S.D.	**Robert S. Dietz**
R.S.F.	**Reino S. Freeman**
R.S.J.	**Richard S. Juvet, Jr.**
R.S.K.	**Robert S. Kapner**
R.S.Ki.	**Robert S. Kirby**
R.S.M.	**Robert S. Mulliken**
R.S.McE.	**Robert S. McEwen**
R.S.N.	**Robert S. Newsham**
R.S.Q.	**Roderick S. Quiroz**
R.S.R.	**Robert S. Ross**
R.S.S.	**Robert S. Sherwood**
R.S.T.	**Robert S. Tikofsky**
R.S.W.	**Ralph S. Wolfe**
R.S.We.	**Richard S. Westfall**
R.S.Y.	**Rosalyn S. Yalow**
R.S.Yo.	**Roland S. Young**
R.T.	**Robert Teasley, Jr.**
R.To.	**Richard Tousey**
R.T.B.	**Richard T. Barber**
R.T.Be.	**Robert T. Beaumont**
R.T.C.	**Robert T. Coupland**
R.T.H.	**Robert T. Hill**
R.T.Ha.	**Richard T. Hanlin**
R.T.J.	**Richard T. Jones**
R.T.L.	**Richard T. Liddicoat, Jr.**
R.T.R.	**Robert T. Ratay**
R.T.S.	**Richard T. Smith**
R.T.W.	**Robert T. Weil, Jr.**
R.T.Wh.	**Richard T. Whitcomb**
R.V.B.	**R. V. Blanden**
R.V.K.	**Robert V. Kesling**
R.V.R.	**Robert V. Ruhe**
R.W.	**Roff Weil**
R.W.B.	**Robert W. Belfit, Jr.**
R.W.Bu.	**Russel W. Burman**
R.W.C.	**Robert W. Carson**
R.W.Ca.	**Richard W. Castenholz**
R.W.Ch.	**R. W. Christie**
R.W.E.	**Ralph W. Engstrom**
R.W.Eg.	**Robert W. Egan**
R.W.F.	**Rhodes W. Fairbridge**
R.W.H.	**Robert W. Holmes**
R.W.Ha.	**Richard W. Harris**
R.W.Hai.	**Roger W. Haines**
R.W.J.	**Richard W. James**
R.W.M.	**Robert W. Mann**
R.W.Mo.	**Robert W. Morse**
R.W.Mu.	**Royce W. Murray**
R.W.N.	**Robert W. Noyes**
R.W.P.	**Robert W. Pennak**
R.W.R.	**Richard W. Robinson**
R.W.Ru.	**Ralph W. Rudolph**
R.W.S.	**Roy W. Simonson**
R.W.Sa.	**Robert W. Sanders**
R.W.Si.	**Richard W. Siegel**
R.W.St.	**Russell W. Strandtmann**
R.W.W.	**Roger W. Wolfe**
R.W.Wo.	**R. W. Woody**
R.W.Y.	**Robert W. Young**
R.W.Ye.	**Ralph W. Yell**
R.Z.	**Rainer Zangerl**
R.Zu.	**Riaz Zuberi**

S

S.A.C.	**Stanley A. Cain**
S.A.K.	**Stephen A. Konz**
S.A.L.	**Simon A. Levin**
S.A.La.	**Sandra A. Larsen**
S.A.M.	**Shelby A. Miller**
S.A.R.	**Scott A. Redhead**
S.A.S.	**Samuel A. Schaaf**
S.A.W.	**Selman A. Waksman**
S.A.Wh.	**Stanley A. White**
S.A.Wi.	**S. A. Williams**
S.B.	**Samuel Baron**
S.Ba.	**Stanley Bashkin**
S.Bo.	**Solomon Bochner**
S.B.E.	**Stuart B. Elston**
S.B.S.	**Steven B. Simon**
S.C.	**Sydney Chapman**
S.Ch.	**Steven Chu**
S.Cho.	**Sudhansu Chokroverty**
S.Cr.	**Sears Crowell**
S.C.C.	**S. C. Coroniti**
S.C.C.T.	**Samuel C. C. Ting**
S.C.E.	**Stephen C. Encleson**
S.C.H.	**Standish C. Hartman**
S.C.K.	**S. Charles Kendeigh**
S.C.L.	**Samuel C. Lee**
S.-C.Li.	**Shao-Chi Lin**
S.C.M.	**Simon Conway Morris**
S.C.R.	**Sydney C. Rittenberg**
S.C.Ri.	**Stephen C. Riser**
S.C.T.	**Samuel C. Tease**
S.C.U.	**Stephen C. Unwin**
S.D.	**Seymour Diamond**
S.D.G.	**Simone Daro Gossner**
S.D.S.	**Seymour D. Silver**
S.DeS.	**Steve deSatnick**
S.E.	**Steven Edwards**
S.E.C.	**Stephen E. Calvert**
S.E.Cr.	**Sheldon E. Cremer**
S.E.M.	**Stephen E. Mudrick**
S.E.W.	**S. E. Weaver**
S.E.Wh.	**Sidney E. White**
S.F.	**Sherman Fried**
S.Fr.	**Sigmund Fritz**
S.F.J.	**Stephen F. Jacobs**
S.F.P.	**Stephen F. Perry**

S.F.S. **S. F. Singer**
S.G. **Sheffield Gordon**
S.Gr. **S. Granick**
S.G.B. **S. Gaylen Bradley**
S.G.S. **Stephen G. Simpson**
S.G.T. **Stanley G. Thompson**
S.G.-W. **Salome Gluecksohn-Waelsch**
S.H. **Sven Horstadius**
S.H.B. **Simon H. Bauer**
S.H.Bl. **Sherman H. Bloomer**
S.H.Bo. **Stefan H. Boshkov**
S.H.C. **Stanley H. Cohn**
S.H.D. **Sven H. Dodington**
S.H.G. **Sidney H. Goodman**
S.H.L. **Sam H. Liu**
S.H.M. **Sergius H. Mamay**
S.H.W. **Samuel H. Wilen**
S.H.Y. **Sulhi H. Yungui**
S.-I.A. **S.-I. Akasofu**
S.I.M. **Sidney I. Miller**
S.I.P. **Shih-I Pai**
S.I.R. **S. I. Rasool**
S.I.S. **Stanley I. Sandler**
S.I.W. **Samuel Isaac Weissman**
S.J. **S. Jafarey**
S.Ju. **Sheldon Judson**
S.J.A. **Stephen J. Andriole**
S.J.C. **S. Joseph Campanella**
S.J.Co. **Steven J. Collins**
S.J.G. **Stephen J. Gould**
S.J.H. **Stanley J. Holt**
S.J.O. **Steven J. Ostro**
S.J.Y. **Samuel J. Yosim**
S.K. **Silve Kallmann**
S.Ka. **Serope Kalpakjian**
S.Ki. **Stanley Kirschner**
S.Kir. **Seth Kirby**
S.Kl. **Susan Kleinmann**
S.Ko. **Seichi Konzo**
S.K.F. **Sheldon K. Friedlander**
S.K.J. **Subodh K. Jain**
S.K.R. **Stanley K. Runcorn**
S.L. **Stephen Lawroski**
S.L.G. **Sol L. Garfield**
S.L.I. **Stanley L. Inhorn**
S.M. **Seymour Melman**
S.Ma. **Stephen Mann**
S.M.F. **Sydney M. Finegold**
S.M.K. **S. Morris Kupchan**
S.M.Ka. **Simon M. Kaplan**
S.M.M. **Stephen M. Matyas**
S.M.R. **S. Meryl Rose**
S.M.S. **Sueo M. Shiino**
S.M.Su. **S. Mark Sumi**
S.N. **Stephen Nygren**
S.N.G. **Stanley N. Gershoff**
S.N.S. **Steven N. Shore**
S.N.U.A.K. **Syed N. U. A. Kirmani**
S.O.R. **Samuel O. Raymond**
S.P. **Sue Povey**
S.Pu. **Seth Putterman**
S.P.H. **Samuel P. Hammer**
S.P.M. **Stephen P. Maran**
S.P.P. **Sybil P. Parker**
S.P.Pa. **S. Peter Pappas**
S.P.Pe. **Sam P. Perone**
S.R. **Sidney Raffel**
S.Ri. **Saul Rich**
S.Ro. **Sidney Ross**
S.R.B. **Sebastian R. Borrello**
S.R.Bo. **Steven R. Bolin**
S.R.C. **Stephen R. Carpenter**
S.R.E. **Steven R. Emerson**
S.R.N. **Suzanne R. Nagel**
S.R.O. **Stanford R. Ovshinsky**
S.S. **Sidney Siggia**
S.Si. **Stanley Singer**
S.Sir. **S. Sircar**
S.S.G. **Samuel S. Goldich**
S.S.H. **Stanley S. Hanna**
S.S.K. **Sudhir S. Kulkarni**
S.S.L.C. **S. S. L. Chang**
S.S.M. **Stephen S. Murray**
S.S.N. **Samuel S. Naistat**
S.S.S. **Shelley S. Sutherland**
S.T. **Shirley Turner**
S.T.A. **S. T. Algermissen**
S.T.P. **Samuel T. Picraux**
S.V.B. **Steven V. Beer**
S.W.B. **Sharon W. Bachinski**
S.W.Ba. **S. W. Bailey**
S.W.MacD. **Samuel W. MacDowell**
S.-Y.C. **Shih-Yuan Chen**

T

T.A. **Tilak Agerwala**
T.Ap. **Thomas Appelquist**
T.A.D. **T. A. Deacon**
T.A.G. **Todd A. Georgi**
T.A.Gl. **Thomas A. Gleeson**
T.A.H. **T. A. Howells**
T.A.Ho. **Thomas A. Horne**
T.A.M. **Terry A. Miller**
T.A.R. **Terence A. Rogers**
T.A.W. **T. A. White**
T.A.Wa. **Thomas A. Waldmann**
T.A.Z. **Thomas A. Zanoni**
T.B. **T. Balderes**
T.Ba. **Theodore Baumeister**
T.Be. **Tor Bergeron**
T.Br. **Thomas Brylawski**
T.B.M. **Thomas B. Mills**
T.B.S. **Thomas B. Sheridan**
T.B.T. **Thomas B. Turner**
T.C. **Thomas Clements**
T.Ca. **Tom Carey**
T.C.B. **Theodore C. Broyer**
T.C.H. **Thomas C. Hoering**
T.C.K. **Thomas C. Kaufman**
T.C.O. **Tobias C. Owen**
T.C.P. **Tram C. Pritchard**
T.C.R. **Theodore C. Ruch**
T.C.S. **Thomas C. Simmons**
T.C.W. **Thomas C. Waddington**
T.D. **Theodore Delevoryas**
T.D.W. **Thomas D. Watkins**
T.E.B. **Thomas E. Bowman**
T.E.L. **Thomas E. Leontis**
T.E.McL. **Thomas E. McLain**
T.E.T. **T. E. Thompson**
T.F.G. **Thomas F. Gallagher**
T.F.N. **T. F. Neubecker**
T.F.S. **Thomas F. Stratton**
T.F.W.B. **T. F. W. Barth**
T.F.Y. **Thomas F. Young**
T.G. **T. Gibson**
T.G.B. **Thomas G. Back**
T.G.G. **Thomas G. Giallorenzi**
T.H. **Thomas Hagler**
T.Hu. **Tomas Hudlicky**
T.H.B. **Theodore H. Bullock**
T.H.D. **Theodore H. Dexter**
T.H.L. **Thomas H. Lee**
T.H.V.A. **Tjeerd H. Van Andel**
T.H.V.H. **Thomas H. Vonder Haar**
T.I. **Teruo Ishihara**
T.I.S. **Tracy I. Storer**
T.J.B. **Thomas J. Bouchard**
T.J.B.H. **Timothy J. B. Holland**
T.J.E. **Thomas J. Eisler**
T.J.M. **Tobin J. Marks**
T.J.Me. **Thomas J. Meade**
T.J.P. **Thomas J. Putz**
T.J.W. **Theodore J. Weiss**
T.K. **Taku Komai**
T.K.M. **Thomas K. Miles**
T.L.F. **Thomas L. Ferrell**
T.L.H. **Thomas L. Hayes**
T.L.P. **Tom L. Phillips**
T.L.S. **Thomas L. Saaty**
T.L.Y. **Terry L. Yates**
T.M. **Theodore Monod**
T.Ma. **Takashi Makinodan**
T.Mo. **Thierry Montmerle**
T.Mor. **Theodore Moreno**
T.M.B. **Thomas M. Buermann**
T.M.Ba. **T. M. Barkley**
T.M.C. **Ted M. Cavender**
T.M.Ch. **T. Ming Chu**
T.M.D. **Todd M. Doscher**
T.M.G. **T. M. Georges**
T.M.M. **Terence M. Murphy**
T.McS. **Tim McSweeney**
T.N.B. **T. N. Bell**
T.N.K. **Tanveer N. Khan**
T. N. T. **Thomas N. Taylor**
T.P. **Thornton Page**
T.Pr. **Tom Pritchard**
T.P.K. **Truman P. Kohman**
T.P.M. **Thomas P. Murray**
T.P.S. **Theodore P. Schilb**
T.R. **Tecwyn Roberts**
T.R.McC. **Thomas R. McCormick**
T.R.S. **Thomas R. Schneider**
T.S. **Tsunemasa Saito**
T.Sn. **Theodore Snook**
T.Su. **Terry Surles**
T.S.P. **Thomas S. Parsons**
T.T.P. **Theodore T. Puck**
T.T.W. **Tai Tsun Wu**
T.U. **Tohru Uchida**
T.V.McE. **Thomas V. McEvilly**
T.W. **Thomas Wartik**
T.Wi. **Therese Wilson**
T.W.A. **Thomas W. Amsden**
T.W.H. **Thora W. Halstead**
T.W.Ha. **Theo W. Hansch**
T.W.M. **Timothy W. Merrill**
T.W.S. **Thomas W. Schoener**
T.W.T. **Theodore W. Torrey**
T.Y. **Tuneo Yamada**
T.Y.L. **Tae Young Lee**

U V

U.L.	**Ulrich Luttge**
U.S.	**Uwe Schumacher**
V.A.	**Vera Alexander**
V.A.B.	**Victor A. Bloomfield**
V.A.F.	**Victoria A. Fromkin**
V.B.S.	**Victor B. Scheffer**
V.D.P.	**Victor D. Petershack**
V.E.B.	**Vladimir E. Bondybey**
V.E.G.	**Victor E. Goldenberg**
V.E.S.	**Vemer E. Suomi**
V.F.	**Val Fitch**
V.Fu.	**Vincent Fulginiti**
V.G.A.	**Vincent G. Allfrey**
V.G.McC.	**V. G. McClusky**
V.H.	**Vladimir Haensel**
V.Hu.	**Vernon Hughes**
V.H.E.	**Van H. English**
V.I.C.	**Vernon I. Cheadle**
V.J.C.	**Valentine J. Chapman**
V.J.W.	**Verner J. Wulff**
V.K.W.	**Vivian K. Walworth**
V.L.S.	**Victor L. Streeter**
V.M.	**Vincent Madison**
V.M.A.	**Vincent M. Altamuro**
V.M.E.	**Victor M. Emmel**
V.N.V.	**Virginius N. Vaughan, Jr.**
V.O.W.	**Virgil O. Wodicka**
V.P.B.	**Victor P. Boyd**
V.R.B.	**Victor R. Baker**
V.R.H.	**Velman R. Huskey**
V.R.S.	**Victor R. Stefanovic**
V.S.	**Vincent Salmon**
V.Sh.	**Virginia Shirley**
V.T.	**Virginia Trimble**
V.T.B.	**Vincent T. Breslin**
V.V.L.	**V. V. Levasheff**
V.W.L.	**Victor W. Laurie**
V.W.U.	**Vincent W. Uhl**

W

W.A.	**William Allan**
W.An.	**Warren Andrew**
W.Ar.	**Walter Aron**
W.A.B.	**William A. Brodsky**
W.A.C.	**William A. Clemens**
W.A.Co.	**William A. Cobban**
W.A.DiM.	**William A. DiMichele**
W.A.E.	**Wilfred A. Elders**
W.A.H.	**Walter A. Harrison**
W.A.Ho.	**William A. Horton**
W.A.L.	**William A. Lanford**
W.A.Le.	**Wayne A. Lea**
W.A.Ly.	**Walter A. Lyons**
W.A.M.	**W. A. Munson**
W.A.N.	**William A. Nierenberg**
W.A.P.	**Wayne A. Pryor**
Wi.A.P.	**William A. Pryor**
W.A.R.	**William A. Rocap, Jr.**
W.A.S.	**William A. Steele**
W.A.V.W.	**W. A. Von Winkle**
W.A.W.	**W. A. Wakeham**
W.A.Wi.	**William A. Wildhack**
W.A.Y.	**W. A. Yost**
W.B.	**William Bradford**
W.Bu.	**Willy Burgdorfer**
W.B.B.	**Warren B. Blumenthal**
W.B.Bo.	**Warren B. Boast**
W.B.Br.	**William B. Brower, Jr.**
W.B.F.	**W. Beall Fowler**
W.B.Fr.	**William B. Fretter**
W.B.N.	**W. B. Nottingham**
W.B.N.B.	**William B. N. Berry**
W.B.R.	**William B. Renfrow, Jr.**
W.B.W.	**Wilse B. Webb**
W.C.	**Wesley Calef**
W.Ch.	**Wallace Chinitz**
W.Cl.	**Walter Clark**
W.C.Co.	**W. Charles Cooper**
W.C.D.	**Willy C. Dekeyser**
W.C.H.	**William C. Hayes**
W.C.Ha.	**William C. Haynes**
W.C.J.	**Woodrow C. Jacobs**
W.C.Jo.	**Walter C. Johnson**
W.C.L.	**W. C. Lineberger**
W.C.Li.	**William C. Livingston**
W.C.McC.	**Walter C. McCrone**
W.C.P.	**Walter C. Pitman, III**
W.C.S.	**Walter C. Sweet**
W.C.Sh.	**William C. Shoemaker**
W.C.W.	**William C. Walter**
W.C.Wo.	**William C. Wonders**
W.D.	**William Dent**
W.Di.	**William Diegel**
W.D.B.	**William D. Billings**
W.D.E.	**W. D. Edson**
W.D.H.	**Willard D. Hartman**
W.D.Hu.	**William D. Hummon**
W.D.J.	**William D. Jackson**
W.D.K.	**Walter D. Keller**
W.D.N.	**William D. Nix**
W.D.R.	**W. David Rust**
W.D.R.-H.	**W. D. Russell-Hunter**
W.D.T.	**W. Dean Trautman**
W.D.W.	**W. Dexter Whitehead**
W.D.Wi.	**W. D. Wilkinson**
W.E.C.	**William E. Cooley**
W.E.D.	**William E. Dossel**
W.E.Go.	**William E. Gordon**
W.E.H.	**William E. Howell**
W.E.J.	**Wendell E. Johnson**
W.E.K.	**William E. Keller**
W.E.Ku.	**Wayne E. Kuhn**
W.E.L.	**Walter E. Lobo**
W.E.N.	**Walter E. Nance**
W.E.S.	**W. E. Sterrer**
W.E.Y.	**Wesley E. Yates**
W.F.	**William Feller**
W.Fi.	**W. Fiers**
W.F.B.	**William F. Bradley**
W.F.Br.	**William F. Brown**
W.F.F.	**W. F. Furter**
W.F.J.	**William F. Jaep**
W.F.M.	**William F. Meggers**
W.F.P.	**Wilson F. Powell**
W.F.S.	**William F. Smith**
W.F.W.	**Warren F. Walker, Jr.**
W.G.	**William Grierson**
W.Gr.	**Walter Greiner**
W.G.A.	**Warren G. Abrahamson**
W.G.B.	**Waldo G. Bowman**
W.G.D.	**W. G. Dukek**
W.G.-F.	**Walter Gibbons-Fly**
W.G.L.	**William G. Lesso**
W.G.M.	**William G. Metcalf**
W.G.McG.	**William G. McGinnies**
W.G.P.	**William G. Pollard**
W.G.V.D.	**William G. Van Dorn**
W.G.W.	**W. Gordon Whaley**
W.H.	**Wilbur Hague**
W.Ha.	**Walter Häntzschel**
W.Hap.	**William Happer**
W.He.	**William Heronemus**
W.Her.	**William Hershleder**
W.H.C.	**Wallis H. Clark, Jr.**
W.H.Cr.	**William H. Crouse**
W.H.G.	**Walter H. Gardner**
W.H.Gi.	**Warren H. Giedt**
W.H.Gr.	**William H. Gross**
W.H.H.	**Walter H. Hartung**
W.H.Ha.	**William H. Hartwig**
W.H.Hu.	**William H. Hunley**
W.H.J.	**W. Hilton Johnson**
W.H.K.	**William H. Konigsberg**
W.H.L.	**Wim H. Laarhoven**
W.H.M.	**William H. Miller**
W.H.P.	**William H. Phillips**
W.H.S.	**Wendell H. Slabaugh**
W.H.Si.	**William H. Simmons**
W.H.U.	**Walter H. Unger**
W.H.W.	**Winona H. Welch**
W.H.Wi.	**William H. Wisely**
W.I.R.	**William Ingersoll Rose, Jr.**
W.I.T.	**Walter I. Thomas**
W.J.A.	**William J. Adelman, Jr.**
W.J.B.	**Walter J. Bock**
W.J.C.	**William C. Clench**
W.J.G.	**William J. Galloway**
W.J.Ge.	**Walter J. Gensler**
W.J.Ger.	**Willis J. Gertsch**
W.J.H.	**Walter J. Hamer**
W.J.Ha.	**William J. Hargis, Jr.**
W.J.K.	**William J. Krefeld**
W.J.L.	**Warren J. Luzadder**
W.J.LeN.	**William J. Le Noble**
W.J.M.	**William J. Mitsch**
W.J.M.M.	**W. J. M. Moore**
W.J.O.	**William J. Oshinski**
W.J.V.W.	**Willem J. Van Wagtendonk**
W.J.W.	**William J. Willis**
W.J.Wi.	**William J. Wisniewski**
W.K.	**Werner Kanzig**
W.Ke.	**William Kerr**
W.Ki.	**Walter Kintsch**
W.Ku.	**Wietse Kuis**
W.K.C.	**William K. Creson**
W.L.	**Willy Ley**
W.Li.	**William Liller**
W.Lu.	**William Luth**
W.Ly.	**Waldo Lyon**
W.Lyo.	**Walter Lyons**
W.L.B.	**William L. Balsam**
W.L.Br.	**William L. Bradford**
W.L.Bro.	**William L. Brown, Jr.**
W.L.H.	**W. Lincoln Hawkins**
W.L.M.	**Walter L. Morgan**
W.L.McC.	**Warren L. McCabe**
W.L.N.	**William L. Nastuk**

W.L.P. **William L. Polhemus**
W.L.R. **William L. Root**
W.L.S. **Waldo L. Schmitt**
W.L.T. **William L. Taylor**
W.L.W. **Willis L. Webb**
W.L.Wi. **W. L. Wilkinson**
W.L.Wo. **William L. Wolfe**
W.M. **William Markowitz**
W.Ma. **Walter Mannheim**
W.Mi. **Wesley Minnis**
W.Mis. **Walter Mischel**
W.Mo. **Walter Moberly**
W.Mos. **William Mosher**
W.My. **Ware Myers**
W.McM. **Wheeler McMillen**
W.M.S. **William M. Sinton**
W.M.Sa. **W. Mark Saltzman**
W.N.S. **Wilson N. Stewart**
W.N.T. **W. Norris Tuttle**
W.O.M. **W. O. Milligan**
W.O.R. **William O. Rainnie, Jr.**
W.P. **William Parrish**
W.P.C. **William P. Clinton**
W.P.M. **Warren P. Mason**
W.P.R. **William P. Rodden**
W.P.S. **Walter P. Siegmund**
W.P.W. **W. P. Wolf**
W.R. **William Rostoker**
W.Ro. **Willard Roth**
W.R.A. **W. Robert Adams**
W.R.B. **William R. Boggess**
W.R.Br. **Winslow R. Briggs**
W.R.E. **William R. Engels**
W.R.El. **Walther R. Ellis, Jr.**
W.R.F. **Walter R. Fehr**
W.R.J. **William R. Judd**
W.R.L. **Werner R. Loewenstein**
W.R.Lo. **Wayne R. Lowell**
W.R.LeP. **W. R. LePage**
W.R.M. **William R. Marshall, Jr.**
W.R.Mi. **Winston R. Miller**
W.R.N. **Walter R. Nagel**
W.R.P. **William R. Philipson**
W.R.R. **William R. Riedel**
W.R.S. **William R. Sistrom**
W.R.Se. **W. Rudolf Seitz**
W.R.Si. **W. Rex Sittner**
W.R.Sm. **William R. Smythe**
W.R.V. **Wyman R. Vaughan**
W.R.W. **William R. Wilcox**
W.S. **Wolfram Saenger**
W.Sc. **William Scher**
W.Sch. **William Schiavoni**
W.Sh. **William Shive**
W.Sp. **William Spackman**
W.St. **Walther Stoeckenius**
W.Stu. **Wilson Sturges**
W.S.A. **Wendall S. Arbuckle**
W.S.B. **William S. Benninghof**
W.S.Ba. **W. Scott Baldridge**
W.S.Br. **Willard S. Bromley**
W.S.C. **W. Storrs Cole**
W.S.Cr. **William S. Cramer**
W.S.G. **Waldo S. Glock**
W.S.J. **William S. Janna**
W.S.L. **W. S. Lyon**
W.S.P. **Wilson S. Pritchett**
W.S.T. **William S. Tolgyesi**
W.S.-Y.W. **William S.-Y. Wang**
W.T.I. **William T. Ingram**
W.T.K. **William T. Keeton**
W.T.McK. **William T. McKinney, Jr.**
W.T.S. **William T. Stuart**
W.V.R. **W. V. Robinson**
W.W. **William West**
W.Wo. **William Wolovich**
W.W.B. **William W. Ballard**
W.W.Be. **Wallace W. Beaver**
W.W.Br. **William W. Bradley**
W.W.Bro. **William W. Bromer**
W.W.E. **W. W. Epstein**
W.W.Ec. **W. Wesley Eckenfelder**
W.W.H. **William W. Hay**
W.W.Ho. **William W. Howells**
W.W.M. **W. Wayne Meinke**
W.W.Mi. **Ward W. Minkler**
W.W.Mo. **W. W. Moyer**
W.W.R. **W. W. Reynolds**
W.W.S. **William W. Seifert**
W.W.Sm. **W. W. Smith**
W.W.Sn. **William W. Snow**
W.W.Sp. **Wesley W. Spink**
W.W.W. **William W. Watson**
W.W.We. **Wesley W. Wendlandt**
W.Y. **Wayne Young**
W.Y.C. **Wai Yiu Cheung**
W.Z. **William Zuk**
W.Z.H. **William Z. Hassid**
W.Z.S. **Wieslaw Z. Stepniewski**

X–Z

X.J.M. **X. J. Musacchia**
Y.E.-M. **Youssef El-Mansy**
Y.I. **Yedy Israel**
Y.W. **Y. Waisel**
Y.W.K. **Yoke Wah Kow**
Z.C.D. **Zbigniew C. Dobrowolski**
Z.R.W. **Zelda R. Wasserman**
Z.S. **Zdenek Sekera**

LIST OF CONTRIBUTORS

The List of Contributors below comprises an alphabetical sequence, according to surname, of all contributors to the Encyclopedia. A brief affiliation is provided for each author. This list may be used in conjunction with the previous section, Contributors' Initials, to fully identify the contributor of each article.

A

Aagaard, Prof. Knut. Department of Oceanography, University of Washington.

Abbott, Dr. Donald P. Professor of Biology, Hopkins Marine Station, Stanford University, Pacific Grove, California.

Abell, Prof. George O. Deceased; formerly, Department of Astronomy, University of California, Los Angeles.

Able, Dr. Kenneth P. Department of Biological Sciences, State University of New York, Albany.

Abrahams, David H. Dexter Chemical Corporation, Bronx, New York.

Abrahams, Prof. Elihu. Department of Physics, Rutgers University.

Abrahamson, Prof. Warren G., II. Department of Biology, Bucknell University, Lewisburg, Pennsylvania.

Abramovitch, Dr. Rudolph A. Department of Chemistry and Geology, Clemson University.

Abzug, Dr. Malcolm J. Senior Staff Engineer, Systems Group, TRW, Inc., Redondo Beach, California.

Ackerman, Dr. Edward A. Deceased; formerly, Carnegie Institution of Washington.

Adair, Dr. Robert K. Department of Physics, Yale University.

Adams, Dr. Chris. Unilever Research, Merseyside, England.

Adams, Prof. Douglas P. Department of Mechanical Engineering, Massachusetts Institute of Technology.

Adams, Dr. E. Dwight. Department of Physics, University of Florida.

Adams, Dr. Elijah. Professor and Head, Department of Biological Chemistry, School of Medicine, University of Maryland.

Adams, Prof. M. W. Department of Crop and Soil Sciences, Michigan State University.

Adams, Dr. Raymond D. Bullard Professor Emeritus of Neuropathology, Massachusetts General Hospital.

Adams, Dr. W. Robert. Director, Clinical Laboratories, College of Medicine, University of Florida.

Adamson, Prof. Arthur W. Department of Chemistry, University of Southern California.

Addicott, Prof. Fredrick T. Department of Botany, University of California, Davis.

Adelberg, Dr. Edward A. Department of Microbiology, Yale University.

Adelman, Dr. William J., Jr. National Institutes of Health, Bethesda, Maryland.

Ader, Prof. Robert. Department of Psychiatry, University of Rochester, New York.

Adkins, George G. Chief (retired), Department of River Basins, Bureau of Power, Federal Power Commission.

Adler, Dr. David. Department of Electrical Engineering, Massachusetts Institute of Technology.

Adler, Dr. Ronald J. Lockheed Palo Alto Research Laboratories, Palo Alto, California.

Agerwala, Dr. Tilak. Manager, Architecture and System Design, IBM T. J. Watson Research Center, Yorktown Heights, New York.

Ahearn, Prof. Gegrory A. Department of Zoology, University of Hawaii, Manoa.

Ahlgren, Dr. Isabel F. Quetico-Superior Wilderness Research Center, Duluth, Minnesota.

Ahlstrom, Dr. Elbert H. Deceased; formerly, Fishery Oceanography Center, U.S. Bureau of Commercial Fisheries, La Jolla, California.

Ajmone-Marsan, Dr. Cosimo. Chief, Branch of Electroencephalography and Clinical Neurophysiology, National Institute of Neurological Diseases and Stroke, National Institutes of Health, Bethesda, Maryland.

Akasofu, Prof. S. I. Geophysical Institute, University of Alaska.

Akesson, Dr. Norman G. Agricultural Engineering Department, University of California, Davis.

Albert, Dr. Henry J. Manager, Physics and Metallurgy Laboratory, Research and Development Department, Engelhard Industries Division, Engelhard Minerals and Chemicals Corporation, Newark, New Jersey.

Albright, Dr. Lyle F. Department of Chemical Engineering, Purdue University.

Alburger, Dr. David E. Brookhaven National Laboratory, Upton, New York.

Aldwinckle, Prof. H. S. Department of Plant Pathology, Cornell University.

Alexander, Dr. Aaron D. Chicago, Illinois.

Alexander, Prof. R. McNeill. Department of Pure and Applied Biology, University of Leeds, England.

Alexander, Dr. Vera. Institute of Marine Science, University of Alaska.

Alexopoulos, Dr. Constantine J. Department of Botany, University of Texas, Austin.

Alger, Prof. Philip L. Deceased; formerly, Consulting Professor of Electrical Engineering, Rensselaer Polytechnic Institute.

Algermissen, Dr. S. T. Aurora, Colorado.

Allan, Roger. Executive Editor, "Electronic Design Magazine," Hasbrouck Heights, New Jersey.

Allan, William. Dean (retired), School of Engineering, City College of the City University of New York.

Allen, Edward W., Jr. Formerly, Chief Engineer, Federal Communications Commission.

Allen, Dr. John A. U.M.B.S.M., Marine Station, Isle of Cumbrae, Scotland.

Allen, Prof. Leland C. Department of Chemistry, Princeton University.

Allen, Prof. Lawrence H. Department of Astronomy, University of California, Los Angeles.

Allen, Dr. L. H., Jr. Department of Soil Sciences, University of Florida.

Allen, Prof. Oscar N. Department of Bacteriology, University of Wisconsin.

Allen, Dr. Robert D. Deceased; formerly, Department of Biological Science, Dartmouth College.

Alley, Prof. Charles L. Department of Electrical Engineering, University of Utah.

Allfrey, Dr. Vincent G. Department of Cell Biology, Rockefeller University.

Allport, H. Burnham. Union Carbide Chemical Company, Cleveland, Ohio.

Allred, Dr. Albert L. Department of Chemistry, Northwestern University.

Almgren, Dr. Frederick J., Jr. Department of Mathematics, Princeton University.

Almond, Dr. Peter R. Department of Biophysics, University of Texas, Houston.

Altamuro, Vincent M. President, Robotics Research, Division of VMA. Inc., Toms River, New Jersey.

Altman, Dr. David. Division Vice President and Technical Director, United Technology Center, Sunnyvale, California.

Alvim, Dr. Paulo de T. Centro de Pesquisas Do Cacau (CEPEC), Itabuna. Bahia, Brazil.

Alvord, Dr. Ellsworth C., Jr. Department of Neuropathology, University of Washington.

Ammons, Dr. Nelle. Retired; Department of Biology, West Virginia University.

Amorocho, Dr. Jaime. Department of Civil Engineering, University of California, Davis.

Amsden, Dr. Thomas W. Geologist, Oklahoma Geological Survey, Norman.

Anders, James E., Sr. President, Hydraulics Associates, Bethlehem, Pennsylvania.

Anderson, Dr. Alfred T., Jr. Department of Geophysical Sciences, University of Chicago.

Anderson, Dr. Everett. Professor of Anatomy and Associate Director of the Laboratory of Human Reproduction and Reproductive Biology, Harvard Medical School.

Anderson, Dr. John D. Jet Propulsion Laboratory, California Institute of Technology.

Anderson, Dr. John R. Vice President, Research and Development, Permutit Research and Development Center, Princeton, New Jersey.

Anderson, Dr. O. Roger. Lamont-Doherty Geological Observatory, Palisades, New York.

Anderson, Dr. Paul M. Department of Electrical and Computer Engineering, Arizona State University.

Anderson, Dr. Robert L. Forest Insect and Disease Management, USDA Forest Service, Asheville, North Carolina.

Andreas, Ronald. Sandia Laboratories, Albuquerque, New Mexico.

Andrew, Dr. Warren. Chairman, Department of Anatomy, School of Medicine, Indiana University.

Andrews, Dr. Lester. Department of Chemistry, University of Virginia.

Andrews, Dr. Peter. Department of Palaeontology, British Museum (Natural History), London.

Andriole, Dr. Stephen J. Department of Information Systems and Systems Engineering, George Mason University, Fairfax, Virginia.

Andrus, Dr. C. F. Agricultural Consultant; formerly, U.S. Department of Agriculture, Charleston, South Carolina.

Anglin, Donald L. Consultant, Automotive and Technical Writing, Charlottesville, Virginia.

Angus, James W. Staff Engineer, Avionics Division, Kollsman Instrument Corporation, Elmhurst, New York.

Ankeney, D. Philip. Naval Weapons Center, China Lake, California.

Anscombe, Prof. Francis J. Department of Statistics, Yale University.

Anthes, Prof. Richard A. Department of Meteorology, College of Earth and Mineral Sciences, Pennsylvania State University.

Antoni, Dr. Charles M. Department of Civil Engineering, Syracuse University.

Anwyl, Robert D. Retired; formerly, Photographic Consultant, Eastman Kodak Company, Rochester, New York.

Apen, John R. Bell Laboratories, Norcross, Georgia.

Apfel, Prof. Robert E. Department of Engineering and Applied Science, Yale University.

Apker, Dr. L. General Electric Research Laboratory, Schenectady, New York.

Appel, Dr. Max J. G. James A. Baker Institute for Animal Health, Cornell University, Ithaca, New York.

Appelman, Dr. Evan H. Chemistry Division, Argonne National Laboratory, Argonne, Illinois.

Appenzeller, Dr. Otto. Department of Neurology, University of New Mexico School of Medicine.

Appleby, Dr. Arnold P. Professor Emeritus of Crop Science, Oregon State University, Corvallis.

Applegate, Charles E. Consulting Engineer, Weston, Maryland.

Applequist, Dr. Thomas. Sloane Laboratory, Department of Physics, Yale University.

Arbuckle, Earl F., III. Engineering Supervisor, WPIX, Inc., New York, New York.

Arbuckle, Prof. Wendell S. Deceased; formerly, Department of Dairy Science, University of Maryland.

Archer, Dr. E. James. Vice President for Academic Affairs, University of Rhode Island.

Ardema, Dr. Mark D. Research Scientist, National Aeronautics and Space Administration, Ames Research Center, Moffett Field, California.

Arey, Prof. Leslie B. Professor Emeritus of Anatomy, Northwestern University.

Arimura, Dr. Kunitaka. Consultant in Electronics, Washington D.C.

Armbruster, Dr. Peter. Gesellschaft fur Schwerionenforschung, MBH, Germany.

Armentrout, Dr. John M. Mobil Dallas Technology Center, Mobil Oil Company, Dallas, Texas.

Armstrong, Dr. Donald. Memorial Hospital, New York, New York.

Armstrong, Dr. Richard L. Institute of Arctic and Alpine Research, University of Colorado.

Arndt, Dr. Kenneth A. Department of Dermatology, Beth Israel Hospital, Boston, Massachusetts.

Arnett, Dr. Ross H., Jr. Editor, Biological Research Institute of America, Inc., Kinderhook, New York.

Arnold, Dr. Chester A. Professor of Botany and Geology and Curator of Paleobotany, University of Michigan.

Arnstein, Dr. Karl. Vice President in Charge of Engineering (retired), Goodyear Aircraft Corporation, Akron, Ohio.

Aron, Dr. Walter. Consultant, Menlo Park, California.

Askey, Dr. Richard. Mathematics Research Center, University of Wisconsin.

Aston, Dr. James. Deceased; formerly, Consulting Metallurgist, A. M. Beyers Company.

Astrom, Prof. Karl J. Department of Automatic Control, Lund Institute of Technology, Lund, Sweden.

Atkins, Dr. Elisha. School of Medicine, Yale University.

Atkins, Dr. P. W. Department of Chemistry, Oxford University, England.

Atkinson, Prof. Gordon. Department of Chemistry, University of Oklahoma.

Atkinson, Dr. Robert d'E. Department of Astronomy, Indiana University.

Atlas, Dr. Ronald M. Department of Biology, University of Louisville, Kentucky.

Atoda, Dr. Kenji. Professor of Zoology, Tohoku University, Sendai, Japan.

Atwood, Dr. Ronald L. Foote Mineral Company, Kings Mountain, North Carolina.

Audley-Charles, Dr. M. G. Department of Geology, Queen Mary College, London, England.

Audus, Dr. Leslie J. Professor of Botany (retired), Bedford College, University of London, England.

Austrian, Dr. Robert. Department of Research Medicine, School of Medicine, University of Pennsylvania.

Averbach, Prof. Benjamin L. Department of Materials Science and Engineering, Massachusetts Institute of Technology.

Ayers, Arthur R. Assistant Professor, Cellular and Development Biology, Harvard University.

Aylmer, Prof. D. Department of Chemistry, Texas A&M University.

Azarnoff, Dr. Daniel L. President, Searle Research and Development, G. D. Searle & Co., Skokie, Illinois.

B

Babcock, Dr. Horace W. Director, Hale Observatories, Pasadena, California.

Bachinski, Dr. Sharon W. Department of Geology, University of New Brunswick, Fredericton, Canada.

Back, Dr. Thomas G. Department of Chemistry, Faculty of Science, University of Calgary, Alberta, Canada.

Bäckstrom, Prof. Gunnar. Institute of Physics, University of Umea, Sweden.

Bada, Dr. Jeffrey L. Scripps Institution of Oceanography, University of California, San Diego-La Jolla.
Badgley, Dr. Catherine. Museum of Paleontology, University of Michigan.
Badgley, Dr. Peter C. Earth Science Division, Office of the Naval Reserve, Arlington, Virginia.
Baer, Prof. Charles J. Department of Mechanical Engineering, University of Kansas.
Baer, Dr. Norbert S. Conservation Center of the Institute of Fine Arts, New York University.
Bagley, Dr. Brian G. Bell Laboratories, Murray Hill, New Jersey.
Bagnall, Dr. Kenneth W. Research Chemist, Atomic Energy Research Establishment, Harwell, England.
Bahcall, Dr. John N. Institute for Advanced Studies, Princeton, New Jersey.
Bailey, Prof. A. Earle. Department of Electrical Engineering, University of Southampton, England.
Bailey, Prof. Catherine H. Department of Horticulture, Rutgers University.
Bailey, Frederick G. Large Steam Turbine-Generator Department, General Electric Company, Schenectady, New York.
Bailey, Prof. Milton E. Department of Food Science and Nutrition, College of Agriculture, University of Missouri-Columbia.
Bailey, Dr. Reeve M. Deceased; formerly, Curator of Fishes, Museum of Zoology, University of Michigan.
Bailey, Dr. S. W. Department of Geology, University of Wisconsin.
Baker, Edward W. Entomology Research Division, Agricultural Research Service, U.S. Department of Agriculture.
Baker, Dr. Herbert G. Department of Botany, University of California, Berkeley.
Baker, Dr. M. Pauline. National Center for Supercomputing Applications, University of Illinois, Urbana.
Baker, Dr. Victor R. Department of Geosciences, University of Arizona.
Bakish, Dr. Robert. Department of Engineering Technology, Fairleigh Dickinson University, Teaneck.
Balabanian, Prof. Norman. Department of Electrical and Computer Engineering, Syracuse University.
Balderes, Dr. T. Engineering Specialist, Grumman Aerospace Corporation, Bethpage, New York.
Baldridge, Dr. W. Scott. Earth and Space Science Division, Los Alamos National Laboratory.
Baldwin, Harry L., Jr. Department of Mathematics, San Diego City College, San Diego, California.
Baliga, Prof. B. Jayant. Department of Electrical and Computer Engineering, North Carolina State University.
Ballard, Prof. William W. Department of Biological Sciences, Dartmouth College.
Bally, Dr. John. AT&T Bell Laboratories, Holmdel, New Jersey.
Balsam, Dr. William L. Division of Natural Sciences, Southampton College, Southampton, New York.
Baltay, Dr. Charles. Department of Physics, Yale University.
Bambach, Dr. Richard A. Department of Geological Sciences, Virginia Polytechnic Institute.
Bamburg, Prof. James R. Department of Biochemistry, Colorado State University.
Banaszak, Prof. Leonard. Department of Biochemistry, University of Minnesota, Minneapolis.
Bandoni, Dr. Robert J. Department of Botany, University of British Columbia, Vancouver, Canada.
Banfield, Dr. A. F. Behre Dolbear and Company, Inc., New York, New York.
Bank, Dr. Arthur. Professor and Acting Chairperson, Department of Human Genetics, and Director, Division of Hematology, Columbia University.
Banks, Dr. Harlan P. Emeritus Professor of Botany, Division of Biological Sciences, Cornell University.
Banner, Dr. Donald W. Banner, Birch, McKie and Beckett, Washington, D.C.
Barash, Prof. David P. Department of Psychology, University of Washington, Seattle.
Barber, Dr. Richard T. Duke University Marine Laboratory.
Barclay, Dr. Harriet G. Professor of Botany, Department of Life Sciences, University of Tulsa.
Barer, Prof. Robert. Department of Human Biology and Anatomy, University of Sheffield, England.
Barghoorn, Prof. Elso S. Department of Biology and Botanical Museum, Harvard University.
Barker, Dr. Fred. Geological Survey, U.S. Department of the Interior, Denver, Colorado.
Barker, Dr. Horace A. Department of Biochemistry, University of California, Berkeley.
Barker, Dr. Joseph W. Chairman of the Board (retired), Research Corporation, New York, New York.
Barkley, Prof. T. M. Division of Biology, Kansas State University.
Barksdale, Dr. Lane. Department of Microbiology, School of Medicine, New York University Medical Center.
Barlow, Prof. George W. Department of Zoology, University of California, Berkeley.
Barlow, Dr. Grant H. Michael Reese Research Foundation, Chicago, Illinois.
Barna, Dr. Gabriel G. Central Research Laboratories, Texas Instruments, Inc., Dallas, Texas.
Barnard, Dr. J. Laurens. Curator of Crustacea, Smithsonian Institution.
Barnes, Dr. Burton V. School of Natural Resources, University of Michigan.
Barnes, Ivan. Water Resources Division, Geological Survey, U.S. Department of the Interior, Menlo Park, California.
Barnes, Dr. Robert K. Research and Development Department, Chemicals and Plastics Operations Division, Union Carbide Corporation, South Charleston, West Virginia.
Barnett, Dr. Roy N. Pathologist-Consultant, Westport, Connecticut.
Barnitz, Edward S. Consultant in Vacuum Engineering, Rochester, New York.
Baron, Dr. Samuel. Department of Microbiology, School of Medicine, University of Texas, Galveston.
Barr, Dr. David. Department of Entomology, Royal Ontario Museum, Toronto, Ontario, Canada.
Barr, Dr. Donald J. S. Principal Research Scientist, Centre for Land and Biological Resources Research, Ottawa, Ontario, Canada.
Barracca, Peter S. Executive Vice President, Merritt Division, Murphy Pacific Marine Salvage, New York.
Barrett, Prof. Harrison H. Optical Sciences Center, University of Arizona.
Barry, Brother B. Austin. Civil Engineering Department, Manhattan College.
Barry, Dr. Patrick E. Department of Electrical Sciences, State University of New York, Stony Brook.
Bars, Dr. Itzhak. Department of Physics, Yale University.
Barth, Prof. T. F. W. Deceased; formerly, Geologic Museum, Oslo.
Barton, Dr. Mark. Brookhaven National Laboratory, Upton, New York.
Barton, Dr. Mark D. Department of Earth and Space Sciences, University of California, Los Angeles.
Bashkin, Dr. Stanley. Department of Physics, University of Arizona.
Basolo, Prof. Fred. Department of Chemistry, Northwestern University.
Bass, Dr. Brenda. Department of Chemistry, University of Colorado.
Bass, Prof. Jack. Department of Physics and Astronomy, Michigan State University.
Batchelder, Howard R. Consulting Chemical Engineer (retired), Battelle Memorial Institute, Columbus, Ohio.
Bateman, Dr. Richard M. Department of Geology, Royal Museum of Scotland, Edinburgh.
Bates, Dr. Robert L. Department of Geology, Ohio State University.
Battan, Prof. Louis J. Deceased; formerly, Associate Director, Institute of Atmospheric Physics, University of Arizona.
Bauer, Prof. Charles L. Department of Metallurgical Engineering and Materials Science, Carnegie-Mellon University.
Bauer, Dr. Simon H. Department of Chemistry, Cornell University.

Baumeister, Prof. Theodore. Deceased; formerly, Consulting Engineer; Stevens Professor of Mechanical Engineering, Emeritus, Columbia University; Editor in Chief, "Standard Handbook for Mechanical Engineers."

Baumgarten, Dr. Alexander. Department of Laboratory Medicine, School of Medicine, Yale University.

Bayfield, Dr. James E. Department of Physics, University of Pittsburgh.

Beaglehole, Prof. David. Department of Physics, Victoria University of Wellington.

Beams, Dr. Jesse. Deceased; formerly, Department of Physics, University of Virginia.

Bean, Howard S. Deceased; formerly, Consultant on Fluid Metering, Liquids and Gases, Sedona, Arizona.

Beard, Dr. Benjamin H. Agronomy and Range Science Extension, University of California, Davis.

Beaty, H. Wayne. Consultant, Fairfax, Virginia.

Beatty, J. Kelly. Consultant, Chelmsford, Massachusetts.

Beatty, Ronald L. Oak Ridge National Laboratory, Oak Ridge, Tennessee.

Beaumariage, Dr. Donald C. Director, Office of Manned Undersea Technology, National Oceanic and Atmospheric Administration, Rockville, Maryland.

Beaumont, Robert T. Director of Meteorological Operations, E. G. G., Inc., Boulder, Colorado.

Beaumont, Prof. Ross A. Department of Mathematics, University of Washington.

Beaver, Dr. Wallace W. Deceased; formerly, Director of Research and Development, Brush Beryllium Company, Cleveland, Ohio.

Beck, Dr. Charles B. Professor of Botany, University of Michigan.

Becken, Eugene D. Vice President, Operations, RCA Global Communications, Inc., New York.

Becker, Dr. Donald E. Head, Department of Animal Science, University of Illinois, Urbana.

Becker, Dr. Elery R. Deceased; formerly, Division of Life Sciences, Arizona State University.

Beckett, Dr. Royce E. Department of Mechanical Engineering, Auburn University.

Beckhorn, Dr. Edward J. Director of Research, Wallerstein Company, New York.

Beckwith, Dr. J. Bruce. Associate Professor of Pathology and Pediatrics, School of Medicine, University of Washington; Pathologist, Children's Orthopedic Hospital and Medical Center, Seattle, Washington.

Bederson, Dr. Benjamin. Department of Physics, New York University.

Beer, Dr. Steven V. Department of Plant Pathology, College of Agriculture and Life Sciences, Cornell University.

Beeson, Dr. Paul B. Nuffield Professor of Clinical Medicine, Oxford University, England.

Beggs, H. L. Consultant, Fluid Processing Division, Selas Corporation of America, Bresher, Pennsylvania.

Beidler, Dr. Lloyd M. Department of Biological Sciences, Florida State University.

Beiler, Dr. A. Clarke. Systems Development Division, Defense and Space Center, Westinghouse Electric Corporation, Baltimore, Maryland.

Belfit, Dr. Robert W. Manager, Quality Standards, Quality Assurance Department, Dow Chemical Company, Midland, Michigan.

Bell, Morton A. American Air Filter Company, Inc., New York.

Bell, Prof. Norman R. Department of Electrical Engineering, North Carolina State University.

Bell, Prof. T. N. Department of Chemistry, Simon Fraser University, Burnaby, British Columbia, Canada.

Bellman, Prof. Richard. Department of Electrical Engineering, School of Engineering, University of California, Los Angeles.

Belrose, Dr. John S. Communications Research Centre, Department of Communications, Ottawa, Ontario, Canada.

Bemis, Dr. David A. Director, Clinical Bacteriology and Mycology Services, University of Tennessee, Knoxville.

Bender, Prof. Myron L. Department of Chemistry, Northwestern University.

Bender, Prof. Paul J. Professor of Physical Chemistry, University of Wisconsin.

Benedict, Dr. Manson. Institute Professor Emeritus, Department of Nuclear Engineering, Massachusetts Institute of Technology.

Benedict, Dr. Robert G. Research Associate Professor of Pharmacognosy, College of Pharmacy, University of Washington.

Benfer, Neil A. Scientific Editor, Barbados Oceanographic and Meteorological Analysis Project.

Benito, Dr. Jesús. Department of Animal Biology I–Zoology of Invertebrates, University Complutense, Madrid, Spain.

Benner, Dr. Frank C. Assistant Director, Research and Development Department, Norton Company, Worcester, Maine.

Bennett, Prof. Michael V. L. Department of Neuroscience, Albert Einstein College of Medicine.

Bennett, Dr. Orus L. Vice President, GRC Panels Unlimited, Leeds, Alabama.

Bennett, Dr. Ralph E. Director, Science Information Department, Squibb Institute for Medical Research, New Brunswick, New Jersey.

Benninghoff, Dr. William S. Department of Botany, University of Michigan.

Benny, Dr. Gerald L. Department of Plant Pathology, University of Florida, Gainesville.

Beranek, Dr. Leo L. Chief Scientist (retired), Bolt Beranek and Newman, Inc., Cambridge, Massachusetts.

Berg, Dr. Clyde. Clyde Berg and Associates, Long Beach, California.

Berger, Dr. France B. Kearfott Division, The Singer Company, Little Falls, New Jersey.

Bergeron, Prof. Tor. Deceased; formerly, Institute of Meteorology, Uppsala, Sweden.

Bergh, Dr. A. A. Bell Telephone Laboratories, Murray Hill, New Jersey.

Bergh, Dr. Berthold O. Department of Plant Sciences, University of California, Riverside.

Berglund, Neil. Manager of Technology Development, Intel Corporation, Aloha, Oregon.

Bergmann, Prof. Peter G. Department of Physics, Syracuse University.

Beringer, Prof. Robert. Physics Department, Yale University.

Berkofsky, Prof. Louis. Institute for Desert Research, Ben Gurion University of the Negev, Beersheba, Israel.

Berner, Prof. Lewis. Acting Director, Division of Biological Sciences, University of Florida.

Bernheimer, Prof. Alan W. Department of Microbiology, New York University Medical Center.

Bernstein, Dr. I. M. Department of Metallurgy and Materials Science, Carnegie-Mellon University.

Bernstein, Dr. Robert I. Professor of Electrical Engineering and Associate Director, Electronics Research Laboratories, Columbia University.

Berry, Dr. William B. N. Museum of Paleontology, University of California, Berkeley.

Bertsch, Dr. George F. Department of Physics, Michigan State University.

Bethe, Prof. Hans A. Floyd R. Newman Laboratory of Nuclear Studies, Cornell University.

Betts, Dr. Russell. Argonne National Laboratory, Argonne, Illinois.

Bever, Prof. Michael B. Department of Materials Science and Engineering, Massachusetts Institute of Technology.

Bewley, Dr. Loyal V. Dean (retired), College of Engineering, Lehigh University.

Biberstein, Dr. Ernst L. School of Veterinary Medicine, University of California, Davis.

Bicknell, Prof. Joseph. Department of Aeronautics and Astronautics, Massachusetts Institute of Technology.

Bidanset, Dr. Jesse. Department of Toxicology, St. John's University.

Biedenharn, Prof. L. C. Department of Physics, Duke University.

Biegel, Prof. John E. Department of Engineering, University of Central Florida.

Bienenstock, Dr. Arthur. Professor of Applied Physics, Stanford University; Director, Stanford Synchrotron Radiation Laboratory.

Bigeleisen, Prof. Jacob. Department of Chemistry, University of Rochester.

Biggs, Dr. Robert H. Department of Fruit Crops, Institute of Food and Agricultural Sciences, University of Florida.

Bignami, Prof. Giovanni F. Istituto per Ricerche in Fisica Cosmica e Tecnologie Relative, Milan, Italy.

Bikales, Dr. Norbert M. Chemical Consultant, N. M. Bikales and Co., Livingston, New Jersey.

Billings, Dr. Bruce H. Special Assistant to the Ambassador of Science and Technology, Embassy of the United States of America, Taipei.

Billings, Dr. Marland P. Geological Museum, Harvard University.

Billings, Dr. William D. Department of Botany, Duke University.

Billington, Dr. Douglas S. Senior Staff Advisor for Materials Science, Metals and Ceramics Division, Oak Ridge National Laboratory, Oak Ridge, Tennessee.

Billmeyer, Prof. Fred W., Jr. Department of Chemistry, Rensselaer Polytechnic Institute.

Bing, Prof. R. H. Department of Mathematics, University of Wisconsin.

Bingham, Dr. Richard. Department of Chemistry, Princeton University.

Bird, Dr. John M. Department of Geology, State University of New York, Albany.

Birkhoff, Prof. Garrett. Department of Mathematics, Harvard University.

Birks, Dr. L. S. Retired; formerly, Radiation Technology Division, Naval Research Laboratory, Washington, D.C.

Birmingham, Jeannette M. Mycology & Botany Department, American Type Culture Collection, Rockville, Maryland.

Biscaye, Dr. Pierre E. Lamont-Doherty Geological Observatory, Palisades, New York.

Bisplinghoff, Raymond L. Dean, School of Engineering, Massachusetts Institute of Technology.

Bjorken, Prof. J. D. Stanford Linear Acceleration Center, Stanford University.

Black, Prof. Clanton C., Jr. Department of Biochemistry, Boyd Graduate Studies Research Center, University of Georgia.

Black, Dr. Harold S. Communications Consultant, Summit, New Jersey.

Black, Prof. Paul H. Department of Mechanical Engineering, Ohio University.

Black, Dr. Robert F. Department of Geology, University of Connecticut.

Blackburn, J. Lewis. Consulting Engineer, Bothell, Washington.

Blad, Dr. Blaine L. Agricultural Meteorology Section, University of Nebraska.

Blair, Dr. John E. Head (retired), Department of Microbiology, Roosevelt Hospital, New York.

Blair, Lewis N. Anaconda Company, Butte, Montana.

Blake, Dr. Daniel B. Department of Geology, University of Illinois, Urbana.

Blake, Neal A. Acting Deputy Associate Administrator for Engineering and Development, Department of Transportation, Federal Aviation Administration.

Blanc, Milton L. Research Climatologist, Office of Climatology, U.S. Weather Bureau, Tempe, Arizona.

Blanden, Dr. R. V. Department of Microbiology, John Curtin School of Medical Research, Canberra City, Australia.

Blandford, Prof. Roger D. Department of Theoretical Astrophysics, California Institute of Technology, Pasadena.

Blankenship, Dr. James E. Department of Physiology and Biophysics, Marine Biomedical Institute, University of Texas, Galveston.

Blanton, J. W. General Manager, Advanced Component Technology Department, General Electric Company, Cincinnati, Ohio.

Blaser, Prof. H. Weston. Department of Botany, University of Washington.

Blass, Dr. Elliott M. Department of Psychology, Johns Hopkins University.

Blatt, Prof. Frank J. Department of Physics, Michigan State University.

Blaxter, Prof. J. H. S. Dunstaffnage Marine Laboratory, The Scottish Association of Marine Science, Scotland.

Bleiweis, John J. COMSAT Labs, Clarksburg, Maryland.

Blencoe, Dr. James G. Chemistry Division, Oak Ridge National Laboratory.

Blew, J. Oscar, Jr. Consultant, Madison, Wisconsin.

Blitz, Dr. Leo. Laboratory for Millimeter-Wave Astronomy, University of Maryland.

Block, Dr. Ira. Department of Textiles-Consumer Economics, University of Maryland.

Bloomer, Dr. Sherman H. Department of Geosciences, Oregon State University, Corvallis.

Bloomfield, Prof. Philip E. Department of Physics, University of Pennsylvania; City College of the City University of New York, Bloomfield.

Bloomfield, Dr. Victor A. Department of Chemistry, University of Minnesota.

Blouin, Dr. Edmour F. College of Veterinary Medicine, Oklahoma State University, Stillwater.

Blumenthal, Prof. Leonard M. Deceased; formerly, Defoe Distinguished Professor of Mathematics, University of Missouri.

Blumenthal, Warren B. Blumenthal-Zirconium, North Tonawanda, New York.

Boast, Dr. Warren B. Anson Marston Distinguished Professor Emeritus of Electrical Engineering, Iowa State University.

Bobart, George F. Manager, Induction Heating and Ultrasonic Cleaning Department, Westinghouse Electric Corporation, Sykesville, Maryland.

Bobbitt, Dr. James M. Department of Chemistry, University of Connecticut.

Bochner, Dr. Salomon. Deceased; formerly, Department of Mathematics, Rice University.

Bock, Dr. Walter J. Professor of Evolutionary Biology, Department of Biological Sciences, Columbia University.

Bocko, Dr. Mark F. Department of Physics and Astronomy, College of Arts and Science, University of Rochester.

Boell, Dr. Edgar J. Department of Biology, Yale University.

Boggess, Prof. William R. Tree Ring Research Laboratory, University of Arizona.

Böhme, Prof. M. Department of Chemistry, Friedrich-Wilhelms University, Germany.

Bohonos, Dr. Nestor. Children's Cancer Research Foundation, Boston, Massachusetts.

Boisvert, Dr. Paul L. Deceased; formerly, Director of Streptococcus Laboratory, School of Medicine, Yale University.

Bok, Dr. Bart J. Deceased; formerly, Professor Emeritus, Department of Astronomy, University of Arizona.

Boke, Dr. Norman H. George Lynn Cross Research Professor of Botany, University of Oklahoma.

Bolender, Dr. Nicole-Fr. Department of Radiology, University of Washington.

Bolin, Dr. Steven R. Research Leader, Virology Cattle Research, Agricultural Research Service, U.S. Department of Agriculture, Ames, Iowa.

Bolt, Prof. J. A. Department of Mechanical Engineering, University of Michigan.

Bolz, Dr. Ray E. Leonard S. Case Professor and Dean of Engineering, Case Western Reserve University.

Bondybey, Dr. Vladimir E. Bell Laboratories, Murray Hill, New Jersey.

Booth, Dr. Grayce M. Manager, Business Communication, Honeywell Bull, Inc., Phoenix, Arizona.

Boothroyd, Dr. Carl W. Department of Plant Pathology, Cornell University.

Borrello, Sebastian R. Central Research Laboratories, Texas Instruments Inc., Dallas, Texas.

Borst, Henry V. Henry V. Borst & Associates, Wayne, Pennsylvania.

Bortman, Dr. Marci L. Waste Management Institute, Marine Sciences Research Center, State University of New York, Stony Brook.

Boshkov, Prof. Stefan H. Henry Krumb School of Mines, Columbia University.
Boston, Dr. Orlan W. Professor Emeritus, College of Engineering, University of Michigan.
Bothwell, Dr. Mark. Biochemical Sciences Laboratory, Princeton University.
Botset, Holbrook G. Professor Emeritus of Petroleum Engineering, University of Pittsburgh.
Bouchard, Prof. Thomas J. Chairman, Department of Psychology, University of Minnesota.
Bouton, George M. Bell Telephone Laboratories, Murray Hill, New Jersey.
Bove, Dr. Joseph R. Department of Laboratory Medicine, Yale University.
Bower, Frank H. Integrated Circuit Engineering Corporation, Palo Alto, California.
Bowler, Dr. Bruce E. Division of Chemistry and Chemical Engineering, California Institute of Technology.
Bowman, Prof. Harry L. Deceased; formerly, Dean, Drexel Institute of Technology, Philadelphia, Pennsylvania.
Bowman, Richard G. Deceased; formerly, Technical Assistant to the President, Republic Aviation Corporation, Farmingdale, New York.
Bowman, Rush A. Consulting Engineer, Memphis, Tennessee.
Bowman, Dr. Thomas E. Division of Crustacea, Smithsonian Institution.
Bowman, Waldo G. Editor (retired), "Engineering News Record."
Boyd, Dr. F. R. Geophysical Laboratory, Washington, D.C.
Boyd, Dr. Richard H. Department of Chemical Engineering, University of Utah.
Boyd, Victor P. Deputy Director, International Electronic Message Systems, U.S. Postal Service, Washington, D.C.
Boyden, Prof. Edward A. Department of Biological Structure, School of Medicine, University of Washington.
Boynton, Dr. Robert M. Department of Psychology, University of California, San Diego.
Boynton, Dr. Robert S. National Lime Association, Washington, D.C.
Bozorth, Dr. Richard M. Research Physicist (retired), Bell Telephone Laboratories, Murray Hill, New Jersey.
Bradfield, Dr. Richard. Emeritus Professor, Cornell University.
Bradford, Dr. William L. Deceased; formerly, School of Medicine and Dentistry, University of Rochester.
Bradley, Dr. Robert L., Jr. Department of Food Science, College of Agricultural and Life Sciences, University of Wisconsin.
Bradley, Prof. S. Gaylen. Department of Microbiology and Immunology, Virginia Commonwealth University, Richmond.
Bradley, Prof. William F. Department of Chemical Engineering, University of Texas, Austin.
Bradley, Prof. William W. Department of Civil Engineering and Institute of Colloid and Surface Science, Clarkson College of Technology.
Bradner, Mead. Technical Director, Foxboro Company, Foxboro, Massachusetts.
Bradsher, Prof. Charles K. Department of Chemistry, Duke University.
Bralla, J. R. Casey. Consultant, Mechanical Engineering, Martinez, Georgia.
Brammer, Dr. Anthony J. Institute for Microstructural Science, National Research Council of Canada, Ottawa, Ontario.
Brand, Dr. Louis. Deceased; formerly, M. D. Anderson Professor of Mathematics, Department of Mathematics, University of Houston.
Brandli, Henry. Satellite Meteorologist, Melbourne, Florida.
Brandt, Dr. John C. Laboratory for Atmospheric and Space Physics, University of Colorado.
Brandt, Dr. Philip W. Department of Anatomy, College of Physicians and Surgeons, Columbia University.
Brantley, Dr. Blake B., Jr. Department of Horticulture, University of Georgia College of Agriculture.
Brazel, Prof. Anthony J. Department of Geography, Laboratory of Climatology, Arizona State University, Tempe.
Breazeale, Prof. M. A. Department of Physics and Astronomy, University of Tennessee.
Breger, Dr. Irving A. Deceased; formerly, Office of Energy Resources, U.S. Geological Survey, Reston, Virginia.
Breier, Dr. Alan. Department of Psychiatry, Yale University School of Medicine, and Connecticut Mental Health Center, New Haven.
Breinan, Dr. Edward M. Materials Science Laboratory, United Technologies Research Center, East Hartford, Connecticut.
Breslin, Dr. Vincent T. Waste Management Institute, Marine Sciences Research Center, State University of New York, Stony Brook.
Breslow, Dr. David S. David S. Breslow Associates, Wilmington, Delaware.
Bressler, Dr. Glenn O. Department of Poultry Science, Pennsylvania State University.
Bretz, Dr. J. Harlen. Deceased; formerly, Department of Geophysical Sciences, University of Chicago.
Breuer, Glenn D. Consulting Engineer, Transmission Systems Development and Engineering Department, General Electric Company, Schenectady, New York.
Brewer, Dr. Charles P. Shell Development Corporation, Emeryville, California.
Brian, Dr. M. V. Institute of Terrestrial Ecology, Furzebrook Research Station, Wareham, England.
Bricker, Dr. Clark E. Department of Chemistry, University of Kansas.
Brickley, Robert L. Marketing Manager, Westinghouse Electric Corporation, Springfield, Massachusetts.
Bridgman, Prof. Percy W. Deceased; formerly, Harvard University.
Briggs, Dr. John D. Department of Entomology, Ohio State University.
Briggs, Dr. Winslow R. Department of Plant Biology, Carnegie Institution of Washington, Stanford, California.
Brigham, Harry S. Executive Vice President, Dixilyn Corporation, New Orleans, Louisiana.
Bright, Prof. James R. Retired; Graduate School of Business, University of Texas, Austin.
Brindley, Dr. George W. Department of Geosciences, Pennsylvania State University.
Brink, Dr. Joseph A., Jr. Deceased; formerly, Professor and Chairman, Department of Chemical and Nuclear Engineering, Washington State University.
Brintzinger, Dr. Hans. Fachbereich Biologie-Chemie, Universitat Konstanz, Germany.
Bristol, Edgar H. Corporate Research Department, The Foxboro Company, Foxboro, Massachusetts.
Broadbent, Carl D. Manager, Mining Engineering, Kennecott Copper Corporation, Salt Lake City, Utah.
Brockhouse, Dr. Bertram N. Institute for Materials Research, McMaster University.
Brockway, Prof. Lawrence O. Deceased; formerly, Department of Chemistry, University of Michigan.
Brodersen, Prof. Klaus. Professor of Inorganic and Analytical Chemistry, University of Erlangen-Nürnberg, Germany.
Brodhag, Dr. Alex E., Jr. Senior Editor, Organic Abstract Editing Department, Chemical Abstracts Service, Columbus, Ohio.
Brodsky, Dr. Frances M. Becton Dickinson Monoclonal Research Center, Inc., Mountain View, California.
Brodsky, Dr. William A. Mount Sinai School of Medicine of the City University of New York; Institute for Medical Research and Studies, New York.
Brody, Aaron L. Mead Packaging, Atlanta, Georgia.
Brofman, Dr. Peter J. Senior Engineering, General Technology Division, IBM Corp., Hopewell Junction, New York.
Bromer, Dr. William W. Research Advisor, Eli Lilly and Company, Indianapolis, Indiana.
Bromley, Prof. D. Allan. Science Adviser to the President of the United States.

Bromley, Willard S. Consulting Forester and Association Consultant, New Rochelle, New York.
Brook, Rollins. Bolt Beranek and Newman Inc., Canoga Park, California.
Brooke, Dr. M. M. Deceased; formerly, Deputy Chief, Licensure and Development Branch, Center for Disease Control, U.S. Public Health Service, Atlanta, Georgia.
Brooks, Dr. Frank D. Department of Physics, University of Capetown, Rondebosch, South Africa.
Brooks, Prof. Philip R. Department of Chemistry, Wise School of Natural Sciences, Rice University.
Brouwer, Prof. Dirk. Deceased; formerly, Director, Observatory, Yale University.
Brower, Prof. William B., Jr. Division of Fluids, Chemical and Thermal Processes, Rensselaer Polytechnic Institute.
Brown, Prof. Frank A., Jr. Morrison Professor of Biology, Northwestern University.
Brown, Prof. Frederick C. Department of Physics, University of Illinois.
Brown, Dr. Glenn H. Professor of Chemistry and Director of the Liquid Crystal Institute, Kent State University.
Brown, Dr. Harold. Lawrence Radiation Laboratory, University of California, Livermore.
Brown, Prof. Harold W. College of Physicians and Surgeons, Columbia University.
Brown, Prof. Herbert C. Department of Chemistry, Purdue University.
Brown, Dr. Robert L. National Radio Astronomy Observatory, Charlottesville, Virginia.
Brown, Dr. William F. Faculty of Medicine, Electromyogram Laboratory, University Hospital of Western Ontario, London, Canada.
Brown, Dr. William L., Jr. Emeritus Professor of Entomology, Cornell University.
Brownell, Dr. Gordon L. Head, Physics Research Laboratory, Massachusetts General Hospital, Boston.
Browning, Dr. Keith A. Chief, Meteorological Office, Radar Research Laboratory, Royal Signals and Radar Establishment, Worcester, England.
Brownlee, Dr. Donald E., II. Department of Astronomy, University of Washington, Seattle.
Broyer, Prof. Theodore C. Department of Soils and Plant Nutrition, University of California, Berkeley.
Brummett, Prof. Anna Ruth. Associate Professor of Biology and Associate Dean, Oberlin College.
Bruner, Prof. Jerome S. Department of Psychology, Harvard University.
Bruno, Michael H. Graphics Arts Consultant, Bradenton, Florida.
Brush, Dr. Alan H. Department of Biology, University of Connecticut.
Bryan, Prof. Leslie A. Director, Institute of Aviation, University of Illinois.
Bryan, Robert J. La Habra, California.
Brylawski, Prof. Thomas. Department of Mathematics, University of North Carolina.
Bryson, Prof. Arthur E., Jr. Chairman, Department of Aeronautics and Astronautics, Stanford University.
Bube, Dr. Richard H. Professor of Electrical Engineering, Department of Material Sciences, Stanford University.
Buchwald, Dr. N. F. Professor of Plant Pathology, Royal Veterinary and Agricultural College, Copenhagen, Denmark.
Buck, Dr. Richard P. Department of Chemistry, University of North Carolina.
Buermann, Thomas M. President, Gibbs & Cox, Inc., New York, New York.
Bullock, Dr. Theodore H. Department of Neurosciences, University of California, San Diego.
Bunnett, Prof. Joseph F. Department of Chemistry, University of California, Santa Cruz.
Bunshah, Dr. Rointan F. Department of Material Science and Engineering, University of California, Los Angeles.
Buntschuh, Dr. Robert F. Astrospace Division, General Electric Company, Princeton, New Jersey.
Burachinsky, Dr. Bohdan V. Consultant, Florham Park, New Jersey.
Burdick, Dr. Donald. Richard B. Russell Agricultural Research Center, Athens, Georgia.
Bureau of Reclamation. U.S. Department of the Interior, Denver, Colorado.
Burgdorfer, Dr. Willy. Scientist Emeritus, Laboratory of Pathology, Rocky Mountain Laboratories, National Institutes of Health, Department of Health and Human Services, Hamilton, Montana.
Burgess, Prof. John C. Department of Mechanical Engineering, University of Hawaii, Manoa.
Burghard, Ronald A. Gould Semiconductors, Pocatello, Indiana.
Burke, Dr. James D. Jet Propulsion Laboratory, California Institute of Technology.
Burman, Russel W. Manager, Refractory Division, AMAX Specialty Metals Corp., Greenwich, Connecticut.
Burnham, Prof. Charles W. Department of Earth and Planetary Sciences, Harvard University.
Burns, Roger G. Department of Earth Science, Massachusetts Institute of Technology.
Bursey, Prof. Maurice M. Department of Chemistry, University of North Carolina.
Bursky, Dave. "Electronic Design Magazine," Sunnyvale, California.
Burte, Dr. Harris M. Chief, Metals and Ceramics Division, Air Force Materials Laboratory, Wright-Patterson Air Force Base, Dayton, Ohio.
Burton, Prof. Ralph A. Chairman, Department of Mechanical Engineering and Astronautical Sciences, Northwestern University.
Burwell, Dr. Robert L. Department of Chemistry, Northwestern University.
Busby, R. Frank. Director, R. Frank Busby Associates, Inc., Arlington, Virginia.
Busch, Prof. Daryle H. Department of Chemistry, Ohio State University.
Busser, Dr. John H. Executive Secretary, Bio-Instrumentation Advisory Council, American Institute of Biological Sciences, Washington, D.C.
Butler, Charles. Consultant, Pescadero, California.
Butler, Dr. Elmer G. Henry Fairfield Osborn Professor of Biology, Emeritus, Princeton University.
Butler, Dr. G. W. Olin Rocket Research Company, Redmond, Washington.
Butler, Dr. Robert W. Department of Psychology, Child Development and Rehabilitation Center, Oregon Health Sciences University, Portland.
Butts, Prof. Allison. Professor Emeritus of Metallurgy and Materials Science, Lehigh University.
Byers, Dr. Breck E. Biological Laboratories, Harvard University.
Byers, Dr. Charles W. Department of Geology, University of Wisconsin.

C

Cable, Prof. Raymond M. Department of Biological Sciences, Purdue University.
Cadle, Dr. Richard D. Retired; formerly, Department of Atmospheric Chemistry, National Center for Atmospheric Research, Boulder, Colorado.
Cady, Dr. Gilbert H. Deceased; formerly, Consulting Coal Geologist, Urbana, Illinois.
Cain, Dr. Stanley A. Director, Institute for Environmental Quality, and Charles Lathrop Pack Professor, Department of Resource Planning and Conservation, University of Michigan.
Caine, Dr. Edsel A. Associate Professor of Marine Sciences, University of South Carolina.
Calderwood, Dr. James H. Department of Electronic Engineering, University College, Galway, Ireland.
Caldwell, Prof. C. Denise. Department of Physics, Yale University.
Calef, Dr. Wesley. Department of Geography, Illinois State University.
Caley, Dr. Earle R. Department of Chemistry, Ohio State University.
Callaham, Dr. Michael. Director, Emergency Medical Services, School of Medicine, University of California, San Francisco.

Callaway, Prof. Joseph. Department of Physics and Astronomy, Louisiana State University.
Calvert, Dr. Donald R. Director, Central Institute for the Deaf, St. Louis, Missouri.
Calvert, Dr. Stephen E. Grant Institute of Geology, University of Edinburgh, Scotland.
Cambel, Dr. Ali B. Executive Vice President for Academic Affairs, Wayne State University.
Cameron, Dr. A. E. Analytical Chemistry Division, Oak Ridge National Laboratory, Oak Ridge, Tennessee.
Cameron, Dr. A. G. W. Associate Director for Planetary Sciences, Center for Astrophysics, Harvard College Observatory.
Camin, Dr. Joseph H. Department of Entomology, University of Kansas.
Camp, Prof. Charles L. Deceased; formerly, Department of Paleontology, University of California, Berkeley.
Campana, Dr. Richard J. Department of Botany and Plant Pathology, University of Maine.
Campanella, Dr. S. Joseph. Retired; formerly, COMSAT Laboratories, Clarksburg, Maryland.
Campbell, Dr. Andrew C. Department of Zoology and Comparative Physiology, Queen Mary College, University of London, England.
Campbell, Harold E. Senior Engineer, Power Distribution Systems Engineering, General Electric Company, Schenectady, New York.
Campbell, Dr. Howard W. National Fisheries and Wildlife Laboratory, Gainesville, Florida.
Campbell, Dr. Laurence J. Low Temperature Physics Group, Los Alamos Scientific Laboratory, Los Alamos, New Mexico.
Cannon, Helen L. U.S. Geological Survey, Denver, Colorado.
Cannon, Dr. Robert H., Jr. Assistant Secretary of Transportation.
Capra, Dr. J. Donald. Department of Microbiology, Southwestern Medical School, University of Texas, Dallas.
Carberry, Dr. James J. Department of Chemical Engineering, University of Notre Dame.
Carew, Dr. H. John. Professor of Horticulture, Michigan State University.
Carey, Prof. Tom. Department of Computing and Information Science, College of Physical Science, University of Guelph, Ontario, Canada.
Carlson, Dr. Jack H. Rhone Meriux, Inc., Athens, Georgia.
Carlson, William C. Tree Physiologist, Weyerhauser Company, Hot Springs, Arkansas.
Carlsson, Dr. D. J. National Research Council of Canada, Ottawa, Ontario.
Carone, Dr. Frank A. Morrison Professor of Pathology, Medical School, Northwestern University.
Carpenter, Dr. Frank M. Department of Biology, Harvard University.
Carpenter, Dr. Stephen, R. Department of Biological Sciences, University of Notre Dame.
Carr, Braxton B. President, American Waterways Operators, Inc., Washington, D.C.
Carroll, Dr. John M. Department of Computer Science, University of Western Ontario, Canada.
Carroll, Robert G., Jr. Manager of Technical Development, Mirafi, Inc., Charlotte, North Carolina.
Carroll, Dr. Robert L. Redpath Museum, McGill University, Montreal, Quebec, Canada.
Carruthers, Dr. Marvin H. Department of Chemistry, University of Colorado.
Carson, Dr. E. R. Department of Systems Science, City University, London, England.
Carson, Robert W. Formerly, Managing Editor, "Product Engineering," New York.
Carter, Archie N. Buan, Carter & Associates, Inc., Minneapolis, Minnesota.
Carter, Dr. George S. Fellow of Corpus Christi College, Lecturer in Zoology (retired), Cambridge University, England.
Carter, Prof. Herbert E. Vice Chancellor for Academic Affairs, University of Illinois.
Carter, Dr. Joseph G. Department of Geology, University of North Carolina.
Casasent, Prof. David. Department of Electrical and Computer Engineering, Carnegie-Mellon University.
Casey, Dr. H. C., Jr. Department of Electrical Engineering, Duke University.
Cassens, Prof. Linda J. Department of Plant Sciences, College of Agriculture, University of Arizona.
Castenholz, Dr. Richard W. Department of Biology, University of Oregon.
Castleman, Prof. A. Welford, Jr. College of Science, Department of Chemistry, Pennsylvania State University.
Catania, Prof. A. Charles. Department of Psychology, University of Maryland.
Catchpole, Dr. John. Admiralty Materials Laboratory, Dorset, England.
Caughlan, Prof. Georgeanne R. Department of Physics, Montana State University.
Cave, Dr. M. Donald. Department of Anatomy, University of Arkansas for Medical Sciences.
Cavender, Ted M. Museum of Zoology, University of Michigan.
Celler, Dr. George K. AT&T Laboratories, Murray Hill, New Jersey.
Cestone, Dr. Joseph A. Formerly, Deep Submergence System Project, U.S. Navy.
Chace, Dr. Fenner A., Jr. Senior Zoologist, Department of Invertebrate Zoology, Smithsonian Institution.
Chaffin, Dr. R. J. Solid State Device Physics, Sandia National Laboratories, Albuquerque, New Mexico.
Chakoumakos, Dr. Bryan C. Department of Geology, Northrup Hall, University of New Mexico.
Chandler, Richard F. Mechanical Engineer, Air Force Missile Development Center, Holloman Air Force Base, New Mexico.
Chang, Dr. M. C. Worcester Foundation for Experimental Biology, Shrewsbury, Massachusetts.
Chang, Prof. S. S. L. Department of Electrical Sciences, State University of New York, Stony Brook.
Chanlett, Dr. Emil T. Department of Environmental Sciences and Energy, University of North Carolina.
Chao, Dr. Edward C. T. Geological Survey, U.S. Department of the Interior, Reston, Virginia.
Chapman, Brian M. Graphic Imaging Systems Division, Eastman Kodak Company, Bedford, Massachusetts.
Chapman, Dr. Carleton A. Department of Geology, University of Illinois.
Chapman, Dr. Clark R. Planetary Science Institute, Tucson, Arizona.
Chapman, Dr. Richard A. Central Research Laboratories, Texas Instruments Inc., Dallas, Texas.
Chapman, Dr. Robert D. Section Head, Laboratory for Solar Physics and Astrophysics, NASA Goddard Space Flight Center, Beltsville, Maryland.
Chapman, Dr. Sydney. Deceased; formerly, High Altitude Observatory, University of Colorado.
Chapman, Dr. Valentine J. Deceased; formerly, University of Auckland, New Zealand.
Chappell, Dr. M. A. University of California, Riverside.
Charig, Dr. Alan J. Chief Curator of Fossil Amphibians, Reptiles and Birds, British Museum, London, England.
Charlton, Dr. Ernest E. Research Engineer (retired), Schenectady, New York.
Charney, Dr. Dennis S. Department of Psychiatry, Yale University School of Medicine, and Connecticut Mental Health Center, New Haven.
Charnock, Dr. Henry. Department of Oceanography, University of Southampton, England.
Chase, Dr. Julia. Department of Biology, Barnard College.
Chaudhari, Praveen. Thomas J. Watson Research Center, IBM, Yorktown Heights, New York.
Cheadle, Prof. Vernon I. Chancellor, University of California, Santa Barbara.
Cheatham, Dr. J. B., Jr. Professor of Mechanical Engineering and Chairman of Department of Mechanical and Aerospace Engineering and Materials Science, Rice University.
Chellis, Robert D. Deceased; formerly, Structural Engineer, Wellesley Hills, Massachusetts.
Chen, Dr. Albert Fu-Tai. Principal Research Chemist, Applied Research Department, Beckman Instruments, Brea, California.

Chen, Prof. C. T. Department of Electrical Science, State University of New York, Stony Brook.
Chen, Dr. Mo-Shing. Director, Energy Systems Research, University of Texas, Arlington.
Chen, Dr. Shih-Yuan. Department of Environmental Sciences, Rand Corporation, Santa Monica, California.
Chenault, Dr. Alice A. Marshall Space Flight Center, Huntsville, Alabama.
Chenault, Dr. Keith. Department of Chemistry, University of Delaware, Newark.
Chenault, Roy L. Chief Research Engineer (retired), Oilwell Division, U.S. Steel Corporation, Dallas, Texas.
Chesnut, Dr. Charles H., III. University Hospital, University of Washington, Seattle.
Cheung, Dr. Wai Yiu. Saint Jude's Children's Research Hospital, Memphis, Tennessee.
Chickering, Dr. Arthur M. Museum of Comparative Zoology, Harvard University.
Chilberg, Guy L. Engineering Director, Outside Plant, American Telephone and Telegraph Company, New York.
Chin, Dr. Gilbert Y. AT&T Bell Telephone Laboratories, Murray Hill, New Jersey.
Chinitz, Dr. Wallace. Department of Mechanical Engineering, Cooper Union.
Chiodo, Prof. Louis A. Chairperson, Department of Pharmacology, Texas Tech University Health Sciences Center, Lubbock.
Chipkin, Dr. Richard E. Department of Pharmacology, Schering Corporation, Bloomfield, New Jersey.
Chodos, Dr. Alan. Department of Physics, Yale University.
Chokroverty, Prof. Sudhansu. Chief, Neurology Service, VA Medical Center, Lyons, New Jersey.
Choueiri, Dr. Edgar Y. Electric Propulsion Laboratory, Engineering Quadrangle, Princeton University.
Christ, Dr. Charles L. Physicist, U.S. Geological Survey.
Christensen, Dr. James. Department of Medicine, University of Iowa.
Christensen, Dr. Martha. Professor Emeritus, Department of Botany, University of Wyoming, Laramie.
Christie, Dr. John M. Department of Geology, University of California, Los Angeles.
Christie, Richard W. Hardesty & Hanover, New York.
Chu, Prof. Benjamin. Department of Chemistry, State University of New York, Stony Brook.
Chu, Dr. Steven. Department of Physics, Stanford University.
Chu, Dr. T. Ming. Roswell Park Memorial Institute, Buffalo, New York.
Chumley, Forrest G. Department of Biology, University of Utah.
Ciegler, Dr. Alex. Southern Regional Research Center, USDA Science and Education Administration, New Orleans.
Clapp, Prof. Leallyn B. Department of Chemistry, Brown University.
Clapp, Philip F. National Weather Service, National Oceanic and Atmospheric Administration, Washington, D.C.
Clark, Dr. David A. Department of Medicine, McMaster University, Ontario, Canada.
Clark, Prof. Francis. Department of Microbiology, University of Illinois.
Clark, Prof. George L. Deceased; formerly, Emeritus Research Professor of Analytical Chemistry, University of Illinois.
Clark, Noel A. Department of Physics, University of Colorado at Boulder.
Clark, Richard F. Division of Physics, National Research Council of Canada, Ottawa, Ontario.
Clark, Prof. Wallis H., Jr. Director, Aquaculture Program, University of California, Davis.
Clark, Dr. Walter. Research Laboratories, Eastman Kodak Company, Rochester, New York.
Clarke, Ellen. Columbia Organic Chemicals Company, Inc., Columbia, South Carolina.
Clarke, Dr. J. Harold. Horticulturist, Clarke Nursery, Long Beach, Washington.
Clarke, Dr. Jeffrey J. U.S. Army Center of Military History, Washington, D.C.
Clarke, Prof. John. Department of Physics, University of California, Berkeley.
Clarke, John H. Engineering Design Consultant, Air Conditioning and Power, Films-Packaging Division, Union Carbide Corporation, Chicago, Illinois.
Clausen, Dr. Harry J. Executive Secretary, Career-Development Review Branch, Division of Research Grants, Department of Health, Education, and Welfare, Bethesda, Maryland.
Clemence, Dr. Gerald M. Deceased; formerly, Observatory, Yale University.
Clemens, Craig B. Geologist, Appalachian Coal Surveys, Pittsburgh, Pennsylvania.
Clemens, Dr. William A. Museum of Paleontology, University of California, Berkeley.
Clement, Prof. Anthony C. Department of Biology, Emory University.
Clements, John A. Director, Corporate Product Evaluation, Gillette Company, Boston, Massachusetts.
Clements, Dr. Leo P. Emeritus Professor of Anatomy, School of Medicine, Creighton University.
Clements, Dr. Thomas. Consulting Geologist, Los Angeles, California.
Cleminshaw, Dr. Clarence H. Deceased; formerly, Director Emeritus, Griffith Observatory, Los Angeles, California.
Clench, Dr. William J. Museum of Comparative Zoology, Harvard University.
Cleve, Dr. Hartwig. Institute for Anthropology and Human Genetics, University of Munich, Germany.
Cleveland, Prof. L. F. Department of Electrical Engineering, Northeastern University.
Clifton, Robert A. Bureau of Mines, U.S. Department of the Interior.
Clinton, William P. Research Fellow, Maxwell House Company, General Food Corporation, Tarrytown, New York.
Cloud, Dr. Preston E., Jr. Department of Geology, University of California, Santa Barbara.
Clowes, Dr. F. A. L. Botany Department, University of Oxford, England.
Cobban, Dr. William A. Geologist, U.S. Geological Survey, Denver, Colorado.
Cobine, Dr. James D. Deceased; formerly, Professor and Senior Scientist, Atmospheric Sciences Research Center, State University of New York, Albany.
Coddington, Dr. Jonathan A. National Museum of Natural History, Smithsonian Institution, Washington, D.C.
Cohart, Prof. Edward M. C.E.A. Winslow Professor of Public Health, School of Medicine, Yale University.
Cohen, Dr. Alan S. Director, Thorndike Memorial Laboratory, Boston City Hospital, Boston, Massachusetts.
Cohen, Charles A. Research Associate (retired), Exxon Research and Engineering Company, Florham Park, New Jersey.
Cohen, Dr. David. National Magnetic Laboratory, Massachusetts Institute of Technology.
Cohen, Dr. David B. Department of Psychology, University of Texas, Austin.
Cohen, Dr. Donald J. Assistant Professor, Child Study Center, Yale University.
Cohen, Dr. E. Richard. Rockwell International Science Center, Thousand Oaks, California.
Cohen, Dr. Gene. National Institute on Aging, National Institutes of Health, Bethesda, Maryland.
Cohen, Dr. Jerome. Department of Psychiatry, Medical School, Northwestern University.
Cohen, Dr. Lisa M. Department of Dermatology, Beth Israel Hospital, Boston, Massachusetts.
Cohn, Stanley H. Department of Industrial Engineering, University of Toronto, Ontario, Canada.
Cole, Dr. Edward H. Department of Immunopathology, Scripps Clinic and Research Foundation, La Jolla, California.
Cole, Dr. W. Storrs. Professor Emeritus, Department of Geology, Cornell University.
Coleal, Dr. Ernest. Formerly, Assistant for Facilities Development, Flight Research Division, Edwards Air Force Base, California.
Coleman, Dr. David C. Department of Entomology, University of Georgia.
Coleman, Dr. Robert E. Staff Scientist, Sugar Crops Production, National

Program Staff, BARC-WEST, Beltsville, Maryland.

Collier, R. J. Bell Telephone Laboratories, Murray Hill, New Jersey.

Collins, Dr. Dean R. Director, CCD Technology Laboratory, Central Research Laboratories, Texas Instruments Inc., Dallas, Texas.

Collins, Prof. J. H. Department of Electrical Engineering, University of Texas, Arlington.

Collins, Dr. Steven J. Department of Clinical Neurosciences, St. Vincent's Hospital, Melbourne, and Department of Pathology, University of Melbourne, Australia.

Collinson, Dr. Charles. Department of Geology, University of Illinois.

Colonna, Richard A. Deceased; formerly, NASA Senior Representative in Australia, United States Embassy.

Colwell, Dr. Rita R. Department of Microbiology, Division of Agricultural and Life Sciences, University of Maryland.

Comeau, Richard. The Foxboro Company, Foxboro, Massachusetts.

Comings, Dr. Edward W. University of Petroleum and Minerals, Dhahran, Saudi Arabia.

Commins, Prof. Eugene D. Department of Physics, University of California, Berkeley.

Compton, Robert D. Electro-Optical Systems Design, Milton S. Kiver Publications, Inc., Chicago, Illinois.

Condon, Prof. E. U. Deceased; formerly, Fellow, Joint Institute for Laboratory Astrophysics, and Professor Emeritus of Physics, University of Colorado.

Confer, Prof. Anthony W. Head, Department of Veterinary, Pathology, Oklahoma State University, Stillwater.

Connors, Dr. James F. Propulsion Aerodynamics Division, National Aeronautics and Space Administration.

Conrad, Prof. Albert G. Dean Emeritus and Professor of Electrical Engineering, College of Engineering, University of California, Santa Barbara.

Contreras, Roland. Laboratory of Molecular Biology, University of Ghent, Belgium.

Conwell, Dr. Esther M. Xerox Webster Research Center, Rochester, New York.

Cook, Dr. Charles G. Research Geneticist, Agricultural Research Service, U.S. Department of Agriculture, Weslaco, Texas.

Cook, Philip H. Plastics Department, Texas Division, Dow Chemical U.S.A., Freeport, Texas.

Cook, Dr. Richard K. Formerly, National Bureau of Standards.

Cooley, Prof. Hollis R. Professor Emeritus of Mathematics, New York University.

Cooley, Dr. William E. Winton Hill Technical Center, Proctor and Gamble Company, Cincinnati, Ohio.

Coombs, Dr. M. M. Imperial Cancer Research Fund, London, England.

Cooper, Dr. Anthony R. Dynapol, Palo Alto, California.

Cooper, Dr. Arthur W. Department of Botany, North Carolina State University.

Cooper, Franklin D. Bureau of Mines, U.S. Department of the Interior.

Cooper, Prof. W. Charles. Department of Metallurgical Engineering, Queens University, Kingston, Ontario, Canada.

Copel, Michel. Acoustronics, Inc., Huntington, New York.

Corben, Dr. Herbert C. Deceased; formerly, Ramo-Wooldridge Corporation, Los Angeles, California.

Core, Prof. Earl L. Professor of Botany, Department of Biology, West Virginia University.

Corliss, Dr. John O. Department of Zoology, University of Maryland.

Coroniti, S. C. Climatic Impact Assessment Program, Office of the Secretary of Transportation, Washington, D.C.

Corson, Almon J. Consulting Engineer (retired), Instrument Department, General Electric Company, West Lynn, Massachusetts.

Corwin, Harold G., Jr. McDonald Observatory at Mount Locke, University of Texas at Austin.

Costerton, Dr. J. William F. Department of Biology, University of Calgary, Alberta, Canada.

Cotterill, Carl H. Chief, Division of Statistical and Technical Services, Bureau of Mines, U.S. Department of the Interior.

Couch, Dr. John C. Matson Navigation Company, San Francisco, California.

Couch, Prof. John N. Department of Botany, University of North Carolina.

Couch, Dr. Richard B. Professor of Naval Architecture and Marine Engineering, Ship Hydrodynamics Laboratory, University of Michigan.

Couclelis, Prof. Helen. Department of Geography, University of California, Santa Barbara.

Coulter, Earl E. Manager, Mechanical Technology, Babcock and Wilcox Company, Barberton, Ohio.

Coupland, Dr. Robert T. Department of Plant Ecology, University of Saskatchewan, Canada.

Court, Dr. Arnold. Retired; formerly, Department of Climatology, California State University.

Cowles, Prof. Philip B. Professor Emeritus of Microbiology, School of Medicine, Yale University.

Cox, Dr. Arthur N. Los Alamos Scientific Laboratory, University of California, Los Alamos, New Mexico.

Cox, Dr. Charles S. Department of Oceanography, University of California, San Diego.

Coxeter, Prof. H. S. M. Department of Mathematics, University of Toronto, Ontario, Canada.

Coyle, Harold P. Harvard-Smithsonian Center for Astrophysics, Harvard College Observatory, Cambridge, Massachusetts.

Coyne, Dr. Michael J. Senior Veterinarian Technical Services, SmithKline Beecham Animal Health, Exton, Pennsylvania.

Crabill, Dr. Ralph E., Jr. Curator and Supervisor of the Division of Myriapoda and Arachnida, Department of Entomology, Smithsonian Institution, National Museum of Natural History.

Crafts, Dr. Alden S. Professor Emeritus, Department of Botany, University of California, Davis.

Craig, Prof. Homer V. Department of Mathematics, University of Texas, Austin.

Cramer, Dr. Friedrich. Max Planck Institut für Experimentelle Medizin, Göttingen, Germany.

Cramer, Dr. William S. Ship Acoustics Department, Naval Ship Research and Development Center, Washington, D.C.

Cranwell, Dr. Lucy M. Department of Geosciences, University of Arizona.

Crawford, Dr. Lionel V. Imperial Cancer Research Fund, London, England.

Cremer, Prof. Sheldon E. Department of Chemistry, Marquette University.

Creson, William K. Consulting Engineer (retired), Ross Gear and Tool Company, Inc., Lafayette, Indiana.

Crittenden, Dr. Charles V. Geographer, Economic Development Administration, U.S. Department of Commerce.

Crocker, Dr. Burton B. Monsanto Company, St. Louis, Missouri.

Crocker, Prof. Malcolm J. Department of Mechanical Engineering, Purdue University.

Cronan, Calvin S. Consulting Editor, "Chemical Engineering" McGraw-Hill Inc., New York.

Cronquist, Dr. Arthur. Director of Botany, New York Botanical Gardens, Bronx, New York.

Cross, Dr. Chester E. Agricultural Experiment Station, University of Massachusetts.

Crouse, William H. Consulting Editor, Automotive Books, McGraw-Hill Book Company, New York.

Crow, Dr. James F. Department of Genetics, University of Wisconsin.

Crow, Dr. Loren W. Consulting Meteorologist, President Consultants, Inc., Denver, Colorado.

Crowe, Prof. Clayton T. Department of Mechanical and Materials Engineering, Washington State University, Pullman.

Crowe, Dr. John H. Department of Zoology, University of California, Davis.

Crowell, Dr. Sears. Department of Zoology, Indiana University.

Crowson, Dr. Roy A. Department of Zoology, University of Glasgow, Scotland.

Cruickshank, Dr. Michael J. Look Laboratory, University of Hawaii.

Crum, Dr. Howard. Herbarium, University of Michigan.

Cuckler, Lee E. Interdivisional Coordinator, Aeronautical and Instrument Division, Robertshaw Controls Company, Anaheim, California.

Cuffey, Dr. Roger J. Department of Geology and Geophysics, Pennsylvania State University.
Culbertson, Dr. Joseph O. Assistant Chief (retired), Oil-seed and Industrial Crops Research Branch, Crops Research Division, U.S. Department of Agriculture, Beltsville, Maryland.
Cumberland, Prof. Kenneth B. Professor of Geography, University of Auckland, New Zealand.
Cummins, Prof. Harold. Professor Emeritus, School of Medicine, Tulane University.
Cunningham, Prof. Burris B. Deceased; formerly, Radiation Laboratory, University of California, Berkeley.
Curdts, Edward B. Deceased; formerly, Consultant; Director of Engineering, James G. Biddle Company, Philadelphia.
Curkendall, Dr. David W. NASA Jet Propulsion Laboratory, Pasadena, California.
Curtin, Dr. Charles B. Department of Biology, Creighton University.
Curtis, Prof. Charles F. Department of Chemistry, University of Wisconsin.
Cutler, Prof. Chapin C. Professor of Applied Physics, Stanford University.
Cutler, Dr. Edward B. Department of Biology, Utica College of Syracuse University.
Cutler, Dr. Leonard S. Director, Physical Research Laboratory, Hewlett-Packard Company, Palo Alto, California.
Cutter, Dr. Elizabeth G. Department of Cryptogamic Botany, University of Manchester, England.
Czamanske, Dr. Gerald K. U.S. Geological Survey, Menlo Park, California.

D

Dahl, Dr. Lawrence F. Department of Chemistry, University of Wisconsin.
Dalitz, Prof. Richard H. Department of Theoretical Physics, Oxford University.
Dallas, Daniel B. Society of Manufacturing Engineering, Dearborn, Michigan.
Dalrymple, Dr. David L. Department of Chemistry, University of Delaware.
Damon, Dr. Albert. Deceased; formerly, Department of Anthropology, Peabody Museum, Harvard University.
Dan, Dr. Bruce. American Medical Association, Chicago.
Danby, Dr. J. M. A. Department of Mathematics, North Carolina State University.
Daniel, James R. Department of Biochemistry, Purdue University.
Daniels, Dr. Farrington. Deceased; formerly, Professor Emeritus, Solar Energy Laboratory, University of Wisconsin.
Daniels, Prof. James M. Department of Physics, University of Toronto.
D'Arcangelo, Prof. Amelio M. Department of Naval Architecture, University of Michigan.
Date, Prof. Hikaru. Acoustics Department, Kyushu Institute of Design, Fukuoka, Japan.
Daubechies, Dr. Ingrid. AT&T Bell Laboratories, Murray Hill, New Jersey.
Daugherty, Prof. Robert L. Deceased; formerly, Division of Engineering, California Institute of Technology.
Davies, Keith. Manager, Propulsion Group, General Electric Astrospace, Princeton, New Jersey.
Davies, Dr. Peter J. Department of Botany, Cornell University.
Davis, Dr. Andrew M. Enrico Fermi Institute, University of Chicago, Illinois.
Davis, D. Dwight. Deceased; formerly, Curator of Vertebrate Anatomy, Chicago Museum of Natural History.
Davis, Dr. George H. Department of Geological Sciences, University of Arizona.
Davis, Prof. Gregory A. Department of Geological Sciences, University of Southern California.
Davis, Dr. Jack. Deceased; formerly, Research and Development, Bright Star Industries, Inc., Clifton, New Jersey.
Davis, Dr. John H. Professor Emeritus of Botany, University of Florida.
Davis, Prof. Kenneth P. Deceased; formerly, School of Forestry, Yale University.
Davis, Lieut. Michael E. P. U.S. Army Armament Material Readiness Command Headquarters, Rock Island, Illinois.
Davis, Dr. Richard A., Jr. Department of Geology, University of South Florida.
Davis, Rowland F. Products Planning Engineer (retired), American Telephone and Telegraph Company, New York.
Dawson, Dr. David M. West Roxbury Veterans Affairs Medical Center, West Roxbury, Massachusetts.
Deacon, T. A. Deceased; formerly, Division of Electrical Science, National Physical Laboratory, Teddington, Middlesex, England.
DeBach, Dr. Paul. Professor of Biological Control, University of California, Riverside.
de Boor, Dr. Carl. Mathematical Research Center, University of Wisconsin.
DeCarli, Dr. P. S. Poulter Laboratory, Stanford Research Institute.
DeClaris, Dr. Nicholas. Division of Emerging Energy Technologies, National Science Foundation, Washington D.C.
de Duve, Dr. Christian. Department of Biochemical Cytology, Rockefeller University.
DeFacio, Prof. Brian. Department of Physics and Astronomy, University of Missouri.
DeGroot, Dr. Leslie J. Department of Medicine, Section of Endocrinology, University of Chicago Medical Center.
Dehmelt, Dr. Hans. Department of Physics, University of Washington.
Dekeyser, Prof. Willy C. Laboratory for Crystallography, Ghent, Belgium.
Dekker, Dr. Adrianus J. Professor of Solid State Physics, University of Groningen, Netherlands.
de la Cruz, Felix. Chief, Mental Retardation and Development Disabilities Branch, National Institute of Child Health and Human Development, Bethesda.
Delahay, Prof. Paul. Department of Chemistry, New York University.
Deland, Dr. Raymond J. Retired; formerly, Department of Meteorology and Oceanography, New York University.
Delevoryas, Prof. Theodore. Department of Botany, University of Texas, Austin.
Del Fava, Dr. Raymond. Chairman, Public Information Committee, American College of Radiology, Reston, Virginia.
DeLong, Dr. Dwight M. College of Biological Sciences, Ohio State University.
Delson, Dr. Eric. Department of Vertebrate Paleontology, American Museum of Natural History.
DeMaggio, Dr. Augustus E. Department of Biological Sciences, Dartmouth College.
Demar, Dr. Robert E. Department of Geology, University of Illinois, Chicago.
Dengler, Dr. Nancy G. Department of Botany, University of Toronto, Ontario, Canada.
Den Hartog, Prof. J. P. Retired; formerly, Department of Mechanical Engineering, Massachusetts Institute of Technology.
Denison, Dr. Robert H. Deceased; formerly, Curator of Fossil Fishes, Field Museum of Natural History, Chicago, Illinois.
Deno, Dr. Richard A. Professor of Pharmacognosy, University of Michigan.
Dent, Dr. William. Department of Physics and Astronomy, University of Massachusetts.
DePaolo, Dr. Donald J. Department of Geology and Geophysics, University of California, Berkeley.
Derge, Prof. Gerhard. Department of Metallurgy and Material Science, Carnegie-Mellon University.
deSatnick, Steve. Vice President, Operations and Engineering, KCET-TV, Los Angeles, California.
Des S. Thomas, Dr. Donovan. Department of Biology, University of Windsor, Ontario, Canada.
Deutscher, Prof. Guy. Department of Physics, Tel Aviv University, Israel.
DeVaney, Fred D. Consulting Metallurgist, Duluth, Minnesota.
de Vaucoulers, Prof. Gerard. Department of Astronomy, University of Texas.

DeVore, Dr. George W. Department of Geology, Florida State University.
DeVries, Dr. Arthur L. Department of Physiology and Biophysics, University of Illinois, Urbana-Champaign.
de Weck, Dr. A. L. Institut für Klinische Immunologic Inselspital, Universitat Bern, Switzerland.
Dexter, Dr. Theodore H. Research Supervisor, Inorganic Chemistry, Hooker Chemical Corporation Research Center, Niagara Falls, New York.
Diamond, Dr. Seymour. Director, Diamond Headache Clinic, Ltd., Chicago, Illinois.
Dick, Prof. B. Gale. Department of Physics, University of Utah.
Dickieson, Alton C. Vice President (retired), Bell Telephone Laboratories, Sedonia, Arizona.
Diegel, William. Marketing Communications Department, E. I. du Pont de Nemours & Co., Wilmington, Delaware.
Dieke, Prof. G. H. Deceased; formerly, Chairman, Department of Physics, Johns Hopkins University.
Dienes, Dr. Louis L. Massachusetts General Hospital, Boston.
Dietrich, Prof. Gunter O. Deceased; formerly, Institute of Oceanography, University of Kiel, Germany.
Dietz, Dr. Robert S. Atlantic Oceanography and Meteorological Laboratories, NOAA, U.S. Department of Commerce, Miami, Florida.
Dill, Dr. Robert F. Ocean Mining Administration, U.S. Department of the Interior.
Dille, Dr. James M. Deceased; formerly, Professor and Chairman, Department of Pharmacology, School of Medicine, University of Washington.
Dillon, J. F., Jr. Bell Telephone Laboratories, Murray Hill, New Jersey.
DiMichele, Dr. William A. Department of Paleobotany, Smithsonian Institution, Washington, D.C.
Dimmel, Prof. Donald R. Research Associate, Wood Sciences, Chemical Sciences Division, Institute of Paper Chemistry, Appleton, Wisconsin.
Dine, Dr. Michael. Physics Department, City College of New York.
Dingle, Prof. Hugh. Department of Entomology, College of Agricultural and Environmental Sciences, University of California, Davis.
Dinsmore, Robertson P. International Ice Patrol, Woods Hole Oceanographic Institution, Woods Hole, Massachusetts.
Dirksen, Dr. Ellen R. Department of Anatomy, School of Medicine, University of California, Los Angeles.
DiSalvo, Dr. Frank J. Bell Laboratories, Murray Hill, New Jersey.
DiTallo, Cynthia. United Technologies, Otis Elevator Company, Farmington, Connecticut.
Dobrowolski, Dr. Zbigniew C. Deceased; formerly, Chief Development Engineer, Kinney Vacuum Company, Canton, Massachusetts.
Dodds, Robert H. Deceased; formerly, Personnel Manager, Gibbs and Hill, Inc., New York.
Dodge, Dr. Donald D. Principal Staff Engineer, Engineering and Research, Ford Motor Company, Dearborn, Michigan.
Dodington, Sven H. International Telephone and Telegraph, New York.
Dodson, Dr. Edward O. Department of Biology, University of Ottawa, Ontario, Canada.
Dogramaci, Prof. Ali. Department of Industrial and Management Engineering, Columbia University.
Doherty, Dr. Michael F. Head, Department of Chemical Engineering, University of Massachusetts.
Dollear, Frank G. Head, Peanut Product Investigation, Southern Utilization Research and Development Division, U.S. Department of Agriculture, New Orleans, Louisiana.
Donaldson, Dr. James A. Professor and Chairman, Department of Otolaryngology, School of Medicine, University of Washington.
Doob, Prof. Joseph L. Department of Mathematics, University of Illinois, Urbana-Champaign.
Doolan, Barry L. Department of Geology, Memorial University of Newfoundland, St. John's, Canada.
Dooley, H. D. Marine Laboratory, Department of Agriculture and Fisheries, Aberdeen, Scotland.
Dorf, Prof. Erling. Department of Geology, Princeton University.
Dorf, Dean Richard C. Division of Extended Learning, University of California, Davis.
Dorfman, Dr. Ralph I. Senior Vice President and Director, Syntex Research Center, Palo Alto, California.
Doscher, Dr. Todd M. Department of Petroleum Engineering, University of Southern California.
Dossel, Dr. William E. Professor and Chairman, Department of Anatomy, School of Medicine, Creighton University.
Dott, Dr. Robert H., Jr. Department of Geology, University of Wisconsin.
Doudoroff, Dr. Michael. Deceased; formerly, Department of Bacteriology, University of California, Berkeley.
Douglass, Prof. David H. Department of Physics and Astronomy, College of Art and Science, University of Rochester.
Douglass, Prof. Raymond D. Department of Mathematics, Massachusetts Institute of Technology.
Dover, Dr. Carl B. Department of Physics, Brookhaven National Laboratory, Upton, New York.
Doviak, R. J. National Severe Storms Laboratory, National Oceanic and Atmospheric Administration, Norman, Oklahoma.
Downey, Dr. R. K. Agricultural Research Station, Saskatoon, Saskatchewan, Canada.
Downton, Dr. A. C. Department of Electronic Systems Engineering, University of Essex, England.
Drake, A. E. Division of Electrical Science, National Physical Laboratory, United Kingdom.
Drake, Prof. Charles H. Department of Bacteriology and Public Health, Washington State University.
Drake, Prof. John C. Department of Geology, University of Vermont.
Draper, Dr. Clifton W. Laser Studies Group, Western Electric Company, Princeton, New Jersey.
Dresden, Prof. Max. Institute for Theoretical Physics, State University of New York, Stony Brook.
Drexter, Dr. Eric. Foresight Institute, Palo Alto, California.
Duby, Prof. Paul. Henry Krumb School of Mines, Columbia University.
Duckworth, Dr. Henry E. Department of Physics, University of Manitoba, Canada.
Dudock, Dr. Bernard S. Department of Biochemistry, State University of New York, Stony Brook.
Duffield, Prof. Roger C. Department of Mechanical and Aerospace Engineering, University of Missouri.
Duguay, Dr. Michel A. Laser Development, Sandia Laboratories, Albuquerque, New Mexico.
Duke, Dr. C. B. Manager, Molecular and Organic Materials Area, Xerox Corporation, Rochester, New York.
Duke, Kenneth L. Department of Anatomy, Duke University Medical Center.
Dukek, W. G. Exxon Research and Engineering Company, Linden, New Jersey.
Dunbar, Prof. Carl O. Deceased; formerly, Professor Emeritus of Paleontology and Stratigraphy, Yale University.
Duncan, Charles C. International Communications Consultant, Manhasset, New York.
Duncombe, Dr. Raynor L. Department of Aerospace Engineering, University of Texas, Austin.
Dunn, Prof. Bruce. Department of Materials Science & Engineering, University of California, Los Angeles.
Dunn, Prof. Floyd. Department of Electrical and Computer Engineering, University of Illinois at Urbana-Champaign.
DuPont, Dr. Herbert L. Medical School, University of Texas, Houston.
Durand, Prof. Loyal, III. Department of Physics, University of Wisconsin.
Duwez, Dr. Pol E. Deceased; formerly, W. M. Keck Laboratory of Engineering Materials, California Institute of Technology.

Dyer, Dr. K. R. Institute of Oceanographic Sciences, Somerset, England.
Dym, Prof. C. L. Department of Civil Engineering, University of Massachusetts.

E

Early, Dr. Louis B. Manager, Engineering Economy, Communications Satellite Corporation, Washington D.C.
Easterbrook, Don J. Department of Geology, Western Washington University, Bellingham.
Eaton, Col. Elbridge P. Booz, Allen, Hamilton, Inc., Arlington, Virginia.
Eckels, Dr. Arthur R. Department of Electrical Engineering, North Carolina State University.
Economou, George A. Formerly, J. W. Fecker Division, American Optical Company, Pittsburgh, Pennsylvania.
Edeskuty, Dr. F. J. Los Alamos National Laboratory, University of California.
Edgerton, Dr. Harold E. Institute Professor, Emeritus, Massachusetts Institute of Technology.
Edson, W. D. Senior Engineer, Federal Railroad Administration.
Edwards, Prof. John O. Department of Chemistry, Brown University.
Edwards, Dr. Russell K. Argonne National Laboratory, Argonne, Illinois.
Edwards, Dr. Steven. Department of Virology, Central Veterinary Laboratory-Weybridge, United Kingdom.
Egan, Patrick. Department of Physics, Yale University.
Egan, Dr. Robert W. Department of Biochemistry of Inflammation, Merck Institute for Therapeutic Research, Rahway, New Jersey.
Egler, Dr. Frank E. Director (retired), Aton Forest, Norfolk, Connecticut.
Ehernberger, L. J. NASA Dryden Flight Research Facility, Edwards, California.
Ehrich, Fredric F. Aircraft Engine Business Group, General Electric Company, Lynn, Massachusetts.
Ehricke, Dr. Krafft A. Formerly, Autonetics Division, North American Rockwell Corporation, Anaheim, California.
Ehrlich, Dr. John. Formerly, Laboratory Director in Antibiotic Research, Parke, Davis and Company, Detroit, Michigan.
Ehrlich, Dr. Paul R. Professor of Biological Sciences and Curator of the Entomological Collections, Stanford University.
Ehrmann, Dr. David A. Department of Medicine, Section of Endocrinology, University of Chicago Medical Center.
Eichhorn, Dr. Dorothy. Institute of Human Development, University of California, Berkeley.
Eichorn, Dr. Gunther L. Gerontology Research Center, National Institutes of Health, Baltimore, Maryland.
Einset, Dr. John. Department of Botany and Plant Sciences, University of California, Riverside.
Eisenbud, Dr. Merril. Institute of Environmental Medicine, New York University Medical Center.
Eisler, Thomas J. National Oceanic and Atmospheric Administration, U.S. Department of Commerce, Rockville, Maryland.
Ek, Dr. Marit. Department of Pathology, University Hospital, Seattle, Washington.
Elder, J. Gordon. Elder Engineering Ltd., King City, Ontario, Canada.
Elderfield, Prof. Robert C. Deceased; formerly, Department of Chemistry, University of Michigan.
Elders, Dr. Wilfred A. Institute of Geophysics and Planetary Physics, University of California, Riverside.
Eldredge, Dr. Niles. American Museum of Natural History, New York.
Elgomayel, Prof. Joseph. School of Industrial Engineering, Purdue University.
Eliel, Prof. Ernest L. Department of Chemistry, University of North Carolina.
Ellett, Dr. C. Wayne. Department of Plant Pathology, Ohio State University.
Ellis, Dr. David E. Research and Development Department, Continental Oil Company, Ponco City, Oklahoma.
Ellis, Dr. Walther R. Division of Chemistry and Chemical Engineering, California Institute of Technology.
El-Mansy, Youssef. Intel Corporation, Aloha, Oregon.
Elston, Dr. Stuart B. Department of Physics and Astronomy, University of Tennessee, Knoxville.
Ely, F. G. Retired; Research and Development Division, Babcock and Wilcox Company.
Emanuel, Dr. Kerry Andrew. Department of Earth, Atmospheric and Planetary Sciences, Massachusetts Institute of Technology.
Emerson, Lewis P. Engineer in Charge of Flow Measurement, Foxboro Company, Foxboro, Massachusetts.
Emerson, Dr. Steven. Department of Oceanography, University of Washington.
Emmel, Prof. Victor M. Professor of Anatomy, School of Medicine and Dentistry, University of Rochester.
Encleson, Stephen C. Technical Editor, Corporate Communications, Dow Chemical Company, Midland, Michigan.
Endicott, Prof. John F. Department of Chemistry, Wayne State University.
Engel, Prof. A. E. J. Department of Geology, University of California, San Diego.
Engels, Dr. William R. Department of Genetics, University of Wisconsin, Madison.
English, Prof. Van H. Department of Geography, Dartmouth College.
Engstrom, Prof. Arne. Head, Department of Medical Physics, Karolinska Institute, Stockholm, Sweden.
Engstrom, Dr. Ralph W. Electro-Optics and Devices, RCA Laboratories, Lancaster, Pennsylvania.
Enkenfelder, Dr. W. Wesley, Jr. Eckenfelder, Inc., Nashville, Tennessee.
Enslow, Prof. Philip H., Jr. School of Information and Computer Science, Georgia Institute of Technology.
hEocha, Colm Ó. Department of Biochemistry, University College, Galway, Ireland.
Epstein, Dr. W. W. Department of Chemistry, University of Utah.
Erasmus, Dr. Baltus J. Head, Onderstepoort Vaccine Factory, Republic of South Africa.
Erb, Dr. Karl A. Oak Ridge National Laboratory, Oak Ridge, Tennessee.
Erdélyi, Prof. A. Professor of Mathematics, University of Edinburgh, Scotland.
Erdman, Dr. Arthur. Department of Mechanical Engineering, University of Minnesota.
Esaki, Leo. Thomas J. Watson Research Center, IBM, Yorktown Heights, New York.
Esau, Prof. Katherine. Department of Botany, University of California, Santa Barbara.
Estes, Dr. Richard. Department of Biology, Boston University.
Ethridge, Dr. Frank G. Department of Earth Resources, Colorado State University.
Ettenberg, Dr. Eugene M. Formerly, Lecturer in Graphic Arts, New York.
Evans, Dr. C. A., Jr. Materials Research Laboratory, University of Illinois.
Evans, Dr. Doyle J. Chief, Bacterial Enteropathogens Laboratory, Veterans Affairs Medical Center, Houston, Texas.
Evans, Dr. F. Gaynor. Department of Anatomy, Medical School, University of Michigan.
Evans, Prof. H. John. Clinical and Population Cytogenetics Unit, Medical Research Council, Western Central Hospital, Edinburgh, Scotland.
Evans, Dr. James B. Department of Microbiology, North Carolina State University.
Evans, Dr. John W. Director, Sacramento Peak Observatory, Air Force Cambridge Research Laboratories, Sunspot, New Mexico.
Evans, Prof. Robley D. Retired; Department of Physics, Massachusetts Institute of Technology.
Evenson, Dr. Paul. Bartol Research Institute, University of Delaware.
Everetts, Dr. John Jr. Professor of Architectural Engineering, Pennsylvania State University.

Ewing, Prof. George E. Department of Chemistry, Indiana University.
Ewing, John. Senior Research Associate, Lamont-Doherty Geological Observatory, Palisades, New York.
Ewing, Dr. Maurice. Deceased; formerly, Lamont-Doherty Geological Observatory, Palisades, New York.
Eyde, Dr. Richard H. Department of Botany, Smithsonian Institution.
Eyring, Dr. Edward M. Department of Chemistry, University of Utah.
Eyring, Prof. Henry. Deceased; formerly, Institute for the Study of Rate Processes, University of Utah.

F

Failla, Dr. Gioacchino. Deceased; formerly, Argonne National Laboratory, Argonne, Illinois.
Fairbank, Prof. Henry A. Department of Physics, Duke University.
Fairbridge, Prof. Rhodes. Department of Geology, Columbia University.
Faller, Dr. John W. Department of Chemistry, Yale University.
Fan, Prof. Hsu Y. Department of Physics, Purdue University.
Fankhauser, Dr. Gerhard. Deceased; formerly, Department of Biology, Princeton University.
Fanta, Prof. Paul E. Department of Chemistry, Illinois Institute of Technology.
Farmer, Dr. David M. Department of the Environment, Pacific Region, Institute of Ocean Sciences, Sidney, British Columbia, Canada.
Faure, Dr. Gunter. Department of Geology, Ohio State University.
Faw, Prof. Richard E. Head, Department of Nuclear Engineering, Kansas State University.
Fawcett, Prof. Don W. Hersey Professor of Anatomy, Harvard Medical School.
Fawcett, Prof. J. J. Department of Geology, University of Toronto, Ontario, Canada.
Fehr, Prof. Walter R. Department of Agronomy, Iowa State University.
Feigenbaum, Mitchell J. Professor, Department of Physics, Cornell University.
Fein, Dr. Jay S. Division of Atmospheric Sciences, National Science Foundation, Washington, D.C.
Feit, Louis. ITT Defense Communication Division, Nutley, New Jersey.
Feldman, Dr. Barry J. Los Alamos Scientific Laboratory, Los Alamos, New Mexico.
Feldman, Cyrus. Analytical Chemistry Division, Oak Ridge National Laboratory, Oak Ridge, Tennessee.
Feldman, Dr. Jerry. Division of Natural Sciences, University of California, Santa Cruz.
Felker, Dr. Peter. Caesar Kleberg Wildlife Research Institute, Texas A&M University.
Fell, Prof. Howard B. Museum of Comparative Zoology, Harvard University.
Feller, Prof. William. Deceased; formerly, Department of Mathematics, Princeton University.
Felver, Prof. Richard I. Designer, Carnegie Institute of Technology.
Fendler, Dr. Janos H. Department of Chemistry, Center for Research in Membranes and Colloid Science, Syracuse University.
Fenske, Prof. Richard F. Department of Chemistry, University of Wisconsin.
Ferguson, Dr. A. John. Retired; formerly, Chalk River Nuclear Laboratories, Chalk River, Ontario, Canada.
Ferguson, Prof. J. Homer. Department of Biological Science, University of Idaho.
Ferguson, Dr. Louise. Extension Pomologist, Kearney Agricultural Center, University of California, Parlier.
Ferm, Dr. John C. Department of Geology, University of South Carolina.
Ferrell, Prof. Thomas L. Health and Safety Research Division, Oak Ridge National Laboratory, Oak Ridge, Tennessee.
Ferrier, Dr. Pierre E. Associate Professor of Pediatrics, University of Washington.
Ferris, Prof. Gordon F. Deceased; formerly, National History Museum, Stanford University.
Ferry, Dr. James F. Department of Biology, Madison College.
Feshbach, Dr. Herman. Department of Physics, Massachusetts Institute of Technology.
Fetz, Dr. Eberhard E. Department of Physiology and Biophysics, University of Washington School of Medicine.
Fiala, Dr. Alan D. U.S. Naval Observatory, Washington, D.C.
Fidecaro, Dr. Giuseppe. EP Division, CERN, Geneva, Switzerland.
Field, Joseph H. Benfield Corporation, Pittsburgh, Pennsylvania.
Fields, Paul R. Senior Chemist, Argonne National Laboratory, Argonne, Illinois.
Fiers, W. Laboratory of Molecular Biology, University of Ghent, Belgium.
Fincham, Prof. J. R. S. Department of Genetics, University of Leeds, England.
Finegold, Dr. Sydney M. Chief, Infectious Disease Section, Veterans Administration Wadsworth Hospital Center, Los Angeles, California.
Fingerman, Dr. Milton. Department of Biology, Tulane University.
Fink, Donald G. Director Emeritus, IEEE; Editor in Chief, "Electronics Engineers' Handbook," McGraw-Hill Book Company, New York.
Finkelstein, Prof. L. Department of Physics, The City University, London, England.
Finks, Dr. Robert M. Department of Geology, Queens College, City University of New York.
Finland, Dr. Maxwell. Epidemiologist, Harvard Medical School, Boston City Hospital.
Fiore, Capt. Alfred E. Maritime Administration, U.S. Merchant Marine Academy, Kings Point, New York.
Firebaugh, Millard S. Naval Sea Systems Command, Washington, D.C.
Firing, Dr. Eric. Joint Institute of Marine and Atmospheric Research, Marine Science Building, University of Hawaii.
Firk, Prof. Frank W. K. Electron Accelerator Laboratory, Yale University.
Fiscarelli, A. Transmission Performance, American Telephone and Telegraph Company, New York.
Fischer, Harold. General Motors Corporation, Flint, Michigan.
Fischer, Henry W. Hardesty & Hanover, New York.
Fischer, Dr. Irene K. Takoma Park, Maryland.
Fisher, Dr. Frederick H. Department of Oceanography, University of California, San Diego.
Fisher, Dr. Jack B. Fairchild Tropical Garden and Research Center, Miami, Florida.
Fisher, Dr. Robert A. Los Alamos Scientific Laboratory, Los Alamos, New Mexico.
Fisher, Dr. Robert L. Geological Research Division, Scripps Institution of Oceanography, La Jolla, California.
Fisher, Prof. Russell A. Department of Physics, Northwestern University.
Fisk, Prof. Nelson S. Department of Civil Engineering, Columbia University.
Fitch, Prof. Val. Department of Physics, Princeton University.
FitzPatrick, James L. G. Dean of Faculty and Professor of Mechanical Technology, Staten Island Community College, City University of New York.
Fitzsimmons, Dr. J. Paul. Professor Emeritus, Department of Geology, University of New Mexico.
Fleischer, Dr. R. L. Physical Science Branch, General Physics Laboratory, General Electric Company, Schenectady, New York.
Fleminger, Dr. Abraham. Research Biologist, Scripps Institution of Oceanography, La Jolla, California.
Flessa, Dr. Karl Walter. Division of Earth Sciences, National Science Foundation, Washington, D.C.
Fleury, Dr. Paul A. Director, Physical Research Laboratory, AT&T Bell Telephone Laboratories, Murray Hill, New Jersey.
Flint, Dr. Richard F. Department of Geology and Geophysics, Yale University.

Floe, Prof. Carl F. Department of Metallurgy, Massachusetts Institute of Technology.

Fogarty, David E. Formerly, Staff Editor, "McGraw-Hill Encyclopedia of Science and Technology," McGraw-Hill Book Company, New York, New York.

Ford, Prof. E. B. Deceased; formerly, Genetics Laboratory, Department of Zoology, Oxford University, England.

Ford, Dr. Floyd E. NASA Goddard Space Center, Greenbelt, Maryland.

Ford, Dr. Karen Grady. Department of Biology, College of Charleston, South Carolina.

Forster, Dr. Denis. Monsanto Company, St. Louis, Missouri.

Forster, Dr. Roy P. Department of Biological Sciences, Dartmouth College.

Fosberg, Dr. F. R. National Museum of Natural History, Smithsonian Institution.

Fosdick, Lloyd D. Department of Computer Science, University of Colorado, Boulder.

Foss, F. E. Evans Energy Company, Denver, Colorado.

Foster, Prof. Jackson W. Deceased; formerly, Professor of Microbiology, University of Texas.

Foster, Dr. Mark G. Retired; Department of Electrical Engineering, School of Engineering and Applied Science, University of Virginia.

Foster, Prof. Merrill W. Department of Geology, Bradley University.

Foster, Dr. Robert J. Department of Physics, San Jose State University.

Fourtner, Dr. Charles R. Department of Biological Science, State University of New York, Buffalo.

Fowkes, Prof. Frederick M. Department of Chemistry, Lehigh University.

Fowler, Prof. Michael. Department of Physics, University of Virginia.

Fowler, Prof. W. Beall. Department of Physics, Lehigh University.

Fox, Prof. Herbert. Chairman, Department of Mechanics and Aeronautics, New York Institute of Technology.

Fox, Dr. Irving. School of Tropical Medicine, University of Puerto Rico.

Fox, Dr. Robert M. Department of Electrical Engineering, College of Engineering, University of Florida.

Fradkin, Dr. David M. Department of Physics, Wayne State University.

Frailey, Dr. Dennis J. Texas Instruments, Inc., Austin, Texas.

Fraiman, Dr. Nelson. Department of Industrial Engineering, Columbia University.

Franc, Dr. Jean-Marie. Laboratoire d'Histologie et Biologie Tissulaire, Université Claude Bernard, Villeurbanne, France.

Francis, Dr. Arthur W. Union Carbide Corporation, Tarrytown, New York.

Frank, Prof. Bernard. Deceased; formerly, Professor of Watershed Management, Colorado State University.

Franzese, Dr. Kenneth. Research and Development, Bright Star Industries, Inc., Clifton, New Jersey.

Frebold, Dr. Hans. Principal Research Scientist (retired), Geological Survey of Canada.

Freedman, Dr. Daniel X. Department of Psychiatry and Behavioral Sciences, University of California School of Medicine.

Freeman, Dr. A. E. Department of Animal Science, Iowa State University.

Freeman, Harry M. Chief, Waste Minimization Branch, Risk Reduction Engineering Laboratory, U.S. Environmental Protection Agency, Cincinnati, Ohio.

Freeman, Prof. Jeremiah P. Department of Chemistry, University of Notre Dame.

Freeman, Prof. Ralph L. Retired; Department of Mechanical Engineering, Iowa State College.

Freeman, Dr. Reino S. Professor of Parasitology, University of Toronto, Ontario, Canada.

Freire, Prof. Ernesto. Director, Biocalorimetry Center, Johns Hopkins University, Baltimore, Maryland.

Fremed, Raymond F. Burson-Marsteller Associates, New York.

French, Dr. J. Bruce. Department of Physics, University of Rochester.

French, Dr. Orval C. Department of Agricultural Engineering, Cornell University.

Frenkel, Dr. Jacob K. Department of Pathology, University of Kansas Medical Center.

Fretter, Prof. William B. Department of Physics, University of California, Berkeley.

Freudenstein, Dr. Ferdinand. Department of Mechanical Engineering, Columbia University.

Freundt, Dr. E. A. Institute for Medical Microbiology, Aarhus University, Denmark.

Frey, Prof. David G. Department of Zoology, Indiana University.

Frey, Herbert. Geophysics Branch, NASA Goddard Space Flight Center, Greenbelt, Maryland.

Frey, Dr. Kenneth J. Department of Agronomy, Iowa State University.

Fridovich, Dr. Irwin. Department of Biochemistry, Duke University Medical Center.

Fried, Dr. George H. Brooklyn College, City University of New York.

Fried, Dr. Sherman. Chemistry Division, Argonne National Laboratory, Argonne, Illinois.

Friedlander, Prof. Sheldon K. Department of Chemical Engineering, University of California, Los Angeles.

Friedman, Prof. Harold L. Department of Chemistry, State University of New York, Stony Brook.

Fritsch, J. Michael. Associate Professor, Department of Meteorology, College of Earth and Mineral Sciences, Pennsylvania State University.

Fritz, Dr. James S. Ames Laboratory, Energy and Mineral Resources Research Institute, Iowa State University.

Fritz, Dr. Sigmund. Chief Space Scientist, National Environmental Satellite Center, National Oceanic and Atmospheric Administration.

Froes, Dr. F. H. Sam. Director, Institute for Materials and Advanced Processes, University of Idaho.

Fromkin, Dr. Victoria A. Dean, Vice Chancellor, Graduate Programs, Department of Linguistics, University of California, Los Angeles.

Frondel, Prof. Clifford. Retired; Department of Geological Sciences, Harvard University.

Fruchter, Dr. Benjamin. Department of Educational Psychology, University of Texas, Austin.

Frumhoff, Dr. Peter. Department of Entomology, University of California, Davis.

Fry, Prof. Edward S. Department of Physics, Texas A&M University.

Fry, Prof. Frederick E. J. Professor of Zoology, University of Toronto, Ontario, Canada.

Fryer, Dr. Geoffrey. Freshwater Biological Association, Cumbria, United Kingdom.

Fudenberg, Dr. H. Hugh. Section of Hematology and Immunology, San Francisco Medical Center, University of California.

Fuhrman, Dr. Ralph E. Black and Veatch, Consulting Engineers, Washington, D.C.

Fulford, Prof. George D. Department of Chemical Engineering, University of Waterloo, Ontario, Canada.

Fulginiti, Dr. Vincent. Department of Pediatrics, University of Arizona Medical College.

Fulks, J. R. Retired; formerly, National Weather Service, Chicago, Illinois.

Fuller, Prof. Dudley D. (Retired) Department of Mechanical Engineering, Columbia University.

Fulton, Dr. James W. Monsanto Company, St. Louis, Missouri.

Furter, Dr. W. F. Dean of Graduate Studies and Research, Royal Military College of Canada.

Fuson, Prof. Reynold C. Deceased; formerly, Department of Chemistry, University of Illinois, Urbana.

Futral, Prof. J. G. Department of Agricultural Engineering, Georgia Agricultural Experiment Station, Experiment, Georgia.

Futuyma, Dr. Douglas J. Section of Ecology and Systematics, Cornell University.

G

Gabelnick, Dr. Henry L. Center for Population Research, National Institutes of Health, Bethesda, Maryland.

Gagne, Dr. Raymond J. Systematic Entomology Laboratory, U.S. Department of Agriculture.
Gaines, Dr. Alan M. National Science Foundation.
Gaines, Dr. George L. Research and Development Center, General Electric Company, Schenectady, New York.
Gall, Dr. Joseph G. Department of Biology, Yale University.
Gallagher, Dr. Thomas F. Deceased; formerly, Institute for Steroid Research, Montefiore Hospital, New York.
Gallant, Dr. Donald M. Department of Psychiatry and Neurology, Tulane University School of Medicine.
Galler, Dr. Bernard A. Computing Center, University of Michigan.
Galloway, Dr. William J. Bolt, Beranek and Newman, Inc., Canoga Park, California.
Galt, Dr. John K. Vice President, Sandia Laboratories, Albuquerque, New Mexico.
Gamborg, Dr. Oluf L. Research Director, International Plant Research Institute, Inc., San Carlos, California.
Gambs, Gerard C. Vice President (retired), Ford, Bacon, & Davis, Inc., New York.
Gandelman, Dr. Ronald Jay. Department of Psychology, Rutgers University.
Gannon, Dr. J. D. Department of Computer Science, University of Maryland.
Garcia, Dr. John. Department of Psychology, University of California, Los Angeles.
Gardiner, Dr. John L. Deceased; formerly, Animal Disease and Parasite Research Division, U.S. Department of Agriculture.
Gardner, Prof. Chester S. Department of Electrical Engineering and Computer Science, University of Illinois, Urbana.
Gardner, Dr. Frank E. Horticulturist, Agricultural Research Service, U.S. Department of Agriculture, Orlando, Florida.
Gardner, Dr. Walter H. Department of Agronomy and Soils, Washington State University.
Garfield, Dr. Sol L. Department of Psychology, Washington University.
Garland, Dr. George D. Department of Physics, University of Toronto, Ontario, Canada.
Garman, Dr. Philip. Connecticut Agricultural Experiment Station, New Haven.
Garrett, Prof. Alfred B. Department of Chemistry, Ohio State University.
Garvey, Dr. Gerald T. Physics Division, Argonne National Laboratory, Argonne, Illinois.
Gastony, Prof. Gerald J. Department of Plant Sciences, Indiana University.
Gatchell, Charles J. U.S. Department of Agriculture Forest Service, Forestry Sciences Laboratory, Princeton, West Virginia.
Gatzke, Irvin H. Surveillance Technology Administrator, Naval Air Systems Command, Washington, D.C.
Gauss, Dr. D. Max-Planck Institut für Experimentelle Medizin, Gottingen, Germany.
Gavlik, Emil L. Council of Reprographic Executives, Southwest Research Institute, San Antonio, Texas.
Gaylord, Prof. Charles N. Deceased; formerly, Chairman, Department of Civil Engineering, University of Virginia.
Gazzaniga, Dr. Michael S. Professor of Psychology and Social Sciences In Medicine, State University of New York, Stony Brook.
Geiger, Prof. Randall L. Department of Electrical Engineering, Texas A&M University.
Gelatt, Dr. C. D., Jr. Watson Research Center, IBM, Yorktown Heights, New York.
Geldard, Prof. Frank A. Stuart Professor of Psychology, Princeton University.
Gelfand, Prof. Erwin W. Chief, Division of Immunology and Rheumatology, Hospital for Sick Children, Toronto, Ontario, Canada.
Geller, Dr. Anne. Department of Pharmacology, Albert Einstein College of Medicine, Yeshiva University.
Gemmell, Dr. Donald S. Physics Division, Argonne National Laboratory, Argonne, Illinois.
Gensler, Prof. Walter J. Deceased; formerly, Department of Chemistry, Boston University.
George, Harvey F. Printing Industry Consultant, East Hampton, New York.
George, Dr. John W. Department of Chemistry, University of Massachusetts.
George, Dr. Melvin R. Agronomy and Range Science Extension, University of California, Davis.
Georges, T. M. Wave Propagation Laboratory, National Oceanic and Atmospheric Administration, Boulder, Colorado.
Georgi, Dr. Todd. Department of Biology, Doane College.
Gergel, Max G. Columbia Organic Chemicals Company, Columbia, South Carolina.
Gergely, Dr. John. Department of Muscle Research, Boston Biomedical Research Institute, Boston, Massachusetts.
Gerjuoy, Dr. Edward. Department of Physics, University of Pittsburgh.
Gerlach, Dr. Arch C. Chief Geographer, Geological Survey, U.S. Department of the Interior.
Gerrity, Dr. Joseph P. National Meteorological Center, National Oceanic and Atmospheric Administration, Camp Springs, Maryland.
Gersh, Prof. Isidore. Research Professor, Department of Animal Biology, School of Veterinary Medicine, University of Pennsylvania.
Gershoff, Prof. Stanley N. Department of Nutrition, Harvard School of Public Health, Boston, Massachusetts.
Gertler, Dr. Janos. Department of Electrical and Computer Engineering, George Mason University, Fairfax, Virginia.
Gertsch, Dr. Willis J. American Museum of Natural History, New York.
Gerwartowski, James W. Bell Telephone Laboratories, Murray Hill, New Jersey.
Ghiorso, Dr. Albert. Department of Chemistry, Lawrence Berkeley Laboratory, University of California, Berkeley.
Giallorenzi, Dr. Thomas G. Superintendent, Optical Science Division, Department of the Navy, Naval Research Laboratory, Washington, D.C.
Giannasi, Dr. David E. Department of Botany, University of Georgia.
Giarman, Prof. Nicholas J. Deceased; formerly, Department of Pharmacology, School of Medicine, Yale University.
Gibbons, Dr. Ian. Kewala Laboratory, University of Hawaii.
Gibbons-Fly, Walter. Deceased; formerly, CBS Television Network, Hollywood, California.
Gibbs, Hyatt M. Optical Science Center, University of Arizona.
Gibbs, Dr. Martin. Institute for Photobiology of Cells and Organelles, Brandeis University.
Gibeling, Dr. Jeffrey C. Department of Materials Science and Engineering, Stanford University.
Giblon, Robert P. President, George G. Sharp, Inc., New York.
Gibson, Harold J. Manager (retired), Research and Development Department, Detroit Laboratories, Ethyl Corporation, Ferndale, Michigan.
Gibson, Dr. John E. Dean of Engineering, Oakland University.
Gibson, Dr. T. Department of Bacteriology, School of Agriculture, University of Edinburgh, Scotland.
Giedt, Prof. Warren H. Department of Mechanical Engineering, University of California, Davis.
Giese, Prof. Arthur C. Department of Biological Sciences, Stanford University.
Giese, Raymond. Department of Mechanical Engineering, University of Minnesota.
Gigg, Roy H. Chemistry Division, National Institute for Medical Research, London, England.
Giguère, Paul A. Emeritus Professor, St. Petersburg Beach, Florida.
Gilardi, Dr. Gerald L. Microbiology Laboratory, North General Hospital, New York, New York.
Gilbert, Barrie. Analog Devices, Inc., Northwest Laboratories, Beaverton, Oregon.
Gilbert, Dr. Edgar N. Deceased; formerly, AT&T Bell Telephone Laboratories, Murray Hill, New Jersey.

Gillespie, Dr. Ronald J. Department of Chemistry, McMaster University, Hamilton, Ontario, Canada.
Gillies, Dr. C. B. School of Biological Sciences, University of Sydney.
Gilliland, Prof. Edwin R. Deceased; formerly, Department of Chemical Engineering, Massachusetts Institute of Technology.
Gingrich, Prof. Newell S. Department of Physics, University of Missouri.
Ginzton, Prof. Edward L. President and Chairman of the Board, Varian Associates, Palo Alto, California.
Giunipero, Dr. Larry C. Management Department, College of Business, Florida State University.
Glasford, Prof. Glenn M. Department of Electrical and Computer Engineering, Syracuse University.
Glass, Prof. H. Bentley. Academic Vice President and Distinguished Professor of Biology (retired), State University of New York, Stony Brook.
Glasscock, Dwight L. Deceased; formerly, Harza Engineering Company, Chicago, Illinois.
Gleeson, Prof. Thomas A. Department of Meteorology, Florida State University.
Glicksman, Dr. Richard. Solid State Division, RCA, Somerville, New Jersey.
Glock, Dr. Waldo S. Institute of Arctic and Alpine Research, University of Colorado.
Gloge, Dr. Detlef C. Crawford Hill Laboratory, Bell Telephone, Holmdel, New Jersey.
Gluckstern, Prof. Robert L. Head, Department of Physics and Astronomy, University of Massachusetts.
Goddu, Dr. Robert F. Manager, Fibers and Film Research Division, Hercules, Inc., Wilmington, Delaware.
Godfrey, Dr. Robert K. Department of Biological Science, Florida State University.
Goebel, Prof. Charles J. Department of Physics, University of Wisconsin.
Goland, Dr. Leonard. Executive Vice President, Dynasciences Corporation, Los Angeles, California.
Gold, Dr. Ronald. Hospital for Sick Children, Toronto, Ontario, Canada.
Goldberg, Prof. David E. Department of General Engineering, University of Illinois, Urbana.
Goldberg, Dr. Edward D. Scripps Institution of Oceanography, La Jolla, California.
Golden, F. B. Semiconductor Products Department, General Electric Company, Auburn, New York.
Golden, Dr. Joseph. Office of Programs, National Oceanic and Atmospheric Administration, Boulder, Colorado.
Goldenberg, Dr. Victor E. Associate Professor in Pathology, University Hospital, Seattle.
Goldfarb, Dr. Marjorie S. School of Oceanography, University of Washington, Seattle.
Goldhaber, Dr. Alfred S. Department of Physics, State University of New York, Stony Brook.
Goldhaber, Dr. Gerson. Lawrence Berkeley Laboratory, University of California, Berkeley.
Goldich, Dr. Samuel S. Department of Geology, Colorado School of Mines, Golden.
Goldstein, Dr. Irving S. College of Forestry Resources, North Carolina State University.
Golley, Dr. Frank B. Institute of Ecology, Athens, Georgia.
Gomer, Prof. Robert. James Franck Institute, University of Chicago.
Gonser, Dr. Bruce W. Battelle Memorial Institute, Columbus, Ohio.
Gonzalez, Prof. R. C. Department of Electrical Engineering and Computer Science, University of Tennessee.
Good, Dr. Irving J. Department of Statistics and Statistical Laboratory, College of Arts and Sciences, Virginia Polytechnic Institute.
Good, Prof. Roland H., Jr. Department of Physics, Pennsylvania State University.
Goodall, Dr. McChesney. Professor of Pharmacology and Professor of Surgery and Physiology, University of Texas Medical Branch, Galveston.
Goodbody, Dr. Ivan. Department of Zoology, University of the West Indies, Jamaica.
Goodchild, Dr. Chauncey G. Candler Professor of Biology, Emory University.
Goodglass, Dr. Harold. Director, Psychology Research, Veterans Administration Hospital, Boston, Massachusetts.
Goodheart, Prof. Clarence F. Department of Electrical Engineering, Union College.
Goodman, Prof. Bernard. Department of Physics, University of Cincinnati.
Goodman, Dr. Joseph W. Stanford Electronics Laboratories, Stanford University.
Goodman, Sidney H. Manager, Materials Products Department, Technology Support Division, Hughes Aircraft Co., Culver City, California; Senior Lecturer, Department of Chemical Engineering, University of Southern California.
Goodnight, Dr. Clarence J. Head, Department of Biology, Western Michigan University.
Goodnight, Marie L. Department of Biology, Western Michigan University.
Goodrich, Dr. Roy G. Department of Physics and Astronomy, Louisiana State University.
Goodstein, David H. Director, Inter/Consult, Cambridge, Massachusetts.
Goodwin, Dr. Richard H. Department of Botany, Connecticut College.
Gordon, Dr. Arnold L. Lamont-Doherty Geological Observatory, Palisades, New York.
Gordon, Dr. Gary D. Program Development, Communication Satellite Corporation, Clarksburg, Maryland.
Gordon, Dr. Isabella. (Retired) Crustacea Section, British Museum (Natural History), London, England.
Gordon, Dr. James P. Bell Telephone Laboratories, Holmdel, New Jersey.
Gordon, Prof. Louis. Deceased; formerly, Case Institute of Technology.
Gordon, Dr. Sheffield. Chemistry Division, Argonne National Laboratories, Argonne, Illinois.
Gordon, Dr. William E. Associate Professor of Physical Chemistry, Pennsylvania State University, and Consultant.
Gore, Edward W., Jr. Vice President, International Market Requirements, Office Products Division, IBM, Franklin Lakes, New Jersey.
Gossard, Prof. Arthur C. Materials Department, University of California, Santa Barbara.
Gossner, Simone Daro. U.S. Naval Observatory, Washington, D.C.
Gots, Dr. Joseph S. Department of Microbiology, School of Medicine, University of Pennsylvania.
Gould, Dr. Richard A. Department of Anthropology, University of Hawaii.
Gould, Dr. Stephen J. Professor of Geology, Museum of Comparative Zoology, Harvard University.
Goulding, Dr. Frederick S. Lawrence Berkeley Laboratory, University of California.
Govindjee, Dr. Department of Botany and Department of Physiology and Biophysics, University of Illinois, Urbana.
Govindjee, Dr. Rajni. Department of Botany, University of Illinois, Urbana.
Gowen, Prof. John W. Department of Radiology and Radiation Biology, Colorado State University.
Grace, Dr. Neville. Biotechnology Division, Department of Scientific and Industrial Research, Palmerston North, New Zealand.
Graham, James A. Business Manager, Industrial Adhesives Division, Lord Corporation, Erie, Pennsylvania.
Gränicher, Prof. H. Laboratory of Solid State Physics, Swiss Federal Institute of Technology, Zurich.
Granick, Dr. S. Department of Biochemistry, Rockefeller University.
Grasshoff, Dr. K. Institut für Meereskunde an der Universität Kiel, Germany.
Grau, Dr. Fred V. Consulting Agronomist, College Park, Maryland.
Graves, Dr. Arthur H. Deceased; formerly, Consultant in Genetics, Connecticut Agricultural Experiment Station.
Graves, Prof. Lawrence M. Professor Emeritus of Mathematics, University of Chicago.
Gray, Dr. Harry B. Division of Chemistry and Chemical Engineering, California Institute of Technology.

Gray, G. Ronald. Director, Syncrude Canada Ltd., Edmonton, Alberta, Canada.
Gray, Dr. Peter. Andrey Avinoff Professor of Biology, University of Pittsburgh.
Green, Dr. Louis C. Department of Astronomy, Haverford College.
Green, Prof. Robert E., Jr. Chairman, Materials Science and Engineering, Johns Hopkins University.
Greenberg, Arnold E. Assistant Chief, Laboratory Services Branch, State of California Department of Health Services, Berkeley.
Greenberg, Prof. David M. Department of Biochemistry, School of Medicine, University of California, San Francisco.
Greenberg, Prof. Jack S. Wright Nuclear Structure Laboratory, Yale University.
Greenberg, Prof. O. W. Division of Mathematical and Physical Science and Engineering, Department of Physics and Astronomy, University of Maryland.
Greening, Nathan R. Director of the Basic Research Department, Cement and Concrete Research Institute, Portland Cement Association, Skokie, Illinois.
Greensfelder, Dr. Bernard S. Deceased; formerly, Director of Oil Research, Shell Development Company, Emeryville, California.
Greenspan, Martin. National Bureau of Standards.
Greenstein, Prof. Jesse L. Hale Observatories, Pasadena, California.
Gregory, Dr. Joseph T. Professor of Paleontology and Curator of Lower Vertebrates, Museum of Paleontology, University of California, Berkeley.
Greiner, Dr. Walter Institut für Theoretische Physik, Universität Frankfurt, Germany.
Grest, Dr. Gary S. Exxon Research and Engineering Company, Annandale, New Jersey.
Grey, Dr. Gene Department of Forestry, Kansas State University.
Grey, Jerry. President, Greyrad Corporation, Princeton, New Jersey.
Grierson, Dr. William. Institute of Food and Agricultural Sciences, University of Florida.
Griffin, Daniel P. Office of the Chief of Military History, U.S. Department of the Army.
Griffiths, Prof. Peter R. Chairman, College of Letters and Science, Department of Chemistry, University of Idaho.
Grim, Dr. Ralph E. Department of Geology, University of Illinois, Urbana.
Grimm, Robert A. AC Spark Plug Division, General Motors Corporation, Flint, Michigan.
Grindlay, Dr. Jonathan E. Center for Astrophysics, Harvard University.
Grobecker, Dr. Alan J. Project Manager, Climatic Impact Assessment Program, Office of the Secretary of Transportation.
Grodzins, Dr. Lee. Physics Department, Massachusetts Institute of Technology.
Grogan, Dr. R. G. Department of Plant Pathology, University of California, Davis.
Groggins, P. H. Deceased; formerly, Chemical Division, Food Machinery and Chemical Corporation.
Groody, Dr. E. Patrick. Integrated Genetics, Framingham, Massachusetts.
Grosch, Prof. Daniel S. Department of Genetics, North Carolina State University.
Gross, Prof. Jonathan L. Department of Mathematics, Columbia University.
Gross, William H. Technical Editor, Corporate Communications, Dow Chemical Company, Midland, Michigan.
Grossman, Dr. Lawrence. Department of Geophysical Sciences, University of Chicago.
Grosswald, Prof. Emil. Deceased; formerly, Department of Mathematics, Temple University.
Groth, Dr. Edward J. Department of Physics, Princeton University.
Groves, Dr. Gordon W. Instituto de Geofisica, Torre de Ciencias, Ciudad Universitaria, Mexico.
Gruen, Dr. Dieter M. Argonne National Laboratory, Argonne, Illinois.
Gruenberger, Prof. Fred J. Department of Computer Science, California State University.
Grunberg, Dr. Emanuel. Director, Department of Chemotherapy, Hoffmann-LaRoche, Inc., Nutley, New Jersey.
Guidry, Dr. Mark R. Fairchild Camera & Instrument Corp., Mountain View, California.
Guildner, Dr. Leslie A. Temperature Division, National Bureau of Standards.
Guirard, Dr. Beverly M. Department of Biochemistry, University of California, Berkeley.
Gulbransen, Dr. Earl A. Department of Metallurgical and Materials Engineering, University of Pittsburgh.
Gunckel, Prof. James E. Department of Botany, Rutgers University.
Gunkler, Dr. Albert A. Chief Process Engineer, Midland Division, Dow Chemical Company, Midland, Michigan.
Gunning, Dr. Brian E. S. Department of Botany, Queen's University, Belfast, Ireland.
Gunning, Dr. Robert C. Department of Mathematics, Princeton University.
Gupta, Dr. Madan M. Intelligent System Research Laboratory, College of Engineering, University of Saskatchewan, Saskatoon.
Gurney, Dr. Ashley B. Systematic Entomology Laboratory, Entomology Research Division, U.S. Department of Agriculture.
Gutsche, Prof. C. David. Department of Chemistry, Texas Christian University, Fort Worth.
Gutzwiller, Dr. Martin C. T. J. Watson Research Center, IBM Corporation, Yorktown Heights, New York.
Guzinski, Dr. Gay M. Chief, Section of Benign Gynecology, Department of Obstetrics and Gynecology, School of Medicine, University of Maryland, Baltimore.

H

Haber, Bernard. Hardesty & Hanover, New York.
Hackl, Ralph R. Manager, R&D Laboratory, Giant Bay Biotech, Burnaby, British Columbia, Canada.
Hader, Rodney N. Secretary, American Chemical Society, Washington, D.C.
Hadler, Jacques B. Webb Institute of Naval Architecture, Glen Cove, New York.
Hadley, Dr. Elbert H. Assistant Dean (retired), College of Liberal Arts and Sciences, and Professor of Chemistry, Southern Illinois University.
Haensel, Dr. Vladimir. Vice President and Director of Research, Universal Oil Products Company, Des Plaines, Illinois.
Hagen, Dr. John P. Department of Astronomy, Pennsylvania State University.
Hagler, Thomas. Senior Systems Engineer, Office of Manned Space Flight, National Aeronautics and Space Administration.
Hagood, Mel A. Irrigated Agriculture Research and Extension Center, Washington State University.
Hague, Wilbur. Oxy Metal Industries Corporation, Warren, Michigan.
Hahn, Dr. John F. Department of Psychology, University of Virginia.
Hailpern, Dr. Brent T. IBM Research Laboratory, Yorktown Heights, New York.
Haines, John E. Deceased; formerly, Vice President, Minneapolis-Honeywell Regulator Company.
Haines, Roger W. Professional Engineer, Laguna Hills, California.
Hainline, Bernard C. Systems Technology, Boeing Commercial Airplane Company, Seattle, Washington.
Halbe, Dr. Douglas N. Homestake Mining Company, Lakewood, Colorado.
Halbouty, Dr. Michael T. Consulting Geologist and Petroleum Engineer, Houston, Texas.
Hale, Dr. Harry P. Department of Mechanical Engineering, Wayne State University.
Hale, Dr. Mason E., Jr. Smithsonian Institution.
Hales, Dr. Milton R. Professor of Pathology, School of Medicine, West Virginia University.
Haley, Dr. Leanor D. Chief, Mycology Training Section, Laboratory Training and Consultation Division, Bureau of Laboratories, Center for Disease Control, Atlanta, Georgia.

Halkias, Prof. Christos C. Chair of Electronics, National Technical University, Athens, Greece.
Hall, Prof. Dennis G. Director, The Institute of Optics, University of Rochester, New York.
Hall, Dr. Richard P. Deceased; formerly, Professor of Zoology, University of California, Los Angeles.
Halpern, Dr. Lawrence M. Department of Pharmacology, University of Washington.
Halstead, Dr. Thora Waters. American Society of Gravitational and Space Biology, Arlington, Virginia.
Halvorson, Dr. Harlyn O. Department of Biochemistry, University of Minnesota.
Hamer, Dr. Walter J. Institute for Basic Standards, National Bureau of Standards.
Hamilton, Prof. C. Howard. Department of Mechanical and Materials Engineering, Washington State University, Pullman.
Hamilton, Dr. Edwin L. U.S. Navy Electronics Laboratory, San Diego, California.
Hamilton, Dr. Howard L. (Retired) Department of Biology, University of Virginia.
Hamilton, Dr. Joseph H. Department of Physics-Astronomy, Vanderbilt University.
Hamilton, Dr. R. I. Research Branch, Agriculture Canada, Vancouver, British Columbia, Canada.
Hammar, Dr. Samuel P. The Diagnostic Specialties Laboratory, Washington.
Hammel, Dr. Edward F. Los Alamos Scientific Laboratory, Los Alamos, New Mexico.
Hammen, Dr. C. S. Department of Zoology, University of Rhode Island.
Hammond, Prof. Edwin H. Department of Geography, University of Tennessee.
Hand, Dr. Cadet. Bodega Marine Laboratory, University of California, Bodega Bay.
Handin, Dr. John. Center for Technophysics, Texas A&M University.
Hanlin, Dr. Richard T. Department of Plant Pathology, University of Georgia, Athens.
Hanna, Prof. Stanley S. Department of Physics, Stanford University.
Hansch, Dr. Theo W. Department of Physics, Stanford University.
Hansen, Arthur G. President, Georgia Institute of Technology.
Hansen, Dr. John. Department of Neurobiology and Anatomy, University of Rochester Medical Center.
Hansen, Dr. Katherine J. Department of Earth Sciences, Montana State University, Bozeman.
Hanson, Prof. Allen L. Department of Chemistry, Saint Olaf College.
Hanson, Dr. Clarence H. Agricultural Research Service, U.S. Department of Agriculture, Beltsville, Maryland.
Hanson, Frederick R. Research Associate, Fermentation Research and Development, Upjohn Company, Kalamazoo, Michigan.
Häntzschel, Dr. Walter. Geologisch-Paläontologisches Institut, Hamburg, Germany.
Happer, Dr. William. Department of Physics, Columbia University.
Haq, Dr. Bilal U. Department of Marine Geology and Geophysics, National Science Foundation, Washington, D.C.
Haratunian, Michael. President, Seelye Stevenson Value & Knecht, New York.
Harborne, Dr. Jeffrey B. Department of Botany, University of Reading, England.
Hardesty, Egbert R. Hardesty & Hanover, New York.
Hardgrove, Ralph M. Sales Engineer, Stock Equipment Company, Cleveland, Ohio.
Hare, Dr. Jonathan. School of Chemistry and Molecular Sciences, University of Sussex, United Kingdom.
Harford, Edward F. Chemical Engineering Consultant, E. I. du Pont de Nemours and Company, Wilmington, Delaware.
Hargis, Dr. William J., Jr. Department of Marine Science, University of Virginia.
Harker, Dr. Robert I. Department of Geology, University of Pennsylvania.
Harleman, Dr. Donald R. F. Department of Civil Engineering, Massachusetts Institute of Technology.
Harley, Prof. J. L. Department of Forestry, Oxford University, England.
Harley, Dr. John H. Department of Environmental Medicine, New York University.
Harned, Dr. Roger L. Research Microbiologist, Commercial Solvents Corporation, Terre Haute, Indiana.
Harper, Dr. Michael J. K. Department of Obstetrics and Gynecology, University of Texas Health Science Center, San Antonio.
Harringon, Roy L. Newport News, Virginia.
Harris, Dr. Cyril M. Professor of Electrical Engineering and Architecture, Department of Electrical Engineering, Columbia University.
Harris, Dr. Forest K. Electrical Measurements Laboratory, National Institute of Standards and Technology, Gaithersburg, Maryland.
Harris, Dr. J. Postgraduate School of Studies in Chemical Engineering, University of Bradford, England.
Harris, Prof. J. Donald. U.S. Navy Submarine Medical Laboratory, Groton, Connecticut.
Harris, Prof. Richard W. Department of Environmental Horticulture, University of California, Davis.
Harrison, David B. Public Relations Department, Goodyear Tire and Rubber, Akron, Ohio.
Harrison, Prof. Edward R. Department of Physics and Astronomy, University of Massachusetts.
Harrison, George R. Deceased; formerly, Dean Emeritus, School of Science, Massachusetts Institute of Technology.
Harrison, Henry T. Consulting Meteorologist, Weaverville, North Carolina.
Harrison, Dr. Walter A. Department of Applied Physics, Stanford University.
Harsh, M. Duffield. Engineering Leader, Display Tube Product Development, RCA Electronic Components, Lancaster, Pennsylvania.
Hartley, Dr. A. M. Department of Chemistry and Chemical Engineering, University of Illinois, Urbana.
Hartman, Dr. Olga. Deceased; formerly, Allan Hancock Foundation, University of Southern California.
Hartman, Dr. Standish C. Department of Chemistry, Boston University.
Hartman, Dr. Willard D. Associate Professor of Biology and Curator in Invertebrate Zoology, Peabody Museum of Natural History, Yale University.
Hartmann, Prof. Hudson T. Pomology Extension, University of California, Davis.
Hartstein, Dr. Jack. Assistant Professor of Clinical Ophthalmology, Washington University School of Medicine; Chairman of Ophthalmology, Missouri Baptist Hospital, St. Louis.
Hartung, Dr. Walter H. Formerly, Professor of Pharmaceutical Chemistry, Medical College of Virginia.
Hartwig, Dr. William H. Department of Electrical Engineering, University of Texas, Austin.
Harvey, Dr. E. Newton. Deceased, formerly, Henry Fairfield Osborne Professor of Biology, Princeton University.
Harvey, Dr. George R. Woods Hole Oceanographic Institution, Woods Hole, Massachusetts.
Harvey, Dr. John A. Oak Ridge National Laboratory, Oak Ridge, Tennessee.
Harvey, Dr. Paul H. Department of Zoology, University of Oxford, United Kingdom.
Hasey, Dr. Janine K. Cooperative Extension, University of California, Yuba City.
Haskins, Dr. Reginald H. Head, Microbiology, Physiology and Biochemistry Section, Prairie Regional Laboratory, National Research Council of Canada, Saskatoon, Saskatchewan.
Hasler, Dr. Arthur D. Laboratory of Limnology, University of Wisconsin.
Hassialis, M. D. Krumb School of Mines, Columbia University.
Hassid, Prof. William Z. Deceased; formerly, Department of Biochemistry, University of California, Berkeley.
Hastings, Prof. J. Woodland. Biological Laboratories, Department of Cellular and Development Biology, Harvard University.
Hauf, Harold D. Consulting Architect, Sun City, Arizona.

Hausner, Henry H. Consulting Engineer, New York.
Haut, Dr. Arthur. Professor of Medicine, Division of Hematology and Oncology, University of Arkansas for Medical Sciences.
Hawkins, Prof. George A. Vice President for Academic Affairs, Purdue University.
Hawkins, Dr. W. Lincoln. Bell Laboratories, Murray Hill, New Jersey.
Hawksworth, Dr. Frank G. U.S. Forest Service, Fort Collins, Colorado.
Hay, Dr. William W. Joint Oceanographic Institutions, Washington, D.C.
Hayes, Dr. Dennis E. Lamont-Doherty Geological Observatory, Columbia University, Palisades, New York.
Hayes, Dr. Donald S. Kitt Peak National Observatory, Tucson, Arizona.
Hayes, Dr. Thomas L. Donner Laboratory, University of California, Berkeley.
Hayes, William C. Editor in Chief, "Electrical World," McGraw-Hill Publications Company, New York.
Hayes-Roth, Dr. Frederick. Executive Vice President, Technology, Teknowledge Inc., Palo Alto, California.
Hayflick, Prof. Leonard. Center for Gerontological Studies, University of Florida.
Haynes, David O. Deceased; formerly, Consulting Industrial Engineer, Tucson, Arizona.
Hazel, Prof. Jeffrey R. Department of Zoology, Arizona State University.
Hazen, Prof. David C. Department of Aerospace-Mechanical Science, Princeton University.
Hazen, Richard. Hazen and Sawyer, Consulting Engineers, New York.
Hazen, Ronald. Consultant (retired), Indianapolis, Indiana.
Hearmon, R. F. S. Formerly, Timber Mechanics Section, Forest Products Research Laboratory, Princes Risborough, Bucks, England.
Hearn, Dr. C. Jack. Research Geneticist, Horticultural Research Laboratory, U.S. Department of Agriculture, Orlando, Florida.
Hecht, Dr. Hans H. Chairman, Department of Medicine, University of Chicago.
Heckel, Dr. P. H. Department of Geology, University of Iowa.
Heckman, Dr. Harry H. Lawrence Berkeley Laboratory, University of California, Berkeley.
Hedberg, Dr. Hollis D. Department of Geology, Princeton University.
Hedgpeth, Dr. Joel W. Marine Science Center, Oregon State University.
Heeger, Dr. Alan J. Institute for Polymers and Organic Solids, University of California, Santa Barbara.
Heezen, Dr. Bruce C. Deceased; formerly, Lamont-Doherty Geological Observatory, Palisades, New York.
Heinrich, Dr. Bernd. Division of Entomology and Parasitology, College of Natural Resources, University of California, Berkeley.
Heinrich, Dr. E. William. Department of Geology and Mineralogy, University of Michigan.
Held, Dr. Gilbert. Director, 4-Degree Consulting, Macon, Georgia.
Heldman, Dr. Julius D. Shell Development Company, Houston, Texas.
Hellemans, Alexander. Formerly, Editor, Encyclopedia Department, Professional and Reference Division, McGraw-Hill Book Company, New York.
Hellmers, Dr. Henry. Botany Department, Duke University.
Helmers, Dr. Carl J. Vice President, Applied Automation, Inc., Bartlesville, Oklahoma.
Helms, Harry L. Formerly, Editor, Professional and Reference Book Division, McGraw-Hill Book Company, New York.
Hendry, Dr. Elaine M. Department of Physics and Astronomy, Georgia State University.
Henry, Dr. Alyson E. Department of Psychology, University of Illinois, Chicago.
Henry, Dr. Charles H. Bell Laboratories, Murray Hill, New Jersey.
Hepburn, Dr. Angus G. Department of Genetics, John Innes Institute, Norwich, England.
Helpler, David S. NASA Goddard Space Center, Greenbelt, Maryland.
Herb, Raymond G. President, National Electrostatics Corporation, Middleton, Wisconsin.
Herber, Prof. Rolfe H. Department of Chemistry, Rutgers University.
Herberman, Dr. Ronald B. Chief, Biological Development Branch, National Cancer Institute, Frederick Cancer Research Facility, Frederick, Maryland.
Hermach, Dr. F. L. National Institute of Standards and Technology, Gaithersburg, Maryland.
Herman, C. J. Consulting Engineer, Advanced Processes and Products, Insulating Materials Department, General Electric Co., Schenectady, New York.
Heronemus, Prof. William. Department of Engineering, University of Massachusetts.
Herrell, Dr. Dennis J. Thomas J. Watson Research Center, Yorktown Heights, New York.
Hersey, Dr. John B. Office of Naval Research, U.S. Department of the Navy.
Hershleder, William. Consulting Construction Engineer, New York.
Herz, Prof. Carl S. Department of Mathematics, McGill University.
Herzberger, Dr. Max J. Deceased; formerly, Consulting Professor, Department of Physics, Louisiana State University.
Hess, Dr. George B. Department of Physics, University of Virginia.
Hesseltine, Dr. C. W. Fermentation Laboratory, Agricultural Research Service, U.S. Department of Agriculture, Peoria, Illinois.
Hewson, Dr. E. Wendell. Chairman, Department of Atmospheric Sciences, Oregon State University.
Heyne, Prof. E. G. Professor of Plant Breeding, Department of Agronomy, Kansas State University.
Hibbard, Robert R. Technical Consultant for Propulsion Chemistry Division, NASA Lewis Research Center, Cleveland, Ohio.
Hickman, Dr. Howard M. Section Manager, Research and Development, Sherex Chemical Company, Dublin, Ohio.
Hicks, Dr. Philip E. President, Hicks & Associates, Consulting Industrial Engineers, Orlando, Florida.
Hieftje, Dr. Gary M. Department of Chemistry, Indiana University.
Higgins, Prof. Richard J. Department of Physics, University of Oregon.
Higgins, Dr. Robert P. Curator, Department of Invertebrate Zoology, National Museum of Natural History, Smithsonian Institution.
Hildebrand, Dr. Joel H. Deceased; formerly, Department of Chemistry, University of California, Berkeley.
Hill, Dr. Armin J. Dean, College of Physical and Engineering Sciences, Brigham Young University.
Hill, Prof. Dorothy. Research Professor, Department of Geology and Mineralogy, University of Queensland, Brisbane, Australia.
Hill, Prof. Edward L. Deceased; formerly, Professor of Physics and Mathematics, School of Physics, University of Minnesota.
Hill, Robert T. Radar Consultant, Bowie, Maryland.
Hill, Dr. Rolla B., Jr. Professor and Chairman, Department of Pathology, Upstate Medical Center, Syracuse.
Hilliard, Dr. Ron. Optomechancis Research Inc., Vail, Arizona.
Hines, Dr. M. E. Vice President, Microwave Associates, Inc., Burlington, Massachusetts.
Hingorani, Dr. Narian G. Electric Power Research Institute, Palo Alto, California.
Hinsch, Dr. Gertrude W. Department of Biology, University of South Florida.
Hipel, Prof. Keith W. Departments of Systems Design Engineering and Statistical Actuarial Science, University of Waterloo, Ontario, Canada.
Hirschberg, Prof. Daniel S. Information and Computer Science, University of California, Irvine.
Hirschfelder, Dr. J. O. Department of Chemistry, University of Wisconsin.
Hobbs, Dr. Herman H. Department of Physics, George Washington University.
Hocking, Dr. Brian. Department of Entomology, University of Alberta, Canada.

Hodge, Dr. Paul W. Department of Astronomy, University of Washington.
Hodge, Prof. Philip. Chair of Polymer Chemistry, Department of Chemistry, University of Manchester, England.
Hodgin, David M. Senior Member, Institute of Electrical and Electronics Engineers, Cedar Rapids, Iowa.
Hodgson, Dr. Robert A. Gulf Research and Development Company, Pittsburgh, Pennsylvania.
Hoering, Dr. Thomas C. Geophysical Laboratory, Carnegie Institution of Washington.
Hoff, Prof. C. Clayton. Department of Biology, University of New Mexico.
Hoff, Nicholas J. Professor Emeritus, Department of Aeronautics and Astronautics, Stanford University.
Hoffman, Dr. Alan J. IBM Corporation, Yorktown Heights, New York.
Hoffman, Dr. Harold L. Gulf Publishing Company, Houston, Texas.
Hoffman, Dr. Paul F. Geological Survey of Canada, Ottawa, Canada.
Hoffman, Richard A. Senior Staff Engineer, Martin Marietta Aerospace, Denver, Colorado.
Hoffman, Prof. Richard L. Department of Biology, Radford College.
Hoffmann, Dr. Friedrich W. Deceased; formerly, Chemical Research and Development Laboratories, Edgewood Arsenal, Maryland.
Hofstadter, Dr. Robert. Department of Physics, Stanford University.
Hogan, Dr. Brigid. Mill Hill Laboratories, Imperial Cancer Research Fund, London.
Hogan, Dr. Neville. Department of Mechanical Engineering, Massachusetts Institute of Technology.
Hogan, Dr. Robert. Department of Psychology, Johns Hopkins University.
Hoge, Harvey J. Manager, Dynamics Technology, Los Angeles Division, North American Rockwell Corporation.
Hogg, Dr. Helen S. David Dunlap Observatory, University of Toronto.
Holdeman, Dr. Louis B. Communications Satellite Corporation (COMSAT), Clarksburg, Maryland.
Holder, Dr. Nigel. Anatomy and Human Biology Group, Biomedical Sciences Division, Kings College, University of London.
Holland, Dr. Heinrich D. Department of Geological and Geophysical Sciences, Harvard University.
Holland, Dr. Timothy J. B. Department of Mineralogy and Petrology, University of Cambridge, England.
Hollowell, Eugene A. Consultant in Agronomy, Port Republic, Maryland.
Hollweg, Prof. Joseph V. Space Science Center, University of New Hampshire.
Holm, Prof. Jens T. Webb Institute of Naval Architecture, Glen Cove, New York.
Holmes, Dr. Jack K. Holmes Associates, Los Angeles, California.
Holmes, Dr. Robert W. Department of Biological Sciences, University of California, Santa Barbara.
Holst, Per A. Foxboro Company, Foxboro, Massachusetts.
Holt, Dr. Donald L. Monsanto Company, St. Louis, Missouri.
Holt, Dr. Perry C. Department of Biology, Virginia Polytechnic Institute.
Hold, Dr. Robert D. Museum of Natural History, University of Kansas.
Holt, Dr. Stanley J. Professor of Cytochemistry, Courtauld Institute of Biochemistry, Middlesex Hospital Medical School, London, England
Holthuis, Dr. Lipke B. Curator, Division of Crustacea, Rijksmuseum van Natuurlijke Historie, Leiden, Netherlands.
Holtje, Malcolm C. General Radio Company, West Concord, Massachusetts.
Holton, Dr. Gerald. Department of Physics, Harvard University.
Holton, Prof. James R. Department of Atmospheric Sciences, University of Washington.
Holzman, Albert G. Professor and Chairman, Department of Industrial Engineering, Systems Management Engineering and Operations Research, School of Engineering, University of Pittsburgh.
Homann, Prof. Peter H. Institute of Molecular Biophysics, Florida State University.
Hong, Prof. Richard. University Hospital and Clinics, University of Wisconsin—Madison.
Hoover, J. Edgar. Deceased; formerly, Director, Federal Bureau of Investigation, U.S. Department of Justice.
Hopper, Bruce. Nematode Investigations, Entomology Research Institute, Canada Department of Agriculture, Ottawa.
Horan, Dr. Michael. National Heart, Lung and Blood Institute, National Institutes of Health, Bethesda, Maryland.
Horen, Dr. D. J. Nuclear Division, Oak Ridge National Laboratory, Oak Ridge, Tennessee.
Horn, Dr. David J. Department of Entomology, College of Agriculture, Ohio State University.
Horne, Thomas A. Aircraft Owners and Pilots Association, Frederick, Maryland.
Horsfall, Dr. James G. Director, Connecticut Agricultural Experiment Station, New Haven.
Horsnail, Dr. R. F. Amax Exploration, Inc., Denver, Colorado.
Horst, Dr. R. K. Department of Plant Pathology, Cornell University.
Horstadins, Prof. Sven. Professor Emeritus of Zoology, University of Uppsala, Sweden.
Horton, G. A. Manager, Engineering Laboratories, Westinghouse Electric Corporation, Cleveland, Ohio.
Horton, Dr. William A. Department of Pediatrics, Medical School, Health Science Center at Houston, University of Texas.
Horzepa, Joseph J. Marketing Director, Visual Communications Services, American Telephone and Telegraph Company, Morristown, New Jersey.
Hotchkiss, Grosvenor. Coordinating Engineer (retired), Western Union Telegraph Company, New York.
Hotchkiss, Prof. Rollin D. Department of Cellular Physiology, Rockefeller University.
Hough, Dr. L. F. Research Professor in Pomology, Department of Horticulture and Forestry, Rutgers University.
Houghton, Dr. David D. Department of Meteorology, University of Wisconsin, Madison.
Hougie, Dr. Cecil. Professor of Pathology, School of Medicine, University of California, San Diego.
Howe, Prof. Carl E. Deceased; formerly, Professor Emeritus of Physics, Oberlin College.
Howell, Dr. Benjamin F., Jr. Department of Geosciences, Pennsylvania State University.
Howell, Prof. F. Clark. Department of Anthropology, University of California, Berkeley.
Howell, William E. NASA Langley Research Center, Hampton, Virginia.
Howells, Dr. T. A Director of Continuing Education (retired), Institute of Paper Chemistry, Appleton, Wisconsin.
Howells, Dr. William W. Peabody Museum of Archeology and Ethnology, Harvard University.
Hoxton, Prof. Llewellyn G. Deceased; formerly, Professor Emeritus of Physics, University of Virginia.
Hromadik, J. J. Director, Amphibious and Harbor Division, U.S. Naval Civil Engineering Laboratory, Port Hueneme, California.
Hrones, Dr. John A. Provost of Science and Technology, Case Western Reserve University.
Hu, Dr. A. S. L. Department of Biochemistry, University of Kentucky.
Hubert, Lester F. National Environmental Satellite Center, U.S. Department of Commerce.
Hudlicky, Dr. Tomas. Department of Chemistry, Illinois Institute of Technology.
Hudson, Dr. Ralph P. Bureau International des Poids et Mesures, Sèvres, France.
Huebler, Dr. Jack. Senior Vice President, Institute of Gas Technology, IIT Center, Chicago, Illinois.
Hughes, Prof. Vernon. Department of Physics, Yale University.
Huizenga, Dr. John R. Nuclear Structure Research Laboratory, University of Rochester.
Hukill, Dr. Peter B. Associate Professor of Pathology, School of Medicine, Yale University.
Hulbary, Dr. Robert L. Department of Botany, University of Iowa.

Hull, Dr. McAllister H., Jr. Department of Physics and Astronomy, State University of New York, Buffalo.
Hultsch, Dr. Roland A. Department of Physics, University of Missouri.
Humes, Dr. Arthur G. Department of Biology, Boston University.
Hummon, Dr. William D. Department of Zoölogical and Biomedical Sciences, Ohio University.
Hundhausen, Dr. Arthur J. High Altitude Observatory, National Center for Atmospheric Research, Boulder, Colorado.
Hunkins, Dr. Kenneth. Lamont-Doherty Geological Observatory, Palisades, New York.
Hunley, Dr. William H. Mechanical Engineer, Alexandria, Virginia.
Hunt, Dr. John M. Department of Chemistry, Woods Hole Oceanographic Institute, Woods Hole, Massachusetts.
Hunten, Dr. Donald M. Lunar and Planetary Laboratory, Department of Planetary Sciences, University of Arizona.
Hunter, Prof. Lloyd P. Department of Electrical Engineering, University of Rochester.
Hunter, Dr. R. H. F. School of Agriculture, University of Edinburgh, Scotland.
Huntington, Prof. H. B. Department of Physics, Rensselaer Polytechnic Institute.
Hurlbut, Prof. Cornelius S., Jr. Professor of Mineralogy, Department of Geological Sciences, Harvard University.
Hurley, Prof. Patrick M. Department of Earth and Planetary Sciences, Massachusetts Institute of Technology.
Hurst, Dr. G. S. Department of Physics, Oak Ridge National Laboratory, Oak Ridge, Tennessee.
Huskey, Dr. Velma R. Information Sciences, University of California, Santa Cruz, California.
Hutcheon, Dr. Ian D. Isotope Sciences Division, Lawrence Livermore National Laboratory, University of California, Livermore.
Hutchison, Dr. Keith. Department of Biology, Brandeis University.
Hyde, Dr. Earl K. Lawrence Berkeley Laboratory, University of California, Berkeley.
Hynek, Dr. J. Allen. Deceased; formerly, Department of Astronomy, Northwestern University.
Hyslop, Dr. Adin. Texas Instruments, Houston.

I

Iachello, Dr. Franco. Department of Physics, Yale University.
Ibrahim, Dr. Aly M. Agricultural Research Service, Western Region, U.S. Department of Agriculture, Salinas, California.
Illis, Alexander. Director, Process Technology, International Nickel Company of Canada, Ltd., Toronto, Ontario.
Ingerson, Dr. Earl. Department of Geology, University of Texas, Austin.
Ingram, Dr. B. Lynn. Department of Geology and Geophysics, University of California, Berkeley.
Ingram, William T. Consulting Engineer, Whitestone, New York.
Inhorn, Dr. Stanley L. Director, Wisconsin State Laboratory of Hygiene; Professor of Preventive Medicine and Pathology, University of Wisconsin Medical Center.
Inman, Dr. Douglas L. Professor of Oceanography, Scripps Institution of Oceanography, La Jolla, California.
Isbin, Prof. Herbert S. Department of Chemical Engineering and Materials Science, University of Minnesota.
Ishida, Dr. Nakao. Department of Bacteriology, Tohoku University School of Medicine, Sendai, Japan.
Ishihara, Prof. Teruo. Department of Mechanical and Aerospace Engineering, Rose Polytechnic Institute.
Israel, Dr. Yedy. Department of Pharmacology, University of Toronto, Ontario, Canada.
Issel, Dr. Charles J. Wright-Markey Professor of Equine Infectious Diseases, Gluck Equine Research Center, University of Kentucky, Lexington.
Iverson, Dr. Brent L. Department of Chemistry and Biochemistry, University of Texas, Austin.

J

Jackson, Donald C. Division of Biology and Medicine, Brown University.
Jackson, Dr. Roscoe G., II. Department of Geological Science, Northwestern University
Jackson, Dr. William D. President, HMJ Corporation, Chevy Chase, Maryland.
Jacobs, Prof. Stephen F. Optical Sciences Center, University of Arizona.
Jacobs, Dr. Woodrow C. Retired; formerly, Senior Scientist, Environmental Data Service, Rockville, Maryland.
Jacobson, A. Walter. Assistant Chief Engineer, Bristol Company, Waterbury, Connecticut.
Jacobson, Prof. Nathan. Department of Mathematics, Yale University.
Jaep, William F. Central Research Department, Experimental Station, E. I. du Pont de Nemours and Company, Wilmington, Delaware.
Jafarey, Prof. S. Behlen Laboratory of Physics, University of Nebraska.
Jaffe, Dr. Julian J. Nuffield Institute of Comparative Medicine, Zoological Society of London, England.
Jahns, Dr. Richard H. Department of Earth Sciences, Stanford University.
Jain, Dr. Subodh K. Department of Agronomy and Range Science, University of California, Davis.
Jakobi, Prof. Hans. Professor of Animal Physiology, Department of Zoology, University of Paraná, Brazil.
James, Dr. Gideon T. Department of Earth Sciences, East Texas State University.
James, John W. Vice President, Research, McDonnell and Miller, Inc., Chicago, Illinois.
James, Dr. Richard W. Head, Forecasting Branch, Oceanographic Prediction Division, U.S. Naval Oceanographic Office.
Janick, Prof. Jules. Department of Horticulture, Purdue University.
Janna, Prof. William S. Department of Mechanical Engineering, Memphis State University, Tennessee.
Jarvik, Dr. Erik. Department of Paleontology, Swedish Museum of Natural History, Stockholm.
Jarvik, Dr. Murray E. Professor of Psychiatry and Pharmacology, Albert Einstein School of Medicine.
Jarvis, Dr. John F. Robotics Systems Research Department, Bell Laboratories, Holmdel, New Jersey.
Jeffries, Dr. R. P. S. Department of Paleontology, British Museum of Natural History, London, England.
Jen, Dr. C. K. Supervisor, Microwave Physics Group, Applied Physics Laboratory, Johns Hopkins University.
Jenkins, Prof. Francis A. Deceased; formerly, Department of Physics, University of California, Berkeley.
Jenkins, Dr. Julian L. Director of Research, Bell Helicopter Company, Fort Worth, Texas.
Jennings, Dr. J. B. Department of Pure and Applied Zoology, University of Leeds, England.
Jensen, Dr. Aldon H. Department of Animal Science, University of Illinois, Urbana.
Jensen, Prof. Arthur R. School of Education, University of California, Berkeley.
Jensen, Dr. Hans L. Department of Bacteriology, State Laboratory for Soil and Crop Research, Lyngby, Denmark.
Jensen, Dr. Howard B. Research Supervisor, Laramie Energy Research Center, Energy Research and Development Administration, Wyoming.
Jespersen, Prof. Neil. Department of Chemistry, St. John's University.
Joel, Amos E., Jr. AT&T Bell Telephone Laboratories, Holmdel, New Jersey.
Johansen, Dr. Kjell. Department of Zoophysiology, University of Aarhus, Denmark.
Johnson, Dr. Donald Lee. Department of Geography, University of Illinois, Urbana.
Johnson, Dr. Edward. Department of Cell Biology, Rockefeller University.

Johnson, Dr. Francis S. Acting President, University of Texas, Dallas.
Johnson, Dr. James G. Department of Geology, Oregon State University.
Johnson, Dr. Jon D. Department of Forestry, University of Florida, Gainesville.
Johnson, Dr. Julian F. Department of Chemistry and Institute of Materials Science, University of Connecticut.
Johnson, K. W. President, Vibra-Grip Corporation, Jamestown, Ohio.
Johnson, Peter K. Metal Powder Industries Federation, American Powder Metallurgy Institute, Princeton, New Jersey.
Johnson, Philip C. Research and Development Department, Chemicals and Plastics Operations Division, Union Carbide Corporation, South Charleston, West Virginia.
Johnson, Dr. Phyllis T. Research Associate, W. M. Keck Engineering Laboratory, California Institute of Technology.
Johnson, Dr. W. Hilton. Department of Geology, University of Illinois, Urbana.
Johnson, Prof. Walter C. Department of Electrical Engineering, Princeton University.
Johnson, Wendell E. Consulting Engineer, McLean, Virginia.
Johnston, Dr. Francis J. Department of Chemistry, University of Georgia.
Johnston, H. Mark. Department of Biology, University of Utah.
Johnston, Prof. Ivan D. A. Department of Surgery, Royal Victoria Infirmary, Newcastle upon Tyne, England.
Johnston, Dr. James D. Senior Research Advisor, Pioneering Research, Ethyl Corporation, Baton Rouge, Louisiana.
Johnston, Prof. Keith. Department of Chemical Engineering, University of Texas, Austin.
Jones, Prof. B. Frank, Jr. Department of Mathematics, Weiss School of Natural Sciences, Rice University.
Jones, Dr. Daniel J. Professor and Chairman, Department of Earth Sciences, California State College, Bakersfield.
Jones, Dr. David R. Department of Zoology, University of British Columbia, Canada.
Jones, Dr. E. Ruffin. Department of Zoology, University of Florida.
Jones, Dr. Edwin C., Jr. Department of Electrical and Computer Engineering, Iowa State University.
Jones, Dr. Hugh D. School of Biological Sciences, University of Manchester, United Kingdom.
Jones, Prof. Lawrence W. Harrison M. Randall Laboratory of Physics, University of Michigan.
Jones, Dr. Norman S. Senior Lecturer, Marine Biological Station, University of Liverpool, England.
Jones, Prof. R. Alan. School of Chemical Sciences, University of East Anglia, England.
Jones, Dr. Richard T. Department of Biochemistry, University of Oregon Medical School.
Jones, Dr. Robert F. Professor of Surgery, School of Medicine, University of Washington.
Jones, Dr. Russell L. Department of Botany, University of California, Berkeley.
Joppa, Prof. Robert G. Department of Aeronautics, University of Washington.
Joravsky, Prof. David. Professor of History, Northwestern University.
Jory, Howard R. Varian Associates, Palo Alto, California.
Joseph, Earl C. President, Planning and Development, Anticipatory Sciences Inc., Minneapolis, Minnesota.
Joseph, Leonard M. Lev Zetlin Associates, New York, New York.
Judd, Dr. Deane B. Deceased; formerly, Office of Colorimetry, National Bureau of Standards.
Judd, Dr. William R. School of Civil Engineering, Purdue University.
Judson, Prof. Sheldon. Chairman, Department of Geological and Geophysical Sciences, Princeton University.
Juliussen, Dr. J. Egil. Senior Member of Technical Staff, Texas Instruments Inc., Dallas, Texas.
Juranek, Dr. Dennis D. Center for Disease Control, Atlanta, Georgia.
Just, Prof. Evan. Department of Mining and Geology, Stanford University.
Juvet, Prof. Richard S., Jr. Department of Chemistry, Arizona State University.

K

Kabler, Dr. Milton N. Naval Research Laboratory, Washington, D.C.
Kado, Prof. Clarence I. Department of Plant Pathology, College of Agricultural and Experimental Sciences, University of California, Davis.
Kaizer, Daniel. Formerly, Staff Editor, "McGraw-Hill Encyclopedia of Science and Technology," McGraw-Hill Book Co., New York.
Kaler, Dr. James B. Department of Astronomy, College of Liberal Arts and Sciences, University of Illinois, Urbana-Champaign.
Kalhammer, Dr. Fritz. Electric Power Research Institute, Palo Alto, California.
Kallio, Dr. R. E. School of Life Sciences, University of Illinois, Urbana.
Kallmann, Silve. Research Director, Ledoux & Company, Teaneck, New Jersey.
Kalpakjian, Prof. Serope. Mechanical and Aerospace Engineering Department, Illinois Institute of Technology.
Kaminow, Dr. Ivan P. Bell Telephone Laboratories, Holmdel, New Jersey.
Kanal, Dr. Laveen N. Department of Computer Science, University of Maryland.
Kan-Mitchell, Dr. June. School of Medicine, University of Southern California.
Kanzig, Prof. Werner. Department of Physics, Massachusetts Institute of Technology.
Kaplan, Prof. Lewis D. Atmospheric and Environmental Research, Inc., Cambridge, Massachusetts.
Kaplan, Dr. Louis. Senior Chemist, Argonne National Laboratory, Argonne, Illinois.
Kaplan, Robert A. Vice President in Charge of Engineering, Automatic Burner Corporation, Chicago, Illinois.
Kaplan, Dr. Simon. Department of Computer Science, University of Queensland, Brisbane, Australia.
Kapner, Robert S. Department of Chemical Engineering, Cooper Union, New York.
Kappus, Peter G. Flight Propulsion Laboratory, General Electric Company, Cincinnati, Ohio.
Kaprelian, Edward K. Vice President and Technical Director, Keuffel and Esser Company, Morristown, New Jersey.
Karady, Dr. George G. Department of Electrical Engineering, Arizona State University.
Karasek, Dr. F. W. Department of Chemistry, University of Waterloo, Kitchener, Ontario, Canada.
Karash, Joseph I. Manufacturing Engineer (retired), Reliance Electric Company, Wickliffe, Ohio.
Karlson, Dr. Peter. Insitut für Physiologische Chemie, Philipps-Universitat, Marburg, Germany.
Karp, Dr. Alan H. IBM Scientific Center, Palo Alto, California.
Katz, Dr. David H. President and Chief Executive Officer, Medical Biology Institute, La Jolla, California.
Katz, Dr. Jack L. Department of Psychiatry, Montefiore Hospital and Medical Center, Bronx, New York.
Katz, Dr. Joseph J. Chemistry Division, Argonne National Laboratory, Argonne, Illinois.
Katz, Dr. Leonard A. Health Care Plan Medical Center, Buffalo, New York.
Kaufman, Dr. Peter B. Department of Botany, University of Michigan.
Kaufman, Dr. Thomas C. Department of Biology, Indiana University.
Kaufmann, R. H. Deceased; formerly, Consultant, Schenectady, New York.
Kayan, Prof. Carl F. Deceased; formerly, Department of Mechanical Engineering, School of Engineering, Columbia University.
Kays, Prof. John M. Animal Industries Department, University of Connecticut.

Keast, David N. Bolt Beranek and Newman, Inc., Cambridge, Massachusetts.
Keating, Richard E. Time Service Division, U.S. Naval Observatory, Washington, D.C.
Keeton, Dr. William T. Deceased; formerly, Professor and Chairman, Section of Neurobiology and Behavior, Langmuir Laboratory, Division of Biological Science, Cornell University.
Keffer, Prof. Frederic. Department of Physics, University of Pittsburgh.
Keith, Dr. David A. Orthopedic Research Laboratories, Children's Hospital Medical Center, Boston, Massachusetts.
Keller, Harold W. Associate Professor, Department of Microbiology and Immunology, Wright State University.
Keller, Dr. Joseph M. Department of Physics, Iowa State University.
Keller, Prof. Walter D. Department of Geology, University of Missouri.
Keller, Dr. William E. Los Alamos Scientific Laboratory, Los Alamos, New Mexico.
Kellison, Prof. R. C. Director, Hardwood Research, Department of Forestry, North Carolina State University.
Kellogg, Dr. Douglas S. Center for Disease Control, Atlanta, Georgia.
Kellogg, Prof. Herbert H. Stanley-Thompson Professor of Chemical Metallurgy, Henry Krumb School of Mines, Columbia University.
Kells, Prof. Lyman M. Deceased; formerly, Professor Emeritus, U.S. Naval Academy.
Kelman, Dr. Arthur. Department of Plant Pathology, University of Wisconsin.
Kempers, Dr. Roger D. Department of Obstetrics and Gynecology, Mayo Foundation, Mayo Graduate School of Medicine, Rochester, Minnesota.
Kendeigh, Dr. S. Charles. Department of Zoology, University of Illinois, Urbana.
Kennedy, Dr. George C. Deceased; formerly, Institute of Geophysics and Planetary Physics, University of California, Los Angeles.
Kensler, Dr. Charles J. Senior Vice President, Life Science Division, Arthur D. Little, Inc., Cambridge, Massachusetts.
Kent, Dr. Ernest. National Bureau of Standards.
Kent, Dr. George C., Jr. Professor Emeritus, Louisiana State University; Authors' Services, Inc., Baton Rouge, Louisiana.
Kerr, Dr. Marilyn S. Department of Biology, Syracuse University.
Kerr, Prof. William. Chairman, Department of Nuclear Engineering, University of Michigan.
Kerst, Prof. Donald W. Department of Physics, University of Wisconsin.
Kesling, Dr. Robert V. Director, Museum of Paleontology, University of Michigan.
Kessler, Prof. Edwin. Director, National Severe Storms Laboratory, Norman, Oklahoma.
Kessler, George W. Vice President, Engineering and Technology, Power Generation Division, Babcock and Wilcox Company, Barberton, Ohio.
Kester, Dr. Dale E. Department of Pomology, College of Agriculture, University of California, Davis.
Ketchum, Dr. Bostwick H. Woods Hole Oceanographic Institution, Woods Hole, Massachusetts.
Ketterson, Dr. J. B. Department of Physics, Northwestern University.
Khan, Dr. Fazlur R. Deceased; formerly, Partner, Skidmore, Owings & Merrill, Chicago, Illinois.
Khan, Dr. Tanveer N. Senior Plant Breeder, Division of Plant Industry, Western Australian Department of Agriculture.
Kharasch, Prof. Norman. Department of Biomedicinal Chemistry, University of Southern California.
Kilbey, Dr. Brian J. Department of Genetics, Institute of Animal Genetics, University of Edinburgh, Scotland.
Kilgore, Dr. Lee A. Retired; formerly, Westinghouse Electric Corporation, East Pittsburgh, Pennsylvania.
Kim, Prof. Chung W. Department of Physics, Johns Hopkins University.
Kimbark, Dr. Edward W. Deceased; formerly, Head, Systems Analysis Group, Bonneville Power Administration, Portland, Oregon.
Kinch, Dr. Michael A. Central Research Laboratories, Texas Instruments Inc., Dallas, Texas.
King, Dr. Donald W., Jr. Richard T. Crane Professor of Pathology, Department of Pathology, University of Chicago.
King, Dr. Elbert A. Department of Geology, University of Houston.
King, Dr. Roger. Computer Science Department, University of Colorado.
Kinnard, Dr. Isaac F. Deceased; formerly, Manager of Engineering, Instrument Department, General Electric Company, Lynn, Massachusetts.
Kinnier, Prof. Henry L. School of Engineering, University of Virginia.
Kinsler, Prof. Lawrence E. Professor of Physics, U.S. Naval Postgraduates School, Monterey, California.
Kinsman, Prof. Blair. College of Marine Studies, University of Delaware.
Kintsch, Prof. Eileen. Department of Psychology, University of Colorado.
Kintsch, Dr. Walter. Department of Psychology, University of Colorado.
Kirby, Robert S. Research Laboratories, NOAA, U.S. Department of Commerce, Boulder, Colorado.
Kirby, Dr. Seth. Dixon Spice Farms, Dixon, California.
Kirmani, Prof. Syed N. U. A. Department of Mathematics, University of Northern Iowa, Cedar Falls.
Kirschner, Prof. Stanley. Department of Chemistry, Wayne State University.
Kissel, Dr. David E. Department of Agronomy, Kansas State University.
Klass, Phillip J. Senior Avionics Editor, "Aviation Week and Space Technology," McGraw-Hill Publications Company, Washington, D.C.
Klein, Cornelius. Department of Geology, University of New Mexico.
Klein, Dr. David H. Department of Chemistry, Hope College.
Klein, Dr. Jan. Max Planck Insitut für Biologie, Abteilung Immungenetik, Tübingen, Germany.
Klein, Prof. Richard G. Department of Anthropology, University of Chicago.
Klein, Prof. Richard M. Department of Botany, University of Vermont.
Kleinmann, Dr. Douglas E. Astronomy Program, Department of Physics and Astronomy, University of Massachusetts.
Kleinmann, Dr. Susan G. Astronomy Program, Department of Physics and Astronomy, University of Massachusetts.
Kleinpell, Dr. Robert M. Department of Paleontology, University of California, Berkeley.
Klemens, Dr. Paul G. Department of Physics, University of Connecticut.
Klerman, Dr. Gerald L. Department of Psychiatry, New York Hospital-Cornell Medical Center, White Plains, New York.
Klick, Dr. Clifford C. Superintendent, Solid Sate Division, U.S. Naval Research Laboratory.
Klieger, Paul. Manager, Concrete Research Section, Research and Development Laboratory, Portland Cement Association, Skokie, Illinois.
Kliewer, Prof. K. L. Department of Physics, Iowa State University.
Kline, Dr. Hibberd V. B., Jr. Chairman, Department of Geography, University of Pittsburgh.
Klock, Dr. Benny L. Defense Mapping Agency, U.S. Department of Defense, Washington, D.C.
Klock, Dr. Harold F. Department of Electrical Engineering, Ohio University.
Knapp, Prof. Anthony W. Department of Mathematics, Cornell University.
Knaysi, Georges. Consultant, Ithaca, New York.
Knebelman, Prof. Morris S. Professor of Mathematics, Bucknell University.
Knight, Dr. C. Arthur. Professor of Molecular Biology and Research Biochemist, Virus Laboratory, University of California, Berkeley.
Knoteck, Dr. M. L. Sandia National Laboratories, Albuquerque, New Mexico.
Knowles, Hugh S. President, Knowles Electronic, Inc., Franklin Park, Illinois.
Knowles, Prof. Paulden F. Department of Agronomy and Range Science, University of California, Davis.
Koch, Leonard J. Deceased; formerly, Senior Engineer, Office of the Director,

Argonne National Laboratory, Argonne, Illinois.

Koch, Dr. Peter M. Department of Physics, State University of New York at Stony Brook.

Kochendorfer, Fred D. Manager, Communications Satellite Program, Philco WDL.

Koerker, Frederick W. Technical Expert (retired), Dow Chemical Company, Midland, Michigan.

Koglenik, Dr. Herwig. Crawford Hill Laboratory, Bell Laboratories, Holmdel, New Jersey.

Kohl, Arthur L. Project Engineer, Advanced Development, Atomics International Division, North American Rockwell, Woodland Hills, California.

Kohls, Dr. Glen M. Sanitarian Director (retired), U.S. Public Health Service, Rocky Mountain Laboratory, Hamilton, Montana.

Kohman, Dr. Truman P. Department of Chemistry, Carnegie-Mellon Institute.

Koller, Dr. Noémie. Department of Physics, Rutgers University.

Komai, Dr. Taku. Professor Emeritus, Kyoto University, Japan.

Konigsberg, Dr. William H. Department of Molecular Biophysics, Yale University.

Konz, Dr. Stephan A. Department of Industrial Engineering, Kansas State University.

Konzo, Prof. Seichi. Department of Mechanical Engineering, University of Illinois.

Koopman, Dr. Karl F. Department of Mammalogy, American Museum of Natural History, New York.

Kopac, Prof. M. J. Department of Biology, New York University.

Koral, Dr. Richard L. Visiting Associate Professor, Pratt School of Architecture; Coordinator, Apartment House Institute, New York City Community College.

Kosikowski, Prof. Frank V. Department of Food Science, Cornell University.

Kosow, Dr. Irving L. Series Editor, Electrical Engineering Technology, John Wiley & Sons, Marietta, Georgia.

Kotter, Dr. F. Ralph. National Bureau of Standards.

Kovar, Dr. Dennis. Hahn-Meitner Institut für Kernsforschung, Berlin, Germany.

Kow, Dr. Yoke Wah. Department of Biochemistry, University of Wisconsin.

Kowal, Dr. Charles T. Department of Astrophysics, California Institute of Technology.

Kraehenbuehl, Prof. John O. Professor Emeritus of Electrical Engineering, University of Illinois, Urbana.

Kramar, Dr. Ernst. Deceased; formerly, Standard Elektrik Lorenz Aktiengesellschaft, Stuttgart, Germany.

Krause, Lloyd N. Consultant, Westlake, Ohio.

Krauss, Prof. George. Colorado School of Mines, Golden.

Krauss, Dr. Lawrence M. Center for Theoretical Physics, Yale University.

Kreek, Dr. Mary Jeanne. Rockefeller University.

Krefeld, Prof. William J. Deceased; formerly, Professor of Civil Engineering, Columbia University.

Kreilik, Dr. Robert. Department of Chemistry, University of Rochester.

Kretchmer, Prof. Richard A. Department of Chemistry, Illinois Institute of Technology.

Krieg, Dr. Noel R. Department of Biology, Virginia Polytechnic Institute.

Krisciunas, Dr. Kevin. Joint Astronomy Centre, Hilo, Hawaii.

Kristofferson, Karl. Public Information Branch, John F. Kennedy Space Center, Cape Canaveral, Florida.

Krombein, Dr. Karl V. Chairman, Department of Entomology, National Museum of Natural History, Smithsonian Institution.

Kroto, Harold W. School of Chemistry and Molecular Sciences, University of Sussex, United Kingdom.

Krous, Dr. Henry F. Department of Pathology and Pediatrics, University of Oklahoma Health Sciences Center.

Krusius, Prof. Matti. Low Temperature Laboratory, Helsinki University of Technology, Finland.

Krzysztofowicz, Dr. Roman. Department of Systems Engineering, University of Virginia.

Kryter, Dr. Karl D. Director, Sensory Sciences Research Center, Stanford Research Institute, Menlo Park, California.

Kuhn, Richard E. V/STOL Consultant, Newport News, Virginia.

Kuhn, Dr. Wayne E. Professional Engineer, Portland, Oregon.

Kuiper, Prof. Gerard P. Deceased; formerly, Lunar and Planetary Laboratory, University of Arizona.

Kuis, Wietse. Department of Molecular and Experimental Medicine, Research Institute of Scripps Clinic, La Jolla, California.

Kulkarni, Sudhir S. Corporate Research Center, UOP Inc., Des Plaines, Illinois.

Kulp, J. Laurence. President, Teledyne Isotopes, Westwood, New Jersey.

Kummel, Dr. Bernhard. Museum of Comparative Zoology, Harvard University.

Kunz, Dr. Kaiser S. Research Professor of Physics, New Mexico State University.

Kuo, Prof. John T. Department of Mining Engineering, Columbia University.

Kupchan, Prof. S. Morris. Deceased; formerly, Department of Chemistry, University of Virginia.

Kupke, Dr. Donald W. Department of Biochemistry, School of Medicine, University of Virginia.

Kurylo, Dr. Michael J. National Measurements Laboratory, National Bureau of Standards.

Kusch, Prof. Polykarp. Department of Physics, Columbia Radiation Laboratory, Columbia University.

Kusters, Dr., N. L. Division of Electrical Engineering, National Research Council, Ottawa, Ontario, Canada.

Kyle, John M. Deceased; formerly, Chief Engineer, Port of New York Authority.

L

Laarhoven, Prof. William H. Department of Organic Chemistry, University of Nijmegen, The Netherlands.

Lach, Dr. Joseph. Fermi National Accelerator Laboratories, Batavia, Illinois.

Lackey, Dr. James B. Emeritus Professor, Department of Environmental Engineering, University of Florida.

Ladd, Dr. Harry S. (Retired) National Museum, Smithsonian Institution.

Lagler, Prof. Karl F. School of Natural Resources, University of Michigan.

Lagowski, Dr. J. J. Department of Chemistry, University of Michigan.

Lagunoff, Dr. David. Department of Pathology, School of Medicine, University of Washington.

Laitinen, Prof. Herbert A. Department of Chemistry, University of Florida.

Lal, Prof. Devendra. Geological Research Division, Scripps Institute of Oceanography, University of California, La Jolla, and Physical Research Laboratory, Navrangpura, Ahmedabad, India.

Lamanna, Dr. Carl. Deputy and Scientific Advisor, Life Sciences Division, Office of the Chief of Research and Development, U.S. Department of the Army, Arlington, Virginia.

Lamar, John E. Illinois State Geological Survey, Urbana.

Lamb, Dr. D. Q. Department of Physics, University of Illinois, Urbana.

Lambert, Dr. Jack L. Department of Chemistry, Kansas State University.

LaMotte, Dr. Clifford E. Botany Department, Iowa State University.

LaMotte, Dr. Robert. Department of Anesthesiology, School of Medicine, Yale University.

Land, Dr. Cecil E. Sandia Laboratories, Kirtland Air Force Base East, Albuquerque, New Mexico.

Landes, Dr. Hugh S. Department of Electrical Engineering, University of Virginia.

Landshoff, Dr. Rolf. Consulting Scientist, Lockheed Missiles and Space Company, Palo Alto, California.

Landweber, Dr. Louis. Iowa Institute of Hydraulic Research, University of Iowa.

Lane, James A. (Retired) Oak Ridge National Laboratory, Oak Ridge, Tennessee.

Lane, Dr. John C. Research and Development Department, Research Laboratories, Ethyl Corporation, Ferndale, Michigan.

Lane, Dr. N. Gary. Indiana University.

Lanford, Prof. William A. Department of Physics, State University of New York, Albany.

Lang, Dr. Anton. Department of Botany, MUS-DOE Plant Research Laboratory, Michigan State University.

Lang, Dr. Karl. Director, Department of Invertebrates, Swedish State Museum of Natural History, Stockholm, Sweden.

Langenberg, Prof. D. N. Department of Physics, University of Pennsylvania.

Langenheim, Prof. Jean H. Division of Natural Sciences, University of California, Santa Cruz.

Langley, Dr. Leroy L. Associate Director for Extramural Programs, National Library of Medicine, Bethesda, Maryland.

Lapedes, Daniel N. Deceased; formerly, Editor in Chief, "McGraw-Hill Encyclopedia of Science and Technology," McGraw-Hill, Inc., New York.

Laport, Edmund A. Director, Communications Engineering, RCA, Princeton, New Jersey.

Lappin-Scott, Hilary. Department of Biology, University of Calgary, Alberta, Canada.

Lapple, Charles E. Consultant (retired), Fluid and Particle Technology, Air Pollution and Chemical Engineering, Los Altos, California.

Larimer, Dr. J. L. Department of Zoology, University of Texas, Austin.

Larsen, Dr. David M. Department of Physics and Applied Physics, University of Lowell, Massachusetts.

Larsen, Prof. Edwin M. Department of Chemistry, University of Wisconsin.

Larsen, Dr. Helge. Professor of Biochemistry, Norway Institute of Technology, University of Trondheim.

Larsen, Dr. R. Paul. Superintendent and Horticulturist, Tree Fruit Research Center, Washington State University.

Larsen, Dr. Robert G. Assistant to the Vice President, Shell Development Company, Emeryville, California.

Larsen, Dr. Sandra A. Centers for Disease Control, Sexually Transmitted Disease Laboratory Program, Atlanta, Georgia.

Larson, Dr. Roy A. Department of Horticulture, North Carolina State University.

Lass, Dr. Harry. Research Specialist, Jet Propulsion Laboratory, California Institute of Technology.

Lassila, Prof. Kenneth E. Department of Physics, Iowa State University.

Latham, Dr. David W. Center for Astrophysics, Cambridge, Massachusetts.

Lattman, Dr. Laurence H. Dean of Mines, University of Utah.

Launay, Dr. Jules D. Consultant, Solid State Division, U.S. Naval Research Laboratory.

Laurie, Prof. Victor W. Department of Chemistry, Princeton University.

Laves, Dr. Friz H. Deceased; formerly, Eidg. Technische Hoschschule, Institut für Kristallographie und Petrographie, Zurich, Switzerland.

Lawler, Frank K. Formerly, Editor in Chief, "Food Engineering," Philadelphia, Pennsylvania.

Lawley, Dr. Alan. College of Engineering, Drexel University.

Lawrence, Dr. J. Dennis. Lawrence Livermore Laboratory, University of California.

Lawrence, Keith W. Acting Chief, Marine Design Center, Department of the Army, Philadelphia District, U.S. Corps of Engineers.

Lawroski, Dr. Stephen. Associate Laboratory Director, Argonne National Laboratory, Argonne, Illinois.

Layzer, Prof. David. Department of Astronomy, Harvard University.

Lazarus, Dr. David. Department of Physics, University of Illinois, Urbana.

Lazarides, Dr. E. Department of Biology, California Institute of Technology.

Lea, Dr. Wayne A. Research Engineer, Speech Communications Research Laboratory, University of Southern California.

Leach, Dr. Richard E. Fellow, Reproductive Endocrinology and Fertility, Department of Obstetrics and Gynecology, Mayo Graduate School of Medicine, Rochester, New York.

Leadon, Dr. Bernard M. Research Professor, Department of Aerospace Engineering, University of Florida.

Lebowitz, Dr. Michael D. Professor of Medicine and Associate Center Director, Respiratory Sciences Center, University of Arizona.

Ledbetter, Dr. Myron C. Biology Department, Brookhaven National Laboratory, Upton, New York.

Lederman, Leon M. Director, Fermi National Accelerator Laboratory, Batavia, Illinois.

Ledig, Dr. F. Thomas. School of Forestry, Yale University.

Ledingham, Dr. George A. Deceased; formerly, Director, Prairie Regional Laboratory, Saskatoon, Saskatchewan, Canada.

Lee, Lin-Nan. Communications Satellite Corporation, Clarksburg, Maryland.

Lee, Low K. Assistant Director, Product Assurance, TRW Systems, Redondo Beach, California.

Lee, Dr. Roberto. Monsanto Company, St. Louis, Missouri.

Lee, Prof. Samuel C. Department of Electrical Engineering, University of Oklahoma.

Lee, Dr. Tae Young. Department of Chemistry, Seoul National University, Seoul, Korea.

Lee, Dr. Thomas H. Strategic Planning Operation, General Electric Company, Fairfield, Connecticut.

Leete, Prof. Edward. Department of Chemistry, University of Minnesota.

Lefkowitz, Prof. Irving. Systems Engineering Department, Case Western Reserve University.

Legget, Prof. Anthony. Department of Physics, University of Illinois, Urbana.

Legget, Robert F. Engineering Consultant, Ottawa, Ontario, Canada.

Lehiste, Prof. Ilse. Department of Linguistics, Ohio State University.

Lehmer, Dr. Derrick H. Department of Mathematics, University of California, Berkeley.

Lehrman, Prof. Daniel S. Deceased; formerly, Institute of Animal Behavior, Rutgers University.

Leiby, Jonathan. Woods Hole Oceanographic Institution, Woods Hole, Massachusetts.

Lein, Dr. Joseph. Panlabs Inc., Fayetteville, New York.

Leland, Dr. Henry. Department of Psychology, Ohio State University.

Lemke, Prof. Paul A. Department of Botany and Microbiology, Auburn University, Auburn, Alabama.

Lemont, Prof. Harvey. Chairman, Department of Podiatric Medicine, Pennsylvania College of Podiatric Medicine, Philadelphia.

Lems, Dr. Kornelius. Deceased; formerly, Associate Professor and Chairman, Department of Biological Sciences, Goucher College.

Lenel, Dr. F. V. Professor Emeritus, Department of Materials Engineering, Rensselaer Polytechnic Institute.

Lennert, Richard. Department of Chemical Engineering, University of Texas, Austin.

Le Noble, Dr. William J. Department of Chemistry, State University of New York, Stony Brook.

Leonard, Dr. Edward F. Chemical Engineering Department, Columbia University.

Leontis, Dr. Thomas E. Magnesium Research Center, Battelle Columbus Laboratories, Columbus, Ohio.

Leopold, A. Starker. School of Forestry and Conservation, University of California, Berkeley.

Leopold, Dr. Luna B. Department of Geology and Geophysics, University of California, Berkeley.

Le Page, Dr. Wilbur R. Department of Electrical Engineering, Syracuse University.

Lepard, Paul E. Mettler Instrument Corporation, Hightstown, New Jersey.

Lerner, Dr. I. Michael. Department of Genetics, University of California, Berkeley.

Lesk, I. A. Director, Central Research Laboratories, Motorola, Inc., Phoenix, Arizona.
Lesso, Dr. William G. Department of Mechanical Engineering, University of Texas, Austin.
Lettau, Prof. Heinz H. Department of Meteorology, University of Wisconsin.
Levasheff, V. V. American Potash and Chemical Corporation, Whittier, California.
Levin, Prof. Simon A. Professor of Applied Mathematics and Ecology, Division of Biological Sciences, Cornell University.
Levine, Dr. I. E. Chevron Research Company, Richmond, California.
Levine, Prof. Ira N. Department of Chemistry, Brooklyn College.
Levine, Dr. Norman D. College of Veterinary Medicine, University of Illinois, Urbana.
Levine, Dr. Rachmiel. City of Hope National Medical Center, Duarte, California.
Lew, Dr. I. Paul. Thornton-Tomasetti, Engineers, New York, New York.
Lewis, Dr. Bernard. Combustion and Explosives Research, Inc., Pittsburgh, Pennsylvania.
Lewis, Prof. D. Department of Botany and Microbiology, University College, London, England.
Lewis, Frank D. Consultant, Lexington, Massachusetts.
Lewis, Dr. James K. Department of Animal Science, South Dakota State University.
Lewis, Prof. M. J. Department of Food Science and Technology, University of California, Davis.
Ley, Dr. Willy. Deceased; formerly, Fairleigh Dickinson University.
Lhermitte, Dr. Roger. Professor of Physical Meteorology, University of Miami School of Marine and Atmospheric Science.
Li, Dr. Ching Chun. Graduate School of Public Health, University of Pittsburgh.
Li, Prof. Choh Hao. Deceased; formerly, Hormone Research Laboratory, University of California Medical Center, San Francisco.
Li, Norman N. Director, Separations Research, UOP Inc., World Headquarters, Des Plaines, Illinois.
Liddicoat, Richard T., Jr. Gemological Institute of America, Los Angeles, California.
Liedholm, George E. Department Head (retired), Petroleum Processing, Shell Development Company, Emeryville, California.
Liener, Dr. Irvin E. Department of Biochemistry, University of Minnesota, St. Paul.
Lieth, Dr. Helmut. Department of Biochemistry, University of North Carolina.
Likens, Prof. Gene E. Director, Institute of Ecosystem Studies, The New York Botanical Garden, Millbrook, New York.
Lill, Gordon G. Coast and Geodetic Survey, Environmental Science Services Administration, U.S. Department of Agriculture.
Liller, Dr. William. Department of Astronomy, Harvard University.
Limpel, Eugene J. Consulting Electrical Engineer, Eagle River, Wisconsin.
Lin, Dr. Shao-Chi. Department of Applied Mechanics and Engineering Sciences, University of California, San Diego.
Lindberg, Prof. Roy A. Department of Mechanical Engineering, University of Wisconsin.
Linde, Dr. Ronald K. President, Envirodyne, Inc., Los Angeles, California.
Linde, Dr. Folke. Flicklaroverket, Halsingborg, Sweden.
Linderorth, Prof. L. Sigfred, Jr. (Retired) Department of Mechanical Engineering, Duke University.
Lindner, Dr. Hans R. Department of Biodynamics, Weizmann Institute of Science, Rehovoth, Israel.
Lindsay, Prof. R. Bruce. Deceased; formerly, Hazard Professor of Physics, Emeritus, Brown University.
Lindsey, J. S. Liquid Carbonic Corporation, Chicago, Illinois.
Lindstrom, Dr. Kirk W. Design Engineer, Hewlett-Packard Company, Optical Communications Division, San Jose, California.
Lineberger, Dr. W. C. Joint Institute for Laboratory Astrophysics, National Institute for Standards and Technology, University of Colorado.
Lines, Malcom E. Bell Telephone Laboratories, Murray Hill, New Jersey.
Linsley, Prof. Ray K. Department of Civil Engineering, Stanford University.
Linthicum, Jack M. Office of Science and Technology, Federal Communications Commission.
Liou, Juhn G. Department of Geology, Stanford University.
Lipkowitz, Prof. Kenneth B. Department of Chemistry, Indiana-Purdue University, Indianapolis.
Lipner, Prof. Harry J. Department of Biological Science, Florida State University.
Litherland, Prof. Albert E. Director, Isotrace Laboratory, Department of Physics, University of Toronto, Ontario, Canada.
Little, Elbert L., Jr. Dendrologist, Forest Service, U.S. Department of Agriculture.
Little, Dr. James N. Research Division, Waters Associates, Milford, Massachusetts.
Liu, Dr. Sam H. Department of Physics, Iowa State University.
Livingston, Dr. Gideon E. Food Science Associates, Rye, New York.
Livingston, Dr. William C. National Optical Astronomy Observatories, National Solar Observatory, Tucson, Arizona.
Ljung, Dr. Donovan A. Fermi National Accelerator Laboratory, Batavia, Illinois.
Lobo, Walter E. Consulting Engineer, New Canaan, Connecticut.
Lochhead, Dr. Allan G. Microbiology Research Institute, Canada Department of Agriculture, Ottawa, Ontario.
Lockhart, Prof. Frank J. Department of Chemical Engineering, University of Southern California.
Lockwood, Dr. Lewis B. Assistant Director of Microbiological Research, Union Division, Miles Laboratories, Elkhart, Indiana.
Loeb, John L. ACLS Corporation, Silver Spring, Maryland.
Loewenstein, Dr. Werner R. College of Physicians and Surgeons, Columbia University.
Loneragan, Prof. J. F. School of Environmental and Life Science, Murdoch University, Perth, Australia.
Long, Prof. Franklin A. Formerly, Department of Chemistry, Cornell University.
Long, Joseph B. Consultant, Columbus, Ohio.
Longacre, Alan. (Retired) Fluor Engineers and Constructors, Irvine, California.
Longmoore, Dr. Kenneth. Department of Chemistry, University of Massachusetts.
Longwell, Dr. Chester R. Department of Geology, Stanford University.
Lonsdale, Dr. Carol J. Infrared Processing & Analysis Center, California Institute of Technology, Pasadena.
Lopiano, Dr. Michael. Chairman, Public Information Committee, American College of Radiology, Reston, Virginia.
Lord, Prof. Richard C. Department of Chemistry, Massachusetts Institute of Technology.
Lorenz, Prof. Oscar A. Department of Vegetable Crops, University of California, Davis.
Lotz, Dr. Martin. Department of Molecular and Experimental Medicine, Research Institute of Scripps Clinic, La Jolla, California.
Loughlin, James E. Chemical Process Pilot Plant, Textile Research Center, Texas Tech University.
Louria, Dr. Donald. Department of Medicine and Community Health, New Jersey Medical School, Newark.
Lowan, Dr. Arnold N. Deceased; formerly, Chairman, Department of Physics, Yeshiva University.
Lowe, Dr. David A. Department of Fermentation Development, Industrial Division, Bristol-Myers Squibb Company, Syracuse, New York.
Lowe, Christopher R. Institute of Biotechnology, Cambridge University, England.
Lowell, Prof. Wayne R. Department of Geology, Indiana University.
Lowenstein, Prof. Otto E. Department of Zoology, University of Birmingham, England.
Lowenthal, Dr. David. Research Associate, American Geographical Society, New York.

Lowin, R. J. Managing Director, Plessey Microwave Ltd.
Luban, Prof. Naomi L. C. Department of Laboratory Medicine, Transfusion Medicine, Children's Hospital, George Washington University Medical Center, Washington, D.C.
Luckenbach, Edward C. Exxon Research and Engineering Company, Florham Park, New Jersey.
Lucker, John T. (Retired) Beltsville Parasitological Laboratory, Agricultural Research Center, Beltsville, Maryland.
Ludlam, Prof. Frank H. Deceased; formerly, Department of Meteorology, Imperial College, London, England.
Ludvik, Dr. George F. Insecticide Application Research, Agricultural Research and Development Department, Monsanto Company, St. Louis, Missouri.
Luebbers, Dr. Ralph H. Department of Chemical Engineering, University of Missouri.
Luanda, Dr. Robert E. Director of Extractive Research and Development, Zinc Smelting Division, St. Joseph Minerals Corporation, Monaca, Pennsylvania.
Luo, Ningyi. Subpicosecond and Quantum Radiation Laboratory, Texas Tech University, Lubbock.
Lusted, Dr. Lee B. Department of Radiology, University of Chicago.
Luth, Dr. William. Department of Earth Sciences, Stanford University.
Luthin, Prof. James N. Department of Civil Engineering, University of California, Davis.
Luttge, Prof. Ulrich. Institut für Botanik, Technische Hochschule Darmstadt, Darmstadt, Germany.
Lutz, Dr. Raymond P. Dean of Graduate Studies and Research, University of Texas, Dallas.
Lazadder, Prof. Warren J. Department of Engineering Graphics, Purdue University.
Lykken, Dr. David T. Department of Psychiatry, University of Minnesota, Minneapolis.
Lyman, Prof. Charles P. Department of Medical Sciences, Harvard University.
Lyman, Dr. John. Department of Oceanography, University of North Carolina.
Lynn, Dr. John W. Director of Technology, Fibers and Fabrics Division, Union Carbide Corporation, New York.
Lyon, Dr. W. S. Analytical Division, Oak Ridge National Laboratory, Oak Ridge, Tennessee.
Lyon, Dr. Waldo. Arctic Submarine Research Laboratory, Naval Undersea Warfare Center, San Diego, California.
Lyons, Prof. Walter. Consultant in Telecommunications, Flushing, New York.
Lyons, Dr. Walter A. Certified Consulting Meteorologist, Forensic Meteorology Associates, Fort Collins, Colorado.

M

McBride, Dr. Earle F. Department of Geology, University of Texas, Austin.
McCabe, Dr. Warren L. Deceased; formerly, Department of Chemical Engineering, North Carolina State University.
McCain, Dr. John W. Department of Plant Pathology, University of Minnesota, St. Paul.
McCann, Dr. Mary B. Nutrition Program, Center for Disease Control, Health Services and Mental Health Administration, U.S. Department of Health, Education and Welfare, Rockville, Maryland.
McCarthy, Dr. Dennis D. Time Service Division, U. S. Naval Observatory, Washington, D.C.
MacCarthy, Donnell D. Manager, Voltage Regulator Engineering, Voltage Regulator Product Section, General Electric Company, Pittsfield, Massachusetts.
McCarthy, Prof. Kathryn A. Professor of Physics, and Dean of Graduate School of Arts and Sciences, Tufts University.
McCarty, Dr. John E. Structure Staff Supervisor, Boeing Company, Renton, Washington.
McClellan, Leslie N. Consulting Engineer, Engineering Consultants, Inc., Denver, Colorado.
McClelland, Richard H. Manager, SNG Project Planning, CNG Energy Company, Pittsburgh, Pennsylvania.
McClung, Dr. Leland S. Professor Emeritus, Department of Microbiology, Indiana University.
McClusky, V. G. Product Planner, High Intensity Discharge Lamps, General Electric Co., Cleveland, Ohio.
McConnaughey, Dr. Bayard H. Department of Biology, University of Oregon.
McCormick, Dr. J. R. D. Research Fellow, Lederle Laboratories, American Cyanamid Company, Pearl River, New Jersey.
McCormick, Dr. Jack. Chairman, Department of Ecology and Land Management, Academy of Natural Sciences, Philadelphia, Pennsylvania.
McCormick, Dr. Thomas R. Department of Biomedical History, School of Medicine, University of Washington.
MacCoull, Neil. Deceased; formerly, Lecturer in Mechanical Engineering, Columbia University.
McCoy, Prof. Barry M. Division of Engineering and Applied Physics, Harvard University.
McCright, Dr. R. D. Corrosion Center, Department of Metallurgical Engineering, Ohio State University.
MacCrone, Prof. Robert K. Department of Physics, Rensselaer Polytechnic Institute.
McCrone, Dr. Walter C. McCrone Research Institute, Inc., Chicago, Illinois.
McCrosky, Dr. Richard E. Smithsonian Astrophysical Observatory, Cambridge, Massachusetts.
McDade, Dr. Joseph E. Center for Disease Control, Atlanta, Georgia.
MacDaniels, Dr. Lawrence H. Professor Emeritus of Floriculture and Ornamental Horticulture, Cornell University.
Macdonald, Dr. Gordon A. Deceased; formerly, Institute of Geophysics, University of Hawaii.
McDonald, Prof. John J. Retired; formerly, Head, Textile Engineering Department, University of Lowell.
MacDonald, Dr. J. M. Small Animal Clinic, Auburn University, Auburn, Alabama.
MacDonald, Prof. Robert C. Department of Biochemistry, Molecular and Cell Biology, Northwestern University.
MacDonald Ruby I. Department of Biochemistry, Molecular, and Cell Biology, Northwestern University.
MacDowell, Dr. Samuel W. Department of Physics, Yale University.
McEvilly, Dr. Thomas V. Department of Geology, University of California, Berkeley.
McEwan, Prof. Robert S. Deceased; formerly, Professor Emeritus, Oberlin College.
McFarland, Forest R. Deceased, formerly, Engineering Consultant, Flint, Michigan.
McGannon, Harold E. Technical Editor, Research and Technology Division, United States Steel Corporation, Pittsburgh, Pennsylvania.
McGaugh, Dr. James L. Department of Psychobiology, University of California, Irvine.
McGinnies, Dr. William G. Professor of Dendrochronology and Arid Land Ecologist, Office of Arid Land Studies, University of Arizona.
McGrain, Preston. Assistant State Geologist, Kentucky Geological Survey, University of Kentucky.
McGuire, Ann B. Aquaculture Program, University of California, Davis.
Macher, Dr. Abe M. Medical Consultant, AIDS Education and Training Centers Programs, U.S. Public Health Service, Rockville, Maryland.
McHugh, Dr. J. L. Marine Sciences Research Center, State University of New York, Stony Brook.
McIntyre, Dr. Andrew. Lamont-Doherty Geological Observatory, Palisades, New York; Queens College of the City University of New York.
McIver, Dr. Richard D. Geochemical Research Inc., Houston, Texas.
McKenna, Dr. Malcolm C. Frick Curator, American Museum of Natural History, New York.
McKenzie, Alexander A. Assistant to the Editor, Editorial Services, Institute of Electrical and Electronic Engineers, New York.
McKenzie, Dr. James. Sandia Laboratories, Albuquerque, New Mexico.
McKetta, Dr. John F. E. P. Schoch Professor, E. P. Schoch Laboratories, University of Texas, Austin.

Mackey, Prof. George W. Department of Mathematics, Harvard University.
McKinney, Dr. William T., Jr. Department of Psychiatry, Medical School, University of Wisconsin.
McClain, Prof. Thomas E. Head, Department of Forest Products, Oregon State University, Corvallis.
McLaughlin, Dr. Patsy A. Shannon Point Marine Center, Western Washington University, Anacortes.
McLellan, Dr. Hugh J. National Science Foundation, Washington, D.C.
McMahon, Dr. John F. Deceased; formerly, New York State Technical Service Program, Alfred, New York.
McMath, Prof. Robert R. Deceased; formerly, Director McMath-Hulbert Observatory, University of Michigan.
McMillen, Wheeler. formerly, Chairman of the Board, Chemurgic Council, Philadelphia, Pennsylvania.
McNamara, Dr. Ken J. Senior Curator, Invertebrate Paleontology, Western Australian Museum, Perth, Western Australia.
McNellis, Dr. Donald. Center for Research of Mothers and Children, National Institute of Child Health and Human Development, National Institutes of Health, Bethesda, Maryland.
McNish, Alvin G. Chief, Meteorology Division, National Bureau of Standards, Chevy Chase, Maryland.
McPartland, Joseph F. Electrical Design and Installation, Englewood Cliffs, New Jersey.
McPherson, Prof. George, Jr. Department of Electrical Engineering, School of Engineering, University of Missouri.
McSweeney, Tim. Director, Technical Services, Screen Printing Association International, Fairfax, Virginia.
McWilliams, Dr. Michael O. Department of Geophysics, School of Earth Sciences, Stanford University.
Maddalon, Dal V. NASA Langley Research Center, Hampton, Virginia.
Maderson, Dr. Paul F. Department of Biology, Brooklyn College.
Madestau, L. Deceased; formerly, Chemicals and Plastics Division, Union Carbide Corp., Bound Brook, New Jersey.
Madin, Dr. Laurence P. Associate Science, Biology Department, Woods Hole Oceanographic Institution, Woods Hole, Massachusetts.
Madison, Dr. Vincent. Department of Medicinal Chemistry, School of Pharmacy, University of Illinois, Chicago.
Magee, Paul T. Department of Microbiology, School of Medicine, Yale University.
Maggenti, Dr. Armand R. Division of Nematology, University of California, Davis.
Mahmoud, Dr. Aly A. School of Engineering, Indiana University-Purdue University of Fort Wayne.
Mahoney, Dr. Lee R. Chemistry Department, Ford Motor Company, Dearborn, Michigan.
Maisel, Dr. Herbert. Computer Science Program, Georgetown University.
Maisey, Dr. John G. Assistant Curator, American Museum of Natural History, New York.
Major, Dr. Jack. Department of Botany, University of California, Davis.
Makinodan, Takashi. Director, Geriatric Research, Education, and Clinical Center, Veterans Administration Medical Center, Los Angeles.
Makulec, Alfred. Product Planning and Application Specialist, Large Lamp Department, General Electric Company, Cleveland, Ohio.
Maldonado-Moll, Dr. José F. Special Assistant for Science and Technology, Sistema Universitario de la Fundación Educative Ana G. Méndez Rio Piedras, Puerto Rico.
Mallet, Prof. Frank McL. Deceased; formerly, Department of Aeronautical and Astronautical Engineering, Ohio State University.
Malloy, Bruce G. Assistant to the President, Wallerstein Company, Division of Baxter Laboratories, New York.
Malm, John G. Argonne National Laboratory, Argonne, Illinois.
Mamay, Dr. Sergius H. U.S. Geological Survey, National Museum.
Mandava, Dr. N. Plant Hormone and Regulators Laboratory, Agricultural Research Service, U.S. Department of Agriculture, Beltsville, Maryland.
Mandel, Dr. Manley. Department of Biology, M.D. Anderson Hospital, Houston, Texas.
Mandler, Dr. George. Department of Psychology, University of California, San Diego.
Mangion, Charles. Technology Laboratory, Science and Technology Division, Systems Group, TRW, Inc., Redondo Beach, California.
Mangum, B. W. Leader, Thermometry Group, Prouss Measurement Division, National Institute of Standards and Technology, Gaithersburg, Maryland.
Mann, Prof. Alfred K. Department of Physics, University of Pennsylvania.
Mann, Dr. Kenneth H. Fisheries Research Board of Canada, Dartmouth, Nova Scotia.
Mann, Prof. Robert W. Department of Mechanical Engineering, Massachusetts Institute of Technology.
Mann, Prof. Stephen. School of Chemistry, University of Bath, United Kingdom.
Mannheim, Prof. Walter. Med. Zentrum für Hygiene, Universität Marburg, Germany.
Manning, Dr. James M. Department of Biochemistry, Rockefeller University.
Manning, Dr. Kenneth V. Professor Emeritus, Pennsylvania State University.
Manning, Dr. Paul D. V. Professor of Chemical Engineering (retired), California Institute of Technology.
Mansfield, Earl C. Western Union Telegraph Company, Upper Saddle River, New Jersey.
Manson, Dr. John A. Deceased; formerly, Department of Chemistry, Lehigh University.
Mansouri, Freydoon. Department of Physics, Yale University.
Mantell, Dr. Charles L. President, C. L. Mantell and Associates, Manhasset, New York.
Manuel, Dr. Oliver K. Department of Chemistry, University of Missouri, Rolla.
Manwell, Dr. Reginald D. Department of Zoology, College of Liberal Arts, Syracuse University.
Maple, M. Brian. Department of Physics, University of California, San Diego.
Maran, Dr. Stephen P. American Astronomical Society, Washington, D.C.
Marcelletti, Dr. John F. Division of Immunology, Medical Biology Institute, La Jolla, California.
Marchalonis, Dr. John J. Chairperson, Department of Biochemistry, Medical University of South Carolina.
Marchand, Prof. Alan P. College of Arts and Sciences, Department of Chemistry, University of Texas, Denton.
Marcus, Prof. Ernesto. Deceased; formerly, Department of Zoology, University of São Paulo, Brazil.
Marcus, Eveline. Department of Zoology, University of São Paulo, Brazil.
Marden, Dr. Albert. School of Mathematics, University of Minnesota.
Margenau, Prof. Henry. Department of Physics, Yale University.
Margon, Dr. Bruce. Space Sciences Laboratory, University of California, Berkeley.
Margrave, Prof. John L. Dean of Research, Rice University.
Marian, Joseph E. Forest Products Laboratory, University of California, Richmond.
Markert, Prof. Clement L. Department of Biology, Yale University.
Markowitz, Dr. William. Oceanographic Center, Nova University; Editor, "Geophysical Surveys."
Marks, Lawrence E. Pierce Foundation, New Haven, Connecticut.
Marks, Peter A. Director, Product Planning and Development, Milford, Ohio.
Marks, Dr. Tobin J. Department of Chemistry, Northwestern University.
Markus, John. Deceased; formerly, Consultant (retired), Sunnyvale, California.
Marshak, Dr. Harvey. National Bureau of Standards.
Marshall, Dr. H. G. Crops Research Division, Agricultural Research Service and Department of Agronomy, Pennsylvania State University.
Marshall, Dr. Max S. School of Medicine, University of California, San Francisco.

Marshall, Peter L. Independent Consultant, Granville, Ohio.
Marshall, Prof. William R., Jr. Associate Dean, College of Engineering, University of Wisconsin.
Marsland, Charles R. Chief Technical Consultant, Handy and Harman, Fairfield, Connecticut.
Martell, Dr. A. E. Department of Chemistry, Texas A&M University.
Martin, Franklin W. Mayaguez Institute of Tropical Agriculture, Mayaguez, Puerto Rico.
Martin, Dr. Peter G. McLennon Physical Laboratories, Canadian Institute for Theoretical Physics, University of Toronto, Canada.
Martinson, C. R. Corporate Engineering Department, Monsanto Company, St. Louis, Missouri.
Marx, Dr. Preston A. California Primate Research Center, University of California, Davis.
Mason, Dr. Basil J. Program Director, Center for Environmental Technology, Imperial College of Science and Technology, London, England.
Mason, Dr. Warren P. Deceased; formerly, Associate Editor, "Journal of the Acoustical Society of America," West Orange, New Jersey.
Mast, Dr. Roy C. Miami Valley Laboratories, Proctor and Gamble Company, Cincinnati, Ohio.
Masters, Dr. Colin L. Head of the National Creutzfeldt-Jakob Disease Registry, Department of Pathology, University of Melbourne, Australia.
Matyas, Stephen M. IBM Systems Communications Division, Kingston, New York.
Mauchline, Dr. John. Scottish Marine Biological Association, Oban, Scotland.
Maul, Prof. George A. Director, Division of Marine and Environmental Systems, Florida Institute of Technology, Melbourne.
May, F. H. Consultant, Kerr-McGee Corporation, Whittier, California.
May, James E. Lockheed-California Company, Burbank, California.
Mayall, Margaret W. American Association of Variable Star Observers, Cambridge, Massachusetts.
Mayall, Robert N. Director, Planning and Research Associates, Boston, Massachusetts.
Mayer, Dr. Manfred M. Deceased; formerly, Department of Microbiology, Johns Hopkins University.
Maynard, Dr. Harold B. President, Maynard Research Council, Inc., Pittsburgh, Pennsylvania.
Maynard, Prof. J. D. Department of Physics, College of Science, Pennsylvania State University.
Mays, Rolland L. Materials Systems Division, Union Carbide Corporation, Tarrytown, New York.
Mazia, Dr. Daniel. Department of Zoology, University of California, Berkeley.
Meade, Prof. M. L. Faculty of Technology, Electronics Discipline, The Open University, Milton Keynes, England.
Meade, Dr. Thomas J. Division of Chemistry and Chemical Engineering, California Institute of Technology.
Mebus, Dr. Charles A. Laboratory Chief, Foreign Animal Disease Diagnostic Laboratory, U.S. Department of Agriculture, Greenport, New York.
Meeuse, Dr. Bastiaan J. D. Department of Botany, University of Washington.
Meggers, Dr. William F. Deceased; formerly, National Bureau of Standards.
Mehra, Prof. Raman K. President, Scientific Systems, Inc., Cambridge, Massachusetts.
Meijer, Prof. Paul H. E. Department of Physics, Catholic University of America.
Meijer, Dr. Roelof J. Assistant Director of Research, Philips Research Laboratories, Eindhoven, Netherlands.
Meiklejohn, Dr. Jane. Senior Research Fellow, University College, Salisbury, Rhodesia.
Meinke, Dr. W. Wayne. Analytical Chemistry Division, National Bureau of Standards.
Meinwald, Dr. J. Department of Chemistry, Cornell University.
Meisel, Dr. Jerome. Department of Electrical and Computer Engineering, Wayne State University.
Meissner, Dr. Hans W. Department of Physics, Stevens Institute of Technology, Hoboken, New Jersey.
Meixner, Dr. Josef. Director, Institute of Theoretical Physics, Aachen, Germany.
Mellichamp, Prof. Duncan. Department of Chemical and Nuclear Engineering, University of California, Santa Barbara.
Mellon, Prof. M. G. Department of Chemistry, Purdue University.
Melman, Prof. Seymour. Department of Industrial and Management Engineering, Columbia University.
Melnick, Dr. Joseph L. Department of Virology and Epidemiology, Baylor College of Medicine.
Meltzer, Herbert Y. Department of Psychology, University of Chicago.
Menard, Dr. Henry W., Jr. Scripps Institution of Oceanography, La Jolla, California.
Menkes, Joshua. Science and Technology Division, Institute for Defense Analyses, Arlington, Virginia.
Menzies, Dr. Robert J. Department of Zoology, Duke University.
Merrill, Dr. E. W. Department of Chemical Engineering, Massachusetts Institute of Technology.
Merrill, Timothy W. Manager, Metallurgical Applications Research, Foote Mineral Company, Exton, Pennsylvania.
Merritt, Frederick S. Consulting Engineer, West Palm Beach, Florida.
Merwin, Earl J. McCormick and Company, Hunt Valley, Maryland.
Messenger, Dr. George C. Vice President and Technical Director, Messenger & Associates, Las Vegas, Nevada.
Meszar, John. Deceased; formerly, Director, Switching Systems Development, Bell Laboratories, Inc., New York.
Metcalf, William G. Woods Hole Oceanographic Institution, Woods Hole, Massachusetts.
Meyer, Dr. Bernard S. Department of Botany, Ohio State University.
Meyer, Dr. Carl H. Advisory Engineer, IBM Systems Communications Division, Kingston, New York.
Meyer, Dr. David B. Department of Anatomy, Wayne State University School of Medicine.
Meyer, Dr. John F. Department of Electric Engineering and Science, College of Engineering, University of Michigan.
Meyer, Dr. Karl F. Deceased; formerly, Professor Emeritus of Experimental Pathology, George Williams Hooper Foundation, San Francisco Medical Center, University of California.
Meyer, Dr. Margaret E. Department of Epidemiology and Preventive Medicine, University of California, Davis.
Meyers, Dr. Edward. Research Group Leader, Department of Microbiology Squibb Institute, New Brunswick, New Jersey.
Miall, Dr. Andrew D. Department of Geology, University of Toronto, Ontario, Canada.
Middlebrook, Dr. Gardner. Professor of International Medicine, School of Medicine, University of Maryland.
Middleton, Dr. Gerard V. Department of Geology, McMaster University, Hamilton, Ontario, Canada.
Midland, Prof. M. Mark. Department of Chemistry, University of California, Riverside.
Miles, Lawrence D. Engineering Consultant, Easton, Maryland.
Miles, Thomas K. Shell Development Company.
Mill, Dr. George S. Research Chemist, Shell Oil Company, New York.
Miller, Barry. Senior Editor, "Aviation Week and Space Technology," McGraw-Hill Publications Company, Los Angeles, California.
Miller, Dr. David A. B. AT&T Bell Laboratories, Holmdel, New Jersey.
Miller, Dr. Frederick R. Department of Soil and Crop Sciences, Texas A & M University.
Miller, Dr. Glenn H. Weapons Effects Division, Sandia Laboratories, Albuquerque, New Mexico.
Miller, John H. Deceased; formerly, Vice President and Chief Engineer, Weston Instrument Division, Dayton, Ohio.

Miller, Dr. Julian C. Department of Horticulture, Louisiana State University.
Miller, Prof. Martin W. Department of Food Science and Technology, College of Agricultural and Environmental Sciences, University of California, Davis.
Miller, Prof. Maynard M. Department of Geology, Michigan State University; Director, Foundation for Glacial and Environmental Research, Seattle, Washington.
Miller, Dr. Milton A. Department of Zoology, University of California, Davis.
Miller, R. A. Engineering Department, Babcock and Wilcox, Barberton, Ohio.
Miller, Dr. Shelby A. Argonne National Laboratory, Argonne, Illinois.
Miller, Prof. Sidney I. Department of Chemistry, Illinois Institute of Technology.
Miller, Terry A. AT&T Bell Laboratories, Murray Hill, New Jersey.
Miller, Dr. William H. Department of Chemistry, University of California, Berkeley.
Miller, Dr. Winston R. Director, Northlands Regional Medical Program, St. Paul, Minnesota.
Milligan, Dr. W. O. Robert A. Welch Foundation, Houston, Texas.
Millman, Dr. Jacob. Department of Electrical Engineering, Columbia University.
Mills, Dr. Harlow B. State Natural History Survey, Urbana, Illinois.
Mills, Thomas B. National Semiconductor Corporation, Santa Clara, California.
Minkler, Ward W. Vice President, TIMET, Pittsburgh, Pennsylvania.
Minnis, Dr. Wesley. Formerly, Assistant to the Director of Research and Development, Allied Chemical Corporation, New York.
Mischef, Prof. Walter. Department of Psychology, Stanford University.
Mitchell, Dr. J. Murray, Jr. Senior Research Climatologist, National Oceanic and Atmospheric Administration, McLean, Virginia.
Mitchell, Dr. Malcolm S. Comprehensive Cancer Center, Kenneth Norris, Jr. Cancer Research Institute, University of Southern California, Los Angeles.
Mitchell, Prof. Rodger. Department of Zoology, Ohio State University.
Mitchison, Dr. J. M. Department of Zoology, University of Edinburgh.
Mitchison, Dr. N. A. Medical Research Council, National Institute for Medical Research, London, England.
Mitsch, Prof. William J. Graduate Program Environmental Science, School of Natural Resources, Ohio State University, Columbus.
Mitterer, Dr. Richard M. Department of Geosciences, University of Texas, Dallas.
Mizell, Dr. David W. Project Leader, University of Southern California/Information Sciences Institute, Marina del Rey.
Moak, Dr. C.D. Oak Ridge National Laboratory, Oak Ridge, Tennessee.
Moberly, Dr. Walter. Deceased; formerly, Department of Biology, University of Oregon.
Mock, Frank C. Deceased; formerly, Bendix Products Division, Bendix Aviation Corporation, South Bend, Indiana.
Moder, Dr. Joseph J. Chairman, Department of Management Science, University of Miami.
Moe, Dr. Richard L. Department of Botany, University of California, Berkeley.
Moffett, Dr. Mark B. Naval Underwater Systems Center, New London Laboratory, New London, Connecticut.
Mohr, Prof. Peter. Department of Physics, Yale University.
Mokhoff, Dr. Nicolas. IEEE Spectrum, New York.
Moll, Dr. Gary A. Director, Urban and Community Forestry, American Forestry Association, Washington, D.C.
Mollenhaurer, Dr. Hilton H. Veterinary Toxicology and Entomology Research Laboratory, College Station, Texas.
Montmerle, Dr. Thierry. Commissariat a l'energie Atomique, Centre d'Etudes Nucléaires de Saclay, Gif-sur-Yvette, France.
Mood, Dr. Alexander M. Director, Public Policy Research Organization, University of California, Irvine.
Moody, Capt. Alton B. Navigation Consultant, La Jolla, California.
Moore, Dr. Alice E. Sloan-Kettering Institute of Cancer Research, New York.
Moore, Prof. Charles A. Department of Civil Engineering, Ohio State University.
Moore, Dr. George P. Department of Biomedical Engineering, University of Southern California.
Moore, Dr. James Alexander. Department of Chemistry and Biochemistry, University of Delaware.
Moore, Dr. Jeffrey S. Department of Chemistry, University of Illinois, Urbana-Champaign.
Moore, Prof. John C. Department of Mathematics, Princeton University.
Moore, Dr. Paul B. Department of the Geophysical Sciences, University of Chicago.
Moore, Dr. Raymond C. Deceased; formerly, Department of Geology, University of Kansas.
Moore, Dr. Richard K. Center for Research in Engineering Science, University of Kansas.
Moore, Dr. W. J. M. Division of Electrical Engineering, National Research Council of Canada, Ottawa, Ontario.
Moo-Young, Prof. Murray. Department of Chemical Engineering, University of Waterloo.
Moreno, Dr. Giuliana. Institute of Cellular Pathology, Paris, France.
Moreno, Dr. Theodore. Vice President, Information Systems Group, Varian Associates, Palo Alto, California.
Morgan, Dr. Anne H. Deceased; formerly, Mount Holyoke College.
Morgan, Dr. Karl Z. Neely Professor, School of Nuclear Engineering, Georgia Institute of Technology.
Morgan, Walker L. Communications Center of Clarksburg, Clarksburg, Maryland.
Morré, Dr. D. J. Department of Biological Sciences, Purdue University.
Morris, Dr. Simon Conway. Department of Earth Sciences, The Open University, Milton Keynes, England.
Morrison, Dr. David. Institute for Astronomy, University of Hawaii, Manoa.
Morrison, Dr. George H. Professor of Chemistry and Director, Analytical Facility of Materials Science Center, Cornell University.
Morrison, Dr. Kenneth J. Cooperative Extension Service, Washington State University.
Morrow, Hugh, III. Supervisor, Technical Information, Climax Molybdenum Company, Greenwich, Connecticut.
Morse, Dr. Robert W. Associate Director and Dean of Oceanographic Studies, Woods Hole Oceanographic Institute, Woods Hole, Massachusetts.
Morse, Dr. Roger A. Department of Entomology, Cornell University.
Morse, Dr. Roy Earl. Food Science Department, Cook College.
Mortenson, Leonard E. Department of Biological Sciences, Purdue University.
Moscowitz, Dr. Albert. Department of Chemistry, University of Minnesota.
Moser, J. F., Jr. Esso Research Laboratories, Humble Oil Refining Company, Baton Rouge, Louisiana.
Mosher, Dr. William. Deceased; formerly, Chairman, Department of Chemistry, University of Delaware.
Mossman, Prof. Harland W. Emeritus Professor, Department of Anatomy, University of Wisconsin.
Motekaitis, Dr. R. J. Department of Chemistry, Texas A&M University.
Mottet, Dr. N. Karle. Professor of Pathology and Director of Hospital Pathology, University Hospital, University of Washington.
Mower, Dr. Richard L. Department of Botany, University of California, Berkeley.
Moyer, Dr. Charles E., Jr. Deceased; formerly, Chemicals and Plastics Division, Union Carbide Corporation, Technical Center, South Charleston, West Virginia.
Moyer, Dr. Kenneth E. Department of Psychology, Carnegie-Mellon University.
Moyer, Dr. William W. Applied Research Laboratory, Pennsylvania State University.
Mrak, Dr. Emil M. Office of Chancellor Emeritus, University of California, Davis.
Mudrick, Prof. Stephen E. Department of Atmospheric Science, College of Agriculture, University of Missouri.

Muffler, Dr. L. J. Patrick. Geologist, Branch of Field Geochemistry and Petrology, Geological Survey, U.S. Department of the Interior, Menlo Park, California.
Mulholland, Dr. J. Derral. Department of Astronomy, University of Texas, Austin.
Müller, Prof. Erwin W. Deceased; formerly Department of Physics, Pennsylvania State University.
Müller, Dr. Richard A. Lawrence Berkeley Laboratory, University of California, Berkeley.
Mulliken, Prof. Robert S. Institute of Molecular Biophysics, Florida State University.
Mulvihill, Michael R. Headquarters, U.S. Army Armament Material Readiness Command, Rock Island, Illinois.
Munger, Prof. H. M. Department of Plant Breeding and Biometry, New York State College of Agriculture and Life Sciences, Cornell University.
Munson, W. A. Consultant, Chatham, New Jersey.
Murdock, Dr. Bennett B., Jr. Department of Psychology, University of Toronto, Ontario, Canada.
Murgatroyd, Dr. R. J. Meteorological Office, Bracknell, England.
Murphy, Dr. Frederick A. Centers for Disease Control, Department of Health and Human Services, Tucker, Georgia.
Murphy, Dr. Glenn. Department of Nuclear Engineering, Iowa State University.
Murphy, Dr. Terence M. Department of Anesthesiology, School of Medicine, University of Washington.
Murray, Dr. Haydn H. Executive Vice President, Georgia Kaolin Company, Elizabeth, New Jersey.
Murray, Dr. Merritt J. Formerly, A. M. Todd Company, Kalamazoo, Michigan.
Murray, Dr. Royce W. Department of Chemistry, University of North Carolina.
Murray, Stephen S. Center for Astrophysics, Harvard College Observatory, Cambridge, Massachusetts.
Murray, Dr. Thomas P. Applied Research Laboratory, United States Steel Corporation, Monroeville, Pennsylvania.
Musacchia, Dr. X. J. Department of Physiology and Space Science Research Center, University of Missouri.
Mushlitz, R. D. Ore Department, American Smelting and Refining Company, New York.
Muther, Richard. Executive Director, Richard Muther & Associates, Inc., Kansas City, Missouri.
Muus, Dr. Bent J. Research Biologist, Danish Institute for Fishery and Marine Research, Charlottenlund, Denmark.
Myer, George H. Department of Geology, Temple University.
Myers, Prof. A. L. Department of Chemical Engineering, University of Pennsylvania.
Myers, Ware. Consultant, Claremont, California.

N

Nace, Dr. Raymond L. Geological Survey, U.S. Department of the Interior, Raleigh, North Carolina.
Nachmansohn, Dr. David. Deceased; formerly, Department of Neurology, College of Physicians and Surgeons, Columbia University.
Nachtrieb, Prof. Norman H. Chairman, Department of Chemistry, University of Chicago.
Natchtsheim, Dr. Philip R. NASA Ames Research Center, Moffett Field, California.
Nagel, Suzanne R. Engineering Research Center, AT&T Bell Laboratories, Princeton, New Jersey.
Nagel, Dr. Walter R. NASA Goddard Space Flight Center, Greenbelt, Maryland.
Nair, Dr. John H. Consultant to Food Industry, Raleigh, North Carolina.
Naistat, Dr. Samuel S. Department of Chemistry, Stephen F. Austin State College Nacogdoches, Texas.
Namias, Jerome. Climate Research Group, Scripps Institution of Oceanography, La Jolla, California.
Nance, Dr. Walter E. Department of Human Genetics, Medical College of Virginia.
Napp-Zinn, Prof. Klaus. Botanical Institute, Cologne, Germany.
Nastuk, Dr. William L. Department of Physiology, College of Physicians and Surgeons, Columbia University.
Naumann, Dr. Robert J. NASA Marshall Space Flight Center, Huntsville, Alabama.
Naylor, Dr. Arch. Department of Electrical Engineering, University of Michigan.
Naylor, Dr. Ernest. Reader in Zoology, University College of Swansea, Wales.
Nedelsky, Prof. Leo. Department of Physical Science, University of Chicago.
Nelsen, Dr. Olin E. Department of Biology, University of Pennsylvania.
Nelson, Elton G. Collaborator (WOC), Crops Research Division, U.S. Department of Agriculture, Beltsville, Maryland.
Nelson, Dr. George C. Department of Mathematics, University of Iowa, Iowa City.
Nelson, James H. Chief (retired), Geomagnetism Division, U.S. Coast and Geodetic Survey, Rockville, Maryland.
Nelson, Prof. Raymond J. Professor of Mathematics and Philosophy, Case Institute of Technology.
Nelson, Dr. Richard B. Chief Engineer, Varian Associates, Palo Alto, California.
Nergaard, Dr. Leon S. Director, Microwave Research Laboratory, RCA Laboratories, Princeton, New Jersey.
Nesbitt, Dr. David J. Joint Institute for Laboratory Astrophysics and Department of Chemistry, University of Colorado.
Nesbitt, Dr. Herbert H. J. Dean, Faculty of Science, Carleton University, Ottawa, Ontario, Canada.
Neter, Dr. Erwin. Deceased; formerly, Department of Pediatrics, State University of New York, Buffalo.
Nettleton, Dr. Lewis L. Consultant, Geophysical Associates International, Inc., Gravity Meter Exploration Company, Houston, Texas.
Neu, Dr. Harold C. Department of Medicine, Columbia Presbyterian Medical Center, New York, New York.
Neubecker, T. F. Manager, Engineering, High Intensity and Quartz Lamp Department, General Electric Company, Twinsburg, Ohio.
Neuhauser, Robert G. Solid State Division, Electro Optics and Devices, RCA, Lancaster, Pennsylvania.
Neumann, Dr. A. Conrad. School of Marine and Atmospheric Sciences, University of Miami.
New, Dr. Ronald. NOAA, U.S. Department of Commerce, Rockville, Maryland.
Newcombe, Dr. Curtis L. Director, San Francisco Bay Marine Research Center, San Francisco State College.
Newell, Dr. Allen. Department of Computer Science, Carnegie-Mellon University.
Newell, Dr. Homer E. National Aeronautics and Space Administration Headquarters, Washington, D.C.
Newsham, Dr. Robert S. Petroleum Economist, Stanford Research Institute, Menlo Park, California.
Newton, Dr. Chester W. National Center for Atmospheric Research, Boulder, Colorado.
Newton, Dr. R. C. Department of the Geophysical Sciences, University of Chicago.
Newton, Prof. Roger G. Department of Physics, Indiana University, Bloomington.
Nichols, Dr. Mark L. Deceased; formerly, U.S. Department of Agriculture, Auburn, Alabama.
Nichols, Richard B. Vice President, AT&T Long Lines, Morris Plains, New Jersey.
Nicholson, Graham G. Department of Plant Biology, University of Hull, England.
Nicodemus, Dr. Fred E. Physicist, Optical Radiation Section, Heat Division, Institute for Basic Standards, National Bureau of Standards.
Nicolai, Col. Leland M. Chief, Unmanned Vehicles Division, Air Vehicles Technology Office, Defense Advanced Research Projects Agency, U.S. Department of Defense, Arlington, Virginia.

Niebel, Benjamin W. Professor Emeritus of Industrial Engineering, Pennsylvania State University; Industrial Engineering Consultant, State College, Pennsylvania.

Niedermair, John C. Technical Director of Preliminary Ship Design (retired), Bureau of Ships, U.S. Department of the Navy; Consultant, Naval Architect.

Niemoeller, Dr. Arthur F. Professional Training, Research Laboratories, Central Institute for the Deaf, St. Louis, Missouri.

Nier, Prof. Alfred O. School of Physics and Astronomy, University of Minnesota.

Nierenberg, Prof. William A. Director, Scripps Institution of Oceanography, University of California, San Diego.

Nieto, Dr. Alberto S. Department of Geology, University of Illinois, Urbana.

Nigrelli, Dr. Ross F. Osborn Laboratories of Marine Sciences, New York Aquarium, New York.

Nippes, Prof. Ernest F. Department of Materials Engineering, Rensselaer Polytechnic Institute.

Nix, Prof. William D. Associate Chairman, Department of Materials Science and Engineering, Stanford University.

Nixon, Dr. Charles W. Consultant, Kettering, Ohio.

Noble, Dr. James A. Geological Sciences, California Institute of Technology.

Noble, R. H. Optical Division, Institute National Astrofis, Puebla, Mexico.

Noll, Dr. Kenneth. Department of Microbiology, University of Illinois, Urbana-Champaign.

Norcross, Prof. Neil L. College of Veterinary Medicine, Cornell University, Ithaca, New York.

Nordell, Eskel. Director of Technical Information, The Permutit Company.

Norris, Dr. H. Thomas. Chairman, Department of Clinical Pathology and Diagnostic Medicine, East Carolina University School of Medicine.

Norris, Robert M. Department of Geology, University of California, Santa Barbara.

North, Dr. Gerald R. Director, Climate System Research Program, Department of Meteorology, Texas A&M University.

Nottingham, Prof. W. B. Deceased; formerly, Professor Emeritus of Physics, Massachusetts Institute of Technology.

Novick, Aaron. Institute of Molecular Biology, University of Oregon.

Novotny, Prof. Milos V. Department of Chemistry, Indiana University.

Noyes, Richard M. Department of Chemistry, College of Arts and Sciences, University of Oregon.

Noyes, Dr. Robert W. Harvard-Smithsonian Center for Astrophysics, Smithsonian Astrophysical Observatory, Harvard College Observatory, Cambridge, Massachusetts.

Nuccitelli, Prof. Richard. Zoology Department, University of California, Davis.

Nygren, Dr. David R. Physics Division, Lawrence Berkeley Laboratory, University of California, Berkeley.

Nygren, Stephen. Bell Telephone Laboratories, Reading, Pennsylvania.

Nylander, Dr. Chaes. Laboratory of Applied Physics, Linköping Institute of Technology, Linköping, Sweden.

Obery, Leonard J. Chief, Plans and Programs Offices, National Aeronautics and Space Administration, Cleveland, Ohio.

O'Brien, Prof. Michael. Department of Agricultural Engineering, University of California, Davis.

Obrist, Dr. Paul A. Deceased; formerly, Department of Psychiatry, Medical School, University of North Carolina.

O'Connor, Dr. Donald J. Department of Civil Engineering, Manhattan College.

O'Connor, Dr. Joseph R. Edward J. Meyer Memorial Hospital, Buffalo, New York.

Odom, I. Edgar. Department of Geology, Northern Illinois University.

O'Dom, Dr. George W. Research Scientist, TRW, Inc., Redondo Beach, California.

O'Donnell, Franklin. Jet Propulsion Laboratory, Pasadena, California.

Odum, Dr. Eugene P. Director, Institute of Ecology, University of Georgia.

Ogburn, Charles B. Cooperative Extension Service, Auburn University.

Ogle, Dr. Kenneth N. Section of Biophysics, Mayo Clinic, Rochester, Minnesota.

Ojemann, Dr. Linda M. Regional Epilepsy Center, Harborview Medical Center, University of Washington.

Olcott, John W. Editor and Publisher, Business and Commercial Aviation Aerospace and Defense Group, McGraw-Hill, White Plains, New York.

Old, Dr. Bruce S. Senior Vice President, Arthur D. Little, Inc., Cambridge, Massachusetts.

Oldenburger, Prof. Rufus. Director, Automatic Control Center, Purdue University.

O'Leary, Dr. John E. Department of Forest Engineering, Oregon State University.

Olins, Dr. Ada L. Deutsches Krebsforshungszentrum, Heidelberg, Germany.

Olins, Dr. Donald E. Deutsches Krebsforshungszentrum, Heidelberg, Germany.

Olive, Lindsay S. University Distinguished Professor of Botany, University of North Carolina.

Oliver, Dr. H. W. Geological Survey, U.S. Department of the Interior, Menlo Park, California.

Olmsted, Leonard M. Utility Consultant and Registered Professional Engineer, South Orange, New Jersey.

Olness, Dr. John W. Brookhaven National Laboratory, Associated Universities, Inc., Upton, New York.

Olsen, Dr. Christopher W. Department of Pathobiological Sciences, University of Wisconsin, Madison.

Olson, Dr. Everett C. Department of Zoology, University of California, Los Angeles.

Olson, Dr. Harry F. Deceased; formerly, Staff Vice President, Acoustical and Electromechanical Research, RCA Laboratories, Princeton, New Jersey.

Olson, Dr. Jerry S. Oak Ridge National Laboratory, Oak Ridge, Tennessee.

Olson, Dr. Norman. Department of Food Science, University of Wisconsin.

O'Neill, Jerome P., Jr. Rumrill-Hoyt, Inc., Rochester, New York.

Oppenheimer, Prof. Jane M. Department of Biology, Bryn Mawr College.

O'Rahilly, Dr. Ronan. Carnegie Embryological Laboratories, University of California, Davis.

Oris, Dr. James T. Department of Zoology, Miami University, Oxford, Ohio.

Orkin, Dr. Frederick K. University of California, San Francisco, The Medical Center at the University of California School of Medicine.

Orrego, Dr. Cristian. Department of Biology, Brandeis University.

Orville, Dr. Harold D. Institute for Atmospheric Science, South Dakota School of Mines, Rapid City, South Dakota.

Osberg, Prof. Philip H. Department of Geological Sciences, University of Maine.

Oshinski, William J. Executive Vice President, Patton Consultants, Inc., Des Plaines, Illinois.

Oster, Prof. Gerald. Mount Sinai School of Medicine, City University of New York.

Ostergaard, Dr. H. L. Department of Biology, University of California, Los Angeles.

Osterman, John A. Government Communications Director, American Telephone and Telegraph Company, New York.

Ostro, Dr. Steven J. Jet Propulsion Laboratory, California Institute of Technology, Pasadena.

Ostrom, Dr. John H. Division of Vertebrate Paleontology, Peabody Museum of Natural History, Yale University.

Othmer, Prof. Donald F. Department of Chemical Engineering, Polytechnic Institute of Brooklyn.

Otto, Dr. Gilbert F. Zoology Department, University of Maryland.
Otto, Dr. Norman. Ford Motor Research Laboratory, Dearborn, Michigan.
Overhauser, Dr. Albert W. Department of Physics, Purdue University.
Overley, Dr. J. C. Department of Physics, University of Oregon.
Oviatt, Dr. Charles D. State Department of Education, Missouri.
Ovshinsky, Stanford R. President, Energy Conversion Devices, Inc., Troy, Michigan.
Owen, Dr. Denis F. Department of Biology, Oxford Polytechnic.
Owen, Prof. Guillermo. Department of Mathematical Sciences, Rice University.
Owen, Dr. Tobias C. Department of Earth and Space Sciences, State University of New York, Stony Brook.

P

Packard, Charles E. Associate Professor of Biology (retired), Randolph-Macon College, Ashland, Virginia.
Page, Prof. Robert M. Department of Biological Sciences, Stanford University.
Page, Dr. Roy C. Research Center in Oral Biology, School of Medicine, University of Washington.
Page, Dr. Thornton. National Aeronautics and Space Administration, Houston, Texas.
Pai, Dr. Shih-I. Institute of Physical Science and Technology, College of Engineering, University of Maryland.
Paine, Dr. Robert. Department of Zoology, University of Washington.
Pake, Dr. George E. Vice President, Xerox Corporation; General Manager, Xerox Palo Alto Research Center, Palo Alto, California.
Palermo, Dr. David S. Department of Psychology, Pennsylvania State University.
Palmer, Dr. Allison R. Department of Earth and Space Sciences, State University of New York, Stony Brook.
Palmer, Lawrence M. Federal Communications Commission, Washington, D.C.
Palmerlee, Prof. Albert S. Deceased; formerly, Professor of Engineering Drawing, School of Engineering and Architecture, University of Kansas.
Palzkill, Dr. David A. Department of Plant Sciences, University of Arizona.
Pangborn, Dr. Jon B. Director, Alternative Energy Systems Research, Institute of Gas Technology, Chicago, Illinois.
Papageorgiou, George, Research Associate, Department of Botany, University of Illinois, Urbana.
Papagiannis, Prof. Michael D. President, IAU Commission 51, Department of Astronomy, Boston University.
Pappas, Dr. S. Peter. Department of Polymers and Coatings, North Dakota State University.
Papuchis, Charles J. Naval Sea Systems Command, Department of the Navy, Washington, D.C.
Paresce, Dr. Francesco. Space Telescope Science Institute, Baltimore, Maryland.
Park, Prof. David. Department of Physics, Williams College.
Parker, Sybil P. Editor in Chief, "McGraw-Hill Encyclopedia of Science and Technology," McGraw-Hill Book Co., New York.
Parkes, Dr. Kenneth C. Curator of Birds, Carnegie Museum, Pittsburgh, Pennsylvania.
Parr, Prof. Robert G. Department of Chemistry, University of North Carolina.
Parrish, Dr. William. Research Staff Member, IBM Research Laboratory, San Jose, California.
Parsons, Prof. Michael G. Department of Naval Architecture, University of Michigan.
Parsons, Dr. Thomas S. Department of Zoology, University of Toronto, Ontario, Canada.
Partridge, Dr. Edward G. Redondo Beach, California.
Pasachoff, Dr. Jay M. Hopkins Observatory, Williams College.
Pask, Dr. Joseph A. Department of Materials Science and Engineering, University of California, Berkeley.
Patel, Dr. C. K. N. Bell Laboratories, Murray Hill, New Jersey.
Paterson, Eric. Iowa Institute of Hydraulic Research and Department of Mechanical Engineering, University of Iowa, Iowa City.
Paterson, Robert A. Professor of Botany, Virginia Polytechnic Institute and State University.
Patrick, Norman W. RCA Corporation, Lancaster, Pennsylvania.
Patton, Dr. Donald J. Department of Geography, Florida State University.
Patton, Dr. Harry D. Chairman, Department of Physiology-Biophysics, School of Medicine, University of Washington.
Patton, Peter C. Minnesota Supercomputer Institute, University of Minnesota.
Pattullo, Dr. June G. Deceased; formerly, Department of Oceanography, Oregon State University.
Paul, Prof. Burton. Department of Mechanical Engineering, University of Pennsylvania.
Paul, Prof. Igor. Department of Mechanical Engineering, Massachusetts Institute of Technology.
Payne, Prof. Joseph N. School of Education, University of Michigan, Ann Arbor.
Peacor, Dr. Donald R. Department of Geology, University of Michigan.
Pearce, Prof. R. H. Department of Pathology, Faculty of Medicine, University of British Columbia, Canada.
Pearson, Allen. Director, National Severe Storms Forecast Center, National Oceanic and Atmospheric Administration, Kansas City, Missouri.
Pearson, Dr. Ralph G. Department of Chemistry, Northwestern University.
Peaslee, Dr. Robert L. Vice President, Stainless Steel Division, Wall Colmonoy Corp., Detroit, Michigan.
Peck, Prof. Raymond E. Department of Geology, University of Missouri.
Pecora, Prof. Robert. Department of Chemistry, Stanford University.
Pedersen, Prof. Niels C. School of Veterinary Medicine, University of California, Davis.
Pederson, Dr. Carl S. Agricultural Experiment Station, State University of New York, Geneva.
Pegg, Prof. David J. Department of Physics and Astronomy, University of Tennessee, Knoxville.
Peirce, Dr. G. R. Engineer, Champaign, Illinois.
Pennak, Dr. Robert W. Department of Biology, University of Colorado.
Penneman, Dr. Robert A. Los Alamos Scientific Laboratory, Los Alamos, New Mexico.
Perch, Michael. Koppers Company, Inc., Monroeville, Pennsylvania.
Perkins, Dr. David D. Department of Biological Sciences, Stanford University.
Perkins, Dr. Jeff. Department of Mechanical Engineering, Naval Postgraduate School, Monterey, California.
Perkins, Kenneth W. Science Writer and Editor.
Perl, Prof. Martin L. Stanford Linear Accelerator Center, Stanford, California.
Perlmann, Dr. Gertrude E. Deceased; formerly, Rockefeller University.
Perone, Prof. Sam P. Department of Chemistry, Purdue University.
Perrin, Arthur M. Deceased; formerly President, National Conveyors Company, Inc., Fairview, New Jersey.
Perry, Astor. Extension Agronomy Specialist, North Carolina State University.
Perry, Stephen F. Exxon Research, Florham Park, New Jersey.
Peters, Prof. Philip C. Department of Physics, University of Washington.
Petershack, Dr. Victor D. Rexnord, Inc., Milwaukee, Wisconsin.
Peterson, Dr. R. L. Department of Botany, University of Guelph, Ontario, Canada.
Petschek, Prof. Albert G. Department of Physics, New Mexico Institute of Mining and Technology, Socorro.
Peyghambarian, Nasser. Optical Sciences Center and Optical Circuitry Cooperative, University of Arizona.

Pfaffmann, Dr. Carl. Department of Psychology, Rockefeller University.

Pfefferkorn, Dr. Hermann W. Department of Geology, University of Pennsylvania.

Pfennig, Prof. Norbert. Institut für Mikrobiologie, Göttingen Universitat, Germany.

Pfleeger, Dr. Charles P. Trusted Information Systems, Inc., Computer and Communications Security, Networking Telecommunications, Glenwood, Maryland.

Phaff, Dr. Herman J. Department of Food Science and Technology, College of Agriculture and Environmental Science, University of California, Davis.

Phelan, Prof. Richard M. Department of Mechanical Systems and Design, Cornell University.

Philip, Dr. Cornelius B. Department of Entomology, California Academy of Sciences, San Francisco.

Philipson, Dr. William R. Department of Botany, University of Canterbury, Christchurch, New Zealand.

Phillips, Dr. James A. CTR-Divison Office, Los Alamos Scientific Laboratory, Los Alamos, New Mexico.

Phillips, Prof. Tom L. Department of Botany, University of Illinois, Urbana.

Phillips, William H. NASA Langley Research Center, Hampton, Virginia.

Phister, Montgomery, Jr. Consultant, Santa Fe, New Mexico.

Picraux, Dr. Samuel T. Supervisor, Ion-Solid Interactions, Sandia Laboratories, Albuquerque, New Mexico.

Pielou, Dr. E. C. Department of Biology, University of Lethbridge, Alberta, Canada.

Pience, Roger D. Watson Technology, Portland, Maine.

Pierce, Prof. Allan D. School of Mechanical Engineering, Georgia Institute of Technology.

Pierce, Prof. E. Lowe. Department of Zoology, University of Florida.

Pierotti, Prof. Robert A. Department of Chemistry, Georgia Institute of Technology.

Pignataro, Prof. Louis J. Department of Transportation Planning and Engineering, Polytechnic Institute of Brooklyn.

Pilgrim, Lt. Hal W. Bureau of Ships, U.S. Department of the Navy.

Pinder, Dr. George F. Department of Civil and Geological Engineering, Princeton University.

Pinkel, Benjamin. Consulting Engineer, Santa Monica, California.

Pipkin, Prof. Francis M. Department of Physics, Harvard University.

Pisoni, Dr. David B. Department of Psychology, Indiana University.

Pitchford, Dr. Leanne C. Sandia National Laboratories, Albuquerque, New Mexico.

Pitman, Waker C., III. Lamont-Doherty Geological Observatory, Palisades, New York.

Platt, Allison M. Manager, Nuclear Waste Technology Department, Battelle Pacific Northwest Laboratories, Richland, Washington.

Platt, Dr. Robert B. Chairman, Department of Biology, Emory University.

Plavec, Dr. Mirek. Department of Astronomy, University of California, Los Angeles.

Plueddemann, Dr. Edwin P. Organic Laboratories, Dow Corning Corporation, Midland, Michigan.

Pogust, Frederick B. Director, Flight Guidance Systems, Eaton Corporation, AIL Division, Farmingdale, New York.

Polhemus, Maj. William L. Polhemus Navigation Sciences, Inc., Burlington, Vermont.

Polking, Dr. John C. Department of Mathematics, Rice University.

Pollack, Dr. Henry N. Department of Geological Sciences, University of Michigan.

Pollack, Dr. James B. Ames Research Center, Moffett Field, California.

Pollard, Dr. Ernest C. Department of Biophysics, Pennsylvania State University.

Pollard, Dr. Maurice G. Engineering Service Manager, Department of Mechanical and Electrical Engineering, British Railways Board, Derby, England.

Pollard, Dr. William G. Oak Ridge Associated Universities.

Polley, Dr. Margaret J. Department of Medicine, Cornell Medical Center, New York

Pollitzer, Dr. Ernest L. Associate Director of Research, Corporate Research Center, Universal Oil Products Company, Des Plaines, Illinois.

Ponikvar, Ade L. Formerly, Editor, Production and Drilling, "Petroleum Week," New York.

Poknivar, Adolph L. Formerly, Manager, Technical Publications, International Lead Zinc Research Organizations, Inc., New York.

Pontecorvo, Dr. Guido. Imperial Cancer Research Fund Laboratories, London, England.

Pope, Michael. Pope, Evans and Robbins, Consulting Engineers, New York.

Popp, Prof. Frank D. Department of Chemistry, Clarkson College of Technology.

Popper, Dr. Daniel M. Department of Astronomy, University of California, Los Angeles.

Porter, Prof. Joseph P. Emeritus Professor, Cornell University.

Porter, Dr. Keith R. Department of Molecular, Cellular and Developmental Biology, University of Colorado.

Porter, Dr. Mark D. Department of Chemistry, Iowa State University of Science and Technology.

Pasada, German. Research Associate, Department of Psychology, State University of New York, Stony Brook.

Postgate, Prof. John R. Department of Microbiology, University of Sussex, England.

Postow, Dr. Elliot. National Institutes of Health, Bethesda, Maryland.

Potter, Prof. M. David. School of Business, San Francisco State College.

Potter, Prof. Norman N. Department of Food Science, Cornell University.

Potts, Dr. Kevin T. Department of Chemistry, Rensselaer Polytechnic Institute.

Poulson, Dr. Donald F. Department of Biology, Yale University.

Povey, Dr. Sue. Galton Laboratory, University College, London, England.

Povh, Dr. Bogdan. Max Planck Institut für Kernphysik, Saupfercheckweg, Germany.

Powell, Dr. Wilson F. Deceased; formerly, Department of Pathology, School of Medicine, Yale University.

Pozar, Prof. David M. Department of Electrical and Computer Engineering, University of Massachusetts, Amherst.

Prescott, Dr. Gerald W. Department of Botany, University of Wisconsin.

Price, Prof. Donald L. Director, Departments of Pathology, Neurology, and Neuroscience, Johns Hopkins University.

Price, Dr. Harvey S. Vice President, Intercomp Resource Developing and Engineering Inc., Houston, Texas.

Price, Dr. P. Buford, Jr. Lawrence Berkeley Laboratory, University of California, Berkeley.

Price, Dr. Peter W. Department of Entomology, Northern Arizona University.

Priester, Gayle B. Consulting Engineer, Baltimore, Maryland.

Priestley, Dr. Robert. Marine Laboratory, Department of Agriculture and Fisheries for Scotland, Aberdeen.

Princen, L. H. Associate Center Director, North Regional Research Center, Department of Agriculture, Peoria, Illinois.

Pringsheim, Prof. Ernest G. Pflanzenphysiologisches Institut, Göttingen, Germany.

Prinn, Prof. Ronald G. Department of Meteorology, Massachusetts Institute of Technology.

Pritchard, Prof. David E. Department of Physics, Massachusetts Institute of Technology, Cambridge.

Pritchard, Dr. Tom. The Nature Conservancy, Bangor, Wales.

Pritchard, Tram C. Lockheed Missiles and Space Company, Sunnydale, California.

Pritchett, Prof. Wilson S. Senior Project Engineer, Noller Control Systems, Inc., Richmond, California.

Pritsker, Prof. A. Alan B. School of Industrial Engineering, Purdue University.

Proctor, Prof. Bernard E. Formerly, Professor and Head, Department of Food Technology, Massachusetts Institute of Technology.

Prosser, Dr. C. Ladd. Department of Physiology, University of Illinois, Urbana.

Protell, Dr. Robert L. Assistant Professor of Medicine, University Hospital, University of Washington.

Prouty, Dr. L. Fletcher. Senior Director, Corporate Communications, National Railroad Passenger Corporation, Washington, D.C.

Pyrke, John K. M. Slocum and Fuller, New York.

Pryor, Prof. Wayne A. Department of Geology, University of Cincinnati.

Pryor, Dr. William A. Department of Chemistry, Louisiana State University.

Puchstein, Albert F. Deceased; formerly, Consulting Engineer, Columbus, Ohio.

Puck, Dr. Theodore T. Department of Biophysics, University of Colorado Medical Center, Denver.

Pullen, Dr. Keats A., Jr. Ballistic Research Laboratories, Aberdeen Proving Ground, Maryland.

Purchon, Dr. Richard D. Department of Zoology, Chelsea College, University of London, England.

Purser, Dr. Kenneth H. Ionex/HEI Corporation, Newburyport, Massachusetts.

Purvis, Dr. Andy. Department of Zoology, University of Oxford, United Kingdom.

Putnam, Prof. Russell C. Professor Emeritus, Case Institute of Technology.

Putterman, Dr. Seth. Department of Physics, University of California, Los Angeles.

Putz, Dr. Thomas J. Deceased; formerly, Westinghouse Electric Corporation, Philadelphia, Pennsylvania.

Q

Quaife, Carol. Department of Pathology, University of Washington.

Qualset, Dr. Calvin O. Department of Agronomy and Range Science, University of California, Davis.

Quate, Prof. C. F. Department of Applied Physics, Stanford University.

Quigg, Dr. Chris. Lawrence Berkeley Laboratory, University of California, Berkeley.

Quimby, Dr. Edith H. Department of Radiology, College of Physicians and Surgeons, Columbia University.

Quirin, Edward J. Consulting Engineer, Besier, Gibble & Quirin, Old Saybrook, Connecticut.

Quiroz, Dr. Roderick S. Retired; formerly, Research Meteorologist, Upper Air Branch, National Oceanic and Atmospheric Administration, U.S. Department of Commerce.

R

Raasch, Dr. Gilbert O. Consulting Paleontologist, Calgary, Alberta, Canada.

Rabb, David D. Mining Engineer-Metallurgist, Tucson, Arizona.

Rabinowicz, Prof. Ernest. Department of Mechanical Engineering, Massachusetts Institute of Technology.

Rabinowitz, Dr. Mario. Senior Scientist, Electrical Systems Division, Electric Power Research Institute, Palo Alto, California.

Radebaugh, Dr. R. National Bureau of Standards, Boulder, Colorado.

Rademacher, Prof. Hans. Deceased; formerly, Professor Emeritus of Mathematics, University of Pennsylvania.

Radforth, Dr. Norman W. Muskeg Research Institute, University of New Brunswick, Canada.

Radke, Dr. Rodney O. Agricultural Product Research Laboratory, Monsanto Company, St. Louis, Missouri.

Rae, James R. Assistant Vice President, American Telephone and Telegraph Company, New York.

Rae, Dr. Peter M. M. Molecular Diagnostics, Inc., West Haven, Connecticut.

Raffel, Dr. Sidney. Department of Medical Microbiology, School of Medicine, Stanford University.

Ragland, Paul G. Department of Geology, University of North Carolina.

Rainnie, William O., Jr. Deceased; formerly, U.S. Department of Commerce, National Space Technology Laboratories, Mississippi.

Raisz, Dr. Erwin. Deceased; formerly, National Atlas Project, U.S. Geological Survey.

Ramberg, Dr. Edward G. RCA Laboratories, Princeton, New Jersey.

Ramberg, Prof. Hans. Institute of Geology and Mineralogy, University of Uppsala, Sweden.

Ramey, Dr. Robert L. Department of Electrical Engineering, University of Virginia.

Ramírez-Angulo, Dr. Jaime. Department of Electrical Engineering, Texas A&M University.

Ramond, Dr. Pierre. Department of Physics, University of Florida.

Ramsdell, Prof. Lewis S. Professor Emeritus, Department of Mineralogy, University of Michigan.

Ramsey, Dr. Norman F. Department of Physics, Harvard University.

Rangan, Dr. P. Venkat. Department of Computer Science and Engineering, University of California, San Diego.

Raper, Dr. John R. Professor of Botany, Harvard University.

Rapp, Prof. George, Jr. Dean, College of Letters and Science, University of Minnesota.

Raski, Dr. Dewey J. Department of Nematology, University of California, Davis.

Rasmussen, Dr. Howard. Department of Biochemistry, School of Medicine, University of Pennsylvania.

Rasmussen, Dr. Norman C. Department of Nuclear Engineering, Massachusetts Institute of Technology.

Rasmusson, Dr. Eugene M. Geophysical Fluid Dynamics Laboratory, Environmental Science Services Administration, Princeton, New Jersey.

Rasool, Dr. S. I. Chief Scientist for Global Change Programs, Laboratoire de Meteorologie Dynamique, Paris, France.

Ratay, Dr. Robert T. Consulting Engineer, Manhasset, New York.

Ratliff, Dr. Floyd. Department of Psychology, Rockefeller University.

Ratnoff, Dr. Oscar D. Department of Medicine, Case Western Reserve University.

Rau, Prof. A. R. P. Department of Physics, Louisiana State University, Baton Rouge.

Rau, James L. Design Engineer, Ross Gear Division, TRW, Inc., Lafayette, Indiana.

Raum, Dr. Donald. Center for Blood Research, Boston, Massachusetts.

Rausch, Dr. Marvin D. Department of Chemistry, University of Massachusetts.

Rawles, Dr. Mary E. Department of Embryology, Carnegie Institution of Washington.

Rawson, Dr. Kenneth S. Department of Biology, Swathmore College.

Raymond, Samuel O. Department of Electrical Engineering and Computer Science, Massachusetts Institute of Technology.

Read, Prof. Clark P. Deceased; formerly, Department of Biology, Rice University.

Reay, Dr. David A. International Research & Development Company, Ltd., Fossway, Newcastle-upon-Tyne, England.

Reay, Dr. Donald T. King County Medical Examiner, Harborview Hospital, Seattle, Washington.

Rebbi, Dr. Claudio. Department of Physics, Brookhaven National Laboratory, Upton, New York.

Recht, Dr. Abram. Joint Center for Radiation Therapy, Boston, Massachusetts.

Reddish, Dr. George F. Deceased; formerly, St. Louis College of Pharmacy and Allied Sciences.

Redhead, Dr. Scott A. Mycologist, Biosystematics Research Centre, Agriculture Canada, Research Branch, Central Experimental Farm, Ottawa, Ontario, Canada.

Reed, Charles A. Department of Anthropology, University of Illinois.

Reed, Dr. R. Dale. NASA Dryden Flight Center, Edwards, California.

Reed-Hill, Dr. Robert E. Department of Metallurgical Science and Engineering, University of Florida.

Rees, Dr. Bryant E. Professor of Biology, Fresno State College.

Reeve, Prof. John. Department of Electrical Engineering, Faculty of Engineering, University of Waterloo, Ontario, Canada.

Reeve, Dr. Roger M. Western Regional Research Laboratory, U.S. Department of Agriculture, Albany, California.
Reich, Prof. Herbert J. (Retired), Department of Engineering and Applied Science, Yale University.
Reichert, Dr. Leo E., Jr. Department of Biochemistry, School of Medicine, Emory University.
Reichmann, Dr. M. E. Department of Microbiology, University of Illinois.
Reid, Prof. Evans B. Professor Emeritus, Department of Chemistry, Colby College.
Reid, Joseph L. Scripps Institution of Oceanography, La Jolla, California.
Reid, Prof. Robert O. Department of Oceanography, Texas A&M University.
Reiling, Dr. Gilbert H. Lighting Research and Technical Services Operation, General Electric Company, Cleveland, Ohio.
Reilly, James D. Vice President, Consolidation Coal Company, Pittsburgh, Pennsylvania.
Reilly, John C., Jr. Naval Historical Center, Navy Department, U.S. Department of Defense.
Reilly, Dr. Norman B. Consultant, Computer Systems, Compata, Inc., Tarzana, California.
Reintjes, John F. Naval Research Laboratory, Washington, D.C.
Reischmann, Rev. Placidus G., O.S.B., Head, Department of Biology, St. Martin's College, Olympia, Washington.
Reiter, Dr. Elmar R. Department of Atmospheric Science, Colorado State University.
Reitz, Dr. Louis P. (Retired) Corps Research Division, Agricultural Research Service, U.S. Department of Agriculture, Beltsville, Maryland.
Remde, Dr. Harry F. Chief, Basic Physics Research Section, Johns-Manville Research and Engineering Center, Manville, New Jersey.
Renfrow, Dr. William B., Jr. (Retired) Department of Chemistry, Oberlin College.
Reynolds, Gardner M. Dames & Moore, Los Angeles, California.
Reynolds, Dr. W. W. Manager, Economic Coordination, Chemical-Economics, Shell Chemical Company, Houston, Texas.
Ribbe, Dr. Paul H. Department of Geological Sciences, Virginia Polytechnic Institute.
Rice, Dr. Mary E. Associate Curator, Division of Worms, Smithsonian Institution.
Rich, Dr. Alexander. Professor of Biophysics, Massachusetts Institute of Technology.
Rich, Prof. Arthur. Department of Physics, University of Michigan.
Rich, Dr. Saul. Connecticut Agricultural Experiment Station, New Haven, Connecticut.
Richards, Dr. Francis A. Department of Oceanography, University of Washington.
Richards, Dr. Frank F. Department of Medicine, Yale University.
Richards, G. N. British Rayon Research Laboratory, Manchester.
Richards, Dr. Oscar W. (Retired) College of Optometry, Pacific University.
Richards, Dr. Paul W. School of Plant Biology, University College of North Wales.
Richardson, Dr. Philip. Department of Physical Oceanography, Woods Hole Oceanographic Institution, Woods Hole, Massachusetts.
Ricksecker, Ralph E. Director of Metallurgy, Chase Brass and Copper Company, Cleveland, Ohio.
Riddle, Dr. Mark. A. Department of Psychiatry and Behavioral Sciences, Johns Hopkins Medical Institutions, Baltimore, Maryland.
Riedel, Dr. Rupert J. Department of Zoology, University of North Carolina.
Riehl, Dr. Herbert. Retired; formerly, Department of Meteorology, Colorado State University.
Ries, Dr. Harold C. Stanford Research Institute, Menlo Park, California.
Riesen, Dr. Austin H. Department of Psychology, University of California, Riverside.
Riggs, Dr. Austen F. Department of Zoology, University of Texas, Austin.
Riggs, Harold C. Manager (retired), Marketing New Product Development, Electric Storage Battery Company, Philadelphia, Pennsylvania.
Riggs, Dr. Lorrin A. Department of Psychology, Brown University.
Rijnsdorp, Prof. Johannes E. University of Twente, Enschede, Netherlands.
Riley, Frank. Senior Vice-President, Bodine Assembly Systems, Bridgeport, Connecticut.
Riley, Dr. Reed F. Quantum Electronics Group, General Telephone and Electronics, Bayside, New York.
Riser, Dr. Stephen C. School of Oceanography, University of Washington, Seattle.
Ritchey, Dr. H. W. President, Thiokol Chemical Corporation, Ogden, Utah.
Ritchie, Dr. J. C. Department of Biology, Trent University, Peterborough, Ontario, Canada.
Rittenberg, Prof. Sydney C. Department of Microbiology, University of California, Los Angeles.
Rizack, Dr. Martin A. Department of Medicine, Rockefeller University.
Robb, D. D. D. D. Robb and Associates, Consulting Engineers, Electric Power Systems, Salina, Kansas.
Robbins, Dr. Elliott. Department of Cell Biology, Albert Einstein Medical School.
Robbins, Nathaniel, Jr. Director of Engineering, Residential Division, Honeywell, Inc., Minneapolis, Minnesota.
Roberts, Dr. Arthur. National Accelerator Laboratory, Batavia, Illinois.
Roberts, J. K. (Retired) Research and Development Department, Standard Oil Company, Chicago, Illinois.
Roberts, Prof. Louis D. Department of Physics, University of North Carolina.
Roberts, Dr. Richard J. Cold Spring Harbor Laboratories, Cold Spring Harbor, New York.
Roberts, Tecwyn. NASA Goddard Space Flight Center, Greenbelt, Maryland.
Robertson, Prof. Burtis L. Professor of Electrical Engineering (retired), University of California, Berkeley.
Robertson, Dr. J. David. Department of Anatomy, Duke University Medical Center.
Robertson, Lemuel C. Newport News Shipbuilding and Dry Dock Company, Newport News, Virginia.
Robinovitch, Dr. Murray. Department of Oral Biology, University of Washington School of Dentistry, Seattle.
Robinson, Prof. Arthur H. Department of Geography, University of Wisconsin.
Robinson, Dr. G. Wilse. Department of Chemistry, Texas Tech University.
Robinson, Prof. Richard W. Department of Seed and Vegetable Sciences, New York State Agricultural Experiment Station, Cornell University.
Robinson, W. V. Processor Technology Research Group, Bell Laboratories, Whippany, New Jersey.
Robison, Dr. G. Alan. Department of Pharmacology, University of Texas Medical School.
Roble, Dr. Raymond G. National Center for Atmospheric Research, Boulder, Colorado.
Rocap, William A., Jr. Printing Consultant, Conover, North Carolina.
Rochkind, Mark M. Bell Laboratories, Holmdel, New Jersey.
Rochow, Prof. Eugene G. (Retired) Department of Chemistry, Harvard University.
Rock, Dr. Irwin. Department of Psychology, University of California, Berkeley.
Rockett, Frank H. Engineering Consultant, Charlottesville, Virginia.
Rodden, Dr. William P. Consulting Engineer, La Cañada-Flintridge, California.
Roddis, Louis H., Jr. President, Consolidated Edison Company of New York, Inc.
Rodgers, Prof. John. Department of Geology, Yale University.
Rodier, Dr. Patricia M. Department of Anatomy, School of Medicine, University of Virginia.
Roe, Kenneth A. President, Burns and Roe, Oradell, New Jersey.
Rogallo, Francis M. (Retired) NASA Langley Research Center, Hampton, Virginia.
Rogers, Prof. R. R. Professor and Chairperson, Department of Meteorology, McGill University, Montreal, Quebec, Canada.
Rogers, D. Terence A. Department of Physiology, Pacific Biomedical Research Center, University of Hawaii.
Rogowski, Prof. Augustus R. Department of Mechanical Engineering, Massachusetts Institute of Technology.

Rohde, Dr. Richard A. Department of Plant Pathology, University of Massachusetts.
Rohlf, Dr. James W. Department of Physics, Boston University.
Rohrlich, Prof. F. Department of Physics, Syracuse University.
Roller, Dr. Duane E. Deceased; formerly, Harvey Mudd College, Claremont, California.
Romer, Prof. Alfred S. Deceased; formerly, Museum of Comparative Zoology, Harvard University.
Roosen-Runge, Prof. Edward C. Department of Biological Structure, School of Medicine, University of Washington.
Root, Dr. William L. Department of Aerospace Engineering, University of Michigan.
Roper, Dr. Clyde F. E. Division of Molluscs, Smithsonian Institution, U.S. National Museum, Washington, D.C.
Roper, Paul J. Department of Geology, University of Southwestern Louisiana.
Rose, Dr. S. Meryl. Department of Anatomy, Tulane University.
Rose, Dr. William Ingersoll, Jr. Department of Geology and Geological Engineering, Michigan Technological University.
Rosen, Prof. Barry. Department of Biological Chemistry, University of Maryland.
Rosenbaum, Prof. Fred J. Department of Electrical Engineering, Washington University.
Rosenberg, Leon T. Senior Consultant, Generator Design, Generation, Installation, and Service, Allis-Chalmers Manufacturing Company, Milwaukee, Wisconsin.
Rosenberg, Prof. Norman J. Department of Agricultural Meteorology, University of Nebraska.
Rosenberger, Harold E. Scientific Instruments Division, Bausch and Lomb, Inc., Rochester, New York.
Ross, Dr. Charles A. Department of Geology, Western Washington State College.
Ross, Dr. Davis A. Woods Hole Oceanographic Institution, Woods Hole, Massachusetts.
Ross, Dr. Edward S. California Academy of Sciences, San Francisco.
Ross, Dr. Herbert H. Illinois State Natural History Survey, Urbana.
Ross, Dr. June R. P. Department of Biology, Western Washington State College.
Ross, Dr. Robert S. Environmental Structures, Inc., Cleveland, Ohio.
Ross, Prof. Sidney. Department of Chemistry, Rensselaer Polytechnic Institute.
Rostoker, Dr. William. Department of Materials Engineering, University of Illinois, Chicago.
Roth, Dr. Louis M. Head, Entomology Group, Food Science Laboratory, U.S. Army Natick Laboratories, Natick, Massachusetts.
Roth, Willard. Engineer, Sunbeam Equipment Corporation, Meadville, Pennsylvania.
Rowe, Dr. Marvin W. Department of Chemistry, Texas A&M University.
Rowell, Prof. A. J. Department of Geology, University of Kansas.
Rownd, Dr. Robert H. Department of Molecular Biology, Northwestern Medical School.
Roy, Dr. Chalmer J. Professor of Geology and Dean, College of Sciences and Humanities, Iowa State University.
Roy, Natalie, U. Director of Recycling and Legislative Affairs, Glass Packaging Institute, Washington, D.C.
Rozeanu, Prof. L. Department of Material Science, Technion, Israel Institute of Technology.
Rozen, Dr. Jerome G., Jr. Department of Entomology, American Museum of Natural History, New York.
Rubis, Prof. David D. Department of Plant Sciences, College of Agriculture, University of Arizona.
Ruch, Prof. Fritz. Department of General Botany, Swiss Federal Institute of Technology, Zurich.
Ruch, Prof. Theodore C. Director, Regional Primate Research Center; Professor, Department of Physiology and Biophysics, School of Medicine, University of Washington.
Ruderman, Dr. Audrey J. Department of Psychology, University of Illinois, Chicago.
Rudnick, Dr. Dorothea. Department of Biology, Albertus Magnus College, Yale University.
Rudolph, Prof. Ralph W. Deceased; formerly, Department of Chemistry, University of Michigan.
Ruhe, Dr. Robert V. Department of Geology, Indiana University.
Rulfs, Dr. Charles L. Department of Chemistry, University of Michigan.
Runcorn, Prof. Stanley K. Head, School of Physics, University of Newcastle-upon-Tyne, England.
Rundman, Prof. Karl. Department of Metallurgical Engineering, Michigan Technological University.
Runham, Dr. Norman. Department of Zoology, University College of North Wales, Bangor.
Runnegar, Dr. Bruce. Department of Earth and Space Sciences, University of California, Los Angeles.
Runyon, Dr. Ernest H. Microbiology Research, Veterans Administration Hospital; College of Medicine, University of Utah.
Ruouff, Dr. Kathryn L. Francis Blake Bacteriology Laboratories, Department of Microbiology and Molecular Genetics, Massachusetts General Hospital, Harvard Medical School, Boston, Massachusetts.
Rusk, Dr. Rogers D. Mount Holyoke College.
Russell, Dr. Allen S. Vice President, Alcoa Laboratories, Aluminum Company of America, Pittsburgh, Pennsylvania.
Russell, Dr. Howard W. Deceased; formerly, Technical Director, Battelle Memorial Institute, Columbus, Ohio.
Russell-Hunter, Prof. W. D. Professor of Zoology, Department of Biology, Syracuse University.
Rust, Dr. David M. Applied Physics Laboratory, Johns Hopkins University, Laurel, Maryland.
Rust, W. David. National Severe Storms Laboratory, National Oceanic and Atmospheric Administration, Norman, Oklahoma.
Ryan, Prof. James J. Professor Emeritus of Mechanical Engineering, University of Minnesota.
Ryland, Prof. John S. Head of Biology, School of Natural Resources, University of the South Pacific, Laucala Bay, Suva, Fiji.

S

Saad, Prof. Michael L. Department of Mechanical Engineering, Santa Clara University, Santa Clara, California.
Saaty, Prof. Thomas L. Graduate School of Business, University of Pittsburgh.
Saenger, Dr. Wolfram. Max Planck-Institut für Experimentelle Medizen, Göttingen, Germany.
Sage, Prof. Andrew P. Founding Dean Emeritus and First American Bank Professor, University Professor, School of Information Technology and Engineering, George Mason University, Fairfax, Virginia.
St. George, Emery, Jr. Charles Stark Draper Laboratory, Inc., Cambridge, Massachusetts.
St. Jean, Dr. Joseph, Jr. Department of Geology, University of North Carolina.
St. Pierre, Prof. George R. Department of Metallurgical Engineering, Ohio State University.
Saito, Dr. Tsunemasa. Lamont-Doherty Geological Observatory, Palisades, New York.
Salam, Prof. Abdus. Director, International Centre for Theoretical Physics, Trieste, Italy.
Salisbury, Dr. Frank B. Plant Science Department, Utah State University.
Salmon, Vincent. Acoustical Consultant, Menlo Park, California.
Saltzman, Dr. W. Mark. Department of Chemical Engineering, G. W. C. Whiting School of Engineering, Johns Hopkins University, Baltimore, Maryland.
Salutsky, Dr. Murrell L. Vice President, Research, Dearborn Chemical Division, W. R. Grace and Company, Lake Zurich, Illinois.

Samaras, Dr. Demetrios G. Air Force Office of Scientific Research, Arlington, Virginia.
Samios, Dr. Nicholas P. Brookhaven National Laboratory, Upton, New York.
Samter, Dr. Max. Institute of Allergy and Clinical Immunology, Grant Hospital of Chicago.
Sandage, Dr. Allan. Mount Wilson and Palomar Observatories, Pasadena, California.
Sanders, Dr. Frederick. Department of Meteorology, Massachusetts Institute of Technology.
Sanders, Dr. Robert W. Department of Zoology, University of Georgia.
Sandler, Prof. Stanley I. Department of Chemical Engineering, University of Delaware.
Sandretto, Brig. Gen. Peter C. Deceased; formerly, Director, Engineering Management, International Telephone and Telegraph Corporation.
Sankey, Dr. Bruce M. Petroleum Products Division, Imperial Oil, Ltd., Sarnia, Ontario, Canada.
Sarachik, Prof. Edward S. Department of Atmospheric Sciences, University of Washington, Seattle.
Sargent, Dr. Anneila I. Owens Valley Radio Observatory, California Institute of Technology, Pasadena.
Sargent, Dr. Murray, III. Max Planck Insitut für Festkoperforschung, Stuttgart, Germany.
Satir, Dr. Peter. Chairman, Department of Anatomy and Structural Biology, Albert Einstein College of Medicine.
Saucer, Prof. Helmut. Department of Biology, Texas A&M University.
Saul, Dr. Everett L. Director of Scientific Liaison, Products Division, Bristol-Myers Company.
Savage, Prof. Blair D. Washburn Observatory, Department of Astronomy, University of Wisconsin.
Savage, Dr. Donald E. Department of Paleontology, University of California, Berkeley.
Savage, Dr. George M. Director, Product Research and Development, Upjohn International Inc., Kalamazoo, Michigan.
Savage, Dr. Jay M. Department of Biological Sciences, University of Southern California.
Sawyer, Dr. Donald. Distinguished Professor, Department of Chemistry, Texas A&M University.
Sawyer, Robert M. Government Communications, American Telephone and Telegraph Company, Washington, D.C.
Sayre, Dr. Albert N. Deceased; formerly, Consulting Groundwater Geologist, Behre Dolbear and Company.
Scalzi, Dr. John B. Program Director, National Science Foundation, Washington, D.C.
Scanlan, Prof. J. O. Department of Electrical and Electronic Engineering, University College, Dublin, Ireland.
Scarlott, Charles A. Manager of Publications Department (retired), Stanford Research Institute, Menlo Park, California.
Schaaf, Prof. Samuel A. Professor of Engineering Science and Chairman, Division of Aeronautical Sciences, University of California, Berkeley.
Schaal, Prof. Klaus P. Oberarzt am Hygiene-Institut der Universität zu Köln, Germany.
Schachter, Dr. Julius. Department of Laboratory Medicine, University of California, San Francisco.
Schaefer, Dr. Henry F., III. Department of Chemistry, University of California, Berkeley.
Schaefer, Dr. Karl E. Deceased; formerly, Chief, Physiological Sciences Division, U.S. Naval Submarine Center, Groton, Connecticut.
Schaeffer, Dr. Bobb. Chairman and Curator, American Museum of Natural History, New York.
Schawlow, Prof. Arthur L. Department of Physics, Stanford University.
Schecter, Dr. Arnold. Department of Preventive Medicine, Upstate Medical Center, State University of New York, Syracuse.
Scheer, Dr. Bradley T. Professor Emeritus, Department of Biology, University of Oregon.
Scheffer, Dr. Victor B. (Retired) Bureau of Commercial Fisheries, Sand Point Naval Air Station, Seattle, Washington.
Schelleng, John C. (Retired) Bell Telephone Laboratories.
Scheltema, Amelie H. Woods Hole Oceanographic Institution, Woods Hole, Massachusetts.
Scher, Dr. William. Mount Sinai Medical School, New York.
Schiavoni, William. Engineering Director, Customer Equipment Systems, American Telephone and Telegraph Company, New York.
Schindler, Dr. Anton. Camille Dreyfus Laboratory, Research Triangle Institute, Research Triangle Park, North Carolina.
Schindler, Dr. James E. Department of Biological Sciences, College of Sciences, Clemson University.
Schmerling, Dr. Louis. Research Associate, Universal Oil Products Company, Des Plaines, Illinois.
Schmid, Dr. Rudolf. Department of Botany, University of California, Berkeley.
Schmidt, Dr. Paul W. Department of Physics, University of Missouri.
Schmidt-Nielsen, Dr. Bodil. Department of Zoology, Duke University.
Schmitt, Dr. Waldo L. Deceased; formerly, Zoologist Emeritus, Department of Zoology, Smithsonian Institution.
Schneider, Dr. Thomas R. Electric Power Research Institute, Palo Alto, California
Schneidewind, Dr. Norman F. Computer Scientist, Computer Research, Pebble Beach, California.
Schoeffler, Dr. James D. Department of Computer and Information Science, Cleveland State University.
Schoener, Dr. Thomas W. Department of Zoology, University of Washington.
Schoenherr, Dr. Karl E. Technical Director (retired), Hydromechanics Laboratory, Naval Ship Research and Development Center, Washington, D.C.
Scholz, Christopher H. Lamont-Doherty Geological Observatory, Palisades, New York.
Schoonmaker, G. R. Retired; formerly, Vice President, Production-Exploration, Marathon Oil Company, Findlay, Ohio.
Schroeder, Dr. Charles A. Department of Botanical Sciences, University of California, Los Angeles.
Schroeder, Dr. Lee S. Lawrence Berkeley Laboratory, Berkeley, California.
Schubert, Dr. Jack. Radiation Health, Graduate School of Public Health, University of Pittsburgh.
Schuchardt, Dr. Lee F. Merck, Sharp, and Dohme, West Point, Pennsylvania.
Schueler, Dr. Donald G. Sandia Laboratories, Albuquerque, New Mexico.
Schule, John J., Jr. Acting Director, Department of Marine Science, U.S. Oceanographic Office.
Schuller, Dr. Ivan K. Physics Department, University of California, La Jolla.
Schulman, Dr. James H. U.S. Naval Research Laboratory.
Schultz, Dr. Jack. Gypsy Moth Research Center, Department of Entomology, College of Agriculture, Pesticide Research Laboratory and Graduate Study Center, Pennsylvania State University.
Schultz, Robert M. General Manager, William Langer Jewel Bearing Plant, Bulova Watch Company, Inc., Rolla, North Dakota.
Schultze, Dr. Hanns-Peter. Curator of Vertebrate Paleontology, Museum of Natural History, University of Kansas.
Schumacher, Dr. Uwe. Max Planck Institut für Plasmaphysik, Munich, Germany.
Schur, Dr. Peter H. Robert B. Brigham Hospital, Boston, Massachusetts.
Schutt, H. C. Deceased; formerly, Consulting Engineer.
Schwab, Frederic L. Department of Geology, Washington and Lee University.
Schwartz, Dr. Herman M. Department of Physics, University of Arkansas.
Schwartz, Dr. Mel M. Rohr Industries, Inc., Chula Vista, California.
Schwartz, Dr. Morton K. Department of Biochemistry, Memorial Hospital, New York.
Schweizer, Ernest. President, Schweizer Aircraft Corporation, Elmira, New York.

Scott, Dave. Industrial Engineering Manager, Modular Computer Systems, Inc., Fort Lauderdale, Florida.
Scott, Dr. David K. National Superconducting Cyclotron Laboratory, Michigan State University.
Scott, Dr. Garland E. Department of Chemical and Materials Engineering, California State Polytechnical University.
Scott, Dr. J. Allen. National Institutes of Health, Bethesda, Maryland.
Scott, Prof. James F. Department of Physics, University of Colorado.
Scott, Dr. John E., Jr. Department of Aerospace Engineering and Engineering Physics, University of Virginia.
Scott, Dr. John P. Director, Center for Research on Social Behavior, Bowling Green State University.
Scott, Prof. Robert L. Department of Chemistry, University of California, Los Angeles.
Seaborg, Dr. Glenn T. Lawrence Berkeley Laboratory, University of California, Berkeley.
Seaman, Dr. R. E. Textile Fibers Department, E.I. du Pont de Nemours and Co., Inc., Wilmington, Delaware.
Searles, Dr. Robert. Rheumatology Division, University of New Mexico School of Medicine.
Sebald, Joseph F. Consulting Engineer and President, Heat Power Products Corporation, Bloomfield, New Jersey.
Seed, Prof. John Richard. Department of Epidemiology, School of Public Health, University of North Carolina, Chapel Hill.
Segal, Dr. Bernard L. Professor of Medicine, Cardiovascular Institute, Hahnemann Medical College and Hospital, Philadelphia, Pennsylvania.
Segeler, C. George. Director, Technical Service, Dave Sage, Inc., New York.
Segrè, Prof. Emilio. Department of Physics, University of California, Berkeley.
Seibel, Dr. Machelle M. Department of Obstetrics and Gynecology, Beth Israel Hospital, Boston, Massachusetts.
Seidelmann, Dr. P. K. Director, Nautical Almanac Office, U.S. Naval Observatory, Washington, D.C.
Seifert, Dr. William W. Professor of Electrical Engineering and Associate Director, Commodity Transportation and Economic Development Laboratory, Massachusetts Institute of Technology.
Seitz, Dr. W. Rudolf. Department of Chemistry, College of Engineering and Physical Sciences, University of New Hampshire.
Sekera, Prof. Zdenek. Deceased; formerly, Department of Meteorology, Institute for Geophysics and Planetary Physics, University of California, Los Angeles.
Selby, Prof. Michael J. Department of Earth Science, University of Waikato, New Zealand.
Sell, Dr. Heinz G. Metals Development Section, Westinghouse Lamp Divisions, Bloomfield, New Jersey.
Sellars, Dr. John R. Manager, Engineering Mechanics Operations, TRW, Inc., Redondo Beach, California.
Sellin, Dr. Ivan A. Department of Physics, University of Tennessee.
Sellmyer, Prof. David J. Behlen Laboratory of Physics, University of Nebraska.
Semmes, Dr. Josephine. Laboratory of Psychology, National Institute of Mental Health, Bethesda, Maryland.
Sessions, Dr. Richard B. School of Chemistry, University of Bristol, England.
Sessler, Dr. Gerhard M. Institut für Ubertragungstechnik Elektroakustik, Technische Hochschule, Darmstadt, Germany.
Shamma, Dr. Maurice. Department of Chemistry, Pennsylvania State University.
Shank, Dr. C. V. Bell Telephone Laboratories, Holmdel, New Jersey.
Shanmugam, Dr. G. Mobil Research and Development Corporation, Research Department, Dallas Research Laboratory, Dallas, Texas.
Shannon, Dr. Hugh F. Products Research Division, Exxon Research and Engineering Company, Linden, New Jersey.
Shannon, L. R. "The New York Times," New York, New York.
Shannon, Prof. Robert R. Optical Sciences Center, University of Arizona.
Shapiro, Hymin. Research and Development Department, Ethyl Corporation, Baton Rouge, Louisiana.
Sharpe, Dr. C. F. Stewart. Falls Church, Virginia.
Sharon, Prof. Nathan. Department of Membrane Research and Biophysics, Weizmann Institute of Science, Rehovot, Israel.
Sharpe, Dr. Ralph H. Gainesville, Florida.
Shaw, Dr. Cheng-Mei. Laboratory of Neuropathy, Department of Pathology, University of Washington School of Medicine.
Shaw, Prof. Gordon L. Department of Physics, University of California, Irvine.
Shaw, Jeffry S. 3M Industrial Tape & Specialties Division, St. Paul, Minnesota.
Shaw, Prof. Peter. Department of Physics, Pennsylvania State University.
Shaw, Dr. Robert A. Department of Chemistry, Birkbeck College, University of London, England.
Shechtman, Dan. Department of Materials Engineering, Israel Institute of Technology, Haifa.
Sheehand, J. J. Outside Plant Supervisor, Operations Department, American Telephone and Telegraph Company, New York.
Sheffield, Dr. Harley G. Laboratory of Parasitic Diseases, National Institutes of Health, Bethesda, Maryland.
Sheft, Dr. Irving. Chemistry Division, Argonne National Laboratory, Argonne, Illinois.
Sheingold, Daniel H. Analog Devices, Inc., Norwood, Massachusetts.
Shellenberger, Dr. John A. Department of Grain Science and Industry, Kansas State University.
Shelton, Dr. Frank H. Chief Scientist, Kaman Sciences Corporation, Colorado Springs, Colorado.
Shepard, Prof. Francis P. Scripps Institution of Oceanography, La Jolla, California.
Sheppard, Dr. Edith M. Senior Lecturer in Zoology (retired), University of South Wales and Monmouthshire.
Sheridan, Prof. Thomas B. Department of Mechanical Engineering, Massachusetts Institute of Technology.
Sheriff, Dr. Robert E. Department of Geology, University of Houston.
Sherman, Dr. P. Dwight, Jr. Research and Development, Chemicals and Plastics, Union Carbide Corporation, South Charleston, West Virginia.
Sherwood, Robert S. Deceased; formerly, Manager of Engineering, Steam Turbine Division, Worthington Corporation, Harrison, New York.
Shigo, Dr. Alex L. Northwest Forest Experiment Station, Durham, New Hampshire.
Shilling, Prof. John J. College of Computing, Georgia Institute of Technology, Atlanta.
Shiino, Dr. Sueo M. Faculty of Fisheries, Prefectural University of Mie, Japan.
Shipman, Dr. Harry L. Department of Physics, University of Delaware.
Shirley, Prof. David A. Deceased; formerly, Department of Chemistry, University of Tennessee.
Shirley, Dr. Virginia S. Lawrence Berkeley Laboratory, University of California, Berkeley.
Shive, Prof. William. Department of Chemistry and Clayton Foundation Biochemical Institute, University of Texas, Austin.
Shively, R. R. Supervisor, Processor Technology Research Group, Bell Laboratories, Whippany, New Jersey.
Shoemaker, Dr. William C. Martin Luther King, Jr./Charles R. Drew Medical Center, Los Angeles, California.
Shore, Dr. Steven N. Laboratory for Astronomy and Solar Physics, NASA Goddard Space Flight Center, Greenbelt, Maryland.
Shortman, Dr. Ken. Head, Lymphocyte Differentiation Unit, The Walter and Eliza Hall Institute of Medical Research, Melbourne, Australia.
Shortreed, Dr. John. Director, Institute for Risk Research, Waterloo, Ontario, Canada.
Shprintzen, Dr. Robert J. Director, Center for Crainofacial Disorders,

Montifiore Medical Center, Albert Einstein College of Medicine, New York, New York.

Shreve, Prof. R. Norris. Professor Emeritus of Chemical Engineering, Purdue University.

Shuman, Dr. Frederick B. Director, National Meteorological Center, National Oceanic and Atmospheric Administration, Washington, D.C.

Shuster, Dr. Carl N., Jr. Ecologist (retired), Bureau of Water Hygiene, U.S. Public Health Service, Cincinnati, Ohio.

Sibley, Dr. Edgar H. Department of Information Systems Management, University of Maryland.

Siddiqi, O. Molecular Biology Unit, Tata Institute of Fundamental Research, Bombay, India.

Sidransky, Dr. Herschel. Department of Experimental Biology, Weizmann Institute, Israel.

Siegbahn, Prof. Kai. Institute of Physics, University of Uppsala, Sweden.

Siegel, Prof. Benjamin M. Department of Applied Physics, Cornell University.

Siegel, Dr. Richard W. Materials Science Division, Argonne National Laboratory, Argonne, Illinois.

Siegman, Dr. A. E. Stanford, California.

Siegmund, C. W. Exxon Research and Engineering Company, Linden, New Jersey.

Siegmund, Dr. Walter P. Director, Research and Development, Schott Fiber Optics, Inc., Southbridge, Massachusetts.

Sieh, Prof. Kerry. Seismological Laboratory, California Institute of Technology, Pasadena.

Sienko, Dr. Michell J. Deceased; formerly, Department of Chemistry, Cornell University.

Siever, Dr. Raymond. Department of Geological Sciences, Harvard University.

Siggia, Prof. Sidney. Department of Chemistry, University of Massachusetts.

Signor, Dr. Philip W. Department of Geology, University of California, Davis.

Silk, Dr. Joseph. Department of Astronomy, University of California, Berkeley.

Silliker, Dr. John H. President, Silliker Laboratories, Chicago Heights, Illinois.

Silva, Dr. Paul C. Department of Botany, University of California, Berkeley.

Silver, Dr. Seymour D. U.S. Department of the Army, Edgewood Arsenal, Maryland.

Silverman, Dr. Joseph. Department of Chemical Engineering, University of Maryland.

Simberloff, Prof. Daniel. Department of Biological Science, Florida State University.

Siminovitch, Dr. David. Ottawa Research Station, Agriculture Canada.

Simmons, Thomas C. Chief, Agents Research Branch, U.S. Department of the Army, Edgewood Arsenal, Maryland.

Simmons, Dr. William H. Department of Biochemistry and Biophysics, Loyola University Medical Center, Maywood, Illinois.

Simon, Dr. Barry. Department of Mathematics, Princeton University.

Simon, Dr. George W. Space Physics Division, Phillips Laboratory (AFSC), National Solar Observatory, New Mexico.

Simon, Dr. Jack, A. Illinois State Geological Survey, Urbana.

Simon, Dr. Steven. Department of Geophysical Science, University of Chicago, Illinois.

Simonsen, Dr. John. Department of Forest Products, Oregon State University, Corvallis.

Simonson, Dr. Roy W. Director (retired), Soil Classification and Correlation, U.S. Department of Agriculture, Hyattsville, Maryland.

Simpson, Dr. F. J. Bacteriologist, Atlantic Regional Laboratory, National Research Council, Halifax, Nova Scotia, Canada.

Simpson, Dr. Joanne. Head, Severe Storms Branch, Goddard Space Flight Center, NASA, Greenbelt, Maryland.

Simpson, Prof. Robert H. Retired; formerly, Director, Experimental Meteorology Laboratory, National Weather Service, Miami, Florida.

Simpson, Prof. Stephen G. Department of Chemistry, Massachusetts Institute of Technology.

Sims, Dr. Chester T. Department of Materials Engineering, Rensselaer Polytechnic Institute, New York.

Sims, Dr. G. K. Agricultural Research Service, U.S. Department of Agriculture, Urbana, Illinois.

Sinclair, Dr. Donald B. Deceased; formerly, President, General Radio Company, Concord, Massachusetts.

Sinclair, Dr. Malcom W. Procurement Executive, Ministry of Defense, Royal Signals and Radar Establishment, Great Malvern, Worcestershire, England.

Singer, Dr. S. F. Formerly, Deputy Assistant Secretary, U.S. Department of the Interior.

Singer, Prof. Stanley. Director, ATHENEX Research Associates.

Singleton, John D. Supervisor of Programming, Radio Division, National Broadcasting Company, New York.

Sinnett, Chester M. Deceased; formerly, RCA, Camden, New Jersey.

Sinton, Dr. William M. Department of Astronomy, University of Hawaii.

Sircar, Dr. S. Senior Research Associate, Air Products and Chemicals, Inc., Allentown, Pennsylvania.

Sisler, C. W. Corporate Engineering Department Monsanto Company, St. Louis, Missouri.

Sisler, Dr. Harry H. Department of Chemistry, University of Florida.

Sistare, George, Jr. (Retired) Consultant, Handy and Harman, Fairfield, Connecticut.

Sistrom, Prof. William R. Department of Biology, University of Oregon.

Sittig, Marshall. Assistant Director, Office of Research Project Administration, Princeton University.

Sittner, Dr. W. Rex. President, Electro Medical Systems, Inc.

Skaar, Dr. Christen. College of Forestry, Syracuse University.

Skilling, Prof. Hugh H. Department of Electrical Engineering, Stanford University.

Skilling, James K. General Radio Company, West Concord, Massachusetts.

Skinner, E. N. Assistant to the Director, Product Research and Development, International Nickel Company, Inc., New York.

Skotland, Dr. C. B. Irrigated Agriculture Research and Extension Center, Washington State University.

Slabaugh, Dr. Wendell H. Deceased; formerly, Department of Chemistry, Oregon State University.

Slichter, Dr. Charles P. Department of Physics, University of Illinois, Urbana.

Slichter, Prof. Louis B. Deceased; formerly, Institute of Geophysics, University of California, Los Angeles.

Slomp, Dr. George. Physical and Analytical Chemistry Division, Upjohn Company, Kalamazoo, Michigan.

Slonecker, Dr. Charles E. Associate Professor of Anatomy, University of British Columbia, Canada.

Sloss, Dr. L. L. Department of Geological Sciences, Northwestern University.

Slowiaczek, Louisa M. Speech Research Laboratory, Department of Psychology, Indiana University.

Smith, Dr. Andrew B. Department of Paleontology, British Museum of Natural History, London, England.

Smith, C. Price. Manager, Power and Electro Optics Products, RCA Corporation, Lancaster, Pennsylvania.

Smith, Carl E. Consulting Radio Engineer, Cleveland, Ohio.

Smith, Prof. David M. Morris K. Jeesup Professor of Silviculture, School of Forestry, Yale University.

Smith, Dr. David W. E. Associate Professor of Microbiology and Pathology, Indiana University.

Smith, Fred W. Western Union Telegraph Company, Rahway, New Jersey.

Smith, Gaylord , Director, High Nickel Alloys and Precious Metals, Product Development and Research Division, International Nickel Company, New York.

Smith, Dr. Guy D. Deceased; formerly, Soil Conservation Service, U.S. Department of Agriculture.

Smith, Prof. Harlan J. Department of Astronomy, University of Texas, Austin.

Smith, Dr. Harold H. Department of Biology, Brookhaven National Laboratory, Upton, New York.

Smith, Dr. Harry. Department of Microbiology, University of Birmingham, England.

Smith, Dr. John Ward. Deceased; formerly, Research Supervisor, Laramie Energy Research Center, Energy Research and Development Administration, Laramie, Wyoming.

Smith, Prof. Julian C. Department of Chemical Engineering, Cornell University.

Smith, Prof. Karl. U. Director, Behavioral Cybernetics Laboratory, and Professor of Psychology, University of Wisconsin.

Smith, Rear Adm. Levering. Director of Strategic Systems Project Office, U.S. Department of the Navy.

Smith, Dr. Lloyd. Lawrence Berkeley Laboratory, Berkeley, California.

Smith, Dr. Nathan J. Professor of Pediatrics, School of Medicine, University of Washington.

Smith, Prof. Paul G. Department of Vegetable Crops, College of Agricultural and Environmental Sciences, University of California, Davis.

Smith, Philip H. Consultant, Sherman, Connecticut.

Smith, Dr. Richard T. Southwest Research Institute, San Antonio, Texas.

Smith, Dr. William F. Department of Mechanical Engineering and Aerospace Sciences, College of Engineering, University of Central Florida, Orlando.

Smith, W. W. Technical Advisor, Legal Department, ESB Inc., Philadelphia, Pennsylvania.

Smuckler, Dr. Edward A. Professor of Pathology, School of Medicine, University of Washington.

Smullyan, Dr. Raymond M. Oscar Ewing Professor of Philosophy Emeritus, Indiana University, Bloomington.

Smythe, Dr. William R. Department of Physics, California Institute of Technology.

Smythies, Dr. John R. Neurosciences Program, Medical Center, University of Alabama.

Snelgrove, Prof. Martin. Department of Electronics, Carleton University, Ontario, Canada.

Snell, Dr. Arthur H. Associate Director, Oak Ridge National Laboratory, Oak Ridge, Tennessee.

Snell, Dr. Esmond E. Department of Biochemistry, University of California, Berkeley.

Snodgrass, Frank E. Institute of Geophysics and Planetary Physics, University of California, San Diego.

Snook, Dr. Theodore. Department of Anatomy, School of Medicine, University of North Dakota.

Snow, William W. Consulting Engineer, William W. Snow Associates, Inc., Woodside, New York.

Soberman, Dr. Robert K. Director, Applied Science Department, Franklin Research Center, Philadelphia, Pennsylvania.

Sohon, Dr. Harry. Deceased; formerly, Moore School, University of Pennsylvania.

Solberg, Dr. Myron. Department of Food Science, Rutgers University, New Brunswick, New Jersey.

Solem, Dr. G. Alan. Curator and Head, Division of Invertebrates, Field Museum of Natural History, Chicago, Illinois.

Somlo, Dr. Peter I. CSIRO Division of Applied Physics, Lindfield, New South Wales, Australia.

Somlyo, Dr. Andrew Paul. Department of Physiology, School of Medicine, University of Virginia.

Sommer, Dr. Alfred. Director, International Center for Epidemiology and Preventive Ophthalmology, Dana Center, Wilmer Institute, Johns Hopkins School of Hygiene and Public Health, Baltimore, Maryland.

Sommer, Dr. Alfred H. Thermo Electron Corporation, Waltham, Massachusetts.

Song, Dr. Pill-Soon. Department of Chemistry, University of Nebraska.

Soost, Dr. R. K. Department of Plant Science, University of California, Riverside.

Sorensen, Dr. Robert M. Coastal Engineering Research Center, Department of the Army, Fort Belvoir, Virginia.

Souder, Dr. Paul A. Department of Physics, Yale University.

Souders, Dr. Mott. Deceased; formerly, Director, Oil Development, Shell Oil Company, Emeryville, California.

Soulen, Robert J., Jr. Naval Research Laboratory, Washington, D.C.

Spackman, Dr. William. Department of Geology, Pennsylvania State University.

Spaulding, Philip F. Nickum & Spaulding Associates, Seattle, Washington.

Spearman, Dr. M. Leroy. NASA Langley Research Center, Hampton, Virginia.

Spedding, Dr. Frank H. Professor Emeritus, Ames Laboratory, Energy Research and Development Administration, Iowa State University.

Speight, Dr. James G. Western Research Institute, Laramie, Wyoming.

Spindler, John C. Anaconda Company, Butte, Montana.

Spink, Dr. Wesley W. Regents, Professor of Medicine, School of Medicine, University of Minnesota.

Spinrad, Dr. Bernard I. Department of Nuclear Engineering, Oregon State University.

Sposito, Dr. Garrison. Department of Plant and Soil Biology, University of California, Berkeley.

Sprague, Dr. Howard B. Agricultural Consultant, Washington, D.C.

Sprankle, Charles H. AT&T Long Lines, Morris Plains, New Jersey.

Spratt, Dr. Nelson T., Jr. Department of Zoology, University of Minnesota.

Sprenkle, Raymond E. Director of Education (retired), Bailey Meter Company, Cleveland, Ohio.

Sprinkle, Dr. James. Department of Geological Sciences, University of Texas, Austin.

Spurgin, Robert A. President, Spurgin & Associates, Irvine, California.

Stahnke, Dr. Herbert L. Director, Poisonous Animal Research Laboratory, Arizona State University.

Staley, Prof. James T. Department of Microbiology, School of Medicine, University of Washington.

Standiford, Ferris C., Jr. W. L. Badger Associates, Inc., Consulting Engineers, Ann Arbor, Michigan.

Stanford, Dennis. Curator, North American Archeology, Austin, Texas.

Stanford, Prof. John L. Department of Physics and Astronomy, Iowa State University, Ames.

Stansbarger, Donald T. Manager, Advanced Composites, Northrop Corporation, Los Angeles, California.

Starr, Dr. Eugene C. Bonneville Power Administration, U.S. Department of the Interior, Portland, Oregon.

Stauffer, Dr. Randy C. Halogens Research Laboratory, Dow Chemical Company, Midland, Michigan.

Stavis, Dr. Gus. Consulting Engineer, Wayne, New Jersey.

Steadman, Dr. David W. Department of Geosciences, University of Arizona.

Stebbins, Prof. G. Ledyard. Department of Genetics, College of Agriculture, University of California, Davis.

Steele, Dr. William A. Department of Chemistry, Pennsylvania State University.

Steenrod, Norman E. Department of Mathematics, Princeton University.

Stefanovic, Dr. Victor R. Electrical Engineering Department, University of Missouri.

Stehle, Dr. Philip M. Department of Physics, University of Pittsburgh.

Stein, Prof. Dale F. Retired; formerly, President Emeritus, Michigan Technological University, Houghton.

Stein, Dr. Nelson. Physics Division, Los Alamos Scientific Laboratory, Los Alamos, New Mexico.

Steinberg, Prof. Arthur G. Department of Biology, Case Western Reserve University.

Steinberg, Dr. Ellis P. Senior Scientist, Argonne National Laboratory, Argonne, Illinois.

Steindler, Dr. Martin J. Chemical Department, Engineering Division, Argonne National Laboratory, Argonne, Illinois.

Steinert, Emil F. (Retired) Arc Welding Division, Westinghouse Electric Corporation.

Steinschneider, Alfred. President, American Sudden Infant Death Syndrome Institute, Atlanta, Georgia.

Stelson, Dr. Paul H. Physics Division, Oak Ridge National Laboratory, Oak Ridge, Tennessee.

Stenderup, Prof. Aksel. Institute of Medical Microbiology, University of Aarhus, Denmark.

Stephens, Dr. Charles R. Director, Chemical Research and Development, Pfizer Inc., Groton, Connecticut.

Stephenson, Dr. R. J. Department of Physics, Wooster College.

Stephenson, Prof. Richard M. Department of Nuclear Engineering, University of Connecticut.

Stepniewski, Wieslaw Z. Aeronautical Consultant, Springfield, Pennsylvania.

Stergis, Dr. Christos G. Department of the Air Force, Los Angeles, California.

Stern, Prof. Fred. Iowa Institute of Hydraulic Research, University of Iowa, Iowa City.

Sternberg, Dr. James C. Applied Research Department, Beckman Instruments, Brea, California.

Sterrer, Dr. W. E. Bermuda Biological Station, St. George's West.

Sterzer, Dr. Fred. RCA Laboratories, Princeton, New Jersey.

Stevens, Dr. Brian. Department of Chemistry, University of South Florida.

Stevens, Dr. Calvin H. Department of Geology, San Jose State University.

Stevens, John E. Director, Advanced Design and Technology, AVCO Systems Division, Wilmington, Massachusetts.

Stevens, Dr. Mary Betty. Division of Rheumatology, Johns Hopkins University School of Medicine, Baltimore, Maryland.

Stevenson, Dr. Edward C. Deceased; formerly, Department of Electrical Engineering, School of Engineering and Applied Sciences, University of Virginia.

Stewart, Dewey. Crops Research Division, Agricultural Research Service, U.S. Department of Agriculture, Beltsville, Maryland.

Steward, Dr. John W. Department of Physics, University of Virginia.

Stewart, Prof. Wilson N. Department of Botany, University of Alberta, Canada.

Stiles, Dr. Martin. Department of Chemistry, University of Kentucky.

Stirton, Dr. Robert I. Vice President (retired), Roger Williams Technical and Economic Services, Inc., Berkeley, California.

Stirton, Prof. Ruben A. Director, Museum of Paleontology, University of California, Berkeley.

Stock, Prof. John T. Department of Chemistry, University of Connecticut.

Stoeckenius, Dr. Walther. Cardiovascular Research Institute, University of California Medical Center, San Francisco.

Stokes, Prof. Jacob L. Department of Bacteriology and Public Health, Washington State University.

Stoloff, Dr. Norman S. Department of Materials Engineering, Rensselaer Polytechnic Institute.

Stone, Prof. A Douglas. Department of Applied Physics, Yale University.

Stone, Prof. Kenneth G. Deceased; formerly, Department of Chemistry, Michigan State University.

Storch, Dr. Henry H. Deceased; formerly, Assistant Professor of Chemistry, New York University.

Storer, Dr. Tracy I. Deceased; formerly, Department of Zoology, University of California, Davis.

Stott, Dr. Jeffrey L. Department of Veterinary Microbiology and Immunology, School of Veterinary Medicine, University of California, Davis.

Stotz, Dr. Elmer H. Department of Biochemistry, School of Medicine and Dentistry, University of Rochester.

Strachan, Dr. Isles. Department of Geology, University of Birmingham, England.

Strand, Dr. Kaj Aa. Retired; formerly, Director, U.S. Naval Observatory.

Strandtmann, Prof. Russell W. Department of Biological Sciences, Texas Technological College.

Stranks, Prof. Donald R. Department of Physical and Inorganic Chemistry, University of Adelaide, Australia.

Stratton, Dr. Robert. Central Research Laboratories, Texas Instruments Inc., Dallas.

Stratton, Dr. Thomas F. Los Alamos Scientific Laboratory, Los Alamos, New Mexico.

Straus, Dr. Reuben. Providence Medical Center, Portland, Oregon.

Strausbaugh, Dr. Perry D. Deceased; formerly, Professor Emeritus of Botany, West Virginia University.

Strausz, Prof. Otto P. Department of Chemistry, University of Alberta, Edmonton, Canada.

Streeter, Prof. Victor L. Department of Civil Engineering, College of Engineering, University of Michigan.

Striker, Dr. Gary. National Institute of Diabetes, Digestive and Kidney Diseases, National Institutes of Health, Bethesda, Maryland.

Strimpel, Dr. Oliver. Computer Museum, Boston, Massachusetts.

Stroke, Dr. George W. Department of Electrical Sciences and Head, Electro-Optical Sciences Center, State University of New York, Stony Brook.

Strominger, Dr. Jack L. Department of Biochemistry and Molecular Biology, Harvard University.

Stroyan, Dr. Henry L. G. Entomology Department, Ministry of Agriculture, Fisheries and Food, Harpenden, England.

Stuart, Dr. Edward G. Deceased; formerly, West Virginia School of Medicine.

Stuart, William T. Deceased; formerly "Electrical Construction and Maintenance," McGraw-Hill Publications Company, New York.

Stubbings, Dr. H. G. Poole, England.

Stuckas, Kenneth J. Engineering Consultation Services, Jacksonville, Florida.

Stumm, Prof. Erwin Charles. Deceased; formerly, Department of Geology and Mineralogy, Museum of Paleontology, University of Michigan.

Stunkard, Dr. Horace W. Research Associate, American Museum of Natural History, New York.

Sturge, Dr. M. D. Bell Laboratories, Murray Hill, New Jersey.

Sturges, Frank C. Deceased; formerly, President, Pennsylvania Drilling Company, Pittsburgh, Pennsylvania.

Sturges, Dr. Wilton. Department of Oceanography, Florida State University.

Sturtevant, Prof. Alfred H. Thomas Hunt Morgan Professor of Biology, Emeritus, California Institute of Technology.

Sturtevant, George R. Manager (retired) Engineering Meter Department, General Electric Company.

Subak-Sharpe, Prof. H. Institute of Virology, University of Glasgow, Scotland.

Sudds, Dr. Richard. Department of Biological Sciences, State University of New York, Plattsburgh.

Suess, Dr. Hans E. Department of Chemistry, University of California, San Diego.

Sullivan, John. Lee Allen Associates, Sunnyvale, California.

Sumi, Dr. S. Mark. Professor of Neuropathology, School of Medicine, University of Washington.

Summers, Dr. B. A. James A. Baker Institute for Animal Health, Cornell University, Ithaca, New York.

Suomi, Prof. Verner E. Department of Meteorology, University of Wisconsin.

Suppe, John. Department of Geological Sciences, Princeton University.

Surles, Dr. Terry. Limnetics, Inc., Milwaukee, Wisconsin.

Suslick, Dr. Kenneth S. Professor, School of Chemical Sciences, Noyes Laboratory, University of Illinois, Urbana.

Suter, Dr. Hans H. Consultant, Calgary, Alberta, Canada.

Sutherland, J. R. Power Transformer Department, General Electric Company, Pittsfield, Massachusetts.

Sutherland, Dr. S. S. Senior Research Officer, Animal Health Laboratories, Department of Agriculture, Western Australia.

Sutton, Dr. B. C. Head of Taxonomic and Identification Services, Mycological Institute, Ferry Lane, Surrey, United Kingdom.

Sutton, George P. Consulting Engineer, Danville, California.

Swanson, Prof. Carl P. Associate Dean for Undergraduate Studies and Professor of Biology, Johns Hopkins University.

Swanson, Dr. R. Lawrence. Director, Waste Management Institute, Marine Sciences Research Center, State University of New York, Stony Brook.
Sweat, F. W. Department of Chemistry, University of Utah.
Sweet, Dr. Walter C. Department of Geology and Mineralogy, Ohio State University.
Swern, Dr. Daniel. Deceased; formerly, Department of Chemistry, Fels Research Institute, Temple University.
Swets, Dr. John A. Bolt Beranek and Newman, Inc., Cambridge, Massachusetts.
Swithinbank, Dr. Charles W. Scott Polar Research Institute, Cambridge, England.
Switzer, Dr. George. Curator, Department of Mineral Sciences, National Museum of Natural History, Smithsonian Institution.
Szalay, Dr. Frederick S. Department of Vertebrate Paleontology, American Museum of Natural History, New York.

T

Tabor, Paul. Soil Conservation Service, Athens, Georgia.
Taft, Dr. Charles K. Department of Mechanical Engineering, University of New Hampshire.
Taft, Prof. Horace D. Department of Physics, Yale University.
Tager, Prof. Morris. Department of Microbiology, Emory University.
Tancreto, Anthony E. Meteorologist in Charge, National Weather Service, National Oceanic and Atmospheric Administration, Logan International Airport, East Boston, Massachusetts.
Tang, Prof. K. Y. Deceased; formerly, Department of Electrical Engineering, Ohio State University.
Tannas, Lawrence E., Jr. Aerojet Electro Systems Company, Azusa, California.
Tanner, Dr. Fred W., Jr. Assistant to Corporate Vice President, Charles Pfizer and Company, New York.
Tapia, Dr. Richard A. Department of Mathematical Sciences, Rice University.
Tappan, Dr. Helen. Department of Geology, University of California, Los Angeles.
Tarling, Dr. D. H. Department of Geophysics and Planetary Physics, University of Newcastle, England.
Tart, Dr. Charles T. Department of Psychology, University of California, Davis.
Tattersall, Dr. Ian M. Peabody Museum of Natural History, Yale University.
Tattersall, Dr. Olive S. Research Zoologist, Hayling Island, England.
Taube, Dr. Henry. Department of Chemistry, Stanford University.
Tauber, Michael E. NASA Ames Research Center, Moffett Field, California.
Taylor, Prof. Angus E. President's Office, University of California, Berkeley.
Taylor, Frank H. Technical Manager of Aviation, Standard Telephones and Cables, Ltd., London England.
Taylor, Dr. Gordon S. Connecticut Agricultural Experiment Station, Windsor.
Taylor, Prof. J. Herbert. Institute of Molecular Biophysics, Florida State University.
Taylor, Jane R. Neurobehavioral Laboratory, Yale University School of Medicine.
Taylor, Dr. Joseph H., Jr. Department of Physics and Astronomy, University of Massachusetts.
Taylor, Dr. R. E. Associate Professor and Director, Radiocarbon Laboratory, University of California, Riverside.
Taylor, Dr. Robert J. Consultant, Microscopy, Falls Church, Virginia.
Taylor, Prof. Thomas N. Department of Biology, University of Kansas, Lawrence.
Taylor, William L. Retired; formerly, National Severe Storms Laboratory, National Oceanic and Atmospheric Administration, Norman, Oklahoma.
Teal, Dr. John M. Department of Biology, Woods Hole Oceanographic Institution, Woods Hole, Massachusetts.
Tease, Samuel C. Pennsalt Chemicals Corporation, Philadelphia, Pennsylvania.
Teasley, Robert, Jr. Cummins Engine Company, Columbus, Indiana.
Tedesco, Dr. Edward F. Jet Propulsion Laboratory, California Institute of Technology.
Tedford, Dr. Richard H. Department of Vertebrate Paleontology, American Museum of Natural History, New York.
Teichert, Dr. Curt. Department of Geological Sciences, University of Rochester.
Teleki, Dr. Geza. Deceased; formerly, Department of Geology, George Washington University.
Teller, Dr. Aaron J. Teller Environmental Systems, Worcester, Massachusetts.
Temin, Dr. Howard M. McArdle Laboratory for Cancer Research, University of Wisconsin Medical Center.
Templeton, Dr. George E. Department of Plant Pathology, University of Arkansas.
Terrill, Dr. Clair E. Chief, Sheep and Fur Animal Research Branch, U.S. Department of Agriculture, Beltsville, Maryland.
Theis, Dr. Douglas J. Computer Systems Department, Aerospace Corporation, Los Angeles, California.
Thelemaque, L. E. Product Family Systems Engineering Center, AT&T Information Systems, Lincroft, New Jersey.
Thiess, Helmut E. Washington, D.C.
Thomas, Dr. David Hurst. American Museum of Natural History, New York.
Thomas, Dr. J. Kerry. Department of Chemistry, University of Notre Dame.
Thomas, Dr. Walter I. Agricultural Experiment Station, Pennsylvania State University.
Thompson, Prof. Gerald L. Professor of Applied Mathematics and Industrial Administration, Graduate School of Industrial Administration, Carnegie-Mellon University.
Thompson, Jack R. U.S. Army Corps of Engineers, Office of the Secretary of the Army, Washington, D.C.
Thompson, Dr. Stanley G. Senior Staff Member, Lawrence Berkeley Laboratory, University of California, Berkeley.
Thompson, Dr. T. E. Department of Zoology, University of Bristol, England.
Thomson, Dr. Keith S. President, Academy of Natural Sciences, Philadelphia, Pennsylvania.
Thomson, Ker Clive. Director, Terrestrial Sciences Laboratory, Hanscom Air Force Base, Bedford, Massachusetts.
Thomson, Prof. Robb M. Chairman, Department of Materials Science, State University of New York, Stony Brook.
Thorn, Dr. Robert J. Divisions of Chemistry and Materials Science, Argonne National Laboratory, Argonne, Illinois.
Thorne, Dr. Gerald G. Department of Plant Pathology, University of Wisconsin.
Thorne, Dr. Kip S. Kellogg Radiation Laboratory, California Institute of Technology.
Throp, Prof. H. Holden. Department of Chemistry, University of North Carolina, Chapel Hill.
Thwaites, Fredrik T. Deceased; formerly, Assistant Professor Emeritus, Department of Geology, University of Wisconsin.
Tigner, Dr. Maury. Laboratory of Nuclear Studies, Cornell University.
Tikofsky, Dr. Ronald S. Department of Radiology, Medical College of Wisconsin, Milwaukee.
Tilling, Dr. Robert I. Branch of Igneous and Geothermal Processes, U.S. Geological Survey, Department of the Interior, Menlo Park, California.
Tilton, Dr. Richard C. Department of Laboratory Medicine, University of Connecticut Health Center, Farmington.
Timoney, Dr. John F. Department of Veterinary Science, University of Kentucky, Lexington.
Ting, Dr. Irwin P. Department of Botany and Plant Sciences, University of California, Riverside.
Ting, Prof. Samuel C. C. Department of Physics, Massachusetts Institute of Technology.
Tobias, Dr. Cornelius A. Donner Laboratories, University of California, Berkeley.

Tobias, Prof. Julian M. Deceased; formerly, Department of Physiology, University of Chicago.
Tolgyesi, Dr. William S. Gillette Research Institute, Rockville, Maryland.
Tomasetti, Richard L. Senior Vice President, Lev Zetlin Associates, Inc. New York, New York.
Tonndorf, Prof. Juergen. Deceased; formerly, Department of Otolaryngology, College of Physicians and Surgeons, Columbia University.
Torrey, Dr. Theodore W. Department of Zoology, Indiana University.
Torres, Dr. Alfonso. Head, Diagnostic Services Section, National Veterinary Services Laboratory, U.S. Department of Agriculture, Greenport, New York.
Toth, Dr. Nicholas. Department of Anthropology, Indiana University.
Tou, Prof. Julius T. Director, Center for Information Research, University of Florida.
Tousey, Dr. Richard. U.S. Naval Research Laboratory, Washington, D.C.
Townes, Prof. Charles H. Department of Physics, University of California, Berkeley.
Toy, Dr. Pearl. San Francisco General Hospital, San Francisco, California.
Tracy, Paul H. Consultant to the Dairy Industry, Deland, Florida.
Tracy, Robert J. Department of Geological Sciences, Harvard University.
Trautman, W. Dean. Manager, Laboratory Administration Research and Development, Brush Beryllium Company, Cleveland, Ohio.
Travis, Elmer W. Swales and Associates, Inc., Beltsville, Maryland.
Treffers, Dr. Henry P. Professor of Pathology, School of Medicine, Yale University.
Treybal, Robert E. Deceased; formerly, Department of Chemical Engineering, New York University.
Triebe, Edward J. Deceased; formerly, Kingsport Press, Inc., Kingsport, Tennessee; Consultant, Franklin Book Programs, Inc.
Trimble, Dr. Virginia. Department of Physics, University of California, Irvine.
Trinkaus, Dr. Erik. Department of Anthropology, University of New Mexico.
Truax, Dr. Robert C. Truax Engineering Company, Saratoga, California.
Trump, Benjamin F. Professor and Chairman, Department of Pathology, University of Maryland Medical Center.
Trump, Prof. John G. Deceased; formerly, Department of Electrical Engineering, Massachusetts Institute of Technology.
Truxal, Dr. John G. College of Engineering and Applied Sciences, State University of New York, Stony Brook.
Tsai, Prof. Chen S. Department of Electrical Engineering, University of California, Irvine.
Tsang, Dr. C. F. Lawrence Berkeley Laboratory, University of California, Berkeley.
Tsaur, Dr. Bor-Yeu. Lincoln Laboratories, Lexington, Massachusetts.
Tsuchiya, Prof. Henry M. Department of Chemical Engineering, University of Minnesota.
Tsutsui, Prof. Minoru. Deceased; formerly, Department of Chemistry, Texas A&M University.
Tukey, Dr. Harold B. Deceased; formerly, Professor Emeritus, Department of Horticulture, Michigan State University.
Tully, Dr. John P. Pacific Oceanographic Group, Fisheries Research Board of Canada, Nanaimo, British Columbia.
Turk, Prof. J. L. Department of Pathology, Royal College of Surgeons of England.
Turnbull, Prof. David. Department of Applied Physics, Harvard University.
Turner, Dr. Charles E. Agricultural Research Service, U.S. Department of Agriculture, Albany, California.
Turner, Prof. C. Donnell. (Retired) Department of Biological Sciences, Duquesne University.
Turner, Dr. Charles W. Deceased; formerly, Professor Emeritus of Dairy Husbandry, University of Missouri.
Turner, Joyce B. Institute for Advanced Study.
Turner, Dr. Monica G. Environmental Sciences Division, Oak Ridge National Laboratory, Oak Ridge, Tennessee.
Turner, Dr. Ruth D. Museum of Comparative Zoology, Harvard University.
Turner, Dr. Shirley. National Institute of Standards and Technology, U.S. Department of Commerce, Gaithersburg, Maryland.
Turner, Dr. Thomas B. Professor of Microbiology and Dean Emeritus, School of Medicine, Johns Hopkins University.
Turok, Dr. Neil. Department of Physics, Joseph Henry Laboratories, Princeton University.
Turrell, Donald L. Marketing Manager, Rochester Instrument Systems, Inc., Rochester, New York.
Turro, Prof. Nicholas J. Department of Chemistry, Columbia University.
Tuteur, Prof. Franz B. Department of Engineering and Applied Sciences, Yale University.
Tuttle, Alan H. Vocational Division, State University of New York Agricultural and Technical Institute, Wellsville.
Tuttle, Douglas E. Pamarco, Roselle, New Jersey.
Tuttle, Dr. W. Norris. Consultant, Concord, Massachusetts.
Tyler, Prof. Albert. Deceased; formerly, Division of Biology, California Institute of Technology.
Tyler, Ivan L. Consultant, Tucson, Arizona.
Tyler, Nathan. Glass Packaging Institute, Washington, D.C.
Tyrer, Dr. Louise B. Medical Affairs Coordinator, Planned Parenthood Federation of America, New York.

U V

Uchida, Dr. Tohru. Professor of Zoology, Hokkaido University.
Uhl, Dr. Vincent W. Department of Chemical Engineering, University of Virginia.
Uhlig, Dr. Ronald P. Bell Northern Research, Ottawa.
Ullman, Dr. John E. Department of Marketing-Management, Hofstra University.
Underwood, Dr. Erwin E. Alcoa Professor, Georgia Institute of Technology.
Unger, Walter H. Anaconda Company, Denver, Colorado.
Untersteiner, Prof. Norbert. Department of Atmospheric Science, University of Washington.
U.S. Army Corps of Engineers. Office of the Secretary of the Army, Washington, D.C.
Unwin, Dr. Stephen C. Owens Valley Radio Observatory, California Institute of Technology, Pasadena.
Van Andel, Dr. Tjeerd H. Department of Geology, Stanford University.
Van Arsdel, Dr. Paul P., Jr. Department of Medicine, Division of Allergy and Infectious Diseases, School of Medicine, University of Washington.
Van Campenhout, Prof. Ernest L. Deceased; formerly Professor of Anatomy, Histology, and Embryology, School of Medicine, University of Louvain, Belgium.
van de Kamp, Dr. Peter. Sproul Observatory, Swarthmore College.
Vandenbergh, Dr. Peter A. Director, Research and Development, Microfile Technics, Sarasota, Florida.
Vander Sande, Prof. John B. Department of Metallurgy and Materials Science, Massachusetts Institute of Technology.
Vanderzee, Dr. Cecil E. Department of Chemistry, University of Nebraska.
Van Dorn, William G. Scripps Institution of Oceanography, La Jolla, California.
Van Dyne, Dr. George. Natural Resource Ecology Laboratory, Colorado State University.
Van Etten, Dr. James P. Navigation Consultant, Nutley, New Jersey.
Van Fleet, Dr. D. S. Department of Botany, University of Georgia.
Van Houten, Prof. Franklyn B. Department of Geology, Princeton University.
Vanick, James S. Consultant, Foundry Castings, Sea Girt, New Jersey.
Váanky, Dr. Kálmán. Botanisches Institut, Universität Tubingen, Germany.
Van Niel, Dr. Cornelius B. Deceased; formerly, Hopkins Marine Station, Pacific Grove, California.

VanPeteghem, Dr. Peter M. Deceased; formerly, Electrical Engineering Department, Texas A&M University.
Van Valen, Dr. Leigh. Committee on Evolutionary Biology, University of Chicago.
Van Valkenburg, Dr. Mac E. Department of Electrical Engineering, University of Illinois, Urbana.
Van Vlack, Prof. Lawrence H. Department of Materials Engineering, University of Michigan.
Van Wagtendonk, Dr. Willem J. Veterans Administration Hospital, Miami, Florida.
Van Wazer, Dr. John R. Department of Chemistry, Vanderbilt University.
Van Winkle, Prof. Quentin. Department of Chemistry, Ohio State University.
Vaughan, Virginius N., Jr. Engineering Director, Data Communications, American Telephone and Telegraph Company, New York.
Vaughan, Prof. Wyman R. Head, Department of Chemistry, University of Connecticut.
Veale, Dr. Chris. A. Zeneca, Inc., Wilmington, Delaware.
Veblen, Dr. David R. Department of Earth and Planetary Sciences, Johns Hopkins University.
Veillon, Dr. Claude. Biophysics Research Laboratory, Peter Bent Brigham Hospital, Harvard Medical School.
Verma, Dr. Pramode K. AT&T Information Systems, Morristown, New Jersey.
Vestal, Prof. Paul A. A. G. Bush Professor of Science, A. G. Bush Science Center and Department of Biology, Rollins College, Winter Park, Florida.
Vestling, Dr. Carl S. Professor and Head, Department of Biochemistry, University of Iowa.
Veverka, Prof. Joseph. Laboratory for Planetary Studies, Cornell University.
Viccione, Dr. Daniel M. Naval Underwater Systems Center, New London Laboratory, New London, Connecticut.
Vidone, Dr. Romeo A. Chairman, Department of Pathology, School of Medicine, Yale University.
Villiger, Peter. Research Associate, Department of Molecular and Experimental Medicine, Research Institute of Scripps Clinic, La Jolla, California.
Viro, Dr. Felix. Kind & Knox, Sioux City, Iowa.
Voelker, Dr. Richard P. Science and Technology Corporation, Columbia, Maryland.
Vogt, Dr. Peter. Department of the Navy, Naval Research Laboratory, Washington, D.C.
Vögtle, Dr. Fritz. Department of Chemistry, Friedrich-Wilhelms University, Bonn, Germany.
Vokac, Charles W. Marketing Administration, Whiting Corporation, Harvey, Illinois.
Volk, Dr. Herbert F. Carbon Products Division, Union Carbide Corporation, Parma, Ohio.
Volkmar, Dr. Fred R. Child Study Center, Yale University.
Vollum, Dr. Howard. Tektronix, Inc., Portland, Oregon.
Vonder Haar, Prof. Thomas H. Department of Atmospheric Sciences, Colorado State University.
Von Doenhoff, Albert E. Deceased; formerly, Staff Scientist, National Aeronautics and Space Administration.
Von Gierke, Dr. Henning E. Consultant, Yellow Springs, Ohio.
von Graevenitz, Dr. Alexander W. C. Institute of Medical Microbiology, University of Zurich, Switzerland.
Vonnegut, Dr. Bernard. Retired; formerly, Atmospheric Sciences Research Center, State University of New York, Albany.
Von Puttkamer, Jesco. Advanced Programs, Office of Space Transportation Systems, National Aeronautics and Space Administration, Washington, D.C.
Von Winkle, William A. Associate Technical Director for Technology, Naval Underwater Systems Center, New London Laboratory, New London, Connecticut.
Voorhees, Dr. Frank R. Department of Biology, Central Missouri State College.
Voss, Dr. Gilbert L. Rosenstiel Institute of Marine and Atmospheric Science, University of Miami.

W

Wackerman, Albert E. Professor Emeritus of Forest Utilization, Duke University.
Waddington, Prof. Thomas C. Deceased; formerly, Department of Chemistry, University of Durham.
Waelsch, Dr. Salome Gluecksohn. Department of Genetics, Albert Einstein College of Medicine.
Wagner, Dr. Frank. Technical Center, Celanese Chemical Company, Corpus Cristi, Texas.
Wagner, Dr. Henry N., Jr. Director, Divisions of Nuclear Medicine and Radiation Health Sciences, Johns Hopkins Medical Institutions.
Wagner, Dr. Robert P. Department of Zoology, University of Texas, Austin.
Wahl, Prof. Floyd M. Professor and Chairman, Department of Geology, University of Florida.
Waidelich, Dr. Donald L. Department of Electrical Engineering, University of Missouri.
Wainwright, Howard W. Coal Research Center, U.S. Bureau of Mines, Morgantown, West Virginia.
Waisel, Prof. Y. Department of Botany, Tel Aviv University, Israel.
Wakeham, Dr. W. A. Department of Chemical Engineering and Chemical Technology, Imperial College of Science and Technology, London, England.
Waksman, Byron H. Department of Pathology, School of Medicine, Yale University.
Waksman, Dr. Selman A. Deceased; formerly, Institute of Microbiology, Rutgers University.
Walczak, Mary M. Department of Chemistry, Iowa State University of Science and Technology.
Waldmann, Katherine S. National Institutes of Health, Department of Health and Human Services, Bethesda, Maryland.
Waldmann, Dr. Thomas A. Chief, Metabolism Branch, National Institutes of Health, Department of Health and Human Services, Bethesda, Maryland.
Waldron, Dr. Robert D. Director, Research Enterprises, Scottsdale, Arizona.
Walker, J. H. Vice President, Portland Cement Association, Skokie, Illinois.
Walker, Lawrence R. Bell Laboratories, Murray Hill, New Jersey.
Walker, Dr. R. M. Department of Physics, Washington University.
Walker, Dr. Warren F., Jr. Department of Biology, Oberlin College.
Walkup, Dr. John T. Yale University School of Medicine, New Haven, Connecticut.
Wall, Dr. Joseph S. Northern Regional Research Center, U.S. Department of Agriculture, Peoria, Illinois.
Wallace, Prof. John F. Division of Metallurgy and Material Science, Case Western Reserve University.
Walsh, Prof. Joseph L. Deceased; formerly, Department of Mathematics, University of Maryland.
Walsh, Dr. Michael A. Department of Biology, Utah State University.
Walsh, Dr. Peter J. Department of Physics, Fairleigh Dickinson University.
Walstedt, Dr. R. E. Bell Laboratories, Murray Hill, New Jersey.
Walt, Dr. Martin. Physical Sciences Laboratory, Lockheed Missiles and Space Company, Palo Alto, California.
Walter, William C. President, Frontier Systems, Rolling Hills, California.
Walton, Prof. Harold F. Department of Chemistry, University of Colorado.
Walton, Dr. Peter D. Department of Plant Science, University of Alberta, Edmonton, Canada.
Walworth, Vivian K. Polaroid Corporation, Cambridge, Massachusetts.
Wamser, Prof. Carl C. Department of Chemistry, California State University.
Wang, Prof. K. K. Sibley School of Mechanical and Aerospace Engineering, Cornell University.

Wang, Prof. William S.-Y. Project On Linguistic Analysis, University of California, Berkeley.
Ware, Dr. Bennie R. Department of Chemistry, Syracuse University.
Warf, Dr. James C. Department of Chemistry, University of Southern California.
Warner, Deborah Jean. Curator, History of Astronomy, National Museum of American History, Washington, D.C.
Warner, Dr. Frank L. Royal Signals and Radar Establishment, Great Malvern, Worcestershire, England.
Warner, Dr. Robert M. Department of Horticulture, University of Hawaii at Manoa.
Warren, Prof. Bertram E. Department of Physics, Massachusetts Institute of Technology.
Wartik, Prof. Thomas. Department of Chemistry, Pennsylvania State University.
Wasserman, Dr. Zelda R. Department of Central Research and Development, E. I. du Pont de Nemours and Company, Wilmington, Delaware.
Waters, Dr. Everett. Department of Psychology, State University of New York, Stony Brook.
Watkins, Dr. Thomas D. Chabot College, Hayward, California.
Watson, William W. Professor Emeritus of Physics, Yale University.
Wattenberg, Ben J. President, Fairfield Publishers, Inc.
Watterson, Dr. Ray L Department of Zoology, University of Illinois, Urbana.
Watts, Dr. Anthony B. Department of Earth Sciences, University of Oxford, England.
Waugh, Prof. John L. T. Department of Chemistry, University of Hawaii.
Wayne, Dr. Andrew W. Department of Haematology, Kings College Hospital, London, England.
Weaver, Prof. David L. Department of Physics, Tufts University.
Weaver, Dr. E. Eugene. Research Scientist, Product Development Group, Ford Motor Company, Dearborn, Michigan.
Weaver, James L. Lockheed Missiles and Space Company, Palo Alto, California.
Weaver, John. National Severe Storms Laboratory, National Oceanic and Atmospheric Administration, Norman, Oklahoma.
Weaver, Dr. Paul. (Retired) Texas A&M College.
Weaver, Prof. Robert J. Department of Viticulture and Enology, College of Agricultural and Environmental Sciences, University of California, Davis.
Weaver, S. E. Northrop Aircraft, Inc.
Webb, Prof. J. E. Professor of Zoology, University College, Ibadan, Nigeria.
Webb, Dr. Richard A. IBM Research Division, Thomas J. Watson Research Center, IBM Corporation, Yorktown Heights, New York.
Webb, Willis L. Deceased; formerly, Chief, Meteorological Satellite TA, Atmospheric Sciences Laboratory, U.S. Army Electronic Command, White Sands, New Mexico.
Webb, Dr. Wilse B. Department of Psychology, University of Florida.
Webber, Bonnie. Department of Computer and Information Science, Moore School, University of Pennsylvania.
Webber, Rear Admiral James H. Naval Sea Systems Command, U.S. Department of the Navy, Washington, D.C.
Webber, David C. Director, Stanford University Libraries.
Weber, Erwin L. Deceased; formerly, Trust Department, National Bank of Commerce, Seattle, Washington.
Weber, Harold C. Chemical Engineer, Boston, Massachusetts.
Webster, Dr. Douglas B. Department of Otorhinolaryngology and Biocommunication, Louisiana State University Medical Center.
Weesner, Prof. Frances M. Department of Zoology, Colorado State University.
Wegner, Dr. Harvey E. Physics Division, Brookhaven National Laboratory, Upton, New York.
Wegner, Prof. Patrick A. Department of Chemistry, California State University.
Weichert, Dr. Charles K. Professor of Zoology and Dean, College of Arts and Sciences, University of Cincinnati.
Weickmann, Dr. Helmut K. Director, Atmospheric Physics and Chemistry Laboratory, Environmental Science Services Administration, Boulder, Colorado.
Weigel, Dr, Fritz. Insitut für Anorganische Chemie, Universität München, Germany.
Weil, Dr. Andrew T. Division of Social Perspectives in Medicine, University of Arizona Health Science Center, Tucson.
Weil, Jonathan F. Staff Editor, "McGraw-Hill Encyclopedia of Science and Technology," McGraw-Hill Book Company, New York.
Weil, Robert T., Jr. Deceased; formerly, Dean, School of Engineering, Manhattan College.
Weil, Rolf. Department of Material and Metallurgical Engineering, Stevens Institute of Technology.
Weiler, Daniel J. Mechanical Engineer, Alexandria, Virginia.
Weinstein, Dr. Marvin J. Director, Microbiology Division, Schering Corporation, Bloomfield, New Jersey.
Weiser, Dr. M. D. Department of Computer Science, University of Maryland.
Weiss, Prof. Gerald. Department of Electrical Engineering, Polytechnic Institute of New York.
Weiss, Dr. Theodore J. Technical Manager, Food Department, Hunt-Wesson Foods, Inc., Fullerton, California.
Weissman, Prof. Samuel Isaac. Department of Chemistry, Washington University.
Weitzenhoffer, Dr. Andre M. Veterans Administration Hospital, Oklahoma City, Oklahoma.
Welch, Arthur A. Engineering Consultant, Purcellville, Virginia.
Welch, Dr. Winona H. Department of Botany and Bacteriology, De Pauw University.
Weller, Dr. J. Marvin. Professor Emeritus, Department of Geophysical Sciences, University of Chicago.
Weller, Prof. Robert A. A. W. Wright Nuclear Structure Laboratory, Yale University.
Wellner, Dr. Daniel. Department of Biochemistry, Cornell University Medical College.
Wells, Dr. Herbert. School of Graduate Dentistry, Boston University.
Wellstead, Dr. Carl F. Doane College, Crete, Nebraska.
Welshimer, Dr. Herbert J. Department of Microbiology, Virginia Commonwealth University.
Wender, Prof. Paul H. Department of Psychiatry, University of Utah.
Wendlandt, Prof. Wesley W. Department of Chemistry, University of Houston.
Wenkert, Prof. Ernest. Department of Chemistry, Indiana University.
Wentorf, Robert H., Jr. General Electric Research and Development, Schenectady, New York.
Wentworth, John W. Director, Educational Development Engineering, RCA, Cherry Hill, New Jersey.
Wenzel, Dr. Elizabeth M. Ames Research Center, National Aeronautics and Space Administration, Moffett Field, California.
West, Dr. William. Eastman Kodak Company, Rochester, New York.
Westfall, Dr. Richard S. Department of History and Philosophy of Science, Indiana University.
Whaley, Dr. W. Gordon. Cell Research Institute, University of Texas, Austin.
Wheeler, Dr. Harry E. Department of Plant Pathology, University of Kentucky.
Wheeler, Prof. John A. Department of Physics, Joseph Henry Laboratories, Princeton University.
Whistler, Prof. Roy L. Department of Biochemistry, Purdue University.
Whitcomb, Dr. Richard T. NASA Langley Research Center, Langley Field, Virginia.
White, Dr. Abraham. Consulting Professor of Biochemistry, Stanford University; School of Medicine; Distinguished Scientist, Institute of Biological Sciences, Syntex Research.
White, Prof. Frank M. Department of Mechanical Engineering, University of Rhode Island.

White, Dr. C. Langdon. Professor of Geography, Stanford University.
White, Prof. Colin. Department of Epidemiology and Public Health, Yale University.
White, Prof. David C. Department of Electrical Engineering, Massachusetts Institute of Technology.
White, Dr. K. Preston, Jr. Department of Systems Engineering, University of Virginia.
White, Dr. Sidney E. Department of Geology and Mineralogy, Ohio State University.
White, Dr. Stanley A. President, Signal Processing and Controls Engineering, San Clemente, California.
White, Dr. T. A. Formerly, Technical Director for Foods, National Starch and Chemical Corporation, Plainfield, New Jersey.
Whitehead, Prof. W. Dexter. Department of Physics, Center for Advanced Studies, University of Virginia.
Whitehouse, Dr. Robert A. Gould-Modicon, Andover, Massachusetts.
Whiting, Frank. Washington Operations Division, McDonnell Douglas Technical Services Co., Inc. Houston, Texas.
Whitney, Dr. Daniel E. The Charles Stark Draper Laboratory, Inc. Cambridge, Massachusetts.
Whitney, Dr. Roy P. Vice President and Dean, Institute of Paper Chemistry, Appleton, Wisconsin.
Whittaker, Dr. Robert H. Department of Ecology, Cornell University.
Whittington, Dr. Harry B. Sedgwick Museum, Cambridge, England.
Wickramasinghe, Dr. H. K. International Business Machines, Yorktown Heights, New York.
Wickstrom, Dr. Conrad. Department of Biological Sciences, Kent State University.
Widder, Prof. David V. Department of Mathematics, Harvard University.
Widom, Dr. Jennifer. IBM Almaden Research Center, San Jose, California.
Wiegand, Dr. Clyde E. Lawrence Berkeley Laboratory, University of California, Berkeley.
Wiegert, Dr. Richard G. Department of Zoology, Franklin College of Arts and Sciences, University of Georgia.
Wiens, Prof. Herold J. Deceased; formerly, Department of Geography, Yale University.
Wienski, Robert M. District Manager, ISDN Architecture Planning, Bell Communications Research, Red Bank, New Jersey.
Wightman, Prof. Arthur S. Department of Physics, Joseph Henry Laboratories, Princeton University.
Wilbur, Donald A., Sr. Emeritus Professor of Entomology, Kansas State University.
Wilcox, Dr. Hugh. State University College of Forestry, Syracuse University.
Wilcox, Dr. Kimerly J. Dight Laboratories, University of Minnesota.
Wilcox, Prof. William R. Department of Chemical Engineering, Clarkson College of Technology.
Wilczek, Prof. Frank. Institute for Advanced Study, Princeton, New Jersey.
Wilder, Dr. George J. Department of Biological Sciences, University of Illinois, Chicago.
Wildhack, William A. Consultant; formerly, Associate Director, Institute for Basic Standards, National Bureau of Standards.
Wilen, Dr. Samuel H. Department of Chemistry, City University of New York.
Wiles, Dr. D. M. National Research Council of Canada, Ottawa, Ontario.
Wilhelm, Dr. Harley A. Principal Scientist, EMRR Institute and Ames Laboratory of U.S. Department of Energy, Iowa State University.
Wilke, Prof. Charles R. Department of Chemical Engineering, University of California, Berkeley.
Wilkinson, Prof. Denys H. Department of Nuclear Physics, Oxford University, England.
Wilkinson, Dr. Michael K. Associate Director, Solid State Division, Oak Ridge National Laboratory, Oak Ridge, Tennessee.
Wilkinson, Dr. W. D. Consultant in Metallurgy and Nuclear Materials, Santa Barbara, California.
Wilkinson, Dr. W. L. Postgraduate School of Studies in Chemical Engineering, University of Bradford, England.
Willett, Prof. Hilda P. Director of Graduate Studies, Department of Microbiology, Duke University Medical Center, Durham, North Carolina.
Willett, Prof. Hurd C. Department of Meteorology, Massachusetts Institute of Technology.
Williams, Prof. Dudley. Department of Physics, Kansas State University.
Williams, Dr. Earle R. Department of Earth, Atmospheric and Planetary Sciences, Massachusetts Institute of Technology, Center for Meteorology and Physical Oceanography.
Williams, Prof. Gary A. Department of Physics, University of California, Los Angeles.
Williams, Dr. Jack M. Chemistry Division, Argonne National Laboratory, Argonne, Illinois.
Williams, Perry F. American Radio Relay League, Inc., Newington, Connecticut.
Williams, Dr. Roger J. Deceased; formerly, Department of Chemistry, University of Texas, Austin.
Williams, S. A. Department of Physics, Iowa State University.
Williamson, Dr. K. D. Los Alamos Scientific Laboratory, Los Alamos, New Mexico.
Willingham, Dr. Mark C. National Institutes of Health, National Cancer Institute, Bethesda, Maryland.
Willis, Homer B. Chief, Engineering Division, Directorate of Civil Works, Office of the Chief of Engineers, Department of the Army, Washington, D.C.
Willis, Dr. William J. European Organization for Nuclear Research (CERN), Geneva, Switzerland.
Wilson, Dr. B. F. Department of Forestry and Wildlife Management, University of Massachusetts.
Wilson, Dr. Curtis B. Department of Immunopathology, Scripps Clinic and Research Foundation, La Jolla, California.
Wilson, Prof. E. Bright, Jr. Department of Chemistry, Harvard University.
Wilson, Dr. James L. Department of Geology and Mineralogy, University of Michigan.
Wilson, Dr. Leonard R. George Lynn Cross Research Professor of Geology and Geophysics, School of Geology and Geophysics, University of Oklahoma.
Wilson, Dr. Therese. Biological Laboratories, Harvard University.
Wilt, Dr. Fred. Department of Zoology, University of California, Berkeley.
Winch, Prof. Ralph P. Department of Physics, Williams College.
Wineland, Dr. David J. Physics Division, Time and Frequency Section, National Institute of Standards and Technology, Boulder, Colorado.
Winick, Dr. Herman. Deputy Director, Stanford Synchrotron Radiation Laboratory, Stanford, California.
Winkler, Dr. Eva M. U.S. Naval Ordnance Laboratory, Silver Spring, Maryland.
Winn, Kevin. American Sudden Infant Death Syndrome Institute, Atlanta, Georgia.
Winston, Dr. Judith E. Assistant Curator, Department of Invertebrates, American Museum of Natural History, New York.
Winter, Prof. Rolf G. Department of Physics, College of William and Mary.
Wise, Edmund M. Consultant; formerly, Assistant to Vice President of Research, International Nickel Company, Inc., New York.
Wisely, Prof. William H. Deceased; formerly, Department of Civil Engineering, University of Virginia.
Wislicenus, Dr. George F. Professor Emeritus; formerly, Director, Garfield Thomas Water Tunnel, and Head, Department of Aerospace Engineering, Pennsylvania State University.
Wisniewski, Dr. William J. Vice President, Engineering, Mutual Broadcasting System, Arlington, Virginia.
Wisnosky, Dennis. WIZdom Systems, Inc., Naperville, Illinois.
Wnek, Prof. Gary E. Director, Polymer Science and Engineering Program, Department of Chemistry, School of Science, Rensselaer Polytechnic Institute.

Wodicka, Virgil O. Consultant in Food Technology and Quality Assurance, Fullerton, California.
Wolf, Dr. Peter. IBM Zurich Research Laboratory, Switzerland.
Wolf, Dr. Werner P. Department of Physics, Yale University.
Wolfe, Dr. Ralph S. Marine Biological Laboratory, Woods Hole, Massachusetts.
Wolfe, Roger W. Plainfield Semiconductor Operation, Burroughs Corporation, Plainfield, New Jersey.
Wolfe, Dr. William L. Optical Sciences Center, University of Arizona.
Wolferz, Alfred H. Chief (retired), Tachometer Systems Development, Weston Instruments Division, Daystrom, Inc., Newark, New Jersey.
Wollersheim, Prof. David. Department of Mechanical and Aerospace Engineering, University of Missouri.
Wolovich, Dr. William. Department of Engineering, Brown University.
Wonders, Dr. William C. Department of Geography, University of Alberta, Canada.
Wood, Dr. Albert E. (Retired) Department of Biology, Amherst College.
Wood, Dr. B. J. Department of the Geophysical Sciences, University of Chicago.
Wood, Dr. Earl H. Mayo Foundation, Graduate School of Medicine, University of Minnesota.
Wood, Dr. Edward J. F. Institute of Marine Science, University of Miami.
Wood, Dr. Frank B. Department of Physics and Astronomy, University of Florida.
Woodall, James E. Division Manufacturing Manager, Westinghouse Electric Corporation, Pittsburgh, Pennsylvania.
Woods, Prof. John M. School of Chemical Engineering, Purdue University.
Woodward, Dr. J. Guy. David Sarnoff Research Center, RCA Laboratories, Princeton, New Jersey.
Woodward, Prof. John B., III. Department of Naval Architecture and Marine Engineering, University of Michigan.
Woody, Dr. R. W. Department of Biochemistry, Colorado State University.
Woram, John M. Woram Audio Associates, New York.
Worley, Dr. J. F. Plant Pathologist, Plant Hormone and Regulators Laboratory, U.S. Department of Agriculture, Beltsville, Maryland.
Wright, Prof. Edward L. Department of Astronomy, University of California, Los Angeles.
Wright, Elliott F. Consulting Engineer (retired), Advanced Products Division, Studebaker-Worthington Corporation, Harrison, New Jersey.
Wright, James A. Strategic Marketing Manager, Motorola, Inc., Boynton Beach, Florida.
Wright, Michael E. National Semiconductor Corporation, Santa Clara, California.
Wu, Prof. Felix F. Department of Electrical Engineering and Computer Sciences, University of California, Berkeley.
Wu, Prof. Tai Tsun. Division of Engineering and Applied Physics, Harvard University.
Wulff, Dr. Verner J. Associate Director of Research, Masonic Medical Research Laboratory, Utica, New York.
Wurm, Dr. M. St. Joseph Hospital, Burbank, California.
Wurtele, Prof. Morgan G. Department of Atmospheric Sciences, University of California, Los Angeles.
Wyant, Prof. James C. Optical Sciences Center, University of Arizona.
Wygodzinsky, Dr. Pedro W. Department of Entomology, American Museum of Natural History, New York.
Wyke, Dr. John A. Wolfson Laboratory for Molecular Pathology, The Beatson Institute for Cancer Research, Glasgow, Scotland.
Wyllie, Dr. Peter J. Chairman, Department of Geophysical Sciences, University of Chicago.
Wyman, Dr. Richard E. Director of Research, Canadian Hunter Exploration, Ltd., Calgary, Alberta.
Wyrtki, Dr. Klaus. Department of Oceanography, University of Hawaii, Manoa.

Y

Yalow, Dr. Rosalyn S. Senior Medical Investigator, Veterans Administration Hospital, New York.
Yamada, Dr. Tuneo. Biology Division, Oak Ridge National Laboratory, Oak Ridge, Tennessee.
Yamaguchi, Prof. M. Department of Vegetable Crops, University of California, Davis.
Yamane, Prof. M. Tokyo Institute of Technology, Japan.
Yang, Prof. C. N. Institute for Theoretical Physics, State University of New York, Stony Brook.
Yang, Dr. Edward. Department of Electrical Engineering, Columbia University.
Yankee, Prof. Herbert W. Mechanical Engineering Department, Worcester Polytechnic Institute.
Yates, Dr. Terry L. Chairperson, Department of Biology, University of New Mexico, Albuquerque.
Yates, Wesley E. Department of Agricultural Engineering, University of California, Davis.
Yatsu, Prof. Eiju. Department of Geography, University of Guelph, Ontario, Canada.
Yeatman, Dr. Harry C. Department of Zoology, University of the South, Sewanee, Tennessee.
Yell, Dr. Ralph W. National Physical Laboratory, Teddington, Middlesex, England.
Yelles, Marvin. Formerly, Editor, "McGraw-Hill Encyclopedia of Science and Technology," McGraw-Hill Book Company, New York.
Yellott, John I. Emeritus Professor, College of Architecture, Arizona State University.
Yochelson, Dr. Ellis L. Geological Survey, U.S. Department of the Interior, Washington, D.C.
Yoeli, Dr. Meir. Deceased; formerly, Department of Preventive Medicine, School of Medicine, New York University.
Yokelson, Dr. Howard B. Research and Development Department, Amoco Chemical Company, Naperville, Illinois.
Yonge, Sir C. Maurice. Department of Zoology, University of Edinburgh, Scotland.
Yosim, Dr. Samuel J. Atomics International Division, North American Rockwell, Canoga Park, California.
Yost, Prof. Don M. Formerly, California Institute of Technology.
Yost, Dr. William A. Director, Parmly Hearing Institute, Loyola University.
Young, Dr. David A. Department of Entomology, North Carolina State University.
Young, Edward M. Deceased; formerly, Associate Editor, "Engineering News Record," McGraw-Hill, Inc., New York.
Young, Prof. N. J. Department of Mathematics and Statistics, University of Lancaster, England.
Young, Dr. Robert W. Associate Editor, "Journal of the Acoustical Society of America," San Diego, California.
Young, Dr. Roland S. Consulting Chemical Engineer, Victoria, British Columbia, Canada.
Young, Prof. Thomas F. Deceased; formerly, Department of Chemistry, of Chicago.
Young, Wayne. Director, Environmental Dredging, Maryland Environmental Service, Annapolis.
Yu, Dr. Ching-Sung. Deceased; formerly, Professor Emeritus of Astronomy, Hood College.
Yuen, Prof. Horace P. Department of Electrical Engineering and Computer Science, Northwestern University.
Yungul, Dr. Sulhi H. Chevron Resources Company, San Francisco, California.

Z

Zaitlin, Dr. Milton. Associate Director, Biotechnology Program, Plant Pathology, Cornell University.
Zander, Dr. Andrew T. Varian Associates, Palo Alto, California.
Zangerl, Dr. Rainer. Chief Curator

(retired), Department of Geology, Field Museum of Natural History, Chicago, Illinois.

Zanoni, Thomas A. Taxonomist, Jardin Botánico Nacional, Santo Domingo, Dominican Republic; and Honorary Research Associate, New York Botanical Garden.

Zapffe, Dr. Carl A. Consultant, Baltimore, Maryland.

Zeilik, Dr. Michael. AstroNet, Santa Fe, New Mexico.

Zelitch, Dr. Israel. Department of Biochemistry and Genetics, Connecticut Agricultural Experiment Station, New Haven.

Ziegler, Dr. Hubert. Technische Universität, München Institut für Botanik, Germany.

Ziemer, Prof. Rodger, E. Chairperson, Department of Electrical and Computer Engineering, University of Colorado, Colorado Springs.

Zifcak, John H. The Foxboro Company, Foxboro, Massachusetts.

Zimmerman, Dr. John R. Department of Mechanical Engineering, Pennsylvania State University.

Zimmerman, Dr. Leroy H. (Retired) U.S. Department of Agriculture, Tucson, Arizona.

Zinder, Dr. Norton D. Department of Microbial Genetics, Rockefeller University.

Zink, Dr. Frank W. Department of Vegetable Crops, University of California, Davis.

Zirin, Dr. Harold. Department of Physics, California Institute of Technology.

Zissis, Dr. George J. Chief Scientist, Environmental Research Institute of Michigan, Ann Arbor.

Zlotnik, Dr. I. Microbiological Research Establishment, Wilts, England.

Zuberi, Dr. Riaz. Department of Immunology, Medical Biology Institute, La Jolla, California.

Zuk, Dr. William. School of Architecture, University of Virginia.

Zuman, Dr. Petr. Department of Chemistry, Clarkson College of Technology.

Zweifel, Dr. Richard G. Curator, Department of Herpetology, American Museum of Natural History, New York.

Zwerman, Prof. Paul J. Department of Agronomy, Cornell University.

Zwolinski, Dr. Bruno J. Department of Chemistry, Thermodynamics Research Center, Texas A&M University.

McGraw-Hill
Concise Encyclopedia of Science & Technology

Fourth Edition

A

A15 phases A series of intermetallic compounds which have a particular crystal structure and the chemical formula A_3B, where A represents a transition element and B can be either a transition or a nontransition element. Many A15 compounds exhibit the phenomenon of superconductivity at relatively high temperatures in the neighborhood of 20 K (−424°F) and in high magnetic fields on the order of several tens of teslas (several hundred kilogauss). High-temperature–high-field superconductivity has a number of important technological applications and is a challenging fundamental research area in condensed-matter physics. *See* SUPERCONDUCTIVITY.

The A15 compounds crystallize in a structure in which the unit cell, the repeating unit of the crystal structure, has the overall shape of a cube. The B atoms are located at the corners and in the center of the cube, while the A atoms are arranged in pairs on the cube faces (see illustration). A special characteristic of the A15 crystal structure is that the A atoms form mutually orthogonal linear chains that run throughout the crystal lattice, as shown in the illustration. The extraordinary superconducting properties of the A15 compounds are believed to be primarily associated with these linear chains of transition-element A atoms. *See* CRYSTAL.

Processes have been developed for preparing multifilamentary superconducting wires that consist of numerous filaments of a superconducting A15 compound, such as Nb_3Sn, embedded in a nonsuperconducting copper matrix. Superconducting wires can be used in electric power transmission lines and to wind electrically lossless coils (solenoids) for superconducting electrical machinery (motors and generators) and magnets. Superconducting magnets are employed to produce intense magnetic fields for laboratory research, confinement of high-temperature plasmas in nuclear fusion research, bending beams of charged particles in accelerators, levitation of high-speed trains, mineral separation, and energy storage. *See* SUPERCONDUCTING DEVICES. [M.B.Ma.]

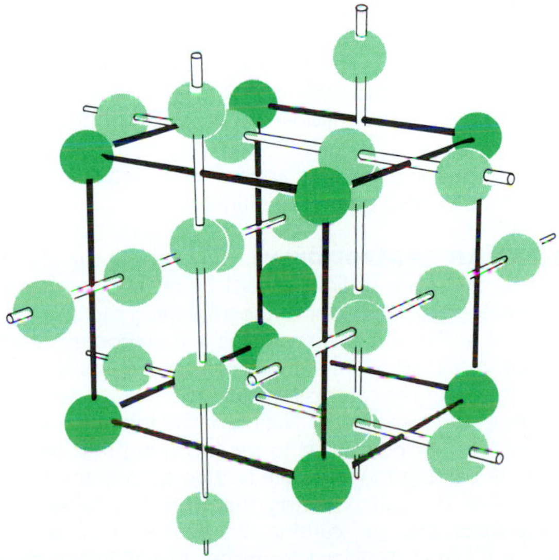

A15 crystal structure of the A_3B intermetallic compounds. The light spheres represent the A atoms; the dark spheres represent the B atoms. The linear chains of A atoms are emphasized.

Aardvark A nocturnal, burrowing, insectivorous mammal of the genus *Orycteropus* in the order Tubulidentata. It averages 6 ft (1.8 m) in length and is covered with short, sparse hair (see illustration).

The cape aardvark (*Orycteropus afer*) ranges from Ethiopia to southern Africa.

The aardvark is structurally modified for its diet of ants. The adult lacks canines and incisors; the only teeth are 20 crushing or cheek teeth in a series of 5 upper and 5 lower teeth on both sides of the jaw. These teeth, which do not have roots or enamel, continually grow throughout the life of the animal, have the appearance of tubelike structures of dentine, and are the basis for the ordinal name Tubulidentata. The short, stout, powerful forelimbs terminate in flat claws, which enable the aardvark to burrow rapidly into termite mounds. The head is long and narrow, ending in a tubular snout. The long tubular tongue extends up to 18 in. (45 cm) from the mouth.

Some disagreement exists as to the number of species of *Orycteropus*. The aardvark is edible and economically important for its thick hide which resembles pigskin. *See* ANTEATER; DENTITION; MAMMALIA; TUBULIDENTATA. [C.B.C.]

Abaca One of the strongest of the hard fibers, commercially known as Manila hemp. Abaca is obtained from the leafstalks of a member of the banana family, *Musa textilis*. The plant resembles the fruiting banana, but is a bit shorter in stature, bears small inedible fruits, and has leaves that stand more erect than those of the banana, and that are slightly narrower, more pointed, and 5–7 ft (1.5–2 m) long. The plant was domesticated long ago in the southern Philippines. *See* ZINGIBERALES.

Abaca prefers a warm climate with year-round rainfall, high humidity, and absence of strong winds. Soils must always be moist but the plant does not tolerate waterlogging. Abaca grows best on alluvial soils in the southern Philippines and northern Borneo below 1500 ft (450 m) elevation. The plant is best propagated by rootstalk suckers. There are about 75 varieties grown in the Philippines, grouped into seven categories, each of which varies slightly in height, length, and quality and yield of fiber.

The fiber ranges 6–14 ft (1.8–4.2 m) in strand length, is lustrous, and varies from white to dull yellow. As one of the longest and strongest plant fibers, resistant to fresh and salt water, abaca is favored for marine hawsers and other high-strength ropes. Abaca is also used in sackings, mattings, strong papers, and handicraft art goods.

Abaca is affected by several diseases, of which the chief are bunchy top, mosaic, and wilt. Bunchy top is caused by a virus spread by the banana aphid (*Pentalonia nigronervosa*). Mosaic is also caused by a virus spread by aphids (chiefly *Rhopalosiphum nymphaeae* and *Aphis gossypii*). Abaca wilt is caused by a soil or water-borne fungus, chiefly attacking plant roots. [E.G.N.]

Abacus The earliest known computing instrument, consisting of beads (counters) strung on wires mounted in a frame (see illustration); technically the term also covers the earlier counting tables and the grooved tablets of the Romans. The abacus is still used in parts of the world.

The Roman hand abacus was a bronze tablet with grooves in which small spherical counters could slide, remarkably like the abacus which came into use later in the Middle and Far East. Although this type of abacus is described in the Chinese literature of the 2d century, it was not until the 13th century that its use became widespread in China. The Chinese abacus had one more counter in each section than the Roman hand abacus, and the counters were mounted on wires in a frame rather than in grooves. It was introduced into Japan in the 15th century, and by the 19th century a modified version had come into wide use there. In China the abacus is called the *suan-pan*, and in Japan the *soroban*. In the Middle East, still another version of the abacus is used, which in Russia is called the *schoty*.

Type of abacus now in use.

By the 19th century the counting-table type of abacus had essentially gone out of use in the Western world, and there the beads-on-shaft type had never been widely used. Conversely, the 19th and 20th centuries, with their economic development, brought an upsurge in the use of the latter type of abacus in the Middle and Far East. By the end of World War II, abacus operation was included in the curriculum of the fourth and upper grades in Japan, and licenses were issued for proficiency in its use. Teachers prefer it over the pocket electronic calculator. In Russia and China, as in Japan, the abacus is still widely used. *See* CALCULATORS. [V.R.H.]

Abalone A gastropod mollusk comprising the single genus *Haliotis*, family Haliotidae, also known as ear shell, ormer, or paua. The abalones are cosmopolitan species of temperate and tropical seas. They are active nocturnally and feed principally on algae.

The shell is flattened, with an enlarged body whorl and reduced spire, and looks like an ear (see illustration). Characteristically, the shell is perforated by a series of small pores which allow for more direct elimination of water from the mantle cavity. Exposed parts of the body are pigmented and show a varying range of colors, such as black, green, or brown. The larvae are pelagic and free-swimming among the plankton in coastal waters for about 2 days, after which they settle down and develop to the adult. *See* ARCHAEOGASTROPODA; CIRCULATION; GASTROPODA; RESPIRATORY PIGMENTS (INVERTEBRATE). [C.B.C.]

Typical abalone ear-shaped shell perforated by pores.

Abdomen A major body division of the vertebrate trunk lying posterior to the thorax; and in mammals, bounded anteriorly by the diaphragm and extending to the pelvis. The diaphragm separates the abdominal or peritoneal cavity from the pleural and pericardial cavities of the thorax. In all pulmonate vertebrates (possessing lungs or lunglike organs) other than mammals, the lungs lie in the same cavity with the abdominal viscera, and this cavity is known as the pleuroperitoneal cavity.

The large coelomic cavity that occupies the abdomen contains the viscera within the peritoneal sac. Connecting sheets of peritoneum from the body wall to the various organs form the mesenteries. Other folds of the peritoneum form the omenta.

The term abdomen is also applied to a similar major body division of arthropods and other animals. [W.J.B.]

Aberration (astronomy) The apparent change in direction of a source of light caused by an observer's component of motion perpendicular to the impinging rays.

To visualize the effect, first imagine a stationary telescope (illustration *a*) aimed at a luminous source such as a star, with photons traveling concentrically down the tube to an image at the center of the focal plane. Next give the telescope a component of motion perpendicular to the incoming rays (illustration *b*). Photons passing the objective require a finite time to travel the length of the tube. During this time the telescope has moved a short distance, causing the photons to reach a spot on the focal plane displaced from the former image position. To return the image to the center, the telescope must be tilted in the direction of motion by an amount sufficient to ensure that the photons once again come concentrically down the tube in its frame of reference (illustration *c*). The necessary tilt angle α is given by $\tan \alpha = vc$, where v is the component of velocity perpendicular to the incoming light and c is the velocity of light. (An analogy illustrating aberration is the experience that, in order for the feet to remain dry while walking through vertically falling rain, it is necessary to tilt an umbrella substantially forward.)

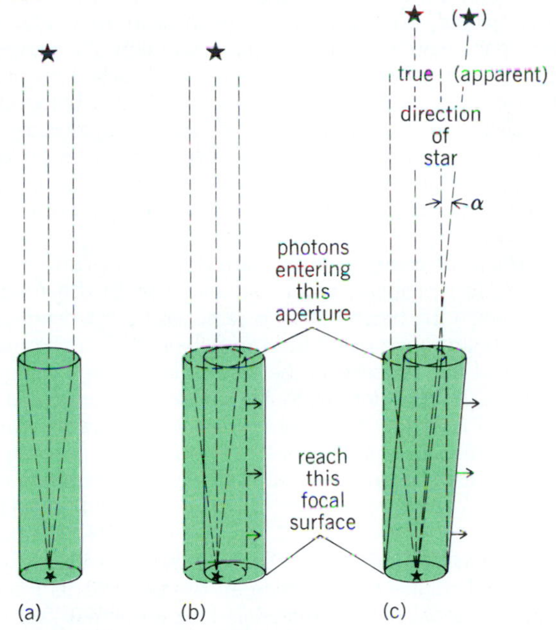

Demonstration of aberration. *(a)* Fixed telescope; photons form image at center of focal plane. *(b)* Moving telescope; image is displaced from center. *(c)* Tilted moving telescope is required to restore image to center.

This discovery provided the first direct physical confirmation of the Copernican theory. A second important application of aberration has been its clear-cut demonstration that, as is axiomatic to special relativity, light reaching the Earth has a velocity unaffected by the relative motion of the source toward or away from the Earth. *See* LIGHT; PARALLAX (ASTRONOMY). [H.J.S.]

Aberration (optics) Deviation from perfect image formation. The deviation arising from the fact that light of different wavelengths follows different paths through an optical system is known as chromatic aberration. The chief monochromatic aberrations are spherical (aperture) aberrations, coma, astigmatism, curvature of field, and distortion. *See* ASTIGMATISM; CHROMATIC ABERRATION; COMA (OPTICS); CURVATURE OF FIELD; DISTORTION (OPTICS); GEOMETRICAL OPTICS; OPTICAL IMAGE; SPHERICAL ABERRATION. [M.J.H.]

Ablastin An antibodylike substance, present in serum as a result of immunization, that inhibits the multiplication of an invading parasite. The best example is that of the serum reactions resulting from *Trypanosoma lewisi* infections in the rat. The ablastin appears a few days after infection, and the reproduction of adult forms of the trypanosome is progressively reduced. The ablastic activity is found in the globulin portion of the serum and can be transferred to normal animals. *See* ANTIBODY; GLOBULIN; SERUM. [H.P.T.]

Abrasive A material of extreme hardness that is used to shape other materials by a grinding or abrading action. Abrasive materials may be used either as loose grains, as grinding wheels, or as coatings on cloth or paper. They may be formed into ceramic cutting tools that are used for machining metal in the same way that ordinary machine tools are used. Because of their superior hardness and refractory properties, they have advantages in speed of operation, depth of cut, and smoothness of finish.

Abrasive products are used for cleaning and machining all types of metal, for grinding and polishing glass, for grinding logs to paper pulp, for cutting metals, glass, and cement, and for manufacturing many miscellaneous products such as brake linings and nonslip floor tile.

The important natural abrasives are diamond, corundum, emery, garnet, feldspar, calcined clay, lime, chalk, and silica, SiO_2, in its many forms—sandstone, sand, flint, and diatomite.

The synthetic abrasive materials are silicon carbide, aluminum oxide, titanium carbide, and boron carbide. The synthesis of diamond puts this material in the category of manufactured abrasives. [J.F.McM.]

Abscisic acid One of the five major plant hormones. It has a number of important functions in plant growth and development. The name abscisic acid (ABA) is derived from the ability of the substance to promote abscission. It is also a potent inhibitor of growth. In this capacity it helps to induce and prolong dormancy of buds. Abscisic acid is a powerful inhibitor of seed germination and has an important role in the closure of stomata. *See* ABSCISSION; DORMANCY.

Abscisic acid is distributed throughout the plant body but is found in highest concentrations in leafy tissues, fruits, and seeds. The principal site of ABA synthesis is the chloroplast. Synthesis of ABA increases markedly when the plant is under stress.

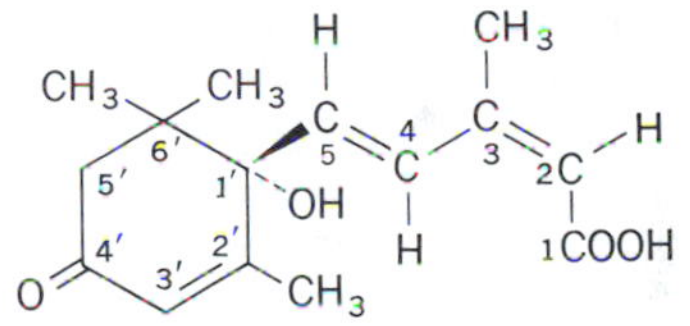

Structural formula for +-abscisic acid.

The chemical structure of ABA is shown in the illustration. ABA is a member of the terpenoid family of chemicals, which includes a number of essential oils, insect hormones, steroids, gibberellins, carotenoids, and natural rubber. It is a weak organic acid, as are two other plant hormones, auxin and gibberellin. Natural ABA is dextrorotatory, hence the "+" in front of the name. *See* AUXIN; GIBBERELLIN; PLANT HORMONES. [F.T.A.]

Abscission The process whereby a plant sheds one of its parts. Leaves, flowers, seeds, and fruits are parts commonly abscised. Almost any plant part, from very small buds and bracts to branches several inches in diameter, may be abscised by some species. However, other species, including many annual plants, may show little abscission, especially of leaves.

Abscission may be of value to the plant in several ways. It can be a process of self-pruning, removing injured, diseased, or senescent parts. It permits the dispersal of seeds and other reproductive structures. It facilitates the recycling of mineral nutrients to the soil. It functions to maintain homeostasis in the plant, keeping in balance leaves and roots, and vegetative and reproductive parts.

In most plants the process of abscission is restricted to an abscission zone at the base of an organ (see illustration); here separation is brought about by the disintegration of the walls of a special layer of cells, the separation layer. The portion of the abscission zone which remains on the plant commonly devel-

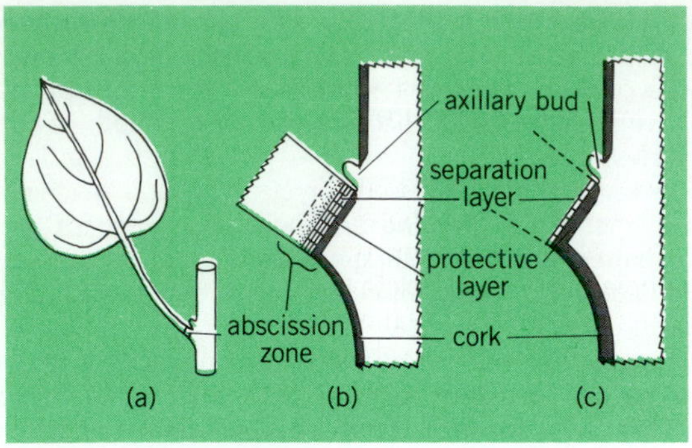

Diagrams of the abscission zone of a leaf. (*a*) A leaf with the abscission zone indicated at the base of the petiole. (*b*) The abscission zone layers shortly before abscission and (*c*) the layers after abscission.

ops into a corky protective layer that becomes continuous with the cork of the stem.

Auxin applied experimentally to the distal (organ) side of an abscission zone retards abscission, while auxin applied to the proximal (stem) side accelerates abscission. The gibberellins are growth hormones which influence abscission. When applied to young fruits or to leaves, they tend to promote growth, delay maturation, and thereby indirectly prevent or delay abscission. Abscisic acid has the ability to promote abscission and senescence and to retard growth. Small amounts of ethylene have profound effects on the growth of plants and can distort and reduce growth and promote senescence and abscission. *See* Abscisic acid; Auxin; Gibberellin. [F.T.A.]

Absolute zero The lowest temperature on the scientific temperature scale. Gross matter, which is in complete thermal equilibrium with all of its subdivided parts, has a property called its temperature, and this is measured in kelvins (K). Like all other similar scalar quantities, the absolute temperature scale starts at zero and has arbitrary but convenient units going on up to high temperatures. A convenient and readily reproducible fixed point on this temperature scale is the triple point of pure water (an equilibrium mixture of ice, water, and water vapor) defined as 273.16 K. There are several convenient thermometers to measure the low temperature range, but none can perform the impossible task of indicating the absolute zero of temperature, 0 K, because that temperature cannot be reached.

The properties of matter at the absolute zero of temperature are always extrapolated from physical measurements made on them at available temperatures such as 1 K or even 0.01 K. Such studies have led to a statement of the third law of thermodynamics. The entropy of a system tends to a constant S_0 as the temperature of the system is made to approach 0 K. One popular error is to regard the absolute zero of temperature as characterized by the complete absence of motion or of energy of the system. The atoms in a solid have considerable energy locked into the lowest allowed energy states of vibration, even at 0 K. *See* Chemical thermodynamics; Cryogenics; Entropy; Kinetic theory of matter; Low-temperature physics; Temperature. [C.F.S.]

Absorption Either the taking up of matter in bulk by other matter, as in the dissolving of a gas by a liquid; or the taking up of energy from radiation by the medium through which the radiation is passing. In the first case, an absorption coefficient is defined as the amount of gas dissolved at standard conditions by 1 cm^3 of the solvent. Absorption in this sense is a volume effect: The absorbed substance permeates the whole of the absorber. In absorption of the second type, attenuation is produced which in many cases follows Lambert's law and adds to the effects of scattering if the latter is present. *See* Attenuation.

Absorption of electromagnetic radiation can occur in several ways. Light is absorbed by atoms of the medium through which it passes; in some cases selected frequencies from a heterochromatic source are strongly absorbed. Electromagnetic radiation can be absorbed by the photoelectric effect, where the light quantum is absorbed and an electron of the absorbing atom is ejected, and also by Compton scattering. Electron-positron pairs may be created by the absorption of a photon of sufficiently high energy. Photons can be absorbed by photoproduction of nuclear and subnuclear particles, analogous to the photoelectric effect. *See* Compton effect; Photoemission.

Sound waves are absorbed at suitable frequencies by particles suspended in the air, where the sound energy is transformed into vibrational energy of the absorbing particles. *See* Sound absorption.

Absorption of energy from a beam of particles can occur by the ionization process, where an electron in the medium through which the beam passes is removed by the beam particles. The finite range of protons and alpha particles in matter is a result of this process. In the case of low-energy electrons, scattering is as important as ionization, so that range is a less well-defined concept. Particles themselves may be absorbed from a beam. Low-energy positrons are quickly absorbed by annihilating with electrons in matter to yield two gamma rays. [McA.H.H.]

Absorption (biology) The net movement (transport) of water and solutes from outside an organism to its interior. The unidirectional flow of materials into an animal from the environment generally takes place across the alimentary tract, the lungs, or the skin, and in each location a specific cell layer called an epithelium regulates the passage of materials.

Absorption across epithelia may occur by several different passive and active processes. Simple diffusion is the net movement of molecules from the apical to basolateral surfaces of an epithelium down chemical and electrical gradients without the requirement of cellular energy sources. Facilitated diffusion across the epithelium is similar to simple diffusion in that energy is not required, but in this process, molecular interaction with protein binding sites (carriers) in one or both membranes must occur to facilitate the transfer. Active molecular transport involves the use of membrane protein carriers as well as cellular energy supplies to move a transported molecule up an electrochemical gradient across the epithelium. Endocytosis and phagocytosis are also examples of active transport because metabolic energy is required, but in these processes whole regions of the cell membrane are used to engulf fluid or particles, rather than to bring about molecular transfer using single-membrane proteins. *See* Cell membranes; Endocytosis; Osmoregulatory mechanisms; Phagocytosis.

Although a wide variety of ions are absorbed by different types of epithelial cells, the mechanisms of Na^+ and Cl^- transport in mammalian small intestine are perhaps best known in detail. Transepithelial transport of these two ions occurs in this tissue by three independent processes: active Na^+ absorption, not coupled directly to the flow of other solutes but accompanied indirectly by the diffusional absorption of Cl^-; coupled NaCl absorption; and cotransport of Na^+ with a wide variety of nutrient molecules. *See* Ion transport.

Net water transport across the epithelium is coupled to net ion transport in the same direction. Pump sites for Na^+ are believed to be located along the lateral borders of epithelial cells. Energy-dependent Na^+ efflux from the cells to the intercellular spaces creates a local increase in osmotic pressure within these small compartments. An osmotic pressure gradient becomes established here, with greatest solute concentrations located nearest the tight junctions. Water flows into the cell across the brush border membrane and out the lateral membranes in response to the increased osmotic pressure in the paracellular spaces. Once water is in the intercellular compartment, a buildup of hydrostatic pressure forces the transported fluid to the capillary network. [G.A.A.]

Abstract algebra A term used synonymously with modern algebra and general algebra to describe the type of algebra which has been developed since the mid-1920s and has become a basic idiom of contemporary mathematics. In contrast with the earlier algebra, which was highly computational and was confined to the study of specific systems generally based on real and complex numbers, abstract algebra is conceptual and axiomatic and deals with systems which are arbitrary sets of elements of unspecified type, together with certain compositions satisfying prescribed lists of axioms.

A good insight into the difference between the older and the present approach can be obtained by comparing the older matrix theory with the more abstract linear algebra. Both deal with roughly the same portion of mathematics, the former a direct perspective which stresses calculations with matrices, the latter from an axiomatic and geometric viewpoint which treats vector spaces and linear transformations as the basic notions, and matrices as secondary to these. *See* LINEAR ALGEBRA; MATRIX THEORY.

Abstract algebra deals with a number of important algebraic structures, such as groups, rings, and lattices. *See* GROUP THEORY; RING THEORY. [N.J.]

Abstract data types Mathematical models which may be used to capture the essentials of a problem domain in order to translate it into a computer program.

Motivation. It is generally understood that people join a line by standing at the end of the line and are serviced from its front. Stepping into the middle of a line violates the rules of line etiquette, as does servicing someone from the middle of the line.

To describe a particular line of people, it suffices to keep track of the first and last person as the line moves. Then, it will always be clear behind whom a new person must stand and who must be served next. Such a description of a line is an example of an abstract data type (ADT). It is a model for representing some sort of structure, with rules for making changes to that structure. The rules may be expressed as operators that perform specific functions.

In the field of computer programming, lines are usually called queues. For a person to enter a queue would be to perform an insertion; to leave the front of a queue would be a deletion. When the rules of line etiquette are broken, that is, an individual enters into or is serviced from the middle, the queue becomes a list. A list is a queue where insertions and deletions may be made at any point. A stack is a queue in which insertions and deletions are both made at the front.

There are many examples of abstract data types, but this formalism is rarely useful in everyday life. However, an ADT queue would be quite helpful in representing a customer line as a computer program. The queue could be used to express certain business operations in terms of a well-understood model for representing information. This would allow the programmer to simplify the problem to that of implementing the queue in a computer language and then using the queue to represent a business function. Techniques for implementing queues are readily available.

Abstract data types allow the use of previously discovered programming techniques. Another benefit of ADTs is that they allow a programmer to encapsulate the rules for altering a model within a set of procedures, and ignore these details when using the given ADT. In other words, the programmer concerned with modeling a business function could forget exactly how an insertion or deletion is made when procedures that perform these functions are used.

ADTs and data structures. An ADT is indeed abstract, in that it must be translated into a programming language. This is done by using program data structures, which are formed from the primitive abstract data types supported directly by the given language. The programmer must also implement operators (for example, insert and delete in the queue example) that follow the rules for manipulating the ADT. In essence, then, ADTs are conceptualizations built of simple data structures and augmented with high-level operators.

Classification of data types. Abstract data types may be grouped according to the ways they arrange information. First, linearly organized data types, such as queues, lists, and stacks, represent information as a one-dimensional series of elements. Second, nonlinear modeling techniques arrange this information so that one element may be logically followed by more than one other element. (This corresponds to a customer line where there is more than one person directly behind any specific person.) Examples of this category include trees, where each element may be logically followed by other elements, called its children; and graphs, which are generalized trees in which each element may have more than one parent. Third, an unordered data type, such as a set, assumes no specific order of the elements of the type. (This corresponds to customers waiting in a chaotic group.) These three data types are examples of models used to represent information in the fast-access, main memory of a computer. Other data types, such as files and data bases, are oriented toward mass storage devices like magnetic tapes and disks. *See* DATABASE MANAGEMENT SYSTEMS; GRAPH THEORY; SET THEORY.

Primitive ADTs. There are also a number of primitive abstract types that are typically implemented directly in a high-level programming language. These include integers and real numbers (with appropriate arithmetic operators), booleans (with appropriate logical operators), text strings (often a programming language provides no special operators for manipulating strings), and pointers. [R.K.]

Acantharia A subclass of Actinopodea. These marine protozoans are related to the Radiolaria and possess a nonliving, organic capsular wall surrounding a central mass of cytoplasm. The intracapsular cytoplasm is connected to the extracapsular cytoplasm by fine cytoplasmic strands passing through pores in the capsular wall. Skeletons are typically constructed of celestite (strontium sulfate) instead of silica. The basic structural elements are 20 rods which pass through the capsule to the center in regular arrangement (polar and equatorial; see illustration). An equatorial rod forms an angle of 90° with a polar rod, and other groups are arranged with similar exactness. This type of skeleton may be modified by addition of a latticework, apparently composed of plates, each fused with a skeletal rod. Some genera show a double latticework, concentric with the central capsule.

Pseudopodia are more or less permanent. Zooxanthellae are of at least two kinds: dinophyceae containing trichocysts in the cytoplasm; and a group of algae characterized, among other things, by numerous discoidal plastids each of which contains an interlamellar pyrenoid.

Myonemes (myophrisks) are significant components of the hydrostatic apparatus and apparently regulate the buoyancy of Acantharia by expanding or contracting portions of the extracapsular cytoplasmic sheath.

Although essentially pelagic, Acantharia may move vertically with the help of their hydrostatic apparatus. Little is known about the ecology or distribution of Acantharia. The Gulf Stream is rich in Acantharia in spring and summer, but rather poor in other seasons.

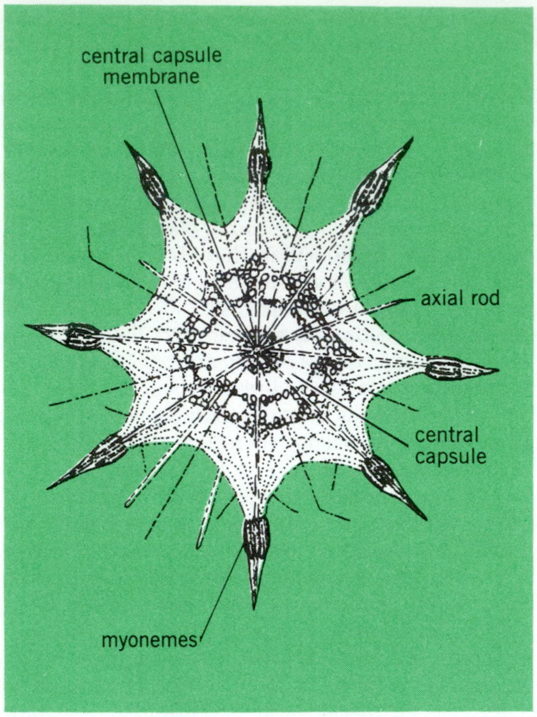

Acantharia: *Acanthometra pellucida.*

The taxonomy of Acantharia is under review, but according to one scheme there are two orders: Acanthometrida, and Acanthophractida. *See* ACANTHOMETRIDA; ACANTHOPHRACTIDA; ACTINOPODEA; PROTOZOA; SARCODINA; SARCOMASTIGOPHORA. [O.R.A.]

Acanthocephala A distinct phylum of helminths, the adults of which are parasitic in the alimentary canal of vertebrates. They are commonly known as the spiny-headed worms. The phylum comprises the orders Archiacanthocephala, Palaeacanthocephala, and Eocanthocephala. Over 500 species have been described from all classes of vertebrates, although more species occur in fish than in birds and mammals and only a relatively few species are found in amphibians and reptiles. The geographical distribution of acanthocephalans is worldwide, but genera and species do not have a uniform distribution because some species are confined to limited geographic areas. Host specificity is well established in some species, whereas others exhibit a wide range of host tolerance. The same species never occurs normally, as an adult, in cold-blooded and warm-blooded definitive hosts. More species occur in fish than any other vertebrate; however, Acanthocephala have not been reported from elasmobranch fish. The fact that larval development occurs in arthropods gives support to the postulation that the ancestors of Acanthocephala were parasites of primitive arthropods during or before the Cambrian Period and became parasites of vertebrates as this group arose and utilized arthropods for food. *See* ARCHIACANTHOCEPHALA; EOCANTHOCEPHALA; PALAEACANTHOCEPHALA.

Adults of various species show great diversity in size, ranging in length from 0.04 in. (1 mm) in some species found in fish to over 16 in. (400 mm) in some mammalian species.

The body of both males and females has three subdivisions: the proboscis armed with hooks, spines, or both; an unspined neck; and the posterior trunk. The proboscis is the primary organ for attachment to the intestinal wall of the host. In most species the proboscis is capable of introversion into a saclike structure, the proboscis receptacle. The proboscis receptacle and neck can be retracted into the body cavity but without inversion. The body cavity, or pseudocoele, contains all the internal organs, the most conspicuous of which are the reproductive organs. There is no vestige of a digestive system in any stage of the life cycle. The reproductive organs of the male consist of a pair of testes and specialized cells, the cement glands. The products of the testes and cement glands are discharged through a penis. Female Acanthocephala are unique in that the ovary exists as a distinct organ only in the very early stages of development and later breaks up to form free-floating egg balls. The eggs are fertilized as they are released from the egg balls and are retained within the ligament sacs until embryonation is complete. The nervous system is composed of a chief ganglion or brain located within the proboscis receptacle. Two nerve trunks pass through the wall of the proboscis receptacle to innervate the trunk wall. Modified protonephridial organs are found closely adherent to the reproductive system, but in most species specialized excretory organs are completely lacking. [D.V.Mo.]

Acanthodii A subclass of importance, including the earliest known jawed fishes or gnathostomes, first appearing in the Lower Silurian and surviving until the Lower Permian. They were usually small, less than 8 in. (20 cm) in length, though a few may have been as much as 100 in. (250 cm) long. The body was fusiform, the mouth terminal or nearly so, the eyes large, and the nasal capsules small. The tail was heterocercal; there were one or two dorsal fins, and all fins except the caudal had a spine on the anterior edge (see illustration). The scales had a square or rhombic crown and typically were nonoverlapping; they grew by periodic additions all around, lacked a pulp chamber, and were composed superficially of dentine or mesodentine and basally of bone, often acellular. Acanthodii are best considered as a subclass of Teleostomi, collateral with Osteichthyes, and can be classified as follows:

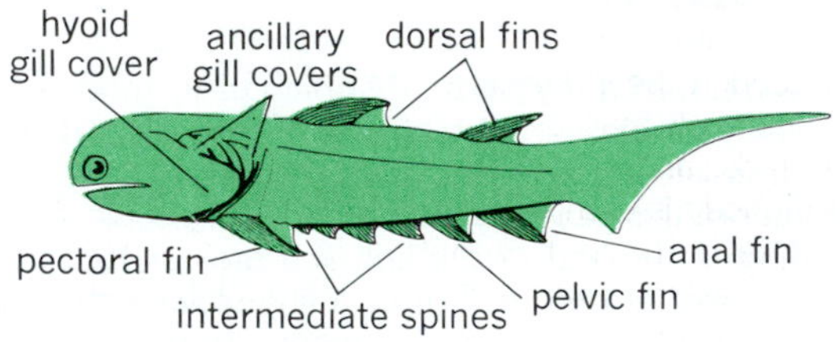

Lateral view of *Climatius reticulatus* (Climatiidae), about 6 in. (15 cm) long.

Subclass Acanthodii
- Order Climatiida
  - Family Climatiidae
  - Family Diplacanthidae
  - ?Family Gyracanthidae
- Order Ischnacanthida
  - Family Ischnacanthidae
- Order Acanthodida
  - Family Acanthodidae

See CHONDRICHTHYES; OSTEICHTHYES; TELEOSTOMI. [R.H.De.]

Acanthometrida An order of Acantharia. These marine protozoans have a skeleton limited to 20 radially arranged rods which extend from the center, forming a characteristic pattern in which angles are quite exact, as in *Acanthometra* (see illustration). The cytoplasm surrounding the spines contains a conical array of contractile microfilaments (myophrisks or myonemes) that act to expand or contract the gelatinous sheath surrounding the cell, a phenomenon that is apparently responsible for changes in the level of flota-

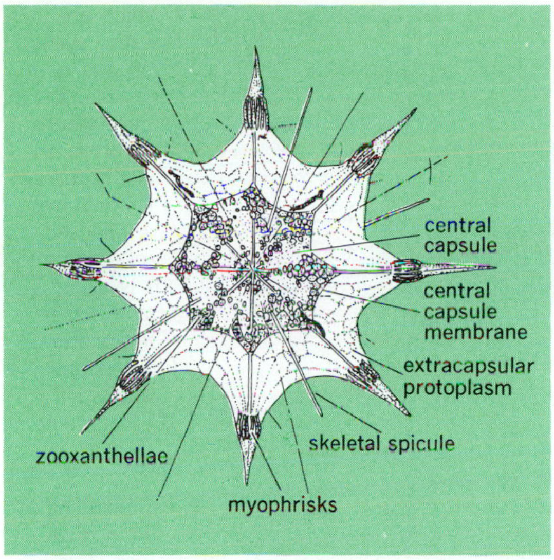

Acanthometra. (After L. H. Hyman, *The Invertebrates*, vol. 1, McGraw-Hill, 1940)

tion. *See* ACANTHARIA; ACTINOPODEA; PROTOZOA; SARCODINA; SARCOMASTIGOPHORA. [O.R.A.]

Acanthophractida An order of Acantharia. In this group of protozoans, skeletons typically include a latticework shell, although the characteristic skeletal rods are recognizable. The latticework may be spherical or ovoid, is fused with the skeletal rods, and is typically concentric with the central capsule. The body is usually covered with a single or double gelatinous sheath through which the skeletal rods emerge. Myonemes extend from the gelatinous sheath to each skeletal rod. These marine forms live mostly below depths of 150–200 ft (45–60 m). The order includes *Coleaspis*, *Diploconus*, *Dorotaspis*, and many other genera. *See* ACANTHARIA; ACTINOPODEA; PROTOZOA: SARCODINA: SARCOMASTIGOPHORA. [R.P.H.]

Acari A subclass of Arachnida, the mites and ticks; also called Acarina. All are small (0.004–1.2 in. or 0.1–30 mm, most less than 0.08 in. or 2 mm in length), have lost most traces of external body segmentation, and have the mouthparts borne on a discrete body region, the gnathosoma. They are apparently most closely related to Opiliones and Ricinulei.

The chelicerae may be chelate or needlelike. Pedipalps are generally smaller than the walking legs, and are simple or sometimes weakly chelate. Legs may be modified by projections, enlarged claws, heavy spines, or bladelike rows of long setae for crawling, clinging to hosts, transferring spermatophores, or swimming. Mites frequently have one or two simple eyes (ocelli) anterolaterally on the idiosoma (occasionally one anteromedially as well). The ganglia of the central nervous system are coalesced into a single "brain" lying around the esophagus. A simple dorsal heart is present in a few larger forms.

Mites are ubiquitous, occurring from oceanic trenches below 13,200 ft (4000 m) to over 20,800 ft (6300 m) in the Himalayas and suspended above 3300 ft (1000 m) in the atmosphere; there are species in Antarctica. Soil mites show the least specialized adaptations to habitat and are frequently well-sclerotized predators. Inhabitants of stored grains, cheese, and house dust are also relatively unspecialized. The dominant forms inhabiting mosses are heavily sclerotized beetle mites (Oribatidae). Flowering plant associates include spider mites (Tetranychidae) and gall mites (Eriophyidae). Fungivores are usually weakly sclerotized inhabitants of moist or semiaquatic habitats, while the characteristic fresh-water mites include sluggish crawlers, rapid swimmers, and planktonic drifters. Mites associated with invertebrate animals include internal, external, and social parasites as well as inactive phoretic stages on Insecta, Crustacea, Myriapoda, Chelicerata, Mollusca, and Parazoa. A similar diversity of species are parasites or commensals of vertebrates. Some parasitic mites are disease vectors. Over 30,000 species have been described, and it is estimated that as many as 500,000 may exist. *See* ARACHNIDA. [D.B.]

Accelerated reference frame A noninertial frame of reference. A point fixed in such a frame will be accelerated with respect to inertial frames and so must be held by a force. An observer in an accelerated frame will see accelerations that are not proportional to the impressed forces, since a force is required to keep the particle at rest. Newton's second law does not hold without modification. *See* FRAME OF REFERENCE; INERTIAL REFERENCE FRAME. [B.G.]

Acceleration The time rate of change of velocity. Since velocity is a directed or vector quantity involving both magnitude and direction, a velocity may change by a change of magnitude (speed) or by a change of direction or both. It follows that acceleration is also a directed, or vector, quantity. If the magnitude of the velocity of a body changes from v_1 ft/s to v_2 ft/s in t, then the average acceleration a has a magnitude given by the following equation. To designate it fully the direc-

$$a = \frac{\text{velocity change}}{\text{elapsed time}} = \frac{v_2 - v_1}{t_2 - t_1} = \frac{\Delta v}{\Delta t}$$

tion should be given, as well as the magnitude. Instantaneous acceleration is the limit of the ratio of the velocity change to the elapsed time as the time interval approaches zero. *See* VELOCITY.

Accelerations are commonly expressed in feet per second per second (ft/s^2), meters per second per second (m/s^2), or in any similar units. Whenever a body is acted upon by an unbalanced force, it will undergo acceleration. [R.D.Ru.]

Accelerator mass spectrometry The use of a combination of mass spectrometers and an accelerator to measure the natural abundances of very rare radioactive isotopes. These abundances are frequently lower than parts per trillion. The most important applications of accelerator mass spectrometry are in archeological and geophysical studies, as, for example, in radiocarbon dating by the counting of the rare carbon-14 (radiocarbon; ^{14}C) isotope. *See* DATING METHODS; GEOCHRONOMETRY; MASS SPECTROSCOPE; PARTICLE ACCELERATOR.

The advantage of counting the radioactive atoms themselves rather than their decay products is well illustrated by radiocarbon dating, which requires the measurement of the number of ^{14}C atoms in a sample. The long half-life of 5730 years for ^{14}C implies that only 15 beta-particle emissions per minute are observed from 1 g (0.035 oz) of contemporary carbon. However, an accelerator mass spectrometer can be used to count the ^{14}C atoms at over 15 per second from a milligram sample of carbon. Consequently, accelerator mass spectrometry can be used to date samples that are a thousand times smaller than those that are dated by using the beta-particle counting method, and the procedure is carried out about 60 times faster. *See* RADIOACTIVITY; RADIOCARBON DATING.

For the study of many rare radioactive atoms, accelerator mass spectrometry also has the important advantage that there can be no background except for contamination with the species being studied. For example, significant interference with the beta-particle counting of radiocarbon from cosmic rays and natural radioactivity occurs for carbon samples about 25,000 years old. In contrast, accelerator mass spectrometer measurements are affected only by the natural contamination of the sample which becomes serious for samples about 50,000 years old.

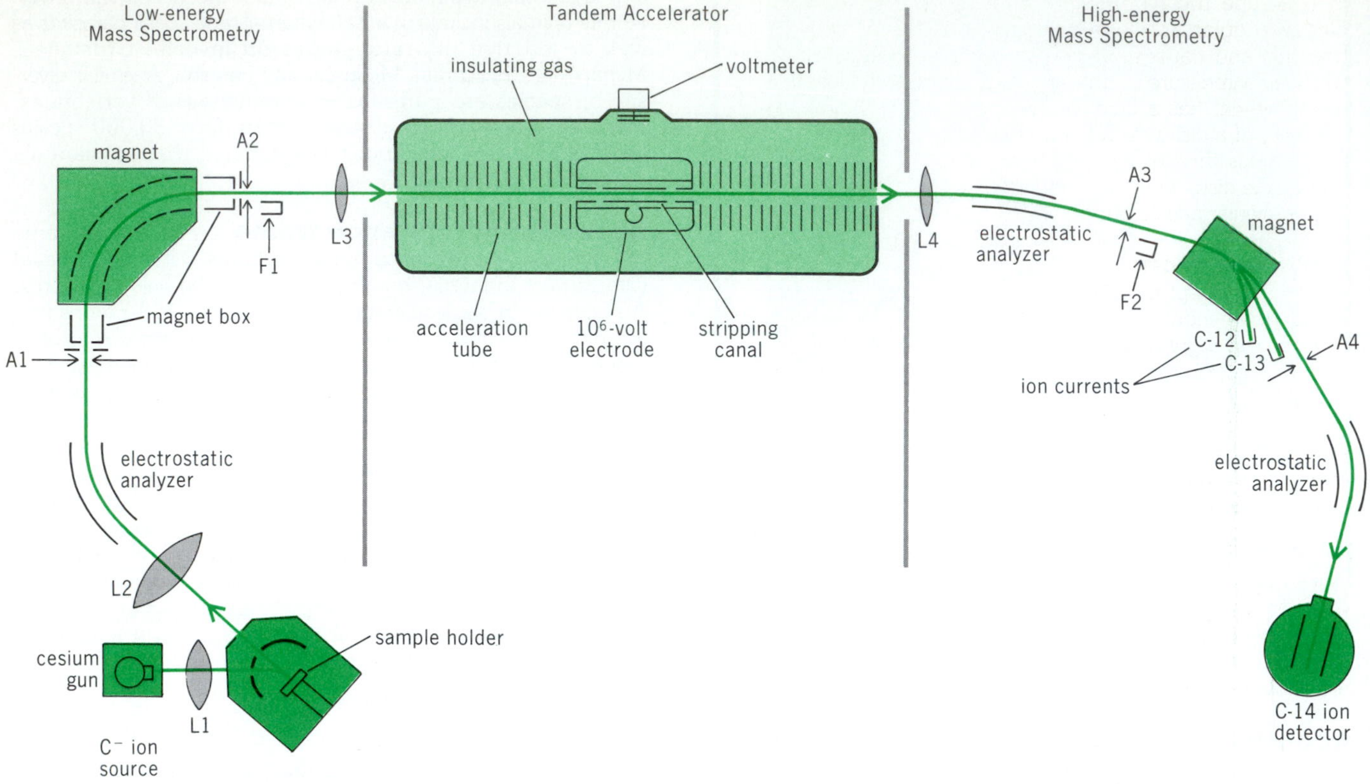

Simplified diagram of an accelerator mass spectrometer used for radiocarbon dating. The equipment is divided into three sections. Electric lenses L1–L4 are used to focus the ion beams; apertures A1–A4 and charge collection cups F1 and F2 are used for setting up the equipment.

Apparatus. The success of accelerator mass spectrometry results from the use of more than one stage of mass spectrometry and at least two stages of ion acceleration. The illustration shows the layout of an ideal accelerator mass spectrometer for radiocarbon studies, divided for convenience into three stages.

The first part of the accelerator mass spectrometer is very similar to a conventional mass spectrometer. In the second stage, a tandem accelerator first accelerates negative ions to the central high-voltage electrode, converts them into positive ions by several successive collisions with gas molecules in a region of higher gas pressure, known as a stripping canal, and then further accelerates the multiply charged positive ions through the same voltage difference back to ground potential. In the third stage, the accelerated ions are analyzed further by the high-energy mass spectrometer.

Distinguishing features. The features that clearly distinguish accelerator mass spectrometry from conventional mass spectrometry are the elimination of molecular ions and isobars from the mass spectrometry.

Elimination of molecular ions. A tandem accelerator provides a convenient way of completely eliminating molecular ions from the mass spectrometry because ions of a few megaelectronvolts can lose several electrons on passing through the region of higher gas pressure in the stripping canal. Molecules with more than two electrons missing have not been observed, so that accelerator mass spectrometry utilizing charge −3 ions is free of molecular interferences.

Elimination of isobars. The use of a negative-ion source, which is necessary for tandem acceleration, can also ensure the complete separation of atoms of nearly identical mass (isobars). In the case of radiocarbon analysis, the abundant stable ^{14}N ions and the very rare radioactive ^{14}C ions are separated completely because the negative ion of nitrogen is unstable whereas the negative ion of carbon is stable. In other cases, it is possible to count the ions without any background in the ion detectors because of their high energy. In many cases, it is also possible to identify the ion. *See* MASS SPECTROMETRY. [A.E.L.]

Accelerometer A mechanical or electromechanical instrument that measures acceleration. The two general types of accelerometers measure either the components of translational acceleration or angular acceleration.

Most translational accelerometers fall into the category of seismic instruments, which means the accelerations are not measured with respect to a reference point. Of the two types of seismic instruments, one measures the attainment of a predefined acceleration level, and the other measures acceleration continuously. In one version of the first type of instrument, a seismic mass is suspended from a bar made of brittle material which fails in tension at a predetermined acceleration level.

Continuously measuring seismic instruments are composed of a damped or an undamped spring-supported seismic mass which is mounted by means of the spring to a housing. The seismic mass is restrained to move along a predefined axis. Also provided is some type of sensing device to measure acceleration.

The type of sensing device used to measure the acceleration determines whether the accelerometer is a mechanical or an electromechanical instrument. One type of mechanical accelerometer consists of a liquid-damped cantilever spring-mass system, a shaft attached to the mass, and a small mirror mounted on the shaft. A light beam reflected by the mirror passes through a slit, and its motion is recorded on moving photographic paper. The type of electromechanical sensing device classifies the accelerometer as variable-resistance, variable-inductance, piezoelectric, piezotransistor, or servo type of instrument or transducer.

There are several different types of angular accelerometers.

In one type the damping fluid serves as the seismic mass. Under angular acceleration the fluid rotates relative to the housing and causes on two symmetrical vanes a pressure which is a measure of the angular acceleration. Another type of instrument has a fluid-damped symmetrical seismic mass in the form of a disk which is so mounted that it rotates about the normal axis through its center of gravity. The angular deflection of the disk, which is restrained by a spring, is proportional to the angular acceleration. [R.C.Du.; T.I.]

Acceptor atom An impurity atom in a semiconductor which can accept or take up one or more electrons from the crystal and become negatively charged. An atom which substitutes for a regular atom of the material but has one less valence electron may be expected to be an acceptor atom. For example, atoms of aluminum, gallium, or indium are acceptors in germanium and silicon, and atoms of antimony and bismuth are acceptors in tellurium crystals. Acceptor atoms tend to increase the number of holes (positive charge carries) in the semiconductor. *See* DONOR ATOM. [H.Y.F.]

Acetal A stable ether usually prepared by acid-catalyzed alcoholation of the carbonyl group of aldehydes. The Y-shaped general formula for these compounds is shown here (R repre-

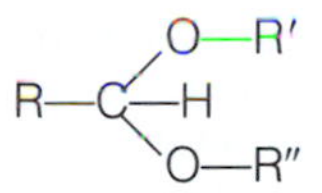

sents an organic radical). Cyclic acetals are formed from diols. Acetals may also be obtained through the Prins reaction of olefins with aldehydes, usually formaldehyde. The acetals are stable toward alkalies, but they hydrolyze in the presence of acids. Their thermal decomposition yields vinyl ethers. Few acetals have any large-scale industrial applications. Their hydrolytic stability to alkali commends the use of some in soaps as fragrances, and their solvency characteristics make them convenient solvents and plasticizers. *See* ALCOHOL; ALDEHYDE; GLYCOL; KETONE. [F.W.]

Acetate One of two types of compounds derived from acetic acid, $HC_2H_3O_2$. One type is obtained by the reaction of acetic acid and bases to give salts containing the negative acetate ion, $C_2H_3O_2^-$. The second type of compound is an ester which is derived from acetic acid and an alcohol, for example, ethyl acetate. *See* ACETYLATION.

Acetate is the official name (Federal Trade Commission) for the textile fiber produced from partially hydrolyzed cellulose acetate and once called acetate rayon.

All the metal acetates are water-soluble except silver acetate. The acetate ion is colorless; it may be identified by heating a sample with sulfuric acid to give the odor of acetic acid. The fruitlike odor of the ester formed when ethyl alcohol and a trace of sulfuric acid are heated with the sample is also characteristic. *See* ACETIC ACID; CARBOXYLIC ACID; ESTER; MANUFACTURED FIBER. [E.E.W.]

Acetic acid A colorless, pungent liquid, CH_3COOH, melting at 16.7°C and boiling at 118.0°C. Acetic acid is the sour principle in vinegar. Concentrated acid is called glacial acetic acid because of its readiness to crystallize at cool temperatures.

Acetic acid is manufactured by three main routes: butane liquid-phase catalytic oxidation in acetic acid solvent, palladium–copper salt–catalyzed oxidation of ethylene in aqueous solution, and methanol carbonylation in the presence of rhodium catalyst. Large quantities of acetic acid are recovered in the manufacture of cellulose acetate and polyvinyl alcohol. Some acetic acid is produced in the oxidation of higher olefins, aromatic hydrocarbons, ketones, and alcohols. *See* OXIDATION PROCESS; WOOD CHEMICALS.

Pure acetic acid is completely miscible with water, ethanol, diethyl ether, and carbon tetrachloride, but is not soluble in carbon disulfide. In a water solution, acetic acid is a typical weakly ionized acid ($Ka = 1.8 \times 10^{-5}$). Acetic acid neutralizes many oxides and hydroxides, and decomposes carbonates to furnish acetate salts, which are used in textile dyeing and finishing, as pigments, and as pesticides; examples are verdigris, white lead, and paris green. *See* VINEGAR. [F.W.]

Acetone A chemical compound, CH_3COCH_3. A colorless liquid with an ethereal odor, it is the first member of the homologous series of aliphatic ketones. Its physical properties include boiling point 56.2°C (133.2°F), melting point −94.8°C (−138.6°F), and specific gravity 0.791.

Acetone is used as a solvent for cellulose ethers, cellulose acetate, cellulose nitrate, and other cellulose esters. Cellulose acetate is spun from acetone solution. Lacquers, based on cellulose esters, are used in solution in mixed solvents including acetone. Acetylene is safely stored in cylinders under pressure by dissolving it in acetone, which is absorbed on inert material such as asbestos. It has a low toxicity. [D.A.S.]

Acetonitrile A nitrogen-containing organic compound having the formula CH_3CN and molecular weight 41.05. It is a flammable liquid and lacrimator with freezing point −48°C (−54°F) and boiling point 81.6°C (179°F). It is miscible with water and with common organic solvents such as alcohols, esters, acetone, alkenes, and chlorinated alkanes, but immiscible with alkanes. Gases such as HCl, SO_2, and H_2S are soluble in acetonitrile.

Because of its high dielectric constant (38.8 at 20°C or 68°F), acetonitrile is useful where a highly polar solvent is required, such as reactions involving ionization; as a solvent in nonaqueous fitrations; or as a solvent for inorganic salts and organic polymers.

Acetonitrile undergoes reactions characteristic of the nitrile group and is used to produce a variety of nitrogen-containing compounds such as amides, amines, isocyanates, and various nitrogen heterocyclic compounds. *See* NITRILE; NITROGEN. [P.E.F.]

Acetyl phosphate The anhydride of acetic and phosphoric acids, with the formula below. The compound occurs in the intermediary metabolism of pyruvic acid and of pentose sugars by some bacteria.

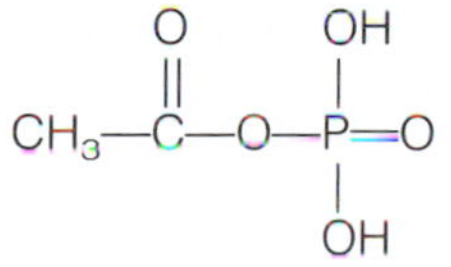

When acetyl phosphate acts as an acetylating agent in the presence of transacetylase, the acetyl group can be transferred reversibly to coenzyme A to yield phosphoric acid and acetyl coenzyme A. In the presence of a kinase, acetyl phosphate can phosphorylate adenosinediphosphate with the formation of acetic acid and of adenosinetriphospate. *See* ADENOSINEDIPHOSPHATE (ADP); ADENOSINETRIPHOSPHATE (ATP); BIOCHEMISTRY; CARBOHYDRATE METABOLISM; METABOLISM. [M.D.]

Acetylation The introduction of an acetyl group

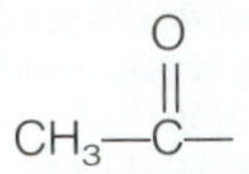

into an organic compound containing the alcoholic or phenolic hydroxyl (—OH) or the amino (—NH_2) and substituted amino groups to yield esters or substituted amides, respectively. Acetylation can be carried out with acetic anhydride, acetyl chloride, or glacial acetic acid (in order of decreasing importance) and with or without an inert solvent, such as benzene or toluene. In general, acetylation reactions are exothermic.

The most important acetates formed by acetylation are cellulose acetate, vinyl acetate, and ethyl acetate. Other important acetylation reactions include production of pharmaceuticals such as aspirin, phenacetin, and derivatives; determination of the hydroxy acid content of fats and oils (acetyl number); preparation of acid anhydrides with acetic anhydride and an organic acid by an exchange reaction; and production of other chemical intermediates. *See* ACID ANHYDRIDE; AMINE. [E.H.H.]

Acetylcholine A chemical compound which plays a key role in the series of reactions that control the permeability of some membranes to ions, a process of paramount importance in nerve excitability and bioelectricity. The action of acetylcholine controls ion permeability in all parts of the excitable membrane, in the axonal, pre- and postsynaptic membranes, and in muscle membranes. The acetylcholine cycle is formed by four proteins: acetylcholine esterase, acetylcholine receptor, acetylcholine storage protein, and choline acetyltransferase. The essential role of the acetylcholine cycle in conduction is well established. Potent inhibitors of acetylcholine esterase and acetylcholine receptor strongly affect and eventually block conduction. All conducting membranes (muscle, myelinated and unmyelinated nerve fibers) are protected by barriers such as myelin, lipid-rich Schwann cells, and connective tissue impervious to quaternary nitrogen derivatives, such as acetylcholine, curare, and neostigmine.

The receptor and acetylcholine esterase proteins are present and functional in nerve terminal membrane. On stimulation, Na^+ ions will enter the axon, and K^+ ions flow into the cleft. Some 20,000,000 to 40,000,000 ions per 1000 molecules of acetylcholine released from acetylcholine storage protein in the membrane move in each direction across the membrane. This high K^+ concentration in the narrow synaptic cleft will decrease the resting-membrane potential to the threshold potential and release acetylcholine from storage protein in the postsynaptic membrane. The action of acetylcholine is limited to the junctions because conducting membranes are protected by barriers. Appearance of acetylcholine at synapses is observed only in the presence of eserine, which prevents the rapid removal by acetylcholine esterase. After degeneration of the nerve terminal, acetylcholine is released by direct stimulation of the muscle. *See* SYNAPTIC TRANSMISSION. [D.N.]

Acetylene A colorless, aromatic gas that burns with a highly luminous flame, sublimes at −83.4°C/760 mmHg (−118°F/260 kilopascals) and is shock-sensitive at low temperatures and partial pressures above 20–30 pounds per square inch (138–207 kPa). Acetylene burns with oxygen to produce the highest achievable flame temperature, over 3300°C (6000°F), of any carbonaceous fuel. Oxyacetylene flames are of major importance in the welding and cutting of metals. *See* TORCH; WELDING AND CUTTING OF METALS.

Acetylene has the formula HC≡CH. The triple bond undergoes addition reactions to produce ethylene and ethane derivatives. The hydrogen atoms of acetylene undergo replacement reactions. Aqueous solutions of copper and silver salts produce explosive metal acetylides. Reactive metals, for example sodium, on contact with acetylene form stable mono- and disodium salts with the generation of hydrogen. Formaldehyde reacts to produce 1,4-butynediol, which can be hydrogenated to the commercially important 1,4-butanediol. Acetylene reacts with carbon monoxide and alcohols in the presence of nickel carbonyl to form acrylate esters.

The major uses for acetylene are in the manufacture of 1,4-butanediol, vinyl chloride, acrylic acid and esters, and vinyl acetate. *See* ALKYNE. [R.K.Ba.]

Acetylide A derivative of acetylene formed by the replacement of one or both (carbides) of the hydrogens by a metal (although carbides such as SiC, B_6C, and WC are not acetylides). Acetylides are prepared by the action of acetylene on active metals or metal compounds, or by the action of metals or metal compounds on carbon at high temperatures (electric furnace). Alkali and alkaline-earth acetylides are relatively stable, but most heavy-metal derivatives are thermodynamically unstable and may explode when dry. They are readily decomposed by dilute acid. Acetylides are salts of a very weak acid and therefore hydrolyze readily. The most important acetylide is calcium carbide, which is used to prepare acetylene. *See* ACETYLENE; CARBIDE. [E.H.H.]

Acid and base Two interrelated classes of chemical compounds, the precise definitions of which have varied considerably with the development of chemistry. Some of these controversies are still unresolved.

Acids initially were defined only by their common properties as substances which had a sour taste, dissolved many metals, and reacted with alkalies (or bases) to form salts. For a time it was believed that a common constituent of all acids was the element oxygen, but gradually it became clear that, if there were an essential element, it was hydrogen, not oxygen. This concept of an acid proved to be satisfactory for about 50 years.

Bases initially were defined as those substances which reacted with acids to form salts (they were the "base" of the salt). The alkalies, soda and potash, were the best-known bases, but it soon became clear that there were other bases, notably ammonia and the amines.

When the concept of ionization of chemical compounds in water solution became established, acids were defined as substances which ionized in aqueous solution to give hydrogen ions, H^+, and bases were substances which reacted to give hydroxide ions, OH^-. These definitions are sometimes known as the Arrhenius-Ostwald theory of acids and bases. Their use makes it possible to discuss acid and base equilibria and also the strengths of individual acids and bases.

A powerful and wide-ranging protonic theory of acids and bases was introduced by J. N. Brönsted in 1923 and was rapidly accepted. Somewhat similar ideas were advanced almost simultaneously by T. M. Lowry and the new theory is occasionally called the Brönsted-Lowry theory. The Brönsted definitions of acids and bases are: An acid is a species which can act as a source of protons; a base is a species which can accept protons. Compared to the water (Arrhenius) theory, this represents only a slight change in the definition of an acid but a considerable extension of the term base. In addition to hydroxide ion, the bases now include a wide variety of uncharged species, such as ammonia and the amines, as well as numerous charged species, such as the anions of weak acids. In fact, every acid can generate a base by loss of a proton. Acids and bases which are related in this way are known as conjugate acid-base pairs, and the table lists examples.

As the table shows, strengths of acids and bases are not independent. A very strong Brönsted acid implies a very weak conjugate base and vice versa. A qualitative ordering of acid

Conjugate acid-base pairs

	Acids	Bases	
Strong acids ↑	H_2SO_4	HSO_4^-	Weak bases ↑
	HCl	Cl^-	
	H_3O^+	H_2O	
	HSO_4^-	SO_4^{2-}	
	$HF_{(aq)}$	F^-	
	CH_3COOH	CH_3COO^-	
	NH_4^+	NH_3	
	NCO_3^-	CO_3^{2-}	
↓ Weak acids	H_2O	OH^-	↓ Strong bases
	C_2H_5OH	$C_2H_5O^-$	

strength or base strength permits a rough prediction of the extent to which an acid-base reaction will go. The rule is that a strong acid and a strong base will react extensively with each other, whereas a weak acid and a weak base will react together only very slightly.

Studies of catalysis have played a large role in the acceptance of a set of quite different definitions of acids and bases, those due to G. N. Lewis: An acid is a substance which can accept an electron pair from a base; a base is a substance which can donate an electron pair. Bases under the Lewis definition are very similar to those defined by Brönsted, but the Lewis definition for acids is very much broader.

Another comprehensive theory was proposed by M. Usanovich in 1939 and is sometimes known as the positive-negative theory. Acids are defined as substances which form salts with bases, give up cations, and add themselves to anions and free electrons. Bases are similarly defined as substances which give up anions or electrons and add themselves to cations. So far, this theory has had little acceptance, quite possibly because the definitions are too broad to be very useful. *See* BASE (CHEMISTRY); BUFFERS (CHEMISTRY); HYDROGEN ION. [F.A.L.; R.H.Bo.]

Acid anhydride One of an important class of reactive organic compounds derived from acids via formal intermolecular dehydration; thus acetic acid, on loss of water, forms acetic anhydride. Anhydrides of straight-chain acids containing from 2 to 12 carbon atoms are liquids with boiling points higher than those of the parent acids. They are relatively insoluble in cold water and are soluble in alcohol, ether, and other common organic solvents. The lower members are pungent, corrosive, and weakly lacrimatory. Anhydrides from acids with more than 12 carbon atoms and cyclic anhydrides from dicarboxylic acids are crystalline solids.

Because the direct intermolecular removal of water from organic acids is not practicable, anhydrides must be prepared by means of indirect processes. A general method involves interaction of an acid salt with an acid chloride.

Large quantities of anhydrides are used in the preparation of esters. Cellulose and acetic anhydride give cellulose acetate, used in acetate rayon and photographic film. The reaction of anhydrides with sodium peroxide forms peroxides (acetyl peroxide is violently explosive), used as catalysts for polymerization reactions and for addition of alkyl halides to alkenes. In Friedel-Crafts reactions, anhydrides react with aromatic compounds, forming ketones such as acetophenone. *See* ACID HALIDE; ESTER; FRIEDEL-CRAFTS REACTION. [P.E.F.]

Acid-base indicator A substance that reveals, through characteristic color changes, the degree of acidity or basicity of solutions. Indicators are weak organic acids or bases which exist in more than one structural form (tautomers) of which at least one form is colored. Intense color is desirable so that very little indicator is needed; the indicator itself will thus not affect the acidity of the solution.

Acid-base indicators are commonly employed to mark the end of an acid-base titration or to measure the existing pH of a solution. Care must be used to compare colors only within the indicator range. A color comparator may also be used, employing standard color filters instead of buffer solutions.

The indicator range is the pH interval of color change of the indicator. In this range there is competition between indicator and added base for the available protons; the color change, for example, yellow to red, is gradual rather than instantaneous. Observers may, therefore, differ in selecting the precise point of change.

The table lists many of the common indicators, their ranges

Common-acid base indicators

Common name	pH range	Color change (acid to base)	pK
Methyl violet	0–2, 5–6	Yellow to blue violet to violet	
Metacresol purple	1.2–2.8, 7.3–9.0	Red to yellow to purple	1.5
Thymol blue	1.2—2.8, 8.0–9.6	Red to yellow to blue	1.7
Tropeoline 00 (Orange IV)	1.4–3.0	Red to yellow	
Bromphenol blue	3.0–4.6	Yellow to blue	4.1
Methyl orange	2.8–4.0	Orange to yellow	3.4
Bromcresol green	3.8–5.4	Yellow to blue	4.9
Methyl red	4.2–6.3	Red to yellow	5.0
Chlorphenol red	5.0–6.8	Yellow to red	6.2
Bromcresol purple	5.2—6.8	Yellow to purple	6.4
Bromthymol blue	6.0–7.6	Yellow to blue	7.3
Phenol red	6.8—8.4	Yellow to red	8.0
Cresol red	2.0–3.0, 7.2–8.8	Orange to amber to red	8.3
Orthocresol-phthalein	8.2–9.8	Colorless to red	
Phenolphthalein	8.4–10.0	Colorless to pink	9.7
Thymolphthalein	10.0–11.0	Colorless to red	9.9
Alizarin yellow GG	10.0–12.0	Yellow to lilac	
Malachite green	11.4–13.0	Green to colorless	

of pH and color change, and pK values. *See* ACID AND BASE; HYDROGEN ION; TITRATION. [A.L.H.]

Acid-fast stain A differential stain used in microbiology as one of the criteria in the identification of the species in the genus *Mycobacterium* and one species in the genus *Nocardia*. The stain technique is a measure of the resistance of stained bacteria to decolorization by acids. The air-dry smear is usually fixed by heating, and stained with a hot solution of carbol fuchsin. This is followed by rinsing with water and exposure to a solution of 3% by volume of concentrated hydrochloric acid in water or in 95% ethanol. The smear is then washed with water and counterstained with methylene blue. Acid-fast substrates appear red, and other substrates blue. Because acid-fast cells are also gram-positive, use of the alcohol hastens decolorization without altering the results. *See* ACTINOMYCETACEAE; GRAM'S STAIN. [G.K.]

Acid halide One of a large group of organic substances possessing the halocarbonyl group,

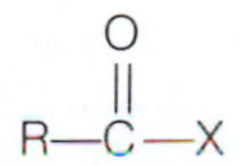

in which X stands for fluorine, chlorine, bromine, or iodine. The terms acyl and aroyl halides refer to aliphatic or aromatic derivatives, respectively.

The great inherent reactivity of acid halides precludes their free existence in nature; all are made by synthetic processes. In general, acid halides have low melting and boiling points and show little tendency toward molecular association. With the exception of the formyl halides (which do not exist), the lower members are pungent, corrosive, lacrimatory liquids that fume in moist air. The higher members are low-melting solids.

Acid chlorides are prepared by replacement of carboxylic hydroxyl of organic acids by treatment with phosphorus trichloride, phosphorus pentachloride, or thionyl chloride.

Although acid bromides may be prepared by these methods, acid iodides are best prepared from the acid chloride treatment with either CaI_2 or HI, and acid fluorides from the acid chloride by interaction with HF or antimony fluoride.

The reactivity of acid halides centers upon the halocarbonyl group, resulting in substitution of the halogen by appropriate structures. Thus with substances containing active hydrogen atoms (for example, water, primary and secondary alcohols, ammonia, and primary and secondary amines), hydrogen chloride is formed together with acids, esters, amides, and *N*-substituted amides, respectively. [P.E.F.]

Acid rain Atmospheric precipitation with a pH below 5.6–5.7; such hydrogen ion concentrations are obtained when pure water is in solution equilibrium with aerial carbon dioxide at 77°F (25°C). Most researchers consider acid rain to be of significance when pH values are at least tenfold more acid than 5.7, that is, 4.7 and below. Alkaline rains are also found, particularly where limestone or alkaline dusts enter the atmosphere. Reports of acid rain are common throughout the Western world and in Japan, with pH values of 3.8–4.5 becoming almost the norm. Snow, too, is now acid, as well as fogs and rime ice formations.

There are several natural sources of acid rain. Among the products of volcanic eruptions are sulfur compounds. The hot springs of Arkansas and the fumaroles in Yellowstone National Park contribute acidity to the air, and considerable amounts of oxides of nitrogen and sulfur are end products of the metabolism of several groups of bacteria. In spite of these natural air pollutants, the pH of glacial ice is close to 5.0, indicating that natural emissions of acid-forming compounds are not major sources of acid rain; anthropogenic sources are the primary contributors.

Burning of fossil fuels for heat and power is the major factor in the generation of oxides of nitrogen and sulfur which are convertible into nitric and sulfuric acids washed down in rain. Wood smoke, too, has significant acid-forming potential. Although gasoline contains only small amounts of nitrogen and sulfur, the modern high-compression engine can produce nitrogen oxides directly from the gaseous nitrogen that makes up about 75% of ordinary air. The combination of air compression in the cylinders, the igniting spark, and the catalytic action of cylinder metal converts N_2 into NO_x, which, in turn, can be converted into HNO_3. Pulp and paper mills use acid bisulfites, sulfurous acid, and sodium sulfide to remove lignin from wood pulp, and the oxidized sulfur compounds are vented into the air. Mining and smelting operations involve the conversion of sulfides and other complexes of heavy metals into pure copper, cadmium, zinc, or iron, with the sulfur oxidized to SO_2. Small but significant quantities of sulfur and nitrogen compounds reach the air from many industries.

All these sources have been active for more years than acid rain has been found. Modern technology is believed to have speeded up the process of acidification of precipitation. Starting about 1950, smokestacks were built to great heights in order to reduce local deposition of acid-forming substances in many steel towns, and communities with coal- and oil-burning power-generating plants. Thus, instead of SO_2 and NO_x or heavy metals dropping within 6 mi (10 km) of the source with only limited airborne time for conversion of oxides into acids, the acid-forming compounds are now carried into the upper atmosphere to be transported for distances known to exceed 600 mi (1000 km). The development of electronic precipitators and filters to remove particulate matter as mandated by various national and local ordinances was also a factor, since fly ash is alkaline and potentially could neutralize acid-forming materials in the stacks. Catalytic converters in automobiles can oxidize sulfur and nitrogen compounds directly into acids.

The effects of acid rain on structures have been amply documented. Exfoliation of marble and limestone monuments and buildings and pitting of granitic stonework are common all over the industrial world. Biological consequences are also evident. Many remote, pristine lakes in northern Europe are virtually sterile, devoid of fish (particularly trout and salmon), and depauperate in invertebrates and most plants. A much more serious potential problem may exist in certain natural terrestrial ecosystems. Research suggests that the interaction of acid rains, fogs, and snows with anthromorphically caused additions of heavy metals (Pb, Cu, Zn) and aluminum can interfere with growth of many important microflora components of forested ecosystems. *See* ECOSYSTEM; pH. [R.M.K.]

Acidolysis A chemical reaction involving the decomposition of a molecule with the addition of the elements of the acid. It is sometimes called acyl exchange. Acidolysis is most commonly an exchange reaction that is catalyzed by concentrated sulfuric acid, zinc chloride, or boron trifluoride. In many cases such reactions are reversible and thus will not go far toward completion unless one of the materials produced is removed as fast at it is formed by distillation or by reaction with a third component. Acidolysis is most commonly applied to the reaction of an organic acid and acetic anhydride to yield an acid anhydride and acetic acid. The acidolysis of esters and acyl halides is a convenient method for the preparation of other esters and acyl halides. It is also a useful reaction to convert neutral esters of dibasic acids to acid esters. The usual method for the preparation of acyl halides from an acid and phosphorus trichloride or thionyl chloride can also be considered to be acidolysis if these reagents are considered the acyl halides of phosphorous acid and sulfurous acid, respectively. *See* ALCOHOLYSIS; HYDROLYSIS. [E.H.H.]

Acinetobacter A genus of nonmotile, strictly aerobic bacteria whose members include gram-negative, nonspore-forming cells which are coccus-shaped and in pairs during the stationary growth phase, and rod-shaped at other times. All members are presently united in a single species, *Acinetobacter calcoaceticus*.

The natural habitat of *Acinetobacter* is soil and water. It has also been found in dairy products, poultry meat, frozen foods, animals, and the hospital environment, and has been isolated from numerous human sources. Sometimes they cause serious or fatal infections, usually in individuals who are predisposed or debilitated, such as patients with burns or malignancies. The most frequent infections are urinary tract infection, wound infection, meningitis, and septic infections with or without endocarditis. *See* INFECTION; MEDICAL BACTERIOLOGY; MENINGITIS; OPPORTUNISTIC INFECTIONS. [G.L.Gi.]

Acipenseriformes An archaic order of actinopterygian fishes, represented by the sturgeons and paddlefishes. The characters include a highly chondrified internal skeleton; fins which are archaic and sharklike in appearance, with more than one ray per pterygiophore; a caudal fin which is strongly heterocercal; scales which are reduced and often form five lengthwise series of bony plates; rostrum produced; and weak jaws (see illustration). Fleshy barbels are present on the lower surface of the snout. The cartilaginous skeleton of the acipenseri-

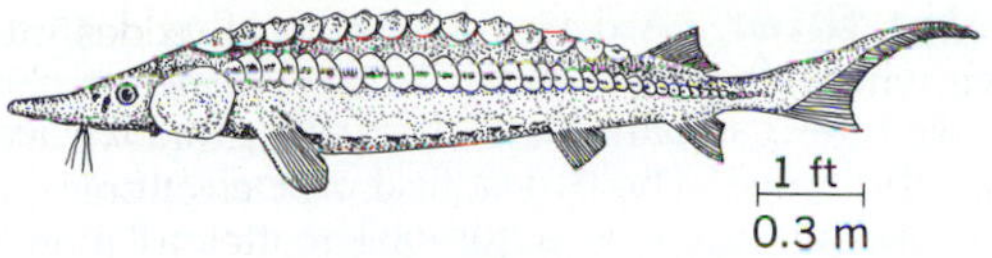

Lake sturgeon (*Acipenser fulvescens*).

form fishes was long regarded as primitive, and was taken as indicative of alliance with the Chondrichthyes. Research reveals, however, that this is secondary, and the true relationship is with the typical bony fishes.

The Acipenseriformes include three families: the extinct Chondrosteidae; the sturgeon family, Acipenseridae, with 4 Recent genera and about 23 species; and the paddlefish family, Polyodontidae, consisting of two Recent species, *Psephurus gladius* of China and *Polyodon spathula* of the eastern United States. *See* ACTINOPTERYGII; OSTEICHTHYES; STURGEON. [R.M.B.]

Ackerman steering Differential gear or linkage that turns the two steered road wheels of a self-propelled vehicle so that all wheels roll on circles with a common center. If a vehicle is to turn without lateral skid of any wheel, the center lines of all wheel axles must intersect, when extended, at every instant in a common center about which the vehicle turns. This requirement is the Ackerman principle of toe-out on turns. It is used universally on wheeled vehicles. For straight, forward motion, the front wheels are substantially parallel, but as the vehicle enters a curve the inner wheel turns more sharply than the outer wheel. The extreme condition occurs when the vehicle is on a curve of its turning radius.

A common configuration that produces Ackerman steering inclines the knuckle arms inwardly and rearwardly (see illustration). The angle of inclination depends on wheelbase and tread

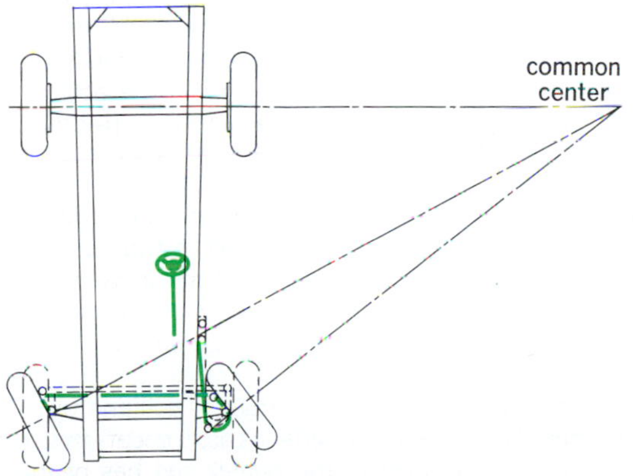

Inwardly inclined steering knuckles cause the wheels to toe-out on the turns.

of the vehicle. Wheelbase is the distance from front to rear wheels, measured between centers of ground contact. Tread is the distance between front or rear wheels, measured from centers of ground contact. *See* FOUR-BAR LINKAGE. [W.K.C.]

Acne A pleomorphic skin disease, commonly referring to acne vulgaris, characterized by blackheads, whiteheads, papules, pustules, and various-sized nodules and scars. The disease involves the sebaceous follicle characterized by large sebaceous glands and ducts and small hairs. *See* SEBACEOUS GLAND; SKIN.

Sebum, a complex lipid liberated by the breakdown of the sebaceous cells, is intimately associated with the development of acne. A perifolliculitis results following rupture of the follicular contents (sebum) into the dermis. This occurs secondary to obstruction of the opening of the sebaceous follicle by a whitehead or blackhead. The inflammatory response eventually heals with scar formation, the ultimate appearance of which is dependent on the depth and exuberance of the inflammatory response.

Other forms of acne are recognized, the most common being acne rosacea, which occurs in older persons. This is seen as reddened inflamed areas of the face, especially the nose and cheeks, and occurs in susceptible individuals following local irritation, faulty digestion, alcoholism, and other predisposing causes. [N.K.M.]

Acoela An order of marine Turbellaria, generally regarded as primitive, and lacking protonephridia, oviducts, yolk glands, permanent digestive cavity, and strictly delimited gonads. The nervous system is epidermal in the most primitive representatives, but it is generally submuscular and consists of three to six

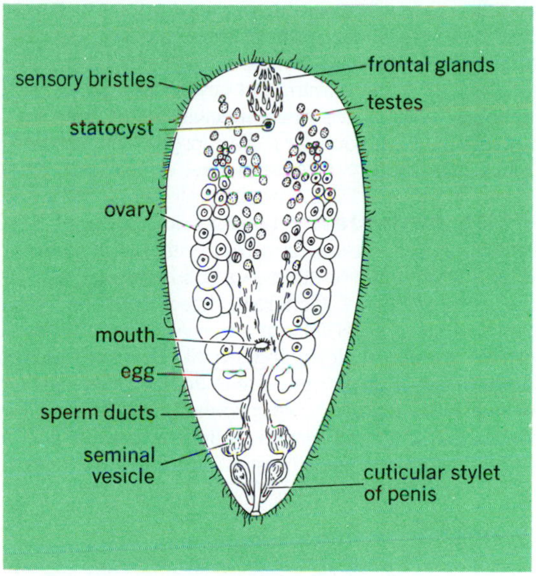

Acoela: *Childia spinosa*.

pairs of longitudinal strands without anterior concentrations recognizable as a brain. The anterior or median ventral mouth opens into a simple pharynx which lacks elaborate musculature and leads directly to the syncytial vacuolated mass of endoderm forming the spongy intestinal tissue and lacking any true lumen.

Eyes are often absent; when present they are of the usual paired turbellarian pigment-cup type, with one to a few pigment and retinal cells. The epidermis, typically, is uniformly ciliated, and locomotion is accomplished by means of rapid, smooth ciliary gliding or swimming.

The acoels (see illustration) are mostly small (one to several millimeters in length). They are virtually worldwide in distribution, living beneath stones, among algae, on the bottom mud, or interstitially, from the littoral zone to deeper waters. A few are pelagic (in tropical and subtropical zones), and others have become symbiotic, living entocommensally in various invertebrates but mainly in echinoderms. *See* COELENTERATA; PLATYHELMINTHES; TURBELLARIA. [J.B.J.]

Aconchulinida An order of the subclass Filosia comprising a small group of naked amebas which form filamentous pseudopodia (filopodia). The genus *Penardia*, possibly the only valid one in the order, includes filopodia-forming amebas with variable shape, an outer zone of clear ectoplasm, and inner

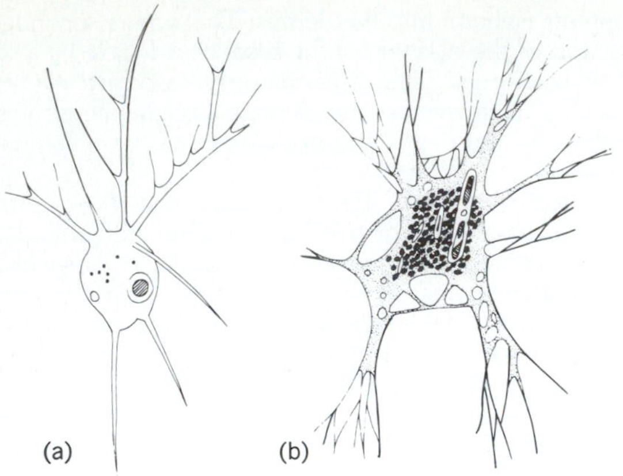

Aconchulinida. (*a*) *Penardia cometa.* (*b*) *P. mutabilis.*

cytoplasm often containing many vacuoles and sometimes zoochlorellae (see illustration). Pseudopodia are sometimes extended primarily at the poles of the organism, but also may fuse laterally to form small sheets of clear cytoplasm. A single nucleus and one water-elimination vesicle are characteristic of these amebas. Size of described species ranges from less than 10 to about 400 micrometers. *See* FILOSIA; PROTOZOA; RHIZOPODEA; SARCODINA; SARCOMASTIGOPHORA. [R.P.H.]

Acoustic emission The phenomenon of transient elastic-wave generation due to a rapid release of strain energy caused by a structural alteration in a solid material; also known as stress wave emission. Generally these structural alterations are a result of either internally or externally applied mechanical or thermal stresses. Depending on the source mechanism, acoustic emission signals may occur with frequencies ranging from several hertz up to tens of megahertz. Often the observed acoustic emission signals are classified as one of two types: burst and continuous (see illustration). The importance of

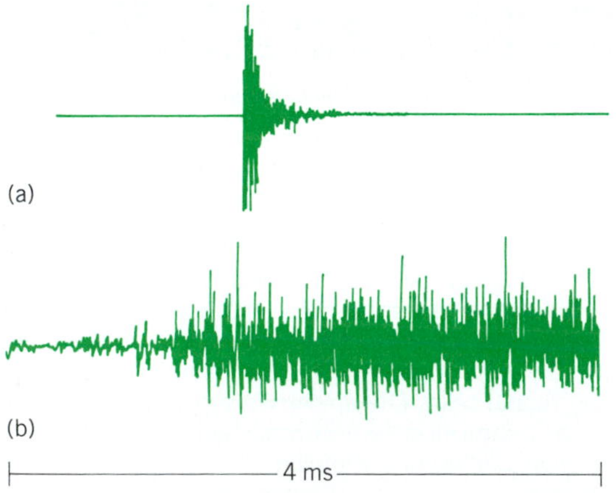

Typical acoustic emission waveforms as detected with piezoelectric transducers: (*a*) burst type; (*b*) continuous type.

acoustic emission monitoring lies in the fact that proper detection and analysis of acoustic emission signals can permit remote identification of source mechanisms and the associated structural alteration of solid materials. This information in turn can augment understanding of material behavior, can be used as a quality-control method during materials processing and fabrication, and as a nondestructive evaluation technique for assessing the structural integrity of materials under service conditions. *See* NONDESTRUCTIVE TESTING; SOUND. [R.E.G.]

Acoustic filter A device for transmitting a given series of waves whose frequencies lie in a preassigned frequency range, called the passband, and attenuating frequencies which lie outside this range. The largest field of application of such filters is in mufflers used on automobiles, exhaust engines, air ducts, and so forth. Since in a muffler a steady or slowly varying stream of gas must be permitted to pass while the more rapid vibrations constituting the undesired noise must be suppressed, a low-pass acoustic filter is required in this case. Such filters in the low-frequency range can be designed in direct analogy to the design of electrical filters. The illustration shows

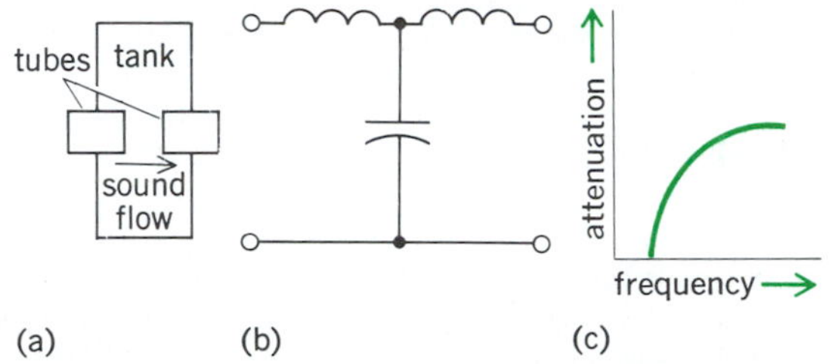

Simple low-pass filter. (*a*) Diagram of filter. (*b*) Electrical analog. (c) Theoretical attenuation.

the design of a simple low-pass filter consisting of two short tubes placed in the ends of a closed tank, with the equivalent electrical analogs and the type of attenuation. Such so-called lumped impedance methods are applicable when the length of the elements is less than an eighth of a wavelength. *See* ACOUSTIC NOISE; ELECTRIC FILTER; MUFFLER. [W.P.M.]

Acoustic impedance At a given surface, the complex ratio of effective sound pressure averaged over the surface to the effective flux (volume velocity or particle velocity multiplied by the surface area) through it. The unit is the $N{\cdot}s/m^5$ (newton-second/meter^5), or the mks acoustic ohm. In the cgs system the unit is the $dyn{\cdot}s/cm^5$ (dyne-second/centimeter^5). *See* SOUND PRESSURE.

Specific acoustic impedance is the complex ratio of the effective sound pressure at a point to the effective particle velocity at a point. The unit is the $N{\cdot}s/m^3$, or the mks rayl. In the cgs system the unit is the $dyn{\cdot}s/cm^3$, or the rayl. The difference between specific acoustic impedance and acoustic impedance is in the specification of impedance at a point, as compared to the average over a surface.

Characteristic acoustic impedance is the ratio of effective sound pressure at a point to the particle velocity at that point in a free, progressive wave. This ratio is equal to the product of the density of the medium times the speed of sound in the medium. The characteristic impedance of a sound wave is analogous to the characteristic electrical impedance of an infinitely long, dissipationless transmission line. It is common in acoustical analyses to represent specific acoustic impedances in terms of their ratio to the characteristic impedance of air.

Acoustic impedance, being a complex quantity, can have real and imaginary components analogous to those in an electrical impedance. In applying this analogy, the real part of the acoustic impedance is termed acoustic resistance, and the imaginary part is termed acoustic reactance. *See* ELECTRICAL IMPEDANCE. [W.J.G.]

Acoustic interferometer A device for measuring the velocity and attenuation of sound waves in a gas or liquid.

It operates by sending an ultrasonic beam of sound waves through the medium to be measured in order to obtain a standing-wave system between the driving crystal and a reflector whose distance from the crystal can be varied by a screw system. An oscillator is attached to the crystal and tuned to the crystal's resonant frequency. The variation in the amplitude of current from the oscillator is measured as the distance between the crystal and the reflector is varied. At half-wavelength intervals, sharp dips occur in the received current, indicating that a high mechanical impedance is impressed on the end of the vibrating crystal by the standing-wave system between the crystal surface and the reflecting plate. By counting the number of half wavelengths occurring in a given displacement of the reflecting plate and by knowing accurately the frequency of vibration, it is possible to determine the velocity of propagation. *See* ULTRASONICS. [W.P.M.]

Acoustic levitation The use of intense acoustic waves to suspend a body which is immersed in a fluid medium without obvious mechanical support.

Levitation can occur in the presence of fluid flow, including the back-and-forth fluid flow produced by the passage of an acoustic wave. Such acoustically generated forces are extremely small in common experience. But intense acoustic waves are nonlinear in their basic character and, therefore, may exert a net acoustic radiation pressure on an object sufficient to balance the gravitational force and thus levitate it. *See* ACOUSTIC RADIATION PRESSURE.

Applications of the levitation of objects in liquids have included measurement of the ultimate tensile strength of liquids, mechanical characterization of superheated and supercooled liquids, the measurement of properties of biological materials, and the study of shape oscillations and interfacial tension of levitated drops. *See* SOUND. [R.E.A.]

Acoustic microscope An instrument for monitoring elastic properties of microscopic scale. It has been used to study structural detail in integrated circuits, the grains and grain boundaries of complex materials, and intercellular detail in biology. It encompasses the domain that is now dominated by the optical microscope and provides a new dimension in microscopy. Acoustic radiation can be used to probe regions inaccessible to optical waves. *See* OPTICAL MICROSCOPE; SCANNING ACOUSTIC MICROSCOPE; SONOSCAN. [C.F.Q.]

Acoustic mine A naval mine which is actuated by acoustic means. Such a mine generally rests on the bottom in fairly shallow water. It is set off by the noise emitted by its intended target as the target passes near by. Since the mine lies on the bottom, it can contain much more explosive charge than a buoyed mine (perhaps 500–1500 lb or 200–700 kg). Very often, the acoustic triggering mechanism is used in conjunction with another triggering device, such as a magnetic sensor. This makes defensive minesweeping more difficult, since the mine will not fire unless both an acoustic and a magnetic signal are received simultaneously. Minesweepers employ high-power underwater acoustic generators in attempting to fire such mines at safe distances. *See* ACOUSTIC TORPEDO. [R.W.Mo.]

Acoustic noise Unwanted sound. Noise control is the process of obtaining an acceptable noise environment for people in different situations. Understanding noise and its control requires a knowledge of the major sources of noise, sound propagation, human response to noise, and the physics of methods of controlling noise. The continuing increase in noise levels from many different human activities in industrialized societies has led to the term noise pollution.

Noise as an unwanted by-product of an industrialized society affects not only the operators of machines and vehicles, but also other occupants of buildings in which machines are installed, passengers of vehicles, and most importantly the communities in which machines, factories, and vehicles are operated. [M.J.Cr.]

Acoustic radiation pressure The net pressure exerted on a surface or interface by an acoustic wave. One might presume that the back-and-forth oscillation of fluid caused by the passage of an acoustic wave will not exert any net force on an object, and this is true for sound waves normally encountered. Intense sound waves, however, can exert net forces in one direction of sufficient magnitude (proportional to the sound intensity) to balance gravitational forces and thus levitate an object in air.

Forces due to acoustic radiation pressure have been used to calibrate acoustic transmitters, to deform and break up liquids, to collect like objects, and to position objects in a sound field, sometimes levitating the sample so that independent studies of the object's properties can be performed. *See* ACOUSTIC LEVITATION; SOUND. [R.E.A.]

Acoustic radiometer A device for measuring the intensity of sound waves in a gas or liquid. The oldest and simplest piece of apparatus for measuring sound intensity is the Rayleigh disk. This consists of a thin disk of radius r set at a 45° angle to the direction of the sound beam. Such a disk experiences a torque τ with magnitude given by the equation below,

$$\pi = {}^4/_3\rho r^3\xi^2$$

where ρ is the density of the medium and ξ the particle velocity. This formula holds for wavelengths that are large compared with the diameter of the disk.

Other radiometers use the pressure of sound waves to deflect a spherical body. Since the radiation pressure is $p = 2E$, where E is the energy of the acoustic wave per unit area, a direct measurement of this pressure will determine the energy of a plane wave. *See* ACOUSTIC RADIATION PRESSURE; SOUND PRESSURE. [W.P.M.]

Acoustic resonator A device consisting of a combination of elements having mass and compliance whose acoustical reactances cancel at a given frequency. Resonators are often used as a means of eliminating an undesirable frequency component in an acoustical system. In other instances resonators are used to produce an increase in the sound pressure in an acoustic field at a particular frequency.

Resonators are useful most often in the control of low-frequency sound. They are of particular value in reducing the noise from sources having constant frequency excitation.

Resonators have also found considerable application in architectural acoustics. It is often difficult to obtain adequate control of reverberation time at low frequencies in a large studio or auditorium using conventional acoustical materials. A number of designs for these spaces have included the construction of resonators behind walls or in the ceiling to obtain increased

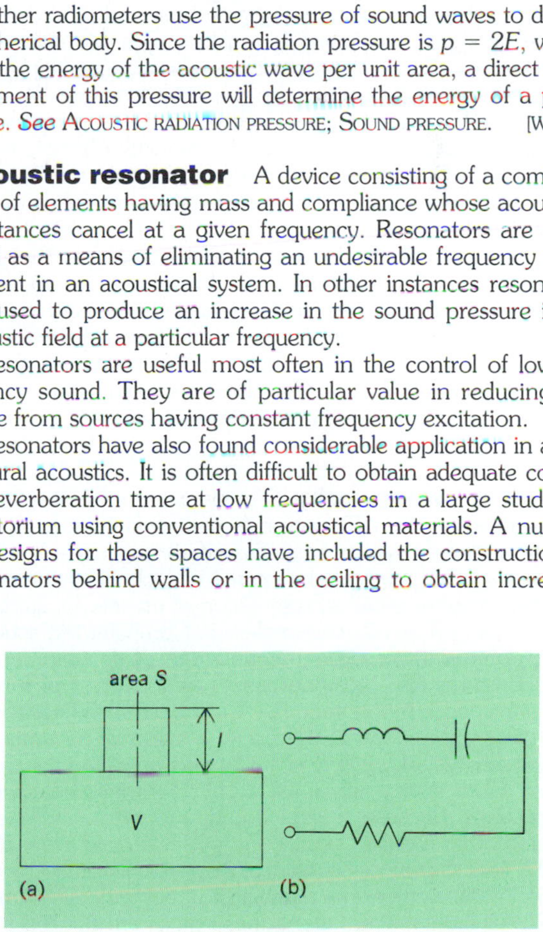

Helmholtz resonator. (*a*) Acoustical unit. (*b*) Electrical analog.

low-frequency absorption and thus provide more satisfactory reverberation characteristics. *See* Architectural acoustics.

The Helmholtz resonator (see illustration) is the simplest and most often utilized acoustical resonator. The unit consists of a straight tube of length l and cross-sectional area S, connected to a closed volume V. This combination is directly analogous to the simple series LC electrical circuit. *See* Acoustic filter; Acoustic impedance; Muffler. [W.J.G.]

Acoustic signal processing A discipline that deals generally with the extraction of information from signals in the presence of acoustic noise. In the undersea, atmospheric, and solid earth environments both signals and noise are propagated as scalar pressure fields which are supported within the medium at audio frequencies ranging from a few hertz to several kilohertz. Because the transport mechanisms vary markedly and in a random way as a function of geometry and of time, the pressure fields are, at best, statistically described. The purpose of signal processing is to implement computational algorithms to produce statistics which can be used to make minimum risk decisions about the measured signal and noise parameters.

The detection, estimation, classification, and extraction of acoustic signals in a random, not necessarily homogeneous, medium are problems of importance in sonar and seismic technologies. In the acoustic environment, the signals, which contain the desired information as well as the noise (or undesired interference), have both spatial and temporal characteristics that are not, in general, appropriately described by simple ideal models. Hence, the treatment of the detection problem must include, explicitly, the physics of the medium, the geometry of the system, and the manner in which coupling to the medium is achieved. *See* Atmospheric acoustics; Seismology; Sonar.

Acoustic signal processing is usually concerned with three basic systems. The active detection system is made up of a transmitter which ensonifies the medium, a target (if one is present), and a receiver situated either at the transmitter (the monostatic case) or at a different location (the bistatic case). The receiver must operate in the presence of reverberation noise (caused by the scattering of transmitted energy) as well as the usual ambient and system noise.

The passive detection system operates usually in the presence of ambient and system noise. In the passive operation (that is, surveillance) the system must detect and classify sources which radiate acoustic energy into the medium.

In a communication system the receiver must first acquire the transmitted information (detection) and then determine the waveform structure of the signal (extraction-estimation). [D.M.V.; W.A.V.W.]

Acoustic tomography An imaging or remote-sensing technique in which information is collected from beams of acoustic radiation which have passed through an object. In general, the word tomography refers to a technique which produces an image or other representation of the interior structure of an object in a two-dimensional slice. Computerized acoustic tomography is in an experimental stage of development in medicine and is also being developed as a method of remote sensing for ocean monitoring. *See* Computerized tomography.

A tomographic array in the ocean consists of a number of acoustic sources and receivers which span an ocean volume of interest. Each source transmits a pulse which is detected at each receiver. Because of refraction due to temperature variations and other (smaller) effects, the pulses do not travel along straight lines but along curved paths, which usually reverse direction in the vertical at a number of upper and lower turning points. Pulse travel times along these paths are the measured data. The information contained in these data is related mainly to the temperature distribution in the ocean, and the monitoring of this distribution has been the primary objective in the development of this technology. [T.J.E.; R.N.]

Ultrasonic pulse echo techniques are also used to obtain clinically useful information about the structure and functioning of tissues and organs. The information is obtained by a mapping technique in which acoustic pulses are emitted from an acoustoelectric transducer, and echoes are received from acoustic impedance discontinuities along the assumed line-of-sight axial propagation path. A number of different modes of operation have emerged, each having areas of usefulness.

The A-mode technique uses acoustic pulse emissions and echo reception along a single line-of-sight axial propagation path. The information is most often displayed on a cathode-ray oscilloscope in which the horizontal axis of the display is a linear time base triggered at the time of the transmitted pulse, and the received echoes are manifested as vertical deflections, with vertical displacement a measure of the amplitude or strength of the returning echo. *See* Oscilloscope.

The M (motion) mode of operation is used to display the movement of time-varying, echo-producing structures by intensity-modulating the trace as it is swept slowly across the oscilloscope screen in a direction at right angles to the fast time-base sweep. This mode of operation is used extensively in diagnosing disorders of the heart. *See* Echocardiography.

For a two-dimensional picture to be obtained, the line-of-sight propagation path must be scanned and the position and direction of the path monitored and used to form a two-dimensional picture. Typically, the B-mode display is formed by moving the transducer so that the line-of-sight path remains in a single plane. The time-base trace of the cathode-ray oscilloscope screen is moved to correspond, in position and direction, to the ultrasonic line-of-sight propagation path, and echoes are displayed as intensity modulations of the trace. The echogram which results is therefore a two-dimensional representation of distribution of acoustic impedance discontinuities in the interrogated object.

C (constant) depth mode operation provides a two-dimensional image display at constant time delay, and presumed constant distance, from the ultrasonic transducer. The scanning is arranged so that the point at constant depth along the propagation path (beam axis) traverses a plane. *See* Ultrasonics. [F.Du.]

Acoustic torpedo A naval torpedo which locates its target by acoustic means. It is often called a homing torpedo. Except for the special method of finding its target, such a torpedo is similar to other types in size, warhead, and method of propulsion. Initially, it is aimed toward the general vicinity of the desired target; when sufficiently close for the acoustic detection system to acquire a fix on the target (perhaps a few hundred meters), the torpedo is steered the remaining distance by servomechanisms. The acoustic detector, which supplies target information for the guidance system, may be of two general types: passive, whereby the torpedo "homes" on sounds emitted by the target vessel, such as propeller noises; or active, in which the nose of the torpedo contains a small sonar device which reflects pulses of sound off the target. *See* Acoustic mine; Sonar; Underwater sound. [R.W.Mo.]

Acoustical holography The recording of sound waves in a two-dimensional pattern (the hologram) and the use of the record to reconstruct the entire sound field throughout a three-dimensional region of space. Portions of the region which show large amplitudes provide a three-dimensional image of the sources of the sound. The technique is one of the most powerful within the general area of acoustical imaging, owing to the enormous increase in the amount of observable information when the two-dimensional hologram is converted into the three-dimensional reconstruction. For example, a digitized acoustical hologram may contain a few thousand data points, but the three-dimensional reconstruction may contain many millions of data points.

Important applications of acoustical holography may be found wherever there is a need to visualize objects or fields that cannot be observed optically, such as objects in an opaque fluid or sound waves in air. Applications include medical imaging, nondestructive evaluation, underwater imaging and surveying, seismic surveying, and noise measurement and control. *See* ACOUSTIC NOISE; GEOPHYSICAL EXPLORATION; MEDICAL IMAGING; NONDESTRUCTIVE TESTING; UNDERWATER SOUND.

Acoustical holography is free from some technical limitations that constrain optical holography. In conventional (optical) holography the extremely rapid oscillations of the light waves can be recorded only through the use of a reference beam of light and the phenomenon of optical interference, and the reconstruction of the three-dimensional images is accomplished through the phenomenon of optical diffraction. For an acoustic wave, the oscillations are much less rapid, and the recording may be accomplished directly with microphones; furthermore, the data can be digitized and then processed with a computer, so there are no intrinsic limitations in frequency or resolution. That a reference beam is necessary to reconstruct a wave field in the acoustic case is a misconception carried over from conventional optical holography. *See* HOLOGRAPHY.

In effect, the recording medium consists of a large array of microphones. The sound source to be studied is positioned just below the microphone array. The microphone signals are routed through a high-speed multiplexer and digitizer to a computer memory, which constitutes the recorded hologram in this case. This hologram is then processed by the computer, and the results are displayed by means of graphic routines, which produce three-dimensional, so-called hidden-line displays, energy flow maps, and so forth. [J.D.Ma.]

Acoustical image If a point source of sound is placed on one side of an extended reflecting surface, it may be considered to have an "image" at an equivalent distance on the other side of the surface along a perpendicular projection, analogous to the familiar optical image. The effects of reflecting surfaces on acoustic waves frequently can be predicted by the use of such images. A source in front of a very reflective wall will have an image of "strength" almost equal to that of the source, vibrating in phase with the source. To obtain the total effect at any point due to the combined action of a source and such a reflecting surface, the effects due to the source itself and another source of equal strength placed at the image point are added. If there is a second reflecting surface present, this so-called first-order image will have an image, called the second-order image of the source, at an equivalent distance on the other side of the second reflecting surface. Similarly, an image of the second-order image is said to be a third-order image, and so on. Although this method is precise only when the reflecting surface is nonabsorptive, it is often of considerable help in investigations of the action of sound waves in rooms. *See* ARCHITECTURAL ACOUSTICS; OPTICAL IMAGE. [C.M.H.]

Acoustics The science of sound, which in its most general form endeavors to describe and interpret the phenomena associated with motional disturbances from equilibrium of elastic media. An elastic medium is one such that if any part of it is displaced from its original position with respect to the rest, as for example by an impact, it will return to its original state when the disturbing influence is removed. Acoustics was originally limited to the human experience produced by the stimulation of the human ear by sound incident from the surrounding air. Modern acoustics, however, deals with all sorts of sounds which have no relation to the human ear, for example, seismological disturbances and ultrasonics.

Basic acoustics may be divided into three branches, namely, production, transmission, and detection of sound. *See* ACOUSTIC NOISE; ARCHITECTURAL ACOUSTICS; ATMOSPHERIC ACOUSTICS; ELECTROACOUSTICS; HYPERSONICS; LOUDSPEAKER; MECHANICAL VIBRATION; MICROPHONE; PHYSIOLOGICAL ACOUSTICS; PSYCHOACOUSTICS; SOUND; SOUND RECORDING; SOUND-REINFORCEMENT SYSTEM; SOUND-REPRODUCING SYSTEMS; ULTRASONICS; UNDERWATER SOUND; VIBRATION; VIBRATION DAMPING; VIBRATION ISOLATION. [R.B.L.]

Acoustooptics The field of science and technology that is concerned with the interactions between acoustic (sound) waves and optical (light) waves, the resulting devices, and their applications. The strain or stress field associated with the acoustic waves propagating in a suitable medium induces a moving optical refractive index grating via the photoelastic effect. For a given medium, the periodicity and the strength of the moving optical grating are, respectively, identical to the wavelength and proportional to the amplitude of the acoustic waves. Under the so-called Bragg condition, discussed below, a significant portion of a light beam incident upon the grating can be diffracted or deflected in a manner similar to that of x-ray diffraction from a crystal lattice. This relatively simple phenomenon has been utilized to study physical properties of materials and, more recently, to construct various types of optical devices of scientific and technological importance. *See* DIFFRACTION GRATING; PHOTOELASTICITY; REFRACTION OF WAVES; WAVE MOTION; X-RAY DIFFRACTION.

Numerous studies were carried out using incoherent light sources and low-frequency acoustic waves, following the prediction of the acoustooptic phenomenon by L. Brillouin in 1922. The simultaneous availability of coherent light sources through the advent of lasers and of very high-frequency acoustic waves through piezoelectric transducer technology stimulated a revival of interest in the subject in the 1960s. The studies then were still limited to bulk-wave acoustooptics, in which both light and sound propagate as unguided (unconfined) columns of waves. Since the early 1970s, numerous studies have focused on guided-wave acoustooptics, in which light propagates as guided optic waves and sound propagates as surface acoustic waves in suitable solid substrates. *See* HYPERSONICS; INTEGRATED OPTICS; LASER; PIEZOELECTRICITY; SURFACE-ACOUSTIC-WAVE DEVICES.

The bulk-wave acoustooptic modulator has three desirable characteristics on which numerous applications are based.

(1) The frequency of the diffracted light is either upshifted or downshifted from that of the incident light by the acoustic frequency. These frequency shifts can be explained in terms of a Doppler shift. This particular characteristic has been utilized to perform frequency modulation and frequency shifting of the lasers.

(2) The angle of deflection with the diffracted light varies linearly with the acoustic frequency. Thus, by electronically tuning the radio-frequency driving signal, the acoustooptic Bragg modulator can be used to perform high-speed light-beam deflection, scanning, and switching.

(3) The intensity of the diffracted light is proportional to the power of the radio-frequency driving signal, provided that the device is not driven into saturation. This characteristic makes the acoustooptic Bragg cell one of the most attractive intensity modulators for lasers. A variety of real-time signal-processing applications such as spectral analysis, correlation, and convolution of wide-band radio-frequency signals have also been realized by using simultaneously the second and the third characteristics of the cell.

Among the variety of waveguide substrates that have been explored for guided-wave acoustooptics, lithium niobate ($LiNbO_3$), an insulator, and gallium arsenide (GaAs), a semiconductor, have so far been found to possess the highest technical potential. Lithium niobate-based Bragg modulators that operate at gigahertz center frequency and octave bandwidth are widely used in integrated optic modules for real-time processing of wide-band radar signals. Also, continuing progress in the fabrication and performance of passive and other active components has significantly advanced the technology for realization of hybrid multichannel integrated optic modules and circuits used in communications, computing, and signal processing. *See* LIGHT; SOUND. [C.S.T.]

Acquired immune deficiency syndrome (AIDS) A profound and eventually fatal deficiency of the immune system caused by a virus. AIDS virus is transmitted by intimate sexual contact, contaminated needles, or, less commonly, by percutaneous inoculation of infectious blood or blood products. The etiologic agent, human immunodeficiency virus, type 1 (HIV-1), formerly called human T-cell leukemia (or lymphotropic) virus type III (HTLV-III) and lymphadenopathy virus, type 1 (LAV-1), is presumed to have a proclivity for attacking and damaging human lymphocytes, culminating in a profound depletion of the helper-inducer T-lymphocyte population. Lacking these critical helper T lymphocytes, the immune system is crippled. In this immunocompromised state, the host becomes susceptible to developing multiple, severe, and recurrent opportunistic parasitic, viral, fungal, and mycobacterial infections, as well as malignancies. *See* CELLULAR IMMUNOLOGY.

The initial manifestations of AIDS can be either an opportunistic infection or Kaposi's sarcoma. Many experience various nonspecific symptoms such as malaise, fevers, anorexia, and weight loss for weeks, months, or years prior to documentation of the first opportunistic infection. Initially individuals may develop shingles and thrush. Extensions of thrush can lead to esophageal erosions, difficulty in swallowing, and a burning sensation behind the sternum. Both primary and recurrent *Herpes simplex* virus infections can also appear. Other conditions derived from opportunistic infections that might occur include a rare mycobacterial infection, pneumocystis pneumonia, neurologic dysfunction, chorioretinitis, and diarrhea.

Kaposi's sarcoma is a vascular tumor of endothelial cell origin and appears as firm red or violet nodules involving the skin and mucous membranes. Since these lesions are often multiple, the body may eventually be covered by these tumors. Some patients develop life-threatening visceral involvement due to massive tumor infiltration of the lung or gastrointestinal tract; bleeding is not uncommon.

Since there is no effective therapy for the underlying acquired immune deficiency, most AIDS patients die from overwhelming infections within 2 to 3 years of the initial appearance of symptoms. *See* ONCOLOGY; OPPORTUNISTIC INFECTIONS. [A.M.Ma.]

Acquired immunological tolerance Failure of immunological responsiveness, brought about by exposure to antigen, in which antigen-sensitive cells become unable to initiate synthesis of a restricted range of antibodies. Through induced tolerance, animals can be prevented from rejecting grafts of foreign cells; rendered more sensitive to infection with pathogens; protected from experimentally induced autoimmune disease; and prevented from developing hypersensitivity to drugs. Tolerance derives its principal practical interest from possible future applications in surgery. It is hoped that permanent survival of transplanted organs will be obtained by inducing tolerance of donor antigen in the host, as can already be done in animal experiments. Other possible applications include treatment of allergy and autoimmune disease. *See* ANTIBODY; ANTIGEN; HYPERSENSITIVITY; IMMUNITY.

The terms immunological tolerance and immunological paralysis can be used interchangeably, although the former is applied mainly to tissue transplants and the latter only to purified antigens. Unresponsiveness to both types of antigen can be acquired in the same way, and probably involves the same cellular mechanism. *See* AUTOIMMUNITY; IMMUNOLOGY; TRANSPLANTATION BIOLOGY. [N.A.M.]

Acrasiomycetes A class of microorganisms considered by some mycologists and many protozoologists to be related to protistans rather than to fungi because of their phagotrophism, mode of sporulation, biochemistry, and probable independent evolution from simpler euprotistans. There is also evidence that the cellular slime molds are polyphyletic in origin. The subclasses are Protostelidae, Acrasidae, and Dictyostelidae. *See* MYXOMYCOTA. [L.S.O.]

Acridine One of a group of organic heterocyclic compounds (also called 9-azaanthracenes) containing benzene rings fused to the 2,3 and 5,6 positions of pyridine. Acridine, with the structure shown, is a typical member of the group. Several

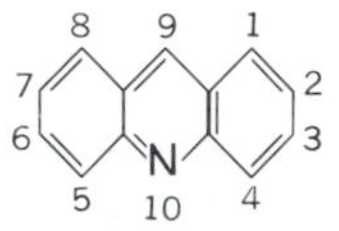

important dyes and medicinals are acridine derivatives.

Acridine is a faintly yellow solid, mp 230–231°F (110–110.5°C), that shows marked fluorescence. The compound, as a tertiary amine, is a weak base (p*Ka* 5.60 at 68°F or 20°C) forming simple and quaternary salts as well as N-oxides. Acridine shows aromatic character. It is stable to heat, alkali, and acid, and it undergoes substitution reactions, such as bromination, nitration, and sulfonation. *See* DYE; HETEROCYCLIC COMPOUNDS. [W.J.Ge.]

Acrothoracica An order of the subclass Cirripedia. All members burrow into shells of mollusks and thoracican barnacles, echinoderm tests, polyzoans, dead coral, and limestone. The normal three pairs of cirriped mouthparts are present: mandibles, maxillules, and maxillae. Four to six pairs of cirri occur, often greatly reduced in size, the first pair close to the oral appendages and the remainder packed together at the posterior end of the body. The sexes are separate; dwarf males are found on the mantle or wall of the burrow of the female. Naupliar larval stages may be omitted, but a cypris larva always occurs in the life cycle. Three families, about eight genera, and 40 species are recognized. *See* CIRRIPEDIA. [H.G.St.]

Acrotretida An order of inarticulate brachiopods whose representatives are known in beds of all ages from Early Cambrian to the present time. Within the order two suborders are recognized, the Acrotretidina and the Craniidina. Acrotretidina includes only species with shells of calcium phosphate. Cambrian and Early Ordovician representatives were typically small, commonly less than 0.08 in. (2 mm) in length, and biconvex or with a subconical pedicle valve. Craniidina probably developed from Acrotretidina in Early Ordovician times. Craniidines are not known to possess a pedicle; all forms are attached by cementing of part or all of the pedicle valve, a process seemingly effected by the periostracum. *See* BRACHIOPODA; INARTICULATA. [A.J.R.]

Acrylonitrile An explosive, poisonous, flammable liquid, boiling at 171°F (77.3°C), partly soluble in water. It may be regarded as vinyl cyanide, and its systematic name is 2-propenonitrile. Acrylonitrile is prepared by ammoxidation of propylene over various sorts of catalysts, chiefly metallic oxides.

Most of the acrylonitrile produced is consumed in the manufacture of acrylic and modacrylic fibers. Substantial quantities are used in acrylonitrile-butadiene-styrene (ABS) resins, in nitrile elastomers, and in the synthesis of adiponitrile by electrodimerization. Smaller amounts of acrylonitrile are used in cyanoethylation reactions, in the synthesis of drugs, dyestuffs, and pesticides, and as co-monomers with vinyl acetate, vinylpyridine, and similar monomers.

Acrylonitrile undergoes spontaneous polymerization, often with explosive force. It polymerizes violently in the presence of suitable alkaline substances. *See* CYANOETHYLATION; NITRILE; POLYMERIZATION. [F.W.]

Actiniaria An order of the Zoantharia known as the sea anemones, which are the most widely distributed of the anthozoans. Usually they are solitary animals which live under the low-tide mark attached to some solid object by a basal expansion or pedal disk. They feed on various prey such as copepods, mollusks, annelids, crustaceans, and fish. The burrowing species lack a pedal disk and bury their elongated bodies in the soft sediment of the oceans.

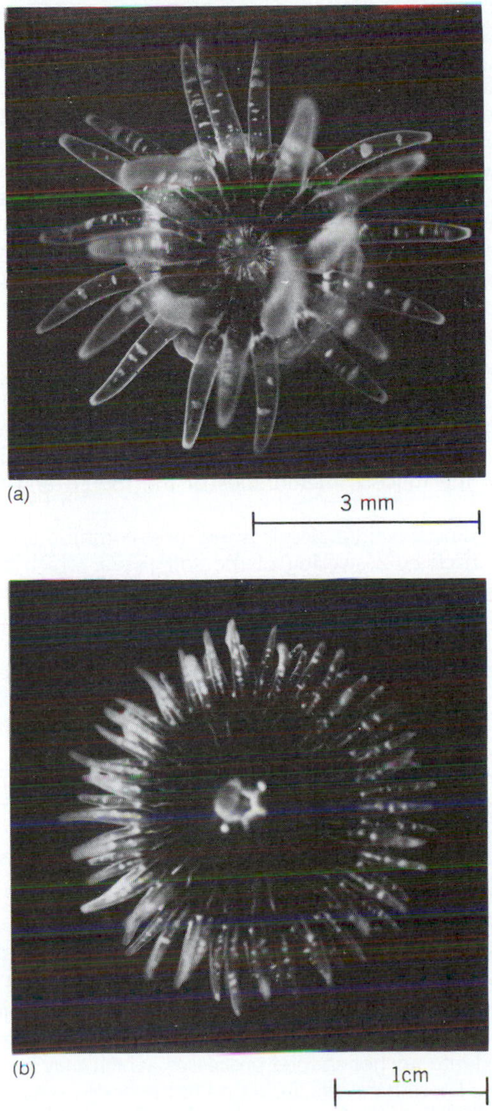

Actiniaria: *Anthopleura* sp. (*a*) Young. (*b*) Adult.

The freely retractile, skeletonless polyp has a cylindrical body, with a thick, tough, rough column wall often bearing rugae, verrucae, tubercles, or suckers (see illustration). The body is often encrusted with sand grains, pebbles, and other detritus. Some species have smooth, thin walls. Nematocysts discharge a toxic substance; however, the human skin is seldom affected by this. The colors of anemones vary with species and many variations occur even among the same species. The tentacles increase in number regularly and are arranged in several cycles. There are 6 primary, 6 secondary, 12 tertiary, 24 quaternary, and so forth in the hexamerous type. The musculature is the most highly developed in the coelenterates.

Most actinians are dioecious. Developing larvae pass through the *Edwardsia* stage then the *Halcampoides* stage. Longitudinal fission frequently occurs as well as budding. Sometimes new individuals result from laceration. *See* COELENTERATA; ZOANTHARIA. [K.At.]

Actinide elements The series of elements beginning with actinium (atomic number 89) and including thorium, protactinium, uranium, and the transuranium elements through the element lawrencium (atomic number 103). These elements have a strong chemical resemblance to the lanthanide, or rare-earth, elements of atomic numbers 57 to 71. Their atomic numbers, names, and chemical symbols are: 89, actinium (Ac), the prototype element, sometimes not included as an actual member of the actinide series; 90, thorium (Th); 91, protactinium (Pa); 92, uranium (U); 93, neptunium (Np); 94, plutonium (Pu); 95, americium (Am); 96, curium (Cm); 97, berkelium (Bk); 98, californium (Cf); 99, einsteinium (Es); 100, fermium (Fm); 101, mendelevium (Md); 102, nobelium (No); 103, lawrencium (Lr). Except for thorium and uranium, the actinide elements are not present in nature in appreciable quantities. The transuranium elements were discovered and investigated as a result of their synthesis in nuclear reactions. All are radioactive and except for thorium and uranium, weighable amounts must be handled with special precautions.

Most actinide elements have the following in common: trivalent cations which form complex ions and organic chelates; soluble sulfates, nitrates, halides, perchlorates, and sulfides; and acid-insoluble fluorides and oxalates. *See* ACTINIUM; LAWRENCIUM; PERIODIC TABLE; PROTACTINIUM; THORIUM; TRANSURANIUM ELEMENTS; URANIUM. [G.T.S.]

Actinium A chemical element, Ac, atomic number 89, and atomic weight 227.0. Actinium was discovered by A. Debierne in 1899. Milligram quantities of the element are available by irradiation of radium in a nuclear reactor. Actinium-227 is a beta-emitting element whose half-life is 22 years. Six other radioisotopes with half-lives ranging from 10 days to less than 1 minute have been identified.

1	2	3	4	5	6	7	8	9	10	11	12	13	14	15	16	17	18
1 H																	2 He
3 Li	4 Be											5 B	6 C	7 N	8 O	9 F	10 Ne
11 Na	12 Mg											13 Al	14 Si	15 P	16 S	17 Cl	18 Ar
19 K	20 Ca	21 Sc	22 Ti	23 V	24 Cr	25 Mn	26 Fe	27 Co	28 Ni	29 Cu	30 Zn	31 Ga	32 Ge	33 As	34 Se	35 Br	36 Kr
37 Rb	38 Sr	39 Y	40 Zr	41 Nb	42 Mo	43 Tc	44 Ru	45 Rh	46 Pd	47 Ag	48 Cd	49 In	50 Sn	51 Sb	52 Te	53 I	54 Xe
55 Cs	56 Ba	71 Lu	72 Hf	73 Ta	74 W	75 Re	76 Os	77 Ir	78 Pt	79 Au	80 Hg	81 Tl	82 Pb	83 Bi	84 Po	85 At	86 Rn
87 Fr	88 Ra	103 Lr	104 Rf	105 Db	106 Sg	107 Bh	108 Hs	109 Mt	110	*111*	*112*	*113*	*114*	*115*	*116*	*117*	*118*

lanthanide series	57 La	58 Ce	59 Pr	60 Nd	61 Pm	62 Sm	63 Eu	64 Gd	65 Tb	66 Dy	67 Ho	68 Er	69 Tm	70 Yb
actinide series	89 Ac	90 Th	91 Pa	92 U	93 Np	94 Pu	95 Am	96 Cm	97 Bk	98 Cf	99 Es	100 Fm	101 Md	102 No

The relationship of actinium to the element lanthanum, the prototype rare earth, is striking. In every case, the actnium compound can be prepared by the method used to form the corresponding lanthanum compound with which it is isomorphous in the solid, anhydrous state. *See* ACTINIDE ELEMENTS; LANTHANUM; NUCLEAR REACTION; RADIOACTIVITY. [S.F.]

Actinobacillus A genus of facultatively aerobic, gram-negative, immotile and nonsporeforming, oval to rod-shaped, often pleomorphic bacteria which occur as parasites or pathogens in mammals (including humans), birds, and reptiles. *Actinobacillus* (*Pasteurella*) *ureae* and *A. hominis* occur in the respiratory tract of healthy humans and may be involved in the pathogenesis of sinusitis, bronchopneumonia, pleural empyema, and meningitis. *Actinobacillus actinomycetemcomitans* occurs in the human oral microflora, and together with anaerobic or capnophilic organisms may cause endocarditis and suppurative lesions in the upper alimentary tract. *See* MEDICAL BACTERIOLOGY. [W.Ma.]

Actinomycetaceae A family of bacteria belonging to the order Actinomycetales. They produce elongated cells, usually filamentous in nature, with a tendency to the development of branches. Some of the cells frequently show swellings or clubbed or irregular shapes. Some species require oxygen and grow best at a high oxygen tension of 20% or more; others grow only in the complete absence of oxygen. Some are parasitic in humans and in animals. These organisms grow readily on artificial media and form well-developed colonies. Some of the organisms are colorless or white, whereas others are pigmented. The family is divided into genera: *Actinomyces*, *Arachnid*, *Bifidobacterium*, *Bacterionema*, and *Rothia*. *See* ACTINOMYCETALES. [S.A.W.]

Actinomycetales An order of bacteria which produce filamentous cells or hyphae. These cells tend to develop branches, giving rise to a true branched mycelium. They are often spoken of as "higher," filamentous, or "moldlike" bacteria. They occur abundantly in nature; some are pathogenic and others are saprophytic. Hyphae do not exceed 1.5 micrometers and are mostly about 1 micrometer or less in diameter. They may produce conidia, special spores, oidiospores, or sporangiospores.

Some species grow best at temperatures between 77–104°F (25–40°C); others grow best at temperatures above 122°F (500°C). Most forms are aerobic, although some are anaerobic. The aerobes require oxygen and grow best at a high oxygen tension of 20% or more, while the anaerobes can grow only in the complete absence of oxygen. They form a variety of both soluble and insoluble pigments.

The order Actinomycetales is divided into eight families: Actinomycetaceae, five genera; Mycobacteriaceae, one genus; Streptomycetaceae, four genera; Actinoplanaceae, ten genera; Frankiaceae, one genus; Dermatophilaceae, two genera; Nocardiaceae, two genera; and Micromonosporaceae, six genera. [S.A.W.]

Actinomycetes A heterogeneous collection of bacteria that form branching filaments. The actinomycetes encompass two groups of filamentous bacteria: the actinomycetes per se and the nocardia/streptomycete complex. The genus *Actinomyces* is more closely related to the genus *Propionibacterium*, which includes bacteria found in pimples and acne, than to the genus *Nocardia*. The actinomycetes have thin hyphae (0.5–1.5 micrometers in diameter) with genetic material coiled inside as free deoxyribonucleic acid (DNA). The cell wall of the hyphae is made up of a cross-linked polymer containing short chains of amino acids and long chains of amino sugars. In general, actinomycetes do not have membrane-bound cell organelles. Actinomycetes are susceptible to a wide range of antibiotics that are used to treat bacterial diseases, such as penicillin and tetracycline. *See* AMINO SUGAR; ANTIBIOTIC; DEOXYRIBONUCLEIC ACID (DNA).

Members of *Actinomyces* are most often found in the mouth and gastrointestinal tract of humans and other animals. *Actinomyces* do not require oxygen for growth and are sometimes referred to as anaerobic bacteria. It is actually the requirement for elevated levels of carbon dioxide rather than the negative effect of oxygen that characterizes *Actinomyces*. When displaced from their normal sites within the mouth or gastrointestinal tract, *Actinomyces* may cause diseases in humans, such as lung abscesses, appendicitis, and lumpy jaw.

The nocardia/streptomycete complex constitutes a continuous spectrum of organisms from those that are most like true bacteria to those that are superficially most like fungi. The nocardiae represent the transition, having members that resemble the bacteria that cause diphtheria (*Corynebacterium*) and tuberculosis (*Mycobacterium*). Members of *Nocardia* require oxygen for growth, are found in soil and water, and have the ability to use a wide range of organic material as a source of energy. Species of *Nocardia* have been used commercially to modify chemical compounds related to cholesterol to make biologically active steroids. A few species of *Nocardia* cause disease in humans. *See* DIPHTHERIA; TUBERCULOSIS.

Streptomycetes have long branching filaments and two types of mycelia. The cell walls are typical bacterial cell walls and do not contain the fatty acids found in nocardiae and mycobacteria. Streptomycetes require oxygen for growth, are found in soil and water, and have the ability to utilize a wide range of organic materials as nutrients. They are particularly important in degradation of dead plant materials in soil. Streptomycetes do not produce disease in humans or animals and are best known for producing many clinically useful antibiotics, including streptomycin, tetracycline, and cephalosporin. *See* BACTERIA; MEDICAL BACTERIOLOGY; SOIL MICROBIOLOGY. [S.G.B.]

Actinomycosis An infectious disease of humans caused by *Actinomyces israelii* and *Aruchnia propionica*. This disease is also seen in cattle, in which the etiologic agent is *Actinomyces bovis*. These microorganisms are normal inhabitants of the mucous membranes of the mouth and tonsillar crypts.

Cervicofacial actinomycosis is the most common form and is characterized by a swelling of the soft tissue, which becomes hard. As the disease process continues, abscess formation and multiple draining sinuses occur. The disease may progress to invade the bone of the lower jaw, causing osteomyelitis. Occasionally the organism may be aspirated into the lungs, resulting in pulmonary abscesses followed by invasion of the chest wall. Abdominal actinomycosis usually results from penetration of the intestinal mucosa.

Actinomyces israelii, *Actinomyces bovis*, and *Arachnia propionica* are sensitive to most antibiotics. Penicillin is the drug of choice in treating this disease in humans. *See* ACTINOMYCETACEAE; ANTIBIOTIC; GRAM'S STAIN; MEDICAL MYCOLOGY. [L.D.H.]

Actinomyxida An order of the protozoan class Myxosporidea (subphylum Cnidospora) characterized by the production of trivalved spores with three polar capsules and one to many sporoplasms. The spore membrane may be extended into anchor-shaped processes, which may have bifurcate tips. These parasites are found in the body cavity or in the intestinal lining of fresh-water annelids and in marine worms of the phylum Sipunculoidea. The life cycle for most species is not well known. *See* CNIDOSPORA; MYXOSPORIDEA; OLIGOCHAETA; PROTOZOA. [R.F.N.]

Actinophage Any of a number of bacteriophages that infect and lyse bacteria of the order Actinomycetales. Actinophages of particular interest are those that include in their host range any organisms of the genus *Streptomyces*, the source of most of the therapeutically useful antibiotics. Contamination of a culture with a specific actinophage may result in lysis and destruction of the bacteria, which obviously halts antibiotic production. The problem of phage contamination is often solved by the isolation and use of a phage-resistant mutant of the antibiotic-producing bacterium. *See* BACTERIOPHAGE; STREPTOMYCETACEAE. [L.B.]

Actinophryida An order of Heliozoia. In these protozoans, a centroplast, a highly organized test, and a capsule are lacking (see illustration). The organisms may be uninucleate, as in *Actinophrys*, or multinucleate, as in *Actinosphaerium* (fresh water) and *Camptonema* (marine). The outer cytoplasm is usually highly vacuolated. In uninucleate genera the nucleus is approximately central. Axopodia are present, and their axial filaments end near the nucleus in uninucleate types. *See* ACTINOPODEA; HELIOZOIA; PROTOZOA; SARCODINA; SARCOMASTIGOPHORA. [R.P.H.]

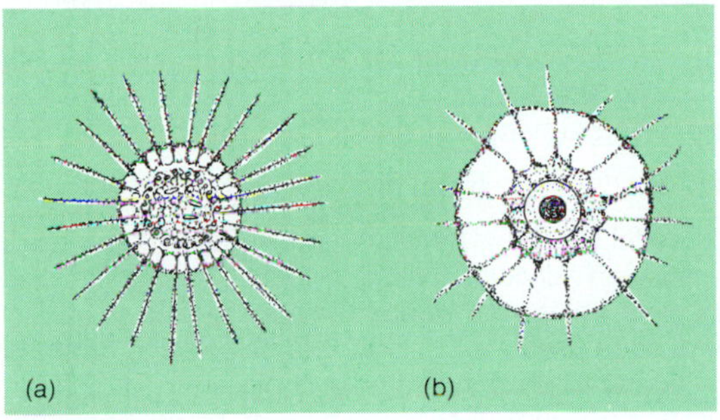

Actinophryida: (a) *Actinosphaerium eichorni*. (b) *Actinophrys pontica*. (After R.P. Hall, *Protozoology*, Prentice-Hall, 1953)

Actinoplanaceae A family of bacteria of the order Actinomycetales, with well-developed mycelium, and spores characteristically formed in sporangia. In many species conidia are formed in moniliform chains, brushlike as in *Penicillium*, or in clusters somewhat as in *Micromonospora*.

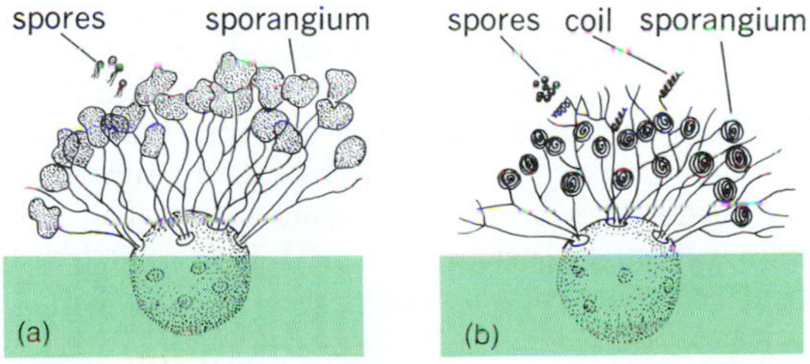

Two genera of Actinoplanaceae, growing on pollen: (a) *Actinoplanes* and (b) *Streptosporangium*.

Of the several genera described (see illustration), the commonest are *Actinoplanes*, having spherical to subspherical planospores with a tuft of flagella, *Ampullariella*, having rod-shaped planospores with polar flagella, *Spirillospora*, having bent-rod to spiral planospores with polar or lateral flagella, and *Streptosporangium*, having spherical to subspherical nonflagellated and nonmotile spores.

This family includes 10 genera: *Actinoplanes*, *Spirillospora*, *Streptosporangium*, *Amorphosporangium*, *Ampullariella*, *Pilimelia*, *Planomonospora*, *Planobispora*, *Dactylosporangium*, and *Kitastoa*. *See* ACTINOMYCETALES; CILIA AND FLAGELLA. [J.N.C.]

Actinopodea A class of Sarcodina. Some of these protozoans have more or less permanent pseudopodia, composed of axial filaments surrounded by a cytoplasmic envelope (axopodia); others have delicate and often radially arranged filopodia (filamentous pseudopodia) or filoreticulopodia (filamentous branched pseudopodia) without axial filaments. Although some are stalked and sessile, most are floating types. There are four subclasses: Radiolaria, Acantharia, Heliozoia, and Proteomyxidia. *See* ACANTHARIA; HELIOZOIA; PROTEOMYXIDIA; RADIOLARIA; SARCODINA. [R.P.H.]

Actinopterygii The rayfin fishes, which form one of three subclasses of the Osteichthyes, or bony fishes. The terms Actinopteri, Teleostomi, and Palaeopterygii, plus Neopterygii of some authors, are equivalent to Actinopterygii. The rayfins are distinguished by the primitive presence of a single dorsal fin, the absence of an epichordal lobe of the caudal fin, and little extension of flesh and skeletal supporting elements into the paired fins. The rayfins lack internal nares, and the swim bladder, if present, usually rises from a dorsal out-pocketing from the gut or is free from it. The rayfins first appeared in the Lower Devonian; they experienced a tremendous differentiation in the Cretaceous and early Cenozoic, since which time they have been the dominant vertebrates of the seas and fresh waters. Approximately 97% of modern fishes are members of this subclass. *See* CROSSOPTERYGII; DIPNOI.

Three centralized structural types may be defined and have been ranked as infraclasses: Chondrostei, Holostei, and Teleostei. *See* CHONDROSTEI; HOLOSTEI; OSTEICHTHYES; TELEOSTEI. [R.M.B.]

Action An integral quantity associated with each possible motion of a system of particles, of fields, or both. For a particle system the action is given by Eq. (1), where the q_j are generalized coordinates, the p_j are their conjugate momenta, and the $\dot{q}_j$ are the derivatives of the generalized coordinates with respect to time. *See* HAMILTON'S PRINCIPLE.

$$S = \int \sum_{j=1}^{N} p_j(t)\dot{q}_j(t)dt \qquad (1)$$

The action permits the following formulation of the dynamical equations, which is similar to but more restrictive than Hamilton's principle: Of all possible motions with a given energy from the initial values q_j, to the final values q_j' the dynamical motion is that for which the action is stationary (not always a minimum despite the name "principle of least action").

When the motion of a degree of freedom q_j, p_j is periodic, the separate integral shown in Eq. (2) holds. Taken over a period of the motion, it is called the action variable J_j. Action variables are useful as adiabatic invariants (quantities insensitive to slow variations in the external parameters of the system). This property was exploited in early quantum theory by assigning to the action variables fixed "quantized" values, namely, integral multiples of Planck's constant. *See* QUANTUM MECHANICS. [B.G.]

$$\int p_j \dot{q}_j \, dt = \int p_j dq_j \qquad (2)$$

Activated carbon A powdered, granular, or pelleted form of amorphous carbon characterized by very large surface area per unit volume because of an enormous number of fine pores. Activated carbon is capable of collecting gases, liquids, or dissolved substances on the surface of its pores.

Adsorption on activated carbon is selective, favoring nonpolar over polar substances. Compared with other commercial adsorbents, activated carbon has a broad spectrum of adsorptive activity, excellent physical and chemical stability, and ease of production from readily available, frequently waste materials. *See* ADSORPTION.

Almost any carbonaceous raw material can be used for the manufacture of activated carbon. Wood, peat, and lignite are commonly used for the decolorizing materials. Bone char made by calcining bones is used in large quantity for sugar refining. Nut shells (particularly coconut), coal, petroleum coke, and other residues in either granular, briqueted, or pelleted form are used for adsorbent products.

Activation is the process of treating the carbon to open an enormous number of pores in the 1.2- to 20-nanometer-diameter range (gas-adsorbent carbon) or up to 100-nm-diameter range (decolorizing carbons). After activation, the carbon has the large surface area (500–1500 m^2/g) responsible for the adsorption phenomena. Carbons that have not been subjected previously to high temperatures are easiest to activate. Selective oxidation of the base carbon with steam, carbon dioxide, flue gas, or air is one method of developing the pore structure. Other methods require the mixing of chemicals, such as metal chlorides (particularly zinc chloride) or sulfides or phosphates, potassium sulfide, potassium thiocyanate, or phosphoric acid, with the carbonaceous matter, followed by calcining and washing the residue. *See* CARBON; CHARCOAL.

[H.B.A.]

Activation analysis A technique in which a neutron, charged particle, or gamma photon is captured by a stable nuclide to produce a different, radioactive nuclide which is then measured. The technique is specific, highly sensitive, and applicable to almost every element in the periodic table.

In neutron activation analysis (NAA), the most widely used form of activation analysis, the sample to be analyzed is placed in a nuclear reactor where it is exposed to a flux of thermal neutrons. Some of these neutrons are captured by isotopes of elements in the sample; this results in the formation of a nuclide with the same atomic number, but with one more mass unit of weight. A prompt gamma ray is immediately emitted by the new nuclide.

Measurement of the induced radioactivities is the key to activation analysis. This is usually obtained from the gamma-ray spectra of the induced radionuclides. Gamma rays from radioactive isotopes have unique, discrete energies, and a device that converts such rays into electronic signals that can be amplified and displayed as a function of energy is a gamma-ray spectrometer. It consists of a detector [germanium doped with lithium, GeLi, or sodium iodide doped with thallium, NaI(Tl)] and associated electronics.

Activation analysis can also be performed with charged particles (protons or He^{3+} ions, for example), but because fluxes of such particles are usually lower than reactor neutron fluxes and cross sections are much smaller, charged-particle methods are usually reserved for special samples. Charged particles penetrate only a short distance into samples, which is another disadvantage. A variant called proton-induced x-ray emission (PIXE) has been highly successful in analyzing air particulates on filters.

Activation analysis has been applied to a variety of samples. It is particularly useful for small (1 mg or less) samples, and one irradiation can provide information on 30 or more elements. Samples such as small amounts of pollutants, fly ash, very pure experimental alloys, and biological tissue have been successfully studied by neutron activation analysis. Of particular interest has been its use in forensic studies; paint, glass, tape, and other specimens of physical evidence have been assayed for legal purposes. In addition, the method has been used for authentication of art objects and paintings where only a small sample is available. *See* FORENSIC CHEMISTRY; NUCLEAR REACTION; PARTICLE DETECTOR; RADIOISOTOPE; TRACE ANALYSIS. [W.S.L.]

Activity (thermodynamics) A function introduced by G. N. Lewis to aid in the thermodynamic treatment of real systems. Like the fugacity, the activity makes possible the correlation of changes in the chemical potential with changes in experimentally measurable quantities, such as concentrations or partial pressures, through relations formally equivalent to those holding for ideal systems. The activity concept retains its usefulness, however, for such cases as condensed phases of low volatility, for which fugacity determinations are impractical. *See* CHEMICAL THERMODYNAMICS; FREE ENERGY; FUGACITY; PHYSICAL CHEMISTRY.

The activity a_i of a constituent i for a given state is defined as Eq. (1), where f_i is the fugacity for the given state and f_i^0 the

$$a_i = f_i/f_i^0 \qquad (1)$$

fugacity of the constituent for a reference state, or standard state, at the same temperature. This definition then requires that Eq. (2) hold, where μ_i and μ_i^0 are the values of the chem-

$$RT \ln a_i = \mu_i - \mu_i^0 \qquad (2)$$

ical potential for the given state and standard state, respectively. Because an actvity is a relative quantity, a numerical value is meaningless unless the standard state involved is known. The standard state used may in principle be selected arbitrarily, but in practice, certain conventional choices are ordinarily made because a maximum of convenience can be obtained thereby.

[P.J.B.]

Acylation The process whereby the acyl group is incorporated into a molecule by substitution. If the acetyl group is incorporated, the process is acetylation; if the benzoyl group is substituted, it is benzoylation. *See* ACETYLATION.

For the process to occur, the acylating agent must attack a molecule containing one or more active or easily replaceable hydrogen atoms. This limits the reaction, for all practical purposes, to simple or polyhydroxy alcohols, phenols, thiols, ammonia, primary and secondary amines, amino acids, esters of malonic acid, β-keto esters, and β-diketones. The common reagents are acid halides, preferably the chloride and occasionally the bromide, and acid anhydrides. For acetylation the highly reactive substance ketene may be used. A related reaction is the Friedel-Crafts acylation, in which the acyl group replaces the hydrogen atom of an aromatic nucleus. *See* FRIEDEL-CRAFTS REACTION.

The most important acylation reaction is acetylation, which is much used industrially. Acetylation of acetylene is accomplished catalytically by the addition of acetic acid to form vinyl acetate, which is an important monomer in the plastics industry. The reaction of cellulose with acetic anhydride in the presence of sulfuric acid gives cellulose triacetate, used in the manufacture of the fiber known as acetate rayon. The important analgesic aspirin, or acetylsalicylic acid, is prepared by the reaction of salicylic acid with acetic anhydride. *See* ACID ANHYDRIDE; ACID HALIDE; AMINE; AMINO ACIDS; ASPIRIN; CELLULOSE. [P.E.F.]

Adamantane The $C_{10}H_{16}$ alicyclic hydrocarbon whose structure has the same arrangement of carbon atoms as does the basic unit of the diamond lattice (see illustration). Adamantane is composed of three interlocking, chair-form cyclohexane rings. The molecular structure is highly symmetrical and compact. All bond angles are found by x-ray crystallography to be nearly 109.5°, and the bonds around all adjacent carbons are ideally staggered. These structural features afford a high degree of stability to the adamantane molecule. The highly symmetrical structure of adamantane also accounts for its extremely high melting point 516°F (269°C), the highest known for aliphatic hydrocarbons, and for the seemingly contradictory ease with which it sublimes. Possibly the most interesting and significant uses of adamantane compounds are

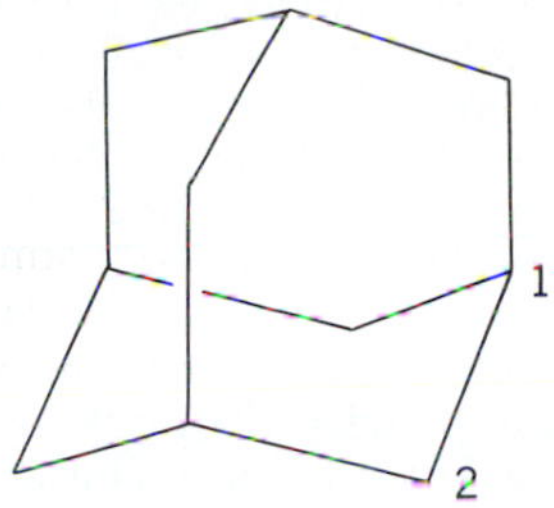

Adamantane, showing bridgehead carbon labeled 1 and alternate position for substitution at carbon 2.

found in the field of pharmacology. Several adamantane derivatives have been found to exhibit antiviral activity, whereas the incorporation of an adamantane moiety into many already useful drugs has been found to significantly enhance their efficacy. *See* ALICYCLIC HYDROCARBON; ALIPHATIC HYDROCARBON. [P.E.F.]

Adaptation (biology) Any characteristic of or process in a living organism which contributes to or furthers the survival of an individual or of the species in a specialized lifestyle (niche) in a peculiar environment or habitat. The term is also used for the process by which an adaptation develops or is accomplished. Permanent, congenital, or inherited adaptations are generally believed to have arisen over long periods of time by means of natural selection of essentially random variations which are inherited.

The term acclimation or adjustment may be used to distinguish from evolutionary adaptations those changes which develop within the lifetime of an individual in response to changes in way of life or habitat, and which are not inherited as such, although their genetic basis may be. The ability to adjust or to become acclimated, possessed somewhat by every living organism, is one of the many adaptive features of life in general. [B.T.S.]

Adaptive control A special type of nonlinear control system which can alter its parameters to adapt to a changing environment. The changes in environment can represent variations in process dynamics or changes in the characteristics of the disturbances. *See* NONLINEAR CONTROL THEORY.

A normal feedback control system can handle moderate variations in process dynamics. The presence of such variations is, in fact, one reason for introducing feedback. There are, however, many situations where the changes in process dynamics are so large that a constant linear feedback controller will not work satisfactorily. For example, the dynamics of a supersonic aircraft change drastically with Mach number and dynamic pressure, and a flight control system with constant parameters will not work well. *See* FLIGHT CONTROLS.

Adaptive control is also useful for industrial process control. Since delay and holdup times depend on production, it is desirable to retune the regulators when there is a change in production. Adaptive control can also be used to compensate for changes due to aging and wear. *See* PROCESS CONTROL. [K.J.A.]

Adaptive optics The science of optical systems that measure and correct wavefront aberrations in real time, that is, simultaneous with the operation of the system. This possibility of improving the performance of an optical system has many applications for both passive imaging systems and laser systems. For example, adaptive optics offers the possibility of removing the effect of atmospheric turbulence on the images of objects viewed through the air; of allowing diffraction-limited imaging of astronomical objects from the ground; of compensating for the degrading effects of the atmosphere in transmitting laser radiation; and of correcting for static or dynamic figure errors in optical surfaces or aberrations in the gain medium of lasers. Active correction and control of optical surfaces allows new methods of construction of large optical telescopes, such as the phasing together of numerous smaller elements or the effective use of thinner, less precise optical surfaces. *See* LASER.

To appreciate the usefulness of an adaptive optics system, it is useful to look at the effect of the atmosphere upon the performance of any Earth-based passive imaging system. Neglecting atmospheric effects, a perfect 1-m-diameter-aperture telescope operating at a wavelength of 0.5 micrometer has a resolution capability that extends out to 10 cycles per arc-second. However, for typical atmospheric conditions the resolution is limited to about 1 cycle per arc-second, that is, the effective aperture diameter is on the order of 0.1 m. Adaptive optics has the potential for making the effects of the atmosphere nearly negligible for many situations. *See* ASTRONOMICAL PHOTOGRAPHY; OPTICS.

Adaptive optics systems consist of three basic components (see illustration). All presently working techniques (other than

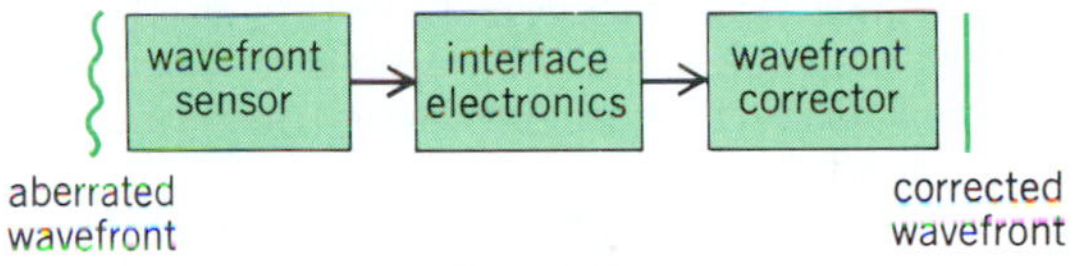

Adaptive wavefront compensation system.

nonlinear phase-conjugation techniques) use a deformable mirror as the wavefront corrector. The main difference between techniques is the wavefront sensor. Three main adaptive optics techniques have been investigated: multidither COAT (coherent optical adaptive techniques); sharpness-function maximization; and phase-conjugate.

The multidither COAT system is used in optical systems that transmit laser radiation and must concentrate this radiation into as small an area as possible. In the sharpness-function technique, the correction is dithered to maximize a given function, such as the integral of the square of the irradiance in the image plane or the amount of energy within a certain region. In operating phase-conjugate systems, the wavefront to be corrected is measured directly by using either a geometric or interferometric test. [J.C.Wy.]

Adaptive sound control Sound field modification, particularly sound cancellation, by electroacoustical means. In active sound cancellation, a control sound source creates sound in a selected region equal in amplitude and opposite in phase to sound that would otherwise exist. In adaptive sound cancellation, active sound cancellation is maintained in the presence of changes in the system. Sound control includes sound reduction and equalization as well as cancellation.

When a person blows across the opening of an empty bottle, a sound can be heard as the cavity resonates. This combination is called a Helmholtz resonator. An example of passive sound modification is the effect that a Helmholtz resonator embedded in the wall of a duct has on a single-frequency plane sound wave traveling along the duct (the incident wave). When the frequency of the incident wave and the resonant frequency of the resonator are the same, fluid moves in such a way (neglecting friction) that no sound travels past the resonator.

The effect is exactly the same as if the resonator were the source of a new sound wave (the control wave) equal in magnitude and opposite in phase to the incident wave. The control wave travels in both directions from the resonator and is superposed on the incident wave: upstream, the incident and control waves travel in opposite directions to create a standing wave;

downstream, they travel in the same direction, 180° out of phase, to create no wave at all. The downstream result is sound cancellation. *See* ACOUSTIC RESONATOR.

The Helmholtz resonator can be replaced with a loudspeaker that is adjusted to create a control wave identical to the incident wave but with opposite phase at the loudspeaker position. The resulting sound field is identical to that achieved with the resonator.

Active (nonadaptive) sound control is equivalent to open-loop control. (This is almost equivalent to driving with one's eyes closed.) Most active sound control realizations use a feed-forward concept. In the duct example, a microphone is placed upstream of the loudspeaker to sense the pressure of the incident-wave sound. Some electronic equipment provides gain and phase control, and one or more loudspeakers create a canceling sound. Active systems require continual manual adjustment, and therefore have little practical value. *See* CONTROL SYSTEMS.

Progress toward the goal of effective active sound control was advanced by two essential developments in the 1960s. One was the microprocessor revolution, which resulted in fast, inexpensive digital hardware. The second was development of practical adaptive signal-processing algorithms. The adaptive echo canceler for telephone networks is a major successful application of adaptive signal processing. *See* ACOUSTIC SIGNAL PROCESSING; MICROPROCESSOR.

While a signal processor works directly on the physical variable to be modified, a control system works via intermediate components and variables. This difference affects the adaptive algorithm. *See* ADAPTIVE CONTROL.

Current successful applications are limited to one-dimensional adaptive systems (with only one error signal), although the acoustical systems may be three-dimensional. Adaptive sound control is most useful in the low-frequency range (up to above 400 Hz), where passive control is often not cost-effective. *See* ACOUSTIC NOISE; SOUND. [J.C.Bu.]

Addictive disorders Any of the human disorders characterized by the chronic use of a drug, alcohol, or similar substance resulting in the development of tolerance (with need for increasing amounts of the substance to achieve the desired effect); physical dependence (characterized by a sequence of well-defined signs and physiological symptoms, for example the withdrawal or abstinence syndrome on cessation of use of the substance); and finally, drug-seeking behavior. Since the early 1960s, research emphasis has shifted from psychological, sociological, and epidemiological studies to studies of the metabolic bases of addiction.

The two major disorders are alcoholism and narcotic (or opiate) addiction. Drug addiction may also occur after chronic use of other types of agents, such as barbiturates and benzodiazepines. Abusive use of other licit as well as illicit drugs, such as cocaine, is common. However, many types of drug abuse do not necessarily result in drug addiction as defined above, since tolerance and physical dependence may not develop despite the presence of drug-seeking behavior.

The finding of a specific opiate receptor in humans in 1973 has been extremely important for many aspects of physiology and pathology. Subsequently it has been determined that there are at least three separate types of opiate receptors—mu receptors, delta receptors, and kappa receptors—all of which meet the accepted criteria of being specific opiate receptors in that: opioid (opiatelike) drugs of different types bind to these receptors; this binding may be displaced by a specific opioid (or narcotic) antagonist, such as naloxone; and this binding leads to some effect.

Subsequent to the discovery of specific opioid receptors, endogenous ligands which bind to these receptors, the so-called endogenous opioids, were discovered. The first of these endogenous opioids to be found were metenkephalin and leuenkephalin. The second class of endogenous opioids to be delineated was beta endorphin, which has been shown to be released from beta lipotropin in equimolar amounts with the release of adrenocorticotropic hormone (ACTH). These peptides are found in the hypothalamic region of the brain, predominantly in the pituitary, but also in other central and peripheral sites, including the gastrointestinal tract and pancreas. The third class of endogenous opioids to be delineated were the dynorphins found in the brain as well as in other sites. It is not known whether or not the endogenous opioids play any role in the addictive diseases, especially narcotic addiction, and alcoholism, although many hypotheses for such a role have been formulated. *See* ALCOHOLISM; BARBITURATES; COCAINE; ENDORPHINS; NARCOTIC; OPIATES. [M.J.Kr.]

Addison's disease A primary failure or insufficiency of the adrenal cortex to secrete the vital cortical hormones, many of which are necessary to life. Causes of adrenal insufficiency include adrenal surgery, infections, tumor formation or infiltration, and other conditions which produce a decrease or cessation of cortical hormone secretion. Occasionally no preceding factors can be demonstrated, so the term idiopathic is used to describe these cases of addisonism.

All four types of adrenocortical hormones are decreased or absent in Addison's disease. Serous fluid and electrolyte imbalances are typical and often accompanied by a peculiar tan skin pigmentation, nausea, vomiting, decreased blood pressure, weakness, weight loss, and other specific and constitutional changes.

Addisonian crisis, or acute insufficiency, may be triggered by a generalized infection or massive stress. In such a crisis rapid deterioration, shock, and death may ensue. *See* ADRENAL GLAND. [E.G.St./N.K.M.]

Addition One of the four fundamental operations of arithmetic and algebra. The term addition is applied to many kinds of objects other than numbers. For example, two vectors $\mathbf{x}, \mathbf{y}$ are added to produce a third vector $\mathbf{z}$ obtained from them by the "parallelogram" law, and two sets A, B are added to form a third set C consisting of all the elements of A and of B. *See* CALCULUS OF VECTORS.

As an operation on pairs of real or complex numbers, addition is associative, Eq. (1), and commutative, Eq. (2); and multiplication is distributive over addition, Eq. (3). There are impor-

$$a + (b + c) = (a + b) + c \quad (1)$$

$$a + b = b + a \quad (2)$$

$$a(b + c) = a \cdot b + a \cdot c \quad (3)$$

tant mathematical structures in which an addition operation is defined that lacks one or more of these properties. *See* ALGEBRA; DIVISION; MULTIPLICATION; NUMBER THEORY; SUBTRACTION. [L.M.Bl.]

Adeno-SV40 hybrid virus A type of defective virus particle in which part of the genetic material of papovavirus SV40 is encased within an adenovirus protein coat (capsid). Adenovirus progeny possesses properties different from those of the original parent adenovirus. It produces tumors in newborn hamsters; it replicates in monkey cell cultures; and although it does not produce infectious SV40, it causes monkey cells in culture to produce a new cellular antigen known to be specifically induced by SV40—the SV40 tumor, or T, antigen. The new virus stock therefore behaves like a hybrid. It has proved to be a population of two distinct kinds of virus particles. One kind of particle is a true adenovirus. The other particle is the adeno-SV40 hybrid, which has an adenovirus coat, but whose genetic material appears to consist of defective adenovirus type 7 DNA, representing about 85% of the adenovirus genome, covalently linked to a portion (about 50%) of an SV40 genome.

In this virus population the particle carrying the SV40 genetic material has been termed PARA (particle aiding replication of adenovirus). The PARA is considered an unconditionally defective virus, since under no known conditions can it reproduce itself except in a cell coinfected with adenovirus. The adenovirus is considered conditionally defective, since it can reproduce independently in human cells but not in monkey cells. *See* ADENOVIRIDAE; ANIMAL VIRUS. [J.L.Me.]

Adenosine diphosphate (ADP) A coenzyme and an important intermediate in cellular metabolism as the partially dephosphorylated form of adenosine triphosphate. The compound is 5′-adenylic acid with an additional phosphate group attached through a pyrophosphate bond. ADP is produced from adenosine triphosphate and reconverted to this compound in coupled reactions concerned with the energy metabolism of living systems. ADP is also produced from 5′-adenylic acid by the transfer of a phosphate group from adenosine triphosphate in a reaction that is catalyzed by an enzyme, myokinase. *See* ADENOSINE TRIPHOSPHATE (ATP); ADENYLIC ACID; METABOLISM. [M.D.]

Adenosine triphosphate (ATP) A coenzyme and one of the most important compounds in the metabolism of all organisms, since it serves as a coupling agent between different enzymatic reactions. Adenosine triphosphate is adenosine diphosphate (ADP) with an additional phosphate group attached through a pyrophosphate linkage to the terminal phosphate group. ATP is a powerful donor of phosphate groups to suitable acceptors because of the pyrophosphate nature of the bonds between its three phosphate radicals. ATP serves as the immediate source of energy for the mechanical work performed by muscle. In its presence, the muscle protein actomyosin contracts with the formation of adenosine diphosphate and inorganic phosphate. ATP is also involved in the activation of amino acids, a necessary step in the synthesis of protein.

In metabolism, ATP is generated from adenosine diphosphate and inorganic phosphate mainly as a consequence of energy-yielding oxidation-reduction reactions. In respiration, ATP is generated during the transport of electrons from the substrate to oxygen via the cytochrome system. *See* ADENOSINE DIPHOSPHATE (ADP); ADENYLIC ACID; COENZYME; METABOLISM. [M.D.]

Adenoviridae A family of viral agents which cause febrile catarrhs and various other respiratory diseases. They have been associated with pharyngoconjunctival fever, acute respiratory disease, epidemic keratoconjunctivitis, and febrile pharyngitis in children. Although most of the illnesses caused by adenoviruses are respiratory, adenoviruses are frequently excreted in stools, and certain adenoviruses have been isolated from sewage. Distinct serotypes are known for simian, bovine, canine, murine, and avian species. *See* ANIMAL VIRUS; KERATOCONJUNCTIVITIS.

Infective virus particles, 70 nm in diameter, are icosahedrons with shells (capsids) composed of 252 subunits (capsomeres). No outer envelope is known. The genome is double-stranded deoxyribonucleic acid (DNA). Three major soluble antigens are separable from the infectious particle—a group-specific antigen common to all adenovirus types, a type-specific antigen unique for each type, and a toxinlike material which also possesses group specificity.

The known types of adenoviruses of humans total at least 33, and previously unrecognized types continue to be isolated. The serotypes are antigenically distinct in neutralization tests, but they share a complement-fixing antigen, which is probably a smaller soluble portion of the virus. Adenoviruses contain four complement-fixing antigens—A, B, C, and P.

The virus does not commonly produce acute disease in laboratory animals but is cytopathogenic in cultures of human tissue. Certain human adenovirus serotypes produce cancer when injected into newborn hamsters. Although complete, infective virus cannot be recovered from adenovirus-induced hamster tumors, a new antigen induced by the virus can be detected by complement fixation or immunofluorescence techniques. This antigen, called tumor or T antigen because of its association with tumor or transformed cells, can also be detected in the cytolytic cycle of the virus.

Live virus vaccines against type 4 and type 7 have been developed and used extensively in military populations. When both are administered simultaneously, vaccine recipients respond with neutralizing antibodies against both virus types. *See* ADENO-SV40 HYBRID VIRUS; ANTIGEN; COMPLEMENT-FIXATION TEST. [J.L.Me.]

Adenylic acid A generic term referring to a group of isomeric nucleotides. Adenylic acid, the phosphoric acid ester of adenosine(9′-β-D-ribosidoadenine), is also known as adenosine monophosphate (AMP). The most common isomer is 5′-adenylic acid (5′-AMP or "muscle adenylic acid"), in which the phosphate group is attached to the fifth carbon atom of the ribose moiety (see illustration).

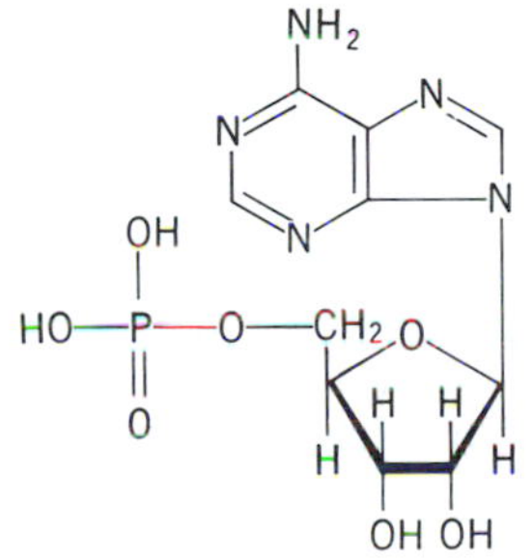

Structural formula for 5′-adenylic acid.

5′-Adenylic acid can be regarded as the parent compound of adenosine diphosphate (ADP) and adenosine triphosphate (ATP). By itself the nucleotide functions as an allosteric effector in the regulation of carbohydrate metabolism. It is also a constituent of several important coenzymes, and in addition plays a role in the incorporation of amino acids into proteins. *See* ADENOSINE DIPHOSPHATE (ADP); ADENOSINE TRIPHOSPHATE (ATP); CARBOHYDRATE METABOLISM; COENZYME.

Other important isomers in this series are 3′,5′-adenylic acid, also known as cyclic AMP, 2′-adenylic acid, 3′-adenylic acid ("yeast adenylic acid"), and 2′,3′-adenylic acid. *See* CYCLIC ADENYLIC ACID; NUCLEIC ACID. [G.A.R.]

Adhesive Substance capable of holding materials together by surface attachment. The term is general and includes, among other things, cement, glue, mucilage, and paste—names which are loosely used interchangeably. The oldest type of adhesives are those of natural origin, such as starch, asphalt, gelatinized collagen, vegetable proteins, plant saps and resins, rubber, and shellac; and most of these continue to find applications. Modified natural products, such as portland cement, dextrins, nitrated and acetylated celluloses or starches, and fatty amides, are widespread in industry. Wholly synthetic adhesives include the emulsion polymers such as polyvinyl acetate, polyvinyl alcohol, and elastomeric solution cements, and the epoxy, polyurethane, and polysulfide condensation adhesives. *See* MUCILAGE. [F.W.]

Adhesive bonding The holding of materials together by surface attachment. This technique of attachment is very ancient. The original glues were based on naturally occurring materials such as coal and pine tars, animal protein, and blood albumin.

Not until the mid-1930s did the chemistry and technology of polymer and resins based primarily on principles of organic chemistry begin to yield new raw materials for new generations of adhesives. The adhesive industry continued to develop new theories of adhesion and appropriate raw materials, based on phenoiic resins, urethanes, polyvinyl esters, and, in the 1950s, the epoxy resins. *See* ADHESIVE.

Among the parameters that must be considered for an adhesive product to have practical applications are contact and transition.

The adhesive must be bought into intimate, continuous contact with the substrate or surface that is to be joined. To accommodate this requirement, the adhesive at some point must be liquid during the formation of the bond in order to wet, spread, and form a continuous film. This requirement is generally fulfilled by application of a solution of solid polymer in a volatile solvent, by subjection of a solid polymer to sufficient heat or pressure, or both, during contact with the substrate to cause a liquid flow, and by application of liquid reactive components that can be made to react chemically in place to form a solid adhesive polymer after wetting and spreading.

Once the adhesive has been brought into intimate, continuous contact with the substrate and has at some point been in a liquid state during contact with the two adherends, the adhesive must pass through a transition to a tough, nonflowing, nonliquid, load-bearing interlayer in the bonded assembly.

In the case of permeable or porous substrates the assembly can be made while the adhesive is still wet with solvent. In the case of nonporous or nonpermeable substrates the transition of the adhesive interlayer from liquid to solid must be accomplished via techniques other than solvent evaporation. There are two methods most commonly used. The method involving heat to flow–cool to set is best represented by thermoplastic adhesives or hot-melt adhesives. The other technique utilizes an in-place chemical reaction. *See* POLYETHER RESINS; POLYMER.

There are many different theories of adhesion. The original theories suggested that a true chemical bond between the adhesive and adherend was necessary and responsible for the formation of the bond. At about the same time, it was theorized that the bond was a result of intermolecular attraction between molecules in close proximity or between the adherend and adhesive. These forces are known as van der Waals or London forces of intermolecular attraction. Most adhesive technologists conjectured that probably combinations of both true chemical bonds (ionic, covalent, or coordinate) and van der Waals forces come into play during the bonding operation. Research involving adhesive bonding continues to refine and expand the various theories of how the bonding occurs. *See* INTERMOLECULAR FORCES. [J.A.G.]

Adiabatic demagnetization The removal or diminution of a magnetic field applied to a magnetic substance when the latter has been thermally isolated from its surroundings. The process concerns paramagnetic substances almost exclusively, in which case a drop in temperature of the working substance is produced (magnetic cooling). Adiabatic demagnetization was employed from 1933 on to produce temperatures below those readily obtainable by using only liquid helium (that is, below 1 K). *See* LIQUID HELIUM; PARAMAGNETISM.

Nuclear magnetic moments are one or two thousand times smaller than their ionic (that is, electronic) counterparts, and the characteristic temperature of their mutual interaction lies in the microkelvin rather than millikelvin region. Successful experiments in nuclear adiabatic demagnetization date from the mid-1950s. [R.P.Hu.]

Adiabatic process A thermodynamic process in which the system undergoing the change exchanges no heat with its surroundings. Reversible adiabatic processes are also isentropic; that is, they take place with no change in entropy. *See* ENTROPY; ISENTROPIC PROCESS.

The events inside an engine cylinder are nearly adiabatic because the wide fluctuations in temperature take place rapidly compared to the speed with which the cylinder surfaces can conduct heat. Similarly, fluid flow through a nozzle may be so rapid that negligible exchange of heat between fluid and nozzle takes place. The compressions and rarefactions of a sound wave are rapid enough to be considered adiabatic. *See* THERMODYNAMIC PROCESSES. [P.E.Bl.]

Adipose tissue A type of connective tissue which is specialized for the storage of lipids (fats). Adipose cells store lipids in the form of neutral triglycerines in intercellular droplets. Adipose tissue is composed largely of adipose cells, with some capillaries, and fibrous connective tissue.

White adipose tissue is the most common and represents stored food reserves and thermal and physical insulation. Layers of white adipose tissue below the skin, especially in aquatic mammals such as seals and whales, serve as heat barriers, preventing heat loss from the body. Adipose tissue also absorbs physical shock, acting like a cushion. The amount of white adipose tissue in an animal is controlled by sex hormones, metabolic hormones, diet, and physical activity. Many animals, especially hibernating and migratory animals, greatly increase their fat reserves in preparation for travel or hibernation.

Brown adipose tissue is mainly found in subscapular, interscapular, and mediastinal areas. It is associated with thermogenesis (heat production) in hibernating animals and newborn mammals and is metabolically much more active than white adipose tissue. It is thought that brown fat may be an endocrine gland. Although its function is unknown, it is thought to trigger arousal from hibernation and to regulate internal temperature in neonates. *See* CONNECTIVE TISSUE; LIPID. [T.A.G.]

Admittance The reciprocal of the impedance of an electric circuit. Admittance is expressed in the unit mho, coined from the inverse spelling of ohm, the unit of impedance. Admittance is used primarily in computations of parallel alternating-current circuits. By using admittance Y, current I can be expressed as $I = EY$, where E is the voltage across the impedance Z. *See* ALTERNATING CURRENT; CONDUCTANCE; SUSCEPTANCE. [B.L.R.]

Adrenal gland A gland of internal secretion in close proximity to the kidneys. It is always composed of two types of cellular elements with dissimilar embryonic origins, hormones, effects, and controls. One element secretes catecholamines—epinephrine (adrenaline) and norepinephrine (noradrenaline); the other secretes steroid hormones. The catecholamine-secreting cells are called chromaffin cells. The steroid-secreting cells are called adrenocortical cells.

The two types of tissue that compose the adrenal gland may occur as discrete, spatially separate bodies as in fishes; they may be intermixed as among amphibia and most reptiles and birds; or the adrenocortical tissue (cortex) may completely envelop the chromaffin tissue (medulla) as in mammals.

Norepinephrine and epinephrine are derivatives of catechol. The ratios of the two amines secreted is variable and species-specific. The catecholamines exert their effects by interacting with specific α-adrenergic or β-adrenergic receptors on their target cells. Norepinephrine exerts predominantly α-adrenergic action, while epinephrine exerts a predominantly β-adrenergic action, but overlap does occur. Excitation of β-adrenergic receptors causes stimulation of the heart, relaxation of branchioles, dilation of arterioles in skeletal muscle, and elevation of blood sugar levels. These effects are usually seen in response to epinephrine. Excitation of α-adrenergic receptors usually results in vasoconstriction of peripheral blood vessels, and contraction of the smooth muscle of the bladder and the sphincters of the gut. Epinephrine is effective in inducing these effects, but norepinephrine is the hormone usually involved even though it is somewhat less potent. *See* EPINEPHRINE.

The adrenal cortex elaborates many steroid hormones, some essential to life, which fall into four general categories: the mineralocorticoids (which affect mineral metabolism), the glucocorticoids (which affect glucose metabolism), the adrenal cortical androgens (which control secondary sex characteristics in males), and the estrogens (estrus-producing hormones).

Secretion of the glucocorticoids is controlled by adrenocorticotropin (ACTH), secreted by the anterior lobe of the pituitary. Secretion of the mineral corticoid is ultimately controlled by the kidney secretion of renin. *See* ANDROGEN; ESTROGEN; PITUITARY GLAND; STEROID. [H.Lip.]

Malfunctions of the adrenal glands may be divided into two major categories, those of the medulla and those of the cortex. The principal disorders of the medulla are tumors. One rare type, the pheochromocytoma, is an actively secreting neoplasm which causes excessive production of the medullary hormones, epinephrine and norepinephrine. Other rare, nonsecreting tumors of the medulla are the ganglioneuroma and the highly malignant neuroblastoma of infancy and childhood.

Congenital defects of the adrenal glands are principally those related to the absence of glandular tissue or to the lack of its normal secretory ability (hypoplasia).

The adrenals are susceptible to certain infections, largely because they are so well vascularized. The resulting inflammations may produce acute reactions and crises, or in lesser cases, may eventually cause scarring and other degenerative changes with or without hormonal alteration.

There are many varieties of neoplasms of the adrenal cortex. They may be either primary or metastatic. Either type may cause alterations of adrenal hormone output, and the classification is often based on clinical findings and sometimes on histologic types. Major clinical categories include Cushing's syndrome, the adrenogenital syndrome, feminizing syndromes, hyperaldosteronism, and combinations of components of any or several of these.

The functional types, in which hormone output is altered, are more readily recognized because of the changes they effect. Nonfunctional tumors may be silent and therefore often offer a problem in diagnosis. The most common metastatic malignancies in the adrenals arise from bronchogenic carcinoma, although other carcinomas may also spread to these glands. *See* ADDISON'S DISEASE. [E.G.St./N.K.M.]

Adsorption The property of an interface between two immiscible phases (solid, liquid, or vapor) to attract and concentrate components of either phase or both phases as an adsorbed interfacial film. Adsorption is a basic thermodynamic property of interfaces, resulting from a discontinuity in intermolecular or interatomic forces. It is also important in nearly all industrial processes and products.

Some definitions that describe adsorption are as follows: The adsorbent is the solid or liquid which adsorbs. The adsorbate is the solid, liquid, or gas which is adsorbed as molecules, atoms, or ions. Physical adsorption or physisorption is reversible adsorption by weak interactions only; no covalent bonds occur between the adsorbent and adsorbate; heats of physical adsorption are usually less than 15–20 kcal/mole (63–84 kilojoules/mole). Chemical adsorption or chemisorption is adsorption involving stronger interaction between adsorbate and adsorbent usually accompanied by rearrangement of atoms within or between adsorbates; reaction occurs between the surface of the adsorbent and the adsorbate; heats of chemisorption are usually in excess of 20–30 kcal/mole (84–126 kilojoules/mole).

Nearly all vapors tend to adsorb onto inorganic solids at temperatures not too much above their boiling point. The intermolecular attractive forces which cause the physical adsorption of vapors are generally dominated by the London dispersion forces, an attraction caused by the perturbation of electron orbits by adjacent atoms. Another attractive force important in vapor adsorption is the interaction of electron-donor (basic) sites of vapor molecules with electron-acceptor (acidic) sites of adsorbents, or vice versa. These short-range attractions are much stronger than dipole interactions. Silica, an acidic adsorbent, adsorbs basic vapors (water, ammonia, and so forth) much more strongly than acidic vapors (chloroform, CO_2, NO_2, and so forth) regardless of the dipole moments.

The adsorption of water is dominated by hydrogen bonding, an intermolecular acid-base interaction onto neutral surfaces such as graphite or polyethylene, except for the acidic or basic sites provided by impurities on these neutral surfaces.

The strong interactions of chemisorption lead to surface compounds with various degrees of covalent bond character. The adsorbed layers are only one molecule thick because covalent bonds exist only between adjacent atoms. Chemisorption occurs on metals and semiconductors and on oxides and sulfides, but is most often observed on transition metals such as silver, nickel, cobalt, platinum, rhodium, and tungsten. Chemisorption is a necessary step in catalysis by these materials. *See* CHEMICAL DYNAMICS; INTERMOLECULAR FORCES.

Heterogeneous catalysis, in which gas or liquid reactants are specifically adsorbed to a dissimilar phase and chemically altered during their brief retention time, is basic to many industrial processes in the petrochemical, polymer, and chemical industries.

Purification by adsorption is perhaps the oldest known application; examples are wine and beer clarification, color removal in sugar processing, industrial wastewater treatment, and toxic gas adsorption in gas masks.

Adsorption is the basic phenomenon of chromatographic separations, which separate and concentrate components of mixtures according to strength of adsorption onto adsorbents in chromatographic columns.

Adsorption of surface-active substances is the key process in the use of soaps, detergents, emulsifiers, wetting agents, dyes, lubricants, and surface treatments. Other industries dependent on adsorption processes include agriculture, mining, petroleum recovery, papermaking, printing, and photography. *See* CATALYSIS; CHEMICAL SEPARATION TECHNIQUES; CHROMATOGRAPHY; RESPIRATOR. [F.M.F.]

Aegyptopithecus A primate that lived during Oligocene times, 30,000,000 years ago, in the Egyptian Fayum (see illustration), and is believed to be the common

Artist's rendering of *Aegyptopithecus*. (Duke University)

ancestor of humans and apes. The names *Aegyptopithecus zeuxis* and "dawn ape" were given by Elwyn Simons.

Aegyptopithecus had an estimated weight ranging from 8 to 12 lb (3.6 to 5.4 kg). The cranial capacity was about 1.8 $in.^3$ (30 cm^3) larger than that of any of its mammalian contemporaries, relative to body size. The teeth were equipped for a vegetarian diet; males had large, fanglike canines, whereas the canines of females were comparatively small. Studies of arm and leg bones lead to the conclusion that *Aegyptopithecus* was a tree dweller. *See* FOSSIL HUMAN; PRIMATES. [S.P.P.]

Aerial photograph A photograph of a portion of the Earth's surface taken from an aircraft or from a satellite. Most often, photographs are taken sequentially and overlap each other, to give complete coverage of the area of interest. Thus they may be viewed stereoscopically to give a three-dimensional view of the Earth's surface. Although the camera in the aircraft may be pointed obliquely to the side of the line of flight, the most common type of aerial photograph is taken with the camera pointed vertically downward beneath the plane.

Aerial photographs have two main uses, the making of planimetric and topographic maps, and interpretation or data-gathering in a variety of specialized fields. All modern topographic (contour) maps are made from stereoscopic aerial photographs, usually black-and-white panchromatic vertical photographs. The science of making accurate measurements and maps from aerial photographs is called photogrammetry. The geometric discrepancy between the distribution of features seen on the photograph and their actual distribution on the Earth's surface depends on several variables, the most important of which is the relief of the surface. The error becomes greater as relief increases. Photogrammetry deals with these errors and allows highly accurate and complete map production. Topographic maps are made from one or more stereoscopic pairs of aerial photographs by means of optical-mechanical plotting machines or mathematical methods (analytical photogrammetry). *See* CARTOGRAPHY; GEOPHYSICAL EXPLORATION; TOPOGRAPHIC SURVEYING AND MAPPING.

The other major use of aerial photographs is photo interpretation, utilizing all types of film. Photo interpretation is the attempt to extract information about the Earth's surface (and sometimes, subsurface) from aerial photographs. The systematic study of aerial photographs, particularly when they are viewed stereoscopically, gives specialists the ability to gather rapidly, record, map, and interpret a great deal of information. In addition, an aerial photograph is a record at a moment in time; photographs taken at intervals give an excellent record of the change with time in surface features of the Earth. [L.H.L.]

Aerodynamic center A position in an airplane wing about which the wing coefficient of pitching moment is unaffected by wing coefficient of lift. Forces acting on a wing, tail, or other aerodynamic surface can be resolved into a lift and a drag acting at the aerodynamic center, and a pitching moment acting about the center (see illustration). The aerodynamic center of an airfoil section at subsonic speeds is usually close to the quarter chord. *See* AIRFOIL; CENTER OF GRAVITY; CENTER OF PRESSURE. [F.H.R.]

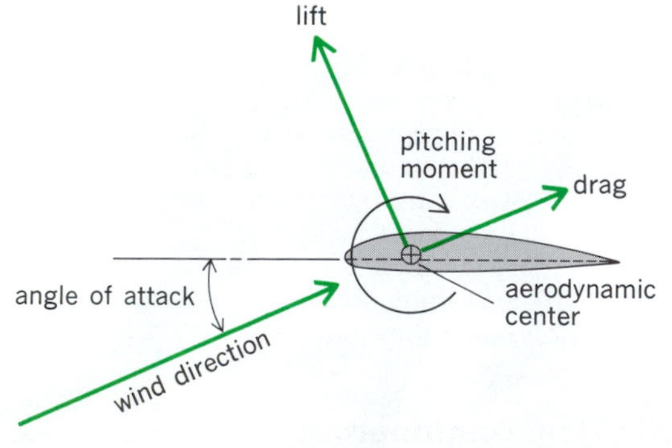

Aerodynamic forces on airfoil section. (*After C.D. Perkins and R.E. Hage, Airplane Performance Stability and Control, John Wiley and Sons, Inc., 1949*)

Aerodynamic force The force between a body and the air caused by their relative motion, such as the wind blowing across the surface of the Earth. Such forces can be described in terms of the change in momentum of the air produced by its interaction with the body. Typical aerodynamic forces are those acting on an aircraft in subsonic flight. These forces are (1) the lift force, perpendicular to the flight path and concerned with the sustentation of an aircraft; (2) the drag force, or resistance acting to the rear along the flight path; and (3) the thrust from the power plant necessary to propel or sustain an aircraft. *See* SUBSONIC FLIGHT; WIND STRESS. [J.Bi.]

Aerodynamic resistance A retarding force that acts upon a body moving through a gaseous medium in a direction opposite to the motion of the body. It is also called aerodynamic drag. The aerodynamic resistance acting on an aircraft in flight is composed of four parts: (1) skin friction drag, which arises from the viscosity of the air; (2) pressure drag, which arises from the pressure field about the body; (3) wave drag, which arises from the compressible nature of air, and which occurs as flight speeds approach and exceed the speed of sound; and (4) induced drag, which is a resistance produced by the generation of lift.

The magnitude of skin friction drag depends upon the surface area of the airplane and the nature of the flow layer that is in immediate contact with it. If the boundary-layer flow is smooth and steady (laminar), the skin friction generated is much less than when this layer becomes turbulent. Unless special devices are employed, such as suction slots to remove the low momentum boundary-layer air before it becomes turbulent, the laminar flow will break down after it has passed over a sufficient length of surface. *See* BOUNDARY-LAYER FLOW.

If sufficient momentum is lost in the boundary layer, the main flow breaks away from the surface, forming a region of separation. Separation results in considerably lower surface pressures than would result in the ideal streamline case. As a consequence, suction pressure on the downstream portions of the body produces pressure drag. This drag is minimized by shaping the body geometrically to reduce the separation region to a minimum by giving it a streamlined shape.

As flight speed of the airplane increases, there comes a point where the flow passing over a surface reaches the local speed of sound. At this speed (Mach 1) pressure waves propagate through the air. This wave propagation forms shock patterns that interact with the boundary layer to cause premature separation and a sudden drag rise. As the speed continues to increase, shock patterns become much stronger until they become analogous to the surface waves created by a ship. Like such waves, at supersonic speeds the generation of these shock waves represents the major contribution to the drag of the vehicle. *See* AERODYNAMIC WAVE DRAG; TRANSONIC FLIGHT.

The final major type of aerodynamic resistance that must be considered arises from the fact that in creating lift the wing must impart a downward momentum to the air flowing past. This downward component of local velocity, termed downwash, rotates the resultant force of wing in such a manner that a force component is directed in the downstream direction. This component, which depends directly upon the magnitude of the lift force, is termed the induced drag. [D.C.H.]

Aerodynamic wave drag The force retarding an airplane, especially in supersonic flight, as a consequence of the formation of shock waves. Although the physical laws governing flight at speeds in excess of the speed of sound are the same as those for subsonic flight, the nature of the flow about an airplane and, as a consequence, the various aerodynamic forces and moments acting on the vehicle at these higher speeds differ substantially from those at subsonic speeds. Basically, these variations result from the fact that at supersonic speeds the airplane moves faster than the disturbances of the air produced by the passage of the airplane. These disturbances are propagated at roughly the speed of sound and, as a result, primarily influence only a region behind the vehicle.

The primary effect of the change in the nature of the flow at supersonic speeds is a marked increase in the drag, resulting from the formation of shock waves about the configuration. These strong disturbances, which may extend for many kilometers from the airplane, cause significant energy losses in the air, the energy being drawn from the airplane. At supersonic flight speeds these waves are swept back obliquely, the angle of obliqueness decreasing with speed. *See* SHOCK WAVE; SUPERSONIC FLIGHT.

The shock waves are associated with outward diversions of the airflow by the various elements of the airplane. This diversion is caused by the leading and trailing edges of the wing and control surfaces, the nose and aft end of the fuselage, and other parts of the vehicle. Major proportions of these effects also result from the wing incidence required to provide lift.

The wave drag at the zero lift condition is reduced primarily by decreasing the thickness-chord ratios for the wings and control surfaces and by increasing the length-diameter ratios for the fuselage and bodies. Also, the leading edge of the wing and the nose of the fuselage are made relatively sharp. With such changes, the severity of the diversions of the flow by these elements is reduced, with a resulting reduction of the strength of the associated shock waves. The wave drag is further reduced by sweeping the wing back. Even with these improvements, aerodynamic efficiencies, or lift-drag ratios, for supersonic airplanes are substantially less than those for subsonic machines. [R.T.Wh.]

Aerodynamics The branch of aeromechanics dealing with the properties and characteristics of, and the forces exerted by, air and other gases in motion. The field of aerodynamics includes the science of a gas itself in motion and the science of bodies immersed in a gas between which there exists a relative motion. *See* GAS DYNAMICS.

Aerodynamics is a broad field with numerous specializations and applications, some of which extend into apparently unrelated fields of science and engineering. Perhaps the most frequently practiced function of the aerodynamicist is the analysis of the forces and moments exerted on a solid body in motion through the air.

Of fundamental significance in the term aerodynamics is the prefix, aero, which refers to the air of the Earth's atmosphere. Until humans can achieve flight within the atmospheres of other planets, aerodynamic flight and the majority of practical considerations will be confined to the limits of the Earth's atmosphere as defined in aerodynamic terms. [W.C.W.]

Aeroelasticity The branch of applied mechanics which deals with the interaction of aerodynamic, inertial, and structural forces. It is important in the design of airplanes, helicopters, missiles, suspension bridges, power lines, tall chimneys, and even stop signs. Variations on the term aeroelasticity have been coined to denote additional significant interactions. Aerothermoelasticity is concerned with effects of aerodynamic heating on aeroelastic behavior in high-speed flight. Aeroservoelasticity deals with the interaction of automatic controls and aeroelastic response and stability. In the field of hydroelasticity, a liquid rather than air generates the fluid forces.

The primary concerns of aeroelasticity include flying qualities (that is, stability and control), flutter, and structural loads arising from maneuvers and atmospheric turbulence. Methods of aeroelastic analysis differ according to the time dependence of the inertial and aerodynamic forces that are involved. For the analysis of flying qualities and maneuvering loads wherein the aerodynamic loads vary relatively slowly, quasi-static methods are applicable, although autopilot interaction could require more general methods. The remaining problems are dynamic, and methods of analysis differ according to whether the time dependence is arbitrary (that is, transient or random) or simply oscillatory in the steady state.

The redistribution of airloads caused by structural deformation will change the lifting effectiveness on the aerodynamic surfaces from that of a rigid vehicle. The simultaneous analysis of the equilibrium and compatibility among the external airloads, the internal structural and inertial loads, and the total flow disturbance, including the disturbance resulting from structural deformation, leads to a determination of the equilibrium aeroelastic state. If the airloads tend to increase the total flow disturbance, the lift effectiveness is increased; if the airloads decrease the total flow disturbance, the effectiveness decreases.

The airloads induced by means of a control-surface deflection also induce an aeroelastic loading of the entire system. Equilibrium is determined as in the analysis of load redistribution. Again, the effectiveness will differ from that of a rigid system, and may increase or decrease depending on the relationship between the net external loading and the deformation.

A self-excited vibration is possible if a disturbance to an aeroelastic system gives rise to unsteady aerodynamic loads such that the ensuing motion can be sustained. At the flutter speed a critical phasing between the motion and the loading permits extraction of an amount of energy from the airstream equal to that dissipated by internal damping during each cycle and thereby sustains a neutrally stable periodic motion. At lower speeds any disturbance will be damped, while at higher speeds, or at least in a range of higher speeds, disturbances will be amplified. *See* FLUTTER (AERONAUTICS).

Transient meteorological conditions such as wind shears, vertical drafts, mountain waves, and clear air or storm turbulence impose significant dynamic loads on aircraft. So does buffeting during flight at high angles of attack or at transonic speeds. The response of the aircraft determines the stresses in the structure and the comfort of the occupants. Aeroelastic behavior makes a condition of dynamic overstress possible; in many instances, the amplified stresses can be substantially higher than those that would occur if the structure were much stiffer. *See* CLEAR-AIR TURBULENCE (CAT); LOADS, DYNAMIC; TRANSONIC FLIGHT. [W.P.R.]

Aeromonas A genus of facultatively anaerobic, polarly flagellated gram-negative rod-shaped bacteria which are of aquatic (not marine) origin and infect chiefly cold-blooded animals, such as fishes, reptiles, and amphibians, and only occasionally warm-blooded animals or humans. The best-known infection is red leg disease in frogs. In humans, wound infections may occur following contact with contaminated water or soil.

There is evidence for a carrier state in the gut. Septicemia has been observed in immunocompromised individuals. The most important disease, however, is diarrhea. *See* DIARRHEA; MEDICAL BACTERIOLOGY. [A.W.C.V.G.]

Aeronautical engineering That branch of engineering which is concerned primarily with the special problems of flight. Historically, aeronautical engineering was an offshoot of mechanical engineering. However, as the technology of flight developed, distinctions were made in the curriculum of the aeronautical engineer. *See* AERODYNAMICS; AIRCRAFT; AIRCRAFT ENGINE; FLUID MECHANICS.

Since 1940 aeronautical engineering has drastically expanded. Flight speeds have increased from a few hundred miles per hour to satellite and space vehicle velocities. The common means of propulsion have changed from propellers to turboprops, turbojets, ramjets, and rockets. *See* JET PROPULSION; ROCKET; SPACE TECHNOLOGY.

The aeronautical engineer now works in fields which were once the exclusive domain of the physicist. Similarly, many jobs which were historically the exclusive domain of the aeronautical engineer are now being done by others who have acquired legitimate interests there.

In general, aeronautical engineering is no longer a definable domain. Rather, the aeronautical engineer with his or her broad training is often most usefully engaged in coordinating the many different disciplines which now enter into the engineering of flight. [J.R.Se.]

Aeronautical meteorology That branch of meteorological study which deals with atmospheric effects on the operation of aircraft—heavier than air and lighter than air—rockets, missiles, and projectiles. Eight major effects are considered: low visibility at terminals, turbulence, clear-air turbulence, upper winds and temperature, jet stream, tropopause, lightning, and icing.

Fog, snow, and rain prevent landings and takeoffs when horizontal visibility at the surface falls below regulatory minimum values for crewed aircraft. The minimum is generally 2400 ft (800 m) in the United States for airplanes equipped to operate under standard instrument conditions. Runway visual range (RVR), an operational concept of visibility, is rapidly supplanting the use of conventional meteorological visibility. It is the distance at which high-intensity lights may be observed down the instrument runway in the direction of landing or takeoff.

Atmospheric turbulence is defined as any deviation from the normal, steady, horizontal flow of air and occurs in two ways: sharp-edged gusts in the horizontal or vertical and large-scale up- and downdrafts which cause aircraft to sink or rise more gradually and for longer periods. [H.T.H.]

The unexpected jolts which an aircraft may receive in clear air turbulence (CAT) add to the danger of this phenomenon, especially with respect to passenger safety and comfort. Forecasting of CAT has only been moderately successful in the past. Since clear-air turbulence is a small-scale phenomenon, the radiosonde network is insufficient in locating or predicting the precise locations of CAT zones. *See* CLEAR-AIR TURBULENCE (CAT).

Since aircraft are part of the air current in which they are embedded, they experience aiding or retarding effects determined by wind direction in relation to the course being flown. Careful planning of long-range flights is completed with the aid of prognostic upper-air charts. These make it possible to select flight tracks and cruising altitudes where more favorable winds and temperature will result in a lower elapsed flight time in spite of added ground distance. This planned procedure is known as minimal time track (least time track, pressure pattern) flying.

The jet stream is a meandering, shifting current of relatively swift windflow embedded in the general westerly circulation at upper levels. Sometimes girdling the globe at middle and subtropical latitudes where the strongest jets are found, this band of strong winds has great operational significance for aircraft flying at cruising levels of 4–10 mi (6–15 km). It is the cause of most serious flight-schedule delays and may result in unscheduled fuel stops. *See* JET STREAM.

The tropopause is the boundary between troposphere and stratosphere and is generally defined in terms of points in the vertical air-mass soundings where the temperature lapse rate becomes less than 5.7°F/mi (2°C/km). Its significance to aviation arises from its frequent association with mild forms of clear-air turbulence and change of vertical temperature gradient with altitude, which has considerable effect on turbine-engine performance. The tropopause also marks the vertical limit of clouds and storms; however, thunderstorms are known to punch through the surface to heights of 14.2 mi (23 km) at times, and lenticular clouds of strong mountain waves have been observed at similar heights. *See* TROPOPAUSE.

Lightning strikes or static discharges, when experienced inside an aircraft, cause a blinding flash and usually a muffled explosive sound audible above the roar of the engines. Atmospheric conditions favorable for strikes follow a consistent pattern: solid clouds or enough clouds for the aircraft to be intermittently on instruments, active precipitation of an icy character, and ambient air temperature close to 32°F (0°C). St. Elmo's fire, radio static, and choppy air often precede the strike.

Flight through a subcooled cloud, freezing rain, or wet snow commonly results in icing. Once one of the major hazards of flying, icing has been reduced to a minor weather factor by effective antiicing and deicing devices for the protection of critical components of aircraft. Icy or deeply snow-covered airport runways cause operational problems for aviation, but these factors have been reduced in importance through the use of modern snow-removal methods at major airports. [H.T.H.]

Aeronautics The principles underlying flight through the air. In the early years of flight, the term aeronautics referred to the art of operating aircraft of all sorts, and included the design of aircraft, which is now usually included as part of aeronautical engineering, and the actual flying of aircraft. However, as the term aeronautical engineering became more widespread, another word was sought to describe the basic scientific knowledge which underlies aeronautical engineering. Aeronautics was adopted to fit this need as being less restrictive than aerodynamics. Thus, for example, the theories of airflow about airfoils and wings may be categorized as aeronautics, although the application of those principles in developing a particular design is more precisely an engineering accomplishment. *See* AERONAUTICAL ENGINEERING; ASTRONAUTICS. [J.R.Se.]

Aeronomy The science of upper atmospheres, especially as influenced by solar ultraviolet energy which dissociates molecules and ionizes molecules and atoms. *See* ATMOSPHERE.

Weather and most human activities are confined to the troposphere. The warm temperatures of the upper stratosphere and lower mesosphere are due to absorption of solar energy by ozone. The hot temperatures of the thermosphere are associated with the generation of the ionosphere. The density, or amount of gas measured in atoms or molecules per unit volume, varies over an enormous range because of the pull of gravity.

On the Earth, oxygen (O_2) molecules are dissociated by radiation with wavelength below 242 nanometers. Some of the atoms remain free for a time; others combine with another O_2 molecule to form ozone, O_3. This molecule in turn absorbs radiation up to about 300 nm, filtering out this radiation which is damaging to many organisms, including humans. Most of the radiation has been absorbed at the 15-mi (25-km) level, putting a near stop to further production at lower altitudes. Although some ozone is transported down by atmospheric motions, it forms a peak, or "layer," in the lower stratosphere. At higher altitudes, especially in the thermosphere, most of the oxygen atoms remain free and come to dominate the upper atmosphere above 95 mi (150 km). *See* AIR POLLUTION; ATMOSPHERIC OZONE; UPPER-ATMOSPHERIC DYNAMICS.

A remarkable property of the stratosphere and higher

regions is extreme dryness: only a few water (H_2O) molecules per million air molecules, in contrast to several per hundred at ground level. Water is partially dissociated above the ozone layer (as are nearly all molecules except nitrogen, N_2), and the fragments participate in the destruction of ozone.

Solar radiation in the far and extreme ultraviolet is able to ionize atmospheric gases (remove an electron, leaving a positive ion). The Earth's ionosphere coincides rather closely with the thermosphere. Its location is determined by depletion of the ionizing solar radiation, which is very strongly absorbed. There is, however, a long tail extending to and even below the mesopause, corresponding to solar wavelengths (mostly x-rays) that are relatively penetrating, and also cosmic rays. All long-distance radio propagation is due to reflection from the ionosphere. *See* IONOSPHERE.

The night and day skies give off a variety of optical emissions characteristic of particular atoms, molecules, and ions. They are of considerable interest because they permit remote sensing from the ground or from satellites. In the daytime the strongest features arise by fluorescence: absorption of solar energy followed by reemission. Some of these phenomena can be followed into twilight. Many others are generated during the recombination of ions with electrons, or oxygen atoms with each other. Generally these processes operate into the night and constitute the night airglow or nightglow. Generally speaking, the dayglow cannot be observed from the ground because of the overwhelming brightness of the sky; the few exceptions involve very delicate experiments. Twilight offers a much easier opportunity for certain emissions, notably those of the alkali atoms sodium, lithium, and potassium. *See* AIRGLOW; ALKALI EMISSIONS; AURORA; SOLAR RADIATION; STRATOSPHERE; ULTRAVIOLET RADIATION. [D.M.Hu.]

Aerosol A suspension of small particles in a gas. The particles may be solid or liquid or a mixture of the two. Aerosols are formed by the conversion of gases to particles, the disintegration of liquids or solids, or the resuspension of powdered material. Aerosol formation from the gas results in much finer particles than disintegration processes (except when condensation takes place directly on existing large particles). Dust, smoke, fume, haze, and mist are terms in common use for aerosols. Dust usually refers to solid particles produced by disintegration, while smoke and fume particles are generally smaller and formed from the gas phase. Mists are composed of liquid droplets. These special terms are helpful but are difficult to define exactly; therefore, the generic term aerosol is preferred.

Aerosol particles range in size from molecular clusters of the order of 1 nanometer to particles as large as 100 micrometers. Only the stable clusters formed by homogeneous nucleation and the small primary particles that compose agglomerates have a significant fraction of their molecules in the surface layer.

Aerosols play an important role in several branches of science and technology. They influence climate both directly and indirectly. They directly affect radiation transfer on the global and regional scales. Indirect effects result from their role as cloud condensation nuclei in changing droplet size distributions that affect the optical properties of clouds and precipitation. The stratospheric aerosol may play a role in ozone destruction. *See* ATMOSPHERIC OZONE; METEOROLOGICAL OPTICS; RADIATIVE TRANSFER.

Undesirable aerosols generated by human activities contribute to local, regional, and global air pollution. Coal combustion and the metallurgical and mineral products industries rank high among industrial sources in total mass emissions. However, much of the submicrometer aerosol in a polluted air mass results from the conversion of pollutant gases such as the oxides of sulfur and nitrogen and organic vapors to aerosols. Aerosols affect human health, and particulate concentrations are regulated both in the atmosphere and in industrial emissions. *See* AIR POLLUTION.

Aerosol processes are used in industry for the production of important commodity products, including titanium dioxide pigment, pyrogenic (fumed) silica, and carbon black. Optical fibers are fabricated by an aerosol process in which a silica fume is deposited on the walls of a quartz tube along with suitable dopants to control the refractive index gradient across the tube diameter. Ceramic and pharmaceutical powdered materials produced by aerosol processes offer particles of controlled size and purity in plant-scale facilities. Aerosol generators have been designed for drug delivery by inhalation; the large surface area of the lung permits rapid uptake of medication by the blood under more benign conditions than by ingestion.

Atmospheric aerosols and aerosols emitted from industrial sources are normally composed of mixtures of chemical compounds. Each chemical species is distributed with respect to particle size in a way that depends on its source and history; hence, different substances tend to accumulate in different ranges of the particle size distribution. This effect has been observed for atmospheric aerosols and for emissions from pulverized coal combustion and municipal waste incinerators, and it undoubtedly occurs in emissions from other sources. Chemical segregation with respect to size has important implications for the effects of aerosols on public health and the environment, because particle transport and deposition depend strongly on particle size. *See* PARTICULATES. [S.K.F.]

Aerospace medicine The special field of medicine that deals with humans in environments encountered beyond the surface of the Earth. It includes both aviation medicine and space medicine and is concerned with humans, their environment, and the vehicle in which they fly. Its objective is to ensure human health, safety, well-being, and effective performance through careful selection and training of flight personnel, protection from the unique flight environment and its physiological and psychological effects, and understanding of the flight vehicle and humans' interaction with it.

The environment encountered in air or space flight is very different from that on Earth. Consequently aerospace medicine must deal with the physics of the atmosphere and space and with the conditions and influences introduced by the flight vehicle. Aerospace medicine, environmental physiology, and psychology are therefore concerned with the effects of changes in barometric pressure, oxygen deprivation at reduced barometric pressure, atmospheric constituents, toxic substances, acceleration, weightlessness, noise, vibration, ionizing radiation, and thermal and other environmental stresses.

All of the environmental hazards encumbent in air travel are present in space flight, but space also has conditions that are unique. The near-vacuum of the space environment can be overcome through engineering design, but the deleterious effects of exposure to high-energy heavy-particle radiation have been avoided only by the low Earth orbit and relatively short duration of space travel. Investigation of the biological response to such radiation is needed prior to long-duration crewed space flight. The condition of near-weightlessness in space has produced significant physiological effects in humans, including motion sickness, bone and muscle loss, reduced red blood cell production, and cardiovascular deconditioning. Aerospace medicine is concerned with developing countermeasures, including physical, psychological, pharmacological, or nutritional means, to prevent or control these physiological effects. Within the confines of the spacecraft, life support devices must provide for the regeneration of oxygen, reclamation of water, and removal of wastes and toxic gaseous materials. *See* SPACE BIOLOGY; WEIGHTLESSNESS. [J.W.H.]

Aerostatics The science of the equilibrium of gases and of solid bodies immersed in them when under the influence only of natural gravitational forces. Aerostatics is concerned with the balance between the weight of the gases and the weight of any object within them. Archimedes' law that an immersed body experiences a buoyancy force equal to the weight of the fluid displaced is the principal law of aerostatics, if the fluid is air, or of hydrostatics, if the fluid is water. Some phases of meteorology and the flight of balloons and dirigibles are based on aerostatics. In meteorology cloud and fog subsidence and simple pressure and temperature relations with altitude are predicted from aerostatic principles. *See* FLUID STATICS; HYDROSTATICS. [J.R.Se.]

Aerothermodynamics Flow of gases in which heat exchanges produce a significant effect on the flow. Traditionally, aerodynamics treats the flow of gases, usually air, in which the thermodynamic state is not far different from standard atmospheric conditions at sea level. In such a case the pressure, temperature, and density are related by the simple equation of state for a perfect gas; and the rest of the gas's properties, such as specific heat, viscosity, and thermal conductivity, are assumed constant. Because fluid properties of a gas depend upon its temperature and composition, analysis of flow systems in which temperatures are high or in which the composition of the gas varies (as it does at high velocities) requires simultaneous examination of thermal and dynamic phenomena. For instance, at hypersonic flight speed the characteristic temperature in the shock layer of a blunted body or in the boundary layer of a slender body is proportional to the square of the Mach number. These are aerothermodynamic phenomena.

Two problems of particular importance require aerothermodynamic considerations: combustion and high-speed flight. Chemical reactions sustained by combustion flow systems produce high temperatures and variable gas composition. Because of oxidation (combustion) and in some cases dissociation and ionization processes, these systems are sometimes described as aerothermochemical. In high-speed flight the kinetic energy used by a vehicle to overcome drag forces is converted into compression work on the surrounding gas and thereby raises the gas temperature. Temperature of the gas may become high enough to cause dissociation (at Mach number ≥ 7) and ionization (at Mach number ≥ 12); thus the gas becomes chemically active and electrically conducting. *See* HYPERSONIC FLIGHT; JET PROPULSION; ROCKET PROPULSION. [S.-Y.C.]

Affective disorders A group of psychiatric states and syndromes, also known as mood disorders, characterized by disturbance of mood, most often depression, but also elation and mania. These states are diagnosed by symptomatic and behavioral criteria that are codified by the American Psychiatric Association, which acknowledges that mood disorders do not make up a homogeneous group as to etiology or treatment; rather, the category includes multiple disorders with varying etiologies.

Clinical syndromes. There is general agreement that the affective disorders include two clinical syndromes, the depressions and the elations and mania.

The term depression has many different meanings. For the neurophysiologist, depression refers to any decrease in the electrophysiological activity of a cell or organ, as in cortical depression. The pharmacologist uses the term to refer to drug actions that decrease the activity of an organ; thus, the central nervous system depressants include the barbiturates and the anesthetics. The psychologist uses the term to refer to a decrease in optimal performance, such as slowing of psychomotor activity or reduction of intellectual functioning. For the clinical psychiatrist, however, depression covers a broad spectrum of changes in mood and affective state, ranging in severity from the normal mood fluctuations of everyday life, sometimes called sadness or despondency, to severe psychotic episodes. *See* ANESTHESIA; BARBITURATES; PSYCHOSIS.

Some manic episodes involve psychotic features which can be specified as mood congruent or mood incongruent. About 75% of manic individuals have more than one manic episode in their lives. Almost all of these individuals also have depressive episodes; however, there is a small minority of people with only recurrent manic episodes. Symptoms of mania are the following: elevated mood (high, elated, euphoric, and ecstatic); irritability, hostility, and belligerence; increased self-esteem; grandiosity; poor judgment; overactivity (motor, social, and sexual); flight of ideas; decreased need for sleep; distractibility; involvement in buying sprees; social intrusiveness; and inappropriate collection of clothes or other objects.

Classification. The four major classes of mood disorders are organic mood disorders, bipolar disorders, depressive disorders, and adjustment disorders with depressed mood.

Both depressions and manic episodes may occur secondary to organic illnesses, including neurologic disorders and systemic medical illnesses, and as adverse effects of drugs commonly used in the treatment of medical conditions. Often these syndromes are indistinguishable symptomatically from bipolar disorders and are recognizable only by careful scrutiny of the individual's history. However, in the case of an organic disorder, careful attention to the treatment of the underlying medical and neurological condition is essential.

The bipolar disorders include a group of conditions whose common features are one or more episodes of mania. Most often, the patient has multiple episodes throughout life, with both manic and depressive features. Cases of recurrent mania without depression are uncommon.

Cyclothymic disorder, previously called cyclothymic personality disorder, refers to a mild form of bipolar disorder in which the intensity of the depressive or manic episodes does not reach full criteria.

Depressive disorders are the most common of the major classes of mood disorders. The epidemiologic evidence indicates that 4–8% of the adult population has a major depression or dysthymia in a given year. Major depression occurs episodically: the majority of patients will have recurrent episodes with a tendency toward relapse. About 15% of patients will become chronic, that is, the depressive symptoms will be unremitting.

Individuals with this disorder are mildly depressed, but chronically distressed. The chronicity of the disorder often results in impairment of social functioning, family relations, and work performance.

Depressive symptoms are quite common in society at large, often occurring in response to some significant life event which serves as a precipitant, or trigger. In these cases, where the relationship between the change in mood and the life event is most clear, the diagnosis of adjustment disorder is appropriate.

Etiology and treatment. Multiple etiologic factors—genetic, biochemical, psychodynamic, and socioenvironmental—may interact in any individual in complex ways. Knowledge of etiology for individuals, however, is uncertain, and clinicians suspect varying combinations of stress, personality, central nervous system changes, and other factors. Researchers have investigated risk factors leading to affective disorders. Sociodemographic factors include sex and gender, age, social class, and marital status; etiologic factors include genetic factors, personality factors, and life events, including childhood loss and separation, and environmental stress.

A variety of treatments are available for depressive disorders and for elations and mania. The most widely used treatments are drug-based, but various forms of psychotherapy have also been developed. Short-term methods of psychotherapy include interpersonal, cognitive behavioral, and family behavior psychotherapy; long-term methods include psychoanalysis and family treatment. In addition, electroconvulsive therapy is still useful in severe cases. *See* PSYCHOTHERAPY.

For treatment of manic states, control of disruptive behavior is usually achieved through the use of neuroleptic drugs, such as haliperidol or the phenothiazines. However, for long-term treatment for bipolar illness, lithium is the preferred drug, since control studies indicate it prevents the recurrence of both manic and depressive episodes and facilitates the individual's personal and social adjustment. [G.L.K.]

Aflatoxin Any of a group of toxic compounds produced during growth of the common mold *Aspergillus flavus* on peanuts, cereal grains, meals, and other food or animal feedstuff. Aflatoxins have been responsible for acute poisoning of fish, poultry, and livestock fed moldy grain and peanut meal containing comparatively high levels of toxin. They can cause liver cancer in livestock, poultry, fish, or domestic animals when ingested in relatively low levels over long periods. In fact, aflatoxin B_1 is the most potent liver carcinogen known. There is considerable indirect evidence that aflatoxins cause liver cancer in humans. [G.E.T.]

Africa A continent of 11,700,000 mi² (30,300,000 km²), Africa is the second largest continent, exceeded only by Eurasia. Despite its great area, Africa is a relatively simple landmass in terms of its climatic and vegetational distributions, its geological structure, the shape of its outline, and the distribution of its surface features. In part this apparent simplicity is the result of its latitudinal extent, which is almost exactly bisected by the Equator, and in part the result of its long degradational history, inasmuch as the greater part of the continent has stood as a land mass of crystalline rocks since Precambrian times. Africa is the most stable portion of the southern, or Gondwana, part of a postulated original continent, Pangaea, from which the Southern Hemisphere continents separated in Cretaceous time. As a result of this remarkable tectonic stability, most of the continent has an elevated rolling plateau surface alternating between broad swells where the ancient rocks crop out and wide basins filled with continental sediments derived from the marginal swells. Because of the rolling surface and the central position of the Equator, Africa displays a notable symmetry in the distribution of climatic types and vegetational categories at equal latitudes north and south of the Equator.

The character of the climates of Africa depends more upon rainfall than upon temperature. Because every part of the continent receives strong insolation throughout the year, and no part of Africa is subjected to the invasion of cold continental air masses from higher latitudes, frost is unknown except in limited areas of high altitude or places so dry that nighttime radiation is very intense. It is at these same dry places that the very highest daytime temperatures occur. The temperature of 136.4°F (58°C) at Azizia, Tripoli, is cited by some authorities as the world's extreme. Nevertheless, temperatures are virtually everywhere within a range that assures vegetational growth throughout the year, and the amount, seasonal distribution, and reliability of rainfall are critical to the life of plants, and therefore broadly to the fauna, including humans.

The continent of Africa is dominated by three types of air masses: tropical maritime, which is warm and moist and has its sources over the oceans north and south of the Equator; tropical continental, which is warm and dry, one mass originating over the Sahara and the other over the Kalahari; and polar maritime, which is cool and moist and invades Africa in two areas, northwest in the Northern Hemisphere wintertime and southwest in the Southern Hemisphere wintertime. The air masses shift in position to conform in general with the shift in the apparent position of the Sun as the seasons pass during the year. *See* AIR MASS.

Africa is a continent much affected by aridity, which in some places endures throughout the year and in others is only seasonal. Contrary to the popular concept of an Africa teeming with vast dark forests of gigantic trees, the greater part is either scrubland, grassland, or desert. The true tropical rainforest is limited to the always rainy areas of humid equatorial climate in the central and western sections of the continent and to a narrow strip along the Guinea coast. All the tropical rainforests and the associated riverine forests and mangrove swamps together cover less than 10% of Africa. Many centuries of slashing and burning by the native inhabitants have reduced most of these forests to degenerate tropical bush.

Poleward from the tropical rainforest position in each hemisphere and also along the higher and drier eastern side of Africa there is a broad horseshoe-shaped band of semideciduous forest, often called dry forest or thorn forest, and of savanna or tall grass with scattered trees and bushes. These vegetation combinations reflect the alternation of precipitation seasons between the wet growing season during "high sun" and the drought during "low sun." A full 40% of Africa experiences the annual changes. Another 45% of Africa is so poorly watered at all seasons that it supports only desert plants and dry savanna of short grasses. This dry zone runs broadly across northern Africa as the Sahara and its margins. It is present in the Southern Hemisphere in the same latitudes, but the continent is narrow there and the dry conditions are not so widely and severely developed. The remainder of Africa displays Mediterranean scrubforest vegetation at its northern and southern extremities, and some middle-latitude and alpine vegetation forms in the highlands of Ethiopia and on the upland areas of eastern and southern Africa.

Africa is sometimes called a plateau because in many places, particularly south of the Equator, the rolling surface of the continent is bounded by steep escarpments which face the sea and drop down to narrow coastal plains. The simple outline of Africa reflects the narrow continental shelf which borders it. Compared to other continents, Africa is a massive, rigid, and stable block, exhibiting folding and faulting in limited zones only, with the basement complex of crystalline rocks exposed over about one-third of its surface in marginal swells.

The surface of the African plateau drops from elevations of 4000–5000 ft (1200–1500 m) in the Southern Hemisphere to about 1000 ft (300 m) in the Sahara. In southern and central Africa there is evidence suggestive of three or four important erosional surfaces, which are presumed to have resulted from periodic uplift of the African block. Locally, the upthrust edges of the block may reach considerable heights, as in the Drakensberg of South Africa where elevations exceed 11,500 ft (3500 m).

The massiveness of the African continent and the simplicity of its outline are expressed in the paucity of islands associated with it. Excluding marshy deltaic islands, the Atlantic shore of Africa has a few islets jutting off points of land and the far more significant line of islands in the Gulf of Guinea which represent extensions of the volcanic peaks found in the Cameroon Mountains. The Indian Ocean coast also has only small close inshore deltaic and headland islands, and coral reefs.

The sole great island is Madagascar, which with an area of 250,000 mi² (650,000 km²) is one of the largest islands of the world. Separated from Africa by the Mozambique Strait and 250 mi (400 km) of open water, Madagascar is almost 1000 mi (1600 km) long. Lying athwart the southeast trade winds, Madagascar has orographic rainfall on its eastern high margin all of the year, whereas the west lies in a rain shadow. Madagascar's long separation from Africa is demonstrated by remarkable faunal and floral differences between the island and the mainland.

One-third of the African continent has no drainage to the surrounding oceans and seas, and most of the remainder is drained by only a few major rivers. Because so much of Africa is a plateau surface of swells and vast basins, most of the great rivers which reach the sea have long middle courses of remarkably low gradients between ungraded headwaters and the waterfalls and cataracts which mark their channels as they fall over the plateau rim to sea level. Although this fact has meant that African rivers were not easy ways of penetrating into the continent in the period of European exploration (nor are they paths of uninterrupted navigation today), the rivers do produce great amounts of energy so that Africa, despite its aridity, has an estimated 23% of the world's potential hydroelectric power.

The only perennial stream which enters the Mediterranean Sea from Africa is the Nile, whose 4000 mi (6400 km) makes it the longest of all African rivers. From the southern edge of the Sahara to its mouth in the Nile Delta, it is a dwindling river, without tributaries to replace water lost to percolation and evaporation and to withdrawal for irrigation. Within Egypt the Nile has been increasingly controlled over the last 2 centuries by engineering structures, culminating in 1971 in the Aswan High Dam which created Lake Nasser, over 300 mi (480 km) long and extending into the Republic of the Sudan. Above the Blue Nile junction the main stream is known as the White Nile. The plain between the two rivers is the Gezira. Upstream the White Nile is joined by the Bahr-el-Ghazel which feeds in water from the left bank. The Bahr-el-Jebel or upper White Nile has drained from the northeast corner of Lake Albert (Lake Mobutu) and the swamps of Lake Kyoga, and these waters have their source in Lake Victoria and the streams which feed it.

The rainy sub-Saharan lands west of the longitude of the Bight of Biafra have many rivers, most of them short, precipitous, and subject to large fluctuations in seasonal volumes. There are only three great river systems: The Senegal River makes a northwestern arc to reach the Atlantic north of Cape Verde and on the south margin of the Sahara. The Niger flows northeastward into ever drier climatic conditions and spreads its waters over a swampy "interior delta" of its own making. Beyond the vicinity of Timbuktu the Niger swings abruptly southeastward, following the valley of a stream that captured the upper reaches, and after a total course of 2600 mi (4160 km) attains the Gulf of Guinea by way of a vast coastal delta. Within the great loop of the Niger lies the Volta system, which collects much of the waters of Ghana and lands to the north and west through stream integration.

The great river of tropical Africa is the Congo. Nearly 3000 mi (4800 km) long, its middle portion provides 1000 mi (1600 km) of low-gradient, navigable waters between Stanley Falls upstream and Stanley Pool downstream. From Stanley Pool to the Atlantic is a descent of over 800 ft (240 m) in rapids and waterfalls to form the premier site for waterpower development in Africa. Below the cataracts, the lower Congo has 80 mi (129 km) of waters navigable by ocean vessels. The upper Congo, or Lualaba, is navigable in stretches.

Because of its confused structure and topography, eastern Africa has developed no great integrated river system other than the Nile. The great lakes occupying troughs in the lift valley complex are minimally integrated. Lakes Albert, Edward, and George are connected to the Nile drainage. Lake Kivu drains to Lake Tanganyika and thence to the Congo system. Lake Malawi drains to the Shire River, which in turn enters the Zambezi and the Indian Ocean. Lake Rudolf and many smaller rift valley lakes have no outlets. [H.V.B.K.]

African horsesickness A highly fatal insect-borne viral disease of horses and mules, and a mild subclinical disease in donkeys and zebras. It normally occurs in sub-Saharan Africa but occasionally spreads to North Africa, the Iberian Peninsula, and Asia Minor.

The African horsesickness virus is an orbivirus (family Reoviridae) measuring 68–70 nanometers in diameter. The outer layer of the double-layered protein shell is ill defined and diffuse and is formed by two polypeptides. The highly structured core consists of five structural proteins arranged in icosahedral symmetry. The viral genome is composed of 10 double-stranded ribonucleic acid (RNA) segments (genes) ranging in size from 240,000 to 2,530,000 daltons. Nine distinct serotypes which can be distinguished by neutralization tests are known. The virus can be cultivated in various cell cultures, in the brains of newborn mice, and in embryonated hen eggs by intravascular inoculation. *See* Animal virus.

African horsesickness is a noncontagious disease that can readily be transmitted by the injection of infective blood or organ suspensions. In nature, the virus is biologically transmitted by midges of the genus *Culicoides*, such as *C. imicola*. The disease has a seasonal incidence in temperate regions (late summer to autumn), and its prevalence is influenced by climatic conditions favoring insect breeding (for example, warm, moist conditions in low-lying areas). Mechanical transmission by large biting flies is possible, but plays a much smaller role than biological transmission in the epidemiology of this disease.

There is no specific treatment for this disease. Infected animals should be given complete rest as the slightest exertion may result in death. Stabling of horses at night during the African horsesickness season reduces exposure to insect vectors and hence the risk of disease transmission. Prophylactic vaccination is the most practical and effective control measure. In epidemic situations outside Africa, the causal virus should be serotyped as soon as possible, allowing the use of a monovalent vaccine. However, in endemic regions it is imperative to use a polyvalent vaccine, which should render protection against all nine serotypes of African horsesickness virus. *See* Disease; Epidemic; Vaccination. [B.E.]

Afterburner A device in a turbojet aircraft engine, between turbine and nozzle, which improves thrust by the burning of additional fuel (see illustration). To develop additional

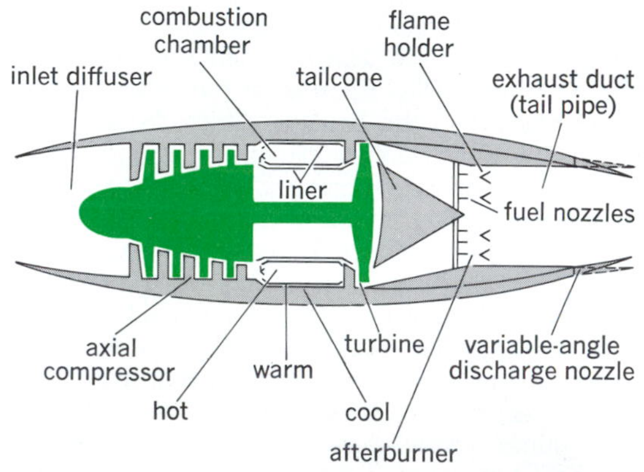

Diagram of turbojet engine showing afterburner.

thrust for takeoff and climb and for periods of dash of military aircraft, it is advantageous to augment the engine thrust. This is done by afterburning, also called reheating, tail-pipe burning, or postcombustion. The augmentation of thrust obtained by afterburning may be well over 40% of the normal thrust and at supersonic flight can exceed 100% of normal thrust. *See* Turbojet. [B.Pi.]

Agammaglobulinemia An immunologic deficiency state characterized by an increased susceptibility to infection. Diagnosis is based on the demonstration of low concentrations or absence of one of several immunoglobulins (gamma globulin), and a poor response of the immune system to infection or other antigenic challenge. Most of the significant agammaglobulinemias are the result of a decreased synthesis of the immune globulins because of a deficiency of plasma cell number or function.

One type of primary (hereditary) agammaglobulinemia is usually transmitted as an autosomal recessive and is termed alymphocytic agammaglobulinemia, characterized by a complete absence of lymphocytes. Affected infants cannot produce a humoral or a cell-mediated immune response and are severely predisposed to infections and usually die within the first year of life. Another type of agammaglobulinemia is one in which the individual is born without a thymus gland (Di George's syndrome). Symptoms include tetany and susceptibility to bacterial but not viral or fungal infections. Another hereditary type of agammaglobulinemia (Bruton's disease) is a sex-linked recessive disorder characterized by a deficiency of all types of immunoglobulins, reflecting a failure of the entire humoral antibody marrow system. In this disease the thymus may be normal, but the lymph nodes and spleen lack lymph cell follicles. *See* BLOOD; HUMAN GENETICS.

Treatment of deficiency disorders of the immunoglobulins has become possible. Infection is a problem, and treatment requires identification of the microorganism and use of specific chemotherapy. Low serum gamma globulin levels can be increased with concentrated human gamma globulin. *See* IMMUNOGLOBULIN. [N.K.M.]

Agar A gelatinous product extracted from several species of red algae, particularly from *Gelidium* species and *Gracilaria lichenoides* (Ceylon moss) and species of *Ahnfeltia* and *Pterocladia*. Agar occurs in the natural state as a cell wall constituent of these plants.

Agar is prepared by boiling the algae in water, after which the product is cooled, purified, and dried. It is an amorphous, translucent material which is used as a gelling agent in culture media (chiefly for bacteria), confectionaries, and canned meat and as a sizing material. It is also used as food roughage, in laxative preparations, and in cosmetic creams and jellies. *See* CULTURE MEDIA. [S.P.P.]

Agate A variety of chalcedonic quartz that is distinguished by the presence of color banding in curved or irregular patterns (see illustration). Most agate used for ornamental purposes is composed of two or more tones or intensities of brownish-red, often interlayered with white, but it is also commonly composed of various shades of gray and white. Since agate is relatively porous, it can be dyed permanently in red, green, blue, and a variety of other colors.

Section of polished agate showing the characteristic banding. (*Field Museum of Natural History, Chicago*)

The term agate is also used with prefixes to describe certain types of chalcedony in which banding is not evident. Moss agate is a milky or almost transparent chalcedony containing dark inclusions in a dendritic pattern. Iris agate exhibits an iridescent color effect. Fortification, or landscape, agate is translucent and contains inclusions that give it an appearance reminiscent of familiar natural scenes. Banded agate is distinguished from onyx by the fact that its banding is curved or irregular, in contrast to the straight, parallel layers of onyx. The properties of agate are those of chalcedony: refractive indices of 1.535 and 1.539, a hardness of $6\frac{1}{2}$ to 7, and a specific gravity of about 2.60. *See* CHALCEDONY; GEM; QUARTZ. [R.T.L.]

Agglutination reaction A clumping of a particulate suspension by a known or by a test reagent. The reactions are usually rapid and easily carried out, and in consequence, have wide application in diagnosis. Several types may be distinguished. In one, natural antigens (bacteria or red blood cells) clump when treated with their specific antibodies. This reaction may be used for the typing of blood for transfusion or the detection of antibodies to typhoid organisms in the sera of suspected carriers. The amount of antibody may be estimated by noting the serum dilution at which agglutination is just perceived. *See* ANTIBODY.

In a second type, a reagent antibody is added to a solution suspected of containing a soluble test antigen; if the antigen is present, the antibody is neutralized and the cell suspension subsequently added will not be agglutinated. The cell suspension has previously been sensitized with the soluble test antigen. This sensitive agglutination-inhibition test is widely used, for example, in the detection of urinary hormones for the determination of pregnancy.

In a third type, certain virus and red cell combinations are agglutinated in a nonspecific manner. However, the reaction may be inhibited, specifically, by antibody to the virus and this permits a high degree of control. This virus hemagglutination-inhibition test may be used to identify the virus antigen or to quantitate the antibody. *See* ANTIGEN; BLOOD GROUPS; VIRUS. [H.P.T.]

Agglutinin A substance that will cause a clumping of particles such as bacteria or erythrocytes. Of major importance are the specific or immune agglutinins, which are antibodies that will agglutinate bacteria containing the corresponding antigens on their surfaces. Analogous reactions involve erythrocytes and their corresponding antibodies, the hemagglutinins. Hemagglutinins to a variety of erythrocytes occur in many normal sera, and their amounts may be increased by immunization. The blood group isoagglutinins of humans and animals are important special cases which must be considered in all proposed blood transfusions lest transfusion reactions result. *See* AGGLUTINATION REACTION; ANTIBODY; BLOOD GROUPS; COMPLEMENT. [H.P.T.]

Aggression A motivational concept relating to the direction of abuse toward a person or property. Aggression is hypothetical for it cannot be directly observed; it must be inferred from the display of certain behaviors. While there is general agreement among psychologists concerning the acts that constitute aggressive behavior (such as killing, wounding, verbal assault, rape, and arson), there is disagreement concerning the nature of aggression and its etiology.

Aggression is typically considered offensive, and involves one organism attacking another or destroying its property. Oftentimes, however, an organism finds itself in a situation in which its self-interests are in jeopardy. Such a situation also requires the display of many of the same behaviors from which aggression is inferred. Aggression thus should be viewed as containing both offensive and defensive components, suggesting that it is not a unitary motivational state.

Based upon evidence derived from both human and nonhuman subjects, many researchers conclude that there are various forms of aggression, each related causally to distinctive physiological and environmental states and each characterized by different behavioral sequences. These forms include territorial aggression, frustration- and isolation-induced aggression, maternal aggression, as well as aggression as a consequence of substance abuse and brain damage. *See* EMOTION; ENDOCRINE SYSTEM (VERTEBRATE). [R.Ga.]

Aging Definitions of aging differ between biologists and behavioral scientists. The biologists regard aging as reflecting the sum of multiple and typical biological decrements occurring after sexual maturation; the behavioral scientists view it as reflecting regular and expected changes occurring in genetically representative organisms advancing through the life cycle under normal environmental conditions. It is difficult to define normal aging, since many changes observed in older adults and previously perceived as concomitants of normal aging are now recognized as effects of disease in later life. The behavioral science view allows for incremental as well as decremental changes with aging. Senescence is not always equated with aging; it is viewed as the increasing vulnerability or decreasing capacity of an organism to maintain homeostasis as it progresses through its life span. Gerontology refers to the study of aging. Geriatrics refers to the clinical science that is concerned with health and illness in the elderly. *See* SENESCENCE.

Theories about the process of aging concern how people age. These biological theories address two sets of factors—those intrinsic and those extrinsic to the organism. Intrinsic factors are influences operating on the body from within, such as the impact of genetic programming. Extrinsic factors are influences on the body from the environment, such as the impact of cumulative stresses.

It is important to differentiate between life expectancy and life span. Life expectancy is the average number of years of life in a given species; it is significantly influenced by factors beyond aging alone, such as famine and disease. Life span is the maximum number of years of life possible for that species; it is more fundamentally linked to the process of aging itself. Over the centuries, life expectancy has increased (due to improved sanitation and health care practices); life span has not. Approximately 115 years appears to be the upper limit of life span in humans. *See* DEATH; HUMAN GENETICS. [G.D.C.]

Agnatha The most primitive class of vertebrates, including the living Cyclostomata (hagfishes and lampreys), as well as extinct ostracoderms of the Ordovician, Silurian, and Devonian periods. They are distinguished by the fact that they have no true jaws. In the absence of jaws Agnatha have fed on minute food particles obtained by suction and trapped by filters or a mucous net, have become nibblers, or have developed a tongue set with horny teeth for rasping flesh, as in living Cyclostomata. The Agnatha are primitive also in their persistent notochord, in the absence of vertebrae, and often in the absence of paired fins.

The class Agnatha may be classified as follows:

Subclass Diplorhina
  Order: Heterostraci
    Thelodonti (or Coelolepida)
Subclass Monorhina
  Superorder Hyperotreti
    Order Myxinoidea
  Superorder Hyperoartii
    Order: Petromyzonida
      Osteostraci
      Anaspida
Subclass Cyclostomata

[R.H.De.]

Agnosia The loss of the ability to recognize objects or other constellations of stimuli. The essential feature of agnosia is the selectivity of the loss. It is confined to impressions received through one sensory channel and occurs despite relative preservation of sensitivity and intellect. The principal varieties of agnosia are visual, auditory, and tactile. Agnosia is caused by cerebral injury or disease. The critical loci of lesions in the various forms of this disorder have not been satisfactorily determined.

The traditional view of agnosia is that sensations of the affected sense modality, although adequate in themselves, fail to arouse the complex of memory images essential for the identification of the object. The failure is said to be due to destruction of associative centers or pathways. It is thought that the lesion should involve the cortex adjacent to the primary projection area, or subcortical tracts issuing from this area. This view of the nature of agnosia, and of the locus of lesion on which it depends, has not been universally accepted. *See* MEMORY; PERCEPTION. [J.Se.]

Agonomycetes Anamorphic (asexual or imperfect) fungi (Deuteromycotina) that not only lack fruit bodies but also fail to product conidia, the thallus consisting of septate hyphae. Somatic structures of propagation or survival, termed mitospores, are varied and include chlamydospores, bulbils, sclerotia, pseudosclerotia, rhizomorphs, strands, cords, and hyphae modified in different ways. About 28 genera (with 30 synonyms) containing 200 species are recognized.

Agonomycetes are generally considered to be combative species which are persistent and long-lived, largely because of the resistant nature of their vegetative structures and their slow and intermittent reproduction or complete absence of reproduction. Whether they are capable of defending captive resources and have good enzymatic competence, which are other features of combative species, is largely unknown. However, they occupy diverse ecological niches, including aquatic habitats (*Beverwykella*), soil, wood in various stages of decay, other decaying plant material, and dung. They also function as root and foliar pathogens, and many cause serious diseases in terms of host damage and economic loss, especially of roots, corms, and bulbs. *See* DEUTEROMYCOTINA; PLANT PATHOLOGY. [B.C.S.]

Agouti A large rodent that resembles the rabbit or hare in size and shape as well as in the elongated hindlegs, which make them well adapted for speed (see illustration). The agouti and the closely related smaller acouchi are inhabitants of clearings in forested areas of the Amazon region. Some range into Central America and as far as the Guianas.

The agouti (*Dasyprocta aguti*), a rodent found in Mexico, South America, and the West Indies.

Thirteen species of agouti have been described, the most common being *Dasyprocta aguti*. Some authorities are of the opinion that all of these are varieties or subspecies of *D. aguti*. The acouchi is represented by two species, the green acouchi (*Myoprocta pratti*) and the red acouchi (*M. acouchy*). *See* RODENTIA. [C.B.C.]

Agricultural aircraft Aircraft adapted or designed for use in agriculture and forestry. Agricultural aircraft are indispensable farm machines in developed countries and are increasingly used in many developing countries as well. In addition to dispersing wet- and dry-type pesticides they perform seeding, fertilizing, crop-drying, and defoliating services.

Agricultural aircraft are a fundamental part of the food production system, but they do not necessarily offer an energy efficiency gain over other methods. Their desirability lies in the rapidity of their operation, in their ability to operate irrespective of field conditions, irrigation, and crop growth, and in their effectiveness as a pest control which is comparable to ground equipment. Aircraft use is limited by: size of fields; proximity to sensitive nontarget crops and human habitat; and biological effectiveness, primarily relating to crop plant coverage obtained by the aircraft application system.

Helicopter or rotary wing aircraft offer significant advantages in agricultural, forestry, and vector control work, due primarily to their maneuverability, low-forward-speed capability, and application precision. However, turbine engines add considerably to the already higher initial price of helicopters, and this, coupled with requirements of higher maintenance and greater pilot skill, has kept helicopter use at less than 10% of the total numbers and area covered by all aircraft. *See* HELICOPTER.

[N.B.A.; W.E.Y.]

Agricultural chemistry The science of chemical compositions and changes involved in the production, protection, and use of crops and livestock. As a basic science, it embraces, in addition to test-tube chemistry, all the life processes through which food and fiber for humans and feed for animals are obtained. As an applied science or technology, it is directed toward control of those processes to increase yields, improve quality, and reduce costs. One important branch of it, chemurgy, is concerned chiefly with utilization of agricultural products as chemical raw materials.

The goals of agricultural chemistry are to expand understanding of the causes and effects of biochemical reactions related to plant and animal growth, to reveal opportunities for controlling those reactions, and to develop chemical products that will provide the desired assistance or control. Agricultural chemistry is not a distinct discipline, but a common thread that ties together genetics, physiology, microbiology, entomology, and numerous other sciences that impinge on agriculture. *See* FERTILIZER; SOIL.

Chemical materials developed to assist in the production of food, feed, and fiber include scores of herbicides, insecticides, fungicides, and other pesticides, plant growth regulators, fertilizers, and animal feed supplements. Chief among these groups from the commercial point of view are manufactured fertilizers, synthetic pesticides (including herbicides), and supplements for feeds. The latter include both nutritional supplements (for example, minerals) and medicinal compounds for the prevention or control of disease. *See* AGRICULTURE; ANIMAL FEEDS; HERBICIDE; PESTICIDE. [R.N.H.]

Agricultural engineering The engineering discipline directly concerned with developing means for providing food and fiber. With the world population explosion and the desire for higher standards of living, the efficiency of food and fiber production must be continually increased. *See* AGRICULTURE.

Nearly all problems that agricultural engineers deal with involve biological materials, systems, and processes. Application of physical and engineering sciences to the solution of such problems is the distinguishing feature of agricultural engineering.

Mechanization of agricultural production has been responsible in large measure for the productivity record of agriculture. While great progress has been made in the application of mechanical power to the production and harvesting of field crops, many crops, such as fruits and vegetables, still require expensive and increasingly scarce hand labor, especially in harvesting operations.

One of the areas of great importance for the future is modification and control of the soil, water, and air environment of plants. As greater production is demanded from an ever-decreasing available land area, more and more attention is being given to modifying the natural environment to produce conditions which are more nearly ideal for plant growth. Controlled environment for animal production within confined housing systems is another area of activity of major concern to agricultural engineers.

The modern livestock farm can be likened to a factory with various inputs for poultry and dairy and meat animals. Mechanization of the modern farmstead has become a reality, and agricultural engineers have been largely responsible. Automation is now the goal that challenges agricultural engineers.

An important phase of agricultural engineering involves the management and beneficial use of soil and water resources. Agricultural engineers are involved in research to develop new techniques and design systems to control soil erosion and soil moisture for crop production, to prevent floods and water pollution, and to provide pure water supplies for human use. *See* SOIL CONSERVATION; WATER CONSERVATION.

Although much agricultural engineering work is involved with the development and design of power machinery for crop production, many engineering problems are encountered after the crops are harvested. Postharvest processing, such as drying of grains and forages, washing, grading, and storage of fruits and vegetables, including controlled-atmosphere systems, and ginning of cotton are typical of on-the-farm applications of agricultural engineering. [O.C.F.]

Agricultural machinery Mechanized systems of food and fiber production that extend from initial tillage of the soil through planting, cultural practices during the growing season, protection from pests, harvesting, conditioning, livestock feeding, and delivery for processing. The trend has been to larger self-propelled special-purpose machines, except for tillage where the trend has been to large four-, six-, or eight-wheel or crawler tractors used to tow high-capacity plows or disks, and also deep rippers used to loosen compacted soils. The use of hydraulic power has made possible highly specialized machinery to perform intricate operations wherever needed, because the fluid power is pumped through flexible hoses to provide linear or rotating motion with push-button control. The most sophisticated technology is used to increase the precision needed in modern agriculture. Lasers are used for laying out irrigation systems; microprocessors are used for sensing and controlling intricate operations, such as controlling feed mixtures for dairy cows and grading fruits and vegetables; electronic devices are used in the automation of many harvesters.

While many implements such as plows, disks, cultivators, and fertilizer spreaders are mounted on tractors, there are many that are too large and are trailed behind and are controlled and operated hydraulically. Some multipurpose machines are used where a high degree of precision is needed. The use of aircraft has revolutionized many farming operations. Rice is sown in flooded fields, and fertilizers and pesticides are applied from the air. Herbicides and insecticides are also applied by air on other crops, such as grains, vegetables, and tree fruits. [M.O'B.]

Agricultural meteorology The study and application of meteorology and climatology (a branch of meteorology) to the specific problems of agriculture. Agriculture is the production of food and fiber in all its forms—crops for human and animal consumption, pasture and range for animal grazing, and crops (including trees) as raw materials for manufactured products. Hence, agricultural meteorology deals with farming, ranching, and forestry as well as with the transportation to the producer of substances required for production—water for irrigation, fertilizer, and agricultural chemicals—and transportation of the products to markets.

Agricultural meteorologists, or agrometeorologists, deal with a variety of problems in cooperation with many other types of specialists. These include prediction of effect of weather on crop yields, water use efficiency, weather-related diseases and pests, the length of the growing season, the timing and severity of frost, predominant speed and direction of wind, and problems associated with severe weather conditions. Some other areas of research and applications in agrometeorology are in turbulence and remote sensing. The Earth's surfaces (terrestrial and marine) exchange radiation and mass with the atmosphere and absorb momentum of the wind. Understanding the mechanisms of these exchanges is the function of micrometeorology.

Agricultural meteorologists have techniques for remotely sensing the condition of the land and the vegetation growing on it. Multispectral scanners on a number of National Oceanic and Atmospheric Administration (NOAA) and National Aeronautics and Space Administration (NASA) satellites and on aircraft provide data on the extent, density, and vigor of vegetation. Agricultural meteorologists have been especially active in conducting ground truth studies, which aid in the interpretation of remotely sensed information. [N.J.R.]

Agriculture The art and science of crop and livestock production. In its broadest sense, agriculture comprises the entire range of technologies associated with the production of useful products from plants and animals, including soil cultivation, crop and livestock management, and the activities of processing and marketing. The term agri-business has been coined to include all the technologies that mesh in the total inputs and outputs of the farming sector. In this light, agriculture encompasses the whole range of economic activities involved in manufacturing and distributing the industrial inputs used in farming; the farm production of crops, animals, and animal products; the processing of these materials into finished products; and the provision of products at a time and place demanded by consumers. The sale of final product of the food and fiber industries in the United States represents about one-fourth of the gross national product and is about five times the value of farm-level production.

Many different factors influence the kind of agriculture practiced in a particular area. Among these factors are climate, soil, topography, nearness to markets, transportation facilities, land costs, and general economic level.

Climate, soil, water availability, and topography vary widely throughout the world. This variation brings about a wide range in agricultural production enterprises. Certain areas tend toward a specialized agriculture, whereas other areas engage in a more diversified agriculture. As new technology is introduced and adopted, environmental factors are less important in influencing agricultural production patterns. Continued growth in the world's population makes critical the continuing ability of agriculture to provide needed food and fiber. [J.J.]

Agroecosystem A model for the functionings of an agricultural system with all its inputs and outputs. Analysis of the basic biology of agriculture in which economic production is explained as only one aspect of biological responses to the environment and to inputs of energy and materials is called agroecosystem research. Unlike the empirical disciplines of agronomy and animal science, which direct primary attention to increasing economic production, agroecosystem research considers all aspects of the biology of an agricultural system. This broader approach to agricultural systems becomes more useful as production methods developed through experimental methods begin to approach the limits of the biological processes. As these limits are approached, it becomes more important to know what biological processes determine yield and to learn how these processes act in determining the biological limits to yields. The chemistry of nutrient cycling must also be considered in order to determine whether these cycles can sustain the high yields of intensive agricultural production that is based on the use of chemicals.

Agroecosystem research uses the methods of ecosystem analysis to measure the material and energy entering plant and animal populations and to explain how these inputs affect the physiological processes determining growth and maintenance. The general diagram for a biologically complete agroecosystem (see illustration) shows three basic components and a pattern of energy flow that is nearly the same in all agroecosystems. All ecosystems are driven by solar energy fixed as organic carbon by green plants. The accumulation of organic carbon fixed by plants in agroecosystems can be divided into three categories: (1) coarse materials, such as fodder, eaten by livestock; (2) grains, fruits, vegetables, and so on, eaten by humans and livestock; and (3) dense stems and leaves which may be used for fuel and constructing shelters or utensils. The patterns of energy and material flow between the plants and the two consumers, livestock and humans, follow simple and consistent patterns that are similar in most agricultural systems. Unlike other ecosystems, which have an immense variety of species with patterns of interaction that form complex food webs, agroecosystems have relatively simple cycles. *See* BIOLOGICAL PRODUCTIVITY; BIOMASS; ECOSYSTEM. [R.Mi.]

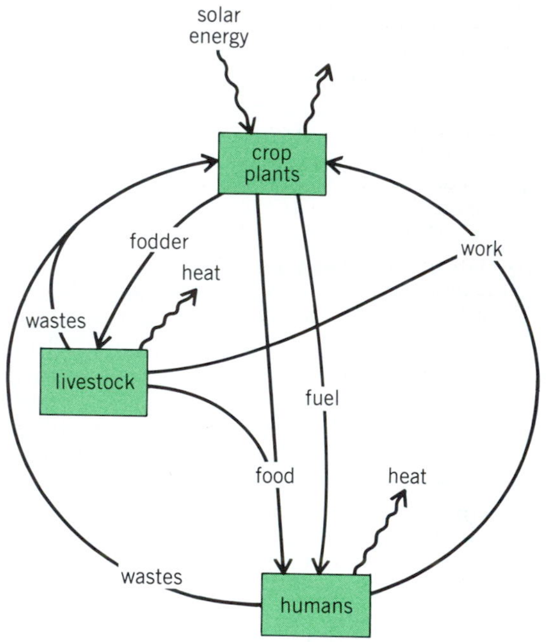

A biologically complete agroecosystem.

Agronomy The principles and procedures of field crop improvement, management, and production. Agronomy is also concerned with the improvement and management of special-purpose plants, such as turf grasses for home lawns, recreational areas, highway embankments, drainage ditches, and waterways. Soil characteristics, properties, uses, and conservation are part of agronomy, and management of soils for efficient production of specific crops is a major interest.

Regarding crops, primary consideration is given to relationship among crop plants as conditioned by their heredity, physiology, and ecology. Soil genesis, classification, and morphology; soil physical, chemical, and biological properties; and water relationships are of prime importance with respect to soils.

[W.I.T.]

Aharonov-Bohm effect The predicted effect of an electromagnetic vector or scalar potential in electronic interference phenomena, in the absence of electric or magnetic fields on the electrons.

The fundamental equations of motion for a charged object are usually expressed in terms of the magnetic field $\vec{B}$ and the electric field $\vec{E}$. The force $\vec{F}$ on a charged particle can be conveniently written as in the equations below, where q is the

$$\vec{F} = q\vec{E} \qquad \vec{F} = q\vec{v} \times \vec{B}$$

particle's charge, $\vec{v}$ is its velocity, and the symbol × represents the vector product. Associated with $\vec{E}$ is a scalar potential V defined at any point as the work W necessary to move a charge from minus infinity to that point, $V = W/q$. Generally, only the difference in potentials between two points matters in classical physics, and this potential difference can be used in computing the electric field. Similarly, associated with $\vec{B}$ is a vector potential $\vec{A}$, a convenient mathematical aid for calculating the magnetic field. *See* CALCULUS OF VECTORS; ELECTRIC FIELD; MAGNETIC FIELD; POTENTIALS.

In quantum mechanics, however, the basic equations that describe the motion of all objects contain $\vec{A}$ and V directly, and they cannot be simply eliminated. Nonetheless, it was initially believed that these potentials had no independent significance. In 1959, Y. Aharonov and D. Bohm discovered that both the scalar and vector potentials should play a major role in quantum mechanics. They proposed two electron interference experiments in which some of the electron properties would be sensitive to changes of $\vec{A}$ or V, even when there were no electric or magnetic fields present on the charged particles. The absence of $\vec{E}$ and $\vec{B}$ means that classically there are no forces acting on the particles, but quantum-mechanically it is still possible to change the properties of the electron. These counterintuitive predictions are known as the Aharonov-Bohm effect.

Surprisingly, the Aharonov-Bohm effect plays an important role in understanding the properties of electrical circuits whose wires or transistors are smaller than a few micrometers. The electrical resistance in a wire loop oscillates periodically as the magnetic flux threading the loop is increased, with a period of h/e (where h is Planck's constant and e is the charge of the electron), the normal-metal flux quantum. In single wires, the electrical resistance fluctuates randomly as a function of magnetic flux. Both these observations, which were made possible by advances in the technology for fabricating small samples, reflect an Aharonov-Bohm effect. They have opened up a new field of condensed-matter physics because they are a signature that the electrical properties are dominated by quantum-mechanical behavior of the electrons, and that the rules of the classical physics are no longer operative. *See* QUANTUM MECHANICS.

[R.A.Web.]

Aileron The hinged rear portion of an aircraft wing, moved differentially on each side of the aircraft to obtain lateral or roll control moments. The angular settings of the ailerons are controlled by the human or automatic pilot through the flight control system. Typical flap- and spoiler-type ailerons are shown in the illustration. *See* FLIGHT CONTROLS.

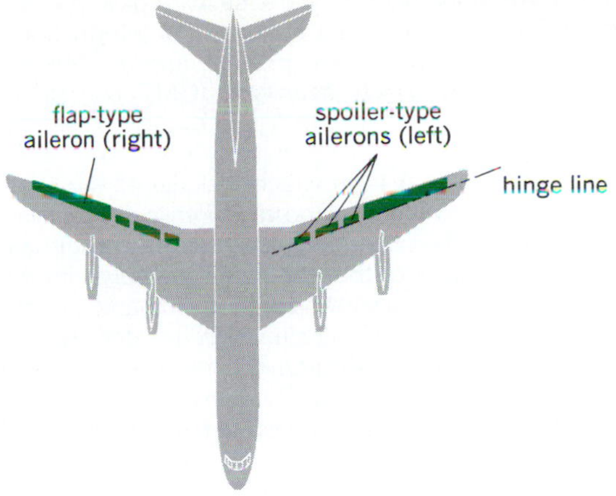

Flap- and spoiler-type ailerons on jet transport airplane.

The operating principles of ailerons are the same as for all trailing-edge hinged control devices. Deflection of an aileron changes the effective camber, or airfoil curvature relative to the wing chord, of the entire wing forward of the aileron. With the trailing edge deflected upward, reduced local flow velocities are produced on the upper wing surface, and increased local flow velocities are produced on the lower wing surface. By Bernoulli's law, this results in a reduction of lift over the portion of the wing forward of the aileron, and on the aileron itself. Conversely, trailing-edge down deflection of a flap-type aileron increases the lift in the same areas. Ailerons are located as close as possible to the wing tips, to maximize rolling moment by increasing the moment arm of the force due to the change in wing lift. In the case of flap-type ailerons, when the trailing edge is raised on one wing, say the left, the trailing edge of the aileron on the opposite or right wing is lowered by about the same amount. The decrease in lift on the left wing is accompanied by a lift increase on the right wing. While the net wing lift remains about the same, a rolling moment or torque about the aircraft's fore-and-aft axis develops in a left, or counterclockwise, direction as seen by the pilot. *See* BERNOULLI'S THEOREM.

Flap-type ailerons are replaced or supplemented by spoiler-type ailerons for a variety of reasons. Spoiler ailerons are usually installed forward of the landing flaps on commercial jet transports, in order to supplement aileron effectiveness during landing approaches, when the landing flaps are extended. Greatly reduced takeoff and landing speeds can be obtained by devoting the trailing edge of the entire wing to high-lift flaps. This is made possible by substituting spoilers for flap-type ailerons.

[M.J.A.]

Air A predominantly mechanical mixture of a variety of individual gases enveloping the terrestrial globe to form the Earth's atmosphere. In this sense air is one of the three basic components, air, water, and land (atmosphere, hydrosphere, and lithosphere), that interblend to form the life zone at the face of the Earth. *See* ATMOSPHERE.

[C.V.C.]

Air Almanac A periodical publication of astronomical statistics useful to, and designed primarily for, air navigation. The U.S. Naval Observatory and H.M. Nautical Almanac Office of Great Britain collaborate on this joint publication available in the United States from the Superintendent of Documents. Several other countries also publish air almanacs. Although designed for air navigators, the *Air Almanac* is used increasingly by mariners, who accept its reduced precision because of its greater convenience compared with the somewhat similar *Nautical Almanac*. *See* CELESTIAL NAVIGATION; NAUTICAL ALMANAC.

The *Air Almanac* is arranged with two "daily pages" (front and back) for each day giving ephemeridal information of the Sun, first point of Aries, three planets, and the Moon at 10-min intervals of Greenwich mean time (GMT) and other useful information. [A.B.M.]

Air armament A term which includes all the equipment through which a combat airplane can release destructive power on a target. Guns, bombs, and guided aircraft rockets (air-to-air missiles) are examples of the armament employed by the U.S. Air Force. Airborne devices used to aim and direct guns, bombs, and rockets are also included in this definition. Steady advances in armament and aircraft have made it possible for fighter-bombers like the F-105 Thunderchief to carry as much destructive power as the heavy bombers of World War II (see illustration).

F-105 Thunderchief. The 1400-mi/h (2250 km/h) aircraft can carry 400 different combinations of weapons, ranging from conventional bombs like these sixteen 750-lb (340-kg) bombs, to nuclear stores, Sidewinder and Bullpup guided missiles, and a 6000-shot-a-minute automatic 20-mm cannon. (*U.S. Air Force photograph*)

Guns mounted in combat airplanes are designed in response to the classical problems of weight and drag. Since aerodynamic drag seriously hinders the operational efficiency of aircraft, aircraft armament must be not only as light as possible but also streamlined in order to offer the least wind resistance. Reliability at extreme altitudes and in frigid temperatures, as well as during the acceleration loads of high-*g* maneuvering, is another consideration. Also, aircraft guns must have a high rate of fire because of the extremely short span of time during which targets may be attacked in aerial combat. *See* AERODYNAMIC WAVE DRAG.

Bombs, the symbol of air armament, represent the ultimate in destructive power. Contrasted with guns and rockets, bombs constitute a more efficient method of destroying a target since, for a given installed weight, they can release more energy than their rivals. Chemical-based bombs are classified explosive, general-demolition, armor-piercing, and fragmentation types.

Devices for aiming guns, bombs, and rockets are called fire-control mechanisms, and include gunsights, rocketsights, and bombsights. Because of the high speed of a combat aircraft, the short time available for acquisition of data and computation of aiming points, and the lack of precise information about targets, the aerial fire-control problem is extremely difficult. Radar is used, as are optical means, to determine exact location and range.

Rockets and missiles used by combat aircraft employ warheads similar to cannon shells, that is, armor-piercing, incendiary, and explosive. They are particularly suited for air armament, where immediate aerodynamic stability is achieved by launching them from a moving platform into a high relative wind. They impart no recoil forces to the airplane structure and can carry a relatively large explosive load, because the warhead is not subjected to the extreme forces of acceleration encountered by gun-launched shells. [D.P.G.; J.J.C.]

Air brake An energy-conversion mechanism used to retard, stop, or hold a vehicle or, generally, any moving element, the activating force being applied by a difference in air pressure. With an air brake only slight effort by the operator quickly applies full braking force.

An air brake performs the energy conversion like other brakes by friction. The feature that distinguishes an air brake is that the friction-producing device, such as disk or drum, is applied by air in contrast to mechanical, hydraulic, or electrical means. In a particular use the choice between an air brake and any other type of brake depends in part on the availability of an air supply and on the method of brake control.

Because air is compressible, air brakes inherently accommodate to wear, such as in brake shoes. Also, because the force with which brakes are applied depends on the area of an air-driven diaphragm as well as on air pressure, large forces can be developed from moderate pressures. Air pressure can be accurately regulated and can also be controlled by other forces besides the operator's hand or foot; consequently air brakes can be made to respond to static and dynamic forces that influence the efficiency of brake operation.

In a typical air brake system an air compressor takes in and compresses air for use by the brakes and usually for other air-operated components of the vehicle. The compressed air is stored in an air brake reservoir until needed. By means of a foot- or hand-operated brake valve the operator controls application of the brake (see illustration). A brake rod from the cylinder diaphragm drives the friction element against the moving surface to provide the braking action. When the operator releases the brake valve, an auxiliary reservoir recharges to full pressure through the triple valve. Also, after the braking action is no longer required, a quick release valve assures rapid discharge of air from the brake cylinder. [F.H.R.]

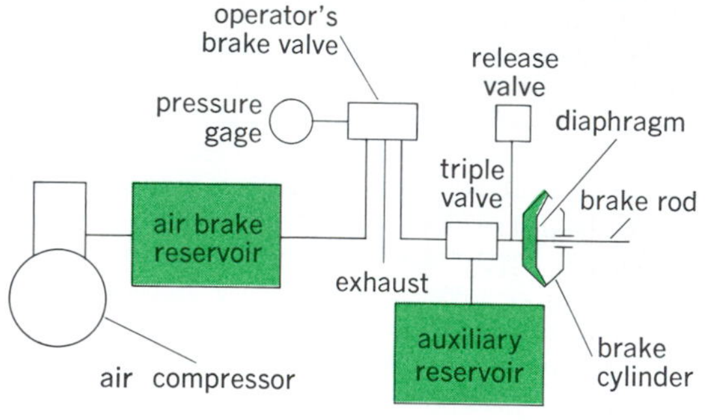

Diagram of air brake. When operator actuates brake valve, pressure to triple valve is reduced. Triple valve then admits compressed air from auxiliary reservoir to brake cylinder, actuating brake rod.

Air-capacitor dosimeter A type of dosimeter most often about the size and shape of a fountain pen and provided with a pen clip; hence, they are sometimes referred to as pen meters or pocket meters. The outer container is usually made of plastic, and the inside chamber wall and central collecting electrode are made of conducting plastic or of aluminum coated with graphite. This inside chamber serves as an air capacitor or ion chamber. This central electrode is insulated from the outer cylinder wall by polyethylene or some other suitable insulating material. An electric charge can be placed on the central electrode by removing an end cap and applying a voltage of 100–200 V between the electrode and the chamber wall.

Such dosimeters, when charged, are essentially air capacitors, and the amount of discharge during use is proportional to the absorbed dose of x- or γ-radiation received. The usual dosimeter is relatively energy-independent for radiation from 300 keV to 2 MeV but often reads high by a factor of 2 to 3 at about 100 keV. [K.Z.M.]

Air conditioning The maintenance of certain aspects of the environment within a defined space to facilitate the intended function of that space. Environmental conditions generally encompassed by the term air conditioning include air temperature and motion, radiant heat energy level, moisture level, and concentration of various pollutants, including dust, germs, and gases. Comfort air conditioning refers to control of spaces inhabited by people to promote their comfort, health, or productivity. Process air conditioning systems are designed to facilitate the functioning of a production, manufacturing, or operational activity.

A comfort air conditioning system is designed to help maintain human body temperature at its normal level without undue stress and to provide an atmosphere which is healthy to breathe. The heat-dissipating factors of temperature, humidity, and air motion must be considered simultaneously. Within limits, the same amount of comfort (or, more objectively, of heat-dissipating ability) is the result of a combination of these factors in a three-dimensional continuum.

In practice most air-conditioning systems for offices and institutions, schools, and other light-work spaces are designed to maintain a temperature at 75±°F (24±°C) year-round with relative humidity maintained at 50±% by dehumidification in summer. In winter relative humidities within the conditioned spaces are usually much lower, on the order of 20–30%, as a result of heating relatively dry outside air, and the impracticability of maintaining higher humidities which would cause excessive condensation on cold surfaces, such as window panes. Both conditions correspond closely to comfort requirements; hence, a single year-round thermostat setting is feasible.

Engineering of an air-conditioning system starts with selection of design conditions; air temperature and relative humidity are principal factors. Next, loads on the system are calculated. Finally, equipment is selected and sized to perform the indicated functions and to carry the estimated loads. Each space is analyzed separately. A cooling load will exist when the sum of heat released within the space and transmitted to the space is greater than the loss of heat from the space. A heating load occurs when the heat generated within the space is less than that lost from it. Similar considerations apply to moisture.

Heat generated within the space consists of body heat, heat from all electrical appliances and lights, and heat from other sources such as gas cooking stoves and industrial ovens. Heat is transmitted through all parts of the space envelope, which includes walls, floor, ceiling, and windows. Whether heat enters or leaves the space depends upon whether the outside surfaces are warmer or cooler than the inside surfaces. The rate at which heat is conducted through the space envelope is a function of the temperature difference across the envelope and the thermal conductance of the envelope. Conductances, which depend on materials of construction and their thicknesses along the path of heat transmission, are a large factor in walls and ceilings exposed to the outdoors in cold winters and hot summers. In these cases insulation is added to decrease the overall conductance of the envelope.

Solar heat loads are an especially important part of load calculation because they represent a large percentage of heat gain through walls and roofs, but are very difficult to estimate because solar irradiation is constantly changing.

Humidity as a load on an air-conditioning system is treated by the engineer in terms of its latent heat, that is, the heat required to condense or evaporate the moisture, approximately 2.3×10^6 J/kg (1000 Btu/lb) of moisture. As with heat, moisture travels through the space envelope, and its rate of transfer is calculated as a function of the difference in vapor pressure across the space envelope and the permeability of the envelope construction. To decrease permeability where vapor pressure differential is large, vapor barriers (relatively impermeable membranes) are incorporated in the envelope construction.

A complete air-conditioning system is capable of adding and removing heat and moisture and of filtering dust and odorants from the space or space it serves. Systems that heat, humidify, and filter only, for control of comfort in winter, are called winter air-conditioning systems; those that cool, dehumidify, and filter only are called summer air-conditioning systems, provided they are fitted with proper controls to maintain design levels of temperature, relative humidity, and air purity.

Air-conditioning systems are either unitary or built-up. The window or through-the-wall air conditioner is an example of a unitary summer air-conditioning system; the entire system is housed in a single package which contains heat removal, dehumidification, and filtration capabilities. When an electric heater is built into it with suitable controls, it functions as a year-round air-conditioning system.

Built-up or field-erected systems are composed of factory-built subassemblies interconnected by means such as piping, wiring, and ducting during final assembly on the building site. Most large buildings are so conditioned.

There are three principal types of central air-conditioning systems: all-air, all-water, and air-water. In the all-air system all return air is processed in a central air-handling apparatus. In the all-water system the principal thermal load is carried by chilled and hot water generated in a central facility and piped to coils in each space; room air then passes over the coils. In an air-water system both treated air and hot or chilled water are supplied to units in each space. *See* Air cooling; Air filter. [R.L.K.]

Air cooling Lowering of air temperature for comfort, process control, or food preservation. Air and water vapor occur together in the atmosphere. The mixture is commonly cooled by direct convective heat transfer of its internal energy (sensible heat) to a surface or medium at lower temperature. In the most compact arrangement, transfer is through a finned (extended surface) coil, metallic and thin, inside of which is circulating either chilled water, antifreeze solution, brine, or boiling refrigerant. The fluid acts as the heat receiver. Heat transfer can also be directly to a wetted surface, such as waterdroplets in an air washer or a wet pad in an evaporative cooler. *See* Air conditioning; Heat transfer.

For evaporative cooling, nonsaturated air is mixed with water. Some of the sensible heat transfers from the air to the evaporating water. The heat then returns to the airstream as latent heat of water vapor. The technique is employed for air cooling of machines where higher humidities can be tolerated; for cooling of industrial areas where high humidities are required, as in textile mills; and for comfort cooling in hot, dry climates, where partial saturation results in cool air at relatively low humidity. *See* Humidity.

In the evaporative cooler the air is constantly changed and the water is recirculated, except for that portion which has evaporated and which must be made up. Water temperature remains at the adiabatic saturation (wet-bulb) temperature. If water temperature is controlled, as by refrigeration, the leaving air temperature can be controlled within wide limits. Entering warm, moist air can be cooled below its dew point so that, although it leaves close to saturation, it leaves with less moisture per unit volume of air than when it entered. An apparatus to accomplish this is called an air washer. It is used in many industrial and comfort air-conditioning systems, and performs the added functions of cleansing the airstream of dust and of gases that dissolve in water, and in winter, through the addition of heat to the water, of warming and humidifying the air.

The most important form of air cooling is by finned coils, inside of which circulates a cold fluid or cold, boiling refrigerant (see illustration). The latter is called a direct-expansion (DX)

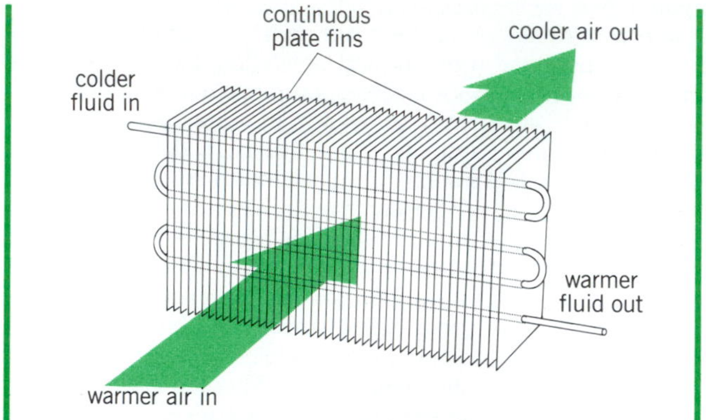

Typical extended-surface air-cooling coil.

coil. In most applications the finned surfaces become wet as condensation occurs simultaneously with sensible cooling. Usually, the required amount of dehumidification determines the temperature at which the surface is maintained and, where this results in air that is colder than required, the air is reheated to the proper temperature. Droplets of condensate are entrained in the airstream, removed by a suitable filter (eliminator), collected in a drain pan, and wasted.

Well water is available for air cooling in much of the world. Temperature of water from wells 30 to 60 ft (10 to 20 m) deep is approximately the average year-round air temperature in the locality of the well. For installations that operate only occasionally, such as some churches and meeting halls, water recirculated and cooled over ice offers an economical means for space cooling. Where electric power is readily available, the cooling function of the ice is performed by a mechanical refrigerator. [R.L.K.]

Air-cushion vehicle A transportation device supported by low-pressure, low-velocity air. It is also known as a ground-effect machine (GEM). Configurations, from which practical applications are developing, grew out of aeronautical experience. While an aircraft is operating close to the ground, as during takeoff or landing, reaction between high-velocity downwash from its wing or rotor and the ground serves to augment lift. By depending predominantly on this reaction, and by designing the underside of the vehicle to retain the delivered air, the vehicle need not be restricted to operation over a prepared roadbed; it can travel equally well over water, marsh, or relatively level land (see illustration).

The air cushion is maintained by several techniques. Some or all of the ground-effect support may be produced by a ram wing operated at high speed near the ground. Support may be produced, as in a helicopter, by a fan that delivers the needed volume of downdraft air against the ground-induced back pressure. Air may be injected at the periphery of the machine in a downward and inward direction. This annular jet creates an air-momentum curtain that reduces escape of supporting air. An apron around the lower portion of the vehicle can form a plenum chamber. For a vehicle that skims over the water, the apron may be fins or thin keels that project into the water, effectively capturing the air as a bubble below the vehicle. For an over-the-ground vehicle the apron may be a flexible skirt of abrasion-resistant plastic. In contrast to an aircraft, the air-

Princeton University experimental 20-ft (6-m) air-cushion vehicle.

cushion vehicle can hover for a protracted period in one location. *See* ANTIFRICTION BEARING; DUCTED FAN; HELICOPTER. [F.H.R.]

Air filter A component of most systems in which air is used for industrial processes, for ventilation, or for comfort air conditioning. The function of an air filter is to reduce the concentration of solid particles in the airstream to a level that can be tolerated by the process or space occupancy purpose. *See* AIR; AIR CONDITIONING; VENTILATION.

Solid particles in the airstream range in size from 0.01 micrometer to objects that can be caught by ordinary fly screens, such as lint, feathers, and insects. The particles generally include soot, ash, soil, lint, and smoke, but may include almost any organic or inorganic material, even bacteria and mold spores. This wide variety of airborne contaminants, added to the diversity of systems in which air filters are used, makes it impossible to have one type that is best for all applications.

Three basic types of air filters are in common use: viscous impingement, dry, and electronic. The principles employed by these filters in removing airborne solids are viscous impingement, interception, impaction, diffusion, and electrostatic precipitation. Some filters utilize only one of these principles; others employ combinations. A fourth method, inertial separation, is finding increasing use as a result of the construction boom throughout most of the Middle East. *See* DRY-AIR FILTER; ELECTRONIC AIR CLEANER; INERTIAL SEPARATOR; VISCOUS-IMPINGEMENT AIR FILTER. [M.A.B.]

Air heater A component of a steam-generating unit that absorbs heat from the products of combustion after they have passed through the steam-generating and superheating sections. Heat recovered from the gas is recycled to the furnace by the combustion air and is absorbed in the steam-generating unit, with a resultant gain in overall thermal efficiency. Use of preheated combustion air also accelerates ignition and promotes rapid burning of the fuel. Air heaters often are used in conjunction with economizers, because the temperature of the inlet air is less than that of the feedwater to the economizer, and in this way it is possible to reduce further the temperature of flue gas before it is discharged to the stack. *See* BOILER ECONOMIZER. [G.W.K.]

Air mass A term applied in meteorology to an extensive body of the atmosphere which approximates horizontal homogeneity in its weather characteristics. An air mass may be followed on the weather map as an entity in its day-to-day movement in the general circulation of the atmosphere. The expressions air mass analysis and frontal analysis are applied to the analysis of weather maps in terms of the prevailing air masses and of the zones of transition and interaction (fronts) which separate them.

The relative horizontal homogeneity of an air mass stands in contrast to the sharp horizontal changes in a frontal zone. The horizontal extent of important air masses is reckoned in millions of square miles. In the vertical dimension an air mass extends at most to the top of the troposphere, and frequently is restricted to the lower half or less of the troposphere. The frontal zones between air masses usually slope in such a manner that the colder air mass underlies the warmer as a wedge. In the vertical direction the properties of an air mass, specifically its content of heat and moisture, may vary between a high degree of stratification and one of homogeneity produced by vertical mixing. *See* FRONT; METEOROLOGY; WEATHER MAP.

The thermodynamic properties of air mass determine not only the general character of the weather in the extensive area covered by the air mass but also, to some extent, the severity of the weather activity in the frontal zone of interaction between air masses. Those properties which determine the primary weather characteristics of an air mass are defined by the vertical distribution of the two elements, water vapor and heat (temperature). The vertical distribution of water vapor determines the presence or absence of condensation forms and, if present, the elevation and thickness of fog or cloud layers. The vertical distribution of temperature determines the relative warmth or coldness of the air mass and, more importantly, the vertical gradient of temperatures, known as the lapse rate. The lapse rate determines the stability or instability of the air mass for thermal convection and, consequently, the stratiform or convective cellular structure of the cloud forms and precipitation. [H.C.Wi.]

Air navigation The process of directing and monitoring the progress of an aircraft between selected geographic points or with respect to some predetermined plan from the moment that it leaves the loading dock to the time that it takes off, attains operating altitude, and reaches its point of disembarkation in safety. Air navigation thus introduces the need for guidance, control, and navigation in the third dimension. The underlying methodology employed is dead reckoning, which in its simplest form is the process of following a planned compass direction at an assumed speed for a specified interval of time. Its fundamental tools are the compass, airspeed meter, altimeter, vertical speed indicator, outside air temperature sensor, clock, chart, and some reference independent of the aircraft, as, for example, a radio navigation aid, by which to verify position. *See* DEAD RECKONING.

All serious air navigation begins with a carefully constructed flight plan, in which the pilot identifies the desired ground track or course, the distance to be flown, the values of magnetic variation and drift due to wind that are assumed to affect the progress of the aircraft, and the compass heading to be steered to make good the desired course. The flight plan includes selection of a flying altitude which is consistent with the performance capability of the aircraft, consideration of the forecast meteorological conditions, and the evaluation of the known or expected hazards to flight, that is, high ground, restricted areas, and obstructions.

The principal air work required of the pilot-navigator is to periodically confirm aircraft position and rate of progress with respect to flight plan and clearance. Each operating region within domestic and international airspace has published performance criteria to be met with respect to allowable deviation from assigned track, altitude, and speed, and these are rigidly enforced for the protection of all users of the airspace. Of particular interest to the flight crew is the validity of estimated time of arrival (ETA) and fuel required to reach destination. The task of confirming progress involves use of dead-reckoning (DR) equipment and procedures and an externally referenced position determination aid.

The dead-reckoning function on long-distance overwater flights may be performed either manually or semi-automatically, depending upon installed equipment. Regardless of equipment employed, the dead-reckoning process is periodically subjected to validation through measurement of position by one or more of the external aids to navigation. These aids can include the pilotage fix, the celestial fix, position derived from one of a variety of radio aids, or position determined from use of a hyperbolic navigation aid. Navigation satellites are becoming an important source of position data as well. *See* CELESTIAL NAVIGATION; ELECTRONIC NAVIGATION SYSTEMS; PILOTAGE; SATELLITE NAVIGATION SYSTEMS.

High ground, artificial obstacles to flight, other aircraft in the sky, turbulence, and icing are avoided through preflight planning, special rules of the road, control of traffic imposed by specially trained ground-based personnel, and detection of potentially dangerous cloud forms by use of radar and special navigation procedures. *See* AERONAUTICAL METEOROLOGY; AIR-TRAFFIC CONTROL.

Operations on the airport surface are under the supervision of ground control. Departure from the ramp and during taxi operations to the duty runway is monitored by this agency. Any assistance in identifying turning points, taxiways, and runup areas that can be obtained from ground control through the pilot's standard airport charts should be sufficient to assist in orientation. Takeoff and the initial stages of climbout are supervised by departure control, a radar-equipped facility, which assists the pilot from start of takeoff roll until the aircraft has joined the en route system. At small airports, or during visual flight rules (VFR) operations, the pilot will generally work with the tower operator in lieu of departure control. The approach and landing phases of flight are the reciprocals of the departure. The departure control facility provides the link between the en route air-traffic control and the airfield until the aircraft is on the ground. At that point the pilot is directed to work with airport ground control as at the start of the flight. [W.L.P.]

Air pocket A concept developed by the early aviators to describe an abrupt loss of altitude. When an airplane encounters a downdraft, the effect is a sudden reduction in the angle of attack of the wings relative to the air. Because the airplane momentarily has less lift than its weight, it drops toward the earth. The acceleration downward allows it to gather a fall velocity that changes the angle of attack back to normal when fall velocity of the airplane equals velocity of the downdraft. With small airplanes the sensation is much what one would imagine to be produced by flying across a pocket devoid of air. [V.L.S.]

Air pollution Alteration of the atmosphere by the introduction of natural and artificial particulate contaminants. Most artificial impurities are injected into the atmosphere at or near the Earth's surface. The atmosphere cleanses itself of these quickly, for the most part. This occurs because in the troposphere temperature decreases rapidly with increasing altitude, resulting in rapid vertical mixing; the rainfall sometimes associated with these conditions also assists in removing the impurities. *See* ATMOSPHERE; TROPOSPHERE.

In the stratosphere, either temperature is constant or it increases with altitude, a condition that characterizes the entire stratosphere as a permanent inversion layer. As a result, vertical mixing in the stratosphere (and hence, self-cleansing) occurs much more slowly than that in the troposphere. Contaminants introduced at a particular altitude remain near that altitude for periods as long as several years. It is a matter of concern that while the turbulent troposphere cleanses itself quickly, the relatively stagnant stratosphere does not. Nonetheless, the pollutants injected into the troposphere and stratosphere have impact on humans and the habitable environment. *See* STRATOSPHERE; TROPOPAUSE.

All airborne particulate matter, liquid and solid, and contaminant gases exist in the atmosphere in variable amounts. Typical natural contaminants are salt particles from the oceans or dust and gases from active volcanoes; typical artificial contaminants are waste smokes and gases formed by industrial, municipal, household, and automotive processes, and aircraft and rocket combustion processes. Another postulated important source of artificial contaminants is certain fluorocarbon compounds (gases) used widely as refrigerants, as propellants for aerosol products, and for other applications. Pollens, spores, rusts, and smuts are natural aerosols augmented artificially by humans' land-use practices. *See* SMOG; SMOKE.

Dispersion of pollution is dependent on atmospheric conditions. Winds transport and diffuse contaminants; rain may wash them to the surface; and under cloudless skies, solar radiation may induce important photochemical reactions.

Wind direction, speed, and turbulence influence atmospheric pollution. Wind direction determines the area into which the pollution is carried. Dilution of contaminants from a source is directly proportional, other factors being constant, to wind speed, which also determines the intensity of mechanical turbulence produced as the wind flows over and around surface objects, such as trees and buildings. Other factors are eddy diffusion by wind turbulence, the primary mixing agency in the troposphere, and thermal turbulence, which occurs in an unstable layer of air. Precipitation, fog, and solar radiation exert secondary meteorological influences. Natural ventilation in the atmosphere is best when the winds are strong and turbulent so that mixing is good, and when the volume in which mixing occurs is large so that dilution of pollution is rapid. As cities have grown in size, air pollution has become more widespread. It has become necessary to think of whole urban complexes as large area sources of pollution. The rate of natural ventilation of an urban area is dependent on two quantities: the wind speed and the mixing volume over the city. Active mixing upward is often limited by a stable layer, perhaps even a very stable inversion layer, aloft. The upward extent of this region of active mixing, known as the mixing height, determines the magnitude of the mixing volume of the city.

Pollution is removed from the atmosphere in such ways as washout, rain-out, gravitational setting, and turbulent impaction. Washout is the process by which contaminants are washed out of the atmosphere by raindrops as they fall through the contaminants; in rain-out the contaminants unite with cloud droplets, which may later grow into precipitation.

Gravitational settling is significant mainly for large particles, those having a diameter greater than 20 micrometers. Agglomeration of finer particles may result in larger ones which settle out by gravitation. Fine particles may also impact on surfaces by centrifugal action in very small turbulent eddies. Gases may be converted to particulates, as by photochemical action of sunlight, which may then be removed by settling or impaction.

The rate of natural cleansing may be slower than the rate of injection of pollutants into the atmosphere, in which case pollution may increase on a global scale. There is evidence that the concentration of atmospheric carbon dioxide has been increasing slowly since the beginning of the century because of combustion of fossil fuels. The tropospheric burden of very small particles and of Freon gas may also be increasing.

Pollution of the stratosphere with nitrogen oxide causes reduction of stratospheric ozone. Ozone reduction in the stratosphere has been linked to biological effects such as skin cancer in two steps: (1) reduced ozone in the stratosphere causes an increase in uv radiation reaching the Earth's surface, and (2) increased uv radiation enhances the normal biological effects of natural uv radiation.

Pollution of the stratosphere also may involve a climate chain of cause-and-effect relations by which aircraft engine effluents, notably sulfur dioxide (SO_2), and to a lesser degree water vapor (H_2O) and nitrogen oxides (NO_x), affect climatic variables such as temperature, wind, and rainfall.

If enough particles larger than 0.1 μm in diameter were added to the stratosphere, they could alter the radiative heat transfer of the Earth-Sun system, and thereby influence climate. Particles of this size are produced by several constituents of engine emissions, in particular those of SO^2.

Among the many effects of tropospheric pollution are damage to human health, animal health, and vegetation, as well as corrosion of materials. Tropospheric pollution may also affect weather in a number of ways. Heavy precipitation at Laporte, Indiana, is attributed to a substantial source of air pollution there, and similar but less pronounced effects have been observed elsewhere. Industrial smoke reduces visibility and also ultraviolet radiation from the Sun, and polluted fogs are more dense and more persistent than natural fogs occurring under similar conditions. See ACID RAIN.

Four main methods of air-pollution control are prevention of pollution by improved equipment design and smokeless fuels; collection of contaminants at the source; containment of pollutants whose noxious characteristics may decrease with time, such as radioactive contaminants from nuclear power plants, or containment with destruction or conversion of the offending substances; and dispersion in industrial processes which can be varied to take advantage of the periods when dispersion conditions are so good that contaminants may be distributed very widely in such small concentrations that they inconvenience no one.

[A.J.G.; S.C.C.; E.W.He.]

Air pollution, indoor The presence of gaseous and particulate contaminants in the indoor environment. Most pollution is due to human (anthropomorphic) sources. Natural sources do exist, including plants, animals, and other living organisms, and water sources that release various chemical aerosols. Contamination can occur from infiltration indoors of atmospheric pollutants generated outdoors, and thus the indoor environment is affected by meteorological conditions.

Natural sources are soils and water that release radon progeny, volatile organic compounds, fungi, and such. Chemical releases (emissions) originate from various types of appliances, combustion sources (including those in garages), building materials and water, and living organisms that release allergens (for example, dander from pets). Carbon dioxide (CO_2) is generated by combustion and living organisms. Other sources involve molds, microorganisms (such as bacteria and viruses), insects and arthropods, and pollen from outdoor and indoor plants. Human activity has been noted to increase air contamination from all of these sources. Some pollutants are individually generated such as tobacco smoke (environmental tobacco smoke, or ETS) or are generated through use of consumer products. *See* PARTICULATES; RADON.

The generation and behavior of pollutants from sources in enclosed environments are also affected by most meteorological factors. Indoor environments are often sealed to the extent that the weather is cold or hot. Infiltration of pollutants is affected also by climate, temperature, humidity, barometric pressure, and wind speed and direction. The small particles and the gases that infiltrate most, that is, NO_2 and volatile organic compounds, are affected by outdoor concentrations, tightness (being sealed, with little air exchange), operation of heating and cooling systems, convection currents, full growth (or lack thereof) of local trees and shrubs, and so forth. Indoor environments also possess individual characteristics of ventilation, dispersion, and deposition. Some gases, for example, sulfur dioxide (SO_2) and ozone (O_3), infiltrating the indoor environment are absorbed readily by materials and exist in high concentrations only when the outdoor concentrations are very high. Chemical-physical mass-balance models are used to estimate pollutant concentrations, as are statistical models. *See* AEROSOL; AIR POLLUTION; OZONE. [M.D.L.]

Air pressure The force per unit area that the air exerts on any surface in contact with it, arising from the collisions of the air molecules with the surface. It is equal and opposite to the pressure of the surface against the air, which for atmospheric air in normal motion approximately balances the weight of the atmosphere above, about 15 pounds per square inch (psi) at sea level. It is the same in all directions and is the force that balances the weight of the column of mercury in the Torricellian barometer, commonly used for its measurement. *See* BAROMETER.

The basic units of pressure, which are the ones mainly used in meteorology, are based on the bar, which is defined as equal to 1,000,000 dynes/cm^2. One bar equals 1000 millibars (mb) equals 100 centibars (cb).

Also widely used in practice are units based on the height of the mercury barometer under standard conditions, expressed commonly in millimeters or in inches. The standard atmosphere (760 mmHg) is also used as a unit, mainly in engineering where large pressures are encountered. The following equivalents show the conversions between the commonly used units of pressure, where $(\text{mmHg})_n$, and $(\text{in. Hg})_n$, denote the millimeter and inch of mercury, respectively, under standard (normal) conditions, and where $(\text{kg})_n$ and $(\text{lb})_n$ denote the weight of a standard kilogram and pound mass, respectively, under standard gravity.

$$\begin{aligned}
1\ \text{mb} &= 1000\ \text{dynes/cm}^2 = 0.750062\ (\text{mmHg})_n \\
&= 0.0295300\ (\text{in. Hg})_n \\
1\ \text{atm} &= 1013.250\ \text{mb} = 760\ (\text{mmHg})_n \\
&= 29.9213\ (\text{in. Hg})_n = 14.6959\ (\text{lb})_n/\text{in.}^2 \\
&= 1.03323\ (\text{kg})_n/\text{cm}^2 \\
1\ (\text{mmHg})_n &= 1\ \text{torr} = 1.333224\ \text{mb} \\
&= 0.03937008\ (\text{in. Hg}) \\
1\ (\text{in. Hg})_n &= 33.8639\ \text{mb} = 25.4\ (\text{mmHg})_n
\end{aligned}$$

Because of the almost exact balancing of the weight of the overlying atmosphere by the air pressure, the latter decreases with height. A standard equation is used in practice to calculate the vertical distribution of pressure with height above sea level. The temperature distribution in a standard atmosphere, based on mean values in middle latitudes, has been defined by international agreement. The use of the standard atmosphere yields a definite relation between pressure and height. This re-lation is used in all altimeters which are basically barometers of the aneroid type. The difference between the height estimated from the pressure and the actual height is often considerable; but since the same standard relationship is used in all altimeters, the difference is the same for all altimeters at the same location, and so causes no difficulty in determining the relative position of aircraft. Mountains, however, have a fixed height, and accidents have been caused by the difference between the actual and standard atmosphere. *See* ALTIMETER.

In addition to the large variation with height, atmospheric pressure varies in the horizontal and with time. The variations of air pressure at sea level, estimated in the case of observations over land by correcting for the height of the ground surface, are routinely plotted on a map and analyzed, resulting in the familiar weather map representation with its isobars showing highs and lows. The movement of the main features of the sea-level pressure distribution, typically from west to east, produces characteristic fluctuations of the pressure at a fixed point, varying by a few percent within a few days. Smaller-scale variations of sea-level pressure, too small to appear on the ordinary weather map, are also present. These are associated with various forms of atmospheric motion, such as small-scale wave motion and turbulence. Relatively large variations are found in and near thunderstorms, the most intense being the low-pressure region in a tornado. The pressure drop within a tornado can be a large fraction of an atmosphere, and is the principal cause of the explosion of buildings over which a tornado passes. *See* ISOBAR (METEOROLOGY); WEATHER MAP.

It is a general rule that in middle latitudes at localities below 1000 m (3280 ft) in height above sea level, the air pressure on the continents tends to be slightly higher in winter than in spring, summer, and autumn; whereas at considerably greater heights on the continents and on the ocean surface, the reverse is true.

The practical importance of air pressure lies in its relation to the wind and weather. It is because of these relationships that pressure is a basic parameter in weather forecasting, as is evident from its appearance on the ordinary weather map.

The large-scale variations of pressure at sea level shown on a weather map are associated with characteristic patterns of vertical motion of the air, which in turn affect the weather. Descent of air in a high heats the air and dries it by adiabatic compression, giving clear skies, while the ascent of air in a low cools it and causes it to condense and produce cloudy and rainy weather. These processes at low levels, accompanied by others at higher levels, usually combine to justify the clear-cloudy-rainy marking on the household barometer. [R.J.D.]

Air register A device attached to an air-distributing duct for the purpose of discharging air into the space to be heated or cooled. These openings are referred to as registers, diffusers, supply outlets, or grills (see illustration). By common acceptance, a register is an opening provided with means for discharging the air in a confined jet, whereas a diffuser is an outlet which discharges the air in a spreading jet. Both registers and diffusers may be placed at a number of locations in a room, including the floor, baseboard, low on the sidewall, window sill, high on the sidewall, or ceiling.

One of the more common diffusers, a round ceiling type. (*Titus Manufacturing Corp.*)

For heating, the preferred location is in the floor, at the baseboard, or at the low sidewall of the outside wall, preferably under a window. For cooling, the preferred location is high on the inside wall or the ceiling. For year-round air conditioning in homes, a compromise location is the floor, baseboard, or low sidewall at the exposed wall, especially if adequate air velocity in an upward direction is provided at the supply outlet. [S.Ko.]

Air temperature The temperature of the atmosphere represents the average kinetic energy of the molecular motion in a small region, defined in terms of a standard or calibrated thermometer in thermal equilibrium with the air. Many different types of thermometer are used for the measurement of air temperature, the most common depending on the expansion of mercury with temperature, the variation of electrical resistance with temperature, or the thermoelectric effect (thermocouple). *See* TEMPERATURE; THERMOMETER.

The temperature of a given small mass of air varies with time because of heat added or subtracted from it, and also because of work done during changes of volume.

The rate at which the temperature changes at a particular point, that is, as measured by a fixed thermometer, depends on the movement of air as well as physical processes such as absorption and emission of radiation, heat conduction, and changes of phase of water involving latent heat of condensation and freezing. The large changes of air temperature from day to day are mainly due to the horizontal movement of air, bringing relatively cold or warm air masses to a particular point, as the large-scale pressure-wind systems move across the weather map. *See* AIR MASS; AIR PRESSURE.

Temperatures near the surface are read at one or more fixed times daily, and the day's extremes are obtained from special maximum and minimum thermometers, or from the trace (thermogram) of a continuously recording instrument (thermograph). The average of these two extremes, technically the midrange, is considered in the United States to be the day's average temperature. The true daily mean, obtained from a thermogram, is closely approximated by the mean of 24 hourly readings, but may differ from the midrange by 1 or 2°F (0.5 or 1°C), on the average. In many countries temperatures are read daily at three or four fixed times, so that their weighted mean closely approximates the true daily mean.

Averages of daily maximum and minimum temperature for a single month for many years give mean daily maximum and minimum temperatures for that month. The average of these values is the mean monthly temperature, while their difference is the mean daily range for that month. Monthly means, averaged through the year, give the mean annual temperature; the mean annual range is the difference between the hottest and coldest mean monthly values. The hottest and coldest temperatures in a month are the monthly extremes; their averages over a period of years give the mean monthly maximum and minimum (used extensively in Canada), while the absolute extremes for the month (or year) are the hottest and coldest temperatures ever observed. The interdiurnal range or variability for a month is the average of the successive differences, regardless of sign, in daily temperatures.

Over the oceans the mean daily, interdiurnal, and annual ranges are slight, because water absorbs the insolation and distributes the heat through a thick layer. In tropical regions the interdiurnal and annual ranges over the land are small also, because the annual variation in insolation is relatively small. The daily range also is small in humid tropical regions, but may be large (up to 40°F or 22°C) in deserts. Interdiurnal and annual ranges increase generally with latitude, and also with distance from the ocean; the mean annual range defines continentality. The daily range depends on aridity, altitude, and noon Sun elevation. *See* ATMOSPHERE; INSOLATION; RADIATION. [R.J.D.]

Air-traffic control A service which promotes the safe and expeditious movement of aircraft operating in the air or on an airport surface. Air-traffic control (ATC) is a component of the aviation system which has evolved to move goods and people through airspace to desired destinations with reliability, efficiency, and minimum infringement on individual intents. The objective of ground-based ATC is to provide pilots of aircraft with information and control services that will enable them to use the airspace efficiently and to conduct operations with maximum safety. Increases in the amount of air traffic using the limited airspace and airports at heavily congested metropolitan terminal areas have profoundly affected traffic control. The challenge to ATC in such an environment is to match system capacity to system demand without compromising safety.

Delays increase with traffic demand when capacity is exceeded. Under saturation demand conditions, delays can be reduced by flow control procedures and additional capacity. Among the possibilities for attaining increased capacity are additional runways at existing airports, additional airports, procedures and equipment that allow operation under conditions of lower visibility, rules to provide segregation of aircraft by speed characteristics, automated techniques to communicate clearances between pilots and controllers, and systems which permit aircraft to operate safely by using reduced spacings in the final approach area. *See* AIRPORT SURFACE DETECTION EQUIPMENT.

Two cardinal rules of flight that apply to the ATC system are instrument flight rules (IFR) and visual flight rules (VFR). IFR governs all flights operating in weather conditions that prevent pilots from seeing well enough to avoid each other; IFR flight prevails when weather conditions fall below the minimum ceiling and visibility prescribed for flight under VFR. All flights receiving ATC separation service in en route airspace must be conducted under IFR rules regardless of weather conditions. All aircraft operating above 18,000 ft (5486 m), where aircraft velocities preclude the pilot from providing visual separation and detecting possible collisions with sufficient lead time to perform an evasive maneuver, must operate within the ATC.

Two broad categorizations of ATC procedures are terminal and en route. Terminal operations pertain to departures and arrivals at airports. En route procedures cover that portion of the flight between the departure and arrival terminals.

Technical devices and equipment of four broad classes are required in order to fly safely and in compliance with the general rules and procedures: navigation systems, communication systems, surveillance systems, and data systems.

To fly within designated airspace, the pilot uses navigational instruments to ascertain aircraft position in three dimensions and to conduct the aircraft over any chosen route from terminal to terminal. The facility used most commonly over the continental United States is known as the very-high-frequency omnirange combined with distance-measuring equipment (VORTAC). It includes a ground radio-transmitting station that provides 360 course radials from the antenna source. The airborne portion of this facility is a receiver and display device, which indicates the magnetic bearing to a particular VOR facility.

Radio voice communication is the means whereby the centralized ground-based ATC entity can learn the position of all aircraft within its jurisdiction, and the means by which safe-flight instructions are transmitted rapidly and unambiguously to all aircraft concerned. Radar is used as a substitute for voice position reporting where available. Digital data-link systems are proposed for future use to increase system capacity and accuracy and to provide relief from voice frequency congestion. *See* RADIO-SPECTRUM ALLOCATIONS.

Surveillance of aircraft in en route and terminal airspace is provided by radar systems. Each primary radar has an ATC radar beacon system (ATCRBS) associated with it which interrogates transponders in equipped aircraft to determine identity code and pressure height data. *See* RADAR.

The ATCRBS is to be expanded to include a discrete mode of interrogation in which each aircraft transponder is called by its discrete identity code. This combined system is called the discrete address beacon system (DABS). The system includes the capability of exchanging data-link messages with the aircraft during surveillance interrogation periods.

The ground-based ATC system assimilates pertinent flight information, such as present position, course, speed, and identification, for each aircraft in its jurisdiction and determines safe, nonconflicting flight procedures for all these aircraft. Computer-based systems which perform computations and drive digital displays have replaced combinations of radar and written information.

ATC and flight advisory services are provided by the ATC command center and associated national flow control facility; air route traffic control centers (ARTCCs); terminal control facilities; and flight service stations. Statistical and real-time information on active and proposed flights, severe weather information, and current and forecast major-airport acceptance-rate information is processed in the national flow control computer facility located at the ARTOC in Jacksonville, Florida. Projected traffic and delay information is forwarded to the ATC command center in Washington, D.C., for dissemination to major airports.

The ARTCC facilities control all traffic operating above a pressure altitude of 18,000 ft or 5486 m. Traffic operating below this altitude may receive control service by filing an IFR flight plan. Terminal control service is provided by the ARTCCs to airports that do not have terminal control facilities and are not satellite airports receiving service from another terminal control facility. The ARTCCs located within the coterminous 48 states have been automated, and automation has been undertaken at centers in Alaska, Hawaii, and Puerto Rico.

Navigation in the terminal area is provided by the VOR in conjunction with DME (VORTAC). When aircraft are about 20–40 mi (32–64 km) from the runway, the air-traffic controllers define the sequence of aircraft arrivals and determine the course, altitude, and speed changes needed to achieve the desired interaircraft spacings on the final approach course to the runway. The final approach guidance is provided by the instrument landing system (ILS), which provides both horizontal and vertical path guidance to the pilot. When development is complete, this service is to be furnished by the microwave landing system (MLS). *See* INSTRUMENT LANDING SYSTEM (ILS).

Control of the aircraft in the terminal area is based primarily on radar position data obtained from the ATCRBS. Many of the major terminals are equipped with an automated radar terminal system (ARTS III) offering automation capabilities equivalent to the en route system. Additional radar terminals are being equipped with a limited automation capability (ARTS II). [N.A.Bl.]

Air transportation The movement of passengers and cargo by aircraft such as airplanes and helicopters. Air transportation has become the primary means of common-carrier traveling. Greatest efficiency and value are obtained when long distances are traveled, high-value payloads are moved, immediate needs must be met, or surface terrain prevents easy movement or significantly raises transport costs. Although the time and cost efficiencies obtained decrease as distance traveled is reduced, air transport is often worthwhile even for relatively short distances. Air transportation also provides a communication link, which is sometimes vital, between the different groups of people being served. [D.V.M.]

Air-velocity measurement The measurement of the rate of displacement of air or gas at a specific location, for example, wind speed, air velocity in the test section of a wind tunnel, airspeed of an aircraft, or air velocity produced by a fan or blower in air conditioning. Velocity indicates the magnitude and direction of the flow rate; the latter is usually measured separately. To obtain the total flow velocity as required in pipes or conduits, integration over the cross section is required, or another method of measurement is used. *See* AIRSPEED INDICATOR; FLOW MEASUREMENT.

Three primary methods are used in measuring air velocity. These methods and the instruments which make use of each are listed in the table. *See* ANEMOMETER; BRIDLED PRESSURE PLATE; KATA THERMOMETER; PITOT TUBE; SHIELDED THERMOCOUPLE; VENTURI TUBE. [H.F.O.]

Methods of measuring air velocity

Method	Instruments
Pressure on an element in the airstream	Pitot-static tube Venturi tube Bridled pressure plate
Speed or number of revolutions of a rotating element in the airstream	Cup anemometer Vane anemometer
Effect of velocity on a heated element in the airstream	Hot-wire anemometer Kata thermometer

Airborne radar Radar equipment carried by aircraft to provide services such as navigation by pilotage, determination of drift, mapping of weather fronts, and determination of ground speed; reconnaissance-system, target-detection, terrain-following, and bombing and gunlaying services provide assistance for tactical missions. *See* RADAR.

To use radar for pilotage, the presentation on the cathode-ray tube (plan position indicator, or PPI) is compared with a chart as in Fig. 1. This method is especially effective when applied to well-mapped terrain that includes bodies of water (bays, rivers, or harbors). *See* PILOTAGE.

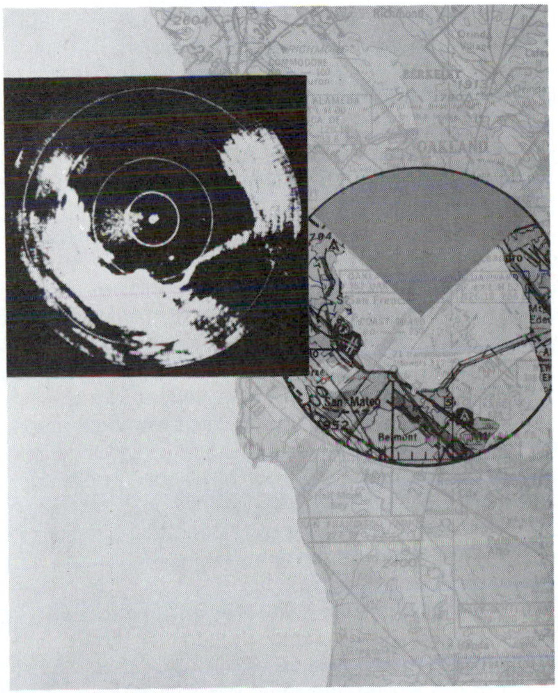

Fig. 1. PPI presentation of an airborne radar over San Francisco Bay. Scope presentation is of the circled area of the aeronautical chart. Note bridge at lower right. (*Radio Corporation of America*)

To determine drift, a transparent disk engraved with parallel lines is mounted on the face of the PPI tube. The disk is then rotated against a scale calibrated in degrees of drift (Fig. 2). The center of the disk coincides with the center of the cathode-ray pattern. A trace on the face of the tube indicates the position of the antenna when it is pointing directly along the fore-and-aft axis of the aircraft. To measure drift, the operator selects a prominent spot on the PPI and rotates the disk until

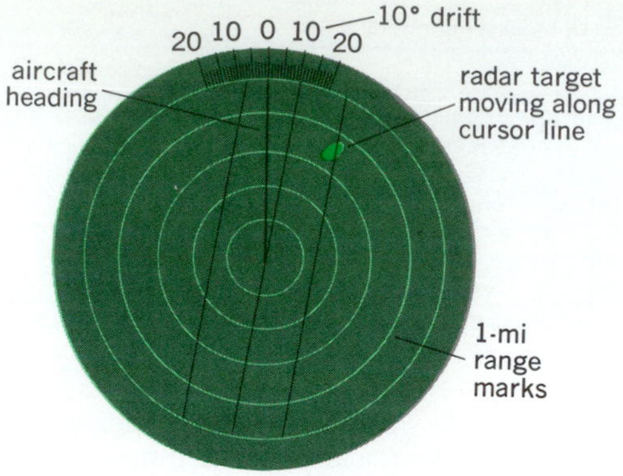

Fig. 2. PPI with cursor and scale for measuring drift. 1 mi = 1.6 km. (*After P.C. Sandretto, Electronic Avigation Engineering, International Telephone and Telegraph Corp., 1958*)

the spot appears to move along or parallel to the lines on the disk. The angle between the center line on the disk and the trace representing the fore-and-aft axis of the aircraft then represents the drift angle. By timing the movement of the spot, ground speed may be estimated.

Airborne radar has been used for determining drift through the Doppler principle. Because of the aircraft movement, the frequency of the reflected signal returning to the aircraft from the ground differs from the frequency of transmission. This difference is proportional to the frequency of the original transmission, twice the aircraft speed, and the cosine of the angle between the direction of aircraft motion and a line to the reflecting point.

One of the most valuable uses of airborne radar is for weather detection. This capability makes for safer and smoother navigation of thunderstorm and precipitation areas. By using radar, a pilot may avoid turbulence associated with these areas. *See* RADAR METEOROLOGY.

Reconnaissance systems, also called ground-mapping systems and side-looking radar, provide data on both fixed and moving ground targets for military reconnaisance purposes. These radars differ from other types in that they use fixed antennas located on both sides of the aircraft fuselage. With this configuration, an antenna system is provided which has an aperture (about 5 ft or 1.5 m) that is much wider than can be provided by any parabolic antenna which would be practical to mount on an aircraft. The wider aperture provides definition of the ground targets that could not otherwise be distinguished. This performance, however, is possible only because of the novel scanning system. Instead of rotating the antenna to scan, the radar beam moves in space because the aircraft carrying the antenna moves.

Terrain following is provided by airborne radar which enables attack aircraft to remain hidden in terrain and thus avoid detection by enemy defenses. Flight must be made below the peaks of the terrain, but there is a limit to how low a level the pilot's own capability may be safely applied. Airborne radar provides range, altitude, and other information that assists the pilot in deciding the proper maneuver which will attain optimal flight.

Airborne radar provides the service which is most essential to the performance of the military aircraft's mission—the detection of targets to be destroyed. These targets are of two general types—strategic and tactical. Strategic targets are those which, when destroyed, inhibit the enemy's capability to wage war. Tactical targets are the weapons and weapon systems which the enemy employs against the opponent.

With a quasi-visual radar PPI presentation of the target area below, a computer is all that is needed to solve the bombing problem. This problem involves the ballistics of the bomb, the range and azimuth offset, and the range closing rate. [P.C.S.]

Aircraft Any vehicle which carries one or more persons and which navigates through the air. The two main classifications of aircraft are lighter-than-air and heavier-than-air. The term lighter-than-air is applied to all aircraft which sustain their weight by displacing an equal weight of air, for example, blimps and dirigibles. Heavier-than-air craft are supported by giving the surrounding air a momentum in the downward direction equal to the weight of the aircraft. *See* AIRPLANE; CONVERTIPLANE; HELICOPTER. [R.G.Bo.]

Aircraft collision avoidance system A system, independent of ground-based systems, that allows the pilot of an aircraft to observe and avoid other aircraft, regardless of weather. As far as civil aviation is concerned, aircraft have been kept separated by the use of communication, navigation, and surveillance systems based on the ground.

Since the mid-1970s, research involving air collision avoidance systems has concentrated on the use of hardware already carried by most aircraft, namely the transponder of the Air-Traffic Control Radar Beacon System (ATCRBS). These transponders reply to interrogations from secondary surveillance radars (SSRs) on the ground. Interrogation is with 1-microsecond pulses at 1030 MHz, and replies are with a train of 0.5-μs pulses at 1090 MHz which can be arranged in 4096 different codes to produce identity and altitude, the latter requiring an additional airborne digitizer. ATCRBS forms the backbone of a ground-based worldwide air-traffic control system.

An independent collision avoidance system has been developed that interrogates these transponders from the air (in addition to continuing to reply to interrogations from the surveillance radar). This system is known as the Traffic Alert and Collision Avoidance System (TCAS). Two features have been added: (1) horizontal directivity, to allow interrogation of one sector at a time; and (2) the so-called whisper-shout technique, which progressively varies the strength of the interrogations to permit selective interrogation of different distances to targets.

In a TCAS-equipped aircraft, replies are fed to a computer which generates two types of information: (1) traffic advisories that tell the pilot there are nearby aircraft, of known distance, altitude, and approximate bearing; and (2) resolution advisories that advise immediate evasive action (for example, "climb" or "descend"). These are displayed to the pilot by various means. In December 1987, the United States Congress passed a law making TCAS II mandatory in airliners carrying 30 or more passengers by the end of 1991. *See* AIR-TRAFFIC CONTROL; SURVEILLANCE RADAR. [S.H.D.]

Aircraft compass system Compass systems are used on aircraft for navigation without reference to ground-based signals. Magnetic compasses use the Earth's magnetic field as the reference; celestial systems use the stars and Sun as the reference; inertial systems use the changes of position and time as references.

The commonest system is the magnetic compass, which at present is found on all aircraft, even when more sophisticated systems are present. Dual magnets are mounted on a float, along with the compass card, to provide the pilot with the direction in which the aircraft is headed. *See* MAGNETIC COMPASS.

Celestial systems constantly track the stars or Sun by precise photodetector-optical systems, and by reference to time and the vertical provide a continuous indication of direction (heading) and position (latitude and longitude).

Inertial systems provide heading and position by the use of high-precision gyroscopes and accelerometers whose signal changes are integrated with respect to time. A lightweight aircraft computer that requires a minimum of service is a critical component of the system. *See* Celestial navigation; Inertial guidance system; Radio compass; Star tracker. [J.W.A.]

Aircraft design The process of designing an aircraft, generally divided into three distinct phases: conceptual design, preliminary design, and detail design. Each phase has its own unique characteristics and influence on the final product. These phases all involve aerodynamic, propulsion, and structural design, and the design of aircraft systems.

Design phases. Conceptual design activities are characterized by the definition and comparative evaluation of numerous alternative design concepts potentially satisfying an initial statement of design requirements. The conceptual design phase is iterative in nature. Design concepts are evaluated, compared to the requirements, revised, reevaluated, and so on until convergence to one or more satisfactory concepts is achieved. During this process, inconsistencies in the requirements are often exposed, so that the products of conceptual design frequently include a set of revised requirements.

During preliminary design, one or more promising concepts from the conceptual design phase are subjected to more rigorous analysis and evaluation in order to define and validate the design that best meets the requirements. Extensive experimental efforts, including wind-tunnel testing and evaluation of any unique materials or structural concepts, are conducted during preliminary design. The end product of preliminary design is a complete aircraft design description including all systems and subsystems. *See* Wind tunnel.

During detail design the selected aircraft design is translated into the detailed engineering data required to support tooling and manufacturing activities.

Requirements. The requirements used to guide the design of a new aircraft are established either by an emerging need or by the possibilities offered by some new technical concept or invention. Requirements can be divided into two general classes: technical requirements (speed, range, payload, and so forth) and economic requirements (costs, maintenance characteristics, and so forth).

Aerodynamic design. Initial aerodynamic design centers on defining the external geometry and general aerodynamic configuration of the new aircraft.

The aerodynamic forces that determine aircraft performance capabilities are drag and lift. The basic, low-speed drag level of the aircraft is conventionally expressed as a term at zero lift composed of friction and pressure drag forces plus a term associated with the generation of lift, the drag due to lift or the induced drag. Since wings generally operate at a positive angle to the relative wind (angle of attack) in order to generate the necessary life forces, the wing lift vector is tilted aft, resulting in a component of the lift vector in the drag direction (see illustration). *See* Aerodynamic force; Airfoil; Wing.

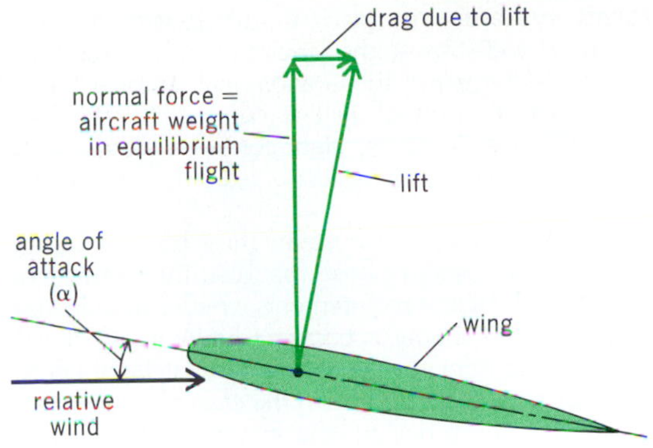

Wing lift and drag due to lift.

Aircraft that fly near or above the speed of sound must be designed to minimize aerodynamic compressibility effects, evidenced by the formation of shock waves and significant changes in all aerodynamic forces and moments. Compressibility effects are mediated by the use of thin airfoils, wing and tail surface sweepback angles, and detailed attention to the lengthwise variation of the cross-sectional area of the configuration. *See* Compressible flow; Shock wave.

The size and location of vertical and horizontal tail surfaces are the primary parameters that determine aircraft stability and control characteristics. Developments in digital computing and flight-control technologies have made the concept of artificial stability practical. *See* Stability augmentation.

Propulsion design. Propulsion design comprises the selection of an engine from among the available models and the design of the engine's installation on or in the aircraft.

Selection of the best propulsion concept involves choosing from among a wide variety of types ranging from reciprocating engine-propeller power plants through turboprops, turbojets, turbofans, and ducted and unducted fan engine developments. The selection process involves aircraft performance analyses comparing flight performance with the various candidate engines installed. In the cases where the new aircraft design is being based on a propulsion system which is still in development, the selection process is more complicated. *See* Aircraft engine; Turbofan; Turboprop.

Once an engine has been selected, the propulsion engineering tasks are to design the air inlet for the engine, and to assure the satisfactory physical and aerodynamic integration of the inlet, engine, and exhaust nozzle or the engine nacelles with the rest of the airframe. The major parameters to be chosen include the throat area, the diffuser length and shape, and the relative bluntness of the inlet lips.

Structural design. Structural design begins when the first complete, integrated aerodynamic and propulsion concept is formulated. The process starts with preliminary estimates of design airloads and inertial loads (loads due to the mass of the aircraft being accelerated during maneuvers).

During conceptual design, the structural design effort centers on a first-order structural arrangement which defines major structural components and establishes the most direct load paths through the structure that are possible within the constraints of the aerodynamic configuration. An initial determination of structural and material concepts to be used is made at this time, for example, deciding whether the wing should be constructed from built-up sheet metal details, or by using machined skins with integral stiffeners, or from fiber-reinforced composite materials.

During preliminary design, the structural design effort expands into consideration of dynamic loads, airframe life, and structural integrity. Dynamic loading conditions arise from many sources: landing impact, flight through turbulence, taxiing over rough runways, and so forth. *See* Loads, dynamic.

Airframe life requirements are usually stated in terms of desired total flight hours or total flight cycles. To the structural designer this translates into requirements for airframe fatigue life. Fatigue life measures the ability of a structure to withstand repeated loadings without failure. Design for high fatigue life involves selection of materials and the design of structural components that minimize concentrated stresses.

Structural integrity design activities impose requirements for damage tolerance, the ability of the structure to continue to support design loads after specified component failures. Fail-safe design approaches are similar to design for fatigue resistance: avoidance of stress concentrations and spreading loads out over multiple supporting structural members. *See* Structural design.

Aircraft systems design. Aircraft systems include all of those systems and subsystems required for the aircraft to operate. Mission systems are those additional systems and subsystems peculiar to the role of military combat aircraft. The major systems are power systems, flight-control systems, navigation and communication systems, crew systems, the landing-gear system, and fuel systems.

Design of these major subsystems must begin relatively early in the conceptual design phase, because they represent large dimensional and volume requirements which can influence overall aircraft size and shape or because they interact directly with the aerodynamic concept (as in the case of flight-control systems) or propulsion selection (as in the case of power systems).

During preliminary design, the aircraft system definition is completed to include additional subsystems. The installation of the many aircraft system components and the routing of tubing and wiring through the aircraft are complex tasks which are often aided by the construction of partial or complete aircraft mock-ups. These are full scale models of the aircraft, made of inexpensive materials, which aid in locating structural and system components. *See* AIRPLANE. [P.L.M.]

Aircraft engine A component of an aircraft that develops either shaft horsepower or thrust and incorporates design features most advantageous for aircraft propulsion. An engine developing shaft horsepower requires an additional means to convert this power to useful thrust for aircraft, such as a propeller, a fan, or a helicopter rotor. It is common practice in this case to designate the unit developing shaft horsepower as the aircraft engine, and the combination of engine and propeller, for example, as an aircraft power plant. In case thrust is developed directly as in a turbojet engine, the terms engine and power plant are used interchangeably.

Air-breathing types of aircraft engines use oxygen from the atmosphere to combine chemically with fuel carried in the vehicle, providing the energy for propulsion, in contrast to rocket types in which both the fuel and oxidizer are carried in the aircraft. *See* INTERNAL COMBUSTION ENGINE; JET PROPULSION; PROPELLER (AIRCRAFT); RECIPROCATING AIRCRAFT ENGINE; ROCKET PROPULSION; TURBINE PROPULSION. [R.Ha.]

Aircraft fuel The source of energy required for the propulsion of airborne vehicles. Aircraft fuel is burned with ambient air and is thereby distinct from rocket propellants, which carry both fuel and oxidant. An important criterion for aircraft fuel is that its energy density, or heat of combustion per unit of weight, be high. This allows reasonable expenditures of fuel during takeoff, efficient performance in flight, and long range of flight duration.

There are two general types of aircraft fuels in conventional use: gasolines for reciprocating (piston) engines, and kerosene-like fuels (called jet fuels) for turbine engines. Because of anticipated limitations in the supply of these crude-oil derived fuels, alternative fuels are being considered for future aircraft.

Piston engine fuels, or aviation gasolines, are special blends of gasoline stocks and additives that produce a high-performance fuel that is graded by its antiknock quality. The gasoline blending stocks are virgin (uncracked) naphtha, alkylate, and catalytically cracked gasoline. In general, the chemical composition of aviation gasoline can be approximated as $C_xH_{1.9x}$ where the number of carbon atoms x is between 4 and approximately 10. Tetraethyllead (Tel) is a common additive used in concentrations of up to 4 ml/gal (1.057 ml/liter) of fuel to increase the antiknock quality of the fuel. *See* GASOLINE; NAPHTHA; TETRAETHYLLEAD.

Turbine engine fuels are distillate hydrocarbon fuels, like kerosenes, used to operate turbojet, turbofan, and turboshaft engines. While all piston engine fuels have the same volatility but differ in combustion characteristics, jet fuels differ primarily in volatility; differences in their combustion qualities are minor. Turbine fuel contains aromatic hydrocarbons; limits are placed on this content owing to concerns about smoke and coke formation. An increasingly important requirement is to provide a fuel that is stable at relatively high temperatures.

Many forecasts of world crude production indicate a peak or plateau in the oil production rate before 2010; hence, the availability of crude-based fuels for aircraft is not certain. Alternative fuels made from coal, oil shale, or solar or nuclear energy plus a suitable raw material are being considered. [J.B.P.]

Aircraft instrumentation A coordinated group of instruments that provide the flight crew with information about the aircraft and its subsystems. These instruments provide flight data, navigation, power plant performance, and aircraft auxiliary equipment operating information to the flight crew, air-traffic controllers, and maintenance personnel. While not considered as instrumentation, communication equipment is, however, directly concerned with the instrumentation and overall indirect control of the aircraft.

Situation information on the operating environment, such as weather reports and traffic advisories, has become a necessity for effective flight planning and decision making. The prolific growth and multiplicity of instruments in the modern cockpit and the growing need for knowledge about the aircraft's situation are leading to the introduction of computers and advanced electronic displays as a means for the pilot to better organize and assimilate this body of information.

Instrumentation complexity and accuracy are dictated by the aircraft's performance capabilities and the conditions under which it is intended to operate. Light aircraft may carry only a minimum set of instruments; an airspeed indicator, an altimeter, an engine tachometer and oil pressure gage, a fuel quantity indicator, and a magnetic compass. These instruments allow operation by a pilotage technique. *See* ALTIMETER; AIRSPEED INDICATOR; PILOTAGE; RATE-OF-CLIMB INDICATOR.

Operation under low visibility and under Instrument Flight Rules (IFR) requires this same information in a more precise form and also requires attitude and navigation data. An attitude-director indicator (ADI) presents an artificial horizon, bank angle, and turn coordination data for attitude control without external visual reference. The attitude-director indicator may contain a vertical gyro within the indicator, or a gyro may be remotely located as a part of a flight director or navigational system. Flying through a large speed range at a variety of altitudes is simplified if the indicated airspeed is corrected to true airspeed for navigation purposes and the Mach number (M) is also shown on the ADI for flight control and performance purposes. Rate-of-climb is provided by an instantaneous vertical-speed indicator (IVSI). Heading data are provided by a directional gyro or data derived from an inertial reference system. *See* GYROSCOPE; INERTIAL GUIDANCE SYSTEM.

Navigation aids include: very-high-frequency omnidirectional radio ranges (VOR) that transmit azimuth information for navigation at specified Earth locations; distance-measuring equipment (DME) that indicates the distance to radio aids on or near airports or to VORs; automatic direction finders (ADF) that give the bearing of other radio stations (generally low-frequency); low-range radio altimeters (LRRA) which by radar determine the height of the aircraft above the terrain at low altitudes; and instrument landing systems (ILS) that show vertical and lateral deviation from a radio-generated glide-path signal for landing at appropriately equipped runways. Some inertial navigation systems include special-purpose computers that provide precise Earth latitude and longitude, ground speed, course, and heading. *See* AIR-TRAFFIC CONTROL; AUTOPILOT; DIRECTION-FINDING EQUIPMENT; DISTANCE-MEASURING EQUIPMENT; ELECTRONIC NAVIGATION SYSTEMS; INSTRUMENT LANDING SYSTEM (ILS); RADIO RANGE.

Engines require specific instruments to indicate limits and efficiency of operation. For reciprocating engines, instruments may display intake and exhaust manifold pressures, cylinder head and oil temperature, oil pressure, and engine speed. For jet engines, instruments display engine pressure ratio (EPR), exhaust gas temperature (EGT), engine rotor speed, oil temperature and pressure, and fuel flow. Vibration monitors on both types of engines indicate unbalance and potential trouble.

Depending on the complexity of the aircraft and the facilities that are provided, there is also an assortment of instruments and controls for the auxiliary systems.

Electronic technology developments include: ring laser gyros, strap-down inertial reference systems, microprocessor digital computers, color cathode-ray tubes (CRT), liquid crystal displays (LCD), light-emitting diodes (LED), and digital data buses. Application of this technology allows a new era of system integration and situation information on the aircraft flight deck and instrument panels. Commercial jet transports will use digital electronics to improve safety, performance, economics, and passenger service. The concept of an integrated flight management system (FMS) includes automatic flight control, electronic flight instrument displays, communications, navigation, guidance, performance management, and crew alerting to satisfy the requirements of the current and future air-traffic and energy-intensive environment.

Effective flight management is closely tied to providing accurate and timely information to the pilot. The nature of the pilot's various tasks determines the general types of data which must be available. The key is to provide these data in a form best suited for use. If the pilot is not required to accomplish extensive mental processing before information can be used, then more information can be presented and less effort, fewer errors, and lower training requirements can be expected. Computer-generated displays offer significant advances in this direction.

The electronic horizontal-situation indicator (EHSI) provides an integrated multicolor map display of the airplane's position, plus a color weather radar (WXR) display. The scale for the radar and map can be selected by the pilots. [B.C.H.]

Aircraft propulsion Flying machines obtain their propulsion by the rearward acceleration of matter. This is an application of Newton's third law: For every action there is an equal and opposite reaction.

In propeller-driven aircraft, the propulsive medium is the ambient air which is accelerated to the rear by the action of the propeller. The acceleration of the air that passes through the engine provides only a secondary contribution to the thrust.

In the case of turbojet and ramjet engines, the ambient air is again the propulsive medium, but the thrust is obtained by the acceleration of the air as it passes through the engine. After being compressed and heated in the engine, this air is ejected rearward from the engine at a greater velocity than it had when it entered. *See* JET PROPULSION.

Rockets carry their own propulsive medium. The propellants are burned at high pressure in a combustion chamber and are ejected rearward to produce thrust. *See* ROCKET PROPULSION.

In every case, the thrust provided is equal to the mass of propulsive medium per second multiplied by the increase in its velocity produced by the propulsive device. This is substantially Newton's second law. *See* FLUID FLOW.

The airplane lift-drag ratio L/D is a primary factor that determines the thrust required from the propulsion system to fly a given airplane. To sustain flight, the airplane lift must be equal to airplane gross weight, and the engine thrust must be equal to the airplane drag. The higher the lift-drag ratio, the more efficient is the airplane. A sharp reduction in L/D occurs with increase in flight Mach number in the vicinity of a Mach number of unity, and this is reflected in a sharp increase in the thrust required for flight. Flight Mach number is the ratio of the airplane speed to the speed of sound in the ambient atmosphere. At standard sea-level conditions, the speed of sound is 773 mi/h (346 m/s).

The competition among nations and among commercial airlines has created a continuing demand for increased flight speed. The reduction in aircraft L/D that accompanies an increase in speed (see illustration) requires an increase in

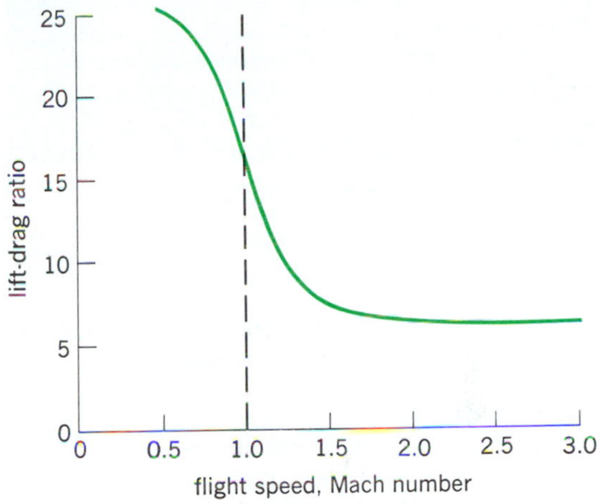

Lift-drag ratios of aircraft. Curve represents envelope of a variety of designs of various wing sweep angles. (*Taken from 27th Wright Brothers Lecture by G. S. Schairer, J. Aircraft, vol. 1, no. 2, 1964*)

engine thrust for an airplane of a given gross weight. For a given engine specific weight, an increase in required engine thrust results in an increase in engine weight and hence a reduction in fuel load and payload that an airplane of a given gross weight can carry. If the engine weight becomes so large that no fuel can be carried by the airplane, the airplane has zero flight range regardless of the efficiency of the engine. At some speed before this point is reached, it becomes advantageous to shift to an engine type that has a lower specific engine weight even at the cost of an increased specific fuel consumption.

At low subsonic flight speeds, the piston-type reciprocating engine, because of its low specific fuel consumption, provides the best airplane performance in terms of payload and flight range. As flight speed increases, specific weight of reciprocating engines increases because of falling propeller efficiency. This effect, coupled with reduction in L/D which accompanies increase in flight speed, results in the weight of reciprocating engines becoming excessive at a flight speed of about 400 mi/h (180 m/s). At about this speed it is advantageous to shift to the lighter-weight turboprops even if the efficiency of the latter is poorer. At about 550 mi/h (245 m/s) it is advantageous to shift from the turboprop to the lighter but less efficient turbojet. Intermediate between the turboprop and turbojet in the spectrum of flight speeds is the turbofan. *See* RECIPROCATING AIRCRAFT ENGINE; TURBOFAN; TURBOJET; TURBOPROP. [B.Pi.]

Aircraft rudder The hinged rear portion of an aircraft vertical stabilizing surface, used to obtain directional or yaw control moments. The angular setting of the rudder is controlled by the human or automatic pilot through the flight-control system. A typical rudder control surface includes aerodynamic balance and tab features, as shown in the illustration. The principles of operation of rudder control surfaces and flutter prevention methods are identical with those for the elevator

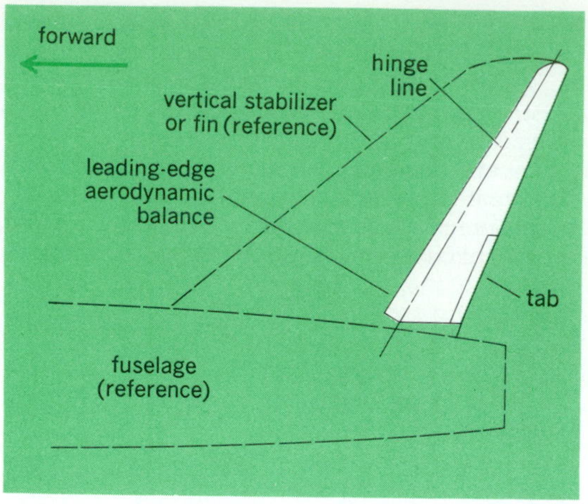

Typical rudder control surface.

and for aileron control surfaces. All-moving vertical stabilizers replace rudders on some supersonic aircraft. [M.J.A.]

Aircraft stability The tendency of forces and moments to be set up that restore an aircraft to its equilibrium position if it is disturbed from this position. Stability is a necessary condition for successful flight. For an airplane to be in equilibrium, the sum of the forces acting at its center of gravity must equal zero, as must the sum of the moments.

To examine the condition of stability, suppose that the equilibrium of the airplane is suddenly disturbed in a manner that causes it to accelerate. This means that its lift coefficient decreases. A stable airplane would immediately be subjected to a nose-up moment which would tend to increase the lift coefficient and reduce the speed. Similarly, a reduction in the speed of a stable airplane should produce a nose-down moment.

Stability is achieved by balancing the force and moment contributions of various portions of the airplane so that the summation is stable. Because the balance is achieved for moments about the center of gravity, the location of this point is important. Its position depends upon the load condition and can change in flight as fuel is consumed or as disposable load is dropped. The airplane must be designed so that it is stable throughout the entire range of center-of-gravity locations encountered in service. [D.C.H.]

Airfoil A body of such shape that the force exerted on it by its motion through a fluid has a larger component normal to the direction of the motion than along the direction of motion. Examples are the wing of an airplane; a tail surface of an airplane; a blade of a propeller, windmill, or helicopter rotor; and a blade of an axial-flow compressor or turbine. The term is sometimes used for a certain cross section of an airfoil. The closed curve defining the external shape of the cross section of an airfoil is called an airfoil profile, airfoil section, airfoil shape, or wing section. Knowledge of the geometry of airfoil profiles is necessary in designing airfoil structures. Both analytical and experimental studies are used in predicting the flow of fluid past an airfoil and therefore in determining the fluid dynamic characteristics of the airfoil. This discussion deals largely with airplane wings, and is mostly applicable to other examples.

The root chord of a wing, c_r (see illustration), is the chord of the cross section at the center of the wing, or at the vertical plane of symmetry of the airplane. The tip chord c_t is the chord of the cross section at the tip of the wing. If the tip is rounded, the rounding should be ignored in determining the tip section. For one of the wing halves, the semispan s is the distance between the root chord and tip chord. For the whole wing the span b is twice the semispan. The leading edge of a wing is the locus of the leading edge points of the cross sections, and the trailing edge of a wing is the locus of the trailing edge points of the cross sections.

Also important in describing the planform shape of a wing are the sweepback angles. The angle Λ_k is the angle between a line in the basic plane perpendicular to the root chord and a line through the points on the chords of the cross sections, which are kc back from the leading edge. The three most used sweepback angles are Λ_0, the sweepback of the leading edge; $\Lambda_{1/4}$, the sweepback of the quarterchord line; and Λ_1, the sweepback of the trailing edge. *See* SUPERCRITICAL WING; SUPERSONIC FLIGHT; WING.

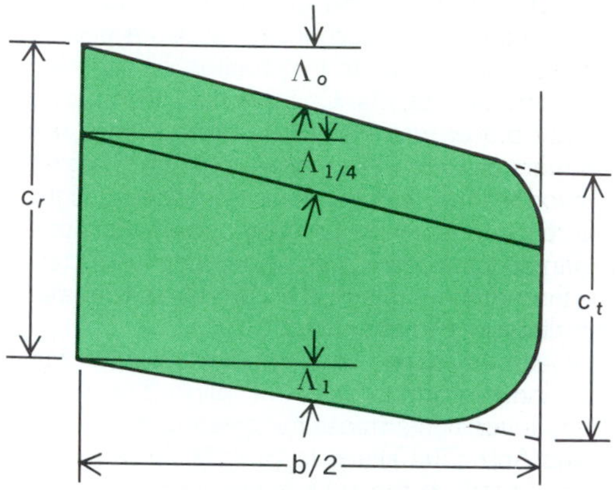

Geometry of a half wing.

In general, the chord of a cross section is a function of spanwise location. The geometric mean chord of a wing $\bar{c}$ is equal to S/b where S is the area of a wing. The aspect ratio of a wing is defined by the equation below. From the formula $\bar{c} = S/b$, it is seen that aspect ratio can also be given by $A = b/\bar{c}$.

$$A = \frac{b^2}{S}$$

The aspect ratio is very important in airplane performance, as it affects most of the performance characteristics.

As in the two-dimensional case, the component of the fluid dynamic force normal to the direction of motion is called lift, and the component along the direction of motion is called drag. It is desirable to design the airfoil so that the lift is as large as possible relative to the drag, that is, for a high lift-drag ratio.

Because the wing is terminated at the wing tips, the flow, especially near the tips, is modified from the two-dimensional pattern. For a wing creating lift the pressure on the lower surface is greater than that on the upper surface, and therefore the air flows up around the tips from the lower side to the upper. Since meanwhile the wing has moved forward, this air continues to move, forming a vortex. This phenomenon is known as the induced flow and represents the chief difference between the two- and three-dimensional cases. *See* SUBSONIC FLIGHT.

Of considerable use are the concepts of mean aerodynamic center and the aerodynamic mean chord of a wing. Consider a wing of general nature, swept, twisted, and tapered. There exists an untwisted rectangular wing with the same area which will have the same lift and moment as the original wing, provided that its fore-and-aft location is properly chosen and that its chord is the right length. The chord length and the fore-and-aft position of the rectangular wing define respectively the aerodynamic mean chord and the mean aerodynamic center of the original wing. The value of the concept is that the analysis is simpler for an untwisted rectangular wing than for a swept, twisted, and tapered wing. *See* Ideal aerodynamics. [F.McL.M.]

Airframe The structure that supports the loads acting upon the airplane. The loads consist of aerodynamic forces exerted by the air flowing around the airplane; hydrodynamic forces, in the case of hydroplanes and flying boats, produced by the water in which the floats or the hull are partially submerged; the weight of the airplane and of the passengers, crew, and stores it is carrying; the dynamic loads caused by accelerations; and the ground loads which appear when the plane is resting or rolling on the ground. The structure must be strong enough to withstand these loads without failure and stiff enough to preclude unduly large deformations; when such deformations appear, they can interfere with the proper functioning of the airplane and can lead to oscillations called flutter which are capable of destroying the structure. *See* Aeroelasticity; Flutter (aeronautics).

Until the end of World War I, all successful planes were built as externally braced monoplanes or biplanes. The externally braced fighters of World War I usually had two spars that supported wooden truss ribs in the shape of the wing section which were covered with canvas. This structure is light in weight, but produces much drag because the wires are exposed to the airflow and the canvas covering cannot maintain the exact aerodynamic shape. In addition, in the biplane the interaction between the flows around the two wings is the source of more drag (induced drag).

The cantilever wing was first used successfully by Hugo Junkers and Anthony Fokker in Germany during World War I. Junkers had found in his wind tunnel that the minimum drag of a thick wing was not much larger than that of the thin biplane wings in general use at that time. Unfortunately this conclusion was correct only at the low speeds of the Aachen tunnel; it was incorrect at the much higher speeds (or Reynolds numbers) of actual flight. Hence in later years designers came to prefer wing sections whose maximum thickness was about 12% of the chord length (which is the distance from the leading edge of the wing to the trailing edge) rather than those with the 20 to 25% proposed by Junkers or the 3 or 4% common in biplanes of World War I. *See* Reynolds number; Wind tunnel.

The predecessor of the modern transport airplane, the six-passenger Lockheed Vega, designed in 1927, had a plywood-covered high wing and a wood monocoque fuselage (*monos* means single, and *coque* means shell). The fuselage of the relatively thick-walled wood Vega was so stable that it needed few internal stiffeners. But the much thinner walls (with thicknesses as low as 0.016 in. or 0.4 mm) of the aluminum alloy fuselages of the planes that followed had to be stiffened to keep them from buckling. The early solution of this design problem, mainly the Reissner-Junkers method of stiffening the skin by corrugating it, was no longer acceptable when the flying speed of the planes increased above 150 mi/h (241 km/h), as the corrugations increased the drag inadmissibly. For this reason many ingenious arrangements of the longitudinal and circumferential stiffeners (stringers and rings) were devised in the late 1920s and early 1930s.

The aluminum-alloy-reinforced monocoque construction described here has been in use since the early 1930s essentially unchanged. But shortly before World War II, the De Havilland company proved that efficient monocoque structures can also be built with a sandwich skin. The sandwich consists of two thin but strong faces kept apart by a thick but weak core. Since the core material is light, the sandwich is light in weight; and since the total sandwich thickness is substantial, the wall is stiff and is not likely to buckle.

The most recent development in airplane structures is the use of composite materials. A composite consists of fibers of very small diameter and very high strength embedded in a matrix of lower strength. Among artificial fibers, thin glass fibers of 0.0001-in. (0.0025-mm) diameter have a failing stress of about 500,000 pounds per square inch (3500 megapascals). When such fibers are spun into a thread, woven into a fabric, and combined with epoxy resin, the average failing stress is reduced to about 55,000 psi (380 MPa), which is the failing stress of the older aluminum alloys. However, the glass fiber–epoxy fabric is lighter than the aluminum alloy, as its specific density is 2.0 (against 2.8 for the aluminum alloy).

Two kinds of fibers of very high strength and rigidity are boron and carbon (or graphite). The main drawback of these materials is their high price. When first produced industrially, boron-epoxy cost as much as gold of the same weight. At first, graphite-epoxy was even more expensive, but its price dropped more rapidly than that of boron-epoxy. Besides its lower price, graphite has the advantage over boron that its minimum fiber diameter, 0.0002 in. (0.005 mm), makes it more flexible than boron, whose minimum fiber diameter is 0.004 in. (0.1 mm). For these reasons, graphite-epoxy composites are used increasingly in aircraft construction, where high strength and low weight are of extreme importance. *See* Composite material. [N.J.H.]

Airglow Weak, nonthermal emissions from the atmosphere above 37 mi (60 km), other than aurora. Subdivisions are dayglow, twilight glow, and nightglow. The energy of the nightglow comes from reactions of atomic oxygen between 43 and 62 mi (70 and 100 km), and from ionic recombination around 190 mi (300 km). Many of the emissions are the direct result of sunlight, scattered by the appropriate species; others are produced by electrons ejected by solar photons, or by enhanced nightglow processes. The twilight period, besides showing the transition between day and night, allows observation of some dayglow emissions from the ground, with much less interference from the overwhelming foreground light of the day sky. *See* Alkali emissions. [D.M.Hu.]

Airplane A heavier-than-air vehicle designed to use the pressures created by its motion through the air to lift and transport useful loads. To achieve practical, controllable flight, an airplane must consist of a source of thrust for propulsion, a geometric arrangement to produce lift, and a control system capable of maneuvering the vehicle within prescribed limits. Further, to be satisfactory, the vehicle should display stable characteristics, so that if it is disturbed from an equilibrium condition, forces and moments are created which return it to its original condition without necessitating corrective action on the part of the pilot. Efficient design will minimize the aerodynamic drag, thereby reducing the propulsive thrust required for a given flight condition, and will maximize the lifting capability per pound of airframe and engine weight, thereby increasing the useful, or transportable, load. *See* Aerodynamic resistance; Aircraft propulsion; Aircraft stability; Airframe; Flight controls. [D.C.H.]

Airport engineering The planning, design, construction, and operation and maintenance of facilities providing for the landing and takeoff, loading and unloading, servicing, maintenance, and storage of aircraft. Inasmuch as an airport may be considered a self-contained community, the subject of airport engineering requires the technical knowledge and diversified skills of many branches of the engineering profession. *See* Aircraft.

To determine the location of an airport, whether or not it should be small or large, what facilities are required, and the stages of airport development involve extensive study and planning. The culmination of the master plan study is expressed in a report and an airport layout plan, which shows in graphic form the existing and planned facilities, airport land uses, and anticipated impact of the airport on the surrounding environment.

The foremost consideration in the design of an airport is the selection of a suitable construction site. Where possible, several potential sites are considered. Adequate size is required to allow for ultimate development requirements, including land acquisition for protection of runway approaches and for buffer zones.

Following site selection, a master plan layout is prepared, showing the relative location of runways, taxiways, aprons, control tower, hangars, and revenue-producing facilities. The ultimate development is often subdivided into suitable stages of construction, such that the completion of each stage will provide a complete and usable facility adequate to meet the growing needs of the area to be served.

Runway lengths are based on the planned category of the airfield and the types of aircraft anticipated. Taxiways usually parallel each runway. They connect at each runway end and at high-speed turnouts at two or more intermediate points. These permit aircraft to exit from the runway as soon as possible after touchdown, thus minimizing headway required between landing aircraft. *See* RUNWAY.

The grading and drainage of an airfield area is important and expensive; however, careful design can achieve major economies. The slope of the runway center line is restricted, and transverse slopes off runways and taxiways must be relatively flat to minimize damage in the event an aircraft performs erratically. Drainage ditches and headwalls are not constructed within the landing strip limits nor immediately off runway ends. The grades in turfed areas slope away from paved surfaces toward drainage structures which conduct the water into underground pipes and discharge it clear of the landing area. *See* SURFACE WATER.

Airport pavements are designed to support aircraft wheel loads and to provide an even, nonskid, lasting surface, free from loose particles. They should require a minimum of maintenance and repair. A trend in runway surfaces is to provide grooves in the pavement to prevent hydroplaning of aircraft during landing. The two basic pavement classifications are rigid and flexible. Rigid pavements are generally composed of rectangular slabs of portland cement concrete interlocked by keys or steel dowels and supported on a well-compacted thickness of granular subbase material. The flexible type consists of bituminous surface courses on compacted granular base and subbase layers. The type selected for a particular site will depend on material and equipment availability, cost, existing soil conditions, and operational considerations. *See* PAVEMENT.

To protect the major investment in a modern airport, a vigorous program of operations and maintenance of the airfield and facilities is essential. Systematic preventative maintenance of the airfield pavements, drainage system, lighting, and navigational aids pays handsome dividends not only in reduction of major repair costs, but also in ensuring that the airfield is in operation for the maximum amount of time. The airport terminal, hangars, cargo buildings, maintenance facilities, fueling system, and many other related facilities and support systems also need the same type of preventative maintenance program to ensure their availability and satisfactory performance for their intended function. *See* AVIATION. [M.H.]

Airport surface detection equipment Equipment used for the observation of the position of aircraft and other vehicles moving on the surface of the airport. This equipment is also known as the airport surface movement indicator (ASMI). *See* AIRPORT ENGINEERING.

With the increase in the density of aircraft movements at the world's major airports, it became evident that control of traffic on the airport surface is as necessary to air safety as is the control in the airspace above the airport. There must be a capability of detecting all vehicles that are on the surface of the airfield, including tractors and trucks. A number of different developments were tried, but it was soon recognized that only high-performance radar could fulfill most of the requirements. *See* AIR-TRAFFIC CONTROL.

Modern ASMI equipment operates in the Q band (36 to 46 gigahertz). The range of these radars is from 200 yd (183 m) to 2.5 mi (4.0 km) from the antenna. At a distance of 1000 yd (914 m), the radar is capable of discriminating by 30 ft (9.1 m) or less in range and 36 ft (11.0 m) or less in azimuth. With this radar it is readily possible to discern the outline of runways and taxiways and detect small vehicles out to distances of 1.5 mi (2.4 km). The lift-off point of the aircraft is easily seen. Some difficult problems in the design of radars of this type are precipitation and high snowbanks. These problems have been reduced by the use of circular polarization and moving-target indicator (MTI) techniques. *See* ELECTRONIC NAVIGATION SYSTEMS; RADAR. [P.C.S.]

Airship A propelled and steered aerial vehicle, dependent on the displacement of air by a lighter gas for lift. An airship, or dirigible balloon, is composed primarily of a streamlined hull, usually a prolate ellipsoid which contains one or more gas cells, fixed and movable tail surfaces for control, a car or cabin for the crew or passengers, and a propulsion system.

Two fundamentally different designs have been successfully used for past airships, the nonrigid and the rigid. A third type, the semirigid, is essentially a variant of the nonrigid type, differing by the addition of rigid keel.

A typical nonrigid airship, or blimp, consists of a flexible envelope, usually fabric, filled with lifting gas that is slightly pressurized. Internal air compartments (ballonets) expand and contract to maintain the pressure in the envelope as atmospheric pressure and temperature vary. Ballonet volume is controlled by ducting air from the prop wash or by electric blowers. The weights of the car structure, propulsion system, and other concentrated loads are supported by catenary systems attached to the envelope.

The nonrigid airships are historically significant for two reasons. First, a nonrigid airship was the first aircraft of any type to achieve controllable flight, in the 1850s. Second, nonrigid airships were the last type to be used on an extensive operational basis; the U.S. Navy decommissioned the last of its nonrigid airship fleet in the early 1960s.

The structure of a rigid airship (see illustration) was usually an aluminum ring-and-girder frame. An outer covering was attached to the frame to provide a suitable aerodynamic surface. Several gas cells were arrayed longitudinally within the frame.

***Akron*, rigid airship of the U.S. Navy.**

These cells were free to expand and contract, thereby allowing for pressure and temperature variations. Thus, despite their nearly identical outward appearance, rigid and nonrigid airships were significantly different in their construction and operation.

Many modern airship vehicle concepts have been proposed. The fully buoyant conventional concepts are modern versions of the classical, fully buoyant, rigid, and nonrigid airship concepts. Fully buoyant means all the lift is provided by displacement. These airships would make extensive use of modern aircraft structural materials, propulsion systems, control systems, and electronics.

Other concepts include partially buoyant, or hybrid, designs in which the buoyant lift is substantially augmented by aerodynamic or propulsive lift. Thus, these vehicles are partly heavier than air and partly lighter than air. Some hybrids are short-takeoff and landing (STOL) vehicles, and others have vertical-takeoff and landing (VTOL) capability. An important advantage of the hybrids is that they promise to alleviate the costly ground-handling requirement of past airship designs. *See* SHORT TAKEOFF AND LANDING (STOL); VERTICAL TAKEOFF AND LANDING (VTOL).

Airships are once again being seriously considered for many civil and military applications. An airship's natural attributes compared with other vehicles draw attention to short-range transportation and to missions requiring high loiter time, that is, patrol and surveillance applications. *See* BALLOON. [M.D.A.]

Airspeed indicator A device that computes and displays speed of an aircraft relative to the air mass in which the aircraft is flying. The commonest type is the indicated airspeed meter, which measures differential pressure between the total ram pressure from the Pitot system and the total static pressure; it then converts this difference into units of speed (mi/h or knots) under standard conditions. Although the indicated values are incorrect above zero altitude, the relationship to the aircraft handling remains essentially unchanged, thus providing a measure of the flyability of the aircraft.

True airspeed indicators are similar but include a more complex mechanism that also senses both the absolute pressure and temperature, and compensates for the change of density of the air mass, thus obtaining true airspeed. This indication is of value in computing course information.

For those aircraft that reach higher speeds (transonic and supersonic), the ratio of the actual speed to the local speed of sound is used. Devices that compute this value are known as Machmeters. *See* MACH NUMBER; PITOT TUBE. [J.W.A.]

Aistopoda An order of extremely elongate, limbless fossil amphibians in the subclass Lepospondyli from Permo-Carboniferous rocks of North America and the British Isles. The order includes three families: Lethiscidae (*Lethiscus*), Ophiderpetontidae (*Coloraderpeton, Ophiderpeton*), and Phlegethontiidae (*Aornerpeton, Phlegethontia, Sillerpeton*).

The skulls of aistopods are fenestrated and exhibit a reduced number of bony elements. The vertebral centra are holospondylous (single-pieced), hourglass-shaped, and fused to their neural arches. Vertebrae can exceed 200 in number and frequently bear foramina for passage of spinal nerves. Ribs typically bear a unique process, which gives them a K-shape in some species.

Most aistopods were presumably aquatic, although rib specializations, and the more gracile proportions of phlegethontiids, suggest a rather snakelike, terrestrial habit. *See* AMPHIBIA; LEPOSPONDYLI. [C.F.W.]

Alanine An amino acid. The biosynthesis of alanine starts with pyruvic acid, by transamination or reductive amination.

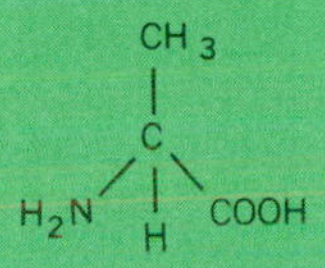

Alanine

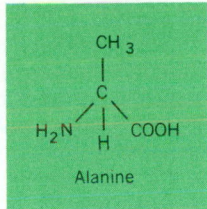

During metabolic degradation, alanine is deaminated to pyruvic acid. *See* AMINO ACIDS; CARBOHYDRATE METABOLISM. [E.A.Ad.]

Alarm systems Systems which operate a warning device after the occurrence of an abnormal or dangerous condition. Alarm systems are used to signal undesirable or dangerous situations such as the presence of an intruder or the existence of a runaway condition in a petroleum refinery process. In factories, these systems signal such things as excessive process temperature or pressure or rapid change in a process condition.

Alarm systems are usually open-loop control systems. A basic alarm system contains two essential components: an alarm detector and an alarm indicator. Frequently, they are remote control systems, that is, the detector is located remotely from the indicator. *See* REMOTE-CONTROL SYSTEM. [D.L.T.]

Albedo A term used to describe the reflecting properties of surfaces. White surfaces have albedos close to 1; black surfaces have albedos close to 0.

Two types of albedos are in common use: the bond albedo determines the energy balance of a planet and is defined as the fraction of the total incident solar energy that the planet reflects back to space. The "normal albedo" of a surface, more properly called the normal reflectance, is a measure of the relative brightness of the surface when viewed and illuminated vertically. Such measurements are referred to as a "perfectly white Lambert surface"—a surface which absorbs no light, and scatters the incident energy isotropically—and are usually approximated by magnesium oxide (MgO), magnesium carbonate ($MgCO_3$), or some other bright powder. *See* PHOTOMETRY; PLANET.

Bond albedos for solar system objects range from 0.76 for cloud-shrouded Venus to values as low as 0.01–0.02 for some asteroids and satellites. The value for Earth is 0.33. The bond albedo is defined over all wavelengths, and its value therefore depends on the spectrum of the incident radiation.

Normal reflectances of materials*

Material	Albedo	Material	Albedo
Lampblack	0.02	Granite	0.35
Charcoal	0.04	Olivine	0.40
Carbonaceous meteorites	0.05	Quartz	0.54
Volcanic cinders	0.06	Pumice	0.57
Basalt	0.10	Snow	0.70
Iron meteorites	0.18	Sulfur	0.85
Chondritic meteorites	0.29	Magnesium oxide	1.00

*Powders; for wavelengths near 0.5 micrometer.

Normal reflectances of some common materials are listed in the table. The normal reflectances of many materials are strongly wavelength-dependent, a fact which is commonly used in planetary science to infer the composition of surfaces remotely. While the bond albedo cannot exceed unity, the normal reflectance of a surface can if the material is more backscattering at opposition than the reference surface. [J.V.]

Albite The sodium feldspar end member $NaAlSi_3O_8$, a common rock-forming mineral component with crystal structure, morphology, and thermochemistry dictated by the degree of Al-Si order in the three-dimensional tetrahedral framework structure. The most ordered triclinic albite occurs frequently as a late-stage mineral associated with gem pocket–bearing lithium-rich pegmatites as the platy variety called cleavelandite. Albite and anorthite are components in the complex plagioclase series, $NaAlSi_3O_8$–$CaAl_2Si_2O_8$. Albite breaks down into jadeite ($NaAlSi_2O_6$) and quartz (SiO_2) at high pressure. *See* FELDSPAR; IGNEOUS ROCKS; PEGMATITE. [P.B.M.]

Albumin A globular protein characterized by its solubility in water and in 50% saturated aqueous ammonium sulfate. Albumins are present in mammalian tissues, bacteria, molds, and plants, and in some foods. Serum albumin, which contains 584 amino acid residues, is the most abundant protein in human serum. This protein is responsible for about 80% of the total osmotic regulation in blood, and it transports fatty acids between adipose tissues.

Another important albumin, ovalbumin, is found in egg white. This protein is about two-thirds the size of serum albumin, and it contains sugar residues in addition to amino acid residues (that is, it is a glycoprotein). *See* PROTEIN. [J.M.M.]

Alcaligenes A genus of bacteria characterized by inert biochemical activity. Its members include gram-negative, nonsporeforming cells that are rod-shaped and motile by means of peritrichous flagella. They are strictly aerobic. Indophenol oxidase and catalase are produced. Colonies are nonpigmented.

There are presently four species: *Alcaligenes faecalis* (*A. odorans*), *A. piechaudii*, *A. xylosoxidans* ssp. *denitrificans* (formerly *A. denitrificans*), and *A. xylosoxidans* ssp. *xylosoxidans* (formerly *Achromobacter xylosoxidans*); additional isolates have been designated *A. faecalis* types I and II. The different species are recognized on the basis of acid production from xylose and extent of reduction of nitrate and nitrite.

The natural habitat of *Alcaligenes* is soil and water. It has also been found in the hospital environment and has been isolated from numerous human clinical specimens. *See* INFECTION; MEDICAL BACTERIOLOGY; OPPORTUNISTIC INFECTIONS. [G.L.Gi.]

Alcohol A member of a class of organic compounds composed of carbon, hydrogen, and oxygen. They can be considered as hydroxyl derivatives of hydrocarbons produced by the replacement of one or more hydrogens by one or more hydroxyl (—OH) groups.

Alcohols may be mono-, di-, tri-, or polyhydric, depending upon the number of hydroxyl groups they possess. They are classified as primary (RCH_2OH), secondary (R_2CHOH), or tertiary (R_3COH), depending on the number of hydrogen atoms attached to the carbon atom bearing the hydroxyl group. Alcohols can also be characterized by the molecular configuration of the hydrocarbon portion (aliphatic, cyclic, heterocyclic, or unsaturated). There are two systems in use for alcohol nomenclature: the common naming system and the IUPAC (International Union of Pure and Applied Chemistry) naming system. The common name is sometimes associated with the natural source of the alcohol or with the hydrocarbon portion (for example, methyl alcohol, ethyl alcohol). The IUPAC method is a systematic procedure and follows agreed-upon rules. Examples of these two systems are given in the table. [J.V.]

Alcohols and their formulas

Common name	IUPAC name	Formula
Methyl alcohol	Methanol	CH_3OH
Ethyl alcohol	Ethanol	CH_3CH_2OH
n-Propyl alcohol	1-Propanol	$CH_3CH_2CH_2OH$
Isopropyl alcohol	2-Propanol	$(CH_3)_2CHOH$
n-Butyl alcohol	1-Butanol	$CH_3(CH_2)_2CH_2OH$
sec-Butyl alcohol	2-Butanol	$CH_3CH_2CHOHCH_3$
tert-Butyl alcohol	2-Methyl-2-propanol	$(CH_3)_3COH$
Isobutyl alcohol	2-Methyl-1-propanol	$(CH_3)_2CHCH_2OH$
n-Amyl alcohol	1-Pentanol	$CH_3(CH_2)_3CH_2OH$
n-Hexyl alcohol	1-Hexanol	$CH_3(CH_2)_4CH_2OH$
Allyl alcohol	2-Propen-1-ol	$CH_2{=}CHCH_2OH$
Crotyl alcohol	2-Buten-1-ol	$CH_3CH{=}CHCH_2OH$
Ethylene glycol	1,2-Ethanediol	$HOCH_2CH_2OH$
Propylene glycol	1,2-Propanediol	$CH_3CHOHCH_2OH$
Trimethylene glycol	1,3-Propanediol	$HOCH_2CH_3CH_2OH$
Glycerol	1,2,3-Propanetriol	$CH_2(OH)CH(OH)CH_2(OH)$

Alcohols have many uses themselves and as intermediates in the synthesis of many chemicals. The lower alcohols are generally employed as solvents, antifreezes, and extractants, whereas the higher alcohols, such as cetyl alcohol, are used as antifoaming agents and evaporation retardants in reservoirs. Their use as chemical intermediates, however, accounts for the largest amount of alcohols produced. By relatively simple chemical reactions, alcohols can be converted into a great number and variety of chemical compounds. Oxidation of primary alcohols (RCH_2OH) produces aldehydes ($RCHO$) and carboxylic acids (RCO_2H); oxidation of secondary alcohols (R_2CHOH) yields ketones (R_2CO). Dehydration of alcohols produces olefins and ethers (RCH_2OCH_2R); treatment with carboxylic acids results in the formation of esters (RCH_2OCOR), a valuable class of organic chemicals.

Esters of the lower monohydric alcohols are used extensively as solvents for lacquer enamels, nail enamel, celluloid products, nitrocellulose, artificial leather, natural and synthetic gums, leather finishes, plastic wood, perfumes, and other natural products. Esters derived from higher alcohols and dibasic acids are used as plasticizers for vinyl, cellulosic and acrylic resins, and synthetic rubber. Their viscosity-temperature coefficients make them useful as lubricants in high-speed applications, such as in jet and gas turbine engines. Alcohols are among the most important industrially produced organic chemicals.

The three most important sources of monohydric alcohols are fermentation of natural products, chemical synthesis based on hydrocarbons from petroleum or natural gas, and chemical treatment of natural fats and oils. Of minor importance are such ancient sources as the destructive distillation of wood and wood rosins.

The fermentation of sugars and starches to produce alcoholic beverages has been employed at least since history has been recorded. The fermentation process is a biological oxidation of a carbohydrate by a highly specialized strain of yeast which produces the desired end products. Fermentation is no longer the major source of *n*-butanol or 2-propanol, but it still accounts for all potable ethanol and two-thirds of worldwide industrial ethanol. Molasses fermentation accounts for about half of the worldwide glycerol production. *See* FERMENTATION; INDUSTRIAL MICROBIOLOGY. [C.E.M./P.E.F.]

Alcohol fuel Any alcohol burned as a fuel. If available, any alcohol may be used as a fuel, but ethanol and methanol are the only ones which are sufficiently inexpensive. Alcohols are useful fuels, even though the oxygen atom in any alcohol molecule reduces its heating value, because the molecular structure increases the combustion efficiency. *See* COMBUSTION; GASOLINE.

Ethanol burns well in engines designed for gasoline and has a high octane rating. However, because of the high cost of the raw materials required, or of their conversion, it costs much more than gasoline. Gasohol is produced by dissolving 5–15% absolute (water-free) alcohol in gasoline. The 5% water fraction in 95% alcohol causes phase separation because of its insolubility, making 95% alcohol unsuitable for use in gasohol. *See* ETHYL ALCOHOL; OCTANE NUMBER.

Carbon in any material can be converted to methanol. Natural gas has been favored as a raw material, and its energy may be delivered at a much lower cost if the gas is converted to methanol than if it is piped directly or if the gas is converted to liquefied natural gas, shipped, and stored at its cryogenic temperature. *See* LIQUEFIED NATURAL GAS (LNG); NATURAL GAS.

Methanol is the most versatile and cheapest liquid fuel that can be made. It is also less flammable than gasoline; accidental fires are extinguished with water instead of being spread as flaming films. When it combusts, it yields neither particulates (soot) nor sulfur oxides, and yields lower quantities of nitrogen oxides than any other fuel. When produced from natural gas or solid fuel, methanol can always be regasified to give substitute natural gas (SNG) with only a small loss of energy. Even when expensive chemical-grade methanol is used in automobiles with relatively insignificant changes, the mileage costs are less for these vehicles than when they use gasoline as a fuel. *See* METHANOL. [D.F.O.]

Alcoholic beverages Potable beverages obtained by distilling an alcohol-containing liquid and further treating the distillate to obtain a beverage of specific character.

Brandies. Brandy is a spirit obtained from the distillation of wine or a fermented fruit juice, usually after aging of the wine in wooden casks. Cognac is a brandy distilled from wines made of grapes grown within the legal limits of Charente and Charente Inférieure Departments, the Cognac region of France. Armagnac is brandy made in the Department of Gers, southeast of Bordeaux. Both types of brandy are aged for many years in oak before bottling. Brandies distilled from grape pomace of the wine press are called *eau de vie de marc* in France and *grappa* in Italy; Spanish brandies are usually distilled from sherry wines. Whereas in Europe pot stills are most common, continuous stills are preferred in California, where most American brandies are produced. Apple brandy, called applejack in the United States and calvados in France, is distilled from completely fermented apple juice and aged in oak barrels for 5–10 years. Other fruits from which brandy is made include black, wild cherries (kirsch or kirschwasser), plums (slivovitz from Hungary or Rumania, and quetsch or mirabelle from France), blackberries, and apricots. When stone fruits are used, some of the stones are broken or crushed and a small amount of the oil is distilled over with the spirit, giving the brandy a more or less pronounced bitter almond flavor.

Whiskeys. These spirits are made by distilling fermented grain mashes and aging the distillate in woods, usually oak.

Scotch whiskey is made primarily from barley. The barley is converted to malt by allowing it to sprout after which it is dried in a kiln over a peat fire. The malt absorbs some of the smoke aroma which is carried over later with the spirit distilled from it. The drying and roasting of the malt is to a large extent responsible for the flavor of the whiskey. After the malt is made into a mash, it is fermented, distilled, and aged in oak casts. A Scotch blended whiskey may contain a certain percentage of grain whiskey besides malt whiskey.

Irish whiskey is made from malt, unmalted barley, and other grains such as wheat, rye, and oats. The malt is not smoke-cured as in Scotland, and as a result the flavor of the resulting whiskey is different from that of Scotch whiskey.

Many types of whiskey are made in the United States. Depending on the principal source of grain, whiskeys are classified as bourbon (corn or grain), rye whiskey (rye), and others. Blended whiskeys are differentiated from straight whiskeys by a certain content of neutral spirits. American whiskeys are aged for a number of years in new charred white-oak barrels. In all types of whiskeys discussed above, the type and quality of the water used in the plant is an important factor in the quality of the finished product.

Gins. These consist essentially of a pure grade of alcohol which has been flavored with an extract of the juniper berry as the chief flavoring agent. The flavor may be imparted by distillation of herbs (distilled gin) or by the addition of essential oils (a compounded gin). There are two principal types of gin: English, or London dry, gin and Dutch gin (jenever).

English gin is made from pure neutral spirits, which are redistilled in the presence of juniper berries and small amounts of other ingredients such as coriander seed, cardamon seed, orange peel, anise seed, cassia bark, and fennel. English gin is not aged.

Dutch gin is made of malt. The flavoring ingredients are mixed directly in the mash, after which it is distilled at a rather low proof. Dutch gins are heavier in body and contain more fusel alcohols and other volatile compounds (congenerics) besides ethyl alcohol than does English gin. Some types of Dutch gins are aged.

Rum. The alcoholic distillate from fermented sugarcane juice or molasses is known as rum. Cuban rum is light-bodied and light-colored. The middle fraction of the distillation, known as aguardiente, is used for making rum. The aguardiente is aged in uncharred oak barrels; it is then decolorized, filtered, and supplied with some caramel to give it the proper color. Occasionally some fruit aroma is added. It is then aged for several more years. Jamaican rum is heavier-bodied and contains more congenerics (fusel alcohol, esters, and aldehydes) than Cuban rum. Arak is a rum that comes from the island of Java. It is a dry, highly aromatic rum. A natural fermentation of the molasses by various species of yeast also contributes to the flavor of this drink.

Vodka. A product originally produced in Russia but now popular in many countries, vodka is usually made from wheat. It is highly rectified during distillation and thus is a very pure neutral spirit without a pronounced taste. It is not aged.

Pulque and tequila. Pulque and tequila are of Mexican origin. A sweet sap obtained from agave is collected and fermented by natural process and is then called pulque. Distilled pulque is tequila.

Liqueurs. Liqueurs or cordials are alcoholic beverages prepared by combining a spirit, usually brandy, with certain flavorings and sugar. In fruit liqueurs, the color and flavor are obtained by an infusion process using the specified fruit and spirit. Sugar is added after the extraction is complete. Plant liqueurs are made by maceration of plant leaves, seeds, or roots with spirit and then distilling the product. Sugar and coloring matter are added after distillation. [E.M.M.; H.J.P]

Alcoholism The continuous or excessive use of alcohol with associated pathologic results. The type of interaction of alcohol with the biological system and the frequency of interaction combine to project the image commonly referred to as alcoholism.

Alcohol is absorbed most rapidly from solutions on the order of 15–30% (30–60 proof) and less rapidly from beverages containing below 10% and over 30%. At the lower concentrations this is to be expected since the rate of absorption depends on the concentration gradient across the mucosal surface. At the higher concentrations ethanol abolishes the rhythmic opening of the pylorus, thus preventing the passage to the intestine, where absorption is faster. The presence of foods in the stomach is also known to delay gastric emptying and thus to slow absorption. Once in the bloodstream, alcohol is distributed evenly in all tissues, according to their water content.

Alcohol can act as a stimulant at lower doses and as a depressant at higher doses. Even at very low doses alcohol can impair the sensitivity to odors and taste. Also, low doses are known to alter motor coordination and time and space perception, important aspects of car driving. Pain sensitivity is diminished with moderate doses. Doses on the order of 0.6 g/kg (about 35 oz of 100-proof per person of 150 lb) are normally lethal by depression of respiratory activity. Alcohol is known to release feelings of self-criticism and to inhibit fear and anxiety.

Continuation of alcohol ingestion produces adaptive changes which act by opposing the effects of alcohol in the brain. This leads to nervous-tissue tolerance. In this new state larger doses of alcohol are required to produce effects similar to those obtained in the initial phases of drinking. Discontinuation of alcohol use leaves an unbalanced biological state which appears to be responsible for the withdrawal syndrome. Characteristic of this state are a marked hyperexcitability, hallucinations, insomnia, sleep fragmentation, and moderate to extreme tremor as seen in delirium tremens. The person is now physically dependent on alcohol.

Since alcohol has a large caloric value, a person drinking 25 oz (750 ml) of a 100-proof beverage per day ingests about 2000 cal (8400 kilojoules), or 70–80% of his total caloric intake as alcohol. However, since alcoholic beverages are normally devoid of essential food factors such as proteins, vitamins, and trace metals, malnutrition is very common in alcoholics. Further, it has been shown that heavy alcohol consumption reduces, both directly and indirectly, the capacity of the intestine to absorb various nutrients such as the vitamins B_1 and B_{12} and some essential amino acids, thus accentuating the malnutrition state. Encephalopathies, beriberi, pellagra, anemia, scurvy, and other hypovitaminoses are very often observed in alcoholics. The encephalopathies are among the most serious complications associated with alcoholism. Among them Wernicke-Korsakoff's encephalopathy is most common. *See* MALNUTRITION.

It was formerly believed that only about 1 out of 10 alcoholics developed cirrhosis. However, it has been shown that the production of cirrhosis is correlated with the amount of ethanol ingested multiplied by the period (years) of ingestion. Thus, an alcoholic that has consumed 200 g of ethanol (18 oz of 100-proof beverage) daily for 20 years has a 50% chance of developing cirrhosis. *See* CIRRHOSIS. [Y.I.]

Alcoholysis A chemical reaction involving the decomposition of a molecule with the addition of the elements of an alcohol. The reaction is most commonly an exchange reaction. The term is frequently applied to the reaction between an alcohol and an ester to produce a different alcohol and a different ester (transesterification). In such cases, higher-molecular-weight alcohols readily replace those of lower molecular weight; tertiary alcohols are more readily replaced than primary alcohols. *See* ALCOHOL.

The term alcoholysis also applies to reactions involving alcohols and a number of other classes of organic compounds, such as nitriles, acid anhydrides, alkyl halides, acyl halides, β-keto esters, lactones, and amides. *See* ACIDOLYSIS; HYDROLYSIS; TRANSESTERIFICATION. [E.H.H.]

Alcyonacea An order of the subclass Alcyonaria. Alcyonacea, the soft corals (see illustration), are littoral anthozoans, which form massive or dendriform colonies with yellowish, brown, or olive colors. Most attach to some solid substratum; however, some remain free in sandy or muddy places. The only skeletal structures are small, elongated, spindle-shaped or rodlike, warty sclerites which are scattered over the mesoglea. The colony is supple and leathery. The polyp body is embedded in the coenenchyme, from which retractile anthocodia protrude in *Alcyonium*. In *Xenia* and *Heteroxenia*, anthocodia are nonretractile. The polyp base is protected by many sclerites and is termed a calyx. *See* ALCYONARIA; COELENTERATA. [K.At.]

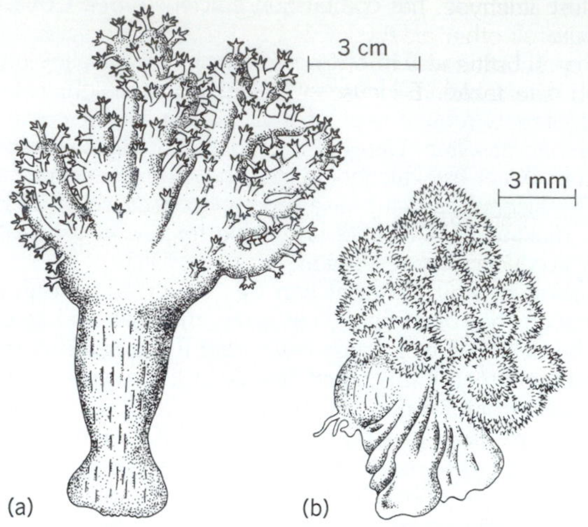

Alcyonaceans. (*a*) *Alcyonium palmatum* (after Y. Delage). (*b*) *Dendronephthya* sp. (after H. Utinomi).

Alcyonaria A subclass of the Anthozoa. These coelenterates are colonial; most are sedentary and littoral, and some live at great depths. The ordinary polyp, the autozooid, has eight motile, contractile, hollow, pinnately branched tentacles. Eight complete mesenteries are present, of which the dorsal or asulcal pair is the most developed and bears the largest, heavily ciliated ectodermal filaments. The remaining mesenteries bear endodermal filaments having many glandular cells. Longitudinal retractor muscles are strongly developed on their ventral or sulcal surfaces. The stomodeum or pharynx, with a single siphonoglyph in its sulcal edge, is lined by a ciliated glandular epithelium. The gastrovascular cavity of each polyp is interconnected by complex canal systems or solenia. These permeate the colonial spiculiferous mesoglea or coenenchyme. The skeleton is formed by a deposition of horny material secreted by ectoderm cells, or by ectodermal scleroblasts which secrete calcareous spicules or sclerites of various shapes. The musculatures are not particularly well developed.

The oral end of the polyp is termed the anthocodium. The basal portion, or anthostele, is embedded in coenenchyme containing numerous ameboid cells and scleroblasts.

The Alcyonaria are characterized by a strongly developed siphonoglyph, which serves to circulate water in a colony. Alcyonaria are either dioecious or monoecious. The gonads ripen in the endodermal mesenteric filaments. The daughter polyp buds asexually from the solenial system or from a parent's body wall. Pennatulacea and some other forms are dimorphic. The siphonozooid is smaller than the autozooid. It lacks tentacles or the tentacles are rudimentary and usually sterile. *See* ANTHOZOA; COELENTERATA. [K.At.]

Aldebaran Alpha Tauri, a red giant star of spectral type K5, temperature 3600 K, and luminosity about 400 times that of the Sun. Aldebaran is at a distance of 20 parsecs (6.2×10^{14} km); it is large enough to show a barely detectable disk, 0″.23 of arc, with the Mount Wilson (California Institute of Technology/Palomar Observatory) stellar interferometer. *See* ASTRONOMICAL PHOTOGRAPHY; STAR. [J.L.Gr.]

Aldehyde One of a class of organic chemical compounds represented by the general formula RCHO. Formaldehyde, the simplest aldehyde, has the formula HCHO, where R is hydrogen. For all other aldehydes, R is a hydrocarbon radical which may be substituted with other groups such as halogen or hydroxyl (see table). Because of their high chemical reactivity,

Aldehydes and their formulas

Compound	Formula
Formaldehyde	HCHO
Acetaldehyde	CH_3CHO
Acetaldol	$CH_3CHOHCH_2CHO$
Propionaldehyde	C_2H_5CHO
n-Butyraldhyde	$CH_3(CH_2)_2CHO$
Isobutyraldehyde	$(CH_3)_2CHCHO$
Acrolein	$CH_2{=}CHCHO$
Crotonaldehyde	$CH_3CH{=}CHCHO$
Chloral	CCl_3CHO
Chloral hydrate	$CCl_3CH(OH)_2$
Benzaldehyde	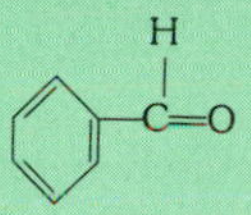
Cinnamaldehyde	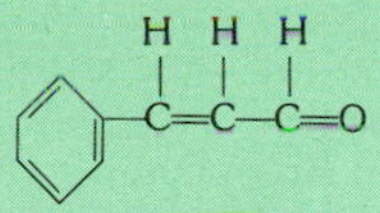

aldehydes are important intermediates for the manufacture of resins, plasticizers, solvents, dyes, and pharmaceuticals.

At room temperature formaldehyde is a colorless gas. The other low-molecular-weight aldehydes are colorless liquids having characteristic, somewhat acrid odors. The unsaturated aldehydes acrolein and crotonaldehyde are powerful lacrimators. The important reactions of aldehydes include oxidation, reduction, aldol condensation, Cannizzaro reaction, and reactions with compounds containing nitrogen.

Because of the importance of aldehydes as chemical intermediates, many industrial and laboratory syntheses have been developed. The more important of these methods include catalytic dehydrogenation of primary alcohols, oxidation of primary alcohols, oxidation of olefins, and hydroformylation of olefins. *See* FORMALDEHYDE. [P.E.F.]

Alder A deciduous tree, *Alnus rubra*, which grows from Alaska to northern California and eastern Idaho. It is recognized by its stalked buds, simple leaves, and dry, conelike, ellipsoid fruit. With the big-leaf maple it shares the role of principal hardwood tree in the Pacific Northwest, where most of the commercially important trees are conifers. The wood is used in furniture. *See* FAGALES. [A.H.G./K.P.D.]

Aldosterone The steroid hormone found in the biologically active amorphous fraction that remains after separation of the various crystalline steroid substances, such as cortisol and corticosterone, from adrenal extracts. In solution, aldosterone exists as an equilibrium mixture of aldo and lactol forms (see illustration).

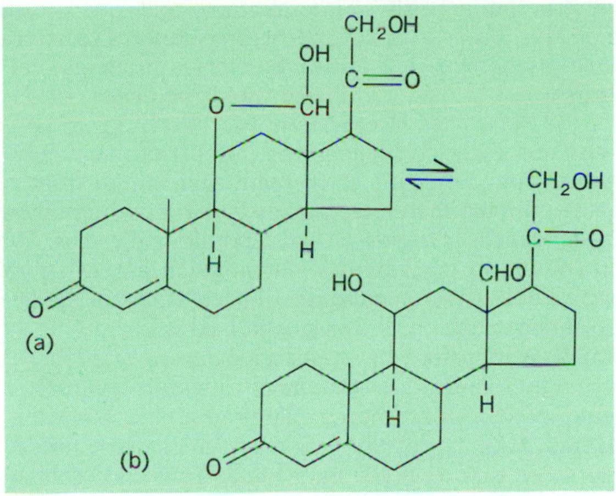

Structures for (*a*) aldo and (*b*) lactol forms of aldosterone in an equilibrium mixture.

The chief function of aldosterone is the regulation of electrolyte metabolism. Aldosterone is the most potent of the hormones which are concerned in this type of metabolism; in adrenalectomized dogs, minute doses on the order of 1–2 micrograms per kilogram per day can maintain the electrolyte balance of the animals and keep them alive. *See* ADRENAL GLAND; HORMONE; STEROID. [C.H.L.]

Alfalfa The world's most valuable forage legume, *Medicago sativa*, known also as lucerne. It is grown for hay, pasture, and silage. Valued highly as livestock feed, alfalfa hay cut at a late bud or early bloom stage (and cured properly) contains 17–24% protein and is a good source of certain vitamins and minerals. *See* ROSALES.

Alfalfa is a herbaceous perennial legume characterized by a deep taproot, which shows a varying degree of branching. Erect or semierect stems, bearing an abundance of leaves, grow to a height of 2–3 ft (0.6–0.9 m). The number of stems arising from a single woody crown may vary from just a few to 50 or more. New stems develop when older ones are mature or have been cut or grazed. Flowers are borne on axillary racemes which vary greatly in size and number of flowers. Flower color is predominantly purple, or bluish-purple, but white, cream, yellow, green, lavender, and reddish-purple occur. The fruit is a legume, or pod, usually spirally coiled in *M. sativa*, but crescent-shaped or straight in *M. falcata*. Seeds are small and the color varies from yellow to brown.

Reproduction in alfalfa is mainly by cross-fertilization. Pollination is effected largely by bees. Seed production is favored by bright sunny days, cool nights, an abundance of bees, and dry weather for harvesting.

Alfalfa is widely adapted to temperate and subtropical climates and soils. It is grown from 40°S in Argentina and New Zealand to 60°N in Canada, Sweden, and Russia. It is not well adapted to humid tropical conditions.

Deep fertile loams are best and good drainage is essential. The water requirement for sustained high yields is great, and moisture deficiency often is a serious limiting factor in areas dependent on natural rainfall.

More than 80 improved cultivars of alfalfa are recognized as eligible for seed certification in the United States. These trace to three basic stocks: *M. sativa*, purple-flowered, narrow-crowned, and erect; *M. falcata*, yellow-flowered, somewhat prostrate, with deep-set crown and branching roots; and an intermediate form, often called variegated, derived from crossing *M. sativa* and *M. falcata*. [C.H.H.]

Alfvén waves Propagating oscillations in electrically conducting fluids or gases in which a magnetic field is present. Magnetohydrodynamics deals with the effects of magnetic fields on fluids and gases which are efficient conductors of electricity. Molten metals are generally good conductors of electricity, and gases can be efficient conductors if they become ionized. A gas that consists of free electrons and ions is called a plasma. Most gases in space are plasmas, and magnetohydrodynamic phenomena are expected to play a fundamental role in the behavior of matter in the cosmos. *See* COSMIC ELECTRODYNAMICS; MAGNETIC FIELD; PLASMA PHYSICS.

Waves are a particularly important aspect of magnetohydrodynamics. They transport energy and momentum from place to place and may, therefore, play essential roles in the heating and acceleration of cosmical and laboratory plasmas. Waves are propagating oscillations and their generation involves restoring forces which tend to return each parcel of the fluid to its equilibrium position. If a magnetic field is present in a conducting fluid there are three restoring forces and, therefore, three magnetohydrodynamic wave modes. However, each restoring force does not necessarily have a unique wave mode associated with it. Put another way, each wave mode can involve more than one restoring force. Thus the usual sound wave, which involves only the thermal pressure, does not appear as a mode in magnetohydrodynamics. The three modes have different propagation speeds, and are named fast mode (F), slow mode (S), and intermediate mode (I). (The intermediate mode is sometimes called the Alfvén wave, but some scientists refer to all three magnetohydrodynamic modes as Alfvén waves.) The magnetohydrodynamic wave modes are analyzed by using the magnetohydrodynamic equations for the motion of a conducting fluid in a magnetic field, combined with Maxwell's equations and Ohm's law. *See* MAXWELL'S EQUATIONS.

It is possible to combine Ohm's law with Faraday's law of induction. The resultant equation is called the magnetohydrodynamic induction equation, which is the mathematical statement of the "frozen-in" theorem. This theorem states that magnetic field lines can be thought of as being frozen into the fluid, with the proviso that the fluid is always allowed to slip freely along the field lines. It is this coupling between the fluid and the magnetic field which makes magnetohydrodynamic waves possible. The oscillating magnetic field lines cause oscillations of the fluid parcels, while the fluid provides a mass loading on the magnetic field lines. This mass loading has the effect of slowing down the waves, so that they propagate at speeds much less than the speed of light (which is the propagation speed of waves in a vacuum.) *See* ELECTROMAGNETIC RADIATION; FARADAY'S LAW OF INDUCTION; LIGHT.

Unfortunately, the basic equations are too difficult to be of much use. The essence of the difficulty is that some of the equations are nonlinear; that is, they contain products of the quantities for which a solution is sought. In order to get solvable equations, scientists accept the limitation of dealing with small-amplitude waves and linearize the equations, so that products of the unknowns are removed. The equations which result have solutions which are harmonic in time and space. *See* HARMONIC MOTION.

The motions in the intermediate mode are pure shears. There is no compression of the plasma. The tension in the magnetic field lines is the only restoring force involved in the propagation of the wave. This mode is therefore closely analogous to the propagation of waves on a string: the tension in the magnetic field lines plays the same role as the tension in the string.

Because these waves channel energy along magnetic fields, they may be responsible for the observed fact that cosmical plasmas are strongly heated in the presence of magnetic fields.

The fast mode is difficult to analyze. However, many cosmical and laboratory plasmas satisfy the strong-magnetic-field case for which the fast mode is more easily understood. Fast waves are compressive, and the magnetic field strength fluctuates as well. Thus fast waves are governed by the two restoring forces associated with the tension and pressure in the magnetic field. The fast mode can propagate energy across the magnetic field.

Like the fast mode, the slow mode is difficult to study in general, and the discussion will again be confined to strong magnetic fields. The slow mode in a strong field is equivalent to sound waves which are guided along the strong magnetic field lines. The strong magnetic field lines can be thought of as a set of rigid pipes which allow free fluid motion along the pipes, but which restrict motion in the other two directions. The slow mode channels energy along the magnetic field. Because the sound speed is small, by assumption, the slow mode transmits energy less effectively than the fast or intermediate modes.

Only small-amplitude waves have been considered. Real waves have finite amplitude, and nonlinear effects can sometimes be important. One such effect is the tendency of waves to steepen, ultimately forming magnetohydrodynamic shock waves and magnetohydrodynamic discontinuities. There is an abundance of such discontinuities in the solar wind. *See* NONLINEAR ACOUSTICS; SHOCK WAVE; SOLAR WIND.

It is also possible that waves can degenerate into turbulence. There are indications that this too happens in the solar wind. *See* MAGNETOHYDRODYNAMICS; TURBULENT FLOW. [J.V.H.]

Algae An informal assemblage of predominantly aquatic organisms that carry out oxygen-evolving photosynthesis but lack specialized water-conducting and food-conducting tissues. They may be either prokaryotic (lacking an organized nucleus) and therefore members of the kingdom Monera, or eukaryotic (with an organized nucleus) and therefore members of the kingdom Plantae, constituting with fungi the subkingdom Thallobionta. They differ from the next most advanced group of plants, Bryophyta, by their lack of multicellular sex organs sheathed with sterile cells and by their failure to retain an embryo within the female organ. Many colorless organisms are referable to the algae on the basis of their similarity to photosynthetic forms with respect to structure, life history, cell wall composition, and storage products. The study of algae is called algology (from the Latin *alga*, sea wrack) or phycology (from the Greek *phykos*, seaweed). *See* BRYOPHYTA; MONERA; PLANT KINGDOM; THALLOBIONTA.

General form and structure. Algae range from unicells 1–2 micrometers in diameter to huge thalli [for example, kelps are often 100 ft (30 m) long] with functionally and structurally distinctive tissues and organs. Unicells may be solitary or colonial, attached or free-living, with or without a protective cover, and motile or nonmotile. Colonies may be irregular or with a distinctive pattern, the latter type being flagellate or nonmotile. Multicellular algae form packets, branched or unbranched filaments, sheets one or two cells thick, or complex thalli, some with organs resembling roots, stems, and leaves (as in the brown algal orders Fucales and Laminariales). Coenocytic algae, in which the protoplast is not divided into cells, range from microscopic spheres to thalli 33 ft (10 m) long with a complex structure of intertwined siphons (as in the green algal order Bryopsidales).

Classification. Sixteen major phyletic lines (classes) are distinguished on the basis of differences in pigmentation, storage products, cell wall composition, flagellation of motile cells, and structure of such organelles as the nucleus, chloroplast, pyrenoid, and eyespot. These classes are interrelated to varying degrees, the interrelationships being expressed by the arrangement of classes into divisions (the next-higher category). Among phycologists there is far greater agreement on the number of major phyletic lines than on their arrangement into divisions.

Superkingdom Prokaryotae
- Kingdom Monera
    - Division Cyanophycota (=Cyanophyta, Cyanochloronta)
        - Class Cyanophyceae, blue-green algae

Division Prochlorophycota (=Prochlorophyta)
Class Prochlorophyceae

Superkingdom Eukaryotae
Kingdom Plantae
Subkingdom Thallobionta
Division Rhodophycota (=Rhodophyta, Rhodophycophyta)
Class Rhodophyceae, red algae
Division Chromophycota (=Chromophyta)
Class: Chrysophyceae, golden or golden-brown algae
Prymnesiophyceae (=Haptophyceae)
Xanthophyceae (=Tribophyceae), yellow-green algae
Eustigmatophyceae
Bacillariophyceae, diatoms
Dinophyceae, dinoflagellates
Phaeophyceae, brown algae
Raphidophyceae, chloromonads
Cryptophyceae, cryptomonads
Division Euglenophycota (=Euglenophyta, Euglenophycophyta)
Class Euglenophyceae
Division Chlorophycota (=Chlorophyta, Chlorophycophyta)
Class: Chlorophyceae, green algae
Charophyceae, charophytes
Prasinophyceae

Placing more taxonomic importance on motility than on photosynthesis, zoologists traditionally have considered flagellate unicellular and colonial algae as protozoa, assigning each phyletic line the rank of order. *See* BACILLARIOPHYCEAE; CHAROPHYCEAE; CHLOROPHYCEAE; CHRYSOPHYCEAE; CRYPTOPHYCEAE; CYANOPHYCEAE; DINOPHYCEAE; EUGLENOPHYCEAE; EUKARYOTAE; EUSTIGMATOPHYCEAE; PHAEOPHYCEAE; PRASINOPHYCEAE; PROCHLOROPHYCEAE; PROKARYOTAE; PROTOZOA; PRYMNESIOPHYCEAE; RAPHIDOPHYCEAE; RHODOPHYCEAE; THALLOBIONTA; XANTHOPHYCEAE.

Characteristics. Prokaryotic algae lack membrane-bounded organelles. Eukaryotic algae have an intracellular architecture comparable to that of higher plants but more varied. Among cell structures unique to algae are contractile vacuoles in some fresh-water unicells, gas vacuoles in some planktonic blue-green algae, ejectile organelles in dinoflagellates and cryptophytes, and eyespots in motile unicells and reproductive cells of many classes. Pyrenoids—proteinaceous bodies around which carbohydrate is stored—occur in the chloroplasts of at least some species of all classes. They have been shown to contain the same carboxylating enzyme that is present in higher plants with C_3 metabolism.

In all classes of algae except Prochlorophyceae, there are cells that are capable of movement. The slow, gliding movement of certain blue-green algae, diatoms, and reproductive cells of red algae presumably results from extracellular secretion of mucilage. Ameboid movement, involving pseudopodia, is found in certain Chrysophyceae and Xanthophyceae. An undulatory or peristaltic movement occurs in some Euglenophyceae. The fastest movement is produced by flagella, which are borne by unicellular algae and reproductive cells of multicellular algae representing all classes except Cyanophyceae, Prochlorophyceae, and Rhodophyceae. Most motile cells have two flagella, inserted apically or laterally, but some have three, four, eight, or a crown of flagella.

Cell division is the process by which unicellular algae multiply and multicellular algae increase in size and complexity. Even within a single class, such as the green algae, details of cell division vary with respect to involvement and interaction of nuclei, microtubules, vesicles, and plasmalemma in wall formation. *See* CELL DIVISION.

Most algae are autotrophic, obtaining energy and carbon through photosynthesis. All photosynthetic algae liberate oxygen and use chlorophyll *a* as the primary photosynthetic pigment. Secondary (accessory) photosynthetic pigments capture light energy and transfer it to chlorophyll *a*. Carotenoids protect the photosynthetic pigments from oxidative bleaching. *See* CAROTENOID; CHLOROPHYLL.

Occurrence. Algae are predominantly aquatic, inhabiting fresh, brackish, and marine waters without respect to size or degree of permanence of the habitat. They may be planktonic (free-floating or motile) or benthic (attached). Benthic marine algae are commonly called seaweeds. Substrates include rocks (outcrops, boulders, cobbles, pebbles), plants (including other algae), animals, boat bottoms, piers, debris, and less frequently sand and mud. Some species occur on a wide variety of living organisms, suggesting that the hosts are providing only space. Many species, however, have a restricted range of hosts and have been shown to be (or are suspected of being) at least partially parasitic. *See* PHYTOPLANKTON.

Fresh-water algae, which are distributed by spores or fragments borne by the wind or by birds, tend to be widespread if not cosmopolitan, their distribution being limited by the availability of suitable habitats. Marine algae, which are spread chiefly by water-borne propagules or reproductive cells, often have distinctive geographic patterns. Many taxonomic groups are widely distributed, but others are characteristic of particular climatic zones or geographic areas.

On the negative side, algae can be a nuisance by imparting tastes and odors to drinking water, clogging filters, and making swimming pools, lakes, and beaches unattractive. Sudden growths (blooms) of planktonic algae can produce toxins of varying potency. In the ocean, toxins produced by dinoflagellate blooms (red tides) can kill fishes and render shellfish poisonous to humans. *See* EUTROPHICATION.

Fossil algae. At least half of the classes of algae are represented in the fossil record, usually abundantly, in the form of siliceous, calcareous, or organic remains, impressions, or indications. Blue-green algae were among the first inhabitants of the Earth, appearing in rocks at least as old as 2.3 billion years. Their predominance in shallow Precambrian seas is indicated by the extensive development of stromatolites. All three classes of seaweeds (reds, browns, and greens) were well established by the close of the Precambrian, 600 million years ago. By far the greatest number of fossil taxa, however, belong to classes whose members are wholly or in large part planktonic. Siliceous frustules of diatoms and endoskeletons of silicoflagellates, calcareous scales of coccolithophorids, and highly resistant organic cysts of dinoflagellates contribute slowly but steadily to sediments blanketing ocean floors, as they have for tens of millions of years. *See* STROMATOLITE. [P.C.Si.; R.L.Moe]

Algebra Classical algebra is a generalization of arithmetic, made possible by the use of symbols, usually letters such as x, y, a, b, α, and β, for unknown numbers. The principal instrument for solving problems is the equation. The answers sought are denoted by letters, and statements in the form of equations are deduced about the quantities in a problem. To solve one or more equations, the value or values must be found which, when substituted for the unknowns, make the equation or equations true. Equations are said to be satisfied by such values. Most of the great developments in algebra were inspired and guided by the search for methods of solving equations, which is the central problem in algebra.

The extension of the number system from the whole numbers and fractions of arithmetic to a complete algebraic number system was influenced by the need for numbers to satisfy equations. Thus, the equations $x^2 - 2 = 0$ and $x^2 + 1 = 0$ have no solutions among whole numbers and fractions. Even the extended number system does not provide solutions for all equations. And even though solutions can be shown to exist, they may be impossible to find. For example, there is no general method for finding the solutions of all fifth-degree equations.

Algebra is also concerned with inequalities, such as $x^2 + y^2 > 2xy$; with arrangements, such as the number of ways a set of elements can be chosen from a larger set (combinations), or the number of ways a set can be ordered (permutations); with the study of special sets of numbers such as the set of coefficients in the expansion of a binomial like $(a + b)^n$; and with manipulations of elements such as matrices, which have properties similar to, but different from, those of numbers. In modern algebra the relationship to arithmetic seems more remote than in classical algebra, and the elements are often combined by rules at variance with those which stemmed from the arithmetic of experience.

The operations of arithmetic—addition, subtraction, multiplication, division, raising to powers, and extracting roots—are the algebraic operations. Applications of these operations to symbols for numbers, known and unknown, produce various kinds of algebraic expressions. A term is formed by multiplication: $4x^2y$. In a term the known multiplier of the literal part is the coefficient. The degree of the term is the sum of the exponents of unknowns. The degree of $4x^2y$ is 3. By multiplication and addition (which includes subtraction because of the rules for combining signed numbers), polynomials, or rational integral expressions, can be constructed. Polynomials are classified according to the number of terms into monomials, binomials, trinomials, and other multinomials. In polynomials the coefficients can be any fixed numbers:

$$5x^2y - 6y^3 + \tfrac{2}{3}x \qquad 3x^4 - \sqrt{2}x^2 + \tfrac{5}{3}$$

A fractional expression permits division by unknowns: $(x^2 + 1)/(x + 1)$. An expression is irrational if it denotes root extraction of an expression involving an unknown.

Whereas the operations in arithmetic are derived from practical experience, their generalizations in algebra accentuate the necessity of axioms. These axioms govern all operations.

The positive and negative whole numbers, zero, and ratios of whole numbers constitute the rational number system. Addition, multiplication, and division are the rational operations. Root extraction is not a rational operation since its results are not usually rational numbers. For example, there is no rational number whose square is 2, nor any rational number whose square is -1. Two extensions of the number system provide irrational numbers and imaginary numbers.

1. A rational number is a number which can be expressed as the ratio of two integers, a/b, with $b \neq 0$. (An integer is a rational number; for example, $2 = 2/1$.)
2. An irrational number is a number which is not rational but which can be approximated as closely as desired by rational numbers. (Thus $\sqrt{2}$ is not rational, but for any $\epsilon > 0$ there are rational numbers a_r such that the difference between a_r and $\sqrt{2}$ is less than ϵ.)
3. The rational numbers and the irrational numbers make up the real number system.
4. A pure imaginary number is a number whose square is negative.
5. A complex number is the sum of a real number and a pure imaginary number.

A pure imaginary number can be expressed as a product $b\sqrt{-1}$, where b is a real number. The symbol i, which replaces $\sqrt{-1}$, is the imaginary unit; hence all pure imaginary numbers have the form bi, and all complex numbers have the form $a + bi$, with a and b both real. The complex numbers include the real numbers (when $b = 0$) and the pure imaginary numbers (when $a = 0$). The term "imaginary" is applied to any complex number $a + bi$, with $b \neq 0$. The complex number system is sufficient for the needs of classical algebra. In modern algebra the number concept is more abstract. The operation of repeated multiplication of like numbers is denoted by exponents and the results are called powers. Thus $2 \cdot 2 \cdot 2 \cdot 2 = 2^4 = 16$; the product of the four 2s is denoted by the exponent 4 applied to the base 2, and 16 is the fourth power of 2.

The inverse operation to forming a power is that of extracting a root and is expressed by a radical sign. For example,

$$\sqrt{4} = 2 \qquad \sqrt[3]{8} = 2$$

In $\sqrt[n]{a}$, is the radicand and n is the index.

An equation is a statement that two expressions are equal. The statement may be false, for example, $x + 2 = x + 1$. It may be true for some values of the unknown or unknowns but not for all possible values. Such an equation is a conditional equation. To solve it is to find the values for which it holds true. For example, $3x - 12 = 0$ is true for $x = 4$ only. Or a statement of equality may be true for all values of the letters for which the expressions have meaning. For example, $3 - 2 - 2x + 6 = x + 4$. Such an equation is called an identity.

Solving conditional equations is the central problem of algebra. Identities are used for transforming and simplifying expressions, mainly to facilitate solving equations. Operations on algebraic expressions, including the factoring of polynomials and the combining of fractions and radicals, produce identities that make possible the reduction of conditional equations to manageable form. *See* ANALYTIC GEOMETRY; BINOMIAL THEOREM; EQUATIONS, THEORY OF. [H.R.C.]

Algebraic geometry A study of solutions of systems of polynomial equations $f_j(x) = 0$ in several variables x_i, thought of as coordinates of a point $x = (x_1, x_2, \ldots x_n)$ in n-dimensional space. Classical studies emphasize properties of an irreducible algebraic plane curve $f(x_1, x_2) = 0$ (f an irreducible polynomial), such as singular and multiple points and genus. In modern studies, the coefficients in the polynomials f_j are chosen from an arbitrary field k, and solutions x are sought in an algebraically closed field K containing k. The set X of all points x at which all the f_j vanish is called an algebraic closed set, or variety. If X is not empty, then the set of all polynomials $f(x)$ that vanish on X form an ideal with a finite basis in the ring of polynomials. Modern algebraic geometry studies these ideals and corresponding varieties. *See* POLYNOMIAL SYSTEMS OF EQUATIONS; RING THEORY. [J.S.F.]

Alginate A major constituent (10–47% dry weight) of the cell walls of brown algae. Extracted for its suspending, emulsifying, and gelling properties, it is one of three algal polysaccharides of major economic importance, the others being agar and carrageenan. The chief sources of alginate are members of the family Fucaceae (rockweeds) and the order Laminariales (kelps), harvested from naturally occurring stands on North Atlantic and North Pacific shores.

Because of its colloidal properties, alginate finds numerous industrial applications, especially in the food, textile, paper, printing, paint, cosmetics, and pharmaceutical industries. About half of the consumption is in the making of ice cream and other dairy products, in which alginate prevents the formation of coarse ice crystals and provides a smooth texture. As an additive to paint, it keeps the pigment in suspension and minimizes brush marks. An alginate gel is used in making dental impressions. *See* AGAR; CARRAGEENAN; ICE CREAM; MILK; PHAEOPHYCEAE. [P.C.Si; R.L.Moe]

Algorithm A precise formulation of a method for doing something. In computers, an algorithm is usually a collection of procedural steps or instructions organized and designed so that computer processing results in the solution of a specific problem.

Algorithms play an important role in computers. It has been suggested that science may be defined as knowledge which is understood well enough to be taught to a computer. In this view, the concept of an algorithm or computer program furnishes an extremely useful test for the depth of knowledge about any particular subject, and the process of going from an art to a science involves learning how to construct an algorithm.

Algorithms are employed to accomplish specific tasks using data and instructions when applying computers. The task may be well definable in either mathematical or nonmathematical terms; it may be either logical or heuristic, and either simple or complex; and it may be either computational or data-processible, or involve sensing and control. In any case, the task must be definable. Then an algorithm can be devised and specified for a computer to perform the task.

Algorithms are further characterized by several properties. Either the data set over which the algorithm will operate or the process of how the computer is to get access to the data must be specifiable. The process required to be performed can be defined with a finite set of operations or actions together with a unique starting point. The sequence of steps, the tree, the list, or the network describing the process is mappable. This, however, does not imply that the path through these steps is known, since in some cases the data or prior process steps will dictate the actual path. That is, for classical computers it must be possible to "program" the algorithm. The algorithm process must terminate, either with the task completed or with some kind of indication that the task (problem) is unsolvable.

Algorithms define the method of operation for performing a task. However, for each task to be performed, there usually are many different mappable methods for the computer to execute it—but they are not of equal desirability. From the above definitions and characterizations it follows that any computer program that does its intended task is also an algorithm. But to be practical an algorithm must perform its task within the time and memory capacity constraints of the system. *See* COMPUTER; DIGITAL COMPUTER. [E.C.J.]

Alicyclic hydrocarbon Organic compounds containing only carbon and hydrogen atoms joined to form one or more rings (see illustration). Synonyms are cycloaliphatic compounds, cycloalkanes, or cycloparaffins. Unsaturated alicyclic hydrocarbons are known as cycloolefins, cycloalkenes, or cycloalkynes, depending on the type of unsaturation.

A number of alicyclic hydrocarbons have industrially important uses. Cyclopropane is an anesthetic. Cyclopentadiene is used in the preparation of the persistent insecticide chlordane. Cyclohexane is the key intermediate in the preparation of nylon-6 from cyclohexanone and caprolactam. Adipic acid, one of the constituents of nylon-66, can be obtained by the oxidation of cyclohexane. The naturally occurring terpenes, limonene and α-pinene, are used in flavors and perfumes and as starting materials in the preparation of synthetic camphor, which is a cyclic ketone, and racemic menthol, which is a cyclic alcohol.

Alicyclic hydrocarbons occur widely in nature. Cyclopentanes and cyclohexanes are found in petroleum and are called naphthenes. An important group of alicyclic hydrocarbons are the cyclic terpenes, such as limonene, derived from citrus fruits, and α-pinene, the principal constituent of turpentine, which is the volatile portion of the gum that exudes from incisions in trunks of living pine trees.

The alicyclic hydrocarbons boil 18–36°F (10–20°C) higher than the corresponding aliphatic hydrocarbons. A comparison of the relationship between the chain length and physical properties of alkanes and cycloalkanes suggests that in rings containing up to 14 carbon atoms, the molecule is roughly spherical in shape, while in the higher rings the sides become parallel and the properties approach those of the straight-chain hydrocarbons. Alicyclic hydrocarbons in general exhibit a stability

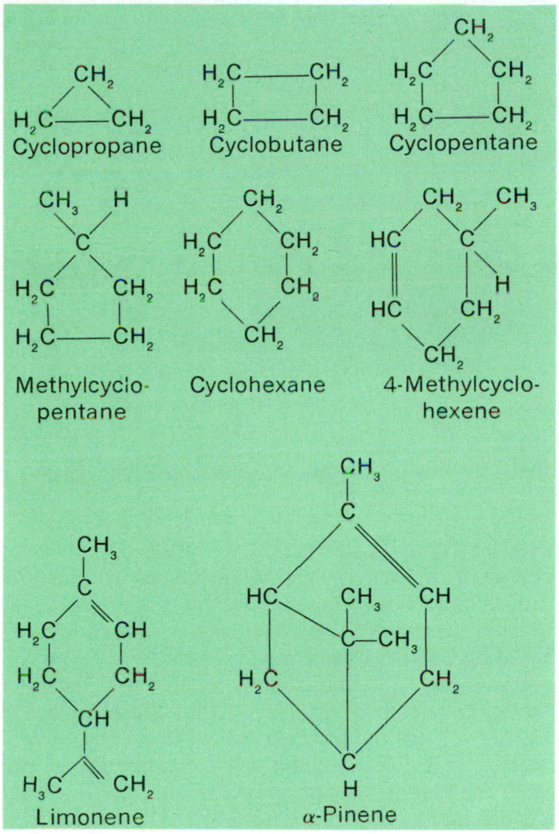

Structural formulas of alicyclic hydrocarbons.

toward heat and chemical attack comparable to that of the corresponding open-chain compounds. Only in cyclopropanes, and in cyclobutanes to a lesser extent, is there any marked ring instability.

Alicyclic hydrocarbons are generally synthesized by three general methods: (1) ring expansions, (2) ring closures, and (3) ring contractions, each followed by removal of any existing functional group. *See* ADAMANTANE; ALIPHATIC HYDROCARBON; ALKENE; AROMATIC HYDROCARBON; HYDROCARBON; TROPOLONE. [P.E.F.]

Alidade An instrument for topographic surveying and mapping by the plane-table method; or the part of the transit which rotates with the telescope about the horizontal circle; or any sighting device of pointer employed for angular measurement. The surveying alidade has a telescope, with attached graduated vertical circle, mounted on a flat base that can be moved about the plane table. The base embodies a straightedge which itself may be a graduated scale for readily plotting the point being sighted (see illustration).

The stadia technique is usually applied in measuring distances and elevation differences between the plane-table point and observed points. The direction to any point is established by sighting the point through the telescope. A direction line is then drawn along the straightedge, and its distance is laid off

An alidade. (*Kern Instruments, Inc.*)

with the scale. *See* PLANE TABLE; STADIA; SURVEYING; TRANSIT (ENGINEERING). [B.A.B.]

Aliphatic hydrocarbon One of a group of hydrocarbons in which the carbon atoms are joined in open chains. Two major classes are the saturated and the unsaturated, the latter including several homologous series depending on the number of double and triple bonds present.

The systematic names adopted by the International Union of Pure and Applied Chemistry (IUPAC) for the aliphatic hydrocarbons are formed by adding a suffix (-ane, -ene, -yne, -adiene, and so forth) indicating the type of compound to a prefix indicating the number of carbon atoms present. The first four prefixes are meth-, eth-, prop-, and but-. The succeeding prefixes are derived from the Greek or Latin word for the number: pent-, hex-, hept-, oct-, and so forth (see table).

The saturated compounds, which are known as paraffin hydrocarbons, or more systematically, as alkanes, fit the empirical formula C_nH_{2n+2}. The members having four or more carbon atoms exist in straight-chain (or normal) and branched-chain isomers. Not counting optical isomers, there are 2 butanes with different carbon skeletons, 3 pentanes, 5 hexanes, 9 heptanes, 18 octanes, and 35 nonanes, all of which are known.

Compounds containing one double bond and having the formula C_nH_{2n}, are olefins (or alkenes). The first two members of the series, ethylene and propene, both exist in one form only. The next higher homolog, C_4H_8 has two straight-chain isomers (1-butene and 2-butene) and one branched-chain isomer (isobutylene or methylpropene, sometimes incorrectly called isobutene). There are two geometrical isomers of 2-butene, namely, *cis*-2-butene and *trans*-2-butene. As the length of the chain increases in both the straight-chain and the branched-chain olefins, the number of isomers formed by a change in the position of the double bond increases rapidly, as shown in the illustration. *See* MOLECULAR ISOMERISM.

There are two homologous series of aliphatic hydrocarbons having the formula C_nH_{2n-2}: (1) acetylenes (or alkynes), which contain one triple bond, and (2) diolefins (or alkadienes), which contain two double bonds. Aliphatic hydrocarbons having three double bonds per molecule are termed triolefins or alkatrienes; those containing a larger number of double bonds are named in analogous fashion. Similarly, the open-chain hydrocarbons containing two triple bonds have the generic names alkadiynes, while those groups of hydrocarbons containing both double and triple bonds are identified by names

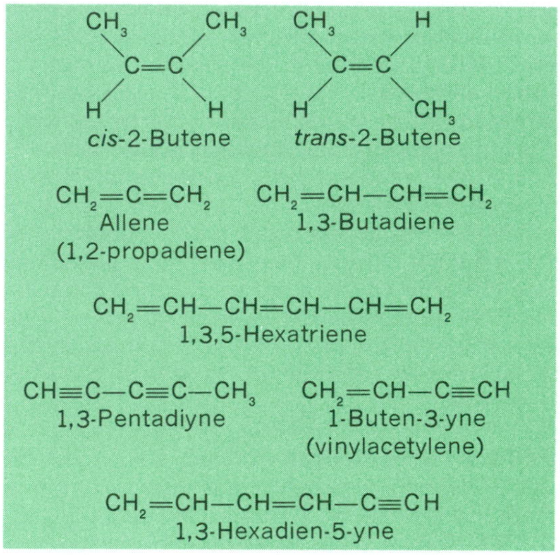

Formulas of some alkenes, alkadienes, and alkynes.

Aliphatic hydrocarbons

Number of carbon atoms	Alkane	Alkene	Alkyne
1	CH_4 Methane	—	—
2	CH_3—CH_3 Ethane	CH_2=CH_2 Ethylene	CH≡CH Acetylene
3	CH_3—CH_2—CH_3 Propane	CH_2=CH—CH_3 Propene	CH≡C—CH_3 Propyne
4	CH_3—CH_2—CH_2—CH_3 Butane	CH_2=CH—CH_2—CH_3 1-Butene	CH≡CH_2—CH_2—CH_3 1-Butyne
	—	CH_3—CH=CH—CH_3 2-Butene	CH_3—C≡C—CH_3 2-Butyne
4	CH_3—CH(CH_3)—CH_3 Isobutane (methylpropane)	CH_2=C(CH_3)—CH_3 Isobutylene (methylpropene)	—

such as alkenynes, alkadienynes, or alkenediynes. *See* ALICYCLIC HYDROCARBON; ALKANE; HYDROCARBON. [L.S.]

Alismatales A small order of flowering plants, division Magnoliophyta (Angiospermae), which gives its name to the subclass Alismatidae of the class Liliopsida (monocotyledons). It consists of three families (Alismataceae, Butomaceae, and Limnocharitaceae) and less than a hundred species. They are aquatic and semiaquatic herbs with a well-developed, biseriate perianth that is usually differentiated into three sepals and three petals, and with a gynoecium of several or many, more or less separate carpels. Each flower is usually subtended by a bract. *Butomus umbellatus* (flowering rush) and species of *Sagittaria* (arrowhead, family Alismataceae) of this order are sometimes cultivated as ornamentals. *See* LILIOPSIDA; MAGNOLIOPHYTA.

The Alismatales and some related orders have often been treated as a single order Helobiae or Helobiales, embracing most of what is here treated as the subclass Alismatidae. *See* ALISMATIDAE; PLANT KINGDOM. [A.Cr.]

Alismatidae A relatively primitive subclass of the class Liliopsida (monocotyledons) of the division Magnoliophyta (Angiospermae), the flowering plants, consisting of 4 orders, 14 families, and scarcely 500 species. Typically they are aquatic or semiaquatic, with apocarpous flowers and nonendospermous seeds. They have trinucleate pollen, and the stomates usually have two subsidiary cells. The orders Alismatales, Hydrocharitales, and Najadales are closely related among themselves and have often been treated as a single order, Helobiae or Helobiales. *See* ALISMATALES; HYDROCHARITALES; LILIOPSIDA; MAGNOLIOPHYTA; NAJADALES; PLANT KINGDOM. [A.Cr.]

Alkali Broadly, any compound having highly basic properties, that is, a compound that readily ionizes in aqueous solution to yield hydroxyl ions so that the pH is high (above 7) or so that litmus paper changes from red to blue. Specifically, an alkali is a hydroxide of one of the alkali metals, although an ammonium salt may also be referred to as an alkali. Alkalies are caustic in that they produce a relatively high concentration of OH ions in solution. Alkalies are soluble in water, tarnish in air, and in concentrated form are corrosive to the touch.

Alkalies are used commercially in soap manufacture, soluble oils, cutting compounds, cleaning solutions, and aluminum etching. In commerce, alkali usually refers to sodium carbonate in the form of soda ash, either from natural deposits or manufactured from salt and either sulfuric acid or ammonium bicarbonate. Other alkali products include lye, potash, caustic soda, potassium hydroxide, water glass, and bicarbonate of soda. *See* ACID AND BASE; ALKALI METALS: ALKALINE-EARTH METALS; HYDROXIDE; SODIUM; SOIL; WATER SOFTENING. [F.H.R.]

Alkali emissions The alkali metals lithium, potassium, and especially sodium can be observed in twilight by the scattering of their resonance lines. Analysis has revealed that the free atoms are concentrated in a layer 50–60 mi (90–100 km) above the Earth's surface. The existence of free atoms can be explained by reactions of the type $NaO+O \rightarrow Na+O_2$. Thus, the alkali layer coincides with the region where oxygen is dissociated. The most likely source of the material may be vaporized meteoric material, but for sodium and perhaps potassium a contribution may also be made by marine salt particles which have been lifted to mesospheric heights. *See* MESOSPHERE.

The sodium abundance (integrated density) has a marked seasonal variation which is also a function of latitude, reversing in the Southern Hemisphere. Sodium dayglow emission has been observed both from the ground and from rockets. The sodium distribution is often observed to be remarkably sharp, with scale height much less than that of the atmosphere. Natural lithium emission can usually be observed by very sensitive equipment, but many artificial enhancements have been measured as well. The behaviors of lithium and sodium are strikingly different, and it seems likely that the lithium enhancements are related to meteoric influx; perhaps certain dust showers are responsible. Or it may be that lithium atoms condense almost irreversibly on the dust particles, whereas the sodium is easily reevaporated. Readily observed with sensitive equipment, potassium emission seems to have about the same intensity at high and low latitudes. *See* ATMOSPHERE. [D.M.Hu.]

Alkali metals The elements of group I in the periodic table (lithium, sodium, potassium, rubidium, cesium, francium). Of the alkali metals, lithium differs most from the rest of the group, and tends to resemble the alkaline-earth metals (group II of the periodic table) in many ways. In this respect lithium behaves as do many other elements that are the first members of groups in the periodic table; these tend to resemble the elements in the group to the right rather than those in the same group. Francium, the heaviest of the alkali-metal elements, has no stable isotopes and exists only in radioactive form.

In general, the alkali metals are soft, low-melting, reactive metals. This reactivity accounts for the fact that they are never found uncombined in nature but are always in chemical combination with other elements. This reactivity also accounts for the fact that they have no utility as structural metals (with the possible exception of lithium in alloys) and that they are used as chemical reactants in industry rather than as metals in the usual sense. The reactivity in the alkali-metal series increases in general with increase in atomic weight from lithium to cesium. *See* CESIUM; ELECTROCHEMICAL SERIES; FRANCIUM; LITHIUM; PERIODIC TABLE; POTASSIUM; RUBIDIUM; SODIUM. [M.Si.]

Alkaline-earth metals Usually calcium, strontium, and barium, the heaviest members of group II of the periodic table (excepting radium). Other members of the group are beryllium, magnesium, and radium, sometimes included among the alkaline-earth metals. Beryllium resembles aluminum more than any other element, and magnesium behaves more like zinc and cadmium. The gap between beryllium and magnesium and the remainder of the elements of group II makes it desirable to discuss these elements separately. Radium is often treated separately because of its radioactivity.

The alkaline earths form a closely related group of highly metallic elements in which there is a regular gradation of properties. The metals, none of which occurs free in nature, are all harder than potassium or sodium, softer than magnesium or beryllium, and about as hard as lead. The metals are somewhat brittle, but are malleable, extrudable, and machinable. They conduct electricity well; the specific conductivity of calcium is 45% of that of silver. The oxidation potentials of the triad are as great as those of the alkali metals.

The elements and their compounds find important industrial uses in low-melting alloys, deoxidizers, and drying agents and as cheap sources of alkalinity. *See* BARIUM; BERYLLIUM; CALCIUM; MAGNESIUM; PERIODIC TABLE; RADIUM; STRONTIUM. [R.F.R.]

Alkaloid One of a group of nitrogenous bases of plant origin, such as nicotine, cocaine, and morphine. Most of the alkaloids show marked physiological activity, and crude extracts of various alkaloid-bearing plants have been used since antiquity because of their curative or poisonous effects. Well over 1000 alkaloids have since been isolated or have been shown to be present in some 97 families of plants. Alkaloid-bearing plants have been found in virtually every habitat in which vascular plants grow.

Alkaloids are found most frequently in the higher seed-bearing plants, and especially in dicotyledons. As more plants are examined, more families are included in the alkaloid-bearing group, for new genera containing alkaloids are reported regularly. The elaboration of alkaloids in plants is not localized but appears to be a characteristic of all organs, including the seeds. However, not all organs of any one species must have alkaloids. When plants elaborate more than one alkaloid, their ratio in the plant is not necessarily the same in all stages of growth. Some alkaloids are virtually absent in the young plant, but increase to isolable amounts as the plant approaches maturity. Furthermore, cultural and climatic conditions exert an effect on the alkaloid content of plants.

The nomenclature of the individual alkaloids has not been systematized, and they draw their names from a variety of sources. A great many important alkaloids have received names derived from those of plants, such as papaverine, quinine, and berberine. A few are named from their physiological action, such as morphine and emetine. Some are named from their physical characteristics, such as hygrine. Only one, pelletierine, has been named for an alkaloid chemist.

The alkaloids are usually crystalline, colorless substances; only a few are liquid. Colored alkaloids are rare. Almost all the known nitrogen-containing ring systems, both saturated and unsaturated, are encountered among the alkaloids. The molecules vary widely in complexity, ranging from the relatively simple systems such as nicotine to the very complex polycyclic structures of morphine and strychnine.

Many alkaloids have great value in medical practice because of specific pharmacological actions. The isolation of pure individual alkaloids on a large scale is sometimes prohibitively expensive. Furthermore, the synthesis of many of the more useful alkaloids is frequently impractical. Therefore considerable effort has been spent on the synthesis of related substitute compounds. In the treatment of malaria, quinine has been largely replaced by synthetic compounds. Synthetic procaine and similar drugs have supplanted naturally occurring cocaine in local anesthesia. *See* ATROPINE; COCAINE; MORPHINE; NICOTINE; QUININE; STRYCHNINE. [S.M.K.]

Alkane A member of the series of saturated aliphatic hydrocarbons having the empirical formula C_nH_{2n+2}. The members of this series are also called paraffinic hydrocarbons or simply paraffins. Alkanes are relatively unreactive compared to unsaturated and aromatic hydrocarbons, and are little used in laboratory synthesis. On the other hand, reactions of alkanes are carried out on a huge scale industrially to convert the hydrocarbons found in natural gas and petroleum to the products needed in modern life.

The alkanes are usually named in accordance with the rules of the International Union of Pure and Applied Chemistry (IUPAC) which adopted the ending "-ane" for saturated hydrocarbons. Branched-chain paraffins are named as derivatives of the longest straight chain in the compound, the location and name of the alkyl radicals in the branches being indicated as prefixes. The carbon atoms in the longest, or parent, chain are numbered so that the positions of the branches have the lower of the two possible values. By this system the two isomeric heptanes having a single branch containing a single carbon atom are named 2-methylhexane and 3-methylhexane, as shown in formulas I–III.

$CH_3CH_2CH_2CH_2CH_2CH_2CH_3$ (I)

Heptane or *n*-heptane

$CH_3CHCH_2CH_2CH_2CH_3$ (II)
|
CH_3

2-Methylhexane or isoheptane

$CH_3CH_2CHCH_2CH_2CH_3$ (III)
|
CH_3

3-Methylhexane (no trivial name)

In an alternative, trivial system of naming alkanes, the straight-chain compound is designated with the prefix "*n*-" (an abbreviation for normal) and the isomer having a methyl branch on the second carbon atom is given the prefix "iso-" (for instance, isobutane, isopentane, isohexane). A selected list of *n*-alkanes is shown in the table. In this connection it should be noted that, although the name isooctane correctly designates 2-methylheptane, it should be avoided because of the unfortunate use of the misnomer "isooctane" in the petroleum industry to represent 2,2,4-trimethylpentane.

The prefix "neo-" indicates the *gem*-dimethyl isomer (two methyl branches on the same carbon atom), but can be used unambiguously only in naming neopentane (dimethylpropane) and neohexane (2,2-dimethylbutane).

For convenience in discussion, the symbol R is used to represent the alkyl group, and X represents a halogen. Thus RH is the general representation of any alkane, and RX is an alkyl halide.

Alkanes containing fewer than 5 carbon atoms are gases at room temperature and atmospheric pressure. The pentanes through the hexadecanes are mostly liquids (with some exceptions). Straight-chain alkanes having more than 16 carbon atoms are waxy solids. The boiling points and the melting points of the normal paraffins increase with increasing molecular weight, and the boiling point decreases with increased branching within a family of isomeric alkanes. The alkanes have lower refractive indices and densities than do the other types of hydrocarbons having the same number of carbon atoms.

The alkanes are soluble in many organic solvents such as alcohol and ether. They are practically insoluble in water.

Natural gas and petroleum contain literally hundreds of alkanes, both straight-chain and branched, and a number of pure alkanes can be isolated from petroleum by careful fractionation. The number of possible isomers increases rapidly with the number of carbon atoms. Beyond the nonanes and decanes, relatively few branched isomers have been isolated from natural sources or prepared synthetically and characterized, since there has been no particular reason for doing so.

Some *n*-alkanes*

No. of carbons	Formula	Name
1	CH_4	Methane
2	C_2H_6	Ethane
3	C_3H_8	Propane
4	C_4H_{10}	Butane
5	C_5H_{12}	Pentane
6	C_6H_{14}	Hexane
7	C_7H_{16}	Heptane
8	C_8H_{18}	Octane
9	C_9H_{20}	Nonane
10	$C_{10}H_{22}$	Decane
11	$C_{11}H_{24}$	Undecane
12	$C_{12}H_{26}$	Dodecane
20	$C_{20}H_{42}$	Eicosane
30	$C_{30}H_{62}$	Tricontane

*From N. L. Allinger et al., *Organic Chemistry*, 2d ed., Worth Publishers, 1976.

The alkanes, particularly the straight-chain isomers, are not readily affected at mild temperature conditions by acids and oxidizing agents, such as sulfuric acid, nitric acid, and potassium permanganate. Because they are saturated hydrocarbons, they react chiefly by substitution of other atoms or groups of atoms for hydrogen atoms. They also undergo reactions involving scission, or splitting, of carbon-carbon or carbon-hydrogen bonds. *See* ALIPHATIC HYDROCARBON; ALKYLATION; ALKYLATION (PETROLEUM); CRACKING; ETHANE; HALOGENATION; HYDROGENATION;

ISOMERIZATION; METHANE; NITRATION; OCTANE; PETROLEUM PROCESSING; PROPANE. [P.E.F.]

Alkene One of the class of acyclic hydrocarbons containing one or more carbon-to-carbon double bonds. Alkenes (also called olefins) and alkynes (also called acetylenes) together constitute the family of organic compounds called unsaturated hydrocarbons, since they contain less than the number of hydrogens found in the corresponding saturated compound, alkane. When the double bond is present in a nonaromatic ring (alicyclic hydrocarbon), the compound is termed a cycloalkene. Hydrocarbons containing more than one double bond are termed dienes, trienes, and so forth, or collectively, polyenes. *See* ALICYCLIC HYDROCARBON; ALKANE; ALKYNE.

In naming alkenes by the system of the International Union of Pure and Applied Chemistry (IUPAC), the longest chain containing the double bond is identified. The presence of the double bond is indicated by changing the "-ane" ending of the alkane having the same number of carbon atoms to "-ene," and the position of the double bond is indicated by a prefixed number. Examples are given in the table, with common or nonsystematic names which are still frequently used given in parentheses.

Alkenes and dienes (common name given in parentheses)

Name	Formula
Ethene (ethylene)	$CH_2{=}CH_2$
Propene (propylene)	$CH_2{=}CHCH_3$
1-Butene	$CH_2{=}CHCH_2CH_3$
2-Butene	$CH_3CH{=}CHCH_3$
2-Methylpropene (isobutylene)	$CH_2{=}C(CH_3)CH_3$ (drawn with CH_3 above the central C: $CH_2{=}\overset{CH_3}{\overset{\vert}{C}}CH_3$)
1,3-Butadiene	$CH_2{=}CHCH{=}CH_2$
2-Methyl-1,3-butadiene (isoprene)	$CH_2{=}\overset{CH_3}{\overset{\vert}{C}}CH{=}CH_2$

The lower alkenes and dienes which have up to five carbon atoms are gases at room temperature and pressure. Higher alkenes are colorless liquids or solids. Like other hydrocarbons, alkenes are insoluble in water. Liquid alkenes have specific gravities well below 1.0. Alkenes may undergo polymerization, cyclization, and addition reactions. A major share of structural and elastic polymers are based on homopolymers or copolymers of alkenes and dienes. Alkenes and dienes cyclize readily under various conditions.

Addition reactions of alkenes are among the most important in the entire field of organic chemistry. Industrially, high-octane gasoline is made by the acid-catalyzed alkylation of the three- and four-carbon alkenes. A variety of alkylated aromatics are made by the alkylation of benzene with olefins.

The commercially important alkenes are produced on a large scale in the petroleum industry by thermal or catalytic cracking processes. In the laboratory, most methods for the preparation of alkenes involve some type of elimination reaction, in which atoms or groups on adjacent carbon atoms are removed with concomitant formation of the carbon-carbon double bond. *See* ALIPHATIC HYDROCARBON; ALKYLATION; ALKYLATION (PETROLEUM); CRACKING; HALOGENATION; HYDROGENATION. [P.E.F.]

Alkoxide An ion having the general formula $(RO)_n^-$; also a salt having one or more alkoxide ions attached to a metal atom at the oxygen atom. The salt may be considered a hydroxide in which the hydrogen is replaced by the alkoxide group. Thus a sodium alkoxide has the general formula NaOR. The physical properties of alkoxides vary depending on the position of the metal in the periodic table. *See* BASE (CHEMISTRY); HYDROXIDE; ION.

Some metal alkoxides can be prepared by a replacement reaction, as shown in the reaction below. Alcohols can react

$$3NaOR + AlCl_3 \rightarrow 3NaCl + \underset{\text{Aluminum alkoxide}}{Al(OR)_3}$$

with active metals, metal amalgams, or any base that is stronger than the alkoxide ion (for example; N or H or $NaNH_2$) to form metal alkoxides. *See* ALCOHOL.

Metal alkoxides are often used as catalysts, for example, sodium or potassium alkoxides in certain condensation or transesterification reactions, particularly those reactions requiring a base stronger than sodium hydroxide or a nonaqueous solvent. *See* CONDENSATION REACTION; ETHER; HYDROLYSIS; TRANSESTERIFICATION. [E.H.H.]

Alkylation A chemical process in which an alkyl radical (C_nH_{2n+1}) is introduced into an organic compound by substitution or addition. By the use of alkylating agents, such as olefins, alcohols, or alkyl halides, alkyl groups are bonded together or to other compounds through carbon, oxygen, nitrogen, sulfur, or metals. Though the general terms addition and substitution cover all alkylation processes, the detailed reactions can be quite dissimilar. Some reactions classified as alkylations are the Williamson ether synthesis, the Friedel-Crafts reaction, and the Wurtz reaction.

Many important chemicals are made by alkylation, some being used directly, others as intermediates, for the manufacture of other chemical compounds. Ethylbenzene, for example, is dehydrogenated to make styrene, important in rubber and plastic manufacture. Ethyl ether is the most widely known anesthetic. Tetraethyllead is an important antiknock additive for gasoline. Many other applications could be mentioned, including some of the synthetic detergents (the alkylaryl sulfonates). *See* ALKYLATION (PETROLEUM); FRIEDEL-CRAFTS REACTION; ORGANIC REACTION MECHANISM; PETROLEUM PROCESSING; SUBSTITUTION REACTION. [H.C.R.]

Alkylation (petroleum) As applied to the petroleum refining industry, the reaction of isoparaffins with olefins to produce higher-boiling isoparaffinic compounds. Specifically, C_3, C_4, and C_5 olefins are reacted with isobutane to form alkylates which are important components of fuels for internal combustion engines and jet aircraft engines. The desirable properties of alkylate include high octane number, excellent stability, high heat of combustion, and low vapor pressure. The illustration shows the alkylation process in a typical plant,

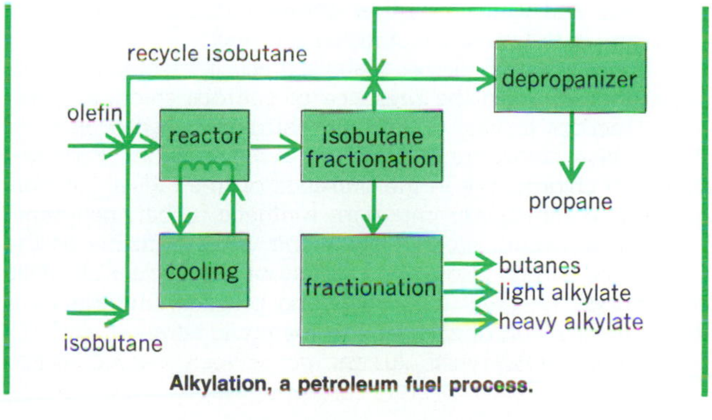

Alkylation, a petroleum fuel process.

including the reaction, isoparaffin separation and recycling, and product fractionation. Cooling is employed because the reaction is highly exothermic.

The effort to change the fuel used in automobiles in the United States to gasoline without the tetraethyllead antiknock additive may change the application of the alkylation process. Although the unleaded octane number of alkylate is higher than that of many other gasoline blending stocks, the refining industry has depended upon the high susceptibility of alkylates to the lead additive to raise the octane number of the final gasoline product.

Successful energy savings in the process have been concentrated at the isobutane fractionation step. A less pure theoretical separation here results in significant operating cost reductions with little influence on product quality. *See* ALKANE; ALKYLATION; ISOMERIZATION; PETROLEUM PROCESSING. [R.S.N.]

Alkyne One of a group of organic compounds containing a carbon-to-carbon triple-bond linkage (—C≡C—). They are termed acetylenes or alkynes. While exhibiting many of the characteristics of alkenes as regards unsaturation, the acetylenes have many unique properties. Since the bonding in alkyne molecules is linear, R—C≡C—R, cis-trans isomerism is not possible. In the simplest alkyne, acetylene (HC≡CH), or in monosubstituted acetylenes, the hydrogen attached to triply bonded carbon is acidic to such a degree that it is replaceable with metals such as sodium. Structural formulas of several alkynes are as follows:

CH≡CH	$CH{\equiv}CH_2—CH_2—CH_3$
Ethyne or acetylene	1-Butyne
$CH{\equiv}C—CH_3$	$CH_3—C{\equiv}C—CH_3$
Propyne	2-Butyne

General methods for the preparation of alkynes depend on dehydrohalogenation of α,β-dihaloparaffins, conversion of aldehydes or ketones to dihaloparaffins with subsequent dehydrohalogenation, and alkylation of metallic acetylides with alkyl halides in liquid ammonia. Grignard reagents of 1-alkynes behave similarly.

The reactions of triply bonded carbon compounds are in general similar to those of compounds containing ethylenic bonds. Addition reactions proceed in two stages to form first a vinyl compound or substituted ethylene, and second the substituted paraffin. In the presence of catalytic quantities of alkoxides, acetylene adds to alcohols and phenols to give vinyl ethers. Reactions of this type, in which addition takes place by replacement of hydrogen with a vinyl group, are termed vinylation. Another form of addition reaction known as ethynylation involves the addition of acetylene to unsaturated compounds. A general reaction of alkynes involves the addition of carbon monoxide and water, or other compounds having an active hydrogen, in the presence of nickel carbonyl.

Polymerization of alkynes may yield acyclic, aromatic, or alicyclic derivatives. In the presence of cuprous chloride, acetylene dimerizes to vinyl acetylene, which adds hydrogen chloride to give chloroprene (2-chloro-1,3-butadiene). Polymerization of chloroprene in the presence of free radical initiators gives the commercially important synthetic rubber, neoprene. Thermal polymerization of acetylene yields benzene as the major product and also a wide variety of polynuclear aromatic compounds. Of great theoretical and practical importance is the polymerization of acetylene to the cyclic tetramer, cyclooctatetraene. *See* ACETYLENE; ALKENE. [C.A.Co./P.E.F.]

All-pass circuit A circuit that is used in an electronic system to compensate for phase or delay distortion. The terms phase distortion and delay distortion refer to the distortion introduced in a signal when its harmonic components are nonuniformly delayed.

An all-pass circuit is characterized by having a constant gain response for all frequencies, and delay and phase responses which can have significantly different values depending on the frequency. Ideally, a system without delay distortion has a constant (flat) delay response. The process of compensating the delay distortion by means of all-pass circuits is called delay equalization. It consists of passing the signal through an all-pass circuit whose delay response is equal to a constant plus a response equal in magnitude but opposite in sign to that of the circuit which is to be compensated. In applications where signals with constant frequencies are used, all-pass circuits also find utilization as delay or phase-shift elements.

A delay equalizer is usually implemented by a cascade (series) connection of basic sections called first- and second-order all-pass circuits. These circuits use operational amplifiers, resistors, and capacitors and belong to the family of *RC*-active circuits. This type of circuit is frequently used in the audio-frequency range and is fabricated as either a discrete or a thick- or thin-film hybrid microcircuit. These circuits require postfabrication adjustment since the values of resistors and capacitors have poor accuracy and this also affects the accuracy of the realizable delay response. *See* AMPLIFIER; OPERATIONAL AMPLIFIER.

More accurate integrated-circuit implementations of all-pass circuits using switched-capacitor circuits have become available. These circuits do not require postfabrication adjustment. Another possible and very accurate integrated-circuit implementation of all-pass circuits uses digital circuits. Integrated-circuit implementations of all-pass circuits are economical only for large-volume applications. *See* DISTORTION (ELECTRONIC CIRCUITS); INTEGRATED CIRCUITS; PHASE (PERIODIC PHENOMENA); SWITCHED CAPACITOR. [J.R.-An.]

Allanite A mineral, also known as orthite, distinguished from other members of the epidote group of silicates by a relatively high content of rare earths, chiefly cerium, lanthanum, and yttrium. Chemical composition is $(Ca,Ce,La,Y)_2(Al,Fe)_3$-$Si_3O_{12}(OH)$. Other lanthanide elements are generally present, as is beryllium in some varieties. Small amounts of thorium and uranium are often present, and the mineral then may be metamict. Allanite is monoclinic in crystallization. The color is black or brownish-black, and the specific gravity ranges from 3.4 to 4.2. *See* CERIUM; EPIDOTE; METAMICT STATE; RADIOACTIVE MINERALS; RARE-EARTH ELEMENTS. [C.Fr.]

Allantois A fluid-filled sac- or sausagelike, extraembryonic membrane lying between the outer chorion and the inner amnion and yolk sac of the embryos of reptiles, birds, and mammals. It is composed of an inner layer of endoderm cells, continuous with the endoderm of the embryonic gut, or digestive tract, and an outer layer of mesoderm, continuous with the splanchnic mesoderm of the embryo. It arises as an outpouching of the ventral floor of the hindgut. The allantois remains connected to the hindgut by a narrower allantoic stalk which runs through the umbilical cord.

The allantois eventually fuses with the overlying chorion to form the compound chorioallantois, which lies just below the shell membranes in reptiles and birds. The chorioallantois is supplied with an extensive network of blood vessels and serves as an important respiratory and excretory organ for gaseous interchange. The allantoic cavity also serves as a reservoir for kidney wastes in some mammals, in reptiles, and in birds. In the latter two groups the allantois assists in absorption of albumin. In some mammals, including humans, the allantois is vestigial and may regress, yet the homologous blood vessels persist as the important umbilical arteries and veins connecting the embryo with the placenta. *See* FETAL MEMBRANE; GERM LAYERS. [N.T.S.]

Allele Any of the different forms of a gene occupying a particular locus on a chromosome. Alleles are symbolized by one or more letters, often with appropriate superscripts; symbols for dominant alleles begin with capital letters. Thus, w^+, w, and w^{cho} represent alleles which respectively cause the eye color of *Drosophila* to be red, white, or chocolate.

The allele most commonly found in a species is called the wild-type allele, designated by the superscript +. A diploid individual may carry two identical or two different alleles of a gene. With a combination of a wild type and a mutant allele, the organism often has a normal phenotype. The wild-type allele is then said to be dominant and the mutant allele recessive. Sometimes mutant alleles are dominant or partially dominant.

Mutant alleles differ from wild type in one or more nucleotide residues. New alleles arise by substitution, addition, deletion, or rearrangement of nucleotides in the wild-type sequence. An average gene contains several hundred nucleotides, and alleles may carry mutational alterations at different sites (heteroalleles) or the same site (isoalleles). In cells containing allelic mutations at nonidentical sites, recombination occasionally gives rise to wild-type alleles. Recombination between alleles can be used to analyze the fine structure of a gene. *See* MUTATION.

The function of an allele may be examined at the level of its primary product or through its effect on the organism's phenotype. Geneticists study interactions between different alleles in order to gain information about gene activity. Alleles that lack gene activity are called amorphic, while those with reduced levels of activity are called hypomorphic. *See* GENE ACTION.

In microorganisms, functionally related genes are often situated on the chromosome in clusters called operons, the genes of which function in a coordinated fashion. Certain mutations of one gene can also block the expression of other genes in the operon. These mutant alleles fail to complement mutations in other genes, which must be considered nonallelic. Clustering of functionally related genes also occurs in higher organisms, where such loci have been described as complex loci. Lack of complementation between apparently different genes has been called pseudoallelism. The distinction between allelism and pseudoallelism is not always easy to make. *See* GENETICS; OPERON. [O.S.]

Allelopathy The harmful effect of one plant or microorganism on another caused by the release of secondary metabolic products into the environment. The effect is usually visible as a partial or complete inhibition of growth or reduction of seed germination. The donor plant releases biochemical toxins into the soil, water, or atmosphere, from which they are absorbed by receptor plants and act as toxins. Such inhibitions can influence species composition of a plant community, ecological succession, and plant productivity. The extent of these influences varies with strength of the allelopathic reaction. If the inhibition is not severe, the effect may be only a slight loss in productivity; stronger reactions may eliminate a species from the plant association.

Inhibition of one species by another is direct allelopathy. Indirect allelopathy is the inhibition of an intermediate organism—often a bacterium, alga, or fungus—upon which the inhibited plant is dependent for nutrients or water. Autoallelopathy is the inhibition of a species by self-produced toxins.

Allelopathy must be distinguished from competition, which involves the removal or reduction of some vital factor—nutrient, moisture, light, space—from the environment by competing species. It must also be distinguished from pollution, which is a general accumulation from a variety of sources of organic or inorganic substances deleterious to many forms of life. Allelopathy is more specific inhibition caused by the metabolic by-products of a particular plant species under conditions which do not necessarily involve crowding, competition, or stagnation. [I.F.A.]

Allergy Altered reactivity in humans and animals to allergens (substances foreign to the body that cause allergy) induced by exposure through injection, inhalation, ingestion, or skin contact. The most common clinical manifestations of allergy are hay fever, asthma, hives, atopic (endogenous) eczema, and eczematous skin lesions caused by direct contact with allergens such as poison ivy or certain chemicals.

A large variety of substances may cause allergies: pollens, animal proteins, molds, foods, insect venoms, foreign serum proteins, industrial chemicals, and drugs. Most natural allergens are proteins or polysaccharides of moderate molecular size (molecular weights of 10,000 to 200,000). Chemicals or drugs of lower molecular weight (haptens) have first to bind to the body's own proteins (carriers) in order to become fully effective allergens.

For the development of the hypersensitivity state underlying clinical allergies, repeated contact with the allergen is required. Duration of the sensitization period is usually dependent upon the sensitizing strength of the allergen and the intensity of exposure. Some allergens (for example, saliva, urine, and hair proteins of domestic animals) are more sensitizing than others. In most instances, repeated contact with minute amounts of allergen is required; several annual seasonal exposures to grass pollens or ragweed pollen usually occur before an overt manifestation of hay fever. On the other hand, allergy to cow milk proteins in infants can develop within a few weeks. When previous contacts with allergens have not been apparent (for example, antibiotics in food), an allergy may become clinically manifest even upon the first conscious encounter with the offending substance.

Besides the intrinsic sensitizing properties of allergens, individual predisposition of the allergic person to become sensitized also plays an important role. Clinical manifestations, such as hay fever, allergic asthma, and atopic (endogenous) dermatitis, occur more frequently in some families. In other clinical forms of allergy, genetic predisposition, though possibly present as well, is not as evident.

Exposure to sensitizing allergens may induce several types of immune response, and the diversity of immunological mechanisms involved is responsible for the various clinical forms of allergic reactions which are encountered in practice. Three principal types of immune responses are encountered: the production of IgE antibodies, IgG or IgM antibodies, and sensitized lymphocytes. *See* ANTIBODY; IMMUNOGLOBULIN.

Diagnosis of allergic diseases encompasses several facets. Since many clinical manifestations of allergy are mimicked by nonallergic mechanisms, it is usually necessary to use additional diagnostic procedures to ascertain whether the person has developed an immune response toward the incriminated allergen. Such procedures primarily consist of skin tests, in which a small amount of allergen is applied on or injected into the skin. If the individual is sensitized, a local immediate reaction ensues, taking the form of a wheal (for IgE-mediated reactions), or swelling and redness occurs after several hours (for delayed hypersensitivity reactions). The blood may also be analyzed for IgE and IgG antibodies by serological assays, and sensitized lymphocytes are investigated by culturing them with the allergen.

Since the discovery of the responsible allergens markedly influences therapy and facilitates prediction of the allergy's outcome, it is important to achieve as precise a diagnosis as possible. Most tests indicate whether the individual is sensitized to a given allergen, but not whether the allergen is in fact still causing the disease. Since in most cases the hypersensitive state persists for many years, it may well happen that sensitization is detected for an allergen to which the individual is no longer exposed and which therefore no longer causes symptoms. In such cases, exposition tests, consisting of close observation of the individual after deliberate exposure to the putative allergen, may yield useful information.

The most efficient treatment, following identification of the offending allergen, remains elimination of allergen from the person's environment and avoidance of further exposure. This form of treatment is essential for allergies caused by most household and workplace allergens. *See* ANTIGEN; HYPERSENSITIVITY. [A.L.deW.]

Alligator Large aquatic reptile of the family Alligatoridae. Common usage generally restricts the name to the two living species of the genus *Alligator*. The American alligator (*A. mississippiensis*) ranges throughout the southeastern United States from coastal North Carolina (southeastern Virginia in historical times) to the Rio Grande in Texas, and north into southeastern Oklahoma and southern Arkansas (see illustration).

American alligator (*Alligator mississippiensis*).

The second species is the Chinese alligator (*A. sinensis*), restricted to the region of the Yangtze River valley in China, where it inhabits burrows in the floodplains and riverbanks. It is also an endangered species and is now protected in China.

The American alligator is by far the larger of the two species, reaching a length in excess of 15 ft (4.5 m). The average length of *A. sinensis* is 4–5 ft (1.2–1.5 m). *See* REPTILIA. [H.W.C.]

Alloeocoela An order of the Turbellaria, mainly marine but with some fresh-water and terrestrial species. They possess a well-defined gut, with a simple, variable (modified dolioform) or plicate pharynx and a diverticulated intestine. The nervous system has three or four pairs of longitudinal nerves, with connecting commissures. The ovary is paired. The testes are follicular, and a well-defined penis papilla is present. There is a pair of protonephridia, often with two or three main branches to each and with a number of nephridiopores. Anterior ciliated pits or grooves are common; eyes and statocysts are less so, but present in some.

Alloeocoels are generally 0.04–0.40 in. (1–10 mm) in length and cylindrical in shape, but sometimes plump or elongate. They are generally grayish white to brown, but may be more brightly colored due to gut contents or yellow to orange lipid reserves. They occur in the littoral zone among algae, on sandy or muddy bottoms, and occasionally in clear rock pools. *See* TURBELLARIA. [J.B.J.]

Allometry The study of changes in the characteristics of organisms with body size. Characteristics such as body parts or timing of reproductive events do not necessarily change in direct proportion to body size, and the ways in which they change relative to body size can often provide insights into organisms' construction and behavior. Large organisms are often not merely magnified small ones.

Many characteristics, ranging from brain size and heart rate to life span and population density, change consistently with body size. These relationships normally fit a simple power function given by Eq. (1),

$$y = km^b \tag{1}$$

where y is the variable under study, m is body mass, k is the allometric constant, and b is the allometric exponent. The use of logarithms makes the equation easier to visualize—the exponent becomes the slope of a straight line when the logarithm of the variable (y) is plotted against the logarithm of body mass (m) [Eq. (2)].

$$\log(y) = \log(k) + b \log(m) \tag{2}$$

Unless the allometric exponent (b) equals 1, the ratio of y/m varies with m. The terms isometry, positive allometry, and negative allometry are sometimes used for $b = 1$, $b > 1$, and $b < 1$, respectively. If the variable of interest is the mass of an organ, under isometry the organ mass is a fixed proportion of body mass; under positive allometry the larger organisms have disproportionately large organs; and under negative allometry the larger organisms have disproportionately small organs. *See* LOGARITHM.

Allometry is perhaps most powerful when known allometric relations are coupled with theory to produce new hypotheses that can be tested. For example, mammal species with large bodies grow more slowly, mature later, have fewer but larger young after longer gestation periods, and live longer than do small species; these traits remain correlated with each other among deviations from the allometric lines. Allometric models predict not only the allometric exponents but also the correlations among the deviations. The models even provide a framework for asking questions vital to understanding scaling, such as what factors affect the evolution of body size itself. *See* BIOPHYSICS. [A.Pu.; P.H.Ha.]

Allosteric enzyme Any one of the special bacterial enzymes involved in regulatory functions. End-product inhibition is a bacterial control mechanism whereby the end product of a biosynthetic pathway can react with the first enzyme of the pathway and prevent its activity. This end-product inhibition is a device through which a cell conserves its economy by shutting off the synthesis of building blocks when too many are present.

A model to explain the action of allosteric enzymes suggests that the enzyme molecule is a complex consisting of identical subunits, each of which has a site for binding the substrate and another for binding the regulatory substance. These subunits interact in such a way that two conformational forms may develop. One form is in a relaxed condition (R state) and has affinity for the substrate and activator; the other form is in a constrained or taut condition (T state) and has affinity for the inhibitor. The forms exist in a state of equilibrium, but this balance can be readily tipped by binding one of the reactants. The substrate and activator are bound by the relaxed form; when this happens, the balance is tipped in favor of that state. Conversely, the inhibitor will throw the balance toward the constrained state. The balance is thus tipped one way or the other, depending on the relative concentrations of substrate and inhibitor. Since the two states require subunit interaction for their maintenance, it can be seen why dissociation of the subunits leads to a simple monomeric enzyme which no longer exhibits allosteric effects. The model also shows how the binding sites may interact in either a cooperative or antagonistic manner. *See* ENZYME. [J.S.G.]

Allotheria One of the four subclasses of Mammalia, containing a single order, the Multituberculata. The Allotheria first appeared in the Late Jurassic and survived well into the Cenozoic, a period of at least 100,000,000 years. Fossils are known from North America, Europe, and Asia.

The diagnostic features of the subclass are in the dentition. There is a pair of enlarged incisors above and below; reduced lateral incisors may persist in the upper jaw. Canines are absent, leaving a diastema between incisors and cheek teeth. Premolars are variable and often reduced. The lower molars have five or more cusps in two parallel longitudinal rows, and the upper molars have two or three parallel rows of cusps; hence the molars are multituberculate. *See* DENTITION; MAMMALIA; MULTITUBERCULATA. [D.D.D./F.S.S.]

Allowance An intentional difference in sizes of two mating parts. With running or sliding fits, allowance is a clearance, usually for a film of oil. In this sense, allowance is the space "allowed" for motion between parts.

With force or shrink fits allowance is an interference of metal; that is, a portion of metal in one part tends to occupy the same space as the adjacent portion of metal in the mating part. In this sense, allowance is the interference "allowed" to produce pressure between parts. *See* FORCE FIT; LOCATION FIT; PRESS FIT; RUNNING FIT; SHRINK FIT; TOLERANCE. [P.H.B.]

Alloy A metal product containing two or more elements (1) as a solid solution, (2) as an intermetallic compound, or (3) as a mixture of metallic phases. Alloys are frequently described on the basis of their technical applications. They may also be categorized and described on the basis of compositional groups. For example, *see* BERYLLIUM ALLOYS; COPPER ALLOYS; IRON ALLOYS.

Except for native copper and gold, the first metals of technological importance were alloys. Bronze, an alloy of copper and tin, is appreciably harder than copper. This quality made bronze so important an alloy that it left a permanent imprint on the civilization of several millennia ago now known as the Bronze Age. Today the tens of thousands of alloys involve almost every metallic element of the periodic table.

Alloys are used because they have specific properties or production characteristics that are more attractive than those of the pure, elemental metals. For example, some alloys possess high strength; others have low melting points; others are refractory with high melting temperatures; some are especially resistant to corrosion; and others have desirable magnetic, thermal, or electrical properties. These characteristics arise from both the internal and the electronic structure of the alloy. An alloy is usually harder than a pure metal and may have a much lower conductivity.

Bearing alloys are used for metals that encounter sliding contact under pressure with another surface; the steel of a rotating shaft is a common example. Most bearing alloys contain particles of a hard intermetallic compound that resist wear. These particles, however, are embedded in a matrix of softer material which adjusts to the hard particles so that the shaft is uniformly loaded over the total surface. The most familiar bearing alloy is babbitt. Bearings made by powder metallurgy techniques are widely used because they permit the combination of materials which are incompatible as liquids, for example, bronze and graphite, and also permit controlled porosity within the bearings so that they can be saturated with oil before being used, the so-called oilless bearings. *See* ANTIFRICTION BEARING; WEAR.

Certain alloys resist corrosion because they are noble metals. Among these alloys are the precious-metal alloys. Other alloys resist corrosion because a protective film develops on the metal surface. This passive film is an oxide which separates the metal from the corrosive environment. Stainless steels and aluminum alloys exemplify metals with this type of protection. The bronzes, alloys of copper and tin, also may be considered to be corrosion-resisting. *See* CORROSION; STAINLESS STEEL.

Dental alloys contain precious metals. Amalgams are predominantly silver-mercury alloys, but they may contain minor amounts of tin, copper, and zinc for hardening purposes. Liquid mercury is added to a powder of a precursor alloy of the other metals. After being compacted, the mercury diffuses into the silver-base metal to give a completely solid alloy. Gold-base dental alloys are preferred over pure gold because gold is relatively soft. The most common dental gold alloy contains gold, silver, and copper. For higher strengths and hardnesses, palladium and platinum are added, and the copper and silver are increased so that the gold content drops. Vitallium and other corrosion-resistant alloys are used for bridgework and special applications. *See* SILVER ALLOYS.

Die-casting alloys have melting temperatures low enough so that in the liquid form they can be injected under pressure into steel dies. Such castings are used for automotive parts and for office and household appliances which have moderately complex shapes. Most die castings are made from zinc-base or aluminum-base alloys. Magnesium-base alloys also find some application when weight reduction is paramount. Low-melting alloys of lead and tin are not common because they lack the necessary strength for the above applications. *See* METAL CASTING.

In certain alloy systems a liquid of a fixed composition freezes to form a mixture of two basically different solids or phases. An alloy that undergoes this type of solidification process is called a eutectic alloy. A homogeneous liquid of this composition on slow cooling freezes to form a mixture of particles of nearly pure copper embedded in a matrix (background) of nearly pure silver.

The advantageous mechanical properties inherent in composite materials have been known for many years. Attention is being given to eutectic alloys as they are basically natural composite materials. *See* EUTECTICS; METAL MATRIX COMPOSITE.

Fusible alloys generally have melting temperatures below that of tin (449°F or 232°C), and in some cases as low as 122°F (50°C). Using eutectic compositions of metals such as lead, cadmium, bismuth, tin, antimony, and indium achieves these low melting temperatures. These alloys are used for many purposes, for example, in fusible elements in automatic sprinklers, forming and stretching dies, filler for thin-walled tubing that is being bent, and anchoring dies, punches, and parts being machined.

High-temperature alloys have high strengths at high temperatures. In addition to having strength, these alloys must resist oxidation by fuel-air mixtures and by steam vapor. At temperatures up to about 1380°F (750°C), the austenitic stainless steels serve well. An additional 180°F (100°C) may be realized if the steels also contain 3% molybdenum. Both nickel-base and cobalt-base alloys, commonly categorized as superalloys, may serve useful functions up to 2000°F (1100°C). Nichrome, a nickel-base alloy containing chromium and iron, is a fairly simple superalloy. More sophisticated alloys invariably contain five, six, or more components; for example, an alloy called René-41 contains Cr, Al, Ti, Co, Mo, Fe, C, B, and Ni. Other alloys are equally complex. A group of materials called cermets, which are mixtures of metals and compounds such as oxides and carbides, have high strength at high temperatures, and although their ductility is low, they have been found to be usable. One of the better-known cermets consists of a mixture of titanium carbide and nickel, the nickel acting as a binder or cement for the carbide. *See* CERMET.

Metals are bonded by three principal procedures: welding, brazing, and soldering. Welded joints melt the contact region of the adjacent metal; thus the filler material is chosen to approximate the composition of the parts being joined. Brazing and soldering alloys are chosen to provide filler metal with an appreciably lower melting point than that of the joined parts. Typically, brazing alloys melt above 750°F (400°C), whereas solders melt at lower temperatures. *See* BRAZING; SOLDERING; WELDING AND CUTTING OF METALS.

Aluminum and magnesium, with densities of 2.7 and 1.75 g/cm^3, respectively, are the bases for most of the light-metal alloys. Titanium (4.5 g/cm^3) may also be regarded as a light-metal alloy if comparisons are made with metals such as steel and copper. Aluminum and magnesium must be hardened to receive extensive application. Age-hardening processes are used for this purpose. *See* ALUMINUM; MAGNESIUM.

Low-expansion alloys include Invar, the dimensions of which do not vary over the atmospheric temperature range, and Kovar, which is widely used because its expansion is low enough to match that of glass. *See* THERMAL EXPANSION.

Soft and hard magnetic materials involve two distinct categories of alloys. The former consists of materials used for magnetic cores of transformers and motors, and must be magnetized and demagnetized easily. For alternating-current applications, silicon-ferrite is commonly used. This is an alloy of iron containing as much as 5% silicon. Permalloy and some comparable cobalt-base alloys are used in the communications industry. Ceramic ferrites, although not strictly alloys, are widely used in high-frequency applications because of their low electrical conductivity and negligible induced-energy losses in the magnetic field. Permanent or hard magnets may be made from steels which are mechanically hardened, either by deformation or by quenching. The Alnicos are also widely used for magnets. Since these alloys cannot be forged, they must be produced in the form of castings. The newest hard magnets are being produced from alloys of cobalt and the rare-earth type of metals. *See* MAGNETIC MATERIALS.

In addition to their use in coins and jewelry, precious metals such as silver, gold, and the heavier platinum metals are used extensively in electrical devices in which contact resistances must remain low, in catalytic applications to aid chemical reactions, and in temperature-measuring devices such as resistance thermometers and thermocouples. The unit of alloy impurity is commonly expressed in karats, where each karat is a $\frac{1}{24}$ part. The most common precious-metal alloy is sterling silver (92.5% Ag, with the remainder being unspecified, but usually copper). The copper is very beneficial in that it makes the alloy harder and stronger than pure silver.

Metallic implants demand extreme corrosion resistance because body fluids contain nearly 1% NaCl, along with minor amounts of other salts, with which the metal will be in contact for indefinitely long periods of time. Type 316 stainless steels resist pitting corrosion but are subject to crevice corrosion. Vitallium and other cobalt-base alloys have orthopedic applications. Titanium alloys gained wide usage in Europe during the early 1970s for pacemakers and for retaining devices in artificial heart valves. While excellent for corrosion resistance, this alloy is subject to mechanical wear; therefore, it is not satisfactory in hip-joint prostheses and applications with similar frictional contacts. *See* PROSTHESIS.

Shape memory alloys have a very interesting and desirable property. In a typical case, a metallic object of a given shape is cooled from a given temperature T_1 to a lower temperature T_2 where it is deformed so as to change its shape. Upon reheating from T_2 to T_1, the shape change accomplished at T_2 is recovered so that the object returns to its original configuration. This thermoelastic property of the shape memory alloys is associated with the fact that they undergo a martensitic phase transformation (that is, a reversible change in crystal structure that does not involve diffusion) when they are cooled or heated between T_1 and T_2. Shape memory alloys are capable of being employed in a number of useful applications. One example is for thermostats; another is for couplings on hydraulic lines or electrical circuits.

Superconducting alloys, with zero resistivity, are of great interest in the design of certain fusion reactors which require very large magnetic fields to contain the plasma in a closed system. The advantage of the use of a material with a resistivity approaching zero is obvious. However, two significant problems are involved in the use of superconducting alloys in large electromagnetics: the critical temperature, and the fact that above a certain critical current density the superconducting materials tend to become normal conductors with a finite resistance. Serious materials problems still have to be solved before these materials can be used successfully. *See* SUPERCONDUCTIVITY.

Thermocouple alloys include Chromel and Alumel. These two alloys together form the widely used Chromel-Alumel thermocouple, which can measure temperatures up to 2200°F (1204°C). Another common thermocouple alloy, constantan, is used to form iron-constantan and copper-constantan couples, employed at lower temperatures. *See* STEEL; THERMOCOUPLE. [L.H.V.V./R.E.R.-H.]

Allspice The dried, unripe fruits of a small, tropical, evergreen tree, *Pimenta officinalis*, of the myrtle family (Myrtaceae). This species is a native of the West Indies and parts of Central and South America. The spice, alone or in mixtures, is much used in sausages, pickles, sauces, and soups. The extracted oil is used for flavoring and in perfumery. Allspice is so named because its flavor resembles that of a combination of cloves, cinnamon, and nutmeg. *See* MYRTALES; SPICE AND FLAVORING. [P.D.St./E.L.C.]

Almanac A book that contains astronomical or meteorological data arranged according to days, weeks, and months of a given year and may also include diverse information of a nonastronomical character. This article is restricted to astronomical and navigational almanacs.

The Astronomical Almanac contains ephemerides, which are tabulations, at regular time intervals, of the orbital positions and rotational orientation of the Sun, Moon, planets, satellites, and some minor planets. It also contains mean places of stars, quasars, pulsars, galaxies, and radio sources, and the times for astronomical phenomena such as eclipses, conjunctions, occultations, sunrise, sunset, twilight, moonrise, and moonset. This volume contains the fundamental astronomical data needed by astronomers, geodesists, navigators, surveyors, and space scientists. *See* ASTRONOMICAL COORDINATE SYSTEMS; EPHEMERIS.

While *The Astronomical Almanac* is basically designed for the determination of positions of astronomical objects as observed from the Earth, *The Nautical Almanac* and *The Air Almanac* are designed to determine the navigator's position from the tabulated position of the celestial object. *The Nautical Almanac* contains hourly values of the Greenwich hour angle and declination of the Sun, Moon, and four planets and the sidereal hour angle and declination of 57 stars for every third day. Monthly apparent positions are tabulated for an additional 173 navigational stars. The positions are tabulated to an angular accuracy of 0.1 minute of arc, which is equivalent to 0.1 nautical mile (0.2 km). This tabular accuracy is sufficient to produce a computed position with an error no greater than 0.3 to

0.4 nmi (0.6 to 0.7 km). *The Air Almanac* gives the positions of the Sun, first point of Aries, three planets, and the Moon at 10-min intervals. As necessary, information is adjusted so that the tabulated data at any given time can be used during the interval to the next entry, without interpolation, to an accuracy sufficient for practical air navigation. While designed for air navigators, *The Air Almanac* is used by mariners who accept the reduced accuracy in exchange for its greater convenience compared with *The Nautical Almanac*. *See* CELESTIAL NAVIGATION.

The *Almanac for Computers*, published annually by the U.S. Naval Observatory since 1977, provides the coefficients for power series and Chebyshev polynomials, which permit the computation for any specific time of the data contained in the nautical, air, and astronomical almanacs.

The data from the publications as well as additional data, such as long-term ephemerides and many star catalogs, are available in machine-readable form. However, in spite of the increased use of computers, there is a continuing need for the printed volumes. [P.K.S.]

Almond A small deciduous tree, *Prunus amygdalus* (also known as *P. dulcis* or *Amygdalus communis*), closely related to the peach and other stone fruits and grown widely for its edible seeds. Almonds are of two general types: the bitter type is a source of prussic acid and flavoring extracts, and the sweet type has various food uses. Almond kernels contain approximately 50% fat or oil, 20% protein, 20% carbohydrate, and a variety of minerals and vitamins. *See* ROSALES.

In the United States, commercial production is limited to California. Spain is the second leading producer, but the amount produced is about one-half that of the United States. Italy has historically been a leading producer, primarily from the Bari and Sicily areas, but production has declined sharply.

Almonds are used in a variety of products. Some are roasted whole and salted to be used as snacks. Others are blanched (the skin is removed) by steam and subjected to slicing, dicing, or halving. These may be roasted, and go into products such as candy bars, bakery products, ice cream, and almond paste, among many other uses. [D.E.K.]

Alpaca A member of the camel family, Camelidae, which belongs to the mammalian order Artiodactyla, the even-toed ungulates. The alpaca (*Lama pacos*) is found at elevations above 12,000 ft (3600 m) along the shores of Lake Titicaca on the boundaries of Peru and Bolivia.

The alpaca's neck and head are elongate, and the upper lip has a deep cleft. The long, slender legs terminate in two toes; the feet are digitigrade, that is, the animals walk on the toes and not on the entire foot or the tip of the digits. The long, fine repellent hair, or wool, ranges in color from black to white and is highly prized for manufacturing cloth, particularly the white wool. Like many breeds of domesticated animals, the alpaca has been bred to produce pure strains for the wool. Although the alpaca is raised chiefly for its wool, its flesh is edible and palatable. *See* ARTIODACTYLA; CAMEL; LLAMA; MAMMALIA; NATURAL FIBER. [C.B.C.]

Alpha Centauri The brightest star in the southern constellation Centaurus (visible only south of geographic latitude +30°). Alpha Centauri is one of a triple system, and is Earth's nearest neighbor in space, at a distance of 1.33 parsecs.

The visual binary of 80-year period contains main-sequence stars of types G2 and K0 separated by about 24 astronomical units, that is, by about the diameter of the solar system. The brighter of the visual binary is almost a twin of the Sun in luminosity, temperature, and mass. There is a faint companion, Proxima Centauri, separated from the other two by about 10,000 astronomical units. *See* STAR. [J.L.Gr.]

Alpha fetoprotein A serum protein that has become associated with the detection of certain types of cancer and of fetal abnormalities. The protein arose in evolution following a reduplication of a primordial gene that coded for a precursor of albumin and alpha fetoprotein some 500 million years ago. The two proteins still share many features, but differ chiefly in that albumin is the major protein component of adult blood plasma while alpha fetoprotein is normally confined to the fetus. The primary structure of alpha fetoprotein is 40% homologous with albumin, but corresponding regions are short and highly dispersed. There is no immunological cross-reactivity between native albumin and alpha fetoprotein.

The function of alpha fetoprotein is unknown. It may be a specialized transport substance for an unidentified ligand. It has been found to bind estrogens in rats and mice, though not in humans, in whom it appears to bind bilirubin. It can also suppress some immunological responses, especially antibody production, probably through the induction of an increased number of suppressor cells. It diminishes the binding of sheep red blood cells to mature T lymphocytes, but not to human ones. It may thus contribute to the maintenance of the mother's tolerance of the fetus, though the significance of this role in the living organism remains undetermined. Both in cancer and in other diseases, the real mechanism responsible for alpha fetoprotein synthesis still remains to be elucidated. *See* ALBUMIN; IMMUNITY; ONCOLOGY. [A.B.]

Alpha particles Helium nuclei which have been ejected at high velocity from atomic nuclei as products of radioactive decay or as products of induced nuclear reactions. Helium nuclei which have been accelerated to high velocities for use as bombarding particles in nuclear reactions also may be called alpha (α) rays or alpha particles. The helium nucleus has a charge of $+2e$, that is, twice the magnitude of the charge e of a proton or of an electron, and a mass of 4.00150 atomic mass units. The velocity of ejection of the alpha ray varies from one radioactive substance to another, but is usually in the domain of 1 to 2×10^9 cm/s. Many radioactive substances emit alpha rays in two or more discrete energy groups, usually in the domain of 4–6 MeV, but occasionally as low as 2 MeV or as high as 10 MeV. The other radiations emitted by radioactive atoms are beta (β) particles [negative or positive electrons] and gamma (γ) rays [high-frequency electromagnetic radiation]. For the theory of radioactive decay *see* RADIOACTIVITY. For alpha-particle detection *see* PARTICLE DETECTOR. *See also* BETA PARTICLES; GAMMA RAYS; NUCLEAR REACTION; NUCLEAR SPECTRA.

As an alpha particle passes through matter, it ionizes many of the atoms along its path, thereby gradually losing its kinetic energy, and is stopped (absorbed). For example, a 5-MeV alpha ray is stopped after traversing 1.4 in. (3.5 cm) of air at 1 atm (10^2 kilopascals) pressure and 59°F (15°C) and engaging in enough collisions with atomic electrons along its path to produce about 150,000 ion pairs.

Along the path of an alpha particle in normal air, some 2000–6000 ion pairs per millimeter are produced, depending upon the velocity of the alpha particle at the point under consideration. The number of ion pairs per unit path length is called the specific ionization. The illustration shows the specific ionization produced by an alpha particle in air for various values of the residual range, that is, at various distances measured back from the end of its range.

The mean range R of a monoenergetic group of alpha particles cannot be derived wholly from theory but must rest on the results of many careful measurements plus interpolation and extrapolation based on theory. [R.D.E.]

Alpha particles are useful in the treatment of inaccessible tumors and vascular disorders. It is clear from the Bragg curve in the illustration that the ionizing power of alpha particles is concentrated near the ends of their paths. Thus they can deliv-

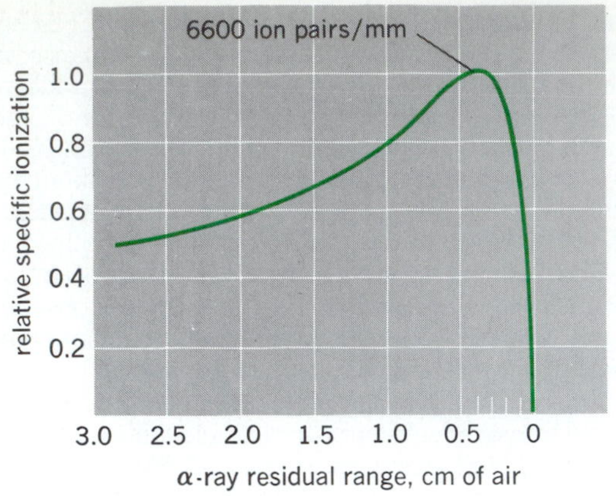

Specific ionization of a single alpha ray in air at 15°C and 760 mmHg (10^2 kilopascals).

er destructive energy to a tumor while doing little damage to nearby healthy tissue. With proper acceleration, positioning, and dosage, the energy can be delivered so precisely that alpha-particle radiotherapy is uniquely suited for treating highly localized tumors near sensitive normal tissue (for example, the spinal cord). *See* RADIOLOGY.

The element-specific energies of backscattered (Rutherford-scattered) alpha particles are used in remote probes to analyze the mineral composition of geological formations. In particular, alpha particles scattered by light elements transfer more energy than those scattered by heavy elements. In another alpha-particle device, the energy from ^{238}Pu alpha decay is reliably harnessed in batteries based on the Brayton cycle, and used to power scientific equipment left on the Moon. *See* BRAYTON CYCLE; ION-SOLID INTERACTIONS; NUCLEAR BATTERY; SPACE POWER SYSTEMS. [V.Sh.]

Alpine vegetation Plant growth forms characteristic of upper reaches of forests on mountain slopes. In such an environment, trees undergo gradual changes that, though subtle at first, may become dramatic beyond the dense forest as the zone of transition leads into the nonforested zone of the alpine tundra. In varying degrees, depending on the particular mountain setting, the forest is transformed from a closed-canopy forest to one of deformed and dwarfed trees interspersed with alpine tundra species. This zone of transition is referred to as the forest-alpine tundra ecotone. The trees within the ecotone are stunted, often shrublike, and do not have the symmetrical shape of most trees within the forest interior. *See* PLANT GEOGRAPHY.

The forest-alpine tundra ecotone is a mosaic of both tree and alpine tundra species; and it extends from timberline (the upper limit of the closed-canopy forest of symmetrically shaped, usually evergreen trees) to treeline (the uppermost limit of tree species) and the exposed alpine tundra. With elevational increases, tree deformation is magnified, tree height is reduced, and the total area occupied by trees becomes smaller as the alpine shrub, grass, and herbaceous perennials become more dominant.

The environment in which these tenacious individuals survive is harsh and involves a complex interaction of many factors, with the major controlling factor often being climate. The climate is characterized by a short growing season, low air temperatures, frozen soils, drought, high levels of ultraviolet radiation, irregular accumulation of snow, and strong winds. The interaction of all these factors produces varying levels of stress within the trees. *See* WIND.

The ultimate cause of the tree deformations and of the eventual complete cessation of tree growth lies in the inability of the tissues of the shoots and the needles to mature and prepare for the harsh environmental conditions. As the length of the growing season decreases with elevation, new needles often do not mature; they have thinner cuticles (the waxlike covering on the needles that protects against desiccation and wind abrasion), and they are less acclimated against low air temperatures. Factors that particularly affect the length of the growing season include air and soil temperatures, and the depth and distribution of snow. *See* AIR TEMPERATURE. [K.J.H.]

Alternating current Electric current that reverses direction periodically, usually many times per second. Electrical energy is ordinarily generated by a public or a private utility organization and provided to a customer, whether industrial or domestic, as alternating current.

One complete period, with current flow first in one direction and then in the other, is called a cycle, and 60 cycles per second (60 hertz) is the customary frequency of alternation in the United States and in all of North America. In Europe and in many other parts of the world, 50 Hz is the standard frequency. On aircraft a higher frequency, often 400 Hz, is used to make possible lighter electrical machines.

When the term alternating current is used as an adjective, it is commonly abbreviated to ac, as in ac motor. Similarly, direct current as an adjective is abbreviated dc.

The voltage of an alternating current can be changed by a transformer. This simple, inexpensive, static device permits generation of electric power at moderate voltage, efficient transmission for many miles at high voltage, and distribution and consumption at a conveniently low voltage. With direct (unidirectional) current it is not possible to use a transformer to change voltage. On a few power lines, electric energy is transmitted for great distances as direct current, but the electric energy is generated as alternating current, transformed to a high voltage, then rectified to direct current and transmitted, then changed back to alternating current by an inverter, to be transformed down to a lower voltage for distribution and use.

In addition to permitting efficient transmission of energy, alternating current provides advantages in the design of generators and motors, and for some purposes gives better operating characteristics. Certain devices involving chokes and transformers could be operated only with difficulty, if at all, on direct current. Also, the operation of large switches (called circuit breakers) is facilitated because the instantaneous value of alternating current automatically becomes zero twice in each cycle and an opening circuit breaker need not interrupt the current but only prevent current from starting again after its instant of zero value.

Alternating current is shown diagrammatically in Fig. 1. In

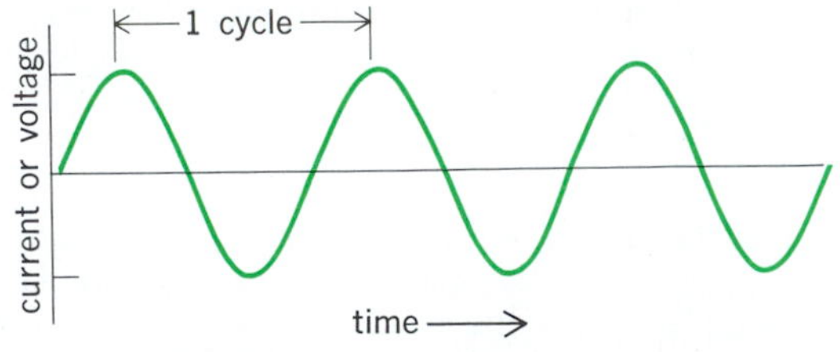

Fig. 1. Diagram of sinusoidal alternating current.

this diagram it is assumed that the current is alternating sinusoidally; that is, the current i is described by the following equation where I_m is the maximum instantaneous current, f is the

$$i = I_m \sin 2\pi ft$$

frequency in cycles per second (hertz), and t is the time in seconds. *See* SINE WAVE.

A sinusoidal form of current, or voltage, is usually approximated on practical power systems because the sinusoidal form results in less expensive construction and greater efficiency of operation of electric generators, transformers, motors, and other machines.

A useful measure of alternating current is found in the ability of the current to do work, and the amount of current is correspondingly defined as the square root of the average of the square of instantaneous current, the average being taken over an integer number of cycles. This value is known as the root-mean-square (rms) or effective current. It is measured in amperes. It is a useful measure for current of any frequency. The rms value of direct current is identical with its dc value. The rms value of sinusoidally alternating current is $I_m/\sqrt{2}$ (see Fig. 1 and the equation). Other useful quantities are the phase difference φ between voltage and current and the power factor. *See* PHASE (PERIODIC PHENOMENA); POWER FACTOR.

The phase angle and power factor of voltage and current in a circuit that supplies a load are determined by the load. Thus a load of pure resistance, such as an electric heater, has unity power factor. An inductive load, such as an induction motor, has a power factor less than 1 and the current lags behind the applied voltage. A capacitive load, such as a bank of capacitors, also has a power factor less than 1, but the current leads the voltage, and the phase angle φ is a negative angle.

Three-phase systems are commonly used for generation, transmission, and distribution of electric power. A customer may be supplied with three-phase power, particularly if a large amount of power is used or the use of three-phase loads is desired. Small domestic customers are usually supplied with single-phase power. A three-phase system is essentially the same as three ordinary single-phase systems, with the three voltages of the three single-phase systems out of phase with each other by one-third of a cycle (120 degrees), as shown in Fig. 2. The three-phase system is balanced if the maximum

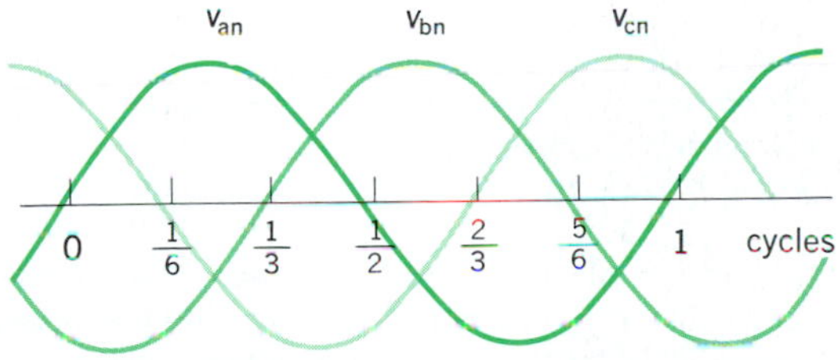

Fig. 2. Voltages of a balanced three-phase system.

voltage in each of the three phases is equal, and if the three phase angles are equal, ⅓ cycle each as shown. It is only necessary to have three wires for a three-phase system (a, b, and c of Fig. 3) plus a fourth wire n to serve as a common return or neutral conductor. On some systems the earth is used as the common or neutral conductor.

Each phase of a three-phase system carries current and conveys power and energy. If the three loads on the three phases of the three-phase system are equal and the voltages are balanced, then the currents are balanced also. The sum of the three currents is then zero at every instant. This means that current in the common conductor (n of Fig. 3) is always zero, and that the conductor could theoretically be omitted entirely. In practice, the three currents are not usually exactly balanced, and either of two situations obtains. Either the common neutral wire n is used, in which case it carries little current (and may be of high resistance compared to the other three line wires), or else the common neutral wire n is not used, only three line wires being installed, and the three phase currents are thereby forced to add to zero even though this requirement results in some imbalance of phase voltages at the load.

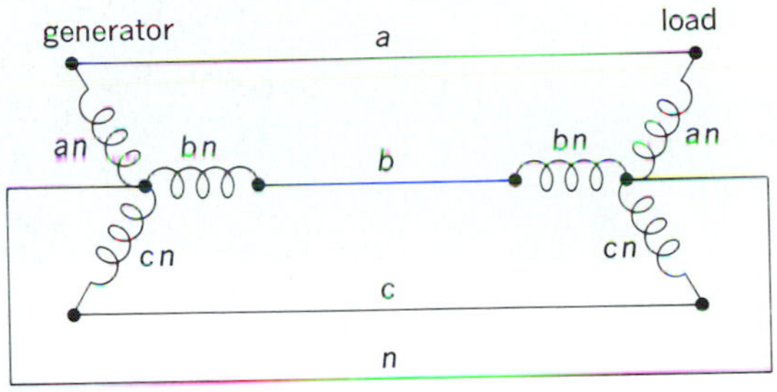

Fig. 3. Connections of a simple three-phase system.

The total instantaneous power from generator to load is constant (does not vary with time) in a balanced, sinusoidal, three-phase system. This results in smoother operation and less vibration of motors and other ac devices. In addition, three-phase motors and generators are more economical than single-phase machines.

AC circuits are also used to convey information. An information circuit, such as telephone, radio, or control, employs varying voltage, current, waveform, frequency, and phase. Efficiency is often low, the chief requirement being to convey accurate information even though little of the transmitted power reaches the receiving end. For further consideration of the transmission of information *see* RADIO; TELEPHONE; WAVEFORM.

An ideal power circuit should provide the customer with electric energy always available at unchanging voltage of constant waveform and frequency, the amount of current being determined by the customer's load. High efficiency is greatly desired. *See* CAPACITANCE; CIRCUIT (ELECTRICITY); ELECTRIC CURRENT; ELECTRIC FILTER; ELECTRICAL IMPEDANCE; ELECTRICAL RESISTANCE; INDUCTANCE; JOULE'S LAW; OHM'S LAW; RESONANCE (ALTERNATING-CURRENT CIRCUITS). [H.H.Sk.]

Alternating-current generator A machine which converts mechanical power into alternating-current electric power. The most common type, sometimes called an alternator, is the synchronous generator, so named because its operating speed is proportional to system frequency. Another type is the induction generator, the speed of which varies somewhat with load for constant output frequency. *See* INDUCTION GENERATOR; SYNCHRONOUS GENERATOR. [L.T.R.]

Alternating-current motor An electric rotating machine which converts alternating-current electric energy to mechanical energy; one of two general classifications of electric motor. Because ac power is widely available, ac motors are commonly used. They are made in sizes from a few watts to thousands of horsepower. *See* DIRECT-CURRENT MOTOR.

Each type of ac motor has special properties. These motors are generally classified by application, construction, principle of operation, or operating characteristics. The most common type of ac motor is the induction motor. Current is induced in a rotor as its conductors cut lines of magnetic flux created by currents in a stator. Where constant speed is essential, a synchronous motor is used. It runs at a fixed speed in synchronism with the frequency of the power supply. For operation from either ac or dc power, the series motor has its field winding connected in series with the armature winding through a commutator and brush arrangement, as in a dc series motor. This universal motor has high starting torque. The repulsion motor

Standard ranges of hp, speeds, and slips at 60 Hz*

No. of phases	Type of motor	Power output, hp†	Synchronous speed, rpm	Slip, %
Single	Capacitor	1–10	1800–1200	0–5
	Split-phase	0–0.5	1800–1200	0–5
	Shaded-pole	0–0.25	1800–900	11–14
	Repulsion-start	0–25	1800–1200	0–5
Three	Induction	0–100,000	3600–450	0–5
Three	Synchronous	20–30,000	3600–80	0

*National Electrical Manufacturers Association.
†1 hp = 0.75 kW.

is also a commutator motor, but the brushes and commutator are short-circuited and not connected in series with the stator. *See* INDUCTION MOTOR; REPULSION MOTOR; SYNCHRONOUS MOTOR; UNIVERSAL MOTOR.

The table gives a comparison of available sizes and characteristics of some common ac motors. For general principles *see* ELECTRIC ROTATING MACHINERY; MOTOR; WINDINGS IN ELECTRIC MACHINERY. [A.F.P.]

The rotor speed of the synchronous motor is a direct function of the number of stator and rotor field poles and the frequency of the ac applied to the stator. Since the number of rotor poles of the polyphase synchronous motor is not easily modified, the change of frequency method is the only way to control synchronous speed of the motor. The speed of the polyphase induction motor is an asynchronous speed varied by one or more of the following methods: (1) by changing the applied frequency to the stator (same method as for a synchronous motor, already discussed); (2) by controlling the rotor slip by means of rheostatic rotor resistance control (used for wound-rotor induction motors); (3) by changing the number of poles of both stator and rotor; and (4) by "foreign voltage" control, obtained by conductively or inductively introducing applied voltages of the proper frequency to the rotor. A number of electromechanical or purely mechanical methods also serve to provide speed control of ac motors.

The principal method of speed control used for fractional hp single-phase induction-type, shaded-pole, reluctance, series, and universal motors is the method of primary line-voltage control. It involves a reduction in line voltage applied to the stator winding (of the induction type) or to the armature of series and universal motors. In the former this produces a reduction of torque and an increase in rotor slip. In the latter it is simply a means of controlling speed by armature voltage control or field flux control or both. The reduction in line voltage is usually accomplished by any one of five methods: autotransformer control, series reactance control, tapped main-winding control, saturable reactor (or magnetic amplifier) control, and silicon-controlled rectifier feedback control.

The development of the thyristor, or silicon-controlled rectifier (SCR), has created unlimited possibilities for control of virtually all types of motors (single-phase, dc, and polyphase). SCRs in sizes up to 1600 amperes (rms) with voltage ratings up to 1600 volts are available. *See* RECTIFIER; SEMICONDUCTOR RECTIFIER. [I.L.K.]

Alternating gradient focusing A configuration of transverse electric or magnetic fields suitable for focusing or confining a charged particle beam. Linear forces are achieved by using a four-pole (quadrupole) geometry (see illustration) with adjacent poles opposite in polarity. Ideally the poles are rectangular hyperbolas in cross section, but adequate linearity can be obtained even with round electrodes. According to Maxwell's equations, a magnetic or electric field which focuses in one transverse direction must be defocusing in the other, in contrast to optical lenses. However, if the polarity alternates in time or along the beam direction (alternating gradient), the net effect is to focus the beam in both directions, because the focusing effect is stronger the farther the particles are from the neutral axis. The net focusing force exceeds that which can be attained by any other external field configuration; thus the configuration is called strong focusing. With excessive focusing, the particle motion becomes unstable. *See* MAXWELL'S EQUATIONS.

This focusing principle was discovered and developed for high-energy physics applications. It became possible to confine high-energy beams circulating in magnetic rings many kilome-

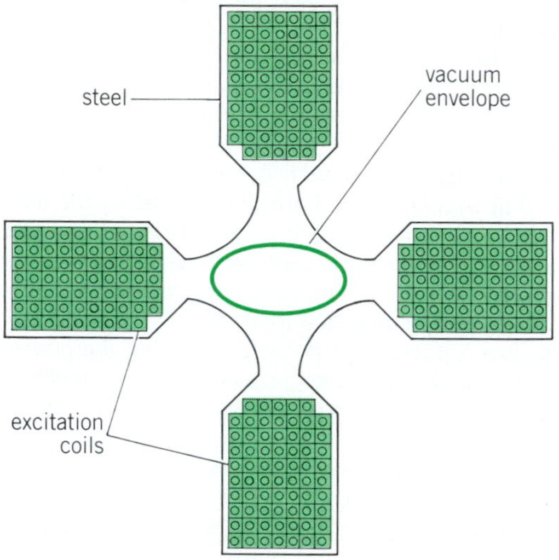

Cross section of a typical magnetic quadrupole for a high-energy physics accelerator.

ters in circumference in vacuum envelopes a few centimeters in transverse dimensions. Thus small, and therefore economically acceptable, magnets could be used. *See* PARTICLE ACCELERATOR.

A newer application of alternating gradient focusing is in a light and compact mass filter well suited to upper atmosphere research and other applications. *See* MASS SPECTROMETRY; MASS SPECTROSCOPE. [L.Smi.]

Alternator A machine that generates an alternating voltage when its rotating portion is driven by a motor, engine, or other means. It is thus a means for converting mechanical power into electric power. The frequency of the generated alternating voltage is exactly proportional to the speed at which the alternator is driven. An alternator generates one cycle of alternating current each time one of its coils passes a pair of magnetic poles in the field structure. *See* ALTERNATING-CURRENT GENERATOR. [J.Mar.]

Altimeter Any device which measures the height of an aircraft. The two chief types are the pressure altimeter, which measures distance above sea level, and the radio altimeter, which measures distance above the ground. *See* PRESSURE ALTIMETER; RADIO ALTIMETER. [J.W.A.]

Altitude In astronomical, navigational, and surveying practice, an angle at the observer, measured positively from 0 to 90° above the observer's horizon along the vertical circle through the object being observed and the observer's zenith. It is also the complement of the angle at the observer along the vertical circle from the observer's zenith to the object. Practitioners often refer to this angle as elevation. *See* ASTRONOMICAL COORDINATE SYSTEMS; CELESTIAL NAVIGATION.

In aeronautics and astronautics, an altitude is the vertical distance (height) of an object such as an aircraft or spacecraft above a reference surface. Aircraft altitude is usually referenced to sea level; spacecraft altitude may be referenced to some arbitrary fiducial surface on the body around which the spacecraft is orbiting. [R.L.Du.]

Alum A colorless to white crystalline substance which occurs naturally as the mineral kalunite and is a constituent of the mineral alunite. Alum is produced as aluminum sulfate by treating bauxite with sulfuric acid to yield alum cake or by treating the bauxite with caustic soda to yield papermaker's alum. Other industrial alums are potash alum, ammonium alum, sodium alum, and chrome alum (potassium chromium sulfate). Major uses of alum are as an astringent, styptic, and emetic. For water purification alum is dissolved. It then crystallizes out into positively charged crystals that attract negatively charged organic impurities to form an aggregate sufficiently heavy to settle out. Alum is also used in sizing paper, dyeing fabrics, and tanning leather. With sodium bicarbonate it is used in baking powder and in some fire extinguishers. *See* ALUMINUM; COLLOID. [F.H.R.]

Aluminate A negative ion usually given the formula AlO_2^- and derived from aluminum hydroxide. Aluminum hydroxide is an amphoteric substance and thus can react with a strong base, such as sodium hydroxide, to form sodium aluminate. It also reacts with acids to form aluminum salts. Solutions of aluminates are strongly basic and the reaction with sodium hydroxide is easily reversed, even by weak acids, to form the insoluble aluminum hydroxide. This is the basis for the commercial use of sodium aluminate in the clarification of water. *See* ALUMINUM; WATER TREATMENT. [E.E.W.]

Aluminum A metallic chemical element, symbol Al, atomic number 13, atomic weight 26.98154, in group III of the periodic system. Pure aluminum is soft and lacks strength, but it can be alloyed with other elements to increase strength and impart a number of useful properties. Alloys of aluminum are light, strong, and readily formable by many metalworking processes; they can be easily joined, cast, or machined, and accept a wide variety of finishes. Because of its many desirable physical, chemical, and metallurgical properties, aluminum has become the most widely used nonferrous metal.

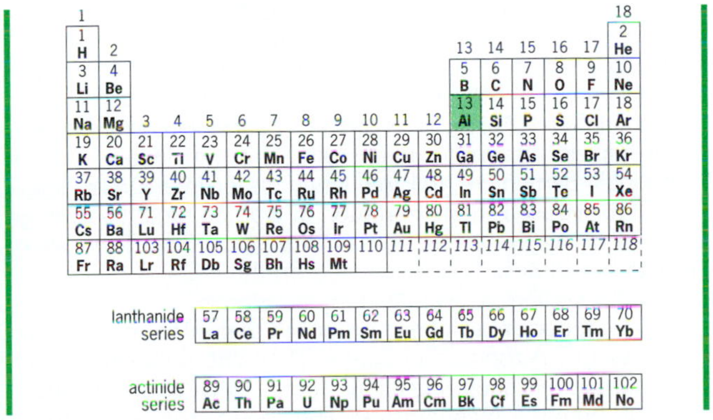
1 18
1 H 2 13 14 15 16 17 2 He
3 Li 4 Be 5 B 6 C 7 N 8 O 9 F 10 Ne
11 Na 12 Mg 3 4 5 6 7 8 9 10 11 12 13 Al 14 Si 15 P 16 S 17 Cl 18 Ar
19 K 20 Ca 21 Sc 22 Ti 23 V 24 Cr 25 Mn 26 Fe 27 Co 28 Ni 29 Cu 30 Zn 31 Ga 32 Ge 33 As 34 Se 35 Br 36 Kr
37 Rb 38 Sr 39 Y 40 Zr 41 Nb 42 Mo 43 Tc 44 Ru 45 Rh 46 Pd 47 Ag 48 Cd 49 In 50 Sn 51 Sb 52 Te 53 I 54 Xe
55 Cs 56 Ba 71 Lu 72 Hf 73 Ta 74 W 75 Re 76 Os 77 Ir 78 Pt 79 Au 80 Hg 81 Tl 82 Pb 83 Bi 84 Po 85 At 86 Rn
87 Fr 88 Ra 103 Lr 104 Rf 105 Db 106 Sg 107 Bh 108 Hs 109 Mt 110 111 112 113 114 115 116 117 118
lanthanide series 57 La 58 Ce 59 Pr 60 Nd 61 Pm 62 Sm 63 Eu 64 Gd 65 Tb 66 Dy 67 Ho 68 Er 69 Tm 70 Yb
actinide series 89 Ac 90 Th 91 Pa 92 U 93 Np 94 Pu 95 Am 96 Cm 97 Bk 98 Cf 99 Es 100 Fm 101 Md 102 No

Aluminum is the most abundant metallic element on the Earth and Moon but is never found free in nature. The element is widely distributed in plants, and nearly all rocks, particularly igneous rocks, contain aluminum in the form of aluminum silicate minerals. When these minerals go into solution, depending upon the chemical conditions, aluminum can be precipitated out of the solution as clay minerals or aluminum hydroxides, or both. Under such conditions bauxites are formed. Bauxites serve as principal raw materials for aluminum production. *See* BAUXITE; CLAY MINERALS; IGNEOUS ROCKS; WEATHERING PROCESSES.

Aluminum is a silvery metal having a density of 1.56 oz/in.^{3} at 68°F (2.70 g/cm^3 at 20°C). Naturally occurring aluminum consists of a single isotope, $^{27}_{13}Al$. Aluminum crystallizes in the face-centered cubic structure with edge of the unit lattice cube of 4.0495 angstroms (0.40495 nanometer). Aluminum is known for its high electrical and thermal conductivities and its high reflectivity.

The electronic configuration of the element is $1s^2 2s^2 2p^6 3s^2$-$3p^1$. Aluminum exhibits a valence of +3 in all compounds, with the exception of a few high-temperature monovalent and divalent gaseous species.

Aluminum is stable in air and resistant to corrosion by seawater and many aqueous solutions and other chemical agents. This is due to protection of the metal by a tough, impervious film of oxide. At a purity greater than 99.95%, aluminum resists attack by most acids but dissolves in aqua regia. Its oxide film dissolves in alkaline solutions, and corrosion is rapid. *See* CORROSION.

Aluminum is amphoteric and can react with mineral acids to form soluble salts and to evolve hydrogen.

Molten aluminum can react explosively with water. The molten metal should not be allowed to contact damp tools or containers.

At high temperatures aluminum reduces many compounds containing oxygen, particularly metal oxides. These reactions are used in the manufacture of certain metals and alloys.

Applications in building and construction represent the largest single market of the aluminum industry. Millions of homes use aluminum doors, siding, windows, screening, and down-spouts and gutters. Aluminum is also a major industrial building product. Transportation is the second largest market. Many commercial and military aircraft have become virtually all-aluminum. In automobiles, aluminum is apparent in interior and exterior trim, grilles, wheels, air conditioners, automatic transmissions, and some radiators, engine blocks, and body panels. Aluminum is also found in rapid-transit car bodies, rail cars, forged truck wheels, cargo containers, and in highway signs, divider rails, and lighting standards. In aerospace, aluminum is found in aircraft engines, frames, skins, landing gear, and interiors, often making up 80% of a plane's weight. The food packaging industry is a fast-growing market.

In electrical applications, aluminum wire and cable are major products. Aluminum appears in the home as cooking utensils, cooking foil, hardware, tools, portable appliances, air conditioners, freezers, and refrigerators, and in sporting equipment such as skis, ball bats, and tennis rackets.

There are hundreds of chemical uses of aluminum and aluminum compounds. Aluminum powder is used in paints, rocket fuels, and explosives, and as a chemical reductant. *See* ALUMINUM ALLOYS; ALUMINUM METALLURGY. [A.S.Ru.]

Aluminum alloys Substances formed by the addition of one or more elements, usually metals, to aluminum. The principal alloying elements in aluminum-base alloys are magnesium, silicon, copper, zinc, and manganese. In wrought products, which constitute the greatest use of aluminum, the alloys are identified by four-digit numbers of the form NXXX, where the value of N denotes the alloy type and the principal alloying element(s) as follows: 1 (Al; at least 99% aluminum by weight), 2 (Cu), 3 (Mn), 4 (Si), 5 (Mg), 6 (Mg + Si), 7 (Zn), 8 (other). *See* ALUMINUM.

Iron and silicon are commonly present as impurities in aluminum alloys, although the amounts may be controlled to achieve specific mechanical or physical properties. Minor amounts of other elements, such as Cr, Zr, V, Pb, and Bi, are added to specific alloys for special purposes. Titanium additions are frequently employed to produce a refined cast structure.

Aluminum-base alloys are generally prepared by making the alloying additions to molten aluminum, forming a liquid solution. As the alloy freezes, phase separation occurs to satisfy phase equilibria requirements and the decrease in solubility as the temperature is lowered. The resultant solidified structure consists of grains of aluminum-rich solid solution and crystals of intermetallic compounds. *See* EUTECTICS.

A decrease in solubility with falling temperature also provides the basis for heat treatment of solid aluminum alloys. In this operation, the alloy is held for some time at a high temperature to promote dissolution of soluble phases and homogenization of the alloy by diffusion processes. The limiting temperature is the melting point of the lowest melting phase present. The time required depends both on temperature and on the distances over which diffusion must occur to achieve the desired degree of homogenization. The solution heat treatment is followed by a quenching operation in which the article is rapidly cooled, for example, by plunging it into cold or hot water or by the use of an air blast. *See* HEAT TREATMENT (METALLURGY).

Casting alloys are significant users of secondary metal (recovered from scrap for reuse). Thus, casting alloys usually contain minor amounts of a variety of elements; these do no harm as long as they are kept within certain limits. The use of secondary metal is also of increasing importance in wrought alloy manufacturing as producers take steps to reduce the energy required in producing fabricated aluminum products. *See* ALLOY. [A.S.Ru.]

Aluminum metallurgy The extraction of aluminum from ore and its subsequent purification and processing. The most widely used technology for producing aluminum involves two steps: extraction and purification of alumina from ores, and electrolysis of the oxide after it has been dissolved in fused cryolite. Bauxite is by far the most used raw material.

In the alumina extraction process, bauxite is crushed and ground, then digested at elevated temperature and pressure in a strong solution of caustic soda. The gibbsite, boehmite, or diaspore in the bauxite reacts with the caustic soda to form soluble sodium aluminate. The residue, known as red mud, contains the insoluble impurities. The red mud is separated from the solution. After cooling, the solution is supersaturated with respect to alumina. It is seeded with recycled synthetic gibbsite (alumina trihydrate) and agitated. A large part of the alumina in solution thus crystallizes out as gibbsite. For metal production, the product is calcined at temperatures up to 2370°F (1300°C) to produce alumina containing about 0.3–0.8% soda, less than 0.1% iron oxide plus silica, and trace amounts of other oxides.

In a smelter, alumina is dissolved in cells (pots)—rectangular steel shells lined with carbon—containing a molten electrolyte (bath) consisting mostly of cryolite. The bath usually contains 2–8% alumina. Carbon anodes are hung from above the cells with their lower ends extending to within about 1.5 in. (3.8 cm) of the molten metal, which forms a layer under the molten bath. The heat required to keep the bath molten is supplied by the electrical resistance of the bath as current passes through it.

The passage of direct current through the electrolyte decomposes the dissolved alumina. Metal is deposited on the cathode, and oxygen on the gradually consumed anode. The smelting process is continuous. Alumina is added, anodes are replaced, and molten aluminum is periodically siphoned off without interrupting current to the cells.

Molten aluminum is siphoned from the smelting cells into large crucibles. From there the metal may be poured directly into molds to produce foundry ingot, or transferred to holding furnaces for further refining or alloying with other metals to form fabricating ingot. As it comes from the cell, primary aluminum averages about 99.8% purity. *See* ALUMINUM; ELECTROMETALLURGY.

In plants not adjacent to a smelter, it is necessary to remelt charges, usually consisting of mill scrap or returned scrap with enough primary metal to provide composition control. Scrap recycling has grown through the years to considerable economic importance and provides a valuable source of aluminum at a much lower energy expenditure than for primary metal.

Although many types of casting methods for ingot have been used historically, most ingot for fabricating is cast by the DC (direct chill) process or some modification thereof. The DC process consists of pouring the metal into a short mold. The base of the mold is a separate platform that is lowered gradually into a pit while the metal is solidifying. The frozen outer shell of metal retains the still-liquid portion of the metal when the shell is past the mold wall. Water is used to cool the mold, and by impingement on the ingot shell also cools the ingot as it is lowered. By sawing the ingot while casting is progressing, the process can be made truly continuous. *See* METAL CASTING.

Aluminum alloys are fabricated by all the methods known to metalworking and are commercially available in the form of castings; in wrought forms produced by rolling, extruding, forging, or drawing; in compacted and sintered shapes made from powder; and in other monolithic forms made by a wide variety of processes. In addition, pieces produced separately can be assembled by common joining procedures to build up complex shapes and structures. *See* METAL FORMING. [A.S.Ru.]

Alunite A mineral of composition $KAl_3(SO_4)_2(OH)_6$. Alunite occurs in white to gray rhombohedral crystals or in fine-grained, compact masses. Alunite is produced by sulfurous vapors on acid volcanic rocks and also by sulfated meteoric waters affecting aluminous rocks. Alunite is used as a source of potash or for making alum. Alum has been manufactured from the well-known alunite deposits at Tolfa, near Civita Vecchia, Italy, since the mid-15th century. In the United States alunite is widespread in the West. *See* ALUM. [E.C.T.C.]

Alzheimer's disease An adult-onset neurological disorder of unknown etiology manifest by loss of memory, impaired thought processes, and abnormal behavior. When the illness begins before the age of 65, the disease is termed Alzheimer's disease; when onset is after 65, the disorder is referred to as senile dementia of the Alzheimer's type.

Affected individuals are, at first, forgetful. As the memory disorder gradually worsens, these individuals, although able to recall occurrences in the distant past, are unable to remember recent events. Subsequently, speech, the ability to calculate, visuospatial orientation, judgment, and social behavior become progressively abnormal. Eventually, the individuals become profoundly demented, and frequently die of intercurrent infection. The disease is more common in women, and the average duration of disease varies from 3 to 8 years. The diagnosis of Alzheimer's disease is usually made on the basis of clinical history and neurological examination.

Several theories have been proposed that attempt to explain the causes of Alzheimer's disease. They cite a genetic defect, aluminum intoxication, infectious agents (such as a slow virus), and autoimmune mechanisms. The theory that Alzheimer's disease may result from a genetic defect has received the greatest attention.

The disease is associated with increasing age but is not an invariable concomitant of aging. Although many cases appear to be sporadic in nature, available evidence suggests that a significant number of affected individuals have inherited a predisposition to develop the disease: in certain families, the disease is inherited as an autosomal dominant; in addition, individuals with Down's syndrome have, if they live into middle life, a very high incidence of Alzheimer's disease. *See* HUMAN GENETICS.

There is no specific treatment for Alzheimer's disease. Because the disease is so common and because of the increasing numbers of individuals reaching the at-risk period of life, Alzheimer's disease has been called an approaching epidemic. [D.L.P.]

Amalgam An alloy of mercury. Practically all metals will form alloys or amalgams with mercury, with the notable exception of iron. Amalgams are used as dental materials, in the concentration of gold and silver from their ores, and as electrodes in various industrial and laboratory electrolytic processes. Amalgams used in dental work require the following composition: silver, 65% minimum; copper, 6% maximum; zinc, 2% maximum; and tin, 25% minimum. These amalgams are prepared by the dentist as needed, and harden within 3–5 min, but may be shaped by carving for 15 min or so. *See* ALLOY; GOLD; MERCURY (ELEMENT); SILVER. [E.E.W.]

Amaranth An annual plant (seldom perennial) of the genus *Amaranthus* (family Amaranthaceae), distributed worldwide in warm and humid regions. Amaranths are botanically distinguished by their small chaffy flowers, arranged in dense, green or red, monoecious or dioecious inflorescences, with zero to five perianth segments and two or three styles and stigmata, and by their dry membranous, indehiscent, one-seeded fruit. *See* FLOWER; FRUIT.

Physiological, genetic, and nutritional studies have revealed their potential economic value. Of particular interest are high rate of productivity as a rapidly growing summer crop, the large amounts of protein in both seed and leaf with high lysine, the overall high nutritional value, and the water use efficiency for the C_4 photosynthetic pathway. Amaranths are important in the culture, diet, and agricultural economy of the people of Mexico, Central and South America, Africa, and northern India. Genetic, ethnobotanical, and agronomic research has been undertaken to develop amaranths as an important food plant in modern agriculture. [S.K.J.]

Amateur radio Two-way radio communications by private individuals as a leisure-time activity. Amateur, or "ham," radio is defined in international treaty as a "service of self-training, intercommunications, and technical investigations carried on by amateurs; that is, by duly authorized persons interested in radio technique solely with a personal aim and without pecuniary interest." *See* RADIO.

There are about 750,000 amateur operators worldwide, and 275,000 in the United States. Licenses in the United States are issued by the Federal Communications Commission in five classes—novice, technician, general, advanced, and amateur extra. Knowledge of the international Morse code and of radio theory and regulations is required for a license of any class. With each succeeding license, greater privileges are granted.

The amateur radio operators' organization is the American Radio Relay League, a nonprofit, nonstock association of radio amateurs in North America. The League serves as the voice for amateurs in regulatory matters; it presents new technical developments and conducts operating contests and other activities to increase the hams' enjoyment of, and skill in, the hobby; it organizes networks to handle messages and coordinates emergency communications training. The League is also headquarters for the International Amateur Radio Union, composed of some 85 similar national radio societies. [P.F.W.]

Amber Most commonly, a generic name for all fossil resins, although it has been restricted by some to refer only to succinite, the mineralogical species of fossil resin making up most of the Baltic Coast deposits. Resins generally are complex mixtures of mono-, sesqui-, di-, and triterpenoids; however, some resins contain aromatic phenols. Among the plants, primarily trees, that produce copious amounts of resin that may fossilize to become amber, two-thirds are tropical.

Although ambers occur throughout the world in deposits from Carboniferous to Pleistocene in age, they have been reported most commonly from Cretaceous and Tertiary strata and often are associated with coal. Amber may contain beautifully preserved insects, spiders, flowers, leaves, and even small animals. The most extensively studied deposits are those from the Baltic Coast, Alaska, Canada, Burma, Dominican Republic, and Mexico.

When amber is used for jewelry, it usually is transparent yellow, reddish-brown, or "amber" color. Translucent or semitranslucent amber is used for pipe stems, decorating small boxes, and a variety of ornamental purposes. The specific gravity varies from 1.05 to 1.10, and hardness from 1 to 3 on Mohs scale.

At one time, chemical studies of amber were mineralogically oriented because the purpose was to describe and classify amber as a semiprecious gem. However, phytochemical studies comparing fossil and present-day resins are providing information regarding the botanical origins of ambers.

The predominantly tropical or subtropical occurrence of amber-producing plants through geologic time has led to evolutionary studies of the natural purpose of resins and their possible defensive role for trees against injury and disease inflicted by the high diversity of insects and fungi in tropical environments. *See* GEM; MINERALOGY; RESIN. [J.H.L.]

Ambergris A fatty substance formed in the intestinal tract of the sperm whale (*Physeter catodon*). Ambergris contains acids, alkaloids, and a fatty substance, its main constituent, called ambrein. Although fresh ambergris is soft and black and has an offensive odor, it hardens into pleasantly fragrant gray or yellow masses when exposed to the air, sun, and sea. Being lighter than water, it is found in lumps floating on tropical seas or cast up on the shores. It is also gathered directly from the abdomens of dead or captured whales. Collecting grounds for ambergris are principally on the shores of China, Japan, Africa, the Americas, tropical islands, and the Bahamas.

Ambergris is valued in the manufacture of perfumes. The ambergris is ground and used in the form of a tincture, dissolved in a dilute solution of alcohol, which when added to perfume acts as a fixative, increasing the duration of the fragrance while adding its own sweet, earthy scent. [S.P.P.]

Amblygonite A mineral consisting of lithium, sodium, and aluminum phosphate $(Li,Na)Al(PO_4)(F,OH)$; generally Li is in excess of Na. Amblygonite crystallizes in the triclinic system. Crystals are short and prismatic. Colors range from white through shades of gray, yellow, green, and brown.

Amblygonite, found at many places in the world, occurs mostly in granitic pegmatites. It is mined in the Black Hills of South Dakota, in southwestern Africa, and elsewhere for its lithium content. [W.R.Lo.]

Amblypygi An order of arachnids, the tailless whip scorpions, comprising about 80 species of flattened, crablike forms from the tropical and warm temperate regions of the world. The somber red or brownish species vary considerably in size, from 0.16–1.80 in. (4 to 45 mm), the largest being *Acanthophyrnus coronatus* of California and Mexico. The pedipalps are long raptorial organs set with many sharp spines that grasp and crush insect prey. The first pair of legs is modified into very long, lashlike whips which are used as sensory feelers. No tail is present on the abdomen. The amblypygids are harmless, nocturnal types that live under stones, in rock fissures and caves, and frequently in houses. No venom or repellent glands are present. *See* ARACHNIDA. [W.J.Ger.]

Ameba The name for a number of species of acellular or unicellular animals of the order Amoebida, which constitutes one of the 11 orders of the class Rhizopodea. Members of this order are naked protozoans; that is, they lack a shell known as a test (see illustration). Characteristically they move by means of protoplasmic waves or protoplasmic extrusions called pseudopods. Amebas are asymmetrical because they do not possess a constant or definite shape. These organisms occur in a variety of habitats. There are fresh-water, marine, and terrestrial species as well as a number which are commensals and parasites (see table).

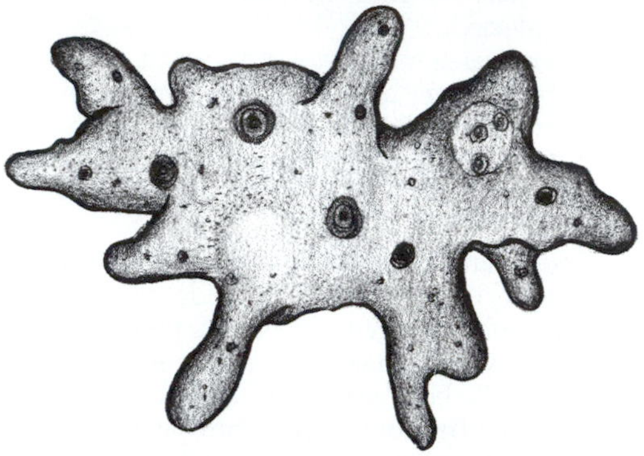

Typical species of ameba showing numerous pseudopods.

Significant features and habitats of some common species of ameba

Species and common name	Significant features	Habitat
Amoeba proteus (ameba)	Used to study ameboid movement and nucleus	Free living in fresh water
Pelomyxa palustris (giant ameba)	Multinucleate; no pseudopods formed	Free living in fresh water
Valkampfia limax	No pseudopods formed	Free living in fresh water
Endamoeba blattae	Cyst with up to 60 nuclei	Commensal in gut of cockroach
Entamoeba histolytica (dysentery ameba)	Quadrinucleate cyst; causes amebic dysentery	Parasite and pathogen in intestine of humans
Entamoeba coli	Octonucleate cyst: nonpathogenic	Commensal in intestine of humans
Iodamoeba bütschlii	Cyst with prominent glycogen vacuole; nonpathogenic	Commensal in intestine of humans
Endolimax nana	Oval, quadrinucleate cyst; nonpathogenic	Commensal in intestine of humans
Dientamoeba fragilis	Usually binucleate; no cyst stage; nonpathogenic	Commensal in intestine of humans
Entamoeba gingivalis	Detritus feeder; no cyst stage; pathogenicity disputed	Probably commensal in mouth of humans

Many species of ameba are uninucleate, whereas others are multinucleate. The type of nucleus is important in differentiating between commensal and parasitic forms. Binary fission, an asexual method of reproduction, is common among the amebas. Under adverse environmental conditions, some amebas encyst or form a heavy cell wall around themselves, and during this period active multiplication of the species occurs. Among the parasitic amebas, this is the infective stage for transmission from one host to another. *See* AMEBIASIS; AMOEBIDA; RHIZOPODEA. [C.B.C.]

Amebiasis A disease of humans caused by the pathogenic ameba *Entamoeba histolytica*. It is characterized by an acute phase (amebic dysentery) and by other clinical pathological manifestations in the intestinal tract and in other organs. *Entamoeba histolytica* has two stages in its life cycle: the resistant and quiescent cyst form and the actively motile, tissue-invading trophozoite. The cysts, discharged in the feces of the sick and the carriers, survive for long periods outside the body. When taken up in contaminated water or with vegetables, they reach the intestinal tract, and the active amebas emerge to divide and to invade the tissues. By their lytic enzymes and active movements, they penetrate the intestinal mucosa to cause typical ulcerations which may lead to perforation of the intestinal wall. The amebas may be carried by the bloodstream to other sites and organs, especially to the liver, to cause dangerous abcesses. On rare occasion they may reach the brain to initiate severe, even fatal complications.

Though amebiasis is cosmopolitan in its distribution, it is in the tropical and subtropical climates that it commonly manifests itself in its severe and acute form and in the liver complications. The diagnosis of an amebic infection is made by microscopic demonstration of the causative *E. histolytica* in the feces of the patient or from liver abcess. A number of drugs have been employed in the treatment of amebiasis. Iodides and bismuth compounds are used, as well as the broad-spectrum antibiotics. Combined treatment with both groups has given the best results. For liver complications, emetine injections and chloroquine by mouth are effective.

Epidemic outbreaks have shown that faulty sewage systems are responsible for contamination of drinking water. Great attention should therefore be given to all city water systems and their regular inspection and supervision. Food handlers in hotels and camps should be examined before employment. Vegetables should be properly cleaned before serving. Tourists going abroad to tropical countries should be advised of the danger of intestinal disorders and the methods of their prevention. *See* MEDICAL PARASITOLOGY. [M.Yo.]

Americium A chemical element, symbol Am, atomic number 95. The isotope ^{241}Am is an alpha emitter with a half-life of 433 years. Other isotopes of americium range in mass from 232 to 247, but only the isotopes of mass 241 and 243 are important. The isotope ^{241}Am is routinely separated from "old" plutonium and sold for a variety of industrial uses, such as 59-keV gamma sources and as a component in neutron sources. The longer-lived ^{243}Am (half-life 7400 years) is a precursor in ^{244}Cm production.

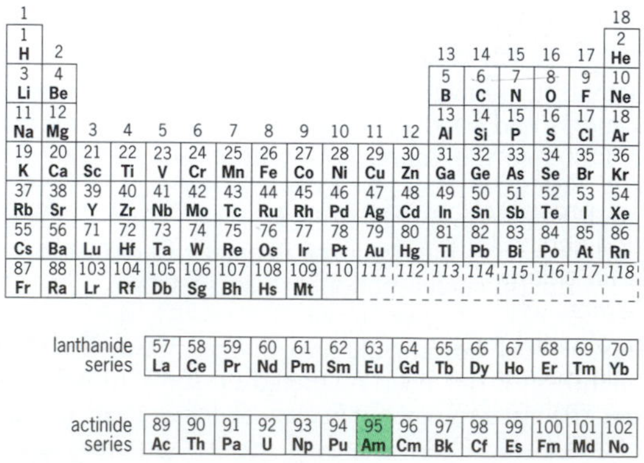
1 H 2 He
3 Li 4 Be 5 B 6 C 7 N 8 O 9 F 10 Ne
11 Na 12 Mg 13 Al 14 Si 15 P 16 S 17 Cl 18 Ar
19 K 20 Ca 21 Sc 22 Ti 23 V 24 Cr 25 Mn 26 Fe 27 Co 28 Ni 29 Cu 30 Zn 31 Ga 32 Ge 33 As 34 Se 35 Br 36 Kr
37 Rb 38 Sr 39 Y 40 Zr 41 Nb 42 Mo 43 Tc 44 Ru 45 Rh 46 Pd 47 Ag 48 Cd 49 In 50 Sn 51 Sb 52 Te 53 I 54 Xe
55 Cs 56 Ba 71 Lu 72 Hf 73 Ta 74 W 75 Re 76 Os 77 Ir 78 Pt 79 Au 80 Hg 81 Tl 82 Pb 83 Bi 84 Po 85 At 86 Rn
87 Fr 88 Ra 103 Lr 104 Rf 105 Db 106 Sg 107 Bh 108 Hs 109 Mt 110 111 112 113 114 115 116 117 118
lanthanide series 57 La 58 Ce 59 Pr 60 Nd 61 Pm 62 Sm 63 Eu 64 Gd 65 Tb 66 Dy 67 Ho 68 Er 69 Tm 70 Yb
actinide series 89 Ac 90 Th 91 Pa 92 U 93 Np 94 Pu 95 Am 96 Cm 97 Bk 98 Cf 99 Es 100 Fm 101 Md 102 No

In its most prominent aqueous oxidation state, 3+, americium closely resembles the tripositive rare earths. The formal analogy to the rare earths is also marked in anhydrous compounds of both tripositive and tetrapositive americium. Americium is different in that it is possible to oxidize Am^{3+} to both the 5+ and 6+ states.

Americium metal has a vapor pressure markedly higher than that of its neighboring elements and can be purified by distillation. The metal is nonmagnetic and superconducting at 0.79 K. Under high pressure the metal has been compressed to 80% of its room-temperature volume and displays the α-uranium structure. *See* ACTINIDE ELEMENTS; BERKELIUM; CURIUM; NUCLEAR REACTION; TRANSURANIUM ELEMENTS. [R.A.Pe.]

Amethyst The transparent purple to violet variety of the mineral quartz. Amethyst is rare in the deep colors that characterize fine quality. It is usually colored unevenly and is often heated slightly in an effort to distribute the color more evenly. Heating at higher temperatures usually changes it to yellow or brown (rarely green), and further heating removes all color. The principal sources are Brazil, Arizona, Uruguay, and Russia. *See* GEM; QUARTZ. [R.T.L.]

Amide A derivative of a carboxylic acid with general formula $RCONH_2$, where R is hydrogen or an alkyl or aryl radical. Amides are divided into subclasses, depending on the number of substituents on nitrogen. The simple, or primary, amides are considered to be derivatives formed by replacement of the carboxylic hydroxyl group by the amino group, NH_2. They are named by dropping the "-ic acid" or "-oic acid" from the name of the parent carboxylic acid and replacing it with the suffix "amide." In the secondary and tertiary amides, one or both hydrogens are replaced by other groups. The presence of such groups is designated by the prefix capital *N* (for nitrogen).

Except for formamide, all simple amides are relatively low-melting solids, stable, and weakly acidic. They are strongly associated through hydrogen bonding, and hence soluble in hydroxylic solvents, such as water and alcohol. Because of ease of formation and sharp melting points, amides are frequently used for the identification of organic acids and, conversely, for the identification of amines.

Commercial preparation of amides involves thermal dehydration of ammonium salts of carboxylic acids. Thus, slow pyrolysis of ammonium acetate forms water and acetamide. *N,N*-dimethylacetamide may be similarly prepared from dimethylammonium acetate.

Amides are important chemical intermediates since they can be hydrolyzed to acids, dehydrated to nitriles, and degraded to amines containing one less carbon atom by the Hofmann reaction. In pharmacology, acetophenetidin is a popular analgesic. However, the most important commercial application of amides is in the preparation of polyamide resins, also called nylons. *See* ACID ANHYDRIDE; POLYAMIDE RESINS. [P.E.F.]

Amiiformes An order of Actinopterygian fishes, also known as the Halecomorphi. The characters include abbreviate heterocercal tail, in some almost symmetrical; usually fusiform body; median fin rays arranged one per pterygiophore; scales with a ganoine surface but typically thin; no spiracle; vascularized swim bladder with a duct; reduced maxilla which is free from the preopercle posteriorly; an orbit which is bordered below and behind by a series of enlarged bones; and, in Recent species, an enlarged gular plate and elongate dorsal fin.

The order, which appeared first in the Triassic, includes six families, of which only the Amiidae survived into the Cenozoic. The single Recent species, *Amia calva*, inhabits sluggish fresh waters of eastern North America. *See* ACTINOPTERYGII; BOWFIN; OSTEICHTHYES. [R.M.B.]

Amination The preparation of amines. Several processes are employed for the preparation of amines. Amines can be prepared by reducing a compound that already has a carbon-nitrogen bond, such as a nitro, nitroso, azoxy, or azo compound. They can also be prepared by treating compounds containing (1) a labile group (—Cl, —OH, —SO_3H), (2) a carbonyl (>C=O) group, or (3) a highly reactive structure, such as oxirane, with ammonia. The preparation of amines from compounds containing a C-N bond is termed amination by reduction; the production of amines by reaction with ammonia or a mixture of ammonia and hydrogen is termed ammonolysis or hydroammonolysis. *See* AMINE. [P.H.G./R.S.K.]

Amine A member of a group of organic compounds which can be considered as derived from ammonia by replacement of one or more hydrogens by organic radicals. Generally amines are bases of widely varying strengths, but a few which are actually acidic are known.

Amines constitute one of the most important classes of organic compounds. The lone pair of electrons on the amine nitrogen enables amines to participate in a large variety of reactions as a base or a nucleophile. Amines play prominent roles in biochemical systems; they are widely distributed in nature in the form of amino acids, alkaloids, and vitamins. Many complex amines have pronounced physiological activity, for example, epinephrine (adrenalin), thiamin or vitamin B_1, and Novocaine. The odor of decaying fish is due to simple amines produced by bacterial action. Amines are used to manufacture many medicinal chemicals, such as sulfa drugs and anesthetics. The important synthetic fiber nylon is an amine derivative.

Amines are classified according to the number of hydrogens of ammonia which are replaced by radicals. Replacement of one hydrogen results in a primary amine (RNH_2), replacement of two hydrogens results in a secondary amine (R_2NH), and replacement of all three hydrogens results in a tertiary amine (R_3N). The substituent groups (R) may be alkyl, aryl, or aralkyl. Another group of amines are those in which the nitrogen forms part of a ring (heterocyclic amines). Examples of such compounds are nicotine, which is obtained commercially from tobacco for use as an insecticide, and serotonin, which plays a key role as a chemical mediator in the central nervous system.

Many aromatic and heterocyclic amines are known by trivial names, and derivatives are named as substitution products of the parent amine. Thus, $C_6H_5NH_2$, is aniline and C_6H_5-NHC_2H_5 is *N*-ethylaniline. *See* ANILINE.

According to the Brönsted-Lowry theory of acids and bases, amines are basic because they accept protons from acids. Stable salts suitable for the identification of amines are in general formed only with strong acids, such as hydrochloric, sulfuric, oxalic, chloroplatinic, or picric.

Commercial preparation of aliphatic amines can be accomplished by direct alkylation of ammonia or by catalytic alkylation of amines with alcohols at elevated temperatures. Reduction of various nitrogen functions carrying the nitrogen in a higher state of oxidation also leads to amines. Such functions are nitro, oximino, nitroso, and cyano. For the preparation of pure primary amines, Gabriel's synthesis and Hofmann's hypohalite reaction are preferred methods. The Bucherer reaction is satisfactory for the preparation of polynuclear primary aromatic amines. *See* AMINO ACIDS. [P.E.F.]

Amino acid dating Determination of the relative or absolute age of materials or objects by measurement of the degree of racemization of the amino acids present. With the exception of glycine, the amino acids found in proteins can exist in two isomeric forms called D- and L-enantiomers. Although the enantiomers of an amino acid rotate plane-polarized light in equal but opposite directions (the D form rotates it to the right and the L form to the left), their other chemical and physical properties are identical. It was discovered by L. Pasteur around 1850, that only L-amino acids are generally found in living organisms, but scientists still have not formulated a convincing reason to explain why life is based on only L-amino acids. *See* AMINO ACIDS, ENANTIOMER. Under conditions of chemical equilibrium, equal amounts of both enantiomers are present ($D/L = 1.0$); this is called a racemic mixture. Living organisms maintain a state of disequilibrium through a system of enzymes that selectively utilize only the L-enantiomers. Once a protein has been synthesized and isolated from active metabolic processes, the L-amino acids are subject to a racemization reaction that converts them into a racemic mixture. Since racemization is a chemical process, the extent of racemization is dependent not only on the time that has elapsed since the L-amino acids were synthesized but also on the exposure temperature: the higher the temperature, the faster the rate of racemization. The rate of racemization is also different for most of the various amino acids.

A variety of analytical procedures can be used to separate amino acid enantiomers; gas chromatography and high-performance liquid chromatography are the most widely used. Since these techniques have sensitivities in the parts per billion range, only a few hundred milligrams of sample material are normally required. Samples are first hydrolyzed in hydrochloric acid to break down the proteins into free amino acids, which are then isolated by cation-exchange chromatography. *See* CHROMATOGRAPHY.

Since the late 1960s, the geochemical and biological significance of amino acid racemization has been extensively investigated. Geochemical uses of amino acid racemization include the dating of fossils or, in the case of known age specimens, the determination of their temperature history. Fossil types such as bones, teeth, and shells have been studied, and racemization has been found to be particularly useful for dating specimens that were difficult to date by other methods. Racemization has also been observed in the metabolically inert tissues of living mammals. Racemization can be studied in certain organisms and used to assess the biological age of a variety of mammalian species; in addition, it may be important in determining the biological lifetime of certain proteins. Fossils have been found to contain both D- and L-amino acids, and the extent of racemization generally increases with geologic age. *See* RACEMIZATION.

The racemization method has been used to calculate the age at death of mummified Alaskan Eskimos, of individuals in a medieval cemetery in Czechoslovakia, and of corpses in forensic cases. The racemization ages have greater accuracy (about $\pm$ 5%) than those estimated by other means, and the method is especially useful in determining the ages of older individuals, whose age at death is often difficult to estimate from anatomical evidence. *See* ARCHEOLOGICAL CHEMISTRY. [J.L.Ba.]

Amino acids Organic compounds possessing one or more basic amino groups and one or more acidic carboxyl groups. Of the more than 80 amino acids which have been found in living organisms, about 20 serve as the building blocks for the proteins. The amino acids are characterized physically by the following: (1) the pK_1, or the dissociation constant of the various titratable groups; (2) the isoelectric point, or pH at which a dipolar ion does not migrate in an electric field; (3) the optical rotation, or the rotation imparted to a beam of plane-polarized light (frequently the D line of the sodium spectrum) passing through 1 decimeter of a solution of 100 g in 100 milliliters; (4) solubility.

All the amino acids of proteins, and most of the others which occur naturally, are α-amino acids, meaning that an amino group (—NH_2) and a carboxyl group (—COOH) are attached to the same carbon atom. This carbon (the α carbon, being adjacent to the carboxyl group) also carries a hydrogen atom; its fourth valence is satisfied by any of a wide variety of substituent groups. In the simplest amino acids, glycine, the substituent group is a hydrogen atom. In all other amino acids, it is an organic radical; for example, in alanine it is a methyl group (—CH_3), while in glutamic acid it is an aliphatic chain terminating in a second carboxyl group (—CH_2—CH—COOH). Chemically, the amino acids can be considered as falling roughly into nine categories based on the nature of R (see table).

Amino acids of proteins, grouped according to the nature of the substituent group (R)

Amino acids*	R
Glycine	Hydrogen
Alanine, valine, leucine, isoleucine	Unsubstituted aliphatic chain
Serine, threonine	Aliphatic chain bearing a hydroxyl group
Aspartic acid, glutamic acid	Aliphatic chain terminating in an acidic carboxyl group
Asparagine, glutamine	Aliphatic chain terminating in an amide group
Arginine, lysine	Aliphatic chain terminating in a basic amino group
Cysteine, cystine, methionine	Sulfur-containing aliphatic chain
Phenylalanine, tyrosine	Terminates in an aromatic ring
Tryptophan, proline, histidine	Terminates in a heterocyclic ring

*See articles on the individual amino acids listed in the table.

Amino acids occur in living tissues principally in the conjugated form. Most conjugated amino acids are peptides, in which the amino group of one amino acid is linked to the carboxyl group of another. This type of linkage is known as a peptide bond; a molecule of water is split out when a peptide bond is formed, and a molecule of water must be added when a peptide bond is broken.

Since each amino acid possesses both an amino group and a carboxyl group, the acids are capable of linking together to form chains of various lengths, called polypeptides. Proteins are polypeptides ranging in size from about 50 to many thousand amino acid residues.

Although most of the conjugated amino acids in nature are proteins, numerous smaller conjugates occur naturally, many with important biological activity. The line between large peptides and small proteins is difficult to draw, with insulin (molecular weight = 7000; 50 amino acids) usually being considered a small protein and adrenocorticotropic hormone (molecular weight = 5000; 39 amino acids) being considered a large peptide. In addition to their role as hormones, peptides often occur in coenzymes, bacterial capsules, fungal toxins, and antibiotics. *See* ANTIBIOTIC; COENZYME.

Free amino acids are found in living cells, as well as in the body fluids of higher animals, in amounts which vary according to the tissue and to the amino acid. The amino acids which play key roles in the incorporation and transfer of ammonia, such as glutamic acid, aspartic acid, and their amides, are often present in relatively high amounts, but the concentrations of the other amino acids of proteins are extremely low, ranging from a fraction of a milligram to several milligrams per 100 g wet weight of tissue. In view of the fact that amino acid and protein synthesis goes on constantly in most of these tissues, the presence of free amino acids in only trace amounts points to the existence of extraordinarily efficient regulation mechanisms.

In addition to the amino acids of protein, a variety of other free amino acids occurs naturally. Some of these are metabolic products of the amino acids of proteins; for example, γ-aminobutyric acid occurs as the decarboxylation product of glutamic acid. Others, such as homoserine and ornithine, are biosynthetic precursors of the amino acids of protein.

At ordinary temperatures, the amino acids are white crystalline solids; when heated to high temperatures, they decompose rather than melt. They are stable in aqueous solution, and with few exceptions can be heated as high as 120°C for short periods without decomposition, even in acid or alkaline solution. Thus, the hydrolysis of proteins can be carried out under such conditions with the complete recovery of most of the constituent free amino acids.

Since all of the amino acids except glycine possess a center of asymmetry at the α carbon atom, they can exist in either of two optically active, mirror-image forms or enantiomorphs. All of the common amino acids of proteins appear to have the same configuration about the α carbon: this configuration is symbolized by the prefix L-. The opposite, generally unnatural, form is given the prefix D-. Some amino acids, such as isoleucine, threonine, and hydroxyproline, have a second center of asymmetry and can exist in four stereoisomeric forms. The prefix allo- is used to indicate one of the two alternative configurations at the second asymmetric center; thus, isoleucine, for example, can exist in the L, L-allo, D, and D-allo forms. *See* STEREOCHEMISTRY.

Another important general property of all amino acids is their ionic state. The basic amino group can bind a proton from solution and become a cation; the acidic carboxyl group can release a proton into solution and become an anion. At the isoelectric point (the pH at which the molecule has no net charge), amino acids exist as dipolar ions or zwitterions, while in strong acid solution, the carboxyl group exists in the undissociated form, and the molecule becomes a cation.

Since most amino acids occur in conjugated form, their isolation usually requires their prior release in free form by acid or alkaline hydrolysis. Hydrolysates of proteins or other polypeptides, or crude extracts of plant, animal, or microbial materials, serve as the starting point for the isolation in pure form of single amino acids. Chromatography has been the method of choice for amino acid isolation ever since its first application in 1941. *See* CHROMATOGRAPHY.

Since amino acids, as precursors of proteins, are essential to all organisms, all cells must be able to synthesize those they cannot obtain from their environment. The biosynthesis of amino acids is relatively complicated and usually requires at least three enzymatic steps. The actual metabolic pathways by which amino acids are synthesized are generally found to be the same in all living cells investigated, whether microbial or animal. Biosynthetic mechanisms thus appear to have developed soon after the origin of life and to have remained unchanged throughout the divergent evolution of modern organisms. There is only one quantitatively important reaction by which organic nitrogen enters the amino groups of amino acids: the reductive amination of α-ketoglutaric acid to glutamic acid by the enzyme glutamic acid dehydrogenase. All other amino acids are formed either by transamination (transfer of an amino group, ultimately from glutamic acid) or by a modification of an existing amino acid.

Most organisms are capable of degrading some amino acids, and metabolic pathways leading to degradation to CO_2 and H_2O are known for each of the common amino acids. These pathways are detailed in the articles on the individual amino acids.

The nutritional requirement for the amino acids of protein can vary from zero, in the case of an organism which synthesizes them all, to the complete list, in the case of an organism in which all the biosynthetic pathways are blocked. There are 8 or 10 amino acids required by certain mammals; most plants synthesize all of their amino acids, while microorganisms vary from types which synthesize all, to others (such as certain lactic acid bacteria) which require as many as 18 different amino acids. *See* NUTRITION; PROTEIN. [E.A.Ad.; P.T.M.]

Amino sugar A sugar in which a nonglycosidic hydroxyl group is replaced by an amino or substituted amino group (see illustration). The most abundantly occurring amino sugar in

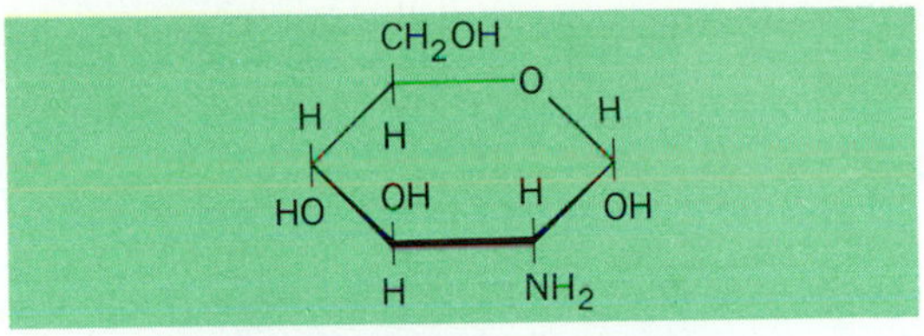

Structural formula of D-glucosamine.

nature is D-glucosamine (2-amino-2-deoxy-D-glucose). This amino sugar is a constituent of glycoproteins, which include the mucins from saliva and from egg albumin. It most frequently occurs as chitin. D-Glucosamine also occurs together with neutral sugars in many bacterial polysaccharides. Other types of amino sugars have been recognized in antibiotics. Uridine diphosphate *N*-acetyl-D-glucosamine occurs in baker's yeast and in a higher plant (mung bean). *See* MONOSACCHARIDE. [W.Z.H.]

para-Aminobenzoic acid A compound also known as PABA, often considered to be a water-soluble vitamin. *p*-Aminobenzoic acid, with the structure below, is widely distributed in

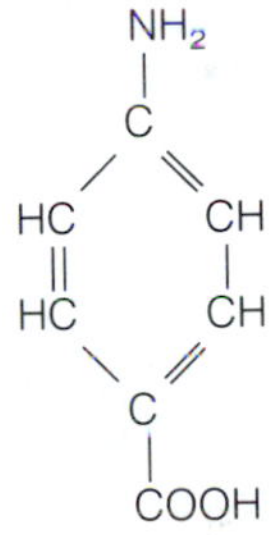

foods and has been isolated from liver, yeast, and other sources rich in vitamin B. *p*-Aminobenzoic acid is a part of the folic acid molecule, and its presence in a folic acid-deficient diet results in increased intestinal synthesis of the folic acid. *p*-Aminobenzoic acid antagonizes the bacteriostatic action of sulfonamides. It is an effective antirickettsial agent and has been used to treat typhus, scrub typhus, and Rocky Mountain spotted fever. *See* FOLIC ACID. [S.N.G.]

para-Aminosalicylic acid A chemical compound (PAS) used in the treatment of tuberculosis. The drug, because of its ability to delay the emergence of resistant tubercle bacilli when used in combination with either isonicotinic acid hydrazide or streptomycin, is one of the mainstays in the combined therapy of tuberculosis. The emergence of strains of organisms resistant to PAS, occasional allergic reactions, and especially gastric irritation from the large quantities necessary in the therapy of tuberculosis, limit its range of usefulness. *See* CHEMOTHERAPY. [N.J.G./E.Gru.]

Ammeter An electrical instrument for the measurement of electric current. In the usual indicating ammeter, an electromechanical system causes a pointer to traverse a calibrated scale, the pointer position indicating the value of the current. Several kinds of ammeter mechanism are used for different kinds of current, different applications, and differing degrees of accuracy. *See* ELECTRODYNAMIC AMMETER; PERMANENT-MAGNET MOVABLE-COIL AMMETER; POLARIZED-VANE AMMETER; SOFT-IRON AMMETER; THERMAL AMMETER. [J.H.M./J.Be.]

Ammine One of a group of complex compounds formed by the coordination of ammonia molecules with metal ions and, in a few instances, such as calcium, strontium, and barium, with metal atoms. Although ammines are formally analogous to many salt hydrates, the general characteristics of the group of ammines differ considerably from those of the hydrates. For example, hydrated Co(III) salts are strong oxidizing agents whereas Co(II) ammines are strong reducing agents. The ammines of principal interest are those of the transition metals and of the zinc family, but even here there is wide variation in stability or rate of decomposition. Ammines are prepared by treating aqueous solutions of the metal salt with ammonia or, in some instances, by the action of dry gaseous or liquid ammonia on the anhydrous salt. *See* AMMONIA; COORDINATION CHEMISTRY. [H.H.S.]

Ammonia The most familiar compound composed of the elements nitrogen and hydrogen, NH_3. It is formed as a result of the decomposition of most nitrogenous organic material, and its presence is indicated by its pungent and irritating odor.

Ammonia has a wide range of industrial and agricultural applications. Examples of its use are the production of nitric acid and ammonium salts, particularly the sulfate, nitrate, carbonate, and chloride, and the synthesis of hundreds of organic compounds including many drugs, plastics, and dyes. Its dilute aqueous solution finds use as a household cleansing agent. Anhydrous ammonia and ammonium salts are used as fertilizers, and anhydrous ammonia also serves as a refrigerant, because of its high heat of vaporization and relative ease of liquefaction. *See* FERTILIZER.

The physical properties of ammonia are analogous to those of water and hydrogen fluoride in that the physical constants are abnormal with respect to those of the binary hydrogen compounds of the other members of the respective periodic families. These abnormalities may be related to the association of molecules through intermolecular hydrogen bonding. Ammonia is highly mobile in the liquid state and has a high thermal coefficient of expansion.

Most of the chemical reactions of ammonia may be classified under three chief groups: (1) addition reactions, commonly called ammonation; (2) substitution reactions, commonly called ammonolysis; and (3) oxidation-reduction reactions.

Ammonation reactions include those in which ammonia molecules add to other molecules or ions. Most familiar of the ammonation reactions is the reaction with water to form ammonium hydroxide. The strong tendency of water and ammonia to combine is evidenced by the very high solubility of ammonia in water. Ammonia reacts readily with strong acids to form ammonium salts. Ammonium salts of weak acids in the solid state dissociate readily into ammonia and the free acid. Ammonation occurs with a variety of molecules capable of acting as electron acceptors (Lewis acids), such as sulfur trioxide, sulfur dioxide, silicon tetrafluoride, and boron trifluoride. Included among ammonation reactions is the formation of complexes (called ammines) with many metal ions, particularly transition metal ions. Ammonolytic reactions include reactions of ammonia in which an amide group ($—NH_2$), an imide group ($=NH$), or a nitride group ($\equiv N$) replaces one or more atoms or groups in the reacting molecule.

Oxidation-reduction reactions may be subdivided into those which involve a change in the oxidation state of the nitrogen atom and those in which elemental hydrogen is liberated. An example of the first group is the catalytic oxidation of ammonia in air to form nitric oxide. In the absence of a catalyst, ammonia burns in oxygen to yield nitrogen. Another example is the reduction with ammonia of hot metal oxides such as cupric oxide.

The physical and chemical properties of liquid ammonia make it appropriate for use as a solvent in certain types of chemical reactions. The solvent properties of liquid ammonia are, in many ways, qualitatively intermediate between those of water and of ethyl alcohol. This is particularly true with respect to dielectric constant; therefore, ammonia is generally superior to ethyl alcohol as a solvent for ionic substances but is inferior to water in this respect. On the other hand, ammonia is generally a better solvent for covalent substances than is water.

The Haber-Bosch synthesis is the major source of industrial ammonia. In a typical process, water gas (CO, H_2, CO_2) mixed with nitrogen is passed through a scrubber cooler to remove dust and undecomposed material. The CO_2 and CO are removed by a CO_2 purifier and ammoniacal cuprous solution, respectively. The remaining H_2 and N_2 gases are passed over a catalyst at high pressures (up to 1000 atm or 100 megapascals) and high temperatures (approx. 1300°F or 700°C). Other industrial sources of ammonia include its formation as a by-product of the destructive distillation of coal, and its synthesis through the cyanamide process. In the laboratory, ammonia is usually formed by its displacement from ammonium salts (either dry or in solution) by strong bases. Another source is the hydrolysis of metal nitrides. *See* AMIDE; NITROGEN. [H.H.S.]

Ammonia maser clock A type of atomic clock which employs maser oscillation of ammonia molecules in a microwave cavity. One pair of energy levels in the ground state of the ammonia molecule is separated by 23,800 megahertz (the inversion line). It is possible to focus ammonia molecules in the upper state into a cavity resonant at the inversion frequency by means of inhomogeneous electric fields. If the cavity losses are low enough and a large enough number of molecules per unit time enter, and if the cavity is long enough so that the interaction time is sufficient, maser oscillator action will occur. Unfortunately the interaction time is not very accurate. The stability for very short times is good, however, since it depends on the power level and frequency of the maser, both of which are relatively high. *See* ATOMIC CLOCK; MASER. [L.S.C.]

Ammonium salt A product of a reaction between ammonia, NH_3, and various acids. The general reaction for formation is $NH_3 + HX \rightarrow NH_4X$. Examples of ammonium salts are ammonium chloride, NH_4Cl, ammonium nitrate, NH_4NO_3, ammonium sulfate, $(NH_4)_2SO_4$, and ammonium carbonate, $(NH_4)_2CO_3$. These compounds are addition products of ammonia and the acid. For this reason, their formulas are sometimes written as $[H(NH_3)]X$.

All ammonium salts decompose into ammonia and the acid when heated. Their stability, however, varies according to the nature of the acid. Salts of weak acids decompose at lower temperatures than do salts of strong acids.

Ammonium chloride is made by absorbing ammonia in hydrochloric acid. This salt, sometimes called sal ammoniac, is used in galvanizing iron, in textile dyeing, and in manufacturing dry cell batteries.

Ammonium nitrate is prepared from ammonia and nitric acid. It is used as a source of nitrous oxide, N_2O, or laughing gas, and in the manufacture of explosives. A mixture of ammonium nitrate and trinitrotoluene is known as amatol.

Ammonium sulfate, obtained from ammonia and sulfuric acid, is prepared commercially by passing ammonia and carbon dioxide, CO_2, into a suspension of finely ground calcium sulfate, $CaSO_4$. Large quantities are also produced as a by-product of coke ovens and coal-gas works. The chief use of ammonium sulfate is as a fertilizer.

Ammonium carbonate may be prepared by bringing ammonia and carbon dioxide together in aqueous solution. It is also obtained by heating a mixture of ammonium sulfate and a fine suspension of calcium carbonate. *See* AMMONIA; FERTILIZER; HYDROLYSIS. [F.J.J.]

Ammonoidea An order of the tetrabranchiate cephalopods possibly derived from the Nautiloidea. They first appeared in the Ordovician, reached their greatest period from the Devonian to the Jurassic, and disappeared at the end of the Cretaceous. The soft parts are unknown; the shells are coiled, have wrinkled septa, more or less highly sculptured exteriors, and complex sutures. It is the largest order of the Cephalopoda. Their morphology and structure are so varied that they are important index fossils. *See* CEPHALOPODA; INDEX FOSSIL; NAUTILOIDEA. [G.L.V.]

Amnesia The pathological loss or impairment of memory. Amnesia may be psychogenic with no detectable organic basis, or it may reflect a gross disturbance of cerebral function.

Psychogenic amnesia varies in degree from the mild lapses of memory to fugue states, in which the subject has no memory for extended periods. Current psychiatric opinion holds that these disorders are the result of emotional conflicts which the amnesia temporarily resolves. Thus the psychiatrist views amnesia as a defense mechanism.

Amnesias associated with cerebral dysfunction may arise from almost any form of injury or disease involving the brain. In cerebral dysfunction of sudden onset, as in acute trauma to the head or in epileptic attacks, there may be amnesia for immediately preceding events (retrograde amnesia), as well as for events subsequent to the apparent restoration of full consciousness (anterograde amnesia). With recovery, the period of complete memory loss may diminish in both directions, sometimes leaving "islands" of memory surrounded by periods of residual memory loss. Amnesia associated with chronic cerebral dysfunction is characterized by continuous difficulty in remembering recent events, whereas early memories are relatively intact. *See* MEMORY. [J.Se.]

Amnion A thin, cellular, extraembryonic membrane forming a closed sac which surrounds the embryo in all reptiles, birds, and mammals and is present only in these forms; hence the collective term amniotes is applied to these animals. The amnion contains a serous fluid in which the embryo is immersed. *See* AMNIOTA; FETAL MEMBRANE.

Typically, the amnion wall is a tough, transparent, nerve-free, and nonvascular membrane consisting of two layers of cells: an inner, single-cell-thick layer of ectodermal epithelium, continuous with that covering the body of the embryo as the outer layer of its skin, and an outer covering of mesodermal, connective, and specialized smooth muscular tissue, also continuous with the mesodermal germ layer of the embryo.

The major function of the amnion and its fluid is to protect the delicate embryo. Thus, developmental stages of terrestrial animals are provided with the same type of cushioning against mechanical shock as is provided by the water environment of aquatic forms. *See* GERM LAYERS. [N.T.S.]

Amniota A collective term for the classes Reptilia, Aves, and Mammalia of the subphylum Vertebrata. Members are characterized by having a series of specialized protective extraembryonic membranes during development. Three of the membranes—amnion, chorion or serosa, and allantois—occur only in this group, but a fourth, the yolk sac, is sometimes present and is found in many anamniotes. The presence of the extraembryonic membranes makes it possible for the embryonic development of the amniotes to take place out of the water. In the most primitive forms the early stages of development take place inside a shell-covered egg that is deposited on land. *See* ALLANTOIS; AMNION; ANAMNIA; CHORION; YOLK SAC. [J.M.S.]

Amoebida An order of Lobosia without protective coverings (tests). Pellicles may be thin, as in *Amoeba proteus*, or thicker and less flexible, as in *Thecamoeba verrucosa*. Pellicular folds may develop during locomotion, particularly in species with thick pellicles. Both flagellate and ameboid stages occur in certain soil amebas.

Locomotion of Amoebida varies somewhat from genus to genus. In some cases movement involves protoplasmic flow of the body as a whole, without typical pseudopodia. There may be several ridged indeterminate pseudopodia, into one of which the organism appears to flow in locomotion. In some species determinate pseudopodia never become large enough to direct locomotion. In some cases the form of the pseudopodia may vary in a single species.

Amebas, normally phagotrophic, usually contain food vacuoles. Certain species contain crystals of apparently differing chemical nature. Other inclusions are globules of different sizes, mitochondria, and stored food reserves. In addition, bacteria or algae may occur in the cytoplasm, changing the color to a gray or green. Nuclei range in number from one to several hundred, as in *Chaos carolinensis*. the giant amebas are visible without a microscope.

Those species found in the digestive tract of invertebrates and vertebrates include relatively harmless species and a few pathogens, such as *Entamoeba histolytica* of humans and *E. invadens* of reptiles. *Entamoeba coli* does not invade human tissues. Also limited to the lumen of the colon are *Endolimax nana*, *Iodamoeba bütschlii*, and *Dientamoeba fragilis*. These four are relatively harmless although sometimes associated with digestive disturbances. Uncooked vegetables from soil fertilized with human feces are a potential source of infection. Standard methods of water purification seem reasonably protective, but it is difficult to control spread by food handlers. *See* AMEBIASIS; LOBOSIA; PROTOZOA; SARCODINA; SARCOMASTIGOPHORA. [R.P.H.]

Amorphous solid A rigid material whose structure lacks crystalline periodicity; that is, the pattern of its constituent atoms or molecules does not repeat periodically in three dimensions. In the present terminology amorphous and noncrystalline are synonymous. A solid is distinguished from its other amorphous counterparts (liquids and gases) by its viscosity: a material is considered solid (rigid) if its shear viscosity exceeds $10^{14.6}$ poise ($10^{13.6}$ Pa · s). *See* CRYSTAL; SOLID-STATE PHYSICS; VISCOSITY.

Oxide glasses, generally the silicates, are the most familiar amorphous solids. However, as a state of matter, amorphous solids are much more widespread than just the oxide glasses. There are both organic (for example, polyethylene and some hard candies) and inorganic (for example, the silicates) amorphous solids. Glasses can be prepared which span a broad range of physical properties. Dielectrics (for example, SiO_2) have very low electrical conductivity and are optically transparent, hard, and brittle. Semiconductors (for example, As_2SeTe_2) have intermediate electrical conductivities and are optically opaque and brittle. Metallic glasses have high electrical and thermal conductivities, have metallic luster, and are ductile and strong. *See* METALLIC GLASSES.

The obvious uses for amorphous solids are as window glass, container glass, and the glassy polymers (plastics). Less widely recognized but nevertheless established technological uses include the dielectrics and protective coatings used in integrated circuits, and the active element in photocopying by xerography, which depends for its action upon photoconduction in an amorphous semiconductor. In optical communications a highly transparent dielectric glass in the form of a fiber is used as the transmission medium. *See* OPTICAL COMMUNICATIONS.

It is the changes in short-range order (on the scale of a localized electron), rather than the loss of long-range order alone, that have a profound effect on the properties of amorphous semiconductors. For example, the difference in resistivity between the crystalline and amorphous states for dielectrics and metals is always less than an order of magnitude and is generally less than a factor of 3. For semiconductors, however, resistivity changes of 10 orders of magnitude between the crystalline and amorphous states are not uncommon, and accompanying changes in optical properties can also be large.

One class of amorphous semiconductors is the glassy chalcogenides, which contain one (or more) of the chalcogens sulfur, selenium, or tellurium as major constituents. These materials have application in switching and memory devices. Another group is the tetrahedrally bonded amorphous solids, such as amorphous silicon and germanium. These materials cannot be formed by quenching from the melt (that is, as glasses) but must be prepared by one of the deposition techniques mentioned above.

When amorphous silicon (or germanium) is prepared by evaporation, not all bonding requirements are satisfied, so a large number of dangling bonds are introduced into the material. These dangling bonds create states deep in the gap which limit the transport properties. The number of dangling bonds can be reduced by a thermal anneal below the crystallization temperature, but the number cannot be reduced sufficiently to permit doping. *See* GLASS SWITCH; SEMICONDUCTOR. [B.G.B.]

Ampère's law A law of electromagnetism which expresses the contribution of a current element of length dl to the magnetic induction (flux density) B at a point near the current. Ampère's law, sometimes called Laplace's law, was derived by A. M. Ampère after a series of experiments during 1820–1825.

Whenever an electric charge is in motion, there is a magnetic field associated with that motion. The flow of charges through a conductor sets up a magnetic field in the surrounding region. Any current may be considered to be broken up into infinitesimal elements of length dl, and each such element contributes to the magnetic induction at every point in the neighborhood. The contribution dB of the element is found to depend upon the current I, the length dl of the element, the distance r of the point P from the current element, and the angle θ between the current element and the line joining the element of the point P (see illustration). Ampère's law expresses the manner of the dependence by Eq. (1). The field near

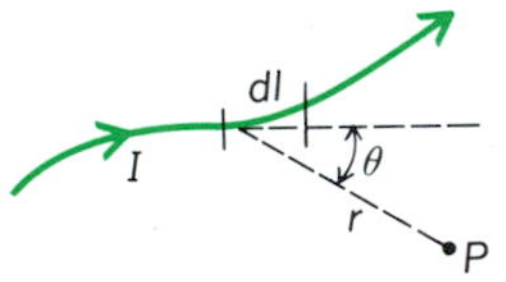

Graphic representation of Ampère's law.

$$dB = k\,\frac{I\,dl \sin\theta}{r^2} \qquad (1)$$

a current may be calculated by finding the vector sum of the contributions of all the various elements that make up the current.

The proportionality factor k depends upon the units used in Eq. (1) and upon the properties of the medium surrounding the current. In the SI system, the factor k is assigned a value of 10^{-7} weber/ampere-meters when the current is in empty space. As in other equations associated with electric and magnetic fields, for example Coulomb's law, it is convenient to replace k by a new factor μ_0 related to k as in Eq. (2). This sub-

$$\mu_0 = 4\pi k \qquad (2)$$

stitution removes the factor 4π from many derived equations in which it would otherwise appear. With this substitution Ampère's law becomes Eq. (3). The factor μ_0 is called the

$$dB = \frac{\mu_0}{4\pi}\,\frac{I\,dl \sin\theta}{r^2} \qquad (3)$$

permeability of empty space. *See* ELECTRICAL UNITS AND STANDARDS; MAGNETIC PERMEABILITY.

The direction of dB at each point may be described in terms of a right-hand rule. If the current element is grasped by the right hand with the thumb pointing in the direction of the current, the fingers encircle the current in the direction of the magnetic induction. [K.V.M.]

Amphetamine A stimulating drug that affects the brain and the body in a variety of ways; also known by the trade name Benzedrine. Chemically, amphetamine is a racemic mixture of the *l* and *d* isomers of α-methyl-ß-phenethylamine. The *l* isomer has more pronounced effects on the body, while the *d* isomer (commonly referred to as Dexedrine) has a greater effect on the brain. On the whole, the pharmacological effects of an amphetamine are to produce an increase in blood pressure, a relaxation of bronchial smooth muscle, a constriction of the blood vessels supplying the skin and mucous membranes, and a variety of alterations in behavior.

The ability of amphetamine to contract blood vessels in the mucous membranes and to relax smooth muscles in the lung make it an efficacious nasal decongestant. In a like manner, the pronounced excitatory effects of amphetamine on the brain have led to several other medical uses. For example, its ability to cause hyperventilation has resulted in its use as an analeptic. The arousing and insomnia-producing effects have been exploited to treat narcolepsy, a disease characterized by an inability to stay awake. Furthermore, the ability of amphetamine to cause a loss of appetite has promoted its use in diet programs as an anorexic; tolerance to this effect limits its usefulness. Finally, the enhanced sense of well-being and mild euphoria that are seen after taking amphetamine have led to its use in certain forms of psychiatric depression. One important use is in the treatment of hyperkinetic children, although the paradox of how a stimulant can do this is not understood.

The typical side effects on the body include dry mouth, heart rhythm alterations (palpitations and arrhythmias), hypertension, stomach cramps, and decreased urinary frequency. The central nervous system side effects include dizziness, dysphoria, headache, tremor, restlessness, insomnia, decreased appetite, increased aggressiveness, anxiety, and paranoid panic states. Extreme overdosage can result in convulsions, cerebral hemorrhaging, coma, and death.

Serious abuse occurs in some individuals who take the drug over long periods. As with other drugs of abuse (such as morphine, alcohol, or barbiturates), tolerance will occur after repeated dosing. Thus, whereas in the past one pill might be sufficient to induce the appropriate high, it becomes necessary to increase that amount in the future to achieve the same feeling. Over time, the users may become "psychologically dependent" on amphetamine. Chronic use of large amounts of amphetamine can have severe effects on personality. It has been shown that large doses of amphetamine can induce a behavioral state in humans that is nearly indistinguishable from paranoid schizophrenia, but can be reversed upon cessation of the drug. The question of whether or not physical dependence occurs with amphetamine is unresolved. It is known that after long-term treatment with amphetamine, removal of the drug results in feelings of anxiety, stomach cramps, headache, and extreme fatigue. However, the mild severity of these symptoms makes it unclear if this is a true withdrawal response. *See* ALCOHOLISM; DRUG ADDICTION; NARCOTIC. [R.E.C.]

Amphibia One of the four classes composing the superclass Tetrapoda of the subphylum Vertebrata. The living amphibians number approximately 2460 species, and are classified in three orders; the Anura or Salientia (frogs and toads, slightly less than 2000 species); Urodela or Caudata (salamanders, 300 species); and Apoda or Gymnophiona (caecilians, about 160 species). The orders in the subclasses Labyrinthodontia and Lepospondyli were in the geologic past and are now extinct. A classification scheme for the Amphibia follows.

- Class Amphibia
  - Subclass Labyrinthodontia
    - Order Ichthyostegalia
    - Order Temnospondyli
    - Order Anthracosauria
  - Subclass Lepospondyli
    - Order Nectridea
    - Order Aistopoda
    - Order Microsauria
  - Subclass Lissamphibia
    - Order Anura
    - Order Urodela
    - Order Apoda

A typical amphibian is characterized by a moist, glandular skin, the possession of gills at some point in its life history, four limbs, and an egg lacking the embryonic membrane called the amnion. *See* AMNION; ANAMNIA.

Amphibians are among the so-called cold-blooded animals; that is, the temperature of the body of an amphibian is not regulated internally to a high level as is that of mammals and birds, but fluctuates with that of the environment. An animal such as an amphibian that burns none of its food energy in keeping warm is able to get along on much less food than a bird or mammal of similar size. This advantage is offset by the inability of amphibians to be active under cold conditions that do not inhibit a warm-blooded animal. Thus the far northern and southern parts of the world which support large populations of birds and mammals are almost devoid of amphibian life. *See* THERMOREGULATION.

The amphibians mark a significant point in the evolution of the vertebrates, the transition from aquatic to terrestrial life. As animals neither divorced from the water nor fully at home on land, they suffer from their intermediate mode of life. Reptiles, and later mammals, came to dominate the land, and fishes the.waters, leaving the amphibians of today as a relatively unimportant but nevertheless highly interesting group of vertebrates. Amphibian history can be divided into two nearly completely separable portions: an older chapter, from Devonian to Triassic, in which the fossils belong to entirely extinct orders, and a later chapter, treating the development of the modern anurans and urodeles. *See* TETRAPODA. [R.G.Z.]

Amphibole A large group of common rock-forming inosilicate (metasilicate) minerals. The amphiboles exhibit a wide range of compositional variation, as indicated by the generalized formulas

$$(Na,Ca)_{2-3}(Mg,Fe^{2+},Fe^{3+},Al)_5(Si,Al)_8O_{22}(OH,O,F)_2$$

$$(Mg,Fe^{2+},Fe^{3+},Al)_7(Si,Al)_8O_{22}(OH,O,F)_2$$

The amphiboles represent a complex series of solid solutions between a variety of idealized end members. The species names of these end members are usually the best known; the more important ones are listed below. *See* SILICATE MINERALS.

Orthorhombic amphiboles	
Anthophyllite	$(Mg,Fe^{2+})_7Si_8O_{22}(OH)_2$
Gedrite	$(Mg,Fe^{2+})_5Al_2(Si_6Al_2)O_{22}(OH)_2$
Monoclinic amphiboles	
Cummingtonite	$(Fe^{2+},Mg)_7Si_8O_{22}(OH)_2$
Tremolite	$Ca_2Mg_5Si_8O_{22}(OH)_2$
Tschermakite	$Ca_2Mg_3Al_2(Si_6Al_2)(OH)_2$
Edenite	$NaCa_2Mg_5(Si_7Al)O_{22}(OH)_2$
Glaucophane	$Na_2(Mg_3Al_2)Si_8O_{22}(OH)_2$
Riebeckite	$Na_2(Fe_3{}^{2+},Fe_2{}^{3+})Si_8O_{22}(OH)_2$
Arfvedspmote	$Na_3Fe_4{}^{2+}Fe^{3+}Si_8O_{22}(OH)_2$
Eckermannite	$Na_3Mg_4(Fe^{3+},Al)Si_8O_{22}(OH)_2$

Each of the listed calcium (Ca) amphiboles has a ferrous iron equivalent and would be indicated by the prefix ferro-, for example, ferrotremolite. Also, some ferric iron and oxygen in the place of the OH groups can occur in the ferroamphiboles and is indicated by the prefix ferri-, for example, ferritremolite; the term oxyhornblende is commonly applied to the ferri-amphiboles. A high content of fluorine is indicated by the prefix fluoro-, for example, fluorocummingtonite. For the Na-Ca amphiboles, solid solution between any of the end members is possible so that a single crystal could be composed of contributions from some 10 different end members. The general term hornblende is usually applied to these solid solutions of the Na-Ca amphiboles. *See* ANTHOPHYLLITE; CUMMINGTONITE; HORNBLENDE.

In detail, the magnesium-rich amphiboles are generally light colored (white, gray, light green) with the color darkening to dark green, dark brown, or black with increasing iron content. The presence of sodium is often indicated by a bluish color, especially in thin sections. The Mg-rich amphiboles commonly develop long needlelike or fibrous crystals, and there is a rough correlation such that, as the iron and aluminum content increases, the individual crystals become progressively more short and stubby.

The amphiboles as a group are present as minor constituents in many volcanic, igneous, and metamorphic rocks and are thus able to form over most of the temperature range observed in the Earth's crust. Amphiboles are present as major constituents in many metamorphic schists and gneisses. The amphiboles can be altered to a variety of decomposition products with talc, antigorite, chlorite, and epidote the most commonly observed. *See* AMPHIBOLITE; ASBESTOS; GLAUCOPHANE; JADE; TREMOLITE. [G.W.DeV.]

Amphibolite A class of metamorphic rocks with one of the amphibole minerals as the dominant constituent. Most of the amphibolites are dark green to black crystalline rocks that occur as extensive layers widely distributed in mountain belts and deeply eroded shield areas of the continental crust. Amphibolite is the main country rock that has been intruded by the large granite masses found in most mountain ranges, with small and large masses of amphibolite present also as inclusions in granites. *See* AMPHIBOLE.

Amphibolites are the products of regional metamorphism and crustal deformation of older materials of appropriate composition. The features of the original rock are obliterated; thus it is difficult and sometimes impossible to determine the premetamorphic rock. Apparent differences in the formation of the bulk composition are used to classify amphibolites as ortho or para. Compositional relations between the minor elements titanium, chromium, and nickel have been used to distinguish the ortho from para amphibolites in some occurrences. *See* METAMORPHIC ROCKS. [G.W.DeV.]

Amphidiscophora A subclass of sponges of the class Hexactinellida, in which the parenchymal microscleres are birotulates and never hexasters. The parenchymal megascleres are always free and unconnected, never fusing to form a rigid network. The sponges are never firmly fixed to a hard substratum, but are anchored to the bottom sediments by a basal tuft or tufts of spicules (see illustration). The recognized orders are Amphidiscosa and Hemidiscosa. Fossils of entire amphidis-

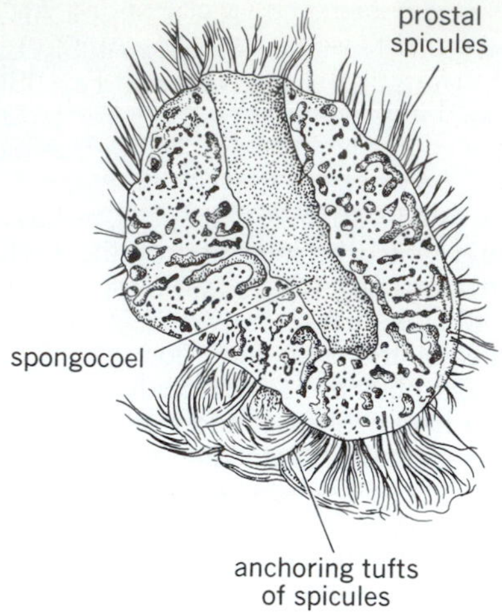

***Pheronema*, sectioned longitudinally. (*After Hyman, 1940, after Schulze, 1887*)**

cophoran sponges are rare. *See* AMPHIDISCOSA; HEMIDISCOSA; HEXACTINELLIDA. [W.D.H.]

Amphidiscosa An order of the subclass Amphidiscophora in the class Hexactinellida. These sponges are distinguished from the order Hemidiscosa in that the birotulates are amphidiscs. Examples of this order are *Pheronema*, *Monorhaphis*, and *Hyalonema*. The record of fossil Amphidiscosa is poor, but goes back to the Carboniferous *Uralonema*. *See* AMPHIDISCOPHORA; HEMIDISCOSA. [W.D.H.]

Amphilinidea An order of tapeworms of the subclass Cestodaria. They have a protrusible proboscis and frontal glands at the anterior end. No holdfast organ is evident. All members of the order inhabit the coelom of sturgeon and other fish. The only life history which is completely known is that of *Amphilina*. The 10-hooked embryos leave the parental uterus through a pore, and if upon escaping into the water they are eaten by an amphipod crustacean, they undergo further development to the procercoid larva. When the parasitized amphipod is eaten by a sturgeon, the larval worm enters its coelom and develops to sexual maturity. *See* CESTODARIA. [C.P.R.]

Amphionidacea An order of the Eucarida comprising a single species, *Amphionides reynaudii*. Because of the similarity of its early larval stages to those found among the caridean shrimps, for more than a century it was classified as an aberrant member of the Caridea. However, the fact that the pleuron of the second abdominal somite never overlaps that of the first in any stage immediately distinguishes this species from all true Caridea. *Amphionides reynaudii* has worldwide distribution, primarily in equatorial regions. Adult females have been collected at depths between 21,000 and 111,000 ft (700 and 3700 m), while most larvae occur in depths of only 90–300 ft (30–100 m). *See* EUCARIDA. [P.A.McL.]

Amphipoda An order of crustaceans in the subclass Malacostraca, which lack a carapace, bear unstalked eyes, and respire by thoracic branchiae, or gills. The abdomen usually bears three pairs of biramous swimmerets (pleopods), three pairs of rather rigid uropods, and a telson which may be lobed or entire. The body is usually flattened laterally, and the pereiopods (walking legs) are elongated so that walking is difficult. The maxillipeds lack epipodites. The sexes are separate, but reproductive and copulatory organs are very simple. The eggs are extruded by the female into a ventral brood pouch composed of setose lamellae attached to the medial bases of the legs. The young hatch as miniature adults, growing usually to a length of 0.12–0.48 in. (3–12 mm), and in exceptional cases to 5.6 in. (140 mm). *See* ISOPODA.

Four suborders are known, the Gammaridea, Hyperiidea, Caprellidea, and Ingolfiellidea. Amphipods are very abundant in the oceans, being represented by 3200 species. More than 600 other species occur in streams, lakes, and subterranean waters and in terrestrial leaf molds and mosses. Many are excellent swimmers. Nonpelagic species of the suborders Gammaridea and Caprellidea live on aquatic bottoms, plants, and epifaunal growths. Predation by amphipods is occasional. Their mouthparts are well adapted for chewing: they either eat aquatic plants, debris, and detritus or swallow mud containing food particles.

Marine species are important food for various stages of many commercial fishes. Hyperiids are the principal food of seals at certain seasons, and also of balaenoid whales at times. One gammaridean genus, *Chelura*, is a minor wood borer, associated with the isopod *Limnoria*.

A few fossil species are known in Tertiary amber deposits. *See* AMBER; CAPRELLIDEA; GAMMARIDEA; HYPERIIDEA; MALACOSTRACA. [J.L.B.]

Amphotericin B An antifungal antibiotic suitable for systemic therapy of deep mycotic infections as well as of superficial infections. It is produced by fermentation with a strain of *Streptomyces nodosus*. In human therapy, desirable therapeutic activity in systemic infections is best shown when the drug is administered intravenously. Oral dosage is apparently not absorbed rapidly enough by humans for the best therapeutic effect in any systemic infections except with very sensitive yeastlike fungi. *See* MEDICAL MYCOLOGY; STREPTOMYCETACEAE.

In therapy of deep mycoses a more soluble mixture of amphotericin B with sodium deoxycholate is dissolved in dextrose solution and infused intravenously. Correct dosages give therapeutic results in many cases of disseminated mycotic infections, including coccidioidomycosis, cryptococcosis, moniliasis, and histoplasmosis. *See* ANTIBIOTIC; COCCIDIOIDOMYCOSIS; CRYPTOCOCCOSIS; HISTOPLASMOSIS. [R.E.B.]

Amplifier A device capable of increasing the magnitude or power level of a physical quantity that is varying with time, without distorting the wave shape of the quantity. The great majority of amplifiers are electronic and depend upon transistors or vacuum tubes for their operation. A small number of electronic amplifiers are magnetic amplifiers, while others take the form of rotating electrical machinery, such as the Amplidyne. A common nonelectrical amplifier is the hydraulic actuator, which is an amplifier of mechanical forces. *See* DIFFERENTIAL AMPLIFIER; DIRECT-CURRENT MOTOR; HYDRAULIC ACTUATOR; OPERATIONAL AMPLIFIER; TRANSISTOR AMPLIFIER; VACUUM-TUBE AMPLIFIER.

Amplifiers of either the transistor or tube variety may be classified in any one of several ways. One is by means of the coupling circuitry, such as an *RC*-coupled (resistance-capacitance-coupled) amplifier. Other categories include classification by purpose, such as hi-fi (high-fidelity audio amplifier), and mode of operation, such as class A. Some of the more common methods of classification follow with a brief description of the operating properties.

A great deal of information about an amplifier is conveyed by merely stating the interstage coupling method used. The *RC*-coupled amplifier is so named because of the coupling capacitor from the load resistor of one stage to the input of the following stage. The transformer-coupled amplifier is so named because two stages are coupled through a transformer. The direct-coupled amplifier, usually called a dc amplifier, allows the amplification of dc as well as ac signals. It has special circuitry, either in the coupling network or in the type of amplifier stages, which eliminates the need for the coupling capacitor.

An important feature of some amplifiers is their ability to amplify only those signals lying within a certain band of frequencies and to reject signals outside this band. Other amplifiers are designed to amplify an extremely wide range of frequencies. A tuned amplifier is one which is designed to amplify signals in a frequency band that is centered on a chosen carrier frequency and not to amplify signals which lie outside this band. Tuned amplifiers form an important part of radio and television receivers in the radio frequency (rf) and intermediate-frequency (i-f) amplifier sections. An untuned amplifier is one in which the coupling circuitry is not a tuned circuit. The audio amplifier and the direct-coupled amplifier are the most important of this type.

The operating modes of vacuum tubes are prescribed by the designations class A, class B, and class C, with certain divisions among the classes. The distinction between the various classes is determined for a sinusoidal signal voltage applied to the input. The position of the quiescent point and the extent of the characteristic that is being used determine the method of operation.

Amplifiers may be classified by the use for which they are designed. Audio-frequency (af) amplifiers are intended to operate over the general range of about 20–20,000 Hz. Radio-frequency amplifiers are used to amplify signals in the range of about 100–1,000,000 kHz. They are usually used as the first amplifier stage in selective radio receivers, including automobile receivers. Because of the additional cost they are usually not used in the average home receiver designed for local reception. Intermediate-frequency amplifiers are used in radio, radar, and television receivers. The majority of broadcast radio receivers employ i-f amplifiers operating at 445 kHz. In frequency-modulated (FM) receivers an i-f of 10.7 MHz is used. Television receivers use 26- or 46-MHz i-f frequencies.

Since, in most applications, one stage of amplification does not provide enough gain, two or more stages are connected together (cascaded) to provide the required gain. For example, many radio receivers have two i-f amplifiers, and the more sensitive receivers have three stages. Similarly, there may be two stages of amplification, and possibly a preamplifier, preceding the power amplifier in an audio amplifier.

The gain of an amplifier is defined as the ratio of the output voltage to the input voltage for a sinusoidal input voltage. Under the usual method for analyzing ac circuits, the gain is a complex number indicating the magnitude of the gain and the phase angle by which the output voltage lags the input voltage in time. Since the gain is a function of frequency, because of the reactive circuit elements, the figure for the gain should be accompanied by the frequency at which it was determined.

An amplifier cannot exactly reproduce the waveform or frequency spectrum of its input signal. The amplifier's ability to faithfully reproduce a signal is measured by either its frequency response or its time response. Special techniques are used to compensate for decreases in gain both above and below the limits of the mid-band region. An amplifier may be called upon to increase the size of a signal whose wave shape is that of a narrow pulse. The definition for pulse rise time is the time required for the pulse to increase from 10% (assuming no overshoot) to 90% of its maximum value. The rise time of the pulse is a function of the frequency response of the amplifier; it is inversely proportional to bandwidth.

There are a number of undesirable conditions that may occur in amplifiers, produced by improper circuit design and by inherent limitations in the physical operation of the devices. In general, good circuit design can reduce all of the undesirable conditions, including those caused by physical limitations, to the point where they are not noticeable. If the output signal from an amplifier is not an exact replica of the input signal, distortion has occurred. In theory it is impossible for an amplifier to avoid introducing distortion. On the other hand, amplifiers have been designed in which the distortion is extremely small. The noise encountered in amplifiers may, in general, be classified into thermal noise and shot noise. Thermal noise is caused by the random motion of electrons inside resistors, conductors, tubes, and transistors. Shot noise is the name given to the noise generated in transistors or vacuum tubes by the random emission of holes or electrons from the emitter or cathode. Degeneration is the loss of gain through unintentional negative feedback. Degeneration can be reduced to a negligible value by having a bypass capacitor with sufficiently small reactance at the signal frequency. Regeneration is an increase in gain through unintentional positive feedback. In its worst form, the amplifier becomes an oscillator. [C.C.H.; C.L.A.]

Amplitude (wave motion) The maximum magnitude (value without regard to sign) of the disturbance of a wave. The term "disturbance" refers to that property of a wave which perturbs or alters its surroundings. It may mean, for example, the displacement of mechanical waves, the pressure variations of a sound wave, or the electric or magnetic field of light waves. Sometimes in older texts the word amplitude is used for the disturbance itself; in that case, amplitude as meant there is called peak amplitude. This is no longer common usage.

If the medium which a wave disturbs dissipates the wave by some nonlinear behavior or other means, then the amplitude will, in general, depend upon position. *See* Displacement (mechanics); Light; Sound; Wave motion. [S.A.Wi.]

Amplitude modulation The process or result of the process whereby the amplitude of a carrier wave is changed in accordance with a modulating wave. This broad definition includes applications using sinusoidal carriers, pulse carriers, or any other form of carrier, the amplitude factor of which changes in accordance with the modulating wave in any unique manner. *See* Modulation.

Practical examples of amplitude modulation (AM) include AM radio broadcasting, single-sideband transmission systems, vestigial-sideband systems, frequency-division multiplexing, time-division multiplexing, phase-discrimination multiplexing, and reduced-carrier systems. *See* Multiplexing; Single sideband; Vestigial-sideband Modulation.

Amplitude modulation is also defined in a more restrictive sense to mean modulation in which the amplitude factor of a sine-wave carrier is linearly proportional to the modulating wave. AM radio broadcasting is a familiar example. At the radio transmitter the modulating wave is the audio-frequency program signal to be communicated; the modulated wave that is broadcast is a radio-frequency, amplitude-modulated sinusoid. *See* Amplitude-modulation radio.

In AM the modulated wave is composed of the transmitted carrier, which conveys no information, plus the upper and lower sidebands, which (assuming the carrier frequency exceeds twice the top audio frequency) convey identical and therefore mutually redundant information. J. R. Carson in 1915 was the first to recognize that, under these conditions and assuming adequate knowledge of the carrier, either sideband alone would uniquely define the message. This eventually led to the development of single-sideband (SSB) and vestigial-sideband (VSB) modulation. Apart from a scale factor, the spectrum of the upper sideband and lower sideband is the spectrum of the modulating wave displaced, respectively, without and with inversion by an amount equal to the carrier frequency. *See* Amplitude-modulation detector; Amplitude modulator; Frequency modulation. [H.S.Bl.]

Amplitude-modulation detector A device for the recovery of the modulating signal from an amplitude-modulated carrier. A detector, like a modulator, must employ a nonlinear device and therefore usually includes either a diode or a nonlinear amplifier. Detectors are sometimes known as

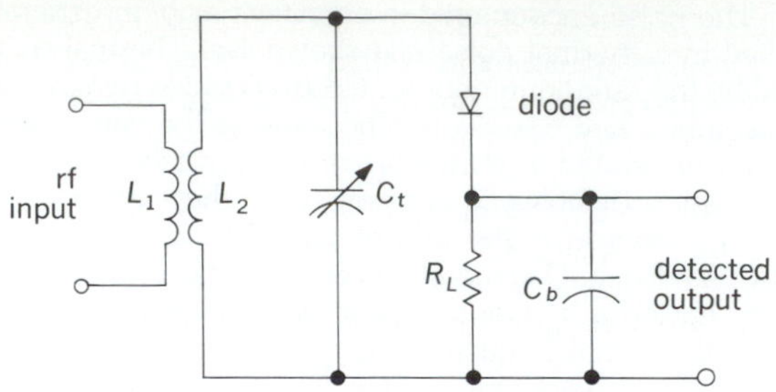

Fig. 1. Typical diode detector circuit.

demodulators. Several types of detector circuits are discussed in this article. *See* MODULATION.

A simple diode detector circuit (Fig. 1) was frequently used as an entire receiver in the early days of radio; it was known as a crystal set. Its radio-frequency (rf) input signal, which was obtained by attaching an antenna and a ground to the input terminals, induced a voltage in a secondary coil L_2 by magnetic induction. Capacitor *Ct* was adjusted to resonate with inductance L_2 at the desired input frequency. Because of the high resonant impedance of the tuned circuit, a voltage was induced in a narrow band of frequencies about the resonant frequency to a high enough level for satisfactory operation of the diode. The desired radio station was thus selected. *See* RADIO.

Modern diode detectors operate similarly. The average value of the modulated input voltage is zero because the negative half cycles have the same amplitude as the positive half cycles. Only high, or radio, frequencies, consisting of the carrier plus sideband frequencies, are present in this input signal; the original modulating frequency appears as the modulation envelope. The diode or rectifier recovers this envelope by allowing current to flow primarily in only one direction, thereby eliminating either the positive or negative half cycles (Fig. 2). Average current through the diode is not zero, but has a component which varies with the amplitude of the input signal and thus contains the original modulating frequencies. Theoretically, nonlinearity produces sum and difference frequencies, as well as harmonics, of the input frequencies. The difference between the carrier frequency and a sideband frequency is the modulating frequency that produced the sideband frequency during modulation. Thus amplitude modulation and detection are basically the same process. *See* RECTIFIER.

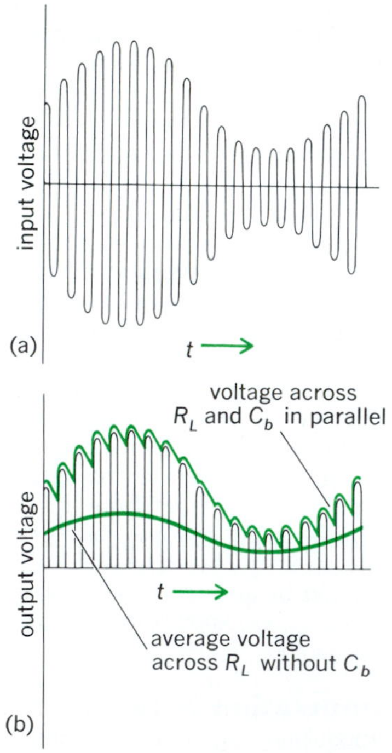

Fig. 2. Input and output waveforms of a detector. (*a*) Modulated rf input. (*b*) Detected output, with modulation envelope recovered by rectification and filtering.

Another important detector is the square-law detector. This device has an output current proportional to the square of the input voltage. Almost all diodes and electronic amplifying devices have this type of characteristic, basically, and linear operation is obtained only by careful selection of the combination of operating point, signal level, driving source resistance, and load resistance. The square-law detector distorts or changes the waveform of the modulation envelope. This distortion is sometimes desirable, as in single-sideband transmission, where one sideband has been eliminated. Single-frequency modulation then produces a nonsinusoidal modulation envelope, but the square-law detector produces a sinusoidal or single-frequency output. In other words, as can be shown mathematically, a square-law detector is necessary for distortionless single-sideband transmission.

A high-gain narrow-band radio receiver can be achieved with a single amplifying detector by providing positive or regenerative feedback from the output to the input. This arrangement is known as a regenerative detector. The feedback adjustment is critical and distortion is high because the feedback must maintain operation on the verge of oscillation.

A four-quadrant multiplier, available as an integrated circuit, may be used as a demodulator or detector. A filter in the output of the multiplier will eliminate high frequencies and leave only the modulating frequency. This type of detector is known as a synchronous detector because the original carrier, without modulation, must be reinserted into the detector. [C.L.A.]

Amplitude-modulation radio Radio communication employing amplitude modulation of a radio-frequency carrier wave as the means of conveying the desired intelligence. In amplitude modulation the amplitude of the carrier wave is made to vary corresponding to the fluctuations of a sound wave, television image, or other information to be conveyed. *See* AMPLITUDE MODULATION; RADIO.

Amplitude modulation (AM), the oldest and simplest form of modulation, is widely used for radio services. The most familiar of these is broadcasting; others include radiotelephony and radiotelegraphy, television picture transmission, and navigational aids.

European and Asian countries use low frequencies in the range 150–255 kilohertz (kHz) for some broadcast services. An advantage of these frequencies is stable and relatively low-attenuation wave propagation. When not limited by atmospheric noise, large areas may be served by one station. In the United States these frequencies are reserved for navigational systems and so are not available for broadcasting.

The frequencies in the range from 535 to 1605 kHz are reserved all over the world for AM (standard) broadcasting. In the Western Hemisphere this band is divided into channels at 10-kHz intervals, certain channels being designated as clear, regional, and local, according to the licensed coverage and class of service. European medium-frequency (mf) broadcasting channels are assigned at 9-kHz intervals.

Small bands of high frequencies between 3000 and 5000 kHz are used in tropical areas of high atmospheric noise for regional broadcasting. This takes advantage of the lower atmospheric noise at these frequencies and permits service under conditions where medium frequencies have only severely limited coverage.

The first radiotelephony was by means of amplitude modulation, and its use has continued with increasing importance. Radiotelephony refers to two-way voice communication. Amplitude modulation and a modified form called single-sideband are used almost exclusively for radiotelephony on frequencies below 30 megahertz. Above 30 MHz, frequency or phase modulation is used almost exclusively, a notable exception being 118–132 MHz, where amplitude modulation is used for all two-way vhf radiotelephony in aviation operations.

The least expensive method known for communicating by telephony over distances longer than a few tens of miles is by using the high frequencies of 3–30 MHz. Furthermore, since radio is the only way to communicate with ships and aircraft, hf AM radiotelephony has remained essential to these operations, except for short distances that can be covered from land stations using the very high frequencies.

Single-sideband (SSB) hf telephony is a modified form of amplitude modulation in which only one of the modulation sidebands is transmitted. In some systems the carrier is transmitted at a low level to act as a pilot frequency for the regeneration of a replacement carrier at the receiver. Since 1933 most transoceanic and intercontinental telephony has been by single-sideband reduced-carrier radio transmission on frequencies between 4000 and 27,000 kHz. In time, SSB will gradually displace AM radiotelephony to reduce serious interference due to overcrowding of the radio spectrum.

Amplitude-modulated radio has a dominant role in guidance and position location, especially in aviation, which is almost wholly under radio guidance.

Amplitude modulation is used everywhere for the broadcasting of the picture (video) portion of television. In England, France, and a few other places amplitude modulation is also used for the sound channel associated with the television picture, but frequency modulation is more commonly used for sound.

Countries of the Western Hemisphere, Japan, Philippines, Thailand, and Iran broadcast television video in an emission band of 4.25 MHz; the English video bandwidth is 3 MHz; the French system, 10 MHz. The rest of continental Europe (except the Soviet Union) use a bandwidth of 5.25 MHz. The carrier frequencies employed are between 40 and 216 MHz, and 470 to 890 MHz. A channel allocation includes the spectrum needed for both sound and picture. Japan and Australia also use 88–108 MHz for television broadcasting. [E.A.L.]

Amplitude modulator A device for amplitude-modulating a carrier signal. The carrier is usually the radio frequency (rf) in a communications system, but it may also be a carrier signal in a multichannel cable communication system, a telemetering system, a control system, or a data-collecting system. The modulator is usually a vacuum tube or semiconductor amplifier, the output power of which controls the output level of another amplifier which modulates the carrier. *See* AMPLITUDE MODULATION; MODULATION.

The goal of a modulator is to vary the amplitude of a carrier in proportion to the modulating voltage. Departure from this linear relation results in modulation distortion. If the carrier being modulated is at a high power level, such as the rf in the final stage of a radio transmitter, the power output capability and efficiency of the modulator and the efficiency of the carrier amplifier become important.

The process of amplitude modulation produces new frequencies known as side frequencies or sidebands; they are the sum and difference of the carrier and modulating frequencies. Because new frequencies are produced only in nonlinear devices, the carrier amplifier in which modulation is to take place must be nonlinear, insofar as the carrier signal is concerned, and must therefore be either a class B or class C amplifier.

The most frequently used method of amplitude modulation employed in radio communications is known as plate or collector modulation. As this name implies, the modulator varies the plate or collector voltage in order to accomplish the modulation.

Low-level amplitude modulation can be produced by applying the modulating voltage to one of the grids of a vacuum tube or to the base of a transistor rf or carrier amplifier. Modulation is accomplished by varying the bias of the grid or base. Low-level modulation can be accomplished by a four-quadrant multiplier. The output of the multiplier modulator contains neither the modulating frequency nor the carrier frequency but only the sidebands. This type of modulator is known as a suppressed carrier or balanced modulator. A four-quadrant multiplier which will handle either positive or negative voltages on both inputs is available as an inexpensive integrated circuit. Regular amplitude modulation, without carrier suppression, may be produced by adding an offset bias voltage to the modulating input of the four-quadrant multiplier. [C.L.A.]

Amygdule A mineral filling (cast) of a vesicle (gas cavity) in congealed lava. A rock containing an amygdule is called amygdaloid, or amygdaloidal lava. Gases dissolved in magma (molten rock) at depth exsolve and form bubbles as lower-pressure regions are reached. When the magma is chilled upon eruption, these gas bubbles are preserved in the lava as open cavities or vesicles (see illustration). In pahoehoe lava, the vesicles typically are spheroidal, but they may be flattened owing to flowage. Vesicles in aa and block lava tend to be more irregular.

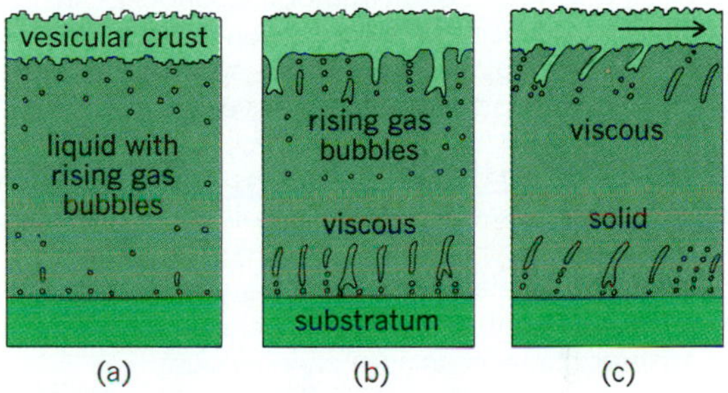

Diagram of spheroidal and pipe vesicles in lava flows. (*a*) Liquid flow with vesicular crust. (*b*) Congealing flow. (*c*) Nearly solidified flow. (*After R. R. Shrock, Sequence in Layered Rocks, McGraw-Hill, 1948*)

Amygdules generally consist of mineral matter deposited by fluids of extraneous origin moving through the vesicular lava after final consolidation. These depositing agents may be hot gases and hydrous solutions rising in volcanic areas, hydrothermal mineralization fluids unrelated to volcanism, or cold solutions in which the water is of meteoric (atmospheric) origin. The material deposited may be derived from remote sources or by alteration of ambient rocks. *See* MAGMA; VOLCANO. [G.A.M./R.I.T.]

Amylase An enzyme which breaks down (hydrolyzes) starch, the reserve carbohydrate in plants, and glycogen, the reserve carbohydrate in animals, into reducing fermentable sugars, mainly maltose, and reducing nonfermentable or slowly fermentable dextrins. Amylases are classified as saccharifying (ß-amylase) and as dextrinizing (α-amylases). *See* CARBOHYDRATE; ENZYME; GLYCOGEN; MALTOSE.

In animals the highest concentrations of amylase are found in the saliva and in the pancreas. Salivary amylase is also known as ptyalin and is found in humans, the ape, pig, guinea pig, squirrel, mouse, and rat. Pig pancreas is rich in amylase, whereas cattle, sheep, and dog pancreases have lower concentrations.

Amylase is also used as a diastase in industry. It is used (1) in brewing and fermentation industries for the conversion of starch to fermentable sugars, (2) in the textile industry for designing textiles, (3) in the laundry industry in a mixture with protease and lipase to launder clothes, (4) in the paper industry for sizing, and (5) in the food industry for preparation of sweet syrups, to increase diastase content of flour, for modification of food for infants, and for the removal of starch in jelly production. [D.N.La.]

Amyloidosis A disorder characterized by the accumulation of an unusual extracellular fibrous protein (amyloid) in the connective tissue of the body. The deposition of amyloid may be widespread, involving major organs and leading to serious clinical consequences, or it may be very limited with little effect on health.

Amyloidosis has been classified clinically as: (1) primary amyloidosis, with no evidence for preexisting or coexisting disease; (2) amyloidosis associated with multiple myeloma; (3) secondary amyloidosis, associated with chronic infections (such as osteomyelitis, tuberculosis, leprosy), chronic inflammatory disease (such as rheumatoid arthritis, ankylosing spondylitis, regional enteritis), or neoplasms (such as medullary carcinoma of the thyroid); (4) heredofamilial amyloidosis, associated with familial Mediterranean fever and a variety of heritable neuropathic, renal, cardiovascular, and other syndromes; (5) local amyloidosis, with local, often tumorlike, deposits in isolated organs without evidence of systemic involvement; (6) amyloidosis associated with aging. There is no specific treatment for amyloidosis, but supportive treatment is very useful. [A.S.Co.]

Anaerobic infections Infections caused by anaerobic bacteria (organisms that are intolerant of oxygen). Many such infections are mixed, involving more than one anaerobe and often aerobic or facultative bacteria as well. Anaerobes are prevalent throughout the body as indigenous flora, and essentially all anaerobic infections arise endogenously.

Anaerobic gram-negative bacilli (*Bacteroides*, *Fusobacterium*) and anaerobic gram-positive cocci (*Peptococcus*, *Peptostreptococcus*) are the most common anaerobic pathogens. *Clostridium* (spore formers) may cause serious infection. The prime pathogen among gram-positive nonsporulating anaerobic bacilli is *Actinomyces*.

In terms of frequency of occurrence, the pleuropulmonary, intraabdominal, and obstetric-gynecologic infections are most important. Foul order is definitive evidence of involvement of anaerobes in an infection, but absence of such odor does not exclude the possibility. Other important clues to anaerobic infection, though not specific, are location near a mucosal surface, bite infection, tissue necrosis, gas formation, prior aminoglycoside therapy, "sulfur granules" (actinomycosis), certain distinctive clinical presentations (such as gas gangrene), septic thrombophlebitis, unique morphology of certain anaerobic organisms on smear, and failure to grow organisms aerobically from specimens.

Therapy includes surgery and antimicrobial agents. *Bacteroides fragilis* is the most resistant anaerobe; only chloramphenicol, clindamycin, and metronidazole are essentially active against all strains. Most other anaerobes are susceptible to penicillin G. *See* ANTIBIOTIC; GANGRENE; INFECTION. [S.M.F.]

Analcime A mineral with a framework structure in which all the aluminosilicate tetrahedral vertices are linked, thus allying it to the feldspars, feldspathoids, and zeolites. Its formula is $Na(H_2O)[AlSi_2O_6]$; in this sense it is a tectosilicate.

The analcime structure type includes several other mineral species. These include high-temperature leucite, pollucite, and wairakite. Crystals are most frequently trapezohedra, and rarely the mineral is massive granular. Hardness is 5–5½ on Mohs scale; specific gravity is 2.27.

Analcime most frequently occurs as a low-temperature mineral in vesicular cavities in basalts, where it is associated with zeolites (particularly natrolite), datolite, prehnite, and calcite. Small grains are frequent constituents of sedimentary rocks and muds in oceanic basins associated with volcanic sources. *See* FELDSPAR; FELDSPATHOID; LEUCITE ROCK; ZEOLITE. [P.B.M.]

Analgesic Any of a group of drugs of diverse chemical structure and physiological effects which are commonly used for the relief of pain. To qualify as an analgesic a drug must selectively reduce or abolish pain without causing impairment of consciousness, mental confusion, incoordination or paralysis, or other derangements of the nervous system.

One class of analgesics are the narcotic alkaloids. The oldest and best-known analgesics are opium, and its most active alkaloid, morphine. Apart from morphine there are also other naturally occurring alkaloids derived from opium; the best known of these is codeine. The narcotic analgesics relieve cough and suppress respiration. They abolish painful spasms of muscles and stimulate smooth muscles. If taken for weeks or months, however, the recipient needs and tolerates larger and larger doses to obtain the desired effects. In other words, the individual becomes habituated. If the drug is then stopped for any reason, very disagreeable withdrawal or abstinence effects are experienced. There is then severe pain, sweating, salivation, hyperventilation, restlessness, and confusion. Thus the narcotic alkaloids are all potentially addictive, and the addicted person will go to extreme measures to obtain the narcotic and avoid the withdrawal symptoms. *See* DRUG ADDICTION; MORPHINE; NARCOTIC; OPIATES.

Because of the strong addictive properties of morphine and related compounds, chemists have synthesized other drugs of similar chemical structure, in the hope of securing analgesia without addiction. This effort has been only partially successful. Methadone, a drug given to addicts as a substitute for morphine, is an effective analgesic when taken orally and is less addictive than morphine. Meperidine (Demerol) is a strong synthetic analgesic but definitely addictive. Other synthetic analgesics are oxycodone (Percodan), levorphanol (levodromoran), propoxyphene (Darvon), and pentazocine (Talwin). The last two of this series cause little or no addiction but, unfortunately, are not strong analgesics.

Another class of analgesic drugs, which are nonnarcotic (nonaddictive), are the salicylates, the most familiar being acetylsalicylic acid (aspirin), and salicylatelike drugs such as phenylbutazone (Butazolidine), indomethacin (Indocin), acetaminophen, and phenacetin. These drugs are most effective in relieving skeletal pain due to inflammation (such as arthritis). Their analgesic properties are not nearly as strong as those of morphine and the synthetic opioids. These drugs also have many other effects, such as reducing fever (antipyrexia) and preventing platelet agglutination. They are the most commonly used of all analgesic medications and are often combined with caffeine or a barbiturate sedative under a variety of trade names and sold for the relief of headache, backache, and so forth. *See* ASPIRIN; CENTRAL NERVOUS SYSTEM; PAIN. [R.D.A.]

Analog computer A computer or computational device in which the problem variables are represented as continuous, varying physical quantities. An analog computer implements a model of the system being studied. The physical form of the analog may be functionally similar to that of the system, but more often the analogy is based solely upon the mathematical equivalence of the interdependence of the computer variables and the variables in the physical system. *See* SIMULATION.

Types. Analog computers can be classified in two ways: in accordance with their use, and in accordance with the type of components used to assemble them. In terms of use, there are general-purpose and special-purpose analog computers. The general-purpose computer is designed so that it can be programmed to solve many kinds of problems or permit the development of different simulated models as needed. The special-purpose computer has a fixed program with a few or no permitted adjustments; it is generally built into or appended to the physical system it serves.

In terms of the components with which computers are made, there are mechanical, hydraulic, pneumatic, optical, electrical, and electronic analog computers serving in various applications. Because of its ease of programming, flexibility of operation, and repeatable results, the general-purpose electronic analog computer has come into wide use. The flexibility of the electronic analog computer has allowed it to be augmented with interface channels to the electronic digital computer, so that during the 1960s a third type of general computer, the hybrid computer, came into being.

Because of the practical difficulties of scaling, interconnecting, and operating an electronic analog computer, the equivalent of its powerful problem-solving capability has been sought and realized in other forms. Since 1965 many digital computer programs have been developed which essentially duplicate the functions of analog computers via digital algorithms. Digital equivalents to the electronic analog computer are free from scaling requirements and are convenient to run; on the other hand, no interaction within the dynamic response is possible, the solution speed is slower by a factor of a hundred times or more, and little or no on-line model building and exploring is possible.

Another type of analog computer is the digital multiprocessor analog system, in which the relatively slow speeds of sequential digital increment calculations have been radically boosted through parallel processing. In this type of analog computer it is possible to retain the programming convenience and data storage of the digital computer while approximating the speed, interaction potential, and parallel computations of the traditional electronic analogs. The computer typically utilizes several specially designed high-speed processors for the numerical integration functions, the data (or variable) memory distributions, the arithmetic functions, and the decision (logic and control) function.

Description and uses. The typical modern general-purpose analog computer consists of a console containing a collection of operational amplifiers; computing elements, such as summing networks, integrator networks, attenuators, multipliers, and function generators; logic and interface units; control circuits; power supplies; a patch bay; and various meters and display devices. The patch bay is arranged to bring input and output terminals of all programmable devices to one location, where they can be conveniently interconnected by various patch cords and plugs to meet the requirements of a given problem. Prewired problem boards can be exchanged at the patch bay in a few seconds and new coefficients set up typically in less than a half hour. Extensive automatic electronic patching systems have been developed to permit fast setup, as well as remote and time-shared operation.

The analog computer basically represents an instrumentation of calculus, in that it is designed to solve ordinary differential equations. This capability lends itself to the implementation of simulated models of dynamic systems. The computer operates by generating voltages that behave like the physical or mathematical variables in the system under study. Each variable is represented as a continuously varying (or steady) voltage signal at the output of a programmed computational unit. Specific to the analog computer is the fact that individual circuits are used for each feature or equation being represented, so that all variables are generated simultaneously. Thus the analog computer is a parallel computer in which the configuration of the computational units allows direct interactions of the computed variables at all times during the solution of a problem.

Components. Manipulations of the signals (voltages) in the analog computer are based upon the fundamental electrical properties associated with circuit components. The interrelation of voltages for representing mathematical functions is derived by combining currents at circuit nodes or junctions. *See* Kirchhoff's laws of electric circuits; Ohm's law.

The simplest arrangement of components for executing addition would be to impress the voltages to be added across individual resistors. The resistors would then be joined to allow the currents to combine and to develop the output voltage across a final resistor. Use of this simple configuration of elements for computation is impractical because of the undesirable interaction between inputs. A change in one input signal (voltage) causes a change in the current that flows through the input resistor; this changes the voltage at the input resistor junction, and the change secondarily causes a different current to flow in the other input resistors. The situation gets more interactive when another computing circuit is attached so that part of the summing current flows away from the summing junction. This interaction effect also prevents exact computing. If, in some way, each voltage to be summed could be made independent of the other voltages connected to the summing junction, and if the required current fed to other circuits could be obtained without loading the summing junction, then precise computation would be possible.

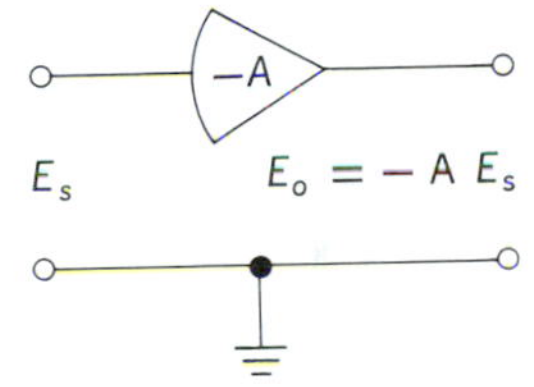

Fig. 1. Symbol for a high-gain direct-current amplifier, with one inverting input referenced to ground.

The electronic analog computer satisfies these needs by using high-gain (high-amplification) dc operational amplifiers. A symbol to represent a dc direct-coupled amplifier is shown in Fig. 1. According to convention, the rounded side represents the input to the amplifier, and the pointed end represents the amplifier output. A common reference level or ground exists between the amplifier input and output, and all voltages are measured with respect to it. The ground reference line is understood and is usually omitted from the symbol. The signal input network (consisting of summing resistors) connects to the inverting (or negative) amplifier input terminal. The noninverting (or positive) amplifier input terminal is normally connected to the reference ground. Generally the inverting input is called the summing junction (SJ) of the amplifier. Internal design of the amplifier is such that, if the signal at the summing junction is positive with respect to ground, the amplifier output voltage is negative with respect to ground. The amplifier has an open-loop voltage gain of $-A$; therefore an input voltage of E_s results in an output voltage of $-AE_s$. Gain of a commercial computing amplifier is typically 10^8; thus, an input voltage of less than 1 microvolt can produce several volts at the amplifier output. *See* Amplifier; Direct-coupled amplifier.

Because the operational amplifier thus inverts the signal (changes its sign or polarity), it lends itself to the use of negative feedback, whereby a portion of the output signal is returned to the input. This arrangement has the effect of lowering the net gain of the amplifier signal channel and of

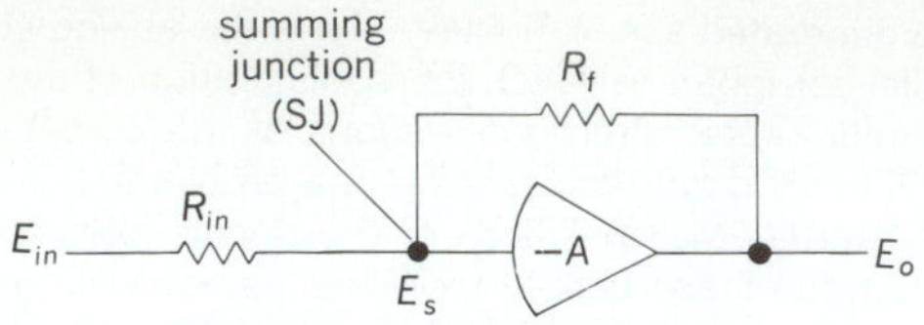

Fig. 2. High-gain amplifier which has been made into an operational amplifier through the inclusion of an input resistor and a feedback resistor tied together at the amplifier's summing junction (*SJ*).

improving overall signal-to-noise ratio and increasing computational accuracy.

Circuit operation (Fig. 2) can be viewed in the following manner. A dc voltage E_{in} applied to input resistor R_{in} produces a summing junction voltage E_s. The voltage is amplified and appears at the amplifier output as voltage E_o (equal to $-AE_s$, where A is voltage gain of the amplifier). Part of output voltage E_o returns through feedback resistor R_f to the summing junction. Because the returned or feedback voltage is of opposite (negative) polarity to the initial voltage at the summing junction, it tends to reduce the magnitude of E_s, resulting in an overall input-output relationship that may be expressed as in the equation below. In fact, the summing junction voltage E_s is

$$\frac{E_o}{E_{in}} = \frac{-A}{A+1} = \frac{-1}{1+1/A} \simeq -1 \qquad A > 10^8$$

so small that it is considered to be at practically zero, a condition called virtual ground.

To illustrate how the operational amplifier serves the needs of the computing network, consider the currents that flow and the corresponding algebraic expressions (Fig. 3). The operational amplifier is designed to have high input impedance (high resistance to the flow of current into or out of its input terminal); consequently the amplifier input current I_s can then be considered to be practically zero. The resulting current equation states that input current I_{in} is equal to feedback current I_f. Since the amplifier has a very high gain, the summing junction voltage is virtually zero. Voltage drop across R_{in} is thus equal to E_{in}; voltage drop across R_f is E_o. The equation in the box is the fundamental relationship for the amplifier. As long as the amplifier has such a high gain and requires a negligible current from the summing junction, the amplifier input and output voltages are related by the ratio of the two resistors and are thus not affected by the actual electronic construction of the amplifier. If several input resistors are connected to the same

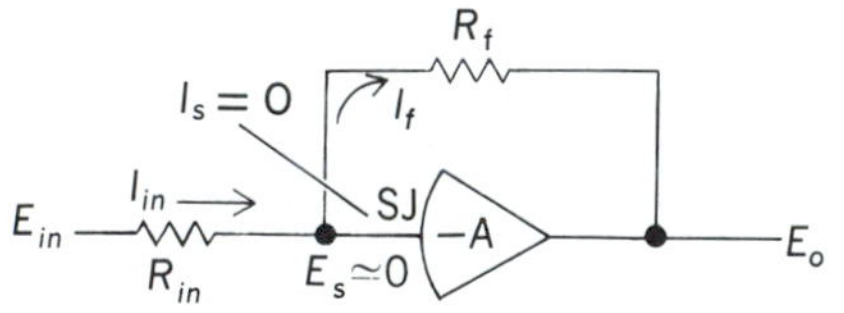

$$I_{in} = I_f \qquad (1)$$

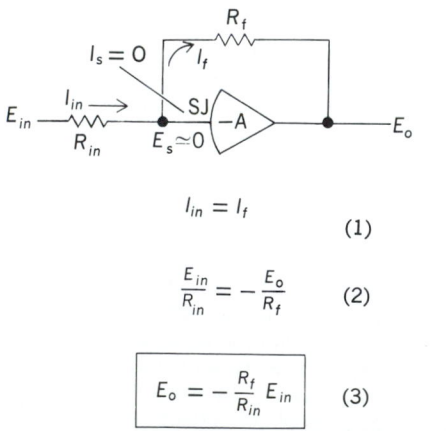

$$\frac{E_{in}}{R_{in}} = -\frac{E_o}{R_f} \qquad (2)$$

$$\boxed{E_o = -\frac{R_f}{R_{in}} E_{in}} \qquad (3)$$

Fig. 3. Summing junction currents into an operational amplifier create the fixed gain function, determined by resistor values, as indicated in the box.

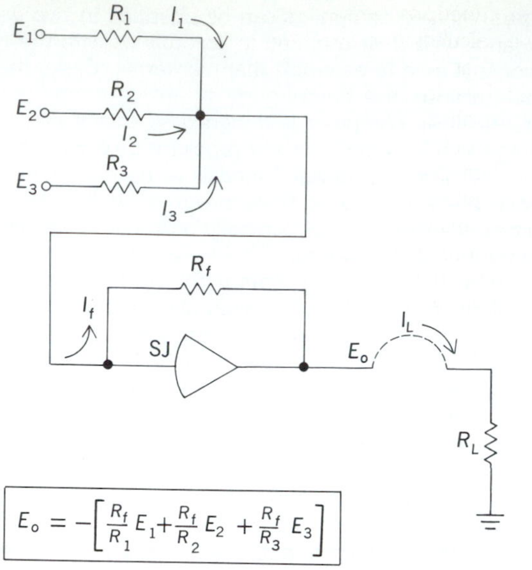

Fig. 4. Operational amplifier which has been made into a summer through the use of several input resistors connected to the summing junction. Output is equal to the inverted weighted sum of the inputs.

summing junction and voltages are applied to them (Fig. 4), then because the summing junction remains at practically zero potential, none of the inputs will interfere with the other inputs. Thus all inputs will exert independent and additive effects on the output.

The input and feedback elements that can be used with an operational amplifier are not restricted to resistors alone. A reactive feedback component such as a capacitor may also be used, resulting in mathematical operations analogous to integration with respect to time. An operational amplifier with a feedback capacitor will generate an output voltage that is the time integral of its input voltage. *See* CALCULUS; CAPACITANCE.

In nature nearly all systems or processes exhibit nonlinear features in the form of constraints, discontinuities, and variable gains. If an analog computer is to be programmed and operated as a realistic system model, it must include devices with programmable nonlinear characteristics. The most versatile computing element for this service is the diode function generator. *See* SEMICONDUCTOR DIODE.

Diodes, in addition to being used to implement simulations of unique nonlinearities such as limits, backlash, and dead zones, can be combined with a network of resistors to produce particular nonlinear functions. A diode function generator (DFG) uses the one-way conducting property of diodes to selectively alter the resultant gain of an operational amplifier in accordance with a desired function of the input voltage. The DFG can be arranged to operate on either positive or negative input signals by choosing the orientation of diodes and reference voltage polarity. Further, by utilizing fully adjustable potentiometers for breakpoint and slope resistances, a variable diode function generator (VDFG) is established, permitting the programming of most analytic, arbitrary, and empirical functions, including functions with one or more inflections.

The principal device for multiplication of one variable by another (to yield a product of variables) is the electronic quarter-square multiplier. This device achieves the multiplication function to frequencies of several thousand hertz by implementing an algebraic identity. The squaring operations are accomplished with high accuracy by two DFG circuits.

An electronic comparator–analog switch combination provides the analog computer with decision-making and preprogrammed automatic signal-rerouting capabilities. These are nonlinear functions preset to take place at some particular point in a calculation run and to cause a change in coefficients or signals, that is, a branching in mathematical procedure or a change in the control of the run.

Programming. To solve a problem using an analog computer, the problem solver goes through a procedure of general analysis, data preparation, analog circuit development, and patchboard programming. Test runs of subprograms may also be made to examine partial-system dynamic responses before eventually running the full program to derive specific and final answers. The problem-solving procedure typically involves eight major steps, listed below:

1. The problem under study is described with a set of mathematical equations or, when that is not possible, the system configuration and the interrelations of component influences are defined in block-diagram form, with each block described in terms of black-box input-output relationships.
2. Where necessary, the description of the system (equations or system block diagram) is rearranged in a form that may better suit the capabilities of the computer, that is, avoiding duplications or excessive numbers of computational units, or avoiding algebraic (nonintegrational) loops.
3. The assembled information is used to sketch out an analog circuit diagram which shows in detail how the computer could be programmed to handle the problem and achieve the objectives of the study.
4. System variables and parameters are then scaled to fall within the operational ranges of the computer. This may require revisions of the analog circuit diagram and choice of computational units.
5. The finalized circuit arrangement is patched on the computer problem board.
6. Numerical values are set up on the attenuators, the initial conditions of the entire system model established, and test values checked.
7. The computer is run to solve the equations or simulate the black boxes so that the resultant values or system responses can be obtained. This gives the initial answers and the "feel" for the system.
8. Multiple runs are made to check the responses for specific sets of parameters and to explore the influences of problem (system) changes, as well as the behavior which results when the system configuration is driven with different forcing functions.

The complementary nature of the input and feedback networks of the operational amplifier gives it one of its most flexible properties. If, for example, a squaring module in an amplifier input circuit is moved to the amplifier feedback circuit, the amplifier will compute the square root. In general, when a function is exchanged between the input and feedback positions of an operational amplifier, the inverse function is obtained.

Because of its fundamental ability to perform continuous integration, the analog computer is particularly well suited to solve ordinary differential equations, typically arising in the problems of calculus. Because of this ability, the analog computer has at times been referred to as the electronic differential analyzer. Significantly, the analog computer integrator will respond correctly to any simple or complex signal waveform presented at its input; it is not restricted to applications involving conventional classes of functions defined by analytic calculus.

Operation. After a program has been connected up on the patchboard in accordance with the foregoing programming procedures, the problem patchboard inserted in the patch bay, and the attenuators adjusted to their proper values, the analog computer is ready for a solution run. The mode of operation of the computer is governed by integrator controls. The input to an integrator amplifier is connected to one of four possible points: potset (PS) for adjusting potentiometers, in which the integrator is effectively short-circuited; the reset or initial condition (IC) mode, in which the capacitor is charged up to the initial value *EIC*; hold (HD), which is a stop and freeze mode mainly for readout purposes; and operate (OP) mode, in which the intended continuous integration with respect to time takes place. The switching between these modes is done with high-speed electronic analog signal switches (older computers use relay contacts) controlled from the computer console or via logic command signals connected on the patchboard. At the beginning of a run, the computer is put in the reset or initial condition mode, energizing a bus circuit so that the integrators assume their initial conditions. Next, the computer is switched to the operate mode, removing the initial condition inputs and connecting the integrand inputs to the integrators. The generated and connected input signals are now integrated by the integrator according to its programmed time scale.

In operating the computer, the user must be prepared to deal with many practical circumstances and to diagnose the troubles or errors that arise from the analog computer program. Mistakes in circuit patching, incorrect attenuator settings, wrong reference polarities, misapplication of ranges of computer units and display devices, and inappropriate time scale factor assignments may occur, as well as a host of details associated with the electronic properties of the computer. A smooth-running program, however, returns an extremely large amount of problem insight and experience, well worth the effort required to get the problem running.

Hybrid computers. The accuracy of the calculations on a digital computer can often be increased through double precision techniques and more precise algorithms, but at the expense of extended solution time, due to the computer's serial nature of operation. Also, the more computational steps there are to be done, the longer the digital computer will take to do them. On the other hand, the basic solution speed is very rapid on the analog computer because of its parallel nature, but increasing problem complexity demands larger computer size. Thus, for the analog computer the time remains the same regardless of the complexity of the problem, but the size of the computer required grows with the problem.

Interaction between the user and the computer during the course of any calculation, with the ability to vary parameters during computer runs, is a highly desirable and insight-generating part of computer usage. This hands-on interaction with the computed responses is simple to achieve with analog computers. For digital computers, interaction usually takes place through a computer keyboard terminal, between runs, or in an on-line stop-go mode. An often-utilized system combines the speed and interaction possibilities of an analog computer with the accuracy and programming flexibility of a digital computer. This combination is specifically designed into the hybrid computer.

In a modern analog-hybrid console, the mode switches in the integrators are interfaced with the digital computer to permit fast iterations of dynamic runs under digital computer control. Data flow in many ways and formats between the analog computer with its fast, parallel circuits and the digital computer with its sequential, logic-controlled programs. Special high-speed analog-to-digital and digital-to-analog converters translate between the continuous signal representations of variables in the analog domain and the numerical representations of the digital computer. Control and logic signals are more directly compatible and require only level and timing compatibility. *See* ANALOG-TO-DIGITAL CONVERTER; BOOLEAN ALGEBRA; DIGITAL-TO-ANALOG CONVERTER.

The programming of hybrid models is a more complex challenge than described above, requiring the user to consider the parallel action of the analog computer interlaced with the step-by-step computations progression in the digital computer. For example, in simulating the mission of a space vehicle, the capsule control dynamics will typically be handled on the analog computer in continuous form, but interfaced with the digital computer, where the navigational trajectory is calculated. *See* COMPUTER. [P.A.H.]

Analog states Certain states belonging to neighboring nuclear isobars and possessing identical structure except for the transformation of one or more neutrons into the same number of protons. Hundreds of examples of analog states (also known as isobaric analog states) have been observed throughout the periodic table, and their existence is considered to be a fundamental property of nuclear structure. *See* ISOBAR (NUCLEAR PHYSICS); NUCLEAR STRUCTURE.

Starting with a nucleus of N number of neutrons and Z number of protons in a particular state of energy, angular momentum, parity, and so forth, an analog state may be constructed in the neighboring isobar with $N - 1$ neutrons and $Z + 1$ protons by replacing a neutron with a proton in the same orbit and then coupling the proton to the remaining nucleons in exactly the same fashion as the original neutron. Under the assumption of charge independence of nuclear forces, the only difference between a state and its analog should arise from the increase in the Coulomb interactions attributed to the increased number of protons in the nucleus. Since the total energy of a state has a Coulomb contribution, the most outstanding difference between analog states is their Coulomb energy. When this difference is taken into account, then the total energies of corresponding analog states become nearly equal. Another, smaller effect which must also be considered is the neutron-proton mass difference of 0.782 MeV. However, even then, all properties of analog states do not become precisely identical because there remain subtle differences in the behavior of neutrons and protons due to electromagnetic effects.

Until about 1961, it was thought that analog states were a property only of light-mass nuclei with mass number A less than about 40. However, in a number of experiments not specifically designed to search for analog states, dramatic evidence was found for their existence throughout the periodic table up to the heaviest known nuclei. The discovery of analog states in heavy nuclei has provided a powerful tool for studying nuclear structure that was previously unforeseen. First, it has enabled the study of Coulomb energies to be extended from low-mass nuclei to the heaviest nuclei; and second, because of their high excitation energy, analog states normally are able to decay by particle emission, unlike their corresponding parent states which usually decay only by γ-ray emission to the ground state. By measuring the properties of the particles that are emitted from the analog state, unique information is derived about the parent state which often cannot be obtained in any other fashion. *See* RADIOACTIVITY. [N.S.]

Analog-to-digital converter A device for converting the information contained in the value or magnitude of some characteristic of an input signal, compared to a standard or reference, to information in the form of discrete states of a signal, usually with numerical values assigned to the various combinations of discrete states of the signal.

Analog-to-digital (A/D) converters are used to transform analog information, such as audio signals or measurements of physical variables (for example, temperature, force, or shaft rotation) into a form suitable for digital handling, which might involve any of these operations: (1) processing by a computer or by logic circuits, including arithmetical operations, comparison, sorting, ordering, and code conversion, (2) storage until ready for further handling, (3) display in numerical or graphical form, and (4) transmission.

If a wide-range analog signal can be converted, with adequate frequency, to an appropriate number of two-level digits, or bits, the digital representation of the signal can be transmitted through a noisy medium without relative degradation of the fine structure of the original signal. *See* COMPUTER GRAPHICS; DATA COMMUNICATIONS; DIGITAL COMPUTER.

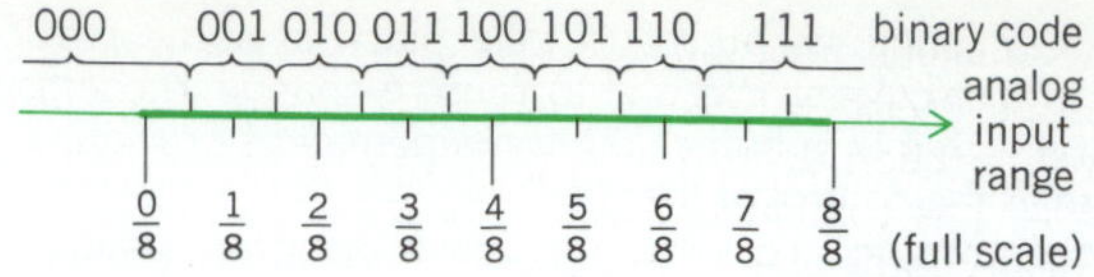

A three-bit binary representation of a range of input signals.

Conversion involves quantizing and encoding. Quantizing means partitioning the analog signal range into a number of discrete quanta and determining to which quantum the input signal belongs. Encoding means assigning a unique digital code to each quantum and determining the code that corresponds to the input signal. The most common system is binary, in which there are $2n$ quanta (where n is some whole number), numbered consecutively; the code is a set of n physical two-valued levels or bits (1 or 0) corresponding to the binary number associated with the signal quantum.

The illustration shows a typical three-bit binary representation of a range of input signals, partitioned into eight quanta. For example, a signal in the vicinity of 3/8 full scale (between 5/16 and 7/16) will be coded 011 (binary 3). *See* NUMBER SYSTEMS. [D.H.S.]

Analysis of variance Total variation in experimental data is partitioned into components assignable to specific sources by the analysis of variance. This statistical technique is applicable to data for which (1) effects of sources are additive, (2) uncontrolled or unexplained experimental variations (which are grouped as experimental errors) are independent of other sources of variation, (3) variance of experimental errors is homogeneous, and (4) experimental errors follow a normal distribution. When data depart from these assumptions, one must exercise extreme care in interpreting the results of an analysis of variance. Statistical tests indicate the contribution of the components to the observed variation. *See* STATISTICS. [R.L.Bri.]

Analytic function A complex-valued function f of a complex variable defined on some region Ω in the complex plane is analytic or holomorphic in Ω if for each point a in Ω the values of the function at all points z sufficiently near a are given by the power series expansion below. This power series

$$f(z) = \sum_{n=0}^{\infty} a_n (z - a)^n$$

must then be the Taylor series expansion of the function f at the point a and must converge in at least the largest circle centered at a and contained in Ω. An analytic function in region Ω is necessarily a continuous function in Ω; moreover, as a function of the two real variables (x,y), an analytic function has partial derivatives of all orders.

There are a number of alternative characterizations of analytic functions, all of which are useful. The most common and perhaps the most surprising characterization, which is often taken as the definition of an analytic function, is that a function f is analytic in Ω precisely when the limit given by the expression below exists at all points a in Ω. This limit is the complex

$$\lim_{\lim z \to a} \frac{f(z) - f(a)}{z - a}$$

analog of the ordinary derivative and is therefore called the complex derivative and written $f'(a)$. *See* CALCULUS.

An analytic function f must satisfy the Cauchy-Riemann equation $\partial f/\partial x + i\,\partial f/\partial y = 0$. Conversely if f is a complex-valued function defined in a region Ω of the complex plane, and if f has continuous first partial derivatives with respect to the real coordinates x and y, and if they satisfy the Cauchy-Riemann equation at all points of Ω, then f is analytic in Ω.

If f is an analytic function in a region Ω of the complex plane, then it follows from the Cauchy-Riemann equation that $\int_\gamma f dz = 0$ for any closed path γ homogeneous to zero in Ω. Conversely Morea's theorem asserts that if f is a continuous complex-valued function in Ω and if $\int_\gamma f dz = 0$ for all closed paths γ homologous to zero Ω, then f is necessarily analytic in Ω.

By analyzing the condition imposed by the Cauchy-Riemann equation it follows that if $f'(a) \neq 0$, if γ_1, and γ_2 are two differentiable paths through the point a, and if the tangent vectors to γ_1, and γ_2 at the point a are at an angle θ apart, then the tangent vectors to the paths $f(\gamma_1)$ and $f(\gamma_2)$ at the point $f(a)$ are also at an angle θ apart. The mapping f thus preserves angles at a, and is therefore said to be conformal at a. Conversely any continuously differentiable conformal mapping is an analytic function with a nonzero complex derivative, thus providing a very geometric characterization of analytic functions. *See* CAUCHY INTEGRAL FORMULA; CONFORMAL MAPPING; ISOLATED SINGULARITY; PARTIAL DIFFERENTIATION. [R.C.G]

Analytic geometry A branch of mathematics in which algebra is applied to the study of geometry; the subject is also called cartesian geometry. The basis for an algebraic treatment of geometry is provided by the existence of a one-to-one correspondence between the elements, "points" of a directed line g, and the elements, "numbers," that form the set of all real numbers. Such a correspondence establishes a coordinate system on g, and the number corresponding to a point of g is called its coordinate. The point O of g with coordinate zero is the origin of the coordinate system. A coordinate system on g is cartesian provided that for each point P of g, its coordinate is the directed distance $\overline{OP}$. Then all points of g on one side of O have positive coordinates (forming the positive half of g) and all points on the other side have negative coordinates. The point with coordinate 1 is called the unit point. Since the relation $\overline{OP} + \overline{PQ} = \overline{OQ}$ is clearly valid for each two points P, Q of directed line g, then $\overline{PQ} = \overline{OQ} - \overline{OP} = q - p$, where p and q are the coordinates of P and Q, respectively.

Choose any two intersecting lines g_1, g_2 of the plane, with cartesian coordinate systems selected on each so that the intersection point has coordinate O in each system. To each point P of the plane an ordered pair of numbers (x,y) is attached as coordinates, where x is the coordinate of the point of intersection of g_1 with the line through P parallel to g_2 and y is the coordinate of the point of intersection of g_2 with the line through P parallel to g_1. The lines g_1 and g_2 are called the x axis and y axis, respectively. It is usually convenient to take the same scale on each axis; that is, the segments joining the unit points on the two axes to the origin are congruent. The notation $P(x,y)$ denotes a point P with coordinates (x,y). If ω denotes the angle made by the positive halves of the two axes, and d the distance between points $P_1(x_1,y_1)$ and $P_2(x_2,y_2)$, application of the law of cosines yields

$$d = [(x_1 - x_2)^2 + (y_1 - y_2)^2 + 2(x_1 - x_2)(y_1 - y_2)\cos\omega]^{1/2}$$

Since $\cos 90° = 0$, this important formula is simplified by taking the axes mutually perpendicular. Though it is occasionally useful to employ oblique axes ($\omega \neq 90°$), the simplifications resulting from a rectangular cartesian coordinate system make it the usual choice. Such a cartesian coordinate system is assumed in what follows. Thus, the distance d between $P_1(x_1,y_1)$ and $P_2(x_2,y_2)$ is given by $[(x_1 - x_2)^2 + (y_1 - y_2)^2]^{1/2}$.

The correspondence between the geometric entity "point" and the arithmetic entity "pair of real numbers," upon which plane analytic geometry is based, results in associating with each geometric locus one or more equations that are satisfied by the coordinates of all those (and only those) points forming the locus (equations of the locus), and in associating with each system of equations in the variables x, y the figure (graph of the equations) whose points are determined by the pairs of numbers satisfying the equations. Thus the algebraic method of studying geometry is balanced by a geometric interpretation of algebra. A central problem in analytic geometry is that of finding equations of certain important figures among curves and surfaces. *See* CURVE FITTING.

By use of the formula for the distance of two points and the definition of a circle, an equation for a circle with center $C(x_0,y_0)$ and radius r ($r \geq 0$) is found to be $(x - x_0)^2 + (y - y_0)^2 = r^2$.

Much of plane analytic geometry deals with a class of curves which (from the way in which they were first studied) are known as conic sections or conics. A conic is the locus of a point P that moves so that its distance from a fixed point F (the focus) is in a constant positive ratio ϵ (the eccentricity) to its distance from a fixed line (the directrix) which is not through F. Let $(c,0)$ be the coordinates of F, $c > 0$, and take the y axis as the directrix. Then $P(x,y)$ satisfies the equation $[(x - c)^2 + y^2]^{1/2} = \epsilon x$; that is, $(1 - \epsilon^2)x^2 + y^2 - 2cx + c^2 = 0$, and it is easily seen that each point whose coordinates satisfy this equation is on the conic. Hence each conic is represented by a second-degree equation in the cartesian coordinates (x,y). A conic is called a parabola, ellipse, or hyperbola accordingly as $\epsilon = 1$, $\epsilon < 1$, $\epsilon > 1$, respectively. *See* CONIC SECTION.

Let cartesian coordinate systems be established on each of three pairwise mutually perpendicular lines of 3-space that intersect in O, the common origin of the systems. Suppose equal scales and call the lines the x axis, y axis, and z axis. To each point P of space an ordered triple (x,y,z) of real numbers is attached as rectangular cartesian coordinates, where x is the coordinate of the foot of the perpendicular from P to the x axis, and y and z are similarly defined. Thus every point of space has unique coordinates, and each ordered triple of real numbers is the coordinates of a point of space. If d denotes the distance between two points $P_1(x_1,y_1,z_1)$ and $P_2(x_2,y_2,z_2)$, then $d = [(x_1 - x_2)^2 + (y_1 - y_2)^2 + (z_1 - z_2)^2]^{1/2}$.

It follows from the definition of a sphere and the formula for distance between two points that $(x - a)^2 + (y - b)^2 + (z - c)^2 = r^2$ is an equation for the sphere with center (a,b,c) and radius r, and by completing the squares of the x, y, and z terms in the equation $x^2 + y^2 + z^2 + 2Dx + 2Ey + 2Fz + G = 0$, it is seen that the locus of such an equation is a sphere with positive or zero radius, or there is no (real) locus.

Any equation in just two of the three coordinates is an equation of a cylinder whose elements are parallel to the axis of the missing variable. Thus the locus in 3-space of $x^2 + y^2 = r^2$ is a (right) circular cylinder whose elements are parallel to the z axis and which intersects the xy plane in the circle $x^2 + y^2 = r^2$, $z = 0$.

Any equation $f(x,y,z)$ 0, with $f(x,y,z)$ homogeneous in x, y, z (for example, $4xy - xz + yz = 0$, $x^3 - xy^2 + z^3 = 0$) has a cone with vertex O as locus.

A surface of revolution is obtained by rotating a plane curve C about a line g of its plane. If $f(x,y) = 0$, $z = 0$ are equations of C, and g is the x axis, the resulting surface of revolution has equation $f(x, \sqrt{y^2 + z^2}) = 0$. Thus the surface generated by revolving the circle $x_2 + (y - b)^2 = a^2$, $z = 0$, about the x axis (the torus or anchor ring, if $b > a$) has the equation $x^2 + (\sqrt{y^2 + z^2} - b)^2 = a^2$.

A quadric surface is the locus of points whose coordinates satisfy an equation of the form $Ax^2 + By^2 + Cz^2 + Dxy + Exz + Fyz + Gx + Hy + Jz + K = 0$, where at least one coefficient of a second-degree term is not zero. Some surfaces obtained by rotating conics about a line belong to this class, for example, spheres, prolate and oblate spheroids (given by rotating an ellipse about its major and minor axes, respectively), hyperboloids and paraboloids resulting from rotations of hyperbolas and parabolas about their axes of symmetry, and right circular cones and cylinders. Cylinders with conics for directrix curves are also members. [L.M.Bl.]

Analytic hierarchy A framework for solving a problem. The analytic hierarchy process is a systematic procedure for representing the elements of any problem. It organizes the basic rationality by breaking down a problem into its smaller constituents and then calls for only simple pairwise comparison judgments, to develop priorities in each level.

The analytic hierarchy process provides a comprehensive framework to cope with intuitive, rational, and irrational factors in making judgments at the same time. It is a method of integrating perceptions and purposes into an overall synthesis. The analytic hierarchy process does not require that judgments be consistent or even transitive. The degree of consistency (or inconsistency) of the judgment is revealed at the end of the analytic hierarchy process.

People making comparisons use their feelings and judgment. Both vary in intensity. To distinguish among different intensities, the scale of absolute numbers in the table is useful.

Scale of relative importance

Intensity of relative importance	Explanation
1 (equal importance)	Two activities contribute equally to the objective
3 (slight importance of one over another)	Experience and judgment slightly favor one activity over another
5 (essential or strong importance)	Experience and judgment strongly favor one activity over another
7 (demonstrated importance)	An activity is strongly favored and its dominance is demonstrated in practice
9 (absolute importance)	The evidence favoring one activity over another is of the highest possible order of affirmation
2, 4, 6, 8 (intermediate values between the two adjacent judgments)	When compromise is needed
Reciprocals of above nonzero numbers (if an activity has one of the above numbers assigned to it when compared with second activity, the second activity has the reciprocal value when compared to the first)	

The analytic hierarchy process can be decomposed into the following steps. Particular steps may be emphasized more in some situations than in others. Also as noted, interaction is generally useful for stimulation and for representing different points of view.

1. Define the problem and determine what knowledge is sought.
2. Structure the hierarchy from the top (the objectives from a broad perspective) through the intermediate levels (criteria on which subsequent levels depend) to the lowest level (which usually is a list of the alternatives).
3. Construct a set of pairwise comparison matrices for each of the lower levels, one matrix for each element in the level immediately above. An element in the higher level is said to be a governing element for those in the lower level since it contributes to it or affects it. In a complete simple hierarchy, every element in the lower level affects every element in the upper level. The elements in the lower level are then compared to each other, based on their effect on the governing element above. This yields a square matrix of judgments. The pairwise comparisons are done in terms of which element dominates the other. These judgments are then expressed as integers according to the judgment values in the table. If element A dominates element B, then the whole number integer is entered in row A, column B, and the reciprocal (fraction) is entered in row B, column A.
4. There are $n(n-1)/2$ judgments required to develop the set of matrices in step 3, where n is the number of elements in the lower level.
5. Having collected all the pairwise comparison data and entered the reciprocals together with n unit entries down the main diagonal, the eigenvalue problem $Aw = \lambda_{max} w$ is solved and consistency is tested, using the departure of λ_{max} from n (see below).
6. Steps 3, 4, and 5 are performed for all levels and clusters in the hierarchy.
7. Hierarchal composition is now used to weigh the eigenvectors by the weights of the criteria, and the sum is taken over all weighted eigenvector entries corresponding to those in the lower level of the hierarchy.
8. The consistency ratio of the entire hierarchy is found by multiplying each consistency index by the priority of the corresponding criterion and adding them together. The result is then divided by the same type of expression, using the random consistency index corresponding to the dimensions of each matrix weighted by the priorities as before. The consistency ratio should be about 10% or less to be acceptable. If not, the quality of the judgments should be improved, perhaps by revising the manner in which questions are asked in making the pairwise comparisons. If this should fail to improve consistency, it is likely that the problem should be more accurately structured; that is, similar elements should be grouped under more meaningful criteria. A return to step 2 would be required, although only the problematic parts of the hierarchy may need revision. *See* DECISION THEORY; SYSTEMS ENGINEERING. [T.L.S.]

Analytical chemistry The science of chemical characterization and measurement. Qualitative analysis is concerned with the description of chemical composition in terms of elements, compounds, or structural units, whereas quantitative analysis is concerned with the measurement of amount.

Instrumental methods, which might better be termed physicochemical methods, make up the vast bulk of both qualitative and quantitative analysis. They include measurements of purely physical characteristics, as well as the use of physical measurements to follow the course of chemical reactions. In the latter category are titrations in which the end points are detected by the measurement of electrical quantities such as electrode potential, conductance, current, impedance, and capacitance; optical quantities such as transmittance, absorbance, emittance, scattering, and reflectance; or other physical quantities such as temperature, density, and surface tension. *See* LIGHT SCATTERING PHOTOMETRY; TITRATION; VOLUMETRIC ANALYSIS.

The scope of analytical chemistry may involve the spatial distribution of elements or compounds within a sample, the distinction between different crystalline forms of a given element or compound, the distinction between different chemical forms (such as the oxidation states of an element), the distinction between a component on the surface or in the interior of particles, and so forth. To permit these more detailed questions to be answered, as well as to improve the speed, accuracy, sensitivity, and selectivity of traditional analysis, a large variety of physical measurements are used. These include methods based on spectroscopic, electrochemical, chromatographic, chemical, and nuclear principles.

The spectroscopic area includes the measurement of emission, absorption, reflection, and scattering phenomena in various regions of the spectrum ranging from gamma rays and x-rays through the ultraviolet, visible, infrared, and microwave. Coupled with these measurements are a wide variety of sample preparation and excitation techniques. Thus the introduction of a sample as either a solid or a solution into a flame, arc, spark, or plasma may be the source of excitation for the measurement of atomic emission, atomic absorption, or atomic fluorescence spectra.

Molecular spectroscopy involves lower-energy forms of excitation such as ultraviolet, visible, or infrared radiation. X-rays are used through emission of characteristic radiation, absorption, or diffraction. Gamma-ray spectroscopy, used to sort out and measure gamma rays according to energy, is usually coupled with nuclear techniques such as neutron activation analysis or photon activation analysis. *See* ACTIVATION ANALYSIS; GAMMA-RAY DETECTORS; MOLECULAR BEAMS; X-RAY DIFFRACTION.

Mass spectrometry is an important method for both inorganic and organic analysis. In the inorganic field, spark-source mass spectrometry permits the detection and estimation of some 70 elements. Secondary-ion mass spectrometry (SIMS) involves the bombardment of a surface with an ion beam, causing the ejection of ions characteristic of the sample. Ion-probe mass spectrometry permits the scanning of a sample to reveal the point-by-point elemental composition of a surface; by gradual erosion of the surface, it determines the composition in the third dimension as well. In the organic field, important structural information is gleaned through high-resolution mass spectrometry, in which molecular fragments are identified through fractional mass differences in atoms of the same principal mass number. *See* MASS SPECTROMETRY; OPTICAL METHODS OF CHEMICAL ANALYSIS; SECONDARY ION MASS SPECTROMETRY (SIMS).

More subtle nuclear phenomena are employed in nuclear magnetic resonance (NMR) spectroscopy, in which the chemical environment of atoms possessing nuclear spins can be probed through the interaction between the nuclear spin and an external magnetic field. *See* NUCLEAR MAGNETIC RESONANCE (NMR).

Several forms of electron spectroscopy are especially useful for surface analysis. The scanning electron microscope (SEM) involves a finely collimated electron beam which sweeps across the surface to produce an image. At the same time, the surface atoms are excited to emit characteristic x-rays which are sorted out according to their energy to reveal point-by-point elemental composition. The same basic technique is used in the electron microprobe. Low-energy electron diffraction (LEED) involves the observation of the diffraction pattern of a reflected electron beam to give structural information about the arrangement of surface atoms.

Techniques used to determine compositional and structural information by measuring the energies of electrons emitted from surface atoms include ultraviolet photoelectron spectroscopy (UPS) and x-ray photoelectron spectroscopy (XPS). The latter technique is sometimes called electron spectroscopy for chemical analysis (ESCA). A related technique, Auger spectroscopy, involves the bombardment of a surface with an electron beam and the observation of the energies of electrons secondarily emitted from surface atoms. *See* ELECTRON DIFFRACTION; ELECTRON PARAMAGNETIC RESONANCE (EPR) SPECTROSCOPY; ELECTRON SPECTROSCOPY.

Methods based on electrochemical principles began with the detection of titration end points through the measurement of electrical quantities. Coulometry utilizes the faraday (96,500 coulombs) as the basis for quantitative measurement, either via a titration with electrochemically generated reagent or via quantitative electrolysis. Potentiometry, the measurement of electrode potentials, is used with various ion-selective electrodes. Various forms of electroanalysis are carried out by applying an electrical signal to an indicator electrode and observing a readout signal. These methods are named according to the measurements; they include cyclic voltammetry, chronopotentiometry, chronoamperometry, and chronocoulometry. Polarography, a special form of diffusion-controlled electrolysis with the dropping mercury electrode, is useful for a wide variety of inorganic and organic analytical applications. *See* COULOMETRIC ANALYSIS; ELECTROCHEMISTRY; ION-SELECTIVE MEMBRANES AND ELECTRODES; POLAROGRAPHIC ANALYSIS.

Chromatographic methods are based on the separation of mixtures by a phase distribution using multiple stages. The stages are not discrete physical entities but are functionally achieved by using a mobile phase, either liquid or gaseous, flowing past a stationary phase, either solid or liquid (in the form of a film on a solid support). In column chromatography the stationary phase is contained in a cylindrical column, whereas in thin-layer chromatography a flat layer is used. In gas chromatography the mobile phase is gaseous, while in liquid chromatography it is liquid. Gel permeation chromatography is a special form of liquid chromatography in which large molecules move faster than small ones because they fail to permeate the structure of the stationary phase. *See* CHEMICAL SEPARATION TECHNIQUES; CHROMATOGRAPHY; GAS CHROMATOGRAPHY; GEL PERMEATION CHROMATOGRAPHY; QUALITATIVE CHEMICAL ANALYSIS; QUANTITATIVE CHEMICAL ANALYSIS. [H.A.L.]

Anamnia Those vertebrate animals, sometimes called Anamniota, which lack an amnion in development. In early classifications the vertebrates were commonly separated on this basis, and the expressions Amniota and Anamniota are still useful in grouping higher and lower vertebrates. Anamnia as a group name includes the Recent classes Agnatha, Chondrichthyes, Osteichthyes, and Amphibia and, by presumption, the class Placodermi, which is known only from fossils. *See* AMNION; AMNIOTA; AMPHIBIA; PISCES (ZOOLOGY). [R.M.B.]

Anaphylaxis An allergic reaction of the immediate hypersensitive type. This can be a local reaction in the skin, if antigen is applied there, or a systemic acute shock syndrome. Both generally begin within seconds after exposure to the appropriate allergen. Anaphylaxis in humans may be of either type.

Sensitization resulting in anaphylaxis and other hypersensitive states requires exposure to an allergen. The induction period for sensitization is usually 7–10 days; this figure is also applicable to the induction period for antibodies.

The antibodies involved in causing anaphylactic reactions share the property of attaching themselves to cell membranes of the sensitive subject and combining with antigen in that locale. Certain cells (mast cells) to which attachment is made contain substances (histamine; "slow-reacting substance," SRS-A; bradykinin; serotonin) which cause contraction of smooth muscle and dilatation of small blood vessels. An eosinophile-attracting substance (ECF-A) and heparin are also released. The antibody-antigen reaction at these cell membranes releases these substances without injury to the cells; these mediators then reveal their activities. *See* HISTAMINE; SEROTONIN.

Useful drugs for the treatment of immediate hypersensitive states are epinephrine, norepinephrine, isoproterinol, certain prostaglandins, histamine and disodium cromoglycate, which act to stabilize mast cell granules. The antihistamine drugs act to block attachment of histamine to receptors on target cells.

The Schultz-Dale reaction provides a method for demonstrating anaphylactic hypersensitivity outside the body. A strip of tissue containing smooth muscle is removed from an animal, suspended in a bath containing a physiologically balanced salt solution, and supplied with oxygen. The tissues commonly used for this purpose are uterine, intestinal, or arterial. With tissue taken from a sensitized animal, addition of the proper allergen to the bath causes contraction of the smooth muscle. The strip may be removed from a normal animal and sensitized by the addition of antibodies to the bath fluid. Again, addition of the appropriate allergen will cause contraction. The test may be useful for studying antigenic relationships of allergenic substances, or for demonstrating antibodies which are too weakly reactive to reveal themselves in test tube reactions. *See* CELLULAR IMMUNOLOGY; HYPERSENSITIVITY; IMMUNOLOGY. [S.R.]

Anaplasia The appearance of cells with an embryonic, immature, or undifferentiated structure at a site where under normal conditions only more mature and differentiated cells are found. Anaplastic cells usually are capable of more rapid growth than normal cells. They appear to have larger, more irregular, darker-staining nuclei than their normal counterparts. There are often several nucleoli and bizarre mitoses in anaplastic cells. The cytoplasm often loses its distinguishing characteristics, for example, the cross striations in muscle. Anaplastic cells are believed by some pathologists to be derived from adult, differentiated cells; it appears more likely that they are derived from the less differentiated intermitotic elements in tissues and under normal conditions would mature as normal cells. Assessment of the degree of anaplasia of a tumor is important because this property usually correlates well with the malignancy of the tumor. *See* ONCOLOGY. [W.F.P./N.K.M.]

Anaplasmosis An infectious arthropod-borne disease of cattle, sheep, goats, and wild ruminants caused by rickettsial organisms of the genus *Anaplasma*. The organism invades red blood cells and causes anemia and sometimes death of the infected animals. *Anaplasma marginale* is the most important species and causes significant losses to the cattle industry throughout tropical and subtropical regions.

Anaplasma belongs to a group of gram-negative bacteria, the rickettsiae, which are obligate intracellular parasites. It is identified in stained blood smears as a round, dense inclusion body measuring approximately 1 micrometer in diameter within red blood cells and comprising a vacuole containing one to eight individual rickettsiae, each of which may invade another red blood cell. The bacteria contain both deoxyribonucleic acid (DNA) and ribonucleic acid (RNA) and multiply by binary fission. *See* RICKETTSIOSES.

The disease is transmitted between animals by blood-feeding arthropods. Mechanical transmission occurs when biting flies (for example, horse flies and deer flies) feed on the blood of an infected animal and quickly move on to another animal. Infection occurs by direct transfer of blood between animals, and the rickettsiae do not develop in the fly. Therefore, many animals may be infected in a short period of time. Mechanical transmission to animals may also occur when needles and instruments contaminated with infected blood are used for vaccinations, ear tagging, and castrations.

Clinical anaplasmosis may not develop for as long as 60 days after infection. In acute anaplasmosis, cattle exhibit depression, loss of appetite, increased temperature, labored breathing, dehydration, and jaundice. During this period, the number of infected red cells increases rapidly. The clinical manifestations follow the rapid removal of large numbers of red blood cells (up to 70%) from circulation by phagocytosis in the reticuloendothelial system. This anemia may become quite severe and death may follow. As immunity develops, rickettsial development stops and the remaining infected red blood cells are removed from circulation. *See* IMMUNITY.

Tetracycline antibiotics are used routinely to treat acute anaplasmosis because the drug inhibits replication of the organisms. Although treated animals remain infected, tetracycline-medicated feeds reduce acute disease symptoms. *See* ANTIBIOTIC; VACCINATION. [E.F.B.]

Anapsida A subclass of reptiles characterized by a roofed temporal region in which there are no temporal fenestrations. Chelonia (turtles), with living representatives, and the extinct Cotylosauria are the two major subdivisions of this subclass. The Mesosauria, an extinct group of aquatic reptiles from the early Permian, have been included, but the assignment is far from certain.

Among the cotylosaurs are found the most primitive known reptiles, which date from the Early Pennsylvanian. These forms and their immediate descendants flourished in the Permian and Triassic periods of the late Paleozoic and early Mesozoic eras, respectively. Turtles first appear as fossils in Triassic rocks and are well represented in the fossil record from that time to the present. Throughout their history anapsids have, for the most part, inhabited areas close to water, with some exceptions among the turtles, and many of them have been semiaquatic in habitat and adaptations. *See* MESOSAURIA; REPTILIA. [E.C.O.]

Anaspida An extinct order of fresh- or brackish-water Agnatha, known from the Upper Silurian of Europe and Canada and from the Upper Devonian of Canada. The Middle Silurian Scottish *Jamoytius* is thought by some to be an anaspid.

Members of this group are small, not exceeding 10 in. (25 cm) in length, and typically have a slender, fusiform body covered with small scales and a rounded, jawless, terminal or subterminal mouth (see illustration). Long paired fins are present,

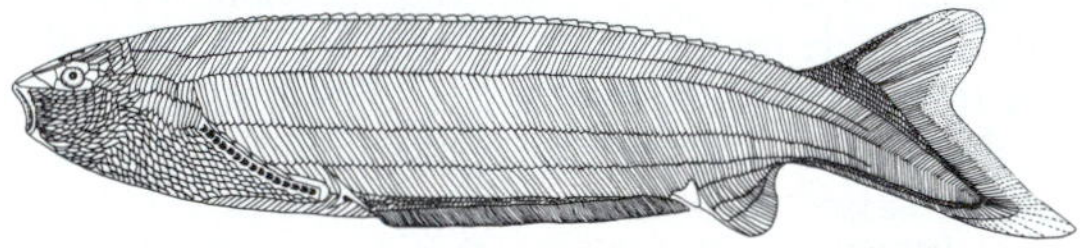

***Pharyngolepsis oblongus* of the Anaspida, reconstruction. (*After A. Ritchie*)**

at least in some genera, and the tail is unusual in having the muscular lobe turned downward. They were probably active, nectonic swimmers, adapted for feeding on minute particles.

A relationship to Osteostraci and Petromyzonida is indicated by the single dorsal nostril lying in front of the pineal eye and between the large paired eyes and by the paired rows of circular gill openings; all three orders are grouped in the superorder Hyperoartii. Anaspida are possibly ancestral to living lampreys. *See* AGNATHA; OSTEOSTRACI; PETROMYZONTIDA. [R.H.De.]

Anaspidacea An order of the crustacean superorder Syncarida, the most primitive of the Eumalacostraca. The families Anaspididae and Koonungidae have living representatives, while the Clarkecarididae are true fossils. The families Gampsonychidae and Palaeocarididae, formerly included in this order, now are assigned to a new order, Palaeocaridacea. The Anaspidacea arose during the Paleozoic Era, and the recent species are now restricted as living fossils to fresh-water habitats in Tasmania and South Australia (not New Zealand). They exhibit a great adaptability to special habitats. The first thoracic somite is incorporated with cephalic tagmata. [H.J.]

Anatomy, regional The detailed study of the anatomy of a part or region of the body of an animal, most commonly applied to regional human anatomy. This is in contrast, but supplementary, to study of organ systems where all the structures pertaining to the system are studied in their continuity.

There are many methods of dividing the body into regions for study, and one such means of classification is shown in the illustration. This system includes only externally visible areas; other systems would include special internal regions as well.

The head, trunk, and extremities are the principal regions. Subdivision of these can be carried out, as illustrated, so that

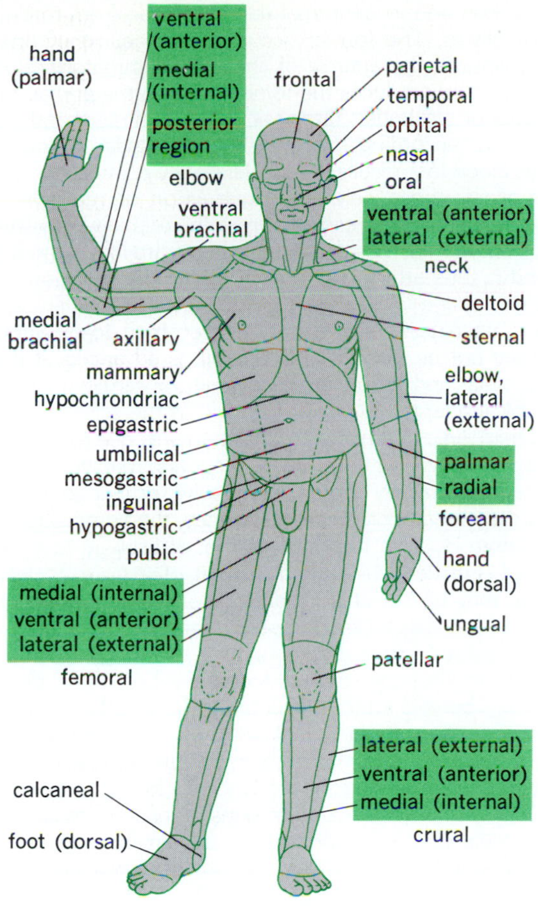

Some of the major regions of the human body seen in ventral view. *(After J. P. Schaeffer, ed., Morris' Human Anatomy, 11th ed., McGraw-Hill, 1953)*

many distinct areas or especially vital regions are indicated. There is no end to the dividing and subdividing a specialist may do to make the task eventually less difficult. [T.S.P.]

Andalusite A nesosilicate mineral, composition Al_2SiO_5, crystallizing in the orthorhombic system. It occurs commonly in large, nearly square prismatic crystals. There is poor prismatic cleavage; the luster is vitreous and the color red, reddish-brown, olive-green, or bluish. Transparent crystals may show strong dichroism, appearing red in one direction and green in another in transmitted light. The specific gravity is 3.1–3.2; hardness is 7½ on Mohs scale, but may be less on the surface because of alteration. *See* SILICATE MINERALS.

Andalusite is abundant in the White Mountains near Laws, California, where for many years it was mined for manufacture of spark plugs and other highly refractive porcelain. Variety chiastolite, in crystals largely altered to mica, is found in Lancaster and Sterling, Massachusetts. Water-worn pebbles of gem quality are found at Minas Gerais, Brazil. [C.S.Hu.]

Andesine A plagioclase feldspar with a composition ranging from $Ab_{70}An_{30}$ to $Ab_{50}An_{50}$, where Ab = $NaAlSi_3O_8$ and An = $CaAl_2Si_2O_8$. In the high-temperature state, andesine has albite-type structure. In the course of cooling, natural material develops a peculiar structural state which, investigated by x-rays, shows reflections that indicate the beginning of an exsolution process, sometimes accompanied by a beautiful variously colored luster (labradorizing). If Fe_2O_3 is present as thin flakes and oriented parallel to certain structurally defined planes, such andesine is called aventurine or sunstone. *See* FELDSPAR; GEM; IGNEOUS ROCKS. [F.H.L.]

Andesite An abundant volcanic rock prevalent in island arcs such as the Aleutians and in ranges of mountains on some continental margins such as the Andes. Most andesite is erupted explosively from steep-sided volcanoes. Andesite has been erupted from most of the world's 400 active volcanoes, including some of the most spectacular mountains (Fuji, Mayon, Shasta, Kliuchevskaya). Four of the United States' national parks encompass andesitic volcanoes (Lassen, Crater Lake, Rainier, Katmai). *See* VOLCANO.

Andesite is a rock with a groundmass of submillimeter-sized crystals. The principal mineral is plagioclase. Pyroxenes, iron oxides, tridymite, cristobalite, and residual glass generally constitute about 40% of the rock. Amphibole or olivine may constitute up to a few percent. Most andesites have 53–62 wt % of SiO_2 and 15–20 wt % of Al_2O_3.

The chemical and mineralogical composition of andesite varies in space and time in a broadly regular way, but with many exceptions. Andesites on continental margins (such as the Alaskan peninsula) are more silicic, in general, than andesites on oceanic islands (such as the Aleutian Islands). Andesites near the ocean and above shallow zones of earthquakes tend to be poorer in alkalies and richer in lime than those farther inland associated with deeper seismicity. Volcanoes formed in the early stage of development of island arcs contain the most iron-rich andesites.

Andesite is similar to continental crust in chemical composition. It is a principal rock type formed at continental margins and along island arcs. The arcs may eventually become attached to the margins of continents; therefore the andesite on island arcs is largely new continental material. Most old andesite now seen on continents probably is also similarly derived. Therefore, continents probably develop by additions of andesite. *See* BASALT; IGNEOUS ROCKS; OCEANIC ISLANDS; PLATE TECTONICS. [A.T.A.]

Androgen One of a class of steroid hormones. Androgens play a major role in the development and maintenance of masculine secondary sexual characters, for example, the seminal vesicle and prostate gland of the male mammal, and the comb, wattles, and spur of the male fowl. They also influence certain other secondary sexual characters, such as hair growth pattern and voice quality in humans. In the fowl, they affect the pattern and seasonal coloration of its feathers, as well as crowing. In addition, androgens affect nitrogen metabolism (anabolic). Androgens are produced in the testis, ovary, adrenal, and most likely, in the placenta. A small portion of the androgen is from corticoids, or adrenal cortex steroids, and from other C_{21} steroids, such as progesterone. *See* HORMONE; OVARY; PROGESTERONE; STEROID; TESTIS. [R.I.D.]

Androgenesis The development of an egg under the influence of the sperm nucleus, without the participation of the egg nucleus. Spontaneous androgenesis occurs as a rare event in nature. It is easily induced experimentally by the following methods: (1) elimination of the egg nucleus from the fertilized egg surgically, with microneedles and pipets; (2) destruction of the egg nucleus in the unfertilized egg by radiation; or (3) by the prolonged refrigeration of freshly fertilized eggs.

Androgenetic development is usually haploid, since the sperm nucleus contains a single, haploid set of chromosomes and genes. Most haploid embryos, in frogs and salamanders at least, become abnormal and retarded at an early stage. *See* GYNOGENESIS; MEROGONY. [G.Fa.]

Andromeda Galaxy The spiral galaxy of type Sb nearest to the Milky Way system. This galaxy is a member of a small cluster of galaxies known as the Local Group. This group contains also the Milky Way system, the Triangulum Nebula (M33), the Large and Small Magellanic Clouds, NGC 6822, and several faint dwarf elliptical galaxies.

The Andromeda Galaxy M31 or NGC 224 is particularly important because it is close enough for its stellar and other content to be studied in great detail. The approximate distance to M31 is known from the apparent luminosities of the cepheid variable stars. The period-luminosity relation for cepheids, combined with the apparent brightness of these stars, gives a distance of about 2,500,000 light-years. All studies show that M31 is typical of other regular spiral galaxies of the Sb class. *See* CEPHEIDS; MILKY WAY GALAXY. [A.Sa.]

Anechoic chamber A room in which essentially an acoustic free field exists. It is sometimes referred to as a free-field or dead room. The word anechoic is derived from the Greek, meaning "without echo." However, a room that is free from echoes (in the usual sense of the word) is not necessarily an anechoic room; in addition, there must be no significant reflections from the boundaries of the room.

Free-field conditions can be approximated when the absorption by the boundaries of the room approaches 100%. To reduce sound reflected by the boundaries to a minimum, the absorption coefficient must be high and the surface areas of the boundaries should be large.

The absorptive material usually installed in such rooms consists of glass fibers or mineral wool held together with a suitable binder. In order to achieve large surface area, parallel blankets of absorptive material are sometimes used to provide a wall construction of considerable thickness. Other types of wall construction include long wedges of absorptive material. [C.M.H.]

Anemia A reduction in the total quantity of hemoglobin or of red blood cells in the circulation. As a matter of convenience and practicality, measures of concentration, rather than the total quantity, are usually employed. However, there are circumstances when, because the plasma volume is less than normal, the concentration of hemoglobin in the circulation will be normal even though the total quantity of hemoglobin, or the red cell mass, is less than normal. In such cases, anemia would be masked. Conversely, in instances where the plasma volume is expanded, the concentration of hemoglobin may be low, even though the total mass of hemoglobin and of red cells is not low. Those cases are examples of spurious or dilutional anemia, which is not really anemia at all. An example is the "anemia" reported in many women during the third trimester of pregnancy when, in fact, the total red cell mass has been increased from the nonpregnant state. *See* BLOOD; HEMOGLOBIN.

There is a normal range over which values of hemoglobin and red cells are distributed among healthy individuals. Values which are less than 2.5 or 3 standard deviations below the mean are indicative of anemia. The mean values are greater for adult males than for females and are greater in adults than in children. Higher altitude of residence results in higher mean values in healthy individuals. The three measures of concentration most often employed are the hemoglobin, the red cell count, and the hematocrit.

The proliferative response to anemia may be insufficient if the production of erythropoietin by the kidneys is defective, or the marrow stem cells themselves are damaged or diminished in number. Ineffective erythropoiesis results when red cell precursors in the marrow fail to mature and survive for delivery to the circulation. Abnormal nuclear maturation results from malabsorption of vitamin B_{12} or of folic acid and from competitive effects of analogs which affect purine or pyrimidine metabolism. Defective synthesis of hemoglobin impairs cytoplasmic maturation. The majority of cases are due to deficiency of body stores of iron and to abnormal release of iron from reticuloendothelial stores. The former occurs in iron-deficiency anemia, and the latter in the anemia of chronic inflammatory diseases. Acute blood loss reduces the total blood volume and produces symptoms of weakness, dizziness, thirst, faintness, and shock, in that order according to increasing magnitude of blood loss. Chronic blood loss results in iron-deficiency anemia.

Hemolysis, the accelerated destruction of red cells, also induces a proliferative response from the marrow. However, it differs from hemorrhage because red cells are lost without plasma, and it thus affects the measures of concentration at the outset.

Pallor, weakness, and fatigue are common to all anemias. They may not be noticed until anemia is advanced, if it is of gradual onset and there has been time for cardiovascular and biochemical adaptation. Faint jaundice in the sclerae is a feature of hemolytic anemia, whereas in anemia due to lack of vitamin B_{12} glossitis and neuropathy occur and may be noted.

The administration of iron will resolve iron-deficiency anemia. In the absence of proved deficiency, the administration of iron, vitamin B_{12} and folic acid are not of value to the anemic individual. Further, the administration of iron, especially by injection, may be harmful. Prednisone and other adrenal corticosteroids are helpful in hemolytic anemias associated with autoantibodies. Hereditary disorders are generally not amenable to therapy, except for those hemolytic diseases which may benefit after splenectomy. In all other cases, the treatment of the anemia is achieved by treating the underlying disease, such as hypothyroidism, rheumatoid arthritis, or leukemia. Blood transfusions are reserved for acute blood loss, or in chronic anemia if there are signs of inadequate cardiovascular or pulmonary compensation and an underlying cause cannot be found or treated. *See* HEMATOLOGIC DISORDERS. [A.H.]

Anemometer A device which measures the magnitude of air velocity. Anemometers are commonly known as the devices which measure wind magnitude, but they are also used to measure the rate of flow of air or of other gases in other applications, for example, in wind tunnels and on aircraft. The most common types are the cup, vane, and hot-wire anemometers. *See* AIR-VELOCITY MEASUREMENT; CUP ANEMOMETER; HOT-WIRE ANEMOMETER; VANE ANEMOMETER. [H.F.]

Anesthesia Loss of sensation with or without loss of consciousness. There are several ways of producing anesthesia, with the choice dependent on the type of surgery and the medical condition and preference of the patient.

During general anesthesia a state of complete insensitivity or unconsciousness is produced when anesthetic gases are inhaled; adjuvant drugs are often given intravenously. Although the mechanism of general anesthesia is unknown, the anesthetics act on the upper reticular formation of neurons in the thalamus and midbrain (neuronal structures necessary for activating the cerebral cortex and maintaining an active, attentive state).

Analgesia, without loss of consciousness, results from injecting a solution of local anesthetic drug either into the cerebrospinal fluid surrounding the spinal cord (spinal anesthesia) or into the epidural space surrounding the cerebrospinal fluid (epidural anesthesia). The local anesthetic acts by blocking the conduction of nerve impulses. Narcotic opioids are injected postoperatively into either the epidural space or cerebrospinal fluid for pain relief.

Analgesia can be localized to a small area, for example, the forearm, by injecting a local anesthetic solution around nerves supplying the area (the bracheal plexus in the upper arm and chest supplies the forearm). Large peripheral nerves may also be blocked individually by using this method.

Acupuncture is an ancient procedure, once used only in

China but now practiced in the United States and elsewhere. It involves inserting needles into specific points around the body, as determined from historical charts, and manipulation of the needles; sometimes electric current is applied. Weak analgesia results through alteration of pain perception. Although sometimes helpful for chronic pain, acupuncture generally has not been found satisfactory for surgical anesthesia. *See* CENTRAL NERVOUS SYSTEM; PAIN. [F.K.O.]

Aneurysm A localized abnormal dilatation of an artery due to a weakening of the vessel wall. Aneurysms are classified as true and false. The wall of a true aneurysm is composed of some layer of the original vessel, while a false aneurysm is a blood-filled space communicating with an artery through an abnormal opening. Among the causes of aneurysms are arteriosclerosis, syphilis, degenerative changes and cystic necrosis of the media of vessels, trauma, bacterial infections, arteritis, and congenital deformities.

Arteriosclerotic and syphilitic aneurysms are the most common types and involve principally the thoracic and abdominal aorta. A less common type of aneurysm is the dissecting aneurysm of the aorta, in which blood enters the wall of the vessel, separating it into two layers, thus creating a new channel. Perforation through the outer layer of the aorta with hemorrhage and sudden death is the usual outcome. Congenital "berry" aneurysms [small spherical dilations, usually 0.2–0.6 in. (0.6–1.5 cm) in diameter] occur in vessels of the brain and are probably developmental in origin, related to defects in the muscular coat of the arteries at their angles of bifurcation. Rupture of these aneurysms accounts for the majority of spontaneous, nontraumatic subarachnoid (intracranial) hemorrhages. *See* HEMORRHAGE; THROMBOSIS. [F.A.C.]

Angina pectoris A clinical complex characterized by various degrees of typical chest pain which occurs in sudden attacks. The chest pain may be accompanied by other symptoms, notably pain or discomfort of the arms, shoulders, or other sites. These symptoms are most often induced by some physical or emotional stress and subside promptly, in most cases, with rest or appropriate therapy.

The most common occurrence preceding an attack of angina is any change which may cause a decrease of blood supply to the heart muscle or sudden extra demands on the heart so that there is a relative inadequacy of blood. Treatment is usually quite effective, and the prognosis is favorable, particularly if the patient develops the proper attitude toward the disorder. *See* HEART DISORDERS. [E.G.St./N.K.M.]

Angle The word angle is commonly used with three different meanings: geometric, arithmetic, and algebraic. Also, there is a distinction made between angles formed by lines (rays) and angles formed by planes.

An angle (in the geometric sense of the term) is a geometric figure formed by two rays that have a common vertex (Fig. 1). The two rays are called the sides of the angle. Two angles are equal if and only if they are congruent. A straight angle is one

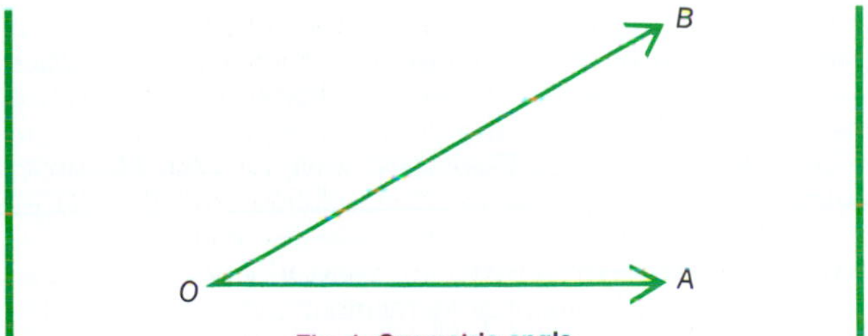

Fig. 1. Geometric angle.

whose sides lie in the same straight line but do not coincide. All straight angles are equal. Three rays having a common vertex form three angles. If one of these angles is a straight angle, the other two angles are called supplementary. If two angles that are supplementary are also equal, each is said to be a right angle. The lines forming the two sides of a right angle are said to be perpendicular or orthogonal or normal to each other. All right angles are equal. If two lines intersect, the pairs of adjacent angles are supplementary and the pairs of vertical angles (opposite angles) are equal.

Just as two points a and b determine a geometric figure $[AB]$ called a line segment, a denominate number $\overline{AB}$ called its length, and an algebraic signed quantity $\overrightarrow{AB}$ called its directed length, so two rays $\uparrow O(A)$ and $\uparrow O(B)$ emanating from a common vertex O determine not only a geometric figure $\angle AOB$, but also both an arithmetic quantity and an algebraic signed quantity, which measure the figure. The word angle, in the second sense, means the number of angular units, commonly between 0° and 180°, assigned as a measure of the sector formed, or of the circular arc intercepted, by the two rays.

In the third sense, the word angle is used to denote a signed measure, which may be positive, zero, or negative, without limit in size, and which, if $\uparrow O(A)$, $\uparrow O(B)$, $\uparrow O(C)$ are any three coplanar rays with common vertex, satisfies the equations

$$\measuredangle AOB = -\measuredangle BOA \qquad \measuredangle AOB + \measuredangle BOC = \measuredangle AOC$$

In this third sense an angle is not determined by its sides alone. An order must be assigned to the sides so that one is the initial side and the other the terminal side; a positive direction of rotation must be assigned in the plane of the angle; and an integer n, positive, negative, or zero, must be assigned such that the angle is greater than, or equal to, n complete revolutions of four right angles but less than $n + 1$ complete revolutions of four right angles. In mathematical texts the usual

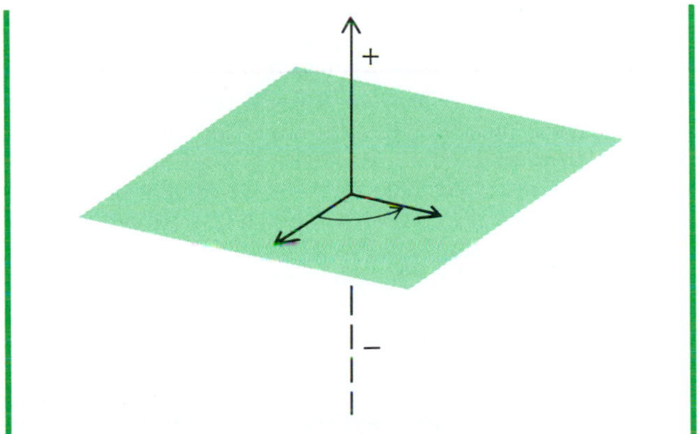

Fig. 2. Directed angle.

method of assigning the positive direction of rotation in a sensed plane is to define it as the counterclockwise direction (opposite to that of the hands of a clock) when viewed from the positive side of the plane (Fig. 2). [J.S.F]

Angle modulation Modulation in which the angle (entire argument) of a sinusoidal carrier is the parameter changed by the modulating wave. This variation in angle may be related to the modulating wave in any predetermined unique manner. Frequency and phase modulation are particular forms of angle modulation. Often the term frequency modulation is used to connote angle modulation. *See* FREQUENCY MODULATION; MODULATION; PHASE MODULATION. [H.S.Bl.]

Anglesite A mineral with the chemical composition $PbSO_4$. Anglesite occurs in white or gray, orthorhombic, tabular or prismatic crystals or compact masses. It is a common secondary mineral, usually formed by the oxidation of galena. Fracture is conchoidal and luster is adamantine. Hardness is 2.5–3 on Mohs scale and specific gravity is 6.38. The mineral does not occur in large enough quantity to be mined as an ore of lead, and is therefore of no particular commercial value. Fine exceptional crystals of anglesite have been found throughout the world. [E.C.T.C.]

Anguilliformes A large order of actinopterygian fishes containing the true eels. This group, also known as the Apodes, now includes the former order Saccopharyngiformes or Lyomeri (gulpers or gulper eels). The Anguilliformes have a ribbonlike, larval stage in development. The chief characters of the Aguilliformes include a pectoral girdle which, when present, is free from the head and suspended from the vertebral column; no symplectic, mesocoracoid, and posttemporal bones; pectoral fin present or absent; absence of a pelvic fin and girdle in recent forms; no fin spines; elongate body with numerous vertebrae; scales present or absent; paired orbitosphenoids; and restricted gill apertures.

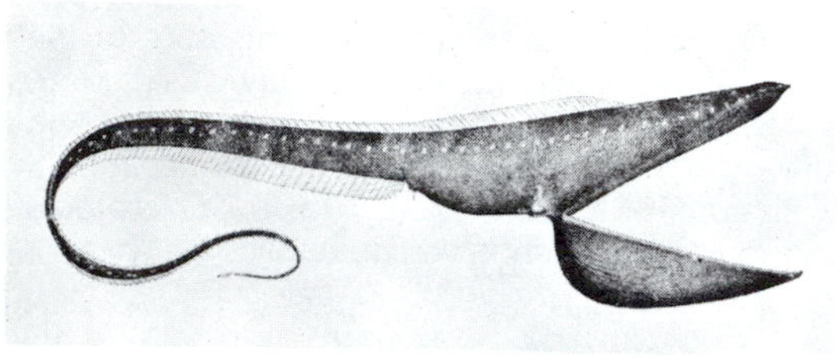

Gulper eel (*Eupharynx bairdi*). (After G. B. Goode and T. H. Bean, *Oceanic Ichthyology, U.S. Nat. Mus. Spec. Bull. no. 2*, 1896)

The typical eels (suborder Anguilloidei) have a swim bladder with a duct, a small opercle, and 6–22 branchiostegals, and the premaxilla, mesethmoid, and lateral ethmoids are fused into a tooth-bearing ethmopremaxillary block that separates the maxillae. Eels, which date from the Upper Cretaceous, are classified in about 20 Recent families and 110 genera, and there are several hundred species. *See* EEL.

The gulpers (suborder Saccopharyngoidei) have degenerative adaptations, including loss of swim bladder, opercle, branchiostegal rays, caudal fin, scales, and ribs. The tremendous mouth is greatly modified (see illustration), the eyes are tiny and placed far forward, the pharynx is enormously distensible, and the tail is slender and tapering. These are rare oceanic fishes, classified in three families, three genera, and nine species. Their fossil history is unknown. *See* ACTINOPTERYGII; OSTEICHTHYES; TELEOSTEI. [R.M.B.]

Angular acceleration A vector quantity representing the rate of change of angular velocity of a body experiencing rotational motion. If, for example, at an instant t_1, a rigid body is rotating about an axis with an angular velocity $\boldsymbol{\omega}_1$, and at a later time t_2, it has an angular velocity $\boldsymbol{\omega}_2$ the average angular acceleration $\bar{\boldsymbol{\alpha}}$ is given by the equation below expressed in radians per second per second.

$$\alpha = \frac{\omega_2 - \omega_1}{t_2 - t_1} = \frac{\Delta\omega}{\Delta t}$$

The instantaneous angular acceleration is given by $\boldsymbol{\alpha} = d\boldsymbol{\omega}/dt$. *See* ACCELERATION; ROTATIONAL MOTION. [C.E.H./R.J.S.]

Angular correlations A technique of nuclear experimentation for measuring spins of nuclear states, the angular momentum mixtures of incoming or outgoing particles, and the multipole mixing of emitted gamma rays. The technique is to measure the dependence of the intensity or the cross section of a nuclear reaction on the directions of two or more radiations. The experiments to which the technique is applicable are reactions excited by a high-energy incident beam, in which the incident beam and an emitted reaction product define two directions, or radioactive decay, in which two emitted radiations, electrons or gamma rays from a decaying nucleus, are measured in coincidence. Angular correlations between an incident and an emitted beam are also called angular distributions, and have been used since the inception of nuclear physics to study the interaction potential between nucleons or nuclei and to investigate direct interaction models. A closely related class of experiments are the polarization–direction correlations, which involve the correlation between an incoming and an outgoing particle or gamma ray coupled with polarization of the incident particle or detection of the polarization of the outgoing radiation, or sometimes both. Here the polarization is equivalent to defining an additional direction in space. *See* ANGULAR MOMENTUM; MULTIPOLE RADIATION; NUCLEAR REACTION; SCATTERING EXPERIMENTS (NUCLEI).

The term angular correlation is normally applied to experiments that can be interpreted in terms of the compound nucleus model. It is assumed that the intermediate state has sharp spin and parity or is a mixture of a small number of overlapping states. This requirement is ordinarily satisfied for experiments with radioactive sources. For bombardment experiments the work must be carried out at a strong resonance. [A.J.F.]

Angular frequency A measure of the rate of oscillation of sinusoidally varying phenomena. If a particle is moving with constant speed v on a circular path of radius r, as indicated in the illustration, it will require a time $t = 2\pi r/v$ to complete one cycle and the frequency of the motion is $f = 1/t$. The angular frequency ω is then defined to be the number of radians traversed per unit time. Since there are 2π radians per cycle, $\omega = 2\pi/T = 2\pi f$.

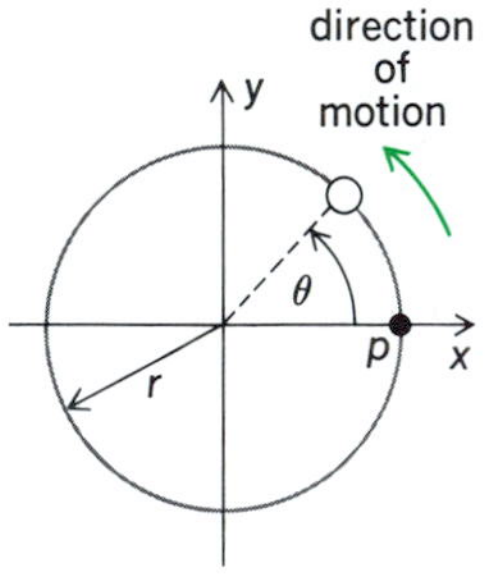

Illustration of angular frequency.

The position of the particle in the diagram can be written in component form as $x = r \cos \theta$, $y = r \sin \theta$. If the particle is arbitrarily taken to have been at the point p of the illustration at the zero of time, $t = 0$, then $\theta = \omega t$ and the position of the particle is given by $x = r \cos \omega t$, $y = r \sin \omega t$. These latter equations, indicating the sinusoidal or harmonic character of the position components, also illustrate the primary role of the angular frequency. *See* FREQUENCY (WAVE MOTION); HARMONIC MOTION. [K.L.K.]

Angular momentum In classical physics, angular momentum is the moment of momentum and is, conceptually, the momentum associated with rotation. A particle of mass m

moving with velocity $\boldsymbol{v}$ and having position vector $\boldsymbol{r}$ with respect to a fixed coordinate origin is said to have an angular momentum $\boldsymbol{L}$ about this origin given by the equation below,

$$\boldsymbol{L} = m\,(\boldsymbol{r} \times \boldsymbol{v}) = \boldsymbol{r} \times \boldsymbol{p}$$

where $\boldsymbol{p}$ is the linear momentum. *See* CALCULUS OF VECTORS; MOMENTUM.

Angular momentum—like linear momentum—is an additive vectorial quantity. The angular momentum of a system of particles is obtained by summing the angular momenta of the individual particles.

The importance of angular momentum in classical mechanics derives from the fact that it is conserved (remains constant in time) if no external torques are applied. Angular momentum conservation and rotation are of key importance, for example, in the structure, shape, and evolution of celestial bodies of all kinds. There is a close connection between conservation laws and invariance (symmetry) transformations (Noether's theorem). To be precise, if Lagrange's equations are invariant for spatial (temporal) translations, then linear momentum (respectively energy) is conserved. Similarly, if Lagrange's equations are invariant to rotations, then angular momentum is conserved. This deep connection between symmetry and conservation laws is not valid for Newton's laws. The requirement that Lagrange's equations apply constitutes a restriction on the forces. *See* CONSERVATION OF MOMENTUM; LAGRANGE'S EQUATIONS; SYMMETRY LAWS (PHYSICS).

This connection between symmetry and conservation laws is valid in quantum physics, since quantum mechanics is based on hamiltonian (and hence lagrangian) equations. *See* QUANTUM MECHANICS. [L.C.B.]

Anharmonic oscillator A system that oscillates with a periodic motion that is not simple harmonic. An oscillator is anharmonic if the restoring force opposing a displacement from the position of equilibrium is a nonlinear function of the displacement. The free motion of such an oscillator may be a complicated function of time, It is periodic, but with a period that depends on the amplitude. A damping nonlinearity in the velocity can also give rise to anharmonicity in an oscillator.

Many oscillators that are approximately harmonic for sufficiently small amplitudes become anharmonic for larger motions. The thermal expansion of solids is attributed to the nonlinear forces between atoms. Small atomic vibrations (at low temperatures) are harmonic, and are centered about the static equilibrium distance. As the temperature rises, the amplitude of atomic oscillations increases, and the oscillations become anharmonic. One consequence is that the resulting motions are unsymmetric, and the mean separation between atoms increases beyond the static equilibrium distance. *See* HARMONIC OSCILLATOR; LATTICE VIBRATIONS; THERMAL EXPANSION. [J.M.Ke.]

Anhydrite A mineral with the chemical composition $CaSO_4$. Anhydrite occurs commonly in white and grayish granular masses, rarely in large, orthorhombic crystals. Fracture is uneven and luster is pearly to vitreous. Hardness is 3–3.5 on Mohs scale and specific gravity is 2.98. Anhydrite is an important rock-forming mineral and occurs in association with gypsum, limestone, dolomite, and salt beds. Under natural conditions, anhydrite hydrates slowly, but readily, to gypsum. It is not used as widely as gypsum. Anhydrite is of worldwide distribution. Large deposits occur in the Carlsbad district, Eddy County, New Mexico, and in salt-dome areas in Texas and Louisiana. *See* GYPSUM; SALINE EVAPORITE. [E.C.T.C.]

Aniline An aromatic primary amine used in the manufacture of urethane polymers, rubber chemicals (antioxidants, accelerators), agricultural chemicals (herbicides, insecticides, fungicides), and dyes, in that order of volume. Its properties are: boiling point 363°F (184°C); melting point 21°F (−6°C); density 1.0215; refractive index 1.5863; solubility about 3 g/100 g of water; basic dissociation constant $K_b = 3.8 \times 10^{-10}$. Aniline is prepared by the catalytic reduction of nitrobenzene in the presence of hydrogen. Aniline readily undergoes the reactions characteristic of primary aromatic amines. In addition, it may be oxidized with reagents such as hydrogen peroxide, chromic acid, or sodium chlorate. Reduction by hydrogen can take place in the presence of a suitable catalyst. The reaction of aniline and phosgene will produce carbonic acid derivatives. *See* AMINE; AROMATIC HYDROCARBON. [L.B.C.]

Animal feeds Substances suitable for the nutrition of animals, mainly livestock, poultry, and pets.

In formulating livestock and poultry diets, animal nutritionists select from a variety of bulk feed ingredients including forages, feed grains, protein concentrates, by-product feeds, and vitamin-mineral supplements. These are formulated into complete rations in modern feed manufacturing plants to provide a balance of energy, protein, vitamins, and minerals to meet the nutrient requirements of various classes of livestock and poultry. Drugs and other additives may also be needed to meet specific needs.

In a broad sense, forages consist of grasses and legumes fed to ruminants (cattle, sheep, goats) and horses. Because of symbiotic microbes (bacteria and protozoa) in the digestive tract of these animals, they are uniquely equipped to utilize nutrients in forages; thus they do not compete with humans for the same food source.

Pasture and rangeland areas are covered with forage that is harvested by the grazing animal. Grazing quality forage has the advantages of providing more digestible protein per acre than most other feeds, eliminating harvesting and feeding costs, and reducing soil erosion that results from the action of water and wind.

Because of variable climatic conditions, forage species do not grow at a uniform rate throughout the growing season, and most farms produce more forage than can be consumed by the livestock. This surplus forage is conserved as hay or silage and either fed or sold at later periods. Hay is dried forage which has been cut and sun-cured in the field to about 10–20% moisture and then baled or otherwise packaged. It is commonly fed to ruminants and horses when grazing is inadequate or when the animals are confined. Silage is fermented forage which is produced by the action of acetic and lactic acid–producing anaerobic bacteria in closed structures called silos. Preserving forage by dehydration is relatively new compared to hay-making or ensiling. Mechanically dehydrated forages usually do not compete with hay or silage as roughages for ruminants. They are used primarily as supplements in swine and poultry rations and pet foods.

Feed grains are any of several grains commonly used as livestock and poultry feeds; they include corn, barley, oats, and grain sorghum (milo). Although wheat is second only to corn as a cereal grain in the United States, wheat is not usually fed to livestock. However, when properly used, wheat is a satisfactory feed for all classes of livestock. Accordingly, when the price of feed grains is high and surplus wheat is available at low price, considerable amounts may be fed to animals. *See* WHEAT.

Protein supplements are feeds which contain more than 20% protein or protein equivalent. They are primarily added to rations containing feed grains or forages which are low in protein. High-protein meals can be made from soybean, cottonseed, sunflower seed, safflower seed, rapeseed, peanut, linseed, and coconut (copra). These meals are by-products of the oilseed-crushing industry; they vary in feeding value (that is, protein quantity and quality), depending upon the amount of hull or seed coat remaining in the meal, conditions of time and temperature during extraction, and constituent amino acids of the protein. Protein supplements of animal origin primarily consist of inedible products derived from meat packing or rendering plants, marine sources, poultry processing plants, and dairies. In general, such products are high in protein and of excellent quality, being well balanced in amino acids and providing vitamins and minerals. Cattle and other ruminants are able to derive a portion of their dietary protein from nonprotein nitrogen via microbial protein synthesis in the rumen. Amino acids, amides, ammonium salts, and other nonprotein nitrogen sources provide nitrogen from which symbiotic rumen bacteria synthesize microbial protein. This microbial protein is then digested by the ruminant animal in the abomasum and gastrointestinal tract.

The absence in the diet of one or more of the required vitamins may prevent the animal from growing or reproducing, or may cause deficiency diseases. If the deficiency is severe, the animals may die. Farm animals fed diets containing high-quality green forage usually receive an adequate supply of most of the required vitamins. To ensure that rations contain sufficient vitamins to meet dietary requirements, manufacturers often fortify feed formulations with one or more vitamins. *See* VITAMIN.

Animals, like plants, require minerals to grow, and if deficiencies occur, livestock become unthrifty, lose weight, or exhibit other deficiency symptoms. Minerals serve many vital functions. To ensure an adequate supply of minerals, feed manufacturers often add mineral supplements. Farm animals may also be permitted free access to a mineral mixture. If this is allowed, care should be taken, particularly with trace minerals, to see that the animals do not consume too much, or toxicity may result.

Feed additives and implants are nonnutritive substances which enhance the utilization of a feed or productive performance of the animal. It is estimated that in the United States 75% of the finishing cattle, finishing lambs, and growing-finishing pigs receive feed additives or implants. Use of these products and guidelines for administration and withdrawal of approved products are regulated and prescribed by the U.S. Food and Drug Administration (FDA). Also, feed-control officials in individual states may regulate use of these substances for livestock marketed intrastate. Prior to receiving approval for marketing a product, the manufacturer must prove its safety and efficacy through a rigorous testing program. Antibiotics, hormones, and hormonelike substances are used. Approval of the use of some has been withdrawn because of undesirable effects on humans consuming the animal products as food. *See* NUTRITION. [D.Bu.]

Animal growth Growth may be most simply defined as increase in mass or dimensions of an organism with time. It is one of the basic characteristics of living things and represents the visible result of a complex and interrelated series of metabolic and developmental events. One of the unique features of biological growth is that the organism changes in size and shape, and to some extent in chemical composition, although it still retains its integrity and its individuality. This is true because growth fundamentally involves synthesis by the organism of more materials like itself.

Growth occurs by two main processes: increase in the number of cells and increase in the size of cells. Living cells, removed from the body and placed in an appropriate culture medium, display growth in its most uncomplicated form. They grow by synthesizing new protoplasm, dividing into smaller cells, and then repeating the process over and over as long as essential nutrients are supplied and accumulation of deleterious waste products is prevented.

In the intact animal, growth begins at, or soon after, the initiation of development. It involves cell division and synthesis of new protoplasm from raw material, either contained in the egg or derived from the environment. However, the process is much more complex than the growth of cells in tissue culture. Except in early embryonic development, growth is normally never dissociated from such processes as differentiation (diversification of cell structure and function) and morphogenesis (change in the form and pattern of the embryo). In later development, growth by cell enlargement is the predominant process. *See* ANIMAL MORPHOGENESIS; CELL DIVISION; DIFFERENTIATION; EMBRYOLOGY; MITOSIS; TISSUE CULTURE.

The vast majority of animals, including most protozoa, require already elaborated organic molecules in order to synthesize new living material during growth. For animals in general, the chemical requirements for growth are (1) inorganic substances, (2) organic substances, especially certain amino acids and fatty acids, and (3) accessory factors or vitamins.

In both invertebrates and vertebrates, silicon, calcium, magnesium, carbon as carbonate, and phosphorus as phosphate are extremely important. Salts of sodium, potassium, calcium, and magnesium are essential components of body fluids. They contribute to the osmotic properties of the body fluids and provide a milieu in which cells, tissues, and organs may function properly. *See* OSMOREGULATORY MECHANISMS.

The organic compounds used by animals as raw materials for growth are carbohydrates, proteins, and fats, or their breakdown products. Carbohydrate apparently is not essential; laboratory animals can grow in the complete absence of carbohydrate. Fat, as such, is probably also nonessential, although some animals need certain unsaturated fatty acids to sustain growth.

Proteins represent the chief organic constituent of living tissues and are the most important raw material for growth. During growth, the proteins stored in the egg or provided as food from outside are digested into their constituent amino acids. These are then resynthesized into the substance of the living cells. The proteins synthesized by the organism have specific characteristics which depend upon the kinds and numbers of amino acids they contain; therefore not all proteins are equally capable of supporting growth. Proteins which fail to induce growth are deficient in one or more essential amino acids (amino acids needed by the organism but which it cannot synthesize).

In addition to the materials used to synthesize the bulk of the protoplasmic system, animal organisms also require certain accessory substances in order to grow. Some of these are now known as vitamins. *See* VITAMIN.

The growth of vertebrates is influenced by a specific growth hormone produced by the anterior lobe of the pituitary gland. Secretions from the thyroid gland, the adrenal cortex, and, in some instances, the gonads also affect growth, but perhaps less directly. Growth in various arthropods is controlled by hormone action. Molting, an essential prelude to growth in insects and crustaceans, is under hormonal control. *See* ENDOCRINE SYSTEM (INVERTEBRATE); ENDOCRINE SYSTEM (VERTEBRATE).

Relative growth is also referred to as heterogony or heterauxesis. Growth determines not only the size of the animal but also its shape and form. As long as an animal grows at the same rate along all its dimensions, it will not change in bodily proportions. However, when the growth rate in certain directions is different from that along others, or when one or more parts of the growing organism develop more rapidly or more slowly than others, progressive changes in form result.

The form changes produced by alterations in body proportion are well illustrated in human development. At the second month of fetal life, the head and neck account for almost one-half the total volume of the fetus; at birth the relative size of the head is only 32% of the body; at maturity it is 10%. Conversely, at the same three stages, the legs comprise, respectively, 2%, 16%, and 29% of the total body volume. During this developmental span the relative size of the trunk remains constant at approximately 50% of body volume. Obviously, the growth of the head during prenatal development after the second fetal month is relatively smaller and that of the legs relatively larger than growth of the body as a whole.

There are many factors involved in the regulation of growth, some intrinsic, some environmental. One important factor is the histological differentiation of the cells of the organism; as differentiation proceeds, the rate of growth declines. Heredity is important in determining the limit and the rate of growth. Tall parents give rise to tall children, and the offspring of toy terriers and mastiffs grow to characteristic size. Hormonal influences also affect growth, as do such environmental factors as nutritive level, temperature, and degree of crowding. In most organisms, growth ceases at maturity, but in some, growth continues throughout life. Complete cessation of growth when the adult stage has been reached occurs mainly in terrestrial animals. [E.J.Bo.]

Animal kingdom One of the two generally accepted major divisions of organisms which live or have lived on the Earth. More than 1,000,000 species of animals are known, and these are divided into various groups (taxa), large and small, according to their relationships and complexity of organization.

Most members of the animal kingdom can be distinguished from those of the plant kingdom by their greater motility, by the more constant form and structure of their bodies, by the placement of the organs, which are internal rather than external, and by the nature of the tissue cells, which are enclosed in delicate membranes rather than rigid walls of cellulose. *See* CELL (BIOLOGY); CELL WALLS (PLANT).

Animals subsist on complex organic materials, derived directly or indirectly from plants, which are chemically altered and metabolized to provide energy for their functions and materials for growth. These processes require oxygen, and the end products are largely carbon dioxide, water, and nitrogenous wastes. Most animals possess a nervous system and exhibit a rapid response to stimuli. However, forms exist which are intermediate between plants and animals in many of these features; for example, certain one-celled organisms such as *Euglena* move about and take food like an animal but contain chlorophyll like a plant. For these reasons, some biologists recognize up to five kingdoms: Monera (bacteria, blue-green algae); Protista (protozoa, chrysophytes); Fungi (slime molds, true fungi); Plantae (algae and higher plants); and Animalia (multicellular animals). Disagreement still exists among zoologists on the limits and relationships of some phyla. *See* ANIMAL SYSTEMATICS; PLANT KINGDOM; TAXONOMIC CATEGORIES. [W.J.B.]

Animal morphogenesis The development of form and pattern in animals. Most animals begin their lives as a single-celled product of the fusion of an egg and a sperm. As the result of repeated cellular divisions, growth, and cellular specialization, this fertilized egg, or zygote, becomes a complicated organism. The number, shape, and arrangement of the cells and the arrangement of their structural products determine the form of the organism. *See* CELL LINEAGE.

The major question is how the specialized cells arise in the proper pattern time after time during normal development. The zygote contains a number of codes within its DNA (deoxyribonucleic acid). These codes can be translated into the different proteins which are synthesized by the cells. It is necessary that only one or a few of the special proteins appear in any one cell; chaos would result if the proteins appeared in the wrong places. For example, the cells of the lens of the eye contain special transparent proteins known as crystallins. If these cells produced opaque materials, vision would be impaired. *See* DEOXYRIBONUCLEIC ACID (DNA); GENE ACTION; GENETIC CODE.

The axes along which cells will communicate during specialization are established in early development. Anterior and posterior poles of an egg are usually established in the ovary. In the eggs of frogs, the dorsoventral axis is determined by the orientation of the egg when it is laid. The middorsal line is the center of a field in which cells will communicate during differentiation. At first and for many hours of development, only quantitative differences are detectable along axes of amphibian eggs. The rates of all activities, such as cellular division, oxygen consumption, and production of RNA (ribonucleic acid), are most rapid in the anterior and dorsal regions and grade off toward the posterior and ventral regions.

Qualitative differentiation begins during gastrulation when the mesoderm invaginates and comes to lie between the ectoderm and endoderm. As the mesoderm glides under the ectoderm, the ectodermal cells are activated. There are many pathways of differentiation open to them, and what they will form depends upon their position with respect to their neighbors. Most cells of the early gastrula are still labile and, when transplanted, can develop in conformity with their new environment. After a day or so the dividing cells begin to differentiate along the anteroposterior and dorsoventral axes. Those in the most anterior position develop more rapidly and become parts of the head, and those in the most dorsal position become parts of the back. Thus, a polarity is established which is focused in the anterior dorsal region. *See* EMBRYONIC DIFFERENTIATION.

As cells differentiate, their surfaces also become different and cellular affinities begin to play a role in the development of pattern. Differences in cellular affinity become apparent as early as the gastrula stage when the embryo is becoming a three-layered sphere. When ectodermal and endodermal cells are removed and cultured together, they segregate, each associating only with its own kind. Mesodermal cells, forerunners of the connective and other tissues, make contact with each other but also make tight junctions with ectodermal and endodermal cells. As differentiation proceeds, more and more different cell surfaces arise.

If two partially developed organs are disaggregated and the dissociated cells are mixed in culture dishes, the dissociated cells move like amebas, coming to rest only when they meet their own kind or a few other kinds with which they can associate. Completely jumbled differentiating kidney tubule cells and precartilage cells segregate and do not come to rest until the kidney cells have made kidney tubules and solid groups of cartilage cells have started to form a cartilaginous matrix. The shape of the group of cells depends on surface patterns of individual cells.

Like differentiating cells, structural molecules secreted by some cells can associate in only one way, depending upon atomic quality and spacing. Under the proper conditions, molecules of collagen join to form long fibers which, in turn, can associate with other fibers in forming tendons and ligaments. Under other suitable conditions, a fibrous substrate produced by fibroblasts serves as a surface upon which another kind of tissue can grow. Under still other conditions, a meshwork of fibers is formed into which other cells will secrete a more rigid matrix, as in the case of cartilage and bone. While molecular forces are important in the origin of tissues, part of the form of the structural tissues depends upon mechanical forces. Sheets of bone, for example, form in the directions which most resist applied pressures.

The form of an organism also depends upon the size of the parts. Normal tissues and organs grow rapidly and then reach an equilibrium, in which there is only enough growth to compensate for losses. For example, if a large part of an adult mammalian liver is cut away, the remainder grows rapidly until the total mass is reconstituted. *See* ANIMAL GROWTH; EMBRYOLOGY. [S.M.R.]

Animal symmetry Animal symmetry relates the organization of parts in animal bodies to the geometrical design that each type suggests. Spherical symmetry is exhibited by

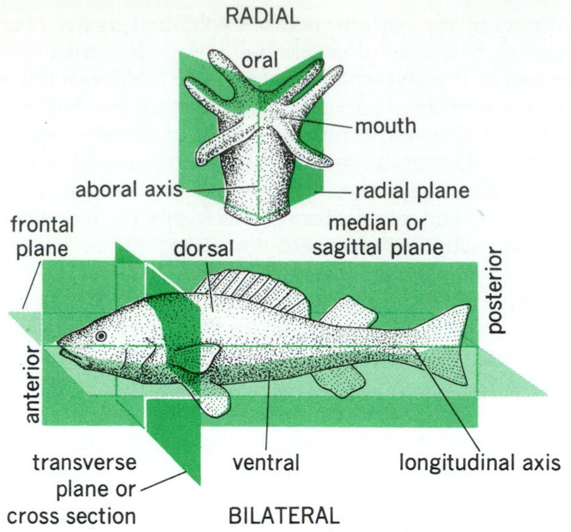

Types of symmetry and the axes, planes, and regions in animal bodies. (*From T. I. Storer and P. L. Usinger, General Zoology, 4th ed., McGraw-Hill, 1965*)

some protozoans, such as the Heliozoia and Radiolaria: the body is spherical with its parts concentrically around, or radiating from, a central point. Radial symmetry is exemplified by the echinoderms and most coelenterates: the body is structurally a cylinder having a central axis named the longitudinal, anteroposterior, or oral-aboral axis (see illustration). Any plane though this axis divides the animal into like halves. Often several planes, from the axis outward, can divide the body into a number of like portions, or antimeres. The ctenophores and many sea anemones and corals possess biradial symmetry, basically radial but with some parts arranged on one plane through the central axis. The great majority of animals have bilateral, or two-sided, symmetry, in which a median or sagittal plane divides the body into equivalent right and left halves, each a mirror image of the other. [T.I.S.]

Animal systematics The science of animal classification. The term had its origin in the systems of classification developed by the early naturalists, such as the *Systema Naturae* (1735) of C. Linnaeus. Animal taxonomy, derived from a word proposed (1813) originally by A. de Candolle for the theory of plant classification, is now generally used interchangeably with animal systematics. The task of naming, characterizing, and cataloging the tremendous number of species has required the cooperative effort of systematists from all parts of the world during the last 200 years, and the job is far from complete, especially in the less conspicuous groups of invertebrates. The more important task of arranging the animals in a hierarchial system of taxa which is designed to reflect their evolutionary relationships has challenged some of the greatest zoologists and has required the use of comparative data derived from many basic fields, including morphology, physiology, ethology, ecology, and genetics. The complex superstructure of classification, into which all newly discovered forms must be made to fit, is being constantly modified and perfected as new knowledge permits. *See* Animal kingdom.

The written record of the study of animals, as with most knowledge, traces back to such early Greek scholars as Hippocrates and Democritus. However, Aristotle (384–322 B.C.) was probably the first to organize the knowledge of animals and to characterize their bodily structures and habits in a comparative manner. His terminology reveals that he recognized animal groups which correspond to many modern phyla and classes and, at a lower level, he utilized the term *genos* to explain a group concept often equivalent to the modern higher categories, and the term *eidos* to designate the individual animal form.

Descriptive phase. For nearly 2000 years these rudiments of an Aristotelian classification were little improved upon, although the knowledge of animal kinds was greatly expanded by the descriptive works of the Renaissance zoologists. It remained for the Swedish naturalist C. Linnaeus (1707–1778) and his immediate predecessors to crystallize emerging concepts, methods, and techniques into a workable system which permitted zoological classification to proceed in an orderly manner and animal systematics to develop as a science. In this system the key position of the species was recognized, a binominal system for their designation was provided, and a framework of classification was set up in which they could be arranged according to degrees of similarity and differences. This ushered in the initial or descriptive phase of animal systematics sometimes referred to as alpha taxonomy. *See* Species concept; Taxonomic categories; Zoological nomenclature.

Phylogenetic phase. The second major step in the development of the science of animal systematics was the concept of organic evolution, which was developed independently by the traveling field naturalists C. R. Darwin (1809–1882) and A. R. Wallace (1823–1913) and brought to a dramatic focus by the publication of Darwin's *Origin of Species* (1859). This ushered in a period of phylogenetic speculation, in which much of the interest of taxonomists shifted from the naming of species to attempts to understand the significance of higher categories and interpret their relationships.

During this period, Ernst Haeckel introduced (1886) the method of representing phylogeny (evolutionary history) by means of treelike diagrams. Modifications of his method are still used by animal systematists to express degrees of presumed relationship in various groups. This level of systematic studies is sometimes referred to as beta taxonomy. *See* Phylogeny.

Species problem. The third stage in the history of animal systematics involved a renewal of interest in the species, with the growth of the genetic approach to problems of speciation and more widespread recognition of the significance of variation. This has been termed the new systematics or gamma taxonomy. Emphasis on the population aspect of the species problem led systematists to give more attention to the geographical distribution of populations within the species, to the ecological requirements of species and smaller populations, and to their genetic composition. The application of this dynamic concept of species to the samples (specimens) of the practicing animal systematist resulted in a sounder and more useful classification, which has permitted more widespread biological generalizations. Renewed interest in evolutionary, morphological, and macroevolutionary mechanisms, combined with the findings of the new systematics, has led to a revival of work on the theoretical foundations of classification and on the practical methods of formulating actual classifications for animal groups. *See* Speciation.

Higher classification. While intensive research continues on the nature of species and species formation, renewed interest has developed in the systematics of the higher categories. This is especially true in the philosophical and basic methodological approaches to classifications. The sources of evidence for classification are comparative studies of observable features of recent and fossil organisms and inferences derived from these studies. Taxa (taxon, singular) or groups of organisms are presumed to be monophyletic (evolving from a common ancestor), and as a general rule, the degree of phenotypical similarity between organisms is closely correlated with the degree of their phylogenetic relationship and time of common origin. The more similar organisms are, the closer is their relationship and the more recent is the time of their common ancestry.

However, many complicating factors, such as different evolution rates, extinction of lineages, convergence, parallelism, and independent origin of similar features, obscure these relationships and render systematic work more an art than an exact science.

The several methods of systematic studies of the higher categories can be grouped under three headings: the phenetic, the cladistic (phylogenetic systematics), and the classical evolutionary approaches. All these approaches are based upon comparative studies of organisms, but differ in basic assumptions and philosophies.

The phenetic approach is based upon the strict assessment of the degree of similarity between organisms in contrast to the strict genealogical or cladistic method. The basic assumption is that a classification may be based on a comparison of phenotypical characters, with the degree of relationship being correlated absolutely with the degree of phenotypical similarity; no other inferences or assumptions are used. It is believed that the problems of convergence and other obscuring factors are overcome by using large numbers of features in the comparison. The family of methods called numerical taxonomy utilizes the evidence based on a series of hypotheses concerning first, the relationships of gene loci to phenotypic characters, that is, observable characters; and second, the sampling of information on these gene loci indirectly by sampling the characters of the phenotype.

In sharp contrast to the phenetic method, which is based solely upon analysis of the degree of similarity between organisms, the cladistic (phylogenetic) flatly rejects similarity as a guide to relationships on the grounds that it is too subjective. W. Hennig, the major proponent of the cladistic method, argues that the only objective approach to classification is on the basis of the forks or branching points between phyletic lineages. In this approach the notion of relationship is restricted to the concept of forks in much the same way as a strict genealogical relationship is in a family. It is argued that basically all branches are dichotomies, and that at least one change in one character occurs at a branch in one of the two lineages, and at least one character change occurs between two forks in a lineage. The two lineages at a branching point represent sister groups, and by definition every taxon has one sister group. Characters are either plesiomorphs (the original character found in the ancestral form) or apomorphs (the derived character found in one of the descendant sister groups). Every monophyletic taxon includes all species that have descended from the stem species of the taxon (the stem species is, by definition, a member of this taxon), and is characterized by the possession by all members of a series of autapomorphous features (those restricted to that taxon).

Both the phenetic approach and the cladistic approach are logical systems as far as they go, but both suffer from the same defect, namely, that they include only one of the two factors that characterize evolutionary relationships between living organisms. These factors are the branching points in the phylogenetic lineages (emphasized strictly by the cladistic school) and the degree of evolutionary modification between groups, or conversely, the degree of similarity between taxa (emphasized strictly by the phenetic school). Because the evolutionary method includes both factors in the analyses of relationships between organisms, it cannot claim or hope to be as simple as either the phenetic or the cladistic methods. Moreover, it is not possible to formulate simple rules for evolutionary systematics, as is the case for the other approaches. However, the evolutionary method has been and still is the prevailing approach by far to classification of living organisms since Darwin. It is used by most taxonomists and is the basis for most classificatory systems used by biologists. Both the phenetic and cladistic methods have been developed and advocated strongly since 1945, partly as a reaction to the very real difficulties in applying the evolutionary approach to the exceedingly complex tangle of relationships between living organisms.

Importance of systematics. Taxonomy is not only the most elementary part of zoology, since animals cannot be discussed or treated in a scientific way until some taxonomy has been achieved, but at the same time it is the most inclusive, since in its various aspects and branches it gathers together, utilizes, summarizes, and implements everything that is known about animals, whether morphological, physiological, psychological, or ecological.

Not only is sound systematics essential to progress in most phases of the basic biological sciences, but it also contributes materially to the applied sciences, in particular, medicine, public health, agriculture, conservation, and natural-resource management. The naming and identification of animal species has provided a filing system for the vast amount of information accumulated about injurious forms. The natural classification has given a means for making predictions and generalizations as to the probable habits, distribution, future importance, and means of control of newly discovered pests, as well as leads and clues as to the endemic source of possible parasites and predators, which might be sought to help suppress populations of introduced pests, especially insects. Applied taxonomy is also the basis of plant quarantine enforcement, in which accurate identification may prevent losses of millions of dollars to farmers, and the basis of agricultural pest control, in which erroneous identification may invalidate costly control procedures or upset the natural balance of populations in an economically disastrous manner. Careful systematic studies have led also to the solution of such public health problems as the cause of the uneven distribution of malaria in Europe and to the concept of species sanitation, a practice which has greatly accelerated the suppression of some arthropod-borne diseases. *See* PLANT TAXONOMY. [W.J.B.]

Animal virus Viruses are the smallest infectious agents, 20–300 nanometers in diameter, containing as their genome (genetic material) a molecule of either RNA (ribonucleic acid) or DNA (deoxyribonucleic acid). The complete infective virus particle is called the virion. In some instances (adenoviruses, papovaviruses, picornaviruses) it may be identical with the nucleocapsid, that is, the capsid together with the enclosed nucleic acid core. More complex virions (herpesviruses, myxoviruses) include the nucleocapsid plus a surrounding envelope. The capsid is the symmetric protein shell which encloses the genome. Often empty capsids are by-products of the viral replicative cycle. The basic building blocks of the capsid are structural units; these may be individual polypeptides. Morphological units known as capsomeres are seen in the electron microscope on the surfaces of isometric virus particles, that is, those with cubic symmetry. These units represent clusters of polypeptides which, when completely assembled, form the capsid.

Viruses are able to propagate only within living cells. The viral genome causes the normal cellular metabolism to be diverted into supplying the raw materials, energy, and synthetic machinery for producing more viral nucleic acid. It codes also for a viral protein shell (capsid), which is necessary for the cell-to-cell transfer of the complete virus particle (virion). Viruses outside living cells are inert particles which neither respire as do bacteria nor possess enzymes customarily associated with metabolic activity. *See* NUCLEIC ACID.

Viruses are generally destroyed by heating at 140°F (60°C) for 30 min and can be preserved by freezing at low temperatures and, in some instances, by lyophilization. *See* LYOPHILIZATION.

Like other microbiological organisms, viruses also mutate. Examples of virus mutations may be seen in nature in the evolution of vaccinia virus and in the almost yearly appearance of new variants of influenza virus. *See* INFLUENZA; MUTATION.

Meaningful genetic studies with animal viruses have been possible only since the early 1970s. Two factors have made these studies possible. The first was the development of the plaque assay for quantitation of virus infectivity. The second factor was the development of stable genetic markers, which, ideally, should be easily recognized and should result from single mutations. Some commonly used markers are specific virus-induced antigens, drug resistance, host range, inability to grow at elevated temperatures, and plaque size. Virus mutants that possess these markers are obtained either after spontaneous mutation or, more frequently, after treatment of the virus population with a mutagen. The use of conditional-lethal mutants provides a powerful tool for the study of the genetics and molecular biology of viruses. Conditional-lethal mutants are mutants which are lethal (in that no infectious virus is produced) under one set of conditions—termed nonpermissive conditions—but which yield normal infectious progeny under other conditions—termed permissive conditions.

Viral multiplication begins with the attachment of the intact virus particle to the host cell. Most of its nucleic acid but little of its protein then enters the cell. The cell nucleic acid disintegrates as a part of the infectious process. The biochemical events in the infected cell are determined by the viral nucleic acid, which not only directs the production of new viral nucleic acid, but also (in separate biochemical steps) governs the synthesis of new viral protein. Complete virus particles are then assembled within the cell, in some cases leaving almost as soon as they are manufactured, but in other cases accumulating and leaving when the cell dies and disintegrates. During viral replication the capsid sometimes encloses host nucleic acid rather than viral nucleic acid. Such particles, pseudovirions, look like ordinary virus particles when observed with the electron microscope, but they do not replicate as do ordinary virus particles. They contain the wrong nucleic acid.

Viruses may be transmitted by (1) direct transmission from person to person by contact, in which droplet or aerosol infections may play the major role (for example, influenza, measles, and smallpox); (2) transmission by means of the alimentary tract as a result of intimate association with carrier, food, and drink (for example, enterovirus infections and infectious hepatitis); (3) transmission by means of dust (for example, Q fever); (4) transmission by bite (for example, rabies); (5) transmission from lower vertebrate to humans by contact or excretions (for example, Q fever and psittacosis); and (6) transmission by means of an arthropod vector (for example, yellow fever, dengue, and equine, St. Louis, and Russian spring-summer encephalitis).

In vertebrates the invasion of most viruses evokes a violent reaction, usually of short duration. The result is decisive. Either the host succumbs or it lives because of the production of antibodies that neutralize or kill the virus. Regardless of the outcome, the sojourn of the active virus is usually short (although latent virus states may occur, as in herpes, adenovirus, and cytomegalovirus infections). In arthropod vectors of a virus, the relationship is usually quite different. The viruses may produce little or no ill effect, and remain active in the arthropod throughout the host's natural life. Thus arthropods, in contrast to vertebrates, act as permanent hosts and reservoirs. In tick-borne infections the virus may be transmitted from the adult arthropod to its offspring by means of the egg (transovarian passage); thus the cycle may continue with or without intervention of a vertebrate host, *See* ADENOVIRIDAE; ARBOVIRAL ENCEPHALITIDES; CHICKENPOX AND SHINGLES; COMMON COLD; COXSACKIEVIRUS; CYTOMEGALOVIRUS INFECTION; ECHOVIRUS; ENTEROVIRUS; EQUINE ENCEPHALITIS; FOOT-AND-MOUTH DISEASE; HERPES; INFECTIOUS MONONUCLEOSIS; INTERFERON; KERATOCONJUNCTIVITIS; MEASLES; MUMPS; MYXOVIRUS; NEWCASTLE DISEASE; PARAINFLUENZA VIRUS; PARAMYXOVIRUS; PHLEBOTOMUS FEVER; PICORNAVIRIDAE; RABIES; RHINOVIRUS; ROSEOLA INFANTUM; SMALLPOX; TICK FEVER; VIRUS; VIRUS INTERFERENCE; WEST NILE FEVER. [J.L.Me.]

Anise One of the earlist aromatics mentioned in literature. The plant, *Pimpinella anisum* (Umbelliferae), is an annual herb about 2 ft (0.6 m) tall and a native of the Mediterranean region. It is cultivated extensively in Europe, Asia Minor, India, and parts of South America. The small fruits are used for flavoring cakes, curries, pastry, and candy. The distilled oil is used in medicine, soaps, perfumery, and cosmetics. *See* APIALES; SPICE AND FLAVORING. [P.D.St./E.L.C.]

Anisomyaria An order containing seven superfamilies of marine and brackish bivalves with byssal attachment at some time in their geological history (the Ostreacea are exceptional, with the lower valve attached by byssal cement at settlement, and with no record of ancestral attachment by byssal threads). The Anisomyaria possess some primitive features and some secondarily simple features: the hinge is edentulous or with inconspicuous teeth (except *Spondylus*); there are no siphons; minimal points of fusion occur between the left and right mantle lobes; the Mytilacea, Pteriacea, Pectinacea, and Anomiacea have filibranch ctenidia. Specialized features include simultaneous hermaphroditism in some Pectinacea, sex reversal and incubation of larvae in some Ostreacea, and protandrous hermaphroditism in others. Few features are common to all Anisomyaria, and the systematic status is debatable. Studies suggest addition of the isomyarian Arcacea and Limopsacea, and placing in an order Pteriomorpha, subclass Lamellibranchia, class Bivalvia. *See* BIVALVIA; MOLLUSCA. [R.D.P.]

Anisotropy (physics) The quality of variation of a physical property with the direction in a body along which the property is measured. For example, the resistivity of certain single crystals measured with the electric field along a particular crystallographic direction may be higher than along directions perpendicular to it. Thus such crystals are anisotropic with respect to resistivity. *See* ELASTICITY; ISOTROPY (PHYSICS). [D.T.]

Ankerite The carbonate mineral $Ca(Fe,Mg)(CO_3)_2$, also commonly containing some manganese. The mineral has hexagonal (rhombohedral) symmetry and has the cation-ordered structure of dolomite. The name is applied only to those species in which at least 20% of the magnesium positions are occupied by iron or manganese; species containing less iron are termed ferroan dolomites. The pure compound, $CaFe(CO_3)_2$, has never been found in nature and has never been synthesized as an ordered compound. *See* DOLOMITE.

Ankerite is commonly white to light brown, its specific gravity is about 3, and its hardness is about 4 on Mohs scale. *See* CARBONATE MINERALS. [A.M.G.]

Ankle The part of the leg consisting of the ankle joint, the tarsal (ankle) bones, and related structures. The joint is formed by two bones of the leg, the tibia and fibula, which articulate with the talus (astragalus), one of the seven tarsal bones. Another prominent tarsal bone is the calcaneus, or heel bone, which bears much of the weight in walking. The inner and outer prominences commonly known as "ankle bones" are the overhanging portions of the tibia and fibula; each prominence is called a malleolus. *See* LEG (ANATOMY).

The tarsal bones are held in place, yet allowed movement, by strong elastic ligaments, as well as by the shapes of the bony surfaces. The ankle joint is a synovial, modified hinge joint, to which movement is imparted by numerous muscles of the leg and foot. *See* JOINT (ANATOMY). [T.S.P.]

Annelida The phylum comprising the multisegmented, invertebrate wormlike animals, of which the most numerous are the marine bristle worms and the most familiar the terrestrial earthworms. The Annelida (meaning little annuli or rings) includes the Polychaeta (meaning many setae); the earthworms and fresh-water worms, or Oligochaeta (meaning few setae); the marine and fresh-water Hirudinea or leeches; and two other marine classes having affinities with the Polychaeta: The Archiannelida (meaning primitive annelids) are small heteromorphic marine worms, and the Myzostomaria (meaning sucker mouth) are parasites of crinoid echinoderms. *See* ARCHIANNELIDA; HIRUDINEA; MYZOSTOMARIA; OLIGOCHAETA; POLYCHAETA.

These five groups share few common characters and little resemblance except that most have a wormlike body. Typically they are bilaterally symmetrical, lack a skeleton, and have a short to long linear body divided into rings or segments (metameres), which are separated from one another by transverse walls or septa. The mouth is an anteroventral or anterior vent at the forward end of the alimentary tract, and the anus is posterodorsal or posterior at the hind end of the gut.

The metameres may be similar throughout, resulting in an annulated cylinder, as in earthworms and *Lumbrineris*. More frequently the successive segments are dissimilar, resulting in regions modified for particular functions. Each segment may be simple or uniannular, corresponding to a metamere, or it may be divided or multiannulate. Segments may have lateral fleshy outgrowths called parapodia (meaning side feet), armed with special secreted bristles or rods, called setae and acicula; they provide protection and aid in locomotion. Setae are lacking in Hirudinea and some polychaetes. The body is covered by a thin to thick epithelium which is never shed.

Annelids have sense organs of many kinds. Most conspicuous are those on the anterior, or head end, and on parapodia. Eyes which function as photoreceptors may be variously developed, simple to complex, small to large, absent to numerous, and obscure to bizarre. Structurally they are simple light-absorbing pigment spots, usually paired on the head, but also on body segments or the anal end. Feelers or tactoreceptors are frequently on the prostomium as antennae; on anterior segments as long filiform or thick fleshy structures called tentacles; and on parapodia as cirri, papillae, scales, or tactile hairs. Cilia in bands or clusters or in grooves occur in specific patterns. The most conspicuous receptors are the large carbuncles (nuchal organs) of amphinomid polychaetes, as simple or complex fleshy, folded, paired organs surrounding the cephalic structures. The nuchal organs of some syllids and phyllodocids, as well as the ciliated ridges of errantiate and sedentary polychaetes, illustrate the diversity and importance of these structures.

Light production is known in many annelids. Some oligochaetes shed luminous mucus through the mouth or anus. Many polychaetes have special photocytes which luminesce. Some *Chaetopterus* colonies shed bright mucus from the palpi and parapodia. *Polycirrus* among the terebellids is highly luminescent. Pelagic spawning stages of *Odontosyllis* (a syllid) and *Tharyx* (cirratulid) flash bright light when disturbed. Some scale worms (polynoids) have light-producing cells on the undersides of elytra. *See* BIOLUMINESCENCE.

The alimentary tract is a straight or sinuous tube, consisting of mouth, pharynx, esophagus, stomach, gut, and pygidium or anus. Its several parts may be further differentiated and named for structure and function. The mouth may be a simple anterior (oligochaetes) or anteroventral pore (many polychaetes) provided with highly complex organs or accessory parts. The upper lip is formed by the prostomium, and the lower lip by the ventrum of the first segment. The mouth is followed by the buccal cavity which may be modified as a proboscis or saclike eversible pouch. The buccal cavity may be followed by a short to long, muscularized, eversible proboscis which captures and breaks up or compresses food particles. Its inner walls may be fortified by papillae or hard gnaths or jaws. A short esophagus leads to the muscular stomach or digestive region, which is followed by the gut. The proctodaeum, or region preceding the anus, expels the wastes as fecal pellets of characteristic form.

The nervous system consists of a dorsal, bilaterally symmetrical, ganglionic mass or brain within or behind the preoral region. The brain is connected to the ventral cord through the circumesophageal connectives which extend about the oral cavity. The ventral cord may be single or paired, nearly smooth or nodular, or ganglionated according to the segmental pattern of the body. The brain sends out lateral branches to the eyes, palpi, antennae, or other structures; the ventral cord has lateral branches to all fleshy parts which receive stimuli.

The circulatory system consists of dorsal and ventral longitudinal, median vessels located above and below the alimentary tract. Lateral branches extend to all parts of the body. Pulsating or propelling contractile portions, sometimes called hearts, are in the anterior dorsal vessel or also at intervals in the ventral vessels. In oligochaetes these are segmental vessels of varying number connecting the dorsal and ventral vessels and surrounding some portion of the alimentary tract. The contained blood is red through the presence of a hemoglobin-like substance, which may be in corpuscles or granular in the fluid, or it is green, through the presence of chlorocruorin. These colors when diluted are yellow or colorless, as in many small annelids. Some annelids lack a closed circulatory system so that blood and coelomic fluids mix freely, resulting in a hemocoel; it may be partial or complete. Many annelids have a special organ called a cardiac or heart body surrounding the pulsating vessel and sometimes visible as a thick brown or red body of spongy tissue; its function is to dispose of circulatory wastes.

The excretory system functions to dispose of wastes in the coelom. The special organs for the function are called nephridia. Nephridia may be few in number and conspicuous, or limited to anterior segments, as in many sedentary families, or may be segmental, a pair to a segment, or dispersed and irregular.

The muscular system consists of an outer circular and an inner longitudinal system of muscles, each varying in extent and density according to species. In addition, an oblique series, between outer and inner layers, is well developed in annelids performing complex lateral movements.

The ability to replace lost parts is highly developed in annelids. Most frequent is the replacement of tail, parapodia, and setae. The anterior end may be replaced, provided the break is postpharyngeal. Least differentiated species regenerate more completely than highly modified ones. Younger individuals repair more easily or quickly than older ones. Fragmentation may result in as many new individuals as there are pieces. The presence of a part of the ventral nerve cord is necessary to develop a new head. *See* REGENERATION (BIOLOGY).

Depending on the species, reproduction may be sexual, asexual, or both. Sexual reproduction may be dioecious, in which male and female are similar, or rarely dissimilar. Individuals may be hermaphroditic, both male and female, but with cross fertilization. Some annelids are protandric hermaphrodites, in which the sexual stages alternate.

Fossil annelids, or segmented worms, are relatively rare, yet sufficiently common to indicate that this large and varied group of invertebrates has been active and abundant for more than 500,000,000 years. [O.H.]

Anomalons High-energy nuclear fragments associated with the fragmentation of projectile (beam) nuclei at relativistic energies that exhibit anomalously short-interaction mean free paths. Because the mean free paths of nuclei in matter are inversely proportional to their reaction cross sections, that is, nuclear sizes, the shortening of the mean free paths of projectile fragments implies the existence of nuclear entities—anomalons—that possess nuclear dimensions beyond those known for conventional nuclei.

Experimental evidence for anomalons was first noted in cosmic-ray experiments in the early 1950s, with subsequent evidence coming from experiments carried out in high-energy heavy-ion accelerators. However, conclusive proof that anomalons exist as a physical entity has yet to be fully established. Still, the possibility that anomalons may exist has far-reaching consequences for theoretical notions involving the basic composition of nuclear matter.

Plausible theoretical speculations on anomalons have ventured into notions of quark structures in nuclei based on quantum chromodynamics, a theory that has had great success in high-energy physics. Nonquark approaches involve the formation of bubble and quasimolecular nuclei and clusters of π^- mesons and neutrons bound to nuclear fragments. *See* NUCLEAR REACTION; NUCLEAR STRUCTURE; QUANTUM CHROMODYNAMICS. [H.H.Hec.]

Anomopoda An order of fresh-water branchiopod crustaceans, formerly included in the Cladocera. Exceptionally these organisms may be as much as 6 mm (0.24 in.) in length, but often are less than 1 mm (0.04 in.). So-called water fleas of the genus *Daphnia* are the most familiar anomopods. *Daphnia* and its close relatives swim freely in open water; however, many anomopods are benthic, predominantly crawling species. Worldwide in distribution, anomopods are among the most abundant and successful of all fresh-water invertebrates. *See* BRANCHIOPODA. [G.Fr.]

Anopla A class of the phylum Rhynchocoela which is divided into the orders Palaeonemertini and Heteronemertini. The simple tubular proboscis lacks stylets and resembles the body wall in structure. The mouth is posterior to the brain. The nervous system lies either immediately below the epidermis or among the musculature of the body wall. The vascular system is well developed. Anoplan rhynchocoelans are common on rocky shores, living beneath stones or interstitially in shell debris or coarser sands. *See* ENOPLA; HETERONEMERTINI; PALAEONEMERTINI; RHYNCHOCOELA. [J.B.J.]

Anoplura A small group of insects usually considered to constitute an order. They are commonly known as the sucking lice. All are parasites living in the covering hair of mammals. About 250 species are now known, and these comprise probably about half of the species in the world, but the group is known in detail to few persons.

These lice are distinguishable from the Mallophaga by their mouthparts and manner of feeding. The mouthparts consist of three very slender stylets which form a tube. When at rest, they are retracted in a pocket which lies just behind the mouth. The antennae are usually five-segmented, rarely three-segmented. The thoracic segments are always very closely fused and lack wings. The claw is one-segmented, and on at least one pair of legs this claw is enlarged and can be folded into a process from the tibia to grasp a hair of the host. In some species two and in others all of the legs are thus modified. The ovipositor is very much reduced. It consists of but little more than two flaps which are able to close around a hair. Eyes are commonly lacking, and in those species in which they do occur, they are reduced to a pair of simple lenses or two light-receptive spots. All the species are quite small; the largest scarcely exceeds 0.2 in. (5 mm) in length and the smallest does not attain 0.04 in. (1 mm).

Feeding is accomplished by thrusting forth the tube formed by the stylets, piercing the skin of the host, and sucking blood by a pump in the throat of the insect. This habit of sucking

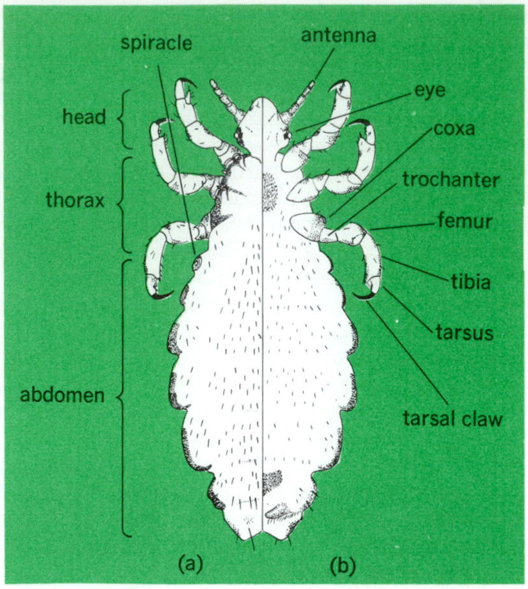

The louse *Pediculus humanus*, female, shown in (*a*) dorsal and (*b*) ventral views.

blood gives the Anoplura a special importance. They take up any disease-producing organisms in the host's blood and may transfer these organisms to another individual. Thus, they transfer the organisms which cause epidemic typhus, relapsing fever, and trench fever. The transfer of a louse, from one host to another, can occur only when the hosts are in close bodily contact, as in the nest or at the time of mating. In the case of humans, the lice may become detached and enter the clothing or the bedding. In times of social disturbance, such as war or when people are crowded together as formerly occurred on ships, or in jails or slums, the opportunities for an exchange of these parasites are enormously increased. The close restriction of each species of lice to its special host is sufficient to account for the fact that diseases are not transferred from one species of mammal to another.

There are two species of lice which occur as common ectoparasites upon humans. One of them is *Pediculus humanus*, the head and body louse which transmits the diseases mentioned above (see illustration). The other species is *Phthirius pubis*, known as the crab louse. It transmits no known disease. The lice of the Old World monkeys are referred to another genus, *Pedicinus*. *See* INSECTA; LOUSE; MALLOPHAGA. [G.F.F./D.M.DeL.]

Anorexia nervosa A psychiatric disorder in which a dramatic reduction in caloric intake consequent to excessive dieting leads to significant bodily, physiological, biochemical, emotional, psychological, and behavioral disturbances.

Anorexia nervosa is typically an illness of adolescent females: 90% of all cases begin in girls who are between 12 and 20 years of age. Nevertheless, this disorder can also occur in males, in prepubertal girls, and in women well into their third decade. Moreover, if the illness becomes chronic, it can persist into mid-life and beyond. Estimates of the incidence of anorexia nervosa are limited but indicate that about 1–5% of adolescent girls living in industrialized cultures are affected, and that the rate is rising.

There is usually a conscious decision made to embark upon a diet. This decision often occurs in the context of some stressful life change situation. In addition, there are usually two personality characteristics: a morbid fear of losing control over one's weight and a distorted body image. The amount of weight lost can vary considerably. The usual criterion for making a diagnosis of anoxeria nervosa is a weight change of at least 25% from premorbid weight (to at least 15% below ideal weight in persons who were overweight at the onset).

In addition to these core disturbances, there is an array of associated symptoms that characterize most persons with anorexia nervosa, such as amenorrhea, increased physical activity, and insomnia. Use of emetics, cathartics, and diuretics also is not uncommon. Difficulty in recognizing satiation is a common complaint, and if the illness becomes chronic, sensations of hunger may also become difficult to discriminate. Obsessional thinking and depression are not unusual in anorectic women.

Perhaps as many as 45–50% of anorectic patients whose illness persists for more than 1–2 years will develop the additional eating disturbance known as bulimia. Bulimia refers to binge eating or compulsive overeating wherein thousands of calories are consumed in a relatively brief period of time. The binge characteristically involves carbohydrates—the very type of food that the anorectic so rigidly avoids the rest of the time—and will usually culminate in self-induced vomiting. There is usually enormous shame and self-loathing associated with bulimia. The precise nature of the relationship of bulimia to anorexia nervosa remains unclear. A major concern about bulimia relates not to the overeating per se but to the subsequent vomiting. The chronic loss of gastric juices can lead to a depletion of the body's potassium, which in turn can seriously affect cardiac function (to the point of death). In general, the onset of bulimia is believed to signify a poor prognosis.

Although there is a broad range of symptoms, personality styles, precipitants, and outcomes that characterize anorexia nervosa, there is no simple explanation of its origins. In one widely accepted conception, the illness is viewed as a desperate struggle by the vulnerable female adolescent to establish a sense of identity separate from that of her domineering, overbearing, controlling, and intrusive mother. As with virtually all psychiatric conditions for which the etiology is unknown and where no single empirically effective treatment exists, the therapeutic approaches to anorexia nervosa are diverse and reflect the different disciplines, training biases, and experiences of their proponents. *See* AFFECTIVE DISORDERS; MALNUTRITION; STRESS (PSYCHOLOGY). [J.L.Ka.]

Anorthoclase The name usually given to alkali feldspars which have a chemical composition ranging from $Or_{40}Ab_{60}$ to $Or_{10}Ab_{90}$ ± up to approximately 20 mole % An (Or, Ab, An = $KAlSi_3O_8$, $NaAlSi_3O_8$, $CaAl_2Si_2O_8$) and which deviate in one way or another from monoclinic symmetry tending toward triclinic symmetry. When found in nature, they usually do not consist of a single phase but are composed of two or more kinds of K- and Na-rich domains mostly of submicroscopic size. In addition, they are frequently polysynthetically twinned after either or both of the albite and pericline laws. It appears that they originally grew as the monoclinic monalbite phase, inverting and unmixing in the course of cooling during geological times. They are typically found in lavas or high-temperature rocks. *See* FELDSPAR; IGNEOUS ROCKS. [F.H.L.]

Anorthosite A rock composed of 90 vol % or more of plagioclase feldspar. Strictly, the rock is composed entirely of crystals discernible with the eye, but some finely crystalline examples from the Moon have been called anorthosite or anorthositic breccia. Scientists have been fascinated with anorthosites because they are spectacular rocks (dark varieties are quarried and polished for ornamental use); valuable deposits of iron and titanium ore are associated with anorthosites; and the massif anorthosites appear to have been produced during a unique episode of ancient Earth history (about 1–2 × 10^9 years ago).

Pure anorthosite has less than 10% of dark minerals—generally some combination of pyroxene, olivine, and oxides of iron and titanium; amphibole and biotite are rare, as are the light minerals apatite, zircon, scapolite, and calcite. Rocks with less than 90% but more than 78% of plagioclase are modified anorthosites (such as gabbroic anorthosite), and rocks with 78–65% of plagioclase are anorthositic (such as anorthositic gabbro). *See* GABBRO.

The structure, texture, and mineralogy vary with type of occurrence. One type of occurrence is as layers (up to several meters thick) interstratified with layers rich in pyroxene or olivine. The second type of occurrence is the massifs type and can have an area up to 11,600 mi^2 (30,000 km^2). Commonly, the massifs are domical in shape and weakly layered. Possibly there is a third group of anorthosite occurrences: extremely ancient bodies of layered rock in which the layers of anorthosite contain calcium-rich plagioclase and the adjacent layers are rich in chromite and amphibole in addition to pyroxene. There are only a few examples of these apparently igneous complexes, in Greenland, southern Africa, and India. However, they appear to be terrestrial counterparts of lunar anorthosites.

By comparison with terrestrial occurrences, most lunar anorthosites are very fine grained, although one rock has crystals up to a centimeter long. Much of the fine grain size results from comminution by meteorite impact, and some of it probably results from rapid crystallization of impact melts. *See* ANDESINE; IGNEOUS ROCKS; LABRADORITE; METAMORPHISM; MOON. [A.T.A.]

Anostraca An order of the subclass Branchiopoda. Commonly called fairy shrimps, they lack a carapace; hence the name, which means "without carapace." About 175 species have been described. The trunk contains 20 or more segments, of which the anterior 11 to 19 bear appendages (see illustration). These are followed by two genital somites and six limbless somites plus a telson. The head carries the five pairs of appendages typical of crustacea: antennules, antennae, mandibles, maxillules, and maxillae. In the male the antennae are modified into claspers. Anostraca feed on the small particles which are strained from the water by the setose appendages and concentrated in a median ventral groove which leads to the mouth. The body is covered by a thin but remarkably impermeable exoskeleton which is molted every 4 to 6 days throughout the adult life of several months. Along with efficient excretory glands, this exoskeleton enables *Artemia* to inhabit very salty waters.

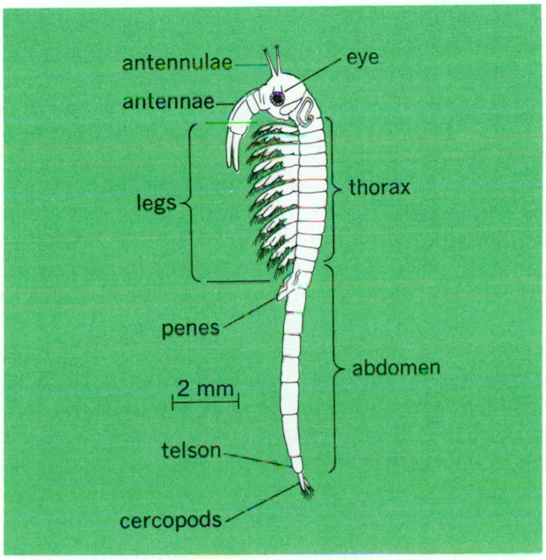

Branchinecta paludosa **of the Anostraca, male, small specimen, lateral aspect.**

In bisexual strains, fertilization is necessary to trigger embryonic development, and copulation occurs after every female molt. However, isolated parthenogenetic types occur, especially in Europe and Asia, in which females give rise to females, and chromosomal polyploidy is common. In favorable environments, nauplius larvae hatch within the ovisac. In unfavorable situations, a copious secretion from shell glands provides a protective cyst wall which helps the embryo to survive years of desiccation if necessary. *See* BRANCHIOPODA. [D.S.G.]

Anoxic zones Oxygen-depleted regions in marine environments. The dynamic steady state between oxygen supply and consumption determines the oxygen concentration. In regions where the rate of consumption equals the rate of supply, seawater becomes devoid of oxygen and thus anoxic. In the open ocean, the only large regions which approach anoxic conditions are between 165 and 3300 ft (50 and 1000 m) deep in the equatorial Pacific and between 330 and 3300 ft (100 and 1000 m) in the northern Arabian Sea and the Bay of Bengal in the Indian Ocean. The Pacific region consists of vast tongues extending from Central America and Peru nearly to the middle of the ocean in some places. In parts of this zone, oxygen concentrations become very low, 15 μmol/liter (atmospheric saturation is 200–300 μmol/liter). Pore waters of marine sediments are sometimes anoxic a short distance below the sediment-water interface. The degree of oxygen consumption in sediment pore waters depends upon the amount of organic matter reaching the sediments and the rate of bioturbation (mixing of the surface sediment by benthic animals). In shallow regions (continental shelf and slope), pore waters are anoxic immediately below the sediment-water interface; in relatively rapid sedimentation-rate areas of the deep sea, the pore waters are usually anoxic within a few centimeters of the interface; and in pore waters of slowly accumulating deep-sea sediments, oxygen may never become totally depleted. *See* MARINE SEDIMENTS.

Restricted basins (areas where water becomes temporarily trapped) are often either permanently or intermittently anoxic. Classic examples are the Black Sea, the Carioca Trench off the coast of Venezuela, and fiords which occupy the Norwegian and British Columbia coasts. Lakes which receive a large amount of nutrient inflow (either from natural or human-produced sources) are often anoxic during the period of summer stratification. *See* BLACK SEA; FIORD.

The chemistry of many elements dissolved in seawater (particularly the trace elements) is vastly changed by the presence or absence of oxygen. Since large areas of the ocean water mass are in contact with oxygen-depleted pore waters, the potential exists for anoxic conditions to have a marked effect on the chemistry of the sea. *See* SEAWATER; SEAWATER FERTILITY. [S.R.E.]

Anseriformes An order of birds comprising two families, the screamers (Anhimidae) of South America and the worldwide waterfowl (Anatidae). They are closely related to the Galliformes, with the screamers being a rather intermediate group. The giant, flightless diatrymids of the early Tertiary are specialized offshoots of the Anseriformes. *See* GALLIFORMES; GASTORNITHIFORMES.

The order Anseriformes is divided into the suborder Anhimae, containing the single family Anhimidae (screamers; 3 species), and the suborder Anseres, including only the family Anatidae (ducks, geese, and swans; 147 species). The waterfowl (Anatidae) are further subdivided into seven subfamilies, namely the primitive Anseranatinae (magpie goose of Australia), Dendrocygninae (tropical tree ducks), Anserinae (swans and geese), Tadorninae (shelducks of the Old World and South America), Anatinae (true ducks), Merginae (mainly Northern Hemisphere sea ducks and mergansers), and Oxyurinae (stiff-tailed ducks).

The South American screamers are turkey-sized, fowllike aquatic birds with a short, heavy, chickenlike beak. The legs are of medium length and heavy, with four toes having only a basal web. They fly slowly but soar well. Screamers live in marshes, walk on mats of floating vegetation, and sometimes swim. They feed on vegetable matter, are gregarious, and nest in solitary pairs.

The waterfowl vary in size from pygmy geese to large swans. They occur worldwide in fresh and marine (coastal) waters. All species have strong legs and feet with webbed toes and a flattened bill with comblike lamellae or teeth. The plumage is waterproof and varies from pure white to multihued to all black; females usually have a brown, cryptic plumage. The tongue is large and fleshy and serves in filter-feeding. Waterfowl feed on both plants and animals, obtained by filtering (many true ducks), grazing (swans and geese), and diving (sea ducks, mergansers, stiff-tailed ducks, and some true ducks). Diving ducks feed on mollusks and water plants, mergansers on fish. All species swim well, and most are strong fliers. Most species are gregarious except at breeding. Waterfowl are monogamous with a strong pair bond (some mate for life) and elaborate courtship.

The waterfowl are of immense economic importance. The mallard (*Anas platyrhynchos*; the common domesticated duck), muscovy duck (*Cairina moschata*), gray-lag goose (*Anser anser*; the common domesticated goose), and swan goose (*Anser cygnoid*) have been domesticated since ancient times for their flesh, eggs, and feathers. *See* AVES. [W.J.B.]

Ant All ants are classified in a single family, Formicidae; this reflects their limited structural variation. They are thought to have evolved from a wasplike ectoparasite of soil insects. Australia contains many primitive species, including the bulldog ants, which are large and fierce and live in small colonies (subfamily Myrmicinae). A common tropical group (Ponerinae) has evolved specialists in group raiding for soil insects, especially preying upon the ubiquitous termites. Also tropical is a set of predacious species (Dorylinae) whose colonies can contain a million workers, but only one queen; as well as hunting in groups, the whole colony roams nomadically through the forests. Other subfamilies are no longer exclusively predacious: in the Formicinae, which include the most conspicuous temperate region ants, plant-sucking bugs are used as a source of honeydew which, though mainly sugary, contains some soluble proteins. Wood ants (*Formica* species) pile dead vegetation into huge mound nests inside which they can retain their body heat. From the nests they establish permanent trackways to aphid-bearing trees, on which they also capture many insect larvae, including those injurious to forest trees. Each tree is part of a large territory which is defended against neighboring colonies. Yellow ants (*Lasius flavus*) culture aphids of many sorts on roots belowground and apparently use their carcasses as meat; in Europe they build soil mounds in old hillside grassland. In the tropics the leaf-nesting ants (*Oecophylla* species), which use larval silk to bind leaves together into a nest, culture plant-bugs in trees as well as prey upon a variety of insects; they can be used to protect fruit bushes from harmful insects.

In subfamily Myrmicinae, though there are many simple, mundane species with mixed diets, some specialists have almost given up predation. Thus grain collectors store and eat seeds; any seeds that germinate are put out and, mixed with rubbish, may start new plants nearer home (surely agriculture in its infancy). Then there are the leaf-cutting ants (tribe Attini) of Central America that collect vegetation and feed it to a species of fungus whose special bodies they then eat. These ants secrete a battery of chemicals for stopping the growth of weed species. Ants of another subfamily (Pseudomyrmicinae) have close mutualistic relations with plants: the plants supply special hollow galls, stems, or swollen thorns and offer nourishing tissues inside; the ants live and feed in these cavities and in exchange protect the plant from phytophagous insects and other enemies.

All female ants are social insects (the males exist only briefly to provide sperm) that live in dense clusters in nests made to protect against weather and enemies and to provide a work surface on which to rear their young. There are two types of female: small, wingless ones (called workers) that construct and defend the nest as well as collect and prepare food for the almost helpless larvae; and large, winged ones (called queens) that fly off to copulate, disperse, and start new nests, and which, after breaking off their wings, not only lay most of the eggs (usually all the female eggs and often the male ones too) but stimulate and organize worker activity. Workers usually have ovaries but no sperm sac, and the few eggs each lays are unfertilized; though haploid, they can produce males by parthenogenesis, for sex is determined by a haplo-diploid mechanism as in most insects of this order (Hymenoptera). *See* HYMENOPTERA; SOCIAL INSECTS. [M.V.B.]

Ant lion Any insect of the family Myrmeleontidae in the order Neuroptera. *Myrmeleon immaculatus* is a common representative (see illustration). These insects occur more commonly in dry areas, such as the South and Southwest, than in other sections of the United States; however, they are not limited to these locales. The larvae construct conical pits by using their heads, which are flat and shovel-shaped. The pits are about 1 in. (2.5 cm) across, and the larvae remain at the bottom as passive predators feeding upon ants and other insects which fall into the pit. A pair of sickle-shaped mandibles, which are sharp, smooth, and curved at the tip, extend from the head of the larvae. There is no mouth. The prey that falls into the pit is grasped by the mandibles. These structures are similar to a hypodermic needle, through which the ant lion can pump its digestive enzymes into the victim. Digestion occurs outside the animal and, since all undigestible material remains outside the larva, solid elimination does not take place; actually, the hindgut of the ant lion is closed. The larval stage may extend from 1 to 3 years, and then pupation occurs with the production of a silky cocoon. *See* NEUROPTERA. [C.B.C.]

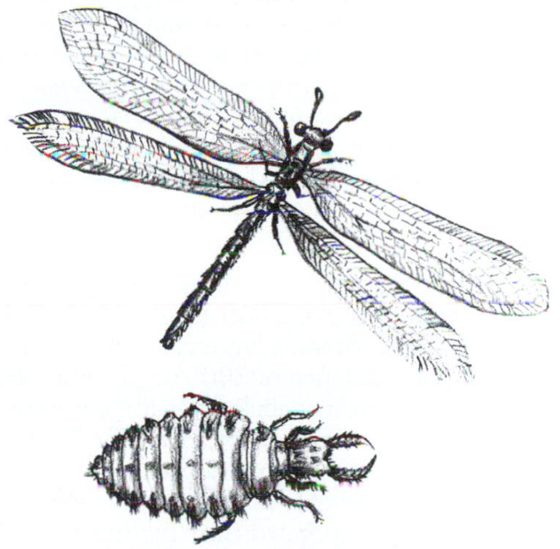

The ant lion (*Myrmeleon immaculatus*) larva (called doodlebug) and winged adult.

Antarctic Circle The parallel of latitude approximately $66\frac{1}{2}°$ (66.°55) south of the Equator and $23\frac{1}{2}°$ from the South Pole; the same angular distance from the Equator as the inclination of the Earth's axis from the plane of the ecliptic. It is the parallel of latitude to which the Sun's rays will extend beyond the South Pole on December 22 (the Northern Hemisphere winter solstice), when that pole is inclined $23\frac{1}{2}°$ toward the Sun. On that day the Sun will be above the horizon for 24 h at the Antarctic Circle. On this same day the Sun's rays at noon will just reach the horizon at the Arctic Circle, $66\frac{1}{2}°$ north.

Although the longest period of continuous sunshine at the Antarctic Circle is 24 h, the long days preceding and following the solstice, and the long periods of twilight, allow a season of about 5 months of continuous daylight. The highest altitude of the noon Sun will be 47° above the horizon on December 22. *See* MATHEMATICAL GEOGRAPHY; SOLSTICE. [V.H.E.]

Antarctic Ocean The three oceans of the Earth, in order of decreasing area are the Pacific, the Atlantic, and the Indian. In the Northern Hemisphere and most of the Southern Hemisphere, the oceans are connected only by narrow, shallow passages which prohibit significant interactions. Such isolation would in time lead to each of the oceans having differing characteristics were it not for a major interconnecting artery farther south. This passage exists between the southern shores of the Australian, South American, and African continents and the coastline of Antarctica. It is a broad, deep, circumpolar ocean belt and is essential in equalizing the characteristics of the three oceans and in maintaining the deep and bottom layers of these oceans as an aerobic, cold environment. Because of the importance of this ocean area and its obvious homogeneity in climate (due to its subpolar geographic position), it is convenient to name it the Antarctic Ocean. This region is not officially recognized as a separate ocean, mainly because of its lack of a northern boundary in the classical sense, that is, a complete (or nearly so) land mass. However, from studying its sea-water characteristics, it becomes possible to set a northern oceanographic limit as the Antarctic Convergence or Subtropical Convergence.

The major flow is the Antarctic Circumpolar Current, or West Wind Drift (see illustration). Along the Antarctic coast is the westward-flowing East Wind Drift. The strongest currents are in the vicinity of the polar front zone and restricted passages such as the Drake Passage and deep breaks in the meridionally oriented submarine ridge systems. Though the magnitude of the current is small, there is little attenuation of flow with depth, which results in the great volume transport of the Antarctic Circumpolar Current. It is estimated that over 7×10^9 ft^3/s (2×10^8 m^3/s) of water goes through the Drake Passage, about three times the flow of the Gulf Stream. *See* OCEAN CIRCULATION.

The extreme cold of the polar regions causes an extensive ice field to form over the southern regions of the Antarctic Ocean. The extent of the ice is seasonal in that during the October-to-March period the area decreases, and it increases during the remaining months. Satellite photographs reveal that the sea ice field is not uniform, but has many large ice-free regions, called polynyas. The sea ice plays an important role in the heat balance since it reflects much more solar radiation (and therefore heat) back into space than would be the case for a water surface. *See* HEAT BALANCE, TERRESTRIAL ATMOSPHERIC; ICEBERG; SEA ICE.

The composition of the sediments is influenced by two factors: the high biological productivity in the nutrient-rich euphotic zone, and the input of glacial debris carried seaward by icebergs. The siliceous ooze (diatoms and radiolaria) is found roughly between the polar front zone and the limit of the pack ice. To the north, the siliceous ooze is replaced by calcareous

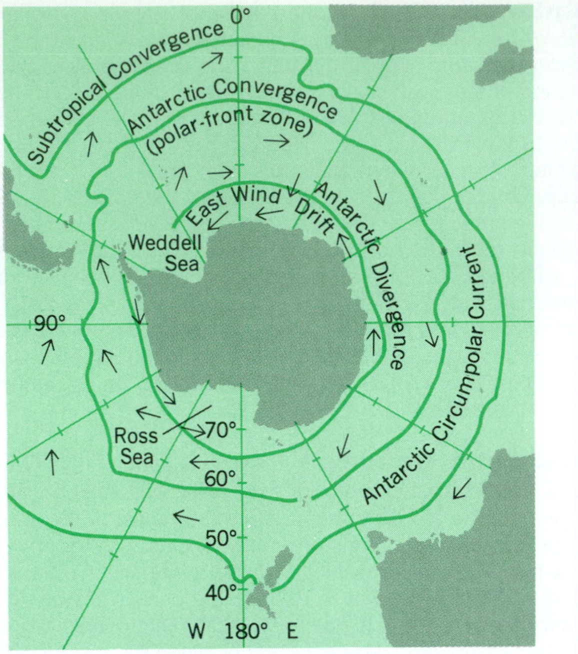

Direction of the surface circulation and major surface boundaries of the Antarctic Ocean.

ooze and red clay in deeper basins. Within the pack ice field, sediments are of glacial marine type. [A.L.G.]

Antarctica A snow- and ice-covered continent roughly centered on the South Pole and surrounded by an ocean consisting of the southern parts of the Atlantic, Pacific, and Indian oceans. It is the fifth largest continent, having an area of 14,000,000 km^2 or 5,400,000 mi^2, which is $1\frac{1}{2}$ times the size of the United States. Antarctica is the coldest and highest of all the continents, and there are no permanent inhabitants. For millions of years snow and ice have built up on the land, burying all but 1% of it.

The continent holds 24,000,000 km^3 or 5,750,000 mi^3 of ice, which is 90% of all the ice on Earth. The mean height of the ice surface is 2000 m (6700 ft) above sea level, giving Antarctica almost three times the mean height of the other continents. The South Pole is 2900 m (9500 ft) above sea level. About 10% of the area is made up of ice shelves, which are ice sheets floating on the sea and commonly 300 m (1000 ft) thick, but in places are over 1900 m (6000 ft) thick. The remainder of the ice sheet rests on rock. If all this ice were to melt, an unlikely event which would take thousands of years, world sea levels would rise some 60 m (200 ft).

As in glaciers, the ice of this huge sheet is moving slowly downhill at speeds varying from a few meters or so to about 900 m (3000 ft) per year. When it reaches the coast, it breaks off to create the tabular icebergs which are characteristic of the Antarctic waters. *See* ANTARCTIC OCEAN.

Antarctica is divided into two parts by one of the world's great mountain chains, the Transantarctic Mountains. Radioactive dating indicates that parts of East Antarctica are at least 1,800,000,000 years old. Similarities of rock types and sequences in other continents indicate that at one time it was part of a single great landmass known as Gondwanaland, including what is now the Indian peninsula, Australia, South America, and Africa.

Antarctica is the coldest region on Earth, the interior being colder than Siberia in winter and colder than the North Pole. At the South Pole there are 6 months (March 21 to September 21) when the Sun remains below the horizon. The length of the period of continuous darkness in winter and continuous daylight in summer decreases with distance from the pole, so that many coastal points have only a few weeks of each. There are considerable differences of climate between coast and interior. The average temperature at the South Pole is −51°C (−60°F), while at Anvers Island off the Antarctic Peninsula it is −3°C (26°F). There is very little precipitation and what there is falls as snow; the high plateau that covers most of Antarctica is the world's largest desert. *See* POLAR METEOROLOGY.

Algae, mosses, and lichens are the characteristic plants of Antarctica; only two genera of flowering plants are known. There are no vertebrate land animals; the largest creature living entirely on land is an insect 0.12 in. (3 mm) long. Four species of seal and four species of flightless bird, the penguins, breed on or near the shores of Antarctica, though they probably spend more than half their lives at sea. Apart from the penguins, only 10 species of birds breed on the coast of Antarctica at higher latitudes than the Antarctic Peninsula. Evidently because of the lack of food, few birds stray inland, though skuas and snow petrels have been seen far inland during the summer months. *See* ALGAE; BRYOPSIDA; LICHENES; OCEANIC BIRDS.

Antarctica was first sighted in 1820 by American and British sealing vessels and by a Russian expedition. But it was not known to be a continent until discoveries made in the period 1838–1843. Extensive land explorations were made by several expeditions during the first 35 years of the 20th century. Since World War II voyages of discovery and great journeys have given way to scientific research on a semipermanent footing. The International Geophysical Year 1957–1958 provided a great stimulus to research in meteorology, glaciology, auroral physics, geomagnetism, ionospherics, seismology, and gravimetry. The Antarctic Treaty of 1959, ratified by 16 countries, encourages cooperation and ensures freedom of scientific investigation in any part of the continent. [C.W.Sw.]

Antares Alpha Scorpii, a bright, red supergiant of spectral type M1, temperature about 3000 K. Antares is variable in light and radial velocity. Antares has a close blue companion of spectral type B5, at a distance of 300 AU (astronomical units). The hot star is involved in nebulosity with an unusual spectrum. The emission lines are of forbidden Fe II and of Si II, without trace of hydrogen, which is normally strong in all other emission nebulae. It is possible that solid interstellar or circumstellar matter is being evaporated near the hot star. This type of circumstellar nebula has been found in many red supergiants, not in the normal spectral region but in the very-far-infrared. Apparently the strong stellar winds, or mass loss, from the red supergiants result in emission-line fluorescence of matter in the immediate surroundings at a rate so large that the nebula then becomes visible. *See* STAR. [J.L.Gr.]

Anteater A name associated with several animals in five different orders of the three major groups of living mammals (see table). They are so named because they are insectivorous, having a diet of ants and termites. The animal most frequently associated with this name is the ground-living *Myrmecophaga tridactyla*, the giant anteater, a member of the family Myrmecophagidae in the order Edentata (see illustration). This family has three other species, *Tamandua longicaudata*, *T. tetradactyla*, and *Cyclopes didactylus*, all of which are arboreal.

All four species are restricted to the tropical regions of South and Central America. *Myrmecophaga tridactyla* prefers the grasslands and more open forested areas. The animal is about 6 ft (1.8 m) long including the tail length, which measures about 2 ft (0.6 m). It is toothless, and the head extends into a

Classification and scientific name of some animals commonly referred to as anteaters

Mammalian order	Scientific name	Common name
Monotremata	*Tachyglossus setosus*	Spiny anteater or Tasmanian echidna
	T. aculeatus	Australian echidna
	Zaglossus bruijini	Bruijn's echidna
	Z. bartoni	Barton's echidna
	Z. bubuensis	Bubu echidna
Marsupialia	*Myrmecobius fasciatus*	Marsupial anteater or banded anteater
	M. rufus	Rusty numbat
Pholidota	*Manis gigantea*	Scaly anteater or giant pangolin
	M. temmincki	Cape pangolin
	M. tricuspis	Tree pangolin
	M. longicaudata	Long-tailed tree pangolin
	M. pentadactyla	Chinese pangolin
	M. crassicaudata	Indian pangolin
	M. javanica	Malayan pangolin
Edentata	*Myrmecophaga tridactyla*	Giant anteater
	Tamandua longicaudata	Long-tailed anteater
	T. tetradactyla	Tamandua
	Cyclopes didactylus	Dwarf anteater
Tubulidentata	*Orycteropus afer*	Aardvark or Cape anteater

long, tubular snout with a small mouth opening. The tongue is long, protrusible, and covered with a viscous mucous material which entraps the insects. The front feet have greatly enlarged claws used for tearing into ant and termite mounds and as defensive weapons. The body is covered with long hair, and in *Cyclopes* the tail is prehensile. Usually a single young is pro-

The giant anteater (*Myrmecophaga tridactyla*).

duced by the female, which she may carry on her back until it is quite large. *See* AARDVARK; EDENTATA; MARSUPIALIA; MONOTREMATA; PHOLIDOTA; TUBULIDENTATA. [C.B.C.]

Antelope The name given to a group of hollow-horned, hoofed ruminants of the order Artiodactyla which are strictly confined to various areas of Africa and Asia. They are assigned to the subfamily Antilopinae in the family Bovidae and include 91 species in 31 genera.

In general, the third and fourth toes are well developed, while the second and fifth toes are reduced or absent. The males and usually the females have a pair of unbranched horns, which differ considerably in appearance among the various species of antelopes. These animals vary in size from the royal antelope (*Neotragus pygmaeus*), which is about 10–12 in. (25–30 cm) high at the shoulder, to the giant eland (*Taurotragus derbianus*), which measures about 6 ft (1.8 m) at the shoulder and weighs about 1 ton (0.9 metric ton). *See* ARTIODACTYLA; MAMMALIA. [C.B.C.]

Antenna (electromagnetism) A portion of a radio system especially designed to couple electromagnetic energy between free space and transmission lines. At a radio transmitter the antenna couples energy from the transmitter circuits into free space, thus serving to radiate the transmitter's output as radio waves. At the radio receiver the antenna couples energy from the radio waves into the receiver's circuits. The same antenna structure functions equally well for transmission or reception. However, in describing antenna operation, it is often more direct to describe the radiation action, with the understanding that reception is the reciprocal. *See* RECIPROCITY PRINCIPLE.

Despite this reciprocity, antennas used for transmission may differ significantly from those used for reception because of the power levels at which they operate. An antenna for transmitting to a major portion of the world or to a space vehicle must handle large powers; the antenna for receiving from radar reflections or from communication satellites handles negligible power. For these reasons antennas are classified by use; they may also be classified by operating frequency. Antennas for very low frequencies may be a mile long, whereas antennas for very high frequencies may be only a few feet long. More basically, if antenna size is measured in units of the wavelength of the frequency at which the antenna operates, structures that differ greatly in size can be described in similar terms.

For an antenna to radiate, it must contain charges that oscillate. Such a charge is the electron shown in Fig. 1. Because no other nearby charges oscillate to counteract the effect of the electron, its field extends infinitely into space. This field contains two components: an electric field E and a magnetic field H. Energy stored in the electric field when the electron is stationary equals energy in the magnetic field when the electron is moving most rapidly. Basically, the changing electric field induces an adjacent magnetic field, and the changing magnetic field induces an adjacent electric field. The result of this continuing energy exchange is radiation.

Figure 1 is an elementary graphical representation of the radiating electric field while a single electron oscillates on a short, thin conductor. At the top of the figure, one line of the electric field terminates at the electron. Successive drawings down the figure show this field line as the electron oscillates on the conductor. Motion of the electron displaces the line. This displacement travels along the field line, moving in the figure to the right, with the velocity of light. After the electron has completed one cycle of oscillation on its conductor, the initial displacement has traveled one wavelength λ from the conductor.

More rigorously, the oscillating electron has an induction field which decreases rapidly with distance. Energy in this field flows alternately away from and toward the electron and never escapes into space. The magnetic component of this field is a maximum when the electric component is zero. In the radiation field, which extends indefinitely into space, energy flows only outward at all instants of time and at all locations in space; magnetic and electric components of this field pass through zero together.

Along with the oscillating negative charge is a positive charge. If all oscillating charges on a short, thin conductor, called a doublet, are considered, the total electromagnetic field is found to contain terms that decrease with distance in the ratios $1/r^3$ and $1/r^2$ in the induction field and $1/r$ in the radiation field. The $1/r$ decrease in radiation field as it spreads out into space from the doublet is due to the greater area of the sphere that bounds it. *See* ELECTROMAGNETIC RADIATION; ELECTROMAGNETIC WAVE TRANSMISSION; RADIO-WAVE PROPAGATION.

The elemental radiator can be either a short electric device or a small current device. If infinitesimally small, their far fields are as given in Fig. 2.

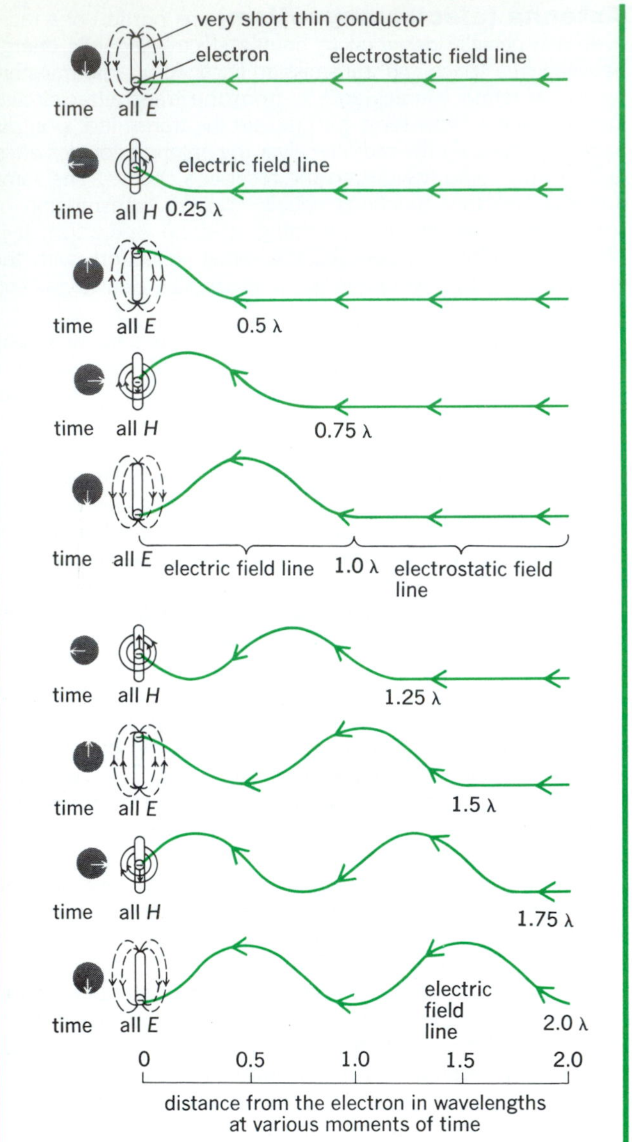

Fig. 1. Graphical representation of electric *E* and magnetic *H* fields of oscillating electron.

As Fig. 2 shows, a small electric or magnetic dipole radiates no energy along its axis, the contour of constant energy being a toroid. This contour in space, called the radiation pattern, is usually the most basic requirement of an antenna. The purpose of a transmitting antenna is to direct power into a specified region; whereas the purpose of a receiving antenna is to accept signals from a specified direction. In the case of a vehicle, such as an automobile with a car radio, the receiving antenna needs a nondirectional pattern so that it can accept signals from variously located stations, and from any one station, as the automobile moves. The antenna of a broadcast station may be directional; for example, a station in a coastal city would have an antenna that concentrated most of the power over the populated land. The antenna for transmission to or from a communication satellite should have a narrow radiation pattern directed toward the satellite for efficient operation, preferably radiating essentially zero power in other directions to avoid interference. *See* DIRECTIVITY; NAVIGATION; RADAR; RADIO BROADCASTING.

The plane of the electric field depends on the direction in which the electric charge moves on the antenna. The electric field is in a plane orthogonal to the axis of a magnetic dipole (Fig. 2). This dependence of the plane of the radiated electromagnetic wave on the orientation and type of antenna is termed polarization. By combining fields from electric and magnetic dipoles that have a common center, the radiated field can be elliptically polarized; by control of the contribution from each dipole, any ellipticity from plane polarization to circular polarization can be produced. A receiving antenna requires the same polarization as the wave that it is to intercept. *See* POLARIZATION OF WAVES.

The input impedance to an antenna is the ratio at its terminals, where the transmission line and transmitter or receiver are connected, of voltage to current. If the antenna is tuned to resonance at the operating frequency, the input impedance will be a pure resistance; otherwise it will also have a reactance component.

Bandwidth of an antenna is related to its input impedance characteristics. Bandwidth may be limited by pattern shape, polarization characteristics, and impedance performance. Bandwidth is critically dependent on the value of Q; hence the larger the amount of stored reactive energy relative to radiated resistive energy, the less will be the bandwidth. *See* Q (ELECTRICITY).

There are two principal approaches to frequency-independent antennas. The first is to shape the antenna so that it can be specified entirely by angles; hence when dimensions are expressed in wavelengths, they are the same at every frequency. Planar and conical equiangular spiral antennas adhere to this principle. The second principle depends upon complementary shapes so that if, for example, an antenna that is cut from a flat conducting sheet is exactly the same shape as the part removed, its input impedance will be frequency-independent. Many log-periodic antennas employ this second principle, and before the structure shape changes very much, when measured in wavelengths, the structure repeats itself. By combining periodicity and angle concepts, antenna structures of very large bandwidths become feasible.

Antennas whose mechanical dimensions are short compared to their operating wavelengths are usually characterized by low radiation resistance and large reactance. This combination results in a high Q and consequently a narrow bandwidth. Current distribution on a short conductor is sinusoidal with zero current at the free end, but because the conductor is so short electrically, typically less than 30° of a sine wave, current distribution will be essentially linear. By end loading to give a

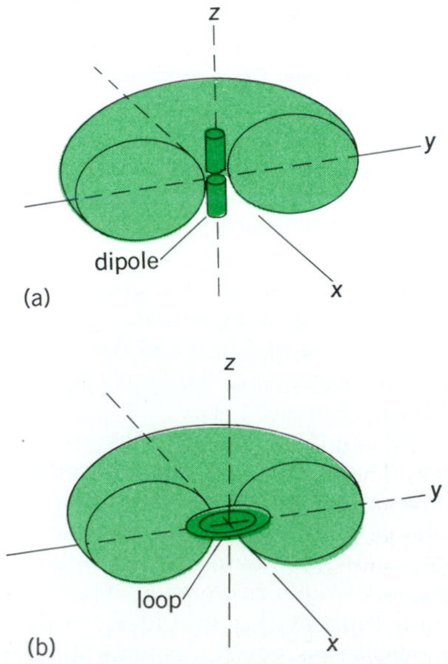

Fig. 2. Cross sections of far-field radiation patterns. (*a*) Small electric dipole. (*b*) Small magnetic dipole.

constant current distribution, the radiation resistance is increased by four times, thus greatly improving the efficiency but not noticeably altering the pattern.

Dipoles and monopoles illustrate resonant antennas which have an approximate sinusoidal current distribution and a pure resistance at their input terminals. However, when the ratio of diameter to length is small, the input impedance varies rapidly and precludes their use as broadband antennas. The impedance bandwidth limitation can be mitigated by increasing the diameter to form a cylinder or by using conical conductors. Long-wire antennas, or traveling-wave antennas, are usually one or more wavelengths long and are untuned or nonresonant.

When they are to be used at short wavelengths, antennas can be built as horns, mirrors, or lenses. Such antennas use conductors and dielectrics as surfaces or solids.

By using reflectors one can achieve high gain, modify patterns, and eliminate backward radiation. A low-gain dipole, a slot, or a horn, called the primary aperture, radiates toward a larger reflector called the secondary aperture. The large reflector further shapes the radiated wave to produce the desired pattern. *See* REFLECTION OF ELECTROMAGNETIC RADIATION.

A beam can be formed in limited space by a two-reflector system. The commonest two-reflector antenna, the Cassegrain system, consists of a large paraboloidal reflector. It is illuminated by a hyperbolic reflector, which in turn is illuminated by the primary feed (Fig. 3).

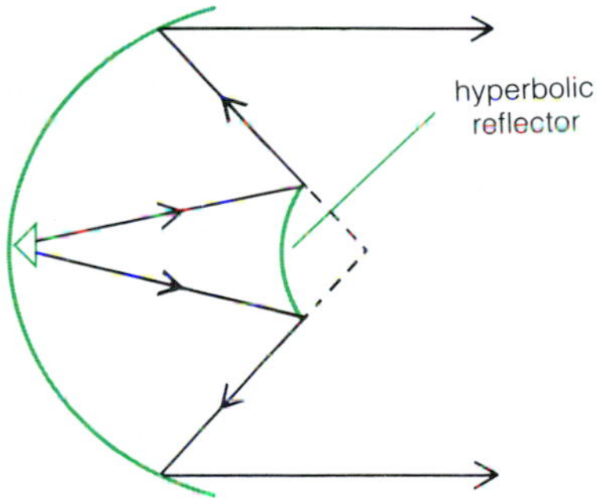

Fig. 3. Cassegrain two-reflector system.

An array of antennas is an arrangement of several individual antennas so spaced and phased that their individual contributions add in the preferred direction and cancel in other directions. One practical objective is to increase the signal-to-noise ratio in the desired direction. Another objective may be to protect the service area of other radio stations, such as broadcast stations. [C.E.Sm.]

Anthocyanin Any of the intensely colored plant pigments responsible for most scarlet, crimson, purple, mauve, and blue colors in higher plants. Unlike the chlorophylls and carotenoids, which are lipid-soluble chloroplast pigments, the anthocyanins are sap-soluble and are located in the cell vacuole. Chemically, they are a class of flavonoids and are particularly closely related, both structurally and biosynthetically, to the flavonols. Their value to the plant lies in the contrasting colors they provide in flower and fruit, against the green background of the leaf, to attract insects and animals for purposes of pollination and seed dispersal. *See* CAROTENOID; CHLOROPHYLL; FLAVONOID.

Anthocyanins are glycosides, and on acid hydrolysis yield sugars and colored aglycones called anthocyanidins. The commonest aglycone is 3,5,7,3′,4′-pentahydroxyflavylium, or cyanidin (see illustration), so called after the cornflower

Structural formula of cyanidin.

(*Centaurea cyanus*), from which it was first isolated. The other 15 known anthocyanidins all have the same basic structure as cyanidin and differ only in having fewer or more hydroxyl groups or in being *O*-methylated. Common anthocyanidins are pelargonidin, peonidin, delphinidin, petunidin, and malvidin. The sugar present in anthocyanins may be glucose, galactose, rhamnose, xylose, or arabinose, a combination of two or three of these monosaccharides, or a combination of a monosaccharide with an aromatic hydroxy acid. *See* GLYCOSIDE.

Anthocyanin colors are most prominent and widespread in flowers in the red, mauve, and blue shades of most cultivated plants, but they are also very effective in fruits such as the plum, raspberry, blackberry, and strawberry. In the leaf, anthocyanin is masked by chlorophyll, except where the pigment concentration is very high, as in the red cabbage or red *Coleus*. Anthocyanin can be present in many other plant tissues, for example, in bean seed, anemone pollen, banana bract, pomegranate juice, potato tuber, and radish root. Anthocyanins frequently occur in plants as mixtures, and a series of related glycosides can often be detected in a single species. Anthocyanins are all but universal in higher plants. They are also present in ferns, pigmenting the juvenile fronds with orange or red coloration, and in a few mosses, notably red sphagnum, but they are not known among the fungi or bacteria.

The primary function of anthocyanins is undoubtedly as flower pigments and, together with scent and the presence of nectar, they attract insect and animal pollinators to plants. The purpose of anthocyanin pigmentation in leaf tissue is more obscure. While permanent leaf coloration due to anthocyanin is not usual, transient anthocyanin synthesis in either young or autumnal leaves is quite common.

Anthocyanins, although they are intensely colored, have never been widely used in commerce for dyeing cloth. Their practical value can only be measured in esthetic terms. Anthocyanin color in foods and in drinks such as red wine is certainly a stimulus to the appetite, and much effort is expended by food scientists in retaining these colors during the canning and processing of fruits and vegetables. [J.B.Ha.]

Anthophyllite The name given to the orthorhombic amphiboles. The anthophyllite amphiboles form a limited solid-solution series between the magnesian end member $Mg_7Si_8O_{22}(OH)_2$ and an approximately 50 mole % ferric iron member. Minerals with more iron than this apparently form the monoclinic mineral cummingtonite, thereby indicating a limited tolerance of the orthorhombic form for iron. Anthophyllite can accept various amounts of aluminum up to the approximate composition $(Mg,Fe)_5Al_2(Al_2Si_6)O_{22}(OH)_2$. Depending on the iron content, it can be white, gray, green, or brown, but usually is colorless in thin sections. Anthophyllite commonly occurs in fibrous masses, sometimes as asbestiform masses and in prismatic needles. *See* AMPHIBOLE.

Anthophyllite is a metamorphic mineral occurring in many schists and gneisses in association with such minerals as tremolite, chlorite, cordierite, garnet, talc, serpentine, spinel, and quartz. Asbestiform varieties are sometimes used as asbestos. *See* CUMMINGTONITE; TREMOLITE. [G.W.DeV.]

Anthozoa A class of the phylum Coelenterata. These organisms are marine, solitary or colonial, and exclusively polypoid coelenterates with no traces of a medusoid stage. Most anthozoans live attached to some firm object of the shore or on the sea bottom; some embed in the soft sediment. Anthozoans have a cylindrical body with an oral disk, mouth, stomodeum, hollow tentacles, endodermal gonad, and cellular mesoglea. The gastrovascular cavity is partitioned longitudinally into radial compartments by endodermal mesenteries or septa whose free edges, particularly, thicken and differentiate into mesenteric or septal filaments. The nervous system is a diffuse network of scattered nerve cells over the ectoderm and the endoderm. No localized sense organs are present.

Both sexual and asexual reproduction occurs. The germ cells are derived from the endoderm, and fertilization occurs either in the female gastrovascular cavity or in the sea. The zygote develops into either a ciliated swimming larva, the planula, or a young polyp.

The class Anthozoa includes the soft, horny, stony, and black corals, the sea pens, and sea anemones. The horny corals include the sea fans, sea whips, and sea feathers. The Anthozoa may be classified as listed below. Separate articles appear on each group.

Class Anthozoa
- Subclass Alcyonaria (Octocorallia)
  - Order: Stolonifera
  - Telestacea
  - Coenothecalia
  - Alcyonacea
  - Gorgonacea
  - Pennatulacea
- Subclass Zoantharia (Hexacorallia)
  - Order: Actiniaria
  - Scleractinia (Madreporaria)
  - Zoanthidea
  - Antipatharia
  - Ceriantharia
  - Rugosa
  - Tabulata

All anthozoans are marine and most are sedentary, except the free-swimming larval stages, while actinians, ceriantharids, and pennatulans are somewhat movable. They are widely distributed over the world, extending from the Arctic to the Antarctic; however, they predominate in the tropic and subtropic areas of the Indo-Pacific Ocean. Actinians also inhabit colder water areas from which deep-sea species of gorgonians, pennatulans, and scleractinians have been collected.

Anthozoans seldom tolerate desiccation or heavy sedimentation. They are so sensitive to reduced salinity that they usually do not live near coastal areas where there is river drainage. Tropical corals are able to endure high temperatures and are adversely affected by low temperatures. Therefore, coral reefs are commonly located in tropic and subtropic regions. *See* COELENTERATA; HYDROZOA; SCYPHOZOA. [K.At.]

Anthracene A colorless crystalline hydrocarbon, $C_{14}H_{10}$, which melts at 421.2°F (216.2°C) and boils at 644°F (340°C). When pure, anthracene shows a blue-violet fluorescence. It is obtained from coal tar, which usually contains about 1% of the hydrocarbon.

Anthracene has had little commercial importance since a direct method for the synthesis of anthraquinone from benzene derivatives was discovered. Anthracene has been used as a light-screening additive for plastics and finds some application in the manufacture of pesticides. *See* AROMATIC HYDROCARBON; POLYNUCLEAR HYDROCARBON. [C.K.B.]

Anthracosauria An order of Carboniferous and Permian labyrinthodont amphibians much less common than their temnospondyl relatives but important as including the ancestors of reptiles. In contrast with temnospondyls, the pleurocentra are retained and developed in the vertebral column, the tabular bones of the skull roof are in contact with the parietals, and the cheek is primitively but loosely attached to the skull roof. *See* AMPHIBIA; LABYRINTHODONTIA; TEMNOSPONDYLI. [A.S.R.]

Anthraquinone pigments Coloring materials which occur in plants, fungi, lichens, and insects. About 50 different derivatives of the parent compound, anthraquinone, have been isolated. Several of the pigments have been used as dyes; others have been utilized as cathartic drugs. The best-known anthraquinone dye is alizarin (see formula), a natural dye

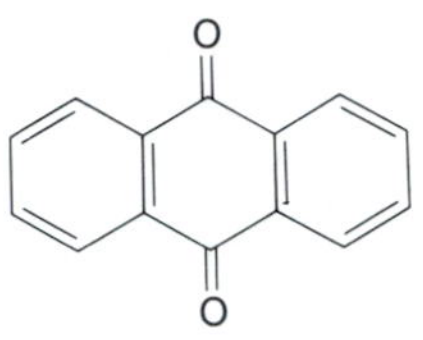

Anthraquinone

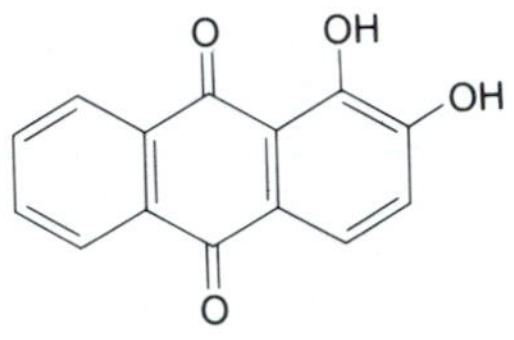

Alizarin

known to the ancient Egyptians and Persians. It occurs in the root of the madder (*Rubia tinctorum*), native to Asia. Another natural anthraquinone pigment is carminic acid. It is the dyeing principle of the cochineal dye, obtained from the dried bodies of the female of the insect *Coccus cacti*. This and other natural anthraquinone dyes have largely given way to better and cheaper synthetic dyes.

Emodin is an anthraquinone pigment with strong cathartic properties and is an active principle of cascara, senna, aloe, and rhubarb. *See* DYE. [S.M.K.]

Anthrax An acute infectious disease, primarily of animals from which humans may be secondarily infected, caused by *Bacillus anthracis*. In animals the disease, known as splenic fever, occurs when spores of *B. anthracis* are eaten with forage. Humans contract the disease by contact with infected animals or animal products, such as bone meal, meat, hide, and fur.

The most susceptible animals are cattle, sheep, pigs, horses, and goats. Usually septicemia, or blood poisoning, occurs. The effects vary from a sudden apoplectic attack to a subacute but eventually fatal illness manifesting fever, an enlarged spleen, and frequently intestinal disturbances. Sometimes local manifestations, which are less often fatal, occur. For example, in cattle and horses circumscribed cutaneous carbuncles may appear, and in swine similar lesions are commonly found in the throat.

The disease in humans occurs almost exclusively from contact with animals or animal products. It takes three main forms: malignant pustule, pulmonary anthrax, and intestinal anthrax.

Malignant pustule (cutaneous anthrax), the commonest form, results from contamination of the skin. Pulmonary anthrax (wool-sorter's disease) is caused by the inhalation of dust containing spores. Intestinal anthrax may follow the eating of infected food. The two latter types of anthrax are rare but almost invariably fatal. Treatment is difficult because of the short period of time between onset of symptoms and death.

Live spores of attenuated virulence form an effective vaccine for cattle and other animals. A cell-free protective vaccine has been produced for use in humans. The disease is diagnosed by microscopic identification of bacteria in the blood and by the Ascoli thermoprecipitin test. In the Ascoli test, a precipitate forms when a boiled saline extract of infected tissue is added to a suitable immune serum. *See* MEDICAL BACTERIOLOGY. [H.Sm.]

Anthropology The observation, measurement, and explanation of human variability in time and space. This includes both biological variability and contemporary behavior in the human species, and these studies are closely allied with the fields of archeology and linguistics as well. Studies range from rigorously scientific approaches, as in work on the demography, physiology, and ecology of traditional hunter-gatherers, to more humanistic research on topics such as symbolism and belief in contemporary religious systems. *See* ARCHEOLOGY.

Research has been emphasizing the integration of these different subfields. Anthropologists increasingly are asking questions about the adaptiveness of various kinds of human behavior in different societies and are approaching these questions in a holistic manner. New subfields such as medical anthropology routinely examine attitudes toward illness and medical practices in non-Western societies with respect to both the social and biological factors involved. *See* PHYSICAL ANTHROPOLOGY.

Most anthropologists are primarily concerned with empirical studies of particular human societies, although there are increasing tendencies to look for generalizing principles that will allow improved predictions of behavior. [R.A.G.]

Anthropometry The measurement of the human body; a basic technique of physical anthropology. Measurement of continuously varying, quantitative traits finds application in studies of evolution, race, genetics, growth and aging, physiology, disease, behavior, constitution (individual anthropology), and equipment design—in fact, in virtually all areas of human biology. *See* ANTHROPOSCOPY; PHYSICAL ANTHROPOLOGY.

The two most important body measurements are height and weight. A ratio of the two, usually in the form height/$\sqrt[3]{\text{weight}}$, measures stockiness-leanness of build, which relates to race, geography, physiological function, and disease. The purpose of an anthropometric study will determine what measurements are required beyond height and weight—body lengths and diameters for studies of race, growth, and machine design; girths for clothing sizing; and skinfolds for nutrition and body composition. Stature generally correlates with body and limb lengths, and weight with body breadths, depths, girths, and skinfolds.

Anthropometry has practical as well as scientific applications. The soldier's physique has always been of concern to military authorities responsible for recruiting and supplying armies. European armies provided information on long-term trends in physique within a country, as well as from one country to another. Military anthropometry in the United States continued with large-scale studies on demobilized soldiers of World Wars I and II and with Air Force research during and since World War II by students of E. Hooton. Military anthropometry has led to industrial anthropometry. Physical anthropologists are currently employed by aircraft and automobile manufacturers and are consulted by designers of other equipment for human use. *See* HUMAN-FACTORS ENGINEERING.

With a shift in emphasis from describing to explaining human variation, new problems and methods have appeared, chiefly interest in body composition and in constitution or anthropology of the individual. Partitioning the body into percentages of water, fat, bone, and muscle helps to establish standards for normal growth and aging, to estimate nutritional status, and to follow the course of medical and surgical treatment. Skinfolds, a pure measure of subcutaneous fat, correlate well with total body fat, determined by measuring specific gravity. Bone density is estimated from x-rays, by comparison with wedges of known density.

Constitutional anthropology, in which the individual person rather than the group is the focus of study, has had a long history dating back to Hippocrates, the father of medicine, about 430 B.C. Relating physique to behavior, physiology, and disease requires more refined assessment of physique than the stocky-lean continuum provides, but at the same time minute details must be avoided. The anthroposcopic classifications of Hippocrates and Ernst Kretschmer, scaled and made more objective by William Sheldon, have indeed proved to be associated with behavior and disease. [A.D.]

Anthroposcopy The description of physical variation in humans by inspection rather than by measurement, as employed in anthropometry. Details of form are rated against standards which can be either subjective or objective. The trend has been to replace subjective with objective ratings, which is easier for some physical features than for others. *See* ANTHROPOMETRY.

Artificial ceramic and fiber standards of skin, hair, and eye color have been virtually supplanted by photometric devices identifying spectral wavelengths. Skin color, for example, can now be partitioned and individuals or groups compared in respect to amounts of melanin, hemoglobin, and carotene. Hair pigment can likewise be described and compared with respect to copper, iron, and traces of other minerals, in addition to melanin.

Other physical measurements, such as hair diameter, period of rotation of the hair shaft, and crystallography, are replacing earlier classification of hair color as a (subjectively estimated) mixture of melanin and red-gold pigment; of hair texture as fine, medium, or coarse, and of hair form as straight, low wave, deep wave, curly, or spiral.

Subjective ratings still have a place in field studies. Ratings have not been replaced for the whole category of specific details such as forehead slope, ear slant, nasal root depression, nasal tip thickness, and the like. The trend here too is toward objectification by photography, but subjectivity remains. The usual standard has been the adult North European male. The usual scale varies from two categories (present and absent) or three categories (small, medium, and large) up to five or six (absent, ssm [subsubmedium], sm [submedium], + [medium], ++ [marked], +++ [very marked]). Clearly one observer's + may be another's sm or ++, and even the same observer's standards may change from one series to another or over time. As long as comparisons are confined to a single observer and a single series, these problems are minimal. The difficulty comes when broader comparison is attempted. *See* PHYSICAL ANTHROPOLOGY. [A.D.]

Anthuridea A suborder of the Isopods. These crustaceans are characterized by slender, elongate, subcylindrical bodies, and by the fact that the outer branch of the paired tail appendage (uropod) arches over the base of the terminal abdominal segment, the telson (see illustration). The uropods of anthurideans attach laterally to the abdomen and together with the telson form a caudal fan.

Marked sexual dimorphism is shown in the first pair of antennae which, in males of many species in both sections, develop brushlike whorls of setae on the flagellum. Female anthurideans carry the developing young beneath the thorax in a brood pouch, formed by overlapping plates originating from the bases of several pairs of legs. *See* SEXUAL DIMORPHISM.

Anthurideans are mostly marine, but some live in brackish or

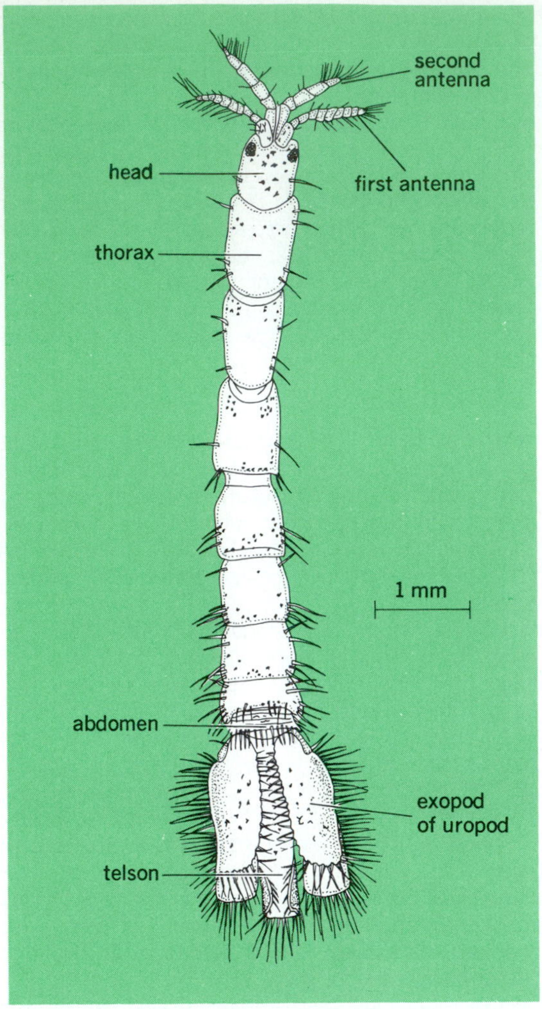

Paranthura infundibulata.

freshwater habitats. Some are facultative ectoparasites of fishes. *See* ISOPODA. [M.A.M.]

Antiarcha An order of highly specialized Placodermi, characterized by heavy armor consisting of a short head shield and a very long, flat-bottomed trunk shield (see illustration). Most antiarchs are small or moderate in size, the largest reaching a length of about 48 in. (120 cm). The eyes and nasal openings occupy an orbital fenestra on top of the skull; in front of them is a premedian plate, unknown in most other placoderms. The mouth is ventral. There is an articulation between the dermal bones of the head and trunk shields, but contrary to the condition in Arthrodira, the condyles are on the head shield, and the glenoid fossae on the trunk shield. The trunk shield has both anterior and posterior median dorsal plates. There are peculiar pectoral appendages, completely covered by small plates, usually jointed near midlength, and articulating in axial fossae on the sides of the trunk shield; they were possibly derived from the pectoral fins of more typical placoderms. The body behind the shield is scaled or naked, has one or two dorsal fins, and a heterocercal caudal fin; pelvic fins are possibly represented by "frills" behind the shield of *Bothriolepsis*.

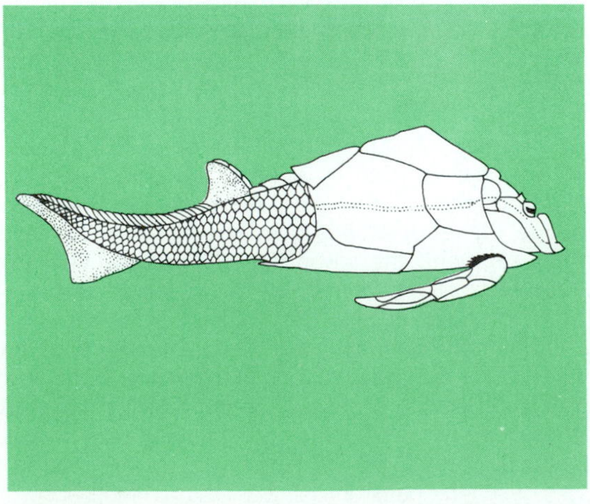

Pterichtyodes (Pterichthys)*, a Middle Devonian antiarch. *(After Traquair)

Antiarcha were predominantly fresh-water stream and lake dwellers, though a few lived in the sea. They were benthic forms, propelling themselves along the bottom by their pectoral appendages, and surely feeding on the bottom. *See* PLACODERMI. [R.H.De.]

Antibacterial agents Synthetic or natural compounds which inhibit the growth or division of bacteria. This article considers gaseous and dye agents. For other antibacterial compounds, *see* ANTIBIOTIC; ANTISEPTIC; SULFONAMIDE.

Gaseous agents serve to disinfect rooms, bedding and other bulky materials, surgical instruments, and valuable heat-labile materials such as viral vaccines which cannot be filter-sterilized. Most widely used is ethylene oxide. Ethylene oxide, boiling point 51.4°F (10.8°C), is commercially available as nonexplosive mixtures with CO_2 or fluorinated hydrocarbons. Special sterilizing chambers are commercially available. Like other potent alkylating agents, it is poisonous and presumably mutagenic. Ethylene oxide has the advantages of quick removal and of not seriously lessening the nutritive value of microbiological culture media. Good ventilation must attend its use. β-Propiolactone, boiling point 311°F (155°C), is 4000 times more effective than ethylene oxide and 50,000 times more effective than CH_3Br (used for fumigating grain elevators). It works best at high humidities, is active even below 50°F (10°C), and presumably is hazardous to humans. Volatile preservatives protect nutrient solutions against microbial action and also preserve samples of sea water for later analysis, especially for vitamins. The preservative is removed on steam sterilization of the culture medium. They probably act as slow alkylating agents and are unsuitable for solutions of compounds having free SH (sulfhydryl) groups. [S.H.H.]

Dye agents are chiefly pyronine (mercurochrome), acridine (neutral acriflavine), and triphenylmethane or rosaniline (gentian violet) dyes. Historically, the discovery of their ability to prevent selectively the growth of and to kill particular kinds of bacteria led to the modern era of search for chemotherapeutic drugs. Except as skin disinfectants, their clinical use has been rapidly displaced by more specific and more effective antibiotics and synthetic drugs with less toxicity for animal tissues. In diagnostic laboratory practice, dyes are among the most valuable substances in bacteriological media for the selective isolation and differentiation of pure culture growth of bacteria from natural sources. *See* CHEMOTHERAPY; ELECTIVE CULTURE. [C.L.]

Antibiotic A chemical substance, produced by microorganisms, that has the capacity in dilute solutions to inhibit the growth of, and even to destroy, bacteria and other microorganisms.

The first antibiotic to receive widespread attention was penicillin, which was developed to meet the need for effective anti-infectious agents during World War II. The action of these substances had been recognized for centuries. Even in 1942 the full importance of antibiotic substances in nature was not recognized. Although penicillin had been discovered in 1928 by Alexander Fleming, the scope of the sulfonamides as anti-infectious agents had not been defined at that time. Gramicidin and tyrothricin had not been discovered, and the fact that antimicrobial agents of natural origin could exert an effect against systemic infections in animals and humans had not been demonstrated. It was not until the pressure of war stimulated research on agents suitable for the control of infectious diseases that Howard Florey and his associates succeeded in demonstrating the potentialities of penicillin therapy. But only after the discovery of streptomycin in 1944 and subsequent observations on the action of bacitracin, neomycin, polymyxin, viomycin, chloramphenicol, chlor- and oxytetracycline, and other antibiotics that were soon isolated did the full import of the antibiotics become apparent.

Penicillin is produced by strains of *Penicillium notatum* and *P. chrysogenum*. Most of the other antibiotics in clinical use are produced by actinomycetes (streptomycetes), a group of organisms that produce a wide variety of such substances. Antibiotics are also produced by bacteria and by fungi.

In general, microorganisms may be subdivided into yeasts, molds (fungi), streptomycetes, bacteria, rickettsiae, and viruses. A large proportion of the infectious diseases of animals and humans is caused by three of these groups, that is, by bacteria, rickettsiae, and viruses. It is against the bacteria and rickettsiae that the antibiotics are most effective. *See* MEDICAL BACTERIOLOGY; MEDICAL MYCOLOGY; RICKETTSIOSES; VIRUS.

Certain antibiotics, for example, penicillin and erythromycin, are primarily effective against gram-positive organisms, while others, such as streptomycin, are also effective against gram-negative and acid-fast organisms. Some antibiotics, such as chloramphenicol and tetracycline, are effective against both gram-positive and gram-negative bacterial species; these are spoken of as broad-spectrum antibiotics because of their more extensive antimicrobial spectrum. It is the broad-spectrum group of antibiotics that has activity against certain of the rickettsiae and large viruses. Several antibiotics are active against pathogenic fungi. Among them may be mentioned amphotericin B, nystatin, griseofulvin, and saramycetin. Griseofulvin is effective against fungi causing skin diseases, whereas the others are active against systemic fungal infections. No antibiotic has so far been found that is effective against small viruses, such as the virus which causes poliomyelitis.

The pioneering observation in 1952 on the antitumor activity of actinomycin sparked an intensive search for antitumor antibiotics. Several were found that received limited use against certain forms of cancer. Among these antibiotics are duanomycin, mitomycin C, chromomycin A_3, and sarkomycin. Unfortunately, the antitumor antibiotics are significantly toxic to the patient, so that great caution must be taken with their use.

Antibiotics are bacteriostatic or bactericidal in action; that is, they either inhibit growth of susceptible microorganisms or actually destroy them. The extent to which they may exert bacteriostatic or bactericidal effects depends upon the concentration of antibiotic present, the number of microorganisms against which it must exert its effect, and various environmental conditions. The exact mechanism by which many antibiotics inhibit microbial growth is not known. Many antibiotics interfere with vital processes of the bacterium, such as protein synthesis, nucleic acid synthesis, cell wall formation, and maintenance of the integrity of the cell membrane.

To be an effective anti-infectious agent, an antibiotic must be capable of exerting its growth inhibitory action in the host against the specific microorganism causing infection without simultaneously damaging to any significant degree the tissues or organs of the host. It must be capable of reaching the site at which the infecting microorganism resides within the body, and it must reach that site in sufficient concentration and remain there for a sufficient period of time to permit it to exert its effects.

Antibiotics may be administered parenterally (that is, by injection), orally, or topically. When administered parenterally or orally, they must be absorbed into the body and transported by the blood and extracellular fluids to the site of the infecting organisms. When administered topically, such absorption is rarely possible, and the antibiotics then exert their effect only against those organisms present at the site of application. [E.Me.]

Antibody A protein found principally in blood serum and characterized by a specific reactivity with the corresponding antigen. Antibodies are important in resistance against disease, in allergy, and in blood transfusions, and can be utilized in laboratory tests for the detection of antigens or the estimation of the immune status. Antibodies are normally absent at birth unless derived passively from the mother through the placenta or colostrum. In time, certain antibodies appear in response to environmental antigens, although the inducers have not been determined for all normal antibodies, particularly the human blood-group isoagglutinins. Antibodies are also induced by artificial immunization with vaccines or following natural infections. The resulting antibody level declines over a period of months, but rapidly increases following renewed contact with specific antigen, even after a lapse of years. This is known as an anamnestic or booster response. *See* ANTIGEN; BLOOD GROUPS; HYPERSENSITIVITY; ISOANTIGEN.

Three principal groups (IgG, IgM, IgA) and two minor groups (IgD, IgE) of antibodies are recognized. These all form part of the wider classification of gamma globulins (immunoglobulins). *See* IMMUNOGLOBULIN. [M.J.Po.]

Anticline A fold in layered rocks in which the strata are inclined down and away from the axes. The simplest anticlines (see illustration) are symmetrical, but in more highly deformed

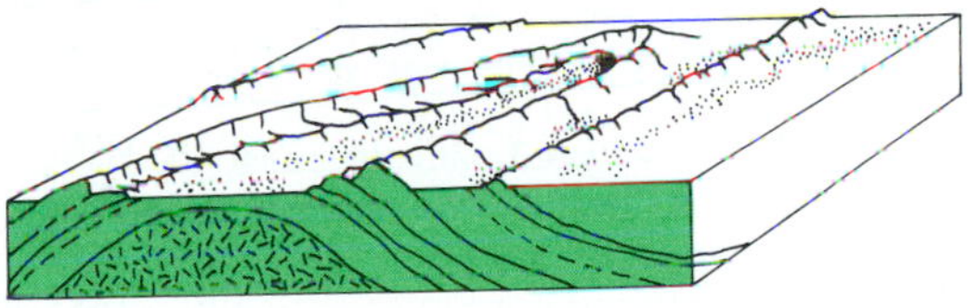

Diagram relating anticlinal structure to topography.

regions they may be asymmetrical, overturned, or recumbent. Most anticlines are elongate with axes that plunge toward the extremities of the fold, but some have no distinct trend; the latter are called domes. Generally, the stratigraphically older rocks are found toward the center of curvature of an anticline, but in more complex structures these simple relations need not hold. Under such circumstances, it is sometimes convenient to recognize two types of anticlines. Stratigraphic anticlines are those folds, regardless of their observed forms, that are inferred from stratigraphic information to have been anticlines originally. Structural anticlines are those that have forms of anticlines, regardless of their original form. *See* SYNCLINE. [P.H.O.]

Antiferromagnetism A property possessed by some metals, alloys, and salts of transition elements in which the atomic magnetic moments, at sufficiently low temperatures, form an ordered array which alternates or spirals so as to give no net total moment in zero applied magnetic field. The most direct way of detecting such arrangements is by means of neutron diffraction. *See* NEUTRON DIFFRACTION.

The transition temperature below which the spontaneous antiparallel magnetic ordering takes place is called the Néel temperature. A plot of the magnetic susceptibility of a typical antiferromagnetic powder sample versus temperature is shown in the illustration. Below the Néel point, which is characterized by the sharp kink in the susceptibility, the spontaneous ordering opposes the normal tendency of the magnetic moments to

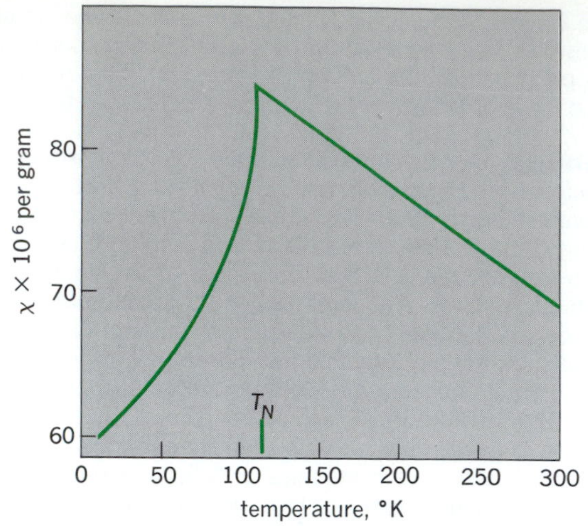

Magnetic susceptibility of powdered manganese oxide. ***(After H. Bizette, C. F. Squire, and B. Tsai, 1938)***

align parallel to the applied field. Above the Néel point, the substance is paramagnetic, and the susceptibility χ obeys the Curie-Weiss law, as in Eq. (1), with a negative paramagnetic

$$\chi = C/(T + \theta) \qquad (1)$$

Curie temperature $-\theta$. The Néel temperature is similar to the Curie temperature in ferromagnetism. *See* CURIE TEMPERATURE; CURIE-WEISS LAW; MAGNETIC SUSCEPTIBILITY.

The cooperative transition that characterizes antiferromagnetism is thought to result from an interaction energy U of the form given in Eq. (2), where $\mathbf{S}_i$ and $\mathbf{S}_j$ are the spin angular

$$U = -2\Sigma J_{ij}\mathbf{S}_i \cdot \mathbf{S}_j \qquad (2)$$

momentum vectors associated with the magnetic moments of neighbor atoms i and j, and J_{ij} is an interaction constant. If all J_{ij} are positive, the lowest energy is achieved with all $\mathbf{S}_i$ and $\mathbf{S}_j$ parallel, that is, coupled ferromagnetically. Negative J_{ij} between nearest-neighbor pairs (i,j) may lead to simple antiparallel arrays; if the distant neighbors also have sizable negative J_{ij}, a spiral array may have lowest total energy. The interaction constant in Eq. (2) probably arises from superexchange coupling. This is an effective coupling between magnetic spins which is indirectly routed via nonmagnetic atoms in salts and probably via conduction electrons in metals. *See* FERROMAGNETISM; HELIMAGNETISM.

The magnetic moments are known to have preferred direction. Anisotropic effects come from magnetic dipole forces and also from spin-orbit coupling combined with superexchange. Some nearly antiparallel arrays such as Fe_2O_3 show a slight bending (called canting) and exhibit weak ferromagnetism. The anisotropy affects the susceptibility of powder samples and is of extreme importance in antiferromagnetic resonance. *See* MAGNETIC RESONANCE. [E.A.; F.Ke.]

Antifreeze (biology) Glycoprotein or protein molecules synthesized by polar and north temperate fishes to enable them to survive in freezing seawater. Similar antifreezes are found in some insects, but relatively little is known about their structure and function.

In a marine fish, the amount of salt and other small molecules in the blood depresses its freezing point to about 30°F (−0.8°C). In the winter, the polar oceans and the nearshore water of north temperate oceans are at the freezing point of seawater 28.6°F (−1.9°C). In the absence of ice, many fishes survive by supercooling, a thermodynamic state of equilibrium in which a solution (the body fluids of the fish in this case) can be in liquid state, in the absence of ice nuclei, at a temperature lower than the equilibrium freezing point. However, polar waters are often laden with ice that can enter the fish by ingestion of seawater. Propagation of ice in the body fluids or tissues of the fish always leads to freezing damage and death. To avoid freezing, many fishes have evolved biological antifreezes that further lower the freezing point of their body fluids to 28°F (−2.2°C), which is 0.6°F (0.3°C) below the freezing point of seawater. *See* COLD HARDINESS (PLANTS); CRYPTOBIOSIS. [A.L.DeV.]

Antifreeze mixture An antifreeze is a substance which is added to a liquid to lower its freezing point. Common usage limits the term to materials added to water in the cooling systems of internal combustion engines. However, it can also include mixtures such as refrigeration brines, snow-melting and deicing agents, aqueous hydraulic fluids, and heat-transfer fluids.

The major criteria for satisfactory antifreeze substances are sufficient miscibility with the solvent to lower freezing point; boiling point somewhat above maximum operating temperatures of the system in which it is used; low viscosity; high degree of chemical stability; low electrical conductivity; good heat-transfer properties; and inertness toward materials of construction in the system used.

The principal automotive antifreeze material is ethylene glycol. Methoxypropanol, methyl and isopropyl alcohols, and diethylene and propylene glycols are also used to a limited extent.

Because of the numerous metals in an automotive cooling system (cast iron, steel, brass, copper, solder, aluminum), commercial antifreeze mixtures also contain corrosion inhibitors. Frequently, an antifoam agent is also added. Electrolytes (inorganic salts), although highly effective, are undesirable for most uses because of their corrosivity and electrical conductivity. Most windshield wiper fluids are solutions of ethanol, methanol, and isopropyl alcohol. Brines for refrigeration of food or making ice are solutions of calcium chloride or propylene glycol. The water content of these solutions is determined by the lowest temperature encountered. *See* ENGINE COOLING; INHIBITOR (CHEMISTRY); REFRIGERATION; SOLUTION. [E.F.Ha.]

Antifriction bearing A machine element that permits free motion between moving and fixed parts. Antifrictional bearings are essential to mechanized equipment; they hold or guide moving machine parts and minimize friction and wear.

In its simplest form, a bearing consists of a cylindrical shaft, called a journal, and a mating hole, serving as the bearing proper. Ancient bearings were made of such materials as wood, stone, leather, or bone, and later of metal. It soon became apparent for this type of bearing that a lubricant would reduce both friction and wear and prolong the useful life of the bearing. Petroleum oils and greases are generally used for lubricants, sometimes containing soap and solid lubricants such as graphite or molybdenum disulfide, talc, and similar substances.

Materials. The greatest single advance in the development of improved bearing materials took place in 1839, when I. Babbitt obtained a United States patent for a bearing metal with a special alloy. This alloy, largely tin, contained small amounts of antimony, copper, and lead. This and similar materials have made excellent bearings. They have a silvery appearance and are generally described as white metals or as Babbitt metals.

Wooden bearings are still used for limited applications in light-duty machinery and are frequently made of hard maple which has been impregnated with a neutral oil. Wooden bearings made of lignum vitae, the hardest and densest of all woods, are still used.

Some of the most successful heavy-duty bearing metals are now made of several distinct compositions combined in one bearing. This approach is based on the widely accepted theory of friction, which is that the best possible bearing material would be one which is fairly hard and resistant but which has an overlay of a soft metal that is easily deformed. Figure 1 shows bearings in which graphite, carbon, plastic, and rubber have been incorporated into a number of designs illustrating some of the material combinations that are presently available.

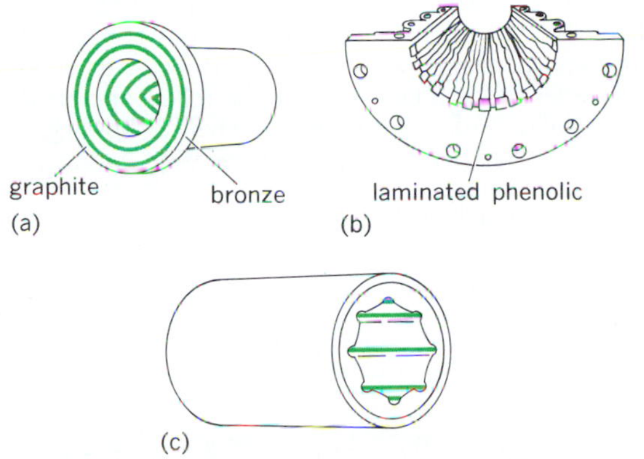

Fig. 1. Bearings with (*a*) graphite; (*b*) wood, plastic, and nylon (*after J. J. O'Connor, ed., Power's Handbook on Bearings and Lubrication, McGraw-Hill, 1951*); (c) rubber.

Rubber has proved to be a surprisingly good bearing material, especially under circumstances in which abrasives may be present in the lubricant. The rubber used is a tough resilient compound similar in texture to that in an automobile tire. Cast iron is one of the oldest bearing materials. It is still used where the duty is relatively light.

Porous metal bearings are frequently used when plain metal bearings are impractical because of lack of space or inaccessibility for lubrication. These bearings have voids of 16–36% of the volume of the bearing. These voids are filled with a lubricant by a vacuum technique. During operation they supply a limited amount of lubricant to the sliding surface between the journal and the bearing. In general, these bearings are satisfactory for light loads and moderate speeds.

Lubricants. The method of supplying the lubricant and the quantity of lubricant which is fed to the bearing by the supplying device will often be the greatest factor in establishing performance characteristics of the bearing. For example, if no lubricant is present, the journal and bearing will rub against each other in the dry state. Both friction and wear will be relatively high. The coefficient of friction of a steel shaft rubbing in a bronze bearing, for example, may be about 0.3 for the dry state. If lubricant is present even in small quantities, the surfaces become contaminated by this material whether it be an oil or a fat, and depending upon its chemical composition the coefficient of friction may be reduced to about 0.1. Now if an abundance of lubricant is fed to the bearing so that there is an excess flowing out of the bearing, it is possible to develop a self-generating pressure film in the clearance space as indicated in Fig. 2. These pressures can be sufficient to sustain a considerable load and to keep the rubbing surfaces of the bearing separated.

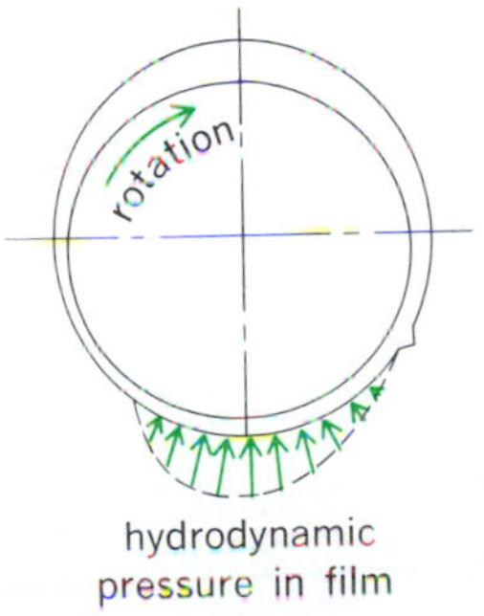

Fig. 2. Hydrodynamic fluid-film pressures in a journal bearing. (*After W. Staniar, ed., Plant Engineering Handbook, 2d ed., McGraw-Hill, 1959*)

The types of oiling devices that usually result in insufficient feed to generate a complete fluid film are, for example, oil cans, drop-feed oilers, waste-packed bearings, and wick and felt feeders. Oiling schemes that provide an abundance of lubrication are oil rings, bath lubrication, and forced-feed circulating supply systems. The coefficient of friction for a bearing with a complete fluid film may be as low as 0.001.

Fluid-film hydrodynamic bearings. If the bearing surfaces can be kept separated, the lubricant no longer needs an oiliness agent. As a consequence, many extreme applications are presently found in which fluid-film bearings operate with lubricants consisting of water, highly corrosive acids, molten metals, gasoline, steam, liquid refrigerants, mercury, gases, and so on. The self-generation of pressure in such a bearing takes place no matter what lubricant is used, but the maximum pressure that is generated depends upon the viscosity of the lubricant. Thus, for example, the maximum load-carrying capacity of a gas-lubricated bearing is much lower than that of a liquid-lubricated bearing. The ratio of capacities is in direct proportion to the viscosity. Gas is the only presently known lubricant that can be used for operation at extreme temperatures. Because the viscosity of gas is so low, the friction generated in the bearing is correspondingly of a very low order. Thus gas-lubricated machines can be operated at extremely high speeds because there is no serious problem in keeping the bearings cool.

The self-generating pressure principle is applied equally as well to thrust bearings as it is to journal bearings. The tilting-pad type of thrust bearing (Fig. 3*a*) excels in low friction and in reliability. A typical commercial thrust bearing (Fig. 3*b*) is made up of many tilting pads located in a circular position. One of the largest is on a hydraulic turbine at the Grand Coulee Dam.

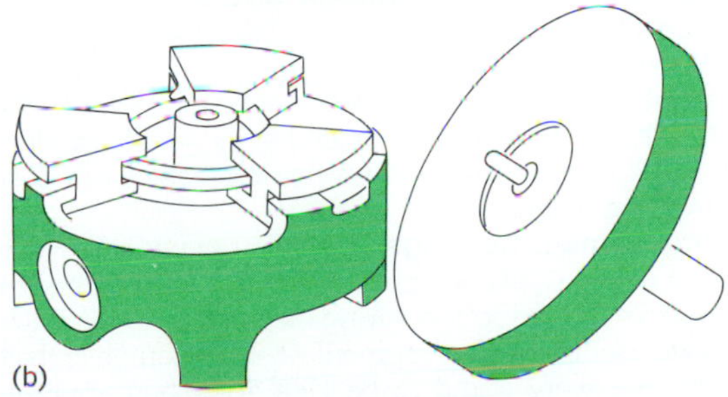

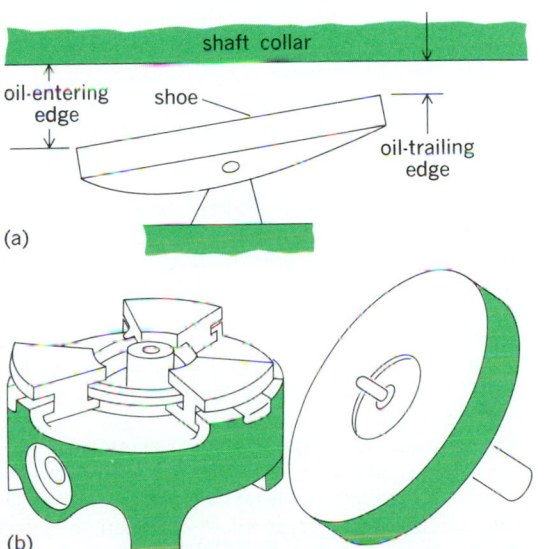

Fig. 3. Tilting-shoe-type bearing. (*a*) Schematic (*after W. Staniar, ed., Plant Engineering Handbook, 2d ed., McGraw-Hill, 1959*). (*b*) Thrust bearing (*after D. D. Fuller, Theory and Practice of Lubrication for Engineers, copyright © 1956 by John Wiley and Sons, Inc.; used with permission*).

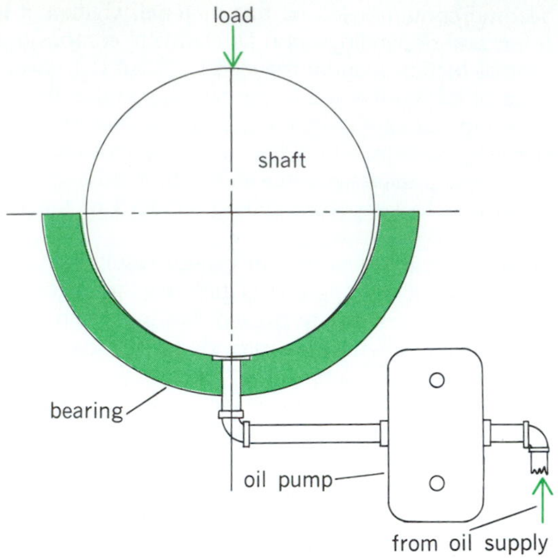

Fig. 4. Fluid-film hydrostatic bearing. Hydrostatic oil lift can reduce starting friction drag to less than one-tenth of usual starting drag. (*After W. Staniar, ed., Plant Engineering Handbook, 2d ed., McGraw-Hill, 1959*)

There, a bearing 96 in. (2.4 m) in diameter carries a load of 2,150,000 lb (9,560,000 newtons) with a coefficient of friction of about 0.0009.

Fluid-film hydrostatic bearings. Sleeve bearings of the self-generating pressure type, after being brought up to speed, operate with a high degree of efficiency and reliability. However, when the rotational speed of the journal is too low to maintain a complete fluid film, or when starting, stopping, or reversing, the oil film is ruptured, friction increases, and wear of the bearing accelerates. This condition can be eliminated by introducing high-pressure oil to the area between the bottom of the journal and the bearing itself, as shown schematically in Fig. 4. If the pressure and quantity of flow are in the correct proportions, the shaft will be raised and supported by an oil-film whether it is rotating or not. Friction drag may drop to one-tenth of its original value or even less, and in certain kinds of heavy rotational equipment in which available torque is low, this may mean the difference between starting and not starting. This type of lubrication is called hydrostatic lubrication and, as applied to a journal bearing in the manner indicated, it is called an oil lift. Hydrostatic lubrication in the form of a step bearing has also been used on various machines to carry thrust.

Large structures have been floated successfully on hydrostatic-type bearings. For example, the Hale 200-in. (5-m) telescope on Palomar Mountain (California Institute of Technology/Palomar Observatory) weighs about 1,000,000 lb (450,000 kg); yet the coefficient of friction for the entire supporting system, because of the hydrostatic-type bearing, is less than 0.000004. The power required is extremely small and a $\frac{1}{12}$-hp (62-W) clock motor rotates the telescope while observations are being made.

Rolling-element bearings. Everyday experiences demonstrate that rolling resistance is much less than sliding resistance. This principle is used in the rolling-element bearing which has found wide use. In the development of the automobile, ball and roller bearings were found to be ideal for many applications, and today they are widely used in almost every kind of machinery.

These bearings are characterized by balls or cylinders confined between outer and inner rings. The balls or rollers are usually spaced uniformly by a cage or separator. The rolling elements are the most important because they transmit the loads from the moving parts of the machine to the stationary supports. Balls are uniformly spherical, but the rollers may be straight cylinders, or they may be barrel- or cone-shaped or of other forms, depending upon the purpose of the design. The rings, called the races, supply smooth, hard, accurate surfaces for the balls or rollers to roll on. Some types of ball and roller bearings are made without separators. In other types there is only the inner or the outer ring, and the rollers operate directly upon a suitably hardened and ground shaft or housing. Figure 5 shows a typical deep-grooved ball bearing, with the parts that are generally used.

These bearings may be classified by function into three groups: radial, thrust, and angular-contact bearings. Radial bearings are designed principally to carry a load in a direction perpendicular to the axis of rotation. However, some radial bearings, such as the deep-grooved bearings shown in Fig. 5,

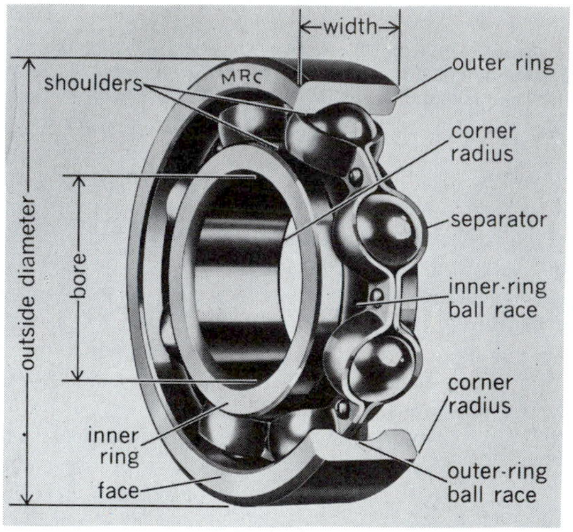

Fig. 5. Deep-groove ball bearing. (*Marlin-Rockwell*)

are also capable of carrying a thrust load, that is, a load parallel to the axis of rotation and tending to push the shaft in the axial direction. Some bearings, however, are designed to carry only thrust loads. Angular-contact bearings are especially designed and manufactured to carry heavy thrust loads and also radial loads.

A unique feature of rolling-element bearings is that their useful life is not determined by wear but by fatigue of the operating surfaces under the repeated stresses of normal use. Fatigue failure, which occurs as a progressive flaking or sifting of the surfaces of the races and rolling elements, is accepted as the basic reason for the termination of the useful life of such a bearing. [D.D.F.]

Antigen A substance that initiates and mediates the formation of the corresponding immune body, termed antibody. Antigens can also react with formed antibodies. Antigen-antibody reactions serve as host defenses against microorganisms and other foreign bodies, or are used in laboratory tests for detecting the presence of either antigen or antibody. *See* ANTIBODY.

With a few exceptions, such as the autoantigens and the isoantigens of the blood groups, antigens produce antibody only in species other than the ones from which they are derived. All complete proteins are antigenic, as are many bacterial and other polysaccharides, some nucleic acids, and some lipids. Antigenicity may be modified or abolished by chemical treatments, including degradation or enzymatic digestion; it may be notably increased by the incorporation of antigen into oils or other adjuvants. *See* BLOOD GROUPS; ISOANTIGEN.

Bacteria, viruses, protozoans, and other microorganisms are important sources of antigens. These may be proteins or polysaccharides derived from the outer surfaces of the cell (capsular antigens), from the cell interior (the somatic or 0 antigens), or from the flagella (the flagellar or H antigens). Gram-negative bacteria also contain the Vi antigen, believed by some investigators to be associated with virulence, and the Boivin antigens, which contain polysaccharide, lipid, and protein. *See* VIRULENCE.

Other antigens are either excreted by the cell or are released into the medium during cell death and disruption; these include many enzymes and toxins, of which diphtheria, tetanus, and botulinus toxins are important examples. The presence of antibody to one of these constituent antigens in human or animal sera is presumptive evidence of past or present contact with specific microorganisms, and this finds application in clinical diagnosis and epidemiological surveys. *See* BOTULISM; DIPHTHERIA; TETANUS.

Microbial antigens, prepared to induce protective antibodies, are termed vaccines. They may consist of either attenuated living or killed whole cells, or extracts of these. Since whole microorganisms are complex structures, vaccines may contain 10 or more distinct antigens, of which generally not more than one or two engender a protective antibody. Examples of these are smallpox vaccine, a living attenuated virus; typhoid vaccine, killed bacterial cells; and diphtheria toxoid, detoxified culture fluid. Several independent vaccines may be mixed to give a combined vaccine, and thus reduce the number of injections necessary for immunization, but such mixing can result in a lesser response to each component of the mixture. *See* VACCINATION.

Haptens, or haptenes, are partial antigens that can inhibit reactions between complete antigens and corresponding antibodies. Some haptens can react visibly with antibodies formed to complete antigens, but the haptens cannot themselves engender the formation of new antibodies. Haptens are usually simple molecules which are small in size, such as degradation products of complete antigens. Some haptens may be converted to complete antigens by combination with a protein carrier.

Allergens are antigens that induce allergic states in humans or animals. Examples are preparations from poison ivy, cottonseed, or horse dander, or simple chemicals such as formaldehyde or picryl chloride. *See* BIOLOGICALS; HETEROPHILE ANTIGEN; HYPERSENSITIVITY; IMMUNITY; IMMUNOLOGY. [M.J.Po.]

Antihistamine Any drug that combats the effects of alleged histamine release following tissue damage by blocking the receptor site at which histamine has its action. They may also act by preventing antigen-antibody reactions and by preventing the release of histamine from mast cells. The antihistaminics are not potent antagonists and fail completely in preventing some of the actions of histamine. They are more potent in preventing the actions of histamines than in reversing these actions once they develop. These drugs may be effective in certain allergies and hypersensitivity reactions but give only symptomatic relief since the reaction will continue, even though suppressed, as long as the inciting agent is present. *See* HISTAMINE; HYPERSENSITIVITY.

Undesirable side effects of the antihistamines vary considerably from one compound to another and according to individual patients as well. Drowsiness, dizziness, dryness of the mouth, nausea, gastrointestinal disturbances, and muscular weakness are not uncommon.

Some antihistamines may be applied locally in solutions, spray, or cream form while others are taken orally or by injection.

When properly used on an individual drug and patient basis, the antihistamines frequently are dramatically effective. However, individual reactions and idiosyncrasies and the variable nature of the disorders treated militate against their indiscriminate or unregulated use. [N.K.M.]

Antiknock agent A substance added to most finished motor and aviation gasolines, and some other hydrocarbon fuels, to increase the resistance of the fuel to knock in spark-ignited engines.

The antiknock quality of a fuel is determined primarily by its hydrocarbon composition, but this quality can be increased by the addition of antiknock agents. These may be organic chemicals such as aromatic amines, or metalloorganic compounds such as tetraethyllead, tetramethyllead, the mixed methylethyl lead alkyls, or methyl cyclopentadienyl manganese tricarbonyl.

The lead alkyls are many times as effective as aromatic amines. The manganese compound, methylcyclopentadienyl manganese tricarbonyl, is also being used as an antiknock agent, principally in the manufacture of lead-free gasolines for vehicles equipped with catalytic converters. A chelate of the rare earth cerium has been patented as an antiknock agent but has not been developed for commercial use.

All metallic antiknock agents vary widely in their effectiveness in different fuels, depending upon constitution of the fuel and the presence of antagonists, such as sulfur compounds. Antiknock agents are believed to function chemically in the engine by decomposing thermally to metals and oxides which interrupt oxidative chain reactions leading to autoignition. *See* GASOLINE; INTERNAL COMBUSTION ENGINE; OCTANE NUMBER; ORGANOMETALLIC COMPOUND; SPARK KNOCK; TETRAETHYLLEAD. [H.J.G.]

Antimatter Substance consisting of atoms which are charge conjugates of atoms found in ordinary matter. Since physicists have demonstrated the existence of the positron, the antiproton, and the antineutron, which are respectively the antiparticles (charge-conjugate particles) of the electron, the proton, and the neutron, it is clear that at least in principle it is possible to have the charge conjugate of any atom. Such an atom would be formed in perfect analogy to the ordinary atom, with each particle replaced by its charge conjugate. Such atoms would constitute antimatter.

An atom of matter and its counterpart of antimatter, if brought in contact, would annihilate each other, giving rise to π-mesons and other particles; but all of these particles created would transform within microseconds into gamma rays, neutrinos, and electrons and their antiparticles.

Antimatter out of contact with ordinary matter would be stable, and there has been speculation about the presence in the cosmos of antiworlds, in which antimatter is prevalent. *See* ANTINEUTRON; ANTIPROTON; ELEMENTARY PARTICLE; POSITRON; SYMMETRY LAWS (PHYSICS). [J.L.]

Antimetabolite An enzyme inhibitor that is structurally related to the substrate and which may form a complex with the enzyme, analogous to the normal enzyme-substrate complex. Products of enzymic reactions have long been known to inhibit the reaction to a greater extent than would be expected by mass action alone, and frequently analogs of the substrate competitively inhibit an enzyme.

Antimetabolite is a term which has been given to analogs of metabolites that inhibit the utilization of their corresponding metabolite. While most antimetabolites are competitive inhibitors, some appear to react with the enzyme in a manner which results in a noncompetitive inhibition. Antimetabolites are frequently termed metabolite antagonists. Structural analogs of most of the known metabolites have been prepared and found to exert inhibitory effects on the utilization of their corresponding metabolites in many biochemical systems; for example, analog antagonists of amino acids, purines, pyrimidines, vitamins, hormones, and other metabolites have been prepared.

Metabolite antagonists have shown some promise, in a few instances, as chemotherapeutic agents. Specificity of the sulfonamides, for example, probably results from the fact that in the host a particular coenzyme is synthesized from folic acid but not from *p*-aminobenzoic acid, whereas in the infective organism this coenzyme is synthesized from *p*-aminobenzoic acid but not from folic acid. Another application of chemotherapy has attempted to produce antitumor agents which will inhibit growing cells without damage to nonproliferating cells.

Inhibitions, involving competition of an analog with a metabolite for an enzyme site, can be used to demonstrate metabolic pathways in living organisms through analysis of the effect of reversing agents other than the metabolite. *See* ENZYME INHIBITION; SULFONAMIDE. [W.Sh.]

Antimicrobial agents A chemical compound biosynthetically or synthetically produced, which either destroys or usefully suppresses the growth or metabolism of a variety of microscopic or submicroscopic forms of life. On the basis of their primary activity, they are more specifically called antibacterial, antifungal, antiprotozoal, antiparasitic, or antiviral agents. Antibacterials which destroy are bactericides or germicides; those which merely suppress growth are bacteriostatic agents. *See* ANTIBACTERIAL AGENTS; FUNGICIDE.

Of the thousands of antimicrobial agents, only a small number are safe chemotherapeutic agents, effective in controlling infectious diseases in plants, animals, and humans. A much larger number are used in almost every phase of human activity: in agriculture, food preservation, and water, skin, and air disinfection. A compilation of some common uses for antimicrobials is shown in the table.

Common antimicrobial agents and their uses

Use	Agents
Chemotherapeutics (animals and humans)	
Antibacterials	Sulfonamides, isoniazid, *p*-aminosalicylic acid, penicillin, streptomycin, tetracyclines, chloramphenicol, erythromycin, novobiocin, neomycin, bacitracin, polymyxin
Antiparasitics (humans)	Emetine, quinine
Antiparasitics (animal)	Hygromycin, phenothiazine, piperazine
Antifungals	Griseofulvin, nystatin
Chemotherapeutics (plants)	Captain, maneb, thiram
Skin disinfectants	Alcohols, iodine, mercurials, silver compounds, quaternary ammonium compounds, neomycin
Water disinfectants	Chlorine, sodium hypochlorite
Air disinfectants	Propylene glycol, lactic acid, glycolic acid, levulinic acid
Gaseous disinfectants	Ethylene oxide, β-propiolactone, formaldehyde
Clothing disinfectants	Neomycin
Animal-growth stimulants	Penicillin, streptomycin, bacitracin, tetracyclines, hygromycin
Food preservatives	Sodium benzoate, tetracycline

The most important antimicrobial discovery of all time, that of the chemotherapeutic value of penicillin, was made after hope of developing its clinical usefulness had been abandoned. The subsequent establishment of penicillin as a nontoxic drug with 200 times the activity of sulfanilamide opened the flood gates of antibiotic research. In the next 20 years, more than a score of new and useful microbially produced antimicrobials entered daily use.

New synthetic antimicrobials are found today by synthesis of a wide variety of compounds, followed by broad screening against many microorganisms. Biosynthetic antimicrobials, although first found in bacteria, fungi, and plants, are discovered primarily in actinomycetes. *See* ACTINOMYCETACEAE; ANTIBIOTIC; PENICILLIN. [G.M.S.]

Antimonate The negative radical in salts derived from antimony pentoxide, Sb_4O_{10}, and bases. Most antimonates contain the $[Sb(OH)_6]^-$ ion in spite of the fact that they were called pyro- or meta-antimonates by analogy with phosphates and arsenates. There is very little similarity between the phosphates and antimonates. The hypothetical acid, $HSb(OH)_6$, is not isolated; instead, a gelatinous precipitate of hydrated antimony pentoxide is obtained when antimony solutions are acidified. Sodium salt, $NaSb(OH)_6$, is one of the very few insoluble sodium salts and can be used as a test for the sodium ion. *See* ANTIMONY; ARSENATE; PHOSPHATE. [E.E.W.]

Antimony A chemical element, symbol Sb, atomic number 51. Antimony is not a naturally abundant element; it is occasionally found native, often in isomorphous mixture with

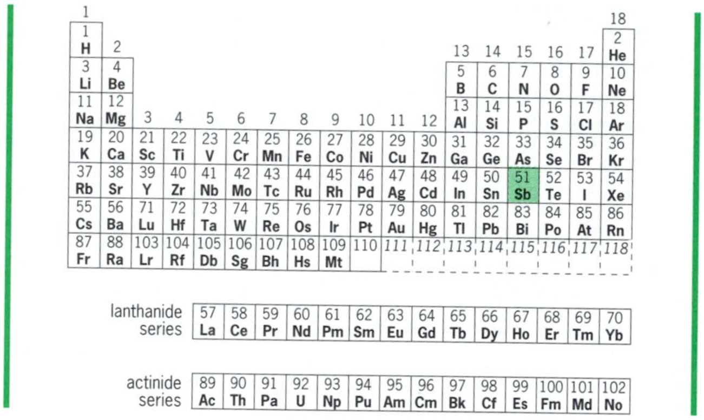

arsenic, as allemonite. The symbol Sb is derived from the Latin name stibium.

The element is dimorphic, existing as a yellow, metastable form composed of Sb_4 molecules, as in antimony vapor and the structural unit in yellow antimony; and a gray, metallic form, which crystallizes with a layered rhombohedral structure. Antimony differs from normal metals in having a lower electrical conductivity as a solid than as a liquid (as does its congener, bismuth). Metallic antimony is quite brittle, bluish-white with a typical metallic luster, but a flaky appearance. Although stable in air at normal temperatures, it burns brilliantly when heated, with the formation of a white smoke of Sb_2O_3. Vaporization of the metal gives molecules of Sb_4O_6, which break down to Sb_2O_3 above the transition temperature.

Antimony occurs in nature mainly as Sb_2S_3 (stibnite, antimonite); Sb_2O_3 (valentinite) occurs as a decomposition product of stibnite. Antimony is commonly found in ores of copper, silver, and lead. The metal antimonides NiSb (breithaupite), NiSbS (ullmannite), and Ag_2Sb (dicrasite) also are found naturally; there are numerous thioantimonates such as Ag_3SbS_3 (pyrargyrite).

Antimony is produced either by roasting the sulfide with iron, or by roasting the sulfide and reducing the sublimate of Sb_4O_6 thus produced with carbon; high-purity antimony is produced by electrolytic refining.

Commercial-grade antimony is used in many alloys (1–20%),

especially lead alloys, which are much harder and mechanically stronger than pure lead; batteries, cable sheathing, antifriction bearings, and type metal consume almost half of all the antimony produced. The valuable property of Sn-Sb-Pb alloys, that they expand on cooling from the melt, thus enabling the production of sharp castings, makes them especially useful as type metal. [J.L.T.W.]

Antineutron The antiparticle (charge-conjugate particle) of the neutron. All electromagnetic properties of a particle are inverted in sign in the charge conjugate; thus the antineutron has a magnetic moment equal and opposite to that of the neutron. Mass, spin, and the decay constant for beta decay are identical for the antineutron and the neutron. Both are fermions of spin 1/2. *See* ELEMENTARY PARTICLE; NEUTRON. [E.G.S.]

Antioxidant An inhibitor which is effective in preventing oxidation by molecular oxygen (autoxidation). Such inhibitors have great commercial significance in the preservation of food and food products and in the prevention of deterioration of petroleum products, rubber, and plastics.

Autoxidations are free-radical chain reactions characterized by the interaction of the radicals with oxygen to yield peroxy radicals, organic peroxides, and a broad spectrum of stable oxygenated products. Autoxidation chains are often long, so that a single initiating event may produce many stable product molecules. Thus only a very small amount of an effective antioxidant need be employed for the protection of a large quantity of a substrate. The role of the antioxidant is to provide an alternate path for oxidation which does not involve the substrate. The antioxidant is destroyed in the process and thus does not function indefinitely.

The major types of antioxidants in use are the phenols, the aromatic amines, sulfur compounds, and a variety of naturally occurring materials. The latter find particular use in the protection of foods and cosmetics from oxidation. Naturally occurring antioxidants include raw seed oils, wheat germ oil, tocopherols, and gums. The activity of the last-named category may often be increased by the use of synergists. These are substances which have little or no activity alone but which enhance the activity of stronger antioxidants. Some effective synergists are phosphoric, citric, and ascorbic acids.

The wide variety of antioxidants available is necessitated by the extreme range of conditions under which protection from oxidation is required. For example, an antioxidant which can delay the development of rancidity in stored butter will seldom prove to be suitable for the protection of hot lubricating oil in the crankcase of an automobile. *See* CATALYSIS; FREE RADICAL; INHIBITOR (CHEMISTRY). [L.R.M.]

Antipatharia An order of the subclass Zoantharia. These animals are the black or horny corals which live in rather deep tropical and subtropical waters and usually form regular or irregularly branching plant-like colonies, often 6.6 or 9.9 ft (2 or 3 m) in height, with thorny, solid lamellar, horny axial skeletons (see illustration). *Stichopathes* forms an unbranching wirelike colony.

The polyp or zooid has six unbranched, nonretractile tentacles with a warty surface due to the presence of nematocysts. Six primary, complete, bilaterally arranged mesenteries occur, of which only two lateral ones bear filaments and gonads. Adjacent zooids are united by a coenenchyme, but their gastrovascular cavities have no connection. The musculature is the most weakly developed in the anthozoans.

The polyps are dioecious. Schizopathidae are dimorphic; the gastrozooid has a mouth and two tentacles, while the gonozooid, the only fertile polyp, lacks a mouth. *See* COELENTERATA; ZOANTHARIA. [K.At.]

Antipathes rhipidion. (After F. Pax)

Antiproton The antiparticle (charge-conjugate particle) of the proton. P. A. M. Dirac's equation of relativistic quantum theory for the electron admits solutions which describe a particle of mass identical to that of the electron but of opposite charge; this is the positron. A generalization of the relativistic equations for fermions of spin 1/2 leads to the prediction that there should be a particle of mass identical to that of the proton but of opposite charge; this is the antiproton. It was first experimentally demonstrated in 1955. *See* ELEMENTARY PARTICLE; RELATIVISTIC QUANTUM THEORY. [E.G.S.]

Antiresonance The condition for which the impedance of a given electric, acoustic, or dynamic system is very high, approaching infinity. In an electric circuit consisting of a capacitor and a coil in parallel, antiresonance occurs when the alternating-current line voltage and the resultant current are in phase. Under these conditions the line current is very small because of the high impedance of the parallel circuit at antiresonance. *See* RESONANCE (ALTERNATING-CURRENT CIRCUITS). [J. Mar.]

Antiseptic A drug used to destroy or prevent the growth of infectious microorganisms on or in the human or animal body, that is, on living tissue. Many chemical substances have been employed as antiseptics.

Iodine is the most important of the halogens used as an antiseptic. Tincture of iodine (iodine in an alcohol solution) has been employed widely as a preoperative antiseptic and in first aid. Tincture of iodine is germicidal by laboratory test in 0.02% concentration, but 2.0% solutions are usually employed in surgery and first aid.

Compounds of mercury were used to prevent infection before the germ theory of disease was established. Because of their high toxicity and severe caustic action, such inorganic mercurials as mercuric chloride, mercuric oxycyanide, and potassium mercuric iodide have been largely replaced by certain organic mercury compounds. Organic mercurial compounds are far less toxic and are nonirritating in concentrated solutions. They are highly bacteriostatic, and in concentrated solutions germicidal as well. They are also nonspecific in antimicrobial activity.

Essential oils have been defined as odoriferous oily substances obtained from such natural sources as plants by steam distillation. Essential oils in alcoholic solutions also were early employed in place of the carbolic acid solution of Lister, and because of the toxic and corrosive action of mercury bichloride, they also replaced this compound. Alcoholic solution of essential oils was first developed in 1881 and was admitted as liquor antisepticus to the U.S. Pharmacopoeia in 1900 and to the National Formulary IV in 1916. Alcoholic solutions of essential oils as represented by liquor antisepticus have proved effective in a wide variety of clinical applications and in first aid.

Silver compounds have been widely used for a variety of purposes. Because of the bland nature of most of these compounds, they have been successfully used in the eyes, nose, throat, urethral tract, and other organs. The most widely used silver compounds are silver nitrate, ammoniacal silver nitrate solution, silver picrate, and certain colloidal silver preparations such as strong protein silver and mild silver protein. These are effective germicides of low tissue toxicity and are not counteracted by organic matter.

Such compounds as ethyl alcohol and isopropyl alcohol are germicidal rather than bacteriostatic and are effective against the vegetative forms of bacteria and virus, but do not kill spores. Ethyl alcohol in 62.5–70% solution is most commonly used, being widely employed for disinfecting the skin before hypodermic injections and other skin punctures. Isopropyl alcohol is equal, if not superior, to ethyl alcohol and is widely used for degerming the skin and for disinfecting oral thermometers. Alcohols are also widely used in other antiseptic preparations, in which they serve to lower the surface tension and to promote spreading and penetration.

Bisphenol compounds such as dichlorophene and tetrachlorophene are essentially bacteriostatic agents and are weaker as germicides. They have proved quite effective as skin-degerming agents, when used in soaps and other detergents, and as mildew-preventing formulations. The halogenated form, such as dichlorophene, tetrachlorophene, hexachlorophene, and bithionol, is most commonly employed. When used repeatedly on the skin, as in soaps and detergents, bisphenols have a tendency to remain for long periods, thus reducing skin bacteria to a significant degree. For this purpose they are especially useful in preoperative hand washing.

Quaternary ammonium compounds have high germicidal activity. Although they are more properly classified as surface-active disinfectants, some of them are employed in certain antiseptic formulations, for instance, Zephiran, especially suited for use on the skin, and Cepacol, for mucous surfaces. Nontoxic and nonirritating, they may be used in place of alcohol after preoperative scrub-up. *See* ANTIMICROBIAL AGENTS; BIOASSAY; FUNGICIDE. [G.F.R./F.C.]

Antitoxin An antibody that will combine with and generally neutralize a particular toxin. When the manifestations of a disease are caused primarily by a microbial toxin, the corresponding antitoxin, if available in time, may have a pronounced prophylactic or curative effect. Apart from this, the other properties of an antitoxin are those of the antibody family (IgG, IgA, IgM) to which it belongs. *See* ANTIBODY; BIOLOGICALS; IMMUNOGLOBULIN.

Antitoxins have been developed for nearly all microbial toxins. Diphtheria, tetanus, botulinus, gas gangrene, and scarlatinal toxins are important examples. Antitoxins may be formed in humans as a result of the disease or the carrier state, or following vaccination with toxoids, and these may confer active immunity. The status of this can be evaluated through skin tests, or by titration of the serum antitoxin level. *See* BOTULISM; DIPHTHERIA; GANGRENE; IMMUNITY; SKIN TEST; TETANUS; TOXIN-ANTITOXIN REACTION. [H.P.T.]

Anura One of the three living orders (sometimes called Salientia) of the class Amphibia, which includes the frogs and toads. About 2400 species of frogs are known. Only the frozen polar regions and remote oceanic islands are without native frogs, and 80% of the species live in the tropics.

Frogs are short-bodied animals with a large mouth and protruding eyes. The externally visible part of the ear, absent in some forms, is the round, smooth tympanum situated on the side of the head behind the eye. There are five digits on the hindfeet and four on the front. Teeth may be present on the upper jaw and the vomerine bones of the roof of the mouth, but are found on the lower jaw of only one species. Often teeth are totally lacking, as in toads of the genera *Bufo* and *Rhinophrynus*. The short vertebral column consists of from 6 to 10 vertebrae, usually 9, and the elongate coccyx. The sacral vertebra precedes the coccyx and bears more or less enlarged lateral processes with which the pelvic girdle articulates. A characteristic feature of frogs is the fusion of the bones in the lower arm and lower leg, so that a single bone, the radioulna in the arm and the tibiofibula in the leg, occupies the position of two in most other tetrapods.

The one character of frogs that comes to the attention of most persons, including many who may never see a frog, is the voice. Most frogs have voices and use them in a variety of ways. In the breeding season great numbers of male frogs may congregate in favorable sites and call, each species giving its own characteristic vocalization. Because no two species breeding at the same time and place have identical calls, it is assumed that the call is important in aiding individuals to find the proper mate. In some species it appears that the female is active in selecting the mate and may be responding to the mating call, but the call may not act in exactly the same way in other species. The mating call is given with the mouth closed. Air is shunted back and forth between the lungs and the mouth, so frogs can call even though submerged. Many species possess one or two vocal sacs, which are expansible pockets of skin beneath the chin or behind the jaws. The sacs (Fig. 1), which may be inflated to a volume as great as that of the frog itself, serve as resonators.

Fig. 1. Toad of the genus *Bufo* giving mating call with vocal sac expanded. (*American Museum of Natural History*)

Other noises made by frogs include the so-called fright scream given with the mouth open, and the warning chirp, which evidently serve as a sex recognition signal when one male contacts another. Some calls evidently serve as territorial signals.

Breeding and development typically take place in the following manner. The male grasps the female about the body with the forelegs, a procedure called amplexus, and fertilizes the eggs externally as they are extruded. The number of eggs may be quite large (up to 20,000 in the bullfrog or 25,000 in a common toad) or may be as few as one in a frog of the West Indies. The larva, called a tadpole, is at first limbless and has external gills and a muscular tail with dorsal and ventral fins (Fig. 2). At hatching there is no mouth opening present, but one soon forms that develops a horny beak and several rows of labial teeth not at all like the true teeth of the adult frog. Shortly after the tadpole hatches, the gills become enclosed within chambers and are no longer visible externally. Except for the gradual development of the hindlimbs, no additional external changes take place as the tadpole grows until the time for metamorphosis. The anterior limbs, which have been forming hidden in the gill chambers, break through the covering skin as metamorphosis begins. The tail dwindles in size as it is absorbed, while the mouth assumes the shape of that of the adult frog. Many other changes are taking place internally, including shortening of the intestine and adapting it to the carnivorous diet of the adult frog.

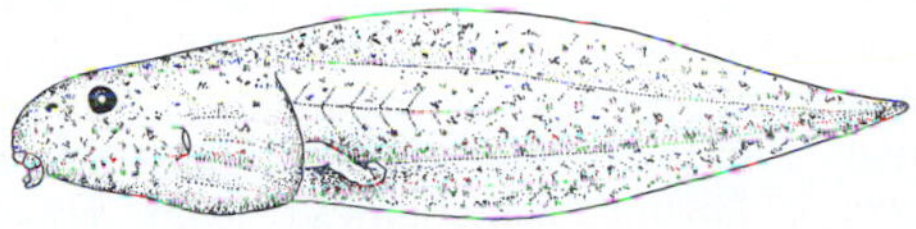

Fig. 2. The tadpole, or larval, stage of the frog *Rana pipiens*. (After W. F. Blair et al., *Vertebrates of the United States*, 2d ed., McGraw-Hill, 1968)

All frogs are carnivorous. The kind of food seems to depend largely upon the size of the frog, whose capacious mouth permits somewhat astonishing feats of swallowing. A large bullfrog, for example, may snap up low-flying bats, ducklings, snakes, and turtles. Insects and other invertebrates form the bulk of the diet of most frogs. The tongue, moistened by a sticky secretion from the intermaxillary gland in the roof of the mouth, is used to catch smaller prey, while larger items of food may bring the front limbs into play. When swallowing, a frog will usually depress the eyeballs into the head to aid in forcing the food down the pharynx. In contrast to transformed frogs, most tadpoles are vegetarian and feed on algae. A few are largely carnivorous or sometimes cannibalistic, and even vegetarian species will scavenge for dead animal matter.

The habitats of frogs are as various as the places where fresh water accumulates. Lakes and streams are tenanted year-round by many species, and others migrate to these places in the breeding season. Any permanent source of water in the desert is likely to support a population of one or more species, and when rainstorms occur, the air around a temporary pool may be filled with mating calls for a few nights, while the frogs take advantage of the water for breeding. As often as not, the pool goes dry before the tadpoles metamorphose, and the adult frogs retreat underground to await another rain. Moist tropical regions provide an abundance of habitats little known to temperate regions, such as the air plants (bromeliads) that hold water and so provide a moist home and breeding site for frogs that may never leave the trees.

Although the majority of frogs fall into fairly well-defined familial categories, the arrangement of the families into subordinal groups by different authorities is not consistent, and there is controversy about the relationships of some smaller groups. *See* FROG. [R.G.Z.]

Anxiety states There is no consensus for a definition of anxiety, and there are few empirical data to distinguish it from fear and other emotional states. Nonetheless, there is agreement that the anticipation of pain, bodily injury, or death usually produces subjective discomfort and physiological changes that are recognizably different from other emotions. It seems reasonable to suppose that the neurobiological substrates associated with normal fear and anxiety in humans also underlie pathological fear, anxiety, or panic, although these disorders may have additional pathophysiologies. Intensive studies in humans and other animals have examined the role of central noradrenergic systems in the production or reduction of anxiety or alarm by using drugs, electrical stimulation, or lesions of the brain. It has been postulated that the major brain noradrenergic system innervated by the nucleus locus coeruleus mediates an alarm function that may include anxiety or fear. *See* NORADRENERGIC SYSTEM. [J.R.Ta.; D.E.Re.]

Anyons Particles obeying unconventional forms of quantum statistics. For many years, only two forms of quantum statistics were believed possible, Bose-Einstein and Fermi-Dirac, but it is now realized that a continuum of possibilities exists. *See* BOSE-EINSTEIN STATISTICS; FERMI-DIRAC STATISTICS.

The more general possibilities for quantum statistics are defined in a mathematically consistent way only for particles whose motion is restricted to two space dimensions. However, many of the most interesting materials in modern condensed-matter physics are effectively two-dimensional, including microelectronic circuits and the copper-oxide layers fundamental to high-temperature superconductivity. The question as to whether the elementary excitations in these systems, the quasiparticles, are anyons can be settled only by a deep investigation of specific cases. The quasiparticles in fractional quantized Hall states are known to be anyons. *See* HALL EFFECT; INTEGRATED CIRCUITS; QUANTUM MECHANICS; QUANTUM STATISTICS; SUPERCONDUCTIVITY. [F.Wil.]

Aorta The main vessel of the systemic arterial circulation arising from the left ventricle of the heart; it is divided into three parts for convenience only. The first portion, the ascending aorta, passes upward under the pulmonary artery; the coronary arteries arise at the base of the ascending aorta behind the aortic valves. The second part, or aortic arch, curves over the hilum of the left lung, giving off the innominate, left carotid, and left subclavian arteries, which supply the neck, head, and forelimbs. The third portion, or descending aorta, continues downward in the thorax on the left side of the vertebral column to the diaphragm, giving off small arteries to the bronchi, esophagus, and other adjacent tissues. Below the diaphragm this vessel, known as the abdominal aorta, descends to the level of the fourth lumbar vertebra where it bifurcates into the two common iliac arteries supplying the hindlimbs.

In the abdomen the major branches of the aorta include the single celiac, superior mesenteric and inferior mesenteric, and the paired renal and internal spermatic (male) or ovarian (female) arteries. In addition, many small branches go to other organs and to the body wall. *See* ARTERY; CIRCULATORY SYSTEM. [W.J.B.]

Apatite The most abundant and widespread of the phosphate minerals, crystallizing in the hexagonal system. The apatite structure type includes no less than 10 mineral species and has the general formula $X_5(YO_4)_3Z$, where X is usually Ca^{2+} or Pb^{2+}, Y is P^{5+} or As^{5+}, and Z is F^-, Cl^-, or $(OH)^-$. The apatite series takes X = Ca, whereas the pyromorphite series includes those members with X = Pb. Three end members form a complete solid-solution series involving the halide and hydroxyl anions. These are fluorapatite, $Ca_5(PO_4)_3F$; chlorapatite, $Ca_5(PO_4)_3Cl$; and hydroxyapatite, $Ca_5(PO_4)_3(OH)$. Thus, the general series can be written $Ca_5(PO_4)_3(F,Cl,OH)$, the fluoride member being the most frequent and often simply called apatite.

The apatite isomorphous series of minerals occurs as grains, blebs, or short to long hexagonal prisms terminated by pyramids, dipyramids, and the basal pinacoid. The minerals are transparent to opaque, and can be asparagus-green (asparagus stone), grayish-green, greenish-yellow, gray, brown, brownish-red, and more rarely violet, pink, or colorless. Apatites are brittle, with hardness 5 on Mohs scale, and specific gravity 3.1–3.2; they are also botryoidal, fibrous, and earthy.

Apatite occurs in nearly every rock type as an accessory mineral. It often crystallizes in regional and contact metamorphic rocks, especially in limestone and associated with chondrodite and phlogopite. It is very common in basic to ultrabasic rocks; enormous masses occur associated with nephelinesyenites in the Kola Peninsula, Russia, and constitute valuable ores which also contain rare-earth elements. Large beds of oolitic, pulverulent, and compact fine-grained carbonate-apatites occur as phosphate rock, phosphorites, or collophanes. Extensive deposits of this kind occur in the United States in Montana and Florida and in North Africa. The material is mined for fertilizer and for the manufacture of elemental phosphorus. *See* FERTILIZER; PHOSPHORUS; PYROMORPHITE. [P.B.M.]

Apes The group of primates most closely related to humans. They include the African great apes, the gorilla and two species of chimpanzee; the Asian great ape, the orangutan; and the lesser apes from Asia, the gibbon and siamang. The apes can be distinguished from the rest of the primates by a number of anatomical and behavioral traits, which indicate their common origin; thus they are termed a monophyletic group called the Hominoidea.

Apes are distinguished from other primates through such obvious features as absence of tail and presence of an appendix. They share a number of specializations of the skeleton, which are useful as diagnostic characters, particularly when it comes to distinguishing fossil apes, because bones and teeth are the most readily preserved parts in the fossil record. The distal end of the humerus is especially useful, both because it is one of the most robust body parts, and therefore readily preserved, and because it is diagnostic of the ape condition, with a large trochlea (ulnar forearm articulation) and a well-developed trochlea ridge. The wrist is also modified for mobility of the joint. There are few synapomorphies of the skull, which in general retains the primitive primate condition except in individual species, but two shared specializations are the deep arched palate and relatively small incisive foramina. The teeth also are generally primitive, except in the broad, low-crowned incisors and enlarged molars.

The fossil history of the apes can be traced back about 20 million years to the early Miocene, for which there is a wealth of fossil evidence from many sites in Kenya and Uganda, yet it is far from certain that apes have been definitely identified. *Proconsul* is the best known of these primates, together with four other genera and a total of 10 species. *Proconsul* mainly shares primitive characters with living apes and is related more closely to living apes than any other group of primates. By the middle Miocene, two groups of apes had differentiated, both of which have apparent affinities with groups of living apes. The best-documented group is the *Sivapithecus-Ramapithecus* group from India, Pakistan, Greece, and Turkey. These are small to medium-size apes that share with the orangutan many of the specializations of the face mentioned earlier as being diagnostic of this species. The similarities of these fossils to humans, however, are of considerably less significance than the many derived characters shared with the orangutan, and there is little doubt that they are related to this ape more closely than to humans. Mention can also be made here of *Gigantopithecus*, which appears to share many of the characters of this group, but there is no evidence for any of the body parts characteristic of the orangutans and so it is not possible to place it exactly. *Gigantopithecus* is a giant ape known from India and China, and like *Sivapithecus* it has thick enameled teeth and a massively built mandibular body. The other group of middle Miocene ape consists of the thin-enameled-teeth species from Europe. It includes the classic *Dryopithecus fontani*, known mainly from France, Germany, Spain, and Hungary, where it has been described under numerous synonyms. There is also evidence from Spain for at least one other species of *Dryopithecus* considerably smaller in size. Finally, brief mention must also be made of *Kenyapithecus* from Kenya, another middle Miocene ape that resembles *Sivapithecus* in having thick-enameled teeth. This led to it being called both *Sivapithecus* and *Ramapithecus* in the past, and at present its exact classification is uncertain. *See* FOSSIL HUMAN; MONKEY; PRIMATES; SKELETAL SYSTEM; SOCIAL ANIMALS. [P.A.]

Aphanite An igneous rock in which the constituents cannot be distinguished by the unaided eye. Aphanitic material may be minutely crystalline, or it may be glassy. Most lavas solidify so rapidly that coarse mineral grains cannot form; thus, volcanic rocks are mostly aphanites. Under conditions of slow cooling, however, such as exist deep below the Earth's surface, molten rock material (magma) solidifies to coarse-grained plutonic rocks. Such visibly crystalline rocks are called phanerites. *See* IGNEOUS ROCKS. [C.A.C.]

Aphasia Impairment in the use of spoken or written language caused by injury to the brain, which cannot be accounted for by paralysis or incoordination of the articulatory organs, impairment of hearing or vision, impaired level of consciousness, or impaired motivation to communicate. The language zone in the brain includes the portion of the frontal, temporal, and parietal lobes surrounding the sylvian fissure and structures deep to these areas. In right-handed persons, with few exceptions, only injury in the left cerebral hemisphere produces aphasia. Lateralization of language function is variable in left-handers, and they are at greater risk for becoming aphasic from a lesion in either hemisphere.

Distinctive recurring patterns of deficit are associated with particular lesion sites within the language zone. These patterns may entail selective impairment of articulation, ability to retrieve concept names, or syntactic organization. Other dissociations affect principally the auditory comprehension of speech, the repetition of speech, or the recognition of written words. The erroneous production of unintended words in speech (paraphasia), oral reading (paralexia), or writing (paragraphia) is a feature of some forms of aphasia.

Mixed forms of aphasia, caused by multiple lesions or lesions spanning anterior and posterior portions of the speech zone, are quite common, and massive destruction of the entire language area results in a global aphasia. Further, individual variations in behavioral manifestations of similar lesions have set limits on the strict assignment of function to structures within the language area.

Preadolescent children suffering aphasia after unilateral injury usually recover rapidly, presumably by virtue of the capacity of the right cerebral hemisphere early in life to acquire the language functions originally mediated by the left hemisphere. Capacity for recovery of function decreases during later adolescence and young adulthood.

Complete recovery in adults after a severe injury is much less common, and severe aphasia may persist unchanged for the duration of the person's life. Many patients are aided by remedial language training, while others continue severely impaired. *See* AGNOSIA; MEMORY. [H.G.]

Aphelion In astronomy, that point at one extremity of the major axis of an elliptical orbit about the Sun where the orbiting body is farthest from the Sun. The Earth passes aphelion on or near July 4, referred to as the time of aphelion passage.

Because the orbit of the Earth is nearly a circle, the Earth is then only some 1.6 × 10⁶ mi (2.5 × 10⁶ km) farther from the Sun than at its mean distance of 9.30 × 10⁸ mi (149.6 × 10⁷ km). *See* Orbital motion. [R.L.Du.]

Aphid One of a group of mostly soft-bodied plant-feeding insects of the suborder Homoptera, superfamily Aphidoidea. The worldwide fauna of over 4000 species is most abundant in north temperate regions. Aphids feed on phloem sap from vascular plants, tapping it through a feeding tube formed from modified mandibles and maxillae called stylets. In so doing they may transmit viruses from plant to plant, spreading serious disease in crops such as potatoes, cereals, sugarbeet, and citrus. Plants sometimes react to aphid feeding by forming galls in which the aphids live protected from drought and enemies. The so-called Chinese gall is valued in commerce for its high tannin content.

Aphids have evolved complex life styles to exploit the changing growth phases of plants. Many divide their yearly cycle by flying between a primary host, on which sexual forms mate and lay winter eggs, and a secondary host, where only parthenogenetic females multiply. Only one generation of males and sexual oviparous females occurs each year, usually in autumn. Most parthenogenetic females are also viviparous, and reproduce very rapidly under favorable conditions. Viviparae are winged or wingless (see illustration). Development of young aphids can be switched toward either wingedness or sexuality by outside factors, such as crowding, decreasing temperature, or shortening days. Aphids in the tropics often remain wholly parthenogenetic. [H.L.G.S.]

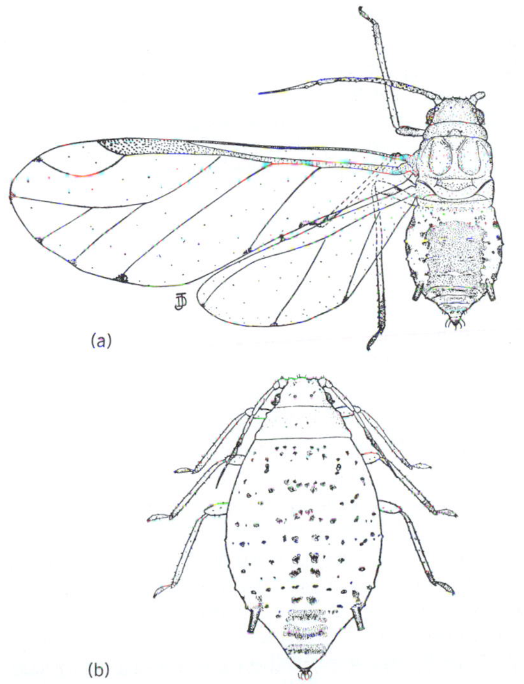

Viviparae of the tulip bulb aphid: (*a*) the winged form and (*b*) the wingless form. (*After J. Davidson, On some aphids infesting tulips, Bull. Entomol. Res., 18:51–62, 1927*)

Apiales An order of flowering plants, division Magnoliophyta (Angiospermae), in the subclass Rosidae of the class Magnoliopsida (dicotyledons). The order (also known as Umbellales) consists of two families, the Araliacea, with about 700 species, and the Umbelliferae, with about 3000. They are herbs or woody plants with mostly compound or conspicuously lobed or dissected leaves, well-developed secretory canals, and separate petals.

The Umbelliferae are mostly aromatic herbs, most numerous in temperate regions. The flowers consistently have an ovary of two carpels, ripening to form a dry fruit that splits into two halves, each containing a single seed. Some common garden vegetables and spice plants, including carrot (*Daucus carota*), parsnip (*Pastinaca sativa*), celery (*Apium graveolens*), parsley (*Petroselinum crispum*), caraway (*Carbum carvi*), and dill (*Anethum graveolens*), belong to the Umbelliferae, as do also such notorious poisonous plants as the poison hemlock (*Conium*) and water hemlock (*Cicuta*). *See* Anise; Carrot; Celery; Fennel; Magnoliopsida; Parsley; Parsnip; Rosidae. [A.Cr.]

Apical meristem Permanently embryonic tissue involved in cell division at the apices of roots and stems, and forming dynamic regions of growth. These apical meristems, usually consisting of small, densely cytoplasmic cells, become established during embryo development. Thereafter they divide to produce the primary plant body of root and shoot. Below the apical meristems, tissue differentiation begins; the protoderm gives rise to the epidermal system, the procambium to the primary vascular system, and the ground meristem to the pith and cortex.

Leaf primordia are formed on the flanks of the shoot apex, in an arrangement or phyllotaxis characteristic of the species. The leaf primordium grows apically at first, but in angiosperms apical growth is short-lived and is followed by intercalary growth; in fern leaf primordia, apical growth is prolonged. Marginal meristems divide to form the leaf blade or lamina; an adaxial meristem may contribute to the thickness of the leaf. In most angiosperms, in addition to leaf primordia the apical meristem, or its derivatives, gives rise to bud primordia. These usually occupy axillary positions (that is, above the junction of the leaf with the stem), and are exogenous (that is, originate superficially), sometimes developing from detached meristems, pockets of meristematic tissue separated from the apical meristem by more mature, differentiated tissue.

Adventitious buds are formed on root (as in raspberry) or leaf (as in *Begonia*) cuttings, and may be exogenous or endogenous in origin.

Primary thickening meristems bring about growth in thickness of many monocotyledons, and a few dicotyledons which have little secondary growth. They consist of files of elongated cells on the "shoulders" of the shoot tip, usually below the leaf primordia; by their division, they add to the width of the shoot tip and sometimes raise the periphery of the apical region so that the apex itself lies in a depression.

Under appropriate environmental stimuli (such as daylength), vegetative shoot apices may be induced to develop as apices or inflorescences or individual flowers. Inflorescence apices may increase markedly in size, become more steeply domed, and finally form floret primordia. Floral apices give rise to a sequence of primordia of floral organs (sepals, petals, stamens, carpels), many of which originate in a manner similar to leaf primordia. [E.G.C.]

Aplacophora A class, also known as Solenogastres, of vermiform mollusks ubiquitous in the deep oceanic basins and trenches to 30,000 ft (9000 m) and common on the continental shelf and slope regions of the world, where they burrow through or creep upon the mud or wrap around alcyonarian corals. Most Aplacophora are less than 10.4 in. (10 mm) in length.

There are two distinct taxa: the subclass Chaetodermomorpha (= Caudofoveata; see illustration); and the subclass Neomeniomorpha (= Ventroplicida; Solenogastres).

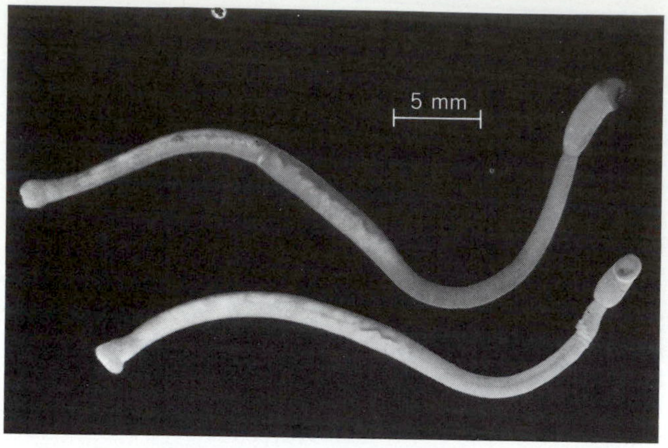

Living *Chaetoderma nitidulum*. (Courtesy of R. Robertson)

Several features are common to all Aplacophora. The ladderlike nervous system consists of paired lateral and ventral cords with many cross-commissures, a buccal ring, and cerebral and suprarectal ganglia. The coelom is restricted to coelomoducts, a spacious pericardium, and gonads that uniquely connect to the pericardium. There is a mantle cavity into which empty the anus and coelomoducts. Other molluscan characters present, but not common to all families, are a radula and its supports; paired gills in the mantle cavity; a mucus-secreting vestigial foot used in creeping; and a style sac containing a mucoid rod which turns against a chitinous gastric shield. *See* MOLLUSCA. [A.H.S.]

Aplite A fine-grained, sugary-textured rock, generally of granitic composition; also any body composed of such rock. This light-colored rock consists chiefly of quartz, microcline, or orthoclase perthite and sodic plagioclase, with small amounts of muscovite, biotite, or hornblende and traces of tourmaline, garnet, fluorite, and topaz. Much quartz and potash feldspar may be micrographically intergrown in cuneiform fashion. *See* GRANITE; IGNEOUS ROCKS.

Aplites may form dikes, veins, or stringers, generally not more than a few feet thick, with sharp or gradational walls. Some show banding parallel to their margins. Aplites usually occur within bodies of granite and more rarely in the country rock surrounding granite. They are commonly associated with pegmatites and may cut or be cut by pegmatites. *See* PEGMATITE. [C.A.C.]

Apodacea A subclass of the Holothuroidea. The tentacles are simple (undivided) or pinnate. The tube feet are markedly reduced or, more usually, not present. There are no retractor muscles; hence the anterior part of the body cannot be retracted as an introvert. The skeletal deposits in the skin may include anchors and anchor plates. There are two orders, Apodida and Molpadida. *See* ECHINODERMATA; HOLOTHUROIDEA. [H.B.F.]

Apodiformes An order of birds consisting of two dissimilar groups, the swifts (Apodi) and the hummingbirds (Trochili). These birds have been placed together because of several anatomical specializations of the wings and feathers. They are excellent fliers but have small, weak feet. The two groups share a unique crossover structure of a neck muscle which cannot be related to their ways of life.

The order Apodiformes is divided into the suborder Apodi, containing the families Aegialornithidae (fossils), Hemiprocnidae (crested swifts; 4 species), and Apodidae (swifts; 83 species), and the suborder Trochili, containing the single family Trochilidae (hummingbirds; 341 species).

Swifts are fast-flying, aerial birds with dull, hard plumage; long, curved, pointed wings; and a short, broad, weak bill and a wide gape, adapted to catching insects in flight. They rest by clinging to cliffs, hollow trees, and other vertical surfaces. Their nest is composed of sticks glued to these surfaces, with the extreme condition being a nest built completely of their gluelike mucus. Swifts are found worldwide except at high latitudes. True swifts never perch crosswise on branches, but crested swifts, found in tropical Asia to New Guinea, are able to perch on branches.

The hummingbirds are small, brightly colored, nectar-feeding birds, found only in the Western Hemisphere. The bill is slender and varies in length and shape, correlated closely with the shape of the flowers utilized by each species. They have a rapid wing beat and flight and are able to hover in front of a flower while feeding or even fly backward. Hummingbirds are attracted to the color red, and flowers specialized on hummingbirds for cross-pollination are red. They are among the smallest birds and include the bee hummingbird (*Mellisuga helenae*) of Cuba, the smallest of all birds. Hummingbirds can hibernate overnight to conserve energy. *See* AVES. [W.J.B.]

Apogee The position most distant from Earth in the orbit of a satellite, as in the orbit of the Moon or of an artificial satellite. The Moon at apogee is 5½% further from Earth than at its mean distance; that is, its orbital eccentricity is 0.055. *See* APHELION; MOON; PERIGEE. [G.P.K.]

Apophyllite A hydrous calcium-potassium silicate containing fluorine. The composition is variable but approximates to $KFCa_4(Si_2O_5)_4 \cdot 8H_2O$. It resembles the zeolites, with which it is sometimes classified, but differs from most zeolites in having no aluminum. It exfoliates (swells) when heated, losing water, and is named from this characteristic; the water can be reabsorbed. It is essentially white, with a vitreous luster, but may show shades of green, yellow, or red. The symmetry is tetragonal, and the crystal structure contains sheets of linked SiO_4 groups; this accounts for the perfect basal cleavage of the mineral. It occurs as a secondary mineral in cavities in basic igneous rocks, commonly in association with zeolites. The specific gravity is about 2.3–2.4, the hardness 4.5–5 on Mohs scale, the mean refractive index about 1.535, and the birefringence 0.002. *See* SILICATE MINERALS; ZEOLITE. [G.W.Br.]

Apoplexy An acute vascular lesion of the brain, also called stroke. This may be the result of hemorrhage from, thrombosis in, or embolism to a cerebral vessel. Cerebral hemorrhage can result from a rupture of the vessel wall, as from a defect in the wall such as a congenital aneurysm, or it may be secondary to death of cerebral tissue as occurs with an infarct.

With cerebral hemorrhage or death of cerebral tissue there is loss of function of the regions involved. One such complication is paralysis of the muscles of the extremities on the side opposite the lesion. Since the brain is enclosed in a nondistensible box, the skull, the increase in size due to the addition of the blood raises intracranial pressure. *See* EMBOLISM; HEMORRHAGE. [R.A.V.]

Aporidea An order of tapeworms of uncertain composition and affinities, found in anseriform birds. The scolex may lack suckers and have only a simple rostellum with hooks, or it may have four large suckers and a complex glandular rostellum with small hooks. The small cylindroid body lacks segmentation, although some species have an internal serial arrangement of reproductive organs. Lack of reproductive ducts and openings to the outside prevents cross fertilization between strobilae. The life cycle is unknown. *See* CESTODA; TAPEWORM DISEASE. [R.S.F.]

Apostomatida A group of ciliates comprising an order of the Holotrichia. The majority occur as commensals on marine crustaceans. Their life histories may become exceedingly complicated, and they appear to bear a direct relationship to the molting cycles of their hosts. Apostomes are particularly characterized by the presence of a unique rosette in the vicinity of an inconspicuous mouth opening and the possession of only a small number of ciliary rows wound around the body in a spiral fashion (see illustration). *See* CILIOPHORA; HOLOTRICHIA; PROTOZOA. [J.O.C.]

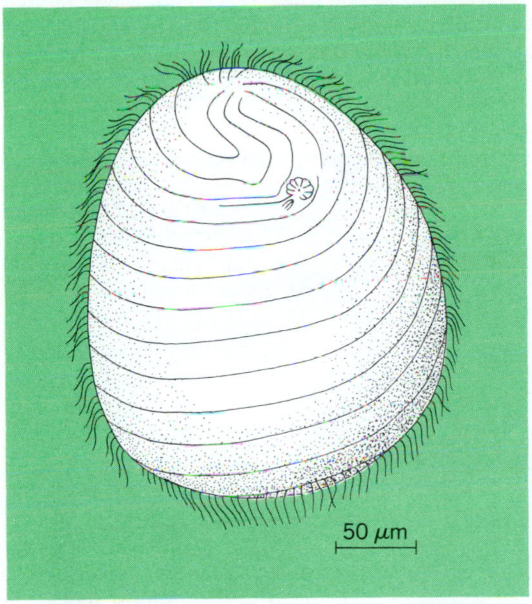

***Foettingeria*, an example of an apostomatid.**

Appendicitis An inflammation of the vermiform appendix. Inflammation may result from bacterial infection following obstruction of the lumen by impacted feces or other causes. The normal bacterial inhabitants of the gastrointestinal tract gain access to the wall of the organ either through penetration of the mucosa or by way of vascular or lymphatic spread. Acute swelling, congestion, and edema ensue if no natural or surgical relief occurs. This may be followed by actual necrosis and gangrene, resulting in a ruptured or perforated appendix and the development of peritonitis, abscesses, or fistulas. *See* APPENDIX (ANATOMY); PERITONITIS.

Occasionally, appendicitis is associated with the presence of worms or other parasites or with inflammatory changes in neighboring tissues.

Chronic appendicitis, as an active low-grade infection, is rare, but scarred and obliterated appendixes are seen quite often. This condition may be the result of previous inflammations or age changes, because the appendix normally loses its lymphoid tissue with an increase in connective-tissue components. [E.G.St./N.K.M.]

Appendicularia A class of the subphylum Tunicata consisting of minute planktonic animals in which the tail, with dorsal nerve cord and notochord, persists throughout life. Two gill slits are commonly present, but an atrium is lacking. The epidermis of the anterior region secretes a gelatinous tunic. In some species this tunic is highly complex and encases the whole body in a "house" equipped with chambers and strainers which enable the inhabitant to filter and concentrate very small plankton from the sea. About 60 species are known; *Oikopleura* and *Fritillaria* are the largest genera. *See* TUNICATA. [D.P.A.]

Appendix (anatomy) A narrow, elongated tube closed at one end, extending from the cecum, a blind pocket off the first part of the large intestine. It is found in only a few mammals. The size and detailed structure of the appendix vary markedly, depending on the species and the age of the individual. Most reptiles, birds, and mammals have a single or a paired cecum at the anterior end of the large intestine, but it is quite rare that this cecum has a thinner projection or true appendix.

In humans the appendix is about 3 in. (7.5 cm) long, but it varies greatly in both size and its specific location in the lower right quarter of the abdomen. The exact function of the human appendix is unknown, and it is considered to be a remnant of a portion of the digestive tract which was once more functional and is now in the process of evolutionary regression. The appendixes presumably function in digestion in forms with larger ones. [T.S.P.]

Apple Apples (genus *Malus*) belong to the family Rosaceae. There are about 30 *Malus* species in the North Temperate Zone. The fruits of most species are edible. More apples are consumed than any other temperate-zone tree fruit. Apples are eaten fresh, processed into jellies or preserves, cooked in pies and pastries, or made into sauces. Apple juice is drunk fresh, or at various stages of fermentation as cider, applejack, or brandy. Apple cider vinegar is popular for use in salads and in many processed foods.

The "European" cultivated apple is now thought to have been derived principally from *M. pumila*, a Eurasian species which occurs naturally from the Balkans eastward to the Tien Shan of central Asia. In the wild, some forms of *M. pumila* approach present cultivars in size and quality. Another Asian species, *M. sylvestris*, whose range extends into western Europe, grows side by side with *M. pumila* and hybridizes with it in the Caucasus Mountains. Thus *M. sylvestris* probably also had some genetic input into the cultivated apple.

Wild apples, mainly the edible *M. baccata*, grow so thickly east of Lake Baikal in Siberia that the region is called Yablonovy Khrebet ("Apple Mountains").

The success of the Delicious and Golden Delicious cultivars may be laid to the demand for better-quality fresh fruit. Both originated on farms as chance seedlings near the end of the 19th century. In spite of the dominance of such "chance" cultivars, cultivars such as Cortland and Idared that were produced by scientific breeding have begun to achieve prominence. Most of the apple breeding programs under way earlier this century have ceased, but the few remaining ones are now introducing some exciting new cultivars. Many of the new cultivars are of excellent quality and in addition are resistant to the most damaging diseases, such as scab, rust, and fire blight. These cultivars promise to revolutionize apple growing in the future. *See* BREEDING (PLANT).

The apple is probably the most widely distributed fruit crop in the world, although it ranks behind grapes, bananas, and oranges in total production. There are substantial apple-growing areas on all temperate-zone continents; the United States, Italy, France, and Germany are leading producers.

Apples can be grown as far north as 60° latitude in the maritime climate of northwestern Europe. In North America, apple culture does not extend much above 50° north latitude. Away from the coasts, the buffering effects of the Great Lakes on temperature extremes facilitate heavy apple production in New York, Michigan, and Ontario. Hardier cultivars are continuing to be developed for use in colder regions. Apples can be grown in the tropics at higher elevations where sufficient chilling to break dormancy is available. Cultivars with lower chilling needs are being developed.

The principal apple-growing regions in North America are, in order of importance: the Northwest (Washington, Oregon, Idaho, and British Columbia), the Northeast (New York, New Jersey, New England, Ontario, Quebec, and Nova Scotia), the Cumberland-Shenandoah area (Pennsylvania, Maryland, Virginia, West Virginia, and North Carolina), Michigan, California, the Ohio Basin (Ohio, Indiana, and Illinois), and Colorado.

Apples may be sold fresh immediately, or after a period of storage, or they may be processed into less perishable products such as canned, frozen, or dried slices or chunks for baking, applesauce, apple juice or cider, and vinegar.

Fresh apples may be sold locally at roadside stands or at farmers' markets, or sold in wholesale quantities to supermarkets. Often, large quantities of fruit are traded by cooperatives or independent buyers for sale in distant markets. Thousands of tons of apples are shipped annually from the state of Washington to the Midwest and the East Coast. Apples can be stored for long periods at low temperatures under controlled atmosphere (CA storage), so that fresh fruits are now available year round.

There is a large international trade in apples, particularly in Europe between the large producing countries of France and Italy and the net-importing countries of Germany, Britain, and Scandinavia, and from the United States to Canada. Apples grown in the Southern Hemisphere (Australia, New Zealand, South Africa, and Argentina) are shipped in large quantities to western Europe during the northern winter. *See* FRUIT, TREE; ROSALES. [H.S.A.; S.V.B.]

Applications satellites Artificial satellites which contribute directly to the well-being of society by providing various services. Satellites were put to practical use almost as soon as the ability to orbit the Earth was attained. The first utilization involved the vantage point of space for observation of the Earth, and the application was in the field of meteorology. Almost as quickly, the unprecedented global communications capability of the satellite was demonstrated, and probably this application has reached the highest state of development. Other areas have rapidly evolved which may be even more important, a situation particularly true of satellites designed to accomplish earth resources surveys. More specialized areas have also received attention, for example, satellite geodesy. *See* COMMUNICATIONS SATELLITE; METEOROLOGICAL SATELLITES; REMOTE SENSING; SATELLITE (SPACECRAFT). [S.I.R.]

Apraxia An inability to perform certain acts despite the absence of complete paralysis of the muscles involved or severe impairment of kinesthesis. Complicated skilled acts are most affected. However, even simple movements, such as protrusion of the tongue or deviation of the eyes, may be impossible except in contexts where they are highly automatic. Apraxia is caused by brain damage; although there has been much speculation about the locus of the causative lesion, the evidence remains inconclusive.

Several forms of apraxia are usually distinguished. The lowest-order apraxia is called limb-kinetic or motor; the intermediate form, ideokinetic or ideomotor (apraxia in the strictest sense); and the highest form, ideational. Limb-kinetic apraxia refers to a loss of coordination usually affecting one upper limb only. Gross movements may be performed fairly well, whereas fine individual movements of the fingers are lost. Ideokinetic apraxia presents the classic feature of this disorder, namely, a dependence of the ability to perform given acts on the evocative situation. For example, the patient cannot make the military salute on command or in imitation, but can raise a hand to the head to ward off a blow. Ideational apraxia is supposedly a loss of the ideational plan governing a series of acts directed toward a goal. The patient may perform the component acts correctly, but the order or the objects utilized may be inappropriate. [J.Se.]

Apricot The stone fruit *Prunus armeniaca* native to China, from which it has been distributed across southern Europe to the United States. Although the tree is slightly hardier than the peach, it blossoms early and is susceptible to damage from spring frost and winter killing. Therefore culture is limited to frost-protected spots and to regions where winter temperatures do not fall below -10 to -15°F (-23 to -26°C). The fruit is nearly smooth, yellow or orange, and excellent for eating raw as well as canned and dried (see illustration). *See* ROSALES.

Apricot (*Prunus armeniaca*) branch showing fruit cluster.

Of the total commercial production in North America, California produces 90%, followed by Washington and Utah. Of this, more than 90% is processed (75% canned). Cultivated varieties are Moorpark, Blenheim, Royal, Tilton, Newcastle, and Wiggins. The Japanese apricot (*P. mume*) is a favorite ornamental and fruit tree in China and Japan, but it is little known in the United States. *See* FRUIT, TREE. [H.B.T.]

Apsides In astronomy, the two points in an elliptical orbit that are closest to, and farthest from, the primary body about with the secondary revolves. In the orbit of a planet or comet about the Sun, the apsides are, respectively, perihelion and aphelion. In the orbit of the Moon, the apsides are called perigee and apogee, while in the orbit of a satellite of Jupiter, these points are referred to as perijove and apojove. The major axis of an elliptic orbit is referred to3 as the line of apsides. *See* ORBITAL MOTION. [R.L.Du.]

Apterygota A subclass of the Insecta characterized by being primitively wingless. Young Apterygota do not change greatly in their metamorphosis to adults, other than for changes in the proportions of structures, and in developing sexual maturity. Although their relationships to each other are not clearly understood, four orders are included in this subclass: Protura, Collembola, Diplura, and Thysanura. *See* INSECTA.

[H.B.M.]

Aqua regia A mixture of one part by volume of concentrated nitric acid and three parts of concentrated hydrochloric acid. Aqua regia was so named by the alchemists because of its ability to dissolve platinum and gold. Either acid alone will not dissolve these noble metals. [E.E.W.]

Aquaculture The cultivation of fresh-water and marine species (the latter type is often referred to as mariculture). The concept of aquaculture is not new, having begun in China about 4000 years ago, and the practice of fish culture there has been handed down through the generations. China probably has the highest concentration of aquaculture operations in a single country today. Yet the science of aquaculture, as a whole, is far behind that of its terrestrial counterpart, agriculture. Aquaculture, though still partly art, is being transformed into a modern, multidisciplinary technology.

Aquacultural ventures occur worldwide. China grows macroalgae (seaweeds) and carp. Japan cultures a wide range of marine organisms, including yellowtail, sea bream, salmonids, tuna, penaeid shrimp, oysters, scallops, abalone, and algae. Russia concentrates on the culture of fish such as sturgeon, salmon, and carp. North America grows catfish, trout, salmon, oysters, and penaeid shrimp. Europe cultures flatfish, trout, oysters, mussels, and eels. Presently, plant aquaculture is almost exclusively restricted to Japan, China, and Korea, where the national diets include substantial amounts of macroalgae.

The worldwide practice of aquaculture runs the gamut from low-technology extensive methods to highly intensive systems. At one extreme, extensive aquaculture can be little more than contained stock replenishment, using natural bodies of water such as coastal embayments, where few if any alterations of the environment are made. Such culture usually requires a low degree of management and low investment and operating costs; it generally results in low yields per unit area. At the other extreme, intensive aquaculture, animals are grown in systems such as tanks and raceways, where the support parameters are carefully controlled and dependence on the natural environment is minimal. Such systems require a high degree of management and usually involve substantial investment and operating costs, resulting in high yields per unit area. Obviously, many aquacultural operations fall somewhere between the two extremes of extensive and intensive, using aspects of both according to the site, species, and financial constraints.

A unique combination of highly intensive and extensive aquaculture occurs in ocean ranching, as commonly employed with anadromous fish (which return from the ocean to rivers at the time of spawning). The two most notable examples are the ranching of salmon and sturgeon. In both instances, highly sophisticated hatchery systems are used to rear young fish, which are then released to forage and grow in their natural environment. The animals are harvested upon return to their native rivers.

While extremely extensive aquacultural operations are continuing, as they have been for generations, in the lesser-developed areas of the world, the more industrialized nations—where land and water resources are limited—are turning increasingly to intensive aquaculture. Intensive aquaculture brings with it high energy costs, necessitating the design of energy-efficient systems. As this trend continues, aquaculture will shift more to a year-round, mass-production industry using the least amount of land and water possible. With this change to high technology and dense culturing, considerable knowledge and manipulation of the life cycles and requirements of each species are necessary. Specifically, industrialized aquaculture has mandated the development of reproductive control, hatchery technology, feeds technology, disease control, and systems engineering.

Regardless of the type of system used, aquacultural products are marketed as are fisheries products, except for some advantages. For one, fisheries products often must be transported on boats and may experience spoilage; whereas cultured products, which are land-based, can be delivered fresh to the various nearby markets. Also, intensively cultured products through genetic selection can result in a more desirable food than those caught in the wild, with uniform size and improved taste resulting from controlled feeding and rearing in pollution-free water.

Aquaculture is following the lead of the development of agriculture, beginning with hunting, moving through stock containment and low-technology farming, and finally arriving at high-density, controlled culture. As the methodology improves, the modern techniques of aquaculture—and of agriculture—will have to be transferred to the lesser-industrialized areas of the world in an effort to offset the predicted food shortages in the coming years. *See* AGRICULTURE; MARINE FISHERIES.

[W.H.C.; A.B.McG.]

Aquarius The Water Bearer, in astronomy, a large zodiacal constellation visible in both summer and autumn. Aquarius is the eleventh sign of the zodiac. To ancients, the constellation resembled a man pouring a stream of water from a jar. Four stars, η, ζ, π, and γ, arranged like a Y, form the head of the Water Bearer (see illustration). The stream of water flows into

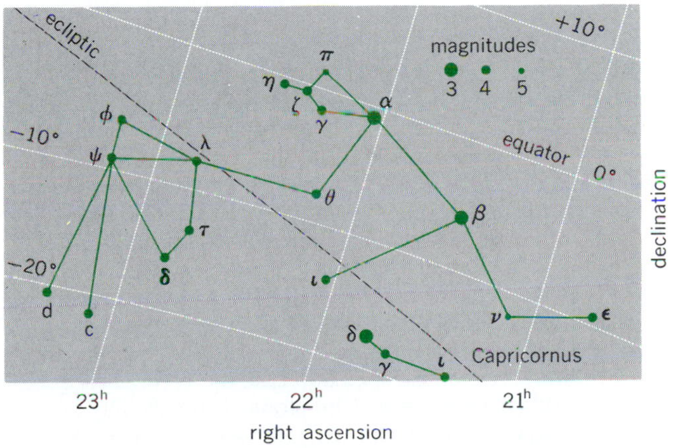

Line pattern of the constellation Aquarius. The grid lines represent the coordinates of the sky. The apparent brightness, or magnitude, of the stars is shown by the sizes of the dots, which are graded by appropriate numbers as indicated.

the Fish's Mouth (Fomalhaut) in the constellation Pisces Austrinus (Southern Fish). Fomalhaut, bright and solitary in this part of the sky, is one of the relatively important navigational stars. From earliest time this constellation has been associated with water, probably because the Sun is seen in Aquarius during the rainy season of February. *See* CONSTELLATION. [C.-S.Y.]

Aquifer A subsurface zone that yields economically important amounts of water to wells. The term is synonymous with water-bearing formation. An aquifer may be porous rock, unconsolidated gravel, fractured rock, or cavernous limestone.

Aquifers are important reservoirs storing large amounts of water relatively free from evaporation loss or pollution. If the annual withdrawal from an aquifer regularly exceeds the replenishment from rainfall or seepage from streams, the water stored in the aquifer will be depleted. This mining of groundwater results in increased pumping costs and sometimes pollution from sea water or adjacent saline aquifers. Lowering the piezometric pressure in an unconsolidated artesian aquifer by overpumping may cause the aquifer and confining layers of silt or clay to be compressed under the weight of the overlying material. The resulting subsidence of the ground surface may cause structural damage to buildings, altered drainage paths, increased flooding, damage to wells, and other problems. *See* ARTESIAN SYSTEMS; GROUNDWATER HYDROLOGY. [R.K.Li.]

Arabinose A plant product which is usually a constituent of polysaccharides such as various gums, hemicelluloses, pectic substances, and some glycosides. It exists in these products mostly in the L form (see illustration), and usually possesses the furanose modification when it is present in combined form as a constituent of an oligosaccharide, polysaccharide, or glycoside. *See* CARBOHYDRATE; GUM; HEMICELLULOSE; MONOSACCHARIDE; POLYSACCHARIDE. [W.Z.H.]

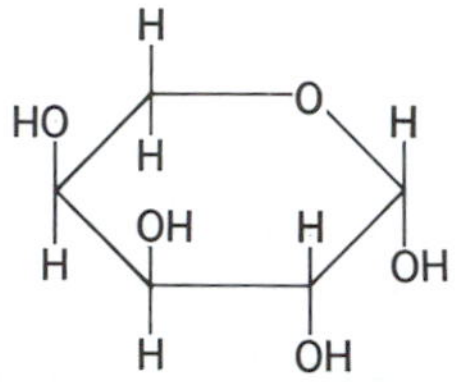

Structural formula of β-L-arabinopyranose.

Arachnida The familiar modern spiders, scorpions, daddy longlegs, ticks, mites, and others, besides a large number of fossil species, are included in this class of the subphylum Chelicerata. The prosoma is commonly called the cephalothorax, and the opisthosoma is called the abdomen. However, as in the scorpions, the abdomen may be differentiated into two parts, and in some forms the whole body is in one piece. The six pairs of appendages on the cephalothorax are the chelicerae, the pedipalps, and four pairs of walking legs. The pedipalps may be leglike, but they often bear large pincers, or chelae. In male spiders they are modified for reproductive purposes. If abdominal appendages are present, they are much reduced and serve for silk spinning. Like the other chelicerates, the arachnids have no jaws. The stomach usually has branching tubular diverticula, in the walls of which the fully digested food is absorbed. The respiratory organs are both book lungs and tracheae. The genital openings of each sex are ventral on the base of the abdomen.

The class Arachnida includes the orders Palpigradida, Solpugida, Scorpionida, Uropygi, Amblypygi, Ricinuleida, Pseudoscorpionida (or Chelonethida), Phalangida, Araneida (spiders), and Acarina (mites and ticks). See separate articles on each order. *See also* ARTHROPODA; CHELICERATA. [J.G.R.]

Araeolaimida An order of nematodes in which the amphids are simple spirals that appear as elongate loops, shepherd's crooks, question marks, or circular forms. The cephalic sensilla are often separated into three circlets: the first two are papilliform or the second coniform, and the third is usually setiform; rarely are the second and third whorls combined. Body annulation is simple. The stoma is anteriorly funnel shaped and posteriorly tubular; rarely is it armed. Usually the esophagus ends in a bulb that may be valved. In all but a few taxa the females have paired gonads. Male preanal supplements are generally tubular, rarely papilloid.

There are three araeolaimid superfamilies: Araeolaimoidea, Axonolaimoidea, and Plectoidea. The distinguishing characteristics of the Araeolaimoidea are in the amphids (sensory receptors), stoma, and esophagus. The amphids are in the form of simple spirals, elongate loops, or hooks. Although araeolaimoids are chiefly found in the marine environment, many species have been collected from fresh water and soil.

The amphids in the Axonolaimoidea are generally prominent features of the anterior end, visible as a single-turn loop of a wide sausage shape. Feeding habits are unknown. All known species occur in marine or brackish-water environments.

Plectoidea comprise small free-living nematodes, found mainly in terrestrial habitats, frequently in moss; some are fresh-water, and a few are marine. Those that inhabit moss cushions can withstand lengthy desiccation. For most, the feeding habit is unconfirmed; where it is known, they are microbivorous. Many are easily raised on agar cultures that support bacteria. *See* NEMATA. [A.R.M.]

Araeoscelida An order of Paleozoic diapsid reptiles including the families Petrolacosauridae and Araeoscelidae. Members of these families resemble primitive modern lizards, such as the green iguana, in size and most body proportions, but are distinguished by their elongate necks and distal limb elements. *Petrolacosaurus*, from the Upper Pennsylvanian of Kansas, is the earliest known diapsid. The skull shows well-developed upper and lateral temporal openings and a suborbital fenestra that are characteristic of the diapsids.

The following derived features distinguish members of the Araeoscelida from other early diapsids: six to nine elongate neck vertebrae; a radius as long as the humerus and a tibia as long as the femur; expanded neural arches; posterior cervical and anterior dorsal neural spines with mamillary processes; a coracoid process; and enlarged lateral and distal pubic tubercles. In contrast to later diapsids, members of the Araeoscelida show no evidence of an impedance-matching middle ear. The stapes is massive, the quadrate is not emarginated for support of a tympanum, and there is no retroarticular process. *See* REPTILIA. [R.L.C.]

Aragonite A mineral species of calcium carbonate; other cations such as strontium and lead may substitute for some of the calcium. Aragonite is much less common than calcite, to which it is frequently or partially transformed. When pure, aragonite is white, but with impurities may assume gray, blue, green, or pink tints. The hardness on Mohs scale is 3.5–4. Specific gravity is 2.947.

Aragonite frequently grows metastably in near-surface secondary deposits in cavities, such as in limestone caverns. It also occurs in precipitates from hot springs, and it forms biogenically in pearls and in the hard parts (shells) of some fossils, pelecypods, and gastropods.

Aragonite is found in many localities throughout the world. In the United States, fine crystals associated with calcite and cerussite are found in the Magdalena district, Socorro County, New Mexico. *See* CALCITE; CARBONATE MINERALS. [R.I.Ha.]

Arales An order of flowering plants, division Magnoliophyta (Angiospermae), in the subclass Arecidae of the class Liliopsida (monocotyledons). It consists of two families, the Araceae, with about 1800 species, and the Lemnaceae, with only about 30.

The Araceae are herbs (seldom woody climbers) with ordinary roots, stems, and leaves (often broad and net-veined) and with the vessels confined to the roots. They have numerous tiny flowers grouped together in a small to very large spadix. They are commonest in forested tropical and subtropical regions. *Anthurium* (elephant ear), *Arisaema* (jack-in-the-pulpit), *Dieffenbachia* (dumb cane), *Monstera*, and *Philodendron* are some well-known members of the Araceae.

The Lemnaceae, or duckweeds, are small, free-floating, thalloid aquatics that are generally conceded to be derived from the Araceae. Their flowers, seldom seen, are much reduced and form a miniature spadix. *Pistia* (water lettuce), a free-floating aquatic (but not thalloid) aroid is seen as pointing the way toward *Spirodela*, the least reduced genus of the Lemnaceae. *See* ARECIDAE; LILIOPSIDA; MAGNOLIOPHYTA; PLANT KINGDOM. [A.Cr.]

Araneae A natural order of the class Arachnida, also called Araneida, commonly known as the spiders. These animals are widespread over most of the land areas of the world and are well adapted to many different habitats. They are known to be one of the oldest of all groups of arthropods, and their remains are known from the Devonian and Carboniferous geological deposits. Through successive geological periods spiders have become adapted to use insects as their chief source of food. On the other hand, certain insects consume the eggs of spiders, others parasitize the eggs, and still others capture adults and place them in their nests for food for their young.

Spiders have only two subdivisions of the body, the cephalothorax and the abdomen, joined by a slender pedicel. All parts of the body are covered by chitinous plates which often extend into curious outgrowths, such as spines, horns, and tubercles. Only simple paired eyes, ocelli, are present, with the number varying from eight, the most common number, to none in a few species inhabiting lightless habitats.

The first of six pairs of appendages are termed the chelicerae or jaws. The second pair of appendages is the six-segmented pedipalps, simple organs of touch and manipulation in the female, but curiously and often complexly modified in males for use in copulation. The four pairs of thoracic legs consist of seven segments each. Just in front of the genital opening in most females there is a more or less specific and often elaborately formed plate, the epigynum. This organ is of importance in the reproductive activities of the female. In most of the true spiders internal breathing tubules occur with ventral openings posterior to the genital apertures. Distinctive paired ventral spinnerets occur near the posterior end of the abdomen. In certain families a sievelike plate lies immediately anterior to the foremost spinnerets. From this plate, the cribellum, a special kind of banded silk is extruded and used in conjunction with a comb on the fourth legs. The spinnerets and cribellum are directly associated with several types of abdominal glands and responsible for the production of the different types of silk characteristic of these animals.

Silk produced by spiders is a scleroprotein which is fine, light, elastic, and strong. It is used industrially only in the making of cross hairs in optical instruments. This use is diminishing as metal filaments and etched glass come into more common usage. In addition to the attractive orb webs, these animals also construct sheet webs, funnel webs, tube webs, and reticular webs. The spider's reliance upon silk extends its use to the making of egg cocoons, sperm webs by males, molting sheets, gossamer threads for ballooning, attachment disks, lining for burrows, hinges for trap doors, binding for captives, retreats, and drag lines.

No general agreement exists at present among araneologists concerning the classification and exact arrangement of the families of spiders. A. Petrunkevitch (1939) recognized 5 suborders: Liphistiomorphae containing only 2 families with a primitively segmented abdomen, and restricted to regions in the Eastern Hemisphere; Mygalomorphae with 8 families; Hypochilomorphae with a single relict family native to the southern Appalachian region; Dipneumonomorphae with 48 families; and finally the Apneumonomorphae with 3 families, which lack the book lungs and represent the most highly modified members of the order. There is now a tendency to increase the number of recognized families. *See* ARACHNIDA.

[A.M.C.]

Arboretum An area set aside for the cultivation of trees and shrubs for educational and scientific purposes. An arboretum differs from a botanical garden in emphasizing woody plants, whereas a botanical garden includes investigation of the growth and development of herbaceous plants as well as trees and shrubs. The largest of the arboretums in the United States is the Arnold Arboretum of Harvard University, founded in 1872. *See* BOTANICAL GARDENS. [E.L.C.]

Arboriculture A branch of horticulture concerned with the selection, planting, and care of woody perennial plants. Knowing the potential form and size of plants is essential to effective landscape planning as well as to the care needed for plants. Arborists are concerned primarily with trees since they become large, are long-lived, and dominate landscapes both visually and functionally.

Plants can provide privacy, define space, and progressively reveal vistas; they can be used to reduce glare, direct traffic, reduce soil erosion, filter air, and attenuate noise; and they can be positioned so as to modify the intensity and direction of wind. They also influence the microclimate by evaporative cooling and interception of the Sun's rays, as well as by reflection and reradiation. Certain plants, however, can cause human irritations with their pollen, leaf pubescence, toxic sap, and strong fragrances from flowers and fruit. Additionally, trees can be dangerous and costly: branches can fall, and roots can clog sewers and break paving. *See* LANDSCAPE ARCHITECTURE.

[R.W.Ha.]

Arborvitae A plant, sometimes called the tree of life, belonging to the genus *Thuja* of the order Pinales (Coniferales). It is characterized by flattened branchlets with two types of scalelike leaves. At the edges of the branchlets the leaves may be keeled or rounded; on the upper and lower surfaces they are flat, and often have resin glands. The cones, about 1/2 in. (1.25 cm) long, have the scales attached to a central axis. *See* PINALES; RESIN; SECRETORY STRUCTURES (PLANT).

The tree is valued both for its wood and as an ornamental. *Thuja occidentalis*, of the eastern United States, is known as the northern white cedar. It occurs in moist or swampy soil from Nova Scotia to Manitoba and in adjacent areas of the United States, and extends south in the Appalachians to North Carolina and Tennessee. Other important species include the giant arborvitae (*T. plicata*); oriental arborvitae (*T. orientalis*); and Japanese arborvitae (*T. standishii*). Among the horticultural forms are the dwarf pendulous and juvenile varieties. *See* FOREST AND FORESTRY; TREE. [A.H.G./K.P.D.]

Arboviral encephalitides A number of diseases, such as St. Louis, Japanese B, and equine encephalitis, which are caused by arthropod-borne viruses (abbreviated as "arboviruses"). In their most severe human forms, the diseases invade the central nervous system and produce brain damage, with mental confusion, convulsions, and coma; death or serious aftereffects are frequent in severe cases. Inapparent infections are common. *See* EQUINE ENCEPHALITIS.

The illness begins 4–21 days after the bite of the infected vector, with sudden onset of fever, headache, chills, nausea, and generalized pains; marked drowsiness and neck rigidity may ensue. Signs and symptoms are often diphasic, the second phase being more severe and involving mental confusion, speech difficulties, convulsions, and coma. Mortality rates vary in different diseases and epidemics. Serious aftereffects such as mental and sensory defects, epilepsy, and paralysis occur.

All arboviruses causing encephalitis require an infected arthropod vector for transmission. The usual natural hosts are mammals or birds, with humans accidentally infected in the cycle. Ticks may serve not only as vectors but also as a natural reservoir, because the virus may pass transovarially to offspring; in other instances birds or mammals are the probable reservoir. In highly endemic regions many persons have antibodies, chiefly from inapparent infections. Each virus is limited to definite geographic areas. The population's immunity to its local virus may cross immunize against closely related viruses.

There is no proved specific treatment. In animals, hyperimmune serum given early may prevent death. Killed virus vaccines have been used in animals and in persons occupationally subjected to high risk; little success has yet been achieved in large field trials in humans. In general, however, control of these diseases continues to be chiefly dependent upon elimination of the arthropod vector. *See* ANIMAL VIRUS. [J.L.Me.]

Arc (mathematics) Either of the two parts of a circle (considered as a curve) intercepted between two distinct radii. If circle C has center O, radius r, and two radii intersect C at points A, B, let $\theta = \measuredangle AOB$ denote the angle, not greater than a straight angle, formed by these radii (see illustration). When

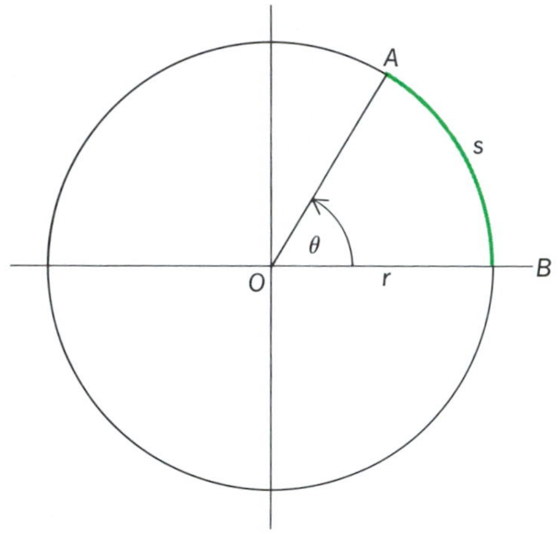

Arc s of a circle.

the length of arc $AB = r$, the angle θ is 1 radian (rad), and consequently there are $2\pi r/r = 2\pi$ rad in one complete revolution. It follows that 2π rad $= 360°$; that is, 1 rad $= 57°17'45''$, to the nearest second. Radius r, angle θ, and corresponding arc length s satisfy the relation $s = r \cdot \theta$. *See* CIRCLE; RADIAN MEASURE. [L.M.Bl.]

Arc discharge A type of electrical conduction in gases characterized by high current density and low potential drop. It is closely related to the glow discharge but has a much lower potential drop in the cathode region, as well as greater current density. *See* GLOW DISCHARGE.

There are many arc devices, and they operate under a wide range of conditions. For example, an arc discharge may be sustained at either high pressure (of the order of atmospheres) or low pressure. The cathode may or may not be heated from an external source. Furthermore, the applied potential difference may be either direct or alternating. Numerous applications have been made of such devices, some having large commercial value. A few of these applications are illuminating devices, high-current rectifiers, high-current switches, welding devices, and ion sources for nuclear accelerators and thermonuclear devices.

Although an arc may be initiated in several ways, it is instructive to consider the transition from a cold-cathode glow discharge (see illustration). In the normal condition, the glow partially covers the cathode, and the voltage drop remains nearly constant as the current is increased. In this condition, the ionization is produced primarily by electron impact. As the

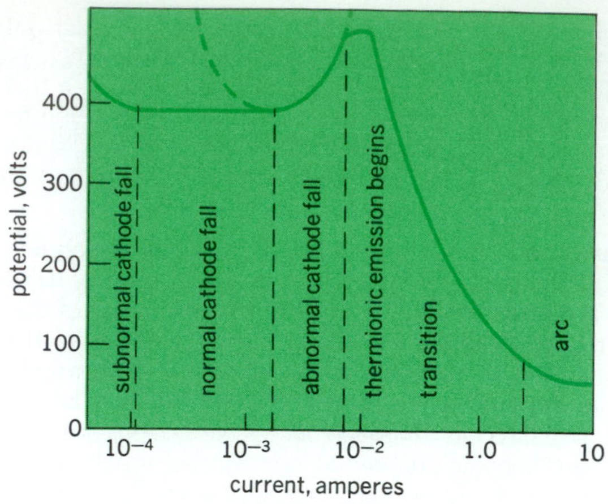

Transition from glow discharge to arc with increase of current. (*After L. B. Loeb, Basic Processes of Gaseous Electronics, University of California Press, 1955*)

current is increased, the cathode glow spreads out, eventually covering the cathode completely. Further increase in current can now be obtained only by an increase in the potential drop across the discharge. The cathode temperature increases in this process. This is the abnormal glow region. As the cathode temperature is increased further, thermionic emission becomes an important factor. At this point the discharge characteristic may acquire a negative slope; that is, a further increase in current may increase the cathode temperature, and the resulting thermionic emission, enough actually to reduce the potential drop. Unless the external resistance is sufficiently great to make the overall resistance positive, the discharge will change suddenly to the arc mode. Typical values in this region are a potential drop of a few tens of volts and a current ranging from amperes to thousands of amperes. *See* ELECTRICAL CONDUCTION IN GASES.

There are three geometrical regions in an arc. These are the cathode fall, the anode rise, and the arc body. The main body of the arc (arc body) is characterized by secondary ionization. This is predominantly a temperature effect. The electrons produce very little ionization by impact, because the electron energy is generally low. However, energy can be imparted to the gas molecules in the form of atomic and molecular excitation. With the many intermolecular collisions that occur, this energy is readily degraded to thermal energy of the gas. Thus a high temperature may be achieved and ionization occur by virtue of intermolecular collisions. *See* ARC HEATING; ARC LAMP. [G.H.M.]

Arc heating The heating of a material by the heat energy from an electric arc. The electric arc is a component of an electric circuit, like a resistor, but with its own peculiar characteristics. Its resistance decreases as its current increases and vice versa. The electric arc is characterized also by its high temperature and high concentration of heat energy. Millions of watts are controllably dissipated in rather confined space, producing temperatures measured in thousands of degrees Celsius. At these temperatures most materials in the Earth's crust melt quickly. The molten materials can be processed under an oxidizing, neutral, or reducing slag or combination of slags to produce a variety of finished materials which are used for many purposes. *See* ARC DISCHARGE; ELECTROCHEMICAL PROCESS.

The electric arc is the heating element in a number of heating, melting, and smelting appliances classified as direct-arc furnaces, submerged-arc furnaces, indirect-arc furnaces, air-arc furnaces, arc welders, arc-machining tools, and arc-boring machines. *See* ARC WELDING; FURNACE.

In iron and steel foundries, in melt shops of mini-steel and large steel mills, and in some nonferrous melt shops, direct-arc furnaces are used for their ability to bring the burden quickly to pour temperature. The direct arcing from electrodes to burden gives the direct-arc furnace its name. In the iron foundry, the direct-arc furnace is used as a batch melter, a duplexer, or continuous melter for making gray iron, malleable iron, nodular iron, and alloy iron castings of closely controlled composition within a wide range of any metallurgical formulation. In the steel foundry and in steel mills, the direct-arc furnace is used almost exclusively as the batch melter. In nonferrous melt shops, the direct-arc furnace is commonly used for either a batch melter or continuous melter for metals, such as nickel, copper, and cobalt for making billets, cakes, ingots, and castings. The direct-arc furnace is used also in vacuum melting of various metals to improve purity and physical properties.

Arcs may be completely submerged under the charge or in the molten bath under the charge. Furnaces operated in this manner are generally smelting furnaces with fixed shells, into which ore granules are fed. The bath is tapped at one, two, or three levels as required for the product, slag, and impurities. Many ores are smelted in the submerged-arc furnace, and the products are numerous.

To use only the radiant heat from an arc, the furnace has a refractory-lined cylindrical shell with its axis horizontal. The shell is mounted on rollers so that it may roll or rock through something less than one revolution. Two electrodes, one from each end, enter the cylinder along the axis and touch and draw an arc inside the furnace. The heat of the arc is radiated to the refractories. The burden in the furnace, as it rocks back and forth, "washes" the heat off the refractories. This indirect heating of the burden gives the indirect-arc furnace its name. It is particularly applicable to melting low-temperature metals and alloys, such as brass, where the difference in the melting points of the burden and the refractory is enough to maintain a high rate of heat transfer, and where the heat of the arc is spread before reaching the burden to prevent large losses of volatile elements, such as zinc, in the burden.

An entirely different application of the arc furnace is to power wind tunnels. The furnace employs a cylindrical shell and a pair of electrodes arranged similarly to those in the indirect-arc furnace; however, the shell is fixed and mounted usually with its axis vertical. There is no refractory; instead, a whorl of water surrounds the arc. Air (or other gas) is blown or pumped through the space between water and arc. The air is superheated to 20,000 K and expanded to emerge at supersonic speeds. [C.W.V.]

Arc lamp A type of electric-discharge lamp in which an electric current flows between electrodes through a gas or vapor. In most arc lamps the light results from the luminescence of the gas; however, in the carbon arc lamp the light is produced by the incandescence of one or both electrodes. The color of the arc depends upon the electrode material and the surrounding atmosphere. Most lamps have a negative resistance characteristic so that the resistance decreases after the arc has been struck. Therefore some form of current-limiting device is required in the electric circuit. For other electric-discharge lamps *see* VAPOR LAMP.

The carbon arc lamp was the first practical commercial electric lighting device, but the use of arc lamps at present is limited. In many of its previous functions, the carbon arc lamp has been superseded by the high-intensity mercury vapor lamp. Arc lamps are now used to obtain high brightness from a concentrated light source, where large amounts of radiant energy are needed, and where spectral distribution is an advantage. Typical uses of arc lamps are in projectors, searchlights, blueprinting, photography, therapeutics, and microscope lighting, and for special lighting in research. *See* CARBON ARC LAMP; FLAME ARC LAMP; METALLIC ELECTRODE ARC LAMP. [J.O.K.]

Arc welding A welding process utilizing the concentrated heat of an electric arc to join metal by fusion of the parent metal and the addition of metal to the joint usually provided by a consumable electrode (see illustration). Electric current for the welding arc may be either direct or alternating, depending upon the material to be welded and the characteristics of the electrode used. The current source may be a rotating generator, rectifier, or transformer and must have transient and static volt-ampere characteristics designed for arc stability and weld performance.

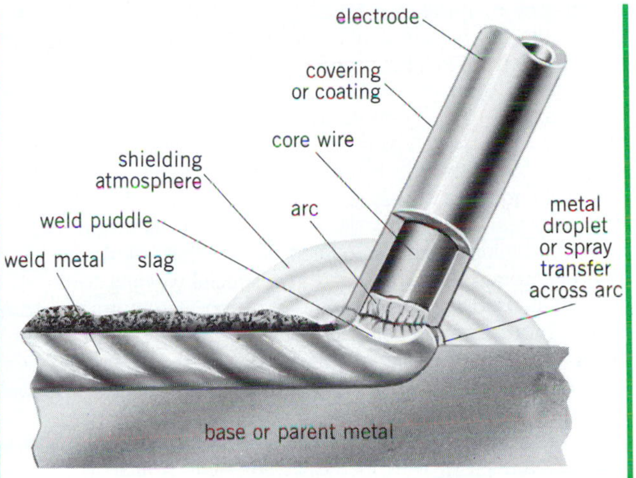

Metallic welding arc.

There are three basic welding methods: manual, semiautomatic, and automatic. Manual welding is the oldest method, and though its proportion of the total welding market diminishes yearly, it is still the most common. Here an operator takes an electrode, clamped in a hand-held electrode holder, and manually guides the electrode along the joint as the weld is made. Usually the electrode is consumable; as the tip is consumed, the operator manually adjusts the position of the electrode to maintain a constant arc length.

Semiautomatic welding is becoming the most popular welding method. The electrode is usually a long length of small-diameter bare wire, usually in coil form, which the welding operator manually positions and advances along the weld joint. The consumable electrode is normally motor-driven at a preselected speed through the nozzle of a hand-held welding gun or torch.

Automatic welding is very similar to semiautomatic welding, except that the electrode is automatically positioned and advanced along the prescribed weld joint. Either the work may advance below the welding head or the mechanized head may move along the weld joint.

There are, in addition to the three basic welding methods, many welding processes which may be common to one or more of these methods. A few of the more common are described below.

Carbon-electrode arc welding is in limited use for welding ferrous and nonferrous metals. Normally, the arc is held between the carbon electrode and the work. The carbon arc serves as a source of intense heat and simply fuses the base materials together, or filler material may be added from a separate source.

Shielded metal arc welding is the most widely used arc-welding process. A coated stick electrode is consumed during the welding operation, and therefore provides its own filler metal.

The electrode coating burns in the intense heat of the arc and forms a blanket of gas and slag that completely shields the arc and weld puddle from the atmosphere. Its use is generally confined to the manual welding method.

Submerged-melt arc welding uses a consumable bare metal wire as the electrode, and a granular fusible flux over the work completely submerges the arc. This process is particularly adapted to welding heavy work in the flat position. High-quality welds are produced at greater speed with this method because as much as five times greater current density is used. Automatic or semiautomatic wire feed and control equipment is normally used for this process.

Tungsten–inert gas welding, often referred to as TIG welding, utilizes a virtually nonconsumable electrode made of tungsten. Impurities, such as thorium, are often purposely added to the tungsten electrode to improve its emissivity for direct-current welding. The necessary arc shielding is provided by a continuous stream of chemically inert gas, such as argon, helium, or argon-helium mixtures, which flows axially along the tungsten electrode that is mounted in a special welding torch. This process is used most often when welding aluminum and some of the more exotic space-age materials. When filler metal is desired, a separate filler rod is fed into the arc stream either manually or mechanically. Since no flux is required, the weld joint is clean and free of voids.

Metal–inert gas welding, often referred to as MIG welding, saw its greatest growth in the 1960s. It is similar to the TIG welding process, except that a consumable metal electrode, usually wire in spool form, replaces the nonconsumable tungsten electrode. This process is adaptable to either the semiautomatic or the automatic method. In addition to the inert gases, carbon dioxide has become increasingly common as a shielding means. *See* WELDING AND CUTTING OF METALS. [E.F.S.]

Arcellinida An order of Lobosia. Characteristically, the shell (test) of these protozoans has a well-defined aperture through which slender fingerlike pseudopodia (lobopodia) extend. The test often has an inner chitinous layer and an outer wall composed of secreted siliceous elements or collected sand grains or the like cemented together by a secretion. Most Arcellinida are uninucleate, but *Arcella* is binucleate. In reproduction certain species prefabricate their tests to some extent. During growth the organism may assemble materials by secretion or ingestion for a new test. During fission such materials pass through the aperture into the extruded protoplasm, where they are moved to the surface and cemented into the outer wall of the new test. The old test is kept by the sister organism. Cyst walls may be formed inside the test; the aperture is often plugged with debris before encystment. The order includes *Arcella*, *Centropyxis*, *Difflugia*, and many others. *See* LOBOSIA; PROTOZOA; SARCODINA; SARCOMASTIGOPHORA. [R.P.H.]

Arch A structure, usually curved, that when subjected to vertical loads causes its two end supports to develop reactions with inwardly directed horizontal components. The designations of the various parts of an arch are given in Fig. 1. The commonest uses for an arch are as a bridge, supporting a roadway, railroad track, or footpath, and as part of a building, where it provides a large open space unobstructed by columns. Although arches are usually built of steel, reinforced concrete, or timber, the Saguenay River Bridge is an all-aluminum arch spanning 290 ft (88 m) from center to center of skewbacks.

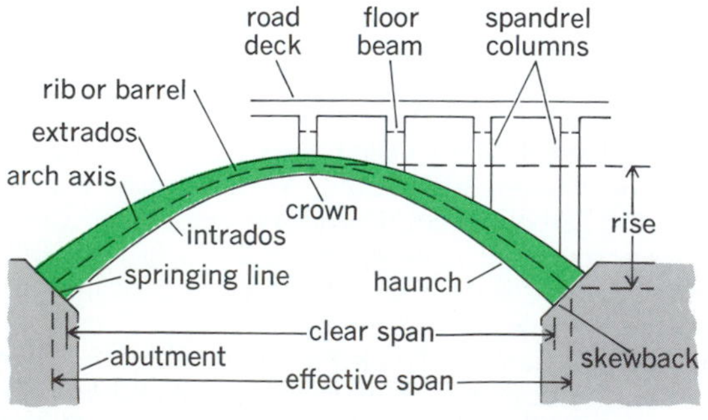

Fig. 1. An open-spandrel, concrete, fixed-arch bridge.

On the basis of structural behavior, arches are classified as fixed (hingeless), single-hinged, two-hinged, or three-hinged (Fig. 2). An arch is considered to be fixed when rotation is prevented at its supports. Reinforced concrete ribs are almost always fixed. For long-span steel structures only fixed solid-rib arches are used. The Rainbow Bridge at Niagara Falls, with a span of 950 ft (290 m), is the longest fixed steel arch in the world. The Henry Hudson Bridge in New York City is a fixed arch of 800-ft (244-m) span. Because of its greater stiffness, the fixed arch is better suited for long spans than hinged arches.

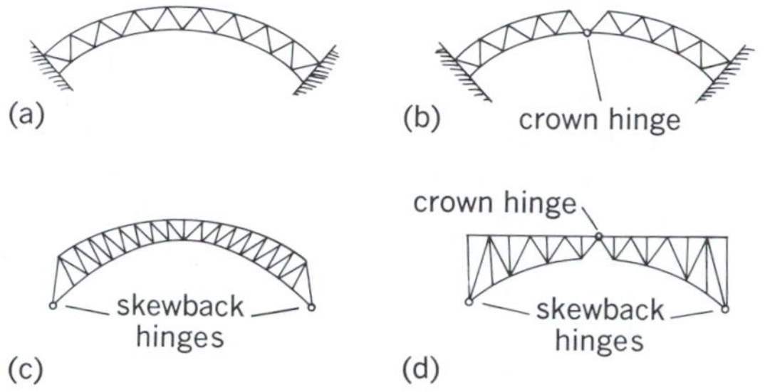

Fig. 2. Various types of bridge arches. (*a*) Fixed. (*b*) Single-hinged. (*c*) Two-hinged. (*d*) Three-hinged. (*After G. A. Hool and W. S. Kinne, Movable and Long-span Steel Bridges, 2d ed., McGraw-Hill, 1943*)

Concrete is relatively weak in tension and shear but strong in compression and is therefore ideal for arch construction. Figure 1 is a sketch of an open-spandrel concrete fixed-arch bridge. Precast reinforced concrete arches of the three-hinged type have been used in buildings for spans up to 160 ft (49 m).

Steel arches are solid-rib or braced-rib arches. Solid-rib arches usually have two hinges but may be hingeless. The braced-rib arch has a system of diagonal bracing replacing the solid web of the solid-rib arch. The world's longest arch spans are both two-hinged arches of the braced-rib type: the Sidney Harbor Bridge in Australia and the Bayonne Bridge at Bayonne, New Jersey, which have spans of 1650 and 1652 ft (502.9 and 503.5 m), respectively. The spandrel-braced arch is essentially a deck truss with a curved lower chord, the truss being capable of developing horizontal thrust at each support. This type of arch is generally constructed with two or three hinges because of the difficulty of adequately anchoring the skewbacks.

Wood arches may be of the solid-rib or braced-rib type. Solid-rib arches are of laminated construction and can be shaped to almost any required form. Arches are usually built up of nominal 1- or 2-in. (2.5- or 5-cm) material because bending on individual laminations is more readily accomplished. Because of ease in fabrication and erection, most solid-rib arches are of the three-hinged type. This type has been used for spans of more than 200 ft (60 m). The lamella arch has been widely used to provide wide clear spans for gymnasiums and auditoriums. The wood lamella arch is more widely used than its counterpart in steel. The characteristic diamond pattern of lamella construction provides a unique and pleasing appearance. *See* BRIDGE; BUILDINGS; TRUSS. [C.N.G.]

Archaebacteria A group of bacteria that is as distantly related to other bacteria as it is to animals and plants. It is one of three groups that define the deepest divisions, or so-called primary kingdoms, in the evolution of organisms from a common ancestor. See ORGANIC EVOLUTION.

Three groups of organisms constitute the archaebacteria: sulfur-metabolizing, extremely thermophilic bacteria; methane-producing bacteria; and extremely halophilic bacteria. Many of these organisms inhabit environments that are hostile to most other organisms.

The sulfur-metabolizing extreme thermophiles have been found in high-temperature environments around the world. They inhabit hot springs in Iceland, the United States (including Yellowstone National Park), Italy, Japan, New Zealand, and Indonesia. They have been isolated from soils and shallow marine sediments heated by nearby volcanoes and from deep-sea hydrothermal vents. Most can grow in anaerobic environments; some grow autotrophically by using only gases and minerals produced by geological action. Thermophiles grow at the highest temperatures that are known to support life.

The methanogenic bacteria also grow in the absence of oxygen. They inhabit the digestive tracts of animals (especially ruminants like cows), decomposing organic matter, sewage sludge digesters, swamps (where they produce marsh gas), and sediments of marine and freshwater environments. They have received renewed interest because of their ability to produce methane, which is a major component of natural gas, from municipal garbage and some industrial wastes. Methanogens are responsible for the production of all the biologically produced methane in the atmosphere.

The last group of archaebacteria comprises those bacteria that require high concentrations of salt in their environment. These organisms live in the Great Salt Lake, the Dead Sea, alkaline salt lakes of Africa, and salt-preserved fish and animal hides. They are also commonly found in pools used to evaporate seawater to obtain salt. These pools turn pink because of the high population of these organisms. Most strains require salt concentrations near the saturation point of sodium chloride. In addition, some grow in highly alkaline brines. See HALOPHILISM (MICROBIOLOGY).

The discovery of the archaebacteria came about as a result of investigations into the origin and early evolution of the cell. One method to determine the ancestral relationships among organisms is to compare the sequences of nucleotides that make up the same kind of nucleic acid from a wide variety of organisms. It was through the analyses of the ribonucleic acids from the ribosomes of many prokaryotes and eukaryotes that the archaebacteria were discovered to be a distinct phylogenetic group. These studies also revealed the primary lines of descent (primary kingdoms) that show the relationships between all living organisms.

This discovery caused a major revision in the understanding of evolutionary history. It had previously been thought that all organisms were either eukaryotes or prokaryotes, that is, the bacteria. Since their cellular organization is simpler, the prokaryotes were assumed to be ancestors of the eukaryotes. The discovery of the relationship of the archaebacteria to the eukaryotes and eubacteria revealed that the prokaryotes do not comprise a monophyletic group since they can be divided into two distinct assemblages. These three groups are the primary kingdoms of life representing the deepest branches of descent. They descended from a common ancestor termed the progenote. The order in which they descended from the progenote is unknown. See BACTERIA; EUKARYOTAE; PROKARYOTAE. [K.M.N.]

Archaeocyatha An extinct group of Early and Middle Cambrian marine animals, considered by some to be an extinct class of sponge and by others an extinct phylum. They are among the oldest known animals to have a mineralized skeleton—in their case, one made of calcium carbonate. Archaeocyathids lived in great abundance on shallow sea bottoms and were able to form reefs by attaching to hard substrates with their holdfasts. These reefs are the oldest constructed by animals.

Most archaeocyathids are shaped like a cup or inverted cone, although flat, saucerlike, annulated, and tubular forms are also encountered (see illustration). The cup of archaeocyathids averages 0.04–1.0 in. (10–25 mm) in diameter and 3.2–6.0 in. (80–150 mm) in height, but specimens with diameters as large as 24 in. (600 mm) have been reported. Although most archaeocyathids are solitary, colonial forms (grouped cups) are occasionally found.

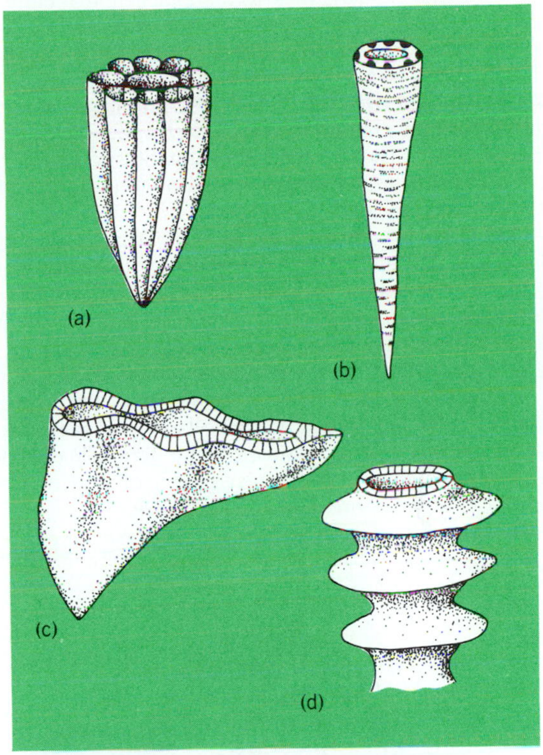

Morphological variations in the Archaeocyatha: (*a*) longitudinally ribbed; (*b*) tubular conical; (*c*) asymmetric conical; and (*d*) annulated.

Because archaeocyathids are among the oldest fossils of multicelled animals, their position in the phyletic system is crucial to understanding the evolution of animals. Yet there is little agreement concerning the phyletic affiliations of this problematic group. Three alternatives are considered the most likely: (1) they are an extinct phylum more primitive than the Porifera (sponges) but more organized than the Protozoa; (2) they are an extinct phylum of multicelled animals on the same organizational level as the Porifera but arising from a different ancestral stock; or (3) they are an extinct class of sponges. For convenience the following classification may be adopted:

Subkingdom Parazoa
  Phylum Archaeocyatha
    Class Regulares—inner wall and radial skeletal elements develop earlier than dissepiments
    Class Irregulares—dissepiments appear earlier than radial skeletal elements or inner wall

Because of uncertainties about the classification of archaeocyathids and the nature of their soft tissue, little is known about the functioning of these enigmatic animals. They were bottom-dwelling marine animals, preferring water depths of 30–100 ft (10–30 m), although estimates of their depth range extend to 330 ft (100 m). They seem to have been confined to shallow environments with substantial water movement because of their dependence on an external flow to propel water through their bodies. In archaeocyathid reef communities, algae and archaeocyathids interacted in much the same way as algae and coral do today. Archaeocyathids formed a skeletal framework which was bound and stabilized by blue-green and red algae. However, unlike modern reef communities, the reefs formed by archaeocyathids were apparently quite tolerant of terrigenous sediments. *See* PARAZOA; PORIFERA. [W.L.B.]

Archaeogastropoda The most primitive order of the subclass Prosobranchia of the class Gastropoda (snails) of the phylum Mollusca. The members retain greatest evidence of the effects of torsion and coiling, which together represent the basic characteristics of the Gastropoda.

Despite their primitive character, the Archaeogastropoda display a wide range of adaptive radiation. The shell is primitively top-shaped, with a rounded, sometimes calcareous, operculum which closes the opening when the animal withdraws. However, the limpet form, with the shell as a widely open cone and the animal permanently attached by a suckerlike foot, has evolved at least three times. Some of the limpets are among the most numerous gastropods of the seashore. For structural reasons, archaeogastropods are generally confined to hard substrates and, with the exception of one freshwater group whose affinities are dubious, to the sea.

Fig. 1. Pleurotomarid slit-shell, *Mikadotrochus hirasei*.

The Archaeogastropoda are divided into well-defined groups. In the Zeugobranchia the shell is coiled (Fig. 1), with openings for expulsion of the water current produced in the midline between the paired, often somewhat asymmetrical gills. The flattened, somewhat coiled *Haliotis* (abalone or ormer) is a widely distributed and highly successful type of a symmetric limpet. In the keyhole limpets (Fissurellidae) the shell is secondarily symmetrical; the gills have regained paired symmetry and expel water through a single apical or marginal opening. The Patellacea include the most common and most enormously successful limpets. They are divided into those with a single gill (Acmaeidae); those in which the gill is replaced by a fringe of secondary growths (Patellidae); and the rarer species from below tidal levels which are without gills (Lepetidae). The Trochacea are coiled "top shells" with a single gill and no shell aperture (Fig. 2). Although they are included in the Archaeogastropoda in most classifications, the Neritacea should probably constitute a separate order. *See* ABALONE; GASTROPODA; LIMPET; MOLLUSCA; PROSOBRANCHIA. [C.M.Y.]

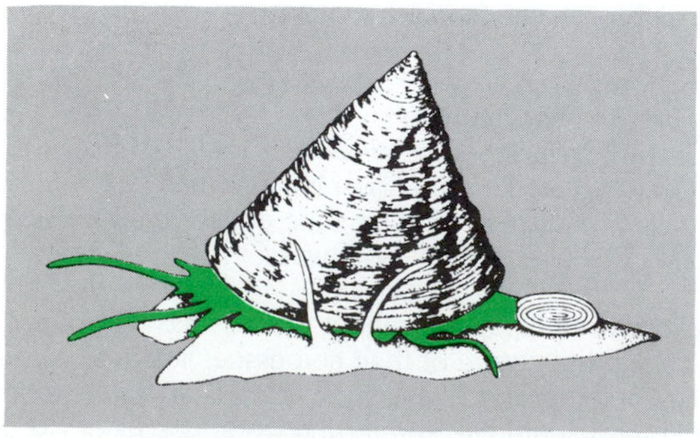

Fig. 2. Top shell, *Calliostoma zizyphinum*, from left side, showing extended head and foot.

Archaeognatha An order of ancestrally wingless insects, commonly known as bristletails, in the subclass Apterygota; about 350 species are known in two families, Machilidae and Meinertellidae. All have slender, spindle-shaped bodies of about 0.4–0.8 in. (1–2 cm) in length, tapered tailward to a long, multisegmented median caudal filament and paired cerci. The relatively thin integument is covered with pigmented scales after the first instar. The antennae are long and filiform, without intrinsic musculature in the flagellum, as in all true insects; and large, contiguous compound eyes are accompanied by a trio of ocelli. The mandibles are exposed and attached to the cranium by a single joint (condyle), and the maxillary palpi are extended, with seven segments, while the labial palpi have only three segments. Each leg has a large coxa, two-segmented trochanter, femur, tibia, and a two- or three-segmented tarsus terminating in a two-clawed pretarsus. The posterior pair or two pairs of coxae and most of the ventral abdominal coxites bear styles, which may represent modified legs. Outwardly projecting vesicles that are apparently water absorbing also occur on most coxites. Internally, the gut is relatively simple, with small crop, digestive ceca, and a dozen or more Malpighian tubules. The nervous system is primitive, with three thoracic and eight abdominal ganglia and twin connectives. Tracheae are well developed.

In mating, the male deposits sperm on a thread or in a spermatophore that is picked up by the female, but in at least one species sperm is placed directly on the female ovipositor. Immature forms are similar to adults in body form, though the first instar may lack cerci. Sexual maturity is attained after eight or more molts.

Bristletails live in a variety of habitats worldwide, often on stones or tree trunks where they find the algae and lichens that appear to be their principal sources of food. Most species are primarily nocturnal feeders. See APTERYGOTA. [W.L.Bro.]

Archaeopteridales An order of spore-bearing plants with woody trunks and simple leaves borne on determinate branch systems that resemble bipinnate fern fronds. The best-known genus is *Archaeopteris*, which is sometimes used as an index fossil of the Upper Devonian. *Archaeopteris* has become one of the best known of Devonian plants. It is no longer classified as a fern, but has been assigned to a new class, the Progymnospermopsida, which is intermediate between ferns and gymnosperms. The Archaeopteridales also include *Actinopodium*, *Svalbardia*, and *Siderella*. *See* INDEX FOSSIL; PINOPSIDA; POLYPODIOPSIDA; PTEROPSIDA. [C.A.A.]

Archaeornithes One of the two subclasses of birds, the ancient birds, containing the single order Archaeopterygiformes, the most primitive taxon of birds. This taxon was established for the seven specimens (one feather impression and six skeletons) of the oldest known fossil bird, *Archaeopteryx lithographica* from the Late Jurassic limestones of Bavaria. All other fossil and living birds are placed in the second subclass, the Neornithes or modern birds.

Archaeopteryx is an outstanding example of an intermediate form demonstrating the pattern of mosaic evolution. It is pigeon-sized with a long feathered tail, a pair of feathers attached to each tail vertebra, and a fully feathered wing. The feathers are completely modern avian feathers. The skull has reptilian jaws with small sharp teeth; cranial kinesis is very similar to that of modern birds. The brain and eyes are somewhat larger than those features in most reptiles. The forelimb is modified to a wing, but with the three clawed fingers still unfused. The pectoral girdle is weak and reptilian except for the presence of a stout furcula (wishbone). A bony, keeled sternum is lacking, but ventral ribs (gastralia) exist. The body is moderately long and flexible. The sacrum is small, and the bones of the pelvic girdle are not fused strongly together. The pubis is not fully reversed. The bones of the tarsometatatsus are not fused together. The long tail is formed of a number of individual vertebrae.

Archaeopteryx probably was aboreal for part of its life, going into trees for hiding, sleeping, and nesting. It could descend to the ground with a controlled glide and most likely foraged for food on the ground. There is no evidence that *Archaeopteryx* had the ability of powered flight or that flight in birds evolved to enable the bird to fly up from the ground. All evidence suggests that active, powered flight in birds evolved from the gliding stage and was to permit the animal to reach the ground from an elevated position in trees. *Archaeopteryx* climbed up tree trunks by using the sharp claws on its hands and feet to grip the tree. It is most likely that *Archaeopteryx* was an obligatory homoiotherm with feathers covering its body to reduce heat loss.

Archaeopteryx has many similarities with small carnivorous dinosaurs, and many workers have argued that birds evolved from this group of dinosaurs. *See* Aves; Dinosaur; Neornithes; Reptilia; Thecodontia. [W.J.B.]

Archean A major division of geologic time between 3.8 and 2.5 billion years before present. The beginning of the Archean corresponds to the age of the oldest preserved continental crust. Before that time, continental crust is believed to have been recycled into the mantle as a result of intense bombardment by meteorites. By the end of the Archean, following an episode of apparent rapid continental growth, approximately half of present continental crust had formed. There is carbon isotopic evidence of microbial activity from the early Archean onward. Major deposits of gold, platinum, nickel, zinc, and copper occur in Archean terrains. *See* Precambrian.

PRECAMBRIAN	1	PROTEROZOIC (ALGONKIAN)	LATE
	2		MIDDLE
	3	ARCHEAN (ARCHEOZOIC)	EARLY
	4		

Archean crust occurs on all continents, forming relatively small but rigid blocks (cratons) flanked by belts of younger deformed rocks. The cratons are rifted fragments of Archean continental crust subjected to many younger cycles of continental fragmentation and reaggregation.

Archean rocks attract attention because of their metallic mineral wealth and what they reveal about early global conditions and processes. The prospects for a better understanding of Archean tectonics have improved greatly since the late 1970s owing to advances in uranium-lead (U-Pb) isotopic dating, which can distinguish Archean rocks differing in age by less than 5 million years. *See* Plate tectonics; Rock age determination. [P.F.H.]

Archeoastronomy The field of science that attempts to determine how much astronomy prehistoric people knew and how it influenced their lives. It involves multiple disciplines: astronomy to chart the heavens, archeology to probe the cultural context, engineering to survey sites, and ethnology to provide clues to the cultural past. Archeoastronomy has prompted valuable insights into the astronomy of the past, even to revolutionizing some of the models of prehistoric cultures. *See* Archeology; Astronomy.

All prehistoric and preliterate peoples have observed the sky. These observations are usually related to keeping a calendar (both solar and lunar) and to regulating sacred time, the sequence of rituals.

The work of archeoastronomers at any site must always be placed in a cultural context. Without that context, finding an astronomical alignment may be no more than coincidence. It is usually possible to find a few astronomical alignments at a site; the real question is whether or not the builders intended them as such. The great danger is the imposition of modern knowledge of astronomy upon an alien culture of the past. While the accuracy of archeoastronomers' ideas about the past can never be known for certain, researchers stand on firmer ground if ethnographic information is available to bolster the archeological evidence.

The kinds of observations of key celestial cycles that can be made without a telescope will be reviewed. Only the Sun and the Moon will be considered.

Sun. Most people are aware that the height of the Sun in the sky at noon changes with the seasons. There is, however, less familiarity with the Sun's seasonal motion along the horizon. On the day of the summer solstice (around June 22), for example, the Sun rises the farthest north of east that it will get for the year (see illus.). On the equinoxes (March 21 and September 23), it rises due east. And on the winter solstice (December 22), it reaches its farthest point south of east. The same occurs, mirror-reflected, at sunset. (This description refers to the Northern Hemisphere; in the Southern Hemisphere, summer and winter are reversed by date.) *See* Equinox; Seasons; Solstice.

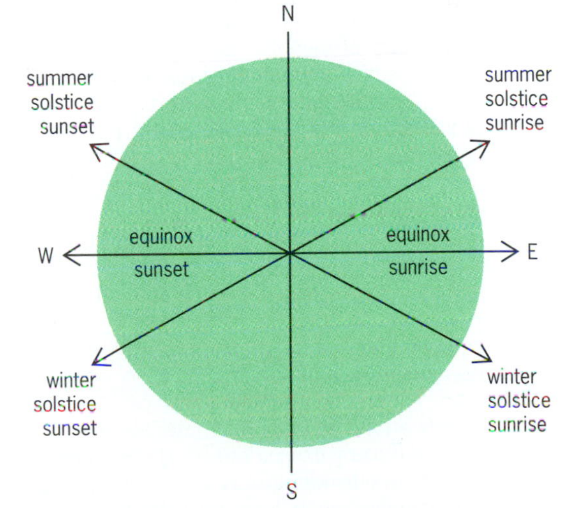

Angular swing of the Sun along the horizon at rising and setting from solstice to solstice. The angle is for a latitude of 36°, that of the North American Southwest.

Moon. The Moon's most obvious change is that of its phases. The month of phases (synodic month), the time from one phase of the Moon to the repetition of that phase, averages 29.5 days. *See* Phase (astronomy).

Suppose the point of moonrise is observed for a month. It would be seen that the moonrise point varies from a point farthest south to one farthest north during the month. In other words, the moonrise motion mimics the sunrise motion but occurs about 12 times as fast. Depending on when the observations are made, the moonrise arc may be larger than, the same as, or smaller than the sunrise arc. This difference results because the Moon's path in the sky with respect to the stars is not the same as the Sun's, but is inclined about 5°, crossing the Sun's path at two points. Thus, the Moon can appear as much as 5° below the Sun's path, 5° above it, or right on it. *See* MOON.

Horizon-marking system. It can now be seen how a simple horizon-marking system is set up. First, a location with a clear view of the horizon must be found, with at least a few prominent features over the angular range of the sunrise (or sunset). Then it is necessary to return to this spot daily and note the rising positions of the Sun throughout the year at significant times: the solstices, the equinoxes, and, perhaps, important times to plant crops. Thus, a basic solar calendar is established. Since the Sun's positions at various dates along the arc remained fixed for a long time, once established the solar calendar will be good for many years.

A solar-horizon calendar keeps reliable track of the seasonal year. To subdivide that interval, people usually rely upon a lunar calendar that follows the phases of the Moon. The essential astronomical problem is that a seasonal cycle and a lunar-phase cycle are incommensurate. The solution can involve complicated, long-term strategies, such as that used in the Gregorian calendar, or more casual, practical ones, such as counting a short month of a few days at the end of each year. *See* CALENDAR. [M.Ze.]

Archeological chemistry The application of chemical theories and experimental procedures, especially the procedures of analytical chemistry, to the solution of problems in archeology. Ordinary analytical procedures may often be applied without modification to the examination of objects and materials from excavations, but frequently some modifications are necessary. Microanalytical methods are especially useful since the amount of material available for examination is often very small. Chemical microscopy is also an important general technique for the examination of ancient objects and materials.

For the examination of valuable objects, which cannot be altered in any way, even to the extent of taking very small samples for analysis, nondestructive methods must be used. Neutron activation analysis is especially useful for the nondestructive testing of metal, glass, and pottery objects. X-ray fluorescence analysis is especially useful for estimating the chemical composition of the surfaces of objects.

The most frequent application of analytical chemistry to archeology is the exact identification of materials found in excavations so that they may be accurately described in archeological reports. Copper is sometimes confused with bronze, bronze with brass, lime mortar with gypsum plaster, solid bitumens with vegetable resins, and so on for other materials that are superficially similar. The problem of making correct identifications of ancient materials from excavations may be very different from identifying the same materials in a fresh condition. In identifying a material found in an excavation, it is usually much less important to determine what it is now than to establish what it was originally, before it underwent extensive chemical change. Though the original nature of inorganic materials may usually be determined with little difficulty, it is otherwise with organic materials, for these are likely to undergo complicated chemical changes during long burial in the ground.

Interpretations of archeological significance that may arise from the exact identification of a particular material may be more important than its identification for the purpose of exact description. Also, adequate chemical analyses of chronologically ordered specimens of materials from a given site or group of sites may yield significant information about technological or economic changes in an ancient civilization.

During long burial in the ground, a great variety of chemical reactions occur between the material of most objects and the surrounding soil and groundwater. To explain fully the appearance of objects found in excavations, to deduce with confidence their probable original appearance, or to decide on proper methods of restoration, an understanding of the nature and course of these chemical reactions is required. The appearance and condition of objects of the same material and of the same age taken from different excavation sites may be very different. The deterioration of buried objects is largely the result of chemical change, and it is therefore reasonable to expect that the restoration of such objects should usually be best brought about by chemical treatment. This principle has been applied with much success to the restoration of metal objects by various electrolytic reduction procedures. Even when objects are of such a nature that they undergo little chemical deterioration during burial, chemical treatment may be required for the effective and safe removal of the foreign matter that encrusts them.

Whether ancient objects are restored by chemical means or not, problems concerning their preservation for exhibition or study often arise that are best solved by the application of chemical principles. For example, the corrosion of an ancient metal object may continue at an appreciable rate in ordinary air unless proper precautions are taken. In addition, certain problems not directly connected with ancient objects, yet important to the archeologist, may sometimes be solved by a physical application of products made available by modern chemistry. The preservation of paper impressions of stone inscriptions is an example. *See* ACTIVATION ANALYSIS; ANALYTICAL CHEMISTRY; ARCHEOLOGY; CHEMICAL MICROSCOPY; NONDESTRUCTIVE TESTING. [E.R.C.]

Archeological chronology The establishment of the temporal sequence of human cultures. Prior to the discovery of nuclear and chemical dating methods, which provide an absolute time scale, archeologists used stratigraphy, lithic and ceramic typology, seriation, index fossils, and a limited range of chemical techniques to establish relative chronologies for cultural remains. The major chemical techniques used in dating of bones in relative sequence have been labeled F-U-N (for fluorine-uranium-nitrogen). In a single site or environment, bones of the same age usually absorb the same amount of fluorine and uranium while losing the same amount of nitrogen.

Since the mid-twentieth century numerous absolute (sometimes called chronometric) dating techniques have been devised by natural scientists. Some of these techniques can give results in calendar years, whereas others yield dates which are expressed in years but which cannot always be correlated precisely with the calendar. The major methods for absolute dating of archeological materials used include radiocarbon dating, dendrochronology, thermoluminescence dating, hydration dating, racemization, potassium-argon dating, lead-210 dating, and archeomagnetism. Many other techniques, for example, fission track dating, have been used to establish archeological chronology, and a host of physical and, to a lesser degree, chemical dating methods have been investigated. *See* DATING METHODS; DENDROCHRONOLOGY; FISSION TRACK DATING; PALEOMAGNETISM; RACEMIZATION; RADIOCARBON DATING; ROCK AGE DETERMINATION; THERMOLUMINESCENCE. [G.R.]

Archeology The scientific study of past material culture. The initial objective of archeology is to construct cultural chronologies, attempting to order past material culture into meaningful temporal segments. The intermediate objective is to breathe life into these chronologies by reconstructing past lifeways. The ultimate objective of contemporary archeology is to determine the cultural processes that underlie human behavior, both past and present.

The material culture of the past is of infinite variety. The scientific study of this evidence is such a broad task that there is no such thing as any single archeological method, although over the past century archeologists have evolved what can be termed an overall archeological approach. By constant confirmation, the archeologist often attempts to establish synchronism with what has already been established historically.

Archeologists use a number of types in order to categorize similar artifacts. Most common is the temporal type, a principle similar to the index fossil concept used by the geologist. A temporal type can be any kind of archeological artifact or feature, but ideally it is some object of common use in which the form is subject to change, due to either the whim of fashion or technological improvement. One example is the simple flint arrowhead with side barbs and central tang. It is typical of the British Bronze Age and was not in fashion earlier or later. Ceramic types have been established by archeologists working around the world, and a thoroughly tested ceramic chronology is invaluable as a temporal ordering device, no matter where the archeologist is working. The nature of the artifact employed as a temporal type is irrelevant, and its use may not even be known. Archeologists also establish other kinds of types. Functional types attempt to group artifacts on the basis of known or presumed functions. Technological types, divisions which reflect the mode of manufacture, are particularly helpful when studying stone tool manufacture.

The concept of culture is used in two different ways by contemporary archeologists. When dealing with cultural chronologies, the archeologist most commonly uses a modal or shared view of culture. It is this normative collection of shared ideas which causes artifacts to change in systematic ways through time, and temporal types can be established on the basis of this shared culture. When attempting to reconstruct lifeways, however, the archeologist can no longer rely on the shared aspects of culture. When transcending temporal associations, contemporary archeologists tend to view culture systematically, as people's extrasomatic (that is, learned) method of dealing with the social and cultural environment.

The principles of stratigraphy are applied to archeology in terms of the law of superposition, which states that, all else being equal, older deposits will tend to be buried beneath younger ones. Mere stratigraphic equivalence, however, does not necessarily indicate contemporaneity, as there can be misleading mixtures of successive occupational debris on one surface. Archeologists must therefore study the processes of cultural deposition in order to recognize the difference between intact and disturbed strata.

Contemporary excavation must be conducted with a plan, a firm research design that attempts to provide answers to definite questions. Archeology is one of the few sciences which destroys its own data in the process of generating them. Archeologists must therefore be extremely careful to make the appropriate observations at the time of excavation.

The task of deciphering meaning from past material culture is so complex that the archeologist is often required to borrow from allied disciplines in the physical and natural sciences, including geology, climatology, paleobotany, paleontology, mineralogy, physics, chemistry, and anthropology. The archeologist must have some understanding of all these sciences to extract from sites and materials every possible piece of information which may lead to a better understanding of prehistory. The archeologist must be able to record and publish every minor fact for the benefit of colleagues and successors, because the writing of prehistory requires the synthesis of all archeological discovery and interpretation. *See* ARCHEOLOGICAL CHEMISTRY; PHYSICAL ANTHROPOLOGY. [D.H.Th.]

Archiacanthocephala An order of the phylum Acanthocephala. The adult worms are parasitic in terrestrial vertebrates. Some common archiacanthocephalans are *Oncicola canis*, *Moniliformis moniliformis*, and *Macracanthorhynchus hirudinaceus*.

Oncicola canis is a short plump acanthocephalan, primarily parasitic in dogs and other Canidae. It occurs also in cats. The body of the adult, 0.24–0.56 in. (6–14 mm) long, is short and heavy with irregular cross furrows. The globular proboscis has six spiral rows of six hooks each. The arthropod intermediate host is unknown. Cystacanths have been found in armadillos and in the esophageal walls of turkeys, which indicates a transport host in the life cycle.

Moniliformis moniliformis is an elongate acanthocephalan which is parasitic in house rats. The females are 4–12 in. (10–30 cm) long, whereas the males measure 2.4–5.2 in. (6–13 cm). The body of both sexes exhibits conspicuous pseudosegmentation except on the extremities. The proboscis is cylindrical with 12–15 rows of 10 or 11 hooks each. Cockroaches (*Periplaneta americana*) serve as the intermediate host; however, in Europe a beetle, *Blaps mucronata*, is the intermediate host. Occasionally infections have been found in humans.

Macracanthorhynchus hirudinaceus is the giant thornheaded worm of hogs and is probably the best known of all acanthocephalans because of its cosmopolitan distribution. Females measure 10–24 in. (25–60 cm) in length and the males 2–4 in. (5–10 cm). The worms are pinkish with a transversely wrinkled body which tapers from a rather broad anterior end to a slender posterior end. The proboscis is globular with six spiral rows of six hooks each. At least 25 species of scarabaeid beetle larvae have been reported as intermediate hosts. In addition to their occurrence in the domestic pig, adults have been reported to occur in squirrels, chipmunks, moles, and occasionally humans. This acanthocephalan is of considerable economic importance to the hog-raising industry. *See* ACANTHOCEPHALA. [D.V.Mo.]

Archiannelida A name applied to a small group of unrelated annelids, probably not primitive as the name implies. Some resemble existing families of polychaetes (see illustration). Most live in marine or brackish water and occur in intertidal and estuarine habitats. They are characterized by their vermiform body and small sizes. Lengths range from less than 0.04 in. (1 mm) to 0.4 in. (10 mm), rarely to 4 in. (100 mm). The number of segments varies from five to many, or segmentation is obscure. Parapodia with setae may be altogether lacking, and the epithelium may be ciliated. Cephalic and anal structures may be modified as holdfast organs to maintain existence in turbulent, intertidal zones and shifting sands. *See* POLYCHAETA.

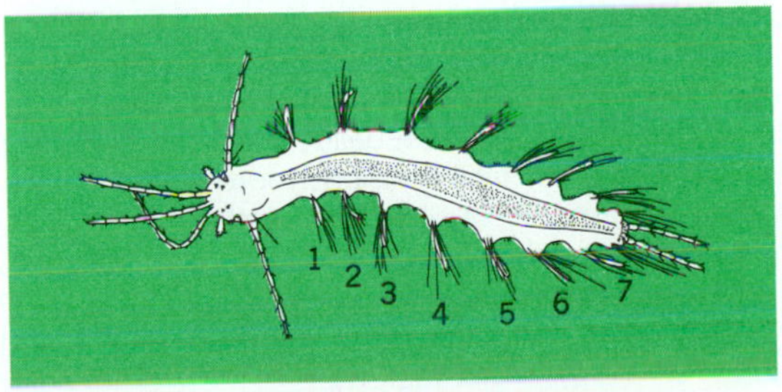

Nerilla antennata **(Nerillidae) with segments 1 to 7. (*After Schlieper, Zool. Anz. Leipzig, 62:233, 1925*)**

Three families with about 82 species in 15 genera are recognized. The Nerillidae have well-developed parapodia and setae;

they are known for 20 species in 7 genera, all but one limited to Europe and western Africa; *Nerilla* is worldwide. The Protodrilidae are known through 47 species in 5 genera; the best known are *Polygordius*, *Protodrilus*, and *Saccocirrus*, which are recorded from cosmopolitan areas. The Dinophilidae are known through 17 species in 3 genera, of which the best known is *Dinophilus*, with 6 or 7 species. *See* ANNELIDA. [O.H.]

Archichlamydeae An artificial group of flowering plants, recognized in the Englerian system of classification, consisting of those families of dicotyledons (Magnoliopsida) that lack petals or have the petals separate from each other, in contrast to the Metachlamydeae, which have the petals united into a sympetalous corolla. *See* MAGNOLIOPSIDA; METACHLAMYDEAE. [A.Cr.]

Archidiidae A subclass of the plant class Bryopsida (mosses). The Archidiidae consists of a single genus, *Archidium*, with 26 species, occurring in ephemeral habitats, especially in wet, grassy places. The small gametophytes consist of erect, simple or forked stems, and elongate, singly costate leaves in numerous rows. The capsules, usually terminal, are globose, immersed, and irregularly rupturing. The capsule wall consists of a single layer of cells; stomota, peristome, and columella are lacking. The spores are few, large (50–310 micrometers), and thick-walled. The calyptra is scarcely differentiated. *See* BRYIDAE; BRYOPHYTA; BRYOPSIDA; BUXBAUMIIDAE; DAWSONIIDAE; POLYTRICHIDAE; TETRAPHIDIDAE. [H.Cr.]

Archigregarinida An order of the protozoan subclass Gregarinia, class Telosporea, subphylum Sporozoa. All gregarines are parasites of the digestive tract and body cavity of invertebrates or lower chordates; their large, mature trophozoites (vegetative stages) live outside the host's cells. The Archigregarinida are primitive gregarines and live in marine worms (annelids and lower chordates—enteropneustids, sipunculids, and ascidians). Their life cycle includes sexual and asexual phases and involves three periods of schizogony (multiple fission). There are only 28 named species in five genera. The most important genus is *Selenidium*, which has 24 species. Its members occur in the intestine of marine polychaete annelids, sipunculids, enteropneustids, and ascidians. *See* GREGARINIA; PROTOZOA; SPOROZOA; TELOSPOREA. [N.D.L.]

Archimedes' principle A body immersed in static fluid is acted upon by a vertical force equal to the weight of fluid displaced, and a body floating in the fluid displaces its own weight of fluid. For example, a balloon ascends because it displaces a volume of air which weighs more than the weight of the balloon. The principle can be proved by determining the difference in vertical components of fluid force acting on the lower and upper curved surfaces of the body. This force, called the buoyant force, acts vertically upward through the centroid of the displaced volume of fluid. *See* BUOYANCY; FLUID STATICS; HYDROSTATICS. [V.L.S.]

Architectural acoustics The science of planning and building an enclosure to ensure the most advantageous flow of properly diffused sound to all listeners. The goal of architectural acoustics is to design a structure in such a manner that it will contribute to speech intelligibility and to the esthetic qualities of music in an environment free from external noise. The primary problems are the selection of the site; the arrangement of the rooms within the building; the design of the shape and size of each room; the selection and placement of absorptive and reflective materials to provide optimum conditions for the growth, decay, and steady-state distribution of sound in the room; and provisions for adequate noise control. *See* ACOUSTIC NOISE.

Sound fields in enclosures. The boundaries of an enclosure alter the distribution of sound emanating from a source by confining the energy that in the open would proceed outward into space. For example, if a source of sound that produces the same sound level at all frequencies at a given point in the open air is placed in a room, the sound pressure at the same distance from the source no longer will be constant with frequency but will be much higher at the resonant frequencies of the enclosure. The resonant frequencies of a room usually are not distributed uniformly with respect to frequency. In some frequency regions the modes "pile up," thereby affecting the acoustic response of the room. The distribution of the normal modes of vibration of a room is one of the considerations that determine the optimum proportions of small rooms whose acoustical characteristics are of importance, such as broadcasting studios. *See* MODE OF VIBRATION.

Wave theory can be used to calculate the transient and steady-state behavior of sound in rectangular rooms and in a few other regular shapes having special distributions of absorptive materials. However, the formulas become so complicated in practical problems in which nonuniform boundaries and many modes are involved that it is not feasible to present a complete set of practical design formulas based on physical acoustics. However, theory can be useful in providing a qualitative explanation of acoustical phenomena, even in complex situations. For example, when a source of sound is turned off in an enclosure, the decay of sound generally does not take place uniformly; instead, the sound pressure fluctuates as it decreases. These fluctuations may be explained by physical acoustics in the following way: The source sets into vibration the normal modes of the room, exciting to the greatest extent modes that have resonant frequencies nearest the frequencies of the source. Each mode has a different sound pressure distribution. When the source is turned off, the sound pressure of each excited mode will decay at its own rate. In so doing, the decaying modes interfere with each other and thus produce fluctuations in pressure. *See* REVERBERATION.

In contrast to physical acoustics, the methods of geometrical, or ray, acoustics assume that sound travels as rays, in a manner similar to that of light rays. Upon reflection from a wall, part of the acoustic energy is absorbed and part reflected; the angle of the reflected ray is equal to the angle of the incident ray if the surface is large compared with the wavelength. *See* REFLECTION OF SOUND.

Sound is said to be perfectly diffuse in a room if its pressure is everywhere the same and if at all points in the room it is equally probable that waves are traveling in every direction. It is impossible (and undesirable from the standpoint of both performers and listeners) to obtain complete diffusion.

Sound absorption. Sound is absorbed by any mechanism which converts incident sound waves into other forms of energy and ultimately to heat. All materials used in building construction absorb some sound, but those especially designed to have relatively high absorption are popularly known as acoustical materials. Most manufactured acoustical materials depend largely on their porosity for their acoustic absorption. The sound waves are converted into heat by propagation through the interstices of the material, and also by vibration of the small fibers of the material. Another important mechanism of absorption occurs when sound waves force a panel into motion; the resulting flexual vibration converts a fraction of the incident sound energy into heat.

When sound waves strike a surface, a fraction of the incident energy is absorbed and a fraction is reflected. The sound absorption coefficient of a surface is the ratio of the energy absorbed by the surface to the energy incident on the surface.

A totally absorbing surface having an area of S square feet is said to have an absorption of S sabins. Thus the sabin, sometimes called a square-foot unit of absorption, is the absorption equivalent of 1 ft^2 (0.093 m^2) of material having an absorption coefficient of unity. Similarly, a totally absorbing surface having an area of S' square meters is said to have an absorption of S' metric sabins. (1 sabin = 0.093 metric sabin.)

A quantity which describes the acoustical properties of a material and which is more fundamental than the absorption coefficient is its acoustic impedance, defined as the complex ratio of sound pressure to the corresponding particle velocity at the surface. *See* ACOUSTIC IMPEDANCE; SOUND ABSORPTION.

Transmission through partitions. The fraction of airborne incident sound energy transmitted through a partition is called its transmission coefficient τ. In rating the noise-insulating value of partitions, windows, and doors, it is convenient to employ a logarithmic quantity, transmission loss *TL*, which is equal to the number of decibels by which sound energy incident on a partition is reduced as a result of transmission through it. The two quantities are related by the equation below.

$$TL = 10 \log \frac{1}{\tau} dB$$

Sound transmission class (STC) is a single number representation of the sound insulation performance of a partition in the frequency range from 125 to 4000 Hz. In this method of rating, the STC value is determined by comparing a curve of the measured transmission loss versus frequency with a set of standard curves.

Airborne sound is transmitted through a so-called rigid partition by forcing the partition into vibration; then the vibrating partition becomes a secondary source and radiates sound to the side opposite the original source. Over a large portion of the audible range, such a partition approximates a mass-controlled system, so that one might expect that its transmission loss would increase by 6 dB each time the weight of the partition is doubled. In most partitions the increase is usually 4–5 dB. A compound-wall construction, for example, a double-leaf wall, can yield relatively high sound insulation with relatively low mass per unit wall area.

Acoustical design of rooms. The floor plan and the shape of the ceiling and walls of a room are important factors in determining its acoustical properties. The optimum length-width ratio is not a fixed number, but varies with the size and shape of the seating area; it also depends on whether a sound-amplification system is used. For most rooms, length-width ratios between 2:1 and 1.2:1 have been found satisfactory.

In order to bring a large audience as close as possible to the stage of an auditorium, it is usually advantageous to design a floor plan with diverging side walls. Reflections from the ceiling and from these walls can aid in the establishment of a higher sound level at the rear of the auditorium. These reflections must be carefully controlled because the reflected sound always travels a greater distance than the sound that reaches the listener by a direct path. If the difference in these path lengths is greater than 65 ft (20 m), the time delay in the arrival of the reflected sound is sufficient to enable the listener to hear it as a separate sound, that is, as an echo. Path length differences of 50–65 ft (15–20 m) produce a blurring quality. The geometrical law of reflection can be used to determine the proper angle for the ceiling and side-wall reflecting surfaces so that they will guide sound to seats where the sound level is not adequate. Some parts of the side walls in very large auditoriums may cause echoes. In such instances these surfaces should not be smooth and reflective but should either be "acoustically rough" to diffuse the sound or be covered with highly absorptive material. *See* ECHO.

Flutter echoes (echoes in rapid succession) frequently occur in rooms having smooth, hard, parallel walls. They can be avoided by the use of diffusing elements on the walls, diverging or tilted walls, or splays. As little as 5/8 in. (16 mm) splay to the running foot will prevent flutter. [C.M.H.]

Architectural engineering That branch of engineering dealing primarily with the properties and characteristics of building materials and components, and with the design of structural systems for buildings rather than the structural design of heavy construction, such as bridges, dams, and highways. The term is also used to indicate a field embracing all of the engineering aspects of building design, including mechanical and electrical equipment, architectural acoustics, illumination, air conditioning, and layout for usage, in addition to the historic primary interest in design of the building structure. When used in this broader sense, architectural engineering connotes building structures as its specialty, since the growing complexity of the other fields generally requires their engineering design to be accomplished by specialists trained in the pertinent branches of engineering. However, the coordination of the structural, mechanical, and electrical systems with each other and with the architectural scheme is often the responsibility of the architectural engineer. *See* ARCHITECTURAL ACOUSTICS; BUILDINGS. [H.D.H.]

Archosauria A subclass of reptiles more easily defined by listing its five component orders, four of them extinct, than by its anatomical characters. The earliest, most primitive archosaurs were the Thecodontia, from which arose independently the Saurischia ("lizard-hipped" dinosaurs), the Ornithischia ("bird-hipped" dinosaurs), the Pterosauria (flying reptiles, including the pterodactyls), and the Crocodilia (including the living crocodiles, alligators, caimans, and gavials). *See* CROCODYLIA; ORNITHISCHIA; PTEROSAURIA; SAURISCHIA; THECODONTIA.

Many Archosauria are certainly bipeds. They have two pairs of openings (fenestrae) in the temporal region of the skull (diapsid condition) and never lose the bony arcades around those openings. Typically there are also one or more antorbital openings in front of the orbit, a character which is virtually confined to this subclass (see illustration).

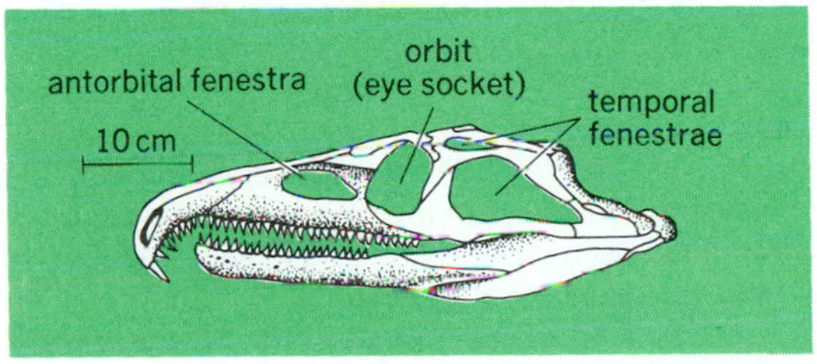

Skull of *Chasmatosaurus*, a primitive archosaur (order Thecodontia), from the Early Triassic of South Africa. (*After Broili and Schroeder*)

The earliest archosaurs are Lower Triassic, with a single species in the uppermost Permian. The Archosauria were undoubtedly a highly successful group, with locomotor systems, sense organs, and probably physiological adaptations far superior to those of any other reptiles. *See* REPTILIA. [A.J.C.]

Arctic and subarctic islands Defined primarily by climatic rather than latitudinal criteria, arctic islands are those in the Northern Hemisphere where the mean temperature of the warmest month does not exceed 50°F (10°C) and that of the coldest is not above 32°F (0°C). Subarctic islands are those in the Northern Hemisphere where the mean temperature of the warmest month is over 50°F (10°C) for less than 4 months and that of the coldest is less than 32°F (0°C). Such islands generally are in high latitudes. Distribution of land and sea masses, ocean currents, and atmospheric circulation greatly modifies the effect of latitude so that it is often misleading to use location relative to the Arctic Circle as a significant criteri-

Size of larger arctic and subarctic islands*

Name	Area mi²	Area km²
Aleutian Is.		
Unimak I.	15,500	40,100
Unalaska I.	10,800	28,000
St. Lawrence I.	18,200	47,100
Nunivak I.	16,000	41,400
Kodiak I.	37,400	96,900
Canadian Arctic Archipelago	500,000	1,295,000
Baffin I.	196,000	507,000
Ellesmere I.	76,000	197,000
Victoria	84,000	217,000
Banks	27,000	70,000
Devon	21,000	55,000
Axel Heiberg	17,000	43,000
Melville	16,000	42,000
Southhampton	16,000	42,000
Prince of Wales	13,000	33,033
Newfoundland	42,734	109,000
Greenland	840,000	2,176,000
Iceland	39,961	102,000
Svalbard (archipelago)	24,100	62,000
Vest-Spitsbergen	15,250	39,000
Franz Josef Land (archipelago)	7,000	18,000
Novaya Zemlya (archipelago)	36,000	93,000
Severny I.	21,000	54,000
Yughny I.	15,000	39,000
Severnaya Zemlya (archipelago)	14,000	36,000
New Siberian Is.	12,000	31,000
Wrangel I.	2,000	5,000
Sakhalin I.	27,000	70,000
Kuriloe Is.	6,000	16,000

*Approximate only in some cases because of incomplete mapping.

on of arctic or subarctic. The largest proportion by area of the islands lies in the Western Hemisphere, primarily in Greenland and in the Canadian Arctic Archipelago. Within this general description, individual islands vary considerably (see table).

Physiographically, the islands include all the varied major landforms found elsewhere in the world, from rugged mountains over 8000 ft (2500 m) high, through plateaus and hills, to level plains only recently emerged from the sea. All have been glaciated except Sakhalin and some of the islands in the Bering Sea sector. Removal of the weight of ice sheets and the resultant crustal rebound has exposed prominent marine beaches and wave-cut cliffs on many of the islands. These now commonly occur at elevations of over 300 ft (150 m) above sea level.

The general climatic pattern of these islands is set by their location relative to the two semipermanent centers of low pressure over the Aleutian Islands and over Iceland. Most of the precipitation is cyclonic in origin. Because they are marine areas, the islands receive more precipitation than they otherwise would, yet even so this is very light for most of the arctic islands removed from the zone of cyclonic activity. Also, because they are marine areas, the islands, regions of low temperatures by definition, are not regions of extreme low temperatures. In general, the larger the island and the closer its proximity to a continental landmass, the higher are the summer temperatures and the lower its winter temperature. *See* POLAR METEOROLOGY.

The climatic differences between arctic and subarctic islands are reflected in their natural vegetation. The arctic islands are treeless. Natural vegetation consists of the tundra—mosses, sedges, lichens, grasses, and creeping shrubs. Bare ground is often exposed and in some places plant growth may be lacking completely except for a few rock-encrusting lichens. In such places the ground surface may consist of frost-shattered rock fragments, tidal mud flats, boulder-strewn fell fields, or snow patches and ice. Permafrost (permanently frozen ground) occurs throughout the Arctic (and in parts of the subarctic) and is reflected in impeded drainage and patterned ground. *See* PERMAFROST; TUNDRA.

The natural vegetation of subarctic islands characteristically is the boreal forest or taiga, composed predominantly of conifers such as spruce, fir, pine, and larch with deciduous trees such as birch, aspen, and willow; the latter are especially common in regrowth of clearings in the forest. Impeded drainage because of permafrost or glaciation gives rise to numerous ponds and muskeg areas. A transitional type of vegetation, the forest-tundra, is recognized on some subarctic islands in sectors where smaller trees are widely spaced and abundant mosses cover the ground. *See* MUSKEG; TAIGA.

The typical soils of the subarctic islands are podzols—the grayish-white surface soil beneath the raw humus layer and highly acidic in nature. The tundra soils of the arctic islands really consist only of a dark-brown peaty surface layer over poorly defined thin horizons, and much of the ground cannot properly be termed soil. [W.C.Wo.]

Arctic Circle The parallel of latitude approximately 66½° (66.55°) north of the Equator, or 23½° from the North Pole. The Arctic Circle has the same angular distance from the Equator as the inclination of the Earth's axis from the plane of the ecliptic. Thus, when the Earth in its orbit is at the Northern Hemisphere summer solstice, June 21, and the North Pole is tilted 23½° toward the Sun, the Sun's rays extend beyond the pole 23½° to the Arctic Circle, giving that parallel 24 h of sunlight. On this same date the Sun's rays at noon will just reach the horizon at the Antarctic Circle, 66½° south. The highest altitude of the noon Sun at the Arctic Circle is on June 21, when it is 47° above the horizon.

At the Arctic Circle the Sun remains above the horizon continuously only 24 h at the longest period. However, with twilight considered, it remains daylight or twilight continuously for about 5 months. Twilight can be considered to last until the Sun drops 18° below the horizon. *See* MATHEMATICAL GEOGRAPHY; SOLSTICE. [V.H.E.]

Arctic Ocean The north polar ocean lying between North Armerica and Asia, extending over about 386,000 mi² (10^6 km²). It is nearly completely covered by 6–9 ft (2–3 m) of ice in winter, and in summer it becomes substantially open only at its peripheries. Its extent has been variably defined, but it is oceanogaphically appropriate to consider it bounded on the south by a line running from northern Greenland through Smith, Jones, and Lancaster sounds, along northwestern Baffin Island to the Canadian mainland, thence to the Alaskan coast, across Bering Strait, along the Siberian coast to Novaya Zemlya, across to Franz Josef Land and Spitsbergen, and over to northern Greenland. This definition omits the Barents, Norwegian, and Greenland seas and Baffin Bay, which have a pronounced North Atlantic character.

The central polar basin, somewhat triangular in shape, is surrounded by continental shelves which are interrupted only by the deep passage running through Fram Strait. The upper 650 ft (200 m) of the Arctic Ocean, referred to as Surface Water or Arctic Water, is characterized by a significant density stratification produced by the strong increase in salinity downward from the surface. This density stratification is of considerable importance, for it prevents a deep-reaching convection from developing within the Arctic Ocean and also prevents the heat of the underlying warm Atlantic Water from reaching the surface. The relatively low salinity at the surface is maintained against the upward diffusion of salt by the addition of fresh water, principally through river outflow. The upper 100–160 ft (30–50 m) of Surface Water tends to be relatively uniform vertically in temperature and salinity. Except for areas which become ice-free in summer, the water will be near the freezing point. Currents in the upper waters tend to be relatively slow (4 in./s or 10 cm/s or less), and they are similar in both speed and direction to the ice motion. The overall circulation in the upper waters has its ultimate cause in the prevailing wind pattern over the Arctic Ocean.

As in other oceans, the current at any instant can vary greatly from the mean condition. The most spectacular example observed in the Arctic Ocean occurs on an occasional basis in the Canadian Basin, consisting of a high-speed current core. *See* OCEAN CIRCULATION; SEAWATER.

Below the Surface Water, the temperature increases to a maximum, which over most of the region is about 33°F (0.5°C) and lies between 1000 and 1500 ft (300 and 500 m). The salinity is nearly uniform, and since at low temperatures the density of seawater depends almost solely on salinity, there is virtually no density stratification beneath the upper waters. Significant deviations from the stated temperature occur only in the southern Eurasian Basin closest to Spitsbergen, for it is there that the warm and saline water (called Atlantic Water) which maintains the temperature maximum throughout the Arctic Ocean first enters. This water has its origin in the North Atlantic. Once into the Arctic Ocean it sinks because of its high salinity and moves eastward along the Eurasian continental slope. Beneath the Atlantic Water lies cold, nearly uniform Bottom Water. These two water masses together constitute over 90% of the volume of the Arctic Ocean. The Bottom Water is formed in the Greenland Sea. [K.A.]

Arcturus Alpha Boötis, a red giant star of spectral type K2. Arcturus is one of the apparently brightest stars in the sky, of apparent magnitude 0.2. Because it is 11 parsecs (2.1×10^{14} mi or 3.4×10^{14} km) distant, its absolute magnitude is -0.04, approximately 100 times brighter than the Sun, with a radius 20 times as large.

An interesting feature of Arcturus is its large space motion with respect to the Sun; it belongs to the high-velocity, or population II, group. The radial velocity is +56 mi/s (+90 km/s), and the observed proper motion in the plane of the sky corresponds to 74 mi/s (120 km/s). W. W. Morgan recognized in Arcturus the spectroscopic features that have since become generally associated with population II. In particular, the bands of CN are weak compared to normal K giants of this luminosity class. [J.L.Gr.]

Area The superficial contents of a geometrical figure of two dimensions. The area of any rectangle or square is the product of two adjacent sides, one of which may be called the base and the other the altitude. In general, any line segment that partially bounds a plane geometric figure may be called a base if its line does not separate the figure, and a perpendicular drawn to the base line from one of its points at greatest distance may be called the altitude. Some area formulas are given in the table. *See* EUCLIDEAN GEOMETRY. [J.S.F.]

Area formulas

Figure	Formula
Triangle	$\frac{hb}{2}$, where h = altitude, b = base; $\sqrt{s(s-a)(s-b)(s-c)}$, where $s = \frac{1}{2}(a+b+c)$, and a, b, and c are sides of the triangle
Rectangle	ab, where a and b are adjacent sides
Square	a^2, where a = side
Parallelogram	$ab \sin \theta$, where a and b are adjacent sides, and θ is the angle between the sides
Trapezoid	$\frac{1}{2}(a+b)h$, where a and b are the parallel sides, and h is the altitude
Quadrilateral	$\frac{1}{2} ab \sin \theta$, where a and b are the diagonals, and θ is the angle between them
Regular polygon	$\frac{1}{4} n l^2 \cot \frac{180°}{n}$, where n is the number of sides, each of length l
Circle	πr^2, where r = radius
Ellipse	πab, where a and b are semiaxes
Sphere	$4\pi r^2$, where r = radius
Spherical triangle	$(A + B + C - \pi)r^2$, where A, B, and C are angles (radians), and r is the radius

Areal velocity The rate at which a line that joins a fixed point and a moving particle sweeps out a surface area is called the areal velocity with respect to the fixed point. In elliptical motion, if the origin is at one focus, the areal velocity is a constant. In astronomy, Kepler's law of areas expresses this characteristic. [R.L.Du.]

Arecales An order of flowering plants, division Magnoliophyta (Angiospermae), of the subclass Arecidae in the class Liliopsida (monocotyledons). The order consists of the single family Arecaceae (Palmae), the palms, with more than 200 genera and about 3500 species, largely confined to tropical and subtropical regions. The order Arecales has also been called Palmales or Principes.

Most palms are trees with an unbranched trunk and a terminal crown of large leaves. The leaf has a blade, petiole, and sheath; the sheath is open or closed, but in any case it fully encircles the stem at the base. The blade as a whole is usually pinnately compound, less often palmately compound or merely lobed. It is pinnately or less often palmately veined, but the individual pinnae or segments have more or less parallel veins and are plicate between the veins. *See* ARECIDAE; CARNAUBA WAX; COCONUT; CYCLANTHALES; DATE; LILIOPSIDA; MAGNOLIOPHYTA; PALM; PLANT KINGDOM; VEGETABLE IVORY. [A.Cr.]

Arecidae A subclass of the class Liliopsida (Monocotyledons) of the division Magnoliophyta (Angiospermae), the flowering plants, consisting of four orders (Arecales, Cyclanthales, Pandanales, and Arales), five families, and about 6400 species. Except for the highly reduced family Lemnaceae (Arales), they have an inflorescence of usually numerous, small flowers, generally subtended by a prominent spathe and often aggregated into a spadix. Except in the Araceae (Arales), the endosperm seldom contains much starch. More than 80% of the species have broad, petiolate leaves that do not have the typical parallel venation commonly associated with monocotyledons, and more than half of the species are arborescent (likewise an unusual character among the monocotyledons). *See* ARALES; ARECALES; CYCLANTHALES; LILIOPSIDA; MAGNOLIOPHYTA; PANDANALES; PLANT KINGDOM. [A.Cr.]

Arenaceous rocks The arenaceous rocks (arenites) include all those classic rocks whose particle sizes range from 0.8 to 0.0025 in. (2 to 1/16 mm), or if silt is included, to 1/256 mm. Some arenites are composed primarily of carbonate particles, in which case they are called calcarenites and grouped with the limestones. Some oolitic iron ores and glauconite beds are properly classified as arenites. But the vast majority of arenites are commonly called sandstones, and the two words are almost synonymous. *See* CALCARENITE; GRAYWACKE; OOLITE; ORTHOQUARTZITE; SANDSTONE; SEDIMENTARY ROCKS; SUBGRAYWACKE. [R.Si.]

Argentite A mineral having composition Ag_2S. Argentite crystals are rare and are cubes or octahedrons. Argentite most commonly occurs in massive form or as coatings with a lead-gray color. Its hardness is 2.5 (Mohs scale) and specific gravity 7.3; it is very sectile and is bright on the fresh surface but becomes dull black on exposure. It occurs in veins associated with other silver minerals and at some places is a major silver ore. Important localities are in Mexico, Peru, Chile, and Bolivia. In the United States it has been extensively mined in Nevada at the Comstock Lode and at Tonopah. *See* SILVER METALLURGY. [C.S.Hu.]

Argillaceous rocks The argillaceous rocks (lutites) include shales, argillites, siltstones, and mudstones; they are clastic sediments whose constituent particles are less than 0.0025 in. or 1/16 mm (if siltstones are included) or less than 0.00015 in. or 1/256 mm (if siltstones are excluded). They are the most abundant sedimentary rock type, varying according to different estimates from 44 to 56% of the total sedimentary rock column. Claystone is indurated clay, which consists dominantly of fine material of which at least a major proportion is clay mineral (hydrous aluminum silicates). Shale is a laminated or fissile claystone or siltstone, in general more consolidated than claystone. Mudstone is a claystone that is blocky and massive. The term argillite is used for rocks which are more indurated than claystone or shale but not metamorphosed to slate. All these argillaceous rocks are consolidated equivalents of muds, oozes, silts, and clays. Loess is a fine-grained, unconsolidated, wind-blown deposit. The term shale has been used by many authors generically to denote all of these types of rock. *See* ARGILLITE; BENTONITE; CLAY; CLAY MINERALS; LOESS; SEDIMENTARY ROCKS; SHALE. [R.Si.]

Argillite A nonfissile argillaceous rock that seems to have formed as the result of incipient metamorphism. The degree of induration is greater than that of shales, and the rock is dense and hard. The term has been used to designate argillaceous rocks in low-grade metamorphic belts that do not have the parting or cleavage of slates. Argillites are normally finely laminated, but the term has also been applied to dense, hard, massive, nonbedded argillaceous rocks. Argillites seem to have a gross mineral and chemical composition similar to that of shales. The argillites have no fissility, for which there is no ready explanation. They may represent conditions under which original settling did not result in preferred orientation of platy minerals, or in which there has been no reorientation due to compaction and pressure. Alternatively, the clay mineral crystal habit in the argillites may be rod-shaped and fibrous rather than platy. *See* ARGILLACEOUS ROCKS; SEDIMENTARY ROCKS; SHALE. [R.Si.]

Arginine An amino acid. Arginine forms a red color when treated with sodium hypochlorite and α-naphthol (Sakaguchi reaction). The amino acid's special functions are: (1) Arginine is an intermediate in the process of urea formation from ammonia and carbon dioxide, by virtue of its position in the urea cycle. (2) Arginine is probably the source of the guanidine groups of streptomycin.

The biosynthesis of arginine starts with glutamic acid and proceeds by way of ornithine and citrulline. The metabolic degradation begins with the hydrolysis of arginine to ornithine and urea. Ornithine can be deaminated to glutamic acid semialdehyde. This compound is oxidized to glutamic acid, which is catabolized via α-ketoglutaric acid and the Krebs cycle. *See* AMINO ACIDS; KREBS CYCLE. [E.A.Ad.]

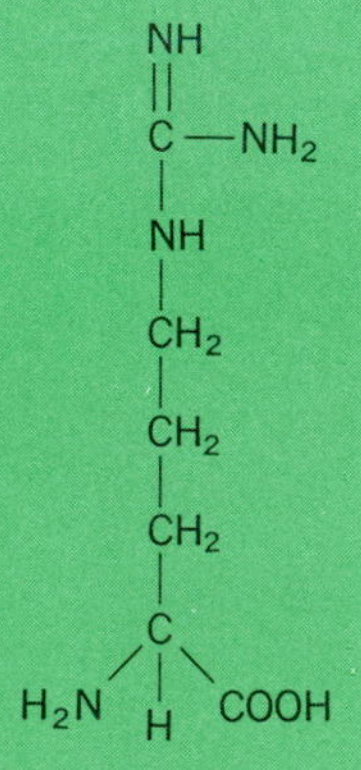

Arginine

Argon A chemical element, Ar, atomic number 18, and atomic weight 39.948. Argon is the third member of group 0 in the periodic table. The gaseous elements in this group are called

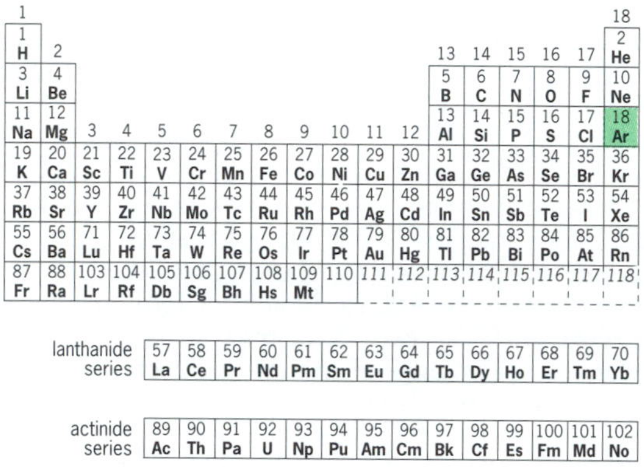

the noble, inert, or rare gases, although argon is not actually rare. The Earth's atmosphere is the only natural argon source; however, traces of this gas are found in minerals and meteorites. Argon constitutes 0.934% by volume of the Earth's atmosphere. Of this argon, 99.6% is the argon-40 isotope; the remainder is argon-36 and argon-38. There is good evidence that all the argon-40 in the air was produced by the radioactive decay of the radioisotope potassium-40. *See* INERT GASES.

Argon is colorless, odorless, and tasteless. The element is a gas under ordinary conditions, but it can be liquefied and solidified readily. Some salient properties of the gas are listed in the table. Argon does not form any chemical compounds in the ordinary sense of the word, although it does form some weakly bonded clathrate compounds with water, hydroquinone, and phenol. There is one atom in each molecule of gaseous argon.

The oldest large-scale use for argon is in filling electric light bulbs. Welding and cutting metal consumes the largest amount of argon. Metallurgical processing constitutes the most rapidly growing application. Argon and argon-krypton mixtures are used, along with a little mercury vapor, to fill fluorescent lamps. Argon mixed with a little neon is used to fill luminous electric-discharge tubes employed in advertising signs (similar to neon signs) when a blue or green color is desired instead of the red color of neon. Argon is also used in gas-filled thyra-

Properties of argon

Property	Value
Atomic number	18
Atomic weight (atmospheric argon)	39.948
Melting point (triple point), °C	−189.4
Boiling point at 1 atm pressure, °C	−185.9
Gas density at 0°C and 1 atm (101.325 kPa) pressure, g/liter	1.7840
Liquid density at normal boiling point, g/ml	1.3998
Solubility in water at 20°C, ml argon (STP) per 1000 g water at 1 atm (101.325 kPa) partial pressure of argon	33.6

trons, Geiger-Müller radiation counters, ionization chambers which measure cosmic radiation, and electron tubes of various kinds. Argon atmospheres are used in dry boxes during manipulation of very reactive chemicals in the laboratory and in sealed-package shipments of such materials.

Most argon is produced in air-separation plants. Air is liquefied and subjected to fractional distillation. Because the boiling point of argon is between that of nitrogen and oxygen, an argon-rich mixture can be taken from a tray near the center of the upper distillation column. The argon-rich mixture is further distilled and then warmed and catalytically burned with hydrogen to remove oxygen. A final distillation removes hydrogen and nitrogen, yielding a very high-purity argon containing only a few parts per million of impurities. [A.W.F.]

Arhynchobdellae An order of the class Hirudinea (the leeches) which do not have an eversible proboscis, but frequently have three jaws armed with sharp teeth. The blood of these annelids contains hemogloblin. They may be divided into the Gnathobdellae, with jaws, and the Pharyngobdellae, without jaws.

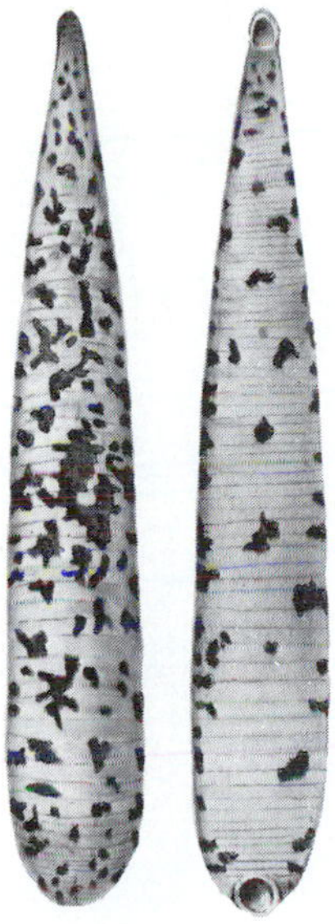

Erpobdella punctata.

Gnathobdellae have bodies which are oval in cross section and have a conspicuous posterior sucker. The anterior sucker does not project beyond the sides of the body but forms a deep cup on the underside of the head in which the jaws can work to make their incision in the host. This group contains most of the important bloodsucking leech parasites of humans and other warm-blooded animals. The land leeches are members of this group. They occur in great numbers on vegetation in swamp and jungle areas and attach themselves to passing warm-blooded animals.

Pharyngobdellae are specialized for carnivorous diets, and in many cases have completely lost the jaws. They have a strong muscular pharynx which extends nearly half the length of the body. *Erpobdella* (see illustration) is common in lakes and streams in the Northern Hemisphere, while *Trocheta* tends to leave the water and forage in moist soil. *See* HIRUDINEA; RHYNCHOBDELLAE. [K.H.M.]

Aries The Ram, in astronomy, a zodiacal and autumnal constellation (see illustration). Among the 12 zodiacal constellations, Aries was considered as the first, because about 2000 years ago when the zodiacal constellations were organized, the Sun was in Aries where it crossed the equator at vernal equinox. Today, because of the precession of the equinoxes, this reference point has moved into the constellation Pisces. However, Aries remains the first sign of the zodiac. *See* CONSTELLATION. [C.-S.Y.]

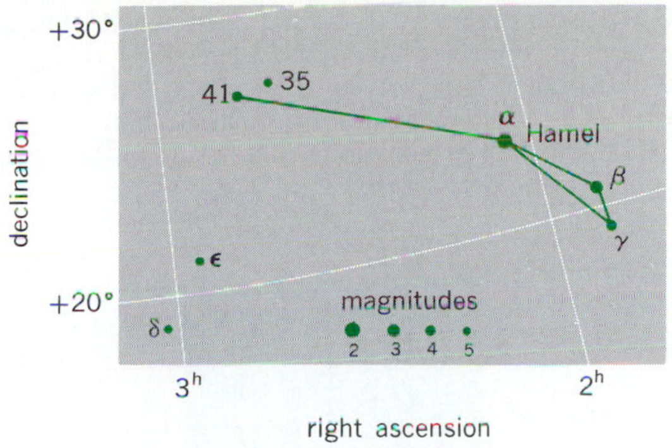

Line pattern of constellation Aries. Grid lines represent the coordinates of the sky. Apparent brightness, or magnitudes, of stars are shown by sizes of the dots, which are graded by appropriate numbers as indicated.

Aristolochiales An order of flowering plants, division Magnoliophyta (Angiospermae), in the subclass Magnoliidae of the class Magnoliopsida (dicotyledons). It contains only the family Aristolochiaceae, with seven genera and about 600 species, most of them in tropical and subtropical regions. Within its subclass the order is marked by the presence of ethereal oil cells, by its uniaperturate or nonaperturate pollen, and especially by its strongly perigynous to epigynous flowers, usually with united carpels, that typically lack petals and have the sepals joined into a highly irregular, corolloid calyx. Many of the species are climbing vines. *Aristolochia* (birthwort or Dutchman's pipe) and *Asarum* are well-known genera of the order. *See* FLOWER; MAGNOLIIDAE; MAGNOLIOPHYTA; MAGNOLIOPSIDA; POLLEN; RAFFLESIALES. [A.Cr.]

Arithmetic A branch of mathematics dealing with numbers, operations on numbers, and computation. Arithmetic is useful in solving many practical problems, such as buying, selling, budgets, sports statistics, and measurement. The usual numbers of arithmetic are whole numbers, fractions, decimals, and percents. Beyond the numbers of arithmetic are negative numbers, rational numbers, and irrational numbers. The rational and irrational numbers together constitute the real numbers.

Whole numbers. The whole numbers include the infinite sequence of counting numbers—one, two, three, four, five, . . . —and the number zero. For numbers to ten, a single symbol is used, and for larger numbers a combination of symbols.

Numbers to ten are designated with a single digit: 0, 1, 2, 3, 4, 5, 6, 7, 8, and 9. Numbers ten and greater are expressed by using a combination of the ten digits, with the place of the digit indicating the value of the digit. This place-value system, named the Hindu-Arabic numeration system, is now used around the world. *See* NUMBER SYSTEMS.

In a multidigit numeral, the value of each place from right to left is a successive power of ten, and the total values for all places are combined or added.

Operations. The basic operations are addition (+) and multiplication (× or ·), with subtraction (−) and division (÷) defined, respectively, by using addition and multiplication.

If two numbers, a and b, are combined or added, the result is a number, c, called the sum. In the example $4 + 6 = 10$, 4 and 6 are addends and 10 is the sum. The whole amount, 10, is the result of combining two parts, 4 and 6.

If a given number, *n*, sets of objects with the same number in each set, *r*, are combined, then multiplication of *n* and *r* is the total number of objects. In 3 × 4 = 12, 3 and 4 are factors and 12 is the product (see illus.).

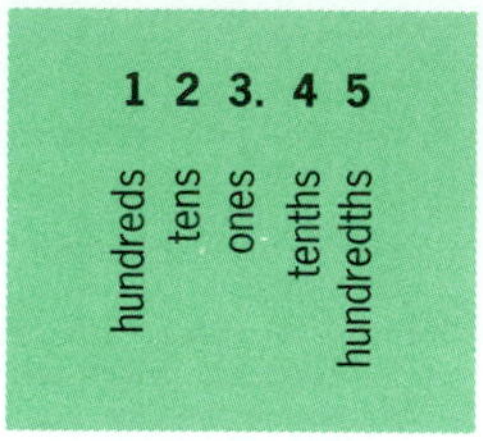

One meaning of multiplication.

Subtraction is the inverse operation to addition, finding an addend when a sum and one addend are known. If there are 18 children on the playground and 10 are boys, then the number of girls is 18 − 10 or 8, illustrating that the whole minus a part leaves the other part. Subtraction is used also to find the difference, for example, to see how many more are in one group of 18 children than in a group of 10 children. Subtraction is checked with addition; 18 − 10 = 8 because 8 + 10 = 18.

Division is the inverse operation to multiplication, that is, finding a factor when a product and a factor are known. In 12 ÷ 3 = 4, 12 is called the dividend, 3 is called the divisor, and 4 is the quotient.

Fractions and decimals. Understanding fractions and decimals, as well as operations on these numbers, is essential for practical uses and for long-term memory.

The initial and most basic idea is that a fraction shows "part of a whole." In the fraction 3/4, read "three-fourths," the 4 shows the number of equal-size pieces in each whole unit as well as the size of one piece, "fourth." The 3 shows the number of equal-size pieces being taken or considered. The top number, 3, "numbers" the parts and is called the numerator. The bottom number, 4, "names" the parts and is called the denominator.

To show decimals less than one, the place value system for whole numbers is extended to the right of the ones place. The value of each place to the right is a successive power of 1/10. The decimal point is needed to designate the ones place because the place on the right is no longer the ones place. The decimal point also separates the whole number from the decimal part.

Percent is another way to express fractions and decimals that show hundredths. For example, 7 hundredths can be expressed as 7/100, as 0.07, or as 7%. All three expressions show the same part of a whole. *See* PERCENT. [J.N.P.]

Arkose An arenaceous rock that contains a high proportion of feldspar in addition to quartz and other detrital minerals. Arkose is also known as feldspathic sandstone. Although there is no universal agreement, many geologists consider a minimum of 25% feldspar a requisite for calling sandstone an arkose. Other geologists accept a lower value. Arkoses may contain a high proportion of other nonquartz detritus, such as igneous and metamorphic rock fragments, micas, amphiboles, and pyroxenes. Frequently the accessory heavy mineral suite consists of a variety of species.

Sedimentary structures of arkoses are similar in kind to those of the orthoquartzites. Cross-bedding, the major feature, may be displayed on a huge scale, some cross-bedded units being many feet thick. Arkoses are associated with a variety of clastic rocks, dominantly conglomerates, and reddish-colored shales. Arkoses also are found with basic lava flows. Most arkoses are found in geosynclinal areas, but the thin, reworked, granite-wash arkoses can be found on stable continental platforms. *See* GEOSYNCLINE.

The granite-wash arkoses appear to have formed as the result of a transgression of the sea over a land area underlain by granite. The fragmented granite in the soil and mantle rock is incorporated in the basal sediment. In some areas the original granite is changed so slightly that the arkose is called recomposed granite and may be almost indistinguishable from the original granite. Since high relief and climatic extremes generally are associated with orogenic movements, arkoses are usually interpreted as sediments that result from tectonically active regions. *See* ARENACEOUS ROCKS; FELDSPAR; GRAYWACKE; ORTHOQUARTZITE; SANDSTONE; SEDIMENTARY ROCKS. [R.Si.]

Arm (anatomy) The upper, or superior, limb, consisting of the upper arm and forearm. Its bony framework (see illustration) consists of the humerus in the upper arm and the two forearm bones, the radius (outer) and ulna (inner). The ulna hinges with the humerus at the elbow, and the radius rotates on the lower (distal) end of the ulna, allowing the wrist to turn.

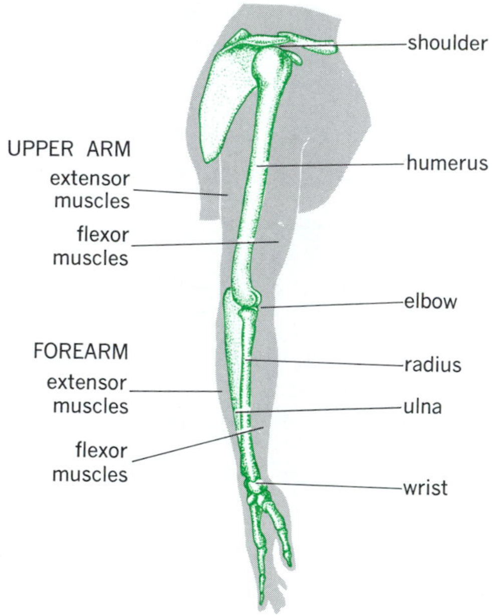

Lateral view of the human right arm showing the bones and major muscles. (*After W. J. Hamilton et al., Textbook of Human Anatomy, Macmillan, 1956*)

Two major muscle groups are found in each portion, the extensors and the flexors. Those of the upper arm act principally on the forearm; those of the forearm act on the wrist, hand, and fingers. Total arm movements are made by means of shoulder and trunk muscles. [T.S.P.]

Armadillo The name for 21 species of mammals of the order Edentata, a group characterized by the lack of enamel on their teeth. They are indigenous to the New World, especially South America.

Armadillos range in size from the lesser pichiciego or fairy armadillo (*Chlamyphorus truncatus*), in which the adult is about 5 in. (7.5 cm) long, to the giant armadillo (*Priodontes giganteus*), which is about 4 ft (1.2 m) in length. The body is covered with horny dermal scales that replace the hair com-

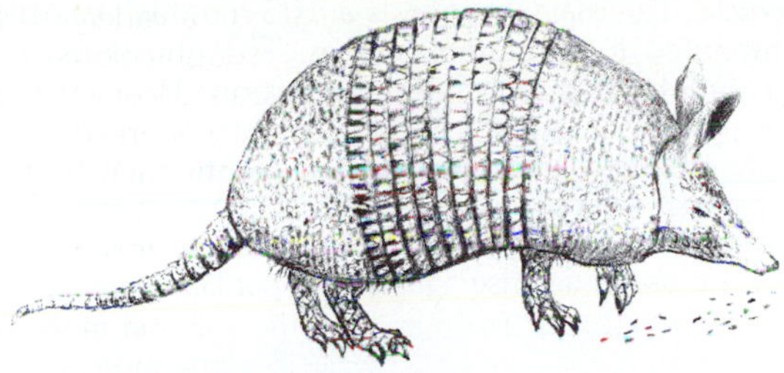

Nine-banded armadillo (*Dasypus novemcinctus*).

mon to most mammals and overlay bony plates. These structures fuse to form rigid shields covering the anterior and posterior ends of the animal, whereas in the midregion they form jointed bands allowing a certain amount of flexibility. The giant armadillo has about 100 teeth, more than any other land mammal. The snout is long, and the tongue is cylindrical and viscous to assist in capturing food. The toes are clawed and are used by the animal to dig into ant and termite colonies for food, as well as for burrowing. When disturbed, many species roll into a ball or wedge themselves into the opening of a burrow.

The nine-banded armadillo (*Dasypus novemcinctus*; see illustration) is the best-known species and ranges from South America to the southwestern and southern United States. It is the only edentate which inhabits the United States, ranging from the Rio Grande area to Oklahoma and eastward along the coast to Louisiana. The nine-banded armadillo has been studied because of its unusual life cycle. Four young are born in a den or chamber at the end of the burrow. The young are always of the same sex and are identical quadruplets. *See* DENTITION; EDENTATA; SCALE (ZOOLOGY). [C.B.C.]

Armature That part of an electric rotating machine which includes the main current-carrying winding. The armature winding is the winding in which the electromotive force (emf) produced by magnetic flux rotation is induced. In electric motors this emf is known as the counterelectromotive force.

On machines with commutators, the armature is normally the rotating member. On most ac machines, the armature is the stationary member and is called the stator. The core of the armature is generally constructed of steel or soft iron to provide a good magnetic path, and is usually laminated to reduce eddy currents. The armature windings are placed in slots on the surface of the core. On machines with commutators, the armature winding is connected to the commutator bars. On ac machines with stationary armatures, the armature winding is connected directly to the line. *See* COMMUTATOR; CORE LOSS; WINDINGS IN ELECTRIC MACHINERY. [A.R.E.]

Armature reaction The reaction of the magnetic field produced by the armature current on the magnetic field produced by the field current. This causes a distorted flux-density distribution in the air gap. Due to the saturation of the armature teeth, the flux density is decreased by a greater amount under one pole tip than it is increased under the other, and therefore the cross-magnetizing armature reaction produces a demagnetizing effect, and the generated voltage or counter-emf will be reduced when the armature is loaded. In a generator this degrades the voltage regulation. In a motor it tends to increase the speed and may cause instability.

For machines subject to heavy overloads, rapidly changing loads, or operation with a weak mainfield, the resultant flux-distribution distortion by excessive cross-magnetizing armature reaction will cause nonuniform distribution of voltage between commutator segments, and may result in flashover between commutator segments. A poleface (or compensating) winding, embedded in slots in the pole face and excited by armature current, is provided to neutralize the cross-magnetizing effect under the pole faces. *See* ELECTRIC ROTATING MACHINERY. [A.R.E.]

Army armament The equipment of a military unit. Army armament covers a broad range of weaponry, from handheld, small-caliber infantry weapons through sophisticated aircraft armament to heavy, self-propelled, large-caliber guns and howitzers. Many weapons have families of ammunition encompassing, for example, high explosives (HE); antipersonnel, armor-penetrating, cannon-launched guided projectiles; and nuclear rounds. Armament also includes numerous devices such as grenades, grenade launchers, mines, and missiles intended to neutralize enemy defenses.

The United States has joined with several of its allies in cooperative efforts to design, develop, produce, and deploy armament under the auspices of rationalization, standardization, and interoperability (RSI) agreements. Rationalization involves increasing the effectiveness of alliance forces through the development of more efficient use of the defense resources committed to the alliance. Standardization means adopting common or compatible technical, logistical, and tactical procedures while developing a new system of interchangeable supplies, components, weapons, and equipment. Interoperability is the ability of allied forces to provide and accept services from each other. Thus, RSI promotes an efficient strategy that reduces defense expenditures while increasing technological advances, and nourishes cooperation in deployment of new weapon systems with the advantage of unifying the alliance defense network.

Army armament is usually divided into categories such as infantry, aircraft armament, artillery, armor, and miscellaneous defense devices.

U.S. Army infantry has several systems. The TOW is a heavy antitank weapon capable of accurate, effective fire against armored vehicles and hard targets. It is a tube-launched, optically tracked, wire-guided missile, launched from a crew-served weapon from either the ground or a suitable vehicle (land or air). The Dragon is a missile weapon system capable of destroying armor vehicles and field fortification. The Dragon is a medium-range antitank/assault weapon, light enough to be carried and deployed by one person. Useful range extends to 3300 ft (1000 m). The missile is optically tracked, with signals from the tracker being transmitted to the missile by means of wires. The Viper is a rocket system intended to replace the M72A2 (LAW) as the Army's light antitank weapon. It is a light-tactical-round, human-portable, shoulder-launch rocket intended for use against tanks and other hard targets. The rocket is unguided after launch, and follows a free ballistic trajectory to the target. The M16A1 rifle (see illustration) is a lightweight, air-cooled, gas-operated weapon which uses a 20- or 30-round magazine. It can be fired from the shoulder or hip in the semi- or full-automatic mode at a cycle rate of 700 to

M16A1 rifle.

940 rounds per minute. The M203 40-mm grenade launcher is a lightweight, compact, breech-loading, pump-action, single-shot, manually operated weapon used with the M16A1 rifle. The squad automatic weapon is a one-person automatic weapon capable of delivering a large rate of automatic, lethal, accurate, and sustained fire to 3300 ft (1000 m). The weapon will supplement and reinforce firepower of other weapons in the rifle squad fire team. The squad automatic weapon will eventually replace the M16A1 in the automatic rifle mission, and may replace one or more of the M60 machine guns in the rifle platoon.

Through the use of Air Force tactical support and Army aviation support, the Armed Forces employ air/ground operations systems (AGOS). Close air support is used to include air attacks from all services against hostile targets which are in close proximity to friendly forces. The Air Force A-10 aircraft, used to sustain close air support, employs a 30-mm GAU-8 seven-barrel gatling gun. The intended mission for GAU-8 is antiarmor. The GAU-8 family consists of three types of cartridges: armor-piercing incendiary (API), high-explosive incendiary (HEI), and target practice (TP). The advanced attack helicopter requires an area weapon subsystem designed to optimize the capability of the aircraft to carry out the Army's area fire mission in support of ground operations. The area weapon system for the AH-64 consists of the following major components: 30-mm chain gun, turret, ammunition storage, and transfer mechanism and control system.

Mine warfare is an integral element in modern land combat, as conflicts since the mid-1960s have clearly demonstrated. Modern tactics and strategies have threatened the effectiveness of conventional mines; therefore, the U.S. Army Armament Research and Development Command (ARRADCOM), in association with the Air Force and Navy, has developed FASCAM (family of scatterable mines). FASCAM provides allied forces with antiarmor and antipersonnel mines with quick-strike emplacement capabilities through air, artillery, and special-purpose ground vehicle delivery techniques, as well as remotely activated and deactivated weapons to counter the enemy's growing strength and technology. Mine fields have the greatest effectiveness when used in conjunction with direct and indirect firepower; therefore, they can be emplaced in advance as an integral element in a well-prepared defense to block, delay, divert, or destroy attackers. During offensive and counterattack operations, FASCAM mine fields can be emplaced to deny escape routes to the enemy.

The field artillery (FA) mission is to destroy, neutralize, or suppress the enemy by cannon, rocket, and missile fire. Field artillery has a target acquisition capability, a responsive and accurate gunnery team, a variety of weapons and ammunition, plus a command and control element. The weapons and ammunition element of field artillery include high- and low-angle cannon fires, rocket and missile fires, scatterable mine delivery, plus appropriate ammunition selection to provide desirable effects on the target. Artillery armament includes both towed (T) and self-propelled (SP) weapons. A variety of field artillery ammunition is available including high-explosive (HE), high-explosive antitank (HEAT), smoke, illumination, white phosphorus (WP), chemical, nuclear, beehive (flechette), and the improved conventional munitions (ICM), both antipersonnel and dual-purpose. Various fusing options such as point-detonating, base-detonating, concrete-piercing, variable-time, and mechanical-time/superquick increase munition lethality. The Honest John, Lance, and strategic weapon missile system augment field artillery capability.

Armor constitutes a basic combat branch of the U.S. Army equipped with tank-mounting artillery and other weapons, and other supporting units and vehicles. Armor includes tank units, armored infantry, armored artillery, and armored engineers. The M60A3 tank is a full-track-laying, heavily armored, combat vehicle. The main weapon is a 105-mm high-velocity cannon mounted to a rotatable turret. The gun is used as an armor-penetrating antimateriel weapon, but can be used against personnel and soft targets. A 7.62-mm machine gun coaxially mounted with the main gun is used as the primary antipersonnel weapon. A .50-caliber machine gun is mounted in the cupola, serving as an antipersonnel and soft target weapon for use in defense against low-performance aircraft. [M.E.P.D.; M.R.M.]

Aromatic A term which describes those organic compounds having physical and chemical properties resembling those of benzene. This quality, aromaticity, is reflected in the stability of these compounds. Although the carbons in aromatic compounds are bonded to three atoms rather than four, aromatic compounds do not behave chemically like unsaturated compounds. Their stability is considered to be the result of a resonating electron structure in which all carbon-to-carbon bonds have the same length. Aromatic compounds show a marked tendency to undergo substitution rather than addition reactions.

There are four main classes of aromatic compounds: benzenoid aromatic compounds, containing one benzene ring; polynuclear benzenoid, containing two or more fused benzene rings; nonbenzenoid, containing planar cyclic carbon rings other than benzene; and those heterocyclic compounds whose properties resemble benzene.

Aromatic compounds owe their name to their original derivation from fragrant natural compounds. *See* AROMATIC HYDROCARBON; BENZENE; RESONANCE (MOLECULAR STRUCTURE). [B.R.]

Aromatic hydrocarbon An organic compound with a chemistry similar to that of benzene. Usually aromatic hydrocarbons have structures that contain at least one benzene ring. Their properties are so different from those of other hydrocarbons that they are considered to be a separate class of compounds, the arenes. Several important aromatic hydrocarbons are benzene, naphthalene, toluene, styrene, biphenyl, and anthracene. Other species usually classed as aromatic hydrocarbons are aromatic ions, for example, the cyclopentadienyl ion; benzologs of benzene such as naphthalene; and a nonbenzenoid polycyclic aromatic hydrocarbon, azulene.

Aromatic hydrocarbons share the aromatic character of benzene and its derivatives in that they exhibit diminished unsaturation, ease of formation, and stability of the nucleus. Benzene's structure has been the focus of attention because the concept of aromatic character has developed in terms of benzene and its derivatives. While the Kekule structure showed a ring consisting of six carbons with three alternate double bonds, the modern view is that all six bonds of benzene are alike and that all are intermediate in nature between double and single bonds. *See* RESONANCE (MOLECULAR STRUCTURE).

Aromatic hydrocarbons, and particularly polycyclic aromatic hydrocarbons, have the ability to form stable molecular complexes with some polynitro compounds. Some of these molecular complexes have proved of value in the isolation and identification of aromatic derivatives. Among the most useful nitro compounds for this are picric acid, trinitrobenzene, and trinitrofluorenone. *See* ANTHRACENE; BENZENE; CARBORANE; HYDROCARBON; NAPHTHALENE; POLYNUCLEAR HYDROCARBON. [C.K.B.]

Aromatization The conversion of any nonaromatic hydrocarbon structures, especially those found in petroleum, to aromatic hydrocarbons. There are numerous routes and means to accomplish this transformation, the simplest and most important of which are: direct dehydrogenation of naphthenes to aromatics; dehydroisomerization of naphthenes to aromatics; dehydrocyclization of aliphatics to aromatics; and high-temperature condensation of hydrocarbons to aromatics. *See* AROMATIC HYDROCARBON.

Beginning in 1948, reforming of naphthas, with catalysts comprising small amounts of platinum on an acidified alumina support, has provided an important means of aromatization that has rapidly displaced earlier processes. It is a major process for making benzene, toluene, and other aromatics from petroleum sources. *See* PETROLEUM PROCESSING.

[B.S.G./M.Sou.]

Arrowroot starch A food product derived from different plant species. Rhizomes of *Maranta arundinacea* of tropical America supply the West Indian arrowroot. Florida arrowroot is the flour made from the rhizomes of the cycad *Zamia floridana*, the Seminole bread plant. Queensland arrowroot is obtained from the edible rhizomes of *Canna edulis*. The tubers of *Curcuma angustifolia* yield the East Indian arrowroot. Arrowroot starch is of no importance in the industries, but because it is very easily digested, it is valued highly as a food for infants and invalids. [P.D.St./E.L.C.]

Arsenate A negative ion having the formula AsO_4^{3-}. Arsenates are derived from orthoarsenic acid, H_3AsO_4. Arsenates are quite similar to the phosphates because the term arsenates also includes metaarsenates and pyroarsenates, as well as three series of salts from orthoarsenic acid.

The various arsenates are also very similar to the phosphates in their solubilities and crystal form. The alkali metal salts are the only soluble tertiary orthoarsenates. Lead and calcium tertiary salts, $Pb_3(AsO_4)_2$, and $Ca_3(AsO_4)_2$, are among those arsenates used in insecticides, all of which are poisonous to humans.

Because of their similarities it is difficult to distinguish between the PO_4^{3-} and AsO_4^{3-} ion. However, silver nitrate gives a chocolate-brown precipitate, Ag_3AsO_4, with arsenate, whereas the corresponding phosphate is yellow. *See* ARSENIC; ARSENITE; PHOSPHATE. [E.E.W.]

Arsenic A chemical element, symbol As, atomic number 33. Arsenic is found widely distributed in nature (approximately 5×10^{-4} of the Earth's crust). It is one of the 22 known elements composed of only one stable nuclide, ${}_{33}^{75}As$; the atomic weight is 74.92158. There are 17 other radioactive arsenic nuclides known.

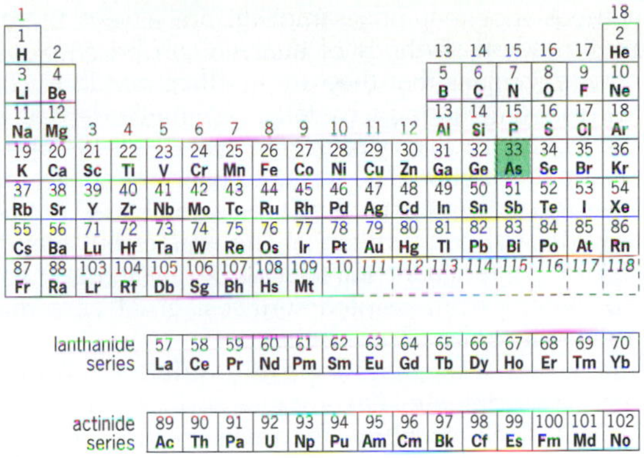

There are three polymorphic modifications of arsenic. The yellow cubic α-form is made by condensing the vapor at very low temperatures. The black β-polymorph is isostructural with black phosphorus. Both these modifications revert to the stable γ-form, gray or metallic, rhombohedral arsenic, on heating or exposure to light. The metallic form is a moderately good thermal and electric conductor and is brittle, easily fractured, and of low ductility.

Arsenic is found native as the mineral scherbenkobalt, but generally occurs among surface rocks combined with sulfur or metals such as Mn, Fe, Co, Ni, Ag, or Sn. The principal arsenic mineral is FeAsS (arsenopyrite, mispickel); other metal arsenide ores are $FeAs_2$ (löllingite), NiAs (nicolite), CoAsS (cobalt glance), NiAsS (gersdorffite), and $CoAs_2$ (smaltite). Naturally occurring arsenates and thioarsenates are common, and most sulfide ores contain arsenic. As_4S_4 (realgar) and As_4S_6 (orpiment) are the most important sulfur-containing minerals. The oxide, arsenolite, As_4O_6 is found as the product of the weathering of other arsenical minerals, and is also recovered from flue dusts collected during the extraction of Ni, Cu, and Sn from their ores; it also results when the arsenides of Fe, Co, or Ni are roasted in air or oxygen. The element may be obtained by roasting FeAsS or $FeAs_2$ in the absence of air or by reduction of As_4O_6 carbon, when As_4 may be sublimed away.

Elemental arsenic has few uses. It is one of the few minerals available in 99.9999+% purity, which is largely used in the laser material GaAs and as a doping agent in the manufacture of various solid-state devices. Arsenic oxide is used in glass manufacture. The arsenic sulfides are used as pigments and in pyrotechnics. Dihydrogen arsenate is used in medicine, as are several other arsenic compounds. Most of the medicinal uses of arsenic compounds depend on their toxic nature. *See* ANTIMONY; PHOSPHORUS. [J.L.T.W.]

Arsenite A negative ion having the formula AsO_3^{3-}. Arsenites are derived from arsenious acid. Arsenious acid is the name given to aqueous solutions of As_4O_6 although the pure acid has never been isolated. These solutions are weakly acidic and can be neutralized by bases to give a variety of *ortho-* and *meta*-arsenites. Paris green, $Cu_2(C_2H_3O_2)(AsO_3)$, is used as an insecticide and Scheele's green, $CuHAsO_3$, is used as a pigment. *See* ARSENATE; ARSENIC. [E.E.W.]

Arsenopyrite A mineral having composition FeAsS and crystallizing in the monoclinic system. Crystals have pseudo-orthorhombic symmetry because of twinning. The Mohs hardness is 5.5–6.0, and the specific gravity is 6.0. The luster is metallic and the color silver-white. Arsenopyrite is the most widespread arsenic-bearing mineral. It is commonly found in veins containing gold (Lead, South Dakota; Deloro, Ontario), tin or tungsten minerals (Bolivia; Cornwall, England), or nickel-cobalt-silver minerals (Cobalt, Ontario; Freiberg, Germany). *See* ARSENIC. [L.Gr.]

Art conservation chemistry The application of chemistry to the technical examination, authentication, and preservation of cultural property. Chemists working in museums engage in a broad range of investigations, most frequently studying the chemical composition and structure of artifacts, their corrosion products, and the materials used in their repair, restoration, and conservation. The effects of the museum environment, including air pollutants, fluctuations in temperature and relative humidity, biological activity, and ultraviolet and visible illumination, represent a second major area of research. A third area of interest is the evaluation of the effectiveness, safety, and long-term stability of materials and techniques for the conservation of works of art. Though analytical techniques appear to dominate, many other areas of chemistry, biology, physics, and engineering, including polymer chemistry, kinetic studies, imaging methodologies, biodegradation studies, dating methods, computer modeling, metallography, and corrosion engineering, play active roles in conservation science.

Methods of examination may be divided into two classes: those that provide an image of the entire object (holistic examination) or a section of it; and those that provide an analysis at a point on the object, with or without sampling. Nondestructive methods, not requiring sampling, are always preferable. However, modern methods of analysis can be employed on such minute samples that they are in effect nondestructive. In some cases, samples must be taken for methods that are in principle nondestructive because the object is too large to fit into a sample chamber. The ability to analyze minute samples introduces the serious concern that the sample may not be representative of the composition of the artifact but may be an inclusion or contaminant introduced by the experimentalist. With specimens from painted surfaces, great care must be taken to identify areas of restoration.

A further concern arises from the differing depths from which signals originate. On a metal surface, ion scattering spectrometry (ISS) would see the initial fraction-of-a-nanometer, predominantly adsorbed species and contaminants. Secondary ion mass spectrometry (SIMS) would begin to penetrate the oxidized area; Auger electron spectrometry (AES) would examine the bulk of the oxidized layer; and x-ray-induced photoelectron spectrometry (XPS) would give data on the bulk sample some 10 nm below the specimen surface. *See* ACTIVATION ANALYSIS; ANALYTICAL CHEMISTRY; AUGER EFFECT; NONDESTRUCTIVE TESTING; SECONDARY ION MASS SPECTROMETRY (SIMS); SURFACE PHYSICS; X-RAY FLUORESCENCE ANALYSIS.

The most commonly employed holistic method is x-ray radiography, where variations in the density and average atomic number of the sample attenuate an x-ray beam, leaving a negative image on film. Other methods, such as ultraviolet and infrared reflectance and fluorescence, are used to show areas of compositional difference indicating restoration or variation in the pigments used by the artist. *See* INFRARED IMAGING DEVICES; INFRARED RADIATION; LUMINESCENCE ANALYSIS; RADIOGRAPHY; ULTRAVIOLET RADIATION.

In the examination of paintings, small samples are taken under the binocular microscope, embedded in transparent resin, and polished to produce a cross section for microscopic examination. This permits a study of the artist's painting technique and shows how several layers may have been built up to achieve a desired effect. Conservation studies of the composition and technique embrace the entire spectrum of modern chemical analysis. *See* CHEMICAL MICROSCOPY; IMAGE PROCESSING; MICROSCOPE.

The separation of the fake from the authentic is a small but often spectacular aspect of the technical examination of artifacts. In some cases, direct age determination (dendrochronology for panel paintings, fission track dating for uranium glass, radiocarbon dating for organic materials, thermoluminescence dating for ceramics) is possible. *See* ARCHEOLOGICAL CHRONOLOGY; DATING METHODS; DENDROCHRONOLOGY; FISSION TRACK DATING; RADIOCARBON DATING.

More commonly, the issue of authenticity turns upon anachronisms in composition or technique when the artifact in question is compared to accepted artifacts of the period. Thus, the greater part of the work in the conservation laboratory concerns the building of databases of analyses of composition, trace-element distributions, and studies of technique.

Many artifacts are sensitive to destructive agents in the museum atmosphere. Rapid changes in relative humidity will cause dimensional changes in wood furniture, polychrome sculpture, and panel paintings, leading to cracking and splitting of the wood with loss of painted surface decoration. High relative humidity can lead to mold growth and foxing on books and prints, while low relative humidity will cause photographic prints and films to become brittle.

Oxidation of iron objects, tarnishing of silver plate, and the development of corrosion products on lead artifacts by the action of formic and acetic acids emitted by wooden display cases have regularly been observed in museums.

The common air pollutants sulfur dioxide (SO_2) and ozone (O_3) have been monitored at elevated levels in museums, libraries, and archives. These pollutants cause the degradation of leather, spotting of photographic prints, and fading of dyes and pigments. Chemical methods of analysis are used to identify degradation products and to study the kinetics of degradation mechanisms. Specialists in air-pollution monitoring use analytical instrumentation to measure ambient pollution levels in museums. [N.S.B.]

Arteriosclerosis The name given a group of degenerative diseases of arteries characterized by thickening and hardening of their walls. The group includes three types of lesions: (1) atherosclerosis involves the aorta and its major branches; (2) medial sclerosis involves the muscular arteries of the legs; and (3) arteriolosclerosis involves the small branches of the arterial tree, called the arterioles.

Atherosclerosis is by far the most common and important form of arteriosclerosis, and the two terms are often used interchangeably. Some degree of atherosclerosis is found at autopsy in persons of all ages, even children, but it increases in severity and extent with advancing years. It is more common in males than in females, especially females before the age of menopause. After menopause the sex incidence and severity are about equal. Atheromas (anatomic lesions) begin with the lining layer of the aorta or its branches and subsequently extend into the middle layer of these vessels. As the lesions enlarge, a plaque forms, and fibrin clots are deposited on the surface, diminishing the diameter of the lumen of the vessel.

Atherosclerosis has global distribution and now occurs in virtually epidemic proportions in the industrialized nations of Western civilization. There is an extremely high prevalence and severity of the lesions in the Northern European countries, the United States, Australia, and New Zealand. There is a relatively low occurrence in economically deprived areas such as India, Africa, and South America. Factors associated with the high incidence and severity of atheromas include a high total caloric intake, high fat intake, sedentary living, aggressive personality, emotional stress, and cigarette smoking. Hypertension (high blood pressure) does not induce arteriosclerosis, but it augments its development and accelerates the progress of the disease if it is present. Although virtually ubiquitous, atherosclerosis fortunately does not cause serious disease symptoms in most individuals. However, in the United States over 60% of the deaths each year are due to cardiovascular disease produced by atherosclerosis and its sequelae. Although the vast majority of cases of atherosclerosis appear to be principally caused by environmental factors, there are some specific genetic defects associated with the genesis of the process. Primary diabetes mellitus patients develop severe arteriosclerosis at an earlier age than nondiabetic individuals do. Familial hypercholesterolemia, homocystinuria, and hypothyroidism are other examples of metabolic defects associated with arteriosclerosis.

Medial sclerosis is an uncommon type of arteriosclerotic lesion of little clinical significance. It affects the arteries of the arms, legs, and genital tracts of both sexes equally. It rarely occurs under the age of 50, and its cause is obscure. The disorder is characterized by ringlike calcifications within the media (middle layer) of affected vessels. The endothelial lining of the vessel is not altered, and the size of the lumen not impinged upon. Therefore, occlusion of the vessel seldom occurs.

Arteriolosclerosis is an increased generalized thickening of the walls of arterioles related to hypertension. The change is often most prominent in the kidneys, although other internal organs may be similarly affected. Patients with diabetes mellitus have an increased incidence of the lesion. [N.K.M.]

Arteritis A general term denoting an inflammatory process in arterial walls; it may be due to various bacterial infections, syphilis, allergic or hypersensitivity states, exceptionally elevated blood pressure, or unknown factors. Arteries situated in regions of localized bacterial infection are frequently involved in the inflammatory process. Such vessels may become thrombosed or weakened to the point of rupture with resulting hemmorhage. Syphilitic arteritis occurs primarily in the proximal or thoracic portion of the aorta. Allergic or hypersensitivity arteritis may be due to drugs or unknown factors. This type of arteritis includes such diseases as periarteritis nodosa, temporal arteritis, lupus erythematosus disseminata, and rheumatic arteritis. This group shows alterations in the vessel wall characterized by inflammation, necrosis, and often thrombosis. Thrombosis with occlusion is a common complication which results in regions of infarction or necrosis of organ tissue due to lack of blood supply. Aneurysms often form in the weakened arteries and frequently rupture, causing fatal hemorrhage. *See* ANEURYSM; HEMORRHAGE; LUPUS ERYTHEMATOSUS; SYPHILIS; THROMBOSIS. [F.A.C.]

Artery Any of the vascular tubes which carry blood away from the heart. The arteries may be divided into four or five categories, depending upon their diameter and proportions of different tissues present in their walls. All arteries have a smooth endothelial lining, which is quite thin in children but gradually thickens with advancing age. *See* ARTERIOSCLEROSIS.

The largest arteries—the aorta and the pulmonary with their major branches—have an ill-defined zone of elastic fibers around the endothelium and its supporting tissue. The middle coat, or tunica media, of these arteries is composed of some smooth muscle, but the predominant tissue is elastic fibers, giving these elastic arteries a yellowish hue in cross section. The outer coat, or adventitia, is a connective tissue containing nerve, collagenous and elastic fibers, and small nutrient blood vessels, the vasa vasorum.

The medium-sized arteries, also known as muscular arteries, have a red-grayish hue in cross section because of the preponderance of smooth muscles in the tunica media. The muscular layer is bounded internally and externally by an elastic membrane. The outer layer, the adventitia, contains a plentiful supply of nerves from the autonomic system that innervate the smooth muscles.

The smaller arteries and arterioles are marked by their small size and progressive loss of first the inner and then the outer elastic membrane, and finally the muscular layer in the smallest vessels. The smallest arterioles, found just before the capillaries, are about 20 micrometers or less in diameter, and consist only of the universal endothelium and a thin layer of smooth muscles. These metaarterioles are still richly innervated, as they are important in the final regulation of blood to the capillary beds. *See* CAPILLARY (ANATOMY).

The terminal ramification of arteries is also connected to veins by direct arteriovenous anastomoses in many parts of the body. These anastomoses contain a well-innervated layer of smooth muscles by which the diameter of their lumen may be regulated. They apparently serve to regulate local blood flow and pressure by allowing blood to bypass the capillary beds and flow directly into the veins. *See* AORTA; BLOOD VESSELS; CIRCULATORY SYSTEM. [W.J.B.]

Artesian systems Groundwater conditions formed by water-bearing rocks (aquifers) in which the water is confined above and below by impermeable beds. Because the water table in the intake area of an artesian system is higher than the top of the aquifer in its artesian portion, the water is under sufficient head to cause it to rise in a well above the top of the aquifer. Many of the systems have sufficient head to cause the water to overflow at the surface, at least where the land surface is relatively low. Flowing artesian wells were extremely important during the early days of the development of groundwater from drilled wells, because there was no need for pumping. Their importance has diminished with the decline of head that has occurred in many artesian systems and with the development of efficient pumps and cheap power with which to operate the pumps. *See* GROUNDWATER HYDROLOGY. [A.N.S./R.K.Li.]

Arthritis A group of diseases affecting joints of their component tissues. Several types of arthritis are recognized, and these can be divided into four groups by their clinical course or pathologic appearance: inflammatory arthritis, degenerative joint disease, nonarticular rheumatism, and miscellaneous arthritis.

Inflammatory arthritis is a group of diseases characterized by inflammation of joint fluid and of synovial tissue and other supporting tissues surrounding diarthrodial and fibrocartilaginous joints. Connective tissue diseases, crystal deposition diseases, infectious arthritis, and spondyloarthropathies are examples of inflammatory arthritis. Connective tissue diseases are a group of acute and chronic diseases characterized by involvement of joints, connective tissue, serosal membranes, and small blood vessels. These diseases are divided into acquired disorders (rheumatoid arthritis, systemic lupus erythematosus, scleroderma, polymyositis, vasculitis) and rare hereditary diseases (Ehlers-Danlos syndrome). Rheumatoid arthritis is the most common variety of inflammatory arthritis occurring in younger and middle-aged persons and is characterized by uninfected inflammation of the synovium, frequently associated with extra-articular manifestations. The systemic disease follows a variable but slowly progressive course, marked by spontaneous remissions and exacerbations. The finger joints are often first involved symmetrically, with the hands, wrists, feet, and other smaller joints later becoming affected. Joint destruction is common with advancing disease, so that the incidence of disability is high. There are three groups of crystal deposition disease based on the type of crystal involvement: monosodium urate (gout), calcium pyrophosphate (pseudo-gout), and hydroxyapatite (calcific tendonitis). Gout, the metabolic disorder of uric acid breakdown, has long been associated with a violent arthritis, particularly of the toes. Infectious arthritis is an inflammatory joint disease caused by the invasion of synovial membrane by living microorganisms, including gonorrheal, streptococcal, and staphylococcal bacteria. These usually result from a generalized infection but may appear following local spread or after trauma. The outcome varies considerably with the patient's age and resistance, adequacy of the treatment, and the nature of the organism. Another group of inflammatory arthritis characterized by involvement of the axial (central) skeleton is spondyloarthropathy. Ankylosing spondylitis, Reiter's syndrome, spondylitis associated with inflammatory bowel diseases, and psoriasis are examples of the spondyloarthropathies. *See* GOUT; LUPUS ERYTHEMATOSUS.

Degenerative joint disease (osteoarthritis) is a ubiquitous joint disease characterized pathologically by deterioration of articular cartilage and new bone formation in the subchondral areas. The disease is thought to be inherent in the aging process, being very common in older persons. In addition to aging, a variety of other factors are involved in pathogenesis such as joint injury, dysplasia, and other forms of underlying arthritis. Degenerative joint disease is marked by a progressive stiffness, loss of function, and destruction of the larger, weight-bearing joints of the body. With advancing age, the continued slow damage causes increasing disability.

The nonarticular rheumatism group includes tendonitis, bursitis, tenosynovitis, fibrositis, and psychogenic rheumatism. Calcific tendonitis and bursitis usually follow local trauma, muscle strain, or excessive exercise and can result from deposition of hydroapatite crystals (crystal deposition disease). Fibrositis is defined as a complex of musculoskeletal complaints frequently involving neck, shoulder girdle, and extremities. Fibrositis and psychogenic rheumatism both are thought to have a psychic origin. *See* RHEUMATISM.

Miscellaneous forms of arthritis include those systemic diseases of unknown or quite different etiology which may produce arthritis or joint destruction. There are neurologic, blood, and endocrine examples of these unusual rheumatic diseases.

These and less common forms of arthritis affect an estimated 11,000,000 persons in the United States. With an increasing survival time in the population, arthritis constitutes one of the greatest medical, social, and economic problems in existence. [R.Se.]

Arthropoda The largest phylum in the animal kingdom, comprising about 75% of all animals that have been described. The estimated number of arthropod species is between 1,000,000 and 2,000,000. Of this number the class Insecta alone contains 750,000–1,000,000 species. Arthropods vary in size from the microscopic mites to the giant decapod crustaceans, such as the Japanese crab with an appendage span of 5 ft (1.5 m) or more. This phylum includes the well-known insects, spiders, ticks, and crustaceans, as well as many smaller groups, some of which are known as fossils.

The adult arthropod typically has a body composed of a series of ringlike segments, muscularly movable on each other. The integument is sclerotized by the formation of hardening substances in the cuticle, and the segmental limbs are many-jointed. These characteristics, taken together, distinguish the arthropods from all other animals. Young stages may be quite different from the adults, and some parasitic species differ very radically from their relatives.

In their evolution the arthropods have diverged structurally into three major groups, the Trilobitomorpha, Chelicerata, and Mandibulata. A classification scheme of the Arthropoda follows:

Phylum Arthropoda
- Subphylum Trilobitomorpha (including Trilobite)
- Subphylum Chelicerata
  - Class: Merostomata
    - Pycnogonida
    - Arachnida
- Subphylum Mandibulata
  - Class: Crustacea
    - Pauropoda
    - Symphyla
    - Diplopoda
    - Chilopoda
    - Insecta (Hexapoda)

Body segmentation, or metamerism, is the most fundamental character of the arthropods, but it is shared by the annelid worms, so there can be little doubt that these two groups of animals are related through some remote, simple, segmented, wormlike ancestor. Modern arthropods possess segmental limbs, highly modified and more specialized than those of any other animals, that serve for walking, feeding, grasping, swimming, silk spinning, egg laying, and sperm transfer. The body segments, corresponding to specialized sets of appendages, tend to be consolidated or united in groups, or tagmata, forming differentiated body regions, such as head, thorax, and abdomen.

The arthropods have all the internal organs essential to any complex animal. An alimentary canal extends either straight or coiled from the subapical ventral mouth to the terminal anus. The nervous system includes a brain and a subesophageal ganglion in the head, united by connectives around the stolmodeum, and a ventral nerve cord of interconnected ganglia. A usually tubular pulsatory heart lies along the dorsal side of the body and keeps the blood in circulation. In some arthropods arteries distribute the blood from the heart; in others it is discharged from the anterior end of the tube directly into the body cavity. The blood reenters the heart through openings located along its sides. Aquatic arthropods breathe by means of gills. Most terrestrial species have either flat air pouches or tubular tracheae opening from the outside surface; some have both. A few small, soft-bodied forms respire through the skin. Excretory organs open either at the bases of some of the appendages or into the alimentary canal. Most arthropods have separate sexes, but some are hermaphroditic, and parthenogenesis is of common occurrence. The genital openings differ in position in different groups and are not always on the same body segment in the two sexes. *See* CHELICERATA; MANDIBULATA; TRILOBITOMORPHA. [J.G.R.]

Arthus reaction An allergic reaction of the immediate hypersensitive type. The classical Arthus reaction in a sensitized rabbit injected with the appropriate allergen is a localized skin reaction characterized by swelling (edema) and inflammation. This reaction differs from other immediate hypersensitive states because larger quantities of circulating antibody (and antigen) are required for its elicitation, and because on passive transfer the recipient becomes immediately reactive, without the incubation period required in passive transfer of anaphylaxis or atopy.

The Arthus reaction becomes visible within 1–2 h after injection of the antigen, reaches its peak in 8–12 h, and may persist for 24 h. If the reaction is sufficiently severe, local damage to blood vessels may lead to necrosis due to deficient blood supply; this degree of damage is never seen in the hive, which is a local anaphylactic reaction. The Arthus reaction also requires the participation of complement, components of which call forth the leukocytes that make up part of the inflammatory and necrotizing reaction.

The sequestration of complexes of antibody, antigen, and complement in the basement membranes of small vessels, such as those of the kidney glomerulus, can also cause systemic disease such as glomerulonephritis and periarteritis nodosa. *See* ANAPHYLAXIS; ATOPIC ALLERGY; AUTOIMMUNITY; HYPERSENSITIVITY; IMMUNOLOGY. [S.R.]

Artichoke *Cynara scolymus*, a herbaceous perennial plant, in the family Compositae; also called globe artichoke. Its

Green Globe artichoke. ***(Burpee Seeds)***

origin is in the Mediterranean region. Artichoke requires a mild winter and cool summer with fog and little bright sunshine. It is a delicacy in Europe, Africa, and North and South America. Artichoke is also a medicinal plant; it is rich in the cynarin and ortho-phenol constituents. In the United States, artichokes are grown in the Pacific Coast area between south San Francisco and Los Angeles, mainly Monterey County.

The marketable portion of the plant, the so-called bud, is actually the immature flower head, made up of numerous closely overlaid bracts or scales (see illustration). The edible portion consists of the tender bases of the bracts, the young flowers, and the receptacle or fleshy base upon which the flowers are borne. The bud can be various shapes, from round to oblong to flat, and the color can be light green to dark green, often with purple or red. [A.M.I.]

Articulata (Brachiopoda) A class of the Brachiopoda, which includes about 92% of the known genera of the phylum. It is presently divided into six orders: Orthida, Terebratulida, Strophomenida, Pentamerida, Rhynchonellida, and Spiriferida. A seventh order, the Kutorginida, is currently unplaced in either the Articulata or the Inarticulata. Although there is no universal agreement on the phylogenetic relationships of the articulate orders, there is little doubt that the Orthida are the oldest stock, appearing in the Early Cambrian, and that ultimately these are the ancestors of all articulate brachiopods. *See* ORTHIDA; PENTAMERIDA; RHYNCHONELLIDA; SPIRIFERIDA; STROPHOMENIDA; TEREBRATULIDA.

One of the characteristic features of the class is its shell microstructure. The basic pattern, found in the earliest orthides and typical of the remainder of the class, with the exception of the strophomenidines, is in two distinct layers of calcite. A thin outer, primary layer secreted extracellularly by a narrow band of columnar epithelium at the mantle margins is underlain by a thicker development of a secondary layer of fibrous calcite, secreted by the remainder of the mantle. The fibers in the latter layer are deposited intracellularly and are bounded by cytoplasmic sheaths in continuity with the cell walls.

Articulation is typically present in the class. A pair of teeth are developed as dorsally directed protuberances, articulating with sockets in the brachial valve. These structures may be replaced by a row of denticles, as in some strophomenidines, or they may be lost completely, as in the majority of the productidines. Even when articulation by teeth and sockets is absent, the musculature develops to allow rotation about a hinge axis.

The valves are closed by contraction of the adductor muscles, opening being affected by contraction of the diductor muscles. Other muscles are typically present which control movement of the shell relative to the pedicle. *See* BRACHIOPODA. [A.J.R.]

Articulata (Echinodermata) The only surviving subclass of the Crinoidea. It differentiated during Triassic times.

The calyx is dicyclic, but considerable reduction of the infrabasals and basals may occur. The uniserial arms bear pinnules and usually branch, and the arm retains its movable articulation with the radial plate, despite the incorporation of the lower brachial ossicles into the calyx. Extant stalked forms with nodal rings of cirri (*Metacrinus*) are included in the order Isocrinida (see illustration). They do not tolerate turbulent waters and live at depths below current action, although they do inhabit shallow water when the conditions are suitable. The feather stars, of the order Comatulida, discard the stem when young, and thereafter remain free, either as swimming animals or as creeping benthic forms. They prefer shallow, clear water, rich in nutrients, and therefore abound on tropical coasts and

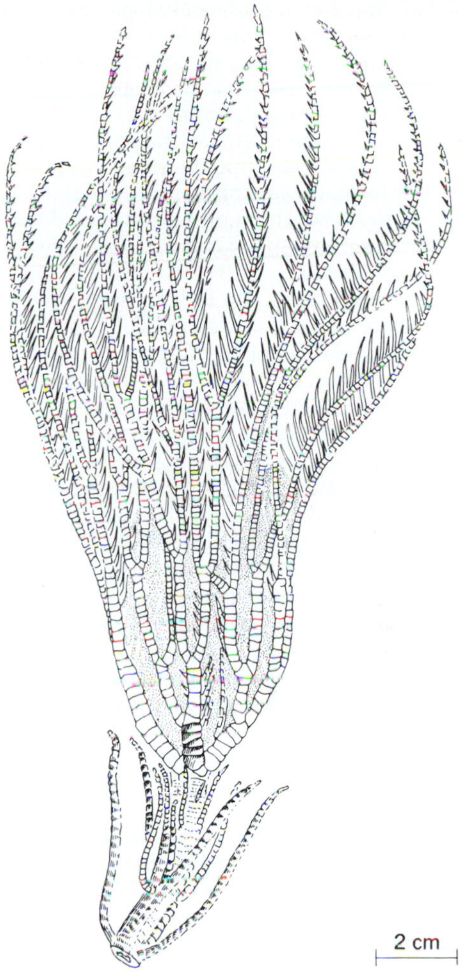

***Metracrinus cyaneus*, a living, articulate stalked crinoid.**

in polar seas rather than in temperate waters. Four other orders have been defined. *See* CRINOIDEA; ECHINODERMATA. [H.B.F.]

Artificial intelligence The subfield of computer science concerned with understanding the nature of intelligent action and constructing computer systems capable of such action. It embodies the dual motives of furthering basic scientific understanding and making computers more sophisticated in the service of humanity.

Many activities involve intelligent action—problem solving, perception, learning, symbolic activity, creativity, language, and so forth—and therein lies an immense diversity of phenomena. Scientific concern for these phenomena is shared by many fields, for example, psychology, linguistics, and philosophy of mind, in addition to artificial intelligence. This implies no contradiction, the phenomena being rich enough to support many starting points and diverse perspectives. The starting point for artificial intelligence is the capability of the computer to manipulate symbolic expressions that can represent all manner of things, including the programs of action of the computer itself.

The approach of artificial intelligence is largely experimental, although it contains some general principles and small patches of mathematical theory. The unit of experimental investigation is the computer program. New programs are created as probes to explore and test new ideas about how intelligent action might be attained.

The foundations of artificial intelligence are divided into representation, problem-solving methods, architecture, and knowledge. The four together are necessary ingredients of any intelligent agent. To work on a task, a computer must have an internal representation in its memory, for example, the symbolic description of a room for a moving robot, or a set of algebraic expressions for a program integrating mathematical functions. The representation also includes all the basic programs for testing and measuring the structure, plus all the programs for transforming the structure into another one in ways appropriate to the task. The representation used for a task can make an immense amount of difference, turning a problem from impossible to trivial.

Given the representation of a task, a method must be adopted that has some chance of accomplishing the task. Artificial intelligence has gradually built up a stock of relevant problem-solving methods (the so-called weak methods) that apply extremely generally: generate-and-test (a sequence of candidates is generated, each being tested for solution-hood); heuristic search (sequences of operations are tried to construct a path from an initial situation to the desired one); hill climbing (a measure of progress is used to guide each step); means-ends analysis (the difference between the desired situation and the present one is used to select the next step); operator subgoaling (the inability to take the desired next step leads to a subgoal of making the step feasible); planning by abstraction (the task is simplified, solved, and the solution used as a guide); and matching (the present situation is represented as a schema to be mapped into the desired situation by putting the two in correspondence). An important feature of all the weak methods is that they involve the process of search.

A general intelligent agent has multiple means for representing tasks and dealing with them. Also required is an operating frame within which to select and carry out these activities. Often called the executive or control structure, it is best viewed as a total architecture (as in computer architecture), that is, a machine that provides data structures, operations on those data structures, memory for holding data structures, accessing operations for retrieving data structures from memory, a programming language for expressing integrated patterns of conditional operations, and an interpreter for carrying out programs. Any digital computer provides an architecture, as does any programming language. Architectures are not all equivalent, and the scientific question is what architecture is appropriate for a general intelligent agent.

In artificial intelligence, this question has taken the form of determining what language is good for programming artificial intelligence systems. Although the question is seemingly about research tools, in reality its investigation has been a search for the properties that make intelligence possible. The main development has been that of list-processing languages, which embody a general homogeneous and flexible notion of symbols and symbolic structure. LISP, one of the early list-processing languages, has evolved until it functions as the common language for artificial intelligence programs.

The basic paradigm of intelligent action is that of search through a space (called the problem space) for a goal situation. The search is combinatorial, each step offering several possibilities, leading to a cascading of possibilities in a branching tree. What keeps the search under control is knowledge, which tells how to choose or narrow the options at each step. Thus the fourth fundamental ingredient is how to represent knowledge in the memory of the system so it can be brought to bear on the search when relevant.

A general intelligent system will have immense amounts of knowledge. This implies a major problem, that of discovering the relevant knowledge as the solution attempt progresses. This is the second problem of search that inevitably attends an intelligent system. Unlike the combinatorial explosion characteristic of the problem space search, this one involves a fixed, though large, data base, whose structure can be carefully tailored by the architecture to make the search efficient. This problem of encoding and access constitutes the final ingredient of an intelligent system.

Three examples will provide some flavor of research in artificial intelligence: (1) All the ingredients of an artificial intelligence program can be seen in game programs. A chess program must have a representation of the current chess position. Besides the bits that describe the position, there must be procedures for analyzing the position and for making the moves to create new legal positions. The basic method used for chess is to generate the moves and test their worth. The test is a form of heuristic search that explores the consequences of a candidate move many potential moves into the future. This search is combinatorial and much too big to accomplish without the aid of knowledge. The main technique for bringing knowledge to bear is the evaluation function, which can be applied to a position to estimate directly its chance of being on a winning path. Many features of the position are calculated, each representing a bit of chess knowledge (for instance, queens are better than pawns). Such value functions are heuristic, only approximating a correct analysis. Evaluation turns the search into hill climbing. Not all the knowledge stems from the evaluation functions; many subtleties of the search allow great improvement. (2) Perception is the formation, from a sensory signal, of an internal representation suitable for intelligent processing. Though there are many types of sensory signals, computer perception has focused on vision and speech. Perception might seem to be distinct from intelligence, since it involves incident time-varying continuous energy distributions prior to interpretation in symbolic terms. However, all the same ingredients occur: representation, search, architecture, and knowledge. Thus, perception by computers is a part of artificial intelligence. (3) Extensive experience enables humans to exhibit expert performance in many tasks, even though no firm scientific or calculational base exists and the knowledge does not exist in any explicit form. So-called expert artificial intelligence programs attempt to exhibit equivalent performance by acquiring and incorporating the same knowledge that the human expert has. An example is MYCIN, developed at Stanford University, which makes judgments on the diagnosis of bacterial infection in patients and proposes antibiotic therapies. [A.N.]

Artificial voice A device, consisting of a small loudspeaker in a baffle, whose shape is proportioned to simulate the acoustical constants of the human head. It is used as a sound source in measuring the characteristics of microphones intended for use in close proximity to the lips of a talker. Since these types of microphone (often referred to as close-talking microphone) are used in the near sound field produced by a person's mouth, it is necessary when measuring their performance to utilize a sound source which approximates the characteristics of the human sound generator. *See* MICROPHONE.

Ideally, an artificial voice should be capable of reproducing speech signals or frequencies without distortion over the range of intensities produced by the human voice. Also, the distribution of the sound field about the source should be the same as that of the human source. Finally, when a microphone is placed in the near sound field of the artificial voice, reaction on the output of the voice and the distortion of the field should be the same as that produced by the same microphone placed in front of a talker's lips. Two types of artificial voice appear to have gained general acceptance: a hemispherical artificial voice and a voice simulator.

The hemispherical artificial voice consists of a loudspeaker driver unit acoustically coupled to a hemispherical baffle. The throat of the artificial voice is acoustically damped to provide a smooth response in the frequency pass band of interest. The voice simulator, an electronic dummy, is a complete replica of the head and torso of the average United States astronaut. The head consists of a rigid skull of polyester-impregnated fiber glass covered with Plastisol which simulates the acoustical and mechanical properties of real flesh. Besides the artificial voice, the head contains two artificial ears. The device is mounted on a base containing amplifiers and power supplies. [M.C.]

Artificially layered structures Manufactured, reproducibly layered structures with layer thicknesses approaching interatomic distances. Modern thin-film techniques are at a stage at which it is possible to fabricate these structures, also known as artificial crystals or superlattices, opening up the possibility of engineering new desirable properties into materials. In addition, a variety of solid-state physics problems can be studied which are otherwise inaccessible. The various possibilities include: the application of negative pressure, that is, stretching of the crystalline lattice; the study of dimensional crossover, that is, the transition from a situation in which the layers are isolated and two-dimensional in character to where the layers couple together to form a three-dimensional material; the study of collective behavior, that is, properties which depend on the cooperative behavior of the whole superlattice; and the effect and physics of multiple interfaces and surfaces. For a discussion of semiconductor superlattices *see* SEMICONDUCTOR HETEROSTRUCTURES. *See also* CRYSTAL.

The preparation techniques can be conveniently classified into two groups: evaporation and sputtering. In the evaporation system, two or more particle sources (thermal or electron beam gun) are aimed at a heated substrate where the artificially layered structure is grown. The sputtering method relies on bombarding targets of the proper materials with an inert gas, such as argon, thus producing the beams of the various elements. *See* MOLECULAR BEAMS; SPUTTERING.

Once the artificially layered structure is prepared, it is necessary to characterize whether the layer structure is stable at the growth temperature. This is of considerable importance, since the interdiffusion of the constituents in many cases eliminates the layered growth. A direct image of the layers in gallium arsenide–aluminum arsenide superlattices, using electron micrography, is shown in the illustration. One of the most successful methods of characterizing layered growth has been x-ray diffraction. *See* ELECTRON MICROSCOPE; X-RAY DIFFRACTION.

Artificially layered structures are especially useful for the construction of mirrors for soft x-rays since there are no suitable, naturally occurring crystals for this purpose. Superlattices with zero temperature coefficient of resistivity are useful as resistor material, and high-critical-field-magnet tapes using superconducting-insulator superlattices have been proposed. *See* ELECTRICAL INSULATION; NEUTRON OPTICS; X-RAYS. [I.K.S.]

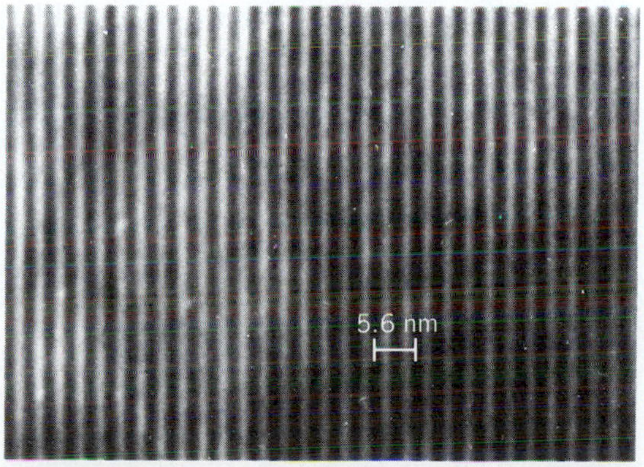

Transmission electron micrograph of ordered superlattice consisting at alternating layers of gallium arsenide (GaAs) and aluminum arsenide (AlAs), each four atomic layers thick, deposited by molecular beam epitaxy. (*From P. M. Petroff et al., Crystal growth kinetics in* $(GaAs)_n$*-*$(AlAs)_m$ *superlattices deposited by molecular beam epitaxy, J. Cryst. Growth, 44:5–13, 1978*)

Artiodactyla An order of mammals which were derived from late Paleocene condylarths and which gradually became one of the most important orders of terrestrial herbivores. By the later Tertiary artiodactyls had forced the previously dominant perissodactyls into a role of secondary importance as herbivores. Living representatives of the artiodactyls include such well-known animals as pigs, peccaries, hippopotamuses, camels, llamas, deer and allies, giraffes, okapi, pronghorn "antelope," bison, buffaloes, antelopes, gazelles, cattle, goats, sheep, and many others. The order also includes numerous extinct species. *See* PERISSODACTYLA.

The ordinal name, Artiodactyla, refers to the fact that most modern and fossil species of the order have an even number of toes. The most characteristic feature of all artiodactyls is that the main axes of the limbs pass between the third and fourth toes, which are strong and usually equal in size (Fig. 1). The first of the five toes is completely reduced, and the second and fifth (the outer toes) when present (lost in advanced forms) are smaller than the third and fourth toes. Probably the most important structural innovation which was responsible for the successful radiation of the artiodactyls is the diagnostic astragalus, the ankle bone. The surface of the astragalus on which the tibia articulates is deeply grooved, restricting lateral motion of the hindfoot, whereas the pulley surface below the astragalus which articulates with the navicular and cuboid (these two bones fuse in several advanced forms), allow swift, free anteroposterior motion with an ability for springing off the ground.

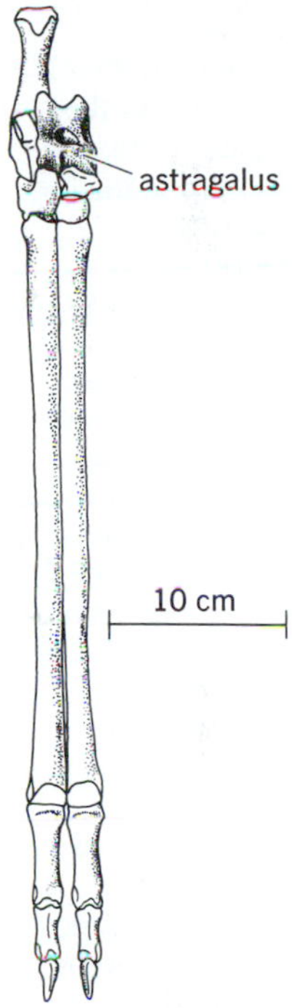

Fig. 1. Right hindfoot of early Miocene camelid, *Oxydactylus longipes*, showing astragalus with its double pulleylike joints. (*After O. A. Peterson, 1904*)

In the living ruminants the stomach is subdivided into three or four chambers, whereas in the more primitive pigs and hippopotamuses the stomach is less advanced, that is, not specialized for rumination. Integumentary glands are common in artiodactyls. Many of these glands are odoriferous and are found on the abdominal and inguinal regions, the back, feet, and tail, and in front of the eyes. As expected, with such variety of scent glands used for marking, artiodactyls have a very keen sense of smell. Thus a great part of artiodactyl biology, from mating and territory marking to predator detection, depends on scent recognition. The essentially terrestrial artiodactyls (no known species is completely arboreal or aquatic) are, as a rule, gregarious. *See* SCENT GLAND; TERRITORIALITY.

A classification scheme for the Artiodactyla is listed below.

Order Artiodactyla
- Suborder Palaeodonta
  - Superfamily Dichobunoidea
    - Family Diacodectidae
    - Family Dichobunidae
    - Family Homacodontidae
    - Family Leptochoeridae
    - Family Achaenodontidae
  - Superfamily Entelodontoidae
    - Family Entelodontidae
    - Family Choeropotamidae
    - Family Cebochoeridae
  - Superfamily Suoidae
    - Family Suidae
    - Family Tayassuidae
  - Superfamily Anthracotherioidea
    - Family Anthracotheriidae
    - Family Hippopotamidae
- Suborder Ruminantia
  - Infraorder Tylopoda
    - Superfamily Merycoidodontoidea
      - Family Merycoidodontidae
      - Family Agriochoeridae
    - Superfamily Cainotheriodea
      - Family Cainotheriidae
    - Superfamily Anaplotherioidea
      - Family Anaplotheriidae
      - Family Xiphodontidae
    - Superfamily Amphimerycoidea
      - Family Amphimerycidae
    - Superfamily Cameloidea
      - Family Oromericidae
      - Family Camelidae
  - Infraorder Pecora
    - Superfamily Traguloidea
      - Family Gelocidae
      - Family Protoceratidae
      - Family Hypertragulidae
      - Family Tragulidae
    - Superfamily Cervoidea
      - Family Palaeomerycidae
      - Family Cervidae
      - Family Giraffidae
    - Superfamily Bovoidea
      - Family Antilocapridae
      - Family Bovidae

Palaeodonta. Of the archaic palaeodonts the dichobunoids are the primitive Eocene and Oligocene families, composed of small- to medium-sized forms with tritubercular to quadritubercular bunodont upper teeth. The entelodontoids, particularly the Oligocene-Miocene entelodonts, are remarkable for their huge skulls, both absolutely and in proportion to the body, enlarged incisors, and large flanges on the zygoma and mandible. These "giant pigs," although not true swine, probably paralleled the pigs in their mode of life.

Of the suoids several forms of pigs and peccaries survive. The exclusively New World peccaries can be traced back to the earliest Oligocene, whereas the pigs, an exclusively Old World assemblage, have a fossil record going back to middle Oligocene times. Both pigs and peccaries are gregarious and have an omnivorous diet, which includes anything from grubs and snails to roots, rats, and snakes. *See* PIG.

Of the anthracotherioids the ubiquitous, long-snouted anthracotheres were once abundant from the middle Eocene into the Pleistocene. The medium- to large-sized anthracotheres ranged throughout Europe, Asia, and Africa, and a few entered North America. Judged by their probable descendants, the living hippopotamuses, anthracotheres also were probably amphibious. Hippopotamuses are now restricted to African rivers and lakes, but during the Pleistocene they were widespread in the Old World tropics. *See* HIPPOPOTAMUS.

Tylopod ruminants. Of the ruminant tylopods the merycoidodontoids (oreodonts and allies) were an exceptionally successful North American group. After oreodonts appeared in the early Oligocene, about 35,000,000 years ago, they became exceptionally abundant, probably outnumbering all other mammals of equivalent size throughout the Oligocene and Miocene. Oreodonts survived well into the Pliocene.

The anaplotheres, xiphodontids, and amphimericids are a late Eocene to middle Oligocene assemblage of primitive ruminants confined to Europe. The European cainotheres, rabbit-sized ruminants, were convergently similar in many features of the skull to the hares and rabbits (Lagomorpha). *See* LAGOMORPHA.

Camels and llamas are the only survivors of these once abundant tylopods. The Camelidae are known as early as the Eocene. The more modern camels, appearing first in the Pliocene, lost the characteristic ungulate hooves, evolving expanded, distal phalanges embedded in the cutaneous pad which formed the sole of the foot. The late Eocene–early Oligocene oromerycids are very camellike, but more primitive in the construction of their limbs and teeth. *See* ALPACA; CAMEL; LLAMA.

Pecoran ruminants. The Pecora, represented by such living forms as deer, gazelles, and sheep, are very successful ruminants with a reduced ulna and usually with antlers (deer), horns (bovids), or deciduous horns (antilocaprids).

Of the traguloids, the most primitive pecorans with large canines, only the chevrotains (Tragulidae) survive. Chevrotain-like hypertragulids were common during the late Eocene in central Asia and throughout the Oligocene and Miocene in North America. The Protoceratidae are spectacular derivatives of the hypertragulids. Some protoceratids have developed a grotesque forked horn on the tip of the snout, as well as an elongated snout. *See* CHEVROTAIN.

The Cervidae (deer) and Giraffidae (giraffe, okapi, and relatives) evolved from a palaeomerycid ancestry sometime during early Miocene or possibly late Oligocene. The varied species of deer range in size from the rabbit-sized Chinese water deer to the giant extinct Irish elk. Unlike bovid horns, deer antlers are supported by permanent bony pedicles and are normally shed and regrown annually. Giraffids are browsers, paralleling the deer in the Old World tropics, but never as successfully. The living giraffe and okapi have large eyes and ears, long thin lips, long extensible tongues, and bony prominences ("horns"), covered by unmodified skin, on the skull (Fig. 2). The *Sivatherium* giraffids of the Pleistocene in Africa and southern Eurasia grew to be giants, and the males developed a variety of large, apparently permanent giraffelike "horns." *See* DEER; GIRAFFE.

Of the bovids the pronghorns (Antilocapridae) appeared in the Miocene. The Antilocapridae is now restricted to North America; only one species, the pronghorn "antelope," survives. Several fossil antilocaprids had strongly branching horns, which were probably permanent, and shed only the horny cover, as does the modern prongbuck. The bovids are the most abundant of the late Tertiary and Recent artiodactyls, and also of ungulates in general. Because bovids are mainly grassland grazers of warm climates, the once abundant species and genera of Europe and Siberia were forced to retreat southward or became extinct as a result of Pleistocene glaciations. A few species, such as the musk-ox, bison, mountain sheep, and

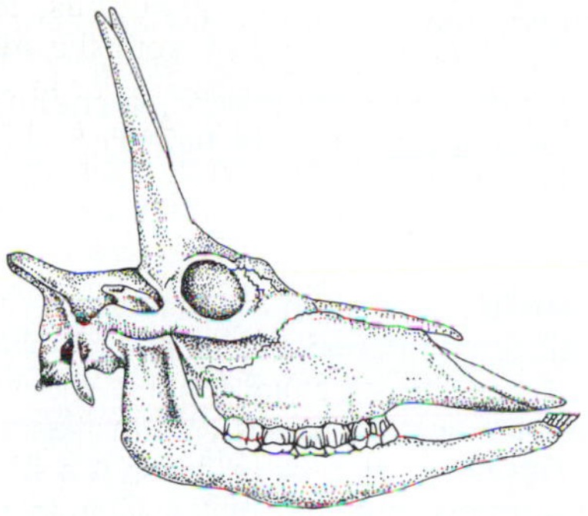

Fig. 2. Skull of *Samotherium*, a Pliocene giraffe, male. *(After Bohlin)*

mountain "goat," adapted to the cold northern climates of the Pleistocene and migrated through the Bering land bridge into the New World. *See* ANTELOPE; BISON; BUFFALO; GOAT; MOOSE; MUSK-OX; PRONGHORN; REINDEER; SHEEP. [F.S.S.]

Asbestos General name for the useful, fibrous varieties of six rock-forming minerals. The value of asbestos derives from its strength as a reinforcing agent, its relative chemical inertness, and the incombustible nature of certain asbestos-containing products. Originally the term applied only to amphibole varieties, but it now applies to all types used in commerce. Over 95% of production is fibrous serpentine.

The fibrous amphiboles that are termed asbestos are generally more heat- and acid-resistant than serpentine asbestos, though not as strong, alkali-resistant, or available. These fibrous amphiboles include amosite, a trade name for a complex iron-magnesium silicate (mostly cummingtonite-grunerite), and crocidolite, a sodium-iron hydrous silicate variety of riebeckite. Of minor importance are tremolite asbestos, anthophyllite asbestos, and actinolite asbestos, a variety of tremolite containing 15–20% ferrous iron. Fibrous serpentine (chrysotile) is usually found associated with altered massive serpentine bodies in olivine-rich ultrabasic rocks, such as dunite and peridotite. Chrysotile is sometimes found in serpentinized dolomitic limestone in association with diabase sills. The more frequent occurrences in either type of host are of the cross-fiber orientation, in which fibers lie perpendicular to the vein walls.

Long fibers are spun into yarn and woven into cloth, either with or without auxiliary fiber material, such as cotton, glass wool, or copper wire. Brake linings, heavy packings and gaskets, electrical insulating materials, and protective clothing require the best fibers. Medium and short fibers are used to make asbestos shingles, sheet siding, pipe, floor tile, less critical packings and gaskets, paper, binders for other heat insulators, and fillers for asphalts, plastics, paints, and greases.

A significant area of asbestos research for the last few years has been concerned with the controversial health aspects of the fibers. Prolonged exposure to high concentrations of airborne asbestos fibers has been demonstrated to be injurious to human health. However, there is disagreement as to the possible hazard from ingestion of asbestos particles. The Environmental Protection Agency (EPA), Occupational Safety and Health Administration (OSHA), and Mine Safety and Health Administration (MSHA) are federal agencies which conduct the enforcement programs in the United States. In their published regulations, the term "asbestos" includes only the fibrous form of one serpentine group and five amphibole group minerals. In late 1976, the U.S. Bureau of Mines established a Particulate Mineralogy Unit at College Park, Maryland, to provide technical information on particulate mineralogy, especially in reference to asbestos; to develop a solid scientific basis for research into particle-related pollution problems; and to facilitate decision making by regulatory bodies. [R.A.Cl.]

Ascariasis The infection of humans by *Ascaris lumbricoides*, a nematode inhabiting the small intestine. The worms may vary in length from about 4 in. (10 cm) in severe infections to over 12 in. (30 cm) in light ones. The severity of the infection may be related to the presence of a larger number of the smaller worms. Large females lay up to 200,000 eggs daily; thus it is easy to diagnose the condition by routine microscopic examination of feces.

Ascariasis may produce toxic, allergic, and mechanical symptoms or damage. The most serious forms are intestinal occlusion or perforation with resulting peritonitis and invasion of the common bile duct, pancreatic duct, or appendix. Death is often the outcome. Chemotherapy is based on oral administration of piperazine compounds. The use of Hetrazan is firmly contraindicated because of possible mechanical complications. *See* FILARIASIS; MEDICAL PARASITOLOGY; NEMATA; PINWORM INFECTION. [J.F.M.]

Ascaridida An order of nematodes in which the oral opening is generally surrounded by three or six labia; in some taxa labia are absent, but the cephalic sensilla are always evident. Usually there are eight cephalic or labial sensilla; the submedians may be fused and then only four sensilla are seen. The stoma varies from being completely reduced to spacious or globose. The esophagus varies from club shaped to nearly cylindrical, never rhabditoid. There may be posterior esophageal or anterior intestinal ceca. The collecting tubules of the excretory system may extend posteriorly and anteriorly. Males generally have two spicules; however, in some taxa there may be none or only one. The gubernaculum may also be present or absent. Though females generally have two ovaries, multiple ovaries do occur. The number of uteri is also variable: two, three, four, or six. Phasmids are sometimes large and pocketlike. Reportedly, the larvae lack a stomatal hook or barb.

The order probably comprises seven superfamilies: Ascaridoidea, Seuratoidea, Camallanoidea, Dracunculoidea, Subuluroidea, Dioctophymatoidea, and Muspiceoidea (*incertae sedis*).

The Ascaridoidea include about 65 genera which comprise large parasitic roundworms whose adult stages usually occur in the stomach or small intestine of terrestrial and aquatic mammals, birds, reptiles, and fishes; the parasitic larval stages of many species occur, either temporarily or indefinitely, in other parts of the host's body.

Many species have a direct life cycle. Others, mainly species with marine mammals, birds, and fishes as definitive hosts, require an intermediate host, such as a fish, amphibian, insect, crustacean, or small mammal. Infestation is typically characterized by pulmonary damage and distress initially, and digestive disturbances later. Damage may also occur during larval migration to other parts of the body, including the liver and brain.

Dracunculoidea, the superfamily of parasitic nematodes, comprise obligate tissue parasites of fishes, reptiles, and mammals. All known species require an intermediate host in order to complete their life cycle, and that host is always a water flea (*Cyclops*). The most widely known example is *Dracunculus medinensis*, the guinea worm.

Ingestion of water containing infective *Cyclops* is the only known source of infection. The encysted nematode larvae are released from *Cyclops* by the digestive juices of the duodenum. Then the larvae burrow through the intestinal wall, and upon reaching the loose connective tissue, they develop to adulthood in 8 months to 1 year.

The gravid females, 28–48 in. (70–120 cm) long, migrate from the site of development to the surface of the skin, and a papule is formed, then a blister, usually on the lower extremities. When the blister comes in contact with fresh water, the uterus bursts through the anterior part of the nematode's body, and the worm also bursts, releasing cloudlike swarms of motile larvae. These larvae are then filtered from the water by *Cyclops* and subsequently ingested.

The formation of the blister and subsequent rupturing of the female produce a profound allergic reaction. This reaction results from the release of large amounts of toxic by-products from the worm. Upon discharge of larvae in fresh water, much of the allergic reaction abates. The reactions and systemic prodromes include erythema, urticarial rash, pruritus, vomiting, diarrhea, and giddiness. Septicemia, suppurating cysts, and chronic abscesses are not uncommonly associated with these infections. The worms can be removed surgically, or in the native manner of winding upon a stick. Chemotherapy is also available.

Control in endemic areas includes keeping infected persons from wading or bathing in water used for drinking purposes, and the education to avoid drinking suspect water. *See* NEMATA. [A.R.M.]

Ascaridina A major group of the Nemata, also known as Ascaridata. They are one of two suborders of Ascaridida. In one modern system of classification, it has essentially the composition and characteristics of Ascaridoidea. In another, it also contains certain groups (Heterakidae, Cosmocercidae, and Kathlaniidae), which the first system assigns to the suborder Oxyurina. *See* ASCARIDIDA; ASCARIDOIDEA; NEMATA; OXYURINA. [J.T.L.]

Ascidiacea A class of Tunicata which occurs as solitary zooids or, by a process of asexual budding, develops into colonies. Zooids vary in length from about 0.1 to 10 in. (a few millimeters to 25 cm). Individuals or colonies are invested by a protective covering, the tunic or test, made of polysaccharide material structurally close to cellulose. Beneath the test is the body wall or mantle. Each zooid has two apertures: inhalant (oral) and exhalant (atrial). Water currents, created by cilia on the margins of stigmata in the pharyngeal wall, draw water into the branchial sac, where it is filtered and passed out through the exhalant aperture. Digestive enzymes are secreted into the stomach, and a pyloric gland, of unknown function, enters at the junction of stomach and intestine. Gonads are hermaphroditic, and may be situated in the loop of the intestine or in the mantle wall.

Three orders of ascidian are recognized. Ascidians occur throughout the seas of the world and at all depths, including the abyssal region. Most species feed on minute particulate matter, but a few are carnivores and engulf small zooplankters. *See* TUNICATA. [I.Go.]

Ascomycotina A subdivision of the Eumycota, known as sac fungi. These fungi are better known under the older name Ascomycetes, which persists as their common name. The distinguishing feature of these fungi is the ascus, a specialized sporangium in which reduction division occurs and in which ascospores are delimited by free-cell formation.

The vegetative body may be unicellular (yeasts) or composed of filaments with cross walls. Nonsexual reproduction is by budding, fission, or conidia (unicellular or multicellular spores borne exposed on conidiophores). In yeastlike forms, sexual reproduction involves fusion of vegetative cells or ascospores which function as gametes. In higher forms, an antheridium, a spermatium, or a conidium fuses with a specialized female sexual organ (ascogonium), and a binucleate stage intervenes between pairing of nuclei in the ascogonium and nuclear fusion in the ascus. Lower Ascomycotina lack fruiting bodies (ascocarps); higher Ascomycetes form asci in ascocarps. *See* YEAST.

The Ascomycotina are important as plant and human pathogens, as food (truffles, morels), in baking, brewing, and wine making (yeasts), in cheese making, in enzyme production, and as sources of such drugs as penicillin and ergot. *See* ASPERGILLOSIS; PENICILLIN. [R.M.P.]

Ascorbic acid A vitamin also known as vitamin C. It is a white crystalline compound that is highly soluble in water. The compound has the structural formula shown below. The stability of ascorbic acid decreases with increases in temperature and pH. This destruction by oxidation is a serious problem in that a considerable quantity of the vitamin C content of foods is lost during processing, storage, and preparation.

While vitamin C is widespread in plant materials, it is found sparingly in animal tissues. Of all the animals studied, only the guinea pig, the red vented bulbul bird, the fruit-eating bat, and

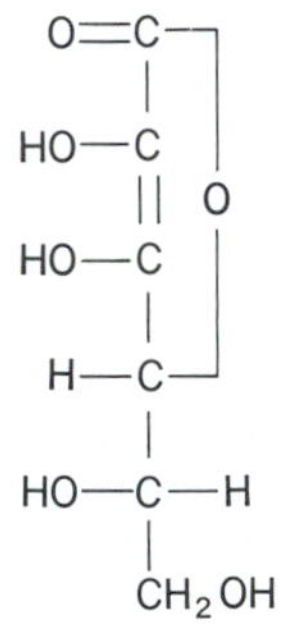

the primates, including humans, require a dietary source of vitamin C. The other species studied are capable of synthesizing the vitamin in such tissues as liver and kidneys.

Vitamin C–deficient animals suffer from defects in their mesenchymal tissues. Their ability to manufacture collagen, dentine, and osteoid, the intercellular cement substances, is impaired. People with scurvy lose weight and are easily fatigued. Their bones are fragile, and their joints sore and swollen. Their gums are swollen and bloody, and in advanced stages, their teeth fall out. They also develop internal and subcutaneous hemorrhages. *See* SCURVY.

There has been great difficulty in establishing the human requirements for vitamin C. The recommended dietary allowances of the Food and Nutrition Board of the National Research Council are 30 mg per day for 1- to 3-month-old infants, 80 mg per day for growing boys and girls, and 100 mg per day for pregnant and lactating women. Many nutritionists believe that the human intake of ascorbic acid should be many times more than that intake level which produces deficiency symptoms. *See* VITAMIN. [S.N.G.]

Ascothoracica An order of the subclass Cirripedia. Members are ecto- or endoparasites of coelenterates and echinoderms. The body is enclosed in a voluminous saclike mantle up to 0.8 in. (20 mm) long, whereas the body is only about one-fourth of this length (see illustration). Ascothoracica are not attached permanently to the host by the antennular region, but the antennules may be modified as a clasping organ. Up to six pairs of thoracic appendages are present, reduced in number and development in the endoparasites. The mouth parts are modified for piercing and sucking. Unlike all other Cirripedia, Ascothoracica retain a more or less segmented abdomen in the adult. There are no cement glands present.

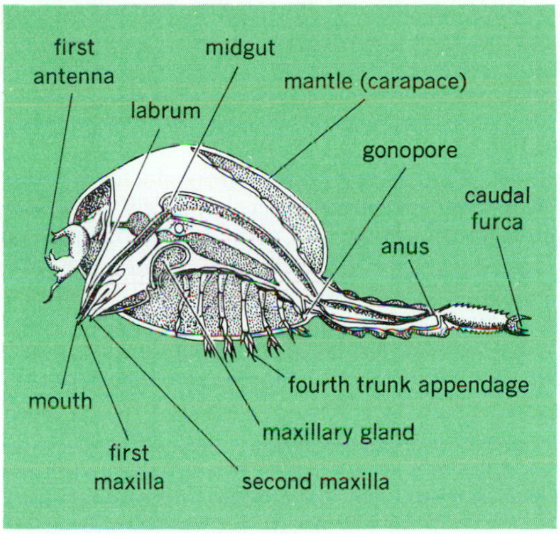

***Ascothorax ophioctenis*, a parasite in the bursae of brittle stars. (After R. D. Barnes, *Invertebrate Zoology*, 2d ed., Saunders, 1968)**

Diverticula of the alimentary canal extend into the mantle. *See* CIRRIPEDIA. [H.G.St.]

Asellota A suborder of the Isopoda containing aquatic species of considerable morphologic and ecologic diversity. These crustaceans are usually divided into three major groups, the Paraselloidea, Aselloidea, and Stenetrioidea. Asellotes such as *Asellus* (see illustration) are found in freshwater streams; *Caecidotea*, in subterranean water; *Caeioniropsis*, in the interstices of marine sands as commensals with other isopods such as *Caecijaera* and *Jaera* and in shallow waters of the seas; and *Macrostylus*, in the greatest depth of the seas. *See* CRUSTACEA; ISOPODA. [R.J.Me.]

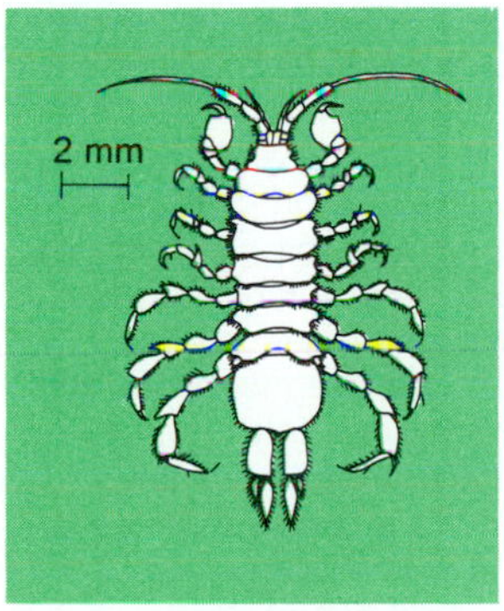

***Asellus communis*. (After H. S. Pratt, *Manual of the Common Invertebrate Animals*, rev. ed., McGraw-Hill, 1951)**

Ash A genus, *Fraxinus*, of deciduous trees of the olive family Oleaceae, order Scrophulariales, which have opposite, pinnate leaflets, except in one species, *F. anomala*, which has only a single leaflet. There are about 65 species in the Northern Hemisphere. This tree occurs in America south to Mexico, in Asia south to Java, and in Europe. *See* SCROPHULARIALES.

The white ash (*F. americana*), of the eastern United States, has stalked leaflets, rusty-colored winter buds, and an erect trunk that is valuable for lumber. The wood is light, strong, but flexible, and is used for oars, baseball bats, furniture, motor vehicle parts, boxes, baskets, and crates. The black ash (*F. nigra*) grows in wet soils in the northeastern United States and Canada and has sessile leaflets and friable outer bark. The wood of black ash is used for the same purposes as that of white ash. The red ash (*F. pennsylvanica*), also of the eastern United States and adjacent Canada, has pubescent (hairy) twigs and leafstalks. The uses of the wood of this species are also similar to those of white ash. Some species of ash are ornamental trees, such as the flowering ash (*F. ornus*) with gray winter buds and white flowers, and the European ash (*F. excelsior*) with black buds and sessile leaflets. *See* FOREST AND FORESTRY; TREE. [A.H.G./K.P.D.]

Asia The largest of the world's continents. With its peninsular extension, commonly called the continent of Europe, it is the major portion of the broad east-west extent of the Northern Hemisphere land masses. In many ways Asia is more a cultural concept than a physical entity. There is no logical physical separation between Asia and Europe, and even Africa is separated from Asia merely by the width of the Suez Canal. For convenience, however, the Eurasian land mass is considered to be divided by the Ural Mountains into Europe in the west and Asia in the east. Thus restricted, Asia has an area of about 17,700,000 mi² (45,800,000 km²), about one-third of the land area of the Earth. In the north, Siberia reaches past the 80th latitude. Southward, India and Sri Lanka (Ceylon) reach nearer than 10°N of the Equator, while the Indonesian islands extend more than 10°S of the Equator. The continental heart of Asia is more than 2000 mi (3200 km) from the nearest ocean. *See* CONTINENT; EUROPE.

Topography. In the topographic framework of Asia, the great mountain systems are the most impressive features. From the central knot of the mighty Pamirs and Kopet Dagh in the heart of the continent originate chains radiating in several directions. In the Peter the First Range there are such heights as Communism Peak, 24,584 ft (7493 m), and Lenin Peak, 23,377 ft (7125 m), above sea level. Running westward through Afghanistan is the Hindu Kush, reaching elevations over 20,000 ft (6100 m). The mountain trendline continues, after a jog northwestward, in the Elburz of northern Iran and thence in the Armenian highlands and the Caucasus, each with elevations reaching 18,000 ft (5500 m), decreasing thereafter to the Pontus and Taurus ranges of northern and southern Turkey. In western and southern Iran are the massive Zagros and Makran ranges.

Southeastward from the Pamir knot run the three most imposing mountain chains on Earth: the Karakorum, which continues the line of the Hindu Kush eastward in an arc convex to the north; the Himalaya in an arc convex to the south; and the shorter Trans-Himalaya, or Nyen-chen Tangla, north of the Himalaya, with higher average elevations but peaks of lesser height. In all of these, the average elevations exceed 4 mi (6400 m), with several scores of peaks reaching a height in excess of 25,000 ft (7600 m) above sea level. Everest, 29,141 ft (8882 m), and Kinchinjunga, 28,146 ft (8579 m), lie in the Himalaya, while the peak designated as K2, 28,250 ft (8611 m), rises in the Karakorum.

In eastern Tibet the Himalaya and Nyen-chen Tangla bend sharply toward the south, and the former is cut through by the gorge of the Brahmaputra River. From the bend zone, great ridges divided by deep gorges run south to form the Burma-China frontiers and the mountain backbones of the Malay peninsula and Vietnam. The Nan-ling system of south China diverges eastward to divide the Yang-tzu (Yangtze) from the Hsi (Si) drainage.

From the western Himalaya, the 11,000-ft (3400-m) Sulaiman Range runs south and, together with the Kirthar Range, divides West Pakistan from Afghanistan.

Beginning at heights over 20,000 ft (6100 m) and branching off from the Karakorum south of Kashgar, the Kuen-lun Mountains run eastward across western China. Genetically they form the longest mountain system of China. With their eastward extensions in the 12,000 ft (3700 m) Ch'in-ling and the lesser Ta-pieh mountains and Huai-yang hills, they reach almost to the Pacific. Together with the northeastward arc of the Altyn Tagh and the Nan Shan branching from it, the Kuen-lun forms the northern wall of the Tibetan plateau. Near the eastern end of the Kuen-lun proper lie the Amne Machin Mountains, with peaks up to 25,000 ft (7600 m) in elevation.

Northeastward of the Pamir knot runs the east-west oriented Tien Shan, over 1000 mi (1600 km) long and maintaining heights of 18,000–20,000 ft (5500–6100 m) over much of its length. Roughly parallel and trending east and west is a series of great ranges to its north, with mutual connections in the west. These include the Altai-Sayan, the Tannu Ola, and the Kentei, which form natural boundaries for Outer Mongolia. They continue the systems of young mountains crossing central Asia; farther northeast, they extend further in the Stanovoi Mountains of Eastern Siberia.

The Asian plateaus are in various stages of erosion and thus present a great variety of landscapes. The Tibetan plateau is a prime example. The western half, because of little rainfall, exhibits a rolling topography with relatively slight local relief except where mountain chains cross it; it is a land of internal drainage basins. Average elevations are over 16,000 ft (4900 m). The eastern half is humid or subhumid and is cut by numerous rivers, producing deep canyons and great ridges. In contrast to this is the Mongolian plateau. This plateau consists mostly of vast, rather level plains 3000–5000 ft (900–1500 m) high, surmounted in places by mountains, and containing broad, shallow basins divided by land swells of low elevation.

Other major topographic units of Asia are blocs of hill lands. Most of southern China and much of southeastern Asia comprise hills which may be roughly defined as slope lands with local relief under 1000–1500 ft (300–450 m) although in absolute elevation they may rise many thousands of feet above sea level. Hilly lands are found to predominate in the northern part of the Indian peninsula and along both flanks of the Indian plateau, where they are called ghats. In southern India are the Nilgiri and Cardomom hills, rising to mountainous elevations of 8000 ft (2400 m). Many parts of different plateaus have hilly regions where erosion has produced uneven local relief, as in the Shan or North Vietnam plateau. Hills are prominent features of southwestern Asia, including eastern Mediterranean regions, such as Israel, Syria, and Lebanon.

The most significant topographic units of Asia are the great alluvial plains and river deltas. The gross drainage pattern of Asia is radial; the rivers flow from the highlands in the heart of the continent and run outward in all directions. Only in the south, east, and north sectors of the continent do the rivers reach the sea. Flowing into the peripheral seas of the Pacific are such mighty rivers as the Mekong, the Hsi, the Yang-tzu, the Huai, the Yellow, and the Amur, each building large, heavily populated plains and, with the exception of the Amur, densely settled deltas. The Yellow Plain (North China Plain), with some 125,000 mi^2 (324,000 km^2) of area, and the Yangtzu Plain, with about 75,000 mi^2 (194,000 km^2), are among the most extensive alluvial plains of the Earth. In the shallow South China, East China, and Yellow seas, the deltas of the first five rivers mentioned above are pushing steadily seaward.

Important sectors of Asia, containing some 200,000,000 people, are completely insular. The most important are the Japanese, Philippine, and Indonesian islands and Taiwan. Almost all of Asia's islands lie in great volcanic arcs bounding large seas off the continent's Pacific coast. At least 160 active volcanoes are found here and in Kamchatka. Few islands lie along the Asiatic coasts of the Indian Ocean, although the Sunda chain of Indonesia has perhaps more of a claim to Indian Ocean frontage than to Pacific frontage. Sri Lanka is the only significant island in the northern part of the Indian Ocean west of Sumatra. In the Persian Gulf off the north coast of Arabia lies the small island Bahrein.

Few islands lie off the alluviated coastlands of northern Siberia. Some moderately large ones are included in the barren and rocky Severnaya Zemlya group, the New Siberian Islands, and Wrangel Island. The Commander Islands and Karaginski Island lie in the Bering Sea only a short distance from the Aleutians.

Climates. Continentality is a strong factor in Asia's climates. Eastward extratropical cyclonic drift, typhoons and westward tropical cyclonic drift, and contrast in rainfall patterns bring regional diversity to the climates of Asia. Describing the climatic patterns very grossly, at least four characteristic types of Asian climatic systems may be distinguished: (1) the monsoonal system of eastern Asia, (2) the monsoonal system of southern Asia, (3) the cyclonic and convectional storm systems of central and northern Asia, and (4) the winter rainfall–summer drought system of southwestern Asia.

Fundamental to the whole problem, however, are the vastness of the unbroken land mass and the long latitudinal stretch from the polar realm to south of the Equator. These are responsible for the great temperature and humidity extremes that occur. The greatest ranges of temperatures in the world have been recorded in interior Asia. Continentality, therefore, is the outstanding feature of climates of interior Asia. In coastal and insular areas of east Asia, however, winds moving over the warm, northward-flowing Japan Current and the western Pacific waters moderate the coastland and island climates. *See* CONTINENTALITY (METEOROLOGY); MARITIME METEOROLOGY; MONSOON METEOROLOGY.

The driest portions of Asia include the vast areas of southern Mongolia, Hsin-chiang, former Soviet Central Asia, and southwestern Asia. Except for small, favored mountain areas, most of this region from the Gobi to the Red Sea gets less than 10 in. (25 cm) of precipitation per year. With the exception of southern Arabia, which is subtropical desert, these are mid-latitude desert and dry steppe regions. Favored with higher rainfall are the Yemen Mountains and the coastal mountains of Turkey, together with Lebanon, Syria, and northern Israel. The highlands of Armenia and the Elburz of Iran are favored also with more abundant rainfall, which may range from 25 to 50 in. (64 to 127 cm) or more per year.

The northeastern Siberian mountains and the Arctic coastal lands also receive meager rainfall, less than 8 in. (20 cm), but are not dry because evaporation is low and the water table is high. Most of Siberia has permafrost below a few feet of surface soil, so that rainwater does not filter far down into the earth. Between the arid belt of central Asia and the northeast Siberian low-precipitation zone, the annual rainfall ranges between 10 and 18 in. (25 and 45 cm).

In eastern Asia the precipitation increases in a southeasterly direction from interior Asia to the coast. The annual maximum seldom exceeds 80 in. (203 cm) in the wetter southeast coastal regions, whereas this drops to less than 30 in. (76 cm) in the North China Plain and less than 15 in. (38 cm) at the Great Wall. In some mountainous parts of Japan and Taiwan, the yearly average may be more than 100 in. (254 cm).

In the Indian subcontinent rainfall is heaviest along the western plateau fringe and in East Bengal, where it may average over 100 in. (254 cm) per year. The interior of the peninsula is relatively dry. Northwestern India and Pakistan share the drought of southwestern Asia. With the exception of the extreme north, Ceylon generally has abundant rainfall.

Southeastern Asia has the heaviest rainfall of the entire Asiatic region. The mainland mountains facing the southwest summer monsoon crossing the Bay of Bengal, and parts of the Vietnamese and Laotian cordilleras facing the humidified northeast winter monsoons of eastern Asia, regularly get average rainfalls of 120–150 in. (305–381 cm) or even more. Equally heavy rainfalls occur in the southwestern half of Sumatra, southwestern Java, the northwestern half of Borneo, and the Pacific fringe of the Philippine Islands. With a few small exceptions, southeastern Asia has no areas that are subject to severe drought.

Vegetation. Asia's vegetation belts and zones follow, in general, the climatic patterns from desert lands through tropical to Arctic margins.

A wide belt of tundra made irregular by topography occupies the entire Arctic lowland of Siberia with widths varying from 250 to 500 mi (400 to 800 km) north and south. It is widest in the extreme northeast and it extends southward and inland with higher elevations. The frozen subsoil permits the growth of little more than mosses, lichens, dwarfed trees, and scrub. *See* PERMAFROST; TUNDRA.

The largest unbroken expanse of forest in the world is the Siberian taiga, a dominantly coniferous forest of larches, spruce, fir, and pines, with such deciduous trees as birch and aspen occurring intermixed with the conifers or taking over as a secondary growth in burnt-over areas. The width of this belt in Siberia is more than 1000 mi (1600 km) and it stretches about 4000 mi (6400 km) from the Sea of Okhotsk to the Urals. *See* TAIGA.

Various admixtures of coniferous and deciduous trees compose the vegetation of mid-latitude mixed forests. In the west Siberian plain there is a narrow zone of mixed taiga and deciduous forests including oaks, maples, ash, and lindens. This zone, with a width of 50–100 mi (80–160 km), lies somewhat south of the parallel of 60°N and fades into the steppelands that form the great spring-wheat region of Siberia. Mixed mid-latitude deciduous and coniferous forest areas of a similar type occupy most of Korea, the northern half of Honshu in Japan, and the hill lands surrounding the Yellow Plain, as well as the Ch'in-ling Mountains. In southern Asia these forests are found chiefly in a narrow belt of mountain land in the outer ranges of the Himalaya. The remaining areas of these mixed forests run from the Elburz Mountains through the Armenian highlands and the Black Sea fringe of Turkey to the Aegean coast, and in southwestern Asia in the Elburz of northern Iran.

From the mixed and deciduous forests of the west Siberian plain southward, an increasingly dry steppeland is encountered. It extends for 400–500 mi (640–800 km) in a belt about 1000 mi (1600 km) long between the Urals and the Altai-Sayan and associated uplands. The northern half of this belt with its higher annual precipitation of 12–16 in. (30–40 cm) is the agricultural heart of the plain. The southern part gradually changes to desert steppe and then to desert along about the 50th parallel. Eastward of Lake Baikal a broadened steppe zone occupies the Trans-Baikal region extending southward to the Gobi Desert of southern Mongolia and eastward to the Great Hsing-an Mountains, where the zone, about 200 mi (320 km) wide, runs southward in Inner Mongolia. The steppe zone in Inner Mongolia widens with the increasing moisture south of the Great Wall to include most of China's loess plateau. Grasses also form the natural vegetation of the Manchurian plain, with tall grass in the eastern portion thinning out to short-grass steppe in the Hsing-an Mountain flanks. The Gobi Desert is flanked by steppelands to its north, east, and south, as well as by mountain steppe zones in the eastern Altai and eastern T'ien Shan.

Mixed evergreen forests appear to be limited mostly to interior southern China and to Japan from the Kwanto Plain southward, South of the Yang-tzu Valley, this forest type extends from the coast at Shanghai to the gorge lands of eastern Tibet. In Asia the characteristic trees of the mixed forest include broad-leafed evergreen trees such as banyans and camphor, and coniferous trees such as pines, cedars, and cypresses, as well as varieties of bamboo.

Tropical and subtropical rainforest is restricted to warm or hot regions of southern and southeastern Asia which get ample rainfall the year round or get so much rain during a large part of the year that a high groundwater table is maintained during the short dry season. The subtropical sectors are found along the southeastern China coast, in Taiwan, and in northern Burma; they merge with the tropical rainforest farther south, where rainfall and temperature increase. *See* RAINFOREST.

Monsoon tropical deciduous forests comprise the tropical parts of Asia which have a moderately high rainfall but a long dry season (usually in the low-sun period or winter). These forests consist mostly of mixed species, but sometimes a single species becomes dominant as a result of selection from frequent burnings.

A large region of savanna grassland surrounds the Thar Desert of northwestern India and occupies most of the Indus Valley, the Punjab, and the Kathiawar peninsula. Much of the drier interior peninsular Deccan of India also has this as a natural vegetation. Other Asian regions with similar cover are found in Yemen and the region in southeastern Arabia from Oman as far westward as the Qatar peninsula; and similar vegetation extends over the Korat plateau of Thailand, lower Thailand west of Bangkok, southern Cambodia, and small areas in interior Borneo and the Philippines. *See* SAVANNA.

Immense areas of central and southwestern Asia have little or no vegetative cover, and bare rock alternates with sand veneering. In places shifting sand dunes are formed. Although the deserts are not necessarily lifeless, the vegetation is so widely spaced that much bare ground is exposed. The tropical desert areas generally receive their meager rainfall in torrential downpours on rare occasions. After such rains numerous herbs may spring to life and flower, while the bunch grass here and there may become green for a short season. [H.J.Wi.]

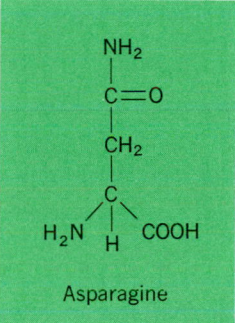

Asparagine An amino acid. Asparagine forms a brown color with ninhydrin. The amide group of asparagine presumably serves as a storage site for nitrogen, especially in plants. Although it is known that asparagine is formed biosynthetically from aspartic acid, it is not certain whether the mechanism involves amide transfer from glutamine, transamination, or ammonia incorporation. Asparagine is deamidated to aspartic acid, which is further catabolized by way of oxaloacetic acid and the Krebs cycle. *See* AMINO ACIDS; ASPARTIC ACID; KREBS CYCLE; TRANSAMINATION. [E.A.Ad.]

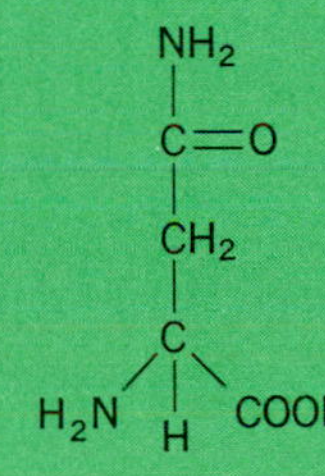

Asparagine

Asparagus A dioecious perennial monocot (*Asparagus officinalis*) of Mediterranean origin belonging to the plant order Liliales. Asparagus is grown for its young shoots or spears, which are canned, frozen, or cooked fresh as a vegetable. These aerial stems arise from rhizomes (underground stems). The rhizomes and the fleshy and fibrous roots constitute the massive underground part of the plant. Blanched or white asparagus is grown by ridging soil over the rows and cutting the spears beneath the soil surface. Chemical weed control is commonly used.

Commercial production is limited to areas where crowns will have a dormant period of 3–5 months each year. Dormancy in the northern states is induced by low temperatures and in

California by withholding irrigation. California, New Jersey, and Washington are important asparagus-producing states. *See* LILIALES. [H.J.C.]

Aspartame A white, crystalline compound, 1-aspartyl-1-phenylalanine methyl ester (AMP). It is slightly soluble in water. Its sweetening properties were discovered accidentally in 1965 when the compound, a dipeptide, was produced as an intermediate in the synthesis of the C-terminal tetrapeptide of gastrin. Aspartame is the L,L-diastereoisomer; the three other possible diastereoisomers are not sweet. The taste of aspartame would not have been predictable based on its component amino acids, aspartic acid and phenylalanine.

The sweetness of aspartame relative to sucrose is a function of the latter's concentration, and is also dependent upon the presence of other flavors and materials. In a number of applications, such as chewing gum and various fruit-flavored products, aspartame favorably extends and enhances the flavor perception, and it shows synergy with other sweeteners. The sweetness perception may also last longer with aspartame than with sucrose or other sweeteners. *See* ASPARTIC ACID; FOOD ENGINEERING; PHENYLALANINE; SUCROSE; TASTE. [D.L.A.]

Aspartic acid An amino acid. The functions are as follows: (1) It is a key intermediate in many transamination reactions and is a product of ammonia incorporation in plants and some microorganisms. (2) It functions as a source (from the amino group) of one of the ring nitrogens of purines and of the amino group of adenylic acid. Aspartic acid also reacts with carbamyl phosphate to form carbamyl aspartate, a precursor of pyrimidines. (3) Aspartic acid can be amidated to form asparagine; it is also the biosynthetic precursor of methionine, threonine, isoleucine, and, in bacteria, of lysine. *See* AMINO ACIDS; ISOLEUCINE; METHIONINE; PYRIMIDINE; THREONINE.

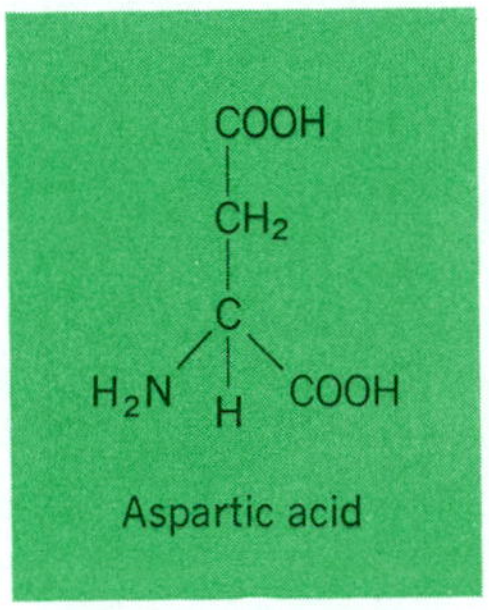

Aspartic acid

Aspartic acid is formed, biosynthetically, by transamination of oxaloacetate or by the addition of ammonia to fumaric acid. The metabolic degradation pathway is by deamination to oxaloacetate, or by decarboxylation. Two decarboxylases are known, one forming α-alanine, the other β-alanine. [E.A.Ad.]

Aspect ratio As originally conceived, the ratio of the span of a wing or airfoil to the chord of a wing, where the span is the maximum cross-stream dimension and the chord is the dimension in the streamwise direction, as illustrated. This definition is unambiguous only in the case of a rectangular wing.

Because early wings were usually nearly rectangular, no confusion resulted from the original definition. Later, when wings were tapered or had complex planforms, another definition became necessary. It was desirable that the new and more general definition correspond to the old definition for the special case of the rectangular wing. The more general definition of geometrical aspect ratio which is now universally used is given in the equation below, where *A* is the aspect ratio and *b* and

$$A = b^2/S$$

S are defined in the illustration. Because *S* is equal to *bc* for a wing of rectangular planform, the definition of aspect ratio given in the equation corresponds to the original idea of the ratio of the span to the chord for a rectangular wing. [A.E.V.D./R.L.Bi.]

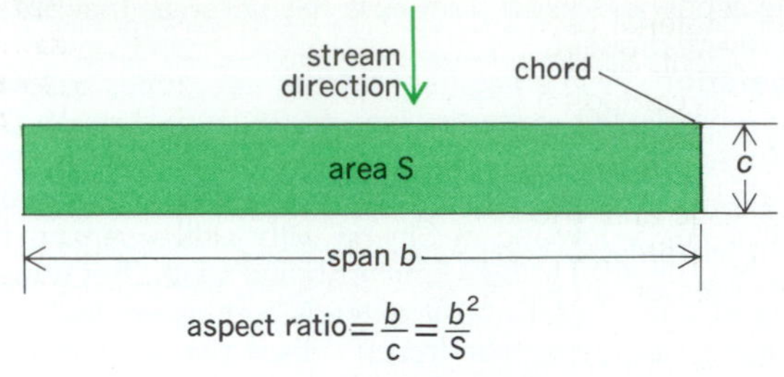

Aspect ratio of a wing.

Aspergillosis A fungus disease of humans caused by several species of *Aspergillus*, particularly *A. fumigatus*, *A. flavus*, and *A. terreus*. Invasive pulmonary and systemic aspergillosis are considered opportunistic diseases, developing in patients whose primary disease may be leukemia, Hodgkin's disease, or a lymphoma. Aspergillosis may develop in patients receiving immunosuppressive drugs for tissue transplantation or cytotoxic and antimitotic drugs. *See* OPPORTUNISTIC INFECTIONS.

Aspergillus species are common airborne contaminants in the laboratory as well as in the general environment; thus, any materials inoculated on laboratory media yielding an *Aspergillus* species must be interpreted carefully. Even repeated isolation of an *Aspergillus* species cannot be considered conclusive proof that aspergillosis exists. Immunodiffusion tests are available and should be used as an additional tool in attempting to diagnose aspergillosis. Treatment of severe forms of aspergillosis requires use of amphotericin B. *See* AMPHOTERICIN B; MEDICAL MYCOLOGY. [L.D.H.]

Asphalt and asphaltite Varieties of naturally occurring bitumen. Asphalt is also produced as a petroleum by-product. Both substances are black and largely soluble in carbon disulfide. Asphalts are of variable consistency, ranging from a highly viscous fluid to a solid, whereas asphaltites are all solid. Asphalts fuse readily, but asphaltites fuse only with difficulty. Asphalts may, moreover, occur with or without appreciable percentages of mineral matter, but asphaltites usually have little or no associated mineral matter. *See* BITUMEN; GILSONITE; IMPSONITE; WURTZILITE.

Many asphalts occur as viscous impregnations in sandstones, siltstones, and limestones. Most such deposits are thought to be petroleum reservoirs from which volatile constituents have been stripped by exposure of the rock. Relatively pure asphalt occurs in Kern, San Luis Obispo, and Santa Barbara counties, California. Occurrences of asphalt are also known in Kentucky and Oklahoma. Although asphalt seeps have long been known in France, Greece, Russia, Cuba, and other countries, the best known and largest are those of Venezuela and Trinidad.

The asphaltites (gilsonite, grahamite, and glance pitch) were probably derived from a saline lacustrine sapropel and owe their variable properties to differences in environment of deposition. These substances occur on a large scale in the Uinta Basin of northeastern Utah, where they are derived from upper Eocene Green River sediments, most of which are oil shales high in carbonate content. *See* OIL SHALE; SAPROPEL. [I.A.B.]

Asphalt is derived from petroleum in commercial quantities

by removal of volatile components. It is an inexpensive construction material used primarily as a cementing and waterproofing agent. *See* PETROLEUM PRODUCTS.

Asphalt is composed of hydrocarbons and heterocyclic compounds containing nitrogen, sulfur, and oxygen; its components vary in molecular weight from about 400 to 5000. It is thermoplastic and viscoelastic; at high temperatures or over long loading times it behaves as a viscous fluid, while at low temperatures or short loading times it behaves as an elastic body.

The three distinct types of asphalt made from petroleum residues are straight-run, air-blown, and cracked. Straight-run asphalt, characterized by a nearly viscous flow, is used in the construction of pavement surfaces for roads and airport runways. Air-blown asphalt is resilient and has a viscosity that is less susceptible to temperature change than that of straight-run asphalt . It is used mainly for roofing, pipe coating, paints, underbody coatings, and paper laminates. Cracked asphalt, with limited applications such as dust laying or as an insulation board saturant, has a nearly viscous flow, and its viscosity is more susceptible to temperature change than straight-run asphalt. [T.K.M.]

Aspidogastrea A group of entoparasites considered to be a subclass or order of the Trematoda. They have strongly developed ventral holdfasts. Two families, Aspidogastridae and Stichocotylidae, are recognized. Aspidogastridae, which are the commonest, occur in various cavities of mollusks and in digestive tracts of fishes and turtles. Elongate Stichocotylidae occur in the digestive tracts of skates. Little is known of the physiology of aspidogastreids, but they appear less host-specific than other trematodes. *See* DIGENEA; TREMATODA. [W.J.Ha.]

Aspidorhynchiformes A small order of specialized holostean fishes which are first recorded from Middle Jurassic deposits of Europe, and probably had a worldwide distribution in the warm seas of the Cretaceous Period. The order contains one family, Aspidorhynchidae, and two genera: *Aspidorhynchus* and *Belonostomus*. These fishes, some of which reached a length of over 3 ft (0.9 m), are characterized by a ganoid scale covering with much deepened scales along the flank, by an elongate fusiform body and head with long slender snout, and by an externally symmetrical tail. All the fins are small and fringing fulcra are reduced or absent. The dorsal and anal fins are positioned opposite one another far back on the body, and the pelvic fins are inserted closer to the anal than to the pectorals (see illustration). *See* HOLOSTEI.

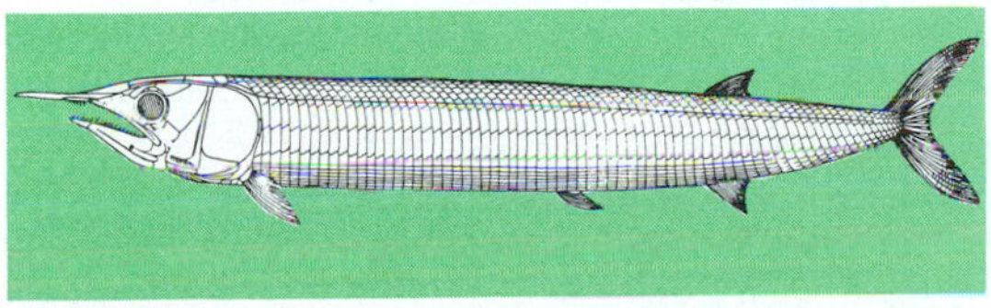

Aspidorhynchus acutirostris*, Upper Jurassic, Bavaria. *(After Assmann)

In the body form, fin position, and elongated snout, the aspidorhynchiforms resemble some of the living teleostean Exocoetoidei (needlefishes and sauries). It seems likely that these two widely separated and unrelated groups of fishes shared a similar mode of life, being predacious open-water forms that utilized their long snouts and strong swimming ability in capturing prey. *See* OSTEICHTHYES: TELEOSTEI. [T.M.C.]

Aspirin The trade name for acetylsalicylic acid. Aspirin is a white, crystalline, weakly acidic substance, with melting point 137°C. It is the most commonly used analgesic and antipyretic against neuralgia, influenza, colds, arthritis, headache, and fever. It is dispensed as tablets or powder in capsules, often mixed with sodium bicarbonate. Soluble aspirin is calcium acetylsalicylate.

Insignificantly affected by digestive juices, aspirin undergoes hydrolysis in the intestinal tract (alkaline) to liberate salicylic acid. The latter is readily absorbed through the intestinal membrane and is responsible for the physiological response. [E.B.R.]

Aspirin is a superior analgesic, but its effect on inflammation or fever is about the same as, and certainly not more than, that of sodium salicylate. Interest in aspirin has increased, in part because of the discovery that it modifies the synthesis of prostaglandins. Prostaglandins participate in a variety of biological reactions and are synthesized on demand and released; in fact, synthesis probably equals release. Aspirin and other peripheral analgesics inhibit prostaglandin synthesis, and the effectiveness of peripheral analgesic drugs and their ability to inhibit prostaglandin synthetase parallel one another.

Intolerance to aspirin is not uncommon. It tends to develop in middle age, showing a classical pattern: it occurs as a "triad" that includes formation of nasal polyps and intermittent (but occasionally progressive) bronchial asthma; some patients have urticaria. Death rarely ensues, because patients rapidly become aware of their intolerance. *See* ANALGESIC. [M.S.]

Assembly machines Machines that take discrete components as they come into an assembly department and bring them together so as to produce a configuration of some practical value. Such machines differ from packaging machinery in two ways: in assembly machinery, components must be inserted in specific sequence and spatial attitude; and they must often be tested functionally as part of the assembly process.

Assembly machinery was originally conceived for situations where volume or hazard of production, parts size, or availability of labor made manual assembly impractical from an operational or economic viewpoint. The early applications of assembly machinery and the majority of modern applications are found in the automotive industry, consumer products, manufacturing, or hazardous assembly.

Assembly machinery can be classified in several ways, including work path (rotary, carrousel, or linear, index-dwell ratios (continuous motion, intermittent motion, or power and free); actuation (mechanical, fluid power, or electronically programmable); work nest configuration (pallet or walking beam); and design (special or standard modular). The engineers who specify assembly machinery will usually select from one or more of these categories based on product size and weight, volume of production, product life cycle, future and present flexibility needs, human resources, and return on investment.

Rotary dial machines have a number of pallets fastened to a rotating dial or ring. Transfer devices are usually mounted to the machine base, while feeders are placed outside the periphery of the dial. Carrousel machines are usually configured in race track shape. Work-holding nests (the machine elements used to hold all of the parts being assembled) are secured to one another (a configuration known as precision link) or fastened to a chain. Linear machines have an open-loop configuration in which pallets transport the product being assembled along one or more linear paths, rather than in a closed circuit such as rotary or carrousel machines.

In any form of automatic assembly, production rates will be controlled by the single longest operation. If new parts can be added to the assembly without the fixture being stopped, the machine can be operated in a continuous motion.

Assembly machines perform several tasks: parts feeding (including orientation, separation, and transfer); parts joining (for example, welding, riveting, and soldering); parts inspection (condition, parameters, presence, and position); functional testing (for example, capacitance, torque, and pressure decay); marking (date coding, model number, and operational characteristics); and ejection (in controlled or uncontrolled positions). Many of the tasks of assembly machinery use commercial units identical to those used on manual lines. The unique task of automatic assembly machinery is that of parts feeding, which consists of accepting the component parts as they come to the assembly area and taking one of these components, separating that component from other parts, changing its spatial orientation to a usable insertion attitude, and transferring it from the orientation device into the work-holding fixture or into a partially completed assembly. *See* MANUFACTURING ENGINEERING; PRODUCTION METHODS. [F.Ri.]

Astatine A chemical element, At, atomic number 85. Astatine is the heaviest of the halogen groups, filling the place immediately below iodine in group VII of the periodic table. Astatine is a highly unstable element existing only in short-lived radioactive forms. About 25 isotopes have been prepared by nuclear reactions of artificial transmutation. The longest-lived

1	2	3	4	5	6	7	8	9	10	11	12	13	14	15	16	17	18
1 H																	2 He
3 Li	4 Be											5 B	6 C	7 N	8 O	9 F	10 Ne
11 Na	12 Mg											13 Al	14 Si	15 P	16 S	17 Cl	18 Ar
19 K	20 Ca	21 Sc	22 Ti	23 V	24 Cr	25 Mn	26 Fe	27 Co	28 Ni	29 Cu	30 Zn	31 Ga	32 Ge	33 As	34 Se	35 Br	36 Kr
37 Rb	38 Sr	39 Y	40 Zr	41 Nb	42 Mo	43 Tc	44 Ru	45 Rh	46 Pd	47 Ag	48 Cd	49 In	50 Sn	51 Sb	52 Te	53 I	54 Xe
55 Cs	56 Ba	71 Lu	72 Hf	73 Ta	74 W	75 Re	76 Os	77 Ir	78 Pt	79 Au	80 Hg	81 Tl	82 Pb	83 Bi	84 Po	85 At	86 Rn
87 Fr	88 Ra	103 Lr	104 Rf	105 Db	106 Sg	107 Bh	108 Hs	109 Mt	110	*111*	*112*	*113*	*114*	*115*	*116*	*117*	*118*

lanthanide series	57 La	58 Ce	59 Pr	60 Nd	61 Pm	62 Sm	63 Eu	64 Gd	65 Tb	66 Dy	67 Ho	68 Er	69 Tm	70 Yb
actinide series	89 Ac	90 Th	91 Pa	92 U	93 Np	94 Pu	95 Am	96 Cm	97 Bk	98 Cf	99 Es	100 Fm	101 Md	102 No

of these is ^{210}At, which decays with a half-life of only 8.3 h. It is unlikely that a stable or long-lived form will be found in nature or prepared artificially. The most important isotope, used for tracer studies, is ^{211}At. Astatine exists in nature in uranium minerals, but only in the form of trace amounts of short-lived isotopes, continuously replenished by the slow decay of uranium, The total amount of astatine in the Earth's crust is less than 1 oz (28 g).

In aqueous solution, astatine resembles iodine except for differences attributable to the fact that astatine solutions are of necessity extremely dilute. Like the halogen iodine, when astatine exists as a free element in solution, it is extracted by benzene. The element in solution is reduced by agents such as sulfur dioxide and is oxidized by bromine. It is more electropositive than the other halogens. It has oxidation states with coprecipitation characteristics similar to those of the iodide ion, free iodine, and the iodate ion. Powerful oxidizing agents produce an astatate ion, but not a perastatate ion. The free state is most readily obtained and is characterized by high volatility and high extractability into organic solvents. *See* HALOGEN ELEMENTS. [E.K.H.]

Asterales An order of flowering plants, division Magnoliophyta (Angiospermae). The order consists of only the single, very large family Compositae (Asteraceae), with about 19,000 species, occurring in nearly all parts of the world but most abundant and conspicuous in areas which are not densely forested.

Most members of the order are herbaceous, but some, such as the sagebrush (*Artemisia tridentata*), are shrubs, and a few

New England aster (*Aster novae-angliae*), a characteristic member of the order Asterales. The numerous apparent petals around the margin of the flower head are corollas of individual flowers, called ray flowers in contrast to disk flowers which occupy the center of the head. (*Courtesy of Alvin E. Staffan, from National Audubon Society*)

of the less familiar tropical species are trees. Many well-known garden ornamentals, such as aster (see illustration), chrysanthemum, dahlia, daisy, sunflower, and zinnia, belong to the Asterales. A few garden vegetables (lettuce and artichoke) and some common weeds (dandelion, thistle, and ragweed) also belong to the order. *See* ARTICHOKE; ASTERIDAE; LETTUCE; MAGNOLIOPHYTA; MAGNOLIOPSIDA; PLANT KINGDOM; SUNFLOWER. [A.Cr.]

Astereognosis In clinical neurology, the loss of recognition of objects by touch in the presence of recognition through another sense, usually vision.

Astereognosis, in the broad usage, may be secondary to gross somatosensory defects caused by lesion of the afferent nerve pathways. It may also occur without gross defects of pressure, pain, and thermal sensitivity, but in association with other sensory changes, such as impaired localization of touch, diminished tactile acuity (two-point discrimination), and defective postural sense. Concomitant difficulty in discriminating the size, weight, texture, and shape of objects by palpation is usually encountered. *See* PERCEPTION; SENSATION.

Astereognosis, in narrow usage, is synonymous with tactile agnosia, sometimes called pure astereognosis. Agnosia literally means lack of knowledge, and is used to indicate a loss which is intellectual in nature yet restricted to one sense modality. Failure to identify objects and their function is thought to be based on a disturbance of associative memory. This view implies a theory of brain function which considers associative activities as based on processes distinct from those involved in sensation or movement, separately localized, and hence selectively vulnerable to brain injury or disease. *See* AGNOSIA; MEMORY. [J.Se.]

Asteridae A large subclass of the class Magnoliopsida (dicotyledons) of the division Magnoliophyta (Angiospermae), the flowering plants, consisting of nine orders, 43 families, and more than 50,000 species. The largest orders of the group are the Asterales (about 19,000 species), Scrophulariales (about

10,000 species), Lamiales (about 7800 species), and Rubiales (about 6500 species). Other orders are the Gentianales, Polemoniales, Plantaginales, Campanulales, and Dipsacales. See individual articles on each order. *See also* MAGNOLIOPHYTA; MAGNOLIOPSIDA; PLANT KINGDOM. [A.Cr.]

Asteroid One of the many thousands of small planets (minor planets) revolving around the Sun, mainly between the orbits of Mars and Jupiter. Newly discovered asteroids are assigned a catalog number and name (such as 433 Eros) only after they are observed often enough to compute an accurate orbit. There are over 2200 cataloged asteroids. *See* PLANET.

The vast majority of asteroids have semimajor axes (mean distances to the Sun; symbolized a) between 2.2 and 3.2 astronomical units (1 AU = distance from Earth to the Sun = 9.30 $\times$ 10^7 mi = 1.496 $\times$ 10^8 km). However, several small asteroids orbit between Earth and Mars, and two groups, the Trojan asteroids, orbit at Jupiter's distance from the Sun.

Two new techniques were used in the 1970s to measure the diameters and albedos of over 130 asteroids: the polarimetric technique, based on an empirical correlation between the albedos of materials and how they polarize light; and the radiometric technique, which measures diameters by comparing the brightness of reflected visible sunlight from an asteroid with the brightness of the asteroid's emitted thermal radiation in the infrared. *See* ALBEDO.

Asteroids are much darker, hence larger, than had been assumed before. There are about 35 asteroids larger than 120 mi (200 km) in diameter; about 75% of them are soot-black (geometric albedos of 3–5%). Asteroids are much more numerous at smaller sizes, generally following a size distribution characteristic of fragmentation processes, as one would expect if the asteroids were smashing into each other. Indeed, there are so many large asteroids confined in the volume of the asteroid belt that collisions sufficient to destroy all but the larger asteroids occur every few billion years, and much more often for smaller ones. Thus the asteroids are mainly collisional fragments, grinding themselves down to dust.

Spectra of sunlight reflected from asteroids have shapes, including absorption bands, characteristic of different rock-forming minerals. Combined with the albedo data from polarimetry and radiometry, the spectra of surfaces of over 270 asteroids show that about 90% fall into two compositional groups: the C-type, having very dark, neutral-colored surfaces of probable carbonaceous composition like the primitive carbonaceous chondritic meteorites; and the S-type, having moderate-albedo, reddish surfaces containing pyroxene and olivine silicates, probably mixed with metallic iron, similar to stony-iron meteorites or ordinary chondrites. Only a few asteroids have spectra clearly indicating a composition like ordinary chondritic meteorites, the most common meteorites to fall on Earth. Since current cosmogonical models for the origin of planets involve accretion from myriads of asteroidlike planetesimals, it is likely that asteroids are a remnant of the planetesimals that failed to accrete into a planet between Mars and Jupiter. Probably bombardment of the asteroid zone by large planetesimals scattered from massive, nearby Jupiter increased the relative velocities of asteroids to the present value of 3.1 mi/s (5 km/s) so that asteroids fragment rather than accrete when they meet each other. Instead of forming a planet, the asteroids have been smashing each other to bits, and those seen now are mostly fragments, except for the "lucky few" that have escaped catastrophic collisions so far.

Evidently some asteroids of primitive, nonvolatile solar composition were heated within the first few hundred million years after the origin of the solar system, perhaps by the solar wind or extinct radionuclides, and they melted. Iron sank to their centers, forming strong stony-iron cores while basaltic lavas floated to their surfaces, producing minor planets like Vesta. The numerous collisions fragmented away the outer crusts and mantles of most of these bodies, with Vesta a prime, intact exception. While the unmelted, weak, C-type asteroids may have been depleted by a large factor by the collisions, most of the strong stony-iron cores of the melted proto-asteroids have survived, perhaps as the S-type asteroids observed today. The asteroids still collide and fragment, occasionally spraying the inner solar system with chips that produce craters or fall as meteorites. [C.R.C.]

Asteroidea A subclass of the class Stelleroidea in the subphylum Asterozoa; asteroids are known as starfish. The arms are not sharply demarcated from the rest of the body. The ambulacral ossicles never fuse to form vertebrae. The tube feet are locomotor organs, usually suctorial, emerging from an open ambulacral groove. The dominant growth gradients are such as to cause the skeletal ossicles to lie in longitudinal rows known as series (for example, adambulacral series, ventrolateral series, and inferomarginal series). A representative asteroid is shown in the illustration.

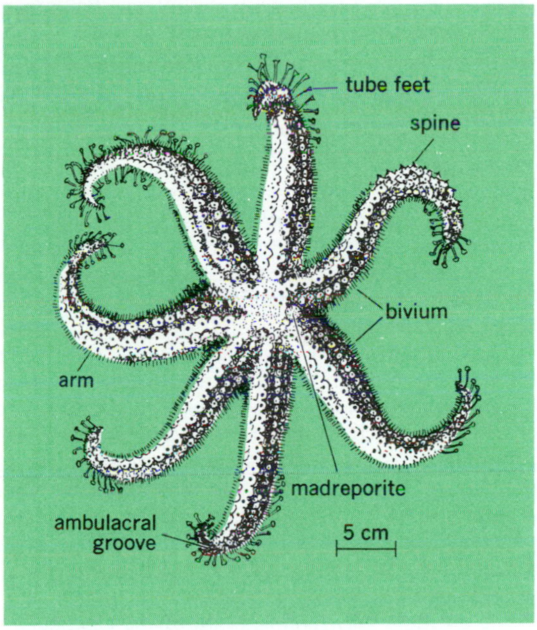

A representative asteroid *(Astrostole scabra)*.

The 1700 or so known fossil and recent species of Asteroidea are now grouped into five orders: Platyasterida, Paxillosida, Valvatida, Spinulosida, and Forcipulatida. Asteroids range in size from about 0.4 to 40 in. (10 mm to 1 m) across. Many starfish are brightly colored and attractive animals, but some are dowdy and cryptic. Like most echinoderms, starfish have a lifespan of about 5 years.

Several species of starfish are regarded as pests, particularly in oyster, mussel, and scallop fisheries. Beyond this they are of little significance, with the notable exception of the coral predator *Acanthaster planci*, which occurs in the Indo-Pacific. When present in large numbers, it may kill extensive areas of reef-building corals. However, large aggregations of this species are occasional phenomena, and coral killed by it is capable of regenerating more quickly than was hitherto supposed.

Starfish inhabit all types of bottom throughout the world's seas and oceans. Some burrow in sand and mud, and others live on rocks and coral reefs, where they are at their most diverse. Most families show well-defined bathymetrical preferences, but there are exceptions, for instance, the offshore fauna of New Zealand. There are 12 families which occur at depths greater than 2 mi (3.2 km), and several genera extend below 4 mi (6.4 km).

The outer surface is coated with a thin layer of ciliated epithelium. Below this lies connective and muscle tissue which forms the body wall and in which the skeletal ossicles are embedded, forming the test. Some of the ossicles, notably the spines, protrude to the exterior. The arms can be slowly moved by the body wall musculature. Although 5 arms are common, up to 12 occur in a number of genera, and more than 12 may occur.

Pedicellariae occur in some starfish. They are small seizing organs used to grip intruders and, according to one report, even to catch food. They probably assist in protecting the delicate epithelium from predators and sediments. The well-developed water-vascular system follows the pattern for the phylum, with the following distinctive features. With the exception of some Paleozoic genera, the madreporite is always on the upper (aboral) side. There is usually only one madreporite situated interradially. The tube feet may be peglike and suckerless or columnar and suckered. The nervous system also follows the pattern for the phylum. Starfish cannot see, but they clearly detect changes in light intensity such as those caused by shadows. They have photosensitive eyespots located at the tip of each arm. The terminal tube feet of each ambulacrum appear to be especially sensitive to water-borne chemicals which may be emitted by prey species, and they are thus used in food detection.

Starfish are voracious feeders. Small food, such as amphipods and young mollusks, may be engulfed whole. *Acanthaster* everts part of its stomach out of the mouth and wraps it over coral polyps, which are thus digested outside the predator's body. Some of the Forcipulatida, for example, *Asterias*, can open large clam and other bivalve shells by applying a continuous force which overcomes the molluscan adductor muscles. The mouth is in the middle of the lower (oral) surface of the disk and leads through a short esophagus to the cardiac stomach. Above the cardiac stomach lies the pyloric stomach, with as many branches as there are arms. Within each arm, the branch forks to give two blind digestive ceca. In most species, a short intestine leads upward to the rectum and anus. This is generally situated near the middle of the upper (aboral) surface. In those starfish which lack an anus, the feces are extruded through the mouth.

Asteroids are generally mature and able to reproduce at 1 year. They normally continue to grow for about 4 years, with growth taking place when food is abundant. The sexes are usually separate, but in a few genera, such as *Fromia*, hermaphrodites occur, the sex changing with age. Regeneration of lost or damaged parts is a characteristic of most genera. In *Linckia* this is most marked, and a whole new individual can be regenerated from a fragment of an arm. *See* ECHINODERMATA. [A.C.C.]

Asteroxylales A small extinct order of the class Lycopsida that bridges the evolutionary gap between the primitive Zosterophyllopsida and relatively advanced Lycopsida; hence the asteroxylaleans are often termed prelycopsids. The best-known asteroxylalean species are of Early Devonian age, although similar forms survived to the Late Devonian. All possessed at least some hydrophytic features; together with a paucity of strengthening tissues, this characteristic suggests that they largely relied on hydrostatic pressure for structural support. *See* LYCOPODIOPHYTA; LYCOPODIOPSIDA.

With the exception of *Asteroxylon*, asteroxylalean fossils are not well preserved. The asteroxylaleans are regarded as the most primitive lycopsids, but they can also be considered as the most advanced zosterophyllopsids; some authorities prefer to disperse the prelycopsids among other taxa.

The pivotal plant in this order is *A. mackiei*, which originated from the remarkable biotic communities petrified in Lower Devonian volcanigenic cherts at Rhynie, Scotland. This species had naked horizontal rhizomes that produced primitive roots and aerial branches up to 20 in. (50 cm) high and 0.4 in (1 cm) in diameter. The radially symmetrical actinostele (a star-shaped protostele) is typical of primitive lycopsids rather than zosterophyllopsids. Vascular traces extended toward, but did not enter, the densely packed clasping tissue outgrowths termed protomicrophylls.

The sporangia were homosporous, kidney shaped, and distributed singly and randomly among the leaves on the more distal axes. Like the zosterophyllopsids, *Asteroxylon* bore sporangia on stalks with a vascular supply that was independent of, and more extensive than, that of the surrounding enations. These reproductive structures are generally regarded as homologous with vegetative lateral branches. [R.M.Ba.; W.A.DiM.]

Asthenosphere A region in the interior of the Earth characterized by less mechanical strength and less resistance to deforming stresses than regions above (lithosphere) and below (mesosphere). This zonation is in contrast to the more familiar subdivision of the interior into crust, mantle, and core, based on discontinuities in seismic-wave velocities. The asthenosphere lies within the seismically defined upper mantle. *See* LITHOSPHERE.

The asthenosphere, because of its ability to flow, has figured prominently in geodynamic theories on the causes of vertical motions observed at the Earth's surface, such as postglacial rebound. Likewise, the asthenosphere plays a prominent role in models of the large horizontal movements of the lithosphere as observed in continental drift and plate tectonics. The asthenosphere is envisioned as the lubricating layer over which the plates glide. *See* CONTINENTAL DRIFT; ISOSTASY; PLATE TECTONICS.

The asthenosphere is thought to be that range of depth where the actual earth temperature is closest to the temperature of incipient melting of mantle rocks. In some places the temperature may actually reach the rock-melting temperature and produce magma, thus giving the asthenosphere another important role in geodynamics—as the source region of many types of igneous rocks.

One consequence of the ability of the asthenosphere to relieve stress through creep and flow is that it appears to be largely earthquake-free. Those intermediate- and deep-focus earthquakes that do occur in the depth range of the asthenosphere are nearly all found in regions where rigid lithosphere has been subducted (recycled) into the Earth's interior and has penetrated the asthenosphere at subduction zones. Deep earthquake activity appears to occur within the subducted lithosphere and not in the surrounding asthenosphere. *See* EARTHQUAKE. [H.N.P]

Asthma A common disease of the bronchioles that is marked by wheezing, breathlessness, and sometimes cough and expectoration of mucus. Typical asthma is intermittent or paroxysmal, but a small proportion of patients suffer continuously if not treated. If severe enough, asthma may become resistant to conventional treatment, and hospitalization is necessary (status asthmaticus).

Asthma may begin at any age, but most cases develop before age 60 and about half before age 10. For some reason childhood asthma is more common in boys than girls. Asthma is a heterogeneous disease, with one common feature: obstruction to air flow which is due to hyperreactive bronchioles. This obstruction may be produced or aggravated by numerous factors. Some are nonspecific, such as dust, tobacco smoke, polluted air, or psychological stress, and may affect most people with asthma.

In about one-third of all cases, including most children, the obstruction has been established to be an allergic reaction. People with allergic asthma have acquired specific sensitivity to common inhaled substances that are ordinarily innocuous, including house dust, animal danders, fungus spores, and pollens. Certain foods may also provoke asthma, especially in early childhood. These inhaled substances and foods are called allergens.

Chronic asthma is often associated with or initiated by bronchopulmonary or sinus infections. Most adult-onset asthma is of this type. Besides infections, other nonallergic factors that may produce or aggravate asthma in some patients, but not others, are exercise, certain occupational dusts, and a few drugs, especially aspirin. Some asthmatic patients, for unknown reasons, become extremely intolerant of aspirin, which has produced fatal reactions. The problem usually develops in adults, especially those with a history of sinus disease and nasal polyps, but occasionally appears in children with allergic asthma as well. An intolerant person may also have trouble with other drugs that are similar to aspirin in their action.

Whatever the type, asthma has typical anatomic features. All cases suffer from spasm of the smooth muscle of the bronchioles and from swelling of the mucous lining. More severe and chronic asthma may be associated with accumulation of inflammatory cells, basement membrane thickening, mucous gland enlargement, and filling of the airway with thick mucus. During asthma attacks, air can enter the lungs more readily than it can leave, so the lungs become hyperinflated.

Asthma can rarely be cured, but almost always can be controlled by a variety of methods. The environment should be altered or the patient relocated to minimize exposure to allergens and irritants. Basic drugs are the bronchodilators: theophylline and modern relatives of adrenaline. For more severe cases, treatment with a glucocorticoid (steroid) drug may be necessary. Immunotherapy (desensitization) is an appropriate way to treat patients with allergic asthma who are unable to avoid allergens that are responsible for their symptoms. Psychotherapy may be necessary for the occasional depressed patient, but is not generally helpful. *See* RESPIRATORY SYSTEM DISORDERS. [P.P.V.A.]

Astigmatism The optical aberration which occurs because a wave surface in general has double curvature. Even for a small circular stop, the rays from an object point do not come to a point focus but intersect a set of image planes in a set of ellipses, the diameters of which are proportional to the distances of the two foci from the image plane under consideration. Such an error exists even on the axis in systems which are not rotation symmetric, such as cylindrical and toric lenses and the astigmatic eye. In systems with rotation symmetry, it exists in general for the rays from an off-axis point going through a small pupil.

Astigmatism on the axis is a common error in the human eye arising from the fact that the refracting surfaces, especially the cornea, can have different powers in different meridians. It can be corrected by a spectacle lens in which at least one surface has different curvatures in different planes through the lens axis. The lens may be a cylinder, a torus, or a surface of second order with double symmetry. *See* ABERRATION (OPTICS); VISION. [M.J.H.]

Astomatida An order of protozoans, subclass Holotrichia, in which all species are mouthless. All species are parasitic in other animals, typically oligochaete annelids. Many astomatids possess an elaborate holdfast organelle. *Anoplophyra* is a typical example. *See* CILIOPHORA; HOLOTRICHIA; OLIGOCHAETA; PROTOZOA. [J.O.C.]

Astrapotheria A relatively small group of extinct South American ungulates, ranging from the late Paleocene to the late Miocene. They are customarily divided into two suborders: the late Paleocene–Eocene Trigonostylopoidea, and the early Eocene–late Miocene Astrapotheroidea.

One of the most spectacular and advanced members of the order was *Astrapotherium* (see illustration). This animal, known from the late Oligocene to the late Miocene, averaged 9–10 ft (2.7–3 m) in length, although some other forms grew even larger. The anterior part of the skull was striking with the huge, persistently growing, curving canines, whose function is unexplained. The retracted, chopped-off appearance of the snout region strongly suggests that this animal had a moderately large trunk. The front legs were somewhat more strongly constructed than the hind ones, and this makes the habits of this animal a puzzle. *See* CONDYLARTHRA; MAMMALIA. [F.S.S.]

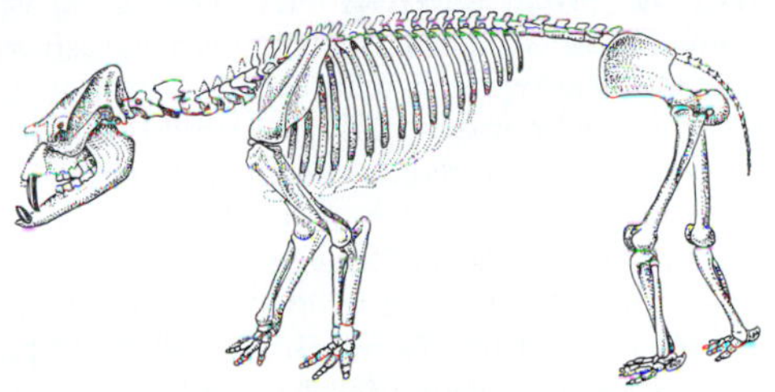

***Astrapotherium magnum* skeleton.**

Astrometry Measurement of space-time relations of celestial objects, primarily as observed on the celestial sphere, that is, perpendicular to the line of sight.

Over the past two centuries the positions of celestial objects have been measured visually with small, stable telescopes designed to measure meridian altitudes with high precision. These transit or meridian-circle telescopes and their variations, in combination with sidereal clocks, are the instrumental backbone of classical knowledge of the positions of celestial objects on the celestial sphere.

While the positions of Sun, Moon, and planets are thus easily obtained, only stars brighter than about the tenth magnitude can be observed in this fashion, because of the limited size of the instruments. Photographs, derived from standardized small telescopes in different locations, have yielded data which includes stars down to the thirteenth magnitude. The positions of stars down to the eighteenth magnitude and fainter are obtained photographically by specially designed powerful telescopes of short focal ratio. Among the most important results of the photographic method is the wealth of newly discovered faint stars of large proper motions, all of which have proved to be comparatively nearby. *See* ASTRONOMICAL PHOTOGRAPHY.

A special branch of astrometry is long-focus photographic astrometry. Accurate stellar positions referred to faint, distant reference stars are measured by means of refractors of focal lengths of 8 m or more, commonly of large apertures of 40 cm or more; that is, with focal ratios up to $f/20$. Originally, long-focus photographic astrometry was developed for the specific problem of stellar parallax, and as a result, a great body of information on the distances of stars now exists. The method has been extended to other astrometric problems requiring high precision, such as the orbital motion of double stars, including the discovery and measurement of unseen companions from the perturbations on visible primary stars. *See* PARALLAX (ASTRONOMY). [P.V.deK.]

The field of radio astrometry is achieving very precise positions (0.005 arc-second) of celestial radio sources. The technique utilizes two radio telescopes configured as an interferometer. One method employs two radio telescopes, situated a few kilometers apart, linked by cables to a receiving system that instantly measures the difference in phase between the radio signals arriving at the telescope. Another technique, very-long-baseline interferometry (VLBI), utilizes a very accurate timing system, usually atomic clocks, to record the signals received at the end of a baseline several thousand kilometers apart. [B.L.Kl.]

Astronautical engineering The engineering aspects of flight and navigation in space, also known as astronautics. Astronautical engineering deals with vehicles, instruments, and other equipment used in space, but not with the sociological or economic aspects of space flight, except as they influence the equipment.

There is a lack of parallelism between astronautic and aeronautic vehicle terminology. An aircraft is a self-contained vehicle, having within its structure essentially all the equipment required to transport its payload from one place to another. A spacecraft, in the more restricted sense, is the container for the payload. Sometimes the word is used to denote the container and payload. Most spacecraft, to date, have had either very limited propulsion or none at all. Since enormous speeds are the hallmark of all astronautic missions, unpowered spacecraft require a "booster," or "launch vehicle," usually a rocket many times as large as the spacecraft. The weight of the spacecraft, in fact, seldom exceeds 5% of the total launch vehicle weight.

It is extremely expensive to put a pound of payload into Earth orbit. Thus designers have been justified in going to great lengths to convert a pound of structure into a pound of payload. Great improvement appears possible in this respect; only the cost of the propellant seems to be irreducible. In view of the high cost of space operations, it is especially important that space vehicles operate long enough to successfully fulfill their missions. A severe reliability requirement is thus imposed upon vehicles and equipment intended for missions, such as journeys to the planets, which may require up to a year or more to accomplish. For complex equipment in space vehicles, operating lifetimes of this order of magnitude are difficult to attain. The requirements for high reliability and low weight add tremendously to the cost of the payloads themselves, to the extent that their cost approaches that of the launch vehicle. All space missions through 1975 used expendable launch vehicles. The space shuttle, a reusable launch vehicle, is expected to reduce the costs of Earth-to-orbit transportation. *See* SPACE SHUTTLE.

Gravity is a dominating influence in the design of space launch vehicles. Despite the fact that the pull of gravity extends to infinity, it is nonetheless possible to escape permanently from the Earth's gravity in the sense of never being drawn back to the ground. The key is speed. Circular velocity is the minimum at which a space vehicle can remain permanently above the Earth. At low altitudes, this velocity is about 25,000 ft/s (7.9 km/s). As the speed is increased above the circular velocity, the path of a vehicle becomes a larger circle or an elongated ellipse. When the speed reaches 37,000 ft/s, or about 7 mi/s (11.2 km/s), the path becomes a parabola and the vehicle will travel along one of the legs to infinity without further propulsion. *See* ESCAPE VELOCITY; SATELLITE (SPACECRAFT).

These velocities are tremendous by any previous standard. To reach them, a vehicle must carry the corresponding amount of energy in the form of propellant.

Even with the most energetic propellants and the lightest structures, it has not yet been possible to reach orbital velocity with a single rocket. To overcome this seemingly insurmountable obstacle, one rocket is carried as the payload of a larger one. When the larger burns out, the second is ignited and adds its velocity to that of the first. This is known as the step-rocket or staging technique. For lunar and planetary missions, lightweight vehicles, powerful propellants, and many stages are used. The lunar orbit rendezvous method required a total of six stages to take the Apollo astronauts to the Moon and back. *See* ROCKET STAGING; SPACECRAFT PROPULSION.

Although propulsion is the key to space flight, other elements are essential and present numerous new problems. One such element is guidance and control. For the ascent phase of space vehicle flight, guidance systems similar to those used for ballistic missiles are employed. Another control requirement of many types of space vehicles is that of maintaining the desired vehicle attitude over long periods of time. Displacement gyroscopes, even excellent ones with very low drift rates, cannot provide an accurate reference for days or weeks. Such devices must be corrected frequently by an external reference.

At least two such references are available: sources of electromagnetic radiation, and the gravitational gradient. The first might be used by such devices as a Sun seeker, a star tracker, or a horizon scanner. In the vicinity of the Earth (or any large celestial body) the difference in the pull of gravity between points on the craft having different distances from the Earth can be usefully employed.

Reaction wheels or other devices capable of storing angular momentum may be used to provide the torque to effect or maintain a given orientation. Such devices are very efficient, both from a weight and an energy standpoint, where disturbing torques on the spacecraft are small, random, and long continued. At the opposite end of the torque spectrum, torques that are large and uncompensating, rocket engines are the most suitable.

Vehicle and payload equipment require electric power. For small amounts of energy, chemical sources, such as batteries or chemically fueled generators, may be used. A great deal more energy can be obtained from a nuclear reactor. Energy also comes continuously from the Sun but at a fairly low density at Earth's distance.

Communications equipment comprises an essential item of nearly all space vehicles. This equipment is designed for light weight, low power consumption, and, usually, long life.

Although a large percentage of the problems of space flight are associated with the vehicles, it would be a mistake to assume that these constitute even a major fraction of the total operating system. Indeed, the cost of overcoming the Earth's gravity is so great that any portion of the total operation which can be performed on the ground should be done there. The supporting ground equipment consists of the preparation and launching equipment, and the tracking, communications, and payload-oriented equipment for turning the received data into usable form. For missions which involve return of space vehicles or booster rockets, recovery equipment may also be required. *See* LAUNCH COMPLEX; SPACECRAFT GROUND INSTRUMENTATION.

In their interaction with the terrestrial and atmospheric environment during reentry, space vehicles resemble ballistic missiles. However, although ballistic reentry techniques have been proved successful, the use of winged vehicles also has certain attractive aspects. There is a basic difference in these two methods in respect to the way atmospheric heat is handled. The ballistic approach absorbs the heat in the reentry body or rejects it back to the air by mass transfer. The winged vehicle dissipates the heat by radiation. Considerable research has been done on compromise reentry vehicles, such as the lifting body approach. The orbital stage of the shuttle is a winged craft designed to land like an airplane. It utilizes a combination of techniques to overcome the reentry heating problems: lift, temperature-resistant materials, and local ablative cooling. *See* ATMOSPHERIC ENTRY.

Astronautical engineering must contend with the unique environment of space outside the Earth's atmosphere. Although gravity is present in space, whenever a vehicle is coasting unpropelled, the shell and everything in it are acted on equally by gravity and therefore appear weightless. Fluids do not flow naturally, but must be confined and extruded. Liquids exposed to the vacuum of space evaporate or freeze. External transfer of heat takes place only by radiation. Metals exposed to the ultraviolet rays of the Sun emit electrons. Small particles of cosmic dust strike external surfaces at fantastic velocities and gradually erode them. Cosmic radiation creates a spectrum of secondary radiation that may reach levels damaging to equipment or personnel. [R.C.Tr.]

Astronautics The application of scientific principles and engineering techniques to flight in space. Astronautics deals with space vehicles in the sense that aeronautics deals with aircraft. The distinguishing feature between astronautics and aeronautics is the extent to which the vehicles are influenced by the Earth's atmosphere. Astronautics encompasses both crewed and uncrewed missions. *See* AERONAUTICS; ASTRONAUTICAL ENGINEERING; SPACE TECHNOLOGY.

The subject matter of astronautics is flight in regions where a vehicle overcomes gravitational attraction and controls its course by reactive propulsion. Aeronautics concerns flight in regions where a vehicle resists gravitational attraction and controls its course by aeromechanical forces. The distinction is convenient but not clear cut. Rockets, by their reaction, assist airplanes to take off. Space vehicles may glide back to Earth. *See* INTERPLANETARY PROPULSION. [R.C.Tr.]

Astronomical atlases Sets of maps of celestial phenomena. Often developed in conjunction with catalogs that list position, brightness, and other features, maps provide a clear picture of the spatial relations between the phenomena.

In the nineteenth century the introduction of larger telescopes, steadier mounts, better graduated circles, and filar micrometers led to more extensive and precise star catalogs and atlases. Premier among these was the *Atlas des nordlichen gestirnten Himmels* (Bonn, 1863), organized by F. W. A. Argelander. These 40 charts showed the positions and magnitudes of 324,198 stars in the northern hemisphere. The charts, along with their companion star catalog, the *Bonner Durchmusterung*, or "B.D.," are still in use today. *See* ASTRONOMICAL CATALOGS.

By the 1880s photography was sufficiently well developed for astronomers to begin considering a photographic atlas of the heavens. In 1949, under sponsorship of the National Geographic Society, was begun the *Palomar Observatory Sky Survey*—the first photographic atlas that showed the sky in two colors. Taken with the 48-in. (1.2-m) Schmidt telescope, it reveals stars brighter than magnitude 20 situated north of −33°. A new *Palomar Observatory Sky Survey* was begun in 1985. This atlas will use improved photographic plates to reach magnitude 22, in three colors. *See* ASTRONOMICAL PHOTOGRAPHY.

In addition to the photographic atlases, printed atlases continue to be published, usually in conjunction with star catalogs. Among the more popular atlases of this type are the *Smithsonian Astrophysical Observatory Star Catalog* and *Star Atlas of Reference Stars and Nonstellar Objects* (Washington, 1966); Antonin Becvar's *Skalnate Pleso Atlas* (Cambridge, 1949) and its successors; *Norton's Star Atlas* (Cambridge, 1979); and W. Tirion's *Sky Atlas 2000.0* (Cambridge, 1982).

Edwin Hubble, known as the founder of modern extragalactic astronomy, charted the distribution of galaxies in space and their morphology. The *Hubble Atlas of Galaxies* appeared in 1961, edited by Allan Sandage. *See* GALAXY, EXTERNAL; MILKY WAY GALAXY; NEBULA.

G. P. Kuiper's *Photographic Lunar Atlas* (Chicago, 1960) contains the best set of Moon plates based on terrestrial observations to date. Lunar probes have generated countless close-ups of the lunar surface. A particularly beautiful photographic atlas of the Moon is the NASA publication *The Moon as Viewed by Lunar Orbiter* (Washington, 1970). As spacecraft explore more of the solar system, atlases of more objects are becoming available. *See* MOON. [D.J.Wa.; C.T.K.]

Astronomical catalogs A list or an enumeration of astronomical data, generally ordered by increasing right ascension of the objects listed. Astronomical catalogs vary a great deal in form and content depending upon their use, which may be purely astronomical, or for navigation, time determination, geodesy, or space science applications. In some catalogs the essential data are stellar positions and motions, while in others astrophysical data, such as magnitudes, spectra, and radial velocities of stars, are important. There are also catalogs of special stellar and of nonstellar objects. *See* ASTRONOMICAL COORDINATE SYSTEMS.

A catalog regarded as the best representation of the celestial coordinate system at the time of its publication is called a fundamental star catalog. The *Fourth Fundamental Catalogue*, designated *FK4*, was published by the Astronomisches Rechen-Institut in Heidelberg, Germany, in 1963. The catalog contains the positions (right ascensions and declinations) and their changes with time (proper motions and precession) of 1535 stars. The successor, the *FK5*, appeared in 1985 with data compiled from more than 250 catalogs with a combined number of 3,000,000 observations. The positions and proper motions of the stars in *FK5* provide a fundamental system for measurements of other star positions and proper motions, which may be carried out for a variety of problems arising in stellar astronomy. *See* FUNDAMENTAL STARS; OPTICAL TELESCOPE.

Moderately bright stars (seventh to ninth magnitude), selected on the basis of one star per square degree of the sky, are related to the fundamental system by meridian circle observations. These stars form a system of sufficient density to serve as position references for photographic observations. Typical of catalogs of such reference stars is the *AGK3R* of some 21,000 stars in the northern celestial hemisphere.

A photographic survey of the northern celestial hemisphere resulted in the *AGK3* (*Dritter Astronomische Gesellschaft Katalog*, 1975). The catalog contains position and proper motions of 183,000 stars. An important photographic catalog covering the entire sky to a limiting magnitude of 13 is the result of international undertaking involving 19 observatories. The catalog, known as the *Carte du Ciel* (*CdC*) or *Astrographic Catalogue* (*AC*), provides the positions of roughly 1,500,000 stars in the form of rectangular coordinates, as measured on the plates. *See* ASTROMETRY.

Among the numerous catalogs of astrophysical data is the monumental *Henry Draper Catalog* (*HD*) of spectral classification, which with its extension includes data for 275,000 stars published in 10 volumes. The general acceptance of the U,B,V photometric system since its inception in the 1950s is shown by the compilation of a general catalog containing data for 53,000 stars (*Astronomy and Astrophysics*, suppl., vol. 29, 1977). *See* ASTRONOMICAL SPECTROSCOPY; MAGNITUDE (ASTRONOMY).

L. E. Dreyer's *New General Catalogue of Nebulae and Starclusters* (1888) contains the objects originally classified as nonstellar, with galaxies included as nebulae. The catalog contains 7840 objects, and was supplemented by two index catalogs in 1895 and 1908 with an additional 5386 objects. The NGC and IC numbers assigned in these catalogs remain the most commonly used designations. Since Dreyer's time, the number of known nonstellar optical objects has increased by an order of magnitude, and the data are scattered through the astronomical literature. R. S. Dixon and B. Sonneborn compiled *A Master List of Nonstellar Astronomical Objects* (1980) with approximately 185,000 listings from 270 catalogs, with multiple listings of objects appearing in several catalogs. *See* GALAXY (EXTERNAL); NEBULA; STAR CLUSTERS. [K.A.S.]

Astronomical coordinate systems Systems of spherical coordinates serving to locate astronomical objects on the celestial sphere, which is the sphere of indeterminate radius, with its center at the observer, on the inside-surface of which astronomical objects may be imagined to be projected. The celestial sphere has no physical existence; it is a concept convenient for specifying directions in space.

The system of astronomical spherical coordinates is analogous to the system of latitudes and longitudes on the Earth, the most important difference being that the Earth is viewed from the outside and the celestial sphere is viewed from the inside. Every such system contains two diametrically opposite poles, a north pole and a south pole, the north pole being the one that is visible from the North Pole of the Earth. There is a primary circle, which is a great circle of the sphere, so placed that each pole is 90° from it. The latitudinal coordinate of an object is its angular distance measured northward or southward from the primary circle. Small circles parallel to the primary circle are called parallels. The longitudinal coordinate of an object is the angular distance measured along the primary circle from a specified origin to the point of intersection with a great circle passing through the object and the two poles: the one of the two intersections that is nearest the object is used. The great circles passing through the poles and perpendicular to the primary circle are sometimes called secondary circles.

The astronomical system most used for locating objects is the equatorial system (see illustration). It is the geometrical extension of the terrestrial system. The poles are the intersections of the Earth's axis of rotation with the celestial sphere, and the primary circle is called the celestial equator, or simply equator. The latitudinal coordinate is called declination, and the parallels are parallels of declination. The longitudinal coordinate is called right ascension. The origin is the vernal equinox, which is the point where the Sun crosses the equator about March 21.

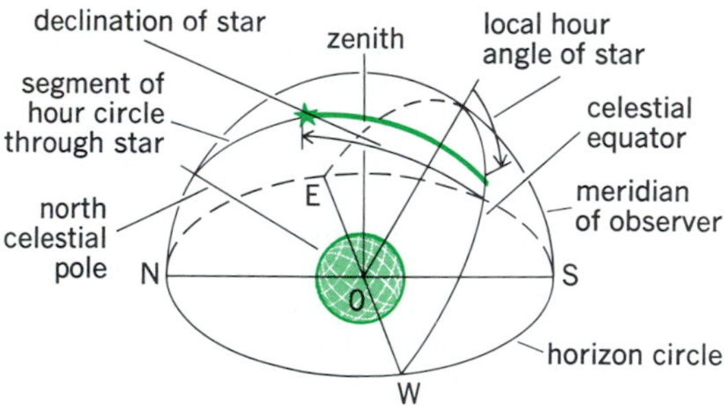

Equatorial astronomical coordinate system.

For some purposes it is convenient to modify the equatorial system by using an origin for the longitudinal coordinate that is fixed with respect to the observer. The origin is then the intersection of the celestial equator with the observer's meridian, and the corresponding coordinate is called the hour angle. Another system much used for theoretical studies of planetary motion is the ecliptic system. The primary circle is the ecliptic which, except for very small deviations, is the path traced out by the Sun on the celestial sphere during a tropical year. A system used for studies of stellar motions is the galactic system. The primary circle is the plane of the galaxy and is called the galactic equator. The coordinates are called galactic latitude and galactic longitude. The horizon system is fixed relative to the observer. The poles are the zenith, which is the point directly overhead, and the nadir. The primary circle is the horizon. The latitudinal coordinate is called altitude if it is measured upward from the horizon, and zenith distance if it is measured downward from the zenith. The longitudinal coordinate is called azimuth. It may be measured in several ways, the commonest origin being the north point of the horizon. *See* LATITUDE AND LONGITUDE. [G.M.C.]

Astronomical observatory A building or group of buildings which houses optical telescopes or the electronics required for radio telescopes or space telescopes, and the astronomers studying the observations. There are over 300 active astronomical observatories, with over 50 of them in the United States. This omits hundreds of amateur telescopes, five or six Earth-orbiting space observatories controlled by radio and computers operated by astronomers on the ground, and at least one high-flying airplane equipped with telescopes. The optical telescopes used at these observatories vary widely in size, from the huge 200-in. (5.1-m) mirror reflector and the 48-in. (1.2-m) Schmidt at Palomar Mountain in California, used mostly to study faint stars, nebulae, and galaxies, to the 10-in. (0.25-m) lens telescopes at Zurich, Switzerland, used to study sunspots. Radio telescopes vary from the 1000-ft (305-m) bowl at Arecibo, Puerto Rico, through the 300-ft (91-m) dish at the National Radio Astronomy Observatory in West Virginia, to arrays of much smaller dishes. The telescopes in orbit outside the Earth's atmosphere are smaller and designed for special purposes, although the Space Telescope, to be put in orbit by NASA in 1985, has an aperture of 95 in. (2.4 m).

The typical optical observatory has a telescope in a building with a hemispherical, rotatable dome in which there is a slot up one side with a cover that can be pulled aside from within, so that the telescope can "see out" (see illustration). Starting about 1890, astronomers took photographs, using the telescope as a large camera, so the observatories have been equipped with darkrooms where photographic plates can be developed, plate files where they are stored, and measuring engines which are used to measure star images on the plates accurately. After 1910 or so, spectrographs were attached to the telescopes to spread the light of a star into a spectrum from red to blue. Photographs of these spectra (spectrograms) are measured on a one-dimensional measuring engine—a microscope mounted on a long, accurate screw. Since the mid-1930s, electronic detectors have been used on optical

Nighttime view of the 200-in. (5.1-m) observatory on Palomar Mountain in California.

telescopes to measure starlight accurately. The electronics in large observatories have become increasingly complex. Since radio and space telescopes are all electronic, modern observatories require electronic technicians and computers. *See* ASTRONOMICAL PHOTOGRAPHY; ASTRONOMICAL SPECTROSCOPY; OPTICAL TELESCOPE; RADIO ASTRONOMY; RADIO TELESCOPE; SCHMIDT CAMERA; TELESCOPE. [T.P.]

Astronomical photography The application of the photographic process to astronomy. Over the past century, photographic observations have made profound contributions to astronomy and astrophysics; and for much of this period, photographs were the best light detectors available for a wide range of applications. Since the mid-1970s, a variety of image detectors utilizing photocathodes or silicon diode arrays have offered detective performance several times better than the best photographic materials. However, it is clear that many astronomers will continue to use photographic techniques for some time to come and for a number of reasons, including low initial cost, relative ease of use, and tradition. In particular, for applications where coverage of a wide field is required, photography is still the detector of choice, because emulsions can be manufactured in much larger formats than electronic detectors.

Nearly all astronomical photography is done with special spectroscopic plates, which are available in several spectral sensitizations. These materials come in several classes of grain sizes. The ultimate accuracy with which photographic photometry, the determination of the amount of light that exposes each picture elements, can be carried out is set by the uncertainties due to graininess. For long exposure times there is a serious breakdown in the reciprocity law that would predict that the photographic sensitivity should depend only on the product of the exposing intensity I and the exposure time t. The loss of sensitivity at low light levels and long exposure times is often called low-intensity reciprocity failure (LIRF). During the 1970s there were major advances in understanding the sources of low-intensity reciprocity failure and in developing various treatments for its reduction. Most of these hypersensitization techniques involve preexposure treatments such as evacuation, baking, bathing, or hypering with hydrogen gas. Many of the procedures work best if the exposures are made in a controlled atmosphere by using special cassettes. Since at least five different mechanisms are involved, it is often beneficial to combine treatments.

Even when they are optimally hypersensitized, astronomical plates have a detective quantum efficiency of only about 2 or 3%. Photocathodes, on the other hand, can have quantum efficiencies as high as 25% or more. A stack of image tubes is often used at the focus of a spectrograph or direct camera, with the image from the final phosphor screen transferred optically onto a film or plate for recording a long exposure. The disadvantages of intensified photography are the distortions, nonuniformities, blemishes, and ion flashes that plague many tubes. Moreover, the tubes must be cooled to reduce the dark current for long exposures. Perhaps the biggest disadvantage for survey work is that the largest photocathodes are rarely 4 in. (100 mm) in diameter. Astronomers have also developed special image tubes which use a nuclear track emulsion to record directly the electrons produced at the photocathode.

Enormous effort has gone into the development of detector systems which give out the image as an electrical signal which can be digitized either for immediate computer processing or magnetic storage for later analysis. These systems divide roughly into two classes. In the first class, a photocathode device such as an image intensifier chain is used to detect the incoming light, while an electronic device such as a television tube or diode array is used to read out the intensifier image. In the second class, silicon diode arrays are illuminated directly.

While the development of new electronic imaging hardware has improved the sensitivity of the big telescopes dramatically, perhaps the greatest advances in astronomical imagery, both photographic and electronic, have come in the area of image processing. Not only are modern high-precision microphotometers commonly available for digitizing photographic images, but elaborate computer schemes for processing and analyzing the images have been developed at several centers. [D.W.L.]

Astronomical reference frame A frame of reference that is fixed relative to celestial objects. The most convenient one for observational purposes is, of course, a geocentric system. The Earth's axis of rotation is used as a coordinate axis. A second reference direction is chosen as the intersection of the Earth's equatorial plane and the plane of the Earth's orbit (the ecliptic). The basic time unit is the Earth's rotation period, the sidereal day. The positions and times of distant astronomical events are determined indirectly. *See* ASTRONOMICAL COORDINATE SYSTEMS.

Reference frames are regarded as more basic as more phenomena appear regular with respect to them. Thus the planetary motions are described naturally in an ecliptic coordinate system centered on the Sun, and the motions of stars as viewed from the Earth appear regular in a coordinate system with its origin at the center of the galaxy. *See* FRAME OF REFERENCE. [B.G.]

Astronomical spectroscopy The use of spectroscopy as a technique for obtaining observational data on the compositions, physical conditions, and velocities of astronomical objects. Astronomical applications of optical spectroscopy from ground-based observatories cover the electromagnetic spectrum from the infrared through the visible to the near-ultraviolet. The advent of space-age techniques—high-altitude balloons, rockets, and satellites—has extended spectroscopic observations through the far-infrared to the domain of radio astronomy, and through the far-ultraviolet to the region of x-rays and gamma rays. *See* GAMMA-RAY ASTRONOMY; RADIO ASTRONOMY; X-RAY ASTRONOMY.

In optical spectroscopy a spectrograph is attached to a telescope, which serves as a light collector. If the telescope is of the reflector type, one distinguishes between prime-focus, newtonian, Cassegrain, and coudé spectrographs, depending on the focus at which they are used. The coudé focus is preferred for high-dispersion work. The image of the celestial body being studied is focused on the spectrograph slit by the telescope objective. The light from the slit passes through the collimator, which is a lens or mirror. The parallel rays of light are dispersed by glass or quartz prisms or, as in most modern spectrographs, by a diffraction grating. The grating is usually "blazed" by shaping its grooves in such a fashion that most of the dispersed light is cast into a chosen order of the spectrum. *See* DIFFRACTION GRATING; OPTICAL TELESCOPE; SPECTROSCOPY.

Until the 1970s, spectra were recorded photographically. When high photographic speeds are required, Schmidt cameras are often used. If only a small field (1 in. or 2–3 cm) in the focal plane suffices, an image tube can be used. The thus intensified image can be registered on a photographic plate or can be scanned electronically, as in the Robinson-Wampler image tube scanner.

Fourier transform spectroscopy, used particularly in infrared work, employs a concept entirely different from the spectrographs described. Instead of being dispersed in a spectrograph, the light of a wide band of wavelengths is passed through a Michelson interferometer with variable spacing of its two apertures. The resulting interferogram, which is an electronic record of the interference signal produced by the interferometer as the separation of the apertures is varied, is converted into a record of intensity versus wavelength by a high-speed computer. *See* INTERFEROMETRY. [L.H.Al.]

Astronomical transit instrument A telescopic instrument adapted to the observation of the passage, or transit, of an astronomical object across the meridian of the observer. The astronomical transit instrument is the classic instrument of positional astronomy, the study of the positions and motions of astronomical objects and the related determination of positions by observation of these astronomical bodies from the Earth (the specific categories of astronomy concerned with these investigations are astrometry and celestial mechanics). The chief variants of the classic design include the meridian circle, the vertical circle, the horizontal transit circle, the broken or prism transit, and the photographic zenith tube.

The modern transit instrument has a telescopic objective with a diameter of 6–10 in. (15–25 cm) and a focal length of 72–90 in. (180–230 cm). The instrument consists of a telescope mounted on a single fixed horizontal axis of rotation. The horizontal axis has a central hollow cube (sometimes a sphere) and two conical semiaxes ending in cylindrical pivots. The objective and eyepiece halves of the telescope are also fastened to the cube of the instrument, perpendicular to the horizontal axis. Rotation of the instrument in its bearings, or wyes, permits the optical axis to sweep only in the plane of the meridian. An accurate clock is the essential ancillary scale by which the transits of the astronomical objects are observed. *See* Astronomical coordinate systems.

The astronomical transit instrument has three interrelated uses. (1) From a known position on the Earth, observations of transits of stars lead to the determination of their right ascension with respect to the astronomical coordinate system, (2) The determination of corrections to the clock may be made by the observation of stars of known position with an instrument situated at a known longitude. (3) Finally, with a knowledge of the positions of the stars observed and of Greenwich time, the longitude of the observer can be computed from observations of the time of transit of a star. *See* Optical telescope. [B.L.Kl.]

Astronomical unit The basic unit of length in the solar system. The astronomical unit (AU) is used to a limited extent for interstellar distances as well through the definition of the parsec (1 pc = 206,265 AU). It is nearly equal to the mean distance a between the center of mass of the Sun and the center of mass of the Earth-Moon system (a = 1.00000023 AU), and for that reason it is often convenient to think of the astronomical unit as the mean distance between the Sun and Earth. *See* Parsec.

The most accurate determination of the length of the astronomical unit in physical units, such as meters, is obtained from phase-modulated continuous-wave (CW) radio signals beamed to other planets. The round-trip travel times of the signals are determined by cross-correlating the returned signal from the planet with the transmitted signal, and as a result, planetary distances are measured directly. CW signals returned from the two Viking orbiters and two landers on the surface of Mars have made possible a determination of the astronomical unit to an accuracy of 65 ft (20 m). With an adopted value of 299,792.458 km/s (186,282.39705 mi/s) for the speed of electromagnetic propagation in vacuum, the value of the astronomical unit from the Viking data is 92,955,807.25 mi (149,597,870.66 km). *See* Doppler effect; Parallax (astronomy). [J.D.A.]

Astronomy The science concerned with the observation and interpretation of the radiation received in the vicinity of the Earth from the component parts of the universe. For the greater part of the history of humanity, the radiation was confined to the visual portion of the electromagnetic spectrum, and the eye was the sole receiver. The invention of the telescope greatly enhanced the power of visual observation of celestial objects, revealing heretofore hidden details of the Sun, Moon, and planets and disclosing the existence of hordes of stars never before seen by the human eye. The increased light-gathering power of the telescope made possible greatly refined observations, which went hand in hand with growing philosophical and theoretical considerations of the nature of the solar system and the universe. These advantages in turn led to the adoption of the heliocentric view of the solar system and the rise of celestial mechanics, which flowered under I. Newton and his followers and gave a detailed physical picture of the motion of celestial objects under gravitational forces.

The telescope made possible the recognition of stars as distant suns through the measurement of their parallaxes and proper motions. Study of the physical nature of the stars became possible when the telescope was subsequently used in conjunction with auxiliary instruments such as the photometer and the spectrograph, designed to measure and analyze the radiation from the stars. The introduction of the photographic plate was the significant step that made possible physical astronomy, or the application of modern physics to astronomy—astrophysics. *See* Astrophysics; Photometry; Spectrography; Star; Sun; Telescope.

The limitation of astronomical observations to the visual portion of the electromagnetic spectrum, imposed by the human eye, was eliminated by the extension of the spectral sensitivity of astronomical instruments and recording devices, including the photographic plate, to the infrared and ultraviolet portions of the spectrum. The most dramatic advance was the development of the radio telescope, which has made radio astronomy a major branch of astronomy.

The observation of the total electromagnetic spectrum of celestial sources came with the lifting of the opaque atmospheric veil by the use of high-altitude balloons, rockets, and craft orbiting well above the Earth's atmosphere. Space probes to the Moon and nearer planets placed astronomical equipment in their vicinity and even upon them. Rockets, high-altitude balloons, and satellites have spawned new branches of astronomy—x-ray and gamma-ray astronomy. The existence of sources of x-rays and gamma rays on the celestial sphere has been revealed in a manner reminiscent of the first detection of discrete radio sources. *See* Celestial sphere; Gamma-ray astronomy; X-ray astronomy; X-ray star; X-ray telescope.

The theoretical and philosophical aspects of astronomy, once concerned with the central position of the human race in a relatively tiny universe, are now concerned with the problems of cosmology and cosmogony. Interpretation of the observed redshift of distant galaxies as evidence for the expansion of the universe is theoretically supported by relativistic cosmology, which presents a picture of the universe in which every galaxy appears to be a center from which all other galaxies are receding with speeds proportional to their distances from that given galaxy. *See* Cosmology; Milky Way Galaxy; Galaxy, external; Infrared astronomy; Pulsar; Redshift; Relativity. [J.A.Hy.]

Astrophysics The application of modern physics to the problems of astronomy. Astrophysics embraces much of the activity of present-day astronomy; specifically excluded are the study of the motions of the planets and satellites (celestial mechanics), the measurement of positions and motions (astrometry), and usually also the structure and dynamics of the Galaxy and other galaxies, although such topics as density wave theory of spiral arms are sometimes consid ered to be part of astrophysics. *See* Astronomy; Milky Way Galaxy.

From an operational point of view, astrophysics differs from terrestrial physics in that it is primarily an observational subject, although certain special problems lend themselves to experimental treatment. The basic problem is often the measurement of the quantity and quality, that is, the energy distribution and polarization, of the electromagnetic radiation of the Sun, stars, planets, and nebulae, and the study of their spectra.

The field of experimental astrophysics involves measurement of shapes and strengths of spectral lines emitted under controlled conditions of temperature and pressure. Techniques involve the use of shock tubes, atomic beams, controlled arcs, beam foils, laser excitation of discrete levels, and so forth. Also involved are measurements of cross sections for atomic and molecular collisional excitation and ionization, low-energy nuclear transformations, and even synthesis of molecular fragments that may be found in interstellar space.

Theoretical astrophysics constitutes a branch of theoretical physics and embraces many subjects that usually are considered the province of the latter. For purposes of discussion, astrophysics may be divided somewhat as follows, according to the types of celestial bodies that are involved.

Solar physics includes all phenomena connected with the Sun and overlaps with geophysics in the consideration of solar-terrestrial relationships, for example, the connection between solar activity and auroras, magnetic storms, and sudden ionospheric disturbances. *See* SUN.

The physics of the solar system includes the nature of planetary atmospheres and interiors and the chemical and physical constitution of comets, meteors, and small particles such as those that constitute the zodiacal cloud. In addition, much attention is paid to interplanetary plasmas (the solar wind) and magnetic fields, and to the magnetosphere of the Earth and Jupiter. *See* COMET; COSMOCHEMISTRY; MAGNETOSPHERE; METEOR; METEORITE; PLANETARY PHYSICS; SOLAR WIND.

The study of stellar atmospheres, including certain aspects of the solar atmosphere, constitutes an important and active field of astrophysics. Much effort is devoted to gaseous nebulae and the interstellar medium. The former includes examples of both thermal excitation, as in Orion and planetary nebulae, and nonthermal excitation, as in the Crab Nebula. The interstellar medium pervades the galactic plasma and extends beyond it as the galactic halo. Stellar structure and evolution trace the development of a star from its condensation from the interstellar medium to its eventual fate as a white dwarf, neutron star, or black hole. *See* BINARY STAR; BLACK HOLE; INTERSTELLAR MATTER; NEBULA; NOVA; STELLAR EVOLUTION; SUPERNOVA; WHITE DWARF STAR.

Radio astronomy data and observations secured from above the Earth's atmosphere have greatly stimulated the development of high-energy astrophysics, particularly the study of the acceleration of charged particles to high energies in supernova remnants, pulsars, and quasistellar sources. Much attention is paid to galaxies, both normal and exotic. *See* GALAXY, EXTERNAL; PULSAR; QUASAR.

Verifications of general relativity are astronomical or astrophysical. In addition to classical tests such as deflection of light by the Sun or advance of the perihelion of Mercury, predictions of dissipation of energy by gravity waves have been verified by observations of a pulsar binary. General relativity has to be applied in interpretation of black holes, which are difficult to observe but are believed to occur in certain binary star systems and nuclei of galaxies. It is also used in constructing large-scale models of the physical universe, in cosmology, and in describing the first moments of the universe following the "big bang." *See* COSMOLOGY; RELATIVITY. [L.H.Al.]

Astrophysics, high-energy The study of astronomical phenomena which generate particles or radiation with energies which range from the order of 10^3 to 10^{21} eV, the highest observed.

Many experiments before World War II showed that cosmic rays are very energetic particles striking the atmosphere, where they produce secondary particles that reach the ground. These cosmic rays were the first indication that high-energy processes are important in astrophysics. The many advances in radio and radar equipment during the war led to a tremendous increase in radio astronomy after 1945. The first radio source identifications were made: Cas A, a supernova remnant in the Milky Way; and Cyg A, a distant peculiar galaxy with enormous radio luminosity. These sources radiate by synchrotron radiation, produced by high-energy electrons spiraling in a magnetic field. While the observed radio photons have very low energy, as low as 1 microelectronvolt, the electrons producing the radio waves must have energies greater than 1 gigaelectronvolt, quite comparable to the energies of cosmic rays, which range from 10^9 to 10^{21} eV. *See* COSMIC RAYS; RADIO ASTRONOMY; SYNCHROTRON RADIATION.

A rocket flight in 1962 discovered a source in the constellation Scorpius, now known as Sco X-1, the brightest nonsolar x-ray source in the sky. This discovery led to the development of the totally new field of x-ray astronomy, which now dominates high-energy astrophysics. *See* X-RAY ASTRONOMY.

Astrophysical sources of high-energy particles or radiation include supernovas, neutron stars, binary systems containing a compact object, supermassive black holes, gamma-ray burst sources, x-ray bursters, and the x-ray background. *See* BINARY STAR; BLACK HOLE; GAMMA-RAY ASTRONOMY; NEUTRON STAR; NOVA; PULSAR; QUASARS; SUPERNOVA; X-RAY STAR.

High-energy astrophysics is a field whose evolution has been driven by the very rapid improvement of instruments and telescopes. The x-ray telescopes used to detect Sco X-1 and other bright sources in the Milky Way were not really telescopes, but collimators. A collimator is like a cardboard mailing tube: it limits the field of view to a particular direction, but it does not concentrate light (see illustration). However, polished metal surfaces will reflect x-rays which arrive at grazing incidence, allowing the construction of true x-ray telescopes. The first satellite using such a detector for objects other than the Sun was the *Einstein Observatory*, flown by NASA, which lasted from 1978 to 1981. This telescope was 1000 times more sensitive than any previous x-ray telescope. *See* SATELLITE ASTRONOMY; SCIENTIFIC SATELLITES; X-RAY TELESCOPE. [E.L.Wr.]

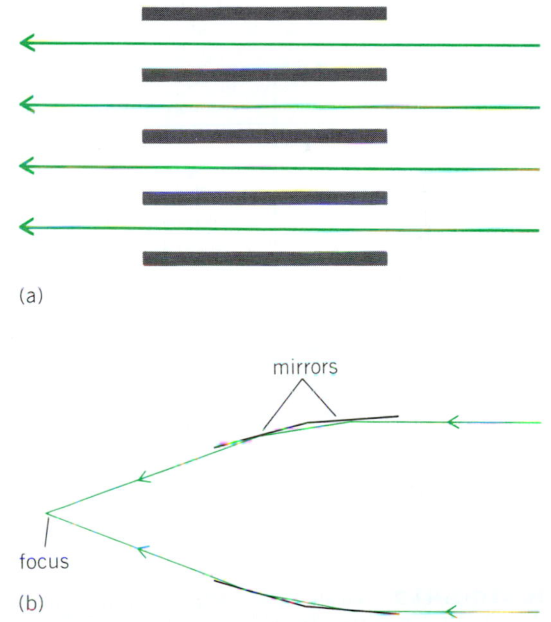

Two types of x-ray instruments: (*a*) stacked x-ray collimators, and (*b*) a grazing-incidence x-ray telescope.

Asymmetric synthesis The chemical synthesis of a pure enantiomer or of an enantiomorphic mixture in which one enantiomer predominates, without the use of resolution. Absolute asymmetric synthesis is achieved by providing dissymmetric conditions by purely physical means such as circularly polarized light or, more commonly, by an optically active reagent. Where a new asymmetric center is produced, the process may be called asymmetric induction, and the selective destruction of one member of an enantiomorphic pair may be called asymmetric decomposition.

Few absolute asymmetric syntheses have been accomplished experimentally, although the natural occurrence of optically active substances presupposes significant numbers of such syntheses early in the evolutionary process. Asymmetric synthesis generally involves introduction of a new asymmetric center into a pure optically active compound; since the original substance is not symmetrical, the probability of producing exactly equal quantities of the new diastereoisomers (epimers) is small. Asymmetric synthesis in the laboratory serves principally for the determination of configurations and for study of reaction mechanisms rather than practical synthesis of optically active products. *See* CONFORMATIONAL ANALYSIS; OPTICAL ACTIVITY; RACEMIZATION; STEREOCHEMISTRY. [W.R.V.]

Asymptote A line that is a limit of lines tangent to a curve as the contact points of those tangents approach infinity along the curve. Thus, an asymptote of a curve is an ordinary line

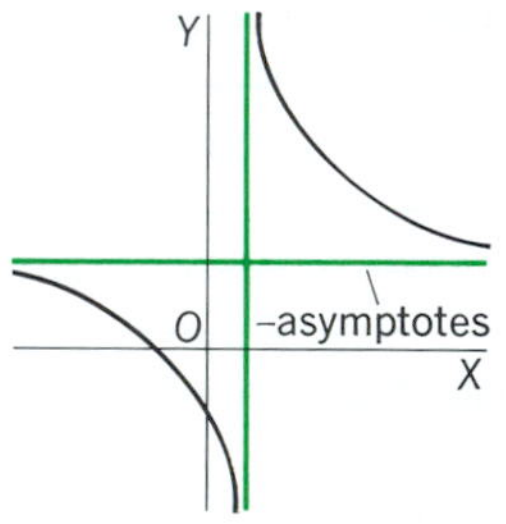

Asymptotes of a hyperbola.

(that is, not the "line at infinity") that is tangent to a curve at the points in which the curve intersects the line at infinity (see illustration). [L.M.Bl.]

Atelostomata A superorder of the Echinoidea, subclass Euechinoidea, characterized by having a rigid, exocyclic test, and lacking a lantern, or jaw, apparatus. The included orders are the Holasteroida and Spatangoida. *See* ECHINODERMATA; ECHINOIDEA; EUECHINOIDEA; HOLASTEROIDA; SPATANGOIDA. [H.B.F.]

Athalamida An order of Granuloreticulosia in which the naked amebas form branched, threadlike interconnected pseudopodia (reticulopodia; see illustration). Species are known from fresh, salt, and brackish water. General characteristics are difficult to select. Heterogeneity may extend even to a genus; described species of *Biomyxa* differ appreciably in morphology and exhibit, for example, uninucleate and multinucleate conditions in different species (although these may represent young and mature stages in life cycles). In addition to *Biomyxa*, the genera *Arachnula*, *Gymnophrys*, and *Pontomyxa* have been assigned to this order. *See* GRANULORETICULOSIA; PROTOZOA; RHIZOPODEA; SARCODINA; SARCOMASTIGOPHORA. [R.P.H.]

***Biomyxa vagans*. (After R. P. Hall, Protozoology, Prentice-Hall, 1953)**

Athecanephria An order of Pogonophora, a group of elongate, tentaculated, tube-dwelling, sedentary, nonparasitic marine worms lacking a digestive system. The coelom in the tentacular region is sac-shaped, with the two ducts set far apart from one another, and their excretory (osmoregulatory) sections lie close to the lateral cephalic blood vessels. The Athecanephria contain two families: Siboglinidae and Oligobrachiidae, with two and five genera respectively. *See* POGONOPHORA; THECANEPHRIA. [E.B.Cu.]

Atheriniformes An order of actinopterygian fishes that includes the flyingfishes, needlefishes, killifishes, silversides, and their allies. Atheriniform fishes date from the Upper Cretaceous. Modern forms are grouped in 3 suborders, 16 families, about 170 genera, and probably 700–800 species. They chiefly inhabit fresh, brackish, and oceanic surface waters of the tropics and subtropics, but some enter the temperate zones, both in fresh water and the sea. They are circumtropical in distribution, abounding in fresh waters of Africa and the Americas and in all warm seas. None lives in deep oceanic waters. Atheriniformes are mostly small, although some needlefishes reach 5 ft (1.5 m) or more in length; most have a single dorsal fin of soft rays, but the silversides have a short anterior spinous dorsal; the pectoral fin is placed high on the side; the pelvic fin, if present, is abdominal, subabdominal, or thoracic and has six (occasionally seven) or fewer rays, one of which may be a spine; scales are usually cycloid, sometimes

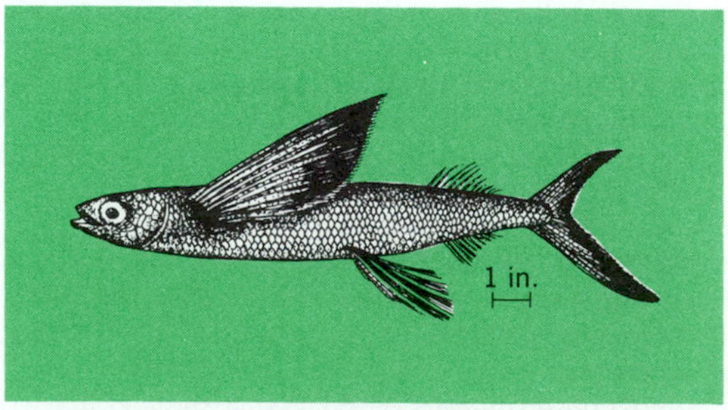

California flyingfish (*Cypselurus californicus*). (After G. B. Goode, Fishery Industries of the U.S., 1884)

ctenoid, and rarely absent. The most familiar families include flyingfishes (see illustration) and halfbeaks (Exocoetidae), the former with stiffened, elongated rays on the paired fins that permit gliding in air; killifishes (Cyprinodontidae); live-bearers (Poeciliidae), among them guppies, swordtails, platyfishes, mollies, and mosquito fishes; and silversides (Atherinidae), that abound on marine beaches and reefs. Viviparity is more frequent and better understood in Atheriniformes than in any other order of bony fishes, occurring in no fewer than eight families. *See* ACTINOPTERYGII; FLYINGFISH; OSTEICHTHYES; TELEOSTEI.

[R.M.B.]

Atlantic Ocean The large body of sea water separating the continents of North and South America in the west from Europe and Africa in the east and extending south from the Arctic Ocean to the continent of Antarctica. The Atlantic is the second largest ocean water body and in area covers nearly one-fifth of the Earth's surface. The two major divisions, North and South Atlantic oceans, have the Equator as the common boundary. The North Atlantic, because of projecting land areas and island arcs, has numerous subdivisions. These include three large mediterranean-type seas, the Mediterranean Sea, the Gulf of Mexico plus Caribbean Sea, and the Arctic Ocean; two small mediterranean-type seas, the Baltic Sea and Hudson Bay; and four marginal seas, the North Sea, English Channel, Irish Sea, and Gulf of St. Lawrence. Parts of the Atlantic are given special names but lack precise boundaries, such as the Bahama Sea, Irminger Sea, Labrador Sea, and Sargasso Sea.

The mean depth of the Atlantic Ocean is 12,960 ft (3868 m), and its volume is 76,300,000 mi^3 (318,000,000 km^3). Broad shelves with depths less than 660 ft (200 m) are found in the region of the North Sea and the British Isles, on the Grand Banks of Newfoundland, and off the coasts of northeastern South America and Patagonia. The Mid-Atlantic Ridge, which extends from the Arctic Ocean to 55°S, is less than 9800 ft (3000 m) beneath the surface and is characterized by a pronounced relief. It separates the east and west Atlantic troughs, both of which have relatively uniform relief.

Three marked east-west ridges—the Greenland-Scotland Ridge in the North Atlantic and the Walvis and Rio Grande Ridges in the South Atlantic—and several less-conspicuous east-west rises separate the two Atlantic troughs into a series of basins including the West European, Canary, and Angola in the eastern Atlantic and the North American, Brazilian, and Argentine basins in the western Atlantic.

Islands in the Atlantic are mostly of volcanic origin. The Bermudas are the northernmost coral reefs, rising from an old submarine volcanic cone. Some islands, such as the British Isles, are continental in character. *See* OCEANIC ISLANDS; WEST INDIES.

The primary circulation of surface winds over the Atlantic Ocean is characterized by a zonal distribution pattern oriented in an east-west direction. The greatest storm frequency, more than 30% in winter, is in the zone of the prevailing westerlies. Air temperatures also follow a zonal pattern of distribution. They are lower in the South Atlantic than in the North Atlantic, and lower in the tropics and subtropics over the eastern Atlantic than they are in the same latitudes over the western Atlantic. Maximum precipitation occurs in the doldrum zone (80 in. or 2000 mm/year). Precipitation also is relatively great in the zone of westerlies but is low in the trade-wind zones.

Sea ice is formed in the northernmost and southernmost parts of the Atlantic Ocean. From these areas drift ice moves equatorward into neighboring regions where it becomes a hazard to sea traffic and limits fishing. Many icebergs drift southward into the sea lanes of the North Atlantic. Most of these have their origin in the valley glaciers of western Greenland. Icebergs generally drift south of the Grand Banks, and some are known to have drifted southeast of Bermuda. In the South Atlantic large, tabular icebergs separate from the Antarctic ice shelf and drift northward. *See* ICEBERG; SEA ICE.

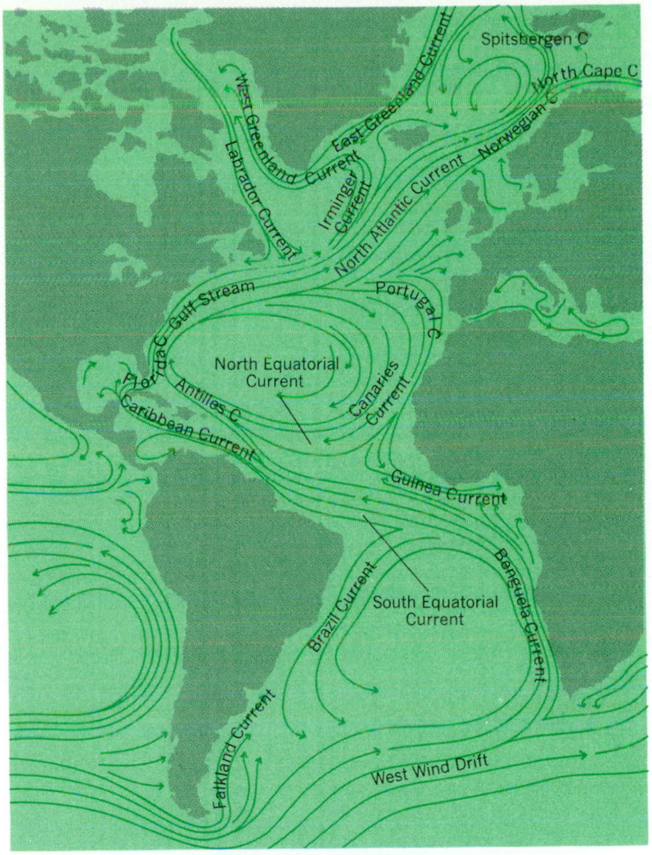

Currents of the Atlantic Ocean. ***(Adapted from J. Bartholomew, Advanced Atlas of Modern Geography, McGraw-Hill, 3d ed.. 1957)***

Surface currents in the Atlantic Ocean flow in much the same direction as the prevailing surface winds (see illustration). Deflections from these directions are caused by the bottom topography and the latitude or increased effect of Coriolis forces. The fairly constant flow of the North and South Equatorial currents is sustained largely by the trade winds. As a result, warm water is piled up along the poleward borders of these currents and on the western sides of the Atlantic Ocean. *See* ANTARCTIC OCEAN; ARCTIC OCEAN; CARIBBEAN SEA; GULF OF MEXICO; GULF STREAM; OCEAN CIRCULATION.

The surface water in certain areas takes on a particularly high density in winter under the influence of climatic conditions. These water masses sink to a depth where the surrounding waters have a corresponding density and then spread out at that level. At the same time they are constantly mixing with the surrounding waters. In this way a multistoried stratification arises. Compared with that of the Indian and Pacific oceans, the deep circulation in the Atlantic Ocean is very vigorous, and the deeper water is therefore rich in oxygen. The abundance of nutrients permits a greater rate of organic production where the nutrient-rich waters nearly reach the surface, as in the Antarctic waters. *See* SEAWATER; SEAWATER FERTILITY.

The semidiurnal tidal form predominates in the Atlantic Ocean. The mean tidal range is about 3.3 ft (1 m) in the open ocean, but it decreases to 6.3 in. (16 cm) off Rio Grande do Sul in southern Brazil and to 3.5 in. (9 cm) off Puerto Rico. Tidal ranges increase beyond broad shelves under favorable physical conditions. The tides of the mediterranean and marginal seas are cooscillations of the tides of the Atlantic Ocean. *See* TIDE.

The Atlantic Ocean, especially the North Atlantic, is by far the most important bearer of the world's sea traffic. Favorable trends include increased transportation capacities for handling bulk goods, regular weather observations for the safety of air and sea traffic by weather ships in selected positions, and the observation and reporting of drifting icebergs by the International Ice Patrol. Communication facilities, including telegraph and telephone cables and radio stations, have been improved and increased in number. [G.O.D.]

Atmosphere The gaseous envelope surrounding a celestial body. The terrestrial atmosphere, by its composition, control of temperature, and shielding effect from harmful wavelengths of solar radiation, makes possible life as known on Earth. The atmosphere, which is retained on the Earth by gravitational attraction and in a large measure rotates with it, is a system whose chemical and physical properties and fields of motion constitute the subject matter of meteorology. The changing atmospheric conditions which affect the environment, particularly temperature, wind, humidity, cloudiness, and precipitation, constitute weather, and the synthesis of these conditions over a period determines the climate at any place. *See* Climatology; Meteorology; Weather.

The average atmospheric pressure at the Earth's surface is about 1013 millibars (101.3 kilopascals) and the density 1.2 $kg \cdot m^{-3}$, and these vary by only a few percent over the globe. They both decrease rapidly and roughly exponentially with height, and at several earth radii the density can be said to have fallen to that of interplanetary space.

The atmosphere is thought to have developed as the result of chemical and photochemical processes combined with differential escape rates from the Earth's gravitational field. Chemical abundances in the atmosphere, therefore, are not directly related to cosmic abundances; in particular, the atmosphere is highly oxidized and contains very little hydrogen.

The atmosphere, apart from its highly variable water-vapor content in the troposphere, its liquid droplets and solid matter in suspension, and its variable ozone content in the stratosphere, is well mixed and constant in composition up to about 62 mi (100 km). This region is termed the homosphere. At higher levels where there is little mixing, diffusive separation tends to take place, with the lighter elements becoming progressively more dominant with height. Moreover, in this region oxygen and the minor constituents, such as carbon dioxide and water vapor, are dissociated by solar ultraviolet radiation. At about 186 mi (300 km) atomic oxygen probably becomes the most important constituent, until about 500 mi (800 km) where helium and hydrogen in turn predominate. This region of highly variable composition is termed the heterosphere. Ionization of the various constituents, also due to absorption of ultraviolet solar radiation, becomes a major factor above about 37 mi (60 km), the base of the ionosphere, which is of major importance to radio communications. At levels above 370–500 mi (600–800 km), collisions between atmospheric particles become so infrequent that some traveling outward may escape from the atmosphere. This region is termed the exosphere. Table 1 gives a summary of the composition of the atmosphere. *See* Air pollution.

A convenient division of the atmosphere is by spherical shells, spheres, each characterized by the way its temperature varies in the vertical, and with tops denoted by pauses, as in the illustration. The troposphere includes the layer closest to the Earth and is the seat of all the important weather phenomena affecting the environment. The general circulation of the troposphere includes wind systems on all scales—prevailing winds, monsoons, long waves, anticyclones and depressions, fronts, hurricanes, thunderstorms, and shower clouds—each system being associated with characteristic weather patterns.

Other divisions of the atmosphere are the stratosphere, mesosphere, and thermosphere. The stratosphere, extending from about 6–10 mi to about 31 mi (10–16 km to 50 km), is generally very stable, with no weather in the popular sense. The mesosphere, extending from 34 mi (55 km) to about 50 mi (80 km), contains the mesopause, a region where noctilucent clouds are sometimes formed, and most of the so-called D layer, a region of low ionization. The thermosphere, extending from 50 mi (80 km) to the edge of the atmosphere, receives energy directly from solar radiation and exhibits phenomena related to solar activity, such as the aurora. *See* Airglow; Aurora; Cloud physics; Mesosphere; Noctilucent clouds; Stratosphere; Weather forecasting and prediction. [R.J.Mu.]

The chemical composition and total weight of the atmosphere have varied through time. Today the atmosphere contains a small number of major components (nitrogen, oxygen, argon) and a very large number of minor components (including carbon dioxide, helium, krypton, xenon, hydrogen, and oxides of nitrogen). Each component exerts a pressure, which

Table 1. Composition of the atmosphere*

Molecule	Fraction by volume near surface	Vertical distribution
Major constituents		
N_2	7.8084×10^{-1}	Mixed in homosphere; photochemical dissociation high in thermosphere
O_2	2.0946×10^{-1}	Mixed in homosphere; photochemically dissociated in thermosphere, with some dissociation in mesosphere and stratosphere
Ar	9.34×10^{-3}	Mixed in homosphere with diffusive separation increasing above
Important radiative constituents		
CO_2	3.1×10^{-4}	Mixed in homosphere; photochemical dissociation in thermosphere
H_2O	Highly variable	Forms clouds in troposphere; little in stratosphere; photochemical dissociation above mesosphere
O_3	Variable	Small amounts, 10^{-8}, in troposphere; important layer, 10^{-6} to 10^{-5}, in stratosphere; dissociated above
Other constituents		
Ne	1.82×10^{-5}	Mixed in homosphere with diffusive separation increasing above
He	5.24×10^{-6}	
Kr	1.14×10^{-6}	
CH_4	1.5×10^{-6}	Mixed in troposphere; dissociated in upper stratosphere and above
H_2	5×10^{-7}	Mixed in homosphere; product of H_2O photochemical reactions in lower thermosphere, and dissociated above
NO	$\sim 10^{-8}$	Photochemically produced in stratosphere and mesosphere

*Other gases, for example, CO, N_2O, and NO_2, and many by-products of atmospheric pollution also exist in small amounts.

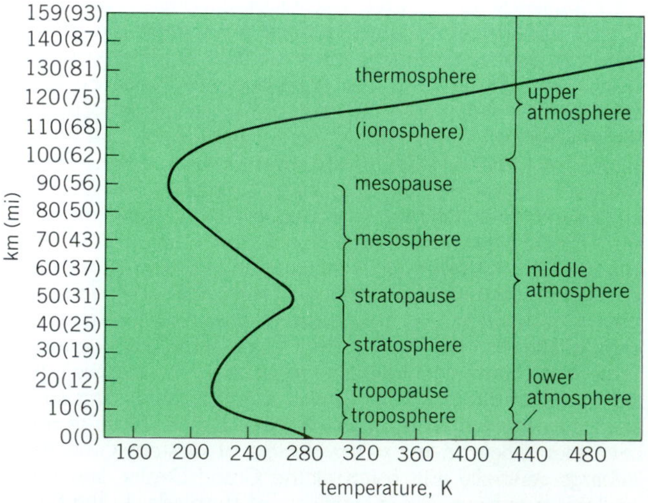

Thermal structure of the atmosphere, showing major divisions.

Table 2. Summary of data on the probable chemical composition of the atmosphere during stages 1, 2, and 3

Components*	Stage 1	Stage 2	Stage 3
Major components: $P > 10^{-2}$ atm ($P > 1$ kPa)	CH_4, H_2 (?)	N_2	N_2, O_2
Minor components: 10^{-2} atm $> P > 10^{-4}$ atm (1 kPa $> P$ 10 Pa)	H_2 (?), H_2O, N_2, H_2S, NH_3, Ar	H_2O, CO_2, A, O_2 (?)	Ar, H_2O, CO_2
Trace components: 10^{-4} atm $> P > 10^{-6}$ atm (10 Pa $> P >$ 0.1 Pa)	He	Ne, He, CH_4, NH_3 (?), SO_2 (?), H_2S (?)	Ne, He, CH_4, Kr

*P = atmospheric pressure. Pa = pascal.

is essentially constant for some components at sea level and which is variable in time and space for other components. At present, gases like ammonia, NH_3, methane, CH_4, and hydrogen can exist in the atmosphere as trace components only, because they are unstable in the presence of the large quantities of oxygen. Conversely, oxygen could not have been a major component of an atmosphere in which ammonia, methane, and hydrogen were abundant.

Free oxygen is somewhat of an anomaly on the Earth. Rocks more than a few feet below the Earth's surface are out of equilibrium with free oxygen and are oxidized in contact with the atmosphere. There are two predominant theories concerning the origin of oxygen in the Earth's atmosphere. The first proposes that atmospheric oxygen has been produced through geologic time by the continuing effect of photosynthesis, during which carbon is effectively separated from oxygen in carbon dioxide. It can be shown that nearly all of the oxygen produced by photosynthesis during a given period of time is lost by plant decay, but that the small amount not lost could account for the present rather large quantity of atmospheric oxygen.

The second theory proposes an alternative way of producing free oxygen. Ultraviolet light from the Sun decomposes water molecules in the upper atmosphere. Most of these recombine, but there is a finite possibility that a given hydrogen atom will manage to escape from the Earth's atmosphere before recombination has taken place. Oxygen atoms, being 16 times as heavy as hydrogen atoms, escape very much more slowly or not at all. The decomposition of water vapor followed by hydrogen escape is therefore a distinctly plausible manner of generating free atmospheric oxygen. A strongly limiting factor for oxygen production by this mechanism is the formation of ozone, O_3, as a by-product of the photodissociation of water. Because it absorbs ultraviolet light very readily, ozone tends to form a screen preventing it from reaching the lower levels of the atmosphere, where water vapor is abundant. The oxygen content of the atmosphere today is therefore only slightly influenced by hydrogen loss from the upper atmosphere; rather, it depends almost exclusively on the operation of a feedback system which links the rate of oxygen production during photosynthesis to the rate of oxygen use by weathering and the decay of organic matter. *See* ATMOSPHERIC OZONE.

Interpretations based on paleontological and geochemical evidence seem to lead quite naturally to the threefold division in Table 2 of the history of the Earth's atmosphere. During the first stage, very shortly after the accretion of the Earth, the atmosphere may have been quite reducing. Reduced gases issuing from volcanoes probably would have given rise to an atmosphere consisting of methane with minor quantities of hydrogen, nitrogen, and ammonia. After metallic iron was removed from the upper mantle, the oxidation state of volcanic gases probably approached its present value, methane was replaced by carbon dioxide, and ammonia was converted to nitrogen. In the atmosphere during this (its second) stage, nitrogen was its dominant component, and carbon dioxide and argon were its most important minor constituents.

The third stage opened when the rate of oxygen production by photosynthesis became sufficiently great so that oxygen became more than a trace component of the atmosphere. The transition from stage 2 to stage 3 may well have occurred about 1.8×10^9 years ago. Since the opening of stage 3, the oxygen pressure has climbed to its present value by a path which was probably simple but which could have been complex in detail. During this period the nitrogen, neon, argon, krypton, and xenon pressures have also gradually climbed to their present value, while the helium and CO_2 pressure have remained reasonably constant, the former suspended between the rate of input and the rate of escape, the latter controlled in large part by reactions which involve carbonate and silicate minerals. [H.D.Ho.]

Atmospheric acoustics The science of sound in the atmosphere. The term is usually reserved for situations where departures of the atmosphere from an ideal homogeneous medium affect propagation. Atmospheric acoustics is concerned with sound outdoors rather than indoors. Infrasound, as well as sound of audible frequencies, is within the scope of the subject.

Audible sound and infrasound with wave periods somewhat shorter than 30 s propagate through the atmosphere primarily along rays. A ray is the trajectory (Fig. 1*a*) of a point that, relative to a person moving with the ambient wind, appears to move with speed *c* (the speed of sound relative to air) normal to a wavefront. Relative to a stationary coordinate system, the

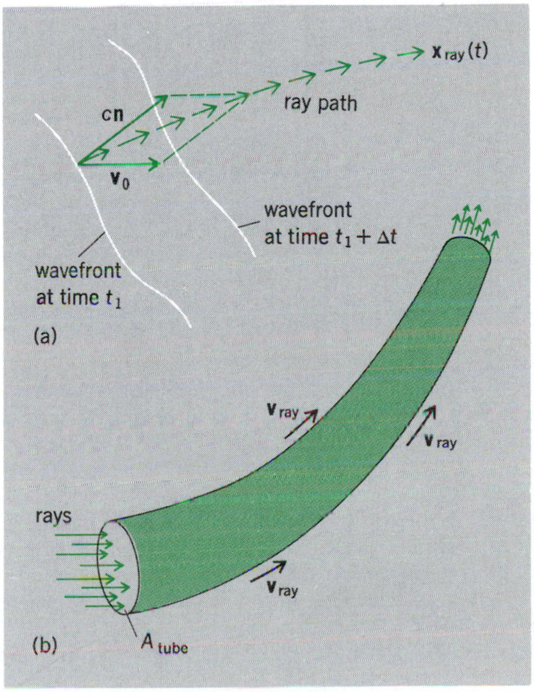

Fig. 1. Ray acoustics. (*a*) Definition of a ray path for a moving medium. (*b*) Ray tube formed by adjacent rays in an inhomogeneous medium.

ray velocity $d\mathbf{x}_{ray}/dt$ is increased by the wind velocity, such that Eq. (1) results, where $\mathbf{n}$ is the unit vector normal to a wave-

$$d\mathbf{x}_{ray}/dt = c\mathbf{n} + \mathbf{v}_0 \quad (1)$$

front and $\mathbf{v}_0$ is the equilibrium fluid velocity. If c and $\mathbf{v}_0$ vary with position, as is typical in the actual atmosphere, then the directions of the normal vector $\mathbf{n}$ and of the ray velocity $d\mathbf{x}/dt$ will in general change along the path; the ray will be bent, or refracted. The general behavior or ray in the atmosphere can be deduced from ray-tracing equations that are derived from Eq. (1). *See* REFRACTION OF WAVES.

General features of long-range ray propagation can be explained qualitatively from a graph of effective sound velocity c_{eff}, sound speed plus wind component in plane of propagation, versus height (Fig. 2). Rays can be trapped in either a

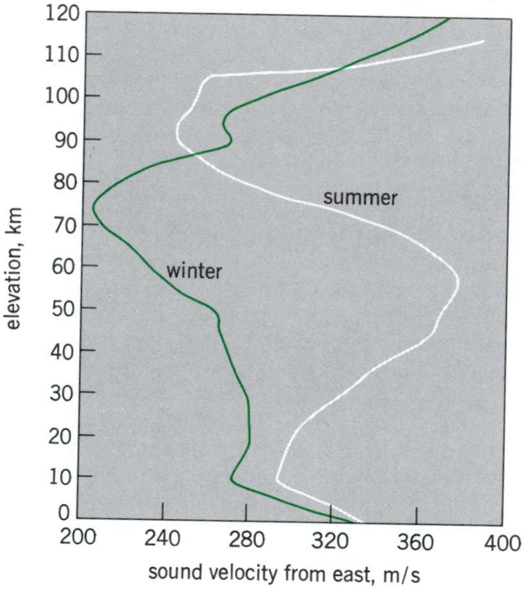

Fig. 2. Model atmospheric profiles of effective sound speed c_{eff} versus height for propagation east to west in northeastern United States. *(After D. Rind and W. E. Donn, Further use of natural infrasound as a continuous monitor of the upper atmosphere. J. Atmos. Sci., 32:1694–1704, 1975)*

low- or an upper-atmosphere sound channel where c_{eff} has a minimum. The greatest height attainable by a ray is its upper turning point, which is also the height for which c_{eff} is maximum along the ray path. If a path has a lower turning point, then at that altitude c_{eff} is the same as at the upper turning point. If there is some altitude at which c_{eff} is larger than at the ground, then it is possible for a ray proceeding obliquely upward from a source at an intermediate altitude to bend downward such that it eventually reaches the ground.

The magnitude and direction of upper atmosphere winds largely determine whether c_{eff} in the stratosphere can exceed its value at the ground. Since winds in the altitude range of 12–50 mi (20–80 km) reverse direction in spring and fall at temperate latitudes, long-range propagation east to west or west to east is markedly different in summer from that in winter. Local meteorological conditions such as inversions (temperature increasing with height near the ground) can also cause a bending back to the ground of rays initially proceeding obliquely upward.

A spectacular consequence of a c_{eff} profile that initially decreases with height, reaches a minimum, then increases to a value higher than that at the ground, is the existence of abnormal zones of audibility. Points at intermediate horizontal distances may receive no sound, while audible sound is received at greater distances.

Refraction can cause the existence of shadow zones within which no ray passes. Since wind velocity near the ground increases with height, shadow zones are frequently found upwind of a source.

Dissipation of acoustic energy in the atmosphere is caused by viscosity, thermal conduction, and molecular relaxation. The last arises because fluctuations in apparent molecular vibrational temperatures lag in phase the fluctuations in translational temperatures. The vibrational temperatures of significance are those characterizing the relative populations of O_2 and N_2 molecules in ground and first-excited vibrational quantum states. Since collisions with H_2O molecules are much more likely to induce vibrational state changes than are collisions with other O_2 or N_2 molecules, the sound attenuation varies markedly with absolute humidity.

Larger-scale turbulence in the atmosphere causes the effective wave speed to fluctuate from point to point, so a nominally smooth wavefront develops ripples. Consequently, the amplitude of the sound at a distant point will fluctuate with time and with small displacements in position.

Infrasound waves with periods longer than a minute are strongly affected by the Earth's gravity. The dominant features of some observed disturbances at stratospheric and ionospheric heights can be interpreted in terms of the model of planar sound waves in an isothermal atmosphere. In this model, there is a forbidden range of angular frequencies ω in which sound waves cannot propagate, namely $\omega_b<\omega<\omega_a$, where ω_a and ω_b, are given by Eqs. (2). (The wave period corresponding to

$$\omega_a = (\gamma/2)g/c \quad (2a)$$

$$\omega_b = (\gamma - 1)^{1/2}g/c \quad (2b)$$

ω_b is of the order of 5 to 10 min in the lower atmosphere.) Possible planar waves are divided into ordinary acoustic waves ($\omega<\omega_a$.) and acoustic-gravity waves ($\omega<\omega_b$).

Acoustic-gravity waves can be channeled in the lower atmosphere and along the ground such that they can carry infrasonic signals over large horizontal distances. Thus, for example, pressure waves generated by large explosions and traveling at slightly less than the sound speed at the ground can be detected at distances greater than halfway around the globe by microbarograph instrumentation. *See* SOUND. [A.D.P.]

Atmospheric electricity The electrical processes constantly taking place in the lower atmosphere. This activity is of two kinds, the intense local electrification accompanying storms, and the much weaker fair-weather electrical activity over the entire globe, which is produced by the many electrified storms continuously in progress over the Earth. The mechanisms by which storms generate electric charge are unknown, and the role of atmospheric electricity in meteorology has not been determined.

Almost all precipitation-producing storms throughout the year are accompanied by energetic electrical activity. The most intense of these are the thunderstorms, in which the electrification attains values sufficient to produce lightning. Electrical measurements show that most other storms, even though they do not give lightning, are also quite strongly electrified. The electric fields of thunderstorms cause three currents to flow, each of a few amperes: lightning, point discharge from the ground beneath, and conduction in the surrounding air. Because the external field and conductivity are greatest over the top of the cloud, most of the conduction current flows to the ionosphere, the upper, highly conductive layer of the atmosphere. *See* THUNDERSTORM.

Fair-weather measurements, irrespective of place and time, show the invariable presence of a weak negative electric field caused by the estimated several thousand electrified storms continually in progress. Together these storms cause a 2000-A current from the earth to the ionosphere that raises the ionosphere to a positive potential of about 300,000 V with respect to the earth. This potential difference is sufficient to cause a return flow of positive charge to the earth by conduction through the intervening lower atmosphere equal and opposite to the thunderstorm supply current. The fair-weather field is simply the voltage drop produced by the flow of this current through the atmosphere. Because the electrical resistance of the atmosphere decreases with altitude, the field is greatest near the Earth's surface and gradually decreases with attitude until it vanishes at the ionosphere.

No importance is presently attached to fair-weather atmospheric electricity except that according to some theories it is responsible for the initiation of the thunderstorm electrification process. *See* CLOUD PHYSICS; LIGHTNING; SFERICS; STORM DETECTION; TORNADO. [B.V.]

Atmospheric entry The entry of a space vehicle into the atmosphere of a planet. Principal problems during entry are deceleration and heating, the severity of which is governed primarily by the type of vehicle and its entry velocity. Entry vehicles are either ballistic or lifting. A ballistic vehicle is acted upon only by gravity and aerodynamic drag forces, while a lifting vehicle also makes use of aerodynamic forces which permit maneuvering. Two types of natural entry bodies are recognized: those coming from interplanetary space (meteoroids and comets), and those thought to come from the Moon (termed tektites).

The maximum deceleration that is caused by aerodynamic forces is determined primarily by the velocity and angle of entry of the vehicle and the density of the atmosphere; it is almost independent of the vehicle size, weight, and shape, although the altitude at which maximum deceleration occurs is not. Maximum deceleration increases rapidly with increasing entry angle. Fairly steep entry angles are required for ballistic vehicles at greater than satellite speed to reduce the centrifugal force that tends to make the vehicle skip out of the atmosphere. Therefore ballistic vehicles experience high deceleration during entry. For crewed vehicles, 10 *g* is presently considered to be the maximum practical deceleration. The required entry angles are so shallow that aerodynamic lift is necessary to balance the centrifugal force and assure capture within the atmosphere.

Since the large kinetic energy of entry vehicles is converted to heat by shock waves and air friction as the vehicle passes through the atmosphere, the air surrounding the vehicle is heated to a temperature of many thousands of degrees. Careful vehicle design can result in more than 99% of this energy remaining in the air. The three means for heat shielding on entry vehicles are heat sinks, mass injection, and reradiation.

The most widely used method of heat shielding is mass injection through natural ablation. A vehicle is coated with an ablative material, frequently a fiber-impregnated plastic; large amounts of heat are absorbed as the ablator is heated and vaporized. In addition, the ablation vapors fend off the hot air, thereby tending to greatly reduce the convective heating. [M.E.T.]

Atmospheric ozone Ozone is found in trace quantities throughout the atmosphere, the largest concentrations being located in a layer in the lower stratosphere between the altitudes of 9 and 18 mi (15 and 30 km). This ozone results almost entirely from the dissociation of molecular oxygen by solar ultraviolet radiation in the upper atmosphere.

Although present in only trace quantities, atmospheric ozone plays a critical role for the biosphere by absorbing the ultraviolet radiation with wavelength λ between 240 and 320 nanometers, which would otherwise be transmitted to the Earth's surface. This radiation is lethal to simple unicellular organisms (algae, bacteria, protozoa) and to the surface cells of higher plants and animals. It also damages the genetic material of cells (DNA) and is responsible for sunburn in human skin. In addition, the incidence of skin cancer has been statistically correlated with the observed surface intensities of the ultraviolet wavelengths between 290 and 320 nm, which are not totally absorbed by the ozone layer. *See* BIOSPHERE.

Ozone also plays an important role in photochemical smog and in heating the upper atmosphere by absorbing solar ultraviolet and visible radiation ($\lambda < 710$ nm) and thermal infrared radiation ($\lambda \cong 9.6$ micrometers). As a consequence, the temperature increases steadily from about 220 K at the tropopause (8–16 km altitude) to about 280 K at the stratopause (50 km altitude). This ozone heating provides the major energy source for driving the circulation of the upper stratosphere and mesosphere.

There has also been concern that industrial production of the chlorofluoromethanes $CFCl_3$ (Freon 11) and CF_2Cl_2 (Freon 12) may be significantly altering the natural chlorine cycle in the atmosphere. These very inert species are principally used as refrigerants, as blowing agents for plastic foams, and as aerosol-can propellants. Once they are released into the atmosphere, their only presently recognized removal mechanism involves photodissociation in the stratosphere. Chlorine atoms released in such reactions can catalytically destroy ozone by conversion to chlorine oxide radicals.

A number of other potential ozone-destroying processes with anthropogenic origins have also been identified. The global measurements of ozone which are required to ascertain the reality of all possible long-term ozone depletions have become possible from satellites. *See* AIR POLLUTION; OZONE. [R.G.Pr.]

Atmospheric waves, upper synoptic Wavelike oscillations in the pattern of wind flow aloft, usually with reference to the stronger portion of the westerly current. The flow is anticyclonically curved in the vicinity of a ridge line in the wave pattern, and is cyclonically curved in the vicinity of a rough line.

Any given hemispheric upper flow pattern may be represented by the superposition of sinusoidal waves of various lengths in the general westerly flow. Analysis of a typical pattern discloses the presence of prominent long waves, of which there are three or four around the hemisphere, and of distinctly evident short waves, of about half the length of the long waves.

Typically, each short wave trough and ridge is associated with a particular cyclone and anticyclone, respectively, in the lower troposphere. These circulations produce the rapid day-to-day weather changes which are characteristic of the climate of the middle latitudes. The long waves aloft do not generally correspond to a single feature of the circulation pattern at low levels. By virtue of their position and amplitude, they can exert an indirect influence on the character of the weather over a given region for a period of the order of weeks. *See* ATMOSPHERE; JET STREAM; STORM; VORTEX; WEATHER FORECASTING AND PREDICTION; WIND. [F.S.]

Atoll An annular coral reef, with or without small islets, that surrounds a lagoon without projecting land area. Most atolls are isolated reefs rising from the deep sea, and vary considerably in size. Small rings, usually without islets, may be less than a mile in diameter, but many atolls have a diameter of about 20 mi (32 km) and bear numerous islets.

The reefs of the atoll ring are flat, pavementlike areas, large parts of which, particularly along the seaward margin, may be exposed at times of low tide. The reefs vary in width from narrow ribbons to broad bulging areas more than a mile (1.6 km) across. The structures form a most effective baffle that robs the incoming waves of much of their destructive power, and at the same time brings a constant supply of refreshing sea water with oxygen, food, and nutrient salts to wide expanses of the reef.

Atolls, like other types of coral reefs, require strong light and warm waters and are limited in the existing seas to tropical and near-tropical latitudes. A large percentage of the world's atolls are contained in an area known as the former Darwin Rise that covers much of the central and southwestern Pacific. Atolls are also numerous in parts of the Indian Ocean and a number are found, mostly on continental shelves, in the Caribbean area. *See* OCEANIC ISLANDS; REEF. [H.S.L.]

Atom The individual structure which constitutes the basic unit of any chemical element. This structure, consisting of a positively charged nucleus surrounded by a number of electrons of total negative charge equal to the positive charge on the nucleus, is essentially identical for all atoms of any one element. The nuclear charge, measured in units of the electronic charge, is called the atomic number and specifies the element. *See* ATOMIC STRUCTURE AND SPECTRA; ISOTOPE; NUCLEAR STRUCTURE. [F.A.J./W.W.W.]

Atom cluster An assembly of atoms (or molecules, in the case of molecular clusters) that are weakly bound together. This assembly may contain as few as two and perhaps as many as several thousand individual components, a unique distinction being that an appreciable number of them are present on the surface at any time. Intense interest has arisen in this configuration because its study promises to serve to bridge the gaps between atomic and molecular science and between condensed matter and surface science.

Characteristically, matter is thought to be in one of three states, solid, liquid, or gas, with possible altered properties if the system is highly ionized (a plasma is sometimes referred to as the fourth state of matter). Clusters are popularly referred to as the fifth, or aggregated, state of matter since they often display properties between those of the gas and the bulk condensed phase. Potential applications of research on clusters include the development of new methods in materials processing, photography, and microelectronics, understanding the physical basis for catalysis, and elucidating the formation of atmospheric aerosols and the nature of small grains in interstellar media.

The gas-phase of clusters depends on suitable methods for effecting the desired degree of aggregation. Adiabatic expansion from high pressure into vacuum is a common technique for producing neutral clusters of atoms and molecules from substances with high vapor pressures. In the case of highly refractory materials, high-temperature ovens or laser vaporization is commonly employed. In the case of cluster ions which comprise atoms or molecules bound to an ion, the most common technique employs thermal ion sources, where clustering is effected through the electrostatic interaction of the ion with other atoms or molecules at controlled reaction temperatures. *See* ION SOURCES.

Studies of metal clusters serve to answer questions concerning size effects on their reactivity, or their approach to the bulk limit of metallic conductivity as a function of their degree of aggregation, as well as the nature of metal bonding.

Ionization potentials of metal clusters often display oscillatory trends, with lower values generally observed for the smaller-sized odd-numbered clusters where an unpaired electron is more easily removed. A number of metal systems correlate reasonably well with a simple theory based on classical electrostatics. *See* ELECTROSTATICS; IONIZATION POTENTIAL.

Another subject of interest is the extent to which adsorption of gases onto small clusters influences their properties, as it does that of bulk condensed matter. For example, studies of oxygen on sodium have shown that an oxygen atom lowers the ionization potential of the metal clusters in a manner analogous to the known effect of impurities on the bulk work function of a metal. *See* ADSORPTION.

Magic numbers are discontinuities in otherwise smooth trends in cluster abundance versus cluster size. The factors influencing magic numbers have been the subject of intense investigation in a wide variety of systems.

Investigations have been carried out of organic molecules embedded within various systems, such as rare-gas aggregates, and on the surface of preformed clusters, such as those composed of ammonia molecules. A combination of photoexcitation and laser-induced fluorescence has enabled a study of the shifts in the electronic states of the organic probe particle as a function of degree of aggregation. *See* ATOMIC STRUCTURE AND SPECTRA; MOLECULAR STRUCTURE AND SPECTRA; SOLID-STATE PHYSICS; SURFACE PHYSICS. [A.W.Ca.]

Atom optics The use of laser light and nanofabricated structures to manipulate the motion of atoms in the same manner that rudimentary optical elements control light. The term refers to both an outlook in which atoms in atomic beams are thought of and manipulated like photons in light beams, and a collection of demonstrated techniques for doing such manipulation. Two types of atom optics elements have existed for some time: slits and holes used to collimate molecular beams (the analog of the pinhole camera), and focusing lenses for atoms and molecules (for example, hexapole magnets and quadrupole electrostatic lenses). However, in the 1980s the collection of optical elements for atoms expanded dramatically because of the use of near-resonant laser light and fabricated structures to make several types of mirrors as well as diffraction gratings. The diffraction gratings are particularly interesting because they exploit and demonstrate the (de Broglie) wave nature of atoms in a clear fashion. *See* LASER.

Diffraction gratings. Diffraction gratings for atoms have been made by using either a standing wave of light or a slotted membrane. The standing light wave makes a phase grating (that is, it advances or retards alternate sections of the incident wavefront but does not absorb any of the atom wave), so that the transmitted intensity is high. This approach requires the complexity of a single-mode laser, and introduces the complication that the light acts differently on the various hyperfine states of the atom. The slotted membrane, however, absorbs (or backscatters) atoms which strike the grating bars but does not significantly alter the phase of the transmitted atoms; it is therefore an amplitude grating. It works for any atom or molecule, regardless of internal quantum state, but with total transmission limited to about 40% by the opacity of the grating bars and requisite support structure. *See* DIFFRACTION GRATING.

Atom interferometers. Atom interferometers have been demonstrated through several different experimental routes, involving both microscopic fabricated structures and laser beams. These interferometers are the first examples of optical systems composed of the elements of atom optics like those discussed above. Atom interferometers, like optical interferometers, are well suited for application to a wide range of fundamental and applied scientific problems. Scientific experiments with atom interferometers divide naturally into three major categories: measurements of atomic and molecular properties, fundamental tests and demonstrations, and inertial effects. *See* ATOMIC STRUCTURE AND SPECTRA; FRAME OF REFERENCE; INTERFERENCE OF WAVES; INTERFEROMETRY; MOLECULAR BEAMS; OPTICS; QUANTUM MECHANICS. [D.E.P.]

Atomic-beam clock A type of atomic clock which makes use of the atomic-beam magnetic resonance technique. In this method a beam of atoms traverses an evacuated space, in which the atoms are irradiated with the electromagnetic energy and subsequently detected. Deflection magnets are arranged so that only those atoms that undergo a transition (caused by the excitation being at their resonance frequency) are deflected to the detector. Beam clocks generally use cesium-133 and have linewidths of the order of 1 part in 10^7 to 1 part in 10^8. Thallium has the advantages over cesium of reduced sensitivity to magnetic fields and a higher frequency (21,300 megahertz as compared to 9192 MHz) but is much harder to detect and deflect. *See* ATOMIC BEAMS; ATOMIC CLOCK; MAGNETIC RESONANCE. [L.S.C.]

Atomic beams Unidirectional streams of neutral atoms passing through a vacuum. These atoms are virtually free from the influence of neighboring atoms but may be subjected to electric and magnetic fields so that their properties may be studied. The method of atomic beams yields extremely accurate spectroscopic data about the energy levels of atoms, and hence detailed information about the interaction of electrons in the atom with each other and with the atomic nucleus, as well as information about the interaction of all components of the atom with external fields. *See* MOLECULAR BEAMS. [P.Ku.]

Atomic bomb A device for suddenly producing an explosively rapid neutron chain reaction in a fissile material such as ^{235}U or ^{239}Pu. In a wider sense, it is any explosive device which derives its energy from nuclear reactions, including not only the foregoing fission bomb but also a fusion bomb, which gets its energy largely from reactions of heavy hydrogen (hydrogen bomb) or other light nuclei, and a fission-fusion bomb which derives its energy in comparable amounts from fission and fusion. Because an atomic bomb derives its energy from nuclear reactions, it is more properly called a nuclear bomb. *See* CHAIN REACTION (PHYSICS); HYDROGEN BOMB; NUCLEAR FISSION; NUCLEAR FUSION; NUCLEAR REACTION.

The first fission bombs released energies of the order of 20 kilotons (1 kiloton is 1000 tons of equivalent chemical high explosive calculated at a conventional energy release of 1000 cal/g; 1 kiloton = 4.18×10^{12} joules), but fractional kiloton yields have since been obtained, overlapping the energy of conventional explosives. On the upper end the yield of fission bombs overlaps that of fusion bombs. Yields of energetic fusion bombs in excess of 100 megatons (4.18×10^{17} J) have been reported.

The energy is communicated by mechanical shock and radiative transport to the surrounding water, earth, or air, ionizing it out to a radius which in the case of explosions in air is known as the fireball radius (150 yd or 135 m about 1 s after the explosion of a conventional or 20-kiloton atomic bomb). Energy goes out from such a fireball into the surrounding relatively transparent air in not very different orders of magnitude in the form of a shock wave and in the form of heat radiation that may continue for a number of seconds. *See* NUCLEAR EXPLOSION; RADIOACTIVE FALLOUT. [J.A.W.]

Atomic clock An electronic clock whose frequency is supplied or governed by the natural resonance frequencies of atoms or molecules of suitable substances. These frequencies are generally in the microwave range from about 1400 to 40,000 megahertz in the clocks presently used or under development. Some work is also being done in the optical and infrared frequency ranges. Atomic clocks are the most precise of all clocks; some of those presently available have errors sufficiently small that they would gain or lose less than 1 s in 30,000 years. The present internationally accepted unit of time, the atomic second, is based on an atomic clock and is defined as that time interval during which occur exactly 9,192,631,770 cycles of the hyperfine resonance frequency of the ground state of the cesium-133 atom. *See* ATOMIC TIME.

Some uses that require the precision of these clocks are: (1) navigation systems such as loran C and Omega, in which position is determined by the relative times of arrival of radio pulses of electromagnetic energy emitted at precisely known times from several transmitters at different locations; (2) deep-space communications and Doppler navigation, both of which require extremely stable frequencies; (3) tests of relativity theory by comparison of rates of clocks carried on aircraft, rockets, or satellites with that of clocks on the ground. *See* CLOCK PARADOX; ELECTRONIC NAVIGATION SYSTEMS; LORAN; OMEGA; RELATIVITY.

The clocks presently in use or under development employ: (1) hyperfine structure (splitting of the spectral lines due to the nuclear magnetic moment) in the ground state of atoms such as cesium, rubidium, hydrogen, or thallium, or in ions such as singly ionized mercury, barium, or magnesium; (2) a molecular line such as the inversion line of ammonia; or (3) a narrow optical or infrared resonance in an atom, molecule, or ion. They may be further classed as passive or active. In a passive clock the atoms or molecules are irradiated with electromagnetic energy close in frequency to their resonance, and some method is provided to sense the difference in frequency and to control the source frequency so as to reduce the error toward zero. The source is usually a stable quartz oscillator at a frequency of a few megahertz followed by a frequency multiplier. Active clocks generate their own signal directly from the atoms or molecules, using the principle of oscillation by stimulated emission of radiation. The microwave frequency signal so generated is low in power, so an oscillator and frequency multiplier chain is usually synchronized to it. In both types some form of frequency synthesis (frequency multiplication by a rational fraction) is necessary to provide an output at a cardinal frequency, such as 1.0 MHz, suitable for clock use. Frequency synthesis into the infrared or optical range is very difficult and requires complex equipment. *See* AMMONIA MASER CLOCK; ATOMIC-BEAM CLOCK; HYDROGEN MASER CLOCK; RUBIDIUM-GAS CELL CLOCK. [L.S.C.]

Atomic energy The energy released in the rearrangement of the particles making up the nucleus of an atom, popularly referred to as atomic energy, but preferably called nuclear energy. Atomic properly refers to phenomena involving the orbital electrons of the atom, but not involving any transformation of the nucleus. *See* NUCLEAR POWER. [J.A.L.]

Atomic mass unit An arbitrarily defined unit in terms of which the masses of individual atoms are expressed. One atomic mass unit is defined as exactly $\frac{1}{12}$ of the mass of an atom of the nuclide ^{12}C, the predominant isotope of carbon. The unit, also known as the dalton, is often abbreviated amu, and is designated by the symbol u. The relative atomic mass of a chemical element is the average mass of its atoms expressed in atomic mass units. *See* ATOMIC WEIGHT; RELATIVE ATOMIC MASS. [J.F.We.]

Atomic nucleus The central region of an atom. Atoms are composed of negatively charged electrons, positively charged protons, and electrically neutral neutrons. The protons and neutrons (collectively known as nucleons) are located in a small central region known as the nucleus. The electrons move in orbits which are large in comparison with the dimensions of the nucleus itself. Protons and neutrons possess approximately equal masses, each roughly 1840 times that of an electron. The number of nucleons in a nucleus is given by the mass number A and the number of protons by the atomic number Z. Nuclear radii r are given approximately by $r = 1.4 \times 10^{-13} A^{1/3}$ cm. *See* NUCLEAR STRUCTURE. [H.E.D.]

Atomic number The number of elementary positive charges (protons) contained within the nucleus of an atom. It is denoted by the letter Z. For an electrically neutral atom, the number of planetary electrons is also given by the atomic number. Atoms with the same Z (isotopes) belong to the same element. The lightest element, hydrogen, has $Z = 1$. The heaviest naturally occurring element, uranium, has $Z = 92$. All elements up to and including $Z = 106$ either occur in nature or have been created artificially. When specifically written, the atomic number is usually placed before and below the elemental symbol, for example, ${}_1H$, ${}_{92}U$. *See* MASS NUMBER; RADIOACTIVITY. [H.E.D.]

Atomic physics The study of the structure of the atom, its dynamical properties, including energy states, and its interactions with particles and fields. These are almost completely determined by the laws of quantum mechanics, with very refined corrections required by quantum electrodynamics. Despite the enormous complexity of most atomic systems, in which each electron interacts with both the nucleus and all the other orbiting electrons, the wavelike nature of particles, combined with the Pauli exclusion principle, results in an amazingly orderly array of atomic properties. These are systematized by the Mendeleev periodic table. In addition to their classification by chemical activity and atomic weight, the various elements of this table are characterized by a wide variety of observable properties. These include electron affinity, polarizability, angular momentum, multiple electric moments, and magnetism. *See* ATOMIC WEIGHT; PERIODIC TABLE; QUANTUM ELECTRODYNAMICS; QUANTUM MECHANICS.

Each atomic element, normally found in its ground state (that is, with its electron configuration corresponding to the lowest state of total energy), can also exist in an infinite number of excited states. These are also ordered in accordance with relatively simple hierarchies determined by the laws of quantum mechanics. The most characteristic signature of these various excited states is the radiation emitted or absorbed when the atom undergoes a transition from one state to another. The systemization and classification of atomic energy levels (spectroscopy) has played a central role in developing an understanding of atomic structure. [B.B.]

Atomic spectrometry A branch of chemical analysis that seeks to determine the composition of a sample in terms of which chemical elements are present and their quantities or concentrations. Unlike other methods of elemental analysis, however, the sample is decomposed into its constituent atoms which are then probed spectroscopically.

In routine atomic spectrometry, a device called the atom source or atom cell is responsible for producing atoms from the sample; there are many different kinds of atom sources. After atomization of the sample, any of several techniques can determine which atoms are present and in what amounts, but the most common are atomic absorption, atomic emission, atomic fluorescence (the least used of these four alternatives), and mass spectrometry.

Most atomic spectrometric measurements (all those just mentioned except mass spectrometry) exploit the narrow-line spectra characteristic of gas-phase atoms. Because the atom source yields atomic species in the vapor phase, chemical bonds are disrupted, so valence electronic transitions are unperturbed by bonding effects. As a result, transitions among atomic energy levels yield narrow spectral lines, with spectral bandwidths commonly in the 1–5-picometer wavelength range. Moreover, because each atom possesses its unique set of energy levels, these narrow-band transitions can be measured individually, with little mutual interference. Thus, sodium, potassium, and scandium can all be monitored simultaneously and with minimal spectral influence on each other. This lack of spectral overlap remains one of the most attractive features of atomic spectrometry. *See* ATOMIC STRUCTURE AND SPECTRA; LINE SPECTRUM; SPECTRUM.

In atomic absorption spectrometry, light from a primary source is directed through the atom cell, where a fraction of the light is absorbed by atoms from the sample. The amount of radiation that remains can then be monitored on the far side of the cell. The concentration of atoms in the path of the light beam can be determined by Beer's law, which can be expressed as the equation below, where P_0 is the light intensity

$$\log \frac{P_0}{P} = kC$$

incident on the atom cell, P is the amount of light which remains unabsorbed, C is the concentration of atoms in the cell, and k is the calibration constant, which is determined by means of standard samples having known concentrations. *See* SPECTROPHOTOMETRIC ANALYSIS.

The two most common kinds of atom cells employed in atomic absorption spectrometry are chemical flames and electrical furnaces. Chemical flames are usually simple to use, but furnaces offer higher sensitivity.

The most common primary light source employed in atomic absorption spectrometry is the hollow-cathode lamp. Conveniently, the hollow-cathode lamp emits an extremely narrow line spectrum of one, two, or three elements of interest. As a result, the atomic absorption spectrometry measurement is automatically tuned to the particular spectral lines of interest.

In atomic emission spectrometry, atomic species are measured by their emission spectra. For such spectra to be produced, the atoms must first be excited by thermal or nonthermal means. Therefore, the atom sources employed in atomic emission spectrometry are hotter or more energetic than those commonly used in atomic absorption spectrometry. Although several such sources are in common use, the dominant one is the inductively coupled plasma. From the simplest standpoint, the inductively coupled plasma is a flowing stream of hot, partially ionized (positively charged) argon. Power is coupled into the plasma by means of an induction coil.

There are two common modes for observing emission spectra from an inductively coupled plasma. The less expensive and more flexible approach employs a so-called slew-scan spectrometer, which accesses spectral lines in rapid sequence, so that a number of chemical elements can be measured rapidly, one after the other. Moreover, because each viewed elemental spectral line can be scanned completely, it is possible to subtract spectral emission background independently for each element. The alternative approach is to view all spectral lines simultaneously, either with a number of individual photo-detectors keyed to particular spectral lines or with a truly multichannel electronic detector driven by a computer. This approach enables samples to be analyzed more rapidly and permits transient atom signals (as from a furnace-based atomizer) to be recorded. *See* EMISSION SPECTROCHEMICAL ANALYSIS.

Elemental mass spectrometry has been practiced for many years in the form of spark-source mass spectrometry and, more recently, glow-discharge-lamp mass spectrometry. However, a hybrid technique that combines the inductively coupled plasma with a mass spectrometer has assumed a prominent place.

At the high temperatures present in an inductively coupled plasma, many atomic species occur in an ionic form. These ions can be readily extracted into a mass spectrometer.

The advantages of the combination of inductively coupled plasma and mass spectrometry are substantial. The system is capable of some of the best detection limits in atomic spectrometry, typically 10^{-3} to 10^{-2} ng/ml for most elements. Also, virtually all elements in the periodic table can be determined during a single scan. The method is also capable of providing isotopic information, unavailable by any other atomic spectrometric method for such a broad range of elements. *See* MASS SPECTROMETRY. [G.M.H.]

Atomic structure and spectra The idea that matter is subdivided into discrete and further indivisible building blocks called atoms dates back to the Greek philosopher Democritus, whose teachings of the 5th century B.C. are commonly accepted as the earliest authenticated ones concerning what has come to be called atomism by students of Greek philosophy. The weaving of the philosophical thread of atomism into the analytical fabric of physics began in the late 18th and the 19th centuries. Robert Boyle is generally credited with introducing the concept of chemical elements, the irreducible units of which are now recognized as individual atoms of a given element. In the early 19th century John Dalton developed his atomic theory, which postulated that matter consists of indivisible atoms as the irreducible units of Boyle's elements, that each atom of a given element has identical attributes, that differences among elements are due to fundamental differences among their constituent atoms, that chemical reactions proceed by simple rearrangement of indestructible atoms, and that chemical compounds consist of molecules which are reasonably stable aggregates of such indestructible atoms. *See* CHEMISTRY.

The work of J. J. Thomson in 1897 clearly demonstrated that atoms are electromagnetically constituted and that from them can be extracted fundamental material units bearing electric charge that are now called electrons. The electrons of an atom account for a negligible fraction of its mass. By virtue of overall electrical neutrality of every atom, the mass must therefore reside in a compensating, positively charged atomic component of equal charge magnitude but vastly greater mass. *See* ELECTRON.

Thomson's work was followed by the demonstration by Ernest Rutherford in 1911 that nearly all the mass and all of the positive electric charge of an atom are concentrated in a small nuclear core approximately 10,000 times smaller in extent than an atomic diameter. Niels Bohr in 1913 and others carried out some remarkably successful attempts to build solar system models of atoms containing planetary pointlike electrons orbiting around a positive core through mutual electrical attraction (though only certain "quantized" orbits were "permitted"). These models were ultimately superseded by nonparticulate-matter wave quantum theories of both electrons and atomic nuclei. *See* QUANTUM MECHANICS.

The modern picture of condensed matter (such as solid crystals) consists of an aggregate of atoms or molecules which respond to each other's proximity through attractive electrical interactions at separation distances of the order of 1 atomic diameter (approximately 10^{-10} m) and repulsive electrical interactions at much smaller distances. These interactions are mediated by the electrons, which are in some sense shared and exchanged by all atoms of a particular sample, and serve as a kind of interatomic glue which binds the mutually repulsive, heavy, positively charged atomic cores together. *See* SOLID-STATE PHYSICS.

The hydrogen atom is the simplest atom, and its spectrum (or pattern of light frequencies emitted) is also the simplest. The regularity of its spectrum had defied explanation until Bohr solved it with three postulates, these representing a model which is useful, but quite insufficient, for understanding the atom.

Postulate 1: The force that holds the electron to the nucleus is the Coulomb force between electrically charged bodies.

Postulate 2: Only certain stable, nonradiating orbits for the electron's motion are possible, those for which the angular momentum is an integral multiple of $h/2\pi$ (Bohr's quantum condition on the orbital angular momentum). Each stable orbit represents a discrete energy state.

Postulate 3: Emission or absorption of light occurs when the electron makes a transition from one stable orbit to another, and the frequency ν of the light is such that the difference in the orbital energies equals $h\nu$ (A. Einstein's frequency condition for the photon, the quantum of light).

Here the concept of angular momentum, a continuous measure of rotational motion in classical physics, has been asserted to have a discrete quantum behavior, so that its quantized size is related to Planck's constant h, a universal constant of nature. *See* ANGULAR MOMENTUM.

Modern quantum mechanics has provided justification for Bohr's quantum condition on the orbital angular momentum. It has also shown that the concept of definite orbits cannot be retained except in the limiting case of very large orbits. In this limit, the frequency, intensity, and polarization can be accurately calculated by applying the classical laws of electrodynamics to the radiation from the orbiting electron. This fact illustrates Bohr's correspondence principle, according to which the quantum results must agree with the classical ones for large dimensions. The deviation from classical theory that occurs when the orbits are smaller than the limiting case is such that one may no longer picture an accurately defined orbit. Bohr's other hypotheses are still valid.

According to Bohr's theory, the energies of the hydrogen atom are quantized (that is, can take on only certain discrete values). These energies can be calculated from the electron orbits permitted by the quantized orbital angular momentum. The orbit may be circular or elliptical, so only the circular orbit is considered here for simplicity. Let the electron, of mass m and electric charge $-e$, describe a circular orbit of radius r around a nucleus of charge $+e$ and of infinite mass. With the electron velocity v, the angular momentum is mvr, and the second postulate becomes Eq. (1). The integer n is called the

$$mvr = n(h/2\pi) \qquad (n = 1, 2, 3, \ldots) \tag{1}$$

principal quantum number. The possible energies of the nonradiating states of the atom are given by Eq. (2). Here ϵ_0 is

$$E = -\frac{me^4}{8\epsilon_0{}^2h^2}\cdot\frac{1}{n^2} \tag{2}$$

the pemittivity of free space, a constant included to give the correct units according to the mks statement of Coulomb's law.

The same equation for the hydrogen atom's energy levels, except for some small but significant corrections, is obtained from the solution of the Schrödinger equation for the hydrogen atom. *See* SCHRÖDINGER'S WAVE EQUATION.

The frequencies of electromagnetic radiation or light emitted or absorbed in transitions are given by Eq. (3) where E' and

$$\nu = (E' - E'')/h \tag{3}$$

E'' are the energies of the initial and final states of the atom. Spectroscopists usually express their measurements in wavelength λ or in wave number σ in order to obtain numbers of a convenient size. The wave number of a transition is shown in Eq. (4). If $T = -E/hc$, then Eq. (5) results. Here T is called the spectral term.

$$\sigma = \frac{\nu}{c} = \frac{E'}{hc} - \frac{E''}{hc} \tag{4}$$

$$\sigma = T'' - T' \tag{5}$$

The allowed terms for hydrogen, from Eq. (2), are given by Eq. (6). The quantity R is the important Rydberg constant. Its

$$T = \frac{me^4}{8\epsilon_0{}^2ch^3}\cdot\frac{1}{n^2} = \frac{R}{n^2} \tag{6}$$

value, which has been accurately measured by laser spectroscopy, is related to the values of other well-known atomic constants, as shown in Eq. (6). *See* RYDBERG CONSTANT.

The effect of finite nuclear mass must be considered, since the nucleus does not actually remain at rest at the center of the atom. Instead, the electron and nucleus revolve about their common center of mass. This effect can be accurately accounted for and requires a small change in the value of the effective mass m in Eq. (6).

In addition to the circular orbits already described, elliptical ones are also consistent with the requirement that the angular momentum be quantized. A. Sommerfeld showed that for each value of n there is a family of n permitted elliptical orbits, all having the same major axis but with different eccentricities. Illustration *a* shows, for example, the Bohr-Sommerfeld orbits for $n = 3$. The orbits are labeled *s*, *p*, and *d*, indicating values of the azimuthal quantum number $l = 0$, 1, and 2. This number determines the shape of the orbit, since the ratio of the major to the minor axis is found to be $n/(l + 1)$. To a first approximation, the energies of all orbits of the same n are equal. In the case of the highly eccentric orbits, however, there is a slight lowering of the energy due to precession of the orbit (illustration *b*). According to Einstein's theory of relativity, the mass increases somewhat in the inner part of the orbit, because of greater velocity. The velocity increase is greater as the eccentricity is greater, so the orbits of higher eccentricity have their energies lowered more. The quantity l is called the orbital angular momentum quantum number or the azimuthal quantum number. *See* RELATIVITY.

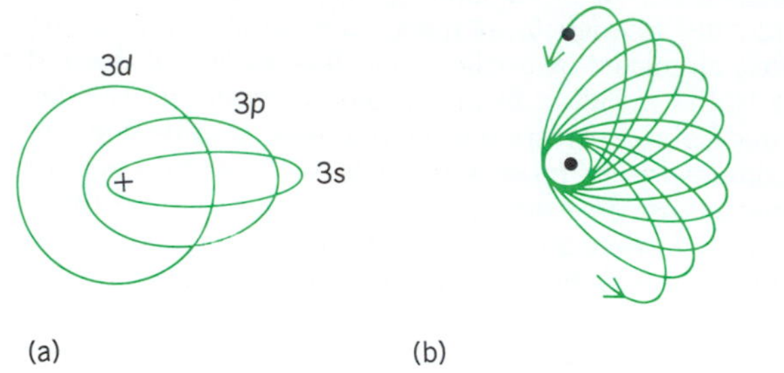

Possible elliptical orbits, according to the Bohr-Sommerfeld theory. *(a)* The three permitted orbits for $n = 3$. *(b)* Precession of the 3s orbit caused by the relativistic variation of mass. *(After A. P. Arya, Fundamentals of Atomic Physics, Allyn and Bacon, pp. 281 and 286, 1971)*

In attempting to extend Bohr's model to atoms with more than one electron, it is logical to compare the experimentally observed terms of the alkali atoms, which contain only a single electron outside closed shells, with those of hydrogen. A definite similarity is found but with the striking difference that all terms with $l > 0$ are double. This fact was interpreted by S. A. Goudsmit and G. E. Uhlenbeck as due to the presence of an additional angular momentum of $\frac{1}{2}(h/2\pi)$ attributed to the electron spinning about its axis. The spin quantum number of the electron is $s = \frac{1}{2}$.

The relativistic quantum mechanics developed by P. A. M. Dirac provided the theoretical basis for this experimental observation.

Implicit in much of the following discussion is W. Pauli's exclusion principle, first enunciated in 1925, which when applied to atoms may be stated as follows: no more than one electron in a multielectron atom can possess precisely the same quantum numbers. In an independent, hydrogenic electron approximation to multielectron atoms, there are $2n^2$ possible independent choices of the principal (n), orbital (l), and magnetic (m_l, m_s) quantum numbers available for electrons belonging to a given n, and no more. Here m_l and m_s refer to the quantized projections of l and s along some chosen direction. The organization of atomic electrons into shells of increasing radius (the Bohr radius scales as n^2) follows from this principle. *See* EXCLUSION PRINCIPLE.

The energy of interaction of the electron's spin with its orbital angular momentum is known as spin-orbit coupling. A charge in motion through either "pure" electric or "pure" magnetic fields, that is, through fields perceived as "pure" in a static laboratory, actually experiences a combination of electric and magnetic fields, if viewed in the frame of reference of a moving observer with respect to whom the charge is momentarily at rest. For example, moving charges are well known to be deflected by magnetic fields. But in the rest frame of such a charge, there is no motion, and any acceleration of a charge must be due to the presence of a pure electric field from the point of view of an observer analyzing the motion in that reference frame. *See* RELATIVISTIC ELECTRODYNAMICS.

A spinning electron can crudely be pictured as a spinning ball of charge, imitating a circulating electric current. This circulating current gives rise to a magnetic field distribution very similar to that of a small bar magnet, with north and south magnetic poles symmetrically distributed along the spin axis above and below the spin equator. This representative bar magnet can interact with external magnetic fields, one source of which is the magnetic field experienced by an electron in its rest frame, owing to its orbital motion through the electric field established by the central nucleus of an atom. In multielectron atoms, there can be additional, though generally weaker, interactions arising from the magnetic interactions of each electron with its neighbors, as all are moving with respect to each other, and all have spin. The strength of the bar magnet equivalent to each electron spin, and its direction in space are characterized by a quantity called the magnetic moment, which also is quantized essentially because the spin itself is quantized. Studies of the effect of an external magnetic field on the stages of atoms show that the magnetic moment associated with the electron spin is equal in magnitude to a unit called the Bohr magneton.

The energy of the interaction between the electron's magnetic moment and the magnetic field generated by its orbital motion is usually a small correction to the spectral term, and depends on the angle between the magnetic moment and the magnetic field, or equivalently, between the spin angular momentum vector and the orbital angular momentum vector (a vector perpendicular to the orbital plane whose magnitude is the size of the orbital angular momentum). Since quantum theory requires that the quantum number j of the electron's total angular momentum shall take values differing by integers, while l is always an integer, there are only two possible orientations for s relative to l: s must be either parallel or antiparallel to l.

For the case of a single electron outside the nucleus, the Dirac theory gives Eq. (7) for the spin-orbit correction to the

$$\Delta T = \frac{R\alpha^2 Z^4}{n^3} \cdot \frac{j(j + 1) - l(l + 1) - s(s + 1)}{l(2l + 1)(l + 1)} \qquad (7)$$

spectral terms. Here $\alpha = e^2/2\epsilon_0 hc \cong 1/137$ is called the fine structure constant.

In atoms having more than one electron, this fine structure becomes what is called the multiplet structure. The doublets in the alkali spectra, for example, are due to spin-orbit coupling; Eq. (7), with suitable modifications, can be applied.

When more than one electron is present in the atom, there are various ways in which the spins and orbital angular momenta can interact. Each spin may couple to its own orbit, as in the one-electron case; other possibilities are orbit–other orbit, spin-spin, and so on. The most common interaction in the light atoms, called *LS* coupling or Russell-Saunders coupling, is described schematically in Eq. (8). This notation indi-

$$\{(l_1,l_2,l_3,\ldots)\,(s_1,s_2,s_3,\ldots)\} = \{L,S\} = J \qquad (8)$$

cates that the l_i are coupled strongly together to form a resultant L, representing the total orbital angular momentum. The s_i, are coupled strongly together to form a resultant S, the total spin angular momentum. The weakest coupling is that between L and S to form J, the total angular momentum of the electron system of the atom in this state.

Coupling of the *LS* type is generally applicable to the low-energy states of the lighter atoms. The next commonest type is called *jj* coupling, represented in Eq. (9). Each electron has its

$$\{(l_1,s_1)\,(l_2,s_2)\,(l_3,s_3)\ldots\} = \{j_1,j_2,j_3,\ldots\} = J \qquad (9)$$

spin coupled to its own orbital angular momentum to form a *ji* for that electron. The various *ji* are then more weakly coupled together to give *J*. This type of coupling is seldom strictly observed. In the heavier atoms it is common to find a condition intermediate between *LS* and *jj* coupling; then either the *LS* or *jj* notation may be used to describe the levels, because the number of levels for a given electron configuration is independent of the coupling scheme.

Most atomic nuclei also possess spin, but rotate about 2000 times slower than electrons because their mass is on the order of 2000 or more times greater than that of electrons. Because of this, very weak nuclear magnetic fields, analogous to the electronic ones that produce fine structure in spectral lines, further split atomic energy levels. Consequently, spectral lines arising from them are split according to the relative orientations, and hence energies of interaction, of the nuclear magnetic moments with the electronic ones. The resulting pattern of energy levels and corresponding spectral-line components is referred to as hyperfine structure. *See* NUCLEAR MOMENTS.

The enormous capabilities of tunable lasers have allowed observations which were impossible previously. For example, high-resolution saturation spectroscopy, utilizing a saturating beam and a probe beam from the same laser, has been used to measure the hyperfine structure of the sodium resonance lines (called the D_1 and D_2 lines). The smallest separation resolved was less than 0.001 cm^{-1} which was far less than the Doppler width of the lines. *See* DOPPLER EFFECT; HYPERFINE STRUCTURE; LASER SPECTROSCOPY.

Nuclear properties also affect atomic spectra through the isotope shift. This is the result of the difference in nuclear masses of two isotopes, which results in a slight change in the Rydberg constant. There is also sometimes a distortion of the nucleus.

It would be misleading to think that the most probable fate of excited atomic electrons consists of transitions to lower orbits, accompanied by photon emission. In fact, for at least the first third of the periodic table, the preferred decay mode of most excited atomic systems in most states of excitation and ionization is the electron emission process first observed by P. Auger in 1925 and named after him. For example, a singly charged neon ion lacking a 1*s* electron is more than 50 times as likely to decay by electron emission as by photon emission. In the process, an outer atomic electron descends to fill an inner vacancy, while another is ejected from the atom to conserve both total energy and momentum in the atom. The ejection usually arises because of the interelectron Coulomb repulsion. *See* AUGER EFFECT. [I.A.S.]

Atomic time Time based on quantum transitions. According to the quantum theory, an atom or a molecule may exist in discrete states, each of which has a distinct energy level. Energy will be absorbed or radiated when the atom jumps between two states. The frequency of the electromagnetic radiation emitted or absorbed is related to the difference in energy: $f = (E_1 - E_2)/h$. Here f is the frequency, E_1 is the initial energy, E_2 is the final energy, and h is Planck's constant. For a selected transition of a given atom, the frequency emitted will remain constant. In contrast, the frequency of oscillation of a pendulum or of a quartz crystal depends upon factors such as size, which may change. *See* CLOCK; QUARTZ CLOCK.

The transition first used to obtain very high precision, in 1955, is the one between the two hyperfine levels of cesium-133 in the ground state. The frequency 9,192,631,770 hertz (Hz) for cesium was determined in 1958. It was adopted in 1967 by the General Conference of Weights and Measures to define the unit of time, the second, in the International System of Units (SI).

A scale of International Atomic Time, or TAI (abbreviation in French), is formed from the operation of a number of cesium-beam atomic clocks, located at various observatories and laboratories about the world. *See* ATOMIC CLOCK. [W.M.]

Atomic weight A number assigned to each chemical element which specifies the average mass of its atoms. Since an element may consist of two or more isotopes having atoms which differ in mass, the atomic weight of such an element depends on the relative proportions of its isotopes. The isotopic composition of all elements as they occur in nature, except those that are the products of natural radioactivity, is practically constant. The atomic weight refers to this natural mixture.

A unit called relative nuclidic mass, defined as one-twelfth the mass of carbon-12, was introduced in 1960. It is designated by the symbol u; thus, $^{12}C = 12$ u. The table of relative atomic weights is now based on the atomic mass of $^{12}C = 12$. *See* RELATIVE ATOMIC MASS.

There are four precision methods of determining atomic weights, all of which have comparable accuracy in the most favorable cases: (1) chemical combining weights, (2) limiting gas densities, (3) mass spectrometer measurements, and (4) nuclear reaction energies. The first two give results on the chemical scale, and the last two on the physical scale. [F.A.J./W.W.W.]

Atomization The mechanical subdivision of a bulk liquid. Other terms are also applied to this process, although the terminology is not standardized. Spraying usually implies the production of coarse drops (100–1000 micrometers in diameter). Sprinkling suggests very coarse drops (greater than 1000 μm). Misting is often applied to the production of fine drops (10–100 μm), and nebulizing to the production of very fine drops (under 10 μm), usually used in inhalation aerosol therapy. Atomization may also be used to subdivide meltable solids, such as certain metals. *See* DUST AND MIST COLLECTION.

Subdivision of a liquid (or a solid) may be desired (1) to distribute materials throughout an area or space; (2) to expose a large surface for increased rates of heat or mass transfer; or (3) to provide desired flow, packing, optical, insulation, deposition, or other properties. The atomization process may be utilized in such operations as spray drying, evaporation, humidification, gas scrubbing, chemical reaction, combustion, fire fighting, spray painting, powder metallurgy, and application of insecticides, herbicides, fungicides, and defoliants. It may also occur inadvertently whenever liquids are handled to produce undesired droplets, for example, entrainment from flash evaporators.

The various techniques by which liquids can be atomized are distinguished either by the geometry of the atomizing device or by the source of the external motivating force employed. Hydraulic, pneumatic, and rotary are the mechanical techniques most widely used in industry, agriculture, and domestic applications. Vibrational and electrostatic techniques are largely in a developmental stage. Ultrasonic nebulizers have been marketed. Much of the interest in electrostatic atomization has been in the application of ion propulsion to rockets in outer space. Explosive techniques have been used in military applications, for example, chemical agent dissemination. Film-bursting and gravitational techniques are prevalent in the natural environment, but not normally capable of atomizing liquids at high rates.

Under normal operating conditions (30–200 psi or 200–1400 kilopascals), hydraulic nozzles produce relatively coarse drops, 100–3000 μm in diameter, the finest ones being produced by small swirl nozzles, wherein pressure is converted into high relative jet velocity. Hydraulic nozzles are exemplified by garden hose nozzles, insecticide spray nozzles, and nozzles in humidification and scrubbing towers. Pneumatic atomizers normally use compressed air (30–100 psi or 200–700 kPa) and produce drops in the 5–100-μm diameter range. They are used in spray painting and fine misting applications, scrubbers or reactors (venturi atomizers), and aircraft application of insecticides. Rotary atomizers (spinning disks) are basically hydraulic atomizers, in which the pump and nozzle are combined, and normally produce drops in the 30–300-μm diameter range. They are widely used in spray drying because of their ability to handle viscous liquids or slurries. Most nebulizers are either pneumatic or ultrasonic in nature and may rely heavily on evaporation to produce the finest drops. In electrostatic spray painting, pneumatic or hydraulic atomization is usually employed, the electrical field merely acting to guide the charged drops to the deposition surface.

The practical application of atomization processes requires that droplets be produced at some predetermined rate, that is, liquid must be supplied at some finite rate and continuously converted into droplets. The kinetics of all such atomization processes involves the sequential steps listed below, although some steps may be negligible or absent under certain circumstances.

1. The extension of a bulk liquid into sheets, jets, films, or streams by accelerating the liquid in some prescribed manner (as through a nozzle or off a rotating disk).
2. The initiation of small disturbances at the liquid surface in the form of local ripples, protuberances, or waves.
3. The formation of short ligaments on the liquid surface, as the result of fluid pressure or shear forces.
4. The collapse of the ligaments into drops, as the result of surface tension.
5. The further breakup of the drops as they move through the gaseous medium by the action of fluid pressure or shear forces. [C.E.La.]

Atopic allergy A type of immediate hypersensitivity which occurs in humans as the result of spontaneous sensitization, usually by inhaled or ingested allergens. Examples of atopic allergy are asthma, hay fever, hives, infantile eczema, and gastrointestinal disturbances. Atopic hypersensitivities differ from the anaphylactic state in these respects: (1) The atopic allergies may be acquired through exposure to minute quantities of antigens, as suggested by the fact that inhalation of hard-shelled grains of pollen is sufficient to induce the state and subsequent reactions. (2) The antibodies associated with human atopic allergy constitute a special class of molecules, one of the five families of immunoglobulin molecules known to exist in human blood. This special class is termed IgE (immunoglobulin E). *See* ANAPHYLAXIS; ASTHMA; IMMUNOGLOBULIN.

The only convenient method available for demonstrating such antibodies is the Prausnitz-Küstner reaction. This is simply a specialized instance of passive transfer of hypersensitivity, carried out in human beings. Serum from a sensitive individual is injected intradermally into a normal subject. Twenty-four hours later, injection of the appropriate allergen into the same site results in a local hive. *See* ANTIBODY; HYPERSENSITIVITY; IMMUNOLOGY. [S.R.]

Atrophy The diminution in the size of a cell, tissue, or organ which was once fully developed and of normal size. Almost all examples of atrophy can be attributed to an alteration in the metabolism of the affected cells, usually through interference with cellular nutrition. It is postulated that increased intracellular activity of proteolytic enzymes is the fundamental cause of the loss of cell substance.

Physiologic atrophy affects certain organs in all individuals as a part of the normal aging process. An example is the atrophy of the thymus and lymphoid tissue that occurs at the time of puberty. In advanced age all organs undergo some atrophy which may in part be physiologic.

Pathologic atrophy refers to a decrease in the size of tissues or organs which is outside the range of normal variability. It is categorized on the basis of apparent clinical cause as disuse atrophy, vascular atrophy, pressure atrophy, exhaustion atrophy, and endocrine atrophy. Disuse atrophy occurs as a result of inactivity or diminished action of a tissue or organ, for example, the muscle atrophy which occurs secondary to immobilization by a plaster cast or secondary to denervation in poliomyelitis. Vascular atrophy occurs when blood supply to an organ or tissue becomes reduced below a critical level. This causes organs to undergo progressive loss of substance. Pressure atrophy results from persistent pressure upon an organ or tissue. It also occurs at a cellular level because of deposition of abnormal substances within tissues. Exhaustion atrophy develops in endocrine organs in conditions of prolonged overwork of the organ. Endocrine atrophy occurs in organs which are dependent upon endocrine stimulation for the maintenance of their normal structure and function. Destruction of the pituitary gland with resultant loss of hormone production leads to atrophy of the endocrine glands which rely upon pituitary hormones for their action. [N.K.M.]

Atropine An alkaloid with the formula $C_{17}H_{23}NO_3$; in pharmacy it is sometimes known as *dl*-hyoscyamine. It occurs in minute amounts in the leaves of *Atropa belladonna*, *A. betica*, *Datura stramonium*, *D. innoxia*, and *D. sanguinea*, as well as many related plants. It melts at 237–241°F (114–116°C) and is poorly soluble in water. The nitrate and sulfate are used in medicine instead of the free base. Atropine is used clinically as a mydriatic (pupil dilator). It is also administered in small doses before general anesthesia to lessen oral and air-passage secretions. *See* ALKALOID. [F.W.]

Attention deficit disorder A common psychiatric disorder of childhood characterized by attentional difficulties and impulsivity. However, attention deficit disorder is actually an example of a syndrome, that is, a group of signs and symptoms which tend to cluster together. The older names for this disorder include minimal brain dysfunction, minimal brain damage, hyperactivity, hyperkinesis, and the hyperactive child syndrome. These names were often misleading.

Some symptoms and signs may be present in all instances of the disorder, while others, though frequently associated, may or may not be present; they may occur in varying combinations, with some signs and symptoms present and others absent. The defining signs of attention deficit disorder are: The children have attentional deficits and are described as having a short attention span. Impulsivity is manifested in such problems as acting before thinking; shifting excessively and rapidly from one activity to another; having difficulty organizing work (not because of intellectual problems); needing much supervision; frequently calling out in class; having difficulty waiting for a turn in games or group activities. Most children with the disorder are hyperactive.

Many children with attention deficit disorder very frequently show an altered response to socialization. They show decreased sensitivity to reward and punishment and are described by their parents by terms such as obstinate, impervious, stubborn, and negativistic. With peers, many affected children are domineering or bullying, and because of this some prefer to play with younger children. Another characteristic often seen is emotional lability; that is, the childrens' moods change frequently and easily, sometimes spontaneously, and sometimes reactively. They may rapidly become upset, sad, or angry. Characteristically, however, they do not remain angry but may calm down rapidly.

Attention deficit disorder, formerly believed to be largely caused by brain damage, and more recently by some to be caused by food allergy, is now believed to be mainly a hereditary disorder. It was formerly believed that attention deficit disorder was outgrown during adolescence, and, in fact, the disorder does often become less severe in adolescence, sometimes gradually disappearing thereafter. In many affected children, some signs of the disorder (such as excessive activity) may diminish or disappear, but other, far more important signs, such as attentional difficulties, impulsivity, emotional lability, and interpersonal problems, may persist. Apparently, the disorder is not uncommon in adults and is frequently overlooked. There seems to be a greatly increased frequency of the residual type, which persists into the third and fourth decades, among young adults with alcoholism, with which the disorder may be genetically associated. *See* ALCOHOLISM.

The treatment of the child or adult with attention deficit disorder involves three steps: evaluation, explanation of the problem to parents and child, and therapeutic intervention. Medication and guidance constitute the mainstay of the treatment of the child with attention deficit disorder. Approximately 70–80% of these children manifest a therapeutic response to one of the major stimulant drugs, such as amphetamines, methylphenidate, and pemoline. Medication is symptomatic rather than curative; that is, it only reduces the signs of the disorder. It must be used until the signs and symptoms decrease with age, usually by adolescence. *See* AFFECTIVE DISORDERS; PSYCHOTHERAPY. [P.H.W.]

Attenuation The reduction in level of a quantity as a function of a characteristic parameter. Usually the quantity is an intensity I, or energy per unit area, and the parameter is the distance from the source x. However, the attenuation of voltage in suitable electrical networks by attenuators should also be mentioned. In the more usual case, the reduction in level follows Lambert's (or Bouguer's) law: $I(x) = I_0 e^{-\alpha x}$, where x is the parameter in question, frequently a distance. The coefficient α is called the attenuation or extinction coefficient.

For unconfined radiation from a point source, the attenuation is inversely proportional to the square of the distance from the source. The attenuation of an electromagnetic wave in a waveguide or a sound wave in a pipe does not follow this inverse-square law. The scattering of radiation and its absorption (that is, loss of energy) are both attenuation processes. *See* ABSORPTION; INVERSE-SQUARE LAW; POINT SOURCE; SCATTERING OF ELECTROMAGNETIC RADIATION. [McA.H.H.]

Attenuation (electricity) The exponential decrease with distance in the amplitude of an electrical signal traveling along a very long uniform transmission line, due to conductor and dielectric losses. If the peak voltage at the sending end of the transmission line is denoted by V_0, the peak voltage at a distance x from the sending end is given by the equation below,

$$V_x = V_0 e^{-\alpha x}$$

where α is the attenuation constant of the line.

Attenuators find numerous applications, typical examples being: in a signal generator, to vary the amplitude of the output signal; and in the input line to a television receiver that is very close to a television transmitter, so that overloading can be avoided. *See* SIGNAL GENERATOR; TELEVISION RECEIVER.

Attenuators for the dc (steady voltage) to very high-frequency (VHF) range (frequencies from 0 to 300 MHz) often contain resistors arranged in T or π configurations.

Piston attenuators (sometimes called waveguide-beyond-cutoff attenuators) are used at both intermediate and microwave frequencies (see illustration). The attenuation is varied by altering the separation between the two coils. The circular tube acts as a waveguide beyond cutoff, and the launching system is designed so that only one mode is excited in it.

A variable waveguide attenuator can be produced by moving

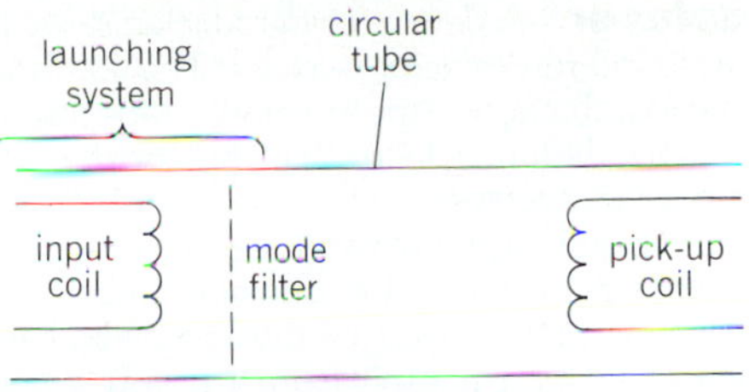

Piston attenuator.

a lossy vane either sideways across the waveguide or into the waveguide through a longitudinal slot.

The rotary vane attenuator is a very popular instrument. At the input end, there is a rectangular-to-circular waveguide taper containing a fixed lossy vane perpendicular to the incident electric vector. The central section contains a lossy vane diametrically across a circular waveguide that can be rotated, and the output section is a mirror image of the input section.

Many different techniques for measuring attenuation have been devised. The power-ratio method is widely used. The simplest configuration requires only a stable well-matched filtered source and a well-matched low-drift power meter. Substitution methods of attenuation measurement are also very popular.

Low values of attenuation can be determined accurately by making reflection coefficient measurements on the device under test with a sliding short behind it. Several bridge techniques for measuring attenuation have been devised.

The attenuation in a waveguide can be found by making Q measurements on resonant sections of different lengths. *See* Q (ELECTRICITY).

When only moderate accuracy (on the order of $\pm$ 0.5 dB) is required over a wide frequency range, a leveled swept source can be connected to the device under test, and the emerging signal can be fed to a diode detector that is followed by a logarithmic amplifier and oscilloscope. *See* AMPLITUDE-MODULATION DETECTOR.

Network analyzers yield both the magnitude and phase angle of the transmission and reflection coefficients of the device under test over a wide frequency range. By using ingenious calibration and computer-correction techniques, high accuracy can be achieved. *See* MICROWAVE TRANSMISSION LINES; TRANSMISSION LINES. [F.L.W.]

Attenuator An arrangement of fixed and variable resistors mounted in a compact package, used to reduce the strength of a radio- or audio-frequency signal by a desired amount without causing appreciable distortion. The attenuator is so designed that its impedance matches that of the circuit in which it is connected, regardless of the amount of attenuation introduced. This characteristic is achieved by properly proportioning the resistance values of the series and shunt elements. [J.Mar.]

Audio amplifier An electronic circuit for amplification of signals within, and in some cases above, the audible range. The term may mean either the complete amplifier, consisting of the voltage-amplifier and power-amplifier stages, or it may mean just one stage. In the usual case, the term implies that the amplifier is not capable of amplifying a direct-current signal; such an amplifier would more likely be called a direct-coupled or dc amplifier. *See* DIRECT-COUPLED AMPLIFIER.

The function of an audio amplifier is to amplify a weak electrical signal, such as from a microphone or a phono pickup, to a level capable of driving a loudspeaker at the desired output sound level. The sound power level may be small in such applications, as in a hearing aid. Audio amplifiers may be constructed with vacuum tubes, but current technology favors transistors and integrated circuits. *See* AMPLIFIER. [H.F.K.]

Audio delayer A device for introducing delay in the audio signal in a sound-reproducing system. The loudspeaker-pipe-microphone combination, the magnetic tape recorder-reproducer, and the digital delayer are the most common. *See* SOUND-REPRODUCING SYSTEMS.

The loudspeaker-pipe-microphone system consists of an equalizer, power amplifier, loudspeaker, pipe, microphone, and additional amplifiers. Delay is introduced by causing sound waves to travel along a pipe from the loudspeaker to the microphone. There is attenuation in the pipe, which increases with frequency. Amplitude compensation with respect to frequency is introduced in the input to the power amplifier to compensate for the frequency discrimination due to attenuation. *See* FREQUENCY-RESPONSE EQUALIZATION.

The magnetic tape recorder-reproducer system consists of an audio magnetic tape recorder and reproducer with multiple spaced heads, and a magnetic tape belt which introduces delay by carrying the recorded signal a suitable distance from the record to the reproduce heads. The system provides a simple means for introducing delay; however, the objection is that the tape belt wears out after a few hours and must be replaced. *See* MAGNETIC RECORDING.

The first element of an audio delay system employing a digital delay is a low-pass filter which eliminates all audio components above 12,000 Hz. A sampling switch operates at a frequency of 30,000 Hz. The sampled signal is converted into the proper digital format by the analog-to-digital converter and fed into the digital shift register serving as a memory. The shift register elements are connected so that each element transfers its digital word to the next stage and reads the word from the previous stage during the sampling interval. In this way each word is entered at the input of the memory and moves down the line relatively slowly. The output of the shift register is fed to the digital-to-analog converter and then to the sampling switch. The output of the sampling switch is smoothed by the low-pass filter. The audio signal output is the same as the audio wave input but with a delay.

One of the major applications for audio delayers is in sound-reinforcement systems. Audio delay of up to 500 milliseconds can be obtained with any of the three systems described above. A signal-to-noise ratio of 60 dB is attainable. *See* SOUND-REINFORCEMENT SYSTEM. [H.F.O.]

Audiometry The quantitative assessment of individual hearing, either normal or defective. Three types of audiometric tests are used: pure tone, speech, and bone conduction tests. Such tests may serve various purposes, such as investigation of auditory fatigue under noise conditions, human engineering study of hearing aids and communication devices, screening of individuals with defective hearing, and diagnosis and treatment of defective hearing. In all of these situations, individual hearing is measured relative to defined standards of normal hearing.

The pure-tone audiometer is the instrument used most widely in individual hearing measurement. It is composed of an oscillator, an amplifier, and an attenuator to control sound intensity. For speech tests of hearing, word lists called articulation tests are reproduced on records or tape recorders. Measurements of detectability or intelligibility can be made by adjusting the intensity of the test words. To make bone conduction tests, sound vibrations from the audiometer activate a vibrator located on the forehead or mastoid bone.

Scientific advance in audiometry demands careful control of all environmental sound. Two types of rooms especially constructed for research and measurement of hearing are the random diffusion, or reverberation, chamber and the anechoic room. In the reverberation chamber, sounds are randomly reflected from heavy nonparallel walls, floor, and ceiling surfaces. In the anechoic room, the fiber glass wedges absorb all but a small percent of the sound.

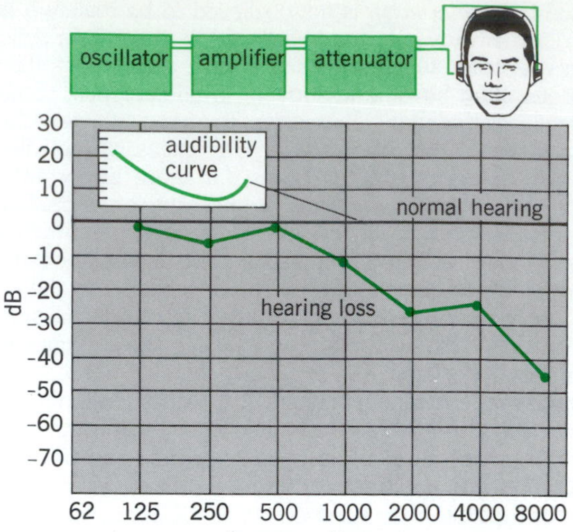

Audiogram for determining the audibility curve for pure-tone hearing loss at various frequency levels.

The measurement of hearing loss for pure tones in defective hearing is represented by the audiogram (see illustration). Sounds of different frequencies are presented separately to each ear of the individual, and the intensity levels of the absolute thresholds for each frequency are determined. The absolute threshold is the lowest intensity which can be detected by the individual who is being tested.

In clinical audiometry the status of hearing is expressed in terms of hearing loss at each of the different frequency levels. In the audiogram the normal audibility curve, representing absolute thresholds at all frequencies for the normal ear, is represented as a straight line of zero decibels. Amount of hearing loss is then designated as a decibel value below normal audibility. The audiogram in the illustration reveals a hearing loss for tones above 500 Hz. Automatic audiometers are now in use which enable individuals to plot an audiogram for themselves.

Articulation tests are speech perception or speech hearing tests used to assess hearing and loss of hearing for speech. The threshold of intelligibility for speech is defined as the intensity level at which 50% of the words, nonsense syllables, or sentences used in the articulation test are correctly identified. The hearing loss for speech is determined by computing the difference in decibels between the individual intelligibility threshold and the normal threshold for that particular speech test. Discrimination loss for speech represents the difference between the maximum articulation score at a high intensity level (100 dB), expressed in percent of units identified, and a score of 100%. The measure of discrimination loss is important in distinguishing between conduction loss and nerve deafness.

Bone conduction audiograms are compared with air conduction audiograms in order to analyze the nature of deafness. Losses in bone conduction hearing generally give evidence of nerve deafness, as contrasted to middle-ear or conduction deafness. *See* EAR; HEARING (HUMAN); HEARING IMPAIRMENT. [K.U.S.]

Auger effect An internal photoelectric process in an atom in which, for example, instead of the emission of a single characteristic x-ray from the filling of a vacancy in the *K* shell of electrons by an electron from a higher shell, an additional electron from the *L*, *M*,... shell is emitted with a kinetic energy equal to the energy of this x-ray minus the binding energy of this ejected electron. Such an electron is called an Auger electron. If the *K*-shell vacancy is in an atom of high atomic number *Z*, the return to the normal state may involve the emission of several x-rays and Auger electrons. That fraction of the vacancies in any electron shell that is filled by this Auger process is called the Auger yield, just as the fraction filled with accompanying x-ray emission is the fluorescence yield. The Auger effect is sometimes called autoionization. *See* X-RAY FLUORESCENCE ANALYSIS. [W.W.W.]

Augite A group of monoclinic calcic pyroxenes which have the general chemical formula $(Ca,Mg,Fe)(Mg,Fe)Si_2O_6$, in which calcium is the dominant cation in the first cation position. Monoclinic pyroxene with substantial iron or magnesium in place of calcium is called pigeonite, and has a different crystal structure from augite. Augite is generally considered a combination of the four end members diopside ($CaMgSi_2O_6$), hedenbergite ($CaFe^{2+}Si_2O_6$), enstatite ($Mg_2Si_2O_6$), and ferrosilite ($Fe_2^{2+}Si_2O_6$), but it almost always has substantial aluminum and minor to substantial amounts of sodium, ferric iron, chromium, and titanium.

Augite occurs in both igneous and metamorphic rocks. It is nearly universal in basalts and gabbros, and occurs somewhat less frequently in less mafic igneous rocks. Magnesium-rich augite is a characteristic mineral in many ultramafic rocks and in rocks of the Earth's mantle. Augite and pigeonite are also rather common constituents of lunar basalts and basaltic meteorites. *See* DIOPSIDE; ECLOGITE; ENSTATITE; PIGEONITE; PYROXENE. [R.J.Tr.]

Aurora A large-scale electrical discharge process which surrounds the Earth. The discharge is powered by the so-called solar wind–magnetosphere generator. A curtainlike luminous feature of the aurora is produced by upper atmospheric atoms and molecules which emit their own characteristic light after colliding with the current-carrying electrons. The most common light of the aurora (the greenish-white light) comes from excited oxygen atoms. Excited and ionized molecular nitrogen adds several band emissions. Since the discharge current flows into and out from the annular region, the aurora appears as a ring-shaped luminosity (called the auroral arc) surrounding both the northern and southern magnetic poles (see illustration). *See* ATMOSPHERIC ELECTRICITY.

Photograph of the aurora taken at the University of Alaska campus. *(Courtesy of V. P. Hessier, University of Alaska)*

The aurora becomes active during geomagnetic storms which occur often about 40 h after an intense solar flare. This is because the efficiency of the solar wind–magnetosphere generator becomes high and variable when the solar wind becomes gusty after a solar flare. During a great magnetic storm, the auroral oval expands from its usual latitude of about 67° to 50° or a little less. It is on such an occasion when the aurora can be seen widely across the continental United States. *See* ATMOSPHERE; ATOMIC STRUCTURE AND SPECTRA; MAGNETOSPHERE; PLASMA PHYSICS; SOLAR WIND. [S.-I.A.]

Australia An island continent in the Southern Hemisphere with a total area of 2,941,526 mi² (7,618,552 km²). It is bounded on the west by the Indian Ocean and on the east by the Pacific Ocean and the Tasman Sea. Numerous small and several large islands lie off the coast, including Tasmania and New Zealand. Australia is generally of remarkably low elevation and moderate relief. Three-fourths of the land mass lies between 600 and 1500 ft (180 and 450 m) in the form of a huge plateau. A cross section from east to west shows first a narrow belt of coastal plain, then the steep escarpments of the eastern face of the Great Dividing Range, stretching 1200 mi (1900 km) from the north of Queensland to the south of Victoria. The descent on the western slope of the Dividing Range is gradual until often elevation in the inland basins is below sea level, rising gradually again across the great plateau until the low ranges of western Australia fringing the plateau are reached, and beyond these lies another coastal plain. With the exception of the Gulf of Carpentaria and Cape York peninsula in the north and the Great Australian Bight in the south, there are few striking features in the configuration of the coast. Australia may conveniently be divided into three great structural and landform regions.

The region called East Australian Highlands consists of a narrow plain extending north and south along the eastern coast. Flanking the plain are the series of ranges and tablelands making up the Great Dividing Range. The East Australian Highlands is the best-watered region in Australia, and some of the river systems are of considerable size. On the flanks of the East Australian Highlands are Australia's principal coal deposits—in the vicinity of Sydney and Newcastle and in the Bowen and Ipswich fields in Queensland. Petroliferous basins at Surat (Roma), flanking the divide in Queensland and off the coast of Victoria in Bass Strait, are Australia's most promising deposits of petroleum and natural gas.

The region known as the Interior Lowland Basins comprises a region of sedimentary rocks that occupy one-third of the continent between the western slope of the eastern highlands and the inner eastern margin of the ancient shield which forms the Western Plateau. Little land is over 500 ft (150 m), and some is below sea level. The rivers of the Murray-Darling Basin, draining the western slopes of the Great Dividing Range, have a marked seasonal variation in flow but never dry up in the lower reaches. South Australia's shallow lakes are more often dry expanses of encrusted white salt than bodies of water—the result of low rainfall and high evaporation. In most parts of the region water from deep artesian wells is available.

The region known as the Western Plateau is the largest area, occupying almost three-fifths of the continent, and is a great shield of ancient rocks standing 750–1500 ft (225–450 m) high. Much of it is buried in desert sand, and only a few ridges of ancient mountains (such as the Macdonnel and Musgrave ranges) break the monotony of the plateau surface. Only in the southwestern corner of the continent and along the northwestern coast is rainfall sufficient to support a sclerophyll forest of eucalypts and a monsoon woodland, respectively. In the north, coastal rivers are of considerable size but change from flooded torrents after rains to a succession of water holes in dry seasons.

Tasmania is a small mountainous island lying 150 mi (240 km) southeast of Australia across Bass Strait, with a total area of 26,383 mi^2 (68,332 km^2). The island is structurally similar to the East Australian Highlands. The dominant feature is the central plateau, falling from a general level of 3500 ft (1070 m) in the northwest toward the southeast. A dense eucalyptus forest covers most of the island except along the wetter west coast, where beech forest predominates. The rivers have short, rapid courses with little seasonal variation in flow. *See* NEW ZEALAND. [K.B.C.]

Australopithecus An extinct genus of the family Hominidae. The existence of this hominid was first demonstrated through the recovery in 1924 of an infant's skull at Taung, northern Cape Province, South Africa. R. A. Dart identified the specimen as humanlike and named it *Australopithecus africanus*. Subsequently, other sites in South Africa yielded further, abundant fossils of this species, as well as numerous remains of a larger, more robust form, usually termed *A. robustus*.

The genus *Australopithecus* is characterized by relatively to absolutely small body size, adaptation of the lower limb for upright posture and bipedal gait, enlarged brain in comparison with the brain of anthropoid apes, large facial skeleton with small, non-apelike anterior teeth (incisors and canines), and large to massive posterior teeth (premolars and molars). The postural and locomotor adaptations and the distinctive anterior dentition clearly ally *Australopithecus* with the family Hominidae rather than with the Pongidae (apes), as was once suggested. However, the relationship of *Australopithecus* to the genus *Homo* has long been debated.

The various discoveries in Africa demonstrate the substantial antiquity of the family Hominidae and of the genus *Australopithecus*. The family Hominidae now appears to have a long evolutionary history, extending back about 7 to 8 $\times$ 10^6 years. However, the fossil record is still too scanty to gain any idea of the skeletal anatomy and adaptation of those very ancient hominids. At least three, and perhaps more, species of *Australopithecus* are known between 3 or 4 and 1 $\times$ 10^6 years ago. The small species *A. africanus* appears to be the oldest, but its evolutionary relationships to the geologically younger robust species *A. boisei* and *A. robustus* still remain unclear. *Australopithecus africanus*, or a similar species, is reasonably regarded as a likely ancestor of *Homo*, but the time and place of origin of that genus are still unresolved. *See* FOSSIL HUMAN. [F.C.H.]

Authigenic minerals Minerals that are formed in place within sediments or sedimentary rocks during or long after their deposition. Their origin distinguishes them from detrital (allogenic) minerals, which are formed elsewhere and transported to the site of deposition, and from biogenic minerals, which compose the hard parts of organisms or are indirectly precipitated by organisms at the site of deposition. Authigenic minerals are among the principal products of diagenetic processes and are frequently the only tangible evidence of these processes. *See* DIAGENESIS.

Authigenic minerals can occur in almost all types of sedimentary rock and can range in abundance from a trace to virtually the total rock. The more common authigenic minerals are shown in the table. Probably the most abundant are calcite and dolomite, the principal minerals composing limestones and dolostones, respectively. The detrital minerals composing sandstones are often cemented by authigenic quartz, calcite, dolomite, or the iron oxides hematite or goethite. Authigenic pyrite and marcasite are frequent constituents of shales and mudstones. Recent marine sediments often contain one or more of the zeolite minerals which have formed from the devitrification of volcanic glass.

Common authigenic minerals

Mineral	Formula
Albite	$NaAlSi_3O_8$
Anatase	TiO_2
Anhydrite	$CaSO_4$
Apatites*	$Ca_5(PO_4)_3(F,Cl,OH)_3$
Aragonite (orthorhombic)	$CaCO_3$
Barite	$BaSO_4$
Boehmite	$AlO(OH)$
Calcite (hexagonal)	$CaCO_3$
Celestite	$SrSO_4$
Chalcedony	SiO_2
Clay minerals	
Chlorites*	$(MgFe^{2+}Fe^{3+})_6(AlSi_3)O_{10}(OH)_8$
Illites*	$K(AlMgFe)_2(AlSi_3)O_{10}(OH)_2$
Koalinite	$Al_2Si_2O_5(OH)_4$
Sepiolite	$Mg_2Si_3O_8 \cdot 2H_2O$
Smectites*	$(NaCa)(AlMgFe)_2(AlSi_3)O_{10}(OH)_2 \cdot nH_2O$
Dolomite	$CaMg(CO_3)_2$
Gibbsite	$Al(OH)_3$
Glauconite*	$K(AlMgFe^{2+}Fe^{3+})_2(AlSi_3)O_{10}(OH)_2$
Geothite	$Fe_2O_3 \cdot n(H_2O)$
Gypsum	$CaSO_4 \cdot 2H_2O$
Halite	$NaCl$
Hematite	Fe_2O_3
Leucoxene	TiO_2
Limonite	$FeO(OH) \cdot n(H_2O)$
Marcasite (orthorhombic)	FeS_2
Microcline	$KAlSi_3O_8$
Muscovite	$KAl_2(AlSi_3)O_{10}(OH)_2$
Opal	SiO_2
Orthoclase	$KAlSi_3O_8$
Psilomelane	$BMn^{2+}Mn_8{}^{4+}O_{16}(OH)_4$
Pyrite (isometric)	FeS_2
Pyrolusite	MnO_2
Quartz	SiO_2
Siderite	$FeCO_3$
Zeolites*	$X_y{}^{1+,2+}Al_x{}^{3+}Si_{1-x}{}^{4+}O_z \cdot nH_2O$

*Group of minerals characterized by considerable chemical variation.

Authigenic minerals may be formed by direct precipitation, by recrystallization or alteration of preexisting minerals, or by structural transformation of one mineral to another. Their formation indicates an attempt to establish an equilibrated mineral assemblage. Critical factors in the formation of authigenic minerals are temperature, pressure, ionic concentration, pH, and Eh (standard oxidation-reduction potential). [I.E.O.]

Autism, infantile A severe neuropsychiatric disorder of early childhood onset, historically regarded as a psychosis of childhood, but now classified as a pervasive developmental disorder of childhood since there is no apparent relationship of autism to adult psychoses such as schizophrenia.

Symptoms of autism generally are apparent within the first 2 years of life and may occasionally be noted from the time of birth. Characteristic disturbances include disruption of social, cognitive, linguistic, motor, and perceptual development. Affected individuals fail to develop appropriate interpersonal relationships. Language may fail to develop or when it does develop, is characterized by pronoun confusion (for example, the use of "you" for "I"), abnormal speech tone or rhythm, and an impaired ability to use abstract terms or communicate symbolic information. Unusual responses to the environment are common and may include resistance to change, exaggerated reactions to sensory stimuli or changes in the environment, ritualistic behavior, and peculiar attachments to inanimate objects. Motor abnormalities include unusual posturing and stereotyped (purposeless and repetitive) movements which may consume much of the individual's time and energy; self-injurious behavior (for example, head banging) is also common. Autistic individuals do not experience delusions and hallucinations, but their metaphorical and bizarre language may mistakenly suggest the kind of thought disturbance found in schizophrenia. *See* SCHIZOPHRENIA.

For the majority of cases of autism, the cause remains unknown. Theoretical explanations have emphasized either a primary psychological or biological vulnerability in the child, the role of environmental factors, and an interaction between an inborn vulnerability and the child's environment. The high incidence of neurological signs, electroencephalographic abnormalities, and the fact that seizures develop in 25% of children during adolescence (especially in lower-IQ children) tend to support the role of a biological vulnerability. The final behavioral expression of the syndrome may be a function of multiple factors.

Various treatment modalities have been used in the management of autistic children. These have included psychotherapy, pharmacotherapy, behavior therapy, various somatic treatments, and educational interventions. However, even with the best of interventions there are no cures and most autistic individuals remain severely impaired. *See* PSYCHOTHERAPY.

[F.R.Vo.; D.J.C.]

Autoclave A pressure vessel for heating and usually agitating mixtures of gases and liquids or liquids and solids. Small-batch autoclaves are used in laboratories and for sterilizing and cooking; large ones in the chemical and petroleum industries for carrying out chemical reactions. The contents may also flow in and out continuously.

Heat may be supplied by steam or hot oil in a jacket or by pipe coils or electric heaters inside. The walls as well as the contents may be heated. Temperatures above a few hundred degrees Fahrenheit may reduce the strength of the wall unless it is of a suitable alloy. At higher temperatures it is preferable to design the autoclave with a cold wall. This is done by heating a space within the vessel cavity with an internal electric heater or with the incoming fluid stream, insulating this space from the inner wall, and cooling the wall with fluid passed through channels in it.

Agitation is provided in various ways. The contents may be stirred by a propeller driven by a shaft which passes through a packing gland, or the entire vessel may be rocked or given a linear reciprocating motion. Agitation in continuous-flow autoclaves may also be provided by baffles inside the vessel.

In batch autoclaves the cover or closure design depends on the use and conditions of operation. If the vessel is seldom opened, a suitable gasket and bolts may suffice. Frequent and ready access requires quick-opening and self-sealing designs. Sight glasses are windows in the wall to view the contents. Breakage has resulted in costly and fatal accidents. A design in which the packing exerts radial pressure on the glass disk increases safety. *See* STERILIZATION. [E.W.C.]

Autogiro A type of aircraft which utilizes a rotating wing (rotor) to provide lift. An autogiro is similar to a helicopter, but uses a conventional engine-propeller combination in addition to the rotor to pull the vehicle through the air like a fixed-wing aircraft (see illustration).

Unlike the helicopter, the autogiro rotor maintains its speed of rotation in the air because of the aerodynamic forces acting upon the rotor blades, and is without direct mechanical drive from the engine (autorotation). All the power required to maintain flight is supplied through the propeller of the autogiro at the front (tractor type) or rear (pusher type) of the fuselage. The rotor is thereby pulled through the air and creates a lifting force.

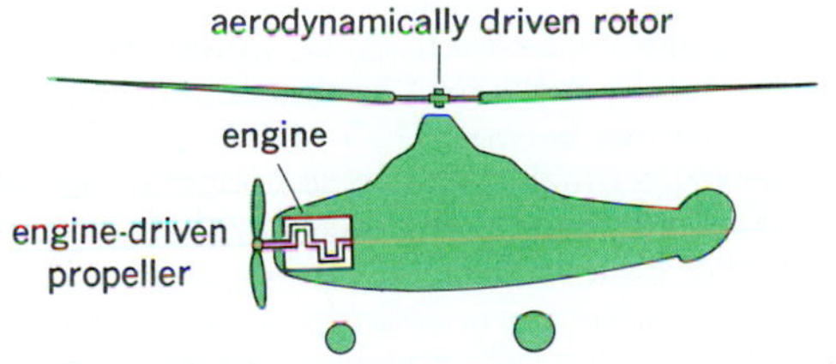

In-flight loaded dynamic components of a typical autogiro.

The autogiro has several inherent advantages. It is capable of takeoff and landing in a shorter distance than the fixed-wing airplane, and is capable of level controlled flight at extremely slow speed, on the order of 20 mi/h (30 km/h). It is mechanically simpler than the helicopter. An autogiro, unlike an airplane, cannot be stalled. Among the disadvantages of the vehicle is that its speed performance is limited as compared to an airplane, although it is capable of flying slightly faster than the helicopter. It cannot hover as can a helicopter. *See* HELICOPTER.

[L.G.]

Autoimmunity The occurrence in an organism of an immune response to one of its own tissues, that is, a response to a self constituent. Foreign substances to which an organism makes a protective immune response are called nonself. Efficient discrimination between self and nonself, the basis of normal immune function, depends upon a function known as immune tolerance, which means an inertness to substances that could be capable of provoking an immune response. Failure of immune tolerance to self constituents results in an autoimmune response which is often, although not invariably, associated with autoimmune disease. Autoimmune disease occurs when the autoimmune response to self constituents has damaging effects of a structural or functional character.

Certain human diseases have been thought to involve immunological mechanisms, especially where any of the following conditions exist: (1) Serological (antibodies, complement) abnormalities appear during the course of the disease; (2) gamma globulin and complement are found at the site of tissue injury (pathology); (3) "delayed hypersensitivity" (cellular immunity) appears; or (4) immunocompetent (lymphocytes and plasma cells) cells predominate at the site of tissue injury.

Most patients with rheumatoid arthritis have a circulating immune complex consisting of a macroglobulin antibody, the rheumatoid factor, and an antigen, the patient's own immunoglobulin G. This rheumatoid factor is therefore considered to be an autoantibody, that is, an antibody directed to a molecule of the host, to which the host, somehow deciding it is foreign, has made an antibody. *See* ARTHRITIS.

All patients with systemic lupus erythematosus have antibodies to nuclear antigens (deoxyribonucleic acid, nucleohistone, and glycoproteins), and many have antibodies to cytoplasmic organelles (mitochondria, lysosomes, and ribosomes). Many patients with lupus develop renal disease, especially those with antibodies to deoxyribonucleic acid. *See* LUPUS ERYTHEMATOSUS.

Patients with poststreptococcal glomerulonephritis also have depression of serum complement levels during acute disease. In addition, streptococcal antigen, gamma globulin, and complement have been found at the site of nephritis.

Pathological studies of acute rheumatic fever have demonstrated gamma globulin and complement on the myocardium. The disease is thought to involve the host making antibodies to some streptococcal antigens during an infection with streptococci which then cross-react with a closely related myocardial antigen. *See* ANTIBODY; ANTIGEN; IMMUNOLOGY. [P.H.S.]

Automata theory In its most general sense, automata theory is concerned with ways in which inputs to a system combine with current states of a system to produce outputs from the system. The subject matter can be divided into (1) abstract theory, (2) structural theory, and (3) self-organizing and adaptive systems. Abstract-theory encompasses general definitions, Turing machines, and language theory. The structural category is concerned more with practical machine construction and includes the propositional calculus, switching algebra (or Boolean algebra), and information theory. Self-organizing systems include neural nets, artificial intelligence, game theory, and pattern recognition. *See* ARTIFICIAL INTELLIGENCE; GAME THEORY; INFORMATION THEORY; PATTERN RECOGNITION; PROPOSITIONAL CALCULUS; TURING MACHINE.

An automaton can be characterized by three sets and two functions: the set A of all input signals, the set S of internal states of a system subjected to the inputs, and a set Z of all output signals (see illustration). At any given time the set of

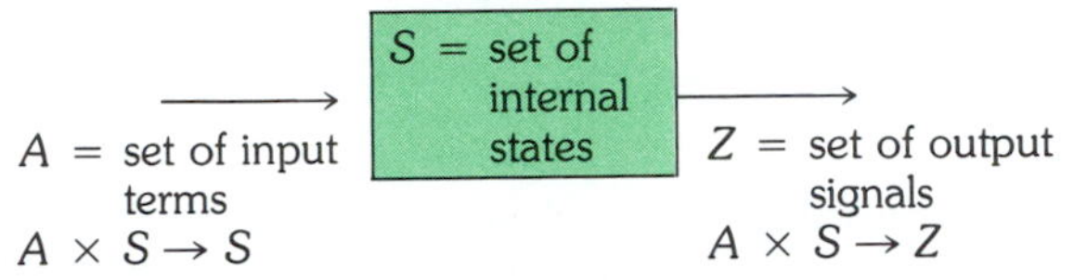

Representation of an automaton.

internal states S is determined from both a combination of the set of possible states and the set of possible input terms, so that $A \times S$ maps into S. Similarly, the output functions, or members of the set Z, are determined from system states and inputs, so that $A \times S$ maps into Z. [N.B.R.]

Automatic frequency control (AFC) The automatic control of the intermediate frequency in a radio, television, or radar receiver, to correct for variations of the frequency of the transmitted carrier or the local oscillator. In high-fidelity broadcast receivers, AFC keeps distortion due to detuning to a low figure. In the reception of long-haul telegraph signals, AFC reduces the error rate due to signal pulse distortion or interference from lower-intensity signals in the same frequency band.

AFC techniques are varied but are mainly of two types. One uses a discriminator to furnish a voltage whose magnitude and polarity are determined by the frequency change. This voltage is used to adjust the frequency of the local oscillator of the receiver, thereby keeping the intermediate frequency constant. The other uses two-polarity pulse accumulation which furnishes a dc potential proportional to frequency error. [W.Lyo.]

Automatic gain control (AGC) The automatic maintenance of a nearly constant output level of an amplifying circuit by adjusting the amplification in inverse proportion to the input field strength, also called automatic volume control (AVC). Almost all radio receivers in use employ AGC. In broadcast receivers, AGC makes it possible to receive incoming signals of widely varying strength, yet have the sound remain at nearly the same volume. In communications receivers a type of AGC circuit called a squelch circuit is used to prevent noise during periods of no transmission, such as in the reception of on-off keying, frequency-shift keying (FSK), and phone. AGC is also useful in accelerating the switching action between receivers in diversity connection. *See* RADIO RECEIVER.

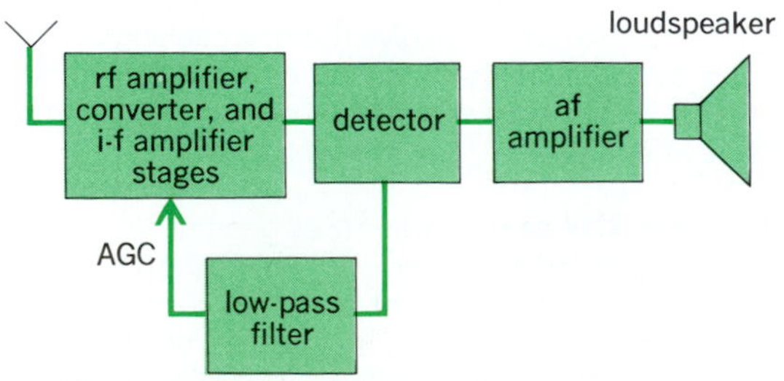

Block diagram of broadcast receiver using AGC.

AGC action depends on the characteristic, possessed by most electronic tubes and transistors, of adjustment of gain by the variation of the applied bias voltage. If the dc voltage applied to the control grid of a vacuum tube is made more negative, the amplification of that stage will be reduced.

In most broadcast receivers the AGC voltage is taken from the detector. This dc voltage, proportional to the average level of the carrier, adjusts the gain of the radio-frequency (rf) and intermediate-frequency (i-f) amplifiers and the converter, as shown in the illustration. AGC tends to keep the input signal to the audio-frequency (af) amplifier constant despite variations in rf signal strength. There are several modifications of this basic circuit.

AGC circuits are also used in dictation recording equipment, public address systems, and similar equipment where a constant output level is desirable. *See* AMPLIFIER. [W.Lyo.]

Automatic horizon A device that provides the pilot with symbols representative of the attitude of an aircraft relative to an artificial horizon. In an automatic or artificial horizon a vertical gyro moves a horizon bar relative to a fixed aircraft index. Motion of the bar simulates changes in pitch or roll of the aircraft. As the pilot views the indicator, the horizon is seen relative to the aircraft as the horizon would appear if the pilot could see it through the windscreen. *See* GYROSCOPE. [J.W.A.]

Automatic landing system The means for guiding and controlling aircraft from an initial approach altitude to a point where safe contact is made with the landing surface. Such systems differ from low-approach systems in three major respects: (1) They furnish not only guidance but control of the aircraft as well. (2) They furnish information on the aircraft's position with respect to the terrain below it, and the rate at which the landing surface is being approached. (3) They do not require the pilot to assume manual control near the ground.

Two automatic landing systems have been developed. One, a radar-beam type, detects the position and rate of change in position of the landing aircraft by means of a radar beam emitted from a ground derived-control complex. The other, a fixed-beam type, derives position and rate of change in position by instrumentation within the landing aircraft, but it makes use of instrument landing system (ILS) type equipment on the ground. In the aircraft are accelerometers (which may be part of an inertial navigation system) and a radio altimeter. Essential to both systems is an autopilot in the aircraft, commanded by a computer on the ground in the radar-beam system and by a computer in the aircraft for the fixed-beam system. *See* INSTRUMENT LANDING SYSTEM (ILS). [J.L.Lo.]

Automatic sprinkler system A safety system used to prevent the spread of fires. An automatic sprinkler system is usually installed inside a building, although sprinklers may also be installed outside for protection against exposure fires.

Common automatic sprinkler systems are wet pipe, dry pipe, preaction, and deluge. In a wet pipe system, the pipes contain water and are connected to an ample water supply. Individual sprinkler heads are normally closed and are preset to open, usually by the melting of an alloy insert, when the local temperature reaches between 135 and 165°F (57 and 74°C) for areas having normal atmospheric temperatures. Thus, the sprinklers discharge only in the area of a fire and in adjacent areas where hot fumes collect.

In a dry pipe system, the sprinkler pipes contain air under pressure. The opening of a sprinkler allows the air to escape and water to enter the system and to discharge from the sprinkler. Dry pipe systems are used in unheated buildings such as warehouses.

In preaction and in deluge systems, separate heat-responsive devices in the same area as the sprinklers admit water to the normally dry sprinkler piping when a fire occurs. The preaction system provides warning before the sprinkler heads operate. In the deluge system, all sprinklers are always open so that water discharges promptly to prevent spread of fire in a hazardous area. *See* FIRE EXTINGUISHER. [F.H.R.]

Automation The concept, development, or use of highly automatic machinery or control systems. *See* CONTROL SYSTEMS; SERVOMECHANISM.

Common concepts of automaton are: production systems so integrated that materials move through the required operations with little or no human assistance; control systems that automatically maintain machinery performance within desired limits; machinery that mathematically manipulates information (usually in the form of numbers), storing, selecting, presenting, and recording input data or internally generated data as required; any machinery or equipment (or changes in product design, materials, or processing technique) that reduces the labor content in a product or service.

INITIATING CONTROL SOURCE	TYPE OF MACHINE RESPONSE		POWER SOURCE	LEVEL NUMBER	LEVEL OF MECHANIZATION
From a variable in the environment	Responds with action	Modifies own action over a wide range of variation	Mechanical (nonmanual)	17	Anticipates action required and adjusts to provide it
				16	Corrects performance while operating
				15	Corrects performance after operating
		Selects from a limited range of possible prefixed actions		14	Identifies and selects appropriate set of actions
				13	Segregates or rejects according to measurement
				12	Changes speed, position, direction according to measurement signal
	Responds with signal			11	Records performance
				10	Signals preselected values of measurement (includes error detection)
				9	Measures characteristic of work
From a control mechanism that directs a predetermined pattern of action	Fixed within the machine			8	Actuated by introduction of work piece or material
				7	Power tool system, remote controlled
				6	Power tool, program control (sequence of fixed functions)
				5	Power tool, fixed cycle (single function)
From man	Variable			4	Power tool, hand control
				3	Powered hand tool
			Manual	2	Hand tool
				1	Hand

Levels of mechanization and their relationship to power and control sources. ***(Harvard Bus. Rev.)***

Seventeen levels of mechanization have been identified. These levels can be arranged in a chart (see illustration) that reveals a distinct evolution. This evolution explains why the effect of increasing automaticity generally is to lower the skill requirements of the operating work force rather than to raise them. It is true, of course, that the increased complexity of automatic machinery and control may increase the amount of maintenance personnel and the degree of maintenance skill required to the point that the increase of skilled personnel in the total work force becomes significant. New types of jobs requiring higher skill on the part of operators also are occasionally created.

By using levels of mechanization as a vertical scale, a mechanization profile can be constructed for any system by plotting the level of each required activity in sequence. Mechanization profiles of so-called automated plants reveal that the spans of mechanization are quite short, that the levels of mechanization fluctuate widely, that very little self-correcting action (levels 12–17) is found in most systems, and that automatic operations are most frequently achieved by mechanization on levels 5–8. Often growth in the span is more significant than an increase in the level of mechanization in achieving automatic operations. [J.R.Br.]

Automobile A generic term for a self-propelled, trackless personal or public carrier which encompasses passenger cars, recreational vehicles, taxis, and buses used to transport people in cities, on highways, or across country. Passenger cars are available in several body styles, with optional components, and they are made in various sizes. The sedan, in two- or four-door models, seats five or six passengers. The hardtop, a metal-roof model that replaced the convertible, offers some of the all-weather advantages of the sedan. The station wagon, a utility vehicle with two or four doors for occupants and a rear door for loading commodities, provides comfortable seating for up to nine passengers. The van is a type of commercial vehicle that has been adapted for personal use. Variations of these body types have been developed for special purposes. *See* AUTOMOTIVE BODY; AUTOMOTIVE BRAKE; AUTOMOTIVE CHASSIS; AUTOMOTIVE ENGINE; AUTOMOTIVE FRAME; AUTOMOTIVE STEERING; AUTOMOTIVE TRANSMISSION; BUS; MOTOR VEHICLE. [P.A.O.]

Automotive air conditioning A system for maintaining comfort of driver and occupants of automobiles, buses, and trucks, limited to air cooling, air heating, and ventilation. Humidification and filtration are not incorporated, but dehumidification is available when refrigeration is employed. Systems built into automobiles by the manufacturers provide a minimum quantity of fresh air to the cab when the vehicle is in motion. The cab is thereby pressurized at 1–2 in. (2.5–5 cm) of water column, and it exhausts by exfiltration through cracks around doors and other body leaks.

Warm air is provided by a coil in the supply airstream; the coil is heated by hot water from the engine cooling system. Cool air is provided by a refrigeration system which consists of a refrigerant compressor, belt-driven by the vehicle engine through a magnetic clutch; a condenser whose configuration is somewhat like the vehicle's radiator and whose location is ahead of the radiator; and a thermostatic expansion valve to meter refrigerant to an evaporator, usually located ahead of the auto's fire wall and across which cab air is recirculated, cooled, and returned to the cab. *See* AIR CONDITIONING; REFRIGERATION. [R.L.K.]

Automotive alternator An alternating-current generator used in automotive vehicles to supply electric current to the vehicle electric system. The typical automotive alternator is a three-phase, Y-wound unit with a six-diode rectifier. Current is supplied to the rotor (rotating field) through slip rings. When the vehicle is inoperative, the field is disconnected from the battery, either by opening the ignition switch or by opening field relay contacts. The six-diode rectifier changes the alternating current to the direct current required by the vehicle electric system. *See* ALTERNATING-CURRENT GENERATOR; ALTERNATOR; AUTOMOTIVE VOLTAGE REGULATOR. [W.H.Cr.]

Automotive body An enclosure mounted on and attached to the frame of an automotive vehicle. The passenger car body is fabricated from sheet metal, glass, interior trim, and upholstery materials for the convenience and protection of the occupants from the weather and from annoyances such as dirt, dust, noise, and smoke.

All-steel body shells are assembled from numerous stampings in automated welding bucks with a minimum of mechanical fasteners in order to reduce noise accountable to moving parts. Roof, quarter panels, doors, deck lid, and hood undergo similar welding operations in fabrication and are then cleaned and undercoated prior to internal and external painting. Body interior, doors, deck lid, and hood are sprayed with sound-deadening materials before door glass, interior trim, upholstery, and floor and deck carpeting are installed; backup lights, windshield, and similar items are added during the final assembly process.

Unitized bodies are a departure from conventional body-frame construction in that shorter side rails are used for a frame that is integrated with the body and floor pan; in conventional construction, full-length standard side rails with fore-and-aft extensions beyond the body shell proper are used.

Commercial bodies are customized to their job to a much greater degree than passenger cars are, with the type of vehicle ranging from light pickup trucks, panel trucks, and vans to huge bus and tractor units for tankers and trailers, many having refrigeration, temperature-control equipment, and so on for critical commodities. *See* MOTOR VEHICLE. [P.A.O.]

Automotive brake A system with mechanical and hydraulic components to slow or stop a vehicle by means of the action of friction upon wheel drums, disks, or drive shafts.

In an automotive power brake assembly, movement of the brake pedal operates a valve which allows one side of a piston to be exposed to atmospheric pressure, while engine-intake-manifold vacuum is applied to the other side. The difference in pressure causes the piston to move. This movement is transmitted to the piston in the brake master cylinder, causing the brakes to be applied.

In an air-brake system, an engine-driven air compressor provides the pressure to the braking system. Operation of the brake pedal admits high-pressure air to brake chambers at each wheel. A diaphragm or piston in the brake chambers is displaced, forcing the brake shoe against the brake drum. *See* AIR BRAKE.

Brakes can also be applied by electromagnets located at each wheel. The driver operates a controller that connects the electromagnet to the vehicle battery. The electromagnet is attracted to disks on the rotating wheels, causing the electromagnet to shift through a limited arc. This movement actuates the brake shoes. Further movement of the controller allows more current to flow in the electromagnets, producing stronger magnetic fields and greater braking action. *See* ELECTROMAGNETIC BRAKE. [W.H.Cr.]

Conventional drum-type service brakes have two steel shoes, faced with heat- and wear-resistant lining, for each wheel, which are forced against the inner brake drum surfaces when pedal pressure is applied by the driver. A dual master cylinder operates the hydraulic systems for front and rear brakes separately to assure braking action in the event that one fails. Power brakes, which are standard in some cars and optional in others, reduce braking effort by means of intake manifold vacuum or atmospheric pressure. Parking brakes are independent of service brakes, but they may be integral with wheel brakes, or used on the drive shaft. Parking brakes are set on rear wheels or the drive shaft by the foot pedal and are released manually. *See* FRICTION BRAKE.

Disk-type brakes are less subject to water fading and more resistant to heat fading due to high speed or repeated stops than drum brakes are, and they are capable of straight-line stops; that is, hard braking effort does not cause them to pull to the right or left. They are available in pressure-plate and fixed- or floating-caliper types. Because high hydraulic operating pressure is necessary for disk brakes, power-assist units must be added. *See* BRAKE. [P.A.O.]

Automotive chassis A bare vehicle (minus body) with frame, power plant, power train, brakes, and wheels, as well as cooling, fuel, exhaust, electrical, lubrication, steering, and suspension systems assembled and operational. For unitized bodies, the chassis components are attached directly to frame members which are integrated with the body shell and which are adapted to serve the functions of conventional frames. In heavy load-bearing commercial vehicles the frames are retained as the foundation for chassis assembly and body attachment. *See* AUTOMOTIVE BODY; AUTOMOTIVE FRAME; MOTOR VEHICLE. [P.A.O.]

Automotive drive axle A theoretical or actual crossbar or assembly which supports a motor vehicle and on which one or more wheels turn. The axle is either a live axle or a dead axle. A live axle, or drive axle, drives the wheels connected to it while supporting part of the weight of the vehicle. A dead axle, or nondrive axle, carries part of the weight of the vehicle but does not drive the wheels. *See* AUTOMOBILE; AXLE.

A drive axle on which the wheels can pivot for steering, such as on the front axle of a four-wheel-drive truck, is a steerable drive axle. The rear axle in most automotive vehicles is a nonsteering drive axle.

The rear drive axle is suspended from the vehicle body or frame by springs attached to the axle housing. The housing encloses the final-drive gears, differential gears, and wheel axle shafts. *See* AUTOMOTIVE SUSPENSION.

Most four-wheel-drive vehicles have a steerable front drive axle that is usually similar in construction and function to the rear drive axle. The principal difference is in the provisions made for steering. *See* AUTOMOTIVE STEERING.

Automobiles with front-engine and front-wheel drive have independent front suspension and do not use a front drive–axle housing. Instead, a separate transaxle combines the functions of the transmission and the drive axle. *See* AUTOMOTIVE TRANSMISSION. [D.L.An.]

Automotive electrical system The system in a motor vehicle that furnishes the electrical energy to crank the engine for starting, recharge the battery after cranking, create the high-voltage sparks to fire the compressed air-fuel charges, and power the headlamps, light bulbs, and electrical accessories.

The vehicle electrical system includes the battery, wiring, starting motor and controls, generator and voltage regulator, electronic ignition, and electronic fuel metering. Also included may be a computerized electronic engine control system, an electronically displayed driver information system, various types of radios and sound systems, and many other electrically operated and electronically controlled systems and devices. *See*

ALTERNATING-CURRENT GENERATOR; BATTERY; COMMUTATION; COMMUTATOR; COMPUTER; CONDUCTOR (ELECTRICITY); CONTROL SYSTEMS; CURRENT MEASUREMENT; DIRECT CURRENT; DIRECT-CURRENT MOTOR; ELECTRIC SWITCH; ELECTRONIC DISPLAY; ELECTRONICS; FUSE (ELECTRICITY); GENERATOR; PARALLEL CIRCUIT; REGULATOR; SPARK PLUG; SPEEDOMETER; STEPPING MOTOR; STORAGE BATTERY; VOLTAGE REGULATOR. [D.L.An.]

Automotive engine An integral major component of and source of power for automotive vehicles. Several types of engines are available for passenger and commercial vehicles, but most vehicles are driven by reciprocating internal combustion gasoline engines. Many large commercial vehicles, as well as some American and imported passenger cars, use diesel engines.

Domestic passenger car engines evolved from an assortment of electrics and steamers, designed with two or four cycles. Present four-, six-, and eight-cylinder engines are four-cycle types of in-line, V-block, slant, or horizontally opposed "pancake" configurations. Formerly popular in-line eights and V-12 and V-16 power plants were discontinued when higher compression ratios were developed. Redesign of six- and eight-cylinder engines provided equivalent horsepower, with incidental reduction in weight.

As vehicles grew in size and type, and greater road speed was demanded, engines tended to increase in size and weight and had to improve in performance. Present cylinder blocks and heads are gray iron castings, as are numerous other engine parts. Weight reduction has been achieved by decreasing cylinder wall sections and substituting stampings for castings, and plastics for metal parts.

Higher engine compression ratios produced greater combustion temperatures, and ignition knocks which could be eliminated only with higher-octane fuel, made possible by introduction of gasoline additives. Such developments created a new family of high-performance engines. This developmental trend was modified by the Clean Air Act of 1970, which mandated specific emission controls and use of nonleaded fuel, and encouraged manufacturers to get more miles per gallon. *See* INTERNAL COMBUSTION ENGINE. [P.A.O.]

Automotive frame The basic structure of an automotive vehicle. Frames are composites of heavy steel stampings which include fabricated side rails, cross members, and mounting brackets assembled into a foundation for the major components and bodies of passenger and commercial vehicles.

Side rails have either channel or box sections, with a swept perimeter conforming to body contours. Heavy cross members support the engine in the front and the loads carried in the rear; additional cross members tie the side rails together through their length; bumpers are attached at both ends. Butyl rubber body mounts insulate occupants from noises or vibrations coming from the power plant and power train. *See* AUTOMOTIVE BODY. [P.A.O.]

Automotive steering Mechanical means by which a driver controls the course of an automotive vehicle. A typical system (Fig. 1) consists of a manually operated steering wheel in the driver's compartment connected by a steering column to the steering gear attached to a connecting rod (pitman arm) from which linkages attach to steerable road wheels mounted at the front of the vehicle. Clockwise rotation of the steering wheel actuates the steering system to turn the vehicle to the right; rotation counterclockwise turns the vehicle to the left. In manual steering the driver provides all the force necessary to turn the steered road wheels; in power steering an auxiliary mechanism assists the driver by supplying part of the steering force. *See* POWER STEERING.

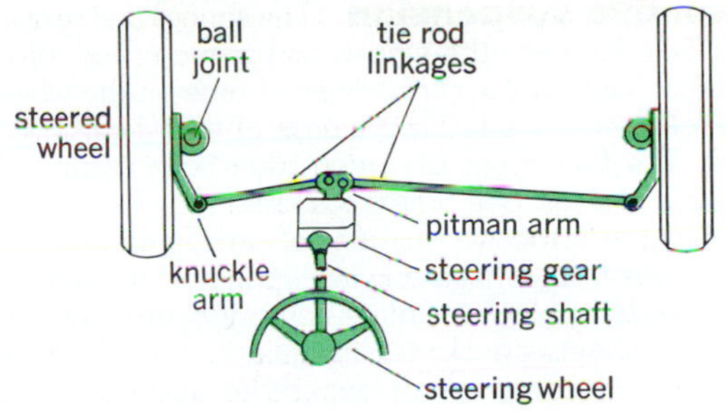

Fig. 1. Components of a typical steering system.

A steering system provides four essential features: (1) sufficient mechanical advantage for a driver to steer the vehicle without excessive physical effort, (2) steering of road wheels without hindering their free movement on the spring suspensions, (3) differential turning of steered wheels so that both roll on circles with a common center, and (4) tendency for steered wheels to return to the straight forward position when the steering wheel is released. *See* ACKERMAN STEERING.

To roll smoothly when the vehicle is directed into the desired course, the steerable wheels are mounted at slight angles to the normal (Fig. 2). This tilt is called camber and is positive if outward. The wheels are given slight positive camber (usually less than a degree) initially so that, when the vehicle is loaded, the wheels become almost vertical.

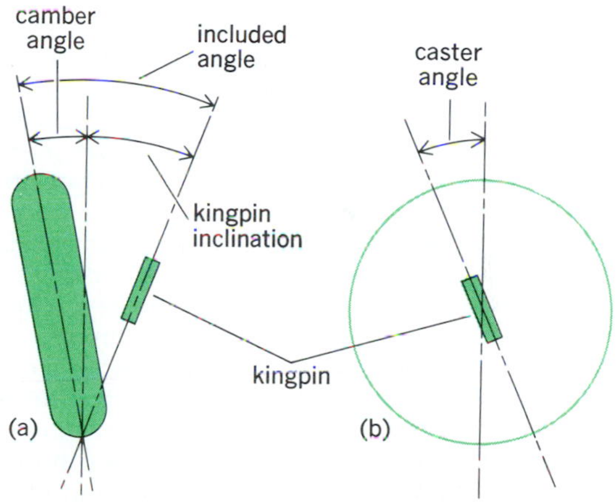

Fig. 2. Front left wheel. *(a)* Viewed from driver's position. *(b)* Viewed from inside hood. Angles exaggerated.

The kingpin, or the axis of the balls on ball suspension, is tilted inward. This kingpin inclination causes the chassis to rise when the wheel turns. Thus the weight of the car tends to keep the wheel turned straight ahead or to return it to that position after a turn. This self-return of a steering system to straight-line travel is called recovery.

Camber and kingpin inclination together are called the included angle. If the apex of this angle lies on the road surface, road resistance on the tire and forward push at the kingpin are along the line of roll. If the apex is not on the road, the tire is forced to toe in or out. To take up play in the front wheel supports, the front wheels are purposely toed in slightly so that the planes of the wheels intersect ahead of the vehicle. Road resistance then forces them to roll parallel and takes up any play in the steering system.

The inclination of the kingpin or its equivalent forward or backward is called caster. Caster is positive if the kingpin inclines backward, negative if it inclines forward, and zero if the kingpin is vertical as viewed along the axis of the front wheels. Positive caster aids in directional stability and recovery. [W.K.C.]

Automotive suspension The springs and related parts intermediate between the wheels and frame of an automotive vehicle that support the car body and frame on the wheels and absorb road shock caused by passage of the wheels over irregularities. The four types of spring have been used: coil, leaf, torsion bar, and air. *See* SPRING (MACHINES).

A typical automotive suspension arrangement uses coil springs at the front wheels and leaf springs at the rear. Weight of the vehicle applies an initial compression to the springs. When the wheels encounter irregularities in the road, the springs further compress or expand to absorb most of the shock. The suspension at the rear wheels is relatively simple in most cars. It is considerably more complex at the front because the front wheels must swing from side to side in steering. This calls for multiple-point attachments that permit the wheel to move up and down and still be swung away from straight ahead for steering. Shock absorbers are used at each wheel to restrain spring movement and prevent prolonged spring oscillations. *See* SHOCK ABSORBER. [W.H.Cr.]

In an air suspension system, suspension units on each wheel contain a cushion of air which supports the sprung weight of the vehicle while maintaining road clearance, alignment of wheels with chassis, and protection of car and passengers from road shock. As in other types of suspension systems, shock absorbers are attached to the frame and suspension unit for additional passenger comfort and prevention of roll on curves and severe shock at all speeds for better steering control and safety. [P.A.O.]

In a torsion-bar system the spring is a bar or rod stressed in torsion about its long axis. In a typical torsion-bar system the bar is rigidly fixed to the car frame at the rear end and is free to rotate at its front end, near which the lower wheel-control arm is splined to the bar. The longitudinal stiffness of the bar aids steering and directional stability. *See* TORSION BAR. [N.MacC.]

Automotive transmission The device for providing different gear or drive ratios between the engine and drive wheels of an automotive vehicle. A principal function of the transmission is to enable the vehicle to accelerate from rest up to a maximum speed through a wide speed range while the engine operates within its most effective range. The transmission in most applications is placed in the vehicle power train between the engine and the propeller shaft. Power passes through the transmission and propeller shaft and is delivered to the differential and drive axles. *See* AXLE; DIFFERENTIAL.

There are two general types of transmission in passenger cars: manual and automatic. Because transmissions in trucks and off-the-road vehicles come in such a variety of designs and number of speeds, only passenger car units will be described here.

Manual transmissions employ a layshaft or countershaft gearing in conjunction with a clutch. This mechanism disengages the engine from the transmission while the sliding of gears in the transmission is controlled manually. Manual transmissions are mainly of the three-speed type with synchronizers in all forward speeds. This type carries a clutch slidably splined on the main shaft to obtain third-speed engagement by shifting to the left, and second speed by shifting to the right. The clutch at the right, also slidably splined on the main shaft, obtains first speed when shifted to the left and reverse when shifted to the right. The two clutches are actuated by forks mechanically connected to the hand shift lever, which may be located on the steering column or on the floor directly over the transmission. Gear ratios are in the order of 1.5:1 for second speed, 2.5:1 for first speed, and 2.7:1 for reverse. Four-speed manual-shift transmissions have enjoyed popularity, particularly in sports cars. Generally they are synchronized in all forward speeds. Reverse generally involves sliding a gear on the main shaft into mesh with the reverse idle by a separate lever. Gear ratios are in the order of 1.28:1 for third speed, 1.64:1 for second speed, 2.2:1 for first speed, and 2.25:1 for reverse. *See* CLUTCH.

In 1974, the first five-speed manual transmission for automobiles to be mass-produced in the United States was introduced. It has overdrive in fifth speed with its ratio either 0.8:1 or 0.84:1. One gear set, for cars with engines under 175 lbf-ft (237 N-m) torque, has ratios of 3.41:1 in first speed, 2.08 in second, 1.40 in third, 1:1 in fourth (direct drive), and 0.8:1 in fifth or overdrive. Reverse is 3.36:1. The other gear set, covering torque up to about 205 lbf-ft (278 N-m) has ratios of 3.10 in first, 1.89 in second, 1.27 in third, 1:1 in fourth, and 0.84:1 in fifth. Reverse is 3.06:1.

In American passenger cars, automatic transmissions generally employ a three-element hydrodynamic torque converter with planetary gear sets of two or three speeds. The torque converter provides a smooth start with a torque (turning effort) multiplication within itself that is continuously variable, depending upon the degree of engine input. The torque may rise as high as 2.0:1 or 2.4:1 at stall speed (full engine torque, the driven member of converter, or turbine, stationary). As car speed increases, the maximum converter torque ratio available decreases until it becomes 1:1 with little slippage. Converters provide an additional advantage in isolating the transmission gear set from torsional vibrations of the engine, which generally make special dampers necessary when engine clutches are used. [F.R.McF.]

Automotive voltage regulator A device in the automotive electrical system to prevent generator or alternator overvoltage. There are two general types: one for direct-current and another for alternating-current generators.

The direct-current voltage regulator contains a set of spring-loaded contacts and a winding shunted across the generator. When the generated voltage exceeds the preset voltage of the regulator, the magnetic flux is sufficient to open the contacts, thereby inserting resistance into the generator field. The generator voltage is reduced, the contacts close, and the cycle is repeated. Included with the voltage regulator are a current regulator and a cutout relay. The current regulator prevents generator overload. The cutout relay opens the circuit when the generator is inoperative to prevent battery discharge through the generator.

The alternating-current voltage regulator has two sets of contacts, lower and upper. At intermediate speeds, the lower contacts open and close, inserting and removing the resistance in the alternator field to provide voltage-limiting action. At higher speeds, the resistance is not sufficient to hold the voltage down. It increases slightly and the upper contacts begin to make and break. When they close, the alternator field is grounded by the contacts. With both ends of the field grounded, additional regulation is achieved. The alternator also has a field relay which connects the alternator field to the battery when the ignition switch is closed and disconnects it when the ignition switch is opened to shut off the engine. This keeps the battery from discharging through the alternator field. Many automotive alternators have integral transistorized voltage regulators which do not use vibrating contacts and do not require adjustment. They prevent excessive alternator voltage by inserting and removing field resistance. *See* AUTOMOTIVE ALTERNATOR; GENERATOR. [W.H.Cr.]

Autonomic nervous system The part of the nervous system which innervates smooth, cardiac, and some striated muscle and glands and viscera; also known as the visceral or involuntary division. The thoracolumbar (sympathetic) and craniosacral (parasympathetic) systems comprise the autonomic system.

The autonomic system is a pure motor system consisting of

efferent neurons. The autonomic system influences the rate of metabolism, muscle tone of the viscera, blood flow, and other aspects of general homeostasis. The craniosacral division and thoracolumbar division are generally antagonistic in their effects. The actions of the autonomic division are duplicated by various hormones, such as epinephrine, and drugs, such as acetylcholine. *See* HOMEOSTASIS; NERVOUS SYSTEM (INVERTEBRATE); NERVOUS SYSTEM (VERTEBRATE); PARASYMPATHETIC NERVOUS SYSTEM; SYMPATHETIC NERVOUS SYSTEM. [W.J.B.]

Autopilot An automatic means for steering an aircraft or other vehicle. An autopilot replaces the human pilot in guided missiles and drone aircraft. In piloted aircraft the autopilot is used to relieve the pilot during cruise, to effect automatic approaches to landing fields, and to control rapid maneuvers, as in tracking a dodging target. For helicopters and vertical takeoff and landing aircraft, the autopilot can provide automatic hovering and transition to forward flight, and may be necessary to provide stable control. Autopilots are also used to steer ships, submarines, torpedoes, and spacecraft, and the automatic control of automobiles is in the developmental stage. This article discusses autopilots as used in aircraft and space vehicles.

Elements of an aircraft autopilot are shown in the illustration. Steering commands may come from a radio receiver,

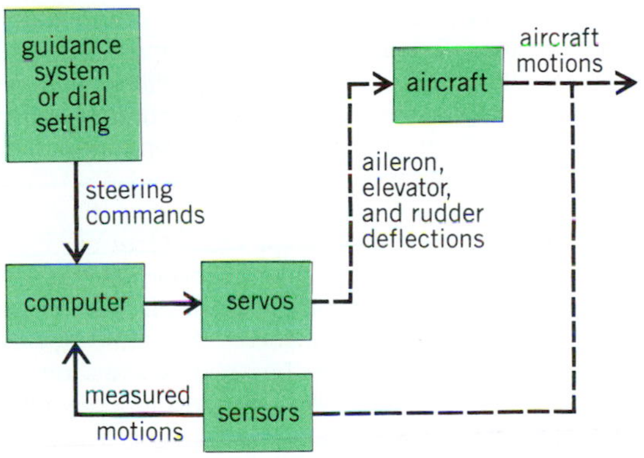

Autopilot system block diagram.

autonavigator, fire-control, or other guidance system, or they may be set by the pilot. The computer compares actual aircraft motion (obtained as signals from the gyroscopes, airspeed indicator, altimeter, and other sensors) with steering commands. Errors between aircraft motion and steering commands are corrected by control-surface servos, which deflect the rudder, ailerons, and elevators to alter the aerodynamic forces on the aircraft and make it roll, pitch, and yaw as necessary to eliminate the errors. *See* FLIGHT CONTROLS; GUIDANCE SYSTEMS.

The oldest form of autopilot simply maintained an airplane in straight and level flight. This is called cruise control. Cruise-control autopilots commonly consist of a heading system and an altitude-control system, in which heading and altitude commands are manually set by the pilot. Guided missiles, drones, and interceptor aircraft can be controlled from remote stations by means of radio signals. Automatic approach to a landing field may be accomplished either by ground command or by furnishing a strong radio beam down which the airplane can be flown. In homing missile or fighter attacks on other aircraft, radar or infrared tracking-error signals from the guidance section feed directly into tight autopilot rate loops to steer the vehicle.

Modern supersonic piloted aircraft are sometimes aerodynamically unstable. They are made flyable by retaining the autopilot rate loops even during manual operation. In such application the autopilot is an essential part of the machine, whereas in the earlier applications described above, it is an assist or substitute for the person in the human-machine system. Hovering helicopter or vertical takeoff and landing (VTOL) aircraft are particularly challenging to control because they are not inherently stable. To assist the pilot, automatic stabilization equipment may be provided during pilot-operated control. In addition, both hover and transition from hover to forward flight may be controlled automatically by using airborne radar sensors to measure vehicle location with respect to the ground. Control is effected by changing blade angles appropriately, as in helicopters; by using gas reaction jets; or by controlling thrust direction, as in deflected-thrust VTOL craft. For piloted aircraft, a mandatory autopilot design feature is that autopilot failure must not endanger life or equipment; that is, the design must be fail-safe. [R.H.C.]

Autoradiography A photographic technique for detecting radioactive substances in almost any object; also known as radioautography. A photographic emulsion is placed in direct contact with the object to be tested and is left for several hours, days, or weeks, depending on the suspected concentration of the radioactive material. The film is then developed, fixed, and washed as in normal photographic processes. At sites where the film was close enough to the radioactive substance it appears dark because of the presence of silver grains. When the number of grains is insufficient to darken the film to the unaided eye, the film may be examined with the aid of a microscope.

Autoradiograms (the picture obtained by autoradiography) can be made with many objects which can be brought in contact with photographic film. The resolution, that is, the precision with which the radioactive substances can be located, is good only when the specimen is flat and in close contact with the film or photographic plate. Whole plants, animals, rocks, or any other objects may be used, but the flat, cut surface of the object provides better contact. However, if the radioactivity is distributed throughout the object being tested, a rather distinct picture results.

The high-resolution autoradiogram is a powerful tool in detecting and measuring small amounts of radioactivity that could not be detected in any other way. Most radioactive isotopes produce radiations that penetrate the film deep enough to produce somewhat out-of-focus autoradiograms, but tritium (radiohydrogen) is an exception. Its very low-energy beta particles are stopped at the surface of the film. Therefore, the grains appear very close to the radioactive site. It has been possible to label chromosomes with a selective label, tritium-thymidine. When they reproduce, each labeled chromosome can be shown to yield one labeled and one unlabeled daughter chromosome. *See* RADIOGRAPHY. [J.H.T.]

Autotomy A process by which appendages of crabs and other crustaceans and tails of some salamanders and lizards may be cast off under stress. The break, in crustaceans, is at a definite site with a preformed internal partition that restricts loss of essential body fluids. In the salamanders and lizards, the center of each tail vertebra is unossified and breaks easily. Autotomy is a protective mechanism. When grasped by an enemy, the part fractures off and the victim escapes. Severed parts usually regenerate. In crustaceans, the replacement increases in size at each successive molt. With salamanders and lizards, repair growth is continuous. [T.I.S.]

Autotransformer A special form of transformer having one winding, a portion of which is common to both the primary and the secondary circuits. The current in the high-voltage circuit flows through the series and common windings (see illustration).

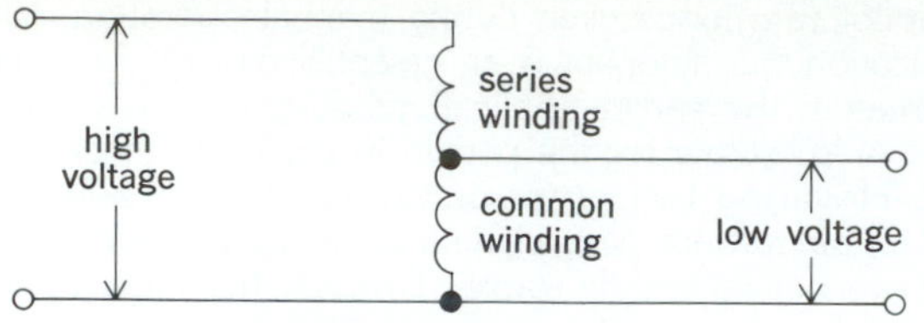

Typical autotransformer circuit.

The current in the low-voltage circuit flows through the common winding and adds vectorially to the current in the high-voltage circuit to give the common winding current. Thus, an electrical connection exists between high-voltage and low-voltage windings. Because of this sharing of parts of the winding, an autotransformer having the same kilovolt-ampere (kVA) output rating is generally smaller in weight and dimensions than a two-winding transformer. One possible disadvantage of autotransformers is that the windings are not insulated from each other and that the autotransformer provides no isolation of the primary and secondary circuits.

Autotransformers of large sizes are used for interconnecting high-voltage power systems. They are used in small sizes for intermittent-duty starting of motors. For this use the motor is connected for a short time to the common winding voltage, and then connected to the full line voltage. Small, variable-ratio autotransformers are used in testing and as components of other apparatus. [J.R.Su.]

Autoxidation The slow, flameless combustion of materials by reaction with oxygen; it is sometimes spelled autooxidation. Autoxidation is important because it is a useful reaction for converting compounds to oxygenated derivatives, and also because it occurs in situations where it is not desired (as in the destructive cracking of the rubber in automobile tires). *See* COMBUSTION; OXIDATION PROCESS.

Although virtually all types of organic materials can undergo air oxidation, certain types are particularly prone to autoxidation, including unsaturated compounds that have allylic hydrogens or benzylic hydrogens; these materials are converted to hydroperoxides by autoxidation.

Autoxidation is a free-radical chain process. Such reactions can be divided into three stages: initation, propagation, and termination. In the initiation process, some event causes free radicals to be formed. For example, free radicals can be produced purposefully by the decomposition of a free-radical initiator, such as benzoyl peroxide. In some cases, initiation occurs by a process that is not well understood but is thought to be the spontaneous reaction of oxygen with a material with a readily abstractable hydrogen. Destructive autoxidation processes also are initiated by pollutants such as those in smog.

Once free radicals are formed, they react in a chain to convert the material to a hydroperoxide. The chain is ended by termination reactions in which free radicals collide and combine their odd electrons to form a new bond. *See* CHAIN REACTION (CHEMISTRY); FREE RADICAL; ORGANIC REACTION MECHANISM.

Autoxidation is a process of enormous economic impact, since all foods, plastics, gasolines, oils, rubber, and other materials that must be exposed to air undergo continuous destructive reactions of this type. All plastics and rubber and most processed foods contain antioxidants to protect them against the attack of oxygen. *See* ANTIOXIDANT; FAT AND OIL (FOOD); FOOD MANUFACTURING; PLASTICS PROCESSING; RUBBER. [Wi.A.P.]

Auxin The generic term originally referring to substances capable of promoting plant growth in a manner similiar to that of the native growth hormone. The currently accepted definition is that of an organic compound which promotes growth along the longitudinal plant axis when applied in low concentrations to shoots freed as far as practical from endogenous growth-promoting substances. Auxins may, and generally do, influence other plant systems, but the effect on plant growth is critical to characterizing an organic compound, synthetic or natural, as an auxin. *See* PLANT GROWTH.

Because of the minute amounts of auxin present in plants, chemical identifications of auxins have been difficult. It was not until 1946 that indole-3-acetic acid (IAA), now considered the most widely occurring auxin, was proved by good chemical characterization techniques to occur in higher plants. It has been isolated from fruits, seeds, pollen, root tips, young leaves, and buds in the flowering plants and from both fungi and bacteria. In general, plant organs, just prior to and during cellular expansion, are good sources of auxin, particularly in actively growing buds and developing ovules in flowers. Many synthetic auxins have been produced as well.

Starting with the first practical use of plant-regulating substances, the stimulation of rooting, the importance of auxins for plant production has steadily increased. In the United States alone over 35 growth-regulating chemicals have been approved for over 100 different agricultural uses. A large portion of the 35 chemicals are either auxins, chemicals that are converted to auxins by certain plants, auxin antagonists, or chemicals that interact with auxins in modifying growth. The largest use of auxins has been as herbicides, particularly 2,4-D (2,4-dichlorophenoxyacetic acid) and 2,4,5-T (2,4,5-trichlorophenoxyacetic acid). Many of the other uses are based on the correlative action of auxin on tissue differentiation: rooting of cuttings; organ abscission, both promotion and inhibition; fruit development; induction of flowering; and alteration of fruiting patterns. *See* PLANT HORMONES. [P.B.K.]

Avalanche In general, a large mass of snow, ice, rock, earth, or mud in rapid motion down a slope or over a precipice. In the English language, the term avalanche is reserved almost exclusively for snow avalanche. Minimal requirements for the occurrence of an avalanche are snow and an inclined surface, usually a mountainside. Most avalanches occur on slopes between 30 and 45°.

Two basic types of avalanches are recognized according to snow cover conditions at the point of origin. A loose-snow avalanche originates at a point and propagates downhill by successively dislodging increasing numbers of poorly cohering snow grains, typically gaining width as movement continues downslope. This type of avalanche commonly involves only those snow layers near the surface. The mechanism is analogous to dry sand. The second type, the slab avalanche, occurs when a distinct cohesive snow layer breaks away as a unit and slides because it is poorly anchored to the snow or ground below. A clearly defined gliding surface as well as a lubricating layer may be identifiable at the base of the slab, but the meteorological conditions which create these layers are complex.

In the case of the loose avalanche, release mechanisms are primarily controlled by the angle of repose, while slab releases involve complex strength-stress problems. A release may occur simply as a result of the overloading of a slope during a single snowstorm and involve only snow which accumulated during that specific storm, or it may result from a sequence of meteorological events and involve snow layers comprising numerous precipitation episodes. Most large snow slides are believed to be caused by an unstable layer of ice grains that develop deep in mountain snow. Called depth hoar by students of avalanche dynamics, these crystals owe their formation to heat from earth and rock which are buried by the snow, and which in late autumn are warmer than the surrounding air. Snow nearest the ground vaporizes, causing growth of angular ice grains that exhibit poor bonding qualities. Gravity combining with the weakness of the depth hoar crystals loosens the upper stable layers. Once the stable layers begin to slide, the depth hoar acts in a manner similar to ball beatings to speed the descent of the slide.

Where snow avalanches constitute a hazard, that is, where they directly threaten human activities, various defense methods have evolved. Attempts are made to prevent the avalanche from occurring by artificial supporting structures or reforestation in the zone of origin. The direct impact of an avalanche can be avoided by construction of diversion structures, dams, sheds, or tunnels. Hazardous zones may be temporarily evacuated while avalanches are released artificially, most commonly by explosives. Finally, attempts are made to predict the occurrence of avalanches by studying relationships between meteorological and snow cover factors. [R.L.A.]

Aves The class of birds, which are characterized by being feathered, warm-blooded, bipedal vertebrates in which the forelimb is modified into a wing. Birds are specialized for flying: all of their features, morphological, physiological, behavioral, and so forth, are clearly modified from the typical terrestrial vertebrate condition. These aspects can even be seen on flightless groups, such as penguins and the ratites, and even in those birds that have become highly specialized for flightlessness. Birds have evolved from some group among the diverse assemblage of archosaurian reptiles (although there is some dispute on their exact ancestral group), whether it was the pseudosuchians or the theropod dinosaurs. All known birds, including flightless groups, are believed to have evolved from flying ancestors, with *Archaeopteryx* being the most primitive known member of the class.

Feathers are the unique feature of birds. These epidermal structures evolved from reptilian scales by elongation and later subdivision of the scale into the barbs of the feather vane. Feathers presumably evolved originally for heat regulation and secondarily specialized as flight surfaces of the wings and tail. *See* FEATHER.

Birds have a higher body temperature than mammals, with high digestive and metabolic rates. Their kidney is simple, as birds excrete uric acid which can be concentrated easily. All species of birds lay eggs which are incubated outside the female's body. Most species provide the needed heat for incubating the eggs with their body; the megapods (Galliformes) use sources of heat from the ground. The young hatch by breaking through the egg shell after an incubation period varying from 10 to over 60 days.

Birds are found over the entire surface of the Earth, being the terrestrial vertebrates found closest to both poles. They live in a diversity of habitats from the open oceans to the deserts, from the tropics to the high latitudes. Aside from nesting, they do not burrow. The ability of flight has permitted birds to travel long distances for food on a daily basis and to migrate great distances between breeding and wintering areas. Flight involves orientation, homing, and navigation; birds possess a magnetic compass and other abilities to determine direction, the home site, and nesting and wintering areas. Birds are able to return to within 100 m (330 ft) of their birth site to breed the following year. *See* FLIGHT; MIGRATORY BEHAVIOR.

Just over 9000 species of living birds are known, and it is doubtful that more than a few hundred species are undiscovered. Much is known about the species taxonomy of birds, including their geographic distribution and variation. The relationships of the higher taxa of birds are far less well known, and there is still considerable disagreement on the limits of avian families and the arrangement of these families into orders and higher categories. The following classification is a conservative consensus and may be subject to considerable revision in future years. Separate articles appear for each subclass, superorder, and order (except those marked by an asterisk; fossil groups are designated by a dagger). *See* VERTEBRATA.

Class Aves
- Subclass Archaeornithes†
  - Order Archaeopterygiformes†*
- Subclass Neornithes
  - Superorder Unknown
    - Order: Palaeocursornithiformes†*
      - Ambiortiformes†*
      - Gansuiformes†*
      - Iberomesornithiformes†*
      - Gobipterygiformes†*
      - Enanthiorniformes†*
      - Alexornithiformes†*
  - Superorder Odontognathae (Odontornithes)†
    - Order: Hesperornithiformes†
      - Ichtyornithiformes†
  - Superorder Neognathae
    - Order: Struthioniformes
      - Gaviiformes
      - Podicipediformes
      - Procellariiformes
      - Sphenisciformes
      - Pelecaniformes
      - Ciconiiformes
      - Phoenicopteriformes
      - Anseriformes
      - Gastornithiformes†
      - Galliformes
      - Falconiformes
      - Gruiformes
      - Charadriiformes
      - Columbiformes
      - Psittaciformes
      - Opisthocomiformes
      - Cuculiformes
      - Strigiformes
      - Caprimulgiformes
      - Apodiformes
      - Coliiformes
      - Trogoniformes
      - Coraciiformes
      - Piciformes
      - Passeriformes

Little question has existed for over 100 years that birds evolved from an ancestor within the great archosaurian radiation of reptiles; crocodiles are the closest living relatives of birds. But it has not been possible to point to the precise ancestral group. Opinion is divided between the pseudosuchians of the order Thecodontia and the theropod dinosaurs of the order Saurischia. The difficulties in resolving this problem lie in a lack of knowledge of the Thecodontia and a lack of fossils bridging the gap between the earliest known bird, *Archaeopteryx*, and its possible reptilian ancestors. *See* DINOSAUR. [W.J.B.]

Avian leukosis A complex consisting of a group of viruses that cause several diseases in fowl. Some viruses in the group cause more than one neoplastic disease. For example, the BAI strain A usually causes myeloblastic leukemia in chickens, but it can also cause visceral lymphomatosis (lymphosarcoma), osteopetrosis (marble bone or big leg), and kidney tumors. Other viruses of the complex induce erythroblastosis, endothelioma, fibrosarcoma, and ovarian carcinoma. Visceral lymphomatosis is of particular economic importance to the poultry industry since it is highly contagious and spreads rapidly in crowded domestic flocks.

These viruses are spherical particles with a diameter of about 100 nanometers. The inner region, or core, of the particle consists primarily of ribonucleic acid (RNA) and protein. The core is surrounded by an outer flexible envelope or membrane which contains lipid and protein. The viruses are released from cells by budding from the cell membrane. *See* ANIMAL VIRUS; TUMOR VIRUSES. [W.Sc.]

Aviation A general term including the science and technology of flight through the air. Aviation also applies to the mode of travel provided by aircraft as carriers of passengers and cargo, and as such is part of the total transportation system. Aviation also describes the employment of aircraft in such fields as military aviation. The world of the airplane, including the people who manufacture, market, and repair aircraft or who work in allied industries, is frequently spoken of as aviation. *See* AIRPLANE; MILITARY AIRCRAFT.

Aviation is broadly grouped into three classes: general aviation, air transport aviation, and military aviation. General aviation comprises all aviation not included in military or air-transport aviation. Military aviation includes all forms of aviation in military activities, and air-transport aviation is primarily the operation of commercial airlines essentially as a public utility for the movement of persons and commodities. *See* GENERAL AVIATION. [L.A.B.]

Avocado A tropical and subtropical fruit tree, *Persea americana*, in the Lauraceae family. It originated in Central America or adjoining regions of North or South America. It has now spread to much of the near-tropical world.

The species is divided into three horticultural races with differing commercial qualities. The so-called West Indian race is least tolerant of cold, the Mexican most tolerant, and the Guatemalan intermediate. This same gradation is found in salt tolerance (West Indian highest) and oil content (West Indian lowest). But in some other respects the West Indian race is intermediate (skin thickness), or one of the races is different from the other two (the West Indian fruit is less tolerant of cold storage; the Mexican has smaller fruit with a unique aniselike odor; the Guatemalan has a smaller seed ratio, and takes twice as long to mature—14 months or more in California).

Mexico is the world's leading producer, followed by Brazil and California, then Colombia and Venezuela, countries of eastern South America, Central America, Caribbean Islands, Florida, Philippines, and Zaire (central West Africa). South Africa and Israel have important export industries, primarily to Europe. Many other countries have begun development. The California industry is expanding rapidly, and the avocado has become one of the state's leading fruit crops. *See* FRUIT; FRUIT, TREE. [B.O.B.]

Avogadro number The number of molecules N_0 in one mole of a substance, 6.02210×10^{23}. Equal numbers of moles of all substances contain the same numbers of molecules. For a perfect gas, the molar volume contains a number of molecules equal to the Avogadro number. *See* MOLE (CHEMISTRY).

Determinations of the Avogadro number by a variety of independent methods give results that agree very closely. The earliest estimates of its value were made during the latter half of the 19th century from the kinetic theory of gases. A modern method that yields the most accurate value of N_0 is obtained from x-ray measurements and density data. [T.C.W.]

Avogadro's law The principle that equal volumes of all gases and vapors, under the same conditions of temperature and pressure, contain identical number of molecules; also known as Avogadro's hypothesis. From Avogadro's law the converse follows that equal numbers of molecules of any gases under identical conditions occupy equal volumes. Therefore, under identical physical conditions the gram-molecular weights of all gases occupy equal volumes. *See* GAS; KINETIC THEORY OF MATTER. [T.C.W.]

Axinellida An order of sponges of the class Demospongiae, subclass Tetractinomorpha, with monactinal or diactinal megascleres or both arranged in plumose tracts bearing more or less spongin (see illustration). The axial skeleton may be dense, with an abundance of spongin especially near the base of the sponge. Axinellidan sponges vary in form from branching or massive to cup-shaped or lamellate. Most occur in shallower waters down to 330 ft (100 m), but a few descend to abysmal depths of at least 14,500 ft (4400 m). *See* DEMOSPONGIAE. [W.D.H.]

***Axinella polycapella*, a representative axinelline sponge.**

Axle A supporting member that carries a wheel. An axle may rotate with the wheel to transmit power to or from it, as does the rear axle of an automobile, or it may allow the wheel to rotate freely, as does the axle of a trailer.

Rear axles in American cars are usually of the unsprung type in which the rear-axle housing carries the right and left rear-axle shafts and the wheels are mounted at the outer end of each shaft. Rear axles in many European and some English cars are of the sprung type, in which the rear-axle gearing housing at the center is mounted to the frame or body, while the wheels and axle shaft outer ends are attached by linkage to the frame, body, or central gear housing. Each type of rear axle has its own characteristics and is selected for suitability to the road conditions found in the countries where the car is to be sold. The unsprung axle is usually simpler and cheaper, permits easier alignment of the rear wheels, and provides an easier design to reduce lowering of the rear end of the car when accelerating or to reduce raising of the rear end when braking. The sprung rear axle permits reducing the unsprung weight at the rear wheels, generally considered an advantage on rough roads. It also permits lowering the tunnel in the floor of the car. *See* AUTOMOTIVE TRANSMISSION; WHEEL AND AXLE. [F.R.McF.]

Aye-aye A rare prosimian primate indigenous to eastern Madagascar. A single living species, *Daubentonia madagascariensis*, makes up the family Daubentoniidae (see illustration). The aye-aye is a nocturnal, arboreal animal. A single young is produced in early spring in a special nest, which is constructed by the female. The hindtoes are opposable, and the fingers are

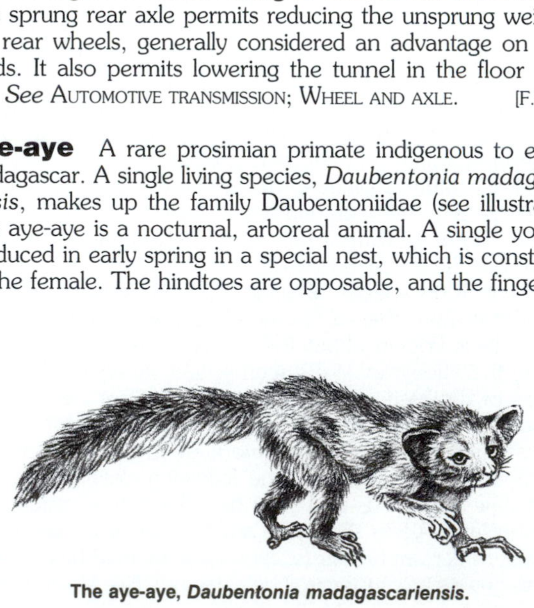

The aye-aye, *Daubentonia madagascariensis*.

quite long and slender, especially the middle one. Using its middle finger or its sharp incisors, the aye-aye digs insect larvae out of tree bark or extracts the contents of sugarcane.

The phylogenetic relationship of this species is obscure; however, the general consensus is that this animal is an aberrant or divergent form of a lemuroid ancestral stock. *See* MAMMALIA; PRIMATES. [C.B.C.]

Azeotropic distillation Any of several processes by which liquid mixtures containing azeotropes may be separated into their pure components with the aid of an additional substance (called the entrainer, the solvent, or the mass separating agent) to facilitate the distillation. Distillation is a separation technique that exploits the fact that when a liquid is partially vaporized the compositions of the two phases are different. By separating the phases, and repeating the procedure, it is often possible to separate the original mixture completely. However, many mixtures exhibit special states, known as azeotropes, at which the composition, temperature, and pressure of the liquid phase become equal to those of the vapor phase. Thus, further separation by conventional distillation is no longer possible. By adding a carefully selected entrainer to the mixture, it is often possible to "break" the azeotrope and thereby achieve the desired separation. *See* AZEOTROPIC MIXTURE; DISTILLATION.

Entrainers fall into at least four distinct categories that may be identified by the way in which they make the separation possible. These categories are: (1) liquid entrainers that do not induce liquid-phase separation, used in homogeneous azeotropic distillations, of which classical extractive distillation is a special case; (2) liquid entrainers that do induce a liquid-phase separation, used in heterogeneous azeotropic distillations; (3) entrainers that react with one of the components; and (4) entrainers that dissociate ionically, that is, salts. *See* SALT-EFFECT DISTILLATION.

Within each of these categories, not all entrainers will make the separation possible, that is, not all entrainers will break the azeotrope. In order to determine whether a given entrainer is feasible, a schematic representation known as a residue curve map for a mixture undergoing simple distillation is created. The path of liquid compositions starting from some initial point is the residue curve. The collection of all such curves for a given mixture is known as a residue curve map (see illustration). These maps contain exactly the same information as the corresponding phase diagram for the mixture, but they represent it in such a way that it is more useful for understanding and designing distillation systems.

Mixtures that do not contain azeotropes have residue curve maps that all look the same. The presence of even one binary azeotrope destroys the structure. If the mixture contains a single minimum-boiling binary azeotrope, three residue curve maps are possible, depending on whether the azeotrope is between the lowest- and highest-boiling components, between the intermediate- and highest-boiling components, or between the intermediate- and lowest-boiling components.

Nonazeotropic mixtures may be separated into their pure components by using a sequence of distillation columns because there are no distillation boundaries to get in the way. The situation is quite different when azeotropes are present, as can be seen from the illustration. It is possible to separate mixtures that have residue curve maps similar to those shown in illus. *a* and *c* by straightforward sequences of distillation columns. This is because these maps do not have any distillation boundaries. These, and other feasible separations for more complex mixtures, are referred to collectively as homogeneous azeotropic distillations. Without exploiting some other effect (such as changing the pressure from column to column), it is impossible to separate mixtures that have residue curve maps like illus. *b*.

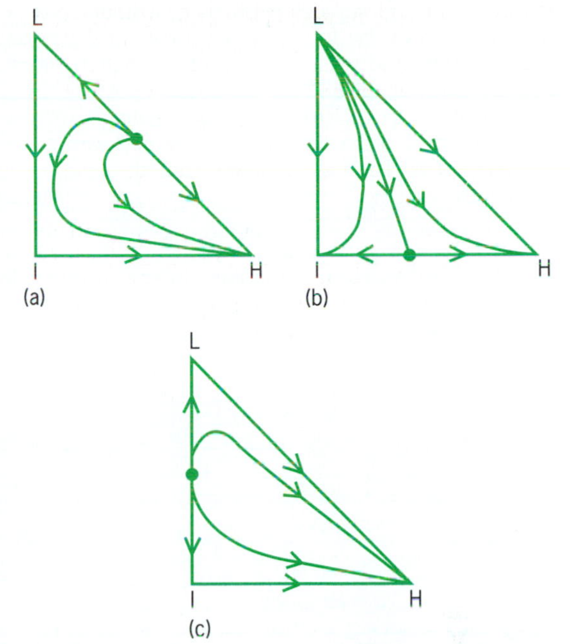

Schematic representation of the residue curve maps for ternary mixtures with one minimum-boiling binary azeotrope. (*a*) Azeotrope between the lowest- (L) and highest-boiling (H) pure components. (*b*) Azeotrope between the intermediate- (I) and highest-boiling components. (*c*) Azeotrope between the intermediate- and lowest-boiling components.

A large number of mixtures have residue curve maps similar to illus. *c*, and therefore the corresponding distillation is given the special name extractive distillation.

Heterogeneous entrainers cause liquid-liquid phase separations to occur in such a way that the composition of each phase lies on either side of a distillation boundary. In this way, the entrainer allows the separation to "jump" over a boundary that would otherwise be impassable. [M.F.D.]

Azeotropic mixture A solution of two or more liquids, the composition of which does not change upon distillation. The composition of the liquid phase at the boiling point is identical to that of the vapor in equilibrium with it, and such mixtures or azeotropes form constant-boiling solutions. The exact composition of the azeotrope changes if the boiling point is altered by a change in the external pressure. A solution of two components which form an azeotrope may be separated by distillation into one pure component and the azeotrope, but not into two pure components. *See* DISTILLATION; SOLUTION. [F.J.J.]

Azide One of several types of compounds containing the —N_3 group and derived from hydrazoic acid, HN_3. The organic azides are represented by two groups of compounds, the alkyl and aryl azides, RN_3, and the acid azides, $RCON_3$. The inorganic azides are important commercially because of the use of lead azide in priming compositions and initial detonating agents. The alkali and alkaline-earth metal azides are water-soluble; hence sodium azide, NaN_3, is usually the starting material for the production of lead azide or other desired salts. *See* NITROGEN. [E.E.W.]

Azimuth An angle measured in the plane of the observer's horizon from one of several arbitrary departure points; also the angle at the observer's zenith between the local meridian and the vertical circle passing through the object that is being observed. *See* ASTRONOMICAL COORDINATE SYSTEMS; CELESTIAL NAVIGATION.

In astronomical and surveying practice, azimuth has been referenced to each of the four cardinal points of the horizon: north, east, south, and west. In current navigation practice, however, azimuth is measured from the north point of the horizon eastward from 0 to 360°. The horizon angle measured from 0° at the north or south horizon point eastward or westward through 90 or 180° is connoted azimuth angle, in contrast to azimuth. [R.L.Du.]

Azole A suffix designating organic compounds with a five-membered *N*-heterocycle containing two double bonds. Formally the term may be, but seldom is, applied to pyrrole. In practice, azole is reserved for diunsaturated rings containing two or more heteroatoms, one of which is nitrogen. Familiar azole systems are given in the table. *See* HETEROCYCLIC COMPOUNDS; IMIDAZOLE; OXAZOLE; PYRAZOLE; THIAZOLE. [W.J.Ge.]

Azole systems

	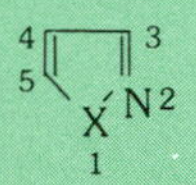	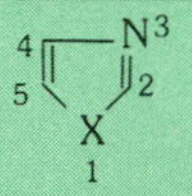
NH	Pyrazole or 1,2-diazole	Imidazole, glyoxaline, or 1,3-diazole
O	Isoxazole or 1,2-oxazole	Oxazole or 1,3-oxazole
S	Isothiazole or 1,2-thiazole	Thiazole or 1,3-thiazole

Azotobacteraceae A family of aerobic gram-negative bacteria capable of vigorous assimilation of elemental nitrogen.

The genus *Azotobacter* has big, bluntly rod-shaped, oval, or spherical cells approximately 2–6 micrometers in diameter, which are usually motile by means of lateral or polar flagella. Some species form thick-walled resting cells known as cysts. The type species *A. chroococcum* is commonly found in soils of neutral and alkaline reaction, especially fertile, cultivated soils.

A group of related organisms, often regarded as a separate genus, is *Beijerinckia*. They differ from *Azotobacter* in the smaller size of their rod-shaped cells with characteristic polar fat inclusions, slower growth rate, formation of tenacious slime, and much greater tolerance of acid reaction (growth between pH approximately 3–9, optimum at pH 4–6). Nitrogen fixation is at least as efficient as in *Azotobacter*, but a narrower range of carbon sources is utilized. The *Beijerinckia* are largely confined to tropical and subtropical soils.

The genus *Azomonas*, found in soil and water, has large ovoid cells that are motile with polar or peritrichous flagella. The type species is *A. agilis*.

A fourth genus, *Derxia*, has been found in soil from India but seems to be rare. A single species, *D. gummosa*, somewhat resembles *Beijerinckia*, but differs in its long rod-shaped cells which lack the typical polar bodies, as well as in its lesser acid tolerance (growth ceases at pH below 5). Nitrogen is fixed with the same efficiency as in *Azotobacter* and *Beijerinckia*. [H.L.J.; M.M.]

Azurite A basic carbonate of copper with the chemical formula $Cu_3(OH)_2(CO_3)_2$. Azurite is normally associated with copper ores and often occurs with malachite. Azurite is monoclinic. It may be massive or may occur in tabular, prismatic, or equant crystals. Invariably blue, azurite was originally used extensively as a pigment. Hardness is 3½–4 (Mohs scale) and specific gravity is 3.8. Notable localities for azurite are at Tsumeb, Southwest Africa, and Bisbee, Arizona. *See* COPPER. [R.I.Ha.]

B

Babinet's principle A principle which states that the diffraction patterns produced by complementary screens are identical. Two screens are said to be complementary when the opaque parts of one correspond to the transparent parts of the other, and vice versa. Babinet's principle is not very useful in dealing with Fresnel diffraction. In Fraunhofer diffraction the pattern due to a disk is the same as that due to a circular hole of the same size. *See* DIFFRACTION. [F.A.J./W.W.W.]

Bacillaceae The family of the rod-shaped, endospore-forming bacteria. Species in this family range from organisms used in retting flax to those pathogenic for animals and humans, such as anthrax. The special features of the group are largely associated with spore production. The spore is formed within the vegetative cell and is therefore of the type known as an endospore. Its outstanding property is a capacity to tolerate conditions which destroy the vegetative stage. The two principal genera of this family are *Bacillus* and *Clostridium*. Other genera include *Sporolactobacillus*, *Desulfotomaculum*, and *Sporosarcina*.

All Bacillaceae appear to have the same mode of life. A large reserve of dormant spores occurs in nature, principally in soil. If a spore is induced to germinate by the presence of food material, rapid multiplication may occur. When the food supply is exhausted, spores may again be formed, and they join the natural reserve, The short multiplication phase becomes possible in the refuse of plant and animal life, in foods, in the bowel of the animal in the case of clostridia, and in the tissues of animals in the case of pathogenic species. *See* BLACKLEG; BOTULISM; TETANUS. [T.G.]

Bacillariophyceae A class of nonflagellate unicellular algae, commonly called diatoms, with boxlike silicified walls. Diatoms range in maximum dimension from 4 micrometers to more than 1 millimeter. The diatom wall or frustule (illustration *a* and *b*) comprises several interlocking, usually elaborately sculptured, lightly or heavily silicified pieces overlying a thin polysaccharide layer. The two largest pieces are the upper and lower valves, which fit together like the top and bottom of a petri dish or shoe box. Between the valves (along the side or girdle of the cell), several smaller pieces—hooplike girdle bands—are intercalated. Depending upon which dimension is larger, breadth or depth, a diatom tends to lie on the valve side or on the girdle side.

Valves are honeycombed by perforate chambers arranged in patterns characteristic of individual species. The overall symmetry of the valves and details of their structure and ornamentation provide the basis for classification. More than 200 genera and 12,000 species of living diatoms have been described. Two main groups are recognized: those in which structural features of the valve are arranged with reference to a central pole (centric valve; illustration *c*) or to two or more poles (gonioid valve); and those in which the features are arranged with respect to a line, often symmetrically (pennate valve).

Most diatoms are photosynthetic. Many, however, are auxotrophic, requiring an external source of certain vitamins. Diatoms are likely to occur wherever there is moisture. They are free-living or attached, solitary or colonial. Marine planktonic diatoms have been important primary producers for at least 100 million years, and over the millennia their frustules have accumulated on the ocean floor. From the abundant fossil record, it is known that centric diatoms evolved first (Cretaceous), followed by pennate diatoms (Paleocene). Uplifted deposits (diatomaceous earth) are mined at several locations. *See* ALGAE; CHRYSOPHYCEAE; DIATOMACEOUS EARTH. [P.C.Si; R.L.Moe]

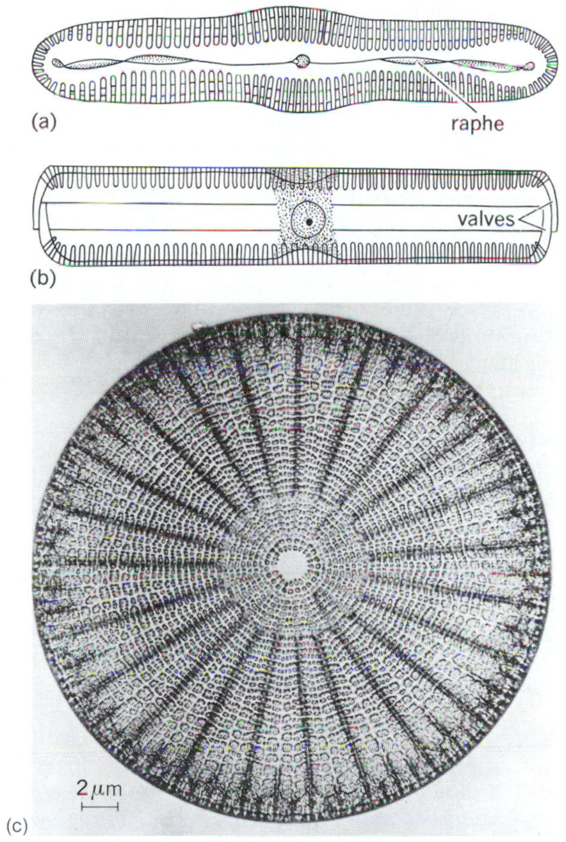

Diatoms. *(a) Pinnularia*, top view; *(b)* side view. *(c) Arachnoidiscus ehenbergii*.

Bacillary dysentery An infectious disease of humans and primates which affects the large bowel and is caused by members of the genus *Shigella*. The organisms enter the body via the fecal-oral route (for example, from unclean hands) or through fecally contaminated water or food, and penetrate the intestinal mucosa only. The incubation time varies from 1 to 7 days. The characteristic symptoms are watery, watery-mucoid, or bloody diarrhea accompanied by fever, dehydration, abdominal cramps, and prostration. Most cases, however, are "subclinical" or mild. Short-term convalescent carriers are an important reservoir; long-term carriers are rare. The main diagnostic tool is the stool culture.

Effective treatment has to rely on the results of sensitivity testing of the strains involved, since some of them have acquired resistance to sulfonamides, ampicillin, and tetracycline. Prevention efforts must stress proper sanitation, cleanliness in handling food and water, and the use of handwashing facilities. *See* MEDICAL BACTERIOLOGY; SALMONELLA. [A.W.C.V.G.]

Bacitracin An antibiotic produced by *Bacillus licheniformis* (*B. subtilis*, strain Tracy). Purified bacitracin is a white to light-brown powder, soluble in water and a variety of aqueous alcohols, including *n*-butanol and *sec*-butanol. It is active against a wide variety of gram-positive bacteria and only a few gram-negative ones. The use of bacitracin is almost entirely limited to topical (including local injection and oral dosage) application because of the toxicity shown after systemic administration. It is therapeutically effective in the local treatment of a wide variety of infections such as furuncles, abscesses, infected wounds, carbuncles, and impetigo caused by staphylococci, streptococci, and certain other gram-positive bacteria. [R.E.B.]

Background count The number of counts that must be subtracted from an observed number of counts in an experiment where atomic or nuclear particles coming from a source are being enumerated. The background count may be due to cosmic rays; to natural radioactivity in the air, in the walls of the room, or in the counter itself; or to the scattered radiation in the vicinity of a source of nuclear particles, such as an accelerator or a nuclear reactor. *See* RADIOACTIVITY. [W.B.Fr.]

Backing wind A wind which changes direction in a counterclockwise sense, for example, a south wind changing to an east wind or an east wind changing to northerly. At a particular place, the wind gradually backs with time when a cyclone passes eastward on a path south of the observer's location. The wind characteristically backs with increasing height, on the west side of an extratropical cyclone. The meaning, in terms of changes of cardinal directions, is reversed in the South Hemisphere. This term is opposite in sense to veering wind. [C.W.N.]

Backward-wave tube A type of microwave traveling-wave tube in which energy on a slow-wave circuit flows opposite in direction to the travel of electrons in a beam. Chief characteristics of backward-wave tubes are regenerative feedback produced by interaction of circuit and beam, and a wide range of electrical tuning, easily produced by changing the beam voltage. Such tubes are useful as voltage-tuned oscillators for signal generators, as power sources for quick tuning transmitters, and as local oscillators in receivers for systems that have quick tuning transmitters. If backward-wave tubes are operated as regenerative amplifiers, they are useful as narrow-band amplifiers in wide-range rapidly tuned receivers. *See* TRAVELING-WAVE TUBE.

An O-type backward-wave oscillator (or O-carcinotron) may be similar in appearance to a forward-wave traveling-wave tube. An electron gun produces an electron beam; this beam is focused longitudinally throughout the length of the tube. A slow-wave circuit interacts with the beam, and at the end of the tube a collector terminates the beam (Fig. 1).

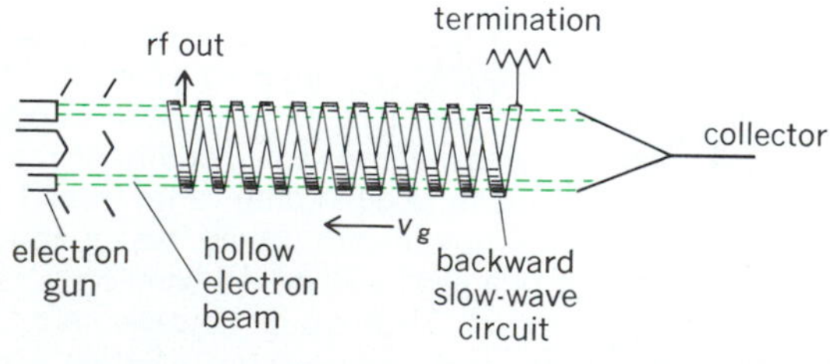

Fig 1. An O-type backward-wave oscillator (or O-carcinotron) uses a helix as the slow-wave circuit and has a hollow cylindrical electron beam.

An M-type backward-wave oscillator (or M-carcinotron) is similar in principle to the O-type, except that focusing and interaction are through magnetic fields, as in magnetrons. In the M-type oscillator (Fig. 2), a transverse magnetic field and a static radial electric field between sole and backward-wave circuit structure confine the beam to the interaction space. Either the voltage connected to the sole or to the slow-wave circuit tunes the frequency. *See* MAGNETRON.

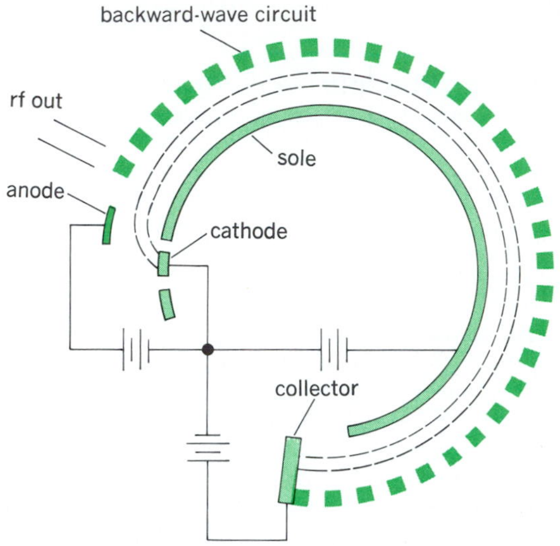

Fig. 2. In an M-type backward-wave oscillator, the beam is focused by a static magnetic field directed into the plane of the page.

When either type tube is operated with currents below the start-oscillation value, narrow-band regenerative amplification is obtained. Amplifier frequency is electrically tunable over a wide range by change of beam voltage. *See* MICROWAVE TUBE. [J.W.Ge.]

Bacteria Extremely small—usually 0.3–2.0 micrometers in diameter—and relatively simple microorganisms possessing the prokaryotic type of cell construction. Although traditionally classified within the fungi as "Schizomycetes," they show no phylogenetic affinities with the fungi, which are eukaryotic organisms. The only group that is clearly related to the bacteria are the blue-green algae. *See* MICROORGANISMS.

Bacteria are found almost everywhere, being abundant in soil, water, and the alimentary tracts of animals. Each kind of bacterium is fitted physiologically to survive in one of the innumerable habitats created by various combinations of space, food, moisture, light, air, temperature, inhibitory substances, and accompanying organisms. Dried but often still living bacteria can be carried into the air. One of the few locations in which bacteria are not usually found is within the cells of other healthy organisms, though even this is subject to exceptions, as there are many bacteria that do live intracellularly in a number of eukaryotic organisms.

As in higher forms of life, each bacterial cell arises either by division of a preexisting cell with similar characteristics, or through combination of elements from two such cells in a sexual process. *See* REPRODUCTION (ANIMAL).

Descriptions of bacteria are preferably based on the studies of pure cultures. Pure cultures are sometimes called axenic, a term denoting that all cells had a common origin and descend from the same cell, without implying exact similarity in all characteristics. Pure cultures can be obtained by selecting single cells, but indirect methods achieving the same result are more common. These involve separating the bacteria from each other and keeping their progeny separate. This is done on a streak plate. Another method is to inoculate the suspension of mixed bacteria into melted nutrient medium, which is cooled to just above the solidification temperature in order not to kill the inoculated cells; the medium is then solidified. *See* PURE CULTURE.

The morphology, that is, the shape, size, arrangement, and internal structures, of bacteria can be distinguished microscopically and provides the basis for classifying the bacteria into major groups. Stains are used to visualize bacterial structures otherwise not seen, and the stain reaction with Gram's stain provides a characteristic used in classifying bacteria. *See* ENDOSPORES (BACTERIA); STAIN (MICROBIOLOGY).

The submicroscopic differences that distinguish many bacterial genera and species are due to structures such as enzymes and genes that cannot be seen. The nature of these structures is determined by studying the metabolic activities of the bacteria. Data are accumulated on the temperatures and oxygen conditions under which the bacteria grow, their response in fermentation tests, their pathogenicity, and their serological reactions. There are also methods for determining directly the similarity in deoxyribonucleic acids between different bacteria. *See* FERMENTATION; PATHOGEN; SEROLOGY.

Interrelationships with other organisms may be close and may involve particular species. Examples are the parasitic association of many bacteria with plant and animal hosts, and the mutualistic association of nitrogen-fixing bacteria with leguminous plants, of cellulolytic bacteria with grazing animals, and of luminous bacteria with certain deep-sea fishes. [R.E.H.]

The study of genetics in bacteria, as in all forms of life, requires the occurrence of readily detectable variations in the properties that characterize a particular type of species. The rapid growth of bacteria (in some cases the number will double in 20 min) and the large populations obtained (as many as 10^9 organisms per cubic centimeter) allow for a wide fluidity in the number of such variations and for detection of any rare change that may occur. These changes, called mutations, are the result of alterations in genes. Depending upon the gene involved, the spontaneous rate of mutation may be as low as one affected bacterium per 10^9 organisms per doubling time or as frequent as one per 10,000. This rate may be increased 100 to 1000 times by the use of mutagenic agents, such as ultraviolet light, x-rays, or a variety of chemicals (nitrogen mustard, acridine dyes, nitrous acid, and purine or pyrimidine analogs). *See* MUTATION.

The genetic characteristics that are most commonly studied are those that occur frequently and are easily detected or selected for. Bacteria produce characteristic colonies when plated on the surface of a solid nutrient agar medium. Morphologic variants may be detected readily by looking for variations in size, shape, color, or texture of such colonies. Biochemical mutants are those in which the change can be related to a specific chemical substance such as a protein, an enzyme, or a unique functional ingredient of the cell. This may show as a difference in the ability to ferment a certain sugar; to survive the inhibitory action of antibiotics, ultraviolet light, or even bacterial viruses; to produce infections in experimental animals; or to produce immunity substances. Mutations that have proved extremely useful to the bacterial geneticists are those which do not allow growth of the bacteria unless a specific growth factor is added to the medium. These are the nutritional or auxotrophic mutants, and they have been of special value in helping to unravel the complex biosynthetic mechanisms of bacteria.

Sexlike mechanisms have been demonstrated in a number of bacterial species by showing that genes can be transferred

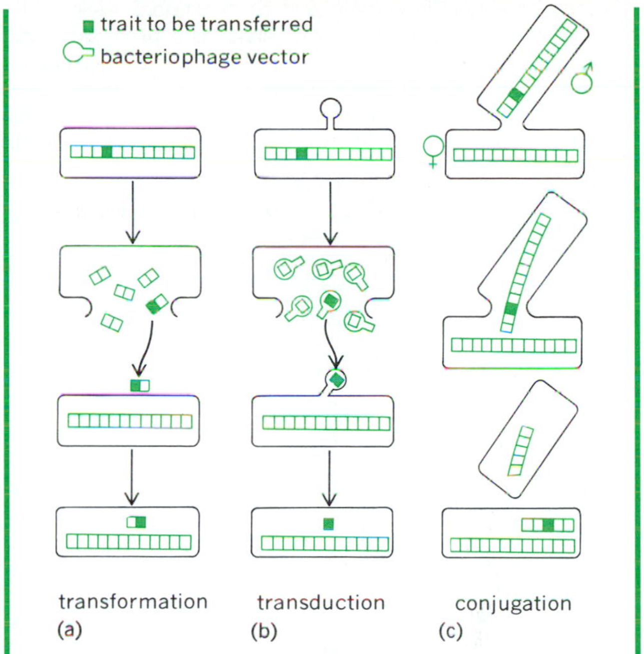

Diagram of three major mechanisms of genetic transfer in bacteria. Trait can be transferred from donor cell to recipient cell by (*a*) transformation using extracted DNA; (*b*) transduction with participation of bacteriophage vector; and (*c*) conjugation involving union of male and female bacteria.

from one bacterial cell to another. Three main types of genetic transfer have been described: transformation, transduction, and conjugation. Transformation involves the transfer of genetic material in the form of virtually pure, highly polymerized DNA. The DNA is extracted from the donor cells or released from them by lysis (illustration *a*). In transduction, small fragments of the bacterial chromosome are transferred by viral vectors, bacteriophages, which pick up the genes in the course of their development in the donor hosts. The host is destroyed by the viral infection; the viruses are liberated and then may be used to infect recipient cells, but this time under conditions favoring the survival of the virus-infected cells (illustration *b*).

Conjugation is the most highly developed form of sexuality in bacteria. It involves a direct union between donor and recipient bacteria. A cytoplasmic bridge is formed, and a considerable part of the single chromosome is transferred (illustration *c*). *See* BACTERIOPHAGE; DEOXYRIBONUCLEIC ACID (DNA); TRANSDUCTION (BACTERIA); TRANSFORMATION (BACTERIA). [J.S.G.]

Bacteriology The science and study of bacteria, and hence a specialized branch of microbiology. It deals with the nature and properties of the bacteria as living entities, their morphology and developmental history, ecology, physiology and biochemistry, genetics, and classification.

The major subjects that have consecutively occupied the forefront of bacteriological research have been the origin of bacteria, the constancy or variability of their properties, their role as causative agents of disease and of spoilage of foods, their significance in the cycle of matter, their classification, and their physiological, biochemical, and genetic features. *See* BACTERIA; MICROBIOLOGY. [C.B.V.N.]

Bacteriophage Any of the viruses that infect bacterial cells. They are discrete particles with dimensions from about 20 to about 200 nanometers. A given bacterial virus can infect only one or a few related species of bacteria; these constitute its host range. Bacteriophages consist of two essential components: nucleic acid, in which genetic information is encoded (this may be either ribonucleic acid or deoxyribonucleic acid), and a protein coat (capsid), which serves as a protective shell containing the nucleic acid and is involved in the efficiency of infection and the host range of the virus.

The description of a bacterial virus involves a study of its shape and dimensions by electron microscopy (see illustration),

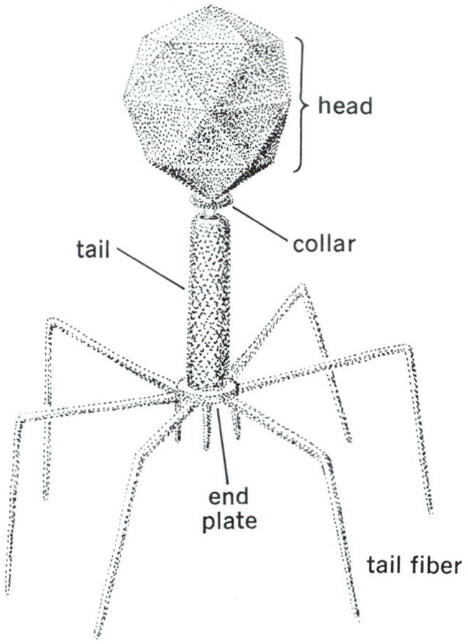

Diagram of a T4 bacteriophage.

its host range, the serological properties of its capsid, the kind of nucleic acid it contains, and the characters of the plaques it forms on a given host. Both the nucleic acid and the capsid proteins are specific to the individual virus; in the case of the capsid proteins this specificity is the basis for serological identification of the virus.

The most striking form of phage infection is that in which all of the infected bacteria are destroyed in the process of the formation of new phage particles. This results in the clearing of a turbid liquid culture as the infected cells lyse. When lysis occurs in cells fixed as a lawn of bacteria growing on a solid medium, it produces holes, or areas of clearing, called plaques. These represent colonies of bacteriophage. The size and other properties of the plaque vary with individual viruses and host cells. *See* ACTINOPHAGE; COLIPHAGE; LYSOGENY; LYTIC INFECTION; VIRUS. [L.B.]

Bacteroidaceae A family of gram-negative anaerobic bacteria. They are obligate parasites normally occurring in the intestinal tracts and mucous membranes of animals and humans. They are sometimes pathogenic, causing calf diphtheria, contagious polyarthritis in mice, and ratbite fever and Haverhill fever in humans. The microorganisms are gram-negative, rod-shaped, obligatorily anaerobic or microaerophilic (need oxygen, but grow best at, or may even require, a reduced oxygen tension of less than 10%), nonsporing, and usually nonmotile. In the eighth edition of *Bergey's Manual of Determinative Bacteriology* (1974), the family Bacteroidaceae is not assigned to an order and is included in Part 9, gram-negative anaerobic bacteria. The family is divided into three genera, *Bacteroides*, *Fusobacterium*, and *Leptotrichia*. [E.A.F.]

Badger The name for a number of species of heavily built omnivorous mammals assigned to the subfamily Melinae of the weasel family, Mustelidae. There are eight species in six genera (see table). *Taxidea taxus*, the American badger, is the only representative in North America. It tends to be more carnivorous with a diet consisting of small rodents, rabbits, prairie dogs, and ground squirrels in addition to vegetation. The American badger is more frequently found in open terrain than is the Eurasian species, which prefers wooded regions. Badgers live in burrows, called sets.

Names and geographic distribution of badgers

Species	Common name	Geographic distribution
Taxidea taxus	American badger	North America, especially United States
Meles meles	Eurasian or common badger	Europe, Asia
Arctonyx collaris	Hog or sand badger	Sumatra, southern Asia
Suillotaxus marchei	Philippines badger	Philippines
Mydaus javanensis	Malay or stinking badger, teledu	Malay Archipelago
Melogale moschata	Chinese ferret, badger	China, especially forested areas
Melogale orientalis	Javanese ferret, badger	Java, Borneo
Melogale personata	Burmese ferret, badger	Malayasia

Badgers are essentially nocturnal animals which have nonretractile claws on each of the five digits. Anal scent glands are present. The badgers walk on their feet and toes, and therefore are plantigrade. They have 38 teeth. Although little is known about the breeding behavior of the animal, the usual litter is three or four, and the gestation period for the Eurasian badger is known to be 7 months. In the colder regions of their range, badgers hibernate for varying periods from October on, with the exception of *Meles meles*, which is active during the winter. *See* CARNIVORA; HIBERNATION. [C.B.C.]

Balance An instrument for measuring weight, or mass. There are many types of balances: spring-type scales, single-lever-deflection types, precise two-lever-arm designs, torsion designs, substitution designs, and complete electronic weighing systems. The application, whether for commercial or scientific usage, determines the type of balance. *See* MASS; WEIGHT.

The two-pan balance, in its crudest form, consists of a rigid bar (lever), two pans suspended from each end, and a pivotal point in the center of the bar (Fig. 1). The center of gravity (cg) of the

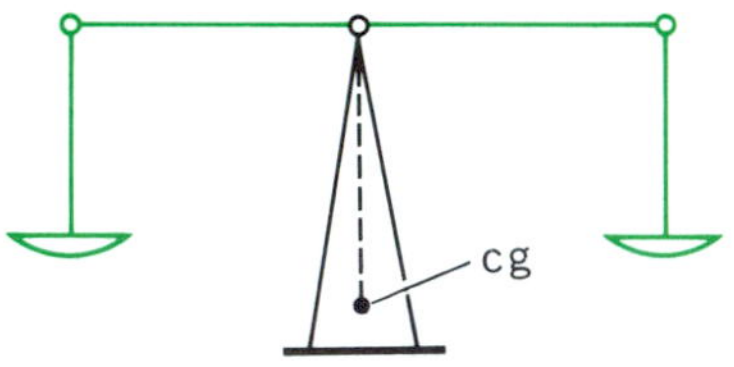

Fig. 1. Basic design of a two-pan balance.

beam system is directly beneath the point of rotation. This design keeps the beam in a horizontal position when unloaded. If, however, a small sample (unknown mass) is placed on one of the pans, the beam will deflect to a specific angle that is proportional to mass. This angle of deflection can be expressed in mass units and read from a graduated scale by means of a pointer. If the unknown mass is so great that the pointer goes off the scale, weights (known standards) are placed in the second pan to bring the balance beam back into a state of equilibrium. Fine differences between the known weights and unknown mass are shown as beam deflection and are read on a graduated scale by means of a pointer. The unknown mass equals the sum of the known weights plus the mass units read from the graduated scale.

The substitution balance is a mechanical balance consisting of one pan, two knife-edges, and one lever arm to compare the known and the unknown mass. A counterweight is rigidly affixed to the beam. Calibrated weights are built into the system and are suspended above the pan. When the unknown sample is placed on the pan, known calibrated weights are removed from the beam until the system equilibrates. In effect, the weights are substituted, giving the relationship between the known and the unknown. Small differences are read as a result of angular deflection of the beam. A simplified schematic of this design is shown in Fig. 2.

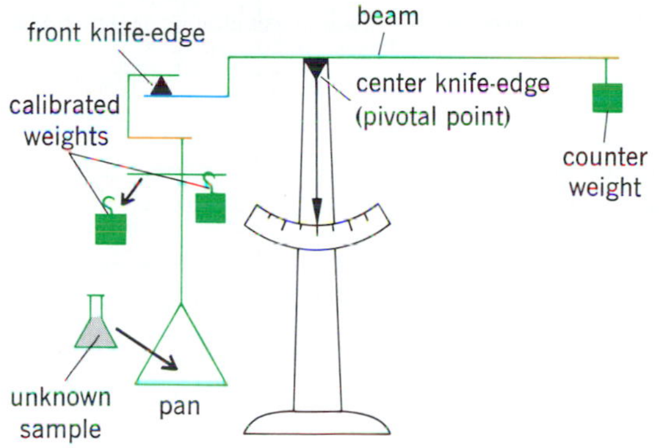

Fig. 2. Basic design of a substitution balance.

Balance manufacturers produce mechanical balances employing the substitution principle in two classifications, top-loading and analytical balances. Analytical balances have a limited capacity but offer better precision and readability than the top-loading counterparts. The top loaders are faster, more rugged, and easier to operate. However, they sacrifice precision and thus readability.

In an electronic weighing system, determination of mass is based on the principle of electronically controlled electrical force compensation. A scanner monitors displacement of the pan support. This modulates the variable gain amplifier, which generates a current that is proportional to the weight of the pan. The current flows through a coil and causes the pan support, located in the force field of the magnet, to return to its zero position. From the signal generated, automatic tare determines the mass value and subtracts any previous stored tare value. The difference is displayed on a digital readout and is also available for further processing at digital output. *See* WEIGHT MEASUREMENT. [P.E.L.]

Balanomorpha A suborder of the Thoracica. This group includes the common acorn barnacles. The mantle secretes a highly calcified test forming the bilaterally symmetrical conical shell, consisting typically of anterior rostrum, posterior carina, and one to three pairs of lateral plates. The broad basis is membranous or calcified. The shell (mantle) opening is closed by paired valves, each formed of two interlocking plates. These animals are hermaphroditic, and cross-fertilization is normal. Dwarf males have been recorded.

Balanomorph barnacles are found on almost every possible site for attachment, inanimate or living. They are most abundant intertidally and in shallow water. Some species occur at great depths. A few have become an economic problem as fouling organisms on ship hulls and in marine installations. *See* BARNACLE; THORACICA. [H.G.St.]

Balantidiasis An intestinal infection of humans caused by the ciliated protozoan *Balantidium coli*. The natural reservoir of the parasite are pigs, in whom *B. coli* causes no clinical signs of disease. The parasite exists in two forms: the resistant cyst and the actively motile trophozoite. Infection occurs by contaminative routes of people in close contact with infective fecal material. The trophozoites, which penetrate the mucosa of the large intestine, cause erosions and ulceration and subsequently the clinical signs of diarrhea and *Balantidium* dysentery. The tetracycline antibiotics have been found to cure this disorder. Prevention is based on personal hygiene and sanitary measures. *See* HETEROTRICHIDA. [M.Yo.]

Ball-and-race-type pulverizer A grinding machine in which balls rotate under pressure to crush materials, such as coal, to a fine consistency. The material is usually fed through a chute to the inside of a ring of closely spaced balls. In most designs the upper spring-loaded race applies pressure to the balls, and the lower race rotates and grinds the coarse material between it and the balls (see illustration). Two or more rings of balls can be cascaded in one machine to obtain greater capacity or output. Counterrotating top and bottom rings also are used to increase pulverizer capacity. Such pulverizers are compact and the power required per ton of material ground is relatively low. *See* CRUSHING AND PULVERIZING. [G.W.K.]

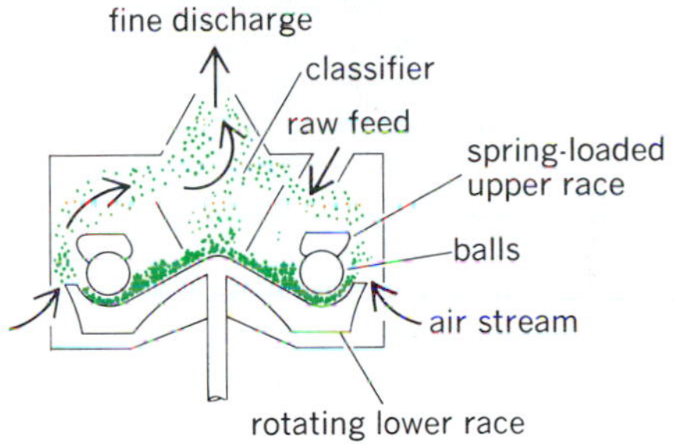

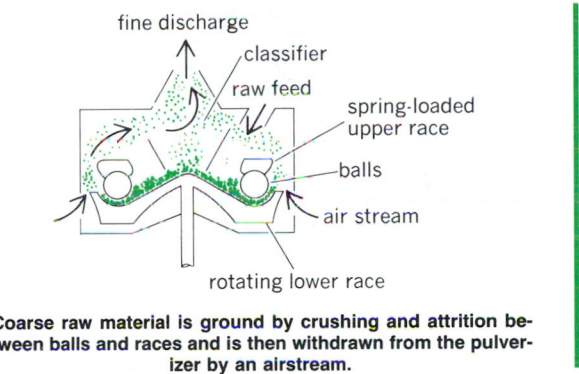

Coarse raw material is ground by crushing and attrition between balls and races and is then withdrawn from the pulverizer by an airstream.

Ballast resistor A resistor that has the property of increasing in resistance as current through it increases and decreasing in resistance as current decreases. Thus it tends to maintain a constant current through it, despite variations in applied voltage or changes in the rest of the circuit. *See* RESISTOR.

The ballast action is obtained by using resistive material that increases in resistance as temperature increases. Any increase in current then causes an increase in temperature, which results in an increase in resistance and reduces the current. Ballast resistors are usually mounted in an evacuated envelope to reduce heat radiation; they are therefore also called ballast tubes.

Ballast resistors have been used to compensate for variations in line voltage or to compensate for negative volt-ampere characteristics of other devices, such as fluorescent lamps and other vapor lamps. *See* VAPOR LAMP; VOLTAGE REGULATOR. [A.A.W.]

Ballistic missile A vehicle capable of guiding and propelling itself in a direction and to a velocity such that it will then follow a ballistic trajectory to a desired point. This trajectory is the nearly elliptic path determined by the momentum gained by self-propulsion, by gravitational acceleration toward the center of mass of Earth, and by the deceleration of aerodynamic drag.

The ballistic missile differs from the guided missile in that the latter applies either aerodynamic or propulsive force, under the direction of its guidance system, to control its trajectory during all or most of its flight. Because aerodynamic forces on a missile in the Earth's atmosphere are much less predictable than the errors of a guidance system, the guided-missile type of design is more commonly chosen for applications that call for a trajectory either entirely or primarily in the Earth's atmosphere. Because aerodynamic forces are essentially absent outside the Earth's atmosphere and because gravitational forces can be predicted with a smaller error than guidance system errors, the ballistic missile type of design is more commonly chosen for applications in which a major portion of the trajectory is exoatmospheric. *See* GUIDED MISSILE; MISSILE.

The characteristics of the ballistic missile fit the requirements of a military weapon to destroy a target located at a known point on the surface of the Earth at ranges of hundreds or thousands of miles. Hence the primary application of ballistic missile principles has been to military weapons, the most widely known of which are the intercontinental ballistic missiles and the submarine-launched ballistic missiles. [L.Sm.]

Ballistic range A long, instrumented enclosure wherein tests of gun-launched projectiles are conducted. Ballistic ranges were originally used for study of projectile flight characteristics such as the rate of velocity loss and the dispersion of trajectories, that is, the imperfect following of the bore sight line of the gun. Ballistic ranges are now used for the measurement of aerodynamic characteristics of projectiles and of scale models of vehicles intended for flight through the atmosphere, including missiles, aircraft, and space-flight capsules.

Ballistic ranges may vary in length from less than 10 ft (3 m), for limited investigations of small arms projectiles, to greater than 1000 ft (300 m), for detailed studies of projectiles from large-bore guns. Instrumentation may be as simple as a series of sheets of paper hung normal to the flight path (yaw cards) in which the projectile, by punching its outline, gives evidence of its vertical and lateral position and its angular orientation at the instant of penetration. The more refined instruments required in modern practice are usually photographic. A spark of extremely short duration (from 1 to 0.1 microsecond) is discharged when the projectile reaches a position between the spark and a piece of film (Fig. 1). The resulting shadow

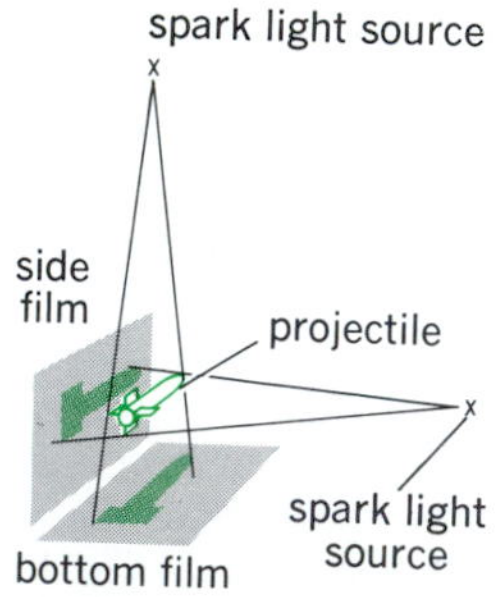

Fig. 1. Simple shadowgraph system.

photograph (shadowgraph) records in projection two components of the linear position of the projectile and one component of angular orientation. Two orthogonal pictures, as in the simple system shown, completely define the position in space of the model. A number of shadowgraph stations placed in sequence along the flight path will define, as a function of distance flown, the history of position and attitude of the projectile in flight. Corollary timing apparatus is used to measure the precise times, frequently accurate within microseconds, at which the pictures are recorded.

The shadowgraph pictures further aid aerodynamic studies by making visible certain details of the airflow about the projectile (Fig. 2). Refraction, or bending, of light rays by the variable-air-density field about the model makes visible the shock

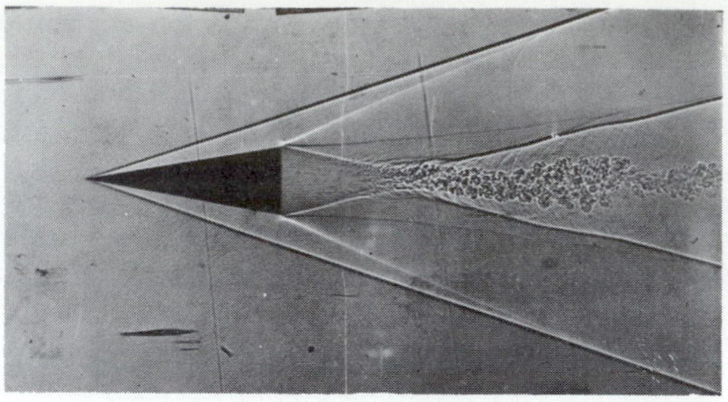

Fig. 2. Shadowgraph of a cone in supersonic flight. (*NASA, Ames Research Center*)

waves, turbulence, boundary layers, and boundaries of expansion fans. The flow visualization is most valuable for the study of aerodynamics.

In modern practice the velocity performance of ballistic ranges has sometimes been enhanced by placing the instrumented range within the test section of a supersonic wind tunnel, which provides a high-velocity airflow opposite in direction to the model's flight direction. The earliest facility of this type used a conventional supersonic wind tunnel with a very long test section, driven by a high-pressure reservoir of air at room temperature. Later devices have been driven by high-temperature air generated in a shock tube. *See* WIND TUNNEL. [A.Se.]

Ballistics That branch of applied physics which deals with the motion of projectiles and the conditions governing that motion. Commonly called the science of shooting, it is, for practical purposes, subdivided into exterior and interior ballistics. Exterior ballistics begins at the instant the projectile leaves the muzzle of the gun barrel; interior ballistics, logically, deals with the events preceding this instant, that is, the events inside the gun barrel. *See* EXTERIOR BALLISTICS; INTERIOR BALLISTICS. [W.L./D.Wo.]

Balloon A nonporous envelope of thin material filled with a lifting gas and capable of lifting any surrounding material and usually a suspended payload into the atmosphere. A balloon which is supported chiefly by buoyancy imparted by the surrounding air is often referred to as an aerostat. The balloon rises because of a differential displacement of air according to Archimedes' principle, which states that the total upward buoyant force is equal to the weight of the air displaced. The upper practical limit for useful ballooning is approximately 34 mi (55 km). Beyond this altitude, the exponential nature of the atmosphere would require balloons of enormous size and delicately thin skin. A record altitude of 32.2 mi (51.8 km) has been recorded. *See* ARCHIMEDES' PRINCIPLE.

Balloons have been configured in many geometrical shapes, but the most common are spheres, oblate spheroids, and aerodynamic configurations. The materials used in the manufacture of the balloon envelope have been paper, rubber, fabric, and various plastics. Several types of lifting gases have been used to inflate balloons, but the most common in use are helium, hydrogen, and heated air.

The many types of balloons in use fall into twin main categories: extensible (expandable) and nonextensible. There are three methods of balloon operation: free balloons which are released into the atmosphere, tethered or moored balloons, and powered or controlled free balloons. The various types of balloons in use are the hot-air balloon, meteorological balloon, zero-pressure balloon, superpressure balloon, tethered balloon, and powered balloon (airship). *See* AIRSHIP; METEOROLOGICAL BALLOON; SUPERPRESSURE BALLOON; TETHERED BALLOON; ZERO-PRESSURE BALLOON. [W.R.N.]

Balsa A fast-growing tree, *Ochroma lagopus*, widely distributed in tropical America, especially in Ecuador. The leaves are simple, angled, or lobed, and the flowers are large and yellowish-white or brownish, and they are terminal on the branches. *See* MALVALES.

With plenty of room for growth in a rich, well-drained soil at low elevations, the wood is very light and soft. However, under adverse conditions, the wood is heavier. Culture is important, for if the trees are injured only slightly, the wood develops a hard and fibrous texture, thereby losing its commercial value. To secure a uniform product the trees must be grown in plantations.

The wood decays easily in contact with the soil and is subject to sap stain if not promptly dried. Seasoned lumber absorbs water quickly, but this can be largely overcome by waterproofing.

Balsa owes most of its present commercial applications to its insulating properties. Balsa also has sound-deadening qualities, and is also used under heavy machinery to prevent transmission of vibrations. The heartwood of balsa is pale brown or reddish, whereas the sapwood is nearly white, often with a yellowish or pinkish hue. Luster is usually rather high, and the wood is odorless and tasteless. [A.H.G./K.P.D.]

Baltic Sea A semienclosed brackish sea located in a humic zone, with a positive water balance relative to the adjacent ocean (the North Sea and the North Atlantic). The Baltic in its present form is a relatively young postglacial sea which developed from the melting Fenno-Scandian glaciers about 14,000 years ago. It is connected to the North Sea by the Great Belt (70% of the water exchange), the Øresund (20% of the water exchange), and the Little Belt. The total area of the Baltic is 147,377 mi^2 (381,705 km^2), its total volume 4,982 mi^3 (20,764 km^3), and its average depth 181 ft (55.2 m). The greatest depth is 1506 ft (459 m), in the Landsort Deep. The topography of the Baltic is characterized by a sequence of basins separated by sills and by two large gulfs, the Gulf of Bothnia (40,000 mi^2 or 104,000 km^2) and the Gulf of Finland (11,400 mi^2 or 29,500 km^2). More than 200 rivers discharge an average of 104 mi^3 (433 km^3) annually from a watershed area of 636,900 mi^2 (1,649,550 km^2). The largest river is the Newa, with 18.5% of the total fresh-water discharge. From December to May, the northern and eastern parts of the Baltic are frequently covered with ice. On the average, the area of maximum ice coverage is 83,000 mi^2 (214,000 km^2). The mean maximum surface-water temperature in summer is between 59 and 63°F (15 and 17°C).

As the Baltic stretches from the boreal to the arctic continental climatic zone, there are large differences between summer and winter temperature in the surface waters, ranging from about 68 to 30°F (20 to −1°C) in the Western Baltic and 57 to 31.7°F (14 to −0.2°C) in the Gulf of Bothnia and the Gulf of Finland. Usually one or more sharp thermoclines develop April through June between 15 and 80 ft (5 and 25 m), separating the brackish winter water from the surface layer. The salt content of the Baltic waters is characterized by two major water bodies: the brackish surface water and the more saline deep water. Total average annual outflow of brackish surface water from the Baltic is 226 mi^3 (942 km^3), and the salt balance of the Baltic is maintained by an average annual inflow of 113 mi^3 (471 km^3). The surface currents of the Baltic are dominated by a general counterclockwise movement and by local and regional wind-driven circulations.

The flora and fauna of the Baltic are those of a typical brackish-water community, with considerably reduced numbers of species compared to an oceanic community. The productivity is relatively low compared to other shelf seas. Other than fish the only major resources that have been exploited are sand and gravel in the Western Baltic. It is believed that the deeper layer under the Gotland Basin contains mineral oil, but so far only exploratory drilling has been carried out in near-coastal regions. Limited amounts of mineral oil have also been located in the Gulf of Kiel. [K.G.]

Bamboo The common name of various perennial, ornamental grasses (Gramineae). There are five genera with approximately 280 species. They have a wide distribution, but occur mainly in tropical and subtropical parts of Asia, Africa, and America, extending from sea level to an elevation of 15,000 ft (4600 m). Their greatest development occurs in the monsoon regions of Asia. Most plants are woody; a few are herbaceous or climbing. The economic uses of bamboo are numerous and varied. The seeds and young shoots are used as food, and the leaves make excellent fodder for cattle. In varying sizes, the stems are used for pipes, timber, masts, bows, furniture, bridges, cooking vessels, buckets, wickerwork, paper pulp, cordage, and weaving. Entire houses are made of bamboo stems. Certain bamboos have been naturalized in California, Louisiana, and Florida. *See* CYPERALES.

[P.D.St./E.L.C.]

Banana A large tropical plant of the family Musaceae; also its edible fruit, which occurs in hanging clusters, is usually yellow when ripe, and is about 6–8 in. (15–20 cm) long. The banana of commerce (*Musa sapientum*), believed to have originated in the Asian tropics, was one of the earliest cultivated fruits. For commercial production the plant requires a tropical climate within the temperature range 50–105°F (10–40°C) and a constant supply of moisture by rainfall or irrigation.

The plant portion above the ground is a false stem (pseudostem) consisting of several concentrically formed leaves, from the center of which develops the inflorescence stalk. The rhizome or true stem is underground. Near the tip of the flower stalk are several groups of sterile male flowers subtended by brilliant purple bracts. The lower female flower clusters on the same stalk give rise to the fruit and contain aborted stamens (male organs). The single fruits are called fingers, a single group of 8–12 fingers is termed a hand, and the several (6–18) hands of the whole inflorescence make up the stem.

The fruit bunch requires 75–150 days to mature and must be removed from the plant to ripen properly. Chilled banana fruits do not soften normally; hence for best edibility the fruit is kept well ventilated at room temperature. Banana fruits of commerce set without pollination, by parthenocarpy, and hence are seedless. When mature, most varieties are yellow, although fine red-skinned types are well known. There are several hundred varieties grown throughout the world. The Cavendish banana (*M. nana*, variety Valery) is becoming important in the American tropics. The more starchy bananas, known as plantains, must be cooked before they can be eaten. *See* FRUIT; FRUIT, TREE; ZINGIBERALES. [C.A.Sch.]

Band spectrum A spectrum consisting of groups or bands of closely spaced lines. Band spectra are characteristic of molecular gases or chemical compounds. When the light emitted or absorbed by molecules is viewed through a spectroscope with small dispersion, the spectrum appears to consist of very wide asymmetrical lines called bands. These bands usually have a maximum intensity near one edge, called a band head, and a gradually decreasing intensity on the other side. In some band systems the intensity shading is toward shorter waves, in others toward longer waves. Each band system consists of a series of nearly equally spaced bands called progressions; corresponding bands of different progressions form groups called sequences.

When spectroscopes with adequate dispersion and resolving power are used, it is seen that most of the bands obtained from gaseous molecules actually consist of a very large number of lines whose spacing and relative intensities, if unresolved, explain the appearance of bands of continua. For the quantum-mechanical explanations of the details of band spectra *see* MOLECULAR STRUCTURE AND SPECTRA. [W.F.M./W.W.W.]

Band theory of solids A quantum-mechanical theory of the motion of electrons in solids which predicts certain restricted ranges, or bands, for the electron energies.

If the atoms of a solid are separated from each other to such a distance that they do not interact, the energy levels of the electrons will then be those characteristic of the individual free atoms, and thus many electrons will have the same energy. As the distance between atoms is decreased, the electrons in the outer shells begin to interact, thus altering their energy and broadening the sharp energy level out into a range of possible energy levels called a band. One would expect the process of band formation to be well advanced for the outer, or valence, electrons at the observed interatomic distances in solids. Once the atomic levels have spread into bands, the valence electrons are not confined to individual atoms, but may jump from atom to atom with an ease that increases with the increasing width of the band.

Although energy bands exist in all solids, the term energy band is usually used in reference only to ordered substances, that is, those having well-defined crystal lattices. In such a case, an electron energy state can be classified according to its crystal momentum $\mathbf{p}$ or its electron wave vector $\mathbf{k} = \mathbf{p}/\hbar$ (where $\hbar$ is Planck's constant h divided by 2π). If the electrons were free, the energy of an electron whose wave vector is $\mathbf{k}$ would be as shown in the equation below, where E_0 is the energy of

$$E(\mathbf{k}) = E_0 + \hbar^2 k^2/2m_0$$

the lowest state of a valence electron and m_0 is the electron mass. In a crystal, however, the electrons are not free because of the effect of the crystal binding and the forces exerted on them by the atoms; consequently, the relation $E(\mathbf{k})$ between energy and wave vector is more complicated. The statement of this relationship constitutes the description of an energy band.

The bands of possible electron energy levels in a solid are called allowed energy bands. There are also bands of energy levels which it is impossible for an electron to have in a given crystal. Such bands are called forbidden bands, or gaps. The allowed energy bands sometimes overlap and sometimes are separated by forbidden bands. The presence of a forbidden band immediately above the occupied allowed states (such as the region A to B in the illustration) is the principal difference in the electronic structures of a semiconductor or insulator and a metal. In the first two substances there is a gap between the valence band or normally occupied states and the conduction band, which is normally unoccupied. In a metal there is no gap between occupied and unoccupied states. The presence of a gap means that the electrons cannot easily be accelerated into higher energy states by an applied electric field. Thus, the substance cannot carry a current unless electrons are excited across the gap by thermal or optical means.

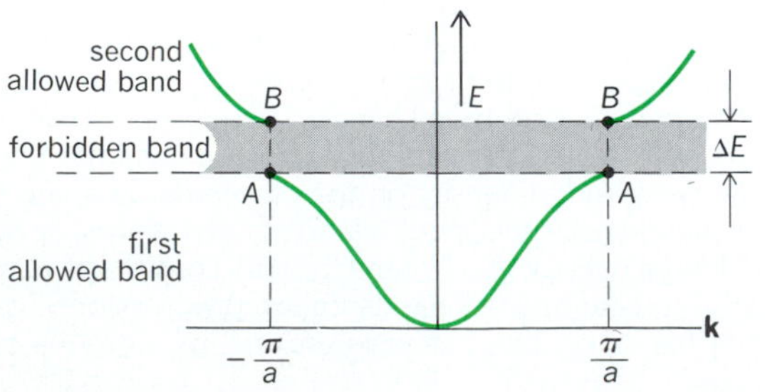

Electron energy *E* versus wave vector k for a monatomic linear lattice of lattice constant *a*. (*After C. Kittel, Introduction to Solid State Physics, 2d ed., 1956*)

Under external influences, such as irradiation, electrons can make transitions between states in the same band or in different bands. The interaction between the electrons and the vibrations of the crystal lattice can scatter the electrons in a given band with a substantial change in the electron momentum, but only a slight change in energy. This scattering is one of the principal causes of the electrical resistivity of metals. *See* ELECTRICAL RESISTIVITY.

An external electromagnetic field (for example, visible light) can cause transitions between different bands. Here momentum must be conserved. Because the momentum of a photon $h\nu/c$ (where ν is the frequency of the light and c its velocity) is quite small, the momentum of the electron before and after collision is nearly the same. Such a transition is called vertical in reference to an energy band diagram. Conservation of energy must also hold in the transition, so absorption of light is possible only if there is an unoccupied state of energy $h\nu$ available at the same $\mathbf{k}$ as the initial state. These transitions are responsible for much of the absorption of visible and near-infrared light by semiconductors. [J.C.]

Bandwidth requirements (communications) The difference between the limiting frequencies of a frequency band containing the useful components of a signal is called the bandwidth, expressed in the unit of frequency, hertz (Hz; the unit was formerly cycles per second, cps). Each communication system requires some optimum bandwidth for the satisfactory transmission of information.

In the case of simple wire telephony, where the speech sound waves cause variation of current in the line, the bandwidth required is the same as the desired bandwidth of the signal to be reproduced. In wire telegraphy, the bandwidth required depends on the signaling speed used.

Radio communications bandwidth requirements are characterized by reference (1) to the bandwidth of the transmitted signal (emission bandwidth) or (2) to the frequency response of the electrical circuits through which the signal passes, including transmitter and receiver. Emission bandwidth is usually stated in terms of the frequency band which contains a specified portion of the total signal power.

The following list shows the approximate bandwidth necessary for several kinds of communications systems.

System	*Bandwidth, Hz*
Telegraphy	
100 words/min	170–400
Telephony	3,000
High-speed data transmission	
1000 bits/s	2,000–3,000
AM radiotelephony	
Commercial	6,000
Broadcasting	8,000–20,000
FM radiotelephony	
Commercial	36,000
Broadcasting	180,000
Radiotelegraphy	
Frequency shift	600
AM facsimile	5,000
FM facsimile	25,000
Television, U.S.	
Broadcasting	6,000,000
Radio relay	25,000,000

When the signal to be transmitted causes the variation of the amplitude, phase, frequency, or pulsing of a carrier wave, the process is known as modulation. Bandwidth requirements then depend on the frequency stability of the carrier wave, the character of the modulation process, and the bandwidth of the modulating signal.

For further information on specific systems *see* FACSIMILE; RADIO; TELEGRAPHY; TELEVISION. [E.W.A.]

Barbiturates A group of drugs widely used for the suppression of anxiety, the induction of sleep, and the control of seizures. Some of them, when injected intravenously, produce a general anesthesia. *See* ANESTHESIA.

Barbituric acid, from which the various barbiturate congeners come, is a malonyl urea. Following the synthesis of this compound, a dozen or more closely related compounds were synthesized by adding or substituting various radicals to the general formula. The names of many of them have become familiar; examples are Phenobarbital, Meberal, Seconal, Nembutal, Amytal, and Pentothal.

These drugs act by suppressing the excitability of all tissues; but all tissues are not equally sensitive. Low dosages induce drowsiness, and high dosages coma and death. The spread between the therapeutic and fatal doses varies with the different barbiturates.

The prolonged use of these drugs results in habituation; and insomnia, agitation, confusional psychosis, and seizures may occur within 24 to 36 hours of withdrawal. Overdose is one of the commonest means of suicide in Western countries, and life can be saved only by admission to a hospital where respiration can be maintained and cerebral anoxia prevented. *See* DRUG ADDICTION. [R.D.A.]

Barite An orthorhombic mineral with chemical composition $BaSO_4$. It possesses one perfect cleavage, and two good cleavages, as do the isostructural minerals. The mineral has a specific gravity of approximately 4.5, and is relatively soft, approximately 3 on Mohs scale. The color ranges through white to yellowish, gray, pale blue, or brown, and a thin section is colorless.

Barite is often an accessory mineral in hydrothermal vein systems, but frequently occurs as concretions or cavity fillings in sedimentary limestones. It also occurs as a residual product of limestone weathering and in hot spring deposits. It occasionally occurs as extensive beds in evaporite deposits. Occurrences of barite are extensive. It is found as a vein mineral associated with zinc and lead ores in Derbyshire, England. Large deposits occur at Andalusia, Spain. Commercial residual deposits occur in limestones throughout the Appalachian states such as Georgia, Tennessee, and Kentucky. It also occurred in substantial amounts in the galena ore deposits in Wisconsin and Missouri.

Since barite is dense and relatively soft, its principal use is as a weighting agent in rotary well-drilling fluids. It is the major ore of barium salts, used in glass manufacture, as a filler in paint, and, owing to the presence of a heavy metal and inertness, as an absorber of radiation in x-ray examination of the gastrointestinal tract. [P.B.M.]

Barium A chemical element, Ba, with atomic number 56 and atomic weight of 137.34. Barium is eighteenth in abundance in the Earth's crust, where it is found to the extent of 0.04%, making it intermediate in amount between calcium and strontium, the other alkaline-earth metals. Barium compounds are obtained from the mining and conversion of two barium minerals. Barite, barium sulfate, is the principal ore and contains 65.79% barium oxide. Witherite, sometimes called heavy spar, is barium carbonate and is 72% barium oxide. *See* BARITE; WITHERITE.

The metal was first isolated by Sir Humphry Davy in 1808 by electrolysis. Industrially, only small amounts are prepared by aluminum reduction of barium oxide in large retorts. These are used in barium-nickel alloys for spark-plug wire (the barium increases the emissivity of the alloy) and in frary metal, which is an alloy of lead, barium, and calcium used in place of babbitt metal because it can be cast.

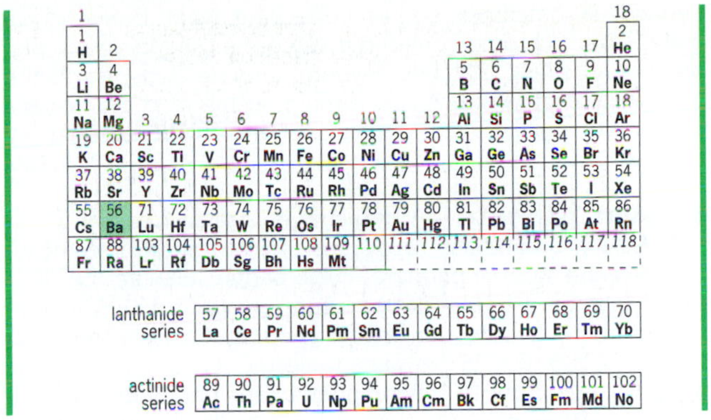

The metal reacts with water more readily than do strontium and calcium, but less readily than sodium; it oxidizes quickly in air to form a surface film that inhibits further reaction, but in moist air it may inflame. The metal is sufficiently active chemically to react with most nonmetals. Freshly cut pieces have a lustrous gray-white appearance, and the metal is both ductile and malleable. The physical properties of the elementary form are given in the table.

For the manufacture of barium compounds, soft (easily crushable) barite is preferred, but crystalline varieties may be used. Crude barite is crushed and then mixed with pulverized coal. The mixture is roasted in a rotary reduction furnace, and the barium sulfate is thus reduced to barium sulfide or black ash. Black ash is roughly 70% barium sulfide and is treated with hot water to make a solution used as the starting material for the manufacture of many compounds.

Lithopone, a white powder consisting of 20% barium sulfate, 30% zinc sulfide, and less than 3% zinc oxide, is widely used as a pigment in white paints. Blanc fixe is used in the manufacture of brilliant coloring compounds. It is the best grade of barium sulfate for paint pigments. Because of the large absorption of x-rays by barium, the sulfate is used to coat the alimentary tract for x-ray photographs in order to increase the contrast. Barium carbonate is useful in the ceramic industry to prevent efflorescence on claywares. It is used also as a pottery glaze, in optical glass, and in rat poisons. Barium chloride is used in purifying salt brines, in chlorine and sodium hydroxide manufacture, as a flux for magnesium alloys, as a water softener in boiler compounds, and in medicinal preparations. Barium nitrate, or the so-called baryta saltpeter, finds use in pyrotechnics and signal flares (to produce a green color), and to a small extent in medicinal preparations. Barium oxide, known as baryta or calcined baryta, finds use both as an industrial drying agent and in the case-hardening of steels. Barium

Properties of barium

Property	Value
Atomic number	56
Atomic weight	137.34
Isotopes (stable)	130, 132, 134, 135, 136, 137, 138
Atomic volume	36.2 cm^3/g-atom
Crystal structure	Face-centered cubic
Electron configuration	2 8 18 18 8 2
Valence	2+
Ionic radius (A)	1.35
Boiling point, °C	1140(?)
Melting point, °C	850(?)
Density	3.75 g/cm^3 at 20°C
Latent heat of vaporization at boiling point, kJ/g-atom	374

peroxide is sometimes used as a bleaching agent. Barium chromate, lemon chrome or chrome yellow, is used in yellow pigments and safety matches. Barium chlorate finds use in the manufacture of pyrotechnics. Barium acetate and cyanide are used industrially as a chemical reagent and in metallurgy, respectively. [R.F.R.]

Barium titanate A dielectric crystalline material which exhibits the anomalous properties typical for ferroelectricity. On lowering the temperature, cubic barium titanate, $BaTiO_3$, transforms to a tetragonal structure at about 130°C (Curie temperature), where the crystal spontaneously acquires an electric polarization (a permanent electric dipole moment). As the polarization can in general point in any one of the originally cubic crystal axes, the crystal consists, in the ferroelectric state, of regions (domains) which differ in the direction of the spontaneous polarization. An applied electric field can align the polarization in the whole crystal and can reverse the polarization direction whereby electric hysteresis occurs.

Barium titanate has a high dielectric constant, typically 2000 for a ceramic sample at room temperature. Thus, capacitors of very small dimensions can be produced for electronics applications. Barium titanate ceramics, when suitably doped with a trivalent oxide such as lanthanum oxide, for example, exhibit a variation of the electrical conductivity by a factor up to 10^6 in the vicinity of the Curie temperature. This effect can be used for sensitive temperature-control devices.

Ceramics of barium titanate and, in particular, of a mixed lead titanate zirconate can be poled by an electric field and retain a substantial remanent polarization. These materials then have a strong piezoelectric effect. In addition, they are mechanically rugged and rather insensitive to temperature and humidity. Therefore, these ceramics find wide applications in devices such as microphones, ultrasonic and underwater transducers, and spark generators. *See* DIELECTRIC CONSTANT; PIEZOELECTRICITY. [H.Gr.]

Bark A word generally referring to the surface region of a stem or a root. Sometimes part or all of the bark is called rind. Occasionally the word bark is used as a substitute for periderm or for cork only. Most commonly, however, it refers to all tissues external to the cambium. If this definition is applied to stems having only primary tissues, it includes phloem, cortex, and epidermis; the bark of roots of corresponding age would contain cortex and epidermis. In the more general usage the term bark is restricted to woody plants with secondary growth.

In most roots with secondary growth, the bark consists of phloem and periderm since the cortex and epidermis are sloughed off with the first cork formation in the pericycle that is beneath the cortex. In stems, the first cork cambium may be formed in any living tissue outside the vascular cambium, and the young bark may include any or all of the cortex in addition to the phloem and periderm. The region composed of the successive layers of periderm and the enclosed dead tissues is called outer bark. The outer bark composed of dead phloem alternating with bands of cork is called, technically, rhytidome. Both stems and roots may have rhytidome. The inner bark is living, and consists of phloem only. *See* CORTEX (PLANT); EPIDERMIS (PLANT); PARENCHYMA; PERICYCLE; PERIDERM; PHLOEM; ROOT (BOTANY); STEM. [H.W.Bl.]

Barkhausen effect An effect, due to discontinuities in size or orientation of magnetic domains as a body of ferromagnetic material is magnetized, whereby the magnetization proceeds in a series of minute jumps. *See* FERROMAGNETISM; MAGNETIZATION.

Ferromagnetic materials are characterized by the presence of microscopic domains of some 10^{12} to 10^{15} atoms within which the magnetic moments of the spinning electrons are all parallel. In an unmagnetized specimen, there is random orientation of the various domains. When a magnetic field is applied to the specimen, the domains turn into an orientation parallel to the field, or if parallel to the field, the domains increase in size. During the steep part of the magnetization curve, whole domains suddenly change in size or orientation, giving a discontinuous increase in magnetization. If the specimen being magnetized is within a coil connected to an amplifier and loudspeaker, the sudden changes give rise to a series of clicks or, when there is a rapid change, a hissing sound. This is called the Barkhausen effect; it is an important piece of evidence in support of a domain theory of magnetism. [K.V.M.]

Barley A cereal grass plant belonging to the family Gramineae and genus *Hordeum*. Most of the modern cultivated barleys are *H. vulgare* (six-rowed) or *H. distichum* (two-rowed; see illustration). All cultivated barleys are annuals, are naturally self-pollinated, and have seven pairs of chromosomes.

Barley spikes: (*a*) six-rowed *Hordeum vulgare*; (*b*) two-rowed *H. distichum*. (USDA)

Barley is grown in nearly all cultivated areas of the temperate parts of the world, and is an important crop in Europe, North and South America, North Africa, much of Asia, and Australia. Barley is the most dependable cereal crop where drought, summer frost, and alkali soils are encountered. In the United States, barley is grown to some extent in 49 states, with the most important production areas in North Dakota, Montana, and California. Principal uses of barley grain are as livestock feed, for the production of barley malt, and as human food. It is also used in some areas for hay and silage.

In the cultivated varieties, a wide diversity of morphological, physiological, and anatomical types are known. There are spring, facultative, and winter growth habits; hulled and naked grain; awned, awnless, and hooded lemmas; black, purple, and white kernels; and also a wide range of plant heights, spike densities, and resistances to a wide range of diseases and insects. There are in excess of 150 cultivars presently commercially grown in the United States and Canada alone, and many additional cultivars are grown in other parts of the world.

Barley can be grown in a variety of rotations with other crops common to a particular region. In areas with long growing seasons, it is possible to double-crop (growing two crops in 12 consecutive months on the same land) with summer crops such as soybeans or cotton. Barley is ideally suited to this purpose because it is usually the earliest maturing of the small grains. *See* CYPERALES; GRAIN CROPS. [D.A.R.]

The term pot barley is applied to the food product which results after the husk has been removed by coarse grinding. Pearled barley is obtained by a more involved, dry-milling process.

For malting purposes, the brewing industries in the United States use different types of barley from those desired throughout the remainder of the world. Outside the United States the demand is for large-kerneled barley of the two-row type which is high in starch and low in nitrogen content. In North America the demand is for the six-row, small-kernel type with higher nitrogen and enzyme content.

There are two systems of malting; flour malting and pneumatic or drum malting. The malting process involves the germination of the barley in a controlled state, followed by drying under conditions that retain enzymatic activity. Industrial value of malt depends on its starch-splitting and protein-degrading enzymes. Brewers' malts are used to produce beer and ale and, to a lesser extent, for malt syrups and evaporated malt extracts. Distillers' malts are employed for starch conversion in the production of distilled beverages and industrial alcohol and for converting starch for the sizing of paper and textiles. [J.A.Sh.]

Barnacle The name popularly applied to two types of Crustacea, subclass Cirripedia, the goose barnacles and acorn barnacles. Both were formerly of importance to seafarers (and acorn barnacles are still so) because of their habit of adhering to ships hulls. Four orders of barnacles are recognized, of which one only, the Thoracica, fits the popular conception of barnacles. *See* CIRRIPEDIA; THORACICA. [H.G.St.]

Baroclinic field A distribution of atmospheric pressure and mass such that the specific volume, or density, of air is a function not solely of pressure. When the field is baroclinic, solenoids are present, there is a gradient of air temperature on a surface of constant pressure, and there is a vertical shear of the geostrophic wind. Significant development of cyclonic and anticyclonic wind circulations typically occurs only in strongly baroclinic fields. Fronts represent baroclinic fields which are locally very intense. *See* FRONT; GEOSTROPHIC WIND; SOLENOID (METEOROLOGY); STORM; WIND. [F.S.]

Barometer An absolute pressure gage specifically designed to measure atmospheric pressure. This instrument is a type of manometer with one leg at zero pressure absolute. *See* MANOMETER.

The common meteorological barometer (see illustration) is a liquid-column gage filled with mercury. The top of the column is sealed, and the bottom is open and submerged below the surface of a reservoir of mercury. The atmospheric pressure on the reservoir keeps the mercury at a height proportional to that pressure. An adjustable scale, with a vernier scale, allows a reading of column height. Aneroid barometers using metallic diaphragm elements are usually less accurate, though often more sensitive, devices, and not only indicate pressure but may be used to record it. *See* PRESSURE MEASUREMENT. [J.H.Z.]

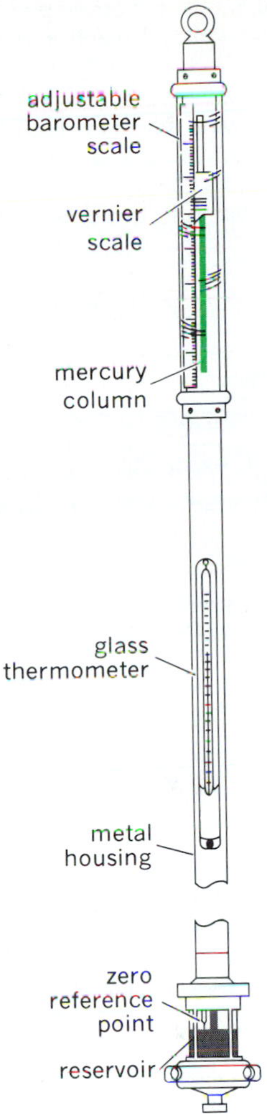

Mercury barometer.

Barotropic field A distribution of atmospheric pressure and mass such that the specific volume, or density, of air is a function solely of pressure. When the field is barotropic, there are no solenoids, air temperature is constant on a surface of constant pressure, and there is no vertical shear of the geostrophic wind. Significant cyclonic and anticyclonic circulations typically do not develop in barotropic fields. Considerable success has been achieved, paradoxically, in prediction of the flow pattern at middle-tropospheric elevations by methods which are strictly applicable only to barotropic fields, despite the fact that the field in this region is definitely not barotropic. *See* BAROCLINIC FIELD; GEOSTROPHIC WIND; SOLENOID (METEOROLOGY); WEATHER FORECASTING AND PREDICTION; WIND. [F.S.]

Barracuda The name for about 20 species of fish found in warm seas throughout the world. All species belong to the genus *Sphyraena* and are members of the order Perciformes. The larger species, such as the great barracuda (*Sphyraena barracuda*), are solitary and not found in schools. Generally, all species are good food fish.

The teeth, which occur both in the jaws and on the roof of the large mouth, are strong and pointed, resembling canine teeth. The jaws are long, with the lower projecting beyond the upper, and the upper incapable of being protracted. The head is large and pointed, and the body is compressed and long, somewhat like the pike. The body is covered with cycloid scales, and the lateral line is well developed. Many of these fish weigh over 100 lb (45 kg) and reach a length of 10 ft (3 m). *See* PERCIFORMES. [C.B.C.]

Barretter A fine wire, having a high temperature coefficient of resistance, which is used as a power-absorbing device in bolometric methods. Its change of resistance is a measure of the power dissipated within it, and it is used either to measure the power directly or to indicate when a comparison dc power is equal to an unknown radio-frequency (rf) power. The overall length of typical barretters (see illustration) is about 1 in. (2.5 cm). *See* BOLOMETER. [D.B.S.]

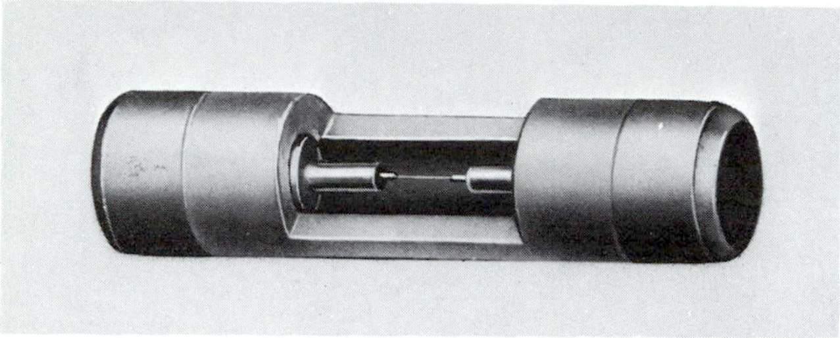

Cutaway view of commercial barretter. (*Sperry Division, Sperry Rand Corporation*)

Barrier islands Elongate, narrow accumulations of sediment which have formed in the shallow coastal zone and are separated from the mainland by some combination of coastal bays and marshes. They are typically several times longer than their width and are interrupted by tidal inlets. Although their origin has been widely discussed, at least three possibilities exist: longshore spit development and subsequent cutting of inlets; drowning of old coastal ridges; and upward shoaling of subtidal sediment accumulations. All three may have occurred; however, the last seems most likely and most prevalent.

Barrier islands must be considered in terms of the adjacent and closely related environments within the coastal system. Beginning offshore and proceeding landward, the sequence of environments crossed is shoreface, beach, dunes, back-island flats or marsh, coastal bay, marsh, and mainland. The barrier island proper consists of the beach, dunes, and back-island flats or marsh; however, of the remaining environments, at least the shoreface is closely integrated with the barrier island in terms of morphology, processes, and sediments. *See* DUNE.

A variety of physical processes exists along the coast. These processes act to shape and maintain the barrier-island system and also to enable the barrier to migrate landward as sea level continues to rise. The most important process in the barrier-island system is the waves, which also give rise to longshore currents. Waves and longshore currents dominate the outer portion of the barrier system, whereas tidal currents are dominant landward of the barrier, although small waves may also be present. Tidal currents are most prominent in and adjacent to the inlets. On the supratidal portion of the barrier island, the wind is the most dominant physical process. *See* NEARSHORE SEDIMENTARY PROCESSES.

Barrier-island sands represent one of the best sources of oil and gas, with the tight organic-rich source rocks being in the form of the bay and shelf muds and the barrier itself being the reservoir rock. These elongate sand bodies have been sought by exploration geologists for decades. The Tertiary sequences of the Texas Gulf coasts are an example of such barrier systems which have been very productive. *See* COASTAL LANDFORMS. [R.A.D.]

Baryon The generic name for any hadronic particle with baryon number $B = +1$. By far the most common baryons are the proton and neutron, the two states of the nucleon doublet $N = (p,n)$, whose intrinsic properties are listed in Table 1. The baryon number of any particular state may be deduced from its production or decay processes, or both, since the total baryon number is conserved (with possible rare exceptions discussed below) and $B = 0$ holds for all mesons and leptons. *See* LEPTON; MESON; NEUTRON; NUCLEON; PROTON.

The scientific view of the hadrons changed greatly during the 1970s. It is now generally accepted that they are composite, consisting of spin-½ quarks (q), corresponding antiquarks ($\bar{q}$), and some number of gluons, the last being the quanta of the intermediate field which binds the quarks and antiquarks to form hadrons. $B = +⅓$ holds for a quark q, $B = -⅓$ for an antiquark $\bar{q}$, while $B = 0$ holds for a gluon. Thus, a baryon consists of three ("valence") quarks, together with some number of quark-antiquark ($q\bar{q}$) pairs (called the quark-antiquark sea) and of gluons. The known quarks are listed in Table 2. They must be assigned fractional charge values, relative to the proton charge. *See* GLUONS; HADRON; QUARKS.

This quark theory of the hadrons has been proposed in a quite specific form, known as quantum chromodynamics

Table 1. Known stable and semistable baryons and their properties

Baryon	Mass, MeV	Spin-parity	Strange-ness (s)	Charm (C)	Lifetime, s	Dominant decay modes	Magnetic moment, n.m.*
p	938.280 ± 0.003	½⁺	0	0	$>10^{36}$	–	2.7928
n	939.573 ± 0.003	½⁺	0	0	917 ± 14	$p\bar{\nu}_e e^-$	−1.9130
Λ	1115.60 ± 0.05	½⁺	−1	0	$2.63 \pm 0.02 \times 10^{-10}$	$p\pi^-$(64%) $n\pi^0$(36%) $p\bar{\nu}_e e^-$(0.081 ± 0.003%)	−0.614±0.005
Σ^+	1189.36 ± 0.06	½⁺	−1	0	$8.0 \pm 0.04 \times 10^{-11}$	$p\pi^0$(52%) $n\pi^+$(48%) $p\gamma$(0.12 ± 0.02%)	2.33±0.13
Σ^0	1192.46 ± 0.08	½⁺	−1	0	$6 \pm 1 \times 10^{-20}$	$\Lambda\gamma$	
Σ^-	1197.34 ± 0.05	½⁺	−1	0	$1.48 \pm 0.01 \times 10^{-10}$	$n\pi^-$ $n\bar{\nu}_e e^-$(0.11 ± 0.005%)	−1.41±0.25
Ξ^0	1314.9 ± 0.6	½⁺	−2	0	$2.9 \pm 0.1 \times 10^{-10}$	$\Lambda\pi^0$	−1.20±0.06
Ξ^-	1321.3 ± 0.15	½⁺	−2	0	$1.64 \pm 0.02 \times 10^{-10}$	$\Lambda\pi^-$ $\Lambda e^-\bar{\nu}_e$(0.028 ± 0.012%)	−1.85±0.75
Ω^-	1672.2 ± 0.3	(3/2⁺?)	−3	0	$0.82 \pm 0.03 \times 10^{-10}$	$\Lambda\bar{K}$(69%) $\Xi^0\pi^-$(23%) $\Xi^-\pi^0$(8%)	
Λ_c	2273 ± 6	(½⁺?)	0	1	$\simeq 7 \times 10^{-13}$	$pK^-\pi^+$(2.2 ± 1%)	

*The abbreviation n.m. denotes the unit $e\hbar 2M_p c$ (nuclear magneton).

Table 2. Properties of established quarks*

Quark type	u (up)	d (down)	s (strange)	c (charmed)	b (bottom)
Charge (Q/e_p)	$\frac{2}{3}$	$-\frac{1}{3}$	$-\frac{1}{3}$	$\frac{2}{3}$	$-\frac{1}{3}$
Mass, GeV	≈0.01	≈0.01	≈0.5	≈1.5	≈4.7
Flavor	$I_3 = +\frac{1}{2}$	$I_3 = -\frac{1}{2}$	$s = -1$	$C = +1$	$b = +1$

*To each quark, there exists an antiquark with the opposite flavor values and with opposite intrinsic parity.

(QCD). It is a gauge theory based on a symmetry hypothesized for the hadronic interactions of the quarks, which says that these interactions are invariant with respect to an SU(3)c group of unitary transformations with modulus unity acting in an abstract three-dimensional space known (whimsically) as color space. Each quark type then has three color states, usually labeled by the suffixes r = red, g = green, and b = blue, corresponding to the three axes of this space. The gauge particle of this symmetry theory is the gluon, a neutral vector particle coupled universally with the currents of color, just as the photon, the gauge particle of quantum electrodynamics (QED), is coupled universally with the electromagnetic current. However, whereas the photon has no charge, the gluon has $(3^2 - 1) = 8$ color components, so that it is a color octet. Consequently, there is a gluon contribution to the color currents, and so the gluon field must interact with itself, introducing a nonlinearity into quantum chromodynamics which has no parallel in quantum electrodynamics. This nonlinearity has important implications for quantum chromodynamics, leading to its asymptotic freedom, the property that the coupling of gluon to the color current approaches zero at short distances, which is essential for even qualitative agreement between quantum chromodynamics predictions and the empirical data on high-energy collision processes. *See* COLOR (QUANTUM MECHANICS); QUANTUM CHROMODYNAMICS; QUANTUM ELECTRODYNAMICS; SYMMETRY LAWS (PHYSICS).

An important element in quantum chromodynamics is the confinement dogma, the assertion that only color singlet states have finite energy. This assertion implies that neither a quark nor a gluon can exist in a free state, since the former is a color triplet and the latter a color octet. Zero mass is expected for a gauge particle (as is the case for the photon), hence for the gluon, and a mass of order 10 MeV is given for the (u,d) quarks in Table 2, but these values refer to the masses effective within hadronic states. In accord with this dogma, no observations of free gluons or quarks have yet been confirmed. Many theoreticians anticipate that this dogma will be deducible from quantum chromodynamics itself but, despite some favorable indications for this expectation, no rigorous proof that the dogma follows from quantum chromodynamics has been given.

Quantitative predictions of the properties of baryonic states are currently made using a simplified quark-quark (qq) potential with the following features: (1) an attractive long-range potential, increasing with separation to ensure confinement, and (2) a spin-dependent potential representing one-gluon exchange, effective at small separation, where the regime of asymptotic freedom holds and perturbation theory is valid. Such predictions have had a great deal of success.

The quark content of the nucleons is given by Eqs. (1).

$$p = (uud) \qquad n = (udd) \tag{1}$$

The replacement of a d quark in the nucleon by an s quark produces a baryon state with spin parity $\frac{1}{2}^+$ and strangeness number $s = -1$, the latter being given by $[n(\bar{s}) - n(s)]$ where $n(q)$ denotes the number of quarks of type q in the system considered. The states thus reached have the quark structures of Eqs. (2), thus the isotriplet Σ and isosinglet Λ states are obtained. If two of the d quarks in Eqs. (1) are replaced by s quarks, the isodoublet Ξ states of Eq. (3) are obtained. The s

$$(\Sigma^+, \Sigma^0, \Sigma^-) = (uus, (ud + du)\, s/\sqrt{2}, dds) \tag{2a}$$

$$\Lambda = (ud - du)\, s/\sqrt{2} \tag{2b}$$

$$(\Xi^0, \Xi^- = (uss, dss) \tag{3}$$

quark has $I = 0$, being unaffected by the SU(2)$_T$ transformations in the (u,d) space. Baryonic states with $s \neq 0$ are collectively termed hyperons. *See* HYPERON; STRANGE PARTICLES.

These eight baryon states (p, n, Σ^+, Σ^0, Σ^-, Λ, Ξ^0, Ξ^-) all have the spin parity $\frac{1}{2}^+$ and the same internal wave functions. They are arrayed in Fig. 1, where the symmetry of their relationship is made evident; for this, it is helpful to use the quantum number $Y = (B + s)$, named hypercharge.

The mass difference $\delta m = (m(s) - m(u,d))$ is quite large, and so SU(3)f symmetry is much more strongly violated than SU(2)$_T$ symmetry. This is most apparent in the baryon mass values, which vary widely over the octet; the leading variation is that proportional to the strangeness s, which counts the s-quark content of each baryon. The approximately 75-MeV difference between the mean Σ mass $\overline{m}$ (Σ) and $m(\Lambda)$ has a more subtle origin, but is well accounted for on the basis of the quark-quark (qq) potential from quantum chromodynamics, described above. The small mass differences within each isospin multiplet are believed due to electromagnetic effects.

The production and reaction processes observed for these baryons are governed by flavor selection rules. For hadronic processes (but not for weak processes) these selection rules are simply Eq. (4) for each flavor $f = B, Q, s, c, \ldots$; that is, each

$$\Delta n(f) = 0 \tag{4}$$

type of quark is independent of the others, and quarks do not change type. Quantum chromodynamics provides a ready understanding of these rules.

The baryon-baryon interactions are of particular interest. Between nucleons, this interaction gives rise to the existence of atomic nuclei, and it has been particularly well studied, both empirically and theoretically, for the NN system. For large separations (greater than 0.8×10^{-15} m), the NN force is due to the exchange of pions and of other known mesons with masses less than about 1 GeV; for small separations (less than 0.4×10^{-15} m), a strong short-range repulsion is observed, possibly arising from the suppressive effects of the Pauli principle for quarks when the quark structures of the two nucleons overlap. At low energies, the outstanding feature of the NN interaction is its strong noncentral tensor component, which is due to one-pion exchange and is a direct consequence of the pseudoscalar nature of the pion. It also has strong spin-orbit interactions, observed in NN interactions at higher energy and of much importance for the shell structure of nuclei.

Many particle-unstable baryon states, with lifetimes in the range 10^{-22} to 10^{-23} s, have become established. All of the resonances now established are consistent with the limitations of the three-quark model.

This qqq model for baryons is far from complete. It provides a good mnemonic for baryon spectroscopy, and even works well quantitatively, but it does not include any account of the quark sea or of multigluon systems, even for the ground states. Other kinds of excitation are also possible; for example, if $s = +1$ resonances are established, they might correspond to mul-

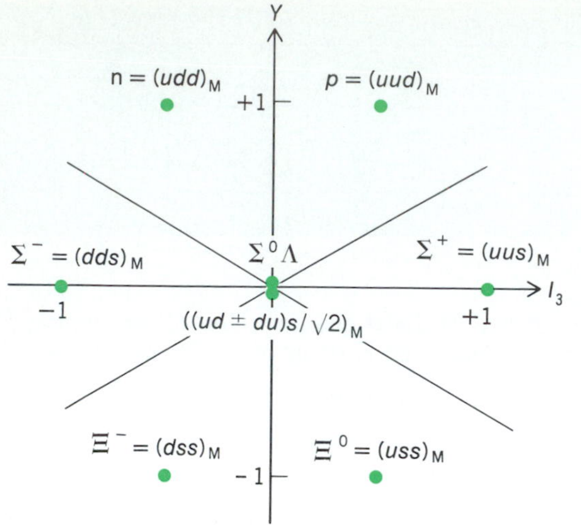

Fig. 1. The baryon octet states, arrayed with respect to I_3 as ordinate and $Y = (B + s)$ as abscissa. The charge number Q is given by $Q = I_3 + Y/2$. There are three axes of symmetry.

tiquark states with the structure ($\bar{q}qqqq$), and there might also exist excited states where the excitations are gluonic. However, there is no data calling for such excitations at present.

Further baryon states can be formed by replacing one or more of the u,d and s quarks of the states discussed above by a c quark. If the s and c quarks both had the same mass as the (u,d) quarks, the states formed would correspond to an SU(4)f symmetry. This is emphasized by Fig. 2, which shows the extensions of the $\frac{1}{2}^+$ baryon octet (Fig. 2a) and the $\frac{3}{2}^+$ baryon decuplet (Fig. 2b) obtained in this way. The lowest plane of each part of the figure consists of the charmless baryon states; those in Fig. 2a are in accord with the octet of Fig. 1.

In reality, the c-quark mass is so large that little quantitative detail of SU(4)f symmetry can survive in the physical situation, and the structure of Fig. 2 has value mainly for general comprehension and for the counting of states. The lowest calculated levels are given in Table 3; the states are labeled as for SU(3)f, the suffix c meaning that the s quark has been replaced by a c quark (which increases the charge by + 1). With these mass values, only the isosinglet Λ_c^+ (2260) will be semistable, and its weak decay will be due to the dominant c-quark transition $c \rightarrow s$.

The existence of a fifth quark b was demonstrated in 1978 in work with the PETRA electron-positron storage ring at the DESY laboratory, Hamburg, and theoreticians anticipate the existence of a sixth quark, to be named t, whose mass must exceed 19 GeV in order to account for the lack of evidence for this quark even at the top energy of PETRA. These quarks will add two further additive quantum numbers with conservation laws of the form of Eq. (4). Of course, there could well exist still further, heavier quarks at present unsuspected. Baryon states built of these heavier quarks, together with the lighter quarks, will certainly exist and will be observed in the course of time; no candidates to be identified with Λb have yet been brought forward. *See* PARTICLE ACCELERATOR.

For every baryonic state mentioned above, there will exist an antibaryon state with opposite flavor quantum numbers, in particular with $B = -1$. The antiproton $\bar{p}$ was first identified in 1954, and antiproton beams have been available at high-energy proton accelerators for many years. Antibaryon decay processes correspond directly to those for the corresponding baryons. They have the same lifetimes, within measurement accuracy.

In grand unified theories (GUTs), which attempt to account for both quarks and leptons, together with their strong, electromagnetic, and weak interactions, transitions $q \rightarrow l$ generally

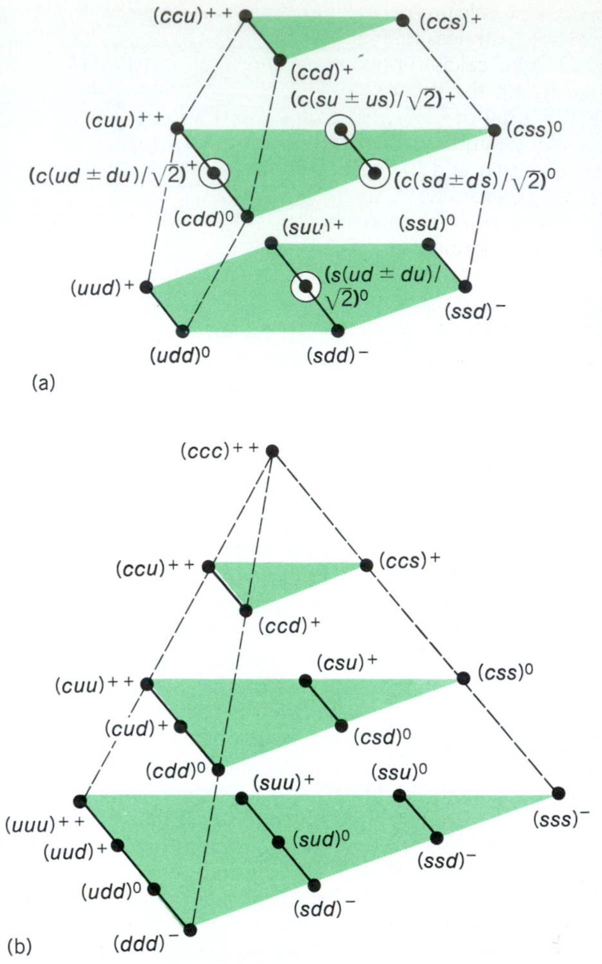

Fig. 2. Arrays of the states of baryons made from three quarks of the types u, d, s, and c, with no internal orbital angular momentum. The quark content and charge of each state is specified. The vertical axis specifies the number of c quarks. (*a*) $\frac{1}{2}^+$ baryons. (*b*) $\frac{3}{2}^+$ baryons.

exist, at some level, since such theories assign leptons and quarks to common multipliers. This violation of baryon conservation opens the possibility that the lightest baryon, the proton, might not be absolutely stable but undergoes decay processes such as $p \rightarrow e^+ \pi^0$ at a very low rate. Empirically, the proton is known to have a half-life exceeding 6×10^{35} s. The simplest grand unified theories under discussion do predict a total proton decay rate of order $10^{-39\pm2}$ s^{-1}, while other theories may predict smaller decay rates, even zero. Since proton decay offers the possibility of discriminating between various grand unified theories, the detection of proton decay, or at least the improvement of the present empirical limits on its rate, is an important subject of investigation. *See* ELEMENTARY PARTICLE; FUNDAMENTAL INTERACTIONS. [R.H.D.]

Basalt An igneous rock characterized by small grain size (less than about 0.2 in. or 5 mm) and approximately equal pro-

Table 3. Mass, spin, and parity values calculated from quantum chromodynamics for the low-lying C = +1 charmed baryonic states

$J^P =$	$\frac{1}{2}^+$	$\frac{1}{2}^+$	$\frac{1}{2}^-$	$\frac{3}{2}^-$
	Λ_c(2260)	Σ_c(2510)	Λ_c(2510)	Λ_c(2590)
	Σ_c(2440)			

portions of calcium-rich plagioclase feldspar and calcium-rich pyroxene, with less than about 20% by volume of other minerals. Olivine, calcium-poor pyroxene, and iron-titanium oxide minerals are the most prevalent other minerals. Most basalts are dark gray or black, but some are light gray. Various structures and textures of basalts are useful in inferring both their igneous origin and their environment of emplacement. Basalts are the predominant surficial igneous rocks on the Earth, Moon, and probably other bodies in the solar system. Several chemical-mineralogical types of basalts are recognized. The nature of basaltic rocks provides helpful clues about the composition and temperature within the Earth and Moon. The magnetic properties of basalts are responsible in large part for present knowledge of the past behavior of the Earth's magnetic field and of the rate of sea-floor spreading. Some meteorites are basaltic rocks. They differ significantly from lunar basalts and appear to have originated elsewhere in the solar system at a time close to the initial condensation of the solar nebula. *See* IGNEOUS ROCKS; METEORITE.

Basalt flows out of fissures and cylindrical vents. Repeated or continued extrusion of basalt from cylindrical vents generally builds up a volcano of accumulated lava and tephra around the vent. Fissure eruptions commonly do not build volcanoes, but small cones of tephra may accumulate along the fissures. *See* VOLCANO; VOLCANOLOGY.

Basalts display a variety of structures mostly related to their environments of consolidation. On land, basalt flows form pahoehoe, aa, and block lava, while under water, pillow lava is formed. Basalt also occurs as pumice and bombs. Commonly, basaltic pumice is called scoria to distinguish it from the lighter-colored, more siliceous rhyolitic pumice.

The mineralogy and texture of basalts vary with cooling history and with chemical composition. As basalt crystallizes, both the minerals and the residual melt change in composition because of differences between the composition of the melt and the crystals forming from it. In basalts, because of the rapid cooling, there is little chance for crystals to react with the residual melt after the crystals have formed. Completely solid basalts generally preserve a record of their crystallization in the zoned crystals and residual glass.

Most basalts contain minor amounts of chromite, magnetite, ilmenite, apatite, and sulfides in addition to the minerals mentioned above. Magnetite in basalts contains a history of the strength and orientation of the Earth's magnetic field at the time of cooling. Therefore, although magnetite is minor in amount, it is probably the most important mineral in terrestrial basalts, because it enables earth scientists to infer both the magnetic history of the Earth and the rate of the production of basaltic ocean floor at the oceanic ridges.

Basalts occur in all four major tectonic environments: ridges in the sea floor, islands in ocean basins, island arcs and mountainous continental margins, and interiors of continents. The principal environment is the deep sea floor. Significant differences exist in the composition of basalt which relate to different tectonic environments (see table). Chemical analyses of basalts are now used instead of, or together with, the textural criteria as a basis of classification. Geologists customarily recast the chemical analysis into a set of ideal minerals according to a set of rules. The result is called the norm of the rock. In general there is rather close correspondence between the normative minerals and the observed minerals in basaltic rocks. *See* PETROLOGY.

Basalt is considered the most widespread material available at the surface of the Earth which is derived from the mantle. Study of the regional distribution of various basalts helps earth scientists formulate some details regarding the tectonic processes affecting their generation. After these factors are taken into account, the differences remain which may be understood in terms of the compositionally inhomogeneous mantle. Some inhomogeneities may relate to earlier episodes of mountain building and possibly to primal planetary differentiation. *See* MAGMA; PETROGRAPHIC PROVINCE. [A.T.A.]

Chemical compositions of basalts, in weight percent

	1*	2	3	4	5	6
SiO_2	49.92	49.20	49.56	51.5	45.90	45.5
TiO_2	1.51	2.32	1.53	1.1	1.80	2.97
Al_2O_3	17.24	11.45	17.88	17.1	15.36	9.69
Fe_2O_3	2.01	1.58	2.78	n.d.	1.22	0.00
Cr_2O_3	0.04	n.d.	n.d.	n.d.	n.d.	0.50
FeO	6.90	10.08	7.26	8.9	8.13	19.7
MnO	0.17	0.18	0.14	n.d.	0.08	0.27
MgO	7.28	13.62	6.97	7.0	13.22	10.9
CaO	11.85	8.84	9.99	9.3	10.71	10.0
Na_2O	2.76	2.04	2.90	4.3	2.29	0.33
K_2O	0.16	0.46	0.73	0.80	0.67	0.06
P_2O_5	0.16	0.23	0.26	n.d.	0.62	0.10
Sum	100.00	100.00	100.00	100.00	100.00	100.02
H_2O	0.4	0.3	n.d.	2	n.d.	0.0
CO_2	0.02	0.01	n.d.	n.d.	n.d.	n.d.
F	0.02	0.03	n.d.	n.d.	n.d.	0.002
Cl	0.02	0.03	n.d.	0.09	0.01	0.0005
S	0.08	0.07	n.d.	0.19	n.d.	0.07

*(1) Average of 10 basalts from oceanic ridges; (2) submarine basalt, Eastern Rift Zone, Kilauea Volcano, Hawaii; (3) average high-alumina basalt of Oregon Plateau; (4) initial melt of Pacaya Volcano, Guatemala; (5) alkali basalt, Hualalai Volcano, Hawaii; (6) average *Apollo 12* lunar basalt.

Base (chemistry) Any chemical species, ionic or molecular, capable of accepting or receiving a proton (hydrogen ion) from another substance. The other substance acts as an acid in giving up the proton. A substance may act as a base, then, only in the presence of an acid. The greater the tendency to accept a proton, the stronger the base. The hydroxyl ion acts as a strong base. Water acts as a weaker base in the reaction $H_2O + H^+ \rightleftharpoons H_3O^+$. This broader definition replaces the older concept of a base as a substance yielding hydroxyl ions in solution. A more general concept of a base is that of a substance which donates a pair of electrons to form a coordinate bond. *See* ACID AND BASE. [F.J.J.]

Basement rock The more resistant, generally crystalline rock beneath layers or irregular deposits of younger, rel-

Basement rock (Precambrian granite here) beneath a sequence of sedimentary strata (Pennsylvanian limestone, sandstone, shale), western front of Sandia Mountains, New Mexico. (*Courtesy of V. C. Kelley*)

atively deformed sedimentary rock. The term is frequently equated with "basement complex," but the latter term should be used only when the composition and structure are truly complex. In much of the world basement rock consists of Precambrian igneous and metamorphic rocks of various types, but locally the basement rock beneath layers of sedimentary strata may be Paleozoic, Mesozoic, or even Cenozoic. It may lie at great depths, beneath hundreds or thousands of feet of sediments, or it may occur at the surface, where it has been exposed by faulting or erosion or both (see illustration). *See* PRECAMBRIAN. [J.P.F.]

Basidiomycotina A division of the fungi; commonly known as Basidiomycetes, the class name used in older classifications. The Basidiomycetes are important as food (mushrooms), as causal agents of plant diseases (rusts, smuts, heart rots of trees), and as causes of dry rots of timbers. The characteristic structure of the Basidiomycetes is the basidium, a specialized sporangium which is the site of nuclear fusion and reduction division. After this division usually four basidiospores are formed in outpocketings in such a way that they appear to be borne on the basidium. There are four classes: Teliomycetes, Phragmobasidiomycetes, Hymenomycetes, and Gasteromycetes. *See* MUSHROOM. [R.M.P.]

Basil An annual herb (*Ocimum basilicum*) of the mint family (Labiatae), grown from seed for its highly scented leaves. It is sold fresh or dried, and its essential oil is used in pharmaceuticals and flavoring. The genus *Ocimum* has approximately 150 species, many of which are also grown as "basil." *See* LAMIALES.

Growth habit varies with type, but it is usually upright and branching from a square central stem up to 24 in. (61 cm) in height. Leaf shape and odor vary considerably from type to type, with "sweet basil," the most popular, having broad, flat, shiny green leaves to 3 in. (75 cm) long. Wild forms of basil are found in South America, Africa, India, and Europe. Cultivated basil differs little from its wild forms, although breeding and selection work have been performed on some of the varieties grown for essential oil production. Some of the more popular cultivated types are: sweet basil (used fresh or dehydrated), lemon-scented basil, opal basil (a purple-leaved type), and large- or lettuce-leaf basil. Many unusual types can be obtained which have a scent similar to camphor, lemon, licorice, or nutmeg. *See* SPICE AND FLAVORING. [S. Kir.]

Basin A low-lying area, wholly or largely surrounded by higher land. Basins vary from small, nearly enclosed valleys to extensive mountain-rimmed depressions, such as the Bighorn Basin of north-central Wyoming and the larger Tarim basin of central Asia.

The term is also applied (sometimes as drainage basin) to the entire area drained by a given stream and its tributaries. An example of this usage is the Mississippi Basin, occupying most of the United States between the Rocky Mountains and the Appalachians, or the Amazon Basin, an area of similar size in northern South America. Depressions that drain entirely inward, without outlet to the sea, are basins of interior drainage.

A geologic basin is an area in which the rock strata are inclined downward from all sides toward the center. Because of erosion and deposition, the surface may not show the form of a depression. [E.H.Ha.]

Basommatophora A superorder of the molluscan subclass Pulmonata containing about 2500 species that are grouped into 11–15 families. Only a few members of the family Ellobiidae are terrestrial, with the other species today being tidal to supratidal or estuarine in habitat. The families Siphonatiidae and Trimusculidae are marine limpets with cap-like shells. The families Otinidae and Amphibolidae also are marine taxa. The remaining families inhabit a great variety of fresh-water situations and are quite varied in shell structure and shape. Because of this great variation in habitat and form, it is difficult to find structures that are common to all taxa and thus diagnostic of the group, which indeed may not have a common origin. The most easily observable characters are having the eyespots normally located at the base of two slender tentacles that are neither contractile nor retractile, and usually having two external genital orifices. Most features of the anatomy are shared with other pulmonate superorders, or are specializations related to major changes in habitat or body form. *See* PULMONATA. [G.A.S.]

Bass The name for a number of fishes assigned to two families in the order Perciformes. Both marine and fresh-water species are included under this common name, and all are highly prized as game fish as well as for food. Members of the family Centrarchidae are commonly referred to as the freshwater or black basses, while the Serranidae are designated as the sea basses. *See* PERCIFORMES. [C.B.C.]

Basswood A member of the linden family in the order Malvales. One species, known as the American linden (*Tilia americana*), is a timber tree of the northeastern quarter of the United States and the adjacent area of Canada. *Tilia* is also an ornamental tree. *Tilia europea*, or lime tree of Europe, is often cultivated along the streets. The lindens are also important as bee trees. The leaves are heart-shaped, coarsely toothed, long, pointed, and alternate. All species of *Tilia* can be recognized by the winter buds, which have a large outer scale that gives the bud a humped, asymmetrical appearance, and by the small, spherical, nutlike fruits borne in clusters.

The wood of basswood, also known as whitewood, is white and soft and is used for boxes, venetian blinds, millwork, furniture, and woodenware. There are about 30 species in the temperate regions of the Northern Hemisphere in North America south to Mexico, but none in the western part. In Asia lindens grow south to central China and southern Japan. *See* MALVALES. [A.H.G./K.P.D.]

Bat The common name for all mammals included in the order Chiroptera. Nearly 850 species are currently recognized; they are grouped in 17 families (see table). They can be combined on the basis of a variety of characters, including size and feeding modifications, into two general groups, recognized as suborders: Megachiroptera, mostly large fruit and flower feeders, with about 150 species; and Microchiroptera, chiefly small insect eaters, with about 700 species. Bats are worldwide in distribution, except for the polar regions and a few isolated oceanic islands.

Small to medium in size, bats are the only mammals that are capable of true flight, with forelimbs highly modified for this function. The bones of the manus, or hand, and the digits are elongate and covered by a membrane; this membrane connects to the body and extends from the forelimb to some position on the lower leg or foot (according to the species). The first digit, or thumb, usually retains a claw, and a membrane runs from its base to the neck. In many species, the tail, if present, is partly or wholly enclosed by the interfemoral membrane, which extends between the small hindlimbs, each of which has five toes that terminate in strong curved claws. The hindlimbs are used by most species for hanging from perches head down while at rest. Bats are principally nocturnal in their activities, and they rest during the daylight hours.

Most bats of the cooler portions of the Northern Hemisphere hibernate during the winter months. Some bats tend to roost singly in a hibernaculum, others in large aggrega-

Nomenclature and distribution of families of Chiroptera

Family and common name	Geographical distribution
Pteropidae Old World fruit bats	Africa, southern Asia, Australasia
Rhinopomatidae Mouse-tailed bats	Northern Africa, southern Asia
Craseonycteridae Bumblebee bat	Thailand
Emballonuridae Sheath-tailed bats	Africa, southern Asia, Australasia, tropical America
Nycteridae Slit-faced bats	Africa, southwestern Asia, Malaysia
Megadermatidae Old World false vampires	Africa, southeastern Asia, Autralia
Rhinolophidae Old World leaf-nosed bats	Africa, Eurasia, Australasia
Noctilionidae Neotropical fish-eating bats	Tropical America
Mormoopidae Ghost-faced bats	Tropical and subtropical America
Phyllostomidae New World leaf-nosed bats (including true vampires)	Tropical and subtropical America
Natalidae Funnel-eared bats	Tropical America
Furipteridae Smoky bats	Central and South America
Thyropteridae New World sucker-footed bats	Central and South America
Myzopodidae Old World sucker-footed bat	Madagascar
Vespertilionidae Common bats	Almost worldwide
Mystacinidae New Zealand short-tailed bat	New Zealand
Molossidae Free-tailed bats	Almost worldwide, in tropics and subtropics

tions. Species may differ markedly in their requirements as regards temperature, humidity, air currents, and other environmental factors. *See* HIBERNATION; THERMOREGULATION.

Bats are one of several different groups of animals that utilize reflected sound for orientation (echolocation). The sense of hearing is indispensable in bat orientation, and bats which cannot transmit sound become disoriented. All Microchiroptera emit high-frequency shortwave sounds, which are used in orientation and often in pursuit of insect prey.

Like other mammals, bats may act as reservoir hosts or as transmitters of pathogens. Most importantly, they can transmit rabies, which may affect not only true vampire bats but also insectivorous species. Some fruit bats can be pests in orchards. On the other hand, bats consume vast quantities of insects, and some are important in seed dispersal and pollination. Bats and their roosts (particularly nursery colonies and hibernacula) should in general be protected, especially since some species have shown alarming declines in numbers in recent years. *See* CHIROPTERA. [K.F.K.]

Batales A small order of flowering plants, division Magnoliophyta (Angiospermae), in the subclass Caryophyllidae of the class Magnoliopsida (dicotyledons). It consists of a single family, with only one genus (*Batis*), of two shrubby, maritime species. The order is somewhat doubtfully placed in the subclass Caryophyllidae, where it is marked by its anatropous ovules, straight embryo, binucleate pollen, four-locular ovary, much reduced flowers borne in unisexual catkins, and the absence of anthocyanins, betalains, endosperm, and perisperm. *See* ANTHOCYANIN; CARYOPHYLLIDAE; FLOWER; MAGNOLIOPHYTA; PLANT KINGDOM. [A.Cr.]

Batholith A geologic name for a body of igneous rock of moderate or large size [40 mi^2 (100 km^2) or more in cross-sectional diameter] emplaced at great or intermediate depth in the Earth's crust. Batholiths are commonly discordant with (cut across) the enclosing rocks, but on a grand scale most of the large batholiths are elongated parallel to the dominant structure of the enclosing crustal rocks. The shape of batholiths in depth is a point of debate. One school argues that batholiths continue downward with increasing size or at least with no reduction in size. The other school argues that large batholiths are composites of many smaller units, each of which maintains a nearly constant cross section, thus suggesting variation in shape with downward development. *See* PLUTON. [J.A.N.]

Bathynellacea An order of syncarid crustaceans found in subterranean waters in central Europe and England. The body is elongate and segmented. Organs and limbs are primitive. Each of the thoracic limbs has two epipodites which function as gills. The thorax and abdomen have a similar form (see illustration). These animals have no metamorphosis. *See* SYNCARIDA. [H.J.]

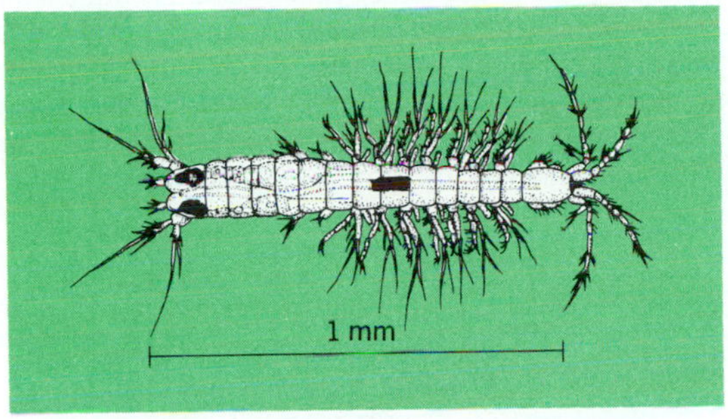

***Bathynella natans*, female.**

Batoidea Skates and rays, constituting one of the two Recent orders of the subclass Elasmobranchii. The order is also known as Rajiformes. The skates and rays differ from the sharks (order Selachii) in having ventral gill slits, the edge of the pectoral fin attached to the side of the head anterior to the gill clefts, and the upper margin of the orbit not free from the eyeball (see illustration).

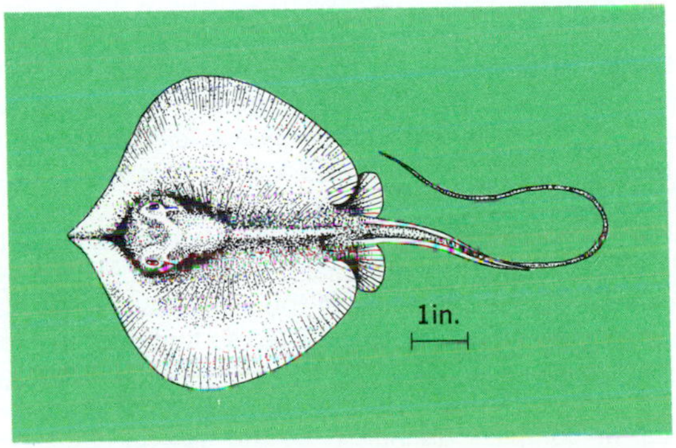

Atlantic stingray (*Dasyatis sabina*). (After G. B. Goode, *Fishery Industries of the United States, 1884*)

Rays have been classified in 5 suborders, 16 families, 47 genera, and 300–350 species. They occur in all oceans, and most are sluggish bottom inhabitants, although the mantas commonly swim at the surface. A few, especially stingrays and sawfishes, penetrate estuaries of tropical rivers or even live far upstream. There are some bizarre skates that live at depths of as much as 1500 fathoms (2700 m); however, none of these is luminescent. The torpedoes have a pair of enlarged electric organs located lateral to the eyes on the expansive pectoral fins; these can deliver a strong and temporarily disabling shock to bathers. See ELECTRIC ORGAN (BIOLOGY).

Internal fertilization is practiced by all rays. Most rays are ovoviviparous, but skates deposit eggs that are protected by horny cases. Rays are carnivorous, feeding on a wide variety of marine worms, mollusks, crustaceans, and other invertebrates, as well as small fishes. See ELASMOBRANCHII. [R.M.B.]

Batrachoidiformes The toadfishes, which make up an order (or suborder) of actinopterygian fishes. This group is also known as the Haplodoci. The first vertebra of Batrachoidiformes is rigidly fused with the broad, flattened cranium; the short posttemporal is not forked and is suturally attached to the skull. There are no epiotics, intercalars, or ribs, and the four or five pectoral radials are elongate. There are only three gill arches, and the gill openings are reduced in size. The pelvic fins are on the throat; the spinous dorsal fin is short, and the soft dorsal and anal fins are long (see illustration).

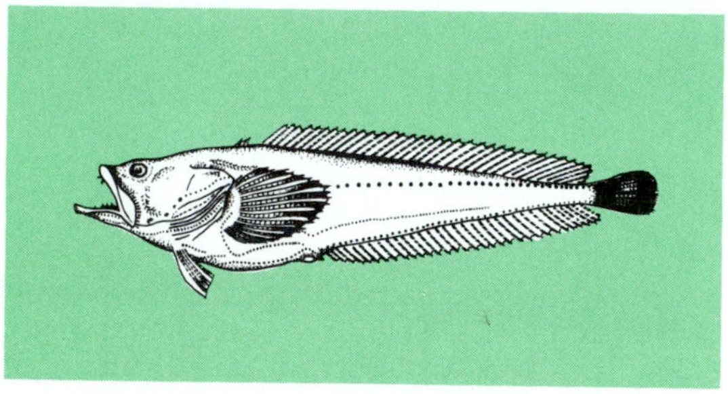

Atlantic midshipman (*Porichthys porosissimus*). (After D. S. Jordan and B. W. Evermann)

The single family, Batrachoididae, known from Miocene to Recent lime, includes 10 genera and about 35 species. Most live in tropical and temperate oceanic shore waters or in moderate depths, but a few ascend tropical rivers. Some species are luminescent; some have hollow opercular and dorsal spines associated with venom glands. Toadfishes have powerfully muscled jaws and strong teeth and are predacious. See ACTINOPTERYGII; BIOLUMINESCENCE; OSTEICHTHYES. [R.M.B.]

Battery A device which transforms chemical energy into electric energy. The term is usually applied to a group of two or more electric cells connected together electrically. In common usage the term battery is also applied to a single cell, such as a flashlight battery.

There are in general two types of batteries, primary batteries and secondary storage or accumulator batteries. Primary types, although sometimes consisting of the same plate-active materials as secondary types, are constructed so that only one continuous or intermittent discharge can be obtained. Secondary types are constructed so that they may be recharged, following a partial or complete discharge, by the flow of direct current through them in a direction opposite to the current flow or discharge. By recharging after discharge, a higher state of oxidation is created at the positive plate or electrode and a lower state at the negative plate, returning the plates to approximately their original charged condition.

Primary and secondary cells may be constructed from several materials. For the more important of these types see PRIMARY BATTERY; STORAGE BATTERY. For other sources of electric energy known as batteries or cells see FUEL CELL; NUCLEAR BATTERY; SOLAR CELL. [H.C.Ri.]

Bauxite A rock name for weathering products composed largely of hydrous aluminum oxides and aluminous laterite. Bauxite, the ore of aluminum, is a mixture and not a mineral species. See ALUMINUM.

Bauxite occurs in a variety of structures, textures, and colors. The color depends on the content of iron oxides and ranges from white to dark red or brown. Commonly bauxite is pisolitic and concretionary, but it may be fine-grained, dense, or vermicular.

Commercial bauxite deposits consist largely of gibbsite, $Al(OH)_3$, or boehmite, AlO(OH), or mixtures of these minerals. The chief impurities are clay minerals and iron oxides, and bauxite may grade into clay or into ferruginous laterite. Some bauxite derived from basaltic rocks in India and Hawaii contains large amounts of titania.

C. S. Fox in 1927 suggested two main groups of bauxite deposits: laterite type and terra rossa type. All deposits are not easily placed in this classification, but bauxite formed more or less in place from crystalline rocks is of the first type, whereas the deposits that overlie limestone or dolomite commonly are referred to as the terra rossa type.

The world production of bauxite has steadily increased. The United States at one time produced one-fourth of the total bauxite used. However, other countries have begun to supply most of the world's needs. Imports are mainly from Jamaica, Australia, and Guinea. Ferruginous bauxite has been explored in Oregon and Hawaii. [S.S.G.]

Bayesian statistics An approach to statistics in which estimates are based on a synthesis of a prior distribution and current sample data. Bayesian statistics is not a branch of statistics in the way that, say, nonparametric statistics is. It is, in fact, a self-contained paradigm providing tools and techniques for all statistical problems. In the classical frequentist viewpoint of statistical theory, a statistical procedure is judged by averaging its performance over all possible data. However, the bayesian approach gives prime importance to how a given procedure performs for the actual data observed in a given situation. Further, in contrast to the classical procedures, the bayesian procedures formally utilize information available from sources other than the statistical investigation. Such information, available through expert judgment, past experience, or prior belief, is described by a probability distribution on the set of all possible values of the unknown parameter of the statistical model at hand. This probability distribution is called the prior distribution. The crux of the bayesian approach is the synthesis of the prior distribution and the current sample data into a posterior probability distribution from which all decisions and inferences are made. This synthesis is achieved by using a theorem proved by Thomas Bayes in the eighteenth century.

The posterior distribution combines the prior information about the unknown parameter θ with the information contained in the observed data to give a composite picture of the final judgments about θ. In order to arrive at a single number as the estimate of θ, it may be necessary to bring in the notion of the loss suffered by the decision maker as a result of estimating the true value θ by the number θ. Depending on the choice of the loss function, the mode, the mean, and the median of the posterior distribution are all reasonable estimates of θ.

The choice of the prior distribution for the unknown parameter θ is of crucial importance in bayesian statistics. The selected prior distribution for θ must be at least a reasonable approximation to the true beliefs about θ. In addition, the prior distribution must be such that the posterior distribution is tractable.

Noninformative and improper priors. The bayesian approach remains applicable even when little or no prior information is available. Such situations can be handled by choosing a prior density giving equal weight to all possible values of θ.

Priors that seemingly impart no prior preference, the so-called noninformative priors, also arise when the prior is required to be invariant under certain transformations. Frequently, the desire to treat all possible values of θ equitably leads to priors with infinite mass. Such noninformative priors are called improper priors.

Criticism. Proponents of the bayesian paradigm claim a number of significant and fairly convincing advantages over the classical non-bayesian approach. However, bayesian statistics is itself criticized for its apparent lack of objectivity. The use of a prior distribution introduces subjectivity into a bayesian analysis so that decision makers having different priors may come to different conclusions even when working on the same problem with the same observed data. This, however, is the price that must be paid in order to incorporate prior information or expert judgment. Moreover, the bayesian approach seems to be the most logical approach if the performance of a statistical procedure for the actual observed data is of interest, rather than its average performance over all possible data. *See* ESTIMATION THEORY; PROBABILITY; STATISTICS. [S.N.U.A.K.]

Bdelloidea A class (formerly order) of the phylum Rotifera. Bdelloid rotifers have a very characteristic appearance; the agile elongate body consists of several (typically 16) false segments or annuli which do not correspond to the true segmentation of the body. Of these, the shorter and smaller head, neck, and foot segments are telescopically retractile into the larger and longer trunk joints. In typical species the corona has the characteristic form of two trochal disks, raised on pedicels, and a single cingulum of smaller cilia. The corona can be completely withdrawn, to reveal the true anterior end of the body. The mouth is large and funnel-shaped, and leads via a ciliated buccal tube into the mastax (a gizzardlike structure) whose ramate trophi are in constant motion. The intestine is syncytial, often colored brown or red by its contents, and digestion appears to be extracellular, rapid, and occurring in an alkaline medium. No males are known, and reproduction is believed to be parthenogenetic. A few genera are viviparous, but most are oviparous.

The Bdelloidea are typically bottom dwellers and crawl over the substratum leech-fashion. They are the most common inhabitants of standing fresh waters throughout the world, occurring in the smallest pools, in the littoral zones of large lakes, and also in mosses, liverworts, and lichens, some of which may be wetted only intermittently. *See* ROTIFERA. [J.B.J.]

Bdellonemertini An order of the class Enopla in the phylum Rhynchocoela. The proboscis is unarmed. The alimentary system comprises the mouth, papillate foregut, sinuous intestine without diverticula, and anus. Members are characterized by the presence of a posterior adhesive disk and dorsoventral flattening of the body, which lacks eyes, cephalic slits, and cerebral organs. The order contains the single genus *Malacobdella*, with three species ectocommensal in the mantle cavity of marine bivalves and one in the pulmonary sac of a freshwater gastropod. *See* ENOPLA; RHYNCHOCOELA. [J.B.J.]

Beam A structural member extending between two reaction points and supporting loads such as floors, walls, and other types of surfacing is a simple beam. The loads are generally placed to produce a bending action about a principal axis of the beam cross section. When the structural member extends over several reaction points, it is referred to as a continuous beam. *See* LOADS, TRANSVERSE. [J.B.S.]

Beam column A structural member subjected to the combined action of axial and transverse loads which produce compression and bending moments in the member. Both compression and bending moments may also be produced by eccentrically applied axial loads. Beam columns exist in frame structures. *See* COLUMN. [J.B.S.]

Beam-foil spectroscopy A method of determining the energies and mean lives of excited electronic levels in monatomic ions and in atoms. Beams of particles (ions) of any element are energized in a particle accelerator and then sent in vacuum through a thin foil, usually of carbon. The particles emerge from the foil in various stages of ionization. Numerous energy levels are excited in each of those stages of ionization. The spontaneous loss of the energy of excitation takes place as the beam moves downstream from the foil. That loss is detected by means of the electromagnetic radiation or the electrons which the ions emit. *See* PARTICLE ACCELERATOR.

The mean life of a level provides sensitive information about the quantum nature of the level itself and the mechanism whereby it connects, by either the absorption or emission of light, to some other level. In beam-foil spectroscopy, this mean life is found from observations on the intensity of a particular spectral line as a function of the separation of the emitter and the exciter foil. Beam-foil spectroscopy is presently the only known way of measuring mean lives for highly ionized systems.

The degree of ionization which can be achieved depends on the atomic number of the ions and their energy. Early work was restricted to a few elements at the low end of the periodic table and to energies of a few megaelectronvolts. Subsequently, more than 60 of the 94 naturally occurring elements have been used, at energies extending to nearly 900 MeV. *See* SPECTROSCOPY. [S.Ba.]

Bean Any of several leguminous plants, or their seeds, long utilized as food by humans or livestock. Some 14 genera of the legume family contain species producing seeds termed "beans" which are useful to humans. Twenty-eight species in 7 genera produce beans of commercial importance, which implies that the bean can be found in trade at the village level or up to and including transoceanic commerce.

The principal Asiatic beans include the edible soybeans, *Glycine* sp., and several species of the genus *Vigna*, such as the cowpea and mung, grams, rice, and adzuki beans. The broad bean (*Vicia faba*) is found in Europe, the Middle East, and Mediterranean region, including the North African fringe. Farther south in Africa occur *Phaseolus* beans, of the *vulgaris* (common bean) and *coccineus* (scarlet runner) species. Some *Phaseolus* beans occur in Europe also. The cowpea, used as a dry bean, is also found abundantly in Nigeria. *See* COWPEA; SOYBEAN.

In the Americas, the *Phaseolus* beans, *P. vulgaris* and *P. lunatus* (lima bean), are the principal edible beans, although the blackeye cowpea, mung bean, and chick pea or garbanzo (*Cicer arietinum*) are grown to some extent. *Phaseolus coccineus* is often grown in higher elevations in Central and South America, as is *Vicia faba*. The tepary bean (*P. acutifolius*) is found in the drier southwestern United States and northern Mexico. *See* ROSALES.

Bean plants may be either bush or vining types, with white, yellow, red, or purple flowers. The seed itself is the most differentiating characteristic of bean plants. It may be white, yellow, black, red, tan, cream-colored, or mottled, and range in weight from 0.0044 to over 0.025 oz (125 to over 700 mg) per seed. Seeds are grown in straight or curved pods (fruit), with 2–3 seeds per pod in *Glycine* to 18–20 in some *Vigna*.

Beans are consumed as food in several forms. Lima beans and snap beans are used as fresh vegetables, or they may be processed by canning or freezing. Limas are also used as a dry bean. Mung beans are utilized as sprouts. Usage of dry beans (*P. vulgaris*) for food is highly dependent upon seed size, shape, color, and flavor characteristics, and is often associated with particular social or ethnic groups. Popular usage includes soups, mixed-bean salads, pork and beans, beans boiled with meat or other vegetables or cereals, baked beans, precooked beans, and powder, and, in the Peruvian Andes, parched or roasted beans. Because of the presence of heat-labile growth retardants, beans must be thoroughly cooked before being used as food. Newer developments are in the areas of cooking time, the manufacture of precooked dry powders that can be readily rehydrated to make a palatable soup, and new research impetus to identify the flatulence factors in dry beans. Since the food grain legumes are an important protein source in human diets, intensive efforts are being made in research and development. *See* LEGUME. [M.W.A.]

Bear The name for a number of species of carnivorous mammals in the family Ursidae (see table). Bears are large and heavy bodied, with short, strong legs and short tails. They are completely plantigrade, walking with their metatarsal and metacarpal regions touching the ground, but are able to stand upright. The foot has five toes which terminate in nonretractile claws. The toes are separate with no membranes between them. The foot soles of the polar bear are covered with hair, lacking in other species, which aids in walking on ice. The eyes of bears are relatively small and vision is rather poor; however, hearing and smell are acute. The rounded ears are small, and the muzzle is elongate.

Common name and geographic distribution of some Ursidae species

Scientific name	Common name	Distribution
Tremarctos omatus	Spectacled bear, Andean bear	Pacific slopes of Andes from northern Chili to Colombia
Selenarctos thibetanus	Asiatic black bear, moon bear, Himalayan bear	Iran, China, Japan, Himalayas
Ursus arctos	Brown bear, many varietal names	Eastern Europe, Asia, North America, Pyrenees
Urses americanus	American black bear	North American forested areas
Thalarctos maritimus	Polar bear	Arctic regions of Northern Hemisphere
Helarctos malayanus	Sun bear, honey bear, bruang, Malay bear	Sumatra, Borneo, southeastern Asia
Melursus ursinus	Sloth bear	Ceylon, southern India

Depending upon the species, the female produces one to three young per lifter; gestation varies from 180 to 250 days. The small, blind, toothless offspring weigh less than 1 lb (0.5 kg). The female cares for the cubs since the male usually does not remain with her.

Bears originated in the Northern Hemisphere and are now widely distributed throughout the world. The brown bear (*Ursus arctos*) is the commonest species, in North America as well as in other areas of the world. The grizzly bear is an example of a species that is on the verge of extinction. It is found in national park areas, Alaska, and the Yukon region. The spectacled bear (*Tremarctos ornatus*) is the only native bear in the Southern Hemisphere, where it is found in the forested foothills of the western slopes of the Andes. The Himalayan bear (*Selenarctos thibetanus*) occurs in China, Japan, Iran, and the Himalayas. The black bear (*Ursus americanus*) is still numerous across North America in forested areas. It is hunted for its fur and as a game animal in many parts of the United States. The sun bear (*Helarctos malayanus*) is found in the tropical forested areas of southern Asia, and thus differs from other bears in its habitat preference. It is the smallest of all bears. The polar bear (*Thalarctos maritimus*) occurs in the polar regions of the Northern Hemisphere. *See* CARNIVORA; HIBERNATION. [C.B.C.]

Beat A variation in the intensity of a composite wave which is formed from two distinct waves with different frequencies. Beats were first observed in sound waves, such as those produced by two tuning forks with different frequencies. Beats also can be produced by other waves. They can occur in the motion of two pendulums of different lengths and have been observed among the different-frequency phonons in a crystal lattice.

One important application of beat phenomena is to use one object with an accurately known frequency to determine the unknown frequency of another such object. The beat-frequency or heterodyne oscillator also operates by producing beats from two frequencies. *See* OSCILLATOR. [B.De.F.]

Beaver The common name for two different and unrelated species of rodents—*Aplodontia rufa*, the mountain beaver, and *Castor canadensis*, the common or true beaver.

The mountain beaver is the only living species of the family Aplodontidae. It is a medium-sized animal and resembles a muskrat. The eyes are small as are the external ears, an adaptation to its fossorial (digging) habits. The tail is a short stump, the body is short and stout, and the limbs are short and terminate in broad feet with five toes and long claws. The mountain beaver is a vegetarian. Although not strictly nocturnal, it is more active at night. The animals form colonies with a single family inhabiting a series of runways. The mountain beaver is found along the Pacific Coast and ranges from British Columbia, Canada, to California.

The common beaver, a member of the family Castoridae, is a large rodent (weighing up to 50 lb or 23 kg). It occurs across Europe, North America from Labrador southward, and Asia, and is a valuable fur-bearing animal. The tail is broad and spatulate and is used, together with the webbed hindfeet, for swimming. The beaver is aquatic and builds its lodge in water by using mud, sticks, and branches interwoven. These beavers are vegetarians. Mating occurs in February, and the average litter of four young is born in May. These beavers appear to be monogamous and mate for life. *See* RODENTIA. [C.B.C.]

Bedbug The name for the approximately 36 recognized species of insects of the family Cimicidae in the order Hemiptera. In the bedbugs true wings are absent; the forewings are represented by two scalelike pads. The mouth parts are modified for piercing and sucking. Bedbugs are small,

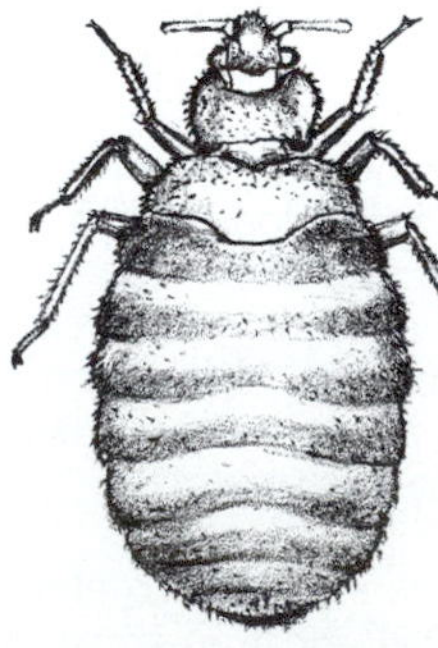

***Cimex lectularius*, a species of bedbug parasitic on humans in many temperate and subtropical regions.**

oval-shaped, reddish-brown insects with a thin, flat body covered with a tough integument. The legs are short; compound eyes are present, and ocelli, or simple eyes, are lacking. These insects are common household pests, having a cosmopolitan distribution. Bedbugs are ectoparasites which feed upon warm-blooded animals, including humans, bats, and birds. Two species attack humans: *Cimex lectularius* and *C. rotundatus*. *Cimex lectularius* (see illustration) is the common species of the temperate and subtropical regions, while the second species is restricted to tropical regions of Asia and Africa. *See* HEMIPTERA. [C.B.C.]

Bee Insects of the superfamily Apoidea in the order Hymenoptera. There are some 3000 species of bees divided into 19 families. Present-day bees subsist almost entirely on pollen as a source of protein, and on the sugar in nectar as an energy source. They are now obligately dependent on flowers, although a few species can sometimes feed on "honeydew" secretions from aphids and scale insects. In turn, many plants have become obligately dependent on bees for pollination. *See* HYMENOPTERA.

Most of the world's bee species are solitary. That is, a female builds a nest cell, provisions the cell with a nectar-pollen mixture, lays an egg on this food, and seals the nest cell, allowing the larva to develop and emerge as much as a year later. In primitively social bees such as some of the sweat bees (family Halictidae) and bumblebees (family Bombidae), an overwintered female (queen) that has mated the previous fall emerges from her underground hibernacula in the spring and attempts to start her own colony. A colony consists initially of sterile female offspring, who are her workers. Near the end of the colony cycle, in late summer or fall, the workers aid the queen in producing large numbers of sexuals, the drones and new queens. Colonies generally consist of several dozen and up to several hundred individuals at the height of the colony cycle in the summer. The workers, drones, and old queens all die off in the fall, and the new queens disperse and hibernate.

In the highly social bees, the honeybee (*Apis mellifera*) and stingless (*Trigona*) species, as many as 50,000 and 150,000 individuals, respectively, may be present in the colonies. In these bees the old queen lives several years, and the colonies are perennial rather than annual. The queens are nearly exclusively concerned with egg laying, and the workers are produced not only during times of food availability in the field, but also before major times of flowering by relying on large stores of pollen and honey. These bees reproduce by swarming, when the old queen leaves the hive accompanied by about half of the workers, which help her to initiate a new colony. *See* SOCIAL INSECTS.

The highly social organization of honeybees and stingless bees is orchestrated by scent, sound, and "dance" signals. The queen emits a chemical (pheromone) which inhibits the workers from developing and laying eggs. Absence or low concentrations of the queen pheromone cause the workers to rear a new queen. The workers also produce pheromones that act as signals to other workers. Honeybees indicate the direction, distance, and quality of rewarding food sources to hive mates by a symbolic dance language. Honeybees also buzz during the dance, which may serve to alert potential followers that then attempt to "read" the dancer's message. Stingless bees also alert hive mates by buzzing. The hive mates follow the scout bee (that has discovered the food) out of the hive; the scout then deposits a trail of scent droplets onto vegetation that aid the recruit to find the food. Bumblebees do not seem to have any way of communicating food found by successful hive mates.

By visiting flowers, bees serve as agents of cross-pollination. They are of inestimable importance in the pollination of crops and in pollination of the natural flora. Their activity is thus vital, not only directly for the human food supply, but also for land, water, and animal resources. *See* BEEKEEPING; POLLINATION. [B.He.]

Beech A genus, *Fagus*, of deciduous trees of the beech family Fagaceae, order Fagales. They can best be distinguished by their long (often more than 1 in. or 2.5 cm), slender, scaly winter buds; their thin, gray bark, smooth even in old trees; and their simple, toothed, ovate or ovate-oblong, deciduous leaves.

The American beech (*F. grandifolia*) is native in the United States east of the Mississippi River and in the lower Mississippi Valley. The hard, strong wood is used for furniture, handles, woodenware, cooperage, and veneer. The small, edible, three-sided nuts, called beechnuts, are valuable food for wildlife. The European beech (*F. sylvatica*) is more popular as an ornamental tree than the American species. Its leaves are smaller, with 5–9 pairs of primary veins compared with 9–14 pairs in the American beech. The leaf teeth are also shorter. Important ornamental varieties are *F. sylvatica purpurea*, the copper or purple beech; var. *incisa*, the cut-leaved or fern-leaved beech; and *F. pendula*, the weeping European beech. *See* FAGALES. [A.H.G./K.P.D.]

Beekeeping The management and maintenance of colonies of honeybees. Although the commonly known honeybee species is native to Europe and Africa only, humans have transported them to other continents, and in most places they have flourished. The natural home for a honeybee colony is a hollow tree, log, or cave. European strains of the honeybee build a nest only in locations which are dry and protected from the wind and sunlight. African bees are less selective and may nest in hollowed-out termite mounds, rock piles, and locations which are less well protected. *See* BEE.

The honey which beekeepers harvest is made from nectar, a sweet sap or sugar syrup produced by special glands in flowers, collected from both wild and cultivated plants. Nectar, the honeybees' source of sugar or carbohydrate, and pollen, their source of protein and fat, make up their entire diet. Nectar contains 50–90% water, 10–50% sugar (predominantly sucrose), and 1–4% aromatic substances, coloring material, and minerals. To transform nectar into honey, bees reduce its moisture content, so that the final honey produced contains between 14 and 19% water, and also add two enzymes which they produce in their bodies.

Scientific beekeeping started in 1851 when an American, L. L. Langstroth, discovered bee space and the movable frame hive. Bee space is the open space which is about 0.4 in. (1 cm) wide and maintained around and between the combs in any hive or natural nest and in which the bees walk. If this space is smaller or larger than 0.4 in. (1 cm), the bees will join the combs. When the combs are stuck together, the hive is not movable, and it is not possible for beekeepers to manipulate a colony or to examine a brood nest. It was found, in 1857, that bees could be forced to build a straight comb in a wooden frame by giving them a piece of wax, called foundation, on which the bases of the cells were already embossed. Bees use these bases to build honeycomb, the cells of which are used for both rearing brood and for storing honey. When a hive of bees is given a frame of foundation, they are forced to build the comb where the beekeeper wants it and not where they might otherwise be inclined to build it. Another discovery, made in 1865, was that honey can be removed from the comb by placing a comb full of honey in a centrifugal force machine, called an extractor. If the beekeeper can return an intact comb to a hive after removing the honey from it, the bees are saved the time and trouble of building a new comb, and the honey harvest is increased. The next discovery, in 1873, was the modern smoker. When bees are smoked, they engorge with honey and become gentle. Without smoke to calm a hive, normal manipulation of the frames would not be possible. By 1880, honey, which had once been a scarce commodity, became abundant.

All beehives used in the industry today are made of wooden boxes with removable frames of comb. Beekeepers have standardized their equipment so that parts will be interchangeable within an apiary and from one commercial operation to another. To be successful in commercial beekeeping, beekeepers must locate in those areas where nectar-producing plants abound. The best-known honey in the United States is clover honey. Alfalfa and oranges are also good nectar-producing plants and are major sources of honey. Bees collect nectar from hundreds of kinds of flowers, and thus there are a great variety of honey flavors. Thousands of colonies of honeybees are rented each year by growers of crops needing cross pollination. As the plants come into bloom, the beekeeper moves the bees, usually at night when all the bees are in the hive, and places them in groups in groves, orchards, and fields. While beekeepers make most of their living producing honey, the real importance of their bees in the agricultural economy is as cross pollinators. Without cross pollination the abundance and variety of food and flowers would not exist. Commercial beekeepers feel they need to own 500 to 2000 colonies to make a living, depending on how they market their honey. Some beekeepers sell their honey on the wholesale market only, while others pack their honey themselves and devote a great deal of time to sales. [R.A.Mo.]

Beet The red or garden beet (*Beta vulgaris*), a cool-season biennial of Mediterranean origin belonging to the plant order Caryophyllales (Centrospermales). This beet is grown primarily for its fleshy root, but also for its leaves, both of which are cooked fresh or canned as a vegetable. Detroit Dark Red strains predominate. Cool weather favors high yields of dark red roots. Wisconsin, New York, and Texas are important beet producing states. *See* CARYOPHYLLALES; SUGARBEET. [H.J.C.]

Beetle A vernacular equivalent of the scientific name Coleoptera. The latter is an order of insects distinguished by the combination of a complete metamorphosis and biting mouthparts with the transformation of the front wings into hard protective elytra, under which the hindwings are folded in repose. Beetle larvae are commonly known as grubs. The beetles include numerous economically important species. A feature of the order is that adults are often long-lived and actively feeding, so that they may be as injurious as the larvae. *See* COLEOPTERA. [R.A.Cr.]

Beggiatoales Formerly recognized as an order of filamentous or unicellular bacteria in the class Schizomycetes. The sulfur bacteria in this order are widely distributed in habitats where sulfur and hydrogen sulfide are present, for example, swamps, sewage, and sulfur mines. The microorganisms move by gliding on a substrate, a method of locomotion also found in many Cyanophyceae (blue-green algae), from which the Beggiatoales differ in their lack of photosynthetic pigments.

In *Bergey's Manual of Determinative Bacteriology* (1957), the Beggiatoales were divided into the Beggiatoaceae, Vitreoscillaceae, Leucotrichaceae, and Achromatiaceae. In the 8th edition of *Bergey's Manual* (1974), Beggiatoaceae, Leucotrichaceae, and Achromatiaceae are redefined and included with seven other families as Gliding Bacteria.

Although many members of the Beggiatoales are able to oxidize hydrogen sulfide and deposit droplets of amorphous sulfur in their cells, the Vitreoscillaceae are not able to do this. There is a certain similarity, morphologically and in the type of locomotion, between members of the Vitreoscillaceae and filamentous Myxobacteria. Further investigations, for example, in cytology, might determine whether or not there is a true relationship. *See* MYXOBACTERALES. [E.G.Pr.]

Behavior genetics The hereditary factors of behavior may be studied in animals and humans. Charles Darwin, who originated the theory that natural selection is the basis of biological evolution, was persuaded by Francis Galton that the principles of natural selection applied equally to behavior and physical characteristics; that is, organisms varied in the expression of certain behaviors. An example would be curiosity—some were very curious, others less so—and curiosity had survival value in some environments. Therefore, more of the organisms which expressed curiosity survived to reproduce in those environments than did those which lacked this trait.

The science of behavior genetics is an extension of these ideas and seeks (1) to determine to what extent the variation of a trait in a population (the extent of individual differences) is due to genetic processes, to what extent it is due to environmental variation, and to what extent it is due to joint functions of these two factors (heredity-environment interactions and correlations); and (2) to identify the genetic architecture (genotypes) that underlies behavior.

The clearest and most indisputable evidence for a hereditary influence on behavior comes from selective-breeding experiments with animals. Behavior genetic research has utilized bacteria, paramecia, nematodes, fruit flies, moths, houseflies, mosquitoes, wasps, bees, crickets, fishes, geese, cows, and dogs. However, although not all of this work involves breeding experiments, enough of it does to demonstrate conclusively the importance of genetic processes on behavior in every species examined. Work with animals allows genetically useful types of relationships, such as half-sibs, to be easily produced. Artificial selection can be used to obtain a population that scores high or low on specific traits. Inbred strains of many animals (populations that are made up of individuals which are nearly identical genetically as a result of inbreeding), particularly rodents, are readily available, and the study of various types of crosses among them can provide a wealth of information. An experimental design using the recombinant inbred-strain method shows great promise for isolating single-gene effects. This procedure derives several inbred strains from the F_2 generation (grandchildren) produced by a cross between two initial inbred strains. Since it is possible to exert a great deal of control over the rearing environments, the experimenter can manipulate both heredity and environment, a feat virtually impossible in human studies but required for precise answers to behavior genetic questions.

Early work with animals was largely directed toward demonstrating genetic variation by comparing the expression of a trait in two different inbred strains or by carrying out studies of artificial selection. Robert Tryon, in a classic study, was able to show that maze-learning ability in rats could be subjected to artificial selection, and populations of maze-bright and maze-dull rats were developed. Other work has often focused on the effects of the environment and genotype-environment interactions. For example, experiments with mice have shown that, with respect to several learning tasks, early environment-enrichment effects and maternal effects were quite small, relative to the amount of normal genetic variation found in the strains of mice tested. Only a few genotype-environment interactions were found. Still other work has shown that early experiences affect later behavior patterns for some strains but not others (a genotype-environment interaction). While it is not always prudent or desirable to generalize from animal results to humans, it is assumed that basic genetic systems work in similar ways across organisms, and it is likely that animal studies will play a key role in elucidating the ways in which environment influences phenotypic variation.

The evidence is now overwhelming that most behavioral traits, whether expressed in humans or in animals, are influenced in an important way by genetic processes. Even the pace and tempo of psychological development appears to be strongly conditioned by genetic factors. In the domain of human behavior, normal or abnormal, there is still considerable controversy about the extent of genetic influence on the expression of various traits, but the existence of genetic influence is widely accepted.

Molecular genetics has made tremendous strides in explaining how genes control and influence biological systems at the biochemical level. The mechanisms can often be precisely specified. With regard to behavior, however, very little is known about the specific mechanisms through which genes influence behavior. For example, it is not known which genes are turning on and off and thereby controlling the spurts and lags in the mental development of twins. It will be necessary to isolate specific genes or groups of genes before the pathways from biochemistry to behavior can be elucidated. *See* MOLECULAR GENETICS. [T.J.B.; K.J.W.]

Behavioral ecology The branch of ecology that focuses on the evolutionary causes of variation in behavior among populations and species. Thus it is concerned with the adaptiveness of behavior, the ultimate questions of why animals behave as they do, rather than the proximate questions of how they behave. The principles of natural selection are applied to behavior with the underlying assumption that, within the constraints of their evolutionary histories, animals behave optimally by maximizing their genetic contribution to future generations. For example, animals must maintain their internal physiological conditions within certain limits in order to function properly, and often they do this by behavior. Small organisms may avoid desiccation by living under logs or by burrowing. Many insects must raise body temperatures to 86–95°F (30–35°C) for effective flight, and achieve this by muscular activity such as the shivering of butterflies in the early morning or by orienting to the Sun. Other adaptive behaviors that are studied may fall in the categories of habitat selection, foraging, territoriality, and reproduction. *See* BEHAVIOR GENETICS; ECOLOGICAL COMMUNITY; ECOLOGICAL INTERACTIONS; ETHOLOGY; MIGRATORY BEHAVIOR; REPRODUCTIVE BEHAVIOR. [H.Di; P.Fr.]

Behavioral psychophysics A branch of psychology concerned primarily with the measurement of the sensory capacities of normal intact animals. In principle, the methods are similar to those of human psychophysics: The animal is "instructed" to observe a stimulus and to "report" what is observed. The stimulus situation, coupled with the past history and present physiological state of the animal, provides the necessary instructions; the animal's behavioral response constitutes the report. *See* PSYCHOLOGY, PHYSIOLOGICAL AND EXPERIMENTAL.

The fundamental measures used are based on the quality, intensity, and spatial and temporal extent of physical stimuli required to elicit some standard response, and include (1) least detectable stimulus intensities (absolute thresholds), (2) least detectable stimulus differences (difference thresholds), and (3) values of sensible stimuli which are not discriminable (points of apparent equality). Sets of such measures, along appropriate dimensions of stimuli, map the range and degree of the animal's sensitivity to those stimuli.

The stimulus-response patterns utilized in these measures may be either innate or learned. Stereotyped unlearned behavior (such as a taxis, orientation of the whole animal in a directional field of stimulation) is particularly useful in the study of lower animals which cannot readily be taught special responses to stimuli. But the use of unlearned behavior is not confined to lower forms; unconditioned reflexes (such as constriction of the pupil in response to light) are useful indices of the effectiveness of stimuli acting on the sense organs of higher animals, including humans. *See* TAXIS.

Such unconditioned responses, even though innately under the control of rather specific stimuli, may be brought under the control of other stimuli by classical (Pavlovian) conditioning, and then used in measuring the animal's sensitivity to these new stimuli. For example, salivary secretion may be conditioned to occur in direct response to a sound by pairing the sound with presentation of food. *See* CONDITIONED REFLEX.

More complex learned discriminative behavior utilizes responses which are instrumental in obtaining some reward. For example, the subject may be required to open one of two doors or to press a lever, in response to some stimulus, to obtain food. Such instrumental behavior may be conditioned to occur in response to almost any sensible stimulus, simply by reinforcing (that is, rewarding) the response only when that stimulus is present. Once such a discriminative response is learned, experiments may be undertaken which are comparable to human psychophysical experiments in their method, scope, and results. [F.R.]

Behavioral toxicology The study of behavioral abnormalities induced by exogenous agents such as drugs, chemicals in the general environment, and chemicals encountered in the workplace. Just as some substances are hazardous to the skin or liver, some are hazardous to the function of the nervous system. In the case of permanent effects, changes in sensation, mood, intellectual function, or motor coordination would obviously be undesirable, but even transient alterations of behavior are considered toxic in some situations. For example, operating room personnel accidentally exposed to very small doses of anesthetic do not exhibit altered performance on an intelligence test, a dexterity test, or a vigilance task. However, significant decrements in performance occur in recognizing and recording visual displays, detecting changes in audiovisual displays, and recalling series of digits.

By comparing the behavior of exposed subjects and control subjects, behavioral toxicologists seek to identify agents capable of altering behavior and to determine the level of exposure at which undesirable effects occur. When the agent under study is one in common use, and there is no evidence of its being hazardous to health, experiments may be carried out on human volunteers, or comparisons may be made from epidemiologic data. More frequently, safety considerations dictate the use of laboratory animals in toxicology research.

Perhaps the best-known example of toxicity in humans is methyl mercury poisoning (Minimata disease), which occurred in epidemic proportions in a Japanese coastal town where the inhabitants ate fish contaminated with mercury from industrial pollution. Although mercury affects a variety of behaviors, the most obvious symptoms are tremors and involuntary movements.

A different set of functional problems is exemplified by the effects of ethyl alcohol, a single agent with direct and indirect, short- and long-term consequences. The short-term, low-dose effects of alcohol include sensory disturbances, motor problems, and difficulties with processing information. Neurologically, alcohol is usually described as a central nervous system depressant which is general, in the sense that it disrupts many functions.

In some individuals, large quantities of alcohol consumed over a long period lead to permanent damage to the nervous system. Behaviorally, individuals with Korsakoff's syndrome exhibit severe memory deficits. Anatomically, their brains are found to have degenerative changes in the thalamus. This syndrome is not just an extension of the short-term effects. In fact, it is thought to arise from alcohol-induced malnutrition rather than as a direct effect of alcohol itself.

Lasting injuries to the nervous system have been reported to occur in children exposed to alcohol before birth. The behavioral problems associated with fetal alcohol syndrome do not appear to be related to either Korsakoff's syndrome or the immediate effects of alcohol. Rather, they constitute a third set of effects, including learning deficits, problems with inhibiting inappropriate behavior, and fine motor dysfunction, along with some visible physical abnormalities. While malnutrition may play a role in this congenital syndrome, the mechanism and locus of damage are not known. *See* ALCOHOLISM; CONGENITAL ANOMALIES; FETAL ALCOHOL SYNDROME.

When toxicity is considered only in terms of direct risk to survival, behavioral toxicity may seem to be of minor importance. However, survival is not the only criterion of good health. In a complex society that places heavy demands on an individual's educability, alertness, and emotional stability, even small deviations in behavior are potentially hazardous. Severe disabilities, as in the gross motor malfunctions of Minimata disease, have drawn attention to behavioral toxicology. Such incidents represent failures of control of toxic substances. Successes are difficult to measure, for they can be seen only in reduction of risk—the ultimate goal of toxicology. *See* TOXICOLOGY. [P.M.R.]

Bejel An infectious disease of humans caused by a spirochete. It is similar to syphilis in character and occurs among seminomadic peoples of the Middle East. It is usually acquired in childhood, either by direct contact or through common drinking utensils. In the latter case, the initial lesion is on the lips. In the generalized stage, lesions are particularly common about mucocutaneous junctions. Bone lesions are common. *See* SYPHILIS. [T.B.T.]

Bel A logarithmic unit expressing the ratios of power, voltage, current, or sound intensity. The number of bels separating two power readings is the logarithm to the base 10 of their ratio (for example, two powers differ by 1 bel when their actual ratio is 10:1), while the number of bels separating two current readings or the sound pressures of an acoustical signal is twice the logarithm of the ratio of the currents.

It is convenient in acoustics to express sound intensity in logarithmic units because of the wide range of pressures to which the ear is sensitive. The strength of a sound is usually specified as the square root of the mean of the squares of the instantaneous pressures measured over a period of time. The sound intensity is proportional to the square of the sound pressure. The measure of the level in bels of a sound is log (I/IR), where I is the intensity of the sound and IR is a specified reference intensity. A smaller unit called the decibel, equal to 1/10 bel, is more commonly used. *See* DECIBEL; NEPER. [K.D.K.]

Belemnoidea An order of extinct dibranchiate cephalopods. These mollusks ranged from the Upper Mississippian through the Cretaceous. The oldest known representative is highly specialized, although the belemnoids are considered to be the most primitive of the dibranchiates. Belemnoids have a chambered shell or phragmocone which fits into a guard or rostrum, the apical portion of the shell (see illustration). *Belemnites* is an important index fossil. *See* CEPHALOPODA; DIBRANCHIA; INDEX FOSSIL. [C.B.C.]

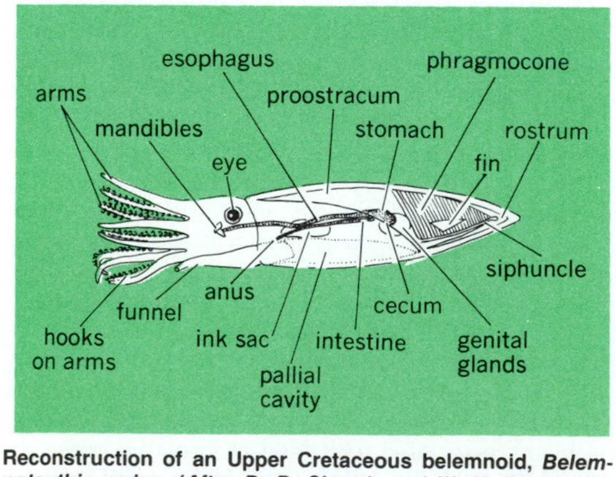

Reconstruction of an Upper Cretaceous belemnoid, *Belemnoteuthis syrica.* (After R. R. Shrock and W. H. Twenhofel, *Principles of Invertebrate Paleontology*, 2d ed., McGraw-Hill, 1953)

Bell A hollow metallic cylinder closed at one end and flared at the other. A bell is used as a fixed-pitch musical instrument or a signaling device, frequently in clocks. A bell is set vibrating by a clapper or tongue which strikes the lip. The bell has two tones, the strike note or key, and the hum tone, which is a major sixth below the strike note. Both tones are rich in harmonics. The finer bells are made for carillons. Small carillons have two full chromatic octaves and large ones have four or more chromatic octaves, the bell with the lowest tone being called the bourdon. All bells of a carillon are shaped for homogeneity of timbre.

Bell metal is approximately four parts copper and one part tin, although zinc, lead, or silver may be used. The shape of a bell—its curves and full trumpet mouth—evolved from experience. A bell can be pitched slightly. The tone is raised by grinding the outer surface, effectively decreasing the diameter, or lowered by grinding the inner surface. Too much tuning of a well-shaped bell is apt to degrade its voice or timbre. [F.H.R.]

Belladonna The drug and also the plant known as the deadly nightshade, *Atropa belladonna*, which belongs to the nightshade family (Solanaceae). This is a coarse, perennial herb native to the Mediterranean regions of Europe and Asia Minor, but now grown extensively in the United States, Europe, and India. During the blooming period, the leaves, flowering tops, and roots are collected and dried for use. The plant contains several important medicinal alkaloids, the chief one being atropine, which is much used to dilate the pupil of the eye. *See* ATROPINE; SOLANALES. [P.D.St./E.L.C.]

Belondiroidea A diverse superfamily of nematodes belonging to the order Dorylaimida. The principal common characteristic is a thick sheath of presumed spiral muscle around the basal swollen portion of the esophagus. The lip region in most species is narrow and varies from rounded to angular. The amphid apertures are slitlike and nearly as wide as the lip region diameter, with a short odontostyle. Flanges are on the odontophore in one genus; in all other species the odontophore is simple and cylindrical. Females may have one or two gonads; if there is only one, it is posteriorly directed. The tails of both sexes are similar. Nothing is known about the feeding habits of this group of nematodes, frequently encountered in soil near plant roots. *See* NEMATA. [A.R.M.]

Belt drive The lowest-cost means for transmitting power between shafts that are not necessarily parallel. Belts run smoothly and quietly, and they cushion motor and bearings against load fluctuations. Belts typically are not as strong or durable as gears or chains. However, improvements in belt construction and materials are making it possible to use belts where formerly only chains or gears would do.

Advantages of belt drive are: They are simple. They are economical. Parallel shafts are not required. Overload and jam protection are provided. Noise and vibration are damped out. Machinery life is prolonged because load fluctuations are cushioned (shock-absorbed). They are lubrication-free. They require only low maintenance. They are highly efficient (90–98%, usually 95%). Some misalignment is tolerable. They are very economical when shafts are separated by large distances. Clutch action may be obtained by relieving belt tension. Variable speeds may be economically obtained by step or tapered pulleys.

Disadvantages include: The angular-velocity ratio is not necessarily constant or equal to the ratio of pulley diameters,

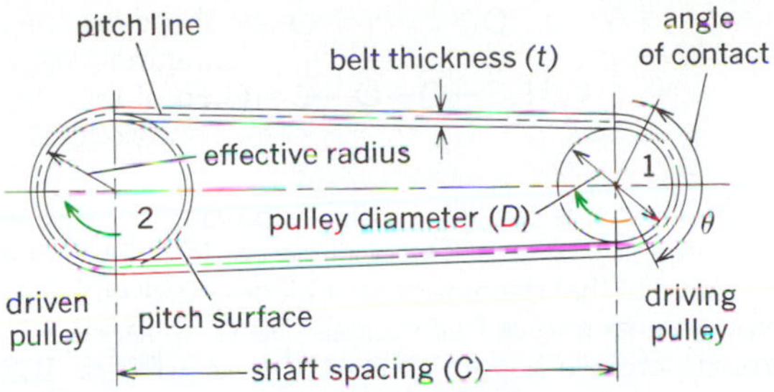

Fig 1. Open belt drive.

because of belt slip and stretch. Heat buildup occurs. Speed is limited to usually 7000 feet per minute (35 meters per second). Power transmission is limited to 370 kilowatts (500 horsepower). Operating temperatures are usually restricted to −31 to 185°F (−35 to 85°C). Some adjustment of center distance or use of an idler pulley is necessary for wear and stretch compensation. A means of disassembly must be provided to install endless belts.

There are four general types of belts: flat belts, V-belts, film belts, and timing belts. Each has its own special characteristics, limitations, advantages, and special-purpose variations for different applications.

Flat belts, in the form of leather belting, served as the basic belt drive from the beginning of the Industrial Revolution. They can transmit large amounts of power at high speeds. Flat belts find their widest application where high-speed motion, rather than power, is the main concern. Flat belts are very useful where large center distances and small pulleys are involved. They can engage pulleys on both inside and outside surfaces, and both endless and jointed construction are available.

V-belts are the basic power-transmission belt, providing the best combination of traction, operating speed, bearing load, and service life. The belts are typically endless, with a trapezoidal cross section which runs in a pulley with a V-shaped groove. The wedging action of the belt in the pulley groove allows V-belts to transmit higher torque at less width and tension than flat belts. V-belts are far superior to flat belts at small center distances and high reduction ratios. V-belts require larger pulleys than flat belts because of their greater thickness. Several individual belts running on the same pulley in separate grooves are often used when the power to be transmitted exceeds that of a single belt. These are called multiple-belt drives.

Film belts are often classified as a variety of flat belt, but actually they are a separate type. Consisting of a very thin strip of material, usually plastic but sometimes rubber, their widest application is in business machines, tape recorders, and other light-duty service.

Timing belts have evenly spaced teeth on their bottom side which mesh with grooves cut on the periphery of the pulleys to produce a positive, no-slip, constant-speed drive. They are often used to replace chains or gears, reducing noise and avoiding the lubrication bath or oiling system requirement.

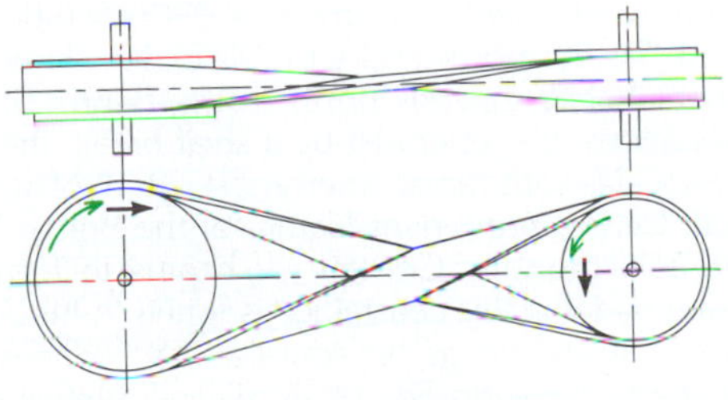

Fig. 2. Cross belt drive.

They have also found widespread application in miniature timing applications. Timing belts, known also as synchronous or cogged belts, require the least tension of all belt drives and are among the most efficient.

The most common belt-pulley arrangement, by far, is the open belt drive (Fig. 1). Here both shafts are parallel and rotate in the same direction. The cross-belt drive of Fig. 2 shows parallel shafts rotating in opposite directions. Timing and standard V-belts are not suitable for cross-belt drives because the pulleys contact both the inside and outside belt surfaces.

Industrial belts are usually reinforced rubber or leather, the rubber type being predominant. Nonreinforced types, other than leather, are limited to light-duty applications.

Belts probably fail by fatigue more often than by abrasion. The fatigue is caused by the cyclic stress applied to the belt as it bends around the pulleys. Belt failure is accelerated when the following conditions are present: high belt tension; excessive slippage; adverse environmental conditions; and momentary overloads caused by shock, vibration, or belt slapping. *See* CHAIN DRIVE; GEAR DRIVE; PULLEY. [A.Erd.; R.G.]

Bending moment The algebraic sum of all moments that are on one side of a cross section of a structural member. For example, in the illustration diagramming a floor joist, a load on the joist tends to bend it at any particular cross section of the joist. At section A-A the bending moment is the reactive upward force of the left wall on the joist times the distance from wall to section A-A. At section B-B the bending moment

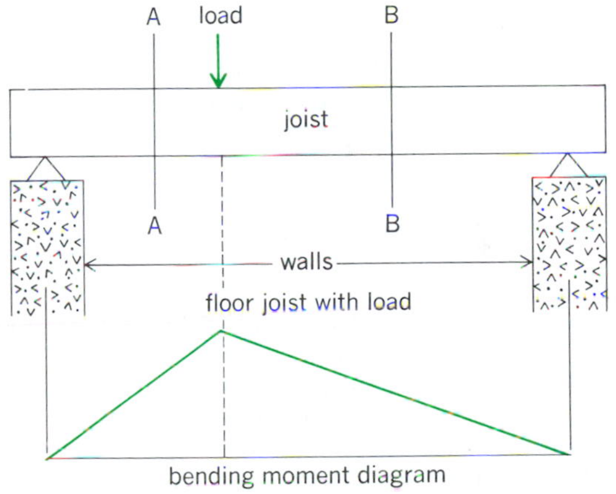

Schematic of bending moment on an end-supported joist with a concentrated load.

is the upward reactive force of the left wall times the distance from wall to B-B plus (in the algebraic sense) the moment produced by the downward load acting at its distance from section B-B. The lower portion of the diagram shows the resultant bending moment at each cross section of the joist. A bending moment that bends a beam convex downward is positive, and that bends a beam convex upward is negative. *See* LOADS, TRANSVERSE; SHEAR. [F.H.R.]

Bentonite The term first applied to a particular, highly colloidal plastic clay found near Fort Benton in the Cretaceous beds of Wyoming. This clay swells to several times its original volume when placed in water and forms thixotropic gels when small amounts are added to water. Later investigations showed that this clay was formed by the alteration of volcanic ash in place; thus, the term bentonite was redefined by geologists to limit it to highly colloidal and plastic clay materials composed largely of montmorillonite clay minerals, and produced by the alteration of volcanic ash in place. The term has been used commercially for any plastic, colloidal, and swelling clays without reference to a particular mode of origin. *See* CLAY; GEL; MONTMORILLONITE.

Bentonites have been found in almost all countries and in rocks of a wide variety of ages. They appear to be most abundant in rocks of Cretaceous age and younger. In the United States, bentonites are mined extensively in Wyoming, Arizona, and Mississippi. England, Germany, Yugoslavia, Russia, Algeria, Japan, and Argentina also produce large tonnages of bentonite. Many bentonites are of great commercial value. They are used in decolorizing oils, in bonding molding sands, in the manufacture of catalysts, in the preparation of oil well drilling muds, and in numerous other relatively minor ways. The properties of a particular bentonite determine its economic use. [R.E.Gr.; F.M.W.]

Benzene A colorless, liquid, inflammable, aromatic hydrocarbon of chemical formula C_6H_6 which boils at 80.1°C (176.2°F) and freezes at 5.4–5.5°C (41.7–41.9°F). In the older American and British technical literature benzene is designated by the German name benzol. In current usage the term benzol is commonly reserved for the less pure grades of benzene.

Benzene is used as a solvent and particularly in Europe as a constituent of motor fuel. In the United States the largest uses of benzene are for the manufacture of styrene and phenol. Other important outlets are in the production of dodecylbenzene, aniline, maleic anhydride, chlorinated benzenes (used in making DDT and as moth flakes), and benzene hexachloride, an insecticide.

The six carbon atoms of benzene, each with a hydrogen atom attached, are arranged symmetrically in a plane, forming a regular hexagon. The hexagon symbol, commonly used to represent the structural formula for benzene, implies the presence of a carbon atom at each of the six angles and, unless substituents are attached, a hydrogen at each carbon atom. Whereas the three double bonds usually included in the formula are convenient in accounting for the addition reactions of benzene, present evidence is that all the carbon-to-carbon bonds are identical.

Nearly all commercial benzene is a product of petroleum technology. The gasoline fractions obtained by reforming or steam cracking of feedstocks from petroleum contain benzene and toluene which can be separated economically. Benzene may also be produced by the dealkylation of toluene.

Benzene is a toxic substance, and prolonged exposure to concentrations in excess of 35–100 parts per million in air may lead to symptoms ranging from nausea and excess fatigue to anemia and leukopenia. *See* AROMATIC HYDROCARBON. [C.K.B.]

Benzoic acid The aromatic carboxylic acid with formula C_6H_5COOH. Benzoic acid melts at 122.4°C (252.3°F), boils at 250°C (482°F), is slightly soluble in water, and is relatively soluble in alcohol and ether. It is found free and combined (gum benzoin; many berries) in nature. The acid of commerce is prepared by hydrolysis of benzotrichloride, $C_6H_5CCl_3$; by monodecarboxylation of phthalic acid; and by oxidation of toluene.

Various derivatives of benzoic acid have applications as miticides, as contrast media in urological and cholecystographic examinations, and in the manufacture of pharmaceuticals. Benzoic acid is also used in curing tobacco; as a preservative (sodium benzoate) of foods (juices and catsup); in dyes; as a mordant in cloth printing (calico); in many esters (benzoates); and as a reference standard in volumetric analysis. *See* CARBOXYLIC ACID. [E.B.R.]

Benzoyl peroxide A chemical compound, sometimes called dibenzoyl peroxide, with the formula below. It is a colorless, crystalline solid, melting point 106–108°C

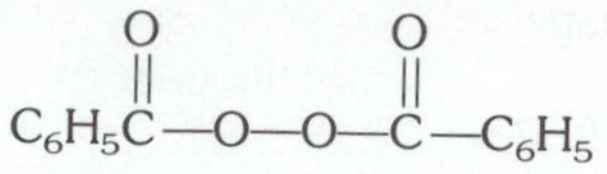

(223–226°F), that is virtually insoluble in water but very soluble in most organic solvents. Benzoyl peroxide is an initiator, a type of material that decomposes at a controlled rate at moderate temperatures to give free radicals. *See* FREE RADICAL.

Benzoyl peroxide has a half-life (10 h at 73°C or 163°F in benzene) for decomposition that is very convenient for many laboratory and commercial processes. This, plus its relative stability among peroxides, makes it one of the most frequently used initiators. Its primary use is as an initiator of vinyl polymerization. It also is the preferred bleaching agent for flour; is used to bleach many commercial waxes and oils; and is the active ingredient in commercial acne preparations. *See* BLEACHING; POLYMERIZATION.

Benzoyl peroxide itself is neither a carcinogen nor a mutagen. However, when coapplied with carcinogens to mouse skin it is a potent promoter of tumor development. All peroxides should be treated as potentially explosive, but benzoyl peroxide is one of the least prone to detonate. *See* MUTAGENS AND CARCINOGENS. [Wi.A.P.]

Beriberi A disorder resulting from a deficiency of vitamin B_1 (thiamine), due usually to idiosyncrasies of diet or excessive cooking or processing of food. Vitamin B_1 is a coenzyme involved in decarboxylation of alpha-keto acids (pyruvic and alpha-ketoglutaric acid) and is important in normal carbohydrate metabolism. It also acts as a factor in maintaining the integrity of the normal conductivity of the nerve fiber for transmission of neural impulses. *See* CARBOHYDRATE METABOLISM; THIAMINE.

Beriberi is characterized by neurologic symptoms, such as tingling of the extremities and muscular weaknesses. Actual inflammation of the nerve fibers occurs and this neuritis, in turn, leads to other alterations in reflexes, in coordination, and in sensory perception, as well as the appearance of occasional psychic states.

When beriberi occurs suddenly, the symptomatology frequently centers about the heart. Derangement of rhythm, decreased blood pressure, the appearance of valvular murmurs, and electrocardiogram changes are typical. These may lead, sometimes precipitously, to cardiac failure and death. In many of these cardiac cases resulting from beriberi, there is associated digestive upset and irregularity and sometimes a paralysis of the diaphragm. A progressive edema of the legs which finally involves the entire body is a common occurrence in severe cases. [N.K.M.]

Bering Sea A water body north of the Pacific Ocean, 875,000 mi^2 (2,268,000 km^2) in area, bounded by Siberia, Alaska, and the Aleutian Islands. The Bering Sea is a biologically productive area, with large populations of marine birds and mammals. An active pollock fishery and a developing bottom-fish industry are evidence of its rich biological resources.

The Bering Sea consists of a large, deep basin in the southwest portion, where depths as great as 9900 ft (3000 m) are encountered. To the north and east, an extremely wide, shallow continental shelf extends north to the Bering Strait. The two major regions are separated by a shelf break, the position of which coincides with the southernmost extent of sea ice in a cold season. Ice is a prominent feature of the Bering Sea shelf during the cold months. Coastal ice begins to form in late October, and by February coastal ice is found in the Aleutians. The sea ice may extend as far south as 58°N. Thus, the ice edge in the eastern Bering Sea advances and retreats seasonally over a distance as great as 600 mi (1000 km). Ice-free conditions can be expected throughout the entire region by early July. *See* SEA ICE.

The main water connections with the Pacific are in the west of the Aleutian Islands, the 6600-ft-deep (2000-m) pass between Attu and Komandorskiye Islands and the 14,000-ft-deep (4400-m) pass between the Komandorskiyes and Kamchatka. Aleutian passes also serve to exchange water. The Bering Sea connection with the Arctic Ocean (Chukchi Sea) is the Bering Strait, 53 mi (85 km) wide and 30 mi (45 m) deep.

Tides in the Bering Sea are semidiurnal, with a strong diurnal inequality typical of North Pacific tides. Three water masses are associated with Bering sea water—Western Subarctic, Bering Sea, and the Alaskan Stream. The general circulation of the Bering Sea is counterclockwise, with many small eddies superimposed on the large-scale pattern. The currents in the Bering Sea are generally a few centimeters per second except along the continental slope, the coast of Kamchatka, and in certain eddies, where somewhat higher values have been found. *See* OCEAN CIRCULATION; TIDE. [V.A.]

Berkelium Element number 97, symbol Bk, the eighth member of the actinide series of elements. In this series the $5f$ electron shell is being filled, just as the $4f$ shell is being filled in the lanthanide (rare-earth) elements. These two series of elements are very similar in their chemical properties, and berkelium, aside from small differences in ionic radius, is especially similar to its homolog terbium. *See* RARE-EARTH ELEMENTS; TERBIUM.

1	2	3	4	5	6	7	8	9	10	11	12	13	14	15	16	17	18
1 H																	2 He
3 Li	4 Be											5 B	6 C	7 N	8 O	9 F	10 Ne
11 Na	12 Mg											13 Al	14 Si	15 P	16 S	17 Cl	18 Ar
19 K	20 Ca	21 Sc	22 Ti	23 V	24 Cr	25 Mn	26 Fe	27 Co	28 Ni	29 Cu	30 Zn	31 Ga	32 Ge	33 As	34 Se	35 Br	36 Kr
37 Rb	38 Sr	39 Y	40 Zr	41 Nb	42 Mo	43 Tc	44 Ru	45 Rh	46 Pd	47 Ag	48 Cd	49 In	50 Sn	51 Sb	52 Te	53 I	54 Xe
55 Cs	56 Ba	71 Lu	72 Hf	73 Ta	74 W	75 Re	76 Os	77 Ir	78 Pt	79 Au	80 Hg	81 Tl	82 Pb	83 Bi	84 Po	85 At	86 Rn
87 Fr	88 Ra	103 Lr	104 Rf	105 Db	106 Sg	107 Bh	108 Hs	109 Mt	110	111	112	113	114	115	116	117	118

lanthanide series	57 La	58 Ce	59 Pr	60 Nd	61 Pm	62 Sm	63 Eu	64 Gd	65 Tb	66 Dy	67 Ho	68 Er	69 Tm	70 Yb
actinide series	89 Ac	90 Th	91 Pa	92 U	93 Np	94 Pu	95 Am	96 Cm	97 Bk	98 Cf	99 Es	100 Fm	101 Md	102 No

Berkelium does not occur in the Earth's crust because it has no stable isotopes. It must be prepared by means of nuclear reactions using more abundant target elements. These reactions usually involve bombardments with charged particles, irradiations with neutrons from high-flux reactors, or production in a thermonuclear device.

Berkelium metal is chemically reactive, exists in two crystal modifications, and melts at 986°C (1806°F). Berkelium was discovered in 1949 by S. G. Thompson, A. Ghiorso, and G. T. Seaborg at the University of California in Berkeley and was named in honor of that city. Nine isotopes of berkelium are known, ranging in mass from 243 to 251 and in half-life from 1 hour to 1380 years. The most easily produced isotope is ^{249}Bk, which undergoes beta decay with a half-life of 314 days and is therefore a valuable source for the preparation of the isotope ^{249}Cf. The berkelium isotope with the longest half-life is ^{247}Bk (1380 years), but it is difficult to produce in sufficient amounts to be applied to berkelium chemistry studies. *See* ACTINIDE ELEMENTS; TRANSURANIUM ELEMENTS. [G.T.S.]

Bermuda grass A long-lived perennial, *Cynodon dactylon*, native to the Old World. Bermuda grass is now common in all tropical and subtropical regions and was introduced to southern United States in colonial days.

Bermuda grass is propagated by surface runners, underground rootstocks, and seed. Common Bermuda grass grows 6–8 in. (15–20 cm) tall; it has short, flat leaves and somewhat wiry stems, and flowers are borne on slender spikes, three to six in a cluster. The grass grows on any soil that is fertile and not too wet, but does best on heavier soils. It requires warm weather for growth, and therefore is adapted to the southern half of the United States. It endures hot dry periods well and thrives under irrigation. Although formerly regarded as a pest in row crops, Bermuda grass is now grown as a valuable forage plant. *See* CYPERALES. [H.B.S.]

Bernoulli's theorem The relation between fluid pressure p and fluid velocity v along a streamline in the steady flow of an incompressible, inviscid fluid. Bernoulli's theorem may be written as the equation below, where H is a constant along a

$$\frac{p}{\rho} + gz + \frac{v^2}{2} = H$$

streamline, ρ is the mass density of the fluid, g is acceleration due to gravity, and z is vertical height. This equation is essentially an expression of conservation of energy because p/ρ is the pressure energy per unit mass of the fluid, gz is the potential energy per unit mass due to gravity, and $v^2/2$ is the kinetic energy per unit mass. It can also be regarded as an integral of momentum. *See* EULER'S MOMENTUM THEOREM. [A.E.Br.]

Beroida An order of the phylum Ctenophora comprising two genera, *Beroë* and *Neis*. These are the only ctenophores that lack tentacles throughout their life, and their origin relative to other ctenophores is uncertain. The conical or cylindrical body ranges in size from a few millimeters to about 40 cm (16 in.) and is strongly flattened in the tentacular plane. Beroids range from clear to bluish, orange, pink, and red in color. Comb rows are well developed in all, and run the entire body length in some species. The mouth is large and extensile, lined with bundles of macrocilia that help to grip and swallow prey.

Beroids are actively swimming predators that feed mainly on other ctenophores. On contact, they rapidly engulf the entire prey or bite off large pieces with the macrocilia. The single species of *Neis* is known only from Australia. Numerous species of *Beroë* have been described from all over the world. Many of these are simply variations in color, and there may be only about a dozen valid species. *See* CTENOPHORA. [L.P.M.]

Beryciformes An order of somewhat intermediate position among actinopterygian fishes. Beryciformes, or Berycomorphi, have fin spines (see illustration) and ctenoid scales, an upper jaw that is bordered by the premaxillae, and a ductless swim bladder and lack a mesocoracoid.

Squirrelfish (*Holocentrus ascensionis*). (After G. B. Goode, *Fishery Industries of the United States*, 1884)

Beryciformes have a rich fossil history extending back to the Upper Cretaceous. Recent representatives are classified in 3 suborders, 12 families, some 30 genera, and about 135 species. They are cosmopolitan in tropical and temperate seas; many are restricted to deep water but some are pelagic and others live in shore waters. Most familiar of the beryciforms are the squirrelfishes and soldierfishes, family Holocentridae, nocturnal animals found in shallow tropical and subtropical reefs. Most are reddish in color. *See* ACTINOPTERYGII; OSTEICHTHYES; TELEOSTEI. [R.M.B.]

Beryl An aluminum beryllosilicate mineral, $Al_2[Be_3Si_6O_{18}]$, crystallizing in the hexagonal system. The hardness is 8 on Mohs scale, specific gravity 2.8, luster vitreous, and color bluish-green, light yellow, and less frequently green, pink, golden yellow, or colorless. It is easily confused with apatite. The rare scandium analog, $Sc_2[Be_3Si_6O_{18}]$, is called bazzite.

Beryl and its gem varieties aquamarine (pale blue), goshenite (colorless), and morganite (pink) most frequently occur in granite-pegmatite druses along with quartz, feldspar, lepidolite, topaz, and tourmaline, where the beryllium ions have concentrated in residual pegmatitic fluids. Beryl crystals up to many tons in weight have been recovered, and the mineral constitutes an important ore of beryllium. Beryl occurs more rarely in mica schists and bituminous limestones (sometimes as emerald, the green gem variety of great value) and nepheline-syenites. Important occurrences include certain pegmatites in New England and California and pegmatites in Brazil, Madagascar, and Siberia. Emeralds occur in certain mica schists in the Urals and in bituminous limestones in Colombia. *See* BERYLLIUM; EMERALD. [P.B.M.]

Beryllium A chemical element, Be, atomic number 4, with an atomic weight of 9.0122. Beryllium, a rare metal, is one of the lightest structural metals, having a density about one-third that of aluminum. Some of the important physical and chemical properties of beryllium are given in the table. Beryllium has a number of unusual and even unique properties.

The largest volume uses of beryllium metal are in the manufacture of beryllium-copper alloys and in the development of beryllium-containing moderator and reflector materials for nuclear reactors. Addition of 2% beryllium to copper forms a nonmagnetic alloy which is six times stronger than copper. These beryllium-copper alloys find numerous applications in industry as nonsparking tools, as critical moving parts in aircraft engines, and in the key components of precision instruments, mechanical computers, electrical relays, and camera shutters. Beryllium-copper hammers, wrenches, and other tools are employed in petroleum refineries and other plants in which a spark from steel against steel might lead to an explosion or fire. *See* ALKALINE-EARTH METALS; BERYLLIUM ALLOYS.

Beryllium has found many special uses in nuclear energy because it is one of the most efficient materials for slowing down the speed of neutrons and acting as a neutron reflector.

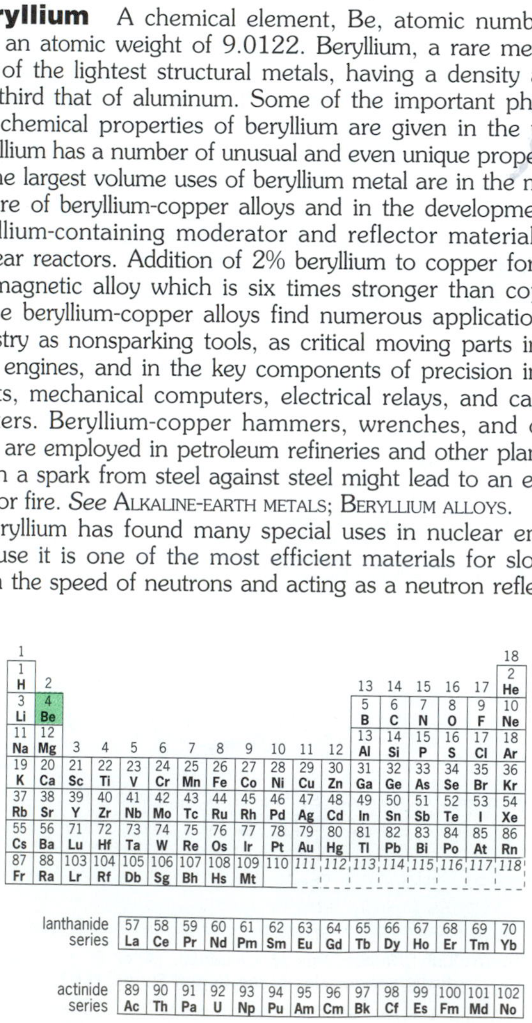

Physical and chemical properties of beryllium

Atomic and mass properties	
Mass number of stable isotopes	9
Atomic number	4
Outer electronic configuration	$1s^2 2s^2$
Atomic weight	9.0122
Atomic diameter	0.221 nm
Atomic volume	4.96 cm^3/mole
Crystal structure	Hexagonal close-packed
Lattice parameters	a=0.2285 nm c=0.3583 nm
Axial ratio	c/a=1.568
Field of cation (charge/radius^2)	17
Density*, 25°, x-ray (theoretical)	1.8477±0.0007 g/cm^3
Density, 1000°, x-ray	1.756 g/cm^3
Radius of atom (Be^0)	0.111 nm
Radius of ion, Be^{2+}	0.034 nm
Ionization energy ($Be^0 \rightarrow Be^{2+}$)	27.4 eV
Thermal properties	
Melting point	1285°C (2345°F)
Boiling point†	2970°C (5378°F)
Vapor pressure (T = K°)	log P (atm)=6.186+1.454 $10^{-4}T$− (16,700/T)
Heat of fusion	250–275 cal/g
Heat of vaporization	53,490 cal/mole
Specific heat (20–100°)	0.43–0.52 cal/(g)(°C)
Thermal conductivity (20°)	0.355 cal/(cm^2)(cm)(s)(°C) (42% of copper)
Heat of oxidation	140.15 cal
Electrical properties	
Electrical conductivity	40–44% of copper
Electrical resistivity	4 microhms/cm (0°C) 6 microhms/cm (100°C)
Electrolytic solution potential, Be/Be†	E^0=−1.69 volts
Electrochemical equivalent	0.04674 mg/coulomb

*Measured values vary from 1.79 to 1.86, depending on purity and method of fabrication.

†Obtained by extrapolation of vapor pressure data, not considered very reliable.

Consequently, much beryllium is used in the construction of nuclear reactors as a moderator and as a support or alloy with the fuel elements. *See* NUCLEAR REACTOR.

The following list shows some of the principal compounds of beryllium. Many of the compounds listed are useful as intermediates in the processes for the preparation of ceramics, beryllium oxide, and beryllium metal. Other compounds are useful in analysis and organic synthesis.

- Acetylacetonate
- Ammonium beryllium fluoride
- Aurintricarboxylate
- Basic acetate, $BeO \cdot Be_3(CH_3COO)_6$
- Basic beryllium carbonate
- Beryllate, BeO_2^{2-}
- Beryllium ammonium phosphate
- Bromide, BeB_2
- Carbide, Be_2C
- Chloride, $BeCl_2$
- Dimethyl, $Be(CH_3)_2$
- Fluoride, BeF_2
- Hydroxide
- Nitrate, $Be(NO_3)_2 \cdot 4H_2O$
- Nitride, Be_3N_2
- Oxide, BeO
- Perchlorate, $Be(ClO_4)_2 \cdot 4H_2O$
- Plutonium-beryllium, $PuBe_{13}$
- Salicylate
- Silicates (emerald)
- Sulfate, $BeSO_4 \cdot 4H_2O$
- Uranium-beryllium, UBe_{13}

Beryllium is surprisingly rare for a light element, constituting about 0.005% of the Earth's crust. It is about thirty-second in order of abundance, occurring in concentrations approximating those of cesium, scandium, and arsenic. Actually, the abundances of beryllium and its neighbors, lithium and boron, are about 10^{-5} times those of the next heavier elements, carbon, nitrogen, and oxygen. At least 50 different beryllium-bearing minerals are known, but in only about 30 is beryllium a regular constituent. The only beryllium-bearing mineral of industrial importance is beryl. Bertrandite substitutes for beryl as a domestic source in the United States. *See* BERYL. [J.Sch.]

Beryllium alloys Dilute alloys of base metals which contain a few percent of beryllium in a precipitation-hardening system. Although beryllium has some solid solubility in copper, silver, gold, nickel, cobalt, platinum, palladium, and iron and forms precipitation-hardening alloys with these metals, the copper-beryllium system and, to a considerably lesser degree, the nickel-beryllium alloys are the only ones used commercially. *See* BERYLLIUM.

In addition to these precipitation-hardening systems, small amounts of beryllium are used in alloys of the dispersion type wherein there is little solid solubility (Al and Mg). Various amounts of beryllium combine with most elements to form intermetallic compounds. Development of beryllium-rich alloys has been chiefly confined to the ductile matrix Be-Al, Be-Cu solid solution alloy with up to 4% Cu, and dispersed-phase-type alloys having relatively small amounts of compounds (0.25–6%), chiefly as BeO or intermetallics, for dimensional stability, elevated temperature strength, and elastic limit control. *See* ALLOY.

Primary applications of beryllium-copper alloys are found in the electronics, automotive, appliance, instrument, and temperature-control industries for electric current–carrying springs, diaphragms, electrical switch blades, and other devices. Applications in structural aerospace and nuclear fields are submarine repeater cable housings for transoceanic cable systems, wind tunnel throats, liners for magnetohydrodynamic generators for gas ionization, and scavenger tanks for propane-Freon bubble chambers in high-energy physics research.

Beryllium intermetallic compounds (beryllides) have high strength at high temperature, good thermal conductivity, high specific heat, and good oxidation resistance. Beryllides are formed with actinide and rare metals, as well as with the transition metals. They are of interest to the nuclear field, to power generation, and to aerospace applications. Evaluation of the intermetallics as refractory coatings, reactor hardware, fuel elements, turbine buckets, and high-temperature bearings has been carried out. [W.W.Be./W.D.T.]

Beryllium metallurgy Beryllium metallurgy involves two processes for the extraction of beryllium oxide or hydroxide from beryl ore. There are also two methods employed to reduce BeO to beryllium metal. The two extraction methods are based on dissolving beryl as either a fluoride or a sulfate. Reduction is accomplished thermally by means of magnesium with beryllium fluoride, and electrolytically with beryllium chloride. Over 90% of the metal is made from the thermal process.

Vacuum-cast beryllium ingots either are produced as raw material for the powder-metal process, which accounts for over 90% of the material being used, or are cast directly for fabrication. Machined parts are usually made from powder-metal products, while ingot beryllium is usually rolled or otherwise worked. *See* BERYLLIUM.

Beryllium can be welded, soldered, brazed, or plastically bonded, with brazing and plastic bonding, as well as mechanical fasteners, being the usual production methods. Surface protection of finished beryllium surfaces can be provided by anodizing, plating, optical polishing, or using conversion coatings. Chemical machining and chemical milling are used to provide patterned and low-damage surfaces.

Beryllium has had a long history of use in atomic energy as neutron sources, reflectors, and moderators in thermal and intermediate reactors. Applications of beryllium to aerospace usually provides weight saving up to 60% of that of a competing material on the same design basis. Extensive use of beryllium in reentry structures depends on high-temperature strength and thermal capacity. The metal's transparency to radiation (as used in x-ray windows) also is important to missile structures, which must withstand electromagnetic pulse conditions created by antimissile tactics. Finally, ability to machine to high tolerance coupled with dimensional stability has created almost exclusive employment of beryllium in inertial guidance navigation and control devices. [W.W.Be./W.D.T.]

Beryllium minerals Minerals containing beryllium as an essential component. Over 50 beryllium minerals have been identified, even though beryllium is a scarce element in the Earth's crust. The unusual combination of low charge (+2) and small ionic radius (0.035 nanometer) of the beryllium ion accounts for this diverse group of minerals and their occurrence in many natural environments. *See* BERYLLIUM.

Nearly all beryllium minerals can be included in one of three groups: compositionally simple oxides and silicates with or without aluminum; sodium- and calcium-bearing silicates; and phosphates and borates. The first group is by far the most abundant; it contains beryl, the most common beryllium mineral, plus the common minerals phenakite, bertrandite, chrysoberyl, and euclase. Of this group, only beryl shows a wide compositional variation.

The beryllium minerals have many structural characteristics similar to the major rock-forming silicate minerals, but are distinguished by containing large quantities of tetrahedrally coordinated beryllium ion (Be^{2+}) in place of, or in addition to, tetrahedrally coordinated aluminum ion (Al^{3+}) and silicon ion (Si^{4+}). *See* SILICATE MINERALS.

Beryllium minerals occur in many geological environments, where they are generally associated with felsic (abundant feldspar ± quartz) igneous rocks and related, metasomatically altered rocks.

Beryl and bertrandite, mined from granitic pegmatites and altered volcanic rocks, are the principal ores of beryllium; deposits of chrysoberyl and phenakite may become economically significant in the future. The colored varieties of beryl (emerald, aquamarine, morganite) are valued gemstones; chrysoberyl, phenakite, and a few of the other minerals are less common gemstones. *See* CHRYSOBERYL; EMERALD; GEM. [M.D.B.]

Bessel functions The solutions of Bessel's differential equation, Eq. (1). Bessel functions, also called cylinder func-

$$z^2\, d^2y/dz^2 + z\, dy/dz + (z^2 - \nu^2)y = 0 \qquad (1)$$

tions, are examples of special functions which are introduced by a differential equation. Bessel functions are of great interest in purely mathematical concepts and in mathematical physics. They constitute additional functions which, like the elementary functions z^n, $\sin z$, e^z, can be used to express physical phenomena.

Applications of Bessel functions are found in such representative problems as heat conduction or diffusion in circular cylinders, oscillatory motion of a sphere in a viscous fluid, oscillations of a stretched circular membrane, diffraction of waves by a circular cylinder of infinite length or by a sphere, acoustic or electromagnetic oscillations in a circular cylinder of finite length or in a sphere, electromagnetic wave propagation in the waveguides of circular cross section, in coaxial cables, or along straight wires, and in the skin effect in conducting wires of circular cross section. In these problems Bessel functions are used to represent such quantities as the temperature, the concentration, the displacements, the electric and magnetic field strengths, and the current density as a function of space coordinates. The Bessel functions enter into all these problems because boundary values on circles (two-dimensional problems), on circular cylinders, or on spheres are prescribed, and the solutions of the respective problems are sought either inside or outside the boundary, or both.

The independent variable z in Bessel's differential equation may in applications assume real or complex values. The para-

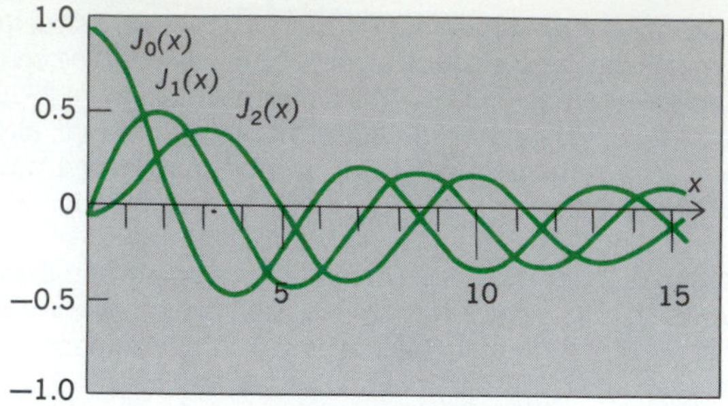

Fig. 1. Bessel functions $J_0(z)$, $J_1(z)$, $J_2(z)$ for $v = 0, 1, 2$ and for positive values of z.

meter v is, in general, also complex. Its value is called the order of the Bessel function. Since there are two linearly independent solutions of a linear differential equation of second order, there are two independent solutions of Bessel's differential equation. They cannot be expressed in finite form in terms of elementary functions such as z^n, sin z, or e^z unless the parameter is one-half of an odd integer. They can, however, be expressed as power series with an exception for integer values of v. The function defined by Eq. (2) is designated as Bessel's

$$J_v(z) = \left(\frac{z}{2}\right)^v \sum_{l=0}^{\infty} \frac{(-z^2/4)l}{l!\Gamma(v + l + 1)} \qquad (2)$$

function of the first kind, or simply the Bessel function of order v. $\Gamma(v + l + 1)$ is the gamma function. The infinite series in Eq. (2) converges absolutely for all finite values, real or complex, of z. Along with $J_v(z)$, there is a second solution $J_{-v}(z)$. It is linearly independent of $J_v(z)$ unless v is an integer n. In this case $J_{-n}(z) = (-1)^n J_n(z)$. *See* GAMMA FUNCTION.

The Bessel function of the second kind, also called the Neumann function, is defined by Eq. (3).

$$Y_v(z) = \frac{\cos v\pi J_v(z) - J_{-v}(z)}{\sin v\pi} \qquad (3)$$

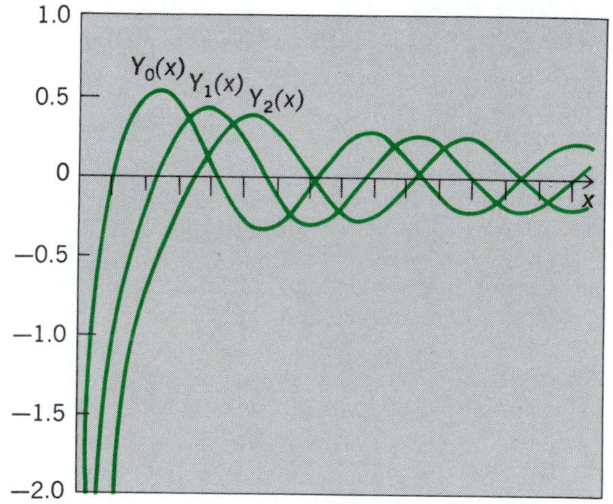

Fig. 2. Neumann functions $Y_0(z)$, $Y_1(z)$, $Y_2(z)$ for $v = 0, 1, 2$ and for positive values of z.

If $v = n$, this expression is indeterminate, and the limit of the right member is to be taken. There are two Bessel functions of the third kind, designated as first and second Hankel functions. They are defined as Eqs. (4).

$$H_v^{(1)}(z) = J_v(z) + iY_v(z) \qquad (4)]$$
$$H_v^{(2)}(z) = J_v(z) + iY_v(z)$$

Figures 1 and 2 give the Bessel and Neumann functions for $v = 0, 1, 2$ and positive values of z. [J.Meix.]

Beta particles Fast, charged particles emitted from certain radioactive nuclei. These particles, which are all identical except for the sign of their charge, are classified as positrons (+) and negatrons (−). The latter class is identical with atomic electrons. The kinetic energies range from zero up to 3–5 MeV. For the theory of beta (β) decay *see* RADIOACTIVITY. *See also* ALPHA PARTICLES; ELECTRON; POSITRON.

Beta particles, often designated beta rays, interact strongly with matter: The particles are generally completely stopped in passing through 0.4 in. (1 cm) of solid material. If various thicknesses of some material are placed across a beam of beta particles, the number of penetrating particles is found to decrease gradually to zero as the thickness increases. There is a definite thickness of absorber which is just sufficient to stop all the beta particles. This quantity, which is called the range of the beta particle, is usually expressed as the product $\rho \times$ (g/cm^2) of density and thickness, since then this "range" will be approximately independent of the material.

The passage of beta particles through matter is macroscopically observable as heat, due to the kinetic energy dissipated. This radiation also may promote certain chemical reactions and cause structural changes in materials, for instance, the discoloring of glass.

The track of a fast electron, as observed in a photographic emulsion, does not follow a straight line until the particle is stopped, but zigzags in a random way because of collisions with the much heavier nuclei. A beta particle loses its energy by excitation and ionization of atoms and at higher energy also by emitting bremsstrahlung, or brake radiation, when scattered. *See* BREMSSTRAHLUNG.

Apart from the stopping process, which proceeds similarly, negatrons and positrons behave differently in matter. Usually after it has been stopped, the positron combines with an atomic negatron and both vanish (annihilate). The two electron masses appear in the form of the electromagnetic energy of two photons, each of energy 0.511 MeV, which are emitted in opposite directions in order to conserve momentum. Before annihilation the positron-negatron pairs form bound systems, called positronium atoms. *See* POSITRONIUM.

Beta detectors may function so as to produce visible tracks of the particles, as in photographic emulsions and cloud chambers, or may give a signal at the advent of each individual particle. *See* PARTICLE DETECTOR. [G.B.]

Beta-ray spectrometer An instrument for measuring the energy or momentum distribution of beta particles. The energy of an electron is in principle inferred from the way it is influenced by a magnetic field. If a charged particle is moving

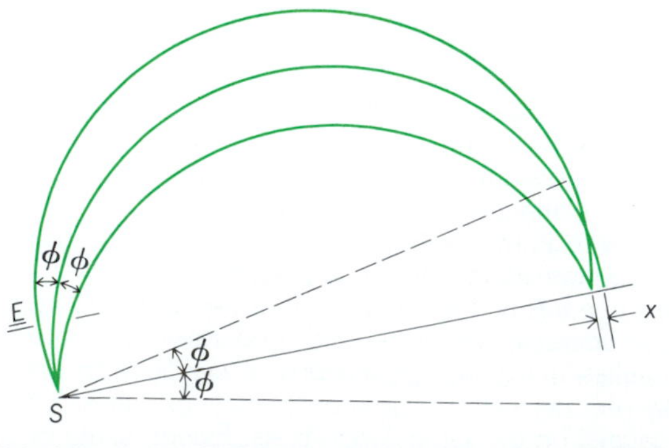

Principle of the semicircular spectrometer. The magnetic field lines are perpendicular to the paper. The broken lines are diameters of the circular orbits.

perpendicularly to the flux of a uniform magnetic field, its orbit is bent into a circle of radius ρ. The momentum is found to be proportional to the product $B\rho$ of magnetic flux density and radius.

A simple and generally used spectrometer is the semicircular type, shown schematically in the illustration. From the electrons that leave the source in all directions, a thin bundle (represented by an angular spread 2ϕ in the illustration) is selected by the entrance slit E to improve resolution. Particles of a definite energy move in congruent circles and are brought to a focus after going through an angle of 180°. Particles of different energies come to a focus at different distances. The particles may be detected by means of a photographic plate along the focal line x, which permits the simultaneous recording of a large energy range. The angular spread 2ϕ allowed by the finite slit width introduces a spread Δx in the position at which the particle strikes the photographic plate. Alternatively a Geiger counter may be located at a fixed distance from the source and the field strength varied to focus one energy after another on a slit in front of the counter. [G.B.]

Betatron A device for accelerating charged particles in an orbit by means of the electric field E from a slowly changing magnetic flux Φ. The electric field is given by $E = -(1/2\,\pi r_o)\,d\Phi/dt$ (in SI or mks units), where r_o is the orbit radius. The name was chosen because the method was first applied to electrons. In the usual betatron both the accelerating core flux and a guiding magnetic field rise with similar time dependence, with the result that the orbit is circular. However, the orbit can have a changing radius as acceleration progresses. For the long path (usually more than 60 mi or 100 km), variations of axial and radial magnetic field components provide focusing forces, while space charge and space current forces due to the particle beam itself also contribute to the resulting betatron oscillations about the equilibrium orbit. In many other instances of particle beams, the term betatron oscillation is used for the particle oscillations about a beam's path.

Collective effects from self-fields of the beam have been found important and helpful in injecting. Circulating currents of about 3 amperes are contained in the numerous industrial and therapeutic betatrons, although the average currents are below 10^{-7} A. *See* Particle accelerator. [D.W.K.]

Betel nut The dried, ripe seed of the palm tree *Areca catechu* (Palmae), a native of Ceylon and Malaya. The nuts, slightly larger than a chestnut, have a faint odor when broken open and a somewhat acrid taste. Betel nuts are chewed by the inhabitants together with the leaves of the betel pepper, *Piper betle* (Piperaceae), and lime. The mixture discolors and eventually destroys the teeth. Frequently, habitual chewers of the nut are toothless by the age of 25. The seeds contain a narcotic that produces some stimulation and a sense of well-being. *See* Narcotic; Palm; Piperales. [P.D.St./E.L.C.]

Betelgeuse Alpha Orionis, a very bright, cool, red supergiant star, spectral type M2, temperature near 3000 K. Typical of the low-temperature end of the supergiant sequence, Betelgeuse varies in light and velocity in periods of about 6 years. Because of high luminosity and low temperature, it has an enormous radius (about as big as the orbit of Mars), 300 times that of the Sun. The angular diameter was measured by the stellar interferometer on Mount Wilson and found to be variable, 0″.034-0″.047 of arc—the largest among the half-dozen stars with measurable disks. The star's mean density is only 5×10^{-7} g/cm^3. Partly because of low surface gravity, the atmosphere is unstable and shows outward streaming of matter at low temperature and density. *See* Star. [J.L.Gr.]

Bevel gears Where shafts intersect, bevel gears transmit the motion. Such gears may be used only to change the shaft axis direction or to change speed as well as direction. Two bevel gears with equal numbers of teeth and running together with their shaft axes intersecting at 90° are called miter gears. Several forms of bevel gears are in use, including straight-tooth, spiral, and skewed bevel gears (see illustration).

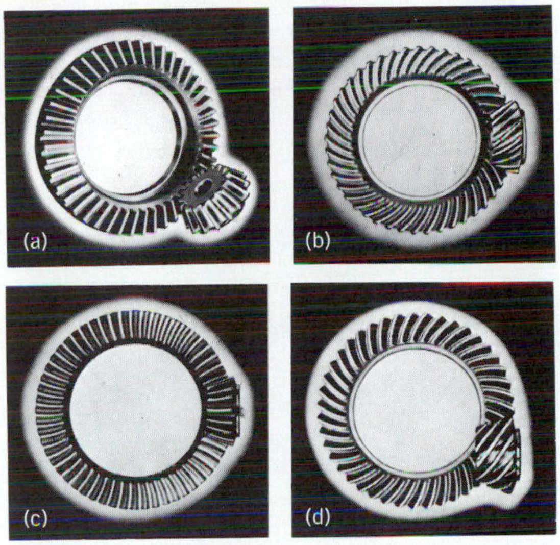

Bevel gears with pinions. (*a*) Straight. (*b*) Spiral. (*c*) Zero. (*d*) Hypoid. (*Gleason Gear Works*)

External bevel gears have pitch angles less than 90°. Internal bevel gears have pitch angles greater than 90°, hence their pitch cones are inverted. A crown gear is one having a pitch angle of 90°. Thus its pitch surface is a plane, and the crown gear corresponds in this respect to a rack in spur gearing.

Straight bevel gears. The simplest form of bevel gear has straight teeth which, if extended inward, would come together at the intersection of the shaft axes. Speeds of the shafts of bevel gears (velocity ratio) are inversely proportional to the numbers of teeth on the gears or to the sines of the pitch angles, but not to the formative number of teeth. Use of straight bevel gears is limited to low-speed operations, ordinarily below 1000 surface feet per minute or, in the case of small gears, 1000 revolutions per minute.

Spiral bevel gears. To provide a gradual engagement, as contrasted to the full line engagement of straight bevel gears, the teeth of spiral bevel gears are curved and oblique. Theoretically the curve is a spiral but, to facilitate manufacture, the curve is actually a circular arc which, within the tooth face width, closely approximates a spiral. This tooth inclination brings more teeth in contact at any one time than with an equivalent straight-tooth bevel gear. The result is smoother and quieter operation, particularly at high speeds, and greater load-carrying ability than with straight bevel gears of the same size. Spiral bevel gears are used in sewing machines, motion picture equipment, machine tools, and other applications where quiet, smooth operation is essential.

Zero bevel gears. A special form of bevel gear has curved teeth with a zero-degree spiral angle. Thus the teeth are not oblique as is the case with spiral bevel gears. Rather, the teeth lie in the same general direction as those of an equivalent straight-tooth bevel gear, and so the gears are usually used in the same types of drives as the straight-tooth gear.

Hypoid gears. To connect nonparallel, nonintersecting shafts, usually at right angles, hypoid gears are used. The shape of the tooth is similar to that of the spiral bevel gear and gives progressive contact across the tooth. In operation these gears run even more smoothly and quietly than spiral bevel gears. To maintain line contact of the teeth with the offset shaft, the pitch surface of the hypoid gear is a hyperboloid of revolution rather than a cone as in bevel gears. The operating smoothness of hypoid gears, along with the lower body lines made possible by the offset pinion shaft, has made them extremely popular for automotive use. [J.R.Z.]

Bicosoecida An order of Zoomastigophorea (Protozoa). They are colorless, free-living cells, each with two flagella, one of which is used for attaching the organism to its exoskeleton (lorica). Although the attachment is normally from the front end, the flagellum emerges alongside the primary or vibrating one (see illustration). The anterior end in many species

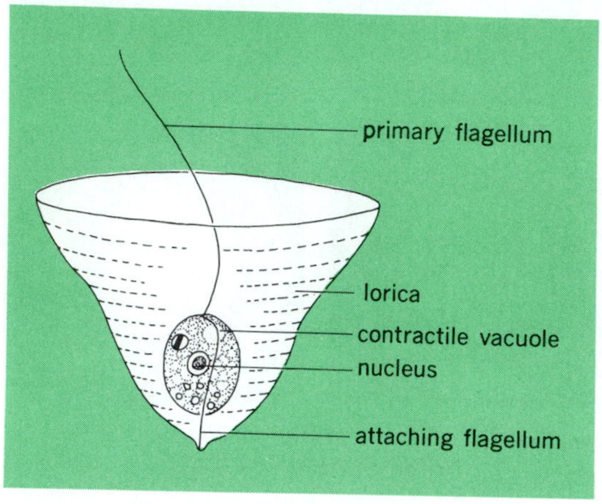

A bicosoecid, *Codomonas annulata.*

(*Codomonas annulata*, *Poteriodendron petiolatum*, and *Stephanocodon stellatum*) is ameboid, or at times appears to be formed into a lip which can turn in and engulf a bacterium.

These Bicosoecida are common in fresh water and often seen attached to desmids or other algae. *Bicosocca mediterranea* is common at times in salt water, where assemblages of it are found on diatoms. The order has very few genera and species. *See* PROTOZOA; ZOOMASTIGOPHOREA. [J.B.L.]

Big bang theory The theory that the universe began in a state of compression to infinite density and has been expanding since some particular instant that marked the origin of the universe. The big bang is the most generally accepted cosmological theory.

Two observations are at the base of observational big bang cosmology. First, the universe is expanding uniformly, with objects at greater distances receding at a greater velocity. Second, the Earth is bathed in an isotropic glow of radiation that has the characteristics expected from the remnant of a hot primeval fireball. *See* HUBBLE CONSTANT.

Tracing the expansion of the universe back in time, it is found that the universe was compressed to infinite density approximately $1.3–2 \times 10^{10}$ years ago. According to the leading theories of cosmology, at that time a "big bang" began the universe. It was the origin of time and space. From the time of the big bang, the universe has been expanding.

In 1917 Einstein found a solution to his own set of equations from his general theory of relativity that predicted the nature of the universe. His universe, though, was unstable: it could only be expanding or contracting. This seemed unsatisfactory at the time, for the expansion had not yet been discovered, so Einstein arbitrarily introduced a special term—the cosmological constant—into his equations to make the universe static. The need for the cosmological constant disappeared with Hubble's discovery of the expansion.

In 1922 A. Friedmann worked out solutions that are at the basis of the cosmological models that are now generally accepted. G. Lemaître in 1927 independently found similar solutions. Lemaître's solutions indicated that the original "cosmic egg" from which the universe was expanding was hot and dense. This is the origin of the current view that the universe began from a hot big bang. Friedmann's solution did not involve the cosmological constant and Lemaître's did, but otherwise the solutions were similar.

Modern theoretical work has been able to trace the universe back to the first instants in time after the big bang. At the earliest instants, the universe may have been filled with exotic elementary particles. Individual quarks may also have been present. By 1 microsecond after the big bang, the exotic particles and the quarks had been incorporated in other fundamental particles. *See* ELEMENTARY PARTICLE; QUARKS.

It is not definitely known why there is an apparent excess of matter over antimatter, though attempts in elementary particle physics to unify the electromagnetic, the weak, and the strong forces show promise in explaining the origin of the matter-antimatter asymmetry. The asymmetry seems to have arisen before the first millisecond after the big bang. *See* ANTIMATTER; FUNDAMENTAL INTERACTIONS.

By 5 s after the big bang, the temperature had cooled to 10^9 K, and only electrons, positrons, neutrinos, antineutrinos, and photons were important. A few protons and neutrons were mixed in, and they grew relatively more important as the temperature continued to drop. The universe was so dense that photons traveled only a short way before being reabsorbed. By the time 4 min had gone by, elements had started to form.

After about a million years, when the universe cooled to 3000 K and the density dropped sufficiently, the protons and electrons suddenly combined to make hydrogen atoms, a process called recombination. Since hydrogen's spectrum absorbs preferentially at the wavelengths of sets of spectral lines rather than continuously across the spectrum, and since there were no longer free electrons to interact with photons, the universe became transparent at that instant. The average path traveled by a photon—its mean free path—became very large. The blackbody spectrum of the gas at the time of recombination was thus released and has been traveling through space ever since. As the universe expands, this spectrum retains its blackbody shape though its characteristic temperature drops. *See* HEAT RADIATION.

As the early universe cooled, the temperatures became sufficiently low for element formation to begin. By about 100 s, deuterium (one proton plus one neutron) formed. When joined by another neutron to form tritium, the amalgam soon decayed to form an isotope of helium. Ordinary helium, with still another neutron, also resulted.

Big bang nucleosynthesis, although at first thought to be a method of forming all the elements, founders for the heavy elements at mass numbers 5 and 8. Isotopes of these mass numbers are too unstable to form heavier elements quickly enough. The gap is bridged only in stars. Thus the lightest elements were formed as direct result of the big bang, while the heavier elements as well as additional quantities of most of the lighter elements were formed later in stars or supernovae. *See* NUCLEOSYNTHESIS.

The two basic possibilities for the future of the universe are that the universe will continue to expand forever, or that it will cease its expansion and begin to contract. It can be shown that the case where the universe will expand forever corresponds to an infinite universe. The term applied is the open universe. The case where the universe will begin to contract corresponds to a finite universe. The term applied is the closed universe. Several lines of evidence agree that the universe is open, including studies of differential velocities of nearby galaxies due to density perturbations, the cosmic abundance of deuterium, and the absence of hot intergalactic gas. The possible existence of neutrino mass is a piece of evidence favoring the opposite conclusion. *See* COSMOLOGY. [J.M.P.]

Bile acid A steroid acid produced in the liver. The bile acids lower surface tension and promote emulsification of fat to aid in digestion, and absorption of fats from the intestine. The bile acids have a five-carbon side chain and occur in the liver in combination, through peptide linkage, with glycine and taurine, forming glycocholic acid and taurocholic acid, respectively (see illustration). Other bile acids are cholic, deoxycholic, lithocholic, and chenodeoxycholic acids. *See* LIVER; STEROID; STEROL. [R.I.D.]

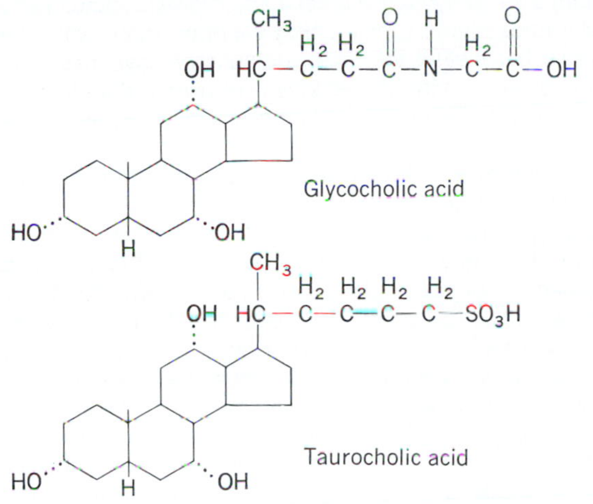

Structural formulas for two common bile acids.

Bilirubin The predominant orange pigment of bile. It is the major metabolic breakdown product of heme, the prosthetic group of hemoglobin in red blood cells, and other chromoproteins such as myoglobin, cytochrome, and catalase. The breakdown of hemoglobin from the old red cells takes place at a rapid rate in the reticuloendothelial cells of the liver, spleen, and bone marrow. The steps in this breakdown process include denaturation and removal of the protein globin, oxidation and opening of the tetrapyrrole ring, and the removal of iron to form the green pigment biliverdin, which is then reduced to bilirubin by the addition of hydrogen. The formed bilirubin is transported to the liver, probably bound to albumin, where it is conjugated into water-soluble mono- and diglucuronides and to a lesser extent with sulfate. *See* LIVER.

In mammalian bile essentially all of the bilirubin is present as a glucuronide conjugate. Bilirubin glucuronide is passed through the liver cells into the bile caniculi and then into the intestine. The bacterial flora further reduces the bilirubin to colorless urobilinogen. Most of the urobilinogen is either reduced to stercobilinogen or oxidized to urobilin. These two compounds are then converted to stercobilin, which is excreted in the feces and gives the stool its brown color. *See* HEMOGLOBIN. [M.K.S.]

Binary number system A system in which numbers are represented as linear combinations of powers of 2. Below is shown the relationship between the binary system and the decimal system.

Decimal notation	*Binary notation*
$3 = 1 \times 2^1 + 1$	$= 11$
$7 = 1 \times 2^2 + 1 \times 2^1 + 1$	$= 111$
$17 = 1 \times 2^4 + 0 \times 2^3 + 0 \times 2^2 + 0 \times 2^1 + 1$	$= 10001$

The binary system has come into importance because of its value in computer applications. *See* NUMBER SYSTEMS. [C.D.O.]

Binary star Two stars held together by their mutual gravitational attraction in a permanent (or at least long-term) association. The term binary system would actually be more appropriate. The two components of the binary system revolve in close elliptical (or circular) orbits around their common center of gravity, which carries them through space so that they also have a common proper motion. Usually treated together with binary stars are multiple systems, consisting of several stars.

The formation of binary stars must be a very common process related to stellar formation itself. According to L. Lucy, the bulk of binary systems are probably formed by successive fragmentations of protostars.

Binary stars play an important role in astrophysics for three reasons: (1) They are a very common phenomenon. It is more likely for a given star to be a member of a binary or multiple system than to be single. (2) Binary systems reveal much more information about stellar properties than do single stars. In particular, masses of stars can be measured only by their gravitational attraction on nearby bodies. The radii, luminosities, and effective temperatures of stars in binary systems can also be obtained, and the mutual relations of these quantities can then be studied. (3) In binary stars where the components interact strongly, new phenomena and new types of stars occur which otherwise would not exist: novae, cataclysmic variables, and binary x-ray sources are examples. *See* HERTZSPRUNG-RUSSELL DIAGRAM; MASS-LUMINOSITY RELATION.

Separations vary over an enormous range from one binary system to another. The upper limit depends critically on the density of stars in space. The lower limit on separations is given by the physical dimensions of the component stars.

Observational techniques. With the enormous range of separations and periods, it is not surprising that different observational techniques must be used. (1) Systems of large separations are often observable directly through a telescope as two separate stars. They are then called visual binary stars. (2) For a closer pair, the orbital velocities may be sufficiently large to be detectable by a periodic shift of spectral lines caused by the Doppler effect. These objects are called spectroscopic binaries. (3) When the two stars are so close to each other that their dimensions are not negligible compared to their separation, they can eclipse each other, provided that the orbital plane is suitably oriented. These objects are then easily detected photometrically from the periodic changes of the light received from the system, and are called eclipsing binaries or photometric binary stars. *See* DOPPLER EFFECT.

Evolution. Stars spend most of their lifetime on the main sequence, generating energy by conversion of hydrogen into helium in their cores. When the hydrogen in the core is exhausted, the stars expand. Single stars encounter no problem in this expansion, but the gravitational field of a nearby companion sets a definite upper limit to the volume available for expansion to a binary star component; it also considerably distorts the shape of the expanded star. Systems in which evolution is substantially affected by the duplicity are called close binary systems. When the expanding component reaches the limiting surface, the evolution of both stars is completely changed. *See* STELLAR EVOLUTION.

The sequence of shapes through which an expanding component passes is shown in the illustration. The curves drawn are cross sections of three-dimensional equipotential surfaces, for a system, with stars of masses M and M', viewed edge-on. When a star is still small with respect to the dimensions of the system, it is very nearly spherical. As it expands, its shape becomes that of a distorted ellipsoid more and more extended toward the companion, until the star reaches the critical surface marked with a heavy line in the illustration. The next

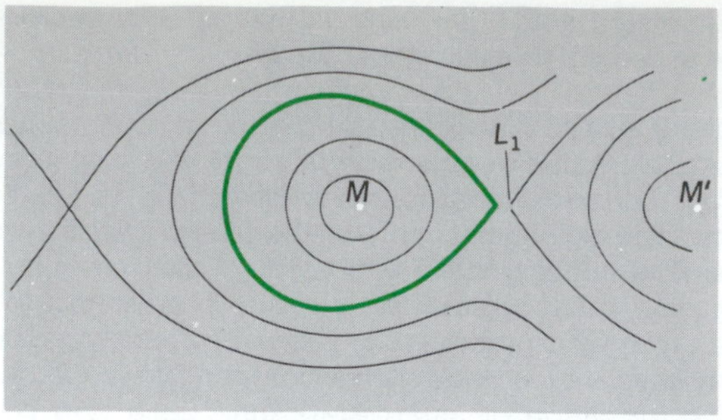

Nature of equipotential surfaces. The Roche limit is marked by a heavy line.

equipotential surface already surrounds both stars. A star that has reached its critical surface will lose mass in the vicinity of the point L_1 (the first lagrangian point) where the effective gravity is zero, and a gas stream will flow from there toward the other star. The stream either impacts directly on the companion or leads to the formation of an accretion disk surrounding it. In either case, the star gains (accretes) mass either directly from the stream or through the disk, in which viscosity makes some particles fall on the star.

X-ray binaries. Since 1970, x-ray satellites have discovered a number of strong sources of x-rays clearly associated with binary systems: the x-ray source undergoes eclipses or the optical counterpart is a binary, or both. The picture which has been developed from study of these systems is as follows. The optically visible star is unstable and loses mass, either by stellar wind or by Roche lobe overflow. The stream flows to the compact star and falls into a very deep "potential well," since the compact star is extremely small. Usually the accretion occurs through a viscous disk, but eventually always a large fraction of the potential energy of the particles is released as radiation. The hot gas reaches temperatures of order 10^8–10^9 °F (50–500 × 10^6 °C) and radiates mostly in x-rays. Some of the x-ray binary sources are among the most powerful energy emitters in the Galaxy. The regular rapid pulses of x-rays are believed to be generated by a rapid rotation of the magnetized neutron star. *See* PULSAR; X-RAY STAR.

The neutron stars found in binary x-ray sources are thought to be remnants of supernova explosions that in the past terminated the normal stellar lifetime of their parental stars. It is also believed that very massive stars leave black holes as the remnants after their supernova explosion, rather than neutron stars. *See* BLACK HOLE.

Cataclysmic variables. Probably all novae are close binary systems. Novae are a special class (conspicuous because of a very high increase in light in an outburst) of cataclysmic variables. Other classes include recurrent novae and dwarf novae. All these objects appear to be binaries of a basically similar nature: a cooler object, usually a red dwarf, is unstable and transfers mass to a hotter and much more compact object, a degenerate white dwarf or a similar body. A viscous disk surrounding the accreting star plays an important role. Variable rates of mass transfer, variable viscosity, and continual interaction between the star, the disk, and the stream lead to instabilities and eruptive events. In the more violent cases, thermonuclear reactions at the surface of the accreting star are believed to be the source of the energy for a nova outburst. *See* NOVA; VARIABLE STAR. [M.Pl.]

Binaural sound A sound-reproducing system in which sound is recorded or transmitted by using two microphones mounted at the ears of a dummy human head. To preserve the binaural effect, the sound must be monitored by a listener wearing a set of earphones identically spaced. In an ideal binaural transmission system, both the amplitude and phase of the sound waves incident on the dummy's ears are duplicated at the listener's ears.

Binaural systems have been made so perfect that the listener is unable to distinguish the monitored sound from the real sound. For example, when a person walks around the dummy head, the listener, upon hearing the footsteps, has the compelling illusion of someone walking around him. No other sound system thus far devised can even approximate such an effect.

Binaural sound-reproducing systems may never become popular because of the inconvenience of having to wear a set of earphones during the rendition of a recording to obtain the true binaural effect. *See* SOUND-REPRODUCING SYSTEMS. [H.F.O.]

Binoculars Optical instruments designed for use with both eyes to give enhanced views of distant objects. The distinguishing performance feature of binoculars, compared to monocular instruments, is the depth perception obtainable.

When an object is viewed with both eyes, the distance of approximately 2.5 in. (6.5 cm) between the eyes requires each to view the object from different angles. To bring the images onto the fovea centralis of each eye, the eyes rotate inward until the lines of sight converge at the object. This convergence is the signal to the brain which stimulates the sense of depth, and its magnitude is the parameter from which judgment of depth is made. The minimum difference between convergence angles which the average adult can detect is about 30 seconds of arc, which is approximately the value corresponding to an object at 500 yd (450 m). The convergence angles for more distant objects are less, and therefore objects beyond 500 yd cannot be resolved stereoscopically.

With binoculars this range can be greatly extended. Binoculars employing prism-erecting systems (see illustration)

Modern prism binocular. (*Bausch and Lomb Optical Co.*)

can be so constructed that the distance between objectives is twice that between the eyepieces. Hence, if the convergence angle at an object subtended by the unaided eyes is α, the angle subtended by the binocular objectives is 2α. This angle in the real field of the binocular becomes $2M\alpha$ in the apparent field, M being the magnification of the binocular. Thus the minimum covergence angle in the real field becomes 30 seconds/$2M$, and the maximum range at which objects can be resolved stereoscopically, using, for example, a seven-power binocular, is extended to about 7000 yd (6500 m). [H.E.R.]

Binomial theorem One of the most important algebraic identities, with many applications in a variety of fields. The binomial theorem, discovered by Isaac Newton, expresses the result of multiplying a binomial by itself any number of times:

$$(a + b)^n = a^n + c_1a^{n-1}b + c_2a^{n-2}b^2 + c_3a^{n-3}b^3 + \cdot \cdot \cdot + cra^{n-r}b^r + \cdot \cdot \cdot + b^n$$

where the coefficients $c_1, c_2, c_3, \ldots, c_r, \ldots$ are

$$c_1 = \frac{n}{1} \qquad c_2 = \frac{n(n-1)}{1 \cdot 2} \qquad c_3 = \frac{n(n-1)(n-2)}{1 \cdot 2 \cdot 3}$$

$$c_r = \frac{n(n-1)(n-2) \cdot \cdot \cdot (n-r+1)}{1 \cdot 2 \cdot 3 \cdot \cdot \cdot r}$$

The standard notation for these coefficients is

$$c_1 = \binom{n}{1},\ c_2 = \binom{n}{2}, \ldots,\ c_r = \binom{n}{r}$$

Here $\binom{n}{r}$ is the coefficient of the term containing b_r in the expansion of $(a + b)n$. It is a fraction with the numerator and denominator each containing r factors; those in the denominator begin with 1 and increase by 1; those in the numerator begin with n and decrease by 1. It is easily shown that

$$\binom{n}{4} = \binom{n}{n-r}$$

Under suitable conditions the binomial formula is valid when n is not a positive integer. In this case the formula does not terminate, but generates an infinite series.

Much of the utility of the binomial theorem stems from the properties of the coefficients. In particular, the coefficient $(n \backslash r)$ gives the number of combinations of n distinct objects taken r at a time. The set of coefficients for any value of n forms a distribution that has fundamental importance in the study of probability and statistics. *See* ALGEBRA; DISTRIBUTION (PROBABILITY); SERIES. [H.R.C.]

Bioacoustics Bioacoustics in its broadest sense treats of the role of sound in the behavior of living things. Hearing and speech of human beings constitute a large segment of this discipline. Yet the acoustical properties of the biological media making up living systems of all kinds wholly aside from the ear and vocal mechanisms have considerable importance. The use of sound (particularly at high frequency) in the diagnosis and treatment of disease has necessitated the acquisition of precise knowledge of the acoustical properties of animal tissue. *See* ACOUSTIC MICROSCOPE; ACOUSTICAL HOLOGRAPHY; HEARING (HUMAN); PHONORECEPTION; PHYSIOLOGICAL ACOUSTICS; PSYCHOACOUSTICS.

Since all mammals and many other animal species can generate and detect sound, the study of the mechanisms responsible for these activities and the way they are used in animal behavior has become of increasing importance. Bats, whales, and porpoises emit sounds which on striking obstacles are reflected; the reflected sound can be detected by the emitting animal and used in the location of the obstacles. This is known as echolocation or animal sonar. It is widely used by the animals in question to avoid obstacles in their motion, to locate prey, and to communicate with other animals of the same species. [R.B.L.]

Bioassay A method for quantitatively determining the concentration of a substance (or class of substances) by its effect on the growth of a suitable animal, plant, or microorganism under controlled conditions. The substance estimated may be stimulatory or essential for growth, such as a vitamin or an amino acid, or it may be inhibitory to growth, for example, the antibiotic penicillin. In its application to the estimation of vitamins, amino acids, and trace elements, the method is an outgrowth of studies of the nutritional requirements of animals and microorganisms. Such studies have revealed that nutritional requirements of exacting microorganisms are similar to those of higher animals. [B.M.G.; E.E.S.]

Biocalorimetry The measurement of the energetics of biological processes such as biochemical reactions, association of ligands to biological macromolecules, folding of proteins into their native conformations, phase transitions in biomembranes, and enzymatic reactions, among others. Two types of instruments have been developed to study these processes: differential scanning calorimeters and isothermal titration calorimeters. Differential scanning calorimeters measure the heat capacity at constant pressure of a sample as a continuous function of temperature. Isothermal titration calorimeters measure directly the energetics (through heat effects) associated with biochemical reactions or processes occurring at constant temperatures. In all cases, the uniqueness of calorimetry resides in its capability to measure directly and in a model-independent fashion the heat energy associated with a process. *See* CALORIMETRY; THERMOCHEMISTRY; TITRATION. [E.Fr.]

Biochemical engineering The development, design, operation, control, and analysis of biological and biochemical processes. Modern biochemical engineering began in the 1940s with the development of techniques to produce penicillin on a mass scale. The goods manufactured by means of biochemical engineering are diverse and include health care products (antibiotics, vaccines), foods and beverages (cheese, vinegar, beer, amino acid components), and chemicals and fuels (organic acids, solvents, enzymes, alcohols). Service industries associated with biochemical engineering include waste treatment, ore leaching, and biomedical support systems. With the development of genetic engineering techniques involving the creation of new or modified life forms, other products and processes have become feasible. *See* FOOD ENGINEERING; GENETIC ENGINEERING; INDUSTRIAL MICROBIOLOGY.

The guiding principles of biochemical engineering are similar to those of its parent discipline, chemical engineering. However, additional knowledge of the life sciences is required for the rational application of these principles to process development, equipment design, and scale-up. A primary task in biochemical engineering is the design, operation, and control of bioreactors which optimize the physical rate processes for mass and heat exchange between the biological agents and their environment. The aim is to ensure that the desired intrinsic (inherent) biokinetics of the bioprocess are satisfied or, if this is not possible, to maximize the global (gross) process kinetics. For the process to be economically feasible, it is crucial that conversion yields from the raw materials are adequate and that the conversion rates are fast enough.

There are numerous types of bioreactors. The common classifications are based on various physical, chemical, or biological criteria as follows: (1) Based on the nature of the biological agent, there are microbial bioreactors (fermentors), subdivided according to bacterial, yeast, and fungal cultures; plant or mammalian tissue-culture bioreactors; and cell-free enzyme bioreactors. (2) Based on the physical and physicochemical environmental requirements, there are psychrophilic (low-temperature), mesophilic (moderate-temperature), and thermophilic (high-temperature) fermentations; acid-tolerant and alkali-tolerant fermentations; aerobic (requiring molecular oxygen) and anaerobic bioreactors; liquid and solid-substrate fermentations; and semi-solid or dense-slurry solid-state bioreactors. (3) Based on the type of mixing energy input, there are mechanically stirred tanks; bubble columns; air-lift bioreactors using multiphase pumps for phase contacting; fluidized beds using free suspensions; packed-beds using fixed-film systems; and immobilized biocatalyst systems using inert carriers for attachment or entrapment of cells or enzymes. (4) Based on mode of flow operation, there are chemostats (continuous-flow steady-state conditions); cyclic-batch systems; and repeated fed-batch systems. *See* CHEMICAL ENGINEERING. [M.M.-Y.]

Biochemistry The study of the substances and chemical processes which occur in living organisms. It includes the identification and quantitative determination of the substances, studies of their structure, determination of how they are synthesized and degraded in organisms, and elucidation of their role in the operation of the organism. Some processes of particular interest are the conversion of foods to energy, respiration, the synthesis of nucleic acids and proteins, and the regulation of the chemical activities of cells and organisms.

Carbohydrates are a class of substances which includes simple sugars such as glucose, large polysaccharides such as cellulose, and starch. Carbohydrates are an important source of energy and structural materials in organisms. *See* CARBOHYDRATE; CELLULOSE.

Proteins are intimately involved in all life processes. Some function as biological catalysts called enzymes, while others may act as carriers of material, as in the case of hemoglobin, which transports oxygen in many animals. Some proteins act as structural material, for instance, as portions of cellular membranes and as the major component of hair, horn, skin, and feathers. Contractile proteins are involved in muscular activity and cell division. *See* AMINO ACIDS; ENZYME; MUSCLE PROTEINS; PROTEIN.

There are two classes of nucleic acids, deoxyribonucleic acid (DNA) and ribonucleic acid (RNA). The sequential array of the nucleotides which compose these large molecules constitutes the form in which genetic information is stored and transferred. *See* DEOXYRIBONUCLEIC ACID (DNA); NUCLEIC ACID; RIBONUCLEIC ACID (RNA).

Lipids are a diverse group of "greasy" substances which are soluble in organic solvents but insoluble in water. Cellular membranes are composed mainly of a class of lipids called phospholipids. The most common lipids are the fats. Some other lipids are cholesterol and steroid hormones. *See* LIPID.

A number of minerals are essential to bodily functions, and trace metals have been found to be constituents of certain enzymes. Vitamins and hormones are substances which are biologically effective in small amounts. *See* CALCIUM METABOLISM; COENZYME; ENDOCRINE SYSTEM (VERTEBRATE); HEMOGLOBIN; HORMONE; PHOSPHATE METABOLISM; VITAMIN.

All of these substances are involved in a variety of biochemical processes. Investigations have sought to explain: the chemical steps involved in the biological breakdown of sugars, fats, and amino acids; the relationship of enzyme structure and catalytic activity; genetics in terms of molecular biology; the biochemical bases of disease; the regulation of biological processes; selective permeability, active transport of materials, and transfer of information across biological membranes; and photosynthesis and nitrogen fixation. *See* CARBOHYDRATE METABOLISM; CELL MEMBRANES; DISEASE; GENETIC CODE; LIPID METABOLISM; MOLECULAR BIOLOGY; NITROGEN CYCLE; PHOTOSYNTHESIS; PROTEIN METABOLISM.

The development of biochemistry closely parallels developments in methods of separately and quantitatively measuring small amounts of substances and of determining the structures of substances. *See* ACTIVATION ANALYSIS; AUTORADIOGRAPHY; CHROMATOGRAPHY; GAS CHROMATOGRAPHY; HISTORADIOGRAPHY; MASS SPECTROMETRY; RADIOCHEMISTRY; SPECTROPHOTOMETRIC ANALYSIS; SPECTROSCOPY; ULTRACENTRIFUGE; X-RAY CRYSTALLOGRAPHY. [A.S.L.H.]

Bioelectric model A conceptual model explaining electrical phenomena of living cells and tissues in terms of physical and structural principles. The phenomena include the excitation and electrical responses of nerves and muscles and the discharge of electricity at relatively high voltage by electric fishes such as the electric eel, *Torpedo*. *See* ELECTRIC ORGAN (BIOLOGY).

The first successful model of this type was proposed for nerves by J. Bernstein, on the basis of W. Nernst's theory of diffusion potentials and on two facts about nerves. First, when one end of an excised nerve is crushed or treated with toxic agents, a potential difference develops between the damaged end and the undamaged portion, the damaged end being negative. Second, the magnitude of this "demarcation potential" depends on the concentration of KCl in the medium bathing the nerve. Bernstein proposed that the demarcation potential is a membrane-diffusion potential dependent on a difference in KCl concentration between the interior and exterior of the undamaged nerve; in the damaged portion of the nerve, this concentration difference is abolished.

Neuron theory, and the theory of electrochemical equilibrium developed by Willard Gibbs and applied to membrane systems by F. G. Donnan, led to a modification of the expression of Bernstein's hypothesis. The nerve axon was seen in neuron theory as a cylinder surrounded by a membrane separating the axoplasm or protoplasm of the nerve cell or neuron from the external saline medium. The potential measured on the whole nerve was seen as the expression of potentials in individual axons. The axon membrane was assumed to be permeable only to potassium ions (K^+) and impermeable to the other major ions of the medium and axoplasm. The demarcation potential results from damage to the membrane, with resultant increase in permeability to other ions. The potential difference across the undamaged membrane is seen as a potassium equilibrium potential, dependent upon the concentrations of K^+ in the external medium and in the axoplasm. Calculations on this basis accurately predict observed changes in potential for given changes in external K^+ concentration. *See* BIOPOTENTIALS AND IONIC CURRENTS; DONNAN EQUILIBRIUM; NEURON.

A. L. Hodgkin and A. F. Huxley developed a new model of the axon membrane: The ion pump is regarded as nonelectrogenic; it is assumed that Na^+ is exchanged for K^+ on a 1:1 basis, and hence no net transfer of charge, or current, is produced by the pump. The electromotive force across the neuron membrane remains, as in the Bernstein model, a diffusion potential resulting from differential distribution of K^+. Excitation, normally initiated by partial depolarization of the membrane, results in a sudden increase in permeability to Na^+. The resulting inflow of Na^+, driven by the electric potential and by the difference in concentration of Na^+, carries enough positive charge across the membrane to reverse the membrane potential locally, and this is accompanied by a delayed increase in permeability to K^+. Electrical and concentration forces now drive K^+ out of the axon, carrying positive charge which restores the membrane potential to its former resting level. During the period of recovery, the normal ion balance, locally disturbed, is restored by the ion pump without change in potential. The action potentials of muscle fibers can also be accounted for by minor modifications of the theory.

The classical picture of the plasma membrane of a cell is that of a lipoid bilayer, a layer of fatty materials two molecules thick. This model is consistent with a large body of evidence, but other evidence has supported the mosaic concept, according to which the lipoid bilagen is interrupted by pores, which may contain molecules of protein and fixed electric charges. The ion pump which exchanges Na^+ for K^+ utilizing metabolic energy has been identified with a form of the enzyme adenosinetriphosphatase (ATPase). [B.T.S.]

Bioelectromagnetics The study of the interactions of electromagnetic energy (usually referring to frequencies below those of visible light) with biological systems. This includes both experimental and theoretical approaches to describing and explaining biological effects. Diagnostic and therapeutic uses of electromagnetic fields are also included in bioelectromagnetics.

The induction of cataracts is commonly associated with exposure of animals to intense microwave fields. Although heating the lens of the eye with electromagnetic energy can cause cataracts, the threshold for cataract production is so high that, if the whole animal were exposed to the cataractogenic level of radiation, it would usually die before cataracts were produced.

In 1961 it was reported that people can "hear" pulsed microwaves at very low averaged power densities (50 microwatts/cm^2). It is now generally accepted that the perceived sound is caused by elastic-stress waves which are created by rapid thermal expansion of the tissue that is absorbing microwaves. Most biological effects of microwaves can be explained by the response of the animal to the conversion of electromagnetic energy into thermal energy within the animal. However, a few experiments yield results that are not readily explained by changes of temperature.

There are reports that microwave irradiation at very low intensities can affect behavior, the central nervous system, and the immune system, but many of these reports are controversial. Animals exposed to more intense electromagnetic fields that produce increases in body temperature of 1.8°F (1°C) or higher (thermal load equal to one to two times the animal's basal metabolic rate) demonstrate modification of trained behaviors and exhibit changes in neuroendocrine levels. Exposure of small animals to weaker fields has been shown to produce some changes in the functioning of the central nervous system and the immune system.

In an industrial society, the effects of stationary electric and magnetic fields, and of extremely low-frequency fields are important because of the ubiquitous nature of electricity. Both strong and weak magnetic fields are being studied—the former to determine if they are hazardous to those who work near cyclotrons, magnetohydrodynamic devices, or isotope-separation facilities; the latter to understand how magnetic fields interact with biological systems, and to appreciate the ecological import. Some bacteria swim northward in stationary magnetic fields as weak as 0.1 gauss (the Earth's magnetic field is about 0.5 gauss at its surface). These bacteria contain iron organized into crystals of magnetite. Magnetite is also found in the brains and Harderian glands of birds that can use local variations in the Earth's magnetic field for navigation and orientation.

In medicine, the therapeutic heating of tissue, diathermy, has been used for many years. Shortwave diathermy provides deeper, more uniform heating than does diathermy at higher frequencies. High-intensity radio-frequency fields have been used to produce hyperthermia in cancer patients. The development of bone tissue (osteogenesis) can be stimulated electrically either with implanted electrodes or by inductive coupling through the skin. This therapy has been used successfully to join fractures that have not healed by other means. Electromagnetic fields were first used for medical diagnosis in 1926 when the electrical resistance across the chest cavity was used to diagnose pulmonary edema. Another diagnostic use of electromagnetic radiation is based on the fact that the spectrum of radiation emitted by any object depends on its temperature (blackbody radiation). Because a tumor is often at a higher temperature than the surrounding tissue, its spectrum of emitted radiation is different. A technique of diagnosing tumors by the radiation they emit (radiometry) is being evaluated. Internally generated fields associated with nerve activity (EEG) and with muscle activity (ECG, MCG) are used to monitor normal body functions. There may be other uses of electric currents or fields in growth differentiation or development which have not yet been explored. *See* ELECTROMAGNETIC RADIATION. [E.P.]

Bioelectronics A discipline in which biotechnology and electronics are joined in at least three areas of research and development: biosensors, molecular electronics, and neuronal interfaces. Some workers in the field include so-called biochips and biocomputers in this area of carbon-based information technology. They suggest that biological molecules might be incorporated into self-structuring bioinformatic systems which display novel information processing and pattern recognition capabilities, but these applications—although technically possible—are speculative.

Of the three disciplines—biosensors, molecular electronics, and neuronal interfaces—the most mature is the burgeoning area of biosensors. The term biosensor is used to describe two sometimes very different classes of analytical devices—those that measure biological analytes and those that exploit biological recognition as part of the sensing mechanism—although it is the latter concept which truly captures the spirit of bioelectronics. Molecular electronics is a term coined to describe the exploitation of biological molecules in the fabrication of electronic materials with novel electronic, optical, or magnetic properties. Finally, and more speculatively, bioelectronics incorporates the development of functional neuronal interfaces which permit contiguity between neural tissue and conventional solid-state and computing technology in order to achieve applications such as aural and visual prostheses, the restoration of movement to the paralyzed, and even expansion of the human faculties of memory and intelligence. The common feature of all of this research activity is the close juxtaposition of biologically active molecules, cells, and tissues with conventional electronic systems for advanced applications in analytical science, electronic materials, device fabrication, and neural prostheses. [C.R.Lo.]

Biogeochemical cycles The more or less circular paths of chemical elements passing back and forth between organisms and environment. The rate at which vital elements become available to biological components of the ecosystem is more important in determining primary and secondary productivity than flow of solar energy. If an essential element or compound is in short supply in terms of potential growth, the substance may be said to be a limiting factor. The productivity of an entire ecosystem is sometimes limited by one material available in least amount in terms of need. Thus water limits the desert ecosystem, and nitrogen or phosphorus often limits ocean ecosystems. However, nature has considerable powers of adaptation and compensation. In many environments, species and varieties have evolved which have low requirements for scarce materials. Also, the amount of one substance which, in itself, may not be limiting often greatly affects the requirement for another substance which is approaching a critical minimum. Consequently, it is usually necessary to consider the interaction of the essential materials if the limiting factors operating in a given situation are to be determined. *See* BIOLOGICAL PRODUCTIVITY; ECOSYSTEM.

Dissolved salts essential to life may be conveniently termed biogenic salts or nutrients. They may be divided into two groups, the macronutrients and the micronutrients. The macronutrients include elements and their compounds needed in relatively large quantities, for example, carbon, hydrogen, oxygen, nitrogen, potassium, calcium, phosphorus, and magnesium. The micronutrients include those elements and their compounds necessary for the operation of living systems but which are required only in minute amounts. At least 10 micronutrients are known to be necessary for primary production: iron, manganese, copper, zinc, boron, sodium, molybdenum, chlorine, vanadium, and cobalt. Several others such as iodine are essential for certain heterotrophs.

From the standpoint of the biosphere as a whole, biogeochemical cycles fall into two groups: the gaseous-type cycles, as illustrated by the nitrogen cycle; and the sedimentary-type cycles involving movement of the more earthbound elements. Cycles of oxygen, carbon, and water resemble the cycle of nitrogen, in that a large gaseous pool is important in the continuous flow between inorganic and organic states. In the nitrogen cycle, the nitrogen of protoplasm is broken down from organic to inorganic form by a series of decomposer bacteria. The nitrogen ends up as nitrate of other form usable by green plants in the synthesis of new organic matter. The air is the great reservoir and safety valve of the system. Nitrogen is continually entering the air and continually returning to the cycle. See NITROGEN CYCLE.

Most biogenic substances are more earthbound than nitrogen, and their cycles follow the pattern of erosion, sedimentation, mountain building, and volcanic activity. Biological activity on land and in the upper layers of water results in local cycles from which there is usually a continual loss downhill and replacement from uphill runoff and from solid matter moving in the air as dust, that is, natural fallout. Humans often disrupt the sedimentary cycles by tending to increase the downhill movement. The phosphorus cycle is a good example. Phosphorus is relatively rare in the surface materials of the Earth in terms of biological demand. At present more phosphorus is apparently escaping to the deep ocean sediments (where it is unavailable to producers) than is being replaced by natural processes. For the time being, the considerable reserves of underground phosphate rock can be mined to make up some of this loss, but eventually a means may have to be found to recover phosphorus from the sea.

Biogeochemical cycles involve elements essential to life. The nonessential elements pass back and forth between organisms and environment and many of them are involved in the general sedimentary cycle. Although they have no known value to the organism, many of these elements become concentrated in tissues, apparently because of similarity to specific vital elements. The ecologist would have little interest in most of the nonessential elements were it not for the fact that atomic bombs and nuclear power operations produce radioactive isotopes of some of these elements, which then find their way into the environment and into food chains. Even a rare element, in the form of a radioactive isotope, can be of biological concern because a very small amount of material from a geochemical standpoint can have marked biological effects. Thus the cycling of such things as strontium, cesium, cerium, ruthenium, and many others is of great concern. See BIOSPHERE.

[E.P.O.]

Biogeochemistry The study of the flux and cycling of chemicals between biological and geological systems. Interest in biogeochemistry has increased since the early 1960s largely because humans have significantly altered biogeochemical cycles on a global scale through activities leading to effects such as erosion of landscapes and pollution of the atmosphere.

Materials are moved across the boundaries of an ecosystem in a variety of ways. The continuous flow of energy, water, and chemicals across the boundaries are input and output fluxes and occur via meteorologic, geologic, and biologic vectors. Meteorologic flux consists of windborne particulate matter, dissolved substances in rain and snow, aerosols, and gases (for example, CO_2). Geologic flux includes dissolved and particulate matter transported by surface and subsurface drainage water and the mass movement of matter by gravity. Biologic flux occurs as matter is transported by animals.

All ecosystems are open in the sense that they are dependent upon the flux of chemicals and energy across their boundaries. Ecosystems may be as small as a pool or a field or as large as the entire biosphere. See BIOMASS; ECOSYSTEM; FOOD WEB; MARINE ECOLOGY; TERRESTRIAL ECOSYSTEM.

Organisms have had an effect on the geochemistry of their environment for about 3 billion years. Although details of the conditions and the types of reactions occurring on the Earth for the first 0.5–1.0 billion years after life appeared are not known, a plausible setting can be presented. Prior to the development of oxygen-releasing photosynthesis, the Earth's surface was anoxic, all organisms were anaerobic, and chemical elements existed in their reduced forms. The gradual release of free oxygen permanently changed the Earth's surface from a reducing environment to an oxidizing one, and this change led to the evolution of eukaryotic organisms and of multicellular life-forms. Elements that had previously existed in reduced form were converted to their oxidized state. As a consequence, the biogeochemical cycles of many elements capable of existing in different oxidation states (such as iron, sulfur, manganese, carbon, and phosphorus) were significantly modified. Key biogeochemical cycles include the carbon cycle, nitrogen cycle, phosphorus cycle, and sulfur cycle. See PHOTOSYNTHESIS.

Over 60 elements occur in the tissues of organisms. Of these, over 30 play a vital role in cellular processes. On an atomic basis, hydrogen (H), the lightest element, constitutes almost 50% of the content of organisms in the biosphere, with carbon (C) and oxygen (O) roughly making up the other half. Taken together, these three elements comprise over 99% of all the atoms in the biosphere. Other important constituents, such as nitrogen (N), phosphorus (P), and sulfur (S), total only about 3 atoms per 1000. This composition reflects the dominance of the woody plants, which are rich in the carbohydrate cellulose, a compound of C, H, and O. Nitrogen and sulfur are important constituents of proteins, while phosphorus is responsible for energy transfer within the cell through formation of adenosinetriphosphate (ATP). The six elements H, C, O, N, S, and P, although present in widely varying amounts, are necessary ingredients of living organisms.

Some essential nutrients are present in trace quantities. Important examples are magnesium in the chlorophyll molecule and iron in hemoglobin. Most of the required elements are needed by both plants and animals, but a few are essential only for one or the other. The necessary constituents are generally those of lower atomic number through zinc (element 30), although some with higher atomic numbers such as iodine and molybdenum also perform vital functions. A number of trace elements, for example, cobalt, copper, selenium, and zinc, have a narrow range of concentrations tolerable for metabolic requirements; greater quantities may be toxic. In some parts of the world, animals have been poisoned by grazing on metal-enriched plants, although the high metal contents did not apparently affect the plant growth. See PLANT MINERAL NUTRITION.

Organisms, through their metabolic activities, continually change the environments of the Earth's surface. Microorganisms, because of their large numbers, biochemical versatility, large surface-to-volume ratio, and rapid growth rate, are the most active agents in these influences. Consequently, numerous reactions affecting the cycling of elements through the biosphere are frequently controlled by microbial processes. Organisms also indirectly influence the chemical behavior of an element or species by altering the concentration of other ions or elements such as H^+ or O_2, which in turn affects equilibrium concentrations of the substance of interest. Mineral surfaces may acquire organic coatings from dissolved organic matter, a process which may affect the rate and extent of equilibration of the mineral with the water. See SOIL MICROBIOLOGY; WEATHERING PROCESSES.

Human beings, of course, are part of the biosphere. Some of their activities have an adverse impact on biogeochemical cycles in many ecosystems by the production of toxic or harmful substances. The major types of environmental pollutants are sewage, trace metals, petroleum hydrocarbons, synthetic organic compounds, and gaseous emissions. See ACID RAIN; AIR POLLUTION; ATMOSPHERIC OZONE; EUTROPHICATION; WATER POLLUTION.

[R.M.M; G.E.Li]

Biogeography The science concerned with distribution of life on the Earth. Plants and animals are irregularly distributed, both on the continents and in the oceans. Some areas have a great abundance and variety of life forms, whereas others are relatively sterile. Biogeographic studies are concerned with learning the manner in which living organisms are arranged on the Earth and the causative factors of this arrangement. All animals are dependent in the final analysis on plants to supply food. Therefore, the distribution of plant life is the basic component of biogeography. Animal life is the secondary or dependent component.

Plant geography is concerned with the climatic conditions and the supply of nutrient substances on which plants feed. Stratification of vegetation is controlled largely by conditions of temperature and moisture, which vary from the flat lowlands to the mountaintop. The nature of the underlying rock, from which plant nutrients are derived by weathering, may strongly influence the growth of plants, but the effect is less striking than that of climate. *See* TERRESTRIAL ECOSYSTEM.

Animal life in turn is stratified in conformance with the zonation of plants. There are typical animals of the desert, the foothills, the forest zone, and the alpine crest. Each of these major communities of plants and animals may be called a life zone, a vegetation zone, a biome, or a climax formation according to various systems of classification. All of these communities are part of the temperate region of the world. To the north is the boreal region, which extends generally from the North Pole to the latitude of southern Canada, and south of the temperate region is the tropical region, extending from the Equator roughly to central Mexico. The Southern Hemisphere may be similarly divided into biotic regions, and these in turn into vegetation zones or communities.

The distribution of plants on the surface of the Earth is constantly changing with shifts in climate, with emergence, submergence, and movement of land masses, and with such surface disturbances as volcanic activity or geologic erosion. As environmental conditions change, floras tend to migrate and in so doing some species become extinct and new species evolve. Certain floristic communities have moved long distances in geologic time. The study of plant geography must consider the historic background of plant communities as well as problems of ecologic adaptation to existing conditions. *See* PALEOBOTANY; POPULATION DISPERSAL.

Terrestrial animal populations are distributed largely in accordance with vegetation patterns. The distribution of individual species of animals often conforms rather closely to the distribution of particular vegetation zones. Some animals make seasonal use of different habitats by migrating. Thus many species of migratory birds nest in northern boreal zones in summer but return southward to temperate or even tropical habitats in winter. On a more local scale, some species of both birds and mammals migrate vertically in mountainous terrain, that is, upward in summer and downward in winter. In geologic time, whole faunas have moved up or down the continental land masses in response to major climatic changes or to shifting of the continents. Animal life in the oceans is distributed as irregularly as on land. The temperature of the water and the concentration of mineral nutrients varies with air temperature, ocean currents, and especially with points of upwelling of enriched waters from the ocean depths. Areas rich in oceanic life are situated as a rule in cold currents and along continental shelves. *See* MARINE ECOLOGY.

The distribution of human populations on the Earth bears a close relationship to the distribution of natural plant and animal life. Throughout history, humans have thrived best in areas that are highly productive of foodstuffs. By and large the most favorable habitats for humans are situated in the temperate regions of the world, where climates are moderate and soils are rich.

Unlike most other animals, humans have altered their environment considerably. Through blind destruction of agricultural soils and exploitation of other resources, many past civilizations have become impoverished or have even destroyed themselves. In the last analysis, humans are still dependent upon many natural features of the environment. *See* ZOOGEOGRAPHY.

Perhaps the outstanding aspect of human biogeography today is the explosive increase of population since 1900. Modern medicine has reduced the normal death rate among people and technological improvements have intensified the processes of resource extraction, permitting, for the moment at least, more people than ever before to live upon the Earth.

[A.S.L.]

Bioherm A lenslike to moundlike structure of strictly organic origin. This term involves two concepts: shape and organic internal composition.

The term shape denotes original topographic relief above the sea floor as well as a three-dimensional quality: crudely conical (sugar loaf-shaped) or ellipsoidal (bread loaf–shaped). Such forms are massive or unbedded, their upbuilding resulting from the very rapid rate of accretion of organic carbonate once it starts in a favorable locality. There are size limitations: bioherms a meter or so in diameter are known, and some rise 300 ft (100 m) or more above the sea floor.

The second concept, organic internal composition, not only embraces sessile, bottom-dwelling organisms forming framebuilding reefy bondstone but also includes piles of organically derived debris replete with organisms which encrust it and cement it in place. Even inorganic cement precipitated from marine and meteoric water is known to play a role in a buildup of massive, moundlike structures. When the internal material is coarse and identifiable, no problem is encountered in applying the term bioherm in its original sense. In some buildups, however, an appreciable amount of lime mud is present, and relatively few organisms are identifiable which could have secreted, bound, encrusted, or trapped carbonate mud (as in some early Carboniferous mounds). In such cases, problems arise in applying that part of the definition based on internal composition. In fact, as a field term, bioherm can hardly ever be completely diagnostic because careful petrographic study is commonly necessary for details of internal composition to be ascertained.

Bioherms may occur on shelves (where they are normally lens-shaped) or in shallow basins, often at the basin margin. In the latter position, they have been called reef knolls or pinnacle reef. *See* BIOSTROME; REEF; STROMATOLITE. [J.L.Wi.]

Bioinorganic chemistry The borderline field between biochemistry and inorganic chemistry. This field, which has also been called inorganic biochemistry, generally involves the application of the principles of inorganic chemistry to biochemical problems. Traditionally biochemistry has been considered "organic," and in fact much of biochemistry involves organic, or carbon, chemistry. It has been known for a long time, however, that living organisms require many other elements, some such as calcium in large quantities, and others such as copper in trace amounts. Most of these elements are metals, though some are nonmetallic. These "inorganic" elements are extremely important in the regulation of many biological phenomena, and it is for that reason that inorganic chemistry and biochemistry have intermingled.

The major achievements of bioinorganic chemistry have consisted of the elucidation of the structures of metal-containing biochemical compounds, and the correlation of structure and function of these compounds. Generally they are complexes of metal ions with organic ligands, and therefore coordination compounds. Much of bioinorganic chemistry can be considered an application of coordination chemistry to biological problems. *See* COORDINATION CHEMISTRY; ION; ORGANOMETALLIC COMPOUND.

Some of the coordination compounds found in living organisms are of relatively low molecular weight; that is, they contain relatively few atoms in comparison to the very large molecules common to living cells. Examples of these small molecules are siderochromes (iron-accumulating and -transporting molecules found in bacteria and fungi), ionophores (substances important in the transport of alkali and alkaline-earth metal ions across biological barriers), porphyrins, chlorophyll, and vitamin B_{12}. *See* CHLOROPHYLL; COENZYME; ENZYME; PHOTOSYNTHESIS; PORPHYRIN; VITAMIN B_{12}.

The large molecules common to living cells, such as proteins, nucleic acids, and lipids, also bind to metal ions, and in many instances the metal ions are important in the biological functioning of the molecules. They include metalloproteins (metal complexes of proteins) and metal enzymes (metalloproteins in the form of enzymes). *See* CATALYSIS; COORDINATION COMPLEXES.

Many of the bioinorganic substances are biological catalysts, transport agents, and regulators. Metal ions are involved in many biological phenomena such as kidney function, muscle contraction, cell division, and nerve impulse transmission, but little is understood about the structures of the molecules to which the metal ions are bound during these processes.

Metal ions are important to the structure not only of individual biological molecules, but also of cell particulates that are molecular aggregates. They are constituents of chromatin, the genetic material in the cell nucleus that is composed of nucleic acids and proteins. Metal ions influence the structure of ribosomes, the cellular particles on which protein synthesis occurs. The structures of these and other cellular particles are strongly affected by metal ions, and unfavorable concentrations of metal ions can distort these structures. Calcium is very important in the structure of bones and teeth. Although metal ions in limited concentration are required for the proper functioning of many biological processes, excesses of metal ions or the presence of undesirable metal ions can produce effects which are deleterious.

Metal complexing agents have been useful for the elimination of toxic metals from the body and in the treatment of diseases characterized by an excess of certain metal ions. Examples of such diseases are Wilson's disease, which is associated with the accumulation of copper, and hemochromatosis, which results from an overload of iron. The toxic effects of metal ions point to relationships between bioinorganic chemistry, toxicology, and studies on environmental effects on the aging process, which is accompanied by changes in the concentration of metal ions in tissues. The biological requirement of inorganic elements, as well as the detrimental effects of excesses of these elements, suggests a relationship to nutrition.

Not all of bioinorganic chemistry is intended to lead to an understanding of the participation of inorganic elements in biological phenomena. Sometimes chemists make use of the reactions of metal ions or metal complexes with large biological molecules to determine the structures of these molecules. The structure of many large biological molecules is elucidated by x-ray diffraction studies that require the formation of metal complexes with the biomacromolecules by a method called isomorphic replacement. *See* BIOCHEMISTRY; GENETIC CODE; INORGANIC CHEMISTRY; ORGANIC CHEMISTRY. [G.L.E.]

Bioleaching The dissolution of metals from their mineral source by certain naturally occurring microorganisms. Dozens of bacterial species have been identified as having bioleaching capabilities; those of commercial interest include species of *Thiobacillus*, *Leptospirillum*, and *Sulfolobus*. *Thiobacillus ferrooxidans* is by far the most widely studied and commercially useful species. It is an aerobic rod-shaped microorganism that derives its energy from the oxidation of various sulfide minerals and soluble ferrous ion (Fe^{2+}). The use of this organism is of considerable interest to the mining industry since many important metals are extracted and refined from sulfide minerals. *Thiobacillus ferrooxidans* has certain other characteristics that makes it a unique microorganism. It thrives in acidic environments of pH 1–3, conditions that would be fatal to most other life forms, and can rapidly oxidize many sulfide minerals. This is of great commercial importance since most sulfide minerals are inert and oxidize very slowly at room temperature and pressure. Also important is the fact that *T. ferrooxidans* can be adapted to withstand high concentrations of dissolved metals. This ability makes it extremely useful in solution mining and other metal recovery operations. *See* SOLUTION MINING.

A typical bioleach reaction, using pyrite as the mineral source, is shown below; here pyrite is oxidized to produce

$$2FeS_2 + {}^{15}\!/_{2}O_2 \rightarrow Fe_2(SO_4)_3 + H_2SO_4$$

soluble ferric sulfate [$Fe_2(SO_4)_3$] and sulfuric acid (H_2SO_4). A considerable amount of oxygen is required in this reaction; it is extracted from the air.

Commercial applications of bioleaching have been developed for the solution mining of copper and uranium from low-grade ores and waste products. Uranium minerals are often found associated with the mineral pyrite. *Thiobacillus ferrooxidans* is used to oxidize pyrite and release the uranium according to the reaction discussed above. The ferric sulfate and sulfuric acid generated in this reaction then dissolve the uranium. *See* URANIUM METALLURGY.

The use of *T. ferrooxidans* to enhance gold and silver recovery from ores that are difficult to treat is considered a major development in bioleaching technology. Bioleaching has gained acceptance because it offers a potentially inexpensive and nonpolluting way to pretreat these ores. The process consists essentially of two stages. In the first step, the ore is bioleached at acidic pH to oxidize the sulfide minerals. In the second step, the oxidized gold-bearing solids are mixed with lime to raise the pH to 10–11, and then subjected to conventional cyanide leaching to dissolve the precious metals. Often, overall gold or silver recovery can be improved to over 90%. *See* GOLD METALLURGY; HYDROMETALLURGY; LEACHING; SILVER METALLURGY; SOLVENT EXTRACTION. [R.Hac.]

Biological clocks Physiological and cellular mechanisms by which organisms measure time. Even the most stable ecological niche shows significant temporal fluctuations in environmental variables such as light and temperatures, which may result in major differences in environmental conditions. Some of these environmental fluctuations are somewhat random and unpredictable, but others show a remarkable degree of regularity. The regular fluctuations, or oscillations, in environmental conditions result from specific planetary cycles, such as the daily rotation of the Earth upon its axis, the monthly revolution of the Moon around the Earth, and the annual revolution of the Earth around the Sun. These daily, tidal, lunar, and annual cycles result in highly predictable and recurrent changes in the Earth's environment; organisms that can modify their internal physiology and behavior in a manner consistent with these changes (that is, that can measure time) are at an advantage.

Although biological rhythmicity has been known and documented for more than a century, it has only been since about 1960 that biologists have recognized the ubiquity of this phenomenon in nearly all groups of plants, animals, and microorganisms. Examples of biological rhythmicity that correspond with each of these environmental cycles have been thoroughly documented: Daily rhythms of sleep and wakefulness in humans, tidal rhythms of plankton migration, lunar rhythms of reproductive behavior in invertebrates, and annual rhythms of flowering in plants and reproduction in animals are but a few of the many biological rhythms corresponding to environmental periodicities which have been studied in depth during recent decades.

One of the most important conclusions from these studies is

Examples of circadian rhythms in various organisms	
Organism	Rhythm
Microorganisms	
Gonyaulax	Bioluminescence, cell division, photosynthesis
Euglena	Photoaxis, cell division, photosynthesis
Paramecium	Mating reactivity, cell division
Neurspora	Conidiation, CO^2 evolution, DNA synthesis, RNA synthesis
Invertebrates	
Cockroach	Locomotor activity
Fruit fly	Adult emergence (eclosion), locomotor activity
Aplysia	Spontaneous neural activity of the optic nerve, locomotor activity
Vertebrates	
Birds and mammals	Locomotory activity, feeding
Human	Sleep, body temperature, drug sensitivity, urine excretion, physical or mental performance

that many biological rhythms are controlled by an internal biological clock, whose periodicity closely matches that of the environmental periodicity and allows the biological rhythm to be coupled to the environment cycle in a manner beneficial to the organism. This internal oscillator has been called a biological clock, or more specifically, a circadian clock. The term circadian comes from the Latin *circa* and *diem*, meaning "about a day," and emphasizes that the biological periodicity in constant laboratory conditions is only approximately equal to 24 h. The use of the term clock signifies that this oscillator provides a timekeeping function for the organism which allows it to determine the time of day at any given moment. Biological clocks that control daily rhythms have been the most thoroughly studied because of the time factors involved in carrying out experiments, and therefore knowledge of these daily clocks is much more extensive than that for other periodicities.

Circadian rhythms have been observed in nearly every major group of organisms except bacteria and blue-green algae (see table). Of particular interest is the fact that in humans a vast number of functions at the biochemical, physiological, and behavioral levels have been observed. *See* MIGRATORY BEHAVIOR; PHOTOPERIODISM; PINEAL BODY; PLANT MOVEMENTS; REPRODUCTIVE BEHAVIOR; SLEEP AND DREAMING. [J.F.]

Biological equilibrium A lifeless object is said to be in equilibrium, or in a state of balance, when all forces acting upon it cancel. This results in a state of rest. In an actively moving animal, internal as well as external forces have to be considered, and the maintenance of a balanced attitude in a body consisting of a number of parts that are loosely connected by movable joints is a complex matter.

The maintenance of equilibrium is relatively easy in limbless animals, such as worms, snails, or starfish, which rest and move on an extensive surface of support. The tactile contact between the normal creeping surface and the substrate furnishes sensory information on the animal's orientation with respect to the plumb line. When such an animal is turned on its back, the lack of contact pressure on the creeping surface and the tactile stimulation of the back initiate muscular movements which return the animal to its normal position. This is known as the righting reflex. The accompanying change in the direction of illumination may also contribute to efficient righting.

Free-swimming and flying animals are often in a precariously poised state of equilibrium, so that the normal attitude can be maintained only by the continuous operation of corrective equilibrating mechanisms. The same applies to a somewhat lesser degree to long-legged quadrupeds, such as many mammals, and to bipeds, such as birds and some primates (including humans). In these animals the center of gravity lies some distance above the area of support which, in some cases, such as in humans, can be relatively small.

The weight of the body is supported by what are known as antigravity muscles, and the coordination of their activity is controlled by sensory information from a number of sources: (1) the eyes; (2) static and dynamic mechanoreceptors incorporated in the various types of statocysts in invertebrate animals and in the vestibular organ or labyrinth of the vertebrates; (3) proprioceptor organs such as muscle spindles, Golgi endings in tendons, Pacinian corpuscles and similar encapsulated endings associated with tendons and joints, and other pressure receptors in supporting surfaces (for example, the soles of feet); and (4) sensory endings in the viscera, capable of being differentially stimulated by changes in the direction of visceral pull on mesenteries, and other structures. All these receptor organs supply the central nervous system with information on the orientation of head, trunk, and limbs in the gravitational field. [O.E.L.]

Biological productivity The amount and rate of production which occur in a given ecosystem over a given time period. It may apply to a single organism, a population, or entire communities and ecosystems. *See* BIOMASS.

Productivity is best considered both in terms of dry matter produced (net production) and in terms of the thermodynamic cost of producing it (gross production, being respiration and heat losses plus net production). The following definitions are useful in calculating production: (1) Gross primary production (GPP) is the total energy fixed by photosynthesis. (2) Net primary production (NPP) is the gross production less losses due to plant respiration. (3) Secondary production is production by heterotrophs (animals, microorganisms), which feed on plant products or other heterotrophs. Note that human-associated terms, such as harvest or yield, refer to just that fraction of production of use to the plant harvester or animal husbander. Thus yield of a grain crop refers solely to the harvested seed portion of the aboveground net primary production. All energy units are presented in kilojoules (kJ); the outmoded, but frequently used, kilocalorie of nutritionists is calculated by kJ/0.2388. *See* ECOSYSTEM; PHOTOSYNTHESIS. [D.C.C.]

Biological specificity Although all living cells contain hundreds of potentially reactive chemical substances, their metabolic and developmental reactions do not occur at random, but follow orderly patterns that give rise to the unique characteristics of the individual and of its species. The guiding principle is termed biological specificity; examples abound in all areas of biology.

Specificity may originate, first, at a chemical level. A most important example is the structural correspondence (presumably in an inverse, complementary sense) between an enzyme and the substrates whose reaction it catalyzes. Analogous relations govern the specific reactions between antigens and antibodies. Again, the specificity of fertilization of an egg by sperm or pollen is universally important in preserving the unique characteristics of animal and plant species, although occasional hybrids can be produced, as exceptions. *See* ANTIGEN; ENZYME; SPECIATION.

In addition to chemical determinants, the specificity of reproduction may also be governed by anatomical or behavioral characteristics of the plant or animal which minimize cross matings between separate species. The relations between two or more widely divergent species may also be precisely ordered, as in parasitism. Some associations may not be unfavorable to either partner. The fertilizations of Smyrna figs, yuccas, and certain orchids require particular insects for pollen transfer. The association of algae and fungi in lichens is also a nonrandom one. *See* LICHENES; PARASITOLOGY. [H.P.T.]

Biologicals Biological products used to induce immunity to various infectious diseases or noxious substances of biological origin. The term is therefore much more restrictive than the words themselves imply, because it is usually limited to immune serums, antitoxins, vaccines, and toxoids that have the effect of providing protective substances of the same general nature that a person develops naturally from having survived an infectious disease or having experienced repeated contact with a biological poison. *See* IMMUNITY.

One major class of biologicals includes the animal and human immune serums. All animals, including humans, develop protective substances (antibodies) in their blood plasma during recovery from many (but not all) infectious diseases or following the injection of toxins or killed bacteria and viruses. These antibodies usually are found in the gamma globulin fraction of the plasma and are specific in the sense that they react with and neutralize only substances identical or closely similar to those that caused them to be formed. *See* SERUM.

Antibody-containing serum from another animal is useful in the treatment, modification, or prevention of certain diseases of humans when it is given by injection. The use of these preformed "borrowed" antibodies is called passive immunization, to distinguish it from active immunization, in which people develop their own antibodies. Serums which contain antibodies active chiefly in destroying the infecting virus or bacterium are usually called antiserums or immune serums; those containing antibodies capable of neutralizing the secreted toxins of bacteria are called antitoxins. Immune serums have been prepared to neutralize the venoms of certain poisonous snakes and black widow spiders; they are called antivenins.

Products used to produce active immunity constitute the other large class of biological products. They contain the actual toxins, viruses, or bacteria that cause disease, but they are modified in a manner to make them safe to administer. Because the body does not distinguish between the natural toxin or infectious agent and the same material when properly modified, immunity is produced in response to injections of these materials in a manner very similar to that which occurs during the natural disease. Vaccines are suspensions of the killed or attenuated (weakened) bacteria or viruses or fractions thereof. Toxoids are solutions of the chemically altered specific bacterial toxins which cause the major damage produced by bacterial infections. Biological products producing active immunity are usually named to indicate the disease they immunize against and the kind of substance they contain: thus typhoid vaccine, diphtheria toxoid, tetanus toxoid, measles vaccine, mumps vaccine, and poliomyelitis vaccine.

Because it is frequently expedient to immunize people simultaneously against several diseases, various immunizing agents have been combined so that they can be given in a single series of two or three doses. The combination of a large number of different immunizing agents within a single immunization dosage is an exceedingly complex problem, because dosage volume must be kept small and the various agents must be compatible.

In addition to products for active and passive immunization, a few other substances of human or animal origin are also classed (mostly for regulatory convenience) as biological products. One group of these is normal human blood plasma and its derivatives. Whole plasma and serum albumin are used to combat hemorrhage and shock, and other fractions are used in the treatment of hemophilia and for their clotting properties. The use of anti-Rh_O, (D) gamma globulin in Rh-negative women at parturition inhibits Rh sensitization and erythroblastosis fetalis in subsequent offspring. *See* BLOOD GROUPS; RH INCOMPATIBILITY.

Another group of biological products consists of a miscellany of reagents used in the diagnosis of infectious diseases. These include immune serums for the serological typing and identification of pathogenic bacteria and viruses, and various antigens for the detection of antibodies in patients' serums as a means of assisting in diagnosis. In addition to these substances used in the laboratory diagnosis of disease, certain extracts of bacteria and fungi have been prepared for injection into the skin to detect allergic hypersensitivity. These skin tests also assist in the diagnosis of various infectious diseases such as tuberculosis and brucellosis. *See* SKIN TEST.

Allergenic products are administered to humans for the diagnosis, prevention, or treatment of allergies to various plant and animal substances. They usually are suspensions of the allergenic substance (pollens, feathers, hairs, or danders) in 50% glycerin. *See* IMMUNOLOGY. [L.F.S.]

Biology A division of the natural sciences dealing with life. Biology is the science of living organisms, in contrast to the physical sciences, which are concerned with inanimate matter. Biology is the broad general field of knowledge concerned with the study of all aspects of living organisms which can be approached by the methods of natural science. Because the field is so broad in scope, it has been divided into two main areas, botany and zoology, each with a large number of specialized branches. No one worker can know the entire field of biology thoroughly. Specialized subdivisions of biology pertaining to both plants and animals include taxonomy, morphology, embryology, physiology, cell biology, genetics, evolution, and ecology. The term biology embraces those principles of widest application to the origin, growth and development, structure, function, evolution, and distribution of plants and animals. *See* BOTANY; CELL BIOLOGY; ECOLOGY; EMBRYOLOGY; GENETICS; MICROBIOLOGY; TAXONOMY; ZOOLOGY. [C.B.C.]

Bioluminescence The emission of light by living organisms, due to an energy-yielding chemical reaction in which a specific biochemical substance, called luciferin, undergoes oxidation, catalyzed by a specific enzyme called luciferase. There are many specific luciferins and luciferases which are chemically different, each involved in some different living luminescent organism.

Bioluminescence does not come from, or depend on, light absorbed by the organism. It is a type of chemiluminescent reaction, in which chemical energy is transformed into light energy. Such reactions represent an unusual kind of chemical reaction, but are not peculiar to living organisms. Bioluminescence is a special type of chemiluminescence because the chemicals are synthesized by living cells and because they are catalyzed by an enzyme, a unique kind of biological catalyst. Since bioluminescence is a type of chemiluminescence, it is not necessary to have a live organism to obtain light emission. The simple preservation of the chemicals involved will suffice. This can be done in some cases by rapidly drying the organism under mild conditions. *See* CHEMILUMINESCENCE.

There is a great diversity of organisms which have the ability to give off light, and they do it in many different ways and for many different reasons. In the plant kingdom only certain mushrooms and some marine algae and bacteria are luminous. The bacteria and mushrooms emit continuously, whereas the algae, which are dinoflagellates (the same group that causes "red tides" in the ocean), give a brief flash when disturbed. Among the animals are many luminous representatives, ranging from the simplest protozoans to fish. Of the approximately 80 different classes of animals, about 30 have light-emitting members. These are mostly marine forms, and include sponges and various jellyfish and related animals, such as comb jellies (sea walnuts). Most of the luminescent annelid worms are marine polychaetes, but a few common terrestrial earthworms possess luminous cells located in the body cavity. Many kinds of mollusks emit light, including clams, snails, and octopuses.

Among arthropods are many luminescent organisms, such as crustaceans (shrimp), millipedes, centipedes, and numerous insects, notably the lampyrid beetles, or fireflies. There are also luminous representations among the brittle stars, acorn worms, and tunicates.

Virtually all of the known or proposed ways in which bioluminescence functions can be classed under three major rubrics: assisting in predation, aiding in escaping predators, and communicating. The first two are interspecific, involving interactions between different species, and the last is intraspecific, that is, involving different individuals of one species.

It seems clear that, unlike many key biochemical molecular species—the hemes, the cytochromes, the histones, or the nucleic acids—luminescent systems have originated and evolved independently more than once. The elegance of the display of luminescence is matched by its diversity—diversity in form, diversity in mode of display, diversity in the chemistry, and diversity in function. In fact, the impressive number of different species possessing bioluminescence, and the various uses for it, is an object lesson on and testimonial to the evolutionary process. [J.W.H.]

Biomagnetism The production of a magnetic field by a living object. The living object presently most studied is the human body, for two purposes: to find new techniques for medical diagnosis, and to gain information about normal physiology. Smaller organisms studied include birds, fishes, and objects as small as bacteria; many scientists believe that biomagnetics is involved in the ability of these creatures to navigate. The body produces magnetic fields in two main ways: by electric currents and by ferromagnetic particles. The electric currents are the ion currents generated by the muscles, nerves, and other organs. For example, the same ion current generated by heart muscle, which provides the basis for the electrocardiogram, also produces a magnetic field over the chest; and the same ion current generated by the brain, which provides the basis for the electroencephalogram, also produces a magnetic field over the head. Ferromagnetic particles are insoluble contaminants of the body; the most important of these are the ferromagnetic dust particles in the lungs, which are primarily Fe_3O_4 (magnetite). Magnetic fields can give information about the internal organs not otherwise available.

These magnetic fields are very weak, usually in the range of 10^{-14} to 10^{-9} tesla; for comparison, the Earth's field is about 10^{-4} T (1 T = 10^4 gauss, the older unit of field). The fields at the upper end of this range, say stronger than 10^{-4} T, can be measured with a simple but sensitive magnetometer called the fluxgate; the weaker fields are measured with the extremely sensitive cryogenic magnetometer called the SQUID (superconducting quantum interference device). The levels of the body's fields, whether they are fluctuating or steady, are orders of magnitude weaker than the fluctuating or steady background fields. They can, however, be measured by using either a magnetically shielded room or two detectors connected in opposition so that much of the background is canceled, or a combination of both methods. The organs producing magnetic fields which are of most interest are the brain, the lungs, and the liver. *See* BIOELECTROMAGNETICS; ELECTROENCEPHALOGRAPHY; MIGRATORY BEHAVIOR; SQUID. [D.C.]

Biomass The mass of living material present in an organism or organisms. Biomass may include some nonliving material, such as the hair and feathers of vertebrates or the heartwood of trees. The intent of the term is to provide a useful quantitative measure of the organism and not to make a sharp distinction between life and nonlife. It is typically referred to or calculated as dry weight. In a model, biomass can be a state variable, into and out of which several flows occur. For example, a heterotrophic organism, whether it is a bacterium, bird, or buffalo, ingests food or nutrients and in turn gives off respiratory gases (carbon dioxide) and waste products, and (at times) makes new body tissue (see illustration). Growth is an increase, and starvation a decrease, in biomass. *See* BIOLOGICAL PRODUCTIVITY.

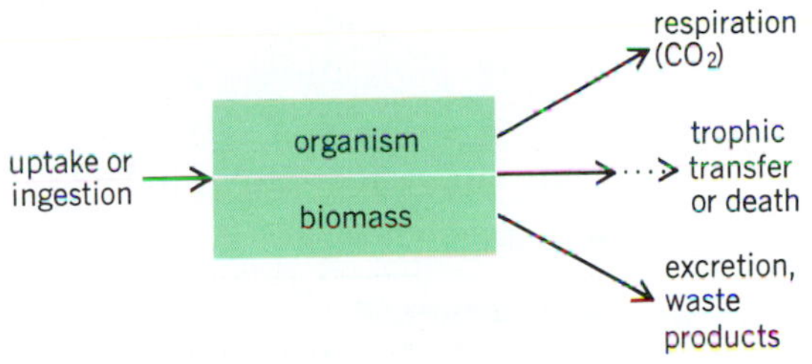

Biomass as a state variable, with flows of energy and nutrients into and out of it.

Due to extensive losses of energy at each step in a food chain, going from primary producers on up the chain, there is a change in the amount of energy flowing to new biomass at each step in the chain. The change in amount of standing crop at each step is often represented as a horizontal bar. With the size of the bar usually decreasing as one ascends the food chain, the shape of the histograms assumes a pyramidal form. This leads to the term biomass pyramid. The terminology has gone out of favor, since there are many instances where one finds small standing crops (as with oceanic phytoplankton) and high turnover rates, so that one can describe inverted pyramids as well. *See* FOOD WEB. [D.C.C.]

Biome A complex biotic community covering a large geographic area and characterized by the distinctive life forms of important climax species of plants and animals. A life form is the common morphological features that characterize a group of organisms. On the land, a biome is identified by the life form of the dominant climax plants, as well as by the distinctive types of vegetation and landscape in which they grow. In the ocean, the life forms of the predominant animals serve as the criterion. The biome incorporates all seral (successional) as well as climax stages of the community, and its distribution is controlled by climate or, in the ocean, by other physical factors of the environment. Principal terrestrial biomes are the temperate deciduous forests, coniferous forest, woodland, chaparral, tundra, grassland, desert, tropical savanna, and tropical broadleaved forest. Principal marine biomes are the pelagic, the barnacle–gastropod–brown algae, the sea urchin–large snail, the bivalve-annelid, the coral reef, and the abyssal-benthic.

Terrestrial biomes are divisible into plant associations, which are distinguished by distinctive combinations of climax dominant species. For instance, the temperate deciduous forest biome includes the beech-maple, oak-hickory, and other associations. Both terrestrial and marine biomes are divisible into animal biociations. A biociation is distinguished by the distinctiveness of the predominant animal species. *See* ECOLOGY. [S.C.K.]

Biomechanics The study of the mechanics of living things. It demands knowledge both of biology and of the various branches of physics and engineering which comprise mechanics.

A major branch of biomechanics is concerned with the properties of biological materials. Particular attention has been paid to bone and blood, because of their medical importance. The aims are to measure the properties and then to explain them in molecular terms. Proteins and polysaccharides are high polymers, like rubber and plastics, and biologists studying their properties have made extensive use of theories and techniques developed by rubber and plastics technologists. Most skeletal materials are composite: bone, for instance, consists of tiny crystals of a calcium salt embedded in a mass of fibers of collagen (the material of which tendons are made). Bone is stronger than tendon and has a higher Young's modulus; that is, it is stiffer (see illustration). That a mass of tiny, brittle crystals

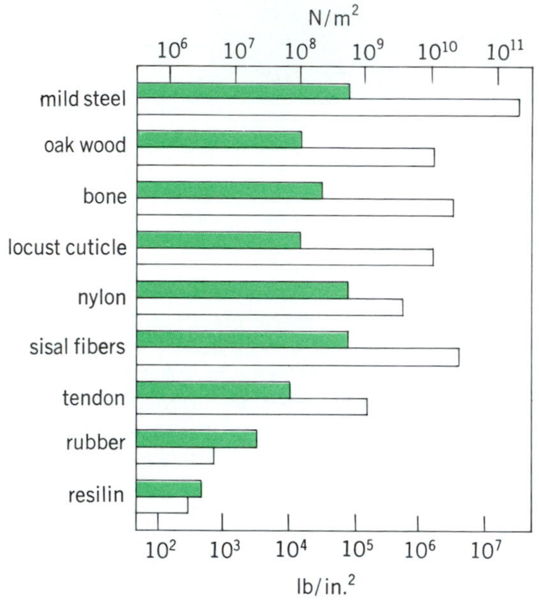

Mechanical properties of some biological materials compared with nylon, steel, and rubber. The top bar of each pair indicates tensile strength, and the lower one indicates Young's modulus of elasticity.

strengthens bone may seem strange; their effect has not been fully explained, but it seems to be comparable to the effect of carbon particles on rubber. Soft rubber is converted to the much stiffer, stronger material of which motor tires are made by incorporating carbon particles in it. *See* YOUNG'S MODULUS.

Other biomedical studies are concerned with surface tension. It is well known that the pressure is greater inside a bubble than immediately outside it, because of surface tension in the wall of the bubble, and that the smaller the bubble, the greater the pressure. The inside of a lung consists largely of very tiny pockets known as alveoli. If the moisture on the walls of the alveoli had a surface tension similar to that of water or of blood plasma, it would have a substantial effect on the pressure in the lung. As one breathed out and the alveoli got smaller, the pressure needed to resist the surface tension and keep the alveoli open would increase, and the alveoli would tend to collapse completely. A strong effort would be needed to inflate them for the next breath. This does not happen, because there is a film of surface-active material on the moist surfaces which makes the surface tension fall to a very low value during expiration. Infants with the respiratory distress syndrome, a dangerous ailment of newborn babies, do not have enough of this material, and are liable to die of exhaustion from the effort of breathing. *See* LUNG; SURFACE TENSION; SURFACTANT.

Aerodynamics, a branch of aeromechanics, has been used in a wide variety of biological investigations, including the study of the falling speeds of winged seeds and the flight of hummingbirds. *See* AERODYNAMICS; BIOPHYSICS. [R.McN.A.]

Biomedical engineering An engineering discipline whose objective is to apply engineering science and technology to devise practical solutions to medical problems. In the broadest sense, biomedical engineering embraces all aspects of the application of technology to living systems. It is an emerging field, and the disciplinary boundaries of biomedical engineering are as yet not well defined. It is distinct from, but closely related to, space medicine, sports medicine, nuclear medicine, biochemistry, and biophysics. *See* AEROSPACE MEDICINE; BIOCHEMISTRY; BIOPHYSICS; NUCLEAR MEDICINE.

Biomedical engineering has had a major impact on all aspects of medicine including: monitoring of a patient's condition and diagnosis of disorders; surgery or treatment to correct or alleviate disorders; rehabilitation to help the disabled patient to participate more fully in society; and fundamental research to gain understanding of the nature and working of living systems.

The first task for a medical practitioner is to determine the current state of health of the patient and to identify or diagnose a malfunction when it occurs. Modern medicine has at its command a number of technological devices to facilitate this task, including those of radiology, ultrasound imaging, electrocardiography, and electroencephalography. *See* ELECTROENCEPHALOGRAPHY; RADIOLOGY; ULTRASONICS.

Biomedical engineering has provided an ever-growing armory of surgical tools and treatment techniques, including ultrasound focusing, iontophoresis, the vibratory saw, and functional electrical stimulation. In addition, biomedical engineering provides an array of implantable devices and artificial organs to assist or replace defective or malfunctioning natural organs. In this area the major problem is biocompatibility. Despite near-insurmountable difficulties, many implantable devices are in common use. The crippling degeneration of the joints which is characteristic of arthritis is commonly treated by replacing the malfunctioning joint by an artificial joint. The development of an implantable pump to replace a failing heart is a major focus of research in biomedical engineering. Internal artificial pacemakers which control the rate at which the heart beats are commonly used. *See* PROSTHESIS.

The major goal of rehabilitation engineering is to make the disabled person as independent as possible by providing assistive devices to alleviate the handicap and compensate for the disability. The toughest problem in rehabilitation engineering is to design the human/machine interface, the means whereby the patient and the assistive device interact with one another. The interface between a person and a machine is treated as two communication pathways: the forward pathway which includes all ways the user commands the machine, and the feedback pathway which includes all ways the user observes the machine. The interface must be designed such that the interactions between the patient and the assistive device will be intimate and unobtrusive. Communication must be achieved without placing further burdens on the patient's limited sensory or motor capabilities, and without drawing unwelcome attention to the disability.

The greatest single impact which biomedical engineering has had on medicine and biology is in the understanding of the underlying phenomena. This is largely due to the fresh perspective which the engineering discipline brings to medical problems. The applicable fundamentals of engineering can be broadly categorized into: (1) observation of fundamental phenomena, making quantitative measurements whenever possible; (2) characterization of phenomena and development of the simplest workable description or mathematical model of the phenomena; (3) quantitative prediction of the consequences of a specific action; (4) design, the creative integration of 1, 2, and 3 to yield new tools and techniques to accomplish a particular purpose. Of these, the development of models shows the value of the engineer's practical-mindedness most clearly. A model is a partial description abstracted from the basic observations.

An engineering background allows the biomedical engineer not only to provide new interpretations of existing data but to devise the correct instrumentation or experimental procedures to obtain the critical data required for the solution of a particular problem. For example, from the engineer's functional standpoint osteoarthritis is the degeneration of a bearing (the interface between two parts in relative motion). The biomedical engineer can draw on the large body of prior knowledge concerning the stresses on a bearing and the modes of failure of a bearing to postulate the causes of osteoarthritis and suggest possible ways to alleviate the condition. Furthermore, engineering experience indicates that the single most informative piece of data about the bearing will be the distribution of pressure (force per unit area) across its surface. Some of the most promising research into the cause of osteoarthritis is being performed by biomedical engineers whose primary objective is to measure the pressure distribution across a living joint. [N.H.]

Bioorganic chemistry The science that describes the structure, interactions, and reactions of organic compounds of biological significance at the molecular level. It represents the meeting of biochemistry, as attempts are made to describe the structure and physiology of organisms on an ever smaller scale, with organic chemistry, as attempts are made to synthesize and understand the behavior of molecules of ever-increasing size and complexity. Areas of research include enzymatic catalysis, the structure and folding of proteins, the structure and function of biological membranes, the chemistry of poly(ribonucleic acids) and poly(deoxyribonucleic acids), biosynthetic pathways, immunology, and mechanisms of drug action. *See* DEOXYRIBONUCLEIC ACID (DNA).

Being at the interface of two disciplines, bioorganic chemistry utilizes experimental techniques and theoretical concepts drawn from both. Important experimental techniques include organic synthesis, kinetics, structure-activity relationships, the use of model systems, methods of protein purification and manipulation, genetic mutation, cloning and overexpression (engineered enhancement of gene transcription), and the elicitation of monoclonal antibodies. Theoretical concepts important to bioorganic chemistry include thermodynamics, transition-state theory, acid-base theory, concepts of hydrophobicity and hydrophilicity, theories of stereocontrol, and theories of adaptation of organisms to selective pressures.

Historically, a major focus of bioorganic research has been the study of catalysis by enzymes. Enzymes have a dramatic ability to increase the rates at which reactions occur. One of the ways that enzymes increase the rates of bimolecular reactions is to overcome the entropic barrier associated with bringing two particles together to form one. *See* ENZYME.

Enzymes also catalyze reactions by facilitating proton transfers. Many of the reactions catalyzed by enzymes, such as the formation and hydrolysis of esters and amides, require the deprotonation of a nucleophile (base catalysis) or the protonation of an electrophile (acid catalysis).

Enzymes may act as preorganized solvation shells for transition states. If the active site of an enzyme has just the right size, shape, and arrangement of functional groups to bind a transition state, it will automatically bind the reactants less well. Selective binding of the transition state lowers the energy of the transition state relative to that of the reactants. Because the energy barrier between reactants and transition state is reduced, the reaction proceeds more rapidly.

The selectivity of enzymes makes them useful as catalysts for organic synthesis. Surprisingly, enzymes are able to catalyze not only the reactions that they mediate in living systems but also similar, selective transformations of unnatural substrates. Because of their selectivity, several enzyme-catalyzed reactions may run simultaneously in the same vessel. Thus, a reactant can undergo several reactions in series without the need for isolation of intermediates. Enzymes can be combined so as to reconstitute within a reaction vessel naturally occurring metabolic pathways or to create new, artificial metabolic pathways. Sequences of up to 12 serial reactions have been executed successfully in a single reaction vessel. *See* CATALYSIS; ORGANIC CHEMICAL SYNTHESIS.

The biological activity of a protein, whether binding, catalytic, or structural, depends on its full three-dimensional or conformational structure. The linear sequence of amino acids that make up a protein constitutes its primary structure. Local regions of highly organized conformation (α-helices, β-pleats, β-turns, and so on) are called secondary structure. Further folding of the protein causes regions of secondary structure to associate or come into correct alignment. This action establishes the tertiary structure of the native (active) protein. A goal of bioorganic chemistry is to achieve an understanding of the process of protein folding. *See* AMINO ACIDS; BIOCHEMISTRY; ORGANIC CHEMISTRY; PROTEIN. [H.K.C.]

Biopolymer A macromolecule derived from natural sources; also known as biological polymer. Some biopolymers are used as structural materials, food sources, or catalysts. Others have evolved as entities for information storage and transfer. Examples of biopolymers include polypeptides; polysaccharides; polypeptide/polysaccharide hybrids; polynucleotides, which are polymers derived from ribonucleic acid (RNA) and deoxyribonucleic acid (DNA); polyhydroxybutyrates, a class of polyesters produced by certain bacteria; and *cis*-1,4-polyisoprene, the major component of rubber tree latex. *See* POLYMER.

Amino acids are the monomers from which polypeptides are derived. Polypeptides alone, as well as multipolypeptide complexes or complexes with other molecules, are known as proteins, and each has a specific biological function. There are numerous examples of biopolymers having a polysaccharide and polypeptide in the same molecule, usually with a polysaccharide as a side chain in a polypeptide, or vice versa. *See* PEPTIDE; POLYSACCHARIDE; PROTEIN. [G.E.W.]

Biophysics A hybrid science involving the overlap of physics, chemistry, and biology. A dominant aspect is the use of the ideas and methods of physics and chemistry to study and explain the structures of living organisms and the mechanisms of life processes. The recognition of biophysics as a separate field is relatively recent, having been brought about, in part, by the invention of physical tools such as the electron microscope, the ultracentrifuge, and the electronic amplifier, which greatly facilitate biophysical research. These tools are peculiarly adapted to the study of problems of great current importance to medicine, problems related to virus diseases, cancer, heart disease, and the like.

The major areas of biophysics are the following:

Molecular biophysics has to do with the study of large molecules and particles of comparable size which play important roles in biology. The most important physical tools for such research are the electron microscope, the ultracentrifuge, and the x-ray diffraction camera. *See* ELECTRON MICROSCOPE; MOLECULAR BIOLOGY; ULTRACENTRIFUGE; X-RAY DIFFRACTION.

Radiation biophysics consists of the study of the response of organisms to ionizing radiations, such as alpha, beta, gamma, and x-rays, and to ultraviolet light. The biological responses are death of cells and tissues, if not of whole organisms, and mutation, either somatic or genetic.

Physiological biophysics, called by some classical biophysics, is concerned with the use of physical mechanisms to explain the behavior and the functioning of living organisms or parts of living organisms and with the response of living organisms to physical forces.

Mathematical and theoretical biophysics deals primarily with the attempt to explain the behavior of living organisms on the basis of mathematics and physical theory. Biological processes are being examined in terms of thermodynamics, hydrodynamics, and statistical mechanics. Mathematical models are being investigated to see how closely they simulate biological processes. *See* BIOMECHANICS; BIOPOTENTIALS AND IONIC CURRENTS; MATHEMATICAL BIOLOGY; MICROMANIPULATION; MICROSCOPE; MUSCLE PROTEINS; MUSCULAR SYSTEM; OXIMETRY; SKELETAL SYSTEM; SPACE BIOLOGY; THERMOREGULATION; THERMOTHERAPY. [M.A.L.]

Biopotentials and ionic currents Biopotentials are voltage differences which are measured between points in living cells, tissues, and organisms. Related to these biopotentials are ionic change transfers, or currents, that give rise to much of the electrical charges occurring in nerve, muscle, and other electrically active cells. Electrophysiology is the science concerned with determining the basic mechanisms underlying the generation of such potentials, and the laws governing the operation of systems contributing to and involving bioelectrical phenomena. Electrophysiology encompasses both the origin and the functional significance of these potentials and their experimental and natural perturbation by a variety of means. Among these, electrical stimulation of cells and tissues has been a long-standing and valuable aspect of electrophysiological methodology.

The potential difference measured with appropriate electrodes between the interior cytoplasm and the exterior aqueous medium of the living cell is generally called the membrane or resting potential (*ERP*). This potential difference is usually on the order of several tens of millivolts and is relatively constant or steady. The range of *ERP* values in various striated muscle cells of animals from insects through amphibia to mammals is about -50 to -100 mV (negative taken inside with respect to outside). Nerve cells show a similar range of *ERP* values in such diverse species as squid, cuttlefish, crabs, lobsters, frogs, cats, and humans. Even in the coenocytic marine algae, *Valonia*, *Halicystis*, and *Nitella*, potentials up to about -140 mV are measured between the cell sap and the outside medium.

Steady potential differences also can be measured across various tissues, such as the frog skin and the gastric mucosa of the stomach, and often have been shown to exist between cells. These potentials are often of complex cellular origins, but it is thought that their origin is predicted from the resting potentials of the cells making up the tissues, or involve mechanisms similar in kind if not in detail to those described here. *See* BIOPHYSICS; CELL (BIOLOGY); CELL MEMBRANES; ELECTRIC ORGAN (BIOLOGY); MUSCLE.

Another type of bioelectric potential is the action potential, a transient change in the cellular transsurface potential which is propagated along the surface of the cell and its processes. Although it has been investigated most extensively in nerve and muscle cells, the action potential probably occurs in some form in all cells; it is even seen in certain plant cells such as *Nitella*, but its function there is unknown. It is the propagated signal or unit of information by which communication is achieved between receptors (sense organs) and effectors (muscles, gland cells) via the central nervous system, between one part of the nervous system and another, and between the nervous system and other organs. It travels along the surface of muscle cells and leads to contraction. It is initiated by a decrease in the resting potential. *See* NERVOUS SYSTEM (VERTEBRATE). [W.J.A.]

Biopyribole A member of a chemically diverse, structurally related group of minerals that comprise substantial fractions of both the Earth's crust and upper mantle. The term was coined by Albert Johannsen in 1911; it is a contraction of biotite (a mica), pyroxene, and amphibole.

The pyroxene minerals contain single chains of corner-sharing silicate (SiO_4) tetrahedra, and the amphiboles contain double chains. Likewise, the micas and other related biopyriboles (talc, pyrophyllite, and the brittle micas) contain two-dimensionally infinite silicate sheets, which result in their characteristic sheetlike physical properties. In the pyroxenes and amphiboles, the silicate chains are articulated to strips of octahedrally coordinated cations, such as magnesium and iron; and in the sheet biopyriboles, the silicate sheets are connected by two-dimensional sheets of such cations. In addition to the classical single-chain, double-chain, and sheet biopyriboles, several biopyriboles that contain triple silicate chains have been discovered. *See* AMPHIBOLE; BIOTITE; MICA; PYROXENE; SILICATE MINERALS.

Pyroxenes, amphiboles, and micas of various compositions can occur in igneous, metamorphic, and sedimentary rocks. Pyroxenes are the second most abundant minerals in the Earth's crust (after feldspars) and in the upper mantle (after olivine). Unlike the pyroxenes, amphiboles, and micas, the wide-chain biopyriboles do not occur as abundant minerals in a wide variety of rock types. However, some of them may be widespread in nature as components of fine-grain alteration products of pyroxenes and amphiboles and as isolated lamellae in other biopyriboles. *See* IGNEOUS ROCKS; METAMORPHIC ROCKS; SEDIMENTARY ROCKS. [D.R.V.]

Biorheology The study of the deformation and flow of biological materials. The application of the principles of rheology to biological systems is a natural consequence of the significant effects which rheological parameters have on physiological functions. Of the many flow systems in the body, there are several in which fluid behavior is not that of a simple liquid such as water. These require more thorough study. The most thoroughly studied biological fluid is blood. Other fluids of interest are mucus and synovial fluid. Characterizing the parameters of natural materials is important not only for the understanding of physiological function in the healthy and diseased state but also for the design of synthetic replacements for diseased or traumatized parts of the body, particularly artificial joints and synovial fluid. *See* RHEOLOGY. [H.L.G.]

Biosphere The thin film of living organisms and their environments at the surface of the Earth. Included in the biosphere are all environments capable of sustaining life above, on, and beneath the Earth's surface as well as in the oceans. Consequently, the biosphere includes virtually the entire hydrosphere and portions of the atmosphere and outer lithosphere. *See* ATMOSPHERE; HYDROSPHERE; LITHOSPHERE.

Neither the upper nor lower limits of the biosphere are sharp. A variety of organisms inhabit the ocean depths. Spores of microorganisms can be carried to considerable heights in the atmosphere, but these are resting stages that are not actively metabolizing. Evidence exists for the presence of bacteria in oil reservoirs at depths of about 7000 ft (2000 m) within the Earth. These are extreme limits to the biosphere; most of the mass of living matter is within the upper 300 ft (100 m) of the lithosphere and hydrosphere, although there are places even within this zone that are too dry or too cold to support much life. Most of the biosphere is within the zone which is reached by sunlight and where liquid water exists.

The biosphere is characterized by the interrelationship of living things and their environments. Communities are interacting systems of organisms tied to their environments by the transfer of energy and matter. Such a coupling of living organisms and the nonliving matter with which they interact defines an ecosystem. An ecosystem may range in size from a small pond to a tropical forest to the entire biosphere. *See* ECOSYSTEM.

The major metabolic processes occurring within the biosphere are photosynthesis and respiration. Green plants, through the process of photosynthesis, form organic compounds composed essentially of C, H, O, and N from CO_2, H_2O, and nutrients, with O_2 being released as a by-product. The O_2 and organic compounds are partially reconverted into CO_2 and H_2O through respiration by plants and animals. Driving this cycle is energy from the Sun; respiration releases the Sun's energy which has been stored by photosynthesis. *See* PHOTORESPIRATION; PHOTOSYNTHESIS.

The fundamental types of organisms engaged in these activities are: producers, or green plants that manufacture their own food through the process of photosynthesis and respire part of it; consumers, or animals that feed on plants directly or on other animals by ingestion of organic matter; and decomposers, or bacteria that break down organic substances to inorganic products. *See* BIOLOGICAL PRODUCTIVITY; ECOLOGICAL INTERACTIONS; FOOD WEB.

Over 60 elements are reported to occur in organisms. Of these, over 30 play a vital role in cellular processes. On an atomic basis hydrogen, the lightest element, constitutes almost 50% of the content of organisms in the biosphere, with carbon and oxygen combined making up the other half. Taken together, these three elements make up over 99% of all atoms in the biosphere. Other important constituents such as nitrogen, phosphorus, and sulfur total only about 3 atoms per 1000.

The carbon cycle (CO_2—HCO_3^-—$CaCO_3$ system) is one of the most important systems affected by organisms because of its effect on the pH balance in the ocean, the content of CO_2 in the atmosphere, and the production of $CaCO_3$ sediment. Despite the dominance of the biosphere in the production of $CaCO_3$, the atmosphere and hydrosphere contain important components of the carbonate system. The inorganic carbon cycle is also an integral part of the larger carbon cycle of the Earth and, in fact, most of the carbon in the cycle is stored in the limestones and dolomites initially precipitated for the most part by organisms. The nitrogen cycle is considerably complex, and portions of it are only incompletely known. While almost all the carbon is contained in the Earth's crust, a significant portion of the nitrogen is stored in the atmosphere, largely as N_2 but also as reactive nitrogen oxides and ammonia. *See* NITROGEN CYCLE; NITROGEN FIXATION.

Organisms, through their metabolic activities, continually influence the processes and reactions occurring in the environments of the Earth's surface. Microorganisms, because of their large numbers, biochemical versatility, large surface-to-volume ratio, and rapid growth rate, are the most active agents in these influences. Consequently, numerous reactions affecting the cycling of elements through the biosphere are frequently controlled by microbial processes. Organisms also indirectly influence the chemical behavior of an element or species by altering the concentration of other ions or elements such as H^+ or O_2, which in turn affects equilibrium concentrations of the substance of interest. Mineral surfaces may acquire organic coatings from the dissolved organic matter, a process which may affect the rate and extent of equilibration of the mineral with the water.

Human beings, of course, are part of the biosphere. Some of their activities have an adverse impact on many ecosystems and on themselves by the addition of toxic or harmful substances to the outer lithosphere, hydrosphere, and atmosphere. Many of these materials are eventually incorporated into or otherwise affect the biosphere. The major types of environmental pollutants are sewage, trace metals, petroleum hydrocarbons, synthetic organic compounds, and gaseous emissions. *See* AIR POLLUTION; ATMOSPHERE; HYDROSPHERE; LITHOSPHERE; WATER POLLUTION. [R.M.]

Biostrome An evenly bedded and generally horizontally layered stratum composed mostly of organic remains, normally considered to be those of sedentary organisms which lived, died, and were buried essentially in place.

The criterion of formation by in-place growth of organisms is subject to some interpretation. Crinoidal limestones obviously resulted from the accumulation of decayed pieces of millions of these stalked echinoderms, often with accompanying detritus of associated fenestrate bryozoans. Usually it is impossible to ascertain whether such debris dropped vertically a few centimeters or meters through the water column as the organisms died and collapsed (an essentially in-place deposit) or whether the layers of crinoidal grainstone were piled mechanically by currents. The same is true of coquinas of many other thin-shelled calcareous tests, such as those of brachiopods, bryozoans, and trilobites.

Biostromal layers need not have been horizontally deposited when they occur as flanking beds around organic buildups. Dips of up to 25 or 30° are possible here. Such biostromes are probably veneers of sessile organisms which lived somewhat above the realm of deposition and were buried as sediment cascaded down the flank of a mound. [J.L.Wi.]

Biosynthesis All those reactions occurring in living systems which lead to the production of chemical compounds from simpler molecules. These biosynthetic reactions, which are catalyzed by enzymes, are divided into two types—primary and secondary. The primary reactions are common to a large number of different species. A good example is photosynthesis, the process by which green plants convert carbon dioxide in the air to carbohydrates such as glucose. Other primary reactions which are common to essentially all species of plants, animals, and microorganisms are those which lead to the production of proteins (from α-amino acids) and nucleic acids (from simple purines, pyrimidines, ribose, and phosphoric acid). The secondary reactions are often unique to a single species or to species belonging to the same genus. The products from these reactions are called secondary natural products, and they can be subdivided into various classes, including alkaloids, terpenes, steroids, flavonoids, fatty acids, porphyrins, and polyketides. The sequence of reactions which proceeds from carbon dioxide (in a green plant) to the ultimate natural product is called a biosynthetic sequence and can involve many steps. *See* ENZYME; FLAVONOID; NUCLEIC ACID; PHOTOSYNTHESIS; PORPHYRIN; PROTEIN; PURINE; PYRIMIDINE; STEROID. [E.Le.]

Biot-Savart law A law of physics which states that the magnetic flux density (magnetic induction) near a long, straight conductor is directly proportional to the current in the conductor and inversely proportional to the distance from the conductor. The field near a straight conductor can be found by application of Ampère's law. The magnetic flux density near a long, straight conductor is at every point perpendicular to the plane determined by the point and the line of the conductor. Therefore, the lines of induction are circles with their centers at the conductor. Furthermore, each line of induction is a closed line. This observation concerning flux about a straight conductor may be generalized to include lines of induction due to a conductor of any shape by the statement that every line of induction forms a closed path. *See* AMPÈRE'S LAW; MAGNETIC INDUCTION. [K.V.M.]

Biotechnology Generally, any technique that is used to make or modify the products of living organisms in order to improve plants or animals, or to develop useful microorganisms. In modern terms, biotechnology has come to mean the use of cell and tissue culture, cell fusion, molecular biology, and in particular, recombinant deoxyribonucleic acid (DNA) technology to generate unique organisms with new traits or organisms that have the potential to produce specific products. Some examples of products in a number of important disciplines are described below.

Recombinant DNA technology has opened new horizons in

the study of gene function and the regulation of gene action. In particular, the ability to insert genes and their controlling nucleic acid sequences into new recipient organisms allows for the manipulation of these genes in order to examine their activity in unique environments, away from the constraints posed in their normal host. Genetic transformation normally is achieved easily with microorganisms; new genetic material may be inserted into them, either into their chromosomes or into extrachromosomal elements, the plasmids. Thus, bacteria and yeast can be created to metabolize specific products or to produce new products. *See* GENE; GENE ACTION; PLASMID.

Genetic engineering has allowed for significant advances in the understanding of the structure and mode of action of antibody molecules. Practical use of immunological techniques is pervasive in biotechnology. *See* ANTIBODY.

Few commercial products have been marketed for use in plant agriculture, but many have been tested. Interest has centered on producing plants that are resistant to specific herbicides. This resistance would allow crops to be sprayed with the particular herbicide, and only the weeds would be killed, not the genetically engineered crop species. Resistances to plant virus diseases have been induced in a number of crop species by transforming plants with portions of the viral genome, in particular the virus's coat protein.

Biotechnology also holds great promise in the production of vaccines for use in maintaining the health of animals. Interferons are also being tested for their use in the management of specific diseases.

Animals may be transformed to carry genes from other species including humans and are being used to produce valuable drugs. For example, goats are being used to produce tissue plasminogen activator, which has been effective in dissolving blood clots.

Plant scientists have been amazed at the ease with which plants can be transformed to enable them to express foreign genes. This field has developed very rapidly since the first transformation of a plant was reported in 1982, and a number of transformation procedures are available.

Genetic engineering has enabled the large-scale production of proteins which have great potential for treatment of heart attacks. Many human gene products, produced with genetic engineering technology, are being investigated for their potential use as commercial drugs. Recombinant technology has been employed to produce vaccines from subunits of viruses, so that the use of either live or inactivated viruses as immunizing agents is avoided. Cloned genes and specific, defined nucleic acid sequences can be used as a means of diagnosing infectious diseases or in identifying individuals with the potential for genetic disease. The specific nucleic acids used as probes are normally tagged with radioisotopes, and the DNAs of candidate individuals are tested by hybridization to the labeled probe. The technique has been used to detect latent viruses such as herpes, bacteria, mycoplasmas, and plasmodia, and to identify Huntington's disease, cystic fibrosis, and Duchenne muscular dystrophy. It is now also possible to put foreign genes into cells and to target them to specific regions of the recipient genome. This presents the possibility of developing specific therapies for hereditary diseases, exemplified by sickle-cell anemia.

Modified microorganisms are being developed with abilities to degrade hazardous wastes. Genes have been identified that are involved in the pathway known to degrade polychlorinated biphenyls, and some have been cloned and inserted into selected bacteria to degrade this compound in contaminated soil and water. Other organisms are being sought to degrade phenols, petroleum products, and other chlorinated compounds. *See* GENETIC ENGINEERING; MOLECULAR BIOLOGY. [M.Z.]

Biotelemetry The use of telemetry techniques on living things, telemetry being metering (or measuring) at a distance. As the word is generally used, the link between the subject and the measurer is understood to be wireless radio waves. In cases where radio waves are ineffective, sound waves, light, magnetic radiations, or electricity in wires is used. *See* TELEMETERING.

The purpose in using biotelemetry is to be able to measure something on, in, or about a living animal or plant that cannot be measured in any other way. Biotelemetry is also a method of tracking animals, as they wander about their home range or as they migrate, that does not necessitate visual contact with the animals. While astronauts and cosmonauts are in space, they need to have several body functions monitored to help them perform their job in safety. Body temperature, breathing rate, heart rate, blood pressure, and alertness are routinely telemetered, first from the space suit to the spacecraft, and then to the ground, where a doctor is on hand to diagnose changes and to anticipate troubles.

Biotelemetry is also used in medicine in the study and treatment of diseases in humans and in patient care. Usually a radio link is used so that the patient is not tied down by wires.

A biotelemetry system has two main parts: a transmitter carried on, or in, the animal or plant subject, and a receiver to pick up the transmitted information. Sound energy is transmitted by a loudspeaker and received by a microphone; light is transmitted by a lamp and received by a photoelectric cell; and magnetic fields are both transmitted and received by coils of wire. A transducer is needed by both the transmitter and the receiver to convert the energy carrying the information to a suitable form. The transmitting transducer must convert electrical energy to a flow of energy which can travel through space. The receiving transducer must accept energy from space and convert it to a current flow in a wire in the receiver. In the case of radio, both the transmitting and the receiving transducers are antennas. The simplest biotelemetry transmitters are those used for tracking.

Sound and ultrasonic waves travel much farther than radio waves underwater. For this reason, tagging of seals and fish for tracking purposes is done with small sonar transmitters, or pingers, fastened to the outside of the animal or fed to it. One of the problems in using sound transmitters on aquatic animals is that many of these animals can hear well up into the ultrasonic range. Carrying such a transmitter may therefore alter the animal's behavior because it can hear its own transmitter.

For extremely short-range work in laboratory animals of all kinds, it is sometimes desirable to use low-frequency magnetic field transmitters. Transmissions at low frequencies will pass through soil, trees, or the body of an animal or fish without being appreciably affected. What is perhaps more important, low-frequency transmissions may not affect the animal as much as the higher frequencies. [J.H.Bu.]

Biotin A vitamin, widespread in nature. It is only sparingly soluble in water; it is stable in boiling water solutions, but can be destroyed by oxidizing agents, acids, and alkalies. Under some conditions, it can be destroyed by oxidation in the presence of rancid fats. Biotin's occurrence in nature is so widespread that it is difficult to prepare a natural deficient diet. Biotin deficiency in animals is associated with dermatitis, loss of hair, muscle incoordination and paralysis, and reproductive disturbances. Biotin deficiency produced in humans by feeding large amounts of egg white resulted in dermatitis, nausea, depression, muscle pains, anemia, and a large increase in serum cholesterol. *See* COENZYME. [S.N.G.]

Biotite The most abundant and widely distributed species of the mica group of minerals, also called black mica or magnesium-iron mica. Its chemical composition is variable:

$$K_2[Fe(II),Mg]_{6-4}[Fe(III),Al,Ti]_{0-2}(Si_{6-5},Al_{2-3})O_{20-22}(OH,F)_{4-2}$$

The color ranges from black to dark brown or green. Unaltered basal cleavage sheets (easy and perfect) are both flexible and elastic. The hardness is 2.5–3 (Mohs scale) and specific gravity 2.8–3.2. Biotite itself is of no commercial importance, but under hydrothermal conditions or when exposed to groundwater action it transforms to vermiculite, which expands (exfoliates) upon heating and has widespread application as lightweight concrete and plaster aggregate, insulation, plant-growing medium, and lubrication. Biotite grades into phlogopite, and no sharp division exists between the two. In a general way, biotite represents the iron-rich half of the biotite-phlogopite series.

Biotite, an important rock-forming mineral, may be an essential constituent of both intrusive and extrusive igneous rocks, ranging in composition from gabbro to granite. It is especially common in granite, granodiorite, and syenite. In porphyritic extrusive rocks biotite appears chiefly as phenocrysts, very rarely in the matrix. Many pegmatites also contain biotite in large amounts in coarse flakes or rough crystals. *See* MICA; SILICATE MINERALS. [E.W.H.]

Biphenyl A colorless aromatic hydrocarbon, with the formula $C_{12}H_{10}$ (melting point 70.0°C or 158°F, boiling point 255.9°C or 492.6°F, and density 1.9896 g/cm^3 or 1.1501 oz/in.^{3}), also called diphenyl or phenylbenzene, which crystallizes as small plates. Although small amounts are present in coal tar, the biphenyl of commerce is obtained by the pyrolysis of benzene at about 750°C (1382°F) or as a by-product in the thermal dealkylation of toluene. Biphenyl is flammable, flash point (open cup) 124°C (255°F).

The high thermal stability and low vapor pressure of biphenyl make it a valuable heat-transfer medium, which can be used without high-pressure equipment. In certain applications it is chlorinated or alkylated to produce agents in some ways superior to the parent hydrocarbon. Unfortunately, polychlorinated biphenyls (PCBs), despite their much prized thermal and chemical stability, are definitely hazardous to health if ingested and have been listed by the Occupational Safety and Health Administration as potential carcinogens. *See* AROMATIC HYDROCARBON; POLYNUCLEAR HYDROCARBON. [C.K.B.]

Birch A deciduous tree of the genus *Betula* which is distributed over much of North America, in Asia south to the Himalaya, and in Europe. About 40 species are known. The birches comprise the family Betulaceae in the order Fagales. The sweet birch, *B. lenta*, the yellow birch, *B. alleghaniensis*, and the paper birch, *B. papyrifera*, are all important timber trees of eastern United States. The yellow and the paper species extend into Canada. The gray birch, *B. populifolia*, is a smaller tree of the extreme northeastern United States and adjacent Canada. Both sweet (black) and yellow birches can be recognized by the wintergreen taste of the bark of the young twigs. A flavoring similar to oil of wintergreen is obtained from the bark of the sweet birch. The paper and gray birches can be easily identified by their white bark. The bark of the paper birch peels off in thin papery sheets, a characteristic not true of the gray birch. The river birch, *B. nigra*, is a less common tree of wet soils and banks of streams and is important as an ornamental and for erosion control. The hard, strong wood of the yellow and the sweet birches is used for furniture, boxes, baskets, crates, and woodenware. The European birches, *B. pubescens* and *B. pendula*, are the counterparts of the paper and gray birches in the United States. European birches are also cultivated in America. *See* FAGALES. [A.H.G./K.P.D.]

Bird Any of about 9022 extant species of warm-blooded vertebrates which make up the class Aves. They are distributed throughout the world and range from the Arctic to the Antarctic.

Birds have certain features, particularly the horny bill and feathers, which distinguish them from other vertebrates. Feathers are specialized epidermal (skin) derivatives, as are many vertebrate structures, such as the hair of mammals, and are nonliving when fully developed. Although most species are aerial, a number of species are incapable of flight.

One group of mainly giant flightless birds are known collectively as the ratites, or running birds. This group represents a monophyletic assemblage, including the ostrich, emu, rhea, kiwi, and tinamous, most of which are large and have lost their ability of flight; they are characterized by the flat keelless sternum. Tinamous can still fly and have a strongly keeled sternum. Many other flightless birds, for example, dodos, rails, coots, and cormorants, are known, some of which also have a flat sternum while others, such as the penguins, still possess a keeled sternum. The bones in most birds are light and not heavy, massive structures found in terrestrial vertebrates and some flightless birds.

The fossil record for birds is poorly known, but is better known for some groups of water and swamp birds. Flying reptiles, the pterosaurs, are not closely related to birds and evolved flight independently. *Archaeopteryx,* the earliest known bird, is an excellent example of an intermediate link. It possessed feathers and a wing, but the hand bones are not fused. However, *Archaeopteryx* retained a long tail, teeth, and a reptilian skull. *See* AVES; BIRDS, UPLAND GAME; BIRDS OF PREY; FEATHER. [W.J.B.]

Birds, upland game A general category for a large number of birds that are hunted for sport, the majority being members of the order Galliformes. Included are the pheasants, quails, grouse, partridges, guinea fowls, turkeys, and the less commonly known brush turkeys and curassows.

All members of the Galliformes have small heads and thickset bodies. The short wings are not adapted to sustained flight, but the long sturdy legs make them good runners; most species are terrestrial and nonmigratory. That they are graminivorous (seedeaters) is apparent by the stout curved bill and strong claws used for scratching soil in search of seeds. Sexual dimorphism is well developed in this group with differences in size, plumage, comb, and wattle. The males are usually polygamous, fierce, and aggressive; the females build nests and rear the young alone, and the young are precocial, that is, they are capable of leaving the nest soon after hatching and feeding by themselves, but are usually brooded and cared for by the female for several weeks. *See* SEXUAL DIMORPHISM.

Natural habitats of these birds have been diminishing for years as a result of human encroachment for a variety of reasons. Extensive efforts have been made to maintain breeding populations in nature. Upland game birds are extensively hunted, and legal restrictions, such as limits on quantity and a season for hunting, have been imposed in most states. Many state game and fisheries commissions raise and release young in selected areas or have farmers incubate eggs with the eventual release of young in selected areas.

The family Meleagrididae includes two species of wild turkeys found in North and Central America. The common turkey (*Meleagris gallopavo*), from which the domestic turkey has been bred, is found along the eastern and southern coasts of the United States from Canada to Mexico, and the ocellated turkey (*Agriocharis ocellata*) is found in Central America and Mexico. A characteristic feature of these birds is the bare head and neck.

The family Phasianidae contains the quails, partridges, peafowls, and true pheasants. The American quail constitute a distinct subfamily unique to the New World, and no other members of the family are native there. Typically, members of the Phasianidae are ground feeders. The partridges belong to two genera, *Alectoris* and *Perdix*, of which the most typical representatives are the common partridge (*P. perdix*) and the red-legged partridge (*A. rufa*). These birds are commonly found in open country living in groups. European quail differ from partridges in having a shorter beak and feathers around the eyes and in being less social. They are among the few species of game birds that migrate, forming flocks only for this activity. The bobwhite (*Colinus virginianus*) is the commonest American quail, found in the southern United States where it is often called a partridge. The principal game bird is probably the ring-necked pheasant (*Phasianus colchicus*), a species that has been introduced into the United States, where it is now common and has partly displaced the bobwhite. Other species include Reeve's pheasant (*Syrmaticus reevesii*), the silver pheasant (*Lophura nycthemera*), and the golden pheasant (*Crysolophus pictus*).

The Tetraonidae include 18 recognized species of ptarmigans and grouse. Members of this family are restricted to the Northern Hemisphere and have a circumpolar distribution. The rock ptarmigan (*Lagopus mutus*) and the willow ptarmigan (*L. lagopus*) are found throughout the colder regions of the Northern Hemisphere. Another ptarmigan, the red grouse (*L. scoticus*), is a species believed by some to be an aberrant form of the willow ptarmigan. A fourth species, the whitetailed ptarmigan, is found in the Rocky Mountains. The largest grouse is the capercaille or wood grouse (*Tetrao urogallus*) with a wing span up to 5 ft (1.5 m). Though once widespread in central Europe, hunting and disease have reduced this species to a more restricted area in some parts of France, Scandinavia, and Russia. The ruffed grouse (*Bonasa umbellus*) is a resident of mixed and deciduous forest areas in Canada and the United States.

The guinea fowl (Numididae) are common in Africa and adjacent islands, where they live in large groups in forest clearings and feed on insects or plants, depending on the season. The common guinea fowl (*Numida meleagris*) has been domesticated for over 2000 years.

The little-known megapodes (Megapodiidae; brush turkeys), native to Australia, are also known as incubator birds, mound builders, and thermometer birds. Curassows (Cracidae) and their relatives range from Texas to Argentina. *See* AVES; BIRD; GALLIFORMES. [C.B.C.]

Birds of prey Birds of prey belong to two unrelated orders: the Falconiformes, which includes the eagles, falcons, hawks, vultures, and buzzards; and the Strigiformes, or owls (see illustration). There are a number of species that are economically important as predators upon mice, rats, and other pests, as well as on game and poultry. Many of these birds are nocturnal while others are diurnal. Members of both orders show similarities in the shape of the bill and have strong claws, adaptations for their mode of life; otherwise, they are dissimilar. Rather dramatic decreases in populations of predatory birds have not been the result of destruction of natural habitats, but are related to the incorporation of chlorinated hydrocarbons into the ecological systems.

Typical birds of prey: (*a*) rough-legged hawk; (*b*) bald eagle; (*c*) great horned owl.

Diurnal species have long, powerful wings for sustained flight and are usually large, fierce, and strong, with a keen sense of vision. The toes on the short, strong legs end in pointed talons. Normally, the birds prey upon warm-blooded vertebrates, but some species are carrion feeders while others prefer fish or amphibians. These birds are monogamous and usually nest in inaccessible areas. The young are helpless for a considerable period of time after hatching.

New World vultures, members of the family Cathartidae, are voiceless species. They include the California condor (*Gymnogyps californianus*), the largest North American bird, which is on the verge of extinction, the Andean condor (*Vultur gryphus*), the king vulture (*Sarcorhamphus papa*), and the turkey buzzard or turkey vulture (*Cathartes aura*).

Old World vultures are large predatory birds, such as kites, harriers, true hawks, and eagles, which are classified as subfamilies in the family Accipitridae. Included are the bald or American eagle (*Haliaëtus leucocephalus*), the golden eagle (*Aquila chrysaëtos*), and the true hawks, exemplified by the goshawk (*Accipiter gentilis*) and the European sparrowhawk (*A. nisus*). Other hawks, grouped together in the genus *Buteo*, include the red-tailed hawk (*B. jamaicensis*), the broad-winged hawk (*B. platyperus*), and the rough-legged hawk (*B. lagopus*).

The family Pandionidae includes a single species, *Pandion haliaëtus*, the osprey or fish hawk. This bird lives near water and feeds on fish.

Falcons are included in the family Falconidae. They usually nest in forested areas and are considered to be the most specialized birds of prey because they destroy large numbers of animal pests, but do not overkill. Birds that have been captured and trained for hunting include the peregrine falcon (*Falco peregrinus*) and the gyrfalcon (*F. rusticolus*).

The owls are frequently referred to as the nocturnal birds of prey. Unlike the diurnal species, owls are silent fliers, mainly because they have soft feathers that extend down to the toes. They differ from most birds in that the eyes are located in a frontal position rather than on the lateral side of the head. Their keen sense of hearing has led many observers to believe that some species hunt by sound. Some of the more common species of owls are *Tyto alba*, the barn owl; *Asio otus*, the long-eared owl; *Bubo virginianus*, the great horned owl; *Otus asio*, the screech owl; *Nyctea scandiaca*, the snowy owl; and *Speotyto cunicularia*, the burrowing owl. *See* AVES; FALCONIFORMES; STRIGIFORMES. [C.B.C.]

Birefringence The splitting which a wavefront experiences when a wave disturbance is propagated in an anisotropic material; also called double refraction. In anisotropic substances the velocity of a wave is a function of displacement direction. Although the term birefringence could apply to transverse elastic waves, it is usually applied only to electromagnetic waves.

In birefringent materials either the separation between

neighboring atomic structural units is different in different directions, or the bonds tying such units together have different characteristics in different directions. Many crystalline materials, such as calcite, quartz, and topaz, are birefringent. Diamonds, on the other hand, are isotropic and have no special effect on polarized light of different orientations. Plastics composed of long-chain molecules become anisotropic when stretched or compressed. Solutions of long-chain molecules become birefringent when they flow. This first phenomenon is called photoelasticity; the second, streaming birefringence. *See* CRYSTAL OPTICS; PHOTOELASTICITY; POLARIZED LIGHT; REFRACTION OF WAVES. [B.H.Bi.]

Birth control Methods of fertility control, including contraception, that are intended to prevent pregnancy, and means of interrupting early pregnancy. Different methods of contraception vary in their effectiveness, and effectiveness under actual conditions of use can be considerably lower than that under ideal conditions. Available methods include oral contraceptives, intrauterine devices, barrier methods, fertility awareness methods, sterilization, and subdermal implants.

Oral contraceptives contain one or both of two compounds (estrogen and progesterone) similar to the hormones that regulate the menstrual cycle. Each monthly series of pills either suppresses ovulation or alters the uterine lining and the cervical mucus, or both. *See* ESTROGEN; MENSTRUATION; PROGESTERONE.

Postcoital contraception is another hormonal method. In emergency situations (for example, rape) high dosages of oral contraception can be used. One dose is given within 72 h after the episode of unprotected intercourse, and an additional dose is given 12 h later. When effective, this treatment renders the uterus inhospitable to implantation of a fertilized ovum.

The main mode of action for the intrauterine device (IUD) is considered to be prevention of fertilization. Of the two commercially available IUDs, one containing copper is designed to remain in place for 4 years; for users over the age of 25, the pregnancy rate is less than 1%. The other one releases a daily dosage of the natural hormone progesterone to suppress the uterine lining and requires annual replacement.

Barrier methods include the condom, diaphragm, cervical cap, vaginal contraceptive sponge, and various chemical preparations. The condom is a sheath of thin latex (sometimes coated with spermicide) or animal tissue that covers the penis. The diaphragm is a shallow rubber cup with a ring rim that fits securely in the vagina to cover the cervix. The cervical cap is a smaller, thimble-shaped latex device that fits over the cervix. The diaphragm and cervical cap are used with spermicides. The vaginal contraceptive sponge is a soft, synthetic, disposable sponge that fits over the cervix and, when moistened, continuously releases spermicide.

Fertility awareness methods enable a woman to estimate when she is fertile so that she can practice abstinence or use a barrier method during those times. Techniques used to determine fertility include cervical mucus observation, and body signs with temperature tracking. Such methods are the least effective in contraception.

Sterilization is the most commonly used method of birth control for women and men both in the United States and worldwide. The procedures do not adversely affect the production of male or female hormones, so that individual sexual characteristics such as sex drive and menses usually remain unchanged. Vasectomy is a minor male surgical procedure that occludes the vas deferens by various means (such as by cautery or suture). In the United States, tubal sterilization in the female is an operation commonly performed through the laparoscope.

The subdermal implant consists of small hollow rods that are placed under the skin of a woman's upper arm and release a low, continuous dose of a progestin. It is more effective than the oral contraceptives and, because it lacks estrogens, does not pose a risk of cardiovascular complications. [L.B.T.]

Birthmark The common term for several types of skin lesions which are present at birth or appear sometime later. Two major types are recognized, the cellular and the vascular nevi. Both are abnormal, benign growths of some tissue component native to an area and result from faulty development. Most are pigmented, and the color depends upon the amount of pigment or blood present and the nearness to the surface of the skin. Pigmented nevi, or common moles, vary from natural flesh color through a range of increasing pigmentation to dark brown or black. *See* MOLE (MEDICINE).

Moles range in size from a few millimeters to several centimeters; they may be flat or elevated, smooth or rough, and other features can be present. Rarely does a tendency toward malignant change occur, so that moles which deepen in color and show increases in size or other changes should be evaluated promptly. The common hairy brown mole is least prone to show malignant changes, especially if present on the legs from childhood.

Vascular birthmarks are either developmental abnormalities of blood vessels or benign neoplasms which arise postnatally. Capillary hemangiomas consist of local overgrowths of small blood vessels. Port wine stains are forms of this lesion which appear as large plaquelike areas of red to blue discoloration on the face or trunk. A strawberry hemangioma is typically a soft, raised nodule of varying size which often regresses spontaneously in childhood. These may be congenital or may appear during the first few weeks of life. Cavernous hemangiomas consist of large dilated vessels which frequently have smaller vessels within them. Although they occur as birthmarks on the skin, they are also found in deeper tissues, such as the liver and brain. *See* ONCOLOGY; TUMOR. [E.G.St./N.K.M.]

Bismuth The metallic element, Bi, of atomic number 83 and atomic weight 208.980 belonging in the periodic table to group V. Bismuth is the most metallic element in this group in both physical and chemical properties. The only stable isotope is that of mass 209. It is estimated that the Earth's crust contains about 0.00002% bismuth. It occurs in nature as the free metal and in ores. The principal ore deposits are in South America. However, the primary source of bismuth in the United States is as a by-product in refining of copper and lead ores.

The main use of bismuth is in the manufacture of low-melting alloys which are used in fusible elements in automatic sprinklers, special solders, safety plugs in compressed gas cylinders, and automatic shutoffs for gas and electric water-heating systems. Some bismuth alloys, which expand on freezing, are used in castings and in type metal. Another important use of bismuth is in the manufacture of pharmaceutical compounds.

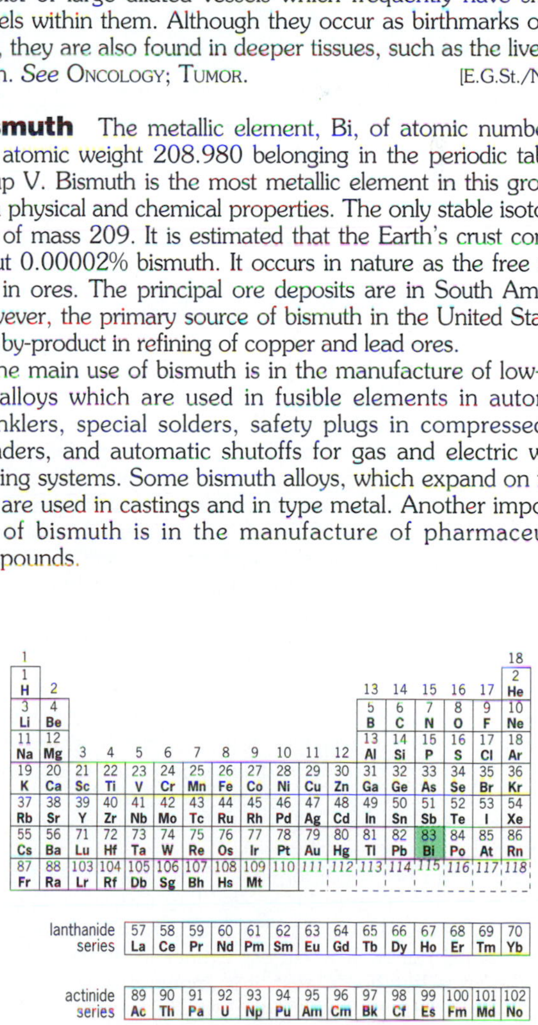

Physical and mechanical properties of bismuth

Property	Value	Temperature
Melting point, °C	271.4	
Boiling point, °C	1559	
Heat of fusion, kcal/mole	2.60	
Heat of vaporization, kcal/mole	36.2	
Vapor pressure, mm Hg	1	917°C
	10	1067°C
	100	1257°C
Density, g/cm^3	9.80	20°(solid)
	10.03	300°(liquid)
	9.91	400°(liquid)
	9.66	600°(liquid)
Mean specific heat, cal/g	0.0294	0–270°C
	0.0373	300–1000°C
Coefficient of linear expansion	13.45×10^{-6}/°C	
Thermal conductivity, cal/(s)(cm^2)(°C)	0.018	100°(solid)
	0.041	300°(liquid)
	0.037	400°(liquid)
Electrical resistivity, μohm-cm	106.5	0°(solid)
	160.2	100°(solid)
	267.0	269°(solid)
	128.9	300°(liquid)
	134.2	400°(liquid)
	145.3	600°(liquid)
Surface tension, dynes/cm	376	300°C
	370	400°C
	363	500°C
Viscosity, centipoise	1.662	300°C
	1.280	450°C
	0.996	600°C
Magnetic susceptibility, cgs units	-1.35×10^{-6}	
Crystallography	Rhombohedral, $a_0 = 0.47457$ nm	
Thermal-neutron absorption cross section, barns	0.032 ± 0.003	
Modulus of elasticity, lb/cm^2	4.6×10^6	
Shear modulus, lb/cm^2	1.8×10^6	
Poisson's ratio	0.33	
Hardness, Brinell	4–8	

Bismuth is a gray-white, lustrous, hard, brittle, coarsely crystalline metal. It is one of the few metals which expand on solidification. The thermal conductivity of bismuth is lower than that of any metal, with the exception of mercury. The table cites the chief physical and mechanical properties of bismuth. Bismuth is inert in dry air at room temperature, although it oxidizes slightly in moist air. It rapidly forms an oxide film at temperatures above its melting point, and it burns at red heat, forming the yellow oxide, Bi_2O_3. The metal combines directly with halogens and with sulfur, selenium, and tellurium; however, it does not combine directly with nitrogen or phosphorus. Bismuth is not attacked at ordinary temperatures by air-free water, but it is slowly oxidized at red heat by water vapor.

Almost all compounds of bismuth contain trivalent bismuth. However, bismuth can occasionally be pentavalent or monovalent. Sodium bismuthate and bismuth pentafluoride are perhaps the most important compounds of Bi(V). The former is a powerful oxidizing agent, and the latter a useful fluorinating agent for organic compounds. [S.J.Y.]

Bison The name for two species of the Bovidae in the mammalian order Artiodactyla, found in North America and Europe. The European bison (*Bison bonasus*) is commonly known as the wisent. The American species (*B. bison*), shown in the illustration, is often called buffalo but should not be confused with the African buffalo.

North American bison (*Bison bison*).

The European bison was originally abundant in the forested areas of western Europe during the late Cenozoic Era. The wisent is a browsing, woodland animal which congregates in relatively small herds. It is almost extinct in its natural range and, although a few herds exist, it is known mainly from a few hundred specimens preserved in zoological gardens and zoos. An intensive effort has been made to preserve the species by breeding and maintaining this animal in captivity.

Although there are differences, the wisent is closely allied to the American species. The wisent has a small head carried in a high position and a short mane, and is more graceful and less massive than the North American species. The hump is less noticeable, the legs are longer, the horns are more slender, and the body is not so shaggy.

Enormous herds of the American bison existed on the Plains area of North America in western Canada and the western United States during the 19th century. It is estimated that there are still about 20,000 bison in Canada and the United States on preserves and national parks. These bovines are massive, with the males attaining a length of 9 ft (2.7 m), a height of 6 ft (1.8 m) at the shoulder, and weight up to 3000 lb (1350 kg). They are herbivorous, and migrated originally when the grass or forage became scarce. The large head is held in a low position, and behind the neck region is a characteristic hump. The forequarters are shaggy with the hair more pronounced in the male than the female. The senses of smell and hearing are well developed, while vision is poor. Both the male and female have horns which are small and set far apart. *See* ARTIODACTYLA; BUFFALO; MAMMALIA. [C.B.C.]

Bit In the pure binary numeration system, either of the digits 0 or 1. The term may be thought of as a contraction of binary digit. *See* NUMBER SYSTEMS.

In a binary notation, bits are used as two different characters. For example, in the American Standard Code for Information Interchange (ASCII), a seven-bit coded character set, the seven bits 1000111 represent the letter G.

Bit is widely used as a synonym for binary element, a constituent element of data that takes either of two values or states (on-off, yes-no, zero-one, and so forth). The brains of animals and registers, memories, and other storage devices of digital computers and electronic calculators store bits (binary elements) as the smallest unit of information.

A byte is a string of bits (binary elements) operated on or treated as a unit and usually shorter than a word. An eight-bit byte comprises eight bits. A word is a string of bits (binary elements) that consists of two or more bytes. The terms halfword, fullword, and doubleword are also used. A computer word is a word stored in one computer location, usually 16, 32, 36, 48, or 64 bits in length, depending on the design of the computer. In microcomputers or in particular applications, a nybble is a string of bits (binary elements) operated as a unit, larger than a bit, and smaller than a byte. *See* DIGITAL COMPUTER.

In information theory, bit is a synonym for the new, preferred term shannon, a unit of measure of information equal to the decision content of a set of two mutually exclusive events. For example, the decision content of a character set of eight characters equals three shannons (the logarithm of 8 to base 2). A shannon also is called an information content binary unit. A hartley is the information content decimal unit. *See* INFORMATION THEORY. [H.E.T.]

Bitumen A term used to designate naturally occurring or pyrolytically obtained substances of dark to black color consisting almost entirely of carbon and hydrogen with very little oxygen, nitrogen, and sulfur. Bitumen may be of variable hardness and volatility, ranging from crude oil to asphaltites, and is largely soluble in carbon disulfide. *See* ASPHALT AND ASPHALTITE. [I.A.B.]

Bivalvia One of the two largest classes in the phylum Mollusca. The class is sometimes termed Pelecypoda or Acephala, or known by the name of its largest subclass, Lamellibranchia. All bivalves are aquatic, living at all depths in the sea and also in brackish and fresh waters. Although the number of living bivalve species (possibly 30,000) is second to the class Gastropoda (over 74,000), the total biomass of marine bivalves is much greater, and certain bivalve species are the most abundant marine benthic animals. The most primitive bivalves are all infaunal, burrowing into soft substrates, but many families are epifaunal, attached to rocks.

Bivalvia are basically bilaterally symmetrical and laterally compressed, with the mantle lobes and thus the shell which they secrete entirely enclosing the body. The head is only the site of the mouth and is without the jaws or radula of other mollusks, but is flanked by pairs of extended lips forming ciliated labial palps. The ciliated molluscan gills (ctenidia) are enlarged, fully occupying the extensive mantel cavity. The shell consists of two valves with a connecting ligament. It is formed of three layers. The ligament is formed of the same three layers similarly produced, but is not calcified. The body is attached to the shell by one or more pairs of pedal retractors with sometimes an additional pair of pedal protractors. The foot is extended by blood pressure, then dilated terminally to form an anchor on which the animal is drawn forward when the retractors contract. A few bivalves, notably scallops and file shells, can swim by jet propulsion, water being suddenly ejected from between the flattened valves.

Bivalves may be hermaphroditic or have separate sexes. Many species show sex change, usually from male to female (protandric hermaphroditism); a few, notably certain oysters, show alternating sexuality.

Various bivalves are economically important. Some, notably oysters, mussels, hard and soft shell clams, and scallops, are widely eaten, many being cultivated. Others are a major food of important food fishers, such as cod, halibut, and plaice. Marine and fresh-water mussels, along with other attached bivalves, may be serious fouling agents, with shipworms (Teredinidae) doing serious damage to wooden vessels and harbor works. [C.M.Y.]

The fossil record of the Bivalvia can be traced to the Lower Cambrian *Fordilla*, a small, laterally compressed shell with a simple, rounded profile. *Fordilla* did not become abundant until the Ordovician Period, several million years later. However, throughout the Paleozoic Era, these bivalves remained second to bivalves of the phylum Brachiopoda in diversity and abundance in most aquatic environments except the ocean margins and fresh-water environments. During the Mesozoic Era, the Bivalvia gradually took over as the dominant marine bivalves. The transition from Mesozoic to Cenozoic time saw the extinction of the rudists and the gradual replacement of many ancient superfamilies with a great variety of newer superfamilies. However, many ancient superfamilies have maintained a high taxonomic diversity to the present day. *See* LAMELLIBRANCHIA; MOLLUSCA; PROTOBRANCHIA; SEPTIBRANCHIA; SHIPWORM. [J.G.C.]

Black dwarf A star that cannot generate thermonuclear energy. Not necessarily invisible, it may radiate its internal heat. None has ever been observed, but as-yet-invisible companions of nearby faint stars have been suspected of having masses within the probable range. Isolated very low masses cannot form by gravitational contraction from interstellar gas and dust. Thermonuclear energy is related if the core temperature (proportional to the mass of a self-gravitating gas sphere) exceeds 2×10^6 K (3.6×10^6 °F). Thus, black dwarf masses should be between 0.007 and 0.07 solar mass. [J.L.Gr.]

Black hole One of the end points of gravitational collapse, in which the collapsing matters fades from view, leaving only a center of gravitational attraction behind. General relativity predicts that, if a star of more than about 3 solar masses has completely burned its nuclear fuel, it should collapse to a configuration known as a black hole. The resulting object is independent of the properties of the matter that produced it and can be completely described by stating its mass and spin. The most striking feature of this object is the existence of a surface, called the horizon, which completely encloses the collapsed matter. The horizon is an ideal one-way membrane; that is, particles and light can go inward through the surface, but none can go outward. As a result, the object is dark, that is, black, and hides from view a finite region of space (a hole). *See* GRAVITATIONAL COLLAPSE; RELATIVITY.

The possible formation of black holes depends critically on what other end points of stellar evolution are possible. There can always be chunks of cold matter which are stable, but their mass must be considerably less than that of the Sun. For masses on the order of a solar mass, only two stable configurations are known for cold, evolved matter. The first, the white dwarf, is supported against gravitational collapse by the same quantum forces that keep atoms from collapsing. However, these forces cannot support a star which has a mass in excess of 1.2 solar masses, this limiting value being known as the Chandrasekhar limit. The second stable configuration, the neutron star, is supported against gravitational collapse by the same forces that keep the nucleus of an atom from collapsing. There is also a maximum mass for a neutron star, estimated to be between 1 and 3 solar masses, the uncertainty being due to the poor knowledge of nuclear forces at high densities. It would appear from the theory that if a collapsing star of over 3 solar masses does not eject matter, it has no choice but to become a black hole. Since there are

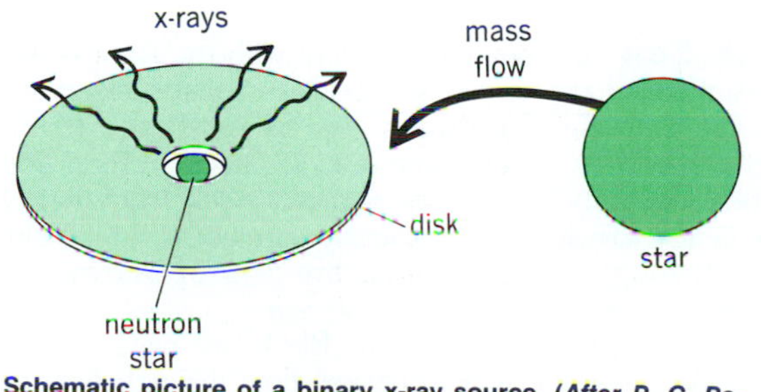

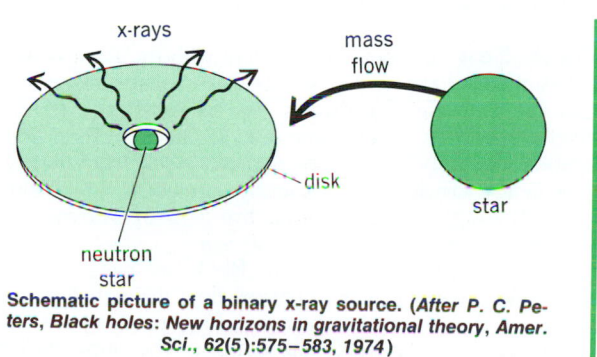

Schematic picture of a binary x-ray source. (*After P. C. Peters, Black holes: New horizons in gravitational theory, Amer. Sci., 62(5):575–583, 1974*)

many stars of over 3 solar masses, and more massive stars evolve more quickly, a considerable number of black holes should have formed.

Because black holes themselves are dark, and therefore essentially unobservable, their existence must be inferred from their effect on other matter. Thus a theoretical model must be constructed to explain the observational data and to show that the most reasonable explanation is that a black hole is responsible for the observed effects. Such is the case with the binary x-ray star system Cygnus X-1. There are a number of binary x-ray systems known. The model which best explains the data is one in which a fairly normal star is in mutual orbit about a very compact object. Because these two are so close, mass flows from the star onto an accreting disk about the compact object, as shown in the illustration. As the mass in the disk spirals inward, it heats up by frictional forces. Because the central body is so compact, the matter heats to a temperature at which thermal x-rays are produced. The only compact objects known that could accomplish this are neutron stars and black holes. The existence of very short time bursts of radiation also points to an object of small diameter, that is, compact. In some of these binary x-ray systems, there is also a regular pulsed component to the x-rays, indicating a rotating neutron star (by reasoning similar to that given for pulsars). In these systems, the compact object could not be a black hole because that would imply a more complicated structure than a black hole would allow. In other systems, however, there are only irregular pulsations or fluctuations; they are candidates for possible black holes. In one of these systems, the Cygnus X-1, a rough mass determination has been made, with the result that the compact object must be more massive than 5 solar masses. According to theory, there is no compact system with this mass other than a black hole. It is in this sense that it can be said that black holes have been observed. *See* Binary star; X-ray astronomy; X-ray star. [P.C.P.]

Black pepper One of the oldest and most important of the spices. It is the dried, unripe fruit of a weak climbing vine, *Piper nigrum*, a member of the pepper family (Piperaceae), and a native of India or Indomalaysia. The fruits are small one-seeded berries which, in ripening, undergo a color change from green to red to yellow. When in the red stage, they are picked, sorted, and dried. The dry, wrinkled berries (peppercorns) are ground to make the familiar black pepper of commerce. White pepper is obtained by grinding the seed separately from the surrounding pulp. *See* Piperales. [P.D.St./E.L.C.]

Black Sea A body of water that with the Azov, Caspian, and Aral seas forms a series of bodies of water extending far into the Eurasian landmass. As the distance from the Mediterranean increases, these seas (or lakes) freshen and the composition of their salts departs more widely from that of the oceans. The Black Sea is tenuously connected with the world ocean through the Bosporus, the Sea of Marmara, the Dardanelles, and the Mediterranean Sea.

Runoff and precipitation on the Black Sea exceed losses by evaporation, causing a net outward flow through the surface layers of the Bosporus and a compensating inflow of Mediterranean water along the bottom. The upper layers of the Black Sea are so diluted that the density increases rapidly with depth, preventing effective vertical mixing. Because of this and the shallowness of the Bosporus (130 to 300 ft or 40 to 90 m), the deep waters of the Black Sea are stagnant.

The surface currents form two counterclockwise gyres with complicated countercurrents between them, and a current sets outward through the Bosporus. The Bosporus undercurrent introduces water denser than that in the Black Sea, but this inflow is so small that it would take 2500 years for it completely to replace the water below a depth of 100 ft (30 m). The tides of the Black Sea are semidiurnal and of small amplitude: about 3.5 in. (9 cm) along the central-eastern and central-western coasts and only 0.8 to 1.2 in. (2 to 3 cm) on the Crimean coast.

The Black Sea is distinctly less saline than the open ocean, but its salt has nearly the same composition as that of ocean salt, indicating that the Black Sea is a part of the ocean system, not a saline lake. River inflow lowers the salinities in the northwestern part of the sea, which allows the formation of ice as much as 70 mi (112 km) offshore in exceptionally severe winters. Surface temperatures vary from the freezing point in winter to over 72°F (22°C) in summer, but at depths greater than 480–660 ft (150–200 m), the temperature is remarkably constant at around 47°F (8.75°C).

The Black Sea has fewer species of plants and animals than the Mediterranean, but more than the Azov, Caspian, or Aral seas. The 1500 animal species are mainly a depleted Mediterranean fauna with a few relict Caspian types and, near the Danube delta, a few fresh-water forms. There are around 170 fish species, of which over 100 are Mediterranean immigrants; 20% of the fishes are of commercial use. Among the most important of these are mackerel, bonito, anchovies, herring, mullets, carp, and sturgeon. *See* Mediterranean Sea; Seawater. [F.A.R.]

Black shale A dark mud rock rich in organic carbon. Black shales are typically very fine-grained and contain pyrite, phosphate, and abnormally large amounts of heavy metals. They commonly display excellent fissility and well-preserved planktonic and nektonic faunas and plant debris. Benthic fossils are rare or absent. Some black shales are sources of hydrocarbons. *See* Shale.

Black shales are enigmatic deposits. Although the large organic carbon content (3–15%) must have required reducing conditions of deposition, there are few unambiguous indicators of the specific environment, most especially of the depth of water. *See* Marine sediments; Sedimentology.

Black shales are typically well laminated on a scale of millimeters. Laminae are produced by variations in the supply of sediment, such as seasonal alternations of clay and planktonic algae. Delicate laminae can be preserved only in the total absence of benthic life, for burrowing animals disrupt lamination, producing bioturbated texture. Hence it is possible to differentiate between totally anaerobic conditions of deposition and marginally oxygenated (dysaerobic) conditions by recognizing laminated or bioturbated fabrics in a shale. *See* Facies (geology). [C.W.By.]

Blackberry Those species of the genus *Rubus*, plant order Rosales, having fruits that persist on the receptacles. The plants are upright or trailing shrubs, usually thorny, with perennial roots and biennial canes (stems). Horticulturally, these fruits are classified as upright and trailing blackberries and black-fruited and red-fruited dewberries, the latter being closely related to the raspberry, with which there are a few hybrids. *See* Fruit.

Blackberries are grown in home gardens over most of the United States, except for the coldest parts of the Great Plains area and the hottest part of the South. Commercial production is extensive in many states. The upright types are grown in the North, whereas the more tender dewberries and trailing blackberries are grown in the South and along the Pacific Coast. States usually leading in blackberry production are Oregon, Texas, California, Washington, Michigan, Arkansas, Oklahoma, Alabama, and North Carolina. Although the fruit is sold fresh or canned for dessert purposes and is made into jelly or jam, quick-freezing is the largest outlet for the crop. *See* Rosales. [J.H.Cl.]

Blackbody An ideal energy radiator, which at any specified temperature emits in each part of the electromagnetic spectrum the maximum energy obtainable per unit time from any radiator due to its temperature alone. A blackbody also absorbs all the energy which falls upon it. The radiation properties of real radiators are limited by two extreme cases—a radiator which reflects all incident radiation, and a radiator which absorbs all incident radiation. Neither case is completely realized in nature. Carbon and soot are examples of radiators which, for practical purposes, absorb all radiation. Both appear black to the eye at room temperature, hence the name blackbody. Often a blackbody is also referred to as a total absorber. *See* HEAT RADIATION. [H.G.S.; P.J.W.]

Blackleg An acute, usually fatal, disease of cattle and occasionally of sheep, goats, and swine. The infection is caused by *Clostridium chauvoei* (*C. feseri*), a strictly anaerobic, sporeforming bacillus of the soil. The disease is also called symptomatic anthrax or quarter-evil. The characteristic lesions in the natural infection consist of crepitant swellings in involved muscles, which at necropsy are dark red, dark brown, or blue black. Artificial immunization is possible; animals surviving an attack of blackleg are permanently immune to recurrence of the disease. *See* IMMUNITY. [L.S.McC.]

Blasting The process of loosening or fragmenting solid materials such as rock, masonry, or hard clays through the use of explosives to permit their excavation. Blasting consists of drilling and cleaning blast holes, loading the holes with a suitable explosive and igniter or detonator, stemming the hole (tamping sand or clay over the charge to help prevent dissipation of its explosive force), and firing or setting off the charge.

The most common explosives are ANFO (ammonium nitrate–fuel oil), water gels, and dynamite. Black powder, the original commercial explosive, is almost never used today. About 80% of all blasting is done with ANFO because of its low cost. Dynamite and water gels are used for special situations.

Dynamite and some water gels are detonated with blasting caps or detonating cord, while other water gels and ANFO require a high-explosive primer detonated with a cap or detonating cord. Electrical blasting caps are detonated with blasting machines, either hand-driven or battery-powered, or from power lines through special switches. Modern nonelectrical systems reduce the possibility of premature detonation because they are not affected by stray electric currents. Explosives in most blasts involving more than a few holes are detonated in sequence to provide better rock breakage and reduce vibration or shock in the earth that could cause damage. The sequential delays between holes are provided by delay blasting caps, delay devices in detonating cord lines, sequential blasting machines with electronic delay circuits, or a combination of these methods.

The blast holes near an exposed face in construction or quarrying are detonated first to move rock out of the way of material which will be moved by explosives further from the face. In a tunnel, holes near the center of the face are detonated first to provide space for the rock nearer the edges to move into. *See* CONSTRUCTION METHODS. [W.Di.]

Blastoidea A class of extinct Pelmatozoa in the subphylum Crinozoa, which arose in the Ordovician and flourished during Carboniferous times; they did not survive the Permian. The symmetrical bud-shaped theca comprised 13 rigid plates arranged in three horizontal rings of three basals, five radials, and five deltoids. There was often a jointed aboral stem. The ambulacral grooves were carried on five lancet plates lying in notches in the radial and deltoid plates. Series of lateral plates margined the lancet plates on either side, and each lateral plate supported a brachiole (see illustration). Each brachiole carried a

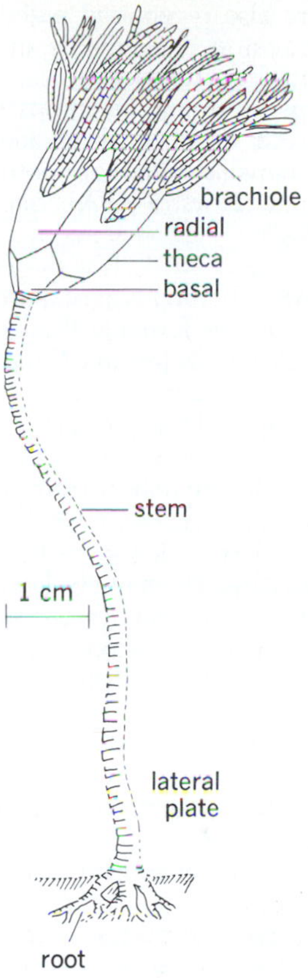

A blastoid, *Pentremites*.

branch of the ambulacral groove. Underneath the ambulacra the radial and deltoid plates were thrown into vertical pleats whose folds hung into the coelom. The reverse folds on the upper side opened to the exterior via pores. *See* CRINOIDEA; CRINOZOA; ECHINODERMATA; PELMATOZOA. [H.B.F.]

Blastomycetes A class of the subdivision Deuteromycotina comprising anamorphic (asexual or imperfect) yeast fungi that lack fruit bodies (conidiomata), have no dikaryophase, and are usually unicellular rather than filamentous. The thallus consists of individual cells. Approximately 22 genera comprising about 300 species are recognized.

The Blastomycetes, like other groups of deuteromycetes, are artificial, composed entirely of anamorphic fungi of ascomycete or basidiomycete affinity. Taxa are referred to as form genera and form species because the absence of sexual, perfect, or meiotic states forces classification and identification by artificial rather than phylogenetic means. Black yeasts are distinguished from anamorphic yeasts by the presence of melanin in the cell walls, abundant production of septate mycelium (filamentous), and aerial dispersal of conidia. Unlike other deuteromycetes, the number of morphological and developmental features for classification of Blastomycetes, although useful, is limited. The emphasis in yeast systematics has therefore been on physiological and biochemical tests, supplemented extensively by serological, electrophoretic, and molecular techniques. *See* ASCOMYCOTINA; BASIDIOMYCOTINA.

Anamorphic yeasts can be recovered from most ecological niches—animals, plants and their surfaces, fresh and marine water, soils, and environments such as manufacturing plants, tanning fluids, and mineral oils. Blastomycetes are of great economic importance in two respects: the production of products and the spoilage of raw materials and products. Selected strains of *Saccharomyces cerevisiae* are used in the baking, brewing, distilling, and wine industries.

Blastomycetes are also recognized pathogens in medicine. Both *Candida*, causing candidiasis or candidosis, and *Cryptococcus*, causing cryptococcosis, are opportunistic pathogens that cause systemic infections only in individuals with lowered resistance. Esophageal candidiasis and cryptococcosis of the central nervous system are both regarded as being particularly strong indicators of AIDS. *See* DEUTEROMYCOTINA; EUMYCOTA; FUNGI; YEAST. [B.C.S.]

Blastomycosis A fungus disease caused by *Blastomyces dermatitidis*. It was formerly thought that this disease was limited to the United States and Canada, but there have been reports of several cases in Africa.

There are two recognized forms of blastomycosis: cutaneous and systemic. In most instances, the cutaneous form is a clinical manifestation of dissemination from some primary site, probably the respiratory tract. However, a large percentage of cases with cutaneous lesions fail to reveal any lesions in the lungs. Systemic blastomycosis may involve the lungs, with dissemination to the viscera, central nervous system, or bones, as well as the skin. Treatment of blastomycosis requires the use of amphotericin B. Another drug, 2-hydroxystilbamidine, may be used if the disease is not severe. *See* AMPHOTERICIN B; ASCOMYCOTINA; MEDICAL MYCOLOGY. [L.D.H.]

Blastulation The formation of a segmentation cavity or blastocoele within a mass of cleaving blastomeres and rearrangement of blastomeres around this cavity in such a way as to form the type of definitive blastula characteristic of each species. The blastocoele originates as an intercellular space which sometimes arises as early as the four- or eight-cell stage. Thus blastulation is initiated during early cleavage stages, and formation of the definitive blastula is thought to terminate cleavage and to initiate gastrulation. Initially the diameter of the blastula is no greater than that of the activated egg; subsequently it increases. *See* CLEAVAGE (EMBRYOLOGY); GASTRULATION.

The blastula is usually a hollow sphere. Its wall may vary from one to several cells in thickness. In eggs which contain considerable amounts of yolk the blastocoele may be eccentric in position, that is, shifted toward the animal pole. The animal portion of its wall is always completely divided into relatively small cells, whereas the vegetative portion tends to be composed of relatively large cells and may be incompletely cellulated in certain species. The blastocoele contains a gelatinous or jellylike fluid, which originates in part as a secretion by the blastomeres and in part by passage of water through the blastomeres or intercellular material, or both, into the blastocoele.

The wall of the blastula is a mosaic of cellular areas, each of which will normally produce a certain structure during subsequent development. In other words, each area of cells in the wall of the blastula has a certain prospective fate which will be realized in normal development. [R.L.W.]

Bleaching The process in which natural coloring matter is removed from a fiber to make it white. The process may be used on fiber, yarn, or fabric. Prior to the bleaching of a fabric, preliminary purification processes should be used. These processes remove applied encrustants (desizing) and natural encrustants (scouring or boil-off) so that the bleaching agent may act uniformly on the material free of impediment. *See* DESIZING; SCOURING.

Bleaching is also classified as a purification process and varies with the content of the substrate or fibrous content of the material. It should not be confused with the stripping process, which is the removal of applied color.

The fabric off the loom is called gray, grey, or greige goods to distinguish it from the partially or completely finished fabric.

The three most prominent commercial bleaching processes are the peroxide, the chlorine, and the chlorite, in that order. In home bleaching the predominant bleach is the chlorine bleach, followed by the use of peroxygen compounds such as persulfates and perborates. The latter two are recommended for minimum care and permanent-press fabrics to preclude the yellowing of the whites and the stripping of colored fabrics, which are always potential problems when chlorine bleaches are used.

Optical bleaching uses organic compounds which are capable of absorbing waves shorter than visual waves and emitting waves within the visible range. The most notable of the short waves absorbed is the ultraviolet light present in daylight. These compounds absorb ultraviolet light which is less than 400 micrometers (0.02 in.) and emit blue light which is in the visible range of 400–700 micrometers (0.02–0.03 in.). This emitted blue light will counteract the yellow on the fabric surface and by the subtractive theory of color will produce a white. These agents are also characterized as brighteners and are extensively used in household detergents.

There are several bleaching systems which are classified as combination bleaches because two different bleaching agents are used together or in tandem. These have been projected to lower the cost, lower the degradation, or shorten the cycle as well as to preclude equipment damage in certain cases. *See* NATURAL FIBER; OXIDIZING AGENT; TEXTILE CHEMISTRY. [J.J.McD.]

Blimp A name originally applied to nonrigid airships and usually implying small size. Early blimps contained about 100,000 ft^3 (2800 m^3) and were only a fraction of the size of the rigid airships of that period. The nonrigid pressure-type airship, however, did not stay small, but each succeeding model built was substantially larger than its predecessor until ships with volumes of approximately 1,500,000 ft^3 (42,000 m^3) were operational. With the advent of these larger sizes, the name blimp tended to be replaced by the more general term airship. However, the principles of construction of these ships are basically the same regardless of the size. *See* AIRSHIP.

The fabric of the main envelope or pressure hull is usually made up of two or three plies of cloth impregnated with an elastomer; at least one of these plies is placed in a bias direction with respect to the others. This tends to make the fabric more resistant to shear loads and results in a stabilized structure. The materials used for the envelope must be lightweight, extremely strong, and resistant to gas diffusion. The airship envelope is a symmetrical airfoil in the form of a body of revolution, circular in cross section and elliptical along its longitudinal axis.

Inside the envelope are catenary curtains (see illustration) that support the weight of the car by distributing the loads imposed by it into the envelope fabric. This suspension system consists of cable systems attached to the car which terminate in fabric curtains, which in turn are cemented and sewed to the envelope proper. The envelope also contains one or more air cells, called ballonets, fastened to the bottom or sides of the envelope, which are used to maintain the required pressure in

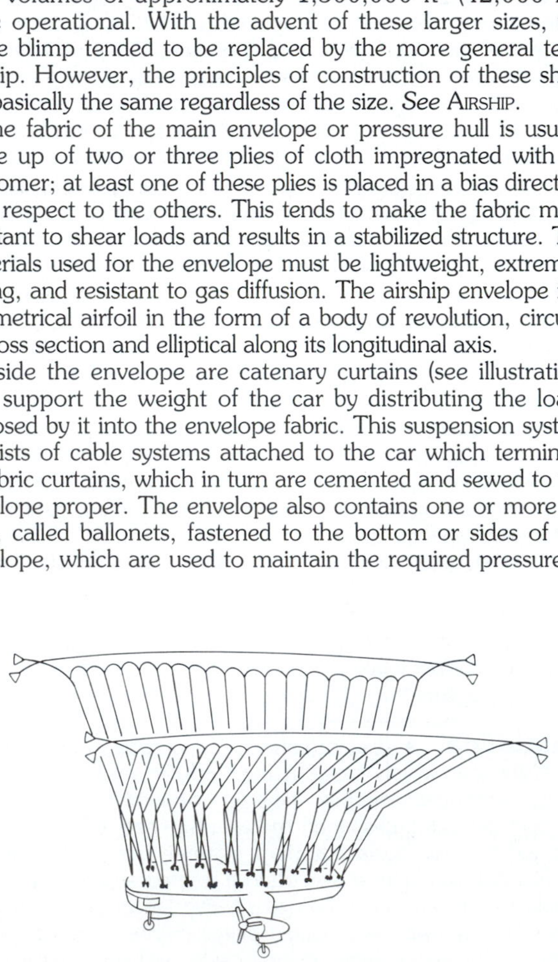

Typical internal rigging of nonrigid airships.

the envelope without adding or valving of gas as the ship ascends or descends.

The only commercial airships operating in the United States are the Goodyear Tire and Rubber Co. advertising blimps of about 150,000 ft^3 (4200 m^3). Fundamentally, they are communications platforms used for public service where airship characteristics are superior to other forms of flight. [R.S.R.]

Bloch theorem The theorem which states that the wave function of an electron moving in a periodic potential (such as in a crystal lattice) has the form of a plane wave, modulated by a function which has the same periodicity as the potential. *See* BAND THEORY OF SOLIDS. [J.C.]

Block and tackle Combination of a rope or other flexible material and independently rotating frictionless pulleys; the pulleys are grooved or flat wheels used to change the direction of motion or application of force of the flexible member (rope or chain) that runs on the pulleys (see illustration). The block

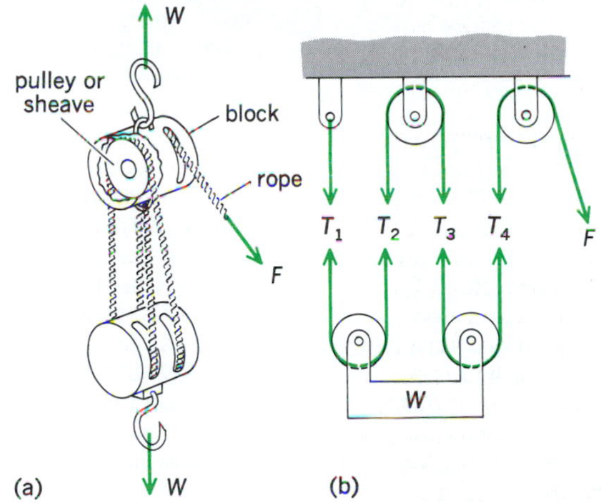

Block and tackle. *(a)* **Actual view.** *(b)* **Schematic. Tension** $T_1 = T_2 = T_3 = T_4$ **= 1/4 weight** W**; applied force** $F = W/4$.

and tackle is used where a large multiplication of the applied forces is desirable. Examples are: lifting weights, sliding heavy machinery into position, and tightening fences. *See* PULLEY; SIMPLE MACHINE. [R.M.Ph.]

Block diagram A diagram consisting of unidirectional, operational blocks that represent the transfer function of the variables of interest in a system. The importance of the cause-and-effect relationship of a transfer function is expressed in diagrammatic form. A block diagram representation of cause-and-effect relationships can be used in engineering and management

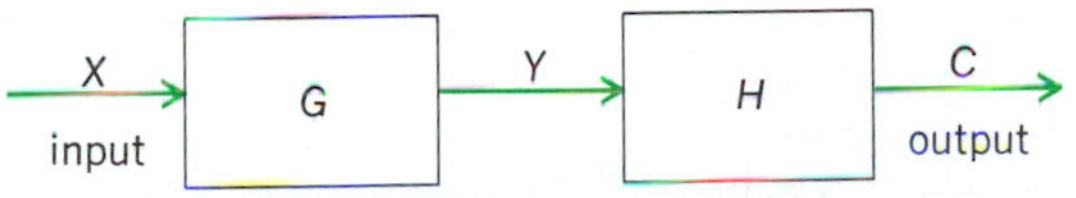

A block diagram with two blocks.

science. An example of a two-block process is shown in the illustration. The first process G acts on the input X to yield an output Y. The output Y of the first block becomes the input to the second block, H. For the total system, X is the input and C is the output. In order to represent a system with several variables, an interconnection of blocks is utilized. *See* CONTROL SYSTEMS; SYSTEMS ENGINEERING. [R.C.D.]

Blocking oscillator A relaxation oscillator that generates a short-time-duration pulse using a single transistor or vacuum tube and associated circuitry. The input and output are coupled together by a transformer in a regenerative feedback arrangement, with the output current being periodically interrupted because of the regenerative action which drives the input far into the reverse-bias or cutoff region. The duration of the period between pulses, which coincides with the period during which the output current is blocked, is dependent upon the value of the RC time constant in the input circuit. The blocking oscillator may be made to operate in either a monostable or astable fashion. When it is operated as a monostable device, considerable control can be exercised over the shape and duration of the pulse, making the circuit useful as a pulse generator. *See* PULSE GENERATOR; RELAXATION OSCILLATOR.

The blocking oscillator may be synchronized with external pulses occurring at a rate slightly faster than its own natural frequency in the same manner as multivibrators. It may be used in the same manner as a frequency divider. *See* MULTIVIBRATOR. [G.M.G.]

Blood The fluid that circulates in the blood vessels of the body. Blood consists of plasma and cells floating within it. The cells are derived from extravascular sites and then enter the circulatory system. They frequently leave the blood vessels to enter the extravascular spaces, where some of them may be transformed into connective tissue cells. The fluid part of the blood is in equilibrium with the tissue fluids of the body. The circulating blood carries nutrients and oxygen to the body cells, and is thus an important means of maintaining the homeostasis of the body. It carries hormones from their sites of origin throughout the body, and is thus the transmitter of the chemical integrators of the body. Blood plasma also circulates immune bodies and contains several of the components essential for the formation of blood clots. Finally, blood transports waste products to excretory organs for elimination from the body. Because of its basic composition (cells surrounded by a matrix), development, and ability to modify into other forms of connective tissues, blood can be regarded as a special form of connective tissue. *See* CONNECTIVE TISSUE.

Formed elements. The cells of the blood include the red blood cells and the white blood cells. In all vertebrates, except nearly all mammals, the red blood cells or corpuscles contain a nucleus and cytoplasm rich in hemoglobin. In nearly all mammals the nucleus has been extruded during the developmental stages.

In normal adult men the blood contains about 5,000,000 red blood corpuscles or erythrocytes per cubic millimeter; in normal adult women, about 4,500,000. Human erythrocytes are about 8 micrometers in diameter and about 2 μm at their thickest and have a biconcave shape. They contain hemoglobin, which imparts to them their color, and possess an envelope. When circulating in the blood vessels, the red blood cells are not evenly dispersed. In the capillaries the erythrocytes are often distorted. In certain conditions they may be densely aggregated. This is known as a sludge. The erythrocytes respond to changes in osmotic pressure of the surrounding fluid by swelling in hypotonic fluids and by shrinking irregularly in hypertonic fluids. Shrunken red blood cells are referred to as crenated cells. The average life of the mature red blood cells is surprisingly long, having a span of about 120 days. *See* HEMATOLOGIC DISORDERS; HEMOGLOBIN.

In humans the white blood cells in the blood are fewer in number. There are about 5000–9000/mm^3. In general, there are two varieties, agranular and granular. The agranular cells include the small, medium, and large lymphocytes and the monocytes (see illustration). The small lymphocytes are spherical, about the diameter of erythrocytes or a little larger, and constitute about 20–25% of the white blood cells. The medium

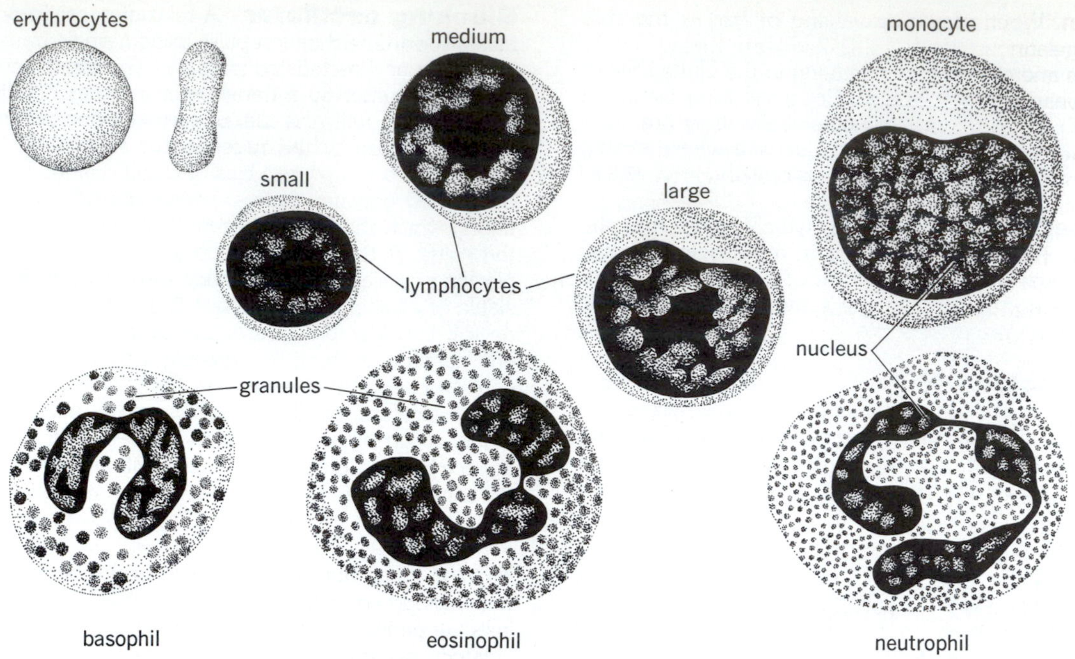

Diagrammatic representation of human blood cells.

and large lymphocytes are relatively scarce. In all lymphocytes the nucleus occupies nearly the whole volume of the cell, and the cytoplasm which surrounds it forms a thin shell. The typical monocyte is commonly as large as a large lymphocyte (12 μm), and constitutes 3–8% of the white blood cells. The nucleus is relatively small, eccentric, and oval or kidney-shaped. The cytoplasm is relatively larger in volume than that in lymphocytes.

The granular leukocytes are of three varieties: neutrophil, eosinophil, and basophil. Their structure varies somewhat in different species, and the following applies to those of humans. The neutrophils make up 65–75% of the leukocytes. They are about as large as monocytes with a highly variable nucleus, consisting of three to five lobes joined together by threads of chromatin. The cytoplasm contains numerous minute granules which stain with neutral dyes and eosin. The eosinophils (also called acidophils) are about the same size as the neutrophils but are less numerous, constituting about 1% of the leukocytes. The nucleus commonly contains but two lobes joined by a thin thread of chromatin. The granules which fill the cytoplasm are larger than those of the neutrophils and stain with acid dyes. The basophils are about the same size as the other granular leukocytes. The nucleus may appear elongated or with one or more constrictions. The granules are moderately large, stain with basic dyes, and are water-soluble.

The functions of the leukocytes while they are circulating in the blood are not known. However, when they leave the blood vessels and enter the connective tissue, they constitute an important part of the defense mechanism and of the repair mechanism. Many of the cells are actively phagocytic and engulf debris and bacteria. Lymphocytes are of two major kinds, T cells and B cells. They are involved in the formation of antibodies and in cellular immunity.

The blood platelets are small spindle-shaped or rodlike bodies about 3 μm long and occur in large numbers in circulating blood. In suitably stained specimens they consist of a granular central portion (chromomere) embedded in a homogeneous matrix (hyalomere). They change their shape rapidly on contact with injured vessels or foreign surfaces and take part in clot formation. The platelets are not to be regarded as cells and are thought to be cytoplasmic bits broken off from their cells of origin in bone marrow, the megakaryocytes. [I.G.]

Plasma. Plasma is the residual fluid of blood left after removal of the cellular elements. Serum is the fluid which is obtained after blood has been allowed to clot and the clot has been removed. Serum and plasma differ only in their content of fibrinogen and several minor components which are in large part removed in the clotting process. *See* SERUM.

The major constituents of plasma and serum are proteins. The total protein concentration of human serum is approximately 7 g/ml, and most other mammals show similar levels. By various methods it can be demonstrated that serum protein is a heterogeneous mixture of a large number of constituents. Only a few are present in higher concentrations, the majority being present in trace amounts. More than 60 protein components have been identified and characterized. Albumin makes up more than one-half of the total plasma proteins and has a molecular weight of 69,000. Because of its relatively small molecular size and its high concentration, albumin contributes to 75–80% of the colloid osmotic pressure of plasma. The immunoglobulins, which represent approximately one-sixth of the total protein, largely constitute the γ-globulin fraction. The immunoglobulins are antibodies circulating in the blood, and therefore are also called humoral antibodies. They are of great importance in the organism's defense against infectious agents, as well as other foreign substances. *See* IMMUNOGLOBULIN.

In addition to the proteins, many other important classes of compounds circulate in the blood plasma. Most of these are smaller molecules which diffuse freely through cell membranes and are, therefore, more similarly distributed throughout all the fluids of the body and not as characteristic for plasma or serum as the proteins. In terms of their concentration and their function, the electrolytes are most important. They are the primary factors in the regulation of the osmotic pressure of plasma, and contribute also to the control of the pH. The chief cations are sodium, potassium, calcium, and magnesium. The chief anions are chloride, bicarbonate, phosphate, sulfate, and organic acids. The circulating blood also contains the many small compounds which are transported to the sites of synthesis of larger molecules in which they are incorporated, or which are shifted as products of metabolic breakdown to the sites of their excretion from the body. [H.Cl.]

Coagulation. When mammalian blood is shed, it congeals rapidly into a gelatinous clot of enmeshed fibrin threads which trap blood cells and serum. Modern theories envision a succession of reactions leading to the formation of insoluble fibrin from a soluble precursor, fibrinogen (factor I). Blood also clots when it touches glass or other negatively charged surfaces, through reactions described as the intrinsic pathway. Several of the steps in this process are dependent upon the presence in blood of calcium ions and of phospholipids, the latter derived principally from blood platelets. The coagulation of blood can also be induced by certain snake venoms which either promote the formation of thrombin or clot fibrinogen directly, accounting in part for their toxicity.

Platelets, besides furnishing phospholipids for the clotting process, help to stanch the flow of blood from injured blood vessels by accumulating at the point of injury, forming a plug. Platelets participate in the phenomenon of clot retraction, in which the blood clot shrinks, expelling liquid serum. Although the function of retraction is unknown, individuals in whom this process is impaired have a bleeding tendency.

Hereditary deficiencies of the function of each of the protein-clotting factors have been described, notably classic hemophilia and Christmas disease, which are disorders of males and clinically indistinguishable. The various hereditary functional deficiencies are associated with a bleeding tendency with one inexplicable exception. Acquired deficiencies of clotting factors, sometimes of great complexity, are also recognized. Therapy for bleeding due to deficiencies of clotting factors often includes the transfusion of blood plasma or fractions of plasma rich in particular substances the patient may lack. *See* HUMAN GENETICS.

Clinical tests of the coagulability of the blood include (1) determination of the clotting time, that is, the time elapsing until shed blood clots; (2) the prothrombin time, the time elapsing until plasma clots in the presence of tissue thromboplastin (and therefore a measure of the extrinsic pathway of clotting); (3) the partial thromboplastin time, the time elapsing until plasma clots in the presence of crude phospholipid (and therefore a measure of the intrinsic pathway of clotting); (4) the enumeration of platelets; and (5) crude quantification of clot retraction and of the various plasma protein-clotting factors.

Heparin, a polysaccharide–sulfuric acid complex found particularly in the liver and lungs, impairs coagulation; its presence in normal blood is disputed. Both coumarin and heparin are used clinically to impede coagulation in thrombotic states, including thrombophlebitis and coronary heart disease. *See* FIBRINOGEN. [O.D.R.]

Blood groups The blood of humans and higher animals has been classified according to the red blood cell antigens. About 200 different blood groups have been identified as part of the surface structure of human red blood cells. Each of these is a unique serologic and biochemical entity, and is one member of a complex genetic polymorphism. Each of the blood groups is identified by its ability to react, selectively and specifically, with a corresponding antibody. In many cases the pattern of reactions with various antibodies not only identifies the blood group, but also places it within the genetic confines of one of the 19 known blood group systems. The blood groups have medical, legal, and anthropologic importance, and are valuable genetic markers in linkage studies. *See* POLYMORPHISM (GENETICS).

Every blood group is a definite and specific structure that is a part of one of the many different blood group antigens. The inheritance of each antigen is genetically controlled, and data support the concept of one gene for each antigen, with the understanding that each of these antigens is a complex serologic (and biochemical) entity. Every blood group antigen has at least several different sites, each of which reacts with a specific blood group antibody. About 200 of these sites have been identified, and each is a separate blood group.

Antibodies directed against blood group antigens are of two types, immune and naturally occurring. Naturally occurring antibodies, anti-A and anti-B, may be found normally in any person who lacks the corresponding antigen (see table). In all probability, these antibodies are formed after contact with immunogens related to, but not identical with, the blood group antigens. Most blood group antibodies are true immune antibodies and do not occur in plasma until after actual immunization with the blood group antigen itself. The immunizing stimulus is almost always transfusion or pregnancy.

ABO blood group system

Blood group	RBC antigens	Possible genotypes	Plasma antibody
A	A	*A/A* or *A/O*	anti-B
B	B	*B/B* or *B/O*	anti-A
O	—	*O/O*	anti-A and anti-B
AB	A and B	*A/B*	—

Blood group systems. The inheritance pattern of many of the 200 known blood groups is well enough understood to place them in one of the clearly established blood group systems. Genetic control for each system is at a single locus, and in most cases the number of allelic genes for each system is greater than two. Since an individual inherits one gene for every blood group system from each parent, the individual's blood group phenotypes are the sum of genetic information from both parents. Almost all blood groups are examples of codominance, and genes are equally well expressed when present in homozygous or heterozygous states. All loci, except for Xg^a, are on autosomal chromosomes.

The four common blood types in the ABO system are A, B, AB, and O. ABO antigens are found not only on red cells but also in soluble form in body fluids of about 80% of the population. Antibodies reacting with the A or B blood factors are always found in the plasma of persons who lack the antigens, and are the most important of the naturally occurring antibodies. No antibody that identifies the product of the *O* gene has been reported.

The inherent complexity of the Rh system has been magnified far out of proportion by a long-standing controversy over the proper genetic interpretation of the data and the choice of terms to use in describing the Rh blood groups. All Rh antigenic determinants are transmitted by a segment on one pair of autosomal chromosomes. At least 36 alleles are known, of which only 8 are common and can be considered basic. Each of these alleles determines an Rh antigen on the red cell. There is almost no biochemical or structural information about the Rh antigens. Of the 30 Rh blood groups now defined, RH_O, or D, is the one of greatest clinical importance, and its presence or absence is the criterion by which bloods are classified as Rh positive or Rh negative. This is because RH_O(D) is the most immunogenic of the Rh blood groups. As a result, individuals who are RH_O(D) negative should, except in emergencies, be transfused with donor blood that is also RH_O(D) negative.

The Lewis blood group system is unique in several ways and, although of relatively little clinical importance, is of great scientific interest. Lewis differs from ABO, and from all other groups, in that the Lewis characteristics of erythrocytes are passively acquired and are not an integral part of the cell membrane.

Xg^a is the only blood group gene not carried on one of the autosomal chromosomes. The frequency of Xg (a+) females is higher than that of Xg (a+) males. In addition, careful family studies have shown without doubt that the gene for the Xg[a] blood group is on the X chromosome. Only one antibody in the Xg[a] system has been found so that, at present, but one allele, Xg[a], and two phenotypes, Xg (a+) and Xg (a −), can be identified. The Xg[a] system is of little importance in transfusion, but, because of the localization of the gene, is of great genetic interest.

Details of other blood group systems are beyond the scope of this article. Some, such as Kell, Duffy, and Kidd, are relatively important to transfusionists, since antibodies to these groups are more frequent and transfusion reactions from incompatibility a distinct hazard.

Significance. Study of blood group factors has made transfusion a safe practice, identified the cause of erythroblastosis fetalis, provided a method for paternity exclusion, and contributed to ethnology, genetics, and medicine.

In most cases only the ABO and RH_O blood groups of donor and recipient are determined before transfusion. The prime consideration in selecting blood for infusion is to avoid the transfusion reaction that will occur when blood containing any of the blood groups is given to a recipient who has developed antibody against that group. Since the antibodies in the ABO system are naturally occurring and always present when the corresponding antigen is lacking from the cells, a transfusion reaction is inevitable if ABO incompatible blood is given. For this reason it is essential to do careful ABO typing on both donor and recipient and to select compatible blood for transfusion. Group O erythrocytes, lacking both A and B antigens, can be used in recipients of other ABO groups without fear of a reaction from ABO incompatibility. In a similar fashion, persons of group AB can be safely transfused with cells of any ABO group. A second consideration in selecting blood for transfusion is to prevent, if possible, the recipient from forming antibodies. Since most blood groups are not strongly immunogenic, it is current practice to disregard all of them except RH_O(D).

Erythroblastosis fetalis is a hemolytic disease of newborn infants that can occur as a result of blood group incompatibility between mother and fetus. In the classifical form, it occurs when an Rh-negative mother carries an Rh-positive fetus. During pregnancy Rh-positive erythrocytes from the baby cross the placenta and immunize the Rh-negative mother, so that she produces anti-Rh. Her antibody, in turn, crosses the placenta and enters fetal circulation, where it reacts with and hemolyzes fetal red cells. In severe cases this leads to intrauterine death; less severely affected infants require treatment by exchange transfusion at birth. Intrauterine transfusion of unborn infants has been used with good results. *See* RH INCOMPATIBILITY.

The inheritance pattern of most blood groups is such that blood group tests of mother, child, and putative father can, at times, exclude paternity. No blood factor can occur in a child unless it is present in at least one of his parents. If both mother and putative father lack a factor that is present on the child's cells, it can be assumed that one of the two is not the real parent. Since maternity is usually not at issue, such findings are interpreted as an exclusion of paternity. Maternity exclusion can also be determined, if indicated, as when it is suspected that babies have been mixed up in a nursery. At present, ABO, Rh, and MN blood group systems are accepted by most courts in paternity trials.

The frequency of certain blood groups in patients with various diseases has been studied at length. Many data have been gathered, and almost every disease has been associated with some blood group. Most of these correlations do not stand close statistical scrutiny, but a few have been so well studied and so commonly found that there is little doubt that they are real. These include the increased incidence of group O in persons with peptic ulcer disease and the increased incidence of group A in cancer of the stomach or pernicious anemia. The mechanism for these relationships is unknown. *See* BLOOD.

[J.R.Bov.]

Blood-plate hemolysis The destruction of red blood cells in an agar medium, such as beef heart infusion agar with rabbit's blood, contained in a petri dish. The cells are destroyed by hemolysin, a toxin produced by such bacteria as streptococcus and staphylococcus. The types of destruction or hemolysis are given Greek letters and usually apply to streptococcus and its classification. Green discoloration (alpha) around a streptococcal colony, or the halo of clearness (beta) contrast with the deep-pink background of growth medium. Absence of change in the red cells and consequent absence of discoloration around the colony represent the gamma type. *See* STAPHYLOCOCCUS; STREPTOCOCCUS.

[P.L.B.]

Blood vessels Tubular channels for blood transport, of which there are three principal types: arteries, capillaries, and veins. Only the larger arteries and veins in the body bear distinct names. Arteries carry blood away from the heart through a system of successively smaller vessels. Capillaries are the smallest but most extensive blood vessels, forming a network everywhere in the body tissues. Veins carry blood from the capillary beds back to the heart through increasingly larger vessels. In certain locations blood vessels are modified for particular functions, as the sinusoids of the liver and the spleen and the choroid plexuses of the brain ventricles. *See* ARTERY; CAPILLARY (ANATOMY); CIRCULATORY SYSTEM; LYMPHATIC SYSTEM; VEIN.

[W.J.B.]

Blowout coil A coil that produces a magnetic field in an electrical switching device for the purpose of lengthening and extinguishing an electric arc formed as the contacts of the switching device part to interrupt the current. The magnetic field produced by the coil is approximately perpendicular to the

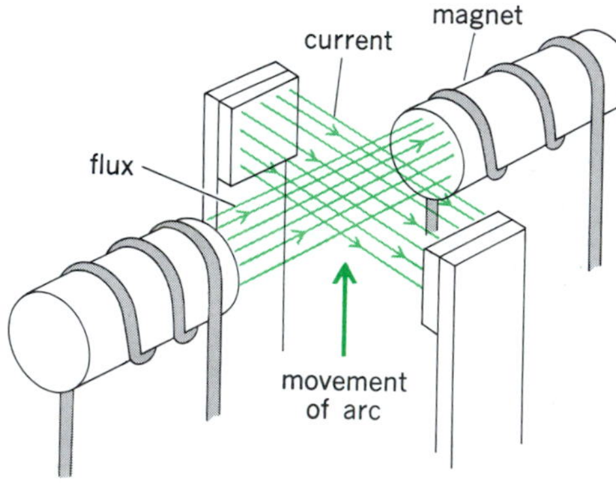

Relation of directions of current, magnetic flux, and movement of arc in a blowout coil.

arc. The interaction between the arc and current and the magnetic field produces a force driving the arc in the direction perpendicular to both the magnetic flux and the arc current (see illustration).

[T.H.L.]

Blowpipe In glass blowing, a long straight tube on which molten glass is gathered and worked, partly by blowing into the tube. The blowpipe is spun to shape the glass object further by centrifugal force, or by a tool, in which case the blowpipe acts as a spindle for turning.

In analytical chemistry, a blowpipe is a small, tapered, and frequently curved tube that directs a jet, usually of air, into a flame to concentrate the flame onto an unknown substance.

Coloration of the flame and other characteristic reactions in the reducing and in the oxidizing portions of the flame created by the blowpipe aid in identifying the substance. Such a blowpipe may be blown directly by the analyst, from a bellows, or from a pressurized line. [F.H.R.]

Blue straggler star A star that is a member of a stellar association and is located at an unusual position on the association's color-magnitude diagram, above the turnoff from the main sequence. *See* Color index; Hertzsprung-Russell diagram; Magnitude (astronomy).

The blue stragglers were discovered by A. Sandage in 1953 in the galactic globular cluster M3 (see illus.). They are located

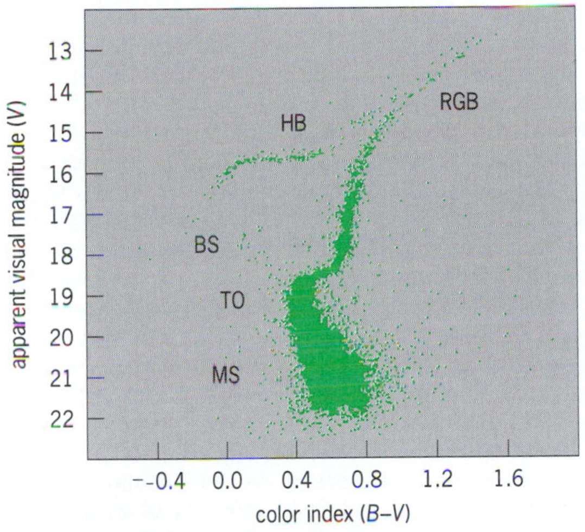

Color-magnitude diagram [apparent visual magnitude (*V*) versus color index (*B*-*V*)] of 10,637 stars in the galactic globular cluster M3 (NGC 5272). The important evolutionary stages are marked by MS (main sequence), TO (turn-off), RGB (red giant branch), HB (horizontal branch), and BS (blue stragglers). (*After R. Buonanno et al., High precision photometry of 10,000 stars in M3, Mem. Soc. Astron. It., 57:391–393, 1986*)

below the horizontal branch and above the turnoff from the main sequence. They form a new sequence extending from the extrapolation of the main sequence to higher luminosities on the left to the red giant branch on the right. While the majority of the cluster members fall rather precisely on the expected isochrones for the approximately 15-billion-year age of the cluster, the 50 or so blue stragglers in M3 are located on much younger isochrones. Approximately 600 objects of this type have been found so far in every known type of stellar association, including dwarf spheroidal galaxies. *See* Galaxy, external.

The simplest explanation of the phenomenon is that the blue stragglers are younger than the rest of the cluster members, perhaps because of a recent burst of star formation. This delayed-formation scenario is the most likely in open clusters and some luminous dwarf spheroidal galaxies. However, there is no evidence at all for recent star formation episodes in the galactic globular clusters. The most plausible explanations for blue stragglers in these systems are internal mixing that supplies new fuel to the core, mass transfer in binary systems in which one component has increased its mass at the expense of the other, and the merging of two low-mass stars to form a more massive star via direct collision. *See* Binary star; Stellar evolution. [F.P.]

Blueberry Several species of the genus *Vaccinium*, plant order Ericales, ranging from low-growing, almost prostrate plants to vigorous shrubs reaching a height of 12–15 ft (3.7–4.6 m). The fruit, a berry, is usually black and covered with bluish bloom, generally occurring in clusters, and has numerous small seeds, a characteristic that distinguishes the blueberry from the huckleberry, which has 10 rather large, gritty seeds. Although there are blueberry species on other continents, all cultivated varieties in the United States are American in origin.

The dryland blueberry (*V. ashei*) is adapted to relatively dry soils and has been brought under cultivation in Florida and Georgia. In the Northeast the lowbush blueberry (*V. lamarckii*) grows wild over thousands of acres of dry hillsides, where it is harvested commercially, especially in Maine, but also in other New England states, Michigan, Minnesota, and a few others. In the Northwest fruit of the evergreen blueberry (*V. ovatum*) is harvested in the wild, and large tonnages of the leafy twigs are shipped for use as florists' greens. The highbush blueberry, represented by *V. australe* and *V. corymbosum*, provides most of the cultivated plants. The highbush blueberry is found in swampy land, usually on hummocks.

The fruit is sold fresh, canned, and frozen. The blueberry is becoming increasingly popular in home gardens, although its cultural requirements are rather exacting. For garden culture, mulching with sawdust or other organic matter is desirable because the roots are shallow. *See* Ericales. [J.H.Cl.]

Bluefish A predatory and voracious species of fish that ranges throughout the tropical and temperate seas of the world, except for the eastern and central Pacific areas. A single species, *Pomatomus saltatrix*, makes up the family Pomatomidae. This fish, also known as the skipjack, is bluish-gray with an average length of 3 ft (0.9 m) and weight of about 5 lb (2.3 kg). The mouth is large with sharp, strong teeth. The bluefish form schools and migrate north along the Atlantic coast, following schools of smaller fish upon which they prey. The bluefish continue to kill and destroy their prey even after feeding. About June they reach the New England coast, where the young can be found in estuaries and bays. *See* Perciformes. [C.B.C.]

Bluegrass A common name applied to several species but particularly to Kentucky bluegrass (*Poa pratensis*) and to Canada bluegrass (*P. compressa*). Kentucky bluegrass is widely grown in temperate humid regions, doing best on fertile soils that are loamy or heavy in texture. Canada bluegrass is more abundant on less fertile soils in cooler regions. Both species are perennial, form sod by underground creeping stems, and have extensive root systems which are finely branched and comparatively shallow.

These grasses are utilized primarily for pasture and lawns. Bluegrasses respond to fertilizers, especially those supplying nitrogen, and the herbage generally is high in the mineral elements, particularly phosphorus. Growth occurs from early spring to late fall, except for near-dormancy during the hot, dry summer period. Kentucky bluegrass is the most important species in permanent pastures, recreation areas, and lawns in the northern half of the United States, where moisture is adequate from rainfall or irrigation. However, in pastures grown in rotation with crops in humid regions, it has been largely replaced by orchard grass and bromegrass. Several varieties have been developed specifically for turf use with special emphasis on disease resistance. *See* Bromegrass; Cyperales. [H.B.S.]

Blueschist metamorphism Regional metamorphism with the highest pressures and lowest temperatures, commonly above 5 kilobars (500 megapascals) and below 752°F (400°C). Metamorphic rocks of the relatively uncommon blueschist facies contain mineral assemblages that record these high pressures and low temperatures. The name "blueschist" derives from the fact that at this metamorphic grade, rocks of ordinary basaltic composition are often bluish because they contain the sodium-bearing blue amphiboles glaucophane or crossite rather than the calcium-bearing green or black amphiboles actinolite or hornblende, which are developed in the more common greenschist- or amphibolite-facies metamorphism.

Blueschist metamorphism developed almost exclusively in young Mesozoic and Cenozoic mountain belts, primarily during the last 5 to 10% of geologic time. During most of geologic time, high-pressure metamorphism has been dominated by high-temperature amphibolite-facies metamorphism, as well as by very high-temperature granulite-facies metamorphism, the latter being uncommon during the last 5 to 10% of geologic time. These phenomena are generally interpreted as being in some way an effect of the cooling of the Earth.

A variety of mineral assemblages are characteristic of blueschist-facies metamorphism, depending upon the chemical composition of the rock and the actual temperature and pressure within the field of blueschist metamorphism. Detailed consideration of these assemblages and their mineral composition allows the determination of the physical conditions of metamorphism.

Blueschist metamorphic rocks are found almost exclusively in the young mountain belts of the circum-Pacific and Alpine-Himalayan chains. The rocks are usually metamorphosed oceanic sediments and basaltic oceanic crust. Previously continental rocks rarely exhibit blueschist metamorphism. The tectonic mechanism for blueschist metamorphism must move the rocks to depths of more than 6 to 12 mi (10 to 20 km) while maintaining relatively cool temperatures (390–750°F or 200–400°C). These temperatures are much cooler than for continental crust at those depths.

What is not well understood is how the blueschist metamorphic rocks return to the surface; it is clear that the mechanism is not simple uplift and erosion of 12–18 mi (20–30 km) of the Earth's crust. Blueschist metamorphic rocks are usually in immediate fault contact with much less metamorphosed or unmetamorphosed sediments, indicating that they have been tectonically displaced relative to their surroundings since metamorphism. *See* METAMORPHIC ROCKS; METAMORPHISM. [J.Sup.]

Bluestem grass The common generic name often applied to the genera *Andropogon*, *Dichanthium*, *Bothriochloa*, and *Schizachyrium* in the grass tribe Andropogoneae. They are adapted to environments with high light intensities and high temperatures during the growing season. The bluestems are medium to tall, warm-season, perennial grasses. The two most important forage species are big bluestem (*A. gerardi*) and little bluestem (*S. scoparius*).

Big bluestem is a tall (usually 3 to 6 ft or 1 to 2 m), deep-rooted grass with strong rhizomes; it occurs throughout the continental United States (except in the extreme western states), in southern Canada, and in northern Mexico. One of the most palatable of all grasses when it is actively growing, its nutritional value declines sharply with maturity. Because of its high palatability and tall growth habit, big bluestem has been reduced or eliminated by heavy grazing over much of its range.

Little bluestem is a medium-height bunchgrass (usually 1 to 2 ft or 30 to 60 cm, but taller in the south) which occurs in the same geographic area as big bluestem but extends farther north in the prairie provinces of Canada to the southern Yukon. It is an excellent grass for grazing and hay when actively growing, but nutritional value declines rapidly with maturity. This is the grass that provides the major aspect of the Flint Hills of Kansas and the Osage Hills of Oklahoma. *See* CYPERALES. [J.K.L.]

Bluetongue An arthropod-borne disease of ruminant species. Its geographic distribution is dependent upon a susceptible ruminant population and climatic conditions that favor breeding of the primary vector, a mosquito (*Culicoides* species).

Bluetongue virus is the prototype of the genus *Orbivirus* (family Reoviridae). The viral genome exists as 10 segments of the double-stranded ribonucleic acid (RNA) that encode for seven structural and three nonstructural proteins. The viral particle has a double capsid, with the outer coat (morphologically poorly defined) being composed of two proteins. Twenty-four serotypes of bluetongue virus have been defined, and their distribution throughout the world is varied. *See* ANIMAL VIRUS.

While multiple ruminant species can become infected, only sheep and deer typically display clinical bluetongue disease. Severity of the disease is dependent upon multiple factors, including virus strain, animal breed, and environmental conditions. Upon infection by a gnat bite, the virus apparently replicates in the local lymphatic system prior to the viral particles moving into the blood (viremia). Viral replication occurs in the endothelial cells of small vessels, resulting in narrowing of the vessel, release of proteinaceous material into the surrounding tissues, and possibly hemorrhage, with the respiratory tract, mucous membranes, cardiac and skeletal musculature, and skin being most affected. Animals experiencing acute clinical symptoms typically die from pneumonia or pulmonary failure; hemorrhage at the base of the pulmonary artery indicates the presence of a vascular lesion.

Control of bluetongue disease requires the application of vaccines and modulation of the farm environment. While bluetongue virus vaccines are available, efficacy is often incomplete and variable, in part because of the multiplicity of serotypes active throughout the world and limited cross-serotype protection. Furthermore, use of polyvalent (multiple-serotype) vaccines in the United States has been discouraged because of potential genetic reassortment between vaccine viruses and wild-type viruses, a process that could possibly lead to pathogenic variants. Relative to environment, elimination of vector breeding sites can also facilitate control of virus transmission. With the multiplicity of serotypes typically active in an endemic area, and the minimal cross-serotype protection observed, administration of vaccine in the face of an outbreak may be of limited value. *See* VACCINATION. [J.L.Sto.]

Boa Name given to reptiles which are members of the family Boidae in the order Squamata. There are about 51 species of boas, none of which is poisonous. These snakes, along with the closely related pythons, are among the largest and are the most primitive of living snakes. They kill their prey by constriction. All species are viviparous, producing living young. Most of the species are found in the New World, being most numerous in the West Indies and Central and South America, although some species are found in Asia and northern and eastern Africa. *See* REPTILIA; SQUAMATA.

Constrictor constrictor, the boa constrictor, is one of the best-known species, living in tropical America from Mexico to Argentina and in the West Indies. It attains a length of 12 ft (3.7 m), is not dangerous to humans, and is easily reared in captivity.

The largest member of the boa family and largest living snake is the giant anaconda (*Eunectes murinus*); 25 ft (7.6 m) is the longest authenticated size. It is arboreal as well as aquatic, in that it may await its prey in trees lining a stream and then drag the captured mammal or bird into the water to kill and devour it.

An arboreal species, the greenish-yellow emerald boa (*Boa canina*), preys on birds and monkeys in forested areas of South America. It grows to about 6 ft (1.8 m) in length. An inhabitant of the arid desert regions of Asia Minor is the sand boa (*Eryx jaculus*), which preys upon small animals. [C.B.C.]

Boat propulsion The action of propelling a boat through water. A boat machinery plant consists principally of a propulsion engine, propulsor (propeller or jet pump), and drive-line components. The engines are almost exclusively of the familiar internal combustion types: gasoline, diesel, or gas turbine. The gasoline engine has traditionally dominated the pleasure-boat field, while the diesel is favored for commercial and military craft. The gas turbine is comparatively rare and is found only in applications where high power from machinery of small weight and volume is essential.

Auxiliary items, such as bilge pumps, domestic water pumps and heaters, and electric generators and switchboards typically are found in the machinery plants of the larger boats. The generator (or alternator) is often driven by a belt from the propulsion engine, as in automotive practice, but is sometimes driven by a separate engine.

The marine gasoline engine appears in inboard and outboard forms. The outboard engine is a unit assembly of engine, propeller, and vertical drive shaft that is usually clamped to the boat transom. It is traditionally the power plant for the smallest motor boats, or an auxiliary for rowboats and sailboats. The inboard form is almost exclusively adapted from one of the mass-produced automotive engines. Inboard engines are compact, light, lower in first cost than any competitor, and familiar to untrained users, and so they predominate among pleasure boats and the smaller fishing boats.

The diesel engine is generally higher in first cost than the gasoline engine and is somewhat heavier for the same power, but it consumes less fuel. Fuel savings make the diesel attractive if the engine is to be used more than a few hundred hours a year or if the boat must have a long cruising range. Its principal market is thus in commercial and military craft, but some of the more compact models are used in pleasure craft.

The gas turbine engine is comparatively light and compact, making it attractive for high-speed, high-power boats. Its disadvantages are high first cost, high fuel consumption, and large exhaust and intake ducts. The last factor is due to its high rate of air consumption; the need for large volumes of air also makes it difficult, in a small vessel, to keep spray from being drawn into the engine, with consequent fouling and corrosion. *See* DIESEL ENGINE; GAS TURBINE; INTERNAL COMBUSTION ENGINE; MARINE ENGINE; MARINE MACHINERY. [J.B.W.]

Bog Plant communities which develop and continue to grow in areas with permanently waterlogged peat substrates. Synonyms such as mire, quagmire, and moor have been used to describe this community. *See* PEAT.

Ecologists have described numerous types of bogs from different parts of the world, and various systems of nomenclature and classification have been adopted. The principal morphological distinction between bog types is concerned with the source of their water: from drainage (rheophilous) or direct precipitation (ombrophilous).

The major features of a bog are the peat substratum and its waterlogged condition, with the water table frequently lying just below, at, or above the peat surface. The peat itself is composed of partially decayed organic matter, originating chiefly from the plants. The reason for the accumulation of this organic matter is the very slow rate of breakdown, because of the lack of oxidative microorganisms in such waterlogged conditions. This adaptive process operates in geographical areas of low temperature and high rainfall.

Bogs are being utilized for an increasing range of purposes. Reclamation for agriculture and forestry and the extraction of peat are accelerating rapidly in some parts of the world. In some regions bogs contribute toward maintaining the hydrological stability of the landscape because they hold large quantities of water which is released slowly. Thus bogs are very important for water conservation and in controlling floods. [T.Pr.]

Bohrium A chemical element, symbol Bh, atomic number 107. Bohrium was synthesized and identified in 1981 by using the Universal Linear Accelerator (UNILAC) of the Gesellschaft für Schwerionenforschung (GSI) at Darmstadt, West Germany, by a team led by P. Armbruster and G. Müzenberg. The reaction used to produce the element was proposed and applied in 1976 by Y. T. Oganessian and colleagues at Dubna Laboratories in Russia. A ^{209}Bi target was bombarded by a beam of ^{54}Cr projectiles.

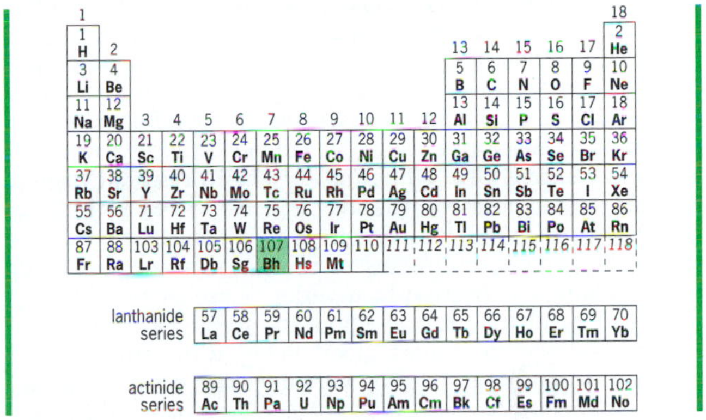
1 18
1 H 2 13 14 15 16 17 2 He
3 Li 4 Be 5 B 6 C 7 N 8 O 9 F 10 Ne
11 Na 12 Mg 3 4 5 6 7 8 9 10 11 12 13 Al 14 Si 15 P 16 S 17 Cl 18 Ar
19 K 20 Ca 21 Sc 22 Ti 23 V 24 Cr 25 Mn 26 Fe 27 Co 28 Ni 29 Cu 30 Zn 31 Ga 32 Ge 33 As 34 Se 35 Br 36 Kr
37 Rb 38 Sr 39 Y 40 Zr 41 Nb 42 Mo 43 Tc 44 Ru 45 Rh 46 Pd 47 Ag 48 Cd 49 In 50 Sn 51 Sb 52 Te 53 I 54 Xe
55 Cs 56 Ba 71 Lu 72 Hf 73 Ta 74 W 75 Re 76 Os 77 Ir 78 Pt 79 Au 80 Hg 81 Tl 82 Pb 83 Bi 84 Po 85 At 86 Rn
87 Fr 88 Ra 103 Lr 104 Rf 105 Db 106 Sg 107 Bh 108 Hs 109 Mt 110 111 112 113 114 115 116 117 118
lanthanide series 57 La 58 Ce 59 Pr 60 Nd 61 Pm 62 Sm 63 Eu 64 Gd 65 Tb 66 Dy 67 Ho 68 Er 69 Tm 70 Yb
actinide series 89 Ac 90 Th 91 Pa 92 U 93 Np 94 Pu 95 Am 96 Cm 97 Bk 98 Cf 99 Es 100 Fm 101 Md 102 No

The best technique to identify a new isotope is its genetic correlation to known isotopes through a radioactive decay chain. These decay chains are generally interrupted by spontaneous fission. In order to apply decay chain analysis, those isotopes that are most stable against spontaneous fission should be produced, that is, isotopes with odd numbers of protons and neutrons. Not only does the fission barrier govern the spontaneous fission of a species produced, but also, in the deexcitation of the virgin nucleus, fission competing with neutron emission determines the final production probability. To keep the fission losses small, a nucleus should be produced with the minimum excitation energy possible. In this regard, reactions using relatively symmetric collision partners and strongly bound closed-shell nuclei, such as ^{209}Bi and ^{208}Pb as targets and ^{48}Ca and ^{50}Ti as projectiles, are advantageous.

Six decay chains were found in the Darmstadt experiment. All the decays can be attributed to ^{262}Bh, an odd nucleus produced in a one-neutron reaction. The isotope ^{262}Bh undergoes alpha-particle decay (10.38 MeV) with a half-life of about 5 ms.

Experiments at Dubna, performed in 1983 using the 157-in. (400-cm) cyclotron, established the production of ^{262}Bh in the reaction ^{209}Bi ^{54}Cr. *See* NUCLEAR FISSION; NUCLEAR REACTION; RADIOACTIVITY; TRANSURANIUM ELEMENTS. [P.Ar.]

Boiler A pressurized system in which water is vaporized to steam, the desired end product, by heat transferred from a source of higher temperature, usually the products of combustion from burning fuels. Steam thus generated may be used directly as a heating medium, or as the working fluid in a prime mover to convert thermal energy to mechanical work, which in turn may be converted to electrical energy. Although other fluids are sometimes used for these purposes, water is by far the most common because of its economy and suitable thermodynamic characteristics.

The physical sizes of boilers range from small portable or shop-assembled units to installations comparable to a multistory 200-ft-high (60-m) building equipped, typically, with a furnace which can burn coal at a rate of 6 tons/min (90 kg/s). Boilers operate at positive pressures and offer the hazardous potential of explosions. Pressure parts must be strong enough to withstand the generated steam pressure and must be maintained at acceptable temperatures, by transfer of heat to the fluid, to prevent loss of strength from overheating or destructive oxidation of the construction materials.

The overall functioning of steam-generating equipment is governed by thermodynamic properties of the working fluid. By the simple addition of heat to water in a closed vessel, vapor is formed which has greater specific volume than the liquid, and can develop an increase of pressure to the critical value of 3208 psia (22.1 megapascals absolute pressure). If the generated steam is discharged at a controlled rate, commensurate with the rate of heat addition, the pressure in the vessel can be maintained at any desired value, and thus be held within the limits of safety of the construction. *See* STEAM.

Addition of heat to steam, after its generation, is accompanied by increase of temperature above the saturation value. The higher heat content, or enthalpy, of superheated steam permits it to develop a higher percentage of useful work by expansion through the prime mover, with a resultant gain in efficiency of the power-generating cycle. *See* SUPERHEATER.

If the steam-generating system is maintained at pressures above the critical, by means of a high-pressure feedwater pump, water is converted to a vapor phase of high density equal to that of the water, without the formation of bubbles. Further heat addition causes superheating, with corresponding increase in temperature and enthalpy. The most advanced developments in steam-generating equipment have led to units operating above critical pressure, for example, 3600–5000 psi (25–34 MPa). Superheated steam temperature has advanced from 500 ± °F (260 ± °C) to the present practical limits of 1050–1100°F (566–593°C). *See* MARINE ENGINEERING; NUCLEAR POWER; STEAM-GENERATING UNIT. [T.Ba.]

Boiler economizer A component of a steam-generating unit that absorbs heat from the products of combustion after they have passed through the steam-generating and super-heating sections. The name, accepted through common usage, is indicative of savings in the fuel required to generate steam.

An economizer is a forced-flow, once-through, convection heat-transfer device to which feedwater is supplied at a pressure above that in the steam-generating section and at a rate corresponding to the steam output of the unit. The economizer is in effect a feedwater heater, receiving water from the boiler feed pump and delivering it at a higher temperature to the steam generator or boiler. Economizers are used instead of additional steam-generating surface because the feedwater, and consequently the heat-receiving surface, is at a temperature below that corresponding to the saturated steam temperature; thus, the economizer further lowers the flue gas temperature for additional heat recovery. *See* BOILER FEEDWATER; THERMODYNAMIC CYCLE.

Generally, steel tubes, or steel tubes fitted with externally extended surface, are used for the heat-absorbing section of the economizer; usually, the economizer is coordinated with the steam-generating section and placed within the setting of the unit. *See* AIR HEATER; BOILER; STEAM-GENERATING UNIT. [G.W.K.]

Boiler feedwater Water supplied to a boiler unit for the generation of steam. Feedwater should be virtually free of impurities that are harmful to the boiler and its associated system. Generally, natural waters are unsuitable for direct use as feedwater because of their contamination by contact with the earth, the atmosphere, or other sources of pollution. These contaminants can be removed or altered by chemical treatment and other means to provide satisfactory feedwater. *See* RAW WATER; WATER TREATMENT. [G.W.K.]

Boiling point The temperature at which the transition from the liquid to the gaseous phase occurs. For pure substances at a fixed pressure, the boiling or vaporization process occurs at a single temperature; as heat is added, the temperature remains constant until all the liquid has boiled.

The normal boiling point is defined as the boiling point at a total applied pressure of 1 atm (101.325 kilopascals), that is, the temperature at which the vapor pressure of the liquid equals 1 atm. The boiling point increases with applied pressure. For substances boiling in the region of room temperature, the rate of change of boiling point with temperature is approximately 0.04°/mmHg or 0.3°/kPa (where the pressure is approximately 1 atm).

The boiling point cannot be raised indefinitely, however. As the pressure is increased, the density of the gas phase increases until it finally becomes indistinguishable from the liquid phase with which it is in equilibrium; this is the critical temperature, above which no distinct liquid phase exists. Helium has the lowest normal boiling point (4.2 K) of any substance, and tungsten carbide has one of the highest (6300 K). *See* DISTILLATION; LIQUID; PHASE EQUILIBRIUM; VAPOR PRESSURE. [R.L.S.]

Bolometer A device for detecting and measuring small amounts of thermal radiation. The bolometer is a simple electric circuit, the essential element of which is a slab of material with an electrical property, most often resistance, that changes with temperature. Typical operation involves absorption of radiant energy by the slab, producing a rise in the slab's temperature and thereby a change in its resistance. The electric circuit converts the resistance change to a voltage change, which then can be amplified and observed by various, usually conventional, instruments.

Although bolometers are useful in studying a variety of systems where detection of small amounts of heat is important, their primary application remains as the instrument of choice for measuring weak radiation signals in the infrared and far infrared, that is, at wavelengths from about 1 to 2000 micrometers, from stars and interstellar material. *See* BARRETTER; INFRARED RADIATION; RADIOMETRY; THERMISTOR. [W.E.K.]

Bolt A rod, usually of metal, with a head at one end and a screw thread on the other. A bolt is used to fasten objects together. A bolt is passed through clearance holes in two or more parts, a nut is engaged on the threaded end, and the parts are drawn together. Bolts can be obtained in a variety of forms, each suited for a specific application (see illustration).

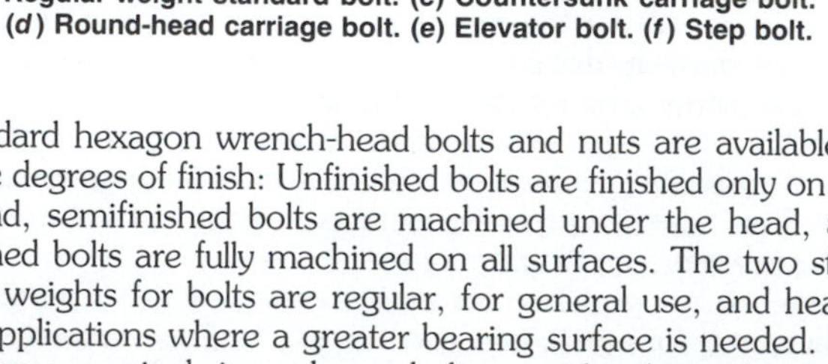

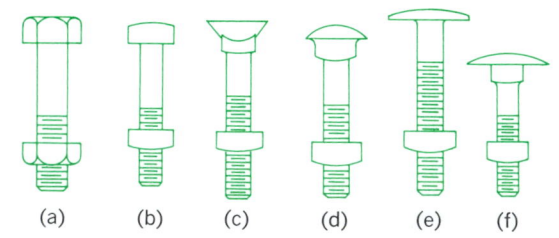

Examples of bolts. **(*a*) Heavyweight standard bolt. (*b*) Regular-weight standard bolt. (c) Countersunk carriage bolt. (*d*) Round-head carriage bolt. (e) Elevator bolt. (*f*) Step bolt.**

Standard hexagon wrench-head bolts and nuts are available in three degrees of finish: Unfinished bolts are finished only on the thread, semifinished bolts are machined under the head, and finished bolts are fully machined on all surfaces. The two standard weights for bolts are regular, for general use, and heavy, for applications where a greater bearing surface is needed. For the same nominal size, a heavy bolt or nut has larger head and nut dimensions than one belonging to the regular series.

Wrench-head bolts are specified by giving the diameter, number of threads per inch, series, class of thread, length, finish, and type of head. Machine bolts are used in the automotive, aircraft, and machinery fields, where a snug full-bodied fit is needed. Stove bolts are used on electrical equipment and household appliances, either in tapped holes or with nuts. The carriage bolt is a round-head type with nut; it is used as a through bolt. Special bolts are elevator, plow, step, strut, tire, T-head, and track bolts. Plow bolts are employed on agricultural equipment. *See* NUT (ENGINEERING); SCREW FASTENER. [W.J.L.]

Bolted joint The assembly of two or more parts by a threaded bolt and nut or by a screw that passes through one member and threads into another (see illustration). A bolted joint can be disassembled more readily than welded or riveted joints. *See* RIVETED JOINT.

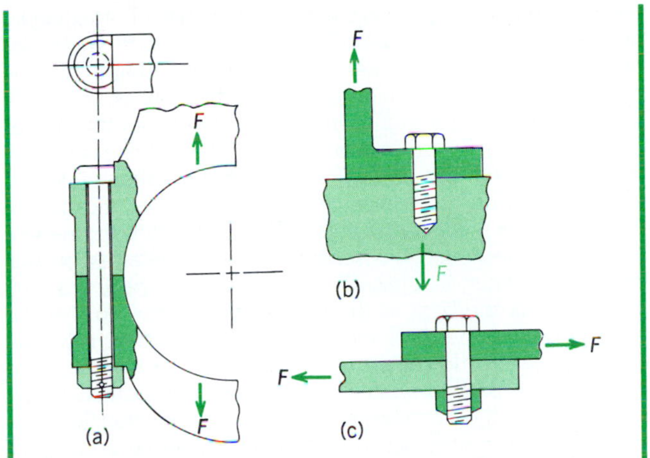

Forms of bolted joint (*F* represents force). (*a*) Direct tension. (*b*) Eccentric loading. (*c*) Single shear.

Structural members, such as I beams, H beams, angles, and plates, may be joined by bolting. When properly tightened, such joints are as strong and reliable as riveted ones. For joints of this type, high-strength alloy-steel bolts and nuts tightened with impact wrenches are generally used. No locking feature is necessary for properly selected and tightened bolts.

The strength of a bolted assembly depends to a great extent on the initial loading and stress placed on the bolt by the assembly torque. Devices such as torque wrenches are used to determine tension as a function of the torque applied to the assembly of nut and bolt or cap screw and machine part despite the uncertain coefficient of friction between nut, bolt, and assembled members. A better measure is to gage the elongation of the bolt as the tightening torque is applied. Bolt and nut assemblies often include special lock washers to prevent accidental loosening of the fastening by vibration. Where critical loads are to be carried and safety is paramount, bolts or screws smaller than ½ in. (1.25 cm) in diameter are dangerous to use because it is easy to overload the bolt or screw in assembly and actually twist it off with standard wrench sizes. *See* JOINT (STRUCTURES); SCREW FASTENER; STRUCTURAL CONNECTIONS; WELDED JOINT. [L.S.L.]

Boltzmann constant A constant occurring in practically all statistical formulas and having a numerical value of 1.3807×10^{-23} joule/K. It is represented by the letter k. If the temperature T is measured from absolute zero, the quantity kT has the dimensions of an energy and is usually called the thermal energy. At 300 K (room temperature) $kT = 0.0259$ electronvolt.

The value of the Boltzmann constant may be determined from the ideal gas law. For 1 mole of an ideal gas Eq. (1*a*)

$$PV = RT \qquad (1a)$$

$$PV = NkT \qquad (1b)$$

holds, where P is the pressure, V the volume, and R the universal gas constant. The value of R, 8.31 J/K mole, may be obtained from equation-of-state data. Statistical mechanics yields for the gas law Eq. (1*b*). Here N, the number of molecules in 1 mole, is called Avogadro's number and is equal to 6.02×10^{23} molecules/mole. Hence, comparing Eqs. (1*a*) and (1*b*), one obtains Eq. (2).

$$k = R/N = 1.3807 \times 10^{-23}\ \text{J/K} \qquad (2)$$

Almost any relation derived on the basis of the partition function or the Bose-Einstein, Fermi-Dirac, or Boltzmann distribution contains the Boltzmann constant. *See* AVOGADRO NUMBER; BOLTZMANN STATISTICS; BOSE-EINSTEIN STATISTICS; FERMI-DIRAC STATISTICS; KINETIC THEORY OF MATTER; STATISTICAL MECHANICS. [M.Dr.]

Boltzmann statistics To describe a system consisting of a large number of particles in a physically useful manner, recourse must be had to so-called statistical procedures. If the mechanical laws operating in the system are those of classical mechanics, and if the system is sufficiently dilute, the resulting statistical treatment is referred to as Boltzmann or classical statistics. (Dilute in this instance means that the total volume available is much larger than the proper volume of the particles.) A gas is a typical example: The molecules interacting according to the laws of classical mechanics are the constituents of the system, and the pressure, temperature, and other parameters are the overall entites which determine the macroscopic behavior of the gas. In a case of this kind it is neither possible nor desirable to solve the complicated equations of motion of the molecules; one is not interested in the position and velocity of every molecule at any time. The purpose of the statistical description is to extract from the mechanical description just those features relevant for the determination of the macroscopic properties and to omit others.

The basic notion in the statistical description is that of a distribution function. Suppose a system of N molecules is contained in a volume V. The molecules are moving around, colliding with the walls and with each other. Construct the following geometrical representation of the mechanical system. Introduce a six-dimensional space (usually called the μ space), three of its coordinate axes being the spatial coordinates of the vessel x, y, z, and the other three indicating cartesian velocity components v_x, v_y, v_z. A molecule at a given time, having a specified position and velocity, may be represented by a point in this six-dimensional space. The state of the gas, a system of N molecules, may be represented by a cloud of N points in this space. In the course of time, this cloud of N points moves through the μ space.

Note that the μ space is actually finite; the coordinates x, y, z of the molecules' position are bounded by the finite size of the container, and the velocities are bounded by the total energy of the system. Imagine now that the space is divided into a large number of small cells, of sizes $w_1, \ldots, w_i, \ldots$. A certain specification of the state of the gas is obtained if, at a given time t, the numbers $n_1(t), \ldots, n_i(t), \ldots$ of molecules in the cells $1, \ldots, i, \ldots$ are given. To apply statistical methods, one must choose the cells such that on the one hand a cell size w is small compared to the macroscopic dimensions of the system, while on the other hand w must be large enough to allow a large number of molecules in one cell. If the cells are thus chosen, the numbers $n_i(t)$, the occupation numbers, will be slowly changing functions of time. The distribution functions $f_i(t)$ are defined by Eq. (1).

$$n_i(t) = f_i(t)w_i \qquad (1)$$

The distribution function f_i describes the state of the gas, and *fi* of course varies from cell to cell. Since a cell i is characterized by a given velocity range and position range, and since for appropriately chosen cells f should vary smoothly from cell to cell, f is often considered as a continuous function of the variables x, y, z, vx, vy, vz. The cell size w then can be written as $dx dy dz dv_x dv_y dv_z$.

Since a cell i determines both a position and a velocity range, one may associate an energy ϵ_i with a cell. This is the energy a single molecule possesses when it has a representative point in cell i. This assumes that, apart from instantaneous collisions, molecules exert no forces on each other. If this were not the case, the energy of a molecule would be determined by the positions of all other molecules.

Most of the physically interesting quantities follow from a knowledge of the distribution function; the main problem in Boltzmann statistics is to find out what this function is. It is clear that $n_i(t)$ changes in the course of time for three reasons: (1) Molecules located at the position of cell i change their positions and hence move out of cell i; (2) molecules under the influence of outside forces change their velocities and again leave the cell i; and (3) collisions between the molecules will generally cause a (discontinuous) change of the occupation numbers of the cells. Whereas the effect of (1) and (2) on the distribution function follows directly from the mechanics of the system, a separate assumption is needed to obtain the effect of collisions on the distribution function. This assumption, the collision-number assumption, asserts that the number of collisions per unit time, of type $(i,j) \rightarrow (k,l)$ (molecules from cells i and j collide to produce molecules of different velocities which belong to cells k and l), called $A_{ij}{}^{kl}$, is given by Eq. (2). Here

$$A_{ij}{}^{kl} = n_i n_j a_{ij}{}^{kl} \qquad (2)$$

$a_{ij}{}^{kl}$ depends on the collision configuration and on the size and kind of the molecules but not on the occupation numbers. Gains and losses of the molecules in, say, cell i can now be observed. If the three factors causing gains and losses are combined, the Boltzmann transport equation, written as Eq. (3), is obtained. Here $\Delta_x f_i$ is the gradient of f with respect to

$$\frac{\partial f_i}{\partial t} + (\mathbf{v}_i \cdot \Delta_x f_i) + (\mathbf{X}_i \cdot \Delta_v f_i) = \sum_{j,k,l} a_{ij}{}^{kl} w_j (f_k f_l - f_i f_j) \qquad (3)$$

the positions, $\Delta_v f_i$ refers similarly to the velocities, and $\mathbf{X}_i$ is the outside force per unit mass at cell i. This nonlinear equation determines the temporal evolution of the distribution function. Exact solutions are difficult to obtain. Yet Eq. (3) forms the basis for the kinetic discussion of most transport processes. There is one remarkable general consequence, which follows from Eq. (3). If one defines $H(t)$ as in Eq. 4, one finds by straight manipulation from Eqs. (3) and (4) that Eqs. (5) hold.

$$H(t) = \sum_i n_i \ln f_i \qquad (4)$$

$$\frac{dH}{dt} \cdots 0 \qquad \frac{dH}{dt} = 0 \quad \text{if } f_j f_i = f_k f_l \qquad (5)$$

Hence H is a function which in the course of time always decreases. This result is known as the H theorem. The special distribution which is characterized by Eq. (6) has the property

$$f_j f_i = f_k f_l \qquad (6)$$

that collisions do not change the distribution in the course of time; it is an equilibrium or stationary distribution.

The form of the equilibrium distribution may be determined from Eq. (6), with the help of conservation laws. For a gas which as a whole is at rest, it may be shown that the only solution to functional Eq. (6) is given by Eqs. (7*a*) or (7*b*). Here A

$$f_i = Ae^{-\beta\epsilon_i} \qquad (7a)$$

$$f(\mathbf{x},\mathbf{v}) = Ae^{(-1/2)\beta mv^2 - \beta U} \qquad (7b)$$

and β are parameters, not determined by Eq. (6), and U is the potential energy at the point x, y, z. Equations (7*a*) and (7*b*) are the Maxwell-Boltzmann distribution. Actually A and β can be determined from the fact that the number of particles and the energy of the system are specified.

The indiscriminate use of the collision-number assumption leads, via the H theorem, to paradoxical results. The basic conflict stems from the irreversible results that appear to emerge as a consequence of a large number of reversible fundamental processes. A careful treatment of the explicit and hidden probability assumptions is the key to the understanding of the apparent conflict. The equilibrium distribution may be thought of as the most probable state of a system. If a system is not in equilibrium, it will most likely (but not certainly) go there; if it is in equilibrium, it will most likely (but not certainly) stay there. By using such probability statements, it may be shown that the paradoxes and conflicts may indeed be removed. A consequence of the probabilistic character of statistics is that the entities computed also possess this characteristic. For example, one cannot really speak definitively of the number of molecules hitting a section of the wall per second, but only about the probability that a given number will hit the wall, or about the average number hitting. In the same vein, the amount of momentum transferred to a unit area of the wall by the molecules per second (this, in fact, is precisely the pressure) is also to be understood as an average. This in particular means that the pressure is a fluctuating entity. The fluctuations in pressure may be demonstrated by observing the motion of a mirror, suspended by a fiber, in a gas. On the average, as many gas molecules will hit the back as the front of the mirror, so that the average displacement will indeed be zero. However, it is easy to imagine a situation where more momentum is transferred in one direction than in another, resulting in a deflection of the mirror. From the knowledge of the distribution function the probabilities for such occurrences may indeed be computed; the calculated and observed behavior agree very well. This clearly demonstrates the essentially statistical character of the pressure. *See* BROWNIAN MOVEMENT. [M.Dr.]

Bomb calorimeter This calorimeter is widely used in determining the heats of combustion of organic compounds and has been developed to yield measurements reproducible to within ±0.01%. Its operation is fairly typical of calorimetric processes carried out near room temperature. *See* CALORIMETRY.

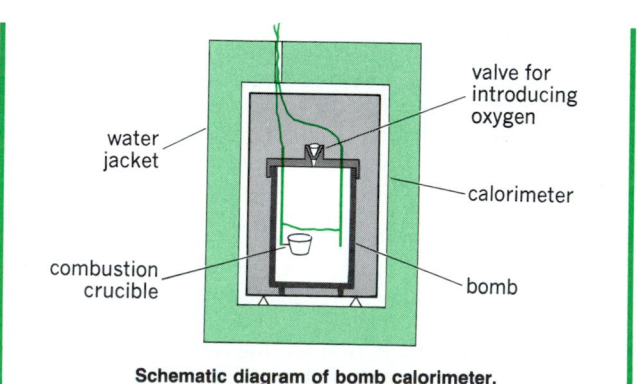

Schematic diagram of bomb calorimeter.

Some of the essential features of the apparatus are shown in the illustration. The bomb, a strong-walled metal container of 0.3–0.4-liter capacity, is constructed of a corrosion-resistant alloy. It is fitted with a screw head, which contains a valve for introducing oxygen to a pressure of about 30 atm (3.0 megapascals). The weighed sample, liquid or solid, is placed in the combustion crucible and ignited by electrical connections which pass through the head.

The bomb is immersed in about 2.5 liters of water in the calorimeter proper, a metal container which is equipped with a platinum resistance thermometer, a stirrer, and a tightly fitting cover to prevent evaporation of its water. It rests on small ivory pegs within the jacket, the two being separated by a 1-cm air space.

This bomb calorimeter is calibrated either electrically or,

more frequently, by burning a sample of standard benzoic acid, supplied by the U.S. National Bureau of Standards. After the heat capacity of the calorimeter has been evaluated, the unknown heat value can be computed from the measured temperature rise. The result so obtained, corrected to 1 atm (101.3 kilopascals) pressure and computed for 1 mole of the substance, represents ΔE, the heat of combustion at constant volume. The corresponding heat effect at constant pressure can be derived by the relation $\Delta H = \Delta E + P\Delta V$. [B.J.Z.]

Bond angle and distance In molecules two atoms are said to be bonded to each other when there are strong attractive forces between them. These two atoms constitute a chemical bond. One important characteristic of a chemical bond is the separation (called the bond length or distance) of the two nuclei of the atoms involved. A given atom may be involved in more than one chemical bond, and another important characteristic is the angle between bonds sharing a common atom. It is also often important to know the angle between two bonds which are connected by a third bond. This angle is called the dihedral or torsional angle. *See* CHEMICAL BONDING.

Bond angles and lengths determine the shape and size of a molecule, and thus determine many physical properties. For example, molecules which are spherical in shape will pack in a crystal in a different manner from long, rodlike molecules, and will likely have a different density and crystal structure. Similarly, the molar volumes in the liquid are likely to be different, as well as the diffusion coefficients. Another example is the observation that the melting points for normal alkanes with an even number of carbons behave differently from those with an odd number. This arises because of their different shapes and the way in which they interact in the solid.

Bond angles and distances are functions of the type of chemical bonding present and give information about the electronic structure of the molecule. For example, bond lengths shorten considerably in going from a single bond to a triple bond, and are thus dependent on the hybridization present. *See* CHEMICAL STRUCTURES.

Since the strength of a bond increases with its multiplicity, there is an inverse relation between the strength of a bond and its length. Related to the strength of a bond is its stretching-force constant, and in fact general relations exist between the stretching-force constant for a bond and its length.

Dimensions and structure are exceedingly important in biological molecules. For example, the functioning of a molecule such as deoxyribonucleic acid depends very precisely on the structure of the individual, small components which make up this very large molecule. Another example is the great specificity of enzymes, which is believed to be associated with the exact shape and size of the enzyme. *See* HYDROGEN BOND. [V.W.L.]

Bonding The act of connecting the various structural metal parts of a metal enclosure or vehicle (as in an aircraft or automobile) so that these parts form a continuous electrical unit. Bonding serves to minimize or eliminate interference, such as that caused by ignition systems. It also prevents buildup of static electricity on one part of the structure, which can, by subsequent discharge to other parts, cause static interference. Bonding is achieved by bolting the parts together in such a way as to achieve good electrical contact or by connecting them with heavy copper cables or straps.

Bonding also refers to the fastening together of two pieces by means of adhesives, as in anchoring the copper foil of printed wiring to an insulating baseboard. *See* ADHESIVE. [J.Mar.]

Bone The hard connective tissue which (along with cartilage) forms the skeleton of vertebrates. *See* SKELETAL SYSTEM.

Bone consists of collagen fibers which are organized in layers and upon which mineral salts, especially calcium phosphate and carbonates, are impregnated. Other organic materials, consisting of blood, bone cells (osteocytes), and cartilage, are also present. *See* CONNECTIVE TISSUE.

The outer portion of most bones is formed of a hard shell-like compacted layer; the inner portions are composed of spongy bone. In long bones of adults the compact outer bone forms the long, hollow shaft surrounding a medullary cavity containing either hematopoietic bone marrow or fatty tissue. The spongy parts of these long bones are present in the ends of the bones where the forces on the bone are distributed over a broad area.

Some bones are formed by transformation of a cartilaginous precursor, while other bones develop directly by deposit of mineral salts in fibrous tissue. The time of ossification in the skeleton follows a definite pattern, providing there is no abnormality of formation.

Aside from furnishing the main support for the body, the bones serve as levers for muscular action and motion of the body. The movement of each bone depends on the structure of the articular surface and associated ligaments. *See* MUSCULAR SYSTEM.

The closely adherent fibrous covering of bones, the periosteum, is osteogenic, that is, bone-producing, especially in younger individuals not yet having attained full growth. The osteogenic properties of the periosteum are vital in the formation of new bone after injury, as in the repair of fractures.

Bone is not an inert substance as commonly believed but participates as a dynamic and powerful factor in general body metabolism, particularly with regard to calcium and other inorganic ions. The influence of vitamin D, pituitary growth hormone, estrogens, and thyroid and parathyroid hormones can readily be seen in many diseases in which these important regulating substances are altered. Moreover, the structure and shape of bones are modified rapidly with changes in the mechanical forces acting upon them. [W.J.B.]

Bonsai The construction of a mature, very dwarfed tree in a relatively small container. Although bonsai is formally an esoteric branch of horticulture, it is firmly based in the social, historical, and cultural ethic of Oriental peoples and its introduction to Western cultures is a development of the twentieth century; indeed, Westerners had not even seen bonsai plants prior to the early decades of this century. At the fundamental level, the word bonsai means merely "plant in a pot," and its origin can be traced back to China's Chou dynasty (900–250 B.C.), when emperors built miniaturized gardens in which soil, rocks, and plants from each province were installed as dwarfed representations of the lands held by the central government.

To a large extent, the technology of bonsai is derived from classical horticultural procedures which are the product of long tradition firmed up with considerations of many areas of plant physiology and ecology. There are, however, some techniques which are unique to bonsai. To maintain shape, not only must growth be controlled, but the branching patterns must be planned. This is done by selective pruning and by pinching out of terminal buds to release axillary buds from apical dominance. Root prunings, done at intervals as short as 2 or 3 months in some species, promote the development of a tight root ball with large numbers of young lateral roots whose ability to take up water and minerals is great. Branch and stem position is controlled by selective pruning and by the wiring techniques developed by the Japanese in the eighteenth century. Weighting or tying is used for the same reasons. [R.M.K.]

Book manufacture A series of professional operations including editing, designing, typography, platemaking, printing, and binding to transform a manuscript and its related illustrations into book form.

Books follow a distinctive technique in assembling and binding pages in sequential order and in fitting and sealing the pages into a separate cover. Those designed for relatively permanent usage are sealed into rigid or semirigid covers (called edition, case, or hardback binding), and those designed for less permanent usage are sealed into paper covers (called soft or paperback binding).

The basic cell of a book's construction is called a signature or section and is composed of multiples of four pages. The folding of signatures printed on a web press is integrated with the printing units. In sheet-fed printing, folding is separate from presswork and is performed on special machines located in a book bindery.

Folding a sheet of paper once produces 4 pages, and each additional fold doubles the number of pages. A sheet folded three times yields a 16-page signature, the first page being the lowest numbered (a page number is called a folio). All right-hand pages are odd numbered, and left-hand pages are even numbered. A book of 160 pages has ten 16-page signatures, the first being folioed 1–16 and the last 145–160.

Signatures, assembled in sequence, are mechanically held together along one folded edge, called the binding edge or spine, by continuous thread sewing through the fold of each signature; or by thread stitching through the side of the book from front to back; or by wire stapling; or by cutting off the binding-edge folds and applying a coating of strong flexible adhesive, sometimes adding an additional locking device of wide mesh textiles or stretch papers.

Purely decorative features include colorful bits of textiles added to the head and tail of the spine (called headbands) and tinted or gilded edges. One spectacular technical breakthrough was the mechanical gilding of book edges; for centuries this had been a closely guarded handicraft process.

The book cover serves to protect the pages against disintegration, to announce the title of the book, and to stimulate the visual interest of the prospective buyer. The degree of usage and permanency influences the selection of raw materials for a book cover (as in reference sets and text and library books), and visual appeal is influenced by the book's price and marketability. Permanent-type covers are made on special equipment in which the cover material and pulp-boards are assembled and joined with special adhesives. In embellishing the cover, the designer has numerous choices of special equipment and processes, such as flat ink, pigment or metal-foil stampings, embossing, silk-screening, and multicolor printing.

Preceding the final assembly of the book into its cover (called "casing in"), strong four-page front and back end-sheets and one or more kinds of hinges (of textiles or paper) have been applied to strengthen the interlocking between book and cover (called "lining up"). The book also has acquired a concave and convex edge and a pair of ridges or joints along the binding edge (called "backing"). These features are added by special equipment and are specifically engineered to securely lock the book into its cover and to transmit the strains of opening the book throughout the entire body instead of only to the first and last pages.

For shelf display, a book is often wrapped in a colorful, eye-catching jacket (called dust wrapper), or in a transparent synthetic if the cover is highly decorative.

Books reach their markets packaged in many different ways. Single copies are packaged and mailed directly to the consumer; sets are assembled into cartons, one set per carton; textbooks and other single titles are bulk-cartoned or skid-packed and shipped to central warehouses. *See* PRINTING. [E.J.T.]

Boolean algebra A branch of mathematics that was first developed systematically, because of its applications to logic, by the English mathematician George Boole, around 1850. Closely related are its applications to sets and probability. Boolean algebra also underlies the theory of relations. A modern engineering application is to computer circuit design. *See* DIGITAL COMPUTER.

Most basic is the use of Boolean algebra to describe combinations of the subsets of a given set I of elements; its basic operations are those of taking the intersection or common part $S \cap T$ of two such subsets S and T, their union or sum $S \cup T$, and the complement S' of any one such subset S. These operations satisfy many laws, including those shown in Eqs. (1), (2), and (3).

$$S \cap S = S \qquad S \cap T = T \cap S$$
$$S \cap (T \cap U) = (S \cap T) \cap U \tag{1}$$

$$S \cup S = S \cup T = T \cup] S$$
$$S \cup (T \cup U) = (S \cup T) \cup U \tag{2}$$

$$S \cap (T \cup U) = (S \cap T) \cup (S \cap U)$$
$$S \cup (T \cap] U) = (S \cup T) \cap (S \cup U) \tag{3}$$

If O denotes the empty set, and I is the set of all elements being considered, then the laws set forth in Eqs. (4) are also fundamental. Since these laws are fundamentals, all other algebraic laws of subset combination can be deduced from them.

$$O \cap S = O \quad O \cup S = S \quad I \cup S = S = S$$
$$I \cup S = I \quad S \cap S' = O \quad S \cup S' = I \tag{4}$$

In applying Boolean algebra to logic, Boole observed that combinations of properties under the common logical connectives *and*, *or*, and *not* also satisfy the laws specified above. These laws also hold for propositions or assertions, when combined by the same logical connectives. *See* LOGIC CIRCUITS.

Boole stressed the analogies between Boolean algebra and ordinary algebra. If $S \cap T$ is regarded as playing the role of st in ordinary algebra, $S \cup T$ that of $s + t$, O of 0, I of 1, and S' as corresponding to $1 - s$, the laws listed above illustrate many such analogies. However, as first clearly shown by Marshall Stone, the proper analogy is somewhat different. Specifically, the proper Boolean analog of $s + t$ is $(S' \cap T) \cup (S \cap T')$, so that the ordinary analog of $S \cup T$ is $s + t - st$. Using Stone's analogy, Boolean algebra refers to Boolean rings in which $s^2 = s$, a condition implying $s + s = 0$. *See* RING THEORY; SET THEORY.

In 1941 M. H. A. Newman developed a remarkable generalization which included Boolean algebras and Boolean rings. This generalization is based on the laws shown in Eqs. (5) and (6). From these assumptions, the idempotent, commutative,

$$a(b + c) = ab + ac(a + b)c = ac + bc \tag{5}$$

$$a1 = 1 \quad a + 0 = 0 + a = a \quad aa' = 0$$
$$a + a' = 1 \tag{6}$$

and associative laws (1) and (2) can be deduced.

Such studies lead naturally to the concept of an abstract Boolean algebra, defined as a collection of symbols combined by operations satisfying the identities listed in formulas (1) to (4). Ordinarily, the phrase Boolean algebra refers to such an abstract Boolean algebra.

The class of finite (abstract) Boolean algebras is easily described. Each such algebra has, for some nonnegative integer n, exactly $2n$ elements and is algebraically equivalent (isomorphic) to the algebra of all subsets of the set of numbers $1, \ldots, n$, under the operations of intersection, union, and complement.

The theory of infinite Boolean algebras is much deeper; it indirectly involves the whole theory of sets. One important result is Stone's representation theorem. Let a field of sets be defined as any family of subsets of a given set I, which contains with any two sets S and T their intersection $S \cap T$, union $S \cup T$, and complements S', T'. Considered abstractly, any such field of sets obviously defines a Boolean algebra. Stone's theorem asserts that, conversely, any finite or infinite abstract Boolean algebra is isomorphic to a suitable field of sets. His proof is based on the concepts of ideal and prime ideal, concepts which have been intensively studied for their own sake. [G.Bi.]

Boötes The Bear Driver, in astronomy, a northern and summer constellation. Boötes is one of the earliest recorded constellations. Arcturus, an orange, first-magnitude navigational star, dominates the constellation. Five prominent stars in this constellation form a pentagon, shaped much like an elongated kite, with Arcturus at the junction of the tail. Boötes is conventionally pictured as a driver of the Bear (Ursa Major) nearby. In a more recent version he is seen seated, smoking a pipe, his feet dangling and with one hand holding the leash of the Hunting Dogs (Canes Venatici). *See* CONSTELLATION. [C.-S.Y.]

Boracite A borate mineral with chemical composition $Mg_3B_7O_{13}Cl$. It occurs in Germany, England, and the United States, usually in bedded sedimentary deposits of anhydrite, gypsum, and halite, and in potash deposits of oceanic type. The chemical composition of natural boracites varies, with Fe^{2+} or Mn^{2+} replacing part of the Mg^{2+} to yield ferroan boracite or manganoan boracite. *See* BORATE MINERALS.

The hardness is 7–7½ on Mohs scale, and specific gravity is 2.91–2.97 for colorless crystals and 2.97–3.10 for green and ferroan types. Luster is vitreous, inclining toward adamantine. Boracite is colorless to white, inclining to gray, yellow, and green, and rarely pink (manganoan); its streak is white; and it is transparent to translucent. It is strongly piezoelectric and pyroelectric and does not cleave. [C.L.Ch.]

Borane One of a class of binary compounds of boron and hydrogen, often referred to as boron hydrides. The term borane is sometimes used to denote substances which may be considered to be derivatives of the boron-hydrogen compounds, such as boron trichloride (BCl_3), and diiododecaborane ($B_{10}H_{12}I_2$).

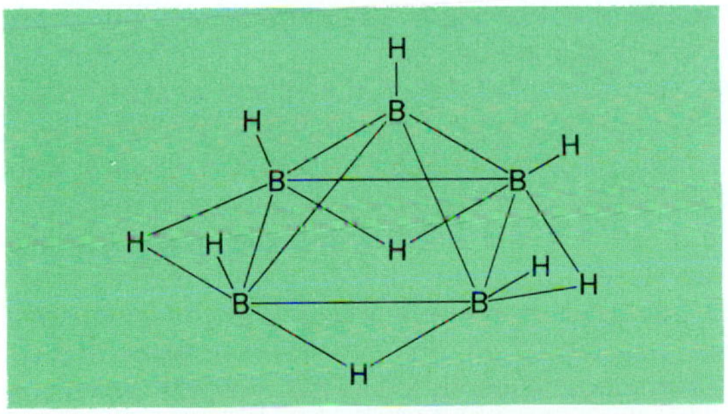

Structure of pentaborane (9).

The simplest borane is diborane (B_2H_6); other boranes of increasingly higher molecular weight are known, one of the least volatile of which is an apparently polymeric solid of composition (BH)*x*. Certain boranes, such as BH_3, and B_3H_7, are not known as such, but can be prepared in the form of adducts with electron-donor molecules.

The most spectacular projected large-scale use of the boranes and their derivatives is in the field of high-energy fuels for jet planes and rockets. The thermal decomposition of diborane (6) [B_2H_6] has been used to produce coatings of pure elementary boron for neutron-detecting devices and for applications requiring hard, corrosion-resistant surfaces. Boranes can also be used as vulcanizing agents for natural and synthetic rubbers, and are especially effective in the preparation of silicone rubbers.

The molecular structures possessed by the boranes are exhibited by no other class of substances. Because of the lack of sufficient electrons for the formation of the requisite number of covalent bonds, normal covalently bonded structures of the hydrocarbon type are not possible. The boranes are sometimes referred to as electron-deficient substances. In no case are the simple chain and ring configurations of carbon chemistry encountered in the more complex boranes. Instead, the boron atoms are situated at the corners of polyhedrons. An example of such a structure is that of pentaborane (9) [B_5H_9], shown in the illustration. Boron nomenclature uses a prefix to designate the number of boron atoms in the molecule and a numeral suffix in parentheses to indicate the number of hydrogen atoms.

As a class, the boranes are quite reactive substances and are generally decomposed, at times explosively, on contact with air. Their reactivities with air and water decrease with increasing molecular weight. Because boranes react readily with air, laboratory investigations are almost invariably carried out in all-glass vacuum apparatus or in inert-atmosphere dry boxes. With the possible exception of decaborane (14) [$B_{10}H_{14}$], the known boranes are not indefinitely stable at room temperature. They decompose more or less rapidly to yield elementary hydrogen and boranes richer in boron.

The known derivatives of the boranes (other than BH_3) are relatively few in number. Several halo, alkyl, and amino boranes have been reported but, in general, these have not been extensively characterized. See BORON; CARBORANE; METAL HYDRIDES. [T.W.]

Borate A generic term referring to salts related to boric oxide, B_2O_3 or commonly to only the salts of orthoboric acid, H_3BO_3. Only orthoboric acid is important in its free state. Some of the others are known only in the form of salts.

A dilute solution of any sodium borate contains mainly $H_2BO_3^-$ and BO_2^- ions. The borates are used in manufacture of borosilicate glasses, ceramic glazes, vitreous enamels, detergents and water-softening agents, flameproofing materials, preservatives, and fluxes. Minerals related to borax are the main source of borates. Research has produced boron nitride, an abrasive harder than diamond, and boron hydrides, which are used as rocket fuels. *See* BORATE MINERALS; BORON. [E.E.W.]

Borate minerals A large and complex group of naturally occurring crystalline solids in which boron occurs in chemical combination with oxygen; hence the term borate (see table). With the exception of avogadrite, (K,Cs)BF_4, and ferrucite, $NaBF_4$, all known boron minerals are borates. Borate minerals may contain silicon, phosphorus, or arsenic. These, specifically, are the borosilicates, the borophosphates, and the boroarsenates. Of these three groups, the first is by far the largest. *See* BORON.

The average amount of boron in the Earth's crust is estimated as 3 ppm. Fortunately, large quantities are concentrated in deposits of hydrated borate minerals that are easily worked, and boron is not, like some minor elements, so completely dispersed that it is hard to find and use. Deposits are found in the United States (Mojave Desert), Turkey, Russia (Inder district), Tibet, Italy, Germany, Argentina, Bolivia, Canada, China, India, Iran, New Zealand, New Guinea, and Syria. The minerals of chief commercial importance are borax, $Na_2B_4O_7 \cdot 10H_2O$, and kernite, $Na_2B_4O_7 \cdot 4H_2O$. Large quantities of these minerals are mined at the Kramer Deposit, Boron,

Borate minerals

Mineral name	Formula
Anhydrous borates	
Suanite	$Mg_2B_2O_5$
Kotoite	$Mg_3B_2O_6$
Nordenskiöldine	$CaSnB_2O_6$
Ludwigite	$(Mg,Fe^{2+})_2Fe^{3+}BO_5$
Boracite	$Mg_3B_7O_{13}Cl$
Hydrated borates	
Sassolite	H_3BO_3
Pinnoite	$Mg(BO_2)_2{\cdot}3H_2O$
Teepleite	$Na_2BO_2Cl{\cdot}2H_2O$
Fluoborite	$Mg_3(BO_3)(OH,F)_3$
Colemanite	$Ca_2B_6O_{11}{\cdot}5H_2O$
Meyerhofferite	$Ca_2B_6O_{11}{\cdot}7H_2O$
Inyoite	$Ca_2B_6O_{11}{\cdot}13H_2O$
Inderite	$Mg_2B_6O_{11}{\cdot}15H_2O$
Borax	$Na_2B_4O_7{\cdot}10H_2O$
Tincalconite	$Na_2B_4O_7{\cdot}5H_2O$
Kernite	$Na_2B_4O_7{\cdot}4H_2O$
Veatchite	$Sr_4B_{22}O_{37}{\cdot}7H_2O$
Probertite	$NaCaB_5O_9{\cdot}5H_2O$
Ulexite	$NaCaB_5O_9{\cdot}8H_2O$
Borosilicates	
Datolite	$CaBSiO_5{\cdot}H_2O$
Reedmergnerite	$NaBSi_3O_8$
Danburite	$CaB_2Si_2O_8$
Tourmaline	$(Na,Ca)(Li,Al)_3$-$Al_6(OH)_4(BO_3)_3Si_6O_{18}$
Borophosphates and boroarsenates	
Lunebergite	$Mg_3B_2(OH)_6(PO_4)_2{\cdot}6H_2O$
Seamanite	$Mn_3(PO_4)(BO_3){\cdot}3H_2O$
Cahnite	$Ca_2B(OH)_4(AsO_4)$

California. Borax is extracted in large amounts from brine at Searles Lake, California. See BORATE; KERNITE. [C.L.Ch.]

Borda mouthpiece A reentrant tube in a hydraulic reservoir. The contraction coefficient of a Borda mouthpiece can be calculated more simply than for other discharge openings.

As a liquid accelerates toward a discharge opening, the streamlines converge (see illustration). This convergence may

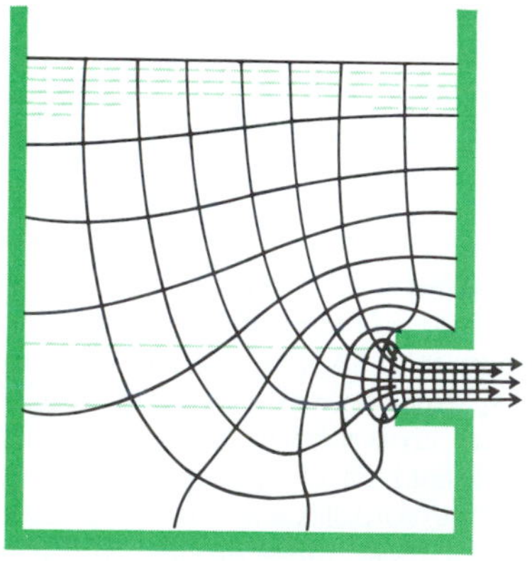

Streamlines around a Borda mouthpiece.

cause the issuing jet to contract to a smaller cross section than that of the opening, the region of the jet of least cross section being called the vena contracta. The ratio of vena contracta area to opening area is the contraction coefficient. The frictional forces in the vicinity of a simple opening alter the accelerations in the liquid, but with a Borda mouthpiece, friction is so reduced near the opening that the contraction coefficient can be calculated from hydrostatic considerations alone with close agreement to measurements. The coefficient thus calculated is 0.5. See METERING ORIFICE; VENTURI TUBE. [F.H.R.]

Bordetella A bacterial genus comprising minute (0.2–0.3 by 0.5–1.0 micrometer), nonsporulating, gram-negative coccobacilli. They grow within 2–5 days at 95°F (35°C) in the presence of oxygen on potato–glycerol–blood agar or in chemically defined media. Litmus milk is made alkaline. Three species are mammalian parasites and one is an avian parasite; they may cause respiratory diseases. A variety of antigenic or toxic factors are secreted or contained in the cell walls. *Bordetella pertussis*, the most fastidious species, causes whooping cough (pertussis) and subacute bronchitis in humans, with high mortality in the young baby. *Bordetella parapertussis* may cause similar but less severe respiratory infections in humans. *Bordetella bronchiseptica* is only rarely found affecting humans, but is a common respiratory parasite in many domestic and wild mammals. *Bordetella avium* causes respiratory infections in birds. See MEDICAL BACTERIOLOGY; PARAPERTUSSIS; VACCINATION; WHOOPING COUGH. [W.Ma.]

Borehole logging The process of analyzing and recording information relating to the character of soil or rock formations penetrated by a borehole, as may be drilled in exploration of minerals, water, oil, and gas, or in geotechnical (subsurface engineering) investigations. See PROSPECTING.

Borehole logs are records of data, compiled on an incremental depth basis, obtained by recording actions of the drill and auxiliary equipment; examination of samples removed from the borehole by drilling; or examination of the walls of the borehole by visual or geophysical methods.

Mineral drilling, as well as most other types of drilling, normally produces samples that provide a basis for logging. Core drilling provides the best samples, comprising essentially unaltered samples from the center of the hole. The rock from the remainder of the borehole is ground up by the core bit while a cylinder in the center remains unabraded. Other types of drilling (destructive drilling) grind up the material from the entire borehole and remove it with the circulating fluid (air, water, or mud). The samples for logging can be collected from the return fluid stream.

In many circumstances, however, an accurate record of the rock penetrated by the hole may not provide all the necessary information. For example, the level and movement of groundwater will not be reflected in a core or sample log. In addition, geophysical logs may be sensitive to parameters not visible to the eye, and they may detect mineralization (such as uranium mineralization) near but not penetrated by the borehole.

The driller's log concentrates on the performance of the drill as the bit advances. It can be a written record made from the driller's notes or a continuous strip chart made from readings gathered by sensors on the drill rig. Of primary interest is the rate of penetration as compared to depth, down (pullback) pressure, rotation speed, and circulation pressure and volume. The driller's log frequently contains a sample or core record delineating the types and thicknesses of units penetrated.

The sample log consists of information garnered from the examination of the samples removed from the borehole (in the form of core, chips, or cuttings). The information can range from that gained from visual examination by a geologist for formation identification to that gained from petrographic microscope examination of thin sections. See PETROGRAPHY.

In many types of geotechnical work, such as mine subsidence investigations, it is important to determine the nature and orientation of fractures in the borehole. Since core recovery is frequently difficult in fractured rock, it is often necessary to examine the borehole wall visually. This can be done, at shallow depths and in very large-diameter (30 in. or 75 cm or more) boreholes, by lowering a geologist into the borehole on a safety chair or ladder. In most cases it is more practical to use borehole video or borehole camera techniques.

Geophysical borehole logging is the technique of electronically measuring properties of the rocks in the borehole wall and of the borehole fluid, and recording these measurements as a series of curves on a graphic strip chart. From the positions of appropriate curves on the charts, geologists can tell the type of rock, rock porosity, the type of fluid in the rock, direction of fluid flow within the borehole, deviation of the borehole, and various other conditions. Geophysical logs are placed in categories depending on the type of property they measure; these include mechanical, electrical, nuclear, thermal, acoustic, and directional.

A mechanical caliper log is made by measuring changes in borehole diameter with spring-loaded mechanical arms. As the tool is drawn up the borehole, the arms move in and out with borehole diameter change, turning a potentiometer in an electrical circuit. The changes in this circuit are recorded as the log curve. *See* CALIPER.

Electrical logs measure electrical properties of the rock and borehole fluid. Wet clays (shales) are good electrical conductors, while sand (quartz), limestone ($CaCO_3$), and coal (carbon) are poor conductors. Resistivity logging is one method to identify lithologic differences. *See* GEOELECTRICITY.

Nuclear logs measure the natural radiation in the borehole wall and the response of the rock in the borehole wall to bombardment from a radioactive source.

The natural gamma log operates on the principle that virtually all material is slightly radioactive. In most sedimentary rocks, the number of gamma-ray counts is directly proportional to the percentage of clay minerals in the rock. The natural gamma log is also useful for uranium exploration. *See* GAMMA-RAY DETECTORS; GEIGER-MÜLLER COUNTER; RADIOACTIVITY.

Thermal logging is the measurement of downhole fluid temperature. In normal circumstances, the temperature of the water will reflect the geothermal gradient, increasing 1–2°F for each 100 ft (2–3.5°C for each 100 m) of depth. Most temperature logs are gradient curves which record temperature versus depth. Temperature logs are useful in groundwater investigations and geothermal exploration as well as oil and gas exploration.

Acoustic logging techniques utilize sound waves. All types of acoustic logs require fluid-filled holes to operate. The fluid is either water or mud. The simplest method is direct velocity testing. This method consists of lowering a geophone down the borehole and measuring the direct transmission time from a sound source (such as a sledgehammer blow or dynamite explosion) on the surface to the tool.

Sonic logging measures sound waves within the borehole. Velocity or delta T logging measures the pressure-wave transit time between a sound source at the bottom of the tool and a geophone at a known distance above the source. Amplitude logging measures the amplitude of the wave at the first geophone. The speed of sound in rock is a function of the modulus (stress or strain relationship) and is a measurement of one parameter of strength. As such, sonic logging is valuable for rock mechanics and geotechnical investigations. *See* ROCK MECHANICS.

The acoustic televiewer maps the wall of the borehole with sonar waves. This technique uses an oscilloscope camera to photograph scan lines from each increment of depth in the borehole. Fractures and other interruptions in the borehole wall form shadows on the scan image. *See* SONAR.

Directional logging techniques determine borehole deviation and orientation of features within the borehole. Verticality logs measure the deviation of a borehole from vertical. Modern verticality logs use gyroscopes or gimbaled electronic compasses which feed directly into a computer in the logging truck. *See* GYROSCOPE.

Other borehole methods include cross-hole and borehole gravity techniques, which make point measurements. They are closely related to logging and often use the same equipment.

Cross-hole techniques place a source of one kind or another in one borehole and a detector in a nearby borehole. Direct seismic (shear wave) transmission is a frequently used cross-hole method for estimating support requirements in tunnels, mines, and excavations.

The borehole gravity technique is a downhole survey technique in which a very sensitive gravity meter measures the gravity at selected points within the borehole, determining the gravity gradient. The average bulk density of the rock can be calculated directly from the gravity readings. *See* GEOPHYSICAL EXPLORATION; GRAVITY METER. [C.B.Cl.]

Boride A binary compound of boron and a metal, formed by heating a mixture of the two elements. Boron does not conform to the usual rules of valence in forming these compounds. The borides, along with the carbides and the nitrides, are called interstitial compounds. These compounds have crystal structures and properties very similar to those of the original metal; hence, it is suggested that the boron is located in the cavities of the metallic lattice.

The borides are very hard and refractory and show high electrical conductivity. Typical borides are Ni_2B, NiB_2, MnB, MnB_2, and Cu_3B_2. Boron nitride is an abrasive harder than diamond. *See* ABRASIVE; BORON; SOLID-STATE CHEMISTRY. [E.E.W.]

Boring A machining operation that increases the size of an existing hole in a workpiece. The usual purpose of boring is to machine a hole to a desired diameter while obtaining required accuracy and finish. In metalworking, boring is performed after making a circular hole in the piece by some method such as coring, drilling, burning, punching, or trepanning. *See* DRILLING MACHINE.

Boring machines may be listed in two categories, horizontal and vertical. The former type holds the workpiece stationary on a movable table. A rotating spindle, holding the cutting tool, can be fed into the work from a vertically adjustable head. On vertical machines the work is revolved on a circular table while tools can be fed in for boring as well as facing operations. [A.H.T.]

Boring bivalves A variety of marine bivalve mollusks which penetrate solid substrata. They represent seven families and vary in the extent to which they are specialized, in the type of substrata they utilize, and in their method of boring.

Unlike other bivalves which use their shells as abrasive "tools," date mussels (*Lithophaga*, Mytilidae) penetrate calcareous rocks, corals, and shells by chemical means, possibly a weak carbonic acid. Larvae of *Botula* (another small, elongate mytilid), which settle in crevices of soft rock, corals, or wood, abrade their way into the substrate by continued movement of their shells and siphons. *Rupellaria* and *Petricola* (Petricolidae) are nonspecialized borers in peat, firm mud, and soft rock. *Platyodon* (Myidae), closely related to the soft-shelled clam, press their valves against the walls of the burrow by engorging the mantle and then abrade the soft rock by continued movement of their unspecialized valves. *Gastrochaena* and *Spengleria* (Gastrochaenidae) are specialized for boring by having a closed mantle cavity, a large pedal gape, and a truncate foot, allowing them to press the foot and shell against the burrow wall. The Pholadacea (Pholadidae and Teredinidae) are worldwide in distribution. The family Pholadidae (common name: piddocks) is composed of 17 genera, of which 5 are restricted to wood. All species of Teredinidae are obligate wood, nut, or plant-stem borers. The major difference between these two families is the presence of accessory plates in the pholads and pallets in the teredinids. *See* MOLLUSCA; SHIPWORM. [R.D.T.]

Boring sponges Sponges that excavate galleries in mollusk shells, corals, limestone, and other calcareous matter. Boring sponges are known in three families of the class Demospongiae; they are also called burrowing sponges. All species of the family Clionidae, order Hadromerida, excavate burrows in calcium carbonate substrata, as do some species of the genera *Anthosigmella* and *Spheciospongia* (family Spirastrellidae, order Hadromerida) and all species of *Siphonodictyon* (family Adociidae, order Haplosclerida).

Water currents are maintained through the tissues of these boring sponges exactly as in other Demospongiae. They do not obtain food from their hosts when the calcareous matter into which they burrow happens to be the shell or skeleton of a living organism. Instead, boring sponges feed on bacteria, flagellates, and other kinds of particulate organic matter. The burrowing process is similar in all genera and involves the excavation of numerous minute particles of calcium carbonate (20–70 micrometers in diameter) that pass out by way of the exhalant canals and oscula. In coral reef environments, up to 30–40% of fine sediment particles may be derived from the activities of burrowing sponges in low-energy environments; lesser proportions (5–22%) characterize sediments in high-energy environments.

Clionids are regarded as a nuisance in oyster beds because their excavations weaken the shells of these mollusks and thus hinder processing in canning factories. Boring sponges are also economically important when they attack limestone breakwaters.

Clionid boring sponges are found in all seas, chiefly in tidal and shallow waters. A few species reach a depth of at most 3000 ft (1000 m). Their excavations are known in fossil mollusk shells at least as far back as the Devonian Period. Burrowing species of the genera *Anthosigmella*, *Spheciospongia*, and *Siphonodictyon* are found on coral reefs at depths of more than 200 ft (60 m). *See* DEMOSPONGIAE. [W.D.H.]

Borneo An island in Sundanaland made up of Kalimantan, Sarawak, Brunei, and Sabah. Borneo has an area of 282,000 mi² (730,000 km²) and is the third largest island of the world. Mount Kinabalu, 13,445 ft (4098 m), in Sabah is the highest mountain on the island.

Crossed by the Equator, Borneo has a tropical climate with little range in temperature, coastal stations varying between 77 and 79°F (25 and 26°C) average monthly temperatures throughout the year. Annual rainfall in the lowlands varies from 90 to 130 in. (2.3 to 3.3 m). This is evenly distributed throughout the year without a distinct wet or dry season, except in the north and southeast, where there is a slight tendency toward a short dry season. A tropical evergreen forest covers large areas of Borneo, but there are also large areas of savanna grass and scrub trees, a result of past farming and fires of human origin. *See* EAST INDIES. [M.G.A.-C.]

Bornite A sulfide of composition Cu_5FeS_4, specific gravity 5.07, and hardness 3 (Mohs scale), commonly occurring as a primary mineral in many copper ore deposits. Crystals are rare; bornite is usually massive or granular. The metallic and brassy color of a fresh surface rapidly tarnishes upon exposure to air to a characteristic iridescent purple, giving rise to the name "peacock ore." Though of lesser importance as an ore than chalcocite or chalcopyrite, masses of bornite have been mined in Chile, Peru, Bolivia, and Mexico and in the United States in Arizona and Montana. *See* CHALCOCITE; CHALCOPYRITE; COPPER. [B.J.Wu.]

Boron A chemical element, B, atomic number 5, atomic weight 10.811, in group III of the periodic table. It has three valence electrons and is nonmetallic in behavior. It is classified as a metalloid and is the only nonmetallic element which has fewer than four electrons in its outer shell. The free element is prepared in crystalline or amorphous form. The crystalline form is an extremely hard, brittle solid. It is of jet-black to silvery-gray color with a metallic luster. One form of crystalline boron is bright red. The amorphous form is less dense than the crystalline and is a dark-brown to black powder. In the naturally occurring compounds, boron exists as a mixture of two stable isotopes with atomic weights of 10 and 11.

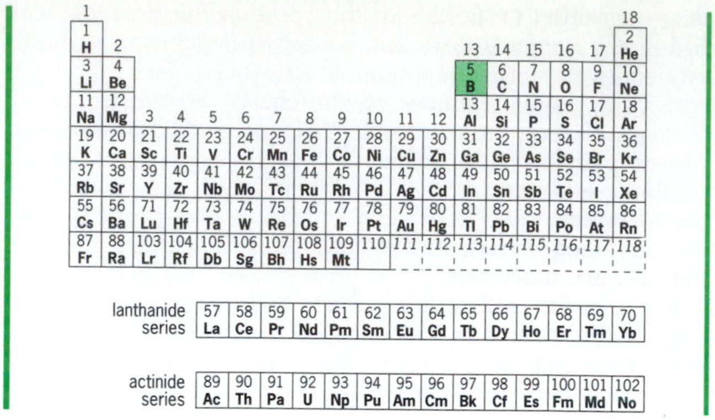

Many properties of boron have not been sufficiently established experimentally as a result of the questionable purity of some sources of boron, as well as of the variations in the methods and temperatures of preparation. A summary of the physical properties is shown in the table.

Physical properties of boron

Property	Temp., °C	Value
Density		
Crystalline	25–27	2.31 g/cm³
Amorphous	25–27	2.3 g/cm³
Mohs hardness		
Crystalline		9.3
Melting point		2100°C
Boiling point		2500°C
Resistivity	25	1.7×10^{-6} ohm-cm
Coefficient of thermal expansion	20–750	8.3×10^{-6} cm/°C
Heat of combustion	25	302.0±3.4 kcal/mole
Entropy		
Crystalline	25	1.403 cal/(mole)(deg)
Amorphous	25	1.564 cal/(mole)(deg)
Heat capacity		
Gas	25	4.97 cal/(mole)(deg)
Crystalline	25	2.65 cal/(mole)(deg)
Amorphous	25	2.86 cal/(mole)(deg)

Boron and boron compounds have numerous uses in many fields, although elemental boron is employed chiefly in the metal industry. Its extreme reactivity at high temperatures, particularly with oxygen and nitrogen, makes it a suitable metallurgical degasifying agent. It is used to refine the grain of aluminum castings and to facilitate the heat treatment of malleable iron. Boron considerably increases the high-temperature strength characteristics of alloy steels. Elemental boron is used in the atomic reactor and in high-temperature technologies. The physical properties that make boron attractive as a construction material in missile and rocket technology are its low density, extreme hardness, high melting point, and remarkable tensile strength in filament form. When boron fibers are used in an epoxy (or other plastic) carrier material or matrix, the resulting composite is stronger and stiffer than steel and 25% lighter than aluminum. Refined borax,

$Na_2B_4O_7 \cdot 10H_2O$, is an important ingredient of a variety of detergents, soaps, water-softening compounds, laundry starches, adhesives, toilet preparations, cosmetics, talcum powder, and glazed paper. It is also used in fireproofing, disinfecting of fruit and lumber, weed control, and insecticides, as well as in the manufacture of leather, paper, and plastics.

Boron makes up 0.001% of the Earth's crust. It is never found in the uncombined or elementary state in nature. Besides being present to the extent of a few parts per million in sea water, it occurs as a trace element in most soils and is an essential constituent of several rock-forming silicate minerals, such as tourmaline and datolite. The presence of boron in extremely small amounts seems to be necessary in nearly all forms of plant life, but in larger concentrations, it becomes quite toxic to vegetation. Only in a very limited number of localities are high concentrations of boron or large deposits of boron minerals to be found in nature; the more important of these seem to be primarily of volcanic origin. *See* BORATE MINERALS. [F.H.M./V.V.L.]

Boron polymers Inorganic polymers of boron compounds. Hexagonal boron nitride, (BN)*n*, is structurally related to graphite. Like graphite, it has lubricating properties, reflecting the relationship between molecular structure and physical properties, but unlike graphite, it is an electrical insulator. Molybdenum disulfide $(MoS_2)n$, with a similar and related structure, is also a solid lubricant. Both graphite and hexagonal boron nitride can be readily machined. Outstanding properties of the latter include high thermal and chemical stability and good dielectric properties. Crucibles and such items as nuts and bolts can be made from this material. See BORON; GRAPHITE; INORGANIC POLYMER. [R.A.S.]

Borrelia A genus of spirochetes. They are motile, helical organisms with 4–30 uneven, irregular coils, and are 5–25 micrometers long and 0.2–0.5 micrometer wide. Their locomotory apparatus consists of 15–22 fibrils coiled around the cell body and situated between the elastic envelope and a cytoplasmic membrane. Motion is forward and backward, laterally by bending and looping, and corkscrewlike. All borreliae are arthropod-borne. Of the 19 recognized species, 17 cause relapsing fevers and similar diseases in human and rodents; of the remaining 2, one is responsible for infections in ruminants and horses, the other for borreliosis in birds. *See* ANTIBIOTIC; BACTERIA; MEDICAL BACTERIOLOGY; RELAPSING FEVER. [W.Bu.]

Bose-Einstein statistics The statistical description of quantum mechanical systems in which there is no restriction on the way in which particles can be distributed over the individual energy levels. This description applies when the system has a symmetric wave function. This in turn has to be the case when the particles described are of integer spin.

Suppose one describes a system by giving the number of particles *ni* in an energy state ϵ_i where the n_i are called occupation numbers and the index *i* labels the various states. The energy level ϵ_i is of finite width, being really a range of energies comprising, say, g_i individual (nondegenerate) quantum levels. If any arrangement of particles over individual energy levels is allowed, one obtains for the probability of a specific distribution Eq. (1*a*). In Boltzmann statistics, this same probability would be written as Eq. (1*b*). *See* BOLTZMANN STATISTICS.

$$W = \prod_i \frac{(n_i + g_i - 1)!}{n_i!(g_i - 1)!} \quad (1a)$$

$$W = \prod_i \frac{g_i^{n_i}}{n_i!} \quad (1b)$$

The equilibrium state is defined as the most probable state of the system. To obtain it, one must maximize Eq. (1*a*) under the conditions given by Eqs. (2*a*) and (2*b*), which express the fact that the total number of particles *N* and the total energy *E* are fixed. One finds for the most probable distribution that Eq. (3) holds. Here, *A* and ß are parameters to be determined from Eqs. (2*a*) and (2*b*); actually, $ß = 1/kT$, where *k* is the Boltzmann constant and *T* is the absolute temperature.

$$\sum n_i i = N \quad (2a)$$

$$\sum \epsilon_i n_i = E \quad (2b)$$

$$ni = \frac{g_i}{\frac{1}{A} e^{ß\epsilon_i} + 1} \quad (3)$$

An interesting and important result emerges when Eq. (3) is applied to a gas of photons, that is, a large number of photons in an enclosure. (Since photons have integer spin, this is legitimate.) The formula for the energy density (energy per unit volume) in a given frequency range is found to be the celebrated Planck radiation formula for blackbody radiation. Thus, blackbody radiation must be considered as a photon gas, with the photons satisfying Bose-Einstein statistics. *See* HEAT RADIATION.

If an ideal Bose gas, consisting of a fixed number of material particles, is compressed beyond a certain point, some of the particles will condense in a zero state, where they do not contribute to the density or the pressure. If the volume is decreased, this curious condensation phenomenon results, yielding the zero state which has the paradoxical properties of not contributing to the pressure, volume, or density. There is now considerable evidence that many of the superfluid properties exhibited by liquid helium are in fact manifestations of an Einstein condensation. *See* FERMI-DIRAC STATISTICS; QUANTUM STATISTICS; STATISTICAL MECHANICS. [M.Dr.]

Botanical gardens A garden for the culture of plants collected chiefly for scientific and educational purposes. Such a garden is more properly called a botanical institution, in which the outdoor garden is but one portion of an organization including the greenhouse, the herbarium, the library, and the research laboratory. *See* HERBARIUM.

It was only in modern Europe, after the foundation of the great medieval universities, that botanical gardens for educational purposes began to be established in connection with the schools. The oldest gardens are those in Padua (established 1533) and at Pisa (1543). The botanical garden of the University of Leiden was begun in 1587, and the first greenhouse is said to have been constructed there in 1599. The Royal Botanical Gardens at Kew, England, were officially opened in 1841. This institution came to be known as the botanical capital of the world.

The first of the great tropical gardens was founded at Calcutta in 1787. The original name, Royal Botanic Garden, was changed in 1947 to Indian Botanic Garden. Another great tropical garden, the Jardin Botanico of Rio de Janeiro, was founded in 1808. The great tropical botanical garden of Buitenzorg (Bogor), Java, which originated in 1817, has an area of 205 acres (83 hectares) with an additional 150 acres (61 hectares) in the Mountain Garden.

The first great garden of the United States was founded by Henry Shaw at St. Louis in 1859, and is now known as the Missouri Botanical Garden. The New York Botanical Garden was chartered in 1891 and the Brooklyn Botanic Garden in 1910. The Jardin Botanique of Montreal, the leading garden of Canada, was opened in 1936. *See* ARBORETUM. [E.L.C.]

Botany That branch of biological science which embraces the study of plants and plant life. Botanical studies may range from microscopic observations of the smallest and obscurest plants to the study of the trees of the forest. One botanist may be interested mainly in the relationships among plants and in their geographic distribution, whereas another may be primarily concerned with structure or with the study of the life processes taking place in plants.

Botany may be divided by subject matter into several specialties, such as plant anatomy, plant chemistry, plant cytology, plant ecology (including autecology and synecology), plant embryology, plant genetics, plant morphology, plant physiology, plant taxonomy, ethnobotany, and paleobotany. It may also be divided according to the group of plants being studied; for example, agostology, the study of grasses; algology (phycology), the study of algae; bryology, the study of mosses; mycology, the study of fungi; and pteridology, the study of ferns. Bacteriology and virology are also parts of botany in a broad sense. Furthermore, a number of agricultural subjects have botany as their foundation. Among these are agronomy, floriculture, forestry, horticulture, landscape architecture, and plant breeding. *See* AGRICULTURE; AGRONOMY; BACTERIOLOGY; BREEDING (PLANT); CELL BIOLOGY; ECOLOGY; FLORICULTURE; GENETICS; LANDSCAPE ARCHITECTURE; PALEOBOTANY; PLANT ANATOMY; PLANT GROWTH; PLANT MORPHOGENESIS; PLANT PATHOLOGY; PLANT PHYSIOLOGY; PLANT TAXONOMY. [A.Cr.]

Bothriocidaroida An order of Perischoechinoidea in which the ambulacra comprise two columns of plates, the interambulacra one column, and the madreporite is placed radially. *Bothriocidaris*, the only known genus, occurs in the Upper Ordovician of Estonia, and is the oldest recognized echinoid. *See* ECHINODERMATA; ECHINOIDEA; PERISCHOECHINOIDEA. [H.B.F.]

Botulism A type of food poisoning due to intoxication by the extremely potent exotoxin produced by the anaerobic bacteria *Clostridium botulinum* and *C. parabotulinum*. Immunization against botulism is generally not necessary since the disease is comparatively rare. Prevention is best accomplished by safe procedures of food preservation.

The disease in humans most commonly follows ingestion of unheated, improperly processed food, with pH above 4.5, in which growth and toxin production has occurred. It is characterized by respiratory failure. In humans, an acute gastroenteritis may or may not precede the principal symptoms of extreme weakness, dry mucosa, loss of accommodation, ocular muscle paralysis, and difficulty in swallowing. These symptoms commonly occur within 48 h after the ingestion of the toxic food. [L.S.McC.]

Boundary-layer flow The flow of that portion of a viscous fluid which is in the neighborhood of a body in contact with the fluid and in motion relative to the fluid. Wherever a viscous fluid flows past a boundary, the layers of the fluid nearest the boundary are subjected to shearing forces, which cause the velocity of these layers to be reduced. As the boundary is approached, the velocity continuously decreases until, immediately at the boundary, the fluid particles are at rest relative to the body. This region of retarded velocity is called the boundary layer, and a graph of the variation of velocity with distance from the wall or boundary describes a boundary-layer profile (see illustration). The primary effects of the viscosity of the fluid are concentrated in this boundary layer, whereas in the outer or free-stream flow the viscous forces are negligible. Because the friction forces due to the viscosity of the fluid are restricted to the boundary layer, this region is also often called the friction layer. *See* FLUID FLOW; NAVIER-STOKES EQUATIONS.

The velocity of the fluid within the boundary layer approaches the local free-stream velocity asymptotically as the distance from the wall is increased; thus the actual thickness of the boundary layer is difficult to determine. An arbitrary boundary-layer thickness δ is usually defined as that distance from the wall or boundary where the velocity of the flow in the boundary layer u reaches 99% of the local free-stream velocity U (see illustration). The thicknesses of boundary layers surrounding

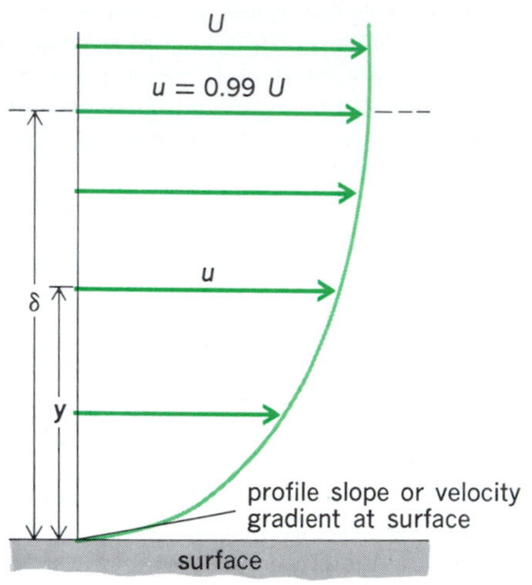

Boundary-layer profile.

vehicles moving through the air range from less than 0.1 in. (2.5 mm) near the front of a high-speed aircraft to more than 10 ft (3 m) at the rear of a dirigible or an airship.

The viscous shearing force created within the boundary layer constitutes a major portion of the resistance or drag experienced by an airplane in flight or by fluids passing through pipes and channels; this shearing force opposes the relative motion of any viscous fluid past a bounding or immersed surface. The boundary-layer thickness generally increases with the extent of the surface over which the fluid has passed. In the flow of fluid through pipes, the boundary layer formed at the walls thickens until the pipe is entirely filled with boundary-layer flow. When such a condition is reached, fully developed pipe flow is said to exist.

In flowing over a surface of changing contour, the velocity and pressure of a fluid vary. As the flow velocity increases, the pressure must decrease. In the free stream, where there are negligible losses to viscosity, the flow is always able to proceed into regions of increasing pressure at the expense of its velocity, the opposing pressure forces being balanced by the forces that result from the change in momentum of the flow. However, within the boundary layer, where the momentum of the fluid has been reduced in overcoming viscous forces, the remaining momentum may be insufficient to allow the flow to proceed into regions of increasing pressure. At the position on the body where this condition exists, the boundary layer no longer continues to follow the contour of the boundary but instead leaves, or separates from, the surface. This phenomenon is known as boundary-layer separation. Beyond the point of separation, between the surface and the separated boundary layer, a region of reversed or upstream flow exists. This reversed flow and the separated boundary layer subsequently coalesce to form a wake of low-velocity fluid behind the body. The separation of the flow in the boundary layer limits the lifting capabilities of wings and causes the stalling of airplanes. The efficiencies of fluid machines and mechanisms, such as turbines, compressors, pumps, and propellers, are decreased by flow separation originating in the boundary layer. *See* AIRPLANE; BERNOULLI'S THEOREM.

The flow in the boundary layer formed on a body exists in

one of two characteristic states: laminar or turbulent. If the flow is laminar, strata or laminae of fluid pass smoothly across the surface, whereas if it is turbulent, the flow is characterized by a rapid churning or mixing between layers of flow having different velocities. The phenomenon of transition from laminar to turbulent flow may occur within the boundary layer after the flow has passed over a surface for some distance.

The natural development of a boundary layer is affected by the contour and surface roughness of the body on which it is formed. Therefore, by the reduction of surface roughness and the choice of surface contours, some amount of control may be exerted upon the development of the boundary layer. Attention to surface roughness, waviness, and continuity is particularly important in the preservation of laminar boundary layers. The overall contour of the body is also an important consideration if laminar boundary-layer flow is to be maintained. Many laminar airfoil sections have been developed for high-speed aircraft. In fact, the term streamlining has come to describe the technique of designing shapes upon which the boundary-layer growth is minimized. The body shape also has a large influence upon the separation of the flow, and, to give low resistance or drag, shapes must be designed which prevent or delay boundary-layer separation. *See* STREAMLINING.

Because the amount of control available by purely geometric means is limited, additional methods for influencing the development of the boundary layer have evolved. These methods in general are intended either to remove the low-momentum flow from the boundary layer or to restore the lost momentum. The method of controlling the boundary layer by the removal of the retarded flow, called suction boundary-layer control, may be applied both to maintain laminar flow and to prevent boundary-layer separation. By sucking away the flow in the lower regions of the boundary layer through slots or perforations in the surface, the development of the boundary layer can be readily influenced. [J.J.C.]

Bovine virus diarrhea A disease of cattle induced by bovine viral diarrhea virus. The virus is common worldwide, infects cattle of all ages, and causes a variety of disease processes.

The mature virion is approximately 50 nanometers in diameter. The viral genome consists of a single strand of positive-sense ribonucleic acid (RNA) that contains approximately 12,500 nucleotides. In the mature virion, viral RNA is packaged in an icosahedral nucleocapsid that is surrounded by a membranous envelope. Cattle are the natural hosts for bovine viral diarrhea virus, but the virus infects most even-toed ungulates, including sheep, swine, goats, deer, and antelope. *See* ANIMAL VIRUS.

Cattle infected with the bovine viral diarrhea virus usually develop an acute disease that is inapparent or clinically mild. The acute form is characterized by fever, low white blood cell counts, mild depression, and a brief loss of appetite. Occasionally, acute bovine viral diarrhea is clinically severe, and then cattle may have diarrhea, develop ulcers in the mouth and intestine, show respiratory distress, hemorrhage internally, and die. Acute bovine viral diarrhea seldom lasts longer than a few days.

The virus replicates in lymphoid cells and induces a transient suppression that affects function of lymphocytes and neutrophils, potentially lowering the host's resistance to other infectious agents. Adverse effects on the fetus are common following infection of pregnant cattle. Embryonic resorption, abortion, stillbirth, and congenital anomalies result from fetal infections with the virus.

Persistently infected cattle are instrumental in the spread of bovine viral disease. These cattle shed the virus in saliva, nasal secretions, urine, milk, and semen. Susceptible cattle become infected within 1 h of direct contact with a persistently infected animal. Because bovine viral diarrhea virus infects several species, wildlife or other farm animals may spread infection. Biting insects, contaminated hypodermic needles, and contaminated farm equipment also spread the virus.

Treatment of bovine viral disease is based on supportive care. Administration of antiserum during the early stages of acute bovine viral disease may be beneficial, but controlling the disease is accomplished by vaccination, and by identification and elimination of persistently infected cattle. [S.R.Bo.]

Bowfin A fish, *Amia calva*, also known as the mudfish, dogfish, and grindle, that is the only living species of the family Amiidae and of the order Amiiformes. This line of fishes originated during the early Mesozoic and became a dominant group during the Jurassic and Cretaceous, when it was represented by many species.

The single extant species has features which are characteristic of extinct primitive fishes, and it is frequently referred to as a living fossil. The group is intermediate between the cartilaginous and bony fishes since the skeletal system is partly cartilaginous, the head shows bony dermal plates, and the scales are cycloid. This fish has a well-developed air bladder which functions as a lung. The lung allows the bowfin to survive out of water for long periods of time, up to a full day.

The bowfin ranges from the Great Lakes through the Mississippi Valley to Florida. This fish is inedible and preys upon game fish. *See* ACTINOPTERYGII; AMIIFORMES. [C.B.C.]

Boyle's law A law of gases which states that at constant temperature the volume of a gas varies inversely with its pressure. This law, formulated by Robert Boyle (1627–1691), can also be stated thus: The product of the volume of a gas times the pressure exerted on it is a constant at a fixed temperature. The relation is approximately true for most gases, but is not followed at high pressure. The phenomenon was discovered independently by Edme Mariotte about 1650 and is known in Europe as Mariotte's law. *See* GAS; KINETIC THEORY OF MATTER. [F.H.R.]

Brachiopoda A phylum of solitary, exclusively marine, coelomate, bivalved animals, with both valves symmetrical about a median longitudinal plane. They are typically attached to the substrate by a posteriorly located fleshy stalk or pedicle. Anteriorly, a relatively large mantle cavity is always developed between the valves, and the filamentous feeding organ, or lophophore, is suspended in it, projecting forward from the anterior body wall (see illustration).

There are two clearly defined groups within the phylum, a division that is particularly marked if only Recent animals are considered. These two groups are regarded as classes; several names have been given to them, but Inarticulata and Articulata are the most widely used and are based on one of the most readily observed differences between them, the presence or absence of articulation between the two valves of the shell. Among the Articulata the valves are typically hinged together by a pair of teeth with complementary sockets in the opposing valve; these hinge teeth are lacking in the Inarticulata, whose valves are held together only by the soft tissue of the living animal. The phylum is currently classified into the following groups:

Class Inarticulata
- Order: Lingulida
- Acrotretida
- Obolellida
- Paterinida

Class Incertae Sedis
- Order Kutorginida

Class Articulata
- Order: Orthida
- Strophomenida
- Pentamerida
- Rhynchonellida
- Spiriferida
- Terebratulida

The pedicle is the only organ protruding outside the valves, while the remainder of the animal is enclosed in the space between them. This space is divided into two unequal parts, a

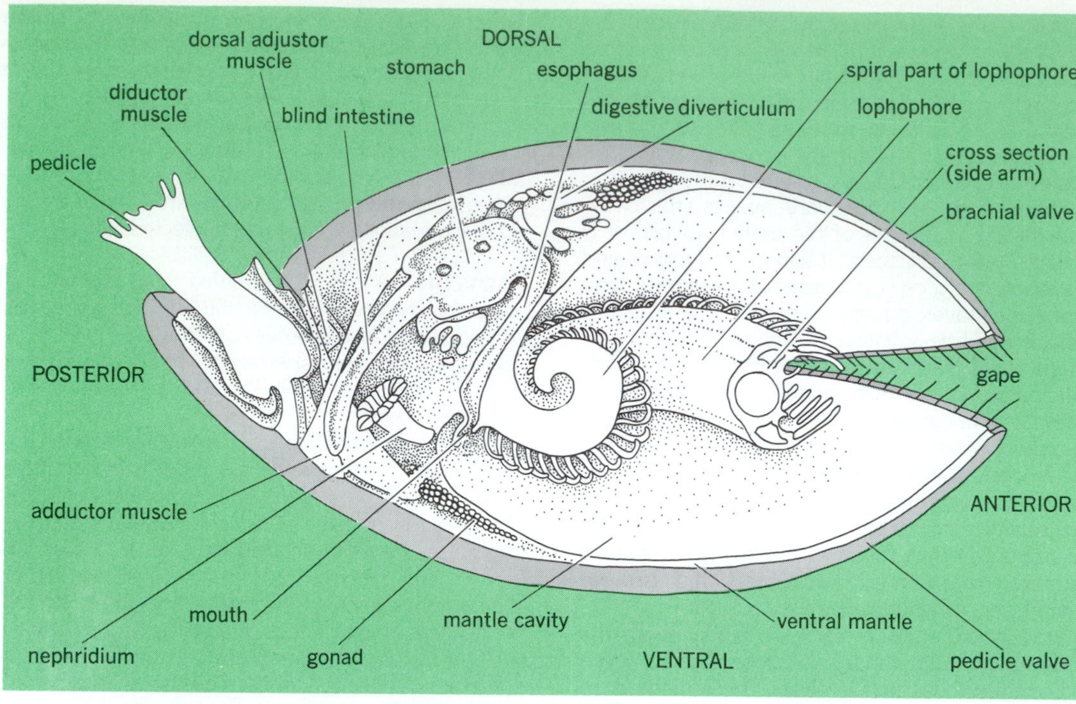

The principal organs of a brachiopod as typified by *Terebratulina*. (*After R. C. Moore, ed., Treatise on Invertebrate Paleontology, pt. H, Geological Society of America, Inc., and University of Kansas Press, 1965*)

smaller posteriorly located body cavity and an anterior mantle cavity. The two mantles approach each other and ultimately fuse along the posterior margin of articulate brachiopods; in contrast, the mantles are invariably discrete in the inarticulates and are separated by a strip of body wall.

The body cavity contains the musculature; the alimentary canal; the nephridia, which are paired excretory organs also functioning as gonoducts; the reproductive organs; and primitive circulatory and nervous systems. Except for the openings through the nephridia, the body cavity is enclosed, but the mantle cavity communicates freely with the sea when the valves are opened. The lophophore is suspended from the anterior body wall within the mantle cavity and is always symmetrically disposed about the median plane. The lophophore consists of a variably disposed, ciliated, filament-bearing tube, with the ciliary beat producing an ordered flow of water within the cavity, flowing across the filaments. The latter trap food particles which are carried along a groove in the lophophore to the medially situated mouth.

All modern brachiopods are marine, and there is little doubt from the fossil record that brachiopods have always been confined to the sea. Recent brachiopods occur most commonly beneath the relatively shallow waters of the continental shelves, which seems to have been the most favored environment, but the bathymetric range of the phylum is large. A few modern species live intertidally and, at the other extreme, a limited number have been dredged from depths of over 16,000 ft (5000 m).

The majority of brachiopods form part of the sessile benthos and are attached by their pedicle during postlarval life. *Glottidia* and *Lingula* are exceptional in being infaunal and making burrows. A commoner modification involves loss of the pedicle, either complete suppression or atrophy early in the life history of the individual. Such forms either lie free on the sea floor, are attached by cementation of part or all of the pedicle valve, or are anchored by spines. The geographic distribution and geological setting of some fossil species suggest that they may have been epiplanktonic, attached to floating weed, but such a mode of life is unknown in modern faunas. *See* ARTICULATA (BRACHIOPODA); INARTICULATA. [A.J.R.]

Bradyodonti An order, perhaps composite in nature, of Paleozoic cartilaginous fishes (Chondrichthyes), presumably derived from primitive sharks. They are mainly represented by dentitions, which appear to have been adapted to the eating of mollusks and other shelled invertebrates; the body (seldom preserved) appears to have been broad and flattened, with large pectoral fins, as in the later skates and rays of similar habits. The first bradyodonts appeared at the end of the Devonian; they flourished during the Carboniferous, but declined and vanished by the end of the Permian, concomitantly with a reduction in the invertebrate fauna at that time. There are four families: Cochliodontidae, Petalodontidae, Psammodontidae, and Copodontidae. *See* CHIMAERIFORMES; CHONDRICHTHYES; ELASMOBRANCHII. [A.S.R.]

Brain That portion of the central nervous system enclosed within the skull. It is composed of the brainstem, or segmental portion of the brain, and the suprasegmental apparatus, comprising a posterior cerebellum and anterior paired cerebral hemispheres (see illustration). *See* CENTRAL NERVOUS SYSTEM.

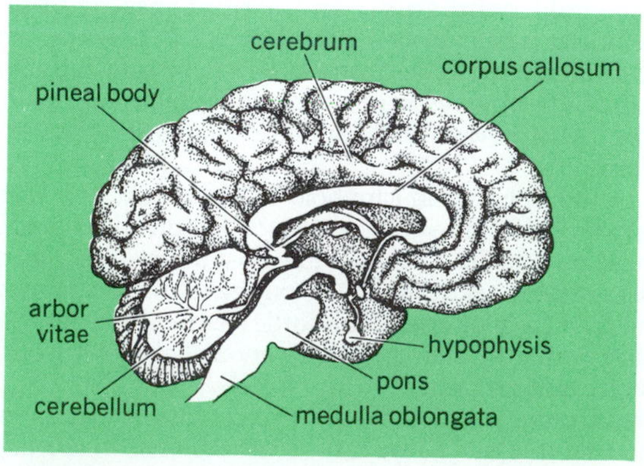

A sagittal section through the human brain. (*After H. E. Walter and L. P. Sayles, Biology of the Vertebrates, 3d ed., Macmillan, 1949*)

The telencephalon is a part of the brain that includes the olfactory bulbs (and paraolfactory bulbs) and cerebral hemispheres. Whereas the olfactory and paraolfactory bulbs are very similarly organized in all vertebrates, the cerebral hemispheres manifest greater diversity of structure among vertebrates than any other part of the brain. The olfactory and paraolfactory bulbs receive and process information from the chemosense organs of the nasal region and transmit this information to the cerebral hemispheres. Not surprisingly, the size of this portion of the telencephalon is proportional to the size and importance of these chemosenses in a given animal. The cerebral hemispheres play an integrative function in all vertebrates. The cerebrum is largest in the vertebrates with the greatest integrative ability, such as humans and whales.

The diencephalon is the anteriormost part of the brainstem. In all vertebrates the diencephalon can be divided into four regions; from dorsal to ventral, these are the epithalamus, dorsal thalamus, ventral thalamus, and hypothalamus. The epithalamus is always relatively small, and its functions poorly understood. It is best developed in vertebrates such as lizards that have pineal or parietal median eyes. The dorsal thalamus is largest in vertebrates with the largest cerebral hemispheres. It is a complex of nuclei relaying information from the spinal cord and other parts of the brain to the cerebral hemispheres. The ventral thalamus is relatively inconspicuous in all vertebrates and functions in motor activity arising from deep portions of the cerebral hemispheres. The hypothalamus, to which the pituitary gland attaches, is a major integrating center for endocrine and visceral activities.

The mesencephalon (midbrain) is divided into a dorsal tectum and a ventral tegmentum. The tectum has large visual and auditory centers and a small somatic sensory center. These sensory centers are integrative in function and send already processed information to both the cerebral hemispheres and other portions of the central nervous system. The tegmentum contains nervous pathways going to and from the forebrain and important motor centers such as the red nucleus.

The dorsal portion of the metencephalon forms the suprasegmental cerebellum. The cerebellum, although it does not direct motor activity, is the all-important organ for coordination of motor activity and maintenance of equilibrium. It is smallest in vertebrates without paired appendages. It is large and well developed in mammals, birds, and active fishes such as sharks. The ventral portion of the metencephalon contains nerve fiber tracts traveling through it, and the central nuclei of several cranial nerves.

Combined with the ventral portion of the metencephalon, the myelencephalon is called the medulla oblongata. In addition to fiber tracts to and from the spinal cord, the medulla oblongata contains the centers for respiration and arousal "reticular" centers. It also has the central nuclei for most of the cranial nerves.

The entire brain has a lumen filled with cerebrospinal fluid, and dilations of the lumen are called ventricles. The size, shape, and number of these ventricles vary greatly among vertebrates. *See* MENINGES; NERVOUS SYSTEM (VERTEBRATE). [D.B.W.]

Brake A machine element for applying a force to a moving surface to slow it down or bring it to rest in a controlled manner. Brakes are used in cars, trains, airplanes, elevators, and other machines. Most brakes are of a frictional type in which a fixed surface is brought into contact with a moving part that is to be slowed or stopped. Brakes in general connect a moving and a stationary body.

The limitations on the applications of brakes are similar to those of clutches, except that the service conditions are more severe because the entire energy is absorbed by slippage which is converted to heat that must be dissipated. The important thing is the rate at which energy is absorbed and heat dissipated. With frictional brakes, if the temperature of the brake becomes too high, the result is a lowering of the friction force, called fading.

There are also electrical and hydrodynamic brakes. The electrical type may be electromagnetic, eddy-current, hysteresis, or magnetic-particle. The hydrodynamic type works somewhat like a fluid coupling with one element stationary. Another type of brake, used on electric trains, is the regenerative brake. The dynamo on this machine can be used as a motor or a generator. As a generator it brakes the train and stores the generated electricity in an accumulator. Another type of brake is the air brakes (flaps) on an airplane. *See* AIR BRAKE; DYNAMO; ELECTROMAGNETIC BRAKE; FRICTION BRAKE. [H.P.H.]

Branch circuit The portion of an electric wiring system that follows the final circuit overcurrent protector and that terminates at the utilization device (lighting fixture, motor, or heater) or a plug receptacle for the connection of portable lamps and appliances.

Branch circuits serving more than one outlet or load are limited by the National Electrical Code to three types: (1) Circuits of 15 or 20 amperes may serve lights and appliances; the rating of one portable appliance may not exceed 80% of the circuit capacity; the total rating of fixed appliances may not exceed 50% of circuit capacity if lights or portable appliances are also supplied. (2) Circuits of 30 A may serve fixed lighting units with heavy-duty lampholders in other than dwellings or appliances in any occupancy. (3) Circuits of 40 or 50 A may serve fixed lighting with heavy-duty lampholders in other than dwellings, fixed cooking appliances, or infrared heating units. [W.T.S.; J.F.McP.]

Branched polymer A polymer chain having branch points that connect three or more chain segments. Examples of branched polymers include long chains having occasional and usually short branches comprising the same repeat units as the main chain (nominally termed a branched polymer); long chains having occasional branches comprising repeat units different from those of the main chain (termed graft copolymers); main chains having one long branch per repeat unit (referred to as comb polymers); and small core molecules with branches radiating from the core (star polymers). Starburst or dendritic polymers are a special class of star polymer in which the branches are multifunctional, leading to further branching with polymer growth. Star, comb, and starburst polymers (see illus.),

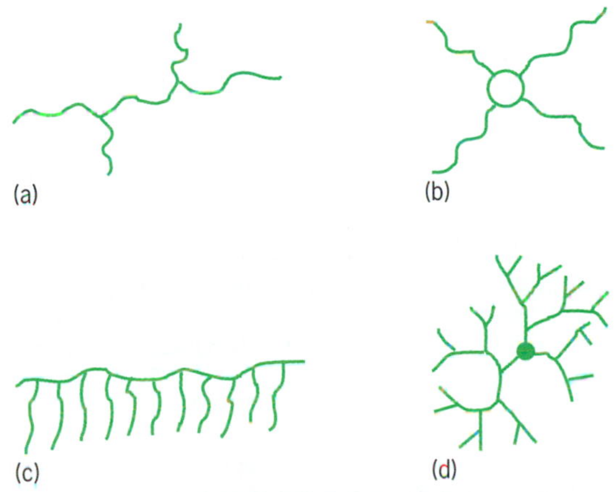

Examples of branched polymers. (a) Branched polymer (if arms are of composition similar to backbone) or graft polymer (if compositions are different). (b) Star polymer. (c) Comb polymer. (d) Dendritic polymer.

especially the last, represent interesting molecular structures that may lead to unusual supramolecular structures (for example, micelles and liposomes) that mimic the functions of complex biomolecules. *See* BIOPOLYMER; POLYMER; POLYMERIZATION. [G.E.W.]

Branchiopoda A class of crustaceans. The Conchostraca consist of two groups which, although superficially similar, differ in so many fundamental features that they have been placed in separate orders, Laevicaudata and Spinicaudata. *See* LAEVICAUDATA; SPINICAUDATA.

Living members of the Branchiopoda range from less than 0.5 mm (0.02 in.) to (exceptionally) 100 mm (4 in.) in length. Form is exceedingly diverse. The trunk may be abbreviated and of probably as few as 5 segments (although segmentation is sometimes obscure) or elongate and of more than 40 segments. Trunk limbs range from 5 to about 70 pairs. A carapace is often present, either as a dorsal shield or as a bivalved structure; anostracans lack a carapace.

Most species are microphagous. Microphagous forms collect their food either by direct scraping from surfaces, which may or may not be followed by filtration, or by abstracting suspended particles by the use of complicated filtering devices.

Reproductive habits are diverse. Parthenogenesis is widespread, and the production of highly resistant resting eggs that can withstand freezing and drying and retain their viability for several years is highly characteristic. Almost all species live in fresh water, but a few are marine and some frequent highly saline situations. Branchiopods have a worldwide distribution. *See* CRUSTACEA. [G.Fr.]

Branchiura A subclass of the Crustacea known as the fish lice. They are ectoparasites of fresh-water and marine fish. *Argulus* is a common genus. They are a small homogeneous group, less than 100 species, with worldwide distribution, and very much alike in appearance. The disklike head and thorax (cephalothorax) is strongly flattened and bears a small unsegmented abdomen bilobed at its distal end. Larger species may exceed 1 in. (25 mm) in length.

The appendages, all on the cephalothorax, consist of two pairs of antennae, the mouthparts which are one pair of mandibles and two pairs of maxillae, and four pairs of swimming legs. The wafer-thin cephalothorax with its appendages appressed to the underside permits fish lice to flatten themselves against a fish's skin. The resulting highly streamlined contour serves to minimize the considerable force of frictional drag found in a dense medium like water.

Other adaptations for holding fast to a mucus-covered swimming fish include numerous spinules and strategically placed hooks on the underside of the body and utilization of specialized portions of the body as suction cups. Despite these elaborate specializations, fish lice never lose the ability to abandon a host and swim. At least some of this activity is associated with mating and spawning.

Branchiurans feed on tissue fluids, especially blood. The mouth, located at the end of a movable proboscis, is applied to the host's skin and the rasping mandibles make the necessary wound. Fish lice usually are no threat to fish populations. In restricted areas such as hatchery ponds, however, an infestation can increase to levels that bring about high fish mortality. Treatment may require such drastic measures as drainage and cleaning before restocking can be successful. *See* CRUSTACEA. [A.Fl.]

Brass An alloy of copper and zinc. In manufacture, lump zinc is added to molten copper, and the mixture is poured into either castings ready for use or into billets for further working by rolling, extruding, forging, or similar process. Brasses containing 75–85% copper are red-gold and malleable; those containing 60–70% are yellow and also malleable; and those containing 50% or less copper are white, brittle, and not malleable. Alpha brass contains up to 36% zinc; beta brass contains nearly equal proportions of copper and zinc. Specific brasses are designated as follows: gilding (95% copper: 5% zinc), red (85:15), low (80:20), and admiralty (70:29, with balance of tin). Naval brass is 59–62% copper with about 1% tin, less than that of lead and iron, and the remainder zinc. The nickel silvers contain 55–70% copper and the balance nickel. Leaded brass is used for castings.

Brass stains in moist air; however, the oxide so formed is sufficiently continuous and adherent to retard further oxidation. When brass is required to remain bright, it is either washed in nitric acid and then coated with clear lacquer, or it is regularly polished and waxed.

Brass is widely used in cartridge cases, plumbing fixtures, valves and pipes, screws, clocks, and musical instruments. *See* ALLOY; COPPER ALLOYS; ZINC. [F.H.R.]

Brassin A lipid complex from rape pollen that is biologically active on a wide range of crop plants. Plants produce hormones which control their growth and behavior. Synthetically prepared growth-regulating substances have been used for many years to supplement endogenous hormones, and they change to some extent the behavior of crop plants to suit the needs of humans. But hormones produced by crop plants themselves so far have not been used in place of synthetic substances for this purpose, mainly because they occur in extremely small amounts and are difficult to extract. Tons of plant material are required to produce a few milligrams of hormones. However, some hormones made by plants eventually may be useful in plant control because sources have been discovered from which relatively large amounts of endogenous regulators can be obtained. Brassins, for example, are hormones that may eventually be useful; they have been discovered in pollen of rape (*Brassica napus*), a widely grown crop plant. Furthermore, endogenous hormones induce some responses that cannot be obtained with synthetic growth-regulating substances. Also of importance for food crops is the fact that these hormones, which are present in plants consumed by humans and animals, are relatively nontoxic substances. *See* PLANT GROWTH; PLANT HORMONES. [J.F.Wo.; N.M.]

Brayton cycle A thermodynamic cycle (also variously called the Joule or complete expansion diesel cycle) consisting of two constant-pressure (isobaric) processes interspersed with two reversible adiabatic (isentropic) processes.

The thermal efficiency for a given gas, air, is solely a function of the ratio of compression. This is also the case with the Otto cycle. For the diesel cycle with incomplete expansion, the thermal efficiency is lower.

The Brayton cycle, with its high inherent thermal efficiency, requires the maximum volume of gas flow for a given power output. The Otto and diesel cycles require much lower gas flow rates, but have the disadvantage of higher peak pressures and temperatures. These conflicting elements led to many designs, all attempting to achieve practical compromises. With the development of fluid acceleration devices for the compression and expansion of gases, the Brayton cycle found mechanisms which could economically handle the large volumes of working fluid. This is perfected in the gas turbine power plant. *See* GAS TURBINE; THERMODYNAMIC CYCLE. [T.Ba.]

Brazil nut A large broad-leafed evergreen tree, *Bertholettia excelsa*, that grows wild in the forests of the Amazon valley of Brazil and Bolivia. The fruit is a spherical capsule weighing 2–4 lb (1–2 kg) which, when mature, consists of an outer hard indehiscent husk enclosing an inner hard-shelled container or pod filled with about 20 rather triangular seeds or nuts.

Although there are a few plantations in Brazil, almost the entire production is gathered from wild trees. The nuts are mostly exported to Europe, Canada, and the United States. About one-fourth of the crop is shelled in Brazil before export.

Brazil nuts have a high oil and protein content and require careful handling and refrigeration to prevent spoilage. The nuts are used in confectionery, baked goods, and nut mixtures. *See* Lecythidales. [L.H.MacD.]

Brazing A method of joining metals, and other materials, by applying heat and a brazing filler metal. The filler metals used have melting temperatures above 840°F (450°C), but below the melting temperature of the metals or materials being joined. They flow by capillary action into the gap between the base metals or materials and join them by creating a metallurgical bond between them, at the molecular level. The process is similar to soldering, but differs in that the filler metal is of greater strength and has a higher melting temperature.

When properly designed, a brazed joint will yield a very high degree of serviceability under concentrated stress, vibration, and temperature loads. It can be said that in a properly designed brazement, any failure will occur in the base metal, not in the joint. There are many design variables to be considered. First among them is the mechanical configuration of the parts to be joined, and the joint area itself. All brazements can be categorized as having one of two basic joint designs: the lap joint or the butt joint. Others are adaptations of these two.

Design considerations should include the informed selection of the base and filler metals. In addition to the basic mechanical requirements, the base metals used in the brazement must retain the integrity of their physical properties throughout the heat of the brazing cycle. No universal filler metal that will satisfy all design requirements is possible, but there are many types available, ranging from pure metals such as copper, gold, or silver to complex alloys of aluminum, gold, nickel, magnesium, cobalt, silver, and palladium.

There are 11 basic brazing processes. In torch brazing, heat is applied by flame, from some type of torch, directly to the base metal. A mineral flux is normally used. The brazing filler metal may be preplaced in the joint, or face-fed into the joint. In induction brazing, brazing temperatures are developed in the parts to be brazed by placing them in or near a source of high-frequency ac electricity. Flux and preplaced filler metals are normally employed. Resistance brazing employs electrodes, which are arranged so that the joint forms a part of an electric circuit. Heat is developed by the resistance of the parts to the flow of the electric current. In dip brazing, the brazing filler metal is preplaced in or at the joint, and the assembly is immersed in a bath of molten salt or flux until the brazing temperature is achieved. In a variation of this process, the assembly is prefluxed and dipped into a bath of molten brazing filler metal. Infrared brazing is a process in which high-intensity quartz lamps are directed on the metals to be joined.

Furnace brazing is a widely used technique, especially useful where the parts to be brazed are machined or formed to their final dimensions, or constitute a complex assembly that has already been lightly joined or fixtured. The atmosphere within a brazing furnace is usually controlled, which permits a great deal of flexibility. An important advantage is that potential distortion of metal, created by heating and cooling, can be predicted and controlled and thereby minimized or eliminated. Also the capacity for automation is facilitated in the furnace brazing process.

Diffusion brazing, unlike furnace brazing, is defined not by the method of heating but rather by the degree of mutual filler-metal solution and diffusion with the base metal resulting from the temperature used and the time interval at heat. In diffusion brazing, temperature, time, in some cases pressure, and selection of base and filler materials are so controlled that the filler metal is partially or totally diffused into the base metal. The joint properties then closely approach those of the base metal.

Other, less used processes include arc brazing, block brazing, flow brazing, and twin carbon arc brazing. [R.L.P.]

Breadboarding Assembling an electronic circuit in the most convenient manner on a board or other flat surface, without regard for final locations of components, to prove the feasibility of the circuit and to facilitate changes when necessary. Standard breadboards for experimental work are made with mounting holes and terminals closely spaced at regular intervals, so that parts can be mounted and connected without drilling additional holes.

Printed-circuit boards having similar patterns of punched holes, with various combinations of holes connected together by printed wiring on each side, are often used for breadboarding when the final version is to be a printed circuit. *See* Circuit (electronics); Printed circuit. [J.Mar.]

Breadfruit The multiple fruit of an Indo-Malaysian tree, *Artocarpus altilis*, of the mulberry family (Moraceae), now cultivated in tropical lowlands around the world. The fruits vary considerably in size and are often borne in small clusters. Breadfruit is a wholesome food for both humans and animals, although it has a high carbohydrate content. It is eaten fresh or baked, boiled, roasted, fried, or ground up and made into bread. There are many varieties, both with and without seeds. *See* Urticales. [E.L.C.]

Breakdown potential That potential difference in a gaseous discharge at which a sudden increase in its electrical conductivity takes place. An alternative term is sparking potential. *See* Electric spark.

In general, the breakdown potential is a complicated function which depends on such factors as electrode material, both macroscopic and microscopic electrode shape, chemical surface states of the electrodes, constitution and pressure of the gas, temperature, electrode separation distance, and time-dependence of applied potential difference.

In certain cases the nature of the influence of one of the above factors may be readily determined. The presence of sharp points at the electrodes, and particularly at the cathode, produces a marked lowering of the breakdown potential. This is because the strong electric field around such a point results in intense ionization once an initial electron enters the region. [G.H.M.]

Breakwater A structure to protect against wave action of the seas and to create comparatively calm water behind for (1) harbors where adjacent land projections and nearby islands are inadequate to give protection, (2) coastal lands subject to damage by the sea, and (3) offshore structures.

The usual breakwater is a mound of stone rubble dumped on the ocean or harbor bed and rising to a height above mean sea level. Sometimes it is capped on the top and sea face with variously shaped concrete blocks. Other common forms of breakwaters are vertical-faced structures built of solid blocks of masonry or concrete, precast-concrete cellular walls floated into place and filled with earth or stone or steel-sheet-piling cells filled with stone (common on the Great Lakes). Masonry and concrete walls are placed on low rubble mounds or on the sea bed if firm. Scouring of the sea bed in front may require a stone protection bed. *See* Coastal engineering; Harbor. [E.J.Q.]

Breast The human mammary gland, usually well developed in the adult female but rudimentary in the male. Each adult female breast contains 15–20 separate, branching glands that radiate from the nipple. During lactation their secretions are discharged through separate openings at the base of the nipple.

In the female, hormonal changes in adolescence cause enlargement of breast tissue, but much of this is connective tissue although some glandular buds form. With the advent of full menstruation ovarian estrogenic hormones influence breast development. If pregnancy ensues, the glandular tissue reaches full development and full lactation begins shortly after birth. After cessation of lactation the breasts regress considerably and once again reflect cyclic regulation. *See* Lactation. [W.J.B.]

Breast disorders may result from congenital or developmental abnormalities, inflammations, hormonal imbalances, and, most important, from tumor formation.

Congenital defects are usually unimportant except for their psychic or cosmetic implications. Supernumerary nipples and breasts or accessory breast tissue are common examples.

Inflammations are not encountered frequently and usually result from a staphylococcal or streptococcal invasion incurred during lactation. A special form of inflammation may result from fat necrosis. Although any age is susceptible, older women show a slightly higher incidence of fat necrosis, the commonest cause of which is injury from trauma. *See* STAPHYLOCOCCUS; STREPTOCOCCUS; SYPHILIS; TUBERCULOSIS.

Hormonal imbalances are believed to be responsible for the variants of the commonest nontumorous breast disorder of women, cystic hyperplasia. The changes are thought to result from exaggeration or distortion of the normal cyclic alterations induced during the menstrual interval. Although a wide range of clinical and pathologic variation is commonplace, three major types or tendencies prevail. The first, called fibrosis or mastodynia, is marked by an increase of connective tissue in the breast, without a proportionate increase in glandular epithelium. The second, cystic disease, is characterized by an increase in the glandular and connective tissues in local areas, with a tendency toward formation of cysts varying in size. The third major type is adenosis, in which glandular hyperplasia is predominant. Each major form of cystic hyperplasia has its own clinical characteristics, ages of highest incidence, and distribution. Each is important because the breast masses which occur require differentiation from benign and malignant tumors. These lesions also have been found to predispose to the subsequent development of carcinoma.

Breast cancer is the most significant lesion of the female breast, accounting for 25,000–30,000 deaths in the United States each year. It rarely occurs before the age of 25, but its incidence increases each year thereafter, with a sharper climb noted about the time of menopause. Early breast cancer may appear as a small, firm mass which is nontender and freely movable. Diagnosis at this time carries a more favorable prognosis than later, when immobility, nipple retraction, lymph node involvement, and other signs of extension or spread are noted. Paget's disease of the nipple is a special form of breast cancer, in which there are early skin changes about the nipple. *See* CANCER (MEDICINE); HORMONE; ONCOLOGY. [E.G.St./N.K.M.]

Breccia A rock composed of fragments, similar to a conglomerate in distribution of sizes of the materials (coarser than 0.08 in. or 2 mm in diameter). Breccias differ from conglomerates in the sharp, angular nature of the fragments. The sedimentary breccias are to be distinguished from volcanic breccias formed by the agglomeration of volcanic materials and tectonic breccias formed by fracturing and disruption due to faulting. Breccias derived from erosion of cliffs or scarps with little or no transportation are called talus breccia or scree. The intraformational conglomerates are frequently called breccias if the pebbles are angular. *See* TALUS.

A type of breccia that has received increased attention is a soft rock deformation product. Such breccias are similar both to flat shale pebble conglomerates and to fault breccias. They develop as a result of gravitational slumping or sliding down a slope where a soft shale bed is interbedded with more competent beds, such as sandstone or siltstone. Very large pieces of the shale bed may become detached and deformed and even squeezed up through overlying layers, giving a picture of badly distorted and in some cases, almost unrecognizable bedding. A characteristic of these breccias is that they are intraformational; that is, higher beds carry across the deformed zone without disturbance. *See* CONGLOMERATE; GRAVEL; SEDIMENTARY ROCKS. [R.Si.]

Breeding (animal) The application of genetic principles to improving heredity for economically important traits in domestic animals. Examples are improvement of milk production in dairy cattle, meatiness in pigs, feed requirements or growth rate in beef cattle, and egg production in chickens. Selection permits the best parents to leave more offspring in the next generation than poorer parents. The science of animal breeding is based on genetic principles; however, the current application of these principles relies heavily on statistical knowledge. Measurement of performance of individuals, such as growth rate or milk production, is essential. Modern computers allow the application of complicated statistical procedures to large volumes of performance data to choose the best individuals as parents. This allows selection to be efficient.

Selection is the primary tool for generating directed hereditary changes in animals. It may be concentrated on one characteristic, may be directed independently on several traits, or may be conducted on an index or total score which includes information on several important traits. In general, the second and third methods are preferable to the first when several important and heritable traits need attention.

Heritability is a key parameter in making decisions in selection. It is the fraction of the total variation in a trait that is due to genetic differences. Managing animals to equalize environmental influences on them, or statistically adjusting for environmental differences between animals, is necessary to choose accurately those with the best inheritance for various traits.

The improvement achieved by selection is directly related to the accuracy with which the breeding values of the subjects can be recognized. Accuracy, in turn, depends upon the heritabilities of the traits and upon whether they can be measured directly upon the subjects for selection (mass selection), upon their parents (pedigree selection), upon their brothers and sisters (family selection), or upon their progeny (progeny testing). For traits of medium heritability, the following sources of information are approximately equally accurate for predicting breeding values of subjects: (1) one record measured on the subject; (2) one record on each ancestor for three previous generations; (3) one record each on five brothers or sisters where there is no environmental correlation between family members; and (4) one record each on five progeny having no environmental correlations, each from a different mate.

Artificial insemination is an effective tool for increasing the intensity of selection, especially in males. It allows one male to breed many more females than with natural mating. It has been most widely applied in dairy cattle, where carefully planned young sire sampling programs are being combined with thoroughly conducted progeny testing schemes to identify genetically superior sires. *See* GENETICS. [A.E.Fr.]

Breeding (plant) The application of genetic principles to improve cultivated plants. New varieties of cultivated plants can result only from genetic reorganization that gives rise to improvements over the existing varieties in particular characteristics or in combinations of characteristics. Thus, plant breeding can be regarded as a branch of applied genetics, but it also makes use of the knowledge and techniques of many aspects of plant science, especially physiology and pathology. Related disciplines, like biochemistry and entomology, are also important, and the application of mathematical statistics in the design and analysis of experiments is essential. *See* GENETICS.

The cornerstone of all plant breeding is selection, or the picking out of plants with the best combinations of agricultural and quality characteristics from populations of plants with a variety of genetic constitutions. Seeds from the selected plants are used to produce the next generation, from which a further cycle of selection may be carried out if there are still differences. Conventional breeding is divided into three categories on the basis of ways in which the species are propagated. First come the species that set seeds by self-pollination; that is, fertilization usually follows the germination of pollen on the stigmas of the same plant on which it was produced. The second category of species sets seeds by cross-pollination; that is, fertilization usually follows the germination of pollen on the stigmas of different plants from those on which it was produced. The third category comprises the species that are asexually propagated; that is, the commercial crop results from planting vegetative parts or by grafting. The procedures used in breeding differ according to the pattern of propagation of the species. Several innovative techniques have been explored to enhance the scope, speed, and efficiency of producing new, superior cultivars. Advances have been made in extending conventional sexual crossing procedures by laboratory culture of plant organs and tissues and by somatic hybridization through protoplast fusion.

The essential attribute of self-pollinating crop species, such as wheat, barley, oats, and many edible legumes, is that, once they are genetically pure, varieties can be maintained without change for many generations. When improvement of an existing variety is desired, it is necessary to produce genetic variation among which selection can be practiced. This is achieved by artificially hybridizing between parental varieties that may contrast with each other in possessing different desirable attributes. This system is known as pedigree breeding, and it is the method most commonly employed, and can be varied in several ways.

Another form of breeding often employed with self-pollinating species involves backcrossing. This is used when an existing variety is broadly satisfactory but lacks one useful and simply inherited trait that is to be found in some other variety. Hybrids are made between the two varieties, and the first hybrid generation is crossed, or backcrossed, with the broadly satisfactory variety which is known as the recurrent parent. Backcrossing has been exceedingly useful in practice and has been extensively employed in adding resistance to diseases, such as rust, smut, or mildew, to established and acceptable varieties of oats, wheat, and barley. *See* PLANT PATHOLOGY.

Natural populations of cross-pollinating species are characterized by extreme genetic diversity. No seed parent is true-breeding, first because it was itself derived from a fertilization in which genetically different parents participated, and second because of the genetic diversity of the pollen it will have received. In dealing with cultivated plants with this breeding structure, the essential concern in seed production is to employ systems in which hybrid vigor is exploited, the range of variation in the crop is diminished, and only parents likely to give rise to superior offspring are retained.

Plant breeders have made use either of inbreeding followed by hybridization or of some form of recurrent selection. During inbreeding programs normally cross-pollinated species, such as corn, are compelled to self-pollinate by artificial means. Inbreeding is continued for a number of generations until genetically pure, true-breeding, and uniform inbred lines are produced. During the production of the inbred lines, rigorous selection is practiced for general vigor and yield and disease resistance, as well as for other important characteristics. To estimate the value of inbred lines as the parents of hybrids, it is necessary to make tests of their combining ability. The test that is used depends upon the crop and on the ease with which controlled cross-pollination can be effected.

Breeding procedures designated as recurrent selection are coming into limited use with open-pollinated species. In theory, this method visualizes a controlled approach to homozygosity, with selection and evaluation in each cycle to permit the desired stepwise changes in gene frequency. Experimental evaluation of the procedure indicates that it has real possibilities. Four types of recurrent selection have been suggested: on the basis of phenotype, for general combining ability, for specific combining ability, and reciprocal selection. The methods are similar in the procedures involved, but vary in the type of tester parent chosen, and therefore in the efficiency with which different types of gene action (additive and nonadditive) are measured.

Special breeding techniques include the following. (1) When cytoplasmic male sterility is utilized, the male sterile lines are pollinated by other, so-called restorer, lines. Using crosses between male sterile lines and restorer lines, hybrid seed supplies can easily be obtained on a large scale. (2) Polyploidy, which can be induced artificially by colchicine treatment and in other ways, produces hybrids that are sterile and thus bear fruit without seeds—obviously a desirable attribute. (3) Multilinear varieties, made up of genetically similar plants containing several lines that differ in having genetically different forms of disease resistance, are advantageous because if the resistance of one of the constituent lines breaks down, production will be maintained by the remaining resistant lines. (4) Means have been discovered of producing essentially normal plants (sporophytes) of several crop species with the haploid chromosome number. The controlled production of such haploids, followed by doubling of the chromosome number using colchicine, enables the immediate fixation of the first products of segregation and recombination. Thus there is a very rapid return to complete homozygosity and homogeneity from hybrids of self-pollinated species. Alternatively, the procedure can give rise to the equivalents of inbred lines in outbreeding species. [R.Ri.]

Plants are also being bred from new sexual crosses, to obtain better disease resistance and crop yield; from somatic cells, to select new disease-resistant cultivars; from somatic hybridization, through protoplast fusion; from selection by genetic complementation, through the use of albino mutants; from cytoplasmic hybridization, which involves the transfer of the male sterility trait; from intergeneric hybridization; through the use of recombinant DNA and the transformation process; and through germ-plasm preservation. [O.L.G.]

Bremsstrahlung The electromagnetic radiation emitted by electrons when they pass through matter. Charged particles radiate when accelerated, and in this case the electric fields of the atomic nuclei provide the force which accelerates the electrons. The continuous spectrum of x-rays from an x-ray tube is that of the bremsstrahlung; in addition, there is a characteristic x-ray spectrum due to excitation of the target atoms by the incident electron beam. The major energy loss of high-energy (relativistic) electrons (energy $\gtrsim$ 10 MeV, depending somewhat upon material) occurs from the emission of bremsstrahlung, and this is the source of the gamma rays in a high-energy cosmic-ray shower. *See* COSMIC RAYS. [C.J.G.]

Brick A building material made usually from clay, molded as a rectangular block, and baked or burned in a kiln. An American building brick measures about $2\frac{1}{4} \times 3\frac{3}{4} \times 8$ in. ($57 \times 95 \times 203$ mm); an English brick is about $2\frac{5}{8} \times 4\frac{3}{8} \times 9$ in. ($67 \times 111 \times 229$ mm). The red color is due to iron oxides in the finished brick. Unselected brick as it comes from the kiln is common brick; brick selected for uniformity is face brick.

Bricks may also be made of other materials. The clay may be mixed with sand or lime, the brick being pressed and steamed. Bricks are commonly fashioned as insulating material, as refractory material termed firebrick, or for structural use. Adobe brick is molded of earth or clay mixed with straw, then hardened by the sun. *See* REFRACTORY. [F.H.R.]

Bridge A structure built to provide ready passage over natural or artificial obstacles, or under another passageway. Bridges serve highways, railways, canals, aqueducts, utility pipelines, and pedestrian walkways. In many jurisdictions, bridges are defined as those structures spanning an arbitrary minimum distance, generally about 10–20 ft (3–6 m); shorter structures are classified as culverts or tunnels. In addition, natural formations eroded into bridgelike form are often called bridges. This article covers only bridges providing conventional transportation passageways.

Bridges generally are considered to be composed of three separate parts: substructure, superstructure, and deck. The substructure or foundation of a bridge consists of the piers and abutments which carry the superimposed load of the superstructure to the underlying soil or rock. The superstructure is that portion of a bridge or trestle lying above the piers and abutments. The deck or flooring is supported on the bridge superstructure; it carries and is in direct contact with the traffic for which passage is provided.

Bridges are classified in several ways. Thus, according to the use they serve, they may be termed railway, highway, canal, aqueduct, utility pipeline, or pedestrian bridges. If they are classified by the materials of which they are constructed (principally the superstructure), they are called steel, concrete, timber, stone, or aluminum bridges. Deck bridges carry the deck on the very top of the superstructure. Through bridges carry the deck within the superstructure. The type of structural action is denoted by the application of terms such as truss, arch, suspension, stringer or girder, stayed-girder, composite construction, hybrid girder, continuous, cantilever, or orthotropic (steel deck plate).

The two most general classifications are the fixed and the movable. In the former, the horizontal and vertical alignment of the bridge are permanent; in the latter, either the horizontal or vertical alignment is such that it can be readily changed to permit the passage beneath the bridge of traffic. Movable bridges are sometimes called drawbridges in an anachronistic reference to an obsolete type of movable bridge spanning the moats of castles.

A singular type of bridge is the floating or pontoon bridge, which can be a movable bridge if it is designed so that a portion of it can be moved to permit the passage of water traffic.

The term trestle is used to describe a series of short spans supported by braced towers, and the term viaduct is used to describe a high structure of short spans, often of arch construction.

Fixed bridges. This type of construction is selected when the vertical clearance provided beneath the bridge exceeds the clearance required by the traffic it spans. For very short spans, construction may be a solid slab or a number of beams; for longer spans, the choice may be girders or trusses. Still longer spans may dictate the use of arch construction, and if the spans are even longer, stayed-girder bridges are used. Suspension bridges are used for the longest spans.

Beam bridges consist of a series of beams, usually of rolled steel, supporting the roadway directly on their top flanges. The beams are placed parallel to traffic and extend from abutment to abutment. Plate-girder bridges are used for longer spans than can be practically traversed with a beam bridge. In its simplest form, the plate girder consists of two flange plates welded to a web plate, the whole having the shape of an I. Box-girder bridges have steel girders fabricated by welding four plates into a box section. A conventional floor beam and stringer can be used on box-girder bridges, but the more economical arrangement is to widen the top flange plate of the box so that it serves as the deck. When this is done, the plate is stiffened to desired rigidity by closely spaced bar stiffeners or by corrugated or honeycomb-type plates. These stiffened decks, which double as the top flange of the box girders, are termed orthotropic. The wearing surface on such bridges is usually a relatively thin layer of asphalt.

Truss bridges, consisting of members vertically arranged in a triangular pattern, can be used when the crossing is too long to be spanned economically by simple plate girders. Where there is sufficient clearance underneath the bridge, the deck bridge is more economical than the through bridge because the trusses can be placed closer together, reducing the span of the floor beams.

The continuous bridge is a structure supported at three or more points and capable of resisting bending and shearing forces at all sections throughout its length. The bending forces in the center of the span are reduced by the bending forces acting oppositely at the piers. Trusses, plate girders, and box girders can be made continuous. The advantages of a continuous bridge over a simple-span bridge (that is, one that does not extend beyond its two supports) are economy of material, convenience of erection (without need for falsework), and increased rigidity under traffic. The disadvantages are its sensitivity to relative change in the levels of supporting piers, the difficulty of constructing the bridge to make it function as it is supposed to, and the occurrence of large movements at one location due to thermal changes.

The cantilever bridge consists of two spans projecting toward each other and joined at their ends by a suspended simple span. The projecting spans are known as cantilever arms, and these, plus the suspended span, constitute the main span. The cantilever arms also extend back to shore, and the section from shore to the piers offshore is termed the anchor span. Trusses, plate girders, and box girders can be built as cantilever bridges. The chief advantages of the cantilever design are the saving in material and ease of erection of the main span. The cable-stayed bridge, a modification of the cantilever bridge which has come into modern use, resembles a suspension bridge. It consists of girders or trusses cantilevering both ways from a central tower and supported by inclined cables attached to the tower at the top or sometimes at several levels.

The suspension bridge is a structure consisting of either a roadway or a truss suspended from two cables which pass over two towers and are anchored by backstays to a firm foundation. If the roadway is attached directly to the cables by suspenders, the structure lacks rigidity, with the result that wind loads and moving live loads distort the cables and produce a wave motion on the roadway. When the roadway is supported by a truss which is hung from the cable, the structure is called a stiffened suspension bridge. The stiffening truss distributes the concentrated live loads over a considerable length of the cable.

Since the development of the prestressing method, bridges of almost every type are being constructed of concrete. Prior to the advent of prestressing, these bridges were of three types: (1) arches, which were built in either short or long spans; (2) slab bridges of quite short spans, which were simply reinforced concrete slabs extending from abutment to abutment; and (3) deck girder bridges, consisting of concrete slabs built integrally with a series of concrete girders placed parallel to traffic. The advent of prestressed concrete greatly extended the utility and economy of concrete for bridges, particularly by making the hollow box-girder type practicable. *See* PRESTRESSED CONCRETE.

Movable bridges. Modern movable bridges are either bascule, vertical lift, or swing; with few exceptions, they span waterways. They are said to be closed when set for the traffic they carry, and open when set to permit traffic to pass through the waterway they cross. Bascule and swing bridges provide unlimited vertical clearance in the open position. The vertical clearance of a lift bridge is limited by its design.

The bascule bridge consists primarily of a cantilever span, which may be either a truss or a plate girder, extending across the channel. Bascule bridges rotate about a horizontal axis parallel with the waterway. The portion of the bridge on the land side of the axis, carrying a counterweight to ease the mechanical effort of moving the bridge, drops downward, while the forward part of the leaf opens up over the channel much like the action of a playground seesaw. Bascule bridges may be either single-leaf, where rotation of the entire leaf over the waterway is about one axis on one side of the waterway, or double-leaf, where the leaves over the waterway rotate about two axes on opposite sides of the waterway.

The vertical-lift bridge has a span similar to that of a fixed bridge and is lifted by steel ropes running over large sheaves at the tops of its towers to the counterweights, which fall as the lift span rises and rise as it falls. If the bridge is operated by machinery on each tower, it is known as a tower drive. If it is driven by machinery located on the lift span, it is known as a span drive.

Swing bridges revolve about a vertical axis on a pier, called the pivot pier, in the waterway. There are three general classes of swing bridges: the rim-bearing, the center-bearing, and the combined rim-bearing and center-bearing. Rim-bearing bridges are supported on circular girder drums on rollers, center-bearing on a single large bearing at the center of rotation.

Vibration. Bridges are generally considered to be statically loaded structures under their own dead load, with dynamic loadings from the live load and from the wind. Vibrations of bridges or individual components of a bridge occur when a resonant frequency of the bridge or a component is excited by one or more of the applied dynamic loadings. Excitation from live loads is more apt to happen on bridges carrying rail traffic than on those carrying highway traffic because of the uniform spacing of railroad cars. Excitation from wind is caused by the repeated formation of eddies or vortices as the wind travels past nonstreamlined members. Light lateral bracing on many spans will vibrate under the passage of live load. *See* VIBRATION.

Substructure. Bridge substructure consists of those elements that support the trusses, girders, stringers, floor beams, and decks of the bridge superstructure. Piers and abutments are the primary bridge substructure elements. Other types of substructure, such as skewbacks for arch bridges, pile bents for trestles, and various forms of support wall, are also commonly used for specific applications.

Both piers and abutments are generally supported on either spread footings or pile footings. A spread footing is usually a concrete pad large enough in area to transmit all superimposed loads and forces directly to the soil on which it is founded. A pile footing is usually a large concrete block supported on piles, so that the superimposed substructure loads and forces are transmitted to the support piles through the footing. The footing piles are used primarily to transmit loads through soil formations having poor supporting properties into or onto formations that are capable of supporting the loads. *See* FOUNDATIONS. [E.R.H.; H.W.F.; R.W.Ch.; B.H.]

Bridge circuit An electric network composed of four impedances forming a diamond-shaped network. These impedances, shown in the illustration as Z_1, Z_2, Z_3, and Z_4, may be combinations of resistors, capacitors, and inductors. Bridge circuits perform numerous functions, depending upon the type of circuit elements used in the arms of the bridge. For example, the bridge circuit can be used to determine the value of an unknown resistance. This is done by making the arm impedances resistances with one of them the unknown. The load impedance Z_L can be that of a measuring instrument, such as a galvanometer. The current through Z_L, and the voltage across it, will be zero when $Z_1Z_4 = Z_2Z_3$. With the proper combination of resistors and capacitors in the bridge arms, the voltage across Z_L will be zero for some frequency. Thus, the bridge acts as a frequency-measuring device. *See* FREQUENCY MEASUREMENT; WHEATSTONE BRIDGE. [H.F.K.]

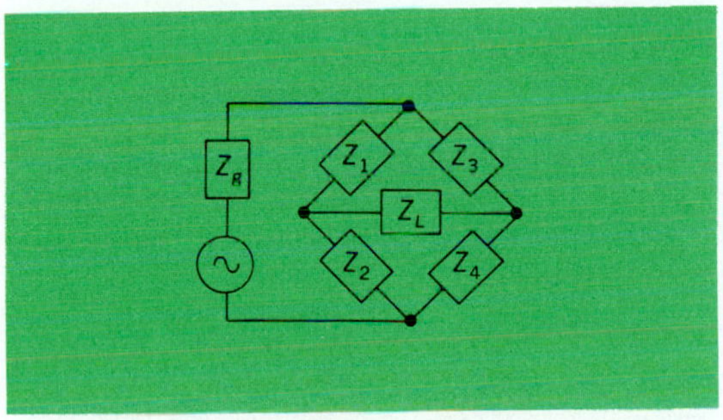

Schematic of bridge circuit. Z_g is the impedance of the seat of the electromotive force.

Bridled pressure plate An instrument that is useful in measuring air velocities in gusts, since it has a greater speed of response than do rotating elements. The velocity pressure on the plate exposed to the wind is balanced by the force of a spring. The deflection is measured by an inductance-type transducer. The signal from the transducer is amplified and transmitted to a recorder. The indication is dependent upon the air density. Simple mechanical designs are also used to measure low air velocities. *See* AIR-VELOCITY MEASUREMENT. [H.Fo.]

Brillouin zone In the propagation of any type of wave motion through a crystal lattice, the frequency is a periodic function of wave vector **k**. This function may be complicated by being multivalued; that is, it may have more than one branch. Discontinuities may also occur. In order to simplify the treatment of wave motion in a crystal, a zone in **k**-space is defined which forms the fundamental periodic region, such that the frequency or energy for a **k** outside this region may be determined from one of those in it. This region is known as the Brillouin zone (sometimes called the first or the central Brillouin zone). It is usually possible to restrict attention to **k** values inside the zone. Discontinuities occur only on the boundaries. If the zone is repeated indefinitely, all **k**-space will be filled. Sometimes it is also convenient to define larger figures with similar properties which are combinations of the first zone and portions of those formed by replication. These are referred to as higher Brillouin zones.

The central Brillouin zone for a particular solid type is a solid which has the same volume as the primitive unit cell in reciprocal space, that is, the space of the reciprocal lattice vectors, and is of such a shape as to be invariant under as many as possible of the symmetry operations of the crystal. *See* CRYSTALLOGRAPHY. [J.C.]

Brittleness That characteristic of a material that is manifested by sudden or abrupt failure without appreciable prior ductile or plastic deformation. A brittle fracture occurs on a cleavage plane which has a crystalline appearance at failure because each crystal tends to fracture on a single plane. On the other hand, a shear fracture has a fibrous appearance because of the sliding of the fracture surfaces over each other. Brittle failures are caused by high tensile stresses, high carbon content, rapid rate of loading, and the presence of notches. Materials such as glass, cast iron, and concrete are examples of brittle materials. [J.B.S.]

Broaching The machine shaping of metal by pushing or pulling a multiple-tooth, barlike tool across a surface or through an existing hole in a workpiece. The peripheries or cutting edges of the broach teeth are shaped to give a desired surface or contour, whereas cutting action results from the fact that each tooth is progressively higher or projects further than the preceding one.

Broaching machines are classed as either horizontals or verticals, depending on the mounting plane of the broach. Most horizontal machines are of the pull type, while vertical models are made to use pull-up, pull-down, push, or surface broaches. [A.H.T.]

Broccoli A cool-season biennial crucifer, *Brassica oleracea* var. *italica*, of Mediterranean origin, belonging to the plant order Papaverales. Broccoli is grown for its thick branching lower stalks which terminate in clusters of loose green flower buds. Stalks and buds are cooked as a vegetable or may be processed in either canned or frozen form. California and Texas are important broccoli-producing states. *See* PAPAVERALES. [H.J.C.]

Bromate A negative ion having the formula BrO_3^- and derived from bromic acid, $HBrO_3$. Bromic acid exists only in aqueous solution. The solid bromates are stable and can be prepared by passing bromine into a warm solution of sodium carbonate.

Since bromine has the oxidation state of 5+ in bromates and these bromates are strong oxidizing agents, these compounds are useful in analytical chemistry. Commercially, they are used in permanent-wave preparations and in flour treatment. *See* BROMINE. [E.E.W.]

Bromegrass A common name designating a number of grasses found in the North Temperate Zone that produce highly palatable and nutritious forage. Of these, smooth bromegrass (*Bromus inermis*) is the most important. Although first widely used in the eastern Great Plains and western Corn Belt regions, improved strains are now grown extensively for hay and rotation pastures north of the Mason-Dixon line, from the Plains to the Atlantic. Smooth bromegrass is a long-lived perennial, spreads by underground creeping stems, and is fairly deep rooted and drought-tolerant. Top growth is used for hay or pasture. Regional strains are available for Canada and the northern two-thirds of the United States. *See* CYPERALES. [H.B.S.]

Bromeliales An order of flowering plants, division Magnoliophyta (Angiospermae) in the subclass Commelinidae of the class Liliopsida (monocotyledons). It consists of the single family Bromeliaceae, with nearly 50 genera and more than 1700 species, occurring chiefly in tropical and subtropical America. They are firm-leaved, terrestrial xerophytes, or very often epiphytes, with six stamens and regular or somewhat irregular flowers that usually have septal nectaries and an inferior ovary. Spanish moss (*Tillandsia*) and the cultivated pineapple (*Ananas*) are familiar members of the Bromeliales, and many others attract attention as unusual houseplants. *See* LILIOPSIDA; MAGNOLIOPHYTA. [A.Cr.]

Bromide A compound which is derived from hydrobromic acid, HBr, and contains the bromine atom in the 1− oxidation state. The properties of bromides are intermediate between those of chlorides and iodides. The bromide ion is a better reducing agent and forms less stable complexes than chloride. *See* CHLORIDE.

The solubilities of bromides are very similar to those of chlorides, and $PbBr_2$, CuBr, AgBr, and Hg_2Br_2 are the common insoluble salts. The bromides are usually produced from bromine, which in turn is obtained from salt brines or sea water.

Bromides are used in medicine as sedatives in treatment of nervous disorders. Silver bromide is used in photographic films and paper. [E.E.W.]

Bromine A chemical element, Br, atomic number 35, atomic weight 79.909, which normally exists as Br_2, a dark-red, low-boiling but high-density liquid of intensely irritating odor. This is the only nonmetallic element that is liquid at normal temperature and pressure. Bromine is very reactive chemically; one of the halogen group of elements, it has properties intermediate between those of chlorine and iodine. *See* HALOGEN ELEMENTS.

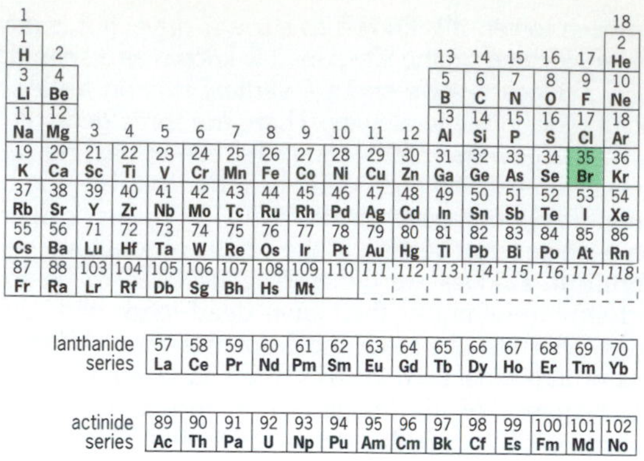

The most stable valence states of bromine in its salts are −1 and +5, although +1, +3, and +7 are known. Within wide limits of temperature and pressure, molecules of the liquid and vapor are diatomic, Br_2, with a formula weight of 159.818. There are two stable isotopes (^{79}Br and ^{81}Br) that occur naturally in nearly equal proportion, so that the atomic weight is 79.909. A number of radioisotopes are also known.

The solubility of bromine in water at 20°C (68°F) is 3.38 g/100 g (3.38 oz/100 oz) solution, but its solubility is increased tremendously in the presence of its salts and in hydrobromic acid. The ability of this inorganic element to dissolve in organic solvents is of considerable importance in its reactions. The table summarizes the physical properties of bromine.

Although it is estimated that from 10^{15} to 10^{16} tons of bromine are contained in the Earth's crust, the element is widely distributed and found only in low concentrations in the form of its salts. The bulk of the recoverable bromine, however, is found in the hydrosphere. Sea water contains an average of 65 parts per million (ppm) of bromine. The other major sources of bromine in the United States are underground brines and salt lakes, with commercial production in Michigan, Arkansas, and California.

While many inorganic bromides have found industrial use, the organic bromides have even wider application. Because of the ease of reaction of bromine with organic compounds and the ease of its subsequent removal or replacement, organic bromides have been much studied and used as chemical intermediates. In addition, any of the bromine reactions are so clean-cut that they can be used for the study of reaction mechanisms without complication of side reactions. The ability of bromine to add into unusual places on organic molecules has added to its value as a research tool.

Bromine and its compounds have found acceptance as disinfection and sanitizing agents in swimming pools and potable water. Certain bromine-containing compounds are safer to use than the analogous chlorine compounds due to certain persis-

Physical properties of bromine

Property	Value
Flash point	None
Fire point	None
Freezing point, °C	−7.27
Density, 20°C	3.1226
Pounds per gallon, 25°C	25.8
Boiling point, 760 mm Hg, °C	58.8
Refractive index, 20°C	1.6083
Latent heat of fusion, cal/g	15.8
Latent heat of vaporization, cal/g, bp	44.9
Vapor density, g/liter, standard conditions (0°C,1 atm)	7.139
Viscosity, centistokes, 20°C	0.314
Surface tension, dynes/cm, 20°C	49.5
30°C	47.3
40°C	45.2
Dielectric constant, 10^5 freq, 25°C	3.33
Compressibility, vapors, 25°C	0.998

Thermodynamic data, cal/(mole K)

	T,K	*Entropy*	*Heat capacity*
Solid	265.9	24.786	14.732
Liquid	265.9	34.290	18.579

tent residuals found in the chlorine-containing materials. Other bromine chemicals are used as a working fluid in gages, as hydraulic fluids, as chemical intermediates in the manufacture of organic dyes, in storage batteries, and in explosion-suppressant and fire-extinguishing systems. Bromine compounds, because of their density, also find use in the gradation of coal and other minerals where separations are effected by density gradients. The versatility of bromine compounds is illustrated by the commercial use of over 100 compounds that contain bromine.

Bromine is almost instantaneously injurious to the skin, and it is difficult to remove quickly enough to prevent a painful burn that heals slowly. Bromine vapor is extremely toxic, but its odor gives good warning; it is difficult to remain in an area of sufficient concentration to be permanently damaging. Bromine can be handled safely, but the recommendations of the manufacturers should be respected. [R.C.S.]

Bronchial disorders A heterogeneous group of diseases of the conducting airways (bronchi and bronchioles) of the lung.

Acute bronchitis is an inflammation of the bronchi that can be caused by irritant gases, bacteria, viruses, and rarely fungi. In the initial stages of the disease there is edema and excessive mucus production. This results in a decreased air flow into the gas exchange units (alveoli) of the lung. In general, acute bronchitis is most commonly seen in debilitated and terminally ill patients and is frequently accompanied by bronchopneumonia. Treatment is aimed at the underlying disease and eliminating the causative agent.

Acute bronchiolitis is an inflammation of the small airways of the lung (bronchioles) usually caused by a variety of viruses, and characteristically occurs in infants and young children. Because of the location of the inflammation, it may be a life-threatening disease. The condition is usually treated with vaporization of water and antibiotics.

Chronic bronchitis is a chronic, widespread, frequently disabling disease which in part is defined clinically by coughing and excessive mucus production for at least 3 months per year for 2 consecutive years. It commonly is associated with emphysema and is a significant cause of morbidity and mortality. The disease is most commonly seen in industrialized large cities. The condition is causally associated with air pollution and, probably more importantly, cigarette smoking. Chronic bronchitis is at least partially reversible if the causitive agents are removed.

Bronchial asthma, a disease of uncertain etiology, is characterized by episodes of incomplete bronchial obstruction (bronchospasm). Asthma is frequently associated with eczema, hay fever, and allergic reactions to a wide variety of substances. If the bronchospasm can be directly related to an allergic reaction to a specific substance, the condition is referred to as extrinsic asthma. In the majority of asthmatics no specific allergen can be demonstrated, and in this instance the disease is called intrinsic asthma. Clinically, persons with bronchial asthma have episodic attacks of difficult breathing, with wheezing and prolongation of the expiratory phase of respiration. The attacks usually last several hours and are followed by coughing, with expulsion of the mucous plugs. Occasionally the asthmatic attacks are unremitting, a condition referred to as status asthmaticus. Bronchial asthma alone rarely causes death. The disease is treated with various drugs such as bronchodilators, mucolytic agents, antihistamines, and antibiotics.

Bronchiectasis is a permanent, abnormal dilatation of bronchi resulting from severe inflammatory injury to their walls. Rarely, congenitally malformed bronchi may provide the basis for secondary damage. In most cases, however, there is a history of severe bronchitis or bronchopneumonia, particularly during childhood. In some instances the disease may follow an inconspicuous clinical course and thus remain undiagnosed.

Cystic fibrosis, also called mucoviscidosis, not only affects the pancreas but is a systemic disease involving the bronchial glands as well. Viscid mucous secretion in the bronchi makes them prone to severe and often intractable bronchitis and bronchiectasis. The infection invariably extends into the lungs.

Bronchial adenoma is a low-grade malignant or potentially malignant tumor of bronchi that appears in relatively young or middle-age adults. In contrast with bronchogenic carcinoma, bronchial adenoma grows slowly but does infiltrate and can metastasize, that is, transfer the disease to a distant focus. Surgical removal results in complete cure in a substantial number of cases. *See* RESPIRATORY SYSTEM DISORDERS. [S.P.H.]

Bronchus At its lower end, the trachea branches into two primary bronchi which connect to each lung. Within the lung, each primary bronchus divides again and again, with the resulting bronchioles forming progressively smaller ducts which finally terminate in the alveoli, where gas exchange occurs. The structure of the large bronchi is similar to that of the trachea, but the smaller bronchioles lack the cartilaginous rings. *See* LUNG; RESPIRATORY SYSTEM. [W.J.B.]

Bronze Usually an alloy of copper and tin. Bronze is used in bearings, bushings, gears, valves, and other fittings both for water and steam.

The properties of bronze depend on its composition and working. Lead, zinc, silver, and other metals are added for special-purpose bronzes. Tin bronze, including statuary bronze, contains 2–20% tin; bell metal 15–25%; and speculum metal up to 33%. Gun metal contains 8–10% tin plus 2–4% zinc. Phosphor bronze is tin bronze hardened and strengthened with traces of phosphorus; it is used for fine tubing, wire springs, and machine parts. Lead bronze may contain up to 30% lead; it is used for cast parts such as low-pressure valves and fittings. Manganese bronze with 0.5–5% manganese plus other metals, but often no tin, has high strength. Aluminum bronze also contains no tin; its mechanical properties are superior to those of tin bronze, but it is difficult to cast. Silicon bronze, with up to 3% silicon, casts well and can be worked hot or cold by rolling, forging, and similar methods. Beryllium bronze (also called beryllium copper) has about 2% beryllium and no tin. The alloy is hard and strong and can be further hardened and strengthened by precipitation hardening; it is one of the few copper alloys that responds to heat treatment, approaching three times the strength of structural steel. *See* ALLOY; COPPER; TIN. [F.H.R.]

Brown dwarf One of the least massive self-gravitating objects that are formed in the fragmentation of an interstellar cloud. Brown dwarfs are less massive than the least massive true stars, the red dwarf stars, and are comparable to or heavier than the most massive planets, that is, the gas-giant planets such as Jupiter.

Brown dwarfs are distinguished from stars in that they are insufficiently massive to sustain long-term nuclear burning of hydrogen, although they undergo a brief period of deuterium (heavy-hydrogen) burning. Accordingly, brown dwarfs cool and fade away over periods of time that are short compared to the age of the Sun (5×10^9 years), while red dwarf stars burn hydrogen continuously at a relatively low rate and should continue to shine over times longer than the present age of the universe (about 20×10^9 years). The current search for brown dwarfs is important because studying them may elucidate the processes by which interstellar clouds break up and condense, and should demonstrate how to distinguish brown dwarfs from extrasolar planets (the planets of stars other than the Sun). *See* PLANET; RED DWARF STAR; STAR; STELLAR EVOLUTION.

Several objects were discovered during the late 1980s that may be brown dwarfs, although no case is proven. Methods used to search for brown dwarfs are astrometry, infrared photometry, infrared speckle interferometry, and radial-velocity monitoring. Astrometry seeks to detect brown dwarfs that are companies of known stars through the slight wobbling that the presence of a brown dwarf must induce in the motion of its accompanying star. Infrared photometry seeks to detect the infrared radiation of a brown dwarf that is the companion of a small hot star such as a white dwarf star. Infrared speckle interferometry is used to attempt to obtain images of infrared radiation sources, such as brown dwarfs, that are close companions of known stars. Radial-velocity monitoring is conducted to detect the orbital motion of a known star around the center of mass of the binary system that it forms with a brown dwarf. *See* ASTROMETRY; ASTRONOMICAL SPECTROSCOPY; BINARY STAR; INFRARED ASTRONOMY; SPECKLE. [S.P.M.]

Brownian movement The irregular motion of a body arising from the thermal motion of the molecules of the material in which the body is immersed. Such a body will of course suffer many collisions with the molecules, which will impart energy and momentum to it. Because, however, there will be fluctuations in the magnitude and direction of the average momentum transferred, the motion of the body will appear irregular and erratic.

In principle, this motion exists for any foreign body suspended in gases, liquids, or solids. To observe it, one needs first of all a macroscopically visible body; however, the mass of the body cannot be too large. For a large mass, the velocity becomes small. *See* KINETIC THEORY OF MATTER. [M.Dr.]

Brucellaceae The family Brucellaceae traditionally was composed of several genera of gram-negative, aerobic bacteria that are pathogenic for humans and animals. They are ovoid in shape and similar in size (0.1–0.5 micrometer by 0.3–1.5 micrometers), and exhibit pleomorphism (variable shape and size) in artificial culture. Many of these organisms are intracellular and, therefore, are resistant to antibiotic therapy. Even though they share this array of characteristics, they are not closely related genetically as determined by guanine-plus-cytosine base ratios and DNA hybridization. Accordingly, this family was deleted from the classification scheme in the eighth edition of *Bergey's Manual* (1974), and its member genera of *Pasteurella*, *Brucella*, *Moraxella*, *Bordetella*, *Haemophilus*, *Actinobacillus*, *Calymmatobacterium*, and *Noguchia* were placed either into other families or into general groups of gram-negative bacteria of uncertain affiliation. [M.E.M.]

Brucellosis An infectious disease of animals and humans caused by bacteria of the genus *Brucella*. The organisms are *B. melitensis*, found in sheep and goats, *B. abortus* in cattle, and *B. suis* in swine. Brucellosis occurs throughout the world and is the most important of the nearly 100 diseases of animals transmissible to human beings.

The brucellae localize in the reproductive organs and mammary glands of animals. This makes the disease of serious economic concern to livestock producers and dairy farmers because it causes abortions and reduces the production of milk. The disease in cattle is known as Bang's disease, or contagious abortion.

Brucellosis in humans, also known as undulant fever or Malta fever, is primarily an occupational disease in which infection through the skin is acquired by contact with the tissues of freshly killed animals, or with the contaminated environment of the animals. The disease is also contracted by drinking unpasteurized milk and by eating fresh cheese prepared from unpasteurized milk. The acute illness is characterized by chills, fever, sweats, and weakness. Chronic disease results in weakness and mental depression. The usual course of the disease is less than 3 months, but antibiotic therapy reduces the length of the illness.

Human brucellosis can be eliminated only by eradicating the disease in animals. The disease is controlled in cattle by eliminating infected animals. Successful vaccination of cattle against the disease has been achieved with a live brucella culture having stable, attenuated virulence. An effective vaccine has been developed for the protection of sheep and goats, but not for swine. [W.W.Sp.]

Brucite A magnesium hydroxide mineral, $Mg(OH)_2$, crystallizing in the trigonal system. It is a member of the important $Cd(OH)_2$ structure type, consisting of hexagonal close-packed oxygen atoms with alternate octahedral layers occupied by Mg. The "brucite layer" is an important structural component in the clay, mica, and chlorite mineral groups. Brucite occurs as tabular crystals and as elongated fibers (as the variety nemalite), hardness 2½ (Mohs scale), color white to greenish, and specific gravity 2.4. Fe^{2+} and Mn^{2+} commonly substitute for Mg^{2+}.

Brucite often occurs in a low-temperature vein paragenesis, usually with serpentine and accessory magnesite. It is also derived by the action of water on periclase, MgO, which results from the thermal metamorphism of dolomites and limestones. Carbonate rocks rich in periclase and brucite are called predazzites. *See* DOLOMITE; MAGNESITE; SERPENTINE. [P.B.M.]

Brush discharge A particular form of corona discharge characterized by strongly ionized streamers. It generally occurs at a field slightly less than that required for complete breakdown. The conditions for a brush discharge are as follows: The pressure should be about atmospheric; one or both of the electrodes should have a radius of curvature small compared to the gap separation; and the gap itself should be large, at least of the order of a few centimeters. The ionized streamers take on a treelike form. The light originates from recombination and radiative transitions from excited states. *See* BREAKDOWN POTENTIAL; CORONA DISCHARGE. [G.H.M.]

Brush-shifting motor A polyphase induction motor whose speed can be controlled similarly to that of a dc shunt motor by use of a regulating winding, a commutator, and a brush-shifting device. Such motors are used for knitting and spinning machines, paper mills, and other industrial services that require controlled variable-speed drive. The primary winding on the rotor is supplied from the line through slip rings. The stator windings are the secondary windings (S_1, S_2, S_3), and the third winding, also in the rotor, is an adjusting winding provided with a commutator (see illustration). Voltages collect-

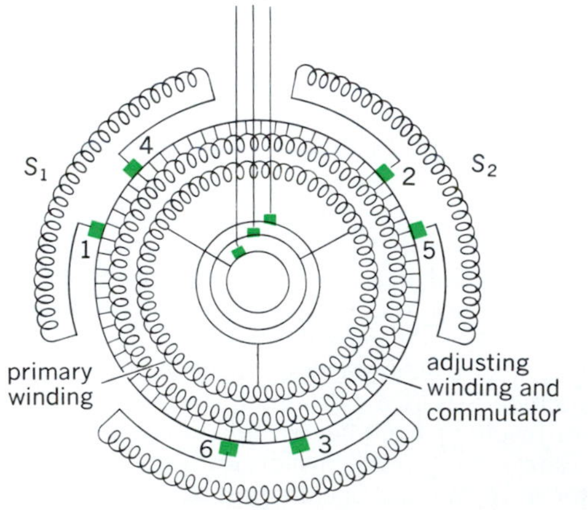

Schematic of adjustable-speed brush-shifting motor. *(After A. E. Fitzgerald and C. Kingsley, Jr., Electric Machinery, McGraw-Hill, 1952)*

ed from the commutator are fed into the secondary circuit. Brushes (1–3, 4–6) are mounted on separate movable yokes. Each set of brushes can be moved as a group. Thus both the spacing between sets of brushes and the angular position of the brushes are adjustable. Brush spacing determines the magnitude of the voltage applied to the secondary. When brush sets are so adjusted that pairs of brushes are in contact with the same commutator segment, the secondary is short-circuited and no voltage is supplied. Under these conditions the motor behaves as an ordinary induction motor. The speed can be reduced by separating the brushes so that secondary current produces a negative torque. The machine can be operated above synchronism by interchanging the position of the brushes, so the voltage collected is in a direction to produce a positive torque. The motor can be reversed by reversing two of the leads supplying the primary. *See* INDUCTION MOTOR. [I.L.K.]

Brussels sprouts A cool-season biennial crucifer (*Brassica oleracea* var. *gemmifera*), which is of northern European origin and belongs to the plant order Capparales. The plant is grown for its small headlike buds formed in the axils of the leaves along the plant stem (see illustration). These buds are eaten as a cooked vegetable. Popular varieties (cultivars) are Half Dwarf and Catskill; however, hybrid varieties are increasingly planted. California and New York are important producing states. *See* CAPPARALES. [H.J.C.]

Brussels sprouts (*Brassica oleracea* var. *gemmifera*), Jade Cross. (*Joseph Harris Co., Rochester, New York*)

Bryales An order of the subclass Bryidae. With 11 families and perhaps 44 genera, it is defined in terms of terminal inflorescences, with rare exceptions, and perfect, double peristomes which are papillose on the outer surface. The capsules are generally inclined and more or less pear-shaped. Erect capsules are associated with reduced peristomes.

These mosses often grow in disturbed places. They are perennial and grow in tufts, with stems erect and simple or forked and often densely covered with rhizoids. The leaves are generally bordered by elongate cells and often toothed. The midrib often ends in a hairpoint. The sporophytes are nearly always terminal. The setae are generally elongate, and the operculate capsules are usually symmetric but generally inclined to pendulous and commonly pyriform owing to the development of a sizable neck. The peristome is normally double, with a well-developed endostome. *See* BRYIDAE; BRYOPHYTA; BRYOPSIDA. [H.Cr.]

Bryidae A subclass of the class Bryopsida. Most genera of true mosses (Bryopsida) belong in the 16 orders of the Bryidae. The most characteristic feature is the peristome consisting of one or two series of teeth, derived from parts of cells rather than whole cells, as in the Tetraphididae, Dawsoniidae, and Polytrichidae. (The Buxbaumiidae have some resemblance in peristome structure to Bryidae.) The stems may be erect and merely forked, or prostrate and freely branched, with sporophytes produced terminally or laterally, respectively. The leaves are inserted in many rows, though sometimes flattened together and appearing two-ranked, but only rarely actually in two rows. The costa may be single or double, sometimes very short, and rarely lacking. The setae are generally present and elongate. The capsules dehisce by means of an operculum except in a few genera that show extreme reduction. *See* ARCHIDIIDAE; BRYOPHYTA; BRYOPSIDA; BRYOXIPHIALES; DAWSONIIDAE; DICRANALES; ENCALYPTALES; FISSIDENTALES; FUNARIALES; GRIMMIALES; HOOKERIALES; HYPNALES; ISOBRYALES; MITTENIALES; ORTHOTRICHALES; POLYTRICHIDAE; POTTIALES; SELIGERIALES; SPLACHNALES; TETRAPHIDIDAE. [H.Cr.]

Bryophyta A relatively small phylum of the plant kingdom including the mosses, Bryopsida (Musci); the liverworts, Marchantiopsida (Hepaticae); and the hornworts, or horned liverworts (Anthocerotae). These plants are found growing on soil, rocks, and upon other plants throughout the world wherever there is sufficient moisture. As individuals these plants are rather inconspicuous, but when closely associated in patches or beds, they present a striking appearance (see illustration). In the tropics many of them live as epiphytes, and in the cold tundra regions vast acreages may be covered by a single species.

Although they lack the true roots, stems, and leaves characteristic of the higher vascular plants, many bryophytes have structures which superficially resemble these organs. Rootlike structures known as rhizoids function in anchorage and absorp-

Moss plant, *Polytrichum juniperinum*. (*General Biological Supply House*)

tion. Since the stemlike and leaflike structures contain chlorophyll and manufacture food, the plants are independently self-sufficient. In structure these are the simplest land plants. Being without specialized conducting and supporting tissues (xylem and phloem), they have remained small. *See* CHLOROPHYLL; LEAF; PHLOEM; PHOTOSYNTHESIS; ROOT (BOTANY); STEM; XYLEM.

The life cycle of bryophytes consists of a regular alternation of generations in which the gametophyte, or gamete-producing generation, alternates with the sporophyte, or spore-producing generation. The gametophyte, though small, is green and independent (self-sustaining). The sporophyte is always attached to the gametophyte and is never entirely independent, although in a few species the sporophyte contains green plastids (chloroplasts) and can manufacture a limited amount of food. In all cases the two generations are entirely different in both appearance and structure. Neither generation produces true flowers, fruits, and seeds. Bryophytes are not considered to be ancestral to any other group. *See* BRYOPSIDA; REPRODUCTION (PLANT); TRACHEOPHYTA. [P.A.V.]

Bryopsida A class of small green plants with some 600 genera and 14,000 species commonly called mosses which make up the largest and most highly developed group of the division Bryophyta. They are worldwide in distribution, occurring in nearly all damp habitats except the ocean. They are abundant on soil, rocks, and other plants in moist woodlands and may form conspicuous carpets of vegetation in arctic and alpine regions. The class is commonly divided into three groups: the peat mosses, the granite mosses, and the true mosses. *See* BRYOPHYTA.

The peat mosses contain only a single genus, *Sphagnum*, which nevertheless has numerous species. In size these are large mosses, being perennial and having practically indefinite growth. The are markedly social, forming extensive masses on boggy soils, often being the dominant plant covering. This habit of growth is responsible for the name sphagnum bog. Peat moss is used as a fuel, as a mulch, and for packing the roots of ornamental seedlings. Brazil is the chief center of commercial distribution of peat moss. *See* BOG.

The granite mosses are represented by only two genera, both having a rather limited distribution. They are restricted to arctic regions or mountaintops, where they are found as blackish masses on siliceous rocks. The general structure resembles that of the true mosses, but in several details they are like the peat mosses.

The true mosses are the largest group of the Bryopsida and include all the common genera. Despite their abundance and wide distribution, they display great uniformity of structure and have a common life history. They differ from the other Bryopsida in having a filamentous protonema, a midrib in their leaflike structures, no pseudopodium, and no differentiation of tissue within the capsule.

Whereas the peat mosses and the granite mosses are relatively limited in geographical distribution, the true mosses are found from the polar regions to the tropics. They may grow on soil or rocks, in bogs, marshes, ponds or streams, or as epiphytes on trees or leaves. Most of them live in moderately moist habitats. Many are pioneer species, colonizing any opening in the vegetation. In this respect they are of considerable importance in forestalling or stopping erosion. [P.A.V.]

Bryopsidales An order of the green algae (Chlorophyceae), also called Caulerpales, Codiales, or Siphonales, in which the plant body (thallus) is a coenocytic filament (tube or siphon). The order comprises six families with about 24 genera. The filaments may be discrete with free or laterally coherent branches, or organized into a dense plexus exhibiting distinctive morphological features. Septa, which are generally infrequent and incomplete, are formed by centripetal deposition of wall material. A large, continuous central vacuole restricts the cytoplasm to a thin layer just beneath the wall. The cytoplasm contains innumerable nuclei, discoid plastids, and other organelles. Vegetative reproduction is common, usually by rhizomes or fragments. Reproductive cells may be formed in unmodified or slightly modified portions of the filament or in special organs. *See* ALGAE; CHLOROPHYCEAE. [P.C.Si.; R.L.Moe]

Bryoxiphiales An order of the class Bryopsida in the subclass Bryidae. The order consists of a single genus and species, *Bryoxiphium norvegicum*, the sword moss. This order is characterized by a swordlike appearance owing to leaves overlapping in two rows. The shiny, rigid leaves are keeled and conduplicate-folded. The apex is long-awned at the stem tip and progressively shorter-pointed downward. The midrib bears at back a low ridge of one to four rows of cells. The leaf cells are smooth and subquadrate within, longer and narrower toward the margins. The plants are dioecious with terminal archegonia. *See* BRYIDAE; BRYOPHYTA; BRYOPSIDA. [H.Cr.]

Bryozoa A phylum of sessile aquatic invertebrates (also called Polyzoa) which form colonies of zooids. Each zooid, in its basic form, has a lophophore of ciliated tentacles situated distally on an introvert, a looped gut with the mouth inside the lophophore and the anus outside, a coelomic body cavity, and (commonly) a protective exoskeleton. The colonies are variable in size and habit. Some are known as lace corals and others as sea mats, but the only general name is bryozoans (sea mosses).

The colony may be minute, of not more than a single feeding zooid and its immediate buds, or substantial, forming masses 3 ft (1 m) in circumference, festoons 1.6 ft (0.5 m) in length, or patches 2.7 ft^2 (0.25 m^2) in area. Commonly the colonies form incrustations not more than a few square centimeters in area, small twiggy bushes up to about 1.2 in. (3 cm) in height, or soft masses up to about 0.3 ft (0.1 m) in the largest dimension. In many colonies much of the bulk consists of the zooid exoskeletons which may persist long after the death of the organism and account for the abundance of fossilized bryozoan remains.

Many bryozoans display polymorphism, having certain zooids adapted in particular ways to perform specialized functions, such as protection, cleaning the surface, anchoring the colony, or sheltering the embryo. The evolution of nonfeeding polymorphs is dependent upon some form of intercommunication between zooids.

Bryozoa is the name of a phylum for which Ectoprocta is generally regarded as a synonym, these names being used by zoologists according to personal preference. Entoprocta (synonym Callyssozoa) is likewise regarded as an independent phylum. A minority regard Ectoprocta and Entoprocta as subphyla within the Bryozoa, while others maintain Ectoprocta and Entoprocta as phyla but link them under Bryozoa as a name of convenience. *See* ENTOPROCTA.

The phylum contains some 20,000 described species, one-fifth of them living. These are distributed among three classes and a somewhat variable number of orders:

- Phylum Bryozoa
  - Class Phylactolaemata
    - Order Plumatellida
  - Class Gymnolaemata
    - Order Ctenostomata
    - Order Cheilostomata
  - Class Stenolaemata
    - Order Cyclostomata
    - Order Cystoporata
    - Order Trepostomata
    - Order Cryptostomata
    - Order Rhabdomesonata
    - Order Fenestrata

Some authorities regard the last two as suborders of an enlarged Cryptostomata. *See* GYMNOLAEMATA; PHYLACTOLAEMATA; STENOLAEMATA.

Fresh-water bryozoans are present on submerged tree roots and aquatic plants in most lakes, ponds, and rivers, especially in clear water of alkaline pH. Most other bryozoans are marine, although some gymnolaemates inhabit brackish water. They are common in the sea, ranging from the middle shore to a depth of over 26,000 ft (8000 m), and are maximally abundant in waters of the continental shelf. Most attach to firm substrata, so that their distribution is primarily determined by the availability of support. Mud is unfavorable and so is sand unless well provided with stone, dead shells, hydroids, or large foraminiferans.

Colony form in bryozoans is to some extent related to habitat. Encrusting and bushy flexible species are adapted to wave exposure; brittle twiglike and foliaceous species are found deeper; some erect branching species tolerate sediment deposition. One group of tiny discoid species lives on sand in warm seas, and in one genus the colonies are so small that they live actually among the sand grains; a few species live anchored in mud. A number of stolonate ctenostomes bore into the substance of mollusk shells; other species are associated only with hermit crabs, and a few are commensal with shrimps or polychaete worms.

Bryozoans have few serious predators. Nudibranch mollusks and pycnogonids (sea spiders) specialize in feeding on zooids but are rarely destructive of entire colonies. Loxosomatids (Entoprocta) and a hydroid (*Zanclea*) are common commensals.

Life spans vary. Small algal dwellers complete their life cycle in a few months. Many species survive a year but have two overlapping generations; others are perennial, with one known to survive for 12 years.

Bryozoans may be a nuisance in colonizing ship hulls and the insides of water pipes, and one species has caused severe dermatitis in fishers. Recently some delicate kinds have been used in costume jewelry, and green-dyed clumps of dried *Bugula* are often sold as "everlasting plants." [J.S.R.]

Fossil Bryozoa have a long geological history from early in the Ordovician Period (500,000,000 years ago) to the Recent. Individual colonies range in size from a few millimeters to several meters in height. A vast array of diverse colony forms record the heterogeneity of this phylum. There are about 1100 known fossil genera and about 15,000 known fossil species. Representatives of the marine orders that secreted calcareous skeletons (Cryptostomata, Cyclostomata, Cystoporata, Fenestrata, Rhabdomesonata, Trepostomata, and Cheilostomata) commonly are abundant in sedimentary rocks deposited in various environments of marine basins where sessile benthic animals flourished. The calcareous skeletons are generally calcite, but some are aragonite or a mixture of calcite and aragonite. The Ctenostomata with only soft membranous or gelatinous colonies are preserved as impressions, excavations, or borings generally in marine shells. The fresh-water Phylactolaemata with soft gelatinous colonies are mostly not preserved except for statoblasts (dormant reproductive bodies) that are reported from Quaternary (2,000,000 years ago) rocks. [J.R.P.R.]

Bubble chamber A particle detector used in elementary particle physics research. Charged particles passing through liquid contained in the chamber leave visible tracks along their paths. These tracks are then photographed for later study. Each curved track is actually a string of small bubbles marking the path of an elementary particle. Bubble chambers vary in size, in the liquid used, and in how fast photographs can be taken. They are particularly useful because they allow direct visualization of all the charged particles involved in a high-energy interaction, and they have been used to discover many new particles, such as the rho (ρ) and omega minus (Ω^-). *See* ELEMENTARY PARTICLE.

The bubble chamber was invented by Donald Glaser in 1952. It was known that a pure liquid in a clean glass container can be raised above its boiling temperature without its starting to boil spontaneously. The boiling must be triggered by something. For example, a broken piece of glass dropped into a superheated liquid can make boiling violently erupt. As an electrically charged particle passes through any material (such as a liquid), it interacts with electrons in the material, leaving atoms ionized. Glaser discovered that in a superheated liquid this ionization energy is enough to trigger boiling along the path of the particle. The boiling bubbles grow rapidly, and so a photograph is quickly taken before the resolution of the individual tracks is destroyed. In order to make the bubble chamber sensitive to the arrival of another charged particle, the boiling must be stopped, and the liquid must be superheated again.

Interest is focused on the nuclear interactions that a beam particle has with the particles in the liquid. Since the interactions occur with particles making up the liquid, the choice of liquid is important to the experimenter. Liquid hydrogen is used to study proton interactions because hydrogen is a simple atom made up of a single proton and electron. Likewise deuterium, with one neutron and one proton, is used to study neutron and proton interactions. Liquid hydrogen and deuterium boil at a very low temperature; thus chambers using these liquids operate at about 49°F (27 K) above absolute zero. *See* PARTICLE DETECTOR. [D.A.L.]

Buckeye A genus, *Aesculus,* of deciduous trees or shrubs belonging to the plant order Sapindales, buckeyes grow in North America, southeast Europe, and eastern Asia to India. The distinctive features are opposite, palmately compound leaves and a large fruit having a firm outer coat and containing usually one large seed with a conspicuous hilum.

The Ohio buckeye (*A. glabra*) is found mainly in the Ohio valley and in the southern Appalachians. It can be recognized by the glabrous winter buds, prickly fruits, and compound leaves having five leaflets. Another important species, the yellow buckeye (*A. octandra*), is native in the Central states, has five leaflets and smooth buds, but differs in its smooth, larger fruit. The horse chestnut (*A. hippocastanum*), which usually has seven leaflets and resinous buds, is a native of the Balkan Peninsula. It is planted throughout the United States and is a beautiful ornamental tree bearing cone-shaped flower clusters in early summer.

The seeds of all species contain a bitter and narcotic principle. The wood of the native tree species is used for furniture, boxes, crates, baskets, and artificial legs. *See* SAPINDALES. [A.H.G./K.P.D.]

Buckwheat A herbaceous, erect annual, the dry seed or grain of which is used as a source of food and feed. It is not a true cereal and is one of the very few plants other than those of the Gramineae family, used for their starchy seed, which is processed as a meal or flour. Buckwheat belongs to the Polygonaceae family. Species of buckwheat that have been commercially grown are *Fagopyrum sagittatum* (= *F. esculentum*), *F. emarginatum*, and *F. tataricum*.

The plant grows to a height of 2–5 ft (0.6–1.5 m) with many broad heart-shaped leaves. It produces a single main stem which usually bears several branches and is grooved, succulent, and smooth except for nodes. Buckwheat is an indeterminate species in response to photoperiod and produces flowers and fruits (so-called seeds) until the beginning of frost. *See* POLYGONALES. [H.G.M.]

Buckwheat milling is similar to wheat milling in principle. The grain is cleaned and crushed between a series of steel rolls, and the flour and feed by-products are separated by sieving. Buckwheat flour is used almost exclusively in pancakes or as one of the ingredients in prepared pancake flour. Sometimes buckwheat flour is milled for groats. Two types of food products are made from buckwheat groats, roasted kernels and a farina, or granular product, used in soup. [J.A.Sh.]

Bud An embryonic shoot containing the growing stem tip surrounded by young leaves or flowers or both, and the whole frequently enclosed by special protective leaves, the bud scales.

The bud at the apex of the stem is called a terminal bud (illustration *a*). Any bud that develops on the side of a stem is a lateral bud. The lateral bud borne in the axil (angle between base of leaf and stem) of a leaf is the axillary bud (illustration *a* and *d*). It develops concurrently with the leaf which subtends it, but usually such buds do not unfold and grow rapidly until the next season. Because of the inhibitory influence of the apical or other buds, many axillary buds never develop actively or may not do so for many years. These are known as latent or dormant buds. Above or beside the axillary buds, some plants regularly produce additional buds called accessory, or supernumerary, buds. Accessory buds which occur above the axillary bud are called superposed buds (illustration *c*), and those beside it collateral buds (illustration *d*). Under certain conditions, such as removal of terminal and axillary buds, other buds may arise at almost any point on the stem, or even on roots or leaves. Such buds are known as adventitious buds. *See* PLANT GROWTH.

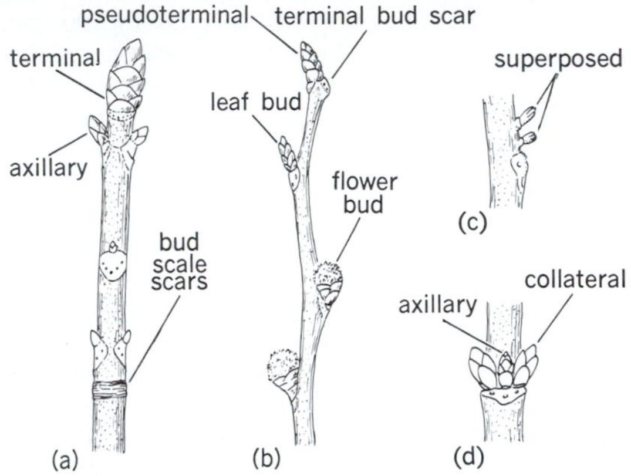

Bud positions. (*a*) Terminal and axillary (buckeye). (*b*) Pseudoterminal (elm). (*c*) Superposed (butternut). (*d*) Collateral (red maple).

Buds that give rise to flowers only are termed flower buds, or in some cases, fruit buds. If a bud grows into a leafy shoot, it is called a leaf bud, or more accurately, a branch bud. A bud which contains both young leaves and flowers is called a mixed bud.

Buds of herbaceous plants and of some woody plants are covered by rudimentary foliage leaves only. Such buds are called naked buds. In most woody plants, however, the buds are covered with modified protective leaves in the form of scales. These buds are called scaly buds or winter buds. In the different species of plants, the bud scales differ markedly. They may be covered with hairs or with water-repellent secretions of resin, gum, or wax. Ordinarily when a bud opens, the scales fall off, leaving characteristic markings on the stem (bud scale scars). *See* LEAF. [N.A.]

Budding A vegetative method of propagation in which a single bud (often a bud sport or bud mutation) serves as a scion and is grafted laterally upon a stock. Bud grafting may be performed throughout the growing season when the bark of the stock is soft and loose, but summer or early fall is usually the most satisfactory time. Success depends upon joining the vascular cambia of the stock and scion. this requires genetic compatibility of stock and scion, pressure to hold the scion firmly against the stock, and protection from desiccation. *See* REPRODUCTION (PLANT). [J.E.Gu.]

Budding and appendaged bacteria The vernacular name given to Part 4 in the organization of bacteria according to the eighth edition of *Bergey's Manual of Determinative Bacteriology* (1974). Typical bacteria divide asexually by binary transverse fission. Some bacteria, that is, many of the bacteria considered in this section, divide by budding which is analogous to budding in yeasts. In this process the daughter bud is formed as a small outgrowth on the surface of the mother cell. It enlarges and eventually separates from the mother cell. In contrast to binary transverse fission, this process involves the formation of the wall layers of the daughter cell (that is, bud) largely from newly synthesized material, with the mother cell retaining essentially all of the components of the cell wall of the original cell.

The appendaged representatives in this group have either stalks or prosthecae extending from the surface of the organism. Stalks are excreted or extracellular appendages, whereas prosthecae are part of the cell, in that they are enclosed by the cell wall of the bacterium that has them.

Most of the bacteria in this group are widely distributed in soil and in fresh waters and marine waters. With rare exception they are all free-living microbes whose major role in nature is the decomposition of the nonliving organic remains of plants, animals, and other microorganisms.

This group is a heterogeneous collection of organisms containing some types that are at best only distantly related to others. The most common genera are:

Group I, budding bacteria with buds formed on tip of prosthecae. Included here are the genera *Hyphomicrobium*, Hyphomonas, and *Rhodomicrobium*.

Group II, prosthecate bacteria whose progeny are not produced on tip of prosthecae. Included here are the genera *Caulobacter*, *Ancalomicrobium*, and *Prosthecomicrobium*.

Group III, budding bacteria which do not have prosthecae; includes the genus *Pasteuria*.

Group IV includes the stalked bacteria, represented by the genus *Gallionella*. [J.T.S.]

Buerger's disease A progressive, chronic inflammation of arteries, veins, and nerves of young to middle-aged men; also known as thromboangiitis obliterans. Buerger's disease is marked by a segmental involvement of larger arteries by inflammation, fibrosis, and a tendency toward formation of thromboses.

The disease is relatively rare and is usually found in men of 25–40 years of age, although a few women and younger males have been affected. A familial susceptibility is sometimes present and the disease shows an increased incidence among Jews.

Although the specific etiologic agent is unknown, almost all patients with Buerger's disease can cause an increase in their symptoms by the excessive use of tobacco or exposure to cold. Conversely, many of those afflicted show beneficial effects or retardation of the process when these two symptom-increasing factors are removed.

Vessels of the lower extremity are usually first affected and the clinical course is marked by muscle pain, aching, and tenderness brought on by exertion and relieved by rest. These symptoms may be accompanied or followed by sensory changes or sensitivity to extremes of temperature. In advanced cases discoloration of the skin, atrophy of the tissues, and even gangrene result from the progressive occlusion of blood vessels.

Progressive invalidism, hastened by the necessity of amputations of affected parts, is commonplace. Death may follow after a long period of illness or may occur during the frequent complications. Although peripheral vessels are most often damaged, the arteries of certain organs may be similarly involved, in which cases the prognosis is much less favorable. *See* CARDIOVASCULAR SYSTEM. [N.K.M.]

Buffalo The name for members of the family Bovidae in the mammalian order Artiodactyla. The buffalo is an Old World species and resembles the oxen in general appearance. The North American bison is often called a buffalo, but is not related to the true buffalo.

The Asiatic buffalo (*Bubalus bubalis*), known as the Indian or water buffalo and also as the carabao, is found as a domestic animal in the Balkans, Asia Minor, and Egypt. These buffalo exist in the wild state in southern Asia and Borneo, where they are considered to be ferocious and dangerous. Water buffalo are stocky, heavy-built animals. They have very short hair and short, splayed horns. Like all buffalo, they have a liking for marshes, where they wallow and become caked with mud that affords protection against insects.

Two other Asiatic species related to, but smaller than, the water buffalo are the tamarau (*Anoa mindorensis*), which is indigenous to the Philippines, and the still smaller anoa (*A. depressicornis*), or wild dwarf buffalo, found in the Celebes.

The African buffalo, classed in the genus *Syncerus*, was very numerous until the turn of the century, when the infectious disease rinderpest caused many deaths. They are still abundant though widely hunted. There are several varieties of African buffalo, and it is thought that all may be subspecies of *S. caffer*, the Cape buffalo. They live in the open country of central, eastern, and southern Africa. Except for its size, this animal is difficult to distinguish from the rare dwarf or forest buffalo (*S. caffer nanus*). It lives in marshy, forested areas of western Africa, where it is known as the bush cow. *See* ARTIODACTYLA; BISON; MAMMALIA. [C.B.C.]

Buffers (chemistry) A solution selected or prepared to minimize changes in hydrogen ion concentration which would otherwise tend to occur as a result of a chemical reaction. In general, chemical buffers are systems which, once constituted, tend to resist further change due to external influences. Thus it is possible, for example, to make buffers resistant to changes in temperature, pressure, volume, redox potential, or acidity. The commonest buffer in chemical solution systems is the acid-base buffer.

Chemical reactions known or suspected to be dependent on the acidity of the solution, as well as on other variables, are frequently studied by measurements in comixture with an appropriate buffer. For example, it may be desirable to investigate how the rate of a chemical reaction depends upon the hydrogen ion activity (pH). This is accomplished by measurements in several buffer systems, each of which provides a nearly constant, different pH. Alternatively, it may be desirable to measure the effects of other variables on a pH-sensitive system, by stabilizing the pH at a convenient value with a particular buffer. *See* PH.

Buffer action depends upon the fact that, if two or more reactions coexist in a solution, then the chemical potential of any species is common to all reactions in which it takes part, and may be defined by specification of the chemical potentials of all other species in any one of the reactions. To be effective, a buffer must be able to respond to an increase as well as a decrease of the species to be buffered. In order to do so, it is necessary that the proton transfer step of the buffer be reversible with respect to the species involved, in the reaction to be buffered. In aqueous solution the proton transfer between most acids, their conjugate bases, and water, is so rapid and reversible that the dominant direct source of protons for a chemical reaction is H_3O^+, the hydronium ion.

Buffers are particularly effective in water, because of the unusual properties of water as a solvent. Its high dielectric constant tends to promote the existence of formally charged ions (ionization). Because it has both an acidic (H) and a basic (O) group, it may form bonds with ionic species leading to an organized sheath of solvent surrounding an ion (solvation). Water also tends to self-ionize to form its own conjugate acid-base system. *See* ACID AND BASE; ACID-BASE INDICATOR; IONIC EQUILIBRIUM; SOLVATION. [A.M.H.]

Buffers (electronics) Electronic circuits whose output voltage exactly replicates the input voltage. An ideal voltage buffer has zero output impedance Z_{out} and infinite input impedance Z_{in}.

Buffers are generally applied in analog systems to minimize performance degradation due to excessive loading of output nodes. This is particularly important in analog integrated circuit design, where impedance levels can exceed 100 megohms. The usual symbol for a buffer is shown in the illustration. Two

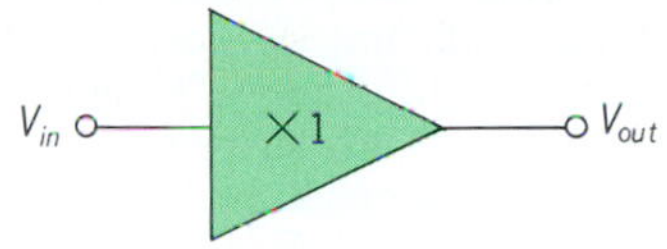

Circuit schematic symbol for a buffer.

circuit configurations are commonly used: the transistor follower and the operational-amplifier-based buffer.

The bipolar emitter follower and the field-effect transistor (FET) source follower are simple circuits, consisting of a single transistor and a bias-current source; they are used in applications where power consumption and circuit area must be reduced to a minimum or where specifications are not too demanding. *See* EMITTER FOLLOWER; TRANSISTOR.

The operational-amplifier-based buffer is essentially an operational amplifier with unity-gain feedback applied to it. The open-loop gain of the operational amplifier should be very high, extending to high frequencies. *See* OPERATIONAL AMPLIFIER. [P.M.VanP.]

Buffing The smoothing and brightening of a surface by an abrasive compound pressed against the work by means of a soft wheel or belt. The abrasive is a fine powder or flour mixed with tallow or wax to form a smooth composition or paste. This is applied as required to the buffing wheel or belt, which is made of a pliable material, such as soft leather, linen, muslin, or felt. Buffing is accomplished on the same types of machines as polishing and frequently both types of wheels are included on one machine. *See* POLISHING. [A.H.T.]

Buhrstone mill A mill for grinding or pulverizing, in which a flat siliceous rock, generally of cellular quartz, rotates against a stationary stone of the same material. Grooves in the stones facilitate the movement of the material. Fineness of the product is controlled by the pressure between the stones and by the grinding speed. A finely ground product is achieved by

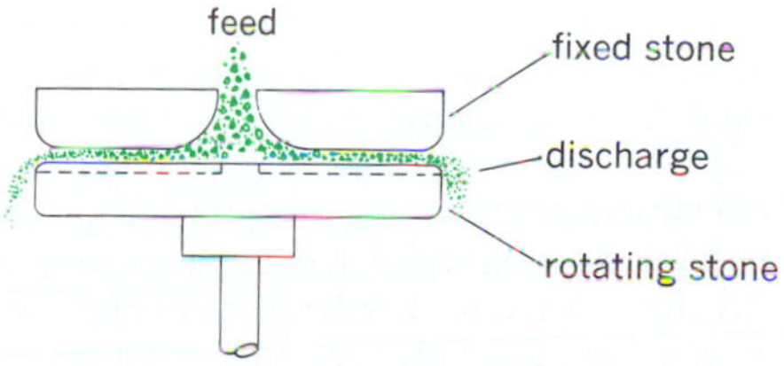

In a Buhrstone mill, material is fed at the center of the fixed stone and moves toward outer edge of the stones where product is discharged.

slowly rotating the stone at a high pressure against the materials and its mate (see illustration). *See* CRUSHING AND PULVERIZING. [G.W.K.]

Buildings Fixed structures for human occupancy and use. The space and structure must be planned to produce the environment and facilities required for the purpose of the building. Items such as light, heat, air conditioning, acoustics, and color must be given careful consideration.

Many governments, states, professional organizations, and municipalities have adopted building codes to establish minimum load capacities and to ensure safety to the occupants under normal use and under common hazards like fire and windstorms. Buildings are usually classified according to occupancy and use as follows: (1) public (auditoriums, churches, theaters), (2) institutional (hospitals, schools, penitentiaries), (3) residential (hotels, apartments, studios), (4) business (stores, factories, office buildings), and (5) storage (garages, warehouses).

The loads that a building supports are called dead load, which includes the weight of walls, columns, partitions, floors, and roof; and live load, which includes the weight of the physical equipment and occupants of the building, Snow load on the roof is considered a live load. Because there is little likelihood of applying the full specified live load all over large areas, codes permit a reduction in the basic design live load when areas exceed a certain minimum. A live load reduction is therefore allowed for supporting members such as girders, columns, piers, and foundations. An allowance for the dynamic effect of all moving loads, such as elevator machinery and cranes, is usually specified as a fixed percentage of the load.

The evaluation of pressures exerted by wind on a building is complex, and code requirements for the wind forces are only rough approximations. In earthquake areas the building code provides for special seismic design requirements. On the basis of predicted earthquake shocks, all components of a building must be designed to take into account the lateral forces that might be induced by such ground motion.

The roof and the floors of the buildings are generally supported by one of the three common framing systems: (1) bearing wall, (2) skeletal framing composed of beams and columns or slabs and columns, and (3) special long-span roof systems. In bearing-wall construction, walls support the ends of the floor and roof beams and slabs in addition to enclosing partitioning space. The bearing wall can be made of bricks, concrete blocks, cast-in-place reinforced concrete, or precast concrete panels. The skeletal frame construction can be either of steel framing or of reinforced concrete. Skeletal frames of reinforced concrete normally have monolithic rigid connections between beams and columns. However, steel skeletal frame construction can be simple, rigid, or semirigid. Long-span construction is used for auditoriums, hangars, gymnasiums, and other structures requiring a wide unobstructed floor area that cannot be economically spanned with standard steel beams or typical reinforced concrete or solid-timber girders. The appropriate forms of construction are plate girders, prestressed girders, trusses, arches, rigid frames, domes, cantilever suspension frames, space frames, and suspended cables. *See* ARCH; CANTILEVER; ROOF CONSTRUCTION; TRUSS; WALL CONSTRUCTION.

It is usually economical to support heavy industrial buildings with a steel framework consisting of a series of trusses supported on columns. Such industrial buildings often attain huge dimensions. Spectacular achievements have been made in the design and construction of ultra-high-rise buildings. Reinforced concrete has been economically and competitively used in a number of structures for both residential and commercial purposes. The newer high-rise buildings, ranging from 50 to 110 stories, are the result of relatively recent innovations and development of new structural systems.

Greater height entails increased column and beam sizes to make buildings more rigid so that under wind load they will not sway beyond an acceptable limit. Excessive lateral sway may cause serious recurring damage to partitions, ceilings, and other architectural details. In addition, excessive sway may cause discomfort to the occupants of the building because of their perception of such motion. New structural systems of reinforced concrete, as well as steel, take full advantage of the inherent potential stiffness of the total building and therefore do not require additional stiffening to limit the sway. [F.R.K.]

Bulk-handling machines A diversified group of materials-handling machines specialized in design and construction for handling unpackaged, divided materials.

Solid, free-flowing materials are said to be in bulk. The handling of these materials requires that the machinery both support their weight and confine them either to a desired path of travel for continuous conveyance or within a container for handling in discrete loads. Wet or sticky materials may also be handled successfully by some of the same machines used for bulk materials. Characteristics of materials that affect the selection of equipment for bulk handling include (1) the size of component particles, (2) flowability, (3) abrasiveness, (4) corrosiveness, (5) sensitivity to contamination, and (6) general conditions such as dampness, structure, or the presence of dust or noxious fumes.

Equipment that transports material continuously in a horizontal, inclined, or vertical direction in a predetermined path is a form of conveyor. The many different means used to convey bulk materials include gravity, belt, apron, bucket, skip hoist, flight or screw, dragline, vibrating or oscillating, and pneumatic conveyors. Wheel or roller conveyors cannot handle bulk materials.

Gravity chutes are the only unpowered conveyors used for bulk material. They permit only a downward movement of material.

Belt conveyors of many varieties move bulk materials. Fabric belt conveyors have essentially the same operating components as those used for package service; however, these components are constructed more ruggedly to stand up under the more rigorous conditions imposed by carrying coal, gravel, chemicals, and other similar heavy bulk materials. Belts may also be made of such materials as rubber, metal, or open wire. Their advantages include low power requirements, high capacities, simplicity, and dependable operation.

An apron conveyor is a form of belt conveyor, but differs in that the carrying surface is constructed of a series of metal aprons or pans pivotally linked together to make a continuous loop. This type of conveyor is suitable for handling large quantities of bulk material under severe service conditions. Apron conveyors are most suitable for heavy, abrasive, or lumpy materials.

Bucket conveyors are constructed of a series of buckets attached to one or two strands of chain or in some instances to a belt. These conveyors are most suitable for operating on a steep incline or vertical path, sometimes being referred to as elevating conveyors. This type of conveyor is most ideal for bulk materials such as sand or coal.

Flight conveyors employ the use of flights, or bars attached to single or double strands of chain. The bars drag or push the material within an enclosed duct or trough. These are frequently referred to as drag conveyors. This type of conveyor is commonly used for moving bulk material such as coal or metal chips from machine tools.

Spiral or screw conveyors rotate upon a single shaft to which are attached flights in the form of a helical screw. When the screw turns within a stationary trough or casing, the material advances. These conveyors are used primarily for bulk materials of fine and moderate sizes, and can move material on horizontal, inclined, or vertical planes.

Vibrating or oscillating conveyors employ the use of a pan

or trough bed, attached to a vibrator or oscillating mechanism, designed to move forward slowly and draw back quickly. The inertia of the material keeps the load from being carried back so that it is automatically placed in a more advanced position on the carrying surface.

Pneumatic, or air, conveyors employ air as the propelling media to move materials. One implementation of this principle is the movement within an air duct of cylindrical carriers, into which are placed currency, mail, and small parts for movement from one point to discharge at one of several points by use of diverters. Pneumatic pipe conveyors are widely used in industry, where they move granular materials, fine to moderate size, in original bulk form without need of internal carriers.

Power cranes and shovels perform many operations moving bulk materials in discrete loads. When functioning as cranes and fitted with the many below-the-hook devices available, they are used on construction jobs and in and around industrial plants. Such fittings as magnets, buckets, grabs, skull-crackers, and pile drivers enable cranes to handle many products. The machines of the convertible, full-revolving type are mounted on crawlers, trucks, or wheels. Specialized front-end operating equipment is required for clamshell, dragline, lifting-crane, pile-driver, shovel, and hoe operations. Specialized equipment for mechanized pit mining has been developed. Power cranes, shovels, and scoops are actively engaged in strip mines, quarries, and other earth-moving operations. *See* CONVEYING MACHINES; ELEVATING MACHINES; HOLSTING MACHINES; INDUSTRIAL TRUCKS; MATERIALS-HANDLING EQUIPMENT; MONORAIL. [A.M.P.]

Bulkhead A structure primarily used to retain earth fill or protect embankments along the shore. Bulkheads sometimes also serve as wharves. They are usually used on comparatively sheltered areas and therefore require less strength against sea action than sea walls. Sheet piling of timber, steel, or concrete is the commonest form of construction, with tiebacks to "deadmen" buried in the earth. Pile-supported platforms behind sheet piling help to relieve the earth pressure against the sheeting and platforms on the water side help to reduce the unsupported height of the sheeting. Masonry or concrete gravity wall construction and stone-filled crib construction are also used. *See* COASTAL ENGINEERING; SEA WALL. [E.J.Q.]

Bulldozer An assembly consisting of a reinforced, curved steel plate or blade and a frame which is attached to the front of a wheeled or crawler tractor for the purpose of pushing dirt, rock, or other materials.

The blade may be straight, angling, or U-shaped. The U blades and straight blades mount perpendicular to the longitudinal axis of the tractor and push material straight ahead; the U blade, because of its shape, can contain and move more material than the straight blade. The angle blade can be set so that either of its ends is ahead of the other; thus it is useful in windrowing material or backfilling trenches. Blades are controlled either by cables or hydraulic units, although the latter are far commoner. The straight and U blades can be raised and lowered, worked with one side lower than the other, and tipped forward or backward. The angle blade cannot be tipped. *See* EARTHMOVER. [E.M.Y.]

Buoy An anchored or moored floating object, other than a lightship, intended as an aid to navigation. Buoys are the most numerous of all artificial aids to navigation.

Buoys are intended to serve as daymarks. Some buoys, particularly those at turning points in channels, are provided with lights for location and identification at night. Some buoys are equipped with apparatus for providing distinctive sounds at intervals so they can be used as aids to navigation in fog and darkness. Some buoys are equipped with radio beacons, and some have reflectors to make them more conspicuous to radar.

Two general systems of buoyage are in use. In the cardinal system buoys are assigned shape, color, and number in accordance with their locations relative to the nearest obstruction. In the lateral system, used in a modified form in United States waters, buoys are assigned shape, color, and number in accordance with their locations relative to navigable waters. *See* PILOTING. [A.B.M.]

Buoyancy The resultant vertical force exerted on a body by a static fluid in which it is submerged or floating. The buoyant force F_B acts vertically upward, in opposition to the gravitational force that causes it. Its magnitude is equal to the weight of fluid displaced, and its line of action is through the centroid of the displaced volume, which is known as the center of buoyancy. *See* AEROSTATICS; FLUID STATICS; HYDROSTATICS.

By weighing an object when it is suspended in two different fluids of known specific weight, the volume and weight of the solid may be determined. *See* ARCHIMEDES' PRINCIPLE.

A body floating on a static fluid has vertical stability. A small upward displacement decreases the volume of fluid displaced, hence decreasing the buoyant force and leaving an unbalanced force tending to return the body to its original position. Similarly, a small downward displacement results in a greater buoyant force, which causes an unbalanced upward force.

A body has rotational stability when a small angular displacement sets up a restoring couple that tends to return the body to its original position. When the center of gravity of the floating body is lower than its center of buoyancy, it will always have rotational stability. Many a floating body, such as a ship, has its center of gravity above its center of buoyancy. Whether such an object is rotationally stable depends upon the shape of the body. [V.L.S.]

Buret An instrument used to deliver variable volumes of a liquid. Burets are usually made from uniform-bore glass tubing in capacities of 5–100 ml, the commonest being 50 ml. The uniform portion is graduated in small divisions, so that volumes of liquid accurate to a small fraction of a milliliter may be measured by reading the volumes in the buret before and after sample has been withdrawn. Most burets are equipped with a stopcock to control the flow of liquid. *See* TITRATION; VOLUMETRIC ANALYSIS. [C.E.B.]

Burgess Shale The Burgess Shale is part of a clay and silt sequence that accumulated beneath a colossal algal reef. The Burgess Shale itself was deposited at the reef base, and various lines of evidence demonstrate that the fauna (and flora) were transferred from a relatively oxygenated environment into deeper anaerobic water in a series of turbidites. *See* TURBIDITE.

The Burgess Shale fauna, occurring in a single locality near Field in southern British Columbia, is probably the finest assemblage of soft-bodied fossils known. It is Mid-Cambrian and thus approximately 530,000,000 years old. The fauna affords a unique glimpse into the state of metazoan evolution comparatively shortly after the major adaptive radiations of the late Precambrian and Early Cambrian.

The importance of the Burgess Shale fauna goes beyond contributing to knowledge of a wide variety of animals otherwise unknown from the fossil record. It must be asked to what extent the Burgess Shale fauna represents a typical Cambrian marine community with "normal" fossil faunas, that is, hard parts only, representing impoverished assemblages decimated by preservational factors such as decay. No firm answer can yet be given, but the scattered occurrence of soft-bodied fossils in other Cambrian rocks strongly suggests that the Burgess Shale is unique only in terms of preservation. *See* CAMBRIAN; FOSSIL; PALEONTOLOGY. [S.C.M.]

Burn An injury to tissues caused by heat, chemicals, electricity, or irradiation effects.

The commonest type of burn is that due to thermal injury, in which some portion of the body surface is exposed to either moist or dry heat of sufficient temperature to cause local and systemic reactions. Clinically, the extent of such a burn is often expressed as first degree, second degree, and so forth. Different systems of classification exist.

First-degree burns result in some redness and swelling of the injured part, without necrosis of any tissue or the formation of blisters. Healing is completed in a few days without scarring.

Second-degree burns show a variable destruction of parts of the epidermis so that blistering occurs. Healing by regeneration in such superficial burns does not necessitate skin grafting, unless secondary infections ensue; no scarring results.

Third-degree burns are marked by complete destruction of the epidermis of a region, including the necrosis of accessory skin structures like hair and sweat glands. A brownish-black eschar marks the destroyed tissue. This is sloughed off and that defect becomes filled with granulation tissue that later consolidates and changes to form a dense, thick scar. Complications may occur without adequate care, and grafting is not unusual, sometimes being required because of contracture of the scar tissue.

In fourth-degree burns, tissue is destroyed to the level of or below the deep fascia lying beneath the subcutaneous fat and connective tissue of the body. Muscle, bone, deeper nerves, and even organs may be injured or destroyed by this severe degree of burn. Healing is usually a slow, involved process, requiring much reparative and reconstructive work by surgical specialists.

Electrical burns result from the amount of heat incident to the flow of a certain amount of electricity through the resistance offered by tissues. From a practical standpoint, most of the resistance offered to the passage of an electric current is that of the skin and the interface between the skin and the external conductor. Therefore, most electrothermal injuries are limited to the skin and immediately subjacent tissues, although deep penetration may follow large voltages.

Most chemical burns result from the action of corrosive agents which destroy tissues at the point of contact. Exposure of the skin, eyes, and gastrointestinal tract are commonest.

[E.G.St./N.K.M.]

Burrowing animals Some terrestrial and aquatic animals are capable of excavating holes in the ground (burrowing) for protection from adverse environmental conditions, as well as for storing food. Burrows vary from temporary structures of simple design (for example, the nesting burrows of some birds) to more permanent underground networks that may be inhabited for several generations (for example, rabbit warrens, badger sets, fox earths, and prairie dog burrows). They vary in structure from blind burrows with a single opening to extensive systems with several openings. Some animals (for example, some species of moles) live permanently underground, and their burrows have no obvious large openings to the surface. Burrows may be shared by a number of species, and abandoned burrows may be used by other species. Animals with limbs usually excavate their burrows by using their legs, but many burrowing animals are limbless and the mechanism of progression is not always obvious.

Worms, slugs, many insects, and many vertebrates live in burrows. Earthworms are important soil organisms because their burrows improve drainage and aeration, their feeding habits enhance leaf decomposition, and their droppings increase soil fertility. Earthworms burrow by contracting circular muscles in their body wall to push forward, and contracting the longitudinal muscles to widen the burrow. Termites (Isoptera) and ants (Hymenoptera) are social insects, most of which live underground. Termites are major consumers of vegetation in warm climates, and many construct extensive underground galleries extending from the mound located on the surface. Most ants are predators or scavengers, but leaf-cutter ants feed on a fungus that grows on the harvested pieces of leaf in carefully tended underground galleries.

Many marine animals, including flatfish, crabs, and shrimps, take temporary refuge or live more permanently in sand or mud by burying themselves just below the surface. Aquatic sand and mud pose several problems for burrowing animals. First, the particles are usually tightly packed together, restricting movement and requiring these organisms to expend 10–1000 times as much energy to move a given distance compared to other forms of locomotion. Second, burrows readily collapse unless reinforced or consolidated. The wall of burrows may simply be consolidated with mucus, but some animals make a more permanent, substantial tube of particles stuck together with mucus. Finally, all burrowing animals must be able to create a current of water through the burrow so that they can breathe. Many animals also feed partly or wholly on particles carried in such currents. [H.D.J.]

Bursa A simple sac or cavity with smooth walls and containing a clear, slightly sticky fluid interposed between two moving surfaces of the body to reduce friction. Subcutaneous bursae are found where the skin stretches around the greater curvature of a joint, as in the elbow or knee, and considerable chafing may occur; they may be single or multiple sacs. These bursae may enlarge as a result of continuous excessive irritation, as in housemaid's knee or miner's elbow. *See* BURSITIS.

Synovial bursae are small closed sacs of fibrous tissue continuous with the joint cavity of a diarthrosis. They are lined with a complex membrane that secretes a clear lubricating fluid, serving to reduce friction between the opposing surfaces of the articulation. *See* JOINT (ANATOMY).

Bursae may exist in the form of elongated sheaths surrounding tendons or ligaments, where these moving bands are in contact with another structure, such as a bone, muscle, or another tendon or ligament. Tendon sheaths are especially common where tendons bend around the ends of two bones at an articulation. *See* MUSCULAR SYSTEM; SKELETAL SYSTEM. [W.J.B.]

Bursitis Any inflammation of a bursa. Inflammatory changes produce acute or chronic swelling, an increase in the fluid contents, and variable degrees of pain and tenderness. Although most bursitis is nonspecific, it may follow the invasion of certain organisms or may result from trauma or physical stress. In most instances the fluid aspirated from an inflamed bursa is sterile, thereby ruling against an infectious etiology. In chronic bursitis the wall becomes thickened, shaggy, and irregular, with calcium deposits commonly present. *See* BURSA; CONNECTIVE TISSUE. [E.G.St./N.K.M.]

Bus A wheeled, self-propelled public carrier available in various capacities and sizes accommodating from 8 to more than 60 people for mass transit of passengers on school, local, intercity, or interstate routes. As an automotive vehicle, the bus retains many passenger car and truck components.

Basic bus design keeps gross weight at a minimum for economic operation and makes maximum utilization of space for entry, aisle, and exit areas; in the United States, seating and safety devices must comply with Federal regulations. Heavy frames perform the same functions as those in other commercial vehicles for load-bearing capacity, rigidity, and resistance to impact. *See* AUTOMOTIVE FRAME.

The chassis assembly includes all the components and systems essential to operation, usually the fundamental package upon which bus bodies built by other manufacturers are mounted. Body styles, passenger capacity, power plant, and equipment are available in great variety. The type of accommodation ranges from three or four bench seats, in suburban units, to units equipped with air conditioning, sanitary facilities, and conveniences for passengers traveling intercity or transcontinental routes. Gasoline and diesel engines are the principal sources of motive power. *See* MOTOR VEHICLE. [P.A.O.]

Bus-bar An electric conductor that serves as a common connection between load circuits and the source of electric power of one polarity in direct-current systems or of one phase in alternating-current systems. Two or more bus-bars which serve all polarities or phases are collectively called a bus.

Bus-bars must be designed to carry the continuous current without overheating. The highest continuous current can be in the order of hundreds of thousands of amperes. Bus-bars must also be designed to withstand the mechanical forces caused by short-circuit currents. Special metallic shields are designed to enclose the bus, thus minimizing the effect of short-circuit forces. Such a bus is known as an isolated phase bus. *See* WIRING. [T.H.L.]

Bushing A removable metal lining, usually in the form of a bearing to carry a shaft. Generally a bushing is a small bearing in the form of a cylinder and is made of soft metal or graphite-filled sintered material. Bushings are also used as cylindrical liners for holes to preserve the dimensional requirements, such as in the guide bearings in jigs and fixtures for drilling holes in machine parts. [J.J.R.]

2,3-Butanediol An organic substance, solid at room temperature, formula $CH_3CH(OH)CH(OH)CH_3$. It is considered a possible source for 1,3-butadiene used for the production of synthetic rubber and a source for a permanent-type antifreeze. Other compounds readily made from 2,3-butanediol are methyl ethyl ketone, methyl vinyl carbinol, and methyl vinyl ketone. The last two compounds are used in the plastics industry.

Preparation of 2,3-butanediol, also referred to as diol or 2,3-butylene glycol, has been by fermentation on a pilot plant basis. It is the major fermentation product of several species of bacteria found in the genera *Aerobacter*, *Bacillus*, *Serratia*, and *Pseudomonas*. Most raw materials suitable for the production of industrial alcohol are also satisfactory for the manufacture of 2,3-butanediol. These fall into three main groups: (1) starchy substrates, including corn, wheat, barley, sweet potatoes, and potatoes; (2) sugars and sugar residues, such as beet and cane molasses; and (3) wood hydrolyzates and pulp-mill residues, including waste sulfite liquor, although the waste sulfite liquor presents some problems. [G.A.L./F.C.]

Butanol Any one of the four isomeric alcohols, liquids at room temperature, having the formula C_4H_9OH and molecular weight 74.12. The butanols are colorless, toxic, flammable materials which are soluble in most organic liquids; their solubilities in water depend upon their respective structures. The chemical properties of the butanols depend on the position of the hydroxyl group in the molecule. *n*-Butyl alcohol and isobutyl alcohol are primary alcohols and, therefore, may be oxidized to the respective aldehydes or carboxylic acids, whereas *sec*-butyl alcohol oxidizes to methyl ethyl ketone; *tert*-butyl alcohol cannot be oxidized without degradation.

The major use of the butanols is as chemical intermediates, although significant quantities are used directly as solvents. Important derivatives include plasticizers and acetate ester solvents for plastics and paints. *See* ALCOHOL. [J.W.L.]

Butter A food fat product made exclusively from milk or cream or both, with or without common salt and added coloring. Butter contains not less than 80% milk fat as well as 1–3% salt, 1% milk solids, and water. It is the second most important food fat. *See* MILK.

On the farm, processing is carried out by natural ripening of the cream, churning, and manually working to the desired consistency. Country butter is more variable in flavor then is creamery butter.

At the creamery, cream having a fat content of 30–40% is skimmed from milk by centrifuging. Pasteurization and spraying, or stripping, under vacuum remove undesirable flavors and odors and reduce action of microorganisms. Butter may be made from sweet or ripened cream. Ripening is carried out at 50–70°F (10–21.1°C) by introduction of a starter prepared from a bacterial culture and pasteurized milk. After ripening overnight, color is added. In the batch process sweet or ripened cream is agitated in large rotary churns; this action converts the emulsion from water in oil to an oil-in-water type and coalesces the fat globules. Churns have internal baffles which lift and drop the contents to provide agitation and splashing. After churning to the desired consistency, the buttermilk is drawn off and the butterfat quickly washed with water. The butter is then worked between grooved rolls or with augurs which impart a kneading action. Salt is added during working, and the moisture content is adjusted. After it is worked, the butter is packaged or formed into prints and wrapped. Butter oil is made by heating butter to break the emulsion and settling or centrifuging to separate the milk serum from the fat. Moisture is reduced to a low level by drying, the butterfat content being over 99%. Butter oil is usually canned, is much more stable than butter, and does not require refrigeration. Ghee, a common food fat in India, is produced from boiled buffalo milk. Its manufacture is similar to that of butter oil. It can be kept for months, or years, without refrigeration but has a more intense flavor than butter or butter oil. [F.G.D.]

Butterfly Insects which make up one of the two divisions of the order Lepidoptera. All butterflies have flattened, scale-like hairs, which cover the wings as well as the body and obscure the venation that is important in species identification. Butterflies have club-shaped antennae, which terminate in a knob, and mouth parts which are modified and adapted to their feeding habits. The mandibles, absent or vestigial in the adults, are well developed in the larvae since they are principally plant feeders. These insects undergo a complete metamorphosis, and most species produce a single generation each year. Generally, the pupal stage is the overwintering stage, although some overwinter as the egg or larval stage. Butterflies have a worldwide distribution, with the larger, more colorful species occurring in tropical regions. They are an interesting group biologically, since the phenomena of mimicry, protective coloration, warning coloration, and other protective adaptations can be studied. *See* CATERPILLAR; LEPIDOPTERA; MOTH; PROTECTIVE COLORATION. [C.B.C.]

Buxbaumiidae A subclass of the class Bryopsida, the true mosses. It consists of a single family with four genera. The Buxbaumiidae is very distinctive in every way and most significantly in the structure of the peristome. The plants are small and occur especially on soil. The gametophyte is greatly reduced (*Buxbaumia*), with no stem and few leaves that are readily disappearing, or better developed with well-formed leaves having a single costa and short cells. The sporophytes are terminal. The capsules are disproportionately large and immersed or elevated on a seta (*Buxbaumia*). They are strongly inclined and asymmetric, tapered to a small mouth from a broad base. The operculum is small and conic. *See* BRYOPHYTA; BRYOPSIDA. [H.Cr.]

Bytownite A member of the plagioclase feldspar solid-solution series with a composition ranging from $Ab_{30}An_{70}$ to $Ab_{10}An_{90}$ (Ab = $NaAlSi_3O_8$ and An = $CaAl_2Si_2O_8$). Bytownite is very abundant in basic igneous rocks where it is the first plagioclase to crystallize under plutonic conditions; it forms the cores of zoned plagioclase phenocrysts in basaltic volcanics, and it sometimes occurs in anorthosites. *See* FELDSPAR. [L.Gr.]

C

Cabbage A hardy, cool-season crucifer (*Brassica oleracea* var. *capitata*) of Mediterranean origin and belonging to the plant order Capparales. Cabbage is grown for its head of overlapping leaves (see illustration), which are generally eaten raw in salads, cooked fresh, or processed into sauerkraut. Because it normally produces seed the second year, cabbage is considered to be a biennial by most authorities. Others regard it a perennial because it will remain vegetative unless subjected to cold weather.

Chinese cabbage is a related annual of Asiatic origin. Two species are grown in the United States, pe-tsai (*B. pekinensis*) and pakchoi (*B. chinensis*). *See* ORIENTAL VEGETABLES.

Cabbage varieties (cultivars) are generally classified according to season of maturity, leaf surface (smooth, savoyed, or wrinkled), head shape (flattened, round, or pointed), and color (green or red). Round, smooth-leaved, green heads are commonest. Varieties differ in their resistance to disease and in the tendency for heads to crack or split in the field.

Cabbage (*Brassica oleracea* var. *capitata*), cultivar Golden Acre 84. (*Joseph Harris Co., Inc., Rochester, New York*)

Texas and Florida are important winter crop producing states; Georgia, Mississippi, and North Carolina produce large acreages in the spring; and New York, North Carolina, and Wisconsin are important for the summer and fall crops. New York and Wisconsin are the important kraut cabbage states. *See* CAPPARALES. [H.J.C.]

Cable television system A system that receives and processes television signals from various sources and retransmits these signals through cables to subscribers' homes. The sources of the signals include broadcast transmissions, satellite-delivered programming, and local television studio productions. The facility that receives, processes, and retransmits the signals is called a headend.

Unlike broadcast television signals, which travel through free space, cable signals travel through coaxial cable or optical fiber, with different programs or channels traveling at different frequencies (much the same as frequency-division multiplex). In effect, the coaxial cable or optical fiber acts as a self-contained, closed, noninterfering frequency spectrum, created inside the cable by the reuse of the spectrum already in use for other purposes. *See* MULTIPLEXING.

Cable television is made possible by the technology of coaxial and optical-fiber cable and is subject to the principles of transmission-line theory. The primary disadvantage of coaxial-cable distribution systems is their relatively high loss or attenuation regarding television signals at the frequencies normally used in cable television systems (50–550 MHz, extending up to 1 GHz in newer systems). Amplifiers are required to overcome this signal loss, and the farther the subscriber is from the cable headend the more amplifiers are needed. Noise and intermodulation distortions created by many cascaded amplifiers limit the practical length of any coaxial cable network. *See* COAXIAL CABLE.

Optical fiber does not have the same high attenuation or loss characteristics as coaxial cable, and optical-fiber networks can therefore be built without amplifiers. Most cable systems that use optical fibers do so in a hybrid fashion. Optical fiber is connected from the headend to some localized node or terminating location. The subscriber is then connected to the optical-fiber node by short distances of coaxial cable. *See* COMMUNICATION CABLES; OPTICAL COMMUNICATIONS; OPTICAL FIBERS.

The system design or architecture is known as a tree-and-branch design. The tree-and-branch architecture is the most efficient way to transmit a package of multiple channels of programming from a headend to all subscribers. *See* CLOSED-CIRCUIT TELEVISION; TELEVISION. [R.D.Pi.]

Cacao *Theobroma cacao*, a small tropical tree (see illustration) that is cultivated for the almond-shaped seeds which are used to make chocolate. The species is native to the rainforest of the Amazon basin, and two regions of distribution in pre-Columbian times are recognized. The crop was first cultivated in Central America and northern South America, the varieties found there being known as Criollos. The second region comprises the Amazon and Orinoco basins, where the cacao populations are known as Amazonian Forastero. The second type is more commonly cultivated, particularly in Brazil, Ivory Coast, Ghana, and Nigeria.

The produce is generally exported in the form of dry beans. The farmers' production is purchased by dealers and exported by registered exporters or government marketing boards. Sales are effected through contracts or futures markets, principally in New York and London. The market distinguishes between bulk

Cacao (*Theobroma cacao*). (*USDA*)

and fine cocoas. The latter have preferred flavor or other characteristics and receive a price premium. *See* THEALES. [P.DeT.A.]

Cachexia Weight loss, weakness, and wasting of the body encountered in certain diseases or in terminal illnesses. Cachexia is most often associated with some malignancy and appears to be the actual cause of death in many cases. Such death, at a specific time, is difficult to explain since often there are no obvious specific events which affect vital organs. *See* ONCOLOGY.

Cachexia predisposes to other disorders, principally infections. There is little resistance to such processes and quite often the emaciated patient dies, apparently as the result of a minor infection which would ordinarily not even be incapacitating. *See* INFECTION.

Extensive irradiation may also cause cachexia. The symptoms may appear at some variable time after exposure, sometimes not becoming pronounced for several months. Recovery may occur or a chronic state of illness may ensue. Usually death follows a gradual deterioration. *See* RADIATION BIOLOGY. [E.G.St./N.K.M.]

Cactus The common name for any member of the cactus family (Cactaceae). There are 120 genera with perhaps 1700 species, nearly all indigenous to America. The cacti are among the most extremely drought-resistant plants, and consequently they thrive in very arid regions. The group is characterized by a fleshy habit, presence of spines and bristles, and large, brightly colored, solitary flowers. There is a great variety of body shapes and patterns, and many of the species are grown as ornamentals or oddities. A few have edible fruits. The saguaro (*Cereus giganteus*) of Arizona and Sonora is the largest of the cacti, attaining a height of 70 ft (21 m). *See* CARYOPHYLLALES. [P.D.St./E.L.C.]

Caddis fly The name for over 3600 species of insects which constitute the order Trichoptera. The adults are mothlike insects (see illustration), small to medium size, dull brown in color, and weak fliers. With one exception, *Enoicyla*, the

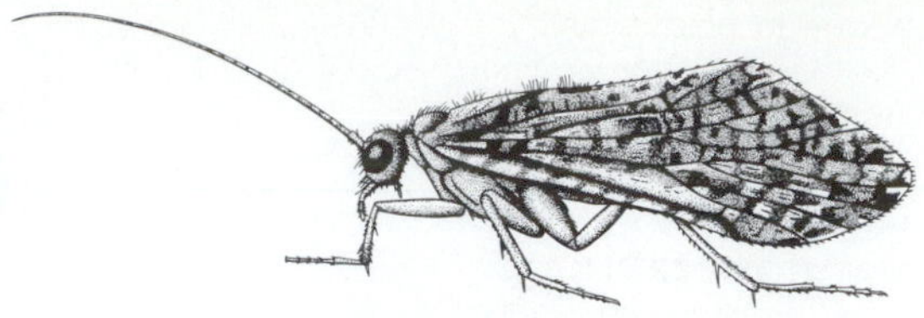

Adult caddis fly.

immature stages are aquatic. The larvae are unusual in that many species construct a silklike case in which they live. These portable cases are of various shapes and are usually strengthened by impregnated material such as small stones, sand grains, bits of snail shells, and vegetation of various types. Other species live in small, silk, netlike structures where several individuals reside. All larvae spin a cocoon before pupation. The larvae live in almost all free-flowing streams and rivers in North America. Species that weave nets may play an important role in keeping the water free of pollutants by filtering out detritus and organic materials. *See* TRICHOPTERA. [C.B.C.]

Cadmium A relatively rare chemical element, symbol Cd, atomic number 48, closely related to zinc, with which it is usually associated in nature. It is a silvery-white ductile metal with a faint bluish tinge. It is softer and more malleable than zinc, but slightly harder than tin. It has an atomic weight of 112.40 and a specific gravity of 8.65 at 20°C (68°F). Its melting point of 321°C (610°F) and boiling point of 765°C (1410°F) are lower than those of zinc. There are eight naturally occurring stable isotopes, and eleven artificial unstable radio isotopes have been reported. Cadmium is the middle member of group II (zinc, cadmium, and mercury) in the periodic table, and its chemical properties generally are intermediate between zinc and mercury. The cadmium ion is displaced by zinc metal in acidic sulfate solutions. Cadmium is bivalent in all its stable compounds, and its ion is colorless. *See* TIN; ZINC.

Cadmium does not occur uncombined in nature, and the one true cadmium mineral, greenockite (cadmium sulfide), is not a commercial source of the metal. Almost all of the cadmium produced is obtained as a by-product of the smelting and refining of zinc ores, which usually contain 0.2–0.4% cadmium. The United States, Canada, Mexico, Australia, Belgium-Luxembourg, and the Republic of Korea are principal sources, although not all are producers.

At one time an important commercial use of cadmium was as an electrodeposited coating on iron and steel for corrosion protection. Nickel-cadmium batteries are the second-largest application, with pigment and chemical uses third. Sizable amounts are used in low-melting-point alloys, similar to Wood's metal, and in automatic fire sprinklers, and relatively smaller uses are in brazing alloys, solders, and bearings. Cadmium compounds are used

1	2	3	4	5	6	7	8	9	10	11	12	13	14	15	16	17	18
1 H																	2 He
3 Li	4 Be											5 B	6 C	7 N	8 O	9 F	10 Ne
11 Na	12 Mg											13 Al	14 Si	15 P	16 S	17 Cl	18 Ar
19 K	20 Ca	21 Sc	22 Ti	23 V	24 Cr	25 Mn	26 Fe	27 Co	28 Ni	29 Cu	30 Zn	31 Ga	32 Ge	33 As	34 Se	35 Br	36 Kr
37 Rb	38 Sr	39 Y	40 Zr	41 Nb	42 Mo	43 Tc	44 Ru	45 Rh	46 Pd	47 Ag	48 Cd	49 In	50 Sn	51 Sb	52 Te	53 I	54 Xe
55 Cs	56 Ba	71 Lu	72 Hf	73 Ta	74 W	75 Re	76 Os	77 Ir	78 Pt	79 Au	80 Hg	81 Tl	82 Pb	83 Bi	84 Po	85 At	86 Rn
87 Fr	88 Ra	103 Lr	104 Rf	105 Db	106 Sg	107 Bh	108 Hs	109 Mt	110	*111*	*112*	*113*	*114*	*115*	*116*	*117*	*118*

lanthanide series	57 La	58 Ce	59 Pr	60 Nd	61 Pm	62 Sm	63 Eu	64 Gd	65 Tb	66 Dy	67 Ho	68 Er	69 Tm	70 Yb
actinide series	89 Ac	90 Th	91 Pa	92 U	93 Np	94 Pu	95 Am	96 Cm	97 Bk	98 Cf	99 Es	100 Fm	101 Md	102 No

as stabilizers in plastics and the production of cadmium phosphors. Because of its great neutron-absorbing capacity, especially the isotope 113, cadmium is used in control rods and shielding for nuclear reactors. *See* ALLOY; CADMIUM METALLURGY. [W.H.]

Cadmium metallurgy Most cadmium occurs in solid solution in the zinc sulfide mineral called sphalerite. Although cadmium may be recovered from some lead and copper ores, it is associated with the zinc which is also found in these ores. Since cadmium is entirely a by-product metal, the supply available is closely aligned with zinc production, averaging about 0.4% of zinc production.

All cadmium recovery processes involve the dissolution of cadmium-bearing feed material, followed by various purification and cadmium displacement steps. Methods of processing can be grouped conveniently into two basic categories, electrolytic and electromotive. In the former case, cadmium is recovered by electrolyzing purified solutions; in the latter case, cadmium in the form of a metallic sponge is displaced from purified solutions by a less noble metal, zinc being used in every known commercial application, and the sponge is melted or distilled, or both.

Major end uses for cadmium are in corrosion-resistant plating and in cadmium compounds for use as pigments in paints, ceramics, and plastics. Cadmium is also used in alloys, plastic stabilizers, batteries, and television picture tube phosphors. In view of the changing world energy situation, uses for cadmium are anticipated in applications such as solar energy cells and energy storage systems. *See* CADMIUM. [R.E.L.]

Caffeine An alkaloid, formerly synthesized by methylation of theobromine isolated from cacao, but now recovered from the solvents used in the manufacture of decaffeinated coffee. It is widely used in medicine as a stimulant for the central nervous system and as a diuretic. It occurs naturally in tea, coffee, and cacao. Caffeine crystallizes into long, white needlelike crystals that slowly lose their water of hydration to give a white solid that melts at 235–237.2°C (455–459.0°F). It sublimes without decomposition at lower temperatures. Caffeine has an intensely bitter taste, though it is neutral to litmus. *See* ALKALOID. [F.W.]

Cage hydrocarbon A compound that is composed of only carbon and hydrogen atoms and contains three or more rings arranged topologically so as to enclose a volume of space. In general, the "hole" within a cage hydrocarbon is too small to accommodate even a proton. The carbon frameworks of many cage hydrocarbons are quite rigid. Consequently, the geometric relationships between substituents on the cage are well defined. This quality makes these compounds exceptionally valuable for testing concepts concerning bonding, reactivity, structure-activity relationships, and structure-property relationships. *See* CHEMICAL BONDING.

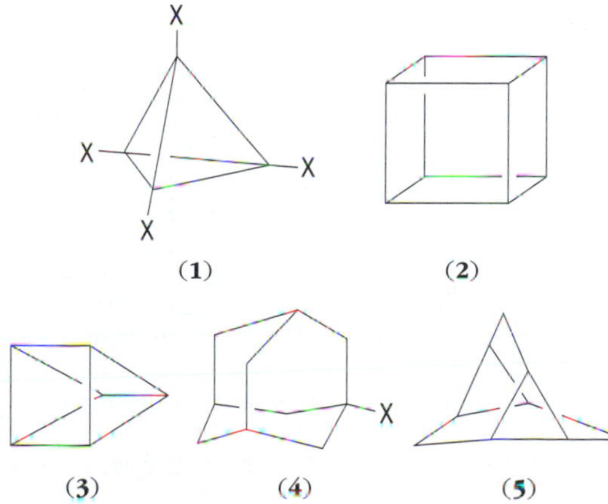

(1) (2) (3) (4) (5)

The carbocyclic analogs of the platonic solids that are tenable are tetrahedrane (structure **1**, where X = −H), cubane (**2**), and dodecahedrane. See ALICYCLIC HYDROCARBON.

An unsubstituted prismane has the general formula of $(CH)_n$, and the carbon atoms are located at the corners of a regular prism. Prismane (**3**), cubane (**2**), pentaprismane, and hexaprismane are the simplest members of this family of cage hydrocarbons.

The monomer of the diamond carbon skeleton is adamantane (**4**, where X = −H).

Amantadine (**4**, where X = $-NH_2$) was developed commercially as the first orally active antiviral drug for the prevention of respiratory illness due to influenza A2-Asian viruses.

The other simple diamondoid hydrocarbons are diamantane and triamantane.

Organic chemists have prepared a wide variety of cage hydrocarbons that do not occur in nature. Among these compounds are triasterane (**5**), iceane or wurtzitane, and pagodane. *See* ORGANIC CHEMICAL SYNTHESIS. [R.K.Mu.]

Caiman A large aquatic reptile of the family Alligatoridae. There are five species of caimans, once often sold as "baby alligators." Caimans differ from alligators in technical details of their internal anatomy and scalation.

The black caiman (*Melanosuchus niger*) of the Amazon Basin of South America strongly resembles the American alligator in superficial appearance and may reach a length approaching 16 ft (4.8 m). The spectacled caiman (*Caiman crocodilus*), with its several subspecies, is the one sold in the pet trade. It is the most widely distributed of the caimans, ranging from southwestern Mexico to Paraguay and Bolivia. One subspecies, *C. c. yacare*, is recognized as endangered. Spectacled caimans reach a length of 8 ft (2.4 m) and can be distinguished from alligators by the presence of a ridge across the snout between their eyes, the nose bridge of the "spectacles."

The broad-nosed caiman (*C. latirostris*) occurs from southern Brazil to Paraguay and northern Argentina and reaches a length of 9 ft (2.7 m; see illustration). The two species of smooth-fronted caiman, *Paleosuchus trigonatus* and *P. palpebrosus*, are the smallest of the family, reaching a maximum length of only 7 ft (2.1 m). They occur over most of northern South America, south to Peru, Bolivia, and southeastern Brazil. *See* ALLIGATOR.

Broad-nosed caiman (*Caiman latirostris*).

Alligators are generally distinguished from crocodiles by their broader, rounded, and more massive snout, the arrangement of their teeth, and other technical anatomical details. The teeth are conical and sharp, equivalent in shape (homodont), and replaced throughout life. Moreover, they possess a functionally four-chambered heart, a secondary palate, a septum or "diaphragm" that separates the lung and peritoneal regions, a compressed tail, webbed feet, and other adaptations for an aquatic existence. *See* CROCODYLIA; REPTILIA. [H.W.C.]

Caisson foundation A permanent substructure that, while being sunk into position, permits excavation to proceed inside and also provides protection for the workers against water pressure and collapse of soil. The term caisson covers a wide range of foundation structures. Caissons may be open, pneumatic, or floating type; deep or shallow; large or small; and of circular, square, or rectangular cross section. The walls may consist of timber, temporary or permanent steel shells, or thin or massive concrete. Large caissons are used as foundations for bridge piers, deep-water wharves, and other structures. Small caissons are used singly or in groups to carry such loads as building columns. Caissons are used where they provide the most feasible method of passing obstructions, where soil cannot otherwise be kept out of the bottom, or where cofferdams cannot be used. *See* BRIDGE; PILE FOUNDATION.

The bottom rim of the caisson is called the cutting edge (see illustration). The edge is sharp or narrow and is made of, or faced with, structural steel. The narrowness of the edge facilitates removal of ground under the shell and reduces the resistance of the soil to descent of the caisson.

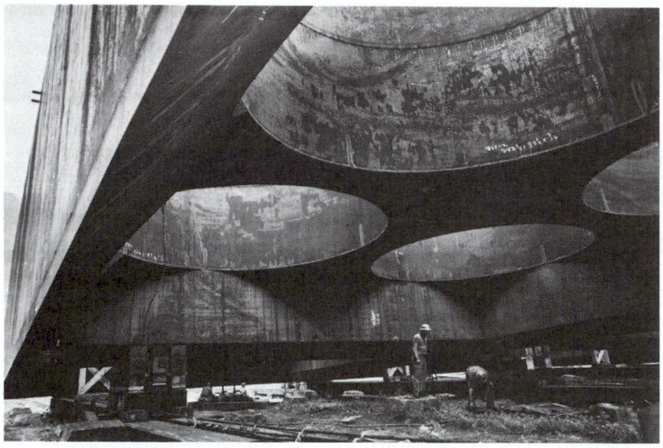

Underside of open caisson for Greater New Orleans bridge over Mississippi River. (*Dravo Corp.*)

An open caisson is a shaft open at both ends. It is used in dry ground or in moderate amounts of water. A pneumatic caisson is like a box or cylinder in shape; but the top is closed and thus compressed air can be forced inside to keep water and soil from entering the bottom of the shaft. A pneumatic caisson is used where the soil cannot be excavated through open shafts or where soil conditions are such that the upward pressure must be balanced. A floating or box caisson consists of an open box with sides and closed bottom, but no top. It is usually built on shore and floated to the site where it is weighted and lowered onto a bed previously prepared by divers. *See* FOUNDATIONS. [R.D.Che.]

Calamine A term that may refer to either a zinc mineral, $Zn_4Si_2O_7(OH)_2 \cdot H_2O$, which is also known as hemimorphite, or to zinc oxide, ZnO, which is used in medicinal or pharmaceutical products and in cosmetics. *See* HEMIMORPHITE. [E.E.W.]

Calanoida An order of Copepoda that includes the larger and more abundant of the pelagic species. Some authorities consider the Calanoida an order of the subclass Copepoda. In the food cycles of the sea these copepods are the most important group of marine animals because of their overwhelming numbers, ubiquitous distribution, and position at the base of the animal food chain. *See* COPEPODA.

The anterior part of the body is cylindrical with five or six segments, and much broader than the posterior part. The first antennae are not for locomotion, but are stabilizers and sinking retarders, and also have an olfactory function. The second antennae and mandibular palps are biramous and create water currents for feeding and slow movement. The five pairs of swimming legs are biramous, but the last pair is sometimes reduced or absent in the female, and the male's right fifth leg may be modified for grasping the female (see illustration).

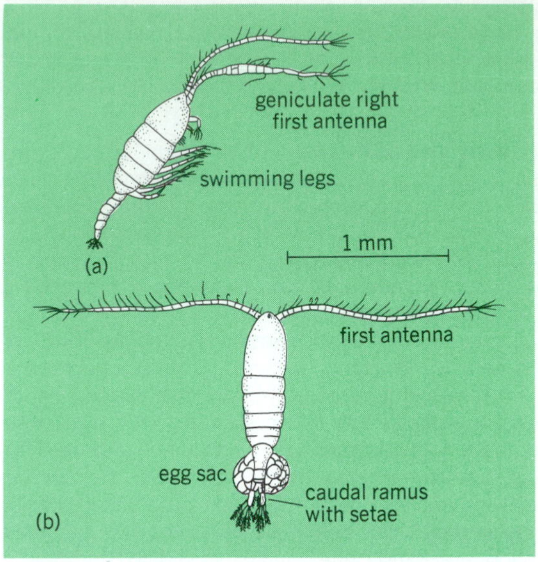

***Diaptomus.* (*a*) Lateral view of male. (*b*) Dorsal view of female.**

Nearly all calanoids are planktonic and, as a group, occur in all parts of the oceans from the surface to abyssal depths. The geographic and bathymetric ranges of many species are, however, restricted by the nature of water currents and the chemical and physical conditions of the water. In their southernmost range the northern species are found at greater depths. [H.C.Y.]

Calcarea A class of the phylum Porifera, including sponges with a skeleton composed of spicules of calcium carbonate. Calcarea vary from radially symmetrical vase-shaped species to colonies made up of a reticulum of thin tubes to irregular massive forms. Calcareous sponges are mostly of small size and inhabit the shallow waters of all seas, from tidal areas to depths of 600 ft (200 m), with a few species extending down to at least 12,000 ft (4000 m).

Primitive calcareous sponges with an ascon grade of construction consist of colonies of upright tubes with unfolded walls made up of an outer epidermis of pinacocytes and an inner lining of choanocytes. Between these layers of cells is a stratum of mesoglea containing amebocytes and spicules. Cells called porocytes, each perforated by a tubular canal, pierce the walls at intervals and allow water to enter the central cavity or spongocoel. Water leaves by way of a terminal osculum.

A somewhat more complicated structure is seen in calcareous sponges of the sycon grade of construction. Syconoid sponges are usually individual vase-shaped forms with a thick wall enclosing a large central spongocoel opening out through a terminal osculum. In the simplest forms the wall is pushed out at intervals into fingerlike projections, called radial canals, in which the choanocytes are localized. Water enters the radial canals directly through pores without the intervention of special inhalant canals. In most syconoid species, however, a dermal membrane made up of pinacocytes and mesenchyme forms a cortex of greater or less thickness which joins the outer ends of the radial canals. Pores or ostia pierce the dermis and open into inhalant canals which are simply the spaces between the radial canals in some cases.

The leuconoid grade of construction has probably evolved independently among the several lines of calcareous sponges. In those with a syconoid ancestry, the radial canals subdivide into many small flagellated chambers which arise as outpocketings of the radial canal wall. [W.D.H.]

Fossils clearly referable to the class Calcarea do not appear before the Carboniferous, later than any other class of sponges. The dominant Calcarea preserved from the late Paleozoic (Carboniferous and Permian periods) are the Sphinctozoa. In the Permian probable Calcinea with the pharetronid type of skeleton first appear. Both the Sphinctozoa and the pharetronids become highly varied and abundant in the Permian and Triassic periods. Calcareous sponges remained fairly abundant in shallow water and reefy deposits until the end of the Cretaceous. After that time the Sphinctozoa became extinct and the pharetronids became gradually less abundant. *See* PORIFERA. [R.M.F.]

Calcarenite A mechanically deposited (clastic) limestone in which the constituent particles of carbonate are sand-grain size. If the grains are larger than 0.08 in. (2 mm), the rock is a calcirudite. If the grains are smaller than 0.0025 in. (0.0625 mm), it is a calcilutite. The calcarenites are normally cemented by clear calcite. The detrital carbonate grains are fossil fragments, broken fragments of calcilutite, oolites, and fecal pellets. Where oolites are dominant, the rock is called an oolite or oolitic limestone.

The fine-grained equivalent of calcarenite, calcilutite, is called lithographic limestone if it is homogeneous, dense, and very fine-grained and breaks with conchoidal or subconchoidal fracture; the name is derived from the use of the rock for lithography. *See* LIMESTONE. [R.Si.]

Calcaronea A subclass of sponges of the class Calcarea, in which the larvae are amphiblastulae. This subclass may be divided into the orders Leucosoleniida and Sycettida. The triradiate spicules in species of this subclass characteristically have one ray longer than the other two. Monaxonid spicules are usually present. The canal system is asconoid, syconoid, or leuconoid. The oldest known fossil member of this group preserved entire is *Leucandra walfordi*, of Jurassic age. *See* CALCAREA; LEUCOSOLENIIDA; SYCETTIDA. [W.D.H.]

Calcichordates Primitive fossil members of the phylum Chordata with a calcite skeleton of echinoderm type. They occur in marine rocks of Cambrian to Pennsylvanian age (530–300 million years old) and, because of their skeletons, have traditionally been placed in the phylum Echinodermata. They are shown to be chordates, however, by many chordate anatomical features. Their calcite skeletons merely confirm an old view—that echinoderms and chordates are closely related. There are three main groups of calcichordates—the Soluta, the Cornuta, and the Mitrata. *See* CHORDATA; ECHINODERMATA. [R.P.S.J.]

Calcinea A subclass of sponges of the class Calcarea, in which the larvae are blastulae. Most species possess at least some triradiate spicules which have equal rays and equal angles between the rays. Monaxonid spicules are present or absent. The canal system is leuconoid. Included in this subclass are two orders: the Clathrinida and Leucettida. *See* CALCAREA; CLATHRINIDA; LEUCETTIDA. [W.D.H.]

This group has a rich fossil record based on the assignment of the pharetronid sponges to the Calcinea. It begins somewhat modestly in the Permian Period, expands in number and variety through the Cretaceous Period, and then slowly contracts to their present obscurity. Some 70 genera of pharetronids are known from the Cretaceous, and about one-seventh as many today. The fossil record of nonpharetronid Calcinea is practically nonexistent. [R.M.F.]

Calcite A mineral with the chemical formula $CaCo_3$; one of the most common and widespread minerals in the Earth's crust. Calcite may be found in a great variety of sedimentary, metamorphic, and igneous rocks. It is also an important rock-forming mineral and is the sole major constituent in limestones, marbles, and many carbonatites. Calcite in such rocks is the main source of the world's quicklime and hydrated, or slaked, lime. It is also widely used as a metallurgical flux to a scavenge siliceous impurities by forming a slag in smelting furnaces. It provides the essential calcium oxide component in common glasses and cement. Limestones and marbles of lower purity may find uses as dimension stone, soil conditioners, industrial acid neutralizers, and aggregate in concrete and road building. Calcite in transparent well-formed crystals is used in certain optical instruments. *See* CRYSTAL OPTICS; GLASS; LIME (INDUSTRY); LIMESTONE; STONE AND STONE PRODUCTS.

When pure, calcite is either colorless or white, but impurities can introduce a wide variety of colors: blues, pinks, yellow-browns, greens, and grays have all been reported. Hardness is 3 on Mohs scale. The specific gravity of pure calcite is 2.7102 ± 0.0002 at 68°F (20°C). Calcite has a very low solubility in pure water (less than 0.001% at 77°F or 25°C), but the solubility increases considerably with CO_2 added, as in natural systems from the atmosphere, when more bicarbonate ions and carbonic acid are formed. The solubility is also increased by falling temperature and rising total pressure. Shallow warm seas are supersaturated with calcite, while enormous quantities of calcite are dissolved in the unsaturated deep oceans. *See* CALCIUM; CARBONATE MINERALS. [R.I.Ha.]

Calcium A chemical element, Ca, of atomic number 20, fifth among elements and third among metals in abundance in the Earth's crust, Calcium compounds make up 3.64% of the Earth's crust. The physical properties of calcium metal are given in the table. The metal is trimorphous and is harder than sodium, but softer than aluminum. Like beryllium and aluminum, but unlike the alkali metals, it will not cause burns on the skin. It is less reactive chemically than the alkali metals and the other alkaline-earth metals.

Occurrence of calcium is very widespread; it is found in every major land area of the world. This element is essential to plant and animal life, and is present in bones, teeth, eggshell, coral, and many soils. Calcium chloride is present in sea water to the extent of 0.15%. *See* CARBONATE MINERALS.

Calcium metal is prepared industrially by the electrolysis of molten calcium chloride. Calcium chloride is obtained either by treatment of a carbonate ore with hydrochloric acid or as a waste product from the Solvay carbonate process. The pure

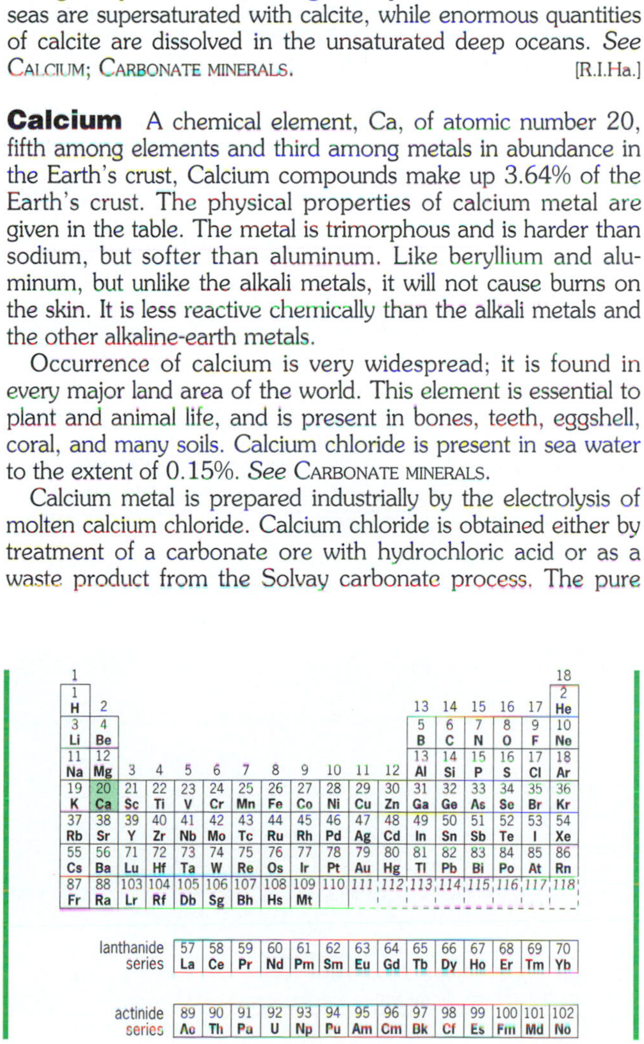

1 18
1 H 2 He
2 13 14 15 16 17
3 Li 4 Be 5 B 6 C 7 N 8 O 9 F 10 Ne
11 Na 12 Mg 3 4 5 6 7 8 9 10 11 12 13 Al 14 Si 15 P 16 S 17 Cl 18 Ar
19 K 20 Ca 21 Sc 22 Ti 23 V 24 Cr 25 Mn 26 Fe 27 Co 28 Ni 29 Cu 30 Zn 31 Ga 32 Ge 33 As 34 Se 35 Br 36 Kr
37 Rb 38 Sr 39 Y 40 Zr 41 Nb 42 Mo 43 Tc 44 Ru 45 Rh 46 Pd 47 Ag 48 Cd 49 In 50 Sn 51 Sb 52 Te 53 I 54 Xe
55 Cs 56 Ba 71 Lu 72 Hf 73 Ta 74 W 75 Re 76 Os 77 Ir 78 Pt 79 Au 80 Hg 81 Tl 82 Pb 83 Bi 84 Po 85 At 86 Rn
87 Fr 88 Ra 103 Lr 104 Rf 105 Db 106 Sg 107 Bh 108 Hs 109 Mt 110 111 112 113 114 115 116 117 118
lanthanide series 57 La 58 Ce 59 Pr 60 Nd 61 Pm 62 Sm 63 Eu 64 Gd 65 Tb 66 Dy 67 Ho 68 Er 69 Tm 70 Yb
actinide series 89 Ac 90 Th 91 Pa 92 U 93 Np 94 Pu 95 Am 96 Cm 97 Bk 98 Cf 99 Es 100 Fm 101 Md 102 No

metal may be machined in a lathe, threaded, sawed, extruded, drawn into wire, pressed, and hammered into plates. *See* ELECTROMETALLURGY.

In air, calcium forms a thin film of oxide and nitride, which protects it from further attack. At elevated temperatures, it burns in air to form largely the nitride. The commercially produced metal reacts easily with water and acids, yielding hydrogen that contains noticeable amounts of ammonia and hydrocarbons as impurities.

The metal is employed as an alloying agent for aluminum-bearing metal, as an aid in removing bismuth from lead, and as a controller for graphitic carbon in cast iron. It is also used as a deoxidizer in the manufacture of many steels, as a reducing agent in preparation of such metals as chromium, thorium, zirconium, and uranium, and as a separating material for gaseous mixtures of nitrogen and argon.

Calcium oxide, CaO, is made by the thermal decomposition of carbonate minerals in tall kilns using a continuous-feed process. The oxide is used in high-intensity arc lights (limelights) because of its unusual spectral features and as an industrial dehydrating agent. The metallurgical industry makes wide use of the oxide during the reduction of ferrous alloys.

Calcium hydroxide, $Ca(OH)_2$, is used in many applications where hydroxide ion is needed. During the slaking process for producing calcium hydroxide, the volume of the slaked lime [$Ca(OH)_2$] produced expands to twice that of quicklime (CaO), and because of this, it can be used for the splitting of rock or wood. Slaked lime is an excellent absorbent for carbon dioxide to produce the very insoluble carbonate. *See* LIME (INDUSTRY).

Calcium silicide, CaSi, an electric-furnace product made from lime, silica, and a carbonaceous reducing agent, is useful as a steel deoxidizer. Calcium carbide, CaC_2, is produced by heating a mixture of lime and carbon to 5432°F (3000°C) in an electric furnace. The compound is an acetylide which yields acetylene upon hydrolysis. Acetylene is the starting material for a great number of chemicals important in the organic chemicals industry.

Pure calcium carbonate exists in two crystalline forms: calcite, the hexagonal form, which possesses the property of birefringence, and aragonite, the rhombohedral form. Naturally occurring carbonates are the most abundant of the calcium minerals. Iceland spar and calcite are essentially pure carbonate forms, whereas marble is a somewhat impure and much more compact variety which, because it may be given a high polish, is much in demand as a construction stone. Although calcium carbonate is quite insoluble in water, it has considerable solubility in water containing dissolved carbon dioxide, because in these solutions it dissolves to form the bicarbonate. This fact accounts for cave formation in which limestone deposits have been leached away by the acidic ground waters. *See* CARBONATE MINERALS.

Properties of calcium metal

Property	Value
Atomic number	20
Atomic weight	40.08
Isotopes (stable)	40, 42, 43, 44, 46, 48
Atomic volume, cm^3/g-atom	25.9
Crystal form	Face-centered cubic
Valence	2+
Ionic radius, nm	0.099
Electron configuration	2882
Boiling point, °C	1487(?)
Melting point, °C	810(?)
Density, g/cm^3 at 20°C	1.55
Latent heat of vaporization at boiling point, kilojoules/g-atom	399

The halides of calcium include the phosphorescent fluoride, which is the most widely distributed calcium compound and which has important applications in spectroscopy. Calcium chloride has in the anhydrous form important deliquescent properties which make it useful as an industrial drying agent and as a dust quieter on roads. Calcium chloride hypochlorite (bleaching powder) is produced industrially by passing chlorine into slaked lime, and has been used as a bleaching agent and a water purifier. *See* BLEACHING; CHLORINE.

Calcium sulfate dihydrate is the mineral gypsum. It constitutes the major portion of portland cement, and has been used to help reduce soil alkalinity. A hemihydrate of calcium sulfate, produced by heating gypsum at elevated temperatures, is sold under the commercial name plaster of paris. *See* PLASTER OF PARIS.

Calcium is an invariable constituent of all plants because it is essential for their growth. It is contained both as a structural constituent and as a physiological ion. Calcium is found in all animals in the soft tissues, in tissue fluid, and in the skeletal structures. The bones of vertebrates contain calcium as calcium fluoride, as calcium carbonate, and as calcium phosphate. *See* CALCIUM METABOLISM. [R.F.R.]

Calcium metabolism Calcium, an essential inorganic element needed by all animals and animal cells, subserves four types of functions in mammals: (1) as an integral part of bone mineral; (2) as an integral part of biomembranes; (3) as the link between excitation and contraction in all forms of muscle, and similarly as the link between excitation and secretion in both exocrine and endocrine glands; and (4) as a regulator, activator, or inhibitor of key enzymes. *See* BIOELECTRIC MODEL; BLOOD; BONE.

Because of the critical importance of calcium in many of these fundamental processes, its concentration in plasma is closely regulated by the organism. However, this regulation is complex because over 90% of the calcium in the body resides in bone. It is further complicated by an intimate relationship with phosphate metabolism. The two needs of ensuring a sufficient supply of both calcium and phosphates are met by the interrelated actions of the parathyroid hormone, thyrocalcitonin, and vitamin D.

A fall in blood calcium, hypocalcemia, leads to tetany, the spontaneous firing of nerves with resultant continuous, uncontrolled muscular contraction. A rise in blood calcium, hypercalcemia, leads to calcification of the kidney with eventual uremia. *See* KIDNEY; PARATHYROID HORMONE; PHOSPHATE METABOLISM; SKELETAL SYSTEM; THYROCALCITONIN; THYROID HORMONE; VITAMIN D. [H.Ras.]

Calculators Portable or desktop devices, primarily electronic, that are used to perform arithmetic and other numerical processing operations at the direction of an operator or a stored program. Early calculators were exclusively mechanical. Although some printing 10-key mechanical calculators are used in offices and small businesses, mechanical calculators have been almost completely replaced by quieter, more efficient, and more capable electronic units.

Inexpensive hand-held calculators employing large-scale integrated circuits as processing elements are in wide use. Inexpensive arithmetic units are available, while more expensive models provide a wide range of complex operations. The rapid evolution of the electronic calculator has been made possible by technological improvements in integrated circuits. An integrated circuit with an area of less than 0.5 $in.^2$ (3 cm^2) can hold all the circuit elements needed to implement the algorithms and the control and timing functions of a calculator. Forty or more algorithms may be held in memory, to be retrieved and used when requested by depression of a function key. A hand-held calculator may use a single integrated circuit to interpret key-switch closures, carry out the requested opera-

Programmable calculator combined with printer. (*Hewlett-Packard Co.*).

tion, and multiplex the individual digits of the result to the display. *See* INTEGRATED CIRCUITS.

Any calculator can be considered a computer in that it contains in nonvolatile memory a fixed set of algorithms for execution of the processing operations in its repertoire. These algorithms are available whenever power is applied to the unit. Programmable calculators are able to accept and act upon a higher level of programming that directs the units through sequences of processing steps without operator intervention other than operand entry. Programming may be done by the operator or, in advance, by the manufacturer, and storage of the program may be volatile or nonvolatile. Programming relieves the drudgery of repeated keystroke entry when a repetitive calculation must be made.

Programming by the operator on the unit to be used for the calculation is known as keystroke programming. The desired sequence of operations is entered into calculator memory by depression of keys as though a calculation were being made, but with the unit in program mode. Each keystroke enters an instruction into the semiconductor memory of the unit. When the unit is returned to the normal run mode, the program sequence can be initiated by the operator through depression of a special function key on the keyboard.

The semiconductor storage in a calculator is usually volatile—program information is lost when power is removed. Some units can accept magnetic tape cartridges or miniaturized magnetic cards, from which the volatile memory can be loaded and to which the memory contents can be written. The magnetic medium provides a means of saving and reentering a program written by the keystroke method. Manufacturers also offer preprogrammed tapes and cards for specialized calculations in a wide range of fields.

Several programmable calculators accept plug-in elements such as memory extender modules, read-only memory modules for specialized computations, magnetic card readers, and printers (see illustration). Programmer assistance in the form of prompting is provided in an alphanumeric display—display capable of showing letters as well as numbers. A calculator with these capabilities could be likened to a numerical processing computer with peripheral devices. As the range of calculator expansion elements increases and calculation capabilities increase, the differences between calculators and computers will become even less distinct. *See* COMPUTER; DIGITAL COMPUTER. [W.W.Mo.]

Calculus The branch of mathematics dealing with two fundamental operations, differentiation and integration, which are carried out on functions. The subject, as traditionally developed in college textbooks, is partly an elementary development of the purely theoretical aspects of these operations and their interrelation, partly a development of rules and formulas for applying calculus to the standard functions which arise in algebra and trigonometry (with exponentials and logarithms included), and partly a collection of applications to problems of geometry, physics, chemistry, engineering, economics, and perhaps a few other subjects.

The fundamental concept of differential calculus is that of the derivative of a function of one variable. The classical physical prototype of this concept is that of instantaneous velocity, which is the derivative of distance as a function of time. The derivative also has a highly significant geometrical realization which depends upon the graphical representation of a function in rectangular coordinates (x,y). If y is a differentiable function of x, perhaps as x increases from x_1 to x_2, the graph of the function is a continuous curve with exactly one y for each x, and at each point the curve has a tangent line which is not parallel to the y axis. If ϕ is the angle, measured counterclockwise, from the positive x direction to the tangent (Fig. 1), then

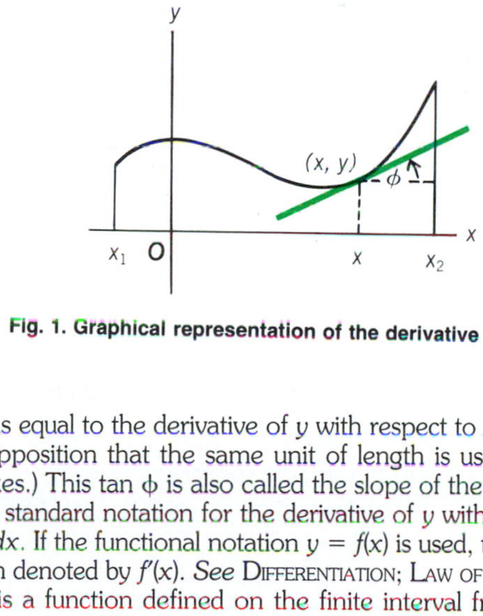

Fig. 1. Graphical representation of the derivative of $f(x)$.

$\tan \phi$ is equal to the derivative of y with respect to x. (This is on the supposition that the same unit of length is used along the two axes.) This $\tan \phi$ is also called the slope of the curve.

The standard notation for the derivative of y with respect to x is dy/dx. If the functional notation $y = f(x)$ is used, the derivative is often denoted by $f'(x)$. *See* DIFFERENTIATION; LAW OF THE MEAN.

If f is a function defined on the finite interval from x_1 to x_2 inclusive, the definite integral of f from x_1 to x_2, denoted by

$$\int_{x_1}^{x_2} (x)\, dx$$

is defined by applying to f a rather intricate process which entails the consideration of what are called approximating sums. When the function f is subjected to certain restrictions, this process culminates in the determination of a number as the limit of the approximating sums, and this number is called the definite integral of f from x_1 to x_2. The integral is not defined unless the approximating sums do converge to a well-defined limit. A sufficient condition that this be so is that the function f be continuous.

There is a geometrical representation of the process of defin-

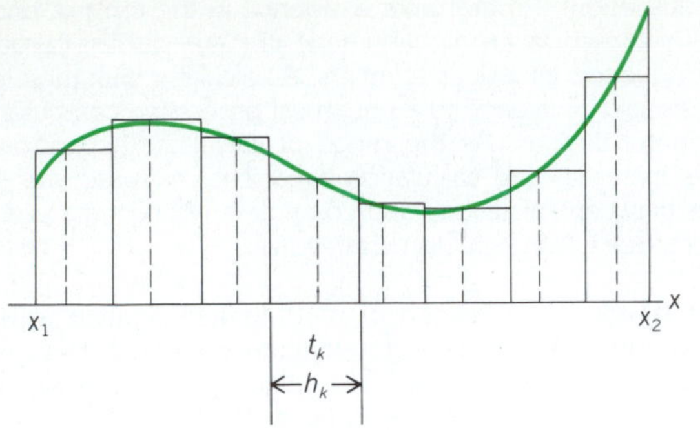

Fig. 2. The definite integral.

ing the definite integral, and it furnishes a plausible argument for the convergence of the approximating sums to a limit. Divide the interval from x_1 to x_2 into a finite number N of not necessarily equal parts. Let the lengths of these parts be h_1, h_2, . . . h_N and let t_k be the value of x in the kth part (Fig. 2). Then the expression

$$f(t_1)h_1 + f(t_2)h_2 + \ldots + f(t_N)h_N$$

is called an approximating sum. In Fig. 2, where the function is continuous and the function values are all positive, each term $f(t_k)h_k$ in the approximating sum is equal to the area of a certain shaded rectangle, and the whole sum is an approximation of the area between the graph of the function and x axis, from x_1 to x_2 inclusive. The limiting process is carried on by the increasing N and making the largest of the h_k's approach 0. It is then intuitively clear that the definite integral is the number which represents the exact area between the x axis and the graph. This geometrical interpretation of the integral is the basis of an important application of integral calculus, to the calculation of areas.

It would be tedious and difficult in practice to compute definite integrals by actually working out the limits of approximating sums. It is therefore fortunate that by purely mathematical reasoning it is possible to demonstrate a theorem which links derivatives and integrals and makes it possible, in many important instances, to compute definite integrals by an easier procedure. *See* INTEGRATION.

The two fundamental theorems of calculus are as follows:

1. For the calculation of

$$\int_{x_1}^{x_2} f(x)\,dx$$

find, if possible, a function F with continuous derivative F' such that $F'(x) = f(x)$ when $x_1 \leqq x \leqq x_2$. Then Eq. (1) can be

$$\int_{x_1}^{x_2} f(x)\,dx = F(x_2) - F(x_1) \tag{1}$$

written. This is one of the two central theorems.

2. Suppose f is continuous, and consider the function F defined by Eq. (2).

$$F(x) = \int_{x_1}^{x_2} f(t)\,dt \tag{2}$$

Then F has a derivative given by $F'(x) = f(x)$. [A.L.Ta.]

Calculus of vectors In its simplest form, a vector is a directed line segment. Physical quantities, such as velocity, acceleration, force, and displacement, are vector quantities, or simply vectors, because they can be represented by directed line segments. Vector analysis is a tool of the mathematical physicist, because many physical laws can be expressed in vector form.

Two vectors **a** and **b** are added according to the parallelogram law (Fig. 1). An equivalent definition is as follows: From

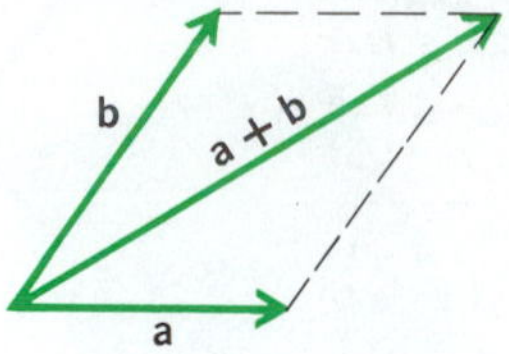

Fig. 1. Addition of two vectors.

the end point of **a**, a vector is constructed parallel to **b**, of the same magnitude and direction as **b**. The vector from the origin of **a** to the end point of **b** yields the vector sum $\mathbf{s} = \mathbf{a} + \mathbf{b}$ (Fig. 2). Any number of vectors can be added by this rule.

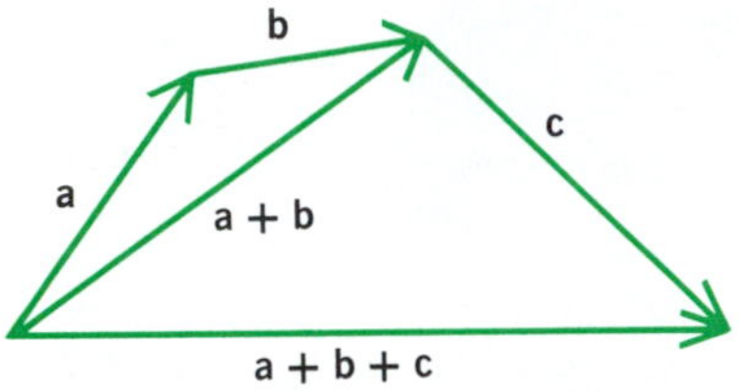

Fig. 2. Addition of three vectors.

Given a vector **a**, a class of vectors can be formed which are parallel to **a** but of different magnitudes. If x is a real number, the vector $x\mathbf{a}$ is defined to be parallel to **a** of magnitude $|x|$ times that of **a**. For $x < 0$, the two vectors **a** and $x\mathbf{a}$ have the same sense of direction, whereas for $x < 0$ the vector $x\mathbf{a}$ is in a reverse direction from that of **a**. The vector $-\mathbf{a}$ is the negative of the vector **a**, such that $\mathbf{a} + (-\mathbf{a}) = \mathbf{0}$, with $\mathbf{0}$ designated as the zero vector (a vector with zero magnitude). Subtraction of two vectors is defined by Eq. (1).

$$\mathbf{a} - \mathbf{b} = \mathbf{b} + (-\mathbf{b}) \tag{1}$$

The cartesian coordinate frame of analytic geometry is very useful for yielding a description of a vector (Fig. 3). The unit vectors **i**, **j**, **k** lie parallel to the positive x, y, and z axes, respectively. Any vector can be written as a linear combination

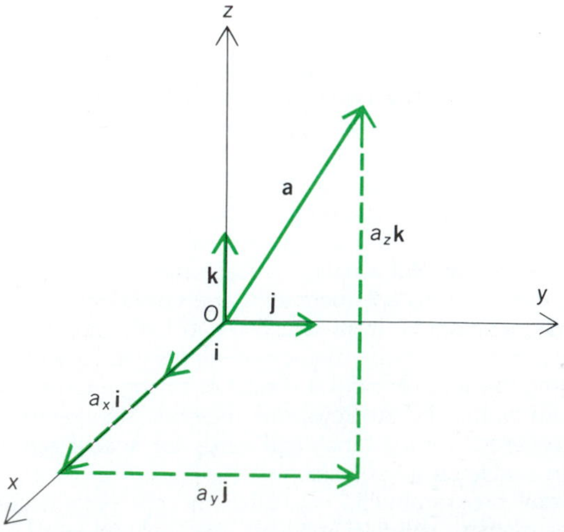

Fig. 3. Vectors in cartesian coordinate system.

of **i**, **j**, **k**. From Fig. 3, it is noted that Eq. (2) holds. Furthermore, the scalars a_x, a_y, a_z are simply the projections

$$\mathbf{a} = a_x\mathbf{i} + a_y\mathbf{j} + a_z\mathbf{k} \qquad (2)$$

of **a** on the x, y, and z axes, respectively, and are designated as the components of **a**. Thus, a_x is the x component of **a**, and so on. If the vector **b** is described by $\mathbf{b} = b_x\mathbf{i} + b_y\mathbf{j} + b_z\mathbf{d}$, then Eq. (3) holds.

$$\alpha\mathbf{a} + \beta\mathbf{b} = (\alpha a_x + \beta b_x)\mathbf{i} + (\alpha a_y + \beta Y_y)\mathbf{j} + (\alpha a_z + \beta b_z)\mathbf{k} \qquad (3)$$

From two vectors **a** and **b**, a scalar quantity is formed from the definition in Eq. (4), where θ is the angle between the two

$$\mathbf{a} \cdot \mathbf{b} = |\mathbf{a}| \cdot |\mathbf{b}| \cos\theta \qquad (4)$$

vectors when drawn from a common origin (Fig. 4).

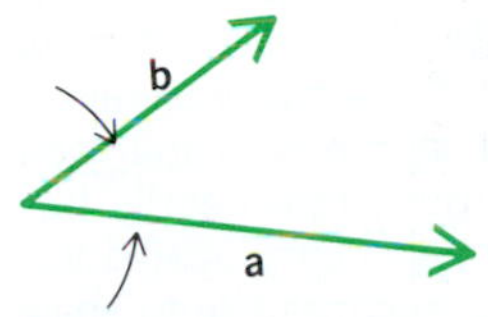

Fig. 4. Scalar product of two vectors.

In three-dimensional space, a vector can be formed from two vectors **a** and **b** in the following manner if they are nonparallel. Let **a** and **b** have a common origin defining a plane, and let **c** be that vector perpendicular to this plane of magnitude $|\mathbf{c}| = |\mathbf{a}|\ |\mathbf{b}| \sin\theta$. If **a** is rotated into **b** through the angle θ, a right-hand screw will advance in the direction of **c** (Fig. 5). Thus Eq. (5) can be written.

$$\mathbf{c} = \mathbf{a} \times \mathbf{b} = |\mathbf{a}|\ |\mathbf{b}| \sin\theta\ \mathbf{n} \qquad (5)$$

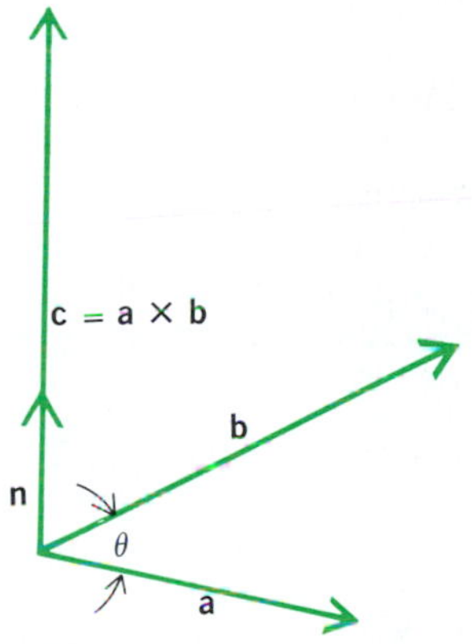

Fig. 5. Vector product of two vectors.

The distributive law can be shown to hold for the vector product so that Eq. (6) holds.

$$(\mathbf{a} + \mathbf{b}) \times (\mathbf{c} + \mathbf{d}) = \mathbf{a}\times\mathbf{c} + \mathbf{a}\times\mathbf{d} + \mathbf{b}\times\mathbf{c} + \mathbf{b}\times\mathbf{d} \qquad (6)$$

It follows from

$$\mathbf{i}\times\mathbf{i} = \mathbf{j}\times\mathbf{j} = \mathbf{k}\times\mathbf{k} = \mathbf{0}$$
$$\mathbf{i}\times\mathbf{j} = \mathbf{k},\ \mathbf{j}\times\mathbf{k} = \mathbf{i},\ \mathbf{k}\times\mathbf{i} = \mathbf{j}$$

that for

$$\mathbf{a} = a_x\mathbf{i} + a_y\mathbf{j} + a_z\mathbf{k} \qquad \mathbf{b} = b_x\mathbf{i} + b_y\mathbf{j} + b_z\mathbf{k}$$

Eq. (7) holds. The expression in Eq. (7) is to be expanded by the ordinary rules governing determinants.

$$\mathbf{a}\times\mathbf{b} = \begin{vmatrix} \mathbf{i} & \mathbf{j} & \mathbf{k} \\ a_x & a_y & a_z \\ b_x & b_y & b_z \end{vmatrix} \qquad (7)$$

It is easy to verify that a reflection of the space coordinates given by $x' = -x$, $y' = -y$, $z' = -z$, reverses the sign of the components of a vector. Under a space reflection, however, the components of the vector $\mathbf{a}\times\mathbf{b}$ do not change sign. This is seen from Eq. (7), for if a_x, a_y, a_z are replaced by $-a_x$, $-a_y$, $-a_z$, and if b_x, b_y, b_z are replaced by $-b_x$, $-b_y$, $-b_z$, the components of $\mathbf{a}\times\mathbf{b}$ remain invariant. Hence, $\mathbf{a}\times\mathbf{b}$ is not a true vector and therefore is given the title pseudovector. In electricity theory the magnetic field vector **B** is a pseudovector, whereas the electric field vector is a true vector provided the electric charge is a true scalar under a reflection of axes.

There are three differentiation processes that are of conceptual value in the study of vectors: the gradient of a scalar, the divergence of a vector, and the curl of a vector.

From the scalar $\phi(x,y,z)$, one can form the three partial derivatives $\partial\phi/\partial x$, $\partial\phi/\partial y$, $\partial\phi/\partial z$, from which the vector in Eq. (8)

$$\left\{\begin{matrix} \text{gradient of } \phi \\ \text{grad } \phi \\ \text{del } \phi \equiv \nabla\phi \end{matrix}\right\} = \frac{\partial\phi}{\partial x}\mathbf{i} + \frac{\partial\phi}{\partial y}\mathbf{j} + \frac{\partial\phi}{\partial z}\mathbf{k} \qquad (8)$$

can be formed. The vector, grad ϕ, has two important properties. Grad ϕ is a vector field normal to the surface $\phi(x,y,z) =$ constant at every point of the surface. Moreover, grad ϕ yields that unique direction such that ϕ increases at its greatest rate.

The del operator defined by Eq. (9) plays an important role

$$\nabla = \mathbf{i}\frac{\partial}{\partial x} + \mathbf{j}\frac{\partial}{\partial y} + \mathbf{k}\frac{\partial}{\partial z} \qquad (9)$$

in the development of the differential vector calculus. For the vector field of Eq. (10), the divergence of **v** is defined by Eq. (11). The divergence of a vector is a scalar.

$$\mathbf{v} = v_1(x,y,z)\,\mathbf{i} + v_2(x,y,z)\,\mathbf{j} + v_3(x,y,z)\,\mathbf{k} \qquad (10)$$

$$\text{div}\,\mathbf{v} = \nabla\cdot\mathbf{v} = \frac{\partial v_1}{\partial x} + \frac{\partial v_2}{\partial y} + \frac{\partial v_3}{\partial z} \qquad (11)$$

By use of the del operator one obtains quite formally Eq. (12). The vector $\nabla\times\mathbf{v}$ is called the curl of **v** (curl **v**). Under

$$\nabla\times\mathbf{v} = \begin{vmatrix} \mathbf{i} & \mathbf{j} & \mathbf{k} \\ \dfrac{\partial}{\partial x} & \dfrac{\partial}{\partial} & \dfrac{\partial}{\partial z} \\ v_1 & v_2 & v_3 \end{vmatrix}$$
$$= \left(\frac{\partial v_3}{\partial y} - \frac{\partial v_2}{\partial z}\right)\mathbf{i} + \left(\frac{\partial v_1}{\partial z} - \frac{\partial v_3}{\partial x}\right)\mathbf{j} + \left(\frac{\partial v_2}{\partial x} - \frac{\partial v_1}{\partial y}\right)\mathbf{k} \qquad (12)$$

a reflection of space coordinates, $\partial v_3/\partial y \rightarrow \partial(-v_3)/\partial(-y) = \partial v_3/\partial y$, and so on, so that the curl of **v** is a pseudovector.

If a closed surface is decomposed into a large number of small surfaces, a vector field normal to the surface can be constructed, each normal element being represented by $d\sigma$. The magnitude of $d\sigma$ is the area of the surface element dS, $d\sigma = \mathbf{N}\,dS$.

If **f** is a vector field defined at every point of the surface, then notation (13) represents the total flux of **f** through the

$$\iint_S \mathbf{f}\cdot d\sigma \qquad (13)$$

surface S. The elements $\mathbf{d}\sigma$ point outward from the interior of S.

The divergence theorem of Gauss states that Eq. (14) is

$$\iint_S \mathbf{f}\cdot d\sigma = \iiint_R (\nabla\cdot\mathbf{f})\,d\tau \qquad (14)$$

true, where R is the region enclosed by S, $d\tau$ a volume element of R. *See* GAUSS' THEOREM.

The line integral of a vector field is described as follows: Let Γ be a space curve, and let $\mathbf{t}$ be the unit vector field tangent to Γ at every point of Γ in progressing from A to B, the initial and end points of the trajectory Γ.

The scalar integral in notation (15) is called the line integral

$$\int_A^B (\mathbf{f} \cdot \mathbf{t})\, ds \qquad (15)$$

of $\mathbf{f}$ along Γ, with arc length s measured along Γ. If $\mathbf{f}$ is a force field, notation (15) represents the work performed by the force field if a unit test particle is taken from A to B along Γ.

The value of the integral of notation (15) will generally depend on the path from A to B. However, if $\mathbf{f} = \nabla\phi$, then

$$\int_A^B (\mathbf{f} \cdot \mathbf{t})\, ds = \int_A^B \nabla\phi \cdot d\mathbf{r} = \int_A^B d\phi = \phi(B) - \phi(A)$$

and the line integral is independent of the path of integration.

The theorem of Stokes states that Eq. (16) holds, where Γ is

$$\oint_\Gamma f \cdot dr = \iint_S (\nabla \times \mathbf{f}) \cdot d\sigma \qquad (16)$$

the boundary of the open surface S. *See* STOKES' THEOREM. [H.La.]

Caldera A large volcanic collapse depression, typically circular to slightly elongate in shape, the dimensions of which are many times greater than any included vent. Calderas range from a few miles to 37 mi (60 km) in diameter. A caldera may resemble a volcanic crater in form, but differs genetically in that it is a collapse rather than a constructional feature. The topographic depression resulting from collapse is commonly widened by slumping of the sides along concentric faults, so that the topographic crater wall lies outside the caldera wall. As originally defined, the term caldron referred to volcanic subsidence structures, and caldera referred only to the topographic depression formed at the surface by collapse. However, the term caldera is now common as a synonym for caldron, denoting all features of collapse, both topographic and structural. *See* PETROLOGY.

Calderas occur primarily in three different volcanic settings, each of which affects their shape and evolution: basaltic shield cones, stratovolcanoes, and volcanic centers consisting of preexisting clusters of volcanoes. These last calderas, associated with broad, large-volume andesitic to rhyolitic ignimbrite sheets, are generally the largest and most impressive, and are those generally denoted by the term. Calderas have been formed throughout much of the Earth's history, ranging in age from Precambrian (greater than 1.4 billion years old) to Holocene (for example, Krakatau in Indonesia, which erupted in 1883). *See* RHYOLITE; VOLCANO.

In addition to Earth, large calderas occur on Mars, Venus, and Jupiter's moon Io. The presence of calderas on four solar system bodies indicates that the underlying mechanisms of shallow intrusion and caldera collapse are basic processes in planetary geology. *See* MAGMA.

Collapse occurs because of withdrawal of magma from an underlying chamber some 2.4–3.6 mi (4–6 km) beneath the surface, resulting in foundering of the roof into the chamber. Withdrawal of magma may occur either by relatively passive eruption of lavas, as in the case of calderas formed on basaltic shield cones, or by catastrophic eruption of pyroclastic material, as accompanies formation of the largest calderas.

Caldera-forming eruptions probably last only a few hours or days. Eruption of pyroclastic material begins as gases (predominantly water) that are dissolved in the magma come out of solution at shallow depths. Magma is explosively fragmented into particles ranging in size from micrometers to meters. An eruption column develops, rising several miles into the atmosphere. This first and most explosive phase of the eruption, known as the Plinian phase, covers the area around the vent with pumice. Caldera subsidence occurs during eruption. As caldera subsidence proceeds and eruption becomes less explosive, the Plinian eruption column collapses. This collapse produces hot, ground-hugging pyroclastic flows that can travel as far as 93 mi (150 km) outward from the vent at speeds of 330 ft/s (100 m/s). Successive collapses of the column produce multiple flow units with an aggregate thickness that may be several hundreds of feet thick near the caldera.

The floors of many of the largest calderas (typically those with diameters exceeding 6 mi or 10 km) have been domed upward, resulting in a central massif or resurgent dome. Resurgence results from the continued or renewed buoyant rise of magma after collapse.

Calderas typically contain or are associated with extensive hydrothermal systems, because of two factors: (1) the shallow magma chambers that underlie them provide a readily available source of heat; and (2) the floors of calderas may be extensively fractured, which, along with the main ring faults, allows meteoric water to penetrate deeply into the crust beneath calderas. Hydrothermal activity related to a caldera system can occur any time after magmas rise to shallow crustal levels, but it is dominant late in caldera evolution.

Many metals, including such base and precious metals as molybdenum, copper, lead, zinc, silver, gold, mercury, uranium, tungsten, and antimony, are mobile in hydrothermal circulation systems driven by the shallow intrusions which underlie and give rise to large calderas. Many economically important ore deposits in the western United States lie within calderas. *See* ORE AND MINERAL DEPOSITS. [W.S.Ba.]

Calendar A list, usually in the form of a table, showing the correspondence between days of the week and days of the month; also, a list of special observances with their dates, especially those that fall on different dates in different years, as a church calendar; also, a set of rules that serves to attach to any day a specific number or name or combination of number and name.

The calendars that have been used at various times and places are numerous and diverse. Most of them represent attempts to divide the synodic month or the tropical year into numbered days. According to whether the emphasis is on month or year, calendars are called lunar or solar.

The calendar used for civil purposes throughout the world, known in Western countries as the Gregorian calendar, was established by Pope Gregory XIII, who decreed that the day following Thursday, October 4, 1582, should be Friday, October 15, 1582, and that thereafter centennial years (1600, 1700, and so on) should be leap years only when divisible by 400 (1600, 2000, and so on), other years being leap years when divisible by four, as previously. The effect of the reform was to restore the vernal equinox (beginning of spring, which governs the observance of Easter) to March 21, and to reduce the number of days in 400 calendar years by 3, making the average number of days in the calendar year 365.2425, instead of 365.25 as formerly, and bringing it into closer agreement with the tropical year, which has 365.24220 days.

The Gregorian calendar is not perfectly adjusted to the tropical year, the error accumulating to 1 day after about 3300 years. Numerous proposals for calendar reform have been made with the object of reducing the error, and there have also been many proposals for altering the succession of days for the purpose of bringing greater regularity into the calendar. [G.M.C.]

Calibration The process of determining, by measurement or by comparison with a standard, the correct value of each scale reading on a meter or the correct value of each setting of a control knob. In a radio receiver, for example, calibration would mean adjusting the tuned circuits in the oscillator to make the readings of the tuning dial correspond exactly to the frequencies of the incoming signals.

With measuring instruments, calibration generally involves adjusting the values of internal components so that the indication is correct at a specified number of points on the indicating scale and approximately correct between these points. With highly accurate instruments, calibration involves the preparation of a graph or table that gives the exact value corresponding to each line on the indicating scale. *See* INSTRUMENTATION. [J.Mar.]

Caliche A term used to describe the crude soda niter of the Chilean nirate deposits. It is a variable admixture of salts, including, in addition to the soda niter, bloedite, anhydrite, gypsum, polyhalite, halite, glauberite, darapskite, and minor amounts of various iodate, chromate, and borate minerals. *See* BORATE MINERALS; SODA NITER.

The term caliche has several other meanings. It is commonly used in Mexico and the southwestern United States to designate gravel, sand, or desert debris cemented by porous, calcium carbonate. Other uses of the term include a thin layer of clayey soil capping auriferous veins; whitish clay in the selvage of veins; a white mineral vein recently discovered; and in placer mining, a bank composed of clay, sand, and gravel. [G.Sw.]

Caliciales An order of the Ascolichenes. This order is characterized by an unusual apothecium. The hymenial layer originates normally, but by the time the spores are mature, the asci and paraphyses have partially disintegrated into a mass of spores and hymenial tissues known as a mazaedium. Two families, Caliciaceae and Roesleriaceae, are so close that some species in either one may or may not lichenize symbiotic algae. The family Cypheliaceae is more typically crustose, with sessile apothecia and a more fully developed thallus. The Sphaerophoraceae are fruticose but the thallus is solid. The apothecia are open or enclosed in a spherical chamber at the tips of branches. The largest genus, *Sphaerophorus*, is widespread in boreal zones and mountains of both hemispheres. [M.E.H.]

Californium A chemical element, Cf, atomic number 98, the ninth member of the actinide series of elements. Its discovery and production have been based upon artificial nuclear transmutation of radioactive isotopes of lighter elements. All isotopes of californium are radioactive, with half-lives ranging from a minute to about 1000 years. Because of its nuclear instability, californium does not exist in the Earth's crust. *See* ACTINIDE ELEMENTS; BERKELIUM; RADIOACTIVITY.

The chemical properties are similar to those observed for other 3+ actinide elements: a water-soluble nitrate, sulfate, chloride, and perchlorate. Californium is precipitated as the fluoride, oxalate, or hydroxide. Ion-exchange chromatography can be used for the isolation and identification of californium in the presence of other actinide elements. Californium metal is quite volatile and can be distilled at temperatures of the order of 1100–1200°C (2010–2190°F). It is chemically reactive and appears to exist in three different crystalline modifications between room temperature and its melting point, 900°C (1600°F).

The most easily produced isotope for many purposes is ^{252}Cf, which is obtained in gram quantities in nuclear reactors and has a half-life of 2.6 years. It decays partially by spontaneous fission, and has been very useful for the study of fission. It has also had an important influence on the development of counters and electronic systems with applications not only in nuclear physics but in medical research as well. *See* TRANSURANIUM ELEMENTS. [G.T.S.]

1	2	3	4	5	6	7	8	9	10	11	12	13	14	15	16	17	18
1 H																	2 He
3 Li	4 Be											5 B	6 C	7 N	8 O	9 F	10 Ne
11 Na	12 Mg											13 Al	14 Si	15 P	16 S	17 Cl	18 Ar
19 K	20 Ca	21 Sc	22 Ti	23 V	24 Cr	25 Mn	26 Fe	27 Co	28 Ni	29 Cu	30 Zn	31 Ga	32 Ge	33 As	34 Se	35 Br	36 Kr
37 Rb	38 Sr	39 Y	40 Zr	41 Nb	42 Mo	43 Tc	44 Ru	45 Rh	46 Pd	47 Ag	48 Cd	49 In	50 Sn	51 Sb	52 Te	53 I	54 Xe
55 Cs	56 Ba	71 Lu	72 Hf	73 Ta	74 W	75 Re	76 Os	77 Ir	78 Pt	79 Au	80 Hg	81 Tl	82 Pb	83 Bi	84 Po	85 At	86 Rn
87 Fr	88 Ra	103 Lr	104 Rf	105 Db	106 Sg	107 Bh	108 Hs	109 Mt	110	*111*	*112*	*113*	*114*	*115*	*116*	*117*	*118*

lanthanide series	57 La	58 Ce	59 Pr	60 Nd	61 Pm	62 Sm	63 Eu	64 Gd	65 Tb	66 Dy	67 Ho	68 Er	69 Tm	70 Yb
actinide series	89 Ac	90 Th	91 Pa	92 U	93 Np	94 Pu	95 Am	96 Cm	97 Bk	98 Cf	99 Es	100 Fm	101 Md	102 No

Caliper An instrument with two legs used for measuring linear dimensions. Calipers may be fixed, adjustable, or movable. Fixed calipers are used in routine inspection of standard products; adjustable calipers are used similarly but can be reset to slightly different dimensions if necessary. Movable calipers can be set to match the distance being measured. The legs may pivot about a rivet or screw in a firm-joint pair of calipers; they may pivot about a pin, being held against the pin by a spring and set in position by a knurled nut on a threaded rod; or the legs may slide either directly (caliper rule) or along a screw (micrometer caliper) relative to each other. *See* MICROMETER.

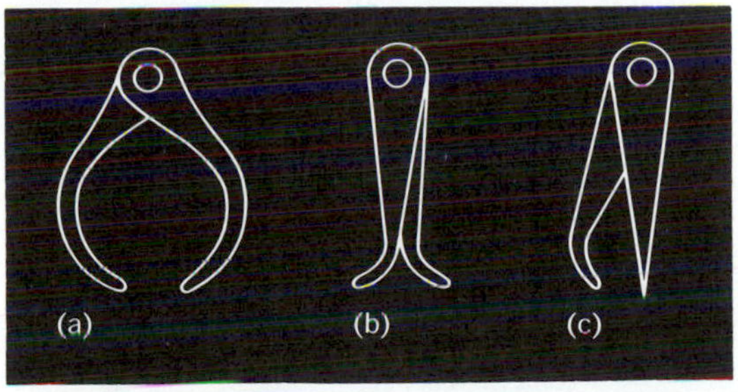

Some typical machinist's calipers. (*a*) Outside. (*b*) Inside. (*c*) Hermaphrodite. (*After R. J. Sweeney, Measurement Techniques in Mechanical Engineering, John Wiley and Sons, Inc., 1953*)

The legs may be shaped to facilitate measuring outside dimensions, inside dimensions, surface dimensions as between points on a plate, or from a surface into a hole as in a keyway (see illustration). *See* GAGE. [F.H.R.]

Callitrichales An order of flowering plants, division Magnoliophyta (Angiospermae), in the subclass Asteridae of the class Magnoliopsida (dicotyledons). The order consists of three small families with about 50 species, most of which are aquatics or small herbs of wet places, and have much reduced vascular systems. The flowers are small and solitary in the axils of leaves or bracts. The perianth is nearly or completely absent. The pistil consists of two carpels united to form a compound, unilocular or four-chambered ovary, or sometimes the pistil appears to be of a single carpel. The Callitrichales are placed in the Asteridae largely on the basis of their embryology and phytochemistry. The ovules are anatropous and tenuinucellular and have a single integument, and the plants generally produce iridoid substances. *See* ASTERIDAE; MAGNOLIOPHYTA; MAGNOLIOPSIDA; PLANT KINGDOM. [T.M.Ba.]

Calobryales An order of liverworts. They are characterized by prostrate, simple or branched, leafless stems and erect, leafy branches of a radial organization. The order consists of a single genus, *Calobryum*, and 12 species, most of them occupying restricted ranges in apparently relic areas indicative of an ancient origin and dispersal. The order is considered primitive in comparison with the Jungermanniales, in which the leafy axis tends to be prostrate and the underleaves reduced. The stems are thick and fleshy, with no differentiated outer layers. Rhizoids are lacking. The leaves may be small or lacking below, larger and more crowded above. They are three-ranked, with those of one rank sometimes more or less reduced. They are broad, unlobed, and entire. *See* BRYOPHYTA; JUNGERMANNIALES; JUNGERMANNIIDAE. [H.Cr.]

Calomel Mercury(I) chloride, Hg_2Cl_2, a covalent compound which is insoluble in water. The substance sublimes when heated. The formula weight is 472.086 and the specific gravity is 7.16 at 20°C (68°F). The material is a white, impalpable powder consisting of fine tetragonal crystals.

Calomel is used in preparing insecticides and medicines. It is well known in the laboratory as the constituent of the calomel reference electrodes which are commonly used in conjunction with a glass electrode to measure pH. *See* CALOMEL ELECTRODE; MERCURY (ELEMENT). [E.E.W.]

Calomel electrode An electrode used widely as a reference electrode of known potential, for example, in electrometric measurements of acidity and alkalinity, in corrosion studies, and in the measurements of the potentials of other electrodes (see illustration). The calomel electrode consists of

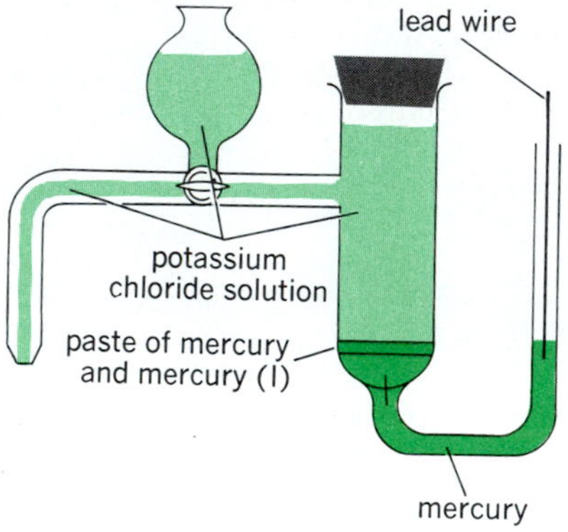

Diagram of calomel electrode.

mercury covered with a layer of paste of mercury(I) chloride in an aqueous solution of potassium chloride. A small amount of mercury is dispersed in the paste of mercury(I) chloride. It is made in three well-known types: the saturated, the normal, and the one-tenth normal. These designations refer to the concentrations of the potassium chloride solutions used in the preparation. The first type is more widely used than the other two because of the ease with which it may be assembled; no analysis of the concentration is required. Crystals of potassium chloride are placed above the layer of mercury(I) chloride in the saturated type. The calomel electrode is commonly referred to as a half-cell because it includes an aqueous phase of potassium chloride as one arm of the cell to be used in the electric circuit. Frequently, calomel electrodes are used in assemblies such as pH meters. *See* ELECTRODE POTENTIAL; HYDROGEN ION; SILVER CHLORIDE ELECTRODE. [W.J.H.]

Calorimetry The measurement of the quantity of heat involved in various processes, such as chemical reactions, changes of state, and formation of solutions, or in the determination of the heat capacities of substances. The fundamental unit for these heat measurements is the joule, but many workers prefer to use the calorie, defined as 4.184 absolute joules.

A calorimeter is an apparatus for measuring these heat quantities. Since the problem involves various kinds of processes and experimental work over a range of temperatures, calorimeters have been developed in great variety.

Nonisothermal calorimeters yield evaluations of the heat quantities by the temperature changes they produce. Many of these use a liquid as a heat reservoir, since uniformity of temperature thus can be realized easily by proper stirring, and the temperature change for the system can be evaluated readily. In contrast, the aneroid type depends on the use of a metal of high thermal conductivity, such as copper, to promote rapid equalization of temperature.

Various isothermal calorimeters also have been developed in which the heat reservoir contains a liquid in equilibrium with its crystalline solid at the melting point or with its vapor at the boiling point. While the isothermal calorimeter is simple, and can yield excellent data, its disadvantage is that it can be operated only at the equilibrium temperature of the two-phase system involved.

All these types of calorimetric apparatus contain (1) the calorimeter proper, or heat reservoir for measuring the amount of heat in the process under study, and (2) a jacket which can be used to control thermal exchange between the calorimeter and its environment. In many cases this jacket is a hollow vessel which may be filled with a liquid, such as water, that can be stirred and accurately regulated in temperature. In the aneroid type the jacket usually is a hollow, cylindrical block of metal with a top and bottom, electrical heating coils, and a suitable thermometer system.

The measurement of temperature is an important factor of calorimetry which has been greatly refined since 1920. Today, resistance thermometers and thermocouples are the instruments most used in high-grade precision measurements. *See* BOMB CALORIMETER; CHEMICAL THERMODYNAMICS; TEMPERATURE MEASUREMENT; THERMOMETER. [B.J.Z.]

Calycerales An order of flowering plants, division Magnoliophyta (Angiospermae), in the subclass Asteridae of the class Magnoliopsida (dicotyledons). The order consists of a single family with about 60 species native to tropical America. The plants are herbs with alternate, simple leaves that do not have stipules. The flowers are borne in involucrate heads with centripetal flowering sequence. The calyx is reduced to small lobes or teeth, and the corolla consists of (4)5(6) fused lobes and is regular or somewhat irregular. The stamens are attached near the summit of the corolla tube, and the filaments are more or less connate. The pistil consists of two united carpels, forming a compound, inferior ovary with a single, pendulous ovule. The order Calycerales is sometimes included within the Dipsacales, and the order has attracted attention because of the overall resemblance of the inflorescence to that of the Asteraceae. *See* ASTERALES; ASTERIDAE; DIPSACALES; MAGNOLIOPHYTA; MAGNOLIOPSIDA; PLANT KINGDOM. [T.M.Ba.]

Calymmatobacterium A genus of encapsulated, gram-negative, facultatively anaerobic bacteria, 1–2 micrometers in length and pleomorphic. It was discovered by C. Donovan in 1905 as the etiologic agent of the tropical granuloma inguinale (often referred to as donovaniosis), a predominantly venereally transmitted human disease. The genus is also referred to as *Donovania*. The only species of the genus known is *Calymmatobacterium granulomatis*. This organism enters the host tissues through epithelial lesions and develops intracellularly, mainly in vacuolic compartments of histiocytes and several types of white blood cells, thereby inducing cell lysis. It may be propagated in the yolk sac of chick embryos, but cultivation in artificial media is extremely difficult. Therefore, *C. granulomatis* has been only poorly characterized, and clinical reports are in general not well documented. *See* GRANULOMA INGUINALE. [W.Ma.]

Cam mechanism A mechanical linkage whose purpose is to produce, by means of a contoured cam surface, a prescribed motion of the output link of the linkage, called the follower. Cam and follower are a higher pair. *See* LINKAGE (MECHANISM).

A familiar application of a cam mechanism is in the opening and closing of valves in an automotive engine. The cam rotates with the cam shaft, usually at constant angular velocity, while the follower moves up and down as controlled by the cam surface. A cam is sometimes made in the form of a translating cam. Other cam mechanisms, employed in elementary mechanical analog computers, are simple memory devices, in which the position of the cam (input) determines the position of the follower (output or readout).

Although many requisite motions in machinery are accomplished by use of pin-jointed mechanisms, such as four-bar linkages, a cam mechanism frequently is the only practical solution to the problem of converting the available input, usually rotating or reciprocating, to a desired output, which may be an exceedingly complex motion. No other mechanism is as versatile and as straightforward in design. However, a cam may be difficult and costly to manufacture, and it is often noisy and susceptible to wear, fatigue, and vibration.

Cams are used in many machines. They are numerous in automatic packaging, shoemaking, typesetting machines, and the like, but are often found as well in machine tools, reciprocating engines, and compressors. They are occasionally used in rotating machinery.

Cams are classified as translating, disk, plate, cylindrical, or drum (see illustration). The link having the contoured surface that prescribes the motion of the follower is called the cam. Cams are usually made of steel, often hardened to resist wear and, for high-speed application, precisely ground.

The output link, which is maintained in contact with the cam surface, is the follower. Followers are classified by their shape as roller, flat face, and spherical face. Followers are also described by the nature of their constraints, for example, radial, in which motion is reciprocating along a radius from the cam's axis of rotation; offset, in which motion is reciprocating along a line that does not intersect the axis of rotation (illustration *b*); and oscillating, or pivoted (illustration *a*). Three-dimensional cam-and-follower systems are coming into more frequent use, where the follower may travel over a lumpy surface. [D.P.Ad.]

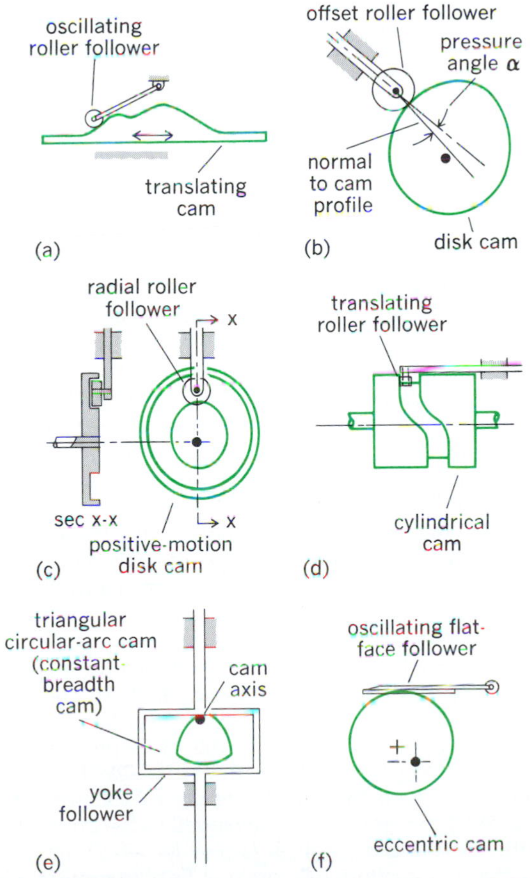

Classification of cams. (*a*) **Translating.** (*b*) **Disk.** (*c*) **Positive motion.** (*d*) **Cylindrical.** (*e*) **With yoke follower.** (*f*) **With flat-face follower.**

Cambrian An interval of time in Earth history (Cambrian Period) and its rock record (Cambrian System). The Cambrian Period spanned about 100,000,000 years and began with the first appearance of marine animals with mineralized (calcium carbonate, calcium phosphate) shells. The Cambrian System includes many different kinds of marine sandstones, shales, limestones, dolomites, and volcanics. There is very little provable record of nonmarine Cambrian environments. Because of the difficulty of finding rocks that can be dated radiometrically in association with rocks that can be dated empirically by fossils, the age in years of most Cambrian deposits is only approximate. The best present estimates suggest that Cambrian time began about 570,000,000 years ago and ended about 475,000,000 years ago. It is one of the longest Phanerozoic periods. *See* PALEOZOIC.

For most practical purposes, rocks of Cambrian age are recognized by their content of distinctive fossils. On the basis of the successive changes in the evolutionary record of Cambrian life that have been worked out during the past century, the Cambrian System has been divided globally into Lower, Middle, and Upper series, each of which has been further divided on each continent into stages, each stage consisting of several zones. Precise intercontinental correlation of series and stage boundaries and of zones is still difficult.

The record preserved in rocks indicates that essentially all Cambrian plants and animals lived in the sea. The few places where terrestrial sediments have been preserved suggest that the land was barren of major plant life, and there are no known records of Cambrian insects or of terrestrial vertebrate animals of any kind. The most abundant remains of organisms in Cambrian rocks are trilobites. They are present in almost every fossiliferous Cambrian deposit and are the principal tools used to describe divisions of Cambrian time and to correlate Cambrian rocks. The next most abundant Cambrian fossils are

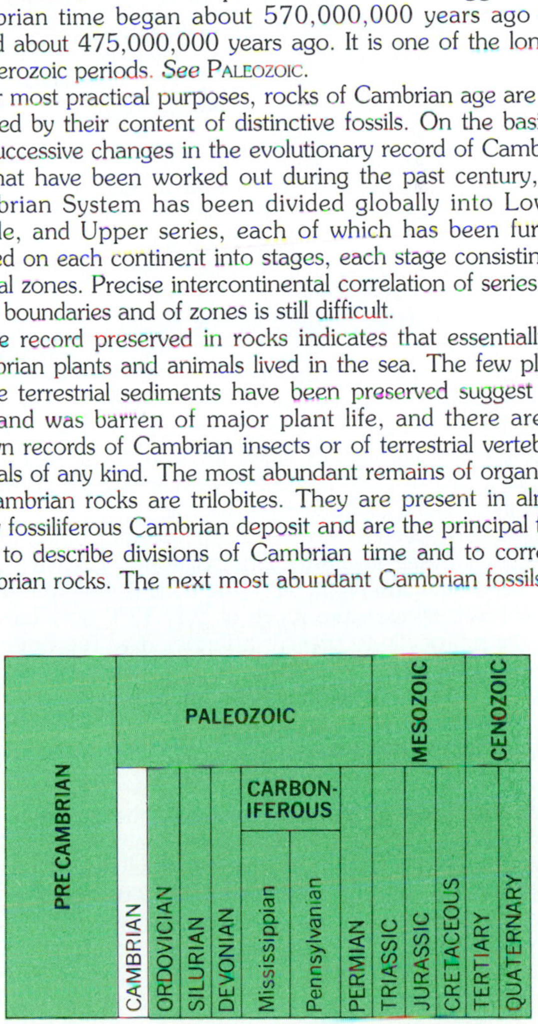

brachiopods. These bivalved animals lived on the sediment surface or on the surfaces of other organisms. Limestones of Early Cambrian age may contain large reeflike structures formed by an association of algae and an extinct phylum of invertebrates called Archaeocyatha. The Cambrian record of mollusks and echinoderms is characterized by many strange-looking forms. Representatives of these phyla, such as cephalopods, clams, and true crinoids, are rare in Cambrian rocks. Snails, however, are found throughout the Cambrian System.

The Cambrian world can be resolved into at least four major continents. These were (1) North America, minus a narrow belt along the eastern coast from eastern Newfoundland to northern Georgia which is most closely related to Cambrian Gondwana; (2) Baltica, consisting of present-day northern Europe north of France and west of the Ural Mountains, but excluding most of Scotland and northern Ireland, which are remains of marginal Cambrian North America; (3) Gondwana, a giant continent whose present-day fragments are Africa, South America, India, Australia and Antarctica, parts of southern Europe, the Middle East, and southeast Asia; (4) Siberia, including much of the northeastern quarter of Asia. Unfortunately, there is not enough reliable information to accurately locate these continents relative to one another on the Cambrian globe. *See* CONTINENT; PLATE TECTONICS. [A.R.P.]

Camel The name given to two species of mammals which are members of the family Camelidae in the order Artiodactyla. These are the Bactrian camel (*Camelus bactrianus*) and the Arabian or dromedary camel (*C. dromedarius*). Both species are domesticated, but a few wild herds of Bactrian camels are still in existence in the Gobi desert. The legs of these animals are long and slender and terminate in two toes. The neck and head are elongate, and there is a cleft upper lip. The period of gestation is about 1 year and the female breeds every second year, producing one young (colt).

The Bactrian camel is stronger and more heavily built than the dromedary and is more suitable as a pack animal. There are two humps of fatty tissue, one over the shoulders and the other atop the hindquarters. This animal is economically important as it provides milk, meat, and leather for the nomads in central Asia.

The Arabian camel is taller than the Bactrian and has a single hump of fatty tissue, which can be used as a food reserve. There are two varieties of this species found in the desert. One is the baggage camel, used as a beast of burden. The other type is the more slightly built racing camel. The species is well-suited to desert life with its broad feet adapted to walking on sand, its ability to close its nostrils completely, and its double row of interlocking eyelashes.

These two species have a most important physiological adaptation in their ability to conserve water. Camels do not store water but conserve it, since the body is well insulated by fur and has a temperature range of over 12°F (7°C) before it perspires sufficiently to prevent a further rise. The camel can lose over 40% of its body water without fear of dehydration. However, although able to survive for long periods without water, it may drink as much as 15 gal (57 liters) when water is available. *See* ARTIODACTYLA. [C.B.C.]

Camel's hair A fine hair known to the American consumer chiefly in the form of high-quality coat fabrics. This textile fiber is obtained from the two-humped Bactrian camel. Camel's-hair fabrics are ideal for comfort, particularly when used for overcoating, as they are especially warm and light in weight. Camel's hair is characterized by strength, luster, and smoothness. The best quality is expensive. It is often mixed with wool to improve the quality of the wool fabric. The price of such a mixed cloth is much less than that of a 100% camel's-hair fabric. *See* CAMEL; NATURAL FIBER; WOOL. [M.D.P.]

Cameo A type of carved gemstone in which the background is cut away to leave the subject in relief. Often cameos are cut from stones in which the coloring is layered, resulting in a figure of one color and a background of another. The term cameo, when used without qualification, is usually reserved for those cut from a gem mineral, although they are known also as stone cameos, The commonly encountered cameo cut from shell is properly called a shell cameo.

Most cameos are cut from onyx or agate, but many other varieties of quartz, such as tiger's-eye, bloodstone, sard, carnelian, and amethyst, are used; other materials used include beryl, malachite, hematite, labradorite, and moonstone. *See* GEM; INTAGLIO (GEMOLOGY). [R.T.L.]

Camera The basic instrument of photography: in its simplest form, a light-tight box with a lens at one end, which forms an image of the subject on light-sensitive film at the other end. A pinhole may be substituted for the lens, but forms an unsharp image. Nearly all modern cameras contain a glass or plastic lens to focus a sharp image of the subject onto the film; a shutter to control the length of time the film is exposed, normally a fraction of a second; a viewfinder, to show the subject being photographed; and a mechanism to hold the film and change film between exposures. Many cameras include a lens aperture control, to regulate image brightness on the film. Advanced cameras may accept a variety of lenses for different photographic effects. Many accessories are commonly used: flash units, exposure meters to measure light, filters to control color rendition, tripods, and close-up focusing attachments.

Cameras designed for still photography include the following types: box and other simple cameras, 35-mm, reflex, view, instant picture, panoramic, underwater, aerial, stereo, medical, and document-copying or microfilm. Some of these types overlap. There are also motion picture cameras, both conventional and high-speed, and television cameras—which are similar to photographic cameras but which use a light-sensitive electronic device instead of film. *See* CINEMATOGRAPHY; PHOTOGRAPHY.

Box and simple cameras usually have no adjustments, making them inexpensive to manufacture and simple to use. The lens focus is fixed, so that reasonably sharp pictures can be taken at any distance beyond a few feet. Lens aperture and shutter speed are factory-set to give correct exposure outdoors in sunshine and indoors with flash.

The 35-mm cameras take perforated film 35 mm wide (about $1\frac{3}{8}$ in.), in cassettes holding 20 or 36 exposures. The normal image size is 24 × 36 mm (1 × $1\frac{1}{2}$ in.), though some cameras make images 18 × 24 mm ($\frac{2}{3}$ × 1 in.) or 24 mm (1 in.) square. The 35-mm cameras can be very compact, and range from simple, inexpensive models to complex camera systems with a great variety of lenses, attachments, and accessories. The two principal kinds of 35-mm cameras are the rangefinder type and the reflex type.

In reflex cameras the light rays that pass through the camera lens are reflected from a mirror onto a horizontal ground-glass screen, where they form an image (illustration *a*). The mirror is at a precise 45° angle, and the distance from mirror to film is exactly equal to the distance from mirror to ground glass, so the image on the ground glass is the same size, focus, and so on, as the image on the film. Since the mirror prevents light from reaching the film, it must be hinged to flip out of the way when the exposure is made. It is also necessary to have the lens shutter and aperture fully open at first, to provide a bright image for viewing and focusing; next, they must close, to shield the film, when the mirror flips up; then they reopen for the set

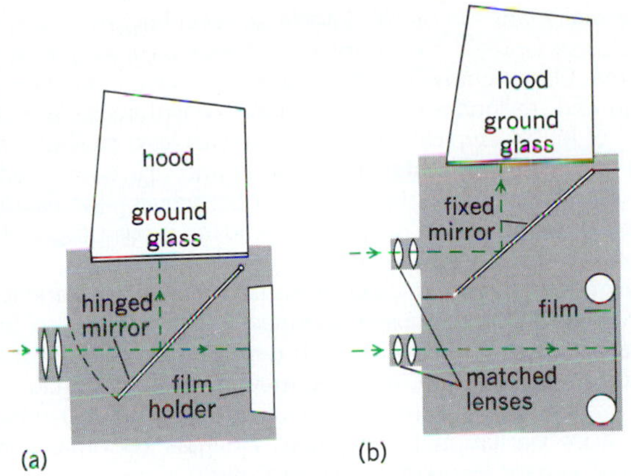

Diagrams of (*a*) single-lens and (*b*) twin-lens reflex cameras.

exposure. Alternatively, cameras can be designed with two identical lenses (illustration *b*). The bottom lens actually takes the picture: the top one is used only as a viewfinder, together with a fixed mirror and a ground glass. The two lenses are coupled so that focusing the image on the ground glass also makes it sharp on the film.

Originally, view cameras were the large cameras used for taking landscapes or views. In present usage the name describes cameras with a range of camera movements: swinging and shifting the front of the camera relative to the back. Other cameras always keep the front parallel to the back, and the center of the lens always aligned with the center of the film. But for special control of perspective and focus, swinging and shifting the camera front and lens independently from the back, and from the film, makes possible photographs which could not be taken any other way. To permit these movements, the view camera body is a flexible bellows, attached to front and rear panels which can swivel on a rigid base.

Polaroid-Land cameras, originally introduced in 1948, were the first cameras to produce a finished print directly. Jellylike developing chemicals, stored in pods attached to the film, quickly process each picture as it is removed from the camera. The photographer can immediately evaluate the print, make any desired changes, and take another.

Panoramic or wide-field cameras are useful for photography of groups of people or in confined spaces, and for taking ultra-wide-angle views. One type uses a wide-angle lens and an elongated film format. Another uses a swinging lens design; some swinging-lens cameras can even photograph a full 360° view.

Underwater cameras follow two general designs. Some are specialized models designed specifically for underwater operation; others are conventional cameras enclosed in waterproof, pressure-resistant housings, with flexible linkages to the camera controls.

Cameras for aerial photography may take photographs as large as 9 × 9 in. (22.5 × 22.5 cm), for maximum sharpness and detail. The pictures overlap by a set amount, so that two photographs can be studied together in a stereo viewer, which gives them the appearance of three dimensions and permits mapping the height of terrain and buildings. Several cameras photographing simultaneously, with special films and color filters, can make spectral analysis photographs to reveal pollution, mineral deposits, and the condition of crops and plant life. Multispectral satellite cameras have provided new information about the Earth's resources. *See* AERIAL PHOTOGRAPH; APPLICATIONS SATELLITES; REMOTE SENSING.

Stereo cameras use two lenses to make two photographs side by side. When properly viewed, the photographs can recreate the feeling of three-dimensional depth in the original scene. Stereo cameras have had periods of popularity in amateur photography and are also used in certain areas of technical and scientific photography.

Special medical cameras can actually photograph inside the body, by means of instruments inserted into the body. A flexible bundle of fiber optics projects light down the probe and relays the image back to the camera. [J.P.O'N.]

Camerata An extinct subclass of stalked Crinoidea comprising about 210 Paleozoic genera, ranging from the Lower Ordovician to the upper Permian. The calyx was composed of a rigid boxlike theca of numerous small, polygonal plates. Together with some of the lower brachial ossicles and some other small interradial and interbrachial plates, these plates were all firmly sutured together (see illustration). The uniserial or biserial arms branched several times. The stem was sometimes unanchored and in such cases served as a prehensile limb. *See* CRINOIDEA; ECHINODERMATA. [H.B.F.]

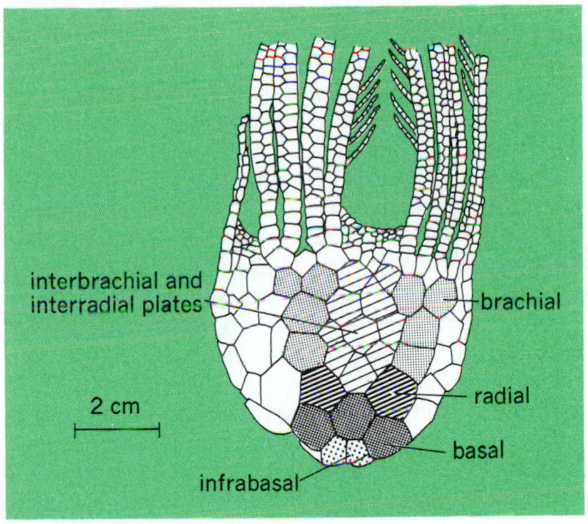

***Thyalocrinus.* (Based on F. Bather and D. P. Oethlert)**

Campanulales An order of flowering plants, division Magnoliophyta in the subclass Asteridae of the class Magnoliopsida (dicotyledons). It consists of 6 families and about 2500 species. The order is distinguished in this subclass by its chiefly herbaceous habit, alternate leaves, inferior ovary, and stamens which are free from the corolla or attached at the base of the tube. About 2000 of the species belong to the single family Campanulaceae. Several familiar ornamentals, including the Canterbury bell (*Campanula medium*) and the cardinal flower (*Lobelia cardinalis*), belong to the Campanulaceae. *See* ASTERIDAE; MAGNOLIOPSIDA; PLANT KINGDOM. [A.CR.]

Camphor A bicyclic, saturated terpene ketone, formula $C_{10}H_{16}O$. It exists in the optically active dextro and levo forms, and as the racemic mixture of the two forms. All of these melt within a degree or so of 178°C (352°F). The principal form is *dextro*-camphor, which occurs in the wood and leaves of the camphor tree (*Cinnamomum camphora*). Camphor is also synthesized commercially on a large scale from pinene which yields mainly the racemic variety. *See* CAMPHOR TREE.

Camphor has a characteristic odor; it crystallizes in thin plates and sublimes readily at ordinary temperatures. It has a wide use in liniments and as a mild rubefacient, analgesic, and antipruritic. It has a local action on the gastrointestinal tract, producing a feeling of warmth and comfort in the stomach. It is also used in photographic film and as a plasticizer in the manufacture of plastics. *See* PINENE; TURPENTINE. [E.L.S.]

Camphor tree The plant *Cinnamomum camphora*, a member of the laurel family (Lauraceae) and a native of China, Japan, and Taiwan. The tree is dense-topped and has shiny, dark, evergreen leaves. All parts of the tree contain camphor, an essential oil which is obtained from the finely ground wood and leaves by distillation with steam. *See* CAMPHOR; MAGNOLIALES. [P.D.St./E.L.C.]

Camptostromatoidea A small class of primitive echinoderms (subphylum Echinozoa) known from the single species *Camptostroma roddyi* based on about 200 specimens from the Early Cambrian (*Bonnia-Olenellus* Zone) in southeastern Pennsylvania. *Camptostroma* was originally described as a hydrozoan or jellyfish, but it was recognized as an echinoderm and the new class Camptostromatoidea was set up for it. *Camptostroma* has been reinterpreted as an early edrioasteroid, and as a "stem echinoderm" ancestral to several other groups including edrioasteroids. Because of *Camptostroma's* puzzling morphology, different reconstructions of it have been made. Its unusual thecal plating is the main feature separating *Camptostroma* from later Edrioasteroidea, which it otherwise resembles, but this genus may also have been ancestral to other classes such as the Eocrinoidea. *See* ECHINODERMATA; ECHINOZOA; EDRIOASTEROIDEA. [J.Sp.]

Canal An artificial open channel usually used to convey water or vessels from one point to another. Canals are generally classified according to use as irrigation, power, flood-control, drainage, or navigation canals or channels. All but the last type are regarded as water conveyance canals.

Canals may be lined or unlined. Linings may consist of plain or reinforced concrete, cement mortar, asphalt, brick, stone, buried synthetic membranes, or compacted earth materials. Linings serve to reduce water losses by seepage or percolation through pervious foundations or embankments and to lessen the cost of weed control. Concrete and other hard-surface linings also permit higher water velocities and, therefore, steeper gradients and smaller cross sections, which may reduce costs and the amount of right-of-way required.

Navigation canals are artificial inland waterways for boats, barges, or ships. A canalized river is one that has been made navigable by construction of one or more weirs or overflow dams to impound river flow, thereby providing navigable depths. Locks may be built in navigation canals and canalized rivers to enable vessels to move to higher or lower water levels. A lock is a chamber equipped with gates at both upstream and downstream ends. Water impounded in the chamber is used to raise or lower a vessel from one elevation to another. The lock chamber is filled and emptied by means of filling and emptying valves and a culvert system usually located in the walls and bottom of the lock. *See* IRRIGATION (AGRICULTURE); TRANSPORTATION ENGINEERING; WATER SUPPLY ENGINEERING. [C.E.; B.R.]

Cancer (constellation) The Crab, in astronomy, a winter constellation and the faintest of the zodiacal groups. Cancer, the fourth sign of the zodiac, is important because during early times it marked the northernmost limit of the ecliptic, when the zodiacal system was adopted. The Tropic of Cancer takes its name from this constellation. Four faint stars form a rough Y outline, which is suggestive of a crab. In the center of Cancer is a hazy object. This is a magnificent cluster of faint stars called Praesepe (the Beehive) or the Manger. *See* CONSTELLATION. [C.-S.Y.]

Cancer (medicine) The common name for a malignant neoplasm or tumor. Neoplasms are new growths and can be divided into benign and malignant types, although in certain instances the distinction is unclear. The most important differentiating feature between a benign and malignant neoplasm is that a malignant tumor will invade surrounding structures and metastasize (spread) to distant sites, whereas a benign tumor will not. Other characteristics used to differentiate between benign and malignant growths include: malignancies are composed of highly atypical cells; malignancies tend to show more rapid growth than benign neoplasms and are composed, in part, of cells showing abnormal mitotic activity, for example, tripolar mitosis; malignant tumors tend to grow progressively without self-limitation. *See* MITOSIS.

Malignant neoplasms which arise from cells of mesenchymal origin (that is, bone, muscle, connective tissue, and so forth) are called sarcomas. Those which develop from epithelial cells and tissues such as skin, mucosal membrane, and glandular tissues are termed carcinomas. Carcinomas usually initially metastasize via lymphatic channels, whereas sarcomas spread to distant organs through the bloodstream.

The cause of most types of cancer is unknown. However, there can be little remaining doubt that cigarette smoking is directly related to cancer of the lung, oral cavity, larynx, and urinary bladder. Likewise, the combination of other substances such as ethyl alcohol or asbestos and cigarette smoking is causally related to other types of cancer. Abundant epidemiologic and experimental data have shown a direct cause-and-effect relationship between various types of industrial exposure and certain neoplasms. Several studies have shown that postmenopausal women taking estrogens have a much higher incidence of cancer of the lining of the uterine cavity. Some malignant tumors seem to be genetically determined or related to chromosomal abnormalities.

The physical changes that cancer produces in the body vary considerably depending on the type of tumor, location, rate of growth, and whether it has metastasized. The American Cancer Society has widely publicized cancer's seven warning signals: (1) change in bowel or bladder habits; (2) a sore that does not heal; (3) unusual bleeding or discharge; (4) thickening or lump in the breast or elsewhere; (5) indigestion or difficulty in swallowing; (6) obvious changes in a wart or a mole; and (7) nagging cough or hoarseness.

The progression, or lack thereof, of a given cancer is also highly variable and depends on the type of neoplasm and its response to treatment. Treatment modalities include surgery, chemotherapy, radiation therapy, hormonal manipulation, and immunotherapy. Often a combination of the various modalities will be used. *See* CHEMOTHERAPY; ONCOLOGY; TUMOR. [S.P.H.]

Cancrinite A family of minerals, related to the scapolite family, characteristically occurring in basic rocks such as nepheline syenites and sodalite syenites. Cancrinite is hexagonal. Four-, six-, and twelve-membered aluminosilicate rings can be discerned in the structure. Large anions such as $[SO_4]^{2-}$ and $[CO_3]^{2-}$ occur in the hexagonal channels of the structure. Four members of the cancrinite family are cancrinite, vishnevite, hauyne, and afghanite. The cancrinite member is white, yellow, greenish, or reddish. It has perfect prismatic cleavage, hardness is 5–6 on Mohs scale, and the specific gravity is 2.45. Localities include the Fen area, southern Norway; the Kola Peninsula, Russia; Bancroft, Ontario, Canada; and Litchfield, Maine, U.S.A. *See* NEPHELINE SYENITE; SILICATE MINERALS. [P.B.M.]

Candidiasis A fungus infection of humans caused by several species of yeasts of the genus *Candida*. The disease may involve the mucous membranes, skin, lungs, central nervous system, and viscera. When the mucous membranes are affected, the disease is known as thrush. The cutaneous form of candidiasis is seen on those areas of the skin where moisture and irritation are common. Amphotericin B and 5-fluorocytosine are used in the treatment of systemic candidiasis. Nystatin is used in the treatment of lesions of the skin and mucous membranes. *See* MEDICAL MYCOLOGY; YEAST. [L.D.H.]

Candlepower Luminous intensity expressed in candelas. The term refers only to the intensity in a particular direction and by itself does not give an indication of the total light emitted. The candlepower in a given direction from a light source is equal to the illumination in footcandles falling on a surface normal to that direction, multiplied by the square of the distance from the light source in feet. The candlepower is also equal to the illumination of metercandles (lux) multiplied by the square of the distance in meters.

The apparent candlepower is the candlepower of a point source which will produce the same illumination at a given distance as produced by a given light source.

The mean horizontal candlepower is the average candlepower of a light source in the horizontal plane passing through the luminous center of the light source.

The mean spherical candlepower is the average candlepower in all directions from a light source as a center. Since there is a total solid angle of 4π (steradians) emanating from a point, the mean spherical candlepower is equal to the total luminous flux (in lumens) of a light source divided by 4π (steradians). *See* LUMINOUS INTENSITY; PHOTOMETRY. [R.C.Pu.]

Canine distemper A fatal viral disease of dogs and other carnivores, with a worldwide distribution. Canine distemper virus has a wide host range; most terrestrial carnivores are susceptible to natural canine distemper virus infection. All animals in the families Canidae (such as dog, dingo, fox, coyote, wolf, jackal), Mustelidae (such as weasel, ferret, mink, skunk, badger, stoat, marten, otter), and Procyonidae (such as kinkajou, coati, bassariscus, raccoon, panda) may succumb to canine distemper virus infection. Members of other Carnivora families, including domestic cats and swine, may become subclinically infected. The virus has also been isolated from large cats (lions, tigers, leopards) that have died in zoological parks in North America, from wild lions in the Serengeti National Park (Tanzania), and from wild javelinas (collared peccaries). *See* CARNIVORA.

Canine distemper virus is classified as a morbillivirus within the Paramyxoviridae family, closely related to measles virus and rinderpest virus of cattle and the phocine (seal) and dolphin distemper virus. The virus is enveloped with a negative-sense ribonucleic acid and consists of six structural proteins: the nucleoprotein and two enzymes in the nucleocapsid, the membrane protein on the inside, and the hemagglutinating and fusion proteins on the outside of the lipoprotein envelope. *See* ANIMAL VIRUS; PARAMYXOVIRUS.

Canine distemper is enzootic worldwide. Aerosol transmission in respiratory secretions is the main route of transmission. Virus shedding begins approximately 7 days after the initial infection. Acutely infected dogs and other carnivores shed virus in all body excretions, regardless of whether they show clinical signs or not.

Great variations occur in the duration and severity of canine distemper, which may range from no visible signs to severe disease, often with central nervous system involvement, with approximately 50% mortality in dogs. The first fever 3–6 days after infection may pass unnoticed; the second peak (several days later and intermittent thereafter) is usually associated with nasal and ocular discharge, depression, and anorexia. A low lymphocyte count is always present during the early stages of infection. Gastrointestinal and respiratory signs may follow, often enhanced by secondary infection.

A specific antiviral drug having an effect on canine distemper virus in dogs is not presently available. Treatment of canine distemper, therefore, is nonspecific and supportive. Antibiotic therapy is recommended because of the common occurrence of secondary bacterial infections of the respiratory and alimentary tracts. Administration of fluids and electrolytes may be the most important therapy for canine distemper because diseased dogs with diarrhea are often dehydrated. [M.J.G.A.; B.A.S.]

Canine parvovirus infection Severe enteritis caused by a small nonenveloped single-stranded deoxyribonucleic acid (DNA) virus that is resistant to inactivation and remains infectious in the environment for 5–7 months. First observed in dogs in 1976, canine parvovirus may have originated by mutation of a closely related parvovirus of cats or wildlife. The original virus was designated as canine parvovirus, type 2 (CPV-2); however, since its discovery the virus has undergone two minor genetic alterations, designated CPV-2a and CPV-2b. These alterations may have enabled the virus to adapt to its new host, replicate, and spread more effectively. *See* ANIMAL VIRUS.

Canine parvovirus is transmitted between dogs by the fecal-oral route. The incubation period is 3–7 days. Virus is first shed in the feces on day 3, and shedding continues for an additional 10 days. Chronically infected dogs that shed virus intermittently are rare. Most naturally occurring infections in dogs are subclinical or result in mild signs of the disease. Dogs with subclinical infections play an important role in the spread of the disease by shedding large amounts of virus into the environment. This shedding, along with the ability of the virus to persist in the environment, contributes to the endemicity of the disease. The development of disease following infection ranges from 20 to 90%, and mortality ranges 0 to 50%.

The goal of treatment is to support the animal until the infection runs its course. There are no specific antiviral therapies available. The intensity of treatment depends on the severity of signs. Dehydrated pups require intensive intravenous fluid therapy. Antimicrobial drugs are useful because of the risk of secondary bacterial infections, and antiemetic drugs help control vomiting and nausea. Good nursing care is essential. All food and water should be withheld until the pup is no longer vomiting, and the pup should be kept warm, clean, and dry. Because of the infectious nature of the disease, pups should be isolated from other dogs. [M.S.Co.]

Canonical transformations Transformations among the coordinates and momenta describing the state of a classical dynamical system which leave the canonical or Hamiltonian form of the equations of motion unchanged. *See* HAMILTON'S EQUATIONS OF MOTION; HAMILTON'S PRINCIPLE. [P.M.S.]

Cantaloupe In the United States the name applied to muskmelon cultivars belonging to *Cucumis melo* var. *reticulatus* of the family Cucurbitaceae. However, this is a misnomer, and the name cantaloupe should be restricted to cultivars of *C. melo* var. *cantalupensis*. The fruits of this group are rough and scaly, with deep vein tracts and a hard rind. Cultivars of the variety *cantalupensis* are grown in Europe and Asia, but seldom in the United States.

The fruits grown in the United States are round to oval; the surface is netted and has shallow vein tracts. At maturity the skin color changes from dark green or gray to light gray or yellow. The flesh is usually salmon-colored, but it may vary from green to deep salmon-orange. When mature the melon is sweet, averages 6–8% sugar, and has a distinct aroma and flavor. The flesh is high in potassium and vitamin C, and, when deep orange, rich in vitamin A. *See* MUSKMELON; VIOLALES. [O.A.L.]

Cantilever A beam supported at one end and supporting a load along its length or at its free end, the upper portion of the beam, if horizontal, being everywhere in tension and the lower portion being everywhere in compression (see illustration). Familiar examples are the symmetrical paired cantilevers

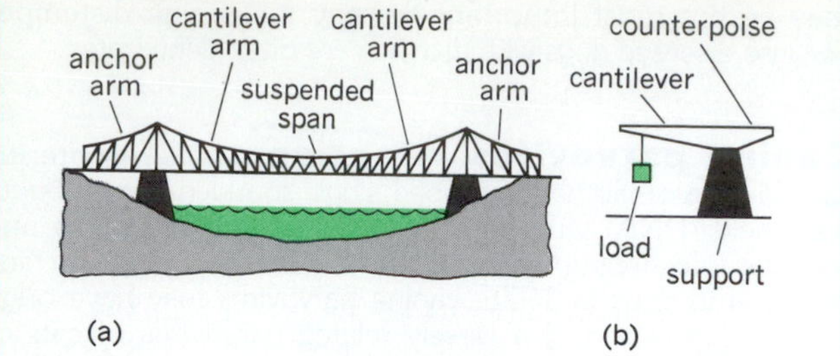

Two applications of the cantilever. (a) Bridge. (b) Hammerhead crane.

of a seesaw or teeterboard, the beam in a chemical balance, and the unsymmetrical cantilever in the overhang of a roof. Cantilever construction permits spanning wide distances without obstructing supports. *See* BEAM. [F.H.R.]

Capacitance The ratio of the charge q on one of the plates of a capacitor (there being an equal and opposite charge on the other plate) to the potential difference v between the plates; that is, capacitance (formerly called capacity) is $C = q/v$.

In general, a capacitor, often called a condenser, consists of two metal plates insulated from each other by a dielectric. The capacitance of a capacitor depends on the geometry of the plates and the kind of dielectric used, since these factors determine the charge which can be put on the plates by a unit potential difference existing between the plates.

In an ideal capacitor, no conduction current flows between the plates. A real capacitor of good quality is the circuit equivalent of an ideal capacitor with a very high resistance in parallel or, in alternating-current (ac) circuits, of an ideal capacitor with a low resistance in series. *See* CAPACITOR; DIELECTRIC MATERIALS. [R.P.Wi.]

Capacitance box An assembly of capacitors and switches which permits adjustments of the capacitance existing at the terminals in nominally uniform steps from a minimum value near zero to the maximum which exists when all the capacitors are connected in parallel. Several arrangements are possible; the most common are those in which each decade consists of four capacitors in the ratios 1:2:2:5 or 1:2:3:4. The switch connects successively the combinations of these four capacitors necessary to provide increasing capacitance at the terminals in uniform steps. The term capacitance decade box is sometimes used. *See* CAPACITOR. [F.R.Ko.]

Capacitance multiplication The generation of a capacitance which is some multiple of that of an actual capacitor. Capacitance multiplication circuits have an input impedance which is capacitive and which is proportional to that of an actual capacitor appearing somewhere in the circuit. In most applications, capacitance multiplication circuits are used to generate an equivalent input capacitance which is much larger than that of the actual capacitor. One scenario where capacitance multiplication might prove useful is where a physical capacitor of a required capacitance may be too large, too expensive, or unavailable. A second is where the capacitor must be reasonably large and capable of handling bidirectional signals, thus precluding the direct use of widely available electrolytic capacitors and hence making practical the utilization of a much smaller nonelectrolytic capacitor in a capacitance multiplication circuit. *See* CAPACITANCE; CAPACITOR.

Capacitance multiplication circuits are often made from operational amplifiers and resistors along with the capacitor that is to be scaled, although transistors and other active devices can also be used. Capacitance multiplication circuits are closely related to classes of circuits termed generalized immittance converters and negative impedance converters. Generalized immittance converters are used to generate an equivalent input impedance that is proportional to products or quotients of specific impedances that appear in the circuit. Generalized immittance converters are often used for capacitance multiplication. Related applications of generalized immittance converters are synthetic inductance simulation and negative-resistance generation. Negative-impedance converters are also used for the generation of negative impedances. *See* ELECTRICAL IMPEDANCE; IMMITTANCE; INDUCTANCE; NEGATIVE-RESISTANCE CIRCUITS; OPERATIONAL AMPLIFIER. [R.L.Ge.]

Capacitor A device for introducing capacitance into a circuit. In general, a capacitor consists of two metal plates insulated from each other by a dielectric. The capacitance of a capacitor depends primarily upon the shape and size of the capacitor and upon the relative dielectric constant of the medium between the plates. In vacuum, and approximately so in air and most gases, k, the dielectric constant, is unity. For other dielectrics, k ranges from one to several hundred. *See* CAPACITANCE; DIELECTRIC CONSTANT.

One classification of capacitors comes from the physical state of their dielectrics. The dielectric may be a gas (or vacuum), a liquid, a solid, or a combination of these. Each of these classifications may be subdivided according to the specific dielectric used.

Capacitors are also classified as fixed, adjustable, or variable. The capacitance of fixed capacitors remains unchanged, except for small variations (caused by temperature fluctuations or vibration, for example); the capacitance of adjustable capacitors may be set at any one of several discrete values; and the capacitance of variable capacitors may be adjusted continuously and set at any value between minimum and maximum limits fixed by construction. Trimmer capacitors are relatively small variable capacitors used in parallel with larger variable or fixed capacitors to permit exact adjustment of the capacitance of the parallel combination.

Made in both the fixed and variable types, air, gas, and vacuum capacitors are constructed with flat parallel metallic plates (or cylindrical concentric metallic plates) with air, gas, or vacuum as the dielectric between plates. Alternate plates are connected, with one or both sets supported by means of a solid insulating material such as glass, quartz, ceramic, or plastic. The purpose of a high vacuum, or a gas under pressure, is to increase the voltage breakdown value for a given plate spacing. Air, gas, and vacuum capacitors are used in high-frequency circuits. Fixed and variable air capacitors of special design are used as standards in electrical measurements.

Solid-dielectric capacitors use one of several dielectrics such as a ceramic, mica, glass, or plastic film. Alternate plates of metal, or metallic foil, are stacked with the dielectric, or the dielectric may be metal-plated on both sides.

The plate assemblies of liquid-dielectric capacitors are mounted in a tank filled with a suitable oil or liquid dielectric. External connection is through insulators or bushings, and provision is made for expansion and contraction of the liquid during temperature changes. They have high capacity and low losses.

A large capacitance-to-volume ratio and a low cost per microfarad of capacitance are chief advantages of electrolytic capacitors. These use aluminum or tantalum plates on which an oxide film forms and acts as the dielectric. The characteristics of the oxide film depend on the polarity of the applied voltage. [W.S.P.]

Capillary (anatomy) The smallest vessel of both the circulatory and lymphatic systems. Lymphatic capillaries are slightly larger in diameter than blood capillaries. The capillary wall is composed of a single layer of flat endothelial cells. The inside diameter of a blood capillary is about 8 micrometers, the size of a red blood cell, but varies slightly in different tissues and organs. Extensive capillary beds result from anastomoses of branching capillaries. These capillary networks are denser in those tissues possessing a high metabolic activity.

Capillaries are the exchange areas of the circulatory system, where passage of materials to and from these vessels and body tissues occurs under the influence of hydrostatic pressure, osmotic pressure, and other factors which affect permeability and rates of diffusion. *See* CIRCULATORY SYSTEM. [W.J.B.]

Capillary viscometer The simplest and oldest device for measuring the rheological properties of liquids. In this instrument, the sample is usually allowed to flow by gravity through a cylindrical capillary, and the transit time is measured. Flow through the capillary is described by Poiseuille's law, Eq. (1),

$$Q = \frac{\pi \Delta p R^4}{8L\eta} \qquad (1)$$

where Q = volumetric flow rate, Δp = pressure drop through the capillary, R = internal radius of capillary, and L = length of capillary.

The shear stress at the wall is $\tau_w = \Delta pR/2L$, for newtonian liquids the shear rate at the wall is $\dot{\gamma}_w = 4Q/\pi R^3$. Although some precautions must be taken to either minimize or correct for errors caused by kinetic and viscous energy entrance and exit effects, variation in the head during gravity flow, heat effects, various wall effects, and so on, the capillary viscometer offers the advantages of a straightforward, inexpensive technique for measuring viscosities. For non-newtonian liquids, the variation in shear rate within the capillary becomes a problem with respect to calculating shear rate. Models such as a Bingham plastic, which has a yield stress and an apparent viscosity given by Eq. (2) or a power law model with $\tau = k\dot{\gamma}^n$,

$$\eta_{\text{apparent}} = \frac{(\tau - \tau_{\text{yield}})}{\dot{\gamma}} \qquad (2)$$

where k is a constant and n is an empirically determined value, must be incorporated into the theoretical equations in order to calculate shear rate. Furthermore, to study the degree of non-newtonian behavior, the overall average shear rate must be changed by using varying capillary sizes or by using pumps to get greater pressure drops. Rheological properties of materials with time-dependency of the same order of magnitude as the transit time cannot be measured with this type of viscometer. *See* VISCOSITY. [H.L.G.]

Capillary wave Capillary waves, or ripples, occur at the interface between two fluids, in which the principal restoring force is controlled by surface tension. Ripples generated by wind at the interface between air and water on oceans and lakes are of importance to the friction of air flowing over water, and to the reflection and scattering of electromagnetic and sound waves. *See* SURFACE TENSION.

In nature, ripples are observed to grow rapidly when the wind blows and to die away rapidly when the wind stops. When the water surface is contaminated, as by an oil film or other surface-active agent, ripples are damped still more rapidly because the contaminated surface acts as an inextensible film against which the water motions due to the ripples must rub. For moderate winds, the increased damping of ripples almost completely inhibits their growth; the surface appears smooth and is called a slick. [C.S.C.]

Capnocytophaga A genus of bacteria comprising fusiform, fermentative, gram-negative rods which require carbon dioxide for growth and show a gliding motility. Three species have been described. All have been isolated from healthy periodontal tissues and as predominant flora from lesions of juvenile periodontitis, periodontitis associated with diabetes mellitus, and the Papillon-Lefèvre syndrome. Septicemia with *C. ochracea* has been observed in individuals with malignancies or granulocytopenia associated with oral ulcerations. *See* BACTERIA; PERIODONTAL DISEASE. [A.W.C.V.G]

Capparales An order of flowering plants, division Magnoliophyta (Angiospermae), in the subclass Dilleniidae of the class Magnoliopsida (dicotyledons). It consists of 5 families and nearly 4000 species.

The families Cruciferae (Brassicaceae), with about 3000 species, and Capparaceae (often spelled Capparidaceae), with about 800 species, form the core of the order. These two families have alternate, usually compound or dissected leaves, and hypogynous flowers with mostly four sepals, four petals, centrifugal stamens, and (at least apparently) two carpels which form a compound ovary with more or less distinctly parietal placentation. Nearly all the Cruciferae and many of the Capparaceae have specialized myrosin cells, which are chiefly but not entirely restricted to this order. Myrosin is an enzyme involved in the formation of mustard oil. Several common vegetables, including cabbage, radish, rutabaga, turnip, water cress, and the various kinds of mustards, belong to the Cruciferae. A number of familiar ornamentals, including candytuft (*Iberis*), stock (*Matthiola*), and sweet alyssum (*Lobularia maritima*) in the Cruciferae, spider flower (*Cleome spinosa*) in the Capparaceae, and mignonette (*Reseda*) in the Resedaceae belong to the Capparales. *See* BROCCOLI; BRUSSELS SPROUTS; CABBAGE; CAULIFLOWER; COLLARD; CRESS; HORSERADISH; KALE; KOHLRABI; MAGNOLIOPSIDA; MUSTARD; RADISH; RAPE; RUTABAGA; TURNIP. [A.Cr.]

Caprellidea A common crustacean suborder in marine and estuarine environments, belonging to the superorder Peracarida, order Amphipoda. As peracarids, female Caprellidea have a ventral brood pouch and eggs develop directly (that is, there are no planktonic larvae); as amphipods, Caprellidea have the second and third thoracic appendages formed into enlarged subchelate claws (gnathopods). The abdomen and abdominal appendages of caprellideans are reduced; thus caprellideans are restricted to a clinging and crawling life-style. Almost all caprellideans use other organisms as substrata. Family distinctions within the Caprellidea are being reevaluated, but the classical divisions are the families Caprellidae and Cyamidae (whale lice). The ecology of both families is poorly understood. *See* AMPHIPODA; CRUSTACEA. [E.A.C.]

Capricornus The Sea Goat, in astronomy, an inconspicuous zodiacal constellation in the southern sky lying between Aquarius and Sagittarius. Capricornus is the tenth sign of the

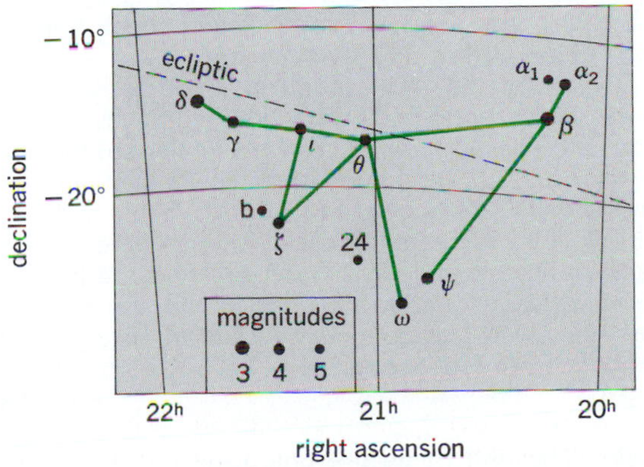

Line pattern of constellation Capricornus. Grid lines represent coordinates of the sky. Apparent brightness, or magnitudes, of stars are shown by size of dots, which are graded by appropriate numbers as indicated.

zodiac. The constellation has been described from the earliest times as a goat, or as a figure that is part goat with the tail of a fish (see illustration). The Tropic of Capricorn originates from this constellation, which marked the southern limit of the ecliptic in ancient times. *See* CONSTELLATION. [C.-S.Y.]

Caprimulgiformes An order of crepuscular or mainly nocturnal birds collectively known as the goatsuckers. The group is apparently most closely related to the owls (Strigiformes) and is found worldwide, mainly in the tropics and warm temperate regions. Species breeding in the Arctic and cooler temperate regions are migratory. *See* STRIGIFORMES.

The order Caprimulgiformes is divided into the suborder Steatornithes, containing the single family Steatornithidae (oilbirds; 1 species), and the suborder Caprimulgi, including the families Podargidae (frogmouths; 13 species), Aegothelidae (owlet-frogmouths; 8 species), Nyctibiidae (potoos; 6 species), and Caprimulgidae (nightjars or goatsuckers; 77 species). The largest family, Caprimulgidae, is found worldwide; Steatornithidae are restricted to northern South America; Podargidae live in the Old World tropics from India to Australia; Aegothelidae are restricted to Australia and surrounding islands; and Nyctibiidae are found in the New World tropics.

The goatsuckers are primarily insectivorous, although the large frogmouths also feed on small vertebrates; the oilbird is unique in feeding on fruits, especially of palms. Goatsuckers have a huge, generally weak mouth with long, stout bristles surrounding it to make an effective trap for insects caught in flight. The wings are well developed, but the feet are weak and serve mainly for perching. The plumage is soft and fluffy, and is mottled and barred brown and gray, serving as a cryptic protective coloration. White patches may exist on the wings, tail, and throat, which are visible only in flight. Goatsuckers are solitary but may migrate in loose flocks (nighthawks). Most species are highly vocal, using the calls to attract mates and defend territories. The English names of a number of species, such as the whippoorwill and chuck-will's-widow, are based on their calls. Goatsuckers nest on the ground or in trees, laying one to five eggs. Young are downy and remain in the nest until they are able to fly. A few species are known to hibernate.

The oilbirds are colonial nesters, placing their nest of seeds and droppings on ledges deep in caves. Paired adults remain together throughout the year, roosting on their breeding ledge. Oilbirds have excellent night vision, but inside their often totally dark caves they find their way by using echolocation based on pulses of sound of about 7000 Hz which are audible to the human ear. *See* AVES; ECHOLOCATION. [W.J.B.]

Captorhinida A moderately coherent group of primitive reptiles constituting an order of the subclass Anapsida. Most members are characterized by a closed cheek (temporal) region. The order is divided into four suborders: Captorhinomorpha, Millerosauria, Procolophonia, and Pareiasauria. Except for the Procolophonia, which continue into the Late Triassic, Captorhinida are confined to the Permo-Carboniferous. They lived in lowlands, where they were associated with amphibians and synapsid reptiles. Some of the smaller animals of the Captorhinomorpha and the Millerosauria fed primarily on insects and small vertebrates, but the larger genera were exclusively herbivorous. Along with the caseid pelycosaurs, they were the dominant consumers of vegetation in the middle Permian ecosystems. *See* ANAPSIDA; REPTILIA; SYNAPSIDA. [E.C.O.]

Carat The unit of weight now used for all gemstones except pearls. It is also called the metric carat (m.c.). By international agreement, the carat weight is set at 200 milligrams. Pearls are weighed in grains, a unit of weight equal to 50 mg, or 1/4 carat. The application of the term carat as a unit of weight must not be confused with the term karat used to indicate fineness or purity of the gold in which gems are mounted. *See* GEM. [R.T.L.]

Caraway An important spice from the fruits of the perennial herb *Carum carvi*, of the family Umbelliferae. A native of Europe and western Asia, it is now cultivated in many temperate areas of both hemispheres. The small, brown, slightly curved fruits are used in perfumery, cookery, confectionery, in medicine, and for flavoring beverages. *See* APIALES. [P.D.St./E.L.C.]

Carbide A binary compound of carbon with an element less electronegative than carbon. Carbon-hydrogen compounds are excluded. Effectively then, carbides are composed of metal-carbon compounds if boron and silicon are included among the normal metals. Essentially no volatile compounds (except AlC or its dimer) are known, because decomposition sets in at higher temperatures before volatilization of the carbide as such.

Most carbides can be prepared by heating a mixture of the powdered metal and carbon, usually to high temperatures, but not necessarily as high as the melting point. Generally, the same result is possible by heating a mixture of the oxide of the metal with carbon ($CaO + 3C \rightarrow CaC_2 + CO$). Some may be prepared by passing a hydrocarbon vapor over the hot metal in the form of an electrically heated filament. *See* CARBON. [R.K.E.]

Carbohydrate A term applied to a group of substances which include the sugars, starches, and cellulose, along with many other related substances. This group of compounds plays a vitally important part in the lives of plants and animals, both as structural elements and in the maintenance of functional activity. Plants are unique in that they alone in nature have the power to synthesize carbohydrates from carbon dioxide and water in the presence of the green plant chlorophyll through the energy derived from sunlight, by the process of photosynthesis. This process is responsible not only for the existence of plants but for the maintenance of animal life as well, since animals obtain their entire food supply directly or indirectly from the carbohydrates of plants. *See* CARBOHYDRATE METABOLISM; PHOTOSYNTHESIS.

The term carbohydrate originated in the belief that naturally occurring compounds of this class, for example, D-glucose ($C_6H_{12}O_6$), sucrose ($C_{12}H_{22}O_{11}$), and cellulose ($(C_6H_{10}O_5)_n$), could be represented formally as hydrates of carbon, that is, $C_x(H_2O)_y$. Later it became evident that this definition for carbohydrates was not a satisfactory one. New substances were discovered whose properties clearly indicated that they had the characteristics of sugars and belonged in the carbohydrate class, but which nevertheless showed a deviation from the required hydrogen-to-oxygen ratio. Examples of these are the important deoxy sugars, D-deoxyribose, L-fucose, and L-rhamnose, the uronic acids, and such compounds as ascorbic acid (vitamin C). The retention of the term carbohydrate is therefore a matter of convenience rather than of exact definition. A carbohydrate is usually defined as either a polyhydroxy aldehyde (aldose) or ketone (ketose), or as a substance which yields one of these compounds on hydrolysis. However, included within this class of compounds are substances also containing nitrogen and sulfur. *See* DEOXYRIBOSE; FRUCTOSE; RHAMNOSE.

The properties of many carbohydrates differ enormously from one substance to another. The sugars, such as D-glucose or sucrose, are easily soluble, sweet-tasting, and crystalline; the starches are colloidal and paste-forming; and cellulose is completely insoluble. Yet chemical analysis shows that they have a common basis; the starches and cellulose may be degraded by different methods to the same crystalline sugar, D-glucose.

The carbohydrates usually are classified into three main groups according to complexity: monosaccharides, oligosaccharides, and polysaccharides. Monosaccharides are simple sugars that consist of a single carbohydrate unit which cannot be hydrolyzed into simpler substances. These are characterized, according to their length of carbon chain, as trioses ($C_3H_6O_3$), tetroses ($C_4H_8O_4$), pentoses ($C_5H_{10}O_5$), hexoses ($C_6H_{12}O_6$), heptoses ($C_7H_{14}O_7$), and so on. Oligosaccharides are compound sugars that are condensation products of two to five molecules of simple sugars and are subclassified into disaccharides, trisaccharides, tetrasaccharides, and pentasaccharides, according to the number of monosaccharide molecules yielded upon hydrolysis. Polysaccharides comprise a heterogeneous group of compounds which represent large aggregates of monosaccharide units, joined through glycosidic bonds. They are tasteless, nonreducing, amorphous substances that yield a large and indefinite number of monosaccharide units on hydrolysis. Their molecular weight is usually very high, and many of them, like starch or glycogen, have molecular weights of several million. They form colloidal solutions, but some polysaccharides, of which cellulose is an example, are completely insoluble in water. On account of their heterogeneity they are difficult to classify. *See* MONOSACCHARIDE; OLIGOSACCHARIDE; POLYSACCHARIDE.

The sugars are also classified into two general groups, the reducing and nonreducing. The reducing sugars are distinguished by the fact that because of their free, or potentially free, aldehyde or ketone groups they possess the property of readily reducing alkaline solutions of many metallic salts, such as those of copper, silver, bismuth, mercury, and iron. The most widely used reagent for this purpose is Fehling's solution. The reducing sugars constitute by far the larger group. The monosaccharides and many of their derivatives reduce Fehling's solution. Most of the disaccharides, including maltose, lactose, and the rarer sugars cellobiose, gentiobiose, melibiose, and turanose, are also reducing sugars. The best-known nonreducing sugar is the disaccharide sucrose. Among other nonreducing sugars are the disaccharide trehalose, the trisaccharides raffinose and melezitose, the tetrasaccharide stachyose, and the pentasaccharide verbascose.

The sugars consist of chains of carbon atoms which are united to one another at a tetrahedral angle of 109°28′. A carbon atom to which are attached four different groups is called asymmetric. A sugar, or any other compound containing one or more asymmetric carbon atoms, possesses optical activity; that is, it rotates the plane of polarized light to the right or left. *See* OPTICAL ACTIVITY. [W.Z.H.]

Carbohydrate metabolism The fields of biochemistry and physiology deal with the breakdown and synthesis of simple sugars, oligosaccharides, and polysaccharides, and with the transport of sugars across cell membranes and tissues. The breakdown or dissimilation of simple sugars, particularly glucose, is one of the principal sources of energy for living organisms. The dissimilation may be anaerobic, as in fermentations, or aerobic, that is, respiratory. In both types of metabolism, the breakdown is accompanied by the formation of energy-rich bonds, chiefly the pyrophosphate bond of the coenzyme adenosinetriphosphate (ATP), which serves as a coupling agent between different metabolic processes. In higher animals, glucose is the carbohydrate constituent of blood, which carries it to the tissues of the body. In higher plants, the disaccharide sucrose is often stored and transported by the tissues. Certain polysaccharides, especially starch and glycogen, are stored as endogenous food reserves in the cells of plants, animals, and microorganisms. Others, such as cellulose, chitin, and bacterial polysaccharides, serve as structural components of cell walls. As constituents of plant and animal tissues, various carbohydrates become available to those organisms which depend on other living or dead organisms for their source of nutrients. Hence, all naturally occurring carbohydrates can be dissimilated by some animals or microorganisms. *See* ADENOSINETRIPHOSPHATE (ATP); CARBOHYDRATE; CELLULOSE; CHITIN; GLYCOGEN; STARCH.

Certain carbohydrates cannot be used as nutrients by humans. For example, cellulose cannot be digested by humans or other mammals and is a useful food only for those, such as the ruminants, that harbor cellulose-decomposing microorganisms in their digestive tracts. The principal dietary carbohydrates available to humans are the simple sugars glucose and fructose, the disaccharides sucrose and lactose, and the polysaccharides glycogen and starch. Lactose is the carbohydrate constituent of milk and hence one of the main sources of food during infancy. The disaccharides and polysaccharides that cannot be absorbed directly from the intestine are first digested or hydrolyzed by enzymes (glycosidases) secreted into the alimentary canal.

The simple sugars reach the intestine or are produced there through the digestion of oligosaccharides. They are absorbed by the intestinal mucosa and transported across the tissue into the bloodstream. This process involves the accumulation of sugar against a concentration gradient and requires active metabolism of the mucosal tissue as a source of energy. The sugars are absorbed from the blood by the liver and are stored there as glycogen. The liver glycogen serves as a constant source of glucose in the bloodstream. The mechanisms of transport of sugars across cell membranes and tissues are not yet understood, but they appear to be highly specific for different sugars and to depend on enzymelike components of the cell.

The degradation of monosaccharides may follow one of several types of metabolic pathways. In the phosphorylative pathways, the sugar is first converted to a phosphate ester. The phosphorylated sugar is then split into smaller units, either before or after oxidation. In the nonphosphorylative pathways, the sugar is usually oxidized to the corresponding aldonic acid. This may subsequently be broken down either with or without phosphorylation of the intermediate products. Among the principal intermediates in carbohydrate metabolism are glyceraldehyde-3-phosphate and pyruvic acid. The end products of metabolism depend on the organism and, to some extent, on the environmental conditions. Besides cell material, the products may include carbon dioxide (CO_2), alcohols, organic acids, and hydrogen gas. In the so-called complete oxidations, CO_2 is the only excreted end product. In incomplete oxidations, characteristic of the vinegar bacteria and of certain fungi, oxidized end products such as gluconic, ketogluconic, citric, or fumaric acids may accumulate. Organic end products are invariably found in fermentations. *See* FERMENTATION.

The principal phosphorylative pathway involved in fermentations is known as the glycolytic, hexose diphosphate, or Embden-Meyerhof pathway (see illustration). This sequence of reactions is the basis of the lactic acid fermentation of mammalian muscle and of the alcoholic fermentation of yeast. The metabolism of simple sugars other than glucose usually involves the conversion of the sugar to one of the intermediates of the phosphorylative pathways described for glucose metabolism.

The oxidative or respiratory metabolism of sugars differs in several respects from fermentative dissimilation. First, the oxidative steps, that is, the reoxidation of DPNH, are linked to the reduction of molecular oxygen. Second, the pyruvic acid produced through glycolytic or other mechanisms is further oxidized, usually to CO_2. Third, in most aerobic organisms, alternative pathways either supplement or completely replace the glycolytic sequence of reactions for the oxidation of sugars. Where pyruvic acid appears as a metabolic intermediate, it is generally oxidatively decarboxylated to yield CO_2 and the two-carbon acetyl fragment which combines with coenzyme A. The

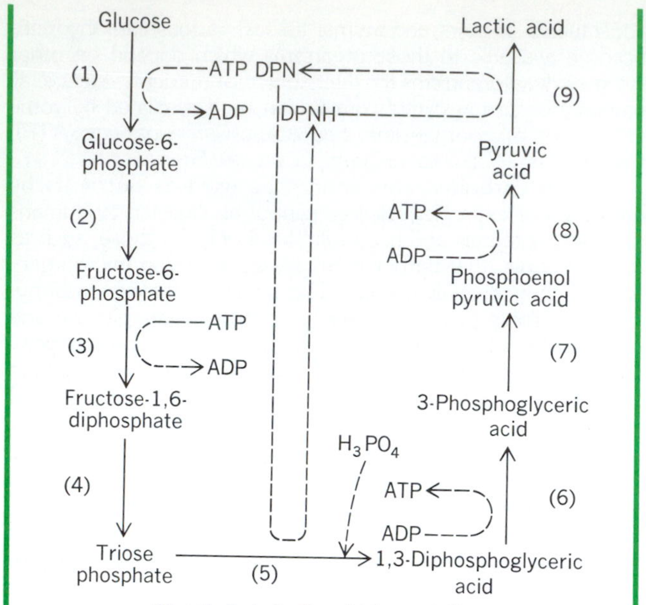

Glycolysis in lactic acid fermentation.

acetyl group is then further oxidized via the Krebs cycle. *See* KREBS CYCLE.

The dissimilation and biosynthesis of the oligosaccharides and polysaccharides are affected through the enzymatic cleavage or formation of glycosidic bonds between simple monosaccharide constituents of the complex carbohydrates. The principal types of enzyme which split or synthesize glycosidic bonds are the hydrolases or glycosidases, phosphorylases, and transglycosylases. The enzymes are generally highly specific with respect to the glycosidic portion, or moiety, and the type of linkage of the substrates which they attack. Lactases, for example, combine the ß-galactosyl portion of the disaccharide lactose with hydroxyl ion of water to yield galactose, liberating glucose. Maltases can hydrolyze maltose to two molecules of glucose. Sucrases or invertases are enzymes which hydrolyze sucrose to yield glucose and fructose. Amylases or diastases are enzymes which hydrolyze (digest) starch. The phosphorylases catalyze the reversible phosphorolysis of certain disaccharides, polysaccharides, and nucleosides by transferring the glycosyl moieties to inorganic phosphate. The transglycosylases interconvert various disaccharides and polysaccharides by the transfer of glycosyl groups. [M.D.]

Carbomycin A macrolide antibiotic isolated from fermentation broths of strains of *Streptomyces halstedii*, *S. hygroscopicus*, and perhaps other species. Macrolide antibiotics are basic antibiotics characterized by a macrocyclic ring structure. Not infrequently, small amounts of a second closely related antibiotic, carbomycin B, occur in the broth. Carbomycin is known commercially as Magnamycin.

The antibacterial spectrum of carbomycin is very similar to that of penicillin; for example, it is principally active against gram-positive bacteria. Unlike penicillin, carbomycin is also active against various rickettsiae. *See* ANTIBIOTIC. [F.W.T.]

Carbon A chemical element, C, with an atomic number of 6 and an atomic weight of 12.01115. Carbon is unique in chemistry because it forms a vast number of compounds, larger than the sum total of all other elements combined. By far the largest group of these compounds are those composed of carbon and hydrogen. It has been estimated that there are at least 1,000,000 known organic compounds, and this number is increasing rapidly each year. Although the classification is not rigorous, carbon forms another series of compounds, classified as inorganic, comprising a much smaller number than the organic compounds. *See* ORGANIC CHEMISTRY.

Elemental carbon exists in two well-defined crystalline allotropic forms, diamond and graphite. Other forms, which are poorly developed in crystallinity, are charcoal, coke, and carbon black. Chemically pure carbon is prepared by the thermal decomposition of sugar (sucrose) in the absence of air. The physical and chemical properties of carbon are dependent on the crystal structure of the element. The density varies from 2.25 g/cm^3 (1.30 oz/in.^{3}) for graphite to 3.51 g/cm^3 (2.03 oz/ in.^{3}) for diamond. For graphite, the melting point is 3500°C (6332°F) and the extrapolated boiling point is 4830°C (8726°F). Elemental carbon is a fairly inert substance. It is insoluble in water, dilute acids and bases, and organic solvents. At elevated temperatures, it combines with oxygen to form carbon monoxide or carbon dioxide. With hot oxidizing agents, such as nitric acid and potassium nitrate, mellitic acid, $C_6(CO_2H)_6$, is obtained. Of the halogens, only fluorine reacts with elemental carbon. A number of metals combine with the element at elevated temperatures to form carbides.

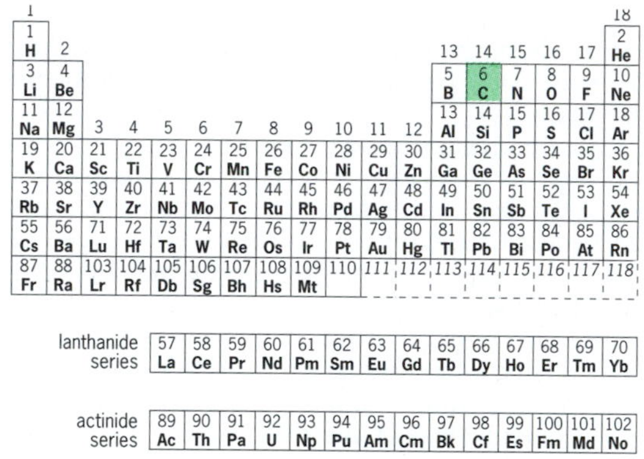

Carbon forms three gaseous compounds with oxygen: carbon monoxide, CO; carbon dioxide, CO_2; and carbon suboxide, C_3O_2. The first two oxides are the more important from an industrial standpoint. Carbon forms compounds with the halogens which have the general formula CX_4, where X is fluorine, chlorine, bromine, or iodine. At room temperature, carbon tetrafluoride is a gas, carbon tetrachloride is a liquid, and the other two compounds are solids. Mixed carbon tetrahalides are also known. Perhaps the most important of them is dichlorodifluoromethane, CCl_2F_2, commonly called Freon. *See* CARBON DIOXIDE; FREON; HALOGENATED HYDROCARBON.

Carbon and its compounds are found widely distributed in nature. It is estimated that carbon makes up 0.032% of the Earth's crust. Free carbon is found in large deposits as coal, an amorphous form of the element which contains additional complex carbon-hydrogen-nitrogen compounds. Pure crystalline carbon is found as graphite and as diamonds.

Extensive amounts of carbon are found in the form of its compounds. In the atmosphere, carbon is present in amounts of up to 0.03% by volume as carbon dioxide. Various minerals such as limestone, dolomite, marble, and chalk all contain carbon in the form of carbonate. All plant and animal life is composed of complex organic compounds containing carbon combined with hydrogen, oxygen, nitrogen, and other elements. The remains of past plant and animal life are found as deposits of petroleum, asphalt, and bitumen. Deposits of natural gas contain compounds that are composed of carbon and hydrogen. *See* CARBONATE MINERALS.

The free element has many uses, ranging from ornamental applications of the diamond in jewelry to the black-colored pigment of carbon black in automobile tires and printing inks. Another form of carbon, graphite, is used for high-temperature crucibles, arc-light and dry-cell electrodes, lead pencils, and as a lubricant. Charcoal, an amorphous form of carbon, is used as an absorbent for gases and as a decolorizing agent. *See* CARBON BLACK; CHARCOAL; DIAMOND; GRAPHITE.

The compounds of carbon find many uses. Carbon dioxide is used for the carbonation of beverages, for fire extinguishers, and in the solid state as a refrigerant. Carbon monoxide finds use as a reducing agent for many metallurgical processes. Carbon tetrachloride and carbon disulfide are important solvents for industrial uses. Freon is used in refrigeration devices. Calcium carbide is used to prepare acetylene, which is used for the welding and cutting of metals as well as for the preparation of other organic compounds. Other metal carbides find important uses as refractories and metal cutters. [E.E.W.]

Carbon arc lamp An arc lamp whose electrodes are pure carbon. The lamp is either open, enclosed, or an intensified arc with incandescence at the electrodes and some light from the luminescence of the arc. The open-arc form of the carbon arc is obsolete. In the enclosed type, a restricted supply of air slows the electrode consumption and secures a life of approximately 100 h. The intensified type uses a small electrode, giving a higher intensity output that is more white in color and more steady than either the open or enclosed types. The electrode life is approximately 70 h. In the high-intensity arc the current may reach values of 125–150 A with positive volt-ampere characteristics, and these may operate directly from the line without a ballast. The brightness of the high-intensity arc is higher and depends markedly on current. Although carbon arc lamps are built for alternating-current operation, direct current produces a steadier and more satisfactory operation. *See* ARC LAMP. [J.O.K.]

Carbon black An amorphous form of carbon produced commercially by thermal or oxidative decomposition of hydrocarbons. It is used principally in rubber goods, pigments, and printer's ink. It is not an inert filler but enhances and reinforces various properties of rubber.

Manufacturing processes may be classed as contact, furnace, or thermal. In the channel (contact) process, natural gas is burned with insufficient air for complete combustion. The smoky flame from individual burners impinges on a cool channel iron, and carbon black deposited on the channel is removed by a scraper blade. In the furnace process, the hydrocarbon and air are fed into a reactor. Combustion of part of the hydrocarbon raises the temperature to 2000–3000°F (1100–1700°C), causing decomposition of the unburned hydrocarbon to carbon black. A water spray quickly cools the hot reaction products, and the finely divided black is recovered by cyclones and bag filters, In the thermal process, natural gas is decomposed to carbon and hydrogen by heated refractories. [C.J.He.]

Carbon dioxide A colorless, odorless, tasteless gas, formula CO_2, about 1.5 times as heavy as air. Under normal conditions, it is stable, inert, and nontoxic. The decay (slow oxidation) of all organic materials produces CO_2. Fresh air contains approximately 0.033% CO_2 by volume. In the respiratory action (breathing) of all animals and humans, CO_2 is exhaled.

Carbon dioxide gas may be liquefied or solidified. Solid CO_2 is known as dry ice. Carbon dioxide is obtained commercially from four sources: gas wells, fermentation, combustion of carbonaceous fuels, and as a by-product of chemical processing. Applications include use as a refrigerant, in either solid or liquid form, inerting medium, chemical reactant, neutralizing agent for alkalies, and pressurizing agent.

Most CO_2 is obtained as a by-product from steam-hydrocarbon reformers used in the production of ammonia, gasoline, and other chemicals; other sources include fermentation, deep gas wells, and direct production from carbonaceous fuels. Whatever the source, the crude CO_2 (containing at least 90% CO_2) is compressed in either two or three stages, cooled, purified, condensed to the liquid phase, and placed in insulated storage vessels. Carbon dioxide is distributed in three ways; in high-pressure uninsulated steel cylinders; as a low-pressure liquid in insulated truck trailers or rail tank cars; and as dry ice in insulated boxes, trucks, or boxcars. [J.S.L.]

Carbon-nitrogen-oxygen cycles A group of nuclear reactions that involve capture of protons by carbon, nitrogen, and oxygen nuclei. These cycles are believed to be the source of energy in main-sequence stars which are more massive than the Sun. Completion of any one of the cycles results in consumption of four protons, synthesis of one helium nucleus (^4He) and two neutrinos, and 26.73 MeV of energy. This energy E reflects the difference in mass m between the four protons and the helium nucleus and is equal to the mass difference times the square of the velocity of light c, as is known from Einstein's statement of mass-energy equivalence, $E = mc^2$. Because the nuclear fuel consumed in these processes is hydrogen, they are referred to as hydrogen burning by means of the carbon-nitrogen-oxygen (CNO) cycles. [G.R.C.]

Carbon star Any of a class of stars with an apparently high abundance ratio of carbon to hydrogen. In normal stars this ratio also prevails, oxygen is more abundant than carbon, and carbon and nitrogen are about equal. In carbon stars, however, carbon is more abundant than oxygen. The majority of the carbon-rich stars are low-temperature red giants of the C class, showing bands of C_2, CN, and CH (the older designation of this is R or N). A few hot carbon-rich stars are also known. The importance of the C/O ratio lies in the theory of the origin of chemical elements, in which it is shown that carbon is easily synthesized at temperatures found in cores of evolving red giants. *See* STAR. [J.L.Gr.]

Carbonate The CO_3^{2-} ion derived from carbonic acid, H_2CO_3, which is a solution of carbon dioxide in water. Because there are two replaceable hydrogen ions in carbonic acid, the salts obtained by neutralizing only one hydrogen are called bicarbonates, for example, sodium bicarbonate, $NaHCO_3$.

Carbonates can be converted into bicarbonates by adding excess carbon dioxide. The reaction can be reversed by heating. Bicarbonates in general are more soluble than carbonates. All the carbonates except those of the alkali metals are insoluble. [E.E.W.]

Carbonate minerals Mineral species containing the carbonate ion as the fundamental anionic unit. The carbonate minerals can be classified as (1) anhydrous normal carbonates, (2) hydrated normal carbonates, (3) acid carbonates (bicarbonates), and (4) compound carbonates containing hydroxide, halide, or other anions in addition to the carbonate. *See* CARBONATE.

Most of the common carbonate minerals belong to group (1), and can be further classified according to their structures. The rhombohedral carbonates are typified by calcite, $CaCO_3$, and by dolomite, $CaMg(CO_3)_2$, The other structural type within this group is that of aragonite, which has orthorhombic symmetry. *See* ANKERITE; ARAGONITE; CALCITE; CERUSSITE; DOLOMITE; MAGNESITE; RHODOCHROSITE; SIDERITE; SMITHSONITE; STRONTIANITE; WITHERITE.

The minerals in groups (2) and (3) all decompose at relatively low temperatures and therefore occur only in sedimentary deposits (typically evaporites) and as low-temperature hydrothermal alteration products. The only common mineral in these groups is trona, $Na_3H(CO_3)_2 \cdot 2H_2O$. *See* SALINE EVAPORATE.

Similarly, the group (4) minerals are relatively rare and are characteristically low-temperature hydrothermal alteration products. The commonest members of this group are malachite, $Cu_2CO_3(OH)_2$, and azurite, $Cu_3(CO_3)_2(OH)_2$, which are often found in copper ore deposits. *See* AZURITE; MALACHITE.

Important occurrences of carbonates include ultrabasic igneous rocks such as carbonatites and serpentinites, and metamorphosed carbonate sediments, which may recrystallize to form marble. The major occurrences of carbonates, however, are in sedimentary deposits as limestone and dolomite rock. *See* DOLOMITE ROCK; LIMESTONE; MARBLE; SEDIMENTARY ROCKS. [A.M.G.]

Carbonatite A carbonate-rich rock of magmatic derivation or descent. Carbonatites are important scientifically because their study provides information (1) on the origin of their associated subsilicic alkalic rocks (nepheline syenites and so forth), (2) on the origin of low-silicate or nonsilicate magmas, (3) on explosive vulcanism, (4) on the origin of kimberlites and subcrustal processes, and (5) on the tectonics of cratonic areas. Their economic significance also is immense. In addition to being the major source of niobium (chiefly as pyrochlore), some carbonatites yield large amounts of rare-earth elements and others are sources or potential sources for phosphate (apatite), barite, thorium, vermiculite, magnetite, titanium, and carbonate rock. One of the world's largest copper deposits (at Palabora, South Africa) is in a carbonatite. The chief carbonate species are calcite, dolomite, and ankerite. Most of the features that characterize silicate igneous rocks also typify carbonatites. There are both intrusive and extrusive types. *See* IGNEOUS ROCKS. [E.W.H.]

Carboniferous A division of later Paleozoic rocks, economically important because of the fossil fuel (coal) deposits they contain. The Carboniferous represents a 65×10^6 year time span that began 345×10^6 years ago and ended 280×10^6 years ago. The name Carboniferous was first applied to the coal-bearing rocks in England: it is now widely used in Europe, Asia, northern Africa, northern South America, and Greenland. In North America the term Carboniferous is not widely used; instead, the time interval and representative rock record are subdivided into two major units, the earlier Mississippian period (345 to 310×10^6 years ago) and the later Pennsylvanian period (310 to 280×10^6 years ago). In southern Africa, southern South America, and Antarctica (the areas referred to as Gondwanaland), the late Paleozoic and early Mesozoic Gondwana System include Carboniferous age rocks. *See* MISSISSIPPIAN; PALEOZOIC; PENNSYLVANIAN.

Although no type locality was designated, the Pennine uplands of Yorkshire and Derbyshire, England, were intended to be used as the typical area of the Carboniferous system of rocks and period of time. They are limited below (earlier) by the Devonian and above (later) by the Permian. The Carboniferous rocks are separated from the overlying and underlying systems by unconformities in many places; however, where there are no physical interruptions, the positions of the lower and upper limits have been based on a comparison of fossil faunas and floras with those from the type Devonian (southwestern England) and the type Permian (European Soviet Union). A widespread and distinct unconformity separates the Lower Carboniferous (Mississippian) rocks from the Upper Carboniferous (Pennsylvanian) rocks throughout much of North America. Recognition of this unconformity, separating the largely non-coal-bearing Mississippian from the coal-bearing Pennsylvanian strata, led to the early disuse of the term Carboniferous in North America. Although Carboniferous rocks are widespread on all continents, they are absent or poorly known in areas of much older rocks, such as the Canadian or African Shield, or in areas where most of the rocks are made up of younger Mesozoic and Cenozoic sediments and volcanics. *See* DEVONIAN; PERMIAN.

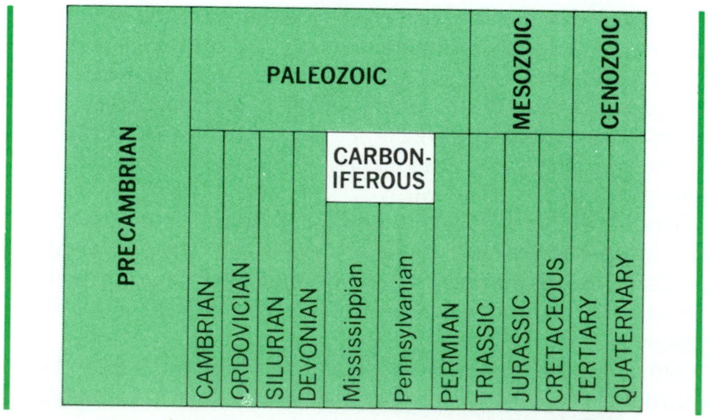

Theories of global tectonics and continental drift propose that at the beginning of Carboniferous time there were three major continental masses moving toward each other; they eventually came together at the end of the Carboniferous to form the supercontinent Pangaea. These three continents were North America–Greenland–western Europe; eastern Europe–northern and eastern Asia; and Gondwanaland (South America, Africa, India, Antarctica, and Australia). North America–western Europe was separated from northern Africa and eastern Europe by a narrow proto-Atlantic, and eastern Europe–northern Asia was separated from northern Africa and Gondwanaland by an eastern widening sea called Tethys. It has been proposed that Gondwanaland was in the region of the South Pole during the Carboniferous and early Permian time. *See* CONTINENT; CONTINENTAL DRIFT; PLATE TECTONICS.

During the Carboniferous the broad epeieric seas supported an abundant diversity of invertebrates, including brachiopods, cephalopods, pelecypods (clams), gastropods (snails), crinoids, corals, and bryozoa. Among the microfossils, fusulinids, ostracods, and conodonts were abundant and diverse. Trilobites were present, but nearing the end of their long history. Corals formed reefs, and calcareous algae formed reeflike mounds in southern Kansas and western and central Texas. Although insect fossils are uncommon, insects were highly diverse and included large dragonflies and giant cockroaches. Among vertebrates there were large sharks, adapted to fresh and marine waters, and amphibians. During the later Carboniferous the first reptiles appeared.

Fossil plants include large trees in widespread swampy, coastal forests throughout many continents. There were also primitive gymnosperms, related to conifers of today, which seem to have grown in somewhat less swampy areas and drier climates. The first seed-bearing plants (pteridosperms), with foliage resembling ferns, appeared during the Middle Carboniferous. *See* PALEOBOTANY; PALEONTOLOGY.

The Upper Carboniferous (Pennsylvanian) is the world's leading source of medium- and high-ranked coal. It is produced in the Appalachian coalfield, the Illinois Basin, and eastern Mid-Continent in the United States, the Maritime Provinces of eastern Canada, England, southern Wales, the Midland Valley of Scotland, the Westphalian coal basin of western Europe, the Silurian coal basin of Poland, the Donets Basin of the southern Soviet Union, and smaller basins in France, Spain, Morocco, Algeria, Norway, and northern China and Korea. Petroleum is produced from Carboniferous rocks in many parts of the central and western United States, western Canada, western Europe, and the North Sea. Oil shales occur in Carboniferous rocks in Scotland. Refractory clays (diaspore and kaolinite) are derived from Lower Pennsylvanian sediments in Pennsylvania, Ohio, eastern Kentucky, northern and western Illinois, eastern Missouri, and the Missouri Ozarks. [W.A.P.]

Carbonyl A functional group found in organic compounds in which a carbon atom is doubly bonded to an oxygen atom:

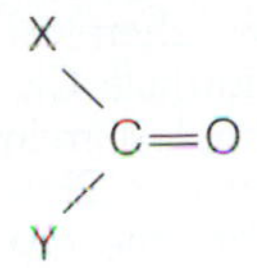

Depending upon the nature of the other groups attached to carbon, the most common compounds containing the carbonyl group are aldehydes (X and Y = H; X = H, Y = alkyl or aryl), ketones (X and Y = alkyl or aryl), carboxylic acids (X = OH, Y = H, alkyl, or aryl), esters (X = O-alkyl or aryl; Y = H, alkyl, or aryl), and amides (X = N—H, N-alkyl, or N-aryl; Y = H, alkyl, or aryl). Other compounds that contain the carbonyl group are acid halides, acid anhydrides, lactones, and lactams. *See* ACID ANHYDRIDE; ACID HALIDE; ALDEHYDE; AMIDE; ESTER; KETONE.

All the compounds containing this functional group are referred to in a general way as carbonyl compounds. It is important, however, to distinguish these compounds from a large group formed from metals and carbon monoxide, which are known as metal carbonyls. In these latter compounds, there is only one group attached to the carbon in addition to the oxygen, and the carbon atom is viewed as triply bonded to the oxygen. *See* METAL CARBONYL. [J.P.Fr.]

Carborane A generic term for a class of chemical compounds composed of boron, carbon, and hydrogen. The term is also the common name for an important member of this class of compounds, $C_2B_{10}H_{12}$. Carborane derivatives have been used to develop new types of polymers which withstand high temperatures. In addition, carboranes have been reacted with transition metals to produce new types of complexes, some of which exhibit catalytic activity.

The carboranes and the boranes have structures which can be rationalized on the basis of a knowledge of triangulated polyhedrons, or fragments of triangulated polyhedrons. Triangulated polyhedrons with 4 to 12 vertexes are especially important in understanding the structures of boranes and carboranes. Three polyhedrons, those with 4, 6, and 12 vertexes, are very symmetrical, and their vertexes are all geometrically equivalent (see illustration). The structures of most boranes and carboranes are derived by placing a B-H unit or a C-H unit at the vertexes of the polyhedrons. For example, the borane anion $B_{12}H_{12}{}^{2-}$ has the icosahedral structure with a B-H unit at each of the 12 vertexes. A carborane exists when a C-H unit replaces a B-H unit at a vertex. Two known icosahedral carboranes are $CB_{10}H_{12}{}^{-}$ and $C_2B_{10}H_{12}$, in which one and two C-H units reside at the vertexes of the icosahedron. There is only a single isomer of $CB_{10}H_{12}{}^{-}$, but all three possible isomers of $C_2B_{10}H_{12}$ have been prepared.

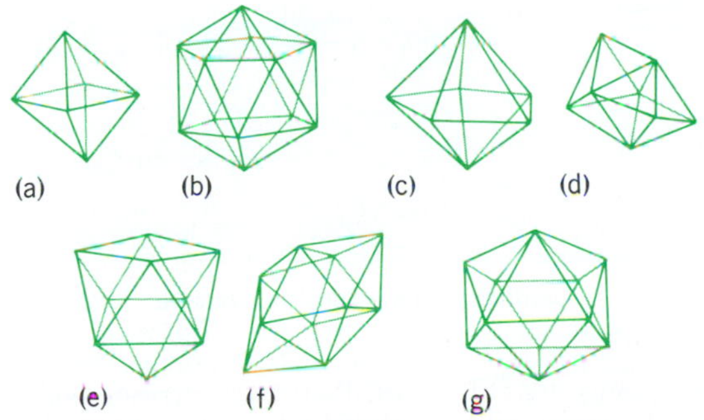

Skeletal structures of polyhedrons. (*a*) Hexahedron. (*b*) Icosahedron. In figures *c–g* the *n* = 7–11 polyhedrons: (*c*) Pentagonal bipyramid. (*d*) Dodecahedron. (*e*) Tricapped trigonal prism. (*f*) Bicapped square antiprism. (*g*) Octadecahedron.

Not all compounds consisting of boron, carbon, and hydrogen are carboranes. To qualify as a carborane, a compound must have at least one carbon atom in the three-dimensional skeleton of the structure. If all the carbon atoms are in side groups, replacing a hydrogen atom in a borane, the compound is not a carborane. For example, $C_2H_5B_{12}H_{11}{}^{2-}$ is a derivative of $B_{12}H_{12}{}^{2-}$ in which a hydrogen atom has been replaced by an ethyl group; it is thus not a carborane even though it is composed of boron, carbon, and hydrogen. *See* BORANE. [P.A.W.]

Carboxylic acid One of a large family of organic substances widely distributed in nature, and characterized by the presence of one or more carboxyl groups (—COOH). These groups typically yield protons in aqueous solution. In the type formula, $R(CXY)_nCOOH$, symbols R, X, and Y can be hydrogen, saturated or unsaturated groups, carboxyl, alicyclic, or aromatic groups, halogens, or other substituents, and *n* may vary from zero (formic acid, HCOOH) to more than 100, provided that the normal carbon covalence of four is maintained.

Physical and chemical properties of carboxylic acids are represented, grossly, by the resultant of the various chemical groupings present in the molecule. A short-chain aliphatic acid, wherein the carboxyl is dominant, is a pungent, corrosive, water-soluble liquid of abnormally high boiling point (because of molecular association), with specific gravity close to 1 (higher for formic and acetic acids). With increasing molecular weight, the hydrocarbon grouping overbalances the carboxyl; sharpness of odor diminishes, boiling and melting points rise, the specific gravity falls toward that of the parent hydrocarbon, and the water solubility decreases. Thus the typical high-molecular-weight saturated acid is a bland, waxlike solid.

Acids are used in large quantities in the production of esters, acid halides, acid amides, and acid anhydrides. They find wide use in the manufacture of soaps and detergents, in thickening lubricating greases (stearate soaps), in modifying rigidity in plastics, in compounding buffing bricks and abrasives, and in the manufacture of crayons, dictaphone cylinders, and phonograph records. The solvent action of acids finds use in manufacture of carbon paper, inks, and in the compounding of synthetic and natural rubber. Because of the stability of saturated fatty acids toward oxidation, these are often used as solvents for carrying out oxidation reactions upon sensitive compounds. [E.B.R.]

Carbuncle An infection of the subcutaneous tissues in which multiple sinuses are created by tissue destruction caused by the spreading microorganism. The commonest agent responsible for carbuncles is *Staphylococcus aureus*, although other microorganisms are sometimes responsible. *See* STAPHYLOCOCCUS.

Most carbuncles begin with the spread of infection from an initial site in a hair follicle or other superficial structure. The invading organisms multiply and penetrate into the deeper regions of the dermis, subcutaneous tissues, and sometimes even below the deep fascia of the body layers. Although the boil is typically a single pocket of pus, the carbuncle differs by being multiloculated and usually deeper in its location. Quite often tissue penetration produces several superficial openings into the carbuncle, all of which may drain pus.

Regional tenderness, pain, and swelling are usual. In older or debilitated persons there is a greater chance of such infections becoming blood-borne, with resultant bacteremia, spread of abscesses to distant sites, and severe damage to such organs as the kidney, brain, and lungs. [E.G.St./N.K.M.]

Carburetor The device that controls the power output and fuel feed of internal combustion spark-ignition engines generally used for automotive, aircraft, and auxiliary services. Its duties include control of the engine power by the air throttle; metering, delivery, and mixing of fuel in the airstream; and graduating the fuel-air ratio according to engine requirements in starting, idling, and load and altitude changes.

A simple updraft carburetor illustrates basic carburetor action (see illustration). Intake air charge, at full or reduced atmospheric pressure as controlled by the throttle, is drawn into the cylinder by the downward motion of the piston to mix with the unscavenged exhaust remaining in the combustion chamber from the previous explosion. The total air-charge weight per unit time is approximately proportional to the square root of its pressure drop in the carburetor venturi throat (with a square root correction, direct for change in air pressure, and inverse for change in air absolute temperature). Also, each individual cylinder air-charge volume follows the intake pressure minus about one-sixth the exhaust pressure, so that the intake manifold pressure and temperature are often taken as an approximate indication of the engine torque; this factor times the engine speed is similarly taken as a measure of the engine power output. Any given fixed part-throttle opening gives higher intake manifold pressure, increased individual-cylinder air charge, and greater engine torque, as the engine speed is decreased.

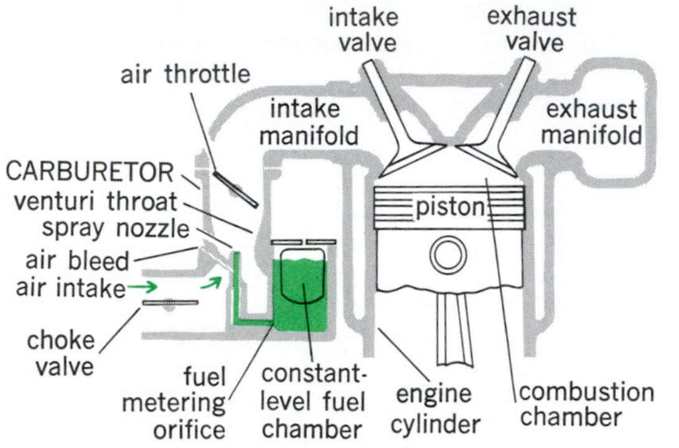

Elements which basically determine air and fuel charges received by an engine through the carburetor.

The fuel is usually metered through a calibrated metering orifice at a differential pressure derived from the pressure drop in a venturi throat in the intake air passage. For normal steady speed, the limits of consistent ignition lie between 0.060 and 0.083 fuel-air by weight. Richer mixtures (more fuel) give higher power, apparently cooler combustion, and less detonation. The weaker mixtures give better fuel economy. A narrowly controlled rich mixture is required for idling because low charge densities and relatively high contamination with residual exhaust make combustion conditions unfavorable. Individual adjustments are required both for the idling fuel feed and for the throttle closure stop to keep the engine firing at minimum speed.

There are two reasons for condensation of atmospheric moisture and formation of ice in the intake system. (1) At partly closed throttle, the adiabatic pressure drop across the throttle orifice generates a temperature drop. When atmospheric humidity is high, ice sometimes partly clogs the throttle orifice, requiring further throttle opening to keep the engine running for a short period after starting and before the engine heat has had time to build up. (2) As fuel evaporates, ice tends to form wherever fuel spray impinges upon an unheated surface. The problem has been largely eliminated either by use of the injection carburetor, which delivers fuel only in the warm parts of the intake system, or by direct injection into the engine cylinders. *See* INTERNAL COMBUSTION ENGINE. [F.C.M./J.A.B.]

Carbyne Either of two chemical systems: a monovalent carbon radical of general formula CX or an allotropic form of elemental carbon. The simplest member of the monovalent carbon radical is methylidyne, CH. Other carbyne radicals which have been studied are the halomethylidynes (CF, CCl, CBr) and carboethoxymethylidyne ($CCO_2C_2H_5$).

Methylidyne is isoelectronic with the nitrogen (N) atom. They differ in structure in that the lowest energy (ground state) of N has three unpaired electrons while that of CH has but one, resulting in a doublet ($^2\pi$) ground state. This configuration is simply described as $1\sigma^2 2\sigma^2 3\sigma^2 1\pi^1$. The ground state of all carbyne radicals conforms to this single spectroscopic state and is unique in carbon chemistry. Studies of the reactions of carbynes relating to the ground-state $^2\pi$ species have shown that the reaction products are themselves highly reactive, and the nature of the products has in general to be inferred. *See* CHEMICAL STRUCTURES.

The normally naturally occurring forms of carbon are graphite and diamond. In 1967, a new allotrope, termed carbyne, supposedly consisting of carbon atoms linked by alternating single and triple bonds, —C≡C—C≡C—C≡, was reported by Soviet workers. Several polymorphs have since been reported to have been found occurring within the natural graphite system, and include the mineral named chaoite. *See* CARBON; DIAMOND; GRAPHITE. [O.P.S.; T.N.B.]

Cardamon The plant *Elettaria cardamomum* (Zingiberaceae), a perennial herb that is a native of India. The small, light-colored seeds, borne in capsules, have a delicate flavor. They are used in curries, cakes, pickles, and in general cooking, as well as in medicine. *See* ZINGIBERALES. [P.D.St./E.L.C.]

Cardinal points The four intersections of the horizon with the meridian and with the prime vertical circle, or simply prime vertical, the intersections with the meridian being designated north and south, and the intersections with the prime vertical being designated east and west (see illustration). The cardinal points are 90° apart; they lie in a plane with each other and correspond to the cardinal regions of the heavens. The four intermediate points, northeast, southeast, northwest, and southwest, are the collateral points. [F.H.R.]

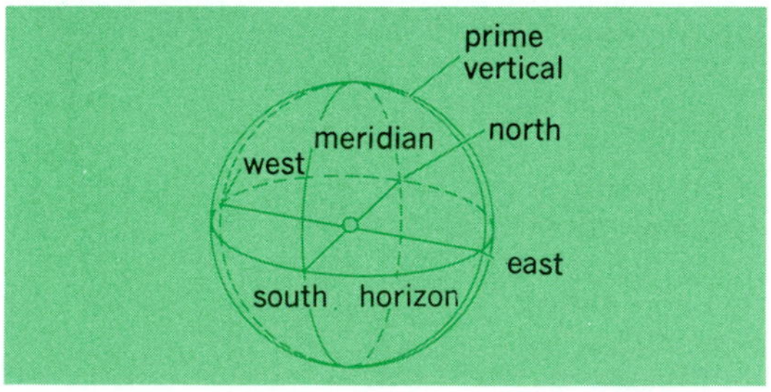

Cardinal points around the horizon.

Cardioid A heart-shaped curve generated by a point of a circle that rolls (without slipping) on a fixed circle of the same diameter. In point-wise construction of the curve, let *O* be a fixed point of a circle *C* of diameter *a*, and *Q* a variable point of *C*. Lay off distance *a* along the secant *OQ*, in both directions from *Q*. The locus of the two points thus obtained is a cardioid (see illustration). If a rectangular coordinate system is chosen with *O* for origin initially and *y* axis tangent to *C* at *O*, the cardioid has equation $(x^2 + y^2 - ax)^2 = a^2(x^2 + y^2)$. The equation in polar coordinates is $\rho = a(1 + \cos\theta)$. Its area is

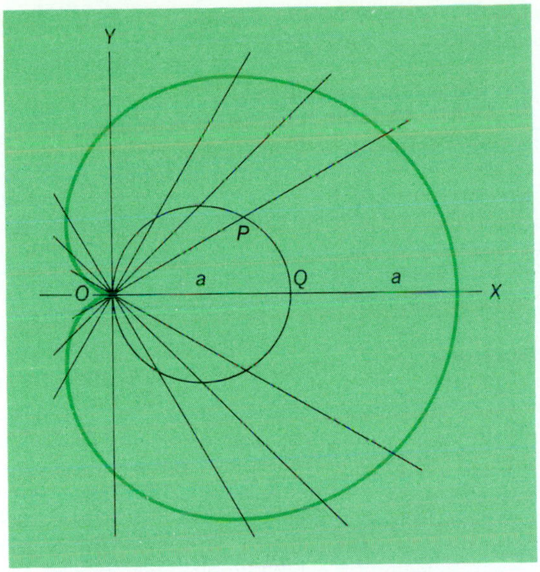

A cardioid (symbols are explained in the text).

$3/2\pi a^2$, or six times the area of C, and its length is $8a$. *See* ANALYTIC GEOMETRY. [L.M.Bl.]

Cardiovascular system Those structures, such as the heart, or pumping mechanism, and the arteries, veins, and capillaries, which provide channels for the flow of blood. The cardiovascular system is sometimes called the blood-vascular system. The circulatory system includes both the cardiovascular and lymphatic systems; the latter consists of lymph channels (lymphatics), nodes, and fluid lymph which finally empties into the bloodstream. *See* ARTERY; BLOOD; CAPILLARY (ANATOMY); HEART (VERTEBRATE); HEMATOPOIESIS; LYMPHATIC SYSTEM; VEIN. [C.K.W.]

Circulatory physiology describes the structure and operation of the circulation in living animals, and enquires as to how or why the circulatory system may have evolved. The circulatory system in all vertebrates has multiple functions, but all functions are involved in regulating the internal environment of the animal (promoting homeostasis). In all vertebrates the circulatory system consists of a central pump, the heart, which drives a liquid transport medium, the blood, continuously around a closed system of tubes, the vascular system. The arterial portion of this system is divided into larger elastic and smaller resistance vessels (arterioles) which distribute blood to specialized regions or organs where transfer of nutrients, oxygen, or waste products takes place across the walls of a fine network of microscopic capillaries. Blood from the capillaries passes through the venules (small venous vessels) into the main vein and returns to the heart. The arterioles, venules, and capillaries make up the microcirculation, which is arguably the most important functional role of the vertebrate circulatory system from a functional point of view.

Cardiovascular system disorders are those disorders which involve the arteries, veins, and lymphatics. *See* ARTERIOSCLEROSIS; ARTERITIS; LYMPHADENITIS; LYMPHANGITIS; PHLEBITIS; VARICOSE VEINS. [D.R.J.]

Caribbean Sea A deep ocean basin roughly 1750 mi (2800 km) long and up to 800 mi (1500 km) wide, bordered by Central and South America and the West Indies island chain. Together with the Gulf of Mexico, it is sometimes called the American Mediterranean Sea. Two major basins, the large Caribbean Basin and the smaller and deeper Cayman Basin, are separated by the Jamaica Rise. The mild and equable climate of the Caribbean Sea may be marred in the summer and early autumn by occasional hurricanes.

The basic circulation is from east to west with North Equatorial Current water entering through passages between the eastern islands and exiting through the Yucatán Channel to the Gulf of Mexico, and eventually through the Florida Straits to become an important portion of the Gulf Stream system. *See* ATLANTIC OCEAN; GULF OF MEXICO; OCEAN CIRCULATION. [W.G.M.]

Carnauba wax Product exuded from the leaves of the wax palm, *Copernicia cerifera*, a native of Brazil and other regions in tropical South America. It is the hardest, highest-melting natural wax and is used in making candies, shoe polish, high-luster wax, varnishes, phonograph records, and surface coating of automobiles. *See* PALM; WAX, ANIMAL AND VEGETABLE. [E.L.C.]

Carnivora One of the larger orders of placental mammals, including fossil and living dogs, raccoons, pandas, bears, weasels, skunks, badgers, otters, mongooses, civets, cats, hyenas, seals, walruses, and many extinct groups organized into 12 families, with about 112 living genera and more than twice as many extinct genera. The subdivision of the order into three superfamilies has long been practiced and the following groups seem appropriate: Miacoidea, Canoidea, and Feloidea. The primary adaptation in this order was for predation on other vertebrates and invertebrates. A few carnivorans (for example, bear and panda) have secondarily become largely or entirely herbivorous, but even then the ancestral adaptations for predation are still clearly evident in the structure of the teeth and jaws. The Carnivora have been highly successful animals since their first appearance in the early Paleocene.

Structural adaptations involve the teeth and jaws. The dentition is sharply divided into three functional units. The incisors act as a tool for nipping and delicate prehension, and the large, interlocking upper and lower canines for heavy piercing and tearing during the killing of prey. The cheek teeth are divided into premolars (for heavy prehension) and molars (for slicing and grinding), which may be variously modified depending on the specific adaptation, but there is a constant tendency for the last (fourth) upper premolar and the first lower molar to enlarge and form longtitudinal opposed shearing blades (the carnassials). In all carnivorans the jaw articulation is arranged in such a manner that movement is limited to vertical hinge motions and transverse sliding. The temporal muscle dominates the jaw musculature, forming at least one-half of the total mass of the jaw muscles.

The earliest fossil records are early Paleocene, but the earliest well-represented material comes from the middle Paleocene of North America. During the Paleocene and Eocene the stem-carnivorans or miacoids underwent considerable diversification in both the Old and New World. At the end of Eocene and beginning of Oligocene time throughout the Northern Hemisphere, a dramatic change took place within the Carnivora; this was the appearance of primitive representatives of modern carnivoran families. *See* BADGER; BEAR; CAT; CIVET; COATI; DOG; FERRET; FISHER; HYENA; MAMMALIA; MARTEN; MINK; MONGOOSE; OTTER; PANDA; PINNIPEDS; RACCOON; SKUNK; WEASEL; WOLVERINE. [R.H.T.]

Carnot cycle A hypothetical thermodynamic cycle used as a standard of comparison for actual cycles. The Carnot cycle shows that, even under ideal conditions, a heat engine cannot convert all the heat energy supplied to it into mechanical energy; some of the heat energy must be rejected.

In a Carnot cycle, an engine accepts heat energy from a high-temperature source, or hot body, converts part of the received energy into mechanical (or electrical) work, and rejects the remainder to a low-temperature sink, or cold body. The greater the temperature difference between the source and sink, the greater the efficiency of the heat engine.

The Carnot cycle (see illustration) consists first of an isentropic compression, then an isothermal heat addition, followed by

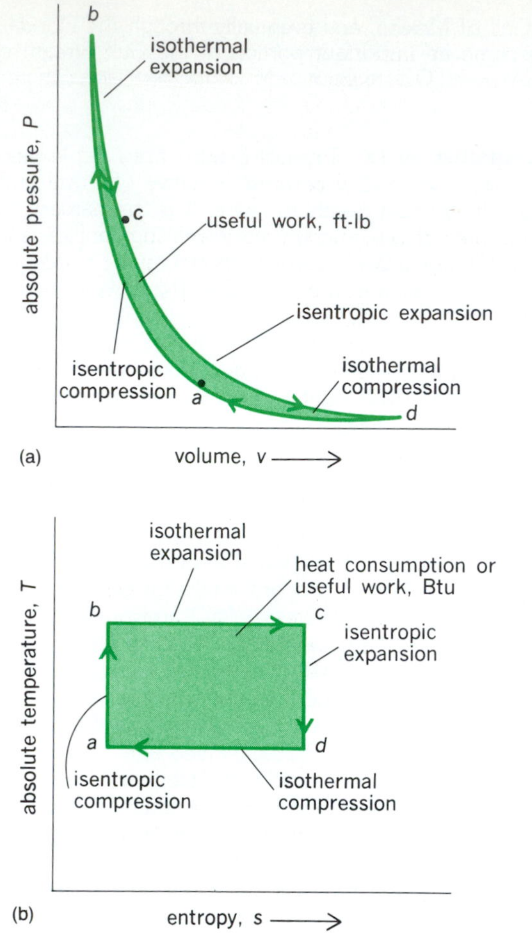

Carnot cycle for air. (*a*) Absolute pressure versus volume. (*b*) Absolute temperature versus entropy.

an isentropic expansion, and concludes with an isothermal heat rejection process. In short, the processes are compression, addition of heat, expansion, and rejection of heat, all in a qualified and definite manner. The net effect of the cycle is that heat is added at a constant high temperature, somewhat less heat is rejected at a constant low temperature, and the algebraic sum of these heat quantities is equal to the work done by the cycle.

A Carnot cycle consists entirely of reversible processes; thus it can theoretically operate to withdraw heat from a cold body and to discharge that heat to a hot body. To do so, the cycle requires work input from its surroundings. The heat equivalent of this work input is also discharged to the hot body. Just as the Carnot cycle provides the highest efficiency for a power cycle operating between two fixed temperatures, so does the reversed Carnot cycle provide the best coefficient of performance for a device pumping heat from a low temperature to a higher one. *See* HEAT PUMP; REFRIGERATION CYCLE.

Good as the ideal Carnot cycle may be, there are serious difficulties that emerge when one wishes to make an actual Carnot engine. The necessarily high peak pressures and temperatures limit the practical thermal efficiency that an actual engine can achieve. Although the Carnot cycle is independent of the working substance, and hence is applicable to a vapor cycle, the difficulty of efficiently compressing a vapor-liquid mixture renders the cycle impractical. *See* POWER PLANT; THERMODYNAMIC CYCLE; THERMODYNAMIC PRINCIPLES. [T.Ba.]

Carnotite A mineral which is a hydrous vanadate of potassium and uranium, $K_2(UO_2)_2(VO_4)_2 \cdot nH_2O$. The water content varies at ordinary temperatures from one to three molecules. Carnotite generally occurs as a powder or as a slightly coherent microcrystalline aggregate. Color ranges from bright yellow to lemon- and greenish-yellow.

In the United States the principal region of carnotite mineralization is the Colorado Plateau and adjoining districts of Utah, New Mexico, and Arizona. Carnotite is found also in Wyoming and in Carbon County, Pennsylvania. Deposits are located at Radium Hill near Olary, Australia, and in Katanga (Zaire). Carnotite is the chief source of uranium in the United States. It is also a source of radium and vanadium. *See* RADIOACTIVE MINERALS; URANIUM; VANADIUM. [W.R.Lo.]

Carotenoid Any of a class of yellow, orange, red, and purple pigments that are widely distributed in nature. Carotenoids are generally fat-soluble unless they are complexed with proteins. In plants, carotenoids are usually located in quantity in the grana of chloroplasts in the form of carotenoprotein complexes. Carotenoprotein complexes give blue, green, purple, red, or other colors to crustaceans, echinoderms, nudibranch mollusks, and other invertebrate animals. Some coral coelenterates exhibit purple, pink, orange, or other colors due to carotenoids in the calcareous skeletal material. Cooked or denatured lobster, crab, and shrimp show the modified colors of their carotenoproteins.

The general structure of carotenoids is that of aliphatic and aliphatic-alicyclic polyenes, with a few aromatic-type polyenes. Most carotenoid pigments are tetraterpenes with a 40-carbon (C_{40}) skeleton. More than 300 carotenoids of known structure are recognized, and the number is still on the rise. *See* TERPENE.

There are several biochemical functions in which the role of carotenoids is well understood. These include carotenoids in the photosynthetic apparatus of green plants, algae, and photosynthetic bacteria, where carotenoids function as a blue light-harvesting pigment (antenna or accessory pigment) for photosynthesis. Thus carotenoids make it possible for photosynthetic organisms more fully to utilize the solar energy in the visible spectral region. *See* CHLOROPHYLL; PHOTOSYNTHESIS.

Another function of carotenoids is to protect biological systems such as the photosynthetic apparatus from photodynamic damage. This is done by quenching the powerful photodynamic oxidizing agent, singlet oxygen, produced as an undesirable by-product of the exposure of pigmented organisms to light.

Perhaps the most important industrial application of carotenoids is in safe coloration of foods, as exemplified in the coloring and fortification of margarine and poultry feedstuff. [P.-S.S.; T.Y.L.]

Carotid body A special sensory organ (glomus caroticum) which is located in the angle between the bifurcation of the common carotid artery into the external and internal carotid arteries. The carotid body is a bilateral ovoid structure which in humans measures approximately 0.2 by 0.3 in. (5 by 7 mm) and is usually embedded within the outer connective tissue layer of the adjacent common carotid artery. The carotid bodies are termed chemoreceptors because they closely monitor the oxygen and carbon dioxide (CO_2) content of the blood. *See* CHEMORECEPTION. [J.Ha.]

Carp The common name for a number of cypriniform fishes of the family Cyprinidae. The carp (*Cyprinus carpio*) is closely related to the goldfish (*Crassius auratus*). The fish originated in China, where for centuries it was raised for food. It was imported into the United States from Europe, where it also has been raised for years as a source of food.

The carp has pharyngeal teeth and a suckerlike mouth. A very long dorsal fin is preceded by a strong spine. The swim bladder of the carp is associated with a group of small bones at the anterior end, the Weberian ossicles. This structural modification enables the carp to perceive sound quite well. *See* CYPRINIFORMES. [C.B.C.]

Carpal tunnel syndrome A condition caused by the thickening of ligaments and tendon sheaths at the wrist, with consequent compression of the median nerve at the palm. Affected individuals report numbness, tingling, and pain in the hand; the discomfort often becomes worse at night or after use of the hand. A physical examination of the injured hand during the early stages of the syndrome often reveals no abnormality. With more severe nerve compression, the individual experiences sensory loss over some or all of the digits innervated by the median nerve (thumb, index finger, middle finger, and ring finger) and weakness of thumb movement.

The incidence of carpal tunnel syndrome is greater among electronic-parts assemblers, frozen-food processors, musicians, and dental hygienists. Highly repetitive wrist movements, use of vibrating tools, awkward wrist positions, and movements involving great force seem to be correlated with the disorder. Awkward and repetitive wrist motions occur in many office tasks, such as typing and word processing.

Carpal tunnel syndrome probably accounts for a minority of the cases of overuse syndrome (cumulative trauma syndrome), which is a common problem in occupational settings. Overuse syndrome symptoms include muscle pain, tendinitis, fibrositis (inflammation of connective tissue in a joint region), and epicondylitis (inflammation of the eminence on the condyle of a bone). Although the causative relationship between the two disorders has not been conclusively proven, the incidence of both carpal tunnel syndrome and overuse syndrome appears to increase in tandem in individuals who are at risk.

Nonsurgical treatment includes avoidance of the use of the wrist, use of a splint to keep the wrist in a neutral position, and anti-inflammatory medications. These treatments are especially useful in individuals with an acute flare-up and in those with minimal and intermittent symptoms. Surgical treatment may be used if conservative approaches fail. The procedure is usually done on an outpatient basis with prognoses of good to excellent in 80% of the cases. Although 40% of the individuals regain normal function, the condition of 5% may worsen. [D.M.D.]

Carpoids An assemblage of enigmatic, rare, Paleozoic echinoderms formerly grouped together in one class (called Carpoidea), but now viewed as comprising four distinct classes which together constitute the subphylum Homalozoa. Like all echinoderms, carpoids had skeletal plates, each of which was made up of a single crystal of calcite; they differ from all other echinoderms in showing no trace of radial symmetry. Bilaterally arranged plates roofed the dorsal side to form a carapace, and similar plates formed a plastron to sheathe the concave ventral side (see illustration). Carpoids range from Early Cambrian to Devonian. *See* ECHINODERMATA; HOMALOZOA. [H.B.F.]

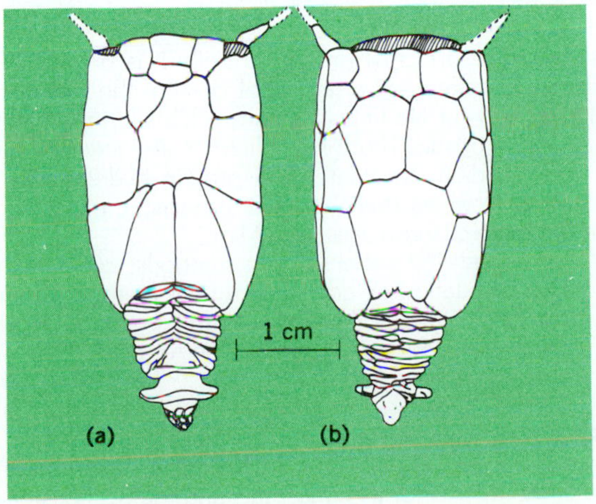

A carpoid, *Enoploura popei*. (*a*) Concave side or plastron. (*b*) Convex side or carapace. (*After K. E. Caster*)

Carrageenan A polysaccharide that is a major constituent of the cell walls of certain red algae (Rhodophyceae), especially members of the families Gigartinaceae, Hypneaceae, Phyllophoraceae, and Solieriaceae. Extracted for its suspending, emulsifying, stabilizing, and gelling properties, it is one of three algal polysaccharides of major economic importance, the others being agar and alginate.

The main sources of carrageenan are *Chondrus crispus* (Gigartinaceae) from the Maritime Provinces of Canada and various species of *Eucheuma* (Solieriaceae) from the Philippines. *Chondrus*, popularly called Irish moss, is harvested from naturally occurring intertidal stands, while *Eucheuma* is successfully grown in mariculture on nets and lines.

About 80% of the refined carrageenan is used in food processing; the dairy industry is the chief consumer. The rest is used in the cosmetic, pharmaceutical, printing, and textile industries. *See* AGAR; ALGINATE; RHODOPHYCEAE. [P.C.Si.; R.L.Moe]

Carrier Any wave to which modulation is subsequently applied. Commonly encountered examples are a sinusoidal wave, a recurrent series of pulses, and a direct current or voltage. It is not always necessary or even desirable to transmit the carrier wave; when not transmitted, it is termed a suppressed carrier. *See* AMPLITUDE MODULATION; ANGLE MODULATION; FREQUENCY MODULATION; PHASE MODULATION. For a definition of pulse carrier *see* PULSE MODULATION; for a definition of subcarrier *see* MODULATION. [H.S.Bl.]

Carrier state The condition of a person who is infected with certain microorganisms and is therefore liable to transmit the disease produced by them, but who does not exhibit signs or symptoms of the infection. The individual may be in such an early phase of the disease that clinical manifestations are not apparent, and in this case is an incubatory carrier. If there has already been a clinical attack, the description is of a convalescent carrier. These carriers sometimes continue to harbor the infectious agent for such an extended period that they are called chronic carriers. It is also possible for a person to carry microorganisms, even without a manifestation of the disease; this association with the germ is an incidental one, and the person is described as a casual carrier. The existence of healthy carriers is a complicating factor in the study of infectious diseases. Since they are without symptoms or signs of disease, they can be identified as carriers only by laboratory investigation. *See* DISEASE; EPIDEMIOLOGY. [C.W.]

Carrion's disease A specific infection, also known as bartonellosis, caused by the microorganism *Bartonella bacilliformis*. The disease presents two clinical types, severe (Oroya fever) and benign (verruga peruana). The disease is endemic to certain regions of South America—Peru, Ecuador, and Colombia—in narrow river valleys and canyons of high altitudes; its distribution parallels the prevalence of certain species of blood-sucking sand flies (*Phlebotomus noguchi* and *P. verrucarum*) which transmit the infection. [M.Yo.]

Carrot A biennial umbellifer (*Daucus carota*) of Asiatic and Mediterranean origin belonging to the plant order Apiales. The carrot is grown for its edible roots which are eaten raw or cooked. Varieties are classified according to length of root (long or short or stump-rooted) and use (fresh market or processing). Popular varieties for fresh market are Imperator and Gold Pak; for processing, Red Cored Chantenay and Royal Chantenay. Texas, California, and Arizona are important producing states. *See* APIALES. [H.J.C.]

Cartilage A specialized kind of connective tissue which is bluish, translucent, and hard but yielding. While it serves as the model for most bones during development, it persists in adults in certain limited regions in three forms: hyaline, or glassy cartilage; elastic cartilage; and fibrocartilage.

Hyaline cartilage is the type found at the ventral ends of ribs, in joints, and in the walls of the larger respiratory passages. The cartilage cells (chondrocytes) occupy about one-third the volume of the cartilage. The remainder is occupied by the matrix, which separates the cells and is thought by many to contain collagen fibrils. As there are no blood vessels in cartilage, exchanges with blood take place over larger distances than in other forms of connective tissue. Hyaline cartilage calcifies during middle age.

Elastic cartilage occurs in mammals in the external ear, the walls of the external auditory and Eustachian tubes, and the epiglottis. It is yellower, more opaque, and more flexible than hyaline cartilage and is rich in elastic fibers.

Fibrocartilage occurs in intervertebral disks, in the symphysis pubis, and in certain tendons near their attachment to bones. It is rich in dense, closely apposed bundles of collagen fibers, and the cells are disposed in columns between the fibers. *See* CONNECTIVE TISSUE. [I.G.]

Cartography The making of maps and charts for the purpose of visualizing spatial distributions over various areas of the face of the Earth. A map can be defined as a conventionalized picture, on a much reduced scale, of the Earth's surface as seen from above. Most maps are laid out on a parallel and meridian system, and have lettering added for identification. Some kinds of maps have long been designated charts, such as aerial and marine navigation charts. A most realistic Earth representation, the globe, is less used than flat maps because of its bulk. *See* GLOBE (EARTH). [E.Ra./W.C.]

Caryophanales An order of bacteria, class Schizomycetes, as defined in the seventh edition of *Bergey's Manual of Determinative Bacteriology* (1957). The order was separated into three families: Caryophanaceae, Oscillospiraceae, and Arthromitaceae. It is not recognized in the classification adopted in the eighth edition (1974). [E.G.Pr.]

Caryophyllales An order of flowering plants, division Magnoliophyta (Angiospermae), in subclass Caryophyllidae, class Magnoliopsida (dicotyledons). It consists of 11 families and about 10,000 species. The names Centrospermae and Centrospermales have often been applied to this order.

The four largest families are the Aizoaceae (about 2500 species), Cactaceae (about 2000 species), Caryophyllaceae (about 2000 species), and Chenopodiaceae (about 1500 species). The Aizoaceae are unarmed leaf-succulents, chiefly of Africa. The Cactaceae are American stem-succulents, mostly spiny and with much reduced leaves. The Caryophyllaceae lack betalains but are otherwise typical of the order. The Chenopodiaceae have reduced, mostly greenish flowers.

Familiar garden plants belonging to the Caryophyllales include beet (*Beta vulgars*) and spinach (*Spinacia oleracea*), both of the Chenopodiaceae; sweet william (*Dianthus barbatus*) and carnation (*D. caryophyllus*), of the Caryophyllaceae; rose moss (*Portulaca grandiflora*), of the Portulacaceae; and the four-o'clock (*Mirabilis jalapa*), of the Nyctaginaceae. *See* BEET; CARYOPHYLLIDAE; SPINACH; SUGARBEET. [A.Cr.]

Caryophyllidae A relatively small subclass of the class Magnoliopsida (dicotyledons) of the division Magnoliophyta (Angiospermae), the flowering plants, consisting of 4 orders, 14 families, and about 11,000 species. Most of these plants contain betalain pigments instead of anthocyanins, and the seeds very often have a perisperm. Most of the families and species of the subclass belong to the order Caryophyllales. The other orders (Batales, Plumbaginales, and Polygonales) have only a single family each. *See* BATALES; CARYOPHYLLALES; MAGNOLIOPSIDA; PLUMBAGINALES; POLYGONALES. [A.Cr.]

Cascode amplifier A transistor amplifier consisting of a common-emitter stage in series with a common-base stage. The input resistance and current gain are nominally equal to the corresponding values for a single common-emitter stage (see illustration). The output resistance is approximately equal

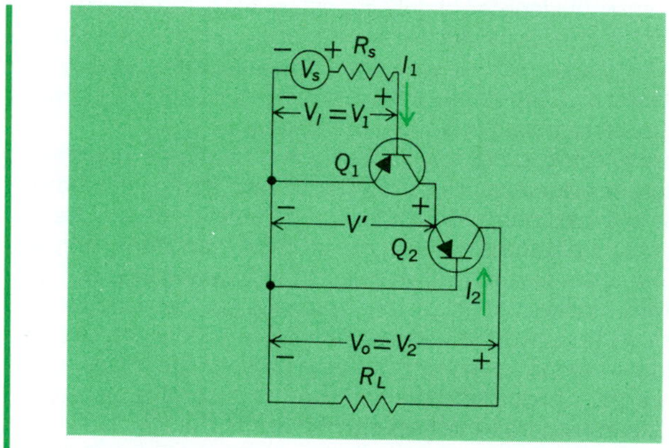

Cascode configuration. Supply voltages are not shown. Q_1, Q_2 = transistors; I_1, I_2 = currents; R_s= input (source) resistance; R_L = output (load) resistance; V_s = source voltage; V_i = V_1 = input voltage; V' = intermediate voltage; V_O = V_2 = output voltage.

to the high output resistance of the common-base stage. The reverse open-circuit amplification parameter h_r is very much smaller for the cascode connection than for a single common-emitter stage. The small value of h_r makes this circuit particularly useful in tuned amplifier design where the collector load is replaced with a tuned circuit.

The cascode amplifier finds its principal application in the amplification of high-frequency signals of 25 MHz and higher. One widespread use for the cascode circuit is in the tuner of a television receiver. *See* TELEVISION RECEIVER. [C.C.H.]

Casein The principal protein of milk. It is present in milk as calcium caseinate and amounts of 80% of the milk protein. Casein has had certain industrial uses, such as in the manufacture of brush bristles, buttons, billiard balls, jewelry, paper, glue, paint, sodium caseinate, and protein hydrolyzates. The competition from synthetic plastics reduced the demand for casein for some of these uses. *See* MILK.

Casein is made by either of two methods: coagulation by rennet or precipitation by acid. The rennet method is used for producing casein for the plastics industry. The acid precipitation method is used for casein production for other commercial uses, including the food industry. [P.H.T.]

Cashew A medium-sized, spreading evergreen tree (*Anacardium occidentale*) native to Brazil, but now grown widely in the tropics for its edible nuts and the resinous oil contained in the shells. The fruit consists of a fleshy, red or yellow, pear-shaped receptacle, termed the apple, at the distal end of which is borne a hard-shelled, kidney-shaped ovary or nut. Although cashew trees are spread throughout the tropics, commercial production is centered in India, which handles 90% of the world trade.

Cashew nut kernels are eaten as nuts and used extensively in the confectionery and baking trade. The cashew shell liquid is a valuable by-product, containing 90% anacardic acid and 10% cardol, and is used in the varnish and plastic industries.

The cashew apples are too astringent for eating without being processed, but when processed may be used for jams, chutney, pickles, and wine. *See* SAPINDALES. [L.H.MacD.]

Cashmere The natural fiber obtained from the Cashmere goat, native to the Himalayan region of China and India. The fleece of this goat has long, straight, coarse outer hair of little value; but the small quantity of underhair, or down, is made into luxuriously soft woollike yarns with a characteristic highly napped finish. Cashmere is a much finer fiber than mohair or wool fiber obtained from sheep. However, it is not as durable as wool. *See* MOHAIR; NATURAL FIBER; WOOL. [M.D.P.]

Cassava The plant *Manihot esculenta* (Euphorbiaceae), also called manioc. It is one of the 10 most important food plants, and the most important starchy root or tuber of the tropics. It originated in Central or South America, possibly Brazil, and was domesticated and widely distributed well before the time of Columbus. Subsequent distribution has established cassava as a major crop in eastern and western Africa, in India, and in Indonesia.

The cassava plant is a slightly woody, perennial shrub. The leaves are deeply palmately lobed; the flowers are inconspicuous, and the prominent capsules are three-seeded and explosive at maturity. The roots (see illustration) are enlarged by the

Tuberous roots from a single cassava plant.

deposition of starch and constitute the principal source of food from the plant. The leaves are also eaten (after cooking), and are noteworthy for their high protein content.

The chief use of cassava is as a boiled vegetable. It is also a source of flour, called *farinha* in Brazil and *gari* in western Africa, and of toasted starch granules, the familiar tapioca. In spite of its popularity, however, cassava root is a poor food. Its protein content is extremely low, and its consumption as a staple food is associated with the protein deficiency disease kwashiorkor. *See* GERANIALES. [F.W.M.]

Cassiduloida An order of exocyclic Euechinoidea which possess five similar ambulacra which form petal-shaped areas (phyllodes) around the mouth. The innermost interambulacral plate between each pair of adjoining phyllodes is swollen and is termed a bourrelet. The five bourrelets and five phyllodes thus form a flower-shaped pattern around the mouth (see illustration). All extant forms live in shallow or moderately deep tropical seas, partly buried in sand. The oldest-known cassiduloids occur in Jurassic strata. The order was most abundant during the Eocene Epoch, after which a gradual decline occurred. In the latest revision of the Cassiduloida, 10 families are recognized. *See* ECHINODERMATA; EUECHINOIDEA. [H.B.F.]

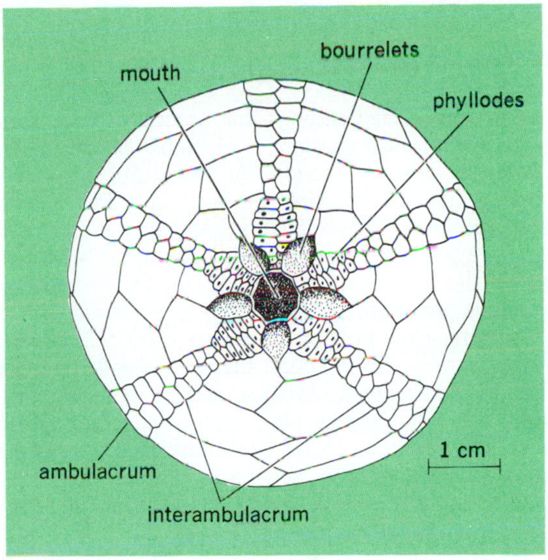

Adoral aspect of the test of Cassiduloida with the mouth at the center.

Cassiopeia A prominent northern circumpolar constellation as seen from the middle latitudes. The five main bright second- and third-magnitude stars of Cassiopeia form the rather distorted W or M by which the constellation is usually identified (see illustration). Cassiopeia and the Big Dipper lie across the opposite sides of the North Celestial Pole. *See* CONSTELLATION; URSA MAJOR. [C.S.Y.]

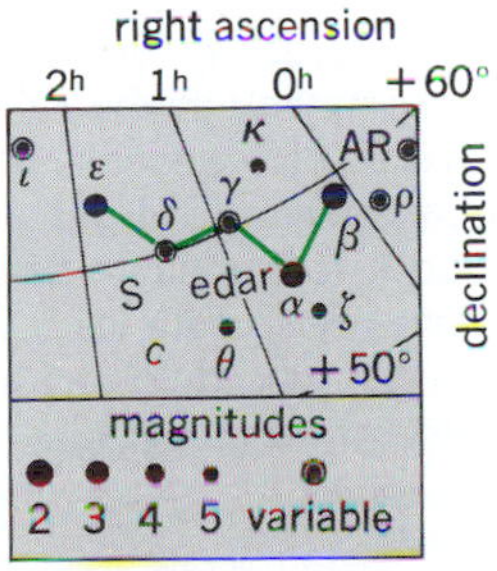

Line pattern of the constellation Cassiopeia. The grid lines represent the coordinates of the sky. Magnitudes of the stars are shown by the sizes of the "dots."

Cassiterite A mineral having the composition SnO_2. It is the principal ore of tin. Cassiterite is usually massive granular, but may be in radiating fibrous aggregates with reniform shapes (wood tin). The hardness is 6–7 (Mohs scale), and the specific gravity is 6.8–7.1 (unusually high for a nonmetallic mineral). The luster is adamantine to submetallic. Pure tin oxide is white, but cassiterite is usually yellow, brown, or black because of the presence of iron.

Cassiterite is most abundantly found as stream tin (rolled pebbles in placer deposits). The world's supply comes mostly from placer or residual deposits in the Malay Peninsula, Indonesia, Zaire, and Nigeria. It is also mined in Bolivia. *See* TIN. [C.S.Hu.]

Cast iron A generic term describing a family of iron alloys containing 1.8–4.5% carbon. Cast iron usually is made into specified shapes, called castings, for direct use or for processing by machining, heat treating, or assembly. In special cases it may be forged or rolled moderately. Generally, it is unsuitable for drawing into rods or wire, although to a limited extent it has been continuously cast into rods and shapes from a liquid bath or swaged from bars into smaller-dimensional units. Silicon usually is present in amounts up to 3%, but special compositions are made containing up to 6% (Silal) and up to 12% (Duriron). Cast iron of the above composition range is often made into blocks or rough shapes and called pig iron. It is an intermediate form of cast iron used for remelting into iron castings. Cast iron may be purchased in several commercial grades called gray iron, chilled iron, mottled iron, white iron, malleable iron, ductile iron, spheroidal graphite iron, nodular iron, and austenitic cast iron. *See* ALLOY; IRON ALLOYS; WROUGHT IRON. [J.S.V.]

Castor plant A plant, *Ricinus communis*, belonging to the spurge family (Euphorbiaceae). Castor seeds are poisonous and also contain allergens.

Current distribution is in the warmer regions of the world, the plant often growing in waste places. The castor oil plant has been of utilitarian value since antiquity. Oil from the seeds is among the world's oldest nonfood products in commerce. Castor oil contains about 85% ricinoleic acid, used in making industrial products such as alkyde resins for surface coatings, blown oil used in plasticizers, cracked oil for production of synthetic perfumes, nylon, sebacic acid, synthetic detergents, drying oils, and special lubricating oils.

Production in the United States, mostly in west Texas, peaked in 1968 but declined to nil because of larger economic return from food and fiber crops. Brazil is the world's largest producer and exporter of castor seed and oil. [L.H.Z.]

Casuarinales An order of flowering plants, division Magnoliophyta (Angiospermae), in the subclass Hamamelidae of the class Magnoliopsida (dicotyledons). The order consists of a single family (Casuarinaceae) and genus (*Casuarina*), with about 50 species. Native to the southwestern Pacific region, especially Australia, they are sometimes called Australian pine. They are trees with much reduced flowers and green twigs that bear whorls of scalelike, much reduced leaves. Some species are grown as street trees in tropical and subtropical regions. *See* HAMAMELIDAE; MAGNOLIOPSIDA. [A.Cr.]

Cat The term used to describe any member of the mammalian family Felidae. More commonly, the term is restricted to the domestic cat and those felids that resemble it in size, shape, and habits. Those that are larger in size are referred to as the big cats. All members of the cat family have a round head, are digitigrade (walk on their toes), have retractile claws (with the exception of the cheetah), and have 30 teeth. All species are carnivorous.

The origin of domestic cats is unknown, but it is established that they have been associated with humans for many centuries. Domestic mixed-breed cats are generally characterized by long, thin tails, straight ears, and short hair of a variety of colorations. The pure breeds, however, are notable exceptions to this general description. There is a long-haired race, apparently developed in Persia, which is represented by the Persian and the Angora. The Abyssinian breed, in contrast to the common domestic cat, is characterized by being ruddy brown in color and having a longer face and ears. There are at least two tailless breeds, of which the best known is the Manx from the Isle of Man. The other tailless variety is found in Japan. One of the most popular breeds is the Siamese cat. Its eyes are deep blue, it has a long, kinky tail, and its fur is short and cream to buff in color.

Wildcats include cats such as the lynx, bobcat, serval, ocelot, puma, leopard, lion, tiger, jaguar, and cheetah. In addition there are a few lesser known forms that are of interest. Among these are the Scottish wildcat and the caracal. *See* CARNIVORA. [C.B.C.]

Cat scratch disease A benign systemic illness in humans characterized by malaise, occasional fever, and a granulomatous lymphadenitis, which may resemble neoplastic disease. Since only a portion of the cases have a history of contact with cats, the term benign lymphoreticulosis may have some justification. The disease is not uncommon, and most cases occur in winter months. Sporadic cases are observed more often in children; occasionally small household outbreaks arise when cats scratch or bite family members.

A few days after a cat scratch has occurred, a persistent scabbed ulcer or papule appears as the primary skin lesion. About 1–3 weeks later the regional lymph nodes on one or both sides become enlarged. Fever, malaise, and symptoms of infection usually develop. The ocular form consists of conjunctivitis, with enlargement of the lymph nodes around the ear. Meningoencephalitis, an inflammation of the brain and the membranes covering it, may complicate the disease. Though benign and self-limiting, the disease may last from 2 weeks to 2 years. [K.F.M.]

Cataclysmic variable A member of a wide class of short-period binary stars, one of whose components is a white dwarf (degenerate) star, capable of irregularly timed but recurrent outbursts of brightness by factors of 2 to 10,000. The other component can be a main-sequence (core-hydrogen-burning) star like the Sun, a red giant (evolved, low-mass) star like that which the Sun will become, or another white dwarf. Classical novae make up the most spectacular, but not the commonest, category of cataclysmic variables. The systems are also called cataclysmic binaries.

Stars with initial masses between 0.8 and 8 solar masses complete hydrogen and helium burning within the present age of the universe, shed their outer layers, which are seen as planetary nebulae, and settle down as white dwarfs. If the star was originally part of a binary system, as about half are, the pair remains bound through this process, leaving the white dwarf in orbit with a normal star. The second star in turn evolves first into a red giant and then into another white dwarf. Such a pair, at any stage, constitutes a minimal cataclysmic variable.

The more interesting manifestations occur when gas flows from the normal star down into the deep gravitational potential well of the compact white dwarf, releasing energy as it goes. Such a transfer can happen in two ways. First, if the stars are close enough together, the normal one fills a critical surface called its Roche lobe, and gas from its surface streams through a neck (called the inner lagrangian point) between the two stars and feeds a disk around the white dwarf. The stream, the disk, and the spot where one hits the other are all hot and radiate with variable brightness, depending on the instantaneous gas flow rate. This accretion-powered luminosity normally exceeds that intrinsic to the stars themselves.

Second, material will be transferred when the companion star becomes a red giant with a strong stellar wind blowing off its surface. Part of the wind hits the white dwarf, heats up, and radiates; but no disk is formed.

Variations in the mass-transfer rate give rise to variable brightness and outbursts. In addition, the transferred hydrogen eventually builds up a layer so thick that its bottom becomes hot and dense enough for nuclear reactions to occur. The hydrogen is degenerate, so that its pressure depends only on density, not on temperature. Thus, when burning starts, the gas heats up but does not expand and cool again, in the way that a normal gas would. Instead, the higher temperature drives more rapid hydrogen burning, and so forth, leading to an explosion called a nova. *See* BINARY STAR; NOVA; STELLAR EVOLUTION; SYMBIOTIC STAR; VARIABLE STAR; WHITE DWARF STAR. [V.T.]

Catalan numbers A series of numbers given by the equation below. They count the number of ways to insert

$$c_n = \frac{1}{n}\binom{2n-2}{n-1}$$

parentheses in a string of n terms so that their product may be unambiguously carried out by multiplying two quantities at a time. Thus $c_4 = 5$, since *abcde* may be computed by $(ab)(cd)$, $((ab)c)d$, $(a(bc))d$, $a((bc)d)$, or $a(b(cd))$. The Catalan numbers also enumerate the ways to place nonintersecting diameters in a convex polygon with $n + 1$ vertices so that only triangular regions result. Further, c_n gives the number of ways to draw in succession all the balls from a container which originally contains $n - 1$ white balls and $n - 1$ black balls, so that at every stage there are at least as many black balls as white balls remaining in the container. Catalan numbers also appear in the theoretical measurement of preferences. *See* COMBINATORIAL THEORY; ENUMERATION OF ARRANGEMENTS. [T.Br.]

Catalysis The phenomenon in which a relatively small amount of foreign material called a catalyst augments the rate of a chemical reaction without itself being consumed. A catalyst is material, and not light or heat. It augments a rate. The term negative catalyst is obsolete. *See* ANTIOXIDANT; INHIBITOR (CHEMISTRY).

If the reaction $A + B \rightarrow D$ occurs very slowly but is catalyzed by some catalyst, Cat, the addition of Cat must open new channels for the reaction. In the very simple case shown below, the two propagation processes, which are fast com-

$$\left.\begin{array}{r} A + \text{Cat} \rightarrow \text{ACat} \\ \text{ACat} + B \rightarrow D + \text{Cat} \end{array}\right\} \quad \text{Chain propagation}$$

$$A + B \rightarrow D \qquad \text{Overall reaction}$$

pared to the uncatalyzed reaction, $A + B \rightarrow D$, provide the new channel for reaction. The catalyst reacts in the first step, but is regenerated in the second step to commence a new cycle. A catalytic reaction is thus a kind of chain reaction.

If a reaction is in chemical equilibrium under some fixed conditions, the addition of a catalyst cannot change the position of equilibrium without violating the second law of thermodynamics. Therefore, if a catalyst augments the rate of $A + B \rightarrow D$, it must also augment the reverse rate, $D \rightarrow A + B$.

Catalysis is conventionally divided into three categories: homogeneous, heterogeneous, and enzyme. Heterogeneous catalysis plays a dominant role in chemical processes in the petroleum, petrochemical, and chemical industries. Homogeneous catalysis is important in the petrochemical and chemical industries. Enzyme catalysis plays a key role in all metabolic processes and in some industries, such as the fermentation industry. *See* ENZYME; HETEROGENEOUS CATALYSIS; HOMOGENEOUS CATALYSIS. [R.L.Bu.]

Catalytic antibody An antibody that can cause useful chemical reactions. Catalytic antibodies are produced through immunization with a hapten molecule that is usually designed to resemble the transition state or intermediate of a desired reaction. Catalytic antibodies combine programmable selectivity with catalytic activity, because a molecule must first bind inside the binding pocket of a catalytic antibody before any reaction takes place. The net result is a high level of control over the catalyzed reaction, control that is reminiscent of enzymes.

Chemical reactions have energy barriers that must be overcome before a reaction can take place and the products be created; the higher the energy barrier, the slower the reaction. A catalyst such as an enzyme accelerates a chemical reaction by somehow lowering the energy barrier. An enzyme, like any true catalyst, is itself not permanently modified during a reaction, so each enzyme molecule can cause the reaction of many substrate molecules, a process referred to as catalytic turnover. *See* CATALYSIS.

Antibodies are the recognition arm of the immune system. They are elicited, for example, when an animal is infected with a bacterium or virus. The animal produces antibodies with binding sites that are exactly complementary to some molecular feature of the invader. The antibodies can thus recognize and bind only to the invader, even in the presence of the animal's complex internal molecular landscape. The antibody binding identifies the invader as foreign, leading to its destruction by the rest of the immune system.

Antibodies are also elicited in large quantity when an animal is injected with molecules, a process known as immunization. A small molecule used for immunization is called a hapten. Ordinarily, only large molecules effectively elicit antibodies via immunization, so small-molecule haptens must be attached to a large protein molecule, called a carrier protein, prior to the actual immunization. Antibodies that are produced after immunization with the hapten-carrier protein conjugate are complementary to, and thus specifically bind, the hapten. *See* ANTIBODY; IMMUNITY.

Catalytic antibodies are produced when animals are immunized with hapten molecules that are specially designed to elicit antibodies that have binding pockets capable of catalyzing chemical reactions. For example, in the simplest cases, binding forces within the antibody binding pocket are enlisted to stabilize transition states and intermediates, thereby lowering a reaction's energy barrier and increasing its rate. This can occur when the antibodies have a binding site that is complementary to a transition state or intermediate structure in terms of both three-dimensional geometry and charge distribution. This complementarity leads to catalysis by encouraging the substrate to adopt a transition state–like geometry and charge distribution. Not only is the energy barrier lowered for the desired reaction, but other geometries and charge distributions that would lead to unwanted products can be prevented, increasing reaction selectivity. *See* REACTIVE INTERMEDIATES. [B.I.]

Catalytic converter An aftertreatment device used for pollutant removal from automotive exhaust. Since the 1975 model year, increasingly stringent government regulations for the allowable emission levels of carbon monoxide (CO), hydrocarbons (HC), and oxides of nitrogen (NO_x) have resulted in the use of catalytic converters on most passenger vehicles sold in the United States. The task of the catalytic converter is to promote chemical reactions for the conversion of these pollutants to carbon dioxide, water, and nitrogen.

For automotive exhaust applications, the pollutant removal reactions are the oxidation of carbon monoxide and hydrocarbons and the reduction of nitrogen oxides. Metals are the catalytic agents most often employed for this task. Small quantities of these metals, when present in a highly dispersed form (often as individual atoms), provide sites upon which the reactant molecules may interact and the reaction proceed.

Two types of catalyst systems, oxidation and three-way, are found in automotive applications. Oxidation catalysts remove only CO and HC, leaving NO_x unchanged. Platinum and palladium are generally used as the active metals in oxidation catalysts. Three-way catalysts are capable of removing all three pollutants simultaneously, provided that the catalyst is maintained in a "chemically correct" environment that is neither overly oxidizing nor reducing. In both oxidation and three-way catalyst systems, the production of undesirable reaction products, such as sulfates and ammonia, must be avoided.

Maintaining effective catalytic function over long periods of vehicle operation is often a major problem. Catalytic activity will deteriorate due to two causes, poisoning of the active sites by contaminants, such as lead and phosphorus, and exposure to excessively high temperatures. To achieve efficient emission control, it is thus paramount that catalyst-equipped vehicles be operated only with lead-free fuel and that proper engine maintenance procedures be followed. *See* AUTOMOTIVE ENGINE; CATALYSIS. [N.Ot.]

Catalytic cracking The major cracking process used throughout most of the world oil industry for the production of high-octane quality gasoline by the conversion of intermediate- and high-boiling petroleum distillates to lower-molecular-weight products. Oil heated to within the lower range of thermal cracking temperatures (850–1025°F; 454–551°C) reacts in the presence of an acidic inorganic catalyst under low pressures (10–35 pounds per square inch gage or 70–240 kilopascals, gage pressure). Gasoline of much higher octane number is obtained than from thermal cracking, a principal reason for the widespread adoption of catalytic cracking. All nonvolatile carbonaceous materials are deposited on the catalyst as coke and are burned off during catalyst regeneration. *See* THERMAL CRACKING.

Although the primary objective of catalytic cracking is the production of maximum yields of gasoline concordant with efficient operation of the process, large amounts of normally gaseous hydrocarbons are produced at the same time. The gaseous hydrocarbons include propylene and butylenes, which are in great demand for chemical manufacture. The other chief product is the material boiling above gasoline, designated as catalytically cracked gas oil. It contains hydrocarbons relatively resistant to further cracking, particularly polycyclic aromatics. The lighter portion may be used directly or blended with straight-run and thermally cracked distillates of the same boiling range for use as diesel and heating oils. Part of the heavier portion is recycled with fresh feedstock to obtain additional conversion to lighter products. The remainder is withdrawn for blending with residual oils to reduce the viscosity of heavy fuel, or else subjected to a final step of thermal cracking.

With the emphasis on conservation of petroleum resources, catalytic cracking has assumed a more important role. The heavier products from a refinery could be replaced by coal-, shale-, or tar-sand-derived products. Thus, there is a need for greater conversion of petroleum to gasoline with emphasis on catalytic cracking as the cheapest major conversion process. *See* CRACKING. [E.C.L.]

Catalytic reforming A process used for upgrading gasoline by improving its antiknock characteristics (increasing the octane number). It is also widely used for the production of aromatic hydrocarbons for the petrochemical industry.

The process utilizes a supported platinum catalyst and involves a number of different reactions. These reactions convert, or reform, the low-octane-number feed components, such as the paraffins, to components with increased octane number. The most important reactions in catalytic reforming are those leading to the formation of aromatic hydrocarbons because these are high-octane components as well as valuable petrochemical intermediates. Aromatics are formed by dehydrogenation of six-membered ring cycloparaffins, rearrangement and dehydrogenation of five-membered ring cycloparaffins, and cyclization (dehydrocyclization) of paraffins.

Liquefied petroleum gas (LPG) consists of propane and butanes and is usually derived from natural gas. In locations where there is no natural gas and LPG is more important than gasoline, or in refineries that require additional isobutane for alkylation, naphtha can be converted to LPG (primarily isobutane) by catalytic reforming. Under suitable conditions it is possible to convert 40% of the naphtha to LPG, the by-product being high-octane gasoline. *See* AROMATIC HYDROCARBON; AROMATIZATION; DEHYDROGENATION; HYDROCRACKING; LIQUEFIED PETROLEUM GAS (LPG); OCTANE NUMBER; PETROLEUM PROCESSING. [E.L.Po.; V.H.]

Cataract Any clouding or opacity of the crystalline lens of the eye. Cataracts vary markedly in degree of density and may be due to many causes, but the majority are associated with aging. Cataracts are the single leading cause of blindness in the world.

Senile cataract occurs with aging and is by far the most common type. Progressively blurred vision is the only symptom. There are a number of other varieties (congenital, metabolic, secondary, or traumatic) and causes (such as a reaction to certain drugs or irradiation). Cataracts may also be associated with systemic diseases such as hypoparathyroidism, myotonic dystrophy, atopic dermatitis, galactosemia, and Lowe's, Werner's, and Down's syndromes.

The treatment for cataracts, when they are sufficiently advanced to impair the vision, is surgical removal. When a cataract is surgically removed, the crystalline lens of the eye is removed much as a lens would be removed from a camera. In order to restore normal vision after surgery, any of a number of methods may be followed: cataract glasses may be prescribed; contact lenses may be fitted; or intraocular lenses made of the plastic polymethylmethacrylate can be inserted into the eye to replace the cataractous lens. *See* EYE (VERTEBRATE). [J.Hart.]

Catastrophe theory A theory of mathematical structure in which smooth continuous inputs lead to discontinuous responses. Water suddenly boils, ice melts, a building crashes to the ground, or the earth unexpectedly buckles and quakes. The French mathematician René Thom conceived and developed an eclectic collection of ideas into catastrophe theory. His idea was to establish a new basis for a more mathematical approach to biology. Connotations of disaster are misleading, since Thom's intention was to emphasize sudden, abrupt changes.

Advanced areas of modern mathematics, including algebraic geometry, differential topology, and dynamical system theory, contributed to the creation of catastrophe theory. A complete mathematical theory exists for the elementary catastrophes, which can be written as the gradient of an energylike function. The physical, chemical, and engineering applications are less developed, although many are known in optics, laser theory, thermodynamics, elasticity, and chemical reaction theory. The Thom classification theorem gives exactly seven elementary catastrophes. Although a theory of generalized catastrophes exists, which extends the theory beyond gradient systems, it is not nearly as well developed mathematically or physically as that of elementary catastrophes. It does include remarkable examples of chaos (or stochastic behavior) in the solutions to nonlinear deterministic equations. These solutions include strange attractors and omega explosions among the examples of nonelementary catastrophes. *See* GEOMETRY; PERIOD DOUBLING; TOPOLOGY.

There are two important aspects of catastrophe theory which are frequently overlooked or misconstrued. One is that as a rigorous mathematical theory the characteristic catastrophe features can be proved. These features include: jumps in the response; hysteresis or a path dependence in the response, representing a storage of energy for some paths; divergence, where a small path change produces a large response change (as if a source or sink were crossed); and type changes in the response, where a smooth response occurs on one path which becomes discontinuous along a nearby path.

All of these features are topological, so that they are independent of the coordinates used to describe the potential. They are, therefore, qualitative features of the solutions. Some critics have concluded that because these aspects were qualitative, they could not be quantitative. This is contradicted by the solid and growing body of quantitative studies in catastrophe theory. (Problems in quantum optics, thermodynamics, and scattering theory have all been clarified by catastrophe theory.) [B.DeF.]

Catechol A naturally occurring dihydric phenol, also known as pyrocatechol, pyrocatechuic acid, and 1,2-dihydroxybenzene, with the empirical formula $C_6H_6O_2$. Catechol is a colorless crystalline substance (melting point 105°C or 221°F) which dissolves readily in water. It resembles phenol in being a severe irritant to skin and in possessing systemic toxicity.

Although catechol and its derivatives are widely distributed among plants, the substance is produced synthetically for use in the manufacture of antioxidants (stabilizers) for rubber and plastics, and of a feed additive for poultry. Catechol also is used in medicinal and photographic preparations. *See* PHENOL. [M.St.]

Catenary The curve formed by an ideal heavy uniform string hanging freely from two points of support. The lowest point *A* (see illustration) is the vertex. The portion *AP* is in

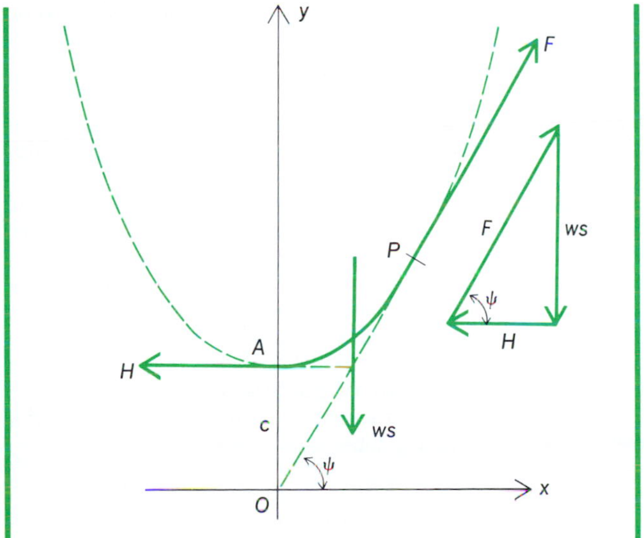

Catenary: force triangle.

equilibrium under the horizontal tension *H* at *A*, the tension *F* directed along the tangent at *P*, and the weight *W* of *AP*. If the weight of the string is *w* per unit length and *s* is the arc *AP*, $W = ws$; and from the force triangle, $\tan\psi = ws/H = s/c$, where $c = H/w$ is called the parameter of the catenary. Thus the catenary has the differential equation (1).

$$dy/dx = s/c \qquad (1)$$

The horizontal line at a distance *c* below the vertex *A* is the directrix of the catenary. With the *x* axis as directrix and the *y* axis through the vertex, the integration of Eq. (1) yields Eqs. (2)

$$y = c\cosh\frac{x}{c} \qquad s = c\sinh\frac{x}{c} \qquad (2)$$

for the ordinate and arc of the catenary. All catenaries are geometrically similar to the hyperbolic cosine curve, $y = \cosh x$.

The surface generated by revolving a catenary about its directrix is a minimal surface, the catenoid. The catenary is an extremal for the problem of finding a curve joining two given points so that the surface generated by revolving it about a given line has minimum area. *See* HYPERBOLIC FUNCTION. [L.Br.]

Caterpillar A term applied to the larval stage of a butterfly or moth and to some scorpion flies and sawflies. In insects that undergo a complete metamorphosis during development, a sequence of stages occurs. As the eggs hatch, a vermiform (wormlike) larval form is released that differs considerably in morphology from the other stages, and several distinct types are recognized.

The caterpillar is an eruciform larva, which is one that has a well-developed head, a cylindrical form, thoracic legs and abdominal prolegs. Most of these larvae are phytophagous. Many caterpillars are cryptically colored and blend in with their environment, with some forms closely resembling twigs. Some larvae are naked, that is, they do not have hairs, while others are covered with hairs or spines. After a period of activity the caterpillar (larva) goes into a quiescent stage, the pupa, from which will emerge the adult. *See* BUTTERFLY; LEPIDOPTERA; MOTH. [C.B.C.]

Catfish The name for a number of fishes which constitute the suborder Siluroidei in the order Cypriniformes. There are about 11 families which include mostly fresh-water species of wide distribution.

The bodies of these fishes do not have scales but may be naked, have plates, or be covered with spines. The common name is derived from the presence of barbels around the mouth. Some species live at great depths and these are usually blind and lack pigmentation.

Catfish are valuable food fish. Common species include the blue catfish, found in the Mississippi drainage, and the European catfish. Some smaller South American species are parasitic and suck blood from the gill filaments of larger fish. *See* CYPRINIFORMES. [C.B.C.]

Cathartics Agents used to produce evacuation of the bowel. They are too frequently used and should probably be administered only after possible organic causes of constipation have been considered. Adding fiber to the diet softens stool, increases daily fecal weight, decreases intestinal transit time, and is often an effective remedy for mild constipation. Bulk can be added medicinally with hydrophilic preparations of psyllium seed, or naturally by ingesting high-fiber foods.

Nonfiber laxatives may be classified in many ways. One useful classification considers them as secretagogues, lubricants, or osmotic cathartics. Secretagogues, such as castor oil, inhibit net transport of water and electrolytes by the intestinal mucosa. Lubricants, such as mineral oil, reduce friction between the stool and the intestinal mucosa. Osmotic cathartics, such as phosphates of magnesium, are poorly absorbed salts that preempt water in the gut lumen because isotonicity must be maintained on both sides of the mucosal membrane. [R.L.Pr.]

Cathode-ray tube An electron tube in which a beam of electrons can be focused to a small area and varied in position and intensity on a surface. In common usage, the term

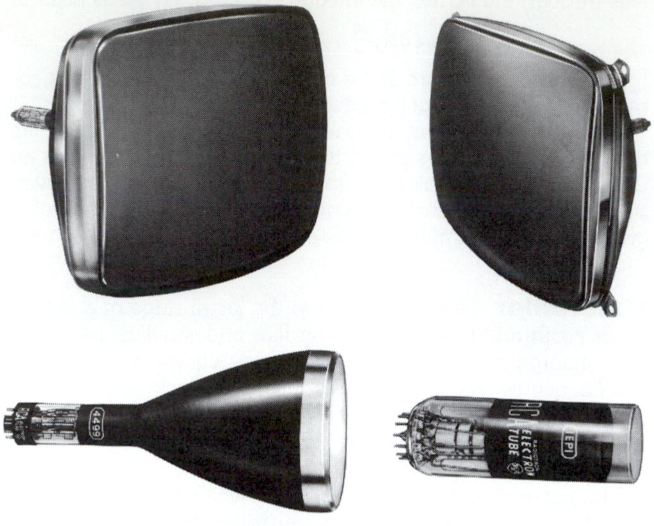

Fig. 1. Typical cathode-ray tubes.

cathode-ray tube is usually reserved for devices in which the surface referred to is cathodoluminescent, that is, luminescent under electron bombardment. In these tubes the output information is presented in the form of a pattern of light. The character of this pattern is related to, and controlled by, one or more electrical signals applied to the cathode-ray tube as input information.

The most familiar form of the cathode-ray tube is the television picture tube, found in home television receivers. Cathode-ray tubes are also used in large quantities in measuring instruments such as oscilloscopes, which have become indispensable in laboratories devoted to experimental studies in the physical and biological sciences and in engineering. The oscilloscope has also been adopted as a vital tool in the fields of automobile, aircraft, and television-receiver maintenance and repair. For navigation the cathode-ray tube is the output device of radars. Large users of cathode-ray tubes are the military services, which employ these devices to display data to guide the employment of weapons systems. Increasingly, cathode-ray tubes are finding use in input/output terminals of digital computers.

The cathode-ray tube is also used in the general field of printed communications, both as an adjunct to, and as a competitor of, traditional means of preparing "hard copy." In the first instance, cathode-ray tubes expose photographic masters that, in turn, are used to prepare offset printing plates. Where only a single copy is required, photosensitive paper is exposed by a cathode-ray tube to deliver printed copy in the business offices from an electronic communication link. *See* ELECTRONIC DISPLAY; OSCILLOSCOPE; STORAGE TUBE; TELEVISION RECEIVER.

The basic elements of a cathode-ray tube are the envelope, electron gun, and phosphor screen. The envelope serves as the vacuum enclosure, the substrate for the phosphor screen, and the support for the electron gun. The envelope of a cathode-ray tube is typically funnel-shaped; its small opening is stopped by the stem. This stem is essentially a disk of glass through which the metallic leads which support and carry applied voltages to the several elements of the gun are sealed. The large end of the funnel is closed by a glass faceplate on the inside of which the phosphor screen is deposited. Some typical cathode-ray tubes are shown in Fig. 1.

The electron gun produces, controls, focuses, and deflects an electron beam. In some cases one or more of the functions enumerated above are performed by components not included in the gun or even in the cathode-ray tube itself. These components are invariably electromagnetic-field-producing devices; a common example is a magnetic solenoid used for beam focusing. A typical electron gun used in a cathode-ray tube for oscillographic applications (Fig. 2) is functionally self-contained because no external components are required to perform beam-intensity control, focusing, or deflection. The phosphor screen converts electrical energy to visible radiation. Materials known as phosphors are said to be luminescent. A phosphor which is excited to luminescence by electron bombardment is described as cathodoluminescent. In the case of phosphors for cathode-ray tubes, luminescence invariably persists after cessation of excitation. *See* CATHODOLUMINESCENCE; LUMINESCENCE.

In cathode-ray tubes built for specialized applications, unique tube characteristics often require distinctive features. Cathode-ray tubes employing multiple independent electron guns are often required for use in oscilloscopes for the simultaneous display of several concurrent signals. Tubes having two guns are common; some have been produced with as many as ten guns. For color television, three guns, one to excite each of the three primary color phosphors, are built into the tube.

For computer input/output terminals, the information to be displayed is in the form of alphanumeric messages. There are a wide variety of electronic techniques available for generating alphanumeric characters. In many instances these techniques require as a display device a cathode-ray tube of conventional design, with due attention to proper choice of electron gun and phosphor screen. However, in some techniques, the combined requirements for high speed and low power consumption require a cathode-ray tube with dual deflection systems.

A character-producing tube is used in some alphanumeric display systems. It employs an electron gun of essentially conventional design to "illuminate," one at a time, characters etched in a thin metal stencil mounted in the envelope neck perpendicular to the tube axis. The electron beam issuing from the stencil, having assumed a cross section corresponding to an appropriate position on the phosphor screen, on which a luminous image of the selected character then appears.

In the simplest printing systems, the image on the face of a conventional cathode-ray tube, usually designed to have a very

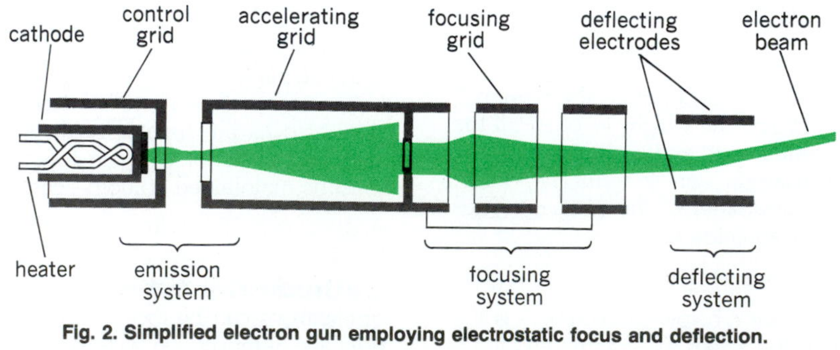

Fig. 2. Simplified electron gun employing electrostatic focus and deflection.

small luminous spot and an optically flat faceplate, is optically projected onto the photosensitive medium page by page. Higher optical coupling efficiency, important where speed of printing must be maximum, can be obtained if the cathode-ray tube faceplate is made of an array of optical fibers, fused together so as to be vacuum-tight. Such a device permits contact printing on the photographic medium. Frequently, to reduce device cost, the fiber optics faceplate takes the shape of a narrow rectangle and the printing is done line by line by moving the paper continuously past the faceplate. Cathode-ray tubes are being made in which the fiber optics assembly is replaced by a window of glass or mica several thousandths of an inch thick. Printing cathode-ray tubes are available in which fine wires, suitably insulated from each other, are brought through the faceplate. These wires conduct charges deposited by the electron beam directly to charge-sensitive paper, which is then passed through a suitable toner material to develop the final printed image. *See* OPTICAL FIBERS; PRINTING.

Television-type color cathode-ray tubes are generally not used in specialized display systems because the dot or line structure does not permit an image of sufficiently high resolution. This limitation is overcome in a color cathode-ray tube of the penetration type. In this device two or more layers of phosphor, each emitting radiance of a different color, are deposited on the inside of the faceplate. If the electron gun is operated at an appropriate overall voltage, only the phosphor layer closest to the gun is excited. At higher voltages a further layer absorbs most of the beam energy and thus its characteristic color dominates the display. By controlling the beam accelerating voltage, therefore, the display can include a range of colors. *See* COLOR TELEVISION. [M.D.H.]

Cathode rays The name given to the electrons originating at the cathodes of gaseous discharge devices. The term has now been extended to include low-pressure devices such as cathode-ray tubes. Furthermore, cathode rays are now used to designate electron beams originating from thermionic cathodes, whereas the term was formerly applied only to cold-cathode devices. *See* CATHODE-RAY TUBE. [G.H.M.]

Cathodoluminescence A luminescence resulting from the bombardment of a substance with an electron beam. The major application of cathodoluminescence is in cathode-ray oscilloscopes and television picture tubes. In these devices a thin layer of luminescent powder is evenly deposited on the transparent glass face plate of a cathode-ray tube. The electron beam originating in the cathode, after undergoing acceleration, focusing, and deflection by various electrodes in the tube, impinges on the luminescent coating. The resulting emission is normally observed through the glass face plate of the tube, that is, from the unbombarded side of the coating. Depending on the type of signal information to be displayed on the tube, phosphors or phosphor blends of various emission colors and persistences are used. *See* LUMINESCENCE; PHOTOLUMINESCENCE. [C.C.K.; J.H.S.]

Cauchy integral formula An integral representation formula due to A. Cauchy which plays a fundamental role in complex analysis and is traditionally the principal tool used in proving the equivalence of some of the alternative characterizations of analytic functions. The formula asserts that if γ is the smooth boundary curve of a finite region Ω in the complex plane and if f is analytic in an open neighborhood of the closed region $\Omega \cup \gamma$ (the union of Ω and γ), then the equation below

$$\frac{1}{2\pi i}\int_{\gamma}\frac{f(\xi)d\xi}{\xi - z} = \begin{cases} f(z) & \text{where } z \text{ is a member of } \Omega \\ 0 & \text{whenever } z \text{ is not a member of } \Omega \cup \gamma \end{cases}$$

is valid. This shows, for example, that the values of the analytic function f in Ω are completely determined by the values of that function on the boundary of Ω. *See* ANALYTIC FUNCTION. [R.C.G.]

Cauliflower A cool-season biennial crucifer (*Brassica oleracea* var. *botrytis*) of Mediterranean origin. Cauliflower belongs to the plant order Capparales. It is grown for its white head or curd, a tight mass of flower stalks, which terminates the main stem. Cauliflower is commonly cooked fresh as a vegetable; to a lesser extent, it is frozen or pickled and consumed as a relish. California and New York are important cauliflower-producing states. *See* CAPPARALES. [H.J.C.]

Causality In classical mechanics causality has been taken to mean that all the dynamical variables of a system can be precisely measured and their evolution in time is strictly determined by the forces. Therefore, classically, one can repeatedly prepare a system in the same specific initial state with arbitrary precision and then measure the same final state after a given lapse in time.

In quantum mechanics there is also a rigorous law of causality. However, there are important differences. All the variables cannot be completely specified: there are inherent uncertainties specified by the Heisenberg uncertainty relation, which is completely built into the structure of quantum mechanics. Hence, one prepares an initial state and specifies a wave function whose (absolute) square is interpreted as a probability distribution. The law of causality in quantum mechanics states that for an isolated system the initial state completely determines, by the dynamics of the Schrödinger equation, the future state, which again leads to a probability distribution. To understand this physically, consider the scattering of protons off a target nucleus. Quantum mechanics can predict only an average distribution for the scattering process: thus it is the average of a large number of measured events that one accurately compares with theory. One also notes that in accord with the uncertainty principle there will always be some spread in momentum for the initial proton states. Furthermore, the detection of the protons disturbs the system (again because of the uncertainty principle), destroying the exact relation between the wave function before and after the measurement. *See* QUANTUM MECHANICS.

Causality has also been used in reference to the principle that an event cannot precede its cause. Consider, for example, a packet of light (or electromagnetic radiation) having a sharp wave front incident on an obstacle. After the incident wave reaches the obstacle or scattering center (located at the origin), scattered light moves out in all directions. Since light travels with the finite velocity, causality requires that no scattered light be observed at any point until there is sufficient time for the incident light to reach that point. This requirement places important restrictions on the behavior of the scattering processes as a function of frequency. [G.L.S.]

Cave A natural cavity in rock beneath the surface of the earth. Most caves have been formed in calcareous rock (limestone and dolomite) by the dissolving action of groundwater circulating along the partings in the rocks. Such partings include existing joints and bedding (stratification) planes. Joints commonly occur in sets as semiparallel and nearly vertical planes, the sets intersecting each other at large angles. Bedding planes usually are horizontal. The intersection of these planes accounts for the three-dimensional pattern observed in caves, but in most caves the horizontal extent greatly exceeds vertical development. *See* JOINT (GEOLOGY).

Caves of no great length may be formed by mechanical erosion of waves against coastal cliffs and by weathering attack on cliffs of jointed rock. Lava flows may, on cooling, develop a stiff upper crust over still molten rock which, continuing to flow after supply at the source has ceased, may leave long caves. *See* WEATHERING PROCESSES. [J.H.B.]

Cavies Rodents comprising the family Caviidae, which includes the guinea pig, rock cavies, mountain cavies, capybara, salt-desert cavy, and mara. All members of the group are indigenous to South America and comprise 15 species in six genera. Cavies have either rounded bodies with large heads and short ears and limbs, or rabbitlike bodies with long limbs and moderately long ears.

The guinea pig (*Cavia aperea*) originated in Peru, where there is still a wild stock. The domestic form (*C. porcellus*) has been produced by selective breeding and is a valuable laboratory animal with a life-span of 3–5 years. Closely related to the guinea pig is the mara or Patagonian cavy (*Dolichotis patagonum*) which resembles a large hare. The salt-desert cavy (*Pediolagus salinicola*) is a smaller species found in the salt deserts of southern Argentina. The largest of all rodents is the capybara or carpincho (*Hydrochoerus hydrochaeris*) which grows to the size of a small pig. It is essentially an aquatic animal which lives in small groups along the banks of lakes and streams in tropical South America. *See* RODENTIA. [C.B.C.]

Cavitation The formation of vapor- or gas-filled cavities in liquids by mechanical forces. If understood in this broad sense, cavitation includes the familiar phenomenon of bubble formation when water is brought to a boil and the effervescence of champagne wines and carbonated soft drinks. In engineering terminology, the term cavitation is used in a narrower sense, namely, to describe the formation of vapor-filled cavities in the interior or on the solid boundaries of vaporizable liquids in motion when the pressure is reduced to a critical value without change in ambient temperature. Cavitation in the engineering sense occurs at suitable combinations of low pressure and high speed in pipelines; in hydraulic machines such as turbines, pumps, and propellers; on submerged hydrofoils; behind blunt submerged bodies; and in the cores of vortices. This type of cavitation has great practical significance because it restricts the speed at which hydraulic machines may be operated and, when severe, it lowers efficiency, produces noise and vibrations, and causes rapid erosion of the boundary surfaces, even though these surfaces consist of concrete, cast iron, bronze, or other hard and normally durable material. [K.E.Sch.]

Cavity resonator An enclosure with conducting walls surrounding a dielectric medium, capable of resonating with electromagnetic waves. It is used mainly as a reservoir of electromagnetic oscillations at microwave or ultra-high frequencies. The conducting walls are usually made of metal such as copper or silver, while the dielectric medium is usually air or vacuum. The enclosure has at least one aperture for coupling the electromagnetic energy between the inside and outside of the cavity.

For an electromagnetic wave to exist inside a cavity, the electric and magnetic field intensities must be solutions of Maxwell's equations, and at the same time they must satisfy the boundary conditions on the inside surfaces of the cavity walls. For all practical purposes it can be assumed that the wall conductivity is so high that the tangential component of the electric field vanishes everywhere along the inside walls. This restriction requires that only waves of certain discrete frequencies be allowed in the cavity. Thus, when the frequency of an exciting wave is varied, the cavity does not extract or contain any energy from the wave unless the exciting frequency happens to be one of the allowed values. When the latter condition holds, then the cavity will resonate strongly with the exciting wave by absorbing a part or all of the incoming energy. The cavity is thus acting as a resonator.

Coaxial cavity resonators are often used as wavemeters, particularly for lower than microwave frequencies. Since cavity resonators are essentially frequency-selective circuit elements for microwaves, they are very useful as components in a microwave filter network. In a radar set, a resonant cavity can be used in a so-called TR box, where a gas-filled cavity breaks down to a short-circuit impedance to protect the receiver from damage when the high-powered transmitter starts working. In high-energy linear accelerators, resonant cavities are used to accelerate charged particles in intense high-frequency electric fields. *See* KLYSTRON; PARTICLE ACCELERATOR; RADAR. [C.K.J.]

Cayley-Klein parameters A set of four complex numbers used to specify the orientation of a body, or equivalently, the rotation R which produces that orientation, starting from some reference orientation. They can be expressed in terms of the Euler angles ψ, θ, and ϕ, as in the equations below.

$$\alpha = \cos\frac{\theta}{2}\, e^{-i(\psi - \phi)/2} \qquad \beta = -i \sin\frac{\theta}{2}\, e^{i(\psi - \phi)/2}$$

$$\gamma = -i \sin\frac{\theta}{2}\, e^{-i(\psi - \phi)/2} \qquad \delta = \cos\frac{\theta}{2}\, e^{i(\psi - \phi)/2}$$

Although these parameters have been used to simplify somewhat the mathematics of spinning top motion, their main use is in quantum mechanics. There they are related to the Pauli spin matrices and represent the change in the spin state of an electron or other particle of half-integer spin under the space rotation R (ψ, θ, ϕ). *See* MATRIX THEORY; SPIN (QUANTUM MECHANICS). [B.G.]

Caytoniales A group of Mesozoic plants. The remains consist of palmately compound leaves with 3–6 lanceolate leaflets previously known as *Sagenopteris phillipsi*, pinnately branched microsporophylls (named *Caytonanthus arberi*) that bore winged pollen in four-chambered microsporangia, and fruit-bearing inflorescences, *Caytonia* (with two species), which bore a dozen or more globular, short-stalked fruits in subopposite rows. The Caytoniales appear related to the pteridosperms. They range from the Triassic to the Cretaceous. *See* PALEOBOTANY. [C.A.A.]

Cedar Any of a large number of evergreen trees having fragrant wood of great durability. Arborvitae is sometimes called northern white cedar. *See* ARBORVITAE.

Chamaecyparis thyoides, the southern white cedar, grows only in swamps near the eastern coast of North America, where it is also known as Atlantic white cedar. The wood is soft, fragrant, and durable in the soil and is used for boxes, crates, small boats, tanks, woodenware, poles, and shingles. The Port Orford cedar (*C. lawsoniana*), also known as Lawson cypress, is native to southwestern Oregon and northwestern California. It is the principal wood for storage battery separators, but is also used for venetian blinds and construction purposes. Alaska cedar (*C. nootkatensis*) is found from Oregon to Alaska. The wood is used for interior finish, cabinetwork, small boats, and furniture. It is also grown as an ornamental tree. Incense cedar (*Libocedrus decurrens*) is found from Oregon to western Nevada and Lower California. Incense cedar is one of the chief woods for pencils, and is also used for venetian blinds, rough construction, and fence posts and as an ornamental and shade tree.

Eastern red cedar (*Juniperus virginiana*) is distributed over the eastern United States and adjacent Canada. The very fragrant wood is durable in the soil and is used for fence posts, chests, wardrobes, flooring, and pencils. Cedarwood oil is used in medicine and perfumes. Cedar of Lebanon (*Cedrus libani*) and Atlas cedar (*C. atlantica*) resemble the larch, but the leaves are evergreen and the cones are much larger and erect on the branches. The cedar of Lebanon is a native of Asia Minor. *See* LARCH.

The cigarbox cedar (*Cedrela odorata*), also known as the West Indian cedar, belongs to the mahogany family, is a broad-leaved tree with pinnate, deciduous leaves, and is related to the *Ailanthus* and sumac. The wood is very durable and fragrant and is valued in the West Indies for the manufacture of cabinets, furniture, and canoes. See PINALES. [A.H.G./K.P.D.]

Celastrales An order of flowering plants, division Magnoliophyta (Angiospermae), in the subclass Rosidae of the class Magnoliopsida (dicotyledons). The order consists of 10 families and about 2300 species, with the families Celastraceae (about 850 species), Aquifoliaceae (about 450 species), Icacinaceae (about 400 species), and Hippocrateaceae (about 300 species) forming the core of the group. The order is marked by its simple leaves and regular flowers, varying from hypogynous (those with the perianth and stamens attached directly to the receptacle, beneath the ovary) to perigynous (those with the perianth and stamens united at the base into a hypanthium distinct from the ovary) and with a single set of stamens which alternate with the petals. Nearly all of the species are woody plants. Various species of holly (*Ilex*, family Aquifoliaceae) and *Euonymus* (Celastraceae) are often cultivated. See FLOWER; HOLLY; ROSIDAE. [A.Cr.]

Celebes An irregular or K-shaped island now called Sulawesi, with a 3000-mi (4800-km) coast and an area of approximately 72,000 mi^2 186,000 km^2. Celebes, part of Wallacea, is a rugged island, with little land under 600 ft (180 m) in elevation; in fact, most of the land is above 1500 ft (450 m). A peak of the Quarles Mountains rises to 11,296 ft (3443 m). Numerous lakes occupy depressions along the major faults thought to be graben. One of these, Lake Poso, is 5000 ft (1500 m) deep. Although at present there are no important mines, indications of nickel, gold, iron, and petroleum have been found. Hot unvarying temperatures, heavy rainfall, and high relative humidity result in a heavy cover of tropical rainforest. See EAST INDIES. [M.G.A.-C.]

Celery A biennial umbellifer (*Apium graveolens* var. *dulce*) of Mediterranean origin and belonging to the plant order Umbellales. Celery is grown for its petioles or leafstalks, which are most commonly eaten as a salad but occasionally cooked as a vegetable. California, Florida, and Michigan are important producing states. See APIALES. [H.J.C.]

Celestial navigation Navigation with the aid of celestial bodies, primarily for determination of position when landmarks are not available. In celestial navigation, position is not determined relative to the objects observed, as in navigation by piloting, but in relation to the points on the Earth having certain celestial bodies directly overhead.

Celestial bodies are also used for determination of horizontal direction on the Earth and for regulating time, which is of primary importance in celestial navigation because of the changing positions of celestial bodies in the sky as the Earth rotates daily on its axis.

The navigator, concerned less with the actual motions of celestial bodies than with their apparent motions as viewed from the Earth, pictures the heavens as a hollow celestial sphere of infinite radius, with the Earth at its center and the various celestial bodies on its inner surface. The navigator visualizes this sphere as rotating on its axis once in about 23 h 56 min—one sidereal day.

Several systems of coordinates are available to identify points on the celestial sphere. The celestial equator system of coordinates is an extension of the equatorial system commonly used on the Earth. The navigator also uses the horizon system of coordinates, in which the primary great circle is the horizon of the observer. See ASTRONOMICAL COORDINATE SYSTEMS.

Position determination in celestial navigation is primarily a matter of converting one set of coordinates to the other. This is done by solution of a spherical triangle called the navigational triangle.

The concept of the spherical navigational triangle is graphically shown in the illustration, a diagram on the plane of the celestial meridian. The celestial meridian passes through the zenith of the observer, and is therefore a vertical circle of the horizon system. Elements of both systems are shown in this illustration, indicating that an approximate solution can be made graphically.

The vertices of the navigational triangle are the elevated pole (P_n), the zenith (Z), and the celestial body (M). The angles at the vertices are, respectively, the meridian angle (t), the azimuth angle (Z), and the parallactic angle (X). The sides of the triangle are the codeclination of the zenith or the colatitude (colat) of the observer, the coaltitude or zenith distance (z) of the body, and the codeclination or polar distance (p) of the body.

A navigational triangle is solved, usually by computation, and compared with an observed attitude to obtain a line of position by a procedure known as sight reduction. With the emergence of electronic computers and hand-held calculators, sight reduction has been performed increasingly with limited use or elimination of tables.

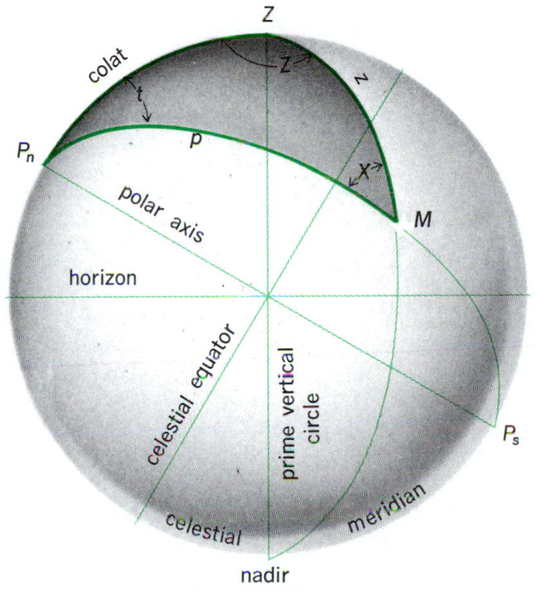

The navigational triangle.

To establish a celestial line of position, the navigator observes the altitude of a celestial body, noting the time of observation. Observation is made by a sextant, so named because early instruments had an arc of one-sixth of a circle. By means of the double reflecting principle, the altitude of the body is double the amount of arc used. The marine sextant uses the visible horizon as the horizontal reference. An air sextant has an artificial, built-in horizontal reference based upon a bubble or occasionally a pendulum or gyroscope. The sextant altitude, however measured, is subject to certain errors, for which corrections are applied. See SEXTANT.

Time is repeatedly mentioned as an important element of a celestial observation because the Earth rotates at the approximate rate of 1 minute of arc each 4 s of time. An error of 1 s in the timing of an observation might introduce an error in the line of position of as much as one-quarter of a mile. Time directly affects longitude determination, but not latitude. The long search for a method of ascertaining longitude at sea was finally solved two centuries ago by the invention of the marine chronometer, a timepiece with a nearly steady rate. *See* NAVIGATION; PILOTING. [A.B.M.]

Celestial sphere A hypothetical sphere of infinite radius with its center at the observer. Because all stars are so far from an Earth-bound observer that they can be located solely by their direction, it is convenient to consider them as located on the inside of a spherical surface centered on the observer. Because the radius of Earth is negligible, it is frequently convenient to consider Earth, rather than the terrestrial observer, centered in the celestial sphere. Stars are designated by their location on this sphere, and planets and other heavenly bodies are located against this background. *See* ASTRONOMICAL COORDINATE SYSTEMS. [F.H.R.]

Celestite A mineral with the chemical composition $SrSO_4$. Celestite occurs commonly in colorless to sky-blue, orthorhombic, tabular crystals. Fracture is uneven and luster is vitreous. Hardness is 3–3.5 on Mohs scale and specific gravity is 3.97. It fuses readily to a white pearl. The strontium present in celestite imparts a characteristic crimson color to the flame.

Celestite occurs in association with gypsum, anhydrite, salt beds, limestone, and dolomite. Large crystals are found in vugs or cavities of limestone. It is deposited directly from sea water, by groundwater, or from hydrothermal solutions. Celestite is the major source of strontium: Although celestite deposits occur in Arizona and California, domestic production of celestite has been small and sporadic. Much of the strontium demand is satisfied by imported ores from England and Mexico. *See* STRONTIUM. [E.C.T.C.]

Cell (biology) One of the microscopic structural units of which living organisms, both plant and animal, are built. Some simple organisms consist of one free-living cell, whereas more complex plants and animals contain from hundreds to billions of cells. A typical cell consists of two major compartments: a nucleus, containing the hereditary material in the form of chromosomes, and a surrounding mass of cytoplasm, which includes a variety of smaller components. *See* CYTOPLASM.

Cells are characterized by certain universal structural features. Almost all cells possess an external limiting membrane, a nucleus (with chromosomes), mitochondria, ribosomes, and a system of internal cytoplasmic membranes. In addition, other components are widely distributed, such as lysosomes, centrioles, cilia, microtubules, and the walls and chloroplasts of plant cells.

The external cell membrane is a very delicate structure less than 10 nanometers in thickness. It is seen in the electron microscope as a three-layered sheet. Other membranes inside the cell display a similar structure, and the name unit membrane has been applied to all these. The cell membrane regulates the exchange of materials in and out of the cell, and it plays an essential role in nervous transmission, contraction, and interaction of cells with one another. *See* CELL MEMBRANES.

Within the cytoplasm is a system of membranes which constitute the endoplasmic reticulum and part of the Golgi zone. The endoplasmic reticulum consists of a series of closed sacs. frequently flattened and stacked in multiple layers. Connections between the sacs establish a complex pattern of membranes and spaces. The outside of the sacs is usually studded with ribosomes. The ribosomes are the sites of protein synthesis in the cells, but their relationship to the membranes of the endoplasmic reticulum is not clear. *See* ENDOPLASMIC RETICULUM; RIBOSOMES.

The Golgi zone also contains thin flattened sacs somewhat like those of the endoplasmic reticulum. However, the sacs of the Golgi zone are smaller and more tightly packed, and lack ribosomes on their surfaces. In addition, this part of the cell regularly contains droplets or granules of various sizes. The exact function of the Golgi zone has been much debated, but it is known that products which are later secreted from the cell receive some kind of processing or packaging in the Golgi zone. *See* GOLGI APPARATUS.

A major role in cell function is played by the mitochondria, which are the respiratory centers of the cell. Mitochondria appear as minute rods or spheres in the cytoplasm, often numbering in the thousands in a single cell. They are surrounded by two unit membranes, the innermost of which sends finger- or sheetlike projections into the interior. Various respiratory enzymes form an integral part of the structure of the membranes. *See* MITOCHONDRIA.

The nuclear envelope is another membranous component of the cell. It was known as the nuclear membrane until the electron microscope showed that it is, in fact, two closely spaced membranes completely surrounding the nucleus. It is pierced by numerous octagonal pores, which connect the cytoplasm with the interior of the nucleus. Within the nucleus are one or more chromosomes, the threadlike structures which control inheritance. Chromosomes are exceedingly long and thin when stretched out completely. In addition to the chromosomes, the nucleus contains one or more nucleoli, which appear as prominent spherical bodies in nondividing cells. Each nucleolus is attached to a chromosome at a point known as the nucleolus organizer. *See* CHROMOSOME; MITOSIS.

Among the widespread but not universal cell components are cell walls and chloroplasts, characteristic of plant cells. The majority of plant cells secrete one or more rigid walls outside the cell membrane. These walls contain cellulose and other strengthening materials and may be extremely thick, as in the woody parts of stems. They are responsible for the characteristically rigid form of plants, as well as for the elasticity and toughness of their tissues. Chloroplasts contain the green pigment chlorophyll and are typical, therefore, of the green plants. They absorb light energy and use it to carry out chemical reactions by the process of photosynthesis. *See* CELL WALLS (PLANT); PLANT CELL.

Many cells contain membrane-bounded vesicles of indefinite form known as lysosomes. These harbor a number of hydrolytic enzymes, one of which, acid phosphatase, is characteristic of this organelle. Lysosomes appear to play a digestive role in the cell. *See* LYSOSOME.

Several cytoplasmic components of widespread occurrence contain thin rods or tubes which are known as microtubules. They may occur free or in bundles in the cytoplasm, where they evidently help to maintain cell shape. The mitotic spindle contains microtubules, which may be concerned with the orderly movements of the chromosomes. [J.G.G.]

Cell biology The name applied to that division of biology which deals with the structure, behavior, growth, and reproduction of cells and the function and chemistry of cell components. Cell biology is therefore the branch of science that explores the properties of living matter, or protoplasm. *See* CELL (BIOLOGY); CYTOCHEMISTRY; MOLECULAR BIOLOGY. [K.R.P.]

Cell cycle In a growing cellular system, a fundamental unit with the dimension of time which stretches from the formation of a cell by the division of its mother cell until it in turn divides to form two daughters. This cycle occurs not only in cultures of isolated cells but also in the growing tissues of multicellular organisms. All components of the cell double during the cycle, and at the end there are the two splitting events of nuclear division (mitosis) and cytoplasmic division (cytokinesis). These two events happen close together in almost all cells.

Although all components of the cell have to double during the cell cycle, deoxyribonucleic acid (DNA) synthesis is not nec-

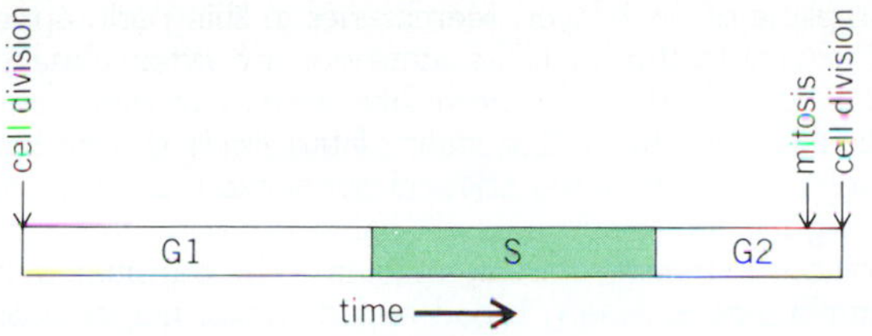

DNA synthesis during the cell cycle of a mammalian cell, showing the period of DNA synthesis (S) and the gap periods (G1 and G2) before and after DNA synthesis.

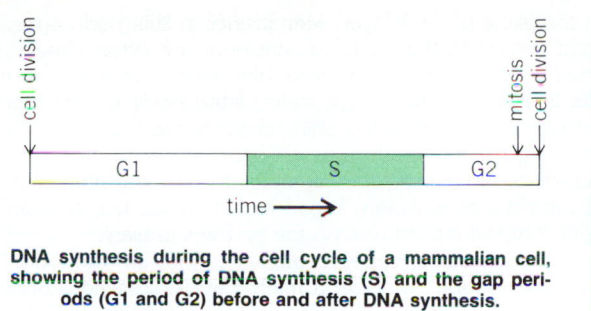

essarily a continuous process (see illustration). There is a gap at the start of the cycle (G1—the first gap) when there is no synthesis of DNA. This is followed by the synthesis period (S), during which DNA doubles, and then by a second gap (G2) until the cycle ends. Proportional division of the cycle into each phase of the DNA cycle varies considerably from one cell type to another although the reason for these variations is uncertain. The synthesis of ribonucleic acid (RNA) stops during mitosis in most eukaryotic cells, except in yeast and some protozoans. Protein synthesis is also reduced. For the rest of the cycle (interphase), synthesis is mostly continuous, though there may be variations in the rate of synthesis. *See* CELL DIVISION; CYTOKINESIS; DEOXYRIBONUCLEIC ACID (DNA); RIBONUCLEIC ACID (RNA). [J.M.Mit.]

Cell differentiation The mechanism by which cells in a multicellular organism become specialized to perform specific functions in a variety of tissues and organs. Specialized cells are the product of differentiation. The process can be understood only from a historical perspective, and the best place to start is the fertilized egg. Different kinds of cell behavior can be observed during embryogenesis: cells double, change in shape, and attach at and migrate to various sites within the embryo without any obvious signs of differentiation. Cleavage is a rapid series of cell cycles during which the large egg cell is divided into a ball of small cells that line the primitive body cavity as a single layer of embryonic cells. This blastula stage is followed by gastrulation, a complex coordinated cellular migration which not only shapes the embryo but segregates the single-cell layer of the blastula into the three germ layers: endoderm, mesoderm, ectoderm. They give rise to specific cell types; for example, skin and nerves from the ectoderm, the digestive tract from the endoderm, and muscle and connective tissue from the mesoderm. *See* BLASTULATION; CELL CYCLE; CLEAVAGE (EMBRYOLOGY); EMBRYOGENESIS; GASTRULATION.

The stable differentiated state is a consequence of multicellularity. A complex organism maintains its characteristic form and identity because populations of specialized cell types remain assembled in a certain pattern. Thus several kinds of cells make up a tissue, and different tissues build organs. The variable assortment of about 200 cell types allows for an almost infinite variety of distinct organisms.

Epithelia, sheets of cells of specific structure and function, cover the outer surface of the vertebrate body and line the lungs, gut, and vascular system. The stable form of a vertebrate is due to its rigid skeleton built from bone and cartilage, forming cells to which the skeletal muscles adhere. All other organs, such as liver and pancreas, are embedded in connective tissue that is derived from fibroblast cells which secrete large amounts of soft matrix material.

Some cells, like nerve cells, are so specialized that they need divide no longer in order to maintain a complex network. Their finite number decreases even during embryonic development. Other cell types are constantly worn out and must be replaced; for example, fibroblasts and pancreas cells simply divide as needed, proving that the differentiated state of cells is heritable, as the daughter cells remember and carry out the same special functions. The renewal of terminally differentiated cells that are unable to divide anymore, such as skin and blood cells, is carried out by stem cells. They are immortal and choose, as they double, whether to remain a stem cell or to embark on a path of terminal differentiation. Most stem cells are unipotent because they give rise to a single differentiated cell type. However, all cell types of the blood are derived from a single blood-forming stem cell, a pluripotent stem cell. A fertilized egg is a totipotent stem cell giving rise to all other cell types that make up an individual organism. *See* EMBRYONIC DIFFERENTIATION; EMBRYONIC INDUCTION; OOGENESIS. [H.Sa.]

Cell division A process by which living cells reproduce their kind. In the larger view, biological increase must be described as an increase in numbers of cells. The only cells which normally will not divide are found in differentiated tissues of multicellular organisms. Cell division includes nuclear division, which occurs most commonly by mitosis, and cytokinesis (fission of cytoplasm). *See* CYTOKINESIS; MEIOSIS; MITOSIS.

Overall growth of biological systems may be described in terms of the cellular growth division cycle. Living systems are composed of cells which have the capacity to synthesize new molecules similar to those already present. As a result, cells grow in mass, but the capacity of an individual cell to grow is limited. Normally, although there are exceptions, a cell can only double its mass before growth ceases. Division refreshes the capacity for growth, because the daughter cells begin to grow immediately after division. They normally cannot divide again until they have grown again. Division makes growth possible, growth makes division possible, and thus biological systems increase by stages of doubling and halving.

Genetic conservation, in this cycle, depends on the duplication of the genetic material before division, followed by its exact qualitative and quantitative distribution to the daughter cells. The distribution is accomplished by mitosis in all plant and animal cells, and in many protozoa, molds, and other microorganisms. Multiplication of cells always involves multiplication of nuclei. On the other hand, there are cases in which cells do grow without division. In these, nuclear divisions parallel the growth, but the division of the cell body is omitted. [D.M.]

Cell lineage A type of embryological study in which the history of individual blastomeres is traced to their ultimate differentiation into tissues and organs. The lineage of blastomers can be readily traced only in certain invertebrate animals, chiefly polyclad flatworms, annelids, mollusks, and ascidians, where the cleavage of the egg follows a rigidly stereotyped pattern. Individual cells are recognizable by differences in size, shape, position, composition, and time of origin. *See* CLEAVAGE (EMBRYOLOGY); OVUM. [A.C.Cl.]

Cell membranes Thin layers, only a few molecules thick, consisting mainly of lipids and proteins. They cover the surface of all cells and are then called plasmalemma, plasma membrane, or cell membrane in the narrower sense. Other membranes similar in composition and appearance to plasma membranes are found in the cytoplasm of most cells, where they create a number of different compartments. Most cell organelles are bounded by such membranes; some of them, for example, mitochondria and chloroplasts, are again subdivided into compartments by internal membranes.

The plasma membrane controls exchange of matter between the interior of the cell and the surrounding medium and enables the cell to maintain concentrations of water, small ions, and larger molecules different from, and within limits independent of, the environment. Plasma membranes are almost impermeable to some ions and molecules; other ions and molecules can diffuse freely through the membrane; and some are transported actively against concentration or electric gradients. Closely related to the permeability properties are the functions of the plasma membrane in excitation and impulse conduction. Bulk transport of fluid (pinocytosis) or solids (phagocytosis) across the cell border occurs through enfoldings of the membrane that close to form vesicles which are detached and move into the cytoplasm. The reverse of this mechanism is used for the secretion of cell products. Membranes also carry enzymes and integrated enzyme systems, especially systems that function as energy transducers. Plasma membranes also carry so-called receptors, which are able to bind specific compounds such as hormones or more complex structures such as virus particles. *See* BIOELECTRIC MODEL; ENDOCYTOSIS; PHAGOCYTOSIS.

Investigations into the molecular structure of cell membranes have for a long time been dominated by the Danielli model, which describes membranes as consisting of a bimolecular layer of lipids 4.0–5.0 nm wide covered with layers of protein on both surfaces to give a total width of 7.5–10.0 nm. In the bilayer the lipid molecules are oriented with their hydrophilic groups facing outward and the hydrocarbon chains facing each other to form the hydrophobic core (see illustration). This

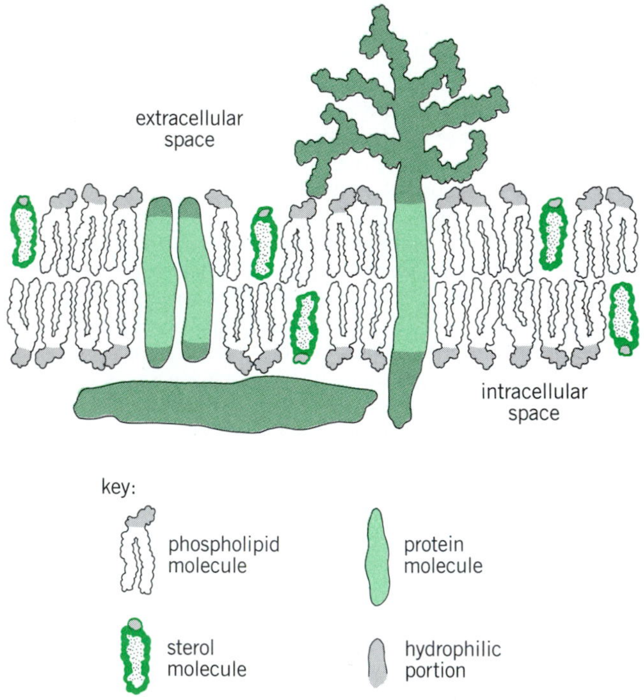

Arrangement of lipids and proteins in membranes. Protein molecules are arranged on the surface of the lipid bilayer or spanning it. The branched structure on the outer surface of the membrane represents the carbohydrate chains attached to the protein.

model may still correctly describe the structure of some membranes, for example, the myelin sheath of nerve fibers, but it must be modified for most other membranes. Electron microscopy, x-ray diffraction, and a variety of spectroscopic techniques have confirmed the existence of a lipid bilayer as the structural backbone of membranes but also show that a large part of the protein is present in the hydrophobic core of the membrane.

The lipids in the bilayer can exist in two states, an essentially solid or gellike state and a liquidlike state which is disordered and in which the lipid molecules have a high degree of mobility in the plane of the bilayer. Membranes in living cells appear to exist usually in the liquidlike state but are often close to, or even in, the transition to the gellike state. The transition from the liquidlike to the gellike state of the lipids can be induced through changes in many physical parameters, for example, by lowering the temperature or changing pH and other ion concentrations in the surrounding medium. The transition in natural membranes is usually broad and depends not only on the lipid composition but also on the proteins present.

The thickness of cell membranes cannot be resolved in the light microscope. In electron micrographs of sectioned cells or tissues they appear as triple-layered structures 7.5–10 nm wide, consisting of two outer dense bands and a central lighter region. This has been termed the unit membrane, because it is the characteristic image of nearly all cell membranes.

Little is known about the formation of new membranes in cells. Most of the direct evidence indicates that membranes grow by insertion of new molecules or molecular complexes into preexisting membranes. It is not known whether areas of undifferentiated membranes are first formed, to which specific functional components are added later, or whether a fully functional membrane piece is assembled and inserted. The Golgi apparatus has been specifically implicated in the synthesis of new cell membranes. The reconstitution of membrane-like structures from disaggregated cellular membranes suggests that formation of new membranes without the requirement of preexisting membrane is also a possibility. *See* CELL WALLS (PLANT); PLANT CELL. [W.St.]

Cell plastids Specialized structures found in the cytoplasm of plant cells, diverse in distribution, size, shape, composition, structure, function, and mode of development. A number of different types are recognized. Chloroplasts occur in the green parts of plants and are responsible for the green coloration, for they contain the chlorophyll pigments. These pigments, along with certain others, absorb the light energy that drives the processes of photosynthesis, by which sugars, starch, and other organic materials are synthesized. Amyloplasts, nearly or entirely colorless, are packed with starch grains and occur in cells of storage tissue. Proteoplasts are less common and contain crystalline, fibrillar, or amorphous masses of protein, sometimes along with starch grains. In chromoplasts the green pigment is masked or replaced by others, notably carotenoids, as in the cells of carrot roots and many flowers and fruits. *See* CAROTENOID; CHLOROPHYLL.

All types of plastids have one structural feature in common, a double envelope consisting of two concentric sheets of membrane. The outer of these is in contact with the cytoplasmic ground substance; the inner with the plastid matrix, or stroma. They are separated by a narrow space of about 10 nanometers.

Another system of membranes generally occupies the main body of the plastid. This internal membrane system is especially well developed in chloroplasts, where the unit of construction is known as a thylakoid. In its simplest form this is a sac such as would be obtained if a balloon-shaped, membrane-limited sphere were to be flattened until the internal space was not much thicker than the membrane itself. It is usual, however, for thylakoids to be lobed, branched, or fenestrated.

The surface area of thylakoids is very large in relation to the volume of the chloroplast. This is functionally significant, for chlorophyll molecules and other components of the light-reaction systems of photosynthesis are associated with these membranes. A chloroplast, however, is much more than a device for carrying out photosynthesis. It can use light energy for uptake and exchange of ions and to drive conformational changes. The stroma contains the elements of a protein-synthesizing system—as much deoxyribonucleic acid (DNA) as a small bacterium, various types of ribonucleic acid (RNA), distinctive ribosomes, and polyribosomes. There is evidence to indicate that much of the protein synthesis of a leaf takes place within the chloroplasts. *See* PHOTOSYNTHESIS; PROTEIN; RIBOSOMES.

One of the most challenging problems in cell biology concerns the autonomy of organelles, such as the plastids. Chloroplasts, for instance, have their own DNA, DNA-polymerase, and RNA-polymerase; can make proteins; and, significantly, can mutate. All this suggests a measure of independence. It is known, however, that some nuclear genes can influence the production of molecules that are normally found only in chloroplasts, so their autonomy cannot be complete. It remains to be seen whether they control and regulate their own morphogenetic processes. [B.E.S.G.]

Cell polarity (biology) The highly organized condition in most cells that is characterized by a distinct apical-basal axis with an asymmetric distribution of cytoplasmic organelles. This phenomenon of polarization is critical for living organisms to function. For example, cells in secretory organs such as the gall bladder generally secrete only at one end where secretory vesicles are localized. Another example would be a cell in the intestinal wall that must collect nutrients on the side near the lumen and transport them through the cell to the opposite end, where they can be delivered to the blood supply for transport to the rest of the body. Clearly, this cell must exhibit a polarized distribution of membrane proteins so that those necessary for sugar uptake are concentrated in the membrane facing the lumen and those needed to move the sugar out of the cell are located at the opposite side.

Most organisms begin life as a rather symmetrical, spherical, single cell called an egg. During early development this egg divides many times and forms an embryo, which exhibits much more intricate patterns (such as the polarized intestine) than were initially expressed by the egg. Determining how the embryo controls the development of such patterns has been an area of active research, and there is evidence that the plasma (outer) membrane can influence the development of cell polarity by driving ionic currents through the cell which, in turn, can influence the polarization process by generating ion concentration gradients within the cell or voltage gradients between cells. *See* BIOPOTENTIALS AND IONIC CURRENTS; CELL (BIOLOGY); DEVELOPMENTAL BIOLOGY. [R.Nu.]

Cell senescence and death The limited capacity of all normal human and other animal cells to reproduce and function. The gradual decline in normal physiological function of the cells is referred to as aging or senescence. The aging process ends with the death of individual cells and then, generally, the whole animal. Aging occurs in all animals, except those that do not reach a fixed body size such as some tortoises and sharks, sturgeon, and several other kinds of fishes. These animals die as the result of accidents or disease, but losses in normal physiological function do not seem to occur. Examples of cells that do not age are those composing the germ plasm (sex cells) and many kinds of cancer cells. These cells are presumed to be immortal.

Although cultured normal human and other animal cells are mortal, they can be converted to a state of immortality. The conversion can be produced in human cells by the SV40 virus and in other animal cells by other viruses, chemicals, and irradiation. This conversion from mortality to immortality is called transformation, and is characterized by the acquisition of many profound abnormal cell properties, including changes in chromosome number and form, and the ability of the cells to grow unattached to a solid surface. These changes, and many more, are characteristic of most cancer cells. *See* CANCER (MEDICINE); ONCOLOGY; SENESCENCE; TUMOR VIRUSES. [L.H.]

Cell walls (plant) The cell wall is the layer of material secreted by the plant cell outside its plasma membrane. With some exceptions among the algae and most of the fungi, plants more complex than prokaryotes (bacteria and blue-green algae) have cell walls that are generally very similar in chemical composition, organization, and development. From an engineering point of view, the plant cell wall can be considered a nonliving system, although its role in the economy of the organism and of the cell is basic to any understanding of the living system. The wall serves as the first point of entry of materials into cells, functions in the movement of water throughout the plant, and is one of the major mechanical strengthening factors. In addition, the wall must be sufficiently flexible and plastic to withstand mechanical stresses while still permitting the growth of the cell. *See* CELL MEMBRANES.

Cellulose forms the skeletal structure of the cell wall. Like starch, it is basically a polymer of glucose, a six-carbon monosaccharide, although the biosynthetic route for cellulose synthesis differs from that of starch. Each chain of cellulose may be as long as 8000 to 12,000 glucose monomers, or up to 4 micrometers long. These are arranged linearly, with no side branching. Cellulose chains are aggregated into bundles of approximately 40 chains each, the cellulose micelles, which are held together by hydrogen bonds. The micelles themselves are embedded in a matrix of other polysaccharides, the hemicelluloses. Micelles, in bundles of variable number, are bound together into the cellulose microfibril, a unit sufficiently large to be seen under the electron microscope; these, in turn, are bound together into macrofibrils which are observable under the light microscope. *See* CELLULOSE. [R.M.K.]

Cellobiose A disaccharide, with the structure shown below, usually prepared from cellulose, in which molecule it

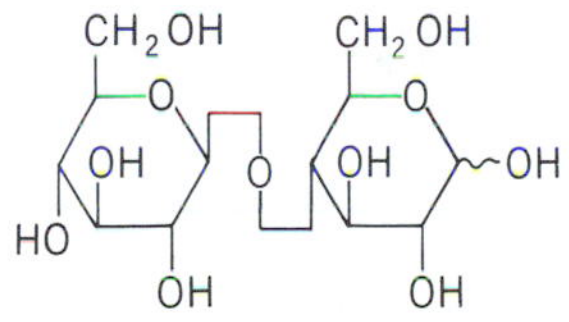

can be viewed as a repeating unit. It is composed of a ß-D-glucopyranosyl group joined by 1→4 glycosidic linkages to D-glucose. Its proper scientific name is 4-*O*-ß-D-glucopyranosyl-D-glucose. The sugar is rarely found free in nature, but its occurrence in pine needles has been reported. *See* CELLULOSE. [R.L.Wh.; J.R.D.]

Cellophane A transparent polycrystalline film produced by regeneration of cellulose in the form of wood pulp, cotton fiber, or another natural source of cellulose. Cellulose nitrate is frequently used as a coating to enhance imperviousness to moisture and receptivity to printing ink.

Cellophane is available as film (thickness less than 0.0098 in. or 0.25 mm) or sheeting (thickness greater than 0.0098 in.). It exhibits excellent transparency, gloss, stiffness, tensile strength, and flexibility, and transmits a high percentage of ultraviolet radiation. Used chiefly in food packaging, it is rapidly being replaced by polypropylene film. However, it has applications in high performance packaging where stiffness and abrasion resistance are primary concerns. Cellophane is considered the most easily processed film in large-scale printing, converting, and in-line packaging operations. *See* CELLULOSE. [F.H.R.]

Cellular immunology The study of the cells of the lymphoid organs, which are the main agents of immune reactions in all the vertebrates. The central cell of immunology is the lymphocyte. Two types of lymphocytes are derived from stem cells in the yolk sac, fetal liver, and bone marrow. T lymphocytes undergo maturation and proliferation in the thymus before entering the "pool" of peripheral immunocompetent lymphocytes. B lymphocytes mature and proliferate in the marrow or fetal liver or (in birds) in a special organ, the bursa of Fabricius. Both cells, when mature, enter the bloodstream.

The two lymphocyte populations, once they enter the peripheral pool, have entirely different roles. The B cell, when adequately stimulated, becomes an antibody-producing and secreting cell. The T cell, on the other hand, becomes the immune effector cell of cell-mediated immunity. Some T cells also play an important role in cooperation, helping B lymphocytes to respond to antigen, and in suppression of excessive or unwanted responses. There are thus several subpopulations of T cells with different functional attributes.

Advances in knowledge of this complex field have depended on development of new materials and techniques. The most important of these has been the use of inbred lines of mice, in which cells and tissues can be freely moved from one animal to another genetically identical recipient. By transfer of different purified cell populations, one can analyze their role in immunity. Also, genetic differences between inbred lines has permitted identification of immune response genes and chromosomal mapping of other significant elements of the immune response.

Immunologists are in the middle of a period of rapid application of the new techniques and new insights to problems of human medicine. A great challenge facing researchers in this field is the use of immunologic methods to alleviate the large range of autoimmune diseases, to permit free grafting of tissues and organs as needed, and to enhance immunity to tumors. *See* IMMUNOLOGY. [B.H.W.]

Cellulose The main polysaccharide in living plants, empirical formula $(C_6H_{10}O_5)_n$. It forms the skeletal structure of the cell wall, hence the name cellulose; wood normally contains 40–50% cellulose. It occurs together with polysaccharides and hemicelluloses derived from other sugars. In the woody parts of plants, cellulose is intimately mixed, and sometimes covalently linked, with lignin. Cellulose is a polymer of ß-D-glucopyranosyl units which are linked together, with elimination of water, to form chains of 2000–4000 units, as shown in the illustration. *See* CELL WALLS (PLANT).

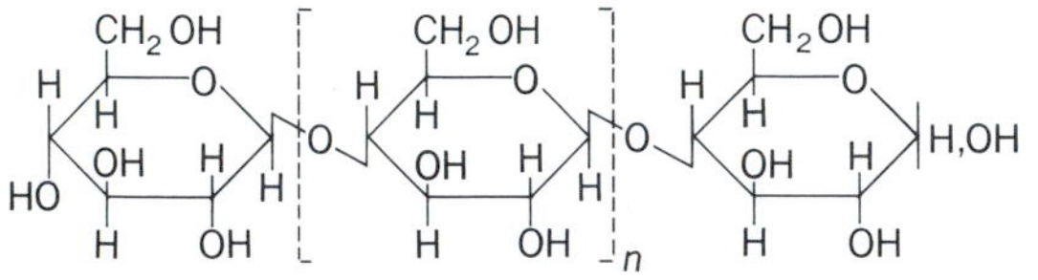

Structure of the cellulose polymer.

Cellulose of high chemical purity for research purposes is prepared from cotton. This cellulose is almost entirely free from other components and is one of the purest readily available polysaccharides. High-molecular-weight cellulose is used for making cellophane, viscose, and acetate rayon. Both high-molecular-weight and lower-molecular-weight celluloses may be used for making other derivatives.

Commercial cellulose may be obtained from cotton linters, but is produced mainly from woods, particularly pine and spruce, by pulping procedures designed to remove lignin and other noncellulosic constituents. In addition to mechanical and semimechanical processes which separate wood fibers for use in paper manufacture, there are three chemical pulping processes in wide commercial use, the kraft, soda, and sulfite processes. *See* CARBOHYDRATE; CELLOBIOSE; HEMICELLULOSE; PAPER; POLYSACCHARIDE. [R.L.Wh.]

Cement Any substance that acts as a bonding agent for materials. In construction and engineering the word almost always means hydraulic cement. For information about other cementing materials *see* ADHESIVE; ASPHALT AND ASPHALTITE; CLAY; EPOXY RESIN; MUCILAGE.

Hydraulic cements are produced by burning an intimate mixture of finely divided calcareous and argillaceous materials and grinding the resulting clinker to a fine powder, usually with gypsum to retard the set. The calcining process produces calcium silicates and calcium aluminates that can react chemically with water to form a hard, stonelike mass. When mixed with sand, coarse aggregate, and water, these cements produce mortars and concretes. *See* CONCRETE; MORTAR. [J.H.W.]

Cenozoic The youngest of the eras or major subdivisions of geologic time, extending from the end of the Mesozoic Era to the present, or Holocene. The term Cenozoic is also applied to all the rocks formed during this time and all the remains of

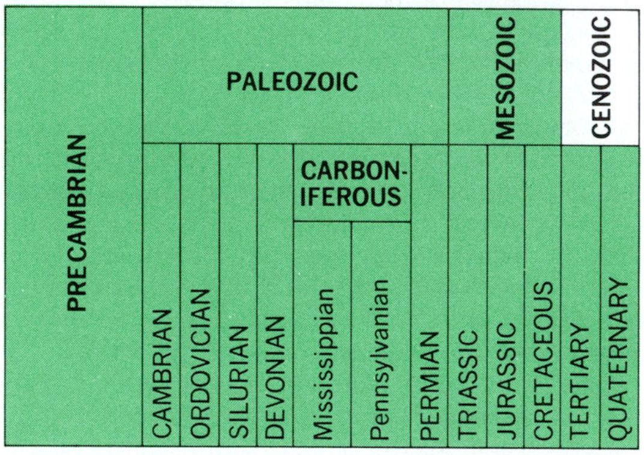

past life which they contain. Although no unit term, such as era, has in the past been used for fossiliferous rock groups of era magnitude, the term erathem has been proposed for such major time-rock units.

Rocks formed during Cenozoic time were deposited mechanically or chemically upon many kinds of older rocks, or in volcanic form extruded to rest also upon older rocks or intruded into or below such older rocks. From time to time these Cenozoic rocks in certain areas of concentrated deposition have been elevated, locally broken, and widely crumpled by stresses operating in the crust of the Earth to produce great mountain ranges such as those around the entire rim of the Pacific Ocean (the Andes, the West Coast ranges of North America, those of Kamchatka, Japan, the East Indies, New Zealand) and the Himalayan chain. Geologically, these are the youngest mountains.

Distribution of land and sea was different at different times during the Cenozoic. Although this was a continuous process of change, with local exceptions to the general trend, three great expansions of the oceans over the continental shelf (with several lesser fluctuations) were interspersed between four extended intervals of great land emergence, uplift, and increased surface relief, all prior to the attainment of geologically modern times. Historically, recognition of subdivisions within the Cenozoic first arrived at a threefold classification (Eocene, Miocene, Pliocene), with intervening gaps filled in only subsequently (Paleocene, Oligocene, "Mio-Pliocene," Pleistocene, Holocene).

Modern landforms and the present configuration of the continental land masses have come into being largely from dynamic interplay of deposition, erosion, mountain-making (orogeny and epeirogeny), and volcanism during Cenozoic time. Eventually continental glaciation became widespread, during a great Ice Age (Pleistocene) that finally began to wane some 25,000 years ago (beginning of the Holocene Epoch).

The entire Cenozoic Era lasted on the order of 70,000,000 years. Customarily, it is formally divided into two larger subdivisions, or periods, of unequal length: the older, or Tertiary, period, and the younger, or Quaternary. *See* QUATERNARY; TERTIARY. [R.M.Kl.]

Centaurus The Centaur, in astronomy; one of the most magnificent of the southern constellations. Two first-magnitude navigational stars, Alpha and Beta Centauri, mark the right and left front feet, respectively, of the centaur. The former is Rigil Kentaurus, or simply Rigil Kent, and the latter Hadar. Rigil Kent is the third brightest star in the whole sky, fainter than only Sirius and Canopus. The line joining Rigil Kent and Hadar points to the constellation Crux (Southern Cross). Thus they are also called the Southern Pointers, in contradistinction to the northern pointers of the Big Dipper. *See* CONSTELLATION; URSA MAJOR. [C.-S.Y.]

Center of gravity A fixed point in a material body through which the resultant force of gravitational attraction acts. The resultant of all forces or attractions produced by the Earth's gravity on a body constitutes its weight. This weight is considered to be concentrated at the center of gravity in mechanical studies of a rigid body. The location of the center of gravity for a body remains fixed in relation to the body regardless of the orientation of the body. If supported at its center of gravity, a body would remain balanced in its initial position. *See* GRAVITY; RESULTANT OF FORCES. [N.S.F.]

Center of mass That point of a material body or system of bodies which moves as though the system's total mass existed at the point and all external forces were applied at the point. The Earth-Moon system moves in the Sun's gravitational field as though both masses were located at a center of mass some 3000 mi (4700 km) from the Earth's geometric center. The function of the center-of-mass concept is to permit analysis of the motion of an entire system as distinguished from that of its individual parts.

Consider a system of mass M composed of n bodies with masses $m_1, m_2, \ldots, m_n$, and radius vectors $r_1, r_2, \ldots, r_n$ measured from some common reference point. Define a point with radius vector R, such that Eq. (1) holds. Then, it is possi-

$$MR = \sum_j m_j r_j \qquad (1)$$

ble to derive Eq. (2), an expression of Newton's second law,

$$\frac{d^2R}{dt^2} = \frac{F}{M} \qquad (2)$$

which states that the center of mass at R moves as though it possessed the total mass of the system and were acted upon by the total external force.

A simplification of the description of collisions can be obtained by using a coordinate system which moves with the velocity of the center of mass before collision. *See* COLLISION (PHYSICS); RIGID-BODY DYNAMICS. [J.P.H.]

Center of pressure A point on a plane surface through which the resultant force due to pressure passes. Such a surface can be supported by a single mounting fixture at its center of pressure if no other forces act. For example, a water gate in a dam can be supported by a single shaft at its center of pressure. *See* RESULTANT OF FORCES. [N.S.F.]

Central force A force whose line of action is always directed toward a fixed point. The central force may attract or repel. The point toward or from which the force acts is called the center of force. If the central force attracts a material particle, the path of the particle is a curve concave toward the center of force; if the central force repels the particle, its orbit is convex to the center of force. Undisturbed orbital motion under the influence of a central force satisfies Kepler's law of areas. [R.L.Du.]

Central heating and cooling The use of a single heating or cooling plant to serve a group of buildings, facilities, or even a complete community through a system of distribution pipework that feeds each structure or facility. Central heating plants are basically of two types: steam or hot-water, The latter type uses high-temperature hot water under pressure and has become the more usual because of its considerable advantages. Steam systems are only used today where there is a specific requirement for high-pressure steam. Central cooling plants utilize a central refrigeration plant with a chilled water distribution system serving the air-conditioning systems in each building or facility.

Advantages of a central heating or cooling plant over individual ones for each building or facility in a group include reduced labor cost, lower energy cost, less space requirement, and simpler maintenance. Central cooling plants, using conventional, electrically driven refrigeration compressors, have the advantage of utilizing bulk electric supply, at voltages as high as 13.5 kV, at wholesale rates. Additionally, their flexible load factor, resulting from load divergency in the various buildings served, results in major operating economies.

The disadvantages of a central heating plant concern mainly the maintenance of the distribution system where steam is used. Corrosion of the condensate water return lines shortens their life, and the steam drainage traps need particular attention. These disadvantages do not occur with high-temperature hot-water installations. *See* AIR CONDITIONING; BOILER; COMFORT HEATING; REFRIGERATION; STEAM HEATING; WARM-AIR HEATING SYSTEM. [J.K.M.P.]

Central nervous system That portion of the nervous system composed of the brain and spinal cord. The brain is enclosed in the skull, and the spinal cord within the spinal canal of the vertebral column. The brain and spinal cord are intimately covered by membranes called meninges and bathed in an extracellular fluid called cerebrospinal fluid. Approximately 90% of the cells of the central nervous system are glial cells which support, both physically and metabolically, the other cells, which are the nerve cells or neurons. *See* MENINGES; NEURON.

Functionally similar groups of neurons are clustered together in so-called nuclei of the central nervous system. When groups of neurons are organized in layers (called laminae) on the outer surface of the brain, the group is called a cortex, such as the cerebral cortex and cerebellar cortex. The long processes (axons) of neurons course in the central nervous system in functional groups called tracts. Since many of the axons have a layer of shiny fat (myelin) surrounding them, they appear white and are called the white matter of the central nervous system. The nuclei and cortex of the central nervous system have little myelin in them, appear gray, and are called the gray matter of the central nervous system. *See* BRAIN; NERVOUS SYSTEM (VERTEBRATE); SPINAL CORD. [D.B.W.]

Centrifugal force A fictitious or pseudo outward force on a particle rotating about an axis which by Newton's third law is equal and opposite to the centripetal force. Like all such action-reaction pairs of forces, they are equal and opposite but do not act on the same body and so do not cancel each other. Consider a mass M tied by a string of length R to a pin at the center of a smooth horizontal table and whirling around the pin with an angular velocity of ω radians per second. The mass rotates in a circular path because of the centripetal force $F_C = M\omega^2R$ which is exerted on the mass by the string. The reaction force exerted by the rotating mass M, the so-called centrifugal force, is $M\omega^2R$ in a direction away from the center of rotation. *See* CENTRIPETAL FORCE.

From another point of view, consider an experimenter in a windowless, circular laboratory that is rotating smoothly about a centrally located vetical axis. No object remains at rest on a smooth surface; all such objects move outward toward the wall of the laboratory as though an outward, centrifugal force were acting. To the experimenter partaking in the rotation, in a rotating frame of reference, the centrifugal force is real. An outside observer would realize that the inward force which the experimenter in the rotating laboratory must exert to keep the object at rest does not keep it at rest, but furnishes the centripetal force required to keep the object moving in a circular path. The concept of an outward, centrifugal force explains the action of a centrifuge. *See* CENTRIFUGATION. [C.E.H./R.J.S.]

Centrifugal pump A machine for moving fluid by accelerating it radially outward. More fluid is moved by centrifugal pumps than by all other types combined. Centrifugal pumps consist basically of one or more rotating impellers in a stationary casing which guides the fluid to and from the impeller or from one impeller to the next in the case of multistage pumps. Impellers may be single suction or double suction. Additional essential parts of all centrifugal pumps are (1) wearing surfaces or rings, which make a close-clearance running joint between the impeller and the casing to minimize the backflow of fluid from the discharge to the suction; (2) the shaft, which supports and drives the impeller; and (3) the stuffing box or seal, which prevents leakage between shaft and casing.

The rotating impeller imparts pressure and kinetic energy to the fluid pumped. A collection chamber in the casing converts much of the kinetic energy into head or pressure energy before the fluid leaves the pump. A free passage exists at all times through the impeller between the discharge and inlet side of the pump. Rotation of the impeller is required to prevent backflow or draining of fluid from the pump. Because of this, only special forms of centrifugal pumps are self-priming. Most types must be filled with liquid, or primed, before they are started.

Every centrifugal pump has its characteristic curve, which is the relation between capacity or rate of flow and pressure or head against which it will pump. At zero pressure-difference, maximum capacity is obtained, but without useful work. As resistance to flow external to the pump increases, capacity decreases until, at a high pressure, flow ceases entirely. This is called shut-off head and again no useful work is done. Between these extremes, capacity and head vary in a fixed relationship at constant rpm. When the required head exceeds that practical for a single-stage pump, several stages are employed. Multistage pumps range from two-stage pumps to pumps built with as many as 20 or 30 stages for high lifts from relatively small-diameter wells. *See* PUMP; PUMPING MACHINERY. [E.F.W.]

Centrifugation A mechanical method of separating immiscible liquids or solids from liquids by the application of centrifugal force. This force can be very great, and separations which proceed slowly by gravity can be speeded up enormously in centrifugal equipment. *See* CENTRIFUGAL FORCE.

Centrifugal force is generated inside stationary equipment by introducing a high-velocity fluid stream tangentially into a cylindrical-conical chamber, forming a vortex of considerable intensity. Cyclone separators based on this principle remove liquid drops or solid particles from gases, down to 1 or 2 μm in diameter. Smaller units, called liquid cyclones, separate solid particles from liquids. The high velocity required at the inlet of a liquid cyclone is obtained with standard pumps. Much higher centrifugal forces than in stationary equipment are generated in rotating equipment (mechanically driven bowls or baskets, usually of metal, turning inside a stationary casing). Rotating a cylinder at high speed induces a considerable tensile stress in the cylinder wall. This limits the centrifugal force which can be generated in a unit of a given size and material of construction. Very high forces, therefore, can be developed only in very small centrifuges.

There are two major types of centrifuges: sedimenters and filters. A sedimenting centrifuge contains a solid-wall cylinder or cone rotating about a horizontal or vertical axis. An annular layer of liquid, of fixed thickness, is held against the wall by centrifugal force; because this force is so large compared with that of gravity, the liquid surface is essentially parallel with the axis of rotation regardless of the orientation of the unit. Heavy phases "sink" outwardly from the center, and less dense phases "rise" inwardly. Heavy solid particles collect on the wall and must be periodically or continuously removed.

A filtering centrifuge operates on the same principle as the spinner in a household washing machine. The basket wall is perforated and lined with a filter medium such as a cloth or a fine screen; liquid passes through the wall, impelled by centrifugal force, leaving behind a cake of solids on the filter medium. The filtration rate increases with the centrifugal force and with the permeability of the solid cake. Some compressible solids do not filter well in a centrifuge because the particles deform under centrifugal force and the permeability of the cake is greatly reduced. The amount of liquid adhering to the solids after they have been spun also depends on the centrifugal force applied; in general, it is substantially less than in the cake from other types of filtration devices. *See* MECHANICAL SEPARATION TECHNIQUES. [J.C.Sm.]

Centriole A cellular organelle of great morphologic complexity, found in metazoans, protozoans, and many lower plants. It is usually near the nucleus in the interphase cell and at the poles of the spindle during mitosis. In differentiating cells, centrioles may give rise to or function as basal bodies, which are necessary to induce the formation of cilia or flagella. *See* CILLA AND FLAGELLA.

The mature centriole is a short, hollow cylinder containing nine groups of three microtubules in a squirrel-cage arrangement (illustration *a*). The centriole is about 300–500 nanometers in length and about 150 nm in diameter. Each microtubule is about 23–27 nm in diameter; its lumen is about 10 nm in diameter (illustration *b*). The wall of a microtubule comprises 13 chains of globular subunits, consisting of a protein called tubulin, and arranged cylindrically down the length of the

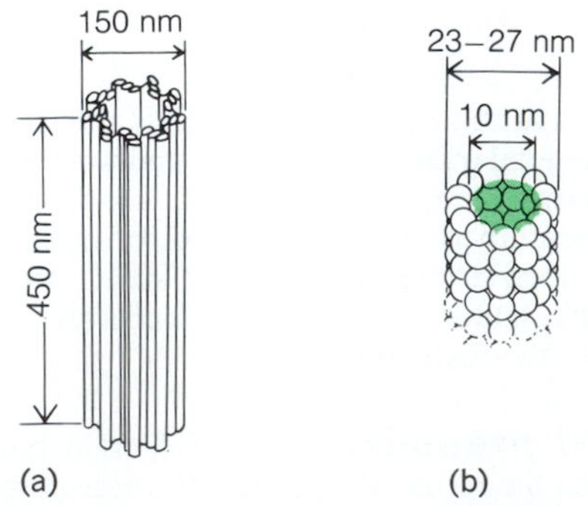

Diagram of mature centriole showing (*a*) arrangement of the microtubules and (*b*) structure of one of the microtubules.

tubule. Each subunit is 4.5–5.0 nm in diameter and 120,000 in molecular weight.

The cellular location of the centriole varies. In certain epithelial cells a pair of centrioles, at right angles to each other and partially surrounded by an extensive Golgi complex, is found near the nucleus. In epithelial cells lining cavities, the centrioles are found just under the membrane of each cell's free surface. In white blood cells the centrioles are found at the center of the cell, surrounded by Golgi cisternae. The location of centrioles also changes with the cell cycle. When the cell is ready to divide, the paired centrioles migrate to opposite sides of the nucleus. Once the cell has divided, a pair of centrioles can again be seen near the nucleus. *See* MITOSIS.

During cell division the positions of the centrioles are related to the plane of cleavage. The orientation of the spindle is established by the migration of the centrioles at the start of mitosis. The cell divides in a plane perpendicular to the main axis of the spindle. It has also been shown that once centrioles function as basal bodies, they induce the formation of cilia and flagella. They are also involved in the formation of the specialized cilia which develop into the rods of the retina and into the specialized processes of other sensory organs.

The centrioles for the first mitotic division of the ovum after fertilization are usually derived from the sperm midpiece, which usually contains two centrioles. Once in the cell, centrioles seem to have morphologic continuity, for they are present at all stages of the mitotic cycle. However, there are instances in which a mature centriole cannot be found in an early developmental stage, but appears at later stages. Centrioles are notably absent from the cells of higher plants. [E.R.D.]

Centripetal force The inward force required to keep a particle or an object moving in a circular path. It can be shown that a particle moving in a circular path has an acceleration toward the center of the circle along a radius. *See* ACCELERATION.

This radial acceleration, called the centripetal acceleration, is such that, if a particle has a linear or tangential velocity v when moving in a circular path of radius R, the centripetal acceleration is v^2/R. If the particle undergoing the centripetal acceleration has a mass M, then by Newton's second law of motion the centripetal force F_C is in the direction of the acceleration. This is expressed by the equation below, where ω is the con-

$$F_C = Mv^2/R = MR\omega^2$$

stant angular velocity and is equal to v/R. From Newton's laws of motion it follows that the natural motion of an object is one with constant speed in a straight line, and that a force is necessary if the object is to depart from this type of motion. Whenever an object moves in a curve, a centripetal force is necessary. In circular motion the tangential speed is constant but is changing direction at the constant rate of ω, so the centripetal force along the radius is the only force involved. [R.J.S.]

Centrode The path traced by the instantaneous center of a plane figure when it undergoes plane motion. If a plane rigid body is constrained to move in its own plane but is otherwise free to undergo an arbitrary translational and rotational motion, it is found that at any instant there exists a point, called the instantaneous center, about which the body is rotating. The path that this instantaneous center traces out in space as the motion unfolds is called the space centrode. The path that it would trace out in a coordinate system which is rigidly attached to the body is called the body centrode. The motion may therefore be specified, when the two centrodes are given, by allowing one curve to roll without slipping along the other. *See* FOUR-BAR LINKAGE; RIGID-BODY DYNAMICS. [H.C.Co./B.G.]

Centrohelida An order of the Heliozoia. There is no central capsule. The usually siliceous skeletal elements are scales and spines secreted by nearly all species. Pseudopodia may be either axopodia or filopodia. A centroplast is typical and the nucleus is consequently eccentric. The order includes *Acanthocystis*, *Heterophrys*, *Actinolophus*, and other genera. A few unusual types are stalked sessile organisms. *See* HELIOZOIA. [R.P.H.]

Centroids (mathematics) Points positioned identically with the centers of gravity of corresponding homogeneous thin plates or thin wires. Centroids are involved in the analysis of certain problems of mechanics, for example, the phenomenon of bending.

The centroid of plane area A is point C (see illustration).

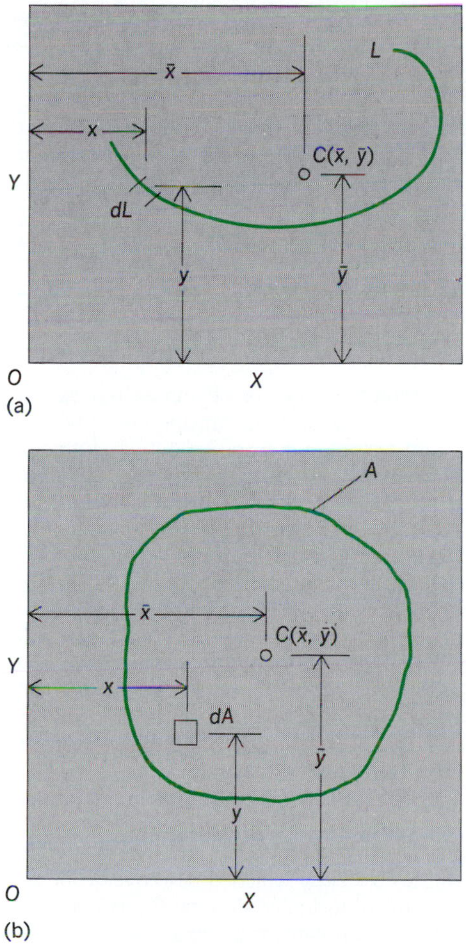

Notation of integral equations for centroids. (a) Centroid of line. (b) Centroid of area.

Coordinates $\bar{x}$ and $\bar{y}$ of C as referred to the indicated X and Y coordinate axes are given by Eq. (1). Similarly, the centroidal

$$\bar{x} = \frac{\int x\,dA}{\int dA} \qquad \bar{y} = \frac{\int y\,dA}{\int dA} \tag{1}$$

coordinates of plane curve L are given by Eq. (2). In these equations $\bar{x}$ and $\bar{y}$ are the coordinate locations of infinites-

$$\bar{x} = \frac{\int x\,dL}{\int dL} \qquad \bar{y} = \frac{\int y\,dL}{\int dL} \tag{2}$$

imal area element dA and infinitesimal line element dL, respectively. [N.S.F.]

Centrosome A poorly staining, optically structureless spherical region which surrounds the centriole, and which may vary in radius from one cell type to another. While it is generally free of other cytoplasmic constituents or organelles, this is not obligatory, and ribosomes, membrane-bounded vesicles, or even mitochondria are sometimes found in what is operationally termed the centrosome, encroaching on the centriole itself. The lack of a precise physical definition of the centrosome within the cell precludes any but the most general characterization of it. Apparently it, or cellular constituents within it, are involved in the fabrication of the mitotic spindle at prophase. *See* CENTRIOLE; MITOSIS. [E.Ro.]

Cephalaspidomorpha The subclass of Agnatha that includes the jawless vertebrates with a single median nostril. The Cephalaspidomorphi, sometimes called Monorhina, includes the superorder Hyperotreti, containing the modern Myxinoidea or hagfishes, and the superorder Hyperoartii, containing the living Petromyzonida or lampreys, as well as the extinct Osteostraci and Anaspida. *See* AGNATHA. [R.H.De.]

Cephalobaenida One of two orders in the class Pentastomida of the phylum Arthropoda. This order includes primitive pentastomids with six-legged larvae. The hooks are simple, lacking a fulcrum, and disposed in trapeziform pattern, with the anterior pair internal to the posterior pair. The mouth is anterior to the hooks. There are two families: Raillietiellidae, which contains two genera (*Raillietiella* and *Cephalobaena*); and Reighardiidae, which contains the single genus *Reighardia* with one species. *See* PENTASTOMIDA. [H.W.S.]

Cephalocarida A group of minute marine crustaceans of great interest to students of arthropod evolution because its members seem to possess many primitive characters and to show relationships not only to various other groups of Crustacea but perhaps also to the trilobites. Their closest apparent relatives are primitive members of other subclasses, such as Notostraca and Lipostraca (including Devonian *Lepidocaris*) within the Branchiopoda, and Leptostraca within the Malacostraca. Nine species are recognized, placed in four genera. They have been found in flocculent surface deposits of mud or silty sand, from the intertidal zone down to depths of 4900 ft (1500 m), on the shores of all continents except Europe. Population densities up to an average of 16 individuals/ft^2 (177 individuals/m^2) have been recorded. *See* BRANCHIOPODA; LEPTOSTRACA; TRILOBITA.

The best-known species is *Hutchinsoniella macracantha*, from the east coast of North and South America, about 0.12–0.16 in. (3–4 mm) long, with a shovel-shaped head and a slender, very flexible body which is not covered by a carapace. Other species are very similar. Among the features that appear to be primitive, most notable is the pronounced serial homology seen in the trunk limbs, the trunk musculature, the ventral nerve cord, and the heart. Other features in the anatomy of *Hutchinsoniella* that seem likely to be primitive are the uniramous, multisegmented antennules, the biramous antennae with both rami multisegmented, the large, flattened pleura on all the limb-bearing somites, and the telson freely articulated with the trunk and bearing a caudal furca. In the larval development, there is a primitive "nauplius" stage, followed by stages showing an unusually regular addition of new somites and limbs.

Hutchinsoniella is a nonselective deposit feeder, subsisting on the organic matter present in its habitat. Its feeding mechanism may well be like that which was used by trilobites and early crustaceans. The animal cannot swim but creeps through the flocculent mud on the bottom, and as it does so, lightweight particles of detritus are stirred up by the antennae and by distal claws on the trunk-limb endopods. *See* CRUSTACEA. [J.H.L.]

Cephalochordata A subphylum of the phylum Chordata comprising the lancelets, including *Branchiostoma* (amphioxus). They are also known as the Leptocardii. Lancelets are small fishlike animals, not exceeding 3.2 inches (80 mm) in length. They burrow in sand on the ocean bed or in estuaries in tropical and temperate regions throughout the world. Only two genera are recognized: *Branchiostoma* (23 species) and *Asymmetron* (6 species).

The structure of the lancelet is based on the same fundamental plan as all other chordates, but there is neither head nor paired fins. The skeletal rod of the back, or notochord, extends the entire length of the body. The animal is thus pointed at both ends and lanceolate in form.

The adults burrow in rather coarse sand or shell gravel and usually lie with the mouth open at the surface of the sand. Large numbers of lancelets congregate in small areas for spawning. The egg develops into a larva which at first is bottom-living, but later becomes planktonic. Metamorphosis takes place after about 11–12 weeks of life. The young adult sinks to the bottom and swims actively until a suitable sand in which it can burrow and remain undisturbed is found.

The importance of lancelets lies chiefly in their being one of the most primitive chordates. However, lancelets are eaten by the Chinese. *See* CHORDATA. [J.E.We.]

Cephalopoda The most highly evolved class of the phylum Mollusca. It consists of squids, cuttlefishes, octopuses, and the chambered nautiluses. The earliest known cephalopods are small, shelled fossils from the Upper Cambrian rocks of northeast China that are 500 million years old. Cephalopods always have been marine, never fresh-water or land animals. Most fossil cephalopods, among them the subclasses Nautiloidea and Ammonoidea, had external shells and generally were shallow-living, slow-moving animals. Of the thousands of species of such shelled cephalopods that evolved, all are extinct except for four species of the only surviving genus, *Nautilus*. All other recent cephalopods belong to four orders of the subclass Coleoidea, which also contains five extinct orders.

Living cephalopods are bilaterally symmetrical mollusks with a conspicuously developed head that has a crown of 8–10 appendages (8 arms and 2 tentacles) around the mouth. These appendages are lined with one to several rows of suckers or hooks. *Nautilus* is exceptional in having many simple arms. The mouth contains a pair of hard chitinous jaws that resemble a parrot's beak and a tonguelike, toothed radula (a uniquely molluscan organ). Eyes are lateral on the head; they are large and well developed. The "cranium" contains the highly developed brain, the center of the extensive, proliferated nervous system. The shell of ancestral cephalopods has become, in living forms, internal, highly modified, reduced, or absent; and is contained in the sac- or tubelike, soft muscular body, the mantle. A pair of fins may occur on the mantle as an aid to locomotion, but primary movement is achieved through jet propulsion in which water is drawn into the mantle cavity and then forcibly expelled through the nozzlelike funnel. Fewer than 1000 species of living cephalopods inhabit all oceans and seas from intertidal pools to depths of over 16,000 ft (5000 m) in the deep sea.

The classification given here concentrates on the living groups and lists only the major fossil groups; see separate articles on each subclass and order.

Class Cephalopoda
- Subclass Nautiloidea
- Subclass Ammonoidea
- Subclass Coleoidea
  - Order Belemnoidea
  - Order Sepioidea
  - Order Teuthoidea
    - Suborder Myopsida
    - Suborder Oegopsida

Order Vampyromorpha
Order Octopoda
Suborder Cirrata
Suborder Incirrata

Species of cephalopods inhabit most marine habitats. Cephalopods inhabit tide pools, rocky patches, sandy bottoms, coral reefs, grass beds, mangrove swamps, coastal waters, and the open ocean from the surface through the water column to depths on the abyssal bottom at over 16,000 ft (5000 m). *See* NERVOUS SYSTEM (INVERTEBRATE).

Cephalopods are high-level, active predators that feed on a variety of invertebrates, fishes, and even other cephalopods. The relatively sluggish nautiluses feed primarily on slow-moving prey such as reed shrimps, and even are scavengers of the cast-off shells of molted spiny lobsters. Cuttlefishes prey on shrimps, crabs, and small fishes, while squids eat fishes, pelagic crustaceans, and other cephalopods. Benthic octopuses prey mostly on clams, snails, and crabs. Salivary glands secrete toxins that subdue the prey and, in octopuses, begin digestion.

To protect themselves from predators cephalopods would rather hide than fight. To this end they have become masters of camouflage and escape. Benthic forms especially (for example, *Sepia* and *Octopus*) have evolved an intricate, complex system of rapid changes in color and patterns via thousands of individually innervated chromatophores (pigment cells) that allow precise matching to the color and pattern of the background. In addition, they regulate the texture of their skins by erecting papillae, flaps, and knobs that simulate the texture of the background. Many midwater oceanic squids camouflage against predation from below by turning on photophores (light organs) that match the light intensity from the surface and eliminate their silhouettes. *See* CHROMATOPHORE; PHOTOPHORE GLAND; PROTECTIVE COLORATION.

Cephalopods have perfected jet propulsion for many modes of locomotion, from hovering motionless, to normal cruising, to extremely rapid escape swimming. Water enters the mantle cavity through an opening around the neck when the muscular mantle (body) expands. The mantle opening seals shut as the mantle contracts and jets the water out through the hoselike funnel, driving the cephalopod tail-first through the water.

The sexes are separate in cephalopods, and many species display complex courtship, mating, spawning, and parental care behavior. At mating, the male of most species transfers the spermatophores to the female with a specially modified arm, the hectocotylus. The spermatophores are implanted into the female's mantle cavity, around the neck, under the eyes, or around the mouth, depending on the species. Incubation takes a few weeks to a few months depending on the species.

Cephalopods are extremely important in the diets of toothed whales (sperm whales, dolphins), pinnipeds (seals, sea lions), pelagic birds (petrels, albatrosses), and predatory fishes (tunas, billfishes, groupers). For example, pilot whales in the North Atlantic feed almost exclusively on one species of squid, *Illex illecebrosus*, that aggregates for spawning in the summer. *See* MOLLUSCA. [C.F.E.R.]

Cephalosporins A group of antibiotics that are effective in eradicating streptococcal, pneumococcal, staphylococcal, *Klebsiella, Neisseria*, and enteric gram-negative rod bacteria that produce pulmonary, skin and soft tissue, bone and joint, endocardial, surgical, urinary, and bacteremic infections. They have been used most often in a preventive or prophylactic fashion at the time of various surgical procedures. All the third-generation cephalosporins penetrate well into tissues, and antibacterially active levels in various body fluids and tissues such as bone are excellent. The toxic potential of the agents, considering their broad antibacterial spectrum, has been minor. Toxicities which are seen are those of bleeding due to vitamin K depletion. *See* ANTIBIOTIC. [H.N.]

Cepheids A class of brightness-variable stars whose prototype is the star Delta Cephei in the constellation Cepheus. While both bluer and redder stars also vary in their intrinsic light, the properties of these ß Cephei, ZZ Ceti, RV Tauri, and Mira variables are much less understood than the yellow-color Cepheids. These yellow stars are known to be pulsating in radius by as much as 10% or more. Their light variations are due to their changing surface temperature. Larger yellow stars are intrinsically brighter because they have more surface area, and they have larger pulsation periods because they have a larger radius. *See* STAR; VARIABLE STAR.

The interest in these stars is twofold: If their intrinsic brightnesses can be inferred from their pulsation period, the brightnesses can be used as indicators of their distance from the Earth. The observed period and a calibrated period-luminosity relation is used to give an intrinsic brightness. The observed distance-dependent apparent brightness then gives the actual distance. The second, and more current, interest in Cepheids is that their pulsation properties reveal their masses and internal structure, which help in understanding how stars age. Thus, Cepheids and the related classes of yellow pulsating stars have been extremely useful in mapping the scale of the universe and in probing the details of stellar interiors. *See* STELLAR EVOLUTION. [A.N.Co.]

Ceractinomorpha A subclass of Demospongiae. Among the Ceractinomorpha, the genus *Halisarca*, lacking skeletal elements, is a primitive form. The larva of *Halisarca* is a diploblastula or parenchymella with an outer layer of flagellated cells and an inner mass of presumptive ectomesenchymal cells. The outer flagellated cells lose their flagella, migrate into the interior, and later differentiate into choanocytes. Other cell types characteristic of the adult sponge differentiate, and inhalant canals begin to form.

In form, ceractinomorph sponges vary from encrustations, thin or massive, to lobate and upright branching colonies. The shallow-water species tend to be more plastic in form than deep-water species, which usually exhibit little intraspecific variation in shape. *See* DEMOSPONGIAE; PORIFERA. [W.D.H.]

Ceramics Nonmetallic, inorganic materials containing high proportions of silicon, silicon oxide, and silicates, and having a wide range of applications. Ceramics are materials which cover a great range in both applications and time. In general, ceramics are hard, brittle, electrical and thermal insulators, require high-temperature processing, and are formed from powders. The major divisions of ceramic technology are similar in processing and in the properties of the materials. However, differences in applications and differences in the behavior of materials during processing require that diverse techniques be used. *See* CERMET; SINTERING.

It is convenient to divide ceramic products into two groups: those, such as pottery and brick, which are shaped or formed before high-temperature treatment, and those, such as glass and cement, which are shaped afterward. Only the first group is discussed here. *See* CEMENT; GLASS; MORTAR; PLASTER.

Structural clay is one of the oldest branches of ceramics and includes building brick, sewer pipe, and decorative ceramic block for walls. To form these products from raw clays, use is made of the plastic forming technique known as extrusion. Extrusion is carried out by forcing a stiff plastic mass through an opening or die in the form of the desired cross section; the continuous ribbon which emerges is cut to the desired lengths. Drying is carried out in conditions of controlled humidity and temperature which prevent the ware from cracking or warping. The dried material is hard and can be broken with hand pressure. After drying, the material is heat-treated to a temperature where the clay is broken up into less complicated molecular structural units. The resulting material is now held together by chemical bonds between glass and oxide compounds, resulting in a hard, brittle material which is resistant to corrosion. This process is known as vitrification. *See* CLAY.

There are two major divisions of whiteware: art ware and consumer ware (tableware, portable lamps, sanitary ware, and so on). To form these products, powders of clay, potter's flint, and feldspar are used. The clay, when sufficiently wet, imparts plasticity or workability to the body. The forming methods used for consumer whiteware production are slip casting and jiggering. Jiggering is a mechanization of the forming process of throwing clay by hand as done by the potter. For large items such as sanitary ware, artware, and portable lamps, slip casting is most often used. Slip casting of clays is done by pouring a water suspension of the body (a slip) into a plaster of paris mold of the desired shape. The porous plaster mold withdraws water from the slip, which results in a buildup of a layer of solid clay next to the mold. Once the materials have been formed into the desired shape, drying and firing are carried out. *See* POTTERY.

Properties that make ceramic products desirable in electrical applications are high resistivity, high dielectric strength, low dielectric loss factor, high dielectric constant, and controllable magnetic properties. The ceramic products used in the electrical industry include porcelains, glasses, steatites, cordierites, titanates, zirconates, carbides, oxides, and ferrites. Ceramic products are used in magnets, electronic tubes, condensers, resistors, transformers, amplifiers, memory devices, transducers, capacitors, and insulators. *See* FERRITE; PORCELAIN.

Ceramics known as refractories are products which thermally insulate the furnaces that produce steel, aluminum, and other metals. They also insulate the furnaces that produce the steam for the generation of electricity, as well as insulate fireplaces in the home. The manufacture of refractories is one of the key industries in the United States. *See* REFRACTORY. [G.E.S.]

Cerargyrite A mineral with composition AgCl. Its structure is that of the isometric NaCl type, but well-formed cubic crystals are rare. The hardness is $2\frac{1}{2}$ on Mohs scale and specific gravity 5.5. Ceragyrite is colorless to pearl-gray but darkens to violet-brown on exposure to light. It is perfectly sectile and can be cut with a knife-like horn; hence the name horn silver. Bromyrite, AgBr, is physically indistinguishable from ceragyrite and the two minerals form a complete series. Both minerals are secondary ores of silver and occur in the oxidized zone of silver deposits. *See* SILVER.

Cercaria The larval generation which terminates the development of a digenetic trematode in the intermediate host, a mollusk or rarely an annelid. It becomes the adult fluke after entering the definitive host, a vertebrate. Access to the vertebrate may be either direct or indirect by entering a second intermediate, or vector, host which is ingested by the vertebrate. The intermediate host is usually aquatic and the cercaria has a tail used in swimming when it escapes from that host, but the tail may be inactive, reduced, or absent and the larva unable to swim. When the tail is absent, the organism is called a cercariaeum. Larvae with unknown life cycles are commonly described and named as species such as *Cercaria micrura* or *Cercariaeum helicis*.

The cercaria is important in taxonomy because that stage of closely related trematodes usually is less variable than the adult, and thus indicates kinships that may be masked in the adult by adaptations to parasitism of the vertebrate host. *See* DIGENEA; TREMATODA. [R.M.C.]

Cereal Any member of the grass family (Gramineae) which produced edible grains usable as food by humans and livestock. Common cereals are rice, wheat, barley, oats, maize (corn), sorghum, rye, and certain millets, with corn, rice, and wheat being the most important. Developed by scientists, triticale is a new cereal derived from crossing wheat and rye and then doubling the number of chromosomes in the hybrid. Occasionally, grains from other grasses (for example, teff) are used for food. Cereals provide more food for human consumption than any other crops.

Four general groups of foods are prepared from the cereal grains. (1) Baked products, made from flour or meal, include breads, pastries, pancakes, cookies, and cakes. (2) Milled grain products, made by removing the bran and usually the germ (or embryo of the seed), include polished rice, farina, wheat flour, cornmeal, hominy, corn grits, pearled barley, semolina (for macaroni products), prepared breakfast cereals, and soup, gravy, and other thickenings. (3) Beverages such as beer and whiskey, made from fermented grain products (distilled or undistilled) and from boiled, roasted grains. (4) Whole-grain products include rolled oats, brown rice, popcorn, shredded and puffed gains, and breakfast foods.

All cereal grains have high energy value, mainly from the starch fraction but also from the fat and protein. In general, the cereals are low in protein content, although oats and certain millets are exceptions. *See* GRAIN CROPS; GRASS. [L.P.R.]

Cerebral palsy In the broadest sense, any nonprogressive motor disorder caused by brain damage incurred during development. Although in common use by both lay people and physicians, the term cerebral palsy is nonspecific and includes many kinds of brain damage from various causes. In fact, few authorities agree on the disorders to be included.

Nonprogressive motor disorders, such as spasticity, paralysis, convulsions, muscular spasms, abnormal reflexes, and delayed or abnormal development of speech, movement, and behavior patterns, are characteristic. Intelligence is not noticeably altered in over two-thirds of the cases.

Possible prenatal causes include certain hereditary and congenital defects, as well as some maternal diseases, such as German measles (rubella) in early pregnancy. Toxemias of pregnancy, abnormal or difficult labor, and insufficient oxygen before, during, or after birth are precipitating causes of brain damage. *See* PREGNANCY DISORDERS; RUBELLA.

Birth injuries are a major cause of cerebral palsy. The infant head is subjected to tremendous stresses during labor, especially in a first pregnancy, which may cause anoxia, edema, hemorrhage, rupture of brain tissue, or other mechanical effects which, in turn, cause degeneration or death of nervous tissue. No regeneration of such tissue takes place and even if nervous tissue damage is slight, there is still the chance that a scar may form and cause later difficulty. Skillful obstetrical techniques have greatly reduced the incidence and severity of damage in many forms of cerebral palsy. [E.G.St./N.K.M.]

Cerenkov radiation Light emitted by a high-speed charged particle when the particle passes through a transparent, nonconducting, solid material at a speed greater than the speed of light in the material. The blue glow observed in the water of a nuclear reactor, close to the active fuel elements, is radiation of this kind. The emission of Cerenkov radiation is analogous to the emission of a shock wave by a projectile moving faster than sound, since in both cases the velocity of the object passing through the medium exceeds the velocity of the resulting wave disturbance in the medium.

Particle detectors which utilize Cerenkov radiation are called Cerenkov counters. They are important in the detection of particles with speeds approaching that of light, such as those produced in large accelerators and in cosmic rays, and are used with photomultiplier tubes to amplify the Cerenkov radiation. These counters can emit pulses with widths of about

10^{-10} s and are therefore useful in time-of-flight measurements when very short times must be measured. They can also give direct information on the velocity of the passing particle. *See* PARTICLE DETECTOR. [W.B.Fr.]

Ceres The first asteroid discovered. It was found serendipitously by G. Piazzi on January 1, 1801. With a diameter of approximately 600 mi (1000 km), Ceres is the largest asteroid but not the brightest since it reflects only 6% of the visual light it receives. Ceres's mass of 1.17×10^{21} kg (that is, 2×10^{-4} that of the Earth) contains approximately 40% of the asteroid belt's total mass. *See* ASTEROID. [E.F.T.]

Ceriantharia An order of the Zoantharia, typified by *Cerianthus*, which lives in sandy marine substrata (illustration *a* and *b*). The animal is enclosed in a sheath formed by mucus secreted from gland cells of the column ectoderm, in which discharged nematocysts, sand grains, and other foreign objects are embedded.

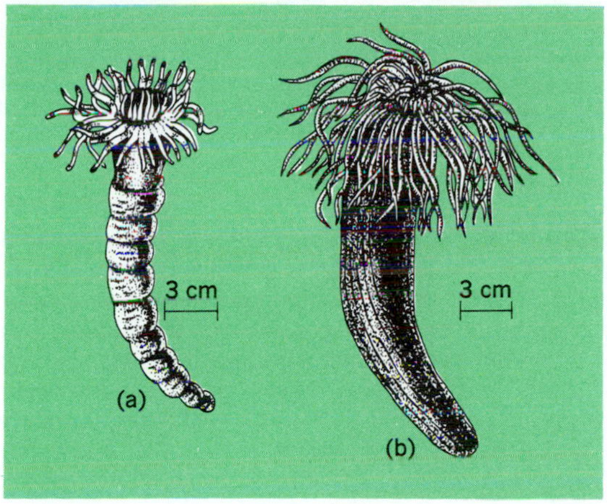

Ceriantharia. (*a*) *Cerianthus solitarius*. (*b*) *Pachycerianthus multiplicatus*.

The polyp is a muscular, skeletonless, elongated, cylindrical body with a smooth wall. Long, slender, unbranching, freely retractile tentacles are arranged in two cycles and consist of smaller labial and larger marginal ones. *See* ZOANTHARIA. [K.At.]

Cerium A chemical element, Ce, atomic number 58, atomic weight 140.12. It is the most abundant metallic element of the rare-earth group in the periodic table. The naturally occurring element is made up of the isotopes ^{136}Ce, ^{138}Ce, ^{140}Ce, and ^{142}Ce. A radioactive α-emitter, ^{142}Ce has a half-life of 5×10^{15} years. Cerium occurs mixed with other rare earths in many minerals, particularly monazite and blastnasite, and is found among the products of the fission of uranium, thorium, and plutonium.

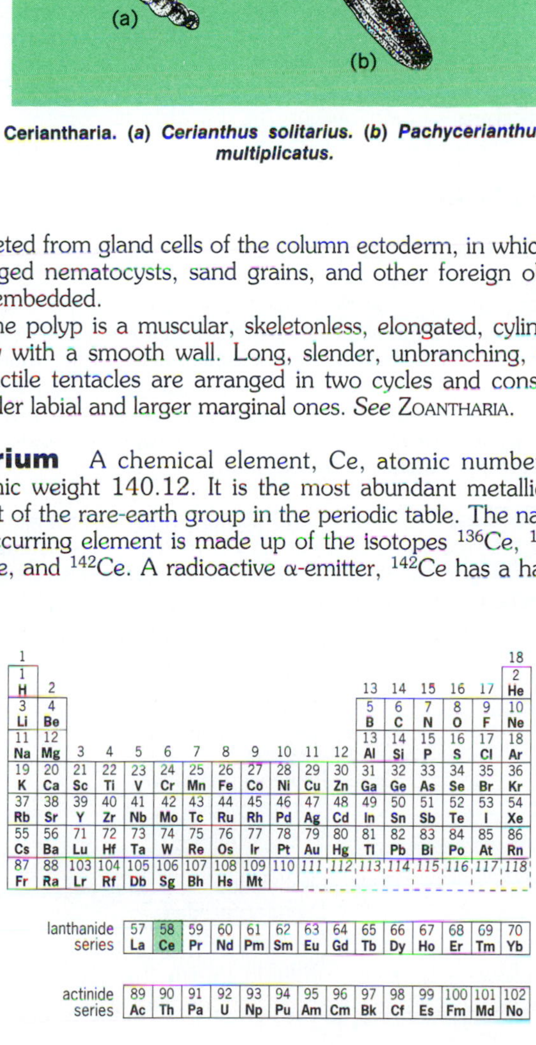
1 H, 2 He
3 Li, 4 Be, 5 B, 6 C, 7 N, 8 O, 9 F, 10 Ne
11 Na, 12 Mg, 13 Al, 14 Si, 15 P, 16 S, 17 Cl, 18 Ar
19 K, 20 Ca, 21 Sc, 22 Ti, 23 V, 24 Cr, 25 Mn, 26 Fe, 27 Co, 28 Ni, 29 Cu, 30 Zn, 31 Ga, 32 Ge, 33 As, 34 Se, 35 Br, 36 Kr
37 Rb, 38 Sr, 39 Y, 40 Zr, 41 Nb, 42 Mo, 43 Tc, 44 Ru, 45 Rh, 46 Pd, 47 Ag, 48 Cd, 49 In, 50 Sn, 51 Sb, 52 Te, 53 I, 54 Xe
55 Cs, 56 Ba, 71 Lu, 72 Hf, 73 Ta, 74 W, 75 Re, 76 Os, 77 Ir, 78 Pt, 79 Au, 80 Hg, 81 Tl, 82 Pb, 83 Bi, 84 Po, 85 At, 86 Rn
87 Fr, 88 Ra, 103 Lr, 104 Rf, 105 Db, 106 Sg, 107 Bh, 108 Hs, 109 Mt, 110, 111, 112, 113, 114, 115, 116, 117, 118
lanthanide series 57 La, 58 Ce, 59 Pr, 60 Nd, 61 Pm, 62 Sm, 63 Eu, 64 Gd, 65 Tb, 66 Dy, 67 Ho, 68 Er, 69 Tm, 70 Yb
actinide series 89 Ac, 90 Th, 91 Pa, 92 U, 93 Np, 94 Pu, 95 Am, 96 Cm, 97 Bk, 98 Cf, 99 Es, 100 Fm, 101 Md, 102 No

Although the common valence of cerium is 3, it also forms a series of quadrivalent compounds and is the only rare earth which occurs as a quadrivalent ion in aqueous solution. Although it can be separated from the other rare earths in high purity by ion-exchange methods, it is usually separated chemically by taking advantage of its quadrivalent state. *See* RARE-EARTH ELEMENTS. [F.H.Sp.]

Cermet A group of composite materials consisting of an intimate mixture of ceramic and metallic components. Cermets can be fabricated by mixing the finely divided components in the form of powders or fibers, compacting the components under pressure, and sintering the compact to produce physical properties not found solely in either of the components. Cermets can also be fabricated by internal oxidation of dilute solutions of a base metal and a more noble metal. When heated under oxidizing conditions, the oxygen diffuses into the alloy to form a base metal oxide in a matrix of the more noble metal. *See* COMPOSITE MATERIAL; CORROSION; POWDER METALLURGY; SINTERING.

The combination of metallic and ceramic components can result in cermets characterized by increased strength and hardness, higher temperature resistance, improved wear resistance, and better resistance to corrosion, each characteristic depending on the variables involved in composition and processing. Friction parts as well as cutting and drilling tools have been successfully made from cermets for many years. Certain nuclear reactor fuel elements, such as dispersion-type elements, are also made as cermets. *See* CERAMICS. [H.H.H.]

Cerussite The mineral form of lead carbonate, $PbCO_3$. Cerussite is common as a secondary mineral associated with lead ores. In the United States it occurs mostly in the central and far western regions. Cerussite is white when pure but is sometimes darkened by impurities. Hardness is 3¼ on Mohs scale and specific gravity is 6.5. Crystals may be tabular, elongated, or arranged in clusters. *See* CARBONATE MINERALS. [R.I.Ha.]

Cesium A chemical element, Cs, with an atomic number of 55 and an atomic weight of 132.905, the heaviest of the alkali metals in group I of the periodic table (except for francium, the radioactive member of the alkali metal family). Cesium is a soft, light, very low-melting metal. It is the most reactive of the alkali metals and indeed is the most electropositive and the most reactive of all the elements.

Cesium reacts vigorously with oxygen to form a mixture of oxides. In moist air, the heat of oxidation may be sufficient to melt and ignite the metal. Cesium does not appear to react with nitrogen to form a nitride, but does react with hydrogen at high temperatures to form a fairly stable hydride. Cesium reacts violently with water and even with ice at temperatures as low as −116°C (−177°F). Cesium reacts with the halogens, ammonia, and carbon monoxide. In general, cesium undergoes some of the same type of reactions with organic compounds as do the other alkali metals, but it is much more reactive. *See* SODIUM.

The physical properties of cesium metal are summarized in the table.

Cesium is not very abundant in the Earth's crust, there being only 7 parts per million (ppm) present. Like lithium and rubidium, cesium is found as a constituent of complex minerals and not in relatively pure halide form as are sodium and potassium. Indeed, lithium, rubidium, and cesium frequently occur together in lepidolite ores, such as those from Rhodesia.

Cesium metal is used in photoelectric cells, spectrographic instruments, scintillation counters, radio tubes, military infrared signaling lamps, and various optical and detecting devices.

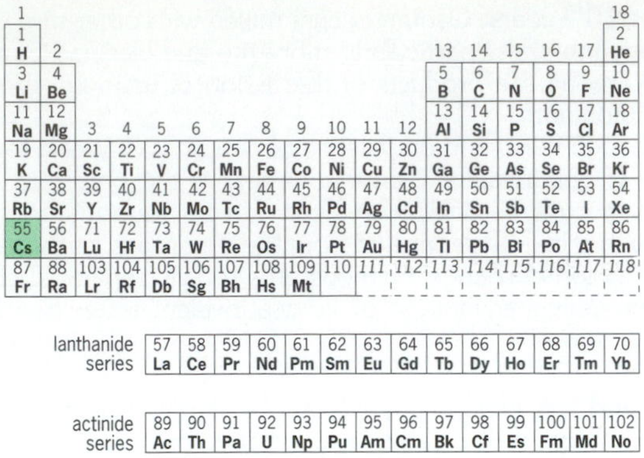

Physical properties of cesium metal

Property	Temp., °C	Valve
Density	20	1.9 g/cm^3
Melting point	28.5	
Boiling point	705	
Heat of fusion	28.5	3.8 cal/g
Heat of vaporization	705	146 cal/g
Viscosity	100	4.75 millipoises
Vapor pressure	278	1 mm
	635	400 mm
Thermal conductivity	28.5	0.044 cal/(s)(cm^2)(°C)
Heat capacity	28.5	0.06 cal/(g)(°C)
Electrical resistivity	30	36.6 microhm-cm

Cesium compounds are used in glass and ceramic production, as absorbents in carbon dioxide purification plants, as components of getters in radio tubes, and in microchemistry. Cesium salts have been used medicinally as antishock agents after administration of arsenic drugs. The isotope cesium-137 is supplanting cobalt-60 in the treatment of cancer. *See* ALKALI METALS. [M.Si.]

Cestida An order of the phylum Ctenophora comprising two genera, *Cestum* and *Velamen*. The morphology of these organisms is unusual; the transparent bodies are flattened in the tentacular plane and greatly elongated in the stomodeal plane so that they have the shape of a belt or ribbon. Cestids are capable of rapid swimming by wriggling the body. *Cestum* is widely distributed in oceanic waters and are among the commonest and most spectacular of epipelagic ctenophores. *See* CTENOPHORA. [L.P.M.]

Cestoda A subclass of tapeworms including most of the members of the members of the class Cestoidea. All species are endoparasites of vertebrates, living in the intestine or related ducts.

Like other members of the class, the cestodes have no digestive tract or mouth. Nutrition presumably occurs by absorption of food through the body surface. The body is usually very elongated and tapelike and frequently divided into segments, or proglottids, with replication of the hermaphroditic reproductive systems. In a few species there is duplication of both male and female organs within a single segment. The anterior end is usually modified into a holdfast organ, the scolex. Since a digestive tract is completely absent, the scolex is of solid construction, typically highly muscular with sucking depressions and hooks (see illustration).

The worms require carbohydrate for growth and reproduction, and this requirement is satisfied only from the host ingesta. On the other hand, nitrogenous nutrients and many micronutrients may be obtained from the body stores of the host; deleting such materials from the host's diet has no appreciable effect on the worms.

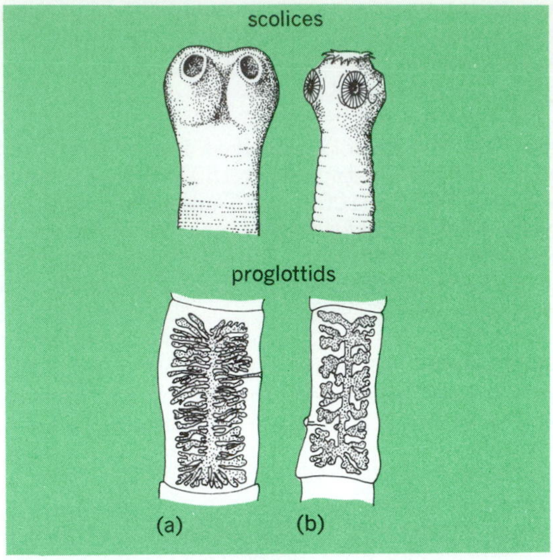

Scolices and mature proglottids of (*a*) *Taenia saginata* and (*b*) *T. solium*. (After T. I. Storer and R. L. Usinger, *General Zoology*, 3d ed. McGraw-Hill, 1957)

Most authorities agree that the tapeworms are ancient parasites, probably evolving as parasites of the earliest fishes. It seems probable that the tapeworms did not evolve from trematodes or other present-day groups of parasitic flatworms. Their ancestry may be directly derived from the acoele or rhabdocoele turbellarians and represents a line of evolution which is completely independent of other parasitic flatworms. These relationships remain obscure in the absence of any fossil record. *See* CESTOIDEA; TURBELLARIA. [C.P.R.]

Cestodaria A subclass of worms belonging to the class Cestoidea. Only a few species are known. All are endoparasites of primitive fishes. The subclass is usually divided into two orders, Amphilinidea and Gyrocotylidea. These worms differ from the other Cestoidea in being unsegmented, in not having the anterior end modified as a holdfast organ, and in frequently occurring as parasites of the coelomic cavity rather than the digestive tract. Some species have the posterior end modified into a holdfast organ. The animals are hermaphroditic and sexual reproduction occurs in the vertebrate host. *See* AMPHILINIDEA; CESTOIDEA; GYROCOTYLIDEA. [C.P.R.]

Cestoidea A class of the phylum Platyhelminthes commonly referred to as tapeworms. All members are endoparasites, usually in the digestive tract of vertebrates. The class has been subdivided as follows:

Class Cestoidea
  Subclass Cestodaria
    Order: Amphilinidea
      Gyrocotylidea
  Subclass Cestoda
    Order: Proteocephaloidea
      Tetraphyllidea
      Lecanicephaloidea
      Trypanorhyncha
    Order: Diphyllidea
      Pseudophyllidea
      Cyclophyllidea
      Nippotaeniidea

Separate articles appear on each group listed in the classification.

In size the tapeworms range from less than 0.04 in. (1.0 mm) to several meters in length. The class is differentially characterized by the presence of a cuticle rather than a cellular epidermis and by the total absence of a mouth and digestive tract. In most species of the class the body is divided into proglottids, each proglottid containing one or two hermaphroditic reproductive systems. The anterior end is usually modified into a holdfast organ, bearing suckers or sucking grooves, and frequently armed with hooks. Early embryonic development occurs in the parental body, usually in a uterus, to a hook-bearing stage, the oncosphere. The oncosphere leaves the parental body through a uterine pore or by liberation of the terminal segment from the main body of the worm. Further development of the worm always occurs within the body of a host, most often an invertebrate, which commonly ingests the larval form.

Food materials are presumed to be absorbed through the external surface. It has been shown that tapeworms have catalytic mechanisms for the active transport into the body of simple sugars and amino acids. The chemical composition of these worms is unique in that polysaccharide and fat represent a much larger proportion of the dry weight than does protein. *See* PLATYHELMINTHES. [C.P.R.]

Cetacea An order of aquatic mammals including about 80 living species, of which the larger ones are called whales and the smaller ones dolphins or porpoises. Cetaceans live in all oceans and in several fresh-water systems. They are remarkably adapted to aquatic life and die within a few hours or days on land. All have a smooth, streamlined body; the forelimbs are flattened and rigid from shoulder to tip, without external fingers; the hindlimbs are absent, though represented by vestiges in the embryo or adult; the tail is flattened horizontally. The external ears are minute pits. The skin is hairless, though some fetal cetaceans have hairs on the snout and some adults have sensory bristles along the body. All cetaceans have a thick layer of blubber (subcutaneous fat) which serves as heat insulation and as food reserve.

So far as known, all cetaceans copulate belly to belly and quickly, near the surface of the sea. The gestation period is 9–16 months. A single, large precocious calf is born tail first, and it rises immediately to the surface for air. The nursing period may extend for 2 years (sperm whale). Cetacean milk is rich, being up to 50% fat, and low in sugars. Cetaceans are generally long-lived.

Specialization of cetaceans for sensing their environment has led to extinction of smell, reduction of vision, and extreme development of hearing. Vibrations serve for communication and echolocation (sonar). While the intelligence of cetaceans is debatable, by most standards (gross weight of brain, weight of brain in relation to cord, ready response to training, social interaction, and playfulness), the species which have been studied in captivity rate nearer the primates than do the dog, cat, rat, and horse. The food of cetaceans is exclusively flesh.

The ancestors of the cetaceans are completely unknown. The earliest fossils are from middle Eocene beds and are clearly whales. Separate lines of evolution have given rise to three suborders: Archaeoceti, Mysticeti, and Odontoceti. *See* DOLPHIN; MAMMALIA; PORPOISE; WHALE. [V.B.S.]

Cetane number A number that indicates the ability of a diesel engine fuel to ignite quickly after being injected into the cylinder. In high-speed diesel engines, a fuel with a long ignition delay tends to produce rough operation. *See* COMBUSTION CHAMBER; DIESEL FUEL. [A.R.R.]

Cetomimiformes An order of oceanic, mostly deep-water fishes that are structurally diverse and rare; most of the 41 species are known from one or a few specimens. Thus, their scientific study has been hindered, the anatomy is imperfectly known, and the relationships are in dispute. Five of the 10 families and 11 of the 21 genera currently placed in the order have been described since World War II. There is no fossil record.

Cetomimiforms are soft-rayed fishes. In most the mouth is large. Many are naked but a few have scales that are thin and deciduous or form an irregular mosaic; a few have the skin spinulose. Pelvic fins may be abdominal, thoracic, or jugular in position or, commonly, absent. In most forms the single dorsal and anal fins are placed rather well back and are opposed; an adipose fin is present in only one species.

Because of the diversity, the five currently recognized suborders may be mentioned separately. Best known are the Cetomimoidei (or Cetunculi) whalefishes, a group of 3 families and 15 rare species of small, red or black deep-sea fishes with whale-shaped bodies and enormous mouths; they are bioluminescent. The Ateleopoidei (or Chondrobrachii) consist of 1 family, 3 genera, and 11 species of elongate fishes in which the long anal fin is continuous with the caudal and there is no dorsal fin. The Mirapinnatoidei (or Miripinnati) are tiny, perhaps larval fishes, all recently described. Three families, 4 genera, and 5 species are included. The Giganturoidei, with 2 families, 3 genera, and 6 species, are small mesopelagic fishes with large mouths and strong teeth; some have telescopic eyes. The Megalomycteroidei, or mosaic-scaled fishes, consist of 1 family, 4 genera, and 4 rare species of small, elongate deep-sea fishes with degenerate eyes and irregularly disposed scales. *See* ACTINOPTERYGII; OSTEICHTHYES; TELEOSTEI. [R.M.B.]

Chabazite A mineral belonging to the zeolite family of silicates. Hardness is 4–5 on Mohs scale; specific gravity is 2.05–2.15. The luster is vitreous and the color usually white but may be yellow, red, or pink. *See* ZEOLITE.

Chabazite is essentially a hydrous calcium sodium aluminum silicate, $(CaNa_2)[Al_2Si_4O_{12}] \cdot 6H_2O$. The ratio of calcium to sodium may vary, and potassium is usually present in small amounts. Chabazite has been the most used natural zeolite for cation exchange and gas sorption, but synthetic zeolites are now widely employed.

Chabazite is a secondary mineral usually found lining cavities in basalt and related rocks. Notable localities are the Giant's Causeway, Ireland; Faeroe Islands; Trentino, Italy; and Oberstein, Germany. In the United States it is found at West Paterson and Bergen Hill, New Jersey; and Table Mountain, Oregon. *See* SILICATE MINERALS. [C.Fr.; C.S.Hu.]

Chaetodermomorpha A subclass of burrowing, vermiform mollusks in the class Aplacophora. They are covered by a spicular integument and recognizable by the presence of a sensory cuticular oral shield, lack of a foot, and presence of paired gills in a posterior mantle cavity. Chaetoderms range in size from less than 0.08 in. (2 mm) to more than 2.8 in. (70 mm) and are found from shelf depths to hadal depths over 22,400 ft (7000 m). There are three families with 10 genera and 84 species worldwide. Chaetoderm species are numerically dominant in certain deep-sea localities. *See* APLACOPHORA; MOLLUSCA. [A.H.S.]

Chaetognatha A phylum of abundant planktonic arrowworms. Their bodies are tubular and transparent, and divided into three portions: head, trunk, and tail. The head possesses one or two rows of minute teeth anterior to the mouth and usually 7–10 larger chaetae, or seizing jaws, on each side of the head. One or two pairs of lateral fins and a caudal fin are present.

Nine genera and about 42 species are recognized by some specialists. Most species belong to the genus *Sagitta*, which can be recognized by the presence of two pairs of teeth and two pairs of lateral fins.

Chaetognaths are cosmopolitan forms which live not only at the surface but also at great depths; however, no one species is found in all latitudes and at all depths. One of the Arctic species, *Eukrohnia hamata*, may extend to the Antarctic by way of deep water across the tropics. A few species are neritic and are not found normally beyond the continental shelf. Their food consists principally of copepods and other small planktonic crustaceans; however, they are very predacious and will even eat small fish larvae and other chaetognaths on rare occasions.

Studies have shown them to be useful as indicator organisms. Certain species appear to be associated with characteristic types or masses of water, and when this water is displaced into an adjacent water mass, the chaetognaths may be used as temporary evidence for such displacement. [E.L.P.]

Chaetonotida An order of the phylum Gastrotricha. These animals usually have two adhesive tubes connected with the distinctive paired, posterior "tail forks" or "toes." Paired protonephridia are present. Parthenogenetic females deposit cuticular-covered, smooth or spinous eggs, and direct development occurs. *Chaetonotus*, *Ichthydium*, *Lepidodermella*, and *Polymerurus* are well-known genera (see illustration), with over 200 species, only about 20 of which have been reported from the United States. See GASTROTRICHA. [C.E.P.]

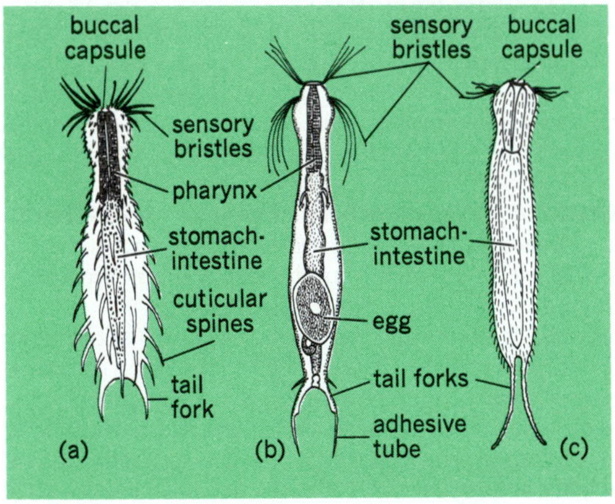

Chaetonotoidea. (a) *Chaetonotus*. (b) *Ichthydium*. (c) *Polymerurus*. (After L. H. Hyman, *The Invertebrates*, vol. 3, McGraw-Hill, 1951)

Chagas' disease An acute and chronic disease, also known as South American trypanosomiasis, caused by the pathogenic hemoflagellate *Trypanosoma* (*Schizotrypanum*) *cruzi* and transmitted to humans by bloodsucking triatoma bugs. The infection originates in wild animals (armadillo and opossum) as a zoonosis.

The acute form of Chagas' disease occurs predominantly in young children and is characterized by fever, lymph node swelling, spleen and liver enlargement, and facial edema. Involvement of the brain occurs often, and convulsions may sometimes end in grave permanent mental damage or even in death. Although the chronic form may be mild, myocarditis due to parasite development in the heart muscles and irreversible tissue damage have not infrequently been described, as well as megalocolon complication.

There is no generally effective treatment. Preventive measures should be applied against the vectors of the disease. See TRYPANOSOMATIDAE. [M.Yo.]

Chain drive A flexible device of connected links used to transmit power. A drive consists of an endless chain which meshes with sprockets located on the shaft of a driving source, such as an electric motor/reducer and a driven source, such as the head shaft of a belt conveyor.

The roller chain (see illustration) meets the demands of heavy-duty oil well drilling equipment, high-production agricultural machinery, construction machinery, and similar equipment. It also meets the precise timing requirements of lighter-duty equipment such as printing, packaging, and vending equipment.

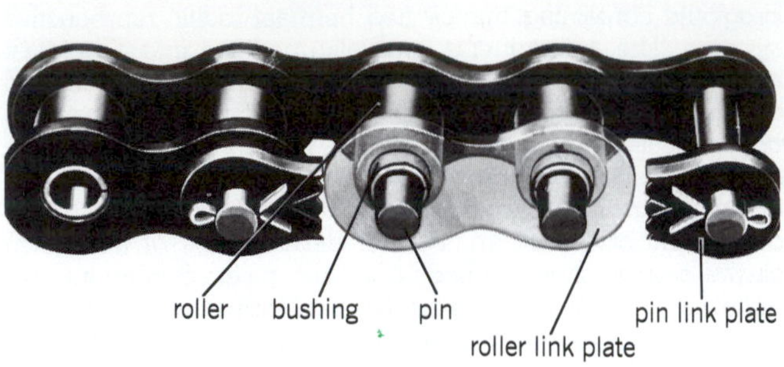

Single-strand roller chain.

Another type of chain, the engineering steel chain, is usually identified by the offset/cranked link sidebar design. Generally, larger pitch sizes as compared to the roller chain and higher-strength chains characterize such chains. A third group used for chain drives is the inverted tooth/silent chains. A familiar application is their use as automotive timing chains in automobile engines.

The use of chains for power transmission rather than another device, such as V-belts or a direct coupling to the power source, is usually based on the cost effectiveness and economy of chains and sprockets. Chains and sprockets offer the following advantages: large speed ratios; sufficient elasticity to absorb reasonable shocks; a constant speed ratio between the driving and driven shaft; long life without excessive maintenance; mechanical understandability regarding installation and functionality; coupling and uncoupling with simple tools; and a simple means to get power from its source to the location where needed. See BELT DRIVE. [V.D.P.]

Chain reaction (chemistry) A chemical reaction in which many molecules undergo chemical reaction after one molecule becomes activated. In ordinary chemical reactions, every molecule that reacts must first become activated by collision with other rapidly moving molecules. The number of these violent collisions per second is so small that the reaction is slow. After a chain reaction is started, it is not necessary to wait for more collisions with activated molecules to accelerate the reaction because the reaction now proceeds spontaneously.

A typical chain reaction is the photochemical reaction between hydrogen and chlorine as described by the following reactions.

$$Cl_2 + \text{light} \rightarrow Cl + Cl$$
$$Cl + H_2 \rightarrow HCl + H$$
$$H + Cl_2 \rightarrow HCl + Cl$$
$$Cl + H_2 \rightarrow HCl + H$$

The light absorbed by a chlorine molecule dissociates the molecule into chlorine atoms; these in turn react rapidly with hydrogen molecules to give hydrogen chloride and hydrogen atoms. The hydrogen atoms react with chlorine molecules to give hydrogen chloride and chlorine atoms. The chlorine atoms react further with hydrogen and continue the chain until some other reaction uses up the free atoms of chlorine or hydrogen. The chain-stopping reaction may be the reaction between two chlorine atoms to give chlorine molecules, or between two hydrogen atoms to give hydrogen molecules. Again the atoms may collide with the walls of the containing vessel, or they may react with some impurity which is present in the vessel only as a trace.

Certain oxidations in the gas phase are known to be chain reactions. The carbon knock which occurs at times in internal combustion engines is caused by a too-rapid combustion rate caused by chain reactions. This chain reaction is reduced by adding tetraethyllead, which acts as an inhibitor.

The polymerization of styrene to give polystyrene and the polymerization of other organic materials to give industrial plastics involve chain reactions. The spoilage of foods, the precipitation of insoluble gums in gasoline, and the deterioration of certain plastics in sunlight involve chain reactions, which can be minimized with inhibitors. *See* CHEMICAL DYNAMICS; PHOTOCHEMISTRY. [F.D.]

Chain reaction (physics) A succession of generation after generation of acts of division (called fission) of certain heavy nuclei. The fission process releases about 200 MeV (3.2×10^{-4} erg = 3.2×10^{-11} joule) in the form of energetic particles including two or three neutrons. Some of the neutrons from one generation are captured by fissile species (^{233}U, ^{235}U, ^{239}Pu) to cause the fissions of the next generations. The process is employed in nuclear reactors and nuclear explosive devices. *See* NUCLEAR FISSION. [N.C.R.]

Chalcanthite A mineral with the chemical composition $CuSO_4 \cdot 5H_2O$. Chalcanthite commonly occurs in blue to greenish-blue triclinic crystals or in massive fibrous veins or stalactites. Fracture is conchoidal and luster is vitreous. Hardness is 2.5 on Mohs scale and specific gravity is 2.28. It has a nauseating taste and is readily soluble in water. It dehydrates in dry air to a greenish-white powder. Although deposits of commercial size occur in arid areas, chalcanthite is generally not an important source of copper ore. Its occurrence is widespread in the western United States. [E.C.T.C.]

Chalcedony A fine-grained fibrous variety of quartz, silicon dioxide. The individual fibers that compose the mineral aggregate usually are visible only under the microscope. Subvarieties of chalcedony recognized on the basis of color differences, some valued since ancient times as semiprecious gem materials, include carnelian (translucent, deep flesh red to clear red in color), sard (orange-brown to reddish-brown), and chrysoprase (apple green). *See* GEM; QUARTZ.

Chalcedony occurs as crusts with a rounded, mammillary, or botryoidal surface and as a major constituent of nodular and bedded cherts. The hardness is 6.5–7 on Mohs scale. The specific gavity is 2.57–2.64.

Crusts of chalcedony generally are composed of fairly distinct layers concentric to the surface. Agate is a common and important type of chalcedony in which successive layers differ markedly in color and degree of translucency. In the most common kind of agate the layers are curved and concentric to the shape of the cavity in which the material formed. *See* AGATE. [R.Si.]

Chalcocite A mineral having composition Cu_2S and crystallizing in the orthorhombic system (below 217°F or 103°C). Crystals are rare and small, usually with hexagonal outline because of twinning. Most commonly, the mineral is fine-grained and massive with a metallic luster and a lead-gray color which tarnishes to dull black on exposure. The Mohs hardness is 2.5–3, and the density 5.5–5.8. Chalcocite is an important copper ore found at Miami, Morenci, and Bisbee, Arizona; Butte, Montana; Kennecott, Alaska; and Tsumeb, South-West Africa. [L.Gr.]

Chalcogenide glasses Amorphous cross-linked inorganic polymers. Probably the best known is arsenic sulfide, $(As_2S_3)_n$, which can be used for infrared transparent windows. Threshold and memory switching are also interesting properties of these glasses. Ultraphosphate glasses resemble glassy organic plastics and can be processed by the same methods, such as extrusion and injection molding. They are used for antifouling surfaces for marine applications and in the manufacture of nonmisting spectacle lenses. See AMORPHOUS SOLIDS; INORGANIC POLYMER. [R.A.S.]

Chalcopyrite A mineral having composition $CuFeS_2$. Crystals are usually small and resemble tetrahedra. Chalcopyrite is usually massive with a metallic luster, brass-yellow color, and sometimes an iridescent tarnish. The Mohs hardness is 3.5–4.0, and the density 4.1–4.3. Chalcopyrite is a so-called fool's gold, but is brittle while gold is sectile. Pyrite, the most widespread fool's gold, is harder than chalcopyrite. *See* PYRITE.

Chalcopyrite is the most widespread primary copper ore mineral. It is commonly found in veins (Braden mine, Chile; Cornwall, England; Butte, Montana; Freiberg, Saxony; Tasmania; Rio Tinto, Spain). Chalcopyrite is also found in contact metamorphic deposits in limestone (Bisbee, Arizona) and as sedimentary deposits (Mansfeld, Germany). *See* COPPER. [L.Gr.]

Chalk A variety of limestone formed from pelagic, or floating, organisms that is very fine-grained, porous, and friable. It is white or very light-colored and consists almost entirely of calcite. The rock is made up of the calcite shells of microorganisms partially cemented by structureless calcite. The most conspicuous are the Foraminiferida, *Globigerina* and *Textularia*. *See* LIMESTONE.

The best-known chalks are those of the Cretaceous, exposed in cliffs on both sides of the English Channel. The Selma (Cretaceous) chalk of Alabama, Mississippi, and Tennessee and the Niobrara chalk of the same age in Nebraska are well-known deposits in the United States. Chalk is used for cements, powders (as soft abrasives and polishers), crayons, and fertilizers. [R.Si.]

Chameleon The name for about 80 species of small-to-medium-sized lizards that make up the family Chamaeleontidae and occur mainly in Africa and Madagascar. The American chameleons (*Anolis*) belong to a different family of lizards, the Iguanidae. *See* IGUANID.

Chamaeleo chamaeleon is the most common species and is a typical example of the group. Its body is flattened from side to side; it has a long, prehensile tail; and both the forelimbs and hindlimbs have two digits that oppose the other three. These feet and the tail make the chameleon well adapted for its arboreal habitat. The eyes are large and can move independently of each other in all directions. The tongue is also prehensile, being extensible for a great distance, about the length of the animal itself, and is a highly efficient organ for capturing insects. The head is triangular in profile and has a pointed crest. Chameleons are noted for their ability to change color. Color changes appear to be related to environmental temperatures as well as other external stimuli.

The chameleon is oviparous. The female digs out a hollow in the ground for a nest where several dozen eggs are laid and then covered with soil. The period of incubation varies inversely with the temperature and may be as short as 4 months or as long as 10 months. Parental care of the young has not been observed. *See* CHROMATOPHORE; LIZARD; REPTILIA; SQUAMATA. [C.B.C.]

Channel electron multiplier A single-particle detector which in its basic form (see illus.) consists of a hollow tube

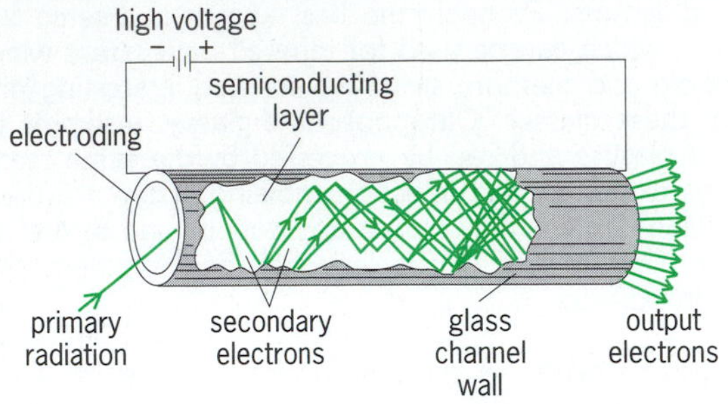

Cutaway view of a straight, single-channel electron multiplier, showing the cascade of secondary electrons resulting from the initial, primary radiation event, which produces an output charge pulse. (*After J. L. Wiza, Microchannel plate detectors, Nucl. Instrum. Meth., 162:587–601, 1979*)

(channel) of either glass or ceramic material with a semiconducting inner surface. The detector responds to one or more primary electron impact events at its entrance (input) by producing, in a cascade multiplication process, a charge pulse of typically 10^4–10^8 electrons at its exit (output). Because particles other than electrons can impact at the entrance of the channel electron multiplier to produce a secondary electron, which is then subsequently multiplied in a cascade, the channel electron multiplier can be used to detect charged particles other than electrons (such as ions or positrons), neutral particles with internal energy (such as metastable excited atoms), and photons as well. As a result, this relatively simple, reliable, and easily applied device is employed in a wide variety of charged-particle and photon spectrometers and related analytical instruments, such as residual gas analyzers, mass spectrometers, and spectrometers used in secondary ion mass spectrometry (SIMS), electron spectroscopy for chemical analysis (ESCA), and Auger electron spectroscopy. *See* AUGER EFFECT; ELECTRON SPECTROSCOPY; MASS SPECTROMETRY; OPTICAL DETECTORS; PHOTOEMISSION; SECONDARY ION MASS SPECTROMETRY (SIMS); SPECTROSCOPY.

A related device is the channel electron multiplier array, often called a microchannel plate. The channel electron multiplier array is usually a disk-shaped device with a diameter between 1 and 4 in. (2.5 and 10 cm) and a thickness of a fraction of a millimeter, and consists of millions of miniature channel electron multiplier devices arranged with channel axes perpendicular to the face of the disk. Channel electron multiplier arrays find application as image intensifiers in night vision devices, and are employed to add either large detection area or imaging capabilities, or both, to charged-particle detectors and spectrometers. *See* IMAGE TUBE (ASTRONOMY); LIGHT AMPLIFIER; PARTICLE DETECTOR. [S.B.E.]

Channeling in solids The steering of energetic charged particles by the atomic rows or atomic planes of a crystalline solid. Solid materials are generally made up of small crystals, and in some cases, materials can be found or made wherein the entire specimen is one single crystal. Within a crystal, the atoms are arranged in a lattice, and various lattice configurations have been observed and classified. In the lattice, the atoms are arranged in orderly rows and columns so that, viewed in some directions, the lattice appears dense, and viewed in other directions, called channels, the lattice has a more transparent appearance. There are planar channels as well as axial channels. Planar channels are the spaces between sheets of atoms. When the atoms in each sheet are along the line of sight, an axial channel will appear. *See* CRYSTALLOGRAPHY.

The phenomenon called channeling occurs when swift particles such as α-particles, protons, electrons, or heavy ions from an accelerator move in a crystal along one of the channel directions. When this occurs, the particle tends to be gently deflected away from the atoms by a series of very-small-angle scatterings; in this way the particle tends to stay in the open spaces of the crystal as it passes through. Such a channeled particle will lose less energy than it would if it moved along some nonchannel path. Since channeled particles have very few close encounters with the atoms, nuclear reactions between particle and atom are almost completely suppressed; the same statement applies to excitation and ionization of inner-shell electrons on the atom, since these processes require close collisions as well.

Channeling has been used as a tool to investigate the properties of solids. For example, when small amounts of a foreign substance exist in a crystal (as in the ion-implantation doping of semiconductor devices), channeling experiments can be used to determine which locations within the crystal are occupied by these added atoms. Another example pertains to the transverse oscillations a channeled particle makes as it passes through a crystal. In this case the exact details of the trajectories of channeled particles provide information about the electrostatic forces acting upon the particle, and these forces relate to the details of the charges distributed through the crystal. *See* ION IMPLANTATION. [C.D.M.]

Chaos System behavior that depends so sensitively on the system's precise initial conditions that it is, in effect, unpredictable and cannot be distinguished from a random process, even though it is deterministic in a mathematical sense.

Throughout history, sequentially using magic, religion, and science, people have sought to perceive order and meaning in a seemingly chaotic and meaningless world. This quest for order reached its ultimate goal in the seventeenth century when newtonian dynamics provided an ordered, deterministic view of the entire universe epitomized in P. S. de Laplace's statement, "We ought then to regard the present state of the universe as the effect of its preceding state and as the cause of its succeeding state."

But if the determinism of Laplace and Newton is totally accepted, it is difficult to explain the unpredictability of a gambling game or, more generally, the unpredictably random behavior observed in many newtonian systems. Commonplace examples of such behavior include smoke that first rises in a smooth, streamlined column from a cigarette, only to abruptly burst into wildly erratic turbulent flow (see illustration); and the unpredictable phenomena of the weather. *See* FLUID FLOW; TURBULENT FLOW.

At a more technical level, flaws in the newtonian view had become apparent by about 1900. The problem is that many newtonian systems exhibit behavior which is so exquisitely sensitive to the precise initial state or to even the slightest outside perturbation that, humanly speaking, determinism becomes a physically meaningless though mathematically valid concept. But even more is true. Many deterministic newtonian-system orbits are so erratic that they cannot be distinguished from a random process even though they are strictly determinate, mathematically speaking. Indeed, in the totality of newtonian-system orbits, erratic unpredictable randomness is overwhelmingly the most common behavior. *See* CLASSICAL MECHANICS; DETERMINISM; STOCHASTIC PROCESS.

One example of chaos is the evolution of life on Earth. Were this evolution deterministic, the governing laws of evolution would have had built into them anticipation of every natural crisis which has occurred over the centuries plus anticipation of every possible ecological niche throughout all time. Nature, however, economizes and uses the richness of opportunity

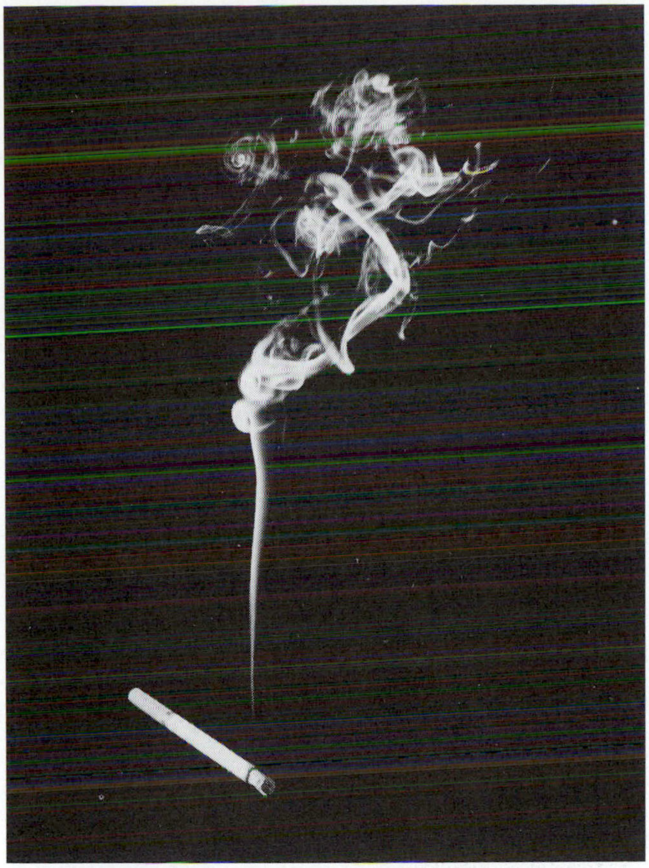

Transition from order to chaos (turbulence) in a rising column of cigarette smoke. The initial smooth streamline flow represents order, while the erratic flow represents chaos.

available through chaos. Random mutations provide choices sufficient to meet almost any crisis, and natural selection chooses the proper one. *See* ORGANIC EVOLUTION.

Another example concerns the problem that the human body faces in defending against all possible invaders. Again, nature appears to choose chaos as the most economical solution. Loosely speaking, when a hostile bacterium or virus enters the body, defense strategies are generated at random until a feedback loop indicates that the correct strategy has been found. A great challenge is to mimic nature and to find new and useful ways to harness chaos. *See* IMMUNITY.

Another matter for consideration is the problem of predicting the weather or the world economy. Both these systems are chaotic and can be predicted more or less precisely only on a very short time scale. Nonetheless, by recognizing the chaotic nature of the weather and the economy, it may eventually be possible to accurately determine the probability distribution of allowed events in the future given the present. At that point it may be asserted with mathematical precision that, for example, there is a 90% chance of rain 2 months from today. Much work in chaos theory seeks to determine the relevant probability distributions for chaotic systems. *See* WEATHER FORECASTING AND PREDICTION.

Finally, many physical systems exhibit a transition from order to chaos, as exhibited in the illustration, and much work studies the various routes to chaos. Examples include fibrillation of the heart and attacks of epilepsy, manic-depression, and schizophrenia. Physiologists are striving to understand chaos in these systems sufficiently well that these human maladies can be eliminated. *See* PERIOD DOUBLING.

Reduced to basics, chaos and noise are essentially the same thing. Chaos is randomness in an isolated system; noise is randomness entering this previously isolated system from the outside. If the noise source is included to form a composite isolated system, there is again only chaos. *See* ACOUSTIC NOISE; ELECTRICAL NOISE. [J.Fo.]

Chaparral A vegetation formation characterized by woody plants of low stature (3.3–10 ft or 1–3 m tall), impenetrable because of tough, rigid, interlacing branches, with small, simple, waxy, evergreen, thick leaves. The term refers to evergreen oak, Spanish *chapparo*, and therefore is uniquely southwestern North American. This type of vegetation has its center in California and occurs continuously over wide areas of mountainous to sloping topography. The Old World Mediterranean equivalent is called maquis or macchie, with nomenclatural and ecological variants in the countries from Spain to the Balkans. Physiognomically similar vegetation occurs also in South Africa, Chile, and southwestern Australia in areas of Mediterranean climates, that is, with very warm, dry summers and maximum precipitation during the cool season. The floras of these five areas with Mediterranean climates are altogether different. *See* MAQUIS.

The characteristic species of the true chaparral of California include *Adenostema fasciculatum*, *Ceanothus cuneatus*, *Quercus dumosa*, *Heteromeles arbutifolia*, *Rhamnus californica*, *R. crocea*, and *Cercocarpus betuloides*, plus a host of endemic species of *Arctostaphylos* and *Ceanothus* and other Californian endemics, both shrubby and herbaceous. These plants determine the formation's physiognomy. It is a dense, uniform-appearing, evergreen, shrubby cover with sclerophyllous leaves and deep-penetrating roots.

Ecologically, chaparral occurs in a climate which is hot and dry in summer, cool but not much below freezing in winter, with little or no snow, and with excessive winter precipitation that leaches the soil of nutrients. The need for water and its supply are exactly out of phase.

Chaparral soils are generally rocky, often shallow, or of extreme chemistry such as those derived from serpentine, and are always low in fertility. In the very precipitous southern Californian mountains, soil erosion rates may be 0.04 in. (1 mm) per year over large watershed areas. [J.Ma.]

Character recognition The technology of using machines to automatically identify human-readable symbols, most often alphanumeric characters, and then to express their identities in machine-readable codes. This operation of transforming numbers and letters into a form directly suitable for electronic data processing is an important method of introducing information into computing systems. *See* DATA-PROCESSING SYSTEMS.

Character recognition machines, sometimes called character readers, print readers, scanners, or reading machines, automatically convert printed alphanumeric characters or symbols into a machine-readable code at high speeds. The output of the character recognition machine may be temporarily stored on magnetic disks or magnetic tape. Alternatively, the recognition system may be operated on line with the data processor. The first commercial application of character recognition was in the banking industry, which adopted magnetic ink character recognition (MICR). In most applications, the printed or typed characters are sensed optically; this process is called optical character recognition (OCR).

Optical mark reading (OMR) refers to the simpler technology of optically sensing marks. The information being read is encoded as a series of marks such as lines or filled-in boxes on a test answer sheet, or as some special pattern such as the Universal Product Code (UPC). The UPC uses a predetermined and reproducible standard pattern which corresponds to each number in a 10-digit code. A relatively simple optical scanner detects the pattern and decodes the number.

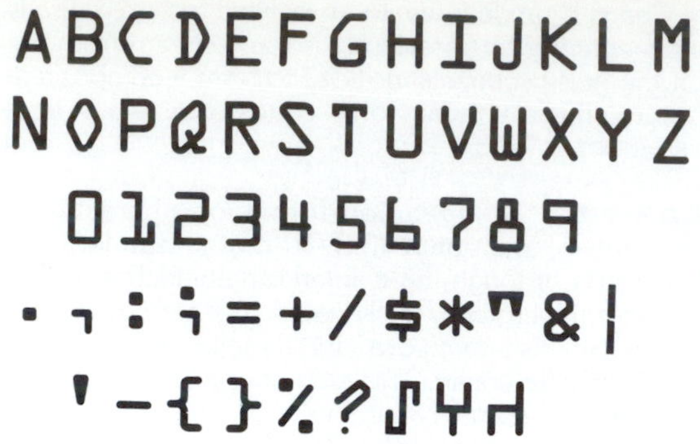

Fig. 1. Character set for use in optical character recognition applications.

Manual alternatives to automatic character recognition equipment for data entry include keypunch, key tape, and key disk systems, in which the data are entered through a keyboard and are recorded on punched cards, magnetic tape, or rotating magnetic disks, respectively. Limited-vocabulary isolated-word voice recognition equipment is available for data entry. However, voice input is also a slow process and would not be competitive with OCR in situations where the data already exist in printed or typed form.

Of all the above application areas, OCR is the single most important commercial application of pattern recognition technology. The technologies used in OCR systems include optical, electronic, mechanical, and computer techniques. In general an OCR system has the following functional systems: input; transport; scanner; preprocessor; feature extraction and classification logic; output.

The classes of OCR applications range from highly controlled stylized fonts to unconstrained multifont typewritten, printed, and handwritten characters encountered by postal address readers. Important elements in rating the performance of an OCR are the error rate, the throughput, the number of fonts capable of being recognized, and the cost, size, and reliability of the machine, The error rate has two components: the undetected substitution rate and the reject rate, that is, the proportion of characters not classified into any character class.

Stylized characters are designed to make automatic recognition easier. One widely used special font is the OCR-A or ISO-A (Fig. 1), which was the first character set standardized by the International Standards Organization. This font is used extensively on turnaround documents such as utility bills and gasoline credit card slips. A common example of stylized font character recognition is the magnetic ink character recognition (MICR) system which is used on bank checks. This special font, adopted as a standard by the American Banking Association, is called E-13B (Fig. 2).

Fig. 2. E-13B font in magnetic ink on the bottom of a typical bank check to permit electronic processing.

Single-stylized-font OCR wands, that is, handheld readers, are being used more and more in inventory and point-of-sale applications. User-programmable hand-held wands in which a microprocessor can be programmed through the wand by reading special codes in the stylized character set have been introduced.

At the opposite end of the scale from stylized font characters is the variability of character fonts used in postal address readers which sort mail in large cities in the United States. With the additional OCR and bar code readers that are to be installed and use of a nine-digit business zip code and special incentives for large-volume business mailers, this equipment should improve service for large-volume mailers. The reading of hand-printed zip codes has been successfully accomplished in Japan by requiring that the numerals be carefully written in boxes preprinted on a standard location on the envelope.

Many data-entry and word-processing applications are aided by the automatic reading of material typed in one of the various popular typewriter fonts. Performance figures similar to those for stylized font characters may be obtained with certain typewriter fonts when typed on a specially aligned typewriter using carbon film ribbon, high-quality paper, and well-specified format conventions. In general, with standard typefaces, error rates are likely to be one or two orders of magnitude higher, depending on the typeface and print quality.

A number of manufacturers offer OCRs for constrained hand-printed numerics, and a few office readers for constrained alphanumeric hand-printed characters. State payroll tax forms, driver's license applications, magazine subscription renewal forms, and other short forms with boxes and directions on how to print the characters give evidence of the business use of constrained hand-printed OCR. The recognition performance depends, among other things, on the number of writers, the number of different character classes, and the training of the writers.

The increasing use of microprocessor technology is aimed at developing low-cost, decentralized OCR devices which can handle a wide variety of fonts and formats. Combinations of word-processing and OCR equipment, and OCR and facsimile store-and-forward communications are likely to lead to many new applications for OCR. *See* COMPUTER; DIGITAL COMPUTER; MICROPROCESSOR; WORD PROCESSING. [L.N.K.]

Characteristic curve A curve that shows the relationship between two changing values. A typical characteristic curve is that which shows how the changes in the control-grid voltage of an electron tube affect the anode current. When three variables are involved, a family of characteristic curves is frequently drawn, with each curve representing one value of the third variable. [J.Mar.]

Charadriiformes A large, diverse, worldwide order of shore and aquatic birds. It may be closely related to the pigeons (Columbiformes) on the one side and to the cranes, rails, and their allies (Gruiformes) on the other. *See* COLUMBIFORMES; GRUIFORMES.

The order Charadriiformes is arranged into three suborders and 17 families as follows: (1) Charadrii, with superfamily Jacanoidea containing the families Jacanidae (jacanas; 8 species) and Rostratulidae (painted snipe; 2 species); superfamily Charadrioidea containing the families Graculavidae, Presbyornithidae, Haematopodidae (oyster catchers; 7 species), Charadriidae (plovers and lapwings; 64 species), Scolopacidae (sandpipers, curlews, phalaropes, and snipe; 86 species), Recurvirostridae (stilts and avocets; 10 species), Dromadidae (crab plovers; 1 species), Burhinidae (thick-knees; 9 species), and Glareolidae (pratincoles; 16 species); and superfamily Chionidoidea with the families Thinocoridae (seed snipe; 4 species) and Chionidae (sheathbills; 2 species).

(2) Lari, with the families Stercorariidae (skuas; 5 species), Laridae (gulls and terns; 88 species), and Rynchopidae (skimmers; 3 species). (3) Alcae, with the single family Alcidae (auks, murres, and puffins; 23 species).

The three suborders of charadiiforms are quite different groups. The Charadrii are the typical shorebirds, usually found in marshy areas and along shores, but some are in dry areas, and a few, the phalaropes, are mainly aquatic. They can run and fly well. Most live in flocks, although most breed solitarily in nests placed on the ground. Most feed on insects and other small animals; the seed snipes are mainly vegetarian. The Lari include the skuas, gulls, terns, and skimmers, predominantly aquatic birds that find their food by flying over the water. They are long-winged, excellent fliers with short legs, but they can walk well. The feet are webbed. Most species breed in large colonies. Alcae include only the alcids, which are true marine, swimming and diving birds found only in the Northern Hemisphere. They have webbed feet and reduced wings. Alcids are excellent divers, and swim underwater by using their wings, which are reduced to rather stiff, paddlelike structures. Alcids feed on fish and other aquatic animals. They breed in large colonies, on rocky ledges or in burrows. *See* AVES. [W.J.B.]

Charales The single order of the class Charophyceae in the division Chlorophyta. These plants, commonly known as the stoneworts, are of macroscopic size and have erect, stemlike axes with nodes and internodes. Whorls of branches ("leaves") arise from the nodes. Two tribes, Nitelleae and Chareae, with seven extant genera and a number of fossil representatives, make up the single family Characeae. *See* CHAROPHYCEAE. [G.W.P.]

Charcoal A porous solid product containing 85–98% carbon produced by heating carbonaceous materials such as cellulose, wood, peat, and coals of bituminous or lower rank at 930–1100°F (500–600°C) in the absence of air.

Chars or charcoals from cellulose or wood are soft and friable. They are used chiefly for decolorizing solutions of sugar and other foodstuffs and for removing objectionable tastes and odors from water. Chars from nutshells and coal are dense, hard carbons. They are used in gas masks and in chemical manufacturing for many mixture separations. Another use is for the tertiary treatment of waste water. Residual organic matter is adsorbed effectively to improve the water quality. *See* ADSORPTION; CARBON. [J.H.Fi.]

Charge-coupled devices Semiconductor devices in which minority charge is stored in a spatially defined depletion region (potential well) at the surface of a semiconductor, and is moved about the surface by transferring this charge to similar adjacent wells. The formation of the potential well is controlled by the manipulation of voltage applied to surface electrodes. Since a potential well represents a nonequilibrium state, it will fill with minority charge from normal thermal generation. Thus a charge-coupled device (CCD) must be continuously clocked or refreshed to maintain its usefulness. In general, the potential wells are strung together as shift registers. Charge is injected or generated at various input ports and then transferred to an output detector. By appropriate design to minimize the dispersive effects associated with the charge-transfer process, well-defined charge packets can be moved over relatively long distances through thousands of transfers.

There are several methods of controlling the charge motion, all of which rely upon providing a lower potential for the charge in the desired direction. When an electrode is placed in proximity to a semiconductor surface, the electrode's potential can control the near-surface potential within the semiconductor. The basis for this control is the same as for metal oxide semiconductor (MOS) transistor action. If closely spaced electrodes are at different voltages, they will form potential wells of different depths. Free charge will move from the region of higher potential to the one of lower potential.

An important property of a charge-coupled device is its ability to transfer almost all of the charge from one well to the next. Without this feature, charge packets would be quickly distorted and lose their identity. This ability to transfer charge is measured as transfer efficiency, which must be very good in order for the structure to be useful in long registers. Values greater than 99.9% per transfer are not uncommon. This means that only 10% of the original charge is lost after 100 transfers.

A second important property of a charge-coupled device register is its lifetime. When the surface electrode is clocked high, the potential within the semiconductor also increases. Majority charge is swept away, leaving behind a depletion layer. If the potential is taken sufficiently high, the surface goes into deep depletion until an inversion layer is formed and adequate minority charge collected to satisfy the field requirements. The time it takes for minority charge to fill the well is the measure of well lifetime. The major sources of unwanted charge are: thermal diffusion of substrate minority charge to the edge of the depletion region, where it is collected in the well; electron-hole pair generation within the depletion region; and the emission of minority charge by traps. Surface-channel charge-coupled devices usually have better lifetime, since surface-state trap emission is suppressed, and the depletion regions are usually smaller.

Once adequate transfer efficiency and sufficient lifetime have been achieved, input and output ports must be established. The port structure of a charge-coupled device depends upon its application. Analog registers require linear inputs and outputs with good dynamic range. Digital registers require precisely metered and detected ones and zeros. Imaging devices use photoemission at the register sites as input with analog-output amplifers. *See* COMPUTER STORAGE TECHNOLOGY; INTEGRATED CIRCUITS; SEMICONDUCTOR. [M.R.Gu.]

Charge-density wave The ground state of a metal in which the conduction-electron charge density is sinusoidally modulated in space. The periodicity of this extra modulation is unrelated to the lattice periodicity. Instead it is determined by the dimensions of the conduction-electron Fermi surface in momentum space.

Coulomb interactions between electrons are usually the cause of a charge-density wave instability. Two such contributions are the exchange energy, an effect of the Pauli exclusion principle, and the correlation energy, an effect of electron-electron scattering. Both energies are reduced in a charge-density wave structure. However, the classical, electrostatic energy opposes the instability and will suppress it unless the positive-ion lattice intervenes. *See* EXCLUSION PRINCIPLE.

A wavelike displacement of this lattice will generate a positive-ion charge density that tends to cancel the electronic charge modulation of the charge-density wave. Typical values of the displacement amplitude are $A \approx 10^{-12}$ to 10^{-11} m. The ion-ion interactions must be small in order to permit this. Consequently charge-density waves are more likely to occur in metals with small elastic moduli. *See* BAND THEORY OF SOLIDS; CRYSTAL; SPIN-DENSITY WAVE. [A.W.O.]

Charles' law A thermodynamic law, also known as Gay-Lussac's law, which states that at constant pressure the volume of a fixed mass or quantity of gas varies directly with the absolute temperature. Conversely, at constant volume the gas pressure varies directly with the absolute temperature. *See* GAS. [F.H.R.]

Charm Ordinary atoms of matter consist of a nucleus composed of neutrons and protons and surrounded by electrons. Over the years, however, a host of other particles with unexpected properties have been found, associated with both electrons (leptons) and protons (hadrons). The hadrons number in the hundreds, and can be explained as composites of more fundamental constituents, called quarks. The originally simple situation of having an up quark (u) and a down quark (d) has evolved as several more varieties or flavors have had to be added. These are the strange quark (s) with the additional property or quantum number of strangeness to account for the unexpected characteristics of a family of strange particles; the charm quark (c) possessing charm and no strangeness, to explain the discovery of the J/ψ particles, massive states three times heavier than the proton; and a fifth quark (b) to explain the existence of the even more massive upsilon (Υ) particles. *See* HADRON; J PARTICLE; QUARKS.

The members of the family of particles associated with charm fall into two classes: those with hidden charm, where the states are a combination of charm and anticharm quarks ($c\bar{c}$), charmonium; and those where the charm property is clearly evident, such as the D^+ ($c\bar{d}$) meson and Λ_c^+ (cud) baryon. Although reasonable progress has been made in the study of charmed states, only a handful of states has been observed, and much work remains to be done. *See* ELEMENTARY PARTICLE.

[N.P.S.]

Charophyceae One of the two classes of the Chlorophyta with one order, Charales. Two tribes, Nitelleae and Chareae, with seven extant genera, including *Nitella* and *Chara*, constitute the order. This group is known as the stoneworts because most species of *Chara* are lime-encrusted. *See* CHARALES; CHLOROPHYCEAE.

The plants have rhizoidal extensions into the substrate and a vertical axis with nodes and internodes. Whorls of short branches bearing sex organs occur at the nodes. An apical cell conducts terminal growth. The axis and "leaves" of *Chara* are usually enclosed by overlaying, elongate, fluting or corticating cells. Plants usually grow in mats and may form extensive bottom meadows in lakes and slow-flowing streams. They are important in the food chain of aquatic animals and birds. *See* CHAROPHYTA; CHLOROPHYCOTA. [G.W.P.]

Charophyta A group of small aquatic plants that have been limited to fresh- and brackish-water environments since the Early Devonian Period. The charophytes live entirely submerged in quiet or slowly moving bodies of alkaline to mildly acidic water. Their calcareous remains are widely distributed in nonmarine shales and limestones, and the group is useful in stratigraphic subdivision and paleoecological interpretation of these strata. Modern forms are a source of food for wildfowl.

Charophyta are generally classified with the green algae, but their reproductive organs, antheridia and sporophydia (unfertilized oogonia), are so specialized that some systematists consider them a separate group. Classification of fossil forms is based chiefly on the gyrogonites (see illustration), the small calcified portions of the sporophydia. *See* ALGAE; CHARALES; PALEOBOTANY. [R.E.P.]

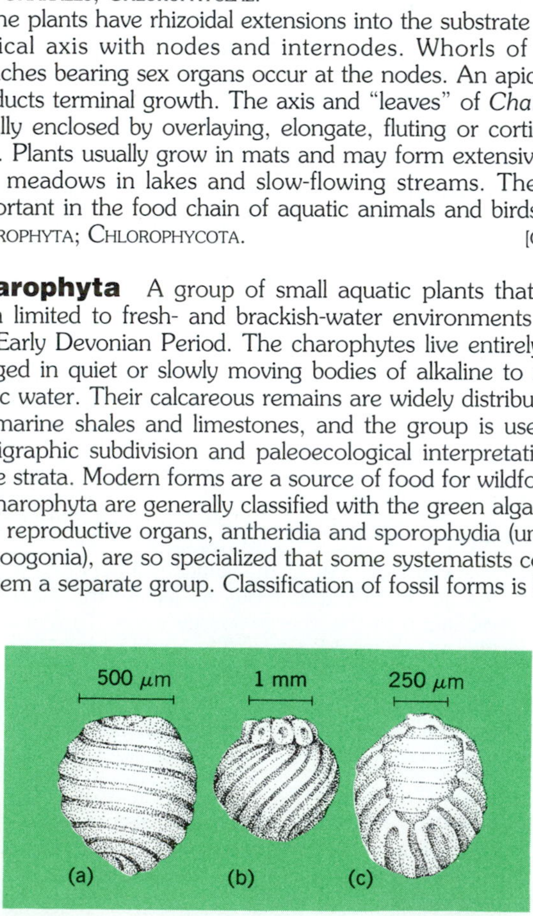

Charophyte gyrogonites, calcified parts of sporophydia. (*a*) Charales. (*b*) Trochiliscales. (*c*) Clavatoraceae.

Cheese A product of milk, selectively concentrated from major milk components. It is generally rich in flavor and contains high-quality nutrients. There are many varieties of cheese, all produced in the following general manner. Raw or pasteurized milk is clotted by acid, rennet, or both. The curd is cut and shaped into the special form of the cheese with or without pressing. Salt is added, or the cheese is brined after pressing.

Acid is produced during manufacture of cheese by fermentation of the milk sugar, lactose. This fermentation is initiated by the addition of a culture of specially selected acid bacteria (starter culture) to the milk. Acid production in cheese curd retards growth of bacteria that cause undesirable fermentations in cheese. Moreover, it favors the expulsion of the whey and the fusion of the curd particles. Fresh cheese (cottage or cream cheese) does not require any ripening, and it is sold soon after it is made. Other varieties of cheese are cured or ripened to obtain the desirable consistency, flavor, or aroma. The flavor and aroma of cheese are obtained by a partial breakdown of mild proteins and fat by the action of microbial, milk, and rennet enzymes. In hard varieties (Cheddar, Gouda, Edam, Emmentaler or Swiss, and provolone) this is done by the microorganisms in the interior of cheese; in semisoft or soft types (Limburger, Camembert, and Roquefort) by the organisms on, or in contact with, the surface of cheese.

Processed cheese is produced from cheeses of different ages by blending. The mixture, melted with the aid of emulsifying salts (citrates and phosphates), is packed in sealed containers (tins, paperboard, foil, or plastic). Few bacteria other than spore-formers survive the heat treatment. No substantial growth of flora occurs in well-preserved process cheese, but spoilage by anaerobic spore-formers may occur. [N.O.]

Key materials for cheesemaking include fresh or precultured milk, cultures, milk-coagulating enzyme preparations, special microorganisms, salt, and beta carotene or annatto color. The amounts used and the manner in which these materials are applied strongly influence the cheese character. Cheese may be made from the milk of the cow, sheep, goat, water buffalo, and other mammals, but the milk of the cow is most widely used despite some limitations. Sheep milk and water buffalo milk generally give more flavor to the cheeses, and the color is uniformly white because of a lack of carotene in such mammalian milks, but they are more expensive to make into cheese.

Two major classes of cheeses exist, fresh and ripened. Fresh cheeses are simpler to make than ripened, are more perishable, and do not develop as intense flavors, but give a mild acid, slightly aromatic flavor and soft, smooth texture. Three basic groups characterize fresh cheese types: group I—ricotta and Broccio; group II—cottage, Neufchatel, and cream; and group III—mozzarella. Most ripened cheeses are contained in one of six basic groups. The characteristic cheeses in each group include: group I—Cheddar and Monterey; group II—Swiss (Emmentaler) and Gruyère; group III—Edam and Gouda; group IV—Muenster, brick, and Limburger; group V—provolone; and group VI—Camembert, Brie, and bleu.

Processed cheese is made from natural types. Nearly any natural cheese can be processed, except that for blue cheese technical difficulties cause blackening of the blue mold due to the high heat. Processed cheese is produced by grinding selected lots of natural cheese and adding cream, color, salt, and emulsifying agents. Processed cheese foods and spreads are made similarly, but include more water and less protein, fat, and other solids. *See* MILK. [F.V.K.]

Cheilostomata An order of ectoproct bryozoans in the class Gymnolaemata. Cheilostomes possess delicate colonies composed of loosely grouped, highly complex, short, uncurved, box- or vase-shaped zooecia, with solid, porous, chitinous, or calcareous walls, and with apertures closed by lid-like opercula. Cheilostome colonies may be encrusting thread-like networks; thin encrusting sheets; tabular, nodular, or globular masses; or erect tufflike, twiglike, frondlike, or trellislike growths. These colonies are not divisible into distinct endozone and exozone regions. Although all lack stolons and monticules or maculae, and some have rhizoids or small coelomic chambers, most cheilostome colonies bear ovicells (brood chambers) developed at the distal ends of the normal autozooecia (feeding zooids). Many cheilostome colonies possess either one or both of two types of heterozooids, avicularia, and vibracula. *See* BRYOZOA; CTENOSTOMATA; GYMNOLAEMATA. [R.J.Cu.]

Chelation A chemical reaction or process involving chelate ring formation and characterized by multiple coordinate bonding between two or more of the electron-pair-donor groups of a multidentate ligand and an electron-pair-acceptor metal ion. The multidentate ligand is usually called a chelating agent, and the product is known as a metal chelate compound or metal chelate complex. Metal chelate chemistry is a subdivision of coordination chemistry and is characterized by the special properties resulting from the utilization of ligands possessing bridged donor groups, two or more of which coordinate simultaneously to a metal ion. *See* COORDINATION CHEMISTRY.

Many of the functional groups of both synthetic and naturally occurring organic compounds can form coordinate bonds to metal ions, producing metal-organic complexes or chelates, many of which are biologically active. Thus chelate compounds are frequently found in an interdisciplinary field of science called bioinorganic chemistry. The biological significance of chelates is demonstrated by the large number of biologically important compounds that are either metal chelates or chelating agents. Included in this group are the alpha amino acids, peptides, proteins, enzymes, porphyrins (such as hemoglobin), corrins (such as vitamin B_{12}), catechols, hydroxypolycarboxylic acids (such as citric acid), ascorbic acid (vitamin C), polyphosphates, nucleosides and other genetic compounds, pyridoxal phosphate (vitamin B_6), and sugars. The ubiquitous green plant pigment, chlorophyll, is a magnesium chelate of a tetradentate ligand formed from a modified porphin compound, and similarly the oxygen transport heme of red blood cells contains an Fe(II) chelate. *See* BIOINORGANIC CHEMISTRY; ORGANOMETALLIC COMPOUND.

The ability of chelating agents to reduce the chemical activity of metal ions has found extensive application in many areas of science and industry. Ethylenediaminetetraacetic acid (EDTA), a hexadentate chelating agent, has been employed commercially for water softening, boiler scale removal, industrial cleaning, soil metal micronutrient transport, and food preservation. Nitrilotriacetic acid (NTA) is a tetradentate chelating agent which, because of lower cost, has taken over some of the commercial applications of EDTA. Many commercially important dyes and pigments, such as copper phthalocyanines, are chelate compounds. Humic and fulvic acids are plant degradation products in lake and sea-water sediments that have been suggested as important chelating agents which regulate metal-ion balance in natural waters. By virtue of its abundance, low toxicity, low cost, and good chelating tendencies for metal ions that produce water hardness, the tripolyphosphate ion (as its sodium salt) is used in large quantities as a builder in synthetic detergents. Both synthetic ion exchangers and the mineral zeolites are chelating ion-exchange resins which are used in analytical and water-softening applications. As final examples, less conventional chelating agents are the multidentate, cyclic ligands, termed collectively crown ethers, which are particularly suited for the complexation of the alkali and alkaline-earth metals. *See* CROWN ETHERS; ETHYLENEDIAMINETETRAACETIC ACID.
[A.E.M.; R.J.Mo.]

Chelicerata A subphylum of the phylum Arthropoda which includes the classes Merostomata, Pycnogonida, and Arachnida. The first pair of ventral appendages, the chelicerae, is modified as pincers. The body usually consists of the cephalothorax and abdomen, except in the Acari in which these body parts are fused. The cephalothorax is unsegmented in all but the solpugids or sun spiders. Four pairs of legs are common to most members of this group. Respiratory structures include book lungs, gills, and tracheae. The sexes are usually separate; there is a single sex opening located anteriorly on the abdomen. Fertilization is usually internal and most species lay eggs. Chelicerates are chiefly terrestrial and predacious. *See* ARACHNIDA; ARTHROPODA; MEROSTOMATA; PYCNOGONIDA.
[C.B.C.]

Chemical bonding The fundamental fact of chemistry is that elements, such as oxygen, hydrogen, carbon, and iron, can combine to form compounds, such as water, methane, and iron oxide, with properties completely different from those of their components. The description of this in terms of atomic theory is that atoms of the elements attach themselves to each other to form molecules. This property that atoms possess, of joining together to form molecules, is known as chemical bonding.

Atoms contain electric charges—a small, positively charged nucleus surrounded by a cloud of moving, negatively charged electrons. All chemical bonding is caused by the mutual attractions and repulsions of these electric charges.

Chemical bonds are very strong. To break one bond in each molecule in a gram mole of material will typically require an energy of many tens of kilocalories per mole. Most of the energy used by humans is chemical energy, derived from changing chemical bonds in food or fuel.

Ionic bonding is the simplest type of bonding, in which one or more electrons are transferred completely from one atom to another, thus converting the neutral atoms into electrically charged ions. These ions are approximately spherical in shape and attract one another because of their opposite charges. The ions are drawn together until their spherical electron clouds sufficiently interpenetrate and repel one another to balance the force of attraction. Molecules can consist of two or more such ions. Many inorganic crystals can be considered as giant molecules made up of ions. Thus common salt, sodium chloride, consists of a lattice of Na^+ ions (positive sodium), each surrounded by six Cl^- ions (negative chlorine) and vice versa. *See* IONIC CRYSTALS.

Covalent bonding is another limiting type of chemical bonding. Here each atom of a bonded pair contributes one electron to form a pair of electrons which move in such a manner as to increase the density of electric charge in the space between the two atoms. The negative charge in the region between the two atoms attracts the two positive nuclei. This type of bond is also called an electron pair bond because its essential feature is the formation of a pair of electrons which spend much time in the region between the two atoms. These two electrons have their spins pointing in opposite directions; in other words, they are paired.

Metallic bonding is a third type of chemical bonding, which is exemplified in the common metals. There are several ways of looking at this bonding, but perhaps the simplest is to consider the crystal as consisting of positive ions of the metallic element immersed in a sea of electrons. The attraction of the positive ions for the electrons holds the crystal together. Some of the electrons are free to move about the whole crystal of the metal, and this is what makes the metal an electrical conductor.

Although many crystals are giant molecules whose atoms are completely linked together by strong ionic or covalent bonds, others consist of discrete molecules, strongly bonded internally, but held to each other by much weaker forces. Most organic crystals are soft and have low melting points. Their discrete molecules are held together by what are called van der Waals or dispersion forces. Although these forces are electrical in nature, they are too weak to be considered as true chemical bonding. *See* CHEMICAL STRUCTURES; INTERMOLECULAR FORCES; MOLECULAR STRUCTURE AND SPECTRA; QUANTUM CHEMISTRY; VALENCE. [E.B.W.]

Chemical compounds Substances composed of two or more elements which do not vary in composition from sample to sample, and which have fixed and definite physical properties, such as density and refractive index. The elements in compounds cannot be separated by simple physical or mechanical means, but only by chemical treatment. When compounds are formed from their elements, heat is generated or absorbed. These properties distinguish them from mixtures. *See* ELEMENTS; MIXTURE.

Most chemical compounds are formed in fixed and definite proportions by weight from their elements, and they obey the laws of chemical combination. However, there are a number of solid compounds, known as nonstoichiometric compounds, that exhibit departures from the law of definite proportions. *See* DEFINITE COMPOSITION, LAW OF; EQUIVALENT WEIGHT; INORGANIC CHEMISTRY; MULTIPLE PROPORTIONS, LAW OF; NONSTOICHIOMETRIC COMPOUNDS. [T.C.W.]

Chemical conversion The term for the chemical change from reactants to products of a chemical industrial process. Preferred for decades in the petroleum industry, the term is being used to an increasing extent for the entire chemical process industries in place of unit processes, since it is a more exact definition of industrial chemical reactions. The following are important chemical conversions:

Alkylation	Hydrogenation, hydrogenolysis
Amination	Hydrolysis, hydration
Ammonolysis	Ion exchange
Aromatization	Isomerization
Calcination	Neutralization
Combustion	Nitration
Condensation	Oxidation
Dehydration	Polymerization
Diazotization and coupling	Pyrolysis, cracking
Esterification (sulfation)	Reduction
Fermentation	Sulfation
Halogenation	Sulfonation
Hydroformylation (oxo)	

All these terms refer to the classes of chemical changes, each embracing many individual reactions or chemical conversions into which chemical industrial processes can be divided. For additional details see the separate articles on individual chemical conversions (unit processes). *See* CHEMICAL ENGINEERING; CHEMICAL PROCESS INDUSTRY; UNIT OPERATIONS. [J.A.Br.]

Chemical dynamics That branch of physical chemistry which seeks to explain time-dependent phenomena, such as energy transfer and chemical reaction, in terms of the detailed motion of the nuclei and electrons which constitute the system.

Reaction kinetics. Although the ultimate state of a chemical system is specified by thermodynamics, the time required to reach that equilibrium state is highly variable. As a consequence, determining the rate of chemical reactions has proved to be important for practical reasons. For example, diamonds are thermodynamically unstable with respect to graphite, but the rate of transformation of diamonds to graphite is negligible. Rate studies have also yielded fundamental information about the details of the nuclear rearrangements which constitute the chemical reaction. *See* CHEMICAL EQUILIBRIUM.

Traditional chemical kinetic investigations of the reaction between species X and Y to form Z and W, reaction (1), sought a rate of the form given in Eq. (2), where $d[Z]/dt$ is the rate of

$$X + Y \rightarrow Z + W \quad (1)$$

$$d[Z]/dt = kf([X], [Y], [Z], [W]) \quad (2)$$

appearance of product Z, f is some function of concentrations of X, Y, Z, and W which are themselves functions of time, and k is the rate constant. Chemical reactions are incredibly diverse, and often the function f is quite complicated, even for seemingly simple reactions such as that in which hydrogen and bromine combine directly to form hydrogen bromide. These are actually complex reactions which proceed through a sequence of simpler reactions, called elementary reactions. For reaction (3*d*), the sequence of elementary reactions is a chain mechanism known to involve a series of steps, reactions (3*a*)–(3*c*). This sequence of elementary reactions was formerly

$$Br_2 \rightarrow 2Br \quad (3a)$$

$$Br + H_2 \rightarrow HBr + H \quad (3b)$$

$$H + Br_2 \rightarrow HBr + Br \quad (3c)$$

$$H_2 + Br_2 \rightarrow 2HBr \quad (3d)$$

known as the reaction mechanism, but in the chemical dynamical sense the word mechanism is reserved to mean the detailed motion of the nuclei during a collision.

An elementary reaction is considered to occur exactly as written. Reaction (3*b*) is assumed to occur when a bromine atom hits a hydrogen molecule. The products of the collision are a hydrogen bromide molecule and a hydrogen atom. On the other hand, the overall reaction is a sequence of these elementary steps and on a molecular basis does not occur as reaction (3*d*) is written. With few exceptions, the rate law for an elementary reaction $A + B \rightarrow C + D$ is given by $d[C]/dt = k[A][B]$. The order (sum of the exponents of the concentrations) is two, which is expected if the reaction is bimolecular (requires only species A to collide with species B). The rate constant k for such a reaction depends very strongly on temperature.

In some instances, particularly for decompositions $AB \rightarrow A + B$, the elementary reaction step is first-order, Eq. (4),

$$d[A]/dt = d[B]/dt = k[AB] \quad (4)$$

which means that the reaction is unimolecular. The species AB does not spontaneously dissociate; it must first be given some critical amount of energy, usually through collisions, to form an excited species AB^*. It is the species AB^* which decomposes unimolecularly.

Molecular dynamics. In principle, it is possible to prepare two reagents in specific quantum states and to determine the quantum-state distribution of the products. In practice, this is much too difficult, and experiments have been limited to preparing one reagent or to determining some aspect of the product distribution. This approach yields data concerning the gross aspects of the dynamics rather than the fine details.

Molecular-beam and luminescence techniques have played a major role in the development of chemical dynamics. Since these techniques largely complement each other, they are illustrated by discussing the results of a single bimolecular reaction, the formation of deuterium chloride (DCl), reaction (5).

$$D + Cl_2 \rightarrow DCl + Cl \quad (5)$$

For infrared chemiluminescence measurements deuterium atoms are made by dissociating D_2 gas in an electrical discharge and injecting them into an observation cell. The reagents mix and react inside a vessel with walls at 77 K (−321°F), which freeze out species hitting the wall. The pressure is kept low to minimize vibrationally deactivating collisions. Infrared (IR) emission is analyzed with an IR spectrometer. By analyzing the spectrum, it is possible to determine which vibration-rotation states are emitting and, as a consequence, which vibration-rotation states are formed in the reaction. *See* CHEMILUMINESCENCE.

Molecules can be isolated in molecular beams, and collisions between these isolated molecules can be observed by crossing two tenuous molecular beams in a region of otherwise high vacuum. Measurements are made of the scattered product intensity and speed at various scattering angles. The speed of the product is high, and corresponds to about half of the reaction exoergicity appearing in translational recoil of the products, with the balance appearing in vibration and rotation of the DCl consistent with the chemiluminescence results. *See* MOLECULAR BEAMS.

Unimolecular reactions of collisionally activated reaction complexes have been studied using both crossed-beam and chemiluminescence techniques. The availability of lasers opened new vistas in chemical dynamics, not only because lasers facilitate conventional measurements, as in detecting reaction products by laser-induced fluorescence, but also because they have uncovered new phenomena. Absorption of photons by molecules has been known for many years, and it has become apparent that under the right circumstances a molecule can absorb not one photon, but so many as to cause the molecule to dissociate or even or ionize. [P.R.B.]

Theoretical methods. The goal of chemical dynamics is to understand kinetic phenomena from the basic laws of molecular mechanics, and it is thus a field which sees close interplay between experimental and theoretical research.

An important question regarding the dynamics of chemical reactions has to do with the product energy distribution in exothermic reactions. For example, because the HF molecule is more strongly bound than the H_2 molecule, reaction (6) re-

$$F + H_2 \rightarrow HF + H \qquad (6)$$

leases a considerable amount of energy. The two possible paths for this energy release to follow are into translation, that is, with HF and H speeding away from each other, or into vibrational motion of HF.

In this case it is vibration, and this has rather dramatic consequences: the reaction creates a population inversion among the vibrational energy levels of HF—that is, the higher vibrational levels have more population than the lower levels—and the emission of infrared light from these excited vibrational levels can be made to form a chemical laser.

The rates of most chemical reactions are increased if they are given more energy. In macroscopic kinetics this corresponds to increasing the temperature, and most reactions are faster at higher temperatures. It seems reasonable, though, that some types of energy will be more effective in accelerating the reaction than others. The general rule of thumb is that vibrational energy is more effective for endothermic reactions (those for which the new molecule is less stable than the original molecule), while translational energy is most effective for exothermic reactions.

Many different theoretical models and methods have been useful in understanding and analyzing kinetic phenomena. Probably the single most useful approach has been the calculation of classical trajectories. Assuming that the potential energy function or a reasonable approximation is known for the three atoms in reaction (6), for example, it is possible by use of electronic computers to calculate the classical motion of the three atoms. It is thus an easy matter to give the initial molecule more or less vibrational or translational energy, and then compute the probability of reaction. Similarly, the final molecule and atom can be studied to see where the energy appears, that is, as translation or as vibration.

It is thus a relatively straightforward matter theoretically to answer the questions and to see whether or not mode-specific excitation leads to significantly different chemistry than simply increasing the temperature under bulk conditions. *See* INORGANIC PHOTOCHEMISTRY; PHOTOCHEMISTRY. [W.H.M.]

Relaxation methods. Since about 1950, considerable use has been made of perturbation techniques to measure rates and determine mechanisms of rapid chemical reactions. These methods provide measurements of chemical reaction rates by displacing equilibria. In situations where the reaction of interest occurs in a system at equilibrium, perturbation techniques called relaxation methods have been found most effective for determining reaction rate constants.

A chemical system at equilibrium is one in which the rate of a forward reaction is exactly balanced by the rate of the corresponding back reaction. Examples are chemical reactions occurring in liquid solutions, such as the familiar equilibrium in pure water, shown in reaction (7). The essence of any of the

$$H_2O \underset{k_b}{\overset{k_f}{\rightleftharpoons}} H^+(aq) + OH^-(aq) \qquad (7)$$

relaxation methods is the perturbation of a chemical equilibrium (by a small change in temperature, pressure, electric-field intensity, or solvent composition) in so sudden a fashion that the chemical system, in seeking to reachieve equilibrium, is forced by the comparative slowness of the chemical reactions to lag behind the perturbation.

The temperature jump (T-jump) relaxation method has proved to be very versatile. The choice of the particular means of effecting the temperature perturbation is dictated only by the requirement that the temperature rise somewhat more rapidly than the time constant of the chemical reaction to be explored, so that a tedious deconvolution can be avoided.

In a situation, such as reaction (7), in which electrically neutral reactant species dissociate into oppositely charged ions, an especially sensitive tool for measuring rate constants of forward and backward reactions is the electric-field jump (E-jump) technique with conductometric detection. In a strong electric field (of the order of 4×10^6 V m^{-1}), a weak acid in solution is caused to dissociate to a greater degree than it would in the absence of the electric field. With a sensitive, high-voltage Wheatstone bridge, the exponential increase with time in the concentration of ions following a precipitous increase in electronic-field strength is readily detected. The measured relaxation time (τ) is clearly that corresponding to the high-electric-field environment, but since the rate constants for these reactions differ little in and out of the electric field, no serious problem is posed.

Two other relaxation methods more widely used than the E-jump technique are pressure jump (P-jump) and ultrasonic absorption. Each relies for its effectiveness on a volume change, $\Delta V°$, occurring in an aqueous sample equilibrium undergoing kinetic investigation. As electrically neutral, weak electrolyte solute species dissociate into ions in aqueous solution, there is an increase in the number of solvent molecules drawn into a highly ordered solvation sheath. The higher the charge density of the ion, the more water molecule dipoles are bound and the greater the change in $V°$ as reactants become products. Unlike the T-jump and E-jump relaxation methods, which usually employ step function perturbations, the ultrasonic absorption techniques are continuous-wave experiments in which the sample chemical equilibrium absorbs a measurable amount of the sound wave's energy when the frequency of the sound wave (f) and the relaxation time of the chemical equilibrium bear the relation to one another given by Eq. (8). *See* ULTRASONICS.

$$\tau^{-1} = 2\pi f \qquad (8)$$

The typical pressure-jump (P-jump) experiment is one in which a liquid sample under about 200 atmospheres (20 megapascals) pressure is suddenly brought to atmospheric pressure by the bursting of a metal membrane in the sample cell autoclave. Relaxation times measured spectrophotometrically or conductometrically are thus accessible if $\tau > 100$ microseconds. This technique has proven particularly useful in the elucidation of micellar systems of great interest for catalysis and for petroleum recovery from apparently depleted oil fields.

The continuous- and stopped-flow techniques antedate somewhat the relaxation techniques described above, and have the sometimes important advantage of permitting kinetic measurements in chemical systems far from equilibrium. The stopped-flow experiment is one in which two different liquids in separate syringes are mixed rapidly in a tangential jet mixing chamber and then the rapid flow of mixed reactants is almost immediately brought to a halt in a spectrophotometric, conductometric, or calorimetric observation chamber. *See* CHEMICAL THERMODYNAMICS. [E.M.Ey.]

Chemical ecology Chemical ecology is concerned with the interaction of organisms mediated by the chemicals they produce. In 1959 scientists coined the word pheromone to designate any substance produced by an organism that serves to induce a behavioral or developmental response in an individual of the same species. Pheromones include sex attractants, alarm substances, trail substances, trail markers, and aggregation-inducing compounds. Animals and plants produce other substances as well, which some behaviorists call allomones and kairomones, that transmit information to individuals of different species.

Allomones are chemicals produced by an organism which induce in a member of another species a behavioral or physiological reaction favorable to the emitter. These chemicals are involved in a variety of mutualistic and antagonistic relationships. The scents of flowers are mutualistic allomones; insects pollinate plants from which they obtain nectar in return. Venoms and repellents, which organisms use in defense against predators, are antagonistic allomones.

Kairomones are chemical substances that benefit the recipient rather than the emitter. These compounds elicit positive responses of predators to their prey, herbivores to their food plants, and parasites to their hosts.

The development and use of synthetic insecticides during this century has saved millions of lives from lethal insect-borne diseases and has helped farmers to feed a larger part of the world's population. At the same time, however, the widespread and often indiscriminate use of toxic chemicals to control agricultural and other pests has created serious problems. Continued use of these toxins has led to the development of insects resistant to insecticides, and insecticide residues have disrupted nontarget ecological systems.

The trend in pest control has shifted toward ecological pest management, usually referred to as integrated control. The goal is not to eradicate a pest but to maintain the pest population below levels at which it causes economic damage or health hazards (insectistasis). Integrated control employs one or a variety of techniques appropriate for a particular species and situation, and among these methods is the use of pheromones.

Now that the chemical structure of many pheromones has been determined and the compounds synthesized, a number of them have been used in integrated control programs. For example, sex and food attractants of certain moths and beetles that destroy stored food products have been used for the early detection of infestations in granaries, mills, and warehouses, and to prevent the increase of pest populations by mass trapping. The use of these pheromone traps minimizes the need for treatment with insecticides. Sex pheromones have been used to disrupt sexual communication among moths and to prevent mating. In this method the sex pheromones are evaporated from many sources to confuse the males and prevent their locating conspecific females. The technique has shown promise in control of the plum fruit moth, codling moth, larch bud moth, and grape moth.

The odorous secretions of mammals have been of interest to perfume makers for ages, but little is known about their role in animal communication, and the chemical structure of the compounds is largely unknown. Studies indicate that mammals do not give an automatic, standardized response to their odor signals in the way that insects react to their pheromones, and the term semiochemicals (chemical or odor signals) has been suggested for mammalian scents.

In mammals, only the male pheromone of the domestic pig has been applied to practical purpose. The saliva and sweat glands of the sexually aroused boar contain two steroids which induce the mating stance in estrous sows. The compounds have been used to facilitate the artificial insemination of pigs. *See* INSECT CONTROL, BIOLOGICAL; SOCIAL INSECTS. [L.M.R.]

Chemical energy In most chemical reactions, heat is either taken in or given out. By the law of conservation of energy, the increase or decrease in heat energy must be accompanied by a corresponding decrease or increase in some other form of energy. This other form is the chemical energy of the compounds involved in the reaction. The rearrangement of the atoms in the reacting compounds to produce new compounds causes a change in chemical energy. This change in chemical energy is equal numerically and of opposite sign to the heat change accompanying the reaction.

Most of the world's available power comes from the combustion of coal or of petroleum hydrocarbons. The chemical energy released as heat when a specified weight, often 1 g, of a fuel is burned is called the calorific value of the fuel.

It is not possible to measure an absolute value for the chemical energy of a compound; only changes in chemical energy can be measured. It is therefore necessary to make some arbitrary assumption as a starting point. One such assumption would be to take the chemical energies of the free atoms as zero and measure the chemical energies of all elements and compounds relative to this standard. If this were done, all chemical energies would then be negative quantities. Heat is given out when all elements and compounds are formed from their atoms. Since chemical energy can be regarded as a form of potential energy, it is interesting that the formation of chemical bonds is always accompanied by a decrease of potential energy. The reason, in qualitative terms at least, lies in the quantum theory. When chemical bonds are formed between atoms, electrons are shared between the two atoms, and this electron sharing produces a lowering of the potential energies of the shared electrons, and thus a lowering of the potential energy of the molecule relative to the free atoms.

Although it would be more fundamental to take the chemical energies of the separated atoms as zero, in practice it is more convenient to take a more arbitrary starting point and to assume that the chemical energies of the elements are zero. For precision, the standard states of the elements at 25°C (77°F) and 1 atm (101.325 kilopascals) pressure are chosen. Thus, in the case of carbon the chemical energy of graphite, not diamond, is said to be zero at 25°C and 1 atm. Diamond then has a definite enthalpy value. *See* CHEMICAL THERMODYNAMICS; ENERGY SOURCES; ENTHALPY. [T.C.W.]

Chemical engineering The branch of engineering serving those industries that chemically convert basic raw materials into a variety of products. Starting with ores, salt, sulfur, limestone, coal, natural gas, petroleum, air, water, and so forth, these industries, by chemical processing techniques, produce widely diversified products, such as aluminum, magnesium, and titanium metals; refined petroleum fractions, such as fuels and solvents; synthetic fertilizers; synthetic fibers, resins, and plastics; antibiotics; wood pulp and paper; and petrochemicals. *See* CHEMICAL CONVERSION; CHEMICAL PROCESS INDUSTRY.

To bring about the conversion of raw materials into such a broad spectrum of products, the chemical engineer starts with basic scientific data on chemical reactions discovered by research chemists. Working first on a small laboratory scale, the engineer develops engineering data for various steps in the process of changing the starting material into a product. Before completing this small-scale investigation, the next step may involve actual operation of various parts of the process in mini- or pilot-scale equipment, simulating what may happen in the commercial-scale process. *See* PILOT PRODUCTION.

In carrying out these development functions, the chemical engineer relies on previous training in scientific and engineering fundamentals and in advanced mathematics to plan the experimental programs and to analyze the resulting data. In many cases the engineer will evolve a mathematical model that interrelates all the process variables in mathematical terms. Then by using electronic computing equipment, the design engineer can attain the optimum process design. Ultimately, when the plant is built, an on-line computer and sophisticated instrumentation will be employed to control the commercial unit. *See* SYSTEMS ENGINEERING.

After completion of the plant, chemical engineering continues to play a vital role in the start-up and operation of the facility. Once in operation, the manufacturing unit becomes a continuing problem in optimization to achieve desired product output and specified quality at lowest cost in the face of constantly shifting variables of raw material supply, availability and quality of personnel, condition of plant equipment, and market conditions.

For their professional group activities, chemical engineers work to a large extent through the American Institute of Chemical Engineers. Headquartered in New York, the AIChE has more than 100 local sections throughout the United States. It also supports an extensive technical publishing program. [C.S.Cr.]

Chemical equilibrium In a dynamic or kinetic sense, chemical equilibrium is a condition in which a chemical reaction is occurring at equal rates in its forward and reverse directions, so that the concentrations of the reacting substances do not change with time. In a thermodynamic sense, it is the condition in which there is no tendency for the composition of the system to change; no change can occur in the system without the expenditure of some form of work upon it. From the viewpoint of statistical mechanics, the equilibrium state places the system in a condition of maximum freedom (or minimum restraint) compatible with the energy, volume, and composition of the system. The statistical approach has been merged with thermodynamics into a field called statistical thermodynamics; this merger has been of immense value for its intellectual stimulus, as well as for its practical contributions to the study of equilibria. *See* CHEMICAL THERMODYNAMICS; STATISTICAL MECHANICS.

Of the three viewpoints, the thermodynamic approach is by far the most powerful and fruitful in treating the quantitative relationships between the position of equilibrium and the factors which govern it. Since thermodynamics is concerned with relationships among observable properties, such as temperature, pressure, concentration, heat, and work, the relationships possess general validity, independent of theories of molecular behavior.

Chemical potential. Thermodynamics attributes to each chemical substance a property called the chemical potential, which may be thought of as the tendency of the substance to enter into chemical (or physical) change. Although the chemical potential of a substance cannot be directly measured (except on a relative basis), differences in chemical potential are measurable. (The units are those of energy per mole.)

The importance of the chemical potential lies in its relation to the affinity or driving force of a chemical reaction. Consider general reaction (1). Let μ_A be the chemical potential per

$$aA + bB \rightleftharpoons gG + hH \qquad (1)$$

mole of substance A, μ_B be the chemical potential per mole of B, and so on. Then, according to one of the fundamental principles of thermodynamics (the second law), the reaction will be spontaneous when the total chemical potential of the reactants is greater than that of the products. Thus, for spontaneous change (naturally occurring processes) notation (2) applies

$$[g\mu_G + h\mu_H] - [a\mu_A + b\mu_B] < 0 \qquad (2)$$

When equilibrium is reached, the total chemical potentials of products and reactants become equal; thus Eq. (3)

$$[g\mu_G + h\mu_H] - [a\mu_A + b\mu_B] = 0 \qquad (3)$$

holds at equilibrium. The difference in chemical potentials in Eqs. (2) and (3) is called the driving force or affinity of the process or reaction; naturally, it is zero when the chemical system is in chemical equilibrium.

For reactions at constant temperature and pressure (the usual restraints in a chemical laboratory), the difference in chemical potentials becomes equal to the free energy change ΔG for the process in Eq. (4). The decrease in free energy

$$\Delta G = [g\mu_G + h\mu_H] - [a\mu_A + b\mu_B] \qquad (4)$$

represents the maximum net work obtainable from the process. When no more work is obtainable, the system is at equilibrium. Conversely, if the value of ΔG for a process is positive, some useful work will have to be expended upon the process, or reaction, in order to make it proceed; the process cannot proceed naturally or spontaneously. (The term spontaneously as used here implies only that a process can occur. It does not imply that the reaction will be rapid or instantaneous. Thus, the reaction between hydrogen and oxygen is a spontaneous process in the sense of the term as used here, even though a mixture of hydrogen and oxygen can remain unchanged for years unless ignited or exposed to a catalyst.)

Since by definition a catalyst remains unchanged chemically through a reaction, its chemical potential does not appear in Eqs. (2), (3), and (4). A catalyst, therefore, can contribute nothing to the driving force of a reaction, nor can it, in consequence, alter the position of the chemical equilibrium in a system. *See* CATALYSIS.

In addition to furnishing a criterion for the equilibrium state of a chemical system, the thermodynamic method goes much further. In many cases, it yields a relation between the change in chemical potentials (or change in free energy) and the equilibrium concentrations of the substances involved in the reaction. To do this, the chemical potential must be expressed as a function of concentration (and other properties of the substance). *See* CONCENTRATION SCALES.

Activity and standard states. It is often convenient to utilize the product fx, called the activity of the substance and defined by $a = fx$. The activity may be looked upon as an effective concentration of the substance, measured in the same units as the concentration x with which it is associated. The standard state of the substance is then defined as the state of unit activity (where $a = 1$) and is characterized by the standard chemical potential $\mu°$. Clearly, the terms $\mu°$, f, and x are not independent; the choice of the activity scale serves to fix the standard state. For example, for an aqueous solution of hydrochloric acid, the standard state for the solute (HCl) would be an (hypothetical) ideal 1 molar (or molal) solution, and for the solvent (H_2O) the standard state would be pure water (mole fraction = 1). The reference state would be an infinitely dilute solution; here the activity coefficients would be unity for both solute and solvent. For the vapor of HCl above the solution, the standard state would be the ideal gaseous state at 1 atm (101.325 kilopascals) partial pressure; the reference state would be a state of zero pressure. (For gases, the term fugacity is used instead of activity.)

It should be noted that the reference state is a limiting state which in many cases can be reached only through an extrapolation from observed behavior. *See* FUGACITY.

Equilibrium constant. If general reaction (1) occurs at constant temperature T and pressure P when all of the substances involved are in their standard states of unit activity, Eq. (4) would become Eq. (5). The quantity $\Delta G°$ is known as the

$$\Delta G° = [g\mu_G^\circ + h\mu_H^\circ] - [a\mu_A^\circ + b\mu_B^\circ] \qquad (5)$$

standard free energy change for the reaction at that temperature and pressure for the chosen standard states. (Standard state properties are commonly designated by a superscript, $\Delta G°$, $\mu°$.) Since each of the standard chemical potentials ($\mu°$) is a unique property determined by the temperature, pressure, standard state, and chemical identity of the substance concerned, the standard free energy change $\Delta G°$ is a constant (parameter) characteristic of the particular reaction for the chosen temperature, pressure, and standard states.

When the system has come to chemical equilibrium at constant temperature and pressure, $\Delta G = 0$, and $\Delta G°$ is given by Eq. (6), where the value of $K°$ is shown as Eq. (7), and the

$$\Delta G° = -RT \ln K° \qquad (6)$$

$$K° = \left[\frac{a_G{}^g a_H{}^h}{a_A{}^a a_B{}^b} \right] \qquad (7)$$

activities are the equilibrium values. The ratio of the activities at equilibrium, $K°$, is called the equilibrium constant or, more precisely, the thermodynamic equilibrium constant. (The terms $K°$ and $Q°$ are written with superscripts to emphasize that they represent ratios of activities.) The equilibrium constant is a characteristic property of the reaction system, since it is determined uniquely in terms of the standard free energy change. The term $-\Delta G°$ represents the maximum net work which the reaction could make available when carried out at constant temperature and pressure with the substances in their standard states.

The kinetic concept of chemical equilibrium introduced by C. M. Guldberg and P. Waage (1864) led to the formulation of the equilibrium constant in terms of concentrations. Although the concept is correct in terms of the dynamic picture of opposing reactions occurring at equal speeds, it has not been successful in coping with the problems of activity coefficients. Conversely, the thermodynamic approach yields no relationship between the driving force of the reaction and the rate of approach to equilibrium. *See* CHEMICAL DYNAMICS.

The influence of temperature upon the chemical potentials, and hence upon the equilibrium constant, is given by the Gibbs-Helmholtz equation, Eq. (8). The derivative on the left

$$\left[\frac{d \ln K°}{dT} \right]_P = \frac{\Delta H°}{RT^2} \qquad (8)$$

represents the slope of the curve obtained when values of $\ln K°$ for a reaction, obtained at different temperatures but always at the same pressure P, are plotted against temperature. The standard heat of reaction $\Delta H°$ for the temperature T at which the slope is measured is the heat effect which could also be observed by carrying out the reaction involving the standard states in a calorimeter at the corresponding temperature and pressure. *See* HEAT CAPACITY; THERMOCHEMISTRY.

Homogeneous equilibria. These involve single-phase systems: gaseous, liquid, and solid solutions. In most cases, solid solutions are so far from ideal that equilibrium constants cannot be evaluated, and such systems are treated in terms of the phase rule. A typical gas-phase equilibrium is the ammonia synthesis shown in reaction (9). A typical liquid-phase equilib-

$$N_2 + 3H_2 \rightleftharpoons 2NH_3 \qquad (9)$$

rium is the dissociation of acetic acid in water, reaction (10).

$$HC_2H_3O_2 + H_2O \rightleftharpoons H_2O^+ + C_2H_3O_2^- \qquad (10)$$

The solvent appears to be inert, since its chemical potential remains practically unchanged over the useful concentration range. As a result of this apparent inertness of the solvent, it is not possible to determine the extent of hydration of any dissolved species from equilibrium studies. Thus, whether the actual ion is H^+, H_3O^+, or $H_9O_4{}^+$, it is the total stoichiometric concentration that is measured and used. *See* IONIC EQUILIBRIUM.

Heterogeneous equilibria. These are usually studied at constant pressure, since at least one of the phases will be a solid or liquid. The imposed pressure may be that of an equilibrium gaseous phase, or it may be an externally controlled pressure.

In describing such systems, the nature of each phase must be specified. In the following example, the terms *s*, *l*, and *g* identify solid, liquid, and gaseous phases, respectively. For solutions or mixtures, the composition is needed, in addition to the temperature and pressure, to complete the specification of the system. If not obvious, the identity of the solvent must be given.

In the equilibrium shown as reaction (11), the relationship

$$H_2O(l) \rightleftharpoons H_2O(g) \qquad (11)$$

of Eq. (12) holds. Here $K° = p/N$, the ratio of the vapor pres-

$$\Delta G° = \mu_g^\circ - \mu_l^\circ = -RT \ln \frac{p}{N} \qquad (12)$$

sure p to the liquid mole fraction N. For pure water, the equilibrium constant is simply the standard vapor pressure $p°$, and the Clausius-Clapeyron equation is just a special case of the Gibbs-Helmholtz equation, Eq. (8). Now when a small amount of solute is added, decreasing the mole fraction of solvent, the vapor pressure p must be lowered to maintain equilibrium (Raoult's law). The effect of the total applied pressure P upon the vapor pressure p of the liquid is given by the Gibbs-Poynting equation, Eq. (13). Here V_l and V_g are the molar

$$\left[\frac{dp}{dP} \right]_T = \frac{V_l}{V_g} \qquad (13)$$

volumes of liquid and vapor. The vapor pressure will increase as external pressure is applied (activity increases with pressure). If the external pressure is applied to a solution by a semipermeable membrane, an applied pressure can be found which will restore the vapor pressure (or activity) of the solvent to its standard state value. *See* OSMOSIS.

Other heterogeneous equilibria are solubility equilibria, reactions involving two immiscible phases, and reactions involving condensed and immiscible phases. *See* EXTRACTION; SOLUBILITY PRODUCT CONSTANT.

[C.E.V.]

Chemical fuel The principal fuels used in internal combustion engines (automobiles, diesel, and turbojet) and in the furnaces of stationary power plants are organic fossil fuels. These fuels, and others derived from them by various refining and separation processes, are found in the earth in the solid (coal), liquid (petroleum), and gas (natural gas) phases.

Special fuels to improve the performance of combustion engines are obtained by synthetic chemical procedures. These special fuels serve to increase the specific impulse of the engine or to increase the heat of combustion available to the engine per unit mass or per unit volume of the fuel. A special fuel which possesses a very high heat of combustion per unit mass is liquid hydrogen. It has been used along with liquid oxygen in rocket engines. Because of its low liquid density, liquid hydrogen is not too useful in systems requiring high heats of combustion per unit volume of fuel ("volume-limited" systems).

A special fuel which produces high flame temperatures of the order of 5000°F (2800°C) is gaseous cyanogen. This is used with gaseous oxygen as the oxidizer. The liquid fuel hydrazine, and other hydrazine-based fuels, with the liquid oxidizer nitrogen tetroxide are used in many space-oriented rocket engines. The boron hydrides, such as diborane and pentaborane, are high-energy fuels which are used in advanced rocket engines.

For air-breathing propulsion engines (turbojets and ramjets), hydrocarbon fuels are most often used. For some applications, metal alkyl fuels which are pyrophoric (that is, ignite spontaneously in the presence of air), and even liquid hydrogen, are being used. *See* METAL-BASE FUEL.

Fuels which liberate heat in the absence of an oxidizer while decomposing either spontaneously or because of the presence of a catalyst are called monopropellants and have been used in rocket engines. Examples of these monopropellants are hydrogen peroxide and nitro-methane.

Liquid fuels and oxidizers are used in most large-thrust rocket engines. When thrust is not a consideration, solid-propellant fuels and oxidizers are frequently employed because of the lack of moving parts such as valves and pumps, and the consequent simplicity of this type of rocket engine. Solid fuels fall into two broad classes, double-base and composites. Double-base fuels are compounded of nitroglycerin (glycerol trinitrate) and nitrocellulose, with no separate oxidizer required. The double-base propellant is generally formed in a mold into the desired shape (called a grain) required for the rocket case. Composite propellants are made of a fuel and an oxidizer. The latter could be an inorganic perchlorate or a nitrate. Fuels for composite propellants are generally the asphalt-oil-type, thermosetting plastics or several types of synthetic rubber and gumlike substances. Metal particles such as boron, aluminum, and beryllium have been added to solid propellants to increase their heats of combustion and to eliminate certain types of combustion instability. *See* HYDROGEN PEROXIDE. [W.Ch.]

Chemical laser A device in which laser action is obtained from the energy released in some fast chemical reactions. Atoms or molecules produced during the reaction are often in excited states. Under special circumstances there may be enough atoms or molecules excited to some particular state for amplification to occur by stimulated emission. Usually the reacting gases are mixed and then infrared output and pulses up to several thousand joules of energy have been obtained in reactions which produce excited hydrogen fluoride molecules. Pulsed laser action in the ultraviolet (193 to 353 nanometers) has been obtained from excimer states of rare gas monohalides (for example, KrF, XeF, KrCl, XeCl, XeBr). *See* LASER. [S.F.J.; A.L.S.]

Chemical microscopy Application of the microscope to the solution of chemical problems. Such problems encompass identification, correlation of structure and performance, troubleshooting, and process control. Several axioms, based on experience, bolster the chemical microscopist's confidence: (1) Apples, oranges, azaleas, or robins are quickly and easily recognized; hence any tiny particles of cement, abrasive, pollen, and so on, magnified microscopically to the size of fruits, flowers, or birds, can also be quickly and easily recognized. (2) The characteristic behavior of adhesives, lubricants, abrasives, and so on is manifest and can be observed at microscopic sites. Such behavior is, then, best investigated, explained, and improved by the chemical microscopist. (3) Any large-scale process, be it baking a cake, cooking meat, solid-state reactions at high temperatures (for example, reaction of sand and carbon to form carborundum), electroplating, or extrusion of a synthetic fiber, can be duplicated and studied on a microscope stage. *See* MICROSCOPE.

The identification of small particles is of interest in air or water pollution, clean-room control, general contamination problems, and criminalistics. The particles may include minerals, fibers, pollens, protozoa, and fly ash or any other industrial dust. Each has characteristic shape and optical properties, and sizes and weights can be identified down to about 1 micrometer (0.001 mm or 1/25,000 in.) and 1 picogram (10^{-12} g); about 10^{15} such particles would weigh approximately 1 kg. [W.C.McC.]

Chemical process industry This term, abbreviated CPI, is now applied to what has long been known as the chemical and allied industries. It includes industries engaged in the manufacture of chemicals both inorganic and organic, petroleum, petrochemicals, rubber, plastics, fertilizers (agrichemicals), sugar and starch, paper, pulp, synthetic fibers, photography, pharmaceuticals, explosives, food additives and flavors, and dyes. *See* CHEMICAL ENGINEERING; PETROLEUM ENGINEERING. [J.A.Br.]

Chemical reactor A vessel in which chemical reactions take place. A combination of vessels is known as a chemical reactor network. Chemical reactors have diverse sizes, shapes, and modes and conditions of operation based on the nature of the reaction system and its behavior as a function of temperature, pressure, catalyst properties, and other factors.

Laboratory chemical reactors are used to obtain reaction characteristics. Therefore, the shape and mode of operation of a reactor on this scale differ markedly from that of the large-scale industrial reactor, which is designed for efficient production rather than for gathering information. Laboratory reactors are best designed to achieve well-defined conditions of concentrations and temperature so that a reaction model can be developed which will prove useful in the design of a large-scale reactor model.

Chemical reactions may occur in the presence of a single phase (liquid or gas), in which case they are called homogeneous, or they may occur in the presence of more than one phase and are referred to as heterogeneous. In addition, chemical reactions may be catalyzed. Examples of homogeneous reactions are gaseous fuel combustion (gas phase) and acid-base neutralization (liquid phase). Examples of heterogeneous systems are carbon dioxide absorption into alkali (gas-liquid); coal combustion and automobile exhaust purification (gas-solid); water softening (liquid-solid); coal liquefaction and oil hydrogenation (gas-liquid-solid); and cake reduction of iron ore (solid-solid).

Chemical reactors may be operated in batch, semibatch, or continuous modes. When a reactor is operated in a batch mode, the reactants are charged, and the vessel is closed and brought to the desired temperature and pressure. These conditions are maintained for the time needed to achieve the desired conversion and selectivity, that is, the required quantity and quality of product. At the end of the reaction cycle, the entire mass is discharged and another cycle is begun. Batch operation is labor-intensive and therefore is commonly used only in industries involved in limited production of fine chemicals, such as pharmaceuticals. In a semibatch reactor operation, one or more reactants are in the batch mode, while the coreactant is fed and withdrawn continuously. In a chemical reactor designed for continuous operation, there is continuous addition to, and withdrawal of reactants and products from, the reactor system.

There are a number of different types of reactors designed for gas-solid heterogeneous reactions. These include fixed beds, tubular catalytic wall reactors, and fluid beds. Many different types of gas-liquid-solid reactors have been developed for specific reaction conditions. The three-phase trickle-bed reactor employs a fixed bed of solid catalyst over which a liquid phase trickles downward in the presence of a cocurrent gas phase. An alternative is the slurry reactor, a vessel within which coreactant gas is dispersed into a liquid phase bearing suspended catalyst or coreactant solid particles. At high ratios of reactor to diameter, the gas-liquid-solid reactor is often termed an ebulating-bed (high solids concentration) or bubble column reactor (low solids concentration). Gas-liquid reactors assume a form virtually identical to the absorbers utilized in physical absorption processes. Solid-solid reactions are often conducted in rotary kilns which provide the necessary intimacy of contact between the solid coreactants. *See* GAS ABSORPTION OPERATIONS; KILN. [J.J.Ca.]

Chemical sampling techniques In analytical chemistry, the operations required to obtain a laboratory sample from a large quantity of raw material. Most commercial materials are not homogeneous; that is, their compositions vary from portion to portion. If the analysis for a constituent is to be significant, the portion used in the laboratory must have the same composition as that of the original material. The problem of obtaining a representative sample may be more difficult than the problems of analysis.

Sampling of a gas is difficult. Special equipment and procedures are required to obtain a representative, homogeneous portion for analysis.

Homogeneous liquids in tanks or barrels are sampled by dipping several portions from different locations or by drawing off portions from several containers. Liquids flowing in open rivers or in pipes are sampled by dipping from the open stream or by using a bypass system on a pipe.

Segregation is a major problem for solids, so usually 0.5–2.0% of the material is taken as the gross sample to be certain of a representative portion. Metallic materials such as steels, brasses, sheet metals, and wires are sampled by drilling, cutting, milling, or filing if reasonably homogeneous, or by taking very fine drillings from several locations if segregation of components is known to exist. Nonmetallic materials, such as coal, ores, rocks, ore veins, and soils, consist of particles of varying size as well as of varying composition. The gross sample comprises particles from various portions of the material that have been ground and mixed together.

Since the final laboratory sample may still be somewhat heterogeneous, a portion several times the amount needed for an analysis is dissolved in a suitable solvent, and the sample solution is diluted to a definite volume. A fraction of this solution is taken for analysis. It is called an aliquot. *See* ANALYTICAL CHEMISTRY. [C.Ru.]

Chemical senses The senses of smell (olfaction) and taste and the so-called common chemical sense make up the three chemical senses. The sense cells of olfaction and taste are specialized receptor neural elements. The receptors of common chemical sense appear to be undifferentiated, free nerve endings. These are distributed throughout the moist mucous membranes of land-dwelling animals and over the skin of aquatic vertebrates. In order of thresholds, olfaction is the most sensitive, taste is intermediate, and common chemical sensitivity is least.

The chemical senses mediate the selection and acceptance of foodstuffs, the avoidance of irritants, and, in lower organisms especially, the detection of enemies and prey and the selection of mates.

The common chemical senses are considered to be sensitivity to mildly irritating chemicals like dilute solutions of alkali, acids, and salts. Such chemical sensitivity is distinct from tactile sensitivity and, in the mouth and nose, is distinct from taste and smell. *See* CHEMORECEPTION; OLFACTION; SENSATION; TASTE. [C.P.]

Chemical separation techniques Methods used in chemistry to purify substances or to isolate them from other substances, for either preparative or analytical purposes. In industrial applications the ultimate goal is the isolation of a product of given purity, whereas in analysis the primary goal is the determination of the amount or concentration of that substance in a sample. There are three factors of importance to be considered in all separations: (1) the completeness of recovery of the substance being isolated, (2) the extent of separation from associated substances, and (3) the efficiency of the separation.

There are many types of separations based on a variety of properties of materials. Among the most commonly used properties are those involving solubility, volatility, adsorption, and electrical and magnetic effects, although others have been used to advantage. The most efficient separation will obviously be obtained under conditions for which the differences in properties between two substances undergoing separation are at a maximum.

The common aspect of all separation methods is the need for two phases. The desired substance will partition or distribute between the two phases in a definite manner, and the separation is completed by physically separating the two phases. The ratio of the concentrations of a substance in the two phases is called its partition or distribution coefficient. If two substances have very similar distribution coefficients, many successive steps may be required for a separation. The resulting process is called a fractionation.

Based on the nature of the second phase, the more commonly used methods of separation are classified as follows:

1. Methods involving a solid second phase include precipitation, electrodeposition, chromatography (adsorption), ion exchange, and crystallization.
2. The outstanding method involving a liquid second phase is solvent extraction, in which the original solution is placed in contact with another liquid phase immiscible with the first.
3. Methods involving a gaseous second phase include gas evolution, distillation, sublimation, and gas chromatography. Mixtures of volatile substances can often be separated by fractional distillation. *See* EXTRACTION; MASS-TRANSFER OPERATION.

[G.H.Mo.]

Chemical structures Much of chemistry is explainable in terms of the structures of chemical compounds. The understanding of the structures of chemical compounds hinges very strongly on the understanding of the electronic configurations of the elements. The union of atoms, and therefore the formation of compounds from the elements, is associated with interactions among the extranuclear electrons of the individual atoms. Electronic interactions among atoms may occur in either of two ways: Electrons may be transferred from one atom to another, or they may be shared by two (or more) atoms. The first type of interaction is called electrovalence and results in the formation of electrically charged monatomic ions. The second, covalence, leads to the formation of molecules and complex ions. *See* CHEMICAL BONDING.

Ions. A number of relationships among the electronic structures of the elements are important in determining the properties of ions.

1. Only the outermost shell of electrons is of significance in compound formation. This is commonly called the valence

shell, and the electrons contained within it are called the valence electrons.

2. The maximum number of electrons encountered in the valence shell of any single isolated atom is eight.

3. In the case of transition elements, the electrons in the *d* subshell of the penultimate (next outermost) shell are often of importance in compound formation and are considered as essentially a part of the valence shell. Consequently, rules 1 and 2 are rigorously true only for the representative elements, not the transition elements.

4. The number of valence electrons and their subshell assignments are given by the arrangement of the elements in the long form of the periodic table. The elements in any vertical family of the periodic table have the same valence-electron configuration. *See* ATOMIC STRUCTURE AND SPECTRA; ELECTRON CONFIGURATION; PERIODIC TABLE. [D.H.B.]

Positive ions are formed by the removal of valence electrons, and negative ions result from the addition of electrons to the neutral atom. The complete analysis of the energy relationships determining the formation of compounds composed of ions is complex; however, a few aspects of the problem are directly related to the electronic structures of the isolated atoms. The energy required for the removal of an electron from a neutral atom is measured by the ionization potential and relates to reaction (1). A second ionization potential

$$M(g) \rightarrow M^{+}(g) + e^{-} \qquad (1)$$

indicates the amount of energy required to convert the unipositive ion into the dipositive ion, and the successive removal of the remaining electrons may similarly be related to ionization potentials. Typical positive ions will be formed by the removal of all of the valence electrons of the alkali and alkaline-earth metallic elements. The resulting ions are called inert-gas-type ions because they have the electronic configurations of the inert gases preceding the respective elements in the periodic system. As a consequence of the increasing amounts of energy (ionization potentials) required to remove successive electrons from ions, monatomic positive ions that are highly charged do not exist among chemical compounds under the usual conditions of temperature and pressure. The most highly charged ions of any apparent significance are tripositive. A number of the actinide elements form tetrapositive monatomic ions, U^{4+}, Th^{4+}, and Pu^{4+}. *See* IONIZATION POTENTIAL.

In addition to positive ions of the inert-gas type, two other electronic configurations recur with some regularity among chemical compounds. The elements of the copper and zinc groups of the transition elements may lose their valence electrons, forming ions having 18 electrons (instead of 8) in their outermost electron shells. These ions are generally called pseudoinert-gas-type ions.

The addition of an electron, as shown in reaction (2), to a

$$M(g) + e^{-} \rightarrow M^{-}(g) \qquad (2)$$

neutral atom with the resulting formation of a negative ion may be associated with the evolution of energy (the electron affinity). It is generally true, however, that the addition of a second electron will require the expenditure of energy. This relationship militates against the formation of highly charged negative ions. The known, monatomic negative ions are formed by the addition of the number of electrons necessary for the atom to attain an inert gas electronic configuration, and no monatomic negative ions are known to have a charge more negative than 3−. *See* ELECTRONEGATIVITY. [R.G.Pe.]

Polyhedral structures. In considering structures more complex than those derived from simple monatomic ions, the next logical step is to consider single polyhedral aggregates of atoms. At this point it is worth noting that the term "structure" now takes on a much more quantitative meaning. In its most precise sense structure is used to denote a knowledge of the bonding distances and angles between atoms in chemical compounds and, in turn, the geometrical arrangements which they form. These atomic arrangements and the associated distances and angles serve uniquely as "fingerprints" of these atoms' spatial configurations, and depend very much on understanding the electronic configurations around atoms because the union of atoms to form compounds occurs via electronic interactions of one atom with another through their valence electrons. The chemical combination of neutral atoms to produce uncharged species results in molecule formation, whereas the similar combination of atoms or ions possessing a net charge results in the formation of complex ions. A basic understanding of the species formed involves the concept of the coordination polyhedron, which allows a simple classification of the structures of many polyatomic molecules and ions. This type of classification is particularly useful because it conveniently explains the packing together of simple chemical molecules or ions in terms of highly symmetrical polyhedra. There is an obvious connection between polyhedra and the structures found in crystalline solids formed from them. Crystal formation often involves the linking of convex polyhedra by the sharing of corners, edges, or faces, ultimately forming space-filling assemblies in which all faces of each polyhedron are in contact with faces of other polyhedra. The most important simple polyhedrons are the tetrahedron, the trigonal bipyramid, the octahedron, the pentagonal bipyramid, and the cube (Fig. 1). The most commonly observed of these polyhedral configurations are the tetrahedron (four faces) and the octahedron (six faces).

The general properties of a coordination polyhedron are summarized in the identification of a central atom, which in many cases resides at the center of mass of the polyhedron, and the determination of the coordination number of the central atom. The coordination number is equal to the number of groups (atoms, ions) directly attached to the central atom. In the simplest cases, the coordination number is also equal to the number of apexes in the coordination polyhedron. It is therefore concluded that the tetrahedron may occur among molecules and ions having a coordination number of 4; the trigonal bipyramid is associated with a coordination number of 5; the

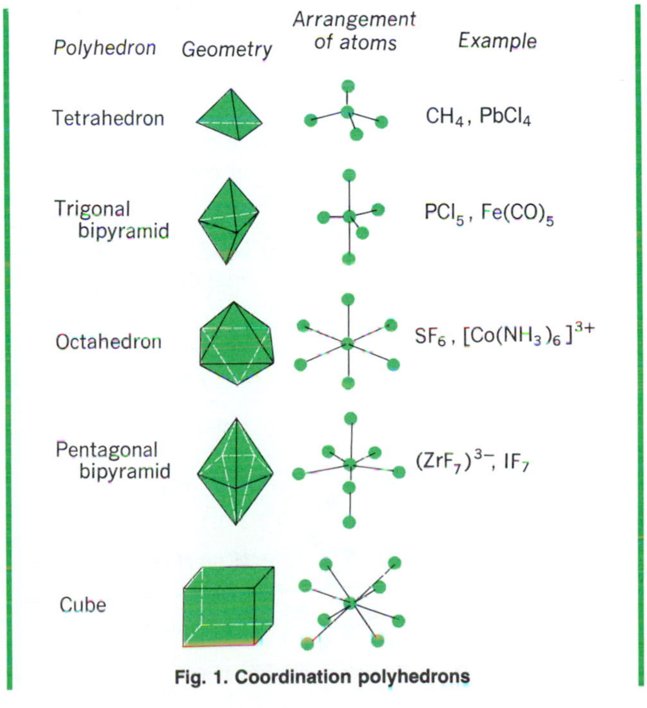

Fig. 1. Coordination polyhedrons

octahedron, with a coordination number of 6; and the pentagonal bipyramid, with a coordination number of 7.

Lower coordination numbers are associated with plane polygons rather than polyhedrons, as shown in Fig. 2. The coordination number of 4 may be associated with either a polygon (square plane) or a polyhedron (tetrahedron). Similarly, alternative structures exist for the coordination numbers 5 (tetragonal pyramid), 6 (trigonal prism), 7 (irregular structures), and 8 (Archimedean antiprism, dodecahedron); however, these alternative structures are rare in occurrence. *See* COORDINATION CHEMISTRY; MOLECULAR ORBITAL THEORY; POLYHEDRON.

The simplest correlative device which accurately summarizes a very large number of such structures and enables the chemist to predict, with a good chance of success, the geometric array of the atoms in a compound of known composition is based on an extreme electrostatic model. This model, or theory, represents the bonds in a purely formal way. The central atom is considered to be a positive ion having a charge equal to its oxidation state. The groups attached to the central atom (the ligands) are then treated either as negative ions or as neutral dipolar molecules.

A knowledge of the electronic structures of the elements, the sizes of the ions, and their relationships within coordination polyhedrons permits prediction of the geometric structures of binary compounds and complex ions for given central atoms and ligands. Such a treatment may be applied to binary halides and oxides and complex oxyanions and haloanions as well as most polyatomic aggregates containing nontransition ions as central elements. In addition, the oxyanions and molecular halides of the transition elements may be treated similarly. Many of the hydrated and ammoniated metallic ions may also be discussed in these terms. Among the particularly useful results of these considerations is the prediction of which compounds have electron pairs that might be shared with molecules or ions having the ability to expand their coordination spheres. In this way the Lewis acids and bases are predicted from their structures. *See* ACID AND BASE.

The most significant limitation of such considerations is the complete lack of concern with the detailed nature of the chemical bonds uniting the central atom to its ligands in a coordination polyhedron. Ions of high charge probably do not exist in chemical compounds, and physical properties of many substances are not explainable in terms of simple electrostatic relationships. The quantum-mechanical theories of valence provide the more detailed descriptions of the manner of bonding required to overcome these limitations. Some classes of compounds cannot be systematized, even from the standpoint of their geometric structures, without the use of these more advanced concepts. Two major theories exist for the application of the idea of the covalent bond to the structures of chemical compounds. One of these is the so-called valence-bond concept.

According to the valence-bond theory, the principal requirements for the formation of a covalent bond are a pair of electrons and suitably oriented electron orbitals on each of the atoms being bonded. The geometry of the atoms in the resulting coordination polyhedron is correlated with the orientation of the orbitals on the central atom. The orbitals utilized depend on the energies of the electrons in them. In general, the order of increasing energy of the electron orbitals is $(n - 1)d < ns < np < nd$. It is concluded that a nontransition atom having one valence electron will form a covalent bond utilizing an *s* orbital. In those cases where an unshared pair of electrons may be assigned to the *ns* orbital, as many as three equivalent bonds may be formed by utilizing the three *np* orbitals of the central atom. *See* CRYSTAL FIELD THEORY; VALENCE.

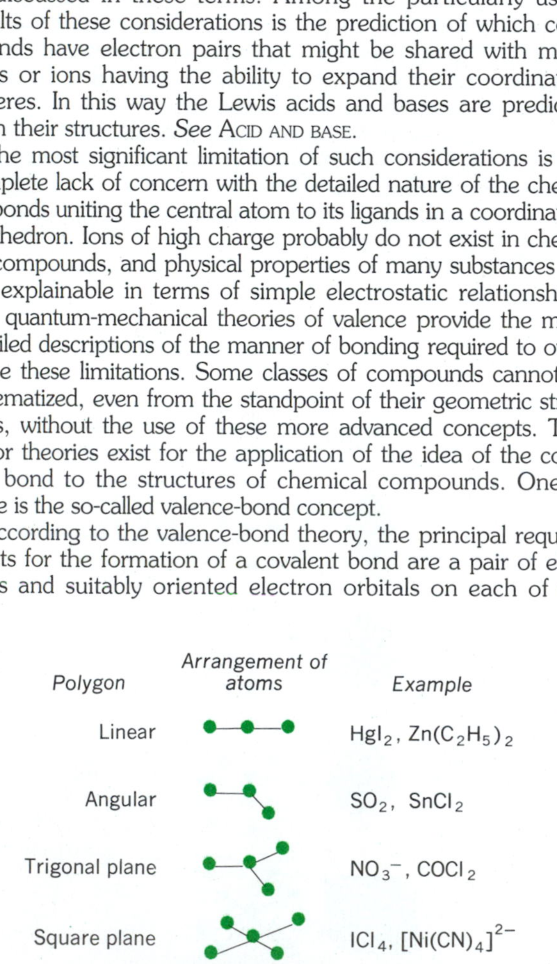

Fig. 2. Coordination polygons.

With regard to the nature of doubly bonded compounds, another problem arises when such structures are viewed from the standpoint of valence-bond theory. In some species nonequivalent bonds are predicted, but this is not found to be true experimentally so long as the similar atoms in the structure are otherwise equivalent. To account for such facts as these, the concept of resonance must be introduced. The resonance method of describing this situation is to say that one of the pictorial structures is inadequate to describe the substance properly but that enough pictorial structures (resonance structures) should be considered to permute the double bond about all the equivalent bonds. The true structure is assumed to be something intermediate to all the resonance structures and more stable than any of them because it exists in preference to any one of them. *See* CONJUGATION (CHEMISTRY); RESONANCE (MOLECULAR STRUCTURE).

The valence-bond theory may be utilized to systematize structures involving unshared pairs of electrons in a manner very similar to that of the resonance method. According to this view, the hybridized orbitals of the central atom determine a total polyhedral structure, but either bonds or unshared electron pairs may make use of the hybridized orbitals.

Condensed polyhedral structures. In the preceding discussions of polyatomic molecules and ions, attention has been confined to those species in which single coordination polyhedrons effectively account for the geometric structures. Understanding of these species is central to the problem of the structures of chemical compounds, but does not provide an adequate basis for the descriptions of the structures of all or even a majority of the known compounds. Among the largest numbers of structures not covered by these concepts are those in which two or more coordination polyhedrons are united to form a more complex total structure. Large numbers of the compounds involving such condensed polyhedrons fall naturally into the subject of solid-state chemistry.

The classic homologous series of compounds in organic chemistry provide useful examples involving the condensation of polyhedrons containing the same central element in the individual units. The series of saturated or aliphatic hydrocarbons, having the general formula C_nH_{2n+2}, contains a large number of compounds extending from the lowest member, methane, CH_4, to polyethylene, a plastic of economic importance in which *n* is a very large number. Two ways exist for the linking together of these tetrahedrons. This gives rise to two molecular forms, both of which are stable, well-known compounds. It is an essential part of these structures that each C—C link is linear (because it is merely a σ-bond); however, when two carbon atoms are linked to a third, the C—C—C angle is essentially determined by the bond angle of the central carbon atom (that is, the other carbons may be treated as ligands to the first). The other familiar homologous series of organic chemistry differ from the saturated hydrocarbons in having at least one unique coordination polyhedron of a different type. The olefins contain two doubly bonded, or unsaturated, carbon atoms, whose polyhedral structures are trigonal planar, and the remainder of the carbons are tetrahedral. As in the case of the aliphatic hydrocarbons, the olefins exhibit an isomerism which is associated with the branching of the chain structure. In addition, the presence of two linked trigonal planar carbon atoms and the fact that the polyhedrons cannot rotate about the double bond give rise to a different kind of isomerism, called cis-trans isomerism. *See* BOND ANGLE AND DISTANCE.

The existence of a predicted isomerism provides one of the most important confirmations of theories of chemical structure. In general, the polyhedral view of molecular structure, as described here, has been thoroughly verified by the discovery of the many types of predicted isomerism. The first really convincing proof of the tetrahedral structures of saturated carbon atoms involved optical isomerism. *See* MOLECULAR ISOMERISM; OPTICAL ACTIVITY.

The property of an element to link to itself, which is so common for carbon, is called catenation. Other elements exhibit this property to a lesser extent. For example, nitrogen forms hydrazine, NH_2—NH_2, the homolog next higher than NH_3; however, the remaining members of the series N_nH_{n+2} are unknown. Similarly, water, H_2O, and hydrogen peroxide, HO—OH, constitute the beginning of a homologous series of which only these two members are known.

A second type of link is commonly involved in joining coordination polyhedrons into more complex structures. This involves the interaction of bridge-forming atoms to link other coordination polyhedrons instead of direct union of the central atoms. The siloxanes provide examples of this behavior. These interesting and useful substances are related to the silicates in structure. The polyphosphates also involve the sharing of corners by PO_4 tetrahedrons through oxygen bridges. [J.M.Wi.]

Electron-deficient structures. A significant number of chemical compounds are known in which the bonding cannot be accurately described by the utilization of a pair of electrons for each bond and between contiguous atoms. Rather, examples exist where the total number of valence-electron pairs is less than the number of bonds. Such molecules are said to have electron-deficient structures which are characterized by the formation of an electron-pair bond between more than two atomic centers. An example is diborane, B_2H_6, the simplest boron hydride. A total of only 12 valence electrons (3 from each boron atom and 1 from each hydrogen atom) is available to form what, under normal circumstances, would be 8 covalent bonds. The bonding in this electron-deficient case can be understood by postulating the existence of a three-center two-electron bond for each B—H—B bridge. That is, the three atoms are joined together by a single electron pair. The formation of such multicenter bonds in response to electron deficiency frequently leads to molecular structures which are cyclic or cagelike. An example involving metal-to-carbon bonds is trimethylaluminum. Electron deficiency leads to a dimeric structure with bridging methyl groups. The bridges are joined by three-center two-electron bonds. *See* BORANE; ORGANOMETALLIC COMPOUNDS. [T.J.M.]

Chemical symbols and formulas

Chemical symbols and formulas A system of symbols and notation for the chemical elements and the combinations of these elements which form numerous chemical compounds. This system consists of letters, numerals, and marks that are designed to denote the chemical element, formula, or structure of the molecule or compound. These symbols give a concise and instantly recognizable description of the element or compound. In many cases, through the efforts of international conferences, the symbols are recognized throughout the scientific world, and they greatly simplify the universal language of chemistry.

Elements. At the present time, 109 chemical elements have been given symbols, usually derived from the name of the element. Examples of names and symbols are chlorine, Cl; fluorine, F; beryllium, Be; aluminum, Al; oxygen, O; and carbon, C. However, symbols for some elements are derived from Latin or other names for the element. Examples are Au, gold (from *aurum*); Fe, iron (from *ferrum*); Pb, lead (from *plumbum*); Na, sodium (from *natrium*); and K, potassium (from *kalium*). The symbols consist of one or, more commonly, two letters. The first letter is a capital, followed by a lowercase letter.

Inorganic molecules and compounds. Simple diatomic molecules of a single element are designated by the element symbol with a subscript 2, indicating that the molecule contains two atoms. Thus the hydrogen molecule is H_2; the nitrogen molecule, N_2; and the oxygen molecule, O_2. Polyatomic molecules of a single element are designated by the element symbol with a subscript corresponding to the number of atoms in the molecule. Examples are the phosphorus molecule, P_4; the sulfur molecule, S_8; and the arsenic molecule, As_4.

Diatomic covalent molecules containing unlike elements are given a similar designation. The formula for hydrogen chloride is HCl; for iodine monochloride, ICl; and for hydrogen iodide, HI. The more electropositive element is always designated first in the formula.

For polyatomic covalent molecules containing unlike elements, subscripts designate the number of atoms of each element that are present in the molecule. Examples are arsine, AsH_3; ammonia, NH_3; and water, H_2O. Again, the more electropositive element is placed first in the formula.

Ionic inorganic compounds are designated by a similar notation. The positive ion is given first in the formula, followed by the negative ion; subscripts denote the number of ions of each element present in the compound. The formulas for several common compounds are sodium chloride, NaCl; ammonium nitrate, NH_4NO_3; and aluminum sulfate, $Al_2(SO_4)_3$.

More complex inorganic compounds are designated in a similar manner. The positive ion is given first, but may contain attached or coordinated groups, and this is followed by the negative ion. Examples are hexammine-cobalt(III) chloride, $[Co(NH_3)_6]Cl_3$; and potassium trioxalatoferrate(III), $K_3[Fe(C_2O4)_3]$. Hydrates of inorganic compounds, such as copper(II) sulfate pentahydrate, are designated by the formula of the compound followed by the formula for water, the number of water molecules being designated by a prefix. Thus the symbol for the last compound is $CuSO_4 \cdot 5H_2O$.

Organic compounds. Because there are many more organic than inorganic compounds, the designation or notation for the organic group becomes complex. Many different types of organic compounds are known; in the case of hydrocarbons, there are aromatic and aliphatic, saturated and unsaturated, cyclic and polycyclic, and so on. The system of notation must distinguish between the various hydrocarbons themselves as well as setting this group of compounds apart from others such as alcohols, ethers, amines, esters, and phenols. *See* CHEMISTRY; COORDINATION COMPLEXES; INORGANIC CHEMISTRY; ORGANIC CHEMISTRY. [W.W.We.]

Chemical thermodynamics

Chemical thermodynamics The application of thermodynamic principles to systems involving physical and chemical transformations in order to (1) develop quantitative relationships among the identifiable forms of energy and their conjugate variables, (2) establish the criteria for spontaneous change, for equilibrium, and for thermodynamic stability, and (3) provide the macroscopic base for the statistical-mechanical bridge to atomic and molecular properties. The thermodynamic principles applied are the conservation of energy as embodied in the first law of thermodynamics, the principle of internal entropy production as embodied in the second law of thermodynamics, and the principle of absolute entropy and its statistical thermodynamic formulation as embodied in the third law of thermodynamics.

The basic goal of thermodynamics is to provide a description of a system of interest in order to investigate the nature and extent of changes in the state of that system as it undergoes spontaneous change toward equilibrium and interacts with its surroundings. This goal implicitly carries with it the concept that there are measurable properties of the system which can be used to adequately describe the state of the system and that the system is enclosed by a boundary or wall which separates the system and its surroundings. Properties that define the state of the system can be classified as extensive and intensive properties. Extensive properties are dependent upon the mass of the system, whereas intensive properties are not. Typical extensive properties are the energy, volume, and numbers of moles of each component in the system, while typical intensive properties are temperature, pressure, density, and the mole fractions or concentrations of the components.

The concept of a boundary enclosing the system and separating it from the surroundings requires specification of the nature of the boundary and of any constraints the boundary places upon the interaction of the system and its surroundings. Boundaries that restrain a system to a particular value of an extensive property are said to be restrictive with respect to that property. A boundary which restrains the system to a given volume is a fixed wall. A boundary which is restrictive to one component of a system but not to the other components is a semipermeable wall or membrane. A system whose boundaries are restrictive to energy and to mass or moles of components is said to be an isolated system. A system whose boundaries are restrictive only to mass or moles of components is a closed system, whereas an open system has nonrestrictive walls and hence can exchange energy, volume, and mass with its surroundings. Boundaries can be restrictive with respect to specific forms of energy, and two important types are those restrictive to thermal energy but not work (adiabatic walls) and those restrictive to work but not thermal energy (diathermal walls).

Changes in the state of the system can result from processes taking place within the system and from processes involving exchange of mass or energy with the surroundings. After a process is carried out, if it is possible to restore both the system and the surroundings to their original states, the process is said to be reversible; otherwise the process is irreversible. All naturally occurring spontaneous processes are irreversible. The first law defines the internal energy as a state function or property of the state of a system, and restricts the system and its surroundings to those processes which conserve energy. The second law establishes which of the permissible processes can occur spontaneously.

According to the first law of thermodynamics, the total energy E of a system is the sum of its kinetic energy T, its potential energy V, and its internal energy U, Eq. (1). If a system has

$$E = T + V + U \qquad (1)$$

constant mass and its center of mass is moving with uniform velocity in a uniform potential, then changes in the total energy of the system δE are equal to changes in its internal energy δU. Chemical thermodynamics concentrates on the internal energy of the system, but kinetic and potential energy changes of the system as a whole can be important for chemical systems. The principle of conservation of energy requires that the change in the internal energy of a system be the result of energy transfer between the system and its surroundings. The internal energy U is a function of the set of extensive variables associated with the various forms of internal energy. Each form of internal energy is manifest by the product of an extensive variable and its conjugate intensive variable.

Thermal energy exchange or heat (that form of energy transferred as a result of temperature differences between a system and its surroundings) plays a central role in thermodynamics, and is singled out from the other forms of energy or work. This is expressed by Eq. (2), where δq is the differential

$$dU = \delta q + \delta w \qquad (2)$$

thermal energy (heat) absorbed by the system from the surroundings and δw is the differential work performed on the system by the surroundings. It is convenient to write Eqs. (3),

$$\delta q = TdS - \delta a \qquad (3a)$$

$$dU = TdS + \delta w - \delta a \qquad (3b)$$

where T is temperature, S is the entropy, and $(-\delta a)$ is a sum of the nonthermal differential work terms. The term δa can be either zero or nonzero. If it is zero, the heat absorbed by the system is equal to TdS. In an adiabatic process δq is zero and $TdS - \delta a$, and hence if δa is nonzero, it must correspond to an internally generated thermal energy. This is frequently referred to as the uncompensated heat of a process, since it does not result from the transfer of heat from the surroundings. *See* HEAT.

The heat capacity of a system is of particular importance in such thermochemical calculations. The heat capacity is the amount of thermal energy that can be absorbed by a system for a unit rise in temperature. This is defined by Eq. (4), where

$$\delta q = C_{\text{process}} dT \qquad (4)$$

C_{process} is the heat capacity of a system for a given type of process. *See* HEAT CAPACITY.

There are many possible and essentially equivalent statements of the second law of thermodynamics. It will suffice to state the empirical result that in all spontaneous processes the uncompensated heat δa in Eqs. (3) is always positive. Equation (3*a*) can be rewritten as Eq. (5), where the term $\delta q/T$ is the

$$dS = \delta q/T + \delta a/T \qquad (5)$$

contribution to the entropy due to heat exchange with the surrounding (d_eS), while $\delta a/T$ is the contribution to the entropy produced as a result of the interconversion of work terms (d_iS). The second law can then be summarized as Eqs. (6), where d_iS greater than zero applies to irreversible pro-

$$dS = d_eS + d_iS \qquad (6a)$$

$$d_iS \geq 0 \qquad (6b)$$

cesses. When $d_iS = 0$, that is, for a reversible process, Eq. (7)

$$dS = \delta q_{\text{rev}}/T \qquad (7)$$

holds. This is the basic equation for establishing the thermodynamic temperature scale based upon the theoretical limits of reversible cycles. The requirement that $d_iS > 0$ for spontaneous processes provides the criteria for examining the specific conditions for spontaneous paths, and the criteria for establishing the equilibrium state of a system. *See* CARNOT CYCLE; TEMPERATURE.

Many chemical systems can be considered closed systems in which a single parameter ξ can be defined as a measure of the extent of the reaction or the degree of achievement of a process. If the reaction proceeds or the process advances spontaneously, entropy must be produced according to the second law and δa must be positive. In terms of the advancement parameter ξ, this uncompensated heat δa can be given by Eq. (8), where $\underline{A}$ is the affinity of the process or reaction. The af-

$$\delta a = \underline{A}\, d\xi = Td_iS \qquad (8)$$

finity is related to internal entropy production by Eq. (9). The

$$\underline{A} = T\, d_iS/d\xi \geq 0 \qquad (9)$$

condition that the entropy production is zero represents equilibrium, and hence $\underline{A} = 0$ is an equivalent condition for equilibrium in a closed system. For spontaneous processes, since the signs of $\underline{A}$ and $d\xi$ must be the same, for positive $\underline{A}$ the process must advance or go in a forward direction in the usual sense of chemical reactions or physical processes, while for negative $\underline{A}$ the process must proceed in the reverse direction.

The affinity of a chemical reaction establishes the spontaneous direction of the reaction, and consequently methods for determining the affinity are important in thermochemical studies. The affinity is simply related to the stoichiometric coefficients of the reaction and the chemical potentials of the reactants and products in the reaction.

Classical equilibrium thermodynamics is primarily concerned with calculations for reversible processes, and deals with irreversibility in terms of inequalities. In the case of irreversible processes in systems slightly removed from equilibrium, the rate of internal entropy production d_iS/dt is related to the fluxes J_i associated with thermal, concentration, or other differences in intensive parameters or potentials X_i. This entropy production is then given by Eq. (10). The fluxes include heat

$$d_iS/dt = \sum_i J_iX_i \geq 0 \qquad (10)$$

conduction, diffusion, electric conduction, and other direct effects.

In addition, a flux of one type may be coupled to a potential difference of another type. For example, a thermal gradient can result in a mass flux (thermal diffusion), or a concentration gradient in any energy flux. Thermal conductivity, thermoosmosis, and thermoelectric effects are all coupled effects.

Far removed from equilibrium, thermodynamics must be formulated somewhat differently and more cautiously. The interplay of thermodynamic stability and kinetics can give rise to macroscopic structures with both temporal and spatial coherence called dissipative structures. Much theoretical effort is being directed to these studies because of their apparent relevance to biological structures, but it is still too early to assess how far-reaching these theories will be in the future. *See* THERMODYNAMIC PRINCIPLES; THERMODYNAMIC PROCESSES. [R.A.Pi.]

Chemiluminescence The type of luminescence wherein a chemical reaction supplies the energy responsible for the emission of light (ultraviolet, visible, or infrared) in excess of that of a blackbody (thermal radiation) at the same temperature and within the same spectral range. Below 930°F (500°C), the emission of any light during a chemical reaction is a chemiluminescence. The blue inner cone of a bunsen burner or the Coleman gas lamp are examples. *See* BLACKBODY.

Many chemical reactions generate energy. Usually this exothermicity appears as heat; whereas, for a visible chemiluminescence to occur, one of the reaction products must be generated in an excited electronic state from which it can undergo deactivation by emission of a photon. Hence a chemiluminescent reaction can be regarded as the reverse of a photochemical reaction.

Only very exothermic, or "exergonic," chemical processes can be expected to be chemiluminescent. Partly for this reason, most familiar examples of chemiluminescence involve oxygen and oxidation processes; the most efficient examples of these are the enzyme-mediated bioluminescences. *See* LUMINESCENCE. [T.Wi.]

Chemiosmosis The coupling of metabolic and light energy to the performance of transmembrane work through the intermediary of electroosmotic gradients. Processes include synthesis of adenosine triphosphate (ATP) by oxidative phosphorylation or by photosynthesis, production of heat, accumulation of small molecules by active transport, movement of bacterial flagella, uptake of deoxyribonucleic acid (DNA) during bacterial conjugation, genetic transformation and bacteriophage infection, and insertion or secretion of proteins into or through membranes.

Mitochondria are the powerhouses of the eukaryotic cell and the site of synthesis of ATP by oxidative phosphorylation. In the oxidation portion of ATP synthesis, reductants, such as reduced nicotinamide adenine dinucleotide (NADH) and succinate, are generated during metabolism of carbohydrates, lipids, and protein. These compounds are oxidized through the series of redox reactions performed by membrane-bound complexes, called electron transport or respiratory chains. *See* ADENOSINE TRIPHOSPHATE (ATP); MITOCHONDRIA.

Bacteria do not contain mitochondria, but many of the functions of the mitochondrial membrane are carried out by the bacterial cytoplasmic membrane. Many bacteria also use respiratory chains. This resemblance to mitochondria is more than chance. The evidence, although mostly circumstantial, suggests that mitochondria, chloroplasts, and perhaps other eukaryotic organelles were originally free-living bacteria. These bacteria and larger proto-eukaryotic cells became mutually symbiotic, so that neither was complete or viable without the other. The animal and plant kingdoms arose from these endosymbiotic events.

Photosynthesis is the conversion of light energy into chemical energy. Overall photosynthetic bacteria and the chloroplasts of eukaryotic plants capture sunlight or other light and use that energy to generate both ATP and a reductant for use in biosynthesis. The mechanism of photophosphorylation, that is, the use of light energy to drive the phosphorylation of adenosine diphosphate (ADP) to ATP, resembles that of oxidative phosphorylation. *See* PHOTOSYNTHESIS.

Oxidative phosphorylation and photophosphorylation are but specialized examples of chemiosmotic energy coupling. Among the forms of useful energy are chemical energy, such as that derived from fossil fuels, and light energy in the case of solar cells. Electricity is transmitted to motors, which couple electrical energy to the performance of work. Bacterial cells, mitochondria, and chloroplasts have protonic generators and protonic motors. Respiratory and photosynthetic electron transport chains are generators of proton currents' proton motive forces, which then drive the various motors of the cell or organelle. When the H^+-translocating ATPase is "plugged in," the proton current drives phosphorylation. There are other motors present in the cell. Most membranes contain specific transport systems for small molecules, such as ions, sugars, and amino acids. Many of these transport systems are protonic; that is, they use the energy of the proton motive force to drive the accumulation or extrusion of their substrate. [B.Ro.]

Chemistry The study of the properties, composition, and structure of matter, the changes in structure and composition which matter undergoes, and the accompanying energy changes. The objective of the chemist is to aid in the interpretation of the universe. Much progress has been made toward meeting this objective because not only has the structure and composition of many of the materials on the Earth been elucidated, but also those of the planets, the satellites, the stars, and the materials of interstellar space.

Matter can be classified in a number of different ways. One method is in terms of the physical state—solid, liquid, and gas; but probably the most useful method is in terms of composition—elements, compounds, and mixtures. An element is a substance which cannot be broken down to simpler substances by chemical reactions. It is also defined as a substance made of one kind of building block (atom) only. There are only a few more than 100 elements known in the entire universe. They can be classified into families or groups; and this classification is called the periodic table. *See* ELEMENTS; PERIODIC TABLE.

The early foundations of chemistry were based on two fundamental laws: the law of conservation of matter and energy and the law of definite composition. The establishment of these laws required much careful research on the quantitative relationship of reactants and products in chemical processes and also of the ratio of weights of the elements on compounds. After the law of conservation of matter had been established, much research was done on the combining weights of elements. These data not only led to the law of definite composition (the composition of a pure substance is always perfectly definite), but also to the atomic theory, to a system of atomic weights, and to the methods of determining formulas of compounds.

To explain several phenomena, including the law of definite composition, which resulted from the observation that elements always combine in the same ratio by weight to form a given compound, John Dalton proposed that matter is made of particles called atoms, that reactions must take place between atoms or groups of atoms, and that atoms of the same element are all alike, but that they differ from atoms of another element. These assumptions are the basis of the accepted modern atomic theory. They are also most useful in explaining another law of chemical combination, called the law of multiple proportions, which states that when two or more elements combine in more than one ratio, they do so in the ratio of small whole numbers.

Atoms are made up of three fundamental particles: protons and neutrons at the heart or center of the atom (the nucleus), and electrons surrounding the nucleus. All the atoms of the same element have the same atomic number (positive charge on the nucleus). Elements whose atoms have the same number of electrons in the outer shell have similar properties; this accounts for the similar properties of the elements in each family of the periodic table. The different atoms tend to react with each other by gaining or losing electrons or sharing pairs of electrons. If the atoms combine by sharing pairs of electrons, molecules are formed. If they combine by gaining and losing electrons, ions are formed. Ions carry positive or negative charges. *See* ATOMIC STRUCTURE AND SPECTRA; MOLECULAR STRUCTURE AND SPECTRA.

Work done in the nucleus of atoms has indicated they contain a series of energy levels similar to the electron levels on the outside of the atom. Work is ongoing on the forces in the nucleus and the explanation of the many fleeting, short-lived particles that have been identified in nuclear reactions. *See* ELEMENTARY PARTICLE.

Most of the elements are made of atoms of different atomic weights: such atoms are called isotopes. More than 1000 isotopes, both stable and radioactive, have been identified. Most of the radioactive isotopes of the elements have been produced synthetically.

Compounds are chemical combinations of elements. They are made of molecules or ions. Several million compounds have been synthesized or have been isolated and identified from products found in plant and animal life or in the minerals of the Earth. The origin of the compounds in plant tissue is the photosynthesis process which takes place in the green leaves. Animals then use the plant tissue to build animal tissue. *See* PHOTOSYNTHESIS.

The six types of chemical bonds that hold atoms to each other are covalent, polar covalent, coordinate covalent, ionic, metallic, and hydrogen bonds. The covalent bonds are formed by sharing pairs of electrons which spin in opposite directions and hence form a strong bond between the atoms, Examples of covalency are H_2, H_2O, NH_3, HCl, and CH_4. The covalent bond is polar if the pair of electrons is located closer to one atom than the other as the result of a difference in attraction for electrons. This gives one end of the molecule a slight positive (+) charge and the other a slight negative (−) charge. The water molecule contains polar covalent bonds; as a result, it has good solvent action for electrovalent compounds. If both electrons are supplied by one atom, the bond is coordinate covalent. This type of bond is formed in acid-base reactions and in the formation of complex ions, for example, $Cu(NH_3)_4^{2+}$.

The ionic bond is formed between ions. The metallic bond is present in all metals and is believed to result from the "sea" of valence electrons free to move through the metal lattice. The hydrogen bond is the result of the polar covalent character of a covalent bond of hydrogen with another element, such as fluorine, oxygen, or sulfur. The boiling point of water is much higher than predicted because the water molecules are polymerized by the formation of hydrogen bonds. It is also important in bonding large organic molecules to each other if they have many OH groups along a chain of carbon atoms. This is particularly true of many large molecules found in plant and animal tissue. *See* CHEMICAL BONDING.

Compounds are sometimes classified as electrolytes, those which dissociate into ions in solution, and as nonelectrolytes, those which do not. Electrolytes can be further classified as acids, bases, and salts. In water solution, acids have a sour taste, change the colors of indicators, and produce H^+ ions in solution. Bases taste bitter or brackish, change the colors of the indicators, and produce OH^- ions in water. A salt is a product of the reaction of an acid with a base. Generally, an acid is an ion or a molecule that can accept a pair of electrons, that is, H^+ and $AlCl_3$. *See* ACID AND BASE.

Abbreviations or symbols are used to designate elements. The symbol of oxygen is O, hydrogen H, helium He, copper Cu, sodium Na, radium Ra, plutonium Pu, and so on. Groups of symbols called formulas are used to designate compounds. The formula for water is H_2O, for carbon dioxide CO_2, for sulfuric acid H_2SO_4 for grain alcohol C_2H_5OH. These symbols and formulas are used as chemists' shorthand to indicate chemical reactions. For example,

Statement:

Water decomposes to form hydrogen and oxygen.

Word equation:

Water → hydrogen + oxygen

Symbol equation:

$$2H_2O \rightarrow 2H_2 + O_2$$

See CHEMICAL SYMBOLS AND FORMULAS.

The total mass of the neutrons and protons in the nucleus along with the electrons gives the mass of the atom. Because the atoms are very tiny, their masses likewise are very small. The weight of an atom of carbon, for example, is 2.0×10^{-23} (or 0.000 000 000 000 000 000 000 020) g. Because this value is so small, a unit of weight known as the atomic weight unit (awu) is often used to express the weights of atoms. It is defined as 1/12 the weight of a carbon atom. Carbon therefore weighs 12 awu and is the chemists' standard of reference for atomic weights. *See* ATOMIC MASS UNIT; ATOMIC WEIGHT; RELATIVE ATOMIC MASS.

The field of chemistry is a very large one. Some of the better-known fields are inorganic chemistry, organic chemistry, physical chemistry, analytical chemistry, biological chemistry, pharmaceutical chemistry, nuclear chemistry, industrial chemistry, colloidal chemistry, and electrochemistry.

The chemical manufacturing industry is the major industry in the United States in terms of value of production. Chemists perform a number of different and important jobs. In the research laboratory, the chemists work on fundamental research, such as synthesizing new compounds and studying the properties of others. The importance of fundamental research is recognized when it is realized that half of the jobs of the nation can be traced to the research laboratory. In the analytical laboratory, chemists develop methods of analysis of elements and of compounds. In industrial chemistry, one of the important jobs is the synthesis or purification of chemicals on a large scale for production purposes. *See* ANALYTICAL CHEMISTRY;

BIOCHEMISTRY: CHEMICAL ENGINEERING; INORGANIC CHEMISTRY; ORGANIC CHEMISTRY; PHYSICAL CHEMISTRY. [A.B.G.]

Chemolithotrophic bacteria Those bacteria capable of generating metabolically useful energy, that is, adenosinetriphosphate (ATP), by the oxidation of inorganic compounds. Molecular hydrogen, elemental sulfur, and salts of NH_4^+, NO_2^-, Fe^{2+}, Mn^{2+}, S^{2-}, and $S_2O_3^{2-}$ can serve as exclusive energy sources for the growth of specific bacteria.

The respiration of NH_4^+, NO_2^-, Fe^{2+} and Mn^{2+} occurs only aerobically, with oxygen as the terminal electron acceptor. Both aerobic and anaerobic respiration of H_2 and reduced sulfur compounds occur. Nitrate, SO_4^{2+} and CO_2 serve as electron acceptors for anaerobic hydrogen utilization, and NO_3^- for anaerobic utilization of the sulfur compounds. The intermediates in the oxidation of NH_4^+ and reduced sulfur compounds, and the components of the electron transport systems, in these oxidation processes are still under debate.

Chemolithotrophic bacteria may be either autotrophic or heterotrophic. There are two classes of autotrophic bacteria: obligate autotrophs growing only in a completely mineral medium (autotrophy) by coupling energy from ammonia oxidation, with the assimilation of CO_2 as the only carbon source; and facultative autotrophs, additionally capable of growing with organic energy and carbon sources. Assimilation of CO_2 is achieved primarily by the reductive pentose (Calvin) cycle, as in green plants. However, chemolithotrophy need not be coupled to CO_2 assimilation. The best-known examples of chemolithotrophic heterotrophs are the desulfovibrios and related sulfate-reducing bacteria. These organisms can derive energy from the oxidation of molecular hydrogen and couple this energy to the assimilation of carbon from organic compounds. They cannot grow autotrophically with H_2 and CO_2.

Chemolithotrophic bacteria are significant geochemical agents. The bacteriological oxidations of ammonia to nitrite and nitrite to nitrate are key steps in the completion of the nitrogen cycle, upon which continued organic productivity depends. The biological oxidations of reduced sulfur compounds to sulfate are key processes in the sulfur cycle. The bacteria oxidizing Fe^{2+} are believed responsible for major deposits of iron oxide ores. The sulfate-reducing bacteria are responsible for vast deposits of iron sulfides and possibly other sulfide minerals. *See* NITROGEN CYCLE.

A major chemolithotrophic process is that of methane production, which occurs extensively in anaerobic marine and fresh-water sediments as well as in regulated human activities such as sewage treatment. Methane production occurs primarily, if not exclusively, by the anaerobic respiration of H_2 with CO_2 as the electron acceptor. [S.C.R.]

Chemoreception The response of most organisms to a change in their chemical environment. Lower forms, such as protozoans, react to chemical stimuli as a consequence of a general property of irritability inherent in all living matter. A similar response to certain obnoxious chemicals is initiated by the small fine nerve endings commonly found in the cornea and mucous membranes of mammals. Such responses are associated with a common chemical sense.

Marine and aquatic invertebrates are highly chemosensitive. Higher invertebrates have discrete chemosensory organs. Small, nonvolatile organic molecules such as amines and amino acids are adequate stimuli for many invertebrate chemoreceptors. Mixtures of such substances interact synergistically and antagonistically to elicit feeding behavior in nature. Different combinations of chemicals are used to recognize different food types. Chemical stimuli also control complex behaviors such as mate recognition and host location between symbionts. [B.W.A.]

The chemical senses play an important role in the nutrition of all forms of animals. This importance is emphasized by the fact that specialized chemical sense organs evolved much earlier than the receptors of the other major senses. Specialized taste receptors are responsive to water-soluble chemicals and are important in the monitoring of all food entering the oral cavity of many animals. Olfactory receptors respond to minute concentrations of odors. Odors are related to many forms of animal behavior. *See* CHEMICAL SENSES.

Specialized chemoreceptors are found within many mammals, and help maintain the internal environment of the organism. The respiratory center of the medulla responds to carbon dioxide of the blood and controls respiration and the oxygen content of the blood. Changes in the bodily constituents of an animal due to a disease may change the taste sensitivity. An animal may tend to compensate for its deficiency to a particular chemical by better detection and greater intake. Thus chemoreception plays an important role in all organisms, particularly in those functions involved with survival and reproduction. *See* CHEMICAL ECOLOGY; OLFACTION; SENSE ORGAN; TASTE. [L.M.B.]

Chemostat A principle for the continuous culture of a bacterial population in a steady state, achieved by restricting the concentration of a nutrilite in the medium. *See* CULTURE.

An apparatus (see illustration) for application of this principle

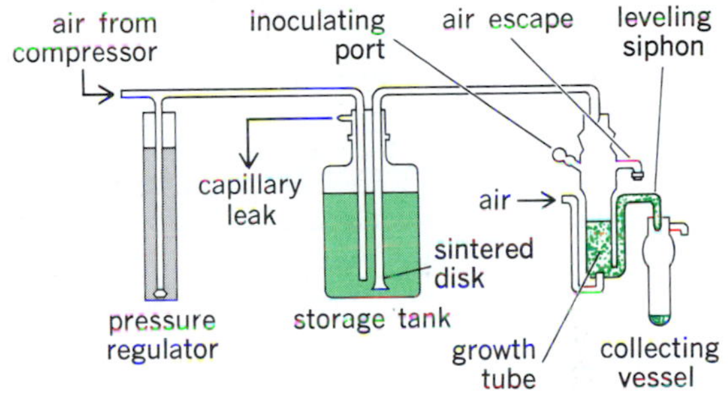

Chemostat apparatus. *(After A. Novick and L. Szilard, Description of the chemostat, Science, 112:715–716, 1950)*

includes a growth tube, provided with an inlet for the entrance of nutrient liquid and an overflow for the exit of bacterial suspension at a rate which keeps constant the volume of bacterial suspension in the tube. Typically air is bubbled through the growth tube for aeration and mixing. Nutrient liquid is fed to the growth tube at a constant rate, generally by an infusion pump.

The entering nutrient contains an excess of all the factors needed for the growth of the bacteria except one, the controlling growth factor, which is present at a selected concentration. Initially rate of growth of the bacteria exceeds rate of removal. The resulting rise in bacterial density (number of organisms per milliliter) causes a fall in the concentration of the controlling growth factor in the tube. Eventually the concentration falls so low as to cause a reduction in the specific bacterial growth rate. When the specific growth rate falls to the level of the dilution rate, a steady state is achieved in which the bacterial density remains constant and the concentration of the controlling growth factor in the growth tube has become independent of its concentration in the incoming nutrient. Because the concentration of all chemical substances in the growth tube remains constant, the bacterial population will grow for long periods of time in a constant physiological state.

The chemostat has been applied to problems in bacterial metabolism, mutations, evolution, and the kinetics of enzyme formation. It is possible to adapt the chemostat to the production of bacterial products. The chemostat method of continuous culture can be of great use in such fields of study as microbiology, physiology, genetics, and biochemistry. [A.No.]

Chemostratigraphy A subdiscipline of stratigraphy and geochemistry that involves correlation and dating of marine sediments and sedimentary rocks through the use of trace-element concentrations, molecular fossils, and certain isotopic ratios that can be measured on components of the rocks. The isotopes used in chemostratigraphy can be divided into three classes: radiogenic (strontium, neodymium, osmium), radioactive (radiocarbon, uranium, thorium, lead), and stable (oxygen, carbon, sulfur). Trace-element concentrations (that is, metals such as nickel, copper, molybdenum, and vanadium) and certain organic molecules (called biological markers or biomarkers) are also employed in chemostratigraphy. *See* DATING METHODS; ROCK AGE DETERMINATION.

Radiogenic isotopes are formed by the radioactive decay of a parent isotope to a stable daughter isotope. The application of these isotopes in stratigraphy is based on natural cycles of the isotopic composition of elements dissolved in ocean water, cycles which are recorded in the sedimentary rocks. *See* ISOTOPE; RADIOISOTOPE.

The elements hydrogen, carbon, nitrogen, oxygen, and sulfur owe their isotopic distributions to physical and biological processes that discriminate between the isotopes because of their different atomic mass. The use of these isotopes in stratigraphy is also facilitated by cycles of the isotopic composition of seawater, but the isotopic ratios in marine minerals are also dependent on water temperature and the mineral-forming processes. *See* SEAWATER.

Certain organic molecules that can be linked with a particular source (called biomarkers) have become useful in stratigraphy. The sedimentary distributions of biomarkers reflect the biological sources and inputs of organic matter (such as that from algae, bacteria, and vascular higher plants), and the depositional environment.

Certain trace metals, such as nickel, copper, vanadium, magnesium, iron, uranium, and molybdenum, are concentrated in organic-rich sediments in proportion to the amount of organic carbon. Although the processes controlling their enrichment are complex, they generally form in an oxygen-poor environment (such as the Black Sea) or at the time of global oceanic anoxic events, during which entire ocean basins become oxygen poor, resulting in the death of many organisms; hence large amounts of organic carbon are preserved in marine sediments. The trace-metal composition of individual stratigraphic units may be used as a stratigraphic marker, or "fingerprint." *See* GEOCHEMISTRY; MARINE SEDIMENTS; STRATIGRAPHY; URANIUM. [B.L.I.; D.J.DeP.]

Chemotaxonomy The use of plant chemistry in taxonomic studies. Plants produce many types of natural products in varying amounts, and quite often the biosynthetic pathways responsible for these compounds also differ from one taxonomic group to another. The distribution of these compounds and their biosynthetic pathways correspond well with existing taxonomic arrangements of plants based on more traditional criteria such as morphology. In some cases, chemical data have contradicted existing hypotheses which necessitates a reexamination of the problem or, more positively, have provided decisive information in situations where other forms of data are insufficiently discriminatory.

Modern chemotaxonomists often divide these natural plant products into two major classes: (1) micromolecules, that is, those compounds with a molecular weight of 1000 or less, such as alkaloids, terpenoids, amino acids, fatty acids, flavonoid pigments and other phenolic compounds, mustard oils, and simple carbohydrates; and (2) macromolecules, that is, those compounds (often polymers) with a molecular weight over 1000, including complex polysaccharides, proteins, and the basis of life itself, deoxyribonucleic acid (DNA).

In studies of existing taxonomic systems, a knowledge of plant chemistry can be a powerful adjunct tool, and in many cases can indicate possible evolutionary relationships between plants more precisely than other evidence alone can. Indeed, the potential of chemistry in plant taxonomy has finally been recognized, resulting in a close synergism between chemist and taxonomist, and in the establishment of the chemotaxonomist, a person trained in both fields. *See* PLANT TAXONOMY. [D.E.G.]

Chemotherapy A term that implies treatment of the sick with chemical substances. It is usually used, however, in a more restricted sense to refer to the treatment, with chemical agents, of diseases caused by parasitic organisms and by cancer cells. This is in essence the use of the term intended by the father of chemotherapy, the German physician Paul Ehrlich. The substances used in chemotherapy include those synthesized by chemists, such as sulfadiazine and chloroguanide, and those which occur naturally, such as emetine and penicillin. All of these substances are called chemotherapeutic drugs, distinguishing them from substances such as antibodies which are produced by the host as its defense against an invading parasite, protein, or other antigenic substance.

The essence of Ehrlich's concept of chemotherapy is specificity of action. He considered a chemotherapeutic compound a chemical agent taken up by parasites in quantities sufficient for lethal action while being tolerated by the organism harboring the parasite. The closest to Ehrlich's ideal came in the remarkable antibiotic penicillin. This drug has a very low inherent toxicity for mammals, except in the sensitized individual, in whom it may produce a wide range of allergic manifestations. Penicillin has the amazing property of preventing the synthesis of the outer wall of certain bacterial cells without influencing the development of mammalian cells. Such a specificity implies that the drug can distinguish subtly between the biochemistry of parasitic and host cells. This ability is decreased in virus infections and in cancer, in which the metabolism of the parasite or neoplasm is intimately associated with the metabolism of the host.

Most of the clinically useful chemotherapeutic agents are not so highly specific that their effects apply only to pathogenic organisms or even to a single kind of microorganism. Thus, these drugs display spectra of activities from the relatively narrow spectrum of penicillin (most specific) to the so-called broad-spectrum antibiotcs such as tetracycline, oxytetracycline, and chlortetracycline. Broad-spectrum antibiotics inhibit the growth of rickettsiae and large viruses as well as numerous types of bacteria. *See* ANTIBIOTIC.

The second great principle of chemotherapy, fixation, led to the formulation of Ehrlich's chemoreceptor, or side-chain, theory. This theory puts forward the concept that the parasitic cell possesses chemical groupings or side chains on its surface, for which the chemotherapeutic drug has an affinity. Once the drug fixes itself to these side chains, the specific lethal action on the parasite is initiated. Such a concept is used today, except that Ehrlich's side chains are now called receptors. It has been a useful working hypothesis which, with a certain extension into the area known as competitive inhibition or metabolite antagonism, has led to the development of a number of drugs useful in malaria, leukemia, and various lymphomata.

The third basic principle proposed by Ehrlich is that a chemotherapeutic drug, when injected into an infected host, must be distributed in effective concentration to the site of the infection. This criterion must be met before fixation to the parasite and the killing action can take place. One of the great advantages of the modern chemotherapeutic approach is that it is possible to attack an occult (hidden) infection by simply introducing the chemotherapeutic agent either directly into the bloodstream or indirectly via the oral route. However, problems emerge of accessibility of the drugs to the site of infection.

Whenever a chemotherapeutic agent is administered to an infected host, a system of interrelationships is established involving three components: the human host, the infecting parasite, and the chemotherapeutic drug. Each of these may act upon, and react to, the other two. The drug itself exerts its lethal action on the parasite, but at the same time produces certain toxic effects upon the infected host, Each of these organisms reacts to the drug in characteristic fashion. The parasite often possesses enzymes which can destroy the drug (such as penicillinase, which destroys penicillin), and in addition a small proportion of the population of parasites may have other mechanisms for resisting the action of the drug. It is these drug-resistant organisms which may, upon continued use of a drug, grow into a drug-resistant population when all susceptible members of the original population have been destroyed by the drug. The host disposes of the drug usually by first metabolically altering it and then excreting it, chiefly with the kidneys. Finally, the host reacts to the presence of the parasite independently of the presence of the drug by marshaling its cellular and immunochemical defense forces to rid itself of the parasite. It is probably the combined action of these forces and the drug which ultimately eradicates the infection. One of the great obstacles in the path of effective cancer chemotherapy may be related to the poorly developed host defense mechanisms against invasion by the malignancy. [N.J.G./E.Gru.]

Cherry The two principal cherries of commerce are the sweet cherry (*Prunus avium*) and the sour cherry (*P. cerasus*). Both are of ancient origin and seem to have come from the region between the Black and Caspian seas. Cherries of minor importance are the dwarf or western sand cherries (*P. besseyi*) of the plains region of North America; the Duke cherries, which are supposedly natural hybrids between the sweet and sour cherry; and the Padus cherries, which bear their small fruits in long clusters or racemes rather than in short fascicles.

Sweet cherries may be divided into two groups: firm-fleshed types known as Bigarreaus, represented by the Napoleon (also called the Royal Anne), and soft-fleshed types known as Hearts, represented by the Black Tartarian. Sour cherries may also be divided into two groups, clear-juice or Amarelle types, represented by the Montmorency, and colored-juice or Morello types, represented by the English Morello.

In North America the principal sweet commercial varieties are the Napoleon (white), Bing, Lambert, Van, Schmidt, and Windsor (dark). The principal commercial variety of sour cherry is the Montmorency. *See* FRUIT; ROSALES. [R.P.L.]

Chert A dense, microcrystalline or cryptocrystalline rock composed of free silica, either in the form of chalcedony and microcrystalline quartz or in the form of opaline silica of various kinds. The composition of cherts varies with age. The older cherts consist of microcrystalline quartz and chalcedony. Common nonsilica components of cherts are calcite and dolomite, detrital quartz, and clay minerals. Flint is synonymous with chert. The latter term is preferred for the rock, although the former is still used for artifacts, such as arrowheads. Other names that have been used for chert are silexite, hornstone, and phthanite. *See* CHALCEDONY.

Chert is found in two forms, as nodules in limestones and dolomites, and as bedded formations. The nodules are flattened parallel to the bedding and vary in size from a few inches to several feet in diameter.

Two theories have been suggested for the origin of chert. One is that of replacement origin of originally calcareous materials. The other calls for primary precipitation of silica from the water of the depositional environment. The relationship of volcanic activity to chert deposition is not wholly clear, for there are many deposits with no apparent link to volcanism; but in those deposits where there is a clear association, high dissolved silica concentrations are thought to have come from hydrolysis of volcanic glass. *See* SEDIMENTARY ROCKS. [R.Si.]

Chestnut Any of several species of large, deciduous, nut-bearing trees native to the Northen Hemisphere. The nuts, usually three in a cluster, are enclosed in a spiny involucre or bur which opens at maturity. Each nut consists of a tough but flexible shell, the ovary wall, enclosing the cotyledons which are covered with the brown seed coats or integuments. Unlike most nuts, the stored food in the cotyledons is mostly starch instead of fat and protein.

The American chestnut (*Castanea dentata*) once was one of the most valuable forest trees in the northeastern United States. In 1904 the chestnut blight, caused by a fungus (*Endothia parasitica*), was recognized as present in the United States. It spread rapidly through the natural range of the chestnut and practically destroyed the species. The European or Italian chestnut (*C. sativa*) is a valuable tree in the Mediterranean region, particularly in the mountain villages of Italy. The nuts, which are larger than those of the American species, are an important export. However, chestnut production in Europe is declining because of the introduction and spread of the chestnut blight disease. The Chinese chestnut (*C. mollissima*) and the Japanese chestnut (*C. crenata*) are both resistant to the blight; they are being used in breeding programs in the United States to secure blight-resistant chestnut types that might replace the American chestnut. Selected varieties of the Chinese chestnut with palatable nuts are being grown as an orchard crop in the milder parts of the eastern United States, but the attempts to produce a good forest tree have not been successful to date.

For the most part, foreign production of chestnuts comes from wild forest trees or trees around farmsteads rather than formally planted orchards. There are some commercial plantings in the United States, but the total production is limited and the many problems of production, propagation, culture, processing, and storage remain troublesome. In the European countries and in parts of China and Japan where the chestnut is native, the nuts are an important staple food source made into flour for bread, roasted, boiled, and used in cookery. In the United States, which imports nuts mainly from Italy, the nuts are boiled and roasted and used in stuffing for poultry and other dishes. *See* FAGALES. [L.H.MacD.]

Chevrel phases A series of ternary molybdenum chalcogenide compounds. They were reported by R. Chevrel, M. Sergent, and J. Prigent in 1971. The compounds have the general formula $M_xMo_6X_8$, where M represents any one of a large number (nearly 40) of metallic elements throughout the periodic table; x has values between 1 and 4, depending on the M element; and X is a chalcogen (sulfur, selenium or tellurium). The Chevrel phases are of great interest, largely because of their striking superconducting properties.

Most of the ternary molybdenum chalcogenides crystallize in a structure in which the unit cell, that is, the repeating unit of the crystal structure, has the overall shape of a rhombohedron with a rhombohedral angle close to 90°. The building blocks of the Chevrel-phase crystal structure are the M elements and Mo_6X_8 molecular units or clusters. Each Mo_6X_8 unit is a slightly deformed cube with X atoms at the corners, and Mo atoms at the face centers. One of these structures, that of $PbMo_6S_8$, is shown in the illustration. *See* CRYSTAL; CRYSTALLOGRAPHY.

Several of the Chevrel-phase compounds have relatively high values of the superconducting transition temperature, T_c, the maximum being about 15 K (−433°F) for $PbMo_6S_8$. The Chevrel-phase $PbMo_6S_8$ has a value of the upper critical mag-

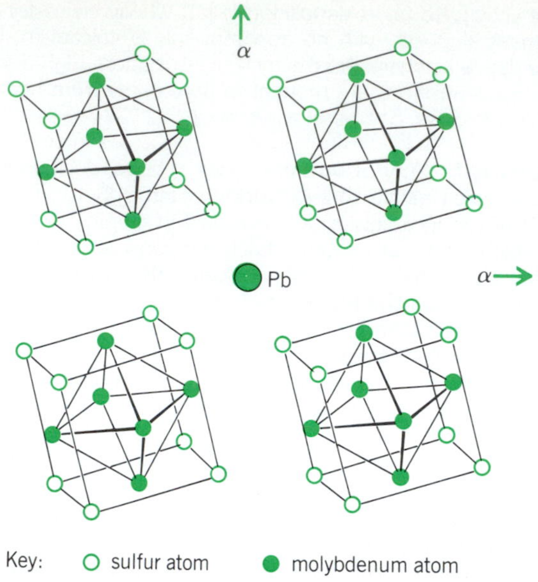

Crystal structure of $PbMo_6S_8$. Each lead atom is surrounded by eight Mo_6S_8 units, the structure of which is shown in the lower right-hand part of the figure. The rhombohedral angle α is indicated.

netic field near absolute zero $H_{c2}(0)$ of about 60 teslas (600 kilogauss), which was the largest value observed prior to the discovery of high-temperature ceramic superconductors in 1986. A number of Chevrel-phase compounds of the form RMo_6X_8, where R is a rare-earth element with a partially filled 4f electron shell and X is S or Se, display magnetic order at low temperatures in addition to superconductivity. *See* SUPERCONDUCTIVITY. [M.B.Ma.]

Chevrotain Any of four species of mammals which constitute the family Tragulidae in the order Artiodactyla. These animals, also known as mouse deer, are the smallest ruminants, growing to a maximum height of 12 in. (30 cm) at the shoulder. The chevrotain lacks horns or antlers. There are two well-developed toes on the feet, and the upper canines of the male are elongate and protrude from the mouth as small tusks. The chevrotain is a shy animal which leads a solitary life except during the breeding season. After a gestation period of 120 days, one or two young are born. The water chevrotain (*Hyemoschus aquaticus*) is found in west-central Africa along the banks of rivers in Sierra Leone through Cameroon to the Congo. The other species are all members of the genus *Tragulus*, and range through the forested areas of Sumatra into Borneo and Java. These are the Indian chevrotain (*T. meminna*), the larger Malay chevrotain (*T. napu*), and the lesser Malay chevrotain (*T. javanicus*). They are differentiated by the pattern of markings (stripes or spots) on their coats. *Tragulus javanicus* has a coat of uniform color.

The Eocene fossil traguloid, *Archaeomeryx*, which was unearthed in Mongolia, shows many general similarities to the modern chevrotains. The main line of evolutionary development of the traguloids occurred in Eurasia. *See* ARTIODACTYLA. [C.B.C.]

Chickenpox and shingles Chickenpox is a mild, highly infectious viral disease characterized by vesicular rash, which occurs chiefly in epidemics among children. It is also known as varicella. Shingles (zoster), a sporadic, severe disease of adults, is characterized by inflammation of certain posterior nerve roots and ganglia, and is accompanied by skin vesicles limited to areas associated with the affected nerves. The viruses of chickenpox and shingles belong to the herpesvirus group. They are so similar morphologically and antigenically that many consider them identical.

In chickenpox, malaise and fever precede the rash, which erupts in successive crops; fever continues after eruption. In shingles, malaise and fever are followed by severe pain and eruption of vesicles in skin areas supplied by sensory nerves associated with the particular dorsal nerve roots and ganglia affected.

Shingles can be the source of epidemics of chickenpox in children, and adults exposed to chickenpox may develop shingles. Infections confer enduring immunity, but shingles may occur in persons who have had chickenpox in the past. This phenomenon may represent latent persistence of virus after chickenpox infection until trauma precipitates development of shingles; or treatment with immunosuppressants, as in organ transplants, may overwhelm the individual's immunity. *See* IMMUNITY. [J.L.Me.]

Chicle A gummy exudate used in the manufacture of chewing gum. It is contained in the bark of a tall evergreen tree, *Achras zapota* (Sapotaceae), a native of Mexico and Central America. The latex is collected and carefully boiled to remove excess moisture. When the water content is reduced to 33%, the chicle is poured off and molded into blocks. The product is an amorphous, pale-pink powder, insoluble in water, and forming a sticky paste when heated. In the manufacture of chewing gum, the chicle is cleaned, filtered, and sterilized, and various flavoring materials and sugar are added. [P.D.St./E.L.C.]

Chicory A perennial herb, *Cichorium intybus* (Compositae), with a long taproot, a coarse branching stem, and a basal rosette of numerous leaves. Although the plant is a native of Europe, it has become a common weed in the United States. It is used as a salad plant or for greens. The roasted root is also used as an adulterant of coffee. *See* CAMPANULALES. [P.D.St./E.L.C.]

Chilopoda An order of the group Myriapoda commonly known as the centipedes. Like all myriapods, centipedes are ground dwellers whose spatial movements are restricted largely by a dependence upon high environmental moisture. Unlike the others, centipedes are evidently exclusively carnivorous and predatory, a way of life reflected in their great agility or fleetness and especially in the extraordinary modification of the first trunk appendages into a pair of usually massive raptorial pincers, the prehensors. Each prehensor contains a poison gland from which venom is conducted through a terminal claw into the victim's body. The poison immobilizes or kills arthropods and even some small vertebrates, but so far as is known it seems only harmlessly painful to human beings.

Excluding the last two or three trunk segments, each segment bears a single pair of functional legs. The genital ducts open at the posterior end of the body. Spiracles open laterally

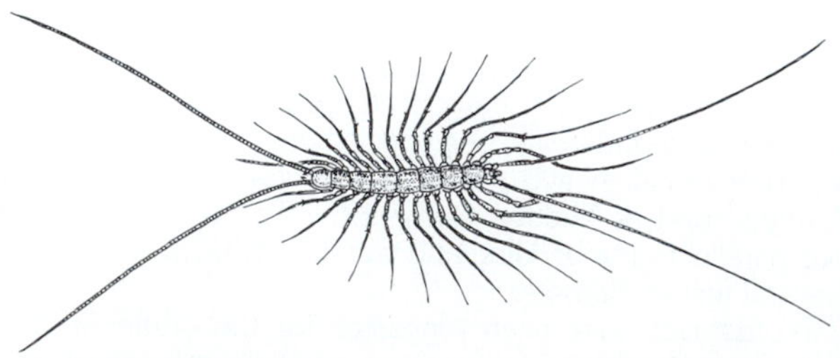

***Scutigera coleoptrata*, the house centipede. (*After R. E. Snodgrass, A Textbook of Arthropod Anatomy, copyright 1952 by Cornell University Press; reprinted with permission*)**

or dorsally; tracheal and circulatory systems are present and often elaborate. The antennae are unbranched; the eyes, when present, are either of the compound type or appear as simple ocelli. The mandibles have a partially free and movable gnathal portion. There are two pairs of maxillae.

In terms of families, genera, and individuals, the Diplopoda, or millipedes, are most abundant in the tropics, but this is not true of centipedes as a group, for whereas certain orders are basically tropical (Scutigeromorpha, Scolopendromorpha), one order, the Lithobiomorpha, is essentially temperate, and the Geophilomorpha is abundant in both zones. In general, centipedes seem less affected by temperature and more sensitive to environmental moisture than are millipedes.

The class is divisible into two subclasses, the Notostigmorphora and the Pleurostigmophora. The notostigmophorous centipedes comprise a single order, Scutigeromorpha. Its members, which are peculiar in embodying primitive as well as highly advanced characters, are signalized by possession of dorsal respiratory openings (stomata), of compound-type eyes, of long flagellate, multisegmental antennae, and of long thin legs with multisegmental tarsi. The order includes the common house centipedes, *Scutigera coleoptrata*, one of which is shown in the illustration.

The Pleurostigmophora, in contrast, have lateral respiratory openings, spiracles. They are reducible into four orders: Lithobiomorpha, Craterostigmomorpha, Scolopendromorpha, and Geophilomorpha. [R.E.Cr.]

Chimaeriformes The only order of the chondrichthyan subclass Holocephali. The chimaeriforms (ratfishes) are a distinctive group of marine bottom-feeding fishes that chiefly inhabit the deeper layers of the Atlantic and Pacific oceans. During the winter months some species move into coastal waters, and a few species may be restricted to this environment.

Like the elasmobranchs, the ratfishes have a cartilaginous skeleton (partly calcified), a urea retention mechanism, and clasper organs in the male for internal fertilization. Unlike the elasmobranchs, however, they have the upper jaw fused to the braincase, a complete hyoid arch that is not involved in jaw suspension, four gill slits opening into a common outer chamber covered by an opercular skin fold, a single branchial opening on each side of the head, no spiracle in the adult, dental plates rather than teeth, a persistent notochord surrounded by calcified centra, and a narrow, tapering tail. *See* ELASMOBRANCHII.

The main line of chimaeriform evolution apparently began with Mississippian to Permian forms called menaspoids. By the beginning of the Jurassic the menaspoids had been replaced by their presumed descendants, the myracanthoids. Modern ratfishes (see illustration) appeared during the Middle Jurassic. They are characterized by a laterally compressed head with a variably developed rostrum, which is frequently exotically shaped, and an ethmoidal canal. The dorsal fin spine is made up almost entirely of lamellar bone. The pectoral fins are dibasal and fairly large, and the caudal fin is tapered to almost whiplike proportions. Placoid scales cover the body, in contrast to the more complicated compound scales of the primitive chimaeriforms.

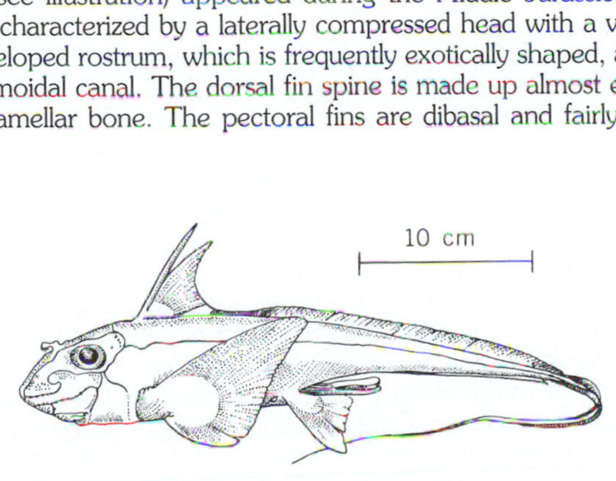

Modern chimaeriform *Chimaera*. (*After H. B. Bigelow and W. C. Schroeder, Fishes of the Western North Atlantic, pt. 2, Sears Foundation for Marine Research, 1953*)

The living ratfishes are divided into three families: the Chimaeridae, the Rhinochimaeridae, and the Callorhinchidae, which can be readily distinguished by the shape of the rostrum. Some 28 species are recognized. Most of them feed on mollusks and crustaceans and occasionally on smaller fishes. *See* CHONDRICHTHYES. [B.S.]

Chimera An individual animal or plant made up of cells derived from more than one zygote or otherwise genetically distinct.

Animals. Although some chimeras do arise naturally, most are produced experimentally, either by mixing cells of very early embryos or by tissue grafting in late embryos or adults. Experimental chimeras have been used to study a number of biological questions, including the origin and fate of cell lineages during embryonic development, immunological self-tolerance, tumor susceptibility, and the nature of malignancy.

Two techniques used to form chimeras by mixing embryo cells are aggregation and injection.

Aggregation chimeras are produced by a technique that involves removing the zonae pellucidae from around 8–16 cell embryos of different strains of mice and pushing the morulae together so that the cells can aggregate. After a short period of laboratory culture, during which the aggregate develops into a single large blastocyst, the embryo is returned to a hormone-primed foster mother. Chimeric offspring are recognized in several ways. If derived from embryos of pigmented and albino strains, they may have stripes of pigmented skin and patches of pigment in the eye. Internal chimerism can be detected by use of chromosomal markers or genetically determined enzyme variants. Chimeras accept skin grafts from the two component strains, but reject grafts from third-party strains.

Injection chimeras are produced by a technique in which a blastocyst of the host mouse strain of mouse embryos is removed from its zona pellucida and held on a suction pipette. Cells of the donor strain are injected through a fine glass needle, either into the blastocoele cavity or into the center of the inner cell mass (the group of cells from which the fetus is derived). After a short period of culture, the blastocyst is returned to a foster mother.

Another kind of cell—the pluripotent stem cell of mouse teratocarcinomas—was found to give rise to normal tissues in adult chimeras after injection into the mouse blastocyst. Teratocarcinomas are tumors consisting of a disorganized mixture of adult and embryonic tissues. They develop spontaneously from germ cells in the gonads of certain mouse strains, or from cells in early embryos transplanted to ectopic sites. All the differentiated tissues in the tumor arise from pluripotent stem cells known as embryonal carcinoma (EC) cells. When embryonal carcinoma cells are injected into a genetically marked host blastocyst, they continue to divide and participate in normal development, and give rise to fully differentiated cells in all tissues of the adult, including skin, muscle, nerve, kidney, and blood. Embryonal carcinoma cells from several sources, including spontaneous and embryo-derived tumors and cultured lines selected to carry specific mutations or even human chromosomes, have contributed to normal chimeras. However, embryonal carcinoma cells from some other sources fail to integrate, but produce teratocarcinomas in the newborn animal or adult. The fact that certain embryonal carcinoma cells give rise to tumors when injected under the skin or into the body cavity, but behave normally in the blastocyst, has been used to support the idea that cancers can develop not only as a result of gene mutations but also as a result of disturbances in environmental factors controlling normal cell differentiation (epigenetic theory of cancer).

Animals that have accepted skin or organ grafts are technically chimeras. Radiation chimeras are produced when an animal is exposed to x-rays, so that blood-forming stem cells in the bone marrow are killed and then replaced by a bone marrow transplant from a genetically different animal. Lymphoid cells in the process of differentiating from stem cells in the donor marrow recognize the recipient as "self " and do not initiate an immune response against the host cells. *See* TRANSPLANTATION BIOLOGY.

Naturally occurring chimeras in humans are not rare and are most easily recognized when some cells are XX and others XY. Such individuals are usually hermaphrodite and probably result from fertilization of the egg by one sperm and the second polar body by another, with both diploid cells then contributing to the embryo (the small polar bodies normally degenerate). Blood chimeras are somewhat more common in animals such as cattle where the blood vessels in placentas of twins fuse, so that blood cells can pass from one developing fetus to the other. [B.Hog.]

Plants. In modern botanical usage a chimera is a plant consisting of two or more genetically distinct kinds of cells. Chimeras can arise either by a mutation in a cell in some part of the plant where cells divide or by bringing together two different plants so that their cells multiply side by side to produce a single individual. They are studied not only because they are interesting freaks or ornamental, but also because they help in the understanding of many of the developmental features of plants that would otherwise be difficult to investigate.

The first type of chimera to be used in this way resulted from grafting. Occasionally a bud forms at the junction of the scion and stock incorporating cells from both, and it sometimes happens that the cells arrange themselves so that shoots derived from the bud will contain cells from both plants forever.

Fig. 1. Variegated *Pelargonium*, a periclinal chimera whose second tunica layer is genetically white and whose corpus is genetically green.

Fig. 2. Variegated *Chlorophytum*, a periclinal chimera whose outer cell layer is genetically green and proliferates at the leaf margins to give green tissue in an otherwise genetically white plant.

Flowering plants have growing points (apical meristems) where the outer cells are arranged in layers parallel to the surface. This periclinal layering is due to the fact that the outer cells divide only anticlinally, that is, by walls perpendicular to the surface of the growing point. In many plants there are two such tunica layers and, because cell divisions are confined to the anticlinal planes, each layer remains discrete from the other and from the underlying nonlayered tissue called the corpus. The epidermis of leaves, stems, and petals is derived from the outer layer of the growing point. *See* APICAL MERISTEM.

With a periclinal chimera it is possible to trace into stems, leaves, and flowers which tissues are derived from each layer in the growing point. For leaves, this can also be done with variegated chimeras where the genetic difference between the cells rests in the plastids resulting from mutation whose effect is to prevent the synthesis of chlorophyll. Tracts of cells whose plastids lack this pigment appear white or yellow. A common form of variegated chimera has leaves with white margins and a green center (Fig. 1). The white margin is derived from the second layer of the tunica, and the green center is derived from inner cells of the growing point. The white leaf tissue overlies the green in the center of the leaf, but does not mask the green color. Chimeras with green leaf margins and white centers are usually due to a genetically green tunica proliferating abnormally at the leaf margin in an otherwise white leaf (Fig. 2).

Since the somatic mutation that initiates chimeras would normally occur in a single cell of a growing point or embryo, it often happens that it is propagated into a tract of mutant cells to form a sector of the plant. If the mutation resulted in a failure to form green pigment, the tract would be seen as a white stripe. Such chimeras are called sectorial, but they are normally unstable because there is no mechanism to isolate the mutant sector and, in the flux that occurs in a meristem of growing and dividing cells, one or other of the two sorts of cells takes over its self-perpetuating layer in the growing point. The sectorial chimera therefore becomes nonchimerical or else a periclinal chimera.

However, in one class of chimera an isolating mechanism can stabilize the sectorial arrangement. This propagates stripes of mutant tissue into the shoot, but because the tunica and corpus are discrete from each other, the plant is not fully sectored and is called a mericlinal chimera. Many chimeras of this type have a single tunica layer; those with green and white stripes in the leaves have the mutant cells in sectors of the corpus. They are always plants with leaves in two ranks, and consequently the lateral growth of the growing point occurs by cell expansion only in the plane connecting alternate leaves. This results in the longitudinal divisions of the corpus cells being confined to planes at right angles to the plane containing the leaves. A mutation in one cell therefore can result in a vertical sheet of mutant cells which, in the case of plastid defect, manifests itself as a white stripe in every future leaf.

The growing points of roots may also become chimerical, but in roots there is no mechanism to isolate genetically different tissues as there is in shoots, and so chimeras are unstable.

Since the general acceptance of the existence of organisms with genetically diverse cells, many cultivated plants have been found to be chimeras. Flecks of color often indicate the chimerical nature of such plants. Color changes in potato tubers occur similarly because the plants are periclinal chimeras. *See* SOMATIC CELL GENETICS. [F.A.L.C.]

Chimney A vertical hollow structure of masonry, steel, or concrete, built to convey gaseous products of combustion from a building. Short, self-supporting steel chimneys are called stacks. A chimney should be high enough to furnish adequate draft and also to discharge the products of combustion without causing excessive local air pollution. For adequate draft, small industrial boilers and home heating systems depend entirely upon the enclosed column of hot gas. A draft is produced because the density of the hot gas inside is less than that of the air outside. The diameter, as well as the height, of a chimney determines the draft. Also, for adequate draft, a chimney in a building should extend at least 3 ft (0.9 m) above the point where it passes through the roof and at least 2 ft (0.6 m) higher than any ridge within 10 ft (3.0 m) of it.

Large power plants usually depend upon forced-draft fans and induced-draft fans to produce the draft necessary for operation, and the chimney is used only for removal of the flue gas. Most modern power plant chimneys are of reinforced concrete, and some reach to the 1000-ft (300-m) level and beyond, thus providing supplementary natural draft. [W.G.B.]

Chimpanzee A Primate indigenous to Central West Africa. There are two species, the pigmy chimpanzee (*Pan paniscus*), also known as the bonobo, and the common chimpanzee (*P. troglodytes*).

These animals lack a tail, and their faces and ears are practically hairless. The brain, the dentition, and the relative length of the limbs resemble those of humans. The forelimbs are about the same length as the legs, and both the thumb and big toe are opposable. Adulthood for these animals is reached at 7 or 8 years and the life-span is about 50 years maximum. One baby is born after a gestation period of about 9 months. The time for menstrual cycle is about 4 weeks. *See* MAMMALIA; PRIMATES.

The chimpanzee is considered to be one of the most intelligent of the apes. It does well in captivity and provides both the biologist and psychologist an opportunity to make comparative studies of this animal and humans. In their native habitat chimpanzees are mainly vegetarians and feed upon fruits, nuts, and other vegetation, but some do eat termites and other animal food. In nature, they have been observed to make crude tools to probe termite nests.

Chimpanzees spend most of their time on the ground but are capable of moving rapidly through the trees. They ascend into the trees at night where they build a platform-type nest on which the entire family huddles together for warmth. Chimpanzees are social animals and move about in family groups consisting of a male and one to several females with the offspring. Mutual grooming is apparently an important aspect in social relationships.

From observations in nature, it has been deduced that two types of communication occur among chimpanzees, gesturing and calls. Intonations vary from low, soft sounds and grunts to screams. These sounds have been associated with definite activities such as greetings, attention to feeding areas, and approach of a new individual or a group or possible predators. [C.B.C.]

Chinchilla The name given to two species of rodents which, together with four species of viscachas, compose the family Chinchillidae. The two species of chinchilla are *Chinchilla brevicaudata* and *C. lanigar*. These animals resemble the squirrel in size and shape and are characterized by long, muscular hindlimbs, with elongate feet bearing four toes, and short forelimbs. Blunt claws occur on the flexible fingers. Chinchillas are gregarious, nocturnally active animals and are found in arid, mountainous regions where they feed principally on vegetation. They often seek shelter in burrows or rock crevices, so that their capture is difficult. The female, which is larger than the male, bears one to six offspring twice each year after a gestation period of 105–111 days.

These animals are native to several areas of South America and are widely bred on farms in North America and Europe for their fur, which is long, fine, and expensive. *See* RODENTIA; VISCACHA. [C.B.C.]

Chinook A mild, dry, extremely turbulent westerly wind on the eastern slopes of the Rocky Mountains and closely adjoining plains. This wind effectively reduces a snow cover by melting or by sublimation. The chinook is a particular instance of a type of wind known as a foehn wind. Foehn winds, initially studied in the Alps, refer to relatively warm, rather dry currents descending the lee slope of any substantial mountain barrier. The dryness is produced by the condensation and precipitation of water from the air during its previous ascent of the windward slope of the mountain range. The warmth is attributable to the previous release of latent heat of condensation in the air mass and to the turbulent mixing of the surface air with the air of greater heat content aloft. *See* WIND; WIND STRESS. [F.S.]

Chipmunk A member of the tribe Marmotini in the rodent family Sciuridae. There are 18 species. The eastern chipmunk (*Tamias striatus*) is found in wooded areas of eastern Canada and the United States. The western species, although quite similar to the eastern form, are included in the separate genus *Eutamias*.

These rodents are intermediate between the squirrels and marmots, having lost the typical bushy tail, tufted ears, and silky fur of the squirrel. They are diurnal animals, active in collecting food such as nuts, grains, and seeds. They fill their large cheek pouches with gathered food to carry it to storage places for the winter.

The animals construct extensive burrows of several chambers at the bottom of a downward sloping entry tunnel, which is about 3 ft (1 m) long. The chambers, used for hoarding food and for nesting, are below the frost line. While chipmunks are not true hibernators, they tend to remain in their underground chambers during the winter months. In the early spring they emerge from the burrows and mating occurs. After a gestation period of 5 weeks six or more young are born, blind and helpless. *See* RODENTIA. [C.B.C.]

Chiroptera An order of mammals (bats) in which the front limbs are modified as wings, thus making the chiropterans the only truly flying mammals. Bats form the second largest order of living mammals (16 families, 171 genera, some 840 species). They range from the limit of trees in the Northern Hemisphere to the southern tips of Africa, New Zealand, and South America, but most species are confined to the tropics. On many oceanic islands they are the only native land mammals.

The wing is formed by webs of skin running from the neck to the wrist (propatagium, or antebrachial membrane), between the greatly elongated second, third, fourth, and fifth fingers (chiropatagium), and from the arm and hand to the body (usually the side) and hindlegs (plagiopatagium). There is also usually a web between the hindlegs (uropatagium, or interfemoral membrane) in which the tail, if present, is usually embedded for at least part of its length (see illustration). The 16 living families may be briefly characterized as follows.

Pteropodidae are in general the most primitive of living bats and are placed in the suborder Megachiroptera, characterized by more primitive ears and shoulder joints. Most still retain a claw on the second digit (absent in all other bats, suborder Microchiroptera), and few have developed an echolocation (sonar) mechanism, found in all Microchiroptera. The teeth, however, are highly modified for eating fruit or nectar. The family, with 38 genera and 149 species, is widely distributed in the tropics and subtropics of the Eastern Hemisphere. While some species are quite small, the family includes the largest of all bats, with wingspreads of up to 5½ ft (1.65 m).

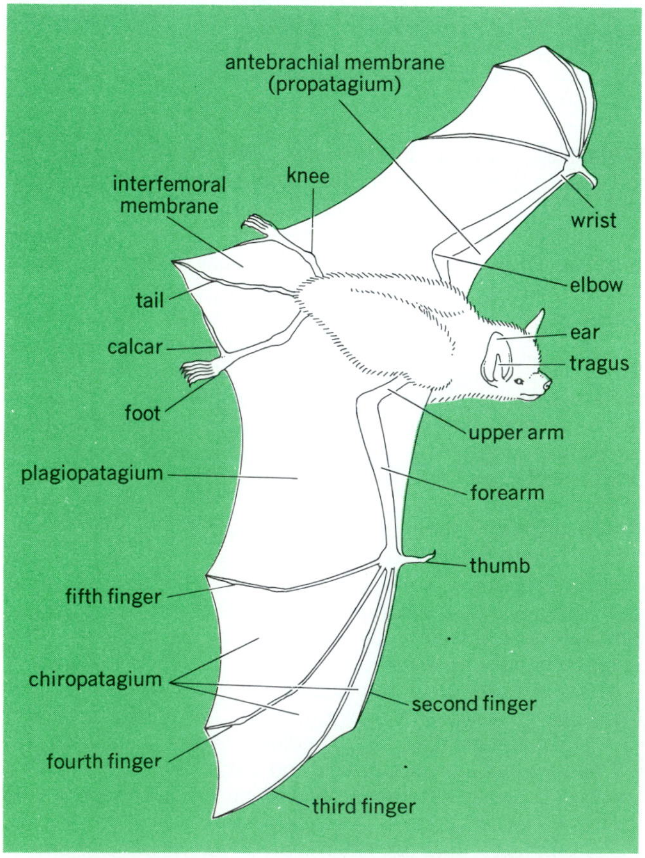

Features of the bat. ***(From R. Peterson, Silently by Night, McGraw-Hill, 1964)***

Rhinopomatidae is an insect-eating family, with one genus and two species, found chiefly in arid regions of northern Africa and southern Asia. These bats are characterized by long wirelike tails and rudimentary nose leaves.

Emballonuridae is an insectivorous family that includes 12 genera with 45 species found in the tropics of both Eastern and Western hemispheres. Like the Pteropodidae, these bats have well-developed bony processes behind the eye sockets. The tail extends only partly across the uropatagium.

Noctilionidae, a tropical American family, is represented by two species in one genus, including a highly specialized fish-eating species. Fish are detected by echolocation and gaffed by the long clawed feet.

Nycteridae are insect eaters, with 1 genus and 12 species found in Africa and southern Asia. These bats have an extensive basin behind the nose, which is partly bridged over by flaps of skin, leaving a mere slit between them.

The four genera of the Megadermatidae (with five species) occur in tropical Africa, southeastern Asia and adjoining islands, and northern Australia. Some species are insect-eating; others feed on small vertebrates, including other bats.

Rhinolophidae are insect-eating bats that are widely distributed in the Eastern Hemisphere. They are remarkable for their extremely complex nose leaves. There are 11 genera represented by 129 species.

Phyllostomatidae includes 120 species (in 50 genera) of tropical and subtropical bats. They possess simple nose leaves and a tremendous variety in structure, reflecting an equal diversity in food habits. Primitively insect-eating, many have become fruit or nectar feeders, and a few are predators on small vertebrates, including other bats.

Desmodontidae are the true vampires, essentially confined to the mainland of tropical America. The teeth and tongue are highly modified for taking of blood. This family includes three genera with three species.

Natalidae (one genus and four species) are tropical American insect-eating species with large funnellike ears.

Furipteridae is an insect-eating family, confined to tropical South America, and represented by two species in two genera. The thumb is reduced in size and largely enclosed in the wing membrane.

Thyropteridae includes two species (in one genus), confined to the tropical American mainland. The large suction disks on the thumbs and hindfeet enable them to roost on the smooth inner surfaces of large rolled-up leaves.

Myzopodidae, with a single species, is confined to Madagascar, eats insects, and has suckers on the thumbs and feet rather like those of the Thyropteridae.

Vespertilionidae, a nearly cosmopolitan family with 279 species in 34 genera, occur almost everywhere that bats occur. Almost all are insect-eating and a few catch fish. In all, the tail is long, extending to the edge of the uropatagium. Most have no special facial modifications. A few have very large ears or small simple nose leaves. In spite of its many species, the family shows little structural diversity. Echolocation, present in all families of bats except the Pteropodidae, is perhaps most highly developed here and is used for catching insects as well as avoiding obstacles.

Mystacinidae, with a single insect-eating species, is confined to New Zealand. It has a short tail, not reaching the edge of the uropatagium, and stout hindlegs and body.

Molossidae is widely distributed in the tropics and subtropics. Feeding on insects, they have long tails which extend beyond their uropatagia. The body and hindlegs are stoutly built. The family includes 81 species in 10 genera.

Bats have a poor fossil record, but have been distinct at least since the Eocene, some 50,000,000 years ago. *See* BAT; MAMMALIA. [K.F.K.]

Chitin A polysaccharide found abundantly in nature. Chitin forms the basis of the hard shells of crustaceans, such as the crab, lobster, and shrimp. The exoskeleton of insects is also chitinous, and the cell walls of certain fungi contain this substance.

Chitin is a long, unbranched molecule consisting entirely of *N*-acetyl-D-glucosamine units linked by ß-1,4 bonds (see illustration). It may be contemplated as cellulose in which the hydroxyl groups on the second carbon are replaced with $NHCOCH_3$ groups. Chitin is considered to be synthesized in nature by an enzyme which is capable of effecting a glycosyl transfer of the *N*-acetyl-D-glucosamine from uridinediphosphate-*N*-acetyl-D-glucosamine to a preformed chitodextrin acceptor, forming the polysaccharide. This stepwise enzymic transfer results in the production of the long chain of ß-*N*-acetyl-D-glucosamine units, which is insoluble chitin. *See* CELLULOSE; OPTICAL ACTIVITY; POLYSACCHARIDE. [W.Z.H.]

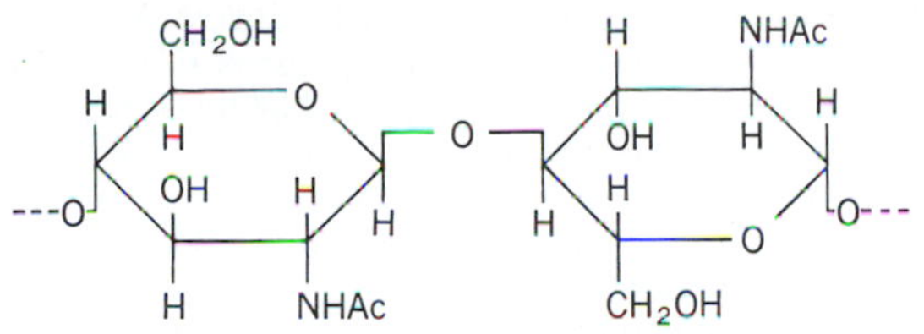

β-*N*-Acetyl-D-glucosamine unit of chitin.

Chiton A member of the class Polyplacophora in the phylum Mollusca. Chitons are also called loricates or coat-of-mail shells. All chitons are marine and, except for a few deep-sea forms, all live in the low intertidal and upper sublittoral, typically on wave-swept rocks. The flattened elliptical body bears eight articulated shell plates dorsally on the mantle and a ventral suctorial foot, features which are clearly adapted for life adhering to hard and uneven surfaces in the lower littoral zone (see illustration). Normally, chitons can resist the strongest surf action but, in the unlikely natural event of a chiton being washed off its rock (or if removed by a human collector), the articulated shell allows it to curl up to protect the soft foot and gills of its underside. Chiton species living on both sides of the Atlantic are relatively small (0.4–1.2 in. or 1–3 cm long) and unobtrusive. Only four genera have living species offshore and in deeper waters (to 20,000 ft or 6000 m), but they may be related to the earliest fossil chitons. *See* MOLLUSCA; POLYPLACOPHORA. [W.D.R.-H.]

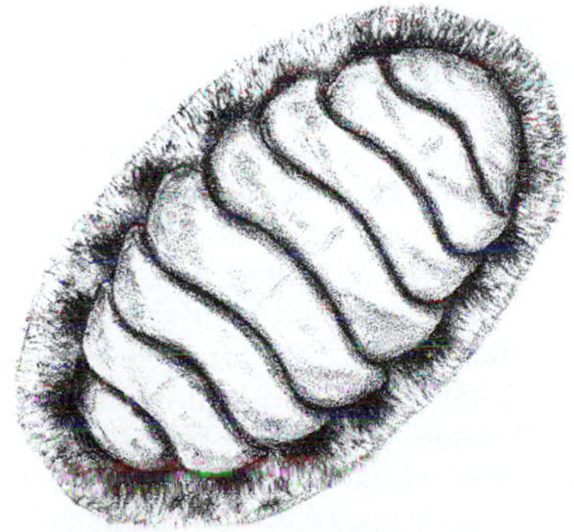

Dorsal view of a mossy chiton (*Mopalia muscosa*) showing the eight articulated plates of the shell and the pallial girdle with calcareous spicules.

Chlamydia A unique genus of bacteria with a growth cycle differing from those of all other microorganisms. They grow only in living cells and cannot be cultured on artificial media. Although capable of synthesizing macromolecules, they have no system for generating energy; the host cell's energy system fuels the chlamydial metabolic processes.

At present three species are recognized within the genus *Chlamydia*: *C. trachomatis*, which is susceptible to sulfonamides and produces inclusions that stain with iodine, because they contain glycogen; *C. psittaci*, which is sulfa-resistant and has inclusions that do not stain with iodine; and *C. pneumoniae*, which must be identified by special tests. *See* CHLAMYDIAL DISEASES. [J.S.]

Chlamydial diseases Diseases caused by *Chlamydia trachomatis*, one of the most common human pathogens. There are 15 serotypes of *C. trachomatis*; four are associated with blinding or endemic trachoma. In many developing countries, trachoma (a serious chronic conjunctivitis which is considered the leading preventable cause of blindness) is hyperendemic. It is most severe in areas of low socioeconomic and poor sanitary conditions.

Three other serotyes cause lymphogranuloma venereum (LGV), a systemic infection characterized by inguinal lymphadenopathy in men. This disease is also found in some developing countries, but is relatively uncommon in industrialized societies. The remaining serotypes are commonly found in industrialized societies, where they cause nongonococcal urethritis (NGU) in men and cervicitis in women. *See* LYMPHOGRANULOMA VENEREUM; NONGONOCOCCAL URETHRITIS (NGU); TRACHOMA.

Many infections of the female genital tract may be transmitted to the newborn infant. Approximately 70% of infants passing through a *Chlamydia*-infected cervix will acquire a chlamydial infection. One-third to one-half of the exposed infants will develop a conjunctivitis called inclusion conjunctivitis of the newborn or inclusion blennorrhea. The most serious consequence of chlamydial infection in the infant appears to be a characteristic pneumonia syndrome, which occurs in infants between 3 weeks and 4 months of age and is characterized by an afebrile course, severe hacking cough, tachypnea, and elevated immunoglobulins. *See* INCLUSION BLENNORRHEA.

Chlamydial diseases have been identified as major public health problems in the United States and other industrialized countries. Although treatment regimens with tetracyclines or erythromycin are readily available for symptomatic men, women, and infants, adequate methods for control of these infections have not been developed. [J.S.]

Chloramphenicol A broad-spectrum antibiotic, also known as Chloromycetin, used to treat serious infectious diseases caused by bacteria, rickettsiae, and certain large viruses.

Crystalline chloramphenicol has an empirical formula of $C_{11}H_{12}Cl_2N_2O_5$, a molecular weight of 323, and a melting point of 150.5–151.5°C. It is moderately soluble in water (2.5 milligrams/milliliter at 25°C), neutral (nonionic but pH 5.5 in water), absorptive of ultraviolet in water solution, and optically active. The biochemical method of manufacture involves aseptic aerobic fermentation of a nutrient broth culture medium seeded with *Streptomyces venezuelae*.

Chloramphenicol is used to treat many infectious diseases, notably typhoid fever and *Hemophilus influenzae* meningitis. It is widely used in systemic infections, including resistant staphylococcal infections, meningitis, septicemia, osteomyelitis, typhus and other rickettsial fevers, brucellosis, tularemia, relapsing fever, plague, pneumonia, whooping cough, psittacosis, endocarditis, peritonitis, and gastrointestinal, genitourinary, eye, ear, nose, throat, and skin infections (see articles on individual diseases). *See* ANTIBIOTIC. [J.E.]

Chlorate A negative ion having the formula ClO_3^- and derived from chloric acid, $HClO_3$. The chlorates are more stable than hypochlorites and chlorites but less stable, and thus better oxidizing agents, than bromates, iodates, and perchlorates. Chlorates are decomposed on heating to liberate oxygen.

Both potassium and sodium chlorates are prepared by electrolysis of a hot chloride solution. Chlorine liberated at the anode is allowed to react with OH^-; hydrogen is liberated at the cathode. Sodium chlorate is used mainly in the production of perchlorates, but also as a weed killer; potassium chlorate is utilized in the manufacture of matches. [E.E.W.]

Chloride A compound which is derived from hydrochloric acid, HCl, and which contains the chlorine atom in the 1− oxidation state. Compounds containing this group may be organic or inorganic. The inorganic chlorides can be divided into three classes: covalent chlorides, ionic chlorides, and complexes. The covalent chlorides can be metallic or nonmetallic and are recognizable because of their low boiling and melting points.

Of the chlorides, sodium chloride is most important commercially. It occurs widely in nature and is the starting point for all sodium, chlorine, and compounds containing these two elements. Calcium chloride is obtained as a by-product in the Solvay soda-ash process. The insoluble silver chloride is used to identify the chloride ion in the absence of bromide ion. Br^-, and iodide ion, I^-. *See* CHEMICAL BONDING; CHLORINE; HALIDE; HALOGENATED HYDROCARBON. [E.E.W.]

Chlorine A chemical element, Cl, of atomic number 17 and atomic weight 35.453. Chlorine exists as a greenish-yellow gas at ordinary temperatures and pressures, It is second in reactivity only to fluorine among the halogen elements, and hence is found free in nature only at the elevated temperatures of volcanic gases. It is estimated that 0.045% of the Earth's crust is chlorine. It combines with metals, nonmetals, and organic materials to form hundreds of compounds. *See* HALOGEN ELEMENTS.

Properties. Naturally occurring chlorine is formed of stable isotopes of mass 35 and 37; radioactive isotopes have been made artificially. The diatomic gas has a molecular weight of 70.906. The boiling point of liquid chlorine (golden-yellow in color) is −34.05°C at 760 mm (101.325 kilopascals), and the melting point of solid chlorine is −100.98°C. The critical temperature is 144°C; the critical pressure is 76.1 atm (7.71 megapascals); the critical volume is 1.745 ml/g; and density at the critical point is 0.573 g/ml. Thermodynamic properties include heat of sublimation at 7370 ± 10 cal/mole at 0 K, heat of evaporation at 4878 ± 4 cal/mole at −34.05°C, heat of fusion at 1531 cal/mole, heat capacity at 7.99 cal/mole at 1 atm (101.325 kilopascals) and 0°C, and 8.2 at 100°C.

Chlorine is one of four closely related chemical elements which have been called the halogen elements. Fluorine is more active chemically, and bromine and iodine are less active. Chlorine replaces iodine and bromine from their salts. It enters into substitution and addition reactions with both organic and inorganic materials. Dry chlorine is somewhat inert, but moist chlorine unites directly with most of the elements.

Manufacture. The first electrolytic process for chlorine production was patented in 1851 by Charles Watt in Great Britain. In 1868 Henry Deacon produced chlorine from hydrochloric acid and oxygen, at 400°C (750°F) with copper chloride absorbed in pumice stone as a catalyst. Modern electrolytic cells may be classified for the most part as diaphragm and mercury types. Both make caustics (NaOH or KOH), chlorine, and hydrogen. The economics of the chlor-alkali industry mainly involves the balanced marketing or internal use of caustic and chlorine in the same proportions as obtained from the electrolytic cell process.

Inorganic compounds. There are many inorganic chlorine compounds, some of which are naturally occurring. Sodium chloride, NaCl, is used directly as mined (rock salt), or as found on the surface, or as brine. It may also be dissolved, purified, and reprecipitated for use in foods or when chemical purity is required. Its main uses are in the production of soda ash and chlorine products. Calcium chloride, $CaCl_2$, is usually obtained from brines or as a by-product of chemical processing. Its main uses are in road treatment, coal treatment, concrete conditioning, and refrigeration. Wet chlorine reacts with metals to form chlorides, most of which are soluble in water. It also reacts with sulfur and phosphorus and with other halogens. *See* INTERHALOGEN COMPOUNDS.

Among the oxides of chlorine is chlorine dioxide, ClO_2, a green gas that has become increasingly important in commercial bleaching of cellulose, water treatment, and waste treatment. Sodium and calcium hypochlorite solutions are in general use for their oxidizing power or bleaching power. The use of sodium chlorite permits controlled bleaching, and it is stable unless heated. Acidified sodium chlorite bleaches are very popular for cellulose and textile processing. Sodium chlorate is used as an intermediate in perchlorate production. It is a weed killer, and is also used in dye preparation, textile and fur processing, and metallurgical operations. Potassium chlorate, $KClO_3$, is used in match manufacture and for fireworks. Potassium perchlorate, $KClO_4$, and ammonium perchlorate, NH_4ClO_4, have become important as fuels for rockets and jet propulsion and as explosives. *See* PERCHLORATE.

The chlorides form a second important group of compounds. Hydrogen chloride, HCl, is used as a strong acid and as a reducing agent. Aluminum chloride, $AlCl_3$ is used principally in anhydrous form. Hydrated and liquid forms are also available, 50% of which are used in drug and cosmetic production. Titanium tetrachloride, $TiCl_4$, is used as a polymerization catalyst and as a smoke screen agent and for sky writing. Phosphorus trichloride, PCl_3, and phosphorus oxychloride, $POCl_3$, are important reagents in organic chemical synthesis. Ferric chloride, $FeCl_3$, is used for the manufacture of salts, pigments, pharmaceuticals, and dyes and for photoengraving, preparation of catalysts, and waste and sewage treatment. *See* CHLORIDE.

Organic compounds. There are many organic chlorine compounds; many of them are of the saturated aliphatic class. Methane may be chlorinated to a variable extent. Methyl chloride, CH_3Cl, is used in production of silicone, tetramethyllead, and butyl rubber. Methylene chloride, CH_2Cl_2, has applications in production of paint remover, plastics, solvent degreasing, aerosol propellants, and other uses. Chloroform, $CHCl_3$, is a colorless, nonflammable, volatile liquid. Fluorocarbon products and resins consume most of the chloroform made, leaving only a small quantity for dye, drug, and other uses. Carbon tetrachloride, CCl_4, is used in the production of chlorofluorohydro-

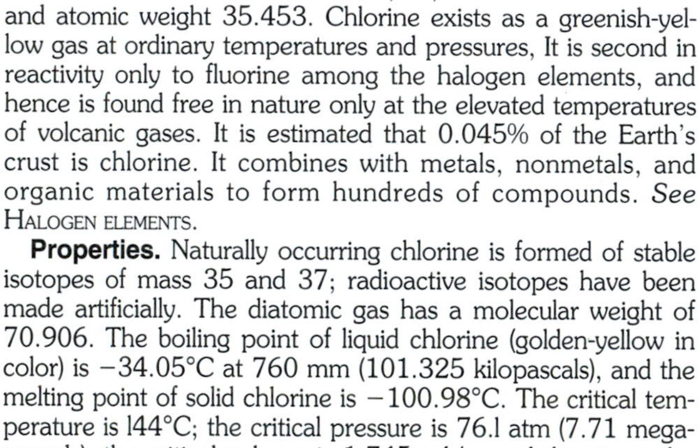
1 H 2 13 14 15 16 17 18 2 He
3 Li 4 Be 5 B 6 C 7 N 8 O 9 F 10 Ne
11 Na 12 Mg 3 4 5 6 7 8 9 10 11 12 13 Al 14 Si 15 P 16 S 17 Cl 18 Ar
19 K 20 Ca 21 Sc 22 Ti 23 V 24 Cr 25 Mn 26 Fe 27 Co 28 Ni 29 Cu 30 Zn 31 Ga 32 Ge 33 As 34 Se 35 Br 36 Kr
37 Rb 38 Sr 39 Y 40 Zr 41 Nb 42 Mo 43 Tc 44 Ru 45 Rh 46 Pd 47 Ag 48 Cd 49 In 50 Sn 51 Sb 52 Te 53 I 54 Xe
55 Cs 56 Ba 71 Lu 72 Hf 73 Ta 74 W 75 Re 76 Os 77 Ir 78 Pt 79 Au 80 Hg 81 Tl 82 Pb 83 Bi 84 Po 85 At 86 Rn
87 Fr 88 Ra 103 Lr 104 Rf 105 Db 106 Sg 107 Bh 108 Hs 109 Mt 110 111 112 113 114 115 116 117 118
lanthanide series 57 La 58 Ce 59 Pr 60 Nd 61 Pm 62 Sm 63 Eu 64 Gd 65 Tb 66 Dy 67 Ho 68 Er 69 Tm 70 Yb
actinide series 89 Ac 90 Th 91 Pa 92 U 93 Np 94 Pu 95 Am 96 Cm 97 Bk 98 Cf 99 Es 100 Fm 101 Md 102 No

carbons, grain fumigation chemicals, and solvents. *See* HALOGENATED HYDROCARBON.

Tetrachloroethane, $Cl_2HCCHCl_2$, is used in the manufacture of the solvents perchloroethylene and trichloroethylene. Ethylene dichloride, $ClCH_2CH_2Cl$, is used in the production of vinyl chloride, ethylenediamine, and solvents. Ethyl chloride, C_2H_5Cl, is used in the manufacture of tetraethyllead, but it also is used in the production of ethyl cellulose, and for refrigerant, anesthetic, and miscellaneous uses. Chloral, CCl_3CHO, became important after World War II with the introduction of newer insecticides.

In addition, there are a number of important unsaturated aliphatic compounds. Perchloroethylene, $Cl_2C[dbd]CCl_2$, is used for dry cleaning, metal degreasing, compounding, and miscellaneous purposes. Vinylidene chloride, $C_2H_2Cl_2$, polymerizes easily and hence is used to make copolymers, such as Saran, Geon, and Velon. Vinyl chloride, C_2H_3Cl, also polymerizes to make polymers, such as Koroseal, Vinylite, Geon, Tygon, Velon, and Saran.

The aromatic chlorine compounds are also important. Chlorobenzene, C_6H_5Cl, at one time used for phenol production, has become the major chemical for phenol manufacture. Dichlorobenzene, $C_6H_4Cl_2$, exists in two isomeric forms. Moth repellents and deodorants are made of *p*-dichlorobenzene. Dye intermediates and miscellaneous items are also produced from this substance. Other aromatic chlorine compounds of interest are benzyl chloride, $C_6H_5CH_2Cl$, benzal chloride, $C_6H_5CHCl_2$, and benzotrichloride, $C_6H_5CCl_3$. These three compounds are starting materials for the synthesis of pharmaceuticals (benzedrine, demerol), and are used as constituents of perfumes and as starting materials for the preparation of insecticides and insect repellents. They are also used as lubricants and as intermediates in dye manufacture. The chlorinated biphenyls find use in electrical insulation, hydraulic mediums, lubricants, and as constituents of paints, varnishes, and plastics. Because these compounds are chemically inert, they have been found to persist in the environment and are being phased out of industrial uses. Dichlorophenol, $Cl_2C_6H_3OH$, is important in the production of another chlorine compound, 2,4-dichlorophenoxyacetic acid (2,4-D). The compound 2,4-D is widely used as a weed killer. *See* HERBICIDE; INSECTICIDE; POLYCHLORINATED BIPHENYLS.

[F.W.Ko.; R.W.B.]

Chlorite One of a group of layer silicate minerals, usually green in color, characterized by a perfect cleavage parallel to (001). The cleavage flakes are flexible but inelastic, with a luster varying from pearly or vitreous to dull and earthy. The hardness on the cleavage is about 2.5. The specific gravity of chlorate varies between 2.6 and 3.3 as a function of composition.

Chlorite is a common accessory mineral in low- to medium-grade regional metamorphic rocks and is the dominant mineral in chlorite schist. It can form by alteration of ferromagnesian minerals in igneous rocks and is found occasionally in pegmatites and vein deposits. It is a common constituent of altered basic rocks and of alteration zones around metallic ore bodies. Chlorite also can form by diagenetic processes in sedimentary rocks. *See* AUTHIGENIC MINERALS; CLAY MINERALS; DIAGENESIS; SILICATE MINERALS.

[S.W.Ba.]

Chloritoid A member of the brittle mica (clintonite) group, which are hydrous aluminum silicates resembling both micas and chlorites. Although they also possess a perfect basal cleavage, cleavage flakes of the clintonites are distinctly more brittle than those of the micas. The composition is [Fe(II),Mg]-[Al,Fe(III)](Si,Al)$O_5(OH)_2$. The variety ottrelite also contains manganese. Margarite, prehnite, and stilpnomelane are other members of the group. Chloritoid has both monoclinic and triclinic modifications which may occur together. It forms rosettes, foliated aggregates, and plates. The color is gray to green, and crystals are weakly pleochroic. The hardness is 6.5, specific gravity 3.5. Inclusions of quartz and magnetite are common. Chloritoid is principally a metamorphic mineral in schists, phyllites, and quartzites, and is associated with quartz, muscovite, and chlorite. *See* MICA; SILICATE MINERALS.

[E.W.H.]

Chlorococcales A large and highly diverse order of green, mostly fresh-water algae in the class Chlorophyceae. These plants are unicellular or colonial. They are characterized by the failure to use customary cell division in vegetative reproduction. Rather, multiplication occurs by the cleavage of cell contents to form small replicas of the parent cell. Eventually these are liberated and enlarge to the adult size. The mother-cell wall sometimes persists as interconnecting strands or as swollen, enclosing membranes.

The order, based primarily on this method of reproduction, brings together a seemingly unrelated variety of genera which comprise eight families varying according to interpretation. The cells have parietal or axial chloroplasts of various forms such as plates, cups, or networks, or they may be stellate. *See* CHLOROPHYCEAE; PHYTOPLANKTON.

[G.W.P.]

Chloroform A colorless, sweet-smelling, nonflammable liquid, of formula weight 119.39 and formula $CHCl_3$. Chloroform has a specific gravity of 1.489 at 20°C (68°F), and a boiling point of 61.2°C (142°F). It is insoluble in water but soluble in organic solvents. In the presence of ultraviolet light, or more slowly in the dark, chloroform tends to react with oxygen of the air to produce poisonous phosgene. Therefore, commercial chloroform contains inhibitors. It is a powerful anesthetic, but prolonged use causes metabolic disturbances and damage to the heart, liver, and kidney, so that it commonly has been replaced by less toxic anesthetics. It is widely used as an extractant and solvent, a chemical intermediate in the production of dyes and drugs, and in pharmaceuticals as an antispasmodic, sedative (particularly in cough medicines), analgesic liniment, and anthelmintic. *See* HALOGENATED HYDROCARBON.

[E.H.H.]

Chloromonadida An order of the class Phytamastigophorea, also known as the Chloromonadina. These poorly known flagellates are grass-green or colorless, somewhat flattened, and have two equal flagella, one anterior, the other trailing. All known genera are free-swimming, although *Reckertia* and *Thaumatomastix* form pseudopodia. They vary in size from 30 to 100 micrometers. Chromatophores are small disks, stigmas are lacking, and fat is the storage product. Trichocysts are found in three genera. The nucleus is large, with nucleoli and chromosomes visible during interphase. One or two anterior vacuoles are present and a reservoir seems to be present in *Trentonia* and *Gonyostomum*. Life cycles are unknown, but longitudinal division occurs. The taxonomic position of the class is poorly defined. *See* PHYTAMASTIGOPHOREA.

[J.B.L.]

Chlorophyceae A large and diverse class of plants, commonly called green algae, in the chlorophyll *a-b* phyletic line (Chlorophycota). Estimated number of taxa varies widely; 560 genera and 8600 species are conservative estimates. *See* CHLOROPHYCOTA.

The green algae exhibit great morphological diversity while sharing fundamental biochemical and ultrastructural features. Their photosynthetic pigments are similar to those in higher plants and include chlorophyll *a* and *b*, α-, ß-, and γ-carotene, and various xanthophylls. Chloroplasts vary in number and shape, but always have two membranes, two to five thylakoids per lamella, and usually one or more pyrenoids. A cell wall, which may be calcified, is usually present, often contains cellulose, hydroxyproline glycosides, xylan, and mannan. The chief food reserve is starch, stored in the chloroplast as granules which often sheathe a pyrenoid. Motile cells usually have two or four apically inserted smooth flagella of approximately equal length. *See* CELL PLASTIDS; CHLOROPHYLL; PHOTOSYNTHESIS.

Almost all somatic cell types known for algae occur among the Chlorophyceae, the exceptions being rhizopodial unicells and complex multicellular thalli differentiated into macroscopic organs. Ultrastructural studies have revealed differences in details of nuclear and cell division, cell coverings, plasmodesmata, pyrenoids, and flagellar structure that are correlated among themselves and with certain traditional characters to a degree that strongly suggests that the ulotrichine line is polyphyletic, with desmoschisis having evolved at least four times. Moreover, these studies suggest that the ultimate ancestral green flagellate was prasinophycean (asymmetrical and covered with scales) rather than *Chlamydomonas*-like. Integration of these studies and inferences suggests four phyletic lines: chlorophycean, ulvophycean, charophycean, and pleurastrophycean.

In addition to the four phyletic lines, each of which includes monads as well as multicellular algae, there is a residual group of primitive green flagellates for which the name Micromonadophyceae has been proposed. These monads are scaly or secondarily naked and have one to four flagella arising from an apical pit or lateral depression. The interzonal mitotic spindle persists during cytokinesis. This group corresponds to the class Prasinophyceae in the present classification. For the sake of consistency, however, Prasinophyceae and Charophyceae are treated elsewhere in this work as separate classes. *See* ALGAE; CHAROPHYCEAE; PRASINOPHYCEAE. [P.C.Si.; R.L.Moe]

Chlorophycota A division of the plant kingdom (also known as Chlorophyta or Chlorophycophyta) comprising all algae that have chlorophyll *a* and *b* except the Euglenophyceae, and that in all other respects are so different as to suggest a separate origin of their photosynthetic pigments. Three classes, Charophyceae (charophytes), Chlorophyceae (green algae), and Prasinophyceae, are recognized. *See* ALGAE; CHAROPHYCEAE; CHLOROPHYCEAE; PRASINOPHYCEAE. [P.C.Si.; R.L.Moe]

Chlorophyll The generic name for the intensely colored green pigments which are the photoreceptors of light energy in photosynthesis. These pigments belong to the tetrapyrrole family of organic compounds.

Five closely related chlorophylls, designated *a* through *e*, occur in higher plants and algae. The principal chlorophyll (Chl) is Chl *a*, found in all oxygen-evolving organisms; photosynthetic bacteria, which do not evolve O_2, contain instead bacteriochlorophyll (Bchl). Higher plants and green algae contain Chl *b*, the ratio of Chl *b* to Chl *a* being 1:3. Chlorophyll *c* (of two or more types) is present in diatoms and brown algae. Chlorophyll *d*, isolated from marine red algae, has not been shown to be present in the living cell in large enough quantities to be observed in the absorption spectrum of these algae. Chlorophyll *e* has been isolated from cultures of two algae, *Tribonema bombycinum* and *Vaucheria hamata*. In higher plants the chlorophylls and the above-mentioned pigments are contained in lipoprotein bodies, the plastids. *See* CAROTENOID; CELL PLASTIDS; PHOTOSYNTHESIS.

Chlorophyll molecules have three functions: They serve as antennae to absorb light quanta; they transmit this energy from one chlorophyll to another by a process of "resonance transfer;" and finally, this chlorophyll molecule, in close association with enzymes, undergoes a chemical oxidation; that is, an electron of high potential is ejected from the molecule; this electron can then be made to do chemical work, that is, reduction of another compound. In this way the energy of light quanta is converted into chemical energy.

The chlorophylls are cyclic tetrapyrroles in which four 5-membered pyrrole rings join to form a giant macrocycle. Chlorophylls are members of the porphyrin family, which plays important roles in respiratory pigments, electron transport carriers, and oxidative enzymes. *See* PORPHYRIN.

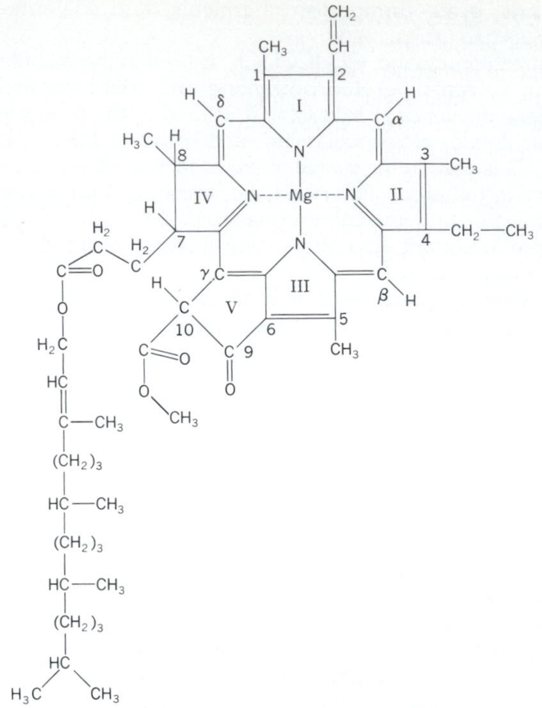

Structure of chlorophyll *a* ($C_{55}H_{72}O_5N_4Mg$).

The structure of chlorophyll *a* is shown in the illustration. The characteristic features of chlorophyll *a* are that it is a magnesium chelate of a dihydroporphyrin with a cyclopentanone ring (V) and is esterified with phytol. The alicyclic 5-membered ring V, which contains a keto C=O group at position 9, is unique to chlorophylls.

The two major pigments of protoplasm, green chlorophyll and red heme, are synthesized along the same biosynthetic pathway to protoporphyrin. Starting from the small building blocks, glycine and succinic acid, they are converted in a series of enzymic steps, which is identical in plants and animals, to protoporphyrin. Here the pathway branches to form (1) a series of porphyrins chelated with iron, as heme and related cytochrome pigments; and (2) a series of porphyrins chelated with magnesium which are precursors of chlorophyll. *See* HEMOGLOBIN.

Chlorophylls reemit a fraction of the light energy they absorb as fluorescence. Irrespective of the wavelength of the absorbed light, the emitted fluorescence is always on the long-wavelength side of the lowest energy absorption band, in the red or infrared region of the spectrum.

The fluorescent properties of a particular chlorophyll are functions of the structure of the molecule and its immediate environment. Thus, the fluorescence spectrum of chlorophyll in the living plant is always shifted to longer wavelengths relative to the fluorescence spectrum of a solution of the same pigment. This red shift is characteristic of aggregated chlorophyll. [G.; S.Gr.; G.P.]

Chloroplatinate One of a group of compounds containing the anion $PtCl_6^{2-}$, which is derived from chloroplatinic acid. This acid is produced when platinum is dissolved in aqua regia. Of the many salts known, the ammonium and potassium chloroplatinates are the most useful because of their low solubility. The formation of a precipitate is used as a qualitative test for the ammonium and potassium ions. *See* AQUA REGIA; PLATINUM. [E.E.W.]

Chlortetracycline A broad-spectrum antibiotic. It was the first tetracycline antibiotic and is effective for many infections. *See* ANTIBIOTIC; TETRACYCLINE.

This crystalline yellow compound is produced by the soil microorganism *Streptomyces aureofaciens*. The original culture has been mutated in order to obtain a strain which produces satisfactory yields of chlortetracycline for commercial production. After several transfers in increasingly larger fermenters, the culture is finally grown for several days in 20,000-gallon (75-m^3) tanks.

The antibiotic is used orally, parenterally, and topically. It is rapidly absorbed from the intestinal tract and is distributed into all body tissues. Chlortetracycline is widely used in both human and veterinary medicine for infections caused by sensitive organisms. It is also used as a feed supplement and as a food preservative. It promotes growth of poultry, hogs, and calves when used in their rations. At very low concentrations it increases the shelf life of perishable foods, such as meats. Commercially, it has been used for the preservation of poultry and fish. [N.Bo.]

Choanoflagellida An order of the class Zoomastigophorea (Protozoa). They are principally small, single-celled or colonial, structurally simple, colorless flagellates distinguished by having the anterior end surrounded by a thin protoplasmic collar, within which is an ingestive area for particulate matter, for example, bacteria. A single flagellum arises from the center of this collar (see illustration). These organisms are widespread in fresh and salt water and are common in inshore waters. *See* PROTOZOA; ZOOMASTIGOPHOREA. [J.B.L.]

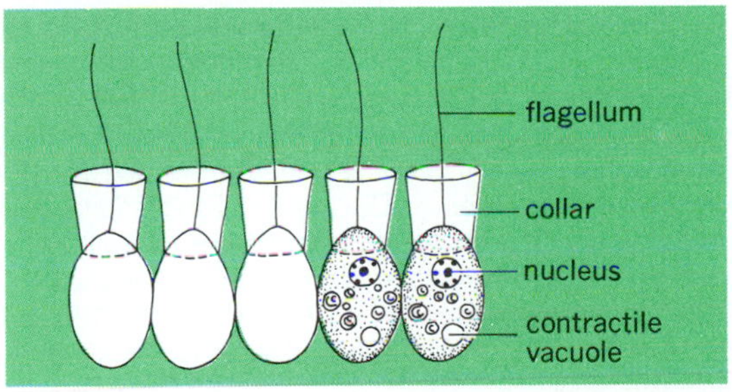

A linear colony of the choanoflagellate *Desmarella moniliformis*. (After McCracken)

Choke (electricity) An inductor used in a low-pass filter in which the useful output is a direct current, as in the filter of a power rectifier. At usual power frequencies a filter choke has an iron core, with an air gap to minimize variation of inductance with direct current. In a choke-input filter (a filter in which the first element is an inductance) some economy is sometimes afforded by designing the choke so that its inductance is allowed to reduce because of saturation at high currents. Such a choke is called a swinging choke. *See* ELECTRONIC POWER SUPPLY. [W.R.LeP.]

Choked flow Fluid flow through a restricted area whose rate reaches a maximum when the fluid velocity reaches the sonic velocity at some point along the flow path. The phenomenon of choking exists only in compressible flow and can occur in several flow situations. *See* COMPRESSIBLE FLOW.

Through varying-area duct. Choked flow can occur through a convergent flow area or nozzle attached to a huge reservoir. Flow exits the reservoir through the nozzle if the back pressure is less than the reservoir pressure. When the back pressure is decreased slightly below the reservoir pressure, a signal from beyond the nozzle exit is transmitted at sonic speed to the reservoir. The reservoir responds by sending fluid through the nozzle. Further, the maximum velocity of the fluid exists at the nozzle throat where the area is smallest.

When the back pressure is further decreased, fluid exits the reservoir more rapidly. Eventually, however, the velocity at the throat reaches the sonic velocity. Then the fluid velocity at the throat is sonic, and the velocity of the signal is also sonic. Therefore, further decreases in back pressure are not sensed by the reservoir, and correspondingly will not induce any greater flow to exit the reservoir. The nozzle is thus said to be choked, and the mass flow of fluid is a maximum. *See* MACH NUMBER; NOZZLE; SOUND; SUPERSONIC DIFFUSER.

With fraction. Choked flow can also occur through a long constant-area duct attached to a reservoir. As fluid flows through the duct, friction between the fluid and the duct wall reduces the pressure acting on the fluid. As pressure is reduced, other fluid properties are affected, such as sonic velocity, density, and temperature. The maximum Mach number occurs at the nozzle exit, and choked flow results when this Mach number reaches 1.

With heat addition. A reservoir with a constant-area duct attached may also be considered in the case that the flow through the duct is assumed to be frictionless but heat is added to the system along the duct wall. *See* FLUID FLOW; GAS DYNAMICS. [W S.J.]

Cholecystitis Inflammation of the gallbladder, a common disease in humans. It is nearly always associated with gallstones and is particularly common in obese middle-aged women. Most cases are thought to be the result of chemical irritation caused by excessively concentrated bile, which is in turn the result of partial or complete obstruction to the outflow of bile. The gallbladder which is irritated and inflamed in this fashion is susceptible to secondary bacterial infection, and this in turn may spread throughout the bile ducts into the liver tissue. Perforation may occur, leading to the development of peritonitis or to abnormal communications with the intestine. Prolonged or recurrent episodes of inflammation result in chronic cholecystitis, characterized by thickening and scarring of the wall, contraction, and impairment of normal function. *See* GALLSTONES; PERITONITIS. [M.R.H.]

Cholera A disease caused by infection and colonization of the small intestine by the bacterium *Vibrio cholerae*. The primary symptom is severe diarrhea. It has been estimated that severe diarrhea accounts for 5 million deaths each year, primarily among children, in developing countries, and cholera is responsible for a significant portion. With proper treatment, however, cholera is fatal to fewer than 1% of the infected individuals. The high death rate in developing countries is primarily due to lack of proper facilities to care for those infected. *See* DIARRHEA. [D.K.]

Cholesterol The chief sterol of vertebrates, with the structural formula

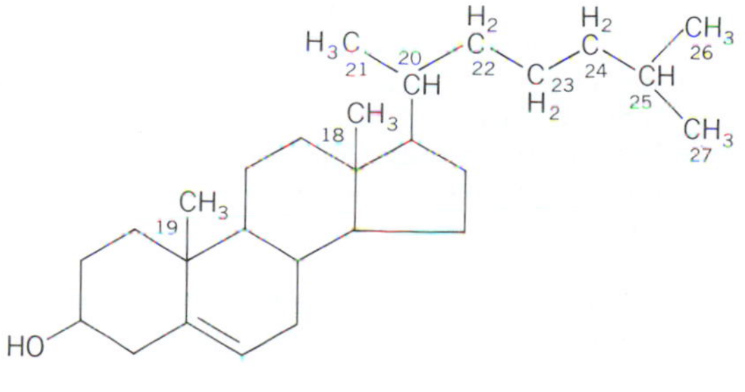

All vertebrate cells at one or another stage of development make cholesterol. In the adult, however, the chief sites for synthesis are the intestine, the liver, and the skin. Nervous tissue is particularly rich in cholesterol. It is a component of myelin, a complex lipoprotein that serves as the sheath of axons and behaves much like an insulator. Cholesterol in tissues and body fluids is always found in combination with proteins (lipoproteins). *See* MYELIN; STEROL.

In humans and other omnivores, and in the carnivores as well, there is a continued intake of cholesterol in food, although the sterol is inefficiently absorbed. Dietary cholesterol supplements biosynthesis. Plant sterols are not absorbed by these species and are absent from tissue sterols. Herbivores can absorb cholesterol and are relatively defenseless against its accumulation in body tissue as a result of excess feeding.

There are two principal paths for removal of cholesterol from the body. The first is by excretion in bile and to a very much lesser degree through the wall of the small intestine. The biliary cholesterol mixes with dietary cholesterol in the intestine; a portion is reabsorbed via enterohepatic circulation and handled indistinguishably from dietary cholesterol. A portion that is not absorbed is metabolized by bacteria in the gastrointestinal tract to a variety of products, chief of which is coprostanol. This latter is not reabsorbed and is eliminated. The second path leads to production of bile acids, which differ with species; these acids are conjugated with compounds such as glycine and taurine and excreted in the bile. They serve as detergents to promote the emulsification of fat in the intestine and are in turn partially reabsorbed in an enterohepatic circulation, together with the products of fat digestion, and are reutilized. *See* LIPID METABOLISM; STEROID.

Cholesterol is a constituent of some gallstones and is also present in the arterial plaques that characterize arteriosclerosis. Many authorities consider elevated plasma cholesterol a causative factor in arteriosclerosis, although others consider its infiltration a consequence of primary damage to vessel walls. All agree that the two factors, a high level of plasma cholesterol and injury to the arterial wall, are intimately associated in formation of arterial plaques. Since arteriosclerosis is a major cause of death in humans, the importance of control of the level of cholesterol is clearly apparent. *See* ARTERIOSCLEROSIS.

Several disorders of cholesterol metabolism in humans are known, especially is the genetically determined elevation of plasma cholesterol known as essential hypercholesterolemia. It is characterized by high plasma cholesterol, with only mild elevation in plasma neutral fat. Other disorders may be associated with very high levels of fats and phospholipids. [T.F.G.]

Choline A compound, trimethyl-ß-hydroxyethylammonium hydroxide, used by the animal organism as a precursor of acetylcholine and as a source of methyl groups. It is a strongly basic hygroscopic substance with the formula

$$(CH_3)_3{\equiv}\overset{+}{N}{-}CH_2{-}CH_2OH$$

Choline deficiency in animals is associated with fatty livers, poor growth, and renal lesions. It is a lipotropic agent. There is no direct evidence of disease in humans due to choline deficiency, although there have been suggestions that some of the liver, kidney, or pancreas pathology seen in various nutritional deficiency states may be related to choline insufficiency. Choline is found in acetylcholine, which is necessary for nerve impulse propagation, and in phospholipids.

Humans eat 50–600 mg of choline per day, but only excrete 2–4 mg. Thus, conventional tests are of no value in studying choline requirements, and no knowledge of human choline requirements exists. *See* ACETYLCHOLINE. [S.N.G.]

Chondrichthyes A class of vertebrates comprising the cartilaginous, jawed fishes. The Chondrichthyes have traditionally included the subclasses Elasmobranchii (sharks, skates, and rays) and Holocephali (ratfishes). A classification scheme for the Chondrichthyes follows; for detailed information see separate articles on each group listed.

Class Chondrichthyes
  Subclass Elasmobranchii
    Order: Cladoselachii
      Pleurocanthodii
      Selachii
      Batoidea
  Subclass Holocephali
    Order Chimaeriformes

A group of Devonian armored fishes, the Placodermi, has usually been regarded as ancestral to the Chondrichthyes, but this derivation is not certain. Another group of primitive jawed fishes called acanthodians, which are considered by many as ancestral to the higher bony fishes, exhibit certain primitive elasmobranch-like features. In any case it is probable that the elasmobranchs and ratfishes arose independently of each other sometime during the Silurian or Early Devonian. *See* ACANTHODII; PLACODERMI.

The most distinctive feature shared by the elasmobranchs and ratfishes is the absence of true bone. In both groups the endoskeleton is cartilaginous; in some cases it may be extensively calcified. Because even calcified cartilage is rarely preserved, the fossil record of the Chondrichthyes is represented mainly by teeth and spines, with only occasional associated skeletons. *See* SKELETAL SYSTEM.

Other characteristics of the Chondrichthyes include placoid scales, clasper organs on the pelvic fins of males for internal fertilization, a urea-retention mechanism, and the absence of an air (swim) bladder. Both groups have primarily always been marine predators, although they have repeatedly invaded fresh water throughout their long history. The elasmobranchs have probably always fed as they do today, on other fishes as well as on soft and hard-bodied invertebrates. The ratfishes have most likely concentrated on invertebrates, although modern forms occasionally also feed on smaller fishes. *See* RAY; SCALE (ZOOLOGY); SHARK; SWIM BLADDER. [B.S.]

Chondrostei The most archaic of three organizational levels (infraclasses) of the subclass Actinopterygii, or rayfin fishes. Chondrosteans of the order Palaeonisciformes appeared first in the Lower Devonian, and by Middle Devonian were abundant contemporaries of the early crossopterygians and lungfishes, representative of the other two subclasses of bony fishes. Palaeonisciformes were diversified and abundant in sturgeons (order Acipenseriformes), which persist in the modern fauna as highly specialized and in some ways degenerate chondrosteans. Even more like the palaeonisciforms than the sturgeons are the surviving African fishes of the order Polypteriformes, which are known only since the Eocene but are probably much older. *See* ACIPENSERIFORMES; PALAEONISCIFORMES; POLYPTERIFORMES.

Primitive chondrosteans were mostly small- or moderate-sized fishes with a heavy armor of rhomboidal, enameled scales. The heterocercal tail was preceded on the midline, above and below, by series of enlarged fulcral scales. There was usually a spiracle behind the eye. The head was heavily plated and relatively inflexible, with the maxilla firmly bound to the cheek bones. Feeding was accomplished largely by biting and gulping, aided by simple conical teeth. *See* ACTINOPTERYGII; SCALE (ZOOLOGY). [R.M.B.]

Chonotrichida An order of the Holotrichia. This is a small group of curious, vase-shaped ciliates, commonly found as ectocommensals on marine crustaceans. They attach to their

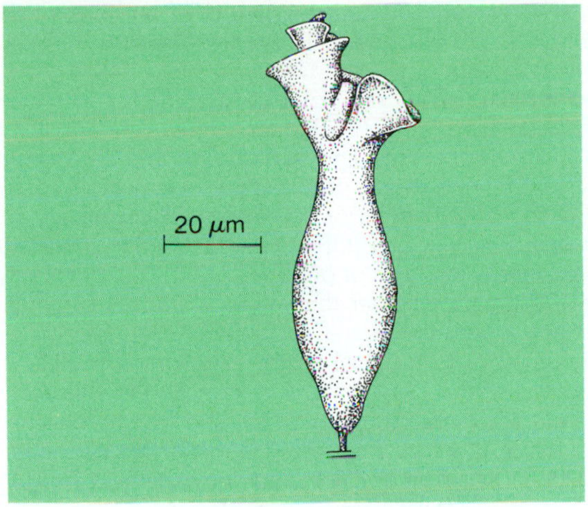

***Spirochona*, an example of a chonotrichid.**

hosts by a short, secreted stalk, and their bodies are practically devoid of ciliature. *Spirochona* (see illustration) is a common example of this order. *See* HOLOTRICHIA; PROTOZOA. [J.O.C.]

Chopping The act of interrupting an electric current, beam of light, or beam of infrared radiation at regular intervals. This can be accomplished mechanically by rotating fan blades in the path of a beam or by placing a vibrating mirror in the path of the beam to deflect it away from its intended source at regular intervals.

A current can be chopped with an electromagnetic vibrator having contacts on its moving armature. A current can also be chopped electronically by passing it through a multivibrator or other switching circuit.

Chopping is generally used to change a direct-current signal into an alternating-current signal that can more readily be amplified. *See* MULTIVIBRATOR; VIBRATOR. [J.Mar.]

Chord A line segment whose end points are points of a circle (see illustration). A chord subtends two (conjugate) arcs of the circle. Equal chords of a circle subtend equal arcs, and of two unequal chords, the greater subtends the greater arc. The chord of maximum length passes through the center of the circle and is called the diameter. A diameter perpendicular to a chord bisects it. Equal chords of a circle are equally distant from the center of the circle, and conversely. An angle formed by two chords that intersect inside the circle is measured by half the sum of the intercepted arcs, and the product of the segments of one chord equals that of the other. *See* ARC (MATHEMATICS); CIRCLE. [L.M.Bl.]

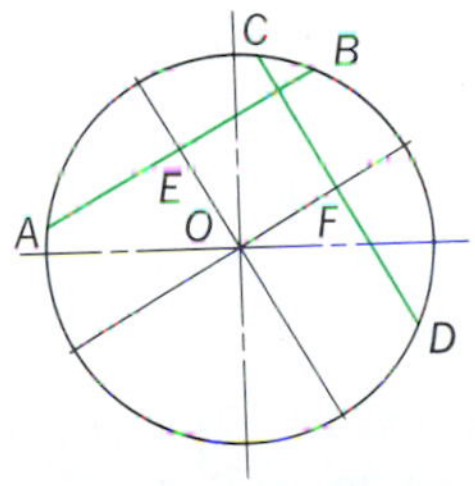

Two intersecting chords *AB* and *CD* bisected by diameters at *E* and *F*, respectively.

Chordata The highest phylum in the animal kingdom, which includes the tunicates (Urochordata), the lancelets or amphioxus (Cephalochordata), and the vertebrates (Craniata), comprising the lampreys, sharks and rays, bony fish, amphibians, reptiles, birds, and mammals. Members of the first two groups, the lower chordates, are small and marine. The vertebrates are free-living, the aquatic ones are basically fresh-water types, and the members include animals of medium size, as well as the largest of all animals, and small ones such as mice, hummingbirds, frogs, and fish. *See* CEPHALOCHORDATA; TUNICATA.

The typical Chordata characteristics are the notochord, the dorsal, hollow nerve cord, and the pharyngeal slits.

The notochord appears in the embryo as a slender, flexible rod filled with gelatinous cells and surrounded by a tough fibrous sheath; it lies above the primitive gut. In lower chordates and the early groups of vertebrates, the notochord persists as the axial support for the body throughout life, but is surrounded and gradually replaced by segmental vertebrae in the higher fish.

The dorsal, hollow nerve cord flows from a specialized band of ectoderm along the middorsal surface of the early embryo by a folding together of two parallel ridges. The anterior end enlarges slightly in larval tunicates and somewhat more in lancelets, but enlarges greatly in the vertebrates to form the brain. Vertebrate evolution is characterized by continual enlargement of the brain. In the craniates or vertebrates, the brain is enclosed by the braincase or cranium, and the spinal cord by the neural arches of the vertebrae. *See* NERVOUS SYSTEM (VERTEBRATE).

Paired slits develop on the sides of the embryonic pharynx, a part of the digestive system, and are retained in all aquatic chordates. Pharyngeal slits originated as adaptations for filter feeding, but soon became the primary respiratory organ, as blood vessels line the fine filaments on the margins of each slit. Water passing over the gills serves for gas exchange in addition to the original filter feeding function, which is soon lost in the vertebrates. Internal gills are lost with the origin of tetrapods; larval and some adult amphibians possess external gills which are quite different structures. The pharyngeal slits in embryonic tetrapods close early in life, with the pharyngeal pouches becoming the site for glandular development, for example, the thyroid and tonsils. *See* RESPIRATORY SYSTEM.

The earliest known vertebrates were fishes of the Ordovician Period, when bone first appeared; thereafter vertebrates are common in the fossil records. Agnathan fish appeared in the Ordovician, and the earliest gnathostomes in the Silurian. The Devonian was a time of vast vertebrate radiation with the sharks and all major subgroups of bony fish and of amphibians (slightly later) appearing. Reptiles originated in the Carboniferous and radiated broadly throughout the Mesozoic Era. The first bird, *Archaeopteryx*, comes from the Jurassic, as do the archaic mammals.

The earliest vertebrates evolved in fresh water and were bottom dwellers with poor swimming abilities. A hard, dermal bony armor served as protection. Better-swimming pelagic groups evolved with gradual reduction of the heavy dermal scales and the origin of paired fins and girdles. Jaws and teeth evolved with feeding on large food particles. Although fish respire via their gills, lungs appeared early in vertebrate history as an accessory respiratory organ using air in stagnant waters. Kidneys serve as a water balance organ. The major line of fish evolution remained in fresh waters, which tetrapods left because of feeding and other advantages on land. Gills were lost and lungs took over respiration, fins became limbs, and kidneys took over nitrogenous waste elimination. Reproduction was still aquatic, with eggs and larval life in the waters. Reptiles were the first true terrestrial tetrapods with a water-impervious skin, specialized extraembryonic membranes, and a dry-shelled egg. Mammals and birds acquired a high sustained body temperature and an insulated body covering of hair or feathers, respectively. *See* VERTEBRATA. [W.J.B.]

Chorion The outermost of the several extraembryonic membranes in amniotes (reptiles, birds, and mammals) enclosing the embryo and all of its other membranes. The chorion, or serosa, is composed of an outer layer of ectodermal cells and an inner layer of mesodermal cells, collectively the somatopleure. Both layers are continuous with the corresponding tissue of the embryo. The chorion arises in conjunction with the amnion, another membrane that forms the outer limb of the somatopleure which folds up over the embryo in reptiles, birds, and some mammals. The chorion is separated from the amnion and yolk sac by a fluid-filled space, the extraembryonic coelom, or body cavity.

In reptiles and birds the chorion fuses with another extraembryonic membrane, the allantois, to form the chorioallantois, which lies directly below the shell membranes. An extensive system of blood vessels develops in the mesoderm of this compound membrane which serves as the primary respiratory and excretory organ for gaseous interchanges. In all mammals above the marsupials, the chorion develops special fingerlike processes (chorionic villi) extending outward from its surface. To a varying degree in different species of mammals, the villous regions of the chorion come into more or less intimate contact with the uterine mucosa, or uterine lining, of the mother, thereby forming the various placental types. *See* ALLANTOIS; AMNIOTA; FETAL MEMBRANE. [N.T.S.]

Choristida An order of sponges of the class Demospongiae, subclass Tetractinomorpha, in which at least some of the megascleres are tetraxons, usually triaenes. Monaxonid spicules may occur as well. Microsclere types are chiefly asters, streptasters, or sigmas. A well-developed cortex comprising an outer gelatinous layer and an inner fibrous layer is often present.

Many choristidan sponges are radially symmetrical; spherical or ellipsoidal shapes are common. Others are encrusting or massive; very few form branching colonies. Some species have basal tufts or mats of long, thin spicules which anchor them in mud or sand on the sea bottom (see illustration). Choristidans are common in tidal regions and shallow waters of all seas, and some species extend down to depths of at least 11,500 ft (3500 m). *See* DEMOSPONGIAE; TETRACTINOMORPHA. [W.D.H.]

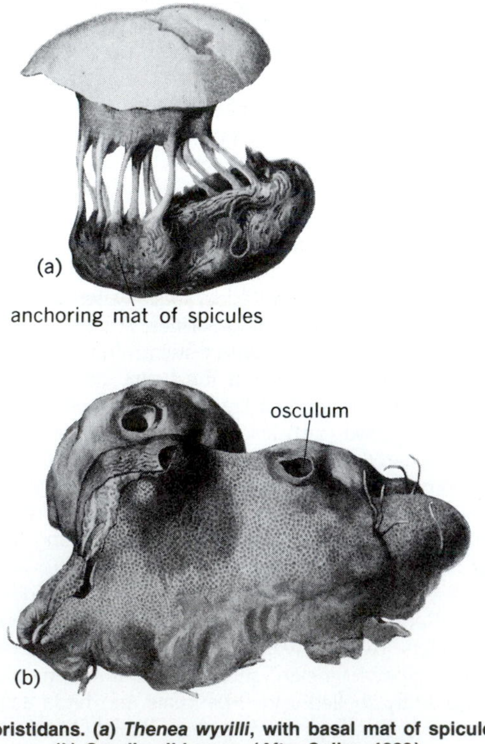

Choristidans. (a) *Thenea wyvilli*, with basal mat of spicules. (b) *Geodia gibberosa*. (After Sollas, 1888)

Chromadorida An order of nematodes in which the amphid manifestation is variable but within superfamilies some constancy is apparent. The various amphids are reniform, transverse elongate loops, simple spirals, or multiple spirals not seen in any other orders or subclasses. The cephalic sensilla are in one or two whorls at the extreme anterior. In all taxa the cuticle shows some form of ornamentation, usually punctations that are apparent whether the cuticle is smooth or annulated. When developed, the stoma is primarily esophastome and is usually armed with a dorsal tooth, jaws, or protrusible rugae. The corpus of the esophagus is cylindrical; the isthmus is not seen; and the postcorpus, in which the heavily cuticularized lumen forms the cresentic valve, is distinctly expanded. The esophagointestinal valve is triradiate or flattened. The females usually have paired reflexed ovaries.

There are four chromadorid superfamilies: Choanolaimoidea, Chromadoroidea, Comesomatoidea, and Cyatholaimoidea. Choanolaimoidea are distinguished by a complex stoma in two parts. The group occupies marine habitats; some species are predaceous, but for many the feeding habits are unknown. Chromadoroidea comprise small to moderate-sized free-living forms that are mainly marine but are also found in fresh water and soil. Known species either are associated with algal substrates or are nonselective deposit feeders in soft-bottom sediments. Comesomatoidea, containing only the family Comesomatidae, are found in marine habitats, but the feeding habits are unknown. Cyatholaimoidea are found in marine, terrestrial, and fresh-water environments. *See* NEMATA. [A.R.M.]

Chromate The CrO_4^{2-} ion derived from the unstable acid H_2CrO_4. Chromic oxide, CrO_3, gives, instead of the expected H_2CrO_4, condensed or polyacids such as $H_2Cr_2O_7$ and $H_2Cr_3O_{10}$ in acid solutions. However, the solid chromates are stable and are the most important salts of the series. The dichromates are also well known. *See* DICHROMATE.

The chromates are easily converted to dichromates by the addition of acid. This is accompanied by a color change from yellow to orange. Most of the solid chromate salts are also yellow, an exception being silver chromate, which is red.

The chromates of calcium, strontium, mercury (I), silver, barium, and lead are insoluble and are used to identify the chromate ion. However, these salts dissolve in strong acid to form dichromates.

Chromates are good oxidizing agents. They are used in paints as pigments and in the tanning of leather. *See* CHROMIUM; OXIDIZING AGENT; PIGMENT. [E.E.W.]

Chromatic aberration The type of error in an optical system in which the formation of a series of colored images occurs, even though only white light enters the system. Chromatic aberrations are caused by the fact that the refraction law determining the path of light through an optical system contains the refractive index, which is a function of wavelength. Thus the image position and the magnification of an optical system are not necessarily the same for all wavelengths, nor are the aberrations the same for all wavelengths. *See* ABERRATION (OPTICS); REFRACTION OF WAVES. [M.J.H.]

Chromatography A separation method whereby individual chemical compounds which were originally present in a mixture are resolved from each other by the selective process of distribution between two heterogeneous (immiscible) phases. The distribution of chemical species to be separated occurs in a dynamic process between the mobile phase and the stationary phase. The stationary phase is a dispersed medium, which usually has a relatively large surface area, through which the mobile phase is allowed to flow. The chemical nature of the stationary phase exercises the primary control over the separation process. The greater the affinity of a particular chemical compound (referred to as the solute) for the stationary medium, the longer it will be retained in the system. The mobile phase can be either gas or liquid; correspondingly, the methods are referred to as gas chromatography and liquid chromatography. *See* COLUMN CHROMATOGRAPHY; FLAT-BED CHROMATOGRAPHY; GAS CHROMATOGRAPHY; LIQUID CHROMATOGRAPHY.

There are four combinations of heterogeneous phase systems, which give rise to four different chromatographic methods: gas-solid, liquid-solid, gas-liquid, and liquid-liquid chromatography. In gas-solid and liquid-solid chromatography, sample molecules are caused to interact physically with the surface of a porous solid by means of a phenomenon called adsorption. Hence, these two methods are also generally referred to as adsorption chromatography. The adsorptive effect of the chromatographic medium for different solutes determines their rates of migration through the medium. In gas-liquid and liquid-liquid chromatography, the liquid stationary phase is held on the surface of an inert solid which serves merely as its support and, ideally, does not participate in the separation process. Primarily, then, the components of a mixture having different solubilities in the stationary phase separate by migrating at different rates. Since the partitioning of sample molecules between the two phases is the basis of these methods, they are also generally referred to as partition chromatography. The rate of migration of a solute can be related to its thermodynamic partition coefficient in a given two-phase system. *See* ADSORPTION.

The most important mode of operation in chromatography is elution. In elution chromatography, the sample (a mixture of components to be separated) is placed as a narrow concentration impulse at the beginning of the chromatographic medium. The mobile phase is then introduced, and the sample components migrate with it through the chromatographic bed. In gas chromatography, in which the mobile phase is most typically an inert inorganic gas, sample molecules are merely transported through the sorption medium. However, in liquid chromatography, the mobile fluid may have an appreciable "extraction" effect on a sample molecule, causing it to migrate more slowly or more quickly; or, alternatively, the mobile-phase molecules can compete with the solute molecules for the available sites on the adsorbent (solid stationary phase).

Chromatography has been used primarily as a separation and isolation method. Unlike classical chemical separation methods (for example, precipitation or crystallization), chromatography is intended to separate many mixture components in a single-step procedure. The use of chromatography in general analytical procedures, in which the separation of mixture constituents can be followed by their direct identifications and quantitative measurements, has been dramatically increasing. Many analytical tasks cannot be adequately dealt with by any other available methods. Chromatographic methods can also be automated for routine analyses. *See* CHEMICAL SEPARATION TECHNIQUES. [M.V.N.]

Chromatophore A cell which contains pigment, especially in a movable form, located in the skin or occasionally in deeper body tissues, and which is responsible for many aspects of the coloration of organisms. Sometimes the word is used to designate the large or complex colored plastids such as chromoplasts or chloroplasts seen in certain plant cells. Usually, animal chromatophores are single cells with specific pigments in their cytoplasm, which may be concentrated in the main cell body or distributed into dendritic protrusions. In cephalopod mollusks, however, chromatophores are miniature organs consisting of a round, saclike pigment cell encircled by a radial arrangement of muscle fibers. Colors and patterns of animals derive from the relative distributions of chromatophores of various types, and color changes may be relatively rapidly effected by the dispersion of pigments within the cells. Depending on the nature and distribution of pigments, light and heat may be absorbed or reflected, so that chromatophores also influence the penetration of radiant energy into the deep body tissues.

This discussion will include only those cells which are capable of effecting color change by movement of their pigments. Also known as pigmentary effectors, they may be found in polychaete worms and leeches, insects and crustaceans, cephalopods, fishes, amphibians, and reptiles. Pigment cells which produce structural color but are only indirectly (if at all) related to color change, notably melanogenic cells in the epidermis, are especially characteristic of birds and mammals, but are also found in other vertebrates. Such cells are not usually called chromatophores. However, it should be noted that "pigmented cells," in the widest, all-inclusive sense, are extremely widespread in many taxa. *See* INTEGUMENT.

Several types of chromatophores are distinguished by their color. The most common are the black or brown melanophores. Others are yellow, red, white, or iridescent and are generally known as xanthophores, erythrophores, leucophores, and iridophores, respectively. No lastingly blue chromatophores are known, but a diffuse water-soluble blue pigment originating in chromatophores of other color appears in the surrounding tissues in some animals, including many crustaceans. In addition, blue, green, or other colors occur as interference effects of light that has been diffracted or refracted by the pigment crystals or intervening films in iridophores or from iridophore colors mingling with the colors of other cells.

A partly different set of chromatophore types is predicated on the chemical nature of their pigment. The predominant type characteristically, as in vertebrates, contains melanin, a class of insoluble black and brown pigments; this type accordingly retains the designation melanophore. Many yellow and red chromatophores are characterized by fat soluble carotenoid pigment ("lipochrome"); they have therefore sometimes been classed as lipophores. Some others were distinguished as allophores because of their different, not fat-soluble, red pigment. The term pterinophore has been introduced for some which owe their yellow or deeper color to pterine, rather than carotenoid, pigment. In the blackish chromatophores of many invertebrates, the pigment has been found to be ommochrome, not melanin.

Environmental conditions that affect the dispersion and concentration of chromatophore pigment chiefly involve light, but in some cases other external factors are decisive, either in themselves or by evoking a nervously excited state.

Induced by mechanical, visual, or other stimuli, which may originate from an enemy or mate, nervous excitement evokes a striking concentration of dark pigment, "excitement pallor," in various reptiles and fishes. A darkening or mottling change, however, is induced in some other reptiles and fishes.

Little general effect on chromatophores results from humidity, but walking-stick insects and frogs darken in moisture and become pale under drier conditions.

Temperature does not often determine chromatophore change when it is in the normal range for vertebrates. But, as a rule, above 86–95°F (30–35°C), or near the extremes of warmth tolerable to the species, melanophore pigment concentration occurs in the cold-blooded vertebrates that have been tested.

In some cases, light affects pigment movement without help from eyes or structures of true visual function. In various crustaceans, light appears to act directly on the chromatophores themselves, which in blinded animals respond as independent effectors. Both white and darker pigments disperse when brightly illuminated and become concentrated in total darkness. Similar responses to light by melanophores have been

Fig. 1. Adaptive changes in skin color. Frogs turned pale in response to a white background and dark to a black background. (*From M. E. Pierce, J. Exp. Zool., 89:293–295, 1942*)

observed in various fishes, amphibians, and reptiles after blinding or in the absence of local nerves or before embryonic development of means of visual mediation.

The chief environmental factor influencing postlarval chromatophore activity is light entering the eyes from the surroundings, or background, of the animal. The responses to this background light depend primarily upon its quantitative relation to the light entering the eyes from overhead or passing to the retinal parts normally exposed to direct light from the sky or main source (Figs. 1 and 2). Though far from universal, the ability to turn paler or darker by such pigmentary responses and thus to conform visibly to the shade of the background is widespread in the classes of chromatophore-possessing animals that have eyes.

A diurnal rhythm of pigment movements that persists for weeks or months independently of the afore-discussed factors characterizes certain marine isopod and decapod crustaceans and such widely representative vertebrates as the lamprey, European minnow, frog, and chameleon. It is evinced particularly under constant darkness; but it may be detected under constant illumination, because by day it slows or reduces and in the night facilitates the chromatophoral adaptation to a white background.

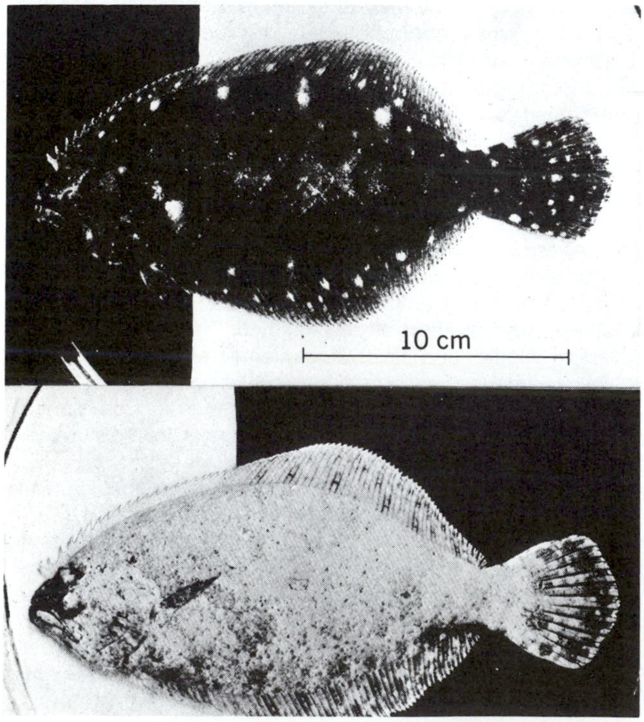

Fig. 2. Darkening and paling reactions of a flounder having only its eyes and its forepart kept over black and white backgrounds, respectively. (*From S. O. Mast, Bull. U.S. Bur. Fish., 821, 1916*)

Chromatophore control systems vary strikingly among different animals. In crustaceans and insects, chromatophore responses to factors affecting them indirectly depend entirely upon the blood's supply of hormones, secreted under visual or other stimulation transmitted through the nervous system. *See* NEUROSECRETION.

Among vertebrates, both bloodborne melanophore hormones and nerve fibers directly innervating the melanophores function prominently in certain elasmobranch dogfishes and probably most bony fishes and reptiles, including horned-toad lizards. Though some evidence of a direct nerve supply to chromatophores has been reported for the lizard *Anolis* and for frogs and toads, color changes in these animals and other amphibians are governed predominantly by hormones. Generally, vertebrate melanophore nerves are pigment-concentrating in action, but in some cases pigment-dispersing nerves are also present. This is true of various bony fishes, including species with partly hormonal systems as well as others with predominantly direct nervous control.

The vertebrate chromatophore hormones include one that is usually called intermedin. It occurs in the intermediate or posterior lobe of the pituitary gland in all species. Intermedin causes melanin dispersal in the many species subject to its action.

The term physiological color change designates any relatively transitory color change that under visual or other stimulation results from pigmentary effector activity, or redistribution of movable pigment already present in existing chromatophores of an animal. In contrast, morphological color change is any intrinsically longer-lasting quantitative change in pigmentation that results from increase or decrease in the total number of chromatophores or from synthesis or destruction of pigment in preexisting chromatophores or from both together.

Color change in certain animals has survival value as camouflage; color adaptation to the background seems to contribute importantly to protection from predators or to capture of prey. In addition, instances of physiological and of morphological color change linked to breeding are found in most of the major groups of chromatophore-possessing animals. Transitory color displays are elements of male mating behavior in some reptiles, fishes, and invertebrates. Bright nuptial coloring, distinctive of sex, develops for the breeding season in many vertebrates, but chromatophores often serve a more fundamental function as screens excluding or absorbing radiation. Thus they can protect inner tissues against harmful illumination. In many reptiles they evidently counteract loss of needed heat by increasing their absorbing area and protect against too high temperature by decreasing that area and affording more reflection. *See* PIGMENTATION; PROTECTIVE COLORATION. [P.F.M.]

Chromite The only important ore mineral of chromium. Chromite crystallizes in the isometric system but crystals are rare and it usually is massive. The hardness is 5.5 (Mohs scale) and the specific gravity 4.6. The luster is submetallic and the color iron-black to brownish-black. Pure chromite with composition $FeCr_2O_4$ is rare, because magnesium usually substitutes for some ferrous iron, and aluminum and ferric iron substitute for chromium. Chromite is a common constituent of peridotites and serpentines and in some is present in large masses. Major producing countries are Turkey, the Union of South Africa, the Soviet Union, the Philippines, and Zimbabwe. *See* CHROMIUM; PERIDOTITE; SERPENTINE. [C.S.Hu.]

Chromium Chemical element, Cr, atomic number 24, atomic weight 51.996, a silver-white metal that is hard and brittle as normally encountered. However, the bulk metal is rel-

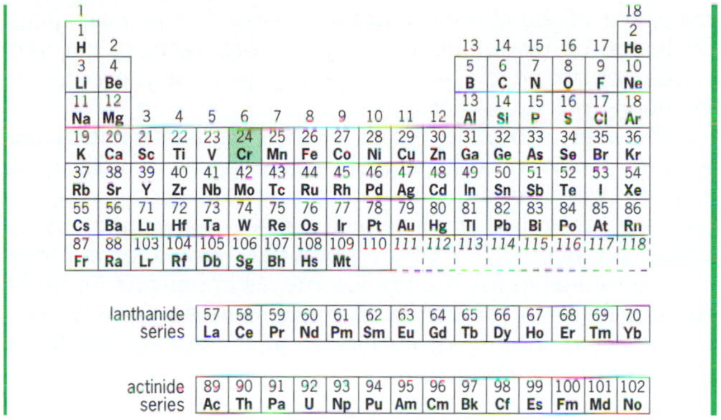

atively soft and ductile when unstressed and either effectively scavenged or extremely pure. Its chief uses are production of noncorrosive, high-strength, heat-resistant characteristics in alloys and as an electroplated coating. Elemental chromium, as such, is not found in nature. The most important mineral occurrence of chromium is in chromite. Of geochemical interest is the fact that basalt found on the Moon contains 0.47 wt % Cr_2O_3, which is 3–20 times as much as that found in representative terrestrial specimens. *See* CHROMITE.

There are four naturally occurring isotopes of chromium, namely, ^{50}Cr, ^{52}Cr, ^{53}Cr, and ^{54}Cr. Various unstable isotopes can be produced by radiochemical reactions. The most important is ^{51}Cr, which emits soft gamma rays and has a half-life of approximately 27 days. Electroplated and polished chromium is bright bluish white. Its reflecting power is 77% that of silver. Chromium has a density of 20°C (68°F) of 7.22 g/cm^3 (0.26 lb/in.^{3}) for a single crystal and 7.14 g/cm^3 (0.259 lb/in.^{3}) for a polycrystalline solid. Chromium has a melting point of 1930 ± 10°C (3505 ± 20°F); its boiling point is 2480°C (4500°F) at 760 mmHg (101.325 kPa).

Mechanical properties reflect the strength and workability of chromium including, among others, its hardness, tensile strength, and forging, rolling, and drawing characteristics. Chromium has relatively poor forging, rolling, and drawing properties. However, when absolutely free of oxygen, hydrogen, carbon, and nitrogen, it is relatively ductile and can be forged and drawn. It is difficult to keep it free from these elements.

Chromium forms three series of compounds with other elements; these series can be represented in terms of the chromium oxides: chromium, valence 2, CrO, chromium(II) or chromous oxide; chromium, valence 3, Cr_2O_3, chromium(III) or chromic oxide; and chromium, valence 6, CrO_3, chromium(VI) or chromic acid anhydride. Chromium is capable of forming compounds with other elements in oxidation states of II, III, and VI. The table lists the typical compounds of each valence state with oxygen.

Valence states of chromium

Compound	Formula	Valence state of Cr
Chromous oxide	CrO	2+
Chromous hydroxide	$Cr(OH)_2$	2+
Chromic tetroxide*	CrO_2	4+
Chromic pentoxide*	CrO_5	5+
Chromic oxide	Cr_2O_3	3+
Hydrous chromic oxid (chromic hydroxide)	$Cr_2O_3{\cdot}xH_2O$	3+
Chromites	$(Cr_2O_4)^{2-}$	3+
Chromic anhydride	CrO_3	6+
Chromates	$(CrO_4)^{2-}$ and $(Cr_2O_7)^{2-}$	6+

*Stable only at high pH.

Peroxides, perchromic acid, and perchromates are also known. The halides (chlorides, fluorides, bromides, and iodides) of chromium are fairly common compounds of this metal. The chloride, for example, has been used to some extent in the production of chromium metal by the hydrogen reduction of chromous chloride, $CrCl_2$. [W.D.Wi.]

Chromomycosis A cutaneous fungus infection of humans also known as chromoblastomycosis. It is usually seen on the legs, feet, and occasionally the arms. Three fungi are most frequently associated with this disease: *Foncecaea pedrosoi*, *F. compactum*, and *Philophora verrucosa*. The lesion begins as an itchy papule, slowly developing into a reddish, warty nodule which eventually becomes pedunculated. As the disease progresses, secondary nodules develop and often coalesce to give rise to a large lesion, cauliflowerlike in appearance. If the lesions are seen in an early stage of development, they may be removed by surgical excision. *See* MEDICAL MYCOLOGY. [L.D.H.]

Chromophycota A division of the plant kingdom (also known as Chromophyta) comprising nine classes of algae: Bacillariophyceae (diatoms), Chrysophyceae (golden or golden-brown algae), Cryptophyceae, Dinophyceae (dinoflagellates), Eustigmatophyceae, Phaeophyceae (brown algae), Prymnesiophyceae, Raphidophyceae, and Xanthophyceae (yellow-green algae). Some of these classes are closely related, while others stand so far apart that they are sometimes assigned to their own divisions. The chief unifying character is the presence of chlorophyll *c* rather than chlorophyll *b* as a complement to chlorophyll *a* (although only chlorophyll *a* is present in Eustigmatophyceae). The chloroplasts are usually brown or yellowish because of large amounts of ß-carotene and various xanthophylls, many of which are restricted to one or more classes. In most classes, motile cells bear two unequal flagella, one of which may be almost completely reduced and at least one of which bears two rows of hairlike appendages. Algae range in size and complexity from unicellular flagellates to gigantic kelps. *See* ALGAE; BACILLARIOPHYCEAE; CHRYSOPHYCEAE; CRYPTOPHYCEAE; DINOPHYCEAE; EUSTIGMATOPHYCEAE; PHAEOPHYCEAE; PRYMNESIOPHYCEAE; RAPHIDOPHYCEAE; XANTHOPHYCEAE. [P.C.Si.; R.L.Moe]

Chromosome One of the cellular organelles which contains the factors of inheritance, or genes, arranged in linear order along its length. The genetic information in a chromosome resides in its deoxyribonucleic acid (DNA), and this linear component is associated laterally with basic proteins (histones) and acidic proteins to give complex, threadlike nucleoprotein structures. The chromosomes are located within, and give rise to, the nucleus in cells of all eukaryotes (plants and animals), but they become clearly visible only under the light microscope when they contract and the nuclear membrane disappears at the time of cell division (mitosis in the soma, meiosis in the gonads). At this time (Fig. 1), they appear as compacted, bipartite, rodlike structures, each consisting of two identical parallel subunits called chromatids. With few exceptions each chromosome normally has a single primary constriction or waist at a fixed position along its length (the centromere or kinetochore), and this structure is involved in separating the two halves of the chromosome at the time of cell division.

New chromosomes originate only by duplication of preexisting ones, and this occurs in the interphase period prior to cell division. During the process of cell division the chromosomes contract, and the chromatids (sister half-chromosomes) separate to opposite poles of the dividing cell. After separation, chromatids swell and uncoil and give rise to a single daughter nucleus in each of the two daughter cells. Each chromosome in

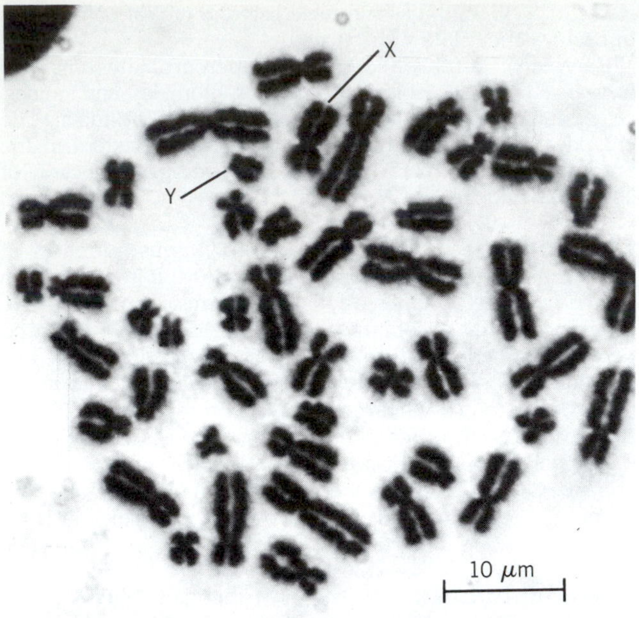

Fig. 1. Peripheral blood leukocyte metaphase from a human male; X and Y chromosomes are indicated.

a newly formed daughter nucleus is therefore represented by a single chromatid which undergoes duplication prior to any succeeding cell division. *See* MEIOSIS; MITOSIS.

Size and number. In the great majority of sexually reproducing organisms the fertilized egg and the cells derived from it contain two sets of chromosomes, one derived from the egg nucleus (maternal set) and the other from the sperm nucleus (paternal set). The gametes themselves contain a single set of haploid (n) number of chromosomes, so that the fertilized egg normally contains two sets or is diploid ($2n$). The minority of organisms, especially among plants, with more than two sets of chromosomes are referred to as polyploids ($>2n$). Thus, in the diploid organism the nucleus in each cell in the soma contains one copy of both the maternally and paternally derived chromosomes. The diploid set therefore contains pairs of chromosomes, and each species has a fixed number of pairs that make up its karyotype or genome. In general, and with the exception of the sex chromosomes, the members of each chromosome pair are identical in shape and size. However, the pairs within a karyotype may differ, and differences in chromosome size and shape are also evident between species.

The DNA that makes up the genetic information is present as a single threadlike molecule in each chromosome so that the bits of information, or genes, in each thread are physically linked together. Each chromosome therefore contains a defined group of genes collectively referred to as a linkage group. Each gene occupies a fixed and specific position within each linkage group, or along the chromosome; and homologous chromosomes, which make up a pair, share common genes. During meiotic cell division in the gonads, homologous chromosomes pair to form bivalent structures and may undergo an exchange of segments. Such reciprocal crossovers (chiasmata) result in a recombination of genes, as between paternally and maternally derived homologs. The number of linkage groups in an organism is thus the same as the number of chromosome pairs, that is, the haploid chromosome number. *See* DEOXYRIBONUCLEIC ACID (DNA); GENE; GENETIC CODE; LINKAGE (GENETICS); RECOMBINATION (GENETICS).

Sex chromosomes. Genes responsible for or associated with sex determination may be located on a special chromosome pair. These are the sex chromosomes or allosomes; the remainder of the chromosomes are referred to as autosomes. Exchange or recombination between segments of chromosomes containing the sex-determining genes is prevented by the nonhomology of their DNAs. The limitation on recombination between allosomes results in their genetic isolation and separate evolution, which in many cases has led to a difference in their size. The larger partner normally carries genes that code for femaleness and is termed the X chromosome, and its smaller partner is the Y chromosome. The genes for maleness may be located in the Y chromosome, as in humans, or in the autosomes, as in the fruitfly. In many species in which maleness is determined by genes located in the autosomes, the Y chromosome may be absent.

An individual's sex is generally determined by the balance of sex genes, that is, by the ratio of the numbers of X to Y chromosomes, or X chromosomes to autosomes. In mammals the female usually has two X chromosomes. In humans, with a normal complement of 46 chromosomes, a normal female has the constitution 46,XX, and all normal human eggs would have the constitution 23,X. The male, being 46,XY, forms two kinds of sperm cells in equal numbers, those carrying an X chromosome (23,X) and those with a Y (23,Y). Random fusion of eggs and sperm at fertilization thus results in equal numbers of XX females and XY males. There are many variations on the basic XX:XY sex chromosome system.

Sex chromosomes, and particularly X chromosomes, not only contain genes concerned with the determination or development of sex, but contain large numbers of genes determining other characters. In humans, for instance, there are genes responsible for color vision, blood clotting, muscular dystrophy, and a host of other characteristics located on the X chromosome. Such genes form part of the linkage group on that chromosome, are associated with the inheritance of sex, and are referred to as sex-linked genes. *See* HUMAN GENETICS; SEX-LINKED INHERITANCE.

Replication. The exact replication of DNA and chromosomes is a basic necessity for the transmission of a complete and unaltered set of genes from parental to daughter cells in the soma and for the familial transmission of heritable traits necessary for the reproductive continuity of the species. The double-helical structure of DNA contains the potential for its exact duplication, since the two strands are complementary. During replication the strands separate, and each acts as a template on which is assembled a complementary strand built from free nucleotides. This results in the formation of two identical double helices, each containing one strand of parental origin (semiconservative replication). Replication takes place during a part of interphase. It is initiated at many hundreds of sites along the chromosome in a temporal pattern characteristic for a given chromosome. Thus some DNA replicates early and some, such as the DNA in heterochromatin, later.

Aberration. An aberration is a change in the normal complement (karyotype) of chromosomes involving an alteration either in chromosome number (heteroploidy) or in chromosome structure (structural changes).

Alterations in chromosome number (heteroploidy) may involve loss, or gain, of entire sets of chromosomes (euploidy) or of one or more individual chromosomes (aneuploidy). There are various causes of heteroploidy, including: cell and nuclear fusion, a process that may be enhanced in animal cells by exposure to virus; successive nuclear divisions without intervening cell division (endomitosis); failure of spindle development and of chromatid separation at mitosis, which can be induced by exposure to drugs like colchicine; failure of meiosis, giving germ cells with a diploid chromosome number (1% of human sperm are diploid); double fertilization, that is, two sperms fertilizing one egg, or a fertilized egg fusing with a polar body. Polyploidy may occur in plant species and even in humans, but polyploid human embryos usually abort although some triploid fetuses may survive to birth. Aneuploidy results from failure of proper separation of chromosomes or chromatids at meiosis or mitosis. It occurs spontaneously, but with increased frequency after mutagen exposure, such as x-rays.

Structural changes within and between chromosomes result from damage to the DNA, leading to breakages across the chromatid or chromosome threads. Breaks may remain unjoined to give fragments which, if acentric (lacking the centromere), may be lost at cell division; alternatively, they may rejoin with other broken ends in the same or different chromosomes to give new structures and gene combinations (Fig. 2). Structural changes can be classified into intra- and

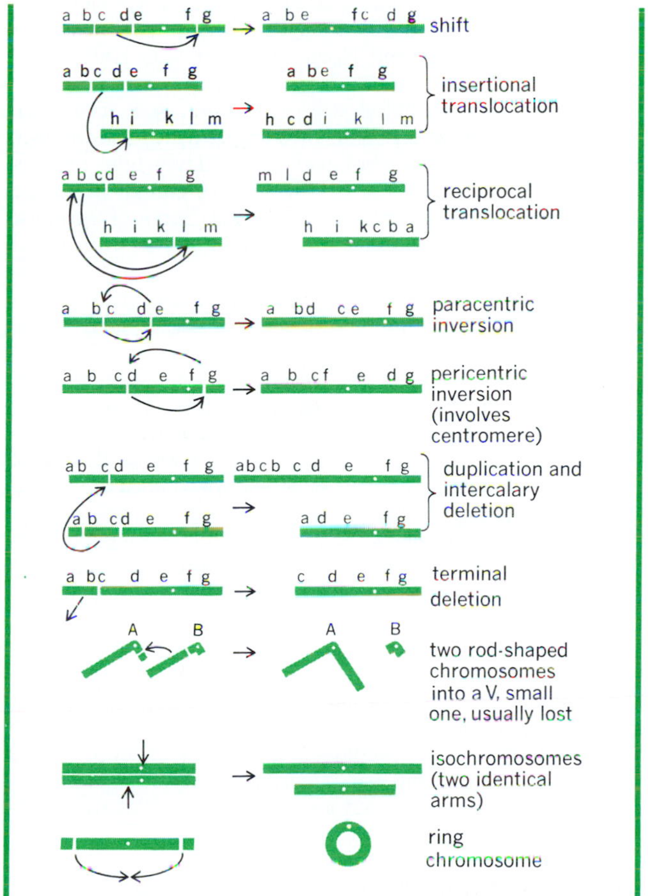

Fig. 2. Common chromosomal alterations.

interchromosomal events and, depending on the unit of breakage and exchange, into chromatid-type and chromosome-type aberrations.

Structural aberrations occur spontaneously, and in humans their incidence increases with age. However, their frequency can be very much increased on exposure to physical mutagens (x-rays, neutrons, ultraviolet light) or chemical mutagens/carcinogens (aflatoxin, burnt products of plant and animal origin, alkylating agents). Chromosomes may also be broken mechanically at cell division if they (abnormally) possess two or more centromeres. Structural rearrangements may be detected genetically, since the linear arrangement of certain genes will be altered, or in favorable cases by microscopic inspection of the chromosomes. *See* Mutation; Radiation biology. [H.J.E.]

Chromosphere A transparent tenuous layer of gas in the atmosphere of the Sun resting on the photosphere. The chromosphere is between 2000 and 3000 mi (3000 and 5000 km) thick. It is a thermal buffer zone between the low temperature at the top of the photosphere (7500°F or 4400 K) and the high temperature of the corona ($2–4 \times 10^6$ °F or $1–2 \times 10^6$ K). The structure of the chromosphere consists of a homogeneous atmosphere, in which are embedded forests of hairy spikes known as spicules. The largest spicules extend 6000 mi (10,000 km) up into the corona above the homogeneous layer. They are believed to carry the energy that maintains the high coronal temperature. *See* Sun. [J.W.E.]

Chronograph A device for recording the epoch of an event. In its older form the astronomical chronograph includes a drum rotated at constant speed and a pen actuated by an electromagnet. A spiral line is traced on paper on the drum. Signals from a clock produce a set of time marks and signals made by an observer produce other marks, so that the epoch of the events observed can be determined.

Modern chronographs are of the digital printing type. In one form rotating type wheels carry numbers. In another form the time is accumulated electronically, and when a signal is given the output is printed by keys. *See* Time-interval measurement. [W.M.]

Chronometer A large, strongly built watch especially designed for precise timekeeping on ships at sea. The name is sometimes loosely applied to any fine watch.

The features that distinguish a chronometer from a watch are (1) a heavy balance wheel, the axis of which is kept always vertical; (2) a balance spring wound in cylindrical shape, instead of a nearly flat helix; (3) a special escapement; and (4) a fusee, by means of which the power of the mainspring is made to work through a lever arm of continuously changing length, being shortest when the spring is tightly wound and longest when it has run down, thus regulating the transmitted power so that it is approximately constant at all times.

Oceangoing ships during a voyage formerly relied completely on chronometers keeping Greenwich mean time as a means for determining longitude. The broadcasting of radio time signals that became widespread in the decade 1920–1930 has made Greenwich mean time available to mariners at almost any time of day, and chronometers are no longer indispensable for determining longitude at sea. *See* Clock. [G.M.C.]

Chrysoberyl A mineral having composition $BeAl_2O_4$ and crystallizing in the orthorhombic system. The hardness is 8.5 (Mohs scale) and the specific gravity is 3.7–3.8. The luster is vitreous and the color various shades of green, yellow, and brown. There are two gem varieties of chrysoberyl. Alexandrite, one of the most prized of gemstones, is an emerald green but in transmitted or in artificial light is red. Cat's eye, or cymophane, is a green chatoyant variety with an opalescent luster. When cut en cabochon, it is crossed by a narrow beam of light. This property results from minute tabular cavities that are arranged in parallel position.

Chrysoberyl is a rare mineral found most commonly in pegmatite dikes and occasionally in granitic rocks and mica schists. Gem material is found in stream gravels in Ceylon and Brazil. The alexandrite variety is found in the Ural Mountains. In the United States chrysoberyl is found in pegmatites in Maine, Connecticut, and Colorado. *See* Beryllium; Gem. [C.S.Hu.]

Chrysocolla A silicate mineral, composition $CuSiO_3 \cdot 2H_2O$. Small acicular crystals have been observed, but it ordinarily occurs in impure cryptocrystalline crusts and masses with conchoidal fracture. The hardness varies from 2 to 4 on Mohs scale, and the specific gravity varies from 2.0 to 2.4. The luster is vitreous and it is normally green to greenish-blue, but may be brown to black when impure. Chrysocolla is a secondary mineral occurring in the oxidized zones of copper deposits, where it is associated with malachite, azurite, native copper, and cuprite. It is a minor ore of copper. *See* SILICATE MINERALS. [C.S.Hu.]

Chrysomonadida An order of the class Phytamastigophorea. Chrysomonads, also known as Chrysomonadina, are usually small flagellates. They are yellow to brown because of phycocrysin in the usual one or two chromatophores, but some lack chromatophores. Many species form diagnostic siliceous cysts. Palmelloid colonies (*Hydrurus*) consist of a tough, gelatinous matrix holding many nonflagellate cells. Starch is not formed, but fats are, and the refractive carbohydrate leucosin is common. The flagella are usually two, rarely three, and are subequal. Nuclei are small. *See* PHYTAMASTIGOPHOREA; SILICOFLAGELLATA. [J.B.L.]

Chrysophyceae A relatively large and diverse class of algae in the chlorphyll *a–c* phyletic line (Chromophycota). In protozoological classification, these organisms constitute an order, Chrysomonadida, of the class Phytomastigophora. Some workers align Chrysophyceae (golden or golden-brown algae) with Bacillariophyceae (diatoms) and Xanthophyceae (yellow-green algae) in the division Chrysophyta. *See* CHROMOPHYCOTA.

Chrysophytes typically are flagellate unicells, either free-living or attached, and solitary or colonial. There are, however, ameboid and plasmodial forms, solitary or colonial nonflagellate cells, filaments, and blades. Fresh-water forms, which are usually found in cold clear water, are more common than marine forms.

Asexual reproduction in unicellular forms is by longitudinal binary fission, autospores (nonflagellate cells that are miniatures of the parent), or zoospores, and in multicellular forms by fragmentation or zoospores. *See* ALGAE; BACILLARIOPHYCEAE; BICOSOECIDA; CHOANOFLAGELLIDA; CHRYSOMONADIDA; XANTHOPHYCEAE. [P.C.Si.; R.L.Moe]

Chytridiomycetes A monophyletic group of true fungi in the subdivision Mastigomycotina, characterized by the presence of a zoosporic state in the life cycle. The zoospore has a single, posteriorly directed, whiplash-type of flagellum; a few species are polyflagellated. The Chytridiomycetes are generally regarded as primitive and are probably representative of the ancestral group for the true fungi.

The thallus may be unicellular, colonial, or filamentous and consists of a single reproductive body (monocentric) or many reproductive bodies (polycentric). The asexual reproductive body comprises a zoosporangium that releases zoospores. Most species are haploid and have no known sexual reproduction. Where sexual reproduction is known, it is highly varied in form. In the least complex species, the thallus consists merely of a naked (having no cell wall) zoosporangium within a host cell. In the most complex species (for example, *Allomyces*), the thallus is polycentric and filamentous with both asexual and sexual reproduction, and with haploid and diploid states. *See* REPRODUCTION (PLANT).

The five orders are classified primarily on differences in zoospore ultrastructure: Chytridiales, Spizellomycetales, Blastocladiales, Monoblepharidales, and Neocallimastigales.

Chytrids are cosmopolitan in distribution and are found in environments from the tundra to the equatorial rainforests, in alkaline or acid soil, in peat bogs, and in fresh or brackish water. Many are saprobes, but some are parasites of microflora and fauna such as algae and rotifers. Some (for example, *Coelomomyces*) are of interest because they destroy mosquitoes, and others are parasites of vascular plants and are of economic concern as vectors of plant viruses. The Neocallimastigales are obligate anaerobes in the rumen and digestive tract of herbivores, where they are important in the breakdown of cellulose and hemicellulose fiber. Many chytrids have been isolated and grown in pure culture, but others are obligate parasites. *See* EUMYCOTA; FUNGI; MASTIGOMYCOTINA. [D.J.S.B.]

Ciconiiformes An order of predominantly long-legged, long-necked wading birds including herons, ibises, spoonbills, storks, and their allies, and also the hawklike New World vultures which were previously placed in the Falconiformes.

The order Ciconiiformes is divided into four suborders and seven families: Ardeae, with the family Ardeidae (herons; 62 species); Balaenicipites, with the family Balaenicipitidae (shoe-billed stork; 1 species); Ciconiae, with the families Scopidae (hammerhead; 1 species), Threskiornithidae (ibises and spoonbills; 33 species), and Ciconiidae (storks; 17 species); and Cathartae, containing the families Teratornithidae (Miocene to Pleistocene of South and North America) and Cathartidae (New World vultures; 7 species).

The herons, shoe-billed stork, hammerhead, ibises and spoonbills, and storks are mainly wading birds with strong legs, living in marshes and other wet areas. They feed on fish, amphibians, and other animals, which are caught in various ways depending on the structure of the bill. Some of the storks are scavengers. Most species are colonial and nest in mixed, often huge, colonies in trees or on the ground, as well as a few on cliff ledges. All are strong fliers, usually having to cover a long distance between the nesting site and feeding areas; some are excellent soarers.

The New World vultures are placed in a separate suborder to emphasize their specialization as scavengers. These vultures are large soaring birds; the Andean condor has the largest wing span of any living land bird. Vultures locate their food from the air either by sight or by smell; these abilities differ in the several species. Vultures may be solitary or hunt in loose flocks; larger concentrations may exist during the winter. They nest solitarily, with the nest placed on the ground or on a cliff ledge. The one to three young remain in the nest until they can fly. *See* AVES. [W.J.B.]

Cidaroida An order of Perischoechinoidea in which the narrow ambulacra comprise two columns of simple plates, and each broad ambulacral plate carries only one primary tubercle. They are unique among echinoids as the only Paleozoic order surviving to the present day. They have remained prominent and beautiful elements in Indonesian and Australasian faunas. *See* ECHINOIDEA; PERISCHOECHINOIDEA. [H.B.F.]

Cilia and flagella Centriole-based, motile cell extensions. These organelles are usually indistinguishable in fine structure as seen with the electron microscope, but quantitatively there are many (several hundred) cilia, and few or fewer (usually one or two) flagella, on one cell. Bacterial or prokaryotic flagella are entirely different organelles that are not considered in this article, which concerns only eukaryotic flagella.

Flagella move with undulatory motion in which successive bending waves progress along the length of the organelle, whereas cilia move with flexural motion consisting of a planar effective stroke, with the organelle extended perpendicular to the cell body, followed by a nonplanar curving recovery stroke, with the organelle pulled parallel to the cell body. Both organelles function to move water past the cell. Their action may bring food and oxygen into an animal, or it may propel the cell to a new environment.

The words cilium (eyelash) and flagellum (whip) are accurate descriptions of the appearance of these cell organelles when they are seen under the light microscope. Cilia are present on protozoa, such as *Paramecium*, and on metazoan cells of many different tissue types. A flagellum is present on human sperm; in fact, the sperm of most animals possess a flagellum,

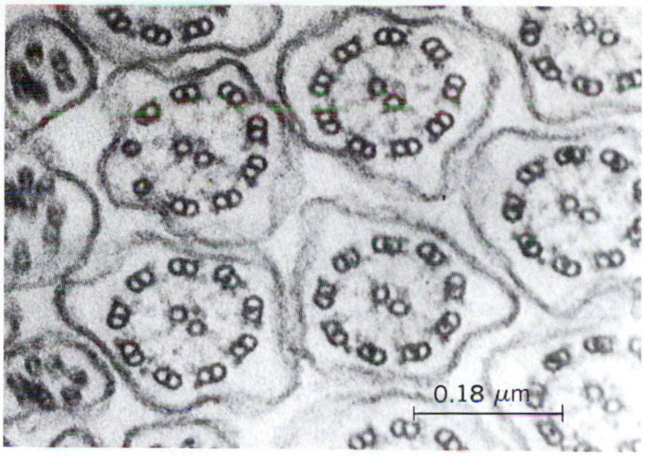

Electron micrograph of cilia showing 9 + 2 pattern of axoneme. (*From P. Satir, Studies on Cilia, II: Examination of the distal region of the ciliary shaft and the role of the filaments in motility, J. Cell Biol., 26:805–834, 1965*)

and, correspondingly, male gametes of many lower plants are flagellated. In addition, many ordinary types of cells of vertebrates, for example, from the thyroid, the kidney, or the pituitary gland, possess modified nonmotile derivatives that resemble cilia. *See* CILIOPHORA; EPITHELIUM; SPERM CELL.

Relative to cell size, both cilia and flagella are very long organelles. Flagella may be over 50 micrometers long. Certain compound cilia, such as the comb plates of ctenophores, are macroscopic structures, visible to the naked eye. Usually, however, cilia range from 10 to 15 μm in length.

The electron microscope reveals that the cilium or flagellum is really an internal organelle since it is bounded by the cell membrane and enclosed at the tip. The main internal structure of the cilium is the axoneme. Under the electron microscope, a single axoneme appears to contain a fixed pattern of microtubules. The microtubules are not simple single units; rather, nine doublet microtubules are found on the periphery of the axoneme surrounding two central elements. This is the so-called 9 + 2 pattern (see illustration). Each peripheral doublet is composed of one complete and one partial microtubule. The microtubules are themselves composed of subunits arranged into microfibers or protofilaments.

At the base of the cilium or flagellum there is a basal body, or kinetosome, that is similar to, and sometimes derived from, a centriole. The basal body may have extensions of various sorts attached to it, notably a basal foot that indicates effective stroke direction and prominent striated rootlet fibers in many cilia. Ordinarily, the ciliary axoneme originates and grows in a membrane protrusion which forms just above the basal body, either at the cell surface or deeper inside the cytoplasm. The basal body remains attached to the cell membrane throughout morphogenesis by a structure that extends from the microtubules to the membrane, where it is seen as a ciliary necklace. *See* CENTRIOLE. [P.S.; I.Gi.]

Ciliatea The single class of the subphylum Ciliophora. This group has the characteristics of those defined for the subphylum. This protozoan class is divided into the subclasses Holotrichia, Peritrichia, Suctoria, and Spirotrichia. *See* CILIOPHORA; HOLOTRICHIA; PERITRICHIA; SPIROTRICHIA; SUCTORIA. [J.O.C.]

Ciliophora A subphylum of the Protozoa. The ciliates are a fairly homogeneous group of highly differentiated, unicellular organisms. Over 5000 species have been described, and many more surely exist but remain to be discovered. Typically, ciliates are larger than most other protozoans, ranging from 10 to 3000 micrometers (about 1/2500 to 1/8 in.). Some larger species are easily visible to the naked eye. The majority of them are free-living forms, found abundantly in a variety of fresh- and salt-water habitats, although a few entire groups live in association with other organisms, generally as harmless ecto- or endocommensals. Their principal value to humans is as experimental animals in a host of biological investigations.

A classification scheme for the ciliates is given below. It is widely recognized as the most reasonable and useful classification currently available. Separate articles appear on each group listed. Subordinal divisions of some orders are commonly recognized by protozoologists.

Subphylum Ciliophora
- Class Ciliatea
  - Subclass Holotrichia
    - Order: Gymnostomatida
      - Trichostomatida
      - Chonotrichida
      - Apostomatida
      - Astomatida
      - Hymenostomatida
      - Thigmotrichida
  - Subclass Peritrichia
    - Order Peritrichida
  - Subclass Suctoria
    - Order Suctorida
  - Subclass Spirotrichia
    - Order: Heterotrichida
      - Oligotrichida
      - Tintinnida
      - Entodiniomorphida
      - Odontostomatida
      - Hypotrichida

The usual ciliate life cycle is fairly simple. An individual feeds and undergoes binary fission, and the resulting filial products repeat the process. Some commensal or parasitic forms have a more complicated life history. Some ciliates, including free-living species, have a cystic stage in their cycle. As in other kinds of Protozoa this stage often serves as a protective phase during adverse environmental conditions, such as desiccation or lack of food. It also may be important in distribution, and thus possibly in preservation, of the species.

Six major characteristics aid in distinguishing the Ciliophora from other protozoan groups. Not all of these are entirely unique, but when taken together they are definitely distinctive of ciliates: mouth, ciliation, infraciliature, nuclear apparatus, fission, and reproduction.

Most Ciliophora possess a true mouth or cytostome often associated with a buccal cavity containing compound ciliary organelles. However, some ciliates are completely astomatous, that is, mouthless. Nutrition is heterotrophic in ciliates.

The Ciliophora possess simple cilia or compound ciliary organelles, often in abundance, in at least one stage of their life cycle. Morphologically, cilia are relatively short and slender hairlike structures, whose ultrastructure is known, from electron microscope studies, to be composed of nine peripheral and two central fibrils. Membranes and membranelles are characteristically associated with the mouth or buccal areas and serve to bring food into the oral opening, although they sometimes aid in locomotion as well. *See* CILIA AND FLAGELLA.

Infraciliature is present, without exception, at a subpellicular level in the cortex. The infraciliature consists essentially of basal bodies, or kinetosomes, associated with cilia and ciliary organelles at their bases, plus certain more or less interconnecting fibrils.

Ciliophora possess two kinds of nuclei, and at least one of each is usually present. The smaller, or micronucleus, contains recognizable chromosomes and behaves much as the single nucleus in cells of metazoan organisms. The larger, or macronucleus, is considered indispensable in controlling metabolic functions, and is recognized as having genic control over all phenotypic characteristics of ciliates.

Ciliophora exhibit a type of binary fission commonly known as transverse division. In ciliates the splitting results in two filial organisms, the anterior or proter and the posterior or opisthe which, geometrically speaking, show homothety with respect to identical structures possessed by each. Thus, homothetogenic is both a broader and most exact descriptive term.

Ciliophora lack true sexual reproduction. Ciliates do not show syngamy, with fusion of free gametes. Processes such as conjugation are considered to be sexual phenomena, since meiosis and chromosome recombination are involved, but not sexual reproduction. In addition to conjugation, certain ciliates exhibit forms of sexual phenomena known as autogamy and cytogamy. *See* Protozoa; Reproduction (animal). [J.O.C.]

Cinematography The process of producing the illusion of a moving picture. Cinematography includes two phases: the taking of the picture with a camera and the showing of the picture with a projector. The camera captures the action by taking a series of still pictures at regular intervals; the projector flashes these pictures on a screen at the same frequency, thus producing an image on the screen that appears to move. This illusion is possible only because of the persistence of vision of the human eye. The still pictures appear on the screen many times a second, and although the screen is dark equally as long as it is lighted by the projected image, they do not seem to be a series of pictures but appear to the viewer to be one continuous picture.

Cameras. Still photographs are taken by motion-picture cameras (movie cameras) at the rate of 24/s, even faster at times. The photographs are sharp and clear, and if they were superimposed on one another would be found to be extremely uniform with respect to position. This last feature is essential in order to have the image appear steady on the screen. A motion-picture camera consists of five basic parts: the lens, the shutter, the gate, the film chamber, and the pull-down mechanism.

The lens on a motion-picture camera collects the light from the scene being photographed and focuses it on the photographic emulsion in the camera aperture. Lenses vary in focal length. Variable-focal-length lenses (zoom lenses) are used extensively not only for filming sport events but also in making feature films, documentaries, and television films. *See* Telephoto lens; Zoom lens.

The shutter in a motion-picture camera rotates and exposes the film according to its angle of opening. When the shutter is open to its widest angle, the film receives its maximum exposure. Many cameras have variable shutters; that is, the angle of opening can be changed while the camera is running.

The aperture called the gate is the passageway through which the film is channeled while it is being exposed. It consists of the aperture plate, which is in front of the film and masks the frame, or picture; the pressure plate, which is in back of the film and holds it firmly against the aperture plate; and the edge guides, which keep the film stable laterally.

The film chamber holds the unexposed film at one end and collects the exposed film at the other. Although some motion-picture cameras have an interior film chamber, it is usually a separate piece of equipment called the film magazine.

The part of the pull-down mechanism called the intermittent movement pulls the film down through the gate of the camera one frame at a time; in a conventional 35-mm camera, each frame is four perforations high. Some wide-screen cameras have frames five or six perforations high. The pull-down claw engages the perforations and pulls the film down into place to be exposed. At the bottom of its stroke, the claw remains stationary for a moment to position the film and then disengages itself and returns to the upper portion of the stroke to pull another frame into place. During the time the claw is returning to the top of its stroke, the film is stationary and the shutter opens and exposes the film.

When sound is being recorded, cameras must be absolutely quiet. If a camera is noisy, it must be put in a sound enclosure called a blimp. Most cameras have to be blimped, although the Mitchell BNC (Blimped News Camera) is self-blimped and is quiet enough to be used on a sound stage (see illustration). The Mitchell BNC, because of its quiet operation and its utmost reliability as well as its ease of operation, has been the standard sound camera the world over. It is used as a model of comparison of all 35-mm professional motion-picture cameras. Several other cameras have been built, however, that compare favorably with the classic Mitchell. These cameras are smaller and lighter than the Mitchell, thereby gaining much of their popularity. They are also reflex cameras, in which the scene being photographed can be viewed directly through the taking lens and does not require a side monitoring finder, which brings about a problem of parallax. Practically all cameras being manufactured today are reflex cameras, including the BNCR, the reflex version of the BNC. These smaller and lighter cameras have enabled production companies to achieve more mobility and to shoot many pictures on actual locations away from studios.

All sound is recorded on ¼-in. (6.35-mm) tape. Since there is no way to accurately control the speed of the tape through the machine, some sort of "sync" pulse must be put on the tape in order to synchronize the recorded sound with the picture when they are put together in a "married" print. The sync pulse is usually an inaudible 1-volt, 60-Hz (in Europe, 50-Hz) pulse put on the tape at the time of recording. There are two types of sound tracks, optical and magnetic, on "release prints," the prints seen by the audience. All 35-mm and practically all 16-mm release prints have optical sound tracks. A 16-mm film may have a magnetic sound track if only one or two prints are to be made, but this is seldom the case. An optical sound track is printed photographically on the film; a magnetic sound track is printed in the same manner as on a tape recorder, after the film is coated with an iron-oxide stripe in the area reserved for the sound track.

Reflexed Mitchell BNC camera with zoom lens on gear head mounted on a crab dolly.

In order to give the camera more mobility and to move quicky from one position to another, 35-mm cameras are almost always used on a four-wheel vehicle, a "crab dolly" (see illustration). This dolly can be steered in the usual manner by turning the wheels for "curve," and it can be put in a mode where all four wheels turn in the same direction so that it can go forward and backward and also "crab" across the stage sideways. A crane is a larger and more versatile version of the crab dolly. Available in a wide range of sizes, some cranes are slightly larger than the crab dolly; others extend to 30 ft (9 m) in the air.

Film. The film used in the standard motion-picture camera is 35 mm (1.378 in.) wide and is perforated along both edges. There are 64 perforations (or sprocket holes) to each foot of film. Motion-picture film consists of a cellulose acetate base approximately 0.006 in. (0.0002 mm) thick and coated with a light-sensitive emulsion. In color film, several layers of emulsion are applied; each of three of the layers is sensitive to one of the three primary colors of light: blue, green, and red.

There are two different types of film: negative film, from which a print is made in order to see the original subject in its true likeness, and reversal film, in which a negative is first formed in the original film and from this a positive is formed in the same piece of film (for example, Kodachrome). In the negative film two pieces of film are necessary to get a picture that can be projected; in the reversal film only one piece of film is required, making it a much less expensive process. In 16 mm or Super 8, as well as regular 8, the reversal film is generally used because of the cost factor. In 16-mm professional work, the original, that is, the film used in the camera, is never used for projection. A "work print" is made in order to edit the film and the original is made to "conform" to the edited work print. The release prints are made from the "conformed" work print and then the sound is added. The original film is all-important and is kept in a vault at the laboratory; it is used only for making prints.

Projectors. The projector in a modern motion-picture theater has four main assemblies: the optical sound head, which reproduces optical (photographically recorded) sound; the projector head, which projects the image onto the screen; the lamphouse, which furnishes the illumination for the picture; and the shutter. On some projectors, there is a second sound head for the reproduction of magnetic (magnetically recorded) sound. The release print of a picture, as contrasted with the negative and the original which are never projected in a theater, is usually mounted on a 2000-ft (600-m) reel. It is placed in the upper film magazine, and the film is threaded through the projector head, down through the optical sound head, and onto the takeup magazine. *See* OPTICAL PROJECTION SYSTEMS.

Wide-screen processes. Until 1952 the motion-picture industry as a whole used film and equipment that was well standardized. Film was 35 mm wide and the camera had a four-perforation pull-down mechanism. These standards had been in effect since early in the 20th century. With the advent of sound in 1926, standard camera speed became established at 24 frames/s and most sound tracks were reproduced optically. Beginning in 1952, however, many new processes were introduced which departed from these standards and utilized a multiplicity of film sizes, aspect ratios, and types of sound tracks. Some processes which continue to employ 35-mm film use a means of getting more picture on the film. Four-track, six-track, and even seven-track magnetic sound has become fairly common. Among the important wide-screen processes have been Cinerama, CinemaScope, Todd-AO, and Dimension 150. [W.G.-F.]

Cinnabar A mineral of composition HgS, crystallizing in the hexagonal system. Crystals are rare, usually of rhombohedral habit and often in penetration twins. Cinnabar most commonly occurs in fine, granular, massive form. It has perfect prismatic cleavage, a Mohs hardness of 2.0–2.5, and a density of 8.09. It has either an adamantine luster and vermilion red color or a dull luster and brownish-red color.

Cinnabar is deposited from hydrothermal solutions in veins and as impregnations near recent volcanic rocks and hot springs. It is the principal ore of mercury. Notable occurrences are Almadén, Spain; Idria, Italy; near Belgrade, Yugoslavia; Kweichow and Hunan provinces, China; Soviet Turkistan; New Almaden and New Idria, California; Terlingua, Texas; and several localities in Utah, Nevada, New Mexico, Oregon, and Idaho. *See* MERCURY (ELEMENT). [L.Gr.]

Cinnamon An evergreen shrub or small tree, *Cinnamomum zeylanicum*, of the laurel family (Lauraceae). A native of Ceylon, the plant is now in cultivation in southern India, Burma, parts of Malaya, West Indies, and South America. The bark is removed from suckers that grow up from the roots, dried, and packaged for shipping. Cinnamon is a very important spice for flavoring foods. It is used in confectionery, gums, incense, dentrifrices, and perfumes. Cinnamon oil is used in medicine and as a source of cinnamon extract. *See* MAGNOLIALES. [P.D.St./E.L.C.]

Circadian rhythms The daily periodic behavior of organisms is commonly governed by the light cycle. However, this cycle serves primarily as a synchronizer or time giver which works upon physiologically based rhythmic processes within the organism, adjusting the periodicity of these processes to the exact 24-h cycle of the environment. Other cyclic features of the environment, such as temperature variations, mechanical disturbances, or noise, may serve as time givers, either alone or in conjunction with a light cycle. With the experimental exclusion of such periodic synchronizing signals, the rhythms of the organisms may continue their characteristic form, but at rates (measured by frequency or period length) which differ slightly from the exact periodicity of the environment. These rhythms are termed circadian rhythms.

Without synchronizing cues, circadian rhythms drift out of phase with the environment so that an organism will perform rhythmic processes at progressively changing times of day. The actual period length of the circadian rhythm, while usually fixed in value for one individual under one set of experimental conditions, still varies from one individual to the next. The range of period lengths for various plants and animals is about 22–28 h, with the majority being fairly close to 24 h. Thus a group of organisms studied simultaneously under conditions which allow the expression of circadian rhythms will drift out of phase with one another as would a group of mechanical clocks each running at a slightly different speed.

The speed of "biological clocks" can be altered by changing the tissue temperature of the organism. Only a few chemicals are known to change the frequency of circadian rhythms. Alcohol slows the rhythm of bean (*Phaseolus*) leaf movements, and this rhythm, as well as a rhythm of swimming activity of the unicellular alga *Euglena*, is slowed by exposure to solutions of heavy water (D_2O). Many of the powerful drugs which either inhibit or enhance metabolic activity are ineffective in altering the frequency of circadian rhythms, although some drugs can change the phase or abolish the measured expression of the rhythm.

The biochemical mechanisms of these rhythms are still unknown. The many rhythmic functions which have been measured appear to be expressions of the underlying timing process of the circadian rhythm without being directly involved in the determination of period length. There is limited evidence suggesting that more than one circadian rhythm may occur within a single organism. *See* PLANT GROWTH. [K.S.R.]

Circle The curve that is the locus of points in a plane with equal distance (radius) from a fixed point (center; see illustra-

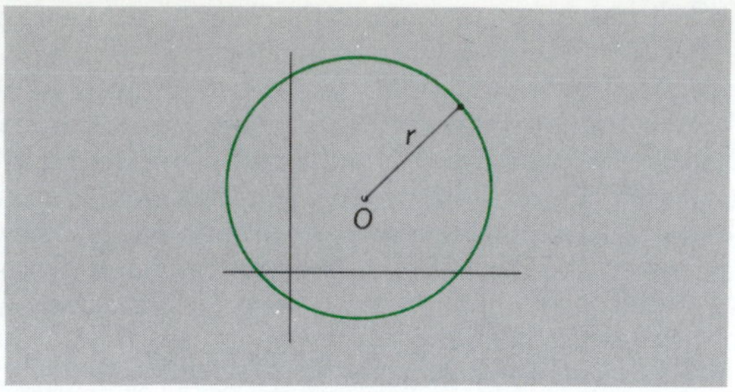

Circle of radius *r* and center *o*.

tion). In elementary mathematics, circle often refers to the finite portion of the plane bounded by a curve (circumference) all points of which are equidistant from a fixed point of the plane, that is, a circular disk. Circles are conic sections and are defined analytically by certain second-degree equations in cartesian coordinates. Electronic computers have calculated π (the ratio of the length of a circle to its diameter) to over 100,000 decimal places. The area of a circle (circular disk) with radius r is πr^2; the length (circumference) is $2\pi r$. The area enclosed by a circle is greater than that bounded by any other curve of the same length. *See* ANALYTIC GEOMETRY; CONIC SECTION. [L.M.Bl.]

Circuit (electricity) A general term referring to a system or part of a system of conducting parts and their interconnections through which an electric current is intended to flow. A circuit is made up of active and passive elements or parts and their interconnecting conducting paths. The active elements are the sources of electric energy for the circuit; they may be batteries, direct-current generators, or alternating-current generators. The passive elements are resistors, inductors, and capacitors. The electric circuit is described by a circuit diagram or map showing the active and passive elements and their connecting conducting paths.

Devices with an individual physical identity, such as amplifiers, transistors, loudspeakers, and generators, are often represented by equivalent circuits for purposes of analysis. These equivalent circuits are made up of the basic passive and active elements listed above.

Electric circuits are used to transmit power as in high-voltage power lines and transformers or in low-voltage distribution circuits in factories and homes; to convert energy from or to its electrical form as in motors, generators, microphones, loudspeakers, and lamps; to communicate information as in telephone, telegraph, radio, and television systems; to process and store data and make logical decisions as in computers; and to form systems for automatic control of equipment.

Electric circuit theory includes the study of all aspects of electric circuits, including analysis, design, and application. In electric circuit theory the fundamental quantities are the potential differences (voltages) in volts between various points, the electric currents in amperes flowing in the several paths, and the parameters in ohms or mhos which describe the passive elements. Other important circuit quantities such as power, energy, and time constants may be calculated from the fundamental variables. For a discussion of these parameters *see* ADMITTANCE; CONDUCTANCE; ELECTRICAL IMPEDANCE; ELECTRICAL RESISTANCE; REACTANCE; SUSCEPTANCE; TIME CONSTANT.

Electric circuit theory is often divided into special topics, either on the basis of how the voltages and currents in the circuit vary with time (direct-current, alternating-current, nonsinusoidal, digital, and transient circuit theory) or by the arrangement or configuration of the electric current paths (series circuits, parallel circuits, series-parallel circuits, networks, coupled circuits, open circuits, and short circuits). Circuit theory can also be divided into special topics according to the physical devices forming the circuit, or the application and use of the circuit (power, communication, electronic, solid-state, integrated, computer, and control circuits). *See* ALTERNATING CURRENT; CIRCUIT (ELECTRONICS); COUPLED CIRCUITS; DIRECT-CURRENT; ELECTRIC TRANSIENT; INTEGRATED CIRCUITS; MAGNETIC CIRCUITS; NEGATIVE-RESISTANCE CIRCUITS; OPEN CIRCUIT; PARALLEL CIRCUIT; SERIES CIRCUIT; SHORT CIRCUIT. [C.F.G.]

Circuit (electronics) An electric circuit in which the equilibrium of electrons in some of the components is upset by a means other than an applied voltage. The means by which electron equilibrium is upset leads to further designation of electronic circuits. In circuits with thermionic tubes, electron equilibrium is upset by thermal emission. In solid-state circuits, electron equilibrium is upset by intentionally introduced irregularities of crystals. Electronic circuits are also classified by their functions. *See* AMPLIFIER; DIGITAL COUNTER; ELECTRONIC POWER SUPPLY; MODULATOR; OSCILLATOR; RECTIFIER; SWEEP GENERATOR; SWITCHING CIRCUIT; WAVE-SHAPING CIRCUITS.

Electronic circuits can consist of discrete elements exclusively, coupled integrated circuits exclusively, or combinations of both kinds. Integrated circuits can consist of groups of lumped components such as resistors, capacitors, and active devices, or of exotic combinations of passive and active components intended to have properties not normally available in lumped discrete-element networks. *See* INTEGRATED CIRCUITS.

Electronic circuits find application in all branches of industry and in the home, both for entertainment equipment and increasingly for control. Because of their low power dissipation and fast response, they are excellent control circuits. Computers, communication systems, and navigation systems use many types of electronic circuits. [K.A.P.]

Circuit breaker A device to open or close an electric power circuit either during normal power system operation or during abnormal conditions. A circuit breaker serves in the course of normal system operation to energize or deenergize loads. During abnormal conditions, when excessive current develops, a circuit breaker opens to protect equipment and surroundings from possible damage due to excess current. These abnormal currents are usually the result of short circuits created by lightning, accidents, deterioration of equipment, or sustained overloads.

Formerly, all circuit breakers were electromechanical devices. In these breakers a mechanism operates one or more pairs of contacts to make or break the circuit. The mechanism is powered either electromagnetically, pneumatically, or hydraulically. The contacts are located in a part termed the interrupter. When the contacts are parted, opening the metallic conductive circuit, an electric arc is created between the contacts. This arc is a high-temperature ionized gas with an electrical conductivity comparable to graphite. Thus the current continues to flow through the arc. The function of the interrupter is to extinguish the arc, completing circuit-breaking action.

In oil circuit breakers, the arc is drawn in oil. The intense heat of the arc decomposes the oil, generating high pressure that produces a fluid flow through the arc to carry energy away. At transmission voltages below 345 kV, oil breakers used to be popular. They are increasingly losing ground to gas-blast circuit breakers such as air-blast breakers and SF_6 circuit breakers.

In air-blast circuit breakers, air is compressed to high pressures. When the contacts part, a blast valve is opened to discharge the high-pressure air to ambient, thus creating a very-high-velocity flow near the arc to dissipate the energy. In SF_6 circuit breakers, the same principle is employed, with SF_6 as the medium instead of air. In the "puffer" SF_6 breaker, the motion of the contacts compresses the gas and forces it to flow through an orifice into the neighborhood of the arc. Both types of SF_6 breakers have been developed for ehv (extra high voltage) transmission systems.

Two other types of circuit breakers have been developed. The vacuum breaker, another electromechanical device, uses the rapid dielectric recovery and high dielectric strength of vacuum. A pair of contacts is hermetically sealed in a vacuum envelope. Actuating motion is transmitted through bellows to the movable contact. When the contacts are parted, an arc is produced and supported by metallic vapor boiled from the electrodes. Vapor particles expand into the vacuum and condense on solid surfaces. At a natural current zero the vapor particles disappear, and the arc is extinguished. Vacuum breakers of up to 242 kV have been built.

The other type of breaker uses a thyristor, a semiconductor device which in the off state prevents current from flowing but which can be turned on with a small electric current through a third electrode, the gate. At the natural current zero, conduction ceases, as it does in arc interrupters. This type of breaker does not require a mechanism. Semiconductor breakers have been built to carry continuous currents up to 10,000 A.

[T.H.L.]

Circuit testing (electricity) The testing of electric circuits to determine and locate any of the following circuit conditions: (1) an open circuit, (2) a short circuit with another conductor in the same circuit, (3) a ground, which is a short circuit between a conductor and ground, (4) leakage (a high-resistance path across a portion of the circuit, to another circuit, or to ground), and (5) a cross (a short circuit or leakage between conductors of different circuits). Circuit testing for complex systems often requires extensive automatic checkout gear to determine the faults defined above as well as many quantities other than resistance.

In cable testing, the first step in fault location is to identify the faulty conductor and type of fault. This is done with a continuity tester, such as a battery and flashlight bulb or buzzer (Fig. 1), or an ohmmeter.

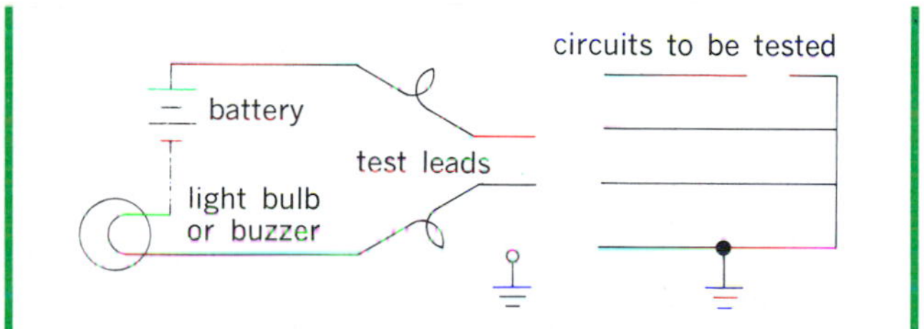

Fig. 1. Simple continuity test setup.

Useful for locating faults in relatively low-resistance circuits, the Murray loop is shown in Fig. 2 with a ground fault in the circuit under test. A known "good" conductor is joined to the faulty conductor at a convenient point beyond the fault but at a known distance from the test connection. One terminal of the test battery is grounded. The resulting Wheatstone bridge is then balanced by adjusting R_B until a null is obtained, as indicated by the detector in Fig. 2. Ratio R_A/R_B is then known. For a circuit having a uniform ratio of resistance with length, circuit resistance is directly proportional to circuit length. Therefore, the distance to the fault is determined from the procedure given by Eqs. (1)–(3). From Eq. (3) and a knowledge

$$R_C \propto l + (l - x) \qquad R_D \propto x \tag{1}$$

$$R_A/R_B = r = R_C/R_D = (2l - x)/x \tag{2}$$

$$x = 2l/(r - 1) \tag{3}$$

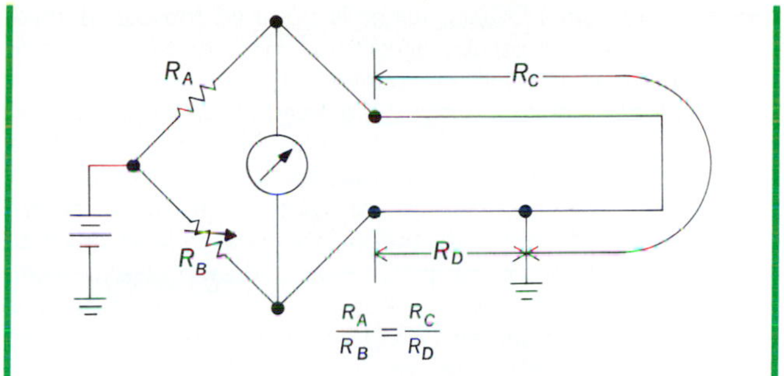

Fig. 2. Murray loop for location of ground fault.

of total length l of the circuit, once ratio r has been measured, the location of the fault x is determined.

The Varley loop test is similar to the Murray loop test except for the inclusion of the adjustable resistance R. The Varley loop is used for fault location in high-resistance circuits.

An alternating-current capacitance bridge can be used for locating an open circuit as shown in Fig. 3. One test terminal

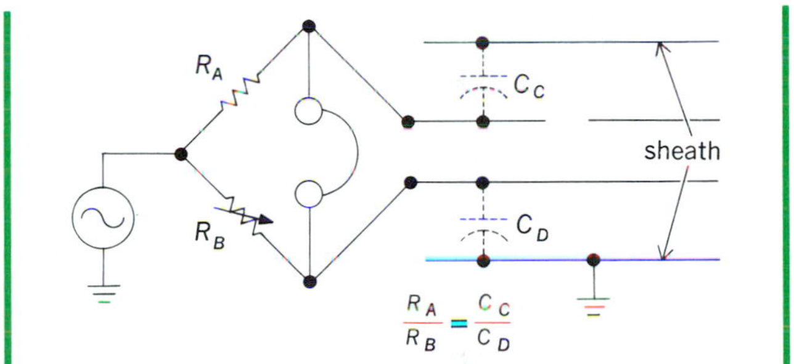

Fig. 3. Alternating-current capacitance bridge used in location of an open circuit in one conductor.

is connected to the open conductor and the other terminal to a conductor of known continuity in the cable. All conductors associated with the test are opened at a convenient point beyond the fault but at a known distance from the test connection. An audio oscillator supplies the voltage to the bridge, which is balanced by adjusting R_B for a null as detected by the earphones. Measured ratio R_A/R_B equals the ratio of capacitances between the lines and the grounded sheath. Because each capacitance is proportional to the length of line connected to the bridge, the location of the open circuit can be determined from Eq. (4).

$$R_A/R_B = C_C/C_D \tag{4}$$

See CIRCUIT (ELECTRICITY).

[C.E.A.]

Circulation Those processes by which metabolic materials are transported from one region of an organism to another. Ultimately, the essential gases, nutrients, and waste products of metabolism are exchanged across cell membranes by diffusion. Diffusion is the movement of material, by random motion of molecules, from a region of high concentration to one of low concentration. The amount of material moved from one place to another depends on the difference in concentrations and on the distance between the two points. The greater the distance, the less movement of material per unit time for a given difference in concentration. Consequently, in all but the smallest animals, convection (or bulk circulation) of materials to the cell must be employed to supplement diffusion.

Protoplasmic movement aids diffusion at the intracellular level. In multicellular animals, however, either the external medium or extracellular body fluids, or both, are circulated. In sponges and coelenterates, water is pumped through definite body channels by muscular activity or, more often, by cilia or flagella on the cells lining the channels.

Coelenterates have a body wall derived from two cell layers; an outer ectoderm is separated from an inner endoderm by a noncellular gelatinous material (mesoglea). All higher animals have bodies consisting of three cell layers, with the ectoderm being separated from the endoderm by a cellular layer of mesoderm. The mesoderm proliferates and separates to develop a fluid-filled body cavity or coelom. The coelom separates the ectoderm (together with an outer layer of mesoderm) from the endoderm (which has an inner layer of mesoderm). Coelomic fluid is moved around by body movements or ciliary activity, but in larger animals this movement is usually inadequate to supply the metabolic requirements of the organs contained within the coelom. These needs are provided for by pumping a fluid, blood, to them through vessels, the blood vascular system. *See* BLOOD.

When the blood is in a separate compartment from the rest of the extracellular fluid, the vascular system is described as closed. The two principal components of such systems are hearts and blood vessels. In such a system, the blood is circulated by a pump, the heart, through special channels, blood vessels; it comes into close association with the tissues only in the capillaries, fine vessels with walls only one cell thick. In some tissues or regions, larger blood spaces may exist, called sinuses. A closed vascular system is found in most annelids (segmented worms and leeches), cephalopod mollusks (squids and octopods), holothurian echinoderms (sea cucumbers), and vertebrates. *See* BLOOD VESSELS; HEART (VERTEBRATE).

In vertebrates, a functional but anatomically closed connection exists between the extracellular spaces (between the cells) and the blood vascular system in the form of lymph channels. Lymph is derived from the noncellular component of blood (plasma), modified in its passage through the tissues, and is conducted to the veins by blind-ending lymphatic vessels, which are separate from blood vessels and coelomic space. *See* LYMPHATIC SYSTEM.

In most arthropods (crustaceans, insects), most mollusks (shellfish), and many ascidians (sea squirts), the extracellular spaces are confluent with the blood system. In these animals, blood is pumped through a limited network of vessels into a body cavity called a hemocoel. After bathing the tissues, blood (called hemolymph in these organisms) collects in sinuses and returns to the heart. This is the open vascular system. In animals with open circulatory systems, the coelom is much reduced. [D.R.J.]

Circulation disorders The function of the circulatory system is to transport and distribute substances either used or produced by cells or both. Excluded are those materials that are discharged directly from sweat glands, digestive glands, and renal tubule cells. Included, however, are nutritive and metabolic substances, hormones, waste products, water, and heat.

Disturbances in the normal pattern of circulation can either result in or from disease conditions. An example is edema, which is an abnormal accumulation of fluid in the cells, tissue spaces, or cavities of the body. There are three main factors in the formation of generalized edema and a fourth which plays a role in the formation of local edema. They are the permeability of the capillary wall, the colloid osmotic pressure of the plasma proteins, and the hydrostatic pressure in the capillaries. The fourth factor, which is of importance in local edema formation, is lymphatic obstruction.

Increased permeability of the capillary walls plays an important role in the formation of inflammatory edema, the edema of severe infections, metabolic intoxications, asphyxia, anaphylactic reactions, secondary shock, and acute nephritis. *See* EDEMA.

A deficiency of circulating blood volume, both cellular elements and fluid, is called oligemia. This may be the result of an acute blood loss or it may be of a chronic nature, such as an anemia combined with dehydration. Anemia, or oligochromemia, is a deficiency of circulating red cell volume or, more specifically, hemoglobin content. It can result from a variety of causes. *See* ANEMIA; BLOOD.

Pancytopenia, or oligocythemia, is a deficiency of all circulating blood cellular elements. This is usually the result of a deficiency of the blood-forming tissue, the bone marrow.

The decrease of blood flow to an organ or tissue is known as ischemia. This can be sudden as when a vessel is ligated, when a thrombus or blood clot forms, or when an embolus comes to lodge in the vessel. A gradual occlusion can follow arteriosclerotic changes in the vessel wall. The effect of a sudden occlusion depends to a great extent on the collateral circulation to the organ or tissue involved. If an adequate collateral circulation comes to be established, the tissue survives; if not, it dies and an infarct results. *See* EMBOLUS; INFARCTION.

An excess of blood is referred to as plethora. This increase may be the result of an increase in the size or the number of red blood cells. The increase in red cell volume may be a polycythemia vera, or true polycythemia, which is a primary increase in the number of red blood cells with no regard to the needs of the organism. Polycythemia, or erythrocytosis, is usually a secondary increase in red cells following conditions of chronic hypoxia. Serous plethora is an increase in the watery part of the blood.

Hyperemia, or congestion, refers to an excess of blood within an organ or tissue. This condition may be localized or generalized. Active hyperemia, caused by an active dilatation of arterioles and capillaries, occurs under certain physiological conditions, such as in the muscle when there is an increased need for blood during exercise. It also occurs in pathological states such as inflammation. Passive hyperemia, a condition which results in an accumulation of blood in the venous system, may be generalized or localized and can result from any obstruction or hindrance to the outflow of blood from the venous circuit. Diseases of the lungs such as emphysema, fibrosis, or pulmonary hypertension of any origin can result in right ventricular failure and generalized congestion. Localized venous congestion results when a main vein from a region is occluded either by a thrombus or some extrinsic pressure such as a tumor or enlarged lymph nodes.

The escape of blood from within the vascular system is hemorrhage. This process can be the result of trauma to, or disease of, the vessel wall. The causes of hemorrhage other than trauma can be divided into three main groups. In the first group are those conditions in which there is a disease process affecting the vessel wall, such as arteriosclerosis or aneurysm formation. In the second group are conditions in which there is an acute process affecting the vessel wall such as in septicemia, poisoning by heavy metals, or even anoxia. The third group consists of those conditions in which there is a defect in the blood itself, which results in hemorrhage. Apoplexy, or stroke, is an acute vascular lesion of the brain. This can be the result of hemorrhage of, thrombosis in, or embolism to a cerebral vessel. *See* HEMORRHAGE.

Thrombosis is the formation of a thrombus, a solid body formed during life and composed of the elements of the blood: platelets, fibrin, red cells, and leukocytes. Thrombosis may occur on a vessel wall anywhere that the endothelium is damaged. However, because the platelets release thromboplastinogen, which activates the clotting mechanism, thrombosis and blood coagulation may occur together. *See* THROMBOSIS.

The sudden blocking of an artery or vein by a clot or other substance which has been brought to its place by the blood current is an embolism. The material carried in the circulation in this process is an embolus. Emboli may be composed of thrombi, fat, air, tumor cells, masses of bacteria or parasites, bone marrow, amniotic fluid, or atheromatous material from the vessel wall. *See* EMBOLISM. [R.A.V.]

Circulatory system The circulatory system of vertebrates actually includes two systems of ramifying tubes, the cardiovascular (blood-vascular) and lymphatic systems. The tubes provide the pathway for the circulating fluids which transport food, water, oxygen, and other necessary materials to the cells and aid in the removal of waste products. *See* BLOOD; CARDIOVASCULAR SYSTEM; LYMPHATIC SYSTEM. [C.B.C.]

Cirque A cliffed rock basin, shaped like half a bowl, at the head of a mountain valley. Cirques may be shallow if glacier erosion is slight, but may be 1600–2600 ft (500–800 m) from the top of the headwall to the cirque floor if erosion is deep. Cirques occur in glaciated mountains all over the world. Some cirques used repeatedly by glaciers over long periods of years are excavated profoundly. Many cirques contain small glaciers today that were inherited from a huge cirque cut by glaciers long ago. Well-formed cirques have steep rock walls and floors that slope down-valley or back toward the base of the headwall. Some cirques are cup-shaped rock basins holding rainwater; a lake formed in them is a tarn. If glacial deposits, such as till or a small moraine on the cirque floor, form a depression for a lake the lake is known as a moraine-dammed lake. [S.E.Wh.]

Cirrhosis Any one of several types of liver disease, all of which are characterized by a real or apparent increased proportion of connective tissue in the organ. Although the term cirrhosis refers to a yellowish or orange color, not all the forms display such discoloration. Common usage has also diverted the term to include scarring or fibrosis and, when used alone, to denote the most common form, that of nutritional (also known as alcoholic, Laennec's, or portal) cirrhosis. *See* LIVER.

In the early stage of cirrhosis the liver is enlarged, loaded with fat, and smooth. Progressive disease eventually leaves a typical nodular, shrunken organ, which is largely scar tissue with an irregular pattern of regenerating liver lobules. Hepatic insufficiency, associated disease, hepatic coma, and death may follow a progressive course in a surprisingly high number of cases which are free of severe symptoms until the terminal stage.

Various classifications of cirrhosis are in use but none is universally accepted, because etiologic factors are sometimes difficult to isolate or to identify. Biliary obstruction and some forms of cardiac disease produce biliary and cardiac cirrhosis, respectively. In addition, liver cell damage from specific infections, chemicals, and other agents may lead to postnecrotic cirrhosis, characterized by massive destruction of liver tissue. Syphilitic cirrhosis is marked by characteristic lesions; other less common types are also recognized. [E.G.St./N.K.M.]

Cirripedia A subclass of the Crustacea, permanently attached when adult, and called barnacles. The carapace forms a complete covering or mantle over the rest of the body and is usually strengthened by calcareous plates. The body within the mantle consists of a mouth region and thorax. The abdomen is usually vestigial. Typically the mouth appendages are paired mandibles with palps, maxillulae, and maxillae. The thorax bears six pairs of biramous appendages (cirri) composed of numerous segments, each with a considerable armament of setae. Compound eyes occur only in the larvae, there is no heart, and typically adults are hermaphroditic. The eggs are incubated in the mantle cavity and hatch into free-swimming nauplius larvae (illustration *a*). By successive molts a further five nauplius larvae follow. The sixth nauplius (or metanauplius) molts into an entirely different larval form, the cypris (illustration *b*), which is characteristic of all Cirripedia. The Cirripedia are divided into four orders: Thoracica, Acrothoracica, Ascothoracica, and Rhizocephala. Only the Thoracica, the goose barnacles, and acorn barnacles are at all conspicuous, and only the latter of any economic significance. *See* ACROTHORACICA; ASCOTHORACICA; BARNACLE; RHIZOCEPHALA; THORACICA. [H.G.St.]

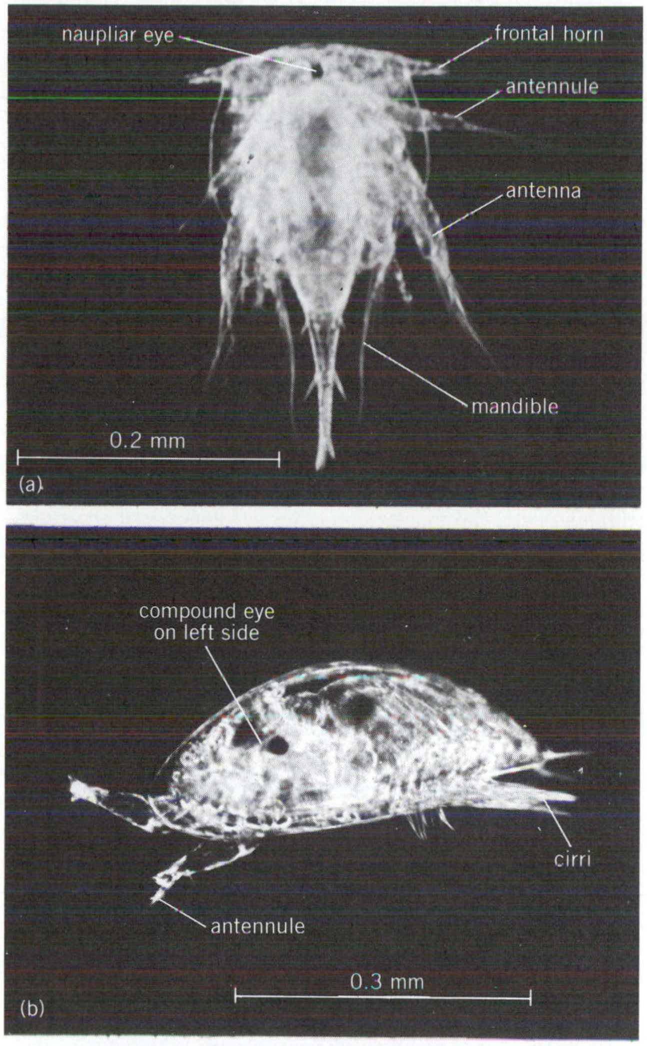

Larval stages of *Balanus*. (*a*) Nauplius. (*b*) Cypris. (*Photographs by D. P. Wilson*)

Citrate A salt or ester of citric acid. Citrates are formed by replacing the acidic hydrogens of citric acid by a metal or an organic radical. Many salts have commercial uses in pharmaceuticals (iron and calcium), cosmetics, and foods; as anticoagulants for blood, saline cathartics (sodium), and effervescent medicines (magnesium); in blueprinting (iron); as plasticizers in plastics; and as resins (simple esters). *See* CITRIC ACID; ESTER. [E.H.H.]

Citric acid An organic acid with formula $C_6H_8O_7 \cdot H_2O$. It is present in almost all natural foods and in the breast milk of nursing mothers. The compound produced commercially by fermentation has a melting point of 153°C (307°F).

Citric acid has many applications in foods, pharmaceuticals, and industry. It is used to enhance flavors, to increase preservative effectiveness, to improve gel strength, and to conserve energy by reducing heat-processing requirements in fruit and vegetable processing. In pharmaceuticals, citric acid can contribute effervescence and improve flavor in certain medications. It also acts to chelate metallic ions in dosage forms and thus lengthens shelf life. In cosmetics and toiletries, citric acid is used in hair rinses, permanent wave neutralizer solutions, low-pH shampoos, lotions, creams, and toothpastes. Citric acid is nontoxic, noncorrosive, and biodegradable. It is used in metal-cleaning applications and as an ingredient in laundry and dishwashing products. In the textile industry it is used in production of soil-release fabrics, carpeting, and dyestuffs. Citric acid is also used in processing a number of minerals, in recovering sulfur from sulfur dioxide emissions in waste gas, as an aid in recovery of secondary oil deposits, in formulating adhesives, and in the concrete and cement industries. [J.W.Fo./R.E.K.]

Citron *Citrus medica*, a species of true citrus. Commercially, citrons are grown almost exclusively in the Mediterranean area, principally in Italy, Sicily, Corsica, Greece, and Israel. The tree is evergreen, as are all citrus, and frost-tender. It is thorny, straggly, shrubby, and tends to be short-lived.

The fruit is scarcely edible fresh, having a very thick skin with little flesh and that lacking in juice. However, it is very fragrant and was valued in ancient times for its aroma and its fragrant peel oil, used in perfumes and as a moth repellent. It is grown commercially only as a source of candied peel for use in cakes and confections. The actual candying is usually done in the importing country, the citron peel being exported in brine.

Confusion sometimes arises due to "citron" also being used for a small, wild, inedible melon in the United States and for lemons (*C. limon*) in France. *See* FRUIT, TREE. [W.G.]

Citronella A tropical grass, *Cymbopogon nardus*, from the leaves of which oil of citronella is distilled. This essential oil is pale yellow, inexpensive, and much used in cheap perfumes and soaps. It is perhaps best known as an insect repellent. A large acreage is devoted to the cultivation of this grass in Java and Ceylon. *See* CYPERALES. [P.D.St./E.L.C.]

Civet Any of 18 species of carnivores assigned to the family Viverridae. Also included in this family are genets, linsangs, and mongooses. *See* MONGOOSE.

Civets are small to medium size with a pointed muzzle, long head, slender body, and long, bushy tail. They have short limbs and nonretractile claws; they are digitigrade, that is, they walk on their toes. The Indian civet (*Viverra zibetha*) and the smaller African civet are two better-known species.

Civets are nocturnal and remain hidden in brush areas during the day. They have well-developed perianal glands, from which a scented substance used in perfumery is secreted. *See* CARNIVORA; SCENT GLAND. [C.B.C.]

Civil engineering The planning, design, construction, and management of all types of works and facilities, including buildings and structures, transportation facilities, water resource development projects, power generation plants, and other facilities for enhancement of the environment. These functions are generally categorized as public works, although the civil engineer also extensively serves industry and private clients.

The civil engineer plans, designs, and constructs residential, industrial, and commercial buildings ranging from warehouses to skyscrapers. Planning involves site selection, allocation of space to building functions, and provision for circulation of people and materials, all in compliance with zoning laws and building codes. Using steel and concrete as the main materials, the civil engineer designs a foundation suited to the geological conditions and loads to be accommodated, and then conceives a frame capable of resisting fire, wind forces, and seismic shocks to support the floors, walls, and roof. *See* BUILDINGS; STRUCTURES (ENGINEERING).

Civil engineering is involved in the creation of a broad spectrum of efficient and economical transportation facilities and terminals, including streets, highways, expressways, and parking facilities; rapid transit systems and subways; transmission systems and pipelines; waterways and harbors; airports; and the launch facilities for vehicles to outer space. *See* HIGHWAY ENGINEERING; TRAFFIC ENGINEERING; TRANSPORTATION ENGINEERING.

Bridges, tunnels, and other special structures appurtenant to highways and railroads are among the most challenging and spectacular projects of the civil engineer. The most creative skills are needed to meet all the parameters of load, topography, geology, social constraints, esthetics, and economy. *See* BRIDGE; TUNNEL.

Water transportation, the most economical method of all, calls upon the civil engineer for channel improvements, canals, locks, dams, ports, harbors, and navigational facilities. Works for controlling erosion by rivers and by ocean tides and currents are also important. *See* COASTAL ENGINEERING; HARBOR; RIVER ENGINEERING.

Air travel involves the civil engineer in airport planning, design, and construction. Site election, runway design, and terminal traffic control for planes, passengers, and baggage must meet demanding constraints of convenience, time-saving, safety, noise control, and esthetics. *See* AIRPORT ENGINEERING.

The vehicle assembly buildings, roadways, pads, and special structures making up the rocket launching complex are demanding the most sophisticated planning, design, and construction techniques of the civil engineer. *See* LAUNCH COMPLEX.

The harnessing of rivers to control floods, produce hydroelectric power, and provide domestic, industrial, and agricultural water supplies is a major area of civil engineering. Groundwater resource development is also important. Efforts toward exploitation of ocean resources for minerals, aquatic life, and power will demand a wide spectrum of civil engineering technology. *See* DAM; HYDRAULICS; SANITARY ENGINEERING.

All large construction projects are planned, directed, and managed by the civil engineer, even though they may require major input by other engineering specialists and architects. Some of the most sophisticated civil engineering works are out of sight beneath the ground, such as complicated foundations and tunnels. Knowledge of geology and soil science is required to solve these problems. *See* ENGINEERING GEOLOGY; FOUNDATIONS; SOIL MECHANICS. [W.H.Wi.]

Cladding An old jewelry art, now employed on an industrial scale to add the desirable surface properties of an expensive metal to a low-cost or strong base metal. In the process a clad metal sheet is made by bonding or welding a thick facing to a slab of base metal; the composite plate is then rolled to the desired thickness. The relative thickness of the layers does not change during rolling. Cladding thickness is usually specified as a percentage of the total thickness, commonly 10%.

Gold-filled jewelry has long been made by this process: the surface is gold, the base metal bronze or brass with the cladding thickness usually 5%. The process is used to add corrosion resistance to steel and to add electrical or thermal conductivity, or good bearing properties, to strong metals. Corrosion-resistant pure aluminum is clad to a strong duralumin base, and many other combinations of metals are widely used in cladding; a development includes a technique for cladding titanium to steel for jet-engine parts.

Cladding supplies a combination of desired properties not found in any one metal. A base metal can be selected for cost or structural properties, and another metal added for surface protection or some special property such as electrical conductivity. Thickness of the cladding can be made much heavier and more durable than obtainable by electroplating.

Cladding can be added to both sides of a sheet or strip of base metal. Tubing can be supplied with a clad surface on inside or outside; round and rectangular wire can be clad similarly (see illustration).

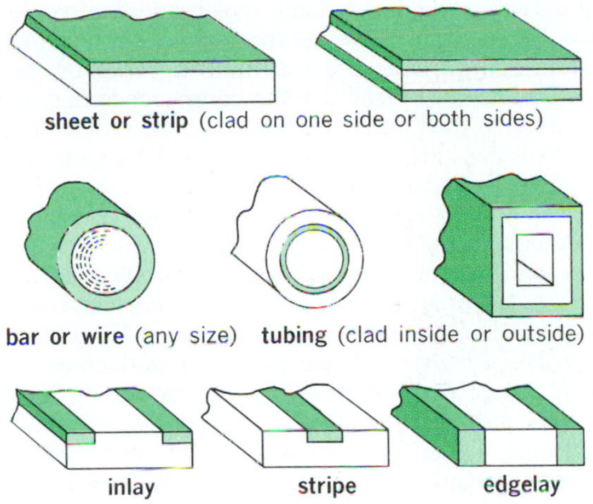

Types of cladding.

For some forms of electrical contacts, the composite materials are bonded side by side, or silver is inset as a stripe on one side or along the edges. This construction can place solid silver just where it is needed to form an electrical contact with no waste of costly metal.

A related form of cladding is found in thermostatic bimetals in which equal thicknesses of low- and high-expansion metals are bonded together. With a change in temperature, differing expansion rates of the two metals cause the composite material to bend and thus operate valves in automobile cooling systems, or electrical contacts in room thermostats.

Clad wires with properly chosen proportions of materials of different thermal-expansion rates can match the thermal expansion of types of glass used for vacuum-tight seals for conductors in lamp bulbs and hermetically sealed enclosures.

In making parts from clad metal, the composite material can be bent, drawn, spun, or otherwise formed just the same as the base metal without breaking the bond. The maximum service temperature is limited by the melting point of the material at the juncture of the two metals. *See* Electroplating of metals; Metal coatings. [R.W.C.]

Cladophorales An order of coarse, wiry filamentous algae in the class Chlorophyceae having a characteristic type of cell division. Some forms are unbranched, as in *Chaetomorpha* and *Urospora*, whereas others are bushily branched, like some species of *Cladophora* and the marine *Spongomorpha*. Most forms are attached and exhibit basal-distal differentiation with rhizoidlike holdfasts. Commonly, thalli become free-floating and *Cladophora* may form globular tangles known as *Cladophora* balls. Some species become symbiotic with sponges. The fresh-water species *Basicladia* occurs on mossback turtles. *See* Chlorophyceae. [G.W.P.]

Cladoselachii An order of extinct elasmobranch fishes including the oldest and most primitive of sharks. Best known is *Cladoselache* (see illustration), of which complete specimens, including even muscle and other soft tissues, have been obtained from shales in the region of Cleveland, Ohio.

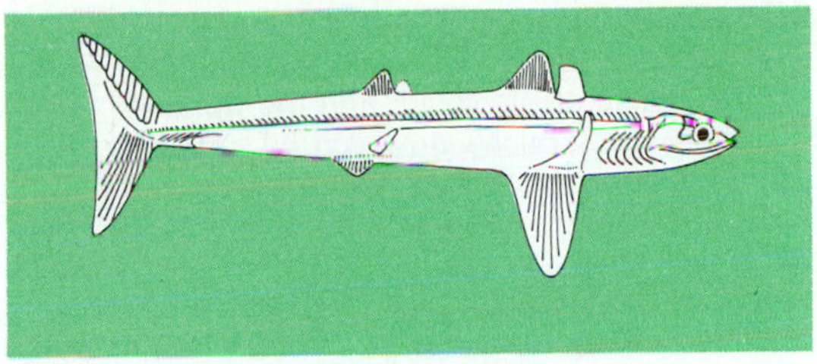

Cladoselache **of the Late Devonian.**

Primitive features are broad-based paired fins, amphistylic jaw support, and absence of claspers. *See* Elasmobranchii. [A.S.R.]

Clamping circuit An electronic circuit that effectively functions as a switch to connect a signal point to a fixed-reference voltage or current level, either at a specific time interval or at some prescribed amplitude level of the signal itself. The circuit is also called, simply, a clamp. Clamps are frequently used to reset the starting level of periodic waveforms, such as sweep generators, and to establish a direct-current (dc) level in a signal which may have lost its dc component because of capacitance coupling. They are also used in many frequency and voltage comparison circuits and in control circuits.

The basic elements of a clamping circuit are shown in the illustration. The clamp between terminals *A* and *B* is shown as

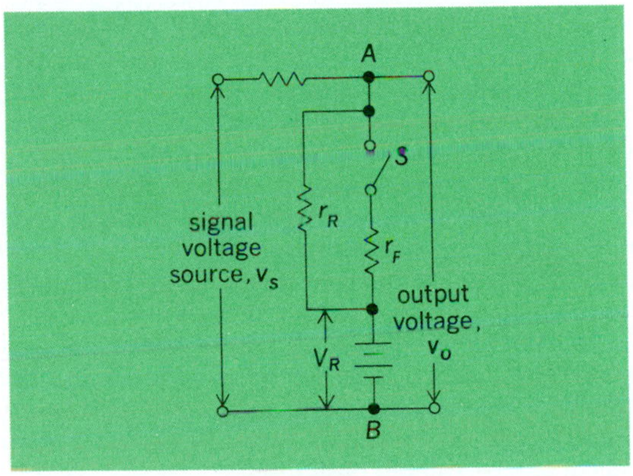

Clamp elements.

a switch *S* in series with reference voltage V_R, having a large resistance r_R when the switch is open and a small resistance r_F when the switch is closed.

If a clamp functions at specific time intervals controllable from separate voltage or current sources rather than from the signal itself, it is a keyed, or synchronous, clamp, and the sources, usually pulses, used to actuate the clamping action are called keying, or clamping, pulses. Such a keyed clamp may be unidirectional but more often it is a bidirectional clamp, defined as a clamp that functions at the prescribed time irrespective of the polarity of the signal source at the time the keying pulses are applied. A three-terminal, high-impedance device such as a field effect transistor (FET) may be used as a keyed clamp to eliminate the need for balanced keying pulses.

A dc restorer is a clamp circuit used to establish a dc reference level in a signal without modifying to any important degree the waveform of the signal itself. For this function to be performed, the signal must have a self-contained reference level which repeats at periodic intervals. A typical example of such a waveform is that comprising the video signal of the television system. *See* Television. [G.M.G.]

Clarification The removal of small amounts of fine, particulate solids from liquids. The purpose is almost invariably to improve the quality of the liquid, and the removed solids often are discarded. The particles removed by a clarifier may be as large as 100 micrometers or as small as 2 micrometers. Clarification is used in the manufacture of pharmaceuticals, beverages, and fiber and film polymers; in the reconditioning of electroplating solutions; in the recovery of dry-cleaning solvent; and for the purification of drinking water and waste water. The filters in the feed line and lubricating oil system of an internal combustion engine are clarifiers.

The methods of clarification include gravity sedimentation, centrifugal sedimentation, filtration, and magnetic separation. Clarification differs from other applications of these mechanical separation techniques by the low solid content of the suspension to be clarified (usually less than 0.2%) and the substantial completion of the particle removal. *See* FILTRATION; MAGNETIC SEPARATION METHODS; MECHANICAL SEPARATION TECHNIQUES; SEDIMENTATION (INDUSTRY); THICKENING. [S.A.M.]

Classical field theory The study of distributions of energy and matter under circumstances where the discrete nature of the latter is unimportant; more commonly known as continuum physics or continuum mechanics. This will generally be the case for a system of a (usually extremely) large number of particles. In this category is the study of the flow of fluids, heat, and other forms of energy; electromagnetic currents and waves, including optical phenomena; and macroscopic theories of the elastic and plastic deformation of solids. Thus, most of applied physical science, such as strength of materials, hydraulics, heat transfer, and aerodynamics, is included.

The quantities involved in these theories, for example, mass and energy densities, fluid velocities, and electric field strengths in a material medium, may be regarded as local averages of corresponding microscopic variables.

Mathematically, field theories are characterized by equations having partial derivatives with respect to position and time. These equations may express the requirements of conservation of energy, or linear and angular momentum, and also the geometrical properties of the particular type of field.

So-called constitutive equations must be added to the general equations. These define ideal materials by restricting the field variables in some way. Simple examples are the assumptions of incompressible or nonviscous fluids, of perfect gases, or of magnetic materials without hysteresis. [B.G.]

Classical mechanics The science dealing with the description of the positions of objects in space under the action of forces as a function of time. Some of the laws of mechanics were recognized at least as early as the time of Archimedes (287?–212 B.C.). In 1638, Galileo stated some of the fundamental concepts of mechanics, and in 1687, Isaac Newton published his *Principia*, which presents the basic laws of motion, the law of gravitation, the theory of tides, and the theory of the solar system. This monumental work and the writings of J. D'Alembert, J. L. Lagrange, P. S. Laplace, and others in the 18th century are recognized as classic works in the field of mechanics. Jointly they serve as the base of the broad field of study known as classical mechanics, or Newtonian mechanics. This field does not encompass the more recent developments in mechanics, such as statistical, relativistic, or quantum mechanics.

In the broad sense, classical mechanics includes the study of motions of gases, liquids, and solids, but more commonly it is taken to refer only to solids. In the restricted reference to solids, classical mechanics is subdivided into statics, kinematics, and dynamics. Statics considers the action of forces that produce equilibrium or rest; kinematics deals with the description of motion without concern for the causes of motion; and dynamics involves the study of the motions of bodies under the actions of forces upon them. For some of the more important areas of classical mechanics *see* BALLISTICS; COLLISION (PHYSICS); DYNAMICS; ENERGY; FORCE; GRAVITATION; KINEMATICS; LAGRANGE'S EQUATIONS; MASS; MOTION; PRECESSION; RIGID-BODY DYNAMICS; STATICS; WORK. [N.S.G.]

Clathrate compounds Well-defined addition compounds formed by inclusion of molecules in cavities existing in crystal lattices or present in large molecules. The constituents are bound in definite ratios, but these are not necessarily integral. The components are not held together by primary valence forces, but instead are the consequence of a tight fit which prevents the smaller partner, the guest, from escaping from the cavity of the host. Consequently, the geometry of the molecules is the decisive factor.

Inclusion compounds can be subdivided into (1) lattice inclusion compounds (inclusion within a lattice which, as such, is built up from smaller single molecules); (2) molecular inclusion compounds (inclusion into larger ring molecules with holes); and (3) inclusion compounds of macromolecules. The best-known lattice inclusion compounds are the urea and thiourea channel inclusion compounds, which are formed by mixing hydrocarbons, carboxylic acids, or long-chain fatty alcohols with solutions of urea. Other representatives of lattice inclusion compounds are the choleic acids, which are adducts of deoxycholic acid with fatty acids, and other lipoic substances. Some aromatic compounds form an open crystal lattice which can accommodate smaller gas and solvent molecules (clathrates in the stricter sense of the word). The gas hydrates are inclusion compounds of gases in a somewhat expanded ice lattice. The gas or solvent molecules are inserted into definite places within the ice lattice and are surrounded by water molecules on all sides.

Crown ether compounds are cyclic or polycyclic polyether compounds capable of including another atom in the center of the ring. In this way, sodium or potassium compounds can be solubilized in organic solvents. Similarly, a series of ionophore antibiotics can complex inorganic cations. *See* CROWN ETHERS.

Some clay minerals are made up of distinct silicate layers. Between these layers some free space may exist in the shape of channels. Smaller hydrocarbon molecules can be accommodated reversibly within these channels. This phenomenon is used in some technical separation processes for separating hydrocarbons (molecular sieves). Furthermore, ion-exchange processes used for water deionization are based on similar minerals. *See* CLAY MINERALS; MOLECULAR SIEVE.

Enzymes are believed to accommodate their substrates in active sites, pockets, or clefts prior to the chemical reaction which then changes the chemical structures of the substrates. These binding processes are identical to those of low-molecular-weight inclusion compounds. [F.Cr.; W.S.; D.G.]

Clathrinida An order of sponges in the subclass Calcinea of the class Calcarea. These sponges have an asconoid structure and lack a true dermal membrane or cortex. The spongocoel is lined with choanocytes. This order contains the family Clathrinidae; *Clathrina*, *Ascute*, and *Dendya* are examples. *See* CALCAREA; CALCINEA. [W.D.H.]

Clavaxinellida An order of sponges of the class Demospongiae, subclass Tetractinomorpha, with monaxonid megascleres arranged in radial or plumose tracts. In the suborder Epipolasina, monactinal or diactinal megascleres are arranged in a radial pattern. Spongin is rare.

Clavaxinellidan sponges vary greatly in shape, from radially symmetrical species which are spherical in shape to encrusting or massive species and upright branching types. They occur in tidal and shallow waters of all seas and extend down to depths of at least 18,000 ft (5500 m). Sponges comparable to existing clavaxinellidans occur scattered through the fossil record from Cambrian strata upward. *See* DEMOSPONGIAE; TETRACTINOMORPHA. [W.D.H.]

Clay A term used both as a rock term and as a particle-size term. As a rock term, it is used for material that has been deposited as a sediment, has formed by hydrothermal action, or is the product of weathering. Chemical analyses of clays show them to be composed primarily of silica, alumina, and water. Iron, alkalies, and alkaline earths are also often present in appreciable quantities. As a rock term clay implies a natural, earthy, fine-grained material which develops plasticity when mixed with a limited amount of water. Plasticity is that property of the moistened material which allows deformation of the material when pressure is applied. In a plastic substance the deformed shape is maintained when the deforming force is removed.

As a particle-size term, clay refers to that fraction of an earthy material containing the smallest particles. In the mechanical analysis of sedimentary rocks and soils, material is normally divided into three size grades, or fractions, on the basis of particle size: sand, silt, and clay. The maximum size of the particles comprising the clay-size grade is defined differently by different observers. A large number of analyses have shown that the clay minerals tend to be concentrated in a size less than 2 micrometers. *See* CLAY MINERALS.

Some types of sedimentary rocks are substantially of the same composition as clay, but differ from clays in their textural characteristics. The expression clay material is often used for any fine-grained, natural, earthy, argillaceous material, when a more precise designation is not possible or desirable. Clay material includes clays, shales, and argillites. Soils would also be included if such materials contained an appreciable quantity of clay-size-grade material and were argillaceous. *See* ARGILLITE; SHALE.

The properties of clay materials are controlled by at least five major factors. These attributes which characterize a clay material are clay-mineral composition, non-clay-mineral composition, organic material, soluble salts and exchangeable ions, and texture. Generally the clay-mineral composition is the most important factor, and sometimes as little as 5% of a particular clay mineral may largely determine the properties of the whole clay. [R.E.Gr.; F.M.W.]

The commercial value of clays is related to their mineralogical and chemical composition, particularly the clay mineral constituents kaolinite, montmorillonite, illite, chlorite, and attapulgite. The presence of minor amounts of mineral or soluble salt impurities in clays can restrict their use. The more common mineral impurities are quartz, mica, carbonates, iron oxides and sulfides, and feldspar. In addition, many clays contain some organic material.

Kaolinitic clays. Clays containing a preponderance of the clay mineral kaolinite are known as kaolinitic clays. Several commercial clays are composed predominantly of kaolinite. These are china clays, kaolines, ball clays, fireclays, and flint clays. The terms china clay and kaolin are used interchangeably in industry. *See* KAOLINITE.

China clays are high-grade white kaolins found in the southeastern United States, England, and many other countries. Many grades of kaolin are used in the manufacture of ceramics, paper, rubber, paint, plastics, insecticides, adhesives, catalysts, and ink. By far the largest consumer of white kaolins is the paper industry, which uses them to make paper products smoother, whiter, and more printable. The kaolin is used both as a filler in the sheet to enhance opacity and receptivity to ink and as a thin coating on the surface of the sheet to make it smoother and whiter for printing. *See* CERAMICS; PORCELAIN; POTTERY.

Ball clays are composed mainly of the mineral kaolinite but usually are much darker in color than kaolin. The term ball clay is used for a fine-grained, very plastic, refractory bond clay. Most ball clays contain minor amounts of organic material and the clay mineral montmorillonite, and are finer grained than china clays. This fineness, together with the montmorillonite and organic material, gives ball clays excellent plasticity and strength. For these reasons and because they fire to a light-cream color, ball clays are commonly used in whitewares and sanitary ware.

The term fireclay is used for clays that will withstand temperatures of 2730°F (1500°C) or higher. Such clays are composed primarily of the mineral kaolinite. Fireclays are generally light to dark gray in color, contain minor amounts of mineral impurities such as illite and quartz, and fire to a cream or buff color. Most fireclays are plastic, but some are nonplastic and very hard; these are known as flint clays. Fireclays are used primarily by the refractories industry. The foundry industry uses fireclay to bind sands into shapes in which metals can be cast. *See* REFRACTORY.

Diaspore clay. This clay is composed of the minerals diaspore and kaolinite. Diaspore is a hydrated aluminum oxide with an Al_2O_3 content of 85% and a water content of 15%. Diaspore clay is used almost exclusively by the refractories industry in making refractory brick. However, after calcination, it is sometimes used as an abrasive material.

Mullite. Mullite is a high-temperature conversion product of many aluminum silicate minerals, including kaolinite, pinite, topaz, dumortierite, pyrophyllite, sericite, andalusite, kyanite, and sillimanite. Mullite is used in refractories to produce materials of high strength and great refractoriness. Mullite does not spall, withstands the shock of heating and cooling exceptionally well, and is resistant to slag erosion. It is used in making spark plugs, laboratory crucibles, kiln furniture, saggars, and other special refractories.

Bentonites. Those clays that are composed mainly of the clay mineral montmorillonite and are formed by the alteration of volcanic ash are known as bentonites. The term bentonite is a rock term, but in industrial usage it has become almost synonymous with swelling clay. Bentonites are used in many industries; the most important uses are as drilling muds and catalysts in the petroleum industry, as bonding clays in foundries, as bonding agents for taconite pellets, and as adsorbents in many industries. *See* BENTONITE; MONTMORILLONITE.

Drilling mud consists of bentonite, water, a weighting material such as barite, and various electrolytes. This mud is circulated in the hole at all times while the drilling is progressing. The functions of the drilling mud are to remove the drill cuttings from the drill hole, to form a filter cake on the wall of the drill hole to keep the drilling fluid in the hole and prevent water in the rock formations from entering, to cool and lubricate the bit, and to prevent blowouts. Bentonite, especially the sodium variety, imparts high viscosity and shear strength to the mud and also imparts a property called thixotropy. A thixotropic body acts as a fluid when agitated and as a thick gel when standing.

Bentonites are used in the foundry industry as bonding clays to bind sands into desired shapes in which metals can be cast. Only 3–5% bentonite is needed to bond the sand grains together. Because of their fine particle size and the nature of their water adsorption, bentonites give the mold a higher green, dry, and hot strength than does any other type of clay.

Bentonites are also used as bleaching clays to remove coloring matter from oils, and as clarifying agents for products such as wines and beers. Bentonite is used as an adsorbent for oils, insecticides, alkaloids, vitamins, proteins, and many other materials. Bentonite suspensions in water are widely used in industry, in addition to that suspension used as a drilling mud by the petroleum industry.

Attapulgite clays. Attapulgite is a hydrated magnesium aluminum silicate with a needlelike shape. Each individual needle is exceedingly small, about 1 micrometer in length and approximately 0.01 micrometer across. Attapulgite is used as a suspending agent, and gives high viscosity because of the interaction of the needles. Some commercial uses are as an oil well drilling fluid, in adhesives as a viscosity control, in oil base foundry sand binders, as thickeners in latex paints, in liquid suspension fertilizers, and as a suspending agent and thickener in pharmaceuticals.

Properties. Most clays become plastic when mixed with varying proportions of water. Plasticity of a material can be defined as the ability of the material to undergo permanent deformation in any direction without rupture under a stress beyond that of elastic yielding. Clays range from those which are very plastic, called fat clay, to those which are barely plastic, called lean clay. The type of clay mineral, particle size and shape, organic matter, soluble salts, adsorbed ions, and the amount and type of nonclay minerals are all known to affect the plastic properties of a clay.

Green strength and dry strength properties are very important because most structural clay products are handled at least once and must be strong enough to maintain shape. Green strength is the strength of the clay material in the wet, plastic state. Dry strength is the strength of the clay after it has been dried.

Both drying and firing shrinkages are important properties of clay used for structural clay products. Shrinkage is the loss in volume of a clay when it dries or when it is fired. Drying shrinkage is high in most very plastic clays and tends to produce cracking and warping. It is low in sandy clays or clays of low plasticity and tends to produce a weak, porous body. Firing shrinkage depends on the volatile materials present, the types of crystalline phase changes that take place during firing, and the dehydration characteristics of the clay minerals.

The temperature range of vitrification, or glass formation, is a very important property in structural products. Vitrification is due to a process of gradual fusion in which some of the more easily melted constituents begin to produce an increasing amount of liquid which makes up the glassy bonding material in the final fired product. The degree of vitrification depends on the duration of firing as well as on the temperature attained.

Color is important in most structural clay products, particularly the maintenance of uniform color. The color of a product is influenced by the state of oxidation of iron, the state of division of the iron minerals, the firing temperature and degree of vitrification, the proportion of alumina, lime, and magnesia in the clay material, and the composition of the fire gases during the burning operation. [H.H.Mu.]

Clay minerals The clay minerals are essentially hydrous aluminum silicates. Alkalies or alkaline earths are present as principal constituents in some of them. Also, in some clay minerals magnesium or iron or both substitute wholly or in part for the aluminum. The ultimate chemical constituents of the clay minerals vary not only in amounts, but also in the way in which they are combined or are present in various clay minerals. They are the major components of clay materials and occur in extremely small particles which are essentially crystalline and are limited in number. *See* CLAY.

While a completely satisfactory classification of the clay minerals has not yet been suggested, the following classification has proved to be a workable one. A major subdivision into amorphous and crystalline groups is made even though the amorphous components are relatively rare and of little importance.

I. Amorphous
  Allophane group
II. Crystalline
  A. Two-layer type (sheet structures composed of units of one layer of silica tetrahedrons and one layer of alumina octahedrons)
    1. Equidimensional
      Kaolinite group: kaolinite, nacrite, dickite, and so forth
    2. Elongate
      Halloysite group
  B. Three-layer types (sheet structures composed of two layers of silica tetrahedrons and one central dioctahedral or trioctahedral layer)
    1. Expanding structure
      a. Equidimensional
        Montmorillonite group: montmorillonite, sauconite, and so forth
        Vermiculite
      b. Elongate
        Montmorillonite group: nontronite, saponite, hectorite
    2. Nonexpanding structure
      Illite group
  C. Regular mixed-layer types (ordered stacking of alternate structural types)
      Chlorite group
  D. Chain-structure types (similar to hornblende—chains of silica tetrahedrons linked together by octahedral groups of oxygens and hydroxyls containing Al and Mg atoms)
      Attapulgite
      Sepiolite
      Palygorskite

The three-layer minerals are divided into the expanding and nonexpanding types in this classification. This subdivision is made even though there may be a continuous gradation between the two types. The expanding attribute is a readily determinable diagnostic property, since it imparts some unique physical properties to the clay materials composed of these minerals. The expanding materials are divided into equidimensional and elongate divisions, because shape is a readily recognizable characteristic that probably reflects structural attributes. *See* CHLORITE; HALLOYSITE; ILLITE; KAOLINITE; MONTMORILLONITE; VERMICULITE.

The atomic structures of the clay minerals consist of combinations of two basic structural units. One unit consists of silica tetrahedrons, in which each tetrahedron is made up of a silicon atom equidistant from four oxygen atoms or hydroxyl ions arranged in the form of a tetrahedron with the silicon atom at the center. The other structural unit consists of two sheets of closely packed oxygen atoms or hydroxyl ions in which aluminum, iron, or magnesium atoms are octahedrally coordinated. Because of the sheetlike nature of their structural units, minerals are tubular or elongate or both and some are fibrous.

The varied properties of the clay minerals are important in that they govern the economic use of clay materials.

The ability of clay minerals to hold certain cations and anions which are readily exchangeable for other cations and anions is most significant. This property of ion exchange and the exchange reaction are of fundamental importance in all fields in which clay materials are used. For example, in soils the retention and availability of potash added in fertilizers depends on cation exchange between the potash salt and the clay mineral in the soil. *See* ION EXCHANGE; SOIL.

Another important property of clay materials is their ability to hold water. This water is of two types, low-temperature water and OH structural water. The low-temperature water is driven off by heating to 212–302°F (100–150°C). This is the water in pores, on the surfaces, and around the edges of the minerals composing the the material.

When clay materials are heated, dehydration occurs. Thus there is a loss of any water (adsorbed, interlayer, or structural OH water) held by the clay minerals. The heating of clay materials also causes changes in the clay-mineral structures. At relatively high temperatures these structural changes facilitate the formation of new mineral phases. These structural modifications are particularly important in the firing of clay materials.

Another property of the clay minerals is their ability to react with organic materials. Clays with a high adsorbing capacity are used in decolorizing oils, while others provide catalysts in the cracking of organic compounds.

Many of the clay minerals are of a hydrothermal origin. Some hydrothermal clay deposits are monomineralic, but most consist of a mixture of clay minerals. The weathering of soils and of different rock types also leads to the formation of clay minerals. The type of clay mineral formed, however, depends upon a number of factors, such as parent rock, climate, topography, vegetation, and time. Clay minerals are abundant in sediments, both recent and ancient. In some cases at least, the character of the clay minerals changes in passing from one environment to another, such as from fresh water to marine, so that the clay mineral composition reflects the geologic history of the sediment. *See* MARINE SEDIMENTS; SEDIMENTARY ROCKS. [R.E.Gr.; M.W.]

Clear-air turbulence Turbulence above the boundary layer but not associated with cumulus convection. The atmosphere is a fluid in turbulent motion. That turbulence of a scale sensed by humans in aircraft is primarily associated with the boundary layer within a kilometer or so of the Earth, where it is induced by the surface roughness, or in regions of deep convection such as cumulus cloud development or thunderstorms. However, aircraft occasionally encounter turbulence when flying at altitudes well above the surface and far from convective clouds. This phenomenon has been given the rather unsatisfactory name of clear-air turbulence (CAT).

What is primarily sensed in CAT by the human is vertical acceleration. This acceleration will depend on the person's location in the plane, the speed of flight relative to the air, and the response characteristics of the airframe. A plane with a wing that generates aerodynamic lift more efficiently or an airframe with less weight per unit wing area will respond more strongly to a given gust magnitude.

CAT is encountered in the atmosphere with a probability depending on flight altitude, geographical location, season of the year, and meteorological conditions. Given this variability and the small scale of the phenomenon, it is difficult to establish reliable statistics on the frequency of its occurrence. Although CAT may be encountered in unexpected meteorological contexts, there are highly favored locations for its occurrence. One is in the vicinity of the jet stream, particularly in ridges and troughs where the wind direction is turning sharply. A second and even more common location of occurrence is in the lee of a mountain range when a strong air flow is distorted by being forced over the range. In this situation a gravity lee wave is generated, which propagates to stratospheric heights. At various altitudes and distances from the mountain, this wave may break, and as many as a dozen or more CAT patches, light to severe, may be formed. Despite knowledge of these favored meteorological areas, it is not possible to forecast with confidence the precise location of a CAT patch. Warning forecasts for substantial portions of routes are typically given to pilots when CAT conditions prevail. *See* JET STREAM.

Aside from the practical implications of CAT for air transport, this phenomenon plays a role of undetermined magnitude in the dissipation of the kinetic energy of the atmosphere. *See* UPPER ATMOSPHERE DYNAMICS. [M.G.W.; L.J.E.]

Cleavage (embryology) The subdivision of eggs into cells called blastomeres. It occurs in eggs activated by fertilization or parthenogenetic agents. Cleavages follow one another so rapidly that there is little opportunity for daughter cells to grow before they divide again. Consequently the size of blastomeres diminishes progressively, although many times unequally, during cleavage. By contrast, the nucleus of each daughter cell enlarges following each cleavage with the result that the ratio of the volume of the nucleus to the volume of cytoplasm (the nucleoplasmic ratio) progressively increases. The cleavage period is said by some authorities to terminate when the nucleoplasmic ratios of various blastomeres attain values characteristic of adult tissues. Cells continue to divide thereafter, but each daughter cell then undergoes a period of growth prior to its division with the result that the nucleoplasmic ratio tends to remain approximately constant for each cell type following termination of cleavage. According to others, cleavage terminates with formation of the definitive blastula. Cleavage appears to be an essential step in development. *See* BLASTULATION.

Cleavage, however, does more than merely subdivide the substance of the egg quantitatively into smaller units, the blastomeres, which are then of such a size that they can readily undergo the subsequent events of blastulation, gastrulation, and interaction that are involved in formation of tissues and organs. Sooner or later cleavage segregates different cytoplasmic areas into different blastomeres, thus subdividing the substance of the egg qualitatively. These qualitative cytoplasmic differences among blastomeres are then sufficient to account for the initial establishment of different lines of differentiation in the progeny of different blastomeres, even though the genetic content of all blastomeres is identical. *See* CELL LINEAGE. [R.L.W.]

Cleft lip and cleft palate Two of the most common congenital anomalies in humans, resulting from incomplete closure of the lip and palate during early embryonic life. During the first trimester of pregnancy, the face and mouth are formed by the fusion of several different parts. The lip is fully fused 6 or 7 weeks after fertilization. The palate, which forms the roof of the mouth and separates the oral cavity from the nasal cavity, is fully fused 10 weeks after fertilization. If fusion fails to occur or breaks down, an opening, or cleft, occurs in the lip, the palate, or both. Although cleft lip and cleft palate are associated with separate embryologic events, that is, are separated by a long time on the embryonic time scale, both often occur in an affected individual.

Clefts of the lip may occur on one side (unilateral), or both sides (bilateral). When they occur in the exact midline of the lip, they are often associated with severe anomalies of the brain. The cleft usually extends through the lip and nostril floor. The underlying dental arch is also typically involved.

Clefts are the only major abnormality in approximately half of newborns who have them. In the other half, they occur as a part of a pattern of multiple anomalies (a syndrome). It has been determined that most clefts are associated with genetic factors. *See* HUMAN GENETICS.

Because of the combination of esthetic, structural, and functional requirements for total repair of clefts, that is, surgical, dental, ear, hearing, and speech, the care of children with clefts has usually been managed by comprehensive teams of specialists, including plastic surgeons, speech pathologists, audiologists, geneticists, otolaryngologists, oral surgeons, orthodontists, and pediatricians. *See* CONGENITAL ANOMALIES. [R.Sh.]

Clevis pin A fastener with a head at one end and a hole at the other used to join a clevis to a rod. A clevis is a yoke with a hole formed or attached at one end of a rod (see illustra-

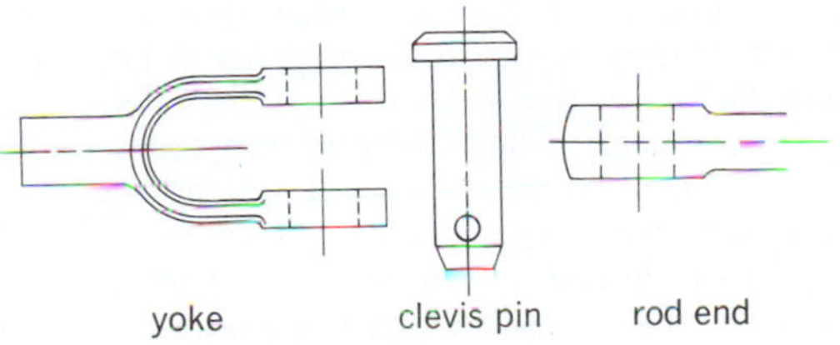

Clevis pin which joins yoke to rod end.

tion). When an eye or hole of a second rod is aligned with the hole in the yoke, a clevis pin can be inserted to join the two. A cotter pin can then be inserted in the hole of the clevis pin to hold it in, yet the fastening is readily detachable. This joint is used for rods in tension where some flexibility is required. *See* COTTER PIN. [P.H.B.]

Climate modeling Construction of a mathematical model of the climate system of the Earth capable of simulating its behavior under present and altered conditions. The Earth's climate is continually changing over time scales ranging from millions of years to a few years. Since the climate is determined by the laws of classical physics, it should be possible in principle to construct such a model. The advent of a worldwide weather observing system capable of gathering data for validation and the development and widespread routine use of digital computers have made this undertaking possible.

The Earth's average temperature is determined mainly by the balance of radiant energy absorbed from sunlight and the radiant energy emitted by the Earth system. About 30% of the incoming radiation is reflected directly to space, and 72% of the remainder is absorbed at the surface. The radiation is absorbed unevenly over the Earth, which sets up thermal contrasts that in turn induce convective circulations in the atmosphere and oceans. Climate models attempt to calculate from mathematical algorithms the effects of these contrasts and the resulting motions in order to understand better and perhaps predict future climates in some probabilistic sense. *See* SOLAR RADIATION; TERRESTRIAL RADIATION.

Climate models differ in complexity, depending upon the application. The simplest models are intended for describing only the surface thermal field at a fairly coarse resolution. These mainly thermodynamical formulations are successful at describing the seasonal cycle of the present climate, and have been used in some simulations of past climates, for example, for different continental arrangements millions of years ago. At the other end of the spectrum are the most complex climate models, which are extensions of the models in weather forecasts. These models aim at simulating seasonal and even monthly averages just shortly into the future, based upon conditions such as the temperatures of the tropical-sea surfaces. Intermediate to these extremes are models that attempt to model climate on a decadal basis, and these are used mainly in studies of the impact of hypothesized anthropogenically induced climate change. *See* WEATHER FORECASTING AND PREDICTION.

Climate models are being used in a large variety of applications that aid in the understanding of Earth history. Many simple climate model simulations have been used to sort out the mechanisms responsible for climate change in the past.

A problem that has received considerable attention is that of the greenhouse effect. Models are being studied to attempt to achieve better understanding of how the increase of atmospheric carbon dioxide and other trace gases from anthropogenic sources are likely to change the climate in the coming decades. The models are being compared to past climates ranging from the ice ages to the records of the last hundred years, for which an instrumental record exists. *See* CLIMATOLOGY; GREENHOUSE EFFECT. [G.R.N.]

Climatology That branch of meteorology concerned with climate, that is, with the mean-physical state of the atmosphere together with its statistical variations in both space and time, as reflected in the totality of weather behavior over a period of many years. Climatology encompasses not only the description of climate but also the physical origins and the wide-ranging practical consequences of climate and of climatic change. Thus it impinges on a wide range of other sciences, including solar system astronomy, oceanography, geography, geology and geophysics, biology and medicine, agriculture, engineering, economics, social and political science, and mathematical statistics.

Like meteorology, climatology is conveniently resolved into subdisciplines in a way that recognizes a hierarchy of geographical scales of climatic phenomena and their governing physics. Macroclimatology refers to the largest (planetary) scale of regimes and phenomena; regional climatology to the scale of continents and subcontinental areas; mesoclimatology to the scale of individual physiographic features such as a mountain, lake, or urban area; and microclimatology to the smallest scale, for example, a house lot or the habitat of an insect. Climatology is also resolved in another way that distinguishes between the theoretical, descriptive, and applied aspects of the science: physical and dynamic climatology, concerned with the governing physical laws; descriptive and synoptic climatology, concerned with comparisons of climatic norms and anomalies, respectively, as concurrently observed in different places; and applied climatology, concerned with the practical utilization of climatological data in engineering design, operations strategy, and activity planning. Other important subdisciplines are climatography, concerned with the comprehensive documentation of climate by means of data summaries, maps, and atlases; and statistical climatology, concerned mostly with the estimation of climatological expectancies, risks of extreme events, and probability distributions of climatic variables (including joint distributions of combinations of variables) as needed to solve problems in applied climatology.

The climate of the Earth as a whole, including its geographical and seasonal variations, is fundamentally prescribed by the disposition of solar radiant energy that is intercepted by the atmosphere. This disposition depends in part on astronomical factors and in part on terrestrial factors, such as atmospheric composition, the distribution of land and sea, and the physiographic nature of the land. At the mean Earth-Sun distance, solar energy arrives in the vicinity of the Earth at the rate of approximately 1.95 g-cal/(cm^2)(min), as measured outside the atmosphere perpendicular to the incident rays. When reckoned per unit horizontal area, however, incoming solar radiation (or insolation) is intercepted at different rates over different parts of the Earth. The rotation of the Earth causes this rate to change locally with time of day. In addition, the shape of the Earth, together with the seasonal changes in orbital distance and axial tilt of the Earth relative to the Sun, causes solar radiation (per unit horizontal area and per unit time) to vary with latitude and with time of year. *See* CONTINENTALITY (METEOROLOGY); GREENHOUSE EFFECT; INSOLATION; TERRESTRIAL RADIATION.

A wide variety of schemes are available for the definition of climatic regions. These schemes fall into two classes, those intended to bring out one or another aspect of relationships and processes within the atmosphere, and those intended to bring out relationships between areal variations in climatic conditions and corresponding variations in phenomena related to climate. It is feasible and instructive to distinguish minor climatic regions, not on a global scale, but within much smaller areas. Where such regions are defined upon the lands, the topographic factor becomes very important. This is especially true where the area is of the order of a few square miles or less. *See* MICROMETEOROLOGY. [J.M.Mi.]

Clinical immunology A branch of clinical pathology concerned with the role of the immune defense system in disease. The subject encompasses diseases where a malfunction of the immune system itself is the basic cause, together with diseases where some external agent is the initiating factor but an excessive response by the immune system produces the actual tissue damage. It also extends to the monitoring of the normal immune response in infectious diseases and to the use of immunological techniques in disease diagnosis. *See* ALLERGY; AUTOIMMUNITY; CLINICAL PATHOLOGY; HYPERSENSITIVITY; IMMUNOLOGICAL DEFICIENCY.

Many features of the immune system make it prone to shift from protecting the body to damaging it. This complex system not only must distinguish between the body's own cells and a foreign invader but must also recognize and eliminate the body's own cells if they are damaged or infected with a virus. The recognition receptors used to make this fine distinction between "self" and "not self" are not encoded in the genes. Rather, they are assembled following random rearrangement of information carried in small gene segments. During their development, immune system cells are subjected to a selective process, those bearing potentially useful receptors being preserved while those bearing dangerous, self-reactive receptors are eliminated. This process is closely balanced, and some potentially self-reactive cells often persist. *See* CELLULAR IMMUNOLOGY.

There are several approaches to suppressing excessive immune reactivity. Desensitization, or modifying the nature of the response by injecting small amounts of the foreign antigen, is sometimes used to treat allergic states. In contrast, there are few therapies for enhancing immune responses. Bone marrow transplantation is used to restore the immune system in some immunodeficiency diseases. Passive transfer of preformed antibody protects against some infections, and transfusion of immunoglobulin is used to treat immunoglobulin deficiencies. However, vaccination or immunization is one of the most effective of all medical procedures. *See* IMMUNOSUPPRESSION. [K.Sh.]

Clinical microbiology The adaptation of microbiological techniques to the study of the etiological agents in infectious disease. Clinical microbiologists determine the nature of infectious disease and test the ability of various antibiotics to inhibit or kill the isolated microorganisms. A contemporary clinical microbiologist is also responsible for a wide range of microscopic and cultural studies in mycology, parasitology, and virology. The consultative skill of the clinical microbiologist is sought by many clinicians. The clinical microbiologist is often the most competent person available to determine the nature and extent of hospital-acquired infections, as well as public-health problems that affect both the hospital and the community. As clinical microbiology encompasses all aspects of infectious disease, the various specialties will be discussed individually. *See* ANTIBIOTIC; HOSPITAL INFECTIONS; MEDICAL BACTERIOLOGY; MEDICAL MYCOLOGY; MEDICAL PARASITOLOGY; VIROLOGY.

Clinical microbiology is a rapidly progressing laboratory science. During the 1970s, clinical microbiologists developed procedures which provide rapid information to the clinician on the causative agents of infectious disease. These mechanisms include direct microscopic evaluation of specimens by using such powerful tools as electron and fluorescent microscopy. Other techniques involve the detection of soluble portions of microorganisms in the patient's body fluids. Detection of these soluble components can be done in a number of ways, including gas chromatography and immunochemical analysis. Many hospitals are using these immunochemical tests to quickly determine the nature of life-threatening diseases. *See* IMMUNOCHEMISTRY.

The clinical microbiology laboratory of the future may be one in which the identification of an infectious disease is made almost instantaneously by direct analysis of the patient's body fluid, and then a rapid computerized appraisal may be made of those antibiotics which might be effective in inhibiting or killing the microbial agents involved. [R.C.T.]

Clinometer A hand-held surveying device for measuring vertical angles; also called an Abney level. It consists of a sighting tube surmounted by a graduated vertical arc with an attached level bubble. A 45° mirror inside the tube enables the observer to see the bubble at the same time that the observer sights a point or a graduated rod with a horizontal wire. Manipulation of the vertical arc brings the bubble to center in coincidence with the sighted point; the vertical angle is indicated on the arc (see illustration).

A typical clinometer. (*Keuffel and Esser Co.*)

The clinometer is used mainly to determine slope angles for reduction of measured slope distance to horizontal distance. If set on a sloping surface (such as a street pavement), the clinometer can be used to give the angle of inclination of the surface. With the arc set at 0°, it can be used as a hand level. *See* SURVEYING. [B.A.B.]

Clipping circuit An electronic circuit that prevents transmission of any portion of an electrical signal exceeding a

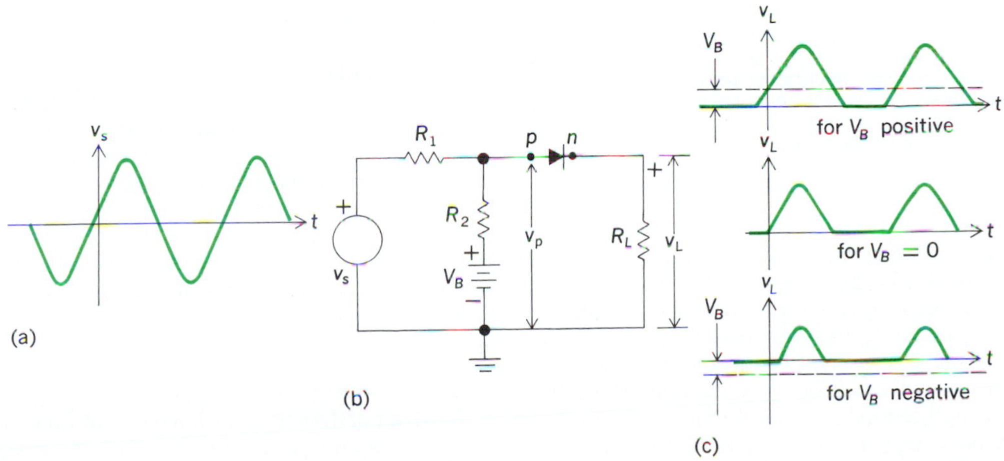

An example of series diode clipping. (*a*) **Waveform of time-varying input voltage.** (*b*) **Circuit diagram.** (*c*) **Waveform of output load voltage v_L when bias voltage V_B is positive, zero, and negative.**

prescribed amplitude. The clipping circuit operates by effectively disconnecting the transmission path for the portion of the signal to be clipped.

A diode may be used as a switch to perform the clipping action. The direct-coupled circuit shown in the illustration represents a typical example. A time-varying signal voltage v_s is applied to the input of the network containing a diode having terminals *p* and *n*. The diode functions as a very low resistance when *p* is positive and as a very high resistance when *p* is negative. Thus, if any portion of the waveform at *p* lies below the common reference or ground level, that portion will be removed, or clipped, and will not appear across output load resistance, R_L. Bidirectional clipping can be achieved by connecting two diodes.

Clipping action may be achieved by using bipolar or field-effect transistors. To do so the bias at the input is set at such a level that output current cannot flow during a portion of the amplitude excursion of the input voltage or current waveform.

[G.M.G.]

Clock A device for indicating the passage of time. Most clocks contain a means for producing a regularly recurring action. This article describes only mechanical clocks. *See* ATOMIC CLOCK; QUARTZ CLOCK; WATCH.

The recurring action of a mechanical clock depends on the swing of a pendulum, or the oscillation of a balance wheel and balance spring or hairspring, or the vibration of a tuning fork, mechanisms capable of repeating their cyclic movements with great regularity. A counting mechanism, consisting of a gear train with calibrated dial and indicating hands, sometimes with a striking mechanism, marks the number of oscillations that have occurred, although the graduations are in seconds, minutes, and hours. A weight or spring ordinarily supplies power to operate the oscillating and the counting mechanisms. However, temperature changes, accelerations, automatic windings, or electricity may provide the power.

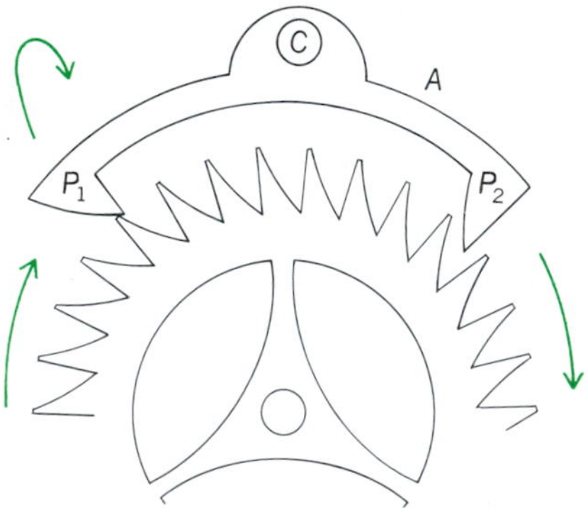

Anchor escapement, common in domestic pendulum clocks. This escapement tends to compensate for changes in amplitude of the swing because of irregularly cut gears and varying lubrication.

Usually an escapement transmits power from the counting mechanism to the oscillating mechanism. The accuracy of a clock depends primarily on the escapement. The illustration depicts an anchor or recoil escapement. Anchor *A* is connected loosely to the pendulum and swings about *C*. At some time after midswing, a tooth of the wheel escapes from pallet P_1 or P_2, giving the pallet a push to maintain oscillation. The other pallet then checks another tooth, the curve of the pallet forcing the wheel slightly backward. The reaction helps to reverse the pendulum swing and to correct for circular error. [R.D'E.A.]

Clock control systems Systems in which a timing device is used to generate the control function; also known as time-controlled systems. A common example of a clock control system is the household alarm clock. In this device the instant of time at which a buzzer sounds an alarm is controlled by a clock.

A clock control system contains two essential components, a timing device and a controlled process, or controlled device. The areas of application of such control systems range from the commonplace sounding of an alarm clock to the control of complex sequences of events in guided missiles and satellite launchers. Clock control systems are usually open-loop control systems, since the control action depends only on the system input, elapsed time, in this case.

There are many different types of timing devices. These devices may be divided into five functional classes: time switches, time-delay relays, interval timers, time-cycle controllers, and time-schedule controllers. *See* CONTROL SYSTEMS. [J.G.Tr.]

Clock paradox The paradox produced by the use of incorrect arguments, concepts, and premises in the relativistic twin clock problem in which two initially synchronized perfect clocks, a and b, are first separated and then reunited. It is the true logical contradiction inherent in the assertion that, when reunited, the two clocks, side by side, are slower than each other. Now, two clocks, side by side, cannot be both slower than each other; those who infer this paradox to be a conclusion have, in some way, introduced premises or concepts incompatible with Albert Einstein's theory of relativity. The clock paradox is *not* a consequence of this theory. When the relativistic twin clock problem is properly handled, no paradox occurs.

"Time dilation" and the twin clock problem must be distinguished. Time dilation is a symmetric artifact of the Lorentz transformations and the Einstein simultaneity convention. The twin clock effect, on the other hand, is an asymmetric physical phenomenon and the physical consequence of the fact that empty space has a metric structure. Time dilation asserts that clocks at rest in different inertial frames symmetrically observe each other to "run slow."

The twin clock effect is completely different from time dilation. The twin clock effect is the natural phenomenon that a clock having a past history of acceleration indicates a total elapsed time less than that indicated by an identical clock having no history of acceleration. It is a physical effect, and the amount by which the accelerated clock is found to be slow is the same regardless of the choice of inertial frame. The physical properties of empty space cause the twin clock effect.

A. H. Bucherer was the first (1909) to implicitly verify the twin clock effect when he experimentally found that high-speed electrons obey relativistic force laws rather than the classical Newtonian laws of motion. In 1971 J. C. Hafele and R. E. Keating flew four automically stabilized clocks around the world and directly observed both the special and general relativistic twin clock effects. This experiment excluded theoretical attempts to "explain" relativistic twin clock effects solely in terms of energy conservation and the principle of equivalence. The Hafele-Keating experiment clearly indicates that both the special and general twin clock effects have as their common physical cause and source the metric structure of the physical field properties of empty space. *See* RELATIVITY; SPACE-TIME.

[R.E.Ke.]

Clonorchiasis An infection of humans and other fish-eating mammals caused by the trematode *Opisthorchis* (*Clonorchis*) *sinensis*, which is usually found in the bile ducts of the infected individuals. The condition is also called Chinese liver fluke disease. Southern China is the classic hyperendemic focus because of the habit there of eating raw fresh-water fish. Endemicity is also associated with the use of night soil as fertilizer. Cats and dogs act as reservoirs.

In the life cycle of the fluke, the minute eggs pass in feces and are ingested by aquatic snails, within which the larval cycle ensues. The larvae emerge and encyst in fish (*Ctenopharyngodon idellus* in particular). When the fish is eaten uncooked, the larvae excyst and arrive in the bile ducts, causing their progressive hypertrophy with resultant cholecystitis and hepatitis. *See* PARAGONIMIASIS. [J.F.M.]

Closed-caption television Method of captioning or subtitling television programs (usually for hearing-impaired or foreign-language viewers) without interfering with the normal television picture, by coding captions as a vertical interval data signal which is decoded at the receiver and superimposed on the television picture. (In open-caption television, captions are part of the normal video picture and are seen by all viewers.) *See* TELEVISION.

Captions are typically prepared at special-purpose workstations optimized for tasks that include preparation and editing of caption texts from the program soundtrack or script; formatting and layout of captions; synchronization of captions to the required video scene; review and correction of the complete captioned program; and storage of caption text and timing information.

When captions are transmitted as part of a teletext data transmission, they are stored in the teletext origination computer and transmitted in synchronism with the required video by interrupting the normal cyclic teletext transmissions as necessary to insert a caption page out of the normal page-number sequence. On the other hand, a dedicated caption transmission system uses a much lower data rate; as a result, the data signal is more rugged, can be successfully received under poorer signal conditions, and can be recorded in encoded form on consumer video recorders, whereas teletext data lie outside the bandwidth of such equipment.

Captions can be decoded only by using a teletext television receiver or adaptor, when they are transmitted as part of a teletext service, or by using a special telecaption adaptor. [A.C.Do.]

Closed-circuit television A video communication system in which the signal is transmitted from the point of origin only to those specific receivers that have access to it by previous arrangement. Closed-circuit television enables visual information to be transmitted, often in parallel with audio information, to a selected audience or between selected groups of people. This private nature of closed-circuit television distinguishes it from broadcast television, in which all receivers in a given area have access to the broadcast signals. *See* TELEVISION.

Closed-circuit television is used for a large number of different types of monitoring applications in which the remote viewing of an ongoing process or activity at a location provides economic, safety, or educational advantages. A growing number of educational institutions and corporations are also using closed-circuit television systems for lectures, presentations, and corporate training applications. Most of these applications consist of one-way television with two-way audio conference facilities to permit audience interaction. Because of nationwide and sometimes worldwide interest in certain sporting or entertainment events, one-way closed-circuit television systems are being used to permit a number of geographically dispersed audiences to view the event simultaneously. Two-way closed-circuit television systems are being used to permit groups of people in geographically separated locations to interact on both a visual and audio basis.

Community antenna television (CATV) or cable television systems are normally considered a type of closed-circuit television. As originally envisioned and implemented, these systems employ standard broadcast television signals which are received by a suitably located antenna and amplified and distributed through coaxial cables. This type of system can be used to provide reliable television reception in a community located beyond the useful range of broadcast television transmitters, or it can be used in an apartment building to avoid having each tenant install a separate antenna.

The range of signal formats and transmission means used in closed-circuit television is great. The signal formats are chosen to suit the particular use, and extend from high-resolution closed-circuit television applications, requiring bandwidths of 5–15 MHz, down to slow-scan television signals which require bandwidths as small as 2.5 kHz. The means of transmission range from broadband radio circuits to standard telephone message circuits.

Technological advances in terminal equipment have broadened the potential applications for closed-circuit television systems. Among these developments are cameras that are capable of operating under very low ambient light conditions, and color cameras that are portable and simple to use. Developments in video tape cassette recording, editing, and playback mechanisms have made the production of video program material easier and less expensive. Developments in video disk technology appear to have a great many potential applications, from mass entertainment to individual instruction in a variety of special fields. [J.J.H.]

Closed-loop control system A system in which the value of some output quantity of the system is continuously or periodically measured and used to manipulate another quantity, an input to the system, in such a way as to make the first quantity or some other dependent output follow a desired pattern of values. At each instant the measured value is compared to the desired value of the measured quantity. The difference is used to change the value of an input to the system. Such systems are also called feedback control systems, because the measured value is fed back, to be used to manipulate an input quantity. *See* CONTROL SYSTEMS. [J.A.Hr.]

Clostridium A genus of bacteria comprising large anaerobic spore-forming rods that usually stain gram-positive. Most species are anaerobes, but a few will grow minimally in air at atmospheric pressure.

The clostridia are widely distributed in nature, and are present in the soil and in the intestinal tracts of humans and animals. They usually live a saprophytic existence, and play a major role in the degradation of organic material in the soil and other nature environments. A number of clostridia release potent exotoxins and are pathogenic for humans and animals. Among the human pathogens are the causative agents of botulism (*Clostridium botulinum*), tetanus (*C. tetani*), gas gangrene (*C. perfringens*), and an antibiotic-associated enterocolitis (*C. difficile*). *See* ANAEROBIC INFECTION; BOTULISM; TETANUS; TOXIN.

Clostridial cells are straight or slightly curved rods, 0.3–1.6 micrometers wide and 1–14 μm long. They may occur singly, in pairs, in short or long chains, or in helical coils. The length of the cells of the individual species varies according to the stage of growth and growth conditions. Most clostridia are motile with a uniform arrangement of flagella. *See* CILIA AND FLAGELLA.

The endospores produced by clostridia are dormant structures capable of surviving for prolonged periods of time, and have the ability to reestablish vegetative growth when appropriate environmental conditions are provided. The spores of clostridia are oval or spherical and are wider than the vegetative bacterial cell. Among the distinctive forms are spindle-shaped organisms, club-shaped forms, and tennis racket–shaped structures:

Clostridia are obligate anaerobes: they are unable to use molecular oxygen as a final electron acceptor and generate their energy solely by fermentation. Clostridia exhibit varying degrees of intolerance of oxygen. Some species are sensitive to oxygen concentrations as low as 0.5%, but most species can tolerate concentrations of 3–5%. The sensitivity of clostridia to oxygen restricts their habitat to anaerobic environments; habitats that contain large amounts of organic matter provide optimal conditions for their growth and survival.

A primary property of all species of *Clostridium* is their inability to carry out a dissimilatory reduction of sulfate. Most species are chemoorganotrophic. The substrate spectrum for the genus as a whole is very broad and includes a wide range of naturally occurring compounds. Extracellular enzymes are secreted by many species, enabling the organism to utilize a wide variety of complex natural substrates in the environment. [H.P.W.]

Cloud Suspensions of minute droplets or ice crystals produced by the condensation of water vapor (the ordinary atmospheric cloud). Other clouds, less commonly seen, are composed of smokes or dusts. *See* AIR POLLUTION; DUST STORM.

If water vapor is cooled sufficiently, it becomes saturated, that is, in equilibrium with a plane surface of liquid water (or ice) at the same temperature. Further cooling in the presence of such a surface causes condensation upon it. In the atmosphere, even in the apparent absence of any surfaces, there are invisible motes upon which the condensation proceeds at barely appreciable cooling beyond the state of saturation. Consequently, when atmospheric water vapor is chilled sufficiently, such motes, or condensation nuclei, swell into minute waterdroplets and form a visible cloud.

The World Meteorological Organization (WMO) uses a classification which divides clouds into low-level (base below about 1.2 mi or 2 km), middle-level (about 1.2–4 mi or 2–7 km), and high-level (4–8 mi or 7–14 km) forms within the middle latitudes. The names of the three basic forms of clouds are used in combination to define 10 main characteristic forms, or "genera."

1. Cirrus are high white clouds with a silken or fibrous appearance.
2. Cumulus are detached dense clouds which rise in domes or towers from a level low base.
3. Stratus are extensive layers or flat patches of low clouds without detail.
4. Cirrostratus is cirrus so abundant as to fuse into a layer.
5. Cirrocumulus is formed of high clouds broken into a delicate wavy or dappled pattern.
6. Stratocumulus is a low-level layer cloud having a dappled, lumpy, or wavy structure.
7. Altocumulus is similar to stratocumulus but lies at intermediate levels.
8. Altostratus is a thick, extensive, layer cloud at intermediate levels.
9. Nimbostratus is a dark, widespread cloud with a low base from which prolonged rain or snow falls.
10. Cumulonimbus is a large cumulus which produces a rain or snow shower. *See* CLOUD PHYSICS. [F.H.Lu.]

Cloud chamber A particle detector in which the path of a fast charged particle is made visible by the formation of liquid droplets on the ions left by the particle as it passes through the gas of the chamber. Cloud chambers are used in research in nuclear physics, and can give detailed information on the particle in addition to the simple fact that the particle passed through. They can be operated at pressures below atmospheric, near atmospheric, or up to 50 atm (5000 kilopascals), depending on the application. They vary in size from a few inches in diameter up to large walk-in chambers containing many metal plates. *See* PARTICLE DETECTOR.

Supersaturated vapor is vapor whose pressure exceeds the saturation vapor pressure at the temperature prevailing. It is unstable, and condensation occurs in the presence of suitable nuclei. The interior of a cloud chamber contains a mixture of a gas such as air, argon, or helium and a vapor such as water or alcohol. When such a mixture is made supersaturated in the vapor component by cooling of the gas, the vapor tends to condense preferentially on charged atoms in the gas. Since a fast-moving charged particle produces many such ions in its passage through the gas, condensation of the vapor leaves a trail of droplets to indicate the path of the particle. When properly illuminated, these droplets are visible and may be photographed. A permanent record of the path of the particle may thus be obtained for examination at a later time.

A cloud chamber can give information on the momentum of the particle if the chamber is placed in a magnetic field. The curvature of the track can be measured on the photograph, and the value of the curvature, together with the magnitude of the field, gives the momentum. Another physical quantity that may be determined in a cloud chamber is the velocity of the particle, since the ionization produced by the charged particle depends on its velocity in a known way. [W.B.Fr.]

Cloud physics The study of the physical and dynamical processes governing the structure and development of clouds and the release from them of snow, rain, and hail (collectively known as precipitation).

The factors of prime importance are the motion of the air, its water-vapor content, and the numbers and properties of the particles in the air which act as centers of condensation and freezing. Because of the complexity of atmospheric motions and the enormous variability in vapor and particle content of the air, it seems impossible to construct a detailed, general theory of the manner in which clouds and precipitation develop. However, calculations based on the present conception of laws governing the growth and aggregation of cloud particles and on simple models of air motion provide reasonable explanations for the observed formation of precipitation in different kinds of clouds.

Clouds are formed by the lifting of damp air which cools by expansion under continuously falling pressure. The relative humidity increases until the air approaches saturation. Then condensation occurs on some of the wide variety of aerosol particles present; these exist in concentrations ranging from less than 1600 particles per cubic inch (100 particles per cubic centimeter) in clean, maritime air to perhaps 1.6×10^7 particles per cubic inch (10^6 particles per cubic inch) in the highly polluted air of an industrial city. These droplets, usually present in concentrations of a few hundreds per cubic centimeter, constitute a nonprecipitating water cloud. Precipitation release follows the production of a relatively few large particles from a large population of much smaller ones, which may be achieved in one of two ways: the coalescence process or the ice crystal process.

Cloud droplets are seldom of uniform size. Droplets arise on nuclei of various sizes and grow under slightly different conditions of temperature and supersaturation in different parts of the cloud. A droplet appreciably larger than average will fall faster than the smaller ones, and so will collide and fuse (coalesce) with some of those which it overtakes.

The second method of releasing precipitation can operate only if the cloud top reaches elevations where temperatures are below 32°F (0°C) and the droplets in the upper cloud regions become supercooled. At temperatures below −40°F

(−40°C) the droplets freeze automatically or spontaneously; at higher temperatures they can freeze only if they are infected with special, minute particles called ice nuclei. As the temperature falls below 32°F (0°C), more and more ice nuclei become active, and ice crystals appear in increasing numbers among the supercooled droplets. Such a mixture of supercooled droplets and ice crystals is unstable. After several minutes the growing crystals will acquire definite falling speeds, and several of them may become joined together to form a snowflake. In falling into the warmer regions of the cloud, however, the snowflake may melt and reach the ground as a raindrop.

The deep, extensive, multilayer-cloud systems, from which precipitation of a usually widespread, persistent character falls, are generally formed in cyclonic depressions (lows) and near fronts. Although the structure of these great raincloud systems, which are being explored by aircraft and radar, is not yet well understood, radar signals from these clouds usually take a characteristic form which has been clearly identified with the melting of snowflakes.

Precipitation from shower clouds and thunderstorms, whether in the form of raindrops, pellets of soft hail, or true hailstones, is generally of greater intensity and shorter duration than that from layer clouds and is usually composed of larger particles. The clouds themselves are characterized by their large vertical depth, strong vertical air currents, and high concentrations of liquid water, all these factors favoring the rapid growth of precipitation elements by accretion.

The development of precipitation in convective clouds is accompanied by electrical effects culminating in lightning. The mechanism by which the electric charge dissipated in lightning flashes is generated and separated within the thunderstorm has been debated for more than 200 years, but there is still no universally accepted theory. However, the majority opinion holds that lightning is closely associated with the appearance of the ice phase, and the most promising theory suggests that the charge is produced by the rebound of ice crystals or a small fraction of the cloud droplets that collide with the falling hail pellets. *See* LIGHTNING.

The various stages of the precipitation mechanisms raise a number of interesting and fundamental problems in classical physics. Worthy of mention are the supercooling and freezing of water; the nature, origin, and mode of action of the ice nuclei; and the mechanism of ice-crystal growth which produces the various snow crystal forms.

The maximum degree to which a sample of water may be supercooled depends on its purity, volume, and rate of cooling. The freezing temperatures of waterdrops containing foreign particles vary linearly as the logarithm of the droplet volumes for a constant rate of cooling. This relationship, which has been established for drops varying between 10 micrometers and 1 centimeter in diameter, characterizes the heterogeneous nucleation of waterdrops and is probably a consequence of the fact that the ice-nucleating ability of atmospheric aerosol increases logarithmically with decreasing temperature.

Measurements made with large cloud chambers on aircraft indicate that the most efficient nuclei, active at temperatures above 14°F (−10°C), are present in concentrations of only about 10 in a cubic meter of air, but as the temperature is lowered, the numbers of ice crystals increase logarithmically to reach concentrations of about 1 per liter at −4°F (−20°C) and 100 per liter at −22°F (−30°C). Since these measured concentrations of nuclei are less than one-hundredth of the numbers that apparently are consumed in the production of snow, it seems that there must exist processes by which the original number of ice crystals are rapidly multiplied, Laboratory experiments suggest the fragmentation of the delicate snow crystals and the ejection of ice splinters from freezing droplets as probable mechanisms.

The most likely source of atmospheric ice nuclei is provided by the soil and mineral-dust particles carried aloft by the wind. Laboratory tests have shown that, although most common minerals are relatively inactive, a number of silicate minerals of the clay family produce ice crystals in a supercooled cloud at temperatures above −4°F (−18°C). A major constituent of some clays, kaolinite, which is active below 16°F (−9°C), is probably the main source of highly efficient nuclei.

The fact that there may often be a deficiency of efficient ice nuclei in the atmosphere has led to a search for artificial nuclei which might be introduced into supercooled clouds in large numbers. In general, the most effective ice-nucleating substances, both natural and artificial, are hexagonal crystals in which spacings between adjacent rows of atoms differ from those of ice by less than 16%. The detailed surface structure of the nucleus, which is determined only in part by the crystal geometry, is of even greater importance.

Collection of snow crystals from clouds at different temperatures has revealed their great variety of shape and form. This multiple change of habit over such a small temperature range is remarkable and is thought to be associated with the fact that water molecules apparently migrate between neighboring faces on an ice crystal in a manner which is very sensitive to the temperature. The temperature rather than the supersaturation of the environment is primarily responsible for determining the basic shape of the crystal, though the supersaturation governs the growth rates of the crystals, the ratio of their linear dimensions, and the development of dendritic forms.

The presence of either ice crystals or some comparatively large waterdroplets (to initiate the coalescence mechanism) appears essential to the natural release of precipitation. Rainmaking experiments are conducted on the assumption that some clouds precipitate inefficiently, or not at all, because they are deficient in natural nuclei; and that this deficiency can be remedied by "seeding" the clouds artificially with dry ice or silver iodide to produce ice crystals, or by introducing waterdroplets or large hygroscopic nuclei.

The science and technology of cloud modification is still in its infancy. It may well be that further knowledge of cloud behavior and of natural precipitation mechanisms will suggest new possibilities and improved techniques which will lead to developments far beyond those which now seem likely. *See* PRECIPITATION (METEOROLOGY); WEATHER MODIFICATION. [B.J.M.]

Clove The unopened flower bud of a small, conical, symmetrical, evergreen tree, *Eugenia caryophyllata*, of the myrtle family (Myrtaceae). The cloves are picked by hand and dried. Cloves, one of the most important and useful spices, are strongly aromatic and have a pungent flavor. They are used as a culinary spice for flavoring pickles, ketchup, and sauces, in medicine, and for perfuming the breath and air. The essential oil distilled from cloves has even more uses. The chief clove-producing countries are Tanzania, Indonesia, Mauritius, and the West Indies. *See* MYRTALES. [P.D.St./E.L.C.]

Clover A common name used loosely to designate the true clovers, sweet clovers, and other members of the plant family Leguminosa.

True clovers. The true clovers are herbaceous annual or perennial plants of the genus *Trifolium*, order Rosales. There are approximately 250 species in the world. Collectively they represent the most important genus of forage legumes in agriculture. Most clovers are highly palatable and nutritious to livestock. The name clover is often applied to members of legume genera other than *Trifolium*. *See* ROSALES.

Clovers are used for hay, pasture, silage, and soil improvement. Certain kinds may be used for all purposes whereas others, because of their low growth, are best suited for grazing. All kinds, when well grown in thick stands, are good for soil improvement. Thoroughly inoculated plants add 50–200 lb of nitrogen per acre (62–252 kg per hectare) when plowed under for soil improvement, the amount added depending on growth, thickness of stand, and length of growing season. *See* Nitrogen cycle.

All the clover species of agricultural importance in the United States are introduced (exotic) plants. Some of the species most widely used are: red clover (*T. pratense*), alsike clover (*T. hybridum*), white clover (*T. repens*), crimson clover (*T. incarnatum*), subclover (*T. subterraneum*), strawberry clover (*T. fragiferum*), persian clover (*T. resupinatum*), and large hop clover (*T. campestre*, or *procumbens*). Other clovers of regional importance, mostly adapted to specific environmental conditions, are rose clover (*T. hirtum*), berseem clover (*T. alexandrinum*), ball clover (*T. nigrescens*), lappa clover (*T. lappaceum*), big-flower clover (*T. michelianum*), and arrowleaf clover (*T. vesiculosum*).

Red clover is composed of two forms, medium and mammoth, producing two and one hay cuts, respectively. The large purplish-red flower heads are round. An upright-growing perennial, red clover generally persists for 2 years in the northern United States, but behaves as a winter annual in the South. There are several varieties and strains of red clover such as Kenland, Pennscott, Lakeland, Dollard, and Chesapeake. These produce higher yields of forage and are more persistent than common red clover.

Alsike clover is an upright-growing species that behaves like a biennial. The growth pattern, seeding methods, mixtures, and uses are similar to those of red clover. The flower heads are much like those of white clover in shape and size, but are slightly more pinkish. Alsike clover is more tolerant of wet, poorly drained soils than red clover and occurs widely in mountain meadows of the West.

White clover, an inhabitant of lawns and closely grazed pastures, is the most important pasture legume in the humid states. The flowers are generally white, but sometimes they are tinged with pink. There are three main types, large, intermediate, and small, with all gradations between. All types are nutritious and are relished by all classes of livestock and poultry. White clover is mainly grown with low-growing grasses, not being tolerant of the tall-growing kinds. For best growth of the clover, grass-clover mixtures should be grazed or cut frequently.

Crimson clover is used principally as a winter annual for pasture and as a soil-improving crop from the latitude of the Ohio River southward, and along the West Coast. Crimson clover is seeded alone, with small grains and grasses, or on grass turf. During the winter it may be grazed, although if it is too heavily grazed, regrowth is slow. The greatest return for soil improvement is obtained when the largest growth is plowed under. There are several varieties of crimson clover, including Dixie, Auburn, Autauga, Chief, Talladega, and many local strains. When used with Bermuda or other perennial summer-growing grasses, the fall growth of the grass must be closely grazed or clipped.

Subclover, a winter annual extensively used for grazing in the coastal sections of the Western states, is the basic pasture crop of the sheep and cattle industry of Australia. The Australian varieties Mount Barker, Tallarook, and Nangeela have proved to be best adapted to most conditions in the United States. Subclover appears to have considerable promise as a pasture legume under many conditions in the southern United States, but better adapted varieties are needed.

Sweet clover. Sweet clover is the common name for all but one species of legumes of the genus *Melilotus*, order Rosales. The exception is sour clover (*M. indica*). There are approximately 20 species of sweet clover. Some of the biennial species have an annual form. Sweet clovers are native to the Mediterranean region and adjacent countries, but several are widely scattered throughout the world, generally by chance introduction.

Sweet clover is used as a field crop in regions of the United States and Canada where the rainfall is 17 in. (42 cm) or more during the growing season, where the soil is neutral, or where limestone and other needed minerals are applied. It is grown either alone or in rotations with small grains and corn, and is used for grazing, soil improvement, and hay. Except for those of certain improved varieties, the plants are somewhat bitter because of the presence of coumarin. [E.A.H.]

Clupeiformes An order of teleost fishes of the subclass Actinopterygii that includes the herrings, sardines, anchovies, and their allies. The Recent clupeiforms are classified in 2 suborders, 3 (by some authorities as many as 6) families, 71 genera, and about 300 species.

The Clupeiformes are generalized teleosts. They are mostly silvery and compressed. A distinctive feature is the extension of the cephalic canal system onto the operculum; the lateral line is undeveloped on the trunk except in *Denticeps*. They lack fin spines, adipose fin, and gular plate. The middle of the belly often bears one or a series of strong scutes. The pelvic fin is abdominal in position, free from the shoulder girdle, and the pectoral fin is placed low on the side. The cycloid scales are usually thin and loosely attached. The upper jaw is bordered by premaxillae and maxillae, but the dentition is usually feeble.

Clupeiform fishes have left a rich fossil record, especially from the Upper Cretaceous to the early Tertiary, but appeared first in the Upper Jurassic. Some herrings and a few anchovies live in lowland rivers and lakes, and others such as the shad and alewife enter rivers to reproduce; but the majority of clupeiforms occur in bays or shore waters of tropical, temperate, or even northern seas, where they commonly make up enormous schools. None inhabits deep water. Most feed on plankton or other minute organisms. Thus they are efficient converters from the base of the food chain to fish flesh.

Some of the great fisheries of the world are based on the conversion of plankton by clupeiform fishes: the tremendous anchovy fishery off western South America; the California fisheries for anchovy and (until depleted) Pacific sardine; the menhaden fisheries of the western Atlantic; and the herring and sardine fisheries of northern seas. The catch is processed variously into oil, fertilizer, or fish meal or is prepared directly for human consumption. *See* Actinopterygii. [R.M.B.]

Clutch A machine element for the connection and disconnection of shafts in equipment drives. If both shafts to be connected can be stopped or made to move relatively slowly, a positive-type mechanical clutch may be used. If an initially stationary shaft is to be driven by a moving shaft, friction surfaces must be interposed to absorb the relative slippage until the speeds are the same. Likewise, friction slippage allows one shaft to stop after the clutch is released.

When positive connection of one shaft with another in a given position is needed, a positive clutch is used. This clutch is the simplest of all shaft connectors, sliding on a keyed shaft section or a splined portion and operating with a shift lever on a collar element. Because it does not slip, no heat is generated in this clutch. Interference of the interlocking portions prevents engagement at high speeds; at low speeds, if connection occurs, shock loads are transmitted to the shafting. Positive clutches may be of the square jaw type (Fig. 1) with two or more jaws of square section meshing together in the opposing clutches, or the spiral jaw type, a modification of the square-jaw clutch that permits more convenient engagement and provides a more gradual movement of the mating faces toward each other.

When the axial pressure of the clutch faces on each other serves to transmit torque instead of the mating shape of their

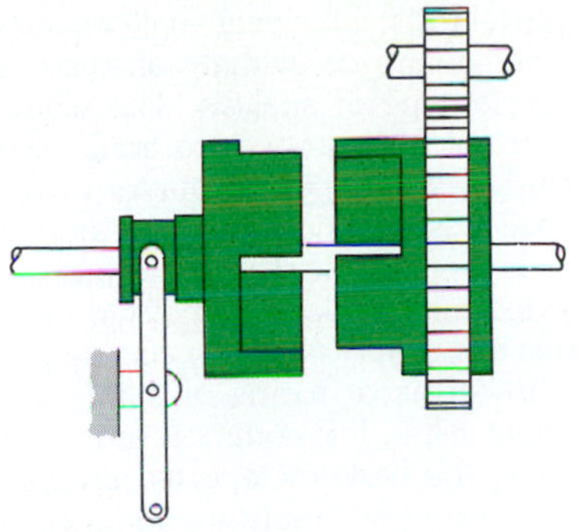

Fig. 1. Square-jaw-type positive clutch.

parts, the clutch operates by friction. This friction clutch is usually placed between an engine and a load to be driven; when the friction surfaces of the clutch are engaged, the speed of the driven load gradually approaches that of the engine until the two speeds are the same. A friction clutch is necessary for connecting a rotating shaft of a machine to a stationary shaft so that it may be brought up to speed without shock and transmit torque for the development of useful work. The three common designs for friction clutches, combining axial and radial types, are cone clutches (Fig. 2), disk clutches, and rim clutches. In a cone clutch, the surfaces are sections of a pair of cones. The disk clutch consists essentially of one or more friction disks connected to a driven shaft by splines. A rim clutch has surface elements that apply pressure to the rim externally or internally.

In the overrunning type of clutch, the driven shaft can run faster than the driving shaft. This action permits freewheeling as the driving shaft slows down or another source of power is applied. Effectively this is a friction pawl-and-ratchet drive, wherein balls or rollers become wedged between the sleeve and recessed pockets machined in the hub (Fig. 3). The clutch does not slip when the second shaft is driven, and is released automatically when the second shaft runs faster than the driver. The centrifugal clutch employs centrifugal force from the speed of rotation. This type of clutch is not normally used because it becomes unwieldy and unsafe with increasing size. Clutch action is also produced by hydraulic couplings, with a smoothness not possible with a mechanical clutch. Automatic transmissions in automobiles represent a fundamental use of hydraulic clutches. *See* FLUID COUPLING; TORQUE CONVERTER.

Magnetic coupling between conductors provides a basis for several types of clutches. The magnetic attraction between a current-carrying coil and a ferromagnetic clutch plate serves to actuate a disk-type clutch. Slippage in such a clutch produces heat that must be dissipated and wear that reduces the life of

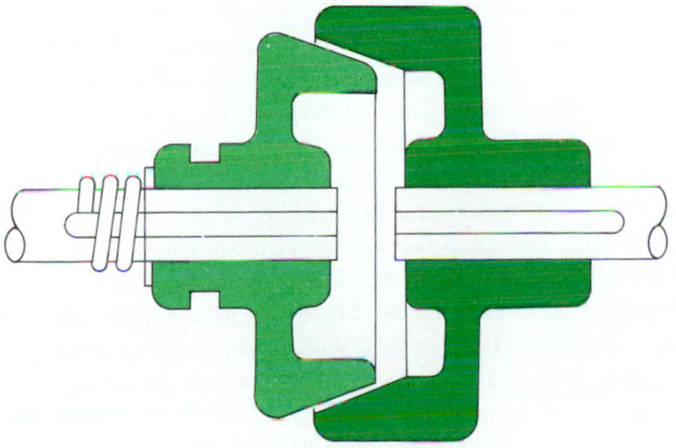

Fig. 2. Cone-type friction clutch.

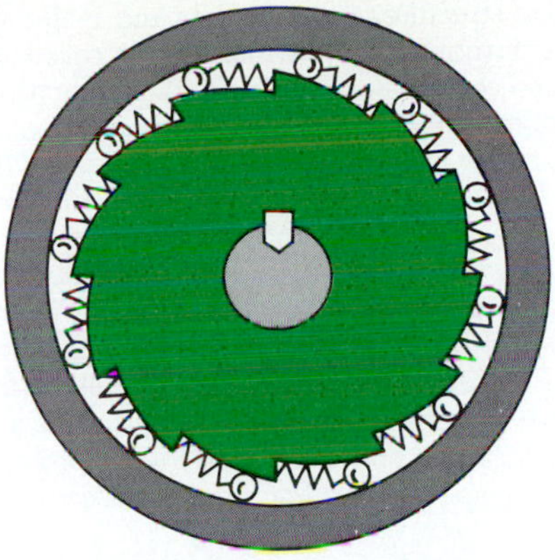

Fig. 3. Overrunning clutch with spring-constrained rollers or balls.

the clutch plate. Thus the electromagnetically controlled disk clutch is used to engage a load to its driving source. *See* BRAKE; COUPLING. [J.E.G.; J.J.R.]

Clypeasteroida An order of exocyclic Euechinoidea which have a monobasal apical system in which all the genital plates fuse together. The ambulacra are petaloid on the aboral side and wider than the interambulacra on the adoral side (see illustration). The order probably arose in the Late Cretaceous.

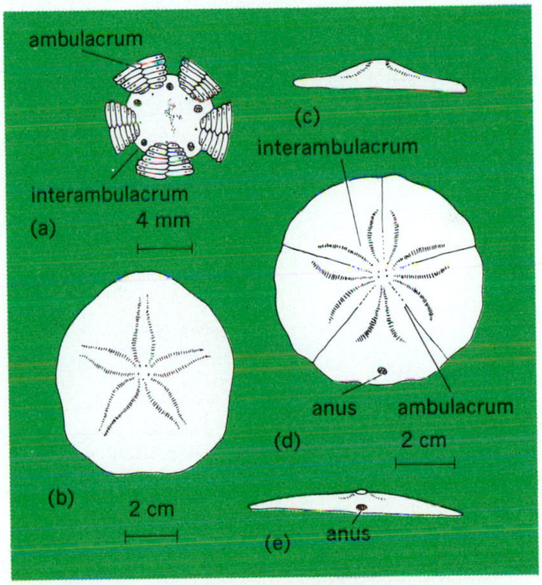

Typical clypeasteroids. *Laganum*, (*a*) monobasal apical system, (*b*) aboral aspect, and (*c*) posterior aspect. *Arachnoides*, (*d*) aboral aspect, and (*e*) posterior aspect.

Tertiary fossils are very common, and extant species range the coastlines of tropical and temperate seas. Relatively short timespans of some fossil species make them valuable as index forms in stratigraphy. Many species are gregarious, and occur in closely packed colonies of perhaps millions of individuals. There are 16 families arranged in four suborders. *See* ECHINODERMATA; EUECHINOIDEA. [H.B.F.]

Cnidospora A subphylum of spore-producing Protozoa containing the two classes Myxosporidea and Microsporidea. Cnidospora are parasites of cells and tissues of invertebrates, fishes, a few amphibians, and turtles. *See* MICROSPORIDEA; MYXOSPORIDEA.

A distinctive structure of this subphylum is the spore, which contains one or more protoplasmic masses called sporoplasms. The spore also contains one or more filaments which lie coiled within the spore proper or within one or more polar capsules.

When the spore is ingested by a new host, the filaments are extruded and at the same time the sporoplasm is released. The sporoplasm, now called an amebula, reaches the specific site of infection directly through the intestinal wall or by way of the bloodstream. The amebula becomes a stage in the life cycle which feeds and grows at the expense of the host. Asexual reproduction by repeated binary fission, multiple fission, internal and external budding, or plasmotomy results in cells which develop into sporonts and eventually into sporoblasts, producing one or more spores. *See* PROTOZOA; SPOROZOA. [R.F.N.]

Coal A brown to black combustible rock that had its origin in the accumulation and physical and chemical alteration of vegetation. Original accumulations of vegetation (mostly woody plants), usually in a swamp or moist environment which reduced decay, resulted in formation of peat, the precursor of coal. Peat is converted to coal after burial and subsequent geological processes, including increased pressure and temperature, which progressively compressed and indurated and otherwise altered the material through a series of coal varieties to an extreme of graphite or graphitelike material. In American terminology, the rank varieties of coal constituting the coalification series consist of lignitic (including brown coal); subbituminous; bituminous (including high-, medium-, and low-volatile); and anthracitic coal (including semianthracite, anthracite, and metaanthracite or graphitic coal). *See* GRAPHITE; LIGNITE; PEAT.

Classification of coals by rank

Class	Group
I. Anthracitic	Metaanthracite Anthracite Semianthracite
II. Bituminous	Low-volatile bituminous coal Medium-volatile bituminous coal High-volatile A bituminous coal High-volatile B bituminous coal High-volatile C bituminous coal
III. Subbituminous	Subbituminous A coal Subbituminous B coal Subbituminous C coal
IV. Lignitic	Lignite A Lignite B

SOURCE: From *Annual Book of ASTM Standards*, pt. 26, copyright 1979 by American Society for Testing and Materials; reprinted with permission.

The term coalification refers to the process of coal metamorphism brought about by increasing weight of overlying sediments, by tectonic movements, by an increase in temperatures resulting from depth of burial, or from close approach to, or contact with, igneous intrusions or extrusions. Increase in pressure affects principally the physical properties of the coal, that is, hardness, strength, optical anisotropy, and porosity. Increase in temperature acts chiefly to modify the chemical composition by increasing the carbon content and decreasing the content of oxygen and hydrogen and increasing the calorific value to a maximum in those coals having 15–30% volatile matter (dry, mineral-matter-free).

The classification and description of coal depend largely upon information supplied by chemical analysis consisting of systematically standardized procedures and the results of a number of empirical tests. Chemical analyses are of two principal kinds: the elementary or ultimate analysis and the proximate or commercial type of analysis. The ultimate analysis is limited to the percent of hydrogen, carbon, oxygen, nitrogen, and sulfur, exclusive of the mineral matter or ash. The proximate analysis, which is empirical in character, is the prevailing form of analysis in North America. In this analysis, percentage values are reported for volatile matter, fixed carbon, moisture, and ash. Calorific and sulfur values are also determined, the latter sometimes in terms of forms of sulfur such as pyritic, organic, and sulfate sulfur. It is common practice to use qualifying terms denoting the basis of reported analysis, such as as-received, moisture-free (mf), moisture-and-ash-free (maf), and dry-mineral-matter-free (dmmf) so-called pure coal, in presenting the results of proximate and ultimate analysis.

Mineral matter is the inorganic material present in coal that is the source of ash when coal is burned. It occurs predominantly as discrete mineral phases, but also as chemical elements combined with organic matter and as dissolved elements in the pore water. Mined coals commonly contain between 5 and 15% mineral matter, although some have greater and lesser amounts.

The term rank refers to the stage of coalification reached in the course of metamorphism. Classification of coal by rank is shown in the table. It is fundamental in coal characterization and generally is based upon the chemical composition, and upon proximate values based upon mineral-matter-free coal obtained from standard face channel samples. The higher-rank coals are classified with respect to fixed carbon on the dry basis, and the low-rank coals according to calorific value on the moist basis. In Europe, coking properties are a more common basis of classification than in North America.

The calorific values of coal are expressed in British thermal units (Btu) per pound in some English-speaking countries, and in calories per gram generally elsewhere. Heat values, determined by a calorimeter, vary for coat material from about 6300 Btu/lb (14.65 megajoules/kg) for California lignite to about 16,300 Btu/lb (37.91 megajoules/kg) for some maf cannel coal. *See* CALORIMETRY.

The lithologic characteristics of coal concern the structural aspects of the coal bed and texture as determined by the megascopic and microscopic physical constitution of the coal itself. Structurally the coal bed (or seam) is a geological stratum characterized by the same irregularities in thickness, uniformity, and continuity as other strata of sedimentary origin. Coal beds may consist of essentially uniform continuous strata or, like other sedimentary deposits, may be made up of distinctly different bands or benches of varying thickness. Like other sedimentary strata, coal beds may be structurally disturbed by folding and faulting so that the originally approximate horizontality of position is lost to the extent that beds may become vertically oriented or even overturned.

The texture of the coal itself is determined by the character, grain, and distribution of its megascopic and microscopic components. In general, banded coals composed of relatively coarse, highly lustrous vitrain lenses are considered coarsely textured. As the thickness of the vitrain bands progressively lessens, the texture becomes fine-banded and then microbanded, with bright laminae composed of fine vitrainlike material. Coals displaying homogeneous texture are regarded as nonbanded or canneloid coals. In general, similar textural units are observed in bituminous and anthracite coals.

Microscopic texture is determined by the physical composition of the lithotypes, or of the anthraxylon and attritus, in terms of the microscopic constituents. The microscopy and petrology of coal are concerned very largely with these constituents.

The uses for coal have varied widely with rank, but three general areas of use may be distinguished. The major use is for combustion to generate electricity and heat (electric utilities, industrial, domestic, and railroads). A significant percentage of the world's coal with suitable properites is used in the manufacture of metallurgical grade coke. Coal was formerly used extensively to manufacture gas, and new research appears to be leading to major developments for conversion of coal to liquid and gaseous fuels to supplement supplies of oil and natural gas. Coal was an important source of a wide range of chemicals before being largely supplanted by oil and gas, but this possible renewed future use is similar to processes of conversion to liquid and gaseous fuels. *See* COAL CHEMICALS; COAL GASIFICATION; COAL LIQUEFACTION.

Coal is found on every continent, in the islands of Oceania, and in the West Indies. Most coal resources of the world are contained in strata from the Carboniferous (Pennsylvanian-Mississippian of the United States) and younger strata, although minor occurrences have been reported in Devonian and older strata. Approximately 95% of the known coal resources and 85% of recoverable reserves of the world lie in the Northern Hemisphere. Estimates of total coal resources of the world are very large, but vary widely depending on assumptions on which the estimates are based. Recoverable coal reserves of the world based on current mining practice have been estimated to total over 8×10^{11} tons. [J.A.Si.]

Coal balls Subspherical masses containing mineral matter and embedded plant material, found in coal seams and overlying beds of the late Paleozoic. The mineral content of the balls usually consists of calcium or magnesium carbonate, or sometimes iron pyrites. In some areas coal balls are found only in coal seams overlain by marine limestones.

All known deposits of coal balls in North America, Europe, and the British Isles are associated with Pennsylvanian and Permian beds of bituminous coal. They usually are present as rounded nodules or irregular masses at the surface of the bed or in the upper part of the coal seam. They vary from small pea-sized nodules to irregularly formed masses weighing tons. Some of the best specimens occur in stream beds in which coal seams have been exposed by erosion. *See* COAL PALEOBOTANY; CONCRETION. [W.N.S.]

Coal chemicals For about 100 years, chemicals obtained as by-products in the primary processing of coal to metallurgical coke have been the main source of aromatic compounds used as intermediates in the synthesis of dyes, drugs, antiseptics, and solvents. Although some aromatic hydrocarbons, such as toluene and xylene, are now obtained largely from petroleum refineries, the main source of others, such as benzene, naphthalene, anthracene, and phenanthrene, is still the by-product coke oven. Heterocyclic nitrogen compounds, such as pyridines and quinolines, are also obtained largely from coal tar. Although much phenol is produced by hydrolysis of monochlorobenzene and by decomposition of cumene hydroperoxide, much of the phenol, cresols, and xylenols are still obtained from coal tar.

Coke oven by-products are gas, light oil, and tar. Coke oven gas is a mixture of methane, carbon monoxide, hydrogen, small amounts of higher hydrocarbons, ammonia, and hydrogen sulfide. Most of the coke oven gas is used as fuel. Although several hundred chemical compounds have been isolated from coal tar, a relatively small number are present in appreciable amounts. These may be grouped as in the table. All the compounds in the table except the monomethylnaphthalenes are of some commercial importance.

The direct utilization of coal as a source of bulk organic chemicals has been the objective of much research and development. Oxidation of aqueous alkaline slurries of coal with oxygen under pressure yields a mixture of aromatic carboxylic acids. Because of the presence of nitrogen compounds and hydroxy acids, this mixture is difficult to refine. Hydrogenation of coal at elevated temperatures and pressures yields much larger amounts of tar acids and aromatic hydrocarbons of commercial importance than are obtained by carbonization. However, this operation is more costly than other sources of these chemicals. *See* COKE; DESTRUCTIVE DISTILLATION; PYROLYSIS. [H.W.W.]

Coal tar chemicals

Compound	Fraction of whole tar, %	Use
Naphthalene	10.9	Phthalic acid
Monomethylnaphthalenes	2.5	
Acenaphthenes	1.4	Dye intermediates
Fluorene	1.6	Organic syntheses
Phenanthrene	4.0	Dyes, explosives
Anthracene	1.0	Dye intermediates
Carbazole (and other similar compounds)	2.3	Dye intermediates
Phenol	0.7	Plastics
Cresols and xylenols	1.5	Antiseptics, organic syntheses
Pyridine, picolines, lutidines, quinolines, acridine, and other tar bases	2.3	Drugs, dyes, antioxidants

Coal gasification The conversion of coal, coke, or char to gaseous products by reaction with air, hydrogen, oxygen, steam, carbon dioxide, or a mixture of these. Products consist of carbon dioxide, carbon monoxide, hydrogen, methane, and some other chemicals in a ratio dependent upon the particular reactants employed and the temperatures and pressures within the reactors, as well as upon the type of treatment which the gases from the gasifier undergo subsequent to their leaving the gasifier. Strictly speaking, reaction of coal, coke, or char with air or oxygen to produce heat plus carbon dioxide might be called gasification. However, that process is more properly classified as combustion, and thus is not included in this coal gasification summary. *See* COMBUSTION.

In most coal gasification processes which are available for use or under development, the reactions are endothermic. Air or oxygen is typically supplied to the gasifier to provide the necessary heat input. From the industrial user's viewpoint, the net result is to produce either low-Btu or intermediate-Btu gas. Each of these gas types has attracted industrial interest. The electric industry has principally investigated the production of clean, low-Btu gas from coal with the object of burning it in a combined-cycle power generation system. The natural gas industry has viewed intermediate-Btu gas production with strong interest since this same gasification product can also be upgraded to synthetic natural gas by using relatively conventional and proved gas-processing steps. Heavy industry has studied the feasibility of using both low- and intermediate-Btu gas for many applications, thereby freeing critical volumes of natural gas and also reducing imported oil consumption. Parts of the chemical industry have examined intermediate-Btu gas as a source of hydrogen and carbon monoxide to offset potentially reduced supplies of these synthetic chemical building blocks. Various projects have also been studied which would involve coproduction of substitute natural gas and intermediate-Btu gas for industrial or chemical use, electric power, coal-derived liquids, and various other products. *See* COAL; COKE; NATURAL GAS. [R.H.McC.]

Coal liquefaction The conversion of most types of coal, anthracite excepted, by direct or indirect technologies primarily to produce synthetic low-sulfur, low-ash liquid or solid fuels to replace some of the fuel coal used by electric utilities. Coal liquefaction can be utilized for making syncrude, a material suitable for use as a refinery feedstock and for petrochemical production.

Although the term liquefaction implies liquid products, several processes also produce gaseous and solid products. The advantage of coal liquefaction is that a wide range of liquid products, especially heavy fuels for electric utilities, distillate fuel oil, and gasoline, can be produced by varying the operating conditions of different processes. The atomic hydrogen: carbon (H:C) ratios of the products are equal or greater than the ratio in the feed coal, depending on the extent of hydrogenation used. Coal liquefaction requires less chemical transformation than gasification, and has an approximate 78% energy-conversion efficiency versus 60–70% for gasification. *See* COAL; COAL GASIFICATION. [F.D.C.]

Coal mining The technical and mechanical job of removing coal from the earth and preparing it for market. Coal, the most abundant and traditionally the most economical source of power in the world, is found in varying amounts throughout the globe. It lies in veins of various thickness and richness beneath the crust of the Earth, resting in giant subterranean sandwiches, shallow or deep, flat or pitched. Obtaining coal in sufficient quantities at a competitive cost and making it the proper grade for the market demand is, in its simplest terms, the science of coal mining. *See* COAL.

Of major importance in coal property development is the accumulation of seam information from borehole drilling. Chemical analysis of the cores will provide details concerning moisture, volatile material, fixed carbon, ash, sulfur, heat content, and fusion temperature of the ash. After preliminary analyses and correlations, the engineering details are planned. Production requirements and the extent of coal reserves generally determine the method of mining and the type and capacity of equipment to be used.

Coal is extracted from the earth by three basic methods: (1) underground, (2) strip, and (3) auger. Approximately two-thirds of America's coal comes from underground mining. The remaining one-third is mined from the surface, either by strip or auger mining.

The systems of underground coal mining generally in use are room-and-pillar, longwall, and, in a few places in Europe, hydraulic. In the room-and-pillar system, tunnels are carved into the seam, leaving pillars of coal for support. In longwall mining, widely spaced tunnels are driven, leaving large blocks of coal. Later, these blocks are completely extracted, allowing the roof material to collapse behind the coal face as it is removed. In hydraulic mining, a stream of water is directed against the coal face with sufficient pressure to dislodge the coal. The water also acts as a transporting medium. *See* UNDERGROUND MINING.

Where the coal seam lies close to the surface, strip mining is an efficient way to mine the coal. It is economical to remove the overburden of earth and rock that covers the coal seam. For this job, power shovels, draglines, or wheel excavators are used. A shovel excavates a "furrow" and casts the overburden parallel to the cut. Draglines sit on the bank above the coal seam and remove the overburden from the seam. Bucket-wheel machines excavate the material from above the coal seam. It then flows in a continuous stream via a transfer to the conveyor system, which in turn transports it to the discharge point. When the seam is exposed, the coal is loaded by smaller power shovels into trucks and hauled to preparation plants or loading bins. *See* SURFACE MINING.

For coal seams which continue under rising land too thick for economical strip mining and where underground mining cannot burrow further to the surface because of the shallow and more treacherous roof conditions, the auger method was developed. The auger miner twists huge drills like carpenter's bits into a hillside coal seam, drawing out the coal to a conveyor which loads it into trucks. Auger recovery gives additional tonnage at minimal cost for equipment and labor and permits recovery of coal that might not be recovered by other methods.

After the coal is mined, it may be loaded directly into transportation facilities for the market if the foreign material in it is not excessive. However, coal from most seams requires preparation to provide a desired and uniform quality. A number of washing devices are used for cleaning coal, all of which operate on the basic principle of the difference of specific gravity between coal and foreign material. Once the coal is ready for market, it is loaded for shipment to the customer by railroad, water transportation, or in some cases belt conveyor systems. [J.D.R.]

Coal paleobotany A special branch of the paleobotanical sciences concerned with the origin, composition, mode of occurrence, and significance of the fossil plant materials that occur in, or are associated with, coal seams. Information developed in this field of science provides knowledge useful to the biologist in attempting to describe the development of the plant world, aids the geologist in unraveling the complexities of coal measure stratigraphy in order to reconstruct the geography of past ages and to describe ancient climates, and has practical application in the coal, coke, and coal chemical industries. *See* PALEOBOTANY.

All coal seams consist of countless fragments of fossilized plant material admixed with varying percentages of mineral matter. The organic and inorganic materials initially accumulate in some type of swamp environment. In some instances, the fossilized plant fragments can be recognized as remnants of a plant of some particular family, genus, or species. When this is possible, information can be obtained on the vegetation extant at the time the source peat was formed, and such data aid greatly in reconstructing paleogeographies and paleoclimatic patterns.

Pollen grains and spores are more adequately preserved in coals than are most other plant parts. Recognition of this fact, coupled with an appreciation of the high degree to which these fossils are diagnostic of floral composition, has led to the rapid development of the paleobotanical subscience of palynology. Fruits, seeds, and identifiable woods also occur as coalified fossils. Occasionally, coal seams contain fossil-rich coal balls. Essentially all types of plant fossil, including entire leaves, cones, and seeds, are encountered in these discrete nodular masses. *See* COAL BALLS; PALYNOLOGY.

Coal seams generally are composed of several superposed sedimentary layers, each having formed under somewhat different environmental conditions. The coal petrologist and paleobotanist recognize these layers because of their distinctive textural appearance and because each consists of a particular association of organic and inorganic materials. Accordingly, each coal seam usually contains several types of coal. These coal types, or lithotypes, possess characteristic suites of physical properties, and knowledge of these properties is very profitably employed in manipulating coal composition in coal preparation and beneficiation plants. [W.Sp.]

Coal Sack An area in one of the brighter regions of the Southern Milky Way which to the naked eye appears entirely devoid of stars and hence dark with respect to the surrounding Milky Way region. Telescopic and spectroscopic observations reveal that the Coal Sack is a cloud of small, solid particles. The cloud not only absorbs about two-thirds of the visible light passing through it but also strongly reddens the transmitted light. The Coal Sack is located just to the southeast of the Southern Cross. *See* INTERSTELLAR MATTER. [W.Li.]

Coastal engineering A branch of civil engineering concerned with the planning, design, construction, and maintenance of works in the coastal zone. The purposes of these works include control of shoreline erosion; development of navigation channels and harbors; defense against flooding caused by storms, tides, and seismically generated waves (tsunamis); development of coastal recreation; and control of pollution in nearshore waters. Coastal engineering usually involves the construction of structures or the transport and possible stabilization of sand and other coastal sediments.

The successful coastal engineer must have a working knowledge of oceanography and meteorology, hydrodynamics, geomorphology and soil mechanics, statistics, and structural mechanics. Tools that support coastal engineering design include analytical theories of wave motion, wave-structure interaction, diffusion in a turbulent flow field, and so on; numerical and physical hydraulic models; basic experiments in wave and current flumes; and field measurements of basic processes such as beach profile response to wave attack, and the construction of works. Postconstruction monitoring efforts at coastal projects have also contributed greatly to improved design practice.

Coastal structures can be classified by the function they serve and by their structural features. Primary functional classes include seawalls, revetments, and bulkheads; groins; jetties; breakwaters; and a group of miscellaneous structures including piers, submerged pipelines, and various harbor and marina structures.

Seawalls, revetments, and bulkheads are structures constructed parallel or nearly parallel to the shoreline at the land-sea interface for the purpose of maintaining the shoreline in an advanced position and preventing further shoreline recession. Seawalls are usually massive and rigid, while a revetment is an armoring of the beach face with stone rip-rap or artificial units. A bulkhead acts primarily as a land-retaining structure and is found in a more protected environment such as a navigation channel or marina. *See* BULKHEAD; REVETMENT.

A groin is a structure built perpendicular to the shore and usually extending out through the surf zone under normal wave and surge-level conditions. It functions by trapping sand from the alongshore transport system to widen and protect a beach or by retaining artificially placed sand. *See* GROIN (ENGINEERING).

Jetties are structures built at the entrance to a river or tidal inlet to stabilize the entrance as well as to protect vessels navigating the entrance channel. *See* JETTY.

The primary purpose of a breakwater is to protect a shoreline or harbor anchorage area from wave attack. Breakwaters may be located completely offshore and oriented approximately parallel to shore, or they may be oblique and connected to the shore where they often take on some of the functions of a jetty. *See* BREAKWATER. [R.M.So.]

Coastal landforms The characteristic features and patterns of land in a coastal zone subject to processes of erosion and deposition. The interactions of various marine processes with the coast and with terrestrial processes produce a great variety of landforms along the coastal zone. Primary terrestrial processes that influence coastal landforms are those which carry sediment to the coast. These include river movement and mass movement by gravity, such as slumping and slides. Marine processes of waves and currents are predominant in controlling the formation of coastal morphology. Another indirectly important process is change in sea level. A passive yet quite important contributing factor in coastal morphology is the nature of the geology and geomorphology of the area adjacent to the coast.

Landforms associated with, and formed along, wide coastal plains are in striking contrast to those which are developed along high-relief, rocky coasts. A classification of coasts based on the modern concepts of global tectonics provides a good framework for consideration of coasts. There are essentially two broad types of plate margins: trailing-edge and leading-edge. Trailing-edge coasts are characterized by coastal plains and adjacent broad shelves with constructional landforms. The Atlantic coast of the United States south of New York is an example. The leading edge coasts are typically high-relief with bedrock at the coast due to structural deformation. The west coast of the United States is a good example. *See* PLATE TECTONICS.

As a rule, coastal plain coasts of trailing-edge plates present a greater diversity of depositional environments than high-relief leading-edge coasts. The primary reason is that submarine morphology generally mimics that of the adjacent subaerial coast. That is, trailing-edge plates show broad gently sloping continental shelves upon which sediment accumulates and is distributed by coastal processes. Leading-edge regions show high-relief submarine morphology and a general lack of a continental shelf, resulting in an absence of sites for accumulation of sediment. *See* COASTAL PLAIN.

Many coastal landform types show no particular association with a specific coastal type. Most, but not all, of these features are relatively small-scale, so-called second-order features. Reefs are probably the most extensive coastal landforms, and their presence has no relation to the type of coast, other than it must lack significant terrigenous runoff. Fiords, like reefs, are climate-related but may develop on a variety of coasts as long as relief is at least modest. These huge glacial excavations are the deepest of the coastal landforms. *See* FIORD; NEARSHORE SEDIMENTARY PROCESSES; REEF. [R.A.D.]

Coastal plain A low-relief area which is bounded by the sea on one side and generally by highlands on the landward side. The continental shelf actually represents a drowned portion of the coastal plain and is geologically linked to the coastal plain. The development of coastal plains ranges widely throughout the world, with the Atlantic and Gulf coasts of the United States being among the most extensive. Development of coastal plains is associated with trailing-edge coasts in the context of global tectonics. These coasts are relatively stable and structurally simple. *See* CONTINENTAL MARGIN.

Existing coastal plains began their development in the Mesozoic Era and have continued to the present. Sediment accumulations are typically some combination of fluvial, deltaic, and marine shelf depositional environments. There is a general thickening of the stratigraphic section in the seaward direction. Down-to-the-coast normal faults and intrusion of salt domes represent the most widespread and significant structural features. Surface topography displays low-relief cuestas along the outcrop belt of the more resistant lithofacies. *See* COASTAL LANDFORMS; GEOMORPHOLOGY; SALT DOME. [R.A.D.]

Coati The name for three species of carnivorous mammals assigned to the raccoon family Procyonidae. The common or ring-tailed coati (*Nasua nasua*) ranges through the forests of Central and South America, the mountain coati (*Nasuella olivacea*) is found only in South America, while *Nasua nelsoni* is confined to Cozumel Island in Central America.

The coatis are characterized by their elongated snouts, body, and tail, which is held erect when they walk (see illustration). They are adept at climbing trees in search of birds and lizards, and their diet also includes fruit, insects, and larvae. *Nasua nasua* roams in bands of females and young during the day in search of food. The males are excluded from the band except during the brief breeding season, when a single male is allowed to join. After mating, the male leaves and the female builds an isolated nest in a tree. Here, following a gestation period of 11 weeks, a litter of about five young are born and are cared for

Common coati (*Nasua nasua*).

by the female away from the group. After about 1 month the female and her young rejoin the band. *See* CARNIVORA. [C.B.C.]

Coaxial cable A two-conductor transmission line with an outer metal tube or braided shield concentric with and enclosing the center conductor. The inner conductor is supported by some form of dielectric insulation, solid, expanded plastic, or semisolid. Semisolid supports are polyethylene disks, helical tapes, or helically wrapped plastic strings. Beads, supporting pins, or periodically crimped plastic tubes are used in some designs.

The significant feature of the coaxial cable is that it is a shielded structure. The electromagnetic field associated with each coaxial unit is nominally confined to the space between the inner and outer conductors. Since alternating current concentrates on the inside of the outer conductor as the frequency of the current increases (skin effect), a coaxial unit is a self-shielded transmission line whose shielding improves at higher frequencies. Unshielded lines, such as the pairs of multipair cable, share the space for the electromagnetic fields. Thus, for equivalent transmission loss, pairs occupy less space than coaxials. The major use of the coaxial cable is for transmitting high-frequency broadband signals. Coaxial cables are rarely used at or near voice frequency since the shielding properties are poor and they are more expensive than twisted pairs having the same transmission loss.

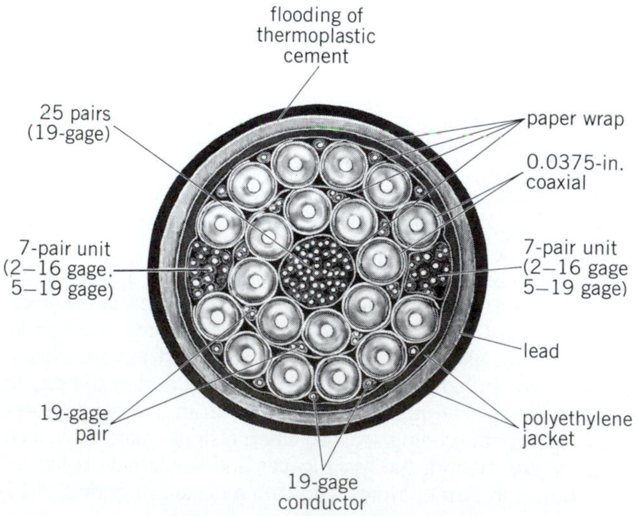

Construction of coaxial transmission line.

Coaxial units are made in three general types for different applications: flexible, semirigid, or rigid. In general, the more rigid the unit, the more predictable and stable its electrical properties. Since loss on a transmission line increases approximately as the square of the frequency, low loss in a coaxial cable is important. Also, physical and electrical irregularities, especially if they are periodic, must be kept to a minimum, or the transmission loss of the coaxial will be very high at certain frequencies. *See* TRANSMISSION LINES.

One variation of the semirigid coaxial, insulated with polyethylene disks enclosed inside a copper tube, is widely used for transcontinental carrier transmission. The disk-insulated coaxial unit has very low loss and is now used in carrier systems in the United States, Japan, and Europe to transmit up to 10,800 two-way voice-frequency channels per pair of coaxial units. The illustration shows a 20-unit coaxial cable with 9 working coaxial pairs and 2 standby coaxials, which automatically switch in if the electronics of the regular circuits fail. [J.R.Ap.]

Cobalt A metallic chemical element, Co, with an atomic number of 27 and an atomic weight of 58.93. Cobalt is similar to iron and nickel in both its free and combined states. It is widely distributed in nature, making up about 0.001% of the igneous rocks of the Earth's crust, compared with 0.02% for nickel. It occurs in meteorites, stars, sea and fresh waters, soils, plants, animals, and in the manganese nodules found on the ocean floor. Traces of cobalt are found in many ores of iron, nickel, copper, silver, manganese, and zinc, but the commercially important cobalt minerals are the arsenides, oxides, and sulfides. Cobalt and its alloys are wear- and corrosion-resistant even at high temperatures. Important commercial applications are in making alloys for use at high temperatures, magnetic alloys, alloys for use in machine tools, glass-to-metal seals, and the dental and surgical alloy Vitallium. Cobalt is required in small amounts by plants and animals. The artificially produced radioactive isotope of cobalt, cobalt-60, is widely used in industry, research, and medicine. *See* RADIOISOTOPE.

1	2	3	4	5	6	7	8	9	10	11	12	13	14	15	16	17	18
1 H																	2 He
3 Li	4 Be											5 B	6 C	7 N	8 O	9 F	10 Ne
11 Na	12 Mg											13 Al	14 Si	15 P	16 S	17 Cl	18 Ar
19 K	20 Ca	21 Sc	22 Ti	23 V	24 Cr	25 Mn	26 Fe	27 Co	28 Ni	29 Cu	30 Zn	31 Ga	32 Ge	33 As	34 Se	35 Br	36 Kr
37 Rb	38 Sr	39 Y	40 Zr	41 Nb	42 Mo	43 Tc	44 Ru	45 Rh	46 Pd	47 Ag	48 Cd	49 In	50 Sn	51 Sb	52 Te	53 I	54 Xe
55 Cs	56 Ba	71 Lu	72 Hf	73 Ta	74 W	75 Re	76 Os	77 Ir	78 Pt	79 Au	80 Hg	81 Tl	82 Pb	83 Bi	84 Po	85 At	86 Rn
87 Fr	88 Ra	103 Lr	104 Rf	105 Db	106 Sg	107 Bh	108 Hs	109 Mt	110	*111*	*112*	*113*	*114*	*115*	*116*	*117*	*118*

lanthanide series	57 La	58 Ce	59 Pr	60 Nd	61 Pm	62 Sm	63 Eu	64 Gd	65 Tb	66 Dy	67 Ho	68 Er	69 Tm	70 Yb
actinide series	89 Ac	90 Th	91 Pa	92 U	93 Np	94 Pu	95 Am	96 Cm	97 Bk	98 Cf	99 Es	100 Fm	101 Md	102 No

Cobalt is ferromagnetic and resembles iron and nickel in hardness, tensile strength, machinability, thermal properties, and electrochemical behavior. The metal is unaffected by air or water under normal conditions, and is rapidly attacked by sulfuric acid, hydrochloric acid, and nitric acid, but only slowly by hydrofluoric acid, ammonium hydroxide, and sodium hydroxide. Cobalt exhibits variable valence and forms complex ions and colored compounds, as do all the transition elements. The table summarizes key properties.

Cobalt(II) chloride, nitrate, and sulfate are all formed by the interaction of the metal, oxide, hydroxide, or carbonate with the corresponding acid. There are three main oxides of cobalt: gray cobaltous oxide, CoO; black cobaltic oxide, Co_2O_3,

Properties of cobalt

Property	Value
Atomic weight	58.93
Atomic number	27
Melting point	1493°C
Specific gravity	8.9 at 20°C
Electronic configuration	$1s^2\ 2s^2\ 2p^6\ 3s^2\ 3p^6\ 3d^7\ 4s^2$
Oxidation states	2+, 3+

formed by heating compounds at a low temperature in excess of air; cobaltosic oxide, Co_3O_4, the stable oxide, formed when salts are heated in air at temperatures which do not exceed 430°F (850°C). Most common cobalt salts are derivatives of cobalt(II), the higher valence state being encountered only in complex coordination compounds. An important, naturally occurring cobalt coordination compound is vitamin B_{12}. Cobalt compounds have a variety of industrial applications, including their use as catalysts, and in agriculture for remedying cobalt deficiency in soil and natural vegetation. *See* TRANSITION ELEMENTS. [R.S.Yo.]

Cobalt toxicity Cobalt in most naturally occurring foodstuffs is largely unabsorbed and is eliminated in the feces. Under natural circumstances cobalt does not produce a toxic syndrome in humans. It takes an unusual set of circumstances to produce cobalt toxicity. Intravenously injected, large doses of cobalt have been observed to cause paralysis and enteritis, and sometimes death. When cobalt is injected into the bloodstream, the amount distributed throughout the tissues is very small; higher concentrations are present in the pancreas, liver, spleen, kidney, and bone. Elimination of cobalt is rapid.

An unusual kind of cobalt toxicity is the "beer drinker's syndrome." During the 1960s, some breweries in Nebraska and Quebec added cobalt to their beer to improve the stability of the foam. Subsequently, numerous fatalities occurred among people who consumed large quantities of beer. Following discontinuance of the cobalt additive, no new cases were reported. Whether the cobalt acted independently or synergistically with other constituents of beer to enhance its toxicity is not known. The principal lesion was found in the heart; autopsies revealed enlarged, flabby hearts that were more than twice normal size. *See* COBALT. [N.K.M.]

Cobaltite A mineral having composition (Co,Fe)AsS. Cobaltite is one of the chief ores of cobalt. It crystallizes in the isometric system, commonly in cubes or pyritohedrons, resembling crystals of pyrite. There is perfect cubic cleavage. The luster is metallic and the color silver-white but with a reddish tinge. The hardness is 5.5 (Mohs scale) and the specific gravity is 6.33. Notable occurrences of cobaltite are at Skutterud, Norway; Lunaberg, Sweden; Ravensthorpe, Australia; and Cobalt, Ontario, Canada. *See* COBALT. [C.S.Hu.]

Cobra Any of several snakes that belong to the reptilian family Elapidae, which also includes kraits, mambas, and coral snakes. There are 10 species of cobra and all are included in the genus *Naja*. They are widely distributed throughout Africa and Asia. These snakes are oviparous, and the spectacled, or Indian, cobra (*Naja naja*) has been known to lay 20 eggs, of which half proved to be fertile. Food preferences include other reptiles, eggs, frogs, and small rodents. The characteristic hood, not pronounced in all species, is formed by a series of ribs, lying against the backbone, which can be extended laterally when the reptile is annoyed. The king cobra (*N. hannah*) is the largest and most dangerous of all poisonous snakes and will attack any animal, including humans, without hesitation. It may grow to a length of over 16 ft (4.8 m). *See* REPTILIA. [C.B.C.]

Coca Shrubs of the genus *Erythroxylum* in the coca family (Erythroxylaceae). The genus contains over 200 species, most in the New World tropics. Two species, *E. coca* and *E. novogranatense*, are cultivated for their content of cocaine and other alkaloids. Leaves of cultivated coca contain 14 alkaloids, chiefly cocaine, and significant amounts of vitamins and minerals. The cocaine content averages 0.5%, rarely exceeding 1.0%.

South American Indians "chew" coca as a mild stimulant by placing dried leaves in the mouth and masticating with various alkalies such as lime or plant ashes to promote the release of the alkaloids, which are swallowed. Coca is important in the folk medicine of Andean Indians, especially as a treatment for gastrointestinal disorders. It may regulate carbohydrate metabolism in a useful way and supplement an otherwise deficient diet. *See* COCAINE; COLA; LINALES. [A.T.W.]

Cocaine The principal alkaloid of coca leaves, a topical anesthetic and stimulant, and popular illicit drug. In 1884 C. Koller demonstrated cocaine's efficacy as an anesthetic in eye surgery, introducing the age of local anesthesia. For the next decade cocaine enjoyed the status of a wonder drug and panacea. It fell into disfavor with increasing reports of acute toxicity and long-term dependence. Today it is used as a topical anesthetic in the eye, nose, mouth, and throat; for injection anesthesia it has been replaced by synthetic drugs with fewer central nervous system effects. *See* COCA.

Cocaine increases heart rate and blood pressure and causes feelings of alertness and euphoria. It does not produce physical dependence, as alcohol and opiates do, but many people find it hard to use in a stable and moderate fashion if they have access to it in quantity. Although it is quite active orally, most users of illicit cocaine take it intranasally by snuffing; few inject it intravenously. Aside from local irritation of the nasal membranes, moderate users suffer few adverse effects. The soluble hydrochloride salt is the common form. Insoluble cocaine free base may be smoked, a practice that may be more harmful. *See* ALKALOID; DRUG ADDICTION. [A.T.W.]

Coccidia A subclass of the class Telosporea. These protozoa are typically intracellular parasites of epithelial tissues in both vertebrates and invertebrates. The group is divided on the basis of life cycles into two orders, the Protococcida (in which there is only sexual reproduction) and the Eucoccida (in which there is both sexual and asexual reproduction). There are only a few species in Protococcida, and all are parasites in marine invertebrates. *See* EUCOCCIDA; PROTOCOCCIDA; TELOSPOREA. [E.R.B./N.D.L.]

Coccidioidomycosis An infectious disease of humans and animals caused by a fungus, *Coccidioides immitis*. This mycotic infection is endemic in the southwestern United States, and is known to occur in Mexico and Central and South America.

This disease has two forms most frequently recognized, primary pulmonary and disseminated. The primary pulmonary form is the one that most uninfected persons coming in contact with *C. immitis* acquire. More than 50% of these individuals do not develop any symptoms. When symptoms do occur, they are usually mild and are often diagnosed as a cold or influenza. This form of the disease, however, may be severe and can be manifested by a wide range of signs and symptoms which include fever, chills, cough, chest pain, and headache. Cavity formation may take place as well as pleural effusions. An allergic reaction may also be noted, characterized by the development of large, tender, reddish nodules on the legs.

Approximately one patient in 1000 with primary pulmonary coccidioidomycosis will have a dissemination of the organisms from the lungs to the skin, bones, viscera, and central nervous system. In this form of the disease, dissemination occurs in Blacks about 15 times and Filipinos about 100 times more frequently than in Caucasians.

Cutaneous coccidioidomycosis in most instances is believed to be secondary to a pulmonary lesion. However, there are cases reported of a primary cutaneous form of this disease acquired by injury to the skin, followed by invasion of the fungus. In these instances the lesions are similar to the sporotrichotic chancre. [L.D.H.]

Coccolithophorida A group of unicellular, biflagellate, golden-brown algae characterized by a covering of extremely small interlocking calcite plates called coccoliths. The Coccolithophorida are mainly classified by the shape of their coccoliths into two groups: the holococcoliths, with simple rhombic or hexagonal crystals arranged like a mosaic, and the heterococcoliths, with complex crystals arranged into boat, trumpet, basket, or collar-button shapes (see illustration). The Coccolithophorida are usually considered plants but possess also some animal characteristics. Botanists assign them to the class Haptophyceae of the phylum Chrysophyta, and zoologists to the class Phytamastigophorea of the phylum Protozoa. *See* PHYTAMASTIGOPHOREA; PROTOZOA.

Chiefly photosynthetic, although epiphytism, phagotrophism, and saprophytism have also been shown, the Coccolithophorida form a significant percentage of the nannoplankton of the tropic through subarctic-subantarctic waters of all oceans; a few brackish and fresh-water species also exist. Together with the diatoms they constitute the primary producers of the open ocean food chain. *See* PHYTOPLANKTON.

Coccoliths preserve well and have a fossil record dating back into the Jurassic, 180,000,000 years ago. Their long and involved evolutionary record makes them useful to geologists for dating ancient sediments. Because they live in surface waters, coccoliths are under direct climatic control. Thus many modern species with relatively long fossil records are excellent temperature indicators. *See* MICROPALEONTOLOGY. [A.McI.]

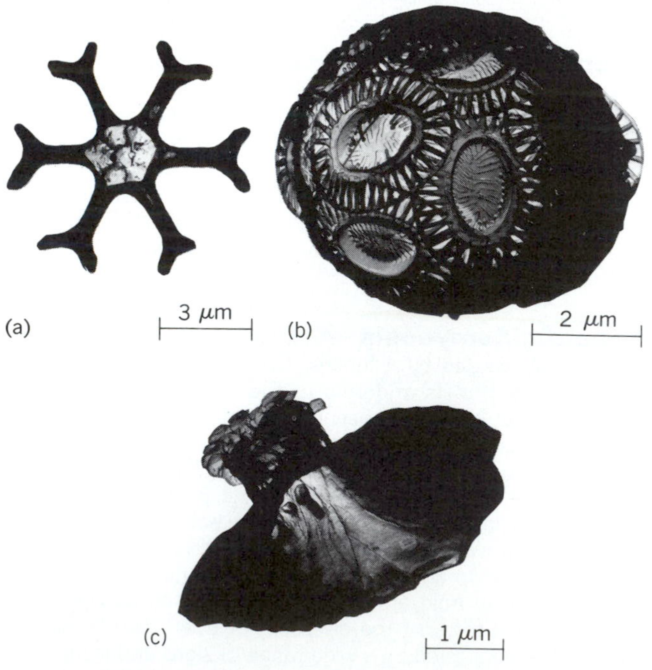

Coccolithophorida: (a) *Discoaster;* (b) a coccosphere; (c) a heterococcolith. (*A. McIntyre, Lamont-Doherty Geological Observatory of Columbia University*)

Cockcroft-Walton accelerator An electrostatic particle accelerator characterized by the method used to obtain the dc accelerating voltage. This type of machine was used in 1932 by J. D. Cockcroft and E. Walton in studying the first nuclear reactions induced by an accelerated ion beam. Since that time, a number of variations have been developed and applied to a wide range of problems in both science and industry. In its most rudimentary form, a Cockcroft-Walton power supply contains a high-voltage transformer and a voltage multiplying circuit (see illustration).

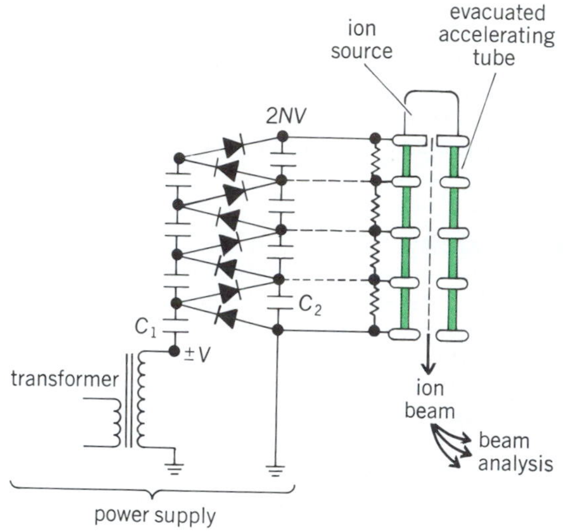

Cockcroft-Walton power supply and accelerator. The multiplier circuit shown has four stages (N = 4). C_1 and C_2 are capacitors. V = peak tranformer output voltage.

The accelerator itself is basically an ion source which injects ions into an evacuated accelerating tube. The power supply voltage is connected across the tube. Either positive or negative ions or electrons may be accelerated depending on the nature of the ion source and the polarity of the accelerating voltage. Voltage gradients along the accelerating tube may be controlled by a resistive bleeder string or by connecting conducting sections of the accelerator tube to intermediate multiplier stages. *See* ION SOURCES.

The physical size of ordinary Cockcroft-Walton accelerators increases with operating voltage, because voltage gradients along the insulating support members and in the surrounding medium must be limited to avoid electrical breakdown (sparking). The size of an accelerator can be vastly reduced by housing it in a vessel filled with high-pressure insulating gas. The size can also be reduced if lower voltages are desired. Tiny accelerator systems have been devised for logging boreholes drilled for oil and mineral prospecting. These accelerators operate at 0. 1 megavolt or less, but this is sufficient for deuterium ions to produce 14-MeV neutrons when they strike a tritium target. *See* PARTICLE ACCELERATOR. [J.C.O.]

Cockle Primarily a member of the bivalve family Cardiidae, with globose shells ornamented with radiating ribs and perhaps spines borne on some ribs posteriorly. The globular shape hinders rapid digging, and cockles occur superficially in sheltered sands where shell ornamentation prevents dislodgement by weak surf. Cockles occur intertidally and subtidally, in temperate and tropical seas. Cockles have a muscular L-shaped foot with which they can "leap" when lying upon the surface of the sand, and so avoid attack by starfish. Young cockles are preyed upon by shore crabs and oyster catchers.

Other applications of the name cockle include the bloody cockle, *Anadara granosa*, of western Malaysia, which resembles the western European *Cardium edule* in appearance, location in shallow water, and commercial exploitation; however, it is a filibranch in the family Arcidae. *See* BIVALVIA; MOLLUSCA. [R.D.P.]

Cockroach Any of a group of insects (approximately 2500 species) of the family Blattidae, order Dictyoptera. This group is among the most primitive of the living and oldest of the fossilized insects, being found in abundance in the Carboniferous age. In general, cockroaches have a flat oval body, a darkened cuticle, and a thoracic shield which extends dorsally over the head. A pair of sensory organs, the whiplike antennae, are located on the head, and a much shorter pair of sensory organs, the cerci, are located at the end of the abdomen. Although some species are wingless, most have two pairs of wings but seldom fly. Most species are nocturnal and primarily live in narrow crevices and under stones and debris. Cockroaches are omnivorous, and a few species have adapted to the human environment, becoming household pests since they can chew foodstuffs, clothing, paper, and even plastic insulation. They also can emit a highly disagreeable odor.

Females deposit egg cases almost anywhere; each case contains several eggs from which the young nymphs hatch and go through several molts as they grow. At the final molt a winged, sexually mature adult will emerge. The time from egg to adult varies from species to species but can be longer than a year.

Cockroaches have been claimed as vectors for numerous human diseases. There is little evidence to support this claim; however, there is evidence that they can produce allergenic reactions in humans. Because of their large size, cockroaches have proved to be very useful animals for scientific research on general problems of insect behavior, physiology, and biochemistry. [C.R.F.]

Coconut A large palm, *Cocos nucifera*, widely grown throughout the tropics and valuable for its fruit and fiber. Usually found near the seacoast, it requires high humidity, abundant rainfall, and mean annual temperature of about 85°F (29°C). Southern Florida, with mean temperature of 77°F (25°C), is at the limit of successful growth.

The fruit, 10 in. (25 cm) or more in length, is ovoid and obtusely triangular in cross section. The tough, fibrous outer husk encloses a spherical nut consisting of a hard, bony shell within which is a thin layer of fleshy meat or kernel. The meat is high in oil and protein and, when dried, is the copra of commerce.

Although many trees grow without special care, the crop lends itself to plantation culture with control of weeds, fertilization, and protection from diseases, insects, and animal pests. Palms begin to bear nuts the sixth year after planting and reach full bearing about the eighth year. Individual nuts mature about a year after blossoming and normally fall to the ground.

The oil from the dried coconut meats (copra) is widely used for margarine, soap, and industrial purposes. High-quality copra may be shredded for confectionery and the baking trade. The residue, after oil removal, is used for animal feed. Coconut husks are an important source of fiber called coir. Various grades of coir are used for ropes, mats and matting, and upholstery filling. *See* COIR.

The coco palm is the most useful of all tropical plants to the native population. An important source of food and drink, it also furnishes building material, thatch, hats, dishes, baskets, and many other useful items. *See* ARECALES. [L.H.MacD.]

Codfish Fish of a subfamily of the Gadidae in the order Gadiformes. Important commercially, these fish are found in cold waters, such as the Baltic and the North Atlantic. *Gadus morhua* is extensively fished off the Newfoundland banks and a circumpolar species, *Boreogadus saida*, the Arctic cod, is found around the ice pack during the summer. *G. macrocephalus*, a related species, occurs in the northern Pacific.

The cod is covered with cycloid scales, has a barbel under the chin, and has pelvic fins on the throat; there are two anal and three dorsal fins. Codfish average about 3 ft (0.9 m) in length and weigh 10–35 lb (4.5–15.7 kg). They live at depths of 100–1500 ft (30–450 m), where they feed on mollusks, small fish, crustaceans, and worms. Spawning occurs from January on to spring. The livers are processed for cod liver oil, which is rich in vitamins, and the swim bladder is made into isinglass. *See* GADIFORMES. [C.B.C.]

Coelenterata That group of the Radiata whose members typically bear tentacles and possess intrinsic nematocysts. The name Cnidaria is also used for this phylum and is preferred by some because the name Coelenterata, as first used, included the sponges (Porifera) and the comb jellies (Ctenophora), as well as the animals called coelenterates. *See* CTENOPHORA; PORIFERA.

The coelenterates are mainly marine organisms and are best known as jellyfish or medusae, sea anemones, corals, the Portuguese man-of-war, small polypoid forms called hydroids, and the fresh-water hydras. Taken together, the phylum is divisible into three classes as follows: (1) Hydrozoa, the hydroids, hydras, and hydrozoan or craspedote jellyfish (hydromedusae); (2) Scyphozoa, the acraspedote jellyfish; and (3) Anthozoa, the sea anemones, corals, sea fans, sea pens, and sea pansies. *See* ANTHOZOA; HYDROZOA; SCYPHOZOA.

It is convenient to recognize two basic body forms in this phylum, the polyp and the medusa, into which all coelenterates can be classified. The polyp and the medusa, however, have many features in common (Fig. 1).

The polyp is a radially, biradially, or radiobilaterally symmetrical individual having a longitudinal oral-aboral axis and is usually sessile. The mouth is at the free end and is surrounded by one to many whorls or sets of tentacles which may be hollow or solid. The aboral end is commonly developed as an adhesive device for attachment and is conveniently referred to as a base. The central body cavity is the gastrovascular cavity, also called the enteron or coelenteron.

The medusa is a tetramerously or polymerously radial individual and is free-swimming. The body is usually bell- or bowl-shaped with the mouth suspended in the center of the underside of the bell on a stalk. Instead of directly surrounding the mouth

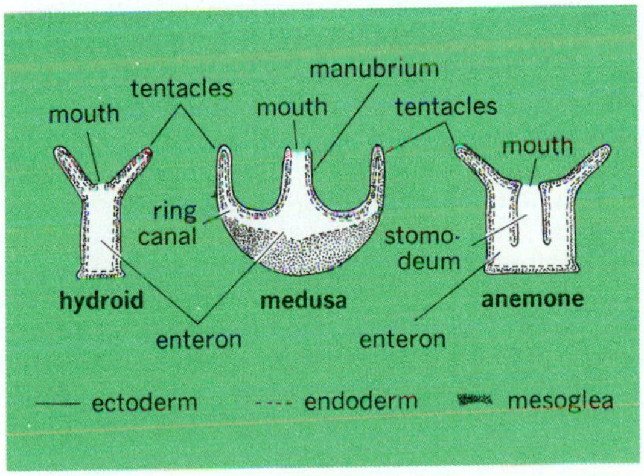

Fig. 1. Comparison of hydroid polyp, medusa (inverted), and anthozoan polyp. (*After T. I. Storer and R. L. Usinger, General Zoology, 3d ed., McGraw-Hill, 1957*)

as in the polyp, the tentacles are located at the margin of the bell. The outer or aboral part of the bell is recognized as the exumbrella and the under or oral part as the subumbrella. The mouth leads to the central stomach which in turn gives rise to four or more radial canals. These radial canals run through the umbrella, on the subumbrellar side, and commonly lead to a ring canal at the margin which is continuous around the margin.

The unique and most distinctive feature of coelenterates is the possession of intracellular, independent effector organelles called nematocysts, but also known as stinging cells or nettle cells. A coiled thread tube in each cell may be rapidly everted under proper stimulation and used for food gathering and for defense against predators, intruders, or enemies (Fig. 2).

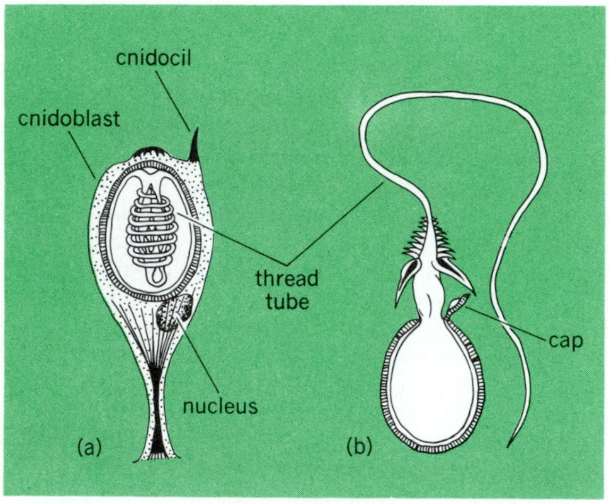

Fig. 2. Penetrant-type nematocyst of *Hydra*. (*a*) Before discharge. (*b*) Discharged. (*After Schulze, from T. I. Storer and R. L. Usinger, General Zoology, 3d ed., McGraw-Hill, 1957*)

Nematocysts are produced within cells, called cnidoblasts, and may be produced either in place or at a location some distance from their definitive site in the coelenterate. The morphologically simplest coelenterates, the Hydrozoa, have nematocysts limited to their outer epidermis whereas the more complex Scyphozoa and Anthozoa bear nematocysts in both the outer epidermis and inner gastrodermis.

The body systems of coelenterates may be divided into the following categories for the purpose of description: epithelial, nervous, muscular, mesogleal, skeletal, digestive, and reproductive. There is no analog of a circulatory system and the functions of respiration and excretion are carried out by each cell.

The outer epithelium or epidermis is in general a single layer of cells, varying from a flattened squamous nature to tall and columnar. The epidermis is usually ciliated in polyps so that the net effect of the ciliary beat is to move material away from the mouth toward the margin of the oral disk and toward the tips of the tentacles. Frequently, some of the epidermal cells have a dual function and the cell base is elongate and possesses contractile fibrils. Such cells are described as epitheliomuscular cells. The gastrodermis, or inner epithelial layer, is also abundantly ciliated. These cilia move food materials about within the gastric cavity and circulate the contained water. One of the prime functions of the gastrodermis is digestion, and essentially every cell of the gastrodermis is capable of ingesting small food particles.

The coelenterate nervous system can be defined as consisting basically of an unpolarized network of bipolar and multipolar neurons. The nervous system is subepithelial in location, both subepidermal and subgastrodermal, and may in some forms consist of two networks in only limited contact with one another and with each specialized either for rapid through conduction or for slower, more general spread of conduction.

The muscular system varies considerably among coelenterates. In simple polyps there is an epidermal, longitudinal muscle sheath of epitheliomuscular cells and a gastrodermal, circular sheath. In medusae and in members of the classes Scyphozoa and Anthozoa, the musculature has become separated from the epithelial tissue as a subepithelial system. In many anthozoans a large sphincter muscle develops in the gastrodermal surface, at the top of the column. In medusae, where the primary swimming muscles are subumbrellar, large muscle bands, both circular and radial, develop.

The mesogleal system is represented by a layer between epidermis and gastrodermis which varies from the thin structureless cementing layer of hydrozoan polyps to the highly complex cellular, fibrous, gelatinous matrix of the scyphomedusans and anthozoans. Where the mesoglea is well developed, as in medusae and some anthozoans, it serves as a type of internal skeleton, against which muscles may act.

Two different types of skeleton occur in the phylum. The first is of an internal nature and is either the mesoglea against which the muscles operate or the contained hydroskeleton of most polyps. Of a quite different nature is the exoskeleton, seen in most hydrozoan polyps, some scyphozoan polyps, and corals, which supports and protects the organisms. Still another skeletal type known in the coelenterates is an axial skeleton composed of sclerified proteins. It is exceedingly tough and flexible and is characteristic of such anthozoans as gorgonians.

The digestive system in coelenterates is developed as a function of the gastrodermis. In hydrozoans, both polyps and medusae, glandular, enzyme-producing cells are abundant throughout the gastrodermis. These cells secrete proteolytic enzymes into the coelenteron which reduce food objects to a fine particulate state. In scyphozoans and anthozoans, although glandular cells are common throughout the gastrodermis, they tend to be most abundant along the free edges of the mesenteries of anthozoans and on the gastric tentacles of scyphozoans. These glandular cells are also the source of proteolytic enzymes which act in the extracellular environment of the coelenteron.

The phylum is characterized by its carnivorous diet, made possible first by the possession of nematocysts which make the predaceous habit successful. After food has been trapped, movements of the tentacles carry it to the mouth, where with the help of ciliary and muscular devices the food is moved to the coelenteron. Here extracellular proteases prepare the way for final intracellular digestion. No herbivorous coelenterates are known.

The reproductive system of coelenterates consists of specialized areas of epithelia, the gonads, which periodically appear and produce gametes. There are no ducts for the sex products or any accessory sexual structures. Fertilization usually occurs in the water surrounding the animal, although a few coelenterates have their eggs fertilized in place and may then brood their young.

The ability to regenerate lost parts is characteristic of coelenterates. Pieces cut from almost any part of polyps will in time grow into new polyps. The regenerative powers of medusae are much less well developed, and not only will the excised piece not develop but it may not even be replaced by the medusa. Gradients of regenerative ability in polyps exist with the ability for a piece to reconstitute a new whole organism decreasing from the mouth to the base. *See* REGENERATION (BIOLOGY). [C.H.]

The Coelenterata have a long and impressive fossil record stretching from the present into the Precambrian, about 700,000,000 years ago. Thus, the known duration of this phylum equals or exceeds that of other animal phyla. Forms with skeletons, primarily the Conulata, Tabulata, Rugosa, and Scleractinia, have left fairly complete fossil records, whereas soft-bodied forms, such as most of the Hydrozoa, Scyphomedusae, and Alcyonaria, are represented by very few fossils. [C.H.St.]

Coelom The mesodermally lined body cavity of most animals above the flatworms and nonsegmented roundworms. Its manner of origin provides one basis for classifying the major higher groups.

Annelids, arthropods, and mollusks have a coelom which develops from solid mesodermal bands. Within the trochophore larva of annelids, a single pole cell proliferates two strips of mesoblast lying on either side of the ventral midline. These bands subdivide transversely into bilateral solid blocks, the somites. Each somite then splits internally to form a hollow vesicle, the cavity of which is the coelom. The somitic vesicles of each side expand until their surfaces become apposed above and below the gut. The mollusks also form bands of mesoderm from a single pole cell, but these bands do not segment. They split internally to form single right and left coelomic sacs, but the cavities are soon reduced and the surrounding mesoblast disperses as separate cells, many of which become muscle. In arthropods paired bands of mesoblast may proliferate from a posterior growth center or may separate inward from a blastoderm, a superficial layer of cells, on the ventral surface of the egg. These bands divide into linear series of somites which then hollow out. Their cavities represent the coelom.

Echinoderms and chordates constitute a second major group, characterized by the origin of the coelom from outpocketings of the primitive gut wall. In echinoderms one pair of bilateral pouches evaginates and separates from the archenteron or primitive digestive cavity. Each pouch constricts into three portions. The protochordates of the groups Hemichordata and Cephalochordata, exemplified by *Balanoglossus* and *Branchiostoma*, respectively, have three coelomic pouches formed by separate evaginations of the archenteral roof. In cephalochordates the head cavity divides into lateral halves. The left side communicates, by a pore, to an ectodermal pit called the wheel organ. The second pair of pouches forms the first pair of mesoblastic somites, and the third pouches subdivide transversely to give rise to the remainder of the linear series of somites. The upper or myotomic portion of each somite remains metameric and forms the segmental muscles. As it enlarges, the coelomic space is displaced ventrally and expands above and below the gut to form the perivisceral cavities and mesenteries, as described for annelids. In vertebrates the mesoderm arises as a solid sheet from surface cells that have been involuted through the blastopore. Lateral to the notochord, beginning at about the level of the ear, the mesoderm subdivides into three parts: (1) the somites; (2) the nephrotomic cord, temporarily segmented in lower vertebrates, which will form excretory organs and ducts; and (3) the unsegmented lateral plate. The coelom arises as a split within the lateral plate. *See* ANIMAL KINGDOM; GASTRULATION. [H.L.H.]

Coelomycetes Anamorphic (asexual or imperfect) fungi (Deuteromycotina) with sporulation occurring inside fruit bodies (conidiomata) that arise from a thallus consisting of septate hyphae. About 700 genera (with 750 synonyms) containing more than 8500 species are recognized.

The Coelomycetes are artificial, comprising almost entirely anamorphic fungi or ascomycete affinity. Some are known anamorphs of Ascomycotina, although there are a few (*Fibulocoela, Cenangiomyces*) with Basidiomycotina affinities because they have clamp connections or dolipore septa. Taxa are referred to as form genera and form species because the absence of a teleomorph (sexual or perfect) state means that they are classified and identified by artificial rather than phylogenetic means. The unifying feature of the group is the production of conidia inside cavities lined by fungal tissue, or by a combination of fungal and host tissue which constitutes the conidioma. *See* ASCOMYCOTINA; BASIDIOMYCOTINA.

Differences in conidiomatal structure traditionally have been used to separate three orders, Melanconiales, Sphaeropsidales, and Pycnothyriales. However, differences in the ways that conidia are produced have been used more recently in classification and identification.

Coelomycetes are known mainly from temperate and tropical regions. They grow, reproduce, and survive in a wide range of ecological situations and can be categorized as either a stress-tolerant or combative species. Coelomycetes are consistently isolated from or associated with disease conditions in all types of vascular plants, often in association with other organisms. *See* PLANT PATHOLOGY. [B.C.S.]

Coenopteridales True ferns which span the Late Devonian through Permian time between the recognizable beginnings of fernlike morphology and the earliest-appearing extant filicalean families (Gleicheniaceae, Osmundaceae). A true fern is a relatively advanced type of vascular land plant with distinct stem, fronds, roots, and foliar-borne annulate sporangia. Coenopterid ferns are mostly small and simple in contrast to late Paleozoic tree ferns of the Marattiales.

There are two major distinct groups. Zygopterid ferns are the most ancient and diverse, and apparently a dead-end evolutionary line; they differ the most from other ferns. Coenopterid ferns are usually placed in the Anachoropteridaceae and Botryopteridaceae, or in one of several extinct families assigned to the Filicales. Coenopterid ferns *sensu stricto* are probably ancestral to, and consequently form an imperceptible transition with, the Filicales.

The Zygopteridaceae appear in Late Devonian time, the Botryopteridaceae in Visean (Mississippian), the Anachoropteridaceae in the lower Westphalian A (Pennsylvanian), and all families extend into the Permian. *See* PALEOBOTANY. [T.L.P.]

Coenothecalia An order of the class Alcyonaria. The Coenothecalia have no spicules but form colonies with a massive skeleton composed of fibrocrystalline aragonite fused into lamellae. The skeleton is perforated by both numerous wide cylindrical cavities occupied by the polyps, and narrow ones containing the solenial systems. The order includes a few genera, of which *Heliopora*, or the blue coral, is often found on coral reefs. *See* ALCYONARIA. [K.At.]

Coenurosis An infestation by a coenurus, the metacestode of *Taenia* (*Multiceps*) sp. The coenurus is most common in sheep, rabbits, and other herbivores. They become infested while browsing on vegetation contaminated with eggs of the parasite. Coenuri occurring in subcutaneous spaces and muscles cause less damage than those in the brain or spinal column. The latter type of infestation, found mainly in sheep, may cause the animal to stagger or circle, and such animals are said to be giddy, from which arises the term gid tapeworm. The infestation is rare in humans; cure requires surgical removal of the coenurus. *See* CYCLOPHYLLIDEA. [R.S.F.]

Coenzyme An organic cofactor or prosthetic group (nonprotein portion of the enzyme) whose presence is required for the activity of many enzymes. In addition, many enzymes need metal ions, such as copper, manganese, and magnesium, for activation. The prosthetic groups attached to the protein of the enzyme (the apoenzyme) may be regarded as dissociable portions of conjugated proteins. The coenzymes usually contain vitamins as part of their structure. Neither the apoenzyme nor the coenzyme moieties can function singly, since dissociation of the two results in inactivation. In general, the coenzymes function as acceptors of electrons or functional groupings, such as the carboxyl groups in α-keto acids, which are removed from the substrate. Some of the well-known coenzymes are the pyridine nucleotides, thiamine pyrophosphate, flavin mononucleotide, hemin, the uridine phosphates, the phosphorylated derivatives of adenylic acid, coenzyme A, and biotin. *See* ADENOSINE DIPHOSPHATE (ADP); ADENOSINE TRIPHOSPHATE (ATP); ADENYLIC ACID; BIOTIN; CYTOCHROME; ENZYME; HEMOGLOBIN; URIDINE DIPHOSPHOGLUCOSE (UDPG). [M.B.McC.]

Coesite Naturally occurring coesite, a mineral of wide interest and the high-pressure polymorph of SiO_2, was first discovered and identified from shocked Coconino sandstone of the Meteor Crater in Arizona in 1960. Since then coesite has been identified from the Wabar (meteorite) Crater in Saudi Arabia, from the Ries Crater in Bavaria in southern Germany, from the Lake Bosumtwi Crater in Ashanti, Ghana, Africa, and from Lake Mien in Sweden. Coesite has also been identified from some Thailand tektites which are considered to have been formed also by an impact cratering process. The finding of natural coesite elevates it as a true mineral species. Because it requires a unique physical condition, extremely high pressure, for its formation, its occurrence is diagnostic of a special natural phenomenon, in this case, the hypervelocity impact of a meteorite. *See* METEORITE.

Synthetic coesite was first produced in the laboratory by L. Coes, Jr., as a chemical compound at pressures of about 35 kilobars (3.5×10^9 pascals) in the temperature range of 500–800°F (246–404°C). Coesite has also been found in synthetic diamonds and has been formed by transformation from alpha quartz by the application of shearing stress. *See* DIAMOND.

Coesite occurs in grains that are usually less than 0.0002 in. (5 micrometers) in size and are generally present in small amounts. The properties of the mineral are known mainly from studies of synthesized crystals. It is colorless with vitreous luster and has no cleavage. It has a specific gravity of 2.915 ± 0.015 and a hardness of about 8 on Mohs scale.

Coesite has as yet no evident commercial use and therefore has no obvious economic value. As a stepping-stone in scientific research, it serves in at least two ways: where it occurs naturally, coesite is diagnostic of a past history of high pressure; the occurrence of coesite from Meteor Crater clearly suggests that shock as a process can transform a low-density ordinary substance to one of high density and unique properties. Coesite has been found in materials ejected from craters formed by the explosion of 500,000 tons (450,000 metric tons) of TNT. Research on the occurrence of coesite from rocks deformed by other energy sources, such as volcanic explosions and deep-seated tectonic movement, is continuing. The study of coesite and craters may be useful in understanding the impact craters on the Earth as well as those on the Moon. [E.C.T.C.]

Coffee Arabica coffee is a shrub or small tree of the genus *Coffea* (Rubiaceae) and a native of Abyssinia. There are 59 species, but only 4 are of commercial importance. Nine-tenths of the world supply of coffee is obtained from the Arabian species, *C. arabica*, of which there are about 15 varieties. Coffee is strictly a tropical crop requiring a moderately cool climate that is moist but not wet. The fruit, called a cherry, is a small spherical drupe containing two seeds, the coffee beans. In the wet method the cherries are picked by hand, then pulped and allowed to ferment, and thus the membranes are removed. In the dry method cherries are harvested and allowed to dry, and the exocarp is milled off. The coffee beans are then graded and packed in burlap bags. Commercially, coffee is the most important caffeine beverage plant in the world. *See* RUBIALES. [P.D.St./E.L.C.]

Coffee beverage is prepared by extraction of roasted and ground coffee with hot water. This can be done in the home by the consumer, or by a coffee manufacturer who prepares coffee extract in a concentrated form and removes the water by spray-drying or freeze-drying.

Green beans are packed and transported in hemp bags. On arrival at the roasting plant, beans from many sources are blended to achieve a particular flavor. The beans are cleaned both before and after roasting. Roasting is the most important operation in coffee processing because it develops the flavor. Systems now employ externally heated gases of combustion, mainly nitrogen, carbon dioxide, and water vapor. In batch roasters, energy at relatively low temperature, 900–1000°F (480–540°C), develops fine flavor and a uniform brown color in about 12 min. In continuous roasters, energy is transferred even more efficiently. Temperature is reduced to about 500°F (260°C) and roasting time to about 5–8 min. Light roasts are attained at bean temperatures of 380–390°F (193–199°C); medium roasts, 400–410°F (204–210°C); and dark roasts, 425–430°F (218–221°C). After roasting, and a second cleaning operation called stoning, the beans are deposited in specially designed bins which prevent segregation as the beans pass to the grinder. Roasted and ground vacuum-packed coffee maintains freshness for at least 2 years.

Soluble or instant coffee manufacturing utilizes about 20% of the green beans imported. Soluble coffee accounts for about one-third of all coffee consumed in the United States, and about 40% of this is freeze-dried. Coffee extract is made by passing water at temperatures up to 350°F (177°C) through coarsely ground coffee in a series of extracting columns. After filtration, the extract can be concentrated in conventional evaporating equipment, followed by spray-drying to remove most of the water. The spray-dried soluble coffee can be processed further by agglomerating the particles with steam to form a granular-appearing product. The extract also can be frozen and then freeze-dried in a low-temperature vacuum system by sublimation from the ice crystal structure to remove most of the water. Most manufacturers of soluble coffees return aroma lost during processing to the package to provide the aroma of a roasted and ground coffee. This requires protection in packaging, and most soluble coffees are packed in a low-oxygen atmosphere by using an inert gas such as carbon dioxide or nitrogen.

Decaffeinated coffee, both regular and instant, is prepared from green beans that have been softened by steam, extracted with low-boiling chlorinated solvents to remove the caffeine, then steamed again to remove residual solvent, and dried. The green decaffeinated coffee then can be processed in the manners described above for a regular or a soluble coffee. [W.P.C.]

Cofferdam A temporary, wall-like structure to permit dewatering an area and constructing foundations, bridge piers, dams, dry docks, and like structures in the open air. A dewatered area can be completely surrounded by a cofferdam structure or by a combination of natural earth slopes and cofferdam structure. The type of construction is dependent upon the depth, soil conditions, fluctuations in the water level, availability of materials, working conditions desired inside the cofferdam, and whether the structure is located on land or in water (see illustration). An important consideration in the design of cofferdams is the hydraulic analysis of seepage conditions, and erosion of the bottom when in streams or rivers.

Where the cofferdam structure can be built on a layer of impervious soil (which prevents the passage of water), the area within the cofferdam can be completely sealed off. Where the soils are pervious, the flow of water into the cofferdam cannot

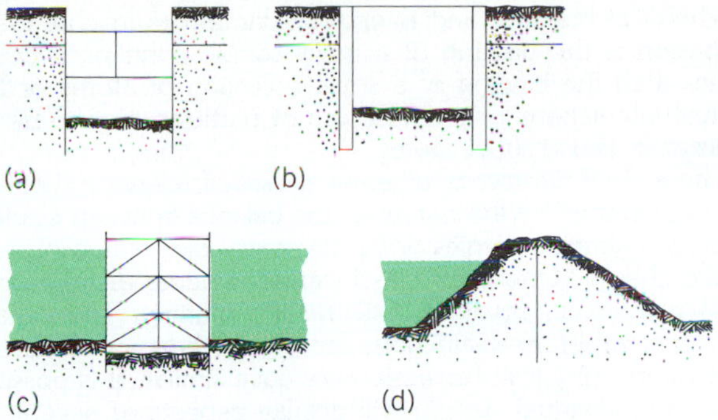

Types of cofferdams for use on land: (*a*) cross-braced sheet piles; (*b*) cast-in-place concrete cylinder; and in water: (*c*) cross-braced sheet piles; (*d*) earth dam.

be completely stopped economically, and the water must be pumped out periodically and sometimes continuously.

A nautical application of the term cofferdam is a watertight structure used for making repairs below the waterline of a vessel. The name also is applied to void tanks which protect the buoyancy of a vessel. [E.J.Q.]

Cogeneration The sequential production of electricity and thermal energy in the form of heat or steam, or useful mechanical work, such as shaft power, from the same fuel source. Cogeneration projects are typically represented by two basic types of power cycles, topping or bottoming. The topping cycle has the widest industrial application.

The topping cycle utilizes the primary energy source to generate electrical or mechanical power. Then the rejected heat, in the form of useful thermal energy, is supplied to the process. The cycle consists of a combustion turbine-generator, with the turbine exhaust gases directed into a waste-heat-recovery boiler that converts the exhaust gas heat into steam which drives a steam turbine, extracting steam to the process while driving an electric generator. This cycle is commonly referred to as a combined cycle arrangement. Combustion turbine-generators, steam turbine-generator sets, and reciprocating internal-combustion-engine generators are representative of the major equipment components utilized in a topping cycle. *See* GENERATOR; STEAM TURBINE; TURBINE.

A bottoming cycle has the primary energy source applied to a useful heating process. The reject heat from the process is then used to generate electrical power. The typical bottoming cycle directs waste heat from a process to a waste-heat-recovery boiler that converts this thermal energy to steam which is supplied to a steam turbine, extracting steam to the process and also generating electrical power. *See* ELECTRIC POWER GENERATION.

Cogeneration for building and district space heating and cooling purposes consists of producing electricity and sequentially utilizing useful energy in the form of steam, hot water, or direct exhaust gases. The two most common heating, ventilating, and air-conditioning cycles are the vapor compression cycle and the absorption cycle. *See* AIR CONDITIONING; CENTRAL HEATING AND COOLING; COMFORT HEATING; DISTRICT HEATING. [C.Bu.]

Cognition A rather loosely defined area that embraces the whole complex system of human mental abilities focusing on perception, attention, learning and memory, the comprehension and use of language, and reasoning. Traditionally, psychologists have regarded cognition in terms of an information-processing system in which a given stimulus is received by the organism and proceeds through a sequence of processing stages until it either drops out of the system or is deposited in long-term memory.

When a stimulus has been perceived, say, a word or letter string, it resides briefly in a preprocessing store called sensory memory, which effectively prolongs its life even when the stimulus is no longer present. Thus, like a fading afterimage, the information remains available for further processing. Sensory memory is relatively unselective in its contents and has a large capacity: most or all of the input information is recorded. At every processing stage, attention is focused on those elements of the input that are relevant for a given purpose.

Short-term memory, unlike sensory memory, has a very small capacity. Only some five to seven pieces of information can be maintained there, and new items quickly displace the old ones. However, quite a bit of information can be retained if it is organized into meaningful chunks—if the letters form words and if the words combine associatively or into sentences.

Short-term memory should more properly be considered the active, conscious, working component of memory, where strict selection processes determine the information relevant for long-term storage. Short-term memory maintains information through rehearsal, reorganizes it in chunks, and relates them to preexisting knowledge. This is what is meant by the encoding process.

Long-term memory is the permanent memory store, the repository for all personal experiences, general world knowledge, and procedures, goals, and interests. Its capacity is virtually unlimited, and forgetting from long-term memory is not so much a loss of information as a retrieval failure.

An important branch of research in cognition has concerned itself with the organization of long-term or semantic memory. Semantic memory is usually defined as general knowledge, information that can be retrieved and used in a variety of contexts. Episodic memory, on the other hand, refers to knowledge of events, how and when something occurred and not so much what it was.

The use of knowledge in reasoning and problem-solving tasks has been another focus of work in cognition. Perceptual aspects appear to be important in a variety of tasks, such as the ability to perceive patterns rather than individual pieces on a chessboard, or to restructure a logical or mathematical problem in such a way that some critical insight is gained. *See* INTELLIGENCE; MEMORY; PERCEPTION; PROBLEM SOLVING (PSYCHOLOGY). [W.Ki.; E.Ki.]

Coherence The attribute of two or more waves, or parts of a wave, whose relative phase is constant during the resolving time of the observer. The concept has been developed most extensively in optics, but is applicable to all wave phenomena.

Consider two waves, with the same mean angular frequency ω, given by Eqs. (1) and (2). It is convenient, and no restriction,

$$\Psi_A(x,t) = A \exp \{i\,[k(\omega)x - \omega t - \delta_A(t)]\} \quad (1)$$

$$\Psi_B(x,t) = B \exp \{i\,[k(\omega)x - \omega t - \delta_B(t)]\} \quad (2)$$

to choose both A and B real. These expressions as they stand could describe de Broglie waves in quantum mechanics. For real waves, such as components of the electric field in light or radio beams, or the pressure oscillations in sound, it is necessary to retain only the real parts of these and subsequent expressions. The frequency spectrum is assumed to be narrow, in the sense that a Fourier analysis of expressions (1) and (2) gives appreciable contributions only for angular frequencies close to ω. This assumption means that, on the average, $\delta_A(t)$ and $\delta_B(t)$ do not change much per period. *See* ELECTROMAGNETIC RADIATION; QUANTUM MECHANICS; SOUND.

Suppose that the waves are detected by an apparatus with resolving time T, that is, T is the shortest interval between two events for which the events do not seem to be simultaneous. For the human eye and ear, T is about 0.1 s, while a fast electronic device might have a T of 10^{-10} s. If the relative phase $\delta(t)$, given by Eq. (3), does not, on the average, change

$$\delta(t) = \delta_B(t) - \delta_A(t) \tag{3}$$

noticeably during T, then the waves are coherent. If during T there are sufficient random fluctuations for all values of $\delta(t)$, modulus 2π, to be equally probable, then the waves are incoherent. If during T there are noticeable random fluctuations in $\delta(t)$, but not enough to make the waves completely incoherent, then the waves are partially coherent. These distinctions are not useful unless T is specified. On the one hand, only waves that have existed forever and that fill all of space can have absolutely fixed frequency and phase. On the other hand, two independent sound waves in which the phases change appreciably in 0.01 s would seem incoherent to the human ear, but would seem highly coherent to a fast electronic device.

The degree of coherence is related to the interference patterns that can be observed when the two beams are combined. *See* INTERFERENCE OF WAVES.

Coherence is also used to describe relations between phases within the same beam. Suppose that a wave represented by Eq. (1) is passing a fixed observer characterized by a resolving time T. The phase δ_A may fluctuate, perhaps because the source of the wave contains many independent radiators. The coherence time Δt_W of the wave is defined to be the average time required for $\delta_A(t)$ to fluctuate appreciably at the position of the observer. If Δt_W is much greater than T, the wave is coherent; if Δt_W is of the order of T, the wave is partially coherent; and if Δt_W is much less than T, the wave is incoherent. These concepts are very close to those developed above.

Extended sources give partial coherence and produce interference fringes with visibility V less than unity. A. A. Michelson exploited this fact with his stellar interferometer, a modified double-slit arrangement with movable mirrors that permit adjustment of the effective separation D' of the slits. It can be shown that if the source is a uniform disk of angular diameter Θ, then the smallest value of D' that gives zero V is $1.22\lambda/\Theta$. The same approach has also been applied in radio astronomy. A different technique, developed by R. Hanbury Brown and R. Q. Twiss, measures the correlation between the intensifies received by separated detectors with fast electronics. *See* INTERFEROMETRY.

Because they are highly coherent sources, lasers and masers provide very large intensities per unit frequency. *See* LASER.

[R.G.Wi.]

Cohesion (physics) The tendency of particles of a condensed phase (a solid or liquid) to stick together, or the processes or forces that lead to this tendency. The cohesive energy is the work necessary to pull the body apart into its constituent particles. For example, the cohesive energy of a metal is the work necessary to break the metal apart into individual atoms in their ground state. The cohesive energy of a molecular solid such as solid carbon dioxide is the energy necessary to separate the solid into isolated carbon dioxide molecules.

The study of cohesion is the study of the competing processes that tend to establish an equilibrium separation between molecules or atoms in a condensed phase. In normal matter all such forces are the result either of the Coulomb interaction between positive and negative charges or of quantum-mechanical kinetic energy. The Coulomb interaction lowers the energy (increases the bond strength) and tends to pull atoms together. This tendency toward higher density and stronger bonding is opposed by the quantum-mechanical kinetic energy. As a quantum-mechanical particle such as an electron is confined to a smaller volume, its kinetic energy increases. The cohesion of matter is a delicate balance between these two effects. *See* COULOMB'S LAW; EXCLUSION PRINCIPLE; UNCERTAINTY PRINCIPLE.

The study of cohesion is the analog in physics of the study of chemical bonding and intermolecular forces in chemistry. Cohesion is the binding of a macroscopic condensed body, rather than the binding of a small collection of atoms or the mutual interactions of a diffuse gas of particles. *See* CHEMICAL BONDING; INTERMOLECULAR FORCES.

There are a number of different classes of cohesion that can be distinguished by the nature of the balance between electrostatic and kinetic energies or by the spatial distribution of electronic charge. One such classification scheme distinguishes molecular, ionic, covalent, and metallic binding. The classes represent arbitrary distinctions among systems in which the microscopic physics of cohesion are quite similar. It is possible that any individual system will display aspects of several of these classes.

In a molecular crystal or liquid some of the atoms are strongly bound together in molecules, and these molecules are relatively weakly bound to the other molecules in a condensed phase. The intermolecular bond is usually due to the electrostatic dipole-dipole interaction.

In an ionic system, such as sodium chloride (NaCl), the constituent particles are ions with a net electronic charge, some positive and some negative. The Coulomb interaction between them leads to an attractive force. The system is kept from collapsing by an increase in kinetic energy due to the compression of charge as the system is compressed.

Covalently bound crystals are a third major class of crystals, in which the atoms are held together by directional bonds. A bond forms because atoms are able to share their electrons. The shared electrons are subject to the Coulomb attraction of more than one nucleus, and thus the system of electrons and nuclei is more tightly bound than the separated atoms.

The fourth class of cohesion is metallic bonding, which is characterized by relatively large cohesive energy and density, and by the fact that metals are good low-temperature electrical conductors in contrast to the other classes which are insulators. The concept of electron pair bonds is not applicable to these systems in which the coordination number, the number of atoms surrounding a given atom, is frequently much larger than the number of valence electrons, the electrons available for bonding. *See* BAND THEORY OF SOLIDS; SOLID-STATE CHEMISTRY; SOLID-STATE PHYSICS.

[C.D.G.]

Coil One or more turns of wire used to introduce inductance into an electric circuit. At power line and audio frequencies a coil has a large number of turns of insulated wire wound close together on a form made of insulating material, with a closed iron core passing through the center of the coil. This is commonly called a choke and is used to pass direct current while offering high opposition to alternating current.

At higher frequencies a coil may have a powdered iron core or no core at all. The electrical size of a coil is called inductance and is expressed in henries or millihenries. In addition to the resistance of the wire, a coil offers an opposition to alternating current, called reactance, expressed in ohms. The reactance of a coil increases with frequency. *See* INDUCTOR; REACTOR (ELECTRICITY).

[J.Mar.]

Coincidence amplifier An electronic circuit that amplifies only that portion of a signal present when another enabling or controlling signal is simultaneously applied. The controlling signal may be such that the amplifier functions during a specified time interval, called time selection, or such that it functions between specified voltage or current levels, called amplitude selection. For example, the amplifier in a transmission gate operates on the basis of time selection where a control voltage of fixed amplitude is applied during a specific time interval. This voltage is the enabling waveform; it allows the amplifier to function normally only during the time it is applied. *See* GATE CIRCUIT.

A linear amplifier used to amplify a section of a waveform above or below specified voltage or current limits, as determined by a control signal, is an example of a coincidence amplifier that operates by amplitude selection. Limiting and clipping circuits that incorporate the function of amplification may be considered coincidence amplifiers. *See* CLIPPING CIRCUIT; LIMITER CIRCUIT. [G.M.G.]

Coining A cold metalworking process in a press-type die. Coining is used to produce embossed parts, such as badges and medals, and for minting of coins. It is also used on portions of a blank or workpiece to form corners, indentations, or raised sections, frequently as part of a progressive die operation. The work is subjected to high pressure within the die cavity and thereby forced to flow plastically into the die details. The absence of overflow of excess metal from between the dies is characteristic of coining and is responsible for the fine detail achieved. *See* METAL FORMING. [R.L.Fr.]

Coir A natural fiber, also known as coco fiber, obtained from the husks of the coconut (*Cocos nucifera*). The outstanding characteristic of coir fiber is its resistance to rot. For example, a coir fiber doormat can stay out in the weather for years. Coir is only intermediate in strength among the vegetable fibers, but its elongation at break is greater than any of the bast or hard fibers. More than 95% of coir exports are from Sri Lanka and India; small amounts are exported from Thailand, Tanzania, Mexico, the Philippines, Malaysia, Kenya, and Trinidad and Tobago.

There are three main types of coir—yarn fiber, bristle fiber, and mattress fiber. Only the finest and longest fiber is suitable for spinning into yarn. It is obtained from the husks of unripe nuts and is the main cash crop rather than a by-product. The bristle fiber and most of the mattress fiber come from mature nuts and are by-products of copra production. Coir fiber ranges in color from light tan to dark brown. Bristles are sorted according to length and graded on the basis of length, stiffness, color, and cleanliness. Long, clean fibers and light color are desirable for any purpose, but stiffness is desirable in bristles and undesirable for yarn.

In developed countries, yarn fiber is used chiefly in mats and mattings; in the United States, for instance, its best-known use is in light-brown tufted doormats. In Asia it is used extensively for ropes and twines, and locally for hand-made bags. The principal outlet for bristle fiber has been in brush making, but the market has declined and most bristle fiber is now used in upholstery padding. Mattress fiber is used chiefly in innerspring mattresses, though it has found uses as an insulating material. *See* COCONUT; NATURAL FIBER. [E.G.N.]

Coke A coherent, cellular, carbonaceous residue remaining from the dry (destructive) distillation of a coking coal. It contains carbon as its principal constituent, together with mineral matter and residual volatile matter. The residue obtained from the carbonization of a noncoking coal, such as subbituminous coal, lignite, or anthracite, is normally called a char. Coke is produced chiefly in chemical-recovery coke ovens, but a small amount is also produced in beehive or other types of nonrecovery ovens.

Coke is used predominantly as a fuel reductant in the blast furnace, in which it also serves to support the burden. As the fuel, it supplies the heat as well as the gases required for the reduction of the iron ore. It also finds use in other reduction processes, the foundry cupola, and house heating.

Coke is formed when coal is heated in the absence of air. During the heating in the range of 660–930°F (350–500°C), the coal softens and then fuses into a solid mass. The degree of softening attained during heating determines to a large extent the character of the coke produced. In order to produce coke having desired properties, two or more coals are blended before charging into the coke oven. In addition to the types of coals blended, the carbonizing conditions in the coke oven influence the characteristics of the coke produced. Oven temperature is the most important of these and has a significant effect on the size and the strength of the coke. In general, for a given coal, the size and shatter strength of the coke increase with decrease in carbonization temperature.

The important properties of coke that are of concern in metallurgical operations are its chemical composition, such as moisture, volatile-matter, ash, and sulfur contents, and its physical character, such as size, strength, and density. The moisture and the volatile-matter contents are a function of manner of oven operation and quenching, whereas ash and sulfur contents depend upon the composition of the coal charged. *See* CHARCOAL; COAL; DESTRUCTIVE DISTILLATION. [M.P.]

Coking (petroleum) A process for thermally converting the heavy residual bottoms of crude oil entirely to lower-boiling petroleum products and by-product petroleum coke. The heavy residual bottoms cannot be catalytically cracked, principally because the metals present are potent catalyst poisons. Traditionally, the residual bottoms have been blended with lighter stocks and marketed as low-quality heavy fuel oils. However, in the United States the demand for heavy fuel oil has remained nearly constant for a number of years, while demands for gasoline, heating oil, and other products, have increased substantially. To meet this shift in demand, refiners have resorted to deeper vacuum flashing, visbreaking (viscosity breaking), solvent deasphalting, and coking. The major products from coking are fuel gas, gasoline, gas oil, and petroleum coke. Generally the gasoline is of poor quality and must be further treated before use in premium fuels. The gas oil is commonly used as catalytic-cracking feed stock. *See* CRACKING; PETROLEUM PROCESSING. [J.F.Mo.]

Cola A tree, *Cola acuminata*, of the sterculia family (Sterculiaceae) and a native of tropical Africa. Its fruit is a star-shaped follicle containing eight hard seeds, the cola nuts of commerce. These nuts are an important masticatory in many parts of tropical Africa. They have a caffeine content twice that of coffee. The nuts also contain an essential oil and a glucoside, kolanin, which is a heart stimulant. Cola nuts, in combination with an extract from coca, are used in the manufacture of the beverage Coca-Cola. Cola is now cultivated in Angola, Jamaica, Brazil, India, and other parts of tropical Asia. *See* COCA; MALVALES. [P.D.St./E.L.C.]

Colchicine An alkaloid contained in the underground stem of the autumn crocus (*Colchicum autumnale*) of the lily family (Liliaceae). In medicine the drug is used principally in the treatment of gout. Colchicine has a specific influence on the division of plant chromosomes, and has been used by geneticists and plant breeders to bring about doubling of chromosomes, thus producing tetraploid plants. Most colchicine comes from the Netherlands, Italy, Yugoslavia, and Hungary. *See* ALKALOID; LILIALES. [P.D.St./E.L.C.]

Cold hardiness (plants) An adaptive capacity of plants enabling them to survive temperatures near and below the freezing point of water. Without this property there would be no forest or fruit trees growing in many regions of the north temperate zone. Furthermore, the limited level of frost hardiness which crops possess is a major limiting factor in the northward extension of agriculture. Even in cultivated areas, sporadic occurrences of unusually low temperatures may result in widespread crop damage with considerable economic loss.

The property of cold hardiness is genetically determined. In plants having the genetic capacity, the property is not always present but is elicited only following a conditioning period which is environmentally evoked. The condition of cold hardiness is one of tolerance of cells and tissues to actual freezing. Since plants do not usually generate large quantities of heat and lack thermoregulation, they assume the temperature of their surroundings. Also, they contain a high percentage of water and freeze solidly when temperatures drop appreciably below the freezing point. Unhardened plants are usually injured or killed under freezing conditions. However, no injury is incurred even when these plants are supercooled well below the freezing point, provided that ice crystallization has not been initiated. If plants or plant tissues whose cellular water has supercooled are quickly rewarmed before ice has formed, no injury will be detected. Also, any plants or plant tissues (such as seeds) which can endure desiccation can, in the desiccated condition, be immersed directly in liquid nitrogen without injury, since seeds contain virtually no free water capable of freezing.

The study of cold hardiness in temperate zone plants includes study of the cellular and environmental factors initiating the cellular phenomenon of cold hardiness, the nature of the injurious effects of ice formation, and the physiological and chemical adaptive mechanisms which have evolved in these plants to render them resistant to these effects. *See* DORMANCY; PLANT GROWTH; PLANT PHYSIOLOGY. [D.Si.]

Cold storage The storage of perishables at low temperatures, usually above freezing, by the use of refrigeration to increase the storage life. In general, the lower the temperature, the longer the storage life. If temperatures are maintained below the freezing point of the product stored, it is called freezer storage. Most fruits and many other products, however, are damaged by freezing and cannot be stored in freezer storage. A cold-storage plant is a large insulated building, with its attendant refrigeration equipment, for storage of commodities at low temperatures. Facilities are often included for quick-freezing fruits, vegetables, meats, and a variety of precooked foods and bakery products for the consumer convenience market. *See* REFRIGERATION. [C.F.K.]

Coleoidea A subclass of the Cephalopoda that appeared in the middle Paleozoic (Early Devonian Period) and presumably evolved from the nautiloid stalk. Of the five orders of fossil coleoids, the belemnites are conspicuous with their fossilized shells termed lightning bolts. All these forms became extinct by the end of the Mesozoic, 65 million years ago. The surviving four orders that developed in the Late Triassic to Late Cretaceous represent all the living cephalopods today, except *Nautilus* (subclass Nautiloidea).

Coleoids are characterized by an internal chitinous or calcareous shell; 8–10 appendages around the mouth (8 arms and 2 tentacles when present) lined with suckers or hooks; one pair of gills; highly developed eyes; a fused, tubelike funnel; an ink sac (lost in some species); chromatophores; and fins on the body (lost in some octopuses).

The living coleoids are the order Sepioidea (cuttlefishes, like *Sepia*; bobtail squids, *Spirula*); the order Teuthoidea (squids—nearshore myopsids, like *Loligo*; open-ocean oegopsids, like *Ommastrephes*); the order Vampyromorpha (the black, deep-sea vampire squid); and the order Octopoda (octopuses, argonauts, deep-sea finned or cirrate octopods). Living coleoids generally are fast, mobile predators with an advanced brain and central nervous system. *See* BELEMNOIDEA; CEPHALOPODA; MOLLUSCA; NAUTILOIDEA; SEPIOIDEA; TEUTHOIDEA; VAMPYROMORPHA. [C.F.E.R.]

Coleoptera An order of Insecta (generally known as beetles) with a complete metamorphosis (with larva and pupa stages), the forewings forming hard protective elytra under which the hindwings are normally folded in repose, and the

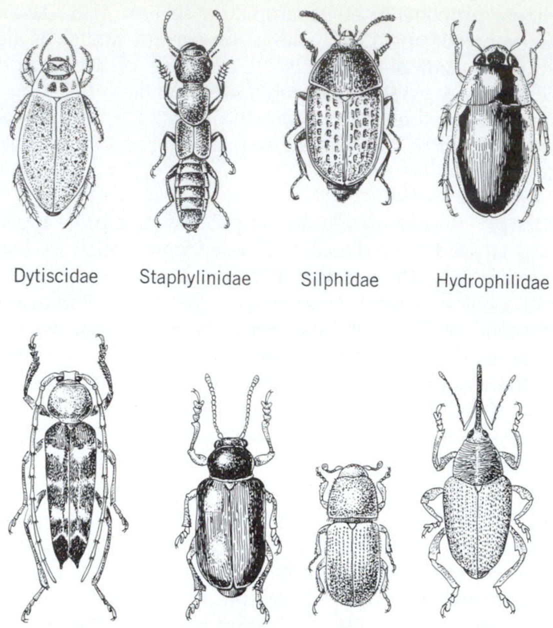

Coleoptera. *(After T. I. Storer et al., General Zoology, 6th ed., McGraw-Hill, 1979)*

indirect flight muscles of the mesothorax having been lost. The head capsule is characterized by a firm ventral (gular) closure, the antennae are basically 11-segmented, and mouthparts are of the biting type with 4-segmented maxillary palpi. The prothorax is large and free. The abdomen has sternite I normally absent, sternite II usually membranous and hidden, and segment X vestigial, with cerci absent. The female has one pair of gonapophyses, on segment IX. There are four or six malpighian tubules, which are often cryptonephric. True labial glands are nearly always absent.

General biological features. The order has well over 250,000 described species, more than any comparable group, showing very great diversity in size, form (see illustration), color, habits, and physiology. Beetles have been found almost everywhere on Earth (except as yet for the Antarctic mainland) where any insects are known, and species of the order exploit almost every habitat and type of food which is used by insects. Many species are economically important.

A common feature is for the exoskeleton to form an unusually hard and close-fitting suit of armor, correlated with a tendency for the adults to be less mobile but longer-lived than those of other orders of higher insects. It seems to be common in a number of groups for adults to survive and breed in 2 or more successive years, a rare occurrence in other Endopterygota. Adult beetles tend to be better runners than most other Endopterygota, and some (such as Cicindelidae) are among the fastest of running insects. The flight of most beetles is not very ready or frequent, often requiring prior climbing up onto some eminence, and in many beetles occurs only once or twice in an average life.

Apart from a tendency for the sclerotized layers to be thicker, the cuticle of beetles differs little from that of other insects in constitution or ultrastructure. The black coloring which is prevalent in many groups of beetles is produced by deposition of melanin in the process of cuticle hardening. In many groups with adults active by day, the black is masked by a metallic structural color produced by interference in the surface cuticle layers. Pigmentary colors, mainly ommochromes, giving reddish to yellowish colors, occur in the cuticle of many species.

Special defenses against predation are common in species

with long-lived adults, particularly the ground-living species of Carabidae, Staphylinidae, and Tenebrionidae. These groups commonly possess defensive glands, with reservoirs opening on or near the tip of the abdomen, secreting quinones, unsaturated acids, and similar toxic substances. In some cases, the secretion merely oozes onto the body surface, but in others may be expelled in a jet, as in the bombardier beetles (Brachinini). Other common defensive adaptations include cryptic and mimetic appearances. The most frequent behavioral adaptations are the drop-off reflex and appendage retraction, in which adults will react to visual or tactile stimuli by dropping off the plant foliage on which they occur, falling to the ground with retracted appendages, and lying there for some time before resuming activity. In death feigning, the appendages are tightly retracted, often into grooves of the cuticle, and held in this position for some time. In such conditions, beetles often resemble seeds and are difficult for predators to perforate or grasp.

Sound production (stridulation) is widespread in beetles and may be produced by friction between almost any counterposed movable parts of the cuticle. Stridulatory organs may show sex dimorphism, and in some species may play an important role in the interrelations of the sexes.

Anatomy. The head of beetles rarely has a marked posterior neck, and the antennal insertions are usually lateral, rather than dorsal as in most other Endopterygota, a feature possibly related to an original habit of creeping under bark. Antennal forms are very various, with the number of segments often reduced, but rarely increased, from the basic 11. Mandible forms are very diverse and related to types of food.

The prothorax commonly has lateral edges which separate the dorsal part from the ventral part of the tergum. The mesothorax is the smallest thoracic segment. Dorsally, its tergum is usually exposed as a small triangular sclerite between the elytral bases, and ventrally, its sternite may have a median pit receiving the tip of the prosternum. The form of the metathorax is affected by far-forward insertions of the hindwings and far-posterior ones of the hindlegs, so that the pleura are extremely oblique and the sternum usually long.

The tarsi of beetles show all gradations from the primitive five-segmented condition to total disappearance. The number of segments may differ between the legs of an individual or between the sexes of a species.

Most beetles have the entire abdomen covered dorsally by the elytra in repose, and ventrally only five or six sternites (of segments III to VII or VIII) exposed; segment IX is retracted inside VIII in repose. Tergites I–VI are usually soft and flexible, those of VII and VIII more or less sclerotized. A feature developed in several groups of Coleoptera, but particularly in Staphylinidae, is the abbreviation of the elytra to leave part of the abdomen uncovered in repose, but usually still covering the folded wings. One advantage of this may be to give greater overall flexibility to the body. Perhaps surprisingly, many beetles with short elytra fly readily.

Internally, the gut varies greatly in length and detailed structure. The foregut commonly ends in a distended crop whose posterior part usually has some internal setae and may be developed into a complex proventriculus; the midgut may be partly or wholly covered with small papillae (regenerative crypts) and sometimes has anterior ceca. The four or six malpighian tubules open in various ways into the beginning of the hindgut and in many Polyphaga have their apices attached to it (cryptonephric). The central nervous system shows all degrees of concentration of the ventral chain, from having three thoracic and eight abdominal ganglia distinct to having all of these fused into a single mass.

The sense organs of beetles, adult or larval, are generally similar in type and function to those of other insects. For vision, some types are adapted for high visual acuity in bright light, others to high sensitivity in poor light, and many show light-dark adaptations of screening pigment. Color vision seems to be restricted to certain groups. In certain beetles, there is evidence of specialized infrared sensitivity. Antennae are mainly receptors for smell and touch, but detection of aerial vibrations (by Johnston's organ in the pedicel) may be important in some. The olfactory sensilla, which commonly are setae, are often concentrated on expanded apical segments, forming a club. Taste, or contact chemoreception, is usually mediated by sensilla with a single large apical opening. These sensilla are concentrated on the palpi and other mouthparts and are often found on tibiae or tarsi of the front legs.

Taste, or contact chemoreception, is usually mediated by sensilla with a single large apical opening. These sensilla are concentrated on the palpi and other mouthparts and are often found on tibiae or tarsi of the front legs.

Specialized sound receptors, other than Johnston's organ, have not been much studied in Coleoptera, though many species produce sounds and in some cases these play a significant part in courtship. A gravitational sense is evident in many burrowing beetles, which are able to sink accurately vertical burrows despite inclinations of the surface started from. A temperature sense is clearly present in some beetles and larvae, but little is known of the sensilla concerned. What might be called a time sense (otherwise known as a biological clock) clearly operates in many beetles that react specifically to changing daylight lengths marking the seasons in nontropical latitudes.

Food specialization. The biting mouthparts of many adult and larval beetles are adaptable to many types of food, and fairly widely polyphagous habits are not uncommon in the order, though most species can be assigned to one or another of a few main food categories: fungivores, carnivores, herbivores, or detritivores. Included in the last category are species feeding on decaying animal or vegetable matter and on dung. Most parasites could be included with carnivores, but those myrmecophiles and termitophiles which are fed by their hosts, and those Meloidae which develop on the food stores of bees, form special categories.

Water beetles. Beetles are unusual among the higher insects in that aquatic habits, where present, usually affect adults as well as larvae. The elytra may have been preadaptive to the invasion of water. Almost any type of fresh or brackish water body is liable to contain some types of water beetles, though very few species can live permanently in full marine salinities. Almost all aquatic beetles maintain an air reservoir, into which the second thoracic and abdominal spiracles open, under their elytra; and some, mainly small, species may have plastrons or physical gills. Aquatic larvae tend to have a last pair of abdominal spiracles that is large and effectively terminal, or tend to develop tracheal gills. Pupae are terrestrial in the large majority of water beetles, and eggs, when deposited underwater, are commonly laid in contact with airspaces in the stems of water plants, or in an air-filled egg cocoon. Most water beetles are confined to shallow waters, and many of them are ready colonists of temporary pools, through adult flight.

Special habitats and adaptations. More or less unusual adaptations and modes of life are manifest in dung and carrion beetles, ambrosia beetles, cave and subterranean beetles, desert beetles, and luminous beetles. Dung and carrion beetles exploit sporadic and very temporary food resources in which competition (particularly with Diptera larvae) tends to be fierce. Many dung beetles avoid the difficulties by burying dung stores in underground cells where their larvae can develop safely, and some beetles (Necrophorinae) adopt a similar strategy for carrion. They may also develop phoretic and symbiotic relations with specific mites preying on fly eggs.

In ambrosia beetles, the adults usually excavate burrows in

wood, carrying with them spores of special fungi, which proceed to develop along the walls of the burrows and are fed on by larvae developing from eggs laid there. This type of relation between beetles, fungi, and trees exists in various forms in a number of families and may be of ecological importance if the fungi concerned are liable to kill trees.

More than one family of beetles contains highly adapted types living exclusively in deeper cave systems (cavernicolous) or the deeper layers of the soil (hypogeous). Such species are usually flightless and eyeless, poorly pigmented, and slow-moving, and of very restricted geographical distribution. Another marginal habitat in which beetles are the principal insect group represented is the desert. Here, too, wings are usually lost, but the cuticle tends to be unusually thick, firm, and black. Also beetles may be found in hot springs. A few water beetles are adapted to underground (phreatic) waters, showing parallel features with the cavernicolous and hypogeous terrestrial ones.

More or less parasitic relations to animals of other groups have developed in a number of lines of Coleoptera. The closest parallels to Hymenoptera-Parasitica are to be seen in the families Stylopidae (Strepsiptera of many authors) and Rhipiphoridae. Stylopid larvae are exclusively endoparasites of other insects, while the adult females are apterous, often legless, and remain in the host's body, while males have large fanlike hindwings and the elytra reduced to halter-like structures. A considerable variety of Coleoptera develop normally in the nests of termites and social or solitary Hymenoptera-Aculeata, some being essentially detritivores or scavengers, but most feed either on the young or the food stores of their hosts. Ectoparasitism on birds or mammals is a rare development in beetles.

The beetles are notable in that some of them manifest the highest developments of bioluminescence known in nonmarine animals, the main groups concerned being Phengodidae, Lampyridae, and Elateridae-Pyrophorini, the glowworms and fireflies. In luminescent beetles, the phenomenon always seems to be manifest in the larvae and often in the pupae, but not always in the adults. In luminous adult beetles, there is often marked sex dimorphism, and a major function of the lights seems to be the mutual recognition of the sexes of a species.

Another possible function of adult luminosity, and the only seriously suggested for that of larvae, is as an aposematic signal. There is definite evidence that some adult fireflies are distasteful to some predators, and the luminous larvae of Phengodidae have dorsal glandular openings on the trunk segments that probably have a defensive function. *See* BIOLUMINESCENCE.

Phylogenetic history. Coleoptera are older than the other major endopterygote orders. The earliest fossils showing distinctively beetle features were found rather before the middle of the Permian Period. By the later Permian, fossils indicate that beetles had become numerous and diverse, and during the Mesozoic Era they appear as a dominant group among insect fossils. Fossils in Triassic deposits have shown features indicative of all four modem suborders, and the Jurassic probably saw the establishment of all modern superfamilies. By early Cretaceous times, it is likely that all "good" modern families had been established as separate lines. In the Baltic Amber fauna, of later Paleogene age (about 40,000,000 years ago), about half the fossil beetle genera appear to be extinct, and the other half have still-living representatives (often in remote parts of the world). Beetle fossils in Quaternary (Pleistocene) deposits are very largely of still-living species.

Geographical distribution. Almost every type of continuous or discontinuous distribution pattern which is known in any animal group could be matched in some taxon of Coleoptera, and every significant zoogeographical region or area could be characterized by endemic taxa of beetles. In flightless taxa, distributional areas are generally more limited than those of comparable winged taxa. Distinct distributional categories can be seen in those small, readily flying groups (in Staphylinidae and Nitidulidae, for example) which are liable to form part of "aerial plankton," and in those wood borers which may survive for extended periods in sea-drifted logs, both of which are liable to occur in oceanic islands beyond the ranges of most other beetle taxa. Climatic factors often seem to impose limits on the spread of beetle species (and sometimes of genera or families), and beetle remains in peats have been found to be sensitive indicators of climatic changes in glacial and postglacial times. *See* INSECTA. [R.A.Cr.]

Coliiformes A small order of birds containing only the family Coliidae with six species (the mousebirds) restricted to Africa. The mousebirds are small, grayish to brownish, with a long tail. The legs are short and the feet strong, with the four toes movable into many positions from all four pointing forward to two reversed backward. Mousebirds perch, climb, crawl, and scramble agilely in bushes and trees. They are largely vegetarian but eat some insects. They are nonmigratory and gregarious, and sleep in clusters, but they are monogamous as breeders. The nest is an open cup in a tree or bush, and the two to four young remain in the nest, cared for by both adults until they can fly. The relationships of the mousebirds to other birds are obscure. Mousebirds have a surprisingly good fossil record from the Miocene of France and Germany. *See* AVES. [W.J.B.]

Coliphage Any bacteriophage able to infect the bacterium *Escherichia coli*; many are able to attack more than one strain of this organism. The T series of phage (T1–T7), propagated on a special culture of *E. coli*, strain B, have been used in extensive studies, from which most of the knowledge of phages is derived. *See* BACTERIOPHAGE; LYSOGENY. [P.B.C.]

Colitis An inflammation of the large bowel, or colon. If the precise causative agent remains unknown, that form of the disorder is called idiopathic colitis. Idiopathic ulcerative colitis is common and may produce severe changes in the bowel. Adults between 20 and 40 years of age are most affected and, at present, multiple contributing factors are blamed for the disease. Small hemorrhages in the lower part of the bowel usually mark the early stages. These develop into tiny abscesses which increase in size, number, and distribution. The symptoms of abdominal pain, bleeding, diarrhea, and those symptoms related to specific complications tend to become chronic or to recur.

Some agents may cause specific forms of colitis, which have very characteristic pathological or clinical features. These include tuberculosis, amebic and bacillary dysentery, typhoid fever, staphylococcal infections, and mercury poisoning.

Various brief, relatively benign cases of colitis occur frequently, probably as a result of mild viral or bacterial infections, food poisoning or intolerances, and emotional stresses. Colitis may also be secondary to other disease or injury, notably as a local response to tumor formation. *See* AMEBIASIS; BACILLARY DYSENTERY; STAPHYLOCOCCUS; TUBERCULOSIS. [N.K.M.]

Collagen The major fibrous protein of many animals. Located in the extracellular connective tissue, collagen is probably the most abundant animal protein in nature. It is estimated that collagen accounts for about 30% of the total human body protein. It is present in all types of multicellular animals, both invertebrate and vertebrate, although in insects and crustaceans chitin replaces collagen as the fibrous supporting matrix in the exoskeleton. Collagen constitutes the fibrillar component of the soft connective tissues (for example skin, ligament, and tendon) and is the major component of the organic matrix of the calcified tissues such as bone and dentine. Besides its structural significance, collagen plays an important role in such events as development and wound healing, and has been implicated in aging and in a number of disease processes.

At least four genetically distinct collagens have been described. The most familiar, type I, is composed of three α-polypeptide chains, each of about 95,000 mol wt. Two chains are identical and are called $\alpha 1$(I); the other is called $\alpha 2$. Type I collagen forms the major portion of the collagen of both soft (skin, tendon, fascia, and so on) and hard (bone and dentine) connective tissue. Type II collagen is the major collagen of cartilage and is composed of three $\alpha 1$(II) chains. Type III collagen is composed of three $\alpha 1$(III) chains and is found in blood vessels, wounds, and certain tumors and appears to be elevated in the tissues of young animals. Reticulin fibers appear to be identified with type III collagen. Basement membrane collagens have been classified as type IV, and other collagen types have been identified. Collagen has a characteristic amino acid composition, containing 33% glycine and about 25% proline and hydroxyproline. The amino acid δ-hydroxy-L-lysine and the amino acid 4-hydroxy-L-proline are almost unique to collagen among the proteins of the vertebrates and invertebrates.

Collagen is recognized by its relative insolubility under mild conditions, by its resistance to most of the proteolytic enzymes of vertebrates, and by various histologic staining properties. Its appearance in the electron microscope is unique, and it has a characteristic banded or striated appearance.

Several heritable disorders of connective tissue have been elucidated at the molecular level. The basic defect has been determined in four types of the Ehlers-Danlos syndrome. People with this condition have "human pretzel" or "India rubber man" characteristics, with hyperextensible skin, hypermobile joints, fragile tissues, and a bleeding diathesis. In three varieties of the syndrome an enzyme defect has been identified, and in the fourth variety it is suggested that the type III collagen is lacking. Several types of osteogenesis imperfecta have also been identified. The individuals exhibit fragile bones, clear or blue sclera, deafness, and loose ligaments. It appears that the bone collagen fails to mature, and there is evidence that the ratio of type I/type III collagen may be abnormal. *See* CONNECTIVE TISSUE; FIBROUS PROTEIN. [D.A.K.]

Collard A cool-season biennial crucifer, *Brassica oleracea* var. *acephala*, similar to nonheading cabbage. Collard is of Mediterranean origin and is grown for its rosette of leaves, which are cooked fresh as a vegetable. Important production centers are in the southern United States, where collards are an important nutritious green, especially during winter months. *See* CABBAGE; CAPPARALES; KALE. [H.J.C.]

Collembola An order of primitive insects, commonly called springtails, belonging to the subclass Apterygota. These tiny insects do not undergo a metamorphosis. They have six abdominal segments, some of which may be ankylosed to give an apparently smaller number (see illustration). Appended to the fourth abdominal segment in most of the species is a spring, or furcula. This consists of a basal piece and two apical parallel structures which end in hooks. Another pair of structures, united at the base, occur ventrally on the third abdominal segment. These bear teeth which engage with the furcula and hold it beneath the body. When released, the spring flies backward, catapulting the insect.

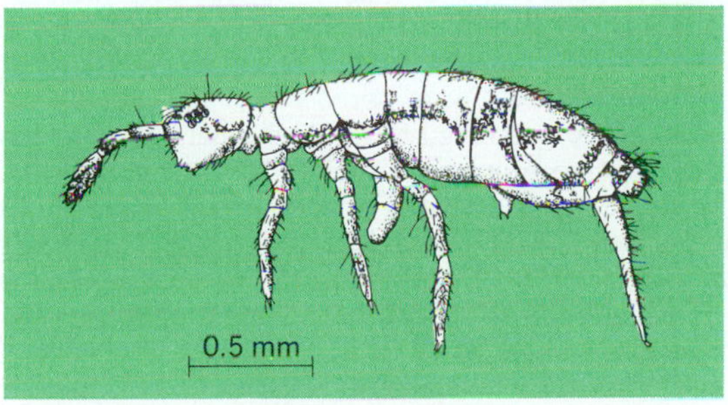

A collembolan, *Entomobrya cubensis*. (After J. W. Folsom, *Proc. U.S. Nat. Mus.*, 72(6), plate 6, 1927)

Collembola live in humid places, often in leaf mold. Some species are active at cool temperatures, hence another common name, snowflea, is used. The earliest known insect fossil, *Rhyniella praecursor*, is thought to belong to this order. *See* INSECTA. [H.B.Mi.]

Collenchyma A primary, or early differentiated, supporting tissue of young shoot parts appearing while these parts are still elongating. It is located near the surface, usually just under the epidermis. When observed in transverse sections, it is characterized structurally by cell walls that are intermittently thickened, generally in the corners or places of juncture of three or more cells. Collenchyma is typically formed in the petioles and vein ribs of leaves, the elongating zone of young stems, and the pedicels of flowers. *See* CELL WALLS (PLANT).

As in parenchyma, the cells in collenchyma are living and may contain chloroplasts and starch grains. The cell wall of a collenchyma cell is its most striking feature structurally and functionally. It is composed of cellulose and pectic compounds plus a very high proportion of water. The cytoplasm is very rich in ribosomes and ribonucleic acids in the early stages of development. Another striking feature of collenchyma cell walls is their plasticity. They are capable of great elongation during the period of growth in length of the plant. The plasticity of collenchyma is associated with a tensile strength comparable to that shown by fibers of sclerenchyma. The combination of strength and plasticity makes the collenchyma effective as a strengthening tissue in developing stems and leaves having no other supporting tissue at that time. *See* CELLULOSE; EPIDERMIS (PLANT); PARENCHYMA; PECTIN. [R.L.Hu.]

Collision (physics) Any interaction between particles, aggregates of particles, or rigid bodies in which they come near enough to exert a mutual influence, generally with exchange of energy. The term collision, as used in physics, does not necessarily imply actual contact.

In classical mechanics, collision problems are concerned with the relation of the magnitudes and directions of the velocities of colliding bodies after collision to the velocity vectors of the bodies before collision. When the only forces on the colliding bodies are those exerted by the bodies themselves, the principle of conservation of momentum states that the total momentum of the system is unchanged in the collision process. This result is particularly useful when the forces between the colliding bodies act only during the instant of collision. The velocities can then change only during the collision process, which takes place in a short time interval. Under these conditions the forces can be treated as impulsive forces, the effects of which can be expressed in terms of an experimental parameter known as the coefficient of restitution. *See* CONSERVATION OF MOMENTUM; IMPACT.

The study of collisions of molecules, atoms, and nuclear particles is an important field of physics. Here the object is usually to obtain information about the forces acting between the particles. The velocities of the particles are measured before and after collision. Although quantum mechanics instead of classical mechanics should be used to describe the motion of the particles, many of the conclusions of classical collision theory are valid. *See* SCATTERING EXPERIMENTS (ATOMS AND MOLECULES); SCATTERING EXPERIMENTS (NUCLEI).

Collisions can be classed as elastic and inelastic. In an elastic collision, mechanical energy is conserved; that is, the total kinetic energy of the system of particles after collision equals the total kinetic energy before collision. For inelastic collisions, however, the total kinetic energy after collision is different from the initial total kinetic energy.

In classical mechanics the total mechanical energy after an inelastic collision is ordinarily less than the initial total mechanical energy, and the mechanical energy which is lost is converted into heat. However, an inelastic collision in which the total energy after collision is greater than the initial total energy sometimes can occur in classical mechanics. For example, a collision can cause an explosion which converts chemical energy into mechanical energy. In molecular, atomic, and nuclear systems, which are governed by quantum mechanics, the energy levels of the particles can be changed during collisions. Thus these inelastic collisions can involve either a gain or a loss in mechanical energy. [P.W.S.]

Colloid A system of which one phase is made up of particles having dimensions of 1–1000 nanometers and is dispersed in a different phase. Some examples of colloidal systems are shown in the table. There are three important types of colloids: small solid particles having the same internal structure as the bulk solid phase (sulfur suspensions in water), aggregates formed from smaller molecules (soaps in water), and large molecules of colloidal size (proteins and high polymers).

The unique properties of colloids are due primarily to the large surface areas of the dispersed phase. One of the results of this large surface area is the adsorption of ions and other materials. The adsorbed ions impart an electric charge to the colloidal particles so that they repel each other. Lyophobic colloids exhibit little affinity between the dispersed phase and the solvent; lyophilic colloids are strongly associated with the solvent. *See* ADSORPTION.

Colloidal particles are so small that they cannot be seen directly in visible light even with an optical microscope. They can be detected by viewing them at an angle to the incident light path. The electron microscope furnishes a means by which the sizes and shapes of colloidal-size particles can be determined. *See* ELECTRON MICROSCOPE; TYNDALL EFFECT.

Types of colloidal system

Dispersed phase	Dispersing phase	Example
Solid	Solid	Some alloys
Solid	Liquid	Suspensions (muddy water)
Solid	Gas	Smoke, airborne dust
Liquid	Solid	Butter, gels
Liquid	Liquid	Emulsions (milk, mayonnaise)
Liquid	Gas	Fog, aerosols
Gas	Solid	Solid foams (marshmallow)
Gas	Liquid	Foams

Because colloids are so small in size, they cannot be separated by filtration through paper or porous porcelain crucibles. They can be separated from solvent and dissolved substances by dialysis and ultrafiltration. Airborne colloidal particles can be separated by special techniques. *See* DIALYSIS; ELECTROSTATIC PRECIPITATOR; ULTRAFILTRATION.

Other more specialized aspects of colloidal properties and surface phenomena are discussed in other articles. *See* DONNAN EQUILIBRIUM; ELECTROPHORESIS; EMULSION; FOAM; GEL; MICELLE; SMOKE; ULTRACENTRIFUGE. [Q.V.W.]

Colloidal crystals Periodic arrays of suspended colloidal particles. Common colloidal suspensions (colloids) such as milk, blood, or latex are polydisperse; that is, the suspended particles have a distribution of sizes and shapes. However, suspensions of particles of identical size, shape, and interaction, the so-called monodisperse colloids, do occur. In such suspensions, a new phenomenon that is not found in polydisperse systems, colloidal crystallization, appears: under appropriate conditions, the particles can spontaneously arrange themselves into spatially periodic structures. This ordering is analogous to that of identical atoms or molecules into periodic arrays to form atomic or molecular crystals. However, colloidal crystals are distinguished from molecular crystals, such as those formed by very large protein molecules, in that the individual particles do not have precisely identical internal atomic or molecular arrangements. On the other hand, they are distinguished from periodic stackings of macroscopic objects like cannonballs in that the periodic ordering is spontaneously adopted by the system through the thermal agitation (brownian motion) of the particles. These conditions limit the sizes of particles which can form colloidal crystals to the range from about 0.01 to about 5 micrometers. *See* BROWNIAN MOVEMENT; KINETIC THEORY OF MATTER.

The most spectacular evidence for colloidal crystallization is the existence of naturally occurring opals. The ideal opal structure is a periodic close-packed three-dimensional array of silica microspheres with hydrated silica filling the spaces not occupied by particles. Opals are the fossilized remains of an earlier colloidal crystal suspension. Another important class of naturally occurring colloidal crystals are found in concentrated suspensions of nearly spherical virus particles, such as *Tipula* iridescent virus and tomato bushy stunt virus. Colloidal crystals can also be made from the synthetic monodisperse colloids, suspensions of plastic (organic polymer) microspheres. Such suspensions have become important systems for the study of colloidal crystals, by virtue of the controllability of the particle size and interaction. *See* COLLOID; CRYSTAL; OPAL; VIRUS. [N.A.C.]

Colon The portion of the intestine that runs from the cecum to the rectum; in some mammals, it may be separated from the small intestine by an ileocecal valve. It is also known as the large intestine. The colon is usually divided into ascending, transverse, and descending portions. In the human a fourth section, the sigmoid, is found. The colon is longer in herbivores and shorter in carnivores, and is about 4 to 6 ft (1.2 to 1.8 m) long in humans. No digestive enzymes are secreted in the colon. Much digestion (for example, all breakdown of cellulose) occurs by bacteria, of which *Escherichia coli* is the most common. Most of the fluid added to the food during digestion is reabsorbed into the body in the colon. All digestive action, water absorption, and so on, is completed before the food materials pass out of the colon into the rectum. *See* DIGESTIVE SYSTEM. [W.J.B.]

Color That aspect of visual sensation enabling a human observer to distinguish differences between two structure-free fields of light having the same size, shape, and duration. Although luminance differences alone permit such discriminations to be made, the term color is usually restricted to a class of differences still perceived at equal luminance. These depend upon physical differences in the spectral compositions of the two fields, usually revealed to the observer as differences of hue or saturation. *See* COLOR VISION.

Color is usually presented to the individual by the surfaces of objects on which a more or less white light is falling. A red surface, for example, is one that absorbs most of the shortwave light and reflects the long-wave light to the eye. The surface colors are easily described by reference to the color solid shown in Fig. 1. The central axis defines the lightness or darkness of the surface as determined by its overall reflectance of white light, the lowest reflectance being called black and the highest, white. The circumference denotes hue, related primarily to the selective reflectance of the surface for particular wavelengths of light.

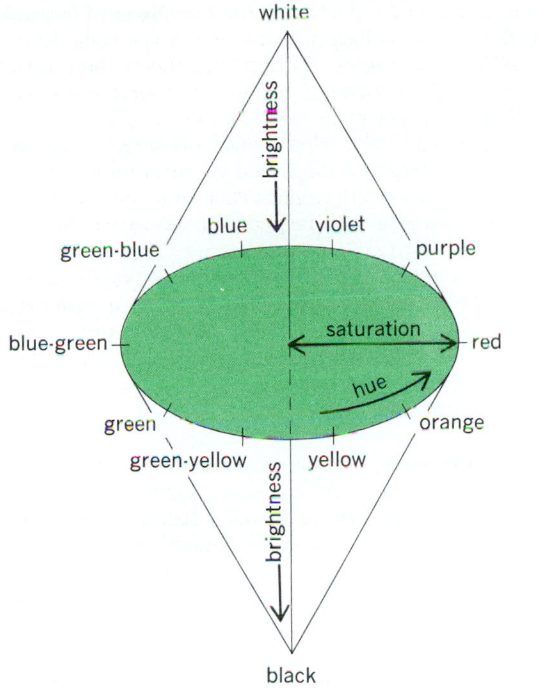

Fig 1. Diagram of a color solid for illustrating the appearance of surface colors.

The color solid is pointed at its top and bottom to represent the fact that as colors become whiter or blacker they lose hue. The distance from central axis to periphery indicates saturation, a characteristic that depends chiefly on the narrowness, or purity, of the band of wavelengths reflected by the surface. At the center of the figure is a medium-gray surface, one which exhibits a moderate amount of reflectance for all wavelengths of light. Colors that are opposite one another are called complementaries, for example, yellow and blue, red and blue-green, purple and green, and white and black.

Two complementaries, when added together in the proper proportions, can oppose or neutralize one another to produce a colorless white or gray. Various contrast effects also attest to the opposition of the complementaries. Staring at a bright-red surface against a gray background results in the appearance of a blue-green border around the red; this is an example of simultaneous contrast. Similarly, when a red light is turned off, there is frequently a negative after-image that appears to have the opposite color, blue-green.

A set of primary colors can be chosen so that any other color can be produced from additive mixtures of the primaries in the proper proportions. Thus, red, green, and blue lights can be added together in various proportions to produce white, purple, yellow, or any of the various intermediate colors. Three-color printing, color photography, and color television are examples of the use of primaries to produce plausible imitations of colors of the original objects. *See* COLOR TELEVISION.

Colors lying along a continuum from white to black are known as the gray, or achromatic, colors. They have no particular hue, and are therefore represented by the central axis of the color diagram in Fig. 1. White is shown at the top of the diagram since it represents the high-brightness extreme of the series of achromatic colors. With respect to the surface of an object, the diffuse, uniform reflectance of all wavelengths characterizes this series from black through the grays to white in order of increasing reflectance. Whiteness is a relative term; white paper, paint, and snow reflect some 80% or more of the light of all visible wavelengths, while black surfaces typically reflect less than 10% of the light.

Gray is the term applied to all the intermediate colors in the series of achromatic colors. Gray may result from a mixture of two complementary colors, from a mixture of all primary colors, or from a fairly uniform mixture of lights of all wavelengths throughout the visible spectrum. Grayness is relative; a light gray is an achromatic color that is lighter than its surroundings, while a dark gray is so called because it is darker than its surroundings.

The opposite extreme from white in the series of achromatic colors is black. Blackness is a relative term applied to surfaces that uniformly absorb large percentages of light of all visible wavelengths. A black object in sunlight absorbs a large percentage of the light, but it may reflect a larger absolute quantity of light to the eye than does a white object in the shade. [L.A.R.]

Colors are often specified in a two-dimensional chart known as the CIE chromaticity diagram, which shows the relations among tristimulus values independently of luminance. Such a diagram is shown in Fig. 2, in which the continuous locus of spectrum colors is represented by the outermost contour. All nonspectral colors are contained within an area defined by this boundary and a straight line running from red to violet. The diagram also shows discrimination data for 25 regions, which plot as ellipses represented at 10 times their actual size. A discrimination unit is one-tenth the distance from the center of an ellipse to its perimeter. Predictive schemes for interpolation to other regions of the CIE diagram have been worked out.

If discrimination ellipses were all circles of equal size, then a discrimination unit would be represented by the same distance in any direction anywhere in the chart. Because this is dramatically untrue, other chromaticity diagrams have been developed as linear projections of the CIE chart. These represent discrimination units in a relatively more uniform way, but never perfectly so.

A chromaticity diagram has some very convenient properties. Chief among them is the fact that additive mixtures of colors plot along straight lines connecting the chromaticities of the colors being mixed. Although it is sometimes convenient to visualize colors in terms of the chromaticity chart, it is important to realize that this is not a psychological color diagram. Rather, the chromaticity diagram makes a statement about the results of metameric color matches, in the sense that a given point on the diagram represents the locus of all possible

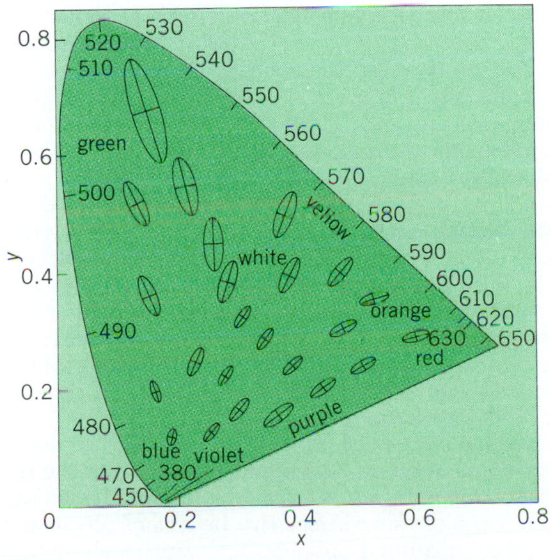

Fig. 2. The 1931 CIE chromaticity diagram showing MacAdam's ellipses 10 times enlarged. (*After G. S. Wyszecki and W. S. Stiles, Color Science, copyright © 1967 by John Wiley and Sons; used with permission*)

metamers plotting at chromaticity coordinates x, y. However, this does not specify the appearance of the color, which can be dramatically altered by preexposing the eye to colored lights (chromatic adaptation) or, in the complex scenes of real life, by other colors present in nearby or remote areas (color contrast and assimilation). Nevertheless, within limits, metamers whose color appearance is thereby changed continue to match.

For simple, directly fixated, and unstructured fields presented in an otherwise dark environment, there are consistent relations between the chromaticity coordinates of a color and the color sensations that are elicited. Therefore, regions of the chromaticity diagram are often denoted by color names, as shown in Fig. 2.

Although the CIE system works rather well in practice, there are important limitations. Normal human observers do not agree exactly about their color matches, chiefly because of the differential absorption of light by inert pigments in front of the photoreceptors. Much larger individual differences exist for differential colorimetry, and the system is overall inappropriate for the 4% of the population (mostly males) whose color vision is abnormal. The system works only for an intermediate range of luminances, below which rods (the receptors of night vision) intrude, and above which the bleaching of visual photopigments significantly alters the absorption spectra of the cones. *See* COLORIMETRY. [R.M.Bo.]

Color (quantum mechanics) A term used to describe a hypothetical quantum number carried by the quarks which are thought to make up the strongly interacting elementary particles. It has nothing to do with the ordinary, visual use of the word color.

The quarks which are thought to make up the strongly interacting particles have a spin angular momentum of one-half unit of h (Planck's constant). According to a fundamental theorem of relativity combined with quantum mechanics, they must therefore obey Fermi-Dirac statistics and be subject to the Pauli exclusion principle. No two quarks within a particular system can have exactly the same quantum numbers. *See* EXCLUSION PRINCIPLE; FERMI-DIRAC STATISTICS.

However, in making up a baryon, it often seemed necessary to violate this principle. The Ω^- particle, for example, is made of three strange quarks, and all three had to be in exactly the same state. O. W. Greenberg is responsible for the essential idea for the solution to this paradox. In 1964 he suggested that each quark type (u, d, and s) comes in three varieties identical in all measurable qualities but different in an additional property, which has come to be known as color. The exclusion principle could then be satisfied and quarks could remain fermions, because the quarks in the baryon would not all have the same quantum numbers. They would differ in color even if they were the same in all other respects. *See* BARYON; ELEMENTARY PARTICLE; MESON; QUARKS. [T.Ap.]

Color blindness A condition of faulty color vision. It appears to be the normal state of animals that are active only at night. It is also characteristic of human vision when the level of illumination is quite low or when objects are seen only at the periphery of the retina. In rare individuals, known as monochromats, there is total color blindness even at high light levels. Such persons are typically deficient or lacking in cone receptors, so that their form vision is also poor.

Dichromats are partially color-blind individuals whose vision appears to be based on two primaries rather than the normal three. Dichromatism occurs more often in men than in women because it is a sex-linked, recessive hereditary condition. One form of dichromatism is protanopia, in which there appears to be a lack of normal red-sensitive receptors. Red lights appear dim to protanopes and cannot be distinguished from dim yellow or green lights. A second form is deuteranopia, in which there is no marked reduction in the brightness of any color, but again there is a confusion of the colors normally described as red, yellow, and green. A third and much rarer form is tritanopia, which involves a confusion among the greens and blues. *See* HUMAN GENETICS.

Many so-called color-blind individuals might better be called color-weak. They are classified as anomalous trichromats because they have trichromatic vision of a sort, but fail to agree with normal subjects with respect to color matching or discrimination tests.

Color blindness is most commonly tested by the use of color plates in which various dots of color define a figure against a background of other dots. The normal eye readily distinguishes the figure, but the colors are so chosen that even the milder forms of color anomaly cause the figure to be indistinguishable from its background. [L.A.R.]

Color centers Atomic and electronic defects of various types which produce optical absorption bands in otherwise transparent crystals such as the alkali halides, alkaline earth fluorides, or metal oxides. They are general phenomena found in a wide range of materials. Color centers are produced by gamma radiation or x-radiation, by addition of impurities or excess constituents, and sometimes through electrolysis. A well-known example is that of the *F*-center in alkali halides such as sodium chloride, NaCl. The designation *F*-center comes from the German word *Farbe*, which means color. *F*-centers in NaCl produce a band of optical absorption toward the blue end of the visible spectrum; thus the colored crystal appears yellow under transmitted light. On the other hand, KCl with *F*-centers appears magenta, and KBr appears blue. *See* CRYSTAL.

Color centers have been under investigation for many years. Theoretical studies guided by detailed experimental work have yielded a deep understanding of specific centers. The crystals in which color centers appear tend to be transparent to light and to microwaves. Consequently, experiments which can be carried out include optical spectroscopy, luminescence and Raman scattering, magnetic circular dichroism, magnetic resonance, and electromodulation. Color centers find practical application in radiation dosimeters; schemes have been proposed to use color centers in high-density memory devices; and tunable lasers have been made from crystals containing color centers.

The illustration shows the absorption bands due to color

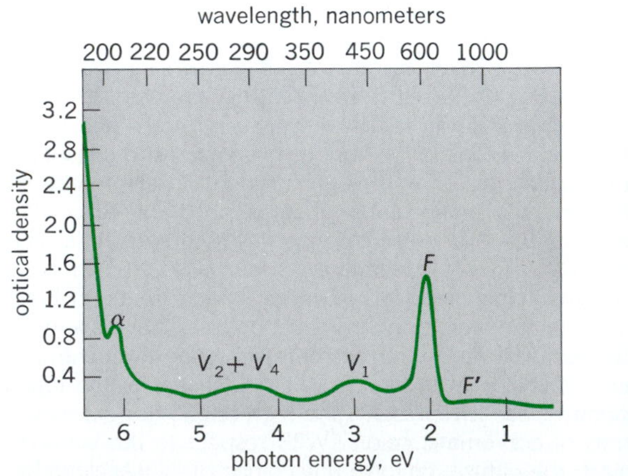

Absorption bands produced in a KBr crystal by exposure to x-rays at 81 K. Bands are designated by letters. Optical density is equal to $\log_{10}(I_0/I)$, where I_0 is the intensity of incident light and I the intensity of transmitted light. Steep rise in optical density at far left is due to intrinsic absorption in crystals (modified slightly by existence of *F*-centers.)

centers produced in potassium bromide by exposure of the crystal at the temperature of liquid nitrogen (81 K) to intense penetrating x-rays. Several prominent bands appear as a result of the irradiation. The *F*-band appears at 600 nanometers and the so-called *V*-bands appear in the ultraviolet.

Color bands such as the *F*-band and the *V*-band arise because of light absorption at defects dispersed throughout the lattice. This absorption is caused by electronic transitions at the centers. On the other hand, colloidal particles, each consisting of many atoms, dispersed through an optical medium also produce color bands. In this case, if the particles are large enough, the extinction of light is due to both light scattering and light absorption. Colloidal gold is responsible for the color of some types of ruby glass. Colloids may also form in alkali halide crystals—for example, during heat treatment of an additively colored crystal with an excess of alkali metal.

Atomically dispersed centers such as *F*-centers are part of the general phenomena of trapped electrons and holes in solids. The accepted model of the *F*-center is an electron trapped at a negative ion vacancy. Many other combinations of electrons, holes, and clusters of lattice vacancies have been used to explain the various absorption bands in ionic crystals.

Impurities can play an important role in color-center phenomena. Certain impurities in ionic crystals produce color bands characteristic of the foreign ion. For example, hydrogen can be incorporated into the alkali halides with resultant appearance of an absorption band (the *U*-band) in the ultraviolet. In this case, the *U*-centers interact with other defects. The rate at which *F*-centers are produced by x-irradiation is greatly increased by the incorporation of hydrogen, the *U*-centers being converted into *F*-centers with high efficiency. [F.C.Br.]

Color filter An optical element that partially absorbs incident light, often called a light filter. The absorption may be selective with respect to wavelength, in which case the color filter purifies the incident light in the sense of making it more nearly like that of a part of the pure spectrum. On the other hand, the absorption may be nonselective, in which case the color filter merely reduces the light flux without change in spectral composition. Color filters are used in photography, optical instruments, and illuminating devices to control the amount and spectral composition of light.

Color filters are made of glass for maximum permanence, of liquid solutions confined to cells with transparent faces for maximum flexibility of control, and of dyed gelatin or plastic for maximum convenience combined with excellent flexibility and often satisfactory permanence. Plastic color filters are supplied cemented between glass sheets for greater toughness than glass and greater scratch resistance than bare plastic. *See* COLOR. [D.B.J.]

Color index A quantitative measure of a star's color. Such color is closely related to temperature. Stars are reddish at around 3000 K (5000°F), orange about 4500 K (7500°F), yellowish about 6000 K (10,300°F), white about 10,000 K (17,500°F), and bluish above 10,000 K. With proper calibration, color provides a means by which the temperature and spectral class can be evaluated. *See* SPECTRAL TYPE; STAR.

Color is defined through the star's magnitude m, which indicates brightness. The original magnitudes were visual, being estimated with the unaided eye, and depended on the eye's response, which is maximized for yellow light. Early photographic emulsions, however, were sensitive primarily to blue light. As a result, blue stars look relatively brighter to the photographic plate than they do the eye, and red stars considerably fainter. It was then necessary to establish a separate magnitude system for photography, called m_{ptg}. The original visual magnitudes were distinguished by calling them m_v. The two magnitudes were set equal for white stars with temperatures of 9300 K (16,280°F). Color can then be quantified by taking the difference between the two magnitudes, the color index becoming $m_{ptg} - m_v$. Blue stars have slightly negative color indices, red stars positive ones. *See* ASTRONOMICAL PHOTOGRAPHY.

Photoelectric photometry allowed the establishment of the more precise UBV system, wherein V (yellow) and B (blue) filters respectively mimic the response of the human eye and the untreated photographic plate, and a U filter is added in the ultraviolet. The traditional color index is then replaced by B − V, and the addition of U allows a second color index, U − B. *See* HERTZSPRUNG-RUSSELL DIAGRAM; INTERSTELLAR EXTINCTION.

The standard system has been expanded into the red and the infrared with other filters. Numerous color indices are then available to examine stars that radiate primarily in the infrared. Other photometric schemes, such as the Strömgren four-color system (uvby, for ultraviolet, violet, blue, and yellow), allow more sophisticated color indices that are responsive not only to temperature but to a variety of other parameters. *See* INFRARED ASTRONOMY. [J.B.Ka.]

Color photography The photographic reproduction of natural colors in terms of two or more spectral components of white light. Red, green, and blue are called additive primary colors because they add to form white light. Their complements are the subtractive primary colors, cyan, magenta, and yellow. Subtractive color processes produce color images in terms of cyan, magenta, and yellow dyes or pigments. *See* COLOR.

Additive color photography. The screen plate system of additive photography uses a fine array of red, green, and blue filters in contact with a single panchromatic emulsion. Reversal processing produces a positive color transparency comprising an array of positive black-and-white image elements in register with the filter elements.

Subtractive color photography. Many types of subtractive color negatives, color prints, and positive transparencies are made from multilayer films with separate red-, green-, and blue-sensitive emulsion layers.

In chromogenic or color development processes, dye images are formed by reactions between the oxidation product of a developing agent and color couplers contained either in the developing solution or within layers of the film. For positive color transparencies, exposed silver halide grains undergo nonchromogenic development, and the remaining silver halide is later developed with chromogenic developers to form positive dye images. Residual silver is removed by bleaching and fixing.

Kodachrome was the first commercially successful subtractive color transparency film. The Kodachrome process forms reversal dye images in successive color development steps, using soluble couplers. Following black-and-white development of exposed grains, the remaining red-sensitive silver halide is exposed to red light, then developed to form the positive cyan image. Next, the blue-sensitive layer is exposed to blue light and the yellow image developed, and finally the green-sensitive emulsion is fogged and the magenta image developed.

Simpler reversal processing is afforded by films with incorporated nondiffusing couplers that produce the three dye images in a single color development step.

One-step color photography. The most dramatic change in the history of color photography was the transition to the single-structure, single-step Polacolor film and Polacolor process from the separate multiple-layer negative and positive and the multiple-step darkroom processing of each of these materials. In the Polacolor process and later one-step, or instant, processes, color prints are produced under ambient conditions within moments after exposure of the film. The processing takes place rapidly, using reagent provided as part of the film unit. *See* PHOTOGRAPHY. [V.K.W.]

Color television Transmission and reception of transient visual images in full color. The technique used for color television broadcasting in the United States is compatible with monochrome, or black-and-white, television. That is, color television signals are sufficiently similar to monochrome signals so that they produce acceptable monochrome pictures on black-and-white receivers while producing full-color pictures on color receivers. Conversely, a standard monochrome signal produces a satisfactory monochrome image on a color receiver. Compatible color television is based on a combination of the primary-color principle, conventional television technology, and special encoding techniques for combining primary-color television signals in such a way that they can be transmitted through a single broadcast channel.

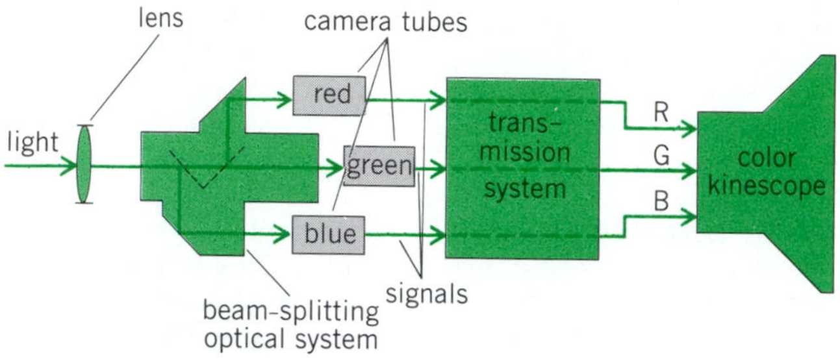

Simplified block diagram of a color television system.

Color television images are formed by combining three primary-color images. The standard set which was proved to be most practical for color television consists of highly saturated red, green, and blue colors. The color kinescope, or picture tube, used to display the picture in a color television receiver can produce independent red, green, and blue images, usually in the form of tiny dots so closely intermingled that they cannot be resolved individually at the normal viewing distance. The three images are thus effectively fused by the visual system of the observer so that only the combined, full-color image is seen. *See* COLOR; COLOR VISION; PICTURE TUBE.

In the simplest type of color television system the three primary-color images in the tricolor picture tube are controlled directly by video signals developed by three separate pickup tubes, adjusted to respond to red, green, and blue light (see illustration). Close registration must be maintained among the three images so that the scanning beams in the three pickup tubes are always at the same relative points in their respective image areas.

Because of the great demand for television broadcast channels in the United States, it is not practical to utilize three separate channels for transmitting the red, green, and blue signals required to control a color kinescope. It is necessary, therefore, to encode the primary-color signals at the transmitter in such a way that they can be transmitted through a single television channel. The encoded color video signal contains three independent components, and is enough like a conventional monochrome signal to produce good images on monochrome receivers. Each color receiver contains decoder circuits (between the second detector and the kinescope) to retrieve from the color signal its red, green, and blue components. *See* TELEVISION; TELEVISION RECEIVER. [J.W.W.]

Color vision The ability to discriminate light on the basis of wavelength composition. It is found in humans, in other primates, and in certain species of birds, fishes, reptiles, and insects. These animals have visual receptors that respond differentially to the various wavelengths of visible light. Each type of receptor is especially sensitive to light of a particular wavelength composition. Evidence indicates that primates, including humans, possess three classes of cone photoreceptors that differ in the photopigments they contain and in their neural connections. In humans, two of these, the R and G cones, are sensitive to all wavelengths of the visible spectrum from 380 to 700 nanometers. The B cones, whose sensitivity peaks at about 440 nm, are not appreciably excited by wavelengths longer than 540 nm. The perception of blueness and yellowness depends upon the level of excitation of B cones in relation to that of R and G cones. No two wavelengths of light can produce equal excitations in all three kinds of cones. It follows that, provided they are sufficiently different to be discriminable, no two wavelengths can give rise to identical sensations. *See* COLOR BLINDNESS; EYE (VERTEBRATE); VISION. [R.M.Bo.]

Colorimetry Any technique by which an unknown color is evaluated in terms of known colors. Colorimetry may be visual, photoelectric, or indirect by means of spectrophotometry. These techniques are widely used in scientific studies involving the appearance of objects and lights, but are of greatest importance in the color specification of the raw materials and finished products of industry.

In visual colorimetry, the unknown color is presented beside a comparison field into which may be introduced any one of a range of known colors from which the operator chooses the one matching the unknown. To be generally applicable, the comparison field must not only cover a sufficient color range but must also be continuously adjustable in color.

In indirect colorimetry, the light leaving the unknown specimen is split into its component spectral parts by means of a prism or diffraction grating, and the amount of each component part is separately measured by a photometer. The quantity evaluated is spectral radiance of a light source, spectral transmittance of a filter (glass, plastic, gelatin, or liquid), or spectral reflectance of an opaque body. *See* SPECTROPHOTOMETRIC ANALYSIS.

In photoelectric colorimetry, the light leaving the specimen is measured separately by three photocells. The spectral sensitivity of these photocells is adjusted, usually by color filters, to conform as closely as possible to the three color-mixture functions for the average normal human eye (CIE standard observer). The responses of the photocells give directly the amounts of red, green, and blue primaries required to produce the color of the unknown specimen for the kind of vision represented by the three photocells.

If two objects have the same color because the light leaving one of them toward the eye is spectrally identical to that leaving the other, any type of colorimetry serves reliably to establish the fact of color match. If, however, the two lights are spectrally dissimilar, they may still color-match for any one observer; such pairs of lights are called metamers. Normal color vision differs sufficiently from person to person so that a metameric color match for one observer may be seriously mismatched for another. On this account, the question of color match of spectrally dissimilar lights can be reliably settled only by the indirect method which uses spectrophotometry combined with a precisely defined standard observer. *See* COLOR; COLOR FILTER; COLOR VISION. [D.B.J./J.L.L.]

Columbiformes An order of birds containing three families, the largest of which is the worldwide pigeons and doves (Columbidae). The members of this order are characterized by an ability to drink water by sucking instead of the sip-and-tilt method of most birds. However, some other groups of birds are able to suck water by various methods.

The order Columbiformes is divided into the suborder Pterocletes, with the single family Pteroclidae (sandgrouse; 16 species), and the suborder Columbae with the families

Raphidae (dodos; 3 species) and Columbidae (pigeons and doves; 303 species). Relationships of Columbiformes appear to be to the Charadriiformes in one direction and to the Psittaciformes in another. Possibly the Columbiformes are a central stock in the evolution of birds. *See* CHARADRIIFORMES; PSITTACIFORMES.

Pigeons are found mainly in the tropics, but a number of species are common in temperate regions. They have a sleek plumage ranging from browns and grays to the brilliant greens, yellows, and reds of the tropical fruit pigeons. They feed on seeds, fruit, and other vegetarian food. Most pigeons live in flocks but breed solitarily. Almost all are nonmigratory.

The sandgrouse live in flocks in dry grasslands and deserts of the Old World, but depend on ponds of water to which flocks come in large numbers. The nest is generally a long distance from water, which is transported to the young by soaking specialized belly feathers. The young drink only from these moistened feathers, and will refuse water in pans placed before them.

The three species of dodos and solitares were found only in the several Macarene Islands and were completely exterminated by the seventeenth century by sailors and released pigs. *See* AVES. [W.J.B.]

Columbite A mineral with composition (Fe,Mn)Nb_2O_6. Tantalum substitutes for niobium in all proportions, and a complete series extends to tantalite (Fe,Mn)Ta_2O_6. Columbite without some tantalum is rare; iron and manganese also vary considerably in their relative amounts. It crystallizes in short prismatic crystals. The hardness is 6 (Mohs scale), and the specific gravity is 5.2 for pure columbite; both increase with increased amounts of tantalum. The luster is submetallic and the color iron-black.

Columbite is the chief ore mineral of niobium (columbium) and is found in small amounts in many pegmatites throughout the world. *See* NIOBIUM; TANTALITE; TANTALUM. [C.S.Hu.]

Column A slender compression member which fails by instability when the applied axial load reaches a critical value. Instability or buckling is said to occur when the system cannot attain an equilibrium condition and the load causes a deformation of an indeterminate amount. The collapse load is a function of the end restraints, the ratio of length to the least radius of gyration (L/r, the slenderness ratio), and the combined action of axial and lateral loads applied to the member. Columns are present in all types of structures, such as bridges, buildings machinery, cranes, and airplanes. Different materials may be used in the various structures, such as steel, concrete, wood, aluminum, and alloy steels of all types.

Column loads and moments are transferred by baseplates to footings and foundations. The plate must be large enough to distribute the load without excessive bearing stresses. The column is connected to the plate by welded or bolted connections. An approximate design procedure assumes uniform pressure and cantilever action of the plate projecting beyond the column boundaries. Anchorage is not required for axially loaded columns or when the resultant pressure due to moment and load falls within the mid-third of the base width. However, bolts are used to position the column. Where lateral forces, such as wind pressures, are involved, the resultant moment requires anchor bolts. [J.B.S.]

Column chromatography A simple chromatographic procedure in which a sample dissolved in a small amount of solvent is placed on the top of a chromatographic column. In the course of the elution process, the different materials in the sample separate out on the column as discrete bands (see illustration). The chemical nature of both the sorbent and the eluting solvent exert primary control over the degree to which individual mixture components are retained on the column. Thus, nonpolar compounds travel more rapidly with the solvent than more polar mixture components do. The rule that "like dissolves (or attracts) like," in chemical terms, is quite generally applicable to chromatography. In addition, some molecular entities may not be completely separated from each other. In order to achieve separation of components that are not resolved, either a different mobile phase or a different stationary phase, or both, should be tried.

The separated compounds can be further isolated in a pure state (if desired, for instance, for additional chemical studies) by drying the column content and subsequently isolating the individual zones mechanically and extracting with an appropriate solvent. Alternatively, the column can be excessively flushed with the mobile phase, and the separated zones recovered into test tubes inserted under the column.

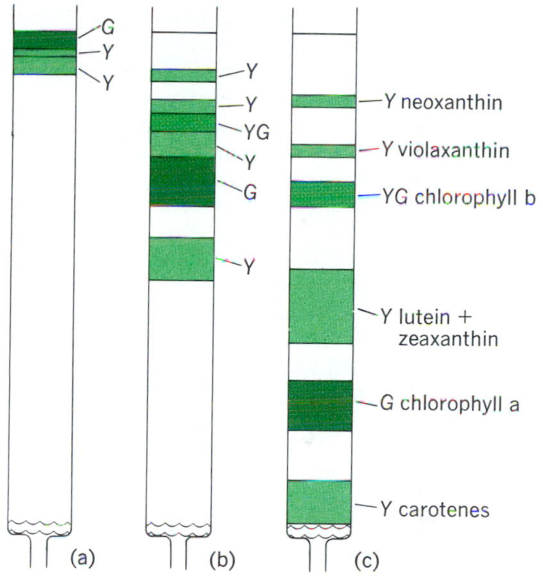

Successive steps a–c in formation of chromatogram of green leaf pigments. *G* signifies green; *Y*, yellow; *YG*, yellow-green.

Although both gas chromatography and modern liquid chromatography utilize these principles, the term column chromatography is universally reserved for the classical system, in which the mobile-phase flow is achieved by means of solvent gravity. In more modern systems, generally applicable to both gas and liquid chromatography, the sample is introduced into the controlled stream of mobile phase at the column top and continuous monitoring of the column effluent is achieved by a suitable detector, followed by a signal recording or, alternatively, further processing of recorded information. *See* CHROMATOGRAPHY. [M.V.N.]

Coma (optics) The popular name for the asymmetric optical aberrations in the image of a point. Coma occurs for one of two reasons: (1) The rays from the object point form a symmetrical image, but its appearance is unsymmetrical because the diaphragm vignettes the rays in an unsymmetrical manner; (2) the rays from the object point form an unsymmetrical image in the absence of vignetting.

The first kind of asymmetry can be easily corrected by shifting the stop (diaphragm) in such a way that the central ray of the imaging bundle goes through its center. The stop in this position is sometimes called the natural stop. The second kind of asymmetry, however, is intrinsic in the design of the system. *See* ABERRATION (OPTICS). [M.J.H.]

Combinatorial theory The branch of mathematics which studies arrangements of elements (usually a finite number) into sets under certain prescribed constraints. Problems combinatorialists attempt to solve include the enumeration problem (how many such arrangements are there?), the structure problem (what are the properties of these arrangements and how efficiently can associated calculations be made?), and, when the constraints become more subtle, the existence problem (is there such an arrangement?). *See* ENUMERATION OF ARRANGEMENTS; LATIN SQUARE. [T.Br.]

Combining volumes, law of The principle that when gases take part in chemical reactions the volumes of the reacting gases and those of the products, if gaseous, are in the ratio of small whole numbers, provided that all measurements are made at the same temperature and pressure. The law is illustrated by the following reactions:

1. One volume of chlorine and one volume of hydrogen combine to give two volumes of hydrogen chloride.
2. Two volumes of hydrogen and one volume of oxygen combine to give two volumes of steam.
3. One volume of ammonia and one volume of hydrogen chloride combine to give solid ammonium chloride.
4. One volume of oxygen when heated with solid carbon gives one volume of carbon dioxide.

The law of combining volumes is similar to the other gas laws in that it is strictly true only for an ideal gas, though most gases obey it closely at room temperatures and atmospheric pressure. Under high pressures used in many large-scale industrial operations, such as the manufacture of ammonia from hydrogen and nitrogen, the law ceases to be even approximately true. *See* GAS. [T.C.W.]

Combustion The burning of any substance, whether it be gaseous, liquid, or solid. In combustion, a fuel is oxidized, evolving heat and often light. The oxidizer need not be oxygen per se. The oxygen may be a part of a chemical compound, such as nitric acid (HNO_3) or ammonium perchlorate (NH_4ClO_4), and become available to burn the fuel during a complex series of chemical steps. The oxidizer may even be a non-oxygen-containing material. Fluorine is such a substance. It combines with the fuel hydrogen, liberating light and heat. In the strictest sense, a single chemical substance can undergo combustion by decomposition, with emission of heat and light. Acetylene, ozone, and hydrogen peroxide are examples. The products of their decomposition are carbon and hydrogen for acetylene, oxygen for ozone, and water and oxygen for hydrogen peroxide.

The combustion of solids such as coal and wood occurs in stages. First, volatile matter is driven out of the solid by thermal decomposition of the fuel and burns in the air. At usual combustion temperatures, the burning of the hot, solid residue is controlled by the rate at which oxygen of the air diffuses to its surface. If the residue is cooled by radiation of heat, combustion ceases.

Liquid fuels do not burn as liquids but as vapors above the liquid surface. The heat evolved evaporates more liquid, and the vapor combines with the oxygen of the air. [B.L.]

Combustion chamber The space at the head end of an internal combustion engine cylinder where most of the combustion takes place. *See* COMBUSTION; INTERNAL COMBUSTION ENGINE.

In the spark-ignition engine, combustion is initiated in the mixture of fuel and air by an electrical discharge. The resulting reaction moves radially across the combustion space as a zone of active burning, known as the flame front. The velocity of the flame increases nearly in proportion to engine speed so that the distance the engine shaft turns during the burning process is not seriously affected by changes in speed. Occasionally a high burning rate, or too rapid change in burning rate, gives rise to unusual noise and vibration called engine roughness. Roughness may be reduced by using less squish or by shaping the combustion chamber to control the area of the flame front. A short burning time is helpful in eliminating knock because the last part of the charge is burned by the flame before it has time to ignite spontaneously. *See* SPARK PLUG.

In compression-ignition (diesel) engines the fuel is injected late in the cycle into highly compressed air; mixing must take place quickly if the fuel is to find oxygen and burn while the piston remains near top center. Rapid mixing is particularly important in the smaller-sized engines, which are usually operated at higher rotative speeds. In simple combustion chambers, mixing can be improved by producing a general swirl in the cylinder so that the air is moving relative to the fuel spray during injection.

In some compression-ignition engines a considerable fraction of the cylinder air is forced into a small auxiliary chamber (prechamber) during the compression stroke. The fuel is injected into the swirl chamber where rapid mixing and combustion take place, and the burning mixture issues into the main chamber where any remaining fuel is burned.

In compression-ignition engines the injected fuel ignites from contact with the hot cylinder air after a short delay. There is no flame front travel to limit the combustion rate, and if mixing of fuel and air is too thorough by the end of the delay period, high rates of pressure rise result, and the operation of the engine is rough and noisy. To avoid this condition, the prechambers of most compression-ignition engines operate at high temperature so that the fuel ignites soon after injection begins. This reduces the amount of fuel present and the degree of mixing at the time ignition takes place. *See* DIESEL ENGINE. [A.R.R.]

Comet One of the major types of objects that move in closed orbits around the Sun. Compared to the orbits of planets and asteroids, comet orbits are more eccentric and have a much greater range of inclinations to the ecliptic (the plane of the Earth's orbit). Physically, a comet is a small, solid body which is approximately a kilometer in diameter, contains a high fraction of icy substances, and shows a complex morphology, often including the production of a tail, as it approaches the Sun. *See* ASTEROID; PLANET.

About 10 comets are discovered or rediscovered each year. On the average, one per year is a bright comet, visible to the unaided eye and generating much interest among the public as well as among comet workers.

Astronomers consider comets to be worthy of detailed study for several reasons: (1) They are intrinsically interesting, involving a large range of physical and chemical processes. (2) They are valuable tools for probing the solar wind. (3) They are considered to be remnants of the solar system's original material and, hence, prime objects to be studied for clues about the nature of the solar system in the distant past.

Comets are nebulous in appearance, and the tail is usually the most visually striking feature. This tail can in some cases stretch along a substantial arc in the sky. An example is given in the illustration, which shows Comet West with tails some 30° in length. Some fainter comets, however, have little or no tail. The coma or head of a comet is seen as the ball of light from which the tail or tails emanate. Within the coma, and not yet seen, is the nucleus, the origin of the material in the tail and coma.

The main components of a comet are the nucleus, the coma, the hydrogen cloud, and the tails. Strong indirect evidence points to the existence of a central nuclear body or nucleus from which all cometary material, both gas and dust,

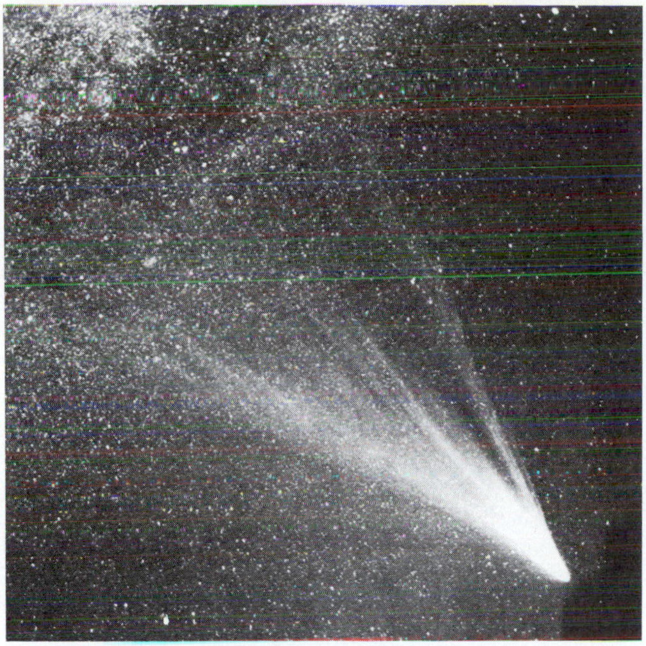

Comet West as photographed on March 9, 1976. ***(S. M. Larson, Lunar and Planetary Laboratory, University of Arizona)***

originates. Unfortunately, no photograph of a nucleus exists. The coma is observed as an essentially spherical cloud of gas and dust surrounding the nucleus. The principal gaseous constituents are neutral molecules. As the gas flows away from the nucleus, the dust particles are dragged along. In 1970, observations of Comet Tago-Sato-Kosaka and Comet Bennett from orbiting spacecraft showed that these comets were surrounded by a giant hydrogen cloud extending to 6×10^6 mi (10^7 km), or a size larger than the Sun. Photographs of bright comets generally show two distinct types of tails (see illustration): the dust tails and the plasma tails. They can exist separately or together in the same comet.

The plasma tails are generally straight and the plasma in these tails is composed of electrons and molecular ions. The dust tails are usually curved and the dust particles are probably silicate in composition.

While not definitely proved, it seems likely that the population of short-period comets can be produced from the population of the long-period comets by their gravitational interaction with Jupiter. The evidence from statistics of cometary orbits indicates an essentially spherical cloud of comets with dimensions in the range 10^{12} to 10^{13} mi (10^4 to 10^5 AU). Gravitational perturbations from passing stars have several effects on the cloud. They limit its size and tend to make the orbits random (as observed). Most importantly, the perturbations continually send new comets from the cloud into the inner solar system, where they are observed. The current consensus on the origin of comets holds that they condensed from the solar nebula at the same time as the formation of the Sun and planets. In other words, although the details are sketchy, comets are probably a natural by-product of the solar system's origin. They may be remnants of the formation process, and their material may be little altered from the era of condensation to the present time. *See* SOLAR SYSTEM. [J.C.B.]

Comfort heating The maintenance of the temperature in a closed volume, such as a home, office, or factory, at a comfortable level during periods of low outside temperature. Two principal factors determine the amount of heat required to maintain a comfortable inside temperature: the difference between inside and outside temperatures and the ease with which heat can flow out through the enclosure.

The first step in planning a heating system is to estimate the heating requirements. This involves calculating heat loss from the space, which in turn depends upon the difference between outside and inside space temperatures and upon the heat transfer coefficients of the surrounding structural members. Outside and inside design temperatures are first selected. Ideally, a heating system should maintain the desired inside temperature under the most severe weather conditions. The design temperature selected depends upon the heat capacity of the structure, amount of insulation, wind exposure, proportion of heat loss due to infiltration or ventilation, nature and time of occupancy or use of the space, difference between daily maximum and minimum temperatures, and other factors. Usually the outside design temperature used is the median of extreme temperatures. The selected inside design temperature depends upon the use and occupancy of the space. Generally it is between 66 and 75°F (19 and 24°C).

In localities where outdoor temperatures are often below 36°F (2°C), it is advisable to provide means for adding moisture in heated spaces to improve comfort. The colder the outside air is, the less moisture it can hold. When it is heated to room temperature, the relative humidity in the space becomes low enough to dry out nasal membranes, furniture, and other hygroscopic materials. This results in discomfort as well as deterioration of physical products. Various types of humidifiers are available. The most satisfactory type provides for the evaporation of the water to take place on a mold-resistant treated material which can be easily washed to get rid of the resultant deposits.

Good insulating material has air cells or several reflective surfaces. A good vapor barrier should be used with or in addition to insulation, or serious trouble may result. Air inside a space in which moisture has been added from cooking, washing, drying, or humidifying has a much higher vapor pressure than cold outdoor air. Therefore, moisture in vapor form passes from the high vapor pressure space to the lower pressure space and will readily pass through most building materials. When this moisture reaches a subfreezing temperature in the structure, it may condense and freeze. When the structure is later warmed, this moisture will thaw and soak the building material, which may be harmful. Good vapor barriers include asphalt-impregnated paper, metal foil, and some plastic-coated papers. Heat loss due to infiltration is the most difficult item to estimate accurately and depends upon how well the house is built. If a masonry or brick-veneer house is not well caulked or if the windows are not tightly fitted and weather-stripped, this loss can be quite large. *See* AIR REGISTER; CENTRAL HEATING AND COOLING; HOT-WATER HEATING SYSTEM; RADIATOR; STEAM HEATING; WARM-AIR HEATING SYSTEM. [G.B.P.]

Commelinales An order of flowering plants, division Magnoliophyta (Angiospermae), which is included in the subclass Commelinidae in the class Liliopsida (monocotyledons). It consists of 4 families and nearly 1000 species, the bulk of which is in the families Commelinaceae (about 600 species) and Xyridaceae (about 270 species). The order is marked by its flowers that ordinarily have the perianth well differentiated into sepals and petals but do not have nectaries or nectar. The wandering Jew (species of *Tradescantia* and *Zebrina* in the Commelinaceae) is a familiar houseplant belonging to this order. *See* COMMELINIDAE. [A.Cr.]

Commelinidae A subclass of the class Liliopsida (monocotyledons) of the division Magnoliophyta (Angiospermae), the flowering plants, consisting of 8 orders, 25 families, and nearly 19,000 species. The orders include Commelinales, Eriocaulales, Restionales, Juncales, Cyperales, Typhales,

Bromeliales, and Zingiberales. For further information see separate articles on each order.

These monocotyledons are syncarpous, (the carpels are united in a compound ovary) or pseudomonomerous (reduced to a single carpel from a syncarpous ancestry). The endosperm is usually starchy, and the perianth is either well differentiated into sepals and petals or more or less reduced and not petal-like. Many of the families have well-developed vessels in all vegetative organs. Several of the orders of Commelinidae have often been treated as a single order Farinosae or Farinales. *See* LILIOPSIDA. [A.Cr.]

Common cold A viral disease of humans, most frequently caused by the rhinoviruses. However, other viruses (including parainfluenza, influenza, and respiratory syncytial viruses and members of the adenovirus, enterovirus, and reovirus groups) also may cause common cold symptoms. *See* ANIMAL VIRUS.

The virus enters via the nose or mouth. After an incubation period of 2–5 days, typical cold symptoms and histopathology of mucous membranes develop, and continue for 4–6 days. Experiments under controlled conditions have not shown that chilling, including the wearing of wet socks, produces the cold itself or increases susceptibility to the virus. Chilliness is an early symptom of the common cold. *See* ADENOVIRIDAE; ENTEROVIRUS; PARAINFLUENZA VIRUS; RHINOVIRUS. [J.L.Me.]

Communication cables An assembly of from 6 to 4200 twisted wire pairs enclosed in a composite metal-plastic sheath. The major use of communication cables is in the telephone industry, to connect customers to their local switching centers (subscriber loops) and to interconnect local and long-distance switching centers (trunk plant). For telephone links, a pair of wires in an aerial or below-ground cable transmit signaling and voice information between the subscriber and the switching center. The instrument most commonly connected to a cable is the telephone set, but teletypewriters, television cameras, and data sets all convert information into electrical signals that can be sent over cables.

Interference between circuits in multipair cables becomes more likely as larger bandwidths are transmitted. Very-high-frequency signals, such as television, high-speed data, or broadband carrier signals, are frequently transmitted over coaxial cables in which one conductor of the pair is a metal tube concentric with and enclosing the other. As higher frequencies are transmitted, the outer conductor becomes more effective in containing radiated electromagnetic energy and preventing one circuit in the cable from interfering with others. However, multipair cable has several advantages when compared with coaxial: a lower loss per circuit cross section; mechanical ruggedness; and rejection of interference from stray power currents. For these reasons, the twisted, balanced pair is the choice for transmission from telegraph frequencies up through a few megahertz. *See* COAXIAL CABLE; CROSSTALK; CONDUCTOR (ELECTRICITY).

Copper is the most commonly used electrical conductor, but aluminum is widely used in England and is becoming more prevalent in the United States. Metallic conductors are commonly insulated with paper pulp or plastics, although paper strip, helically or longitudinally wrapped around the wire, which was introduced in 1890, is still used. Sheaths protect the transmission properties of the cable pairs by providing mechanical protection, electrical protection from lightning-induced currents, and shielding. In paper-insulated cables sheaths also function to keep out moisture. *See* TELEPHONE SERVICE; TELEPHONE SYSTEMS CONSTRUCTION.

Communication cables or telephone cables are divided into four general classes. Local and exchange area cable is placed aerially or underground in ducts, or is directly buried. It is used to connect one switching office to another or to connect the switching office and the subscriber. It has pulp paper, plastic, or expanded plastic insulation, with either copper or aluminum conductors. Cables filled with petroleum jelly, that is, waterproofed, are rapidly supplanting unfilled cables in buried subscriber applications.

Toll cable is used to connect metropolitan areas and to bring toll circuits through cities to toll switching offices. It is generally designed to have superior transmission characteristics compared to exchange area designs. Frequently, this cable carries many voice or data channels multiplexed onto high-frequency carrier systems. Older versions still being manufactured have helically applied paper insulation. Newer versions have expanded plastic insulation. Sheaths are lead, lead-plastic composites, or special multilayer metal-plastic composites. Heavy-duty sheaths and elaborate gas-pressurization schemes are used to protect and monitor the integrity of the cable sheaths because of the large number of circuits.

Switchboard cable is intended to be used as equipment cabling in switching offices, business premises, and office buildings. Polyvinyl chloride (PVC) insulation has largely supplanted older insulations.

Examples of special cables are coaxial cables used for high-frequency carrier systems on heavy toll routes, video pairs for baseband video transmission, stub cables used to connect apparatus such as loading coils and repeater cases to main cables, terminating cables from the central office cable vault to the equipment, and building cables used to distribute telephone service within commercial buildings. Also included in this group are composite cables having various combinations of coaxials, video pairs, and regular pairs. *See* ELECTRIC DISTRIBUTION SYSTEM; ELECTRIC POWER SYSTEMS. [J.R.Ap.]

Communications satellite A spacecraft placed in orbit around the Earth to intercept and retransmit radio signals. The communications satellite may also amplify, sort, or route these signals. The communications function is similar to a ground microwave repeater facility but with greatly increased coverage. Whereas a ground repeater relays communications signals between two fixed locations, a communications satellite interconnects many locations, both fixed and mobile over a wide area. *See* MICROWAVE; SATELLITE (SPACECRAFT).

There are two types of communications satellites of practical interest: passive and active. There is no amplification capability in a passive satellite; it simply reflects incident signals. Passive communications satellite systems fall into two types: One uses large discrete structures, such as the Echo satellite; the other uses the West Ford approach, in which a large volume of space is filled with a large number of tiny passive satellites. The general disadvantage of passive systems is the limitation on message capacity resulting from the extremely low level of reradiated energy.

In all active satellite systems complete electronic equipment for processing and transmission of signals must be carried. Also, means must be provided for directing antennas and maintaining desired satellite orientation and orbital position. With on-board electronic processing capability, an active satellite can amplify, modify, and retransmit the signals as appropriate. Active satellites can also be considered as two basic types: the store-and-forward design and the real-time repeater. In the store-and-forward type an Earth station transmits a message to a satellite passing overhead. The message is stored in an on-board recorder and later played back through the satellite transmitter when the satellite passes over the appropriate Earth receiving stations. The only communications satellite now considered for practical use is the real-time repeater. Operation of this instantaneous repeater requires the transmitting and receiving stations to be simultaneously in view of the spacecraft.

The first satellite with the primary purpose of communications was *Score*, launched December 18, 1958. It received messages from Earth stations in the United States and recorded them on tape for later transmission when the satellite was over the receiving stations. *Echo 1*, launched August 12, 1960, was the first artificial satellite whose primary purpose was to provide real-time relay of messages. It was a metallic-coated plastic balloon 100 ft (30 m) in diameter which could only reflect the signals directed toward it. The American Telephone and Telegraph Company developed and built *Telstar*, which was launched by NASA on July 10, 1962. It was the first active communications satellite with a real-time repeater. *Telstar* provided the first live television between the United States, Europe, Japan, and South America. On July 26, 1963, NASA launched the first successful Syncom into a synchronous orbit. This satellite proved that the synchronous equatorial orbit was achievable and that a satellite in this orbit was suitable for relaying telecommunication signals for uninterrupted service.

To exploit commercial utilization of communications satellites for international communications, an international consortium was formed in August 1964. By 1979 more than 100 nations were members of this consortium (Intelsat). *Early Bird*, or *Intelsat 1*, was launched from Cape Kennedy (Cape Canaveral) on April 6, 1965. This satellite with its 240 two-way voice channels was designed to relay all types of communications traffic, including television between North America and Europe. Three successors to *Early Bird* were launched in 1967 to expand the commercial communications satellite system. These satellites, called *Intelsat 2*, have more powerful signals that enable them to operate with Earth stations in both the Northern and Southern hemispheres. The *Intelsat 2* that extended full-time commercial service to the Pacific region was launched January 11, 1967. The development of the *Intelsat 3* satellite began in 1966, and the first was launched December 18, 1968. The *Intelsat 4*, the fourth-generation satellite, was first launched in 1971. The *Intelsat 5*, with a design capacity of 12,000 voice channels plus TV, went into service in 1980.

The Soviet Union also developed and launched communications satellites for its domestic use. The first launch was in April 1965, with inaugural service in November 1967. The satellite (*Molniya*) was placed in a highly elliptical orbit to provide maximum visibility over the Soviet Union with 8–10 h of continuous communication every 24 h.

The United States and Canada also have separate domestic systems (equatorial orbits) serving their own communication needs, while many countries lease satellite capacity from Intelsat to interconnect their Earth stations.

A separate maritime communications system, Marisat, provides reliable, high-quality, full-time communications connecting properly equipped ships and offshore equipment such as oil drilling platforms all over the globe with the world's telephone systems. [L.B.E.]

Communications scrambling The methods for ensuring the privacy of voice, data, and video transmissions. Various techniques are commonly utilized to perform such functions.

Analog voice-scrambling methods typically involve splitting the voice frequency spectrum into a number of sections by means of a filter bank and then shifting or reversing the sections for transmission in a manner determined by switch settings similar to those of a combination lock; the reverse process takes places at the receive end. Digital methods first convert the analog voice to digital form and then scramble or encrypt the digital voice data by one of the methods discussed below. *See* ANALOG-TO-DIGITAL CONVERTER; ELECTRIC FILTER.

A simple data-scrambling method involves the addition of a pseudorandom number sequence to the data at the transmit end. Devices using this method are known as stream ciphers. A second method partitions the data into blocks. Data within a block may be permutated bit by bit or substituted by some other data in a manner determined by the switch setting, which is often called a key. Devices using this method are known as block ciphers. *See* CRYPTOGRAPHY.

Typical video scrambling devices used for cable television applications involve modifying the amplitude or polarity of the synchronization signals, thereby preventing the normal receiver from detecting the synchronization signals. A more sophisticated technique, used in satellite transmission, introduces a random delay to the active video signal on a line-by-line basis. An even more advanced technique called cut-and-rotate has been proposed. Video signals can also be digitized by a number of coding techniques and then scrambled by any of the data-scrambling techniques discussed above to achieve high security. *See* CLOSED-CIRCUIT TELEVISION; TELEVISION. [L.-N.L.]

Communications systems protection The protection of wire communication systems equipment and services. This includes the electrical protection of lines and station equipment and inductive coordination, or the protection against interference from adjacent electric power lines and equipment. *See* ELECTRIC PROTECTIVE DEVICES.

Wire communications systems generally require protection against voltages and currents from lightning, from accidental energization by power lines and, under some conditions, from earth potentials from magnetic storms caused by sunspots. The basic methods of protection are: (1) to reduce the likelihood of energization through shielding or arranging the lines so they will be less subject to lightning strokes or contact with power wires, (2) to use adequate insulation so that physical contact may occur with power wires without energizing the communication plant, (3) to increase the insulation of the wires and the conductivity of cable sheaths to minimize damage from power contacts, and (4) to install protective devices that prevent damage to persons or equipment.

Protective devices for communications systems generally take the form of protector blocks and fuses. The protector blocks usually consist of two small carbon electrodes arranged to be separated by a few thousandths of an inch. These electrodes are paired together in a suitable mounting, so as to connect each set of protector blocks between the conductor to be protected and the grounded sheath of a cable or a grounding electrode at the equipment location.

To protect the user, the premises, and the communications equipment, telephone sets require protection when the lines serving them are subject to contact with power lines or lightning. The protector unit consists of carbon protector blocks connected between each wire and ground (usually the metallic water pipe serving the building). Sometimes a fuse on the incoming wire side of the protector blocks is also necessary.

Inductive coordination applies to measures which reduce the effects of potentials and currents induced in the communication plant by paralleling power facilities or power lines. When power and telephone lines on separate structures are parallel for long distances, a fault current to ground on the power circuit creates, in effect, a one-turn transformer primary. The induced voltages may be high enough to result in an electric-shock hazard to telephone personnel or interference to service on the communication plant. Remedial measures to reduce the induction or its effects include: (1) installing effectively grounded shield wires on either or both power and telephone plants, (2) using grounded metal-sheathed cable on either or both lines, (3) moving either line for greater separation to reduce the coupling between them, (4) using faster circuit breakers on the power line to reduce the duration of the fault, and (5) installing suitable devices in the communication plant to reduce the effect of the induced voltages on the services transmitted. *See* LIGHTNING AND SURGE PROTECTION. [J.J.Sh.]

Commutation The process of current reversal in the armature windings of a direct-current (dc) electric rotating machine to provide dc at the brushes. *See* COMMUTATOR.

The basic process of commutation may be explained with the aid of the elementary motor diagram illustrated; *N* and *S* indicate the north and south magnetic poles; the force on conductor *A* is f_A, and on conductor *B* is f_B. With the rotor position shown at *a*, the battery will force current into conductor *B*

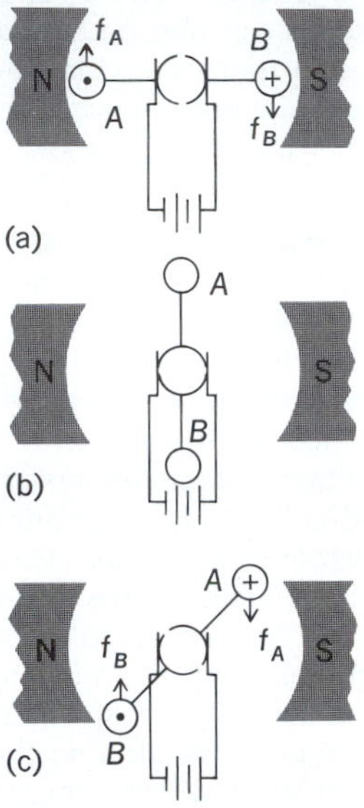

Views of an elementary motor with two-segment commutator; *a*, *b*, and *c* explained in text. The symbol f_A indicates the force on conductor *A*, and f_B the force on conductor *B*.

and out of conductor *A* to produce a clockwise torque and rotation. The generated electromotive force (emf) in the conductors is opposite to the direction of current. With the rotor position of *b*, the conductors are not cutting lines of flux; no emf is induced in the coil; no tangential forces are produced in the coil; and the torque is zero. With the rotor position of *c*, the battery will force current into conductor *A* and out of conductor *B*. The torque is again clockwise; clockwise rotation will continue, and the generated emf will have a polarity opposite to the conductor current. Note that the coil current and the coil emf are reversed from *a* to *c*. The switching is accomplished by the commutator. Practical machines have many coils and many commutator segments on the armature. *See* DIRECT-CURRENT GENERATOR; DIRECT-CURRENT MOTOR; ELECTRIC ROTATING MACHINERY. [A.R.E.]

Commutator That part of a dc motor or generator which serves the dual function, in combination with brushes, of providing an electrical connection between the rotating armature winding and the stationary terminals, and of permitting the reversal of the current in the armature windings. For explanation of the necessity of this function *see* COMMUTATION.

A commutator (see illustration) is composed of copper bars assembled to form a drumlike cylinder which is concentric with the axis of rotation. Insulation, commonly mica, to provide exceptional mechanical and electrical stability, is placed between commutator bars and between the bars and the shaft. Conducting brushes, commonly carbon, sufficient in size and number to carry the current, are spaced at intervals of 180 electrical degrees about the surface of the commutator and held in contact with the surface of the commutator by spring

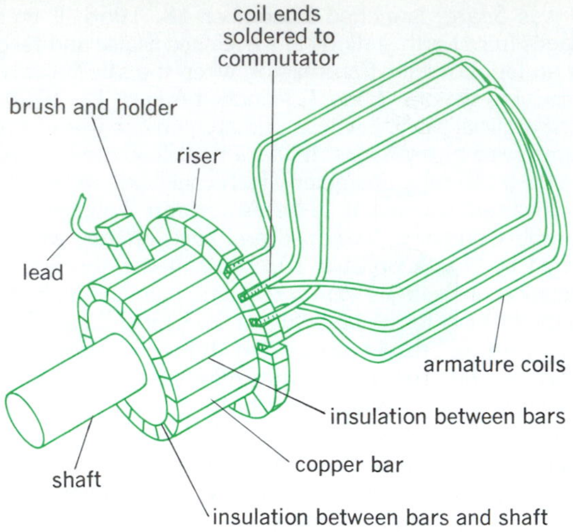

Commutator and brush assembly with coil connections for lap winding.

tension. *See* DIRECT-CURRENT GENERATOR; DIRECT-CURRENT MOTOR; ELECTRIC ROTATING MACHINERY. [A.R.E.]

Compact disk A system of audio or video recording or data storage in which digitally encoded information in the form of microscopic pits on a rotating disk is accessed by optical readout. The implementation of these ideas in the audio compact disk will be described, along with their extension to current developments in video disks and optical computer storage.

Optical readout. The acoustic signal waveform is recorded on the disk in the form of a binary code, as a series of 0s and 1s. This is done by forming pits along spiral tracks on a transparent plastic disk, overlaying this with a reflective coating, and then covering this coating with a protective layer. The light from a semiconductor laser is focused onto the pits from below (see illustration), and the reflected light, picked up by a photodetector, converts the presence or absence of the pits into a binary electrical signal. *See* LASER.

With the light focused at the level of the pits, the surface of the transparent disk is well out of focus; hence the presence of dust, finger marks, and minor scratches on this surface is so blurred as not to affect the detection process. Even more important, the use of an optical pickup means that there is no mechanical contact and hence no wear.

Because the focused laser spot size is so small, the amount of information that can be packed onto the surface of the disk is very great. Adjacent tracks of the spiral of pits need be only 1.6 μm apart, and hence 20,000 such tracks are available on a 120-mm-diameter (5-in.) audio compact disk.

Digitally encoded waveform. The human ear is insensitive to sounds whose frequency is higher than about 20,000 Hz. C. Shannon's sampling theorem guarantees that a waveform can be reproduced exactly from samples taken at a uniform rate of at least twice that of the highest frequency. To allow a small margin of error, the standard sampling rate for audio compact disks is 44.1 kHz. The amplitude of each of these samples is then approximated to the nearest of 2^{16} (65,536) equally spaced levels (a process known as quantization). Hence each sample amplitude can be represented by a 16-bit binary word. With two separate stereo channels, the number of bits per second required to be stored on the disk is given by Eq. (1).

$$2 \times 16 \times 44{,}100 = 1{,}411{,}200 \text{ bits/s} \tag{1}$$

See INFORMATION THEORY; PULSE MODULATION.

Both the dynamic range, defined as the ratio of the amplitude of the largest undistorted signal to the amplitude of the

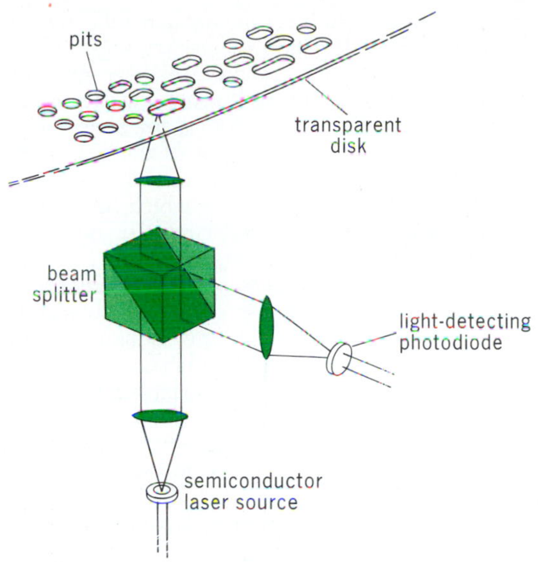

Optical readout system for audio compact disk.

noise, and the signal-to-noise ratio are just 2^{16}, and are given in decibels by Eq. (2). This is very close to the accepted

$$\text{Dynamic range} = 10 \log_{10}(2^{16})^2 = 92 \text{ dB} \qquad (2)$$

figure of 100 dB for the dynamic range of a live orchestra. As a comparison, the best figure of dynamic range for stylus disks and professional magnetic tapes is about 70 dB, and acceptable signal-to-noise ratios for conventional audio systems are in the region of 50–60 dB.

Signal processing. Since the information that is read off the disk is in digital form, as a sequence of 0s and 1s, it can be processed in many more ways than were possible with analog systems. To enable this, information can be stored for as long as is desired, and then output at a rate that is controlled by the player's quartz-crystal (oscillator) clock, hence eliminating the wow and flutter of conventional systems entirely.

Video disks and computer storage. The video disk and the compact disk for computer storage have been slow to develop for both economic and technical reasons. A major difficulty in video recording is the large bandwidths required. The main difficulty in using video and compact disks for recording computer data involves the allowable error rate. For data this must be effectively zero, whereas for both audio and video recording a small but finite error rate is tolerable.

Erasable compact disks are also produced. The information is stored magnetooptically. A bit is erased by using a higher-power laser to reverse the direction of magnetization. *See* Computer storage technology; Disk recording; Video disk recording. [R.H.Cl.]

Comparator A device used to inspect a gaged part for deviation from a specific dimension. The term comparator is also used to identify control-system and analog-computer devices which compare two information signals for such characteristics as simultaneity, size, direction, or rate of change. *See* Analog-to-digital converter; Comparator circuit; Control systems.

To check a gaged part, a comparator is usually preset to the basic critical dimension to be inspected, for example, to 1.505 in the dimension 1.505 ± 0.005 or 2.000 in 2.000 ± 0.002. A master having this dimension is used for presetting the comparator. A comparator may also be set by two masters, one for the maximum limit and one for minimum limit. Two classes of comparator are in common use: one in which the comparison is made visually or optically, the other in which comparison is made by contact. The three common types of contact comparators are mechanical, electrical, and pneumatic. *See* Inspection and testing.

A general-purpose comparator of the contact type usually consists of (1) a fixed contact point or surface; (2) an indicating contact or measuring head; and (3) a means for physically maintaining the distance between. This last means takes the form of a frame in a gage making external measurements, or the form of a plug in a gage making internal measurements. The indicator or meter, which records the results, may be an integral part of the measuring head or a separate unit. A general-purpose comparator can usually be set with gage blocks, as well as with a master. Movement in mechanical comparators is amplified usually by a rack, pinion, and pointer, or by a parallelogram arrangement. Movement in electrical comparators is amplified by several methods. The most popular method consists of a floating core in a solenoid attached to the contact point. *See* Gage.

Pneumatic comparators sense the surface of the part by means of an airstream. A precision orifice is brought so close to the surface to be measured that the flow of air from the orifice is restricted. The closer the orifice comes to the surface, the greater the restriction; a measurement of this restriction in flow is, in effect, a measurement of this distance.

Visual or optical comparators of the general-purpose type consist of (1) a light source, (2) a stage on which to place the part being inspected, (3) a series of lenses and mirrors for magnifying and projecting the shadow of the object, and (4) a screen on which the image is projected. The screen is usually a frosted glass. [R.A.Bo.]

Comparator circuit An electronic circuit that produces an output voltage or current whenever two input levels simultaneously satisfy predetermined amplitude requirements. A comparator circuit may be designed to respond to continuously varying (analog) or discrete (digital) signals, and its output may be in the form of signaling pulses which occur at the comparison point or in the form of discrete direct-current levels.

A linear comparator operates on continuous, or nondiscrete, waveforms. Most often one voltage, referred to as the reference voltage, is a variable dc or level-setting voltage and the other is a time-varying waveform. One common application of the comparator is in a linear time-delay circuit. Inputs consist of a sawtooth waveform of linearly increasing magnitude (ramp function) and a variable dc reference voltage. The reference voltage can be calibrated in units of time, as measured from the beginning of the sawtooth. A clipper (usually called a pick-off diode for this application) and a coincidence amplifier, together with a resistance-capacitance (RC) differentiating circuit, can perform the function of comparator.

Multivibrators can be used in several ways directly as comparators without need for the pick-off diodes; such comparators sense the required coincidence accurately and introduce little additional delay. These are called regenerative comparators. A simple type is the direct-coupled bistable circuit, sometimes known as the Schmitt circuit.

High-gain dc operational amplifiers ("op-amps") operated in the nonfeedback mode are often used to perform the comparator function, and many such amplifiers are classified as comparators because they are specifically designed to meet the needs for accurate voltage comparison applications. Such "op-amps" have two inputs, the output being inverting with respect to one and noninverting with respect to the other.

The term "digital comparator" has historically been used when the comparator circuit is specifically designed to respond to a combination of discrete level (digital) signals, for example, when one or more such input signals simultaneously reach the reference level which causes the change of state of the output.

Among other applications, such comparators perform the function of the logic gate such as the AND, OR, NOR, and NAND functions. More often, however, the term "digital comparator" is used to describe an array of logic gates designed specifically to determine whether one binary number is less than or greater than another binary number. Such digital comparators are sometimes called magnitude comparators or binary comparators. [G.M.G.]

Complement The complement system comprises a group of proteins in the blood serum and other body fluids that play an important role as mediators of immune and allergic reactions. When activated by antigen-antibody complexes, or by other agents, complement kills bacteria, fungi, protozoans, and cells of higher organisms by damaging the lipid bilayer of their membranes. Complement also destroys certain viruses or interferes with their infectivity. In addition, activation of complement generates permeability-enhancing and chemotactic factors, as well as ligands, which participate in a variety of immune and allergic processes. See COMPLEMENT-FIXATION TEST. [M.M.M.]

Complement-fixation test A sensitive reaction used in serology for the detection of either antigen or antibody, as in the diagnosis of many bacterial, viral, and other diseases, including syphilis. It involves two stages: Stage I is the binding or fixation of complement if certain antigen-antibody reactions occur, and stage II is detection of residual unbound complement, if any, by its hemolytic action on the sensitized erythrocytes subsequently added. This is shown in the illustration.

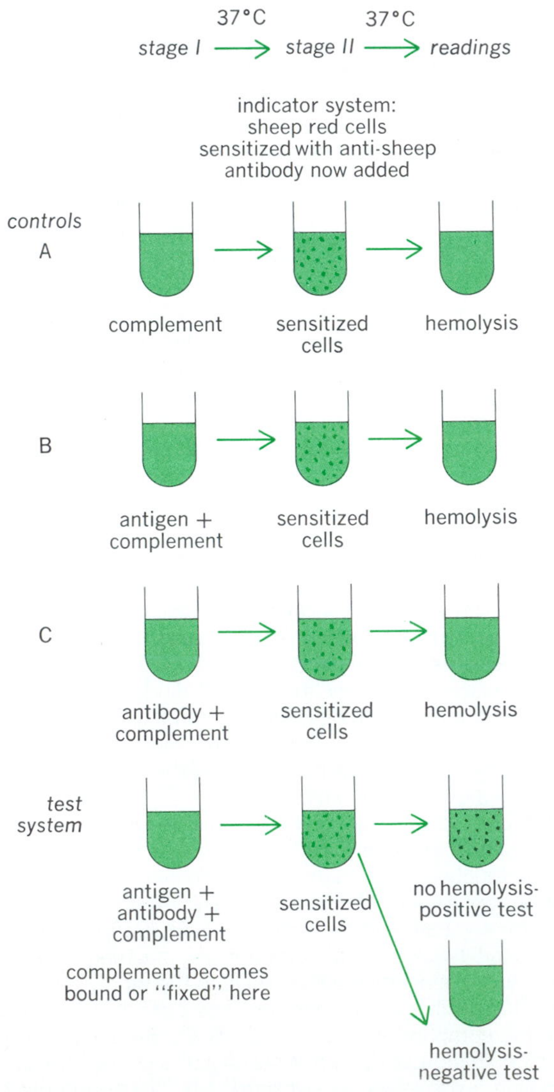

Diagram of the complement-fixation reaction.

In the first stage either the antigen or the antibody must be supplied as a reagent, with the other of the pair as the test unknown. Fresh guinea pig serum is normally used as a complement source. Sheep erythrocytes which are coated with their corresponding antibody (amboceptor or hemolysin) are used in the second stage. See COMPLEMENT.

The controls A, B, and C in the diagram demonstrate that sufficient complement is present to effect hemolysis of the sensitized indicator cells and that neither antigen or antibody added alone will interfere with this by binding complement. In the test system the combination of a suitable antigen and antibody in the presence of complement will bind the complement to the complex so that the complement becomes unavailable for the hemolysis of the indicator cells added in stage II. If either antigen or antibody is added as a reagent in stage I, then the presence of the other in the test unknown added can be detected through its ability to complete the antigen-antibody system. A lack of hemolysis denotes that the complement is bound. See IMMUNOLOGY; LYTIC REACTION. [H.P.T.]

Complementation (genetics) The complementary action of different genetic factors. The term usually implies two homologous chromosomes or chromosome sets, each defective because of mutation and unable by itself to promote the normal development or metabolism of the organism, but able to do so jointly when brought together in the same cell. See CHROMOSOME; MUTATION.

S. Benzer proposed the term cistron for the unit within which mutants do not complement each other. The word gene is often used in the same sense. The usual biochemical function of a cistron, or gene, is to determine the structure of a specific polypeptide component of a protein. Full complementation between different genes is the rule except when, as sometimes in bacteria, the genes form part of a functionally coordinated complex (operon). Allelic mutants (mutants within one gene) show limited complementation in some cases, for example, when certain pairs of mutant polypeptides correct each other's defects through coaggregation in a complex protein. See GENETICS; OPERON. [J.R.S.F.]

Complex numbers A natural and extremely useful extension of the familiar real numbers. They can be introduced formally as follows. Consider the two-dimensional real vector space consisting of all ordered pairs (a_1,a_2) of real numbers. Geometrically this space can be identified with the ordinary euclidean plane, viewing the real numbers a_1,a_2 as the coordinates of a point in the plane.

The addition of vectors is defined by $(a_1,a_2) + (b_1,b_2) = (a_1 + b_1, a_2 + b_2)$ and is just the usual addition of vectors by the parallelogram law. The multiplication of a vector (a_1,a_2) by a real number c is defined by $c(a_1,a_2) = (ca_1,ca_2)$, and is just the uniform dilation of the plane by the factor c. It may be asked whether it is possible to define a multiplication of one vector by another in such a manner that this multiplication is linear and satisfies the same formal rules as multiplication of real numbers. There does exist such a multiplication, given by the equation below, and it is essentially unique in the sense that any

$$(a_1,a_2) \cdot (b_1,b_2) = (a_1b_1 - a_2b_2, a_1b_2 + a_2b_1)$$

multiplication satisfying all the desired properties can be reduced to this equation by a suitable choice of coordinates in the plane. The plane with the ordinary addition and scalar multiplication of vectors and with the vector multiplication given above is the complex number system.

The above multiplication law can most easily be interpreted geometrically by introducing polar coordinates in the plane and writing $(a_1,a_2) = (r \cos \theta, r \sin \theta)$, where r is the length of the vector (a_1,a_2), or the modulus of the complex number (a_1,a_2), defined by $r^2 = a_1^2 + a_2^2$, and θ is the angle between the vector (a_1,a_2) and the first coordinate axis, or the argument of the complex number (a_1,a_2), defined by $\tan \theta = a_2/a_1$. Then multiplication of complex numbers amounts to multiplying their moduli and adding their arguments (see illustration).

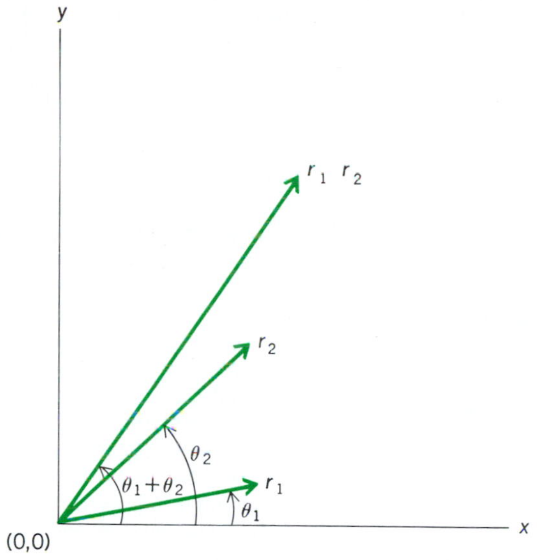

Multiplication of complex numbers.

Introducing the basis vectors $I = (1,0)$ and $i = (0,1)$, any vector (x,y) can be written uniquely as the sum $(x,y) = xI + yi$. It follows readily from the multiplication law that $I \cdot I = I$, $I \cdot i = i \cdot I = i$, and $i \cdot i = -I$, so that in particular the vector I is the identity element for the multiplication of complex numbers. The complex number I can be identified with the ordinary real number 1, and reflecting this identification the notation can be simplified by writing a complex number $z = (x,y) = xI + yi$ merely as $z = x + iy$. The component x is called the real part of the complex number z, and the component y is called the imaginary part. The modulus of the complex number z is denoted by $|z|$. *See* ANALYTIC FUNCTION. [R.C.G.]

Composite beam Composite beams are so called because they provide beam action of two materials joined together in such a way as to act as a unit. In civil engineering the term is most commonly given to the beam action developed by a concrete slab resting on a steel beam, usually an I or a wide flange shape, the slab and the beam being made to act together by shear connectors.

In noncomposite design the structural function of the concrete slab is only to span transversely across a series of parallel steel beams. In composite design the concrete slab performs its normal function and in addition increases to a considerable extent the capacity of the structure to resist bending in the direction of the steel beams. The unit stresses in the concrete slab are not added algebraically from the dual function the slab performs, because the flexural stresses are in mutually perpendicular directions. Thus the concrete slab thickness and area of reinforcing steel are not necessarily greater for this type of design than for the noncomposite. *See* STRESS AND STRAIN.

Shear connectors are devices that provide unified action between the concrete slab and the steel beam by resisting the horizontal shear developed in beam bending. They take the form of steel projections welded to the top flange of the steel beam and around which the concrete slab is poured. The connectors perform a necessary function in the development of composite beam action, for without them the two elements, the concrete slab and the steel beam, would act separately as two beams and the beam action would be analogous to two unconnected wooden planks placed one on top of the other and slipping on each other when bent under a transverse load. There are many types of shear connectors, but three commonly used types are stud bolts, a helical coil of steel rod, and short lengths of a standard channel shape. *See* SHEAR.

Composite beams are used both in building and highway bridge construction and can be used in any type of construction where it is appropriate to use a concrete slab resting on steel beams or stringers. Definite and considerable economy is achieved in composite construction when the spans are of sufficient length to offset the increased cost of fabricating the shear connectors with the saving from the lesser beam size required. [H.L.K.]

Composite material A material that results when two or more materials, each having its own, usually different characteristics, are combined in order to provide the composite with useful properties for specific applications. Each of the input materials should serve a specific function in the composite, which in turn should show distinctive new or improved characteristics. Advanced composites are a class of structural high-performance, fiber-reinforced plastic or metallic materials.

The development and use of composite materials goes back to antiquity. Combining two or more high- and low-carbon irons or steels by skillfully hand-forging them together into a samurai sword with a lightweight, tough blade which could keep its edge is a typical example. Modern technology applies them to a growing variety of functions from fishing poles to lightweight automobiles or aircraft, and from machine tool bits to the thermal protection systems for reentry from space. The expanding activity in composites is stimulated both by the demands of such applications and by the potential embodied in the outstanding properties which have been achieved in substances such as high-strength glass filaments or stiff, refractory graphite filaments. The challenge with such filaments, for example, is to combine them with a suitable supporting, load-transferring, environmental protective material such as an epoxy polymer or a graphitizable pitch to produce efficient structural or refractory composite materials.

Attempts are often made to differentiate composite materials from conventional materials on the basis of the heterogeneous or multiphase nature of the former, but many materials which are considered conventional by generally accepted usage, such as high-strength structural alloys, are also multiphase. Such familiar materials as concrete, plywood, or carbon black–reinforced rubber may or may not be considered to be composites. [H.M.B.]

Advanced composites are rapidly emerging as the primary material for use in aircraft systems because they provide greater structural efficiency at lower weights than equivalent metallic structures. Advanced composites consist of basically two distinct materials, fibers and matrix, which are combined in a preimpregnated form and subsequently bonded under heat and pressure to achieve the properties of high strength and stiffness with low density. On a specific basis all the advanced composite materials are much stronger and stiffer than steel or aluminum. Other significant advantages of advanced composites are their resistance to fatigue and corrosion and their tendency to dampen vibrations more rapidly than less stiff materials.

The matrix material can be thermosetting resins, such as epoxies, polyesters, or polyimides, which usually cure (harden) in the presence of a catalyst, heat, or pressure. Or they can be of thermoplastic resins, such as polycarbonate or polysulfones that are also cured in the presence of heat and pressure but, unlike the thermosetting resins, can be softened and reshaped by heat. For metal matrixes the most commonly used materials are aluminum and titanium. Of all of these, expoxies are the matrix materials most commonly used for advanced composites. High-performance fibers available for such applications are boron, graphite, and the Aramid fiber known as Kevlar. *See* EPOXY RESIN; POLYMER; POLYMERIC COMPOSITE.

The use of advanced composites in military applications has progressed from the basic replacement of metal in selected secondary components to include primary structures (wing, fuselage, empennage) that are critical to flight safety. As a result, the F-14, F-15, F-16, F/A-18, and the AV8-B military aircraft use advanced composites in various degrees, ranging 3–20% of the structural weight. In addition to military applications, the commercial and general aviation industries use advanced composites because the weight savings and increased performance result in significant increases in payload or decreases in fuel consumption.

The technology developed by the aerospace industry provided a tremendous baseline and spin-off to nonaerospace applications. Golf shaft producers have manufactured graphite and boron shafts for clubs. Manufacturers also introduced composite tennis rackets, offering lighter weight at an increased stiffness. Other sports applications include fly rods and fishing rods, archery equipment, hang gliders, and skis. Substantial work in the medical field has led to use of advanced composites in external and implanted prosthetic devices, radiological equipment, and lightweight mechanical supports and equipment for orthopedic and handicapped patients. The automotive industry has initiated evaluations of graphite-reinforced composites as a possible means of reducing weight in automobiles. [D.St.]

Composition board A wood product in which the grain structure of the original wood is drastically altered. Composition board may be divided into several types. When wood serves as the raw material for chemical processing, the resultant product may be insulation board, hardboard, or other pulp product. When the wood is broken down only by mechanical means, the resultant product is particle board. Because composition board can use waste products of established wood industries and because there is a need to find marketable uses for young trees, manufacture of composition board is one of the most rapidly developing portions of the wood industry. *See* PAPER.

Fiberboard is produced from wood chips. Synthetic resin may be added as a binder before the board is formed. After the board is formed, it may be impregnated with drying oils and heated in a kiln until the oils are completely polymerized to produce tempered board. If insulating board is required instead of hardboard, the material is less compacted, the degree of compaction being described by the specific gravity of the finished board.

When formed from wood particles that retain their woody structure, the product is termed particle board. Properties of such boards depend on the size and orientation of the particles, which may be dimensioned flakes, random-sized shavings, or splinters. After the particular type of particles are produced, they are screened to remove fines and to return oversizes for further reduction. Graded particles are dried, mixed with synthetic adhesive and other additives such as preservatives, and delivered to the board-forming machine.

Development of adhesives specifically for composition boards is extending their utilitarian value, and variety of textures is increasing their esthetic appeal. *See* WOOD PRODUCTS. [F.H.R.]

Compressible flow Flow in which the fluid density varies. In aerodynamic phenomena, when the flow velocity is large, it is necessary to consider that the fluid is compressible rather than to carry over from classical aerodynamics the assumption that the fluid has a constant density. Under this condition the speed of sound becomes an important factor. At relatively low speeds the changes of temperature and density of a fluid caused by the motion of a body in the fluid are almost negligible. However, if the body moves at high speed through the fluid, the motion can cause pronounced changes in density and temperature of the fluid. Hence, consideration of phenomena of this type involves not only classical fluid mechanics but thermodynamics as well. *See* AEROTHERMODYNAMICS.

The essential difference between an incompressible fluid and a compressible fluid is in the speed of sound. In an incompressible fluid the propagation of pressure change is essentially instantaneous; in a compressible fluid the propagation takes place with finite velocity. For example, if one strikes the surface of an incompressible fluid, the effect observed at great distances is, of course, less than at a smaller distance, but it reaches even infinite distance in essentially zero time. In a compressible fluid the effect propagates at finite velocity. A small disturbance propagates at the velocity of sound. [J.E.Sc.]

Compression ratio In a cylinder, the piston displacement plus clearance volume, divided by the clearance volume. This is the nominal compression ratio determined by cylinder geometry alone. In practice, the actual compression ratio is appreciably less than the nominal value because the volumetric efficiency of an unsupercharged engine is less than 100%, partly because of late intake valve closing. In spark ignition engines the allowable compression ratio is limited by incipient knock at wide-open throttle. *See* COMBUSTION CHAMBER; INTERNAL COMBUSTION ENGINE; VOLUMETRIC EFFICIENCY. [N.MacC.]

Compressor A machine that increases the pressure of a gas or vapor, typically air, by increasing the gas density and delivering the fluid against the connected system resistance. The resistance may be on the suction side, as with vacuum pumps, or on the discharge side, as with air compressors.

Compressors are typically applied to the operation of pneumatic tools and rock drills, conveying systems, furnace blast systems, ventilation systems, and the inflation of automobile tires. Gas and vapor compressors are represented in the refrigeration, air conditioning, and heat pump fields, where they handle a wide variety of refrigerants near the condensation point, such as Freon, ammonia, carbon dioxide, and sulfur dioxide. Compressors are used in industrial process operations such as nitrogen fixation and gas liquefaction. They are used for repressuring and pumping on natural-gas transmission lines.

The reciprocating compressor is suited to the highest-pressure services. Rotary compressors are suitable for direct connection to high-speed drivers such as automotive engines and electric motors.

Free-compression devices include fans and blowers, of the centrifugal or axial-flow type, limited to pressure ratios so small that the change in density of the fluid on passage through the unit is negligible.

Jet compressors are free of moving parts. They are especially suitable for vacuum service when steam is used as the actuating jet for entrainment of the noncondensible gas. [T.Ba.]

Compton effect The increase in wavelength of electromagnetic radiation, observed mainly in the x-ray and gamma-ray region, on being scattered by material objects. This increase in wavelength is caused by the interaction of the radiation with the weakly bound electrons in the matter in which the scattering takes place. The Compton effect illustrates one of the most fundamental interactions between radiation and matter and displays in a very graphic way the true quantum nature of electromagnetic radiation. Together with the laws of atomic spectra, the photoelectric effect, and pair production, the

Compton effect has provided the experimental basis for the quantum theory of electromagnetic radiation. For information on these and related topics *see* ANGULAR MOMENTUM; ATOMIC STRUCTURE AND SPECTRA; ELECTRON-POSITRON PAIR PRODUCTION; LIGHT; PHOTOEMISSION; QUANTUM MECHANICS; UNCERTAINTY PRINCIPLE.

The Compton effect has played a significant role in several diverse scientific areas. Compton scattering (often referred to as incoherent scattering, in contrast to Thomson scattering or also Rayleigh scattering, which are called coherent scattering) is important in nuclear engineering (radiation shielding), experimental and theoretical nuclear physics, atomic physics, plasma physics, x-ray crystallography, elementary particle physics, and astrophysics, to mention some of these areas. In addition the Compton effect provides an important research tool in some branches of medicine, in molecular chemistry and solid-state physics, and in the use of high-energy electron accelerators and charged-particle storage rings.

Perhaps the greatest significance of the Compton effect is that it demonstrates directly and clearly that in addition to its wave nature with transverse oscillations, electromagnetic radiation has a particle nature and that these particles, the photons, behave quite like material particles in collisions with electrons. This discovery by A. H. Compton and P. Debye led to the formulation of quantum mechanics by W. Heisenberg and E. Schrödinger and provided the basis for the beginning of the theory of quantum electrodynamics, the theory of the interactions of electrons with the electromagnetic field. [E.N.H.]

Compton wavelength A convenient unit of length that is characteristic of any particle. By definition, the Compton wavelength λ_C of a particle of rest mass m is $\lambda_C = h/mc$, where h is Planck's constant and c is the velocity of light; this definition is analogous to that of the quantum-mechanical de Broglie wavelength. *See* DE BROGLIE WAVELENGTH; PLANCK'S CONSTANT.

The Compton wavelength of the electron is 2.42631×10^{-12} m; of the proton 1.32141×10^{-15} m; of the muon 1.1734×10^{-14} m; and of the pion 8.883×10^{-15} m. The last is, by definition, the range of the strong nuclear force; in general, the Compton wavelength provides a convenient scale length in any given quantum-mechanical situation. *See* ELEMENTARY PARTICLE; QUANTUM FIELD THEORY.

The so-called reduced Compton wavelength, $\lambda\!\!\!^{-}_C = \lambda_C/2\pi$, is very frequently used instead of the Compton wavelength itself, See QUANTUM MECHANICS. [D.A.B.]

Computational chemistry A branch of theoretical chemistry that uses a digital computer to model systems of chemical interest. In this discipline, the computer itself is the primary instrument of research. The use of computers for analysis of experimental data, and for the storage and display of results obtained with other tools, is distinct from computational chemistry. The latter permits calculation of quantities which can be measured experimentally, such as molecular geometries of ground and excited states, heats of formation, and ionization potentials. Alternatively, quantities not readily accessible by existing experimental techniques, such as geometries of transition states and detailed structure of liquids, may be evaluated. *See* DIGITAL COMPUTER.

Because of the increasing power and availability of computers, and the simultaneous development of well-tested and reliable theoretical methods, the use of computational chemistry as an adjunct to experimental research has increased rapidly. Calculations ranging from a few seconds to many hours of computer time can serve as a guide to exclude less favorable reactions or unstable products, or to select several more fruitful procedures from the many possible ones. In addition, modeling of chemical systems with a computer enables the researcher to examine them on a scale of space or time as yet unmeasurable by experimental techniques. This can give insight into a chemical system beyond that provided by experiment. Examples are examination of the dynamics of a chemical reaction or of detailed changes in conformation of a polymer in solution. Examination of the molecular orbitals occupied by the electrons of the molecule can provide insight into chemical bonding and the electronic interactions which determine specific geometric configurations. Thus computational chemistry can yield information which may not be experimentally available. *See* CHEMICAL DYNAMICS.

Computational chemistry may be the application of existing theory and numerical methods to new molecules, or it may be the development of new computational methods. The latter may include incorporating more physics into the mathematical model in order to provide a better theoretical description of the system being studied, for example, inclusion of interactions between individual electrons in molecular orbital calculations. These more complete studies, for "large" molecules, usually need to be performed on a supercomputer, or on a mid-size computer with an array processor. Another approach is to devise simpler methods which will approximate the accuracy of more complex calculations. These include semiempirical methods in which values of hard-to-calculate terms are derived from experiment. Such an approach allows fruitful work with a mid-size or even a desk-top computer, avoiding the need for expensive computer resources. *See* MICROCOMPUTER; SUPERCOMPUTER. [Z.R.W.]

Computational fluid dynamics The numerical approximation to the solution of mathematical models of fluid flow and heat transfer. Computational fluid dynamics is one of the tools (in addition to experimental and theoretical methods) available to solve fluid-dynamic problems. With the advent of modern computers, computational fluid dynamics evolved from potential-flow and boundary-layer methods and is now used in many diverse fields, including engineering, physics, chemistry, meteorology, and geology. The crucial elements of computational fluid dynamics are discretization, grid generation and coordinate transformation, solution of the coupled algebraic equations, turbulence modeling, and visualization.

Discretization. Numerical solution of partial differential equations requires representing the continuous nature of the equations in a discrete form. Discretization of the equations consists of a process where the domain is subdivided into cells or elements (that is, grid generation) and the equations are expressed in discrete form at each point in the grid by using finite difference, finite volume, or finite element methods. The finite difference method requires a structured grid arrangement (that is, an organized set of points formed by the intersections of the lines of a boundary-conforming curvilinear coordinate system), while the finite element and finite volume methods are more flexible and can be formulated to use both structured and unstructured grids (that is, a collection of triangular elements or a random distribution of points). *See* FINITE ELEMENT METHOD.

Turbulence modeling. There are a variety of approaches for resolving the phenomena of fluid turbulence. The Reynolds-averaged Navier-Stokes (RANS) equations are derived by decomposing the velocity into mean and fluctuating components. An alternative is large-eddy simulation, which solves the Navier-Stokes equations in conjunction with a subgrid turbulence model. The most direct approach to solving turbulent flows is direct numerical simulation, which solves the Navier-Stokes equations on a mesh that is fine enough to resolve all length scales in the turbulent flow. Unfortunately, direct numerical simulation is limited to simple geometries and low-Reynolds-number flows because of the limited capacity of even the most sophisticated supercomputers. *See* TURBULENT FLOW.

Visualization. The final step is to visualize the results of the simulation. Powerful graphics workstations and visualization software permit generation of velocity vectors, pressure and velocity contours, streamline generation, calculation of secondary quantities (such as vorticity), and animation of unsteady calculations. Despite the sophisticated hardware, visualization of three-dimensional and unsteady flows is still particularly difficult. Moreover, many advanced visualization techniques tend to be qualitative, and the most valuable visualization often consists of simple x-y plots comparing the numerical solution to theory or experimental data. See COMPUTER GRAPHICS.

Applications. Computational fluid dynamics has wide applicability in such areas as aerodynamics, hydraulics, environmental fluid dynamics, and atmospheric and oceanic dynamics, with length and time scales of the physical processes ranging from millimeters and seconds to kilometers and years. Vehicle aerodynamics and hydrodynamics, which have provided much of the impetus in the development of computational fluid dynamics, are primarily concerned with the flow around aircraft, automobiles, and ships. See AERODYNAMIC FORCE; AERODYNAMICS; CLIMATE MODELING; DYNAMIC METEOROLOGY; FLUID FLOW; FLUID-FLOW PRINCIPLES; HYDRAULICS; HYDRODYNAMICS; OCEAN CIRCULATION; RIVER; SIMULATION; WATER POLLUTION. [E.Pa.; F.Ste.]

Computer A device that receives, processes, and presents information. The two basic types of computers are analog and digital. Although generally not regarded as such, the most prevalent computer is the simple mechanical analog computer, in which gears, levers, ratchets, and pawls perform mathematical operations—for example, the speedometer and the watt-hour meter (used to measure accumulated electrical usage). The general public has become much more aware of the digital computer with the rapid proliferation of the hand-held calculator and a large variety of intelligent devices, ranging from typewriters to washing machines.

An analog computer uses inputs that are proportional to the instantaneous value of variable quantities, combines these inputs in a predetermined way, and produces outputs that are a continuously varying function of the inputs and the processing. These outputs are then displayed or connected to another device to cause action, as in the case of a speed governor or other control device. Small electronic analog computers are frequently used as components in control systems. If the analog computer is built solely for one purpose, it is termed a special-purpose electronic analog computer. General-purpose electronic analog computers are used by scientists and engineers for analyzing dynamic problems. See ANALOG COMPUTER.

In contrast, a digital computer uses symbolic representations of its variables. The arithmetic unit is constructed to follow the rules of one (or more) number systems. Further, the digital computer uses individual discrete states to represent the digits of the number system chosen. A digital computer works with a symbolic representation of variables; consequently, it can easily store and manipulate numbers, letters, or graphical information represented by a symbolic code. Through the use of the stored program, the digital computer achieves a degree of flexibility unequaled by any other computing or data-processing device.

The most prevalent special-purpose mechanical digital computers have been the supermarket cash register, the office adding machine, and the desk calculator. Each of these is being widely replaced by electronic devices allowing much greater increased speed. For example, in suitably equipped retail stores, the Universal Product Code (UPC) is scanned by a light-sensitive device, bringing information about each product into the point-of-sale (POS) terminal that has replaced the mechanical cash register.

Most digital computers are occupied with performing applications related to bookkeeping, accounting, engineering design, or test data reduction. On the other hand, applications previously considered esoteric are beginning to lead to industrial applications, such as the use of robots on manufacturing assembly lines. Another area of major societal impact is the growing application of digital computers to word processing, including the entire concept of office automation. See ARTIFICIAL INTELLIGENCE; DATA-PROCESSING SYSTEMS; DIGITAL COMPUTER; ROBOTICS; WORD PROCESSING. [B.A.G.]

Computer-aided design and manufacturing The application of digital computers in engineering design and production. Computer-aided design (CAD) refers to the use of computers in converting the initial idea for a product into a detailed engineering design. The evolution of a design typically involves the creation of geometric models of the product, which can be manipulated, analyzed, and refined. In CAD, computer graphics replace the sketches and engineering drawings traditionally used to visualize products and communicate design information. See COMPUTER GRAPHICS.

Engineers also use computer programs to estimate the performance and cost of design prototypes and to calculate the optimal values for design parameters. These programs supplement and extend traditional hand calculations and physical tests. When combined with CAD, these automated analysis and optimization capabilities are called computer-aided engineering (CAE). See COMPUTER-AIDED ENGINEERING; OPTIMIZATION.

Computer-aided manufacturing (CAM) refers to the use of computers in converting engineering designs into finished products. Production requires the creation of process plans and production schedules, which explain how the product will be made, what resources will be required, and when and where these resources will be deployed. Production also requires the control and coordination of the necessary physical processes, equipment, materials, and labor. In CAM, computers assist managers, manufacturing engineers, and production workers by automating many production tasks. Computers help to develop process plans, order and track materials, and monitor production schedules. They also help to control the machines, industrial robots, test equipment, and systems which move and store materials in the factory.

CAD/CAM is more expensive than traditional manufacturing technology and requires a more highly trained work force. Wisely applied, however, CAD/CAM can improve productivity, product quality, and profitability. Computers can eliminate redundant design and production tasks, improve the efficiency of workers, increase the utilization of equipment, reduce inventories, waste, and scrap, decrease the time required to design and make a product, and improve the ability of the factory to produce different products. Today most manufacturers employ CAM/CAM to varying degrees. See PRODUCTIVITY.

The fact that CAD, CAE, and CAM work best together has led to the breakdown of many of the traditional barriers between functional and manufacturing units. The goal of computer-integrated manufacturing (CIM) is a database, created and maintained on a factory-wide computer network, that will be used for design, analysis, optimization, process planning, production scheduling, robot programming, materials handling, inventory control, maintenance, and marketing. Although many technical and managerial obstacles must be overcome, computer-integrated manufacturing appears to be the future of CAD/CAM. See COMPUTER-INTEGRATED MANUFACTURING; DATABASE MANAGEMENT SYSTEMS; FLEXIBLE MANUFACTURING SYSTEM; MATERIALS HANDLING; ROBOTICS. [K.P.W.]

Computer-aided engineering The use of computer-based tools to help solve engineering problems. Computer-aided engineering can be considered as a set of four interrelated problem-solving aids: computer data bases and communications; computer graphics and modeling; computer simulations and analyses; and data acquisition and control of physical prototypes and production processes. The applications of computer-aided engineering range up to final production, at which time similar computer applications are usually described as computer-aided manufacturing (CAM) or computer-integrated manufacturing (CIM). *See* COMPUTER-AIDED DESIGN AND MANUFACTURING; COMPUTER-INTEGRATED MANUFACTURING.

The complexity of many products outstrips the capacity of any single engineer to remember all the relevant information at any point in design. Access to the collective experience of the engineering team, as well as the experience embodied in references and standards, requires efficient organization of a project team, which may include hundreds of engineers. Engineering information that companies have organized into computer data bases include test results, warranty reports, drawings, analysis results, cost and materials data, codes, design standards, component specifications, and documentation. *See* DATABASE MANAGEMENT SYSTEMS.

The language of design engineers is in the form of graphical renderings, that is, sketches, drawings, plots, histograms, maps, vectors, schematics, graphs, and projections. Computer graphic systems, which have made it possible for engineers to create and manipulate graphic representations much faster, are almost invariably tied to an alphanumeric data-base program, which names drawings and parts for retrieval. A computer-aided drawing system might be considered to be a power tool to replace manual drafting. *See* COMPUTER GRAPHICS; DRAFTING; ENGINEERING DRAWING.

Engineers are trained to model real-world systems with more or less analogous equations. This use of simulations can be thought of as building a mathematical prototype, which is often faster and cheaper to create than building and testing the real thing. An added advantage of mathematical simulation is that it forces a better understanding of the system under investigation. Furthermore, it is relatively easy to try many variations in search of an optimal solution. Finite element analysis techniques are an excellent example of the analytical applications of computer-aided engineering. The technique can be applied to any field problem, such as heat transfer, electrodynamics, and the behavior of structures. *See* FINITE ELEMENT METHOD; NUMERICAL ANALYSIS; SIMULATION.

Computer tools have emerged to build prototypes faster through numerical control and computer-aided manufacturing techniques and to gain more insight into product performance through computer-aided testing techniques. In addition to general-purpose data acquisition and control programs, specialized testing techniques are well developed in the electronics industry and in the area of structural and vibrations testing.

Computer-assisted engineering applications use hardware similar to that in data processing, with input devices, central processors, main and auxiliary memory, and output devices. Some notable developments for engineering applications include very high-resolution graphic displays, a variety of easy-to-use graphic input devices, and specialized output devices such as plotters and machine controls. Robots might also be considered as specialized computer peripherals. Another significant advancement has been the development of superminicomputers, which approach the performance of mainframe computers but at a much lower cost. Similarly, individual 32-bit microcomputers have been designed that approach the power of superminicomputers, but in desk-top work stations which can be networked to share resources and information. The result is often a trend away from highly centralized computing to decentralized computing. *See* COMPUTER SYSTEMS ARCHITECTURE.

Programs that are used to solve engineering problems are often written in structured programming languages such as Ada and C, making them easier to maintain and upgrade. Attention has also been given to simplifying and easing user interactions with user-friendly help messages, step-by-step command menus, and so on. To the extent that artificial-intelligence programming methods become feasible, engineers can expect that working with a computer system will become increasingly like a dialog with a colleague. *See* ARTIFICIAL INTELLIGENCE; NONLINEAR PROGRAMMING; PROGRAMMING LANGUAGES; SOFTWARE ENGINEERING. [P.A.M.]

Computer graphics Communication between people and computers using imagery as opposed to text and numbers. Computer graphics enables people to channel communication with computers into pictures, thus employing the highly evolved human ability to recognize patterns. Trends portrayed in a graph or the relationship between objects shown on a map are absorbed much faster and more firmly than the same information presented in numbers or words.

Computers can generate all types of images, from simple line drawings to realistically shaded pictures that resemble photographs. These can be totally synthesized by the computer or can be enhanced or manipulated natural images. *See* IMAGE PROCESSING.

Graphics-display hardware. Computers must be equipped with special hardware in order to display images. There are two types of graphics: vector and raster.

Vector images are specified as a series of lines. The computer stores a list of coordinates for the starting and end points of each line, called a display list. To display the image, the computer passes through the list, redrawing each line on a cathode-ray tube from 10 to 30 times a second, depending on the number of vectors in the image. *See* CATHODE-RAY TUBE; STORAGE TUBE.

Raster images are continuous smoothly shaded pictures, composed of rows of square or rectangular picture elements, called pixels. The quality of a raster image is determined by its spatial and contrast resolution. The spatial resolution, or sharpness, is determined by the number of rows and columns of pixels. The contrast resolution refers to the number of different shades of gray (in a monochrome image) or different colors that a pixel can assume. A raster image is stored as a bit map in a special, high-speed video memory within the computer. The memory contents are read out and passed to the video display 30 times a second, fast enough to create a flicker-free image. Raster graphics has largely supplanted the line drawings of vector graphics because it offers smoothly shaded images in continuous tones. *See* INTEGRATED CIRCUIT; SEMICONDUCTOR MEMORIES.

The most common display device for raster images is the cathode-ray tube. Plasma panels and liquid-crystal displays are commonly used in lap-top computers. *See* ELECTRONIC DISPLAY; LIQUID CRYSTALS.

Hardcopy output. Computer-generated images can be recorded on paper in different ways. Vector drawings can guide pen plotters, in which electric motors, under computer control, move pens over paper. Extremely precise plotters are suitable for the creation of engineering drawings, integrated circuit masks, and maps. *See* DRAFTING.

Raster images can be recorded on film by cameras that photograph images displayed on a small, extremely high-resolution monitor. Raster images can also be printed by such devices as electrostatic plotters and ink-jet and dot matrix printers. Electrostatic plotters, which use a technology similar to that found in photocopy machines, are faster than pen plotters and are approaching the same degree of precision. *See* DATA-PROCESSING SYSTEMS; PHOTOCOPYING PROCESSES.

Input devices. Line drawings can be entered into a computer by tracing them with a special stylus or cursor over a tablet embedded with a fine grid of wires. Entire images can be input at a stroke by means of such devices as scanning digitizers and charge-coupled-device video cameras.

Interactive graphics. Interactive graphics enables users to create, modify, and respond to graphical objects in real time. Graphical commands are mediated by a pointing or locator device, such as a mouse, trackball, joystick, or stylus with tablet. These devices convert the physical movement of the user's hand into the movement of a cursor or an object on the computer display screen. The mouse, which is the most widely used device, is a hand-held box with a ball on the underside coupled to rollers that respond to movement in two orthogonal directions. The stylus with tablet is used for precise sketching, being easier to control for fine movements than the mouse.

Image synthesis. A collection of surfaces can be used to model three-dimensional objects. This is called geometric or shape modeling. One commonly used method is to create a mesh of polygons, a connected set of flat surfaces. Another technique approximates the surface of objects with a set of patches, each of which is described by a set of three bicubic equations. Unlike geometric modeling, solid modeling uses three-dimensional solid primitives (cubes, cones, spheres, ellipsoids, and cylinders) for representing objects.

The creation of a geometric or solid model is only the first step toward the realistic rendering of an object. The first improvement that can be made is to provide some depth cuing by eliminating all edges or surfaces that would be obscured by visible surfaces. The next step is to shade the visible surfaces, taking into account the light sources, surface properties, and positions and orientations of the surfaces. The computation of shading tends to be very demanding of computer time, and many graphics-display systems incorporate special hardware for shading.

Further refinements include the modeling of shadows, the refraction of light by transparent objects, and the interplay between objects and their environments. The most realistic and costly technique of modeling shadows, reflections, and refractions is ray tracing, in which a separate light ray for each pixel is traced backward from the viewpoint through the pixel to its origin.

Models of nature. Realism in computer image synthesis can depend on the convincing generation of natural structures. The models are usually grounded in physical or biological concepts, but simplify or alter reality for computational convenience. Clouds, plants, and mountain ranges have been simulated by using fractals. *See* FRACTALS. [O.B.R.S.]

Computer-integrated manufacturing

A system in which individual engineering, production, and marketing and support functions of a manufacturing enterprise are organized into a computer-integrated system. Functional areas such as design, analysis, planning, purchasing, cost accounting, inventory control, and distribution are linked through the computer with factory floor functions such as materials handling and management, providing direct control and monitoring of all process operations.

Computer-integrated manufacturing (CIM) may be viewed as the successor technology which links computer-aided design (CAD), computer-aided manufacturing (CAM), robotics, numerically controlled machine tools (NCMT), automatic storage and retrieval systems (AS/RS), flexible manufacturing systems (FMS), and other computer-based manufacturing technology. Computer-integrated manufacturing is also known as integrated computer-aided manufacturing (ICAM). Autofacturing includes computer-integrated manufacturing, but also includes conventional machinery, human operators, and their relationships within a total system. *See* COMPUTER-AIDED DESIGN AND MANUFACTURING; ROBOTICS; TOOLING.

Optimization of the CIM factory is based on specific criteria. By considering efficiency and flexibility and effectiveness, the CIM factory enables economy of scope. In economy of scope, flexibility is at least as important as efficiency, because the factory is seen to be an ever-changing dynamic environment which must always respond quickly to the needs of the marketplace, if not actually lead the need. The traditional factory with both rigid transfer-line technology and a rigid organizational structure cannot satisfy this objective; it strives for maximum volume of each product. In the ideal case, the CIM factory provides for profit with an order quantity of one, that is, when no two products are identical.

The CIM factory is made up of a parts fabrication center, a component assembly center, and a product assembly center. Centers are subdivided into work cells, cells into stations, and stations into processes. Processes comprise the basic transformations of raw materials into parts which will be assembled into products. In order for the factory to achieve maximum efficiency, raw material must come into the factory and move smoothly and continuously through the factory to emerge as a product. No part must ever be standing; each part is either being worked on or is on its way to the next work station. In conventional factories, a typical part is worked on about 5% of the time. There is no storage of work in process and no warehousing in the CIM factory of the future. To accomplish this objective, parts movement is handled by robots of various types. *See* AUTOMATION; ROBOTICS. [D.Wis.]

Computer peripheral devices

Any device connected internally or externally to a computer and used in the transfer of data. A personal computer or workstation processes information and, strictly speaking, that is all the computer does. Data (unprocessed information) must get into the computer, and the processed information must get out. Entering and displaying information is carried out on a wide variety of accessory devices called peripherals, also known as input/output (I/O) devices. Some peripherals, such as keyboards, are only input devices; other peripherals, such as printers, are only output devices; and some are both. *See* DIGITAL COMPUTER; MICROCOMPUTER.

The monitor is the device on which images produced by the computer operator or generated by the program are displayed on a cathode-ray tube (CRT). Electron guns—one in a monochrome monitor, three in a color monitor—irradiate phosphors on the inside of the vacuum tube, causing them to glow. The flat-panel displays on most portable computers, known as liquid-crystal displays (LCDs), use two polarizing filters with liquid crystals between them to produce the image. *See* CATHODE-RAY TUBE; ELECTRONIC DISPLAY; LIQUID CRYSTALS.

The computer keyboard, based on the typewriter keyboard, contains keys for entering letters, numbers, and punctuation marks, as well as keys to change the meaning of other keys. The function keys perform tasks that vary from program to program. *See* TYPEWRITER.

The mouse is a device that is rolled on the desktop to move the cursor on the screen. A ball on the bottom of the mouse translates the device's movements to sensors within the mouse and then through the connecting port to the computer. There are also mice that substitute optical devices for mechanical balls, and mice that use infrared rather than physical connections.

The trackball is essentially an upside-down mouse, with the ball that is used to move the cursor located on the top rather than on the bottom.

The joystick is a pointing device used principally for games.

The light pen performs the same functions as a mouse or trackball, but it is held up to the screen, where its sensors detect the presence of pixels and send a signal through a cable to the computer.

The graphics, or digitizing, tablet is a pad with electronics beneath the surface which is drawn upon with a pointed device, called a stylus. The shapes drawn appear on the monitor's screen.

The most common input, or storage, device in personal computers or workstations is a hard disk drive, a stack of magnetized platters on which information is stored by heads generating an electrical current to represent either 1 or 0 in the binary number system. The device is called hard because the platters are inflexible, and is called a drive because it spins at 3600 revolutions or more a minute, within a sealed case. Diskettes, made of flexible film like that used in recording tape, are usually stored within a hard shell and are spun by their drives at about 360 revolutions per minute. *See* COMPUTER STORAGE TECHNOLOGY.

Data can also be stored and retrieved with light, the light of a laser beam reading a pattern of pits on an optical disk. The most familiar type of optical disk is the CD-ROM (compact disk-read only memory). Another kind of optical disk is the WORM (write once read many times). *See* COMPACT DISK; MULTIMEDIA TECHNOLOGY; OPTICAL RECORDING.

As a consequence of their greater efficiency and speed, disk drives have quickly replaced tape drives as the primary means of data and program storage. Tape drives are still in use for backup storage, copying the contents of a hard disk as insurance against mechanical failure or human error. *See* MAGNETIC RECORDING.

The scanner converts an image of something outside the computer, such as text, a drawing, or a photograph, into a digital image that it sends into the computer for display or further processing. The image is viewed as a graphics image, not a text image, so it can be altered with a graphics program but cannot be edited with a word-processing program, unless the scanner is part of a character-recognition system. To digitize photographs, a scanner may dither the image (put the dots a varying amount of space apart), or use the tagged image file format (TIFF), storing the image in 16 gray values. Some scanners can use standard video cameras to capture images for the computer. *See* CHARACTER RECOGNITION; IMAGE PROCESSING.

The printer puts text or other images produced with a computer onto paper or other surfaces. Printers are either impact or nonimpact devices.

Daisy-wheel or thimble printers are so called from the shape of the elements bearing raised images of the characters. Their speed, perhaps 30 characters per second, is now considered unacceptably slow. Dot-matrix printers produce their images by striking a series of wire pins, typically, 9, 18, or 24, through the ribbon in the pattern necessary to form the letter, number, line, or other character.

Ink-jet printers carry their ink in a well, where it is turned into a mist by heat or vibration and sprayed through tiny holes to form the pattern of the character on paper. Laser printers are similar to photocopying machines. The quality of laser-printer output is the highest generally available. *See* PHOTOCOPYING PROCESSES; PRINTING.

The modem connects one computer to another, ordinarily through the telephone lines, to exchange information. *See* MODEM. [L.R.S.]

Computer security

Computer security The process of ensuring confidentiality, integrity, and availability of computers, their programs, hardware devices, and data. Lack of security results from a failure of one of these three properties. The lack of confidentiality is unauthorized disclosure of data or unauthorized access to a computing system or a program. A failure of integrity results from unauthorized modification of data or damage to a computing system or program. A lack of availability of computing resources results in what is called denial of service.

An act or event that has the potential to cause a failure of computer security is called a threat. Some threats are effectively deflected by countermeasures called controls. Kinds of controls are physical, administrative, logical, cryptographic, legal, and ethical. Threats that are not countered by controls are called vulnerabilities.

Encryption. Encryption is a very effective technique for preserving the secrecy of computer data, and in some cases it can also be employed to ensure integrity and availability. An encrypted message is converted to a form presumed unrecognizable to unauthorized individuals. The principal advantage of encryption is that it renders interception useless. *See* CRYPTOGRAPHY.

Access control. Computer security implies that access be limited to authorized users. Therefore, techniques are required to control access and to securely identify users. Access controls are typically logical controls designed into the hardware and software of a computing system. Identification is accomplished both under program control and by using physical controls.

Typically, access within a computing system is limited by an access control matrix administered by the operating system or a processing program. All users are represented as subjects by programs executing on behalf of the users; the resources, called the objects of a computing system, consist of files, programs, devices, and other items to which users' accesses are to be controlled. The matrix specifies for each subject the objects that can be accessed and the kinds of access that are allowed.

Access control as described above relates to individual permissions. Typically, such access is called discretionary access control because the control is applied at the discretion of the object's owner or someone else with permission. With a second type of access control, called mandatory access control, each object in the system is assigned a sensitivity level, which is a rating of how serious would be the consequences if the object were lost, modified, or disclosed, and each subject is assigned a level of trust.

Access control is not necessarily as direct as just described. Unauthorized access can occur through a covert channel. One process can signal something to another by opening and closing files, creating records, causing a device to be busy, or changing the size of an object. All of these are acceptable actions, and so their use for covert communication is essentially impossible to detect, let alone prevent.

Security of programs. Computer programs are the first line of defense in computer security, since programs provide logical controls. Programs, however, are subject to error, which can affect computer security.

A computer program is correct if it meets the requirements for which it was designed. A program is complete if it meets all requirements. Finally, a program is exact if it performs only those operations specified by requirements.

Simple programmer errors are the cause of most program failures. Fortunately, the quality of software produced under rigorous design and production standards is likely to be quite high. However, a programmer who intends to create a faulty program can do so, in spite of development controls. *See* SOFTWARE ENGINEERING.

A salami attack is a method in which an accounting program reduces some accounts by a small amount, while increasing one other account by the sum of the amounts subtracted. The amount reduced is expected to be insignificant; yet, the net amount summed over all accounts is much larger.

Some programs have intentional trapdoors, additional undocumented entry points. If these trapdoors remain in operational systems, they can be used illicitly by the programmer or discovered accidentally by others.

A Trojan horse is an intentional program error by which a program performs some function in addition to its advertised use. For example, a program that ostensibly produces a formatted listing of stored files may write copies of those files on a second device to which a malicious programmer has access.

A program virus is a particular type of Trojan horse that is self-replicating. In addition to performing some illicit act, the program creates a copy of itself which it then embeds in other, innocent programs. Each time the innocent program is run, the attached virus code is activated as well; the virus can then replicate and spread itself to other, uninfected programs. Trojan horses and viruses can cause serious harm to computing resources, and there is no known feasible countermeasure to halt or even detect their presence.

Security of operating systems. Operating systems are the heart of computer security enforcement. They perform most access control mediation, most identification and authentication, and most assurance of data and program integrity and continuity of service.

Operating systems structured specifically for security are built in a kernelized manner, embodying the reference monitor concept. A kernelized operating system is designed in layers. The innermost layer provides direct access to the hardware facilities of the computing system and exports very primitive abstract objects to the next layer. Each successive layer builds more complex objects and exports them to the next layer. The reference monitor is effectively a gate between subjects and objects. *See* OPERATING SYSTEM.

Security of databases. Integrity is a much more encompassing issue for databases than for general applications programs, because of the shared nature of the data. Integrity has many interpretations, such as assurance that data are not inadvertently overwritten, lost, or scrambled; that data are changed only by authorized individuals; that when authorized individuals change data, they do so correctly; that if several people access data at a time, their uses will not conflict; and that if data are somehow damaged, they can be recovered.

Database systems are especially prone to inference and aggregation. Through inference, a user may be able to derive a sensitive or prohibited piece of information by deduction from nonsensitive results without accessing the sensitive information itself. Aggregation is the ability of two or more separate data items to be more (or less) sensitive together than separately. Various statistical methods make it very difficult to prevent inference, and aggregation is also extremely difficult to prevent, since users can access great volumes of data from a database over long periods of time and then correlate the data independently. *See* DATABASE MANAGEMENT SYSTEMS.

Security of networks. As computing needs expand, users interconnect computers. Network connectivity, however, increases the security risks in computing. Whereas users of one machine are protected by some physical controls, with network access, a user can easily be thousands of miles from the actual computer.

Furthermore, message routing may involve many intermediate machines, called hosts, each of which is a possible point where the message can be modified or deleted, or a new message fabricated. A serious threat is the possibility of one machine's impersonating another on a network in order to be able to intercept communications passing through the impersonated machine. Another impediment to security is the use of exposed communication media, typically telephone lines.

Finally, a network is a larger, more complex structure than a single computer. The reliability of single resources or vital links has a direct effect on the availability of the network resources, and hence, on the many users of the network.

The principal method for improving security of communications within a network is encryption. Messages can be encrypted link or end to end. With link encryption, the message is decrypted at each intermediate host and reencrypted before being transmitted to the next host. End-to-end encryption is applied by the originator of a message and removed only by the ultimate recipient. *See* LOCAL-AREA NETWORK; WIDE-AREA NETWORK. [C.P.P.]

Computer storage technology The techniques, equipment, and organization for providing the memory capability required by computers in order to store instructions and data for processing at high electronic speeds. In early computer systems, memory technology was very limited in speed and high in cost. Since the 1970s, the advent of high-speed random-access memory (RAM) chips has significantly reduced the cost of computer main memory by more than two orders of magnitude. Chips are no larger than 1/4 in. (6 mm) square and contain all the essential electronics to store thousands of bits of data or instructions. Traditionally, computer storage has consisted of three or more types of memory hierarchy storage devices (for example, magnetic core memories, disks, and magnetic tape units).

Memory hierarchy. Memory hierarchy refers to the different types of memory devices and equipment configured into an operational computer system to provide the necessary attributes of storage capacity, speed, access time, and cost to make a cost-effective, practical system. The fastest-access memory in any hierarchy is the main memory in the computer. In most computers manufactured after the late 1970s, RAM chips are used because of their high speed and low cost. Magnetic core memories were the predominant main memory technology in the 1960s and early 1970s prior to the RAM chip. The secondary storage hierarchy usually consists of disks (both movable-arm and head-per-track). Significant density improvements have been achieved in disk technology so that disk capacity has doubled every 3–4 years. Between main memory and secondary memory hierarchy levels, however, there has always been what is called the memory gap. It is characterized by a wide gap in both performance and cost. The last, or bottom, level (sometimes called the tertiary level) of storage hierarchy is made up of magnetic tape transports and mass storage tape systems.

Memory organization. The efficient combination of memory devices from the various hierarchy levels must be integrated with the central processor and input-output equipment, making this the real challenge to successful computer design. The resulting system should operate at the speed of the fastest element, provide the bulk of its capacity at the cost of its least expensive element, and provide sufficiently short access time to retain these attributes in its application environment. Another key ingredient to a successful computer system is its operating system (that is, software) that allows the user to execute jobs on the hardware efficiently. Operating systems are available which achieve this objective to a reasonably high level of optimization.

The computer system hardware and the operating system software must work integrally as one resource. In many computer systems, the manufacturer provides what is called a virtual memory system. It gives each programmer automatic access to the total capacity of the memory hierarchy without specifically moving data up and down the hierarchy and to and from the central processing unit. *See* DIGITAL COMPUTER; MULTIACCESS COMPUTER.

Main memory. There are two basic types of RAMs, static and dynamic. The differences are significant. Dynamic RAMs are those which require their contents to be refreshed periodically. They require supplementary circuits to do the refreshing and to assure that conflicts do not occur between refreshing and normal read/write operations. Even with those extra circuits, dynamic RAMs still require fewer on-chip components per bit than do static RAMs (which do not require refreshing). The fact that they require fewer components makes it possible to achieve higher densities with dynamic RAMs than are possible with static RAMS. These higher densities also lead to lower costs per bit. Static RAMs are easier to design, and compete well in applications in which less memory is to be provided, since their higher cost then becomes less important.

There is another trade-off to be made with semiconductor random-access memories in addition to the choice between static and dynamic types, namely that between MOS and bipolar chips. Bipolar devices are faster, but have not yet achieved the higher densities (and hence the lower costs) of MOS.

Microcomputers have evolved their own special set of semiconductor memory chips to suit their application needs. Whereas large, medium-size, and "mini" computers primarily use only RAMs that have read/write capability, microcomputers have found significant use for read-only memory (ROM) chips and programmable read-only memory (PROM) chips. Because microcomputers are typically used in dedicated applications and not in general-purpose programming installations, their data and program storage can be separately allocated between RAMs and ROMs. ROMs are cheaper and provide protection, since the contents are fixed or hard-wired. During microcomputer program development, PROMs are used because they can be rewritten to modify the contents. Typically, PROMs are cleared by putting the PROM chip under an ultraviolet light to zero out its contents and then using a PROM programmer to write in the new bit pattern for the modified program. After the program has completed final test and acceptance, the final bit pattern can be put into ROMs (using a chip mask) for making production units. There are also chip devices called electrically alterable programmable read-only memories (EPROMs); they have an electrical switch to allow one to rewrite new contents into them.

Memory gap technologies. The wide gulf between fast, expensive main memory devices and slow, inexpensive secondary memories can be seen in the following memory gap performance/cost ranges: access time 1 microsecond–30 milliseconds; capacity 10^7–10^9 bits; price per bit 0.1–0.001¢. The memory gap for access time is bounded on one side by a 1-μs typical computer main memory cycle time and on the other by the 30-ms typical disk-drive access time. Capacity is bounded by a reasonable main-frame memory capacity of about 1 megabyte and by the typical disk capacity of 100 megabytes. The price is then determined by these existing technologies, that is, 0.1¢ per bit for RAMs and 0.001¢ per bit for disks.

The semiconductor RAM and disk technology continues to improve in performance and price, narrowing the gap. It is unlikely, however, that the proved technologies will spread far enough to completely close it. Charge-coupled devices (CCDs) of 64K bits and 256K bits per chip have been used integrally in a few commercial computer systems as bulk store between main memory and secondary disks. Some disk replacement units have been designed by using CCDs. Bubble memory units are on the slow-access-time side of the memory gap. Bubble memories are most promising as low-capacity disk replacements and for special applications. *See* CHARGE-COUPLED DEVICES; MAGNETIC BUBBLE MEMORY.

Secondary memories. Secondary memories consist of disk units which vary in capacity from the small 250-kilobyte floppy disks (used with microcomputers), through 5-megabyte cartridge disks (used with minicomputers), to large-capacity (up to 600-megabyte) disks used with mini, medium, and large-scale computers. Disk memories are also characterized by access times in the millisecond region (versus access time of 1μs to hundreds of nanoseconds for RAMS).

All the disk units have a movable arm which can be positioned to any track. This device minimizes the cost of the unit, because there are less read/write head electronics to be amortized over the large capacity, at the expense of longer access times to position the arm. Disk memories store information along tracks or cylinders traced in a circular fashion by a transducer which is held in a fixed position during the write or read process. The density of the data is referred to as the number of bits packed along the track in the recording process.

Many significant advances in the capabilities of commercially available disk memory equipment were made during the late 1970s. One of the major trends was toward larger-capacity disks from 100 megabytes to 600 megabytes per disk drive. The majority of technology improvements to provide higher capacities, lower costs, and better reliabilities resulted from what is known as the Winchester technology. Prior to this technology, removable disk packs were established as the most advanced technology for large-capacity disks. The Winchester technology is characterized by nonremovable or sealed disk packs, which inherently provide more reliability due to less handling and make more stringent alignment tolerances practical. The extremely narrow Winchester track has been achieved by a combination of factors. The Winchester head weighs only 0.25 g and floats above the surface. The iron oxide particles on the disk surface have also been magnetically oriented so that both the track and the bits along the track are more precisely defined. Another improvement which characterizes the new technology is a lubricated surface that allows the lightweight head to rest on the disk during start and stop operations. The in-contact start/stop capability of Winchester drives has largely eliminated the "head crash" problems encountered with the earlier technologies.

Central to the advance to increase the disk packing density was the reduction in the flying height of the read/write head. Because flux spreads with distance, reduced separation between the read/write head and the magnetic surface will decrease the area occupied by an information bit. All movable media memories utilize an air-bearing mechanism known as the slider or flying head to space the transducer in close proximity to the relatively moving media. Head-to-medium spacings, commonly up to 250 microinches (6 μm) in 1965, were 50 μin. (1.2 μm) in 1975 and 20 μin. (0.5 μm) in 1980. *See* TRANSDUCER.

The head-per-track disk unit can make available many data tracks on the disk without physically moving the read/write head and thus can provide faster access times. The disk storage medium is nonremovable. These types of disks are called swapping disks because different user programs can be more rapidly exchanged than by using moving-arm disks. One prime application for head-per-track disks is in the communications equipment portion of data-processing systems.

Removable storage cartridges containing single-disk packs are used extensively with minicomputer systems. The price and performance features of cartridge disks compete favorably with the more expensive, larger-capacity disk pack storage equipment.

Floppy disks have become the most widely used secondary memory for microcomputer systems with random-access capability. Floppy disks derive their name from the recording media itself, which is an oxide-coated flexible disk (similar to, but more flexible than, plastic 45-rpm music records) enclosed with a protective envelope. The envelope contains alignment holes for spindle loading, head contact, sector/index detection, and write protect detection. The capacity of this easily transportable medium is 3 megabits laid out on 77 tracks.

Optical or video disks have the potential to provide very inexpensive mass-produced read/write mass memory capability. These rotating devices are used to provide read-only television storage on easily replicable disks. The pressed disk has a spiral groove at the bottom of which are submicrometer-sized depressions. These depressions are read out either through a capacity pickup in a needle-in-the-groove or through a light-beam technique. The storage capacity is equivalent to 10^{10} bits per disk. *See* VIDEO DISK RECORDING.

Magnetic-tape units. In magnetic-tape units the tape is held in intimate contact with the fixed head while in motion, allowing high-density recording. The long access times to find user data on the tape are strictly due to the fact that all intervening data have to be searched until the desired data have been found. This, of course, is not true of rotating disk memories or RAM word addressable main memories. The primary use of tape storage is for seldom-used data files and as backup storage for disk data files. Half-inch (12.7-mm) tape has been the industry standard since it was first used commercially in 1953. Half-inch magnetic-tape drive transports are reel-to-reel recorders with extremely high tape speeds (up to 200 in. or 5 m per second), and fast start, stop, reverse, and rewind times (on the order of 1 ms).

The most frequently used magnetic-tape memory devices for microcomputer systems are cassette and cartridge tape units.

Both provide very-low-cost storage, although their access times are high and their overall throughout performance is not as great as that of floppy disks. Cassette and cartridge units both use quarter-inch-wide (6.35-mm) magnetic tape, but the packing densities used vary considerably between the two types of units.

Mass storage systems. With the gradual acceptance of virtual memory and sophisticated operating systems, a significant operational problem arose with computer systems, particularly the larger-scale installations. The sheer expense and attendant delays and errors of humans storing, mounting, and demounting tape reels at the command of the operating system began to become a problem. Mass-storage facilities are designed to alleviate this problem. All such configurations mechanically extract from a bin, mount on some sort of tape transport, and replace in a bin, following reading or writing, a reel of magnetic tape. In practice, most users have been satisfied with tapes and disks, and have not chosen the middle ground of mass-storage systems. *See* COMPUTER; DATA-PROCESSING SYSTEMS; DIGITAL COMPUTER. [D.J.T.]

Computer systems architecture The discipline that defines the conceptual structure and functional behavior of a computer system. It is analogous to the architecture of a building, determining the overall organization, the attributes of the component parts, and how these parts are combined. It is related to, but different from, computer implementation. Architecture consists of those characteristics which affect the design and development of software programs, whereas implementation focuses on those characteristics which determine the relative cost and performance of the system. The architect's main goal has long been to produce a computer that is as fast as possible, within a given set of cost constraints. Over the years, other goals have been added, such as making it easier to run multiple programs concurrently or improving the performance of programs written in higher-level languages.

A computer system consists of four major components (see illustration): storage, processor, peripherals, and input/output (communication). The storage system is used to keep data and programs; the processor is the unit that controls the operation of the system and carries out various computations; the peripheral devices are used to communicate with the outside world; and the input/output system allows the previous components to communicate with one another.

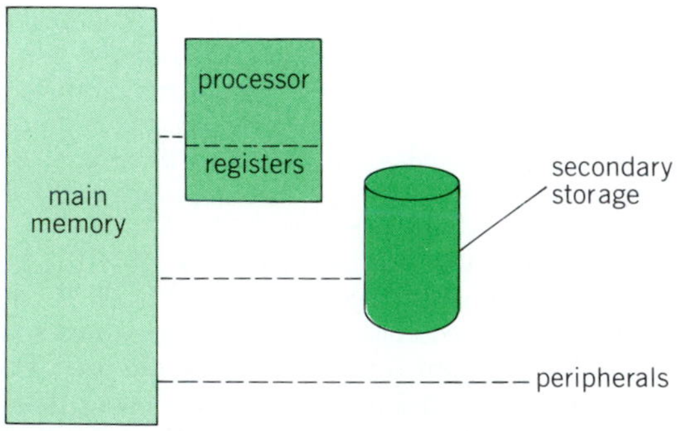

Overview of a computer system. Storage is made up of registers, main memory, and secondary storage. Broken lines indicate input/output.

Storage. The storage or memory of a computer system holds the data that the computer will process and the instructions that indicate what processing is to be done. In a digital computer, these are stored in a form known as binary, which means that each datum or instruction is represented by a series of bits. Bits are conceptually combined into larger units called bytes (usually 8 bits each) and words (usually 8 to 64 bits each). A computer will generally have several different kinds of storage devices, each organized to hold one or more words of data. These types include registers, main memory, and secondary or auxiliary storage. *See* BIT.

Registers are the fastest and most costly storage units in a computer. Normally contained within the processing unit, registers hold data that are involved with the computation currently being performed.

Main memory holds the data to be processed and the instructions that specify what processing is to be done. A major goal of the computer architect is to increase the effective speed and size of a memory system without incurring a large cost penalty. Two prevalent techniques for increasing effective speed are interleaving and cacheing, while virtual memory is a popular way to increase the effective size. Interleaving involves the use of two or more independent memory systems, combined in a way that makes them appear to be a single, faster system. With cacheing, a small, fast memory system contains the most frequently used words from a slower, larger main memory.

Virtual memory is a technique whereby the programmer is given the illusion of a very large main memory, when in fact it has only a modest size. This is achieved by placing the contents of the large, "virtual" memory on a large but slow auxiliary storage device, and bringing portions of it into main memory, as required by the programs, in a way that is transparent to the programmer.

Auxiliary memory (sometimes called secondary storage) is the slowest, lowest-cost, and highest-capacity computer storage area. Programs and data are kept in auxiliary memory when not in immediate use, so that auxiliary memory is essentially a long-term storage medium. There are two basic types of secondary storage: sequential and direct-access. Sequential-access secondary storage devices, of which magnetic tape is the most common, permit data to be accessed in a linear sequence. A direct-access device is one whose data may be accessed in any order. Disks and drums are the most commonly encountered devices of this type.

Memory mapping is one of the most important aspects of modern computer memory designs. In order to understand its function, the concept of an address space must be considered. When a program resides in a computer's main memory, there is a set of memory cells assigned to the program and its data. This is known as the program's logical address space. The computer's physical address space is the set of memory cells actually contained in the main memory. Memory mapping is simply the method by which the computer translates between the computer's logical and physical address spaces. The most straightforward mapping scheme involves use of a bias register. Assignment of a different bias value to each program in memory enables the programs to coexist without interference.

Another strategy for mapping is known as paging. This technique involves dividing both logical and physical address spaces into equal-sized blocks called pages. Mapping is achieved by means of a page map, which can be thought of as a series of bias registers. *See* COMPUTER STORAGE TECHNOLOGY.

Processing. A computer's processor (processing unit) consists of a control unit, which directs the operation of the system, and an arithmetic and logic unit, which performs computational operations. The design of a processing unit involves selection of a register set, communication paths between these registers, and a means of directing and controlling how these operate. Normally, a processor is directed by a program, which consists of a series of instructions that are kept in main memory.

Although the process of decoding and executing instructions is often carried out by logic circuitry, the complexity of instruction sets can lead to very large and cumbersome circuits for this purpose. To alleviate this problem, a technique known as microprogramming was developed. With microprogramming, each instruction is actually a macrocommand that is carried out by a microprogram, written in a microinstruction language. The microinstructions are very simple, directing data to flow between registers, memories, and arithmetic units.

It should be noted that microprogramming has nothing to do with microprocessors. A microprocessor is a processor implemented through a single, highly integrated circuit. *See* MICROPROCESSOR.

Peripherals and communication. A typical computer system includes a variety of peripheral devices such as printers, keyboards, and displays. These devices translate electronic signals into mechanical motion or light (or vice versa) so as to communicate with people.

There are two common approaches for connecting peripherals and secondary storage devices to the rest of the computer: the channel and the bus. A channel is essentially a wire or group of wires between a peripheral device and a memory device. A multiplexed channel allows several devices to be connected to the same wire. A bus is a form of multiplexed channel that can be shared by a large number of devices. The overhead of sharing many devices means that the bus has lower peak performance than a channel; but for a system with many peripherals, the bus is more economical than a large number of channels.

A computer controls the flow of data across buses or channels by means of special instructions and other mechanisms. The simplest scheme is known as program-controlled input/output (I/O). Direct memory access I/O is a technique by which the computer signals the device to transmit a block of data, and the data are transmitted directly to memory, without the processor needing to wait.

Interrupts are a form of signal by which a peripheral device notifies a processor that it has completed transmitting data. This is very helpful in a direct memory access scheme, for the processor cannot always predict in advance how long it will take to transmit a block of data. Architects often design elaborate interrupt schemes to simplify the situation where several peripherals are active simultaneously. *See* COMPUTER; DIGITAL COMPUTER. [D.J.F.]

Computer vision The use of digital computer techniques for extracting, characterizing, and interpreting information in visual images of a three-dimensional world; it is also known as machine vision. Visual sensing technology is receiving increased attention as a means to endow machines with the capability of exhibiting a greater degree of "intelligence" in dealing with their environment. Thus, a robot or other machine that can "see" and "feel" should be easier to train in the performance of complex tasks while requiring less stringent control mechanisms than preprogrammed machines. A sensory, trainable system is also adaptable to a much larger variety of tasks, thus achieving a degree of universality that ultimately translates into lower production and maintenance costs.

The computer vision process may be divided into five principal areas: sensing, segmentation, description, recognition, and interpretation. These categories are suggested to a large extent by the way in which computer vision systems are generally implemented.

Visual information is converted to electrical signals by the use of visual sensors. The most commonly used visual sensors are vidicon cameras and solid-state diode arrays. Illumination of a scene is an important factor affecting the complexity of vision algorithms. When control of the illumination is possible, the lighting system should illuminate the scene so that the complexity of the resulting image is minimized, while the information required for analysis is enhanced. For instance, the so-called structured lighting approach used in some industrial vision systems projects points, stripes, or grids onto the scene. The way in which these features are distorted by the presence of an object simplifies the computer interpretation of the scene. Two examples of this approach are shown in the illustration.

Segmentation is the process that breaks up a sensed scene into its constituent parts or objects. Segmentation algorithms are generally based on one of two basic principles: discontinuity and similarity. The principal approach in the first category is edge detection. The principal approaches in the second category are thresholding and region growing. Thresholding is by far the most widely used approach for segmentation in industrial applications of computer vision.

The description problem in computer vision is one of extracting features from an object for the purpose of recognition. Ideally, these features should be independent of object location and orientation and should contain enough discriminatory information to uniquely identify between objects. Descriptors for computer vision are based primarily on shape and amplitude (for example, intensity) information. Shape descriptors attempt to capture invariant geometrical properties of an object. Approaches for shape analysis and description are generally either global-oriented (region-oriented) or boundary-oriented.

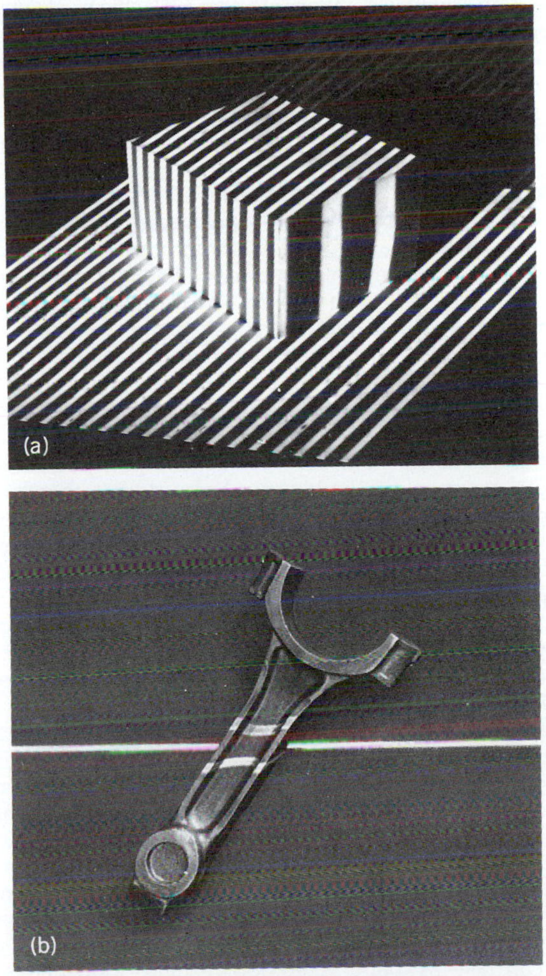

Two examples of scene illumination by structured lighting. (*a*) Illumination of a block by a series of parallel strips of light (*from W. Myers, Industry begins to use visual pattern recognition, Computer, 13(5):21–31, 1980*). (*b*) Illumination of a workpiece by two strips of light which coincide at a background surface (*from F. Rocher and E. Keissling, Methods for analyzing three-dimensional scenes, Proceedings of the 1975 International Conference on Artificial Intelligence, American Association for Artificial Intelligence, pp. 669–673*).

Recognition is basically a labeling process; that is, the function of recognition algorithms is to identify each segmented object in a scene and to assign a label (for example, wrench, seal, or bolt) to that object. Recognition approaches presently in use may be subdivided into two principal categories: decision-theoretic and structural. Decision-theoretic techniques are based on the use of decision (discriminant) functions. These methods deal with patterns on a quantitative basis and largely ignore structural interrelationships among pattern primitives. Structural methods of pattern recognition attempt to describe fundamental relationships among pattern primitives via discrete mathematical modes. Here, the most widely used method is syntactic pattern recognition in which concepts and results from formal language theory provide the basic mechanisms for handling structural descriptions. *See* DECISION THEORY.

Interpretation can be viewed as the process which endows a vision system with a higher level of conception about its environment. Computer vision techniques may be divided into three basic levels of processing: low-, medium-, and high-level vision. Low-level vision techniques attempt to extract "primitive" information from a scene. Medium-level vision refers to procedures which use the results from low-level vision to produce structures that somehow carry more meaning than the elements extracted by the low-level vision process. Medium-level vision processes include edge linking, segmentation, description, and recognition of individual objects. High-level computer vision may be viewed as the process that attempts to emulate cognition. At this level of processing, the present knowledge and understanding of a suitable model is considerably more vague and speculative. *See* ARTIFICIAL INTELLIGENCE; CHARACTER RECOGNITION; COMPUTER GRAPHICS; ROBOTICS. [R.C.Go.]

Computerized tomography An imaging technique in which the x-ray beam of a computerized tomography scanner passes through a human body or an object and is collected with an array of detectors; the beam is rotated to produce the equivalent of a "slice" through the area of interest. The x-ray information collected during the rotation is then used by a computer to reconstruct the "internal structures," and the resulting image is displayed on a television screen. This technique represents a noninvasive way of seeing internal structures, and has in many ways revolutionized diagnostic approaches. In the brain, for example, computerized tomography can readily locate tumors and hemorrhages, thereby providing immediate information for evaluating neurological emergencies. Equally important is the fact that, in a great majority of cases, computerized tomography offers an alternative approach to the more problematic invasive procedures such as pneumoencephalography and angiography. The reduction in scanning time made it possible to do whole-body scans; abdominal tumors, hemorrhages, and infections can be detected without the need of exploratory surgery.

Basically the scanner gantry is composed of an x-ray tube, an array of detectors opposite the tube, and a central aperture in which the person (or object) is placed. X-rays are generated in short bursts, usually lasting 2–3 ms; the x-ray beam contains an "invisible image" of the internal structures. The role of the detectors is to collect this information, which is then fed into a computer. The computer reconstructs the image from the information collected by the detectors. In order to obtain enough information to calculate one image, the newer scanners can take as many as 90,000 readings (300 pulses and 300 detectors). *See* RADIOGRAPHY. [N.-F.B.]

Concave grating A widely used form of reflection-type diffraction grating with which a spectrograph can be formed that has no auxiliary optical parts except a slit and a camera. Being ruled on a concave mirror, this type of grating can both collimate and focus the light that falls upon it. It is made by spacing straight grooves equally along the chord (rather than the arc) of a spherical or paraboloidal mirror surface. Light which passes through a slit and falls on such a grating is dispersed by it into spectra which are in focus on the Rowland circle, a circle drawn tangent to the face of the grating at its midpoint, having a diameter equal to the radius of curvature of the grating surface. *See* DIFFRACTION GRATING. [G.R.H.]

Concentration scales Concentration is a very important property of mixtures, because it defines the quantitative relation of the components. In solutions the concentration is expressed as the mass, volume, or number of moles of solute present in proportion to the amount of solvent or of total solution.

The simplest scale to measure is percentage; hence it is often used for medicinal or household solutions. Weight percent is the number of parts of weight of solute per hundred parts of solution (total). For example, a 10% saline solution contains 10 g of salt in 90 g of water, that is, 100 g total weight. Gaseous mixtures, being difficult to weigh, are often expressed as volume percent. Thus, air is said to contain 78% nitrogen by volume. Solutions of liquids in liquids (say, alcohol in water) may also be expressed in volume percent.

To the chemist, the number of moles of solute is of more significance than the number of grams. The molarity (abbreviated *M*) is the number of moles of solute per liter of total solution. Thus, 12 *M* HCl means that the solution contains 12 formula weights (12 × 36.5), or 438 g, of HCl/liter. (As used here, the mole is an amount of substance whose weight in grams is numerically the same as the molecular weight.)

Molality (abbreviated *m*) relates the number of moles of solute to the weight of solvent rather than to the volume of solution. This scale indicates the number of moles of solute/1000 g of solvent. Thus 34.2 g of sucrose ($C_{12}H_{22}O_{11}$, mol wt 342), if dissolved in 200 g of water, has the concentration of 0.5 mole of sucrose/1000 g of water, and hence is 0.5 *m*.

When it is important to know the reactive capacities of reagents, as in volumetric analysis, the normality scale is used. Normality (abbreviated *N*) is found by multiplying molarity by the number of active units in the formula.

In recent years many chemists have sought to avoid the molarity scale lest it imply that electrolyte solutes exist as molecules rather than ions. The formality scale (abbreviated *F*) represents formula weights per liter. Its values are identical with the molarities of un-ionized solutes.

Many properties of solutions (for example, vapor pressure of one component) are dependent on the ratio of the number of moles of solute to the number of moles of solvent, rather than on the ratios of respective volumes or masses. The mole fraction (abbreviated N_A or X_A for component *A*) is the ratio of the number of moles of solute to the total number of moles of all components. Thus for 16 g of methanol (0.5 mole) dissolved in 18 g of water (1 mole), the mole fraction of methanol is 0.5/1.5, or 1/3; the mole percent is 33.3. For gases the mole percent is identical with the volume percent. *See* GRAM-MOLECULAR WEIGHT; SOLUTION; TITRATION. [A.L.H.]

Conch The name given to many species of the Strombidae, a mainly tropical family of large, imposing, marine gastropods with massive shells. They are found on sand or mud bottoms in the Caribbean and Indo-Pacific, in shallow waters, from just below the littoral zone down to depths of 200 ft (60 m). In all strombids the inner nacreous (or mother-of-pearl) layer of the shell is heavily developed with the sheen of porcelain, and it is often colored a rich pink or orange.

Strombids are all herbivores, some species grazing specifically on epiphytic red seaweeds attached to larger brown algae and to turtle grass (*Thalassia*). The mantle cavity in strombids is in the typical mesogastropod condition, with a single pectinibranch gill and simplified asymmetric pallial and cardiac structures. The flesh of conchs is eaten throughout the Antilles and in several areas of the Pacific, and their shells have long been articles of commerce. *See* GASTROPODA; PROSOBRANCHIA; SNAIL; WHELK. [W.D.R.-H.]

Concrete A versatile engineering material consisting of a hydraulic cementing substance (usually portland cement), aggregate, water, and often controlled amounts of entrained air.

Concrete is initially a plastic, workable mixture which can be molded into a wide variety of shapes. Strength is developed in the hydration reaction between the cement and water. The products, mainly calcium silicates, calcium aluminate, and calcium hydroxide, are relatively insoluble and bind the aggregate in a hardened matrix.

Concrete is considerably stronger in compression than in tension. For structures required to carry only compressive loads, such as massive gravity dams and heavy foundations, no reinforcement is required, and the concrete is consequently called plain concrete. When the structure will involve tensile stresses, steel bars are embedded in the concrete, which is then called reinforced concrete. An extension of the reinforcement principle is prestressing, in which high-tensile-strength steel wires or wire strand embedded while in tension induces a precompression in the concrete so as to increase its resistance to tensile stress. *See* PRESTRESSED CONCRETE; REINFORCED CONCRETE.

Concrete has a wide variety of applications. In buildings it is used in footings, foundations, columns, beams, girders, wall slabs, roof units—in short, all important building elements. Concrete not only meets structural demands but also lends itself readily to architectural treatment. Other important applications of concrete are in road pavements, airport runways, and bridges. Normally, concrete for these structures is plain concrete, containing embedded steel only in order to resist the small volume changes caused by changes in temperature and moisture content. Concrete is widely used in conservation projects such as dams, irrigation canals, water diversion structures, sewage treatment plants, and water distribution pipelines. A considerable amount of concrete is used in the form of masonry units manufactured in equipment utilizing vibration and pressure to achieve compaction. Examples are building blocks and brick. *See* CEMENT. [P.Kl.]

Concretion A loosely defined term used for a sedimentary mineral segregation that may range in size from inches to many feet. Concretions are usually distinguished from the sedimentary matrix enclosing them by a difference in mineralogy, color, hardness, and weathering characteristics. Some concretions show definite sharp boundaries with the matrix, while others have gradational boundaries. Most concretions are composed dominantly of calcium carbonate, with or without an admixture of various amounts of silt, clay, or organic material. Coal balls are calcareous concretions found in or immediately above coal beds. Concretions are normally spherical or ellipsoidal; some are flattened to disklike shapes. Frequently a concretion is dumbbell-shaped, indicating that two separate concretionary centers have grown together. *See* COAL BALLS; SEDIMENTARY ROCKS. [R.Si.]

Concurrent processing The simultaneous execution of several interrelated computer programs. A sequential computer program consists of a series of instructions to be executed one after another. A concurrent program consists of several sequential programs to be executed in parallel. Each of the concurrently executing sequential programs is called a process. Process execution, although concurrent, is usually not independent. Processes may affect each other's behavior through shared data, shared resources, communication, and synchronization.

Concurrent programs can be executed in several ways. Multiprogramming systems have one processing unit and one memory bank. Concurrent process execution is simulated by randomly interleaving instructions of the sequential programs. All processes have access to a common pool of data. In contrast, multiprocessing systems have several processing units and one memory bank. Processes are executed in parallel on the separate processing units while sharing common data. In distributed systems, or computer networks, each process is executed on its own processor with its own memory bank. Interaction between processes occurs by transmission of data from one process to another along a communication channel. *See* DISTRIBUTED SYSTEMS (COMPUTERS); MULTIPROCESSING.

One of the first uses of concurrent processing was in operating systems. If the computer is to support a multiuser environment, the operating system must employ concurrent programming techniques to allow several users to access the computer simultaneously. The operating system should also permit several input/output devices to be used simultaneously, again utilizing concurrent processing. *See* MULTIACCESS COMPUTER; OPERATING SYSTEM.

Concurrent programming is also used when several computers are joined in a network. An airline reservation system is one example of concurrent processing on a distributed network of computers. *See* LOCAL-AREA NETWORK; WIDE-AREA NETWORK.

A simple example of a task that can be performed more efficiently by concurrent processing is a program to calculate the sum of a large list of numbers. Several processes can simultaneously compute the sum of a subset of the list, after which these sums are added to produce the final total.

Concurrent programs can be created explicitly or implicitly. Explicit concurrent programs are written in a programming language designed for specifying processes to be executed concurrently. Implicit concurrent programs are created by a compiler that automatically translates programs written in a sequential programming language into programs with several components to be executed in parallel. *See* DATA-FLOW SYSTEMS; PROGRAMMING LANGUAGES. [J.Wi.]

Concussion A state following injury in which there is temporary functional impairment without physical evidence of damage to the impaired tissues. The term usually refers to cerebral concussion produced by any type of trauma.

From a clinical point of view cerebral concussion is produced by a head injury which causes temporary unconsciousness but with complete recovery within 24 h. This temporary alteration is believed to result from one of several mechanisms. In all of these a sudden acceleration or deceleration appears to be a prerequisite. The sudden movement is thought to cause an unequal shifting of tissues of different specific gravities within the skull, between skull and brain, or between different brain tissues. *See* BRAIN. [E.G.St./N.K.M.]

Condensation reaction One of a class of chemical reactions involving a combination between molecules or between parts of the same molecule. A relatively small molecule such as water or alcohol is often eliminated in the process. The conversions of oxygen to ozone and of diethyl pimelate to 2-carbethoxycyclohexanone are examples, as seen in reactions (1) and (2).

$$3O_2 \rightarrow 2O_3 \qquad (1)$$

$$C_2H_5O_2C(CH_2)_5CO_2C_2H_5 \rightarrow$$

$$\begin{array}{l} CH_2CH_2C{=}O \\ | \qquad\quad\;\; | \\ CH_2CH_2CHCO_2CH_5 \end{array} + C_2H_5OH \qquad (2)$$

The term condensation reaction is used very loosely by chemists. Some authors define the term to include all reactions which result in formation of carbon-carbon bonds. General usage, however, does not classify all reactions resulting in formation of carbon-carbon bonds as condensation reactions. Furthermore, the reaction commonly referred to as condensation polymerization involves formation of carbon-oxygen or carbon-nitrogen bonds. *See* POLYMERIZATION. [W.B.R.]

Condensed-matter physics The fundamental science of very large numbers of strongly interacting particles. It includes the familiar solid and liquid states but covers as well dense plasmas, liquid crystals, glasses, polymers, gels, and so forth. It may be regarded as generalizing the more traditional field of solid-state physics in several directions. Whereas solid-state physics has provided a quantitative and predictive understanding of the structures and properties of well-ordered (crystalline), homogeneous systems in three dimensions under nearly equilibrium conditions, condensed-matter physics also treats disorder in both space and time, and heterogeneous compositions in both one and two, as well as three dimensions. It seeks to provide an understanding of macroscopic physical (for example, electric, magnetic, mechanical, and optical) properties in terms of microscopic interactions and phenomena at the atomic level. This requires a thorough understanding and innovative application of the laws and equations of quantum mechanics. Many of the phenomena in condensed-matter physics, such as superconductivity, magnetism, and superfluidity, cannot be understood on the basis of classical physics alone. *See* AMORPHOUS SOLID; COLLOID; GEL; GLASS; LIQUID CRYSTALS; MAGNETISM; PLASMA PHYSICS; POLYMER; QUANTUM MECHANICS; SUPERCONDUCTIVITY; SUPERFLUIDITY. [P.A.F.]

Condenser (electricity) A device for introducing capacitance into a circuit. In modern usage, the word capacitor replaced the word condenser since nothing is really condensed in this device. Hence, for a discussion of this electric circuit component *see* CAPACITOR. *See also* CAPACITANCE. [R.P.Wi.]

Conditioned reflex A learned response performed by a trained animal to a signal that was previously associated with an event of consequence for that animal. Conditioned reflex (CR) was first used by the Russian physiologist I. P. Pavlov to denote the criterion measure of a behavioral element of learning, that is, a new association between the signal and the consequential event, referred to as the conditioned stimulus (CS) and unconditioned stimulus (US), respectively. In Pavlov's classic experiment, the CS was a bell and the US was sour fluid delivered into the mouth of a dog restrained by a harness; CS was followed by US regardless of the dog's response. After training, the CR is manifested when the dog salivates to the sound of the bell.

Ideally certain conditions must be met to demonstrate the establishment of a CR according to Pavlov's classical conditioning method. Before conditioning, the bell CS should attract the dog's attention or elicit the orienting reflex (OR), but it should not elicit salivation, the response to be conditioned. That response should be specifically and reflexively elicited by the sour US, thus establishing its unlearned or unconditional status. After conditional pairings of the CS and the US, salivation is manifested prior to the delivery of the sour US. Salivation in response to the auditory CS is now a "psychic secretion" or the CR.

Pavlov's methods provide important guidelines for basic research upon brain mechanisms in learning and memory. Classical conditioning is considered a general biological or psychobiological phenomenon which promotes adaptive functioning in a wide variety of physiological systems in various phylogenetic settings.

The CR is the criterion of learning, but the nature and origin of the CR are still at issue. In his original experiment, Pavlov circumvented many ambiguities by his selection of an auditory CS and a salivary CR. His CR resembled the unconditioned reflex (UR), and was clearly differentiated from the dog's natural OR to the auditory CS. Early investigators were guided by the UR in their search for classically conditioned CRs. Although the UR may not be a sure guide in all situations, it is still a powerful analytical device.

Methodological complexities and conceptual confusions arise when an investigator selects a CS which, before training, has a natural tendency to elicit reactions similar to the CR or any reaction which is similar to that elicited by the chosen US. For example, before training, an electric shock US causes a startle UR in a subject; the soft sound of a bell elicits the OR but not the startle UR, and thus the bell seems to be a valid CS and the startle reaction a valid CR. When the bell CS and the electric shock US are paired, the bell quickly elicits the startle response, indicating that a new pathway between the auditory and startle mechanisms has been presumably established in the brain of the subject. But the loud clang of a bell will elicit the startle reflex without training in a second subject, indicating that connection with a low threshold of sensitivity may have existed in the brain of this subject before training. Presumably the loud clang was intense enough to activate this pathway, but the soft CS was not. Furthermore, administration of shock alone to a third subject sensitizes the subject to any sound, so that the subject starts at the soft sound of the bell CS, even though it has never been paired with the shock US. Because there has been no explicit CS-US pairing during training, the third subject is said to be pseudoconditioned. For the same reason, sensitization and pseudoconditioning are often called nonassociative effects.

Nonassociative effects cause some practical and theoretical problems in the interpretation of an apparent CR. Practically, the problems become formidable when the subject is a member of an exotic species, such as a marine invertebrate whose thresholds of sensitivity and response repertoires are unknown. Theoretically, nonassociative effects raise questions about the associative definition of learning, and perhaps any purely behavioral definition of learning. [J.G.]

Conductance For a series circuit the real, or in-phase, part of the complex representation for the admittance Y of a circuit. This is shown in Eq. (1), where B is the imaginary part,

$$Y = G \pm jB \tag{1}$$

called the susceptance. Since $Y = 1/Z = 1/(R + jX)$, where R is the resistance and X is the total reactance $X_L - X_C$, Eq. (2) holds, and conductance G is given by Eq. (3). This is the general expression for conductance and shows that conductance is

$$Y = \frac{R}{R^2 + X^2} - j\frac{X}{R^2 + X^2} \tag{2}$$

$$G = \frac{R}{R^2 + X^2} \tag{3}$$

a function involving both resistance and reactance.

If reactance is negligible, as in a dc circuit, then $G = R/R^2 = 1/R$, or conductance is the reciprocal of resistance. *See* ADMITTANCE. [A.A.W.]

Conducting polymer An organic polymer exhibiting the electronic properties of a metal. Interest in conducting polymers began with the discovery in 1976 that polyacetylene could be doped with electron acceptors and electron donors to conductivity levels approaching those of some metals and comparable to those obtained with organic charge-transfer crystals. The latter is especially important since, compared to the single-crystal charge-transfer complexes, doped polyacetylene is highly disordered and would probably provide a tortuous path for carrier transport. The work on electrochemical doping of polyacetylene led to the application of conducting polymers in rechargeable batteries. The soliton concept was then developed and applied toward an understanding of the experimental results in polyacetylene, and it was soon realized that the strong coupling of electronic excitations to structural distortions (solitons, polarons, bipolarons) is a general feature of these linear conjugated polymers. *See* POLARON; SOLITON.

The discovery that poly(*p*-phenylene) could be doped to conductivity levels quite comparable to those obtained in the polyacetylene system was significant in that it demonstrated that the polyacetylene system was not unique. This paved the way for the discovery of a number of polyaromatic-based conducting polymer systems, including poly(*p*-phenylene sulfide), polypyrrole, polythiophene, and polyquinoline. By changing the aromatic constituents and by inserting bridging atoms or groups between rings, one can alter the electronic bandwidth, change the ionization potential (or electron affinity), or vary the energy gap. Thus a wide range of electronic properties is potentially available in this class of organic polymer conductors. *See* BAND THEORY OF SOLIDS; ELECTRON AFFINITY; IONIZATION POTENTIAL.

Research in conducting polymers is directed toward combining the electronic properties of metals with the materials properties (especially processibility) of plastics. There are a number of potential applications involving the replacement of metals, one of the most attractive being electromagnetic shielding. A promising feature of doped polymers is that the conductivity can be varied in a controlled manner over many orders of magnitude. *See* ELECTROCHEMISTRY; MAGNETIC MATERIALS; MAGNETISM; POLYMER; SOLID-STATE PHYSICS; SUPERCONDUCTIVITY.

[F.Wu.; A.J.He.]

Conduction (electricity) Electrical conduction may be defined as the passage of electric charge. This can occur by a variety of processes. In metals the electric current is carried by free electrons. These are not bound to any particular atom and can wander throughout the metal. In semiconductors (germanium, silicon, and so on) there are a limited number of free electrons and also "holes," which act as positive charges, available to carry current. Aqueous solutions of ionic crystals readily conduct electricity by means of the positive and negative ions present, for example, Na^+ and Cl^- in an ordinary solution of sodium chloride. A strong electric field ionizes gas molecules, and thereby permits a flow of current through the gas in which the ions are the charge carriers. If sufficient ions are formed, there may be a spark. Electric current can flow across a vacuum, for example, in a vacuum tube. The charge carriers are electrons emitted by the filament. *See* ELECTRIC SPARK; ELECTRICAL CONDUCTION IN GASES; ELECTROLYTIC CONDUCTANCE; ELECTRON EMISSION; IONIC CRYSTALS; SEMICONDUCTOR; SUPERCONDUCTIVITY.

[J.W.St.]

Conduction (heat) The flow of thermal energy through a substance from a higher- to a lower-temperature region. Heat conduction occurs by atomic or molecular interactions. Conduction is one of the three basic methods of heat transfer, the other two being convection and radiation. *See* CONVECTION (HEAT); HEAT RADIATION; HEAT TRANSFER.

Steady-state conduction is said to exist when the temperature at all locations in a substance is constant with time, as in the case of heat flow through a uniform wall. Examples of essentially pure transient or periodic heat conduction and simple or complex combinations of the two are encountered in the heat-treating of metals, air conditioning, food processing, and the pouring and curing of large concrete structures. Also, the daily and yearly temperature variations near the surface of the Earth can be predicted reasonably well by assuming a simple sinusoidal temperature variation at the surface and treating the Earth as a semi-infinite solid. The widespread importance of transient heat flow in particular has stimulated the development of a large variety of analytical solutions to many problems. The use of many of these has been facilitated by presentation in graphical form.

For an example of the conduction process, consider a gas such as nitrogen which normally consists of diatomic molecules. The temperature at any location can be interpreted as a quantitative specification of the mean kinetic and potential energy stored in the molecules or atoms at this location. This stored energy will be partly kinetic because of the random translational and rotational velocities of the molecules, partly potential because of internal vibrations, and partly ionic if the temperature (energy) level is high enough to cause dissociation. The flow of energy results from the random travel of high-temperature molecules into low-temperature regions and vice versa. In colliding with molecules in the low-temperature region, the high temperature molecules give up some of their energy. The reverse occurs in the high-temperature region. These processes take place almost instantaneously in infinitesimal distances, the result being a quasi-equilibrium state with energy transfer. The mechanism for energy flow in liquids and solids is similar to that in gases in principle, but different in detail.

[W.H.Gi.]

Conduction band The electronic energy band of a crystalline solid which is partially occupied by electrons. The electrons in this energy band can increase their energies by going to higher energy levels within the band when an electric field is applied to accelerate them or when the temperature of the crystal is raised. These electrons are called conduction electrons, as distinct from the electrons in filled energy bands, which, as a whole, do not contribute to electrical and thermal conduction. In metallic conductors the conduction electrons correspond to the valence electrons (or a portion of the valence electrons) of the constituent atoms. In semiconductors and insulators at sufficiently low temperatures, the conduction band is empty of electrons. Conduction electrons come from thermal excitation of electrons from a lower energy band or from impurity atoms in the crystal. *See* BAND THEORY OF SOLIDS; ELECTRIC INSULATOR; SEMICONDUCTOR; VALENCE BAND.

[H.Y.F.]

Conductivity A measure of the ability of a material to conduct electric current. It is the reciprocal of resistivity. Conductivity is commonly expressed as siemens (mhos) per meter, since the unit of resistivity is the ohm-meter. The conductivity of metallic elements varies inversely with absolute temperature over the normal range of temperatures, but at temperatures approaching absolute zero the imperfections and impurities in the lattice structure of a material make the relationship more complicated.

The conductivity associated with conduction electrons in a semiconductor is known as *n*-type conductivity; that associated with the holes in an impurity semiconductor (equivalent to positive charges) is known as *p*-type conductivity. *See* CONDUCTION (ELECTRICITY); SEMICONDUCTOR.

[J.Mar.]

Conductor (electricity) Metal wires, cables, rods, tubes, and bus-bars used for the purpose of carrying electric current. (The most common forms are wires, cables, and bus-bars.) Although any metal assembly or structure can conduct electricity, the term conductor usually refers to the component parts of the current-carrying circuit or system.

Wires employed as electrical conductors are slender rods or filaments of metal, usually soft and flexible. They may be bare or covered by some form of flexible insulating material. They are usually circular in cross section; for special purposes they may be drawn in square, rectangular, ribbon, or other shapes. Conductors may be solid or stranded, that is, built up by a helical lay or assembly of smaller solid conductors.

Insulated stranded conductors in the larger sizes are called cables. Small, flexible, insulated cables are called cords. Assemblies of two or more insulated wires or cables within a common jacket or sheath are called multiconductor cables.

Bus-bars are rigid, solid conductors and are made in various shapes, including rectangular, rods, tubes, and hollow squares. Bus-bars may be applied as single conductors, one bus-bar per phase, or as multiple conductors, two or more bus-bars per phase. The individual conductors of a multiple-conductor installation are identical.

Most wires, cables, and bus-bars are made from either copper or aluminum. Copper, of all the metals except silver, offers the least resistance to the flow of electric current. Both copper and aluminum may be bent and formed readily and have good flexibility in small sizes and in stranded constructions. Aluminum, because of its higher resistance, has less current-carrying capacity than copper for a given cross-sectional area. However, its low cost and light weight (only 30% that of the same volume of copper) permit wide use of aluminum for bus-bars, transmission lines, and large insulated-cable installations.

For overhead transmission lines where superior strength is required, special conductor constructions are used. Typical of these are aluminum conductors, steel reinforced, a composite construction of electrical-grade aluminum strands surrounding a stranded steel core. Other constructions include stranded, high-strength aluminum alloy and a composite construction of aluminum strands around a stranded high-strength aluminum alloy core.

For extra-high-voltage transmission lines, conductor size is often established by corona performance rather than current-carrying capacity. Thus special "expanded" constructions are used to provide a large circumference without excessive weight. Typical constructions use helical lays of widely spaced aluminum strands around a stranded steel core. The space between the expanding strands is filled with paper twine, and outer layers of conventional aluminum strands are applied.

[H.W.Be.]

Condylarthra A mammalian order of extinct, primitive, hoofed herbivores with five-toed plantigrade to semidigitigrade feet. Condylarths are not far removed from carnivorous ancestors. The astragalus is carnivorelike and known skulls are carnivorelike in outline, although the dentition has been modified for an omnivorous or herbivorous diet. While the teeth are usually low-crowned and superficially piglike, one group possesses selenodont teeth. Condylarths are best represented in the early Cenozoic deposits of Europe and of North and South America, appearing first in the North American Cretaceous. The order is near the ancestry of all living hoofed mammals. Condylarthra is subdivided into seven families: Arctocyonidae, Hyopsodontidae, Phenacodontidae, Didolodontidae, Periptychidae, Meniscotheriidae, and Mesonychidae. *See* CARNIVORA.

[M.C.McK.]

Cone The solid of revolution obtained by revolving a right triangle about one of its shorter sides is called a cone, or more precisely a right circular cone (see illustration). More generally, the term cone is used in solid geometry to describe a solid bounded by a plane and a portion of one nappe of a conical surface. In analytic geometry, however, the term cone refers not to a solid but to a conical surface. This is a surface generated by a straight line which moves so that it always intersects a given plane curve, called the directrix, and passes through a point, called the vertex, not in the plane of the directrix. The generating line in each of its positions is called an element of the cone. The vertex divides the surface into two parts, called nappes.

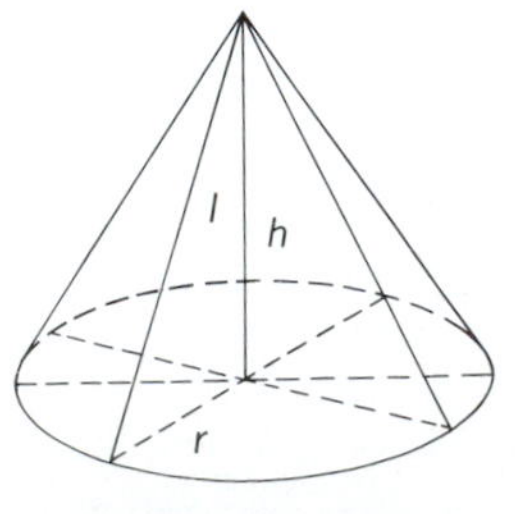

Right circular cone.

If the elements of a cone make equal angles with a line through the vertex, the cone is called a cone of revolution. Plane sections of a cone of revolution are called conic sections.

The volume of a solid cone is $V = Bh/3$, where B is the base area included within the directrix, and h is the altitude (or height) measured from the vertex to the plane of the base. The volume V and surface area S of a right circular cone are $V = \pi r^2 h/3$ and $S = 2\pi rl + 2\pi r^2$, where r denotes the radius of the base and $l = \sqrt{r^2 + h^2}$ denotes the slant height, measured from vertex to base along an element of the cone. *See* CONIC SECTION; EUCLIDEAN GEOMETRY; SURFACE AND SOLID OF REVOLUTION.

[J.S.F.]

Confocal microscopy A technique that creates high-resolution images of very small objects but differs from conventional optical microscopy in that it uses a condenser lens to focus the illuminating light from a point source into a very small, diffraction-limited spot within the specimen, and an objective lens to focus the light emitted from that spot onto a small pinhole in an opaque screen. Located behind the screen is a detector capable of quantifying how much light passes through the hole at any instant. Because only light from within the illuminated spot is properly focused to pass through the pinhole and reach the detector, any stray light from structures above, below, or to the side of the spot is filtered out. The image quality is therefore greatly enhanced.

Only the smallest possible spot is illuminated at any one time, and so a coherent image must be built up by scanning point by point over the desired field of view and recording the intensity of the light emitted from each spot. Scanning can be accomplished in several ways, but the most common system is laser scanning.

[R.J.Ta.]

Conformal mapping A special operation in mathematics in which a point set in one coordinate system is mapped or transformed into a corresponding point set in another coordinate system.

A mapping or transformation of a point set E in the xy plane is merely a correspondence $u = u(x,y)$, $v = v(x,y)$ which makes each point of E correspond to some point (u,v) of the uv plane. If distinct points (x,y) are transformed into distinct points (u,v), the mapping is one-to-one. A mapping is conformal if it is one-to-one and if small triangles are transformed into small triangles approximately similar to the original triangles, the approximation to similarity approaching perfection as the size of the triangles diminishes indefinitely. Since any small figure can be cut up into triangles, it follows that any small figure can be transformed into an approximately similar figure.

The transformation defined by an analytic function whose derivative is different from zero is conformal. It can also be shown that, conversely, a conformal map must be defined by an analytic function with nonvanishing derivative. The term conformal applies also to the transformation of any surface onto another, such as the mapping of a portion of the Earth's surface onto a portion of a plane. The Mercator and stereographic projections are conformal in this same sense. Small regions are essentially unchanged in shape but not in relative size. *See* ANALYTIC FUNCTION; MAP PROJECTIONS.

When the map of a sector of the complex plane $0 < \arg z < \alpha$ is studied under the transformation $w = z^2$, it is seen that for

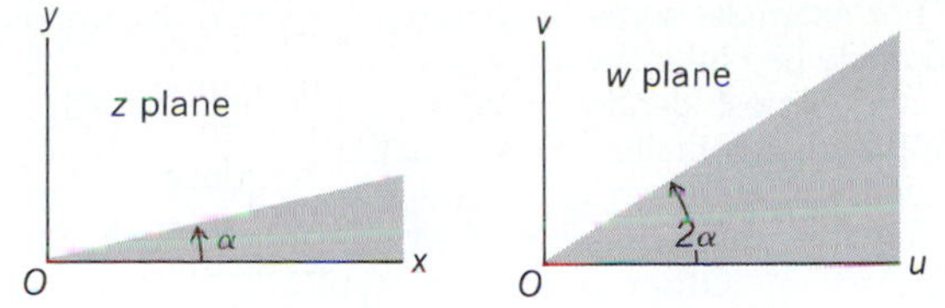

Formation of a Riemann surface.

small α the image is the sector $0 < \arg w < 2\alpha$, and that as α increases the image sector also increases (see illustration). When α passes through the value π, the boundary of the image sector passes through the value 2π and continues to increase beyond 2π, say, in a new sheet over the w plane, in the sense that as new points z occur it is natural to continue to adjoin new points w. But when the sector in the z plane reaches the angle 2π, it abuts on values of z already considered, so it is natural to "heal" the boundaries of the sector of angle 4π in the w plane by passing one sheet through another (say, along the positive half of Ox) and joining the edges. There is thus formed a Riemann surface over the w plane, which is useful for studying the behavior of $z = \sqrt{w}$ as a function of w with reference to the two branches of the function, their continuity, and their passing (along various w curves) continuously from one into the other.

There exists more generally a Riemann surface of n sheets for the inverse of any irreducible polynomial function $w = a_0 z^n + a_1 z^{n-1} + \cdots + a_n$. Moreover, the exponential function $w = e^z = e^{x+iy}$ can be defined as $e^x(\cos y + i \sin y)$ for complex z, whose analyticity for all z can be verified by the Cauchy-Riemann equations. The inverse function, given below, is

$$z = \log w = \log [r(\cos\theta + i \sin\theta)] = \log r + i\theta$$

infinitely many valued; its Riemann surface consists of infinitely many sheets over the w plane, each connected (for instance) along the positive half of Ox to its predecessor along one edge and to its successor along the other. Riemann surfaces may be defined abstractly also. *See* TOPOLOGY. [J.L.W.]

Conformational analysis The determination of the arrangement in space of the constituent atoms of a molecule that may rotate about a single bond. These rotations produce changes in conformation. Conformational analysis also includes the study of the chemical and physical differences between the conformations of a molecule. A particular conformation is chemically significant only when capable of finite existence.

Conformations may be treated in a manner analogous to the treatment of geometrical isomers, which are differentiated by the large energy barrier to rotation imposed by the carbon-carbon double bond. However, the barrier to rotation about a carbon-carbon single bond is usually considerably lower, and consequently different conformations are more readily interconvertible.

Ethane theoretically possesses an infinite number of conformations; but if the hydrogen atoms are not differentiated, there are but two significant conformations. Viewed along the axis of the carbon-carbon bond, one conformation has the hydrogen atoms of one methyl group exactly lined up behind those of the other methyl group, and the other has the hydrogens of one methyl group failing midway (60° from the former position) between those of the other methyl group. The former is an eclipsed (hydrogens opposed to each other) conformation, and the latter a staggered conformation. The eclipsed conformation represents a free-energy maximum and the staggered conformation a free-energy minimum, presumably due to repulsions, or lack thereof, between the electrons of the carbon-hydrogen bonds.

If the ethane carbons each carry two substituents other than hydrogen, three staggered conformations are possible, and the most stable one is determined by the relative magnitudes of steric repulsions between the substituent groups on the central carbons. The staggered conformations for the (±) configurations (I*a*), (I*b*), and (I*c*), and the meso configurations (II*a*), (II*b*), and (II*c*) of stilbene dibromide are shown here.

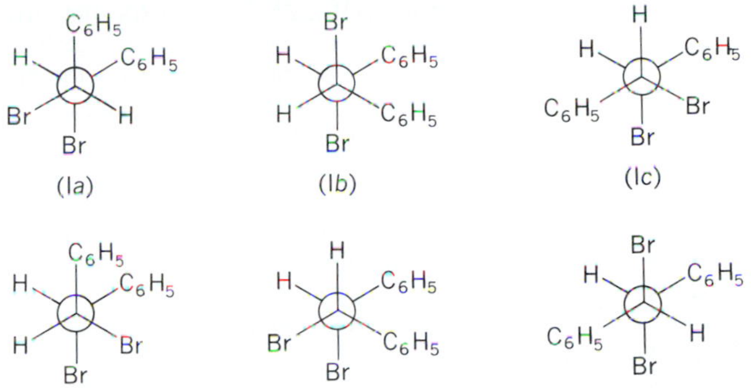

(Ia) (Ib) (Ic)

(IIa) (IIb) (IIc)

The extension of conformational analysis to cyclic and polycyclic systems leads to many useful conclusions concerning the structure, configuration, and mode of reaction of such important natural products as the steroids, terpenes, and alkaloids. *See* STEREOCHEMISTRY. [W.R.V.]

Congenital anomalies Abnormalities of any portion or organ of the human body that develop during pregnancy. These structural changes are often called birth defects. They may be minor deformities, such as supernumerary toes, which do not affect the health of the individual, or they may be major defects of internal organs that lead to serious physical disability or even death. Congenital anomalies may be single or associated with other anomalies, or multiple defects that may involve several organ systems. Birth defects may be inherited or produced by harmful external factors; however, the cause of most anomalies is unknown.

Certain common abnormalities show a significant regional or ethnic difference. Thus anencephaly, a defective development of the brain, is relatively uncommon in persons of African descent but is seen quite often in India. Polydactyly, extra toes or fingers, is more common in Africans than in other groups. In the United States, multiple congenital anomalies occur more frequently in whites than in blacks.

Congenital defects are found in a high proportion of abortions and stillbirths and so are responsible for considerable pregnancy wastage. Defects that cause death of the embryo, with spontaneous abortion, represent the severe end of the biologic spectrum. Similar defects of a milder degree may cause impairment that becomes manifest in childhood or adult life, while very slight defects may cause no clinical manifestations. One mechanism that causes syndromes of multiple malformations and also spontaneous abortions is gross chromosomal abnormality. An abnormal chromosome constitution has been found in approximately 25% of cases of spontaneous human abortion.

While most abnormalities occur as often in males as in females, several common ones display distinct sex differences. Pyloric stenosis, an abnormality of the stomach outlet that causes severe vomiting in infants, is found five times more frequently in boys than in girls. The common clubfoot abnormality is twice as common in males, while anencephaly is twice as common in females. Congenital dislocation of the hip occurs six times more often in females. By studying twins with malformations, the genetic component can be further identified. For cleft palate, congenital dislocation of the hip, clubfoot, and pyloric stenosis, the monozygotic (single-egg) twin partner will have the malformation 5 to 10 times more often than the dizygotic (two-egg) twin partner. On the other hand, anencephaly is extremely rare in both identical twins, indicating a minimal genetic contribution.

Most defects of the neural tube are more common in older mothers or in children born later in the birth order. This relationship of abnormality with increasing maternal age is also well known for Down syndrome and the two other autosomal trisomy syndromes. In congenital dislocation of the hip, the opposite effect is true, with more cases occurring in the firstborn than would be expected. A possible explanation for hip deformities developing in firstborns is poor positioning of the fetus within the uterus during the last trimester of pregnancy. *See* DOWN SYNDROME; TRISOMIC SYNDROME.

Malformations of structures derived from the neural tube include severe conditions like anencephaly, hydrocephaly (increased fluid in skull), spina bifida, and meningocele (exteriorization of membranes). Approximately 50% of all deaths due to malformations are caused by such neural defects. Congenital heart disease includes many separate abnormalities, several of which are not apparent at birth. Large defects in the septum separating the two sides of the heart, however, may cause blood to be shunted from right to left, the unoxygenated blood causing cyanosis ("blue baby"). Defects of the esophagogastrointestinal tract may occur at any level. One common defect is the tracheoesophageal fistula, a connection between the esophagus and major airway, which may lead to aspiration with subsequent suffocation or pneumonia. In imperforate anus, seen more often in males, the large intestine ends blindly, leading to intestinal obstruction. Diaphragmatic hernia is caused by incomplete formation of the diaphragm, permitting abdominal organs to enter the thoracic cavity, with secondary pressure effects on the heart and lungs. Harelip and cleft palate are defects which often coexist and which may vary in severity from a slight dimple in the lip to extreme lack of fusion of palatal and facial bones. Talipes (clubfeet) are common malpositions of the feet which develop during the last trimester of pregnancy due to intrauterine mechanical factors, as does congenital dislocation of the hip. Other common malformations of the extremities include polydactyly, syndactyly (fusion of the fingers or toes, "webfeet"), and reduction deformities of various descriptions. Defects of the genitourinary system encompass abnormalities of the external genital organs which may confuse the assignment of the proper sex to the newborn (intersex genitalia). If the infant has genital organs resembling those of the opposite sex, it may suffer false sex assignment and be categorized as a pseudohermaphrodite. Abnormalities of the urinary tract and kidneys may lead to failure to produce adequate volumes of urine, to obstruction and degeneration of the kidneys, or to other severe complications.

Syndromes are associations of two or more anomalies in one individual. The finding of several abnormalities does not necessarily mean that they are produced by the same etiologic mechanism; several anomalies may be present in any person by coincidence. Furthermore, certain abnormalities may have diverse etiologies. As an example of this principle, polydactyly may be caused by a simple dominant mutant gene, may be a part of a malformation syndrome caused by a single gene (Laurence-Moon-Biedl syndrome), may be a chromosomal abnormality (D_1, trisomy syndrome), or may be of complex (multifactorial) genetic contribution. Most cases of polydactyly, however, are sporadic and of unknown etiology.

Another important principle of congenital abnormalities is the concept of variable expressivity, where some cases of a certain condition have few malformations while others have numerous ones. This principle is sometimes referred to as the biologic spectrum of a disease; it can be applied to syndromes with genetic causes and those produced by environmental factors. For example, some women who received the tranquilizer thalidomide had infants with severe limb deformities, while others, under almost identical conditions, bore infants with few or no apparent anomalies. Some malformation syndromes are characterized by the presence of few anomalies in the majority of cases, for example, the XXX trisomy syndrome (an extra X chromosome). Other syndromes typically have numerous anomalies, such as the XO syndrome (absence of a sex chromosome), with a dozen or more anomalies in most cases. *See* TERATOLOGY. [S.L.I.]

Conglomerate The consolidated equivalent of gravel. Conglomerates are aggregates of more or less rounded particles greater than 0.08 in. (2 mm) in diameter. Frequently they are subdivided on the basis of size of particles into pebble (fine), cobble (medium), and boulder (coarse) conglomerates. The common admixture of sand-sized and gravel-sized particles in the same deposit leads to further subdivisions, into conglomerates (50% or more pebbles), sandy conglomerates (25–50% pebbles), and pebbly or conglomeratic sandstones (less than 25% pebbles). The pebbles of conglomerates are always somewhat rounded, giving evidence of abrasion during transportation; this distinguishes them from some tillites and from breccias, whose particles are sharp and angular (see illustration).

Conglomerates fall into two general classes: the well-sorted, matrix-poor conglomerates with homogeneous pebble lithology, and the poorly sorted, matrix-rich conglomerates with het-

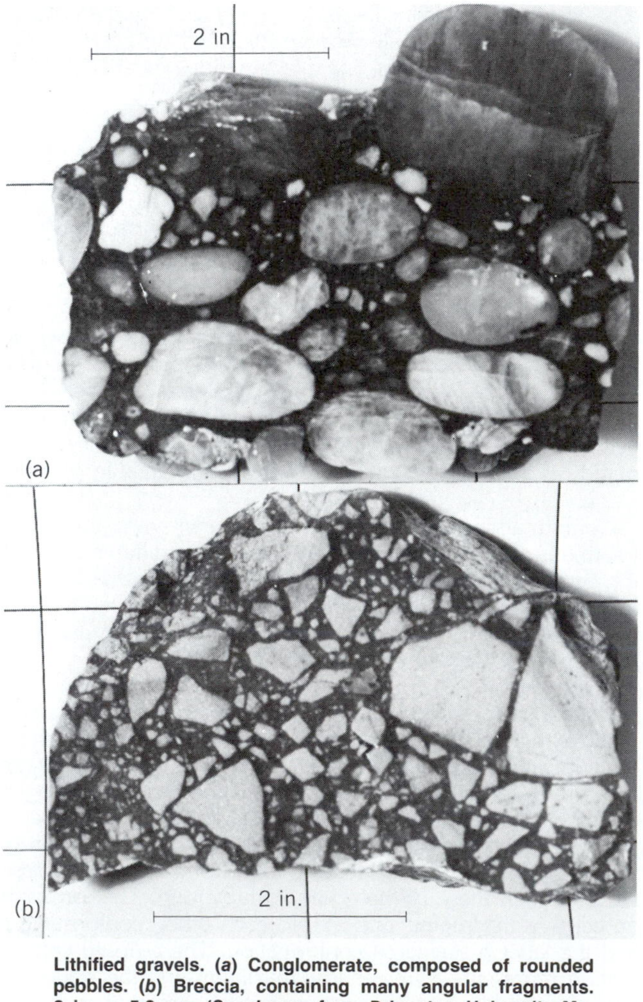

Lithified gravels. (*a*) Conglomerate, composed of rounded pebbles. (*b*) Breccia, containing many angular fragments. 2 in. = 5.2 cm. (*Specimens from Princeton University Museum of Natural History; photo by Willard Starks*)

erogeneous pebble lithology. The well-sorted class includes quartz-pebble, chert-pebble, and limestone-pebble conglomerates which tend to be distributed in thin, widespread sheets, normally interbedded with well-sorted, quartzose sandstones. The poorly sorted conglomerates include many different types, all related in having very large amounts of sandy or clayey matrix and pebbles of many different rock classes. The graywacke conglomerates are the outstanding representatives. All poorly sorted conglomerates tend to occur in fairly thick sequences, and some of them, typically the fanglomerates (conglomerates formed on alluvial fans) are wedge-shaped accumulations. *See* Sedimentary rocks.

Special types of conglomerates, such as volcanic conglomerates and agglomerates and some intraformational conglomerates composed of shale pebbles or deformed limestone pebbles, do not seem to fall easily into either class. *See* Breccia; Gravel; Graywacke; Till. [R.Si.]

Conglutination A term used in serology to describe the completion or enhancement of an incomplete agglutinating system by the addition of certain substances. Some bacteria or erythrocyte suspensions do not exhibit the visible agglutination ordinarily expected after they have been coated with their specific antibodies and complement. Further addition of a conglutinating agent—normal bovine serum—however, initiates visible agglutination. Complement is an essential component of the system. The conglutination reagent is itself without effect in the absence of antibody and complement. Similar overall actions, such as agglutination of Rh^+ cells, occur when human serum, gelatin, or bovine serum albumin are added to the incomplete Rh antibodies found in some sera, but since this enhancing effect occurs also in the absence of complement, these reactions probably are to be distinguished in mechanism from that of the traditional conglutination. The conglutination reaction can be used for a variety of serological diagnostic reactions in bacterial and viral infections. *See* Agglutination reaction; Blood groups; Complement; Complement-fixation test. [H.P.T.]

Conic section One of the class of curves in which a plane may cut a cone (surface) of revolution. The section is a parabola if the plane is parallel to an element of the cone, an ellipse or circle if the plane cuts all elements of one nappe (but does not go through the apex), and a hyperbola if the plane cuts elements of both nappes (for example, the plane parallel to the cone's axis of revolution) and does not go through the apex (see illustration).

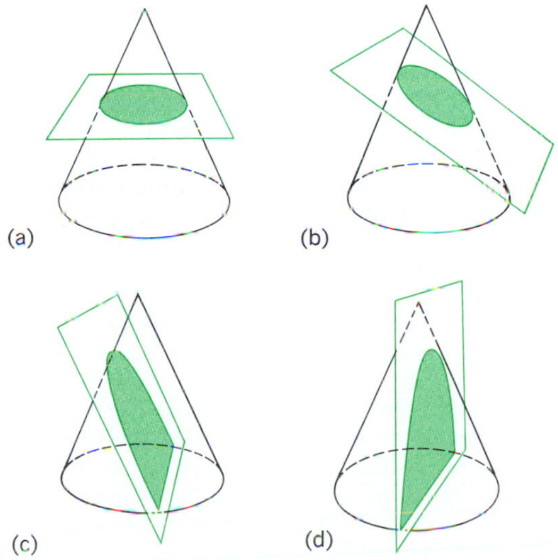

Four conic sections illustrated. (*a*) Circle. (*b*) Ellipse. (*c*) Parabola. (*d*) Hyperbola.

After the advent of analytic geometry, the synthetic method used by the Greeks to develop the properties of conics gave way to algebraic procedures which, using other definitions, divorced these curves from their original relationship to cones and regarded them as graphs of second-degree equations in cartesian coordinates x,y. The third phase in the study of conic sections began with the development of projective geometry. *See* Analytic geometry; Cone; Ellipse; Hyperbola; Parabola; Projective geometry. [L.M.Bl.]

Coniconchia A fossil group that has never seen general usage and probably should be abandoned. The hyoliths make up about half of the Coniconchia. The systematic position of Hyolitha continues to be troublesome. B. Runnegar and others suggest it constitutes an extinct phylum, whereas L. Mareck and E. L. Yochelson consider it as an extinct class of Mollusca. Studies of shell structure by J. Carter inferentially support assignment to the mollusks. Final systematic placement is still uncertain. It would appear that Tentaculitida and Dacryoconarida constitute classes within an extinct phylum. Possibly these fossils were lophophorate "worms." *See* Mollusca. [E.L.Y.]

Conjugation (chemistry) An arrangement of bonds in a molecule that provides a framework consisting of alternating carbon-carbon single and multiple (that is, double or triple) bonds. The single bond lies directly between two multiple bonds or between a multiple bond and a group containing a lone π-electron, a π-electron pair or quartet, or a π-electron vacancy. Molecules possessing such a framework show unusual chemical behavior and physical properties. Such molecules and such multiple bonds are called conjugated. Simple examples are 1,3-butadiene ($H_2C{=}CH{-}CH{=}CH_2$) and allylamine ($H_2C{=}CH{-}\ddot{N}H_2$).

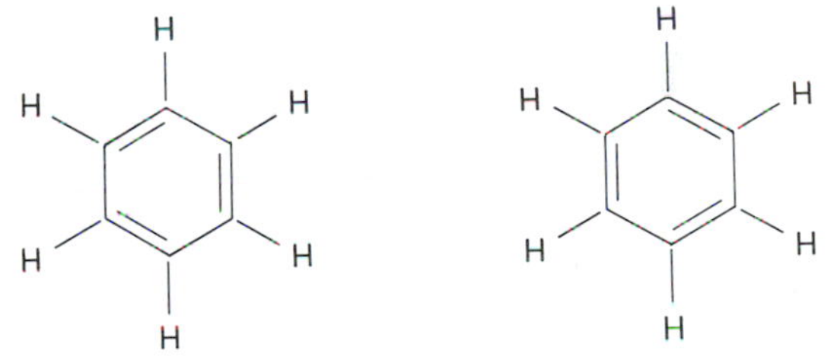

Isovalent structures of benzene.

The unusual properties of conjugated molecules can be understood theoretically, according to quantum mechanics, in terms of a bond structure in which, in the two examples mentioned, small proportions of $H_2\dot{C}{-}CH{=}\dot{C}H{-}CH_2$ or $H_2C^-{-}CH{=}N^+H_2$, respectively, "resonate with" the main bond structures given above. These examples are considered to exhibit sacrificial conjugation, since in the minor resonance structures there is one less bond, or else much energy is required to transfer a charge. *See* Chemical bonding; Resonance (molecular structure).

Stronger resonance effects occur in conjugated molecules in which alternative structures with equal numbers of bonds can be written (isovalent conjugation), for example, in benzene (see illustration) or the allyl ion ($H_2C{=}CH{=}C^+H_2$ and $H_2C^+{-}CH{=}CH_2$) or radical ($\dot{H}_2C{=}CH{-}CH_2$ and $H_2\dot{C}{-}CH{=}CH_2$). Conjugation brings about energy stabilization, but much more so in isovalent than in sacrificial conjugation. [R.S.M.]

Conjunctivitis An inflammation of the thin transparent membrane that covers the posterior surfaces of the eyelids and the anterior surface of the white of the eye. It is the most common eye disease in the Western Hemisphere and varies in severity from a mild redness (hyperemis) with tearing, such as hay-fever conjunctivitis, to a severe destructive process.

Infectious agents that may cause conjunctivitis are bacteria, viruses, and rickettsia. Fungal infection is rare. Allergic conjunctivitis may take the form of a hay-fever conjunctivitis or a seasonal type (vernal conjunctivitis). Chemical inflammation of the conjunctiva may result from eye drops, drugs, acids, or other chemicals. Conjunctivitis may also be associated with systemic diseases, such as thyroid disease or gout.

The important signs of conjunctivitis include redness, tearing, secretions, a drooping of the eyelids, and balloonlike swelling of the conjunctival tissue. Enlargement of the glands located immediately in front of the ear indicates a virus-type conjunctivitis. [J.Hart.]

Connecting rod A link in several kinds of mechanisms. Usually one end of a connecting rod is intended to follow a circular path, while the other end follows a path along a straight line or curve of large radius. The term is sometimes applied, however, to any straight link that transmits motion or power from one linkage to another within a mechanism. The illustration shows some conventional arrangements.

The connecting rod of the four-bar linkage, often called the coupler, has special significance. The motion of its plane can now be synthesized to furnish desired paths for points, or desired positions of the entire connecting-rod plane. The connecting rod is then not primarily used for transmission of force or motion from input to output crank, but the entire mechanism is employed to impart to the connecting-rod plane certain displacements, and sometimes velocities and accelerations. *See* FOUR-BAR LINKAGE; MECHANISM. [D.P.Ad.]

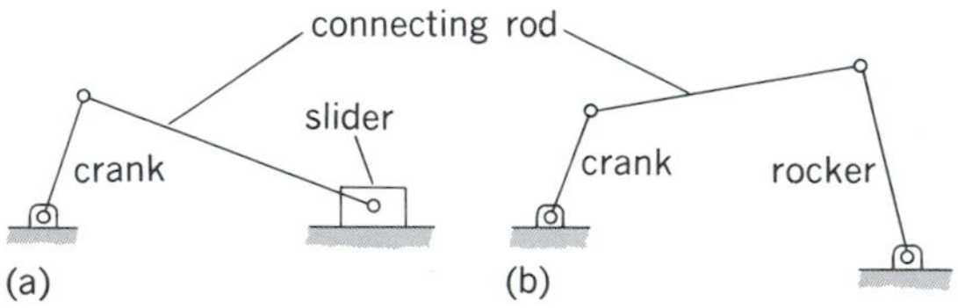

Connecting rod in (*a*) slider-crank mechanism and (*b*) crank-and-rocker mechanism.

Connective tissue One of the four primary tissues of the body. It differs from the other three tissues in that the extracellular components (fibers and intercellular substances) are abundant. It cannot be sharply delimited from the blood, whose cells may give rise to connective tissue cells, and whose plasma components continually interchange with and augment the ground substance of connective tissue. Bone and cartilage are special kinds of connective tissue.

The functions of connective tissues are varied. They are largely responsible for the cohesion of the body as an organism, of organs as functioning units, and of tissues as structural systems. The connective tissues are essential for the protection of the body both in the elaborate defense mechanisms against infection and in repair from chemical or physical injuries. Nutrition of nearly all cells of the body and the removal of their waste products are both mediated through the connective tissues. Connective tissues are important in the development and growth of many structures. Constituting the major environment of most cells, they are probably the major contributor to the homeostatic mechanisms of the body so far as salts and water are concerned. They act as the great storehouse for the body of salts and minerals, as well as of fat. The connective tissues determine in most cases the pigmentation of the body. Finally, the skeletal system (cartilage and bones) plus other kinds of connective tissue (tendons, ligaments, fasciae, and others) make motion possible.

The connective tissues consist of cells and extracellular or intercellular substance (see illustration). The cells include many varieties, of which the following are the most important: fibroblasts, macrophages (histiocytes), mast cells, plasma cells, melanocytes, and fat cells. Most of the cells of the connective tissue are developmentally related even in the adult; for example, fibroblasts may be developed from histiocytes or from undifferentiated mesenchymal cells.

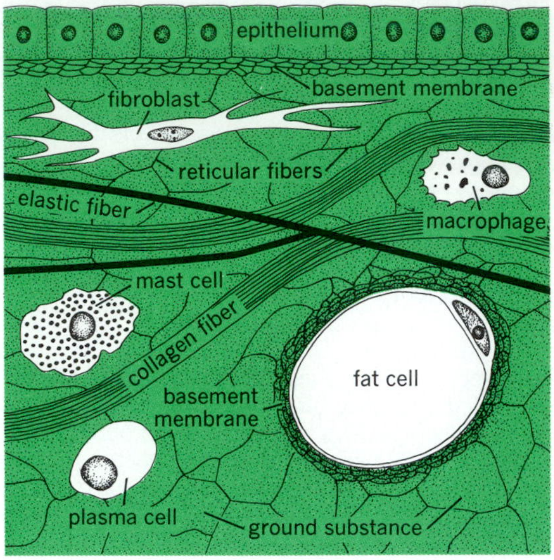

Components of connective tissue.

The extracellular components of connective tissues may be fibrillar or nonfibrillar. The fibrillar components are reticular fibers, collagenous fibers, and elastic fibers. The nonfibrillar component of connective tissues appears amorphous with the light microscope and is the matrix in which cells and fibers are embedded. It consists of two groups of substances: (1) those probably derived from secretory activity of connective tissue cells including mucoproteins, protein-polysaccharide complexes, tropocollagen, and antibodies; and (2) those probably derived from the blood plasma, including albumin, globulins, inorganic and organic anions and cations, and water. In addition, the ground substance contains metabolites derived from, or destined for, the blood.

All the manifold varieties of connective tissue may contain all the cells and fibers discussed above in addition to ground substance. They differ from each other in the relative occurrence of one or another cell type, in the relative proportions of cells and fibers, in the preponderance and arrangement of one or another fiber, and in the relative amount and chemical composition of ground substance. They are classified as:

1. Irregularly arranged connective tissue—which may be loose (subcutaneous connective tissue) or dense (dermis). The dominant fiber type is collagen.
2. Regularly arranged connective tissue—primarily collagenous—with the fibers arranged in certain patterns depending on whether they occur in tendons or as membranes (dura mater, capsules, fasciae, aponeuroses, or ligaments).
3. Mucous connective tissue—ground substance especially prominent (umbilical cord).
4. Elastic connective tissue—predominance of elastic fibers or bands (ligamentum nuchae) or lamellae (aorta).

5. Reticular connective tissue—fibers mostly reticular, moderately rich in ground substance, frequently numerous undifferentiated mesenchymal cells.

6. Adipose connective tissue—yellow or brown fat cells constituting chief cell type, reticular fibers most numerous.

7. Pigment tissue—melanocytes numerous.

8. Cartilage—cells exclusively of one type, derived from mesenchymal cells.

9. Bone—cells are predominantly osteocytes, but also include fibroblasts, mesenchymal cells, endothelial cells, and osteoclasts.

See Adipose tissue; Blood; Bone; Cartilage; Collagen; Histology; Ligament; Tendon. [I.G.]

Connective tissue disease Any of a group of diseases involving connective tissue; formerly termed collagen vascular disease. These diseases are clinically and pathologically discrete from each other but have overlapping features. The group includes lupus erythematosus, systemic vasculitis, polymyolitis, scleroderma (and systemic sclerosis), and Sjögren's syndrome. Each of the diseases can involve multiple organ systems and is often coupled with various immunologic abnormalities. *See* Connective tissue.

The set of diseases known as lupus erythematosus includes limited, primarily cutaneous disorders (discoid lupus and subacute cutaneous lupus) and a diffuse systemic illness (systemic lupus erythematosus), all of which are of unknown cause. Also, a lupuslike reaction known as drug-induced lupus can be caused by certain therapeutic agents. *See* Lupus erythematosus.

Systemic vasculitis comprises a series of different clinical illnesses that are all characterized by intense inflammation in the walls of blood vessels, especially arteries (that is, arteritis). These illnesses differ clinically, pathologically, and with respect to therapeutic responsiveness and outcome. Even the age and sex of the typical affected person varies from one to another. The clinical pattern of illness varies with the sites of blood vessel involvement and the character of vascular injury. Tissue death in the vessel wall can lead to narrowing, or even blockage, of the vessel or weakening of the wall with formation of an aneurysm. Impaired circulation and altered blood flow result. Symptoms, therefore, can arise not only from the tissue inflammation itself but also from the lack of adequate blood supply. *See* Aneurysm; Circulation disorders.

Polymyositis involves intense inflammation in skeletal muscle that, if untreated, can lead to destruction of muscle fibers. When typical skin lesions are present, the term dermatomyositis is applied. The cause of polymyositis, or dermatomyositis, is not known in most cases. These conditions can occur alone or in association with other connective tissue disorders (that is, systemic lupus, scleroderma, and Sjögren's syndrome). In older persons, they sometimes accompany malignant tumors. If diagnosed early, before loss of muscle fibers and replacement by fibrous scarring is extensive, myositis is reversible; recovery with minimal residual loss of strength is to be expected in almost all patients.

The term scleroderma means "hard skin," designating a disorder in which increased deposition of collagen fibers in the deeper dermis leads to thickened, leathery, bound-down skin. When organ involvement is associated with such skin changes, the term systemic sclerosis is used. Raynaud's phenomenon, a hyperreactivity to cold exposure with blanching and discoloration of the fingers and toes, is the common initial manifestation and may precede sclerodermatous skin changes by decades. Systemically, the gastrointestinal tract is most commonly involved. Difficulty in swallowing and mid-chest discomfort result from loss of peristalsis in the esophagus. Diarrhea, inability to absorb nutrients, abdominal pain, and distention can occur with bowel involvement and lead to weight loss and wasting. Scarring of the lungs and decreased pulmonary function occur frequently. The most serious lesions are those of the heart and kidney, which may result from vascular abnormalities related to Raynaud's phenomenon. Joint inflammation is mild and uncommon; but advanced skin changes, especially in the hands, may restrict joint mobility

Sjögren's syndrome is characterized primarily by dryness of the membranes due to excretory gland failure, especially dryness of the eyes (xerophthalmia) and mouth (xerostomia) from loss of tears and saliva, respectively. This sicca (dryness) syndrome reflects the infiltration of lacrimal and salivary glands by immunologically competent cells (lymphocytes). Sjögren's syndrome may be primary or secondary; that is, superimposed on another disorder like rheumatoid arthritis or systemic lupus. *See* Immunology; Inflammation. [M.B.S.]

Conodont A member of a group of extinct animals, the Conodonta, which were abundant in the seas from early in the Cambrian Period almost to the end of the Triassic Period. Conodonts are found as fossils in most marine sedimentary rocks, in which they are represented by tiny cone-, bar-, or platform-shaped skeletal elements (see illustration). These range from less than 0.04 in. (1 mm) to somewhat more than 0.16 in. (4 mm) in size and, in unaltered form, are translucent and amber in color. Conodont elements contain minor amounts of organic substances, but consist largely of phosphatic mineral matter which approaches the carbonate apatites in composition.

Most conodont elements are collected as discrete entities

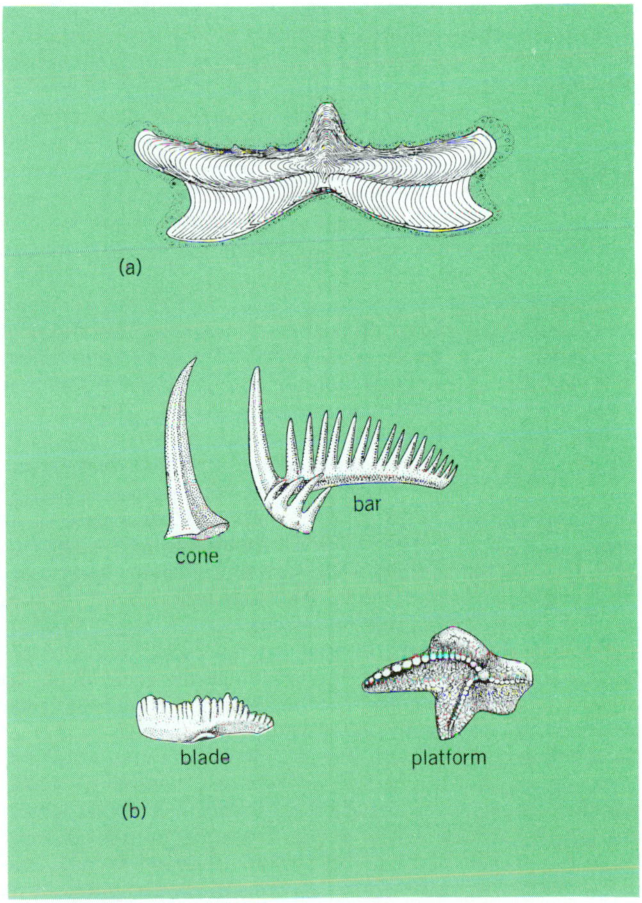

Conodont elements: (*a*) Internal structure, (*b*) major shape categories. (*After W. H. Hass, Conodonts, in Treatise on Invertebrate Paleontology, pt. W, University of Kansas Press and Geological Society of America, 1962*)

because the soft tissues of the conodont decayed rapidly upon the death of the animal. However, a few clusters of elements (called element clusters or natural assemblages) have been found on the surfaces of black-shale slabs. These elements are thought to be more or less undisturbed skeletons of individual conodonts. Analyses of the components of such clusters (and of others that are generated from large collections of discrete elements by various grouping techniques) reveal that a complete conodont skeleton may have consisted of 20 or more separate elements, representing as many as seven different shape categories.

Most present-day students of conodonts base species on recurrent groups of differently shaped conodont elements (rather than on elements of different shape), and then group these species into genera and higher taxonomic categories on the basis of overall composition of the skeletal apparatuses. In this so-called multielement taxonomy careful attention must be paid not only to the composition of the element group but also to such characters as the shape, microscopic surface ornamentation, and internal structure of individual elements. More than 300 genera of conodonts have been named, but there is little agreement as to how they should be grouped together into families or larger taxonomic units that can be interpreted to be genetically interrelated. [W.C.S.]

Conservation laws (physics)

Principles which state that the total values of specified quantities remain constant in time for an isolated system. Conservation laws occupy enormously important positions both at the foundations of physics and in its applications.

Realization in classical mechanics. There are three great conservation laws of mechanics: the conservation of linear momentum, often referred to simply as the conservation of momentum; the conservation of angular momentum; and the conservation of energy.

The linear momentum, or simply momentum, of a particle is equal to the product of its mass and velocity. It is a vector quantity. The total momentum of a system of particles is simply the sum of the momenta of each particle considered separately. The law of conservation of momentum states that this total momentum does not change in time. *See* CONSERVATION OF MOMENTUM; MOMENTUM.

The angular momentum of a particle is more complicated. It is defined by the vector product of the position and momentum vectors. The law of conservation of angular momentum states that the total angular momentum of an isolated system is constant in time. *See* ANGULAR MOMENTUM.

The conservation of energy is perhaps the most important law of all. Energy is a scalar quantity, and takes two forms: kinetic and potential. The kinetic energy of a particle is defined to be one-half the product of its mass and the square of its velocity. The potential energy is loosely defined as the ability to do work. The total energy is the sum of the kinetic and potential energies, and according to the conservation law it remains constant in time for an isolated system.

The essential difficulty in applying the conservation of energy law can be appreciated by considering the problem of two colliding bodies. In general, the bodies emerge from the collision moving more slowly than when they entered. This phenomenon seems to violate the conservation of energy, until it is recognized that the bodies involved may consist of smaller particles. Their random small-scale motions will require kinetic energy, which robs kinetic energy from the overall coherent large-scale motion of the bodies that are observed directly. One of the greatest achievements of nineteenth-century physics was the recognition that small-scale motion within macroscopic bodies could be identified with the perceived property of heat. *See* CONSERVATION OF ENERGY; ENERGY; KINETIC THEORY OF MATTER.

Position in modern physics. As physics has evolved, the great conservation laws have likewise evolved in both form and content, but have never ceased to be important guiding principles.

In order to account for the phenomena of electromagnetism, it was necessary to go beyond the notion of point particles, to postulate the existence of continuous electric and magnetic fields filling all space. To obtain valid conservation laws, energy, momentum, and angular momentum must be ascribed to the electromagnetic fields. *See* ELECTROMAGNETIC RADIATION; MAXWELL'S EQUATIONS; POYNTING'S VECTOR.

In the special theory of relativity, energy and momentum are not independent concepts. Einstein discovered perhaps the most important consequence of special relativity, that is, the equivalence of mass and energy, as a consequence of the conservation laws. The "law" of conservation of mass is understood as an approximate consequence of the conservation of energy. *See* CONSERVATION OF MASS; RELATIVITY.

A remarkable, beautiful, and very fruitful connection has been established between symmetries and conservation laws. Thus the law of conservation of linear momentum is understood as a consequence of the homogeneity of space, the conservation of angular momentum as a consequence of the isotropy of space, and the conservation of energy as a consequence of the homogeneity of time. *See* SYMMETRY LAWS (PHYSICS).

The development of general relativity, the modern theory of gravitation, necessitates attention to a fundamental question for the conservation laws: The laws refer to an "isolated system," but it is not clear that any system is truly isolated. This is a particularly acute problem for gravitational forces, which are long ranged and add up over cosmological distances. It turns out that the symmetry of physical laws is actually a more fundamental property than the conservation laws themselves, for the symmetries remain valid while the conservation laws, strictly speaking, fail.

In quantum theory, the great conservation laws remain valid in a very strong sense. Generally, the formalism of quantum mechanics does not allow prediction of the outcome of individual experiments, but only the relative probability of different possible outcomes. One might therefore entertain the possibility that the conservation laws were valid only on the average. However, momentum, angular momentum, and energy are conserved in every experiment. *See* QUANTUM MECHANICS; QUANTUM THEORY OF MEASUREMENT.

Conservation laws of particle type. There is another important class of conservation laws, associated not with the motion of particles but with their type. Perhaps the most practically important of these laws is the conservation of chemical elements. From a modern viewpoint, this principle results from the fact that the small amount of energy involved in chemical transformations is inadequate to disrupt the nuclei deep within atoms. It is not an absolute law, because some nuclei decay spontaneously, and at sufficiently high energies it is grossly violated. *See* RADIOACTIVITY.

Several conservation laws in particle physics are of the same character: They are useful even though they are not exact because, while known processes violate them, such processes are either unusually slow or require extremely high energy. *See* ELEMENTARY PARTICLE. [F.WIL.]

Conservation of energy

The principle of conservation of energy states that energy cannot be created or destroyed, although it can be changed from one form to another. Thus in any isolated or closed system, the sum of all forms of energy remains constant. The energy of the system may be interconverted among many different forms—mechanical, electrical, magnetic, thermal, chemical, nuclear, and so on—and as time progresses, it tends to become less and less available; but within the limits of small experimental uncertainty, no change in total amount of energy has been observed in any situation in which it has been possible to ensure that energy has not entered or left the system in the form of work or heat. For a system that is both gaining and losing energy in the form of work and heat, as is true of any machine in operation, the energy principle asserts that the net gain of energy is equal to the total change of the system's internal energy. *See* THERMODYNAMIC PRINCIPLES.

There are many ways in which the principle of conservation of energy may be stated, depending on the intended application. Of particular interest is the special form of the principle known as the principle of conservation of mechanical energy which states that the mechanical energy of any system of bodies connected together in any way is conserved, provided that the system is free of all frictional forces, including internal friction that could arise during collisions of the bodies of the system.

J. P. Joule and others demonstrated the equivalence of heat and work by showing experimentally that for every definite amount of work done against friction there always appears a definite quantity of heat. The experiments usually were so arranged that the heat generated was absorbed by a given quantity of water, and it was observed that a given expenditure of mechanical energy always produced the same rise of temperature in the water. The resulting numerical relation between quantities of mechanical energy and heat is called the Joule equivalent, or is also known as mechanical equivalent of heat.

In view of the principle of equivalence of mass and energy in the restricted theory of relativity, the classical principle of conservation of energy must be regarded as a special case of the principle of conservation of mass-energy. However, this more general principle need be invoked only when dealing with certain nuclear phenomena or when speeds comparable with the speed of light (1.86×10^5 mi/s or 3×10^8 m/s) are involved. [D.E.R./L.N.]

Conservation of mass The notion that mass, or matter, can be neither created nor destroyed. According to conservation of mass, reactions and interactions which change the properties of substances leave unchanged their total mass; for instance, when charcoal burns, the mass of all of the products of combustion, such as ashes, soot, and gases, equals the original mass of charcoal and the oxygen with which it reacted.

The special theory of relativity of Albert Einstein, which has been verified by experiment, has shown, however, that the mass of a body changes as the energy possessed by the body changes. Such changes in mass are too small to be detected except in subatomic phenomena. Furthermore, matter may be created, for instance, by the materialization of a photon (quantum of electromagnetic energy) into an electron-positron pair; or it may be destroyed, by the annihilation of this pair of elementary particles to produce a pair of photons. *See* ELECTRON-POSITRON PAIR PRODUCTION; RELATIVITY. [L.N.]

Conservation of momentum The principle that, when a system of masses is subject only to forces that masses of the system exert on one another, the total vector momentum of the system is constant. Since vector momentum is conserved, in problems involving more than one dimension the component of momentum in any direction will remain constant. The principle of conservation of momentum holds generally and is applicable in all fields of physics. In particular, momentum is conserved even if the particles of a system exert forces on one another or if the total mechanical energy is not conserved. Use of the principle of conservation of momentum is fundamental in the solution of collision problems. *See* COLLISION (PHYSICS); MOMENTUM. [P.W.S.]

Conservation of resources Conservation is concerned with the utilization of resources—the rate, purpose, and efficiency of use. Many natural resources must be utilized to maintain the quality of human life, yet some are needlessly wasted or destroyed, diminishing the heritage of future generations. Direct effects on humans are not the only standard for conservation decisions, but actions affecting any natural system eventually affect human systems because of the interrelations of all parts of the biosphere. Historically, conservation measures came into being after an adequate supply of material goods was available to an organized population; until that time, the meeting of material needs took precedence, and the long-term health of the environment was considered secondary to human needs. For treatment of individual resources *see* SOIL CONSERVATION; WATER CONSERVATION; WILDLIFE CONSERVATION.

Universal natural resources are the land and soil, water, forests, grassland and other vegetation, fish and wildlife, rocks and minerals, and solar and other forms of energy. Some natural resources, such as metallic ores, coal, petroleum, and stone, are called fund or stock resources. They usually are referred to as nonrenewable natural resources because extraction from the stock depletes the usable quantity remaining and, even if some is being formed, the rate of formation is too slow for practical meaning. Other natural resources, such as living organisms and their products and solar and atomic radiation, are called flow resources. They usually are referred to as renewable natural resources because they involve organic growth and reproduction or because they are relatively quickly recycled, or renewed in nature, as in the case of water in the hydrologic cycle and certain atmospheric phenomena. Some natural resources are difficult to fit into such a simple system. Soil, for example, is commonly thought of as a renewable, as erosion and nutrient depletion can in some cases be rather quickly corrected, but if the upper layers of the soil are removed or bedrock is exposed, renewal may take thousands of years. Water also is commonly renewable, but rapid extraction of water by wells from deep aquifers may be equivalent to mining minerals.

Human resources are of two types: the people themselves (their numbers, qualities, knowledge, and skills) and their culture (the tools and institutions of society). The three broad classes of resources correspond to the economic factors of production: land (natural resources), labor (personal human resources), and capital (cultural tools and institutions). The natural resources have meaning only as there is human ability to make use of them.

Conservation has received many definitions because it has many aspects. Conservation receives impetus from the social conscience aware of an obligation of future generations and is viewed differently according to one's social and economic philosophy. To some extent, the meaning of conservation changes with the time and place. It is understood differently when approached from the natural sciences and technologies, and when it is approached from the social sciences.

No definition of conservation exists that is satisfactory to all elements of the public and applicable to all resources. In its absence, an operational or functional definition can be arrived at by considering a series of conservation measures.

1. Preservation is the protection of nature from commercial exploitation to prolong its use for recreation, watershed protection, and scientific study. It is familiar in the establishment and protection of parks and reserves of many kinds.

2. Restoration, another widely familiar conservation measure, is essentially the correction of past willful and inadvertent abuses that have impaired the productivity of the resources base. This measure is familiar in modern soil and water conservation practices applied to agricultural land.

3. Beneficiation is the upgrading of the usefulness or quality of something, for instance, the utilization of ores that were formerly of uneconomic grade. Modern technology has provided many examples of this type of conservation.

4. Maximization includes all measures to avoid waste and increase the quantity and quality of production from resources.

5. Reutilization, in industry commonly called recycling, is the reuse of waste materials, as in the use of scrap iron in steel manufacture or of industrial water after it has been cleaned and cooled.

6. Substitution, an important conservation measure, has two aspects: the use of a common resource instead of a rare one when it serves the same end and the use of renewable rather than nonrenewable resources when conditions permit.

7. Allocation concerns the strategy of use—the best use of a resource. For many resources and products from them, the market price, as determined by supply and demand, establishes to what use a resource is put, but under certain circumstances the general welfare may dictate usage and resources may be controlled by government through the use of quotas, rationing, or outright ownership.

8. Integration in resources management is a conservation measure because it maximizes over a period of time the sum of goods and services that can be had from a resource or a resource complex such as a river valley; this is preferable to maximizing certain benefits from a single resource at the expense of other benefits or other resources. This is one of the meanings of multiple use, and integration is a central objective of planning.

A generalized definition that fits many but not all meanings of conservation is "the maximization over time of the net social benefits in goods and services from resources."

National resources policy in the United States is framed by acts of Congress and in the states by acts of legislatures. The language of specific acts, however, usually permits some freedom for administrative decisions and also permits different interpretations that must be settled in the courts of law. As the country has developed and conditions have changed, a sequence of laws has been passed to deal with the exigencies of natural resource conditions. Also, there has occurred some evolution of political philosophy to meet the changing conditions. In addition to legislative and court actions, some natural resources are subject to treaties and other international agreements. [S.A.C.]

Constellation A name used in astronomy originally to designate any one of the star groups that were imagined to form configurations in the sky. Now the term constellation also refers to any one of the definite areas of the sky. The ancients originated most of the constellations visible in northern latitudes. Ptolemy, the Alexandrian astronomer, recorded 48 of them 2000 years ago, as listed in the table; they are still used. The ancients living in northern latitudes were unable to see the southern sky; hence it was not mapped. In 1603 J. Bayer charted the southern regions and named 13 new southern constellations, and in 1690 Hevelius added nine more. Finally in 1763 N. L. de Lacaille completed the list by adding 14 names to new constellations. These constellations, together with Coma Berenices (Berenices Hair) whose origin is uncertain, make up a total of 88 groups, now recognized and universally accepted.

Constellation boundaries were irregular. When only bright stars were considered, there was no ambiguity in locating them. But when fainter and fainter stars were considered and new and more constellations were added, the sky areas became so crowded that uncertainty and confusion naturally resulted at the boundaries. To remedy the confusion and to secure a general agreement on the boundaries, the International Astronomical Union in 1928 instituted a much-needed reform by remapping all the constellations. Replacing the irregular and uncertain outlines, the new boundaries now run only from north to south and from east to west, though they zigzag considerably to contain as many stars as possible in their former constellations. *See* ZODIAC. [C.S.-Y.]

Traditional constellations

Ancient name	Modern name
Andromeda	Andromeda
Aquarius*	Water Bearer
Aquila	Eagle
Ara	Altar
Argo†	Ship
Aries*	Ram
Auriga	Charioteer
Boötes*	Bear Driver
Cancer*	Crab
Canis Major	Greater Dog
Canis Minor	Lesser Dog
Capricornus*	Sea Goat
Cassiopeia*	Cassiopeia
Centaurus*	Centaur
Cepheus	Cepheus
Cetus	Whale
Corona Australis	Southern Crown
Corona Borealis	Northern Crown
Corvus	Crow
Crater	Cup
Cygnus*	Swan
Delphinus	Dolphin
Draco	Dragon
Equuleus	Little Horse
Eridanus	River Po
Gemini*	Twins
Hercules	Hercules
Hydra	Snake
Leo*	Lion
Lepus	Hare
Libra*	Balance
Lupus	Wolf
Lyra*	Lyre
Ophiuchus	Serpent Carrier
Orion*	Warrior
Pegasus*	Winged Horse
Perseus*	Perseus
Pisces*	Fishes
Pisces Austrinus	Southern Fish
Sagitta	Arrow
Sagittarius*	Archer
Scorpius*	Scorpion
Serpens	Serpent
Taurus*	Bull
Triangulum	Triangle
Ursa Major*	Great Bear
Ursa Minor*	Little Bear
Virgo*	Maiden

*See separate article.

†The large Ptolemy constellation Argo (Ship) is now divided into four smaller ones, namely, Carina (Keel), Puppis (Stern), Pyxis (Compass), and Vela (Sail).

Constipation The passage of hard, dry stools. The term has come to indicate the irregularity or absence of bowel movements for any period of time considered abnormal by an individual. However, the character of the stool is generally of more significance than the interval between bowel movements or the quantity of feces evacuated. Atonic constipation is that form not accompanied by signs of abdominal griping or distress, whereas hypertonic constipation is marked by spasms, spasticity of the bowel, and abnormal activity of its movements.

Common causes of constipation include faulty nutrition, inadequate water intake, poor general health, and emotional stresses or tensions. The use and abuse of laxatives is a common finding in cases of irregularity and bowel complaints. Local disease of the colon such as inflammation, fistulas, tumors, polyps, and obstructions may produce constipation, sometimes alternating with diarrhea. Certain systemic diseases such as hyperthyroidism, circulatory disturbances, stomach ailments, and infections may be accompanied by constipation. [E.G.St./N.K.M.]

Constraint A restriction on the natural degrees of freedom of a system. If n and m are the numbers of the natural and actual degrees of freedom, the difference $n - m$ is the number of constraints. In principle $n = 3N$, where N is the number of particles, for example, atoms. In practice n is determined by the number of effectively rigid components.

A holonomic system is one in which the n original coordinates can be expressed in terms of m independent coordinates and possibly also the time. It is characterized by frictionless contacts and inextensible linkages. The new coordinates are called generalized coordinates. *See* LAGRANGE'S EQUATIONS.

Nonholonomic systems cannot be reduced to independent coordinates because the constraints are not on the n coordinate values themselves but on their possible changes. For example, an ice skate may point in all directions but at each position it must point along its path. *See* DEGREE OF FREEDOM (MECHANICS). [B.G.]

Construction engineering A specialized branch of civil engineering concerned with the planning, execution, and control of construction operations for such projects as highways, buildings, dams, airports, and utility lines. Planning consists of scheduling the work to be done and selecting the most suitable construction methods and equipment for the project. Execution requires the timely mobilization of all drawings, layouts, and materials on the job to prevent delays. Control consists of analyzing progress and cost to ensure that the project will be done on schedule and within the estimated cost. *See* CONSTRUCTION EQUIPMENT; CONSTRUCTION METHODS. [W.Her.]

Construction equipment A wide variety of relatively heavy machines which perform specific construction (or demolition) functions under power. The power plant is commonly an integral part of an individual machine, although in some cases it is contained in a separate prime mover, for example, a towed wagon or roller. It is customary to classify construction machines in accordance with their functions such as hoisting, excavating, hauling, grading, paving, drilling, or pile driving. The selection of a machine for a specific job is mainly a question of economics and depends primarily on the ability of the machine to complete the job efficiently, and secondarily on its availability.

Hoisting equipment is used to raise or lower materials from one elevation to another or to move them from one point to another over an obstruction. The main types of hoisting equipment are derricks, cableways, cranes, elevators, and conveyors. *See* BULK-HANDLING MACHINES; HOISTING MACHINES.

The two main types of derricks are the guy and the stiff-leg. The former has a mast that is held in a vertical position by guy wires and a boom that can rotate with the mast 360°; the latter has the mast tied to two or more rigid structural members, with rotation of the boom limited by the position of these members. The derrick is practical only where little mobility is required.

The cableway is a combination hoist and tram system comprising a trolley that runs on main load cables stretched between two or more towers which may be stationary or tiltable. Its use in construction is generally restricted to dams and other special cases.

The construction elevator, like the passenger elevator, consists of a car or platform that operates within a structural framework and is raised and lowered by cables. It is limited to moving relatively light materials on jobs of small areas.

A crane is basically a fast-moving boom mounted on a frame containing the power supply and mechanisms for moving the boom and for raising and lowering the load-bearing cables that run through sheaves at the top of the boom. Mobile cranes are predominant, but tower cranes and climbing cranes have become popular throughout the United States and the world for high-rise building construction where mobility requirements are negligible.

Occasionally a conveyor is used to raise materials needed for construction of low buildings. This machine, which has a movable endless belt mounted on a frame, can carry only light loads such as bricks and cement sacks and is not practical for buildings of more than one or two floors. *See* CONVEYING MACHINES.

Excavating equipment is divided into two main classifications: standard land excavators and marine dredges, each of which has many variations. Standard land excavators are machines that dig earth and rock and place it in separate hauling units, as well as those that pick up and transport the materials. Among the former are power shovels, draglines, backhoes, cranes with a variety of buckets, front-end loaders, excavating belt loaders, trenchers, and the continuous bucket excavator. The second group includes such machines as bulldozers, scrapers of various types, and sometimes the front-end loader. Combination excavators and haulers include bulldozers, carrying scrapers, self-loading or elevating scrapers, and sometimes front-end loaders.

Hauling equipment is used to move excavated materials great distances. The most common conveyances are the self-propelled rubber-tired rear-dump trucks. Conveyors, while not commonly used on construction jobs for hauling earth and rock great distances, have been used to good advantage on large jobs where obstructions make impractical the passage of trucks.

Graders are high-bodied, wheeled vehicles that mount a leveling blade between the front and rear wheels. The principal use is for fine-grading relatively loose and level earth. Commonly considered a maintenance unit, the grader is still an important machine in highway construction.

Pavers place, smooth, and compact paving materials. They may be mounted on rubber tires, endless tracks, or flanged wheels. Asphalt pavers embody tamping pads that consolidate the material; concrete pavers use vibrators for the same purpose.

Drilling equipment is used for drilling wells and for blasting, grouting, and exploring. Drills are classified according to the way in which they penetrate rock, namely, percussion, rotary percussion, and rotary.

Among the most common specialized construction equipment are augers, compactors, and pile hammers. Augers are used for drilling wells, dewatering purposes, and cutting holes that can be filled with concrete for foundations. Compactors are machines designed solely to consolidate earth and paving materials to sustain loaders greater than those sustained in the uncompacted state. Actual compaction may be achieved by heavily loaded rubber tires or steel rollers, or by a vibratory action induced into the compacting units so that compaction is achieved by impact force rather than sheer weight. Pile hammers are machines used to install bearing piles for foundations, or sheet piles for cofferdams and retaining walls. Conventional hammers contain pistons that are raised and then either allowed to fall freely or to be driven downward to impart an impact to the pile. Vibratory hammers have been developed which vibrate piles into place.

Steam is seldom used as a source of power today except in some marine applications and rarely for pile driving. Gasoline and diesel engines of the piston type are the most common source of power for construction machines, with the diesel attaining an increasingly greater prominence for at least two major reasons. First, the economy of operation and maintenance of the diesel has come to far outweigh its greater initial cost. Second, manufacturers have been able to build lightweight economical diesel engines in very small sizes, making practical their application in small units.

Electricity is sometimes, though not often, used as a primary source of power for some construction machines. Electricity is making its appearance as a secondary source of power on several types of mobile construction machinery. A more common type of secondary power, however, is the hydraulic motor activated by a gasoline or diesel-powered hydraulic pump. [E.M.Y.]

Construction methods The procedures and techniques utilized during construction. Construction operations are generally classified according to specialized fields. These include preparation of the project site, earth-moving, foundation treatment, steel erection, concrete placement, asphalt paving, and electrical and mechanical installations. Procedures for each of these fields are generally the same, even when applied to different projects, such as buildings, dams, or airports. However, the relative importance of each field is not the same in all cases. For a description of tunnel construction, which involves different procedures, *see* TUNNEL. [W.Her.]

Consumer electronics A variety of consumer electronic products are available which simulate human characteristics of listening, talking, amusing, and educating. Among these microprocessor-controlled devices are hand-held language translators, programmable personal computers, educational talking toys, intelligent television receivers, and portable remote terminals that utilize the telephone system to gain access to various computer data bases. These electronic devices are made possible through electronic microminiaturization called large-scale integration (LSI). Large-scale integration allows more features to be implemented in compact, relatively inexpensive products that were bulky and expensive only a few years ago. *See* INTEGRATED CIRCUITS.

Portable language translators register keyboard entry of phrases in one language and speak the translated phrase in another. Each language for translation is preserved in memory modules called read-only-memory (ROM), which are basically tables storing corresponding words in both languages.

Hand-held computers have been developed that fit into a briefcase which has receptacles for acoustically coupling the computer with any telephone. A variety of information can thereby be accessed while the user is out of reach of data bases. *See* MICROCOMPUTER.

Two types of electronic games exist: those that measure users' dexterity and those that challenge their intelligence. Both types of games can be played alone against the computer or between two players.

Decoders are being developed that will enable text and graphics to be retrieved on the home television screen. The information is received by the television set as part of the regular broadcast but cannot be seen by the viewer unless a decoder is used. The decoder transforms information taken from the unseen (blanking) portion of the screen to the visible portion.

A more sophisticated system, called viewdata, is an outgrowth of the business community. Here the television again serves as the receiver of data and graphics and displays them when the decoder is activated; but the information reaches the set through the telephone. By modulating the voice line on the telephone, digital bit streams are sent through these circuits to the decoder. As a result, different data bases can be accessed by dialing their respective codes. In teletext a remote keypad is used to access the pages as they are broadcast. The switching process takes a few seconds and limits the viewer's participation in the service. In viewdata, on the other hand, the keypad is used as a two-way transmitter and can be used to engage in a two-way conversation with the system. [N.Mo.]

Contact aureole A halo or zone of alteration surrounding a body of igneous rock (such as granite or granodiorite) and presumably formed by heat and emanations from the igneous mass. Aureoles are usually not more than a few thousand feet wide. They may be detected by mineralogical or textural changes in the surrounding (country) rocks, which become progressively more intense as the igneous contact is approached. The aureole, thus, represents a shell of changed or metamorphosed rock. *See* IGNEOUS ROCKS.

Aureoles are generally most pronounced where developed in previously unmetamorphosed or weakly metamorphosed rocks. They may not be detected where impressed upon intensely metamorphosed rocks. Shale and limestone generally show a more conspicuous aureole than does sandstone. The metamorphic effect around small igneous bodies is generally less extensive and intensive than that around large ones. The country rock against small dikes and sills may only be baked or indurated, whereas against stocks, large sheets, and laccoliths it is usually recrystallized. *See* PLUTON.

The inner boundary of the aureole (against the plutonic rock) may be sharp or transitional. Along some contacts abundant fragments (inclusions) of country rock appear to have been ripped off and enclosed in the granite, and apophyses (offshoots) of granite extend well into the country rock. Along other contacts, notably where granite comes against well-bedded or foliated rocks, migmatites (mixed rocks) have developed. *See* MIGMATITE. [C.A.C.]

Contact condenser A device in which a vapor is brought into direct contact with a cooling liquid and condensed by giving up its latent heat to the liquid. In almost all cases the cooling liquid is water, and the condensing vapor is steam. Contact condensers are classified as jet, barometric, and ejector condensers. In all three types the steam and cooling water are mixed in a condensing chamber and withdrawn together. Noncondensable gases are removed separately from the jet condenser, entrained in the cooling water of the ejector condenser, and removed either separately or entrained in the barometric condenser. The jet condenser requires a pump to remove the mixture of condensate and cooling water and a vacuum breaker to avoid accidental flooding. The barometric condenser is self-draining. The ejector condenser converts the energy of high-velocity injection water to pressure in order to discharge the water, condensate, and noncondensables at atmospheric pressure. *See* VAPOR CONDENSER. [J.F.Se.]

Contact potential difference The potential difference that exists across the space between two electrically connected metals; also, the potential difference between the bulk regions of a junction of two semiconductors. In illustration *a* consider two metals M_1 and M_2 with work functions ϕ_1, and ϕ_2, respectively. When the metals are brought in contact, the Fermi levels, that is, the levels corresponding to the most energetic electrons, must coincide, as in illustration *b*. Consequently, if $\phi_1 > \phi_2$, metal M_1 will acquire a negative and metal M_2 a

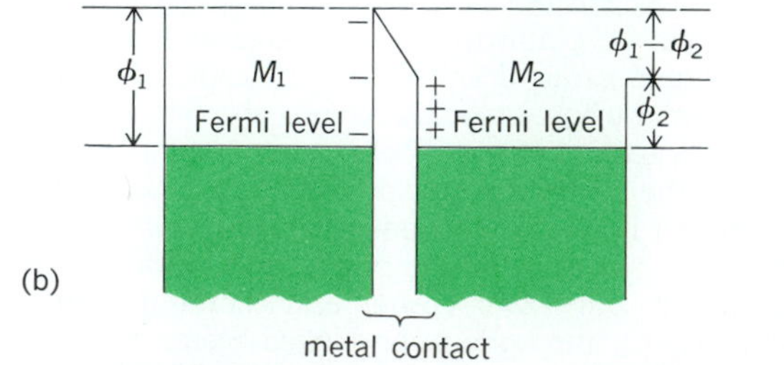

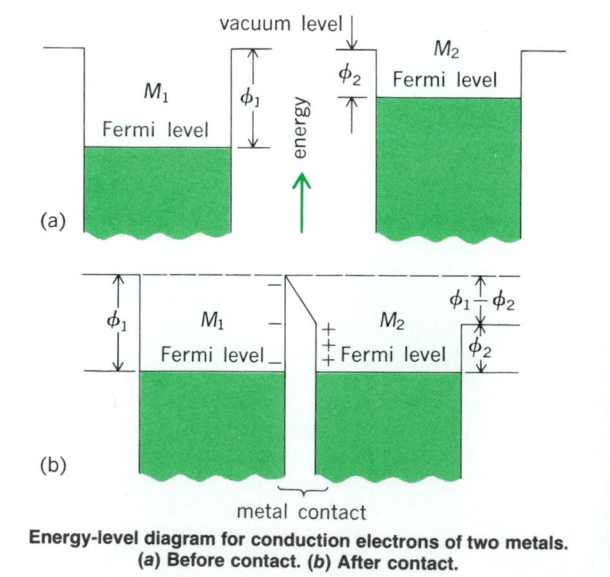

Energy-level diagram for conduction electrons of two metals. (*a*) Before contact. (*b*) After contact.

positive surface charge at the contact area. Contact potentials must be taken into account in analyzing physical electronics experiments. [A.J.D.]

Continent A protuberance of the Earth's crustal shell with an area of several million square miles and with sufficient elevation above neighboring depressions (the ocean basins) so that much of it is above sea level. There are six continents: Eurasia (Europe, China, and India are parts of this largest continent), Africa, North America, South America, Australasia (including Australia and New Guinea), and Antarctica.

The term was originally applied to the most extensive continuous land areas of the globe. The great majority of maps now in use imply that the boundaries of continents are their shorelines. From the geological point of view, however, the line of demarcation between a continent and an adjacent ocean basin lies offshore, at distances ranging from a few to several hundred miles, where the gentle slope of the continental shelf changes somewhat abruptly to a steeper declivity. This change occurs at depths ranging from a few to several hundred fathoms at different places around the periphery of various continents. *See* CONTINENTAL MARGIN.

All continents have similar structural features but display great variety in detail. Each includes a basement complex (shield) of metamorphosed sedimentary and volcanic rocks of Precambrian age, with associated igneous rocks, mainly granite. Originally formed at considerable depths below the surface, this shield was later exposed by extensive erosion, then largely covered by sediments of Paleozoic, Mesozoic, and Cenozoic age, chiefly marine limestones, shales, and sandstones. In at least one area on each continent, these basement rocks are now at the surface (an example is the Canadian Shield of North America). In some places they have disappeared beneath a sedimentary platform occupying a large fraction of the area of each continent, such as the area in the broad lowland drained by the Mississippi River in the United States. In each continent there are long belts of mountains in which thick masses of sedimentary rocks have been compressed into folds and broken by faults. *See* BASEMENT ROCK; OROGENY.

Continents are the less dense, subaerially exposed portion of the large and not-so-large plates that make up the Earth's lithosphere, or outer shell of rigid rock material. As such, continents together with part of the ocean's floor are intimately joined portions of the lithospheric plates. As plates rip apart and migrate horizontally over the Earth's surface, so too do continents rip apart and migrate, sometimes colliding with other continental segments scores of millions of years later.

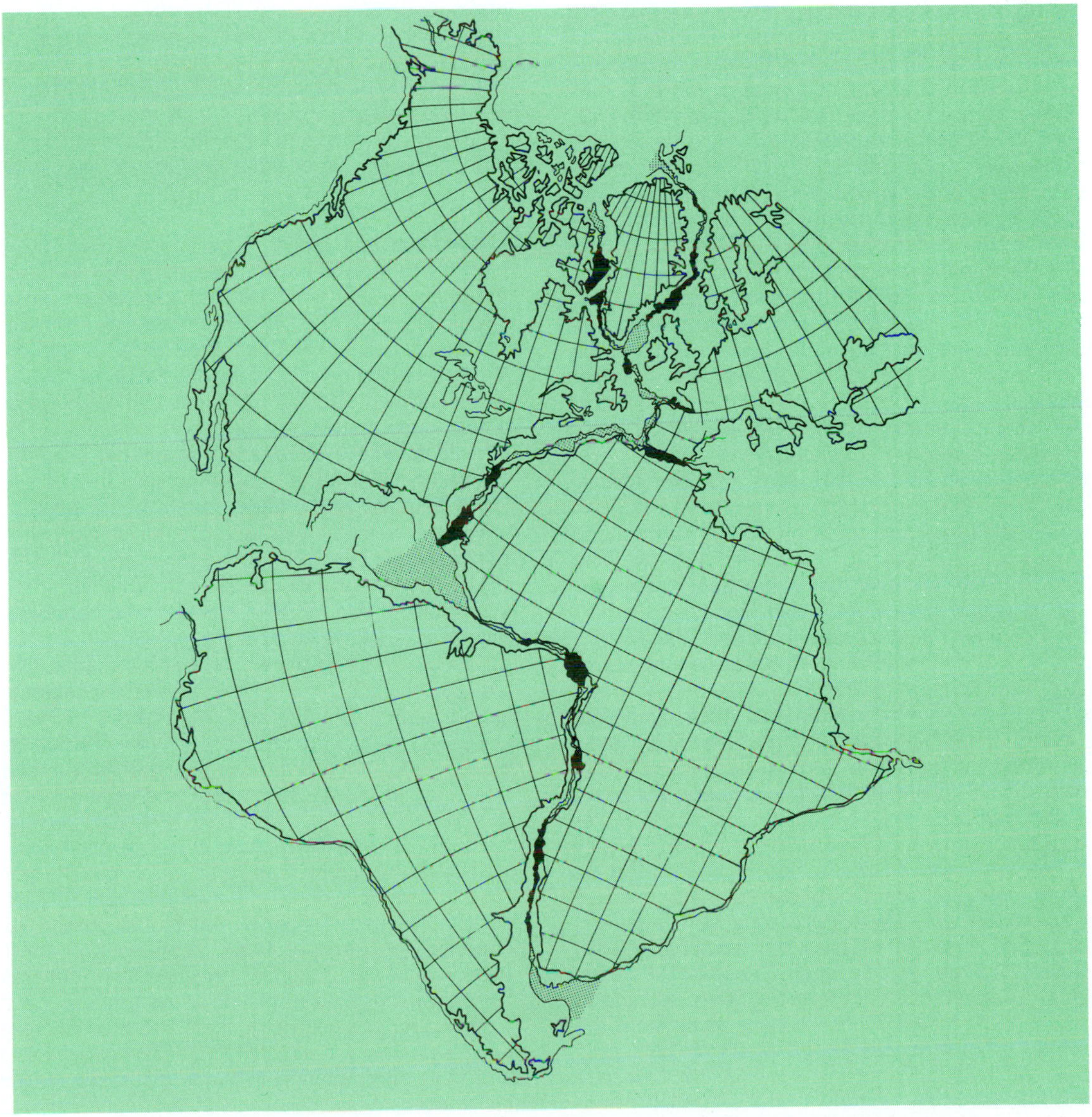

Map showing how the Atlantic margins of the Americas, Greenland, Europe, and Africa can be fitted together. The fit is made along the continental slope at the 500-fathom (3000-ft) contour line. Regions where the continents overlap are shown in black, gaps in the fit by shaded areas.

Such continental collision and accretion are believed to have occurred throughout most of the Earth's history. [D.L.J.]

The continents, or more accurately, the protocontinents, began to form in the final stages of the origin of the Earth. But because the precise processes contributing to the origin of the Earth are speculative, the initial stages of continent formation also remain enigmatic. It is known that the Earth accreted from clouds and chunks of solar matter about 4.6×10^9 years ago. What is not known in detail is the actual rate, or kind of accretionary process involved, or the precise composition of the particles. The Earth's accretion is assumed to have occurred over a period of perhaps 0.25 to 1×10^8 years. The crust of the primordial Earth, and especially the protocontinents, evolved during the final stages of the Earth's origin, essentially as the last stage of its accretionary growth.

Studies of the nature, and the ages, of continental rocks clearly indicate that the rate of continent formation has varied with time. The oldest continental rocks discovered to date occur in Minnesota, in southwestern Greenland, and in southern Rhodesia. In these places granitic rocks, as well as metamorphosed sedimentary rocks formed in ancient seas, range from 3.5 to almost 4×10^9 years old. These ancient sgments of the continents are almost as old as the oldest definitively dated lunar crust.

If the view that the Earth accreted inhomogeneously is correct, the early fragments of the continents were probably quite large, conceivably covering the entire Earth. If this conclusion is accepted, then the continents have decreased in area but have greatly increased in thickness and become more granitic as they have aged. Hence the more striking changes in the formation of the contemporary continents are the shape, area, and distribution over the Earth. Less dramatic but critical changes have occurred in the structure and composition of the continents, and in their relationships to the Earth's other crusts: the oceanic crust and the island arcs.

Generally, the continents are coupled with large fragments of oceanic crust and ride piggyback on the entire outer, cooler lithosphere. Great fragments, or plates, of the lithosphere appear to jostle and move over the softer upper mantle of the Earth. A crude analogy is that of the movements of great plates of Arctic and Antarctic ice in response to the currents in the underlying ocean. The most recent major episode of continental drift that has given the continents their present form and distribution began about 2×10^8 years ago. Immediately prior to that time, the Americas, India, Australia, and Antarctica were joined to Africa and Europe and formed one vast megacontinental cluster sometimes called Pangea. The reconstruction of events involved in the breakup of Pangea, and of one of its largest fragments, called Gondwana, which resulted in the opening of new oceans, especially the Atlantic and Indian oceans, has preoccupied earth scientists for decades (see illustration). *See* CONTINENTAL DRIFT.

Studies of the oceans have revealed many of the major features of the Earth's crusts and their origins. One of the most imposing and significant features of the sea floor is the "ridge and rise complex." This is a vast, largely submarine mountain chain dotted with numerous volcanoes. The complex is worldgirdling, extending some 37,000 mi (60,000 km) largely through the median parts of ocean basins. In the Pacific Ocean, however, one great segment of the complex, the East Pacific Rise, converges with the western edge of Central and North America. The ridge-rise complex is a tensional feature, with great central rifts from which emerge floods of a unique basaltic magma. These eruptions of basalt continuously add to the existing ocean crust, which is spreading away from the ridge-rise complex as it forms. *See* MARINE GEOLOGY.

The evidence for sea-floor spreading is severalfold. One indication of spreading lies in the patterns of the stripelike masses of basalt episodically welded to the crust. These basaltic stripes, which crudely parallel the ridges, have frozen into them the orientation of the Earth's magnetic field extant at the time of volcanic eruption. In places, the ocean floor is striped for about 600 mi (1000 km) or more from the zones of spreading and on both sides of this zone. The result is a series of symmetrical magnetic patterns, indicative of the rate of sea-floor spreading. Further corroboration of sea-floor spreading and continental drift is found in the patterns of sediments formed on the sea floor, and in the trench–island arc systems that almost encircle the Pacific, offshore from the continents. *See* ROCK MAGNETISM.

A tantalizing question is the fate of the striped ocean crust that is being conveyed away from the ridge-rise spreading centers, presumably by convection in the Earth's mantle. This process is a dynamic one that forces the great platelike masses of the lithosphere, with its overlying continents and ocean crust, against other plates. Patterns of plate movements, that is, plate tectonics, can be defined by their pushing and frictional movements relative to one another, because the movements of adjoining or colliding plates produce most of the world's earthquakes. A world map of recorded earthquakes defines the outlines and patterns of the great lithospheric plates active today throughout the world. Plate outlines and relative motions are further defined both by the direction of slip during the earthquakes and by the relative displacements of the magnetic strips frozen into the basalts of the sea floor. *See* PLATE TECTONICS.

One of the most fascinating relationships that emerges is the appearance of essentially all the world's recent mountain belts, the evolving island arcs, and the great trenches of the sea floor at sites where moving lithospheric plates are converging and either colliding or overriding one another. The worldwide pattern is completely consistent with the patterns of rift and drift of the continents and with the emergence of the Atlantic, Indian, and smaller oceans. *See* OCEANIC ISLANDS; OROGENY.

[A.E.J.E.]

Continental drift The concept that the world's continents once formed part of a single mass and have since drifted into their present positions. Although it was outlined by Alfred Wegener in 1912, the idea was not particularly new. Paleontological studies had already demonstrated such strong similarities between the flora and fauna of the southern continents between 300,000,000 and 150,000,000 years ago that a huge supercontinent, Gondwana, containing South America, Africa, India, Australia, and Antarctica, had been proposed. However, Gondwana was thought to be the southern continents linked by land bridges, rather than contiguous units.

Wegener's ideas were almost universely rejected in 1928; the fundamental objection was the lack of a suitable mechanism. Almost simultaneously with the temporary eclipse of Wegener's theory, Arthur Holmes was considering a mechanism that is still widely accepted. Holmes conceived the idea of convective currents within the Earth's mantle which were driven by the radiogenic heat produced by radioactive minerals within the mantle. At that time, Holmes's ideas, like those of Wegener, were largely ignored. Nonetheless, several geologists, particularly those living in the Southern Hemisphere, continued to believe the theory and accumulate more data in its support.

By the 1950s, convincing evidence had accumulated, with studies of the magnetization of rocks, paleomagnetism, beginning to provide numerical parameters on the past latitude and orientation of the continental blocks. Early work in North America and Europe clearly indicated how these continents had once been contiguous and had since separated. The discovery of the midoceanic ridge system also provided further evidence for the geometric matching of continental edges, but the discovery of magnetic anomalies parallel to these ridges and their interpretation in terms of sea-floor spreading finally led to almost universal acceptance of continental drift as a reality. *See* PALEOMAGNETISM; PLATE TECTONICS; TECTONOPHYSICS.

[D.H.T.]

Continental margin The submerged fringes of a continents, comprising three characteristic zones: the continental shelf, the continental slope, and the continental rise. The continental shelf is the zone around the continent, extending from the low-water line to the depth at which there is a marked increase in slope to greater depth. The continental slope is the declivity from the edge of the shelf extending down to great ocean depths. The shelf and slope comprise the continental terrace, which is the submerged fringe of the continent, connecting the shoreline with the 2½-mi-deep (4-km) abyssal ocean floor.

The continental shelf is a comparatively featureless plain, with an average width of 45 mi (72 km), sloping gently seaward at about 10 ft/mi (1.9 km). At a depth of about 70 fathoms (128 m) there generally is an abrupt increase in declivity called the shelf break, or the shelf edge. This break marks the limit of the shelf, the top of the continental slope, and the brink of the deep sea. For some purposes, especially legal, the 100-fathom (200-m) line is conventionally taken as the limit of the shelf. Characteristically, the shelves are thinly veneered with clastic sands, silts, and silty muds, which are patchily distributed. Geologically, the shelf is an extension of, and in unity with, the adjacent coastal plain. The position of the shoreline is geologically ephemeral, being subject to constant prograding and retrograding, so that its precise position at any particular time is not important.

The drowned edges of the low-density "granitic" or sialic continental masses are the continental slopes. The continental plateaus float like icebergs in the Earth's mantle with the slopes marking the transition between the low-density continents and the heavier oceanic segments on the Earth's crust. The continental slopes are the most imposing escarpments on the Earth. The slope is comparatively steep. Most slopes resemble a straight mountain front but are highly irregular in detail; in places they are deeply incised by submarine canyons, some of which cut deeply into the shelf. Usually the slope does not connect directly with the sea floor; instead there is a transitional area, the continental rise, or apron, built by the shredding of sediments from the continental block. *See* MARINE GEOLOGY; SUBMARINE CANYON. [R.S.D.]

Continentality (meteorology) The attributes of notably greater temperature ranges with associated weather characteristics that make inland and continental regions distinctly different from marine areas or areas bordering large bodies of water.

The reasons are associated with the greatly different thermal behavior of water and land surfaces. The soil surfaces of the continents can be heated or cooled comparatively rapidly because the thermal effects are limited to a shallow surface layer. The heat received at the ground warms the top layers, which consequently become hot in summer. In winter, by contrast, the surface becomes notably cold since there is little conduction of heat from subsurface depths. As a result of the responsiveness of the surface air temperature to the temperature of the underlying soil surface, the air in contact with the ground undergoes temperature fluctuations of large amplitude, lagging slightly behind the annual and the diurnal periods of insolation.

On the other hand, the water surfaces which make up the oceans are heated or cooled much less rapidly, partly because the thermal effects extend to greater depths and partly because much of the solar energy absorbed at the sea surface is utilized for evaporation rather than heating. As a result, the upper layers of the ocean are not heated as strongly as the land surfaces in summer or cooled as strongly by radiation in winter. It follows that large annual and diurnal ranges in temperature with hot summers and cold winters are characteristic of the continental interiors of middle and high latitudes. In contrast, the summers over the oceans are cool, while the winters are comparatively mild. [W.C.J.]

Continuous-wave radar A radar in which the transmitter output is uninterrupted, in contrast to pulse radar, where the output consists of short pulses. The chief disadvantage of a continuous-wave (CW) radar is that it is difficult to use a single antenna for both reception and transmission, because both the transmitter and receiver are on all the time. This difficulty has been overcome by means of the use of isolation circuitry, which gives the receiver protection from a transmitter output up to 200 watts. Among the advantages of CW radar is its ability to measure velocity with extreme accuracy by means of the Doppler shift in the frequency of the echo. *See* DOPPLER RADAR; RADAR.

In order to measure the range of targets, some form of frequency modulation of the CW output must be used. In one very effective form of modulation, the carrier frequency of the transmitted signal is varied at a uniform rate. Range is determined by comparing the frequency of the echo with that of the transmitter, the difference being proportional to the range of the target that produced the echo. Systems in which this is done are known as FM-CW radars. A modified form of FM-CW radar employs long, but not continuous, transmission. This might be regarded as the same as transmitting extremely long pulses on a frequency-modulated carrier. Systems of this type are referred to as pulse compression radars. *See* FREQUENCY MODULATION.

The principal use of CW radar is in short-range missile guidance. Typically the missile's course is tracked from the ground while the missile is simultaneously illuminated. CW radar, in some cases the same radar, can be used both for tracking and illumination, although it is more common for pulse Doppler radar to be used for tracking.

A variant of CW radar is used to illuminate persons under surveillance by techniques employing semiconductor tracer diodes. These devices are either secreted on the subject's person without his or her knowledge or else concealed in protected objects the subject has stolen. Despite the fact that pulsed X-band sources would provide the necessary high power levels more conveniently, the requirement to reduce clutter makes it essential to use CW power. [J.M.C.]

Contour The locus of points of equal elevation used in topographic mapping. Contour lines represent a uniform series of elevations, the difference in elevation between adjacent lines being the contour interval of the given map. Thus, contours represent the shape of terrain on the flat map surface (see illustration). Closely spaced contours indicate steep ground; sparseness or absence of contours indicates gentle slope or flat ground. Contours do not cross each other unless there is an overhang. *See* TOPOGRAPHIC SURVEYING AND MAPPING. [R.H.Do.]

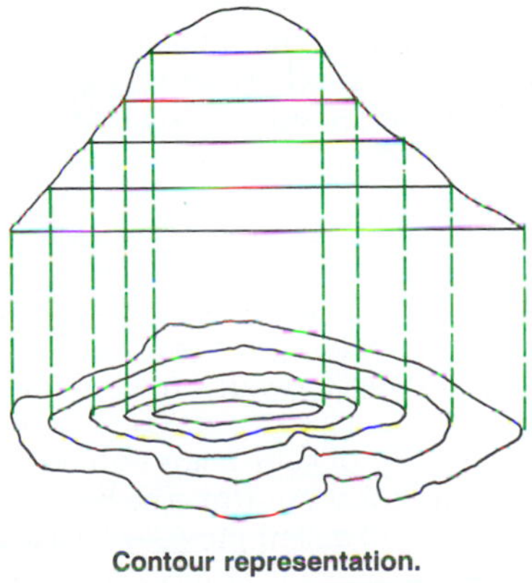

Contour representation.

Control charts Charts for the analysis and presentation of data obtained from production processes and research investigations. They are also used to display data in other commercial applications, such as labor turnover, office efficiency, costs, and accident rates. In the simplest charts, results are plotted in sequence over time, so that trends and changes may be identified and acted upon. In a more advanced form, limits calculated by statistical methods are placed on the chart. Provided that results stay within these limits, no action is taken, but if any results fall outside the limits, some action is indicated. Such statistical control charts are known as Shewhart charts.

Due to natural variations in raw materials and processes, it is rare to find any two manufactured units which are exactly alike. There is also a tendency for industrial processes to change over time (due to tool wear and changes in external conditions, for example). In order to detect such changes, quality-control personnel take samples from the process output and make measurements on important characteristics of the units in the sample. In isolation the results of individual samples are not very informative, but when plotted on a control chart they provide an effective cumulative picture of process behavior. Together with a history of changes made to the process, they provide an invaluable diagnostic tool to production supervision.

Even when a process is running satisfactorily there is still some variation in results about the target value. Such variations are due to a system of chance causes, and provided they can be tolerated in the finished product, it is unnecessary and usually uneconomical to try and reduce them. Under these conditions the process is said to be in a state of statistical control. However, changes which move the process average away from its target must be identified and dealt with. These changes are due to assignable causes, as distinct from chance causes. The object of Shewhart control charts is to provide rules for determining whether changes are more likely due to assignable causes or chance variation.

The statistical theory on which Shewhart charts are based depends on the assumption that the underlying data are normally distributed, and this must be kept in mind when interpreting charts which are constructed by other methods, such as average characteristics, fraction defective, or defects per unit. Misleading indications can result from the assumptions being violated, and therefore a thorough analysis of the data by skilled statisticians may be necessary. Custom-designed charts may be required in some cases. *See* CONTROL SYSTEMS; INDUSTRIAL COST CONTROL; INSPECTION AND TESTING; QUALITY CONTROL.

[J.A.Cl.]

Control systems Interconnections of components forming system configurations which will provide a desired system response as time progresses. The steering of an automobile is a familiar example. The driver observes the position of the car relative to the desired location and makes corrections by turning the steering wheel. The car responds by changing direction, and the driver attempts to decrease the error between the desired and actual course of travel. In this case, the controlled output is the automobile's direction of travel, and the control system includes the driver, the automobile, and the road surface, The control engineer attempts to design a steering control mechanism which will provide a desired response for the automobile's direction control. Different steering designs and automobile designs result in rapid responses, as in the case of sports cars, or relatively slow and comfortable responses, as in the case of large autos with power steering.

Open- and closed-loop control. The basis for analysis of a control system is the foundation provided by linear system theory, which assumes a cause-effect relationship for the components of a system. A component or process to be controlled can be represented by a block. Each block possesses an input (cause) and output (effect). The input-output relation represents the cause-and-effect relationship of the process, which in turn represents a processing of the input signal to provide an output signal variable, often with power amplification. An open-loop control system utilizes a controller or control actuator in order to obtain the desired response (Fig. 1).

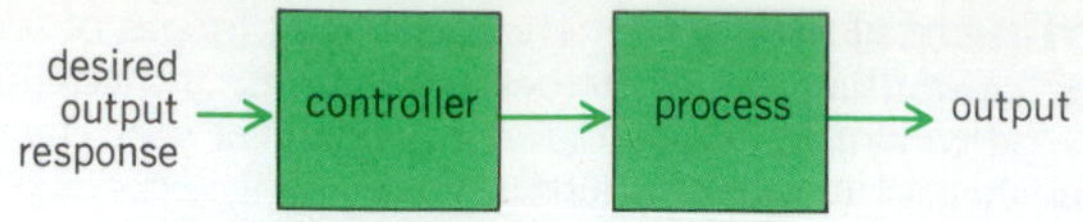

Fig. 1. Open-loop control system.

In contrast to an open-loop control system, a closed-loop control system utilizes an additional measure of the actual output in order to compare the actual output with the desired output response (Fig. 2). A standard definition of a feedback control system is a control system which tends to maintain a prescribed relationship of one system variable to another by comparing functions of these variables and using the difference as a means of control. In the case of the driver steering an automobile, the driver uses his or her sight to visually measure and compare the actual location of the car with the desired location. The driver then serves as the controller, turning the steering wheel. The process represents the dynamics of the steering mechanism and the automobile response.

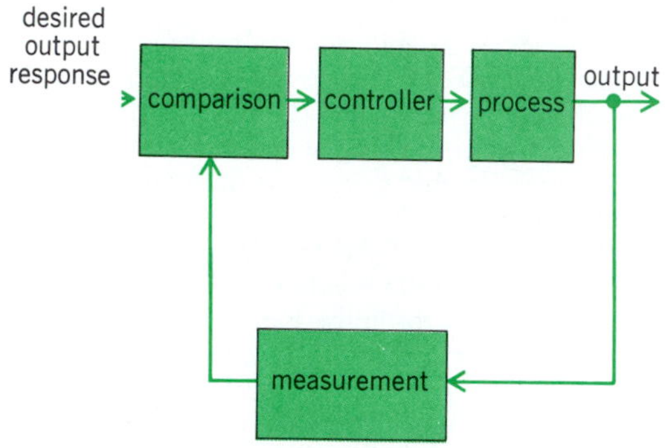

Fig. 2. Closed-loop control system.

A feedback control system often uses a function of a prescribed relationship between the output and reference input to control the process. Often, the difference between the output of the process under control and the reference input is amplified and used to control the process so that the difference is continually reduced. The feedback concept has been the foundation for control system analysis and design.

Applications for feedback systems. Familiar control systems have the basic closed-loop configuration. For example, a refrigerator has a temperature setting for desired temperature, a thermostat to measure the actual temperature and the error, and a compressor motor for power amplification. Other examples in the home are the oven, furnace, and water heater. In industry, there are controls for speed, process temperature and pressure, position, thickness, composition, and quality, among many others. Feedback control concepts have also been applied to mass transportation, electric power systems, automatic warehousing and inventory control, automatic control of agricultural systems, biomedical experimentation and biological control systems, and social, economic, and political systems. *See* BIOMEDICAL ENGINEERING; ELECTRIC POWER SYSTEMS; MATHEMATICAL BIOLOGY; SYSTEMS ANALYSIS; SYSTEMS ENGINEERING.

Advantages of feedback control. The addition of feedback to a control system results in several important advantages. A process, whatever its nature, is subject to a changing environment, aging, ignorance of the exact values of the process parameters, and other natural factors which affect a control process. In the open-loop system, all these errors and changes result in a changing and inaccurate output. However, a closed-loop system senses the change in the output due to the process changes and attempts to correct the output. The sensitivity of a control system to parameter variations is of prime importance. A primary advantage of a closed-loop feedback control system is its ability to reduce the system's sensitivity.

One of the most important characteristics of control systems is their transient response, which often must be adjusted until it is satisfactory. If an open-loop control system does not provide a satisfactory response, then the process must be replaced or modified. By contrast, a closed-loop system can often be adjusted to yield the desired response by adjusting the feedback loop parameters.

A second important effect of feedback in a control system is the control and partial elimination of the effect of disturbance signals. Many control systems are subject to extraneous disturbance signals which cause the system to provide an inaccurate output. Feedback systems have the beneficial aspect that the effect of distortion, noise, and unwanted disturbances can be effectively reduced.

Costs of feedback control. While the addition of feedback to a control system results in the advantages outlined above, it is natural that these advantages have an attendant cost. The cost of feedback is first manifested in the increased number of components and the complexity of the system. The second cost of feedback is the loss of gain. Usually, there is open-loop gain to spare, and one is more than willing to trade it for increased control of the system response. Finally, a cost of feedback is the introduction of the possibility of instability. While the open-loop system is stable, the closed-loop system may not be always stable.

Stability of closed-loop systems. The transient response of a feedback control system is of primary interest and must be investigated. A very important characteristic of the transient performance of a system is the stability of the system. A stable system is defined as a system with a bounded system response. That is, if the system is subjected to a bounded input or disturbance and the response is bounded in magnitude, the system is said to be stable.

The concept of stability can be illustrated by considering a right circular cone placed on a plane horizontal surface. If the cone is resting on its base and is tipped slightly, it returns to its original equilibrium position. This position and response is said to be stable. If the cone rests on its side and is displaced slightly, it rolls with no tendency to leave the position on its side. This position is designated as neutral stability. On the other hand, if the cone is placed on its tip and released, it falls onto its side. This position is said to be unstable.

The stability of a dynamic system is defined in a similar manner. The response to a displacement, or initial condition, will result in either a decreasing, neutral, or increasing response.

Design. A feedback control system that provides an optimum performance without any necessary adjustments is rare indeed. Usually one finds it necessary to compromise among the many conflicting and demanding specifications and to adjust the system parameters to provide a suitable and acceptable performance when it is not possible to obtain all the desired optimum specifications.

It is often possible to adjust the system parameters in order to provide the desired system response. However, it is often not possible to simply adjust a system parameter and thus obtain the desired performance. Rather, the scheme or plan of the system must be reexamined, and a new design or plan must be obtained which results in a suitable system. Thus, the design of a control system is concerned with the arrangement, or the plan, of the system structure and the selection of suitable components and parameters. For example, if one desires a set of performance measures to be less than some specified values, one often encounters a conflicting set of requirements. If these two performance requirements cannot be relaxed, the system must be altered in some way. The alteration or adjustment of a control system, in order to make up for deficiencies and inadequacies and provide a suitable performance, is called compensation.

In redesigning a control system in order to alter the system response, an additional component or device is inserted within the structure of the feedback system to equalize or compensate for the performance deficiency. The compensating device may be an electric, mechanical, hydraulic, pneumatic, or other-type device or network, and is often called a compensator.

Digital computer systems. The use of a digital computer as a compensator device has grown since 1970 as the price and reliability of digital computers have improved dramatically.

Within a computer control system, the digital computer receives and operates on signals in digital (numerical) form, as contrasted to continuous signals. The measurement data are converted from analog form to digital form by means of a converter. After the digital computer has processed the inputs, it provides an output in digital form, which is then converted to analog form by a digital-to-analog converter. *See* ANALOG-TO-DIGITAL CONVERTER; DIGITAL-TO-ANALOG CONVERTER.

There are more than 1,000,000 control systems using a computer in industry, although the computer size and power may vary significantly. If only computer control systems of a relatively complex nature are considered, such as chemical process control or aircraft control, the number of computer control systems is approximately 100,000. Automatic handling equipment for home, school, and industry is particularly useful for hazardous, repetitious, dull, or simple tasks. Machines that automatically load and unload, cut, weld, or cast are used by industry in order to obtain accuracy, safety, economy, and productivity. Robots are programmable computers integrated with machines. They often substitute for human labor in specific repeated tasks. Some devices even have anthropomorphic mechanisms, including what might be recognized as mechanical arms, wrists, and hands. Robots may be used extensively in space exploration and assembly. They can be flexible, accurate aids on assembly lines. *See* ROBOTICS. [R.C.D.]

Controlled rectifier

Controlled rectifier A three-terminal semiconductor junction device with four regions of alternating conductivity type (*p-n-p-n*), also called a thyristor. This switching device has a characteristic such that, once it conducts, the voltage in the circuit in which it conducts must drop below a threshold before the controlled rectifier regains control. Such devices are useful as high-current switches and may be used to drive electromagnets and relays.

The principle of operation can be understood by referring to the illustration. The central junction is reverse-biased (positive

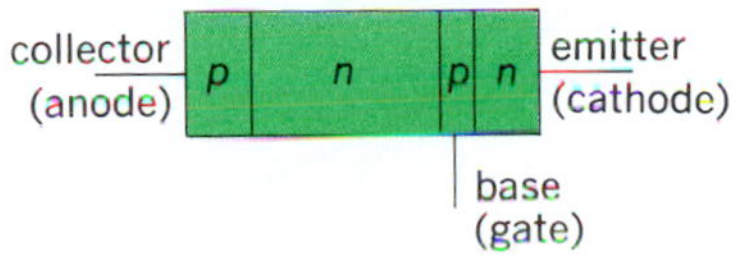

Diagram of a controlled rectifier.

collector, grounded emitter). The wide *n* region between collector and base regions prevents holes injected at the collector junction from reaching the collector-to-base barrier by diffusion. The junction between emitter and base is the emitter. When operated as a normal transistor, this device shows a rapid increase of current gain with collector current. This effect may be due to a field-induced increase of transport efficiency across the floating *n* region, or to increased avalanching in the high-field barrier region, or to increased injection efficiency at the two forward-biased junctions, or to a combination of these phenomena.

If this device is operated as a three-terminal device, the switching between the nonconducting and conducting states can be controlled by the base. *See* SEMICONDUCTOR RECTIFIER. [L.P.H.]

Conularida A small group of extinct invertebrates showing a fourfold symmetry and a narrow pyramidal shape; the cross section commonly is square. Specimens are chitinous and frequently impregnated with calcium phosphate. The four side walls are characteristically ornamented by numerous transverse lines. The posterior tip may bear an attachment disk or may be broken off and sealed with a septum. Internal structures are rare; the few known show fourfold symmetry. Conularida are regarded as a subclass of the Coelenterata because of the characteristic symmetry. *See* COELENTERATA.

Specimens are worldwide in occurrence and are known from rocks of Cambrian through Triassic age. They have been found in sandstone, shales, and limestones. About 20 genera and 150 species are known. Except for a few local occurrences, Conularida are rare fossils. [E.L.Y.]

Convection (heat) The transfer of thermal energy by actual physical movement from one location to another of a substance in which thermal energy is stored. A familiar example is the free or forced movement of warm air throughout a room to provide heating. Technically, convection denotes the nonradiant heat exchange between a surface and a fluid flowing over it. Although heat flow by conduction also occurs in this process, the controlling feature is the energy transfer by flow of the fluid—hence the name convection. Convection is one of the three basic methods of heat transfer, the other two being conduction and radiation. *See* CONDUCTION (HEAT); HEAT RADIATION; HEAT TRANSFER.

Natural convection is exemplified by the cooling of a vertical surface in a large quiescent body of air of temperature t_∞. The lower-density air next to a hot vertical surface moves upward because of the buoyant force of the higher-density cool air farther away from the surface. At any arbitrary vertical location *x*, the actual variation of velocity *u* with distance *y* from the surface will be similar to that in the illustration *b*, increasing from zero at the surface to a maximum, and then decreasing to zero as ambient surrounding conditions are reached. In contrast, the temperature *t* of the air decreases from the heated wall value *t's* to the surrounding air temperature. These temperature and velocity distributions are clearly interrelated, and the distances from the wall through which they exist are coincident because, when the temperature approaches that of the surrounding air, the density difference causing the upward flow approaches zero.

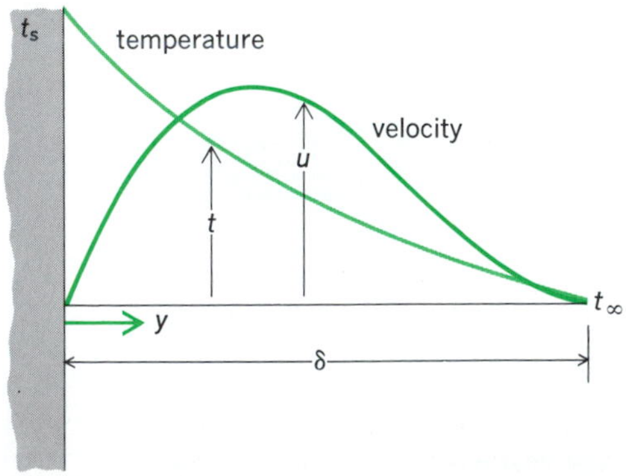

Temperature and velocity distributions in air near a heated vertical surface at arbitary vertical location. The distance δ is that distance at which the velocity and the temperature reach ambient surrounding conditions.

The region in which these velocity and temperature changes occur is called the boundary layer. Because velocity and temperature gradients both approach zero at the outer edge, there will be no heat flow out of the boundary layer by conduction or convection. *See* BOUNDARY-LAYER FLOW.

When air is blown across a heated surface, forced convection results. Although the natural convection forces are still present in this latter case, they are clearly negligible compared with the imposed forces. The process of energy transfer from the heated surface to the air is not, however, different from that described for natural convection. The major distinguishing feature is that the maximum fluid velocity is at the outer edge of the boundary layer. This difference in velocity profile and the higher velocities provide more fluid near the surface to carry along the heat conducted normal to the surface. Consequently, boundary layers are very thin.

Heat convection in turbulent flow is interpreted similarly to that in laminar flow. Rates of heat transfer are higher for comparable velocities, however, because the fluctuating velocity components of the fluid in a turbulent flow stream provide a macroscopic exchange mechanism which greatly increases the transport of energy normal to the main flow direction. Because of the complexity of this type of flow, most of the information regarding heat transfer has been obtained experimentally. *See* LAMINAR FLOW; TURBULENT FLOW.

Convection heat transfer which occurs during high-speed flight or high-velocity flow over a surface is known as aerodynamic heating. This heating effect results from the conversion of the kinetic energy of the fluid as it approaches a body to internal energy as it is slowed down next to the surface. In the case of a gas, its temperature increases, first, because of compression as it passes through a shock and approaches the stagnation region, and second, because of frictional dissipation of kinetic energy in the boundary layer along the surface.

The phenomena of condensation and boiling are important phase-change processes involving heat release or absorption. Because vapor and liquid movement are present, the energy transfer is basically by convection. Local and average heat-transfer coefficients are determined and used in the Newton cooling-law equation for calculating heat rates which include the effects of the latent heat of vaporization. [W.H.Gi.]

Convective instability A state of fluid flow in which the distribution of body forces along the direction of the net body force is unstable and will thus break down. Fluid flows are subject to a variety of instabilities, which may be broadly viewed as the means by which relatively simple flows become more complex. Instabilities are an important step in the transition between smooth and turbulent flow, and in the atmosphere they are responsible for phenomena ranging from thunderstorms to low- and high-pressure systems. Meteorologists and oceanographers divide instabilities into two broad classes: convective and dynamic. *See* DYNAMIC INSTABILITY.

In the broadest terms, convective instabilities arise when the displacement of a small parcel of fluid causes a force on that parcel which is in the same direction as the displacement. The parcel of fluid will then continue to accelerate away from its initial position, and the fluid is said to be unstable. In most geophysical flows, the convective motions that result from convective instabilities operate very quickly compared with the processes acting to destabilize the fluid; the result is that such fluids seem to be nearly neutrally stable to convection.

The simplest type of convective instability arises when a fluid is heated from below or cooled from above. Warm air rises and cold air sinks; thus a fluid whose temperature decreases with altitude is convectively unstable, while one in which the temperature increases with height is convectively stable. A fluid at constant temperature is said to be convectively neutral. *See* ARCHIMEDES' PRINCIPLE.

The above description assumes that density depends on temperature alone and that density is conserved in parcels of fluid, so that when the parcels are displaced their density does not change. However, in the Earth's atmosphere, neither of these assumptions is true. In the first place, the density of air depends on pressure and on the amount of water vapor in the air, in addition to temperature. Second, the density will change when the parcel is displaced because both its pressure and its temperature will change. Because of these conditions it is convenient to define a quantity known as virtual potential temperature (θ_v) that both is conserved and reflects the actual density of air. This quantity is given by the equation below, where T is the

$$\theta_v \equiv T\left(\frac{1 + (r/0.622)}{1 + r}\right)\left(\frac{1000}{p}\right)^{0.287}$$

temperature in kelvins, p is the pressure in millibars, and r is the number of grams of water vapor in each gram of dry air. When θ_v decreases with height in the atmosphere, it is convectively unstable, while θ_v increasing with height denotes stability.

In the oceans, density is a function of pressure, temperature, and salinity; convection there is driven by cooling of the ocean surface by evaporation of water into the atmosphere and by direct loss of heat when the air is colder than the water. It is also driven by salinity changes resulting from precipitation and evaporation. In many regions of the ocean, a convectively driven layer exists near the surface in analogy with the atmospheric convective layer. This oceanic mixed layer is also nearly neutral to convection.

Convective instabilities are also responsible for convection in the Earth's mantle, which among other things drives the motion of the plates, and for many of the motions of gases within other planets and in stars. [K.A.Em.]

Converter A device for processing alternating-current (ac) or direct-current (dc) power to provide a different electrical waveform. The term converter denotes a mechanism for either processing ac power into dc power (rectifier) or deriving power with an ac waveform from dc (inverter). Some converters serve both functions, others only one. *See* ALTERNATING CURRENT; DIRECT CURRENT; RECTIFIER.

Converters are used for such applications as (1) rectification from ac to supply electrochemical processes with large controlled levels of direct current; (2) rectification of ac to dc followed by inversion to a controlled frequency of ac to supply variable-speed ac motors; (3) interfacing dc power sources (such as fuel cells and photoelectric devices) to ac distribution systems; (4) production of dc from ac power for subway and streetcar systems, and for controlled dc voltage for speed-control of dc motors in numerous industrial applications; and (5) transmission of dc electric power between rectifier stations and inverter stations within ac generation and transmission networks. *See* ALTERNATING-CURRENT MOTOR; DIRECT-CURRENT TRANSMISSION.

The introduction of the thyristor (silicon-controlled rectifier) in the 1960s had an immediate effect on converter applications because of its ruggedness, reliability, and compactness. Power semiconductor devices for converter circuits include (1) thyristors, controlled unidirectional switches that, once conducting, have no capability to suppress current; (2) triacs, thyristor devices with bidirectional control of conduction; (3) gate turn-off devices with the properties of thyristors and the further capability of suppressing current; and (4) power transistors, high-power transistors operating in the switching mode, somewhat similar in properties to gate turn-off devices. Thyristors are available with ratings from a few watts up to the capability of withstanding several kilovolts and conducting several kiloamperes. *See* SEMICONDUCTOR RECTIFIER. [J.Re.]

Convertiplane An aircraft combining vertical takeoff and landing capabilities of the helicopter with forward-flight effectiveness and high-speed potentials of the airplane. In forward flight a convertiplane relies, at least partially, on the fixed wing, while for vertical takeoffs, landings, and hovering, a separate vertical thrust generator is provided. Between vertical and forward-flight regimes, the aircraft goes through a conversion.

Of many systems suitable as vertical-thrust generators, those applied to practical convertiplanes will probably be either of the direct type, in which combustion products are used directly for thrust generation (turbojets and rockets), or of the indirect type, in which all or a part of the available combustion energy is used to drive a mechanical system, functioning as an actuator disk to accelerate the ambient air. The actuator disks may be either free (helicopter rotors and propellers) or shrouded (shrouded propellers and ducted fans).

The principle of the convertiplane can be applied to many types of aircraft. The first practical attempts were directed toward rotary-wing aircraft. The presently used scheme consists of transferring as much lift as possible in forward flight from the rotor (which is either put into autorotation, or slowed down) to more efficient fixed wings, while a propeller (or propellers) provides the forward thrust. The MacDonnell XV-1 (Fig. 1) and the Fairey Rotodyne are examples of the pioneering efforts.

Fig. 1. In McDonnell XV-1 helicopter rotor serves for hovering, and fixed wing and propellers provide lift and propulsion in forward flight.

The so-called X-wing design represents another approach to the high-speed problem of helicopter-based convertiplanes. Here, the circulation-controlled, reaction-driven, four-bladed, rigid rotor is stopped in flight to form an X-shaped wing. Jet propulsion should assure high-subsonic-speed capabilities for this aerodynamically clean aircraft (Fig. 2).

In the fixed-wing, free-airscrew group, the original Bell XV-3 of the 1950s paved the way for the development of a flight research aircraft with 373 mi/h (600 km/h) speed capabili-

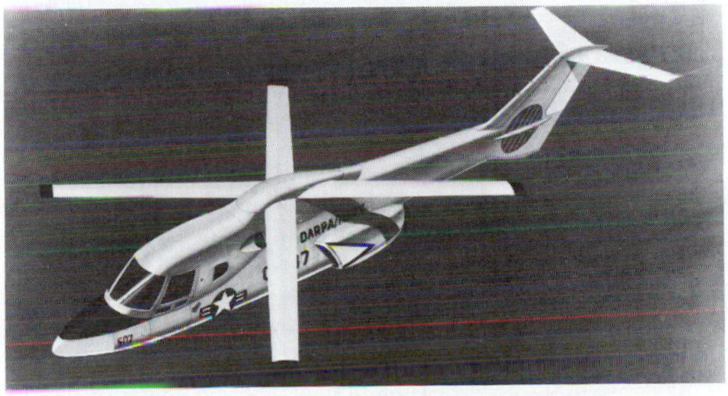

Fig. 2. An early version of the X-wing aircraft, as proposed by Lockheed. (*Lockheed Aircraft Corp.*)

Fig. 3. NASA-Army-Bell XV-15 in airplane configuration after conversion from helicopter mode. (*Bell Helicopter Textron*)

ties—the NASA-Army-Bell XV-15, which in July 1979 went through complete conversion (Fig. 3).

The Hawker Siddeley Harrier, with Bristol Siddeley Pegasus vectored-thrust turbofan engines, is a fighter in quantity production for the military forces (including the U.S. Marines) of several Western countries (Fig. 4), while, similar in concept, the

Fig. 4. Hawker Siddeley Harrier. Close ground support and reconnaissance are provided through use of vectored-thrust turbofans.

Yakovlev single-seat, Carrier-based combat aircraft is in service in Russia. *See* ORNITHOPTER; VERTICAL TAKEOFF AND LANDING (VTOL). [W.Z.S.]

Conveying machines A family of materials-handling machines designed to move individual articles of solid, free-flowing bulk materials over a fixed horizontal, inclined, declined, or vertical path of travel with continuous motion. The actuating medium may be the force of gravity, air, vibration, or power-operated components of the machine such as continuous belts, chains, or cables. Of the many varieties, some are used almost exclusively to convey bulk materials. The kinds described in this article fall largely in the category popularly referred to as package conveyors, and consist of gravity and powered belt, chain, and cable conveyors. *See* BULK-HANDLING MACHINES.

Gravity conveyors provide the most economical means for lowering articles and materials. Chutes depend upon sliding friction to control the rate of descent; wheel and roller conveyors use rolling friction for this purpose.

Gravity chutes may be made straight or curved and are fabricated from sheet metal or wood, the latter being sometimes covered with canvas to prevent slivering. The bed of the chute can be shaped to accommodate the products to be handied. In spiral chutes centrifugal force is the second controlling factor. Spirals with roller beds or wheels provide smooth descent of an article and tend to maintain the position of the article in its original starting position. Rollers may be constructed of metals, wood, or plastic and can be arranged in an optimum position, depending upon the articles to be carried.

To move loads on level or inclined paths, or declining paths that exceed the angle of sliding or rolling friction of the particular material to be conveyed, powered conveyors must be employed. There are various types of powered conveyors. Belt conveyors move loads on a level or inclined path by means of power-driven belts. Belt conveyors with rough-top belts make possible inclines up to 28°; cleated belts are limited on degree of incline only by the position of the center of gravity of the conveyed item.

Live-roller conveyors move objects over series of rollers by the application of power to all or some of the rollers. The power-transmitting medium is usually belting or chain.

There are three basic types of chain conveyors: (1) those that support the product being conveyed, (2) those that carry actuating elements between the two chains, and (3) those that operate overhead or are set into the floor. In the sliding chain conveyor either plain links or links with such special attachments as lugs are made up into the conveyors to handle cases, cans, pipes, and other similar products.

Slats or apron conveyors are fitted with slats of wood or metal, either flat or in special shapes, between two power chains. The slats handle such freight as barrels, drums, and crates. Push-bar conveyors are a variation of the slat conveyor, in which two endless chains are cross-connected at intervals by pusher bars which propel the load along a stationary bed or trough of the conveyor. [A.M.P.]

Cooling tower A tower- or building-like device in which atmospheric air (the heat receiver) circulates in direct or indirect contact with warmer water (the heat source) and the water is thereby cooled (see illustration). A cooling tower may serve

Counterflow natural-draft cooling tower at Trojan Power Plant in Spokane, Washington. (*Research–Cottrell*)

as the heat sink in a conventional thermodynamic process, such as refrigeration or steam power generation, or it may be used in any process in which water is used as the vehicle for heat removal, and when it is convenient or desirable to make final heat rejection to atmospheric air. Water, acting as the heat-transfer fluid, gives up heat to atmospheric air, and thus cooled, is recirculated through the system, affording economical operation of the process.

Two basic types of cooling towers are commonly used. One transfers the heat from warmer water to cooler air mainly by an evaporation heat-transfer process and is known as the evaporative or wet cooling tower. Evaporative cooling towers are classified according to the means employed for producing air circulation through them: atmospheric, natural draft, and mechanical draft. The other transfers the heat from warmer water to cooler air by a sensible heat-transfer process and is known as the nonevaporative or dry cooling tower. Nonevaporative cooling towers are classified as air-cooled condensers and as air-cooled heat exchangers, and are further classified by the means used for producing air circulation through them. These two basic types are sometimes combined, with the two cooling processes generally used in parallel or separately, and are then known as wet-dry cooling towers.

Evaluation of cooling tower performance is based on cooling of a specified quantity of water through a given range and to a specified temperature approach to the wet-bulb or dry-bulb temperature for which the tower is designed. Because exact design conditions are rarely experienced in operation, estimated performance curves are frequently prepared for a specific installation, and provide a means for comparing the measured performance with design conditions. [J.F.Se.]

Coordinate systems Schemes for locating points in a given space by means of numerical quantities specified with respect to some frame of reference. These quantities are the coordinates of a point. To each set of coordinates there corresponds just one point in any coordinate system, but there are useful coordinate systems in which to a given point there may correspond more than one set of coordinates.

A coordinate system is a mathematical language that is used to describe geometrical objects analytically; that is, if the coordinates of a set of points are known, their relationships and the properties of figures determined by them can be obtained by numerical calculations instead of by other descriptions. It is the province of analytic geometry, aided chiefly by calculus, to investigate the means for these calculations.

The most familiar spaces are the plane and the three-dimensional euclidean space. in the latter a point P is determined by three coordinates (x,y,z). The totality of points for which x has a fixed value constitutes a surface. The same is true for y and z so that through P there are three coordinate surfaces. The totality of points for which x and y are fixed is a curve and through each point there are three coordinate lines. If these lines are all straight, the system of coordinates is said to be rectilinear. If some or all of the coordinate lines are not straight, the system is curvilinear. If the angles between the coordinate lines at each point are light angles, the system is rectangular.

A cartesian coordinate system is one of the simplest and most useful systems of coordinates. It is constructed by choosing a point O designated as the origin. Through it three intersecting directed lines OX, OY, OZ, the coordinate axes, are constructed. The coordinates of a point P are x, the distance of P from the plane YOZ measured parallel to OX, and y and z, which are determined similarly (Fig. 1). Usually the three axes are taken to be mutually perpendicular, in which case the system is a rectangular cartesian one. Obviously a similar construction can be made in the plane, in which case a point has two coordinates (x,y).

A polar coordinate system is constructed in the plane by choosing a point O called the pole and through it a directed straight line, the initial line. A point P is located by specifying the directed distance OP and the angle through which the initial line must be turned to coincide with OP in position and direction. The coordinates of P are (r,θ). The radius vector r is the directed line OP, and the vectorial angle θ is the angle through which the initial line was turned, + if turned counterclockwise, − if clockwise.

Spherical coordinates are constructed in three-dimensional euclidean space by choosing a plane and in it constructing a polar coordinate system. At the pole O a polar axis OZ is constructed at right angles to the chosen plane. A point P, not on OZ, and OZ determine a plane. The spherical coordinates of P are then the directed distance OP denoted by ρ, the angle θ through which the initial line is turned to lie in ZOP and the angle $\phi = ZOP$ (Fig. 2).

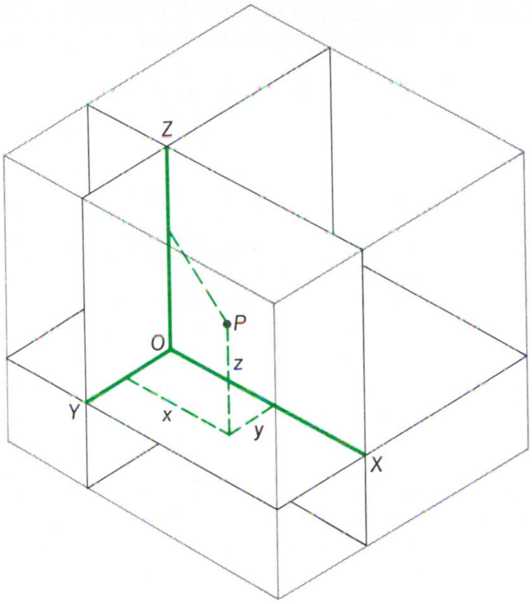

Fig. 1. Cartesian coordinate system.

Cylindrical coordinates are constructed by choosing a plane with a pole O, an initial line in it, and a polar axis OZ, as in spherical coordinates. A point P is projected onto the chosen plane. The cylindrical coordinates of P are (r,θ,z) where r and θ are the polar coordinates of Q and $z = QP$ (Fig. 2).

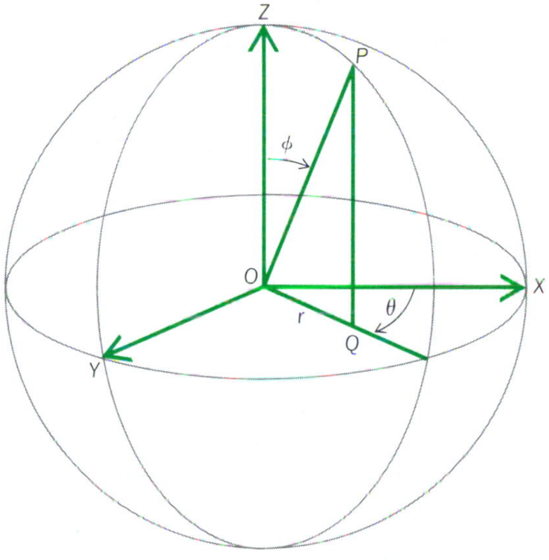

Fig. 2. Spherical coordinate system.

By means of a system of equations the description of a geometrical object in one coordinate system may be translated into an equivalent description in another coordinate system. *See* ANALYTIC GEOMETRY; CALCULUS; CONFORMAL MAPPING; CURVE FITTING; SPHERICAL HARMONICS. [M.S.Kn.]

Coordination chemistry A field which, in its broadest usage, is acid-base chemistry as defined by G. N. Lewis. However, the term coordination chemistry is generally used to describe the chemistry of metals and metal ions in their interactions with other molecules or ions. For example, reactions (1)–(3) show acid-base-type reactions; the products formed

$$Mg^{2} + 6H_2O \rightarrow Mg(H_2O)_6^{2+} \quad (1)$$

$$Ni + 4CO \rightarrow Ni(CO)_4 \quad (2)$$

$$Fe^{2+} + 6CN^- \rightarrow Fe(CN)_6^{4-} \quad (3)$$

are coordination ions or compounds, and this area of chemistry is known as coordination chemistry.

Thus, it follows that coordination compounds are compounds that contain a central atom or ion and a group of ions or molecules surrounding it. Such a compound tends to retain its identity, even in solution, although partial dissociation may occur. The charge on the coordinated species may be positive, zero, or negative, depending on the charges carried by the central atom and the coordinated groups. These groups are called ligands, and the total number of attachments to the central atom is called the coordination number. Other names commonly used for these compounds include complex compounds, complex ions, Werner complexes, coordinated complexes, chelate compounds, or simply complexes. *See* ACID AND BASE; CHELATION.

Experimental observations as early as the middle of the 18th century reported the isolation of coordination compounds, but valence theory could not adequately account for such materials. The correct interpretation of these compounds was given by Alfred Werner in 1893. He introduced the concept of residual or secondary valence, and suggested that elements have this type of valence in addition to their normal or primary valence. Thus, platinum(IV) has a normal valence of 4 but a secondary valence or coordination number of 6. This then led to the formulaton of $PtCl_4 \cdot 6NH_3$ as $[Pt(NH_3)_6]^{4+}$, $4Cl^-$ and of $PtCl_4 \cdot 5NH_3$ as $[Pt(NH_3)_5Cl]^{3+}$, $3Cl^-$. The compound with five ammonias has only three ionic chlorides, the fourth is inside the coordination sphere, and therefore is not readily precipitated upon the addition of silver ion. Although the exact nature of the coordinate bond between metal and ligand remains the subject of considerable discussion, it is agreed that the formulations of Werner are essentially correct.

Three theories have been used to explain the nature of the coordinate bond. These are the valence bond theory, the electrostatic theory, including crystal field corrections, and the molecular orbital theory. Currently, the theory used almost exclusively is the molecular orbital theory. The valence bond theory for metal complexes considers that the pair of electrons on the ligand enter the hybridized atomic orbitals of the metal and that the bond is either essentially covalent or essentially ionic. Several of the properties of these substances can be explained on the basis of this theory. The electrostatic theory, plus the crystal field theory for the transition metals, assumes that the metal-ligand bond is caused by electrostatic interactions between point charges and dipoles and that there is no sharing of electrons. In addition to explaining the structure and magnetic properties, the crystal field theory affords an adequate interpretation of the visible spectra of metal complexes. The molecular orbital theory assumes that the electrons move in molecular orbitals which extend over all the nuclei of the metal-ligand system. In this manner, it serves to make use of both the valence bond theory and crystal field theory. The molecular orbital theory is therefore the best approximation to the nature of the coordinate bond because it is sufficiently flexible to permit both covalent and ionic bonding as well as the splitting of *d* orbitals into various energy levels. *See* CRYSTAL FIELD THEORY; MOLECULAR ORBITAL THEORY.

The stability of metal complexes depends both on the metal ion and the ligand. In general the stability of metal complexes increases if the central ion increases in charge, decreases in size, and increases in electron affinity. Several characteristics of the ligand are known to influence the stability of complexes: (1) basicity of the ligand, (2) the number of metal-chelate rings per ligand, (3) the size of the chelate ring, (4) steric effects, (5) resonance effects, and (6) the ligand atom. Since coordination compounds are formed as a result of acid-base reactions where the metal ion is the acid and the ligand is the base, it follows that generally the more basic ligand will tend to form the more stable complex. The size of the chelate ring is likewise an important factor. For saturated ligands such as ethylenediamine, five-membered rings are the most stable for chelates containing one or more double bonds.

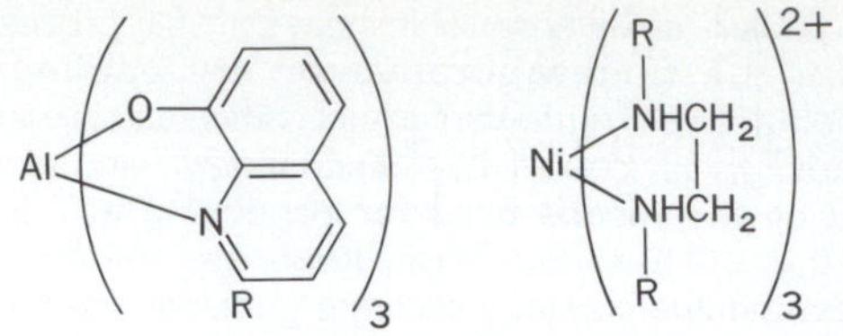

Structural formulas of metal complexes which are affected by steric factors.

Steric factors often have a very large effect on the stability of metal complexes. This is most frequently observed with ligands having a large group attached to the ligand atom or near it. Thus complexes, of the type shown in the illustration, with alkyl groups R in the position designated are much less stable than the parent complex where R = H. This results from the steric strain introduced by the size of the alkyl group on or adjacent to the ligand atom. In contrast to this, alkyl substitution at any other position results in the formation of more stable complexes because the ligand becomes more basic, and the bulky group is now removed from a position near the coordination site.

Finally, the ligand atom itself plays a significant role in controlling the stability of metal complexes. For most of the metal ions, the smallest ligand atom with the largest electron density will form the most stable complex.

Often the most stable complex is also the least reactive or most inert. Several factors, such as the electronic configuration of the central metal ion, its coordination number, and the extent of chelation, all have a marked effect on the rate of reaction of a given compound. *See* CHEMICAL BONDING; MAGNETOCHEMISTRY; SOLID-STATE CHEMISTRY; STEREOCHEMISTRY.

[F.Ba.]

Coordination complexes A group of chemical compounds in which a part of the molecular bonding is of the coordinate covalent type. For a discussion of the nature of the coordinate bond, and the stability and reactivity of complex compounds *see* CHELATION: COORDINATION CHEMISTRY.

Coordination complexes contain a central atom or ion and a group of ions or molecules surrounding it. Many simple hydrates, such as $MgCl_2 \cdot 6H_2O$, are best formulated as $[Mg(H_2O)_6]Cl_2$ because it is known that the 6 molecules of water surround the central magnesium ion. Therefore, $[Mg(H_2O)_6]^{2+}$ is a complex ion, and $[Mg(H_2O)_6]Cl_2$ is a complex compound. The charge on this complex ion is +2, because this is the charge on the magnesium ion and the coordinated water molecules are neutral. However, if the coordinated groups are charged, then the charge on the complex is represented by the sum of the charge on the metal and that of the coordinated ions. See, for example, the progression of charges on the platinum(IV) complexes listed below.

$[Pt(NH_3)_6]Cl_4$	Hexaammineplatinum(IV) chloride
$[Pt(NH_3)_5Cl]Cl_3$	Chloropentaammineplatinum(IV) chloride

$[Pt(NH_3)_4Cl_2]Cl_2$	Dichlorotetraammineplatinum(IV) chloride
$[Pt(NH_3)_3Cl_3]Cl$	Trichlorotriammineplatinum(IV) chloride
$[Pt(NH_3)_2Cl_4]$	Tetrachlorodiammineplatinum(IV)
$K[Pt(NH_3)Cl_5)]$	Potassium pentachloroammineplatinate(IV)
$K_2[PtCl_6]$	Potassium hexachloroplatinate(IV)

Thus the charge is +4 for $[Pt(NH_3)_6]^{4+}$ because Pt is +4 and NH_3 is neutral. But the charge is −2 for $[PtCl_6]^{2-}$ because of the 6 for Cl^-, that is, $+4 - 6 = -2$.

Metal complexes exhibit various types of isomerism. In many ways, inorganic stereochemistry is similar to that observed with organic compounds. Geometrical isomers are common among the inert complexes of coordination numbers 4 and 6. *See* STEREOCHEMISTRY.

The synthesis of metal complexes containing only one kind of ligand generally involves just the reaction of the metal salt in aqueous solution with an excess of the ligand reagent, as shown below.

$$[Ni(H_2O)_6](NO_3)_2 + 6NH_3 \rightarrow [Ni(NH_3)_6](NO_3)_2 + 6H_2O$$

The desired complex salt can then be isolated by removal of water until it crystallizes, or by addition of a water-miscible organic solvent to cause it to separate. For the inert complexes (those slow to react), prolonged treatment at more drastic conditions is often necessary. The preparation of geometrical isomers is much more difficult, and in most cases, the approach used is rather empirical. Generally, reactions yield a mixture of cis-trans products, and these are separated on the basis of their differences in solubility. *See* AMMINE; HYDRATE. [F.Ba.]

Coordination number The number of nearest neighbors of a point in a space lattice, of an atom or an ion in a structure, or of an anion or cation in a complex ion. The figure shows that the coordination number in the three cubic lattices *P*, *I*, and *F* is respectively 6, 8, and 12. In closely packed structures this number is 12.

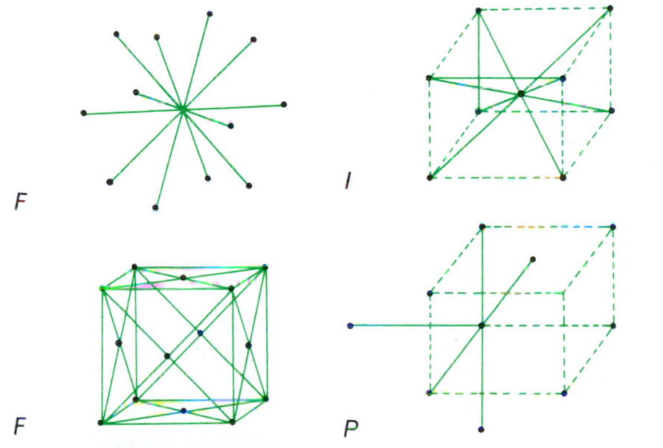

The coordination number in the three cubic lattices *P*, *I*, and *F* is shown to be 6, 8, and 12, respectively.

In ionic structures each ion is surrounded by a number of ions of opposite sign. This coordination is respectively 8/8, 6/6, 4/4, 8/4 in the cesium chloride, sodium chloride, zinc sulfide, and calcium fluoride structures. *See* CRYSTAL.

The anions are, in nearly all cases, larger than the cations. For any given coordination the distribution of the anions is as regular as possible. Each cation is in the center of a coordination polyhedron. The structures can be described in terms of the way in which the anion polyhedrons are packed. For example, in the rock salt structure, each cation is in the center of an octahedron formed by six anions. Two neighboring octahedrons have a common edge. [W.C.D.]

Copepoda A subclass of Crustacea. Most are free-living in aqueous environments, though many are parasitic or symbiotic with other aquatic animals. The free-living forms are the most abundant of all animals in the sea and directly constitute the main food for vast numbers of fishes, many invertebrates, and also at times for baleen whales.

The structure of copepods, although similar in basic pattern, varies greatly. In free-living forms the body is composed externally of a series of small chitinous cylinders, or segments, some of which may be more or less fused. The head bears two pairs of antennae. The feeding appendages are one pair of mandibles, two pairs of maxillae, and one pair of maxillipeds. The thorax, consisting of six segments, bears the swimming feet. The abdomen consists of one to five segments and bears no appendages except a pair of terminal processes known as caudal rami. The sexes are separate and sexual dimorphism is frequently evidenced. Free-living species usually range in size from less than 0.02 in. to about 0.4 in. (0.5 mm to about 10 mm) in length.

Seven orders (with over 6000 species) of copepods are usually recognized. Three of these, Calanoida, Cyclopoida, and Harpacticoida, are wholly or primarily free-living and because of overwhelming numbers are of greatest general interest. The remaining orders (Notodelphoida, Monstrilloida, Caligoida, and Lernaepodoida) are parasitic at least part of their lives. Their numbers are relatively small, but they may be conspicuous and important as "fish lice," burrowing deep into the flesh or feeding externally upon the skin and gills. *See* CALANOIDA; CYCLOPOIDA; HARPACTICOIDA; MONSTRILLOIDA.

Copepod habitats are extremely varied, but the vast majority of species live in the sea. They are mostly planktonic throughout their lives, some occurring only in coastal waters while others live in the open sea. Many frequent only the upper layers to about 660–990 ft (200–300 m) depth. Others are bathypelagic at great depths. The harpacticoids, however, are primarily a benthic, or bottom-dwelling, group, though many are truly planktonic.

The role of copepods in the economy of the sea consists mainly of converting plant into animal substance. Many species are adapted to graze directly upon the chief synthesizers of organic material, the microscopic diatoms and dinoflagellates that constitute the great pasturage of the sea. Some species are predatory on other copepods, fish larvae, and other small animals, such as protozoans. Commonly copepods constitute 70% of the zooplankton, and at times their numbers are so great as to impart a pink color to the sea. [H.C.Y.]

Copolymer A macromolecule in which two or more different species of monomer are incorporated into a polymer chain. The properties of copolymers depend on both the nature of the monomers and their distribution in the polymer. Thus, monomers A and B can polymerize randomly to form ABBAABA; they can alternate to give ABAB; they can form blocks AAABBB; or one monomer can be grafted onto a polymer of the other:

AAAA
B
B

Copolymers can be prepared by all the known methods of polymerization: addition polymerization of vinyl monomers (by free-radical, anionic, cationic, or coordination catalysis), ring-opening polymerization, or condensation polymerization. *See* POLYMERIZATION.

In the polymerization of two monomers, M_1 and M_2, the monomers can add to a growing chain ending in either monomer, designated as M_1^* and M_2^*. To a first approximation, only the terminal group of the growing chain is important, so two reactivity ratios can be defined: r_1 is the relative reactivity of chain end M_1^* to monomers M_1 and M_2, and r_2 is the relative reactivity of chain end M_2^* to monomers M_2 and M_1. The reactivity ratios can be determined experimentally by analyzing polymer compositions at different monomer feeds. *See* ACRYLONITRILE; POLYACRYLONITRILE RESINS; POLYMER. [D.S.Br.]

Copper A chemical element, Cu, atomic number 29, one of the transition elements and an important nonferrous metal. Its usefulness is accounted for by its combination of chemical, physical, electrical, and mechanical properties and its fairly abundant supply. Copper is one of the first metals to have been used by humans. *See* TRANSITION ELEMENTS.

By far the greater part of the world's copper is obtained from the sulfide ores chalcocite, covellite, chalcopyrite, bornite, and enargite. Oxidized ores include cuprite, tenorite, malachite, azurite, chrysocolla, and brochantite. Native copper, once widespread in the United States, is now mined in quantity only in Michigan. The grade of ore used for copper production has been going steadily downward as the richer ores have become exhausted and the demand for copper has grown. There are vast amounts of copper in the ground, available for future use if ores of still lower grades are utilized, and there is no prospect of exhaustion for a long time to come.

1	2	3	4	5	6	7	8	9	10	11	12	13	14	15	16	17	18
1 H																	2 He
3 Li	4 Be											5 B	6 C	7 N	8 O	9 F	10 Ne
11 Na	12 Mg											13 Al	14 Si	15 P	16 S	17 Cl	18 Ar
19 K	20 Ca	21 Sc	22 Ti	23 V	24 Cr	25 Mn	26 Fe	27 Co	28 Ni	29 Cu	30 Zn	31 Ga	32 Ge	33 As	34 Se	35 Br	36 Kr
37 Rb	38 Sr	39 Y	40 Zr	41 Nb	42 Mo	43 Tc	44 Ru	45 Rh	46 Pd	47 Ag	48 Cd	49 In	50 Sn	51 Sb	52 Te	53 I	54 Xe
55 Cs	56 Ba	71 Lu	72 Hf	73 Ta	74 W	75 Re	76 Os	77 Ir	78 Pt	79 Au	80 Hg	81 Tl	82 Pb	83 Bi	84 Po	85 At	86 Rn
87 Fr	88 Ra	103 Lr	104 Rf	105 Db	106 Sg	107 Bh	108 Hs	109 Mt	110	*111*	*112*	*113*	*114*	*115*	*116*	*117*	*118*

lanthanide series	57 La	58 Ce	59 Pr	60 Nd	61 Pm	62 Sm	63 Eu	64 Gd	65 Tb	66 Dy	67 Ho	68 Er	69 Tm	70 Yb
actinide series	89 Ac	90 Th	91 Pa	92 U	93 Np	94 Pu	95 Am	96 Cm	97 Bk	98 Cf	99 Es	100 Fm	101 Md	102 No

Copper is the first element in the transition-element subgroup of the periodic table which also includes the other coinage metals, silver and gold. The copper atom has the electronic structure $1s^2 2s^2 2p^6 3s^2 3p^6 3d^{10} 4s^1$. The low ionization potential of the $4s^1$ electron results in easy removal to give the copper(I), or cuprous ion, Cu^+; and the copper(II), or cupric ion, Cu^{2+}, is readily formed by removal of one electron from the $3d$ shell. The atomic weight of copper is 63.546. It has two stable natural isotopes, ^{63}Cu and ^{65}Cu. There are also known nine unstable (radioactive) isotopes. Copper is characterized by low chemical activity. The element combines chemically in one of three valences. The most common valence is 2+ (cupric), but 1+ (cuprous) is also frequent; the valence 3+ occurs in only a few unstable compounds.

A comparatively heavy metal, pure solid copper has a density of 8.96 g/cm^3 at 20°C (68°F). That of commercial copper varies with method of manufacture, averaging 8.90–8.94. The melting point of copper is 1083.0 ± 0.1°C (1981.4 ± 0.2°F). Its normal boiling point is 2595°C (4703°F). Copper is nonmagnetic; or more precisely, it is slightly paramagnetic. The thermal and electrical conductivities of copper are very high. Copper is one of the strongest pure metals. It is moderately hard, extremely tough, and wear resistant. The strength of copper is accompanied by high ductility. The mechanical and electrical properties of a metal are strongly dependent on physical condition, temperature, and grain size of the metal.

Of the hundreds of copper compounds, only a few are manufactured industrially on a large scale. The most important is copper(II) sulfate pentahydrate or blue vitriol, $CuSO_4 \cdot 5H_2O$. Others include bordeaux mixture, $3Cu(OH)_2CuSO_4$; paris green, a complex of copper metaarsenite and acetate; cuprous cyanide, CuCN; cuprous oxide, Cu_2O; cupric chloride, $CuCl_2$; cupric oxide, CuO; basic cupric carbonate; and copper naphthenate, the most widely used agent for prevention of rotting in wood, fabric, rope, and fishing nets. Leading uses of copper compounds are in agriculture, especially as fungicides and insecticides; as pigments; in electroplating solutions; in primary cells; as mordants in dyeing; and as catalysts. *See* COPPER ALLOYS; COPPER INTOXICATION; COPPER METALLURGY. [A.Bu.]

Copper alloys Solid solutions of one or more metals in copper. Many metals, although not all, alloy with copper to form solid solutions. Some insoluble metals and nonmetals are intentionally added to copper alloy to enhance certain characteristics. *See* SOLID SOLUTION.

Copper alloys form a group of materials of major commercial importance because they are characterized by such useful mechanical properties as high ductility and formability and excellent corrosion resistance. Copper alloys are easily joined by soldering and brazing. Like gold alloys, copper alloys have decorative red, pink, yellow, and white colors. Copper has the second highest electrical and thermal conductivity of any metal. All these factors make copper alloys suitable for a wide variety of products.

Copper and zinc melted together in various proportions produce one of the most useful groups of copper alloys, known as the brasses. Brasses containing 5–40% zinc constitute the largest volume of copper alloys. One important alloy, cartridge brass (70% copper, 30% zinc), has innumerable uses, including cartridge cases, automotive radiator cores and tanks, lighting fixtures, eyelets, rivets, screws, springs, and plumbing products.

Lead is added to both copper and the brasses, forming an insoluble phase which improves machinability of the material. Free cutting brass (61% copper, 3% lead, 36% zinc) is the most important alloy in the group. It is machined into parts on high-speed (10,000 rpm) automatic screw machines for a multiplicity of uses. Increased strength and corrosion resistance are obtained by adding up to 2% tin or aluminum to various brasses.

Alloys of copper, nickel, and zinc are called nickel silvers. Nickel is added to the copper-zinc alloys primarily because of its influence upon the color of the resulting alloys; color ranges from yellowish-white, to white with a yellowish tinge, to white. Because of their tarnish resistance, these alloys are used for table flatware, zippers, camera parts, costume jewelry, nameplates, and some electrical switch gear.

Three copper-base alloys containing 10%, 20%, and 30% nickel, with small amounts of manganese and iron added to enhance casting qualities and corrosion resistance, have gained commercial importance. These alloys are known as cupronickels and are well suited for application in industrial and marine installations as condenser and heat-exchanger tubing because of their high corrosion resistance and particular resistance to impingement attack.

Copper-tin alloys (3–10% tin), deoxidized with phosphorus, form an important group known as phosphorus bronzes. Tin increases strength, hardness, and corrosion resistance. These alloys are widely used for springs and screens in papermaking machines. *See* ALLOY; COPPER. [R.E.R.]

Copper intoxication Copper and its salts are remarkably nontoxic to mammalian tissues. It is necessary to ingest a large quantity of a soluble copper salt such as copper sulfate to produce intoxication. Nausea, vomiting, diarrhea, abdominal cramps, melana, coma, and death can ensue. The copper solution not only produces vascular congestion of the gastrointestinal tract but also acts as an irritant producing focal necrosis, microscopic thrombi in the capillaries of the intestinal wall, and liver necrosis. Death is usually a result of vascular collapse (shock). Ingestion of smaller amounts of copper salts dissolved from the walls of pots and vessels by the action of citrus juices seldom produces symptoms and does not represent a threat to life. The symptoms that occur are mild (nausea and vomiting) and are related to acute gastroenteritis. Inhalation of dust, fumes, or mists of copper salts can cause congestion and the irritation of the upper airway, but generalized symptoms do not ordinarily ensue.

Chronic copper poisoning does not represent a major health hazard. Chronic human toxicity is extremely rare except in those individuals who inherit a set of autosomal recessive genes producing an abnormal increase in body copper (a syndrome called hepatolenticular degeneration or Wilson's disease). A normal individual absorbs a scant amount of copper, sufficient to meet the body's essential needs, and excess copper is readily eliminated. Copper, in small amounts, is a trace element essential for the activity of such mammalian metalloenzymes as ceruloplasmin, cytochrome *c* oxidase, dopamine, and tyrosinase. *See* COPPER. [N.K.M.]

Copper loss The power loss due to the flow of current through copper conductors. When an electric current flows through a copper conductor (or any conductor), some energy is converted to heat. The heat, in turn, causes the operating temperature of the device to rise. This happens in transformers, generators, motors, relays, and transmission lines, and is a principal limitation on the conditions of operation of these devices. Excessive temperature rises lead to equipment failure.

Much research has been undertaken to reduce the loss in machines and thus to increase the output capability and efficiency for a given amount of conductor. For large machines and for transmission lines, the concept of superconductivity appears to hold much promise for substantial improvements in operating efficiencies. *See* SUPERCONDUCTING DEVICES. [E.C.Jo.]

Copper metallurgy There are many variations in the production processes, and improvements are constantly being made. Only the most common methods can be indicated here because of space limitations.

The ores are first concentrated, usually by flotation, to yield concentrates containing from 20 to over 40% copper. *See* ORE DRESSING.

Roasting has usually been the next step, but is omitted in plants treating the richer concentrates. Roasted concentrates are mixed with raw concentrates before smelting to obtain a sulfur content that will yield the desired grade of matte in the reverberatory furnace.

Smelting in blast furnaces has been replaced, except in a few instances, by reverberatory smelting. The reverberatory furnace has the advantage of being able to smelt finely divided flotation concentrates without sintering, and it is not dependent on coke as fuel. Whatever fuel is cheapest in the locality of the smelter can be used. The basis of the process is that reactions between copper oxide and iron sulfide in the molten charge form copper sulfide and iron oxide. With excess iron sulfide, the copper sulfide forms a molten solution called copper matte. The iron oxide formed in the reactions, together with that added as flux, unites with silica in the ore to form a slag.

The matte is taken to a converter where the sulfur in the matte is oxidized to SO_2, and passes off in the gases, leaving copper in the converter. When cast, the copper forms cakes with a surface roughened and blistered by the escape of gases during freezing, and hence is called blister copper. It is usually 98–99% Cu.

The blister copper is next partly refined in a furnace and cast into anodes containing about 99.0–99.3% Cu. *See* PYROMETALLURGY.

The anodes are hung in electrolytic tanks, spaced alternately with cathodes, in an electrolyte of copper sulfate and free sulfuric acid, contaminated with soluble impurities. Insoluble impurities fall to the bottom of the tank. Soluble impurities enter the electrolyte and do not plate out at the cathode. Some of the impurities form insoluble compounds. Thus the copper is freed of nearly all impurities. The cathodes are usually remelted for casting into shapes such as wirebars, cakes, billets, or ingots. Some of the cathodes are sold without remelting. *See* ELECTROMETALLURGY.

Continuous casting of oxygen-free copper is practiced on a large scale, the copper being protected from air while molten.

Most copper ores contain gold and silver. The recovery of these and other metals as by-products is an important part of the metallurgy of copper.

A considerable amount of copper is produced by hydrometallurgy. The ore is leached without being concentrated, dilute sulfuric acid being the solvent most often used. The process is applied chiefly to oxidized ores. A further limitation of leaching is that the solvents employed do not dissolve gold or silver. *See* HYDROMETALLURGY.

In another proposed method of copper recovery, low-grade ore in the ground would be shattered by a nuclear energy explosion, then leached in place with dilute sulfuric acid; the solution would next be pumped to the surface and the copper cemented out on scrap iron. *See* COPPER; METALLURGY. [A.Bu.]

Copulatory organ An organ utilized by certain male vertebrates for insemination, that is, to deposit spermatozoa directly into the female reproductive tract. In fish, internal fertilization is restricted to certain groups. The pelvic fin of male elasmobranchs and holocephalians is modified for the transmission of sperm and is known as the clasper or clasping organ. A pair of anterior claspers (illustration *a*) and a frontal clasper which protrudes from the head occur among holocephalians in addition to the modified pelvic clasper of elasmobranchs. The anal fin of teleosts, in which copulation occurs, may be elongated to form a gonopodium.

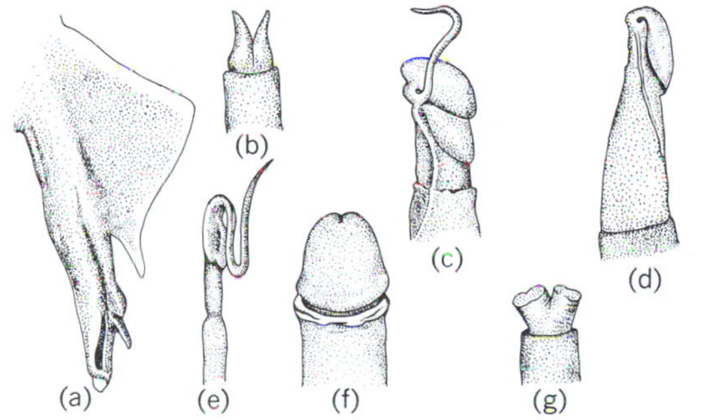

Vertebrate copulatory organs. (*a*) Clasper of dogfish (*Squalus*). Glans penis of (*b*) opossum, (*c*) ram, (*d*) bull, (*e*) short-tailed shrew, (*f*) man, (*g*) Echidna.

Hemipenes which can be everted during copulation are common to both snakes and lizards; however, these structures lack erectile tissue. Turtles and crocodiles possess a single penis with associated erectile tissue, the corpora cavernosa, which becomes distended with blood. A few species of birds have a penis; among these are the ostriches and anseriforms. The penis is present in all mammals.

The copulatory organ (illustration *b–g*) is variously modified morphologically. Among certain mammals such as rodents, bats, whales, some carnivores, and lower primates, a penis bone, the os priapi or baculum, occurs, increasing the rigidity of the penis. *See* PENIS. [C.B.C.]

Coquina A calcarenite or clastic limestone whose detrital particles are chiefly fossils, whole or fragmented. The term is most frequently used for an aggregate of large shells more or less cemented by calcite. If the rock consists of fine-sized shell debris, it is called a microcoquina. Some coquinas show little evidence of any transportation by currents. *See* CALCARENITE; LIMESTONE. [R.S.]

Cor pulmonale A cardiomyopathy consisting of a right ventricular hypertrophy caused by hypertension in the pulmonary blood vessels, as the result of primary diseases within the lungs or lung blood vessels. Almost any long-standing lung disease may lead to cor pulmonale, including chronic obstructive pulmonary diseases (chronic bronchitis, emphysema, pneumonconioses, interstitial fibrosis, and so on). Vascular obstructions of the pulmonary blood vessels (thrombi or emboli within the pulmonary vessels) similarly produce hypertension within the pulmonary blood vessels and lead to cor pulmonale. *See* HEART (VERTEBRATE); HEART DISORDERS. [N.K.M.]

Coraciiformes A diverse order of land birds that is found mainly in the tropics. The relationships of these families have been disputed, with no resolution; the closest affinities may be with other land birds, such as the Piciformes and Passeriformes. *See* PASSERIFORMES; PICIFORMES.

The Coraciiformes are divided into four suborders and families: Alcedines, including the families Alcedinidae (kingfishers; 91 species), Todidae (todies; 5 species), and Momotidae (motmots; 9 species); Meropes, with the single family Meropidae (bee-eaters; 24 species); Coracii, containing the families Coraciidae (rollers; 11 species), Brachypteraciidae (ground-rollers; 5 species), Leptosomatidae (cuckoo-rollers; 1 species), Upupidae (hoopoes; 1 species), and Phoeniculidae (wood hoopoes; 8 species); and Bucerotes, with the single family Bucerotidae (hornbills; 45 species). Several of these families, such as the different groups of rollers, are sometimes merged into a single family, and several families, such as the kingfishers and hornbills, are divided into subfamilies.

Coraciiform birds are characterized by a syndactyl foot in which the three anterior toes are joined together at the base, although the cuckoo rollers have a zygodactyl foot. Most groups are tropical or warm temperate and are brilliantly colored. Except for bee eaters and wood hoopoes, coraciiforms are generally solitary. Most temperate species are migratory; the tropical ones are permanent residents. Most coraciiforms feed on insects and other animal prey, including fish; hornbills are omnivorous and some feed mainly on fruit. All nest, either solitarily or in colonies in holes burrowed in earthern banks or in tree cavities. *See* AVES. [W.J.B.]

Corallimorpharia An order of the subclass Zoantharia or Hexacorallia. These sea anemones resemble corals in their weak musculature, capitate (spherically tipped) tentacles, and complex nematocysts. They do not possess a skeleton, however. The tentacles are arranged in radial rows, not cycles, on the disk, and the base has no musculature. The group occurs in shallow water from the temperate zone to the tropics. The species are solitary, although they may occur in large, asexually produced aggregations. Examples are *Corallimorphus* and *Corynactis*. *See* ANTHOZOA. [C.H.]

Corallinales An order of red algae (Rhodophyceae), commonly called coralline algae. Only one family is recognized, the Corallinaceae, which formerly was assigned to the order Cryptonemiales. These algae are distinguished by the impregnation of cell walls with calcite, a form of calcium carbonate, which causes the thallus to be stony or brittle. *See* RHODOPHYCEAE.

Coralline algae, comprising about 40 genera and 500 species, are widespread, abundant, and ecologically important. They are divisible into two groups on the basis of the presence or absence of uncalcified, moderately flexible joints (genicula) between calcified segments (intergenicula). The most simple nonarticulated coralline algae are individual crusts of varying extent and thickness (to 8 in. or 20 cm thick). These often become confluent and cover large expanses of substrate. Many crusts bear rounded or pointed, branched or unbranched protuberances that can break off and continue to grow as free nodules known collectively as maerl.

Coralline algae are exclusively marine, although some species can tolerate a reduction in salinity to 13 parts per thousand. Some species thrive only where light is intense, as at the crest of a coral reef, while others grow only in shaded habitats or in deep water. Most species require constant immersion.

The ecosystem in which coralline algae are most important is the coral reef, where they are primary producers, adding carbon to the ecosystem, adding new material to the reefs, and cementing together other calcareous organisms. They have been engaged in similar activities through the millennia, with modern genera recognizable in limestones at least as old as 150 million years (Jurassic). [P.C.Si.; R.L.Moe]

Corbino disk A disk of conducting material with inner and outer concentric-ring electric contacts. When subjected to a magnetic field perpendicular to the plane of the disk, the current between the inner and outer ring contacts tends to spiral rather than flow straight out, thus introducing a considerable magnetoresistance. *See* CORBINOTRON; MAGNETORESISTANCE. [L.P.H.]

Corbinotron The combination of a corbino disk, made of high-mobility semiconductor material, and a coil arranged to produce a magnetic field perpendicular to the disk. This device can be used as a switch in which the current in the coil controls the resistance of the disk and hence the radial current flowing in the disk. *See* CORBINO DISK.

If the coil is connected in series with the disk, the resulting two-terminal device exhibits rectification. If the coil is connected in parallel with the disk, the device exhibits a negative-resistance characteristic for one direction of current. [L.P.H.]

Cordaitales An extensive and for the most part natural grouping of forest trees of the late Paleozoic. With inclusion of certain presumed relatives, the stratigraphic range of the Cordaitales extends from the Upper Devonian to the Lower Triassic. The order may be divided into three families: Pityaceae, Cordaitaceae, and Poroxylaceae. The degree of relationship between the three families is quite uncertain, and the bulk of knowledge and significance of the group resides in the Cordaitaceae, abundant remains of which occur in Carboniferous sediments in most parts of the Earth. *See* PALEOBOTANY. [E.S.B.].

Cordierite An orthorhombic magnesium aluminosilicate mineral of composition $Mg_2[Al_4Si_5O_{18}]$. The crystal structure is related to beryl. Limited amounts of Fe^{2+} may substitute for Mg^{2+}, and Fe^{3+} for Al^{3+}. The hardness is 7 (Mohs scale); specific gravity 2.6; luster vitreous; cleavage poor; and color greenish-blue, lilac blue, or dark blue, often strongly pleochroic colorless to deep blue. Transparent pleochroic crystals are used as gem material. The disordered cordierite structure, $Mg_2[(Al,Si)_9O_{18}]$, is called indialite and is hexagonal, isotypic with beryl. Osumilite, $KMg_2Al_3[(Al,Si)_{12}O_{30}] \cdot H_2O$, is also related but has a structure built of double six-membered rings of tetrahedrons. Cordierite, indialite, and osumilite are difficult to distinguish. Pale colored varieties are often misidentified as quartz, since these minerals have many physical properties in common. *See* BERYL; SILICATE MINERALS.

Cordierite possesses unusually low thermal expansion, and synthetic material has been applied to thermal-shock-resistant materials, such as insulators for spark plugs and low-expansion concrete.

Cordierite frequently occurs associated with thermally metamorphosed rocks derived from argillaceous sediments. It may occur in aluminous schists, gneisses, and granulites; though usually appearing in minor amounts, cordierite occurs at many localities throughout the world. [P.B.M.]

Cordilleran belt A mountain belt or chain which is an assemblage of individual mountain ranges and associated plateaus and intermontane lowlands. A cordillera is usually of continental extent and linear trend; component elements may trend at angles to its length or be nonlinear.

The term cordillera is most frequently used in reference to the mountainous regions of western South and North America, which lie between the Pacific Ocean and interior lowlands to the east. Farther north, the extensive and geologically diverse mountain terrane of western North America is formally known as the Cordilleran belt or orogen. This belt includes such contrasting elements within the United States as the Sierra Nevada, Central Valley of California, Cascade Range, Basin and Range Province, Colorado Plateau, and Rocky Mountains. *See* MOUNTAIN SYSTEMS.

Cordilleras represent zones of intense deformation of the Earth's crust produced by the convergence and interaction of large, relatively stable areas known as plates. Mountain belts have been analyzed in terms of different modes of plate convergence. Cordilleran-type mountain belts, such as the North American Cordillera, are contrasted with collision-type belts, such as the Himalayas. The former develop during long-term convergence of an oceanic plate toward and beneath a continental plate, whereas the latter are produced by the convergence and collision of one continental plate with another or with an island arc. Characteristics of cordilleran-type mountain belts include their position along a continental margin, their widespread volcanic and plutonic igneous activity, and their tendency to be bordered on both sides by zones of low-angle thrust faulting directed away from the axis of the belt. *See* OROGENY. [G.A.D.]

Core drilling The boring of a hole in the earth by drilling a circular groove and leaving a central core. It is usually done to obtain the core as a sample for visual examination and to test its physical and chemical properties. But core drilling is also done for making the hole itself whenever it is more economical to drill the annular groove and remove the core than to drill the entire cross-sectional area of the hole.

The greatest use for core drilling is in prospecting for mineral deposits. It is usually done with a diamond core drill, so called because it utilizes the extreme hardness of the diamond to cut the annular groove in rock. Diamond core drilling is also widely used in civil engineering work to determine the strength of the rocks that must support the tremendous weights of buildings, bridges, and dams.

In shaft sinking by the core-drill method, and in core drilling other large-diameter holes, chilled steel shot is used as a cutting medium instead of diamonds. Other cutting mediums sometimes used are hard alloy-steel teeth, sintered carbides, and corundum grit. *See* PROSPECTING. [F.C.St.]

Core loss The rate of energy conversion into heat in a magnetic material due to the presence of an alternating or pulsating magnetic field. It may be subdivided into two principal components, hysteresis loss and eddy-current loss. *See* EDDY CURRENT; MAGNETIC HYSTERESIS.

The energy consumed in magnetizing and demagnetizing magnetic material is called the hysteresis loss. It is proportional to the frequency and to the area inside the hysteresis loop for the material used. Most rotating machines are stacked with silicon steel laminations, which have low hysteresis losses. The cores of large units are sometimes built up with cold-reduced, grain-oriented, silicon iron punchings having exceptionally low hysteresis loss, as well as high permeability when magnetized along the direction of rolling. *See* MAGNETIZATION.

Induced currents flow within the magnetic material because of variations in the flux; this is called eddy-current loss. For 60-cycle rotating machines, core laminations of 0.014–0.018 in. (0.35–0.45 mm) are usually used to reduce this eddy-current loss. *See* ELECTRIC ROTATING MACHINERY. [L.T.R.]

Coriander An ancient flavoring material. The plant, *Coriandrum sativum* (Umbelliferae), is a strong-scented perennial herb. The small, spherical, brown fruits have an unpleasant odor when fresh, but are pleasingly aromatic when dried. They are used as a flavoring in cookery. The oil of coriander is used in medicine and for flavoring beverages. *See* APIALES. [P.D.St./E.L.C.]

Coriolis acceleration An acceleration which arises as a result of motion of a particle relative to a rotating system. Only the components of motion in a plane parallel to the equatorial plane are influenced. Coriolis accelerations are important to the circulation of planetary atmospheres, and also in ballistics. *See* ACCELERATION; BALLISTICS.

Newton's second law of motion is valid only when the motions and accelerations are those observed in a coordinate system that is not itself accelerating, that is, an inertial reference frame. In order to utilize familiar concepts in mathematical treatment, the Earth is commonly treated as if it were fixed, as it appears to one observing from a point on the surface, and the Coriolis force is introduced to balance the acceleration observed by virtue of the observer's motion in the rotating frame. As with the influenced components of motion, the Coriolis force is directed perpendicularly to the Earth's axis, that is, in a plane parallel to the equatorial plane. Since the direction of its action is also perpendicular to the particle velocity itself, the Coriolis force affects only the direction of motion, not the speed. This is the basis for referring to it as the deflecting force of the Earth's rotation.

A simple illustration of a Coriolis effect in the Northern Hemisphere is afforded by a turntable in counterclockwise rotation, and an external observer who moves a marker steadily in a straight line from the axis to the rim of the turntable. The trace on the turntable is a right-turning curve, and obviously this is also the nature of the path apparent to an observer who rotates with the table. Now consider the contrasting case of an air parcel near the North Pole that moves directly south (away from the Earth's axis of rotation) so that its motion to an observer on the Earth is in a straight line. To a nonrotating observer in space, this same motion appears curved toward the east because of the increased linear velocity of the meridian at lower latitudes. The force necessary to produce the eastward acceleration in the inertial frame is equal to the Coriolis force and would be produced by a gradient of air pressure from west to east, and shown by north-south–oriented isobars. In the absence of the pressure gradient force, the Coriolis force would cause the trajectory of the southward-moving air to curve westward on the Earth's surface. Then an air parcel moving uniformly away from the North Pole along a line which appears straight to an observer in space would appear to earthbound observers to curve westward. *See* GEOSTROPHIC WIND; ISOBAR (METEOROLOGY). [E.Ke.]

Corn *Zea mays*, a grain crop of the order Cyperales grown for its edible seeds (technically fruits). *See* CYPERALES.

Corn is a cross-pollinated plant; the staminate (male) and pistillate (female) inflorescences (flower clusters) are borne on separate parts of the same plant (see illustration). Plants of this type are called monoecious. The staminate inflorescence is the tassel; it produces pollen that is carried by the wind to the silks produced on the ears.

A corn plant in full tassel (above) and silk (below). (*Courtesy of J. W. McManigal*)

As a crop. Corn occupies a larger area than any other grain crop in the United States, where 60% of the world production is grown. Production throughout the United States reaches its greatest concentration in the states of Iowa, Illinois, Indiana, Ohio, Nebraska, Minnesota, Wisconsin, Michigan, and Missouri. This area, called the Corn Belt, is characterized by moderately high temperature, fertile, well-drained soils, and normally adequate rainfall. The Corn Belt accounts for approximately 80% of corn production in the United States. Although corn is grown in the United States primarily for livestock feed, about 10% is used for the manufacture of starch, sugar, corn meal, breakfast cereals, oil, alcohol, and other specialized products. In many tropical countries, corn is used primarily for human consumption.

The origin of corn is still unsettled, but the most widely held hypothesis assumes that corn developed from its wild relative teosinte (*Z. mexicana*) through a combination of favorable mutations, recognized and selectively propagated by early humans. Corn migrated from its presumed center of origin in Mexico or Central America and was being cultivated by the Indians as far north as New England when the first European colonists arrived, whose survival was due largely to the use of corn as food.

Each form (botanical variety) of corn is conditioned by fairly few genetic differences, and each may exhibit the full range of differences in color, plant type, maturity, and so on, characteristic of the species. All types have the same number of chromosomes (10 pairs), and all may be intercrossed to produce fertile progeny. Dent corns (*Z. m. indentata*) are the most important in the United States. Sweet corn is grown more extensively in the United States than in any other country. It is eaten as fresh corn or canned or frozen. In other countries, flint (*Z. m. indurata*), dent, or flour (*Z. m. amylaceae*) corns may be eaten fresh, but at a much more mature stage than the sweet corn eaten in the United States. The commercial production of popcorn p(*Z. m. everta*) is almost exclusively American.

The development of varieties and strains of corn made possible the extension of its culture under diverse soil and climatic conditions. However, modern research methods, used for its further improvement, led to the present widespread use of hybrid corn. Hybrid corn is the first generation of a cross involving inbred lines. Inbred lines are developed by controlled self-pollination. When continued for several generations, self-pollination leads to reduction in vigor but permits the isolation of types which are genetically pure or homozygous. Intense selection is practiced during the inbreeding phase to identify and maintain genotypes having the desired characteristics. Crosses involving any two unrelated lines will exhibit heterosis, that is, yields above the means of the two parents. *See* BREEDING (PLANT); HETEROSIS.

Planting dates depend upon temperature and soil conditions. Germination is very slow at soil temperatures of 50°F (10°C), and seedling growth is limited at temperatures of 60°F (16°C) or below. Because of these temperature relations, planting begins in Florida and southern Texas in early February and is completed in New York and New England by late May or early June. Planting rates are influenced by water supply, soil type, and fertility and by the maturity characteristics of the hybrid grown. With planting rates above 16,000 plants per acre (40,000 per hectare), drilling in rows 24–36 in. (60–90 cm) apart has become common practice. The use of nitrogen fertilizer has increased greatly; lesser amounts of phosphorus and potash are applied as needed.

The corn plant grows vegetatively until about silking, after which all weight increase is in the form of grain. Almost the entire grain yield results from photosynthesis during the grain growth period, which runs from silking to maturity. Contrary to popular opinion, grain yields are highest under cool conditions, when the lengthened grain growth period more than compensates for the slower growth rate.

Relationships among solar radiation, temperature, growing-season length, soil moisture, day length, soil fertility, and corn genotype in producing grain yields are complex and not well understood. Attempts to study the system as a whole, using simulation models on digital computers, may add considerably to knowledge of the subject. [W.G.Du.]

Processing. The high starch content and caloric value of corn have resulted in extensive production of animal feeds, human food, alcoholic beverages, and industrial products from the grain. Corn grain consists of three main parts: endosperm (horny and floury), germ, and hull, which differ in composition. The dry hard kernel of common dent corn is excellent for storage and handling but requires suitable processing for use. The whole grain may be ground, or alternatively, it may be degermed and dehulled during dry milling. It may also be wet-milled to further separate endosperm into starch and protein.

New World Indians discovered that cooking corn in solutions of lime or wood ashes (lye) rendered it soft and easier to grind or dehull. Lime-treated corn is used for preparation of tortillas and other popular Mexican-style foods, whereas lye-treated corn is widely sold as canned hominy. Modern steel mills are used to grind most corn today.

About 90% of the corn produced in the United States is used for animal feeds. Because of its nutritional inadequacy, the grain is generally supplemented with protein sources such as soybean meal and with minerals and vitamins.

The high unsaturated oil content of corn germ limits the

shelf life and restricts the use of whole ground corn. Most dry-milled corn is produced by degermer processes that fractionate grain into various-size endosperm particles, bran, and germ. Endosperm particles, after separation according to size, are classified into grits, meal, and flour. Grits serve as breakfast porridge, in corn flakes, and as a brewing substrate. Meal is a constituent of corn bread or muffin recipes and snack items. Corn flour is used in pancakes, soup thickeners, and dusting flours. Corn flour may be cooked on heated rolls or in extruders to pregelatinize its starch for use in baby foods and in industrial binders and adhesives. *See* FAT AND OIL (FOOD).

Oil is expelled from the germ and then refined. The germ fraction may also be highly purified and defatted by solvent extraction. This process yields additional oil and a germ flour food product.

Demand for purified starch and corn sugars has resulted in continued growth of the corn wet-milling industry in the United States. In wet milling, the cleaned corn is steeped in solutions of sulfurous acid in large tanks, 45 h at 122°F (50°C). The extracted solids in the steep liquor are then concentrated. The corn is next passed through an attrition mill that tears the kernels apart. After washing and purification, starch and gluten, dewatered and dried, are the products.

Of the starch produced in the United States, 40% is used directly or chemically modified in food products and industrial applications, and the remainder is converted to corn syrup. High-fructose corn syrup is manufactured by treatment of the syrup solution with purified bacterial-derived isomerase enzyme, which converts half the glucose to fructose to yield a syrup with sweetness equivalent to sucrose.

Oil is recovered from clean dry germ and, after refinement, is sold as table oil. Valuable protein-rich feed by-products are also obtained, including defatted corn germ meal and 60% protein gluten meal.

For preparing alcoholic beverages and alcohol, whole ground corn is cooked, and the starch is hydrolyzed with malt enzymes to provide substrate for whiskey fermentation. Cooked grits treated with malt are used to produce beer by yeast fermentation. Corn sugar syrups are also widely used in brewing. Industrial ethyl alcohol may be distilled from the yeast fermentation broth of enzyme-treated cooked whole grain or wet-milling by-products. Production of grain alcohol for automotive fuel supplements has increased. *See* ALCOHOLIC BEVERAGES; MALT BEVERAGE. [J.A.Sh.]

Cornales An order of flowering plants, division Magnoliophyta (Angiospermae), in the subclass Rosidae of the class Magnoliopsida (dicotyledons). The order consists of 6 families and about 250 species. More than 200 of the species belong to only 2 of the families, the Rhizophoraceae (120) and the Cornaceae (90).

Within its subclass the order is marked by its usually woody habit, simple leaves, flowers with the perianth attached to the surface of the ovary and appearing to grow from the top of it, usually well-developed endosperm, and fleshy fruits that remain closed at maturity (indehiscent). One of the common kinds of mangrove (*Rhizophora mangle*) and the various species of dogwood (*Cornus*) are well-known members of the order, as is the sour gum (*Nyassa sylvatica*, family Nyssaceae). *See* DOGWOOD; ROSIDAE; TUPELO. [A.Cr.]

Corner reflector antenna A directional antenna consisting of the combination of a reflector comprising two conducting planes forming a dihedral angle and a driven radiator or dipole which usually is in the bisecting plane (see illustration). It is widely used both singly and in arrays, gives good gain in comparison with cost, and covers a relatively wide band of frequency. The distance S from the driven radiator D to

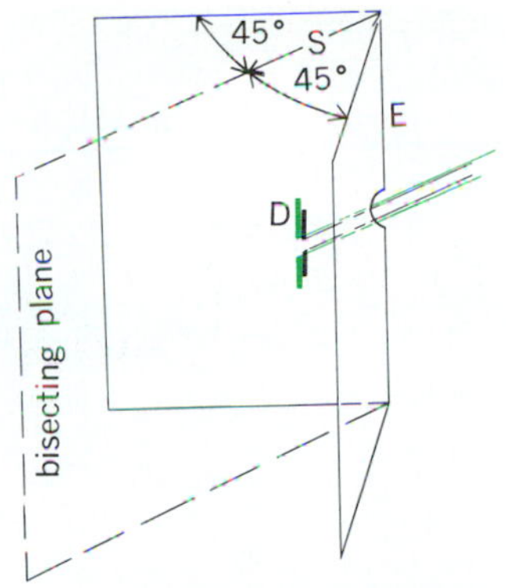

A 90° corner reflector antenna.

edge E need not be critically chosen with respect to wavelength. *See* ANTENNA (ELECTROMAGNETISM). [J.C.S.]

Corona discharge A type of electrical conduction that generally occurs at or near atmospheric pressure in gases. A relatively strong electric field is needed. External manifestations are the emission of light and a hissing sound. The particular characteristics of the discharge are determined by the shape of the electrodes, the polarity, the size of the gap, and the gas or gas mixture. *See* ELECTRICAL CONDUCTION IN GASES.

In some cases corona discharge may be desirable and useful, whereas in others it is harmful and attempts are made to minimize it. The effect is used for voltage division and control in direct-current nuclear particle accelerators. On the other hand, the corona discharge that surrounds a high-potential power transmission line represents a power loss and limits the maximum potential which can be used. Because the power loss due to Joule heating decreases as the potential difference is increased, it is desirable to use the maximum possible voltage.

In the potential current characteristic, the corona region is found above the dark current region and is field-sustained. Near the upper end it goes into a glow discharge or a brush discharge, depending on pressure. Higher pressure favors the brush discharge. *See* BRUSH DISCHARGE; GLOW DISCHARGE.

For still higher potential difference, breakdown occurs and a continuously ionized path forms. *See* ELECTRIC SPARK. [G.H.M.]

Coronagraph A specialized astronomical telescope used for observation of the corona of the uneclipsed Sun. This observation is difficult because the brightness of the corona is only 10^{-6} that of the Sun. The stray light surrounding the solar image in a conventional telescope is about 100 times brighter, and its glare drowns the corona completely. The coronagraph is designed to minimize this glare by artificially eclipsing the solar image itself with an occulting disk. Further, it suppresses the diffracted light from the edge of the primary objective and the doubly reflected light from its center by forming an image of the objective on an annular screen which occults these areas. In addition to such matters of design, success depends on an objective which produces the least possible scattered light. *See* SOLAR CORONA. [J.W.E.]

Coronatae An order of the class Scyphozoa which includes mainly abyssal species. The exumbrella is divided by a circular or coronal furrow into two parts, an upper central disk and a lower coronal part. The central disk is usually domelike, but in *Periphylla* (see illustration) it often narrows toward the top. The coronal part has gelatinous thickenings (pedalia) situated on the radii running from the center of the umbrella to the tentacles and the sensory organs. The pedalia are separated from one another by clefts, running down the radii, that are half-way between each tentacle and rhopalium (a sensory

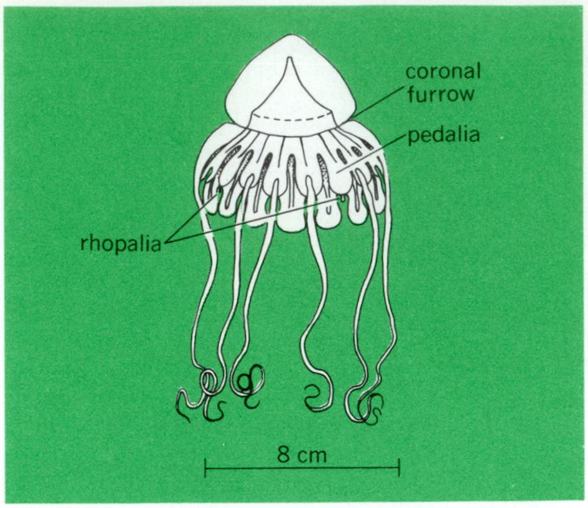

***Periphylla*. (After L. H. Hyman, *The Invertebrates*, vol. 1, McGraw-Hill, 1940)**

organ). Abyssal forms of this group are dark brown or reddish-purple. The littoral species of *Nausithoë* is reported to show alternation of generations, but the life history of abyssal forms remains unknown. *See* SCYPHOZOA. [T.U.]

Coronoidea A small class of "arm"- and brachiole-bearing, stemmed echinoderms in the subphylum Crinozoa, based on five genera known from the Middle Ordovician to Late Silurian of Europe and North America. Coronoids have a crested theca or body with well-developed pentameral symmetry and plate arrangement very similar to that found in blastoids. A thin stem supported the theca above the sea floor, allowing coronoids to live as attached, low- to medium-level suspension feeders. Coronoids had previously been assigned to the blastoids, eocrinoids, or crinoids by different researchers, but they have been elevated in rank to a separate class. Coronoids appear to be most closely related to the Blastoidea and may have been the ancestors of this class. *See* BLASTOIDEA; CRINOZOA; ECHINODERMATA; EOCRINOIDEA. [J.Sp.]

Correspondence principle A fundamental hypothesis according to which classical mechanics can be understood as a limiting case of quantum mechanics; or conversely, many characteristic features in quantum mechanics can be approximated on the basis of classical mechanics, provided classical mechanics is properly reinterpreted. This idea was first proposed by N. Bohr in the early 1920s as a set of rules for understanding the spectra of simple atoms and molecules.

The classical motions in simple dynamical systems can be understood as composed of independent partial motions, each with its own degree of freedom. Each degree of freedom accumulates its own classical action-integral. The frequency of the classical motion for any particular degree of freedom is given by the partial derivative of the energy function with respect to the corresponding action. Bohr noticed that this classical result yields the correct quantum-theoretical result for the light frequency in a transition from one energy level to another, provided the derivative is replaced by the difference in the energies. Moreover, precise information about the possibility of such transitions and their intensities is obtained by analyzing the related classical motion. This information becomes better as the quantum numbers involved become larger. The apparent inconsistencies in Bohr's quantum theory are thereby overcome by a set of rules that came to be called the correspondence principle. *See* ACTION.

After 1925, the success of the new quantum mechanics, particularly wave mechanics, reduced the correspondence principle to a somewhat vague article of faith among physicists. However, the appeal to classical mechanics is still convenient for some rather crude estimates such as the total number of levels below a given energy. Such estimates help in finding the approximate shape of large atoms and large nuclei in the Thomas-Fermi model.

The correspondence principle, however, has assumed a more profound significance. Experimental techniques in atomic, molecular, mesoscopic, and nuclear physics have improved dramatically. High-precision data for many thousands of energy levels are available where the traditional methods of quantum mechanics are not very useful or informative. However, the basic idea behind the correspondence principle must still be valid: Quantum mechanics must be understandable in terms of classical mechanics for the highly excited states, even in difficult cases like the three-body problem, where the overall behavior seems unpredictable and chaotic. The wider application of Bohr's correspondence principle allows many basic but difficult problems to be seen in a new light. *See* ATOMIC STRUCTURE AND SPECTRA; CHAOS; MESOSCOPIC PHYSICS; MOLECULAR STRUCTURE AND SPECTRA; QUANTUM MECHANICS. [M.C.G.]

Corrosion The destruction, degradation, or deterioration of material due to the reaction between the material and its environment. Generally, the reaction is chemical or electrochemical, but there are often important physical and mechanical factors in the corrosion process.

Around 40% of corrosion problems are due to stress-corrosion cracking, hydrogen embrittlement, and corrosion fatigue. These corrosion phenomena have a mechanical, as well as a chemical, aspect to the failure. They constitute some of the worst corrosion problems because failure of the structure occurs often without warning and with sometimes catastrophic results. Other forms of corrosion attack include fretting corrosion, liquid-metal embrittlement, and corrosion by gases.

Stress-corrosion cracking results from the conjoint action of mechanical stress and chemical reaction of the structure with the environment. Cracks form which propagate through the structure, rendering it useless for supporting a load or containing a gas or liquid. Although stress corrosion is usually thought to be a problem in metallic structures, glasses and ionic solids exhibit a formation of cracks which is called crazing. In most cases involving stress corrosion, the metal is not severely attacked by the environment. The metal usually exhibits passivity under the circumstances; in fact, it is this passive behavior which makes the material useful for the particular application. Because stress corrosion commonly occurs when the metal is passive, one explanation for its occurrence is that the action of the stress and the environment is to rupture the protective passive films and expose the underlying metal to severe local attack. While other mechanisms have been suggested to explain stress-corrosion cracking, the film rupture mechanism does explain most of the observed phenomena.

Closely related to stress-corrosion cracking is damage to the metal or alloy caused by the entry of hydrogen. In many environments, the principal cathodic reaction is the evolution of hydrogen gas from hydrogen ions (protons). It is believed that hydrogen enters the metal in the form of a proton or as a single atom. Because of the small size of these particles, hydrogen is able to diffuse readily in many metals, particularly ferrous alloys. Iron, mild steels, alloy steels, and stainless steels with body-centered cubic structure are most prone to hydrogen damage. The damage takes the form of blisters in the low-strength and more ductile materials, and the form of embrittlement in the higher-strength steels.

Corrosion fatigue is the synergistic action of corrosion and fatigue, and results in crack formation. Corrosion fatigue and stress corrosion share many similarities. In corrosion fatigue, the load is applied cyclically (tension-compression). However, while stress corrosion occurs only in the presence of rather specific environmental species, corrosion fatigue may occur in a much wider range of environments. Corrosion fatigue occurs in components such as gears, propellers, impellers, turbines, and the shafts, housings, and attachments to these components that are subjected to cyclic stress and a corrosive atmosphere. Another example is failure of metal implants in the human body. These bone and joint substitutes are exposed to severe load cycling in an environment whose salinity is much like that of sea water.

Fretting corrosion is a phenomenon which occurs between mated surfaces which are stressed and in small relative motion or "slip" to one another. Mechanical vibration is often sufficient to cause the effect; fretting is frequently a problem in engine and machine components. Fretting damage can be minimized by: (1) eliminating the motion or vibration between parts; (2) lubricating the surfaces to lower the friction and prevent oxygen entry; (3) using a gasket material such as rubber or Teflon; (4) increasing the load at the surface to discourage relative motion between parts; and (5) increasing the hardness of one or both components.

Liquid-metal embrittlement is the rapid loss in mechanical properties (fracture stress and ductility) of a metal or alloy due to the contact with certain liquid metals. As with stress corrosion, the alloy–liquid-metal combination is very specific. Occurrence of liquid-metal embrittlement results from joining operations, such as soldering or brazing, and from coating operations, such as galvanizing and tin dipping. Further, the use of liquid-metal coolants (such as liquid sodium in nuclear reactors) can possibly endanger the construction material of the heat exchanger. Another source of problems is mercury contamination of aluminum pipelines. Also, such metals as zinc, cadmium, tin, and lead can embrittle certain steels after they have been soldered or electroplated onto the steel. *See* EMBRITTLEMENT. [R.D.McC.]

Chemical reactions between metals and gases produce two effects on the metal: Metal is consumed, and the metal's properties are changed. Although corrosion by gases is similar to corrosion by liquids, two important differences are found: First, in gases the products of corrosion usually remain on the surface, forming in some systems a protective film, and second, local electrochemical cell reactions are suppressed. Four types of reactions are observed. (1) The gas is dissolved in the metal as molecules, atoms, or ions, and solution is followed by diffusion into the metal. (2) A solid reaction product is formed on the surface, in which case reaction continues by diffusion of the gas or metal atoms or ions through the surface film. (3) A volatile reaction product is formed. (4) A liquid reaction product is formed. Most corrosion reactions involve two or more of these types. Rapid corrosion occurs if liquid or volatile corrosion products are formed or if the solid compound is nonadherent (spalling). For these conditions, new metal is continually being exposed to the gaseous environment. [E.A.G.]

Cortex (plant) The mass of primary tissue in roots and stems extending inward from the epidermis to the phloem. The cortex may consist of one or a combination of three major tissues: parenchyma, collenchyma, and sclerenchyma (see illustration); in roots the cortex almost always consists of parenchyma. The cortex is bounded, more or less distinctly, by the hypodermis (exodermis) on the periphery and by the endodermis on the inside.

Cortical parenchyma is composed of loosely arranged thin-walled living cells. Prominent intercellular spaces usually occur in this tissue. In stems the cells of the outer parenchyma may appear green due to the presence of chloroplasts in the cells. This green tissue is sometimes called chlorenchyma, and it is probable that photosynthesis occurs in it.

In some species the cells of the outer cortex are modified in aerial stems by an irregular deposition of hemicellulose as an additional wall substance, especially in the corners or angles of the cells. This tissue is called collenchyma, and the thickening of the cell walls gives mechanical support to the shoot.

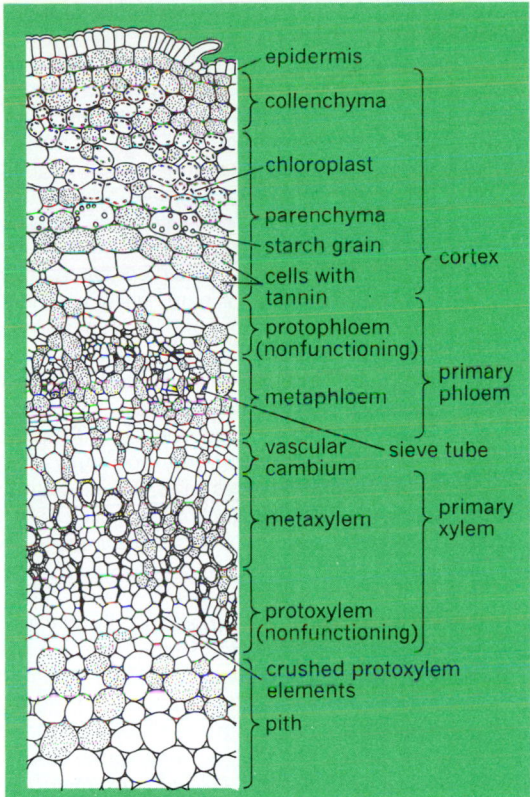

Transverse section of the *Prunus* stem showing the cortex which is composed of collenchyma and parenchyma. (*After K. Esau, Plant Anatomy, 2d ed., 1965*)

Lignifed fibers and sclereids, as well as other lignified cells, also occur in the cortex of many plants. The tissue composed of such hard-walled cells is often referred to as cortical sclerenchyma.

The cortex makes up a considerable proportion of the volume of the root, particularly in young roots, where it functions in the transport of water and ions from the epidermis to the vascular (xylem and phloem) tissues. In older roots it functions primarily as a storage tissue.

In addition to being supportive and protective, the cortex functions in the synthesis and localization of many chemical substances; it is one of the most fundamental storage tissues in the plant. The kinds of cortical cells specialized with regard to storage and synthesis are numerous. The localization of enzyme systems apparently results in a remarkable differentiation of functional cell types.

Because the living protoplasts of the cortex are so highly specialized, patterns and gradients of many substances occur within the cortex, including starch, tannins, glucosides, organic acids, crystals of many kinds, and alkaloids. Oil cavities, resin ducts, and laticifers (latex ducts) are also common in the midcortex of many plants. *See* COLLENCHYMA; ENDODERMIS; HYPODERMIS; PARENCHYMA; ROOT (BOTANY); SCLERENCHYMA; SECRETORY STRUCTURES (PLANT); STEM. [D.S.V.F.]

Corundum A mineral with the ideal composition Al_2O_3. It is one of a large group of isostructural compounds including hematite (Fe_2O_3) and ilmenite ($FeTiO_3$), all of which crystallize in the hexagonal crystal system, trigonal subsystem. Corundum has the high hardness of 9 on Mohs scale and is therefore commonly used as an abrasive, either alone or in the form of the rock called emery, which consists principally of the minerals corundum and magnetite. Crystals occurring in igneous rocks usually have an elongated barrellike shape, while crystals from metamorphic rocks are generally tabular. The specific gravity is approximately 3.98. *See* HEMATITE; ILMENITE.

Pure corundum is transparent and colorless, but most specimens contain some transition elements substituting for aluminum, resulting in the presence of color. Substitution of chromium results in a deep red color; such red corundum is known as ruby. The term "sapphire" is used in both a restricted sense for the "cornflower blue" variety containing iron and titanium, and in a general sense for gem-quality corundums of any color other than red. Star ruby and star sapphire contain tiny needies of the mineral rutile. *See* RUBY; SAPPHIRE.

Corundum occurs as a rock-forming mineral in both metamorphic and igneous rocks, but only in those which are relatively poor in silica, and never in association with free silica. Igneous rocks which most commonly contain corundum include syenites, nepheline syenites, and syenite pegmatites. Both contact and regionally metamorphosed silica-poor rocks may contain corundum. *See* IGNEOUS ROCKS; METAMORPHIC ROCKS.

[D.R.P.]

Coryneform bacteria A large and diverse group of nonspore-forming gram-positive rod-shaped aerobic bacteria which, at some time in their growth cycle on artificial media, yield cells with somewhat irregular outlines or predominant granules. These granules (polyphosphate) often have a metachromatic appearance; that is, they appear red when stained with a blue dye. For the sake of convenience, these organisms are classified together with the "Actinomycetes and Related Organisms," which constitute Part 17 of the eighth edition of *Bergey's Manual*. The rationale for this arrangement, aside from the morphological resemblance to some of the bacteria included in the Actinomycetes, is the finding of chemical components in the cell walls of members of *Corynebacterium* which are also found in *Mycobacterium* and *Nocardia*. In addition, a type of branched lipid which has been found to be peculiar to *Mycobacterium* and to *Nocardia* has also been found in *Corynebacterium*. These mycolic acids and corynemycolic acids have the general structure shown here.

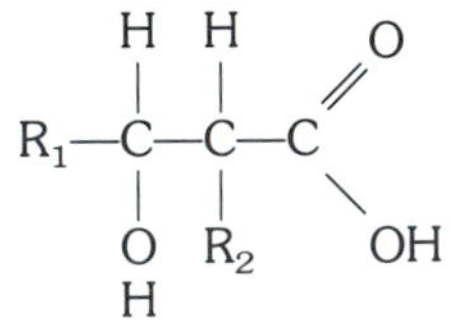

The coryneform group of bacteria loosely corresponds to the family Corynebacteriaceae, which was recognized in the sixth edition of *Bergey's Manual* and broadened in the seventh edition of that treatise. The group, as currently envisioned, includes the genera *Corynebacterium*, *Arthrobacter*, *Cellulomonas*, and *Kurthia*. In addition, a number of other genera are included as *genera incertae sedis* (of uncertain standing in the nomenclature); their representatives may most likely be assigned to *Arthrobacter* when adequate studies of *Brevibacterium* and *Microbacterium* have been completed.

[M.M.]

Cosmic background radiation A nearly uniform flux of microwave radiation that is believed to permeate all of space. The discovery that "empty" space is filled with microwave radiation has had a profound impact on understanding the nature and history of the universe. Although the discovery of the radiation was made in 1965 by A. Penzias and R. W. Wilson, it was only in the late 1970s that its origin as the remnant fireball from the "big bang" became nearly universally accepted by astrophysicists. The correct prediction of the entire spectrum by P. J. E. Peebles, R. H. Dicke, R. G. Roll, and D. T. Wilkinson, after the observation of the intensity at only one frequency, was one of the great triumphs of cosmological theory.

Cosmic microwave radiation is the strongest evidence available for a hot, compressed early universe. In the theory of the big bang, as originally detailed by G. Gamow, H. Alpher, and R. Hermann, the universe began with an explosion about 15×10^9 years ago. This was not an explosion of matter into empty space, but an explosion of space itself. In the big bang theory the entire early universe is assumed to be filled with dense, hot, glowing matter. There is no region of space which is free of matter or radiation. The explosion of space increases the volume of the matter and radiation, and thus also reduces the density and the temperature. The initial temperature is so high that, even after 500,000 years, the temperature of the universe is still several thousand kelvins, the temperature of the surface of the present-day Sun. The matter of the universe is in the form of a plasma of photons, electrons, protons, and alpha particles (the nuclei of helium). About this time the expansion of the universe causes the temperature to drop to the point that the electrons and protons combine to form neutral hydrogen. The universe, which was previously opaque to visible light, suddenly becomes transparent, thus decoupling matter and radiation. This happens because the visible photons are much less strongly absorbed by neutral hydrogen than by the plasma. From that time until now the cosmic blackbody photons have been traveling virtually unscattered, carrying information about the nature of the universe at the time of the decoupling. *See* BIG BANG THEORY.

To an observer moving with the plasma, the photons have a blackbody spectrum with a characteristic temperature of a few thousand kelvins. Although the glow from the plasma is in the visible region, the radiation is redshifted from the visible by a factor on the order of 1000 into the microwave region, with a characteristic temperature of 3 K (−454°F). When one measures the radiation, one is really "looking at" the shell of matter which last scattered the radiation. *See* HEAT RADIATION; REDSHIFT.

This microwave radiation is coming from the most distant region of space ever observed, and was emitted earlier in time than any other cosmological signal. The radiating shell of matter forms the spatial background in front of which all other astrophysical objects, such as quasars, lie. Until methods are devised to detect the neutrinos or gravity waves that were emitted earlier, there will be no direct means of viewing beyond this background.

In the late 1970s, in a series of experiments flown aboard a balloon gondola which had risen above 99.9% of the Earth's atmosphere, P. L. Richards and D. Woody observed the full spectrum of the radiation and verified that the shape was essentially as predicted in the big bang theory.

A small anisotropy in the radiation had been observed. The smooth "cosine" behavior of the data suggests that its cause is not cosmological but local: the motion of the Earth relative to the radiation. From the maximum amplitude of the anisotropy, which is about 10^{-3} of the intensity of the 3 K signal, one can deduce that the velocity of the Earth is 10^{-3} of the speed of light, or about 300 km/s. *See* COSMOLOGY. [R.A.Mu.]

Cosmic electrodynamics The science concerned with electromagnetic phenomena in ionized media encountered in interstellar space, in stars, and above the atmosphere. Because these ionized materials are excellent electrical conductors, they are strongly linked to magnetic-field lines; they can travel freely along but not across the field lines. Phenomena treated under cosmic electrodynamics include acceleration of charged particles to cosmic-ray energies; collisions between galaxies; the correlation of magnetic fields with galactic structure; sunspots and prominences; magnetic storms; aurora; and Van Allen radiation belts. *See* IONOSPHERE; MAGNETOHYDRODYNAMICS. [R.L.]

Cosmic rays Electrons and the nuclei of atoms—largely hydrogen—that impinge upon Earth from all directions of space with nearly the speed of light. These nuclei with relativistic speeds are often referred to as primary cosmic rays, to distinguish them from the cascade of secondary particles generated by their impact against air nuclei at the top of the terrestrial atmosphere. The secondary particles shower down through the atmosphere and are found all the way to the ground and below.

The primary cosmic rays provide the only direct sample of matter from outside the solar system. Measurement of their composition can aid in understanding which aspects of the matter making up the solar system are typical of the Galaxy as a whole and which may be so atypical as to yield specific clues to the origin of the solar system. Cosmic rays are electrically charged; hence they are deflected by the magnetic fields which are thought to exist throughout the Galaxy, and may be used as probes to determine the nature of these fields far from Earth. Outside the solar system the energy contained in the cosmic rays is comparable to that of the magnetic field, so the cosmic rays probably play a major role in determining the structure of the field. Collisions between the cosmic rays and the nuclei of the atoms in the tenuous gas which permeates the Galaxy change the cosmic-ray composition in a measurable way and produce gamma rays which can be detected at Earth, giving information on the distribution of this gas.

Cosmic-ray detection. All cosmic-ray detectors are sensitive only to moving electrical charges. Neutral cosmic rays (neutrons, gamma rays, and neutrinos) are studied by observing the charged particles produced in the collision of the neutral primary with some type of target. At low energies the ionization of the matter through which they pass is the principal means of detection. A single measurement of the ionization produced by a particle is usually not sufficient both to identify the particle and to determine its energy. However, since the ionization itself represents a significant energy loss to a low-energy particle, it is possible to design systems of detectors which trace the rate at which the particle slows down and thus to obtain unique identification and energy measurement.

At energies above about 500 MeV per nucleon, almost all cosmic rays will suffer a catastrophic nuclear interaction before they slow appreciably. An ionization measurement is commonly combined with measurements of physical effects which vary in a different way with mass, charge, and energy. Cerenkov detectors and the deflection of the particles in the field of large superconducting magnets or the magnetic field of the Earth itself provide the best means of studying energies up to a few hundred GeV per nucleon. Detectors employing the phenomenon of x-ray transition radiation promise to be useful for measuring composition at energies up to a few thousand GeV per nucleon. *See* CERENKOV RADIATION; SUPERCONDUCTING DEVICES.

Above about 10^{12} eV, direct detection of individual particles is no longer possible since they are so rare. Such particles are studied by observing the large showers of secondaries they produce in Earth's atmosphere. These showers are detected either by counting the particles which survive to strike ground-level detectors or by looking at the flashes of light the showers produce in the atmosphere with special telescopes and photomultiplier tubes. *See* PARTICLE DETECTOR; PHOTOMULTIPLIER.

Atmospheric cosmic rays. The primary cosmic-ray particles coming into the top of the terrestrial atmosphere make inelastic collisions with nuclei in the atmosphere. When a high-energy nucleus collides with the nucleus of an air atom, a number of things usually occur. Rapid deceleration of the incoming nucleus leads to production of pions with positive, negative, or neutral charge. A few protons and neutrons (in about equal proportions) may be knocked out with energies a significant fraction of that of the incoming nucleus. They are called knock-on protons and neutrons.

All these protons, neutrons, and pions generated by collision of the primary cosmic-ray nuclei with the nuclei of air atoms are the first stage in the development of the secondary cosmic-ray particles observed inside the atmosphere. Since several secondary particles are produced by each collision, the total number of energetic particles of cosmic-ray origin will increase with depth, even while the primary density is decreasing.

The uncharged π^0 mesons decay into two gamma rays with a life of about 8×10^{-17} s. The two gamma rays each produce a positron-electron pair. Upon passing sufficiently close to the nucleus of an air atom deeper in the atmosphere, the electrons and positrons convert their energy into bremsstrahlung. The bremsstrahlung in turn create new positron-electron pairs, and so on. This cascade process continues until the energy of the initial π^0 has been dispersed into a shower of positrons, electrons, and photons with insufficient individual energies (≤ 1 MeV) to continue the pair production. The electrons and photons of such showers are referred to as the soft component of the atmospheric (secondary) cosmic rays. *See* ELECTRON-POSITRON PAIR PRODUCTION.

The $\pi^{\pm}$ mesons produced by the primary collisions have a life of about 2.5×10^{-8} s before they decay into muons. Most low-energy $\pi^{\pm}$ decay into muons before they have time to undergo nuclear interactions. The muons will not interact with nuclei, and are too massive (207 electron masses) to produce bremsstrahlung. They can lose energy only by the comparatively feeble process of ionizing an occasional air atom as they progress downward through the atmosphere. Because of this ability to penetrate matter, they are called the hard component.

The high-energy nucleons—the knock-on protons and neutrons—produced by the primary-particle collisions and a few pion collisions proceed on down into the atmosphere. They produce nuclear interactions of the same kind as the primary nuclei, though of course with diminished energies. This cascade process constitutes the nucleonic component of the secondary cosmic rays.

Solar modulation. The cosmic-ray intensity is lower during the years of high solar activity and sunspot number, which follow an 11-year cycle. This effect has been extensively studied with ground-based and spacecraft instruments.

The primary cause of solar modulation is the solar wind, a highly ionized gas (plasma) which boils off the solar corona and propagates radially from the Sun at a velocity of about 250 mi/s (400 km/s). The wind is mostly hydrogen, with typical density of 80 protons per cubic inch (5 protons per cubic centimeter). This density is too low for collisions with cosmic rays to be important. Rather, the high conductivity of the medium traps part of the solar magnetic field and carries it outward.

In addition to the bulk sweeping action, another effect of great importance occurs in the solar wind, adiabatic deceleration. Because the wind is blowing out, only those particles which chance to move upstream fast enough are able to reach Earth. However, because of the expansion of the wind, particles interacting with it lose energy. Thus, particles observed at Earth with energy of 10 MeV per nucleon actually started out with several hundred MeV per nucleon in nearby interstellar space, and those with initial energy of only 100–200 MeV per nucleon probably never reach Earth at all.

Composition of cosmic rays. Nuclei ranging from protons to lead have been identified in the cosmic radiation. The relative abundances of the elements may be compared with the best estimate of the "universal abundances" obtained by combining measurements of solar spectra, lunar and terrestrial rocks, meteorites, and so forth. Most obvious is the similarity between the two distributions. However, a systematic deviation is quickly apparent: the elements lithium-boron and scandium-manganese as well as most of the odd-charged nuclei are vastly overabundant in the cosmic radiation. This effect has a simple explanation: the cosmic rays travel great distances in the galaxy and occasionally collide with atoms of interstellar gas—mostly hydrogen and helium—and fragment. This fragmentation, or spallation as it is called, produces lighter nuclei from heavier ones but does not change the energy per nucleon very much. Thus the energy spectra of the secondaries are similar to those of the primaries. Calculations involving reaction probabilities determined by nuclear physicists show that the overabundances of the secondary elements can be explained by assuming that cosmic rays pass through an average of about 1 oz per square inch (5 g per square centimeter) of material on their way to Earth.

When spallation has been corrected for, differences between cosmic-ray abundances and solar-system or universal abundances still remain. The most important question is whether these differences are due to the cosmic rays having come from a special kind of material (such as would be produced in a supernova explosion), or simply to the fact that some atoms might be more easily accelerated than others.

Cosmic-ray electron measurements pose other problems of interpretation, partly because electrons are nearly 2000 times lighter than protons, the next lightest cosmic-ray component. Protons with kinetic energy above 1 GeV are about 100 times as numerous as electrons above the same energy, with the relative number of electrons decreasing slowly at higher energies. But it takes about 2000 GeV to give a proton the same velocity as a 1-GeV electron. Viewed in this way electrons are several thousand times more abundant than protons. It is thus quite possible that cosmic electrons have a different source entirely from the nuclei.

Age. Another important result which can be derived from detailed knowledge of cosmic-ray isotopic composition is the "age" of cosmic radiation. Certain isotopes are radioactive, such as beryllium-10 with a half-life of 1.6×10^6 years. Since Be is produced entirely by spallation, study of the relative abundance of ^{10}Be to the other Be isotopes, particularly as a function of energy to utilize the relativistic increase in this lifetime, will yield a number related to the average time since the last nuclear collision. Measurements show that ^{10}Be is nearly absent at low energies and yield an estimate of the age of the cosmic rays of approximately 10^7 years. An implication of this result is that the cosmic rays propagate in a region in space which has an average density of 1.5–3 atoms per cubic inch (0.1–0.2 atoms per cubic centimeter). This is consistent with some astronomical observations of the immediate solar neighborhood.

Origin. Although study of cosmic rays has yielded valuable insight into the structure, operation, and history of the universe, their origin has not been determined. The problem is not so much to devise processes which might produce cosmic rays, but to decide which of many possible processes do in fact produce them.

It is thought that cosmic rays are produced by mechanisms operating within galaxies and are confined almost entirely to the galaxy of their production, trapped by the galactic magnetic field. The intensity in intergalactic space would only be a few percent of the typical galactic intensity, and would be the result of a slow leakage of the galactic particles out of the magnetic trap. [P.E.]

Cosmic spherules Solidified droplets of extraterrestrial materials that melted either during high-velocity entry into the atmosphere or during hypervelocity impact of large meteoroids onto the Earth's surface. Cosmic spherules are rounded particles that are millimeter to microscopic in size and that can be identified by unique physical properties. Although great quantities of the spheres exist on the Earth, they are ordinarily found only in special environments where they have concentrated and are least diluted by terrestrial particulates. *See* METEOR.

The most common spherules are ablation spheres produced by aerodynamic melting of meteoroids as they enter the atmosphere. Typical ablation spheres are produced by melting of submillimeter asteroidal and cometary fragments that enter the atmosphere at velocities ranging from 6.5 to 43 mi (11 to 72 km) per second. The spheres are formed near 48 mi (80 km) altitude, where deceleration, intense frictional heating, melting, partial vaporization, and solidification all occur within a few seconds. During formation, the larger particles can be seen as luminous meteors or shooting stars.

Ablation spheres fall to Earth at a rate of one 0.1-mm-diameter sphere per square meter per year, and every rooftop contains these particles. Unfortunately they are usually mixed in with vast quantities of terrestrial particulates, and they are very difficult to locate. They can, however, be easily found in special environments, such as the ocean floor, that do not contain high concentrations of terrestrial particles that could be confused with cosmic spheres larger than 0.004 in. (0.1 mm) in diameter. *See* MARINE SEDIMENTS.

Impact spheres constitute a second and rarer class of particles that are produced when giant meteoroids impact the Earth's surface with sufficient velocity to produce explosion craters that eject molten droplets of both meteoroid and target materials. They are very abundant on the Moon, but they are rare on the Earth, and they have been found in only a few locations. Impacts large enough to produce explosion craters occur on the Earth every few tens of thousands of years, but the spheres and the craters themselves are rapidly degraded by weathering and geological processes. Meteoritic spherules have been found around a number of craters. Silica-rich glass spheroids (microtektites) are found in thin layers that are contemporaneous with the conventional tektites. Microtektites are believed to be shock-melted sedimentary materials that were ejected from large impact craters. They were ejected as plumes that covered substantial fractions of the surface of the Earth. Microspherules of a different composition have been found in the thin iridium-rich layer associated with the global mass extinctions at the Cretaceous–Tertiary boundary. *See* TEKTITE.

Cosmic spherules are of particular scientific interest because they provide information about the composition of comets and asteroids and also because they can be used as tracers to identify debris resulting from the impact of large extraterrestrial objects. In general, cosmic spherules can be confidently identified on the basis of their elemental and mineralogical compositions, which are radically different from nearly all spherical particles of terrestrial origin. [D.E.Br.]

Cosmic string A hypothetical object that may account for the origin of large-scale structure in the universe. One of the most active areas of research in physics and astronomy is the search for a theory of the formation of such structure. The simplest possibility is that the structure grew by the gravitational collapse of small initial fluctuations in the matter density.

An important idea, which emerged from particle physics in the early 1980s, was that the process that broke the symmetry between the particles and forces, which forms the basis for the standard model for particle physics, might break the spatial symmetry of the universe and produce the fluctuations. The process is analogous to the freezing of ordinary liquids. *See* ELEMENTARY PARTICLE; STANDARD MODEL; SYMMETRY BREAKING.

Most liquids are homogeneous and isotropic. However, when a liquid freezes, the crystal structure of the solid picks a particular direction; but different directions are chosen in different regions. Thus the solid is full of defects where there is a mismatch between neighboring crystalline regions. *See* CRYSTAL; GRAIN BOUNDARIES.

Cosmic strings are very similar to these defects. They are predicted to occur by some grand unification theories and superstring theories. In the very early universe, at high temperature the fields in these theories are random, just as the atoms of a liquid are. As the universe cools below a certain temperature, some of the fields freeze. As they do so, defects are formed, just as in solids. *See* GRAND UNIFICATION THEORIES; SUPERSTRING THEORY.

In different theories, the defects may be at points, along lines, or in sheets. Cosmic strings are the linelike defects; they have special properties that make them ideally suited to forming structure later in the universe. For topological reasons they cannot have ends; they must form closed loops or continue on forever. They are very thin, about 10^{-15} the radius of a proton, but very massive. One meter of cosmic string weighs about 10^{20} kg. *See* COSMOLOGY; UNIVERSE. [N.T.]

Cosmochemistry The science of the chemistry of the universe, particularly that beyond the Earth. As currently practiced, cosmochemistry is concerned primarily with inferences on pre-solar-system events, solar nebular processes, and early planetary processes as deduced from minerals in meteorites and from chemical and isotopic compositions of meteorites and their parts.

Minerals in meteorites. Meteorites provide a great deal of otherwise unobtainable information about the formation and early history of the solar system. The solar system formed from the solar nebula, a large cloud of gas and dust. The chemical composition of the solar system, which can be assumed to be the same as that of the solar nebula, is known in detail from spectroscopic analysis of the Sun, which contains 99.9% of the system's mass. From a combination of this knowledge, an estimate of temperature and pressure from hydrodynamic models, and thermodynamic data for all possible gas, liquid, and solid species, it can be determined which minerals would have become stable as the gas cloud underwent gravitational collapse, spun down into a disk, and cooled. Pressure in the solar nebula was probably so low that condensation occurred at temperatures below the melting points of the minerals, and they condensed as solids. Different minerals became stable at different temperatures, giving rise to a condensation sequence, which can be calculated. *See* ASTRONOMICAL SPECTROSCOPY; HYDRODYNAMICS; PROTOSTAR; SUN; THERMODYNAMIC PRINCIPLES.

Detailed chemical information, however, requires samples, and the Earth is too geologically active to provide samples dating to its formation. Meteorites are therefore studied because, in general, they have undergone much less modification than terrestrial rocks.

A certain type of meteorite, the carbonaceous chondrite, contains millimeter-sized rocklets, or refractory inclusions, that are composed of the very minerals (especially hibonite, perovskite, melilite, spinel, and fassaite) believed to have been among the first to condense from the solar nebula. This finding indicates that assumptions used to calculate the condensation sequence are reasonably accurate. *See* CORUNDUM; MELILITE; PEROVSKITE.

Some meteorites also contain trace amounts of grains of carbon-rich minerals (silicon carbide, diamond, and graphite) whose condensation requires a gas with a carbon/oxygen ratio greater than 1, whereas that of the Sun is 0.6. This finding, along with very large isotopic anomalies for many elements, provides strong evidence that these grains are interstellar in origin, predating the Sun by as much as about 2.4 billion years. These grains should have disappeared by reacting with the solar nebular gas, however; why they did not and how they survived long enough to be incorporated into meteorites are not yet understood. *See* DIAMOND; INTERSTELLAR MATTER; SILICON. [S.B.S.]

Chemical compositions of meteorites. The chemical composition of meteorites and independently formed objects within them have taught much about pre-solar-system events, solar nebular processes, and early planetary processes.

As mentioned above, several types of interstellar grains have been separated from primitive meteorites. While chemistry alone has not unambiguously identified any of these grains as presolar, chemistry does help to understand the formation conditions of objects whose isotopic composition proves a presolar origin. A good example is silicon carbide, which is believed, on the basis of isotopic arguments, to come from the circumstellar envelopes of asymptotic giant branch stars. *See* GIANT STAR.

Chemistry has been used to infer a number of important processes in the solar nebula. These include high-temperature condensation and vaporization, melting and crystallization of millimeter- to centimeter-sized objects in the solar nebula, separation of metal from silicate grains, and reaction of condensed grains with solar nebular gas. Investigation of these processes helps determine important properties of the solar nebula, such as its temperature, how long it lasted, and what the planets, which formed from the nebula, are made of.

A number of important planetary processes, such as separation and differentiation of a metallic core and differentiation of the silicate portion of a planet into mantle and crust, can be studied through the chemical compositions of meteorites. [A.M.D.]

Isotopic compositions of meteorites. Studies of the isotopic compositions of meteorites are a unique source of information on the age of the solar system, the time scales for formation of the first solid bodies in the solar system and the growth and evolution of small planets, the nucleosynthesis of solar material, and the interaction of solar and galactic cosmic rays with matter.

The age of the solar system is most reliably determined by the application of radioactive dating techniques based on the long-lived parent-daughter systems samarium-neodymium, rubidium-strontium, and uranium-thorium-lead. The oldest and most precisely dated meteoritic material is refractory, calcium-aluminum-rich inclusions from the Allende meteorite. The age of these objects, 4.559 ± 0.004 billion years, indicates the time of formation of the first solid bodies in the solar system. The basaltic achondrite meteorites crystallized within a few tens of millions of years of this time. The tight clustering of meteorite ages reflects a period of rapid growth, as millimeter- to centimeter-sized objects collided and coalesced to form meter- to kilometer-sized planetesimals. Younger ages of some meteorites indicate disturbances due to shock and thermal metamorphism beginning as early as 4.48 billion years and continuing for 100–200 million years. *See* EARTH.

An unusual group of basaltic meteorites, the SNC group, is much younger, with crystallization ages of approximately 1.3 billion years. The young igneous ages suggest an origin on a body larger than the Moon, capable of sustaining volcanic activity for at least 3.2 billion years. It is now widely believed that the SNC meteorites come from Mars, an argument strengthened by discoveries that the isotopic compositions of hydrogen, nitrogen, and noble gases in the SNC meteorites and the Martian atmosphere are strikingly similar and distinct from the Earth and other meteorites. *See* MARS.

The short-lived, or extinct, radionuclides have half-lives shorter than about 100 million years and are no longer present in meteorites, having completely decayed. Evidence that they were present at the earliest epoch of solar system history, about 4.56 billion years ago, occurs as excesses of the daughter isotope of the extinct radionuclide. For example, large excesses of ^{26}Mg, produced by the decay of ^{26}Al (half-life 720,000 years), are found in aluminum-rich, magnesium-poor minerals in refractory inclusions in carbonaceous chondrites.

Evidence of extinct radionuclides is found in all classes of meteorites. This widespread evidence of short-lived nuclides requires a late addition of fresh nucleosynthetic material followed by formation of solid bodies (some of which survive intact) within a few million years of nucleosynthetic production.

Excluding the radiogenic isotope variations discussed above, for the most part there is a close similarity between the isotopic compositions of bulk meteorites, the Moon, and the Earth. For many years this apparent homogeneity was taken as evidence that the solar nebula was chemically and physically well mixed prior to the formation of solid bodies. However, the development of new analytical techniques to determine the isotopic compositions of many elements in small constituents of meteorites has led to numerous discoveries of important differences in isotopic composition not explainable in terms of radioactive decay or cosmic-ray interactions. These isotope anomalies in meteorites provide unequivocal evidence that dust derived from disparate stellar sources was not fully mixed in the solar nebula, and offer a window through which the nucleosynthesis of solar system material may be examined in considerable detail.

Research on isotope anomalies and presolar grains has yielded important clues to nucleosynthesis and the types of stars contributing material to the nascent solar nebula. The hydrogen and the bulk of the helium in the solar system are relicts of big bang nucleosynthesis, while the elements carbon through iron were produced dominantly in charged particle reactions in massive stars. The radionuclides ^{129}I, ^{232}Th, ^{235}U, ^{238}U, and ^{244}Pu are all produced in rapid neutron capture (r-process) nucleosynthesis in supernovae. Determination of their relative abundances in meteorites 4.55 billion years ago suggests that r-process nucleosynthesis began 10-15 billion years ago and ended about 150 million years prior to meteorite formation. The synthesis of the stable elements constituting most of the solar system most plausibly occurred on a similar time scale. Late addition, only 5–10 million years prior to meteorite formation, of fresh nucleosynthetic material is required to account for the abundances of short-lived nuclides ^{26}Al, ^{53}Mn, ^{60}Fe, and ^{107}Pd. *See* Big bang theory; Meteorite; Solar system; Supernova. [I.D.H.]

Cosmogenic nuclides Nuclides produced by the interaction of cosmic rays with matter. Cosmic rays are composed of high-energy particles of nuclear matter; they fragment target nuclear material and produce a variety of stable and radioactive nuclides known as cosmogenic nuclides. Protons are the most abundant nuclei in the cosmic radiation, constituting more than 90% of all nuclei above a given kinetic energy per nucleon. Helium and heavier nuclei in roughly the same proportion as in the solar system material account for the rest of the nuclei. Primary cosmic rays and even the secondary particles produced in their interactions are very effective in breaking up nuclear matter. *See* Cosmic rays; Cosmochemistry; Proton; Solar system.

The isotopic composition of any matter exposed to cosmic radiation thus gets altered because of nuclear interactions; in this process, the composition of the cosmic-ray beam itself also gets altered. Cosmogenic nuclides have been studied in diverse forms of extraterrestrial and terrestrial matter and in the cosmic-ray beam with a view toward understanding the evolutionary history of matter. Studies of cosmogenic nuclides in the cosmic-ray beam itself give information on the characteristics of propagation of cosmic-ray particles in the galaxy (for example, the amount of matter traversed and the travel time). The information thus obtained is generally not available from other studies. *See* Isotope.

Cosmogenic nuclides have received increasing emphasis and importance since higher-sensitivity techniques for detection of nuclides have become available. This aspect is of great significance in cosmic-ray studies since the cosmic-ray flux is small.

Cosmogenic nuclides in meteorites have proven useful principally for the study of (1) exposure ages of meteorites, (2) temporal variations in the flux of galactic cosmic radiation during the past 10^9 years, and (3) the characteristics of low-energy charged particles (mostly protons) accelerated by the Sun (the solar cosmic radiation) during the recent and early history of the solar system. The latter records are found in grains within the gas-rich meterorites. Availability of documented lunar samples since 1969 has made it possible to study in detail the temporal variations in the long-term average flux of solar cosmic-ray protons during the past 5 million years. *See* Meteorite; Moon.

The development of accelerator mass spectrometry has made it possible to detect most of the long-lived radionuclides at levels of approximately 10^6 atoms, corresponding to a gain in sensitivity of detection by a factor of about 10^6 over the earlier methods, which were based on detection of the decay product. The gain in sensitivity was particularly important for studies of both meteorites and rare terrestrial material, for example, samples of old ice.

In the field of terrestrial studies, use of accelerator mass spectrometry was extended to include studies of nuclides produced in place in solids, for example, rocks, tree rings, and ice. These studies have made it possible to consider a wide range of geophysical processes which could not be studied earlier. *See* Accelerator mass spectrometry; Radioisotope. [D.Lal]

Cosmology The study of the large-scale structure and the evolution of the universe, and including, in its modern connotation, that part of cosmogony that deals with the origin of the universe and of the chemical elements. *See* Universe.

Structure of the universe. Beginning about 1924, E. P. Hubble developed ways of estimating distances to remote galaxies, and also studied their distribution in space. Hubble concluded that on the large scale the distribution of galaxies is homogeneous and isotropic. Hubble's finding was the first observational evidence for the cosmological principle, which states that at any instant of cosmic time the universe is, on the large scale, the same as seen by all hypothetical observers; that is, it is the same everywhere (is isotropic and homogeneous). *See* Galaxy, external.

On the small scale, however, there is a tendency for galaxies to cluster. The Milky Way Galaxy, in fact, is a member of a system of about two dozen galaxies, most of them much smaller than the Milky Way, spread over a region of space with a linear diameter of about 3×10^6 light-years (1 LY = 5.88×10^{12} mi = 9.46×10^{12} km); the system is called the Local Group. The distribution of galaxy images in photographic surveys of the sky is compatible with a statistical model of complete clustering. Occasional clusters, the great clusters, contain thousands of member galaxies each, and have radii of $5–20 \times 10^6$ LY. On the small scale, even the clusters are clumped into larger aggregates, called superclusters; typical superclusters are $1-3 \times 10^8$ LY across. Observations suggest that the superclusters are the end of the hierarchy. *See* Superclusters.

In summary, the structure of the universe is observed to be very lumpy on scales of up to a few hundred million light-years, but on larger scales appears to be the same everywhere. The cosmological principle, therefore, is a starting assumption for all theories of cosmology that are widely considered today.

Gravitation in the universe. The theory of gravitation embodied in Einstein's general theory of relativity leads to field equations that, in their simplest form, do not permit a static stable universe even if it is infinite. In general relativity the field equations describe the curvature of space-time by matter; it is in curved space-time that all objects, material bodies as well as photons of light, move along unaccelerated paths (geodesics). Imbued with the idea that the universe should be static, Einstein, in his effort to apply the field equations to cosmology, modified them with the introduction of a term now called the cosmological constant. The cosmological constant, if positive, implies a repulsive force that is greater in proportion to distance. By choosing just the correct value for the constant, the force can be made to balance gravitation on the large scale, and permit a static universe. *See* GRAVITATION; RELATIVITY; SPACE-TIME.

Expanding universe. In the 1920s it occurred to others that the universe need not necessarily be static, and that the field equations in their simplest form (zero cosmological constant) can apply perfectly well to a uniformly expanding or contracting universe. Indeed, observations that could test the hypothesis were begun about 1912 by V. M. Slipher; these concerned the spectra of the "nebulae" that were later found to be galaxies. Slipher found the features in the spectra of the objects to be displaced from their normal wavelengths by the Doppler effect, indicating that the nebulae are moving away from the Earth at high speeds. *See* DOPPLER EFFECT; REDSHIFT.

Now, if the cosmological principle is applied to an expanding universe, the universe must expand uniformly; otherwise, irregularities would be generated that would make the universe inhomogeneous. In a uniform expansion, the separations of all pairs of objects must increase by the same factor in the same time, which means that widely separated objects, having further to move apart, must separate at a greater speed than objects close together, which have less far to move to increase their separations by the same relative amount. In other words, every observer, everywhere in an expanding universe, will see every other object moving away from him or her at a speed that is proportional to its distance; it is a necessary and unique condition for a uniform expansion. Of course, since all observers see the same effect, this fact gives no information about the observer's location in the universe, nor of any "center." In an infinite expanding universe, the concept of "center" has no meaning.

In 1931 Hubble and M. L. Humason jointly published a comparison of the distances and radial velocities of galaxies receding at speeds up to 15 times those of the most remote galaxies observed by Slipher. The velocities and distances were in direct proportion; the expansion of the universe had been firmly established. The observed relation between the velocities and distances of galaxies is now called the Hubble law.

Cosmological models. Because the cosmological principle requires that the universe expand uniformly, its evolution can be described uniquely with a simple change in scale. Thus the actual distance $r(t)$ between two remote objects in space at time t can be expressed as the product of a suitably normalized time-varying scale factor, usually denoted as $R(t)$, and the fixed distance u between those objects at some arbitrarily chosen time. According to the Hubble law, the radial velocity of a galaxy is given by $V_r = Hr(t)$, where the constant of proportionality, H, is called the Hubble constant. The Hubble "constant" is not really constant over time, but has a value that changes as the universe evolves and the scale $R(t)$ changes with time. Even if all galaxies maintained their present speeds, H would diminish as the scale grows. In fact, however, the mutual gravitation of the objects in the universe must slow its expansion, so that galaxies, as they recede from each other, are gradually reducing their relative speeds. A major goal of observational cosmology is to determine how strongly the gravitation of the universe decelerates its expansion. *See* HUBBLE CONSTANT.

In general relativity, photons and material objects flow unaccelerated along geodesics in space-time, and that space-time is curved by the presence of matter. In a random distribution of massive objects the curvature can be very complex, but for the universe at large, adoption of the cosmological principle results in an enormous simplification. Except for small-scale local effects, the curvature is the same everywhere. If the mean universal density of matter and energy is high enough, the curvature is positive, in which case the universe is unbounded but finite, for information can never be received from beyond a particular volume of space. Such a universe is said to be closed. If the universal density is below a critical value, however, the curvature is negative, and the universe is open and infinite. The dividing line between these two classes of space-time, requiring the critical density that does not quite close the universe, is flat space-time, in which the universe is infinite and the geometry is euclidean.

A closed universe is closed not only to radiation but to matter, for it implies that the geodesics of particles of matter and of photons are closed; such a universe will eventually cease expanding and start to contract. The contraction will continue until all of the matter and radiation approaches a singularity at the origin of a universal black hole—the "big crunch," as J. A. Wheeler described it.

Alternatively, the open universe will expand forever into greater and greater infinities; neither radiation nor material objects will ever come together again. Closed and open universes are analogous to rockets fired from the surface of the Earth at respectively less than and greater than the velocity of escape. A rocket launched with greater than the escape velocity is always slowed by the Earth's gravitation, but it has enough energy that it never stops; similarly, in the open universe the mean density of matter and radiation is too low to provide enough gravitation to ever stop the expansion.

The flat universe has the critical density to barely stop the expansion, analogous to a rocket launched with precisely the escape velocity. It is mathematically equivalent to a model by Einstein later modified by the astronomer W. de Sitter, and so is also called the Einstein–de Sitter universe.

All general relativistic models of the universe with zero cosmological constant have a unique beginning, usually called the big bang, and a finite age. The maximum possible age for any of these models is that of the open universe with so low a mean density that gravitation can be ignored, so that the galaxies continue to recede at constant speed. In this case, the age of the expansion is simply the reciprocal of the Hubble constant. With the range of current estimates for the Hubble constant, that maximum possible age lies in the range $1\text{–}2 \times 10^{10}$ years. [G.O.A.]

Cotter pin A cotter pin is a split pin (as illustrated) formed by doubling a piece of wire semicircular in cross section to

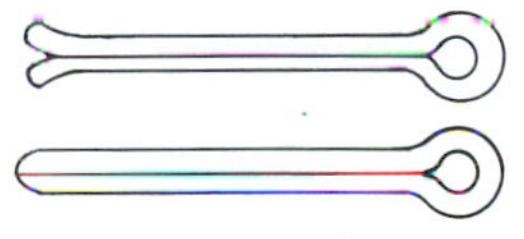

Two of the common forms of cotter pin.

form a loop at one end. After insertion in a hole or a slot in a nut and through a mating crosswise hole in a bolt, the ends of the pin are separated and bent to hold it in place. This fastener is widely used in locations where service inspection is difficult and where failure would be disastrous. Cotter pins are similarly used in many other applications to prevent relative rotation or sliding. *See* NUT (ENGINEERING). [P.H.B.]

Cotton A fiber obtained from the cotton plant *Gossypium*, of the order Malvales. It has been used for more than 3000 years, and is the most widely used natural fiber, because of its versatility and comparatively low cost to produce and to manufacture into finished products. Cotton traditionally has been used alone in textile products, but blends with artificial fibers are also important. Chemically, cotton is essentially pure cellulose. The natural waxes occurring on the original fiber surface act as a finish which facilitates spinning. *See* CELLULOSE; MALVALES; NATURAL FIBER; TEXTILE.

To mature, cotton requires about 180 days of continuously warm weather with adequate moisture and sunlight; frost is harmful and may kill the plant. The plants begin to bloom about 4 months after planting. When the flowers fall off, seed pods grow and become cotton capsules or boils, which must be protected against the boll weevil or other insects. When fully grown, the boils burst, exposing the fleecy white fiber. When the raw cotton is harvested, it contains seeds, leaf fragments, and dirt that must be removed before baling. Cotton seeds alone constitute approximately two-thirds of the weight of the picked cotton.

After harvesting, the fiber, or lint as it is then called, is separated from the seed by gins. The lint is compressed into bales, covered with jute bagging, and bound with metal bands. After baling, the lint is sampled, graded, and sent to cotton mills, where the lint is blended, cleaned, carded, and spun into yarns. In the oil mill processing industry, the cottonseed is separated into fuzz or linters, hulls, oil, and protein cake. Each of these products is converted to several subproducts.

The cotton plant is one of the world's major agricultural crops. It is grown in all countries lying within a wide band around the Earth. The limits of this band in the New World are at about 37°N latitude and at about 32°S latitude. In addition to the effects of latitude, suitability of climate for the growth of cotton is also regulated by elevation, wind belts, ocean currents, and mountain ranges. The regions of more intensive culture comprise the Cotton Belt of the United States, the northern valleys of Mexico, India, Pakistan, eastern China, central Asia, Australia, the Nile River valley, eastern Africa, South Africa, northeastern and southern Brazil, northern Argentina, and Peru. Within these regions and outlying districts, 44 countries report data on cotton production. [E.G.N.]

Cotton effect The characteristic wavelength dependence of the optical rotatory dispersion curve or the circular dichroism curve or both in the vicinity of an absorption band.

When an initially plane-polarized light wave traverses an optically active medium, two principal effects are manifested: a change from planar to elliptic polarization, and a rotation of the major axis of the ellipse through an angle relative to the initial direction of polarization. Both effects are wavelength dependent. The first effect is known as circular dichroism, and a plot of its wavelength (or frequency) dependence is referred to as a circular dichroism (CD) curve. The second effect is called optical rotation and, when plotted as a function of wavelength, is known as an optical rotatory dispersion (ORD) curve. In the vicinity of absorption bands, both curves take on characteristic shapes, and this behavior is known as the Cotton effect, which may be either positive or negative (see illustration). There is a Cotton effect associated with each absorption process, and hence a partial CD curve or partial ORD curve is associated with each particular absorption band or process. *See* OPTICAL ROTATORY DISPERSION; POLARIZED LIGHT.

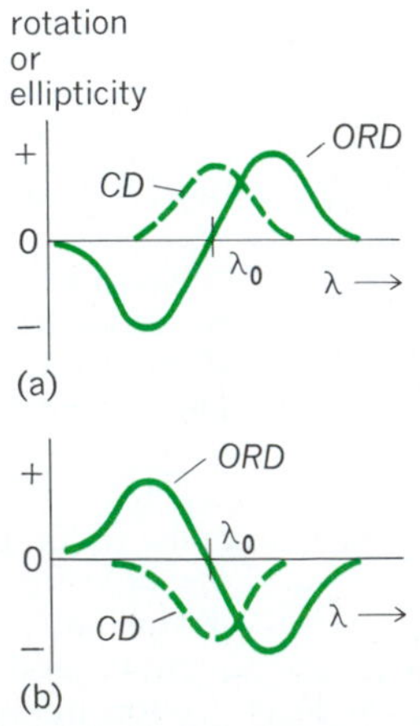

Behavior of the ORD and CD curves in the vicinity of an absorption band at wavelength λ_0 (idealized). (*a*) Positive Cotton effect. (*b*) Negative Cotton effect.

The rotational strengths actually observed vary over quite a few orders of magnitude; this variation in magnitude is useful in stereochemical interpretation of molecular structure. [A.Mo.]

Cotton-Mouton effect A magnetooptic effect concerned with the double refraction of light in a liquid when the liquid is placed in a transverse magnetic field. It is analogous to the electrooptical Kerr effect and is observed in liquids with complicated molecular structure. If the molecule has a magnetic moment, the field tries to orient the molecule, but the themal motion tends to oppose this action. There is thus a degree of orientation, which is dependent on the temperature. If the molecule itself is optically anisotropic, the liquid also will be anisotropic and will exhibit double refraction. *See* KERR EFFECT.

The Cotton-Mouton effect is observed chiefly in nitrobenzene and aromatic organic liquids. Aliphatic compounds have a considerably smaller effect. *See* MAGNETOOPTICS. [G.H.Di.; W.W.W.]

Coulomb excitation A process in which two atomic nuclei that approach each other undergo transitions to excited states that are caused by the long-range Coulomb forces acting between the nuclei. The process can occur even when the nuclei do not come sufficiently close to allow the strong short-range nuclear forces to act.

That nuclei can be excited by this process is not so important in itself. What has proved to be enormously fruitful is the fact that experimental results based on this process have provided accurate values for electrical or electromagnetic properties of nuclei. To take an easily visualized example, Coulomb excitation measurements were instrumental in showing that many nuclear species are not spherical but are shaped like a football. *See* COULOMB'S LAW; NUCLEAR STRUCTURE.

Coulomb excitation can be measured either by the direct analysis of the energy spectrum of the scattered projectiles or by the analysis of the gamma rays subsequently emitted by the excited nuclei. Both methods are used extensively. [P.H.St.]

Coulomb explosion A process in which a molecule moving with high velocity strikes a solid and the electrons that bond the molecule are torn off rapidly in violent collisions with the electrons of the solid; as a result, the molecule is suddenly transformed into a cluster of charged atomic constituents that then separate under the influence of their mutual Coulomb repulsion. *See* COULOMB'S LAW.

Coulomb explosions are most commonly studied using a particle accelerator, normally employed in nuclear physics research (Van de Graaff generator, cyclotron, and so forth), to produce a beam of fast molecular ions that are directed onto a solid-foil target. The Coulomb explosion of the molecular projectiles begins within the first few tenths of a nanometer of penetration into the foil, continues during passage of the projectiles through the foil, and runs to completion after emergence of the projectiles into the vacuum downstream from the foil. Detectors located downstream make precise measurements of the energies and charges of the molecular fragments together with their angles of emission relative to the beam direction. The Coulomb explosion causes the fragment velocities to be shifted in both magnitude and direction from the beam velocity. *See* PARTICLE ACCELERATOR; VAN DE GRAAFF GENERATOR.

Coulomb explosion experiments serve two main purposes.

First, they yield valuable information on the interactions of fast ions with solids. For example, it is known that a fast ion generates a polarization wake that trails behind it as it traverses a solid. This wake can be studied in detail by using diatomic molecular-ion beams, since the motion of a trailing fragment is influenced not only by the Coulomb explosion but also by the wake of its partner. Second, Coulomb-explosion techniques can be used to determine the stereochemical structures of molecular-ion projectiles. *See* ELECTRON WAKE; MOLECULAR STRUCTURE AND SPECTRA; STEREOCHEMISTRY. [D.S.Ge.]

Coulomb's law For electrostatics, Coulomb's law states that the direct force F of point charge q_1 on point charge q_2, when the charges are separated by a distance r, is given by $F = k_0 q_1 q_2/r^2$, where k_0 is a constant of proportionality whose value depends on the units used for measuring F, q, and r. It is the basic quantitative law of electrostatics. In the rationalized meter-kilogram-second (mks) system of units, $k_0 = 1/4\pi\epsilon_0$, where ϵ_0 is called the permittivity of empty space and has the value $\epsilon_0 = (1/4\pi \times 8.98776 \times 10^9)$ coulomb^2/newton-meter^2 $\cong 8.85 \times 10^{-12}$ coulomb^2/newton-meter^2. Thus, Coulomb's law in the rationalized mks system is as in the equation below,

$$F = \frac{1}{4\pi\epsilon_0} \frac{q_1 q_2}{r^2}$$

where q_1 and q_2 are expressed in cou-lombs, r is expressed in meters, and F is given in newtons. *See* ELECTRICAL UNITS AND STANDARDS.

The direction of F is along the line of centers of the point charges q_1 and q_2, and is one of attraction if the charges are opposite in sign and one of repulsion if the charges have the same sign. For a statement of Coulomb's law as applied to point magnet poles *see* MAGNETOSTATICS.

Experiments have shown that the exponent of r in the equation is very accurately the number 2. Lord Rutherford's experiments, in which he scattered alpha particles by atomic nuclei, showed that the equation is valid for charged particles of nuclear dimensions down to separations of about 10^{-12} cm. Nuclear experiments have shown that the forces between charged particles do not obey the equation for separations smaller than this. *See* ELECTROSTATICS. [R.P.Wi.]

Coulometer Electrolytic cell for the measurement of quantities of electricity by quantitative determination of chemical changes; sometimes known as a voltmeter. Coulometers are used for the determination of quantities of electricity in standardization procedures and some electroanalytical measurements.

The quantity of substance produced or consumed in a coulometer is generally determined by weighing, titration, or volume measurements. The deposition of silver, copper, or mercury, the electrolysis of water with hydrogen and oxygen evolution, and the oxidation of iodide to iodine are the electrode reactions generally selected for coulometers. In the silver coulometer the quantity of electricity is determined by weighing the amount of silver plated upon the platinum cup cathode during passage of the current. One coulomb of electricity will deposit 0.0011180 gram of silver. The passage of 1 coulomb of electricity through the iodine coulometer will liberate 0.001315 gram of iodine at the anode. [P.De.]

Coulometric analysis An electroanalytical chemistry technique in which the amount of a substance is determined quantitatively by measuring the total amount of electricity required to deplete a solution of this substance. The total electricity or charge consumed is related to the concentration by Faraday's law. The faraday, F, is 96,487 coulombs per equivalent.

Coulometric analyses are classified as primary or secondary. Primary coulometric analyses are those in which the substance to be determined is electrolyzed directly. Secondary coulometric analyses are those in which an electrolytically generated intermediate reacts stoichiometrically with the substance to be determined. When the solution is completely depleted, the total charge used is related directly to the amount of starting material. Secondary coulometric processes are often referred to as coulometric titrations. However, any coulometric analysis is a titration with electrons and is therefore a coulometric titration.

A condition of 100% current efficiency must exist for a coulometric analysis; that is, all the current must be used either directly or indirectly to deplete the desired substance. For a coulometric analysis either the applied current or the applied potential can be controlled. The controlled-current method is seldom used in primary coulometric analysis, for, as the reaction proceeds, the electroactive substance is depleted and is unable to maintain the required current. The controlled-potential method, however, is frequently used for primary coulometric analyses. *See* COULOMETER; ELECTROCHEMICAL TECHNIQUES. [G.W.O'D.]

Countercurrent exchange (biology) Engineers have known for decades that efficient, almost complete heat or other exchange could be achieved between two fluids flowing in opposite directions in separate tubes. Such countercurrent systems have evolved numerous times in living organisms for all types of exchange function. They are most commonly found in the circulatory, respiratory, and excretory (kidney) systems, serving in heat, oxygen, and ion exchange. Biological countercurrent systems can be classified into two main types: downhill exchanges and hairpin multipliers. In both cases, the basic mechanism is the same—exchange of substance between fluids flowing in opposite directions—but the consequences are very different.

Downhill exchange systems are commonest in the circulatory system where their morphological structure is a rete (network) of closely oppressed sets of small arteries and veins. They are also found in gills of fish and in the minute air tubules of the avian lung. In downhill exchanges, fluids flow in opposite directions in separate tubes with the possibility of exchange, for example, heat flow or diffusion of oxygen, between them. The fluid entering one tube is warmest at that end, while that entering the second tube is coolest at the other end. Heat flows from higher to lower temperature. Although the temperature differential between the two fluids is small at any point along the length of the countercurrent system, almost all the heat contained in the warmer tube is transferred to the cooler tube. Exchange of heat or oxygen occurs by passive diffusion. Most of the heat that entered the countercurrent system at one end leaves the system at the same end.

Retia of blood vessels thus serve as thermal isolating mechanisms within the body. Downhill exchange systems in the gills of fish and in the air tubules of birds permit maximum exchange of oxygen from the environment into the blood. Blood in respiratory capillaries flows against the water or air current and thus can pick up most of the oxygen contained in the external fluid. The advantage of downhill exchangers is that they achieve greater efficiency without extra energy cost simply by arranging flow in a countercurrent rather than in a concurrent fashion.

Hairpin multiplier systems take their name from the structure of the tubes, which have a hairpin turn between the afferent (descending) and the efferent (ascending) limbs. Hairpin countercurrent systems are found in the nephron (the loop of Henle) of the kidney and in the capillary system of the gas gland in the swim bladder of many fish. In contrast to downhill systems, which operate by passive transport, hairpin multipliers must employ active transport of materials. These are always materials pumped out of the efferent limb of the system. *See* CIRCULATORY SYSTEM; KIDNEY; RESPIRATORY SYSTEM; SWIM BLADDER. [W.J.B.]

Countercurrent transfer operations Industrial processes in chemical engineering or laboratory operations in which heat or mass or both are transferred from one fluid to another, with the fluids moving continuously in very nearly steady state or constant manner and in opposite directions through the unit. Other geometrical arrangements for transfer operations are the parallel or concurrent flow, where the two fluids enter at the same end of the apparatus and flow in the same direction to the other end, and the cross-flow apparatus, where the two fluids flow at right angles to each other through the apparatus.

In heat transfer there can be almost complete transfer in countercurrent operation. The limit is reached when the temperature of the colder fluid becomes equal to that of the hotter fluid at some point in the apparatus. At this condition the heat transfer is zero between the two fluids. Most heat transfer equipment has a solid wall between the hot fluid and the cold fluid, so the fluids do not mix. Heat is transferred from the hot fluid through the wall into the cold fluid. Another type of equipment does use direct contact between the two fluids—for example, the cooling towers used to remove heat from a circulating water stream. *See* COOLING TOWER; HEAT BALANCE; HEAT EXCHANGER; HEAT TRANSFER.

Mass transfer involves the changing compositions of mixtures, and is done usually by physical means. A material is transferred within a single phase from a region of high concentration to one of lower concentration by processes of molecular diffusion and eddy diffusion. In typical mass transfer processes, at least two phases are in direct contact in some state of dispersion, and mass (of one or more substances) is transferred from one phase across the interface into the second phase. Mass transfer takes place between two immiscible phases until equilibrium between the two phases is attained. In mass transfer there is seldom an equality of concentration in the two equilibrium phases. This means that a component may be transferred from a phase at low concentration (but at a concentration higher than that at equilibrium) to a second phase of greater concentration. The approach to equilibrium is controlled by diffusion transport across phase boundaries.

Although the two phases may be in concurrent flow or cross-flow, usual arrangements have the phases moving in countercurrent directions. The more dense phase enters near the top of a vertical cylinder and moves downward under the influence of gravity. The less dense phase enters near the bottom of the cylinder and moves upward under the influence of a small pressure gradient. *See* ADSORPTION; CHEMICAL SEPARATION TECHNIQUES; DIFFUSION IN GASES AND LIQUIDS; DISTILLATION; ELECTROPHORESIS; EXTRACTION; LEACHING. [F.J.L.]

Counting circuit One of several types of functional circuits used in switching or digital data-processing systems, such as dial telephone systems and digital computers. The general function of a counting circuit is to receive and count repeated current or voltage pulses, which represent information arriving from some other circuit within the same switching system or from a source external to the system. Counting circuits may be devised with a variety of switching circuit devices, such as electromagnetic relays, vacuum tubes, and transistors. *See* SWITCHING CIRCUIT; SWITCHING SYSTEMS (COMMUNICATIONS).

The basic requirement of a counting circuit is to detect each incoming pulse and to establish within itself a unique state for each successive appearance of such a pulse. The number of pulses that have arrived can then be indicated to another circuit by an output which reflects the particular state of the counting circuit.

The simplest counter is one that counts a single pulse. With electromagnetic relays the counting of a single pulse requires at least two relay actions, one to record the arrival of the input signal, and a second to record its disappearance.

A frequently used counting circuit configuration is a two-pulse, or binary, counter which recycles or returns to its original state after the second pulse. Several single-stage binary counters may be connected so that the output of one counter stage acts as the input to the next counter stage. A two-stage circuit of this type can count four pulses; a three-stage, eight pulses, and so on. [J.Mes.]

Couple A system of two parallel forces of equal magnitude and opposite sense. Under a couple's action a rigid body tends only to rotate about a line normal to the couple's plane. This tendency reflects the vector properties of a couple.

The total force of a couple is zero. The total moment **C** of a couple is identical about any point. Accordingly, **C** is the moment of either force about a point on the other and is perpendicular to the couple's plane. *See* RESULTANT OF FORCES; STATICS.

The moment of a couple about a directed line is the component of its total moment in the line's direction. Couples are equivalent whose total moments are equal. [N.S.F.]

Coupled circuits Two or more electric circuits are said to be coupled if energy can transfer electrically or magnetically from one to another. If electric charge, or current, or rate of change of current in one circuit produces electromotive force or affects the voltage between nodes in another circuit, the two circuits are coupled.

Between coupled circuits there is mutual inductance, resistance, or capacitance, or some combination of these. The concept of a mutual parameter is based on the loop method of analysis. A mutual parameter can be one that carries two or more loop currents; such a network has conductive coupling because electricity can flow from one circuit to the other.

Also, there can be purely inductive coupling, which appears if the magnetic field produced by current in one circuit links the other circuit. A two-winding transformer is an application of inductive coupling, with energy transferred through the magnetic field only.

It is also possible to have mutual capacitance, with energy transferred through the electric field only. Examples are the mutual capacitance between grid and plate circuits of a vacuum tube, or the capacitive interference between two transmission lines, as a power line and a telephone line, that run for a considerable distance side by side.

There are several ways to show the relative polarities of inductive coupling. Figure 1 shows two coils wound on the

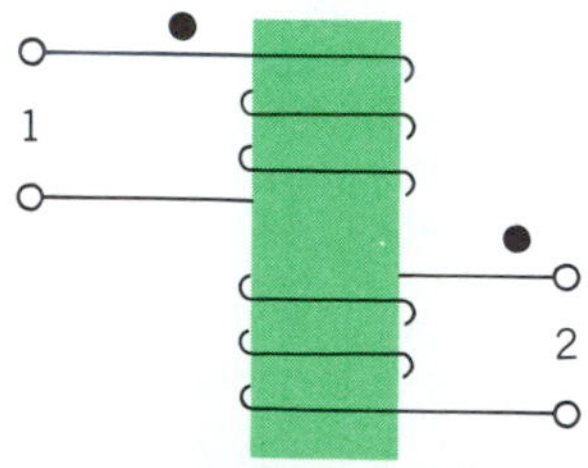

Fig. 1. Polarity of coils.

same core. Current flowing into the upper end of coil 1 would produce magnetic flux upward in the core, and so also would current flowing into the upper end of coil 2. For this reason the upper ends of the two coils are said to be corresponding

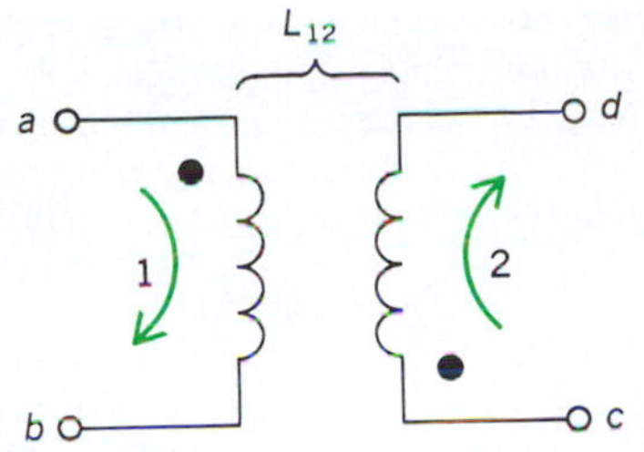

Fig. 2. Mutual inductance L_{12}. Letters *a*, *b*, *c*, and *d* label terminals.

ends. Dots are placed on a diagram at the corresponding ends of coupled coils. Such dots are shown in Fig. 1, though they are not needed in this figure. Dots are also shown in Fig. 2, where they give the only means of identifying corresponding ends of the coils shown. [H.H.Sk.]

Coupling The mechanical fastening that connects shafts together for power transmission, often to form long sections of shafting or to connect the shaft of a machine to an external drive. Couplings are classified with respect to the kind of alignment and the centerline position of the connecting shafts. Rigid couplings are used where the axes of the shafts are directly in line, flexible couplings where the axes may be at a slight angle and slightly displaced, and universal joint couplings for large angularity or large displacement.

A rigid coupling is suitable for shafts in close alignment or held in alignment without inducing destructive forces in shafts or bearings. For commercial shafting, the coupling may be a sleeve with the shafts pressed into each end, or it may be a clamping sleeve. The sleeve on each shaft end may have an external flange with bolt holes. Couplings for large power machines are bolted together to hold the shafts rigidly; hence the shafts must be accurately aligned before assembly.

Flexible coupling provides the mechanical connection for shafts at a slight angle and displaced somewhat, particularly where power machines may set up vibratory forces. Such couplings are used when alignment is difficult to obtain, when a nonrigid support makes alignment impossible to maintain, or if shaft position changes because of operation. Flexible couplings consist of symmetrical hubs on each end of the joined shafts with a flexible connecting member between. This member may take such forms as the floating gear, the embedded spring, the flexible pin, and the flexible disk. Flexible couplings not only provide for lateral and angular misalignment, but also reduce the transmission of shock loads and change the vibration characteristics and critical speeds of the shafts. Oldham's coupling is the original flexible coupling, and presents the basic principle from which most flexible couplings are derived.

A flexible coupling for connecting shafts that have appreciable permanent angularity is termed a universal joint. *See* CLUTCH; FLUID COUPLING; SHAFTING; UNIVERSAL JOINT. [J.J.R.]

Covellite A mineral having composition CuS and crystallizing in the hexagonal system. It is usually massive or occurs in disseminations through other copper minerals. The luster is metallic and the color indigo blue. The hardness is 1.5 (Mohs scale), the specific gravity 4.7. Covellite is a common though not abundant mineral in most copper deposits. It is a supergene mineral and is thus found in the zone of sulfide enrichment associated with other copper minerals, principally chalcocite, chalcopyrite, bornite, and enargite, and is derived by their alteration. *See* BORNITE; CHALCOCITE; CHALCOPYRITE; COPPER; ENARGITE. [C.S.Hu.]

Cover crops Crops grown for the express purpose of preventing and protecting a bare soil surface. Many crops serve this purpose, and some give a high direct economic return. Other crops of lower return value are mainly important as a part of a system of good soil and water conservation and soil productivity maintenance, and it is these that are ordinarily considered cover crops. Basic benefits of cover crops are in the prevention of soil detachment by wind, rainfall, and splash erosion and in the return of cover crop roots and tops to replenish and add to the supply of soil organic matter.

Cover crops can be readily grown in humid climates and in arid and semiarid climates where irrigation provides sufficient water so that cover crops do not rob the soil of needed moisture. Reseeded rangeland is an outstanding example of the use of cover crops. Many special areas such as road banks, ditch banks, and recreation areas need the protection of permanent vegetative cover of high quality. A variety of grasses and legumes may be used, depending upon soil and climatic conditions. *See* SOIL CONSERVATION. [P.J.Z.]

Cowpea *Vigna unguiculata* ssp. *unguiculata*, a legume of such ancient cultivation that its country of origin, either in Asia or Africa, is uncertain. Large-scale production occurs mainly in the southern United States, Africa, the Mediterranean region, and Australia. Cowpeas are grown primarily in the United States as a vegetable crop and are also known as southern peas and blackeye peas, the latter referring to a group of varieties. There are at least 200 varieties in the United States, but only about 20 of these are of commercial and home garden importance. It is also grown as a hay and forage crop. In the tropics and subtropics, cowpeas provide food for millions of people and feed for a large number of livestock. *See* LEGUME. [B.B.B.]

Coxsackievirus A large subgroup of the enteroviruses, in the picornavirus group. The coxsackieviruses produce various human illnesses, including aseptic meningitis, herpangina, pleurodynia, and encephalomyocarditis of newborn infants. At least 23 antigenically distinct types in group A are now recognized, and 6 in group B. Several types (B1–6, A9) grow well in tissue culture. *See* ENTEROVIRUS; PICORNAVIRIDAE.

After incubation for 2–9 days, during which the virus multiplies in the enteric tract, clinical manifestations appear which vary widely. Diagnosis is by isolation of virus in tissue culture or in infant mice. Stools are the richest source of virus. Neutralizing and complement-fixing antibodies form during convalescence and are also useful in diagnosis. *See* ANTIBODY; COMPLEMENT-FIXATION TEST; NEUTRALIZING ANTIBODY.

The coxsackieviruses have worldwide distribution. Infections occur chiefly during summer and early fall, often in epidemic proportions. Spread of virus, like that of other enteroviruses, is associated with family contact and contacts among young children. *See* ANIMAL VIRUS. [J.L.Me.]

CPT theorem A fundamental ingredient in quantum field theories, which dictates that all interactions in nature, all the force laws, are unchanged (invariant) on being subjected to the combined operations of particle-antiparticle interchange (so-called charge conjugation, *C*), reflection of the coordinate system through the origin (parity, *P*), and reversal of time, *T*. In other words, the *CPT* operator commutes with the hamiltonian. The operations may be performed in any order; *TCP*, *TPC*, and so forth, are entirely equivalent. If an interaction is not invariant under any one of the operations, its effect must be compensated by the other two, either singly or combined, in order to satisfy the requirements of the theorem. *See* QUANTUM FIELD THEORY.

The *CPT* theorem appears implicitly in work by J. Schwinger in 1951 to prove the connection between spin and statistics. Subsequently, G. Lüders and W. Pauli derived more explicit proofs, and it is sometimes known as the Lüders-

Pauli theorem. The proof is based on little more than the validity of special relativity and local interactions of the fields. The theorem is intrinsic in the structure of all the successful field theories. *See* QUANTUM STATISTICS; RELATIVITY; SPIN (QUANTUM MECHANICS).

CPT assumed paramount importance in 1957, with the discovery that the weak interactions were not invariant under the parity operation. Almost immediately afterward, it was found that the failure of *P* was attended by a compensating failure of *C* invariance. Initially, it appeared that *CP* invariance was preserved and, with the application of the *CPT* theorem, invariance under time reversal. Then, in 1964 an unmistakable violation of *CP* was discovered in the system of neutral *K* mesons. *See* PARITY (QUANTUM MECHANICS).

One question immediately posed by the failure of parity and charge conjugation invariance is why, as one example, the π^+ and π^- mesons, which decay through the weak interactions, have the same lifetime and the same mass. It turns out that the equality of particle-antiparticle masses and lifetimes is a consequence of *CPT* invariance and not *C* invariance alone. *See* ELEMENTARY PARTICLE; MESON; SYMMETRY LAWS (PHYSICS). [V.F.]

Crab The name applied to arthropods in sections Anomura and Brachyura of the reptantian suborder of the order Decapoda, class Crustacea. The term crab is also sometimes used for two species of sucking lice (order Anoplura) which prey upon humans. *Phthirus pubis* is the crab louse which inhabits the pubic region.

Section Anomura includes over 1400 species in 12 different families. Commonly called hermit, king, sand, or mole crabs, the anomurans have lobsterlike abdomens which bend beneath the cephalothorax in crablike manner, setting them apart in a somewhat ill-defined group.

Section Brachyura encompasses the so-called true crabs, with reduced abdomens folded snugly beneath the cephalothorax. More than 4500 species making up 28 families have been grouped into subsections. In addition to subsections Dromiacea, Gymnopleura, and Oxystomata, the Hapalocarcinidea has one family of the same name containing three genera of small crabs with elongated abdomens. Subsection Brachygnatha is the largest group of brachyurans and includes the most typical crabs. It has 20 families encompassing over 3700 species which occupy a variety of habitats, undergo different types of development, and exhibit contrasting patterns of behavior. *See* CRUSTACEA; DECAPODA (CRUSTACEA). [M.S.K.]

Crab Nebula The Crab Nebula in Taurus (see illustration) is the most remarkable known gaseous nebula. Observed optically, it consists of an amorphous mass that radiates a continuous spectrum and is involved in a mesh of delicate filaments which radiate a bright line spectrum characteristic of typical gaseous nebulae. This nebula has been identified as the expanding remnant of a supernova that appeared in the year 1054. Presumably it comprises only material from the star not yet mixed with the interstellar medium. *See* SUPERNOVA.

The Crab Nebula radiates strongly in the radio range, and the infrared, ultraviolet, x-ray, and gamma-ray spectral regions. Furthermore, the radiation of the central amorphous mass is strongly polarized, even in the x-ray region—indicating that the source of the radiation must be the synchrotron emission of electrons accelerated in a field on the order of 10^{-8} tesla. Energy must be supplied to these electrons at a rate equivalent to about 30,000 times the solar power output. The source of energy for the Crab Nebula appears to be a remarkable, rapidly spinning object known as a pulsar, or neutron star, which is presumed to be the residue of the supernova that created the nebula. Its period of variability (which is associated with its rota-

Crab Nebula, in the constellation Taurus, emitter of strong radio waves and x-rays.

tion period) is 0.033 s. Energy output variations are observed in the radio, infrared, optical, and x-ray range. *See* NEBULA. [L.H.Al.]

Crabapple A fruit, represented commercially by such varieties as Martha, Hyslop, and Transcendent, comprising hybrids between *Malus baccata* (Siberian crabapple) and *M. domestica* (cultivated apple).

Trees are very hardy. At one time the fruit of the crabapple was esteemed because of its high pectin content and its usefulness in jam and jelly manufacture. With the introduction of commercial pectin preparations, demand for the crabapple declined sharply. Except for its use as a pickled product, there is little commercial interest in this fruit. *See* APPLE; FRUIT, TREE; PECTIN; ROSALES. [H.B.T.]

Cracking A process used in the petroleum industry to reduce the molecular weight of hydrocarbons by breaking molecular bonds. Cracking is carried out by thermal, catalytic, or hydrocracking methods. Increasing demand for gasoline and other middle distillates relative to demand for heavier fractions makes cracking processes important in balancing the supply of petroleum products.

Thermal cracking depends on a free-radical mechanism to cause scission of hydrocarbon carbon-carbon bonds and a reduction in molecular size, with the formation of olefins, paraffins, and some aromatics. Side reactions such as radical saturation and polymerization are controlled by regulating reaction conditions. In catalytic cracking, carbonium ions are formed on a catalyst surface, where bond scissions, isomerizations, hydrogen exchange, and so on, yield lower olefins, isoparaffins, isoolefins, and aromatics. Hydrocracking is based on catalytic formation of hydrogen radicals to break carbon-carbon bonds and saturate olefinic bonds. It converts intermediate- and high-boiling distillates to middle distillates high in paraffins and low in cyclics and olefins. *See* CATALYTIC CRACKING; HYDROCRACKING; THERMAL CRACKING. [E.C.L.]

Cranberry The large-fruited American cranberry, *Vaccinium macrocarpon*, a member of the heath family, Ericaceae, is a native plant of open, acid peat bogs in northeastern North America. Well-tended cranberry bogs continue to produce annual crops for a century or more without replanting. Selections from the wild have been cultivated since the early 19th century. It is an evergreen perennial vine producing runners and upright branches with conspicuous terminal flower buds.

Commercial cranberry growing is confined to Massachusetts, New Jersey, Wisconsin, Washington, and Oregon and to several provinces in Canada. Most of the fruit destined for processing is harvested in flood waters, either by machines which pick and deliver the fruit into towed plastic boats or by water reels which detach the berries to float and be driven by wind to shore where they are elevated into bulk trucks. Berries for juice manufacture are usually vine-ripened and deep-frozen for a month or more prior to thawing and extraction.

Good-quality fresh cranberries can be stored for several months, refrigerated, with very little loss to decay. Good-quality cranberries can be kept in deep-freeze storage for several years with only minor moisture loss. Frozen berries on thawing are soft and juicy, unlike the firm fresh berry, and must be utilized promptly.

The special requirements of the cranberry plant for low fertility, acid soil, and winter protection make it a poor choice for home garden cultivation. [C.E.C.]

Crane hoist A mobile construction machine built principally for lifting loads by means of cables. A crane hoist consists of three principal elements: (1) an undercarriage on which the unit moves, (2) a cab or house which envelops the main frame and which contains the power units and controls, and (3) a movable boom over which the cables run and to which miscellaneous working attachments are connected.

A crane is usually typed according to its undercarriage. The crawler crane is a self-propelled unit mounted on continuous tracks similar to those on army tanks. The truck crane is a unit consisting of a crane house and boom mounted on a truck chassis (see illustration). The wheeled crane is self-propelled and rides on a rubber-tired chassis that is an integral part of the unit. The rail or locomotive crane is mounted on a railroad flatcar or a special chassis with flanged wheels for use on standard- or special-gage tracks. The floating crane has a barge or scow for an undercarriage. It is used for water work and for work on waterfronts and generally lacks its own motive power, being towed from place to place by tugboats. See HOISTING MACHINES. [E.M.Y.]

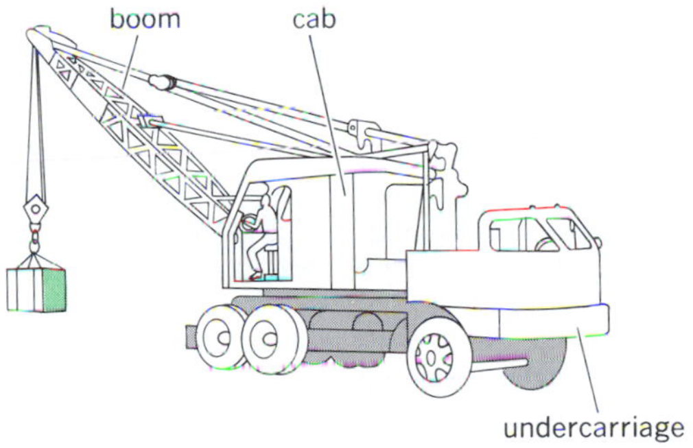

Typical truck crane. (*Link-Belt Co.*)

Cranial nerve Any peripheral nerve which has its central nervous system connection with the brain, as opposed to the spinal cord, and reaches the brain through a hole (foramen) in the skull. Nerve fibers are sensory if they carry information from the periphery to the brain, and motor if they carry information from the brain to the periphery. Sensory fibers are classified as (1) somatic sensory (ss) if they come from the skin or muscle sense organs, (2) visceral sensory (vs) if they come from the viscera, and (3) special sensory (sp.s) if they come from special sense organs such as the eye and ear. Motor fibers are classified as (1) somatic motor (sm) if they carry information to somatic striated muscles, (2) general visceral motor (gvm) if they carry information to glands, smooth muscle, or cardiac muscle, and (3) special visceral motor (svm) if they carry information to visceral striated muscle. A cranial nerve may have only one fiber type or several; cranial nerves with several fiber types are called mixed nerves.

Cranial nerves of vertebrates

No.	Name (type)	Peripheral origin or destination
—	Terminal (ss)	Anterior nasal epithelium
I	Olfactory (sp.s)	Olfactory mucosa
—	Vomeronasal (sp.s)	Vomeronasal mucosa
II	Optic (sp.s)	Retina of eye
III	Oculomotor (sm)	Four extrinsic eye muscles
IV	Trochlear (sm)	One extrinsic eye muscle
V	Trigeminal (svm)	Muscles of mandibular arch derivative
	(ss)	Most of head
VI	Abducens (sm)	One extrinsic eye muscle
—	Anterior lateral line (sp.s)	Lateral line organs of head
VII	Facial (svm)	Muscles of hyoid arch derivative
	(gvm)	Salivary glands
	(ss)	Small part of head
	(vs)	Anterior pharynx
	(sp.s)	Taste, anterior tongue
VIII	Vestibulocochlear (sp.)	Inner ear
—	Posterior lateral line (sp.s)	Lateral line organs of trunk
IX	Glossopharyngeal (svm)	Muscles of third branchial arch
	(gvm)	Salivary gland
	(ss)	Skin near ear
	(vs)	Part of pharynx
	(sp.s)	Taste, posterior tongue
X	Vagus (svm)	Muscles of arches 4–6
	(gvm)	Most viscera of entire trunk
	(vs)	Larynx and part of pharynx
	(sp.s)	Taste, pharynx
XI	Spinal accessory (svm)	Some muscles of arches 4–6
XII	Hypoglossal (sm)	Muscles of tongue and anterior throat

In mammals, 12 pairs of cranial nerves, numbered I through XII, are usually described. However, mammals also have an anterior unnumbered nerve, the terminal nerve. Many vertebrates also have two pairs of lateral line nerves (unnumbered), and lack discrete nerves XI and XII. The table summarizes the cranial nerves of vertebrates. See BRAIN; NERVOUS SYSTEM (VERTEBRATE). [D.B.W.]

Crank In a mechanical linkage or mechanism, a link that can turn about a center of rotation. The crank's center of rotation is in the pivot, usually the axis of a crankshaft, that con-

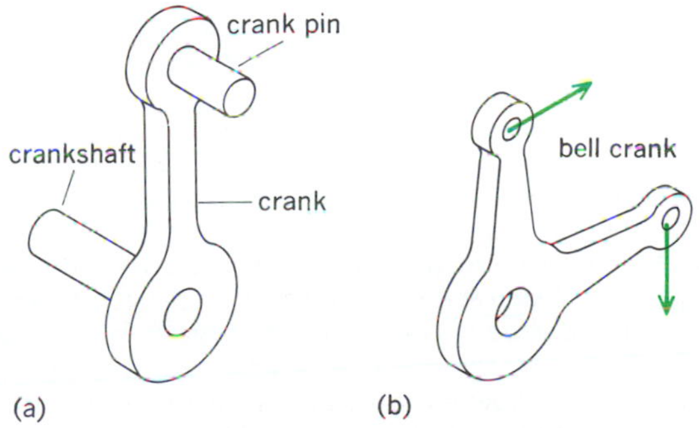

Cranks (*a*) for changing radius of rotation, and (*b*) for changing direction of translation.

nects the crank to an adjacent link. A crank is arranged for complete rotation (360°) about its center; however, it may only oscillate or have intermittent motion. A bell crank is frequently used to change direction of motion in a linkage (see illustration). *See* LINKAGE (MECHANISM). [D.P.Ad.]

Crayfish Fresh-water decapod crustaceans which are related to lobsters and classified with them in the section Astacura of the suborder Reptantia (see illustration); also called crawfish or crawdads. They are generally cryptic and nocturnal, but other habits and habitats of crayfish vary. They live in streams, rivers, ponds, and lakes and demonstrate various extents of burrowing. Crayfish are scavengers and eat detritus and dead animals; they also eat each other, small fish, and probably any other small living animal they can catch; some eat algae and other forms of vegetation. Crayfish are eaten by some fishes, amphibians, reptiles, wading or diving birds, and mammals including humans.

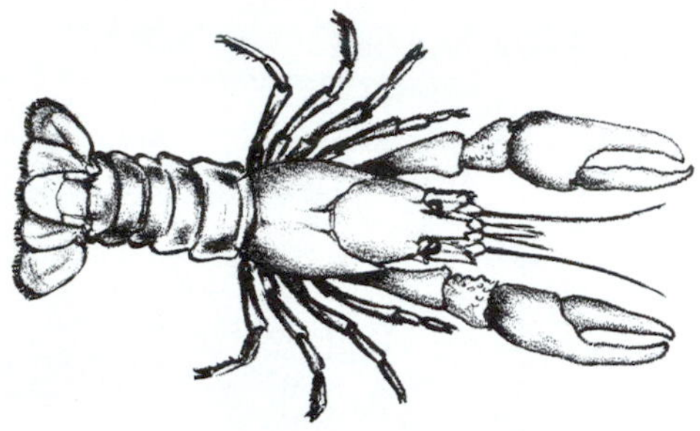

***Cambarus*, a crayfish common to eastern North America.**

Crayfish anatomy and physiological processes, including growth by molting, regeneration, and autotomy, are generally like those of their marine relatives. Crayfish undergo abbreviated development, having no larval stages characteristic of marine crustaceans. Hatchlings resemble adults except that they lack uropods, broad appendages of the terminal abdominal segment. Baby crayfish use their chelipeds to hang onto the mother's pleopods until they undergo the molt after which their uropods are formed and they swim off on their own.

The largest of fresh-water crustaceans, long-lived and abundant in many locales, crayfish are integral components of their ecosystems. *Astacus* species have been products of commerce in Europe since before the Middle Ages. Crawfish farming, usually alternating a rice crop with a crawfish crop in what were originally rice ponds, has prospered in Louisiana and has been undertaken in several other states. In other areas, catches of natural populations are shipped to dealers in live fishing-bait, or are processed and frozen for export to European food markets. *See* CRUSTACEA; DECAPODA (CRUSTACEA); LOBSTER. [M.S.K.]

Creep (materials) The time-dependent strain which occurs when solids are subjected to an applied stress. *See* STRESS AND STRAIN.

Materials exhibit different kinds of creep phenomena. For the most common kind of creep response, following the loading strain, the creep rate is high but decreases as the material deforms during the primary creep stage. At sufficiently large strains, the material creeps at a constant rate. This is called the secondary or steady-state creep stage. Ordinarily this is the most important stage of creep since the time to failure is determined primarily by the secondary creep rate. In the case of tension creep, the secondary creep stage is eventually interrupted by the onset of tertiary creep, which is characterized by internal fracturing of the material, creep acceleration, and finally failure. At low temperatures of applied stresses the time scale can be thousands of years or longer while under different conditions the entire creep process can occur in a matter of seconds. Another kind of creep response is the sort of strain-time behavior observed when the applied stress is partially or completely removed in the course of creep. This results in time-dependent or anelastic strain recovery.

Creep of materials often limits their use in engineering structures. The centrifugal forces acting on turbine blades cause them to extend by creep. In nuclear reactors the metal tubes that contain the fuel undergo creep in response to the pressures and forces exerted on them. In these examples the occurrence of creep is brought about by the need to operate these systems at the highest possible temperatures. Creep also occurs in ordinary structures. For example, when a bolt is tightened to fasten two parts together, the tension stress in the bolt can be quite high. Over a long period of time the bolt creeps, with the result that the force holding the two parts together diminishes. Another example of stress relaxation caused by creep is found in prestressed concrete beams, which are held in compression by steel rods that extend through them. Creep and stress relaxation in the steel rods eventually leads to a reduction of the compression force acting in the beam, and this can result in failure.

There are many examples of creep that are not related to high technology. The window panes in very old homes or cathedrals are often thicker at the bottom than they are at the top due to creep under the force of gravity. Similarly, when a glacier flows down the side of a mountain the ice undergoes extensive creep deformation. *See* PLASTICITY; RHEOLOGY.

[W.D.N.; J.C.G.]

Creeping flow Fluid flow in which the velocity of the flow is very small. For creeping flow, the Reynolds number $R_e = Ud/\upsilon$ is small (less than unity). Here U is the reference velocity, d is the reference length, and υ is the kinematic viscosity of the fluid. For low Reynolds number, the inertial force is negligible and the nonlinear terms in Navier-Stokes equations are neglected. The resultant flow is known as Stokes flow. *See* FLUID FLOW; NAVIER-STOKES EQUATIONS; REYNOLDS NUMBER; VISCOSITY. [S.-I.P.]

Creosote Either of two products of tar distillation: commercial creosote is a toxic substance obtained by high-temperature carbonization of soft coal, and pharmaceutical creosote is an aromatic mixture obtained by redistillation of wood tar.

Commercial creosote is obtained from creosote oil, which amounts to about 30–35% of coal tar. It contains numerous tar acids such as cresols, xylenols, naphthols, and higher homologs, as well as various tar bases such as picolines and collidines, and hydrocarbons. It is primarily used for wood treatment. Small amounts are consumed in the manufacture of lampblack, in disinfectant sprays for fruit trees, and as melting auxiliaries for pitch and bitumen.

Hardwood creosote is prepared from settled hardwood tar, chiefly derived from beechwood. It is the source of pharmaceutical-grade creosote. *See* COAL CHEMICALS; CRESOL; WOOD PRESERVATION.

[F.W.]

Cresol Any one of the three methylphenols (see illustration). The name is also often used to describe a mixture of the three. At room temperature the ortho and para isomers are colorless solids when highly purified, whereas the meta isomer is a liquid, as is a mixture of the three. The cresols are moderately soluble in water and readily soluble in alcohol and ether. They are only slightly less acidic than phenol, and they also resemble the parent substance in toxicity.

Solutions of cresol have long been used as disinfectants. Large quantities are utilized in the manufacture of synthetic resins, antioxidants, surface-active agents, plasticizers, and

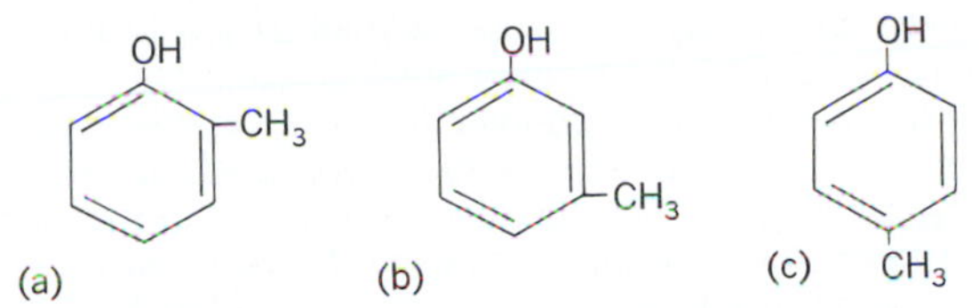

Structural formulas for the three isomeric methylphenols: (*a*) *ortho*-cresol, (*b*) *meta*-cresol, (*c*) *para*-cresol.

antibacterial agents. The phosphate ester of cresol is used as an additive for fuel and lubricating oils. *See* PHENOL. [M.St.]

Cress A prostrate hardy perennial crucifer of European origin belonging to the plant order Capparales. Watercress (*Nasturtium officinale*) is generally grown in flooded soil beds and used for salads and garnishing. Virginia is an important producing state. Garden cress (*Lepidium sativum*) is a cool-season annual crucifer of western Asian origin grown for its flavorful leaves. Of lesser commercial importance is upland or spring cress (*Barbarea verna*), a biennial crucifer of European origin. *See* CAPPARALES. [H.J.C.]

Cretaceous The latest system of rocks or latest period of the Mesozoic Era. It includes that part of the Earth's history

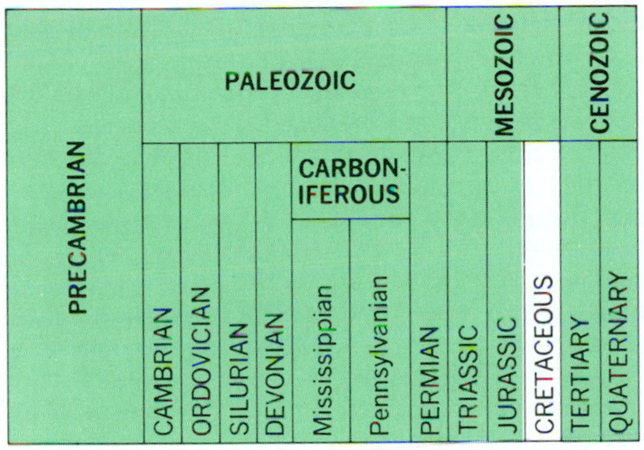

from about 127,000,000 to 64,000,000 years ago—the time of the greatest flooding of the Earth's surface during the Mesozoic Era. It was also the time of extensive chalk deposition over much of the Northern Hemisphere. *See* MESOZOIC.

The Cretaceous marks the rapid rise and spread of deciduous trees over the landscape. Its close was marked by the withdrawal of the seas from the continents, and by the extinction of the dinosaurs, flying reptiles and huge marine reptiles, and the ammonites and certain other marine mollusks that had been so conspicuous throughout the Mesozoic Era.

Much of the Earth's present land surface was submerged by shallow seas during the Cretaceous. Two seas covered much of Europe most of the time: A northern sea extended from the British Isles eastward across Germany and Poland to Russia, and a southern sea, Tethys, covered most of the present Mediterranean Sea and the surrounding parts of southern Europe and northernmost Africa.

According to concepts of continental drift, at the beginning of the Cretaceous the Earth consisted of four closely spaced landmasses and a vast Pacific Ocean. The landmasses consisted of South America united to Africa, an India landmass lying to the east, North America united to Europe by way of Greenland, and Antarctica united to Australia. The northern landmass (North America-Eurasia) is known as Laurasia, and the rest of the landmasses are known collectively as Gondwana. *See* CONTINENT.

Two broad climatic belts were present in the Cretaceous: a southern Mediterranean or Tethyan belt and a northern or boreal belt. The Mediterranean belt, extending from Central America eastward across southern Europe and northern Africa to India and the East Indies, was characterized by much deposition of limestone and other calcareous rocks. Characteristic fossils include large thick-shelled gastropods, certain ammonites, and sessile bivalves known as rudistids which formed reefs. The boreal belt was characterized by an extinct group of squids known as belemnites, certain bivalves (such as aucellas), and numerous ammonites. Temperatures were warmer than at present and probably more uniform. Palms grew as far north as Alaska and Greenland.

The Cretaceous is generally divided into two epochs, Lower Cretaceous and Upper Cretaceous. These are subdivided into

Stages of the Cretaceous

Epoch	Stage	Source of name
Upper Cretaceous	Maestrichtian	Maastricht, Netherlands
	Senonian	Sens, France
	Turonian	Touraine, France
	Cenomanian	Cenomanum, Latin for Mans, France
Lower Cretaceous	Albian	Aube, France
	Aptian	Apt, France
	Neocomian	Neocomum, Latin for Neuchâtel, Switzerland

stages whose names derive from localities in western Europe, where the Cretaceous was first studied intensively (see table). [W.A.Co.]

Crinoidea A class of radially symmetrical Crinozoa in which the body is flower-shaped and either is carried on top of a long, anchored stem (see illustration), as in sea lilies, or is stemless and free to move about, as in comatulids or feather stars. The flowerlike appearance is the result of a circlet of arms that range like petals around the edge of a cup-shaped plated structure called the calyx. The skeleton of the calyx is first formed of two circlets of five plates each. The free-moving arms, usually five in number, occur around the upper margin of the calyx. Each of the five arms is usually divided into one or more branches, for a total of 10 to 30 branches. The stem is attached to the aboral pole of the calyx. It may be long and slender with a rooflike organ attaching it to the sea floor, or it may carry regular whorls of appendages, called cirri, arranged in groups of five at points on the stem termed nodes. The cirri on the distal part of the stem also serve to anchor the crinoid on the sea floor.

Crinoids are the only group of Crinozoa to have survived the Paleozoic Era, which ended 200,000,000 years ago, and thus they are of great importance in the interpretation of the abundant fossil remains of other Crinozoa.

About 690 living species are known, and of these, 80 are stemmed forms. More than 5500 fossil species, representing about 800 genera, have been described. All living crinoids are referred to one subclass, the Articulata. The three other subclasses of crinoids, the Inadunata, Camerata, and Flexibilia, are known only from fossils.

No crinoids are of direct economic importance to humans because they are not used for food, nor do they appear to be eaten by other animals. None are venomous.

Crinoids occur in all seas except the Black and the Baltic. Shallow-water comatulids are most abundant and varied around tropical coral reefs. Many are nocturnal, coming out of crevices in the reefs to perch on coral heads at night to feed. The arms are arrayed in a fan-shaped filtration net to catch

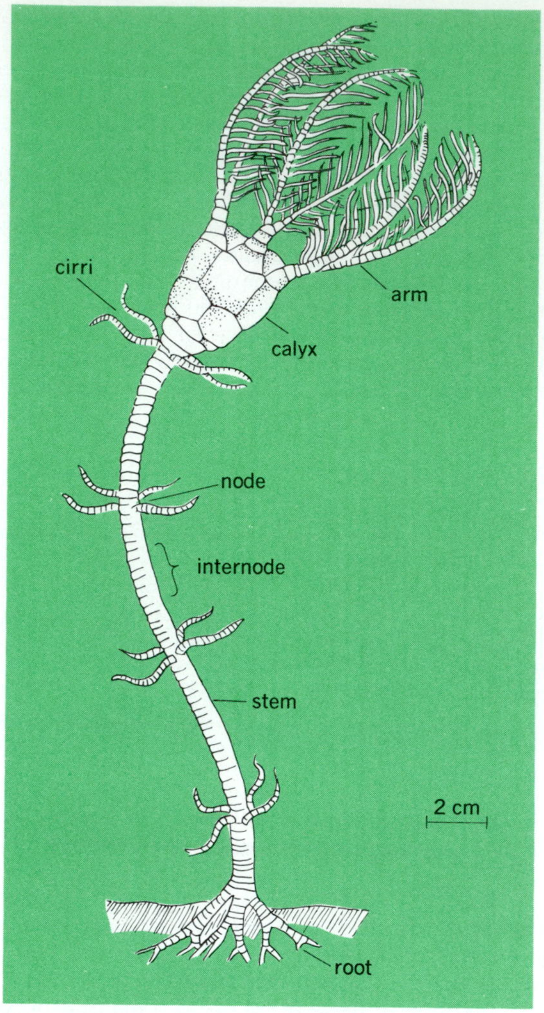

Schematic diagram of a crinoid.

suspended food, which consists mainly of zooplankton. Crinoids prefer areas with moderate currents so that food can be carried through the feeding fan. All living stemmed crinoids inhabit water more than 330 ft (100 m) deep, where they and some comatulids are widespread. Sea lilies use the cirri on the lower end of the stem to anchor themselves on sea-floor sediments; and comatulids, which also have cirri on the centrodorsal plate, use them for temporary attachment. *See* ARTICULATA (ECHINODERMATA); CRINOZOA; ECHINODERMATA. [N.G.L.]

Crinozoa A subphylum of the Echinodermata including radially symmetrical echinoderms showing a partly meridional pattern of growth. This tends to produce an aboral cup-shaped theca and a partly radially divergent pattern of growth, forming appendages, called brachioles or arms and bearing ambulacral grooves. Included in the subphylum are the blastoids, cystoids, edrioblastoids, eocrinoids, parablastoids, and paracrinoids. *See* BLASTOIDEA; CRINOIDEA; ECHINODERMATA; EOCRINOIDEA. [H.B.F.]

Critical care medicine The treatment of acute, life-threatening disorders, usually in intensive care units. Critical care medicine has been practiced informally for many decades in trauma centers, postanesthesia recovery rooms, coronary care units, delivery rooms, emergency rooms, and postoperative areas. The facilities and trained personnel available in the intensive care unit (ICU) permit extensive monitoring of physiological variables, organization of complex, multidisciplinary diagnostic and therapeutic plans, administration of therapy to predetermined goals, and expert nursing care.

Critical care thus runs counter to the traditional division of specialties by organ or organ system. Specialists in critical care undergo training beyond a primary qualification (internal medicine, surgery, anesthesia, or pediatrics), and must be able to manage acute respiratory, cardiovascular, metabolic, cerebral, and renal problems, as well as infections. The patients may be newborns, children, or adults suffering from trauma or acute life-threatening disease. Patients having failure of multiple organs, complicated medical problems, disorders falling into several medical specialities, or a need for 24-h care often become the responsibility of the critical care specialist.

The intensive care unit is the most labor-intensive, technically complex, and expensive part of hospital care. The intensive care unit, however, may be crucial to the patient's survival. Perhaps the single most useful function of the intensive care unit is to provide life-support systems for desperately ill patients who would not survive without them. Such systems include mechanical ventilation; cardiopulmonary resuscitation; peritoneal dialysis and hemodialysis; circulatory support with intraaortic balloon pumping; and extracorporeal membrane oxygenation.

Another aspect of critical care is the provision of life-sustaining therapy. Components include administration of intravenous fluids, provision of nutritional support, and control of infections. [W.C.Sh.]

Critical mass That amount of fissile material (^{233}U, ^{235}U, or ^{239}Pu) which permits a self-sustaining chain reaction. The critical mass is increased by the presence of neutron absorptive materials. It is reduced by a moderator, such as graphite or heavy water, which slows down the neutrons, inhibits their escape, and indirectly increases their chance to produce fission. *See* CHAIN REACTION (PHYSICS); NUCLEAR FISSION; REACTOR PHYSICS. [J.A.W.]

Critical path method (CPM) A diagrammatic network-based technique, similar to the program evaluation and review technique (PERT), that is used as an aid in the systematic management of complex projects. The technique is useful in: organizing and planning; analyzing and comprehending; problem detecting and defining; alternative action simulating; improving (replanning); time and cost estimating; budgeting and scheduling; and coordinating and controlling. It has its greatest value in complex projects which involve many interrelated events, activities, and resources (time, money, equipment, and personnel) which can be allocated or assigned in a variety of ways to achieve a desired objective. It can be used to complete a multifaceted program faster or with better utilization of resources by reassignments, trade-offs, and judiciously using more or less assets for certain of the individual activities composing the overall project. *See* PERT.

Applications of CPM include: research and development programs; new product introductions; facilities planning and designs; plant layouts and relocations; construction projects; equipment installations and start-ups; major maintenance pro-

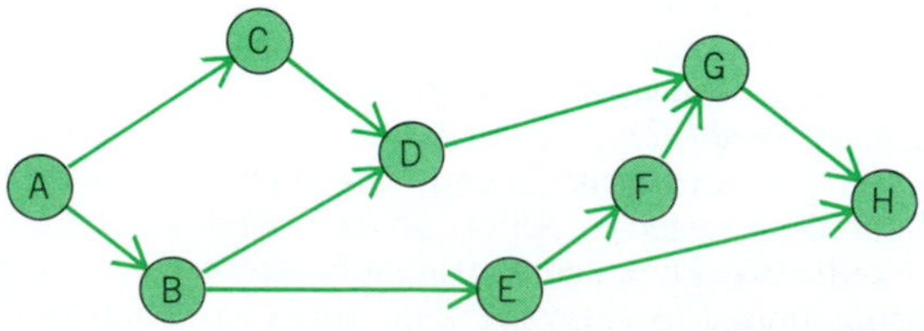

Network of relationship between events (circles) and their activities (arrows).

grams; medical and scientific researches; weapons systems developments; and other programs in which cost reduction, progress control, and time management are important. It is best suited to large, complex, one-time or first-time projects rather than to repetitive, routine jobs.

The illustration shows a critical path diagram constructed for a comparatively simple project, designing a personal home-use computer. The table lists the required events. The critical path is defined as the longest route from the first event in the project to the final outcome. This is the total time that the project can be expected to take. Analysis may shorten the critical path by identifying activities whose time can be shortened. A schedule is constructed showing the earliest and latest permissible start and completion dates of each activity.

CPM events table

A = Authorization to start received
B = Computing circuits designed
C = Video circuits designed
D = Keyboard and cabinets designed
E = Programming completed
F = Operating systems completed
G = Testing and debugging finished
H = Design specifications and programs finalized

The use of CPM adds another step to a project, and it requires continual updating and reanalysis as conditions change, but experience has shown that the effort can be a good investment in completing a project in less time, with less resources, with more control, and with a greater chance of on-time, within-budget completion. *See* INDUSTRIAL ENGINEERING. [V.M.A.]

Crocco's equation A relationship between vorticity and entropy gradient for the steady flow of an inviscid compressible fluid. Crocco's equation, given below, pertains to

$$\mathbf{v} \times \boldsymbol{\omega} = -T \text{ grad } s$$

isoenergetic flow, which is a common type of flow where the total energy, or stagnation enthalpy, of the fluid per unit mass is constant throughout the fluid. In the equation $\mathbf{v}$ is the fluid velocity vector, $\boldsymbol{\omega} = \text{curl } \mathbf{v}$ is the vorticity vector, T is the fluid temperature, and s is the entropy per unit mass of the fluid. Thus entropy gradients can occur only at right angles to $\mathbf{v}$ and $\boldsymbol{\omega}$. An irrotational flow is one where $\boldsymbol{\omega} = 0$ throughout the flow, and it follows from Crocco's equation that it must also be an isentropic flow, that is, one with constant entropy throughout. *See* FLUID FLOW; ISENTROPIC FLOW; LAPLACE'S IRROTATIONAL MOTION. [A.E.Br.]

Crocodile The common name used for 14 species of large reptiles included in the family Crocodylidae (order Crocodylia) which also includes the alligators, caimans, and the gharial (also known as the gavial). Like all crocodilians, the crocodiles are primarily distributed throughout the tropical regions of the world. Species occur in both saltwater and freshwater habitats. Crocodiles are generally omnivorous, feeding on invertebrates, fish, other reptiles and amphibians, birds, and mammals—practically any animal they can overpower. A few, very narrow-snouted species are believed to subsist primarily on fish. Crocodiles are primarily aquatic and nocturnal, leaving the water only to bask by day or to build their nests. Some species construct burrows into the banks of rivers or lakes where they spend part of their time.

These animals are powerful predators with large teeth and strong jaws. Large adults of some species may exceed 20 ft (6 m) in length and are capable of overpowering and eating large grazing mammals, even occasionally humans. The webbed feet, flattened tail, and placement of the nostrils, eyes, and ears on raised areas of the head are adaptations for an aquatic existence. The raised nostrils, eyes, and ears allow the animals to float almost completely submerged while still monitoring their environment.

Reproduction in crocodiles is the most elaborate of the reptiles. Courtship and mating occur in the water. The female digs a hole in the soil for the 30 or so eggs. Nests are often guarded by the female. The young may remain together as a pod with the female for a year or more. Hearing and vocal communication are well developed in the crocodiles, and a variety of bellows, snarls, and grunts are utilized in their elaborate social behavior.

Crocodiles are considered very valuable for the leather obtained from their hides, and all species are becoming very rare owing to hunting and to the loss of their habitats through land development for other uses. Most countries with native crocodile populations are implementing conservation measures, and international efforts are being made to regulate trade in crocodile products. *See* ALLIGATOR; CROCODYLIA; REPTILIA. [H.W.C.]

Crocodylia An order of the class Reptilia (subclass Archosauria) which is composed of large, voracious, aquatic species which include the alligators, caimans, crocodiles, and gavials. The group has a long fossil history from Late Triassic times and its members are the closest living relatives of the extinct dinosaurs and the birds. The 21 or 22 living species are found in tropic areas of Africa, Asia, Australia, and the Americas. One form, the salt-water crocodile (*Crocodylus porosus*), has traversed oceanic barriers from the East Indies as far east as the Fiji Islands. *See* ARCHOSAURIA.

The order is distinguished from other living reptiles in that it has two temporal foramina, an immovable quadrate, a bony secondary palate, no shell, a single median penis in males, socketed teeth, a four-chambered heart, and an oblique septum that completely separates the lung cavities from the peritoneal region. Certain of these unique features and other salient characteristics of the Crocodylia are intimately associated with their aquatic life. For example, there is a special pair of fleshy flaps at the posterior end of the mouth cavity which form a valvular mechanism which separates the mouth from the region where the air passage opens into the throat. This complex arrangement allows crocodilians to breathe even though most of the head is under water, or the mouth is open holding prey or full of water.

During the breeding season male crocodylians set up territories on land which they defend against intruders of the same species. Fertilization is internal and the hard-shelled eggs are deposited in excavations in the sand or in large nests of decaying vegetation, depending upon the species.

The living species are placed in two families and eight genera. The family Crocodylidae contains two subgroups: the true crocodiles, Crocodylinae, including the genera *Crocodylus* found in all tropic areas, *Osteolaemus* in central Africa, and the false gavial (*Tomistoma*) in Malaya and the East Indies; the alligators and caimans, Alligatorinae, including the genera *Alligator* of the southeastern United States and near Shanghai, China, the *Caiman* from Central and South America, and *Melanosuchus* and *Paleosuchus* of South America. The gavial (*Gavialis gangeticus*) of India and north Burma is the only living member of the family Gavialidae. Crocodiles differ most obviously from alligators and caimans in head shape and in the position of the teeth, although other technical details also separate them. *See* ALLIGATOR; CROCODILE; GAVIAL.

Crocodylians first appear in deposits of Late Triassic or possibly Early Jurassic age in North America and South Africa. They formerly were much more widely distributed in temperate latitude than today; abundant paleobotanical evidence confirms a warmer climate in mid-latitudes at this time, It seems probable that the restriction of crocodilians to the tropics was a direct result of cooling climate in the late Cenozoic. *See* REPTILIA. [J.T.G.]

Crocoite A mineral with the chemical composition $PbCrO_4$. Crocoite occurs in yellow to orange or hyacinth red, monoclinic, prismatic crystals with adamantine to vitreous luster; it is also massive granular. Hardness is 2.5–3 on Mohs scale and specific gravity is 6.0. Streak, or color of the mineral powder, is orangish-yellow. It fuses easily.

Crocoite is a secondary mineral associated with other secondary minerals of lead such as pyromorphite and of zinc such as cerussite. It has been found in mines in California and Colorado. *See* LEAD. [E.C.T.C.]

Cro-Magnon people The name which refers to the earliest representatives of anatomically modern *Homo sapiens* in western and central Europe. These fossil humans, who are associated with the upper Paleolithic archeological period and date from about 35,000 to about 10,000 years ago, were physically within the ranges of variation of modern Europeans (see illustration). They differed only in their tendency to be ruggedly built, especially the males.

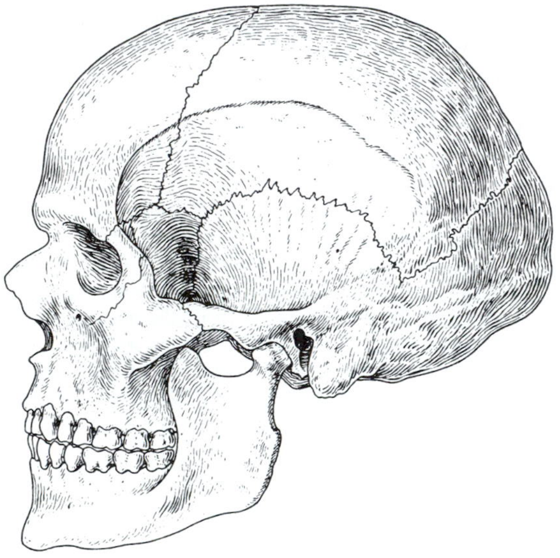

Skull of a Cro-Magnon male. ***(After M. F. Ashley Montagu, An Introduction to Physical Anthropology, 2d ed., Charles C. Thomas, 1951)***

The name Cro-Magnon refers specifically to the remains of at least five individuals found in the Cro-Magnon rock shelter near Les Eyzies, Dordogne, France, in 1868 by E. Lartet and H. Christy. As these were the first well-known fossils of this period, their name has been employed for most European fossils of the same time period. Other notable sites which have yielded human remains of the Cro-Magnon type are Combe Capelle and La Madeleine in France, Grimaldi in Italy, and Mladeč in Czechoslovakia. *See* FOSSIL HUMAN. [E.T.]

Cromwell Current The strong subsurface current that flows eastward along the Equator in the Pacific Ocean. The Cromwell Current, also known as the Pacific Equatorial Undercurrent, flows from north of New Guinea in the western Pacific to the Galápagos Islands in the eastern Pacific. Its length, about 8700 mi (14,000 km), is larger than one-third the circumference of the Earth. The current is typically 120–240 mi (200–400 km) wide and 300–900 ft (100–300 m) thick. Its core of highest speed, usually 1.5–4.5 ft/s (0.5–1.5 m/s), is found at depths of 150–600 ft (50–200 m).

The Cromwell Current is usually a continuous flow along the Equator from about 140°E to 91°W. It may at times originate further to the west, and its terminus in the east sometimes extends to the South American coast, where it turns south into the Peru-Chile Undercurrent. The high-speed core is found in the tropical thermocline, typically near 68°F (20°C). The extensive westerly winds associated with the El Niño phenomenon can destroy the east-west slope of the thermocline, and with it the Cromwell Current.

The Cromwell Current is indirectly forced by the southeast tradewinds that blow westward along the Equator over most of the Pacific. The winds blow the surface waters to the west as the South Equatorial Current. Also, because the vertical component of the Coriolis force is zero at the Equator, the existence of the Cromwell Current can be explained without reference to the Earth's rotation. However, the Coriolis force is important in the dynamics and structure of the current. *See* CORIOLIS ACCELERATION; OCEAN CIRCULATION; PACIFIC OCEAN. [E.Fi.]

Crop micrometeorology The science concerned with the interaction of crops and their immediate physical environment. Especially, it seeks to measure and explain net photosynthesis (photosynthesis minus respiration) and water use (transpiration plus evaporation from the soil) of crops as a function of meteorological, crop, and soil moisture conditions. These studies are complex because the intricate array of leaves, stems, and fruits modifies the local environment and because the processes of energy transfers and conversions are interrelated. As a basic science, crop micrometeorology is related to plant anatomy, plant physiology, meteorology, and hydrology. Expertise in radiation exchange theory, boundary-layer and diffusion processes, and turbulence theory is needed in basic crop micrometeorological studies. A practical goal is to provide improved plant designs and cropping patterns for light interception, for reducing infestations of diseases, pests, and weeds, and for increasing crop water-use efficiency. Shelter belts are modifications that have been used in arid or windy areas to protect crops and seedlings from the harshness. *See* AGRICULTURAL METEOROLOGY; MICROMETEOROLOGY. [L.H.A.]

Crossing-over (genetics) The process whereby one or more gene alleles present in one chromosome may be exchanged with their alternative alleles on a homologous chromosome to produce a recombinant (crossover) chromosome which contains a combination of the alleles originally present on the two parental chromosomes. Genes which occur on the same chromosome are said to be linked, and together they are said to compose a linkage group. In eukaryotes, crossing-over may occur during both meiosis and mitosis, but the frequency of meiotic crossing-over is much higher.

Crossing-over is a reciprocal recombination event which involves breakage and exchange between two nonsister chromatids of the four homologous chromatids present at prophase I of meiosis; that is, crossing-over occurs after the replication of chromosomes which has occurred in premeiotic interphase. The result is that half of the meiotic products will be recombinants, and half will have the parental gene combinations. Using maize chromosomes which carried both cytological and genetical markers, H. Creighton and B. McClintock showed in 1931 that genetic crossing-over between linked genes was accompanied by exchange of microscopically visible chromosome markers. *See* RECOMBINATION (GENETICS).

In general, the closer two genes are on a chromosome, that is, the more closely linked they are, the less likely it is that crossing-over will occur between them. Thus, the frequency of crossing-over between different genes on a chromosome can be used to produce an estimate of their order and distances apart; this is known as a linkage map. *See* ALLELE; CHROMOSOME; GENE; GENETIC MAPPING; LINKAGE (GENETICS). [C.B.G.]

Crossopterygii An infraclass of the bony fishes (class Osteichthyes), also known as fringe-finned fishes, that forms one of the two major divisions of the lobe-finned fishes (Sarcopterygii). The group first appeared as fossils in the Early Devonian; in the Paleozoic they were mostly small to medium-sized carnivorous fish living in shallow tropical seas, estuaries, and fresh waters. There were two principal groups: a diverse set of fishes termed Rhipidista and the Coelacanthini. Their principal radiations were in the Devonian, and by the Mississippian they were in sharp decline. The Rhipidista were wholly extinct by the Middle Permian, but the coelacanths underwent a second, smaller, Mesozoic radiation and managed to survive to the present day as the most famous lobe-fin of all, the living species *Latimeria chalumnae*. Crossopterygii are characterized by a unique hinge in the skull that allowed the front portion to be raised and lowered during feeding and respiratory movements. *See* SARCOPTERYGII.

Members of the order Rhipidistia were principally fusiform, fast-swimming carnivores that flourished in the rivers and lakes of the Late Devonian. They could breathe air, and the use of lungs as well as gills gave them an advantage in warm, shallow-water environments where dissolved oxygen was often low. Members of the order Coelacanthini are characterized by a special trifid tail and scales ornamented with tubercles. They are not thought to be close to the ancestors of tetrapods. *See* OSTEICHTHYES. [K.T.]

Crossover network A selective network used to divide the audio-frequency output of an amplifier into two or more bands of frequencies. The frequency separation is employed to feed two or more loudspeakers, each operating in a restricted frequency band and thereby operating more efficiently and with less distortion. Most crossover networks used in high-fidelity audio systems are designed for crossover frequencies somewhere between 400 and 2000 cycles. *See* LOUDSPEAKER; SOUND-REPRODUCING SYSTEMS. [J.Mar.]

Crosstalk A term originated for the sound produced at a receiving terminal of one voice-frequency circuit by speech signals carried by one or more other such circuits. The term is used commonly for the induction of speech signals in one circuit from a neighboring circuit. It is also used for cross modulation. Crosstalk is unwanted; it impairs communication service to a degree determined by the tolerance of the receiving system or the listener, by the frequency of occurrence, and by the duration of high ratios of crosstalk to normal-speech signal intensity. If it causes loss of presumptive secrecy, the condition is intolerable. *See* DISTORTION (ELECTRONIC CIRCUITS).

Several forms of crosstalk are often distinguished. Intelligible crosstalk is sufficiently understandable under pertinent circuit and room noise conditions that meaningful information can be obtained by the more sensitive observers. Nonintelligible crosstalk cannot be understood regardless of its received volume, but because of its syllabic nature is more annoying subjectively than thermal-type noise. Near-end crosstalk is that type whose energy travels in the opposite direction to the speech signal in the disturbing circuit. Far-end crosstalk is that type whose energy travels in the same direction as the speech signal in the disturbing circuit.

With open-wire circuits, coupling control is effected most practically by transposing (interchanging) positions of the wires of each pair in accordance with a systematic pattern seeking to oppose the coupling of one interval by that of another. *See* TELEPHONE SERVICE. [A.F.]

Crown ethers The crown ethers are macrocyclic polyethers (see structure) which are large enough to encircle metal

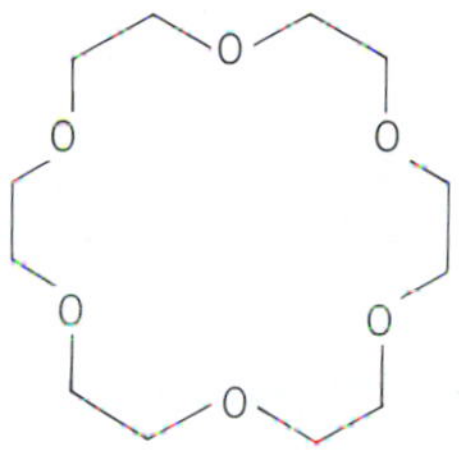

18-crown-6

cations. These compounds possess cavities which can encapsulate metal cations. The crown ethers have the capability of forming stable and selective complexes with monoatomic or molecular ions and, as such, are potential receptor, carrier, or catalyst molecules.

Crown ethers have numerous applications in chemistry for solubilization of salts, anion activation, stabilization of unusual species, analytical techniques, and isotope separation. The ability of the crown ethers to complex metal ions selectively suggests possibilities as drugs for use in detoxification of organisms contaminated with heavy or radioactive metals. [R.B.S.]

Crown gall A neoplastic disease of primarily woody plants, although the disease can be reproduced in species representing more than 90 plant families. The disease results from infection of wounds by the free-living soil bacterium *Agrobacterium tumefaciens*. At 1 or 2 weeks after infection, swellings and overgrowths take place in tissue surrounding the site of infection, and with time these tissues proliferate into large tumors (see illustration). If infection takes place around the main stem or trunk of woody hosts, continued tumor prolif-

Crown gall on peach.

eration will cause girdling and may eventually kill the host. Crown gall is therefore economically important, particularly in nurseries where plant material for commercial use is propagated and disseminated.

Analyses made of crown gall genetic material (DNA) have revealed the presence of certain unique plasmidlike DNA and genetic products. This evidence has strengthened the hypothesis that the bacterial cells introduce part of their genetic information into plants. *See* PLANT PATHOLOGY. [C.I.K.]

Crushing and pulverizing The reduction of materials such as stone, coal, or slag to a suitable size for their intended uses such as road building, concrete aggregate, or furnace firing. Reduction in size is accomplished by five principal methods: (1) crushing, a slow application of a large force; (2) impact, a rapid hard blow as by a hammer; (3) attrition, a rubbing or abrasion; (4) sudden release of internal pressure; and (5) ultrasonic forces. The last two methods are not in common use.

Crushing and pulverizing are processes in ore dressing needed to reduce valuable ores to the fine size at which the valueless gangue can be separated from the ore. These processes are also used to reduce cement rock to the fine powder required for burning, to reduce cement clinker to the very fine size of portland cement, to reduce coal to the size suitable for burning in pulverized form, and to prepare bulk materials for handling in many processes. *See* MATERIALS-HANDLING EQUIPMENT.

Equipment suitable for crushing large lumps as they come from the quarry or mine cannot be used to pulverize to fine powder, so the operation is carried on in three or more stages called primary crushing, secondary crushing, and pulverizing. The three stages are characterized by the size of the feed material, the size of the output product, and the resulting reduction ratio of the material. The crushing-stage output may be screened for greater uniformity of product size.

There are four principal types of primary crushers. The Blake jaw crusher uses a double toggle to move the swinging jaw and is built in a variety of sizes from laboratory units to large sizes having a feed inlet 84 by 120 in. (213 by 305 cm). The Dodge jaw crusher uses a single toggle or eccentric and is generally built in smaller sizes. The Gates gyratory crusher has a cone or mantle that does not rotate but is moved eccentrically by the lower bearing sleeve. The Symons cone crusher also has a gyratory motion, but has a much flatter mantle or cone than does the gyratory crusher. The top bowl is spring-mounted. It is used as a primary or secondary crusher.

Secondary crushers include the single-roll crusher and the double-roll crusher which have teeth on the roll surface and are used mainly for coal. Smooth rolls without teeth are sometimes used for crushing ores and rocks. The hammer crusher is the type of secondary crusher most generally used for ore, rock, and coal. The reversible hammer mill can run alternately in either direction, thus wearing both sides of the hammers.

In open-circuit pulverizing, the material passes through the pulverizer once with no removal of fines or recirculation. In closed-circuit pulverizing, the material discharged from the pulverizer is passed through an external classifier where the finished product is removed and the oversize is returned to the pulverizer for further grinding.

Ball and tube mills, rod mills, hammer mills, and attrition mills are pulverizers operating by impact and attrition. In ball race and roller pulverizers, crushing and attrition are used. *See* BALL-AND-RACE-TYPE PULVERIZER; BUHRSTONE MILL; PEBBLE MILL; TUMBLING MILL. [R.M.H.]

Crustacea A class of animals with mandibles and jointed feet, in the subphylum Mandibulata of the phylum Arthropoda. Most people are familiar with species of Crustacea such as the shrimp, prawn, crab, or lobster. However, there are many more with less common vernacular names such as the water fleas, beach fleas, sand hoppers, fish lice, wood lice, sow bugs, pill bugs, barnacles, scuds, slaters, and krill or whale food.

The Crustacea are one of the most difficult animal groups to define because of their great diversity of structure, habit, habitat, and development. No one character or generalization will apply equally well to all. The adults of some aberrant or highly specialized parasitic species are not recognizable as crustaceans, and evidence of their kinship rests on a knowledge of their life histories. Virtually every form of marine animal life as well as a number of fresh-water animals have their crustacean parasites.

In common with the two other classes of the Mandibulata, the Myriapoda and the Hexapoda or Insects, the Crustacea have segmented, chitin-encased bodies; articulated appendages; mouthparts known as mandibles during some stage of their life, however modified they may be for cutting, chewing, piercing, sucking, or licking; and two pairs of accessory feeding organs, the maxillae. In Crustacea the first pair are the maxillulae, and the second are the maxillae. One or the other pair of maxillae is sometimes vestigial or may be lacking. The Crustacea typically have two pairs of antennae: the first pair, or antennules, and the second pair, the antennae proper. The latter, moreover, are almost always functional at some stage of every crustacean's life, although they disappear in the sedentary adult barnacles and in some lower shrimplike forms. In some of the same forms the antennules may also become reduced or vestigial.

Taxonomy. Although all questions regarding the subclasses of the Crustacea have not been resolved, eight subclasses are recognized: Cephalocarida, Branchiopoda, Ostracods, Mystacocarida, Copepoda, Branchiura, Cirripedia, and Malacostraca. See articles on each group.

General morphology. The true body segments, the somites or metameres, are usually somewhat compressed or depressed. The somite's dorsum, or upper side, is the tergum or tergite; the lower or ventral surface, the sternum or sternite; and either lateral portion, the pleuron or epimeron. Between the embryonic anterior or prostomial element and the telson at the posterior end, the linear series of somites making up the body of a crustacean are more or less distinctly organized into three regions, the head, thorax, and abdomen. Where regional organization of the postcephalic somites is not clearly marked, they collectively form the trunk. The somites are variously fused with one another in diagnostic combinations in different groups of the Crustacea.

The head region of the simplest type is found in the characteristic crustacean larva, the nauplius. When the nauplius hatches, it has only three pairs of appendages, the antennules, antennae, and mandibles, which correspond to the three fused somites of the primary head of the lower Crustacea. In more advanced forms, and at later stages, two additional somites bearing the maxillules and maxillae become coalesced with this protocephalon to form the most frequently encountered crustacean head, or cephalon. Thoracic somites frequently are fused with the cephalon to form a cephalothorax. A dorsal shield or carapace of variable extent arises from the dorsum of the third cephalic somite and covers the cephalothorax. In some crustaceans it is in the form of a bivalved shell; in the barnacles it is modified as a fleshy mantle reinforced by calcareous plates. The carapace reaches its greatest development in the malacostracan Decapoda.

Body. The chitinous cuticle covering the crustacean body is its external skeleton (exoskeleton). The chitin is flexible at the joints, in foliaceous appendages, and throughout the exoskeletons of many small and soft-bodied species, but it is often thickened and stiff in others. It becomes calcified in many species as a result of the deposition of lime salts, and figuratively can be as hard as a rock in some crabs.

The paired appendages are typically biramous and consist of two branches, the endopodite and exopodite. Either the endopodite or exopodite or both may be reduced or suppressed, or both may be coalesced as in the foliaceous limbs of lower Crustacea. The endopodite is definitely segmented in the higher Crustacea. The endopodites are variously modified to serve a variety of functions and needs such as sensory perception, respiration, locomotion, prehension and comminution of food, cleansing, defense, offense, reproduction, and sex recognition and attraction. If retained in the adult, the exopodites may remain leaf- or paddlelike, or become flagellated structures, facilitating swimming or aiding respiration.

The alimentary tract is a relatively straight tube that curves dorsally from the mouth. In the alimentary tract three regions are recognizable, the foregut, midgut, and hindgut. The anterior part of the foregut or stomodeum is esophageal in nature, whereas the posterior and greater part is modified as a stomach, gullet, or gizzard. The mechanical preparation of food occurs here because this region (called the gastric mill) contains chitinous ridges or "teeth" and setae for the trituration and straining of food. The midgut or mesenteron in most cases is provided with several ceca or diverticula which produce digestive secretions and serve as organs for the absorption of food. When these ceca are present in considerable number, they become organized into a sizable gland, the hepato-pancreas or "liver" of higher Crustacea. The hindgut or proctodeum is the rectal portion which terminates, with few exceptions, in the anal opening on the underside of the telson.

Most crustaceans have a heart perforated by openings which admit venous blood from the pericardial sinus in which it is located. The heart may be elongated and tubular and extend through the greater part of the body, but generally it is a more compact organ. It pumps blood through an arterial system or through connecting sinuses or lacunae within the body tissues. The blood of the great majority of the crustaceans is bluish because it contains the respiratory pigment hemocyanin. A few crustaceans have red blood as a result of the presence of erythrocrurion. *See* RESPIRATORY PIGMENTS (INVERTEBRATE).

Crustacea have appropriately been described as water-breathing insects of the sea, for the vast majority are aquatic arthropods. They take up oxygen by means of gills, the general body surface, or special areas of it. Some of the few species that have become more or less terrestrial in their habits have developed modifications of their branchial apparatus which when sufficiently moist enable them to breathe air. Some sow or pill bugs have special tracheal developments in their abdominal appendages for the same purpose.

The nervous system is distinctly arthropodan in character. A supraesophageal ganglion somewhat larger than the other ganglia is considered to be the brain. It is connected by circumesophageal commissures to a double ventral nerve chord with segmentally arranged ganglia.

The organs of special sense are the eyes, the antennules, and the antennae, or feelers in a very real sense. The antennules have also sense of taste of a kind, and organs of equilibration, the statocysts, in their basal segments. Crustacea have setae variously located on their bodies which are sensitive to various external stimuli.

Those glands recognized as definite excretory organs are the maxillary and antennal glands in the adult crustaceans. The antennal gland is best developed in the Malacostraca. It is often called the green gland and opens externally on the underside of the coxal segment of the antennal peduncle. Many crustaceans, especially deep-sea forms, have phosphorescent or luminous organs. The barnacles and some other crustaceans have cement glands. This may also be true of tube-building species which construct their homes of extraneous materials. The most studied crustacean gland is the sinus gland of the eyestalks. This gland is part of a neurosecretory system producing hormones which control color change and pattern, molting cycle, oogenesis, and development of eggs within the ovary. *See* ENDOCRINE SYSTEM (INVERTEBRATE); NEUROSECRETION.

Reproduction. The sexes are separate in most Crustacea and usually can be differentiated from each other by secondary sex characters. Chief among these characters are the size and shape of the body, appendages, or both, and placement of the genital apertures. Hermaphroditism is the rule in the sessile Cirripedia (barnacles), in isolated cases in the lower crustaceans, and in certain parasitic forms. Parthenogenesis (eggs developing and hatching without prior fertilization) occurs frequently in some of the lower crustaceans such as the Branchiopoda and Ostracoda, which have what might be called an alternation of generations. The parthenogenetic generations (and there are usually very many of them in succession) alternate with a generation produced by fertilized eggs. *See* PARTHENOGENESIS.

The eggs of most crustaceans are carried attached to the female until hatched. Some females develop brood pouches in which the young are retained for a time. A nutrient secretion which sustains the young until they are released is produced in some species having a brood chamber. Penaeid shrimp and a few of the lower Crustacea deposit their eggs in the medium in which they live, in some cases attaching them to aquatic vegetation.

Larvae. The nauplius larva is characteristic of Crustacea. Its oval, unsegmented body has three pairs of appendages indicative of three embryonic somites. These are the antennules which are uniramous in the nauplius, the biramous antennae, and mandibles. Besides subserving other functions such as sensory and feeding, all three are the organs of locomotion of this free-swimming larva. There is a single dorsal, median, naupliar eye composed of three closely united, similar parts which form a relatively simple compound eye and a frontal organ, probably sensory, evidenced externally by a very minute pair of projections. The integument of the upper surface of the nauplius body has the appearance of a dorsal shield, as in most crustaceans. Typical as this first larval stage may be, it does not appear in all crustaceans. It is common in the lower forms, but in many of the higher forms it occurs during development in the egg, and the young are hatched as a different and more advanced larva or, as in many Malacostraca, in a form similar to the adult. After the naupliar stage very remarkable larval stages follow in some groups: the free-swimming cypris larvae of barnacles; the phyllosomae of the spiny lobsters; the erichthus and alima larvae of the mantis shrimps (Stomatopoda); and the zoeae and megalopa of crabs.

Molting (ecdysis). This process involves several steps: (1) preparation, which includes some degree of resorption of the old cuticle; (2) the formation of a new, temporarily soft and thin one within it; (3) the accumulation and storing of lime salts in the hepato-pancreas or as lenticular deposits (gastroliths), one of which occurs on either side of the stomach or gizzard in certain forms such as crayfishes and lobsters. The preparatory period is less involved in the thinly chitinous forms. The actual molt follows. The old shell or cuticle splits at strategic and predetermined places, permitting the crustacean within, already enclosed in new but still soft exoskeleton, to withdraw. In the postmolt period the crustacean is quite helpless and, if unsheltered, is at the mercy of its enemies. It is in the soft-shelled stage immediately after molting that the East Coast blue crab (*Callinectes*) becomes a much-sought seafood delicacy. Molting takes place quite frequently in the larval stages when growth is rapid, but becomes less frequent as the animal ages. Hormones from the sinus gland play an important role in both initiating and inhibiting molting. *See* GASTROLITH.

Autotomy and regeneration. The mechanisms of autotomy and regeneration are developed in the crustaceans to minimize injury or loss to an enemy. When an appendage is broken, it is cast off or broken at the fracture or breaking plane. This sacrifice often enables the victim to escape. Even more remarkable is the fact that crustaceans, by voluntary muscular contraction, can part with a limb which may be injured. *See* AUTOTOMY.

Crustacea also have the ability to regenerate lost parts. Although the regenerated parts are not always the same size as the original in the first molt after injury, increase in size in successive molts soon restores a lost limb to virtually its former appearance. *See* REGENERATION (BIOLOGY).

Phylogeny. The geologic history of crustaceans dates from the earliest fossiliferous rock formations, the Cambrian, estimated to be about 350,000,000 years of age. Many lower forms of crustaceans have been discovered in those strata, including some thought to be distantly related, morphologically, to present-day Malacostraca, but the state of preservation of these early fossils, especially of the appendages, leaves much to be desired. It is not unlikely that because of their small size and fragility many crustaceans of bygone ages never did become recognizably fossilized.

Bionomics and economics. Crustacea are ubiquitous. They live at almost all depths and levels of the sea, in fresh waters at elevations up to 12,000 ft (3.7 km), in melted snow water, in the deepest of the sea's abysses more than 6 mi (10 km) down, and in waters of 0°C (32°F) temperature. They live on land, trees, and mountains, although most of these species must descend to salt water areas again to spawn their young. Some live in strongly alkaline waters and others in salt water which is at the saturation point, still others in hot springs with temperatures of 55°C (131°F).

All crustaceans of sufficient size are eaten by humans somewhere in the world. Very few properly prepared have proved to be poisonous so far as is known. Myriads of small fry are the sustenance of larger aquatic animals, chiefly fish and whales.

Most crustaceans are omnivorous, and essentially scavengers. Many are filter feeders and screen particulate life, plankton, and organic detritus from the waters in which they live; others are largely carnivorous, still others vegetarian. Among the vegetarians are the grazers of the ocean meadows which convert the microscopic plant life (diatoms) into flesh and food for larger animals which in turn are harvested as food for humans. *See* ARTHROPODA. [W.L.S.]

Crux The Southern Cross in astronomy, the most celebrated of the constellations of the far south. The four principal bright stars of the group, α, γ, β, and δ form the figure of a cross, giving the constellation its name (see illustration). The brightest star in Crux is Alpha Crucis at the foot of the cross, which has received the artificial name Acrux. This is a navigational star. *See* CONSTELLATION; CYGNUS. [C.-S.Y.]

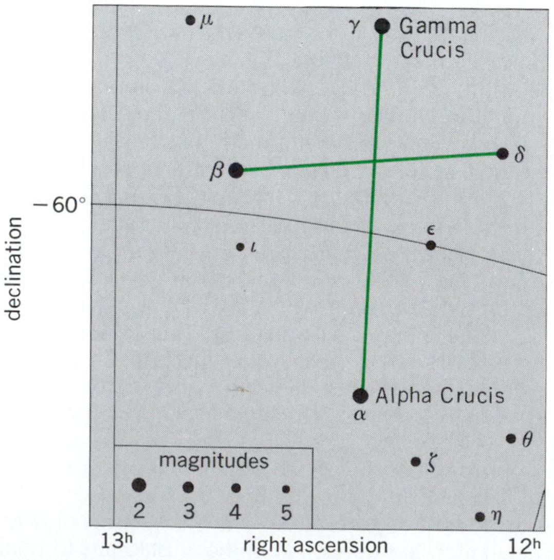

Line pattern of the constellation Crux. The grid lines represent the coordinates of the sky. The apparent brightness, or magnitude, of the stars is shown by the sizes of dots, graded by appropriate numbers as indicated.

Cryobiology The use of low-temperature environments in the study of living plants and animals. The principal effects of cold on living tissue are destruction of life and preservation of life at a reduced level of activity. Both of these effects are demonstrated in nature. Death by freezing is a relatively common occurrence in severe winter storms. Among cold-blooded animals winter weather usually results in a comalike sleep that may last for a considerable length of time.

The extreme cold of liquid nitrogen (boiling at −320°F or −196°C) can cause living tissue to be destroyed in a matter of seconds or to be preserved for an indefinite time, certainly for years and possibly for centuries, with essentially no detectable biochemical activity. The result achieved when heat is withdrawn from living tissue depends on processes occurring in the individual cells. Basic knowledge of the causes of cell death, especially during the process of freezing, and the discovery of methods which circumvent these causes have led to practical applications both for long-term storage of living cells or tissue (cryopreservation) and for calculated and selective destruction of tissue (cryosurgery).

The biochemical constituents of a cell are either dissolved or suspended in water. During the physical process of freezing, water tends to crystallize in pure form, while the dissolved or suspended materials concentrate in the remaining liquid. In the living cell, this process is quite destructive. In a relatively slow freezing process ice first begins to form in the fluid surrounding the cells, and the concentration of dissolved materials in the remaining liquid increases. The equilibrium between the fluid inside the cell and the fluid surrounding the cell is upset. A concentration gradient is established across the cell wall, and water moves out of the cell in response to the osmotic force. As freezing continues, the cell becomes quite dehydrated. Salts may concentrate to extremely high levels. In a similar manner the acid-base ratio of the solution may be altered during the concentration process. Dehydration can affect the gross organization of the cell and also the molecular relationships, some of which depend on the presence of water at particular sites. Cellular collapse resulting from loss of water may bring in contact intracellular components normally separated to prevent destructive interaction. Finally, as the ice crystals grow in size, the cell walls may be ruptured by the ice crystals themselves or by the high concentration gradients which are imposed upon them.

By speeding the freezing process to the point that temperature drop is measured in degrees per second, some of these destructive events can be modified. However, most of the destructive processes will prevail. To prevent dehydration, steps must be taken to stop the separation of water in the form of pure ice so that all of the cell fluids can solidify together. The chief tools used to accomplish this are agents that lower the freezing point of the water. Typical materials used for industrial antifreeze agents are alcohols or glycols (polyalcohols). Glycerol, a polyalcohol which is compatible with other biochemical materials in living cells, is frequently used in cell preservation. Besides the antifreeze additive, refrigeration procedures are designed to control the rate of decline in temperature to the freezing point, through the liquid-solid transition, and below, to very low temperatures.

In the case of preservation of sizable aggregates of cells in the form of whole organs, the problems are significantly increased. While there are signs of progress, whole-organ preservation is not yet a normally successful procedure.

Cryopreservation. The earliest commercial application of cryopreservation was in the storage of animal sperm cells for use in artificial insemination. In the frozen state sperm may be transported great distances. This permits the spread of high-quality breeding stock to all parts of the Earth at much lower costs than would be involved in the shipment of an entire herd.

In an entirely similar manner the microorganisms used in cheese production can be frozen, stored, and transported without loss of lactic acid–producing activity. Pollen from various plants can be frozen for storage and transport. In plant-breeding experiments it is sometimes desired to crossbreed plants that do not grow in similar environments. In such instances the transport of frozen pollen can be utilized.

Among the most valuable applications of cryopreservation is the storage of whole blood or separated blood cells. Frozen storage permits the buildup of the large stocks of blood required in a major catastrophe. It is useful, too, in storing the less common types of blood. Frozen blood storage is also useful in long-term diagnostic studies.

Cryosurgery. The causes of cellular destruction from freezing can be effected deliberately to destroy tissue as a surgical procedure. A technique for controlled destruction of tissue in the basal ganglia of the brain was developed for relief from tremor and rigidity in Parkinson's disease.

One of the significant advantages of cryosurgery is that the apparatus can be employed to cool the tissue to the extent that the normal or the aberrant function is suppressed; yet at this stage the procedure can be reversed without permanent effect. When the surgeon is completely satisfied that he or she has located the exact spot to destroy, the temperature can be lowered enough to produce irreversible destruction. This procedure is of particular assistance in neurosurgery.

A second major advantage of cryosurgery is that the advancing front of reduced temperatures tends to cause the removal of blood and the constriction of blood vessels in the affected area. This means that little or no bleeding results from cryosurgical procedures.

A third major advantage of cryosurgery is that cryosurgery equipment currently employs a freezing apparatus (which is about the size of a large knitting needle) that can be placed in contact with area to be destroyed with a minimum incision to expose the affected area. *See* CRYOGENIC ENGINEERING. [A.W.F.]

Cryogenic engineering A branch of engineering specializing in technical operations at very low temperatures. Cryogenic engineers combine competence in the techniques of low-temperature physics and chemistry with capability in conventional branches of engineering, such as mechanical, chemical, electrical, and nuclear. For discussions of other aspects of low-temperature technology *see* LIQUEFACTION OF GASES; LOW-TEMPERATURE THERMOMETRY.

An important military application of cryogenics has been the development of oxygen-breathing apparatus for crews of high-altitude, long-range aircraft. By storing oxygen as a dense liquid in lightweight, well-insulated containers from which it can be vaporized as needed, significant weight reductions are effected over the older method of high-pressure, low-density gas storage in heavy steel cylinders. Cryogenic engineers have also been active in the aerospace field where low-molecular-weight fuels are highly desirable. These, however, are all cryogenic liquids.

Gases employed in the chemical industry usually require extensive pretreatment because their source is most often a gas mixture—either the atmosphere, natural gas, or a by-product stream from some other chemical operation. In consequence the components must first be separated, then purified. Both these processes involve cryogenics. Depending upon the mixture, one or more of the following low-temperature techniques must be used: fractional distillation, scrubbing, and selective adsorption. The last technique has proved particularly applicable for the separation and purification of the noble, or inert, gases.

Metallurgical applications for pure gases are increasing as new alloy steels are created and as the production of hitherto rare or seldom-refined metals is undertaken in order to meet the unusual demands of the nuclear and the space age. Oxygen, nitrogen, argon, and hydrogen, all of which are produced cryogenically, are the gases most commonly used.

Although still in the developmental stage, the application of superconductivity and associated cryogenic systems to electric power generation and transmission appears promising. Development of superconducting magnets to provide the magnetic fields needed for the generation of electric power by magnetohydrodynamics (MHD) has been undertaken. Superconducting magnets will also be needed for the magnetic confinement of plasmas in thermonuclear (nuclear fusion) power generation. *See* ENERGY STORAGE; MAGNETOHYDRODYNAMIC POWER GENERATOR; NUCLEAR FUSION; SUPERCONDUCTIVITY.

Cryogenic engineering is applied to several modes of transportation. Superconducting magnets are used in the development of magnetically levitated, high-speed trains in which speeds of 300 mi/h (500 km/h) or more are anticipated. Cryogenic (or liquid) hydrogen fuel appears promising for the propulsion of jet aircraft, and has also been investigated for use as a fuel in both rail and automotive transportation.

Several elementary superconducting circuits have been developed to produce binary memory elements, switching devices, and multivibrators for high-speed computers. The small space requirements, negligible power inputs, increased speed, and high reliability of these units combine to promise significant advantages over other types of computer components. *See* CRYOTRON.

A high-speed vacuum pump has been developed for specialized applications, capable of producing an extremely low ultimate vacuum and requiring less power than conventional diffusion-type pumps. The principle employed is well known: lowering the pressure within a gas-tight enclosure by freezing out condensible gases on a cold surface. In the cryopump a low-temperature refrigeration unit circulates helium at −487°F (20 K) through coils in plate condensers upon which the gas to be pumped condenses. *See* VACUUM PUMP.

Gases whose critical temperatures lie below ambient temperature and which, therefore, are not liquefiable by pressure alone have conventionally been transported and stored in heavy-walled steel cylinders at high pressures. In the case of hydrogen, for example, about 100 lb of steel must be used to contain each pound of gas at high pressure. In contrast, current cryogenic technology yields a corresponding weight ratio of only 3 for liquid containment. Although this ratio is not quite so small for heavier gases, the dead-weight savings are still sufficient to justify liquid storage of nitrogen and oxygen in almost all large-scale industrial operations.

Efforts have been undertaken to develop new and better superconductors that have improved mechanical properties and that can carry higher currents at higher temperatures. The development of Nb_3Ge with a transition temperature of over 23 K (−418°F) has opened the possibility of operation of superconducting devices using liquid hydrogen coolant. There has also been extensive research into the low-temperature mechanical properties of structural materials, including epoxy resins and composites.

Low temperatures are finding increasing use in basic biological studies as well as in practical medical procedures. *See* CRYOBIOLOGY.

Large quantities of liquid nitrogen are employed in the preservation of food by rapid freezing. This is carried out by direct immersion of the food product in liquid nitrogen or by direct contact with a liquid-nitrogen spray or cold vapors. *See* CRYOGENICS. [K.D.W.; F.J.E.; W.E.K.]

Cryogenics The science of producing and maintaining very low temperatures. From a practical standpoint, any given object may be cooled to and maintained at a low temperature most simply by placing it in thermal contact with a suitable liquefied gas held at a constant pressure. In principle, a liquefied gas can be used to provide constant bath temperatures from its triple point to its critical point. Bath temperature is varied by changing the pressure above the liquid, and, within the liquid range, any desired bath temperature can be maintained by removing precisely that amount of gas vaporized by heat leak into the bath liquid. In general, because heat leak cannot be excluded entirely, baths will eventually boil away unless replenished periodically.

Liquefied gases do not exist over the entire low-temperature range. The lowest temperature at which it is practical to use a liquid bath is about 0.3 K (−459.1°F), and gaps exist between 5 and 14 K (−451 and −434°F) and between 44 and 55 K (−380 and −361°F). If it is necessary to extend the accessible temperature range or realize certain ranges more conveniently, other methods of cold production are available.

Refrigerators can, in principle, be designed to produce and maintain a predetermined temperature at any point along the temperature scale. For refrigerators operating cyclically, the low temperature produced is not strictly constant because of the periodic withdrawal of heat from the low-temperature reservoir, but the temperature variation can be large or extremely small, depending on the cycle characteristics. *See* REFRIGERATION CYCLE.

Refrigerator cycles operating above about 1 K (−458°F) generally involve compression and expansion of appropriately chosen gases. Because at lower temperatures it is impractical or impossible to use gases, liquid or solid working materials are employed along with different techniques to produce cooling. Adiabatic demagnetization of paramagnetic ions in solids, involving electron-spin paramagnets, is useful in the range 1 to 0.003 K (−458 to −459.665°F), but could be applied at temperatures up to 300 K (80°F) or more by appropriate choice of magnetic material. *See* ADIABATIC DEMAGNETIZATION.

The same basic principles of refrigeration may be used to explain gas liquefaction, the periodic temperature reduction of the refrigerator working substance, and the adiabatic demagnetization process. Cold production is related directly to the concept of temperature, which in turn is intimately connected with entropy S defined as $\int(C/T)\,dT$, where C is the specific heat of the substance being cooled. Entropy is a measure of the intrinsic thermal disorder associated with a substance, and, at absolute zero, the entropy of every substance in internal equilibrium is zero. Entropy is increased either by the introduction of thermal energy or by appropriately varying some other parameter upon which S depends. A conjugate relation exists such that, provided all other parameters are held constant, a change in T results in a change of S of the same sign. The significance of S for low temperatures stems from the concept of idealized processes, called isentropic (or constant S), which result in the maximum possible decrease in T for a given spontaneous parametric change. Because an isentropic process must be both adiabatic (thermally isolated) and reversible, in practice it can be only approximated. Nevertheless, the isentropic process provides a criterion against which the efficiency of any actual process may be gaged. From the illustration, in which the variation of S and T for an idealized substance is shown at two different constant values of a thermodynamic parameter X such as pressure or magnetization, it is seen that an isentropic parameter change from $A \rightarrow B$ produces the maximum reduction in T. (Spontaneous processes resulting in a decrease in S with a change in X are impossible.) *See* THERMODYNAMIC PRINCIPLES.

Once a temperature lower than ambient has been produced, it may be maintained indefinitely (provided the region itself contains no heat sources) if flow of heat to the cooled region can be prevented. Because no perfect thermal insulators exist, the problem resolves itself into minimizing the flow of heat to the refrigerated region. Devices designed to provide low-temperature environments in which operations may be carried out under controlled conditions are called cryostats. *See* CRYOGENIC ENGINEERING; LOW-TEMPERATURE PHYSICS. [F.J.E.; K.D.W.; W.E.K.]

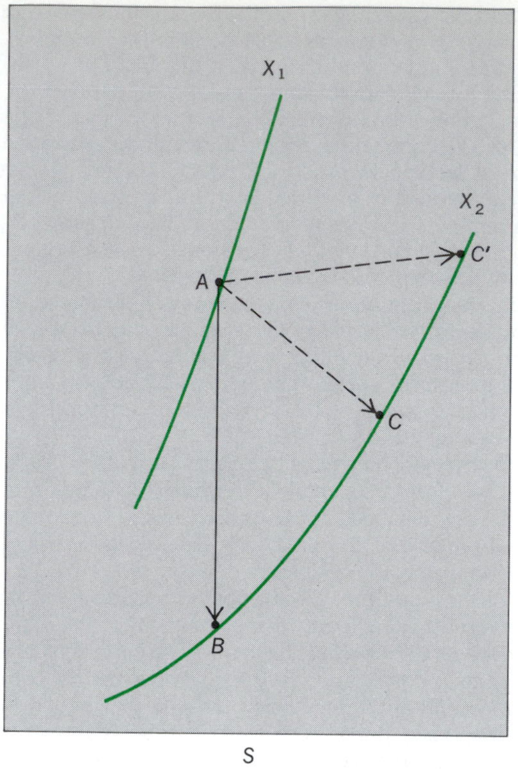

Variation of entropy with temperature for two different constant values of thermodynamic parameter *X*.

Cryolite A mineral with chemical composition Na_3AlF_6. It crystallizes in the monoclinic system. Hardness is 2½ on Mohs scale and the specific gravity is 2.95. Crystals are usually snow-white but may be colorless and more rarely brownish, reddish, or even black. The mean refraction index is 1.338, approximately that of water, and thus fragments become invisible when immersed in water.

Cryolite was once used as a source of metallic sodium and aluminum, but now is used chiefly as a flux in the electrolytic process in the production of aluminum from bauxite. *See* ALUMINUM. [C.S.Hu.]

Cryotron A current-controlled switching device based on superconductivity for use primarily in computer circuits. The early version has been superseded by the tunneling cryotron, which consists basically of a Josephson junction. In its simplest form the device has two electrodes of a superconducting material (for example, lead) which are separated by an insulating film only about 10 atomic layers thick. Switching is accomplished by a magnetic field generated by sending a current through the control line on top of the junction. The tunneling cryotron is still in the research stage, but various computer circuits have already been realized. *See* SUPERCONDUCTING DEVICES; SUPERCONDUCTIVITY. [P.Wo.]

Cryptobiosis A state of life in which the metabolic rate of an organism is reduced to an imperceptible level. The several kinds of cryptobiosis ("hidden life") include anhydrobiosis (life without water), cryobiosis (life at low temperatures), and anoxybiosis (life without oxygen). The most is known about anhydrobiosis.

States of anhydrobiosis occur in early developmental stages of various organisms, including seeds of plants, spores of bacteria and fungi, cysts of certain crustaceans, and larvae of certain insects; they occur in both developmental and adult stages of certain soil-dwelling micrometazoans (rotifers, tardigrades, and nematodes), mosses, lichens, and certain ferns.

A central question in the study of anhydrobiosis has been whether metabolism actually ceases. Available evidence strongly suggests that dry anhydrobiotes are ametabolic. In that case, a philosophical question immediately arises concerning the nature of life. This philosophical quandary can be avoided by applying the definition of life adopted by most students of anhydrobiosis: an organism is alive, provided its structural integrity is maintained. When that integrity is violated, it is dead. *See* METABOLISM. [J.H.Cr.]

Cryptococcales An order in the class Fungi Imperfecti created to accommodate the yeasts which do not form sexual spores. This group, also known as Torulopsidales, comprises the genera *Cryptococcus*, *Rhodotorula*, *Torulopsis*, *Pityrosporum*, *Candida*, *Brettanomyces*, *Kloeckera*, *Trigonopsis*, and *Trichosporon*. Some species, such as *Torula utilis*, are used in industrial fermentation to make food and fodder yeast; other species are pathogenic for humans and animals (*Pityrosporum ovale* is usually found associated with dandruff; *Candida albicans* causes thrush in infants; and *Cryptococcus neoformans* causes cryptococcosis). *See* CRYPTOCOCCOSIS.

Vegetative reproduction may be by budding, by fission, by formation of a pseudomycelium or true mycelium, or combinations thereof. The Cryptococcales are separated from the related order Moniliales by their lack of an aerial mycelium and aerial asexual reproductive structures. *See* YEAST. [H.J.P.]

Cryptococcosis An infection of humans, primarily of the central nervous system, having as its etiological agent an encapsulated yeast, *Cryptococcus neoformans*. Skin, lungs, bones, and viscera may also be invaded by this organism. *Cryptcoccus neoformans* is widespread in nature and worldwide in distribution. Diagnosis of the disease is made by cultural and histological methods. Serologic tools are available for detection of antibodies and antigens. Treatment for this disease includes the use of amphotericin B and 5-fluorocytosine. *See* AMPHOTERICIN B; MEDICAL MYCOLOGY; YEAST. [L.D.H.]

Cryptography The various methods for writing in secret code or cipher. As society becomes increasingly dependent upon computers, the vast amounts of data communicated, processed, and stored within computer systems and networks often have to be protected, and cryptography is a means of achieving this protection. It is the only practical method for protecting information transmitted through accessible communication networks such as telephone lines, satellites, or microwave systems. Cryptographic procedures can also be used for message authentication, personal identification, and digital signature verification for electronic funds transfer and credit card transactions. *See* DATA COMMUNICATIONS; DATABASE MANAGEMENT SYSTEMS; DIGITAL COMPUTER; ELECTRICAL COMMUNICATIONS.

Cryptography helps resist decoding or deciphering by unauthorized personnel; that is, messages (plaintext) transformed into cryptograms (codetext or ciphertext) have to be able to withstand intense cryptanalysis. Transformations can be done by using either code or cipher systems. Code systems rely on code books to transform the plaintext words, phrases, and sentences into ciphertext code groups. To prevent cryptanalysis, there must be a great number of plaintext passages in the code

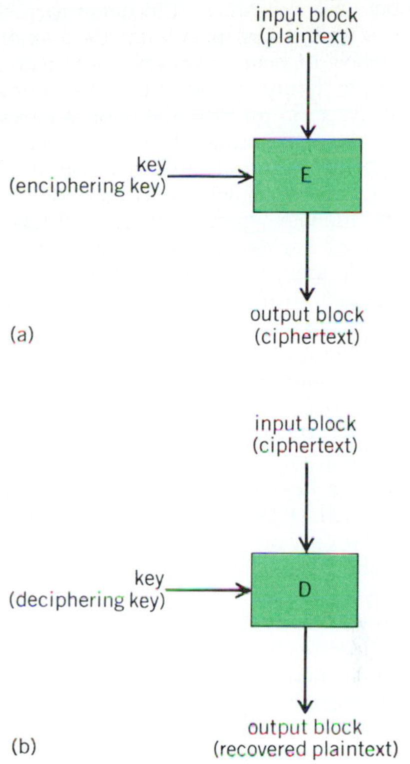

Fig. 1. Block cipher. (*a*) Enciphering. (*b*) Deciphering.

book and the code group equivalents must be kept secret, making it difficult to utilize code books in electronic data-processing systems.

Cipher systems are more versatile. Messages are transformed through the use of two basic elements: a set of unchanging rules or steps called a cryptographic algorithm, and a set of variable cryptographic keys. The algorithm is composed of enciphering (**E**) and deciphering (**D**) procedures which usually are identical or simply consist of the same steps performed in reverse order, but which can be dissimilar. The keys, selected by the user, consist of a sequence of numbers or characters. An enciphering key (Ke) is used to encipher plaintext (X) into ciphertext (Y) as in Eq. (1), and a deciphering key (Kd) is used to decipher ciphertext (Y) into plaintext (X) as in Eq. (2).

$$\mathbf{E}_{Ke}(X) = Y \qquad (1)$$

$$\mathbf{D}_{Kd}[\mathbf{E}_{Ke}(X)] = \mathbf{D}_{Kd}(Y) = X \qquad (2)$$

Algorithms are of two types—conventional and public-key. The enciphering and deciphering keys in a conventional algorithm either may be easily computed from each other or may be identical [Ke = Kd = K, denoting $\mathbf{E}_K(X) = Y$ for encipherment and $\mathbf{D}_K(Y) = X$ for decipherment]. In a public-key algorithm, one key (usually the enciphering key) is made public, and a different key (usually the deciphering key) is kept private. In such an approach it must not be possible to deduce the private key from the public key.

When an algorithm is made public, for example, as a published encryption standard, cryptographic security completely depends on protecting those cryptographic keys specified as secret.

Unbreakable ciphers. Unbreakable ciphers are possible. But the key must be randomly selected and used only once, and its length must be equal to or greater than that of the plaintext to be enciphered. Therefore such long keys, called one-time tapes, are not practical in data-processing applications. To work well, a key must be of fixed length, relatively short, and capable of being repeatedly used without compromising security. In theory, any algorithm that uses such a finite key can be analyzed; in practice, the effort and resources necessary to break the algorithm would be unjustified.

Strong algorithms. Fortunately, to achieve effective data security, construction of an unbreakable algorithm is not necessary. However, the work factor (a measure, under a given set of assumptions, of the requirements necessary for a specific analysis or attack against a cryptographic algorithm) required to break the algorithm must be sufficiently great. Included in the set of assumptions is the type of information expected to be available for cryptanalysis. For example, this could be ciphertext only; plaintext (not chosen) and corresponding ciphertext; chosen plaintext and corresponding ciphertext; or chosen ciphertext and corresponding recovered plaintext.

A strong cryptographic algorithm must satisfy the following conditions: (1) The algorithm's mathematical complexity prevents, for all practical purposes, solution through analytical methods. (2) The cost or time necessary to unravel the message or key is too great when mathematically less complicated methods are used, because either too many computational steps are involved (for example, in trying one key after another) or because too much storage space is required (for example, in an analysis requiring data accumulations such as dictionaries and statistical tables).

To be strong, the algorithm must satisfy the above conditions even when the analyst has the following advantages: (1) Relatively large amounts of plaintext (specified by the analyst, if so desired) and corresponding ciphertext are available. (2) Relatively large amounts of ciphertext (specified by the analyst, if so desired) and corresponding recovered plaintext are available. (3) All details of the algorithm are available to the analyst; that is, cryptographic strength cannot depend on the algorithm remaining secret. (4) Large high-speed computers are available for cryptanalysis.

Digital signatures. Digital signatures authenticate messages by ensuring that: the sender cannot later disavow messages; the receiver cannot forge messages or signatures; and the receiver can prove to others that the contents of a message are genuine and that the message originated with that particular sender. The digital signature is a function of the message, a secret key or keys possessed by the sender of the message, and sometimes data that are nonsecret or that may become nonsecret as part of the procedure (such as a secret key that is later made public).

Digital signatures are more easily obtained with public-key than with conventional algorithms. When a message is enciphered with a private key (known only to the originator), anyone deciphering the message with the public key can identify the originator. The latter cannot later deny having sent the message. Receivers cannot forge messages and signatures, since they do not possess the originator's private key.

Since enciphering and deciphering keys are identical in a conventional algorithm, digital signatures must be obtained in some other manner. One method is to use a set of keys to produce the signature. Some of the keys are made known to the receiver to permit signature verification, and the rest of the keys are retained by the originator in order to prevent forgery.

Data Encryption Standard. Regardless of the application, a cryptographic system must be based on a cryptographic algorithm of validated strength if it is to be acceptable. The Data Encryption Standard (DES) is such a validated conventional algorithm already in the public domain. This procedure enciphers a 64-bit block of plaintext into a 64-bit block of ciphertext under the control of a 56-bit key. The National Bureau of Standards accepted this algorithm as a standard, and it became effective on July 15, 1977.

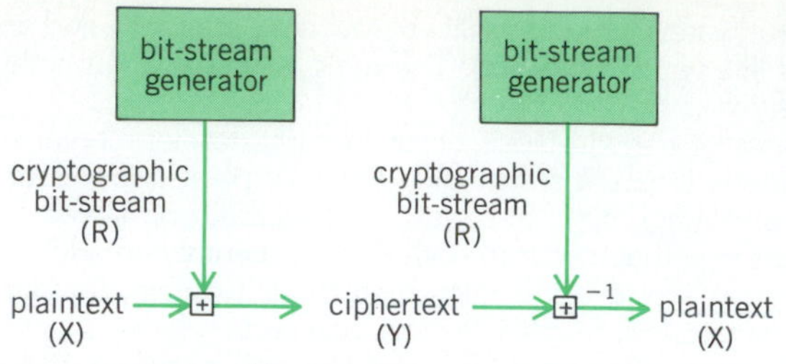

Fig. 2. Stream cipher concept. *(After C. H. Meyer and S. M. Matyas, Cryptography: A New Dimension in Computer Data Security, 1980).*

Block ciphers. A block cipher (Fig. 1) transforms a string of input bits of fixed length (termed an input block) into a string of output bits of fixed length (termed an output block). In a strong block cipher, the enciphering and deciphering functions are such that every bit in the output block jointly depends on every bit in the input block and on every bit in the key. This property is termed intersymbol dependence.

Stream ciphers. A stream cipher (Fig. 2) employs a bit-stream generator to produce a stream of binary digits (0's and l's) called a cryptographic bit stream, which is then combined either with plaintext (via the ⊞ operator) to produce ciphertext or with ciphertext (via the $⊞^{-1}$ operator) to recover plaintext.

[C.H.M.; S.M.M.]

Cryptomonadida

Cryptomonadida An order of the class Phytamastigophorea, also known as the Cryptomonadina. Cryptomonads are considered to be protozoans by zoologists and algae by botanists. Most species occur in fresh water. They are olive-green, blue, red, brown, or colorless. All flagellated members have two subequal flagella inserted in a gullet opening through

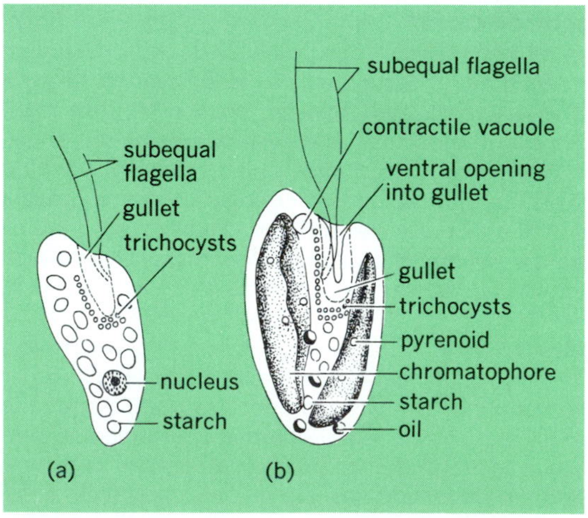

Examples of cryptomonads. (a) ***Chilomonas paramecium.*** **(b)** ***Cryptomonas erosa.***

an obliquely truncate anterior end (see illustration). Although few in genera and species, Cryptomonadida form extensive marine and fresh-water blooms. *See* PHYTAMASTIGOPHOREA; PROTOZOA.

[J.B.L.]

Cryptophyceae

Cryptophyceae A small class of biflagellate unicellular algae (cryptomonads) in the chlorophyll *a–c* phyletic line (Chromophycota). In protozoological classification, these organisms constitute an order, Cryptomonadida, of the class Phytomastigophora. Cryptomonads are 4–80 micrometers long, ovoid or bean-shaped, and dorsiventrally flattened. The cell is bounded by a moderately flexible periplast comprising the plasmalemma and underlying rectangular or polygonal proteinaceous plates. A tubular invagination (gullet, groove, or furrow) traverses the ventral cytoplasm and opens just below the apex of the cell. A pair of subequal flagella, which are covered with hairs and small scales, arise from the center or apical end of the gullet. Cryptomonads may be photosynthetic, osmotrophic, or phagotrophic. In photosynthetic species, chloroplasts occur singly or in pairs.

Photosynthetic pigments include chlorophyll *a* and *c*, α-carotene and ß-carotene (the former being predominant, an unusual ratio for algae), and red and blue phycobiliproteins closely related to those in red algae but occurring in the intrathylakoidal space rather than forming phycobilisomes. The color of the chloroplasts, and thus the color of the cryptomonad, depends upon pigment composition and may be green, olive, brown, yellow, red, or blue. Pigment composition is affected by environmental conditions.

Cryptomonads are found in fresh, brackish, marine, and hypersaline bodies of water, sometimes in great abundance. Several species are endosymbionts of marine ciliates and invertebrates, and of fresh-water dinoflagellates. *See* ALGAE; CHROMOPHYCOTA; CRYPTOMONADIDA. [P.C.Si.; R.L.Moe]

Cryptostomata An extinct order of bryozoans in the class Stenolaemata. Cryptostome colonies have delicate, flattened branches of bifoliate construction in which the short, tubular, solid-walled, moderately complex zooecia are tightly packed back to back and open out onto the two opposite sides of the branches. *See* BRYOZOA; STENOLAEMATA.

Possibly evolving from early cystoporates or from an ancestor they shared with them, cryptostomes first appeared in the sea at the start of the Middle Ordovician. They became moderately common immediately, and remained so, until their extinction during latest Permian time.

Bifoliate cryptostomes are also termed ptilodictyoids or timanodictyoids. The order Cryptostomata was much more broadly defined in the past, and included the Fenestrata and Rhabdomesonata. Moreover, although formerly suggested as possibly ancestral to cheilostomes, the cryptostomes may have died out without modern descendants. *See* CHEILOSTOMATA; CYSTOPORATA. [R.J.Cu.]

Crystal This term, as used in science and technology, usually denotes a single crystal. A single crystal is a solid throughout which the atoms or molecules are arranged in a regularly repeating pattern. In electronics the term crystal is usually restricted to mean a single crystal which is piezoelectric. Examples of single crystals are most gems, piezoelectric quartz crystals used in controlling the frequencies of radio transmitters, and single crystals of galena (lead sulfide) used in crystal radios. *See* SINGLE CRYSTAL.

Most crystalline solids are made up of millions of tiny single crystals called grains and are said to be polycrystalline. These grains are oriented randomly with respect to each other. Any single crystal, however, no matter how large, is a single grain. Single crystals of metals many cubic centimeters in volume are relatively easy to prepare in the laboratory. [H.H.Ho.]

Structure. Knowledge of the precise ways in which atoms and ions are distributed in crystals is of prime importance in solid-state physics, chemistry, metallurgy, mineralogy, geochemistry, and other fields. Knowledge of crystal structure has been obtained mainly from x-ray diffraction data, although electron diffraction and neutron diffraction have become important tools in crystal analysis. *See* CRYSTALLOGRAPHY; ELECTRON DIFFRACTION; IONIC CRYSTALS; NEUTRON DIFFRACTION; SOLID-STATE CHEMISTRY; X-RAY CRYSTALLOGRAPHY; X-RAY DIFFRACTION.

A three-dimensional, indefinitely extended array of points, each of which is surrounded in an identical way by its neighbors, is known as a lattice or space lattice. The space lattice of a crystal is the representation of the periodicity with which matter is distributed in it. It is essential to distinguish a lattice from a crystal structure; a crystal structure is formed by associating with every lattice point an assembly of atoms identical in composition, arrangement, and orientation. The space-lattice concept explains the special geometric properties of crystal polyhedrons. It was originally postulated that an elementary unit, having all the properties of the crystal, should exist, or conversely that a crystal was built up by the juxtaposition of such elementary units. If mathematical points forming the vertices of a parallelepiped $OABC$ (defined by three vectors $\overline{OA}$, $\overline{OB}$, $\overline{OC}$ are considered (Fig. 1), a space lattice is obtained by

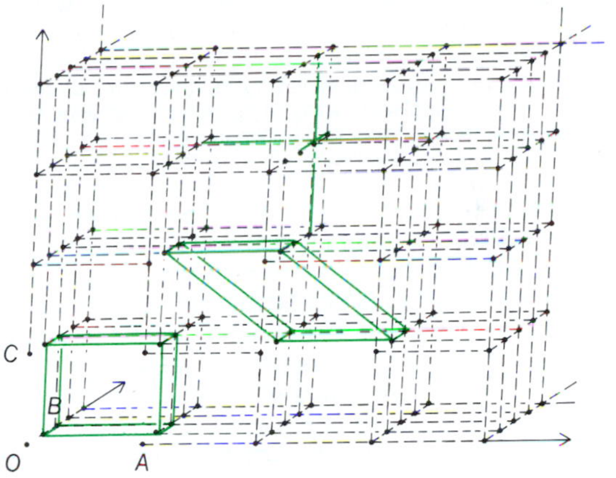

Fig. 1. A space lattice, two possible unit cells, and the environment of a point.

translations parallel to and equal to $\overline{OA}$, $\overline{OB}$, $\overline{OC}$. The parallelepiped is called the unit cell. The vector **r** joining O to any lattice point can be written as $\mathbf{r} = m\overline{OA} + n\overline{OB} + r\overline{OC}$, where m, n, and r are integers.

The Bravais lattices are the 14 different possible space lattices obtained on the basis that two lattices are different when the environment of their points is different. If considered as solids, the combination of symmetry elements they exhibit can be determined. Seven point groups which are respectively the most symmetrical of each system are found. Accordingly the unit cells, and by extension the lattices, can be divided into seven groups corresponding to the seven crystallographic systems. The external symmetry of a crystal is due to the periodically repeated arrangement of its atoms or ions. The Bravais lattices give the possible periodicities. A further step must consist in finding the number of possible periodic arrangements, two such arrangements being considered as different when they give rise to a different symmetry.

Two-dimensional structures are very well suited to illustrate this. In Figs. 2 and 3 a numeral 9 is chosen as the object; it is fully asymmetric and no false symmetry can therefore be introduced. It is multiplied by a periodic array of mirror planes and fourfold axes. Several interpenetrating identical lattices are formed. The two arrangements have the same Bravais lattice but not the same symmetry. In three-dimensional space the problem is similar, but other symmetry elements are considered.

Screw axes combine the rotation of an ordinary symmetry axis with a translation parallel to it and equal to a fraction of the unit distance in this direction. If screw axes are present in crystals, it is clear that the displacements involved are of the order of a few angstroms and that they cannot be distinguished

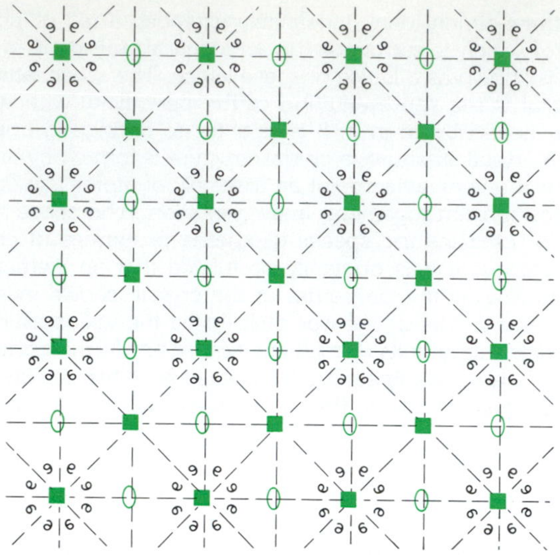

Fig. 2. Periodic pattern obtained by multiplication of asymmetric object by indicated array of symmetry elements. Both pattern and lattice have same symmetry elements. Mirror planes are indicated by broken lines.

macroscopically from ordinary symmetry axes. The same is true for glide mirror planes.

Glide mirror planes combine the mirror image with a translation parallel to the mirror plane over a distance which is half the unit distance in the glide direction. These new symmetry elements must be taken into account when the number of possible periodic arrangements is considered. A total of 230 arrangements or space groups is possible; of these, 32 can be distinguished macroscopically.

The forces which link the atoms together in metallic crystals are nondirectional. This means that each atom tends to surround itself by as many others as possible. This results in a dense packing, similar to that of spheres of equal radius, and yields three distinct systems: close-packed (face-centered) cubic, hexagonal close-packed, and body-centered cubic.

Simple crystal structures are usually named after the compounds in which they were first discovered (diamond or zinc sulfide, cesium chloride, sodium chloride, and calcium fluoride). Many compounds of the type A^+X^-, $A^{++}X_2^-$ have such structures. They are highly symmetrical, the unit cell is cubic, and the atoms or ions are disposed at the corners of the unit cell and at points having coordinates which are combinations of 0, 1, 1/2, or 1/4.

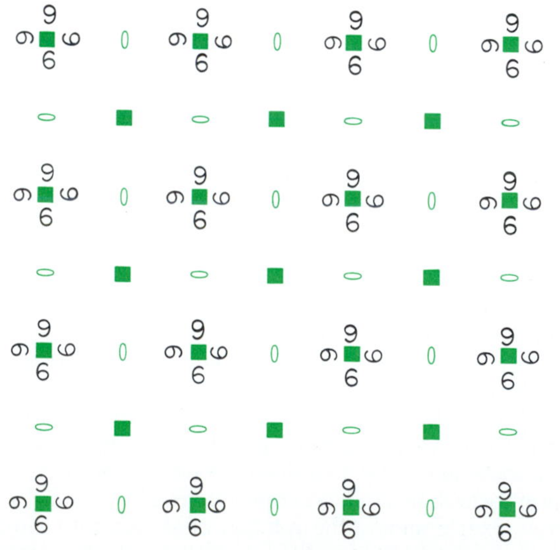

Fig. 3. Arrangement in which the periodic pattern has fewer symmetry elements than the lattice. This shows that the symmetry of the Bravais lattices can be lowered by the structure of the lattice points.

The sodium chloride structure is an arrangement in which each positive ion is surrounded by six negative ions and vice versa. Systematic study of the dimensions of the unit cells of compounds having this structure has revealed that each ion can be assigned a definite radius and each ion tends to surround itself by as many others as possible of the opposite sign because the binding forces are nondirectional.

In the cesium chloride structure each of the centers of the positive and negative ions forms a primitive cubic lattice whose centers are mutually shifted.

In the diamond structure each atom is in the center of a tetrahedron formed by its nearest neighbors. The 4-coordination follows from the well-known bonds of the carbon atoms.

The zinc blende structure is basically similar to the diamond structure. Each zinc atom is in the center of a tetrahedron formed by sulfur atoms (large circles) and vice versa.

In the calcium fluoride structure the unit cell is divided into eight equal cubelets, and calcium ions are situated at corners and centers of the faces of the cell. The fluorine ions are at the centers of the eight cubelets. [W.C.D.]

Growth. The growth of crystals, of which all crystalline solids are composed, generally occurs by means of the following sequence of processes: (1) diffusion of the molecules of the crystallizing substance through the surrounding environment (or solution) to the surface of the crystal, (2) diffusion of these molecules over the surface of the crystal to special sites on the surface, (3) incorporation of molecules into the crystal at these sites, and (4) diffusion of the heat of crystallization away from the crystal surface. The rate of crystal growth may be limited by any of these four steps. The initial formation of the centers from which crystal growth proceeds is known as nucleation. *See* CRYSTALLIZATION; NUCLEATION.

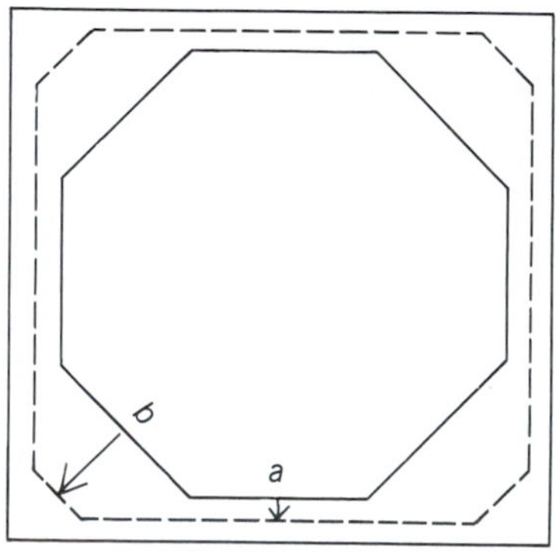

Fig. 4. Schematic representation of cross section of crystal at three stages of growth.

During its growth into a fluid phase, a crystal often develops and maintains a definite polyhedral form which may reflect the characteristic symmetry of the molecular pattern of the crystal. The bounding faces of this form are those which are perpendicular to the directions of slowest growth. How this comes about is illustrated in Fig. 4, in which it is seen that the faces *b*, normal to the faster-growing direction, disappear, and the faces *a*, normal to the slower-growing directions, become predominant.

The molecular binding sites on the surface of a crystal can be of several kinds. Thus a molecule must be more weakly bound on a perfectly developed plane of molecules at the crystal surface than at a ledge formed by an incomplete plane one molecule thick. Therefore, the binding of molecules in an island monolayer on the crystal surface will be less per molecule than it would be within a completed surface layer. The potential energy of a crystal is most likely to be minimum in forms containing the fewest possible ledge sites. According to the screw dislocation theory of F. C. Frank, growth is sustained by indestructible surface ledges which result from the emergence of screw dislocations in the crystal face. [D.T.]

Defects. Departures of a crystalline solid from a regular array of atoms or ions are referred to as defects. A "perfect" crystal of NaCl, for example, would consist of alternating Na^+ and Cl^- ions on an infinite three-dimensional simple cubic lattice, and a simple defect (a vacancy) would be a missing Na^+ or Cl^- ion. There are many other kinds of possible defects, ranging from simple and microscopic, such as the vacancy and other structures illustrated in Fig. 5, to complex and macroscopic, such as the inclusion of another material, or a surface.

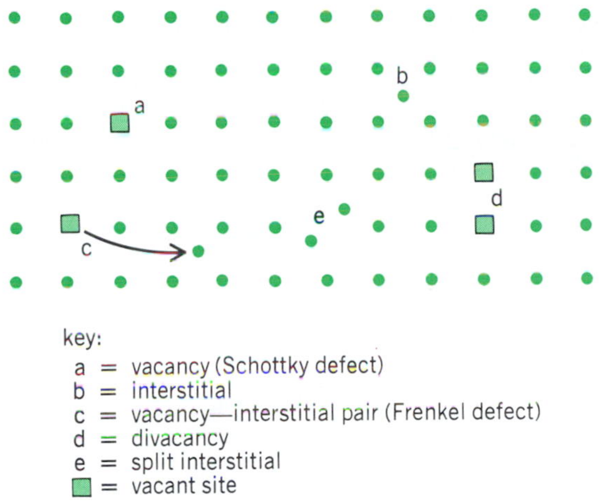

Fig. 5. Some simple defects in a lattice.

Natural crystals always contain defects, often in abundance, due to the uncontrolled conditions under which they were formed. The presence of defects which affect the color can make these crystals valuable as gems, as in ruby (Cr replacing a small fraction of the Al in Al_2O_3). Crystals prepared in the laboratory will also always contain defects, although considerable control may be exercised over their type, concentration, and distribution.

The importance of defects depends upon the material, type of defect, and properties which are being considered. Some properties, such as density and elastic constants, are proportional to the concentration of defects, and so a small defect concentration will have a very small effect on these. Other properties, such as the color of an insulating crystal or the conductivity of a semiconductor crystal, may be much more sensitive to the presence of small numbers of defects. Indeed, while the term defect carries with it the connotation of undesirable qualities, defects are responsible for many of the important properties of materials, and much of solid-state physics and chemistry and materials science involves the study and engineering of defects so that solids will have desired properties. A defect-free silicon crystal would be of little use in modern electronics; the use of silicon in devices is dependent upon small concentrations of chemical impurities such as phosphorus and arsenic which give it desired electronic properties.

An important class of crystal defect is the chemical impurity. The simplest case is the substitutional impurity, for example, a zinc atom in place of a copper atom in metallic copper. Impurities may also be interstitial; that is, they may be located where atoms or ions normally do not exist. In metals, impurities usually lead to an increase in the electrical resistivity. Impurities in semiconductors are responsible for the important electrical properties which lead to their widespread use. The energy levels associated with impurities and other defects in nonmetals may also lead to optical absorption in interesting regions of the spectrum.

Even in a chemically pure crystal, structural defects will occur. These may be simple or extended. One type of simple defect is the vacancy, but other types exist (Fig. 5). The atom which left a normal site to create a vacancy may end up in an interstitial position, a location not normally occupied. Or it may form a bond with a normal atom in such a way that neither atom is on the normal site, but the two are symmetrically displaced from it. This is called a split interstitial. The name Frenkel defect is given to a vacancy-interstitial pair, whereas an isolated vacancy is a Schottky defect.

The simplest extended structural defect is the dislocation. An edge dislocation is a line defect which may be thought of as the result of adding or subtracting a half-plane of atoms. A screw dislocation is a line defect which can be thought of as the result of cutting partway through the crystal and displacing it parallel to the edge of the cut. Dislocations are of great importance in determining the mechanical properties of crystals. A dislocation-free crystal is resistant to shear, because atoms must be displaced over high-potential-energy barriers from one equilibrium position to another. It takes relatively little energy to move a dislocation (and thereby shear the crystal), because the atoms at the dislocation are barely in stable equilibrium. Such plastic deformation is known as slip. *See* PLASTICITY.

For both scientific and practical reasons, much of the research on crystal defects is directed toward the dynamic properties of defects under particular conditions, or defect chemistry. Much of the motivation for this arises from the often undesirable effects of external influences on material properties, and a desire to minimize these effects. Examples of defect chemistry abound, including one as familiar as the photographic process, in which incident photons cause defect modifications in silver halides or other materials. Properties of materials in nuclear reactors are another important case. [W.B.F.]

Crystal counter A detector of radiation in which the sensitive material is a high-resistivity crystal. The crystals are diced into wafers 0.02 to 0.08 in. (0.5 to 2 mm) in thickness and 0.006 to 0.15 in. (4 to 100 mm^2) in surface area. Various techniques are used to contact the larger surfaces. A field of about 2.5×10^4 V/in. (1×10^4 V/cm) is used to collect the charge liberated by the radiation. Crystal and junction counters are very similar, but differ fundamentally in the way the high-resistivity material is created. In a junction counter, a reverse-biased junction creates a depletion region which is the high-resistivity region. At the same time, the reverse bias gives high charge-collection field. In the crystal counter, the high resistivity comes from the basic crystalline material. Cadmium telluride (CdTe) and mercury iodide (HgI_2) are considered the most preferable crystals. *See* JUNCTION DETECTOR.

When a charged particle enters the crystal, it loses energy by creating electron-hole pairs. A gamma ray entering the crystal interacts with an atom of the crystal. Then, depending upon the interaction process—photoelectric, Compton, or pair production—the gamma ray will transfer all or part of its energy to an electron or to an electron-positron pair. These charged particles, in turn, dissipate their energy by creating electron-hole pairs. The number of electron-hole pairs is proportional to the energy deposited in the crystal. *See* GAMMA-RAY DETECTORS; PARTICLE DETECTOR; SCINTILLATION COUNTER. [J.McK.]

Crystal field theory An essentially ionic approach to chemical bonding which is often used with coordination compounds. These compounds consist of a central transition metal ion that is surrounded by a regular array of coordinated atoms or ligands. Accordingly, the ligands are assumed to be sources of negative charge which perturb the energy levels of the central metal ion. In this respect the ligands subject the metal ion to an electric field which is analogous to the electric or crystal field produced by the regular distribution of nearest neighbors within an ionic crystalline lattice. For example, the crystal field produced by the Cl ion ligand in octahedral $TiCl_6^{3-}$ is considered to be similar to that produced by the octahedral array of the six Cl ions about each Na ion in NaCl. The Na ion with its rare-gas configuration has an electronic charge distribution which is spherically symmetric both within and without the crystal field. The paramagnetic Ti(III) ion, which possesses one 3*d* electron (d^1), has a spherically symmetric charge distribution only in the absence of the crystal field produced by the ligands. The presence of the ligands destroys the spherical symmetry and produces a more complex set of energy levels within the central metal ion. The crystal field theory allows the energy levels to be calculated and related to experimental observation. *See* COORDINATION CHEMISTRY.

To illustrate the results of a typical crystal field calculation, assume that the single *d* electron in the Ti(III) ion will experience a coulombic repulsion with each of the six nearest neighbor Cl ions which are taken as point negative charges. This model for coulombic repulsion may be described mathematically as the summation $eq\Sigma r^{-1}$, where *e* and *q* are the electronic and ligand charges, respectively, and *r* is the distance between the electron and the ligand. The summation extends over all the ligands. A detailed quantum-mechanical calculation can then be made. Fortunately, it is possible to arrive at identical results by a very qualitative procedure. This method considers the spatial orientation of *d* orbitals with relation to the ligands when both are viewed within the same coordinate system.

Crystal field theory calculations have enabled inorganic chemists to understand certain natural phenomena; for example, why some coordination compounds containing a given metal ion exhibit full paramagnetism, while others containing the metal in exactly the same formal oxidation state show either a much weaker paramagnetism or none at all. A striking example of this is provided by paramagnetic CoF_6^{3-}, which possesses six unpaired *d* electrons, and the diamagnetic $Co(NH_3)_6^{3+}$, in which none of the *d* electrons are unpaired. Both contain Co(III).

The many successes of crystal field theory in the interpretation of phenomena associated with transition metal compounds should not lead one to the conclusion that the bonding within these compounds can be truthfully represented by a strictly ionic model. Crystal field theory is essentially a very specialized form of the more complete molecular orbital theory, and some aspects of transition metal behavior can only be interpreted in terms of covalent bonding. *See* CHEMICAL BONDING; MOLECULAR ORBITAL THEORY. [R.A.D.W.]

Crystal optics The study of the propagation of light, and associated phenomena, in crystalline solids. For a simple cubic crystal the atomic arrangement is such that in each direction through the crystal the crystal presents the same optical appearance. The atoms in anisotropic crystals are closer together in some planes through the material than in others. In anisotropic crystals the optical characteristics are different in different directions. In classical physics the progress of an electromagnetic wave through a material involves the periodic displacement of electrons. In anisotropic substances the forces resisting these displacements depend on the displacement direction. Thus the velocity of a light wave is different in different directions and for different states of polarization. The absorption of the wave may also be different in different directions. *See* DICHROISM; TRICHROISM.

In an isotropic medium the light from a point source spreads out in a spherical shell. The light from a point source embedded in an anisotropic crystal spreads out in two wave surfaces, one of which travels at a faster rate than the other. The polarization of the light varies from point to point over each wave surface, and in any particular direction from the source the polarization of the two surfaces is opposite. The characteristics of these surfaces can be determined experimentally by making measurements on a given crystal.

In the most general case of a transparent anisotropic medium, the dielectric constant is different along each of three orthogonal axes. This means that when the light vector is oriented along each direction, the velocity of light is different. One method for calculating the behavior of a transparent anisotropic material is through the use of the index ellipsoid, also called the reciprocal ellipsoid, optical indicatrix, or ellipsoid of wave normals. This is the surface obtained by plotting the value of the refractive index in each principal direction for a linearly polarized light vector lying in that direction (see illustration). The different indices of refraction, or wave velocities

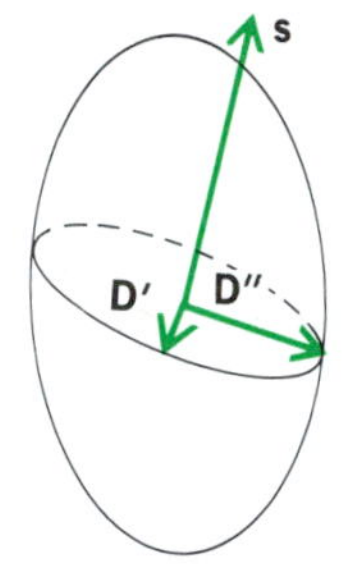

Index ellipsoid, showing construction of directions of vibrations of *D* vectors belonging to a wave normal *s*. (*After M. Born and E. Wolf, Principles of Optics, Pergamon, 1959*)

associated with a given propagation direction, are then given by sections through the origin of the coordinates in which the index ellipsoid is drawn. These sections are ellipses, and the major and minor axes of the ellipse represent the fast and slow axes for light proceeding along the normal to the plane of the ellipse. The length of the axes represents the refractive indices for the fast and slow wave, respectively. The most asymmetric type of ellipsoid has three unequal axes. It is a general rule in crystallography that no property of a crystal will have less symmetry than the class in which the crystal belongs.

Accordingly, there are many crystals which, for example, have four- or sixfold rotation symmetry about an axis, and for these the index ellipsoid cannot have three unequal axes but is an ellipsoid of revolution. In such a crystal, light will be propagated along this axis as though the crystal were isotropic, and the velocity of propagation will be independent of the state of polarization. The section of the index ellipsoid at right angles to this direction is a circle. Such crystals are called uniaxial and the mathematics of their optical behavior is relatively straightforward.

In crystals of low symmetry the index ellipsoid has three unequal axes. These crystals are termed biaxial and have two directions along which the wave velocity is independent of the polarization direction. These correspond to the two sections of the ellipsoid which are circular. *See* CRYSTALLOGRAPHY.

The normal to a plane wavefront moves with the phase velocity. The Huygens wavelet, which is the light moving out from a point disturbance, will propagate with a ray velocity. Just as the index ellipsoid can be used to compute the phase or wave velocity, so can a ray ellipsoid be used to calculate the ray velocity. The length of the axes of this ellipsoid is given by the velocity of the linearly polarized ray whose electric vector lies in the axis direction. *See* PHASE VELOCITY.

The refraction of a light ray on passing through the surface of an anisotropic uniaxial crystal can be calculated with Huygens wavelets in the same manner as in an isotropic material. For the ellipsoidal wavelet this results in an optical behavior which is completely different from that normally associated with refraction. The ray associated with this behavior is termed the extraordinary ray. At a crystal surface where the optic axis is inclined at an angle, a ray of unpolarized light incident normally on the surface is split into two beams: the ordinary ray, which proceeds through the surface without deviation; and the extraordinary ray, which is deviated by an angle determined by a line drawn from the center of one of the Huygens ellipsoidal wavelets to the point at which the ellipsoid is tangent to a line parallel to the surface. The two beams are oppositely linearly polarized. [B.H.Bi.]

Crystal whiskers Single crystals that have grown in a filamentary form. Such filamentary growths have been known for centuries; however, great interest in them developed only after it was discovered at Bell Telephone Laboratories in the 1950s that the strength exhibited by some whiskers in bend tests approaches that expected theoretically for perfect crystals. *See* SINGLE CRYSTAL.

Whiskers have been grown by spontaneous extrusion or out of a supersaturated medium, the medium being supersaturated by physical methods or by chemical reaction. In all instances whiskers apparently form from singularities in the medium. [D.T.]

Crystallization The formation of a solid from a solution, melt, vapor, or a different solid phase. Crystallization from solution is an important industrial operation because of the large number of materials marketed as crystalline particles. Fractional crystallization is one of the most widely used methods of separating and purifying chemicals. In fractional crystallization it is desired to separate several solutes present in the same solution. This is generally done by picking crystallization temperatures and solvents such that only one solute is supersaturated and crystallizes out. By changing conditions, other solutes may be crystallized subsequently. Repeated crystallizations are necessary to achieve desired purities when many inclusions are present or when the solid solubility of other solutes is significant. For a discussion of crystallization in glass, a supercooled melt, *see* GLASS. Polymer crystals obtained from solutions are used to study the properties of these crystals, while crystallization of polymer melts dramatically influences polymer properties; *see* POLYMER. For methods of preparing large crystals *see* SINGLE CRYSTAL. For crystallization from vapors *see* SUBLIMATION. For solubility and other relationships between solid and liquid phases *see* PHASE EQUILIBRIUM; SOLUTION. *See also* CRYSTAL; CRYSTALLOGRAPHY; NUCLEATION. [W.R.W.]

Crystallography The branch of science that deals with the geometric description of crystals, their internal arrangement, and their properties. Long a part of mineralogy, crystallography is now a vast, loosely defined area between physics, chemistry, physical metallurgy, mineralogy, and even biology.

The anisotropic nature of crystals results in a freely growing crystal, bounded by flat faces. The following discussion presents a formalism for describing crystals exactly in terms of these faces.

Faces which are parallel to the same edge are said to form a zone; the direction of the edge is the zone axis. The number and relative importance of the faces bounding crystals of the same chemical compound can be very different. The crystals are then said to have a different habit. Faces and edges which are present as well as those which are geometrically possible must therefore be taken into account, and in the following discussion the term

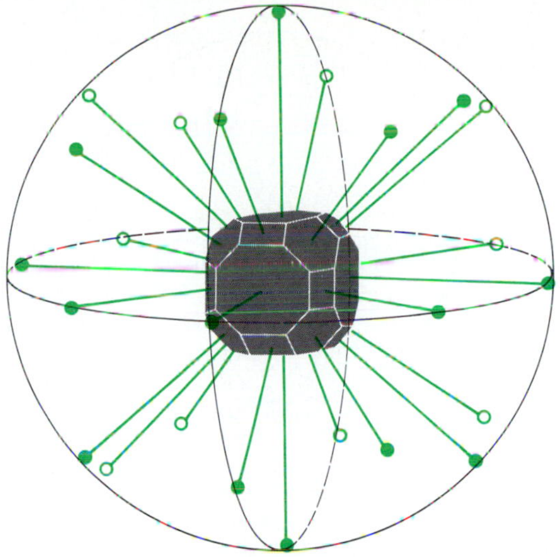

Fig. 1. Replacement of the constellation of faces by a constellation of poles on a sphere. Faces parallel to the same edge, that is, forming a zone, have their poles on a great circle called the zone circle.

face is extended to any plane determined by two edges. An edge is, in the same sense, the intersection of two faces.

The angle between similar faces is constant for different-sized crystals of the same substance. This means that a crystal grows by the parallel displacement of its faces. Therefore, only the direction of a normal to a face is of importance. As a consequence, the crystal is considered to be in the center of a sphere (Fig. 1). The normals to its faces from the center cut the sphere in points or poles whose position can be fixed by two coordinates φ and ρ (Fig. 2*a*). The constellation of faces is in this way replaced by a constellation of points on a sphere. The zone axes are replaced by great circles or zone circles; they are the intersections of the sphere with planes passing through the center of the sphere, perpendicular to the zone axis. In practice, φ and ρ values are obtained by "measuring" the crystal; two-circle goniometers are used for the purpose. All problems which can arise are connected with angles between faces and between zone axes; they can in principle be solved by spherical trigonometry once the φ and ρ values are known. Graphical constructions are, however, most helpful, and gnomonic and stereographic projections are used in such constructions. These projections are shown in Fig. 2. In stereographic projection, circles are projected as circles, and the angles between them are not altered. In gnomonic projection, great circles are projected as straight lines.

Symmetry is a basic property of crystals. A crystal polyhedron (or better, the constellation of the normals) can be brought into successive indistinguishable positions by nontrivial operations. These are inversion, rotation around an axis over a specific angle, reflection into a plane, or combinations of these. It is, however, convenient to consider all operations as rotations, and two types of axis can be considered: (1) a rotation or symmetry axis involving a rotation over an angle of $360°/n$ (n is the multiplicity of the axis) and (2) an inversion axis which combines the preceding operation with an inversion with respect to a point of the axis. It is found that only multiplicities of 1, 2, 3, 4, and 6 occur.

The seven groups obtained when only the dominant symmetry elements of the classes are taken into account are called crystal systems, In order of decreasing symmetry, they are called cubic, hexagonal, tetragonal, trigonal, orthorhombic, monoclinic, and triclinic. Sometimes the hexagonal and trigonal systems are considered as a single system.

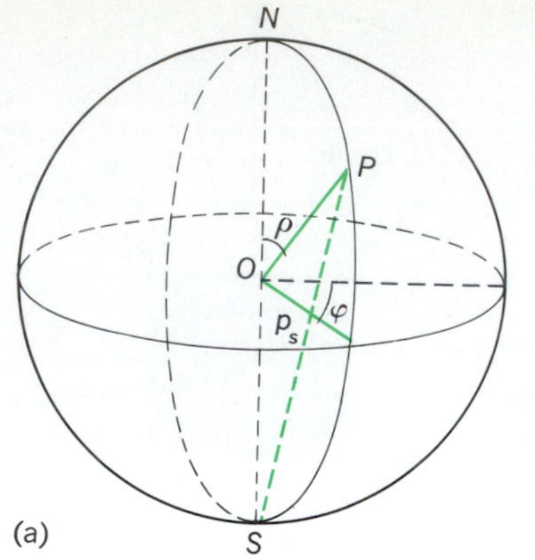

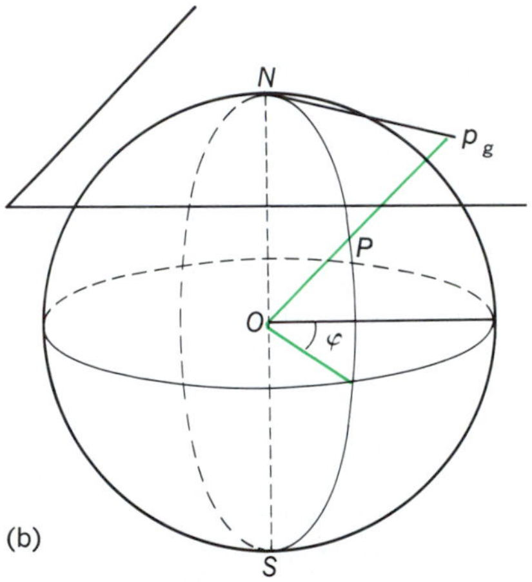

Fig. 2. Graphical constructions used in description of crystals. (*a*) Stereographic projection of *P*. Line joining pole *P*, defined by φ and φ, to south pole *S* cuts equator at p_s. (*b*) Gnomonic projection of *P*. Line *OP* cuts the plane tangent to the sphere at *N* at point p_g.

Miller indices are used for the analytical description of crystal faces, and are shown in Fig. 3. A specially well-suited reference system used for that purpose consists of three nonparallel possible edges of the crystal which are chosen as axes and a possible plane of the crystal which intersects all

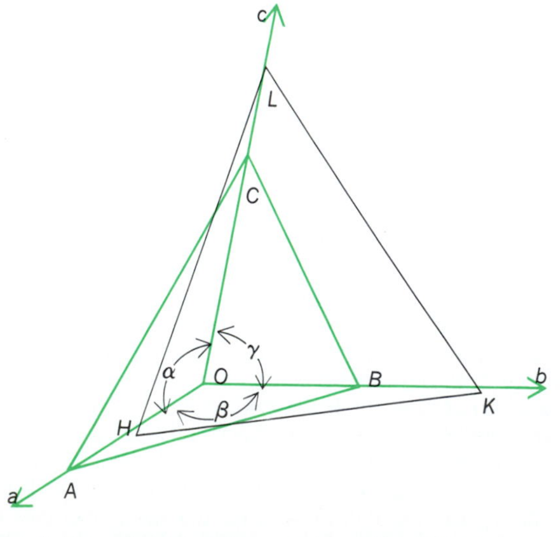

Fig. 3. The reference system used to define Miller indices; *ABC* is the unit plane.

three axes respectively at *A*, *B*, *C*. It is defined by the three angles α, ß, γ and the ratio *OA*:*OB*:*OC*. Another plane, *HKL*, of the same crystal is defined by the ratio *OH*:*OK*:*OL*. The lengths *OH*, *OK*, *OL* are respectively measured with $OA = a$, $OB = b$, $OC = c$ as units, that is, by the ratios a/OH, b/OK, c/OL. In the equation below, *h*, *k*, *l* are rational numbers which are

$$(a/OH)/(b/OK)/(c/OL) = h/k/l$$

small when the axes are well chosen (the law of rational indices). A different unit is used along each axis to take the anisotropy of the crystal into account. When, for instance, the temperature is raised, *a*, *b*, and *c* expand differently, but the ratios a/OH, b/OK, c/OL remain constant. The ratios $h{:}k{:}l$ define the plane *HKL*; *h*, *k*, *l* are called its Miller indices, and by convention, this is written (*hkl*), or ($\overline{hkl}$) when the indices are negative numbers. *See* CRYSTAL. [W.C.D.]

Ctenophora A phylum of exclusively marine organisms, formerly included in the jellyfish and polyps as coelenterates. These animals, the so-called comb jellies, possess a biradial symmetry of organization, They are characterized by having eight rows of comblike plates as the main locomotory structures. Most of these animals are pelagic, but a few genera are creeping. Many are transparent and colorless; others are faintly to brightly colored. These organisms are hermaphroditic. Development is biradially symmetrical, with a cydippid larval stage. Two classes and five orders constitute this phylum as follows:

Phylum Ctenophora
  Class Tentaculata
    Order: Cydippidea
      Lobata
      Cestida
      Platyctenea
  Class Nuda
    Order Beroida

The body is gelatinous and extremely fragile; its form may be globular, pyriform, or bell- or helmet-shaped (see illustration). Some species resemble a ribbon. Their size ranges from about 0. 1 in. to 20 in. (3 mm to 50 cm). Comb plates are the most characteristic structure of the ctenophores, possessed by all members of the phylum. Eight meridional comb-plate rows or ribs stretch from the aboral pole on the surface of the body. Each plate consists of a great number of very long related cilia. With the exception of Beroida, all ctenophores possess two highly extensile tentacles bearing many lateral branches (tentilla). Tentacles and tentilla are thickly covered with adhesive cells (colloblasts) which are exclusive to the phylum. When a prey collides with a tentillum, some colloblasts burst and their granules stick the prey to the helical threads. These threads apparently act as shock absorbers of the captured prey's movements rather than as springs projecting the head.

Fertilization usually occurs in sea water. Pelagic ctenophores are self-fertile, but cross-fertilization might also take place inside a swarm of ctenophores. The earliest larva formed in the development of all types of ctenophores is the cydippid, with a globular body, a pair of primary tentacles (or rudiments), and eight costae of comb plates. In the Lobata, Cestoida, and Platyctenea, the larva undergoes secondary changes in later development.

Almost all ctenophores are luminescent. This capacity, linked to a calcium-dependent photoprotein, is assumed by specialized cells underlying the comb plates.

The ctenophores feed on zooplankton. Prey are trapped either with the tentacles (Cydippidea, Cestida) or with oral extensions of the body (Lobata), or are engulfed directly

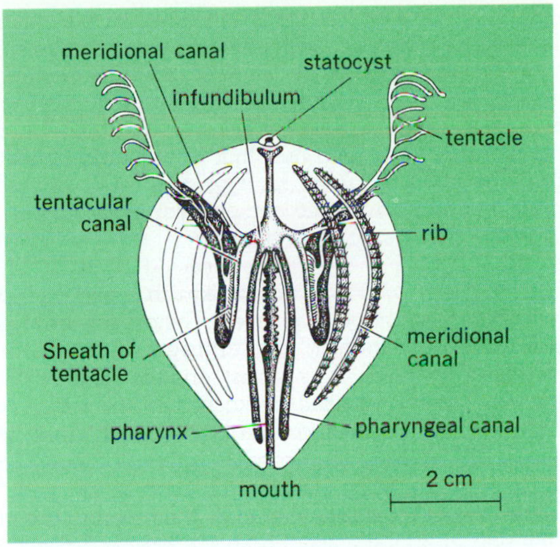

Structure of a cydippid ctenophore.

through the widely opened mouth (Beroida). Extracellular digestion occurs first in the pharynx, where gland cells secrete enzymes. The partially digested material is then carried throughout the gastrovascular system, where it is phagocytized by digestive cells.

Ctenophores have high powers of regeneration. Asexual reproduction (in a few platyctenid genera) is by regeneration of an entire organism from a small piece of the adult body.

Ctenophores are themselves important plankton organisms. They are quite common and have a worldwide distribution in all seas, where they can appear in enormous swarms. Because of their voracity as predators of zooplankton, they can play an important role in plankton equilibrium and in fisheries.

The Ctenophora constitute a well-defined phylum. Among its orders, the Cydippidea are undoubtedly the most primitive, with a radiating evolution of the other orders. The ancestor of the phylum is possibly to be found among the Trachylina (Cnidaria, Hydrozoa). However, because of such forms as *Coeloplana* (Platyctenea), ctenophores have also been suspected of affinity with the polyclad Turbellaria. *See* POLYCLADIDA; TRACHYLINA. [J.-M.F.]

Ctenopoda An order of branchiopod crustaceans formerly included in the order Cladocera. The body is up to about 4 mm (0.2 in.) in length. Although superficially similar to the large order Anomopoda in the protection of the trunk and its limbs by a functionally bivalved carapace, in the nature of the eye, in the use of antennae for swimming, and in the possession of a somewhat similar postabdomen, ctenopods differ in various ways. *See* ANOMOPODA.

The six trunk limbs are all constructed on a similar plan but differ in detail among themselves. They beat with a metachronal rhythmn and by so doing draw in suspended food particles. Some species have become planktonic; *Holopedium* surrounds itself by a sphere of jelly to assist flotation. Others are sedentary; *Sida* attaches itself by a sucker at the back of the head.

Reproduction is mostly by parthenogenesis. Most species occur in fresh water, where they are widely distributed, but *Penilia* is marine. *See* BRANCHIOPODA. [G.Fr.]

Ctenostomata An order of bryozoans in the class Gymnolaemata. Ctenostomes have inconspicuous delicate colonies made up of relatively isolated, simple, short, straight, vase-shaped zooecia with solid chitinous walls. *See* GYMNOLAEMATA.

Most ctenostome colonies are encrusting threadlike networks of stolons or proximal portions of autozooecia, with the distal portions of the autozooecia arising erect from the substrate at regular intervals. A few other ctenostome colonies are tuft-like or are gelatinous masses. Though sometimes having stolons or rhizoids, ctenostome colonies lack ovicells or heterozooids (avicularia, vibracula, and so on). The colonies cannot be divided into distinct endozone and exozone regions.

Each ctenostome zygote develops into only one larva. After attaching to a substrate, each larva metamorphoses into an ancestrular zooid. Additionally, the few fresh-water (but not marine) ctenostomes produce irregularly shaped resistant resting bodies (hibernacula) by asexual budding when unfavorable environmental conditions threaten.

First appearing early in the Early Ordovician, ctenostomes are the oldest known bryozoans and may possibly have been the stem group from which the other ectoproct groups evolved, the stenolaemates early in the Paleozoic and the cheilostomes and possibly phylactolaemates late in the Mesozoic. *See* BRYOZOA; CHEILOSTOMATA; PHYLACTOLAEMATA; STENOLAEMATA. [R.J.Cu.]

Ctenothrissiformes A small order of teleostean fishes that are of particular interest because they seem to be on or near the main evolutionary line leading from the generalized, soft-rayed salmoniforms to the spiny-rayed beryciforms and perciforms, dominant groups among higher bony fishes. Ctenothrissiformes lack fin spines but they share many characters with primitive beryciforms. The few fishes confidently assigned to the Ctenothrissiformes, *Ctenothrissus* (Ctenothrissidae; see illustration) and *Aulolepis* and *Pateroperca*

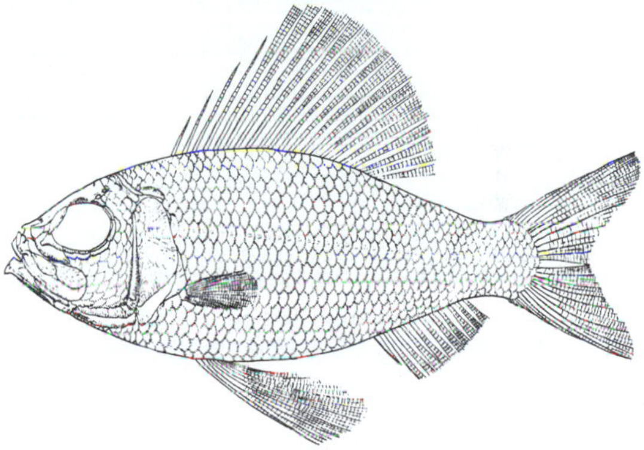

Cretaceous ctenothrissid, *Ctenothrissus radians*. (After C. Patterson, *Phil. Trans. Roy. Soc. London, Ser. B, Biol. Sci.*, no. 739, vol. 247, 1964)

(Aulolepidae), are known as Upper Cretaceous marine fossils from Europe and southwestern Asia. Two little-known genera of Recent oceanic fishes, *Macristium* and *Macristiella* (Macristiidae), are assigned by some authors to the order but some doubt attends this placement. *See* BERYCIFORMES; PERCIFORMES; SALMONIFORMES. [R.M.B.]

Cube A parallelepiped whose six faces are all squares. The cube is one of the five regular solids known to the ancient Greeks. A cube has eight vertices and twelve edges. Each vertex is on three edges and three faces, each edge is on two vertices and two faces, and each plane is on four vertices and four edges. *See* PARALLELEPIPED; POLYHEDRON; REGULAR POLYTOPES. [L.M.Bl.]

Cubeb The dried, nearly ripe fruit (berries) of a climbing vine, *Piper cubeba*, of the pepper family (Piperaceae). This species is native to eastern India and Indomalaysia. The crushed berries are smoked, and the inhaled smoke produces a soothing effect in certain respiratory ailments. Cubebs are used in medicine as a stimulant, expectorant, and diuretic. Medicinal properties of cubeb are due to the presence of a volatile oil which formerly was thought to stimulate healing of mucous membranes. *See* PIPERALES. [P.D.St./E.L.C.]

Cubomedusae An order of the class Scyphozoa. Their distribution is mostly tropical. The umbrella is cubic, and a pedalium, a gelatinous leaf-shaped body, is present at the base of each ridge of the exumbrella (see illustration). A well-developed tentacle, or a cluster of tentacles, arises from each pedalium. No connecting stage between polyp and medusa is known. *See* SCYPHOZOA. [T.U.]

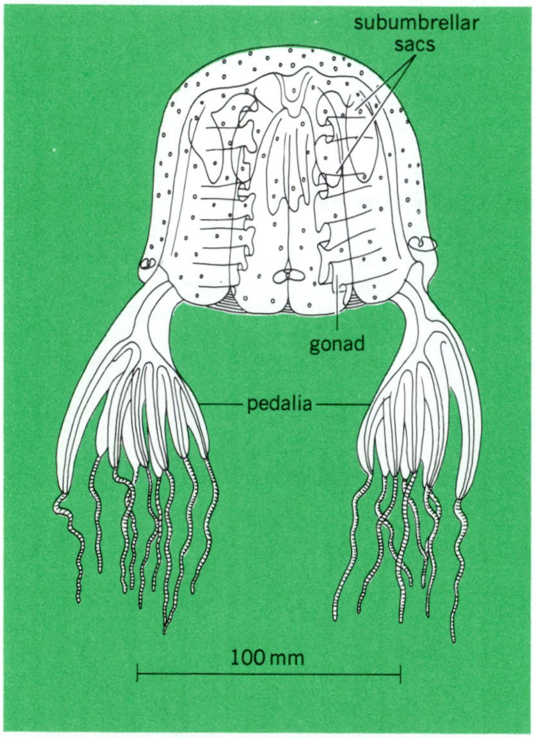

Cubomedusan, *Chiropsalmus quadrumanni*. (After L. Hyman, *The Invertebrates, vol. 1*, McGraw-Hill, 1940)

Cuckoo The name for about 130 species of the family Cuculidae, in the order Cuculiformes, which are primarily arboreal and widely distributed throughout the Old World. The arrangement of the toes is zygodactyl with two in front and two behind, an arrangement that is well suited for these essentially perching birds which fly from tree to tree.

There are about 50 species which are social parasites in their breeding habits, and they are widely represented in the Old World, ranging from northern Europe to Australia. Essentially, the parasitic species lay their eggs in the nests of other birds for incubation and for feeding of young after hatching. Egg mimicry makes these cuckoos well adapted to parasitism with the eggs being about the same size and color as those of the host species. The eggs of the hosts are removed from the nest by the cuckoo to make room for her own. The eggs of the cuckoo hatch before those of the host, or shortly after. The cuckoo nestlings attempt to push the host's eggs out of the nest, or to rid the nest of the nestlings. If they are unsuccessful in the latter, they beg for food with such intensity that they are fed first and more frequently. The female cuckoo occupies a definite territory during this season, even though she exhibits no interest in the offspring. *See* TERRITORIALITY.

In the nonparasitic species, both parents incubate the eggs and care for the young in a nest that is frequently constructed of twigs. The clutch is two to six eggs, and when hatched the young are more or less naked and are nidicolous; that is, they remain in the nest. *See* CUCULIFORMES. [C.B.C.]

Cuculiformes An order of birds with zygodactyl feet in which the outer fourth toe is reversed. The order is divided into the suborder Musophagi, containing the single family Musophagidae (touracos; 18 species), and the suborder Cuculi, with the lone family Cuculidae (cuckoos; 129 species).

Touracos are African woodland birds with a loose green, gray, or brown plumage with patches of bright yellow, red, or violet. The nest is a bulky platform placed in a tree, with incubation and care of the chicks shared by the parents. The young leave the nest at an early age (about 10 days) and crawl about the trees before being able to fly.

Cuckoos are diverse. They are mostly arboreal, but some are terrestrial, and some, such as the American roadrunners (*Geococcyx*) are fast runners with top speeds of about 15 mi/h (25 km/h) and fly rarely. Most cuckoos are secretive, sulking in heavy vegetation. They are strong fliers but do not fly often. Temperate species are migratory. Cuckoos feed on insects, small vertebrates, other animals, and rarely on fruit. Most are solitary, but the anis (Crotophaginae) live in flocks and build a cooperative nest housing several females. Many Old World species (Cuculinae) are nest parasites. *See* AVES. [W.J.B.]

Cucumber A warm-season annual cucurbit (*Cucumis sativus*) native to India and belonging to the plant order Campanulales. The cucumber is grown for its immature fleshy fruit which is used primarily for pickling and for slicing as a salad. The development of gynoecious (bearing only female flowers) and hybrid varieties (cultivars) with increased disease resistance is resulting in a continuing change in the varieties of both pickling and slicing cucumbers. Important producing states are Florida and South Carolina for fresh market and Michigan and Wisconsin for pickling. *See* CAMPANULALES. [H.J.C.]

Culture A growth of living cells in a controlled artificial environment. It may be a culture of microorganisms, such as a bacterial culture, or one of animal or plant cells, such as a tissue culture. Cultures require appropriate sources of food and energy, provided by the culture medium, and a suitable physical environment. Tissue cultures can themselves become a culture medium for viruses, which grow only with live cells.

Cultures of only one kind of cells are known as pure cultures, as distinguished from mixed or contaminated cultures. Isolating microorganisms in pure culture for identification or study is a basic microbiologic procedure. Cultures are used commercially as "starters" in preparing several dairy products, in fermentation industries, in bread making, and in other instances in which seeding from a culture initiates further controlled massive growth to produce desired products, ranging from cottage cheese to antibiotics. *See* CULTURE MEDIA; EMBRYONATED EGG CULTURE; PURE CULTURE; TISSUE CULTURE. [M.S.M.]

Culture technique involves the procedures for isolation, cultivation, and manipulation of microorganisms, including viruses and rickettsia, and for propagation of plant and animal cells and tissues. A relatively minute number of cells, the inoculum, is introduced into the culture medium. The medium in a suitable vessel is protected by cotton plugs or loose-fitting covers with overlapping edges so as to allow diffusion of air, yet prevent access of contaminating organisms from the air or from unsterilized surfaces. The inoculation is usually done with the end of a flamed, then cooled, platinum wire. Sterile swabs may also be used and, in the case of liquid inoculum, sterile pipets. The aqueous solution of nutrients may be left as a liquid medium or be solidified by incorporatingg a nutritionally inert substance, commonly agar or silica gel. Special gas requirements may be provided in culture vessels closed to the atmosphere, as for anaerobic organisms. Inoculated vessels are held at desired temperature in an incubator or water bath. Liquid culture media may be mechanically agitated during incubation. Maximal growth, visible as a turbidity or as masses of cells, is usually attained within a few days, but some organisms may require weeks to reach this stage. *See* CHEMOSTAT; INOCULATION. [J.W.F./R.E.K.]

Culture media Combinations of nutrients used for the growth of living cells in cultures. Water is the chief constituent of most media. A medium may consist of natural materials, such as enzymatic digests, extracts of yeast or beef, milk, potato slices, or chick embryos. Artificial media are prepared by mixing various ingredients according to particular formulas. A complex medium contains at least one crude ingredient derived from a natural material, hence of unknown chemical composition. A chemically defined, or synthetic, medium is one in which the chemical structure and amount of each component are known.

All media furnish minerals, sources of nitrogen, carbon, and energy, and sometimes vitamins or other growth factors. An ingredient furnishing carbon—usually the chief source—is called the substrate for the utilizing organism.

A selective medium usually contains an individual organic compound as the sole source of carbon, nitrogen, or sulfur for growth of an organism. An agar medium of a given composition permits selection, through development of colonies, of organisms capable of utilizing the specific sources of carbon, energy, nitrogen, or sulfur available. A colony on such a medium is prima-facie evidence that such organisms have been selected, but this must be confirmed by restreaking (subculturing on agar medium) on a medium of the same composition.

Enrichment media are liquids of a given composition which permit preferential emergence of certain organisms that initially may have made up a relatively minute proportion of a mixed inoculum. This enrichment of the population in the desired type expedites isolation by direct plating on the corresponding selective solid medium.

Indicator media are usually solid, and contain substances capable of undergoing a color change in the vicinity of the colony which has effected a particular chemical change, such as fermenting a certain sugar. Thus, organisms with distinctive types of metabolism may be readily detected by the color change in the medium, even when the surface of the medium contains many other colonies, Indicators may be used in liquid media to characterize a pure culture of organisms. *See* ELECTIVE CULTURE. [J.W.Fo./R.E.K.]

Cumacea An order of the class Crustacea, subclass Malacostraca. All have a well-developed carapace which is fused dorsally with at least the first three thoracic somites and which overhangs the sides to enclose a branchial cavity. The telson may be free or may appear lacking because it is fused with the last abdominal somite. Eyes are sessile and, except for a few genera, are coalesced in a single median organ or wholly wanting. There are eight pairs of thoracic appendages. There is no free larval stage. In the few species whose life history has been investigated there are five to seven molts after release from the brood pouch. There is always some sexual dimorphism in adults, but immature males resemble females until the later stages.

The order is rather uniform and classification is based on trivial characters. Seven families are recognized. Affinities are probably closer to the Mysidacea than to the Tanaidacea. A few fossils are now known.

Cumacea are found in all seas and in some estuaries and areas of brackish water, including the Caspian Sea. Most inhabit water shallower than 660 ft (200 m), but many occur in great depths. They normally burrow in sand or mud with the front protruding. They feed on organic detritus or very small organisms and may themselves form part of the diet of various fishes. *See* MALACOSTRACA. [N.S.J.]

Cummingtonite The name given to the calcium-poor monoclinic amphiboles that form a limited solid-solution series between the iron end member grunerite, $Fe_7Si_8O_{22}(OH)_2$, and a magnesian end member of approximately $Mg_5Fe_2Si_8O_{22}(OH)_2$. More magnesium-rich minerals apparently form the closely related orthorhombic form, anthophyllite. Cummingtonite is usually yellow-brown to brown and is slightly colored in thin sections. It forms elongate prismatic to fibrous crystals. Grunerite commonly occurs as brown fibrous crystals with a silky luster and is usually found associated with iron formation.

Cummingtonite is a metamorphic mineral that occurs in various schists, in contact metamorphic rocks, and in some ore deposits. It is found associated with such minerals as hornblende, biotite, garnet, plagioclase, quartz, and anthophyllite. *See* AMPHIBOLE; ANTHOPHYLLITE. [G.W.DeV.]

Cup anemometer An instrument widely used to measure horizontal wind speed, independent of direction. The use of three hemispherical cups mounted on a vertical shaft is standard in the United States (see illustration). Commonly the cups rotate a worm gear which operates an electric contact every mile of wind, or fraction thereof. These contacts are recorded on a chart driven at constant speed. From the number of contacts in a selected time interval, the wind speed is deduced. Since the number of cup revolutions per mile varies with wind speed, particularly at low speeds, a correction must be made. [H.Fo.]

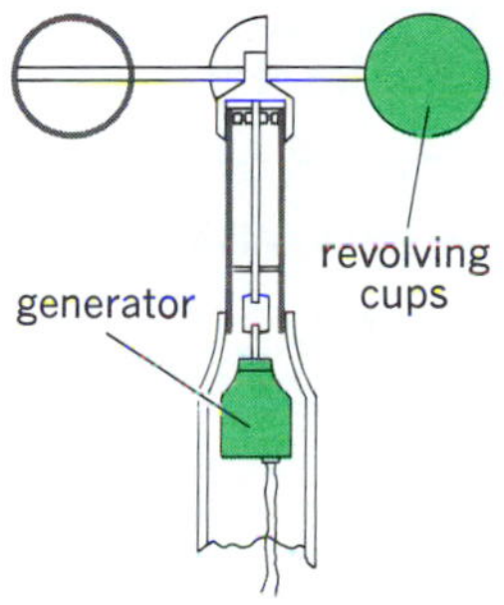

Cutaway diagram of revolving-cup electric anemometer. (After D. M. Considine, *Process Instruments and Controls Handbook*, 2d ed., McGraw-Hill, 1974)

Cuprite A mineral having composition Cu_2O and crystallizing in the isometric system. Cuprite is commonly in crystals showing the cube, octahedron, and dodecahedron. It is various shades of red and a fine ruby red in transparent crystals which have a metallic to adamantine luster. The hardness is 3.5–4 (Mohs scale) and the specific gravity is 6.1.

Cuprite is a widespread supergene copper ore. Fine crystals have been found at Cornwall, England, and Chessy, France. It has served as an ore in the Congo, Chile, Bolivia, and Australia. In the United States it has been found at Clifton, Morenci, Globe, and Bisbee, all in Arizona. *See* COPPER. [C.S.Hu.]

Curie temperature The critical or ordering temperature for a ferromagnetic or a ferrimagnetic material. The Curie temperature Tc is the temperature below which there is a spontaneous magnetization M in the absence of an externally applied magnetic field, and above which the material is paramagnetic. In the disordered state above the Curie temperature, thermal energy overrides any interactions between the local magnetic moments of ions. Below the Curie temperature, these interactions are predominant and cause the local moments to order or align so that there is a net spontaneous magnetization.

In the ferromagnetic case, as temperature T increases from absolute zero, the spontaneous magnetization decreases from M_0, its value at $T = 0$. At first this occurs gradually, then with increasing rapidity until the magnetization disappears at the Curie temperature. In ferrimagnetic materials the course of magnetization with temperature may be more complicated, but the spontaneous magnetization disappears at the Curie temperature.

In antiferromagnetic materials the corresponding ordering temperature is termed the Néel temperature. Below the Néel temperature the magnetic sublattices have a spontaneous magnetization, though the net magnetization of the material is zero. Above the Néel temperature the material is paramagnetic. *See* ANTIFERROMAGNETISM.

The ordering temperatures for magnetic materials vary widely. The ordering temperature for ferroelectrics is also termed the Curie temperature, below which the material shows a spontaneous electric moment. *See* FERROMAGNETISM; PYROELECTRICITY. [J.F.Di.]

Curie-Weiss law A relation between magnetic or electric susceptibilities and the absolute temperature which is followed by ferromagnets, antiferromagnets, nonpolar ferroelectrics and antiferroelectrics, and some paramagnets. The Curie-Weiss law is usually written as the following equation, where χ is the

$$\chi = C/(T - \theta)$$

susceptibility, C is a constant for each material, T is the absolute temperature, and θ is called the Curie temperature. Antiferromagnets and antiferroelectrics have a negative Curie temperature. The Curie-Weiss law refers to magnetic and electric behavior above the transition temperature of the material in question. It is not always precisely followed, and it breaks down in the region very close to the transition temperature. Often the susceptibility will behave according to a Curie-Weiss law in different temperature ranges with different values of C and θ. *See* CURIE TEMPERATURE; ELECTRIC SUSCEPTIBILITY; MAGNETIC SUSCEPTIBILITY. [E.A.; F.Ke.]

Curing A process in which polymers or oligomers (low-molecular-weight polymers) are chemically cross-linked into polymer networks. Commercially important applications include coatings, adhesives, elastomers, molding compounds, and thermosetting plastics. Cross-linking is generally required to impart solvent resistance as well as other properties related to durability of the cured product. An ideal cured composition consists of one giant molecule. *See* POLYMER.

The system must not cure prior to application (since it would become intractable) and, therefore, must exhibit a period of latency adequate for storage, mixing, and application. Following application, curing may be induced by heat (thermosetting), light (ultraviolet curing), electron beams (EB curing), or exposure to an atmospheric component such as oxygen (air cure) or water (moisture cure).

The formation of polymer networks requires the utilization of multifunctional oligomers. This principle may be illustrated by the reaction of an isocyanate and an alcohol to produce a urethane [reaction (1)].

$$\mathrm{ROH + R'N{=}C{=}O \rightarrow R'NH{-}CO{-}OR} \qquad (1)$$

Reaction of a difunctional alcohol (diol) and diisocyanate yields a linear polyurethane [reaction (2)].

$$\mathrm{HO{-}R{-}OH + O{=}C{=}N{-}R'{-}N{=}C{=}O \rightarrow} \\ \mathrm{+O{-}R[{-}O{-}CO{-}NH{-}R'{-}NH{-}CO+_n} \qquad (2)$$

Formation of a polyurethane cross-linked network requires utilization of triols or triisocyanates, which provide a third functionality for network formation depicted graphically in reaction (3) with a triol.

$$\mathrm{HO{-}\underset{\displaystyle OH}{\underset{|}{R}}{-}OH + O{=}C{=}N{-}R'{-}N{=}C{=}O \rightarrow}$$

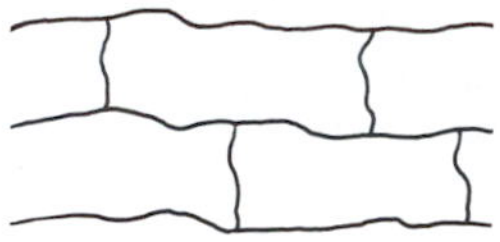

(3)

See ISOCYANATE; POLYURETHANE RESINS; URETHANE. [S.P.Pa.]

Curium A chemical element, Cm, in the actinide series, with an atomic number of 96. Curium does not exist in the terrestrial environment, but may be produced artificially. The chemical properties of curium are so similar to those of the typical rare earths that, if it were not for its radioactivity, it might easily be mistaken for one of these elements. The known isotopes of curium include mass numbers 238–250. The isotope ^{244}Cm is of particular interest because of its potential use as a compact, thermoelectric power source, through conversion to electrical power of the heat generated by nuclear decay.

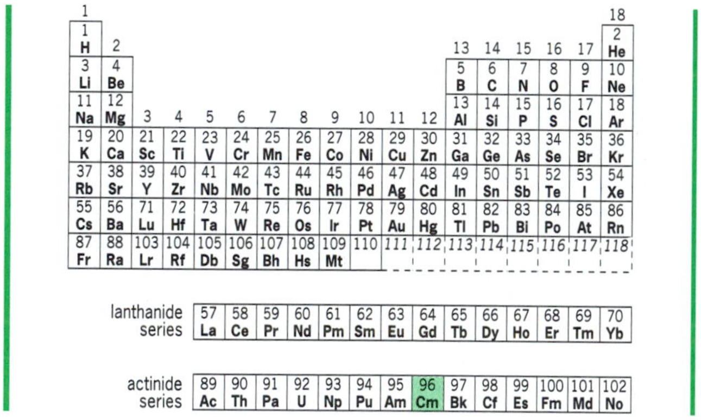

Metallic curium may be produced by the reduction of curium trifluoride with barium vapor. The metal has a silvery luster, tarnishes in air, and has a specific gravity of 13.5. The melting point has been determined as 2444 ± 72°F (1340 ± 40°C). The metal dissolves readily in common mineral acids with the formation of the tripositive ion.

A number of solid compounds of curium have been prepared and their structures determined by x-ray diffraction. These include CmF_4, CmF_3, $CmCl_3$, $CmBr_3$, CmI_3, Cm_2O_3, and CmO_2. Isostructural analogs of the compounds of curium are observed in the lanthanide series of elements. *See* ACTINIDE ELEMENTS; TRANSURANIUM ELEMENTS. [G.T.S.]

Currant A shrubby, deciduous plant of the genus *Ribes*, order Rosales, related to the gooseberry. The fruits are berries, borne in clusters. The black currant (*R. nigrum*), popular in Europe, is not cultivated in the United States. Red currants are derived from five species, mostly European. There are also white currants which are variants of the red type. Only the red varieties are produced commercially. The greatest commercial production occurs in New York, Washington, and Michigan, but production has been declining in recent years. *See* GOOSEBERRY; ROSALES. [J.H.Cl.]

Current balance An apparatus with which force is measured between current-carrying conductors (fixed and movable coils; see illustration). The purpose of the measurement is to assign the value of the ampere; therefore the relative locations of the conductors are prescribed so that the force between them can be computed from their geometry and the current they carry. Experimentally this force is measured in terms of the gravitational force on a known mass; thus the ampere may be said to be weighed. Three types of current balance have been used in ampere determinations. In two of these, the Rayleigh and the Ayrton-Jones balances, the fixed and movable coils are coaxial, and the motion is linear along the coil axis. In the third type, the Pellat balance, the axes of the fixed and movable coils are at right angles, and the movable coil turns about its center, the pivot point of the weighing balance. *See* CURRENT MEASUREMENT. [F.K.H.]

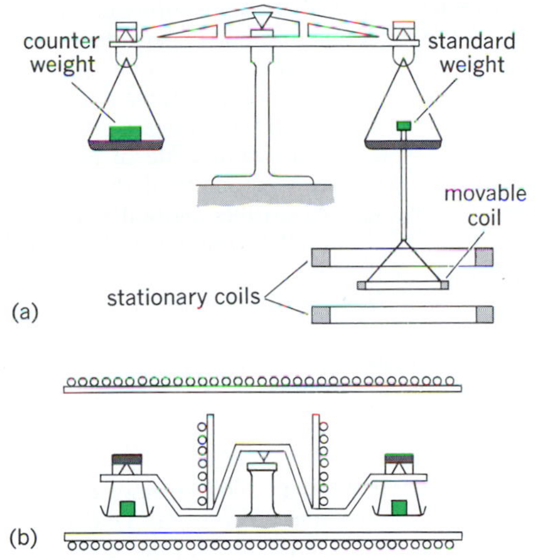

Current balances. (*a*) Rayleigh. (*b*) Pellat.

Current comparator An instrument for determining the ratio of two currents, based on Ampère's laws. Many electrical measurements are made by using voltage ratio transformers to compare the unknown to a standard. The current comparator provides the means for similar measurements to be performed with current ratio techniques. Moreover, the application of the current comparator, unlike the voltage ratio transformer, is not limited to alternating currents but can be used with direct currents as well. *See* INSTRUMENT TRANSFORMER.

The current comparator is based on Ampère's circuital law, which states that the integral of the magnetizing force around a closed path is equal to the sum of the currents which are linked with that path. Thus, if two currents are passed through a toroid by two windings of known numbers of turns, and the integral of the magnetizing force around the toroid is equal to zero, the current ratio is exactly equal to the inverse of the turns ratio. *See* AMPÈRE'S LAW; MAGNETIC FIELD.

For alternating-current comparators, the ampere-turn unbalance is given by the voltage at the terminals of a uniformly distributed detection winding on a magnetic core. Direct-current comparators use two magnetic cores which are modulated by alternating current in such a way that the dc ampere-turn unbalance is indicated by the presence of even harmonics in the voltage. However, neither method is an exact measure of the integral of the magnetizing force, and various design features are added to overcome this deficiency. The most important of these are the magnetic shields, which are configured as hollow toroids. *See* CURRENT MEASUREMENT; ELECTRICAL MEASUREMENTS. [W.J.M.M.; N.L.K.]

Current density A vector quantity equal in magnitude to the instantaneous rate of flow of electric charge, perpendicular to a chosen surface, per unit area per unit time. If a wire of cross-sectional area A carries a current I, the current density J is I/A. The units of J in the rationalized meter-kilogram-second system are amperes per square meter. [J.W.St.]

Current measurement The measurement of the flow of electric current. The current balance is tedious to operate, and is ordinarily used only by the national laboratories of the various countries for the determination of the absolute ampere, whereby simpler (but less accurate) current-measuring devices may be calibrated. In the practical determination of current value the current-carrying wires are formed into circular coils. The moving coil may move up or down and may rotate in jeweled bearings against the countertorque of a spiral spring. If a pointer is attached to the moving system rotating over an appropriately marked scale, calibrated in terms of the absolute current determinations of an ampere balance, the result is an indicating electrodynamometer ammeter. Such instruments are widely used as laboratory standards for both direct- and alternating-current measurements and for the calibration of secondary instruments. *See* CURRENT BALANCE.

Direct current may also be measured by placing the movable coil in the magnetic field of a permanent magnet. This is the construction of the conventional type of dc ammeter. *See* AMMETER.

For ordinary switchboard or test indications of alternating current, the fixed-coil, moving-iron type of instrument is widely used. A fixed iron vane and a similar vane on a staff are magnetized by the current in the coil, the force between the vanes acting to rotate the staff against a spring as a function of the rms value of the current.

In the thermoammeter a short metal strip or wire is heated by the current in question; its temperature rise is measured by a contacting thermocouple connected, in turn, to a sensitive dc meter which deflects in proportion to the temperature rise.

When direct currents of over a fraction of 1 ampere are to be measured, the bulk of the current is diverted around the instrument through a shunt or bypass circuit. Large alternating currents, on the other hand, are connected to meters rated at 5 A through current stepdown transformers, or current transformers. Extremely heavy direct currents, above 10^4 A, can be measured with a transductor, essentially an ac transformer with the iron core surrounding the dc bus. [J.H.M./J.B.]

Small direct currents in low-resistance circuits can be measured with a standard-resistance shunt and a potentiometer. In high-resistance circuits, vacuum-tube devices are now usually preferred to electrometers. Currents as low as 10^{-15} A can be measured with shunts of 10^{11} or 10^{12} ohms, and lower values can be measured by observing the rate at which the input capacitance is charged by the flow of current. *See* ELECTROMETER. [W.N.T.]

Current sources and mirrors A current source is an electronic circuit that generates a constant direct current which flows into or out of a high-impedance output node. A current mirror is an electronic circuit that generates a current which flows into or out of a high-impedance output node, which is a scaled replica of an input current, flowing into or out of a different node.

Most specifications of analog integrated circuits depend almost uniquely on the technological parameters of the devices, and on the direct or alternating current that flows through them. The voltage drop over the devices has much less impact on performance, as long as it keeps the devices in the appropriate mode of operation (linear or saturation). High-performance analog integrated-circuit signal processing requires that currents be generated and replicated (mirrored) accurately, independent of supply voltage and of those device parameters that are least predictable (such as current gain ß in a bipolar transistor). Hence, current sources and mirrors occupy a large portion of the total die area of any analog integrated circuit. They are also used, but less often, in discrete analog circuits. *See* INTEGRATED CIRCUITS; TRANSISTOR. [P.M.VanP.]

Curry A mixture of plant spices used with meats and fish and other seafoods, each requiring its own specific combination. One recipe calls for a mixture of turmeric, coriander, cinnamon, cumin, ginger, cardamon, cayenne pepper, pimiento, black pepper, long pepper, cloves, and nutmeg. [P.D.St./E.L.C.]

Curvature of field The optical aberration which occurs because the best image of a plane object sometimes lies on a curved surface. An image formed on a flat screen will then be subject to the error known as curvature of field. If R_κ designates the curvature of the κth optical surface, the quantity shown in the equation, where n'_κ and n_κ are the refractive indices before and after the κth surface, is called the Petzval curvature or Petzval sum of the system.

$$R = \sum R_\kappa(1/n'_\kappa - 1/n_\kappa)$$

When all the image-forming errors are corrected and the field is small, the Petzval curvature gives the axial curvature of the image of a plane object. The vanishing of R indicates that the Petzval condition is fulfilled.

The discovery of the significance of the Petzval sum enabled photographic lenses with plane fields to be constructed. These lenses were called anastigmats to distinguish them from aplanats, in which the meridional and sagittal fields were merely balanced against each other, one field having a positive curvature and the other a negative of nearly equal size. *See* ABERRATION (OPTICS); LENS (OPTICS). [M.J.H.]

Curve fitting A procedure in which the basic problem is to pass a curve through a set of points, representing experimental data, in such a way that the curve shows as well as possible the relationship between the two quantities plotted. It is always possible to pass some smooth curve through all the points plotted, but since there is assumed to be some experimental error present, such a procedure would ordinarily not be desirable. *See* INTERPOLATION.

The first task in curve fitting is to decide (1) how many degrees of freedom (number of unspecified parameters, or independent variables) should be allowed in fitting the points, and (2) what the general nature of the curve should be. Since there is no known way of answering these questions in a completely objective way, curve fitting remains something of an art. It is clear, however, that one must make good use of any background knowledge of the quantities plotted in order to answer the two questions. Thus, if one knew that a discontinuity might occur at some value of the abscissa, one would try to fit the points above and below that value by separate curves.

Against this background knowledge of what the curve should be expected to look like, one may observe the way the points fall on the paper. One may even find it advantageous to make a few rough attempts to draw a reasonable curve "through" the points.

A knowledge of the accuracy of the data is needed to help answer the question of the number of degrees of freedom to be permitted in fitting the data, If the data are very accurate, one may use as many degrees of freedom as there are points. The curve can then be made to pass through all the points, and it serves only the function of interpolation. At the opposite extreme when the data are very rough, one may attempt to fit the data by a straight line representing a linear relation, Eq. (1),

$$y = ax + b \tag{1}$$

between y and x. Using the above information and one's knowledge of the functions that have been found useful in fitting various types of experimental curves, one selects a suitable function and tries to determine the parameters left unspecified. At this point there are certain techniques that have been worked out to choose the optimum value of the parameters.

One of the most general methods used for this purpose is the method of least squares. In this method one chooses the parameters in such a way as to minimize the sum, Eq. (2),

$$S = \sum_{i=1}^{n} [y_i - f(x_i)]^2 \tag{2}$$

where y_i is the ordinate of ith point and $f(x_i)$ the ordinate of the point on the curve having the same abscissa x_i as this point. *See* LEAST-SQUARES METHOD. [K.S.K.]

Cutaneous sensation The sensory quality of skin. The skin consists of two main layers, the epidermis and the dermis. Sensory receptors in or beneath the skin are peripheral nerve-fiber endings that are differentially sensitive to one or more forms of energy. The sensory endings can be loosely categorized into three morphological groups: endings with expanded tips, such as Merkel's disks found at the base of the epidermis; encapsulated endings, such as Meissner's corpuscles (particularly plentiful in the dermal papillae), and other organs located in the dermis or subcutaneous tissue, such as Ruffini endings, Pacinian corpuscles, Golgi-Mazzoni corpuscles, and Krause's end bulb; and bare nerve endings that are found in all layers of the skin (some of these nerve endings are found near or around the base of hair follicles).

There is a remarkable relationship between the response specificities of cutaneous receptors and five primary qualities of cutaneous sensation, the latter commonly described as touch-pressure (mechanoreceptors), cold and warmth (thermoreceptors), pain, and itch. Each quality is served by a specific set of cutaneous peripheral nerve fibers. More complex sensations must result from an integration within the central nervous system of information from these sets of nerve fibers. Exploration of the skin surface with a rounded metal point reveals that there exist local sensory spots on the skin, stimulation of which evokes only one of the five qualities of sensation. Thus, there may be plotted maps of pressure, warm, cold, pain, or itch spots. *See* MECHANORECEPTORS; NOCICEPTORS; PAIN; SKIN; SOMESTHESIS; TEMPERATURE SENSES; TOUCH. [R.LaM.]

Cyanamide A term used to refer to the free acid hydrogen cyanamide, H_2NCN, or commonly to the calcium salt of this acid, calcium cyanamide, $CaCN_2$. Calcium cyanamide is manufactured by the cyanamide process, in which nitrogen gas is passed through finely divided calcium carbide at a temperature of 1832°F (1000°C). Most calcium cyanamide is used in agriculture as a fertilizer; some is used as a weed killer, as a pesticide, and as a cotton defoliant.

Hydrogen cyanamide is prepared from calcium cyanamide by treating the salt with acid. It is used in the manufacture of dicyandiamide and thiourea. [E.E.W.]

Cyanate A compound containing the —OCN group and derived from cyanic acid, HOCN. Cyanates are isomeric with fulminates which have the same atoms arranged —ONC.

Cyanic acid may exist in two forms, HO—C≡N and H—N=C=O, the latter of which is called isocyanic acid. The alkali and alkaline-earth metal cyanates contain an anion which may also be written in two forms: $[N{\equiv}C{-}O]^-$ and $[O{=}C{=}N]^-$. These salts are water-soluble.

The heavy metal cyanates are more covalent in nature and are water-insoluble. The main use of cyanates is in the synthesis of organic compounds.

Ammonium cyanate rearranges easily to urea: $NH_4OCN \rightleftharpoons NH_2CONH_2$. *See* CYANIDE; THIOCYANATE. [E.E.W.]

Cyanide A compound containing the —CN group, for example, potassium cyanide, KCN; calcium cyanide, $Ca(CN)_2$; and hydrocyanic (or prussic) acid, HCN. Chemically, the simple inorganic cyanides resemble chlorides in many ways. Organic compounds containing this group are called nitriles, for example, acrylonitrile, CH_2CHCN. *See* ACRYLONITRILE.

HCN is a weak acid. In the pure state, it is a highly volatile liquid, boiling at 26°C (78.8°F). HCN and the cyanides are highly toxic to animals and humans.

The cyanide ion forms a variety of coordination complexes with transition-metal ions, a property responsible for several of the commercial uses of cyanides. The cyanide process T is the most widely used method for extracting gold and silver from the ores. In silver-plating, a smooth adherent deposit is obtained on a metal cathode when electrolysis is carried out in the presence of an excess of cyanide ion. *See* ELECTROPLATING OF METALS; GOLD METALLURGY; SILVER METALLURGY.

$Ca(CN)_2$ is extensively used in pest control and as a fumigant in the storage of grain. In finely divided form, it reacts slowly with the moisture in the air to liberate HCN.

In case hardening of metals, an iron or steel article is immersed in a bath of molten sodium or potassium cyanide containing sodium chloride or carbonate. The cyanide decomposes at the surface, forming a deposit of carbon which combines with and penetrates the metal. *See* COORDINATION CHEMISTRY. [F.J.J.]

Cyanobacteria A group of one-celled to many-celled aquatic organisms formerly known as blue-green algae. Superficially they appear similar to many of the smaller algae (for example, green algae), but since the mid-1950s it has become evident to many scientists that their cellular structure, including lack of bounded nucleus and chloroplasts, is comparable to that of bacteria (for example, photosynthetic bacteria) rather than algae and more complex plants. *See* ALGAE.

Cyanobacteria are photosynthetic and produce molecular oxygen (O_2) in the process, although some can shift to a nonoxygen-producing type of photosynthesis when exposed to hydrogen sulfide. The cellular pigments include chlorophyll *a* (green), ß-carotene and xanthophylis (orange to yellow), c-phycocyanin (blue), and allophycocyanin (blue-green). Phycoerythrin (red) also occurs. *See* CAROTENOID; CHLOROPHYLL.

The diameter of individual cells varies among species from less than 1 micrometer to about 60 micrometers. The genetic material consists of deoxyribonucleic acid (DNA) arranged as a continuous double-stranded thread that is greatly folded. The unfolded molecule forms a circle as in other prokaryotic organisms. *See* DEOXYRIBONUCLEIC ACID (DNA).

The pigments are located in and attached to internal membrane foldings called thylakoids. These are not packaged into membrane-bound chloroplasts, the universal unit in all true algae and complex plants. *See* CELL PLASTIDS.

Cells have a rigid wall and reproduce by a simple division called binary fission, giving rise to two daughter cells. Daughter cells may separate or adhere (even after several divisions), giving rise to aggregates of cells called colonies or to filaments which occur when cell divisions proceed in one plane of division only. A few species may reproduce when the large parent cell cleaves internally into many small cells (baeocytes) which are eventually released. Usually all the cells of a colony or filament have the potential for division. In some species the filaments taper, however, and divisions are restricted to a certain region, mainly the thicker basal area. No sexual conjugation of cells is known, but some genes may be transferred from one cell to another, perhaps by specific viruses (cyanophages).

Cyanobacteria occur almost anywhere that there is at least dim light. Cyanobacteria occur as marine and fresh-water plankton, often forming blooms (dense populations) in tropical seas and in tropical and temperate lakes, particularly eutrophic lakes. Other species are attached to rocks, sediment, and plants in marine and fresh water. A few penetrate below the crust of aquatic rocks; some are symbiotic partners in association with cells of terrestrial fungi (for example, many lichens) or with terrestrial or aquatic vascular green plants (*Gunnera*, cycads, and the aquatic fern *Azolla*) and nonvascular plants and algae (such as liverworts, hornworts, and diatoms). In the main, the symbiotic cyanobacteria are those with the ability to fix nitrogen from the atmosphere or water, thus providing the host organism with a source of nitrogen to draw on which is directly available only to the cyanobacterial partner. *See* NITROGEN FIXATION.

Many free-living cyanobacteria also live in terrestrial habitats, usually on damp wood, soil, or rocks, but others thrive during the short wet periods on desert floors and within the crusts of porous desert rocks. Hot springs are inhabited by several species of cyanobacteria above a temperature of 113°F (45°C), with one species occurring as high as 165°F (74°C). Eukaryotic algae (that is, those with a nucleus) are very rare above 113°F (45°C). The photosynthetic organisms of hypersaline lagoons and lakes are also often predominantly cyanobacteria or photosynthetic bacteria of other types.

A major branch of filamentous cyanobacteria produce special cells called heterocysts. They develop from normal photosynthetic cells when environmental supplies of fixed nitrogen (ammonium and nitrate) are depleted. The heterocyst becomes a thick-walled cell that is converted to a "factory" for fixing gaseous nitrogen into cell material (eventually protein). The interior of the cell is maintained as an anaerobic site for this process. Another common specialized cell of some cyanobacteria is the thick-walled resting spore (akinete) which is more resistant to desiccation and extreme temperatures than the active cells. *See* BACTERIA; MONERA. [R.W.Ca.]

Cyanocarbon A derivative of hydrocarbon in which all of the hydrogen atoms are replaced by the —C≡N group. However, the term cyanocarbon has been applied to compounds which do not strictly follow the above definition: tetracyanoethylene oxide, tetracyanothiophene, tetracyanofuran, tetracyanopyrrole, tetracyanobenzoquinone, tetracyanoquinodimethane, tetracyanodithiin, pentacyanopyridine, diazomalononitrile, and diazotetracyanocyclopentadiene.

Tetracyanoethylene (with the formula below), the simplest

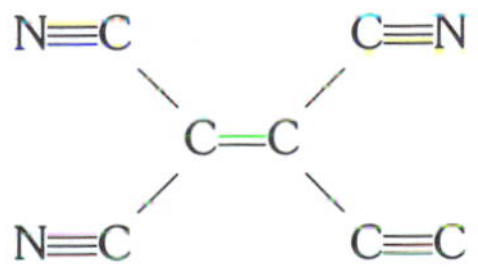

olefinic cyanocarbon, is a colorless, thermally stable solid, having a melting range of 198–200°C (388–392°F), and readily forming stable complexes with most aromatic systems. Cyanocarbon acids are among the strongest protonic organic acids known and are usually isolated only as the cyanocarbon anion salt. [D.W.W.]

Cyanoethylation A chemical reaction involving the addition of acrylonitrile ($CH_2{=}CH{-}CN$) to compounds carrying a reactive hydrogen, to introduce the ß-cyanoethyl grouping ($-CH_2CH_2CN$). The reaction is represented by

$$ZH + CH_2{=}CH{-}CN \rightleftharpoons ZCH_2CH_2CN$$

where Z is CN or a C=O group.

Cyanoethylation can be effected with HBr, HCl, HCN, and with compounds containing the functional groups

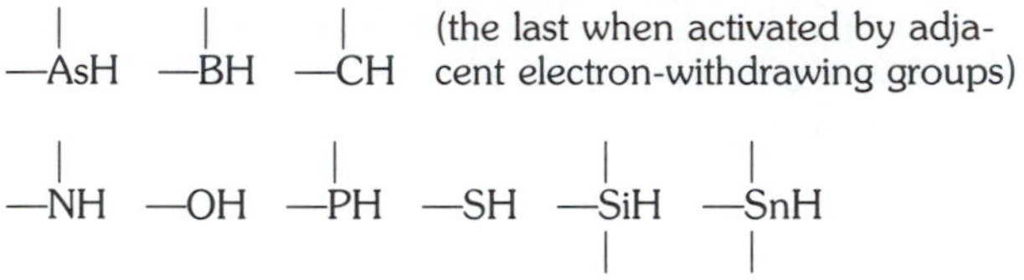

Because of its versatility, cyanoethylation has assumed great importance in synthesis, as shown by the fact that over 1200 compounds have been used as reactants. Introduction of the cyanoethyl group tends to increase hydrophobic character and resistance to biological attack. Natural products, notably cellulose, have been cyanoethylated to take advantage of these properties.

The nature of the catalyst is used to classify substrates into three groups:

(1) No catalyst:

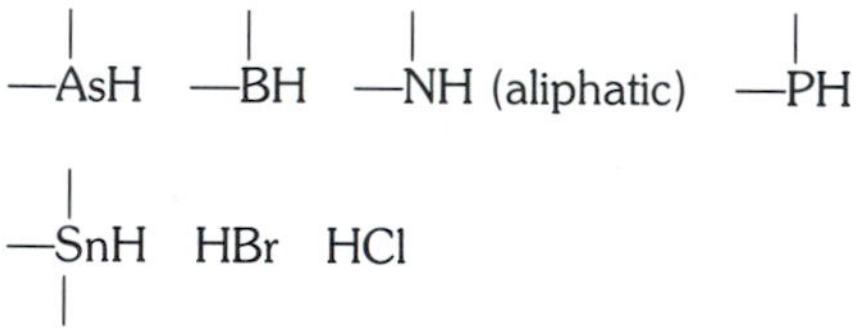

(2) Basic catalyst (for example, NaOH, $NaOCH_3$, benzyltrimethylammonium hydroxide):

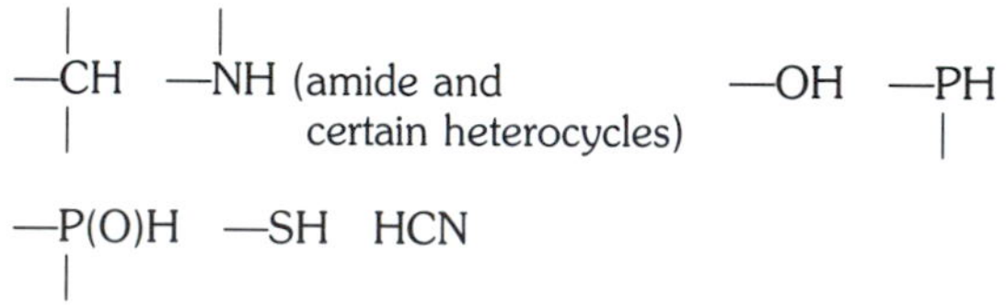

(3) Acid catalyst (for example, acetic acid with or without a copper salt):

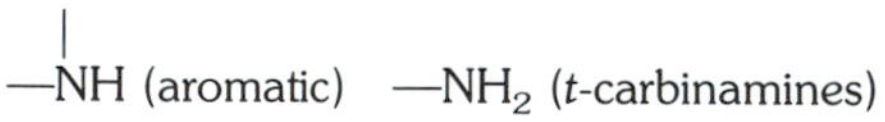

Yields in cyanoethylation are generally high. The reaction is strongly exothermic and is usually carried out at moderate temperatures in a solvent such as dioxane or *t*-butanol.

Since the extent of cyanoethylation can be varied by choice of conditions, a variety of products can be obtained from a given material. These may be further modified by reactions of the cyanoethyl group, for example, to give polyelectrolytes.

Partially cyanoethylated cellulose, in which one-sixth of the hydroxyl groups have reacted, is rot- and termite-proof; it has improved resistance to degradation by heat and by acids, altered dyeing characteristics and greater hydrophobicity. Utility is in marine nets and ropes, electrical insulation, industrial fabrics, structural wood, and dimensionally stable paper.

Cyanoethylation is also used in the preparation of amino acids, chemotherapeutic agents, dyes, blowing agents, long-chain diamines, special-purpose silicones, food stabilizers, and liquids to effect separations by gas chromatography. *See* ACRYLONITRILE; TEXTILE CHEMISTRY. [N.M.B.]

Cyanogen A colorless, highly toxic gas having the molecular formula C_2N_2. Structurally, cyanogen is written $N{\equiv}C{-}C{\equiv}N$. Cyanogen belongs to a class of compounds known as pseudohalogens, because of the similarity of their chemical behavior to that of the halogens. Liquid cyanogen boils at −21.17°C (−6.11°F) and freezes at −27.9°C (−18.2°F) at 1 atm (101.325 kilopascals).

Cyanogen reacts with hydrogen at elevated temperatures in a manner analogous to the halogens, forming hydrogen cyanide, HCN. With hydrogen sulfide, H_2S, cyanogen forms thiocyanoformamide or dithiooxamide. Cyanogen burns in oxygen, producing one of the hottest flames known from a chemical reaction. It is considered to be a promising component of high-energy fuels. *See* CYANIDE. [F.J.J.]

Cyanophyceae The single class of the division Cyanophycota. Sometimes also known as Cyanophyta, Myxophyceae, or Schizophyceae, they are commonly called blue-green algae. There are about 1500 species of blue-green algae, which are arranged into orders, the Chroococcales, Pleurocapsales, Stigonematales, and Nostocales.

Blue-green algae occur free-floating in fresh and salt water and in moist subaerial habitats throughout the world. They are more abundant in fresh than in salt water, and more abundant near the surface than at depths of more than a few feet. Spray-dampened cliffs, dripping ledges, and temporary pools are likely to be especially rich in blue-green algae. They are also commonly present in and about hot springs, sometimes at temperatures not much below boiling.

Blue-green algae occur as single cells of microscopic size, as small cell colonies of various shapes, and as multicellular filaments which may form a tangled mat or a gelatinous ball. The individual cell consists of a prokaryotic protoplast surrounded by a cell wall, which in turn is usually covered by a gelatinous sheath composed of pectic compounds. The cell wall, like that of bacteria, consists chiefly of mucopolymers composed of glucosamine, amino acids, and muramic acid. Some genera also have cellulose in the cell wall, but the cellulose is not usually organized into the sort of fibrils found in other algae. Blue-green algae are completely without flagella or other obvious means of locomotion, but some of the filamentous ones move nevertheless. The whole filament may move slowly forward and backward, or the end of the filament may wave slowly back and forth, as in *Oscillatoria*. The mechanism of these movements is not yet clearly understood. No sexual processes are known in any blue-green algae.

The blue-green algae resemble the bacteria in the prokaryotic organization of their protoplasm. In general, prokaryotic organisms do not have a highly organized, vesicular nucleus in the cell, nor do they have complex cytoplasmic organelles such as mitochondria, dictyosomes, and membrane-bounded chloroplasts. Like other algae, the blue-green algae carry on photosynthesis and release oxygen as a by-product. In this respect they differ sharply from the bacteria. Those few bacteria which are photosynthetic carry on a more primitive type of photosynthesis which does not result in the release of oxygen. Phycobilin pigments play an essential (though accessory) role in the photosynthesis of blue-green algae. *See* PHYCOBILIN.

All groups of algae are ultimate sources of food for fish and other aquatic animals, but the blue-green algae are less important than groups such as the diatoms and green algae. Like the eukaryotic photosynthetic organisms, the blue-green algae play an integral role in the balance of nature by using up carbon dioxide and contributing to the supply of atmospheric oxygen, which is necessary for human and other animal life. Some blue-green algae, like some bacteria, can also use atmospheric nitrogen in the formation of complex organic compounds, thus playing a critical role in the nitrogen cycle. Photosynthetic activity of many blue-green algae in water containing dissolved bicarbonates tends to cause the deposition of calcium carbonate on and about the cells, and this action by blue-green and other algae is believed to have been an important factor in the formation of many or most limestone deposits. Blue-green algae sometimes become so abundant in reservoirs and lakes as to color the water and make it offensive to the smell and taste, especially in water that has in effect been fertilized by inflow of detergents or sewage. *See* ALGAE; LIMESTONE; NITROGEN CYCLE; PHOTOSYNTHESIS; PHYTOPLANKTON. [A.Cr.]

Cycadales An ancient order of the plant division Pinophyta (gymnosperms), having a few living representatives (living fossils). These plants are popularly but incorrectly called palms. The tuberous or columnar stems bear a crown of large, usually pinnate leaves. *Cycas revoluta*, which is the best-known cycad, is called sago palm (see illustration). *See* PALM.

Leaves of the Sago palm (*Cycas revoluta*). (*New York Botanical Garden*)

The cycads appeared, along with the Cycadeoidales (or Bennettitales), in the Upper Carboniferous and made their greatest display in the Cretaceous. In the Mesozoic the distribution was worldwide, but the living members of the order are confined to the tropical and subtropical regions, where they occur in scanty patches. Today, there is but 1 family with 9 genera and approximately 100 species. *See* CYCADEOIDALES; PINOPHYTA; PTERIODOSPERMAE. [P.D.St./A.Cr.]

Cycadeoidales An order of extinct plants that formed a conspicuous part of the landscape during the Triassic, Jurassic, and Cretaceous periods. The Cycadeoidales (or Bennettitales) had unbranched or sparsely branched trunks with a terminal crown of leaves (family Cycadeoidaceae; see illustration), or they were branched, at times profusely (family Williamsoniaceae).

No known relatives of the Cycadeoidales exist at the present time, although members of the order Cycadales show some resemblances to the extinct group. *See* CYCADALES; EMBRYOBIONTA. [T.D.]

Reconstruction of a plant of the genus *Cycadeoidea*, showing terminal crown of leaves, persistent leaf bases, and positions of the fruiting structures. (*Courtesy of T. Delevoryas*)

Cycadicae A subdivision of the plant division Pinophyta (gymnosperms), consisting of three classes, the Lyginopteridopsida (pteridosperms or seed ferns), Cycadopsida (cycads), and Bennettitopsida (cycadeoids). Only the Cycadopsida, with the single small order Cycadales, has any living species; the rest are represented by fossils chiefy from the Mesozoic and late Paleozoic eras. *See* PINOPHYTA.

The Cycadicae are gymnosperms with relatively large leaves that are often pinnately compound. The stems are stout and usually unbranched or with only a few relatively large branches. The seeds are borne on ordinary leaves (in many of the seed ferns) or on more or less strongly modified leaves (called megasporophylls) that are often aggregated into a simple strobilus (cone) at the end of the stem. *See* CYCADALES; CYCADEOIDALES; PTERIDOSPERMAE. [A.Cr.]

Cycadopsida A class of the division Pinophyta (gymnosperms), consisting of six orders: Lagenostomales (=Lyginopteridales, pteridosperms or seed ferns), Trigonocarpales (=Medullosales), Cycadales (cycads), Bennettitales (cycadeoids or bennettites), Gnetales (gnetum), and Welwitschiales (welwitschia). The Cycadales, Gnetales, and Welwitschiales are the only orders with living species.

The Cycadopsida are gymnosperms with radially symmetrical seeds and bilateral or radial cupules. Stems are simple (unbranched) to pinnate or dichotomous-branched. Leaves are usually compound (except in Gnetales and Welwitschiales). The seeds are borne on leaves (seed ferns), or on modified leaves (megasporophylls) that may be aggregated into a simple strobilus (cone) at the end of the stem or in clusters of strobili (Welwitschiales) or on linear axes (Gnetales). *See* CYCADALES; GNETALES; PINOPHYTA; PLANT KINGDOM; WELWITSCHIALES. [T.A.Z.]

Cyclanthales An order of flowering plants, division Magnoliophyta (Angiospermae), sometimes called Synanthae or Synanthales, in the subclass Arecidae of the class Liliopsida (monocotyledons). The order consists of the single tropical American family Cyclanthaceae, with about 180 species. They are herbs or, seldom, more or less woody plants, with characteristic leaves that have a basal sheath, a petiole, and an expanded, usually bifid (cleft into two equal parts) blade which is often folded lengthwise (plicate) like that of a palm leaf. One of the species, *Carludovicia palmata*, is a principal source of fiber for panama hats. *See* ARECIDAE. [A.Cr.]

Cyclic adenylic acid A nucleotide, also known as adenosine 3′,5′-monophosphate or cyclic AMP, with the structure shown. Widely distributed in nature, cyclic AMP may be

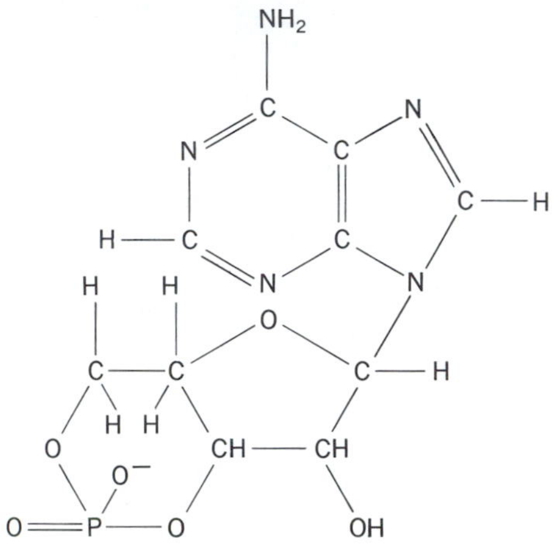

regarded as a versatile regulatory agent which acts to control the rate of a number of cellular processes (see table). It has become especially important in endocrinology because of the evidence that many hormones and neurohormones produce a number of their effects by changing the concentration of cyclic AMP within the cells of their target tissues.

Cyclic AMP and the enzymes that produce and destroy it have been found in all nucleated animal cells that have been studied, as well as in bacteria and other unicellular organisms. The regulatory function of cyclic AMP seems to have emerged at a very early stage in the evolutionary process. However, cyclic AMP appears not to exist in green plants. A second cyclic nucleotide, cyclic guanylic acid, has been found to exist in most animal cells, but less is known about it than about cyclic AMP. Evidence suggests that both nucleotides may play important roles during growth and development and may also be involved in regulating the immune response.

Cellular processes known to be influenced by cyclic AMP

Enzyme or process affected	Changes in activity or rate
Phosphorylase	Increased
Glycogen synthetase	Decreased
Phosphofructokinase	Increased
Fructose 1,6-diphosphatase	Decreased
Steroidogenesis	Increased
Lipolysis	Increased
Gluconeogenesis	Increased
Ketogenesis	Increased
Amino acids → liver proteins	Decreased
Release of protein from ribosomes	Increased
Acetate → liver cholesterol	Decreased
Release of amylase (salivary glands)	Increased
Release of insulin (pancreas)	Increased
HCl secretion (gastric mucosa)	Increased
Permeability (kidney)	Increased
Tone of smooth muscle	Decreased
Pigmentation (amphibia and reptiles)	Increased
Cell aggregation (cellular slime mold)	Increased

Defects in the formation or action of cyclic AMP have been implicated in the etiology and pathogenesis of a number of disease processes, as for example, bronchial asthma and certain other allergic disorders. It is also known that abnormally high levels of cyclic AMP may be dangerous. For example, many of the metabolic abnormalities seen in diabetic patients are the result of an excess of glucagon over insulin, leading to excessive cyclic AMP formation in the liver and adipose tissue. *See* ADENYLIC ACID; CYCLIC NUCLEOTIDES; HORMONE. [G.A.R.]

Cyclic nucleotides Mononucleotides in which the phosphate moiety is attached to both the third and the fifth carbon atoms of the ribose component of the molecule, thus forming a cyclic diester. They are more correctly referred to as 3′,5′-mononucleotides. The only cyclic nucleotides which have been unequivocally shown to exist in living tissues are 3′,5′-adenylic acid and 3′,5′-guanylic acid, also known as cyclic AMP and cyclic GMP, respectively. They are important because they influence a large variety of essential biological processes, and sometimes the two nucleotides may act in opposition to each other. A third cyclic nucleotide, 3′,5′-cytidylic acid (cyclic CMP), has been reported to exist in some cells, but its role is poorly understood. *See* ADENYLIC ACID; CYCLIC ADENYLIC ACID; NUCLEIC ACID. [G.A.R.]

Cycloamylose Any of a group of cyclic oligomers of glucose in the normal C-1 conformation in which the individual glucose units are connected by 1,4 bonds. They are called Schardinger dextrins (after the discoverer), cyclodextrins (to emphasize their cyclic character), or cycloamyloses (to emphasize both their cyclic character and their amylose origin). The most common ones are cyclohexaamylose (which has six glucose units in a cyclic array) and cycloheptaamylose (which has seven glucose units in a cyclic array). The toroidal shape of these molecules has been determined by x-ray analysis (see illustration). The molecular weights are around 1000.

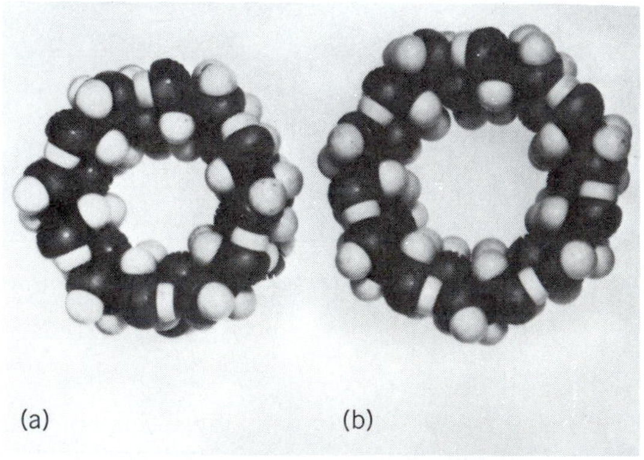

Molecular models of cycloamyloses viewed from the secondary hydroxyl side of the torus. (*a*) Cyclohexaamylose. (*b*) Cycloheptaamylose. (*From M. L. Bender and M. Komiyama, Cyclodextrin Chemistry, Springer, 1978*)

In water or in a mixture of water and dimethyl sulfoxide, cycloamyloses form 1:1 complexes with many organic molecules and most benzenoid derivatives. X-ray and spectroscopic evidence indicates that the guest molecule (usually a small organic molecule) is bound tightly in the cavity of the host molecule (the cycloamylose) just like a key fitting into a lock. The result is an inclusion complex, which is what is formed by enzymes when they first bind the molecules whose reactions they subsequently catalyze. Inclusion complex formation is probably the most important similarity between cycloamyloses and enzymes.

Cycloamyloses have been used to mimic many enzymes, such as ribonuclease, transaminase, and carbonic anhydrase. They have also been used to effect selectivity in several chemical reactions. Cycloamyloses are the premier example of chemical compounds that act like enzymes in inclusion complex formation followed by reaction. *See* CATALYSIS; CLATHRATE COMPOUNDS; ENZYME; STEREOSPECIFIC CATALYST. [M.L.Be.]

Cyclocystoidea A class of small, disk-shaped echinozoans in which the lower surface of the body probably consisted of a suction cap for adhering to substrate, and the upper surface was covered by separate plates arranged in concentric rings. The mouth lay at the center of the upper surface, and the anus, also on the upper surface, lay some distance between the margin and the mouth. Little is known of the habits of cyclocystoids. They occur in rocks of middle Ordovician to middle Devonian age in Europe and North America. *See* ECHINODERMATA; ECHINOZOA. [H.B.F.]

Cycloid A curve traced in the plane by a point on a circle that rolls, without slipping, on a line. If the line is the x axis of a rectangular coordinate system, at whose origin O the moving point P touches the axis, parametric equations of the cycloid are $x = a(\theta - \sin\theta)$, $y = a(1 - \cos\theta)$, when a is the radius of the rolling circle, and the parameter θ is the variable angle through which the circle rolls (see illustration). *See* ANALYTIC GEOMETRY. [L.M.Bl.]

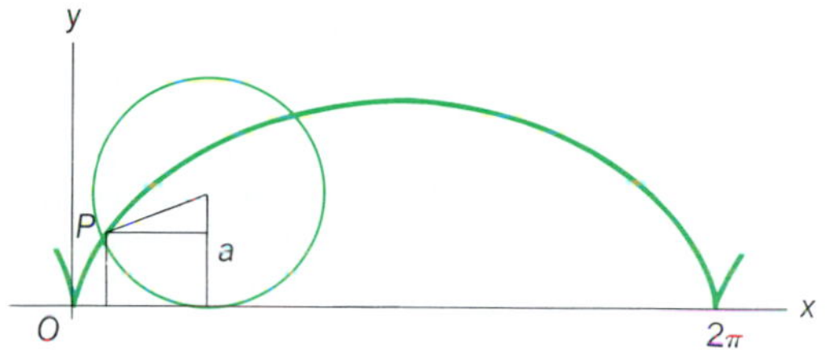

Diagram of a cycloid.

Cyclone An atmospheric circulation system in which the sense of rotation of the wind about the local vertical is the same as that of the Earth's rotation. Thus, a cyclone rotates clockwise in the Southern Hemisphere and counterclockwise in the Northern Hemisphere. In meteorology the term cyclone is reserved for circulation systems with horizontal dimensions of hundreds (tropical cyclones) or thousands (extratropical cyclones) of kilometers. For such systems the Coriolis force due to the Earth's rotation, which is directed to the right of the flow in the Northern Hemisphere, and the pressure gradient force, which is directed toward low pressure, are in opposite directions. Thus, there must be a pressure minimum at the center of the cyclone, and cyclones are sometimes simply called lows. *See* AIR PRESSURE; CORIOLIS ACCELERATION.

Extratropical cyclones are the common weather disturbances which travel around the world from west to east in mid-latitudes. They are generally associated with fronts, which are zones of rapid transition in temperature. Extratropical cyclones arise due to the hydrodynamic instability of the upper-level jet stream flow. *See* FRONT; JET STREAM.

Tropical cyclones, by contrast, derive their energy from the release of latent heat of condensation in precipitating cumulus clouds. Over the tropical oceans, where moisture is plentiful, tropical cyclones can develop into intense vortical storms (hurricanes and typhoons), which can have wind speeds in excess of 200 mi/h (100 $m \cdot s^{-1}$). *See* HURRICANE; PRECIPITATION (METEOROLOGY); STORM; WIND. [J.R.H.]

Cyclone furnace A water-cooled horizontal cylinder in which fuel (coal, gas, or oil) is fired and heat is released at extremely high rates. When firing coal, the crushed coal is introduced tangentially into the burner at the front end of the cyclone (see illustration). About 15% of the combustion air is used as primary and tertiary air to impart a whirling motion to the particles of coal. The whirling, or centrifugal, action on the fuel is further increased by the tangential admission of high-velocity secondary air into the cyclone.

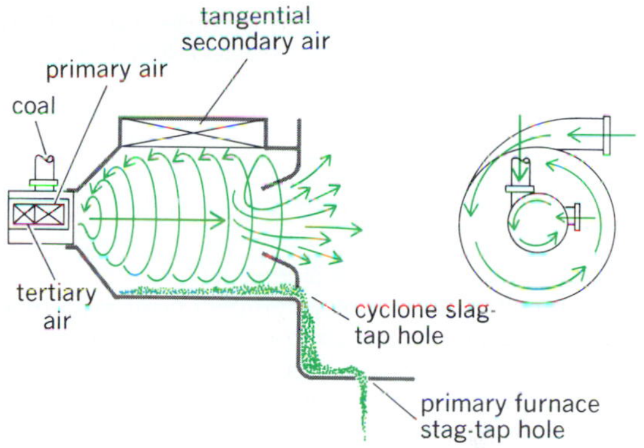

Schematic diagram of cyclone furnace. (*From T. Baumeister, ed., Standard Handbook for Mechanical Engineers, 8th ed., McGraw-Hill, 1978*)

The products of combustion are discharged through a water-cooled reentrant throat at the rear of the cyclone into the boiler furnace. Essentially, the fundamental difference between cyclone furnaces and pulverized coal–fired furnaces is the manner in which combustion takes place. In pulverized coal–fired furnaces, particles of coal move along with the gas stream; consequently, relatively large furnaces are required to complete the combustion of the suspended fuel. With cyclonic firing, the coal is held in the cyclone and the air is passed over the fuel. Thus, large quantities of fuel can be fired and combustion completed in a relatively small volume, and the boiler furnace is used to cool the products of combustion. *See* BOILER; STEAM-GENERATING FURNACE; STEAM-GENERATING UNIT. [G.W.K.]

Cyclophane A molecule composed of two building blocks: an aromatic ring (most frequently a benzene ring) and an aliphatic unit forming a bridge between two (or more) positions of the aromatic ring. Those cyclophanes that contain heteroatoms in the aromatic part are generally called phanes, as "cyclo" in cyclophanes more strictly means that only benzene rings are present in addition to aliphatic bridges. Phane molecules containing hetero atoms in the aromatic ring (for example, nitrogen or sulfur) are called heterophanes, while in heteraphanes they are part of the aliphatic bridge.

Cyclophanes usually are not at all planar molecules; they exhibit an interesting stereochemistry (arrangement of atoms in three-dimensional space); molecular parts are placed and sometimes fixed in unusual orientations toward each other, and often the molecules are not rigid but (conformationally) flexible. Ring strain and, as a consequence, deformation of the benzene rings out of planarity are often encountered. Electronic interactions between aromatic rings fixed face to face can take place. In addition, influence on substitution reactions in the aromatic rings results; that is, substitutents in one ring induce transannular electronic effects in the other ring, often leading to unexpected products. *See* CHEMICAL BONDING; CONFORMATIONAL ANALYSIS; STEREOCHEMISTRY.

Cyclophanes have become important in host-guest chemistry and supramolecular chemistry as they constitute host compounds for guest particles because of their cavities. Recognition processes at the molecular level are understood to mean the ability to design host molecules to encompass or attach selectively to smaller, sterically complementary guests in solution—by analogy to biological receptors and enzymes. For this reason, water-soluble cyclophanes were synthesized; they have the ability to include nonpolar guest molecules in aqueous solution. *See* CAGE HYDROCARBON; MOLECULAR RECOGNITION; POLAR MOLECULE.

Cyclophanes also exist in nature as alkaloids, cytotoxic agents, antibiotics, and many cyclopeptides. With the possibility of locating groups precisely in space, cyclophane chemistry has provided building units for molecular niches, nests, hollow cavities, multifloor structures, helices, macropolycycles, macrohollow tubes, novel ligand systems, and so on. Cyclophane chemistry has become a major component of supramolecular chemistry, molecular recognition, models for intercalation, building blocks for organic catalysts, receptor models, and variation of crown ethers and cryptands. *See* MACROCYCLIC COMPOUND; SUPRAMOLECULAR CHEMISTRY. [F.Vo.; M.Boh.]

Cyclophyllidea The order which includes most tapeworms that inhabit the gut of warm-blooded vertebrates. They are frequently referred to as the Taenioidea. Each worm has a head, or scolex, and a segmented body, called the strobila. The scolex typically has an apical rostellum, or muscular pad and hooks, and two pairs of lateral suckers. New segments are produced immediately posterior to the scolex, so that the oldest segment is at the hind end. As a segment is pushed further from the scolex, the male and female reproductive organs mature and open on the lateral margin. Ova, fertilized by sperm from the same or another segment, gather in a uterus. The ripe terminal segments containing infectious eggs are shed into gut contents of the host.

A number of tapeworms occur in humans and domestic animals, either as adults or immature stages, the metacestodes, but rarely in both stages. *See* COENUROSIS; CYSTICERCOSIS; ECHINOCOCCOSIS; PLATYHELMINTHES; TAPEWORM DISEASE. [R.S.F.]

Cyclopoida An order of small copepod Crustacea. The abundant fresh-water and salt-water species form an important link in the food chains of aquatic life, consuming tiny plants, animals, and detritus and, in turn, furnishing food for small fish, some large fish, and other organisms. Some species are important intermediate hosts for human parasites.

The front part of the body is oval and sharply separated from the tubular hind end, which bears two caudal rami with distinctly unequal setae (see illustration). Usually 10 body segments are present in the male, and 9 in the female because of fusion of two to form a genital segment.

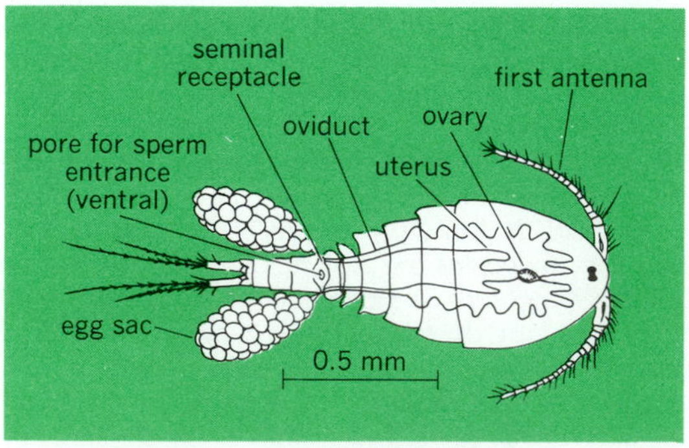

***Cyclops vernalis*, female, dorsal view.**

Cyclopoids are intermediate hosts of the parasitic guinea worm (*Drucunculus medinensis*), and sometimes the fish tapeworm (*Dibothriocephalus latus*). Most fresh-water species live in shallow water, swimming from plant to plant, but salt-water species are generally water-treaders. Food is not secured by filtration, but is seized and eaten with the biting mouthparts. Many species have a world wide distribution. *See* COPEPODA. [H.C.Y.]

Cycloserine A broad-spectrum antibiotic produced during the growth of certain species of *Streptomyces*. It is effective in humans in the treatment of tuberculosis, including cases resistant to other drugs such as streptomycin, isoniazid, and *p*-aminosalicylic acid. It is also effective in urinary tract infections caused by gram-negative bacteria which are often resistant to other antibiotics. *See* PARA-AMINOSALICYLIC ACID; STREPTOMYCIN; TUBERCULOSIS. [R.L.H.]

Cyclostomata (Bryozoa) An order of bryozoans in the class Stenolaemata. Cyclostomes tend to have delicate colonies composed of relatively isolated, loosely bundled, or tightly packed, comparatively simple, long, slender, and tubular zooecia, with thin, highly porous, calcareous walls. Cyclostome colonies most commonly are small and delicate, but some are moderately large and massive. They may be inconspicuous encrusting threadlike networks, thin encrusting sheets, nodular masses, or erect tufflike, twiglike, or frondlike growths. Individual cyclostome zooecia are straight to slightly curved long tubes. The walls of adjacent zooecia may be either distinctly separate or fused together.

Exclusively marine, cyclostomes (occasionally termed Stenostomata) are known first from the Late Ordovician, when they may possibly have evolved from cystoporates (which were formerly included in the Cyclostomta). Rare and inconspicuous throughout the rest of the Paleozoic and Triassic, cyclostomes became moderately common in Jurassic and Cretaceous seas, but have since (including today) been less numerous. *See* BRYOZOA; CYSTOPORATA; STENOLAEMATA. [R.J.Cu.]

Cyclostomata (Chordata) The simplest and most primitive of living vertebrates. Cyclostomes, also known as Cyclostomi and Marsipobranchii, are the modern representatives of the class Agnatha. The Cyclostomata constitute a subclass that is descended at least in part (lampreys—Petromyzonida or Petromyzontiformes) from the subclass Monorhina (Cephalaspidomorphi).

Cyclostomes differ notably from all other Recent fishes. The absence of jaws and the presence of a single median nostril are distinctive. In addition, they have an uncalcified cartilaginous skeleton and lack true teeth and paired fins and girdles.

Cyclostomes are eellike in form and have tough, more or less slimy skin. They are oviparous and may be either bisexual or hermaphroditic. Development may involve a protracted larval stage (ammocoetes) followed by profound structural change at metamorphosis, as in lampreys, or it may be direct, as in hagfishes. *See* AGNATHA; PETROMYZONIDA. [R.M.B.]

Cyclothem A specific sequence of different kinds of rocks, one of which is generally coal. These sequences, each on the order of tens of feet in thickness, occur repeatedly, one above the other, in coal-bearing strata. Cyclothems have been recognized mainly in rocks of late Paleozoic age in the eastern United States but also have been described in late Mesozoic coal-bearing strata of the Rocky Mountain region.

Because there has been no real agreement about factual knowledge concerning cyclothems, there is even less about their origin. Some believe that the eroded surface represents an episode of uplift of the land, and the succeeding deposits a period of subsidence, resulting first in nonmarine deposition and then in marine deposition. Others believe that the land surface may have remained stable but that the sea level rose and fell in response to the growth and melting of continental ice sheets in the Southern Hemisphere. Some support for this notion is found in the late Paleozoic glacial deposits recognized in portions of South Africa and Australia. At the time when cyclothems were being formed, these now isolated areas are believed to have been joined in a single continent designated Gondwana. *See* STRATIGRAPHY. [J.C.F.]

Cydippida The largest order of the phylum Ctenophora, comprising 5 families (Bathyctenidae, Haeckelidae, Lampeidae, Mertensidae, and Pleurobrachiidae) and 11 genera. The body of cydippids is usually globular or cylindrical in shape; sizes range from a few millimeters to about 15 cm (6 in.). Most cydippids are colorless and transparent, but some deep-sea species are pigmented dark red. Adult cydippids retain the larval morphology common to all ctenophores (except Beroida) and are usually thought to be the most primitive order. All forms have two main tentacles. Cydippids catch prey on the outstretched tentacles, contracting them to bring it to the mouth. The well-known genus *Pleurobrachia* is widely distributed in temperate coastal waters and estuaries, where it can be a major predator of copepods and larval fish. Many other species of cydippids occur in oceanic waters, from the surface to mesopelagic depths. *See* BEROIDA; CTENOPHORA. [L.P.M.]

Cygnus The Swan, in astronomy, is a conspicuous northern summer constellation. The five major stars of the group, α, γ, ß, ϵ, and δ, are arranged in the form of a cross (see illustration). Hence Cygnus is often called the Northern Cross, to distinguish it from the Southern Cross of the constellation Crux. The constellation is represented by a swan with widespread wings flying southward. The bright star Deneb is the tail of the swan: it lies at the head of the cross. Albireo, a beautiful double star of contrasting orange and blue colors, is the head of the swan. The whole constellation lies in, and parallel to, the path of the Milky Way. *See* CONSTELLATION; CRUX. [C.-S.Y.]

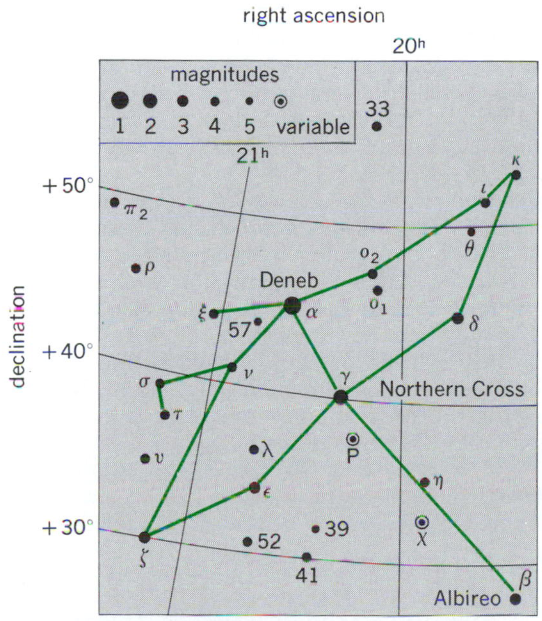

Line pattern of constellation Cygnus. Grid lines represent coordinates of sky. Apparent brightness, or magnitudes, of stars is shown by sizes of "dots," which are graded by appropriate numbers as indicated.

Cylinder The solid of revolution obtained by revolving a rectangle about a side is called a cylinder, or more precisely a right circular cylinder. More generally the word cylinder is used in solid geometry to describe any solid bounded by two parallel planes and a portion of a closed cylindrical surface. In analytic geometry, however, the word cylinder refers not to a solid but to a cylindrical surface (see illustration). This is a surface generated by a straight line which moves so that it always intersects a given plane curve called the directrix, and remains parallel to a fixed line that intersects the plane of the directrix. Cylinders whose right sections are circles, ellipses, parabolas, or hyperbolas are called circular cylinders, elliptic cylinders, parabolic cylinders, or hyperbolic cylinders, respectively. All these are quadric cylinders. *See* EUCLIDEAN GEOMETRY. [J.S.F.]

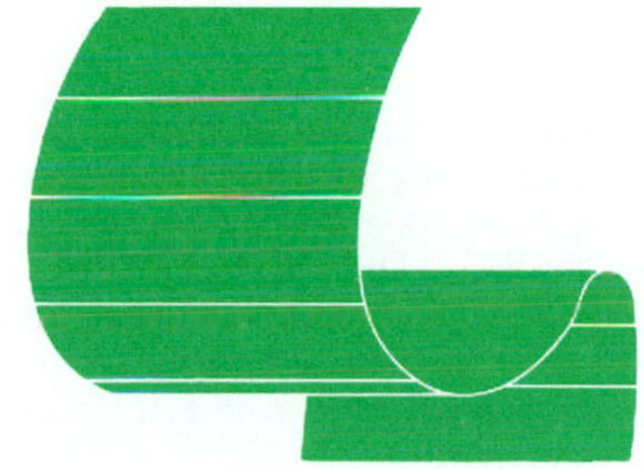

Cylindrical surface.

Cyperales An order of flowering plants, division Magnoliophyta (Angiospermae), in the subclass Commelinidae of the class Liliopsida (monocotyledons). The names Gluminflorae, Graminales, and Poales have also been used for this order. There are only two families, the Gramineae (Poaceae) with about 8000 species, and the Cyperaceae, with nearly 4000. The Cyperales are Commelinidae with reduced, mostly wind-pollinated or self-pollinated flowers that have a unilocular, two- or three-carpellate ovary bearing a single ovule. The flowers are arranged in characteristic spikes or spikelets representing reduced inflorescences. The perianth is represented only by a set of bristles or tiny scales, or is completely missing. *See* COMMELINIDAE.

The Cyperaceae, or sedge family, includes the bulrushes (*Scirpus*) and the papyrus (*Cyperus papyrus*) of Egypt, as well as the sedges (*Carex*). The Gramineae, or grass family, embraces all true grasses, including bamboos and the cereal grains such as wheat, maize, oats, and rye. The two families differ in a number of more or less consistent technical characters of the inflorescence, fruits, stems, and leaves. *See* BAMBOO; BARLEY; CEREAL; CORN; FLOWER; GRASS; MILLET; OATS; RICE; RYE; SORGHUM; SUGARCANE; TIMOTHY; WHEAT. [A.Cr.]

Cypress The true cypress (*Cupressus*), which is very close botanically to the cedars (*Chamaecyparis*). All of the species of *Cupressus* in the United States are western and are found from Oregon to Mexico. The Arizona cypress (*Cupressus arizonica*) of the southwestern United States and the Monterey cypress (*Cupressus macrocarpa*) of California are medium-sized trees and are chiefly of ornamental value. The Italian cypress (*Cupressus sempervirens*) and its varieties are handsome ornamentals, but usually do well only in the southern parts of the United States. Other trees are also called cypress, such as the Port Orford cedar (*Chamaecyparis lawsoniana*) known also as the Lawson cypress, and the Alaska cedar

(*Chamaecyparis nootkatensis*), known also as the Nootka cypress or cedar.

The bald cypress (*Taxodium distichum*) is an entirely different tree that is found in the swamps of the South Atlantic and Gulf coastal plains and in the lower Mississippi Valley. The soft needlelike leaves and short branches are deciduous; hence, they drop off in winter and give the tree its common name. Also known as the southern or red cypress, this tree yields a valuable decay-resistant wood used principally for building construction, especially for exposed parts or where a high degree of resistance to decay is required as in ships, boats, greenhouses, railway cars, and railroad ties. *See* CEDAR; PINALES.

[A.H.G./K.P.D.]

Cypriniformes An order of actinopterygian fishes which ranks second in size only to the Perciformes. The order Cypriniformes (or Eventognathi) includes the characins, minnows, and their allies. The Weberian apparatus of the Cypriniformes involves only the first four vertebrae. The body is typically invested in cycloid scales, although these are sometimes lost. In further contrast to catfishes, the Cypriniformes have parietal, suboperculum, symplectic, and intercalar bones; they lack a pectoral spine (present in most catfishes); and they have intermuscular bones.

The 8 families of the Cypriniformes, numbering about 3000 species, may be arranged in 3 suborders: Characoidei, Gymnotoidei, and Cyprinoidei. The suborder Characoidei includes the single family Characidae (split by some authors into several smaller families) or characins. They are primarily fresh-water fishes that abound in South America, where they have undergone an extensive adaptive radiation resulting in perhaps 850 species. The Gymnotoidei includes a single family, Gymnotidae, consisting of about 16 genera and perhaps 40 or 45 species that inhabit fresh waters of South and Central America. They are eel-shaped fishes that seem to be capable of producing an electric shock. The electric eel is the largest species. Cyprinoidei includes about 6 families of primarily fresh-water fishes. The Cyprinidae, including minnows and carps, are the largest family of fishes, with about 275 genera and more than 1500 species. The center of abundance and diversity of form of this group is in southeastern Asia. The family is well represented in Africa, Europe, Asia, and North America (about 230 species), but there are none in Central or South America, Madagascar, Australia, or in that part of the East Indies lying east of Lombok.

In the mountainous regions of southeastern Asia are found 3 other small families of the Cyprinoidei: the Homalopteridae, the Gastromyzontidae with 16 genera, and the Gyrinocheilidae. All are adapted to life in torrential streams. *See* ACTINOPTERYGII; CARP; CATFISH; EEL; MINNOW; OSTARIOPHYSI; SILURIFORMES.

[R.M.B.]

Cyrtosoma A subphylum of the Mollusca comprising members of the classes Monoplacophora, Gastropoda, and Cephalopoda. Each class has, in turn, been the most diverse molluscan group. In the Ordovician the cephalopods succeeded the monoplacophorans as the most diverse molluscan class, and they held this position until the close of the Mesozoic. They were followed by the gastropods, which by then had diversified into a great variety of aquatic and terrestrial habitats.

Primitive members of the Cyrtosoma have, or had, a conical univalved shell, often twisted into a spiral. Although primitive cyrtosomes, including the only living shelled cephalopod, *Nautilus*, are shell-bearing animals, many advanced cyrtosomes have either lost or greatly reduced their shells (examples include slugs, nudibranchs, squids, and octopuses). Consequently, it is not any particular shell form which unites all members of the Cyrtosoma; rather, it is a knowledge of their evolutionary history. *See* CEPHALOPODA; GASTROPODA; MOLLUSCA; MONOPLACOPHORA.

[B.Ru.]

Cysteine An amino acid with the formula shown. In microorganisms, cysteine probably is formed from serine and

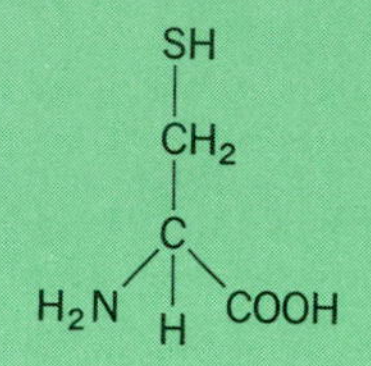

Cysteine

inorganic sulfate or sulfide. In animals, cysteine is formed by transsulfuration of serine with the homocysteine from dietary methionine. A variety of metabolic degradation pathways are known, of which the principal ones are the following: (1) degradation to pyruvate, hydrogen sulfide (H_2S), and ammonia (NH_2); (2) oxidation to cysteic acid and decarboxylation of this compound to taurine; (3) transamination (as cysteinesulfinic acid) with α-ketoglutarate to form ß-sulfinylpyruvate, which breaks down spontaneously to pyruvate and sulfite. Cysteine furnishes the sulfhydryl group of coenzyme A and, in microorganisms, of homocysteine, the precursor of methionine. It is a constituent of the tripeptide coenzyme, glutathione, and of the amino acid cystine. *See* AMINO ACIDS; CYSTINE; METHIONINE; TRANSAMINATION.

[E.A.Ad.]

Cysticercosis The infestation by cysticerci of the genus *Taenia*. The cysticerci are rarely larger than a pea, and damage usually results from mechanical pressure on a variety of delicate tissues. The intermediate host becomes infested with eggs from the feces of the definitive host, which usually is a carnivore.

Cysticercus cellulosae is the old name for the metacestode of *T. solium*, the pork tapeworm. Humans acquire the adult worm by eating inadequately cooked pork containing the cysticerci. Not only do humans harbor the adult worm, but the much more dangerous cysticercus can develop in humans as well. The cysticercus has a predilection for nervous tissue, frequently in the brain or eye, causing serious pathology. Anthelmintics can eliminate the tapeworm, but surgical removal is the only known cure for cysticercosis. *See* CYCLOPHYLLIDEA; TAPEWORM DISEASE.

[R.S.F.]

Cystine An amino acid with the formula shown. Cystine can be reversibly reduced to cysteine. It is formed biosyntheti-

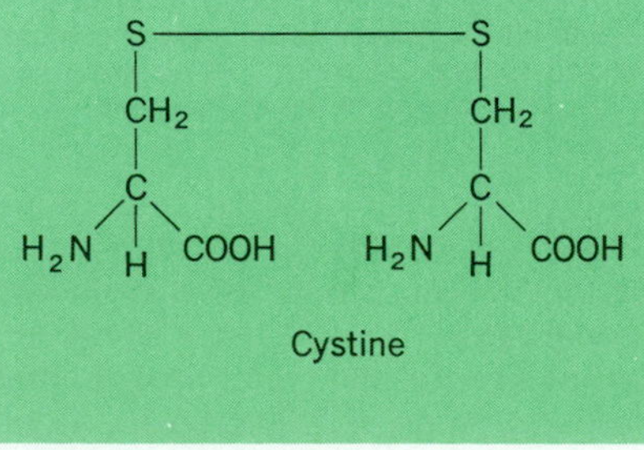

Cystine

cally from cysteine, by oxidation. In addition to the metabolic degradation pathway of cysteine, cystine may also be oxidized directly to cystine disulfoxide, which is decarboxylated and then further oxidized to taurine. *See* CYSTEINE. [E.A.Ad.]

Cystitis Inflammation of a fluid-filled organ. As employed in clinical medicine, cystitis is assumed to refer to inflammation of the urinary bladder. When other viscera are affected, they are identified more accurately, for example, cholecystitis being inflammation of a gallbladder. Principal symptoms of urinary cystitis are pain and frequency of urination. The usual cause is bacterial infection, particularly by the coliform group of microorganisms. Cystitis often occurs in association with infection of the kidneys. [P.B.B.]

Cystoporata An extinct order of bryozoans in the class Stenolaemata. Cystoporates tend to have robust colonies composed of relatively simple, long, slender, and tubular zooecia, separated by blisterlike vesicles stacked upon one another. *See* BRYOZOA; STENOLAEMATA.

Cystoporate colonies range from small and delicate to large and massive; they can be thin encrusting sheets, tubular or nodular masses, or erect twiglike or frondlike growths.

The cystoporates may possibly have evolved from the earliest ctenostomes. Cystoporates first appeared late in the Early Ordovician, somewhat earlier than representatives of other stenolaemate orders; thus, the cystoporates may possibly have in turn given rise to those orders, although all these forms may simply share a common ancestor further back in time. Apparently marine, the cystoporates became moderately common by Mid-Ordovician time—contributing to building small reefs, as well as to level-bottom communities—and remained so until they died out in the latest Permian. *See* BRYOZOA; STENOLAEMATA. [R.J.Cu.]

Cytochalasin A class of lipophilic antibiotics produced by fungi. The cytochalasins elicit in animal and plant cells a puzzling diversity of membrane phenomena. There is evidence that numerous chemicals, including cytochalasins, interact directly with plasma membrane components, modulate activity of membrane-bound enzymes, and often produce changes in membrane structure.

In animals, cytochalasins inhibit cytokinesis, cell movement, and embryonic morphogenesis, as well as intracellular movement such as the transport of melanin granules. In addition, nuclear extrusion is induced; lymphocyte-mediated destruction of target cells is inhibited; and there is selective "pulverization" of certain chromosomes, which are converted to the unraveled, interphase form, while the other chromosomes in the cell remain in the condensed, metaphase form.

In plants, intracellular movements such as cytoplasmic streaming and chloroplast movements are inhibited. Cytochalasins also inhibit root growth and water uptake in onion seedlings; cells become more spherical in shape.

The great value of the cytochalasins as research tools is that they appear to achieve their reversible impact on cell behavior with a minimum of undesirable side effects such as inhibition of respiration or protein synthesis. Cytochalasin is extensively applied as a chemical "scalpel" to enucleate mammalian cells rapidly, precisely, and efficiently in studies of nuclear-cytoplasmic relations and in cell hybridization and nuclear transplant work. Another major application is in examining the consequences of arrested cytoplasmic movement. [D.D.S.T.]

Cytochemistry A science concerned with the chemistry of cells and cell components, primarily with the location in cells of chemical constituents and enzymes. Cytochemical investigations fall into two main categories, visualization and biochemical analysis.

Visualizing methods involve the application of various microscopic or staining techniques to locate chemical constituents within cells. The techniques of tissue preparation and staining for this purpose are closely related to those used for histochemistry. In cytochemistry they must be selected and used in such a way as to achieve the greatest possible accuracy of localization, particularly when the investigator wishes to take advantage of the great resolving power of the electron microscope. *See* HISTOCHEMISTRY.

Biochemical methods make use of techniques whereby the outer membranes of cells in a tissue sample are mechanically ruptured, followed by centrifuging to separate the various cell constituents so released into different fractions. [S.J.H.]

Cytochrome A complex protein occurring within the cells of a wide variety of animals and plants. An integral part (prosthetic group) of a cytochrome is a heme (iron tetrapyrrole) moiety. The heme imparts a color, generally red, to the cytochrome, as well as characteristic absorption bands which are used for the spectroscopic observation and identification of these intracellular hemoproteins. The iron atom of the heme may be reduced and oxidized by appropriate substances in the cell, thus constituting the function of the cytochromes as electron carriers and as essential components of biological oxidation systems. *See* MITOCHONDRIA. [E.H.St.]

Cytokine Any of a group of soluble proteins that are released by a cell to send messages and that act on the same cell (autocrine), on an adjacent cell (paracrine), or on a distant cell (endocrine). The cytokine binds to a specific receptor and causes a change in function or in development of the target cell. Cytokines are involved in reproduction, growth and development, normal homeostatic regulation, response to injury and repair, blood clotting, and host resistance (immunity). Unlike cells of the endocrine system, many different types of cells can produce the same cytokine, and a single cytokine may act on a wide variety of target cells. Further, several cytokines may produce a similar effect on a target, so that the loss of one type of cytokine has few consequences for the organism; this effect is called redundancy. Finally, the response of a target cell may be altered by the context in which it receives a cytokine signal. Thus has developed the concept of cytokines as letters of an alphabet combining to spell words that make up a molecular language.

Cytokines may be divided into five groups: interleukins, colony-stimulating factors, interferons, tumor necrosis factors, and growth factors.

Interleukins are proteins that are produced by one type of lymphocyte or macrophage and that act on another leukocyte. At least 13 types of this important class, with varying origin and function, exist. *See* INTERLEUKIN.

Colony-stimulating factors are produced by lymphoid and nonlymphoid cells. These factors provide a mechanism whereby cells that are distant from bone marrow can call for increased production of different types of hemopoietic progeny.

Interferons classically interfere with virus replication mechanisms in cells. Alpha-interferon (produced by leukocytes) and beta-interferon (produced by fibroblasts) activate natural killer–type cells. Gamma-interferon is a potent activator of macrophages and has effects on other cells as well. *See* INTERFERON.

Tumor necrosis factor-alpha is produced by a variety of cell types, but activated macrophages provide the dominant source. This factor activates natural killer cells, enhances cytotoxic T-cell generation where there is inhibition by transforming growth factor-beta, and activates natural killer–type cells (which then produce gamma-interferon). Tumor necrosis factor-alpha also acts on vascular endothelial cells to promote inflammation and thrombosis. Tumor necrosis factor-beta is a product of a subset of T cells with helper and delayed-type hypersensitivity properties.

Transforming growth factors have the ability to promote proliferation of cells which otherwise have a benign phenotype. There are two factor groups: transforming growth factor-alpha, produced by a variety of cells including tumor cells, and epidermal growth factor. Transforming growth factor-beta is a peptide that synergizes with transforming growth factor-alpha or epidermal growth factor (which is similar to transforming growth factor-alpha) in promoting colony formation of some cell types such as normal rat kidney fibroblast cells. It has potent and pleiotropic effects on a wide variety of tissues and is a potent fibrogenic and immunosuppressive agent. *See* GROWTH FACTOR.

Cytokines are also involved in the process of wound healing. They ensure that the restorative sequences are carried out in the appropriate order by signaling the blood cells to coagulate and fill a wound opening, signaling macrophages and neutrophils to engulf microbes, and guiding the protective skin cells over the wound. If the damage is more extensive, cytokines stimulate the production of new skin cells, blood vessels, connective tissue, and bone. [D.A.C.]

Cytokinesis The partitioning of cells, following nuclear division, or karyokinesis. As a result, each nucleus is contained in a portion of cytoplasm and is isolated from other nuclei by a cell membrane. Where cytokinesis does not take place following nuclear division, a syncytium results, that is, a cytoplasmic continuum containing more than one nucleus and correspondingly larger than the parent cell. Syncytial organisms and tissues are fairly common, but cytokinesis may be considered a normal terminal step in cell reproduction. Syncytia may vary in form from shapeless plasmodial masses, such as are seen in slime molds, to highly elongated cells containing a chain of nuclei, such as are found sometimes among the bacteria. *See* CELL DIVISION; MITOSIS.

The timing of cytokinesis with respect to the mitotic cycle is variable; the one rule that can be made is that it begins only after anaphase movement is completed. It may coincide with telophase changes in the chromosomes or may come later. The extreme case is that in which cytokinesis is delayed while many nuclear divisions proceed. The syncytium so produced may later be partitioned off into cells, usually providing a cell per nucleus. This delayed cytokinesis is observed in the growth of plant endosperm and in spore formation and gamete formation in many lower organisms. [D.M.]

Cytokinins Plant hormones (phytohormones) that act together with other phytohormones, such as auxins, in regulating virtually every aspect of plant growth and development. Naturally occurring and synthetic compounds are operationally defined as cytokinins if they promote growth and cell division (hence the name cytokinin, from cytokinesis) of excised tissues from plants of tobacco, soybean, or carrot when these substances are added to a nutrient medium containing auxin, vitamins, mineral salts, and sucrose. Compounds active in these tissue culture bioassays have other effects on plants as well. *See* AUXIN; BIOASSAY.

One of the most important uses of cytokinins for both practical and experimental purposes involves its ability to regulate development and organ formation in tissue cultures. With tobacco callus cultures, for example, the formation of roots or shoots depends on the relative concentrations of auxin and cytokinin added to the medium. Practically speaking, this cytokinin effect enables agriculturalists to produce large numbers of genetically identical plants using tissue cultures. *See* TISSUE CULTURE.

Not only are cytokinins involved in the normal processes of plant development, but they also seem to be important as a cause of abnormal growth associated with certain plant diseases. Cytokinin effects also include promotion of seed germination, breaking of dormancy of some buds, stimulation of the mobilization of various metabolites to the point at which cytokinin has been applied, retardation of senescence, increased anthocyanin and flavanoid synthesis, and stimulation of the opening of stomates. *See* CROWN GALL; DORMANCY; PLANT HORMONES. [J.Ei.]

Cytolysis An important immune function involving the dissolution of certain cells. There are a number of different cytolytic cells within the immune system that are capable of lysing a broad range of target cells. The most thoroughly studied of these cells are the cytotoxic lymphocytes, which appear to be derived from a wide range of cells and may employ a variety of different lytic mechanisms. Cytotoxic cells are believed to be essential for the elimination of oncogenically or virally altered cells, but they can also play a detrimental role by mediating graft rejection. *See* CELLULAR IMMUNOLOGY.

Large granular lymphocytes that spontaneously lyse certain tumor cells are called natural killer cells and are probably essential for immune surveillance. That is, these cells may be responsible for lysing spontaneously occurring tumors in the body. They are unique in that no previous sensitization is required for them to kill.

Another killer cell, called the lymphokine-activated killer cell, appears to be derived from natural killer cells. They lyse any target cell, including cells from freshly isolated tumors, and are being employed in cancer therapy.

The last group of cytotoxic cells is the cytotoxic T lymphocyte. These are T cells that can lyse any target cell in an antigen-specific fashion. These are truly immune cells in that they require prior sensitization in order to function. These cells are thought to mediate graft rejection, mount responses against viral infections, and play a major role in tumor destruction. *See* ANTIGEN; HISTOCOMPATIBILITY.

Many models have been proposed to explain the lytic mechanism. The dominant model is that cytotoxic cells release the contents of cytotoxic granules after specific interaction with the target cell. After target cell binding, the Golgi apparatus and the granules containing the cytolytic molecules orient toward the target cell in preparation for secretion. Vectorally oriented granule exocytosis then occurs, with the release of cytolytic molecules into the intercellular space between the killer and target cells. This results in the leakage of salts, nucleotides, and proteins from the target cell, leading to cell death. *See* GOLGI APPARATUS.

If primary cytotoxic T lymphocytes do not use the degranulation mechanism, they must have another way of killing. In the programmed cell death model, a cytotoxic T lymphocyte merely triggers an endogenous suicide pathway within the pathogenically altered cell so that it goes on to kill itself by a mechanism that produces nuclear disintegration and deoxyribonucleic acid (DNA) fragmentation prior to plasma membrane damage. The target cell plays an active role in its own demise, which is consistent with much of the data available on cytotoxic T lymphocyte-mediated killing. It would make sense for all of the cells in the body to have a built-in self-destruct mechanism so that if a cell became pathogenically altered, it could be easily triggered to die. *See* IMMUNOLOGY. [H.L.O.]

Cytomegalovirus infection A viral illness also known as cytomegalic inclusion disease. The virus, a member of the herpesvirus group, produces enlargement of cells in the affected organs or in tissue cultures, as well as large, intranuclear inclusion bodies and sometimes intracytoplasmic inclusions.

In young infants clinical cases are characterized by jaundice, enlargement of liver and spleen, and blood and circulatory disturbances. Cytomegalovirus infection appears in its most severe, and sometimes fatal, form in infants under 2 years of age, who became infected before birth or when newborn. In the congenital infection the virus may persist even in the presence of antibody, an unusual situation similar to that found in congenital rubella infection. Mild or inapparent infection is common during late childhood and adolescence; it is recognized by the detection of the virus, which may be excreted in the urine for months. The virus may remain hidden for years, but its reappearance is brought on by immunosuppressive drugs, such as those given to patents receiving organ transplants. *See* ANTIBODY; TRANSPLANTATION BIOLOGY. [J.L.Me.]

Cytoplasm That portion of living cells bordered externally by the plasma membrane (cell membrane) and internally by the nuclear envelope. In the terminology of classical cytology, the substance in living cells and in living organisms not compartmentalized into cells was called protoplasm. It was assumed at the time that the protoplasm of various cells was similar in structure and chemistry. Results of research on cell chemistry and ultrastructure after about 1960 showed that each cell type had a recognizably different "protoplasm." Primarily for that reason, the term protoplasm gradually fell into disuse in contemporary biology. The terms cytoplasm and nucleoplasm have been retained and are used descriptively; they are used almost synonymously with the terms cytosome (body of cytoplasm) and nucleus, respectively.

Many cells, especially the single-celled organisms or protistans, have regional cytoplasmic differentiation. The outer region is the cortex or ectoplasm, and the inner region is the endoplasm. In many cases the cortical layer is a gel made up of a meshwork of cytoskeletal fibers.

Cytoplasm contains mostly water, from 80 to 97% in different cells, except for spores and other inactive forms of living material, in which water may be present in lesser amounts. The dry mass of cells consists mainly of macromolecules: proteins, carbohydrates, nucleic acids, and lipids associated with membranes. The small molecules present in cells are mainly metabolites or metabolic intermediates. The principal ions other than the hydrogen and hydroxyl ions of water are the cations of potassium, sodium, magnesium, and calcium, and the anions chloride and bicarbonate. Many other elements are present in cytoplasm in smaller amounts. Iron is found in cytochrome pigments in mitochondria; magnesium is present in chlorophyll in chloroplasts; copper, zinc, iodine, bromine, and several other elements are present in trace quantities. *See* CYTOCHEMISTRY.

Sedimentation of cells by centrifugation shows that organelles and inclusions can be separated from the ground cytoplasm, the fluid phase of the cytoplasm in which they are suspended. The ground cytoplasm in turn has been shown to consist of a cytoskeletal network and the cytosol, the fluid in which the cytoskeleton is bathed. The cytoskeleton consists of several biopolymers of wide distribution in cells. Microtubules have been observed in electron micrographs of a vast number of different cell types. They consist of the protein tubulin, and are frequently covered by a fuzzy layer of microtubule-associated proteins. *See* CYTOSKELETON.

In most cells the smaller particles exhibit Brownian motion due to thermal agitation. In some cells lacking extensive cytoskeletal structure, particles can be moved freely around the cell by Brownian motion. In others they are restricted by their surrounding cytoskeletal elements. Particles of various types may also undergo saltatory motions which carry them farther than Brownian motion possibly could. Such excursions result from the interaction of a particle with an element of the cytoskeleton such as one or more microtubules or microfilaments. *See* CELL (BIOLOGY). [R.D.Al.]

Cytoskeleton A system of filaments found in the cytoplasm of cells and responsible for the maintenance of and changes in cell shape, cell locomotion, movement of various elements in the cytoplasm, integration of the major cytoplasmic organelles, cell division, chromosome organization and movement, and the adhesion of a cell to a surface or to other cells.

Three major classes of filaments have been resolved on the basis of their diameter and cytoplasmic distribution: actin filaments (or microfilaments) each with an average diameter of 6 nanometers, microtubules with an average diameter of 25 nm, and intermediate filaments whose diameter of 10 nm is intermediate to that of the other two classes. The presence of this system of filaments in all cells, as well as their diversity in structure and cytoplasmic distribution, has been recognized only in the modern period of biology.

A technique that has greatly facilitated the visualization of these filaments, as well as the analysis of their chemical composition, is immunofluorescence applied to cells grown in tissue culture. *See* IMMUNOFLUORESCENCE.

Actin is the main structural component of actin filaments in all cell types, both muscle and nonmuscle. Actin filaments assume a variety of configurations depending on the type of cell and the state it is in. They extend a considerable distance through the cytoplasm in the form of bundles, also known as stress fibers since they are important in determining the elongated shape of the cell and in enabling the cell to adhere to the substrate and spread out on it. Actin filaments can exist in forms other than straight bundles. In rounded cells that do not adhere strongly to the substrate (such as dividing cells and cancer cells), the filaments form an amorphous meshwork that is quite distinct from the highly organized bundles. The two filamentous states, actin filament bundles and actin filament meshworks, are interconvertible polymeric states of the same molecule. Bundles give the cell its tensile strength, adhesive capability, and structural support, while meshworks provide elastic support and force for cell locomotion.

Microtubules are slender cylindrical structures that exhibit a cytoplasmic distribution distinct from actin filaments. Microtubules originate in structures that are closely associated with the outside surface of the nucleus known as centrioles. The major structural protein of these filaments is known as tubulin. Unlike the other two classes of filaments, microtubules are highly unstable structures and appear to be in a constant state of polymerization-depolymerization. *See* CENTRIOLE; MITOSIS.

Intermediate filaments function as the true cytoskeleton. Unlike microtubules and actin filaments, intermediate filaments are very stable structures. They have a cytoplasmic distribution independent of actin filaments and microtubules. In the intact cell, they anchor the nucleus, positioning it within the cytoplasmic space. During mitosis, they form a filamentous cage around the mitotic spindle which holds the spindle in a fixed place during chromosome movement. [E.L.]

D

Dacite Aphanitic (very finely crystalline or glassy) rock of volcanic origin, composed chiefly of sodic plagioclase (oligoclase or andesine) and free silica (quartz or tridymite) with subordinate dark-colored (mafic) minerals (biotite, amphibole, or pyroxene). If alkali feldspar exceeds 5% of the total feldspar, the rock is a quartz latite. As quartz decreases in abundance, dacite passes into andesite. Thus, dacite is roughly intermediate between andesite and quartz latite. *See* ANDESITE. [C.A.C.]

D'Alembert's paradox A theorem in fluid mechanics which states that no forces act on a body moving at constant velocity in a straight line through a large mass of incompressible, inviscid fluid which was initially at rest (or in uniform motion). This seemingly paradoxical theorem can be understood by first realizing that inviscid fluids do not exist. If such fluids did exist, there would be no internal physical mechanism for dissipating energy into heat; hence there could be no force acting on the body, because work would then be done on the fluid with no net increase of energy in the fluid. *See* FLUID FLOW. [A.E.Br.]

D'Alembert's principle The principle that the resultant of the external forces **F** and the kinetic reaction acting on a body equals zero. The kinetic reaction is defined as the negative of the product of the mass m and the acceleration **a**. The principle is therefore stated as $\mathbf{F} - m\mathbf{a} = 0$. While D'Alembert's principle is merely another way of writing Newton's second law, it has the advantage of changing a problem in kinetics into a problem in statics. The techniques used in solving statics problems may then provide relatively simple solutions to some problems in dynamics; D'Alembert's principle is especially useful in problems involving constraints. *See* CONSTRAINT. [P.W.S.]

Dalitz plot Pictorial representation in high-energy nuclear physics for data on the distribution of certain three-particle configurations. Many elementary-particle decay processes and high-energy nuclear reactions lead to final states consisting of three particles (which may be denoted by a, b, c, with mass values m_a, m_b, m_c). Well-known examples are provided by the K-meson decay processes, Eqs. (1) and (2), and by the K- and $\bar{K}$-meson reactions with hydrogen, given in Eqs. (3) and (4).

$$K^+ \rightarrow \pi^+ + \pi^+ + \pi^- \quad (1)$$

$$K^+ \rightarrow \pi^0 + \mu^+ + \upsilon^- \quad (2)$$

$$K^+ + p \rightarrow K^0 + \pi^+ + p \quad (3)$$

$$K^- + p \rightarrow \Lambda + \pi^+ + \pi^- \quad (4)$$

For definite total energy E (measured in the barycentric frame), these final states have a continuous distribution of configurations, each specified by the way this energy E is shared among the three particles. (The barycentric frame is the reference frame in which the observer finds zero for the vector sum of the momenta of all the particles of the system considered.) *See* ELEMENTARY PARTICLE.

If the three particles have kinetic energies T_a, T_b, and T_c (in the barycentric frame), Eq. (5) is obtained. As shown in the

$$T_a + T_b + T_c = E - m_a c^2 - m_b c^2 - m_c c^2 = Q \quad (5)$$

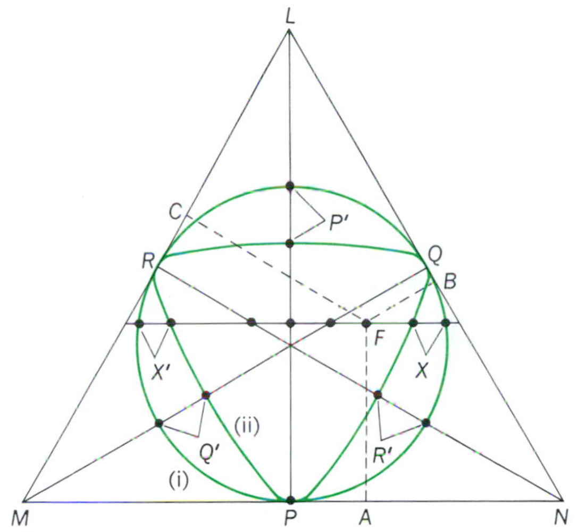

Configuration of a three-particle system (*abc*) in its barycentric frame is specified by a point *F* such that the three perpendiculars *FA*, *FB*, and *FC* to the sides of an equilateral triangle *LMN* (of height *Q*) are equal in magnitude to the kinetic energies T_a, T_b, T_c, where *Q* denotes their sum.

illustration, this energy sharing may be represented uniquely by a point F within an equilateral triangle LMN of side $2Q/\sqrt{3}$, such that the perpendiculars FA, FB, and FC to its sides are equal in magnitude to the kinetic energies T_a, T_b, and T_c. The most important property of this representation is that the area occupied within this triangle by any set of configurations is directly proportional to their volume in phase space.

Not all points F within the triangle LMN correspond to configurations realizable physically, since the a, b, c energies must be consistent with zero total momentum for the three-particle system. With nonrelativistic kinematics and with equal masses m for a, b, c, the only allowed configurations are those corresponding to points F lying within the circle inscribed within the triangle, shown as (i) in the illustration. More generally, with relativistic kinematics, the limiting boundary is distorted as illustrated by the boundary curve (ii), drawn for the $\omega \rightarrow 3\pi$ decay process, where the final masses are equal. [R.H.D.]

Dallis grass A general term for a genus of grasses of which the most important species is the deeply rooted perennial *Paspalum dilatatum*. Dallis grass is widely used in the southern United States, mostly for pasture, and remains productive indefinitely if well managed. Dallis grass does best on fertile soils and responds to lime and fertilizer. On heavier soils it remains green throughout the winter unless checked by heavy frosts. Ergot-bearing seed heads are very toxic to livestock, and Dallis grass must be so managed as to prevent consumption of infected heads by livestock. *See* CYPERALES; ERGOT. [H.B.S.]

Dalton's law The total pressure that a mixture of gases exerts is equal to the sum of the separate pressures which each of the gases would exert if it alone occupied the whole volume.

This law was observed by John Dalton (1766–1844) and is also called the law of additive pressures. It may be stated mathematically as $p_m = p_a + p_b + p_c + \ldots$, where p_m is the pressure produced by the mixture and $p_a, p_b, p_c, \ldots$ are the partial pressures of the several gases in the mixture. The partial pressure is the pressure exerted by a constituent gas that occupies the whole volume occupied by the mixture at the same temperature and pressure in the absence of the other gases. *See* AVOGADRO'S LAW; GAS; THERMODYNAMIC PRINCIPLES. [G.A.H.]

Dam A structure that bars or detains the flow of water in an open channel or watercourse. Dams are constructed for several principal purposes. Diversion dams divert water from a stream; navigation dams raise the level of a stream to increase the depth for navigation purposes; power dams raise the level of a stream to create or concentrate hydrostatic head for power purposes; and storage dams store water for municipal and industrial use, irrigation, flood control, river regulation, recreation, or power production. A dam serving two or more purposes is called a multiple-purpose dam. Dams are commonly classified by the material from which they are constructed, such as masonry, concrete, earth, rock, timber, and steel. Most dams are built either of concrete or of earth and rock.

Concrete dams may be typed as gravity, arch, or buttress. Gravity dams depend on weight for stability against overturning and for resistance to sliding on their foundations. An arch dam may have a near-vertical face or, more usually, one that curves concave downstream. The dam acts as an arch to transmit most of the horizontal thrust from the water pressure against the upstream face of the dam to the abutments of the dam. Buttress dams depend on the weight of the structure and of the water on the dam to resist overturning and sliding.

Principal forces acting on a concrete dam are (1) vertical forces from weight of the structure and vertical component of water pressure against the upstream and downstream faces of the dam, (2) uplift pressures under the base of the structure, (3) horizontal forces from the horizontal component of the water pressure against the upstream and downstream faces of the dam, (4) forces from earthquake accelerations in regions subject to earthquakes, (5) temperature stresses, (6) pressures from silt deposits and earth fills against the structure, and (7) ice pressures.

The earth embankment is the most common dam. Earth dams may be built of rock, gravel, sand, silt, or clay in various combinations. Most earth dams are constructed with an inner impervious core, with upstream and downstream zones of more pervious materials, sometimes including rock zones. Earth dams limit the flow of water through the dam by use of fine-grained soils. Where possible, these soils are formed into a relatively impervious core. In some cases where pervious soils are scarce, the entire dam may be a homogeneous fill of relatively impervious soil. Downstream pervious drainage blankets are provided to collect seepage passing through, under, and around the abutments of the dam.

A spillway releases water in excess of storage capacity so that the dam and its foundation are protected against erosion and possible failure. All dams must have a spillway, except small ones where the runoff can be safely stored in the reservoir without danger of overtopping the dam. Ample spillway capacity is of particular importance for large earth dams, which would be destroyed or severely damaged by being overtopped. Failure of a large dam could result in severe hazards to life and property downstream. Spillways are of two general types: the overflow type, constructed as an integral part of the dam; or the channel type, located as an independent structure discharging through an open chute or tunnel. Either type may be equipped with gates to control the discharge.

Reservoir outlet works are used to regulate the release of water from the reservoir; they consist essentially of an intake and an outlet connected by a water passage, and are usually provided with gates. Outlet works usually have trashracks at the intake end to prevent clogging by debris. Bulkheads or stop logs are commonly provided to close the intakes so that the passages may be unwatered for inspection and maintenance. A stilling basin or other type of energy dissipator is usually provided at the outlet end.

A penstock is a pipe that conveys water from a forebay, reservoir, or other source to a turbine in a hydroelectric plant. It is usually made of steel, but reinforced concrete and woodstave pipe have also been used. Pressure rise, or water hammer, and speed regulation must be considered in the design of a penstock.

The foundation of a dam must support the structure under all operating conditions. For concrete dams, following removal of unsatisfactory materials to a sound foundation surface, imperfections such as adversely oriented rock joints, open bedding planes, localized soft seams, and faults lying on or beneath the foundation surface receive special treatment. In addition, the foundation of an earth dam must safely support the weight of the dam, limit seepage of stored water, and prevent transportation of dam or foundation material away or into open joints or seams in the rock by seepage. Earth-dam foundation treatment may include removal of excessively weak surface soils to prevent both potential sliding and excessive settlement of the dam, excavation of a cutoff trench to rock, and grouting of joints and seams in the bottom and downstream side of the cutoff trench.

Because failure of a dam may result in loss of life or property damage in the downstream area, it is essential that dams be inspected systematically both during construction and after completion. The design of dams should be reviewed to assure competency of the structure and its site, and inspections should be made during construction to ensure that the requirements of the design and specifications are incorporated in the structure. After completion and filling, inspections may vary from cursory surveillance during day-to-day operation of the project to regularly scheduled comprehensive inspections. The objective of such inspections is to detect symptoms of possible distress in the dam at the earliest time. Detection of any such symptoms of distress should be followed by an investigation of the causes, probable effects, and remedial measures required. Inspection of a dam and reservoir is particularly important following significant seismic events in the locality. Systematic monitoring of the instrumentation installed in dams is essential to the inspection program. [J.R.T.]

Damping A term broadly used to denote either the dissipation of energy in, and the consequent decay of, oscillations of all types or the extent of the dissipation and decay. The energy losses arise from frictional (or analogous) forces which are unavoidable in any system or from the radiation of energy to space or to other systems. For sufficiently small oscillations, the analogous forces are proportional to the velocity of the vibrating member and oppositely directed thereto; the ratio of force to velocity is $-R$, the mechanical resistance. For the role of damping in the case of forced oscillations, where it is decisive for the frequency response, *see* FORCED OSCILLATION; RESONANCE (ACOUSTICS AND MECHANICS). *SEE ALSO* HARMONIC MOTION; MECHANICAL VIBRATION; OSCILLATION; VIBRATION.

An undamped system of mass m and stiffness s oscillates at an angular frequency $\omega_0 = (s/m)^{1/2}$. The effect of a mechanical resistance R is twofold: It produces a change in the frequency of oscillation, and it causes the oscillations to decay with time. If u is one of the oscillating quantities (displacement, velocity, acceleration) of amplitude A, then Eq. (1) holds in the damped case, whereas in the undamped case Eq. (2)

$$u = Ae^{-\alpha t} \cos \omega_d t \qquad (1)$$

$$u = A \cos \omega_0 t \qquad (2)$$

holds. The reciprocal time $1/\alpha$ in Eq. (1) may be called the damping constant.

The damped angular frequency ω_d in Eq. (1) is always less than ω_0. According to Eq. (1), the amplitude of the oscillation decays exponentially; the time required for the amplitude to decrease to the fraction $1/e$ of its initial value is equal to $1/\alpha$.

A common measure of the damping is the logarithmic decrement δ, defined as the natural logarithm of the ratio of two successive maxima of the decaying sinusoid. If T is the period of the oscillation, then Eq. (3) holds. Then $1/\delta$ is the number

$$\delta = \alpha T \tag{3}$$

of cycles required for the amplitude to decrease by the factor $1/e$ in the same way that $1/\alpha$ is the time required.

The Q of a system is a measure of damping usually defined from energy considerations. The Q is π times the ratio of peak energy stored to energy dissipated per cycle and is equal to π/δ.

If α in Eq. (1) exceeds ω_0 then the system is not oscillatory and is said to be overdamped. If the mass is displaced, it returns to its equilibrium position without overshoot, and the return is slower as the ratio α/ω_0 increases. If $\alpha = \omega_0$ (that is, $Q = 1/2$), the oscillator is critically damped. In this case, the motion is again nonoscillatory, but the return to equilibrium is faster than for any overdamped case. [M.Gr.]

Daphniphyllales An order of flowering plants, division Magnoliophyta (Angiospermae), in the subclass Hamamelidae of the class Magnoliopsida (dicotyledons). The order consists of a single family with but one genus, *Daphniphyllum*, containing about 35 species. All are dioecious trees or shrubs native to eastern Asia and the Malay region. The plants produce a unique type of alkaloid (daphniphylline group); they often accumulate aluminum and sometimes produce iridoid compounds. The leaves are simple and entire, alternate or sometimes closely clustered at the ends of the branches. The flowers are small and inconspicuous, unisexual, regular, and hypogynous. Usually there are 2–6 sepals, or sometimes the sepals are absent; petals are lacking. *Daphniphyllum* has sometimes been included in the Euphorbiales of the subclass Rosidae, but structural details such as the very tiny embryo make it highly aberrant there. *See* HAMAMELIDAE; MAGNOLIOPSIDA; PLANT KINGDOM. [T.M.Ba.]

Dark current An ambiguous term used in connection with both gaseous-discharge devices and photoelectric cells or tubes. In gaseous-conduction tubes it refers to the region of operation known as the Townsend discharge. When applied to photoelectric devices, the term refers to background current. This is current which may be present as the result of thermionic emission or other effects when there is no light incident on the photosensitive cathode. *See* ELECTRICAL CONDUCTION IN GASES; TOWNSEND DISCHARGE. [G.H.M.]

Darter Any of 93 species of small, slender, fresh-water fishes of the family Percidae. They occur in a variety of habitats, including swift- and slow-moving streams and lakes and riffles with stony bottoms, and they are widely distributed throughout the various drainage systems of the United States.

Darters are typical bony fishes. There is a single external gill opening on each side which is covered with an opercle or bony flap. Pelvic fins are present, and the body is scaled and lacks an adipose fin. The body is semitranslucent in some species and most are brilliantly colored. These fishes are not of commercial value. *See* OSTEICHTHYES; PERCIFORMES. [C.B.C.]

Dasheen Common name for the plant *Colocasia esculenta*, including the variety *antiquorum* (taro). These plants are among the few edible members of the aroid family (Araceae). Native to southeastern Asia and Malaysia, the plants supply the people with their most important food. A main dish in the Polynesian menu is poi, a thin, pasty gruel of taro starch, often fermented, which is frequently formed into cakes for baking or toasting. The raw corms (underground stems) are baked or boiled to eliminate an irritating substance present in the cells. *See* ARALES. [P.D.St./E.L.C.]

Dasycladales An order of green algae (Chlorophyceae) in which the plant body (thallus) is composed of a nonseptate axis, attached by rhizoids and bearing whorls of branches. The walls utilize a partially crystalline mannan (polymer of mannose) as the skeletal component and usually are impregnated with aragonite, a form of calcium carbonate, causing these algae to be easily fossilized. Among green algae, only the Dasycladales have a fossil record sufficient to permit meaningful phylogenetic speculation. About 140 extinct genera have been described from limestones as old as the Ordovician, with three peaks of abundance and diversity—Carboniferous, Jurassic-Cretaceous, and Eocene. Only about 11 genera comprising 50 species are extant, with 3 of the genera extending from the Cretaceous. All are confined to warm marine waters except *Batophora*, which occurs as a relict in brackish sinkholes in New Mexico in addition to having a normal Caribbean distribution.

The extant Dasycladales are usually placed in a single family, Dasycladaceae, but those forms in which fertile and sterile whorls alternate or only one fertile whorl is produced may be segregated as the Acetabulariaceae (or Polyphysaceae). *See* ALGAE; CHLOROPHYCEAE. [P.C.Si.; R.L.Moe]

Data communications The function of electronically conveying digitally encoded information from a source to a destination where the source and destination may be displaced in space (distance) and time. Encoded information can represent alphabetic, numeric, and graphic forms. Such information may originate as keystrokes at a terminal, as images on a page, as motion of a hand-held stylus, as a signal from some electrical or mechanical measurement device, or as output from a computer. The data communications function includes management and control of connections as well as the actual transfer of information.

Digitally encoded information typically takes the form of a series of 1s and 0s called bits (contraction for binary digit). Groups of eight bits are often referred to as bytes. Standard codes (such as ASCII and EBCDIC) have been adopted to establish a correspondence between bytes (patterns of eight bits) and alphabetic, numeric, and commonly used graphic and control characters. These codes are used for the interchange of information among equipments made by different manufacturers. *See* BIT.

Data communications usually do not refer to the conveyance of voice and video, although these too may be digitally encoded and then conveyed in digital form just as data are conveyed. The increased use of computers in the design and implementation of communications systems and the efficiencies realized from the handling of digitally encoded information have markedly broadened the range of data communications, and tend to mask old distinctions between data communications and other forms of information transfer.

Data communications has evolved from a function serving simple terminal-to-terminal communications (telegraphy) to one dominated by communications with and among computers. Terminal-to-computer communications remains a sizable component of traffic, but the migration of intelligence into terminal systems (an aspect of distributed processing) is causing terminals to take on more of the communications characteristics of computers. Networks of terminals and computers are emerging to serve the needs of business, education, government, industry, medicine, recreation, and other forms of human endeavor. These systems, both publicly and privately owned and operated, manage and direct the movement of data between stations (terminals, computers, and other sources of digitally encoded information) and across networks. Where two stations are geographically separated, data communications manages the transfer of information as a displacement in space. Where there is value in delaying the transfer of information, data communications manages the transfer also as a displacement in time. *See* COMPUTER; MULTIACCESS COMPUTER; TELEGRAPHY.

Data communications applications fall into three principal categories: transaction-oriented, message-oriented, and batch-oriented. Such applications exhibit different needs in terms of total bytes transferred, delay, and special features. Consequently, they impose different demands upon the functional components of data communications: media conversion, communications processing, and transmission.

Transaction-oriented applications involve the transfer of small amounts of data in the range of tens to hundreds of bytes and frequently require minimal delay in the transfer process. Transaction-oriented applications are typically bidirectional or conversational. Information is exchanged between source and destination stations. Message-oriented applications involve data transfers in the range of hundreds to a few thousand bytes or characters and are usually unidirectional; information flows from source to destination. Message applications frequently take advantage of time displacement (store and forward) as well as space displacement, and they exhibit data transfer options which go beyond the simple movement of information between two points. Batch-oriented applications involve the transfer of thousands or even millions of bytes of data and are usually point-to-point (place-to-place) and computer-to-computer.

Media conversion achieves the conversion of some input, such as a keystroke, graphic material, a signal representing some measurement, and information recorded on magnetic media such as tape or disk, into a digitally encoded signal which may be transmitted. Likewise, it achieves the conversion of a digitally encoded signal into similar forms of useful output.

Communications processing provides the management and control in a data communications system. In the data terminal or media conversion system, it must at least manage the protocol (rules) used for intelligibly exchanging information with other stations including the administration of techniques designed to detect errors in transmission. Depending on the application, it may be necessary to perform code conversion and to multiplex or concentrate inputs from several terminals into a single output stream.

Transmission accomplishes the physical transfer of data. It includes the provision of transmission channels, the signal conversions necessary to use such channels, and any required switching among channels. Since it is generally impractical to install privately owned cables beyond single buildings or campus environments and because transmission distances of more than a few miles are commonly of interest, the conventional approach is to use the telephone network for transmission facilities. To support applications where communications among multiple stations is required, data communications systems offer switching beyond that available in the dial switching capabilities of the telephone network. Circuit switching, packet switching, and message switching are the common forms. *See* ELECTRICAL COMMUNICATIONS; SWITCHING SYSTEMS (COMMUNICATIONS).
[M.M.R.]

Data compression The process of transforming information from one representation to another, smaller representation from which the original, or a close approximation to it, can be recovered. The compression and decompression processes are often referred to as encoding and decoding. Data compression has important applications in the areas of data storage and data transmission. Besides compression savings, other parameters of concern include encoding and decoding speeds and workspace requirements, the ability to access and decode partial files, and error generation and propagation.

The data compression process is said to be lossless if the recovered data are assured to be identical to the source; otherwise the compression process is said to be lossy. Lossless compression techniques are requisite for applications involving textual data. Other applications, such as those involving voice and image data, may be sufficiently flexible to allow controlled degradation in the data.

Data compression techniques are characterized by the use of an appropriate data model, which selects the elements of the source on which to focus; data coding, which maps source elements to output elements; and data structures, which enable efficient implementation.

Models. Information theory dictates that, for efficiency, fewer bits be used for common events than for rare events. Compression techniques are based on using an appropriate model for the source data in which defined elements are not all equally likely. The encoder and the decoder must agree on an identical model. *See* INFORMATION THEORY.

A static model is one in which the choice of elements and their assumed distribution is invariant. For example, the letter "e" might always be assumed to be the most likely character to occur. A static model can be predetermined with resulting unpredictable compression effect, or it can be built by the encoder by previewing the entire source data and determining element frequencies. The benefits of using a static model include the ability to decode without necessarily starting at the beginning of the compressed data.

An alternative dynamic or adaptive model assumes an initial choice of elements and distribution and, based on the beginning part of the source stream that has been processed prior to the datum presently under consideration, progressively modifies the model so that the encoding is optimal for data distributed similarly to recent observations. Some techniques may weight recently encountered data more heavily. Dynamic algorithms have the benefit of being able to adapt to changes in the ensemble characteristics. Most important, however, is the fact that the source is considered serially and output is produced directly without the necessity of previewing the entire source.

In a simple statistical model, frequencies of values (characters, strings, or pixels) determine the mapping. In the more general context model, the mapping is determined by the occurrence of elements, each consisting of a value which has other particular adjacent values. For example, in English text, although generally "u" is only moderately likely to appear as the "next" character, if the immediately preceding character is a "q" then "u" would be overwhelmingly likely to appear next.

The use of a model determines the intended sequence of values. An additional mapping via one coding technique or a combination of coding techniques is used to determine the actual output. Several data coding techniques are in common use.
[D.S.Hi.]

Digitized audio and video signals. The information content of speech, music, and television signals can be preserved by periodically sampling at a rate equal to twice the highest frequency to be preserved. This is referred to as Nyquist sampling. However, speech, music, and television signals are highly redundant, and use of simple Nyquist sampling to code them is inefficient. Reduction of redundancy and application of more efficient sampling results in compression of the information rate needed to represent the signal without serious impairment to the quality of the remade source signal at a receiver. For speech signals, redundancy evident in pitch periodicity and in the format (energy-peaks) structure of the signal's spectrum along with aural masking of quantizing noise is used to compress the information rate. In music, which has much wider bandwidth than speech and far less redundancy, time-domain masking and frequency-domain masking are principally used to achieve compression. For television, redundancy evident in the horizontal and vertical correlation of the pixels of individual frames and in the frame-to-frame correlation of a moving picture, combined with visual masking that obscures quantizing noise resulting from the coding at low numbers of bits per sample, is used to achieve compression. *See* SPEECH; TELEVISION.

Compression techniques may be classified into two types: waveform coders and parametric coders. Waveform coders replicate a facsimile of a source-signal waveform at the receiver with a level of distortion that is judged acceptable. Parametric coders use a synthesizer at the receiver that is controlled by signal parameters extracted at the transmitter to remake the signal. The latter may achieve greater compression because of the information content added by the synthesizer model at the receiver.

Waveform compression methods include adaptive differential pulse-code modulation (ADPCM) for speech and music signals, audio masking for music, and differential encoding and sub-Nyquist sampling of television signals. Parametric encoders include vocoders for speech signals and encoders using orthogonal transform techniques for television. [S.J.C.]

Data-processing systems Electronic, electromechanical, or mechanical machines for transforming information into suitable forms in accordance with procedures planned in advance. The term data-processing system is also applied to the scheme, or procedure, that prescribes the sequence of operations to be performed in processing the information.

Typical business applications of data-processing systems include record keeping, financial accounting and planning, processing personnel information (including payroll processing), sales analysis, inventory control, production scheduling, operations research, and market research. Data-processing systems are also used to process correspondence and other information in business offices. These systems are called word processors. *See* WORD PROCESSING.

In science and engineering, data-processing systems are used in data reduction, statistical analysis of experimental data, planning and design of engineering projects, and the display in tabular and graphic form of the results of research and development. Special-purpose computer systems, which also fall within the definition of data-processing systems, are used to monitor and control such things as chemical processes, machine tools, typesetting equipment, and power generation plants. *See* DATA REDUCTION.

Data-processing systems can be broadly classified as manual, semiautomatic, or automatic, depending on the degree of human effort required to control and execute the procedures. Automatic systems are usually built around electronic digital computers and termed electronic data-processing (EDP) systems. *See* DIGITAL COMPUTER.

Data-processing functions. Virtually every data-processing system, regardless of the degree of automation, consists of six basic functions: recording, transmission, manipulation, reporting, storage, and retrieval—collectively termed the data-processing cycle.

Before information can be processed, it must be recorded in some form that is meaningful to a person or a machine. At one time the principal medium used to enter information into computers was the punched card. Subsequently keyboards linked to magnetic tapes (key to tape) or to magnetic disk (key to disk) systems became widely used. These recording systems are being replaced in turn by key entry devices that are linked directly to the computer via communication lines.

Once information has been recorded, it will usually need to be transmitted to another location for processing. Where fast response is desirable and economically feasible, the data can be transmitted at high speeds by means of telephone or telegraph circuits, microwave links, or satellite channels. *See* DATA COMMUNICATIONS.

Manipulation is the stage in which most of the actual processing is performed. The operations involved can range from simple to highly complex. The most common types of operations are classifying, sorting, calculating, and summarizing. More specifically, the manipulation stage frequently involves (1) arrangement of the information into a sequence that will facilitate further processing; (2) determination of the exact procedure to be followed in processing each item of information; (3) references to files containing data that must be associated with the current information; (4) arithmetic operations upon the current information or file data or both; and (5) updating (changing) of the file data to reflect the current information.

After information has been processed, it is usually necessary to report the results in a meaningful form. Reports are the people-oriented products of the data-processing activity. They should be timely, complete, understandable, and in a convenient format.

Some or all of the processed information will need to be stored for future reference or retrieval. Depending on the system, the information may be stored manually (as in ledgers or notebooks), electromechanically (as in punched cards or

Input/output devices

Unit	Medium	Input or output	Typical approximate transfer rate in characters per second
Disk pack unit	Magnetic disks housed in interchangeable cartridges	Both	200,000–2,000,000
Diskette unit	Flexible magnetic disks	Both	Tens of thousands
Magnetic tape unit	Plastic tape with magnetizable coating, housed on reels	Both	Thousands–1,250,000
Magnetic tape cartridge unit	Plastic tape with magnetizable coating, housed in cartridges	Both	Tens of thousands
Card reader	Punched cards	Input	20–2000
Card punch	Punched cards	Output	20–700
Paper tape reader	Perforated paper or plastic tape	Input	10–1000
Paper tape punch	Perforated paper or plastic tape	Output	10–200
Printer	Continuous paper forms	Output	10–20,00
Optical character reader	Paper or card documents	Input	10–3000
Magnetic ink character reader	Paper or card documents	Input	200–3000
Display unit	Cathode-ray tube or other display medium	Output*	10–100,000
Computer output microfilmer	Microfilm	Output	2000–60,000

*Input capabilities are usually provided by an associated keyboard or light pen.

punched paper tape), or electronically (as in disk packs or magnetic tape).

The importance of the contents of storage and the availability of transmission facilities have resulted in many automatic data-processing systems providing immediate retrieval of stored information at work stations. Questions regarding the contents of storage can be posed at remote locations, and responses can be prepared and transmitted to these locations in a few seconds. This is called an on-line query facility.

Electronic data-processing (EDP) systems. EDP systems take advantage of the great speed and versatility of stored-program digital computers to process large volumes of data with little or no need for human intervention. There is no fundamental difference between the computers used for business data processing and those used for scientific calculations; in fact, a single computer is often used for both types of applications. Scientific computers, however, tend to emphasize high computational speeds, while computers used for business data processing often place primary emphasis on fast, flexible input and output equipment.

Every computer system has four basic functional parts: input equipment to permit data and instructions to be entered into the system; a storage unit to permit data and instructions to be stored until called for; a control unit, called the central processing unit (CPU), to interpret the stored instructions and direct their execution; and output equipment to permit the processed data to be removed from the system. The common types of input and output devices, their recording media, and their typical speeds are summarized in the table. Magnetic tape and disk can also be viewed as storage media that supplement the storage available in the computer itself. They have been termed auxiliary storage. *See* COMPUTER STORAGE TECHNOLOGY.

Disk pack units. The disk pack has become the preeminent high-speed computer input/output medium. A disk pack typically consists of a stack of from 1 to 20 round stainless-steel plates mounted on a vertical spindle. Data are magnetically recorded on some or all of the surfaces of the disks, The disk pack can be conveniently mounted on a drive unit that contains magnetic read/write heads mounted on a comblike access mechanism that moves horizontally between the disks in order to reach and record information on them. The key advantage of disk packs that accounts for their preeminence is their rapid-access capability; any record stored in a disk pack can be located and read into the computer in a fraction of a second, whereas it may take as long as several minutes to locate a particular record stored in a reel of magnetic tape.

One development in magnetic disk technology deserves special mention. Flexible plastic disks, called diskettes or floppy disks, are in widespread use for applications in which low cost and ease are more important than high speeds and large data-storage capacities. Most diskettes are $3\frac{1}{2}$ or 5 in. (9 or 13 cm) in diameter, and may be housed in thin plastic envelopes; the envelope remains stationary while the drive unit spins the disk.

Magnetic tape units. Magnetic tape is an important input/output medium for EDP systems because it permits large quantities of information to be stored in a highly compact, economical, and easily erasable form. The tape is usually made of plastic with a magnetizable oxide coating on which data can be recorded in the form of magnetic spots.

As an alternative to high-performance magnetic tape units, many EDP systems include smaller, less costly tape transports in which the data are recorded on shorter lengths of narrow magnetic tape housed in conveniently interchangeable cartridges. The cartridge tape units have comparatively slow read/write speeds and small data-storage capacities; their offsetting advantages are economy and convenience of use.

Magnetic tape is widely used to provide back-up for disk units and to permit the exchange of large volumes of information by sending reels of tape from one place to another. For example, computer programs that are purchased are often delivered on one or more reels of magnetic tape.

Card reader and punch. The fact that most early computers replaced or augmented punched-card systems gave the punched card a strong head start as a computer input/output medium, and it is still widely used. The punched card is a highly flexible medium that has many advantages for use as a source document and as a storage medium for permanent records. In EDP systems its principal drawbacks are the difficulty of correcting errors in the punched data, the fixed upper limit on the amount of data a card can hold (usually 80 or 96 characters), and the comparatively low speeds of even the fastest card readers and punches (see table).

Paper tape readers and punches. Paper tape can be punched and read by relatively simple, inexpensive equipment, making it practical to produce tape records as a direct by-product of the normal operations of many business machines, such as teletypewriters and cash registers.

Though generally called paper tape, the tape may be made of paper, plastic, metal, or laminated combinations of these materials. Data are recorded by punching round holes into the tape and read by sensing the holes either mechanically or photoelectrically. As with punched cards, the principal disadvantage of paper tape is the low speeds of the tape readers and punches available for use with computers.

Printers. Electromechanical printing devices are the primary means for making information processed by computers available to people. However, display units are overtaking printers in this regard. Computer printers can be broadly classified as either serial or line printers and as either impact or nonimpact printers. A serial unit prints one character at a time in the manner of a conventional typewriter, whereas a line printer plants a full line, usually consisting of 80 to 132 characters, at the same time. An impact printer uses direct mechanical force to produce character images on the paper, whereas nonimpact printers utilize electrical or chemical processes to form the characters.

Matrix printers may be impact or nonimpact. The most successful are nonimpact and use thermal-chemical or electrostatic processes to create an image. Typically a 5-by-7 matrix of dots is used to print each character. The set of dots that is printed provides an image of the character. A matrix printer that is widely used, especially in offices using word processing, is the ink jet printer.

A laser can be used to write the printed output on the light-sensitive surface of a drum. The image on this surface is then transferred to paper. Speeds of tens of thousands of characters per second can be realized by this type of printer. *See* LASER.

Character reader. Optical character recognition (OCR) has been heralded as the solution to the great computer input bottleneck because it promises to reduce, and in some cases eliminate, the manual keystroke operations which are normally required in preparing the input data for machine processing. Machines capable of reading characters printed on paper documents by typewriters, cash registers, computer printers, and many other business machines are in use in applications such as utility billing and credit-card processing. Moreover, machines that can read hand-printed characters are on the market, and extensive development work is in progress on machines to recognize ordinary handwriting. The principal barriers to wider use of optical character readers have been reliability problems and comparatively high costs, and significant progress has been made in both areas. *See* CHARACTER RECOGNITION.

Optical recognition techniques can also be used effectively to read handmade pencil marks or printed bar codes. Optical mark readers and bar-code readers are considerably less costly than character readers and are widely used in applications such as test scoring, survey analysis, credit-card processing, and supermarket checkout.

Displays. Display units are widely used as "electronic blackboards" that provide rapid access to data stored in computer systems and facilitate human-machine communication. The great majority of current display units use cathode-ray tubes, but other technologies such as light-emitting diodes, liquid-crystal displays, and plasma (gaseous) displays are also used. The display units may be connected directly to the computer or located remotely and connected by a data communications link. The displayed information may consist of characters or lines on the face of a cathode-ray tube 6–22 in. (15–56 cm) in diameter. Most display units can also be used for input to the computer by a keyboard or hand-held light pen which the operator focuses upon particular points on the screen. *See* ELECTRONIC DISPLAY. [H.M.]

Data reduction The transformation of information, usually empirically or experimentally derived, into corrected, ordered, and simplified form.

The term data reduction generally refers to operations on either numerical or alphabetical information digitally represented, or to operations which yield digital information from empirical observations or instrument readings. In the latter case data reduction also implies conversion from analog to digital form either by human reading and digital symbolization or by mechanical means. *See* ANALOG-TO-DIGITAL CONVERTER; DIGITAL COMPUTER.

In applications where the raw data are already digital, data reduction may consist simply of such operations as editing, scaling, coding, sorting, collating, and tabular summarization.

More typically, the data reduction process is applied to readings or measurements involving random errors. These are the indeterminate errors inherent in the process of assigning values to observational quantities. In such cases, before data may be coded and summarized, the most probable value of a quantity must be determined. Provided the errors are normally distributed, the most probable (or central) value of a set of measurements is given by the arithmetic mean or, in the more general case, by the weighted mean. *See* STATISTICS.

Data reduction may also involve operations of smoothing and interpolation, because the results of observations and measurements are always given as a discrete set of numbers, while the phenomenon being studied may be continuous in nature. [R.J.Ne.]

Database management systems Special data-processing systems or parts of data-processing systems which are developed to aid in the storage, manipulation, reporting, and managing of data. They may be considered as building blocks constructing a data-processing system which also acts as a mechanism for the effective control of the data. The database management system provides a means for management of data by making available data to a large range of authorized users, while preserving the integrity of the data and improving the overall system performance.

It has been estimated that a medium-sized corporation may need to retain 10^{10} characters of current important data in high-speed automated storage. By comparison, a book of 300 pages, 40 lines per page, and 12 words (each of about 6 characters) per line, contains about 10^6 characters (including spaces between words) and thus the 10^{10}-character "database" could be stored in 10^4 books, which is indeed a good-sized library. However, the organization of a library makes it relatively easy to find any book, but imagine a library that had no catalog, no librarians and no one putting the books back on the shelves. The cost of a library staff is similar to the cost of organizing data in a corporate database: it is a necessary cost if data must be found and used effectively. Indeed the old systems, without well-organized data, suffered from the problem that people knew data existed, but could not find it easily and became frustrated. One of the reasons for the development of database management systems was to solve this need.

Data become truly available when a large cross section of potential users know of their existence, when their location and format are known, and when there is a mechanism for retrieving the data. If the mechanism to locate its format is used by humans, the location is normally termed a directory, while the format is stored in a dictionary. If the locating device is embedded in the database system, the mechanism is termed a schema.

A schema is therefore a machine-readable version of the location and format of the stored data. The difference between the schema and its "instances" or "occurrences" is shown in Fig. 1. This simple schema contains information about a person: name, date of birth, and salary. The representation is: name 10 characters in length, with blanks for shorter names; date of birth as year, month, and day, with two decimal digits representing each; and salary of up to $9999.99 per month. The "model" is defined in the schema, while the data are recorded in each instance.

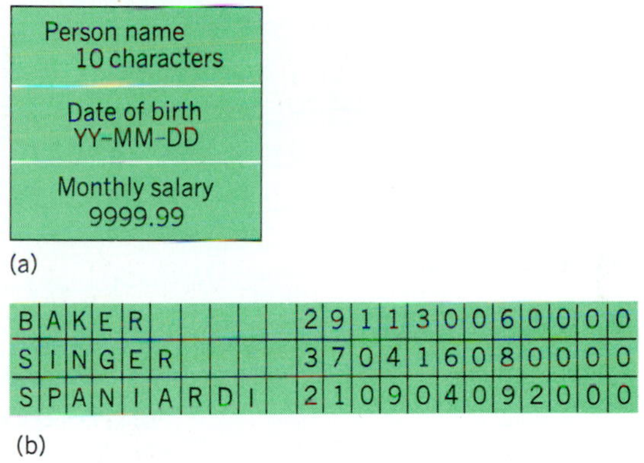

Fig. 1. Simple schema representing (*a*) a very simple record structure and (*b*) its stored data base of three instances.

In order for the database management system to retrieve or store data about a person, it must consult the schema to determine the form of the data as well as its location in machine storage. This means that a database management system with a "query language" interface can respond to a request like FOR PERSON, FIND MONTHLY-SALARY, WHERE PERSON-NAME = "SINGER," or FOR PERSON, FIND PERSON-NAME AND DATE-OF-BIRTH, WHERE SALARY GREATER THAN 700.00

This powerful capability of the database management system enables a wide variety of potential users to obtain information that would otherwise be unavailable. However, this also creates some social problems, because access to the data may become too easy. Typically, security may be enforced by insisting that users of the database management system identify themselves both by name or employee number and by some password.

Because the data are often stored in a device with moving parts they can be destroyed through mechanical or other failures. Data may also be lost through user carelessness or program malfunction. To protect the integrity of the data, several defensive steps may be taken. One major precaution is to take dumps, that is, copies of the data. This action is taken at regular intervals if the data are frequently changed. In order to be able to restore a database after it has been destroyed, it is merely necessary to get a copy of the last dump and update it.

The somewhat simplified architecture of a database management system and its environment is shown in Fig. 2. It illustrates how a data request is satisfied. For example, consider

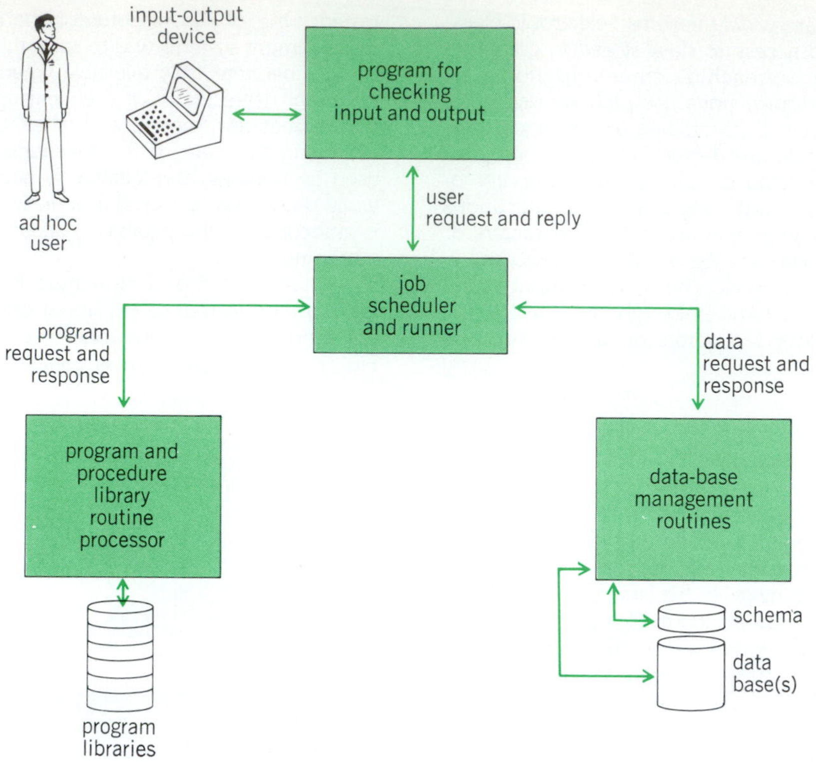

Architect of a database management system.

the request for person-name and date of birth in the PERSON database. An ad hoc user (that is, one who can query the system without having recourse to a previously coded program) goes to an input/output device with a keyboard and/or cathode-ray-tube combination. The person types in SIGNON JONES PASSWORD 51Z5QP, and then (after some response) AD HOC DBMS FOR PERSON, FIND PERSON-NAME AND DATE-OF-BIRTH WHERE SALARY GREATER THAN 700.00.

The following operations will be performed: (1) The word SIGNON is transmitted by the checking program to the scheduler program which requests the SIGNON program from its library. (2) The SIGNON program is run by the scheduler program to check that Jones has given the correct password. (3) The AD HOC program is called from the library and starts to look for more inputs. (4) The word PERSON is taken by this program as the name of the schema. (5) The scheduler program requests the database routine to load the PERSON schema from storage. (6) The next sentence is now checked to see whether PERSON-NAME, DATE-OF-BIRTH, and SALARY are all valid words in the PERSON schema. (7) A search of the instances is made to determine which have a value of more than 700.00 in the SALARY item. In the above example, two people qualify. (8) The names and dates of birth of these two people are displayed for the user. In this example, the authority of the user (Jones) to view any element (for example, date of birth) could be checked at step 6, if necessary.

Different database management systems model the data in different ways. The so-called relational system represents data as tables. No entry in the table contains more than one elementary value (for example, it would be impossible to have two salaries for BAKER). The hierarchical system allows more structure. In this type of system, groupings of elements are connected by parent-child relationships. Here, a particular person may have one or several different educational categories (high school, bachelor's degree, and so forth), as well as several job histories (office boy, clerk, manager, president). The network system has more structure and complexity, because it uses named relationships. Because these three types of systems are structurally different, most current commercial database management systems service only one type; future systems are expected to be able to allow the users to choose the best system to suit their special needs. *See* DATA-PROCESSING SYSTEMS; DIGITAL COMPUTER. [E.H.S.]

Dataflow systems An alternative to conventional programming languages and architectures, in which values rather than value containers are dealt with, and all processing is achieved by applying functions to values to produce new values. These systems can realize large amounts of parallelism (present in many applications) and effectively utilize very-large-scale-integration (VLSI) technology.

Dataflow systems use an underlying execution model which differs substantially from the conventional one. The model deals with values, not names of value containers. There is no notion of assigning different values to an object which is held in a global, updatable memory location. A statement such as $X: = B = C$ in a dataflow language is only syntactically similar to an assignment statement. The meaning of $X: = B = C$ in dataflow is to compute the value $B = C$ and bind this value to the name X. Other operators can use this value by referring to the name X, and the statement has a precise mathematical meaning defining X. This definition remains constant within the scope in which the statement occurs. Languages with this property are sometines referred to as single assignment languages. The second property of the model is that all processing is achieved by applying functions to values to produce new values. The inputs and results are clearly defined, and there are no side effects. Languages with this property are called applicative. Value-oriented, applicative languages do not impose any sequencing constraints in addition to the basic data dependencies present in the algorithm. Functions must wait for all input values to be computed, but the order in which the functions are evaluated does not affect the final results. There is no notion of a central controller which initiates one statement at a time sequentially. The model described above can be applied to languages and architectures.

The computation specified by a program in a dataflow language can be represented as a data dependence graph, in which each node is a function and each arc carries a value. Very efficient execution of a dataflow program can be achieved on a stored program computer which has the properties of the dataflow model. The machine language for such a computer is dependence graph rather than the conventional sequence of instructions. There is no program counter in a dataflow computer. Instead, a mechanism is provided to detect when an instruction is enabled, that is, when all required input values are present. Enabled instructions, together with input values and destination addresses (for the result), are sent to processing elements. Results are routed to destinations, which may enable other instructions. This mode of execution is called data-driven.

Dataflow systems can overcome many of the disadvantages of conventional approaches. In principle, all the parallelism in the algorithm is exposed in the program and this the programmer does not have to deal with parallesism explicitly.

Several important problems remain to be solved in dataflow systems. The handling of complex data structures as values is inefficient, but there is no complete solution to this problem yet. Dataflow computers tend to have long pipelines, and this causes degraded performance if the application does not have sufficient parallelism. Since the programmer does not have explicit control over memory, "separate garbage collection" mechanisms must be implemented. The space-time overheads of managing low levels of parallelism have not been quantified. Thus, though the parallelism is exposed to the hardware, it has not been demonstrated that it can be effectively realized. *See* DIGITAL COMPUTER: PROGRAMMING LANGUAGES. [T.A.]

Date The fruit of the evergreen date palm (*Phoenix dactylifera*). Date fruits are dry, semidry, or soft, depending on the moisture content at maturity of the particular variety. Soft types tend to spoil or sour more readily than dry types. Most varieties contain about 65% sugar as glucose or fructose. Palms are grown mainly in the semiarid and desert regions of Asia and Africa. Commercial date culture in the United States is limited to areas of high heat (100–120°F; 38–49°C) and low atmospheric humidity, such as the low-elevation deserts of California and a small part of Arizona. *See* ARECALES; FRUIT; FRUIT, TREE. [C.A.Sch.]

Dating methods Methods and techniques used in archeology, biology, and geology to fix dates, assign periods of time, and determine age. The uranium-lead, rubidium-strontium, potassium-argon, and carbon-14 methods have yielded reliable results on suitable samples and have permitted the construction of absolute geologic history in many areas of the world. *See* GEOLOGICAL TIME SCALE; METEORITE; RADIOCARBON DATING; ROCK AGE DETERMINATION.

For descriptions of other methods and techniques used to establish the sequence of events, the succession of strata, and relative chronologies *see* ARCHEOLOGY; DENDROCHRONOLOGY; FISSION TRACK DATING; GEOCHRONOMETRY; INDEX FOSSIL; ROCK MAGNETISM; STRATIGRAPHY; TREE-RING HYDROLOGY. [D.E.F.]

Datolite A mineral nesosilicate, composition $CaBSiO_4(OH)$, crystallizing in the monoclinic system. It usually occurs in crystals showing many faces and having an equidimensional habit. It may also be fine granular or compact and massive. Hardness is 5–5½ on Mohs scale; specific gravity is 2.8–3.0. The luster is vitreous, the crystals colorless or white with a greenish tinge. Datolite is found in the Harz Mountains, Germany; Bologna, Italy; and Arendal, Norway. In the United States fine crystals have come from Westfield, Massachusetts; Bergen Hill, New Jersey; and various places in Connecticut. In Michigan, in the Lake Superior copper district, datolite occurs in fine-grained porcelainlike masses which may be coppery red because of inclusions of native copper. *See* SILICATE MINERALS. [C.S.Hu.]

Dawsoniidae A subclass of the true mosses (Bryopsida) largely limited to the South Pacific. The subclass consists of a single genus, *Dawsonia*, of nine species. The plants are coarse and rigid in loose dark green or brownish tufts. They are simple or occasionally forked, grow erect from a procumbent base, and sometimes achieve a height of 26 in. (65 cm), much more than any other mosses. The stems may have a central strand of homogeneous fiberlike cells or both xylemlike and phloemlike cells. The leaves are long-linear or long-lanceolate from a relatively short, oblong-sheathing base. The limb, which may be as long as 1.6 in. (40 mm), is concave-tubulose and spreading or flat and spirally twisted when dry. It is coarsely toothed at the margins and at back. The inflorescences are dioecious and terminal. Male plants repeatedly form new growth through the perigonial bud. *See* BRYOPHYTA; BRYOPSIDA; POLYTRICHIDAE. [H.Cr.]

Day A unit of time equal to the period of rotation of Earth. Different sorts of day are distinguished, according to how the period of rotation is reckoned with respect to one or another direction in space.

The apparent solar day is the interval between any two successive meridian transits of the Sun. It varies through the year, reaching about 24 h 30 s of ordinary clock time in December and about 23 h 59 min 39 s in September. The mean solar day is the interval between any two successive meridian transits of an imagined point in the sky that moves along the celestial equator with a uniform motion equal to the average rate of motion of the Sun along the ecliptic. Ordinary clocks are regulated to advance 24 h during a mean solar day.

The sidereal day is the interval between any two successive meridian transits of the vernal equinox. Similarly, as for the solar day, a distinction is made between the apparent sidereal day and the mean sidereal day which, however, differ at most by a small fraction of a second. A mean sidereal day comprises 23 h 56 min 4.09054 s of a mean solar day.

The mean solar day and the sidereal day vary together in consequence of variations in the speed of rotation of Earth, which are of three sorts: seasonal, irregular, and secular. *See* TIME. [G.M.C.]

Dead reckoning A form of navigation that determines position of a craft by advancing a previous position to a new one on the basis of assumed distance and direction moved.

The parameters of dead reckoning are direction of motion and distance traveled. Several directions are involved in the motion of a craft, as shown exaggerated in the illustration. The intended direction of travel is called the course. The direction steered, which may differ somewhat because of anticipated offset due to wind or current, is also called course (or course steered) by the marine navigator, and heading by the air navigator. The direction actually made good between two points is called course made good. Air navigators often refer to this direction as the track, although this expression is used also to refer to the path followed or intended to be followed. The direction in which the craft is pointed at any moment is called heading by both air and marine navigators.

A compass is used to indicate direction. Distance is usually determined indirectly by measurement of speed and time, but it may be measured directly.

Air and land navigators, and some marine navigators, use the best estimate of direction and distance in their dead reckoning. Many marine navigators, however, prefer to use course

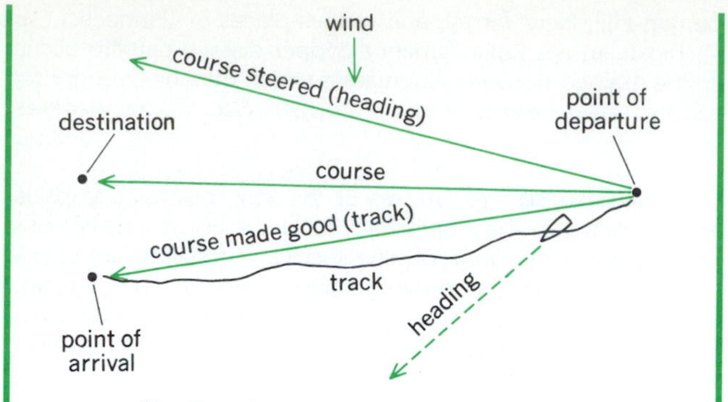

Directions involved in the motion of a craft.

steered and estimated speed through the water (without allowance for the effect of wind) for their dead reckoning; they consider positions determined by allowance for estimated effects of wind and current as estimated positions.

The uncertainty of dead-reckoning positions, however determined, increases with time and perhaps also with distance traveled. From time to time an independent determination of position is made by celestial navigation or piloting. When a reliable position, called a fix, is so obtained, a new dead reckoning is started from this point.

The average combined effect of wind, current, and other sources of error since the last fix can be determined by comparison of a fix with the position the craft would have occupied had there been no disturbing force. It is for this reason, in part, that many marine navigators prefer to use course steered and speed at which the vessel is propelled by its primary source of power for their dead reckoning. For. convenience, they consider the entire offset effect as the result of current. The direction from the dead-reckoning position to a fix at the same time is called the set of the current. The distance between these two positions, divided by the time since the last fix, is called the drift (speed) of the current. Air navigators consider the entire offset effect as the result of wind, determined by comparison of a no-wind position, sometimes called an air position, with a fix at the same time.

Whatever other forms of navigation may be employed, dead reckoning is generally considered essential to safe navigation. It provides a continuous position which is valuable in the evaluation of all other navigational information received. It is always available in some form.

Aboard ship, distance or speed is measured by means of a log or by an engine revolution counter. The pitometer log, now in wide use, uses a pitot-static tube. The Forbes log uses a small rotor in a tube projecting below the bottom of the vessel. An electromagnetic log has a sensing element which produces a voltage directly proportional to speed through the water.

In aircraft, speed through the air is measured by means of an airspeed indicator or Mach meter. The latter provides an indication of speed in units of the speed of sound, which varies with density of the atmosphere. For measurement of air speed a pitot-static tube is generally used with a suitable registering device. *See* AIR-VELOCITY MEASUREMENT.

Two sophisticated systems for performing the dead-reckoning process have been developed. One of these determines course and speed of an aircraft over the surface by measuring the Doppler shift of echoes of radio signals directed downward at an angle. Ships use sound signals in the same manner. The other system, employed both in aircraft and in ships, uses accelerometers and gyroscopes, with a good heading reference, to determine accelerations in various directions. These are converted by integration to components of speed and distance in each direction.

As speed of aircraft and volume of air traffic have both increased, and the need for following an assigned track to maintain established separation standards has developed, a need has arisen to reduce the pilot workload and decrease the time lag associated with the traditional dead-reckoning process. This need is being met by the development of light, compact, reliable digital computers suitable for use in aircraft. Errors can be reduced significantly by the use of a Doppler or inertial navigator, or a combination of both. *See* ELECTRONIC NAVIGATION SYSTEMS; PILOTING. [A.B.M.]

Death Cessation of life functions. This can involve the whole organism (somatic death), individual organs (organ death), individual cells (cellular death), and individual parts of cells (organelle death). Although the next smaller level of organization, the macromolecules which make up the cell organelles, may also cease to function, their disintegration is ordinarily not spoken of as death.

Pathogenesis. A normal animal, cell, or organelle exists in a normal state of homeostasis, or a normal "steady state." Following a disturbance of the normal steady state, referred to as an "injury," the living system can be assumed to undergo a change in steady-state level, that is, to exhibit an altered steady state. Such altered steady states can be regarded as having increased, decreased, or unaltered levels of ability to maintain homeostasis. An altered steady state with increased homeostatic ability might include the hypertrophy of a muscle cell resulting from continued exercise; and altered steady state with decreased homeostatic ability is the atrophy or shrinking of cells that occurs following disuse or lack of stimulation by hormones or suitable nutrients. In the case of injuries that do not result in the death of the living system, the system may persist in one of the new or altered steady states for long periods up to many years; that is, although the system differs significantly from the normal steady state, it is able to continue its existence and to maintain homeostasis. On the other hand, certain severe injurious stimuli result in an altered state that ultimately passes a point beyond which recovery of homeostatic ability is impossible even if the injurious stimulus or situation is removed. This point may be considered as a point of no return or the point of death of the living system. Following this point, the system undergoes exponential decay, leading to complete equilibrium with its environment. This period of decay, which is characterized by reactions resulting in the breakdown of structural components of the system, is referred to as necrosis. *See* HOMEOSTASIS.

Somatic death. In the biological sense, somatic death refers to the cessation of characteristic life functions. In higher organisms it can be assumed to result from an interplay between the genetic background of the individual and the effects of the environment. Death is, in a sense, the price of differentiation; in another sense, protoplasm is immortal and all life springs from preexisting life. The undifferentiated amebas need never die, only divide. It would appear that death is inevitable for higher organisms such as mammals and, in a sense, life is a series of little deaths.

Developments in medicine and associated technology involving supportive therapy, resuscitation, hypothermia, extracorporeal circulation, and organ transplantation have reopened serious questions regarding the definition of when an individual is "dead." The vagueness and inconsistency in previous legal and medical definitions have led to serious dilemmas. These are of particular importance in the selection of donors for transplantation of vital organs such as the heart.

Medical techniques have shown that individuals can be sustained for prolonged periods without spontaneous respiratory or cardiac movements as well as without electrical evidence of central nervous system activity. These techniques have also achieved the resuscitation of individuals who would formerly have been considered dead and have clearly shown the survival of transplanted organs from cadavers. Just as organs may survive after the death of the organism, the organism, properly supported by mechanical or other means, can survive the death of organs, including the brain.

The following are changes that occur in the body after somatic death: 1. Cooling (algor mortis) of the body occurs postmortem. 2. Rigidity (rigor mortis) or stiffening of the musculature usually begins within 5–10 h and begins to disappear after 3 or 4 days. 3. Staining (livor mortis) occurs, that is, there is reddish-blue discoloration in the dependent portions of the body resulting from the gradual gravitational flow of unclotted blood. 4. Blood begins to clot shortly after death and sometimes prior to cessation of cardiac function. 5. Autolysis (necrosis) occurs in the cells of the body following somatic death. 6. Putrefaction, bacterial and enzymatic decomposition, begins shortly after death, caused by microorganisms which enter the body.

Organ death. Following somatic death, all organs do not die at the same rate. Length of survival can be prolonged by reducing the temperature of the organs to near 32°F (0°C), which appreciably retards metabolic function and extends survival time. This survival of organs after the death of the host forms the basis for organ transplantation from cadavers.

Cellular death. Death and subsequent necrosis of cells form an important part of many reactions of cells to injury in disease states. The death and disintegration of cells within a living animal incite a vascular reaction called an inflammatory reaction on the part of the host. It is characterized by a series of vascular phenomena which include the migration of leukocytes (white blood cells) from the blood capillaries into the area of tissue damage with subsequent digestion and phagocytosis of the cellular debris resulting from cellular necrosis. [B.F.T.]

Issues. The advancements of modern medicine have brought significant changes in the ways of life and death, and challenge former concepts and definitions of death. As the determination of death becomes increasingly complex in a technological society, the definition of death becomes correspondingly more complex. The meaning of death is a philosophical concept which lies outside the scientific point of view. Although dying is a process which may continue over a considerable period of time, for all practical purposes death occurs at that point where there is an irreversible loss of one's essential human characteristics. Such loss indicates the death of the organism as a whole, the death of a person.

An interdisciplinary committee at the Harvard Medical School examinined possible new criteria for determining death. In 1968 they cited four criteria as determinative of a permanently nonfunctioning brain: unreceptivity and unresponsivity to externally applied stimuli and inner need; absence of spontaneous movements of breathing; absence of reflexes other than spinal column reflexes, and fixed dilated pupils; and a flat electroencephalogram (EEG) to be used as a confirmatory but not mandatory test.

These tests were to be repeated after 24 h, and in the absence of hypothermia or depressant drugs, were to be considered a confirmation of brain death. It has since been conclusively demonstrated that when the flow of blood to the brain has been interrupted for approximately 15 min there is a uniform necrosis and liquefaction of the brain. This definition appears to be a major milestone in determining new criteria for making the determination that death has occurred.

With regard to the issue of organ transplantation, it is contrary to moral intuitions that a dying person might be prematurely defined as dead in order that someone else might harvest the organs. A moral obligation exists to treat the living as such until death occurs. To avoid a conflict in the role of the physician, it has become common practice that the physician for the recipient of an organ donation is not party to the judgment that the donor has died. The needs of a patient for a viable organ must not contaminate the judgment that another person has died.

The ability to define death with more precision is important in cases involving the comatose patient. It is generally accepted as both a moral and legal right that competent persons have the right to refuse medical treatment. Increasingly, dying patients claim a more active role in self-determination regarding medical treatment in the final stage of life.

One of the more radical issues is that of actively ending the life of a suffering person or helping another to die through a positive action. Some ethicists see no moral distinction between removing life-support systems or life-saving medications which allow a patent to die and taking a direct action which causes a patient to die, since the intention and the consequence are the same in both cases. Others disagree, claiming that there is indeed a morally significant difference if a person dies of the underlying disease and is not made to endure still greater suffering by prolonging the dying process than if the person dies because another has caused the death through an intentional and fatal action. [T.R.McC.]

De Broglie wavelength The wavelength $\gamma = h/p$ associated with a beam of particles (or with a single particle) of momentum p; $h = 6.61 \times 10^{-27}$ erg-second is Planck's constant. The same formula gives the momentum of an individual photon associated with a light wave of wavelength γ. This formula, along with the profound proposition that all matter has wavelike properties, was first put forth by Louis de Broglie in 1924, and is fundamental to the modern theory of matter and its interaction with electromagnetic radiation. *See* QUANTUM MECHANICS. [E.G.]

Decapoda (Crustacea) One of the more highly specialized orders of the class Crustacea. This order includes the shrimps, lobsters, hermit crabs, and true crabs. The order is so diverse that satisfactory definition is difficult, but a few characters are common to nearly all decapods. The most obvious is a head shield, or carapace, which covers and coalesces with all of the thoracic somites and which overhangs the gills on each side. The first three of the eight pairs of thoracic appendages are specialized as maxillipeds and closely associated with the true mouthparts. The gills are usually well developed and arranged in several series.

Decapods vary in size from less than 1/2 in. (1.25 cm) in length to that of the giant Japanese crab, the largest living arthropod, a spider crab which may span more than 12 ft (3.6 m) between the tips of the outstretched claws. Although most decapods are found in the sea, they are by no means restricted to that habitat. Crayfishes are well-known inhabitants of freshwater streams and ponds, as are several kinds of shrimps and some true crabs. A number of crabs and hermit crabs have become well adapted to a terrestrial existence far from water; they return to the sea only seasonally to hatch their eggs. Many crayfishes burrow in the ground, and one species of crab spends its entire life in the tops of lofty trees.

In a general way, the decapods may be divided into two groups: the long-tailed and the short-tailed forms. The long-tailed species, such as the shrimps and lobsters, have a more or less cylindrical, or laterally compressed, carapace that often bears a head spine or rostrum. The large, muscular abdomen permits the shrimp, or lobster, to dart quickly backward out of danger, but the succulence of this tail makes the animals prey to numerous enemies, including humans. In the short-tailed species, the crabs, the carapace is often broadened and flattened dorsally and usually does not form a rostrum. The much-reduced and feeble abdomen is tucked under the thorax, where it serves the female as a brood pouch for eggs.

Physiology. The nervous system is somewhat variable. Although usually fewer, there may be as many as 11 ganglia, 5 thoracic and 6 abdominal, in the ventral nerve chain posterior to the subesophageal ganglion. The sense organs include eyes, statocysts, olfactory filaments, and tactile setae. The eyes are well developed in most decapods, but they may be reduced or entirely absent in species living in the deep sea, in caves, or in burrows. Luminous organs of various kinds occur in a number of deep-sea decapods, and some shrimps have the ability to eject a luminous fluid.

As in most crustaceans, the alimentary tract is a more or less straight tube divided into three parts: the foregut, the midgut, and the hindgut. The first and last have a chitinous, and sometimes calcified, lining. The foregut, in most decapods, is dilated to form a stomach. Its anterior, or cardiac, portion is at least partially lined with movably articulated, calcified ossicles, controlled by a complex system of muscles called the gastric mill. The posterior, or pyloric, part is provided with hairy ridges which combine to form a straining apparatus. The midgut is variously furnished with blind tubes, and a mass of minutely ramified tubules, known as the hepatopancreas or liver, empties through one or more ducts into the anterior end of the midgut.

The heart is short, polygonal, and situated under the posterior part of the carapace. It has three or more pairs of venous ostia, and two pairs of arteries lead from it, in addition to the median anterior and posterior arteries. The blood collects in a venous sinus in the midventral line, passes to the gills, and then back to the pericardial sinus surrounding the heart.

In practically all decapods, respiration is accomplished by a series of gills, attached to the lateral walls of the thoracic somites and to the basal segments of the thoracic appendages. A continuous current of water is drawn through the branchial chamber and over the gills by the fluttering motion of the outer lobe of the second maxilia. The current usually moves forward, but this direction may be periodically reversed in burrowing species. In those crabs best adapted for life on land, the action of the gills is supplemented by the lining membrane of the branchial chamber. This chamber is covered with minute villi and is unusually well supplied with blood vessels, and thus functions as a lung.

The antennal gland is the chief excretory organ in all decapods. Maxillary glands are often present in the larval stages, but they never persist in the adult. The antennal gland is compact and more complex than in any other crustacean group. It consists of three divisions: the saccule, which is usually partitioned or ramose; the labyrinth, a spongy mass with a complex system of canals; and the bladder with a duct leading to the exterior.

The sexes are separate in most decapods, although protandric hermaphroditism occurs in some shrimps. The testes most often lie partly in the thorax and partly in the abdomen, and they are usually connected across the midline. The ovaries resemble the testes in shape and position.

Development. In only one group of shrimps are all of the presumably primitive larval stages represented. The first stage is a typical crustacean nauplius with an unprotected oval body bearing a median eye and three pairs of appendages. This stage is followed by a metanauplius in which four more appendage rudiments appear. The third pair of the original appendages becomes less of a swimming organ and more like a pair of primitive mandibles. In the third stage, the protozoea, the seven pairs of appendages become more highly developed, a carapace covers the anterior part of the body, the abdomen is clearly formed though unsegmented at first, the rudiments of paired eyes appear, and the heart is formed. In the next stage, the zoea, the eyes become movable and the carapace develops a rostrum. All thoracic appendages are present, at least in rudimentary form, and those of the abdomen appear, especially the uropods. In the last larval stage, the mysis, the well-developed thoracic appendages replace the antennae as the chief swimming organs. The abdomen increases markedly in size and takes on a form similar to that of the adult.

Phylogeny. No true decapod fossils are known with certainty from Paleozoic deposits, but decapod shrimps are not uncommon in Triassic and, especially, Jurassic rocks. Lobsterlike forms related to Recent deep-sea species also appeared in the Triassic; true lobsters are known from the Jurassic, and crayfishes from the Cretaceous. Crabs are well known as fossils, with primitive groups appearing in the Jurassic; several more specialized groups are found in the Cretaceous, and all of the major existing groups were represented in the Tertiary.

The evolution of the decapods has not been satisfactorily worked out and may never be known completely. The presence of a nauplius larva in the development of some shrimps suggests a link with the primitive crustaceans, but it can hardly be doubted that the Decapoda, as a group, are highly specialized. They are certainly closely related to the Euphausiacea, and fusion of the two groups has been suggested by some authors.

Classification. The classification of the 8500 living species of decapods presents a difficult problem that has not yet been entirely solved. The chief disparity seems to lie between the shrimps (Natantia) and all of the other decapods (Reptantia):

Order Decapoda
- Suborder Natantia
  - Section Penaeidea
  - Section Caridea
  - Section Stenopodidea
- Suborder Reptantia
  - Section Macrura
  - Section Anomura
  - Section Brachyura

Penaeidea. The primitive section of the Decapoda, penaeideans, is distinguished by the following characters: The pleura of the first abdominal somite overlap those of the second; the third legs are nearly always chelate, but are no stouter than the first pair; the first pleopods of the male bear a complicated flaplike appendix, the petasma; and the females have a spermatophore receptacle, the thelycum, on the ventral surface of the posterior thoracic somites. The best-known penaeideans, including most of the commercially important shrimps or prawns of the warmer seas (Penaeidae), live on muddy bottoms in shallow or moderate depths. However, they also occur both in the deep sea and as pelagic organisms in the mid-depths. There are more than 300 species of living penaeideans.

Caridea. This is the largest and most diverse group of shrimps and prawns. The pleura of the second abdominal somite overlap those of the first; the third legs are never chelate; there is no petasma or thelycum. The sexes can usually be distinguished by the presence, in the male, of two stylets on the inner edge of the inner branch of the second pleopods. Carideans occur in all parts of the sea, often in association with other marine animals. Members of at least two families, Atyidae and Palaemonidae, are also widespread in fresh water. The edible shrimps and prawns of northern Europe and of northwestern North America, of the genera *Crangon* and *Pandalus*, belong to the Caridea, as do the large fresh-water prawns of the tropical genus *Macrobrachium*. Most shrimps and prawns are favorite foods of fishes. Some carideans maintain "service stations" where they remove parasites from the outer skin, mouths, and gills of the fish. More than 1500 species of Caridea are known.

Stenopodidea. This is a small group of shrimps which superficially resemble the Penaeidea. The third pereiopods are chelate but are much longer and stouter than the first pair; the pleura of the second abdominal somite do not overlap those of the first; and there is no petasma or thelycum. All of the 20 or more living species are marine; some are closely associated with other animals, such as sponges.

Macrura. This long-tailed group of the Reptantia includes the deep-water and fossil eryonids, the spiny lobsters, the true lobsters, and the mud shrimps (Fig. 1). The abdomen is extended and bears a well-developed tail fan. There are about 700 known living species.

The Eryonidea are rather thin-shelled, blind inhabitants of the depths of the sea. The carapace is flattened dorsally and considerably expanded laterally. The first four, or all five, pairs of pereiopods are chelate, with the first pair much elongated. There are about 40 living species; the group was probably more numerous in ancient seas.

The Scyllaridea include the spiny lobsters or langoustes (Palinuridae) and the Spanish, or shovel-nosed, lobsters (Scyllaridae). They are heavily armored like the true lobsters but are distinguished by the absence of a rostrum and of chelae, except occasionally on the last pereiopod of the female. They are abundant in shallow and moderate depths of warm and temperate seas, where they are often of considerable commercial importance. There are about 85 Recent species.

The true lobsters and crayfishes, Nephropidea, also have a firm shell but are also characterized by a rostrum and by chelae on the first three pairs of pereiopods, with the first pair being noticeably larger than the others. Most lobsters (*Nephropidae*) are found in cool seas or in the cool, off-shore waters of the tropics. The most familiar lobsters (*Homarus*) are those found along the Atlantic coasts of Europe and North America. The Norway lobster of Europe (*Nephrops*) is also of some commercial importance but is smaller and less meaty. The crayfishes

Fig. 1. Reptantian decapods. (*a*) Spanish lobster, *Scyllarides aequinoctialis*. (*b*) Mud shrimp, *Callianidea laevicauda*. (c) Eryonid, *Polycheles crucifer*. (*d*) Spiny lobster, *Panulirus interruptus*. (e) Crayfish, *Orconectes limosus*. (Smithsonian Institution)

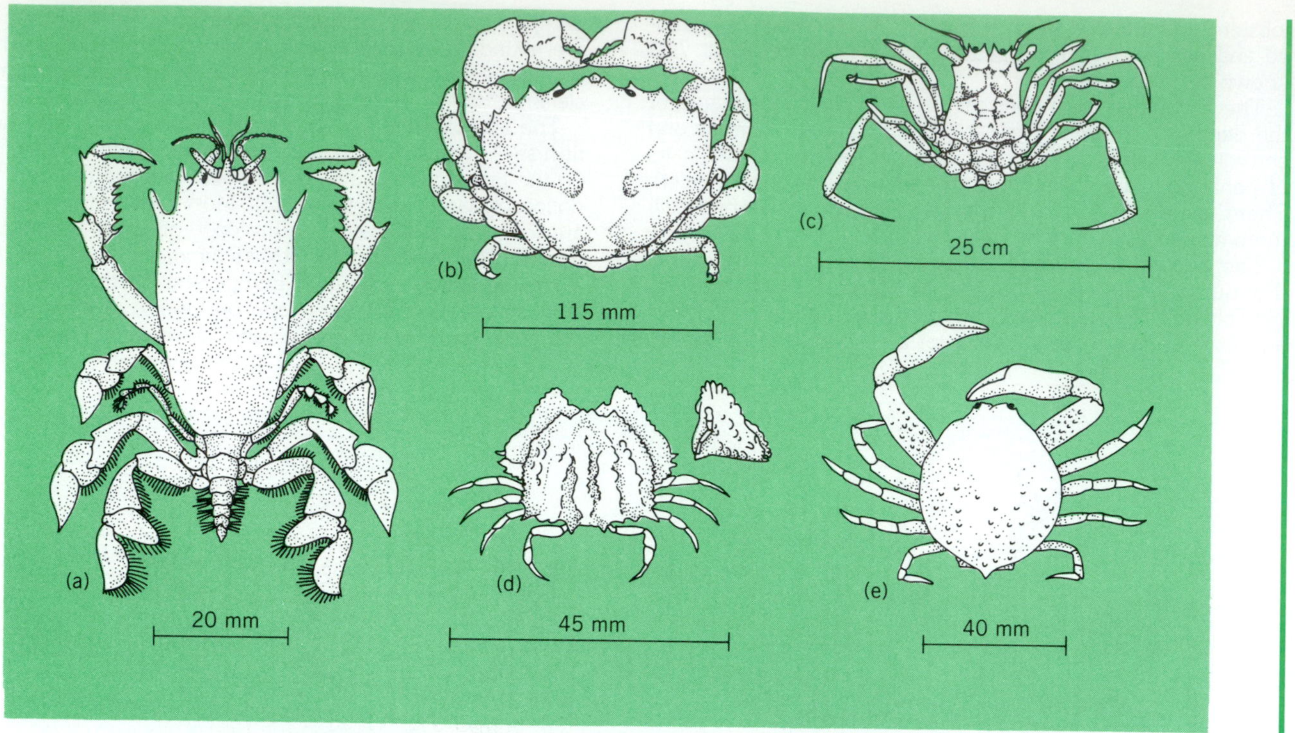

Fig. 2. Common brachyuran crabs. (*a*) Gymnopleuran crab, *Raninoides louisianensis.* (*b*) Dromilid crab, *Dromia erythropus.* (c) Mask crab, *Ethusa mascarone americana.* (*d*) Box crab, *Calappa sulcata.* (e) Purse crab, *Persephona punctata.* (*Smithsonian Institution*)

(Cambaridae, Astacidae, Parastacidae) are widespread in fresh waters of the temperate regions of all continents except Africa. They are of commercial importance in southern Europe and Australia. More than 300 living species of lobsters and crayfishes are known, more than half of them being from the fresh waters of the United States.

The mud shrimps, Thalassinidea, are usually thin-shelled, burrowing crustaceans with large, chelate or subchelate first pereiopods, and no chelae on the third pereiopods. More than 250 living species are found in shallow and deep seas throughout the world, especially in tropical and warm temperate regions.

Anomura. This intermediate, and possibly unnatural, group lies between the Macrura and the Brachyura. It includes the galatheids, the porcelain crabs or rock sliders, the hermit crabs, the king crabs, and the mole crabs or hippas. In nearly all of these diverse crustaceans, the first, and sometimes the last, pereiopods are chelate or subchelate. The abdomen is usually bent forward ventrally or is asymmetrical, soft, and twisted. There are about 1300 Recent species.

Brachyura. These are the true crabs (Fig. 2). The abdomen is symmetrical, without a tail fan, and bent under the thorax. The first pereiopods are always chelate or subchelate. The true crabs are as numerous as all other decapods combined, numbering nearly 4500 Recent species. They are divided into four subsections.

The Gymnopleura include about 30 species of primitive burrowing crabs; their carapaces are more or less trapezoidal or elongate, the first pereiopods are subchelate, and some or all of the remaining pereiopods are flattened and expanded for burrowing.

The Dromiacea is a primitive group of about 200 species. The first pereiopods are chelate; the last pair is dorsal in position and modified for holding objects, such as sponges, tunicates, and bivalve mollusk shells, over the crab.

In Oxystomata the first pair of pereiopods are chelate and the last pair are either normal or modified as in the Dromiacea. The oxystomes include the mask crabs (Dorippidae), the box crabs (Calappidae), and the purse crabs (Leucosiidae). There are nearly 500 Recent species.

Most of the crabs, more than 3700 living species, belong to the Brachygnatha, in which the last pereiopods are rarely reduced or dorsal in position. To this group belong the swimming crabs (Portunidae). Also worthy of mention are freshwater crabs (Potamidae, Pseudothelphusidae, and Trichodactylidae), found in tropical and some temperate regions, usually in areas not inhabited by crayfishes. The crabs of the genus *Cancer* are large and abundant enough to be of commercial importance in northern Europe and North America. The rock, or Jonah, crabs of New England and the Dungeness crab of the Pacific Coast are familiar edible crabs of this genus. The ubiquitous little mud crabs (Xanthidae) are well known to every visitor to rocky shores. One of them, the stone crab (*Menippe*), is highly esteemed by connoisseurs of seafood. The pea crabs (Pinnotheridae) are often found in oysters and other bivalve mollusks. The square-backed crabs (Grapsidae) are characteristic of warm, marshy areas but are not restricted to that habitat; they mark the trend toward the true land crabs (Gecarcinidae), the depreciations and intrusions of which are familiar to all who live in the tropics. The ghost crabs (*Ocypode*), which scuttle almost unseen over sandy beaches, and the fiddler crabs (*Uca*) of muddy shores, are closely related. The end of the list is reached with the spider, or decorator, crabs (Majidae), slow-moving animals that often conceal themselves by attaching seaweeds and various sessile animals to their carapaces. One of them (*Macrocheira*) is the largest arthropod now alive. *See* ARTHROPODA; CRAB; CRAYFISH; CRUSTACEA; LOBSTER; SHRIMP. [F.A.Ch.]

Decca A hyperbolic navigation system which establishes a line of position from measurement of the phase difference between two continuous-wave signals. The intersection of the

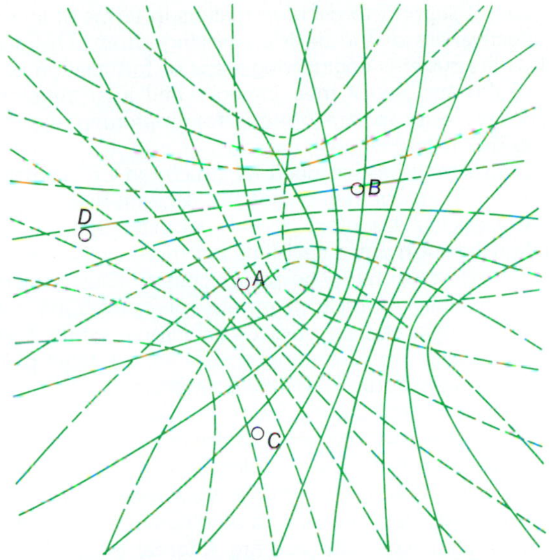

Decca station location and patterns. *A* **is master station and** *B, C,* **and** *D,* **slave stations. (***After P. C. Sandretto, Electronic Avigation Engineering, International Telephone and Telegraph Corp., 1958***)**

two lines of position from two pairs of transmitting stations establishes a navigational fix, or location. Charts surprinted with Decca hyperbolas of constant phase difference are used to show the navigator's position. Moving charts upon which a pen traces the track of the craft can be used to show position and provide a permanent record.

The two lines of position required for a fix can be provided by one "master" and two "slave" transmitters. A third slave located at the apex of a triangle in which the master is centered is also used (see illustration). Since continuous waves are employed, each station must use a different radio frequency. *See* HYPERBOLIC NAVIGATION SYSTEM. [A.A.McK.]

Decibel A logarithmic unit used to express the magnitude of a change in level of power, voltage, current, or sound intensity. A decibel (dB) is 1/10 bel.

In acoustics a step of 1 bel is too large for most uses. It is therefore the practice to express sound intensity in decibels. The level of a sound of intensity I in decibels relative to a reference intensity I_R is given by notation (1). Because sound

$$10 \log_{10} \frac{I}{I_R} \qquad (1)$$

intensity is proportional to the square of sound pressure P, the level in decibels is given by Eq. (2). The reference pressure is

$$10 \log_{10} \frac{P^2}{P_R{}^2} = 20 \log_{10} \frac{P}{P_R} \qquad (2)$$

usually taken as 0.0002 dyne/cm^2 or 0.0002 microbar or 20 micropascals. *See* BEL; SOUND PRESSURE.

The neper is similar to the decibel but is based upon natural (napierian) logarithms. One neper is equal to 8.686 dB. *See* NEPER; VOLUME UNIT (VU). [K.D.K.]

Deciduous plants Plants that regularly lose their leaves at the end of each growing season. Dropping of the leaves occurs at the inception of an unfavorable season characterized by either cold or drought or both. Most woody plants of temperate climates have the deciduous habit, and it may also occur in those of tropical regions having alternating wet and dry seasons. Many deciduous trees and shrubs of regions with cold winters become evergreen when they are grown in a warm climate. Conversely, such trees as magnolias, evergreen in warm areas, become deciduous when they are grown in colder climates. [N.A.]

Decimal number system The system, used in ordinary arithmetic, in which numbers are represented as linear combinations of powers of 10. The base 10 has as its origin the fact that humans have a total of 10 digits on both hands. Other than this, the decimal number system has little to recommend it over other systems. *See* NUMBER SYSTEMS. [M.Ye.]

Decision analysis An applied branch of decision theory. Decision analysis offers individuals and organizations a methodology for making decisions; it also offers techniques for modeling decision problems mathematically and finding optimal decisions numerically. Decision models have the capacity for accepting and quantifying human subjective inputs: judgments of experts and preferences of decision-makers. Implementation of models can take the form of simple paper-and-pencil procedures or sophisticated computer programs known as decision aids or decision systems.

The methodology is rooted in postulates of rationality—a set of properties which preferences of rational individuals must satisfy. One such property is transitivity: if an individual prefers action a to action b and action b to action c, he or she should prefer a to c. From the rationality postulates, principles of decision-making are derived mathematically. The principles prescribe how decisions ought to be made, if one wishes to be rational. In that sense, decision analysis is normative.

The methodology is broad and must always be adapted to the problem at hand. An illustrative adaptation to a class of problems known as decision-making under uncertainty (or risk) is outlined in the illustration and consists of seven steps:

1. The problem is structured by identifying feasible actions, one of which must be decided upon; possible events, one of which occurs thereafter; and outcomes, each of which results from a combination of decision and event. Problem structuring can be facilitated by displays such as decision trees and decision matrices.

2. At the time of decision-making, the event that will actually occur cannot be predicted perfectly. The degree of certainty about the occurrence of an event, given all information at hand, is quantified in terms of the probability of the event. *See* PROBABILITY.

3. Preferences are personal: the same outcome may elicit different degrees of desirability from different individuals. The subjective value that a decision-maker attaches to an outcome is quantified and termed the utility of outcome.

4. The preceding steps conform to the principle of decomposition: probabilities of events and utilities of outcomes must be measured independently of one another. They are next combined in a criterion for evaluating decisions. The utility of a decision is defined as the expected utility of the outcome. The optimal, or the most preferred, decision is one with the maximum utility.

5. The probability encodes the current state of information about a possible event. Often, additional information can be acquired in the hope of reducing the uncertainty. The monetary value of such information is computed before purchase and compared with the cost of information. Thus, one can determine whether or not acquiring information is economically rational.

6. The source of information may be a real-world experi-

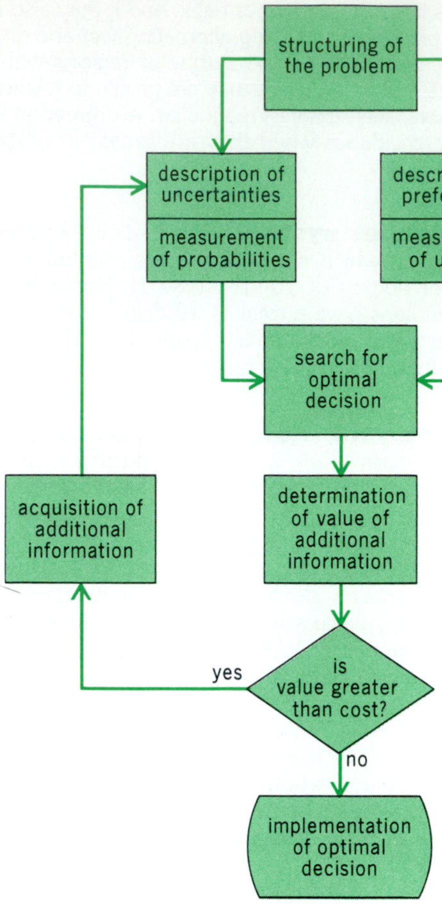

A methodology of decison analysis.

ment, a laboratory test, a mathematical model, or the knowledge of an expert. The informativeness of the source is described in terms of a probabilistic relation between information and event. This relation, known as the likelihood function, makes it possible to revise the prior probability of the event and to obtain a posterior probability of the event, conditional on additional information. The revision is carried out via Bayes' rule.

7. Given the additional information, prior probabilities can be replaced by posterior probabilities, and the analysis can be repeated from step 4 onward. Steps 4–6 may be cycled many times, until the cost of additional information exceeds its value, at which moment the optimal decision is implemented.

Measurement of probability and utility functions is guided by principles of decision theory, statistical estimation procedures, and empirical laws provided by behavioral decision theory—a branch of cognitive psychology. *See* COGNITION; DECISION THEORY. [R.Krz.]

Decision support system A system that supports technological and managerial decision making by assisting in the organization of knowledge about ill-structured, semistructured, or unstructured issues. A structured issue has a framework comprising elements and relations between them that are known and understood. Structured issues are generally ones about which an individual has considerable experiential familiarity. A decision support system (DSS) is not intended to provide support to humans about structured issues since little cognitively based decision support is generally needed.

Emphasis in the use of a decision support system is upon provision of support to decision makers in terms of increasing the effectiveness of the decision-making effort. This support involves the systems engineering steps of formulation of alternatives, the analysis of their impacts, and interpretation and selection of appropriate options for implementations. *See* SYSTEMS ENGINEERING.

Decisions may be described as structured or unstructured, depending upon whether or not the decision-making process can be explicitly described prior to its execution. Generally, operational performance decisions are more likely than strategic planning decisions to be prestructured. Thus, expert systems are usually more appropriate for operational performance and operational control decisions, while decision support systems are more appropriate for strategic planning and management control. *See* EXPERT SYSTEMS.

The primary components of a decision support system are a database management system (DBMS), a model-base management system (MBMS), and a dialog generation and management system (DGMS). An appropriate database management system must be able to work with both data that are internal to the organization and data that are external to it. Model-base management systems provide sophisticated analysis and interpretation capability. The dialog generation and management system is designed to satisfy knowledge representation, and control and interface requirements. *See* DECISION THEORY. [A.P.Sa.]

Decision theory A broad spectrum of concepts and techniques which have been developed to both describe and rationalize the process of decision making, that is, making a choice among several possible alternatives. In the broadest sense, one can differentiate between prescriptive decision theory, which formulates how decisions should be made, and descriptive decision theory, which deals with how people actually make decisions. Behavioral and social scientists and psychologists are trying to discover more elaborate descriptive models of the decision process in order to provide mathematicians, economists, strategists, business managers, and others with more sophisticated prescriptive decision-making procedures.

The information necessary for a rational decision is usually summarized in a decision matrix, as shown in the illustration. The alternate pure (independent from each other) or mixed (combinations of pure strategies) strategies (S_1 to S_n) are listed in the first column. Across the top all possible states of nature (N_1 to N_m) which might affect the outcome of the various strategies are enumerated. The probabilities of occurrence of various states of nature (p_1 to p_m) must be estimated and listed with the states of nature. Since these values will critically influence the decision, they should be as reliable as possible. In value theory one distinguishes between the subjective probability of occurrence which reflects the decision maker's opinion on what the probability is and the objective probability which is derivable mathematically or as a result of observations.

Corresponding to each interaction of a strategy (S_i) and a state of nature (N_j), one determines a consequence or outcome (O_{ij}) in terms of some value or utility parameters. The outcome may sometimes be easily computed, for example, the interest cost on a business loan, but quite often must be based on insufficient data, past experience, or questionably valid assumptions, such as would be the case in trying to assess the outcome of a limited war. The generation of reliable outcome data can be expensive and time-consuming, and a probability of accuracy is often associated with each outcome. The question of how to assess the relative value or utility of mixed parameters to present a single parameter for the outcome is still a paramount question in decision and utility or value theory.

States of nature		N_1	N_2	— —	N_j	— — — — — —	N_m	Expected value
Probabilities of occurrence		p_1	p_2	— —	p_j	— — — — — —	p_m	
Alternate strategies	S_1	O_{11}	O_{12}	— —	O_{1j}	— — — — — —	O_{1m}	EV_1
	S_2	O_{21}	O_{22}	— —	O_{2j}	— — — — — —	O_{2m}	EV_2
	|	|	|		|		|	|
	|	|	|		|		|	|
	S_i	O_{i1}	O_{i2}	— —	O_{ij}	— — — — — —	O_{im}	EV_i
	|	|	|		|		|	|
	|	|	|		|		|	|
	|	|	|		|		|	|
	S_n	O_{n1}	O_{n2}	— —	O_{nj}	— — — — — —	O_{nm}	EV_n

Form of decision matrix. O_{ij} = outcome of *j*th state of nature (N_j) intersecting with the *i*th strategy (S_i).

The information in the decision matrix is used to generate the last column which describes the overall expected outcome, or expected value (*EV*), or expected utility (*EU*), or expected profit for each strategy. Depending upon the confidence one places in the prediction of the probabilities of the states of nature, the decision process is classified as one of three types.

In the case of decision making under certainty (DMUC) the state of nature which will interact with the strategies can be defined with sufficient certainty so that its probability is close to 1. This case reduces the decision matrix to one column showing the outcomes (profit, cost, and utility) of each strategy with the state of nature. Decision theory indicates that, assuming the alternate strategies are equal in all other respects, the one which will maximize the utility (maximize profit or minimize cost) should be selected. Since in many cases the dependent variable (outcome) is continuously related to the independent variable (strategy), one can obtain relationships which can be used to analytically find the strategy which will correspond to maximum utility. The process is normally termed optimization, which in its broadest interpretation refers to decision making under certainty. *See* OPTIMIZATION.

Decision making under risk (DMUR) implies that, although a number of possible states of nature may interact with each strategy, the probability of occurrence of each of these is known with some degree of certainty. The expected value or utility for each strategy is obtained as a weighted average sum of the individual possible outcomes or, in mathematical terms,

$$EV_i = \sum_{j=1}^{m} p_j O_{ij}$$

The condition of dominance is usually used to eliminate all strategies which are dominated. A dominant strategy has more desirable outcomes under any of the possible states of nature than the dominated strategy.

If the probabilities of the states of nature cannot be predicted, then the decision has to be made under conditions of uncertainty (decision making under uncertainty; DMUU). Decision theory provides a number of techniques to deal with this situation. Which of these will work best in a given situation is questionable, and their application depends primarily on the philosophical outlook of the decision maker.

1. Equal probability criterion: This is also known as the LaPlace criterion and assumes that all states of nature considered are equally likely (that is, $p_j = 1/m$). The decision matrix is then used just as in DMUR.
2. Maxim criterion: This represents a pessimistic outlook in that it prescribes the strategy which will maximize the minimum profit.
3. Maximax criterion: The optimist will like this one since it chooses the strategy which will maximize the maximum possible profit (the best of the best outcomes in each row).
4. Hurwicz criterion: A coefficient of optimism is defined as $\alpha = 1$ corresponding to a complete optimist, and $\alpha = 0$ corresponding to a complete pessimist. The maximum possible profit for each strategy is then multiplied by α and the minimum profit by $(1 - \alpha)$ to find the best expected value.
5. Regret criterion: Also known as the Savage principle, this criterion is used to construct a regret matrix in which each outcome entry represents a regret defined as the difference between the best possible outcome and the given outcome. The matrix is then used as in DMUR with expected regret as the decision-determining quantity. *See* GAME THEORY. [I.P.]

Declinometer A geomagnetic instrument for measuring magnetic declination, sometimes called variation of the compass, which is defined as the angle between true or geographic north and magnetic north—the direction of the magnetic meridian. The instrument is suited for determining the direction in which a perfect compass needle would point and for comparing this with the known true azimuth of a fixed reference point or a celestial object such as the Sun or the North Star.

The compass declinometer employs a thin compass needle 6 in. (15 cm) long, supported on a sapphire bearing and steel pivot of high quality. Peep sights serve for aligning the compass box on an azimuth mark. The accuracy attained is 2–3 minutes of arc in middle latitudes.

The transit declinometer, a surveyor's transit, built to exacting specifications with respect to freedom from traces of magnetic impurities and quality of the compass needle, has a 17-power telescope for sighting on a mark and for making solar and stellar observations to determine true directions. A microscope mounted on the side of the telescope permits a more accurate alignment of the instrument axis with the magnetic needle (see illustration).

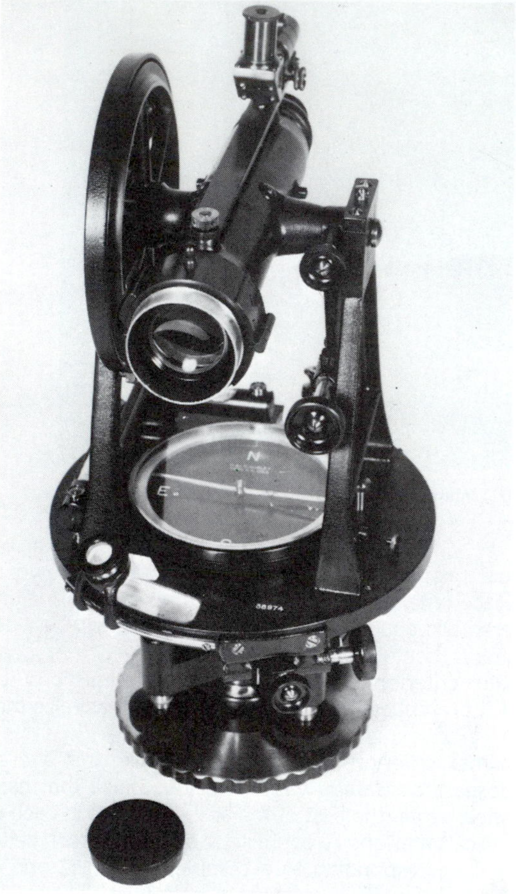

Transit declinometer. ***(U.S. Coast and Geodetic Survey)***

For magnetic fieldwork of highest accuracy and for standard calibrating instruments in a magnetic observatory, the precision declinometer (sometimes called a magnetometer) employs a cylindrical permanent magnet suspended in a horizontal position by a fine gold ribbon and equipped either with a collimating lens and cross hair or with a mirror. The position of the magnet is observed with a suitable telescope and scale, which are integral parts of the equipment. The accuracy for an observatory declinometer is 0.1–0.2 minute of arc. *See* MAGNETOMETER. [J.H.Ne.]

Decomposition potential The electrode potential at which the electrolysis current begins to increase appreciably. Decomposition potentials are used as an approximate characteristic of industrial electrode processes. *See* ELECTROCHEMICAL PROCESS; ELECTROLYSIS.

Decomposition potentials are obtained by extrapolation of current-potential curves. Extrapolation is not precise because there is a progressive increase of current as the electrode potential is varied. The decomposition potential, for a given element, depends on the range of currents being considered. The cell voltage at which electrolysis becomes appreciable is approximately equal to the algebraic sum of the decomposition potentials of the reactions at the two electrodes and the ohmic drop, or voltage drop, in the electrolytic cell. The ohmic drop term is quite negligible for electrolytes with high conductance. For a discussion of current-potential curves *see* OVERVOLTAGE. *See also* ELECTROLYTIC CONDUCTANCE. [P.De.]

Decompression illness Symptoms in humans which result from a sudden reduction in atmospheric pressure. It is also called dysbarism, caisson disease, the bends, and compressed-air illness. It is most commonly seen in two groups of subjects, (1) those who rapidly ascend in nonpressurized airplanes to altitudes in excess of 18,000 ft (5500 m), and (2) divers, scuba divers, sandhogs, and professional workers in hyperbaric chambers who work under increased ambient pressures and are decompressed. *See* DIVING.

This condition is caused by the formation of nitrogen bubbles in the tissues and blood vessels. These bubbles plug vessels and expand in tissue spaces producing the characteristic symptoms and signs of this illness. Helium, when present in high concentrations utilized by divers or found in some spacecraft, can also produce a similar condition.

The onset of symptoms may occur at any time from a few minutes to several hours after decompression. The most common manifestation is pain in the joints and muscles. However, skin, respiratory, and neurologic symptoms are not uncommon. The skin manifestations are itching, discoloration, and edema (swelling). Respiratory symptoms are coughing and dyspnea (difficulty in breathing). The neurologic symptoms vary from mild paresthesia (sensations of tingling, crawling, or burning of the skin) and weakness to total paralysis; loss of bladder and rectal sphincter control is common. The severe forms of this illness are followed by circulatory failure, paralysis, coma, and death.

Treatment is recompression, followed by gradual decompression to normal atmospheric pressure. Prognosis is generally good except in those subjects who show central nervous system damage. [McC.G.]

Decontamination of radioactive materials The removal of radioactive contamination which is deposited on surfaces or may have spread throughout a work area. Personnel decontamination is also included.

Decontamination methods follow two broad avenues of attack, mechanical and chemical. Commonly used mechanical methods are vacuum cleaning, sand blasting, blasting with other abrasives, flame cleaning, scraping, ultrasonic radiation, and surface removal (for example, removal of concrete floors with an air hammer). The principal chemical methods of decontamination are water washing, steam cleaning, and scrubbing with detergents, acids, caustics, and solvents.

Another important method of handling contamination is to store the contaminated object, or temporarily abandon the contaminated space. This can be done when the use of the material or space is not necessary for a period of time and the half-life of the contaminant is relatively short.

Personnel decontamination methods differ from those used for materials primarily because of the possibilities of injury to the person being decontaminated. Soap and water (sequestrants and detergents) normally remove more than 99% of the contaminants. If it is necessary to remove the remainder, chemical methods which remove the outer layers of skin upon which the contamination has been deposited can be used. These chemicals—citric acid, potassium permanganate, and sodium bisulfite are examples—should be used with caution and preferably under medical supervision, because of the increased risk of injury to the skin surface. It is very difficult to remove radioactive material once it is fixed inside the body, and the ensuing hazard depends very little on the method of entry into the body. *See* MONITORING OF IONIZING RADIATION. [K.Z.M.]

Deep inelastic collisions A type of nuclear reaction in which the two nuclei interact strongly while their surfaces overlap. Deep inelastic collisions of one nucleus with another represent a class of reaction mechanism with interaction times intermediate between direct and compound nucleus processes. They often occur in collisions of heavy nuclei at center-of-mass energies less than 5 MeV per nucleon above the Coulomb barrier. During the brief encounter of the two nuclei, energy and mass flow from one nucleus into the other. After a short time, corresponding to the time it takes for the intermediate dinuclear complex to make a partial rotation (10^{-22} to 5×10^{-21} s), the two nuclei separate again. Of the characteristic features of these collisions, two of the most important properties are the sizable amounts of energy dissipation and nucleon (neutron and proton) exchange. *See* NUCLEAR REACTION. [J.R.Hu.]

Deep-sea trench A long, narrow, very deep, and asymmetrical depression of the sea floor, with relatively steep sides. Oceanic trenches characterize active margins at the ocean-basin–continent or ocean-basin–island-arc boundaries. They contain the greatest oceanic depths and are associated with the most active volcanism, largest negative gravity anomalies, most frequent shallow seismicity, and almost all of the intermediate and deep-focus earthquake activity. As the surface expression of the widely accepted process of subduction by which oceanic crustal material is returned to the upper mantle, they are key elements in current models of plate tectonic evolution on Earth and possibly on Venus. *See* PLATE TECTONICS; VOLCANOLOGY.

Deep-sea trenches are the signature relief form of the Pacific; in a counterclockwise direction, they occur from southern Chile to just northeast of North Island, New Zealand. A secondary or outer branch trends southward from near Tokyo Bay in a festoon of arcs to south of Palau. The principal gaps in the circum-Pacific chain are from Baja California to southeastern Alaska, and off the northern coast of New Guinea. From eastern New Guinea to southern Vanuatu the trenches lie southwest, or "inside," the island chains; otherwise their characteristics are like those facing the Pacific. The Indian Ocean contains only the very long contorted Sunda Trench that appears near the northwestern end of Sumatra and extends southeast and east past Timor, to curve north and west near Aru and end adjacent to Buru. In the Atlantic, the Puerto Rico–Antillean trench system extends outside the island arc from eastern Hispaniola around to Trinidad; but south of 14°N, off Barbados, the trench is filled with sediment. In the far South Atlantic a typical island-arc–trench complex extends from near South Georgia through the South Sandwich Archipelago.

A series of pioneering gravity observations with pendulum instruments on Dutch submarines during the 1920s and 1930s established that the East Indian trenches, and several others, were characterized by a belt of negative gravity anomalies of 150–200+ milligals, that is, values 150–200 parts per million less than normal, interpretable as deficiency of mass near and at their axes.

It was established that oceanic crust is thin, that crust under island arcs is thicker, and its layers display different sound transmission velocities, indicating different composition. Shipboard studies in the Middle America, Tonga, Cedros, Aleutian, Peru-Chile, and Sunda trenches established that the characteristic oceanic crustal layer [that is, 6.8–7.0 km/s (4.1–4.2 mi/s) compressional wave velocity] does not end or thin under the trench; rather, it may thicken slightly but does deepen steeply as it passes beneath the island arc or continental slope by the process of subduction. *See* EARTH, GRAVITY FIELD OF; EARTH, HEAT FLOW IN; EARTH CRUST; FAULT AND FAULT STRUCTURES; OCEANIC ISLANDS; SEISMOLOGY; SUBDUCTION ZONES. [R.L.Fi.]

Deer The name for 41 species of even-toed ungulates (artiodactylids) that comprise the distinct and homogeneous family Cervidae. The males have antlers, which are bony outgrowths on their heads that do not contain horn, are generally branched, and are shed yearly. The feet have four toes and, as such, are less developed than those of other ruminants. The third and fourth digits are well developed, while the second and fifth are small and do not touch the ground.

Musk deer occur in the mountains of central Asia at high altitudes in Tibet, Korea, Mongolia, and Siberia. They are solitary animals and are active nocturnally. The male has a scent gland, known as a pod or musk gland, in the skin of the abdomen in front of the navel that secretes musk, used as a fixative in the perfume industry.

A number of species of deer that occur throughout Eurasia are called European deer. The red deer (*Cervus elaphus*), one of the best-known species, is still common throughout Europe and also is found in southwestern Asia and northwestern Africa. These deer are closely related to the wapiti of North America.

One of the best-known species of American deer is the white-tailed deer (*Odocoileus virginianus*), which is also known as the Virginia deer. This animal is found to range in North America and into South America as far as Peru, where it prefers woods and thickets to heavily forested areas. In the western states the mule deer (*O. hemionus*) may be distinguished from the white-tailed deer by the long ears, black-tipped tail, and greater tendency to occur in herds. These animals range from Alaska to New Mexico and prefer more open habitats. The wapiti or North American elk (*Cervus canadensis*) is more common in Canada than in the United States. The Key deer of the Florida Keys (*O. virginianus clavium*) is in danger of extinction because of the destruction of its habitat. *See* ARTIODACTYLA; MOOSE; REINDEER. [C.B.C.]

Definite composition, law of The law that a given chemical compound always contains the same elements in the same fixed proportions by weight. Thus, whatever its source, silver chloride always contains 100 g (3.53 oz) of silver to every 32.85 g (1.159 oz) of chlorine. If a compound is formed by the union of m atoms of one element, each weighing a, with n atoms of another element, each weighing b, the composition by weight of one molecule of the compound is in the ratio $ma{:}nb$. This must be the composition of any mass of the compound, provided that all atoms of the same kind have the same weight. It is now known that this is not usually the case but that the atoms of an element may consist of a number of isotopes, having different masses. However, as long as any sample of the element always contains the same relative proportions of the isotopes, the law still holds. Much more widespread and serious departures from the law of definite composition occur in a large variety of solid compounds (the nonstoichiometric compounds). *See* ATOMIC WEIGHT; ISOTOPE; NONSTOICHIOMETRIC COMPOUNDS; STOICHIOMETRY. [T.C.W.]

Defoliant and desiccant Defoliants are chemicals that cause leaves to drop from plants; defoliation facilitates harvesting. Desiccants are chemicals that kill leaves of plants; the leaves may either drop off or remain attached; in the harvesting process the leaves are usually shattered and blown away from the harvested material. Defoliants are desirable for use on cotton plants because dry leaves are difficult to remove from the cotton fibers. Desiccants are used on many seed crops to hasten harvest; the leaves are cleaned from the seed in harvesting. Defoliants and desiccants have also been used during war to destroy vegetation. [A.S.C.]

Degaussing Neutralization of the magnetization of a ship by properly located and oriented current-carrying coils which produce a magnetic field of desired strength and direction. *See* DEMAGNETIZATION.

A steel ship has structural components of many different ferromagnetic characteristics. Many parts become magnetized during the construction, and retain that magnetization for a long time. Other parts are "soft" iron and do not retain a magnetized condition permanently but become magnetized by induction in the magnetic field of the Earth. The ship then has a magnetic field. This magnetization of the ship causes deviation of the magnetic compass and may trigger magnetic mines or other explosive devices when the ship passes near them. *See* MAGNETIZATION.

One method for neutralizing the magnetic field of the ship is to install coils in which currents are maintained to produce components of the field that will neutralize the field due to the magnetization of the ship. These coils are called degaussing coils. [K.V.M.]

Degeneracy (quantum mechanics) A term referring to the fact that two or more stationary states of the same quantum-mechanical system may have the same energy even though their wave functions are not the same. In this case the common energy level of the stationary states is degenerate. The statistical weight of the level is proportional to the order of degeneracy, that is, to the number of states with the same energy; this number is predicted from Schrödinger's equation. In quantum mechanics and in other branches of mathematical physics, the term degeneracy is employed also to characterize the eigenvalues of operators other than the energy operator. *See* EIGENVALUE (QUANTUM MECHANICS). [E.G.]

Degenerative cardiomyopathy Cardiomyopathy characterized by degenerative changes and fibrosis within the myocardium. This may occur as a secondary change associated with generalized disease processes. Metabolic disorders such as nutritional deficiencies (protein, calorie or vitamin deficiencies such as beriberi), anemia, endocrine disorders, hypersensitivity, lupus erythematosus, and amyloidosis may induce secondary cardiomyopathy. Primary cardiomyopathies are often caused by excessive alcohol consumption and toxic chemicals such as chloroform, carbon tetrachloride, arsenic, and phosphorus. Many of the degenerative cardiomyopathies are of unknown etiology. *See* HEART DISORDERS. [N.K.M.]

Degree-day A unit used in estimating energy requirements for building heating and, to a lesser extent, for building cooling. It is applied to all fuels, district heating, and electric heating. Origin of the degree-day was based on studies of residential gas heating systems. These studies indicated that there existed a straight-line relation between gas used and the extent to which the daily mean outside temperature fell below 65°F (18°C).

The number of degree-days to be recorded on any given day is obtained by averaging the daily maximum and minimum outside temperatures to obtain the daily mean temperature. The daily mean so obtained is subtracted from 65°F and tabulated. Monthly and seasonal totals of degree-days obtained in this way are available from local weather bureaus.

A frequent use of degree-days for a specific building is to determine before fuel storage tanks run dry when fuel oil deliveries should be made. Number of Btu which the heating plant must furnish to a building in a given period of time is

$$\text{Btu required} = \text{heat rate of building} \times 24 \times \text{degree-days}$$

where "Btu required" is the heat supplied by the heating system to maintain the desired inside temperature. "Heat rate of building" is the hourly building heat loss divided by the difference between inside and outside design temperatures. When the estimating procedure is applied to buildings with high levels of internal heat gains, as in a well-lighted office building, then degree-day data on other than a 65°F basis are required. *See* AIR CONDITIONING; COMFORT HEATING; PSYCHROMETRICS. [C.G.S.]

Degree of freedom (mechanics) Any one of the number of independent ways in which the space configuration of a mechanical system may change. A material particle confined to a line in space can be displaced only along the line, and therefore has one degree of freedom. A particle confined to a surface can be displaced in two perpendicular directions and accordingly has two degrees of freedom. A particle free in physical space has three degrees of freedom corresponding to three possible perpendicular displacements. A system composed of two free particles has six degrees of freedom, and one composed of N free particles has $3N$ degrees. Any requirement which diminishes by one the degrees of freedom of a system is called a holonomic constraint. *See* CONSTRAINT. [R.A.Fi.]

De Haas–van Alphen effect An oscillatory behavior of the magnetic moment of a pure metal crystal with changes in the applied magnetic field B, at very low temperatures. It is named after its discoverers, W. J. de Haas and P. M. van Alphen. This effect has its origin in the Bohr-Sommerfeld quantization of the orbits of conduction electrons under the influence of the magnetic field.

A measurement of the temperature dependence of the oscillation amplitude permits a determination of the cyclotron frequency, or equivalently the electron mass. In a metal, as in a semiconductor, electrons behave with an effective mass m^* rather than the free electron mass, and the de Haas–van Alphen effect thus allows a measurement of m^* in metals. *See* KINETIC THEORY OF MATTER.

Application of the Bohr-Sommerfeld quantization rules shows that the "period" of the de Haas–van Alphen oscillations (it is a period in magnetic field, not time) is given by the equation below. Here a_p is the area of the orbit in momentum

$$\Delta\left(\frac{1}{B}\right) = \frac{2\pi\hbar e}{a_p}$$

space of an electron at the Fermi level, that is, the cross section of the Fermi surface; e is the charge of the electron; and $\hbar$ is Planck's constant divided by 2π. By studying the dependence of a_p on the direction B of the magnetic field, sufficient information can be obtained to construct the detailed shape of the Fermi surfaces of a metal.

Virtually all metals available in sufficient purity have been studied by the de Haas–van Alphen effect. The effect is also the most powerful probe of Fermi surface properties in alloys and intermetallic compounds. *See* BAND THEORY OF SOLIDS; FERMI SURFACE. [J.B.K.]

Dehumidifier Equipment designed to reduce the amount of water vapor in the atmosphere. There are three methods by which water vapor may be removed: (1) the use of sorbent materials, (2) cooling to the required dew point, and (3) compression with aftercooling. *See* DEW POINT.

Sorbents are materials which are hygroscopic to water vapor. Solid sorbents include silica gels, activated alumina, and aluminum bauxite. Liquid sorbents include halogen salts such as lithium chloride, lithium bromide, and calcium chloride, and organic liquids such as ethylene, diethylene, and triethylene glycols and glycol derivatives.

Solid sorbents may be used in static or dynamic dehumidifiers. Bags of solid sorbent materials within packages of machine tools, electronic equipment, and other valuable materials subject to moisture damage constitute static dehumidifiers. A dynamic dehumidifier for solid sorbent consists of a main circulating fan, one or more beds of sorbent material, reactivation air fan, heater, mechanism to change from dehumidifying to reactivation, and aftercooler.

The liquid-sorbent dehumidifier consists of a main circulating fan, sorbent-air contactor, sorbent pump, and reactivator including contactor, fan, heater, and cooler. This unit will control the effluent dew point at a constant level because dehumidification and reactivation are continuous operations with a small part of the sorbent constantly bled off from the main circulating system and reactivated to the concentration required for the desired effluent dew point.

A system employing the use of cooling for dehumidifying consists of a circulating fan and cooling coil. The cooling coil may use cold water obtained from wells or a refrigeration plant, or may be a direct-expansion refrigeration coil. In place of a coil, a spray washer may be used in which the air passes through two or more banks of sprays of cold water or brine, depending upon the dew-point temperature required.

Dehumidifying by compression and aftercooling is used when the reduction of water vapor in a compressed-air system is required. This is particularly important, for example, if the air is used for automatic control instruments or cleaning of delicate machined parts. The power required for compression systems is so high compared to power requirements for dehumidifying by either the sorbent or refrigeration method that the compression system is not an economical one if dehumidifying is the only end result required. *See* DRYING OF GASES. [J.E.]

Dehydrogenation A reaction in which hydrogen is detached from a molecule. The reaction is strongly endothermic, and therefore heat must be supplied to maintain the reaction temperature. When the detached hydrogen is immediately oxidized, two benefits accrue: (1) the conversion of reactants to products is increased because the equilibrium concentration is shifted toward the products (law of mass action); and (2) the added exothermic oxidation reaction supplies the needed heat of reaction. This process is called oxidative dehydrogenation. On the other hand, excess hydrogen is sometimes added to a dehydrogenation reaction in order to diminish the complete breakup of the molecule into many fragments.

The primary types of dehydrogenation reactions are vapor-phase conversion of primary alcohols to aldehydes, vapor-phase conversion of secondary alcohols to ketones, dehydrogenation of a side chain, and catalytic reforming of naphthas and naphthenes in the presence of a platinum catalyst. All four of these types of dehydrogenation reactions are of major industrial importance. They account for the production of billions of pounds of organic compounds that enter into the manufacture of lubricants, explosives, plastics, plasticizers, and elastomers. *See* HYDROGENATION; OXIDATION PROCESS. [J.W.Fu.]

Delay line A transmission line (as nearly dissipationless as possible) or an electric network approximation of it which, if terminated in its characteristic impedance, will reproduce at its output a waveform applied to its input terminals with little distortion but at a time delayed by an amount dependent upon the electrical length of the line. If the delay line is not terminated in its characteristic impedance, there is multiple reflection back and forth along the line.

Delay lines are used for establishing a time sequence for the occurrence of events. A delay line with a total length equal to the greatest time delay required in a system may be used as a basic element. Pulses occurring at intermediate times may be obtained from taps at various points along the line. A specific application is found in the synchronizing signal generator of the television system. Also the lumped-circuit delay line is an essential element of the wide-band distributed amplifier.

When a signal is digital in nature, or consists of a series of pulses, the series of pulses may be delayed by using a shift register, which might, for example, consist of a chain of cascaded type D flip-flops. If the register has n stages, the pulse series will appear at the output delayed by a time $(n - 1)T$, where T is the periodicity of the pulses. The same function can be realized by using an array of charge-coupled devices (CCDs). *See* TRANSMISSION LINES. [G.M.G.]

Delayed neutron A neutron emitted spontaneously from a nucleus as a consequence of excitation remaining from a preceding radioactive decay event. Analogously, delayed emission of protons and alpha particles is also observed, but the known delayed neutron emitters are more numerous, and some of them have practical implications. In particular, they are of importance in the control of nuclear chain reactors.

In a ^{235}U nuclear reactor, about 0.7% of the neutrons are delayed, the others being prompt. In a conventional, moderated reactor, the prompt neutrons are born, slowed down, and reabsorbed to produce the next fissions in a cycling time of about 1 millisecond. (In a fast-neutron reactor, the time is much shorter.) Consequently, if the reactor were to become overcritical (more neutrons generated per millisecond than are absorbed or leak out), the chain reaction would exponentiate or "run away," and the reactor might overheat itself and possibly cause a dangerous accident unless the control rods could respond within a few milliseconds to correct the situation. The fortunate presence of the delayed neutrons eases the situation, because so long as the reactor operates within the margin of 0.7% ("delayed critical"), the control rods can take as long as several seconds to respond, and thus the chain reaction comes within the range of easy and leisurely control. *See* NEUTRON; NUCLEAR FISSION; REACTOR PHYSICS. [A.H.Sn.]

Deliquescence The absorption of atmospheric water vapor by a crystalline solid until the crystal eventually dissolves into a saturated solution. This behavior is well known for certain salts such as hydrated calcium chloride, $CaCl_2 \cdot 6H_2O$, and zinc chloride, $ZnCl_2$, but it is a property of all soluble salts in air of sufficiently high humidity.

Thermodynamically, the condition for deliquescence is that the partial pressure of the water vapor in the air exceed the vapor pressure (aqueous tension) of the water in the saturated solution of the salt. The speed at which the process takes place depends upon the rate of diffusion of water vapor into the crystal lattice, crystal size, and other factors. The process will stop when the water vapor in the atmosphere is depleted to the point at which its partial pressure equals that of the saturated solution.

Crystalline solids also may absorb water by increasing their water of hydration if the dissociation pressure of the hydrated species to be formed is less than the partial pressure of the water vapor. It is this process, not deliquescence, which is the opposite of efflorescence.

Deliquescent substances can be used to remove water vapor from air, although they have no special advantage over substances which merely add water of hydration and remain crystalline. *See* DESICCANT; DRYING OF GASES; EFFLORESCENCE; VAPOR PRESSURE. [L.S.]

Delocalization A phenomenon in which the most loosely held bonding electrons of some molecules serve to bind not two but several atoms. This contrasts with localization, a characteristic of ordinary single bonds, as in the normal paraffins, for example, methane (CH_4); in ethane, (C_2H_6); or in the molecules water (H_2O) and ammonia (NH_3). *See* CHEMICAL BONDING.

The prototype completely delocalized system is the ideal crystalline metal, in which the valence electrons are spaced uniformly over a periodic lattice of positive-ion cores. This confers a special stability to the metal, and it also accounts for its high electrical conductivity and other metallic properties. *See* METAL.

The aromatic and conjugated molecules of organic chemistry contain delocalized electronic systems. The benzene molecule (C_6H_6) is considered the archetypical aromatic molecule. Benzene possesses an underlying single-bonded planar framework $(C_6H_6)^{6+}$ plus six additional electrons. Available for these electrons are six $2p$ orbitals, one on every carbon atom, each perpendicular to the molecular plane. The six electrons, called pi electrons, are ascribable to the six carbon atoms in such a way that neither structure (I) nor structure (II) is an accurate representation, but instead

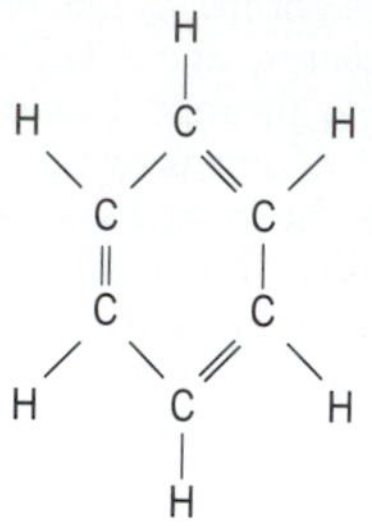

(I)

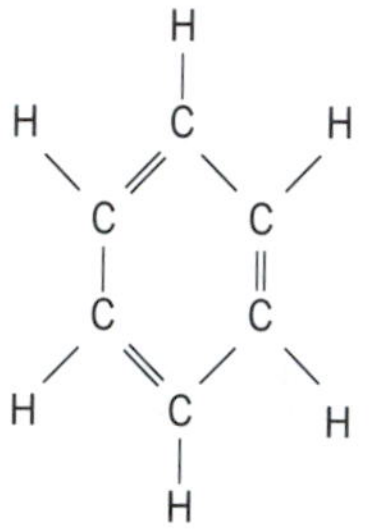

(II)

a structure that can be described in the valence bond language as a resonance hybrid of the two, schematically written as (III), where the circle stands for the

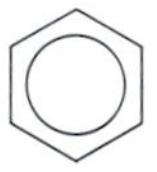

(III)

delocalized electrons. All molecules containing such rings possess an extra stability associated with the delocalization phenomenon. As a consequence, they also have a low propensity for chemical reactivity. *See* AROMATIC; MOLECULAR ORBITAL THEORY; RESONANCE (MOLECULAR STRUCTURE). [R.G.P.]

Delta A deposit of sediment at the mouth of a river or tidal inlet. It is also used for storm washovers of barrier islands and for sediment accumulations at the mouths of submarine canyons. *See* FLOODPLAIN.

The shape and internal structure of a delta depend on the nature and interaction of two forces: the sediment-carrying stream from a river, tidal inlet, or submarine canyon, and the current and wave action of the water body in which the delta is building. This interaction ranges from complete dominance of the sediment-carrying stream (still-water deltas) to complete dominance of currents and waves, resulting in redistribution of the sediment over a wide area (no deltas). This interaction has a large effect on the shape and structure of the delta body.

Most of the sediment carried into the basin is deposited when the inflowing stream decelerates. If there is little density contrast, this deceleration is sudden and most sediment is deposited near the mouth of the river. If the inflowing water is much lighter than the basin water, for example, fresh water flowing into a colder sea, the outflow spreads at the surface over a large distance away from the outlet. If the inflow is very dense, for instance, cold muddy water in a warm lake, it may form a density flow on or near the bottom, and the principal deposition may occur at great distance from the outlet.

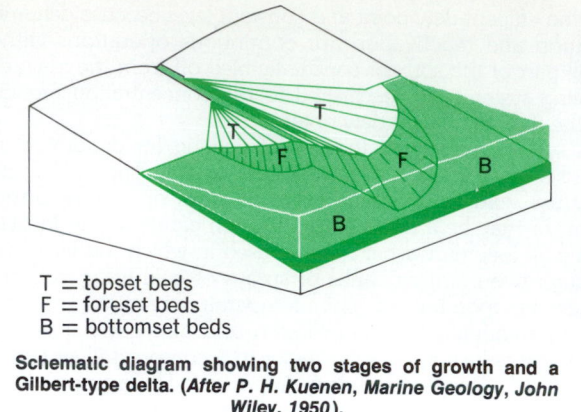

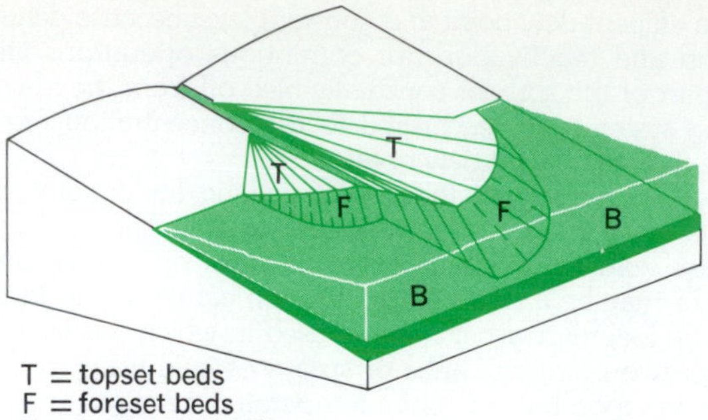

Schematic diagram showing two stages of growth and a Gilbert-type delta. (*After P. H. Kuenen, Marine Geology, John Wiley, 1950*).

Three principal components make up the bodies of most deltas in varying proportions: topset, foreset, and bottomset beds (see illustration). As defined for most deltas, the topset beds comprise the sediments formed on the subaerial delta: channel deposits, natural levees, floodplains, marshes, and swamp and bay sediments. The foreset beds are those formed in shallow water, mostly as a broad platform fronting the delta shore, and the bottomset beds are the deep-water deposits beyond the deltaic bulge. In marine deltas the fluviatile influence decreases and the marine influence increases from the topset to the bottomset beds.

In a different way, deltas can be viewed as being composed of three structural elements: (1) a framework of elongate coarse bodies (channels, river-mouth bars, levee deposits), which radiate from the apex to the distributary mouths (sand fingers); (2) a matrix of fine-grained floodplain, marsh, and bay sediments; and (3) a littoral zone, usually of beach and dune sands which result from sorting and longshore transport of river-mouth deposits by waves, currents, tides, and wind. The relative proportions of these components vary widely. The Mississippi delta consists almost entirely of framework and matrix; its rapid seaward growth is the result of deposition of river-mouth bars and extension of levees, and the areas in between are filled later with matrix. This gives the delta its characteristic bird-foot outline. A different makeup is presented by the Rhone delta, where the supply of coarse material at the distributary mouths is slow, and dispersal by wave action and longshore drift fairly efficient, so that nearly all material is evenly redistributed as a series of coastal bars and dunes across a large part of the delta front. This delta advances as a broad lobate front, while the present Mississippi delta grows at several localized and sharply defined points.

Despite difficult engineering problems, many cities, such as Calcutta, Shanghai, Venice, Alexandria (Egypt), and New Orleans, were constructed on deltas. These problems include shifting and extending shipping channels; lack of firm footing for construction except on levees; steady subsidence; poor drainage; and extensive flood danger. Moreover, in certain deltas the tendency of the main flow to shift away to entirely different areas, with resulting disappearance of the main channels for water traffic, is a constant problem that is difficult and costly to counter. *See* ESTUARINE OCEANOGRAPHY. [T.H.V.A.]

Delta electrons Energetic electrons ejected from atoms in matter by the passage of ionizing particles. In every primary ionizing collision between a charged particle and an atom, one or more electrons are ejected. Delta electrons are, by definition, that small fraction of these emitted electrons having energies which are large compared to the ionization potential. The name is a traditional one—comparable to alpha particles, for energetic helium nuclei, and beta particles, for energetic electrons emitted in radioactive decays. *See* ALPHA PARTICLES; BETA PARTICLES; IONIZATION.

Delta electrons are responsible for the "hairy" appearance of charged particle tracks when they are observed in cloud chambers or in photographic emulsions. In studies of super-high-energy particles in cosmic radiation and from the highest-energy accelerators, observation of the number of delta electrons per centimeter of path length has been shown to lead to a reliable determination of the charge of the energetic particle. [D.A.B.]

Delta resonance A member of a class of subatomic particles called baryons, which exists in four electric charge states and has a total spin of $J = 3/2$. In the underlying quark model, the delta resonance (Δ) consists of three quarks whose intrinsic spins of $1/2$ are lined up in the same direction. The Δ is closely related to the more familiar nucleon constituents of atomic nuclei, the neutrons (n) and protons (p). *See* NUCLEON; QUARKS.

The Δ was first observed as a resonant interaction of a beam of pi mesons (π) with a proton target. The probability of a scattering interaction between the π and the proton is strongly dependent on energy, attaining a maximum at the Δ mass of 1236 MeV/c^2 (where c is the speed of light). The formation of the very short-lived Δ (with a lifetime on the order of 10^{-23} s) followed immediately by its decay back into pion and nucleon. *See* MESON; SCATTERING EXPERIMENTS (NUCLEI).

The Δ plays an important role in a wide variety of nuclear phenomena, even under conditions of low energy and momentum transfer. The study of these phenomena reveals much about the presence of pions in nuclei, in addition to neutrons and protons. *See* BARYON; ELEMENTARY PARTICLE; NUCLEAR STRUCTURE. [C.D.]

Deltatheridia An order of mammals that includes the dominant carnivores of the early Cenozoic. While all deltatheridians have been persistently primitive in most respects, such as a small brain and the pattern of carotid circulation, the teeth are specialized in diverse ways for eating invertebrates or vertebrates. In the carnivorous forms, the molars, not the last upper premolar and the first lower molar, become specialized slicing teeth. The paracone and metacone of the upper molars are close together or even fused, and the terminal phalanges are fissured for the insertion of the claws. *See* DENTITION.

The ancestral family, the Palaeoryctidae, was insectivorous and contained some of the most primitive of the known placental mammals. From this family evolved the Oxyaenidae and the Hyaenodontidae. These groups evolved a carnivorous habit separately from the order Carnivora, which contains all the terrestrial carnivores among recent mammals. *See* CARNIVORA.

The tenrecs and chrysochlorids of Africa may possibly be surviving insectivorous members of the Deltatheridia rather than belonging to the Insectivora. A few aberrant middle Cenozoic genera may also be deltatheridians. *See* INSECTIVORA; MAMMALIA. [L.V.V.]

Delusion A conviction based upon faulty perceptions, feelings, and thinking. Delusions in adults are signs of mental abnormality and are accompanied by severe behavioral difficulties; yet, at times it is difficult to differentiate delusions from fixed, faulty, and rigid ideas, particularly if they are shared by groups.

Delusions seen in psychotic patients are usually classified according to content. The main groups are as follows: (1) persecutory or paranoid delusions; these may be systematized and have a logic of their own, as in paranoia, or may be unsystematic and incoherent; (2) delusions of influence related to paranoid delusions; one of the most frequent forms is the delusion of being hypnotized or under a spell; (3) delusions of grandeur in which the patient imagines he or she is an extraordinary person or has extraordinary powers; (4) self-deprecatory delusions occurring primarily in depressed patients; and (5) delusions of body changes and of unreality which may be a feeling of not being one's self, of having changed bodily, or of having gross abnormalities and diseases; these are also referred to as hypochondriacal delusions. *See* PARANOIA; PSYCHOSIS; PSYCHOSOMATIC DISORDERS. [F.C.R.]

Demagnetization A process for reducing a magnetized material to a neutral state, that is, to a state when both the magnetic induction B and the magnetic field strength H are equal to zero.

The standard demagnetization process involves applying a magnetizing force above a critical value for the material used, and successively reversing H in direction as the value of H is steadily reduced to a value less than those to be used in subsequent experiments. If the final value of H is essentially zero, the specimen is demagnetized.

The adiabatic demagnetization of paramagnetic salts is a technique used to produce temperatures very near the absolute zero. *See* ADIABATIC DEMAGNETIZATION. [K.V.M.]

Demeclocycline A broad-spectrum antibiotic formerly known as demethylchlortetracycline. The antibacterial activity of demeclocycline in the blood is markedly more prolonged than that of tetracycline, chlortetracycline, or oxytetracycline. *See* CHLORTETRACYCLINE; OXYTETRACYCLINE; TETRACYCLINE.

Demeclocycline has a wide range of clinical usefulness, corresponding generally to the range of usefulness of the other tetracyclines. It is rapidly absorbed from the intestinal tract and is distributed into all body tissues. The drug is used in the treatment of a wide range of diseases, including pneumonia, genitourinary infections, brucellosis, acute childhood infections, pustular dermatoses, gonorrhea, lymphogranuloma venereum, and granuloma inguinale. *See* ANTIBIOTIC. [J.R.D.McD.]

Dementia Any one of many diseases of the brain that cause intellectual deterioration (forgetfulness: these involve inability to reason clearly, to use words, and to perform arithmetic calculations) in an individual who has previously been mentally normal. These conditions, which are loosely called the dementias, take many forms and are increasingly frequent during the senium. In the later period of life, two diseases occur with notable frequency—Alzheimer's disease (called also senile dementia in the aged) and occlusions of many of the arteries of the brain (multi-infarct dementia). As much as a third of the patient population of psychiatric hospitals has these two types of diseases. *See* ARTERIOSCLEROSIS; SENESCENCE. [R.D.A.]

Demodulator The stage in a radio, television, radar, or other receiver at which demodulation of the received signal takes place. Thus, in a tuned radio-frequency (rf) receiver the demodulator separates the audio-frequency (af) signal from the amplified incoming rf carrier signal. A demodulator is often called a detector. In a superheterodyne receiver the af signal is separated from the carrier signal at the second detector, because the converter or first detector merely serves to change the modulated rf carrier signal to a modulated intermediate-frequency (i-f) carrier signal. In a frequency-modulation receiver the demodulator converts carrier frequency changes into corresponding audio signals. In a color television receiver the demodulator extracts the color difference signals from the incoming modulated carrier signal. [J.Mar.]

Demospongiae A class of the phylum Porifera, including sponges with a skeleton of one- to four-rayed siliceous spicules or of spongin fibers or both. Several genera lack a skeleton. The Demospongiae constitute the most abundant and widely distributed group of sponges, occurring in the sea from the tidal zone down to abyssal depths (at least to 18,000 ft or 5500 m). One family has invaded fresh water. The species vary in size from thin encrustations several centimeters in diameter to huge cake-shaped forms which may measure up to as much as 6.5 ft (2 m) in diameter. *See* PORIFERA.

Comparative studies of the embryology and early attached stages of sponges of the class Demospongiae suggest at least two evolutionary lines within this group: the subclass Ceractinomorpha and the subclass Tetractinomorpha. *See* CERACTINOMORPHA; TETRACTINOMORPHA. [W.D.H.]

Dendritic macromolecule A large molecule having a well-defined three-dimensional structure. Dendritic macromolecules play a crucial role in the chemistry of living systems. In contrast to the high level of structural precision that characterizes many biologically active macromolecules, the sizes and shapes of macromolecules made by polymer chemists are usually far less controlled. Most synthetic polymers are best described as statistical mixtures. However, chemists have sought to develop ways to prepare large molecules with more control over their architecture. If properly designed, such molecules might be capable of performing chemical or physical functions reminiscent of the macromolecules found in living systems.

Dendritic macromolecules are characterized by a highly branched molecular connectivity, whereby each repeat unit forms a branch juncture. The monomers used to prepare dendrimers possess three or more functional groups, and are of the type AB_2, AB_3, and so forth, where A and B represent a functionality (a site of chemical activity) that can combine to form a new covalent bond. Monomer chemistry is thus similar to that used to make condensation polymers except that the functionality is higher (there are more sites of chemical activity). *See* POLYMER.

The architectures of dendritic macromolecules are dramatically different from those of conventional macromolecules. Variations in molecular size, shape, and flexibility are all possible, depending on the monomer's chemical structure and geometry, the branch-point multiplicity, and the number of repetitive cycles carried out. The architecture most typical of dendritic macromolecules is a globular shape where segment density increases in going from the core to the periphery. The unusual structure of dendritic macromolecules, yet to be fully investigated, results in rheological and solubility characteristics that are dramatically different from linear macromolecules. *See* ORGANIC CHEMICAL SYNTHESIS. [J.S.Mo.]

Dendroceratida A small order of sponges of the class Demospongiae. Members of this order either have a skeleton of spongin fibers or lack a skeleton. The spongin fibers, when present, are typically dendritic in form, seldom anastomosing to form a network, and arise from a basal plate of spongin adherent to the substratum. The fibers, which in most genera lack foreign inclusions such as sand grains, are made up of concentric layers of spongin, new layers apparently being added throughout the life of the sponge. The flagellated chambers are large and sac-shaped and open directly into the exhalant canals without the intervention of a special channel. Dendroceratid sponges occur chiefly in tidal and shallow coastal regions of all seas. *See* DEMOSPONGIAE. [W.D.H.]

Dendrochronology The science of dating historical and environmental events by comparative studies of the annual variations in the width of growth rings of trees and wood specimens.

Dendrochronology developed by using the well-defined annual rings that are visible on stem cross sections of both coniferous and nonconiferous tree species growing in temperate and subpolar climates. Annual rings are formed by cell division in the cambial tissues located just inside the bark. When diameter growth begins in the spring, large, thin-walled cells are produced. These cells form the earlywood. Later in the growing season, the newly formed cells are smaller, more dense, and have thick walls. This part of the ring is called latewood. The ring boundary is formed by the contrasting structures of the dark-colored, dense latewood on the outside of one ring against the light-colored, low-density earlywood of the next. *See* STEM; XYLEM.

Tree growth is related to both genetic and environmental factors. Genetic factors determine the range of environmental conditions that an individual tree can tolerate, as well as its response to those factors. The environment supplies essential nutrients, moisture, solar radiation, and atmospheric gases. Growth is related to either an abundance or scarcity of any or all of these constituents. In most temperate regions, especially the American Southwest, precipitation is the limiting factor for diameter growth, hence the width of annual growth rings. Narrow rings are formed when moisture is scarce, and wide rings when it is abundant. Near the alpine and arctic timberlines, temperature is usually the controlling factor.

The principle of cross dating is basic to all dendrochronological work. The procedure of matching ring patterns between trees or wood specimens is used to identify and assign a calendar date to each annual ring. Unfortunately, all rings may not represent annual growth layers. In addition, many species inherently produce multiple rings and are of little use to dendrochronologists. However, those who become experienced in working with species in a given area often learn to recognize ring patterns that are associated with certain years, and can cross-date some specimens with relative ease. *See* TREE-RING HYDROLOGY.

Cross dating provides the basis for developing precisely dated ring sequences known as chronologies. Those developed for a specific location may be combined with chronologies from other sites, assuming that they cross-date, to form a master chronology for a region. A site or master chronology forms the base for dating older material, such as wood or charcoal, whose ages overlap those of the chronology. This is done by matching ring patterns of the unknown sample with those of the chronology. If cross dating is established, each ring of the unknown sample is assigned a calendar date. These records can then be incorporated in the master chronology and thus extend it farther back in time.

Dendrochronology has become a useful tool in many areas of research. As a result of intensive tree-ring research carried out in the Southwest, the chronological prehistory of the area is perhaps better known than in any part of the world. From its beginning in the American Southwest, dendrochronological dating has spread to other parts of the United States and countries throughout the world.

In the absence of instrumented records, tree-ring series can be used to determine past climatic patterns (dendroclimatology). The reconstruction procedure is complicated and requires the understanding and use of advanced computer techniques and statistical methods. These methods and techniques are still being refined and may provide even more precise information on climatic variation.

The same methods used for climatic reconstructions can be applied to hydrologic events such as streamflow—the amount of water flowing in a stream or river. Tree-ring chronologies can serve as a proxy data source to reconstruct past streamflows and to determine whether modern records are typical or represent periods of above- or below-normal flows (dendrohydrology).

Analysis of tree rings for trace elements can often provide a chronological record of pollution from industrial and other sources. Tree rings may also be used to date catastrophic events such as landslides, earthquakes, floods, volcanic eruptions, and other phenomena that cause an identifiable change in the growth rate. These are only a few of the interesting and unusual applications of tree rings. Other uses will undoubtedly be found as research continues. [W.R.B.]

Dendrology The division of forestry concerned with taxonomy of trees and other woody plants. Dendrology, called forest botany in some countries, usually is limited to taxonomy of trees but may also include shrubs and woody vines. This basic subject in the training of foresters teaches how trees are named (nomenclature), described (morphology), and grouped (classification); how to find the name of an unknown tree and recognize important forest species (identification); and where trees occur both by geographic ranges of species and by forest types (distribution). Forest stands of similar composition, appearance, and structure are grouped together into areas characterized by major forest types or formation, and are named from the predominant or characteristic species. *See* Forest and forestry; Plant taxonomy. [E.L.L.]

Dengue fever An acute viral disease with fever, rash, prostration, and lymphadenopathy. Inapparent infections are frequent; complications or deaths are rare. The virus is a member of arbovirus group B. *See* Arboviral encephalitides.

The human infection cycle is from mosquito (*Aedes aegypti*) to human. Monkeys and *A. albopictus* possibly form the jungle reservoir. Outbreaks occur chiefly in Africa, India, the Far East, and also in Hawaii, the Philippine and Caribbean islands, and the eastern Mediterranean area. Destruction of *A. aegypti* in the American tropics may be eliminating dengue there. *See* Animal virus. [J.L.Me.]

Density The mass per unit volume of a material. The term is applicable to mixtures and pure substances and to matter in the solid, liquid, gaseous, or plasma state. Common units of density are grams per cubic centimeter, and slugs or pounds per cubic foot. The specific gravity of a material is defined as the ratio of its density to the density of some standard material, such as water at a specified temperature, for example, 60°F (15.56°C), or, for gases the basis may be air at standard temperature and pressure. Another related concept is weight density, which is defined as the weight of a unit volume of the material. *See* Mass; Weight. [L.N.]

The density of all substances depends on temperature; in the case of gases, on temperature and pressure. The temperature used as a base for determining or reporting values of density is not the same for all substances. For solids 32°F (0°C) is the preferred temperature; for many liquids the reference temperature is 60°F (15.56°C); and for gases 32°F and a pressure of 29.921 in. Hg (or 0°C and 760 mmHg of mercury or 101.325 kilopascals) are used for most scientific work and for tables of gas data.

In the case of a solid, if the sample is of regular shape, such as a cube or a cylinder, its volume may be determined by linear measurement. The mass of the sample is determined by weighing it on a suitable scale or balance; then this weight divided by the volume gives the density. Ordinarily the weighing is done in air, and the density value is the density in air, or apparent density. By adjusting for the buoyant effect of the air upon the weight of the sample (and also upon the weights if a balance is used), the real density is obtained.

A second procedure, applicable to irregularly as well as regularly shaped samples, is to weigh the sample in air and then to suspend it in a liquid of known density. The volume of the sample is equal to its loss of weight in the liquid divided by the density of the liquid. This is the method of hydrostatic weighing. Conversely, the density of a liquid may be determined by weighing in the liquid a weight of known mass and volume.

One method to determine the density of a gas is to completely evacuate a light but strong vessel of suitable size, the interior volume of which is known. The evacuated vessel is weighed, filled with a sample of the gas, and then weighed again. Of course the pressure and temperature of this sample of gas must be obtained.

A densitometer or gravitometer may be used to indicate and record the density or specific gravity of a flowing stream of a liquid or a gas. [H.S.B.]

The increasing demand for accurate values of the density of phases and systems has generated much effort toward improving the technology of the necessary measurements. Two methods, the oscillator or vibrator method and the magnetic method, have emerged which eliminate most of the past difficulties associated with rapid and accurate determinations on liquid systems.

In the oscillator method the density of a sample is related to the change in resonance frequency f of a laterally vibrating tube (made of special glass or quartz). This frequency is inversely proportional to the square root of the mass m of the tube and its contents.

The instruments utilizing the magnetic method are called magnetic densimeters. This densimeter is a device whereby a tiny (<0.4 in.^{3} or 10 mm^3) ferromagnetic cylinder, encased in a glass or plastic jacket, is held at a precise height within a medium by virtue of a solenoid controlled by a servo system in circuit with a height sensor. The jacket and ferromagnetic material constitute a buoy or float. [D.W.Ku.]

Density matrix A matrix which is constructed as the most general statistical description of the states of a many-particle quantum-mechanical system. The state of a quantum system is described by a normalized wave function $\psi(x,t)$ [where x stands for all coordinates of the system, and t for the time], which satisfies the Schrödinger equation, Eq. (1), where H is

$$H\psi(x,t) = \frac{h}{i}\,\frac{\partial\psi(x,t)}{\partial t} \tag{1}$$

the hamiltonian of the system, and $\hbar$ is Planck's constant divided by 2π. Furthermore, $\psi(x,t)$ may be expanded in terms of a complete orthonormal set $\{\varphi(x)\}$, as in Eq. (2). Then, the

$$\psi(x,t) = \sum_n a_n(t)\varphi_n(x) \tag{2}$$

density matrix is defined by Eq. (3), and this density matrix

$$\rho_{mn}(t) = a_n^*(t)a_m(t) \tag{3}$$

describes a pure state. Examples of pure states are a beam of polarized electrons and the photons in a coherent beam emitted from a laser. *See* Laser; Quantum mechanics.

In quantum statistics, one deals with an ensemble of N systems which have the same hamiltonian. If the αth member of the ensemble is in the state ψ^α in Eq. (4), the density matrix is

$$\psi^\alpha(x,t) = \sum_n a_n^\alpha(t)\varphi_n(x) \tag{4}$$

defined as the ensemble average, Eq. (5). In general, this den-

$$\rho_{mn}(t) = \frac{1}{N}\sum_\alpha [a_n^\alpha(t)]^* a_m^\alpha(t) \tag{5}$$

sity matrix describes a mixed state, for example, a beam of unpolarized electrons or the photons emitted from an incoherent source such as an incandescent lamp. The pure state is a special case of the mixed state when all members of the ensemble are in the same state. *See* STATISTICAL MECHANICS. [S.H.L.]

Dental caries A disease in which the mineralized tissues of the tooth undergo progressive destruction from the surface of the tooth. It is caused by bacteria that colonize the tooth surface and, under certain conditions, produce sufficient acids to demineralize the enamel covering of the tooth crown or the cementum covering the root, and then the underlying dentin. As the destruction of the dentin progresses, along with breakdown of the organic components, the bacteria invade the dead tissue and enter the pulp chamber. The pulpal tissue becomes infected and the typical toothache may ensue. The infection can ultimately destroy the pulpal tissue and extend through the apical openings of the roots and into the surrounding periodontal tissues.

The sites of caries development have been correlated with the presence of dental plaque containing one of a number of mutans streptococci—notably in humans. *Streptococcus mutans* and *S. sobrinus*. These microorganisms are cariogenic because of their ability to generate a considerable amount of acid as a result of their metabolism of carbohydrates, and to survive in an acid environment. The acid environment is only slowly neutralized since its dilution is hindered by the fact that the acid is within the dental plaque and thereby shielded from the saliva. Each time a carbohydrate-rich substance is ingested, acid is formed. Therefore, the frequency of ingestion and physical consistency of fermentable carbohydrates are important factors in caries formation. *See* MICROBIOLOGY; TOOTH.

Dental caries can be prevented by making the tooth less susceptible to acid attack, removing cariogenic bacteria from the teeth, and limiting ingestion of cariogenic substrate.

The most effective and least expensive means of rendering the tooth less susceptible to the caries attack is through fluoridation, either by systemic means such as water fluoridation, or by topical application of fluoride to the tooth surfaces. Water fluoridation (approximately 1 part per million) by itself reduces the incidence of caries by 50–60%. Topical fluoridation can be instituted by many means, ranging from professional applications in the dental office to self-applied fluoride in the form of toothpastes, mouth rinses, and gels. The mode of action of fluoridation is not fully understood, but the benefit to the tooth structure is probably a result of a number of actions. It is believed that fluoride, when incorporated into the mineral phase of enamel and the other hard tissues, results in a better crystalline structure that is less susceptible to acid dissolution. Alternatively, the anticaries effect of fluoride may be due to its ability to inhibit demineralization and promote remineralization. Other means of improving the tooth's resistance to dental caries include the application of plastic pit and fissure sealants which eliminate retentive areas on the biting surfaces of the posterior teeth that are more susceptible to carious lesions. [M.R.R.]

Dentistry An autonomous branch of biomedical science that is concerned with the prevention, diagnosis, and treatment of diseases and abnormalities of the teeth, jaws, oral cavity, and adjacent structures.

Dental caries (tooth decay) is one of the most prevalent diseases affecting humans, and the greatest portion of the dentist's time and efforts is expended on treating dental decay and its consequences. In addition to caries, teeth can be damaged by trauma, erosion, and abrasion. Restorative dentistry encompasses efforts to conserve and restore decayed, defective, missing, and traumatically injured teeth.

Significant advances have been made in the practice of restorative dentistry. Development of high-speed, air-driven turbines combined with rotary cutting instruments fashioned from diamonds and ultrahard steel permits the rapid removal of tooth structure with little discomfort to the patient. In addition, many new materials for restorations and impression taking have become available. Especially important is the availability of composite resins which have sufficient strength to withstand biting and chewing pressures.

Seven branches of specialization are recognized by the American Dental Association: oral surgery, orthodontics, pedodontics, periodontics, prosthodontics, oral pathology, and public health dentistry. Other subspecialties such as oral medicine, dental radiology, and periodontal prosthetics exist but are not recognized.

Oral surgery treats diseases and abnormalities of the maxillofacial region by surgical means. Oral surgeons treat a wide variety of problems by removing teeth, reducing bone fractures, removing cysts, tumors, and growths, and correcting congenital anomalies and malformation of the structures of the maxillofacial region.

Orthodontists deal with abnormalities in tooth position and jaw relationships that result in facial disharmony and malfunction. The objective of orthodontic treatment is to establish normal occlusion and facial harmony. The teeth are repositioned and the jaws modified through the use of mechanical force applied with fixed or removable appliances. Successful treatment results in normal shape and expression of the mouth and lips, aids in enunciation and the sounding of words, and permits proper mastication.

Pedodontics is the branch of dentistry concerned with the detection, prevention, and treatment of oral and dental diseases and abnormalities in children. The deciduous or primary teeth are very small and have shapes which differ from those of adult teeth; special procedures and materials are required for their conservation and restoration.

Periodontics is the branch of dentistry devoted to the study, prevention, diagnosis, and treatment of diseases of the tissues supporting the teeth: gingiva (gum tissue), alveolar bone, periodontal ligament, and cementum. Periodontal diseases include gingivitis, periodontitis (sometimes called pyorrhea), primary and secondary occlusal traumatism, gingival hyperplasia, and periodontal atrophy. Several types of anaerobic gram-negative microorganisms are thought to be associated with chronic periodontitis. *See* ANAEROBIC INFECTIONS; PERIODONTAL DISEASE.

Prosthodontics is the branch of dentistry devoted to the construction and replacement of oral structures with artificial substitutes. The replacement of teeth and other oral structures is necessitated by congenital abnormalities, loss of teeth from disease or trauma, and destruction of teeth or jaws or other parts of the mouth by surgical management of neoplasms or trauma.

Oral pathology is concerned with the detection and diagnosis of the diseases of the teeth, oral cavity, and jaws, and also with the oral manifestations of systemic diseases. *See* PATHOLOGY.

Public health dentistry is defined as the science and art of preventing and controlling dental diseases and promoting dental health through organized community efforts. It comprises research, education, prevention, diagnosis, prescription, treatment of problems related to dentistry, and evaluation of community dental care. *See* TOOTH DISORDERS. [R.C.P.]

Dentition The arrangement, type, and number of teeth which are variously located in the oral or in the pharyngeal cavities, or in both. Teeth are found in areas where there is an underlying supporting structure of cartilage or bone and where stomodeal ectoderm is present.

Attachment of teeth is variable among vertebrates. They may be inserted in sockets in the jawbones (thecodont condition), fused to the edge of the bone proper (acrodont condition), or attached to the inner surface of the jawbone (pleurodont condition). In polyphyodont animals, teeth may be constantly replaced. The polyphyodont condition is characteristic of the fishes and the lower tetrapods in which the teeth are replaced as they are worn out. Most mammals have two sets of teeth during their lifetime, which is the diphyodont condition. These sets are the deciduous, or milk, teeth and the permanent dentition. Monophyodont dentition is the development of only one set of teeth.

Teeth may have a similar form in all regions where they occur (homodont), as distinct from those that are variable in shape (heterodont). The homodont condition is characteristic of nonmammalian vertebrates, although the anterior teeth may differ from those lying in the cheek region of the jaws. A heterodont dentition began to evolve in the reptilian ancestors of mammals, but became fully developed in the mammals. The mammalian heterodont condition is frequently described by a dental formula expressing both the number and kind of teeth in each half jaw, both upper and lower (see table). For humans the formula is

Dental formulas of some mammals

	Teeth				
Animal	I	C	Pm	M	Total
Human	2/2	1/1	2/2	3/3	32
Cony	3/3	1/1	4/4	4/4	48
Beaver	1/1	0/0	1/1	3/3	20
Cat	3/3	1/1	3/2	1/1	30
Dog	3/3	1/1	4/4	2/3	42
Sheep	0/3	0/1	3/3	3/3	32
Lynx	3/3	1/1	2/2	1/1	28
Rat	1/1	0/0	0/0	3/3	16
Horse	3/3	1/1	4/4	3/3	44
Mole	3/3	1/1	4/4	3/3	44
Squirrel	1/1	0/0	2/1	3/3	22
Reindeer	0/3	0/1	3/3	3/3	32
Pig	3/3	1/1	4/4	3/3	44
Common seal	3/2	1/1	4/4	1/1	34
Skunk	3/3	1/1	3/3	1/2	34
Raccoon	3/3	1/1	4/4	2/2	40
Bear	3/3	1/1	4/4	2/3	42

I 2/2 C 1/1 Pm 2/2 M 3/3. This can be read for either the right or left side of the jaw; there are 2 upper and 2 lower incisors, 1 upper and 1 lower canine, and 2 premolars and 3 molars in both the upper and lower jaws. According to this formula, humans have a total of 32 teeth, 8 located in each half of the upper jaw and 8 in each half of the lower jaw. *See* TOOTH. [F.S.S.]

Deoxyribonucleic acid (DNA) The main carrier of genetic information in all living organisms except viruses that contain ribonucleic acid. Most of the deoxyribonucleic acid (DNA) of the cell occurs in the nucleus as a component of the chromosomes. Small amounts also occur in mitochondria, plastids, and other cytoplasmic organelles.

Structure and function. The molecules of DNA are extremely long polymers, built up of deoxyribonucleotide repeating units. Each unit comprises a sugar (2-deoxyribose), phosphate, and a purine or pyrimidine base. The deoxyribonucleotide units are linked together by the phosphate groups, joining the 3′ position of one sugar to the 5′ position of the next. The alternate sugar and phosphate residues form the backbone of the molecule, and the purine and pyrimidine bases are attached to the backbone via the 1′ position of the deoxyribose. This sugar-phosphate backbone is the same in all DNA molecules; what gives each DNA its individuality is the sequence of the purine and pyrimidine bases. *See* DEOXYRIBOSE; PURINE; PYRIMIDINE.

Most DNA molecules are double stranded, consisting of two DNA chains wound round each other, the two chains running in opposite senses, or antiparallel (Fig. 1). The structure is a right-handed helix in which the bases stack on top of each other like a stack of coins (Fig. 2). The two chains are linked

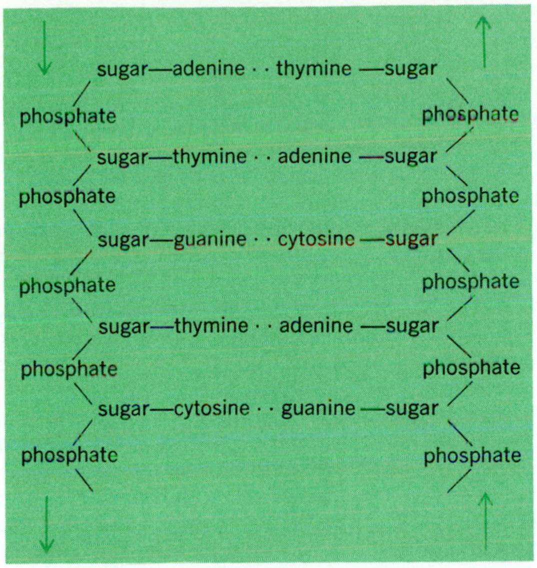

Fig. 1. Schematic diagram of the DNA molecule. The two chains are antiparallel, as shown by arrows. The dotted lines between the bases represent hydrogen bonding. Chains are drawn flat; however, they are actually wound around each other in the molecule.

together by a large number of weak bonds (hydrogen bonds) formed between complementary bases. The complementary base pairs are adenine-thymine and guanine-cytosine, that is, always a purine (adenine or guanine) with a pyrimidine (cytosine or thymine). This base pairing is extremely important not only in the structure of DNA but also in the performance of its two functions, replication and transcription. The information carried in the DNA is encoded in the sequence of purine and

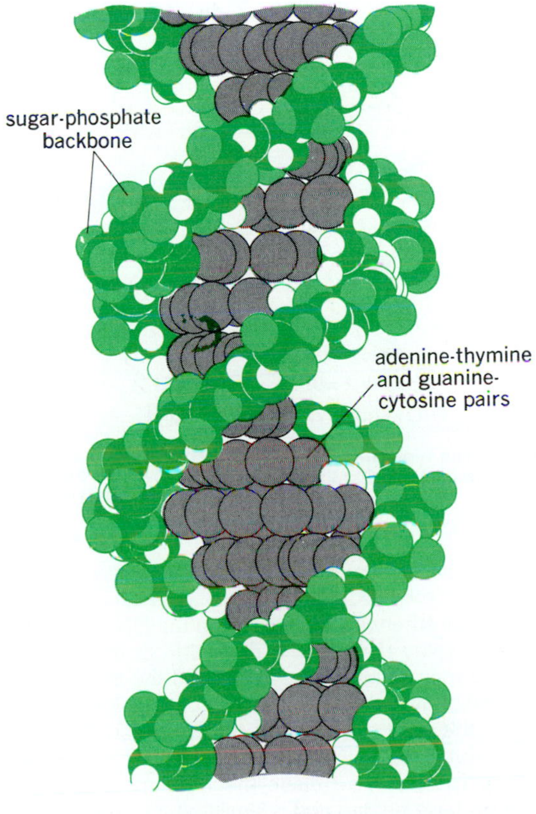

Fig. 2. Watson-Crick model of DNA structure. (*Upjohn*)

pyrimidine bases in the molecule, and perpetuation of this information involves production of two exact copies of each parental DNA molecule (replication). Much of this information is expressed in the form of protein and the products of protein action. Expression takes place in several stages: first, transcription of the DNA to give rise to ribonucleic acid (RNA); second, RNA processing at both 3′ and 5′ ends and by splicing within the molecule to remove intervening sequences; third, translation of the resultant messenger RNA to protein. Each amino acid in the protein chain is represented by a group of three bases in the RNA derived from a group of three corresponding bases in the DNA. Some of the information in DNA is expressed directly via transcription, as RNA, for example, ribosomal RNA and transfer RNA. *See* Genetic code; Protein; Ribonucleic acid (RNA); Ribosomes.

Replication. The structure of the DNA double helix immediately suggests the mechanism by which replication might take place. The base sequence of either strand defines the structure of the whole molecule. The mechanism of DNA replication involves the separation of the complementary strands over a short distance, followed by the synthesis of complementary strands on each of the parental strands (Fig. 3). Errors in this process occur very rarely, probably less than one per million bases incorporated. Once an error has occurred, it may be corrected, but otherwise the change in the DNA is perpetuated in all the progeny DNA that is derived from this altered molecule. Errors may so alter the sequence of the protein coded for by the DNA that the activity of the protein is lost or altered. A change in the base sequence of the DNA constitutes a mutation. *See* Mutation.

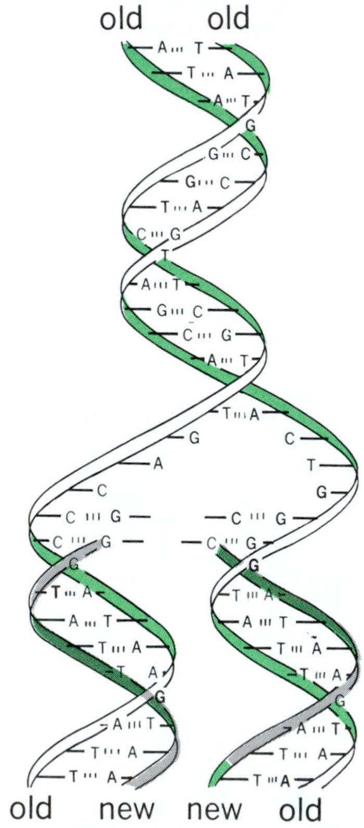

Fig. 3. The replication of DNA: A = adenine, C = cytosine, G = guanine, and T = thymine. (*After J. D. Watson, Molecular Biology of the Gene, W. J. Benjamin, 1965*)

Transcription. The basic mechanism of transcription depends on base pairing and is similar in a formal sense to that of replication, except that only one of the strands of the DNA is copied and the copy differs from the template. RNA differs from DNA in that it contains ribose rather than deoxyribose and in that thymine is replaced by uracil as one of the two pyrimidine bases. This latter change does not alter the basic pattern of base pairing except that adenine now pairs with uracil. As in the case of DNA synthesis the building blocks for RNA synthesis are the nucleoside 5′ triphosphates, each losing two of its phosphates in the process of being attached to the growing end of the nucleic acid chain. Transcription occurs by attachment of an enzyme (DNA-primed RNA polymerase) to specific sites on the DNA at the beginning of operons. At these sites the strands of the DNA separate locally, and one strand, which is rich in pyrimidines, is selected as the template strand. Synthesis of the RNA then commences with the complementary purine deoxyribonucleoside triphosphate and continues along the DNA molecule, inserting complementary bases in the sequence dictated by the template DNA strand, until reaching a stop signal at the end of an operon. The strands of the DNA rejoin as soon as the wave of RNA synthesis has passed, thus forming again the original structure.

Variation. In addition to the main chromosomal DNA, many cells contain other DNA molecules. In bacteria these include plasmids, which are of considerable importance since they carry genes for fertility and resistance to drugs and antibiotics, and many other characters. The plasmids may replicate independently or, in some cases, become integrated into the bacterial chromosome. From here they can later be excised and return to their autonomous state. The small size of plasmid DNAs has made it possible to examine their structure in detail and to manipulate them in various ways. DNA sequences from other sources can be inserted into such plasmids to form hybrid DNA molecules capable of indefinite replication.

Recombinant DNA technology uses bacteriophages and plasmids as vectors for DNA from other sources. It makes use of the ability of restriction endonucleases to cut DNA at specific sequences, ligases to join the fragments, and DNA transformation to reinsert the hybrid molecules into recipient organisms. This form of genetic engineering has great potential in the elucidation of the structure and function of the DNA of higher organisms. The practical advantages of being able to propagate specific regions of such DNAs are also considerable. The many potential applications include insertion of the genes for human proteins into plasmids and their production in bacteria. The hormone insulin, used in treatment of diabetes, and the antiviral, antitumor agent interferon may both be produced commercially in this way. *See* Genetic engineering; Hormone; Interferon.

Plasmid DNAs, many organelle DNAs, and the DNAs of some small viruses all have circular structures. The importance of circular structure may simply be that DNA strands without free ends are less subject to chemical or enzymic degradation. Other viruses of animals and bacteria have unusual DNAs which are single- rather than double-stranded. Unusual DNA compositions are also found in some bacterial viruses, cytosine being completely replaced by methyl cytosine or hydroxymethyl cytosine. Thymine may be replaced by uracil or hydroxymethyl uracil. These alterations do not affect overall DNA structure and base pairing in the double helix, but they may affect its stability. *See* Animal virus; Bacteriophage; Nucleic acid; Virus. [L.C.]

Deoxyribose A nucleic acid constituent (see illustration) of all animal, microbial, and plant cells; also known as 2-D-deoxyribose. Deoxyribose is enzymically formed in living cells by reduction of ribonucleoside di- or triphosphate. The four deoxyribose nucleotides, containing adenine, guanine, cytosine, and thymine, are the major constituents of the deoxyribonucleic acids (DNA), which control the hereditary character-

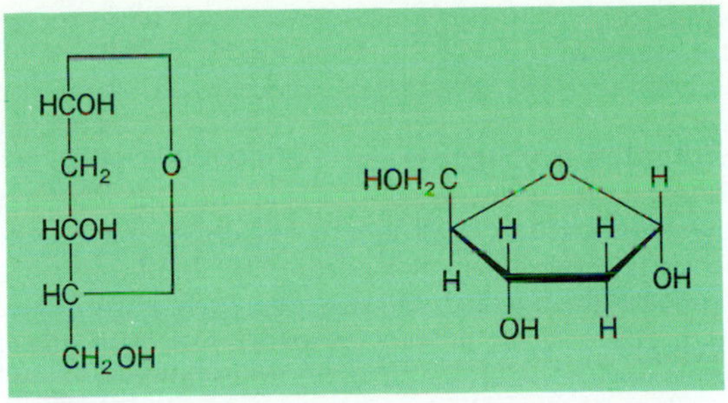

Formulas of 2-D-deoxyribose (α-D-2-deoxyribofuranose).

istics of every living organism. *See* DEOXYRIBONUCLEIC ACID (DNA); NUCLEIC ACID; RIBOSE. [W.Z.H.]

Dependovirus Any of a group (genus) of defective viruses which seem unable to reproduce without help from adenoviruses. They were formerly known as adeno-associated viruses and belong to the family Parvoviridae. These 20-nanometer virus particles were found during electron microscopic studies (see illustration) in several preparations of adenovirus from both human and monkey sources, and have since been observed in many adenovirus stocks. *See* ADENOVIRIDAE.

Like adenovirus, the dependovirus has a deoxyribonucleic acid (DNA) core. However, the dependovirus, although dependent upon adenovirus for its growth, does not appear to be structurally related to its "helper." Its genetic content, as well as its size, is much smaller than that of adenovirus, and its protein coat is completely different from that of adenovirus. These particles contain single-stranded DNA and a protein coat with icosahedral symmetry. The single-stranded DNA has been shown to be present within the dependovirus virion as either plus or minus complementary strands in separate particles. Upon extraction, the minus and plus strands unite to form a double-stranded helix.

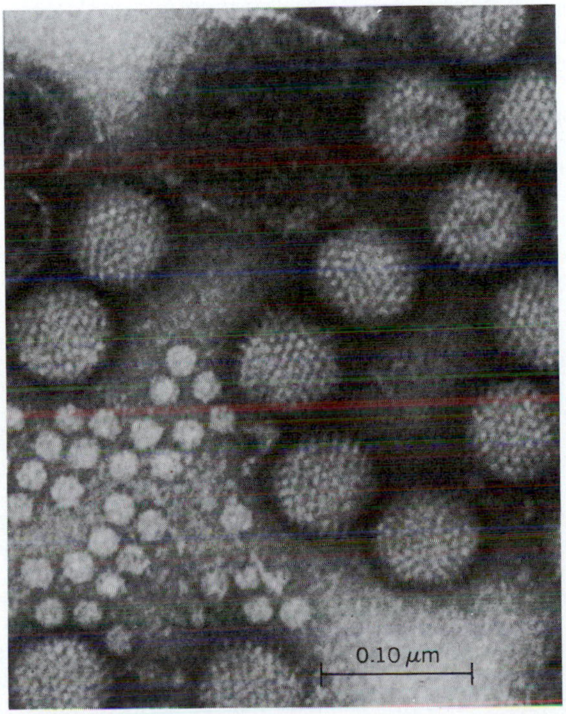

Electron micrograph showing adenovirus particles (70-nm diameter; 0.10 μm = 100 nm) and dependoviruses (20-nm diameter).

The genetic material of dependovirus not only can persist in cultured cells for long periods without giving evidence of its presence but also can survive within human beings in a latent state, becoming detectable only in specimens taken when the person is concurrently infected with adenovirus. This characteristic permits survival of a defective virus in nature, even though it cannot regularly replicate or be passed in infective form from host to host. *See* VIRUS CLASSIFICATION. [J.L.Me.; M.E.Re.]

Depolymerization The reversion of a polymer to a monomer. Depolymerization normally is not employed as a distinct process in the petroleum industry, but may occur incidentally during catalytic or thermal cracking operations if the feedstocks contain polymers. Depolymerization by heat alone will proceed by a thermal cracking mechanism. In both cases the product may contain substantial quantities of polymer fragments differing from the original monomer. *See* CRACKING; POLYMERIZATION. [B.S.G./M.Sou.]

Depression (psychology) Along with mania, depression belongs to the general area of affective disorders. There are 10 major models of depression, reflecting five dominant schools of thought. For those in the psychoanalytic tradition, depression represents the introjection of hostility resulting from the loss of an ambivalently loved object, or a reaction to separation from a significant object of attachment. More recent ego-psychology approaches focus on helplessness, lowered self-esteem, and negative cognitive set. For the behaviorist, depression is a set of maladaptive behavioral responses—elicited by uncontrollable aversive stimuli or by loss of reinforcement—that are additionally maintained by the rewards of the sick role. Each theoretical framework has generated its own therapeutic modalities, ranging from psychoanalysis to behavior therapy, from sociopolitical activism to pharmacotherapy and electric convulsive therapy (ECT).

The distinction between unhappiness and depressive illness is important because almost all people experience mood swings in association with the demands of everyday life. However, clinical depression involves sustained states of deep dejection characterized by dysphoric mood (depressed, blue, despondent, hopeless, discouraged, fearful, worried, irritable). Frequently appearing symptoms include insomnia or hypersomnia, loss of energy, slowing down of mental and physical activities, decreased libido, lowered self-esteem and increased self-reproach, poor concentration, slow thinking, or confused thoughts. Sometimes there are recurrent preoccupations with death or suicide.

Of the different symptoms of depression, suicide and suicidal attempts are the most alarming and must be considered in all serious depressions. An increasing amount of psychiatric research is being done to try to determine the high risk factors for suicide attempts and to develop ways to intervene usefully.

It is increasingly possible to treat and even prevent some, although certainly not all, specialized forms of depression. Lithium for the treatment and prophylaxis of manic-depressive disorders has made it possible for many manic-depressives to avoid the frequent ups and downs that attend their illness. Other drugs are being used for the treatment of different kinds of depressive disorders, with reasonable success in properly selected patients. For the milder forms of depression, researchers have acquired new knowledge about the most effective psychotherapeutic or behavioral approaches that can be used. *See* MANIC-DEPRESSIVE PSYCHOSIS. [W.T.McK.]

Dermaptera An order of slender insects (commonly called earwigs) having incomplete metamorphosis, chewing mouthparts, short forewings, and cerci—the paired sensory appendages, on the last abdominal segment—that are shaped like forceps (illustration *a*). They have hard, shiny bodies and

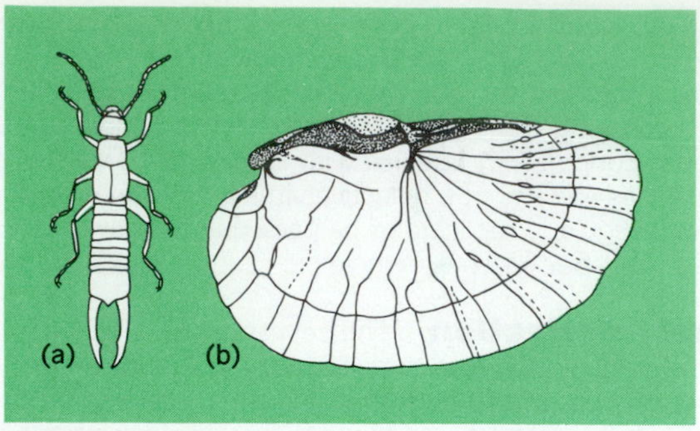

Some features of Dermaptera. (*a*) Adult of *Labidura*. (*b*) Hindwing of *Forficula*. (After C. P. Brues, A. L. Melander, and F. M. Carpenter, Classification of Insects, Harvard University Press, 1954)

running legs. The forewings (elytra) extend only to the end of the metathorax; the hindwings are large and semicircular and at rest they fold up under the elytra (illustration *b*). Most species fly only rarely and many are apterous.

Earwigs are terrestrial and mainly nocturnal; they live in moist, shady regions and during the day hide under stones, logs, or the bark of trees. They are omnivorous feeders; a few are destructive to tender foliage and flowers, and others, mostly tropical species, are carnivorous. The order is a relatively small one, including about 1000 species; it is a primitive group extending back to the Jurassic Period. *See* INSECTA. [F.M.C.]

Dermatitis Inflammation of the skin with redness and scaling, or if acute, with blisters, edema, and formation of a crust; also known as eczema. Of the several distinct types of dermatitis, each has unique etiology, clinical characteristics, pathology, and treatment.

Atopic dermatitis. Atopic dermatitis usually begins in infancy and may continue into adult life. The eruption is characterized by red patches accompanied by intense itching. The lesions often become secondarily infected, leading to moist discharge and crusting. Although any anatomic site may be affected, the classic locations in infants are the extremities, face, and scalp. In older children and adults, the inside of the elbows, the back of the knees, wrists, eyelids, and neck are most often involved.

Treatment of atopic dermatitis includes oral antihistamines, emollients, topical corticosteroids (particularly ointments), and for more severe cases, systemic corticosteroids. Avoidance of irritants (wool, chemicals, harsh soaps and detergents, perfumes), extremes in temperature and humidity, overbathing, and certain foods is imperative. Mild cleansers are preferred. *See* ANTIHISTAMINE; STEROID.

Contact dermatitis. Contact dermatitis is a reaction that causes acute itching, redness, swelling, large blisters, and in chronic cases, red, scaly papules (raised bumps) and plaques (abnormal flat areas). The two forms are irritant contact dermatitis and allergic contact dermatitis. In irritant contact dermatitis the eruption is caused by a nonallergic reaction resulting from exposure to an irritating substance. Allergic contact dermatitis is an immunologic reaction in persons who have been previously exposed (sensitized) to the allergen. The lesions occur 24–48 h after exposure to an irritant or allergen. The key to treatment is removal of the offending agent. Antihistamines, emollients, and topical and systemic corticosteroids are helpful. *See* ALLERGY.

Seborrheic dermatitis. Seborrheic dermatitis is characterized by reddish to yellow greasy scale that frequently forms on the face (eyebrows, nasolabial folds, forehead), ears, scalp, upper back, central chest, and anogential region. In contrast to atopic and contact dermatitis, itching is uncommon.

Topical corticosteroids are the mainstay of treatment for seborrheic dermatitis. In cases in which *Pityrosporum ovale* is involved, topical antifungal agents (such as ketoconazole cream) are useful. Topical or oral antibiotics are required in cases with secondary bacterial infection. In generalized or recurring cases, systemic corticosteroids may be necessary.

Exfoliative erythroderma. Exfoliative erythroderma is characterized by widespread, warm redness and scaling. Nail degeneration and loss, hair loss, fever, chills, and enlargement of the lymph nodes may also occur. The eruption may be due to an underlying primary skin disorder, such as dermatitis (seborrheic, contact, or atopic) or psoriasis; drug allergy; or leukemia or lymphoma, particularly cutaneous T-cell lymphoma. *See* LEUKEMIA; LYMPHATIC SYSTEM; LYMPHOMA.

Individuals with exfoliative erythroderma are best managed in a hospital setting. Because heat, water, and protein losses occur through damaged skin, careful monitoring of fluid balance, protein losses, and body temperature is imperative. The application of wet dressings or topical corticosteroids and administration of oral antihistamines are important aspects of treatment. *See* INFECTION. [L.M.Co.; K.A.A.]

Flea allergy dermatitis. Flea allergy dermatitis, sometimes called flea-bite dermatitis, affects the skin of small pets (dogs, cats, and ferrets), and results from an allergic reaction to protein substances deposited on (or under) the surface of the skin at the time of flea feeding. It is by far the most common cause of allergic dermatitis in pet animals and a major cause of itchy skin in geographic regions where fleas are found. [J.M.MacD.]

Dermatophytosis An infection of the skin of humans and animals caused by fungi which live in the keratinized tissues but which are unable to invade the subcutaneous or deeper tissues. Classification of the infection is based upon its clinical appearance rather than upon the etiologic agents. For example, ringworm of the scalp may be caused by any one of several species of *Microsporum* or *Trichophyton*. Thus, fungus infection of the scalp and hair is tinea capitas; of the feet, tinea pedis. Fungi causing dermatophytoses are also included in the genus *Epidermophyton*.

Some patients develop an allergy to the fungus which causes their dermatophytosis; this is manifested as an eruption secondary to the primary infection. The allergic reaction clears up as the primary lesions heal. *See* HYPERSENSITIVITY; MEDICAL MYCOLOGY. [L.D.H.]

Dermoptera An ancient order of somewhat primatelike herbivorous (plant-eating) and frugivorous (fruit-eating) gliding mammals, now confined to the extreme southeast comer of Asia and parts of the East Indies and the Philippines. Colugos, or flying lemurs, as they are often called, are the only extant members of the group. Modern dermopterans are usually placed in a single genus, though two genera may exist.

The patagium, a flap of skin used in gliding, extends from behind the ears to the hands, then to the feet, and finally to the tip of the tail. The extent of the patagium and the gliding ability, of course, are not known in fossil forms, nor are certain features of the soft anatomy. Skeletal features not related to their gliding adaptations point to distant connections with primitive primatelike insectivores that were no doubt living in southeastern Asia as long ago as the Cretaceous. Colugos probably were restricted to southeastern Asia by the contraction of the tropics that occurred during the Oligocene and later Cenozoic. *See* EUTHERIA; INSECTIVORA; MAMMALIA. [M.C.McK.]

Derrick A hoisting machine consisting usually of a vertical mast, a slanted boom, and associated tackle. Derricks have a wide variety of forms. The mast may be no more than a base for the boom; it may be a tripod, an A-frame, a fixed column, and so on. Fixed stays may guy it in place. The boom may be fixed, it may pivot at the base of the mast, it may swing horizontally from near the top of the mast, or it may be omitted.

The derrick may be permanently fixed, temporarily erected, or mobile on a cart or truck.

Derricks are widely used in construction, in cargo handling, and in shops. A manual or powered winch provides the lifting action by coiling in the running tackle, the load swinging free as it rises. *See* CRANE HOIST; HOISTING MACHINES. [D.O.H./F.H.R.]

Descriptive geometry A mathematical-graphical procedure for the visualization of structures and their exact representation in drawings. After analysis of the structure, each element is shown in the drawing in its exact geometrical relation to the other elements. There are two basic methods of descriptive geometry: the projection method and the direct method. The two methods differ as regards the attitude of mind toward the structure and toward the drawing that represents the structure.

Projection method. In the projection method the horizontal projection plane H and the vertical projection plane V intersect in the line GL, which is called the ground line (Fig. 1). These two projection planes divide space into four quadrants, or angles, as numbered in Fig. 1. Point A, in the first quadrant, is projected onto the horizontal plane at a_h by means of a projection line perpendicular to the H plane, and onto the vertical plane at a_v by means of a projection line perpendicular to the V plane. The projections of the points B, C, and D in the other quadrants are located in a similar manner. Right-angle projection, as described above, is called orthographic projection.

To represent horizontal and vertical projections on a flat sheet of paper, the planes are conceived as being hinged along the ground line and brought together by closing the second and fourth quadrants. Projections of A, B, C, and D then appear in a single plane (Fig. 2). The H and V projections of a point are always in the same perpendicular to the ground line.

There are two general types of views, perspective and orthographic (Fig. 3). A perspective view of an object is observed from a fixed station point, or point of view, by means of converging rays of light that meet at the eye of the observer. An orthographic view of an object is observed in a chosen direction by means of parallel rays of light.

Direct method. In the direct method the attention is focused on the visualized structure or object. Each view of the object is obtained by looking at the object in a definite direction. The view is orthographic. A view never is considered as two-dimensional or as projected or drawn on a plane. This is the way the engineer who makes and reads the drawing thinks of views.

Orthographic views may be classified into three types: principal views, auxiliary views, and oblique views. The object can be viewed from any direction around three rings—horizontal, frontal, and profile (Fig. 4). The rings rep-

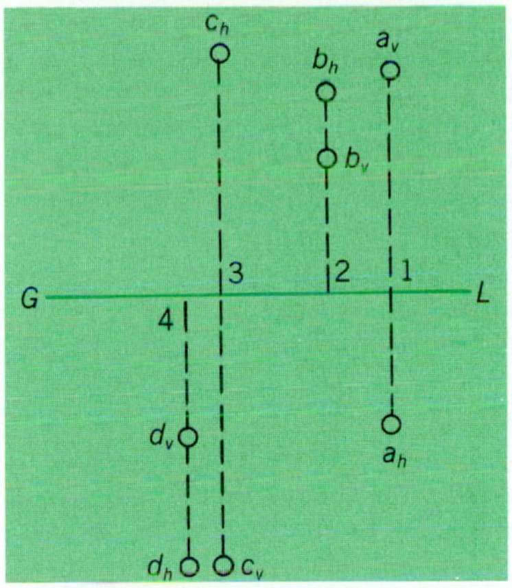

Fig. 2. Projection of points onto projection planes. *(After G. J. Hood and A. S. Palmerlee, Geometry of Engineering Drawing, 4th ed., McGraw-Hill, 1958)*

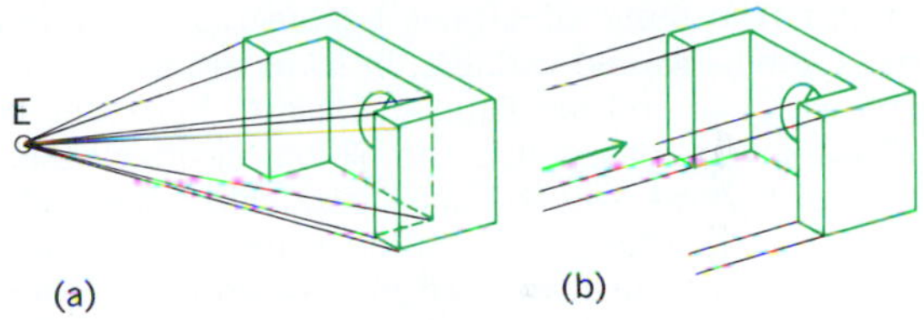

Fig. 3. Methods of viewing objects. (*a*) Perspective viewpoint (converging rays). (*b*) Orthographic viewpoint (parallel rays). *(After G. J. Hood and A. S. Palmerlee, Geometry of Engineering Drawing, 4th ed., McGraw-Hill, 1958)*

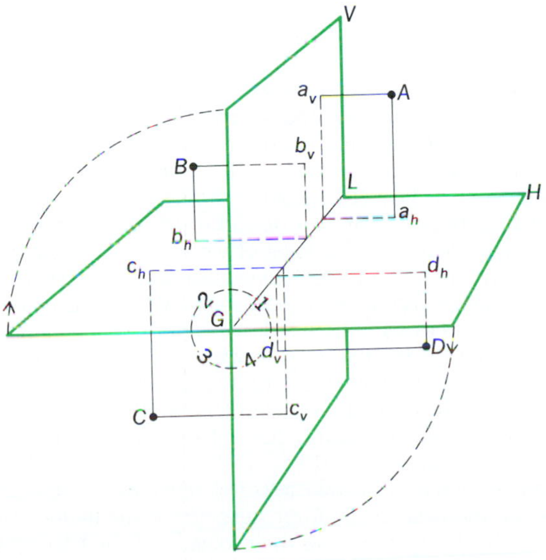

Fig. 1. Planes of projection. *(After G. J. Hood and A. S. Palmerlee, Geometry of Engineering Drawing, 4th ed., McGraw-Hill, 1958)*

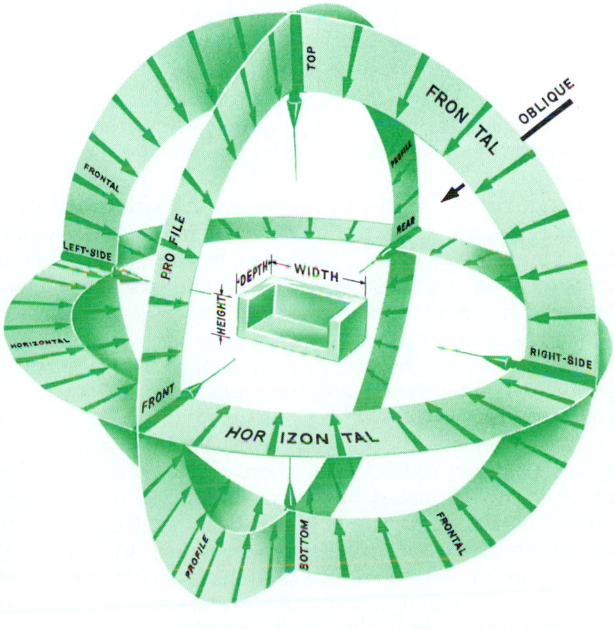

Fig. 4. A series of viewing positions for auxiliary views. *(After G. J. Hood and A. S. Palmerlee, Geometry of Engineering Drawing, 4th ed., McGraw-Hill, 1958)*

resent three mutually perpendicular planes. The intersections of the rings define three mutually perpendicular directions from which six principal views are observed: front and rear, top and bottom, right and left sides. An auxiliary view can be observed around any ring in a direction perpendicular to one, and only one, of the directions in which principal views are observed. All views other than principal or auxiliary views are oblique. In Fig. 4 a single arrow, marked oblique, indicates one of the infinite number of directions in which oblique views are observed. *See* DRAFTING; ENGINEERING DRAWING. [A.S.P./C.J.B.]

Desert No precise definition of a desert exists. From an ecological viewpoint the scarcity of rainfall is all important, as it directly affects plant productivity which in turn affects the abundance, diversity, and activity of animals. It has become customary to describe deserts as extremely arid where the mean precipitation is less than 2.5–4 in. (60–100 mm), arid where it is 2.5–4 to 6–10 in. (60–100 to 150–250 mm), and semiarid where it is 6–10 to 10–20 in. (150–250 to 250–500 mm). However, mean figures tend to distort the true state of affairs because precipitation in deserts is unreliable and variable. In some areas, such as the Atacama in Chile and the Arabian Desert, there may be no rainfall for several years. It is the biological effectiveness of rainfall that matters and this may vary with wind and temperature, which affect evaporation rates. The vegetation cover also alters the evaporation rate and increases the effectiveness of rainfall. Rainfall, then, is the chief limiting factor to biological processes, but intense solar radiation, high temperatures, and a paucity of nutrients (especially of nitrogen) may also limit plant productivity, and hence animal abundance. Of the main desert regions of the world, most lie within the tropics and hence are hot as well as arid. The Namib and Atacama coastal deserts are kept coot by the Benguela and Humboldt ocean currents, and many desert areas of central Asia are cool because of high latitude and altitude.

The diversity of species of animals in a desert is generally correlated with the diversity of plant species, which to a considerable degree is correlated with the predictability and amount of rainfall. There is a rather weak latitudinal gradient of diversity with relatively more species nearer the Equator than at higher latitudes. This gradient is much more conspicuous in wetter ecosystems, such as forests, and in deserts appears to be overridden by the manifold effects of rainfall. Animals, too, may affect plant diversity: the burrowing activities of rodents create niches for plants which could not otherwise survive, and mound-building termites tend to concentrate decomposition and hence nutrients, which provide opportunities for plants to colonize.

Each desert has its own community of species, and these communities are repeated in different parts of the world. Very often the organisms that occupy similar niches in different deserts belong to unrelated taxa. The overall structural similarity between American cactus species and African euphorbias is an example of convergent evolution, in which separate and unrelated groups have evolved almost identical adaptations under similar environmental conditions in widely separated parts of the world. Convergent structural modification occurs in many organisms in all environments, but is especially noticeable in deserts where possibly the small number of ecological niches has necessitated greater specialization and restriction of way of life. The face and especially the large ears of desert foxes of the Sahara and of North America are remarkably similar, and there is an extraordinary resemblance between North American sidewinding rattlesnakes and Namib sidewinding adders. *See* ECOLOGY; PHYSIOLOGICAL ECOLOGY (PLANT); PRECIPITATION (METEOROLOGY). [D.F.Ow.]

Desertification The spread of desertlike conditions in arid and semiarid areas, due to human influence or climatic change, particularly as a result of the interaction of naturally recurring drought with unwise land-use practices. Natural vegetation is being lost over increasingly larger areas because of overgrazing, especially in Africa and southwestern Asia. In many areas the deserts seem to be spreading, with an apparent speed of about one or more kilometers per year.

The world distribution of dry climates depends mainly on the subsidence associated with the subtropical high-pressure belts, which migrate poleward in summer and equatorward in winter. These migrations are connected with the atmospheric general circulation. There results a threefold structure in the arid zone: a Mediterranean fringe, with rains occurring only in winter; a desert core (in about 20–30° latitude), with litfle or no rain; and a tropical fringe, with rains mainly in the high sun season. Throughout the arid zone, rainfall variability is high. Clear skies and low humidities in most regions give the dry climates very high solar radiation (averaging 200 W/m^2, which leads to high soil temperatures. The light color and high reflectivity of many dry surfaces cause large reflectional losses, and long-wave cooling is also severe. Hence net radiation incomes are relatively low.

Climatic variation takes place on many time scales. The world's deserts and semideserts are very old, although they have shifted in latitude and varied in extent during geological history. The modern phase of climate began with a major change about 10,000 years before present, when a rapid warming trend removed most of the continental ice sheets. The Sahara and Indus valleys were at first moist, but since about 4000 years before present, when severe natural desiccation took place, aridity has been profound.

Recent climatic variations, such as the Sahelian drought, are natural in origin, and are not without precedent. Prolonged desiccation, lasting a decade and more, is common and often ends abruptly with excessive rainfall. This persistence suggests that feedback mechanisms may be operating, whereby drought feeds drought and rain feeds rain.

A common mechanism of desertification includes the following steps: (1) expansion and intensification of land use in marginal dry lands during wet years, including increased grazing, plowing, and cultivation of new lands, and wood collection around new camps or settlements; and (2) wind erosion during the next dry year, or water erosion during the next maximum rainstorm.

This desertification process has the following climatological implications: (1) Increased grazing during wet years tends to compact the soil near water holes, and increased livestock numbers cause pressure on perennial plants during dry seasons. The result is to expose surface soil to erosion by wind. Runoff may be increased, and so may albedo (that is, reflectivity of the ground with respect to solar radiation). (2) Increased cultivation during wet years greatly improves the chances of wind erosion of fine soil materials during the dry season, and possibly increases evapotranspiration (in the case of water-demanding crops). (3) Removal of wood increases direct solar heating and may considerably decrease evapotranspiration. (4) In the ensuing dry years, the acceleration of wind erosion by the above processes further reduces water storage capacity by removal of topsoil. The loss of some or all perennial plant cover lowers infiltration rates and hence the potential percolation. During subsequent rains, surface runoff is increased, with an attendant loss of water for subsequent use by shrubs, herbage, or crops.

More research is needed into the relationship between climate and the desertification process. Experimentation is needed that specifically examines the problems of the general circulation of the Earth's atmosphere, and even of the smaller-scale circulation, over the dry land areas of the subtropics. There is also a need for more detailed study of the physical climatology and bioclimatology of dry land surfaces, and particularly for a closer synthesis of climatology with geomorphology, soil science, hydrology, and ecology. *See* DROUGHT. [L.Be.]

Desiccant A substance (adsorbant) used to withdraw moisture from other materials. Although the removal of large quantities of water is done by evaporation, aided by moving air currents and by elevated temperature, the last traces of moisture are often held very tightly and do not evaporate readily. Furthermore, evaporation ceases when the moisture content of the material is reduced to that of the drying-air current. For final drying, a desiccant is used. It may react with water chemically or retain water through capillarity of adsorption. The drying agent is placed directly into the gas or liquid to be dried; solid materials are placed in a desiccator, a closed vessel in which moisture diffuses to the desiccant through the dry desiccator atmosphere. A desiccant loses potency as it takes on water; often it can be renewed by heating. Desiccants which form hydrates can be selected to maintain certain levels of low humidity in a closed vessel. *See* ADSORPTION; DELIQUESCENCE.

Among the more important types of solid desiccants are silica gel, activated alumina, anhydrous calcium sulfate, magnesium perchlorate, oxides (of barium and calcium), and activated carbon. [A.L.H.]

Design standards Specifications of materials, physical measurements, processes, performance of products, and characteristics of services rendered. Design standards may be established by individual manufacturers, trade associations, and national or international standards organizations. The general purpose is to realize operational and manufacturing economies, to increase the interchangeability of products, and to promote uniformity of definitions of product characteristics.

Individual firms often maintain extensive and detailed standards of parts that are available for use in their product designs. Usually the standards have the effect of restricting the variety of parts to certain sizes and materials. In this way the production lots required for inventory purposes are increased, and production economies may thereby be realized through the wider use of mass production. However, even if the larger quantities needed of the relatively few sizes do not in themselves lead to a cheaper manufacturing process, the costs of carrying inventory and setting up for production runs are reduced. A further development of this design approach may lead to the modulization of the entire product line, by reducing it to certain major subassemblies that are common to as many products as possible. Special jobs then typically require only a few added features, and cost savings may be realized.

A possible disadvantage is that the extensive use of general-purpose parts may jeopardize the space parts business, especially where outside manufacturers can skim off the market for the more commonly used and profitable spare parts once the original patents, if any, have expired, and then leave the more complex and slow-moving spares to the original manufacturer. However, it is precisely this aspect of standardization which is often welcomed by the users of the product.

Standardization also determines the nature of design practice. Especially when the specifications also give data on strength and performance as well as the usual dimensions, it is only necessary to compute loads approximately and then select the nearest standard sizes. Much design effort is thereby saved, especially on detail drawings, bills of material, and so forth. This approach also simplifies programming when computer-aided design is used. *See* COMPUTER-AIDED DESIGN AND MANUFACTURING.

Trade associations are the principal sources of American industrial standards. These involve standardization over an entire product line. In general, their scope is considerably less than that within firms with extensive standardization programs, but the technical and policy considerations in the two levels of standardization are quite similar. Trade standards are primarily concerned with specifying overall dimensions, so that products of different manufacturers may be used interchangeably; with performance, so that customers know what they are buying; and with certain design features, such as major materials, in order to assure proper function. In some cases, dimensional standards particularly must be related to standards in other industries; for instance, an American butter dish must accommodate the standard 4-oz (113.6-g) sticks in which butter is packed. Like national standards to which they are closely related, trade association standards should be established on the basis of as broad a consensus as possible within the industry. If standards were established such that any required burden of retooling and product change would fall in a discriminatory fashion upon only certain members of the industry, legal remedy would certainly be sought under the American antitrust laws.

The principal industrial countries have official agencies that approve, consolidate, and in some cases establish standards. Among them are the British Standards Institution (BSS), German Institute for Norms (DIN), and the American National Standards Institute (ANSI, formerly ASA). The national standardization agencies are members of a wide variety of international groupings and United Nations agencies. The principal ones are the International Organization of Standardization (ISO) and the International Electrotechnical Commission (IEC). These attempt to coordinate national activities and promote cooperation in the area of standardization. Several of the more than 50 organizations deal with weights and measures; others engage in transnational or international activities in the standardization of many products or cover specific regional issues and requirements. Some, like the European Economic Community (EEC), the International Telecommunications Union (ITU), the International Civil Aviation Organization (ICAO), or the administration of the General Agreement on Tariffs and Trade (GATT) mainly have other political, economic, and scientific concerns but must necessarily take note of standardization as part of their work. *See* ENGINEERING DESIGN; PHYSICAL MEASUREMENT. [J.E.U.]

Desizing A process that removes sizing agents (slasher compounds) which have been added to the warp yarns (lengthwise yarns) prior to weaving to protect them against the abrasive action of loom parts during the formation of gray fabric. Cellulosics are slashed primarily with starches which may be removed with enzymes of the malt or bacteriostatic types and are called amylolytic enzymes. Other effective agents are dilute mineral acids at high temperatures and sodium bromite in cold solution. Protein warps are sized or slashed or pretreated with proteinaceous agents such as gelatin, casein, albumin, or glue and require the use of proteolytic enzymes for their removal. Synthetics are slashed with water-soluble agents such as polyvinyl alcohol or sodium salts of polyacrylic acids which are removed by hot detergent solutions. Use of polyvinyl alcohols and carboxymethyl celluloses is increasing with cotton and other cellulosic fiber materials because of their biodegradable characteristics and as a means of decreasing stream pollution. Elimination of the desizing operation creates insurmountable problems in subsequent processes, including bleaching. *See* BLEACHING; TEXTILE CHEMISTRY. [J.J.McD.]

Desmodorida An order of nematodes in which the variable stoma may be armed with a dorsal tooth which may be opposed by subventral denticles. The variable amphids range from reniform-to-elongate loops to simple or multiple spirals. The cephalic sensilla are generally in three whorls; however, the second and third whorls may be combined. Distinguishing the order is the cephalic capsule or helmet and the conspicuous somatic annuli. In some groups, anterior and posterior adhesion tubes are utilized in locomotion.

There are five desmodorid superfamilies: Ceramonematoidea comprise marine forms with distinctive cuticular ornamentation that takes the form of crested annuli; nothing is known of the feeding habits. Most species of Desmodoroidea are found in the marine environment. Draconematoidea are an unusual superfamily of marine forms; when relaxed, the body is dorsally and then ventrally arched into a shallow sigmoid shape. Epsilonematoidea and Monoposthioidea comprise marine forms. *See* NEMATA. [A.R.M.]

Desmoscolecida An order of nematodes in which the taxa are readily recognized by their conspicuous body annulation. The annuli may be covered with concretion rings, or the cuticle may be ornamented with scales, warts, or bristles. The cephalic sensilla reportedly are reduced in number. The internal whorl is absent, the second whorl is papilliform, and the four sensilla of the third circlet are setiform. The vesiculate amphids are oval to circular and occupy much of the cephalic region. The somatic setae are tubular, and the open distal ends are often elaborate. The posterior esophagus is only slightly expanded. When pigment spots or ocelli are present, they are just posterior to the esophagus. Females are amphidelphic, and the ovaries are generally outstretched. The male spicules are generally accompanied by a gubernaculum. Both sexes have three caudal glands.

There are two desmoscolecid superfamilies: Most species of Desmoscolecoidea are found in the marine environment, although some species have also been collected from brackish waters. Greeffielloidea are free-living nematodes found primarily in marine or, rarely, estuarine habitats. *See* NEMATA. [A.R.M.]

Desmostylia An extinct order of large, hippopotamuslike, amphibious, gravigrade, shellfish-eating mammals that frequented shallow bays and coastal margins during the Oligocene and Miocene (see illustration). Their main distribution was northern trans-Pacific, but discoveries in Florida indicate a much broader distribution. There are two families, the Desmostylidae and the Paleoparadoxidae. These animals probably descended from an ancestral stock that also gave rise to the Proboscidea and Sirenia. *See* MAMMALIA; PROBOSCIDEA; SIRENIA. [G.T.J.]

A restoration of *Desmostylus* from the Miocene of California. (*After R. A. Stirton, 1959*)

Desmothoracida An order of Heliozoia. The shell (test) of these protozoans is typically perforate and chitinous (often with appreciable siliceous impregnation, but a completely siliceous wall seems to be rare). The body is commonly spherical. Some species are normally floating (illustration *a–c*); others are stalked sessile forms (illustration *d–e*). There is no centroplast, and there is a more or less central nucleus. The pseudopodia are typically delicate filamentous types (filopodia), which may branch. Little is known about most of the described species. *See* HELIOZOIA; PROTOZOA. [R.P.H.]

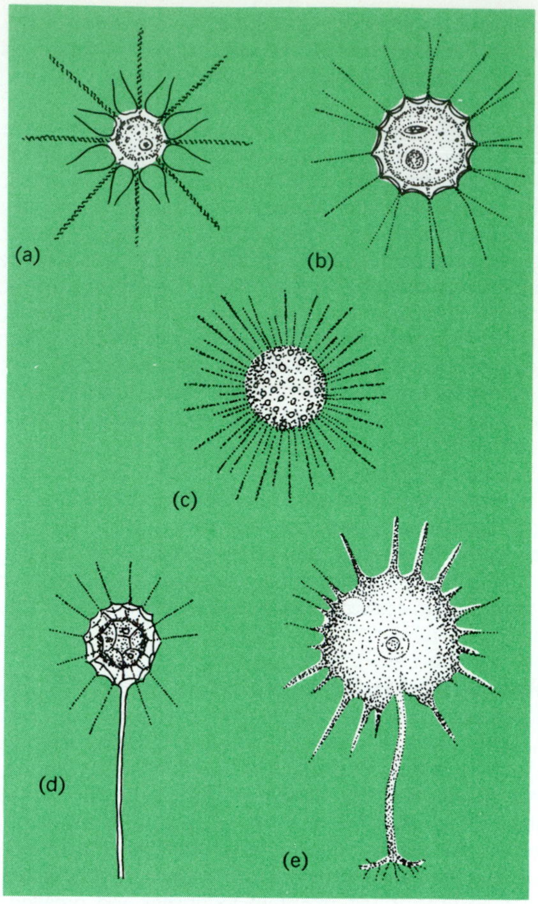

Examples of Desmothoracida. (*a*) *Choanocystis lepidula*; (*b*) *Clathrella forelli*; (*c*) *Elaster greefi*; (*d*) *Hedriocystis reticulata*. (*e*) *Clathrulina elegans*. (*After R. P. Hall, Protozoology, Prentice-Hall, 1953*)

Desorption A process in which atomic and molecular species residing on the surface of a solid leave the surface and enter the surrounding gas or vacuum. In stimulated desorption studies, species residing on a surface are made to desorb by incident electrons or photons. Measurements of these species provide insight into the ways that radiation affects matter, and are useful analytical probes of surface physics and chemistry. In thermal desorption studies, adsorbed surface species are caused to desorb as the sample is heated under controlled conditions. These measurements can provide information on surface-bond energies, the species present on the surface and their coverage, the order of the desorption process, and the number of bonding states or sites.

Stimulated desorption. Stimulated desorption from surfaces is initiated by electronic excitation of the surface bond by incident electrons or photons. The classical model of desorption is an adaptation of the theory of gas-phase dissociation, in which desorption results from excitation from a bonding state to an antibonding state.

Another model which is more applicable to the phenomenon of ion desorption was first observed in studies of the desorption of positively ionized oxygen (O^+) from the surface of titanium(IV) oxide (TiO_2). Here it is found that O^+ is desorbed not by valence level excitation, but by ionization of the titanium and oxygen core levels. These levels, of course, have little to do with bonding. Furthermore, the fact that the oxygen is desorbed as an O^+ ion (whereas it is nominally at O^{2-} on the surface) implies a large (three-electron) charge-transfer preceding desorption. This mechanism for desorption can also be effective for covalently bonded surface species.

Stimulated desorption studies are finding wide use. First, they can show the ways in which radiation affects the structure of solids. This will have important applications in the areas of radiation-induced damage and chemistry. Second, as an analytical tool, they offer a unique new way to study the physics and chemistry of atoms on surfaces which, when combined with the many other surface techniques based largely on electron spectroscopy, can provide new insight. Finally, models of the surface bond are put to a much sterner test in attempting to explain desorption phenomena. *See* RADIATION DAMAGE TO MATERIALS.

An additional important discovery is that ion angular distributions from stimulated desorption are not isotropic, but show that ions are emitted in relatively narrow cones which project along the nominal ground-state bond directions. Thus this technique provides a direct display of the surface-bonding geometry.

Thermal desorption. Thermal desorption mass spectroscopy is possibly the oldest technique for the study of adsorbates on surfaces. Three primary forms of the thermal desorption experiment involve measurement of (1) the rate of desorption from a surface during controlled heating (temperature-programmed thermal desorption), (2) the rate of desorption at constant temperature (isothermal desorption), and (3) surface lifetimes and diffusion under exposure to a pulsed beam of adsorbates (molecular-beam experiments). Of the three, temperature-programmed thermal desorption is by far the most widely applied. The most straightforward information provided is the nature of the desorbed species from mass analysis, and a determination of the absolute coverage by the adsorbate, which is very difficult to obtain with other techniques. The technique can also provide important kinetic parameters of the desorption process.

While the thermal desorption techniques are among the simplest of surface probes, they remain indispensable because of their directness and the variety of information they convey. Thus while surface science moves to detailed methods involving extremely sophisticated apparatus, the simple thermal desorption methods remain an important part of the overall picture. *See* SURFACE PHYSICS. [M.L.K.]

Destructive distillation The primary chemical processing of materials such as wood, coal, oil shale, and some residual oils from refining of petroleum. It consists in heating material in an inert atmosphere at a temperature high enough for chemical decomposition. The principal products are (1) gases containing carbon monoxide, hydrogen, hydrogen sulfide, and ammonia, (2) oils, and (3) water solutions of organic acids, alcohols, and ammonium salts.

Crude shale oil may be obtained by destructive distillation of carboniferous shales. It may be subjected to a destructive, or coking, distillation to reduce its viscosity and increase its hydrogen content. Residual oils from petroleum refinery operations are subjected to coking distillation to reduce the carbon content. The coke is used for the manufacture of electrode carbon. The main product of the destructive distillation of wood is 40–45% charcoal used in metallurgical processes in which the low content of ash, sulfur, and phosphorus is important. *See* CHARCOAL; COAL CHEMICALS; COKE; COKING (PETROLEUM); OIL SHALE; PYROLYSIS; WOOD CHEMICALS. [H.H.St./H.W.W.]

Detergent A substance used to enhance the cleansing action of water. A detergent is an emulsifier, which penetrates and breaks up the oil film that binds dirt particles, and a wetting agent, which helps them to float off. Emulsifier molecules have an oillike nonpolar portion which is drawn into the oil, and a polar group that is water-soluble; by bridging the oil-water interface, they break the oil into dispersible droplets (emulsion). As a surfactant, a detergent decreases the surface tension of water and helps it penetrate soil.

Soap, the sodium salt of long-chain acids, was the principal detergent until superseded in 1954 by synthetic detergents (syndets) which, unlike soap, do not form insoluble products with the calcium in hard water. Most syndets are of the anionic type, that is, sodium salts of alkyl sulfates or sulfonates. Alkyl benzene sulfonates (ABS) with branched carbon chains were found to persist in wastewater and have been replaced by linear alkyl benzene sulfonates (LAS), which are biodegradable by bacterial action. Anionic detergents are best for water-absorbing fibers such as cotton, wool, and silk. Nonionic detergents are polyethers made by combining ethylene oxide with a 12-carbon lauryl alcohol. They are used for water-repelling "permanent press" fabrics, and their low-foaming property is desirable for automatic washers. Cationic syndets are quarternary base compounds. They are more expensive, but some are germicidal; some are used as fabric softeners and as good metal cleaners.

Detergents must contain alkaline "builders" to bind dissolved metal ions and support emulsification. Sodium pyrophosphate or polyphosphate were preferred because of low cost and high cleansing effectiveness. However, when discharged with laundry wastewater, these compounds supply nutrient to phosphate-deficient lakes and streams and thus lead to eutrophication, and their use is now banned by law. Less harmful, but less effective, builders such as sodium carbonate are now widely used in detergents. *See* EUTROPHICATION.

Many additives are used in detergents to provide scent, brightening (usually through fluorescent action), or bleaching action. Biodegradability is essential for detergents; it ensures that components of detergents will be broken down by bacterial action before undesirable aftereffects can occur. Nonbiodegradable detergents can prevent effective bacterial action in septic tanks and sewage treatment plants, and can cause undesirable persistent foaming in rivers. *See* SOAP. [A.L.H.]

Determinant The theory of determinants had its origin in the solution of linear systems of equations. Determinants occur in virtually all branches of mathematics and related fields. *See* LINEAR SYSTEMS OF EQUATIONS; MATRIX THEORY; POLYNOMIAL SYSTEMS OF EQUATIONS.

The concept of a determinant can best be explained with reference to a matrix A.

$$\text{Let} \qquad A = \begin{pmatrix} a_{11} & a_{12} & \cdots & a_{1n} \\ a_{21} & a_{22} & \cdots & a_{2n} \\ \cdots & \cdots & \cdots & \cdots \\ a_{n1} & a_{n2} & \cdots & a_{nn} \end{pmatrix}$$

be an n by n square array of numbers. Such an array is called an n by n square matrix, and the numbers a_{ij} in the array are called elements of the matrix. The determinant of the matrix A, as indicated by Eq. (1), is a number which is the value of a certain function of the elements of A.

$$|A| = \begin{vmatrix} a_{11} & a_{12} & \cdots & a_{1n} \\ a_{21} & a_{22} & \cdots & a_{2n} \\ \cdots & \cdots & \cdots & \cdots \\ a_{1} & a_{n2} & \cdots & a_{nn} \end{vmatrix} \qquad (1)$$

The determinant of Eq. (1) is called a determinant of order n, and it can be defined by induction on n. For explanatory purposes, determinants of orders $n = 1$, 2, and 3 are defined as follows:

For $n = 1$, $|a_{11}| = a_{11}$

For $n = 2$, $\begin{vmatrix} a_{11} & a_{12} \\ a_{21} & a_{22} \end{vmatrix} = a_{11}|a_{22}| - a_{12}|a_{21}| = a_{11}a_{22} - a_{12}a_{21}$

$$\text{For } n = 3, \begin{vmatrix} a_{11} & a_{12} & a_{13} \\ a_{21} & a_{22} & a_{23} \\ a_{31} & a_{32} & a_{33} \end{vmatrix}$$

$$= a_{11}\begin{vmatrix} a_{22} & a_{23} \\ a_{32} & a_{33} \end{vmatrix} - a_{12}\begin{vmatrix} a_{21} & a_{23} \\ a_{31} & a_{33} \end{vmatrix} + a_{13}\begin{vmatrix} a_{21} & a_{22} \\ a_{31} & a_{32} \end{vmatrix}$$

$$= a_{11}\,(a_{22}a_{33} - a_{23}a_{32}) - a_{12}(a_{21}a_{33} - a_{23}a_{31}) + a_{13}\,(a_{21}a_{32} - a_{22}a_{31})$$

The determinants of order 2 and 3 are defined in terms of determinants of order 1 and 2, respectively. Suppose now that a determinant of order $n - 1$ has been defined. In Eq. (1), denote by A_{1j}, $j = 1, 2,\ldots, n$, the determinant of order $n - 1$ formed by deleting from Eq. (1) the first row and the jth column. Equation (2) then, by definition, holds.

$$|A| = a_{11}A_{11} - a_{12}A_{12} + a_{13}A_{13} + + (-1)^{n-2}a_{1n-1}A_{1n-1} + (-1)^{n-1}a_{1n}A_{1n} \qquad (2)$$

This definition is seen to agree with the definitions for determinants of order 2 and 3 and, by the principle of mathematical induction, defines a determinant of order n for every n. There are many other definitions of a determinant of order n which can be proved to be equivalent to the definition given here. *See* EQUATIONS, THEORY OF. [R.A.B.]

Determinism The principle that nature follows exact laws, so that what will happen in the future is a necessary consequence of the state of the world at any given moment in the past. This view, if fully adopted, implies that events which seem to occur by chance would be fully understood if more was known about them, and that apparently free thoughts and choices are explainable and in principle predictable in terms of neuroscience. In a looser sense, determinism refers to claims that mental freedom is much more restricted than people are inclined to suppose.

The question of determinism in physical science cannot be considered apart from the philosophical problem; this gives it added importance and forces its consideration in a very critical spirit.

The idea that the world is composed of atoms moving under the influence of certain forces according to certain laws can be traced back to the Greek philosopher Leucippus. Deterministic philosophy was prominent in the work of the seventeenth-century thinker René Descartes, and became widely known through his influence. Isaac Newton carried out a large part of the cartesian program. His theory explained so many natural processes that it began to appear that the universe since the time of Creation might actually have run its course in a deterministic fashion like a machine, without divine intervention. A century after Newton, Pierre Simon de Laplace argued that an Omniscient Calculator, provided with exact knowledge of the state of the universe at present, would be able to predict the entire future. *See* CLASSICAL MECHANICS; NEWTON'S LAWS OF MOTION.

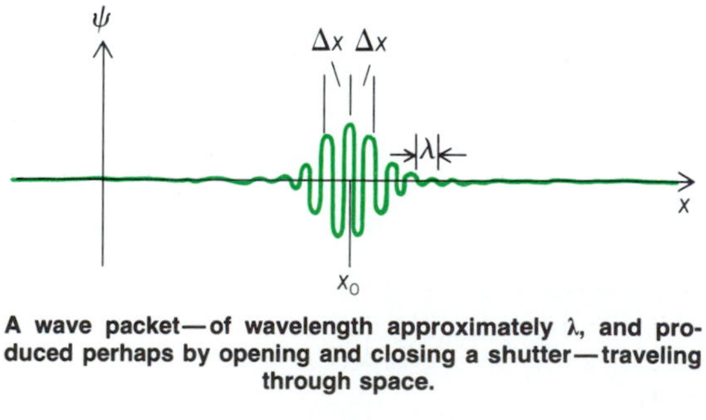

A wave packet—of wavelength approximately λ, and produced perhaps by opening and closing a shutter—traveling through space.

The quantum mechanics of the 1920s introduced the paradox of particles which are, at the same time, waves. A wave of length λ was supposed to accompany, or describe, a particle of momentum $p = h/\lambda$, where h is Planck's constant. The probability of finding a free particle is expressed by a (complex) wave packet (see illustration), and the particle has appreciable probability of being found only where the wave function is large; that is, within a region of size roughly Δx on each side of x_0 in the illustration. A theorem of Fourier analysis shows that Δp, the range of momenta present in the wave packet, is related to Δx as shown below. This means that the more pre-

$$\Delta x \, \Delta p \geq h(2\pi)$$

cisely the specification of a particle's position x is attempted by means of a localized wave function, the less precisely can its momentum be specified. W. Heisenberg's principle of uncertainty, or indeterminacy, states that it is impossible to measure, and therefore to know, x and p any more accurately than is allowed by this relation, and there are also other pairs of dynamical variables similarly related. *See* FOURIER SERIES.

There remains the question as to whether Heisenberg's principle is merely an unfortunate limitation on an experimenter's ability to know or whether it goes deeper. The general opinion of physicists is that of N. Bohr: the principle expresses a limitation of the precision with which concepts such as position and momentum can be applied at all. Therefore Laplace's Omniscient Calculator cannot predict the future. *See* QUANTUM MECHANICS; UNCERTAINTY PRINCIPLE.

Perceptive mathematicians have warned that determinism is not as obvious a consequence of newtonian physics as it might appear. A series of mathematical results have been proven whose general effect is that for the vast majority of dynamical systems any error in the initial conditions, however small, will be amplified, in general exponentially, and so quickly that the predicted result will soon bear no relation to reality. Thus unless it is assumed that initial conditions are known with perfect accuracy, and perfectly accurate computation takes place thereafter, the Omniscient Calculator will wind up getting everything wrong.

Very few people now think that all events in the natural world are exactly determined. Experiments suggest that some human and animal behavior can reliably be predicted and controlled, but nobody knows the limits within which this can be done. [D.P.]

Detonator A device used to initiate the explosion of a high explosive. Detonators employ a very sensitive primary explosive which detonates readily when set afire by the primer. Primary explosives in common use are mercury fulminate and lead azide. In addition, the detonator or blasting cap usually contains a small charge of secondary explosive, such as PETN (pentaerythritol tetranitrate), to transmit a more powerful shock wave than the primary explosive alone. An electric detonator is ignited by a fuse wire which serves to touch off the primer. The explosion of the detonator readily sets off a dynamite charge. *See* FUSE (EXPLOSIVE). [W.E.Go.]

Deuterium The isotope of the element hydrogen with atomic weight 2.0144 and symbols ^2H or D. The terrestrial natural abundance of deuterium is 1 part in 6700 parts of ordinary hydrogen (protium). Small variations in natural sources are found as a result of fractionation by geological processes.

Deuterium is a gas (D_2) at room temperature. It is prepared from heavy water, D_2O, either by electrolysis or by reaction of D_2O with metals such as zinc, iron, calcium, and uranium. It is also prepared directly by the fractional distillation of liquid hydrogen.

Deuterium is used mainly in the form of heavy water. In the uncombined state it finds uses as a research tool. Liquid deuterium is used in bubble chambers to study the reactions of elementary particles with the deuterium nucleus, the deuteron. Deuterons are frequently accelerated in cyclotrons to study their reactions with other nuclei and also to produce radioactive nuclides. Deuterium gas is used in the direct synthesis of organic compounds for tracer studies. *See* DEUTERON; HEAVY WATER; HYDROGEN; NUCLEAR FUSION; NUCLEAR REACTION; TRITIUM. [J.Big.]

Deuteromycotina A heterogeneous group of anamorphic (asexual or imperfect) fungi in which sporulation may occur on separate hyphae or composite fruit bodies (conidiomata). The diagnostic feature of the group is the lack of a teleomorphic (sexual or perfect) state. There are more than 1700 genera containing 18,700 species.

The Deuteromycotina are also artificial inasmuch as the group consists of fungi or states of fungi reproducing vegetatively by mitosis, not meiosis. Some organisms, such as *Aspergillus, Penicillium*, and *Fusarium* have a parasexual process in which reciprocal exchange of chromatid segments between homologous chromosomes during mitotic propagation of the diploid nucleus occurs. Instead of the regular sequence of events which obtain in the meiotic cycle, the various processes may all be occurring at one time in one mycelium. For deuteromycetes, parasexuality brings the advantage of genetic recombination. *See* MITOSIS; RECOMBINATION (GENETICS).

Traditionally, Deuteromycotina are separated into two classes: Hyphomycetes, which are mycelial forms bearing conidia on separate hyphae or aggregations of hyphae but not within discrete conidiomata; and Coelomycetes, which are forms that produce conidia in pycnidial, pycnothyrial, acervular, cupulate, or stromatic conidiomata. A third class is also now recognized, the Agonomycetes (Mycelia Sterilia); these are forms that lack conidiomata and fail to produce conidia but form somatic survival propagules. *See* AGONOMYCETES; COELOMYCETES; HYPHOMYCETES.

Deuteromycetes are ubiquitous and occupy most conceivable ecological niches, whether they are aquatic, terrestrial, or artificial. Deuteromycetes have been used extensively in manufacturing processes of different kinds. All fungal infections of humans are opportunistic, and mycoses are almost always the result of abnormal host immunity. Even before the association of the HIV virus with acquired immune deficiency syndrome (AIDS), criteria used in defining the syndrome relied heavily on the existence of certain opportunistic infections in the absence of any of the established predisposing conditions. A significant number of these deep-seated infections are caused by deuteromycetes such as *Aspergillus, Candida, Cryptococcus*, and *Histoplasma*. *See* ACQUIRED IMMUNE DEFICIENCY SYNDROME (AIDS); MEDICAL MYCOLOGY; MYCOTOXIN; OPPORTUNISTIC INFECTIONS; PLANT PATHOLOGY. [B.C.S.]

Deuteron The nucleus of the atom of heavy hydrogen, ^2H (deuterium). The deuteron *d* is composed of a proton and a neutron; it is the simplest multinucleon nucleus. Its binding energy is 2.227 MeV; that is, this is the amount of energy which must be added to a deuteron for it to dissociate into a proton and a neutron. Deuterons are much used as projectiles in nuclear bombardment experiments. *See* DEUTERIUM; NUCLEAR REACTION. [H.E.D.]

Developmental biology The current and appropriate term for a large field of investigation that has taken shape by a coalescence of several paths of research. It was in the interdisciplinary spirit induced by the maturing of the biologically oriented natural sciences that a small group of experimentalists interested in organic development met for a symposium in 1939. Professionally, their fields included biochemistry, botany, cytology, embryology, genetics, mathematics, medical (cancer) research, microbiology, and zoology. They continued, as the Society of the Study of Development and Growth, to meet annually. In 1959 a journal entitled *Developmental Biology* was organized; it has become the official organ of the society, which a few years later changed its name to the Society for Developmental Biology. At present, there is also an International Society of Developmental Biologists. Individual investigators in the field work with all varieties of organism, at many levels, but in any case keep in mind the picture of cell function that is accumulating at the molecular level. A developmental view must also consider the life history of the organism at hand.

In multicellular animals the sexual cycle is simple in outline. An adult form, with a double (diploid) chromosome set, produces gametes by two specialized cell divisions (meiosis). Sperm and egg have single-set (haploid) chromosomes. Fertilization restores the diploid state, with added genetic variability. The zygote undergoes a developmental history, which may be direct or may be interrupted by one or more metamorphoses from a larval state. All cell divisions in the history will be mitotic, until germ cells are produced. *See* OOGENESIS; REPRODUCTION (ANIMAL); SPERMATOGENESIS.

In multicellular plants the haploid phase is prolonged. Spores, which may be of one type or may be distinguished as micro- or macrospores, are produced by meiotic divisions. The haploid spores undergo mitotic divisions to form gametophytes. In mosses the gametophyte is the principal, green plant. In seed plants the male and female gametophytes are very reduced in size; the female is sessile within a parental structure, to which the male pollen grain is transported. Sperm and egg differentiate in the haploid plants, then unite to develop into the plant as commonly known, which is the sporophyte generation. *See* REPRODUCTION (PLANT).

Embryogenesis. Normally insemination follows release of the ovum, triggering activation of the egg. After the sperm enters the egg, male and female pronuclei reconstruct as they approach the center of the egg and unite, while the mitotic spindle assembles. The early development that follows the initial mitosis includes cleavage, leading to gastrulation, whereby three germ layers move to final position. The cells of the animal region remain external; vegetal cells become enclosed, as innermost entoderm and as mesoderm between the two layers.

Cleavage is a period of successive mitoses, subdividing the zygote into numerous cells, with almost no alteration of cytoplasmic localization or polarity, or change in mass or shape. The mitotic apparatus is the immediate mechanism, modified by the cytoplasmic structure, which may produce unequal divisions or restrict mitosis to one area, as in the case of the bird egg. Cleavage patterns differ among phyla and classes, morphologically and developmentally. In some groups, including vertebrates, isolated cells or parts of the cleaving zygote can form a complete organism, if both animal and vegetal cytoplasm are present in proportions related to the polar structure. By contrast, in phyla having so-called mosaic development, even the first cleavage is differential. *See* CLEAVAGE (EMBRYOLOGY).

In echinoderms, vetebrates, and related phyla, cleavage ceases as a fluid-filled hollow blastula stage is reached, with some hundreds of cells displaying rudimentary differentiation. The future ectoderm is a coherent epithelium. Gastrulation involves invagination of mesodermal and entodermal areas, while the ectoderm stretches to cover the entire structure, all without altering the original spherical shape. The involuting cells retain contact with one another and with the ectoderm by long pseudopodia if not by the entire surface.

Gastrulation is a critical period, known as a stage when incompatible hybrids die. Primitive organogenesis progresses from anterior to posterior. The dorsal ectoderm develops longitudinal folds that meet, pinching off a tube that will become the central nervous system; the entodermal layer, similarly, produces the tube that will line the digestive tract. Mesoderm, except for its anteriormost region, which remains relatively unorganized, forms the axial notochord (the foundation of the vertebral column), while adjacent mesoderm on either side assumes a segmental, repetitive pattern that will impose itself on the entire dorsal axis. The remaining mesoderm spreads laterally between ectoderm and entoderm, to meet ventrally, forming the heart and the body cavity. *See* GASTRULATION.

Cell differentiation, involved in these rearrangements, is at first reversible, becoming stabilized as the germ layers find their relative positions. Contacts between cells with different previous histories appear to be decisive factors in stabilization. Such contacts have been given the general term of embryonic induction; they are mutual, not one-way, reactions, eventually leading to a fixed cellular program not reversible by experimental methods so far employed. The primary induction in vertebrate development is that of the central nervous system by the dorsal mesoderm during gastrulation. In the absence of mesodermal lining, the ectoderm can retain only an undifferentiated epithelial structure. Different foreign molecules can regularly induce three recognizable regional levels of the nervous system: anteriormost (forebrain-nose-eye-midbrain); secondary (hindbrain-ear-spinal cord); and posterior (tail). Since these three levels, through gastrular movements, become underlain by structurally different mesodermal domains, regional differences of signal are not unexpected. *See* CELL LINEAGE.

After the original induction, cell differentiation progresses stepwise by new syntheses, or intensification of established ones. The process may depend on mitoses, migration, new contacts, and even cell fusion, as well as hormonal signals. Visible differentiation takes many forms, some well analyzed at molecular levels. Selective protein synthesis is basic; it may characterize cell types, such as red cells, keratin-producing epidermis, or, as an extreme example, the elaborately assembled fibers of striated muscle. Many cell types are characterized by elaboration and export of specific molecules or subunits: exocrine or endocrine secretions, matrix, or the various neurotransmitters and neurohormones that are continually being discovered. The plasma membrane is probably specifically structured in every cell type, or perhaps in each cell clone, as seems the case in the immune system. In neurons the plasma membrane not only is the functional messenger of impulses but, in response to a growth factor, is critical in production and guidance of the polarized outgrowing fibers that establish the incredibly complex structure of nervous tissue. *See* EMBRYONIC DIFFERENTIATION; EMBRYONIC INDUCTION.

Plant development. In seed plants the preparatory functions for early development are not concentrated in a single oocyte cell. Meiosis is completed and a haploid gametophyte formed in a polarized capsule structure within a modified leaf; in flowering plants this is the pistil base, connected to the parental vascular tissue. Upon fertilization, the zygote commences syntheses; the cell soon divides. Cleavage of the zygote, unlike that in the animal, at first forms a linear cell series in the polar direction, later taking, successively, globular and bilateral shapes, while continually enlarging by mitosis. The growth pattern first emphasizes the storage structures of the embryo itself. Later a centrally located area of rapidly dividing cells indicates the future plant axis, with the shoot apex directed toward the nutritive pole, and the root tip and cap opposite. *See* PLANT GROWTH.

The plant axis is organized on the basis of cell and tissue differentiation in an elegant geometry pattern within a cylindrical structure. Growth is partly by addition, in layers, of axially aligned cells from actively dividing meristem apices. The walled, individually immovable cells, as they are distanced from the meristem, elongate, thus adding to axial growth. The cells differentiate, following a radial pattern of position within the cylinder. The basic pattern consists of an outer epidermis, an adjacent cortex of supporting or storage tissue, and a vascular ring, with or without a central core. These landmarks become variously elaborated according to species, region, and developmental stage. Radial growth derives from dividing tissue (cambium) organizing between the outer phloem and inner xylem layers of the vascular ring.

Much cell differentiation relates to wall structure. Tracheids and xylem tubes, phloem sieve tubes, sclerenchyma, epidermal walls with interspersed apertures such as lenticels and stomata, and depots of lignin and cork are examples. Final differentiation may result from death of the cell, as in xylem vessels. Parenchymal cells, distinguished by chloroplasts or starch grains in the cytoplasm, are typically thin-walled, with intercellular spaces. All living cells are interconnected by protoplasmic strands, passing through wall apertures.

It is not known to what extent differentiation is due to intercellular relations; the precision of the pattern suggests considerable local signaling. It is certain that translocation of growth substances plays an important role. Several molecular types have been identified. Three of these substances are organic acids: auxins, gibberellins, and abscisic acid. Another family of substances, the purine-related cytokinins, are involved in seed germination and in releasing meristems from inhibition. Other transported substances, such as flower-inducing factors, have resisted isolation. *See* ABSCISIC ACID; AUXIN; CYTOKININ; GIBBERELLIN.

Developmental effects of growth substances often appear contradictory, probably because response of cultured cells is very sensitive to concentrations, and is also species- and stage-specific. Cytokinins and gibberellins are present in gametophyte tissues; abscisic acid is involved in the transfer of endosperm-stored carbohydrate to cotyledon tissue. *See* PLANT MORPHOGENESIS.

Cellular totipotency. Not only is cell differentiation in vascular plants less diverse than in animals, it is reversible. Adventitious embryo formation is known in many species. Modern culture methods have demonstrated that many, perhaps all, living cells can be induced to dedifferentiate and to undergo embryo formation and even development to reproductive stages. Haploid embryos have been produced from isolated microspores, although, as in animal parthenogenesis, regulation to the diploid state is the rule.

Although differentiated animal cells may lose some specialized structures when cultured, nothing like a reversion to the zygote stage has been observed. In amphibians the nucleus of many types of differentiated cells can, if transplanted to oocyte cytoplasm, collaborate in the development of a reproductive adult, thus demonstrating a complete genome. It is known that in the oocytes of many animal phyla a cytoplasmic factor or factors, localized in the vegetal region, determine a germ cell lineage to be isolated until the gonad is organized.

Morphogenesis. Although much is known about cellular aspects of organ and organism structure in both plants and animals, the molecular basis of these larger developmental problems is far from clear. In plants growth patterns appear to depend on mitosis and on axial and lateral distribution of growth substances. The suggestion, in geriatric investigation, that cells have genetically limited powers of mitosis might, if applicable to apical meristems, explain pattern and size limitations. A molecular counting device would be required; this might be an aspect of the biological clock problem that still remains to be solved.

In animals several levels of control must be understood. Anteroposterior polarity, originating in the oocyte as animal-vegetal differential, is transmitted, directionally unaltered, to blastula, gastrula, and embryo, as regards external tissue. In amphibian development a dorsoposterior pole is added in the zygote, recognizable by pigment migration in the cortex. Gastrulation movements follow these two axial directives, the mesoderm first migrating from posterior to anterior until the anterior pole is reached. At this point the anteroposterior direction takes over all layers, and the mesoderm spreads dorsoventrally. Regional differentiation must depend on establishment of subordinate centers; cell surfaces would be decisive in grouping, moving, or otherwise interacting to produce organ primordia.

In all cases of size or direction control, centers producing radiating or linear cellular gradients of communication appear to furnish a reasonable model. A spectacular case of reversal of the principal axis is found in the bilateral, elongated insect egg. In certain fly species, destruction of small bilateral areas of the anterior cytoplasm produces bicaudate adults: that is, the head and thorax are replaced by a reversed abdomen, mirror-imaging the original undamaged abdominal segments. It is difficult to interpret such organized reversal other than on a gradient pattern.

Specific size-controlling substances, with promotion or inhibition of mitosis as a target, have been identified or suspected in several cases of individual organs or tissues. These must be attributed to some species-specific spatial control. Something of the sort is strikingly illustrated in tetraploid amphibians, which, unlike plants, grow no larger than their diploid relatives. The large cell size, proportional to the enlarged nucleus, does not affect the size of functioning units; for example, the kidney tubules are no larger than normal, but are composed of fewer cells. [D.Ru.]

Developmental genetics A field that deals with problems of gene expression and its control on various levels from gene transcription to the phenotype. Studies in this field serve to identify genes concerned with processes of development and differentiation, mechanisms of action of such genes and their control, and mechanisms of development. Genes involved in developmental processes include structural as well as regulatory genes of various types, for example, controlling and modifying genes. Mutations causing abnormalities of development and differentiation expose the existence of the respective normal genes, and thus serve as tools for the genes' identification. Gene analysis can be performed on both plants and animals.

Concepts and mechanisms pertaining to the fields of genetics and of developmental biology have guided analytical approaches to problems of developmental genetics. The genetic concepts of pleiotropy, phenocopies, penetrance, and expressivity are there. Two concepts of development that are of concern in developmental genetics are determination and inductive interaction.

The concept of determination in development refers to the progressive restriction of the multitude of developmental potentials of a cell to a single one serving the differentiation of one specific cell type only. After having undergone the process of determination, the cell is "programmed" to carry out its specific fate. This particular determined fate is perpetuated by cell hereditary. The development of vertebrates is characterized by inductive interactions between different cells and tissues; these interactions are responsible for progressive differentiation. Mutations in mice are known to interfere with induction processes, thus indicating a genetic basis for the responsible mechanisms. *See* DEVELOPMENTAL BIOLOGY.

Studies of developmental genetics involve a variety of systems, which include the following:

1. Studies of the developmental genetics of individual enzymes serve to identify genetic mechanisms of enzyme differentiation. The normal differentiation of an enzyme requires the interaction of structural genes, controlling genes, and regulatory genes of various types. These are involved in temporal as well as spatial aspects of enzyme differentiation, that is, time of expression of the enzyme and integration into subcellular compartments.

2. Mutations in the mouse provide particularly suitable model systems for studies of mammalian developmental genetics. The effects of such mutations have been described at any stage between fertilization and birth. The majority of developmental mutations act as embryonic lethals. Included among organs and tissue systems affected by developmental mutations are pigment cells, blood cells, germ cells, kidneys, cartilage and bone, muscle, nervous system, and subcellular ultrastructure of liver and kidney cells. Mutations affecting enzyme and protein differentiation in prenatal stages include those causing specific liver and kidney enzyme deficiencies.

3. The T-locus is one of the most interesting and most thoroughly studied systems in mammalian developmental genetics. Several mutations in this complex are dominant mutations, whereas the larger portion is occupied by recessive alleles which seem to represent quantitative and qualitative changes of the respective chromosome region. The presence of T-locus mutations is expressed in effects on the tail. Embryos homozygous for certain recessive alleles die in early stages of development. The T-complex is a prime example of a genetic system involved in the regulation and control of early embryonic differentiation.

4. Lethal albino deletions in the mouse provide a system of deletion mutations affecting biochemical and morphogenetic differentiation and thus serving the analysis of the genetic control of differentiation. They include a series of overlapping deletions, all of which are lethal in homozygotes, either in early embryonic stages or perinatally, causing phenotypic abnormalities of ultrastructural morphogenesis of subcellular organelles, of embryonic membrane differentiation, of early cleavage stages of the zygote, and of the developmental pattern of certain proteins.

5. The genetic control of muscle differentiation has been studied with the help of a recessive mutation in the mouse. This mutation affects specifically the skeletal striated muscle of homozygotes, whereas cardiac and smooth muscle differentiate normally, thus demonstrating the existence of a separate genetic control of differentiation for one of the three types of muscle cells.

6. Several mutations in the mouse are known to affect fetal red blood cell differentiation and have thus served the analysis of the genetic control of blood development. Studies of development and regulation of gene expression find a particularly suitable system in mammalian hemoglobins and their development, both in humans and in mice. At least seven genes coding for hemoglobin polypeptide chains in humans and an approximately equal number in mice participate in the formation and the development of six different hemoglobins characteristically present at different stages between early embryonic and adult life.

7. A variety of mutations affect the nervous system in its morphological and functional development. They have served the identification of some of the normal processes involved in neural development and their genetic control.

8. The mammalian female has two X chromosomes. However, in any one adult female cell, only one X chromosome is active. The mechanism which governs and assures activity of a single X chromosome is unknown, but it may provide a model system for the study of activation and inactivation of genes in general.

9. Gene expression, gene control, and gene interaction during mammalian differentiation may be analyzed with the help of a powerful experimental system, that of experimental

mouse chimeras. Various developmental processes and their genetic control may be studied with this method. In the case of lethal mutations it has been possible to "rescue" lethal homozygotes by fusing them with normal early embryos, with resulting survival of the fusion products. Mouse chimeras may also be obtained by injecting single cells from various embryonic donors into early mouse embryos. The injected cells become incorporated into the host embryo, providing opportunities for studies of differentiation. *See* CHIMERA; GENE; GENE ACTION; GENETICS; MUTATION. [S.G.-W.]

Developmental psychology The study of the course of behavior in human beings, and other animals, across their life spans. Although early studies of behavioral development were made with infants and children, child psychology is not synonymous with developmental psychology—child psychology, adolescent psychology, and the psychology of infancy or old age refer to subdivisions of developmental psychology that deal with behaviors during a particular age range and changes in behaviors during that period.

With the rapid expansion of information about behaviors and their development, developmental psychologists have become increasingly specialized. Some study only visual development during infancy, or prosocial behaviors (such as helping) in different cultures, while others concentrate on memory in old age. Nevertheless, the ultimate aims of the discipline are to describe behavioral development across the entire life span, and to explain such development, that is, to determine how and why it occurs, and hence to contribute to the remediation and prevention of behavioral disorders and to the promotion of healthy development.

Developmental psychology encompasses all the forms of behavior studied in other branches of psychology, such as perception, cognition, psychomotor skills, personality, and social interactions. It also uses the same methods and techniques.

Behavioral research may be experimental in which an independent variable is manipulated by the researcher and the effect on a dependent variable is measured. Other research is based on naturalistic observations, in which behaviors occurring in natural settings are recorded. Often experimental and naturalistic procedures are combined, as when an experimenter devises naturalistic situations in which to observe helping behavior but systematically varies the conditions, such as the number of other persons present. Developmental psychologists use all these approaches. Similarly, they use all the kinds of assessment techniques used by other psychologists, such as reaction time, tests, interviews, and ratings. Most developmentalists assume that behavioral development is influenced by social and biological variables and their interactions, and therefore many also draw on the techniques and data provided by other disciplines, such as anthropology, genetics, physiology, and sociology.

A major issue for developmental science is the polarity of continuity and change. On the one hand, even a casual observer notes that 2-year-olds differ in appearance and abilities from newborns, 6-year-olds differ from 12-year-olds, and 16-year-olds differ from 66-year-olds. But developmental psychologists address the questions of whether the child is the "father of the man," if changes are gradual or abrupt, if all characteristics change throughout life, and what influences make for continuity or change.

Societies with a history of emphasis on rational thought think in terms of antecedents and consequences, so the concept that the child is the "father of the man" has been characteristic of most of modern western society's ways of thinking about development. There have been, however, sharp disagreements about the factors underlying continuity or change. One position is represented by the construct of maturation, which is based on the assumption that the developmental patterns are determined primarily by genetic endowment, with only minor variations among healthy individuals resulting from individual differences in experiences. The other extreme accepts species differences as genetically determined but regards the individual within a species as a "tabula rasa" (blank page) whose course of development is shaped primarily by the physical and social environments to which it is exposed and particular experiences within those environments. Although a complex interactionist view is now most common, this "nature-nurture" issue still arises frequently and in many guises in interpretations of development.

Neither general psychology nor developmental psychology has a theory that applies to all forms of behavior. Instead there are several theories of cognitive development, of personality development, of perceptual development, and so on, which have little to say about other aspects of behavior. Also, most developmental theories deal with only a portion of the life span, such as infancy through adolescence or middle through old age. All theories have been contradicted to at least some extent by experimental studies designed to test predictions from them or by systematic observations of naturally occurring behavior. Nevertheless, theories have been, and will continue to be, useful in guiding the kinds of observations that are made and the way in which they are made. Major changes have occurred since the mid-1960s in the way behavioral development is conceptualized and measured. Much greater emphasis is placed on understanding how development occurs rather than simply describing what occurs. Instead of counting vocabularies, for example, developmentalists are analyzing linguistic development, such as the acquisition of grammar and the ways children communicate. Learners are no longer viewed as passive recipients of stimuli, reinforcements, and punishments, but as active participants who employ strategies and monitor their own skills and progress. [D.E.]

Devonian The fourth period of the Paleozoic Era, encompassing an interval of geologic time between about 410,000,000 and 345,000,000 years before present, based on radiometric data. The Devonian System encompasses all sedimentary rocks deposited, and all igneous and metamorphic rocks formed, during the Devonian Period. It is conventional

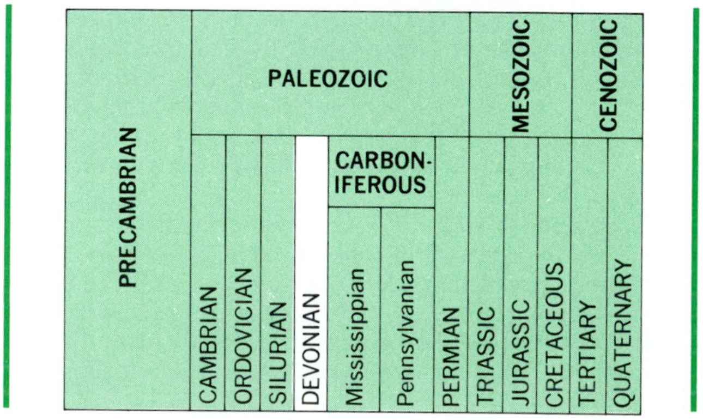

that recognition of Devonian time is determined by the definition of Devonian rocks. *See* PALEOZOIC.

The base of the Devonian System has been fixed, by international agreement, at an actual outcrop of sedimentary rocks in Czechoslovakia, where it corresponds to the base of the *Monograptus uniformis* graptolite zone. The top of the Devonian System, corresponding to the base of the Carboniferous System, has not yet been similarly fixed, but many stratigraphers use the base of the *Siphonodella sulcata* conodont zone for this boundary.

The Devonian is customarily divided into Lower, Middle, and Upper series and their corresponding epochs. These, in turn, have been divided into stages and their corresponding ages, of which several sets are in use (see illustration). Determination of time equivalency (correlation) and age dating in the Devonian

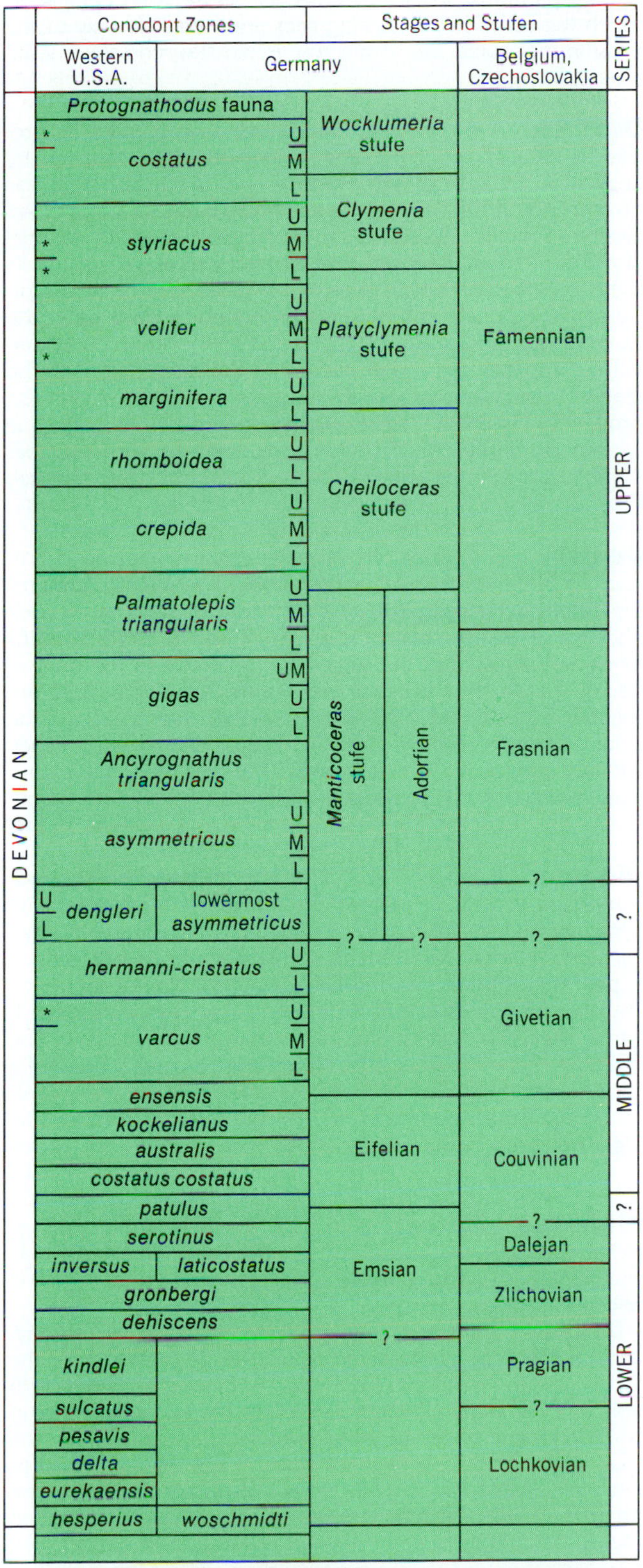

Conodont Zones
Stages and Stufen
SERIES
Western U.S.A.
Germany
Belgium, Czechoslovakia
Protognathodus fauna
costatus
styriacus
velifer
marginifera
rhomboidea
crepida
Palmatolepis triangularis
gigas
Ancyrognathus triangularis
asymmetricus
dengleri
lowermost asymmetricus
hermanni-cristatus
varcus
ensensis
kockelianus
australis
costatus costatus
patulus
serotinus
inversus
laticostatus
gronbergi
dehiscens
kindlei
sulcatus
pesavis
delta
eurekaensis
hesperius
woschmidti
Wocklumeria stufe
Clymenia stufe
Platyclymenia stufe
Cheiloceras stufe
Manticoceras stufe
Adorfian
Eifelian
Emsian
Famennian
Frasnian
Givetian
Couvinian
Dalejan
Zlichovian
Pragian
Lochkovian
UPPER
MIDDLE
LOWER
DEVONIAN

*Not yet identified in western U.S.A.

Biostratigraphic zones based on conodonts plotted to show approximate correspondence to the chronostratigraphic stages and series which compose the Devonian System. Stufen are groups of biostratigraphic zones based on ammonoids.

are usually accomplished biostratigraphically, by utilizing zones based on the evolutionary successions in individual fossil groups, or on a composite zonal framework based on several fossil groups.

Devonian fossils are found to be distributed in three major realms. Within each realm there is taxonomic similarity, which indicates that there was reproductive interchange among members of the same phyletic groups, but between each two realms there are various degrees of taxonomic dissimilarity, which indicates that there were various degrees of reproductive isolation among members of the same phylogenetic groups. The largest realm covered Australia, Asia, Europe, western North America, and the Morocco-India fringe of Gondwana, and is termed the Old World Realm. The Appalachian Realm covered most of eastern North America and the Colombia-Venezuela-Amazon part of northern South American Gondwana, which was adjacent to Appalachian North America during the Devonian. The Malvinokaffric Realm covered central Gondwana, including southern South America, southern Africa, and Antarctica.

Devonian mountain building was particularly noticeable along the margins of Euramerica. The Acadian orogeny formed mountainous highlands accompanied by a chain of granitic intrusions from Nova Scotia to Pennsylvania during much of Devonian time. Simultaneously, along the Arctic margin of Canada, the Ellesmerian orogeny was producing folded mountains. In Late Devonian time, the Ellesmerian orogeny was joined by the Antler orogeny, which formed a narrow, incipient mountain belt along the ancient continental margin from northern British Columbia to Nevada. *See* OROGENY; PLATE TECTONICS.

Among the marine invertebrates, trilobites were much less abundant than during the Cambrian. The planktic members of the extinct graptolites died out during the Early Devonian, at about the same time as the pelagic ammonoid cephalopods first evolved. The brachiopods were at their greatest diversity. Lime-secreting corals and stromatoporoids were important and widespread in warm-water environments, and formed reefs during the Middle and Late Devonian. The extinct microfossil group known as conodonts was abundant, widespread, and rapidly evolving during the Devonian, so that conodont fossils are now regarded as the principal tools to be used for international correlation and relative age determination (see illustration). *See* BRACHIOPODA; CONODONT; GRAPTOLITHINA; MICROPALEONTOLOGY; STROMATOPOROIDEA; TRILOBITA.

Placoderm fish were successful predators in Devonian waters. Bony fishes, or Osteichthyes, were represented in the Devonian by the primitive acanthodians, but more modern groups of bony fishes appeared in the Early Devonian. Lobe-finned bony fishes, or Sarcopterygii, include both the lungfish (Dipnoi) and crossopterygians in the Devonian. The oldest known amphibian, *Ichthyostega*, which evolved from the rhipidistian crossopterygians, occurs in strata thought to be high Upper Devonian. *See* CROSSOPTERYGII; DIPNOI; OSTEICHTHYES; PLACODERMI; SARCOPTERYGII.

Land plants began to flourish near the beginning of Devonian time, and were exemplified by the vascular genus *Psilophyton* of the phylum Psilopsida. The latter gave rise in the Devonian to the Lycopsida (scale trees), Sphenopsida, and Pteropsida (true ferns); the pteropsids remain important in the world flora of today. *See* LYCOPODIOPHYTA; PALEOBOTANY; PSILOTOPHYTA; PTEROPSIDA. [J.G.J./P.H.H.]

Dew Drops or films of water formed by condensation of water on outside exposed surfaces, especially of vegetation. Hoarfrost is the corresponding phenomenon at temperatures below freezing. Dew forms on clear nights when there is cooling by radiation, provided exposed surfaces cool below the dew point of the air. Moisture then condenses on the exposed surfaces. If the ground is wet, moisture is evaporated, which adds to the supply and increases the deposit of dew on vegetation. *See* Dew point; Humidity; Vapor pressure. [J.R.F.]

Dew point The temperature at which air becomes saturated when cooled without addition of moisture or change of pressure. Any further cooling causes condensation; fog and dew are formed in this way.

Frost point is the corresponding temperature of saturation with respect to ice. At temperatures below freezing, both frost point and dew point may be defined because water is often liquid (especially in clouds) at temperatures well below freezing; at freezing (more exactly, at the triple point, +.01°C) they are the same, but below freezing the frost point is higher. For example, if the dew point is − 9°C, the frost point is − 8°C. Both dew point and frost point are single-valued functions of vapor pressure. *See* Dew; Evaporation; Fog; Humidity; Vapor pressure. [J.R.F.]

Dewar flask A vessel having double walls, the space between being evacuated and the surfaces facing the vacuum being heat reflective. It was invented in 1892 by James Dewar as a container for liquid oxygen.

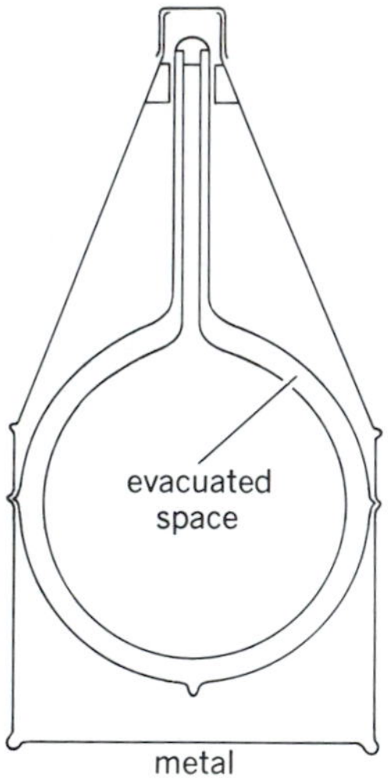

A typical Dewar container.

Dewar's original flasks were made of glass with a coating of mirror silver; this type is still used in the laboratory. But for shipment and storage of liquid gases, metal vacuum vessels are used (see illustration). Thermos Bottle is a trademark for a Dewar vessel for hot and cold foods. [H.W.Ru./G.R.H.]

Dewaxing of petroleum The process of separating hydrocarbons which solidify readily (waxes) from petroleum fractions. Removal of wax is usually necessary to produce lubricating oil which will remain fluid down to the lowest temperature of use. The wax removed may be purified further to produce commercial paraffin or microcrystalline waxes.

Most commercial dewaxing processes utilize solvent dilution, chilling to crystallize the wax, and filtration. Wax crystals are formed by chilling through the walls of scraped surface chillers, and wax is separated from the resultant wax-oil-solvent slurry by using fully enclosed rotary vacuum filters. In a process modification, most of the chilling is accomplished by multistage injection of very cold solvent into the waxy oil with vigorous agitation, resulting in more uniform and compact wax crystals which filter faster.

Complex dewaxing requires no refrigeration, but depends upon the formation of a solid urea–*n*-paraffin complex which is separated by filtration and then decomposed. The catalytic dewaxing process is based on selective hydrocracking of the normal paraffins; it uses a molecular sieve-based catalyst in which the active hydrocracking sites are accessible only to the paraffin molecules. *See* Petroleum; Petroleum processing; Wax, petroleum. [S.F.P.]

Dextran A polyglucose biopolymer characterized by preponderance of α-1,6 linkage, and generally produced by enzymes from certain strains of *Leuconostoc* or *Streptococcus*. While formerly its principal utility was as blood plasma substitute, dextran is also employed as packing material in column chromatography and as pharmaceutical agents; its average molecular weight determines usage to a great extent. Dextran's chemical and physical properties depend upon the strain of microorganism employed and the environmental conditions imposed upon the bacterium during growth, or the reaction conditions where an enzymatic method of dextran production is employed. *Leuconostoc* and *Streptococcus* species primarily convert sucrose to dextran and fructose. *Acetobacter* species convert dextrin to dextran. *See* Dextrin; Polymerization; Polysaccharide. [H.M.Ts.]

Dextrin A polymer of D-glucose which is intermediate in complexity between starch and maltose. The dextrins are usually obtained by hydrolysis of starch with diastase (amylases). The higher dextrins resemble starch, while the lower dextrins more nearly resemble the sugars. Compared with the original starch, the dextrins produce less viscous solutions. They are soluble in water but insoluble in alcohol. Dextrins may be obtained from starch by controlled hydrolysis with acids. They are used commercially as adhesives. Tapioca, waxy maize, and sweet potato starch represent the best material for their manufacture. *See* Glucose. [W.Z.H.]

Diabase A fine-textured, dark-gray to black igneous rock composed mostly of plagioclase feldspar (labradorite) and pyroxene and exhibiting ophitic texture. It is commonly used for crushed stone. Its resistance to weathering and its general appearance make it a first-class material for monuments. Diabase is a common and extremely widespread rock type. It forms dikes, sills, sheets, and other small intrusive bodies. The Palisades of the Hudson, near New York City, are formed of a thick horizontal sheet of diabase. As defined, diabase is equivalent to the British term dolerite. The British term diabase is an altered diabase in the sense defined here. *See* Igneous rocks. [C.A.C.]

Diabetes A condition in which excessive amounts of some substance are excreted from the body. The term may refer to either of two unrelated diseases, diabetes mellitus and diabetes insipidus.

Diabetes mellitus. This is a disease of abnormal carbohydrate metabolism in which the sugar glucose cannot enter the body's cells to be utilized and therefore remains in the blood in high concentrations. In diabetes mellitus the excess of sugar in the blood (hyperglycemia) leads to the excretion of sugar in the urine (glycosuria), one of the cardinal diagnostic signs of this disease. Glycosuria in turn causes the excretion of large amounts of urine (polyuria), which results in dehydration and intense thirst (polydipsia). Although blood glucose is high, glucose cannot enter the appetite-regulating cells of the hypothalamus, and hunger is therefore great, so that the diabetic person tends to eat constantly (polyphagia). But because glucose cannot enter and nourish the cells, body tissues are subjected to the equivalent of starvation conditions, and rapid weight loss occurs. *See* Carbohydrate metabolism.

Diabetes mellitus appears in two varieties, each with its own cause: juvenile-onset diabetes, caused by a shortage of the pancreatic hormone insulin (whose chief function is to promote the entry of glucose into cells), and maturity-onset diabetes, in which insulin is available but the body's cells do not respond to it normally. Both juvenile-onset and maturity-onset diabetes tend to produce a number of debilitating and life-threatening complications despite the use of insulin or oral hypoglycemic agents. These include damage to the eyes and kidneys, premature atherosclerosis, and nervous system disturbances (collectively, diabetic neuropathy). It is primarily these complications, not the risk of ketoacidosis, that account for the heavy burden this disease inflicts on society, and make diabetes the third leading cause of death in the United States. *See* INSULIN; PANCREAS.

Three factors believed important in the causation of diabetes are heredity, virus infections, and immunological damage to the pancreas. Heredity does operate in determining one's risk of diabetes, but only as a predisposing factor, not an absolute determinant. The two types of diabetes mellitus have entirely different patterns of inheritance. Virus infection is strongly implicated as one causative factor in the beta-cell destruction that characterizes juvenile-onset diabetes, but it is not involved in maturity-onset diabetes. Autoimmune reactions, wherein the body's immune defense system attacks its own pancreatic tissue as though it were a foreign substance, are also suggested as a cause for the beta-cell destruction of juvenile-onset diabetes. *See* AUTOIMMUNITY; HUMAN GENETICS.

Dietary management has always been part of diabetic therapy, including weight reduction for maturity-onset diabetes and carbohydrate restriction for both types of diabetes mellitus. Other dietary measures are now believed beneficial for maturity-onset diabetics. The inclusion of foods high in fiber content, for example, has been shown to reduce hyperglycemia, despite the fact that high-fiber foods are also high-carbohydrate foods. Also, foods with a high content of the mineral element chromium seem to enhance insulin's effect on body tissues; such foods include whole-grain products, mushrooms, bran, wheat germ, and brewer's yeast.

For juvenile-onset diabetics, the traditional diet is high in protein and quite restricted in carbohydrate, with only about 45% of the total daily caloric intake being carbohydrate-derived. There is a trend toward relaxing somewhat the carbohydrate restriction on diets for juvenile-onset diabetics, as evidence accumulates that diets containing even as much as 80% carbohydrate can be compatible with excellent control of diabetes, so long as the carbohydrate is complex (that is, whole grains, legumes, potatoes, bread) rather than refined (that is, sugars, candy, soft drinks, desserts, and so on).

Diabetes insipidus. This is a condition in which a large amount of very dilute urine, sometimes as much as 6.5 gal (25 liters) per day, is excreted. Diabetes insipidus may be classified as primary, secondary, and nephrogenic. Both primary and secondary diabetes insipidus result from failure of the posterior lobe of the pituitary gland to secrete antidiuretic hormone (ADH), which encourages the reabsorption of water by kidney tubules. Both conditions can be treated by administration of ADH. In nephrogenic diabetes insipidus adequate ADH is present, but the kidney tubules are genetically defective and do not respond to its signal to reabsorb water from the urine. Nephrogenic diabetes insipidus is a sex-linked recessive disorder. All forms of diabetes insipidus are characterized by extreme polyuria and polydipsia, but polyphagia is not present. *See* PITUITARY GLAND. [A.C.]

Diadematacea A superorder of Euechinoidea having a rigid or flexible test, perforate tubercles, aulodont lantern, and branchial slits. The group arose in the Late Triassic and differentiated into four stocks, the Diadematoida, Pedinoida, Echinothurioida, and Pygasteroida. *See* DIADEMATOIDA; ECHINOIDEA; ECHINOTHURIOIDA; PEDINOIDA; PYGASTEROIDA. [H.B.F.]

Diadematoida An order of Diadematacea with hollow primary radioles and diademoid ambulacral plates. The tubercles are normally crenulate and the anus remains within the apical system. Three families compose this order: Diadematidae, Micropygidae, and Aspidodiadematidae. *See* DIADEMATACEA. [H.B.F.]

Diagenesis All the chemical, biochemical, and physical changes that sediments undergo from the time of deposition until the stage of metamorphism is reached. Diagenetic changes are gradational, with metamorphism at elevated temperatures or pressures, and with atmospheric weathering effects when sedimentary rocks become exposed at the surface. Sandstones, shales, and carbonate sediments are particularly susceptible to diagenetic modifications. *See* CARBONATE MINERALS; SANDSTONE; SHALE.

Diagenetic processes include purely physical ones that involve rearrangement of the sediments such as compaction, slumping, bioturbation by organisms, infiltration, and soft sediment deformation; biochemical or organic processes such as particle accretion, flocculation, boring, and decomposition; and physiochemical processes such as cementation, authigenesis (formation of new minerals), inversion, recrystallization, grain growth, replacement, and interstratal solution. Most of these processes involve a reduction in porosity and permeability, which are two important sediment properties in considering the migration of subsurface fluids and the accumulation of oil and gas and certain types of mineral deposits in subsurface rock units. However, one diagenetic process, interstratal solution, is of major importance in creating secondary porosity. *See* AUTHIGENIC MINERALS.

There have been attempts to group diagenetic processes into phases (or stages) in order to develop a comprehensive model for diagenetic evolution. The boundary limits of phases are a function of chemical as well as physical conditions and include pH, oxidation potential (Eh), ionic adsorption phenomena, temperature, pressure, depth of burial, and geologic time. In one model, three diagenetic phases are recognized: syndiagenesis, anadiagenesis, and epidiagenesis. Syndiagenesis includes sediment modifications that take place during and immediately following deposition. Anadiagenesis refers to the diagenesis processes that are characterized by expulsion and upward migration of connate water and other fluids, such as petroleum. Epidiagenesis includes those sediment-modifying processes that take place during and after uplift and emergence. *See* SEDIMENTARY ROCKS. [F.G.Et.]

Diakoptics A piecewise approach to the solution of large-scale interconnected systems. First the large system is torn into several small pieces or subdivisions by removing the interconnections. Then the subdivisions are solved separately. Finally, the effect of interconnection is determined and added to each subdivision to yield the complete solution of the system. The advantage of the approach is that the subdivisions, being smaller, are easier to solve.

Diakoptics was conceived by G. Kron in the 1950s and developed by H. H. Happ and others. The motivation behind the development of diakoptics came from electric power system problems. There are many electric power companies, and most of their systems are interconnected. The operation of one system affects the others; the entire network represents the problem to be solved. An ideal method for power system analysis is one that, as a first step, solves the problem as if all individual power companies were isolated from each other, and then, in subsequent steps, modifies the solutions to take the interconnections into consideration. This is the central idea of diakoptics. *See* ELECTRIC POWER SYSTEMS.

The development of powerful sparse-matrix techniques in the late 1960s for handling large-scale system problems made the use of diakoptics unnecessary in most cases, although sparse-matrix techniques can be combined with diakoptics. Then, in the 1970s, the use of computers with multiple processors to solve large-system problems led to renewed interest in diakoptics, since it is ideally suited to a parallel-processing solution. *See* MULTIPROCESSING.

Diakoptics has been applied to a number of large-scale system problems. Three power systems applications are as follows:

1. Simulation of power system steady-state and dynamic responses. The steady-state behavior of a power system is described by a set of nonlinear equations, called load-flow equations. Iterative schemes are used in solving the load flow. At each iteration, a linear equation that preserves the topological relation of the system is solved. A diakoptic approach can be applied at each iteration. To study the transient stability after a fault, a system is described by a set of nonlinear algebraic and differential equations. Again, iterative methods are used in the solution, and diakoptics can be applied at each iteration.

2. Economic dispatch. Economic dispatch is the determination of the most economic combination of generation in a system to supply the load demand. A quadratic formula for the system losses is used. A diakoptic approach can be applied to multiarea economic dispatch.

3. Automatic generation control. Automatic generation control is the coordination of generation to maintain constant-frequency operation and scheduled tie-line interchanges. The idea of diakoptics can be applied to multiarea automatic generation control. [F.F.W.]

Dialysis A process of selective diffusion through a membrane by dissolved solutes in liquid solution. As dialysis is usually carried out, the membrane permits the diffusion of low-molecular-weight solutes (crystalloids) but prevents the passage of colloidal and high-molecular-weight solutes (macro-molecules). Membranes suitable for this purpose include vegetable parchment, animal parchment, goldbeater's skin (peritoneal membranes of cattle), fish bladders, dialyzing cellophane (Visking sausage casing), and collodion (nitrocellulose deposited from alcohol-ether solution).

The solution is contained within such a membrane. The low-molecular-weight solutes are removed by placing pure solvent outside the membrane. This solvent is changed periodically or continuously until the concentration of diffusible solutes in the solution is reduced to near zero. The technique is used extensively in separating and purifying macro-molecules of biological origin. *See* ELECTRODIALYSIS. [Q.V.W.]

Diamagnetism That branch of magnetism which treats of diamagnetic phenomena and of the properties of diamagnetic bodies. Diamagnetism is a property exhibited by substances with a negative magnetic susceptibility, that is, by substances which magnetize in a direction opposite to that of an applied magnetic field. A diamagnetic substance has a magnetic permeability less than 1, and is repelled when placed near a magnet. The magnetization of diamagnetic substances is associated with the currents induced on application of a magnetic field. *See* MAGNETIC PERMEABILITY; MAGNETIC SUSCEPTIBILITY.

Although all matter exhibits diamagnetism, only those substances in which paramagnetism is absent are referred to as diamagnetic. This is because paramagnetism, if present, usually predominates, and the gross magnetic response of the material is paramagnetic. Important exceptions are the alkali and alkaline earth metals. The condition for pure diamagnetism is that all electronic spins be paired and all orbital moments either be zero or effectively cancel one another. *See* PARAMAGNETISM.

As stated previously, the diamagnetic response of a substance is small; only a very small fraction of the applied magnetic field is shielded from the interior of the substance by the induced diamagnetic currents. There is one case, however, in which the inducing field is completely shielded (except for small surface effects). This is the perfect diamagnetism exhibited by superconductors, and is known as the Meissner effect. *See* MEISSNER EFFECT; SUPERCONDUCTIVITY. [E.A.; F.Ke.]

Diameter Any chord of a circle that contains the center of the circle. It is also a chord of maximum length. (This property furnishes a basis for extending the concept to more general figures; for example, the diameter of any point set is the least upper bound of the distances of two points of the set.) *See* CIRCLE. [L.M.Bl.]

Diamond A mineral composed entirely of the element carbon crystallized in the isometric system. Gem diamonds have a density of 3.53, but the tough, black, cokelike aggregates of microscopic crystals sold as carbons and known to the lay person as black diamond may have a density as low as 3.15. It is the hardest known substance. Diamond slowly burns to carbon dioxide in oxygen at a temperature as low as 1650°F (900°C) and slowly inverts to graphite at temperatures as low as 1830°F (1000°C). Diamond has the highest thermal conductivity of any known substance; at room temperature it is approximately five times that of copper. The points of diamonds used as cutting tools do not become hot owing to this very high thermal conductivity. All except a few diamonds are nonconductors of electricity, but all are excellent heat conductors, superior to iron and steel.

Occurrence and recovery. Diamonds crystallize directly from rock melts rich in magnesium, at depths of 93 mi (150 km) or more in the Earth. The melts from which they crystallize are essentially saturated in carbon dioxide gas at exceedingly high pressures; temperatures in excess of 2550°F (1400°C) are required. Thus rocks which contain diamonds as a natural component are samples of the deep mantle of the Earth and worldwide are remarkably similar in composition. Only garnet-bearing lherzolite and garnet-pyroxene eclogites are known to contain diamond. *See* ECLOGITE.

Diamond pipes are commonly called volcanic necks in the literature. However, this terminology is incorrect, for there is no evidence that the material in diamond pipes that has reached the surface is a liquid or is related to volcanoes. They are more properly called gas-driven diatremes. These dikes and pipes of diamond-bearing hydrated magnesium silicates form primary diamond deposits. Productive dikes have been mined in Sierra Leone, the Orange Free State, the Transvaal, the Kimberley district of Cape Province, Tanzania, the Republic of the Congo, and the United States at Murfreesboro, Arkansas. In addition, a substantial number of these pipes have been found in Yakutsk, Siberia, and are now being actively mined. Several hundred dikes and pipes of kimberlite that contained diamonds have also been found. Only a few have been profitable to mine at depth. All pipes have an enriched surface layer which has been developed by erosional processes. Initially the upper few feet of many pipes may be profitably mined, although they may not warrant mining at depth.

Most diamonds have been produced from secondary (placer) deposits, called alluvials. Most alluvial diamonds are recovered from modern gravels, all of which are stream gravels except the beach gravels of the Atlantic Ocean. Diamonds also have been found as irregular microcrystalline clumps up to a millimeter or so in size and as tiny crystals in meteorites. These meteoritic diamonds were formed from graphite nodules as the iron meteorites were subjected to intense high pressures by the shock of the meteorite impacting on the surface of the Earth.

Of the world's output of diamonds, 95% comes from Africa. Most of the production is by large mining companies, but a significant amount comes from individual operations from streambed deposits in Sierra Leone, Ghana, and the Union of South Africa. These small operations recover diamonds by a method similar to the panning of gold from placers.

Diamond cutting. After recovery, diamonds are referred to as rough. Those of gem quality are called cuttable rough, and all others are classed as industrial rough. In cutting diamonds, the objective is to obtain the maximum price for the finished gems, not the maximum weight of the finished stones. The most important step in diamond cutting is the decision as to how the stone will be cut.

In subdividing an irregularly shaped crystal or eliminating flaws, the stone may be either cleaved or sawed, but only in certain crystallographic directions. There are four directions in which a diamond may be cleaved parallel to any octahedron face, and nine directions in which it may be sawed—three parallel to any cube face and six parallel to any dodecahedron face. If the finished diamond is to be round or oval in shape, it goes to a person who is known in the trade as a cutter. The process of putting on the facets is known as polishing.

The weight of the finished gems is 50–60% of the weight of the rough stones if they are well-formed crystals with few flaws. The metric carat, 0.200 g, is the unit of weight by which both gem and industrial diamonds are sold. A point is 1/100 of a carat and is used only in reference to gem diamonds.

Industrial diamonds. Industrial diamonds vary from the better grades, which are identical with inferior gems, to crushing bort, which is suitable only for crushing to grit and powder sizes. The better qualities are made into shaped diamond-cutting tools or wire-drawing dies. However, shaped diamond tools have been supplanted in many of their former uses by sintered tungsten carbide. For truing and shaping grinding wheels of alumina or silicon carbide, diamond crystals, in the shapes in which they are mined, are used.

Drill bort consists of crystals in their original form as mined which are mounted in the end of a cylinder called a bit or crown. When the rotating cylinder is forced against a rock surface, it wears its way into the rock. The mining industry is the largest user of drill bort, although this method of coring has been adapted to the testing of concrete and the foundations of buildings, dams, and bridges.

Synthetic and imitation diamonds. The synthesis of diamond has been effected in various ways: by static crystallization from certain molten metals or alloys at pressures upward of about 50 kilobars (5 gigapascals) and temperatures over 2240°F (1500 K); by shock conversion from graphite at transient pressures of about 300 kilobars (30 GPa) and temperatures of about 1880°F (1300 K); and by static conversion from graphite at pressures more than 130 kilobars (13 GPa) and transient temperatures more than about 5480°F (3300 K). Synthetic diamonds are identical with natural diamonds in fundamental properties but differ in those characteristics that depend on the process of manufacture, such as impurities, size, and shape.

Imitation diamonds are not made of carbon crystallized in its densest form. Most imitation diamonds are yttrium-aluminum-garnet or strontium-titanate. The distinction between synthetic gems and imitation gems is not always made. Synthetic gems are chemically and crystallographically identical to the natural stone, whereas imitation gems are made of some other component, such as glass, or another crystal. *See* CARBON; GEM.

[C.Fr./G.C.K.]

Diapensiales An order of flowering plants, division Magnoliophyta (Angiospermae), in the subclass Dilleniidae of the class Magnoliopsida (dicotyledons). The order has only a single family, the Diapensiaceae, with about 18 species of herbs and dwarf shrubs in temperate and arctic regions of the Northern Hemisphere. Within its subclass the order is marked by sympetalous flowers (flowers with the petals joined to each other by their margins, at least toward the base, forming a basal tube, cup, or saucer), separate pollen grains, and ovules that have a single integument (unitegmic). *Shortia* and *Galax* genera are sometimes cultivated. *See* DILLENIIDAE; FLOWER. [A.Cr.]

Diapir An anticline or dome in which a core of mobile rock has broken through the overlying strata. In the process of concentric folding, the rocks of the core occupy increasingly more limited space as the curvature of the fold becomes smaller. If the rocks of the core are extremely plastic, they may exert an upward pressure sufficient to cause expulsion of the core through the crest. Diapiric structures may also form as a result of unequal loading by the sediments overlying a plastic layer. *See* SALT DOME.

[P.H.O.]

Diapsida A subclass of reptiles characterized by two pairs of temporal openings and a suborbital fenestra, or features derived from this condition. This subclass includes lizards, snakes, tuatara, and crocodiles of the modern fauna, as well as the Mesozoic ancestors of these groups—dinosaurs, flying pterosaurs, aquatic nothosaurs, plesiosaurs, and placodonts—and their ancestors from the late Paleozoic Era. Birds evolved from bipedal theropod dinosaurs and are part of the diapsid radiation, but are placed in a separate class, Aves. *See* AVES; DINOSAUR.

The earliest adequately documented diapsid is *Petrolacosaurus* from the Upper Pennsylvanian Series of Kansas. It is about 8 in. (20 cm) in length, with a relatively long neck and distal limb elements. Aside from these differences in proportions and the presence of the diagnostic temporal openings, the skeleton of Petrolacosaurus strongly resembles that of the most primitive amniotes of the family Protorothyridae.

Fossils from the upper Permian and Lower Triassic are examples of the early stages of a great radiation that has continued to the present. Some Permian and Mesozoic groups, including the coelurosauravids (gliding reptiles from the upper Permian and Lower Triassic series) and the aquatic thalattosaurs, placodonts, and possibly the ichthyosaurs, may have evolved from reptiles at the beginning of the diapsid radiation. All other advanced diapsids can be grouped in two large infraclasses, the Lepidosauria and the Archosauria, represented in the modern fauna by the lizards and snakes on one hand and the crocodiles (and birds) on the other. *See* ARCHOSAURIA; LEPIDOSAURIA.

[R.L.C.]

Diarrhea The passage of loose or watery stools, usually at more frequent than normal intervals. Diarrhea is a symptom of many diseases and may be accompanied by nausea, vomiting, griping, tenesmus, and other general or specific indications of a disease.

The more common specific disorders which may produce diarrhea include intestinal infections, such as dysentery, cholera, typhoid fever, food poisonings, and parasitic infestations; food sensitivities; drug and chemical irritation; and vitamin deficiency states.

Emotional and psychic disturbances frequently produce diarrhea and other visceral derangements. The poorly understood entities of regional enteritis and ulcerative colitis are perhaps related to these disturbances, as are other psychosomatic disorders.

Diarrhea is a common symptom in gastrointestinal obstruction or in inflammations from local infections or tumor invasion. *See* BACILLARY DYSENTERY; FOOD POISONING; MEDICAL PARASITOLOGY.

[E.G.St./N.K.M.]

Diastem A temporal break between adjacent geologic strata that represents nondeposition or local erosion but not a change in the general regimen of deposition (in contrast to unconformity). Diastems may be produced by the scouring action of shifting submarine currents which temporarily interrupt deposition on the continental shelf, or by a shifting river within the deposits of its floodplain. Or they may be produced simply by nondeposition in either environment of deposition where the absence of sediment reflects normal shifting of currents rather than an overall change in conditions. *See* MARINE SEDIMENTS; UNCONFORMITY. [J.R.]

Diastereoisomer One of a pair of optical isomers which are not mirror images of each other. A given diastereoisomer (or diastereomer) may not be optically active, in which case it is an optically inactive meso form; or it may be optically active, in which case with its nonsuperimposable mirror image it constitutes an enantiomorphic pair of optical isomers. Thus *meso*-tartaric acid is a diastereoisomer of both the (+) and (−) tartaric acids, and it can form two optically active monoesters, each of which is a diastereoisomer of monoesters of the (+) and (−) acids. *See* OPTICAL ACTIVITY. [W.R.V.]

Diastrophism The process of deformation of the Earth's crust. The most spectacular and conspicuous products of diastrophism are the mountain systems within continents. These consist of regionally extensive belts of folded and buckled layers, igneous intrusions, volcanic outpourings, metamorphosed strata, and tremendous faults.

Diastrophism contributes enormously to the architecture of the Earth's crust. Although the term diastrophism has fallen out of widespread usage among earth scientists, the subject of deformation of the Earth's crust is of significant and growing interest and practical importance. Scientists study the products of diastrophism out of a fundamental historical interest and curiosity, but also as a means to carry out difficult operational tasks such as predicting earthquakes and volcanic eruptions, identifying places of concentration of metals and oil and gas, and analyzing the feasibility of safe disposal of radioactive waste in the Earth's interior.

Diastrophism is attributed to multiple origins, most of which are intimately linked to major crustal movements related to seafloor spreading, transform faulting, and the collision, underthrusting (subduction), and general interference of shifting lithospheric plates which compose the Earth's outer shell. The fundamental plate configurations which establish the settings for diastrophism include the subduction of oceanic crust under oceanic crust at sites of deep ocean trenches; subduction of oceanic crust under continental margins; collision of two continents; and the movement of one plate past another along high-angle transform fault zones. *See* MOUNTAIN SYSTEMS; OROGENY; PLATE TECTONICS; STRUCTURAL GEOLOGY. [G.H.D.]

Diatomaceous earth Earth consisting of a friable, porous silica deposit made up of the opaline silica tests (shells) of diatoms. It is the dry, relatively unconsolidated equivalent of diatom ooze found on some parts of the sea floor. It is usually white or cream-colored. Because this material is inert chemically and has unusual physical properties, it is admirably suited for many scientific and industrial purposes, such as filtering agent; insulator against heat, cold, and sound; catalyst carrier; absorbent; filler; building material; abrasive; ingredient in pharmaceutical preparations; and stratigraphic indicator.

Diatomite is the indurated equivalent of diatomaceous earth; the pores are partially or completely filled with silica. *See* CHERT. [R.Si.]

Diazotization The reaction between a primary aromatic amine and nitrous acid to give a diazo compound. Diazotization is important in organic chemical synthesis. Its most striking use is in the large-scale manufacture of the important azo class of dyes, but it has likewise been invaluable in general synthesis in both chemical manufacturing and in research.

The light sensitivity of aromatic diazo compounds has led to their use in photocopying processes. Azo dyes derived by coupling diazonium salts to aromatic amines and phenols remain an important class of dyes, although the use of some members of this class in food has been restricted in recent years because of carcinogenic properties observed in experiments with laboratory animals. *See* ANILINE; DYE. [M.St.]

Dibranchia A subclass of the Cephalopoda which contains the extinct Belemnoidea and all living cephalopods (Decapoda, Octopoda, and Vampyromorpha) with the exception of *Nautilus*. All members of this group possess two gills and, when present, the shell is internal. The belemnites arose in the Upper Mississippian and disappeared at the close of the Cretaceous. *Belemnites* is an index fossil of Jurassic and Cretaceous rocks throughout the world. *See* BELEMNOIDEA; INDEX FOSSIL; OCTOPODA; VAMPYROMORPHA.

Most dibranchiates emit ink from a special gland. Contrary to popular belief, in most species the ink discharged is not a smoke screen but coagulates in the water as an object approximately the size of the animal. This is seized by the enemy while the colorless cephalopod darts away. *See* CEPHALOPODA. [G.L.V.]

Dichroism In certain anisotropic materials, the property of having different absorption coefficients for light polarized in different directions. There are few natural materials which exhibit strong dichroism. One of the first to be discovered was tourmaline. Light transmitted by thin plates of dark forms of tourmaline is almost completely polarized. *See* POLARIZED LIGHT.

If the absorption in a dichroic material is different for different linear states of polarization, the material is termed linear dichroic. If it is different for right and left circularly polarized light, it is termed circular dichroic. Similarly, there can be elliptically dichroic crystals. [B.H.Bi.]

The study of dichroism allows conclusions as to the submicroscopic fine structure of cells. In visible light only a few cellular components, such as chloroplasts, show absorption. An absorption can, however, be produced by staining. The dichroic staining of plant fibers is especially simple. The elongate stain particles of benzidine dyes, for example, congo red, are deposited in an oriented manner in the spaces between the microfibrils and produce an intrinsic dichroism of the fiber: colored for a vibration plane parallel, colorless for a plane perpendicular to the stain particles and fibrils. Therefore, the direction of strongest absorption indicates the course (parallel or helical) of the microfibrils in the fiber.

Ultraviolet dichroism gives direct information as to the orientation of the absorbing molecules or molecular groups in cell structures. The method has been especially helpful for studies of orientation of deoxyribonucleic acid in nuclei and chromosomes. Lignifed plant cell walls show ultraviolet dichroism. It is pure form dichroism, a fact which eliminates the possibility that lignin is in an anisotropic state in the wall.

By irradiation with ultraviolet light, various compounds of the cell are caused to fluoresce. The fluorescent light is polarized if the object is anisotropic. This phenomenon, called difluorescence, is observable in lignifed cell walls, and leads to the same conclusions as to lignin deposition as emerge from dichroism studies. [F.Ru.]

Dichromate The $Cr_2O_7^{2-}$ ion which is derived from dichromic acid, $H_2Cr_2O_7$. This acid and other polyacids are obtained when chromium trioxide, CrO_3, is dissolved in water.

The dichromates are converted to chromates under basic conditions.

Most dichromates are water-soluble and give orange-colored solutions. Dichromates are good oxidizing agents, and this property is utilized in chemical processes, analytical chemistry, explosives, and safety matches. Dichromates are also used in electroplating, tanning, photography, and pigments. *See* CHROMATE. [E.E.W.]

Dicranales An order of mosses. The stems are erect, dichotomously branched, and densely foliated. The costate, narrow leaves vary from subulate to lanceolate, and in some species are falcate-secund. The seta is elongated. The single peristome consists of 16 teeth, usually two-parted into linear or awllike divisions from the apices to the middle or even beyond. *See* BRYOPSIDA. [W.H.W.]

Dictyoceratida An order of sponges of the class Demospongiae which includes the bath sponges of commerce. Dictyoceratida are mostly of considerable size, and form massive, lobate, or branching colonies. Some are leaflike in shape or vase-shaped. A reticulate skeleton of spongin fibers is always present. In many species some or all the fibers enclose foreign bodies, such as sand grains, shell fragments, or spicules of other sponges. There are three families, Spongiidae, Thorectidae, and Dysideidae.

Dictyoceratid sponges are most abundant in tidal and shallow waters of tropical and subtropical regions. Relatively few species occur in Arctic and Antarctic seas, and most occur above the continental slope. A few species are found down to 3300 ft (1000 m). *See* DEMOSPONGIAE. [W.D.H.]

Dicyemida An order of Mesozoa comprising minute, wormlike parasites of the renal organs of cephalopod mollusks. They are composed of a single layer of large, ciliated epithelial cells enclosing one or more elongate axial cells (see illustration).

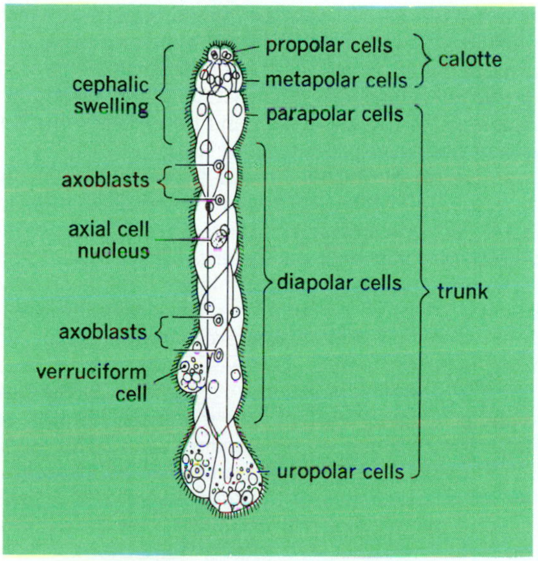

Morphology of a young dicyemid.

Each axial cell contains, in addition to its own nucleus, reproductive cells and developing larvae. These larvae, when fully developed, escape from the parent organism.

There are two reproductive phases termed the nematogens and rhombogens, respectively. During the nematogen phase vermiform larvae are formed asexually from the germ cells. They increase the infection in the same cephalopod. This is succeeded by the rhombogen phase in which infusoriform, or swarm, larvae are liberated. These are discharged into the sea with their host's urine and their fate is unknown. Zoologists are not in agreement regarding many points in the life cycle, particularly in respect to the interpretation of the infusorigen and its products. *See* MESOZOA. [B.H.McC.]

Didymelales An order of flowering plants, division Magnoliophyta (Angiospermae), in the subclass Hamamelidae of the class Magnoliopsida (dicotyledons). The order contains a single family with but one genus, *Didymeles*, with two species. These dioecious, evergreen trees are restricted to Madagascar. The wood has vessels with scalariform perforations, which is a putatively primitive feature. The leaves are alternate, simple, and entire. The flowers are small; the staminate ones are in open clusters, without perianth, with two stamens, and the pollen grains have three germinal furrows. The pistillate flowers are commonly paired in spikelike clusters; they are hypogynous and sometimes have 1–4 scalelike sepals but no petals. The pistil has but one carpet. This curious genus has often been included in the Hamamelidales; however, the primitive nature of the wood and of the pistil make it anomalous there. *See* HAMAMELIDAE; MAGNOLIOPSIDA; PLANT KINGDOM. [T.M.Ba.]

Dielectric constant For a given dielectric material, the ratio of electrical capacitance of a dielectric-filled capacitor to a vacuum capacitor of identical dimensions. This is defined by the equation below, where C is the capacitance of the

$$\kappa = \frac{C}{C_0}$$

dielectric-filled capacitor and C_0 is the capacitance of the empty capacitor.

The dielectric constant κ is also known as the specific inductive capacity or as the relative permittivity. It is perhaps most familiar as the proportionality constant in Coulomb's law of electrostatics. For a given charge distribution, the dielectric constant expresses the ratio of electric field strength in vacuum to that in a dielectric, the latter field being reduced by the polarization of the dielectric medium. *See* CAPACITANCE; CAPACITOR; COULOMB'S LAW; DIELECTRIC MATERIALS; ELECTRIC FIELD; PERMITTIVITY.

The values of κ for low frequency or static fields range from 1 to more than 10,000 for typical dielectrics. The dielectric constants of gases are only slightly greater than unity, while high values occur for many polar liquids and certain ionic solids.

The experimental methods of measuring dielectric constants depend on the frequency range under investigation. For frequencies below about 10^9 Hz, the permittivity or impedance of a dielectric sample inserted in a parallel-plate capacitor may be measured in suitable circuits; a Schering bridge arrangement is commonly employed up to 10^7 Hz, and resonant circuits in the range 10^4–10^9 Hz. For frequencies above 10^8 Hz, the dielectric constant may be determined by measuring the interaction of electromagnetic waves with the medium. From about 10^8 to 10^{11} Hz, the material is usually inserted in wave guides or coaxial lines, and the standing-wave patterns measured. At still higher frequencies, optical techniques involving reflection and transmission measurements are employed.

Because the dielectric constant is related to chemical structure, it can be used for both qualitative and quantitative analysis. From measurements of the dielectric constant and refractive index taken together, it is possible to compute the dipole moment of a species. [R.D.W.]

Dielectric heating The heating of a nominally electrical insulating material due to its own electrical (dielectric) losses, when the material is placed in a varying electrostatic field.

The material to be heated is placed between two electrodes

(which act as capacitor plates) and forms the dielectric component of a capacitor (see illustration). The electrodes are connected to a high-voltage source of 2–90-MHz power, produced by a high-frequency vacuum-tube oscillator.

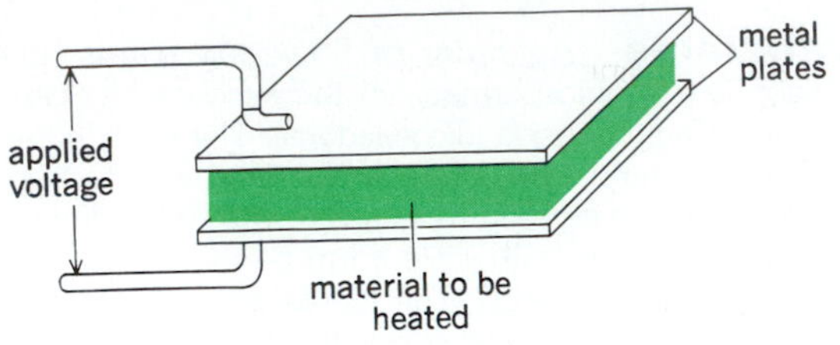

Basic assembly for dielectric heating.

The resultant heat is generated within the material, and in homogeneous materials is uniform throughout. Dielectric heating is a rapid method of heating and is not limited by the relatively slow rate of heat diffusion present in conventional heating by external surface contact or by radiant heating.

This technique is widely employed industrially for preheating in the molding of plastics, for quick heating of thermosetting glues in cabinet and furniture making, for accelerated jelling and drying of foam rubber, in foundry core baking, and for drying of paper and textile products. Its advantages over conventional methods are the speed and uniformity of heating, which offset the higher equipment costs. Because of the absence of high thermal gradients, an improved end-product quality is usually obtained. [G.F.B.]

Dielectric materials Materials which are electrical insulators or in which an electric field can be sustained with a minimum dissipation of power. In a more general sense, dielectrics include all materials except condensed states of metals. *See* ELECTRIC INSULATOR.

Dielectrics are employed as insulation for wires, cables, and electrical equipment, as polarizable media for capacitors, in devices used for the propagation or reflection of electromagnetic waves, and for a variety of dielectric devices, such as rectifiers and semiconductor devices, piezoelectric transducers, dielectric amplifiers, and memory elements.

The electrical response of a normal dielectric can be described by its dielectric or breakdown strength, conductivity or dielectric loss, and dielectric constant. The behavior of nonlinear dielectrics depends also on the amplitude and time variation of the electric field.

Dielectric strength is defined as the maximum electric field which can be applied to a dielectric without causing breakdown, the abrupt irreversible drop in resistivity at high fields often accompanied by destruction of the material. Dielectric breakdown is caused by an enormous increase in the number of charge carriers because of collisions or thermal ionization and field emission. *See* FIELD EMISSION.

Dielectric loss is the power dissipated in a dielectric because of conduction processes. This power loss results from thermal dissipation of the electrical energy expended by the field. It is caused by molecular collisions.

The dielectric constant or permittivity relative to vacuum is important in many applications. Materials with high dielectric constants are desirable for capacitors, since they permit a reduction in size for a given capacitance, while low dielectric constants are usually preferred for cable and transformer insulation. *See* DIELECTRIC CONSTANT.

Vacuum or gaseous dielectrics other than air have had relatively little dielectric application except in electron tube devices, voltage regulators, and lightning arresters. Dielectric liquids are principally employed for impregnating porous insulation in high-voltage cables and capacitors and as insulating media for transformers and circuit breakers. In the latter applications, the heat transfer properties of the liquid are also important. Mineral oils, halogenated hydrocarbons, and silicone oils are the most important commercial dielectric liquids. Solid dielectrics are employed for the vast majority of commercial applications. Important solid dielectrics include many ceramics and glasses; plastics and rubber; minerals such as quartz, mica, magnesia, and asbestos; and paper and fibrous products. The mechanical and thermal properties as well as the electrical response are important in the choice of a dielectric for a particular product. [R.D.W.]

Dielectric measurements Measurements of the dielectric properties of a material, which are characterized by its complex relative permittivity ϵ_r. For all materials except ferroelectrics, this quantity does not depend on applied field: the general behavior is linear, and so voltage of any convenient magnitude can be used for measurement. *See* FERROELECTRICS; PERMITTIVITY.

Bridge methods. The most commonly used apparatus for measuring ϵ_r is the alternating-current (ac) bridge. These bridges are readily available in the operating range $10–10^6$ Hz, and sometimes outside it; ultralow-frequency bridges can go as low as 10^{-3} Hz. Most specimen holders for solids are essentially parallel-plate capacitors with the specimen filling all the space between the plates; for liquids, a test cell with cylindrical electrodes is usually employed. The bridges most commonly used are of the Wheatstone type, the most versatile for dielectric measurements being the Schering bridge.

Resonance methods. Resonance methods, useful for frequencies greater than 1 MHz, involve the injection of voltage or current by one of several methods into an *LC* (inductance-capacitance) resonant circuit. Measurements over a range of frequencies may be made by using coils with different inductance values, but ultimately the inductance required becomes impracticably small, and in the range $10^8–10^9$ Hz reentrant cavities are often used. These are hybrid devices in which the plates holding the specimen still form a lumped capacitor, but the inductance and capacitances are distributed along a coaxial line. At higher frequencies, the wavelength is comparable to the dimensions of the apparatus, and transmission methods in coaxial lines and waveguides must be used.

Transmission methods. Coaxial lines are used in the frequency range 300 MHz–3 GHz, and waveguides in the range 3–30 GHz. The transmission characteristics are determined by the complex permittivity of the material filling the line or guide. Many different measurement techniques have been devised, but all derive values of the complex relative permittivity ϵ_r from its relationship to the complex propagation factor γ. *See* MICROWAVE TRANSMISSION LINES; TRANSMISSION LINES; WAVEGUIDE.

In practice, traveling waves are rarely used as the basis of measurement, except for high-loss materials. Usually, reflections from terminations set up standing waves, the amplitude of which in the case of a liquid-filled line can be measured by a suitable probe. The ratios of the field magnitudes at adjacent maxima, and the distance between them, give the information required. *See* ATTENUATION (ELECTRICITY); WAVELENGTH.

Submillimeter measurements. Dielectric measurements are difficult to carry out in the frequency range 30–300 GHz, for which λ_0 is in the range 1 cm–1 mm, but for λ_0 less than 1 mm, methods related to infrared spectroscopy are used. Broadband continuous spectra result from Fourier transform spectroscopy, which in its simplest form is equivalent to normal infrared spectroscopy, with the specimen in one of the two passive arms of the interferometer, between the beam divider and either the source or the detector. In the more sophisticated dispersive Fourier transform spectroscopy, the specimen is in one of the active arms, that is, between the beam divider and either mirror. Discrete-point spectra also may be obtained by the use of a Mach-Zehnder interferometer and a laser source. By using interferometric techniques, the frequency range can be extended up to about 5 THz. *See* INFRARED SPECTROSCOPY; INTERFEROMETRY; SPECTROSCOPY.

Time-domain methods. If a constant direct-current (dc) voltage is suddenly applied to a dielectric specimen, in principle the charging current is related through the Fourier integral transformation to the steady-state ac current which would flow if the applied voltage were sinusoidal at any particular frequency. If the dc voltage is suddenly removed, a similar relationship holds between the discharge current and ac current. Thus the variation of complex permittivity with frequency can in principle be derived from a transient signal in the time domain. Because of various limitations, the method is not capable of giving results of an accuracy at all frequencies comparable to those obtainable from a single frequency measurement. Nevertheless, with the aid of computer analysis, the response over a large frequency range can be obtained much more quickly than would be possible by using point-by-point measurement methods. [J.H.Ca.]

Diels-Alder reaction The 1,4-addition of an alkene (the dienophile) to a conjugated diene. The reaction, also known as the diene synthesis, is one of the most valuable and versatile methods for the preparation of compounds containing a six-membered ring, and proceeds most rapidly when the dienophile is substituted by electron-attracting groups. An example is in the illustration. The Diels-Alder reaction does not require a catalyst, nor is the reaction retarded by the presence of oxidation inhibitors with which dienes are commonly treated to prevent formation of peroxides.

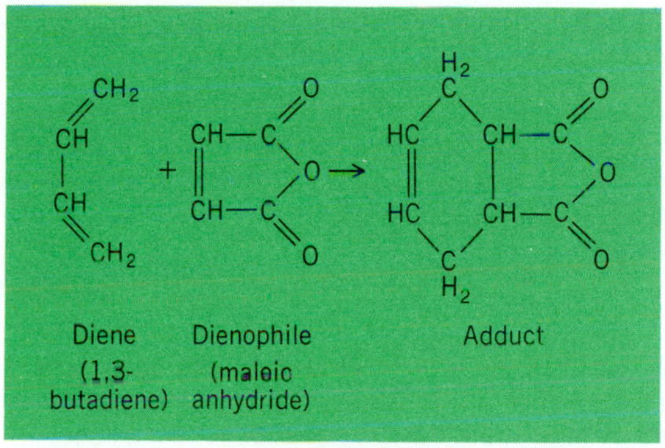

Example of Diels-Alder reaction.

Industrially the diene synthesis is used in the production of the insecticides aldrin and dieldrin. The adduct of butadiene and maleic anhydride is used in the synthesis of the important fungicide captan. [P.E.F.]

Diesel cycle An internal combustion engine cycle in which the heat of compression ignites the fuel. Compression-ignition engines, or diesel engines, are thermodynamically similar to spark-ignition engines. The sequence of processes for both types is intake, compression, addition of heat, expansion, and exhaust. Ignition and power control in the compression-ignition engine are, however, very different from those in the spark-ignition engine. *See* THERMODYNAMIC CYCLE.

Usually, a full-unthrottled charge of air is drawn in during the intake stroke of a diesel engine. A compression ratio between 12 and 20 is used, in contrast to a ratio of 4 to 10 for the Otto spark-ignition engine. This high compression ratio of the diesel raises the temperature of the air during the compression stroke. Just before top center on the compression stroke, fuel is sprayed into the combustion chamber. The high temperature of the air ignites the fuel, which burns almost as soon as it is introduced, adding heat. The combustion products expand to produce power, and exhaust to complete the cycle.

In an actual engine with a given compression ratio, the Otto engine has the higher efficiency. However, fuel requirements limit the Otto engine to a compression ratio of about 10, whereas a diesel engine can operate at a compression ratio of about 15 and consequently at a higher efficiency.

In addition, heat can be added earlier in the cycle by injecting fuel during the latter part of the compression process. This mode of operation is the dual-combustion or semidiesel cycle. With most of the heat added near peak compression, semidiesel efficiency approaches Otto cycle efficiency at a given compression ratio. *See* DIESEL ENGINE. [T.Ba.]

Diesel engine An internal combustion engine operating on a thermodynamic cycle in which the ratio of compression of the air charge is sufficiently high to ignite the fuel subsequently injected into the combustion chamber. The engine differs essentially from the more prevalent mixture (Otto) engine in which an explosive mixture of air and gas or air and the vapor of a volatile liquid fuel is made externally to the engine cylinder, compressed to a point some 200°F (110°C) below the ignition temperature, and ignited at will as by an electric spark. The diesel engine utilizes a wider variety of fuels with a higher thermal efficiency and consequent economic advantage under many service applications.

The true diesel engine, as projected by R. Diesel and as represented in most low-speed engines, such as about 300 rpm, uses a fuel-injection system where the injection rate is delayed and controlled to maintain constant pressure during combustion. Adaptation of the injection principle to higher engine speeds, such as 1000–2000 rpm, has necessitated departure from the constant pressure specification because the time available for fuel injection is so short (milliseconds). Combustion proceeds with little regard to the constant-pressure specification. High peak pressures may be developed. Yet nonvolatile (distillate) fuels are burned to advantage in these engines which cannot be rigorously identified as true diesels but which properly should be called commercial diesels. In ordinary parlance all such engines are classified as diesels. *See* DIESEL CYCLE.

Distillate fuel prevails with locomotive, truck, bus, and automotive applications. Lower-speed engines (stationary and motorship service) burn heavier fuels. Alternative fuels are burned in dual-fuel and gas diesel engines for stationary service. The main fuel is typically natural gas (90–95%) with oil (5–10%) used to control burning and to stabilize ignition. In the more prevalent liquid-fuel-injection system, the technical problems are numerous and embrace such elements as pumps, spray nozzles, and combustion chambers for the delivery, atomization, and burning of the fuel in the hot compressed air. There must be accurate timing (measured in milliseconds) for the entire process to give clean, complete combustion without undue excess air. *See* DIESEL FUEL.

Small size (<200 horsepower) engines are conveniently started by an electric motor and storage battery. Larger engines use compressed air introduced through valves in the cylinder head. Starting, with engine-driven generator sets, may be accomplished by motoring the generator.

Cooling systems use water at 120–180°F (49–82°C) with radiators, cooling towers, and cooling ponds employed for conservation and reclamation. Lubrication costs can become prohibitive with inadequate engine maintenance. *See* INTERNAL COMBUSTION ENGINE. [T.Ba.]

Diesel fuel A broad class of petroleum products which includes distillate or residual materials (or blends of these two) from the refining of crude oil. Diesel fuel generally has a distillation range between 374 and 716°F (190 and 380°C) and a specific gravity (59°F or 15°C) range between 0.760 and 0.935. In addition to these two primary criteria, other properties which are used to define diesel fuel are: viscosity (1.4 to 26.5 mm^2/s), sulfur content (usually <1.0% by weight), cetane number (30 to 60), ash content (usually <0.10% by weight), water and sediment content (usually <0.5% by volume), flash point (usually >131°F or 55°C), and cloud point (that temperature at which wax crystals begin to precipitate as the oil is cooled under prescribed conditions) or pour point (the lowest temperature at which the oil will flow when cooled under prescribed conditions). The properties of diesel fuel greatly overlap those of kerosine, jet fuels, and burner fuel oils; thus all these products are generally referred to as intermediate distillates. *See* FUEL OIL; JET FUEL; KEROSINE.

The diesel engine uses a method of ignition which requires that the fuel ignite spontaneously and quickly (within 1 to 2 milliseconds in a high-speed engine). The time lag between the initiation of fuel injection and the initiation of combustion is called ignition delay characterized by two major factors: a mechanical factor which is influenced by such things as compression ratio, motion of the charge air during injection, and ability of the injector to atomize the fuel; and a chemical factor which is influenced by such things as the fuel's autoignition temperature, specific heat, density, thermal conductivity, surface tension, and coefficient of friction. *See* DIESEL ENGINE.

The fuel's effect on the chemical portion of ignition delay is expressed by a quantity called the cetane number. Cetane, which has a high-ignition quality (short chemical ignition delay), has arbitrarily been assigned a cetane number of 100, whereas heptamethylnonane has been assigned a cetane number of 0. The cetane number of diesel fuel is determined by comparing it to a blend of cetane and heptamethylnonane which has the same ignition quality. The cetane number is the percentage by volume of cetane in the blend which has an ignition quality equal to the test fuel. *See* CETANE NUMBER; COMBUSTION CHAMBER.

[R.T.]

Difference equation A relationship between one or more independent variables, one or more dependent variables, and differences of those variables. Difference equations arise in the analysis of discrete systems (for example, a string loaded along its length with small masses), in the solution of differential equations by means of digital computers, in the implementation of digital filters, and in the discrete-time control of systems. An ordinary difference equation expresses a relationship between an independent variable t and one or more dependent variables, $y(t)$, $w(t)$, and so forth, and any successive differences of y, w . . . . The first forward difference of y, relative to the increment h, is defined by Eq. (1). The second forward

$$\Delta y(t) = y(t+h) - y(t) \qquad (1)$$

difference can be obtained by applying the above definition to itself, as in Eq. (2).

$$\begin{aligned}\Delta^2 y(t) &= \Delta[\Delta y(t)] \\ &= \Delta y(t+h) - \Delta y(t) \\ &= [y(t+2h) - y(t+h)] \\ &\quad - [y(t+h) - y(t)] \\ &= y(t+2h) - 2y(t+h) + y(t)\end{aligned} \qquad (2)$$

A difference equation is linear if it is of the first degree with respect to all the quantities $\Delta y(t)$, $\Delta^2 y(t)$, $\Delta^3 y(t)$, ... The general solution of difference equations is possible only for the case where they are linear with constant coefficients. However, nonlinear difference equations arise naturally in the solution of nonlinear differential equations. Likewise, the numerical solution of a partial differential equation leads to a partial difference equation. Methods exist for approximating governing equations for linear continuous-time systems by difference equations.

Linear constant-coefficients equations. The independent variable of a difference equation will usually have discrete values. Then the notation can be simplified by making the substitution in Eq. (3). In this case, the arguments of the independent

$$t_k = t_0 + kh \qquad k = 0, 1, 2, \ldots \qquad (3)$$

variables, if more than one, can be replaced by the subscript k. Then, the forward difference operators given by Eqs. (1) and (2) can be written as Eqs. (4), where the shifting operator E has

$$\begin{aligned}\Delta y_k &= y_{k+1} - y_k = Ey_k - y_k \\ \Delta^2 y_k &= E^2 y_k - 2Ey_k + y_k\end{aligned} \qquad (4)$$

been defined, incrementing the index on the dependent variable by 1 each time it is applied.

The general homogeneous, linear, constant-coefficient difference equation of order n can be written in terms of the shifting operator as Eq. (5). From the solution of linear, constant-

$$E^n y_k + A_1 E^{n-1} y_k + \cdots + A_{n-1} E y_k + A_n y_k = 0 \qquad k \geq 0 \qquad (5)$$

coefficient differential equations, the solution is assumed to be of the form $e^{rk} = \beta^k$, where r is a suitably chosen constant and $\beta = e^r$. From the definition of the shifting operator, Eqs. (6)

$$\begin{aligned}E\beta^k &= \beta^{k-1} = \beta(\beta^k) \\ E^2\beta^k &= \beta^{k-2} = \beta^2(\beta^k)\end{aligned} \qquad (6)$$

follow, as well as similar equations for higher powers of E. When these equations are substituted into Eq. (5), the result is Eq. (7).

$$\beta^n + A_1\beta^{n-1} + \cdots + A_{n-1}\beta + A_n = 0 \qquad (7)$$

Hence, $y_k = \beta^k$ will satisfy Eq. (5) if β is a root of Eq. (7). If all roots are real and distinct, the most general expression satisfying Eq. (5) is Eq. (8), where $\beta_1, \beta_2, \ldots, \beta_n$ are the roots to

$$y_k = c_1\beta_1^k + c_2\beta_2^k + \cdots + c_n\beta_n^k \qquad k \geq 0 \qquad (8)$$

Eq. (7) and $c_1, c_2, \ldots, c_n$ are constants to be determined by the initial conditions. *See* ALGEBRA.

Nonhomogeneous linear equations. The general solution of a nonhomogeneous, linear, constant-coefficient difference equation, given by Eq. (9), is of the form given by Eq. (10),

$$E^n y_k + A_1 E^{n-1} y_k + A_{n-1} E y_k + A_n y_k = x_k \qquad k \geq 0 \qquad (9)$$

$$y_k = y_{k,h} + y_{k,p} \qquad (10)$$

where $y_{k,h}$ is the solution of the homogeneous equation [$x_k = 0$ in Eq. (9)], and $y_{k,p}$ is the particular solution of Eq. (9). With constant coefficients on Eq. (9), the method of undetermined coefficients can be used to find the particular solution. The solution of constant-coefficient, linear difference equations is facilitated greatly by use of the z transform. *See* DIFFERENTIAL EQUATION; DIGITAL FILTER; INTERPOLATION; LAPLACE'S DIFFERENTIAL EQUATION; LINEAR SYSTEM ANALYSIS; NUMERICAL ANALYSIS. [R.E.Z]

Differential A mechanism which permits a rear axle to turn corners with one wheel rolling faster than the other. An automobile differential is located in the case carrying the rear-axle drive gear (see illustration).

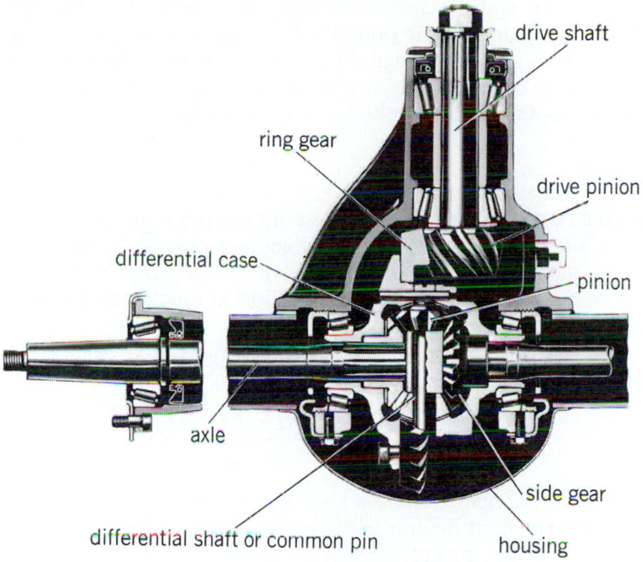

A rear-axle differential. (*Chrysler*)

The differential gears consist of the two side gears carrying the inner ends of the axle shafts, meshing with two pinions mounted on a common pin located in the differential case. The case carries a ring gear driven by a pinion at the end of the drive shaft. This arrangement permits the drive to be carried to both wheels, but at the same time as the outer wheel on a turn overruns the differential case, the inner wheel lags by a like amount.

Special differentials permit one wheel to drive the car by a predetermined amount even though the opposite wheel is on slippery pavement; they have been used on racing cars for years and are now used by a number of car manufacturers. *See* AUTOMOTIVE TRANSMISSION. [H.Fi.]

Differential amplifier An electronic circuit that generates an output voltage proportional to the voltage difference between two high-impedance input nodes. This output voltage can be defined either as the voltage difference between the output node and ground (single-ended output), or the voltage difference between a pair of output nodes (fully differential amplifier).

Two types of differential amplifiers are fabricated. In differential-in, single-ended-out amplifiers, the most widely used type, the output signal is referenced to local ground. Differential-in, differential-out or fully differential amplifiers are applied when maximum dynamic range or signal-to-noise ratio is required in the presence of noisy power supplies or other strong sources of interference, such as mixed analog-digital integrated circuits. *See* SIGNAL-TO-NOISE RATIO. [P.M.VanP.]

Differential equation A relationship between a function and its derivatives.

If there is one independent variable, the differential equation is called an ordinary differential equation. The general form of such an equation is shown in Eq. (1), where t is the

$$F(t,u,u',u'',\ldots,u^{(n)}) = 0 \qquad (1)$$

independent variable; u is a function of t; $u' = du/dt$, $u'' = d^2u/dt^2,\ldots,u^{(n)} = d^nu/dt^n$ are the derivatives of u; and $F(t,u_0,u_1,\ldots,u_n)$ is a given function of the $n + 2$ variables $t,u_0,\ldots,u^n$. The positive integer n is called the order of the differential equation; that is, the order of the differential equation is the order of the highest derivative that occurs in the equation. As an example, consider Eq. (2), which is an ordi-

$$u'' + u = 0 \qquad (2)$$

nary differential equation of order 2. A function u is said to be a solution to Eq. (1) if, when u and its derivatives up to order n are substituted into F, the identity expressed in Eq. (1) is valid for t in some interval $a < t < b$. In Eq. (2), $u = \cos t$ and $u = \sin t$ are solutions for $-\infty < t < \infty$, as can be immediately verified.

If there are two or more independent variables, the equation is called a partial differential equation. Again, the order of the equation is the order of the highest partial derivative that occurs in the equation. Thus, the general form of a partial differential equation of order 2, with two independent variables, is given by Eq. (3), where x and y are the independent vari-

$$F(x,y,u,u_x,u_y,u_{xx},u_{xy},u_{yy}) = 0 \qquad (3)$$

ables; $u = u(x,y)$ is a function: $u_x = \partial u/\partial x$, $u_y = \partial u/\partial x$, $u_{xx} = \partial^2u/\partial x^2$, $u_{xy} = \partial^2u/\partial x\partial y$, and u_{yy}, $= \partial^2u/\partial y^2$ are the derivatives of u; and $F(x, y, u, p, q, r, s, t)$ is a given function of eight variables. For example, Eq. (4) is a partial differential equation

$$u_{xy} = 0 \qquad (4)$$

of order 2. It is easily verified that $u = h(x) + g(y)$ is a solution of Eq. (4) for any choice of functions h and g.

A differential equation is linear if the function u is nonlinear otherwise.

Partial differential equations occur with more than two independent variables. Another possibility is a system of differential equations. This consists of two or more equations involving one or more unknown functions which are to be solved simultaneously. The Cauchy-Riemann equations, Eqs. (5), are a

$$u_x - v_y = 0 \qquad (5)$$
$$u_y + v_x = 0$$

first-order linear system of partial differential equations. *See* COMPLEX NUMBERS. [J.C.P.]

Differential geometry A branch of mathematics that deals with intrinsic properties of curves and surfaces in three-dimensional euclidean space. The intrinsic properties are those which are independent of the geometrical objects orientation or location in space. The subject is also concerned with nets of curves and families of surfaces, these having wide application in the arts.

The space is referred to a rectangular cartesian coordinate system (x,y,z). A space curve may be defined by a pair of independent equations, $f(x,y,z) = 0$ and $g(x,y,z) = 0$, or, more meaningfully, by parametric equations, $x = x(t)$, $y = y(t)$, $z = z(t)$. In this case the arc length between two points t_0 and t is given by Eq. (1).

$$s = \int_{t_0}^{t} \sqrt{\left(\frac{dx}{dt}\right)^2 + \left(\frac{dy}{dt}\right)^2 + \left(\frac{dz}{dt}\right)^2}\, dt \qquad (1)$$

Obviously, if s is chosen as parameter, Eq. (2) holds. In this

$$\left(\frac{dx}{ds}\right)^2 + \left(\frac{dy}{ds}\right)^2 + \left(\frac{dz}{ds}\right)^2 = 1 \qquad (2)$$

case dx/ds, dy/ds, dz/ds at a point P are the direction cosines of the tangent to the curve at P.

Consider three nearby points P, P_1, P_2. Through them, in general, one plane may be constructed. The limiting position of this plane as P_1 and P_2 approach P is the osculating plane of the curve at P. The limiting position of the circle through P, P_1, P_2 as P_1 and P_2 approach P is the osculating circle. Its center is the center of curvature and its radius is the radius of curvature ρ. The reciprocal of the radius of curvature κ is the curvature; its value is shown in expression (3). The perpendic-

$$\sqrt{\left(\frac{d^2x}{ds^2}\right)^2 + \left(\frac{d^2y}{ds^2}\right)^2 + \left(\frac{d^2z}{ds^2}\right)^2} \qquad (3)$$

ular to the tangent at P in the osculating plane is the principal normal and the perpendicular at P to the osculating plane is the binormal.

A surface in three-dimensional euclidean space may be given by $f(x,y,z) = 0$ or, more conveniently, by Eqs. (4). For a

$$x = x(u,v) \qquad y = y(u,v) \qquad z = z(u,v) \qquad (4)$$

fixed value of v, Eqs. (4) describe a curve in the surface, a u line, and similarly for a fixed value of u. Thus (u,v) are curvilinear coordinates of a point in the surface. The most important quantity in the study of surfaces is the arc length of a curve. This is given by Eq. (5), where E, F, G are functions of

$$ds^2 = E\,du^2 + 2F\,du\,dv + G\,dv^2 \qquad (5)$$

partial derivatives of (x,y,z). In the real domain E, G, and $EG - F^2$ are nonnegative and may be zero only at singular points of the surface, or at points where the matrix of the partial derivatives of (x,y,z) is of rank less than two. The right-hand side of Eq. (5) is called the first fundamental form of the surface, where E, F, G are the first fundamental quantities of the surface. All applicable surfaces have the same first fundamental form. Thus any cylinder or cone which is developable (applicable to a plane) has the fundamental form $du^2 + dv^2$, where (u,v) are rectangular cartesian coordinates. Under an arbitrary transformation of the surface coordinates (u,v) the fundamental quantities E, F, G transform linearly so that many problems in the study of surfaces reduce to the question whether a coordinate system exists on the surface in which E, F, G satisfy desired conditions. *See* COORDINATE SYSTEMS; RIEMANNIAN GEOMETRY. [M.S.Kn.]

Differential transformer An iron-core transformer with movable core. A differential transformer produces an electrical output voltage proportional to the displacement of the core. It is used to measure motion and to sense displacements. It is also used in measuring devices for force, pressure, and acceleration which are based on the conversion of the measured variable to a displacement.

Various available configurations, some translational and others rotational, all employ the basic circuit shown in the illustration: a primary winding, two secondary windings, and a movable core. The primary winding is energized with alternating voltage. The two secondary windings are connected in series opposition, so that the transformer output is the difference of the two secondary voltages. When the core is centered, the two secondary voltages are equal and the transformer output is zero. This is the balance or null position. When the core is displaced from the null point, the two secondary voltages are no longer alike and the transformer produces an output voltage. With proper design, the output voltage varies linearly with core position over a small range. Motion of the core in the opposite direction produces a similar effect with 180° phase reversal of the alternating output voltage. *See* TRANSDUCER; TRANSFORMER. [G.W.]

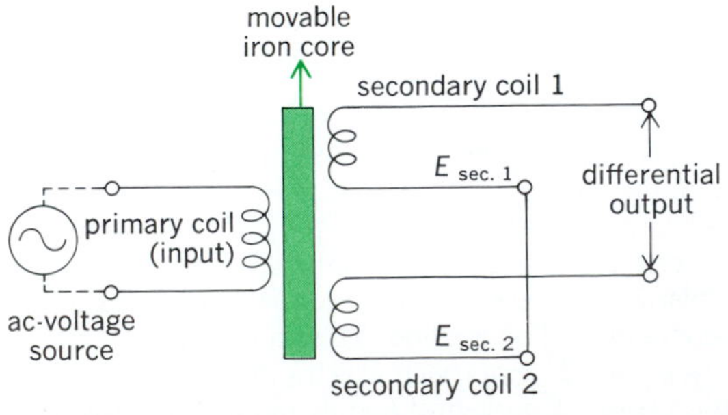

Basic circuit of differential transformer. *E* refers to E pickoff (from E-shaped iron core).

Differentiation A mathematical operation performed on a function to determine the effect of a change in the value of the independent variable. Functions of one variable are considered in this article. For differentiation of functions of several variables *see* PARTIAL DIFFERENTIATION.

If f is a function of x, defined on an interval containing x_0, the derivative at x_0 is, by definition, that shown in Eq. (1). For

$$f'(x_0) = \lim_{x \to x_0} \frac{f(x) - f(x_0)}{x - x_0} \qquad (1)$$

generalities about derivatives and calculus *see* CALCULUS.

In the quotient on the right, in the definition of $f'(x_0)$, x is restricted to the interval on which f is defined. If $y = f(x)$, $f'(x_0)$ is also denoted by

$$\left(\frac{dy}{dx}\right)_{x = x_0}$$

The limit defining $f'(x_0)$ may not exist. If it does, f is called differentiable at x_0.

The derivative of $f'(x)$, called the second derivative, is denoted by

$$f''(x) \text{ or } \frac{d^2y}{dx^2}$$

A function f is called continuous at x_0 if x_0 is in the domain of f and

$$\lim_{x \to x_0} f(x) = f(x_0)$$

The precise formulation of the limit concept used here is the following: Let g be a function and let A be a number. Then Eq. (2) means that to each positive number ϵ corresponds to

$$\lim_{x \to x_0} g(x) = A \qquad (2)$$

some positive number δ such that $|g(x) - A| < \epsilon$ whenever x is a number in the domain of g such that $x \neq x_0$ and $|x - x_0| < \delta$. It is required of g that its domain shall contain numbers x as close to x_0 as desired, but different from x_0. The domain of g may also contain x_0, but this is irrelevant.

It is a theorem that, if f is differentiable at x_0, then f is continuous at x_0. However, a function can be continuous at x_0, but not differentiable there. An example is $f(x) = |x|$ at $x_0 = 0$. It is even possible for a function to be continuous on an interval and yet not differentiable at any point of this interval.

The chief elementary applications of differentiation are (1) in expressing rates of change (velocity, acceleration), and in solving problems where, through functional relationship, the rate of change of one variable is calculated when the rate of change of another variable is known; (2) in studying graphs of functions and, more generally, in studying curves in the plane or in space of three dimensions: (3) in expressing scientific laws or principles in the form of differential equations; and (4) in the expression of various extensions and applications of the law of the mean, including such topics as l'Hospital's rule and Taylor's formula or series.

The general technique of differentiation is built upon the rules for differentiating combinations of differentiable functions. If $u = f(x)$ and $v = g(x)$ are functions differentiable for the same values of x, then $u + v$ and uv are differentiable and so is u/v if $v \neq 0$. The formulas are shown in Eqs. (3). If u is

$$\frac{d}{dx}(u+v) = \frac{du}{dx} + \frac{dv}{dx}$$

$$\frac{d}{dx}(uv) = u\frac{dv}{dx} + v\frac{du}{dx} \qquad (3)$$

$$\frac{d}{dx}\left(\frac{u}{v}\right) = \frac{v\dfrac{du}{dx} - u\dfrac{dv}{dx}}{v^2}$$

constant in value, then $du/dx = 0$. A very powerful instrument of technique is furnished in the chain rule for composite functions. If y is a differentiable function of u and u is a differentiable function of x, then y is a differentiable function of x, and

$$\frac{dy}{dx} = \frac{dy}{du}\,\frac{du}{dx}$$

In functional notation, if $y = f(u)$ and $u = h(x)$, then $y = F(x)$, where $F(x) = f[h(x)]$, and then $F'(x) \cdot f'[h(x)]h'(x)$. [A.E.Ta.]

Diffraction The bending of light, or other waves, into the region of the geometrical shadow of an obstacle. More exactly, diffraction refers to any redistribution in space of the intensity of waves that results from the presence of an object that causes variations of either the amplitude or phase of the waves. Most diffraction gratings cause a periodic modulation of the phase across the wavefront rather than a modulation of the amplitude. Although diffraction is an effect exhibited by all types of wave motion, this article will deal only with electromagnetic waves, especially those of visible light. For discussion of the phenomenon as encountered in other types of waves *see* ELECTRON DIFFRACTION; NEUTRON DIFFRACTION; SOUND.

Diffraction is a phenomenon of all electromagnetic radiation, including radio waves; microwaves; infrared, visible, and ultraviolet light; and x-rays. The effects for light are important in connection with the resolving power of optical instruments. *See* RADIO-WAVE PROPAGATION; X-RAY DIFFRACTION.

There are two main classes of diffraction, which are known as Fraunhofer diffraction and Fresnel diffraction. The former concerns beams of parallel light, and is distinguished by the simplicity of the mathematical treatment required and also by its practical importance. The latter class includes the effects in divergent light, and is the simplest to observe experimentally.

To illustrate the difference between the methods of observation of the two types of diffraction, Fig. 1 shows the experimental arrangements required to observe them for a circular hole in a screen S. The light originates at a very small source O, which can conveniently be a pinhole illuminated by sunlight. In Fraunhofer diffraction, the source lies at the principal focus of a lens L_1 which renders the light parallel as it falls on the aperture. A second lens L_2 focuses parallel diffracted beams on the observing screen F, situated in the principal focal plane of L_2. In Fresnel diffraction, no lenses intervene. The diffraction effects occur chiefly near the borders of the geometrical shadow, indicated by the broken lines. An alternative way of distinguishing the two classes, therefore, is to say that Fraunhofer diffraction concerns the effects near the focal point of a lens or mirror, while Fresnel diffraction concerns those effects near the edges of shadows. Photographs of diffraction patterns are shown in Fig. 2.

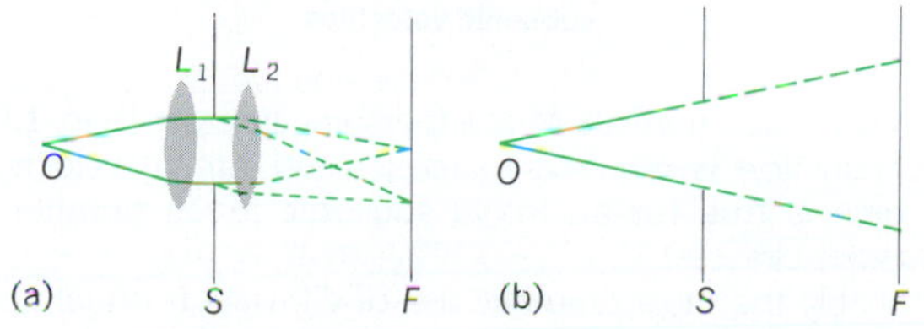

Fig. 1. Observation of the two principal types of diffraction, in the case of a circular aperture. (*a*) Fraunhofer and (*b*) Fresnel diffraction.

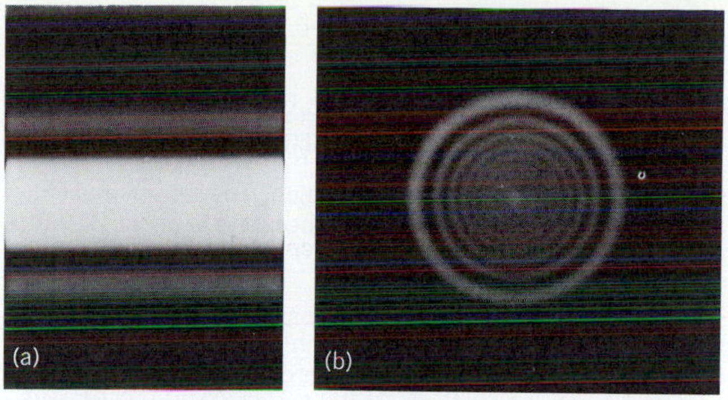

Fig. 2. Diffraction patterns, photographed with visible light. (*a*) Fraunhofer pattern, for a slit; (*b*) Fresnel pattern, circular aperture.

Fraunhofer diffraction. This class of diffraction is characterized by a linear variation of the phases of the Huygens secondary waves with distance across the wavefront, as they arrive at a given point on the observing screen. At the instant that the incident plane wave occupies the plane of the diffracting screen, it may be regarded as sending out, from each element of its surface, a multitude of secondary waves, the joint effect of which is to be evaluated in the focal plane of the lens L_2. The analysis of these secondary waves involves taking account of both their amplitudes and their phases. The simplest way to do this is to use a graphical method, the method of the so-called vibration curve, which can readily be extended to cases of Fresnel diffraction. *See* HUYGENS' PRINCIPLE.

The vibration curve results from the addition of a large (really infinite) number of infinitesimal vectors, each representing the contribution of the Huygens secondary waves from an element of surface of the wavefront. If these elements are assumed to be of equal area, the magnitudes of the amplitudes to be added will all be equal. They will, however, generally differ in phase, so that if the elements were small but finite each would be drawn at a small angle with the preceding one, as shown in Fig. 3*a*. The resultant of all elements would be the vector A. When the individual vectors represent the contributions from infinitesimal surface elements (as they must for the Huygens wavelets), the diagram becomes a smooth curve, the vibration curve, shown in Fig. 3*b*. The intensity on the screen is then proportional to the square of this resultant amplitude. In this way, the distribution of the intensity of light in any Fraunhofer diffraction pattern may be determined.

The intensity distribution for Fraunhofer diffraction by a slit as a function of the angle θ from the center may be simply calculated by the method of the vibration curve. The intensity at any angle is given by Eq. (1), where I_0 is the intensity at the

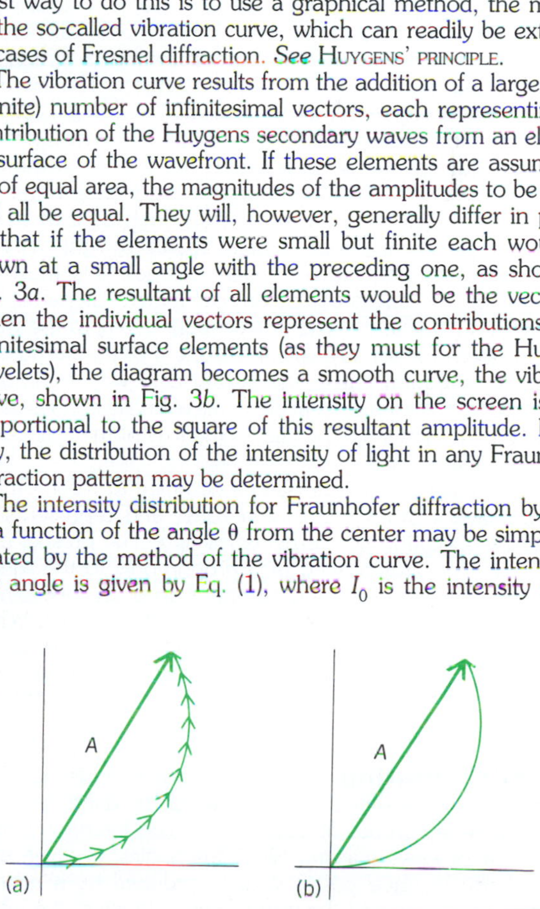

Fig. 3. Vibration curves. (*a*) Addition of many equal amplitudes differing in phase by equal amounts. (*b*) Equivalent curve, when amplitudes and phase differences become infinitesimal.

$$I = I_0 \frac{\sin^2 ß}{ß^2} \tag{1}$$

center of the pattern, and ß is given by Eq. (2), where b is the

$$ß = (\pi b \sin \theta)\lambda \tag{2}$$

width of the slit and λ is the wavelength. The central maximum is twice as wide as the subsidiary ones, and is about 21 times as intense as the strongest of these. A photograph of this pattern is shown in Fig. 2*a*.

Fraunhofer diffraction by a circular aperture determines the resolving power of instruments such as telescopes, cameras, and microscopes, in which the width of the light beam is usually limited by the rim of one of the lenses. The method of the vibration curve may be extended to find the angular width of the central diffraction maximum for this case. An exact construction of the curve or, better, a mathematical calculation shows that the extreme phase differences required are ± 1.220π, yielding Eq. (3) for the angle θ at the first zero of

$$\sin \theta \approx \theta = \pm \frac{1.220\lambda}{d} \tag{3}$$

intensity. Here d is the diameter of the circular aperture. This pattern has circular symmetry and consists of a diffuse central disk, called the Airy disk, surrounded by faint rings. The angular radius of the disk, given by Eq. (3), may be extremely small for an actual optical instrument, but it sets the ultimate limit to the sharpness of the image, that is, to the resolving power. *See* RESOLVING POWER (OPTICS).

Fresnel diffraction. The diffraction effects obtained when the source of light or the observing screen are at a finite distance from the diffracting aperture or obstacle come under the classification of Fresnel diffraction. This type of diffraction requires for its observation only a point source, a diffracting screen of some sort, and an observing screen. The latter is often advantageously replaced by a magnifier or a low-power microscope. The observed diffraction patterns generally differ according to the radius of curvature of the wave and the distance of the point of observation behind the screen. If the diffracting screen has circular symmetry, such as that of an opaque disk or a round hole, a point source of light must be used. If it has straight, parallel edges, it is desirable from the standpoint of brightness to use an illuminated slit parallel to these edges. In the latter case, it is possible to regard the wave emanating from the slit as a cylindrical one. For the purpose of deriving the vibration curve, the appropriate way of dividing the wavefront into infinitesimal elements is to use annular rings in the first case, and strips parallel to the axis of the cylinder in the second case.

The zone plate is a special screen designed to block off the light from every other half-period zone, and represents an interesting application of Fresnel diffraction. The Fresnel half-period zones are drawn, with radii proportional to the square roots of whole numbers, and alternate ones are blackened. The drawing is then photographed on a reduced scale. When light from a point source is sent through the negative, an intense point image is produced, much like that formed by a lens. [F.A.J./W.W.W.]

Diffraction grating

Diffraction grating An optical device consisting of an assembly of narrow slits or grooves, which by diffracting light produces a large number of beams which can interfere in such a way as to produce spectra. Since the angles at which constructive interference patterns are produced by a grating depend on the lengths of the waves being diffracted, the waves of various lengths in a beam of light striking the grating will be separated into a number of spectra, produced in various orders of interference on either side of an undeviated central image. By controlling the shape and size of the diffracting grooves when producing a grating and by illuminating the grating at suitable angles, a beam of light can be thrown into a single spectrum whose purity and brightness may exceed that produced by a prism. Gratings can now be made with much larger apertures than prisms, and in such form that they waste less light and give higher intrinsic dispersion and resolving power. *See* DIFFRACTION.

Transmission gratings consist of a large number of narrow transparent and opaque slits alternating side by side in regular order and with uniform separation, through which a beam of light will appear as a series of spectra in various orders of interference. Reflection gratings, either plane or concave, are used in most spectrographs. Such a grating may consist of an original ruling or of a metal-coated replica from an original. Large grating replicas can now be made which are practically indistinguishable in performance or permanence from an original.

Gratings are engraved by highly precise ruling engines, which use a diamond tool to press into a highly polished mirror surface a series of many thousands of fine shallow burnished grooves. If a grating is to give resolution approaching the theoretical limit, its grooves must be ruled straight, parallel, and equally spaced to within a few tenths of the shortest incident wavelength. Scattered light and false images may arise from local spacing error and groove shape variations of only a few hundredths of the diffracted wavelength.

A grating spectroscope usually consists of a slit, a lens or mirror to collimate the light sent through the slit into a parallel beam, a transmission or reflection grating to disperse the light, a lens or mirror to focus the light into spectrum lines (which are monochromatic images of the slit in the light of each wavelength passing through it), and an eyepiece for viewing the spectrum. If a camera is substituted for the telescope, the instrument becomes a grating spectrograph. If a photoelectric cell, a thermocouple, or other radiation-detecting device is used instead of a camera or telescope, the device becomes a grating spectrometer. *See* CONCAVE GRATING; ECHELETTE GRATING; ECHELLE GRATING; INFRARED SPECTROSCOPY. [G.R.H.]

Diffuser

Diffuser A device to convert a high-velocity low-pressure stream of fluid (usually air) into a low-velocity high-pressure flow. In doing so, a diffuser converts the velocity or kinetic energy into pressure or potential energy. An efficient diffuser operates by decelerating the high-velocity stream with as little loss in total energy as possible.

The shape of a diffuser depends on the Mach number of the entering fluid. If a supersonic stream is to be decelerated, the diffuser cross-sectional area must be progressively decreased (ideally to just sonic velocity) and then increased until the flow reaches the desired velocity, as indicated in the illustration. If

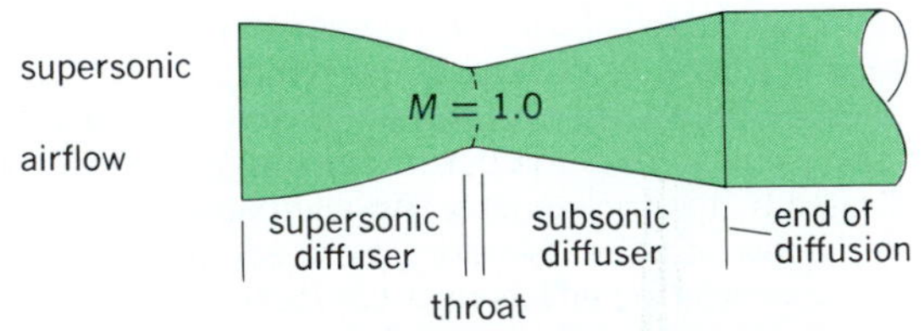

Idealized diffuser for decreasing supersonic velocities to subsonic velocities.

the initial Mach number M is supersonic (greater than 1.0), an area reduction is required to reduce the stream velocity; the converse is true for an initial subsonic Mach number. *See* SUPERSONIC DIFFUSER.

Probably the most dramatic use of diffusers is on jet aircraft. Diffusers supplying air to the engine have had inlets in the nose, at the wing root, along the airplane fuselage, and even in the tail section. [J.F.C.; L.J.O.]

Diffusion in gases and liquids The spreading or scattering of matter under the influence of a concentration gradient. For a related discussion *see* DIFFUSION IN SOLIDS.

Consider a fluid confined in a space of dimensions which are large compared to the mean free path of the fluid molecules. For gases at atmospheric pressure and room temperature, the mean free path is on the order of 10^{-5} cm and varies inversely with pressure. Liquids in general have much smaller free paths. Assume that one of the components of the fluid exists initially at different concentrations, or more rigorously at different activities, in two or more different locations in the confining space. At constant temperature, and in the absence of external forces, there will be a spontaneous movement, that is, diffusion, of the component in the direction of establishing a uniform concentration of that component in all parts of the enclosure.

In a very qualitative way the cause of this spontaneous mixing may be interpreted as follows. As a consequence of thermal agitation, molecules of a fluid are in constant motion in all directions. The number of molecules of a given kind moving in any given direction at a particular point in the fluid is proportional to the number of these molecules present per unit volume. In the absence of a concentration gradient, on the average as many molecules per unit time leave any hypothetical plane in a given direction as return to the plane from the opposite direction. However, if the number of molecules per unit volume decreases in a given direction, more molecules on the average move into the region of lower concentration than return from it. This results in a net transfer of molecules of that kind toward the region of lower concentration.

In general, the diffusion of one component of a mixture will be accompanied by diffusion of one or more other components in the opposite direction to maintain the volume of fluid constant in the region under consideration. This may not be the case when special restrictions are placed on the system, for example, in the selective absorption of one component from a gas by a liquid acting as a boundary of the confining space, or for liquids when there is a volume change on mixing.

Fluid motion may be viscous or turbulent. In viscous flow, mass transfer resulting from a concentration gradient occurs by molecular diffusion. In turbulent flow, transfer of material occurs by the more rapid process of mixing of the swirling eddies of fluid, a process termed turbulent diffusion or eddy diffusion.

S. Chapman demonstrated in 1916 as a consequence of the kinetic theory that a temperature gradient in a mixed gas might give rise to a flow of one constituent relative to the mixture as a whole. His finding was verified experimentally by L. Chapman and F. W. Dootsen, who placed various mixtures in a bulb which was heated at one end and cooled at the other. A similar effect had been observed for liquids by Ludwig in 1856 and by J. L. Soret in 1879, and is usually called the Soret effect. The phenomenon as it pertains to fluids generally is called thermal diffusion.

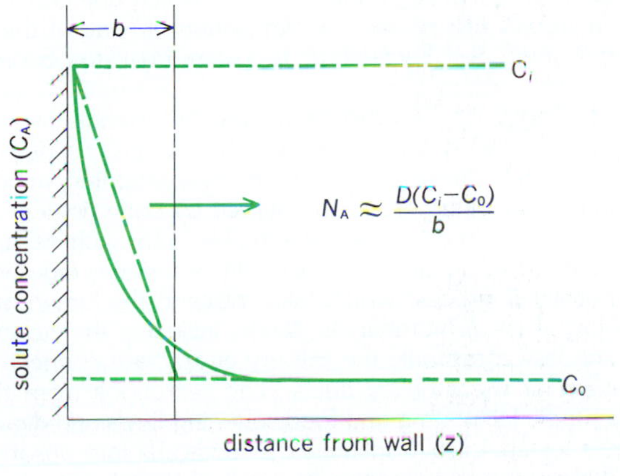

Concentration profile for mass transfer from pipe wall into turbulent fluid.

An external force acting upon each of a group of molecules may induce molecular movement analogous to diffusion. Such movements, which may result from gradients of pressure within a fluid, or from electric or magnetic fields, may be termed forced diffusion. An important example is the process of electrolytic migration of ions under an applied potential.

Mass transfer through a turbulent fluid to an interface formed by some other fluid with which it is immiscible or by a solid surface is commonly encountered in processes of industrial importance. For such situations it is necessary to resort to empirical equations to describe mass-transfer rates. As an example, consider the dissolution of a slightly soluble solute, such as benzoic acid, from the wall of a pipe into water in turbulent flow. Solute concentrations in the fluid in the vicinity of the wall are shown qualitatively in the illustration. At the interface the water is saturated with solute at concentration C_i corresponding to the solubility of the solute. The concentration falls progressively from the wall until a value corresponding essentially to the value in the bulk liquid, C_0, is reached. In the steady state the mass-transfer flux at a given point is commonly expressed by a mass-transfer coefficient, k_C, as in Eq. (1)

$$N_A = k_C(C_i - C_0) \tag{1}$$

Similar concepts can be applied to both gases and liquids and for surfaces other than pipe walls. Various mass-transfer coefficients may be defined corresponding to the units used for mass-transfer rates, concentrations, and distance.

An alternate method of describing the rate of mass transfer is to define an equivalent film b of stagnant fluid which offers the same resistance to transfer (that is, requires the same total drop in concentration between the interface and bulk liquid) as the actual fluid region which is not completely stagnant (see illustration). Integration of Eq. (1) for steady-state diffusion gives the relation between the mass-transfer flux and the film thickness for equivolume diffusion, Eq. (2).

$$N_A = \frac{D_A(C_i - C_0)}{b} \tag{2}$$

See MASS-TRANSFER OPERATION. [C.R.W.]

Diffusion in solids The actual transport of mass, in the form of discrete atoms, through the lattice of a crystalline solid is called diffusion. Diffusion processes are fundamental to a large class of solid-state effects: solubility, phase transformations, nucleation, annealing, devitrification, recrystallization, oxidation, corrosion, creep, welding, sintering, tempering, the photographic process, and ionic conductivity. Almost any change in a solid—in its form, strength, or even its color—is usually associated with mass motion within the lattice. *See* IONIC CRYSTALS.

Scientific interest in diffusion has centered on elucidating the basic atomic mechanisms which permit diffusion to occur without violating the solidity of the lattice. The jump rates of individual atoms necessary to explain observed bulk effects is formidable: Well below the melting point, each atom in a solid must jump from one "fixed" lattice site to another over a million times per second. A large number of models have been proposed, and with precise radioactive tracer measurements it has finally been possible to define the basic atomic mechanisms. It is clear that diffusion occurs only by virtue of lattice imperfections. A perfect lattice would permit no mass motion. At temperatures below about half the absolute melting temperature, diffusional motion is largely confined to surfaces and regions of inherent disorder: dislocations, grain boundaries, and paths disarranged by the passage of high-energy particles. True bulk diffusion, where atoms migrate through the whole volume, predominates at high temperatures. In most solids it has been shown that the principal defect responsible for bulk diffusion is the vacancy—a missing lattice atom; in a few cases the defect is the interstitial—an atom in an extra nonlattice site. These defects can be formed externally at free surfaces or internally along jogs on dislocations. In order to maximize the entropy, they will tend toward a random distribution by migration through the lattice. Results of high-pressure measurements show that vacancies and interstitials are "many-body" rather than pure "point" defects, since their sizes are clearly different from that of a single atom. This implies a considerable relaxation of the lattice around a defect, severely complicating theoretical attempts to calculate defect energies from first principles. [D.Laz.]

Digenea A group of parasitic flatworms or flukes constituting a subclass or order of the class Trematoda, in the phylum Platyhelminthes. The name Digenea refers to the two types of generations in the life cycle: (1) the germinal sacs which parasitize the intermediate host (a mollusk or rarely an annelid) and reproduce asexually; and (2) the adult which is primarily endoparasitic in vertebrates and reproduces sexually. The adult usually is hermaphroditic, but many of the blood flukes and a few others are dioecious. Vertebrates of all classes, except the Cyclostomata, serve as definitive hosts. Those feeding on aquatic plants and animals harbor the greatest variety of digenetic trematodes, but several species occur in strictly terrestrial hosts. Effects on the vertebrate vary from no apparent harm to severe and even fatal injury. Control of flukes is by preventive measures, including interruption of their life cycles and treatment with drugs. *See* TREMATODA.

The adult differs from other trematodes in several respects, the most obvious being the absence of a posterior adhesive organ with sclerotized hooks, plates, or multiple suctorial structures. Usually the mouth is anterior and opens into an oral sucker. A second sucker, the ventral one, is at the posterior end of amphistomes, on the ventral surface of distomes, or absent in monostomes. Adult digenetic trematodes occur in any part of the vertebrate that provides egress for their eggs and also in the circulatory system, from which eggs work through the tissues and escape in the feces or urine.

There is no general agreement concerning the classification of the Digenea. The scheme of G. La Rue proposes two superorders, the Anepitheliocystidia and Epitheliocystidia. *See* PLATYHELMINTHES. [R.M.C.]

Digestive system That system of structures in which food substances are digested; that is, complex food materials are subjected to the action of digestive enzymes throughout the system, in the presence of water, and broken down into simpler chemical compounds. Absorption into the cells and tissues is then possible. The digestive system of invertebrates varies among the various phyla. Many protozoans have no permanent structures; such animals as sponges have intracellular digestion while in coelenterates digestion is partly extracellular in the body cavity; most animals have a complete digestive system in that there is a mouth and anal opening. Among the vertebrate classes, the structures of the digestive system are essentially similar, although they may be highly modified according to the food habits of the animal.

Anatomy. The major digestive organs form the posterior portions of the digestive tract, while the mouth and pharynx are the reception point for food and have numerous specialized structures and glands.

The mouth is the expanded and structurally complex upper end of the digestive tract. When lips and cheeks are present, as in humans, the space anterior to the teeth and gums is termed the vestibule or buccal cavity. Behind this is the much larger oral cavity, in the floor of which lies the tongue; the roof of this cavity is termed the palate. The mouth is continuous posteriorly with the pharynx. Although the mouth is primarily concerned with feeding and varies accordingly, in terrestrial vertebrates it also participates importantly in the production and modulation of sound, a function which attains its highest development in the voice and articulate speech of humans.

The pharynx is a relatively short segment of the digestive tract lying between the posterior part of the oral cavity and the upper end of the esophagus. It is an important region because its passage is shared by both the digestive and respiratory systems. Below the pharynx the digestive tract proper (or gut) consists of an elongated muscular tube extending from the mouth to the anus. Four successive segments of the gut are distinguished on the basis of their structural and functional specialization (see illustration): the esophagus, for rapid conveyance of food from the mouth to the stomach; the stomach, where the food is held for a time and digestion begins; the small intestine, where digestion and absorption are virtually completed; and the large intestine, into which undigested and waste materials are passed for elimination by way of the rectum and anus. *See* ESOPHAGUS; INTESTINE; PHARYNX; STOMACH. [V.M.E.]

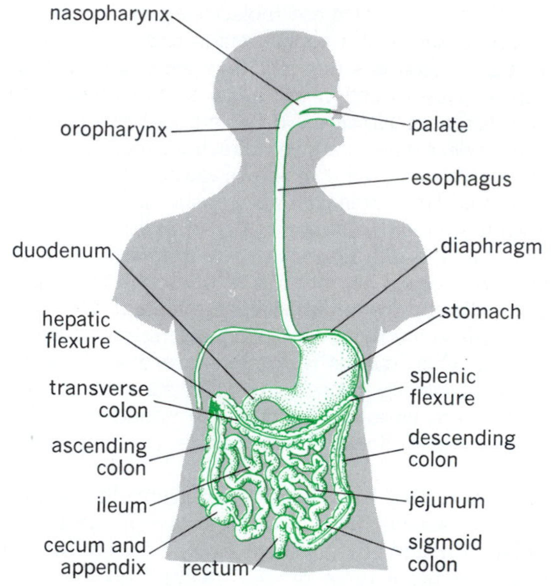

Digestive tract in humans. (*After L. L. Langley, E. Cheraskin, and R. Sleeper, Dynamic Anatomy and Physiology, 2d ed., McGraw-Hill, 1963*)

Physiology. The foodstuffs utilized by animals to satisfy their energy and growth requirements consist of nutrients (complex carbohydrates, proteins, and fats), vitamins, minerals, and water. The nutrients, to be utilized by body tissues, must first be digested into smaller, absorbable units such as simple sugars, amino acids, and fatty acids. In vertebrates digestion is an extracellular process which takes place in the lumen of the alimentary tract. A number of glands, including the pancreas, liver, and less commonly the salivary glands, are connected to the lumen of the tube by ducts. The secretions from these glands supply lubricating and food-softening fluids and digestive enzymes for chemically splitting the foodstuffs into absorbable units. Other glands lodged in the wall of the alimentary tube also supply fluids, acids, and other important aids to the digestive process. Digestion is a highly complex, highly integrated, and delicately controlled physiological process. Integration and control of the functioning of the alimentary tract are achieved by means of an interlocking system of nervous networks and hormones.

In the tract food is masticated, triturated, mixed with hydrochloric acid, digestive enzymes and fluid, and propelled toward the anus by peristaltic action. The digestive enzymes have similar physiological actions throughout the vertebrates. The amounts of each enzyme produced vary with the species and appear to be functionally adapted to differences in dietary habits.

Mouth. In vertebrates the primary functions of the mouth in digestion are the reception and swallowing of food. Other important functions subserved by it in some, but not all, vertebrates are the mastication and moistening of whole foods. Actual digestion in the mouth is of limited importance. The functions of the mouth in the digestive process, then, are mainly mechanical; that is, prehension, tearing, cutting and grinding of food, and mixing it with a fluid medium.

Salivary glands. The fluids which bathe the oral cavity are produced by the salivary glands, which are found in many groups of animals, invertebrates as well as vertebrates. Saliva performs a number of different functions. During mastication it helps to form food into a convenient bolus for mastication and swallowing. It also facilitates swallowing by keeping the mouth and esophagus moist. Saliva aids the taste mechanism by dissolving the substances in the food which stimulate the taste buds. Finally saliva helps keep the oral tissues and teeth clean of food debris.

In humans there are three pairs of large salivary glands: the parotid, the submaxillary, and the sublingual. There are also numerous, smaller glands lodged in the submucosal tissue of the lips, cheeks, tongue, and palate. Secretion by the salivary glands is under the control of the autonomic nervous system, especially the parasympathetic branch. The presence of food in the mouth causes a reflex secretion of saliva. Salivary secretion is easily conditioned and the sight, smell, and in humans even the thought of food cause increased salivary secretion (thus the familiar expression "makes the mouth water"). Saliva is secreted in much greater volume than is generally appreciated; the daily secretion for humans is probably about 37 in.^{3} (600 milliliters).

Teeth. Along with their role as biologic weapons for offense and defense, the teeth serve in the prehension, cutting, and mastication of food. It would seem logical to suppose that mastication, because it increases the surface area of the food exposed to digestive action, would be an aid to digestion. This is definitely the case in ruminant forms. On the other hand, the food swallowed whole by snakes or wolfed by carnivores is clearly subject to complete digestion. *See* TOOTH. [H.W.]

Tongue. The tongue is composed principally of skeletal muscle and is enveloped by mucous membrane. The mucous membrane on the undersurface forms a fold, termed the frenulum linguae or more commonly the frenulum, which binds the tongue, with the exception of the tip, to the floor of the mouth. Normally the upper surface of the tongue presents a moist, pink appearance. Under other conditions it may appear yellow and dry because of the accumulation of shed epithelial cells, remains of food, and organisms. The surface of the tongue is characterized by the presence of papillae. The taste buds are scattered over the upper surface of the tongue.

The taste buds are chemoreceptors and as such are attuned to the chemical composition of the oral cavity. There are specific substances which will stimulate only one of the four types of taste buds. It is common experience that more than four basic tastes can be easily differentiated. This ability depends upon two factors: utilization of more than one type of taste bud in a blend, so to speak, and derivation of clues from activation of extragustatory receptors such as those for temperature, pressure, touch, and olfaction. For example, if an individual were to chew a piece of uncooked potato it could be distinguished from a raw carrot or an apple. Yet is it clear that none of these foods can be described as truly acid or sour, bitter, or sweet. The texture of the substance, its moisture content, the temperature, the relative proportion of the various types of taste buds stimulated, and finally the sense of smell all contribute. It is easy to demonstrate that the taste of a substance depends on contrast. For example, coffee consumed after a very sweet dessert does not taste as sweet as the same beverage does in the morning when nothing else has preceded it. *See* CHEMORECEPTION; TASTE; TONGUE.

Deglutition. The act of swallowing is termed deglutition. It is a complex process, involving both voluntary and reflex mechanisms, and occurring in three stages.

In the first stage of deglutition the food is ground and rolled into a mass, called a bolus, and thoroughly soaked with saliva. The tongue then directs the bolus to the back of the mouth and forces it to enter the pharynx. All these acts are purely voluntary. Once the bolus is forced into the pharynx, however, the voluntary phase of swallowing has ended, and all subsequent mechanisms are purely reflex. During the second stage of deglutition, the bolus passes through the pharynx to enter the esophagus.

In the third phase of deglutition the food traverses the esophagus to enter the stomach. The force of gravity assists in transporting the bolus through the esophagus, but it is not indispensable to the swallowing act. The esophagus is composed chiefly of muscles which contract in wavelike fashion along the length of the tube. These are called peristaltic waves, or movements. Thus the bolus is moved through the esophagus to enter the stomach. [L.L.L.]

Esophagus. The esophagus extends from the pharynx to the stomach in mammals. It transfers swallowed material from the pharynx to the stomach, and prevents the return of gastric contents to the esophagus and pharynx. The esophagus contains four functionally distinct regions: the upper esophageal sphincter, the part of the esophageal body composed of striated muscle, the part of the esophageal body composed of smooth muscle, and the lower esophageal sphincter.

The upper esophageal sphincter is a segment, 1.2–1.6 in. (3–4 cm) long in humans, between the pharynx and the esophageal body, in which the lumen is completely closed at rest but opens during swallowing. The opening of the esophageal lumen in swallowing occurs immediately after pharyngeal contraction, so that the lumen offers low resistance to the movement of the swallowed bolus. After the brief relaxation, resting tonic contraction is resumed.

The striated muscle of the proximal esophageal body is flaccid at rest. After a swallow, an occlusive contraction begins at the most proximal level as an extension of the closure of the upper esophageal sphincter, and moves as a front throughout the length of the organ. The peristaltic contraction of the esophageal body generated by swallowing passes smoothly from the striated muscle to the smooth muscle, but the mechanism of the contraction in the smooth muscle is different.

The lower esophageal sphincter, 1.2–1.6 in. (3–4 cm) long in humans, separates the esophageal body from the stomach. In this segment the lumen is closed at rest but opens during swallowing. In humans, this segment lies in an opening through the diaphragm, the esophageal hiatus of the diaphragm. The stable contraction of the muscle at rest is due to a special property of the muscle itself, although there may also be some contribution from constantly active nerves to the region. After a swallow, luminal occlusion is transiently obliterated through relaxation of the muscle. The relaxation persists until esophageal peristalsis has been accomplished.

The temporal interrelationships of these events are maintained by a swallowing center in the brainstem. The swallowing center not only activates the somatic nerves to the striated muscle but also triggers the autonomic nerves, which initiate contractions in the esophageal body smooth muscle and relaxation of the lower esophageal sphincter. Also, peristalsis in the smooth muscle segment can be produced by esophageal distension. Thus, there are neural sensory receptors in the esophageal wall which can, through nerve reflex pathways, modify the process. [J.Ch.]

Stomach. Broadly speaking, there are three kinds of stomach found among the vertebrates: the simple stomach, as found in humans, the dog, and the great majority of higher and lower vertebrates; the complex stomach, as found in the ruminants; and the stomachs found in animals with a gizzard, such as birds and crocodiles. The basic functions are similar in all simple stomachs: storage, secretion of digestive fluids, and mixing.

Simple stomachs. The simple stomach, as found in humans, is a bag-shaped organ into which food passes from the esophagus. The partially digested food is afterward released piecemeal to be dealt with by the digestive processes in the small intestine. The main function of the stomach is to initiate the digestion of proteins. This digestion is only partial, but sufficient to disrupt the cells of foodstuffs and to release the intracellular contents, rendering them susceptible to digestion in the small intestine. The gastric secretions include mucus, which moistens the food and helps protect the gastric mucosa from the action of the digestive juices and from the action of hydrochloric acid and pepsinogen.

Hydrochloric acid is secreted in a surprisingly concentrated form. The fresh secretion has a pH of about 1.0, but dilution of the acid with other gastric contents gives an overall pH in the stomach of about 2.0 (which is only one-tenth as acid). This extreme acidity is important in several respects; it not only kills any live food, but also prevents bacterial putrefaction if a large meal is stored in the stomach for a considerable time. The low pH also denatures and precipitates soluble proteins, which then remain longer in the stomach. The most essential role of the gastric acid, however, is in providing suitably acid conditions for pepsin, a hydrolytic enzyme which splits proteins into smaller units but does not carry the digestion all the way to single amino acids.

Digestion of carbohydrates in the stomach is negligible, but sometimes a large starchy meal can form a dense mass in the stomach, in which the action of salivary ptyalin can continue for some time before the acid gastric juices can penetrate and deactivate the ptyalin. Digestion of fats proceeds to a small extent. The stomach secretes a small amount of lipase which is adequate to work on the type of fat found in egg yolk, for example. Additional active lipase is frequently regurgitated from the duodenum.

The mixing action of the stomach comes about through waves of powerful contraction which push the food toward the pyloric sphincter. If the sphincter remains closed, the inner part of the food mass in the antrum is pushed back into the body and thereby mixed thoroughly. When the stomach contents are digested enough to be largely fluid and if the first part of the small intestine is empty (or nearly so), the pyloric sphincter relaxes and the antral contraction empties that portion of the stomach. Contrary to what might be expected, the antral muscles do not eject the stomach contents in vomiting. This is actually accomplished by compression of the stomach against the liver and diaphragm by the muscles of the abdominal wall.

The stomach plays an extremely important role in the absorption of vitamin B_{12}, which is essential to the normal manufacture of red blood cells. The stomach secretes Castle's intrinsic factor, which combines with dietary B_{12} to form a complex that is resistant to digestion in the stomach and most of the small intestine. This complex is absorbed into the blood much further down in the intestine. *See* VITAMIN B_{12}.

Birds. The characteristic modification of the stomach in birds is a gizzard derived from the posterior portion of the stomach. The anterior portion, the proventriculus, is glandular and similar in function to the human stomach. In most birds there is probably a minimum of digestion in the proventriculus because the food reaches it in an undivided state and it is too small for prolonged storage. The food is rapidly passed on to the gizzard, where it is triturated for 1–12 min, depending on the consistency of the food. This mixes the food thoroughly with the digestive enzymes previously secreted in the proventriculus; digestion continues in the duodenum. In birds the biliary and pancreatic ducts enter the duodenum at its distal end, so that the full length of the duodenum is available for gastric digestion.

The gizzard is most highly developed in seed-eating birds, less so in birds of prey. Birds do consume grit and small rocks, which are retained in the gizzard. The gizzard is an advantage to birds because it enables them to utilize the high nutrient value of plant seed. Mammals do not derive the full benefit from this food source unless the seeds are first ground or cooked.

Ruminants. The ruminants include cows (and wild relatives), sheep, goats, deer, musk oxen, camels, and llamas. Of all the vertebrates, the ruminants show the most spectacular modification of the stomach. The anterior portion is enlarged to form a fermentation vat in which microorganisms break down the otherwise indigestible cellulose of the animal's vegetable diet, and make the products of the fermentation available for the nutrition of the animal.

The rumen and reticulum together form a large fermentation vat in which the ingesta are held in a very liquid suspension. Their total volume may be 80 gal (300 liters) in a large cow. The pH of the contents is about 6.0–7.5. Fermented material from the rumen-reticulum passes through the omasum, where it is subjected to squeezing and trituration, and then the omasum discharges into the abomasum, which is generally similar to the simple stomach of other mammals and is the only part of the ruminant's complex stomach that contains digestive glands.

When grazing, ruminants swallow the grass rapidly and it remains in the rumen for some time. Later, boli of rumen contents are regurgitated and chewed more thoroughly before being returned to the stomach. This permits fermentation to begin disrupting the cellulose structure before it is chewed. It is also likely that this practice of rumination has survival value in the wild state, where the animal can rapidly eat its fill on the open plain and then work it over in some safe retreat. [T.A.R.]

Small intestine. Digestion in the small intestine (that is, in the duodenum, jejunum, and ileum) depends on the secretions of the liver, pancreas, and intestinal glands, which provide liquid for dilution and solution of the food, a reducer of surface tension for the emulsification of fat, and enzymes for the breakdown of food into smaller molecules. Digestion also depends on a great deal of metabolic activity within the columnar cells lining the small intestine and on the movement of the intestine which mixes the food with the secretions and propels it in a caudal direction toward available absorptive surfaces.

The mixture delivered to the duodenum is strongly acid. One of the first functions of the duodenum is to bring the contents to, or close to, neutrality in order to facilitate the activity of pancreatic hormones. Alkaline duodenal secretions, bile, and pancreatic juice accomplish neutralization quickly. Once neutrality and osmotic equilibrium have been established, digestion proceeds rapidly.

Pancreatic secretion. Pure pancreatic juice is a colorless, clear or opalescent, rather viscous fluid, with a pH of 8.4–8.9. The inorganic constituents include sodium, potassium, and bicarbonate ions; the bicarbonate renders the juice alkaline. The organic constituents consist of various proteins which contain the enzymes of the juice. The three main enzyme systems in pancreatic secretions are proteolytic, amylolytic, and lipolytic. *See* PANCREAS.

Trypsin is the main proteolytic enzyme formed by the pancreas. It is secreted in an inactive form, trypsinogen. Trypsinogen does not become active until it enters the duodenum. This is a safety mechanism to prevent autodigestion of the pancreas. Chymotrypsinogen is another proteolytic enzyme that is secreted in an inactive form in the pancreatic juice. It is activated by trypsin to form chymotrypsin. This enzyme clots milk and hydrolyzes casein and gelatin.

Carbohydrates are acted upon by pancreatic amylase, or amylopsin, which acts like salivary amylase to hydrolyze starch to dextrin and then to maltose. A maltase capable of splitting maltose into glucose has been identified in pure pancreatic juice.

Fats are subjected to the action of pancreatic lipase, or steapsin, which hydrolyzes neutral fat into glycerol and fatty acids. The digestion of fat is enhanced considerably by the concomitant action of bile salts on the duodenal contents. Bile salts lower the surface tension between water and neutral fat so that the fat becomes emulsified into minute globules. Fat also aids its own digestion. Liberated fatty acids unite with alkali in the bowel to form sodium salts or soaps that have the same emulsifying action as the bile salts.

Bile. Bile is an orange alkaline fluid. The principal components of bile are bile salts, bile pigments (bilirubin and biliverdin), cholesterol, and inorganic salts. The bile salts and cholesterol have a digestive function, while the other components are waste products. The bile salts promote the emulsification of fat in the intestine and also keep cholesterol and lecithin in solution. Of the bile salts, 90% are reabsorbed in the small intestine and remetabolized by the body. *See* BILE ACID; BILIRUBIN.

Absorption. The average daily intake of protein in humans is about 1.5 oz (45 grams). This is broken down and absorbed completely. Gastrointestinal secretions and desquamated intestinal cells are important sources of intestinal protein and can exceed the daily oral intake in amount. The small intestine receives proteins after they have been broken down in the duodenum to amino acids and polypeptides. Products of the digestion of protein are absorbed in the first part of the jejunum under normal conditions. *See* AMINO ACIDS; PROTEIN METABOLISM.

The initial splitting of carbohydrates into disaccharides occurs in the upper gastrointestinal tract. More amylase is added to the duodenal contents by the pancreas. The disaccharides maltose, lactose, and sucrose are the products of carbohydrate digestion which enter the small intestine. Further digestion occurs in the columnar cells of the intestine; maltose is broken down to glucose, sucrose becomes glucose and fructose, and lactose is split to glucose and galactose. The sugars are absorbed rapidly and completely in the first part of the jejunum. *See* CARBOHYDRATE METABOLISM.

The absorption of dietary fat is almost complete, and very small amounts appear in the feces. Fat is absorbed mainly in the terminal part of the duodenum and upper jejunum, but the reserve capacity of the intestine to absorb fat is considerable. Pancreatic lipase hydrolyzes fat in the intestine first to diglycerides and then to glycerol and free fatty acids.

The gastrointestinal membrane plays little part in the normal control of the body's composition of fluid and electrolytes, even though the daily transfer across the intestinal membrane is very great. Gastric secretion, bile, pancreatic juice, and intestinal secretions provide 2 gal (8 liters) of fluid for the small intestine to deal with daily, yet only 6 in.^{3} (100 milliliters) of fluid is actually excreted in the feces each day.

Electrolytes pass both ways across the intestinal membrane in health, and the rate of this transfer can be demonstrated easily by isotopes. [I.D.A.J.]

Colon. The colon or large intestine may be described as a reservoir where water and food material that have escaped digestion and absorption in the small intestine may be stored and concentrated until the time for elimination. Although there is no evidence of any secretion of digestive enzymes by the colon, those enzymes that pass from the small intestine with the undigested food materials into the colon continue digestion for a time.

The colon and cecum of all mammals provide an excellent environment for the growth and development of bacteria. These bacteria act effectively on protein and carbohydrates that have escaped digestion and produce a number of volatile compounds and gases that may be either absorbed or excreted. They can also synthesize certain vitamins, especially the B vitamins.

The principal secretion of the cells of the walls of the cecum, colon, and rectum is mucin, which lubricates the accumulating feces and thus counteracts the loss of water that is absorbed from the colon. The normal water content of the passed feces varies considerably in different animals. Likewise, the volume and weight of feces passed per day vary from animal to animal and are related to the amount of undigestible cellulose consumed. The composition of fecal material varies with the nature of the diet consumed and the completeness of the digestion and absorption in the small intestine. Such substances as mucin, live and dead bacteria, and sloughed-off epithelial cells make up a considerable percentage of the feces.

For effective absorption of water and certain end products of digestion, it is necessary that the contents be well churned, squeezed, and moved to the walls of the colon and cecum. Following this, the concentrated material must be propelled down toward the rectum where it can be temporarily stored and finally expelled. [J.C.]

Digital computer Any device for performing mathematical calculations on numbers represented digitally; by extension, any device for manipulating symbols according to a detailed procedure or recipe.

This class of digital computers includes microcomputers, conventional adding machines and calculators, digital controllers used for industrial processing and manufacturing operations, store-and-forward data communication equipment, and electronic data-processing systems. *See* CALCULATORS; COMPUTER.

In this article emphasis is on electronic stored-program digital computers. These machines store internally many thousands of numbers or other items of information, and control and execute complicated sequences of numerical calculations and other manipulations on this information in accordance with instructions also stored in the machine.

A digital system can be considered from many points of view. At the lowest level it is a network of wires and mechanical parts whose voltages and positions convey coded information. At another level it is a collection of logical elements, each of which embodies certain rules, but which in combination can carry out very complex functions. At a still higher level, a digital system is an arrangement of functional units or building blocks which read (input), write (output), store, and manipulate information.

Codes. Numbers are represented within a digital computer by means of circuits that distinguish various discrete electrical signals on wires inside the machine. In practice, the most reliable and economical circuit elements distinguish between only two signal levels. These two-valued signals make it necessary to represent numbers and symbols using a corresponding base-two or binary system.

Data are stored and manipulated within a digital computer in units called words. The binary digits (called bits), which make up a word, may represent either a binary number or a collection of binary-coded alphanumeric characters.

Logical circuit elements. Two kinds of logical circuits are used in the design and construction of digital computers: decision elements and memory elements. A typical decision element provides a binary output as a function of two or more binary inputs. The AND circuit, for example, has two inputs and an output which is 1 only when both inputs are 1. A memory element stores a single bit of information and is set to the 1 state or reset to the 0 state, depending on the signals on its input lines. And because such a circuit can be caused alternately to store 0's and 1's from time to time, a memory element is commonly called a flip-flop. These two basic logical elements are all that are required to construct the most elaborate and complex digital arithmetic and control circuits.

Physical components. The actual physical components which were used to realize conceptual gates and flip-flops in some specific piece of equipment are dependent on the status of electronic technology at the time the equipment was designed. In the 1950s the earliest commercial computers used vacuum tubes, resistors, and capacitors as components. A flip-flop typically required a dozen or more such components in these first-generation computers. Between the late 1950s and middle 1960s, solid-state transistors and diodes replaced the vacuum tubes, and the resulting second-generation systems were considerably more reliable than their first-generation predecessors; they were also smaller and consumed less power. But the number of electronic components per conceptual logical component remained about the same—a dozen or more for a flip-flop.

Since the mid-1960s the integrated circuit (IC) has been the principal logical building block for digital systems. Integrated circuit technology has consistently improved, and typical large-scale-integration (LSI) circuits now contain thousands of flip-flops and gates. *See* INTEGRATED CIRCUITS.

System building blocks. On a completely different conceptual level, a digital computer can be regarded as being composed of functional, system building blocks, containing (among other things) subassemblies of the fundamental logical components. A computer viewed at this level may be described in an oversimplified fashion by the diagram of Fig. 1. The computation and control block (often called the central processing unit, or CPU) is constructed entirely of logical elements of the kind described above. The main memory, which may store from a few thousand to several million binary digits, and the input/output and auxiliary memory devices (the so-called peripheral equipment) are specialized devices available over a range of speeds and operating characteristics.

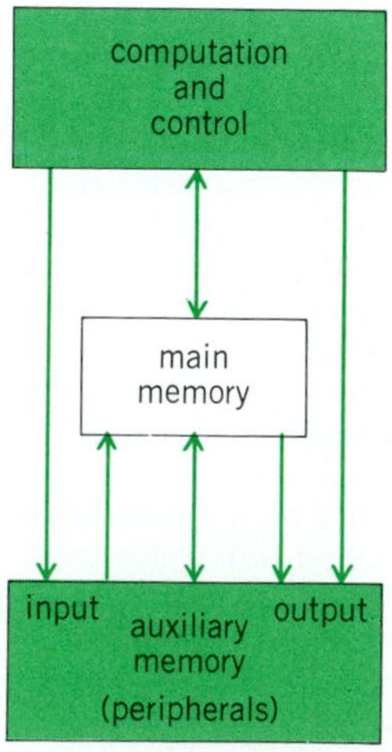

Fig. 1. Block diagram of a digital computer.

Main memory is a building block capable of storing data or instructions in bulk for use by the computation and control portion of the computer. The important characteristics of a memory are capacity, access time, and cost. Starting in the early 1970s, the integrated circuit memory was introduced, and is now the most widely used technology. *See* COMPUTER STORAGE TECHNOLOGY.

Input/output and auxiliary memory peripherals represent the other major computer building blocks. Equipment is now available, and has been so from the beginning, for feeding information to the computer from paper tape and punched cards, and for receiving data from the computer and printing it, or punching it on tape or cards. But in the intervening years, designers have provided additional output devices which record computer data on microfilm, or plot data on graphs, or use data to control physical devices such as valves or rheostats. They have also designed input equipment which feeds the computer data from laboratory instruments, and from devices which scan documents and "read" printed characters. Data can be transmitted to and from the computer over ordinary telephone lines, and a wide variety of devices, generally called terminals, make it possible for people to send data to, or receive requested data from, a computer system located hundreds or thousands of miles away. *See* CHARACTER RECOGNITION; COMPUTER GRAPHICS; DATA COMMUNICATIONS.

Stored program computer. The concept of the stored-program computer is simple and can be described with reference to Fig. 1. Main memory contains, in addition to data and the results of intermediate computations, a set of instructions (or orders, or commands, as they are sometimes called); these specify how the computer is to operate in solving some particular problem. The computation and control section reads these instructions from the memory one by one and performs the indicated operations on the specified data. The instructions can control the reading of data from input or auxiliary memory peripherals, and (when the prescribed computations are completed) can send the result to auxiliary memory, or to output devices where it may be printed, punched, displayed, plotted, and so forth. The feature that gives this form of computer organization its great power is the ease with which instructions can be changed; the particular calculations carried out by the computer are determined entirely by a sequence of instructions stored in the computer's memory; that sequence can be altered completely by reading a new set of instructions into the memory through the computer input equipment. [M.Ph.]

Each word in storage is assigned an address in order to refer to it, in a manner analogous to postal addresses. Each instruction contains two basic parts: (1) a coded number that dictates what operation is to be performed, for example, ADD, MULTIPLY, and STORE; and (2) the address of the word (data) on which to perform that operation.

The stored-program computer gains additional power from two additional concepts. The first concept is derived naturally in that the words of information referenced by an instruction can be other instruction words; thus, a computer can manipulate its own instructions and alter them dynamically, particularly in their address portions, during the execution of a program. The second concept concerns the ability to branch, based on the condition of the accumulator at any given moment. Basically, the machine can interrogate the accumulator and, on the basis of its contents being negative, zero, or positive, call for operating on the instruction counter so that the next executed instruction is not at the location that is one more than the last executed location, but is any other desired location. This branch of control may also be called for arbitrarily, that is, not based on the condition of the accumulator. For the avail-

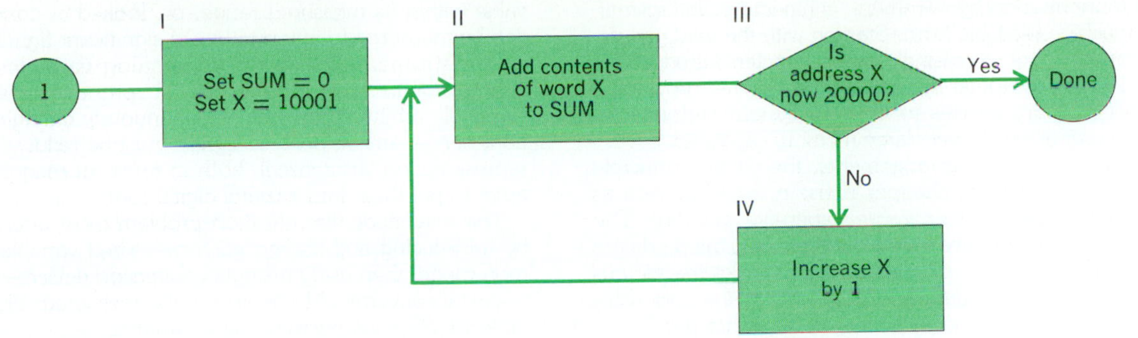

Fig. 2. Flow chart for summing the contents of 10,000 words.

able branch instructions the address portion is then the location of the next instruction rather than that of a data word.

The logic of any problem situation can be expressed graphically in a flow chart. For the problem of summing the contents of 10,000 words, the flow chart of Fig. 2 applies. This is an example of the programmed loop technique.

Programmers prefer to work in languages that are at a higher level than the machine language. The use of mnemonic operation codes and the replacement of all absolute machine addresses by symbols greatly speed up the work of the programmer at the modest cost of an extra computer run whose sole function is the translation back to the machine language. The translating program is called an assembler, and the language used is assembly language. Generally, assembly language follows the format of the machine language and is one-for-one with it; that is, for every instruction written in assembly language, one machine language instruction will be produced.

Most programming is done in the language of compilers, for example, FORTRAN or COBOL. Compiler language permits a format that fits the problems rather than the machine, and is many-to-one. As with assemblers, a separate machine run is needed to translate from the compiler language to the language of the machine; the translating program is called a compiler.

The looping technique shown in Fig. 2 is one of the two basic building blocks of the programmer. The other is the closed subroutine. Both techniques have the purpose of allowing a small number of instructions to be executed many times. The instructions of this program can be stored once, and simple arrangements can be made in the main program to cause their linkage and return at each of the points at which the operation is needed. [F.J.G.]

Computer characteristics. A computer installation is complex. Consequently it is difficult to describe a system or to compare the characteristics of two systems without listing their instruction types and describing their modes of operation at some length. Nevertheless, certain important descriptors are commonly used for comparison purposes and are shown in the table, where salient characteristics of two typical systems are shown. Memory cycle time is the time required to read a word from main memory. Add time is the time required to perform an addition, including the time necessary to extract the addition instruction itself and the operand from memory.

Typical computer characteristics

Characteristics	Typical systems	
	Large	Small
Memory cycle time	0.08 μs	0.90 μs
Add time	0.17 μs	2.84 μs
Main memory storage capacity	750,000 words	128,000 words
Word length	64 bits	16 bits

Evolution of capabilities. The process by which new circuit and peripheral equipment technologies led to the development of a series of generations of computers was discussed above. But simultaneous with the changes in technology, there came changes in the structure or architecture of computers. These changes were introduced to improve the capability and efficiency of systems, as designers came to understand how computers were actually used.

In the first generation of computer equipment only one activity could be carried out at a time. Between jobs the computer was idle while an operator made ready for the next task. When the operator was ready, the program was read into the computer from some input device and the input data were then loaded. The program operated upon the data and performed necessary calculations. When the calculations were complete, the computer printed out answers, and the operator took steps to set up the next problem.

This series of operations was inefficient, and the designers of second-generation equipment removed some of the inefficiency by arranging input and output operations to be performed directly between the input/output peripherals and the computer memory without interfering with computations. As a result, second-generation computers were able to perform computations while reading in data and printing out replies, and efficiency was greatly enhanced.

First-generation equipment was most efficient while performing tedious and lengthy computations. The input/output capabilities of the second generation made them useful in applications where large volumes of data had to be handled with relatively little computation—applications such as billing, payroll, and inventory control. At the same time, the great capability and increased reliability of second-generation systems encouraged engineers to apply them to situations where the computer acts as a control element.

In the mid-1960s a new set of trends in computer applications was becoming apparent, and a third generation of computer systems became available to cope with those trends. The usefulness and flexibility of the stored-program computer, together with its improved cost-performance ratio, made it apparent that the computer had the basic power to perform a great variety of small and large tasks simultaneously. A computer system which serves a number of users in this way is called a time-sharing system. To perform in such complex applications, third-generation computers required elaborations of the features found in second-generation computers.

The operating system is the set of supervisory programs which manages the system, keeps it operating efficiently, and takes over many of the scheduling and monitoring functions which had previously been the function of the computer operator.

Computer memory has become increasingly important with the passing generations, as users have found it useful to store more and more business, government, engineering, and scientific data in machine-readable form. Starting with the third generation of computers, but increasingly with computers introduced in the early 1970s, computer-system architects have provided a hierarchy of memory devices to improve system performance and give the user access to very large memory capacities.

In all these hierarchical arrangements, the guiding principle is that the larger, slower, cheaper memory supplies data as needed to the smaller, faster, more expensive memory. The various levels of memory are invisible to the user; the hardware and, where necessary, the operating system make the various hierarchical levels deliver data and programs to the user without the necessity for any special action on his or her part.

The versatility of the digital computer has led to its application in a wide variety of industries and activities. The evolution of the integrated circuit has made it possible for designers to provide even cheaper systems, and as a result the computer has become economical for use even in small organizations. The development of the microcomputer—the stored-program computer on an integrated-circuit chip—has made possible the introduction of the personal computer, which can supply entertainment, educational, record-keeping, and computing capabilities in the home at a relatively low cost. *See* MICROCOMPUTER; MICROPROCESSOR. [M.P.]

Digital control The use of digital or discrete technology to maintain conditions in operating systems as close as possible to desired values despite changes in the operating environment. Traditionally, control systems have utilized analog components, that is, controllers which generate time-continuous outputs (volts, pressure, and so forth) to manipulate process inputs and which operate on continuous signals from instrumentation measuring process variables (position, temperature, and so forth). In the 1970s, the use of discrete or logical control elements, such as fluidic components, and the use of programmable logic controllers to automate machining, manufacturing, and production facilities became widespread. In parallel with these developments has been the accelerating use of digital computers in industrial and commercial applications areas, both for logic-level control and for replacing analog control systems. The development of inexpensive mini- and microcomputers with arithmetic and logical capability orders of magnitude beyond that obtainable with analog and discrete digital control elements has resulted in the rapid substitution of conventional control systems by digital computer-based ones. With the introduction of microcomputer-based control systems into major consumer products areas (such as automobiles and video and audio electronics), it is clear that the digital computer will be widely used to control objects ranging from small, personal appliances and games up to large, commercial manufacturing and production facilities. *See* MICROCOMPUTER; PROGRAMMABLE CONTROLLERS.

The object that is controlled is usually called a device or, more inclusively, process. A characteristic of any digital control system is the need for a process interface to mate the digital computer and process, to permit them to pass information back and forth.

Measurements of the state of the process often are obtained naturally as one of two switch states; for example, a part to be machined is in position (or not), or a temperature is above (or below) the desired temperature. Control signals sent to the process often are expressed as one of two states as well; for example, a motor is turned on (or off), or a valve is opened (or closed). Such binary information can be communicated naturally to and from the computer, where it is manipulated in binary form. For this reason the binary or digital computer/process interface usually is quite simple.

Process information also must be dealt with in analog form; for example, a variable such as temperature can take on any value within its measured range, or, looked at conceptually, it can be measured to any number of significant figures by a suitable instrument. Furthermore, analog variables generally change continuously in time. Digital computers are not suited to handle arbitrarily precise or continuously changing information; hence, analog process signals must be reduced to a digital representation (discretized), both in terms of magnitude and in time, to put them into a useful digital form.

The magnitude discretization problem most often is handled by transducing and scaling each measured variable to a common range, then using a single conversion device—the analog-to-digital converter (ADC)—to put the measured value into digital form. *See* ANALOG-TO-DIGITAL CONVERTER.

Discretization in time requires the computer to sample the signal periodically, storing the results in memory. This sequence of discrete values yields a "staircase" approximation to the original signal, on which control of the process must be based. Obviously, the accuracy of the representation can be improved by sampling more often, and many digital systems simply have incorporated traditional analog control algorithms along with rapid sampling. However, newer control techniques make fundamental use of the discrete nature of computer input and output signals. Analog outputs from a computer most often are obtained from a digital-to-analog converter (DAC), a device which accepts a digital output from the computer, converts it to a voltage in several microseconds, and latches (holds) the value until the next output is converted. Usually a single DAC is used for each output signal. *See* DIGITAL-TO-ANALOG CONVERTER.

In order to be used as the heart of a control system, a digital computer must be capable of operating in real time. Except for very simple microcomputer applications, this feature implies that the machine must be capable of handling interrupts, that is, inputs to the computer's internal control unit which, on change of state, cause the computer to stop executing some section of program code and begin executing some other section. The ability to initiate operations on schedule and to respond to process interrupts in a timely fashion is the very basis of real-time computing; this feature must be available in any digital control system.

Computer control systems for large or complex processes may involve complicated programs with many thousands of computer instructions. Several routes have been taken to mitigate the difficulty of programming control computers. One approach is to develop a single program which utilizes data supplied by the user to specify both the actions to be performed on the individual process elements and the schedule to be followed. Another approach is to develop a rather sophisticated operating system to supervise the execution of user programs, scheduling individual program elements for execution as specified by the user or needed by the process. *See* DIGITAL COMPUTER.

Many applications, particularly machining, manufacturing, and batch processing, involve large or complex operating schedules. Invariably, these can be broken down into simple logical sequences. Some applications—in the chemical process industries, in power generation, and in aerospace areas—require the use of traditional automatic control algorithms.

Attempts to expand the digital control medium through development of strictly digital control algorithms is an important and continuing trend. Such algorithms typically attempt to exploit the sampled nature of process inputs and outputs, significantly decreasing the sampling requirements of the algorithm. *See* CONTROL SYSTEMS. [D.A.Me.]

Digital counter An instrument which, in its simplest form, provides an output that corresponds to the number of pulses applied to its input. Counters may be categorized into two types: the Moore machine or the Mealy machine. The simpler counter type, the Moore machine, has a single count input (also called the clock input or pulse input), while the Mealy machine has additional inputs that alter the count sequence. Digital counters take many forms, including geared mechanisms (tape counters and odometers are examples), relays (old pinball machines and old telephone switching systems), vacuum tubes (old test equipment), and solid-state semiconductor circuits (most modern electronic counters).

Most digital counters operate in the binary number system, since binary is easily implemented with electronic circuitry. Binary allows any integer (whole number) to be represented as a series of binary digits, or bits, where each bit is either a 0 or 1 (off or on, low or high, and so forth). *See* NUMBER SYSTEMS.

Digital counters are found in much modern electronic equipment, especially equipment that is digitally controlled or has digital numeric displays. A frequency counter, as a test instrument or a channel frequency display on a radio tuner, consists simply of a string of decade counters that count the pulses of an input signal for a known period of time, and display that count on a seven-segment display. A digital voltmeter operates by using nearly the same idea, except that the counter counts a known frequency for a period of time proportional to the input voltage. *See* ANALOG-TO-DIGITAL CONVERTER.

A digital watch contains numerous counter/dividers in its large-scale integration (LSI) chip, usually implemented with complementary metal oxide semiconductor (CMOS) technology.

Digital computers may contain counters in the form of programmable interval timers that count an integral number of clock pulses of known period, and then generate an output at the end of the count to signal that the time period has expired. Most of the counters in a microprocessor consist of arithmetic logic units that add one many-bit number to another, storing the results in a memory location. The program and data counters are examples of this kind of counter. *See* DIGITAL COMPUTER; MICROPROCESSOR.

Counters have progressed from relays to light-wavelength-geometry very-large-scale-integrated circuits. There are several technologies for building individual digital counters. Single counters are available as integrated circuit chips in emitter-coupled logic, transistor-transistor logic, and CMOS. [M.E.W.]

Digital filter Any digital computing means that accepts as its input a set of one or more digital signals from which it generates as its output a second set of digital signals. While being strictly correct, this definition is too broad to be of any practical use, but it does demonstrate the possible extent of application of digital-filter concepts and terminology.

Capabilities. Digital filters can be used in any signal-manipulating application where analog or continuous filters can be used. Because of their utterly predictable performance, they can be used in exacting applications where analog filters fail because of time- or other parameter-dependent coefficient drift in continuous systems. Because of the ease and precision of setting the filter coefficients, adaptive and learning digital filters are comparatively simple and particularly effective to implement. As digital technology becomes more ubiquitous, digital filters are increasingly acknowledged as the most versatile and cost-effective solutions to filtering problems.

The number of functions that can be performed by a digital filter far exceeds that which can be performed by an analog, or continuous, filter. By controlling the accuracy of the calculations within the filter (that is, the arithmetic word length), it is possible to produce filters whose performance comes arbitrarily close to the performance expected of the perfect models. For example, theoretical designs that require perfect cancellation can be implemented with great fidelity by digital filters.

Linear difference equation. The digital filter accepts as its input signals numerical values called input samples and produces as its output signal numerical values called output samples. Each output sample at any particular sampling instant is a weighted sum of present and past input samples, and past output samples. If the sequence of input samples is x_n, x_{n+1}, x_{n+2}, . . . , then the corresponding sequence of output samples would be y_n, y_{n+1}, y_{n+2}, ... *See* DIFFERENCE EQUATION.

From this simple time-domain expression, a considerable number of definitions can be constructed. If the filter coefficients (the a's and the *b*'s) are independent of the *x*'s and *y*'s, this digital filter is a linear filter. If the a's and *b*'s are fixed, this is a linear time-invariant (LTI) filter. The order of the filter is given by the largest of the subscripts among the *a*'s and *b*'s, that is, the larger of *M* and *N*. If the *b*'s are all zero (that is, if the output is the weighted sum of present and past input samples only), the digital filter is referred to as a nonrecursive (having no feedback) or finite impulse response (FIR) filter because the response of the filter to an impulse (actually a unit pulse) input is simply the sequence of the "*a*" coefficients. If any value of *b* is nonzero, the filter is recursive (having feedback) and is generally an infinite impulse response (IIR) filter.

If the digital filter under consideration is not a linear, time-invariant filter, the transfer function cannot be used.

Transfer functions. Although the time-domain difference equation is a useful description of a filter, as in the continuous-domain filter case, a powerful alternative form is the transfer function. The information content of the transfer function is the same as that of the difference equation as long as a linear, time-invariant system is under consideration. A difference equation is converted to transfer-function form by use of the *z* transform. The *z* transform is simply the Laplace transform adapted for sampled systems with some shorthand notation introduced. *See* LAPLACE TRANSFORM; Z TRANSFORM.

Adaptive filters. So far only LTI filters have been discussed. An important class of variable-parameter filter change their coefficients to minimize an error criterion. These filters are called adaptive because they adapt their parameters in response to changes in the operating environment. An example is an FIR digital filter whose coefficients are continually adjusted so that the output will track a reference signal with minimum error. The performance criterion will be the minimization of some function of the error. *See* ELECTRIC FILTER. [S.A.Wh.]

Digital navigation system An electronic digital device that provides an aircraft, spacecraft, or surface vehicle with frequently updated positional information in two or three dimensions in relation to some reference source and translates this information into a form usable either by human operators or directly by vehicle control systems. The principal difference between digital and analog navigation systems is in the way that they manage information about the position of the vehicle.

Both analog and digital systems derive input from one or more reference sources which provide information about the position, location, velocity, and direction of travel of the vehicle. Both system types then calculate the expected, revised position of the vehicle based on this information and produce information which may be used to direct the vehicle toward a selected path of travel. In the case of analog systems, the navigation device creates an electrical analog of the information about the state of the vehicle and then converts this into a display for the operator or into control signals which are fed to actuators. The digital navigation system, on the other hand, converts all the information from electrical analog to digital numeric data and then uses these numeric data to calculate position status and control information. This "digitization" permits operation under the control of a stored digital program, which provides the system with the ability to evaluate multiple inputs automatically, without human intervention. The digital program can also include a planned flight profile for an aircraft which contains routing, economical airspeed and altitude planning, avoidance of weather problems, and a timed arrival at a prespecified destination. The actual computer activity consists of a series of comparisons between the actual and the desired (programmed) state of the aircraft until arrival at the programmed destination. *See* DIGITAL COMPUTER.

When combined with other digital aircraft or vehicle control systems (such as in surface ships or submarines), digital navigation systems permit fully automatic control of these vehicles. In these fully automatic systems, changes in vehicle position are accomplished by altering the program of the digital navigation system, rather than by manipulation of the vehicle controls. This changes the role of the vehicle operator from active participant in the control process to manager and monitor of the progress of the system. *See* NAVIGATION. [F.Wh.]

Digital recording The application of digital techniques to the recording and reproduction of music on disk records. During recording, the output of the microphone amplifier is converted to a stream of digital bits. Typically the sampling rate lies between 44,000 and 55,000 samples per second with 16 bits per sample. The digitized program is recorded on magnetic tape. Special encoding schemes involving the insertion of additional bits permit correction in playback of errors due to loss or misinterpretation of bits. Thus the digital signal may be recorded through many generations without deterioration of quality. Editing and mixing are done with the signals in digital form. *See* ANALOG-TO-DIGITAL CONVERTER.

After editing and mixing, the digital signals are transferred from the tape to a disk rotating on the disk-recorder turntable. The recorded signal is carried in the form of microscopic indentations in a spiral track on the surface of the disk. Typically the indentations are less than 1 micrometer across and 0.15 micrometer deep. The disk format and the recording techniques are those developed for video-disk recording of television signals. The master disk is used to form stampers for pressing vinyl records following the same processes used in the production of conventional phonograph records.

Players for the digital records resemble, and in some cases may be identical to, video-disk players. Depending on the record format offered by the manufacturers, the pickup in the player may be a capacitive, mechanical, or optical type. The digital output of the pickup and player circuits is error-corrected and stored temporarily in a memory. It is removed from the memory at a bit rate controlled by a precise quartz crystal, and is converted to analog audio signals to drive loudspeakers. Because the output bit rate is precisely controlled, any effects due to small fluctuations in speed of the tape recorders or of the disk turntables are removed from the analog output. *See* DIGITAL-TO-ANALOG CONVERTER. [J.G.W.]

Digital-to-analog converter A device for converting information in the form of combinations of discrete states of a signal, often representing binary number values, to information in the form of the value or magnitude of some characteristics of a signal, in relation to a standard or reference. Most often, it is a device which has electrical inputs representing a parallel binary number, and an output in the form of voltage or current.

Digital-to-analog (D/A) converters (sometimes called DACs) are used to present the results of digital computation, storage, or transmission, typically for graphical display or for the control of devices that operate with continuously varying quantities. D/A converter circuits are also used in the design of analog-to-digital converters that employ feedback techniques, such as successive-approximation and counter-comparator types. In such applications, the D/A converter may not necessarily appear as a separately identifiable entity. *See* COMPUTER GRAPHICS; DATA COMMUNICATIONS; DIGITAL COMPUTER.

The fundamental circuit of most D/A converters involves a voltage or current reference; a resistive "ladder network" that derives weighted currents or voltages, usually as discrete fractions of the reference; and a set of switches, operated by the digital input, that determine which currents or voltages will be summed to constitute the output.

The output of the D/A converter is proportional to the product of the digital input value and the reference. In many applications, the reference is fixed, and the output bears a fixed proportion to the digital input. In other applications, the reference, as well as the digital input, can vary; a D/A converter that is used in these applications is thus called a multiplying DAC. It is principally used for imparting a digitally controlled scale factor, or "gain," to an analog input signal applied at the reference terminal. *See* AMPLIFIER; ANALOG COMPUTER. [D.H.S.]

Digitalis A genus of the figwort family (Scrophulariaceae) ranging from the Canary Islands to central Asia. Foxglove (*Digitalis purpurea*), a native of western Europe, is the source of the important drug digitalis, much used in the treatment of heart disorders. The active ingredient, digitalin, slows and regulates the heartbeat, improving the tone and rhythm, and making the contractions more effective. *See* SCROPHULARIALES. [P.D.St./E.L.C.]

Dihedron A geometric figure formed by two half planes that are bounded by the same straight line. This line is called the edge of the dihedron. Planes perpendicular to the edge cut

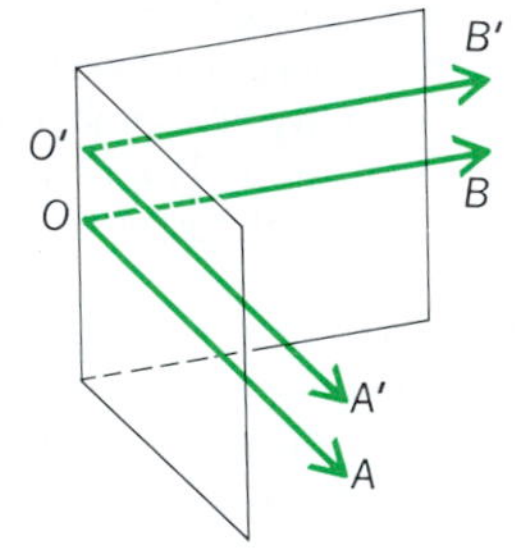

Dihedron and dihedral angle.

the dihedron in equal plane angles *AOB* and *A′O′B′* whose common measure is called the dihedral angle of the dihedron (see illustration). *See* ANGLE. [J.S.F.]

Dill A small annual or biennial herb, *Anethum graveolens* (Umbelliferae), of high repute among ancient peoples and now cultivated in Europe, India, and the United States. In the United States dill is used mainly as a flavoring for pickles. In France, India, and other countries it is used in soups and sauces, stews, and other dishes. Both the fruits and oil are used in medicine. *See* APIALES. [P.D.St./E.L.C.]

Dilleniales An order of flowering plants, division Magnoliophyta (Angiospermae), subclass Dilleniidae, class Magnoliopsida (dicotyledons). It consists of 3 families: the Dilleniaceae, the Paeoniaceae, and Crossosomataceae. Within its subclass the order is marked by separate carpels; numerous stamens; and seeds that have a well developed endosperm and are provided with an aril (a partial or complete covering or appendage, usually derived from the funiculus of the ovule). The garden peony (*Paeonia lactiflora*) is a familiar member of the order. *See* DILLENIIDAE. [A.Cr.]

Dilleniidae A large subclass of the class Magnoliopsida (dicotyledons) of the division Magnollophyta (Angiospermae), the flowering plants, consisting of 12 orders, 69 families, and nearly 24,000 species. The subclass is morphologically ill-defined, but most of the orders and species have the carpels united to form a compound pistil. The petals are either separate or joined into a sympetalous corolla. The stamens, when numerous, are initiated in centrifugal sequence. Many of the species have numerous ovules on parietal placentas; that is, the placentas are borne along the walls of an ovary which is usually with a single chamber.

The largest orders included in the subclass are the Violales, Capparales, Malvales, Theales, and Ericales. Other orders include the Dilleniales, Lecythidales, Sarraceniales, Salicales, Diapensiales, Ebenales, and Primulales. See separate articles on each order. *See* MAGNOLIOPHYTA. [A.Cr.]

Dimensional analysis A technique that involves the study of dimensions of physical quantities. Dimensional analysis is used primarily as a tool for obtaining information about physical systems too complicated for full mathematical solutions to be feasible. It enables one to predict the behavior of large systems from a study of small-scale models. It affords a convenient means of checking mathematical equations. Finally, dimensional formulas provide a useful cataloging system for physical quantities.

All the commonly used systems of units in physical science have the property that the number representing the magnitude of any quantity (other than purely numerical ratios) varies inversely with the size of the unit chosen. Thus, if the length of a given piece of land is 300 ft, its length in yards is 100. The ratio of the magnitude of 1 yd to the magnitude of 1 ft is the same as that of any length in feet to the same length in yards, that is, 3. The ratio of two different lengths measured in yards is the same as the ratio of the same two lengths measured in feet, inches, miles, or any other length units. This universal property of unit systems, often known as the absolute significance of relative magnitude, determines the structure of all dimensional formulas. *See* UNITS OF MEASUREMENT.

In defining a system of units for a branch of science such as mechanics or electricity, certain quantities are chosen as fundamental and others as secondary, or derived. The choice of the fundamental units is always arbitrary and is usually made on the basis of convenience in maintaining standards. In mechanics the fundamental units most often chosen are mass, length, and time.

It can be proved that every secondary quantity which satisfies the condition of the absolute significance of relative magnitude is expressible as a product of powers of the primary quantities. Such an expression is known as the dimensional formula of the secondary quantity. There is no requirement that the exponents be integral.

The technique of dimensional analysis has several important applications. It is intuitively obvious that only terms whose dimensions are the same can be equated. The equation 10 kg = 10 m/s, for example, makes no sense. A necessary condition for the correctness of any equation is that the two sides have the same dimensions. This is often a help in the verification of complicated analytic expressions. Of course, an equation can be correct dimensionally and still be wrong by a purely numerical factor.

The application of dimensional analysis to the derivation of unknown relations depends upon the concept of completeness of equations. An expression which remains formally true no matter how the sizes of the fundamental units are changed is said to be complete. Assume a group of n physical quantities $x_1, x_2, \ldots, x_n$, for which there exists one and only one complete mathematical expression connecting them, namely, $\phi(x_1, x_2, \ldots, x_n) = 0$. Some of the quantities $x_1, x_2, \ldots, x_n$ may be dimensional constants. Assume further that the dimensional formulas of the n quantities are expressed in terms of m fundamental quantities $a, \alpha, \beta, \gamma, \ldots$. Then it will always be found that this single relation ϕ can be expressed in terms of some arbitrary function F of $n - m$ independent dimensionless products $\pi_1, \pi_2, \ldots, \pi_{n-m}$, made up from among the n variables, as in the equation below. This is known as the π

$$F(\pi_1, \pi_2, \ldots, \pi_{n-m}) = 0$$

theorem. It was first rigorously proved by E. Buckingham. The main usefulness of the π theorem is in the deduction of the form of unknown relations. The procedure is particularly useful in hydraulics and aeronautical engineering, where detailed solutions are often extremely complicated.

A further application of dimensional analysis is in model design. Often the behavior of large complex systems can be deduced from studies of small-scale models at a great saving in cost. In the model each parameter is reduced in the same proportion relative to its value in the original system. *See* MODEL THEORY.

Dimensional formulas also provide a convenient shorthand notation for representing the definitions of secondary quantities and are helpful in changing units from one system to another. [J.W.St.]

Dimensioning Assigning of dimensions on a mechanical drawing. A dimension on a drawing is a labeled measure in a straight line of the breadth, height, or thickness of a part, the angular position of a line, or the location of a detail such as a hole or boss. Dimensions specify the size and shape of the part as required by the designer and aid the worker in constructing the part.

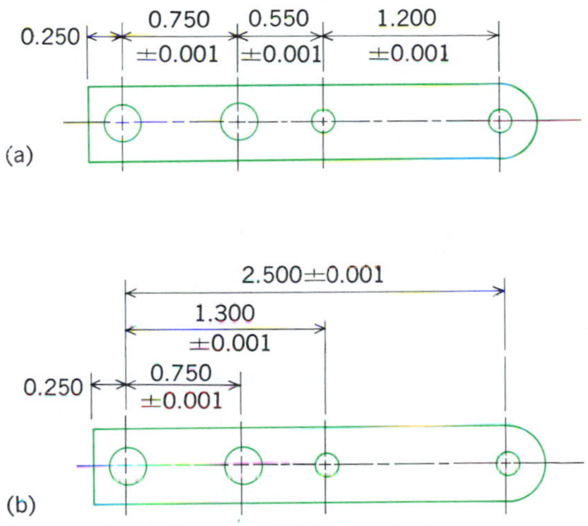

Two standard methods for marking dimensions on mechanical drawings. (*a*) Point-to-point dimensioning. (*b*) Datum dimensioning.

Two plans of dimensioning are used on mechanical drawings, as illustrated. One is called point-to-point or chain dimensioning; each length is dimensioned from the end of the preceding one. These dimensions are usually taken directly from the designer's sketch; however, tolerances are cumulative and, rather than averaging out, usually add or subtract so that intermediate clearances may be disturbed. In the other plan, called datum dimensioning or the reference-line method, all dimension lines in each direction extend from a datum or reference line or plane which is usually the first machined surface. Here, tolerances are not cumulative and a mating part will be within the specified limits. *See* DRAFTING; TOLERANCE. [P.H.B.]

Dimensionless groups A dimensionless group is any combination of dimensional or dimensionless quantities possessing zero overall dimensions. Dimensionless groups are frequently encountered in engineering studies of complicated processes or as similarity criteria in model studies. A typical dimensionless group is the Reynolds number (a dynamic similarity criterion), $N_{Re} = VD\rho/\mu$. Since the dimensions of the quantities involved are velocity V: $[L/\theta]$; characteristic dimension D: $[L]$; density ρ: $[M/L^3]$; and viscosity μ: $[M/L\theta]$ (with M, L, and θ as the fundamental units of mass, length, and time), the Reynolds number reduces to a dimensionless group and can be represented by a pure number in any coherent system of units. *See* DYNAMIC SIMILARITY.

Many important problems in applied science and engineering are too complicated to permit completely theoretical solutions to be found. However, the number of interrelated variables involved can be reduced by carrying out a dimensional analysis to group the variables as dimensionless groups. *See* DIMENSIONAL ANALYSIS.

The advantages of using dimensionless groups in studying complicated phenomena include:

1. A significant reduction in the number of "variables" to be investigated; that is, each dimensionless group, containing several physical variables, may be treated as a single compound "variable," thereby reducing the number of experiments needed as well as the time required to correlate and interpret the experimental data.
2. Predicting the effect of changing one of the individual variables in a process (which it may be impossible to vary much in available equipment) by determining the effect of varying the dimensionless group containing this parameter (this must be done with some caution, however).
3. Making the results independent of the scale of the system and of the system of units being used.
4. Simplifying the scaling-up or scaling-down of results obtained with models of systems by generalizing the conditions which must exist for similarity between a system and its model.
5. Deducing variation in importance of mechanisms in a process from the numerical values of the dimensionless groups involved; for instance, an increase in the Reynolds number in a flow process indicates that molecular (viscous) transfer mechanisms will be less important relative to transfer by bulk flow ("inertia" effects), since the Reynolds number is known to represent a measure of the ratio of inertia forces to viscous forces. *See* FROUDE NUMBER; KNUDSEN NUMBER; MACH NUMBER; REYNOLDS NUMBER; UNIT OF MEASUREMENT.

[J.Ca.; G.D.F.]

Dimensions (mechanics) Length, mass, time, or combinations of these quantities serving as an indication of the nature of a physical quantity. Quantities with the same dimensions can be expressed in the same units. For example, although speed can be expressed in various units such as miles/hour, feet/second, and meters/second, all these speed units involve the ratio of a length unit to a time unit; hence, the dimensions of speed are the ratio of length L to time T, usually stated as LT^{-1}. The dimensions of all mechanical quantities can be expressed in terms of L, T, and mass M. The validity of algebraic equations involving physical quantities can be tested by a process called dimensional analysis; the terms on the two sides of any valid equation must have the same dimensions. *See* DIMENSIONAL ANALYSIS; UNITS OF MEASUREMENT. [D.Wi.]

Dimethyl sulfoxide A versatile solvent (formula C_2H_6OS) abbreviated DMSO, used industrially and in chemical laboratories as a medium for carrying out chemical reactions. Its uses have been extended to that of a chemical reagent where DMSO itself is involved in a chemical change. It is the simplest member of a class of organic compounds which are typified by the polar sulfur-oxygen bond represented in the resonance hybrid shown below. The molecule is pyramidal in

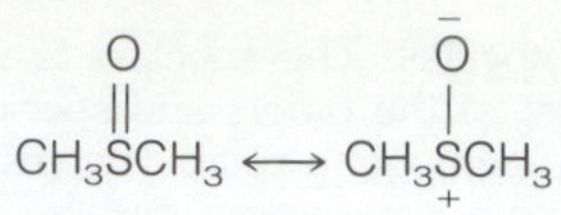

shape with the oxygen and the carbons at the corners.

DMSO is a colorless, odorless (when pure), and very hygroscopic stable liquid (bp 189°C, mp 19.5°C). It is manufactured commercially by reacting the black liquor from digestion in the kraft pulp process, with molten sulfur to form dimethyl sulfide which is then oxidized with nitrogen tetroxide. This highly polar aprotic solvent is water- and alcohol-miscible and will dissolve most polar organic compounds and many inorganic salts.

As an aprotic solvent, DMSO strongly solvates cations, leaving a highly reactive anion. Thus in DMSO, basicity and nucleophilicity is enhanced, and it is a superior solvent for many elimination, nucleophilic substitution, and solvolysis reactions in which nucleophile and base strength are important. Nucleophilic substitution reactions in which halogens or sulfonate esters are displaced by anions such as cyanide, alkoxide, thiocyanate, azide, and others are accelerated 1000 to 10,000 times in DMSO over the reaction in aqueous alcohol. Dimethyl sulfoxide is also superior to protic solvents as a media for elimination reactions. Its high dielectric constant makes DMSO a useful solvent for dissolving resins, polymers, and carbohydrates. It is employed commercially as a spinning solvent in the manufacture of synthetic fibers.

The apparent low toxicity and high skin permeability have led to extensive studies in numerous biological systems, including humans. Inorganic salts or small-molecular-weight organic compounds dissolved in DMSO can be transported across skin membrane, indicating a potential hazard in commercial use.

[W.W.E./F.W.Sw.]

Dinocerata An extinct order of large herbivorous mammals, often called uintatheres, from early Cenozoic deposits of North America and northern Asia. Members of this group have semigraviportal limbs, that is, adapted to bearing considerable weights, with hoofed, five-toed feet (see illustration). A saber-

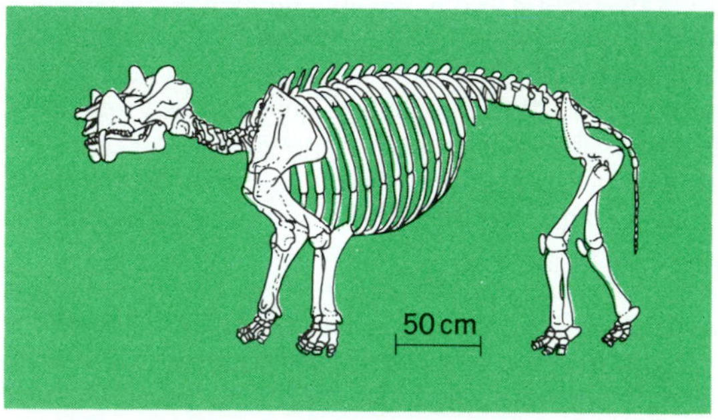

Skeleton of *Uintatherium*, a middle Eocene member of the Dinocerata.

like canine tooth and protective lower jaw flange are present in all forms except the aberrant *Gobiatherium*. Horns are absent or very small on the most primitive forms. Middle and late Eocene uintatheres in North America developed an imposing array of six horns. One pair was on the tips of the nasal bones, another was above the root of the saberlike canine tooth, and a third pair above the ear region.

The order Dinocerata consists of one family, the Uintatheriidae, which is divided into three subfamilies: Prodinoceratinae, Uintatheriinae, and Gobiatheriinae. The Dinocerata left no descendants. They are believed to have arisen from the arctocyonid condylarths but from a different subfamily than that which gave rise to the order Pantodonta. *See* CONDYLARTHRA; MAMMALIA; PANTODONTA. [M.C.McK.]

Dinoflagellida An order of the class Phytamastigophorea; also known as the Dinoflagellata. Although primarily marine, some dinoflagellates occur in fresh water. Some possess brown chromatophores, some are variously colored, and others are colorless. All types of nutrition exist, and some species are parasitic. Two flagella emerge laterally from a longitudinal depression or sulcus. An encircling flagellum in a groove or girdle divides the body into epicone and hypocone; the other, extending backward, propels the organism forward. The nucleus is very large.

Ceratium hirundinella blooms in fresh water, causing tastes and odors. Marine blooms of *Gymnodinium breve* produce the fish-killing red tides along the Gulf Coast of the southern United States. *See* PHYTAMASTIGOPHOREA. [J.B.L.]

Dinophyceae A large and extremely diverse class of biflagellate algae (dinoflagellates) in the chlorophyll *a–c* phyletic line (Chromophycota). In protozoological classification, these organisms constitute an order, Dinoflagellida, of the class Phytomastigophora. Many taxonomists emphasize the distinctness of dinoflagellates by placing them in a separate division (Pyrrophyta or Pyrrhophyta) or even in a separate kingdom (Mesokaryota). More than 1200 species are known, classified into 18 orders and 54 families. Most are microscopic, but a few reach a diameter of 2 mm (0.08 in.). Cell shape is highly variable, with many planktonic species having elaborately modified surfaces. Dinoflagellates occur in marine, brackish, and fresh waters, frequently producing algal blooms. They may be benthic as well as planktonic, and a few are colonial. Ameboid, palmelloid, coccoid (with or without a gelatinous sheath), and filamentous forms are also known. *See* DINOFLAGELLIDA.

Dinoflagellates span the spectrum of nutritional diversity. About half of the species are photosynthetic, and some of these are facultatively osmotrophic (absorbing nutrients) or phagotrophic (engulfing food). Symbiosis involving dinoflagellates is a common occurrence in marine environments. The phenomenon is most important in coral reefs, where up to half of the carbonate in calcified structures is derived by way of photosynthesis carried out by endosymbiotic dinoflagellates.

Several dinoflagellates contain substances that are toxic to other organisms. Blooms of some of these dinoflagellates cause "red tides" that are lethal to fishes or invertebrates. *See* ALGAE. [P.C.Si.; R.L.Moe]

Dinosaur The term Dinosauria, meaning "terrible lizard," was coined in 1841 by Richard Owen in reference to huge fossil bones then being discovered in England. These bones were recognized as reptilian, but of extinct species much larger than any living reptile. The term is no longer used in a formal sense, the dinosaurs now being classified in two separate orders, Saurischia and Ornithischia, under the subclass Archosauria. The diagnostic distinction between representatives of these two orders lies in the arrangement of the three pelvic bones, either in a triradiate pattern (saurischian), as in modern crocodilians and lizards, or in a tetraradiate arrangement (ornithischian), superficially "birdlike." The name dinosaur is obviously appropriate, for among these reptiles were the largest land animals that ever lived. Some were so large, in fact, that it has been theorized that they could not have supported their great weight without the buoyant effect of water in swamps or lakes. And it still is a great mystery as to how such giants could have obtained adequate food to grow to, and to sustain, such enormous bulk. *See* ARCHOSAURIA.

Not all dinosaurs were gigantic, however. One variety, *Compsognathus*, was no larger than a pigeon, and several others were chicken- or turkey-sized. Dinosaurian habits and life styles can only be inferred from their skeletal anatomy, but that evidence clearly indicates a wide variety of living habits, many of which were quite different from those of modern animals. As a whole, dinosaurs were almost universal in distribution, having been found on all continents except Antarctica, and were dominant among land animals throughout the Mesozoic Era, a time span of approximately 150,000,000 years. There were many kinds, some of which apparently existed in great numbers. But at the end of the Cretaceous, dinosaurs became extinct as a consequence of unknown, but probably multiple, causes.

Both dinosaurian orders are believed to have evolved from advanced members of the Triassic order Thecodontia known as pseudosuchians, probably from two separate kinds, but perhaps from a common ancestor. Some advanced pseudosuchians resemble some of the later saurischians, with marked tendencies toward upright posture and bipedal locomotion. This upright posture previews one of the distinctive conditions of all dinosaurs, where the limbs are largely or completely beneath the body, rather than sprawled out to the side as in lizards, turtles, and crocodilians. Erect posture permits longer strides and more rapid and efficient locomotion. Development of erect carriage was accompanied in all dinosaurs by enlargement of the sacrum, strengthening the attachment of the pelvic bones to the backbone, and rearrangement of the limb skeleton and musculature. Most dinosaurs were quadrupedal, but the trend toward bipedal gait in some pseudosuchians was perfected in certain saurischians and ornithischians (see Figs. 1–4). In the ornithischians (Ornithopoda), bipedality seems to have been facultative only, with a four-legged pose still possible for activities like grazing. Bipedal saurischians (Theropoda), however, became obligatory bipeds, losing all potential for assuming a quadrupedal stance. Such highly perfected bipedality suggests a high level of agility and activity—and perhaps a higher physiologic level than is typical of modern reptiles. *See* ORNITHISCHIA; SAURISCHIA; THECODONTIA.

Fig. 1. Restored bipedal carnivorous saurischian *Allosaurus*, about 40 ft or 12 m long, from the late Jurassic, North America.

Because all dinosaurs possessed certain skeletal features which also occur in all living reptiles, they have been classified as reptiles. Consequently, it is natural that they have been interpreted as reptilian in their habits, physiology, and levels of activity. This view of dinosaurs as cold-blooded, sluggish, and generally inactive animals has been challenged, and various lines of evidence have been cited as indicative that they were probably warm-blooded and avianlike or mammallike in their metabolism and activity levels.

Because dinosaurs are classified as reptiles, it has generally been assumed that they deposited eggs like modern crocodilians, turtles, and lizards rather than giving live birth. The

Fig. 2. Restoration of the duck-billed dinosaur *Anatosaurus* (= *Trachodon*) from the Late Cretaceous of North America; these were 30–40 ft (9–12m) long.

1922 discovery in Mongolia of clutches of eggs associated with skeletons of *Protoceratops* confirmed that some dinosaurs, at least, were egg layers rather than live bearers. Several other kinds of dinosaur eggs have been found elsewhere in Mongolia, France, Brazil, Argentina, and western North America, indicating that most, if not all, types of dinosaurs laid eggs. In 1978 a nest of a dozen or more very young hadrosaurs, associated with eggshell fragments, was discovered in central Montana. The fact that all the young had worn teeth shows that they were not fresh hatchlings, yet they were all crowded together in what appears to have been a nest. This suggests that they may have had some parental care.

At the end of the Mesozoic Era, the dinosaurs and many other kinds of reptiles, including the flying reptiles, marine mosasaurs, ichthyosaurs, and plesiosaurs, became extinct. Various other forms of life also died out at about the same time. The causes of such widespread extinction of such a large number of different kinds of organisms, all at about the same time, are among the most perplexing problems in paleontology. Inability to adapt to changing environments is the usual cause of extinction. Under conditions of change, those animals with the narrowest specializations are most apt to be wiped out. Contrarily, broad adaptations to a variety of foods and habitats may enable species to survive such changing conditions. *See* EXTINCTION (BIOLOGY).

Extinction of the dinosaurs has been attributed to many causes, including diseases, evolutionary senility, exploding supernovae and cosmic radiation, egg-eating mammals, falling temperatures, rising mountains, drifting continents, receding sea level, and rising temperatures. None appears adequate to account for the extinction of such a wide variety of animal kinds, and at the same time to explain why other kinds were little, or not, affected. It is known that the end of the Mesozoic

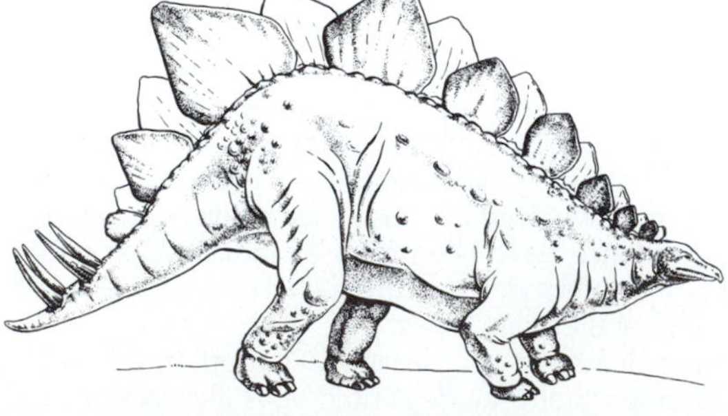

Fig. 3. Restoration of the plated Jurassic dinosaur *Stegosaurus*, about 20 ft or 6 m long. (*After C. W. Gilmore*)

Fig. 4. Restoration of the armored Cretaceous dinosaur *Euoplocephalus* (= *Ankylosaurus*), about 20 ft or 6 m long.

was a time of many changes, including sea-level and climatic changes, mountain building, and continental drifting, but no one of these phenomena alone is judged capable of eliminating the dinosaurs. The most logical explanation is that dinosaurian extermination resulted from a combination of factors, including some of those listed here as well as others that are still unknown to scientists. *See* REPTILIA. [J.H.O.]

Dioctophymoidea A group of nematodes characterized by the peculiar structure of the copulatory bursa of the male. Various species are parasites of fish, aquatic birds, and mammals. This group is considered to constitute either an order or superfamily, according to the viewpoint of the specialist.

Dioctophyme renale, the so-called giant kidney worm, is the best-known species in the group. It is a fairly common parasite of dogs and a wide variety of mammals, especially the mink. The adult worms are among the longest nematodes, the females measuring up to 1 m (3 ft). The size, blood-red color, and characteristic location in the kidney led to the common name. They live in the kidney for 1–3 years, digesting its substance until it is only a shell. The infection may be fatal but is often symptomless.

Other genera include *Eustrongylides* and *Hyrstrichis*, which inhabit the proventriculus of aquatic birds and are apparently transmitted by fish. Species of *Soboliphyme* are parasites in the intestine of foxes, cats, and the wolverine. *See* NEMATA. [J.A.S.]

Diode A two-terminal electron device exhibiting a nonlinear current-voltage characteristic. Although diodes are usually classified with respect to the physical phenomena that give rise to their useful properties, in this article they are more conveniently classified according to the functions of the circuits in which they are used. This classification includes rectifier diodes, negative-resistance diodes, constant-voltage diodes, light-sensitive diodes, light-emitting diodes, and capacitor diodes.

A circuit element is said to rectify if voltage increments of equal magnitude but opposite sign applied to the element produce unequal current increments. An ideal rectifier diode is one that conducts fully in one direction (forward) and not at all in the opposite direction (reverse). This property is approximated in junction and thermionic diodes. Processes that make use of rectifier diodes include power rectification, detection, modulation, and switching. *See* RECTIFIER.

Negative-resistance diodes, which include tunnel and Gunn diodes, are used as the basis of pulse generators, bistable counting and storage circuits, and oscillators. *See* NEGATIVE-RESISTANCE CIRCUITS; OSCILLATOR; TUNNEL DIODE.

Breakdown-diode current increases very rapidly with voltage above the breakdown voltage; that is, the voltage is nearly independent of the current. In series with resistance to limit the current to a nondestructive value, breakdown diodes can therefore be used as a means of obtaining a nearly constant reference voltage or of maintaining a constant potential difference between two circuit points, such as the emitter and the base of a transistor. Breakdown diodes (or reverse-biased ordinary junction diodes) can be used between two circuit points in order to limit alternating-voltage amplitude or to clip voltage peaks. *See* Limiter circuit.

Light-sensitive diodes, which include phototubes, photovoltaic cells, photodiodes, and photoconductive cells, are used in the measurement of illumination, in the control of lights or other electrical devices by incident light, and in the conversion of radiant energy into electrical energy. Light-emitting diodes (LEDs) are used in the display of letters, numbers, and other symbols in calculators, watches, clocks, and other electronic units. *See* Light-emitting diode; Photoconductive cell; Photodiode; Photoelectric devices; Phototube; Photovoltaic effect.

Semiconductor diodes designed to have strongly voltage-dependent shunt capacitance between the terminals are called varactors. The applications of varactors include the tuning and the frequency stabilization of radio-frequency oscillators. *See* Junction diode; Microwave solid-state devices; Point-contact diode; Semiconductor diode; Varactor. [H.J.R.]

Diopside The monoclinic pyroxene mineral which in pure form has the formula $CaMgSi_2O_6$. Pure diopside melts congruently at 1391°C (2536°F) at atmospheric pressure. Diopside has no known polymorphs. Its structure consist of chains of SiO_4 tetrahedrons in which each silicon ion shares an oxygen with each of its two nearest silicon neighbors. These chains are linked together by Ca and Mg ions in octahedral coordination.

Diopside forms gray to white, short, stubby, prismatic, often equidimensional, crystals. Small amounts of iron impart a greenish color to the mineral. Pure diopside is common and occurs as a metamorphic alteration of impure dolomites in medium and high grades of metamorphism.

Natural diopsidic pyroxenes which show extensive solid solution with jadeite and to a lesser extent with acmite are called omphacite. Omphacite is a principal constituent of eclogites, rocks of basaltic composition which have formed at high pressure. *See* Dolomite; Eclogite; Pyroxene. [F.R.B.]

Diopter A measure of the power of a lens or a prism. The diopter (also called dioptrie) is usually abbreviated D. Its dimension is a reciprocal length, and its unit is the reciprocal of 1 m (3.28 ft). *See* Focal length; Lens (optics).

The dioptric power of a prism is defined as the measure of the deviation of a ray going through a prism measured at the distance of 1 m. A prism that deviates a ray by 1 cm in a distance of 1 m is said to have a power of one prism diopter. *See* Optical prism.

Spectacle lenses in general consist of thin lenses, which are either spherical, to correct the focus of the eye for near and far distances, or cylindrical or toric, to correct the astigmatism of the eye. An added prism corrects a deviation of the visual axis. The diopter thus gives a simple method for prescribing the necessary spectacle for the human eye. [M.J.H.]

Diorite A phaneritic (visibly crystallized) plutonic rock having intermediate SiO_2 content (53–66%), composed mainly of plagioclase (oligoclase or andesine) and one or more ferromagnesian minerals (hornblende, biotite, or pyroxene), and having a granular texture. Diorite is the plutonic equivalent of andesite (a volcanic rock). This dark gray rock is used occasionally as a building stone and is known commercially as black granite. *See* Igneous rocks.

Gray or white plagioclase feldspar is the dominant mineral. Rocks with more calcic plagioclases and more abundant ferromagnesian minerals are gabbros. Rocks with greater proportions of alkali feldspar are called monzonite. Those with more quartz are called quartz-diorite or tonalite.

The texture of diorites is notably variable. Most often diorites are equigranular, with coarse, partly or mostly anhedral plagioclase and hornblende crystals, subordinate biotite, and interstitial quartz and orthoclase.

Diorite is found as isolated small bodies such as dikes, sills, and stocks, but it is also found in association with other plutonic rocks in batholithic bodies. It is closely associated with convergent plate boundaries where calc-alkalic magmatism and mountain building are taking place. *See* Magma. [W.I.R.]

Dioxin Any of a family of chlorinated aromatic hydrocarbons, also known as polychlorinated dibenzo-*para*-dioxins (PCDDs). Dioxins are highly toxic, lipophilic contaminants found in a number of chemical products; they are also found in the food chain in fish, meat, eggs, poultry, and milk. Dioxins are chemically and toxicologically similar to the chlorinated dibenzofurans and polychlorinated biphenyls (PCBs; see illustration). The compound 2,3,7,8-TCDD, or simply TCDD, is the most toxic of the dioxins in many species, including monkeys, rats, mice, rabbits, and guinea pigs.

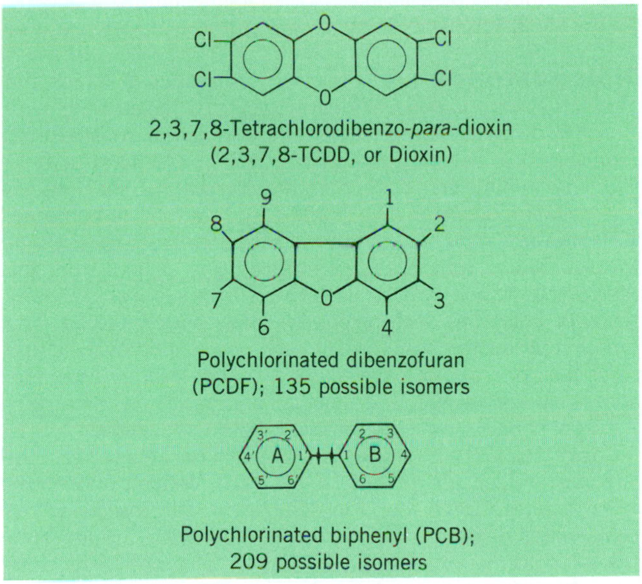

The chemical structure of a dioxin, furan, and polychlorinated biphenyl.

PCDDs are found as impurities or contaminants in chlorinated phenols, which are frequently used as wood or paper preservatives, in the herbicide 2,4,5-trichlorophenoxyacetic acid (2,4,5-T), and in hexachlorophene, an antibacterial agent. These compounds have also been reported in fly ash as well as in stack effluent from municipal incinerators. PCB and polychlorinated benzene-containing electrical transformers can release furans and dioxins when involved in fires by pyrolytic conversion of PCBs to furans and of the chlorinated benzenes to dioxin isomers.

The finding of trace amounts, in the parts-per-million range, of 2,3,7,8-TCDD in Agent Orange, a mixture of 2,4-D and 2,4,5-T, used extensively as a defoliant in Vietnam during the 1960s and early 1970s, has caused great concern about health effects of dioxins. *See* Poison; Toxicology. [A.Sch.]

Diphtheria A communicable disease of humans caused by the growth of the bacterium *Corynebacterium diphtheriae* on the mucous membranes or, less commonly, in cutaneous tissues. Both toxinogenic and nontoxinogenic strains of *C. diphtheriae* are capable of inducing the formation of a spreading, pseudomembranous exudate which may cause serious obstruction in the larynx and trachea. The classic pattern of the disease, caused only by the toxinogenic strains, involves the spread of diphtheria toxin to target tissues far from the superficial site of the infection.

Although diphtheria has been of worldwide occurrence, its incidence has declined sharply in countries where there has been a general immunization of children with diphtheria toxoid. The immune status of an individual, with respect to diphtheria toxin, may be determined through the use of the modified Schick test. Approximately 0.0008 microgram of toxin protein is injected into the skin of the forearm, the test site. An equal amount of toxoid is injected at a control site. Necrosis at the test site indicates nonimmune state, while immunity, the presence of circulating antibody, is indicated by a lack of reaction at either site. Immunity complicated by allergy to diphtherial proteins results in a delayed inflammatory reaction at both sites. *See* IMMUNITY; SKIN TEST. [L.B.]

Diphyllidea An order of tapeworms of the subclass Cestoda. Species of this order belong to a single genus and live in the intestine of elasmobranch fishes. The scolex has a large muscular rostellum armed with hooks and two fused pairs of suckers; the neck is armed with T-shaped hooks. Larval stages have been found parasitizing marine mollusks and crustaceans, but the life history is not known. *See* CESTODA. [C.P.R.]

Diplogasterida An order of nematodes in which the labia are seldom well developed; however, a hexaradiate symmetry is distinct. The external circle of labial sensilla may appear setose, but they are always short, never long or hairlike. The stoma may be slender and elongate, or spacious, or any gradation between. The stoma may be armed or unarmed; the armature may be movable teeth, fossores, or a pseudostylet. The corpus is always muscled and distinct from the postcorpus, which is divisible into an isthmus and glandular posterior bulb. The metacorpus is almost always valved. The female reproductive system may have one or two ovaries, and males may or may not have caudal alae; however, a gubernaculum is always present. The male tail commonly has nine pairs of caudal papillae; three are preanal and six are caudal.

There are two superfamilies: The members of Diplogasteroidea are predators, bacterial feeders, and omnivores; they are often found in association with insects or the fecal matter of herbivores. Cylindrocorporoidea include both free-living forms and intestinal parasites of amphibians, reptiles, and certain mammals. *See* NEMATA. [A.R.M.]

Diplomonadida An order of the class Zoomastigophorea, phylum Protozoa. These are small, colorless flagellates, some of which are free-living, some parasitic. The body is bilaterally symmetrical, composed of two mirror halves, each with a nucleus and a full set of kinetic organelles. There are four flagella to a side, not all of the same length.

Trepomonas rotans is probably the most common of the free-living species (see illustration), and usually occurs in water of high organic, low (or no) oxygen content, such as sewage. It occurs in ocean water also. This genus also is found as a parasite or symbiont in other animals, such as amphibians. *Hexamita* is another genus containing both free-living and parasitic representatives; *Spironucleus* is parasitic; and *Giardia*

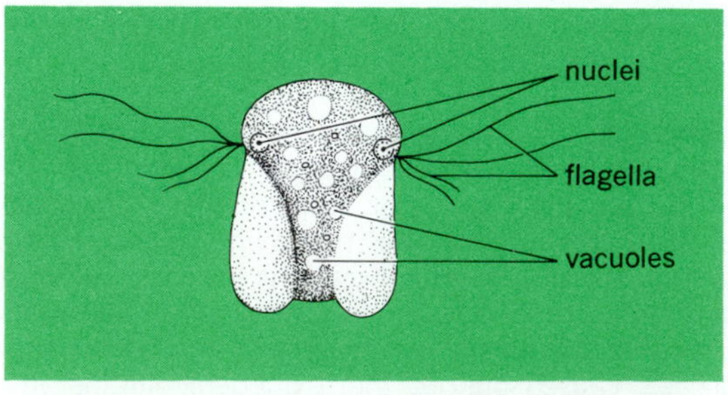

A diplomonad, *Trepomonas rotans*. (After McCracken)

has several species which parasitize vertebrates. In humans giardiasis results in severe dysentery; however, adequate medication is available. *See* ZOOMASTIGOPHOREA. [J.B.L.]

Diplopoda A class of terrestrial, tracheate, oviparous arthropods. Diplopods (millipeds) are largely cryptic in habits, are saprophytic feeders, and are characterized by the development of a compact head with a pair of short, simple, eight-jointed antennae, and powerful mandibles. The body is not differentiated into thorax and abdomen, but is composed of a variable number of similar, cylindrical diplosomites, each of which (except the first two or three) bears two pairs of walking legs (see illustration). The body wall is chitinous, with a thick

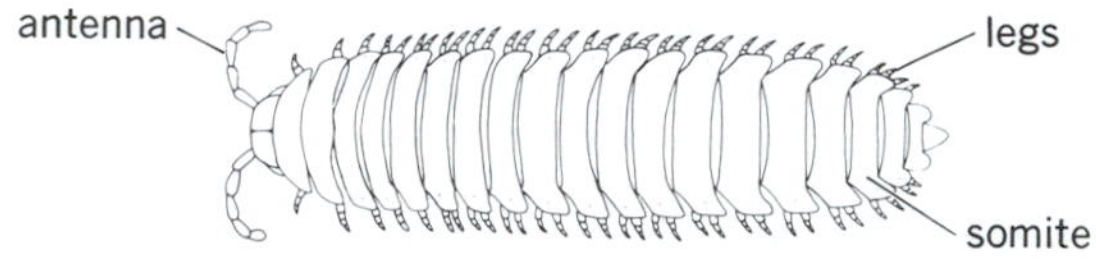

A diplopod. (*After R. E. Snodgrass, A Texbook of Arthropod Anatomy, Cornell University Press, 1952*)

impregnation of calcium carbonate in the majority of species. Most millipeds are variously adapted for rolling into a closed spiral or nearly perfect sphere when threatened.

The sexes are separate and fertilization is internal, following prolonged clasping behavior. Eggs may be laid in a cluster and "brooded" by the mother, scattered singly in the humus environment, or enclosed in an igloo-shaped mud nest built by the mother. Postembryonic development is gradual, without major changes in appearance, and may require a year or more for completion.

More than 8000 species have been described, although so far only the fauna of Europe is well known. In general, about 11 orders and more than 111 families are recognized. Probably as many as 25,000 species will eventually be recognized. Classification is based to a large extent upon shape of the male gonopods, which are quite constant and characteristic for each species.

Most species are local in distribution, because millipeds generally remain close to the parental habitat, and some may be restricted to a few square miles. Even genera are limited in distribution, and only a few occur on more than one continent.

The diplopods have a long geological history; they arose in the Early Devonian and were well developed by the Late Pennsylvanian. *See* ARTHROPODA. [R.L.Ho.]

Diploporita An extinct class of Echinodermata in which the thecal canals were associated in pairs; the canals ran perpendicularly to the surface and, when exposed in fossils by superficial abrasion, appear as paired pores termed diplopores. Exposure of solitary canals produces haplopores. These so-called pores are, of course, artifacts. The theca was ovoid or pear-shaped, built up of numerous small polygonal plates arranged without order. There was usually no stem. Diploporita probably gave rise to blastoids through transitional forms similar to *Asteroblastus*. None of them survived beyond the Devonian. *See* BLASTOIDEA; ECHINODERMATA. [H.B.F.]

Diplura An order of primarily wingless insects, measuring normally 0. 12–0.4 in. (3–10 mm) but occasionally larger, up to 2 in. (50 mm) in length. There are four families, given higher rank by some authors: Procampodeidae, Campodeidae, Projapygidae, and Japygidae. There are approximately 450 described species in this order, but many more are still to be found.

These insects generally lack pigment, and only a few campodeids have scales. The bodies are elongate, cylindrical, and

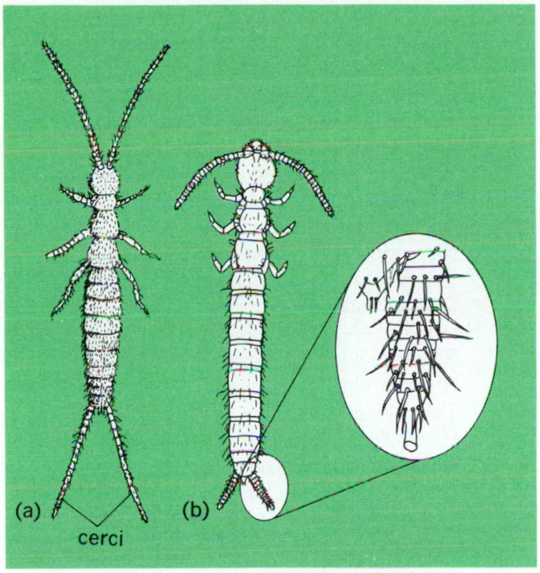

Diplura. (a) *Campodea folsomi Silvestri* (Campodeidae). (b) *Anajapyx vesiculosus* Silvestri (Projapygidae). (*After E. O. Essig, College Entomology, Macmillan, 1942*)

highly flexible, in accordance with their subterranean habits (see illustration). The mandibles and maxillae are retracted into the head capsule when at rest; maxillary and labial palps are obsolescent. The antennal flagellum is provided with intrinsic musculature. Eyes and ocelli are always absent. Well-developed external genitalia are absent. All are subterranean and can be found in the soil, detritus, rotten wood, and under bark or rocks. Diplura are of no importance to the economy of humans. Campodeids and projapygids feed on detritus and fungal mycelia and spores; japygids attack and devour small soil arthropods. *See* APTERYGOTA; INSECTA. [P.W.W.]

Dipnoi The lungfishes, one of the three subclasses of Osteichthyes. In comparison with Devonian lungfishes, extant species have reduced the number of median fins, changed the shape of the tail, and decreased ossification of the braincase and other endochondral bones.

The recent family Lepidosirenidae includes five species from Africa and South America. Lepidosirenids have eellike bodies, filamentous paired fins, two lungs, tooth plates with three cutting blades, a greatly reduced skull roof, and an elongated ceratohyal important in suction feeding and respiration. Lepidosirenids are obligate air breathers, and the heart, arterial tree, and gills are highly specialized. To breathe air, the fish protrudes its snout from the water surface, gulps air into the mouth, and then forces it into the lung by closing the mouth and raising the floor of the oral cavity.

The family Ceratodontidae is represented by the single species *Neoceratodus forsteri* from Australia. They are stout-bodied with large scales, leaf-shaped paired fins, and tooth plates with flattened crushing surfaces. Adults reach more than 3 ft (1 m) in length. The diet includes plants, crustaceans, and soft-bodied invertebrates. [W.Be.]

Early dipnoans were marine; fresh-water adaptations occur in Paleozoic and most post-Paleozoic dipnoans. Burrowing habits developed independently in late Paleozoic gnathorhizids and extant lepidosirenids. Dipnoans were common worldwide from the Early Devonian to the end of the Triassic, but are rare since. They are restricted to the southern continents in the Cenozoic. *See* OSTEICHTHYES. [H.P.S.]

Dipole Any object or system that is oppositely charged at two points or poles, such as a magnet or a polar molecule. The properties of a dipole are determined by its dipole moment, that is, the product of one of the charges by their separation directed along an axis through the centers of charge.

An electric dipole consists of two electric charges of equal magnitude but opposite polarity ($-Q$ and $+Q$ in the illustration), separated by a short distance (d); or more generally, a

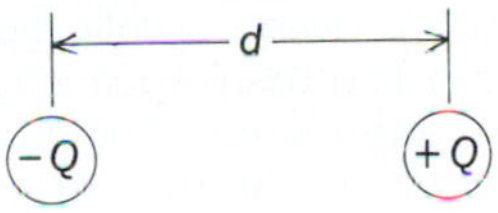

Electric dipole with moment $\mu = Qd$.

localized distribution of positive and negative electricity without net charge whose mean positions of positive and negative charge do not coincide. For a discussion of electric dipole radiation *see* ELECTROMAGNETIC RADIATION.

Molecular dipoles which exist in the absence of an applied field are called permanent dipoles, while those produced by the action of a field are called induced dipoles. *See* POLAR MOLECULE. [R.D.W.]

The term magnetic dipole originally referred to the fact that a magnet has two poles and, because of these two poles, experiences a torque in a magnetic field if its axis is not along a magnetic flux line of the field. It is now generalized to include electric circuits which, because of the current, also experience torques in magnetic fields. *See* MAGNET. [R.P.Wi.]

Dipole-dipole interaction The interaction of two atoms, molecules, or nuclei by means of their electric or magnetic dipole moments. This is the first term of the multipole-multipole series of invariants. More precisely, the interaction occurs when one dipole is placed in the field of another dipole. The interaction energy depends on the strength and relative orientation of the two dipoles, as well as on the distance between the centers and the orientation of the radius vector connecting the centers with respect to the dipole vectors. The electric dipole-dipole interaction and magnetic dipole-dipole interaction must be distinguished.

The center of the negative charge distribution of a molecule may fail to coincide with its center of gravity, thus creating a dipole moment. An example is the water molecule. If such molecules are close together, there will be a (electric) dipole-dipole interaction between them. Atoms do not have permanent dipole moments, but a dipole moment may be induced by the presence of another atom nearby; this is called the induced dipole-dipole interaction. Induced dipole-dipole forces between atoms and molecules are known by many different names: van der Waals forces, London forces, or dispersion forces. These induced dipole-dipole forces are responsible for cohesion and surface tension in liquids. They also act between unlike molecules, resulting in the adsorption of atoms on macroscopic objects. *See* ADSORPTION; COHESION (PHYSICS); INTERMOLECULAR FORCES; SURFACE TENSION; VAN DER WAALS EQUATION.

The magnetic dipole-dipole interaction is found both on a macroscopic and on a microscopic scale. Two compass needles within reasonable proximity of each other illustrate clearly the influence of the dipole-dipole interaction. In quantum mechanics, the magnetic moment is partially due to a current arising from the motion of the electrons in their orbits, and partially due to the intrinsic moment of the spin. The same interaction exists between nuclear spins. Magnetic dipole-dipole forces are particularly important in low-temperature solid-state physics, the interaction between the spins of the ions in paramagnetic salts being a crucial element in the use of such salts as thermometers and as cooling substances. *See* ADIABATIC DEMAGNETIZATION; DIPOLE; ELECTRON; LOW-TEMPERATURE THERMOMETRY; MAGNET; MAGNETIC MOMENT; MAGNETIC THERMOMETER; NUCLEAR MOMENTS. [P.H.E.M.]

Dipsacales An order of flowering plants, division Magnoliophyta (Angiospermae), in the subclass Asteridae of the class Magnoliopsida (dicotyledons). The order consists of 5 families and about 1100 species. Within its subclass the group is marked by its inferior ovary; usually opposite, exstipulate (without stipules, the paired basal appendages found on many leaves) leaves; often anisomerous stamens (fewer in number than the petals); and frequently irregular corolla. The flowers are sometimes grouped in heads suggesting those of the Asterales, but then the ovules are either pendulous (instead of erect from the base) or more than one in each ovary. Elderberry (*Sambucus*) and honeysuckle (*Lonicera*), in the Caprifoliaceae, and teasel (*Dipsacus*) are familiar members of the Dipsacales. *See* ASTERIDAE; MAGNOLIOPSIDA. [A.Cr.]

Diptera An order of the class Insecta known as the true flies and so named because they possess only two wings. This characteristic, together with a pair of balancers or halteres, distinguishes flies from all other orders of the Insecta. The three names for members of the order, known commonly as flies, gnats, or midges, form a part of the common names of most families, genera, and species of the order. The term fly also forms a part of the compound names of the insects in many other orders, such as butterfly, mayfly, and chalcid fly, but when used alone it is correctly applied only to the members of the Diptera. The forms most commonly known as maggots are actually dipterous larvae, and keds (Hippoboscidae) are parasitic forms of flies that have lost their wings. The Diptera are the most important group of insects considered in medical entomology. Many, especially the mosquitoes, are the vectors for numerous parasites and diseases.

In number of species, it is the third largest order; only the Coleoptera and Lepidoptera are larger. There are approximately 100,000 kinds of flies in the world, of which over 18,500 inhabit America north of Mexico.

The order is ubiquitous and is more widespread than any other of the insects. Diptera are found on every continent and most islands of the world, except for the coldest parts of the Arctic and Antarctic, and at the tops of extremely high mountain ranges. They occur in almost all available ecological niches. Specimens have been taken at altitudes of many thousands of feet by specially constructed traps attached to airplanes.

Taxonomy. For convenience, the families of Diptera may be divided into the Nematocera and the Brachycera, and the latter group further divided into the Orthorrhapha and the Cyclorrhapha. Adults of the Nematocera are characterized by their slow flight and long antennae. These are the more primitive Diptera, and the major groups are represented in fossil deposits as old as Jurassic and Cretaceous age.

The Brachycera are generally swift fliers, and their antennae are no longer than the head. In the Orthorrhapha, the adult fly emerges from the pupa through a T-shaped anterior opening, while in the Cyclorrhapha it emerges from a circular opening at the anterior end of the puparium.

Following is a list of scientific and vernacular names of the more common families of flies.

Nematocera
- Trichoceridae (winter crane flies)
- Ptychopteridae (phantom crane flies)
- Tipulidae (crane flies)
- Psychodidae (moth flies and sand flies)
- Blepharoceridae (net-winged midges)
- Chironomidae (midges)
- Ceratopogonidae (biting midges)
- Simuliidae (black flies)
- Chaoboridae (phantom midges)
- Culicidae (mosquitoes)
- Dixidae (dixid midges)
- Anisopodidae (wood gnats)
- Bibionidae (march flies)
- Scatopidae (scatopsid flies)
- Mycetophilidae (fungus gnats)
- Sciaridae (black fungus gnats)
- Cecidomyiidae (gall midges)

Brachycera—Orthorrhapha
- Stratiomyidae (soldier flies)
- Rhagionidae (snipe flies)
- Tabanidae (horse flies and deer flies)
- Bombyliidae (bee flies)
- Asilidae (robber flies)
- Empididae (dance flies)
- Dolichopodidae (long-legged flies)

Brachycera—Cyclorrhapha
- Lonchopteridae (spear-winged flies)
- Phoridae (hump-backed flies)
- Platypezidae (flat-footed flies)
- Syrphidae (syrphid flies)
- Pipunculidae (big-headed flies)
- Lonchaeidae (lonchaeid flies)
- Lauxaniidae (lauxaniid flies)
- Drosophilidae (pomace or fruit flies)
- Ephydridae (shore flies)
- Anthomyiidae (anthomyiid flies)
- Muscidae (house flies)
- Calliphoridae (blow flies)
- Sarcophagidae (flesh flies)
- Oestridae (bot flies)
- Tachinidae (tachinid flies)
- Hippoboscidae (louse flies)
- Micropezidae (stilt-legged flies)
- Diopsidae (stalk-eyed flies)
- Sciomyzidae (marsh or snail-killing flies)
- Sepsidae (black scavenger flies)
- Heleomyzidae (heleomyzid flies)
- Sphaeoceridae (small dung flies)
- Agromyzidae (leaf-miner flies)
- Tethinidae (tethinid flies)
- Chloropidae (frit flies)
- Conopidae (thick-headed flies)
- Tephritidae (fruit flies)
- Piophilidae (skipper flies)

Morphology. Adult flies vary from somewhat less than 1 mm (0.04 in.) to over 25 mm (1 in.) in body length. The head is vertical, usually with three ocelli at the dorsal vertex and the mouthparts ventral. In some cases, however, the head is distinctly longer than high when viewed laterally. The compound eyes are usually large and prominent.

The mouthparts are modified for either lapping or piercing, but never with the mandibles apposable and capable of chewing. The structure of the mouthparts varies greatly throughout the order so that their homologies are often rather difficult to determine. Throughout the Orthorrhapha there tends to be a retention of mandibles for piercing, especially in those flies that require blood meals. In the Cyclorrhapha, on the other hand, mandibles tend to be reduced or lost entirely, and with few exceptions the work of the mouthparts is largely sucking, with the maxillae and labium taking over the large share of work. Since all adult flies imbibe fluids only, the presence of a pharyngeal "pump" to transfer fluids from the external substrate to the gut is characteristic.

Each antenna typically has two basal segments and one or more additional segments called the flagellum. In the Nematocera the flagellum consists of a variable number of quite similar segments. In all remaining flies that are three distinct antennal segments. The third segment, representing the flagellum, is usually longer than the two basal ones and sometimes complex.

The prothorax and metathorax are small and closely united with the large mesothorax which contains the musculature for the single pair of wings. The legs are usually alike, except in species in which the prothoracic pair are raptorial and species in which the fore, mid, or hind legs of the males are modified by secondary sexual characters.

Only the first pair of wings, the mesothoracic pair, is developed. The second pair is reduced to a pair of club-shaped organs, or halteres, which serve as balancing organs in flight. They are present in most species, even when the mesothoracic wings are absent. In many species, especially those of the Nematocera, abundant scalelike setae clothe the veins and sometimes the margins of the wings. The wings of some of the higher Diptera bear yellow, brown, or black markings that aid in species identification.

In certain families of Diptera, some of the setae covering the head and body are greatly enlarged and are used for identification. The location and relative size of these bristles are often characteristic of species, genera, or families.

Life cycle. The adult stage in the life history may be regarded as a reproductive and dispersal phase, during which these insects increase both in number and often in geographic range. Females select suitable oviposition sites and lay a variable number of eggs, after which they usually die. The eggs, after a period of incubation of varying length, give rise to larvae, which represent a growth phase. During this period, the insect does almost nothing but nourish itself, thus providing the tissues necessary for a later transformation into an adult. This growth period is divided into stages, or instars, each of which terminates with the molting of the larval skin to allow increase in size for the stage that follows. In the Diptera there are four larval instars. In most of the Orthorrhapha, especially those with aquatic immature forms, all four instars are active; but in the Cyclorrhapha, at the third larval molt, a hard puparium is formed inside which a quiescent fourth larval stage occurs. At the conclusion of the final larval instar, there is a dramatic reorganization in tissue structure initiated by the action of appropriate hormones on the so-called "imaginal buds," a process that eventually results in the formation of an adult insect. During this time, the insect is called a pupa because its external form is now altered and any external activity virtually ceases.

When the adult tissues are almost fully formed, the fly emerges from its pupal case and spends some time in drying, hardening, and expanding its wings, attaining color, and reaching sexual maturity. It is then ready to start its cycle once again.

Economic importance. Diptera have probably more economic importance than any other insect order. This importance comes from the relatively few species that affect domestic animals and plants. The vast majority of forms have no direct importance although a number are beneficial in one way or another.

Only a relatively few dipteran species cause severe economic loss to humans. Perhaps of most concern is the role played by them in disease transmission. About 70 species of *Anopheles* mosquitoes transmit an estimated 500,000 cases of malaria each year. Yellow fever is transmitted principally by a single mosquito, *Aedes aegypti*. Dengue, or breakbone fever, a usually nonfatal disease of worldwide distribution that leaves its victims debilitated for several weeks, is transmitted by *Aedes aegypti* and *A. albopictus*. Filariasis, primarily a disease of peoples of Africa, the Orient, and the Pacific islands, is caused by a minute roundworm whose larvae are transmitted by a few species of *Anopheles*, *Mansonia*, *Culex*, and *Aedes*. With modern methods used by bacteriologists and virologists, a large number of virus diseases known to be transmitted by mosquitoes have been found in humans. Many of these, such as Sindbis virus of Egypt, may not produce clinical symptoms of disease, but others, such as the equine encephalitides, which include western, eastern, St. Louis, Japanese, and others, may be quickly fatal in humans. With the exception of the viruses, none of the diseases mentioned above are transmitted by any insects other than members of the order Diptera.

A few less well-known diseases are transmitted by other flies. Black flies transmit *Onchocerca volvulus*, the causative agent of onchocerciasis, a disease affecting the eyes of natives of Central and parts of South America. Sandflies of the genus *Phlebotomus* transmit organisms that cause kala azar, Oriental sore, pappataci fever, and Oroya fever. Even today large parts of Africa remain underdeveloped because of the presence of sleeping sickness, a fatal disease transmitted by tsetse flies. Species of the genus *Chrysops* (Tabanidae) are instrumental in carrying tularemia, primarily a disease of rodents.

Domestic animals, hides, meat, and dairy products are affected by disease transmission or by direct attack by flies. Anthrax, tularemia, botulism, many virus diseases, and nagana, a form of sleeping sickness, are some of the diseases transmitted by members of the Diptera that take an annual toll amounting to millions of dollars. Some flies wreak their damage by direct attack. The primary screwworm fly deposits eggs on hides of animals, and the larvae, upon hatching, burrow through the skin and into the flesh. The secondary screwworm gains entrance through holes, often infected, already present on the skin surface. *Hypoderma lineata* and *H. bovis* are botflies of special importance. Eggs of both species are laid on hairs of cattle, and the hatching larvae bore through the skin into the connective tissue. During their development they wander through the tissues of the animal, and when mature they escape through holes which they make along the spine of their host and drop to the ground to pupate. Horses are afflicted by a species of *Gasterophilus*. They lick the eggs from their bodies, and the larvae settle and dwell in their stomachs, often in large numbers. Sheep are victims of the sheep botfly, whose larvae live in their nasal passages and sinuses. All of these insects affect the meat, milk, and wool production of animals through debilitation and irritation, and hides may be rendered completely useless by their entrance and exit holes. An interesting dipteran is the bee louse, which is a wingless, ectoparasite of the honeybee.

Economic losses as a result of damage by flies to crops are perhaps not as great as those caused by other insects, yet their presence has necessitated the expenditure of large sums of money for control. Among these, perhaps the most important are members of the family Tephritidae, the larvae of which feed upon and ruin the succulent flesh and seeds of their hosts. The most important of these are the European and American cherry fruit flies; their relatives in the genus *Rhagoletis* that attack walnuts in North America; six or seven species of the genus *Anastrepha* which attack many kinds of fruit in the New World; the Mediterranean fruit fly which is a limiting factor in the production of many fruits, especially citrus; and several species of the genus *Dacus*, which attack olives, citrus fruits, many kinds of vegetables, and other edible plants and plant parts in Europe, the Orient, and Pacific islands.

Larvae of some flies mine the leaves of ornamentals thereby defacing them, reducing their growth potentials, and affecting their production and sale by nurseries. The larvae of some species of Muscidae are known as root maggots and burrow into the underground parts of plants with considerable loss to growers of truck crops over the world. *See* ARTHROPODA; INSECTA. [R.J.G.]

Direct broadcasting satellite systems Systems for distributing television signals, by transmitting them directly from a satellite to the home over superhigh-frequency radio waves. Both performance and economic advantages may be realized by this direct method.

A direct broadcasting satellite (DBS) system consists of a television repeater station in geostationary orbit, a ground station that transmits program signals to the spacecraft, and the consumer terminals that receive the signals from the satellite and convert them to a format compatible with existing television sets.

To reduce the cost and installation problems of the user equipment, commonly known as homesat terminals, the satellite transmitter provides as much as 100 times the signal strength used by previous satellite transmission systems. As a result, the homesat terminal antenna (a dish type) can be made as small as 2 ft (0.6 m) in diameter as compared with the 10–15-ft (3–4.5-m) diameters of existing earth terminal equipment. The consumer cost for equipment and its installation is reduced to affordable levels through this reduction in antenna size.

The increase in transmitter power does, however, limit the number of channels that can be transmitted from a particular spacecraft. A typical spacecraft can provide 3 such higher-power transmission channels compared with 24 lower-power channels. This limitation requires that many such satellites be collocated to provide the 32 channels available in the designated frequency band for each orbital location. Alternatively, larger satellites can be employed. The grouping or clustering of spacecraft at such an orbital location allows the consumer's antenna to be fixed in direction and receive the maximum number of channels available.

Problems encountered in the implementation of a DBS system include rain attenuation, impact of high-power radio-frequency satellites, initial implementation cost, and establishment of a broad user base. *See* COMMUNICATIONS SATELLITE; TELEVISION. [R.F.Bun.]

Direct-coupled amplifier A device for amplifying signals with direct-current components. There are many different situations where it is necessary to amplify signals having a frequency spectrum which extends to zero frequency. Some typical examples are amplifiers in electronic differential analyzers (analog computers), certain types of feedback control systems, and some medical instruments such as the electrocardiogaph. Amplifiers which have capacitor coupling between stages are not usable in these cases, because the gain at zero frequency is zero. Therefore, a special form of amplifier, called a dc or direct-coupled amplifer, is necessary. These amplifers will also amplify alternative-current (ac) signals. *See* AMPLIFIER.

Some type of coupling circuit must be used between successive amplifer stages to prevent the relatively large supply voltage of one stage from appearing at the input of the following stage. These circuits must pass direct-current signals with the least possible amount of attenuation.

Transistor direct-coupled amplifiers have a number of advantages and disadvantages relative to vacuum-tube amplifiers. Interstage direct coupling is much easier with transistors because of the availability of both *pnp* and *npn* transistors and zener diodes. The main disadvantage of this amplifier is the temperature dependence of the transistor parameters.

It is generally recognized that the differential amplifier is the most stable direct-coupled amplifier circuit available. This is true because in this circuit the performance depends on the difference of the device parameters, and transistors can be manufactured using the planar epitaxial technique with very close matching of their parameters.

A method of amplifying direct-current (or slowly varying) signals by means of ac amplifiers is to modulate a carrier signal by the signal to be amplified, amplifying the modulated signal, and demodulating at the output.

A chopper-stabilized amplifier is composed of two amplifiers: One amplifier is a carrier direct-coupled amplifier employing a chopper to modulate and demodulate a portion of the signal to be amplified; the second amplifier is a straight direct-coupled amplifier. The success of this circuit in reducing to a very low value the amount of drift appearing at the output depends upon the fact that the drift is a very-low-frequency phenomenon. [C.C.H.]

Direct current Electric current which flows in one direction only through a circuit or equipment. The associated direct voltages, in contrast to alternating voltages, are of unchanging polarity. Direct current corresponds to a drift or displacement of electric charge in one unvarying direction around the closed loop or loops of an electric circuit. Direct currents and voltages may be of constant magnitude or may vary with time.

Direct current is used extensively to power adjustable-speed motor drives in industry and in transportation. Very large amounts of power are used in electrochemical processes for the refining and plating of metals and for the production of numerous basic chemicals.

Direct current ordinarily is not widely distributed for general use by electric utility customers. Instead, direct-current (dc) power is obtained at the site where it is needed by the rectification of commercially available alternating-current (ac) power to dc power. *See* DIRECT-CURRENT TRANSMISSION; ELECTRIC POWER SYSTEMS. [D.D.R.]

Direct-current generator A rotating electric machine which delivers a unidirectional voltage and current. An armature winding mounted on the rotor supplies the electric power output. One or more field windings mounted on the stator establish the magnetic flux in the air gap. A voltage is induced in the armature coils as a result of the relative motion between the coils and the air gap flux. Faraday's law states that the voltage induced is determined by the time rate of change of flux linkages with the winding. Since these induced voltages are alternating, a means of rectification is necessary to deliver direct current at the generator terminals. Rectification is accomplished by a commutator mounted on the rotor shaft. Carbon brushes, insulated from the machine frame and secured in brush holders, transfer the armature current from the rotating commutator to the external circuit. *See* COMMUTATION; ELECTRIC ROTATING MACHINERY; GENERATOR; WINDINGS IN ELECTRIC MACHINERY.

The field windings of dc generators require a direct current to produce a magnetomotive force (mmf) and establish a magnetic flux path across the air gap and through the armature. Generators are classified as series, shunt, compound, or separately excited, according to the manner of supplying the field excitation current.

In the separately excited generator, the field winding is connected to an independent external source. Separately excited generators are among the most common of dc generators, for they permit stable operation over a very wide range of output voltages.

Using the armature as a source of supply for the field current, dc generators are also capable of self-excitation. Residual magnetism in the field poles is necessary for self-excitation. Series, shunt, and compound-wound generators are self-excited, and each produces different voltage characteristics. The armature winding and field winding of a series generator are connected in series. The field winding of a shunt generator is connected in parallel with the armature winding. A compound generator has both a series field winding and a shunt field winding. Both windings are on the main poles with the series winding on the outside. The shunt winding furnishes the major part of the mmf. The series winding produces a variable mmf, dependent upon the load current, and offers a means of compensating for voltage drop. [R.T.W.]

Direct-current motor An electric rotating machine energized by direct current and used to convert electric energy to mechanical energy. It is characterized by its relative ease of speed control and, in the case of the series-connected motor, by an ability to produce large torque under load without taking excessive current. *See* ELECTRIC ROTATING MACHINERY.

The principal parts of a dc motor are the frame, the armature, the field poles and windings, and the commutator and brush assemblies. The frame consists of a steel yoke of open cylindrical shape mounted on a base. Salient field poles of sheet-steel laminations are fastened to the inside of the yoke. Field windings placed on the field poles are interconnected to form the complete field winding circuit. The armature consists of a cylindrical core of sheet-steel disks punched with peripheral slots, air ducts, and shaft hole. These punchings are aligned on a steel shaft on which is also mounted the commutator. The commutator, made of hard-drawn copper segments, is insulated from the shaft. Segments are insulated from each other by mica. Stationary carbon brushes in brush holders make contact with commutator segments. Copper conductors placed in the insulated armature slots are interconnected to form a reentrant lap or wave style of winding. *See* COMMUTATION; WINDINGS IN ELECTRIC MACHINERY.

Rotation of a dc motor is produced by an electromagnetic force exerted upon current-carrying conductors in a magnetic field. For basic principles of motor action *see* MOTOR.

Direct-current motors may be categorized as shunt, series, compound, or separately excited.

The field circuit and the armature circuit of a dc shunt motor are connected in parallel. The field windings consist of many turns of fine wire. The entire field resistance, including a series-connected field rheostat, is relatively large. The field current and pole flux are essentially constant and independent of the armature requirements. The torque is therefore essentially proportional to the armature current. Typical applications are for load conditions of fairly constant speed, such as machine tools, blowers, centrifugal pumps, fans, conveyors, wood- and metal-working machines, steel, paper, and cement mills, and coal or coke plant drives.

The field circuit and the armature circuit of a dc series motor are connected in series. The field winding has relatively few turns per pole. The wire must be large enough to carry the armature current. The flux of a series motor is nearly proportional to the armature current which produces it. Therefore, the torque of a series motor is proportional to the square of the armature current, neglecting the effects of core saturation and armature reaction. An increase in torque may be produced by a relatively small increase in armature current. Typical applications of this motor are to loads requiring high starting torques and variable speeds, for example, cranes, hoists, gates, bridges, car dumpers, traction drives, and automobile starters.

A compound motor has two separate field windings. One, generally the predominant field, is connected in parallel with the armature circuit; the other is connected in series with the armature circuit. The field windings may be connected in long or short shunt without radically changing the operation of the motor. They may also be cumulative or differential in compounding action. With both field windings, this motor combines the effects of the shunt and series types to an extent dependent upon the degree of compounding. Applications of this motor are to loads requiring high starting torques and somewhat variable speeds, such as pulsating loads, shears, bending rolls, plunger pumps, conveyors, elevators, and crushers. *See* DIRECT-CURRENT GENERATOR.

The field winding of a separately excited motor is energized from a source different from that of the armature winding. The field winding may be of either the shunt or series type, and adjustment of the applied voltage sources produces a wide range of speed and torque characteristics. Small dc motors may have permanent-magnet fields with armature excitation only. Such motors are used with fans, blowers, rapid-transfer switches, electromechanical activators, and programming devices. [L.F.C.]

Direct-current transmission The conduction of electric power by means of electric currents which flow in one direction only. After the almost complete replacement of direct current (dc) by alternating current (ac), interest in dc power transmission began increasing in 1954. One reason is that the cost of a dc line is about two-thirds that of the corresponding ac line. This is because the usual dc bipolar line has two conductors, compared with three conductors of an ac single-circuit three-phase line. *See* DIRECT CURRENT.

The dc line requires additional apparatus at its terminals, not required for an ac line, which increases the cost. These are the converters for changing ac to dc at the sending end and dc to ac at the receiving end. Considering the combined costs of the line and of the terminal equipment, the dc link is thus more expensive than ac for a short line but less expensive for a long line. However, the choice between ac and dc usually depends on other factors.

Direct-current lines are used to perform functions for which ac lines are unsuited, such as transmitting power across a wide body of water, using submarine cables. Cables have a much larger shunt capacitance per unit of length than overhead lines have. When an ac voltage is applied between the two conductors of a cable, an alternating "charging" current flows even if the distant end is not connected to a load. When the length of the cable exceeds some distance—about 30 mi (50 km)—the thermal limit is reached, even though no power is transmitted. This was a significant factor in the choice of dc for connecting the island of Gotland to the mainland of Sweden, Denmark to Sweden, England to France, and the South Island of New Zealand to the North Island.

Direct-current lines are also used to provide interconnections of two ac systems of different nominal frequencies, for example, 50 Hz and 60 Hz, as was done in Japan, or interconnections between two ac systems having the same nominal frequency but different frequency controls. An ac tie of the latter type might require a much higher power rating than the greatest power which was desired to be interchanged. In some cases, the two converters are installed in the same station, so that really there is no dc line. Such stations are called frequency-changer stations if there are two different nominal frequencies, and asynchronous ties if the nominal frequencies are equal, as in New Brunswick, Canada, or Steagle, Nebraska. [E.W.K.]

Direction-finding equipment Equipment used by aircraft and ships as an aid to navigation or used to obtain information for other purposes such as plotting the flight of meteorological balloons, locating storm fronts, discovering the location of illicit transmitters, locating the source of galactic noise, and determining source and characteristics of signals for military purposes. The equipment attempts to determine the direction of arrival of a radio wave by measuring the orientation of the wavefront or of the magnetic or electric vector of a radio wave.

Generally speaking, all radio direction finders consist of three components: (1) the directional antenna or collector system; (2) the radio receiver that detects the emission from which the directional information will be derived; and (3) the display and indicator system which transforms the output of the two previous components into usable information for presentation to the operator.

The directional antenna or collector systems are either signal-minimum (null) or signal-maximum types.

The common airborne direction finder (ADF) was derived from the right-left radio compass. The radio compass employs both a loop antenna and a nondirectional antenna. The loop antenna is mounted rigidly to the aircraft, with its plane at right angles to the longitudinal axis of the aircraft. By proper phasing within the receiver, the receiving antenna pattern is a cardioid which can be reversed so that its maximum moves from the right to the left side of the aircraft as the connection to the loop antenna is reversed. The output of the compass receiver is rectified and connected to a differential meter. The meter reads the relative output for the two loop-antenna connections. Therefore, it shows the relative strength of the signals received from one side of the aircraft as compared with those received from the other side. If the radio station is directly ahead, the reading of the meter will be zero. In actual practice the switching is accomplished by electronic methods as frequencies ranging from 30 to 90 Hz.

Automatic direction finders for marine services would employ solid-state components used in circuits consistent with modern technology. *See* ELECTRONIC NAVIGATION SYSTEMS. [P.C.S.]

Directional coupler A waveguide network of four guide terminals A, B, C, and D (see illustration) such that there

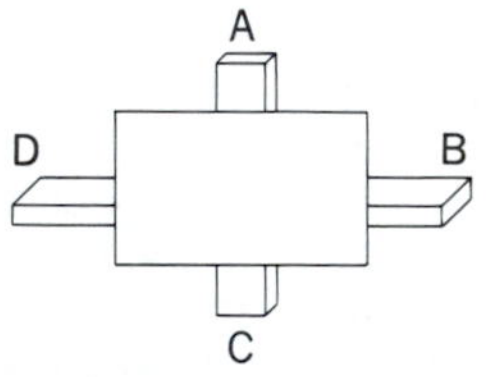

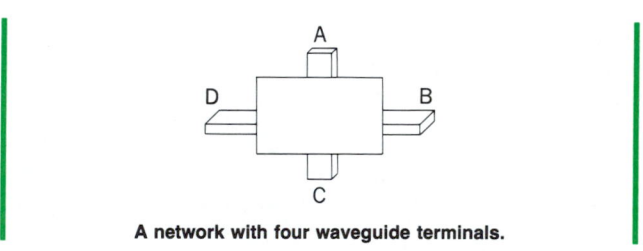

A network with four waveguide terminals.

is a complete isolation between A and C and between B and D, but no isolation between the two terminals of any other combination.

A directional coupler is most useful in selectively measuring either the forward (incident) or the backward (reflected) wave. In the case of a small hole coupling (large decibel rating), this objective can be achieved with a negligible attrition of the main wave. [C.K.J.]

Directivity The general property of directional discrimination displayed by systems that receive or emit waves. Thus, loudspeakers, microphones, radio antennas, underwater sound projectors, hydrophones, and even telescopes all have the common property that their effectiveness depends upon the direction from which the wave is either emitted or received. The manner in which a sender or receiver is directional depends largely upon its geometrical shape and, in particular, upon its dimensions compared to those of the wavelength involved.

Directivity is a desirable property of a receiver because it permits the identification of the direction from which a signal comes and because noise from other directions is eliminated. It is desirable in a sender because the available energy can be concentrated in a given direction. A simple example of directivity is furnished by the megaphone, which effectively increases the size of the emitting area, thus increasing directivity. [W.S.Cr.]

Dirigible A lighter-than-air craft equipped with means of propelling and steering for controlled flight in all three dimensions of the atmosphere. As a general designation, the term dirigible has gradually become associated with the term airship and has come to indicate primarily that type of airship with the hull of rigid structure. *See* AIRSHIP. [K.Am.; R.S.R.]

Disasteroida A paraphyletic grouping of extinct irregular echinoids, members of which gave rise to the Holasteroida and Spatangoida in the Lower Cretaceous. Disasteroids have a split apical disc; the posterior two ocular plates and their associated ambulacra are separated from the remainder of the apical disc (and the anterior three ambulacra) by intercalated interambulacral plates. Disasteroids have a strong bilateral symmetry. Larger pore pairs, arranged in phyllodes around the mouth, suggest that disasteroids were deposit feeders with penicillate feeding tube feet. Most species probably lived infaunally, though not burrowing deeply.

Five families are included: Acrolusiidae, Collyritidae, Disasteridae, Pygorhytidae, and Tithoniidae, separated principally on apical disc plating. Approximately 75 species have been named, arranged in 17 genera. The oldest species is late Bajocian (Middle Jurassic) and the youngest Albian (Lower Cretaceous). The group achieved its greatest diversity during the Upper Jurassic. *See* ATELOSTOMATA; ECHINODERMATA. [A.B.S.]

Discomycetes A class in the subdivision Ascomycotina, commonly known as cup fungi. They form a well-developed mycelium which bears the sexual (ascus) and asexual (conidium) states. The ascocarp (apothecium) is generally cup or saucer shaped, but other shapes occur as well. In general, the asci are exposed when they mature, except in one group (truffles) that forms the ascocarps underground (hypogeous ascocarps). The ascocarp arises as coiled or intertwined initials that can be distinguished as male (antheridium) and female (ascogonium) in some species. Layers of hyphae develop around these initials to form the tissue of the ascocarp. As this development occurs, the initial coil forms a system of ascogenous hyphae that transform into asci. The asci form in a broad layer (hymenium) on the apothecium. They have a single rigid wall (unitunicate), with two basic types of apex, operculate and inoperculate. Operculate asci have a hinged cap (operculum) at the tip that opens to allow discharge of ascospores. In inoperculate asci, the apex has a pore or a small slit through which the spores are discharged. In most species, ascospores are forcibly discharged. Their shape is variable, but they are usually one celled and hyaline.

Discomycetes occur on soil and on living and dead plants, either as saprobes or as parasites. The operculate discomycetes include the largest of all ascomycetes: the edible morels (*Morchella*), prized for their flavor; those with large, showy red or orange apothecia (*Cookeina, Sarcoscypha, Scutellinia*); and truffles (*Tuber*). Among the inoperculate discomycetes are numerous plant pathogens, such as *Monilinia fructicola* (brown rot of peach), which forms small, brown, cup-shaped ascocarps with long stalks, and the needle cast fungus (*Lophodermium pinastri*), which forms narrow, elongate ascocarps that open by a slit on pine needles. *See* ASCOMYCOTINA; EUMYCOTA; FUNGI. [R.T.Ha.]

Discriminant For a polynomial $f(x) = a_n x^n + a_{n-1}x^{n-1} + \cdots + a_1 x + a_0$, the discriminant is given by the expression $D = a_n^{2n-1}(x_1 - x_2)^2(x_1 - x_3)^2 \cdots (x_1 - x_n)^2(x_2 - x_3)^2 \cdots (x_2 - x_n)^2 \cdots (x_{n-1} - x_n)^2$, where $x_1, x_2, \ldots, x_n$ are the roots of the equation $f(x) = 0$.

The importance of the discriminant D lies in the fact that D vanishes if, and only if, the equation $f(x) = 0$ has equal roots. Since the value of the discriminant is unchanged if any two letters x_i and x_j are interchanged, it is a symmetric function of the roots, and can be expressed in terms of the coefficients of $f(x)$. The discriminant of a quadratic polynomial $f(x) = a_2x^2 + a_1x + a_0$ is $D = a_1^2 - 4a_2a_0$. The discriminant of a cubic polynomial $f(x) = a_3x^3 + a_2x^2 + a_1x + a_0$ is

$$D = 18a_3a_2a_1a_0 - 4a_2^3a_0 + a_2^2a_1^2 - 4a_3a_1^3 - 27a_3^2a_0^2$$

See EQUATIONS, THEORY OF; POLYNOMIAL SYSTEMS OF EQUATIONS. [R.A.B.]

Disease A deleterious set of responses which occurs at the subcellular level, stimulated by some injury, and which is often manifested in altered structure or functioning of the affected organism. With advances in understanding and the development of sensitive probes, it has become clear that the fundamental causes of diseases are based on biochemical and biophysical responses within the cell. These responses are now being categorized and, slowly, the mechanisms are being understood.

The term homeostasis refers to functional equilibrium in an organism and to the processes that maintain it. There is a range of responses that is considered normal. If cells are pushed to respond beyond these limits, there may be an increase, a decrease, or a loss of normal structure or function. These changes may be reversible or irreversible. If irreversible, the cells may die. Thus, subcellular changes may be reflected in altered tissues, organs, and consequently organisms, and result in a condition described as diseased. *See* HOMEOSTASIS.

Lesions are the chemical and structural manifestations of disease. Subjective manifestations of a disease process such as weakness, pain, and fatigue are called symptoms. The objective measurable manifestations such as temperature, blood pressure, and respiratory rate changes are called signs or physical findings. Changes in the chemical or cellular makeup of an organ, tissue, or fluid of the body or its excretory products are called laboratory findings. To make a diagnosis is to determine the nature of the pathologic process by synthesizing information from these sources evaluated in the light of the patient's history and compared with known patterns of signs and symptoms. In common usage, the term disease indicates a constellation of specific signs and symptoms attributable to altered reactions in the individual which are produced by agents that affect the body or its parts.

Etiology is the study of the cause or causes of a disease process. Although a disease may have one principal etiologic agent, it is becoming increasingly apparent that there are several factors involved in the initiation of a disease process. Susceptibility of the individual is an ever present variable. The etiologic factors can conveniently be divided into two categories (see table). One group consists of endogenous (internal; within the body) factors, and may originate from errors in the genetic material. The other category of etiologic factors is exogenous (environmental). These account for the majority of disease reactions. Exogenous factors include physical, chemical, and biotic agents.

Pathogenesis refers to the mechanisms by which the cell, and consequently the body, responds to an etiologic agent. It involves biochemical and physiological responses which are reflected in ultrastructural, microscopic, or gross anatomic lesions. There are a limited number of ways in which cells respond to injury. The nature of the response is modified by the nature of the agent, dose, portal of entry, and duration of exposure, as well as many host factors such as age, sex, nutritional state, and species and individual susceptibility.

The diseases which are important in causing human death have changed in the last 80 years. In 1900 six of the ten leading causes of death in the United States were infectious (biotic) agents. At present, only one of the ten leading causes of death in the United States, influenza and pneumonia, is due to biotic agents. While most of the biotic causes of diseases were being

Common exogenous and endogenous causes of disease

Causative agent	Disease
EXOGENOUS FACTOR	
Physical	
Mechanical injury	Abrasion, laceration, fracture
Nonionizing energy	Thermal burns, electric shock, frostbite, sunburn
Ionizing radiation	Radiation syndrome
Chemical	
Metallic poisons	Intoxication from methanol, ethanol, glycol
Nonmetallic inorganic poisons	Intoxication, from phosphorous, borate, nitrogen dioxide
Alcohols	Intoxication from methanol, ethanol, glycol
Asphyxiants	Intoxication from carbon monoxide, cyanide
Corrosives	Burns from acids, alkalies, phenols
Pesticides	Poisoning
Medicinals	Barbiturism, salicylism
Warfare agents	Burns from phosgene, mustard gas
Hydrocarbons (some)	Cancer
Nutritional deficiency	
Metals (iron, copper, zinc)	Some anemias
Nonmetals (iodine, fluorine	Goiter, dental caries
Protein	Kwashiorkor
Vitamins:	
A	Epithelial metaplasia
D	Rickets, osteomalacia
K	Hemorrhage
Thiamine	Beriberi
Niacin	Pellagra
Folic acid	Macrocytic anemia
B_{12}	Pernicious anemia
Ascorbic acid	Scurvy
Biological	
Plants (mushroom, fava beans, marijuana, poison ivy, tobacco, opium)	Contact dermatitis, systemic toxins, cancer, hemorrhage
Bacteria	Abscess, scarlet fever, pneumonia, meningitis, typhoid, gonorrhea, food poisoning, cholera, whooping cough, undulant fever, plague, tuberculosis, leprosy, diphtheria, gas gangrene, botulism, anthrax
Spirochetes	Syphilis, yaws, relapsing fever, rat bite fever
Virus	Warts, measles, German measles, smallpox, chickenpox, herpes, roseola, influenza, psittacosis, mumps, viral hepatitis, poliomyelitis, rabies, encephalitis, trachoma
Rickettsia	Rocky Mountain spotted fever, typhus
Fungus	Ringworm, thrush, actinomycosis, histoplasmosis, coccidiomycosis
Parasites (animal)	
Protozoa	Amebic dysentery, malaria, toxoplasmosis, trichomonas vaginitis
Helminths (worms)	Hookworm, trichinosis, tapeworm, filariasis, ascariasis
ENDOGENOUS FACTOR	
Hereditary	Phenylketonuria, alcaptonuria, glycogen storage disease, Down's syndrome (trisomy 21), Turner's syndrome, Klinefelter's syndrome, diabetes, familial polyposis
Hypersensitivity	Asthma, serum sickness, eczema drug idiosyncrasy

brought under control, continued population growth (in large part, a consequence of the control of infectious disease) and the remarkable growth of industrialization have been associated with an increased prevalence of diseases caused by physical and chemical agents. These include cancer, cirrhosis, and cardiovascular disease.

General principles of the organism's response to toxic substances, some of which occur naturally in the environment, have evolved from a great number of investigations of agent-host interaction. They are: (1) All substances entering the organism are toxic; none is harmless. Dose rate of exposure and route of entry into the body determine whether a toxic response will occur or not. (2) All agents evoke multiple responses. (3) Most of the biological responses are undesirable, leading to the development of pathological changes. (4) A given dose of an agent does not produce the same degree of response in all individuals. Thus, when disease is viewed as interaction between the environment and the individual, the control of disease is largely the management of the environmental causes of disease. [N.K.M.; C.Qu.]

Disk recording The process of inscribing suitably transformed acoustical or electrical signals on a flat circular plate that may be played back at a subsequent time. Virtually all modern disk recorders and reproducers are used to record or reproduce sound signals, mainly music and voice.

Monophonic system. A monophonic disk recording system consists of a disk record rotated by a turntable mechanism and a cutter for producing undulations in a groove in the disk corresponding to the sound signals. A monophonic disk reproducing system consists of a pickup and mechanism for rotating the disk record by means of which the recorded undulations in the disk record are converted into electrical signals of approximately like form.

The elements in a complete monophonic disk recording system are shown in Fig. 1. The first element is the acoustics of the studio. The output of each microphone is amplified and fed

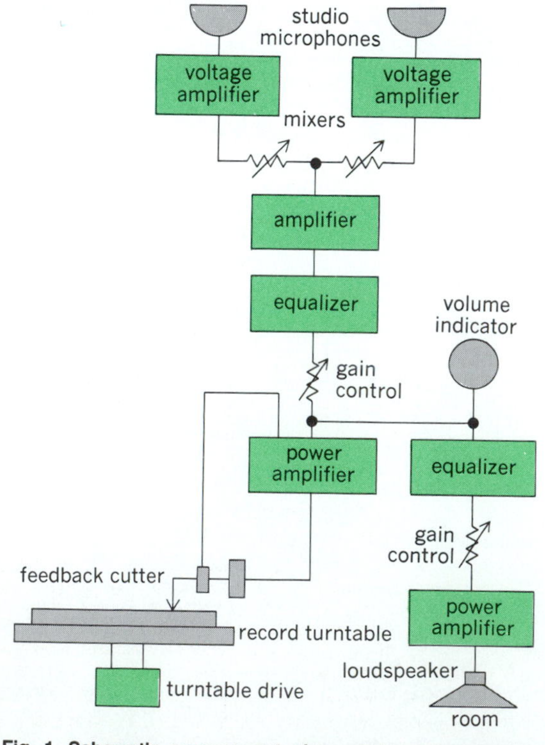

Fig. 1. Schematic arrangement of apparatus in a complete monophonic disk recording system.

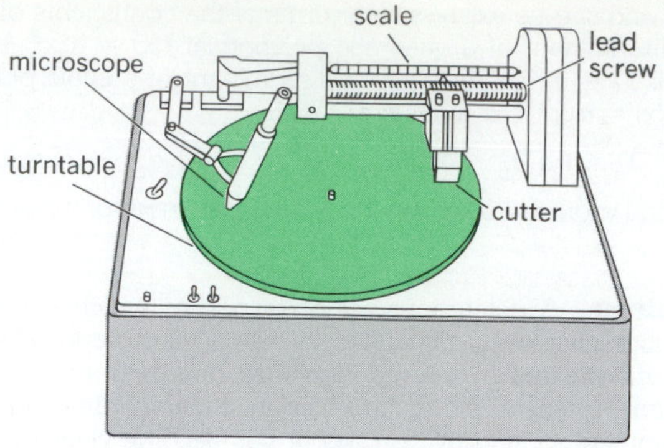

Fig. 2. Disk phonograph recorder. Microscope is used for periodic inspection of groove. (*After H. F. Olson, Acoustical Engineering, Van Nostrand, 1957*)

to a mixer, a device having two or more inputs and a common output. If more than one microphone is used (for example, when an orchestra accompanies a soloist, there is one microphone for the orchestra and one for the singer), the outputs of the two microphones may be adjusted for the proper balance by means of the mixers. An electronic compressor is used to reduce a large amplitude range to that suitable for reproduction in the home. A corrective electrical network called an equalizer provides the recording characteristics. The attenuator, or gain control, provides a given control on the overall level fed to the power amplifier. The cutter, actuated by the amplifier, cuts a wavy path in the groove of the revolving record corresponding to the undulations in the original sound wave striking the microphone. A monitoring system consisting of a volume indicator, complementary equalizer, attenuator or gain control, power amplifier, and loudspeaker is used to control the recording operation.

A phonograph recorder (Fig. 2) is an instrument for transforming acoustical or electrical signals into motion of approximately like form and inscribing such motion in an appropriate medium by cutting or embossing. The lacquer disk used in recording the master record is placed on the recording turntable. The turntable is heavy, to ensure against spurious rotational motions. A suitable mechanical filter is placed between the driving motor and the turntable so that uniform rotational motion of the turntable will be obtained. The lead screw drives the cutter in a radial direction so that a spiral groove is cut in the record. Lead screws of different pitches are used, ranging from 100 to 500 grooves per inch (4 to 20 per millimeter). In some recordings a variable pitch is used. In this procedure the spacing between the grooves is made to correspond to the amplitude—small spacing for small amplitudes and large spacing for large amplitudes. Under these conditions the maximum amount of information can be recorded on a record.

The electromechanical transducers used as cutters can be of either the lateral or the vertical type. In the lateral type of disk recording, the undulations are cut in a direction parallel to the surface of the record and perpendicular to the groove. The vibrating system is of the dynamic type with two coils, one the driving coil and the other the sensing (feedback) coil, wound on a common cylinder. The cutting stylus is attached to the coil cylinder.

The cutting stylus consists of a sapphire, synthetic ruby, or other hard material fashioned in the form of a pointed chisel. The stylus is heated in recording and thereby imparts a smooth sidewall to the groove. This expedient results in considerable reduction in noise in reproduction.

The original recording of disk records is made on a lacquer disk. The lacquer disk consists of a coating of an acetate plastic on two sides of an aluminum disk. The grooves are cut in the plastic.

In the vertical (hill-and-dale) type of disk recording, the undulations of the groove are cut in a direction perpendicular to the surface of the record and perpendicular to the groove. The vertical disk phonograph system is used to a limited extent in broadcasting stations, but it is not used in home disk phonograph systems.

In the mass production of disk phonograph records, the original lacquer disk, termed the original, is metalized and then electroplated. The plating is separated from the lacquer and reinforced by backing with a solid metal plate. The assembly, called the master, is electroplated. This plating is separated from the master and reinforced by backing with a solid plate. The resulting assembly, the mother, is electroplated and reinforced by a solid metal plate, forming an assembly termed the stamper. Several stampers are made from each mother. One stamper containing a sound selection to be placed on one side of the final record is mounted in the upper jaw, and another stamper containing a sound selection to be placed on the other side of the record is placed in the lower jaw of a hydraulic press equipped with means for heating and cooling the stampers. A preform, or biscuit, of thermoplastic material such as a shellac compound or vinyl compound is placed between the two stampers.

The stampers are heated, and the jaws of the press are closed to bring the two stampers against the thermoplastic material. When an impression of the stampers has been obtained in the thermoplastic material, the stampers are cooled, thus cooling and setting the plastic record. The jaws of the hydraulic press are opened, and the record is removed from the press. The modulated grooves in the record correspond to those in the original lacquer disk. The stamping procedure is repeated until the desired number of records is obtained.

Stereophonic system. Two-channel disk phonograph sound reproduction was commercialized in 1958.

In a two-channel disk phonograph dynamic-type feedback cutter the two vibrating systems are arranged at right angles; therefore, the two channels in the groove are recorded at right angles. The same type of recorder described in the beginning of this article may be used in the recording of stereophonic records. *See* STEREOPHONIC SOUND.

Quadraphonic system. There are two approaches to quadraphonic sound reproduction by means of the disk record: first, the matrixing and encoding from four channels, to recording on a two-channel disk, and reproducing from a two-channel disk and decoding and matrixing to four channels; and, second, the recording and reproducing of four discrete channels by means of the disk record. *See* QUADRAPHONIC SOUND SYSTEM.

Distortion and noise. Distortion is any undesired change in the waveform of a signal in a sound-reproducing system. Noise is any foreign, erratic, intermittent, or statistically random signal produced within a sound-reproducing system. Distortion and noise are the two major problems that must be solved to achieve high-fidelity performance in disk sound-reproducing systems.

The recording and reproducing of a phonograph record constitute a complicated process, with many sources of nonlinear distortion. The record does not present an infinite mechanical impedance to the stylus. As a consequence, the vibrating system of the pickup is shunted by the effective mechanical impedance of the record at the stylus. Nonlinear distortion will be introduced if this impedance of the record is variable. Other sources of distortion are tracking error and tracing distortion. A nonlinear distortion due to a deviation in tracking is commonly termed tracking error. A form of distortion in lateral disk-record reproduction known as tracing distortion is a function of the diameter of the stylus, the lateral velocity, and the linear groove velocity. This distortion is due to the fact that there is not a one-to-one correspondence between the shapes of the cutting and reproducing styli.

The record surface noise, in the absence of any signal, is one of the factors which limits the volume range and the frequency range of shellac phonograph records. The noise of Vinylite records is extremely low and, in general, is not a problem. Adequate volume ranges can be obtained with Vinylite and similar plastics. *See* SOUND-REPRODUCING SYSTEMS. [H.F.O.]

Dispersion (radiation) The separation of a complex of electromagnetic or sound waves into its various frequency components. For example, a beam of white light can be separated into its monochromatic components by virtue of the different velocities of rays of different wavelength of the beam as it passes through a prism. The dispersion of a material, such as glass or water, at a given wavelength in the electromagnetic spectrum is defined as the rate of change of refractive index with wavelength at the wavelength in question. [W.W.]

Dispersion relations Relations between the real and imaginary parts of a response function. A response function relates a cause and its effect through an integral equation. The term dispersion refers to the fact that the index of refraction of a medium is a function of frequency. In 1926 H. A. Kramers and R. Kronig showed that the imaginary part of an index of refraction (that is, the absorptivity) determines the real part (that is, the refractivity); this is called the Kramers-Kronig relation. The term dispersion relation is now used for the analogous relation between the real and imaginary parts of the values of any response function. *See* COMPLEX NUMBERS. [C.J.G.]

Displacement (mechanics) When an object is moved from one position to another, it is said to be displaced, and the linear distance from the initial to the final position, regardless of the length of path followed, is called the displacement. The displacement is always in a particular direction, and consequently displacement is a vector quantity involving direction as well as magnitude. In rectilinear motion the magnitude of the displacement is also the length of path or the distance traversed, but since length of path does not involve direction, it is a scalar quantity.

When a body is rotated about any axis, it is said to undergo angular displacement. Angular displacement is commonly measured in radians or degrees. *See* ROTATIONAL MOTION. [R.D.Ru.]

Displacement current The name given by J. C. Maxwell to the term $\partial \mathbf{D}/\partial t$ which must be added to the current density $\mathbf{i}$ to extend to time-varying fields A. M. Ampère's magnetostatic result that $\mathbf{i}$ equals the curl of the magnetic intensity $\mathbf{H}$. In integral form this result is given by the equation below,

$$\oint \mathbf{H} \cdot d\mathbf{s} = \int_S \left(\mathbf{i} = \frac{\partial \mathbf{D}}{\partial t} \right) \cdot \mathbf{n}\, dS$$

where the unit vector $\mathbf{n}$ is perpendicular to the surface dS. The concept of displacement current has important consequences for insulators and for free space where $\mathbf{i}$ vanishes. For conductors, however, the difference between the above equation and Ampère's result is negligible. *See* AMPÈRE'S LAW; MAXWELL'S EQUATIONS.

If one defines current as a transport of charge, the term displacement current is certainly a misnomer when applied to a vacuum where no charges exist. If, however, current is defined in terms of the magnetic fields it produces, the expression is legitimate. [W.R.Sm.]

Displacement pump A pump that develops its action through the alternate filling and emptying of an enclosed volume. There are five basic types: reciprocating, direct-acting steam, rotary, vacuum, and air-lift.

Positive-displacement reciprocating pumps have cylinders and plungers or pistons with an inlet valve, which opens the cylinder to the inlet pipe during the suction stroke, and an outlet valve, which opens to the discharge pipe during the discharge stroke. Reciprocating pumps may be power-driven through a crank and connecting rod or equivalent mechanism, or direct-acting, driven by steam or compressed air or gas. Power-driven reciprocating pumps are highly efficient over a wide range of discharge pressures. Except for some special designs with continuously variable stroke, reciprocating power pumps deliver essentially constant capacity over their entire pressure range when driven at constant speed.

A reciprocating pump is readily driven by a reciprocating engine; a steam or power piston at one end connects directly to a fluid piston or plunger at the other end. Steam pumps can be built for a wide range of pressure and capacity by varying the relative size of the steam piston and the liquid piston or plunger. The delivery of a steam pump may be varied at will from zero to maximum simply by throttling the motive steam, either manually or by automatic control. Reciprocating pumps are used for low to medium capacities and medium to highest pressures. They are useful for low- to medium-viscosity fluids, or high-viscosity fluids at materially reduced speeds. Specially fitted reciprocating pumps are used to pump fluids containing the more abrasive solids.

Another form of displacement pump consists of a fixed casing containing gears, cams, screws, vanes, plungers, or similar elements actuated by rotation of the drive shaft. Most forms of rotary pumps are valveless and develop an almost steady flow rather than the pulsating flow of a reciprocating pump.

Although vacuum pumps actually function as compressors, displacement pumps are used for certain vacuum pump applications. Simplex steam pumps with submerged piston pattern fluid ends are used as wet vacuum pumps in steam heating and condensing systems. Sufficient liquid remains in the cylinder to fill the clearance volume and drive the air or gas out ahead of the liquid. *See* VACUUM PUMP.

In handling abrasive or corrosive waters or sludges, where low efficiency is of secondary importance, air-lift pumps are used. The pump consists of a drop pipe in a well with its lower end submerged and a second pipe which introduces compressed air near the bottom of the drop pipe. The mixture of air and water in the drop pipe is lighter than the water surrounding the pipe. As a result, the mixture of air and water is forced to the surface by the pressure of submergence. *See* COMPRESSOR; PUMP. [E.F.W.]

Distance-measuring equipment An internationally standardized navigation system which allows an aircraft to measure its distance from a selected ground-based beacon. Such beacons are used throughout the world by all airliners, most of the military aircraft of the West, and a large number of general-aviation aircraft. The range of service is line-of-sight up to 300 mi (480 km) and system accuracy is usually 0.1 mi (0.16 km) but precision equipment, intended for use during landing, has accuracy of 100 ft (30 m).

The airborne equipment, called an interrrogator, transmits pulses of 1 kW peak power on 1 of 126 frequencies. These are in the 1025–1150-MHz band and are spaced 1 MHz apart. Each pulse is of 3.5 microseconds duration and is paired with another, spaced 12 or 36 microseconds later. The combination of frequencies and pulse spacings therefore provides 252 operating channels.

The beacon on the ground, called a transponder, receives these pulses, delays them by 50 μs, and then retransmits them, usually with a power of 1 kW, on 252 frequencies lying between 962 and 1213 MHz. The pulse-pair spacing is 12 μs on those frequencies not used by the interrogator, and 30 μs on those frequencies shared with the interrogator. The transponder transmission is called the reply. The frequency difference between interrogation and reply is always 63 MHz. This arrangement allows each transmitter frequency to act as the local oscillator for its associated superheterodyne receiver, the intermediate frequency of which is 63 MHz. For landing purposes, some transponders have powers as low as 100 W.

In the aircraft the replies to its own interrogations are recognized by their phase coherence with their own transmissions, and by the elapsed time measured between transmissions and reception (minus the 50-μs transponder delay), usually by means of a crystal clock. This elapsed time is about 12 μs for each nautical mile (7 μs for each kilometer), and is displayed in the cockpit on a digital meter, which is usually calibrated in miles and tenths of miles. *See* ELECTRONIC NAVIGATION SYSTEMS; INSTRUMENT LANDING SYSTEM (ILS); RADIO RANGE; RHO-THETA SYSTEMS; TACAN. [S.H.D.]

Distillate fuel A broad term for any one of the wide variety of fuels obtained from fractions boiling above gasoline in the distillation of petroleum. The most important distillate fuels are kerosine, furnace oils, and diesel fuels. *See* DIESEL FUEL; KEROSINE; PETROLEUM PRODUCTS. [M.Sou.]

Distillation The process of producing a gas or vapor from a liquid or solid. However, the term sublimation is used ordinarily to describe the vaporization of a solid. Heat is generally supplied to the liquid during the distillation, although in special cases the latent heat required for the vaporization may be obtained from the internal energy of the liquid.

The main purpose of distillation is either the separation of volatile components from nonvolatile materials, or the separation of a mixture of volatile components. The separation of volatile components from nonvolatile materials is carried out by a simple distillation in which the material is placed in a still and heated, and the vapor removed and condensed. Simple distillation is similar to the process of evaporation, but the former term usually describes the operation in which the volatile material is a desired product, whereas evaporation generally is applied to aqueous solutions of nonvolatile materials in which the nonvolatile material is the desired product. Simple distillation is frequently used for high-boiling organic compounds; to prevent thermal degradation of the product, the operation is usually carried out either at reduced pressure, termed simple vacuum distillation, or with the addition of steam, termed steam distillation. *See* EVAPORATION; EVAPORATOR; HEAT TRANSFER; SUBLIMATION.

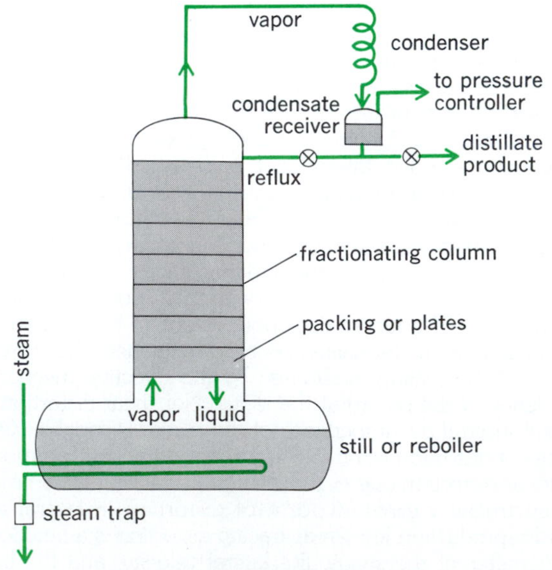

Batch fractional distillation system.

Although simple distillation can be applied to mixtures of volatile components, the separation obtained is usually not complete, particularly if the components have boiling points that are close to each other. To obtain greater separations in such cases, fractional distillation is employed. In this process, the vapors from the still are permitted to come in contact with a portion of the condensate in a countercurrent or stepwise countercurrent operation. Because of its lower operating and capital costs, this countercurrent type of operation has completely replaced the multiple distillation-condensation method formerly employed. This process has been widely adopted for both laboratory and industrial operations because it is usually the most effective method of separating mixtures of miscible volatile liquids. It is so effective and efficient that it is also frequently employed to separate mixtures which are not normally liquids, such as fractional distillation of liquid air.

Separation of two liquids is possible only when the composition of the vapor is different from that of the liquid from which it was produced. The design of distillation equipment is usually based on vapor-liquid equilibrium compositions. The vapor-liquid equilibrium data needed for distillation work either obtained experimentally or estimated from physical chemistry relationships. *See* BOILING POINT; GAS; LIQUID; PHASE EQUILIBRIUM; VAPOR PRESSURE.

In steam distillation, steam is introduced directly into the liquid in the still. The method is usually limited to those cases in which the solubility of the steam in the liquid is low at the operating temperature and pressure. It is employed with relatively high-boiling organic materials which would decompose if they were distilled directly at atmospheric pressure or with liquids that have such poor heat-transfer characteristics that excessive local overheating would result with indirect heating. By steam distillation, a volatile material can be separated from nonvolatile impurities, or mixtures can be separated with results about equivalent to those obtained with simple distillation. *See* HEAT EXCHANGER; VAPOR CONDENSER.

In fractional distillation, the vapor produced in the still is brought into contact with a portion of the condensate in a countercurrent or stepwise countercurrent system. The operation can be batch or continuous. The unit consists of a still to which a vertical column is attached (see illustration). This column is filled with some type of packing or plate construction which will permit the descending liquid added at the top to come into contact with the vapor rising from the still. The vapor from the top of the column is condensed and a portion of it is removed as product; the remaining liquid is returned to the top of the column as liquid, called reflux. The flow rates of liquid and vapor are adjusted so that, at every place within the column, the liquid has a higher concentration of the more volatile components than corresponds to equilibrium with the vapor with which it is in contact. As a result, the more volatile components pass from the liquid to the vapor and the less volatile components pass in the reverse direction. The vapor becomes progressively more enriched in the volatile components as it flows up the column to the condenser, and the liquid becomes more concentrated in the less volatile components as it flows down the column to the still. The separation obtainable by such a system is much greater than that of a simple distillation. [E.R.G]

Distortion (electronic circuits) Any undesired change in the waveform of an electric signal passing through a circuit, including the transmission medium. In the design of any electronic circuit one important problem is to modify the input signal in the required way without producing distortion beyond an acceptable degree. Amplifier and loudspeaker systems are examples where maximum effort has been expended to produce a design for faithful amplification of speech and music input signals. There are four general types of distortion: amplitude, frequency, phase, and cross modulation.

Amplitude distortion is generally considered to mean distortion produced by a nonlinear relationship between the input and output amplitudes of a device. Amplitude distortion is usually introduced by a transistor or vacuum tube. The change in collector current of a transistor is nearly proportional to the change in signal voltage only over a small range of signal voltage amplitude. For further increase, the collector current change begins to depart from proportionality. *See* TRANSISTOR.

Frequency distortion is an inherent feature of all amplifiers but can be minimized by proper design. It occurs because the reactive elements and inherent reactances in the amplifier circuit do not allow the same amplification for all frequencies, and therefore some components of a signal are amplified more than others. Furthermore, in the case of audio amplifiers, the loudspeaker and enclosure characteristics affect the load presented to the amplifier in a manner which depends upon frequency.

Like frequency distortion, phase distortion is caused by the reactive elements in the circuit producing a phase shift that is not the same for each frequency component of the input signal. It is possible for an input signal to have frequency components with approximately constant magnitude amplification but not with a phase shift for each component which is proportional to the frequency. As a result, the output is not an amplified replica of the input. Fortunately, the ear is more tolerant of phase distortion than frequency distortion.

Cross modulation, also called intermodulation, is caused by nonlinear device characteristics. If two signals of different frequencies are applied to the input of a nonlinear transistor or vacuum tube stage, the output will contain the fundamental and harmonic components of each signal, frequency components equal to the sum and difference of the input signal frequencies, and sums and differences of the harmonics of the two input signals. Therefore, if the input is a signal composed of several frequencies, the nonlinearity will produce new frequencies not integrally related to those of the input signal. The distortion, if bad enough, is generally more noticeable than harmonic distortion. *See* FEEDBACK CIRCUIT. [C.C.H.]

Distortion (optics) The optical aberration arising from the variation in magnification over the field of an optical system. Distortion that is positive, the magnification increasing with field angle, is called pincushion distortion because the image of a square has concave sides and thus looks like a pincushion (see illustration). The opposite type, negative distortion, is called barrel distortion because the image of a square has bulging sides. It is possible to balance distortion in an optical system by balancing higher-order distortion through introducing some third-order distortion of opposite sign. [M.J.H.]

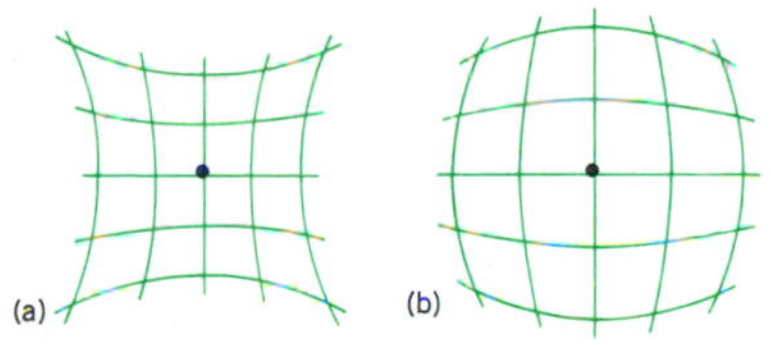

Images of a rectangular object screen shown with (*a*) pincushion distortion and (*b*) barrel distortion. (*After F. A. Jenkins and H. E. White, Fundamentals of Optics, 3d ed., McGraw-Hill, 1957*)

Distributed systems (computers) Systems that implement related information-processing functions in two or more programmable devices, connected so that data can be exchanged, or that store segments of a database on two or more computer systems. The subject of distributed systems thus encompasses both distributed processing and distributed databases. Distributed processing is sometimes referred to as distributed data processing (DDP).

Distributed processing. Distributed processing systems are distinguished from other types of computer-based systems by three characteristics. First, information-processing (application) functions are distributed. This excludes systems which consist of one information-processing computer surrounded by a network with programmable concentrators, network switches, or similar network-control devices.

The second important characteristic is programmability in the distributed devices. Hard-wired terminals or terminal controllers, which provide management of only the terminal devices, do not fit within this definition.

The final important characteristic is connection for data exchange. Two independent computer systems operated by the same organization, but without any linking network, are an example of decentralized, not distributed, processing.

At about the same time that minicomputers became generally available (in the late 1960s and early 1970s), the growing experience with data networks made these practical for use in more organizations. These two conditions came together to make possible the early attempts at distributed processing. Not long thereafter, the introduction of the microprocessor-based personal computer made it still more attractive to distribute functions rather than retaining and controlling them all centrally. *See* DATA COMMUNICATIONS; LOCAL-AREA NETWORK; MICROCOMPUTER; WIDE-AREA NETWORK.

Although distributed systems are, in practice, extremely diverse, there are only two distributed processing structures, or ways in which to organize the functions within a distributed system. These can be used independently or can be combined.

Hierarchical systems represent the earliest form of distributed processing. In these systems, processing functions are distributed outward from a central computer, called the host system, to departmental systems or personal computers. A system of this type is sometimes called a host-centered structure. A host processor is typically located at some central site, where it serves as a focal point for the collection of data and often for the provision of services which cannot economically be distributed (for example, the management of large databases, or simultations). *See* DATA-PROCESSING SYSTEMS; DATABASE MANAGEMENT SYSTEMS; DIGITAL COMPUTER.

Horizontal systems form the other major structure for distributed processing. In this structure, two or more computers which are logically equivalent are connected. The term logically equivalent means that there is no connotation of hierarchy or master/slave relationship, even though the systems may not be physically equal in their functional capability or processing capacity. This form is also called a peer structure or a host/host connection.

Distributed databases. Distributed processing is often accompanied by a distributed database. A distributed database is a collection of logically related data, segments of which are stored on two or more computer systems. In addition, the distributed systems must allow users to access multiple segments of the distributed database.

A typical example of a distributed database is a bank that maintains its records of customer accounts in two geographically separate locations, attached to two separate host systems. Because customers travel, it must be possible to check an account in one host system when the customer wants to cash a check at a branch served by the other host.

This type of distributed database is called a partitioned distributed database, and ideally each partition contains only unique data. In practice, some data elements that are accessed frequently but changed infrequently are maintained in parallel in all partitions.

Such parallel data storage is characteristic of the other type of distributed database, a replicated distributed database. A typical example of such a database is a manufacturing management system, in which the production schedule for the entire factory is maintained by the host, while the schedule for each department is copied (replicated) on the departmental system for that part of the factory. *See* DIGITAL COMPUTER. [G.M.Bo.]

Distributed systems (control systems)

Collections of modules, each with its own specific function, interconnected to carry out integrated data acquisition and control. Typical modules include data acquisition modules, direct digital controllers, operator consoles, and the supervisory computer.

Industrial control systems have evolved from totally analog systems through centralized digital computer-based systems to multilevel, distributed systems. In an analog system the individual process variables were each controlled by a single analog feedback controller. Analog control systems had many advantages: simplicity due to the separate nature of the realizations of the control loops; reliability from the independence of the analog control devices (separate transducers, transmitters, and controllers); and graceful degradation in the face of a device failure, since the failure of one element or loop has a limited effect on the rest of the control system. The major disadvantage of analog control is the lack of integrated process information display to the process operator and the difficulty in implementing complex control strategies. Supervisory control using digital computers alleviated these problems. This mode of control is called supervisory control because the computer itself is not directly involved in the dynamic feedback. Here the analog control system is maintained as before, but a digital computer is added which periodically scans, digitizes, and inputs process variables to the computer. *See* DIGITAL COMPUTER.

Direct digital control evolved from supervisory control and later merged with supervisory control to produce the best available system prior to the introduction of distributed control. Direct digital control replaces the analog control with a periodically executed equivalent digital control algorithm carried out in the central digital computer. The advantages of direct digital control are the ease with which complex dynamic control functions can be carried out and the elimination of the cost of the analog controllers themselves. The cost reduction which resulted from the introduction of direct digital control was offset by a number of disadvantages. The most notable of these were the decrease in reliability and the total loss of graceful degradation. Failure of a sensor or transmitter had the same effect as before, but failure of the computer itself threw the entire control system into manual operation. Hence it was necessary to provide analog controllers to back up certain critical loops which had to function even when the computer was down.

Increasing demand for higher and higher levels of supervisory control, coupled with increasing installation cost, highlighted two disadvantages of the centralized digital computer control of processes. First, process signals were still being transmitted from the process sensor to the central control room in analog form, meaning that separate wires had to be installed for every signal going to or from the computer. Cost of installing this cable is very high. Second, the digital computer system itself evolved into a very complex unit because of the number of devices attached to the computer and because of the variety of different programs needed to carry out the myriad control and management functions. The latter resulted in the need for an elaborate real-time operating system for the computer. Design, coding, installation, and checkout of centralized digital control systems were so costly and time-consuming that application of centralized digital control was limited.

Low-cost electronic hardware utilizing large-scale integrated (LSI) circuits provided the technology to solve both of these problems while retaining the advantages of centralized direct digital and supervisory control. The solution involved distributing control functions in such a way that software complexity was eliminated and process signals could share communication paths to the central computer. *See* INTEGRATED CIRCUITS.

Distributed systems have a number of valuable modern applications. Chemical processing plants, and electrical energy generating plants in particular, have stringent control requirements related to process quality and process safety which demand very reliable, very fast, and very complex control systems. These needs are being met by distributed systems. *See* CONTROL SYSTEMS; DIGITAL CONTROL. [J.D.S.]

Distribution (probability) The results of a series of independent trials, random variables, or errors often occur in fairly regular and predictable patterns that can be expressed mathematically. The most important of them are called the binomial, normal, and Poisson distributions.

Consider n independent trials, each of which results in success S or failure F, with corresponding probabilities p and $q = 1 - p$. Denote by S_n the number of successes. Because there are $\binom{n}{k}$ possible ways to select k places for S and $n - k$ places for F, the probability distribution of the random variable S_n is given by

$$P\{S_n = k\} = \binom{n}{k} p^k q^{n-k}$$

where $k = 0, 1, \ldots, n$. This is the binomial distribution. Its expectation is np, its variance npq. *See* PROBABILITY.

The standard normal density is defined for $-\infty < x < \infty$ by

$$\phi(x) = (2\pi)^{-1/2} e^{-x^2/2}$$

The standard normal distribution function or error function $\Phi(x)$ is its integral from $-\infty$ to x. The normal distribution function with mean m and variance s^2 is $\Phi[(x - m)/s]$; its density is $(2\pi)^{-1/2} s^{-1} e^{-(x-m)^2/(2s^2)}$. As x goes from $-\infty$ to ∞, the function increases from 0 to 1. It plays an important role in many fields. In particular, $u(t,x;\xi) = 2\pi^{-1/2} t^{-1/2} e^{-(x-\xi)^2/4t}$ is the fundamental solution of the heat (or diffusion) equation $u_t = u_{xx}$ and represents the heat distribution on the x axis at time t caused by a unit heat source initially concentrated at $x = \xi$. Probabilistically, this represents the transition probabilities in the Wiener process. A random variable whose distribution is normal is called normal or Gaussian. Many empirical quantities (for example, the amount of water in a reservoir, certain inherited characteristics such as height, and the experimental error of physical measurements) represent the cumulative effect of many small components, and the statistical fluctuations of such quantities may be expected to follow the normal distribution.

The Poisson distribution with parameter λ is the probability distribution of a random variable assuming the values 0, 1, 2,...with probabilities $p_k(\lambda) = e^{-\lambda}\lambda^k/k!$ Both its expectation and variance equal λ. This is one of the most important distributions; it plays a basic role in the theory of stochastic processes and in many applications.

Consider a large number n of independent trials, each of which results in success or failure with probabilities p and $q = 1 - p$. Ordinarily, interest is restricted to the case where p is very small, but the expected number of successes $np = \lambda$ is of moderate size. Typical examples may be obtained by considering centenarians, color-blind people, or triplets in a large population. The number of successes in the n trials is a random variable with the binomial distribution, but under the present circumstances the binomial is close to the Poisson distribution and may be replaced by it. In fact, the probability of no success is $q^n = (1 - \lambda/n)^n$, which is close to $e - \lambda$, the first term of the Poisson distribution.

In other circumstances the Poisson distribution appears, not as an approximation, but as the exact expression of a law of nature. This is true in particular of processes where certain events, such as radioactive disintegrations, mutations, power failures, and accidents, occur in time in such a way that (1) the probability that an event occurs during any given time interval of length dt is, asymptotically, λdt and (2) there is no interaction or aftereffect between nonoverlapping time intervals. A similar argument applies to random distributions of points in space, with t interpreted as volume; typical examples are stars, flaws of material, raisins in a cake, and animal litters in a field. *See* ANALYSIS OF VARIANCE; STATISTICS. [W.F.]

Distributor A rotary switch that directs the high-voltage ignition current in the proper firing sequence to the various cylinders of an internal combustion engine. In automotive practice the distributor housing usually contains, in addition, apparatus for timing the ignition to occur when each piston is at optimum position in the cycle. Timing is also varied with engine load, by the movement of a diaphragm exposed to the pressure in the engine intake manifold. *See* IGNITION SYSTEM. [A.R.R.]

District heating The supply of heat, either in the form of steam or hot water, from a central source to a group of buildings. Most district heating systems in the United States rely on separate steam generation facilities close to load centers. In some cities, notably New York, high-pressure district steam is used extensively to feed turbines that drive pumps and refrigerant compressors. Although some district heating plants serve detached residences, the cost of underground piping and the small quantities of heating service required makes this service generally unfeasible. District heating, apart from utility-supplied systems in downtown areas of cities, is accepted as efficient practice in many colleges, universities, and government complexes.

In Europe, a substantial part of the heat distribution is by hot water from plants combining power and heat generation. Hot water distribution systems have the advantage of high thermal mass (hence, storage for peak periods) and freedom from problems of condensate return. On the other hand, pumping costs are high and two mains are required, supply and return. *See* COMFORT HEATING. [R.L.K.]

Disulfide One of a group of organosulfur compounds, RSSR′, that may be symmetrical (R = R′) or unsymmetrical (R and R′ different). They are of great biochemical interest, since the S-S link occurs in natural products. The cleavage of the S-S bond in proteins, for example, has important biological and industrial interests in the dehairing of hides, in the preparation of cysteine from wool, in wavesetting of hair, and in petroleum refining. Disulfides are also of interest in the manufacture of polysulfide rubbers and polymers. *See* ORGANOSULFUR COMPOUND. [N.K.]

Diverticulosis A degenerative process in the gastrointestinal tract that results in saccular outpouchings of the colonic mucosa which extend to the serosa. These most frequently occur where the blood vessels penetrate the muscular wall to supply the mucosa. It has been now well documented that the presence of a diet low in bulk is one of the leading causes of this disease. As with other lesions of the gastrointestinal tract which become acutely inflamed (diverticulitis), the lesion may perforate, causing peritonitis, or may bleed, causing massive gastrointestinal hemorrhage. [H.T.N.]

Diving Skin diving, scuba diving, saturation diving, and "hard hat" diving are techniques used by scientists to investigate the underwater environment. Skin diving is usually without breathing apparatus and is done with fins and faceplate. The diver's underwater observation is limited to the time that breath can be held (1–2 min). Diving with scuba (self-contained underwater breathing apparatus) and "hard hat" provide the diver with a breathable gas, thus expanding the submerged time and the depth range of underwater observations. This type of diving is limited by human physiology and the diver's reaction to the pressure and nature of the breathing gas. Saturation diving permits almost unlimited time down to depths of 100 ft (30 m).

Scuba diving. Scuba is used by trained personnel as a tool for direct observation in marine research and underwater engineering. This equipment is designed to deliver through a demand-type regulator a breathable gas mixture at the same pressure as that exerted on the diver by the overlying water column. The gas which is breathed is carried in high-pressure cylinders (at starting pressures of 2000–3000 psi or 14–20 megapascals) worn on the back.

Scuba can be divided into three types: closed-circuit, semiclosed-circuit, and open-circuit. In the first two, which use pure oxygen or various combinations of oxygen, helium, and nitrogen, exhaled gas is retained and passed through a canister containing a carbon dioxide absorbent for purification and then recirculated to a bag worn by the diver. During inhalation additional gas is supplied to the bag by various automatic devices from the high-pressure cylinders. These two types of equipment are much more efficient than the open-circuit system, in which the exhaled gas is discharged directly into the water after breathing. Most open-circuit systems use compressed air because it is relatively inexpensive and easy to obtain. Although open-circuit scuba is not as efficient as the other types, it is preferred because of its safety, the ease in learning its use, and its relatively low cost.

For physiological reasons scuba diving is limited to about 165 ft (50 m) of water depth. Below this depth when using compressed air as a breathing gas, the diver is limited, not by equipment, but by the complex temporary changes which take place in the body chemistry while breathing gas (air) under high pressure.

Saturation diving. This type of diving permits long periods of submergence (1–2 weeks). It allows the diver to take advantage of the fact that at a given depth the body will become fully saturated with the breathing gas and then, no matter how long the submergence period, the decompression time needed to return to the surface will not be increased. Using this method, the diver can live on the bottom and make detailed measurements and observations, and work with no ill effects. This type of diving requires longer periods of decompression in specially designed chambers to free the diver's body of the high concentration of breathing gas. Decompression times of days or weeks (the time increases with depth) are common on deep dives of over 200 ft (60 m). [R.F.Di.]

Physiology. When a skin diver descends into the sea, the pressure of the surrounding water increases rapidly and compresses the gas-filled space of the lungs. At some depth between 100 and 300 ft (30 and 90 m), depending on the individual's lung capacity and shift of blood into the chest, a threshold is reached where the thorax and lungs cannot withstand this pressure and collapse. With the aid of diving equipment, helmet or scuba, air is supplied to the diver under high pressure, thus preventing the lungs from collapsing. The effects of increased pressure observed on divers in a water environment are different from those found in caisson workers, who are exposed to a dry, pressurized environment in which hydrostatic pressure differences are not present.

A diver breathing air at a pressure equivalent to that at 300 ft (90 m) of water is subject to certain changes in personality and performance, ascribed to nitrogen narcosis. The subjective symptoms and mental reactions consist of euphoria, overconfidence accompanied by dulling of mental ability, and difficulty in assimilating facts and making quick and accurate decisions. These symptoms are similar to alcohol intoxication. J. Y. Cousteau called it "rapture of the deep."

The use of nitrogen-oxygen gas mixtures limited the maximal depth of diving to about 300 ft (90 m) because of nitrogen narcosis and resistance to breathing. Helium was found superior to nitrogen as an inert gas diluent for oxygen, and it is used for divers below 150 ft (45 m). It extended practical diving operations to 500 ft (150 m). Helium is lighter than nitrogen, thus reducing the resistance to breathing. Moreover, helium is less soluble in the tissues than nitrogen. Therefore, smaller quantities of helium are absorbed in the body during a dive, and less gas is eliminated from the tissues during decompression to result in bubble formation. Nevertheless, in spite of these advantages, helium-oxygen diving has not lessened incidents of bends during decompression. Diving with helium-oxygen atmospheres requires special equipment and special decompression tables.

Inhalation of oxygen concentration above 65% at 1 atmospheric pressure has deleterious effects in most warm-blooded animals; it leads eventually, even after days or weeks, to death due to pulmonary oxygen toxicity. Inhalation of oxygen at approximately twice the atmospheric pressure results in generalized convulsions. With increasing depth the toxicity symptoms develop after shorter exposure times. Under higher partial pressures oxygen toxicity manifests itself primarily in the central nervous system. No permanent damages resulting from oxygen convulsions in humans have been reported.

With properly designed diving equipment and sufficient ventilation, one would expect the danger of CO_2 accumulation to be minimal. However, because of increased resistance to breathing under higher pressures, CO_2 retention does occur in diving operations. Inhaling CO_2 can cause symptoms of dyspnea, headaches, dizziness, warmth, restlessness, or increased motor activity.

The tissues of a diver who has been exposed to high pressure for a long period will have accumulated a large amount of dissolved nitrogen. If the diver were brought back to the surface immediately, the result would be bubble formation, which can cause minor or severe damage, depending on the quantity of bubbles and the organs in which they form. It is fortunate that the blood and tissues can hold a certain amount of gas in supersaturation without significant bubble formation. If bubbles are liberated from supersaturated tissues, they can exert pressure on nerves and result in local pain to muscles, joints, or bones. Bubbles formed in the spinal cord can produce paralysis. If gas bubbles develop in the brain, they may cause dizziness, blindness, paralysis, unconsciousness, or convulsions, depending on quantity and location. Bubbles formed in the blood can interfere with the pulmonary blood flow and result in asphyxia and the "chokes." The first symptom is often pain below the sternum during deep breathing, followed by shallow and rapid breathing. The onset of symptoms varies greatly, occurring within 1–18 h following decompression. *See* DECOMPRESSION ILLNESS.

Aside from narcosis, toxicity, and decompression sickness, the scuba, helmet, and breath-hold divers are subject to the effects of unequal pressure differences (barotrauma) across their air-containing structures, such as middle ears, sinuses, lungs, and gastrointestinal tract. Swimmers and divers often have difficulty in equalizing pressure in the middle ear during descent because of blockage of the Eustachian tube. They will experience pain and upon further descent will rupture the eardrum. If the sinuses do not equalize during descent, transudation and hemorrhage can develop.

During ascent the air contained in the lungs of the diver expands in relation to the diminishing ambient pressure. If the diver uses a breathing apparatus and breathes normally during ascent, the excess of expanding lung air is discharged through the exhaust valve. But if the diver panics and holds his breath, overexpansion of lung tissues may cause air embolism from gases entering the pulmonary blood vessels from adjoining tissues. Air embolism frequently results in death due to blockage of circulation. [K.E.S.]

Diving animals Truly aquatic animals have no need for a specialized diving response, because they are able to obtain oxygen directly from the water and thus sustain themselves indefinitely in the aquatic medium. Terrestrial or land animals, on the other hand, have in many cases gone back to the aquatic mode of living in varying degrees. In all of these has arisen the problem of obtaining the necessary gas exchange in an aquatic environment with only the physical attributes of a terrestrial animal. Many of these terrestrial animals have adapted so well as to achieve almost perfect freedom in the water.

All classes of terrestrial vertebrates (birds, mammals, and reptiles) have species which spend part of their lives under water or seek food or shelter by becoming submerged for a period of time. One example, a fresh-water mammalian species, is the North American beaver. One of the first changes noted in these and other animals that normally dive beneath the surface of the water is a reduction in heart rate, called bradycardia. However, those animals which traditionally utilize an aquatic habitat show a more pronounced reduction in heart rate in a much shorter time. In the other two classes of animals which contain divers, the reptiles and birds, some excellent examples reflect reductions in heart rate similar to the rates seen in mammals.

Warm-blooded animals show a more rapid response to submergence than do reptilian species. Those mammals, such as seals, which are traditionally found in open-ocean environments show the most dramatic responses of all. The reptiles in general utilize less oxygen than do warm-blooded animals and have slower resting heart rates, so that their response would be expected to be less spectacular. Also, reptilian reaction times are generally slower than those seen in mammals, and are usually modified by the temperature of the water in which they dive, so that a slower response to the diving stimulus is expected.

Specifically, the diving response enables the animal to remain underwater for longer periods of time. Not only does the heart rate slow down, but blood flow to specific muscles is decreased or stopped altogether. Lactic acid produced as a result of anaerobic glycolysis in muscles thus isolated is less likely to enter the systemic circulation and alter the functions of vital organs such as the brain and heart. This response also reduces the likelihood that hydrogen ions will build up in the blood and stimulate chemoreceptors (receptors which respond to changes in blood carbon dioxide levels and hydrogen ion concentration) which control breathing. Blood flow thus remains constant and unrestricted to the most essential organs, such as brain, heart, and lungs, during the dive, and arterial pressure to these organs remains at the normal level, so that perfusion of the capillaries with blood is not altered.

Other changes in the bodies of divers are also apparent, but are seen as changes in total body chemistry. For instance, glycogen and adenosinetriphosphate, which are the principal energy sources of muscles, are in more than ample supply in divers, and appear to have been stored for that time when they are most needed. Oxygen stores are essential to the length of time that a diver may stay submerged, and diving animals have concentrations of myoglobin in their muscles which gives them a remarkable ability to store oxygen. *See* DIVING. [J.H.F.]

Division In arithmetic and algebra, the process of finding one of two factors of a number (or polynomial) when their product and one of the factors are given. The symbol ÷ is used in elementary English and American arithmetics to denote division. Division is more often symbolized by the double dot :, the bar —, or the solidus /; thus $x{:}y$, $\frac{x}{y}$, or x/y indicates division of a number x by a number y. Considered as an operation inverse to multiplication, x/y is a symbol denoting a number whose product with y is x. Another way to base division upon multiplication is provided by the concept of the reciprocal of a number. If y is any number (real or complex) other than 0, there is a number, denoted by $1/y$ and called the reciprocal of y, whose product with y is 1. Then x/y is the symbol for the product of x and $1/y$. This view of division furnishes a means of extending the concept to objects other than real or complex numbers. *See* ALGEBRA; MULTIPLICATION. [L.M.Bl.]

Docodonta One of the most primitive mammalian orders known, found in the Jurassic of North America and England and possibly in Rhaetic deposits. In docodonts the main jaw articulation was formed by the dentary and squamosal bones, but the articular and quadrate bones formed a secondary jaw articulation. Early members of other Mesozoic orders possessed the same double articulation, but by Late Jurassic time only the docodonts still retained this transitional reptile-mammal condition. *See* MAMMALIA. [M.C.McK.]

Dog Any breed of domestic or wild dogs, and related species, that belong to the family Canidae. The origin of domestic dogs is obscure, but they seem to be most closely related to the wolf. In many respects the dog is structurally primitive and shows a genetic plasticity which accounts for the many varieties.

Domestic breeds. There are more than 100 breeds of domestic dogs, and their classification is based principally on their uses. The breeds are the sporting breeds, hounds, terriers, working breeds, and toy breeds.

Sporting breeds are trained and employed for retrieving or finding game. The largest group in this class is the spaniels, of which there are about 10 types. The spaniels have been trained to retrieve and to flush game, and they hunt both birds and fur-bearing animals. The Brittany spaniel is the only member of the group which points. Retrievers are specially trained to locate and return game to hunters and are used most commonly in hunting waterfowl. There are four varieties: the curly, the Chesapeake Bay, the golden retriever, and the Labrador. With the exception of the golden retriever, the Labrador is the origin of the others by crossbreeding with spaniels and setters. Pointers and setters, used for hunting upland game birds, range ahead of the hunter, point the game until the hunter arrives, and retrieve the fowl after it has been flushed and shot. Among the varieties that are used in these pursuits are the Weimaraner, the English and Irish setters, and the German short-haired pointer.

The hound group includes the basset, the bloodhound, the whippet (produced by crossing a terrier with an Italian greyhound and then backbreeding to the English greyhound), the dachshund, the wolfhound, and the beagle. The greyhound is built for speed with its thin body and long legs. The bloodhounds and foxhounds are used for hunting, mostly by scent. The beagle, now more a pet than a hunter, can also follow a scent and is easier to follow on foot.

The terriers originally were bred for hunting burrowing animals, such as the badger and fox. The Boston terrier is the only breed to have originated in the United States. The fox terrier was originally bred for fox hunting but is an excellent ratter. Other terriers include the Airedale and the Cairn, Yorkshire, Scottish, and Skye terriers.

Most of the working breeds are large animals used as draught animals, for police work, for herding, and as guide dogs for the blind. Draught breeds are the Alaskan malamute, the Eskimo, the Samoyed, and the St. Bernard. Among the animals used as guard dogs and for police work are the

Doberman, the German shepherd, and the Great Dane. The collies, Belgian sheep dogs, and English sheep dogs are outstanding sheep herders. The bulldog is now more of a pet and house dog than a guard dog. The poodle can be trained as a gun dog and was originally used for duck shooting.

Toy breeds are all quite small, some being miniatures of the larger breeds. The chihuahua is the smallest. Some of the more popular varieties are the Pomeranian, Pekingese, and pug.

Wild species. The wild species of the family, numbering about 36 and having a wide distribution, include several wild dogs, wolves, coyotes, foxes, and jackals. There are a number of wild dogs which have never been domesticated, unlike the dingo of Australia (*Canis dingo*) that is believed to have been a domestic dog introduced into Australia during prehistoric times and then reverted to the wild state. The Asiatic wild dog occurs throughout Asia, Java, and Sumatra and is considered by some authorities to be three distinct species. The Cape hunting dog (*Lycaon pictus*) ranges throughout the grasslands of eastern and southern Africa but has become reduced in numbers.

The common European wolf (*C. lupus*) is the species that once ranged throughout the temperate forested regions of Europe, Asia, and North America. While they occur in some European countries such as Italy, Spain, and the Balkans, they are plentiful only in the Scandinavian countries. The gray wolf or timber wolf, originally extremely common in North America, is now restricted to Alaska and the subarctic regions of Canada.

The coyote (*C. latrans*), sometimes called the prairie wolf, is a close relative of the true wolf, although it is smaller. Coyotes inhabit the prairies, open plains, and desert areas of North America.

Jackals are scavengers as well as menaces to domestic poultry. The oriental jackal (*C. aureus*) has spread from southeastern Europe and northern Africa through Asia as far south as Burma. It prefers higher elevations in contrast to the black-backed jackal (*C. mesomelas*), which is found in the grasslands of eastern and southern Africa. The most common enemy of the jackal is the leopard.

Foxes have relatively short legs, long bodies, and long bushy tails. The Old World red fox (*Vulpes vulpes*) is closely related to, but is smaller than, the American red fox (*V. fulva*), which is found throughout North America. The American species has undergone many color phases and mutations and includes other varieties such as the silver fox and the cross fox. *See* CARNIVORA; MAMMALIA. [C.B.C.]

Dogwood

A tree, *Cornus florida*, also known as flowering dogwood, found in the eastern half of the United States and in southern Ontario, Canada. When this tree is in full flower, the four large, white, notched bracts or petallike growths surrounding the small head of flowers give an ornamental effect that is unequaled by any native tree. Pink, rose, and cream-colored varieties are commonly planted. The wood is very hard and is used for roller skates, carpenters' planes, and other articles in which hardness is desired.

The Pacific dogwood (*C. nuttallii*) grows in Idaho and from southwestern British Columbia to southern California. The Japanese dogwood (*C. kousa*) is a similar small tree that blooms in June. Other shrubby species of dogwood are used as ornamentals. *See* CORNALES. [A.H.G./K.P.D.]

Dolomite

The carbonate mineral $CaMg(CO_3)_2$. Often small amounts of iron, manganese, or excess calcium replace some of the magnesium; cobalt, zinc, lead, and barium are more rarely found. Dolomite is normally white or colorless with a specific gravity of 2.9 and a hardness of 3.5–4 on Mohs scale. It can be distinguished from calcite by its extremely slow reaction with cold dilute acid. Dolomite is a very common mineral, occurring in a variety of geologic settings. It is often found in ultrabasic igneous rocks, notably in carbonatites and serpentinites, in metamorphosed carbonate sediments, where it may recrystallize to form dolomite marbles, and in hydrothermal veins. The primary occurrence of dolomite is in sedimentary deposits, where it constitutes the major component of dolomite rock and is often present in limestones. *See* DOLOMITE ROCK; LIMESTONE; SEDIMENTARY ROCKS. [A.M.G.]

Dolomite rock

A limestone whose carbonate fraction contains more than 50% of the mineral dolomite. Most rocks called dolomite have more than 90% dolomite. Newer classifications of carbonate rocks based on textural elements introduce the amount of dolomite as a secondary criterion. *See* DOLOMITE; LIMESTONE.

Dolomites show gross textures similar to those of limestones, but the finer details differ. Most coarsely crystalline dolomite appears to be secondary. An original clastic texture may be revealed only by scattered quartz grains floating in a mosaic of dolomite crystals. The original outlines of fossils or oolites may show as faint dust outlines in the dolomite, but the original structure has been lost. A sharply contrasting type of dolomite is finely crystalline, shows little or no evidence of replacement origin, and preserves many sedimentary structures such as weak current bedding.

Dolomite and dolomitic limestone are known from rocks of all ages but are more common in older rocks, particularly the Paleozoic. Dolomite is most often found in association with limestone, with which it may be interbedded or laterally gradational. Modern dolomite has been found in supratidal flats in carbonate depositional areas such as the Bahamas and in a variety of hypersaline lagoons or arms of the sea in warm climates. The deposits are associated with precipitates of calcite, aragonite, and magnesian calcite as well as gypsum and anhydrite. *See* SEDIMENTARY ROCKS. [R.Si.]

Dolphin

The name for about 33 species of cetacean mammals included in the family Delphinidae. These animals may be distinguished from the porpoises by the shape of the mouth, which is seen as a pronounced beak in the dolphin while the mouth of the porpoise is blunt. *See* PORPOISE.

The common dolphin (*Delphinus delphis*) is found in the North Sea and the Mediterranean and in all other warm and temperate waters. This dolphin is one of the swiftest cetaceans and swims in schools as large as 10,000. Dolphins surface frequently to breathe air which they take in through the dorsally located "blowhole," which has a single aperture.

The bottle-nosed dolphin (*Tursiops truncatus*), restricted to the Mediterranean and North Atlantic, is known in the United States as a porpoise and is found performing tricks in most seaquariums. Other species of dolphins are *T. nuuana* and *T. gilli*, both closely related to the bottle-nosed dolphin and found along the North Pacific coast. The Amazonian white dolphin (*Sotalia fluviatilis*) is an estuarine species found in many of the rivers of South America, while other species of this genus occur in Africa and Asia. They are known collectively as the long-beaked river dolphins. *See* CETACEA; MAMMALIA. [C.B.C.]

Domain (electricity and magnetism)

A region in a solid within which elementary atomic or molecular magnetic or electric moments are uniformly arrayed.

Ferromagnetic domains are regions of parallel-aligned magnetic moments. Each domain may be thought of as a tiny magnet pointing in a certain direction. A ferromagnet is generally composed of many such domains, pointing in many different directions. *See* FERROMAGNETISM.

Antiferromagnetic domains are regions of anti-parallel-aligned magnetic moments. They are associated with the presence of grain boundaries, twinning, and other crystal inhomogeneities.

Ferroelectric domains are electrical analogs of ferromagnetic domains. *See* FERROELECTRICS. [E.A.; F.Ke.]

Dominance The development of the same phenotype in homozygotes with two doses of an allele and heterozygotes with one dose of the allele and one dose of another allele which in homozygotes produces a different phenotype. *See* MENDELISM.

Although complete dominance is common, there are exceptions in which the heterozygote is intermediate in character between the two parents. An example of this kind of exception is found in the plant *Mirabilis*: a red-flowered strain crossed with a white-flowered strain produces a pink-flowered hybrid. This is a case of semidominance.

When an inherited character is detectable at the protein or enzyme level, it is sometimes found that the heterozygote contains two slightly different polypeptides. Several aspects of dominance are well illustrated by sickle-cell anemia in humans; the homozygote sickle cell H^SH^S is lethal before the individual reaches the age of maturity. The heterozygote H^AH^S is almost normal. Therefore, insofar as the criterion of viability is concerned, the normal is dominant to the sickle cell. The hemoglobins in the heterozygote, however, are of two types, normal and sickle cell, which differ in one amino acid. At this level, both the H^A and H^S genes are expressed; this is an example of codominance. Similar codominance is found in the blood group antigens ABO, in which A and B are codominant to one another but A and B are both dominant to 0, which lacks the antigen altogether. *See* BLOOD GROUPS.

Another expression of the heterozygote which has been classified as a type of dominance occurs when the heterozygote transcends both parents. This has been called overdominance. Examples of overdominance are rare and difficult to establish, but, again, sickle cell provides the best illustration. In an environment of falciparum malaria the heterozygote is more fit then either of the homozygotes: It is greatly resistant to this persistent malaria; the homozygote H^AH^A is sensitive, the homozygote H^SH^S is virtually lethal.

These examples have been chosen not only to show the different kinds of dominance but also to demonstrate that dominance is a phenomenon of the character (phenotype) and not of the gene (genotype). Since characters are the result of interactions between genotype and environment, in some cases, the type of dominance, as in the overdominance of sickle cell, is expressed only in one environment. R. Goldschmidt called this type of dominance conditioned dominance. Despite the many factors affecting dominance, the type of dominance found in a gene system can be a powerful indication of the type of gene action in the system. *See* ALLELE; GENE ACTION; HETEROSIS. [D.Le.]

Donkeys and allies Donkeys, asses, and mules are included in the family Equidae along with the horses and zebras. These are odd-toed ungulates and therefore belong to the mammalian order Perissodactyla.

Donkeys (*Equus asinus*) originated in Africa, and their only relatives in the wild state are found in Ethiopia and Somaliland. They have been domesticated and are used principally as pack animals throughout Asia and Europe. They are hardy animals but better suited to hot climates since they are sensitive to cold.

The wild ass of Asia (*E. hemionus*) is known by a variety of common names, and because of its extensive distribution it has been divided into a number of races or subspecies. In the central regions of Mongolia is the kulan (*E. h. hemionus*), the onager (*E. h. onager*) occurs in Persia and India, while the kiang (*E. h. kiang*), largest of all the races, is restricted to Tibet and the Sikkim. The Syrian wild ass (*E. h. hemippus*) is extremely rare and possibly extinct. The Asian wild asses are the most abundant of the wild Equidae after the zebras. The African wild ass (*E. asinus*) is also becoming rare in its natural habitat, being found only between the Sudan and Somaliland.

The mule is a hybrid resulting from the cross of the male ass (*E. asinus*) and the mare or female horse (*E. caballus*). When the female ass is bred to a stallion, the cross is called a hinny. These crosses do not occur naturally and both mule and hinny are usually sterile. *See* PERISSODACTYLA. [C.B.C.]

Donnan equilibrium The particular equilibrium set up when two coexisting phases are subject to the restriction that one or more of the ionic components cannot pass from one phase into the other. This equilibrium was first recognized by F. G. Donnan in 1911. Commonly, the restriction is caused by a membrane which is permeable to the solvent and small ions but impermeable to colloidal ions or charged particles of colloidal size. The presence of a membrane is not essential, since the restriction of movement on the charged colloid can be provided by a centrifugal or gravitational field or by gel coherence.

An immediate consequence of such a restriction in a system is the uneven distribution of diffusible ions at equilibrium. This is apparent in the following example.

$$\begin{array}{cc|cc} Na^+ & R^- & Na^+ & Cl^- \\ c_1 & c_1 & c_2 & c_2 \end{array}$$

Let the initial state of a system be a solution of the ions Na^+ and R^- of concentration c_1 separated from a solution of sodium chloride of concentration c_2 by a membrane freely permeable to all but the R^- ions. Then at equilibrium, a certain concentration of chloride ions, x, will have diffused through the membrane, accompanied by the same number of sodium ions in order to preserve electrical neutrality on both sides of the membrane, and the final equilibrium will be:

$$\begin{array}{ccc|cc} Na^+ & R^- & Cl^- & Na^+ & Cl^- \\ (c_1 + x) & & x & (c_2 - x) & (c_2 - x) \\ & (1) & & (2) & \end{array}$$

It can be shown thermodynamically that at equilibrium the product of the concentrations, or more strictly, the activities of the sodium and chloride ions, will be the same on both sides of the membrane. Hence may be formed

$$[Na^+]_1[Cl^-]_1 = [Na^+]_2[Cl^-]_2$$

or

$$(c_1 + x)x = (c_2 - x)^2$$

Obviously, the diffusion of the chloride ions (and an equal number of sodium ions) through the membrane has been hindered by the presence of the nondiffusible ion R^-. Calculations based on this equation show that sodium chloride is almost completely prevented from diffusing through the membrane if it is present in small concentration relative to the concentration of the nondiffusible ion R^-. As the relative concentration of sodium chloride is increased, more of it diffuses through the membrane. Finally, when the salt concentration is very high relative to that of R^-, an even distribution of sodium and chloride ions on either side of the membrane is approached. A similar equilibrium is also attained when the diffusible salt has no ion common to the colloidal electrolyte. *See* COLLOID; DIALYSIS; ION-SELECTIVE MEMBRANES AND ELECTRODES. [G.S.M.; W.O.M.]

Donor atom An impurity atom in a semiconductor which can contribute or donate one or more conduction electrons to the crystal by becoming ionized and positively charged. For example, an atom of column 15 of the periodic table substituting for a regular atom of a germanium or silicon crystal is a donor because it has one or more valence electrons which can be detached and added to the conduction band of the crystal. Donor atoms thus tend to increase the number of conduction electrons in the semiconductor. *See* ACCEPTOR ATOM; SEMICONDUCTOR. [H.Y.F.]

Dopamine A catecholamine neurotransmitter that is synthesized by certain neurons in the brain and interacts with specific receptor sites on target neurons. *See* BRAIN.

Dopamine is manufactured inside dopamine neurons in a controlled manner from the amino acid precursor L-tyrosine, which mammals obtain through the normal diet. Dopamine is then stored in vesicles within the nerve terminals, which may fuse with the cell membrane to release dopamine into the synapse.

The release of neurotransmitter is controlled by a variety of factors, including the firing rate of the dopamine nerve cell (termed impulse-dependent release) and the release- and synthesis-modulating presynaptic dopamine receptors located on the dopamine nerve terminals. Once released, dopamine also acts at postsynaptic receptors to influence behavior. The actions of dopamine in the synapse are terminated primarily by the reuptake of neurotransmitter into the presynaptic terminal by means of an active dopamine transporter. *See* SYNAPTIC TRANSMISSION.

The midbrain dopamine neurons which project to a variety of forebrain structures are critically involved in normal behavioral attention and arousal; abnormalities in the normal functioning of these systems have been implicated in a variety of disorders. For example, Parkinson's disease involves a degeneration of the midbrain dopamine neurons. This condition is often successfully treated by providing affected individuals with L-dopa, which is readily converted to dopamine in the brain. *See* ATTENTION DEFICIT DISORDER; PARKINSON'S DISEASE. [L.A.C.]

Doppler effect A change in the observed frequency of sound, light, or other waves, caused by motion of the source or of the observer. A familiar example for sound waves is the increase (decrease) in pitch of a train whistle as the train approaches (passes). The optical phenomenon is shown in the altered frequencies of spectral lines in the light emitted from a moving star. If the star and the Earth are moving closer to one another, more light pulses are received in a given time interval, and the color emitted from the star appears to be shifted toward the violent end of the spectrum. When the distance between the Earth and the star is increasing, the observed light is shifted toward the red end of the spectrum. The color shifts of remote galaxies are taken as evidence that the universe is expanding. *See* REDSHIFT. For an important application of the Doppler effect *see* DOPPLER RADAR.

Three fundamental differences exist between the acoustical and the optical Doppler effects.

1. The optical frequency change does not depend upon whether it is the source or the observer that is moving with respect to the other, whereas the acoustical frequency change is different in the two cases.
2. No effect is observable in the acoustical case when the source, or the observer, moves at right angles to the line connecting the source and the observer. An optical frequency change is observable under such conditions.
3. The motion of the medium through which the waves are propagated does not affect the observed optical frequency, whereas it does affect the observed acoustical frequency. [G.W.S.]

Doppler radar A radar system used to measure the relative velocity of the system and the radar target. The operation of these systems is based on the fact that the Doppler frequency shift in the target echo is proportional to the radial component of target velocity. *See* DOPPLER EFFECT.

Airborne systems are used to determine the velocity of the vehicle relative to the Earth for such purposes as navigation, bombing, and aerial mapping, or relative to another vehicle for fire control or other purposes. Ground or ship equipment is used to determine the velocity of vehicular targets for fire control, remote guidance, intercept control, traffic control, and other uses.

Doppler navigation radar is a type of airborne Doppler radar system for determining aircraft velocity relative to the Earth's surface. It is generally used with a navigation computer. The signal from a single beam can provide only the velocity component in the direction of that beam. Complete velocity determination requires, therefore, the use of at least three beams. Most systems use four beams for symmetry.

A preferred technique for obtaining coherent detection is to employ sinusoidal frequency modulation. A sideband of the detected beat between echo and transmitter signal is used. Modulation index and rate and the sideband order are chosen such that echoes from nearby objects are rejected, while those from distant objects are accepted. Leakage noise is reduced at the expense of lowered efficiency. [F.B.B.]

Pulse Doppler radars are useful tools for the observation of the movements of precipitation particles. The Doppler frequency shift associated with the velocity of atmospheric targets, such as precipitation particles or artificial chaff, is always a very small fraction (10^{-6} to 10^{-8}) of the radar operating frequency. The observation and measurement of such small frequency shifts require excellent radar system frequency-stability characteristics that are not usually found in conventional radars but can be added without a drastic increase in equipment cost. *See* RADAR. [R.Lh.]

Doppler VOR A vhf omnidirectional radio range (VOR) employing the Doppler principle. If VOR beacons are installed in the vicinity of obstructions, or when aircraft using VOR signals fly over mountainous terrain, the bearing accuracy is deteriorated by reflections (site and enroute errors). The Doppler VOR solves this problem by two fundamental principles maintaining full compatibility of the radiated information with existing airborne receivers. It uses (1) a wide-base antenna array for suppressing the effects of multipath propagation and (2) the Doppler principle for determination of bearing. *See* DOPPLER EFFECT; RADIO RANGE.

Disturbances by multipath propagation can be reduced by wide-aperture systems because of integration or averaging between the information transferred over the direct and the reflected propagation path. Wide-base systems, however, have the drawback that their directional information becomes ambiguous as soon as the aperture is larger than half a wavelength, approximately.

By applying the Doppler principle, one can overcome these limitations. A dipole rotating eccentrically on an orbit periodically changes the distance between transmitter and receiver, thus modulating the frequency of the radiated carrier by a deviation f_D according to the equation below. Here D is the diameter of the array and $f_{rot.}$ is the number of rotations, or revolutions, per second.

$$f_D = \frac{\text{velocity}}{\text{wavelength}} = \frac{\pi D f_{rot.}}{\lambda}$$

The resulting Doppler effect produces a sinusoidal frequency modulation of the carrier frequency; the phase of this FM contains the bearing information (variable phase).

The Doppler VOR presents two essential advantages over conventional VOR: the improvement of absolute accuracy in good and poor terrain, and the spectacular radial stability and smoothness even over mountains. [E.Kr.]

Dormancy A periodic phase in the development of many plants characterized by their inability to grow, even in the presence of favorable environmental conditions. Dormancy is often considered a phase in the plant's life history during which the usual processes contributing to orderly growth and development are inhibited. Viewed in this restricted sense, the fact that a succession of important morphological and physiological stages precede dormancy and contribute to its development is overlooked.

This article is concerned primarily with dormancy and its regulation in buds and seeds of woody plants. While consideration is given only to these two dormancy systems, the dormancy process is widespread throughout the plant kingdom and many phenomena common to buds and seeds also may be active in regulating dormancy in other systems.

Buds. A summary of the events occurring during the seasonal cycle of many woody plants illustrates the complexity of the dormancy process and emphasizes that both internal and external factors actively regulate specific phases of the developmental process. *See* BUD.

For most woody plants growing in the temperate zone, the termination of shoot extension and the appearance of buds in summer often provide the first visible evidence of dormancy development. During the initial weeks following bud formation, growth activity can be renewed by subjecting the plant to a regime of long days or favorable temperatures or both. This reversible phase of dormancy development has been termed summer dormancy. These observations suggest that the changes leading to the dormant state do not take place immediately but usually develop progressively during the autumn months. Ultimately, growth activity can no longer be stimulated by manipulating the environmental conditions and winter dormancy, or true dormancy, develops.

Many changes take place within the bud during the time of dormancy development. In some species of plants the shoot apex becomes active and produces all the leaves for the next season's growth. In other plants only a part of the total complement of leaves is produced before dormancy develops, the remainder being initiated during the growing season. For some time after the primordia have formed, they continue to develop within the bud. The bud remains in this condition throughout the dormant period of winter.

Release from dormancy may be brought about by a variety of factors, including an adequate period of low temperature, lengthening photoperiod, and the appropriate chemical stimulation. The bud is now exposed to the environmental conditions of spring, which promote growth. During spring and early summer, the bud scales open, the shoot elongates rapidly, leaf growth occurs, and in some species the shoot apex actively produces new leaves. Then, long before the low temperatures and short days of winter arrive, shoot extension ceases and dormant buds are produced, thus completing the seasonal cycle of bud development. *See* PHOTOPERIODISM; PLANT HORMONES.

Seeds. Most seeds are shed in the mature condition and are morphologically complete and apparently physiologically capable of germinating. Those which germinate immediately or within a brief period after exposure to favorable environmental stimuli are termed quiescent and are distinguished from dormant seeds which will not germinate, even under optimal conditions, until they have undergone a rest period, that is, a period of afterripening. Afterripening, defined as the time during which the synthetic machinery of the seed is prepared for germination and growth, is considered to be necessary for releasing dormancy. The major causes of seed dormancy are those attributed to the seed coat, the embryo, and the presence of various physiological inhibitors. *See* SEED.

In a large number of seeds, dormancy is imposed by the presence of a hard seed coat which is impermeable to water and gases or is mechanically restrictive to embryo growth. Impermeable seed coats having chemical and structural properties preventing water absorption are characteristic of many families of plants, particularly the Leguminosae. Water imbibition is an important factor in that it is responsible for reactivation of metabolic processes leading to seed germination, and treatments which alter the chemical or structural barriers to water absorption can be effective agents for overcoming dormancy.

Probably the most common type of dormancy in seeds is associated either with immature embryos or with fully mature embryos which are dormant. In both cases a rest period is required by the seeds before germination takes place. For seeds with immature embryos, afterripening may be necessary before further morphological and physiological developments can occur. Low temperatures are usually necessary for embryo maturation and subsequent seed germination. In addition, there is a very close correlation between growth of the embryo and loss of stored materials from the endosperm. This suggests that low temperature stimulates the transfer of nutrients from endosperm to embryo and promotes growth. Once the embryo has matured, germination follows rapidly.

Inhibitors of seed germination are known to exist in a variety of fleshy and nonfleshy fruits and in seeds. They may occur in the pericarp, endosperm, seed coat, or embryo; when present in high concentrations, they prohibit embryo growth and seed germination. Removal of the tissue surrounding the embryo or extensive leaching of the seed or embryo often overcomes dormancy associated with inhibitors. *See* PLANT GROWTH. [A.E.DeM.]

Dorylaimida An order of nematodes in which the labia are generally well developed; however, many taxa exhibit a smoothly rounded anterior. The labial region is often set off from the general body contour by a constriction. The cephalic sensilla are all located on the labial region. When there is no constriction, the labial region is defined as that region anterior to the amphids. The amphidial pouch is shaped like an inverted stirrup, and the aperture is ellipsoidal or a transverse slit. The stoma is armed with a movable mural tooth or a hollow axial spear. The anterior portion of the tooth or spear is produced by a special cell in the anterior esophagus. The esophagus is divided into a slender, muscular anterior region and an elongated or pyriform glandular/muscular posterior region. There are generally five esophageal glands with orifices posterior to the nerve ring. In some taxa there are three glands, and in others seven have been reported. The esophagointestinal valve is well developed. The mesenteron is often clearly divided into an anterior intestine and a prerectum. Females have one or two reflexed ovaries; when there is only one, the vulva may shift anteriorly. Males have paired equal spicules that are rarely accompanied by a gubernaculum. The males often have the ventromedial preanal supplements preceded by paired adanal supplements.

There are seven dorylaimid superfamilies: Actinolaimoidea, Belondiroidea, Diphtherophoroidea, Dorylaimoidea, Encholaimoidea, Nygolaimoidea, and Trichodoroidea. *See* NEMATA. [A.R.M.]

Dosimeter A meter used for measuring the dose of ionizing radiation received by the person wearing it. There are many types of dosimeters, the most common being the air-capacitor dosimeter and the thermoluminescent dosimeter. *See* AIR-CAPACITOR DOSIMETER; FILM BADGE; MONITORING OF IONIZING RADIATION; THERMOLUMINESCENT DOSIMETER. [K.Z.M.]

Doublet flow In hydrodynamics, a doublet is the combination of a source and a sink of equal strength which are allowed to approach each other in such a manner that the product of their strength and the distance between them remains constant in the limit. Doublets have directional properties, the line drawn from the sink toward the source being the axis of the doublet. The strength of a doublet is proportional to the product of strength of source and distance between source and sink before the limit is taken. *See* SINK FLOW; SOURCE FLOW.

The doublet is a flow element that is used in combination with other elements to build up special flow cases. For example, a uniform flow superposed on a two-dimensional doublet so that the axis of the doublet is directed upstream yields the flow case of uniform flow around a circular cylinder. In three-dimensional flow, uniform flow and a doublet directed upstream result in the case of uniform flow around a sphere. [V.L.S.]

Douglas-fir A large coniferous tree, *Pseudotsuga menziesii*, also known as red fir. This tree grows in the Pacific Coast region and the Rocky Mountains of the United States and differs from true fir in having short leaf stalks, elliptical leaf scars, and pendant cones. The wood is hard and strong and is used throughout the United States for construction timber, lumber, and plywood veneer. It is also used for cooperage, mine timbers, mill work, railway car construction, flooring, furniture, ships, and ladders. In addition, it is used as a shade tree, an ornamental, and for shelter belts. Many cultivated varieties are planted in the eastern United States and Europe. *See* PINALES. [A.H.G./K.P.D.]

Down syndrome A type of developmental disability historically known as mongolism because of the characteristic appearance of the eyes. This condition is due to an abnormality in either the number or structure of the chromosomes. The most common type of abnormality in Down syndrome is due to an extra chromosome. Such persons have 47 chromosomes (instead of 46).

The characteristic physical features include almond-shaped eyes and a rounded, brachycephalic skull with a flattened occipital region; an enlarged, fissured tongue; broad hands with stubby fingers; hypotonic muscle development; a short nose and depressed nasal bridge; thick, everted, and cracked lips; dry, rough skin; subnormal height; and infantile genitalia. All these physical signs are not present in every case.

The degree of mental defect may be directly related to the number and gravity of the physical signs; however, few Down children are classified as severely retarded. Most cases are moderate to mild, and are often educable; the rest are usually trainable by educational definition.

Down syndrome does not appear to be a hereditary disease following a typical genetic pattern. The major factor which appears with consistency in studies of the etiology is that the probability of a Down syndrome birth increases with maternal age after age 35, and in advanced paternal age. Other than the certainty that Down syndrome is a manifestation of a defect in embryonic development and chromosomal abnormality, etiology is still in doubt. Prenatal determination of the presence of a Down syndrome fetus can be ascertained through amniocentesis. *See* HUMAN GENETICS; MENTAL DEFICIENCY. [H.Le.]

Drafting The making of drawings of objects, structures, or systems that have been visualized by engineers, scientists, or others. Such drawings may be executed in the following ways: manually with drawing instruments and other aids such as templates and appliqués, freehand with pencil on paper, or with automated devices.

A drafting machine has two straightedges which, although fixed at right angles to each other, maintain a preset angular relationship to the work when moved from place to place over the drafting surface. Engineering drawings often require making numerous parallel lines. Drafting machines simplify the making of these and other types of drawings and substantially reduce the worker-hours required.

On most drafting machines the straightedges can be pivoted to any desired angle, then locked. Two types of mechanisms are in use to allow the straightedges to move over the drafting surface. One consists of vertical and horizontal tracks or bars (see illustration). The other consists of mechanical linkages

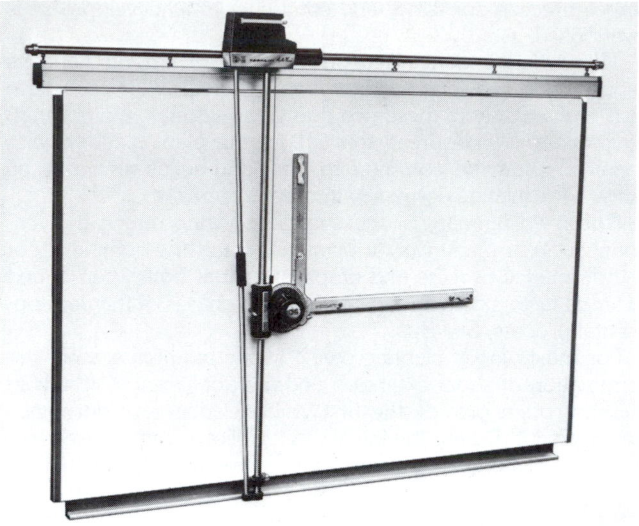

Drafting machine using vertical and horizontal bars. (*Keuffel and Esser Co.*)

which may take the form of a double set of parallel four-bar linkages having a common member at the elbow, or two sets of equal-diameter pulleys tightly coupled by metal bands. *See* FOUR-BAR LINKAGE.

As the complexity of designs has increased and the use of standard parts has become widespread, drawings have become increasingly stylized. Templates carry frequently used symbols, from which the symbols in the required positions on the drawing can be quickly traced.

Where the design procedures from which drawings are developed are repetitive, computers can be programmed to perform the design and to produce their outputs as instructions to automatic drafting equipment. Essentially, automated drafting is a method for creating an engineering drawing or similar document consisting of line delineation either in combination with, or expressed entirely by, alphanumeric characters. *See* COMPUTER GRAPHICS; ENGINEERING DRAWING. [T.C.P.]

Dragonfly A member of the suborder Anisoptera of the order Odonata of the class Insecta. The adults are large, attractive insects characterized by four similar-appearing elongate, membranous wings with numerous characteristic veins; very large compound eyes; tiny antennae; chewing mouthparts; and a long, slender abdomen terminating in a short pair of cerci. Adults are strong, agile fliers and are all predacious. In flight, their legs are held to form a "basket" which entraps small insects. Dragonflies are considered very beneficial since they consume large numbers of mosquitoes, flies, gnats, and similar pests. Dragonflies are most commonly found near permanent water. Eggs are laid in the water or on the banks. The nymphs are all aquatic and predacious. *See* ODONATA. [F.R.V.]

Drawing of metal An operation wherein the workpiece is pulled through a die, resulting in a reduction in outside dimensions. This article deals only with bar and wire drawing and tube drawing. *See* Sheet-metal forming.

Among the variables involved in the drawing of wires and bars are properties of the original material, percent reduction of cross-sectional area, die angle and geometry, speed of drawing, and lubrication. The operation usually consists of swaging the end of a round rod to reduce the cross-sectional area so that it can be fed into the die; the material is then pulled through the die at high speeds. Most wire drawing involves several dies in tandem to reduce the diameter to the desired dimension. Die materials are usually alloy steels, carbides, and diamond. Diamond dies are used for drawing fine wires. The purpose of the die land is to maintain dimensional accuracy (see illustration).

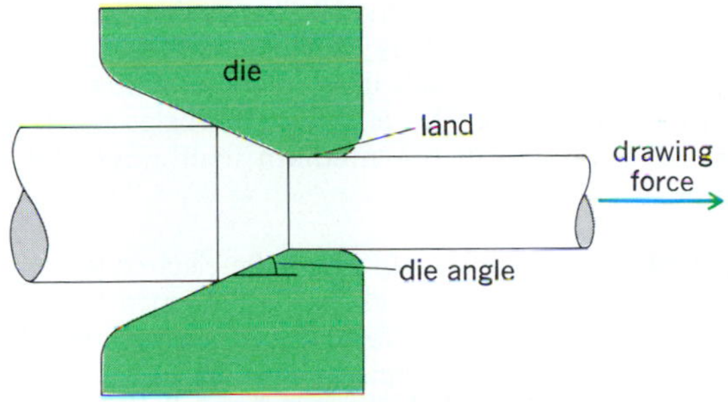

Cross section of drawing die.

Tubes are also drawn through dies to reduce the outside diameter and to control the wall thickness. The thickness can be reduced and the inside surface finish can be controlled by using an internal mandrel (plug). Various arrangements and techniques have been developed in drawing tubes of many materials and a variety of cross sections. Dies for tube drawing are made of essentially the same materials as those used in rod drawing. [S.Ka.]

Dredge A digging machine installed on a floating hull used to excavate underwater and discharge into a containment vessel or a transport pipeline. A dredge may be either mechanical or hydraulic. Mechanical dredges pick up material with a scoop, shovel, or bucket and deposit it in a barge for transport. They include the revolving crane with clamshell or dragline bucket, the backhoe, the front-opening dipper, and the endless-chain bucket dredge. Hydraulic dredges pick up material with a pump. The solids are carried up the pipe with the flow of water. They include the trailing suction hopper dredge, the cutterhead or pipeline dredge, and the dustpan dredge. *See* Construction equipment. [K.W.L.]

Drier (paint) A material added to paint formulations to facilitate the oxidation of oils. Driers are salts of metals. Acids used to render metals soluble in oil, and therefore useful as driers, may be fatty acids, rosin, or naphthenic or octoic acids derived from petroleum. Cobalt, the most reactive of drier metals, is generally regarded as a surface drier, and it is widely used as the only additive in thin-film paint formulations. Lead, while less reactive than cobalt, may have increasing restrictions in its applications because of ecological requirements. Numerous other metals, including cerium and vanadium, have been used occasionally, and are effective driers. Certain organic compounds also catalyze the drying of oils and have been used for this purpose when freedom from all metallic contamination is required.

In general, small amounts of drier are essential to the formation of a satisfactory paint film within a reasonable time. Addition of large amounts of drier, however, can lead to premature embrittlement and failure of the paint film. *See* Drying oil; Paint. [C.R.Ma.; C.W.Si.]

Drilling machine A motor-driven device fitted with an end cutting tool that is rotated with sufficient power either to create a hole or to enlarge an existing hole in solid material. One or more flutes or grooves in the drill tool conduct coolant to the cutting lips and also provide chip relief. Spot-facing to finish the area around a hole, counterboring to enlarge the diameter over part of the depth, and countersinking to chamfer edges of a hole are operations frequently performed during drilling setups. *See* Boring; Reamer. [A.H.T.]

Drone A pilotless aircraft capable of performing any nondestructive mission; when used destructively, the device is properly termed a missile. There are three basic types of drones: the preprogrammed drone, the smart drone, and the remotely piloted vehicle. *See* Guided missile; Missile.

A preprogrammed drone responds to an on-board timer or scheduler on board the aircraft and has no sensor contact with the ground. The drone follows a set routine of maneuvers, altitude changes, speed changes, and course changes that are programmed through an autopilot to the drone's control surfaces and engine throttle. The drone is usually recovered by a parachute at the end of the mission.

A smart drone has any number of sensor contacts and is equipped with an on-board computer. The ability of a smart drone to make decisions governing course and altitude changes is limited only by its computer and sensor capacity. *See* Guidance systems; Navigation.

The remotely piloted vehicle, probably the most common type of drone, is under the constant control of an operator or pilot through radio links. The pilot or pilots can be located on the ground, in other aircraft, or on boats. [R.D.R.]

Drought A general term implying a deficiency of precipitation of sufficient magnitude to interfere with some phase of the economy. Agricultural drought, occurring when crops are threatened by lack of rain, is the most common. Hydrologic drought, when reservoirs are depleted, is another common form. The Palmer index is used by agriculturalists to express the intensity of drought as a function of rainfall and hydrologic variables.

The meteorological causes of drought are usually associated with slow, prevailing, subsiding motions of air masses from continental source regions. These descending air motions, of the order of 660–1000 ft (200 or 300 m) per day, result in compressional warming of the air and therefore reduction in the relative humidity. Since the air usually starts out dry, and the relative humidity declines as the air descends, cloud formation is inhibited—or if clouds are formed, they are soon dissipated. [J.N.]

Drug addiction The state wherein a person is unable to live without the drug in question and is compelled to take it. The most dangerous addictive drugs are the amphetamine-like drugs, opiates, and cocaine. Various mechanisms of action for the addictive drugs have been proposed. It is considered that these drugs induce addiction by a direct action on the basic conditioning mechanisms of the limbic system so as to either block transmitters that reduce reward or increase punishment or to potentiate transmitters that have the opposite effect. In addition, addictive drugs induce withdrawal effects, but this appears to play a less important part in the establishment of the craving for the drug. The interference with the chemical mechanisms of conditioning causes the patterns of behavior that led to taking the drugs to become so powerfully reinforced, and the behavior-directing mechanism of the brain originating in the limbic circuits to become so altered, that these patterns become too powerful to be broken by an act of "will," that is, by cortically originated directives to the altered limbic system. It is well known that monkeys become more quickly addicted to morphine if they make the injection themselves; that is, there must be some behavior emitted that can be reinforced. *See* BARBITURATES; NARCOTIC. [J.R.Sm.]

Drug delivery systems The engineering of physical, chemical, and biological components into systems for delivering controlled amounts of a therapeutic agent over a prolonged period, thereby maintaining plasma or tissue drug levels at a constant level. Controlled drug delivery systems can take a variety of forms, such as mechanical pumps, polymer matrices or microparticulates, and externally applied transdermal patches. *See* PHARMACOLOGY; TOXICOLOGY.

The most obvious approach for controlled drug delivery involves miniaturization of the familiar infusion system. Mechanical pumps, either totally implantable or requiring percutaneous catheters, have been used to deliver insulin, anticoagulants, analgesics, and cancer chemotherapy. Because pumps deliver precisely controlled amounts of therapeutic agents, they may find additional important applications, particularly when coupled with implanted biosensors for feedback control. *See* BIOELECTRONICS.

Many disadvantages of controlled drug delivery via pumps can be avoided by using controlled-release polymers as delivery vehicles. Techniques for the production of polymer implants that slowly release drug molecules have become available. Nondegradable polymers, biodegradable polymers, swellable polymers, and biopolymers have been used as the basis of controlled drug delivery systems. For a specific therapeutic agent, the rate, pattern, and duration of drug release can be modified by selecting the appropriate biocompatible polymer and method of device fabrication. *See* POLYMER. [W.M.Sa.]

Drug resistance Reduction or lack of activity of particular synthetic or natural substances, upon living organisms. Drug resistance can occur for viruses, bacteria, fungi, parasites, and even mammalian cells such as tumor cells. The best-understood form of drug resistance is that of viruses and bacteria. Resistance in these microorganisms is due to one of four principal mechanisms: failure of the drug to reach a receptor; destruction of the drug by the microorganism; failure of the drug to bind to a receptor; or use of an alternative pathway by the microorganism to produce an essential component.

Many fungi resist antimicrobial agents which do not damage the walls of fungi or inhibit fungal protein synthesis. The situation is similar for most parasites. *Plasmodium falciparum*, the most lethal cause of malaria, has become resistant to most antimalarial drugs and to agents such as dichlorodiphenyltrichloroethane (DDT) by changing its metabolism.

Human cancer cells sometimes are not killed by chemotherapeutic agents. The most common reason is that the cells change their metabolic requirements and prevent the agents from inhibiting DNA synthesis. *See* CHEMOTHERAPY.

Drug resistance has been most effectively overcome for bacteria, but far less so for fungi, parasites, and cancer cells. Molecular modification of existing molecules or the discovery of new molecules has provided a wide array of agents to overcome spreading bacterial resistance. *See* ANTIBIOTIC. [H.C.N.]

Drumlin A streamlined, oval-shaped hill which has been shaped by flowing glacial ice. The long axis is parallel to the direction of ice flow, the up-glacier slope is usually steeper than the lee slope, and composition includes a variety or combination of materials—till, outwash, or bedrock. Drumlins are highly localized, but where present, they occur in large numbers. Some drumlins are clearly erosional in origin, but in others till deposition appears to have been synchronous with drumlin formation. Thus, one or both processes must be operative at some time in the subglacial environment where drumlins form. [W.H.J.]

Dry-air filter The broadest category of air filters in terms of the variety of designs, sizes, and shapes in which they are manufactured. The most common filter medium is glass fiber. Dry filters employ the principles of interception, in which particles too large to pass through the filter openings are literally strained from the airstream; impaction, in which particles strike and stick to the surfaces of the glass fibers because of natural adhesive forces, even though the fibers are not coated with a filter adhesive; and diffusion, in which molecules of air moving in a random pattern collide with very fine particles of airborne solids, causing the particles to have random movement as they enter the filter media. It is this random movement which enhances the likelihood of the particles coming in contact with the fibers of filter media as the air passes through the filter. Through the process of diffusion a filter is able to separate from the airstream particles much smaller than the openings in the medium itself. *See* AIR FILTER; MOLECULAR ADHESION. [M.A.B.]

Dry cell A primary cell in which the electrolyte is absorbed in a porous medium, or is otherwise restrained from flowing. Common practice limits the term dry cell to the Leclanché cell, which is the major commercial type. Other dry cells, not discussed in this article, include the mercury cell, the alkaline-zinc-manganese dioxide cell, and the air-depolarized cell. These cells all use aqueous electrolytes immobilized in absorbent materials or gels. By order of the Federal Trade Commission, "leakproof" must not be printed on any dry cell. *See* PRIMARY BATTERY.

The Leclanché cell is made in a variety of sizes in either round or flat shapes. The negative electrode is usually solid zinc. The carbon positive electrode is embedded in a black mixture of manganese dioxide and carbon black. The carbon is either a rod located in the center of a cylindrical cell or a coating on the back of the flat zinc electrode in the flat cell. The separator between the black mix and the zinc electrode consists of a paper barrier coated with cereal or methyl cellulose. The electrolyte is a solution of ammonium chloride and zinc chloride in water. The cell enclosure consists of a top seal in the cylindrical cell or thin plastic wrappings for the flat cell.

The service capacity of dry cells is not a fixed number of ampere-hours, but varies with current drain, operating schedule, cutoff voltage, operating temperature, and storage conditions prior to use. Most cells are tailor-made for their rated end use. The higher the temperature during discharge, the greater is the energy output. Conversely, the lower the temperature, the lower is the output. At −9.4° (−23°C) the battery is virtually inoperative. However, shelf life is influenced in the reverse direction by environmental temperatures. Shelf life is the period of time that a battery can be stored before it drops to 90% of its capacity when tested fresh at 70°F (21°C) and 50% relative humidity. In general, shelf life decreases as the cell size becomes smaller. Tests show that, when batteries are frozen, they suffer no deterioration for about 10 years. They must, however, be allowed to reach room temperature before use.

It is possible to recharge dry-cell batteries for five or six cycles provided the following precautions are taken: The battery should be subject only to shallow discharges between cycles, used immediately after charging, charged over a 10- to 15-h period at constant current, and not overcharged. [J.D.; K.F.]

Dry ice A solid form of carbon dioxide, CO_2, which finds its largest application as a cooling agent in the transportation of perishables. It is nontoxic and noncorrosive and sublimes directly from a solid to a gas, leaving no residue. At atmospheric pressure it sublimes at −109.6°F (−78.7°C). Slabs of dry ice can easily be cut and used in shipping containers for frozen foods, in refrigerated trucks, and as a supplemental cooling agent in refrigerator cars. *See* CARBON DIOXIDE; REFRIGERATION. [C.F.K.]

Drydocking A technique used to remove a ship from the water so that the underwater portion may be inspected, repaired, maintained, or altered. Occasionally underwater repairs may be undertaken while a ship is afloat; however, at regular intervals, or as dictated by emergency, it may be necessary to expose all of the underwater portion, regardless of whether the ship is a small harbor tug or a large transoceanic liner.

The four types of dry docks are known as marine railways, floating dry docks, graving docks, and mechanical lift docks. The size of the ship usually determines which type is used.

The marine railway consists of a cradle of wood or steel with rollers on which the ship may be hauled out of the water along a fixed inclined track leading up the bank of a waterway. The advantages of a marine railway lie in the economy of the original construction and the relative low cost of maintenance. A marine railway is ideal for ships up to 5000 tons.

The floating dry dock may be constructed of wood, steel, or concrete. The dock is submerged, to provide the required depth of water over the keel blocks, by partially filling its tanks with water. The ship to be drydocked is then positioned within, the tanks of the dock are rapidly pumped out by powerful pumps located within the dock walls, and the ship is lifted out of the water.

The graving dock consists of an excavation in the ground with a thick concrete base supported, if necessary, by piling and surrounded on three sides by earth held back by timbers, stone, cement, or steel supports, or a combination of these materials. The entrance, or seaward end of the dock, is usually closed by a caisson of the pontoon type which, when flooded, is trimmed down into position. The dock is flooded, the caisson is floated, and the ship enters the dry dock and is positioned over the keel blocks. The caisson is then replaced and submerged, the dock is pumped out, and the ship settles on the keel blocks. This process is reversed when the ship is ready to leave the dock.

The mechanical lift dock is somewhat similar in action to the floating dry dock. The vessel, after taking up on the keel and bilge blocks in the dock, is bodily lifted clear of the water. The mechanical platform dock has much more flexibility than other types and has increased greatly in size and use. [L.C.R.]

Drying of gases The removal of 95–100% of the water vapor in air or other gases. Gases having a dew point of −40°F (−40°C) are considered commercially dry. The more important reasons for the removal of water vapor from air are: (1) comfort, as in air conditioning; (2) control of the humidity of manufacturing atmospheres; (3) protection of electrical equipment against corrosion, short circuits, and electrostatic discharges; (4) requirement of dry air for use in chemical processes where moisture present in air adversely affects the economy of the process; (5) prevention of water adsorption in pneumatic conveying; and (6) as a prerequisite to liquefaction.

Gases may be dried by the following processes: (1) absorption by use of spray chambers with such organic liquids as glycerin, or aqueous solutions of salts such as lithium chloride, and by use of packed columns with countercurrent flow of sulfuric acid, phosphoric acid, or organic liquids; (2) adsorption by use of solid adsorbents such as activated alumina, silica gel, or molecular sieves; (3) compression to a partial pressure of water vapor greater than the saturation pressure to effect condensation of liquid water; (4) cooling below dew point of the gas with surface condensers or cold-water sprays; and (5) compression and cooling, in which liquid desiccants are used in continuous processes in spray chambers and packed towers—solid desiccants are generally used in an intermittent operation that requires periodic interruption for regeneration of the spent desiccant.

The mechanical methods of drying gases, compression and cooling and refrigeration, are used in large-scale operations, and generally are more expensive methods than those using desiccants. Such mechanical methods are used when compression or cooling of the gas is required. *See* DESICCANT; GAS ABSORPTION OPERATIONS. [W.R.M.]

Drying of solids An operation in which a liquid, usually water, is removed from a wet solid in equipment termed dryers. The use of heat to remove liquids distinguishes drying from mechanical dewatering methods such as centrifugation, decantation or sedimentation, and filtration, in which no change in phase from liquid to vapor is experienced. Drying is preferred to the term dehydration, which is sometimes used in connection with the drying of foods. Dehydration usually implies removal of water accompanied by a chemical change.

Drying is a widespread operation in the chemical process industries. It is used for chemicals of all types, pharmaceuticals, biological materials, foods, detergents, wood, minerals, and industrial wastes. The materials dried may be in the form of thin solutions, suspensions, slurries, pastes, granular materials, bulk objects, fibers, or sheets. Drying may be accomplished by convective heat transfer, by conduction from heated surfaces, by radiation, and by dielectric heating. In general, the removal of moisture from liquids (that is, the drying of liquids) and the drying of gases are classified as distillation processes and adsorption processes, respectively. *See* ADSORPTION; CENTRIFUGATION; DISTILLATION; DRYING OF GASES; EVAPORATION; FILTRATION; SEDIMENTATION (INDUSTRY).

Equipment for drying solids may be conveniently grouped into three classes on the basis of the method of transferring heat for evaporation: direct dryers, indirect dryers, and radiant heat dryers. Batch dryers are restricted to low capacities and long drying times. Most industrial drying operations are performed in continuous dryers. The large numbers of different types of dryers reflect the efforts to handle the large numbers of wet materials in ways which result in the most efficient contacting with the drying medium. Thus, filter cakes, pastes, and similar materials, when preformed in small pieces, can be dried many times faster in continuous through-circulation dryers than in batch tray dryers. Similarly, materials which are sprayed to form small drops, as in spray drying, dry much faster than in through-circulation drying. [W.R.M.]

Drying oil An oil that readily undergoes autoxidation and polymerization to form a hard, dry film on exposure to air. Drying oils are relatively highly unsaturated; that is, they are composed of triglycerides constructed from unsaturated fatty acids. The best drying oils contain several nonconjugated double bonds per molecule.

Raw (untreated) drying oils are not suitable for paints and varnishes because they polymerize too slowly, and various methods have been introduced to improve the polymerization process. One method involves boiling the oil after addition of soluble resin-acid salts. Boiled oil dries in approximately one-fifth the time in which raw oil dries. Blown oil, produced by blowing air through the oil (to which driers have been added) at about 120°C (248°F), is said to have superior wetting or surface-covering properties. Stand oil has been partially polymerized, with admixture of driers, by heating to 260–280°C (500–536°F). This material is used extensively in antifouling paints, printing inks, and linoleum, as well as in varnishes and enamels. Linseed oil is the most widely used drying oil in paints and varnishes. *See* DRIER (PAINT). [E.B.R.]

Dubnium A chemical element, symbol Db, atomic number 105. It was synthesized and identified unambiguously in March 1970 at the heavy-ion linear accelerator (HILAC) at the Lawrence Radiation Laboratory, Berkeley, University of California. The discovery team consisted of A. Ghiorso and colleagues.

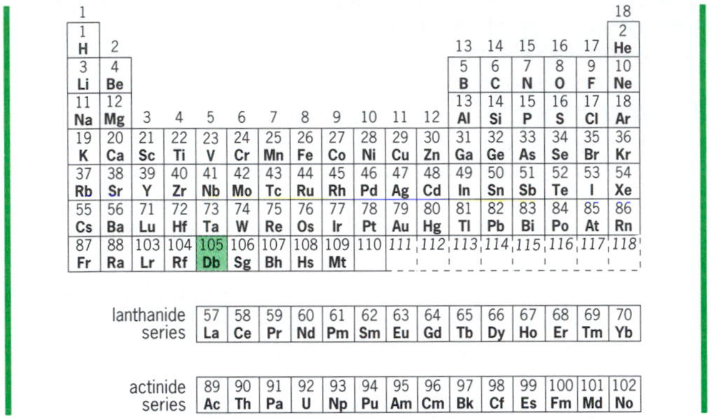

The dubnium isotope, with a half-life of 1.6 s, decayed by emitting alpha particles with energies of 9.06 (55%), 9.10 (25%), and 9.14 (20%) MeV. It was shown to be of mass 260 by identifying lawrencium-256 as its daughter by two different methods.

Previous work on dubnium was reported in 1968 by G. N. Flerov and colleagues at Dubna Laboratories in Russia. They claimed to have discovered two isotopes of dubnium produced by the bombardment of ^{243}Am by ^{22}Ne ions. However, the Lawrence Radiation Laboratory work did not confirm these findings (due to energy and decay differences). *See* NUCLEAR CHEMISTRY. [A.Gh.]

Ducted fan A propeller or multibladed fan inside a coaxial duct or cowling, also called a ducted propeller or a shrouded propeller, although in a shrouded propeller the ring is usually attached to the propeller tips and rotates. The duct serves to protect the fan blades from adjacent objects and to protect objects from the revolving blades, but more importantly, the duct prevents radial flow of the fluid at the blade tips. Fan efficiency remains high over a wider speed range with a properly shaped duct than without. However, fan efficiency is sensitive to duct shape at off-center design conditions. *See* DUCTED FLOW; FAN.

Ducted fans are used in axial-flow blowers or compressors of several stages for turbine engines. A ducted fan engine is a gas turbine arranged to move a larger mass of air than passes through the turbine, the additional air leaving at lower exit velocity and hence higher jet propulsion efficiency for moderate-speed aircraft than obtainable with a simple turbojet. *See* GAS TURBINE; TURBOFAN; TURBOJET. [F.H.R.]

Ducted flow Fluid flow with zero velocity at the boundary relative to the boundary. This condition is distinguished from jet flow, in which the boundary is a fluid, either liquid or gas, and does not remain stationary.

Under certain conditions a propeller has a housing or shrouding around it to control the fluid approaching it, as well as to control the induction of fluid into the jet downstream from the propeller. Shrouding a propeller may be used on a ship to decrease interference of propeller and hull, or it may be used to protect the propeller. *See* DUCTED FAN; JET FLOW. [V.L.S.]

Dumortierite A nesosilicate mineral with composition $Al_8BSi_3O_{19}(OH)$. Dumortierite crystallizes in the orthorhombic system but well-formed crystals are rare; the mineral usually occurs in parallel or radiating fibrous aggregates. The hardness is 7 on Mohs scale, and the specific gravity is 3.26–3.36. The mineral has a vitreous luster and a color that varies not only from one locality to another but in a single specimen. It may be pink, green, blue, or violet. In the United States it occurs at Dehesa, California, and at Rochester, Nevada, where it has been mined for the manufacture of high-grade porcelain. *See* SILICATE MINERALS. [C.S.Hu.]

Dune A mobile mound, hill, or ridge of windblown material, usually sand. Dunes develop where the velocity of sand-driving winds is checked or reduced, often as a result of barriers or obstructions. Dunes require not only a suitable site for development but an adequate sand source as well. Consequently, most dunes are found downwind from sandy lake or ocean beaches; dry stream beds, particularly in the desert; and barren, sandy, glacial outwash plains. *See* DESERT.

No commonly accepted classification of dunes exists, partly because of dune complexity but chiefly because numerous intermediate and special forms must be accommodated in any classification. Nevertheless, there are dune types that are widely recognized.

Draa are very large dune forms, found only in the North African desert, whose dimensions are measured in hundreds of meters and whose spacing is in kilometers.

Transverse dunes include three main types: barchan, aklé, and transverse ridge. Crescent-shaped barchan dunes are those which develop on nearly flat, barren desert surfaces where wind direction is constant and sand supply moderate (Fig. 1). Aklé dunes are sinuous transverse ridges of sand somewhat intermediate in form between the crescentic barchan and the simple transverse ridge. Transverse ridges are nearly straight in ground plan, and are oriented more or less at right angles to the sand-driving winds. Transverse ridges are most common in coastal localities where vegetation affects sand movement, but they may occur in desert regions as well.

Fig. 1. Barchans, or crescent-shaped dunes, near Biggs, Oregon. These riverbed dunes move from left to right with steep side away from the wind. (*USGS*)

Parabolic or blowout dunes are hairpin or spoon-shaped with horns pointing toward the wind, opposite to those of the barchan.

Longitudinal or self dunes are sometimes hundreds of kilometers long, as in the Sahara or Australian deserts. Barchan chains pass gradually into self ridges as winds change from unidirectional to multidirectional. As sand-driving winds come

Fig. 2. Seif dune or linear ridge, located in the Algodones dunes, Imperial County, California.

from increasingly divergent directions, the seif form is lost altogether (Fig. 2). [R.M.N.]

Dunite An ultrabasic rock consisting almost solely of olivine. The type locality is Dun Mountain, situated in the so-called Mineral Belt in New Zealand. Besides a magnesium-rich olivine, chromite and picotite are common constituents of dunites. Dunites are therefore important sources of chromium. Spinel, magnetite, ilmenite, pyrrhotite, and native platinum also occur in some dunites. Serpentinization of olivine in dunites is widespread. Some dunites have changed completely to serpentine bodies. *See* SERPENTINE; SERPENTINITE.

Dunites may form by fractional crystallization of basic magma or by replacement of rocks such as limestone and dolomites; metamorphic differentiation has also been suggested as a mechanism of formation. It is also possible that some dunites are directly derived from a deep-seated basic layer in the earth. *See* MAGMA; METAMORPHISM. [H.Ra.]

Dust and mist collection The physical separation and removal of solid or liquid particles from a gas in which they are suspended. Such separation is required for one or more of the following purposes: (1) to collect a product which has been processed or handled in gas suspension, as in spray-drying or pneumatic conveying; (2) to recover a valuable product inadvertently mixed with processing gases, as in kiln or smelter exhausts; (3) to eliminate a nuisance, as a fly-ash removal; (4) to reduce equipment maintenance, as in engine intake air filters; (5) to eliminate a health, fire, explosion, or safety hazard, as in bagging operations or nuclear separations plant ventilation air; and (6) to improve product quality, as in cleaning of air used in processing pharmaceutical or photographic products.

All particle collection systems depend upon subjecting the suspended particles to some force which will drive them mechanically to a collecting surface. The known mechanisms by which such deposition can occur may be classed as gravitational, inertial, physical or barrier, electrostatic, molecular or diffusional, and thermal or radiant. There are also mechanisms which can be used to modify the properties of the particles or the gas to increase the effectiveness of the deposition mechanisms. For example, the effective size of particles may be increased by condensing water vapor upon them or by flocculating particles through the action of a sonic vibration. Usually, larger particles simplify the control problem. To function successfully, any collection device must have an adequate means for continuously or periodically removing collected material from the equipment.

Devices for control of particulate material may be considered, by structural or application similarities, in seven principal categories as follows: gravity setting chamber, inertial device, packed bed, cloth collector, scrubber, electrostatic precipitator, and air filter. *See* AIR FILTER; CHEMICAL SEPARATION TECHNIQUES; ELECTROSTATIC PRECIPITATOR; MECHANICAL SEPARATION TECHNIQUES; UNIT OPERATIONS. [C.E.La.]

Dust storm A strong, turbulent wind carrying large clouds of dust. In a large storm, clouds of fine dust may be raised to heights well over 10,000 ft (3000 m) and carried for hundreds or thousands of miles (1 mi = 1.6 km).

Sandstorms differ by the larger mass, more rapid setting speeds of the particles involved, and the stronger transporting winds required. The sand cloud seldom rises above 3.3–6.6 ft (1–2 m) and is not carried far from the place where it was raised.

Dust storms cause enormous erosion of the soil, as in the dust bowl disasters of 1933–1937 in the Great Plains of the United States. Besides causing acute physical discomfort, they present a severe hazard to transportation by reducing the visibility to very low ranges. Conditions required are an ample supply of fine dust or loose soil, surface winds strong enough to stir up the dust, and sufficient atmospheric instability for marked vertical turbulence to occur.

Small dust particles increase scattering of light, mainly in short (blue) wavelengths. The Sun often appears a deep orange or red when seen through a dust cloud; however, optical effects are variable. Large particles are effective reflectors, and an observer in an aircraft above a dust storm may see a solid sheet with an apparent dust horizon. [C.W.N.]

Dwarf star The most common type of star in the Galaxy, also known as a main-sequence star. The Sun is a typical dwarf, with surface temperature of 5750 K, radius of 6.96×10^5 km, mass of 2×10^{30} kg, and luminosity of 4×10^{33} ergs/s (that is, 4×10^{23} kW). These numbers are fairly typical as to radius and mass. However, in luminosity, main-sequence stars occur over a range from O and B stars, up to more than 10^4 times brighter than the Sun, with surface temperatures of 35,000–15,000 K, down to the faintest M dwarfs, 10^{-4} as bright as the Sun, with a temperature of only 2500 K. Dwarfs form a single-parameter family in which the significant variable is mass; members range in mass from 20 Suns down to less than 0. 1 Sun. The M or red dwarfs are the most common stars in space, and because of their low luminosity have the longest life. Almost all other types of stars have evolved from main-sequence stars. *See* STAR. [J.L.Gr.]

Dwarfism and gigantism Underdevelopment and overdevelopment of the skeleton, respectively. Skeletal growth is a complex process and can be distributed in many ways. For example, overstimulation by excessive growth-hormone production during childhood can produce gigantism. This is usually due to a pituitary tumor. Insufficient stimulation of skeletal growth, resulting from hormonal, metabolic, or nutritional disturbances, leads to reduced height with normal body proportions. The shortness of stature depends on the degree of the disturbance; the designation proportionate dwarfism is used in severe cases. Proportionate dwarfism may or may not be genetic. *See* PITUITARY GLAND.

Commonly, the term dwarfism is applied to short individuals with abnormal body proportions resulting from disturbances of bone growth itself. The bones of either limbs or trunk can be most affected. The disorders, called chondrodysplasias, number well over 100 and show a wide variation in clinical features: some appear at birth, others become evident in later childhood. The severity ranges from conditions in which infants die at birth to conditions compatible with normal life. In most instances, however, the abnormal skeleton creates medical problems.

Almost all of the chondrodysplasias are inherited as mendelian or single-gene traits. However, because of reduced fertility and a high incidence of new mutations for certain traits, most individuals with these disorders are born to parents of average stature. *See* HUMAN GENETICS; SKELETAL SYSTEM DISORDERS. [W.A.Ho.]

Dyadic A mathematical abstraction corresponding to an expression of the type $\boldsymbol{\beta}\boldsymbol{\gamma} + \boldsymbol{\delta}\boldsymbol{\epsilon} + \cdots$, in which the elements (dyad symbols) consist of two vector symbols in juxtaposition without the intervention of either the dot (·) or cross (×). Essentially, a dyad is an ordered pair of vectors subject to certain rules of operation. The first symbolic factor in a dyad ($\boldsymbol{\beta}$ in $\boldsymbol{\beta}\boldsymbol{\gamma}$, for example) is called the antecedent and the second the consequent. *See* CALCULUS OF VECTORS. [H.V.C.]

Dye A colored substance, also called a dyestuff, which imparts more or less permanent color to other materials. *See* DYEING.

Customarily, colored water-insoluble substances are called pigments. Dyes are generally water-soluble, although some are soluble only during application, after which they become insoluble. *See* PIGMENT.

The mechanism by which soluble colored substances enter the internal structure of fibers and there become fixed has been variously explained in terms of the physical and chemical concepts of the times when the explanations were given. It is said to be an adsorption phenomenon, a salt formation, a quasi-chemical union caused by hydrogen bonding, or an ether linkage, and in some cases it is considered to be a true solution effect. The end result, however, is that the dye has imparted a color (not necessarily that of the solid dye itself) to the fiber which is more or less resistant to washing or removal by similar mechanical operations. The dye is said to be fixed on and to have affinity for the material it has colored. The material is designated as the substrate. If the color is quite resistant to washing and light, it is called a fast color; if the color is easily removed or fades quickly, it is a fugitive dye.

Dyestuffs may be classified in various ways: according to color (blue, red, and so on); origin (natural—from vegetable and animal matter—or synthetic); chemical structure (the most precise); kinds of material to which they are applied (cloth, paper, leather, plastic, food, biological specimens, etc.); and method of application (used most frequently by the practical dyer). [W.Mi.]

Dyeing The application of color-producing agents to material, usually fibrous or film, in order to impart a degree of color permanence demanded by the projected end use. True dyeing covers mechanisms in which molecules of material to be dyed become involved by various means with the molecules of the coloring matter, or small aggregates thereof. There is some overlapping between true dyeing and other methods of coloring, which are called dyeing in the industry. Products which are commonly dyed include textile fibers, plastic films, anodized aluminum, fur, wood, paper, leather, and some foodstuffs. *See* DYE.

The broad term affinity is used to describe the various types of attraction between the material to be colored and the dye. Affinity may be caused by attraction between charged dye particles and oppositely charged dye sites on the material, by various types of chemical attraction, and by formation of solid solutions of dye in the material.

When affinity is involved, dyeing is an exothermic process. More dye will be absorbed by a fiber or film at low temperatures than at higher temperatures. However, the rate of dyeing increases geometrically, whereas maximum dye absorption decreases only arithmetically with rise in temperature. Since maximum dye absorption is rarely necessary or desirable, the trend in modern practice is toward dyeing at high temperatures and short times.

Dyeing is accomplished by dissolving or dispersing the colorant in a suitable vehicle (usually water) and bringing this system into contact with the material to be dyed. Although many dye molecules or aggregates may adhere to the material surface when they meet, dyeing does not occur until the adhering dye particles migrate within the fibers or films. All dyeing processes are designed to accomplish ultimately penetration of the undyed substance by the colorant.

Assistants are materials which do not impart color to the product to be dyed but, as the name suggests, promote or retard dyeing. Usually, but not always, they affect the dye molecule.

Swelling agents are assistants which open up the structure of the fiber temporarily so that dye molecules or aggegates may enter more freely and reach otherwise inaccessible dye sites.

Carriers are agents (often solvents of low water solubility) which accelerate dyeing by breaking up or dissolving dye aggregates and bringing them to the fiber-water interface in a size small enough to be absorbed by the material.

Dye retarders are a class of dyeing assistants, usually inorganic or organic salts, which slow up the dyeing process by forming evanescent compounds with the dye, by buffering or depressing the ionization of an acid assistant, or by temporarily occupying the more active or more accessible dye sites on the fiber, later to be dislodged therefrom by the dye.

Aftertreating agents are salts, resins, or other products (more frequently applied to cellulosic fibers) to render the colored fabric more resistant to the effects of washing, perspiration, or fading by ozone or combustion gases. More often than not, their application causes a loss in light fastness of the dyed material.

Textiles. Cellulose fibers, such as cotton and rayon, are most commonly dyed by immersion of the fibers in a solution of direct dyes using an electrolyte such as common salt as assistant and then boiling this dyebath. Such dyeings usually exhibit only commercial (minimum) resistance to washing. Treatment of the properly dyed fibers with resins and copper, for example, increases the resistance to washing with minimum loss of light resistance.

The dyeing of wool, silk, and fur (felt) involves the formation of salts by reactive positively charged groups on the fiber and negatively charged color groups on anionic dyes. Level-dyeing (acid or anionic) dyes, which require a relatively large amount of strong acid to force the dye onto the fiber, are used for carpet yarns, felt hats, and wherever even dyeing and penetration instead of washing fastness is a paramount requirement. Milling dyes, which demand less acid, are faster to washing and are used for blankets and sweaters. Where a very high resistance to light and washing is required, acid-dyeing or neutral-dyeing metallized dyes are employed.

Synthetic fibers, such as cellulose acetates and triacetate (Arnel), are dyed in a supension of solvent-soluble dyes by immersion. Polyamide synthetic fibers are dyed like wool with acid, metallized acid, neutral metallized, or fiber-reactive mordant dyes, azoics, and selected direct dyes from an acid bath. Special processes have also been developed for acrylic, polyester, and propylene fibers. *See* TEXTILE CHEMISTRY.

There are two fundamental types of textile-dyeing processes—the dye liquor is pumped through stationary material, or the material to be dyed moves through the dye liquor.

Kiers are tanks which hold loose fiber or packaged yarn and the dye liquor is pumped through the fiber at temperatures up to the boiling point or (if the fiber can withstand it) under pressure at temperatures up to 270°F (132°C). In continuous dyeing, cloth is impregnated with dye and then passed through a series of developing, washing, and drying zones to a final take-up roll.

Nontextile materials. Dyeing anodized aluminum was formerly believed to be accomplished by the formation of aluminum salts of dyes by the electrically deposited aluminum oxide coating. However, many dyes having no salt-forming groups were found to dye this film. Anodized aluminum is readily dyed by many textile dyes. Light and weather resistance undreamed of in textile applications of some of these same dyes is achieved.

Paper pulp is usually dyed in the paper beater by dyes normally employed for cotton; on occasion, it is tinted by wool dyes, and it is frequently tinted by addition of pigments to the beater. Finished paper is also colored by passing it over rollers which supply dye or colored coatings to its surface (calender staining).

Wood is normally stained with solutions of dyes or dispersions of pigments in water, in solvents, or in lacquers. Wood, however, is frequently dyed by the application of water solutions of dyes at high temperature under pressures of 100 psi (6.9×10^5 pascals) or more. Freshly felled trees have been dyed by force-pumping dye solution through the length of the log. The dye solution replaces the natural moisture and sap.

Leather is dyed at low temperatures with the classes of dyes normally used for wool and cotton. Formic acid is normally used to exhaust the dye. For dress gloves, leather is usually colored by applying the dye on the grain surface, leaving the flesh side undyed. Leather is also dyed with natural dyes such as logwood, fustic, and quercitron. Leather fresh from tanning and containing considerable moisture is dyed in Europe by tumbling with dry water-soluble dye.

Most food products which are artificially colored are not actually dyed. Maraschino cherries, however, are dyed for several hours with food dyes, then washed and placed in flavored syrup. Glazed citron, citrus, or watermelon rinds are colored by long immersion followed by drainings in a series of food-dye-colored, increasingly strong, sugar syrups. Pistachio nuts in the shell are dyed by applying food-dye solutions with or without salt during roasting operations.

It is not generally realized that human hair is, in the final analysis, a class of textile fiber. Cosmetic companies, appreciating this fact, have hired textile experts to do research on dyeing, waving, and washing hair. The dyeing of hair is of course limited to processes employing relatively low temperatures. Food dyes (also carbon black) mixed with citric or malic acid are used for so-called rinses. Other rinses employing basic dyes and salt or textile dyes diluted with salts, detergents, and malic, citric, or tartaric acids are on the market. Products selected, as a rule, from among the safer fur-dye intermediates are used for more permanent and intense coloring techniques; these employ hydrogen peroxide for developing. Natural dyes, such as the age-old henna and some metallic salts, are still encountered in some hairdyeing establishments. Rinses which are used to mask the yellowish tinge in gray or white hair frequently use blue or violet food or textile dyes and contain optical bleaches—substantially colorless compounds which receive invisible ultraviolet light and reflect it to the eye as a blue or violet fluorescence.

Many plastic materials may be dyed by processes similar to those employed for textiles. Nylon, cellulose acetate, polyethylene, polypropylene, and polyester resins are dyeable with dyes which color these materials in yarn form. *See* MANUFACTURED FIBER; NATURAL FIBER. [J.E.Lo.]

Dynamic braking A technique of electric braking in which the retarding force is supplied by the same machine that originally was the driving motor. Dynamic braking is effective only in high inertial systems wherein the kinetic energy of the motor rotor with its connected load is converted into electrical energy and dissipated mainly as I^2R losses or returned to the source.

The commonest type of dynamic braking will be explained for a direct-current (dc) motor. To accomplish braking action, the supply voltage is removed from the armature of the motor but not from the field. The armature is then connected across a resistor. The electromotive force generated by the machine, now acting as a generator driven by the kinetic energy of the rotating system, forces current in the reverse direction through the armature. Thus a torque is produced to oppose rotation, and the load decelerates as its kinetic energy is dissipated, mostly in the external resistor but to some extent in core and copper losses of the machine.

Electric braking can also be accomplished by causing the kinetic energy of the rotating system to be converted in the armature to electrical energy and then returned to the supply lines. This mode of operation, called regenerative braking, occurs when the counter electromotive force exceeds the supply voltage.

Interchanging two of the lines supplying a three-phase alternating-current (ac) induction motor also produces braking. In this case, called plugging, the direction of the electromagnetic torque on the rotor is reversed to cause deceleration. Both the kinetic energy of the system and the energy drawn from the supply lines are expended in copper and core losses in the machine. *See* ALTERNATING-CURRENT GENERATOR; DIRECT-CURRENT MOTOR; INDUCTION MOTOR; SYNCHRONOUS MOTOR. [A.R.E.]

Dynamic instability A state of fluid flow in which the distribution of mass and momentum is unstable. Fluid flows are subject to a variety of instabilities which generally cause the flow to become more complex and which often lead to turbulent, chaotic flow. Instabilities are responsible for a variety of phenomena in natural flows, including cyclones, hurricanes, and thunderstorms in the atmosphere; mantle convection in the Earth's interior; and granules and supergranules in stellar atmospheres. A great deal of research in the geophysical and astrophysical sciences has focused on flow instabilities; and instability plays an important role in engineering problems ranging from naval engineering and aeronautics to the design of efficient heating and cooling systems. *See* FLUID FLOW; TURBULENT FLOW.

Fluid instabilities may be broadly divided into two classes: convective and dynamic. While convective instabilities can be understood rather easily in terms of forces acting on displaced parcels of fluid, dynamic instabilities are more varied and more challenging to understand. This broad class of fluid instabilities is responsible for the cyclones and anticyclones that dominate the weather in middle and high latitudes, as well as meanders and rings in ocean currents such as the Gulf Stream. These instabilities are ultimately responsible for the lack of predictability of complex flows such as are found in the atmosphere. *See* CONVECTIVE INSTABILITY; CYCLONE; WIND.

Many, but not all, dynamic instabilities can be understood with the aid of a vorticity principle and an invertibility principle. The vorticity of a fluid can be thought of as the rate of rotation of a rigid paddle wheel embedded in the flow (see illus.). [Mathematically, it is the curl of the vector velocity field.] The rotation can be brought about by shear (a change in the speed of the flow in the direction across the flow) and by the curvature of the flow. Vorticity is important for two reasons. First, in two-dimensional flows, vorticity is conserved; that is, if the pad-

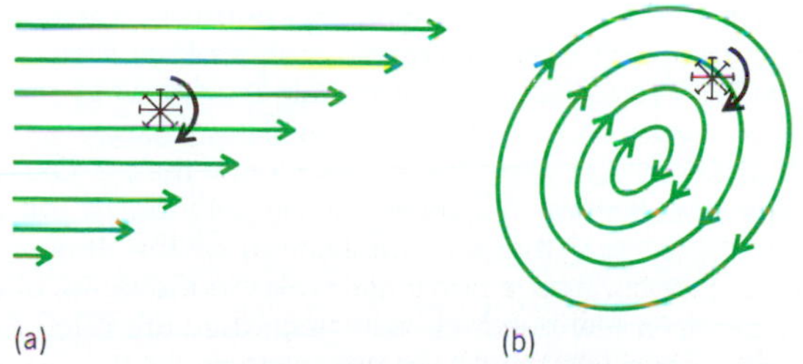

The vorticity of a fluid is related to the rate at which a rigid paddle wheel would rotate if placed in the flow. (*a*) The paddle wheel rotates because the velocity of the fluid varies in a direction perpendicular to the flow. (*b*) Rotation is generated by the curvature of the flow.

dle wheel is followed around it is observed that its rate of rotation remains constant. Second, in such flows, it is possible to work backward from knowledge of the vorticity at every point in the fluid to find the flow field at every point. This is called the invertibility principle. Basically, a clockwise vorticity at a point in the fluid will induce a clockwise rotation of the fluid nearby; the total velocity field in the fluid can then be found by adding the velocities induced by all the vorticities at each point. *See* CALCULUS OF VECTORS.

The flow of the Earth's atmosphere is three-dimensional. Under this circumstance, vorticity is not conserved because fluid converging on or diverging from a point in the fluid causes it to spin faster or slower. (The principle at work here is the conservation of angular momentum.) Fortunately, the Earth's atmosphere is usually stably stratified; that is, a parcel displaced upward will be colder than its environment and will accelerate downward toward its original position. In this sense the atmosphere is convectively stable. It is possible to identify a quantity that reflects the air's density and at the same time is conserved as long as no heat is added to or subtracted from the air. This quantity is called the potential temperature (θ) and is related to temperature and pressure by the expression below, where T

$$\theta = T\left(\frac{1000}{p}\right)^{0.287}$$

is the temperature in kelvins and p is the pressure in millibars. Generally, θ increases upward in the atmosphere: slowly in the troposphere (a layer of air that extends upward to roughly 6 mi or 10 km) and more rapidly in the stratosphere (a layer of air between about 6 and 18 mi, or 10 and 30 km). *See* STRATOSPHERE; TROPOSHERE.

The criterion for the dynamic instability of large-scale three-dimensional flows is analogous to barotropic instability of two-dimensional flows. The necessary condition for this kind of instability is that potential vorticity must be a maximum or minimum somewhere on a surface of constant θ. The type of instability that results from maxima or minima of potential vorticity in a fluid is called internal baroclinic instability. This type of instability can draw on both the kinetic energy of the original flow and the potential energy that occurs when there are horizontal variations of density.

Observations of the Earth's atmosphere show that the potential vorticity of the troposphere in middle and high latitudes is nearly uniform. But when a gently sloping θ surface is followed into the stratosphere, a large increase of potential vorticity occurs. Even so, there is no localized maximum or minimum of potential vorticity, but just a large gradient of the quantity, concentrated near the boundary between the troposphere and stratosphere. (Many meteorologists define the stratosphere as a region of large potential vorticity.) For this reason, internal baroclinic instability seldom occurs in the atmosphere. *See* ATMOSPHERE. [K.A.Em.]

Dynamic meteorology The study of those motions of the atmosphere that are associated with weather and climate. Atmospheric motions span an enormous range of spatial and temporal scales; dynamic meteorology concentrates mainly on large-scale and mesoscale motions. Large-scale motions are those with horizontal scales in excess of a few hundred kilometers and time scales longer than a day. Such motions are strongly influenced by the rotation of the Earth and by the vertical thermal stratification of the atmosphere. Mesoscale motions have horizontal scales in the range of a few kilometers to a few hundred kilometers; they are often associated with convective clouds and precipitation.

The mechanics and thermodynamics of the unsaturated atmosphere are governed by three fundamental conservation laws: (1) conservation of mass, expressed by the mass continuity equation; (2) conservation of momentum, expressed by Newton's second law of motion; and (3) conservation of thermodynamic energy, expressed by the first law of thermodynamics. For saturated conditions, conservation of water substance must also be considered. *See* CONSERVATION OF MASS; CONSERVATION OF MOMENTUM; CORIOLIS ACCELERATION; NEWTON'S LAWS OF MOTION; THERMODYNAMIC PRINCIPLES.

Statics. The vertical structure of the atmosphere is determined by the equation of state for an ideal gas and by the hydrostatic relationship. The former expresses the relationship among pressure, density, and temperature at any point; the latter expresses the balance between the upward-directed component of the pressure-gradient force (associated with the approximate exponential decrease of pressure with height) and the downward-directed gravity force. The equation of state and hydrostatic equation may be combined to form the hypsometric equation (1), which

$$Z_2 - Z_1 = \frac{R}{g}\int_{p_2}^{p_1} T\, d\ln p \qquad (1)$$

relates the geopotential height differences ($Z_2 - Z_1$) between two pressure surfaces p_2 and p_1 to the mean temperature T in the layer between the two surfaces; and where R (the gas constant for dry air) = 287 J · kg^{-1} · K^{-1} and g (the acceleration of gravity) = 9.81 m · s^{-2}. Thus, pressure decreases more rapidly with height in cold air than in warm air. *See* GAS; HYDROSTATICS.

Except in regions of active precipitation, where the heating rate due to latent heat of condensation is large, temperature changes following the motion of individual parcels of air are controlled primarily by adiabatic expansion and compression as the air parcels move to lower or higher pressure. The thermodynamic state of such parcels can be characterized by the potential temperature θ, as in Eq. (2), where p_0

$$\theta = T\left(\frac{p_0}{p}\right)^{R/C_p} \qquad (2)$$

[= 10^5 pascals (1000 millibars)] is a reference pressure and c_p (= 1004 J · kg^{-1} · K^{-1}) is the specific heat at constant pressure. The potential temperature is the temperature that a parcel of air at pressure p and temperature T would acquire if it were moved adiabatically to pressure p_0. When all diabatic heat sources can be neglected, θ remains constant in time for each air parcel. Normally, θ increases with altitude in the atmosphere, so that an air parcel displaced upward (downward) has a value of θ less (greater) than that of its environment, and hence experiences a net buoyancy force that tends to return it to its equilibrium level. The atmosphere is then said to be statically stable. When θ is constant with height, a condition that occurs when the temperature decreases with height at a rate 10°C · km^{-1} (30°F · mi^{-1}), the atmosphere is said to be neutrally stable; if θ decreases with height, the atmosphere is absolutely unstable and convective motion develops spontaneously. If an air parcel is saturated, upward displacement causes water vapor to condense and release its latent heat of condensation; the potential temperature is then no longer conserved.

The hydrostatic relationship implies that pressure decreases monotonically with altitude. Pressure may thus be substituted for height as the independent vertical coordinate. If the atmosphere is everywhere statically stable so that θ increases monotonically with height, potential temperature can also be used as a vertical coordinate. Potential temperature coordinates are useful for analysis of adiabatic motions, since in that reference frame prediction of adiabatic flow is reduced to a two-dimensional problem of following the motion on θ surfaces. Isobaric coordinates, in which pressure is the vertical coordinate, have the advantage of eliminating any explicit reference to the density field. These are the most commonly used vertical coordinates in dynamic meteorology. *See* CONVECTIVE INSTABILITY; DYNAMIC INSTABILITY; HYDROSTATICS; ISOBAR (METEOROLOGY).

Baroclinic instability. Baroclinic energy conversion processes are responsible for the growth and maintenance of most large-scale weather disturbances.

In midlatitudes in the troposphere, potential temperature normally decreases from Equator to pole on isobaric surfaces. This decrease does not occur uniformly, but tends to be concentrated in the jet stream, a narrow band of strong westerly winds in the upper troposphere that encircles the globe in midlatitudes. *See* JET STREAM.

When the shear of the zonal wind is sufficiently strong so that the meridional gradient of potential vorticity on a constant potential temperature surface is locally reversed, or when there is a nonvanishing gradient of potential temperature at the surface of the Earth, the linearized equations have solutions in the form of exponentially growing disturbances. These baroclinically unstable modes have growth rates, structures, and scales similar to those observed in developing extratropical cyclones. Baroclinic instability provides a mode whereby infinitesimal disturbances may amplify into large-amplitude storms. However, in many cases it appears that storms in the atmosphere grow through nonlinear interactions involving preexisting disturbances. Such processes must generally be studied by numerical simulations on high-speed computers.

Mesoscale convective systems. For horizontal scales less than several hundred kilometers, the major energy source is not baroclinic instability; it is latent heat release by cumulonimbus clouds. The convective storms associated with such clouds can occur only when the atmosphere is conditionally unstable [see Eq. (2)], sufficient moisture is present, and there is an initial disturbance strong enough to lift air parcels high enough to release the conditional instability. Mesoscale convective systems take a variety of forms. Among these are hurricanes, squall lines, and mesoscale convective complexes. *See* HURRICANE; MESOMETEOROLOGY; SQUALL LINE.

Numerical weather prediction. In current global weather prediction models, a mixture of techniques is often employed. Nearly all models use finite difference representations for the vertical coordinate and time; the horizontal variation is generally represented either by a network of grid points at uniform intervals of latitude and longitude or by a finite set of spherical harmonics. *See* SPHERICAL HARMONICS.

To predict the weather dynamically, it is necessary to solve the dynamics equations by integrating in time, starting from an initial state of the atmospheric variables determined from observations. In practice, observational errors and poor data coverage, particularly over the oceans, make it impossible to exactly determine the initial state of the atmosphere. Moreover, the state determined from observations generally does not properly represent the true dynamical balance among the pressure and wind fields. This imbalance introduces noise that is interpreted as high-frequency inertia-gravity waves. If a prediction is attempted from such an initial state, the noise rapidly dominates the true, slowly evolving weather disturbances. In order to prevent the growth of such spurious noise, the analysis must be initialized by processing it in a manner that assures a dynamical balance between the pressure and wind fields while preserving the actual observations as closely as possible. However, even if the initial state were known perfectly, there still would be a limit beyond which errors would dominate the forecast. *See* WEATHER FORECASTING AND PREDICTION.

Global climate modeling. In addition to their role in weather prediction, dynamical models can also be used to simulate global climate. Climate is the study of the average state of the atmosphere and its seasonal and interannual variability. Climate is determined by the joint influence of energy sources and sinks at the Earth's surface, and the transformation and transport of energy in the atmosphere and the oceans. Models that simulate these processes are usually referred to as general circulation models. Such models must contain accurate representations of all the important physical processes that influence the circulation. *See* ATMOSPHERE; CLIMATE MODELING; METEOROLOGY; WIND.

[J.R.H.]

Dynamic nuclear polarization

Dynamic nuclear polarization The creation of assemblies of nuclei whose spin axes are not oriented at random, and which are in a steady state that is not a state of thermal equilibrium. Under commonly occurring conditions, the spin axes of nuclei (with nonzero spin) are oriented at random; where this is not so, the nuclei are said to be polarized. Assemblies of polarized nuclei are not in a state of thermal equilibrium except under rather extreme conditions (for example, temperatures below 10 millikelvins and magnetic fields greater than several teslas), and therefore schemes have been devised to produce polarized assemblies, in a steady state which is not a state of thermal equilibrium, under less extreme conditions of temperature and so forth. Such schemes constitute dynamic nuclear polarization.

Among the many applications of polarized nuclei are the following. Nuclear forces are spin-dependent, and although the spin-dependent part can be found by using unpolarized assemblies, the experiments are simpler and their interpretation is clearer if polarized nuclei are used. Assemblies of polarized nuclei have a lower geometrical symmetry than assemblies of randomly oriented nuclei, and so these have been used to investigate the fundamental symmetries of nature. Polarized nuclei have been used to enhance the signal in free precession magnetometers and similar instruments, and the use of an assembly of polarized nuclei as a gyroscope has also been suggested. *See* MAGNETOMETER; NUCLEAR ORIENTATION; PARITY (QUANTUM MECHANICS); SPIN (QUANTUM MECHANICS). [J.M.D.]

Dynamic programming

Dynamic programming A mathematical technique for the study of systems that can be represented by certain types of multistage processes. In such processes the output at each is a function of that at the preceding stage (the state), an action taken by a decision maker (the control), and possibly a random event. An example occurs in inventory control, in which the inventory of a given item depends on the previous day's inventory (the state), the amount produced (the control), and the external demand (random event). This gives rise to a constraint space of very high dimension. The technique of dynamic programming reduces this sequence to that of one-dimensional problems, which are generally much simpler to solve. [G.Ow.]

Dynamic similarity

Dynamic similarity A relationship existing between two homologous fluid-flow systems such that corresponding parts of the systems experience similar net forces. Dynamically similar flows about geometrically similar bodies will themselves be geometrically similar. Consequently, this concept is basic to the meaningful extrapolation of model results to full-scale performance. However, geometrically similar flows are not necessarily dynamically similar.

Dynamic similarity between two flow systems will occur if certain nondimensional parameters formed from the flow variables have the same values for both systems (see table).

In practice, it is often difficult to establish equality of all similarity parameters simultaneously for two flows. Equality of

Dynamic similarity parameters

Nondimensional parameter*	Name	Physical effect
$\rho VL/\mu$	Reynolds number	Viscosity
V/c	Mach number	Compressibility
V^2/Lg	Froude number	Gravity
x_m/L	Knudsen number	Pressure
$\rho V^2L/\sigma$	Weber number	Surface tension
$c_p\mu/\kappa$	Prandtl number	Heat conduction
$\beta TgL^3\rho^2/\mu^2$	Grashof number	Free convection

*The reference variables are defined as follows: L, length; μ, coefficient of viscosity; c, speed of sound; g, acceleration of gravity; σ, surface tension; κ, coefficient of thermal conductivity; β, coefficient of thermal expansion; T, temperature; x_m, mean free path of molecule; and c_p, specific heat at constant pressure.

parameters corresponding to dominant flow properties is usually sufficient. *See* DIMENSIONAL ANALYSIS; FLUID MECHANICS; FROUDE NUMBER; MACH NUMBER; MODEL THEORY. [A.G.H.]

Dynamical time A time scale based on the orbital motions of celestial bodies. It is defined as the independent variable T in the differential equations of motion of bodies subject to Newton's laws of dynamics and gravitation as modified by relativity. Dynamical time scales can differ in epoch and rate, but cannot have a relative secular acceleration.

The atomic time scale obtained from atomic clocks is not corrected for periodic relativistic terms, and thus differs from dynamical time by their sum, P_R. Universal Time One (UT1), based on the Earth's rotation, differs from dynamical time because of secular and other variations in rotational speed. *See* ATOMIC TIME; EARTH.

In 1967 the General Conference of Weights and Measures defined the second, the unit of time in the International System of Units (SI), as the duration of a certain number of periods of cesium-133 radiation. This is the second of International Atomic Time (TAI). Standard time differs from TAI by integral numbers of hours and seconds.

In 1976 the International Astronomical Union defined two time scales to meet needs for increased accuracy of ephemerides and to utilize the high accuracy and availability of atomic time, TAI. Barycentric Dynamical Time (French abbreviation, TDB) is a dynamical scale. It is the independent variable in the differential equations of motion, referred to the barycenter (center of mass) of the solar system (near the center of the Sun). Terrestrial Dynamical Time (TDT) is an auxiliary, atomic scale, defined as TDT = TAI + 32.184 s. TDT differs from TDB only by the relativistic terms, P_R, included in TDB. P_R is calculable, and has a maximum value of 0.0017 s, but it depends somewhat on the relativity theory used to form TDB (although the differences are orders of magnitude less than the maximum). *See* RELATIVITY; TIME. [W.M.]

Dynamics That branch of mechanics which deals with the motion of a system of material particles under the influence of forces, especially those which originate outside of the system under consideration. From Newton's third law of motion, namely, to every action there is an equal and opposite reaction, the internal forces cancel in pairs and do not contribute to the motion of the system as a whole, although they determine the relative motion, if any, of the several parts.

Particle dynamics refers to the motion of a single particle under the influence of external forces, particularly electromagnetic and gravitational forces. The dynamics of a rigid body is the study of the motion, under given forces, of a system of particles, the distances between which are postulated to be constant throughout the motion.

In classical dynamics the basic relation that enables the motion to be determined once the force is known is Newton's second law of motion, which states that the resultant force on a particle is equal to the product of the mass of the particle times its acceleration. For a many-particle system it becomes impracticable to write and solve this equation for each individual particle and, in general, the motion may be computed only on a statistical basis (that is, by the methods of statistical mechanics) unless, as for a few particles or a rigid body, the number of degrees of freedom is sufficiently small. *See* DEGREE OF FREEDOM (MECHANICS); KINEMATICS; KINETICS (CLASSICAL MECHANICS); NEWTON'S LAWS OF MOTION; RIGID-BODY DYNAMICS; STATISTICAL MECHANICS. [H.C.Co./B.G.]

Dynamitron accelerator A particle accelerator that utilizes a steady high-voltage direct-current (dc) potential to accelerate charged particles. The process is similar to that used in electrostatic accelerators; however, the high-voltage potential of the Dynamitron accelerator is produced by electrical means rather than by the mechanical charge transport system utilized in electrostatic accelerators. The high voltage is produced by a large number of dc voltage sources, all of which are connected in series. Each of these dc sources receives its power by means of individual parallel capacitive coupling to a radio-frequency (rf) power system operating at 100 kHz. This high voltage can then be used for the acceleration of particles in a manner similar to that used in any other kind of dc accelerator. Although other accelerators, for example, the Cockcroft-Walton machines, use various types of cascade voltage-multiplying circuits, the Dynamitron principle makes possible considerably higher operating particle currents. The practical limit of size for a Dynamitron-type accelerator is probably 8–10 MV; the largest machines constructed are rated at 5 MV. The advantages of the Dynamitron arise from its use of high frequency and low capacitance; other systems use low frequency and high capacitance, the high-energy storage of which sometimes leads to component failure. Invented by M. R. Cleland in 1956, Dynamitron accelerators are used in industrial, medical, and basic research applications. *See* COCKCROFT-WALTON ACCELERATOR; VAN DE GRAAF GENERATOR.

A smaller, less powerful version of the Dynamitron tandem accelerator called a Tandetron. These tandem accelerators are made in different sizes varying from 1 to 3 MV and have a power supply capable of only 1 mA, in contrast to the 100 mA or more of a Dynamitron. The Tandetron is used as an analytical system by the semiconductor industry and in ^{14}C analysis as an ultrasensitive mass spectrometer. *See* PARTICLE ACCELERATOR. [H.E.W.]

Dynamo An electric machine for the conversion of electrical energy into mechanical energy or, conversely, mechanical energy into electrical energy. It is called a generator if it converts mechanical into electrical energy, and it is called a motor if it converts electrical into mechanical energy. *See* ELECTRIC ROTATING MACHINERY; GENERATOR; MOTOR. [A.R.E.]

Dynamometer A special type of electric rotating machine used to measure the output torque or driving torque of rotating machinery. Most dynamometers consist of a direct-current (dc) machine with the stator cradle-mounted in antifriction bearings. The rotor is connected to the rotor of the machine under test. The field current is introduced through flexible leads. The stator is constrained from rotating by a radial arm of known length to which is attached a scale for measuring the force required to prevent rotation. The torque of the connected machine is found from the product of the lever arm length and the scale reading, after correcting the scale reading by the amount of the zero torque reading. *See* ELECTRIC ROTATING MACHINERY; TORQUE. [A.R.E.]

Dyschondroplasia A deforming nonhereditary disease of early youth also referred to as Ollier's disease or enchondromatosis. The fingers and the long bones of the extremities are most commonly affected, while skull, ribs, and pelvis are rarely involved. The disease is progressive and involves the growing portions of the affected bone.

In dyschondroplasia the growth rate of the epiphyseal cartilage is retarded and formation of new bone is defective, leaving irregular residual islands of abnormal cartilage scattered along the length of the shaft. These two factors result in an abnormally short bone and irregular nodular swellings along the shaft of the bone. The former results in arms or legs of unequal length; the latter may cause such irregular enlargement of fingers that function is severely impaired. The deformities produced increase steadily until skeletal maturity is reached, at which time the process slows down or stops entirely, always leaving some degree of residual disability. No cure is known. [W.R.A.]

Dysprosium A metallic rare-earth element, Dy, atomic number 66 and atomic weight 162.50. The naturally occurring element is composed of seven stable isotopes. Dysprosium forms a white oxide, Dy_2O_3, which dissolves in acid to give a yellowish-green solution. For properties of the metal *see* RARE-EARTH ELEMENTS.

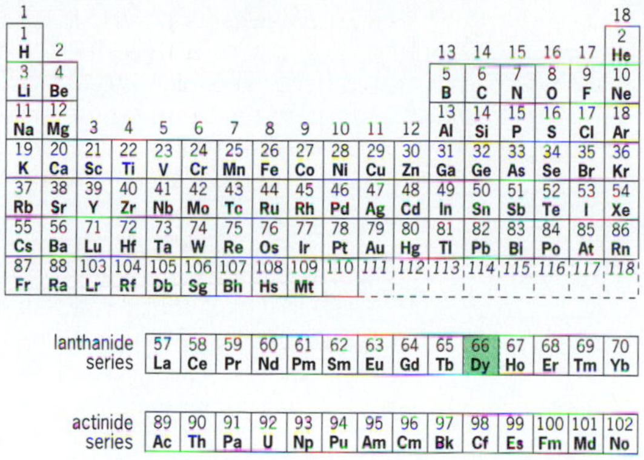

The metal is attacked readily by air at high temperatures, but at room temperatures, in massive blocks, it is fairly stable in the atmosphere and remains shiny for long periods of time. Dysprosium is paramagnetic, but as the temperature is lowered, it becomes antiferromagnetic at the Néel point (178 K or −139°F) and ferromagnetic at the Curie point (85 K or −306.4°F). At very low temperatures, the metal shows strong anisotropic magnetic properties. [F.H.Sp.]

E

e (mathematics) The number e is usually defined as the limit approached by the expression

$$\left(1 + \frac{1}{n}\right)^n$$

as n approaches infinity. If the given expression is expanded by the binomial theorem, and if one uses the theorem that the limit of the quotient of two polynomials of equal degree as the variable tends to infinity is equal to the ratio of the coefficients of highest degree, one obtains the expansion in relation (1). It can be shown by elementary methods that e is

$$e = 1 + \frac{1}{1} + \frac{1}{1 \cdot 2} + \frac{1}{1 \cdot 2 \cdot 3} + \cdots + \frac{1}{1 \cdot 2 \cdot 3 \cdots n} + \cdots \qquad (1)$$

irrational; that is, it cannot be represented as the quotient of two integers. Furthermore, e is transcendental; it does not satisfy any algebraic equation with integral coefficients.

By the method outlined above, it may be shown that the limit of

$$\left(1 + \frac{x}{n}\right)^n$$

as n tends to infinity is e^x, and moreover that relation (2) holds.

$$e^x = 1 + \frac{x}{1} + \frac{x^2}{1 \cdot 2} + \cdots + \frac{x^n}{1 \cdot 2 \cdot 3 \cdots n} + \cdots \qquad (2)$$

The function e^x is of great importance in mathematical analysis and is encountered in numerous problems in applied mathematics. *See* Binomial theorem; Calculus; Logarithm.

[A.N.L./S.Bo.]

Ear The organ which is responsible for sending both auditory information and space orientation information to the brain.

Ear structure. The basic ear structures common to all vertebrate ears are those of the internal ear, embedded in the ear capsule. The basic sensory units in the inner ear are hair cells. These structures are all morphologically similar but have either auditory or vestibular functions, depending upon the superstructure that surrounds them. The superstructure around the auditory receptor must facilitate the transmission of vibrations from the environment to the hair cells. For the vestibular apparatus, the superstructure must block external vibratory energy but must sensitize the hair cells to the pull of gavity and to acceleratory and deceleratory movements of the head.

The other parts of the ear, namely, the middle ear and external ear, are strictly auditory in function and vary considerably among vertebrate groups. They are lacking among fishes; amphibians, reptiles, birds and mammals have an external auditory meatus; and mammals have three middle-ear ossicles and usually an external pinna.

Fishes. In bony fishes the ear is embedded in the otic bones of the skull. An elaborate closed system of canals within these bones (or cartilages, in the case of cartilaginous fish), termed the bony labyrinth, is filled with perilymph. Within this fluid, its shape closely corresponding to that of the bony labyrinth, is the membranous labyrinth; it too is hollow but is filled with endolymph, a fluid of different chemical composition. Connective tissue attaches the membranous labyrinth to the walls of the bony labyrinth at several points, but most of its outer surface is in contact with the perilymphatic fluid only. The fish's membranous and bony labyrinths correspond to the structures of the same name in other vertebrates, and thus collectively are termed the inner ear, even though there are no middle or external ears in fishes. *See* Osteichthyes.

For many years fishes were assumed to be deaf. However, since early in the 20th century, there has been strong behavioral evidence to the contrary. Some fishes can respond to frequencies as low as 16 Hz and as high as 7000 Hz.

The members of one prominent group of bony fishes, the Ostariophysii (including minnows, carp, and catfish), relay acoustical information to the ear relatively directly, and there is behavioral and physiological evidence that they hear better than do other fish. In all bony fishes a large gas-filled chamber, the swim bladder, is located in the dorsal part of the body cavity, where it helps control the fish's buoyancy. In the Ostariophysii the anterior part of this gas bladder is connected by a set of tiny bones (Weberian ossicles) to the membranous labyrinth. Vibratory energy causes deformation of the swim bladder, and the Weberian ossicles carry energy from this vibrating bladder to the inner ear. *See* Swim bladder.

Tetrapoda. Many structural and functional features of the fish ear are also found in the tetrapod ear (the Tetrapoda include birds). The inner ear is embedded in the otic bones of the skull, with the membranous labyrinth attached to the bony labyrinth by connective tissue but suspended in perilymphatic fluid. There are three semicircular canals, with cristae, and a sacculus and utriculus, with maculae; there are usually also a lagena and a lagenar macula (Fig. 1). In their morphology and physiology the vestibular parts of fish and tetrapod ears are nearly the same; the tetrapod maculae, however, function as vestibular organs only, rather than playing an auditory role as they do in fish.

Amphibians. The tympanic membrane of frogs and toads is located on the lateral surface of the head. Attached to its inner aspect is a small rodlike bone, the stapes, or columella, which runs through the air space of the middle-ear cavity and plugs a small hole, the oval window, beyond which are the inner ear fluids. The frog's tympanic membrane collects sound energy and transmits it through the columella to the inner ear fluids. In the lagenar portion of the amphibian's membranous labyrinth are two areas of hair cells, the lagenar macula and a sensory structure not found in fish ears, termed the basilar papilla. In addition, amphibians have a second unique area of sensory epithelium for hearing, the amphibian papilla. Vibratory energy enters the inner ear at the oval window, passes through the basilar and amphibian papillae causing them to vibrate, and then dissipates at the round window.

Birds and reptiles. In most reptiles and birds the tympanic membrane lies not on the surface of the head but internally, at the end of the tube called the external auditory meatus. A mid-

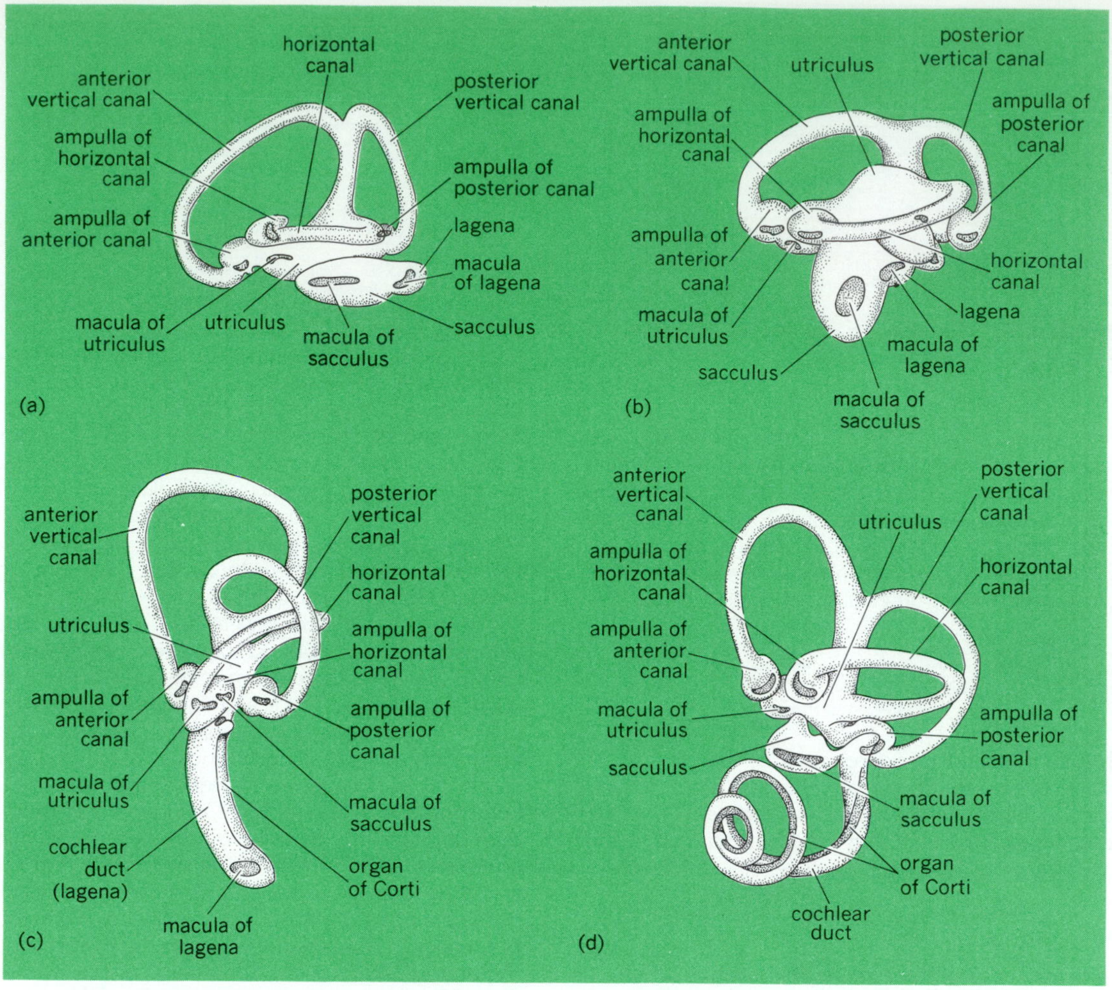

Fig. 1. Left-ear external views of membranous labyrinths of (*a*) the teleost, (*b*) the frog, (*c*) the bird, and (*d*) the mammal. (*After Retzius, from A. S. Romer, The Vertebrate Body, 3d ed., Saunders, 1962*)

dle-ear cavity (with its Eustachian tube to the mouth) lies medial to the tympanic membrane. A single ossicle, the columella, crosses this cavity from the tympanic membrane to the oval window at the inner ear.

The lagena of birds and reptiles, sometimes termed a cochlea, is elongated; there is a small maculae lagena at its tip. Along most of its length there is an elongated basilar papilla, sometimes termed an organ of Corti, which is usually somewhat shorter in reptiles than in birds but which in both classes exhibits considerable species variation. Because of its drawn-out structure, the bird and reptile lagenae contain greater numbers of hair cells than does the amphibian structure.

Mammals. In mammals the ear consists of three parts: the external ear, which receives the sound waves; the middle ear, which transmits the vibrations by a series of three small bones; and the inner, or internal, ear, a complex bony chamber placed deep in the skull (Fig. 2). The external auditory meatus plus the newly evolved pinna, a cartilaginous structure projecting from the ear, compose the external ear. The shape and size of the pinna vary greatly; in elephants it is a huge, flapping structure; in whales and moles it is completely absent; and in other species it falls between the two extremes. Its auditory function is similarly varied; it is an accurate and acute collector of sound waves in animals such as the deer, whereas in humans it plays little or no auditory role.

As in other tetrapods, the middle-ear cavity communicates with the pharynx by way of the Eustachian tube. In other tetrapods this tube is permanently open; in mammals it is usually closed, and in order to equalize the pressure between the inside of the middle-ear cavity and the environment, the tube must be opened by yawning or swallowing. This is the "popping" in the ears which is experienced when one drives rapidly up a mountain or rides in an elevator.

Instead of the single columella of other tetrapods, the mammalian middle ear has three bones, closely articulated with one another. The innermost is the stapes, fitting into the oval window of the inner ear and homologous with the columella. Attached to the tympanic membrane is the malleus, and lying between the malleus and stapes is the incus.

In the mammalian inner ear the vestibular apparatus is much like that of other tetrapods; the auditory portion, however, is considerably different. The lagena, greatly elongated and coiled into a snail shape, is now termed the cochlea; in its center, like the center pole of a spiral staircase, is the bony modiolus. The epithelium of the basilar papilla, more differentiated in mammals than in other tetrapods, is termed the organ of Corti. The number of turns in the cochlea varies, from less than one in the generalized monotremes of Australia, to two and a half in humans, to five in the guinea pig. At the base of the cochlea is the oval window, which carries sound energy into the inner ear, and the round window, where this energy is dissipated after traveling in the cochlea.

The mammalian auditory ear is structurally more complex than that of reptiles, birds, amphibians, and fishes—for instance, in its larger size and greater differentiation of cochlear cell types. It is probable that the overall hearing ability

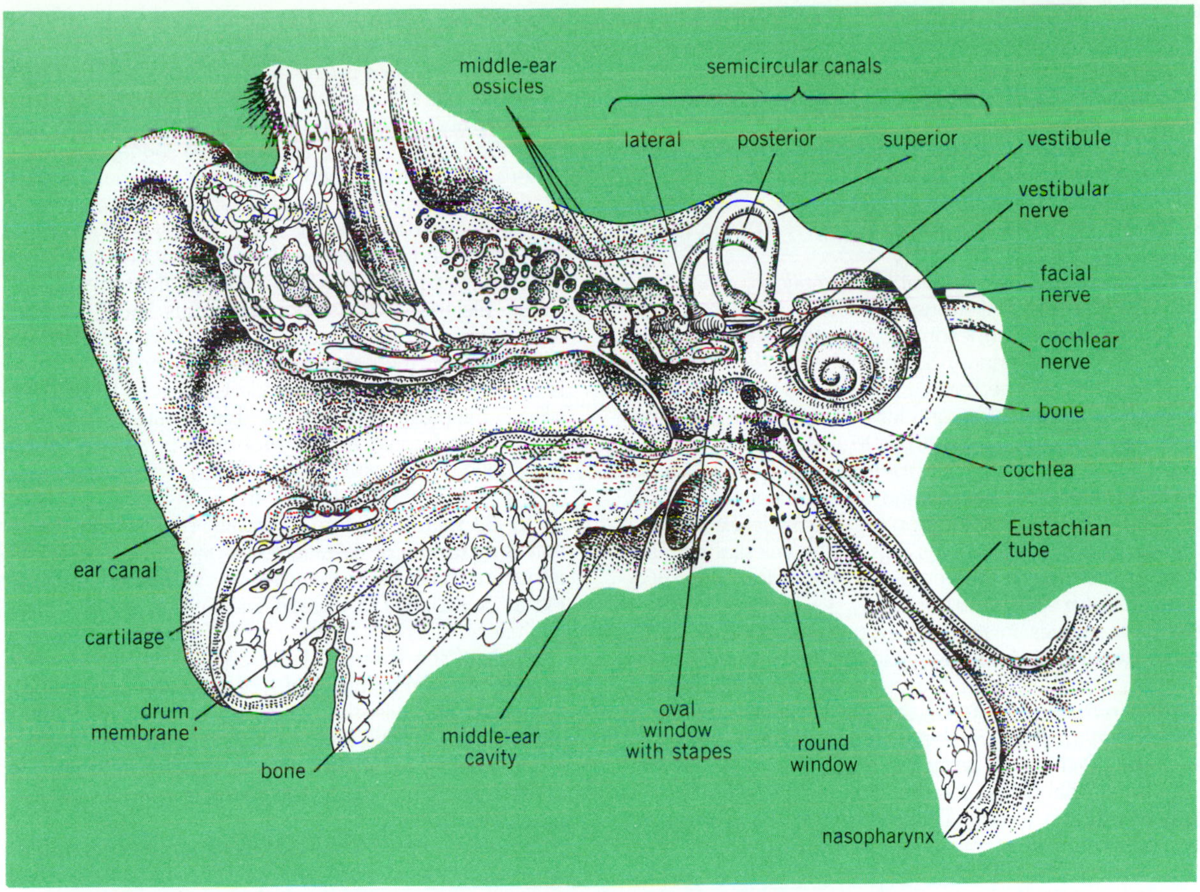

Fig. 2. Schematic drawing of the human ear. (***After drawing by M. Brödel, Three Unpublished Drawings of the Anatomy of the Human Ear, Saunders***)

of mammals is superior to that of other vertebrates and that this is due to their large cochleae with more greatly differentiated cell types. *See* HEARING (HUMAN); PHONORECEPTION. [D.B.W.]

Ear protectors Devices used to protect the human ear from noise that may be injurious to hearing. There are situations in industry, military services, and commercial aviation where persons must be exposed to sounds having levels in excess of those which may damage hearing.

There are three general classes of ear protectors: (1) earplugs that fit into the ear canal; (2) earmuffs or cushions that fit over the external ear, under which earplugs may be worn; and (3) rigid helmets that fit over the entire head, under which earplugs and earmuffs may be worn.

The amount of protection afforded by an ear protector, or combination of protectors, is measured in terms of the number of decibels it attenuates or reduces the intensity of a sound transmitted through it. In general, the higher-frequency components in a noise are attenuated more than are the low-frequency components. [K.D.K.]

Earphones A class of energy transducers capable of receiving alternating current and generating acoustic waves resembling very closely the characteristics of that current. The movement of an element (diaphragm) is accomplished by magnetic attraction, electrostatic attraction, or the piezoelectric effect (the expansion or contraction of certain crystalline substances in response to electric charges). Earphone systems include the driver element with its diaphragm and arrangements for magnetic flux, or electrostatic or direct electric charge, plus a casing, one or more acoustic cavities and ports, acoustic damping and insulation, and some arrangement for coupling the driver to the human ear.

In a magnetic earphone, a permanently magnetized diaphragm is moved in and out by an electromagnet energized by alternating current. Dynamic earphones are actually small dynamic loudspeakers. In some, a small coil fed by the sound source is bonded to the membranous diaphragm. In another configuration, the coil is relatively large and is attached to the diaphragm only at its edge. A miniature magnetic earphone is a small unit whose output port fits snugly into a plastic olive in the ear canal. These are widely used with small radios, with the more powerful hearing aids, by television commentators, in business transcription devices, and in many other communications situations. *See* HEARING AID.

Efficiency is increased if the mass of the diaphragm is reduced to a minimum. The electrostatic earphone utilizes a thin metallized plastic film on which a large constant electrostatic charge is placed by an auxiliary unit. The motion of the diaphragm is controlled by the audio signal impressed on perforated wire mesh plates on either side.

The diaphragm of a dynamic-electrostatic (or orthodynamic) earphone is a permanently polarized electret of fluorocarbon. Consequently the need for an added source of polarization voltage (a drawback inherent to electrostatic earphones) is eliminated. *See* ELECTRET TRANSDUCER.

Certain crystalline substances expand and contract when alternating voltage is applied. In some piezoelectric earphones, a crystal element is coupled mechanically to the center of a small cone. Such earphones can be lightweight and cheap and may be acceptable for speech communication. *See* PIEZOELECTRICITY. [J.D.H.]

Earth The third planet, of nine, from the Sun. The Earth is unique, so far as is known, in having life, although statistics strongly suggest that many similar planets exist, and some probably also have life. The Earth is orbited by the Moon and many artificial satellites. *See* MOON.

The Earth's nearest neighbors in space, other than the Moon, are the planets Venus, which is about 108 × 10⁶ km (67 × 10⁶ mi) from the Sun, and Mars, 227 × 10⁶ km (141 × 10⁶ mi) from the Sun. Earth is 150 × 10⁶ km (93 × 10⁶ mi) from the Sun.

Motions. There are two types of Earth motions, orbital and rotational. The Earth's orbit around the Sun is an ellipse, with the Sun at one focus. Perihelion, or the closest approach to the Sun, occurs about January 3, and aphelion, farthest from the Sun, about July 4. The Earth's velocity varies systematically, being greatest at perihelion and least at aphelion. The average velocity is 107,000 km/h (66,600 mi/h) or 29.6 km/s (18.5 mi/s). The Earth's orbital period is the year, whose length is determined by the average distance between Earth and Sun. The calendar year or tropical year is defined as the time between two successive crossings of the celestial equator by the Sun at the time of the vernal equinox. It is exactly 365 days 5 h 48 min 46 s. *See* CALENDAR.

As the Earth moves in its orbit around the Sun, its north spin axis, or geographic pole, points in the direction of the star Polaris, making it the North Star or polestar. One obvious result is that different parts of the Earth receive differing amounts of sunlight; this is the primary cause of seasons. Because of the gravitational attraction of the Moon and Sun on the Earth's equatorial bulge, the direction in which the Earth's axis points moves very slowly. This motion is called precession, and a complete cycle requires about 25,800 years. *See* PRECESSION OF EQUINOXES.

The Earth's period of rotation is the day, and the day has been used to define the second, which is the basic unit of time. The length of the solar day varies because the Earth's orbital velocity changes predictably with distance from the Sun, and less regularly owing to tidal friction, changes in the Earth's core, and seasonal atmospheric circulation. To avoid the obvious problems in determining the length of the mean solar day, the second is now defined atomically. *See* DAY; TIME.

Structure. The Earth is not quite a sphere, but has an equatorial bulge caused by its rotation. Its shape therefore is an oblate spheroid. The equatorial radius is 3963.205 mi (6378.160 km), and the polar radius is 3949.917 mi (6356.775 km).

The Earth's mass is 2.108 × 10²⁶ oz (5.975 × 10²⁷ g), giving it an average density of 3.20 oz/in.³ (5.53 g/cm³). This figure is about twice the density of the common rocks that form the Earth's surface and strongly suggests that the interior is denser than the surface. Seismic studies have confirmed that the Earth's interior has a layered structure (see illustration).

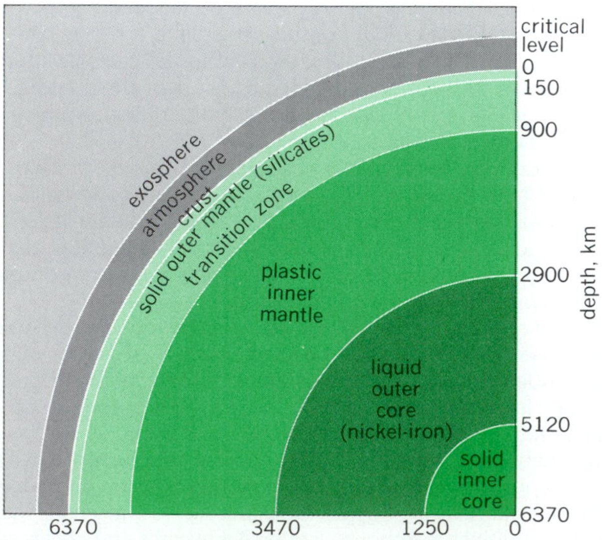

The principal layers of the Earth. 1 km = 0.621 mi.

The deepest layer is the core, which is divided into a solid inner core and a liquid outer core. The core is believed to be composed of nickel-iron and probably lighter elements such as sulfur. Electric currents moving in this liquid conducting core are believed to be the origin of Earth's magnetic field. The layer between the crust, or outermost rock layer, and the core is called the mantle. The mantle is believed to be composed of silicate minerals. The horizon that separates the mantle from the crust is known as the Mohorovičić seismic discontinuity. *See* MOHO (MOHOROVIČIĆ DISCONTINUITY).

The crust is the thinnest of the rock layers, and is composed of 19–37 mi (30–60 km) of granitic rocks under the continents and 3 mi (5 km) of basaltic rocks under the oceans. The continents are much older than the oceans, and deformed rocks as old as 3,900,000,000 years indicate an eventful history. *See* CONTINENT.

Much of the surface of the continents is covered by a thin veneer of sedimentary rocks. Where the underlying rocks of the plains and hills that make up most of the continents are exposed, the rocks and their structures are similar to those found in most mountain ranges. This leads to the theory that continents are formed by the deeply eroded remnants of earlier mountain ranges.

The topographic features underlying the oceans are similarly diverse and reveal more evidence of a dynamic Earth. The continental shelf, an area covered by shallow water, generally less than 150 m (500 ft) deep, surrounds the continents at most places. Such areas are generally underlain by continental, granitic rocks, and are submerged parts of the continents. Continental slopes are the transition between the continental shelf and the ocean floors. *See* CONTINENTAL MARGIN.

The ocean floors are the most widespread surface feature of Earth. Interrupting the ocean floor at many places are submarine mountains formed by basalt volcanoes. Some of these volcanoes are very large and form oceanic islands such as the Hawaiian Islands. The ocean floors rise gradually to the midocean ridges, a more or less continuous feature through all the oceans with some branches and offsets. Lines of parallel volcanoes and steep scarps mark the midocean ridges. *See* MARINE GEOLOGY; VOLCANO.

Atmosphere. The Earth's temperature and gravitation are such that an atmosphere is present. The major constituents are nitrogen and oxygen. The atmosphere, especially oxygen, and the presence of water, both at the surface and in the atmosphere, make life possible. A thin ozone layer in the atmosphere shields the Earth from lethal ultraviolet radiation from the Sun. *See* ATMOSPHERE; ATMOSPHERIC OZONE; METEOROLOGY.

Age. The Earth with the rest of the solar system is believed to have formed about 4,500,000,000 years ago. This age is determined by dating radioactive isotopes in meteorites. Meteorites are believed to be fragments produced by collisions among small bodies formed by the same process that created the solar system. *See* COSMOLOGY; DATING METHODS; GEOCHRONOMETRY; METEORITE; PLANET; SOLAR SYSTEM. [R.J.F.]

Earth sciences A group of overlapping areas of study that consider all aspects of the earth and its place in the universe. The principal Earth sciences are geology, oceanography, meteorology, astronomy, geophysics, and geochemistry. Each of these is subdivided into several disciplines. Geography, the study of all aspects of the earth's surface, including political and economic as well as scientific aspects, and especially physical geography, are included in the earth sciences by many. *See* ASTRONOMY; GEOCHEMISTRY; GEODESY; GEOGRAPHY; GEOMORPHOLOGY; GEOPHYSICS; METEOROLOGY; OCEANOGRAPHY. [R.J.F.]

Earth tides Cyclic motions of the Earth, sometimes over a foot or so in height, depending on latitude, caused by the same lunar and solar forces which produce tides in the sea. These forces also react on the Moon and Sun, and thus are significant in astronomy in evaluations of the dynamics of the three bodies. For example, the secular spin-down of the Earth due to lunar tidal torques is best computed from the observed acceleration of the Moon's orbital velocity. In oceanography, earth tides and ocean tides are very closely related. *See* GEODESY; TIDE.

By far the most widely used earth tide instruments are the tiltmeter and the gravimeter. Both instruments have the merits of portability, high potential precision, and low cost. Thus they are able to advance economically an important mission—the global mapping of earth tides and ocean tides. [L.B.S./J.T.K.]

Earthmover Any of a variety of construction machines designed to move or transport earth. Earthmovers include heavy-duty trucks with high-sided dump bodies, self-propelled or towed scrapers, wagons, and bulldozers. The bulldozer mounted on a wheeled or crawler tractor is suitable for moving large quantities of earth for distances of several hundred feet. Scrapers and wagons are efficient for moving earth over relatively level terrain for distances up to 1 or 2 mi (2 or 3 km). For longer distances or for grades in excess of 5%, trucks are the most practical. *See* BULK-HANDLING MACHINES; CONSTRUCTION EQUIPMENT. [E.M.Y.]

Earthquake A phenomenon during the occurrence of which the Earth's crust is set shaking for a period of time. The shaking is caused by the passage through the Earth of seismic waves—low-frequency sound waves that are emanated from a point in the Earth's interior where a sudden, rapid motion has taken place. It is more proper today to use the term earthquake to refer to the source of seismic waves, rather than the shaking phenomenon, which is an effect of the earthquake.

Earthquakes vary immensely in size. Although thousands of earthquakes occur every day, and have for billions of years, a truly great earthquake occurs somewhere in the world only once every 2 or 3 years. When a great earthquake occurs near a highly populated region, tremendous destruction can occur within a few seconds.

Cause. Earthquakes are not distributed randomly over the globe but tend to occur in narrow, continuous belts of activity. These earthquake belts link up so that they encircle large seismically quiet regions, which are known as plates. The plates are in continuous motion with respect to one another at rates on the order of centimeters per year; this plate motion is responsible for most geological activity. *See* PLATE TECTONICS.

As the plates move past each other, the motion at their boundaries does not occur by continuous slippage but in a series of rapid jerks. Each jerk is an earthquake. This happens because, under the pressure and temperature conditions of the shallow part of the Earth's lithosphere, the frictional sliding of rock exhibits a property known as stick-slip, in which frictional sliding occurs in a series of jerky movements, interspersed with periods of no motion—or sticking. In the geologic time frame, then, the lithospheric plates chatter at their boundaries, and at any one place the time between chatters may be hundreds of years.

The periods between major earthquakes is thus one during which strain slowly builds up near the plate boundary in response to the continuous movement of the plates. The strain is ultimately released by an earthquake when the frictional strength of the plate boundary is exceeded. *See* FAULT AND FAULT STRUCTURES.

Classification. Most great earthquakes occur on the boundaries between lithospheric plates and arise directly from the motions between the plates. Although these may be called plate boundary earthquakes, there are many earthquakes, sometimes of substantial size, that cannot be related so simply to the movements of the plates.

Near many plate boundaries, earthquakes are not restricted to the plate boundary itself, but occur over a broad zone—often several hundred miles wide—adjacent to the plate boundary. These earthquakes, which may be called plate boundary–related earthquakes, do not reflect the plate motions directly, but are secondarily caused by the stresses set up at the plate boundary.

While most earthquakes occur on or near plate boundaries, some also occur, although infrequently, within plates. These earthquakes, which are not related to plate boundaries, are called intraplate earthquakes, and can sometimes be quite destructive. Intraplate earthquakes are probably caused by the same convective forces which drive the plates; their immediate cause is not understood.

In addition to the tectonic types of earthquakes, some earthquakes are directly associated with volcanic activity. These volcanic earthquakes result from the motion of underground magma that leads to volcanic eruptions.

Sequences. Earthquakes often occur in well-defined sequences in time. Tectonic earthquakes are often preceded, by a few days to weeks, by several smaller shocks (foreshocks), and are nearly always followed by large numbers of aftershocks. Foreshocks and aftershocks are usually much smaller than the main shock. Volcanic earthquakes often occur in flurries of activity, with no discernible main shock. This type of sequence is called a swarm.

Size. Earthquakes range enormously in size, from tremors in which slippage of a few tenths of an inch occurs on a few feet of fault, to the greatest events, which may involve a rupture many hundreds of miles long, with tens of feet of slip. Accelerations as high as 1 *g* (acceleration due to gravity) can occur during an earthquake motion. The velocity at which the two sides of the fault move during an earthquake is only 1–10 mi/h (0.5–5 m/s), but the rupture front spreads along the fault at a velocity of nearly 5000 mi/h (2200 m/s).

The size of an earthquake is in terms of a scale of magnitude based on the amount of seismic waves generated. Magnitude 2.0 is about the smallest tremor that can be felt. Most destructive earthquakes are greater than magnitude 6; the largest shock known measured 8.9. The scale is logarithmic, so that a magnitude 7 shock is about 30 times more energetic than one of magnitude 6, and 30×30, or 900 times, more energetic than one of magnitude 5.

The intensity of an earthquake is a measure of the severity of shaking and its attendant damage at a point on the surface of the Earth. The same earthquake may therefore have different intensities at different places. The intensity usually decreases away from the epicenter (the point on the surface directly above the onset of the earthquake), but its value depends on the many factors in addition to earthquake magnitude. Intensity is considered more a measure of the earthquake's effect on humans than an innate property of the earthquake. *See* SEISMOLOGY. [C.H.S.]

East Indies The Indonesian islands and, for convenience of presentation, New Guinea and the Bismarck Archipelago, lying between Asia and Australia. They make up the territory of Indonesia, Brunei, part of Malaysia, and regions formerly under Australian trusteeship. Together with the intervening seas, the East Indies form a region of great relief and of geological diversity. The islands extend from western Sumatra to eastern New Guinea, a distance of about 4000 mi or 6400 km.

Geographically, three basic divisions are recognized: (1) Sundaland, or the islands west of Macassar and the Lombok Straits; (2) Papualand, or the islands east of the Aru Islands and New Guinea; and (3) Wallacea, or the islands between Sundaland and Papualand and south of the Philippines. Each division is an important biological and geomorphological region.

During the Quaternary, a lowering of the general level of the world's oceans is thought to have resulted in the emergence of dry land between the western edge of Wallacea and Asia, and on the east in exposed portions of the Sahul Shelf beneath the Arafura Sea. The islands of Wallacea, however, remained separated by deep water, which was an effective barrier to the migration of plants and animals. Thus, the flora and fauna of Sundaland are closely related to the flora and fauna of Asia, while the flora and fauna of Papualand show a relationship to those of Australia and a separation from Asia since early Cretaceous. Because of this isolation and the great variation of topography, soils, rainfall, humidity, and altitude within short distances, there have existed conditions favorable for the development of numerous species of plants. *See* ASIA; AUSTRALIA; ISLAND BIOGEOGRAPHY; NEW ZEALAND.

Since almost all the islands lie in a belt within 10° of the equator, an equatorial climate prevails, with high temperatures throughout the year, except at higher elevations. The diurnal variations are greater than the range in mean temperatures of the hottest and coldest months. Monsoons, representing an interplay of air masses between Asia and the Southern Hemisphere, control the seasonal change in winds and the variation in precipitation. *See* BORNEO; CELEBES; JAVA; NEW BRITAIN; NEW GUINEA; NEW IRELAND; SUMATRA; TIMOR; TROPICAL METEOROLOGY. [M.G.A.-C.]

Ebenales An order of flowering plants, division Magnoliophyta (Angiospermae) in the subclass Dilleniidae of the class Magnoliopsida (dicotyledons). The order consists of 5 families and about 1700 species, the Sapotaceae (about 800 species) and Ebenaceae (about 450 species) being the largest and most familiar families. The Ebenales are woody, chiefly tropical, sympetalous plants (those with flowers have the petals joined by their margins, at least toward the base, forming a basal tube cup, or saucer) with usually twice as many stamens as corolla lobes.

Chicle (from *Manilkara sapota*, in the Sapotaceae) and ebony (*Diospyros ebeneum*) are obtained from members of the Ebenales. *See* CHICLE; DILLENIIDAE; EBONY. [A.Cr.]

Ebony A genus, *Diospyros*, of the ebony family, containing more than 250 species. Some species are important for their succulent fruits, such as date plum, kaki plum, and persimmon, and several for their timber, particularly the heartwood, which is the true ebony of commerce. *See* EBENALES.

Although it is popularly supposed to be a black wood, most species have a heartwood that is only streaked and mottled with black. The heartwood is very brittle, and is difficult to work, but it has long been in demand. The sapwood is white, becoming bluish or reddish when cut.

Black ebony is used for knife handles, piano keys, finger boards of violins, hairbrush backs, inlays, and marquetry. Some of the woods called ebony, however, belong to different families, especially the pulse family, Leguminosae.

Persimmon (*D. virginiana*), of the southeastern United States, is one of numerous tropical or subtropical species. The species in tropical America are too small or rare to be of economic value, although several of them have black heartwood used locally for making walking sticks, inlays, and miscellaneous articles of turnery and carving. [A.H.G./K.P.D.]

Ebriida An order of flagellate Protozoa, subphylum Sarcomastigophora, class Phytamastigophorea. Two genera, *Ebria* and *Hermesinum*, remain of the once numerous order, as determined by fossil remains. They possess an internal solid siliceous skeleton forming a shallow flattened or slightly arched structure that is enclosed by clear cytoplasm. Skeleton form and structure are distinctive. These phytoflagellates do not have chromatophores.

The organisms are common in the inshore waters of the Atlantic Ocean and the Gulf of Mexico. However, numbers sufficient to influence the general ecology have not been observed. Their only occurrence seems to be marine, although somewhat lowered salinities are certainly tolerated. *See* PHYTAMASTIGOPHOREA; PROTOZOA. [J.B.L.]

Ecdysone The molting hormone of insects. It is a derivative of cholesterol. The most striking physiological activity of ecdysone is the induction of puffs (zones of gene activity) in giant chromosomes of the salivary glands and other organs of the midge *Chironomus*. The induction of puffs has been visualized as primary action of the hormone, indicating that ecdysone controls the activity of specific genes. It has been shown that ecdysone stimulates the synthesis of messenger RNA, among which is the messenger for dopa decarboxylase. This enzyme is involved in the biosynthesis of the sclerotizing agent *N*-acetyl-dopamine. *See* INSECTA. [P.K.]

Echelette grating A diffraction grating that has coarse groove spacing and is designed for the infrared region, the grooves being so shaped and of such size that most of the radiation is concentrated by reflection into a small angular coverage. Radiation of any given wavelength is thus concentrated largely into one order by shaping the point of the ruling diamond to give grooves with comparatively flat sides and by choosing a groove separation which minimizes diffraction effects. *See* DIFFRACTION GRATING. [G.R.H.]

Echelle grating A type of diffraction grating designed for use in high orders and at angles of illumination greater than 45° to obtain high dispersion and resolving power by the use of high orders of interference. Echelles have properties lying midway between those of plane gratings of the ordinary type and interferometers of the reflection echelon type, orders of interference ranging from 100 to 1000 being used. Overlapping orders are separated by using crossed dispersion. *See* DIFFRACTION GRATING. [G.R.H.]

Echinacea A superorder of Euechinoidea, having a rigid test, the periproct within the apical system, keeled teeth, a complete perignathic girdle, and branchial slits. J. Durham and R. Melville (1957) include five orders in this group. These were formerly distributed among the Stirodonta and Camarodonta in the classification of R. Jackson (1912). *See* ECHINODERMATA; ECHINOIDA; EUECHINOIDEA; HEMICIDAROIDA; PHYMOSOMATOIDA; SALENIOIDA; TEMNOPLEUROIDA. [H.B.F.]

Echinococcosis The infestation by the hydatid (cyst) of *Echinococcus granulosus*, also known as hydatidosis. The adult worm occurs in the gut of dogs and other canines, and the hydatid occurs primarily in the liver and lungs of herbivores, and occasionally in humans.

Human infestations are found throughout the world in nontropical areas but are particularly common in certain areas of South America, Australia, New Zealand, and Africa, where dogs, sheep, and cattle are abundant and in intimate association. The normal cycle of the tapeworm involves the dog and the herbivore, but humans may become infested directly from the dog feces or from fecally contaminated food or water.

Human infestations are rare in North America, except in the Far North. Here the parasite normally cycles from moose and other wild herbivores to wolves, but dogs become infested when eating offal from the herbivores. *See* CYCLOPHYLLIDEA. [R.S.F.]

Echinocystitoida An extinct order of Perischoechinoidea which arose in the Ordovician and seems to have inhabited calm, shallow lagoons. There was no perignathic girdle, the number of columns of plates was variable, the ambulacral plates overlapped adorally, and the interambulacral plates overlapped one another toward the apex and also overlapped the adjacent ambulacrals (see illustration). The shape of the test

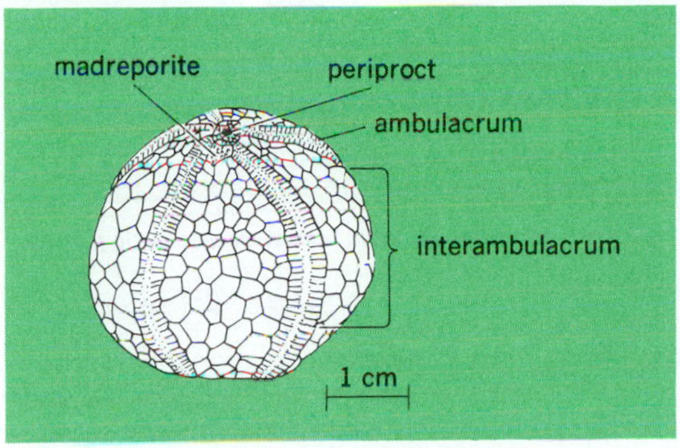

***Aulechinus grayae*, an echinocystitoid example from the Upper Ordovician of Scotland.**

varied from spherical (*Echinocystis*) to flattened pentagonal (*Hyattechinus*). The largest forms (*Fournierechinus*) reached a diameter of 30 cm (12 in.). All were extinct by the close of the Permian, but before extinction they had given rise to the Archaeocidaridae, from which all surviving echinoids probably stem. *See* ECHINOIDEA; PERISCHOECHINOIDEA. [H.B.F.]

Echinodermata A unique group of exclusively marine animals with peculiar body architecture. They are headless with a fivefold radial symmetry. The body wall contains the endoskeleton, made of numerous independent calcareous plates which frequently support spines. The plates may be tightly interlocked or loosely associated. The spines may protrude through the outer epithelium, and are often used for defense. The skeletal plates of the body wall, together with their associated muscles and connective tissue, form a tough and sometimes rigid test which encloses the large coelom. There is a unique multipurpose water-vascular system involved in locomotion, respiration, food gathering, and sensory perception. This shows outside the body as rows of fluid-filled tube feet lying in conspicuous double lines or ambulacra. Within the body wall lie the ducts and fluid reservoirs necessary to protract and retract the tube feet by hydrostatic pressure. The nervous system consists of a ring around the mouth with connecting nerve cords associated with each ambulacrum. There may also be diffuse nerve plexuses lying below the outer epithelium. The coelom houses the alimentary canal and associated organs and in most groups the reproductive organs. The larvae are usually planktonic with a bilateral symmetry, but the adults are almost always sedentary and benthic. They inhabit all seas and oceans of the world, ranging from the shores to the ocean depths.

The phylum comprises about 6000 existing species and many fossils, providing a good fossil record. Echinoderms first appeared in the Early Cambrian and have been evolving over 600,000,000 years. During this vast time several divergent patterns have arisen. The surviving groups show few resemblances to the original stock. The existing representatives fall into three subphyla: Crinozoa, Asterozoa, and Echinozoa. The fourth subphylum, Homalozoa, has no living representatives. Following is the outline of classification for the phylum.

Phylum Echinodermata
- Subphylum Homalozoa
- Subphylum Crinozoa
  - Class: Cystoidea
    - Diploporita
    - Rhombifera
    - Blastoidea
    - Eocrinoidea
    - Paracrinoidea
    - Crinoidea
      - Subclass: Inadunata
        - Camerata
      - Subclass: Flexibilia
        - Articulata
- Subphylum Echinozoa
  - Class: Edrioasteroidea
    - Helicoplacoidea
    - Ophiocistioidea
    - Cyclocystoidea
    - Holothuroidea
      - Subclass Dendrochirotacea
        - Order: Dactylochirotida
          - Dendrochirotida
      - Subclass Aspidochirotacea
        - Order: Aspidochirotida
          - Elasipodida
      - Subclass Apodacea
        - Order: Molpadida
          - Apodida
  - Class Echinoidea
    - Subclass Perischoechinoidea
      - Order: Bothriocidaroida
        - Echinocystitoida
        - Palaechinoida
        - Cidaroida
    - Subclass Euechinoidea
      - Superorder Diadematacea
        - Order: Diadematoida
          - Pedinoida
          - Echinothurioida
          - Pygasteroida
      - Superorder Echinacea
        - Order: Salenioida
          - Hemicidaroida
          - Phymosomatoida
          - Arbacioida
          - Temnopleuroida
          - Echinoida
      - Superorder Gnathostomata
        - Order: Holectypoida
          - Clypeasteroida
          - Cassiduloida
      - Superorder Atelostomata
        - Order: Holasteroida
          - Spatangoida
- Subphylum Asterozoa
  - Class Stelleroidea
    - Subclass Somasteroidea
    - Subclass Asteroidea
      - Order: Platyasterida
        - Hemizonida
        - Phanerozonida
        - Paxillosina
        - Valvatina
      - Suborder Valvatina

Order: Spinulosida
Euclasterida
Forcipulatida
Subclass Ophiuroidea
Order: Stenurida
Oegophiurida
Phrynophiurida
Ophiurida

See CRINOZOA; ECHINOZOA; HOMALOZOA. [A.C.C.]

The origin of the echinoderm invertebrates surely belongs some time before the beginning of the Paleozoic Era, for Lower Cambrian deposits contain such divergent branches of the phylum as Homalozoa, Helicoplacoidea, Edrioasteroidea, and Eocrinoidea. These are primitive sorts of echinoderms. Cystoids, crinoids, and blastoids, as well as all recognized main groups of asterozoans and echinozoans, appear in Ordovician strata.

It is evident from their already well-marked development of characteristic structures that a long period of antecedent evolution must have operated to produce such widely divergent forms. For example, each main division of the Paleozoic crinoids (Inadunata, Flexibilia, Camerata), although seeming to appear very suddenly, is fully differentiated in the Middle Ordovician. One can only guess as to reasons for the absence of these echinoderms in older deposits. During the Paleozoic numerous well-marked evolutionary trends are discernible in nearly all echinoderm groups, including free-moving forms (especially echinoids) as well as crinozoans. All groups of modern echinoderms have their origin in early Paleozoic stocks, and the lines of their phylogeny are mostly indicated by the fossil record. Echinoids predominate in Mesozoic and Cenozoic echinoderms. [R.C.Mo.]

Echinoida An order of Echinacea with a camarodont lantern, smooth test, imperforate noncrenulate tubercles, ambulacral plates of echinoid type, and shallow branchial slits. There are numerous tropical and temperate species, some of them remarkably adapted to living on coral reefs. The four included families (Parasaleniidae, Echinidae, Strongylocentrotidae, and Echinometridae) are principally distinguished by characters of the pedicellariae. *See* ECHINACEA; ECHINODERMATA; ECHINOIDEA. [H.B.F.]

Echinoidea A class of Echinozoa known as the sea urchins. These animals have a compact body enclosed in a hard shell, or test, formed from regularly arranged plates which bear movable spines (see illustration). There are no arms, but radii are represented by five double rows of tube feet arranged as meridians between the upper and lower poles of the body.

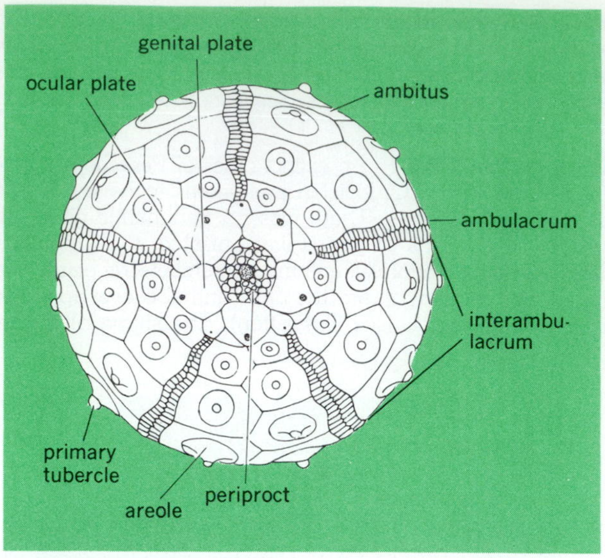

Structure of naked echinoid test of *Goniocidaris parasol*.

There are about 850 living species, and some 500 fossil species have been recorded, included in 225 genera. They are classified in 18 orders, grouped in 2 subclasses. Sea urchins range in size from a few tenths of an inch (few millimeters) across the test to 8 in. (20 cm). They differ from other echinoderms in possessing echinochromes in the pigmentation complex. Although many species are dull or dark in color, some are brilliant shades of purple, red, green, or orange. Others have particolored striped spines, and deep-sea forms may be white.

The test is globular, or nearly so, in those forms in which the anus lies within the apical system. These are often called Regularia. In exocyclic forms the test tends to assume a secondary bilateral symmetry. The radial symmetry is always evident, however, and is five-part (pentamerous). *See* IRREGULARIA; REGULARIA.

The mouth lies on the lower (oral) side of the test, surrounded by soft skin, the peristome. In endocyclic forms the mouth is central, but in exocyclic forms it may suffer displacement into the radius and lie opposite the interamb which contains the anus. The radius is then termed anterior. In endocyclic forms and in some of the exocyclic forms the mouth is furnished with a ring of five powerful jaws, each with one large tooth. The jaws and teeth collectively comprise the so-called lantern, first described by Aristotle. The water-vascular system is highly developed and greatly influences the form of the ambulacral plates.

Small grasping organs, pedicellariae, are well developed in echinoids, in which they take the form of a beak carried on a stalk. The beak is made up of three (sometimes only two) movable jaws, operated by muscles and sometimes provided with venom glands. They respond to tactile stimuli and seize any small organisms or particles which may touch the skin.

The nervous system follows the same pattern as the water-vascular system. The tube feet evidently serve as tactile and taste organs. A few sea urchins have photosensitive eyespots on the upper surface of the test, scattered in the ectoderm. The ocular pores are not sensitive to light. Minute spherical stalked bodies attached to the skin in some echinoids are believed to be organs of balance.

The alimentary canal is tubular. It lies in the coelom, attached to the wall of the test by mesenteries. The stomach runs from the esophagus in a counterclockwise coil (as viewed from above), and the intestine retraces the route in reverse. The rectum passes upward to the anus.

Like starfishes, sea urchins seem to be sexually mature after 1 year, but continue to grow for several years. The life span is unknown, but may average 5–6 years in medium-sized species. The sexes are normally separate, although fertile hermaphrodites are known. Spines and some other organs are regenerated after injury. Autotomy and autoevisceration are unknown.

Sea urchins occur in all seas from low-tide level downward. *Pourtalesia* reaches a depth of more than 4 mi (7250 m) in the Banda Trench. The rounded (or regular) urchins feed mainly on algae, often hiding by day under stones held over the test by tube feet, and emerging at night or at high tide. The heart urchins and other exocyclic forms live buried in mud or sand, feeding on organic matter in the mud or on selected detritus.

Sea urchins move slowly, using muscles at the bases of the spines to swing the spines like stilts. The suctorial tube feet are used to ascend steep surfaces and as anchors. *See* ECHINODERMATA; EUECHINOIDEA; PERISCHOECHINOIDEA. [H.B.F.]

Echinothurioida An order of Diadematacea with solid or hollow primary radioles, diademoid ambulacral plates, noncrenulate tubercles, and the anus within the apical system. The extant members of the two included families, Pedinidae and Echinothuriidae, are all deep-water forms. The Pedinidae have solid radioles and simple diademoid plates. The Echinothuriidae have a large flexible test which collapses into a disk at atmospheric pressure, and the middle element of the diademoid plates is much larger than the other two elements. Some species carry venomous spines. *See* DIADEMATACEA; ECHINODERMATA; ECHINOIDEA. [H.B.F.]

Echinozoa A subphylum of free-living echinoderms in which the body is essentially globoid with meridional symmetry. They lack arms, brachioles, or other appendages, and do not at any time exhibit pinnate structure. The Echinozoa range in time from the Lower Cambrian to the present day. There are six classes: Helicoplacoidea (Lower Cambrian); Edrioasteroidea (Lower to Middle Paleozoic); echinozoans, in which the mouth and anus were both directed upward and ambulacra (three to five in number) served as food-collecting areas; Echinoidea (sea urchins); Holothuroidea (sea cucumbers); Ophiocistoidea (Paleozoic); and Cyclosteroidea (Middle Ordovician to Middle Devonian).

The ancestry of the Echinozoa is unknown, but the oldest known members suggest forms similar to extant plated holothurians, the Dendrochirotacea, and may have arisen from cystoids or similar crinozoans. *See* CYCLOCYSTOIDEA; ECHINODERMATA; ECHINOIDEA; EDRIOASTEROIDEA; HELICOPLACOIDEA; HOLOTHUROIDEA. [H.B.F.]

Echiurida A small group of wormlike animals, once linked with the Sipuncula and Priapulida under the term Gephyrea, but now regarded as a separate phylum, with affinities to the annelid worms. They range from the topical to the polar seas, living buried in the sea floor from the intertidal area to depths of 30,000 ft (9000 m).

Their classification has always presented difficulties. In the classification of W. Fisher three orders—Echiuroinea, Xenopneusta (4 species), and Heteromyota (1 species)—are recognized, with one problematic species attached. By far the largest number of species are included in the Echiuroinea, which is divided into the families Bonellidae and Echiuridae. The classification of the Bonellidae is the most uncertain and now includes 21 genera with about 35 species. The Echiuridae are more stable with only 8 genera but more than 70 species.

The body is saclike or sausage-shaped and often highly colored. Anteriorly there is a prostomium which is readily detached. This may be very long and cleft at the tip (Bonellidae), or short and flaplike (Echiuridae; see illustration). The mouth, located at the base of the prostomium, leads into the gut, which is divided into several distinct regions. At the posterior end of the body are two very characteristic structures, the anal vesicles, for which an excretory function has been proposed.

The sexes are separate and similar in the Echiuridae. In the Bonellidae they are separate and very dissimilar, the male being minute and parasitic within or on the female. [M.E.Ri.]

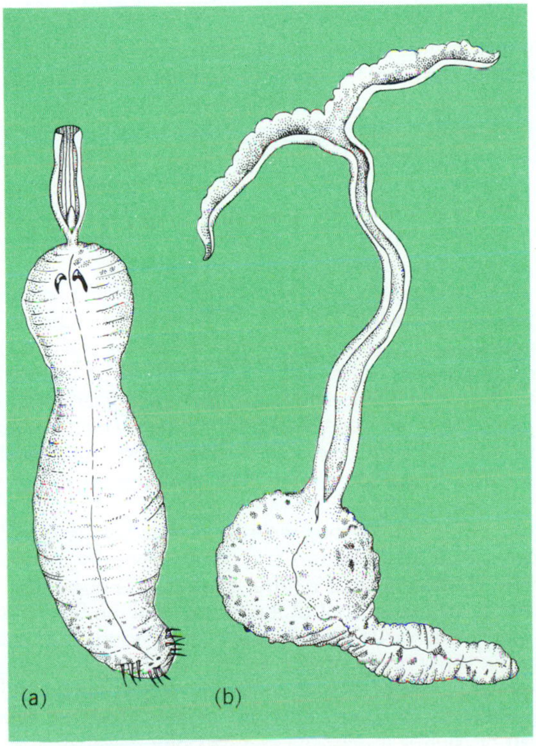

Echiurida. (*a*) *Echiuris* and (*b*) *Bonellia*, half size.

Echo A sound wave which has been reflected or otherwise returned with sufficient magnitude and time delay to be perceived in some manner as a sound wave distinct from that directly transmitted. Multiple echo describes a succession of separately indistinguishable echos arising from a single source. When the reflected waves occur in rapid succession, the phenomenon is often termed a flutter echo. Echoes and flutter echoes are generally detrimental to the quality of the acoustics of rooms. They may be minimized through the proper selection of room dimensions, room shape, and distribution of sound-absorbing materials.

Echoes have been put to a variety of uses in measurement problems. For example, the distance between two points can be measured by timing the duration required for a direct sound originating at one location to strike an object at the other point and to return an echo to the location of the initial source. Ultrasonic echo techniques have achieved considerable success in nondestructive testing of materials. *See* ECHO SOUNDER; ECHOCARDIOGRAPHY; NONDESTRUCTIVE TESTING; REFLECTION OF SOUND; SONAR; SOUND; ULTRASONICS. [W.J.G.]

Echo box A device used to check the output power and spectrum of a radar transmitter. It consists of a low-loss, tunable, resonant cavity connected to the antenna feed line through a fixed coupling circuit so that the fraction of output power supplied to the cavity is always constant. The signal level within the cavity depends on the strength of the portion of the transmitter output spectrum lying within the cavity's narrow passband, which can be tuned to traverse the entire frequency range of interest. A microammeter connected through a crystal rectifier to a loop within the cavity permits reading the signal level. The spectrum can be measured as the cavity is tuned to different frequencies. *See* RADAR. [R.I.B.]

Echo sounder A marine instrument used primarily for determining the depth of water by means of an acoustic echo. A pulse of sound sent from the ship is reflected from the sea bottom back to the ship, the interval of time between transmission and reception being proportional to the depth of the water. An echo sounder is really a type of active sonar. It consists of a transducer located near the keel of the ship which serves (in most models) as both the transmitter and receiver of the acoustic signal; the necessary oscillator, receiver, and amplifier which generate and receive the electrical impulses to and from the transducer; and a recorder or other indicator which is calibrated in terms of the depth of water.

Echo sounders, sometimes called fathometers, are used by vessels for navigational purposes, not only to avoid shoal water, but as an aid in fixing position when a good bathymetric chart of the area is available. Some sensitive instruments are used by commercial fishers or marine biologists to detect schools of fish or scattering layers of minute marine life. Oceanographic survey ships use echo sounders for charting the ocean bottom. *See* SCATTERING LAYER; SONAR; UNDERWATER SOUND. [R.W.Mo.]

Echocardiography A noninvasive diagnostic procedure in clinical and investigative cardiology that uses a transducer held against the chest to send high-frequency sound waves which pass into the heart; as they strike structures within the heart, they are reflected back to the transducer and recorded on an oscilloscope. *See* ULTRASONICS.

Mitral stenosis, idiopathic hypertrophic subaortic stenosis, atrial myxoma, and pericardial effusion are disorders that can be accurately diagnosed by echocardiography. If used in conjunction with the clinical, electrocardiographic, and roentgenologic findings, echocardiography is useful in the evaluation of numerous cardiac disorders. When combined with phonocardiography and cardiac catheterization, echocardiography has extended knowledge of the genesis of heart sounds, murmurs, and left ventricular function. In addition, the technique is useful in measuring the internal dimensions of the ventricles, atria, and outflow vessels. These measurements can be used with reasonable accuracy to estimate ventricular volumes and ejection fractions in normal subjects. However, in patients with grossly dilated ventricles or in those exhibiting localized contraction abnormalities, the correlation with ventriculography is unreliable. *See* PHONOCARDIOGRAPHY.

Other techniques, such as compound-B ultrasonography, which produces a two-dimensional cross-sectional image of intracardiac structures, and multiscan echocardiography, may provide a more accurate anatomic display of cardiac chambers and outflow vessels. Unique features of the multiscan technique are the capacity to display substantial segments of cardiac structures and to show them in continuous, real-time motion. *See* HEART DISORDERS. [B.L.S.]

Echolocation A form of acoustic navigation used by porpoises, bats, and certain other animals that live in dark or murky environments. Such animals locate objects with echoes: they produce sounds, usually with the larynx, tongue, or teeth, and listen for echoes reflected from objects in their path.

Microchiropteran bats, which include all the New World and European bats and many of the smaller African and Asian species, all echolocate. Many show prominent anatomical specializations for echolocation, such as enormous ears that aid in the detection of insects, or bizarre arrays of cartilaginous structures on the face, nose, or external ear canal which are thought to enhance the ear's directional sensitivity. Others have large mouths shaped like parabolic horns which focus the outgoing cries into a narrow beam of sound—an acoustic flashlight of sorts. For bats, echolocation is primarily important for the detection of objects at close range.

Several species of dolphins (bottle-nosed, common, and Amazon River) and two of the toothed whales (beluga and killer) are known to echolocate, and it is presumed that related cetaceans also use sonar. The heads of echolocating cetaceans show considerable modification for the emission and reception of sonar signals. On the right side, the dorsal structures are greatly enlarged, leading to a marked asymmetry of the skull. The right nasal passage is modified for phonation, while the left nostril is used only for respiration. Sound emission and focusing probably involve the large fatty structure in the forehead known as the melon.

Blind humans sometimes show a remarkable ability to avoid obstacles by using echoes of sounds such as their own footsteps or snapping fingers. Interestingly, such people experience this detection as a tactile sensation around the face; hence, it has been termed facial vision. Under experimental conditions, however, it can be shown that this is acoustic orientation. Attempts to enhance human echolocation by specially devised sound-generating mechanisms have been largely unsuccessful.

Cave-dwelling birds, including the South American oilbird *Steatornis* and a half dozen Asian swifts, use echolocation to navigate in and out of the cave to their nests. In none of these cave-dwelling fliers does echolocation appear to be used in general orientation as it is in bats and cetacea. Echolocation has been demonstrated in only three terrestrial mammals—rats, shrews, and tenrecs—and has also been reported in fishes and reptiles, suggesting that this form of navigation is far more widespread than was previously suspected. *See* CETACEA; CHIROPTERA; PHONORECEPTION; SONAR; ULTRASONICS. [J.Cha.]

Echovirus One of the divisions of the enterovirus subgroup, within the picornavirus group of viruses. The name is derived from the term enteric cytopathogenic human orphan virus. More than 34 antigenic types exist. Only certain types have been associated with human illnesses, particularly with aseptic meningitis and febrile disease. Their epidemiology is similar to that of other enteroviruses. Echoviruses resemble polioviruses and coxsackieviruses in size (about 28 nanometers) and in many other properties. Diagnosis is made by isolation and typing of the viruses in tissue culture. Antibodies form during convalescence. *See* ANIMAL VIRUS; ENTEROVIRUS; PICORNAVIRIDAE. [J.L.Me.]

Eclipse The darkening of one celestial body by the shadow of another. A solar eclipse occurs when the Moon passes directly between Earth and the Sun, hiding at least a part of the Sun from view. The Moon is eclipsed when it goes into Earth's shadow. This article deals only with these two kinds of eclipses, but there are other similar phenomena. *See* ECLIPSING VARIABLE STARS; JUPITER; OCCULTATION: TRANSIT (ASTRONOMY).

The Moon's orbit around Earth is inclined 5° to the plane of Earth's orbit around the Sun. The latter is called the ecliptic, because eclipses can occur only when the Moon is on or near it.

The two points where the Moon's orbit crosses the ecliptic are called nodes. The line connecting them and passing through Earth's center is the line of nodes. When this line points to the Sun, the Moon's shadow strikes Earth, producing a solar eclipse. About 2 weeks before or after that event, the Moon can go through Earth's shadow, producing a lunar eclipse. This interval during which the Earth-Sun line is approximately along the line of nodes is called an eclipse season. Another one occurs about 6 months later.

The Moon's orbit is not fixed in position, being disturbed by the gravitational pull of the Sun. This causes the line of nodes to move in a direction opposite to that of Earth's orbital motion. This effect is called the regression of the nodes. It makes the eclipse seasons occur earlier each year by about 20 days, resulting in an interval of 346.62 days which is called the eclipse year. At least one solar eclipse must occur during an eclipse season, making a minimum of two solar eclipses in a year.

Solar eclipses. A total eclipse of the Sun is one of the greatest spectacles of nature and results from one of the most fortunate coincidences of nature. In round numbers, the Sun's diameter is 400 times that of the Moon, and the Sun's distance from Earth is also 400 times that of the Moon. Both subtend nearly the same angle of about 0.5°, so that the Moon's disk can appear just large enough to cover the Sun's disk.

Since the Moon and Earth are smaller than the Sun, the shadow of each is a cone having its point directed away from the Sun. All sunlight is geometrically excluded within this region, which is called the umbra. The Moon's umbra is formed by tangents to the Sun and Moon, which intersect on or near Earth's surface. Surrounding it is the larger inverted cone of the penumbra, formed by tangents intersecting between the Sun and Moon. A total eclipse can be observed within the umbra, and a partial eclipse within the penumbra.

The length of the Moon's umbra is very nearly $\frac{1}{400}$ of its distance from the Sun. At new moon it averages about 374,000 km (232,000 mi), and it varies about 6400 km (4000 mi) either way. Since the Moon's distance ranges from about 357,000 km (221,000 mi) to 407,000 km (253,000 mi), its umbra is sometimes long enough to reach Earth, but more often falls short. In the first case an observer located on the line of centers of the Sun and Moon can see a total solar eclipse. In the second case the Moon appears smaller than the Sun and the observer sees an annulus or ring of sunlight around the Moon. Hence this is called an annular eclipse.

In the most favorable case of a total eclipse, the diameter of the circular umbra where Earth's surface cuts it squarely is 269 km (167 mi). This occurs near the Equator. The path of totality traced out by the umbra is so narrow compared with Earth's diameter that the average interval between total eclipses at any one place on Earth is 360 years.

Under the most favorable conditions a total solar eclipse can last at a given point near the Equator for about $7\frac{1}{2}$ min. The maximum possible duration of an annular eclipse is nearly $12\frac{1}{2}$ min.

Since the Moon moves its own diameter in about 1 h, the complete phenomenon of a total solar eclipse lasts about 2 h: 1 h each in covering and uncovering the Sun and not more than a very few minutes for the total phase.

With the arrival of totality, the pearly white corona appears (see illustration). It is the Sun's outer atmosphere, extending in

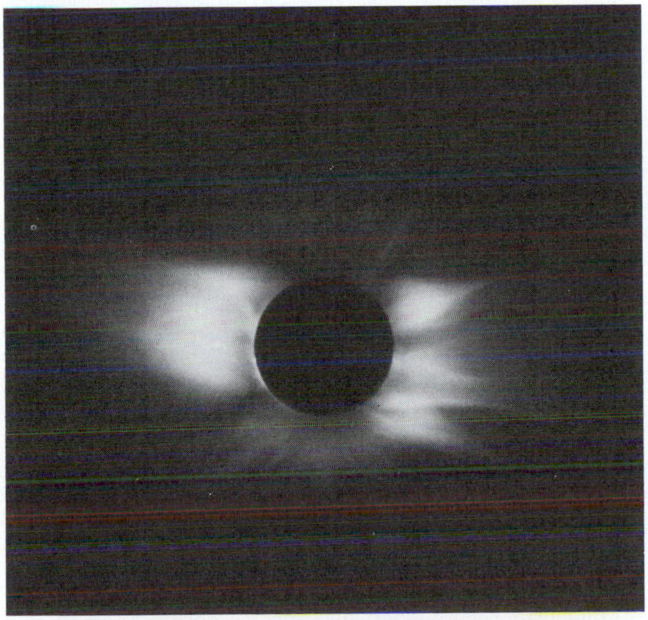

Solar corona of June 30, 1973, photographed from a jet aircraft flying over Africa at an altitude of 37,800 ft (11.5 km). This picture is a composite of a 0.5-s exposure with no filter and a 1-s exposure with a radial-gradient neutral-density filter. (*Los Alamos National Laboratory, University of California*)

all directions for millions of miles. It is bright at its inner edge and fades out in irregular streamers. Flamelike prominences are often visible to the unaided eye and usually with binoculars. They are protuberances from the red chromosphere, which is the layer of gases just above the Sun's visible surface. The brighter stars and planets are usually visible to the unaided eye.

Lunar eclipses. Although lunar eclipses do not occur as often as solar eclipses, they are seen by many more people. They occur during full moon, which is opposite the Sun and can be observed at any one instant by all of the night half of Earth. Also, since a total lunar eclipse can last for more than an hour, it can be seen from a little more than half of Earth's surface. On the other hand, solar eclipses are visible even in the partial phases from limited areas of Earth.

When the Moon passes through the center of Earth's shadow, it takes about 5 h 40 min. The first hour is spent in the penumbra. No darkening is noticeable until near the end of that period, because all of the Moon's side facing Earth is still receiving some direct sunlight. Then as the Moon enters the umbra, the edge of the Moon appears nearly black by contrast with the bright parts of the lunar surface. The second hour is required for all of the Moon to get into the umbra.

The diameter of the umbra where the Moon crosses it is about $2\frac{2}{3}$ times the Moon's diameter. The Moon can remain totally eclipsed for 1 h 40 min. The next hour is spent in moving out of the umbra, and the final hour in passing through the penumbra.

If Earth had no atmosphere, the Moon would disappear from view while in the umbra. However, the Earth's atmosphere acts like a lens and bends the sunlight into the umbra. The longer waves of red light penetrate the atmosphere better than the shorter waves of blue light, which are scattered to form the blue of the sky. An observer on the Moon would see Earth's disk surrounded by a thin ring of bright sunset colors. This accounts for the usual reddish color of the totally eclipsed Moon. [C.H.C.]

Eclipsing variable stars Double star systems in which the two components are too close to be seen separately but which reveal their duplicity by periodic changes in brightness as each star successively passes between the other and the Earth, that is, eclipses the other. Studies of the light changes and the radial velocity changes of each component permit the computation of the radii, masses, and densities of the components—important quantities that cannot be measured directly in single stars. In addition, these close double stars are useful in studies of mass loss and of stellar evolution. Since eclipsing stars are variable in light, they are included in general variable star catalogs under the same system of nomenclature. *See* BINARY STAR; VARIABLE STAR.

It is now believed that all explosive variables (novae, recurrent novae, and so forth), with the exception of supernovae, are members of close binary systems. At least some of the x-ray sources are close binaries in this state. In a few of the eruptive variables, particularly those known as dwarf novae, rapid scintillation is found, presumably in each system from a hot spot where the transferring mass collides violently with a circumstellar disk of relatively low-density material revolving around the collapsed star; the scintillation stops periodically when the spot is eclipsed by the other component. *See* NOVA; STELLAR EVOLUTION. [F.B.W.]

Ecliptic The path in the sky traced by the Sun in its apparent annual journey as Earth revolves around it. The ecliptic is a great circle on the celestial sphere, inclined about 23.5° to the celestial equator, the angle of inclination being called the obliquity of the ecliptic. *See* ASTRONOMICAL COORDINATE SYSTEMS; SUN. [G.M.C.]

Eclogite A very dense rock composed of red-brown garnet and the grape-green pyroxene omphacite. Eclogites possess basaltic bulk chemistry, and their garnets are rich in the components pyrope, almandine, and grossular, while the pyroxenes are rich in jadeite and diopside.

Eclogite occurrences may be subdivided into three broad categories: Group a eclogites are found as layers, lenses, or boudins in schists and gneisses seemingly of the amphibolite facies. Quartz, together with zoisite or kyanite, commonly occurs in these rocks. Amphibole of barroisitic composition may also be present. Group b eclogites are found as inclusions in kimberlites and basalts. They are frequently accompanied by xenoliths of garnet peridotite. Group c eclogites are found as blocks and lenses in schists of the glaucophane-schist facies. Such eclogites do not contain kyanite, rarely contain quartz, but bear amphibole, epidote, rutile, or sphene.

Eclogite is the name given to the highest-pressure facies of metamorphism; the critical mineral assemblage defining this facies is garnet + omphacite, together with kyanite or quartz in rocks of basaltic composition. Where sedimentary and granitic rocks have been metamorphosed under eclogite facies conditions, they result in spectacular omphacite + garnet + quartz-bearing mica schists and metagranitic gneisses such as are found in the Sezia-Lanzo zone of the western Alps. *See* METAMORPHIC ROCKS.

The high density of eclogite, together with its elastic properties, makes it a candidate for upper mantle material. Large quantifies of basaltic oceanic crust are returned to the mantle through the process of subduction, where prevailing high pressures convert it to eclogite. The quantity and distribution of eclogite within the mantle is not known; that it occurs is known from the nodules brought up in kimberlite pipes and in basalts. [T.J.B.H.]

Ecological community An assemblage of living organisms that occur together in an area. The nature of the forces that knit these assemblages into organized systems and those properties of assemblages that manifest this organization have been topics of intense debate among ecologists since the early years of this century. On the one hand, there are those who view a community as simply consisting of species with similar physical requirements, such as temperature, soil type, or light regime. The similarity of requirements dictates that these species are found together, but interactions between the species are of secondary importance and the level of organization is low. On the other hand, there are those who conceive of the community as a highly organized, holistic entity, with species inextricably and complexly linked to one another and to the physical environment, so that characteristic patterns recur, and properties arise that one can neither understand nor predict from a knowledge of the component species. In this view, the ecosystem (physical environment plus its community) is as well organized as a living organism, and constitutes a superorganism. Between these extremes are those who perceive some community organization but not nearly enough to invoke images of holistic superorganisms.

The crux of this debate is whether communities have emergent properties, a concept that itself is sufficiently confused, so that part of the ecological debate may be semantic. Some denote by emergent property any property of a group that is not a property of the component individuals. Others call any group trait a collective property, and reserve as emergent only those collective properties that do not derive trivially from properties of component individuals or from the very definition of the group. An emergent property, in this conception, represents a new level of organization, and cannot be predicted even with complete knowledge of all properties of individuals in the group. Thus, the number of species in a community is a property that is not defined for any particular species, but once one knows all the species in a community one can easily measure the property "number of species" for that community. In the restricted view of emergent property, the number of species in a community is a collective property, but not an emergent one. Whether any properties of communities are emergent is debatable, but all ecologists agree that communities have collective properties that are significant both biologically and practically.

A description of an ecological community may include such key characteristics as (1) its size in terms of number of species, (2) a list of species, (3) the spatial distribution patterns of species, (4) the pattern of succession, (5) the functional organization of species, (6) the productivity of species, and (7) the ability of species to reproduce. *See* BIOLOGICAL PRODUCTIVITY; BIOSPHERE; ECOLOGICAL INTERACTIONS; ECOLOGICAL SUCCESSION; ECOSYSTEM; PHYSIOLOGICAL ECOLOGY (PLANT). [D.Sim.]

Ecological interactions Relationships between members of different species belonging to a particular ecological community. Ecologists generally classify interactions according to the effect that members of one species have on the population growth rate of a second species. If an increase in one species' population increases the growth rate of a second species' population, the effect is positive or beneficial (denoted by +). If an increase in one species' population decreases the growth rate of a second species' population, the effect is negative or harmful (denoted by −). If neither occurs, the effect is neutral (denoted by 0). Effects may be indirect, in which case they involve one or more additional species in the community. For example, earthworms increase plant growth and thereby provide more food for caterpillars. Alternatively, effects may be direct; for example, injuries or death may result from territorial fighting between members of different species.

Interspecific competition occurs when two species negatively affect one another. Two principal kinds of competition are recognized. Exploitative competition between species occurs when individuals of one species, by consuming some resource, usually food, deprive individuals of another species of that resource and thereby lower the second species' population growth rate. Interference competition is more direct; it occurs when individuals of one species directly harm individuals of a second species. The degree to which such an effect is harmful can be quite varied. Organisms may kill other organisms by producing toxins or by fighting; fighting may also produce injuries. More subtle types of interference competition occur when individuals interact with one another in such a way as to use energy that might have been used for the production of offspring, or take up time that might have been used to gather additional energy for the production of offspring.

Predation occurs when individuals of one species, the predator, consume part or all of individuals of another species, the prey. In this interaction, the predator benefits and the prey is harmed. Broadly defined, the predator and prey may be animal or plant, and the predator may be a parasite. Examples of predation include (1) a hawk eating a sparrow, (2) a lizard eating an insect, (3) a sparrow eating a seed, (4) a sheep eating grass, (5) a Venus' flytrap eating an insect, (6) a tick drawing blood from a dog, and (7) a pathogenic bacterium afflicting a human. Some ecologists exclude plants from the definition, in which case examples 3–5 do not apply, or exclude parasitism, in which case examples 6 and 7 do not apply.

Mutualism occurs when two species positively affect one another. The degree to which the interaction is necessary for the survival of one or both species is sometimes emphasized; when the interaction is not necessary for either species, it is called protocooperation. Several well-known examples of mutualistic interactions that are necessary for the survival of both species exist. Various animals, such as termites and ruminants, that consume woody plant parts are dependent on the protozoa in their guts to digest cellulose. Various bacteria live in nodules on the roots of leguminous plants such as peas and clover. These bacteria are able to extract nitrogen directly from air spaces in the soil and make it available to the plant, which otherwise would have to acquire the nitrogen from nitrates. A third example is the lichen, which is composed of two kinds of plants, algae and fungi, living in a mutualistic association. The algae can photosynthesize and so provide various organic compounds. The fungi are believed to be able to dissolve nutrients from the rock on which the lichen lives; they also provide a structural matrix. This relationship is usually but not always obligatory. *See* LICHENES; NITROGEN CYCLE.

Commensalism occurs when one species positively affects another while being little affected itself. Several examples meet this criterion to various degrees. Plants such as bromeliads and orchids grow on other plants such as trees; the former are benefitted, while the latter are not usually seriously affected. Cattle egrets, which are insectivorous heronlike birds, have become adapted to following cattle and consuming the insects that the cattle disturb. The egrets are benefitted, and their actions seem to bother the cattle little. Certain fish (*Amphiprion percula*) are specialized to live among the poisonous tentacles of sea anemones. The anemones feed on other kinds of fish, and the *Amphiprion* feed on the leftovers from the anemones' meals. *See* ECOLOGY; ECOSYSTEM. [T.W.S.]

Ecological modeling The conceptualization and implementation of computer simulations of the behavior of living systems. In contrast to systems ecology, with which it somewhat overlaps, ecological modeling focuses on the predictions derived from models of particular systems rather than on the general behavior of entire classes of systems. The living systems that are modeled range from a single population to an entire ecosystem, including models of global processes. The models may be causal or correlative. Causal models, also known as mechanistic models, are constructed by incorporating hypothesized rules of interaction. The resulting behavior of such models can be compared to the behavior of real systems by way of validation. Correlative, or empirical, models, which are constructed by fitting observed system behavior with some appropriate mathematical functions, are more properly the domain of statistical ecology. They have great utility as predictors, provided an adequate database is available, but they offer no explanation of why the system behaves as it does.

Most ecological model builders begin by defining the system of interest: the boundaries must be delineated, the abstract system structure must be described, and the specific functional components to be modeled must be added.

Traditionally, ecological models consist of a set of nonlinear, often discontinuous differential equations. The discontinuities that result from the incorporation of threshold behavior preclude global analytical solutions to these sets of equations, although the large number of interactions and nonlinear processes that are present in all but the simplest ecological systems would make analytical solutions impossible in any case. Ecological modelers are fortunate in that the systems they simulate normally are relatively stationary or else oscillate around some long-term mean. Thus errors in the numerical integration of sets of differential equations tend to cancel each other, with the only result being decreases in the predicted peaks and increases in the predicted troughs. *See* SYSTEMS ECOLOGY. [R.G.W.]

Ecological succession A gradual process brought about by the change in the number of individuals of each species of a community and by the establishment of new species populations which may gradually replace the original inhabitants. Succession depends upon several major factors: (1) the floristic and faunistic resources of the general region, that is, what organisms are in a position to invade a site; (2) the rate of change of the habitat and its receptivity to these potential invaders; and (3) chance factors which may influence interactions.

The term sere is used to describe all of the temporary communities which occur during a successional sequence on a given site. Eventually succession ends, changes in species composition of the community cease or fluctuate within some bounds, and the result is a climax community.

Successional sequences may be classified on the basis of starting habitat. Thus, hydroseres are communities in which pioneer plants invade open water, eventually forming some kind of soil such as peat or muck; xeroseres are communities on dry, sterile ground, such as rock, sand, or clay.

If an open habitat has never been occupied before, the way is opened for primary succession. If the habitat has been disturbed, leaving vestiges of soil, seeds, and other organic debris from previous occupancy, a secondary succession will begin.

As an area is invaded and taken over by successive populations of plants and animals, the physiognomy of the community changes. Isolated pioneer plants eventually give way to a consolidation of the plant cover. With continued exploitation of the habitat, new species take over which have life forms able to utilize more fully the resources of the habitat than previous life forms until finally the climax community is established. Such a community can perpetuate itself indefinitely, at least in theory, because each species reproduces and maintains a stable number of individuals.

It was traditional to divide succession into pioneer, consolidation, subclimax, and climax stages on the basis of physiognomy. However, such discrete stepwise changes are not always shown by close study of the populations, because there is a continuously changing array of species. Therefore, the subdivision of a successional continuum must depend on local conditions and purposes of study.

An understanding of succession is important for effective human management of communities. Human activities generally have the effect of stopping succession at an early stage. Fire was used by most primitive people to keep forests suitable for hunting by clearing away the underbrush or to clear land for agriculture. Since almost all major crop plants are annuals of the pioneer stage, regular plowing is needed to hold down succession. Other methods of checking succession include mowing and grazing, which favor sod-forming grasses and kill most tree seedlings, and selective weed control with chemicals.

Succession can be either retarded or accelerated by manipulation of drainage. Such manipulation may be used to establish a cranberry bog or a stand of productive timber. Many timber trees occur naturally as successional species rather than in climaxes. Thus the forest must be managed so as to permit these species to grow free from the competition of the climax species which will eventually replace them. Fire, herbicides, and cutting practices are all useful in regulating the growth and composition of forest stands so that early successional stages are maintained. *See* CONSERVATION OF RESOURCES; ECOLOGY. [A.W.C.]

Ecology A study of the relation of organisms to their environment, or in more simple terms, environmental biology. Ecology is concerned especially with the biology of groups of organisms and with functional processes on the lands, in the oceans, and in fresh waters. Ecology is the study of the structure and function of nature (humans being considered part of nature).

Ecology is one of the basic divisions of biology which are concerned with principles or fundamentals common to all life. Since biology may also be divided into taxonomic divisions which deal with specific kinds of organisms, ecology is not only a basic division of biology but also an integral part of any and all of the taxonomic divisions. Both approaches are profitable. It is often productive to restrict certain studies to taxonomic groups, insects, for example, because different kinds of organisms require different methods of study. (Some groups of organisms are of greater interest to humans than others.) It is also important to seek and to test unifying principles which may be applicable to nature as a whole.

Approaches to ecology. From the standpoint of general ecology, an instructive subdivision is in terms of the concept of level of organization. For convenience, a biological spectrum can be visualized as follows: protoplasm, cells, tissues, organs, organ systems, organisms, populations, communities, and ecosystems. Ecology is concerned largely with the levels beyond that of the individual organism. In ecology the term of population, originally used to denote a group of people, is broadened to include groups of individuals of any one species of organism. Community in the ecological sense, sometimes designated as biotic community, includes all of the species populations of a given situation. The community or individual and the nonliving environment function together as an ecological system or ecosystem. The Earth as an ecosystem is conveniently designated as the biosphere. The term autecology is often used to refer to the study of environmental relations of individuals or species, whereas the term synecology refers to the study of groups of organisms such as communities. Mathematical concepts have permeated the basic foundations of ecology and have provided new means of exploring qualitative and quantitative relations at all levels. This new brand of ecology is known as systems ecology. *See* BIOSPHERE; SYSTEMS ECOLOGY.

For convenience, subdivisions of ecology may also be based on the kind of environment or habitat to be considered or studied. Marine ecology, fresh-water ecology, and terrestrial ecology are the three broad divisions from this point of view; estuarine ecology, stream ecology, or grassland ecology represent more restricted interests. *See* FRESH-WATER ECOSYSTEM; MARINE ECOLOGY; TERRESTRIAL ECOSYSTEM.

Some attributes, obviously, become more complex and variable during the procession from cells to ecosystems; however, it is often overlooked that other attributes may become less complex and less variable from the small to the large unit. The reason is that homeostatic mechanisms, which may be defined as checks and balances or forces and califorces, operate all along the line to produce a certain amount of integration, and smaller units function within larger units. For example, the rate of photosynthesis of a whole forest or a whole cornfield may be less variable than that of the individual trees or corn plants within the communities, because when one individual or species slows down, another may speed up in a compensatory manner. *See* BIOGEOCHEMICAL CYCLES; ECOSYSTEM; HOMEOSTASIS.

[E.P.O.]

Applied ecology. Applied ecology involves the application of ecological principles to the solution of human problems and the maintenance of a quality life. It is assumed that humans are an integral part of ecological systems and that they depend upon healthy, well-operating, and productive systems for their continued well-being. For these reasons, applied ecology is based on a knowledge of ecosystems, and the principles and techniques of ecosystem ecology are used to interpret and solve specific environmental problems and to plan new management systems in the biosphere. Although a variety of management fields, such as forestry, agriculture, wildlife management, environmental engineering, and environmental design, are concerned with specific parts of the environment, applied ecology is unique in taking a view of the whole system, and attempting to account for all inputs to and outputs from the system—and all impacts. In the past, applied ecology has been considered as being synonymous with the above applied sciences. *See* ENVIRONMENTAL PROTECTION. [F.B.Go.]

Economic geology

Commercial application of the general knowledge of geology. A large number of geologists receive special training for service in solving problems met in the mining of metals and nonmetals, in discovering and producing petroleum and natural gas, and in engineering projects of many kinds. *See* ENGINEERING GEOLOGY; MINING; PETROLEUM GEOLOGY. [C.R.L.; P.J.Wy.]

Ecosystem

A functional system which includes the organisms of a natural community together with their environment. The term is a contraction of ecological system. All natural communities, or assemblages of organisms which live together and interact with one another, are closely related to their environments. It is consequently appropriate to conceive of community and environment as forming a single, complex whole—the ecosystem. *See* COMMUNITY; ECOLOGY; ENVIRONMENT.

The term ecosystem can be applied to any size or level of environment chosen for study. Communities of small and distinctive environments, such as the organisms inhabiting an animal's intestine, a decaying log, or the foliage of a particular plant, are sometimes referred to as microcommunities, and their environments, as microcosms or microenvironments. These are generally parts of larger ecosystems, such as a forest and its environment or a lake and its organisms, for which the terms macrocommunity and macroenvironment are appropriate. At the opposite extreme from microcommunities, all the organisms of the world together may be regarded as a world community which, with its environment, forms a single world ecosystem. The living part of this world ecosystem is often termed the biosphere. The transfer and concentration of materials in the world ecosystem by both organisms and physical processes are the subject of biogeochemistry. Since humans are part of the world ecosystem, their place and effects on it are part of the study of human ecology. *See* BIOSPHERE.

Any organism lives in a natural context which includes an environment and a community of other organisms. The natural context of the life of an organism is thus an ecosystem. Two aspects of an organism's relation to the ecosystem in which it lives are often distinguished, though they cannot be sharply separated. Its location, as described in terms of physical and chemical factors, and often, kind of community, constitute its functional habitat. Its place within the community, including relations to other organisms, is its niche.

Within an ecosystem certain major fractions, or divisions, may be recognized. A terrestrial ecosystem is often thought of as having four or five such subdivisions: (1) the physical environment, such as climate; (2) the soil, if distinguished from physical environment; (3) the vegetation or plant community; (4) the animal community; and (5) the saprobe community, which consists of bacteria and fungi. These same divisions may be recognized in a marine or fresh-water ecosystem, with the substitution of substrate or bottom material for soil. Aquatic ecosystems, however, are often divided in a different way. These divisions are the communities and environments of the open water, those of the shores, and those of the deeper bottom. *See* FRESH-WATER ECOSYSTEM; MARINE ECOLOGY; TERRESTRIAL ECOSYSTEM. [R.H.Wh.]

Eczema

Any skin disorder characterized by redness, thickening, oozing from blisters or papules, and occasional formation of fissures and crusts. The words eczema, eczematous, and chronic eczematous are difficult to define and use. It is suggested that these terms be avoided in favor of the specific type of dermatitis. Various authorities differ as to the specific skin lesions which are considered under this catchall heading, but it is most commonly used to mean acute dermatitis, while some also include chronic dermatitis. Few eczemas represent pure forms if they have been present for a while.

Infantile eczema may be a form of contact dermatitis, atopic or neurodermatitis, or combinations of these with others of known or unknown cause.

Contact dermatitis occurs in many individuals as the result of either chronic irritation or sensitization to some external substance. Poison ivy, industrial dermatoses, and cosmetic sensitivities are examples.

Neurodermatitis may be either localized or general in its manifestations. An initial itching, or pruritis, is followed by a thickening of the skin due to repeated scratching. There is often a prior history of psychic disturbance or a family history of allergy, such as hay fever or asthma.

Stasis dermatitis is primarily the result of poor circulation and is found principally on the legs and ankles. Older patients or those with an occupation that requires prolonged periods of standing appear to be especially susceptible.

The senile eczema seen in elderly people is thought to be caused by various factors, notably dryness of the skin, soap sensitivity, poor hygiene or diet, or a preceding neurodermatitis. *See* SKIN DISORDERS. [E.G.St./N.K.M.]

Eddy current An electric current induced within the body of a conductor when that conductor either moves through a nonuniform magnetic field or is in a region where there is a change in magnetic flux. It is sometimes called Foucault current. Although eddy currents can be induced in any electrical conductor, the effect is most pronounced in solid metallic conductors. Eddy currents are utilized in induction heating and to damp out oscillations in various devices.

It is possible to reduce the eddy currents by laminating the conductor, that is, by building the conductor of many thin sheets that are insulated from each other rather than making it of a single solid piece. The laminations do not reduce the induced emfs, but if they are properly oriented to cut across the paths of the eddy currents, they confine the currents largely to single laminae, where the paths are long, making higher resistance. *See* CORE LOSS. [K.V.M.]

Edema An abnormal accumulation of fluid in the cells, tissue spaces, or cavities of the body; also known as dropsy. An excess of fluid in the pleural spaces is referred to as hydrothorax; in the pericardial sac as hydropericardium; and in the peritoneal cavity as ascites. Anasarca is a generalized subcutaneous edema.

There are three main factors in the formation of generalized edema and a fourth which plays an important role in the formation of local edema. They are (1) permeability of the capillary wall, (2) colloid osmotic pressure of the plasma proteins, (3) hydrostatic pressure in the capillaries, and (4) lymphatic obstruction. [R.A.V.]

Edentata A mammalian order consisting of many strange fossil and living animals found exclusively in the Western Hemisphere. Included is the suborder Xenarthra. Xenarthrans include Gravigrada (sloths), Vermilingua (anteaters), and Cingulata (the armored animals, armadillos and glyptodonts). Edentate, meaning toothless, directly applies only to the infraorder Vermilingua ("worm tongue," the South American true anteaters). Other edentates have simple prismatic teeth with no enamel, except in a few extinct genera and in the embryos of living armadillos. There is little or no specialization of the dentition.

Some edentates were, or are, no larger than rats; others (the megatheriine ground sloths, which ranged from Patagonia to the southern United States) rivaled large elephants in size. All animals in this order are clawed (some were secondarily hoofed) and quadrupedal. Some live in trees; others are ambulatory-terrestrial; and others are burrowers. Many edentates can be termed omnivorous; some are more strictly insectivorous or herbivorous. *See* ANTEATER; ARMADILLO; SLOTH.

Armor seems to have been a very early development in xenarthrans, and there is some evidence to suggest that the families of sloths may have been derived from an armored armadillolike stock. South America was the center of early history and diversification of the Xenarthra.

Xenarthran edentates may have originated in North America in an early Paleocene or Late Cretaceous stock of small clawed quadrupedal eutherian mammals without dermal armor and with generalized skeleton. Such a group would most likely be included in the order Insectivora. *See* EUTHERIA; INSECTIVORA; MAMMALIA. [D.E.S.]

Ediacaran fauna A widely distributed group of soft-bodied marine animals (metazoans) that are preserved as fossils in rocks of the latest Precambrian age (600–550 million years ago). It represents the oldest known animal fossils, and identifies a geological period known as the Ediacaran (or Vendian), which precedes the appearance of animals with mineral skeletons. The name Ediacaran is taken from Ediacara, an abandoned mining area about 380 mi (600 km) north of Adelaide, South Australia.

Although the first report was from South-West Africa (Namibia) in the 1930s, it was the discovery in 1946 of abundant fossil "jellyfish" at Ediacara that created international interest in this fauna. Subsequently, similar or identical fossils were found in central England (1957), southeastern Newfoundland (1967), the Soviet Union (1960s and 1970s), North Carolina (1966), northwestern Canada (1979), and elsewhere. As some of these sites could not have been less than about 6000 mi (10,000 km) apart, no matter how the continents were arranged at the time, there is no doubt that the Ediacaran fauna was a globally distributed marine biota.

All known occurrences of Ediacaran fauna are found either in strata that underlie those containing the earliest skeletal fossils (archaeocyaths, trilobites, mollusks, brachiopods) or in locations where other evidence indicates a latest Precambrian age. None of the distinctive Ediacaran fossils has been found associated with typical Cambrian fossils; thus it appears that this biota became extinct before the beginning of the Cambrian. *See* CAMBRIAN.

At almost all sites the fossils are preserved as impressions in some kind of detrital sedimentary rock. Commonly, they are found on the bases of sandstone beds (South Australia), within sandstone beds (Namibia), or below volcanic ashes swept into deep water by wind and turbidity currents (Newfoundland). The animals appear to have lived in continental shelf to continental slope environments and are normally preserved in sediments that were deposited under fairly quiet conditions below normal wave base.

There are between 20 and 30 distinct genera that have been referred to a number of different higher taxa. Scientists who have worked in South Australia have generally attempted to relate the Ediacaran animals to the phyla of living invertebrates. Others, however, recognized that the Namibian fossils cannot be placed in any modern group, and in one case, therefore, a new phylum (Petalonamae) was proposed for them. *See* COELENTERATA.

The marks left in soft sediments by otherwise unknown animals provide another source of knowledge of late Precambrian animal life. Animals capable of directed, muscular, gliding motion (like that of a garden snail) coexisted with more typical members of the Ediacaran fauna. In a similar fashion, strings of fecal pellets demonstrate the existence of animals with one-way guts, and closely meandering marks imply an ability for systematic grazing. There is little evidence for vertical burrowing in rocks of this age.

The Ediacaran fossils are the remains of sizable (up to 3 ft or 1 m) sessile or weakly mobile animals that had large surfaces compared with their volumes. It is not clear how they obtained their food, but it has been suggested that their bodies housed symbiotic, photosynthetic microorganisms like modern reef corals and other kinds of tropical marine animals. Alternatively, the high ratios of surface to volume may represent adaptations to an atmosphere and hydrosphere relatively low in oxygen. *See* PALEONTOLOGY. [B.Ru.]

Edrioasteroidea A class of extinct Echinozoa which arose early in Cambrian times and died out before the end of the Carboniferous Period. They resemble the other Echinozoa in having the ambulacral radial areas bordered by tube feet which emerge through pores in or between the ambulacral plates. They differ from extant Echinozoa in having the mouth and anus on the upper side of the theca. Hinged ambulacral plates bordered the groove on either side and could be erected over it to convert it into a tube (see illustration). The animals in some cases were cemented to the substrate by the lower surface; in other cases they were free and rested upon the sea bed. Brachioles are lacking. About 30 genera have been described and grouped into 7 families. *See* ECHINODERMATA; ECHINOZOA. [H.B.F.]

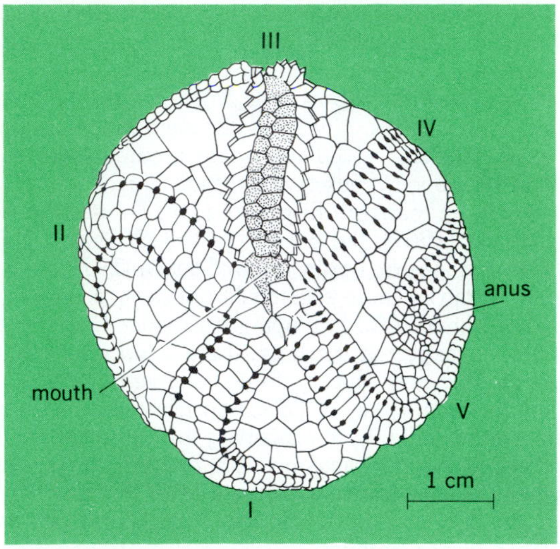

An edrioasteroid (*Edrioaster*) seen from above. The five ambulacra are numbered I to V. In IV and V the hinged plates are removed, leaving the adambulacral plates exposed, and showing the pores. (*Modified from F. Bather, 1900*)

Edrioblastoidea A very small class of early echinozoan echinoderms known from the single genus *Astrocystites*, with several species occurring in the Middle Ordovician of eastern Canada, Virginia, Oklahoma, and Australia, and in the Late Ordovician of Sweden. Edrioblastoids have a globular theca or body with fairly regular rows of plates, including basals, radials, infradeltoids, and large parabolic-shaped deltoids that bear one-half of an ambulacral groove with pores on each limb. The theca was attached to the sea floor by a primitive multipart stem that lacked disklike columnals. Edrioblastoids were apparently attached, low- to medium-level suspension feeders that used tube feet in their ambulacra to catch small food particles drifting by the theca. The distinctive pores in their covered ambulacra resemble those in other echinozoans, suggesting that edrioblastoids are an edrioasteroidlike offshoot that converged on blastoids in their thecal shape and way of life. *See* CRINOZOA; ECHINODERMATA. [J.Sp.]

Eel The name for a number of unrelated fish included in the orders Anguilliformes and Cypriniformes.

The true eels are members of the Anguilliformes, which is also known as the Apodes (see table). There are several hundred species, most of which are marine. They are most common in the shallower waters of tropical and subtropical seas, although a few species do occur in colder waters or at considerable depths. *See* ANGUILLIFORMES.

Scientific and common names of some species of true eels

Scientific name	Common name
Muraena helena	Moray eel
Phisodonopis boro	Indian eel
Simenchelys parasiticus	Pug-nosed eel
Anguilla rostrata	American eel
Anguilla vulgaris	European eel
Conger conger	Conger eel

The family Gymnotidae of the order Cypriniformes contains about 32 species which are found in fresh waters of Central and South America. The electric eel (*Electrophorus electricus*), the largest and best-known species of the family, may grow to a length of 8 ft (2.4 m) and weigh 60 lb (27 kg). It can produce an electric shock from its electric organs, which extend almost the length of the body. The bodies of the members of this family are eellike and may be naked or scaled. *See* CYPRINIFORMES; ELECTRIC ORGAN (BIOLOGY). [C.B.C.]

Effective dose 50 A term used chiefly to characterize the potency of a drug by the amount required to produce a response in 50% of the subjects to whom the drug is given; also known as ED_{50} or median effective dose.

Median effective dose is most commonly applied in connection with drugs, but it may be used when various other sources of stimuli, for example, x-rays, are under consideration. The response must be of the kind known as quantal, or all-or-nothing, where the investigator is simply able to report that the response either was or was not elicited.

The median effective dose is a well-defined quantity only when the test animal, the end response, and such factors as the route of injection of a drug and the state of nutrition of the animal are specified. [C.W.]

Effector systems Those organ systems of the animal body which mediate overt behavior. Injury to an effector system leads to loss or to subnormal execution of behavior patterns mediated by the system, conditions termed paralysis and paresis, respectively.

Overt behavior consists of either movement or secretion. Movement results from contraction of muscle. Secretion is a function of glands. Neither muscular contraction nor glandular secretion is autonomous but is regulated by an activating mechanism which may be either neural or humoral. In neurally activated systems the effector organ, whether muscle or gland, is supplied by nerve fibers originating from cell bodies situated in the central nervous system or in peripherally located aggregates of nerve cell bodies known as ganglia.

In other effector systems (humeromuscular and humeroglandular) the activating agent is normally a blood-borne chemical substance produced in an organ distant from the effector organ. For example, uterine smooth muscle is uninfluenced by the uterine nerve activity but contracts vigorously when the blood contains pitocin, a chemical substance elaborated by the posterior lobe of the hypophysis.

Finally, some effector systems are hybrid in the sense that both nerves and humors regulate their functions. The smooth muscle of arterioles contracts in response to either nerve stimulation or epinephrine. Secretion of hydrochloric acid by the gastric mucosa is increased by activation of the vagus nerve or by the presence in the blood of histamine. Effector systems with both neural and humoral regulation are never completely paralyzed by denervation but may be deficient in reaction patterns when the quick integrated activation provided by neural regulation is essential. *See* NERVOUS SYSTEM (VERTEBRATE). [T.C.R.; H.D.P.]

Efficiency The ratio, expressed as a percentage, of the output to the input of power (energy or work per unit time). As is common in engineering, this concept is defined precisely and made measurable. Thus, a gear transmission is 97% efficient when the useful energy output is 97% of the input, the other 3% being lost as heat due to friction. A boiler is 75% efficient when its product (steam) contains 75% of the heat theoretically contained in the fuel consumed. All automobile engines have low efficiency (below 30%) because of the total energy content of fuel converted to heat; only a portion provides motive power, while a substantial amount is lost in radiator and car exhaust. [F.R.E.C.]

Efflorescence The spontaneous loss of water (as vapor) from hydrated crystalline solids. The thermodynamic requirement for efflorescence is that the partial pressure of water vapor at the surface of the solid (its dissociation pressure) exceed the partial pressure of water vapor in the air. A typical efflorescent substance is Glauber's salt, $Na_2SO_4 \cdot 10H_2O$. The spontaneous loss of water normally requires that the crystal structure be rearranged, and consequently, efflorescent salts usually go to microcrystalline powders when they lose their water of hydration. *See* PHASE EQUILIBRIUM; VAPOR PRESSURE. [R.L.S.]

Egg (fowl) A single, large, living, female sex cell enclosed in a porous, calcareous shell through which gases may pass. Although they vary in size, shape, and color, the eggs of chickens, ducks, geese, and turkeys are essentially the same in structure and content (see illustration). Inward from the shell are the outer and inner shell membranes which are also permeable to gases. The membranes are constructed to prevent rapid evaporation of moisture from the egg but to allow free entry of oxygen, which is necessary for life. Air begins to penetrate the shell soon after the egg is laid, and it tends to accumulate in a space between the two membranes at the large end of the egg. *See* CELL (BIOLOGY); RESPIRATION.

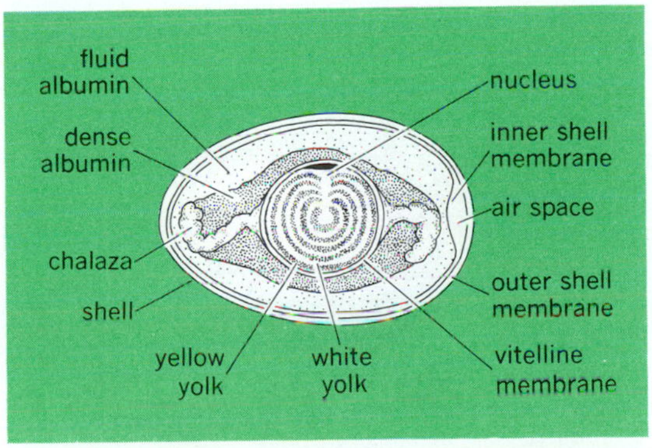

Structure of the egg of a bird.

The inner shell membrane surrounds a mass of fluid albumin which, in turn, encloses a body of dense albumin; these two types of protoplasm constitute the so-called egg white. The central part of the egg is occupied by the yolk, which contains the vital egg nucleus and its associated parts. The yolk consists of alternating layers of yellow and white yolk. The yolk, enclosed by the vitelline membrane, is held in place by the chalaza which is anchored at each end of the egg and prevents undue mechanical disturbance. *See* YOLK SAC. [J.F.F.]

Eggplant A warm-season vegetable (*Solanum melongena*) of Asiatic origin belonging to the plant order Polemoniales (formerly Tubiflorales). Eggplant is grown for its usually egg-shaped fleshy fruit and is eaten as a cooked vegetable. Popular purple-fruited varieties (cultivars) are Black Beauty and a number of hybrid varieties; fruits of other colors, including white, brown, yellow, and green, are used chiefly for ornamental purposes. Florida and New Jersey are important eggplant-producing states. *See* SOLANALES. [H.J.C.]

Eicosanoids A family of naturally occurring substances derived from polyunsaturated fatty acids, such as arachidonic acid, and including prostaglandins, thromboxanes, and leukotrienes. Several are physiologically active, and the free radicals formed during their syntheses control the biosynthetic potential and can lead to the cooxidation of xenobiotics.

Although the actions of prostaglandins and leukotrienes in normal physiology and in pathophysiology have been studied in detail, their extensive and frequently antagonistic actions preclude the identification of a specific role for these agents. Some of their more important actions include the effects of these eicosanoids on smooth muscle tone, cyclic nucleotide levels, platelet aggregation, and cell migration. The mutually antagonistic actions of sets of prostaglandins such as PGE_2 and $PGF_{2\alpha}$ or PGI_2 and thromboxane A_2 (TXA_2) make control of biosynthesis at a level subsequent to endoperoxide synthase extremely important. Most inhibitors such as nonsteroidal anti-inflammatory agents block this primary enzyme and, thereby, depress all prostaglandins equally. An exception is aspirin, which inhibits irreversibly by acetylation with preference for the platelet, where protein resynthesis cannot occur and TXA_2 is the major product. Platelets are, therefore, unable to form prostaglandin products for the remainder of their circulating lifetime. *See* ASPIRIN.

A host of tissues are influenced by eicosanoids as a result of their ability to mediate vascular tone and alter cyclic adenosinemonophosphate concentrations. These effects are observed on reproductive, cardiovascular, respiratory, renal, gastrointestinal, and endocrine tissues and skin, as well as the hematological system. As a result of these actions, imbalances in prostaglandin or leukotriene levels have been implicated in a variety of disorders, including inflammation, pain, ulcers, atherosclerosis, renal failure, asthma, psoriasis, and gout. In most of these disorders, contractile and aggregatory prostaglandins would be considered causative, whereas those that relax smooth muscle would be palliative. Total inhibition of prostaglandin biosynthesis may, therefore, not be most advantageous. On the other hand, inhibition of synthesis of sulfidopeptideleukotrienes or antagonism of their effects should be generally beneficial, especially in asthma, where sulfidopeptideleukotrienes play a significant role in the human disease. *See* ARTERIOSCLEROSIS; ASTHMA; GOUT; PAIN; ULCER. [R.W.Eg.]

Eigenvalue (quantum mechanics) If an equation containing a variable parameter possesses nontrivial solutions only for certain special values of the parameter, these solutions are called eigenfunctions and the special values are called eigenvalues.

The eigenfunction-eigenvalue relation is of particular importance in quantum mechanics because of its prominence in the equations which relate the mathematical formalism of the theory with physical results. *See* QUANTUM MECHANICS. [D.P.]

Eikenella A genus of bacteria containing only one species, *E. corrodens*. These are small rod-shaped (0.3 by 1–3 micrometers), nonflagellated, and nonsporulating gram-negative bacteria which can be cultivated on blood agar at reduced atmospheric oxygen and increased carbon dioxide tensions.

Eikenella corrodens appears to be a normal inhabitant of

the human upper respiratory tract and the oral cavity. Occasionally it is also isolated from mucous membrane surfaces of healthy intestinal and genitourinary tracts. It is, however, also found as an opportunistic pathogen in pure culture or together with other bacteria in wound infections, soft tissue abscesses, and sinusitis; it may be even involved in endocarditis, meningitis, septic arthritis, and osteomyelitis. *See* MEDICAL BACTERIOLOGY; OPPORTUNISTIC INFECTIONS. [W.Ma.]

Einsteinium A chemical element, Es, atomic number 99, a member of the actinide series in the periodic table. It is not found in nature but is produced by artificial nuclear transmutation of lighter elements. All isotopes of einsteinium are radioactive, decaying with half-lives ranging from a few seconds to about 1 year. *See* ACTINIDE ELEMENTS; RADIOACTIVITY.

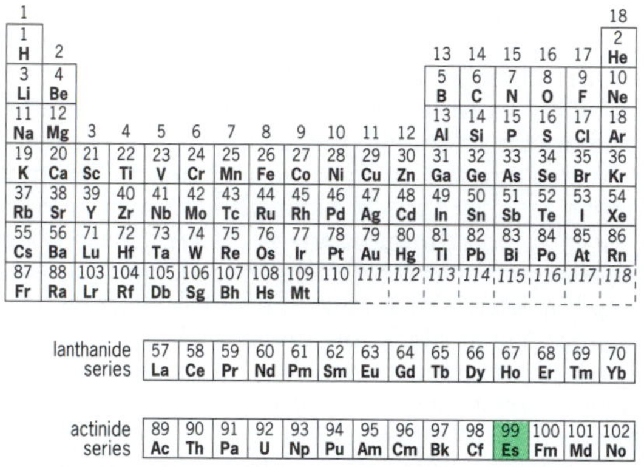

Einsteinium is the heaviest actinide element to be isolated in weighable form. The metal is chemically reactive, is quite volatile, and melts at 860°C (1580°F); one crystal structure is known. *See* TRANSURANIUM ELEMENTS. [S.G.T.; G.T.S.]

El Niño An interannual rise and fall in the sea surface temperature of the equatorial Pacific Ocean near the west coast of South America. It is a component of a phenomenon known as the El Niño–Southern Oscillation (ENSO), an irregular alternation of anomalous warm and cold temperatures and concomitant rainfall variations across the vast span of the tropical Pacific Ocean. ENSO has been measured, explained, and predicted a year in advance.

The tropical Pacific Ocean normally has very warm surface water at the western end and cooler water at the eastern end. Superimposed on this east-to-west temperature difference is an annual cycle that has east Pacific sea surface temperatures coolest in August and warmest in April, in contrast to midlatitudes, which have January and July as the extreme months. In general, the warm pool in the west moves northwest during northern summer and southeast during northern winter. Throughout this complicated annual cycle, the basic pattern is maintained, with warm water in the west and its concomitant heavy rainfall, low pressure, and upward atmospheric motion.

Every 4 years or so, the temperature pattern of the tropical Pacific changes. It is as if all the processes that normally maintain the climatic conditions in the tropical Pacific no longer occur, and the tropical Pacific reverts to what would be expected in the absence of winds: a uniformly warm tropical Pacific. Occasionally, the opposite also occurs: the eastern Pacific becomes cooler than normal, rainfall decreases still more, atmospheric surface pressure increases, and the westward winds become stronger. This irregular cyclic swing is ENSO. *See* PACIFIC OCEAN.

Numerical models that couple the atmosphere to the ocean have been used to successfully predict the sea surface temperature of the tropical Pacific a year or so in advance. Predicting the evolution of ENSO requires (1) a coupled model of the tropical Pacific atmosphere-ocean system capable of accurately simulating the normal climate (that is, mean and annual variation) of the system; (2) a measurement system that provides data, in real time, on the state of the atmosphere and ocean near the ocean surface; (3) a method of using the data to provide the best estimate of the current state of the coupled atmosphere-ocean system; (4) coupled-model computer calculations starting from the estimated current state and proceeding to predicting the state of the coupled system a year in advance; and (5) a method for validating the predictions and improving the prediction system in response to errors in the predictions. *See* CLIMATE MODELING.

The basic reason that the cycle is predictable is that ENSO evolves slowly and regularly. ENSO prediction offers the possibility of predicting rainfall around the tropical Pacific basin (the rainfall is tightly tied to the sea surface temperature—the warmer the sea surface temperature, the more the rainfall). Such rainfall forecasts have economic value for Australia, Indonesia, Peru, Chile, and the Pacific islands. Known correlations between ENSO and rainfall not near the direct domain of ENSO (for example, in India, in Brazil, on the northwest coast of North America) increase the value of ENSO forecasts even more. *See* CLIMATOLOGY; METEOROLOGY; SEAWATER; TROPICAL METEOROLOGY; UPWELLING. [E.S.S.]

Elasipodida An order of aspidochirotacean holothurians in which there are no respiratory trees and the mesentery of the posterior loop of the intestine is attached in the right dorsal interradius. The order is composed almost exclusively of deep-sea forms, in which bilateral symmetry is often extremely conspicuous (see illustration). They are common in deep water and

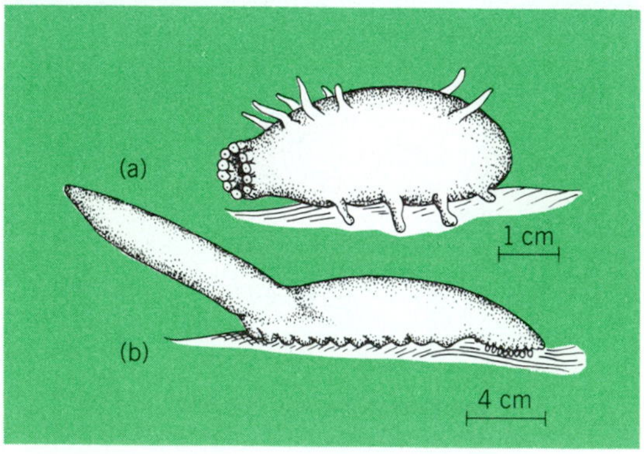

Two examples of bottom-dwelling Elasipodida. (*a*) *Elpidia*. (*b*) *Psychropotes*.

especially common in the faunas of deep trenches, ranging down to at least 5000 fathoms (9000 m). This distribution probably occurs because they are able to extract nutriment from mud in an environment almost devoid of other sources of food. *See* ECHINODERMATA; HOLOTHUROIDEA. [H.B.F.]

Elasmobranchii A subclass of the class Chondrichthyes, the cartilaginous fishes. The elasmobranchs are distinguished by separate gill openings, amphistylic or hyostylic jaw suspension, and sensory ampullae (of Lorenzini) in the head region. Characters shared with members of the other subclass (Holocephali) include a variably calcified cartilaginous endoskeleton, placoid scales, urea-retention mechanism, clasper organs in the male for internal fertilization, and the absence of an air (swim) bladder. *See* BATOIDEA; CHONDRICHTHYES; CLADOSELACHII; SELACHII; SHARK. [B.S.]

Elastic limit The largest stress that a material can sustain without permanent deformation or strain. The proportional limit, it may be noted, is the maximum stress at which the stress is directly proportional to the strain.

Actual materials are not perfectly elastic upon first application of load because they lack homogeneity and have residual stresses produced by the manufacturing process. Upon subsequent loadings, the material will exhibit elastic properties. *See* ELASTICITY; STRESS AND STRAIN. [J.B.S.]

Elasticity The property whereby a solid material changes its shape and size under the action of opposing forces, but recovers its original configuration when the forces are removed. The theory of elasticity deals with the relations between the forces acting on a body and the resulting changes in configuration, and is important in many branches of science and technology, for instance, in the design of structures, in the theory of vibration and sound, and in the study of the forces between atoms in crystal lattices.

The forces acting on a body are expressed as stresses and measured as force per unit area. Thus if a bar *ABCD* of square cross section (see illustration *a*) is fixed at one end and subjected to a force *F* uniformly distributed over the other end *DC*, the stress is $F/(DC)^2$. This stress causes the bar to become longer and thinner and to assume the shape $A'B'C'D'$. The strain is measured by the ratio (change in length)/(original length), that is, by $(B'C' - BC)/(BC)$. According to Hooke's law, stress is proportional to strain, and the ratio of stress to strain is therefore a constant, in this case the Young's modulus, denoted by *E*, so that $E = F(BC)/(DC)^2\,(B'C' - BC)$. *See* HOOKE'S LAW; STRESS AND STRAIN; YOUNG'S MODULUS.

Poisson's ratio σ is the ratio of lateral strain to longitudinal strain so that $\sigma = BC(DC - D'C')/DC(B'C' - BC)$. The bar of illustration *a* is in a state of tension, and the stress is tensile; if the force *F* were reversed in direction, the stress would be compressive. Stresses of this type are called direct or normal stresses; a second type of stress, known as tangential or shear stress, is shown in illustration *b*. In this case, the configuration *ABCD* becomes $ABC'D'$, with the shear forces *F* acting in the directions *AB* and *CD*. The shear strain is measured by the angle θ, and if the body is originally a cube, the shear stress is $F/(DC)$. The ratio of stress to strain, $F/(DC)^2\,\theta$, is the shear or rigidity modulus *G*, which measures the resistance of the material to change in shape without change in volume.

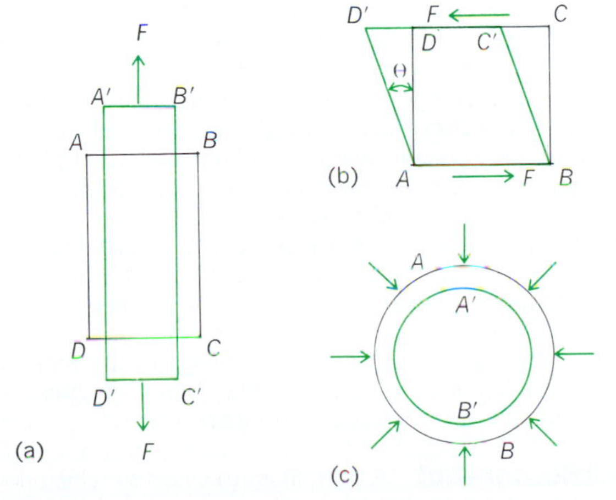

Stresses on a bar. (a) Direct or normal stress. (*b*) Tangential or shear stress. (c) Change in volume with no change in shape. (All deformations are exaggerated.)

A further elastic constant, the bulk modulus *k*, measures the resistance to change in volume without changes in shape, and is shown in illustration *c*. The original configuration is represented by the circle *AB*, and under a hydrostatic (uniform) pressure *P*, the circle *AB* becomes the circle $A'B'$. The bulk modulus is then $k = Pv/\Delta v$, where $\Delta v/v$ is the volumetric strain. The reciprocal of the bulk modulus is the compressibility.

The elastic constants may be determined directly in the way suggested by their definitions; for instance, Young's modulus can be determined by measuring the relative extension of a rod or wire subjected to a known tensile stress. Less direct methods are, however, usually more convenient and accurate. Prominent among these are the dynamic methods involving frequency of vibration and velocity of sound propagation. The elastic constants can be expressed in terms of frequency of (or velocity in) regularly shaped specimens, together with the dimensions and density, and by measuring these quantities, the elastic constants can be found. The elastic constants can also be determined from the flexure and torsion of bars. *See* ULTRASONICS.

In practice, stress is only proportional to strain, and the strain is only completely recoverable within certain limits called the elastic limits of the material. Above the elastic limits, the material is subject to time-dependent effects, and as the stress is further increased, the ultimate strength of the material is approached. *See* ELASTIC LIMIT; PLASTICITY; STRENGTH OF MATERIALS. [R.F.S.H.]

Elaterite A light-brown to black, naturally occurring carbonaceous substance having specific gravity 0.90–1.05. Elaterite is insoluble in carbon disulfide and infusible. It is moderately soft and elastic, and on heating leaves only 2–5% fixed carbon. Elaterite is thought to be of algal origin and occurs in small quantities in Derbyshire County, England, in the Coorong District of South Australia, and in Turkestan. [I.A.B.]

Elective culture A type of microorganism grown selectively from a mixed culture by culturing in a medium and under conditions selective for only one type of organism. No single culture medium or environment will support growth of more than a small proportion of the known types of microorganisms. Each organism has characteristic nutritional and environmental requirements. Thus, in a culture composed of more than one type of organism, the medium and conditions are selective for only one type of organism present and the other types are suppressed. These selective, or elective, culture principles and techniques are invaluable for the detection and isolation of specified organisms from natural sources. Elective culture techniques are more effective for obtaining bacteria from natural sources than for obtaining fungi, actinomycetes, and other groups; the faster growth rates, larger numbers, and greater variety of bacteria usually confer a competitive advantage upon them. In general, elective culture procedures are applicable to organisms capable of utilizing particular sources of energy, carbon, nitrogen, and other nutrients in a given physicochemical environment. *See* CULTURE MEDIA. [J.W.Fo./R.E.K.]

Electret A solid dielectric possessing persistent dielectric polarization. An electret is the analog of a magnet. Electrets are made by cooling suitable dielectrics from elevated temperatures in strong electric fields. A special class called photoelectrets is produced by the removal of light from an illuminated photoconductor in an electric field. *See* DIELECTRIC MATERIALS; FERROELECTRICS; MAGNET; POLARIZATION OF DIELECTRICS. [R.D.W.]

Electret transducer A device for the conversion of acoustical or mechanical energy into electrical energy, and vice versa, which utilizes a quasi-permanently charged dielectric material (electret). Examples are certain microphones, earphones, and phonograph cartridges.

In the simplest implementation, such a transducer consists of a metal backplate (first electrode) covered by a mechanically tensioned diaphragm. The diaphragm is a foil electret carrying a metal coating (second electrode) on the side facing away from the backplate. Provisions are made to maintain a shallow air gap between electret and backplate. The air gap is occupied by an electrostatic field originating from the electret charges. Upon acoustical or mechanical deflection of the diaphragm, such a device generates an electrical output signal between its two electrodes; similarly, application of an electrical signal results in diaphragm deflections. Electret devices are therefore self-biased electrostatic or condenser transducers. They thus exhibit all the advantages of this transducer class, such as wide dynamic range and flat response over a frequency range of several decades, without requiring the external bias necessary in conventional transducers of this kind. *See* ELECTRET; ELECTROSTATICS; TRANSDUCER.

The basic component of all modern electret transducers is a "foil electret" consisting of a thin film of insulating material that has been electrically charged to produce an external electric field. Strongly insulating materials capable of trapping charge carriers, such as the halocarbon polymers, in particular polyfluoroethylenepropylene (Teflon), are best suited for this purpose. Before charging, the material is either metallized on one side or backed up with a metal electrode.

A simple implementation of the most widely used electret transducer, the foil-electret microphone, is shown in the illustration. The diaphragm, typically 12- or 25-micrometers Teflon metallized on one surface, is charged to 10–20 nanocoulombs per square centimeter, corresponding to an external bias of about 200 volts. The nonmetallized surface of this foil electret is placed next to a backplate, leaving a shallow air layer, the thickness of which (about 20 micrometers) is controlled by ridges or raised points on the backplate surface. The stiffness of the air layer can be decreased (and thus the sensitivity of the microphone can be improved) by connecting the air layer to a larger cavity by means of small holes through the backplate. The backplate is either a metal disk or a metal-coated dielectric with a thermal expansion coefficient about equal to that of the diaphragm. The electrical output of the microphone is taken between the backplate, which is insulated from the outer case, and the metal side of the foil. The output is fed into a high-impedance preamplifier. *See* MICROPHONE.

Compared with conventional electrostatic transducers, electret microphones have the following advantages: they do not require a dc bias; they have three times higher capacitance per unit area, resulting in a better signal-to-noise ratio; and they are not subject to destructive arcing between foil and backplate in humid atmospheres and under conditions of water condensation.

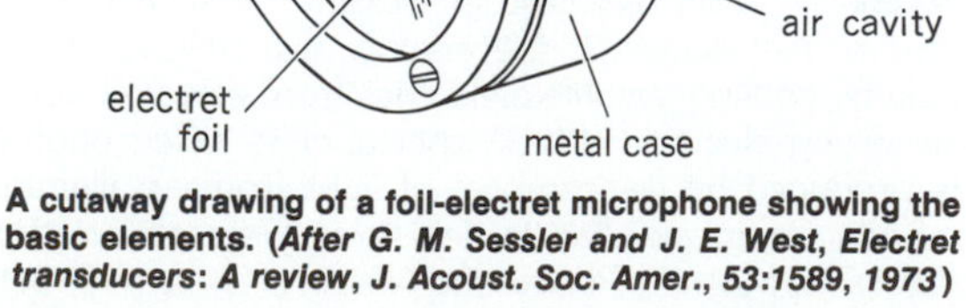

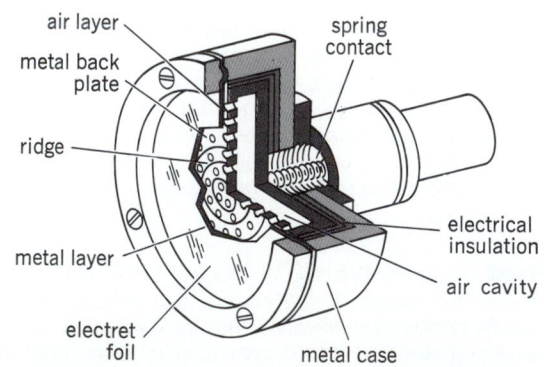

A cutaway drawing of a foil-electret microphone showing the basic elements. (*After G. M. Sessler and J. E. West, Electret transducers: A review, J. Acoust. Soc. Amer., 53:1589, 1973*)

Another group of electret transducers of interest are electret headphones. The above microphone design can also be utilized in this case. They produce a certain amount of nonlinear distortion due to the quadratic dependence of the force per unit area, exerted on the diaphragm, on the applied ac voltage. This drawback can, however, be minimized in a well-designed system, and high-voltage sensitivities have been achieved by making the air-gap thickness small. The distortion can be eliminated by the use of push-pull transducers and monocharge electrets.

Apart from its use in electroacoustic transducers, such as microphones and earphones, the electret principle has been applied to electromechanical transducers. Examples are phonograph cartridges, touch or key transducers, and impact transducers.

Because of their favorable properties, simplicity, and low cost, electret transducers have been used in many applications, both as research tools and in the commercial market.

Among the research applications are microphones for use in acousto-optic spectroscopy, applied to the detection of air pollution and to the study of reaction kinetics of gases and optical absorption of solids. Other applications of electret microphones have been reported in aeronautics and shock-tube studies, in which the low vibration sensitivity of these transducers is crucial. The wide frequency range of electret transducers, discussed above, made possible their application in infrasonic atmospheric studies and in ultrasonic investigations of liquids and solids. In addition, ultrasonic arrays of electret microphones have been used in acoustic holography. *See* ACOUSTICAL HOLOGRAPHY; PHOTOACOUSTIC SPECTROSCOPY; ULTRASONICS.

Of all commercial applications of electret devices, the high fidelity electret microphone for amateur, professional, studio, and tape recorder use is most prominent. Although the electret microphone was introduced only about 1970, tens of millions of the instruments are produced annually, accounting for about one-half of the entire output of high fidelity microphones.

[G.M.Se.]

Electric charge A basic property of elementary particles of matter. One does not define charge but takes it as a basic experimental quantity and defines other quantities in terms of it.

According to modern atomic theory, the nucleus of an atom has a positive charge because of its protons, and in the normal atom there are enough extranuclear electrons to balance the nuclear charge so that the normal atom as a whole is neutral. Generally, when the word charge is used in electricity, it means the unbalanced charge (excess or deficiency of electrons), so that physically there are enough "nonnormal" atoms to account for the positive charge on a "positively charged body" or enough unneutralized electrons to account for the negative charge on a "negatively charged body."

In line with this usage, the total charge q on a body is the total unbalanced charge possessed by the body. For example, if a sphere has a negative charge of 1×10^{-10} coulomb, it has 6.24×10^{8} electrons more than are needed to neutralize its atoms. The coulomb is the unit of charge in the meter-kilogram-second (mks) system of units. *See* COULOMB'S LAW; ELECTRICAL UNITS AND STANDARDS; ELECTROSTATICS. [R.P.Wi.]

Electric contact A part, in an electrical switching device, made of conducting material, for the purpose of closing, opening, or changing the conductive path of an electrical circuit. To open or close a circuit, an electric contact is made to come in contact with or separate from its mating part. Devices embodying contacts for these purposes are electric switches, relays, contactors, and circuit breakers. Contacts may be actuated directly or through a linkage that is driven either manually, mechanically, electromagnetically, hydraulically, or pneumatically.

An electric contact is also an essential part of a variable resistor. In such an application the contact or wiper provides a moving connection to the resistive element. *See* RHEOSTAT.

[T.H.L.]

Electric current The net transfer of electric charge per unit time. It is usually measured in amperes. The passage of electric current involves a transfer of energy. Except in the case of superconductivity, a current always heats the medium through which it passes. *See* JOULE'S LAW.

On the other hand, a stream of electrons or ions in a vacuum, which also may be regarded as an electric current, produces no local heating. Measurable currents range in magnitude from the nearly instantaneous 10^5 or so amperes in lightning strokes to values of the order of 10^{-16} ampere, which occur in research applications.

All matter may be classified as conducting, semiconducting, or insulating, depending upon the ease with which electric current is transmitted through it. Most metals, electrolytic solutions, and highly ionized gases are conductors. Transition elements, such as silicon and germanium, are semiconductors, while most other substances are insulators. *See* ALTERNATING CURRENT; CONDUCTION (ELECTRICITY); DIRECT CURRENT; DISPLACEMENT CURRENT; ELECTRIC INSULATOR; SEMICONDUCTOR; SUPERCONDUCTIVITY. [J.W.St.]

Electric-discharge machining A metal-removal process (also called electric-spark machining) in which materials that conduct electricity can be removed by an electric spark. The spark is a transient electric discharge through the space between the tool (cathode) and the workpiece (anode). It is used to form holes of varied shape in materials of poor machinability and to form cavities in steel dies. [O.W.B.]

Electric distribution system That part of an electric power system that supplies electric energy to the individual user or consumer. The distribution system includes the primary circuits and the distribution substations that supply them; the distribution transformers; the secondary circuits, including the services to the consumer; and appropriate protective and control devices. The four general classes of individual users are residential, industrial, commercial, and rural.

The three-phase, alternating-current (ac) system is practically universal, although a small amount of two-phase and direct-current systems from early days are still in operation. Three-phase transmission and subtransmission lines require three wires, termed phase conductors. Most of the three-phase distribution systems consist of three phase conductors and a common or neutral conductor, making a total of four wires. Single-phase branches (consisting of two wires) supplied from three-phase mains are used for single-phase utilization in residences, small stores, and farms. Loads are connected in parallel to common supply circuits. *See* ELECTRIC POWER SYSTEMS.

The distribution substation is an assemblage of equipment for the purpose of switching, changing, and regulating the voltage from subtransmission to primary distribution. More important substations are designed so that the failure of a piece of equipment in the substation or one of the subtransmission lines to the substation will not cause an interruption of power to the load. *See* ELECTRIC POWER SUBSTATION.

The primary system leaving the substation is most frequently in the 11,000–15,000 volt range. A particular voltage used is 12,470-volt line-to-line and 7200-volt line-to-neutral (conventionally written 12,470Y/7200 volts). Some utilities use a lower voltage, such as 4160Y/2400 volts. Secondary voltages are derived from distribution transformers connected to the primary system and they usually correspond to utilization voltages. Residential and most rural loads are supplied by 120/240-volt single-phase three-wire systems. Commercial and small industrial needs are supplied by either 208Y/120-volt or 480Y/277-volt three-phase four-wire systems. The secondary voltage is usually used to supply multiple street lights, in addition to service to consumers.

Service continuity is the providing of uninterrupted electric power to the consumer; therefore, good continuity is doing this a high percentage of the time. This is accomplished for large industrial and commercial loads by use of some form of duplicate power supply. Downtown commercial areas are supplied from three-phase 208Y/120-volt grid networks. The system is arranged so that the failure of a primary feeder will not cause a loss of load on the secondary. Commercial buildings and shopping centers are often served by spot networks. All of the transformers and protectors are at the same location. Residential and rural loads are usually supplied by a radial system. Good continuity for them is obtained by sectionalizing the system with fuses, circuit breakers, and manual switches to reduce the extent of an outage due to a failure.

The majority of distribution systems already built in residential, industrial, and rural areas has been overhead construction. For aesthetic reasons, underground distribution systems are being installed in most new residential developments. *See* TRANSMISSION LINES. [H.E.C.]

Electric energy measurement The measurement of the integral, with respect to time, of the power in an electric circuit. The absolute unit of measurement of electric energy is the joule, or the charge in coulombs times the potential difference in volts. The joule, however, is too small (1 watt-second) for use in commercial practice, and the more commonly used unit is the watt-hour (3.6×10^3 joules). The most common measurement application is in the utility field.

Electric energy is one of the most accurately measured commodities sold to the general public. Many methods of measurement, with different degrees of accuracy, are possible. Basically, measurements of electric energy may be classified into two categories, direct-current power and alternating-current power. The fundamental concepts of measurement are, however, the same for both.

There are two types of methods of measuring electric energy: electric instruments and timing means, and electricity meters.

Electric instruments and timing means make use of conventional procedures for measuring electric power and time. Typical methods are listed below. *See* ELECTRIC POWER MEASUREMENT.

1. Measurement of energy on a direct-current circuit by reading the line voltage and load current at regular intervals over a measured period of time. *See* CURRENT MEASUREMENT; VOLTAGE MEASUREMENT.
2. Measurement of energy on a direct-current circuit by controlling the voltage and current at constant predetermined values for a predetermined time interval.
3. Measurement of energy on an alternating-current circuit by reading the watts input to the load at regular intervals over a measured period of time.
4. Measurement of energy on an alternating-current circuit by controlling the voltage, current, and watts input to the load at constant predetermined values.
5. Measurement of energy by recording the watts input to the load on a linear chart progressed uniformly with time.

Electricity meters are the most common devices for measuring the vast quantities of electric energy used by industry and the general public. The same fundamentals of measurement apply as for electric power measurement, but in addition the electricity meter provides the time-integrating means necessary for electric energy measurement. A single meter is sometimes used to measure the energy consumed in two or more circuits. However, multistator meters are generally required for this purpose.

Watt-hour meters are generally connected to measure the losses of their respective current circuits. These losses are extremely small compared to the total energy being measured and are present only under load conditions. Watt-hour meters used for the billing of residential, commercial, and industrial loads are highly developed devices. *See* ELECTRICAL MEASUREMENTS. [W.H.Ha.]

Electric field A condition in space in the vicinity of an electrically charged body such that the forces due to the charge are detectable. An electric field (or electrostatic field) exists in a region if an electric charge at rest in the region experiences a force of electrical origin. Since an electric charge experiences a force if it is in the vicinity of a charged body, there is an electric field surrounding any charged body.

The electric field intensity (or field strength) **E** at a point in an electric field has a magnitude given by the quotient obtained when the force acting on a test charge q' placed at that point is divided by the magnitude of the test charge q'. Thus, it is force per unit charge. A test charge q' is one whose magnitude is small enough so it does not alter the field in which it is placed. The direction of **E** at the point is the direction of the force **F** on a positive test charge placed at the point. Thus, **E** is a vector point function, since it has a definite magnitude and direction at every point in the field, and its defining equation is Eq. (1).

$$\mathbf{E} = \mathbf{F}/q' \qquad (1)$$

Electric flux density or electric displacement **D** in a dielectric (insulating) material is related to **E** by either of the equivalent equations shown as Eqs. (2), where **P** is the polarization of the

$$\mathbf{D} = \epsilon_0\mathbf{E} + \mathbf{P} \qquad \mathbf{D} = \epsilon\mathbf{E} \qquad (2)$$

medium, and ϵ is the permittivity of the dielectric which is related to ϵ_0,by the equation $\epsilon = k\epsilon_0$, k being the relative dielectric constant of the dielectric. In empty space, $\mathbf{D} = \epsilon_0\mathbf{E}$.

In addition to electrostatic fields produced by separations of electric charges, an electric field is also produced by a changing magnetic field. *See* ELECTRIC CHARGE; ELECTROMAGNETIC INDUCTION; POTENTIALS. [R.P.Wi.]

Electric filter A transmission network used in electrical systems for the selective enhancement (or reduction) of specified components of an input signal. The filtering is accomplished by selectively attenuating those frequency components of the input signal which are undesired, relative to those which it is desired to enhance.

A typical filter network having two ports is shown in Fig. 1. Since the output is generally smaller than the input, it is customary to specify the amplitude information as the amount of reduction of the input signal; the loss or attenuation is measured in decibels (dB). *See* DECIBEL.

An optical filter transmits, or passes, certain colors in the light spectrum and rejects, or stops, others. Electrical filters

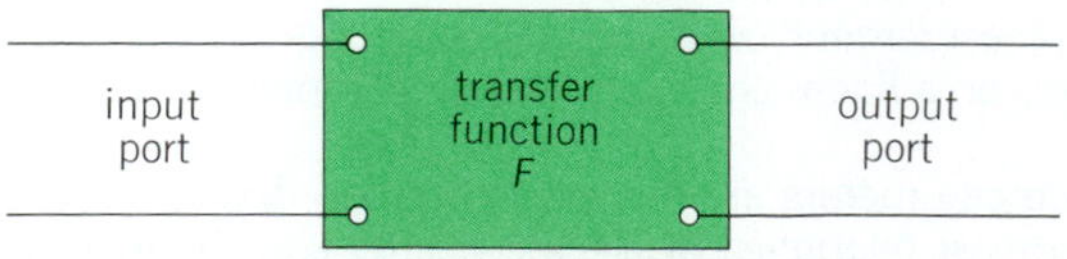

Fig. 1. Two-port filter network.

perform in a similar way. A band of frequencies which is rejected or greatly attenuated is called a stopband; a band which is transmitted with little attenuation is called a passband. Ideally, there should be zero attenuation in the passband and infinite attenuation in the stopband. More practically, a small amount of attention can be tolerated in the passband, and a noninfinite attenuation in the stopband.

Applications. Electrical filters are incorporated in almost all systems of electrical communications and control. Radio and television receivers contain filters in order to pass the particular channel selected and to stop others. The telephone network incorporates millions of filters in order to pass each conversation, which is carried by a different band of frequencies, to the appropriate telephone receiver and to stop all others. Bandwidths vary from a fraction of a hertz in control systems, to 3 kHz in telephone systems, to 20 kHz in AM radio systems, to 6 MHz in television systems. *See* AMPLITUDE-MODULATION DETECTOR; CONTROL SYSTEMS; FREQUENCY-MODULATION DETECTOR; RADIO RECEIVER; TELEVISION RECEIVER.

Narrow-band band-rejection filters are often used to reduce interference from extraneous electrical noise, which may come from power lines or from industrial equipment. One example of such a null network is the twin-tee filter shown in Fig. 2. *See* ELECTRICAL INTERFERENCE.

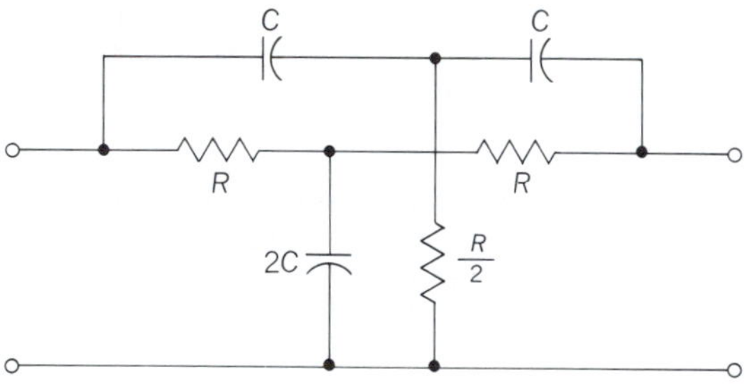

Fig. 2. *RC* twin-tee null filter. *R* = resistance in ohms; *C* = capacity in picofarads.

A common use of a filter is to smooth the output from a rectifier to make it more closely constant by attenuating the alternating-current component. The power supply in every radio and television receiver has such a low-pass filter. Another common use of a low-pass filter is as a decoupling network when several amplifier stages share the same power supply. A low-pass decoupling filter is placed between the power supply and each amplifier stage, thereby reducing interaction between the stages.

Special filters. Special categories of filters can be classified in accordance with either their application or the components used in their construction. Piezoelectric crystalline materials such as quartz have the property of electrical resonance when mechanically stressed along certain axes. This property can be used to replace electrical components in a filter. The resulting crystal filters are almost always used in band-pass applications. The incoming (and outgoing) electrical signals must be coupled to the mechanical vibrations of the crystal through transducers, which are themselves electrical elements. Ceramic materials can also be used in place of crystals; the result is a ceramic filter.

Mechanical filters have some characteristics in common with crystal filters, in that the resonant elements are mechanical. But they differ from crystal filters in that the coupling mechanisms between mechanical vibrations and electrical signals (the transducers) are also mechanical. Ferrites and iron-nickel alloys whose elastic modulus is constant over a temperature range are common materials from which mechanical filters are constructed.

Microwave filters operate at microwave frequencies, generally considered to lie above 1 GHz. However, they can also be used in the much lower high-frequency (hf) range of 3–30 MHz. The components are no longer lumped elements, but consist of sections of transmission lines (called stubs), strip lines, coupled lines, and waveguide structures such as irises and posts. *See* MICROWAVE TRANSMISSION LINES; WAVEGUIDE.

Distributed RC filters are made of thin-film integrated circuits or various large-area semiconductor devices.

A filter that is matched to the characteristics of an input signal of known time dependence is called a matched filter. The term is also applied to describe an impedence match between a filter and its termination. A matched filter has the property that, when the input consists of noise in addition to the desired signal, the output signal-to-noise ratio is the maximum that can be obtained by any linear filter. Accordingly, such filters are used (in radar and similar systems) to optimize the detection of weak signals.

Switched filters are similar to other electrical filters except that a switch is incorporated into one of the branches. The switch is operated (turned on or off) in a periodic manner so that the switched filter is a time-varying linear network. Switched filters find particular application in modulators used in communications systems, both in frequency-division multiplexing (FDM) and in time-division multiplexing (TDM).

Digital filters. An increasing amount of signal processing is being performed in digital form. A digital signal is one which takes on only a limited number of discrete values and only at discrete and equally spaced intervals of time. Signals may already be in digital form (for example, sent between digital computers and most business machines), or a continuous signal may be converted to digital form by sampling it at uniform intervals and digitally coding its values, for ease of transmission or some other reason. This digital signal is then processed by a digital filter in accordance with a set of instructions in order to regain the original information. [N.B.]

Electric furnace An enclosed space heated by electric power. The furnace may be in such forms as a refractory crucible, a large tiltable refractory basin with a capacity of 100 tons (91 metric tons) and a removable roof, or a long insulated chamber equipped with a continuous conveyor. Heat is provided by an arc to the charge or melt (direct-arc furnace), by an arc between electrodes (indirect-arc furnace), or by an arc confined for concentrated heating by an electromagnetic field (plasma-arc furnace). Heat may also be produced by current flowing in the melt. *See* ELECTROMETALLURGY; PYROMETALLURGY.

Because the source of heat is nonchemical, electric furnaces are especially desirable in melting alloys of controlled composition. Temperature is also readily controlled. The arc furnace may be used to smelt ores or to refine metals or alloys. Induction furnaces are widely used to melt alloys for castings. Because electric furnaces can be enclosed, they are used for operations that require controlled or inert atmospheres, such as growing crystals or annealing. When sealed and evacuated, they are used in degassing metals. Furnaces with hearth resistors are used for operations below melting temperatures, such as annealing, and with infrared heat lamps, for drying paints or setting glues. *See* ARC HEATING; ELECTRIC HEATING; HEAT TREATMENT (METALLURGY); KILN; REFRACTORY. [F.H.R.]

Electric heating Methods of converting electric energy to heat energy by resisting the free flow of electric current. Electric heating has several advantages; it can be precisely controlled to allow a uniformity of temperature within very narrow limits; it is cleaner than other methods of heating because it does not involve any combustion; it is considered safe because it is protected from overloading by automatic breakers; it is quick to use and to adjust; and it is relatively quiet. For these reasons, electric heat is widely chosen for industrial, commercial, and residential use. There are four major types of electric heaters: resistance, dielectric, induction, and electric-arc. *See* ARC HEATING; DIELECTRIC HEATING; INDUCTION HEATING; RESISTANCE HEATING.

In the past, electricity has not been a popular choice for heating houses. Power plants lose 60–65% efficiency in generating electricity, and another 10% in transmission and distribution. With only a 35% efficiency rate, electricity has long been overlooked in favor of the direct use of gas and oil for heating homes. As these fossil fuels diminish, however, electric heating is a more attractive option. Also, if nuclear power becomes more widespread, so will electric heating because it will make electric heating far more economical than the direct use of gas and oil.

One type of electric heater for houses is the heat pump. The heat pump can be used alone or to supplement the output of direct resistance heating (heat strips). The heat pump works on a reversed refrigeration cycle; the pump pulls the heat from the cold outside air and forces it into the house. Since the heat pump performs both heating and cooling and is more efficient than other forms of electric heat, it can become the most economical way to heat houses in the future. *See* HEAT PUMP.

Common electric heating systems in houses are: central heating employing an electric furnace with forced air circulation; central heating employing an electric furnace with forced water circulation; central heating using radiant cables; electrical duct heaters; space (strip) heaters which use radiation and natural convection for heat transfer; and portable space heaters. [M.-S.C.]

Electric insulator Ideal insulators are substances into which static electric fields penetrate uniformly. Substances for which Ohm's law is valid are often classed as insulators if their resistivities exceed about 10^{14} ohm-cm. Insulators used primarily as media for the storage of electrostatic energy, for example, in capacitors, are called dielectrics. Important electrical properties of insulators are resistivity, dielectric constant, and breakdown voltage, the applied voltage above which a material ceases to act as an insulator. *See* DIELECTRIC CONSTANT; DIELECTRIC MATERIALS.

At high pressures gases are excellent insulators, with higher breakdown voltages than most solids or liquids or high vacuum. *See* ELECTRICAL CONDUCTION IN GASES.

In solids electric current is carried by electrons or ions. Liquids show many of the features of solids in their conducting mechanisms. *See* BAND THEORY OF SOLIDS. [B.G.D.]

Electric organ (biology) An effector organ found in six different groups of fishes; output is an electric pulse. Voltages large enough to aid in prey capture or predator deterrence are produced by various strongly electric fishes. These include the electric eel (*Electrophorus electricus*) from South America; the electric catfish (*Malapterurus electricus*) from Africa; the family of electric rays, Torpedinidae, which are widely distributed in the world's oceans; and possibly the stargazer genus, *Astroscopus*, of the western Atlantic. Weakly electric fishes emit a lower voltage, the energy source for an active electrosensory system that monitors electrical impedance in the environment. These weak signals also serve in intra- and interspecific communication. There are three groups of weakly electric fishes. First, the South American knifefishes, the Gymnotiformes, are a large and diverse group of several families that also include *Electrophorus*. Second, the electrically active African Mormyriformes are composed of the numerous species of the family Mormyridae and the single species *Gymnarchus niloticus* in the family Gymnarchidae. Finally, many species of skates and rays of the family Rajidae occur in marine waters around the world.

The strongly electric fishes are remarkable for the high voltage and power of their discharges. For example, the electric eel generates pulses in excess of 500 V. In contrast, a large torpedo generates a smaller voltage, about 50 V in air, but the maximum current is larger, and the peak pulse power can exceed 1 kW (or about 1 hp). The electric organs may constitute a substantial fraction of the body mass. The weakly electric fishes have much smaller organs emitting pulses of a few tenths of a volt to several volts; such amplitudes that are still large compared to those recorded outside ordinary, nonelectric fishes.

Electric organ discharge is explicable in terms of the properties of ordinary excitable cells, that is, those of nerve and muscle. The single generating cells of electric organs are called electrocytes. They are modified striated muscle fibers that have lost the ability to contract, although they retain muscle filaments to varying degrees, and they produce electric signals as muscle fibers do.

A question often asked about the strongly electric fishes is how they keep from electrocuting themselves. When immersed in water, they do not appear to be at all affected by their own discharges, although their electroreceptors are undoubtedly activated. In air, when the organ is not loaded down by the conductivity of the water, they often do twitch when they discharge their organs, but as would be expected, they are much more resistant than other fishes. One factor is that their nerves and central nervous system are well insulated by many layers of connective and fatty tissue, Whether their brains are more resistant to stunning by a given current density has not been determined; certainly the excitability of individual cells is no different from that in animals in general. [M.V.L.B.]

Electric power generation The production of bulk electric power for industrial, residential, and rural use. Although limited amounts of electricity can be generated by many means, including chemical reaction (as in batteries) and engine-driven generators (as in automobiles and airplanes), electric power generation generally implies large-scale production of electric power in stationary plants designed for that purpose. The generating units in these plants convert energy from failing water, coal, natural gas, oil, and nuclear fuels to electric energy. Most electric generators are driven either by hydraulic turbines, for conversion of failing water energy; or by steam or gas turbines, for conversion of fuel energy. Limited use is being made of geothermal energy, and work is progressing in the use of solar energy in its various forms. Electric power generating plants are normally interconnected by a transmission and distribution system to serve the electric loads in a given area or region. *See* ELECTRIC DISTRIBUTION SYSTEM; GENERATOR.

An electric load is the power requirement of any device or equipment that converts electric energy into light, heat, or mechanical energy, or otherwise consumes electric energy as in aluminum reduction, or the power requirements of electronic and control devices. The total load on any power system is seldom constant; rather, it varies widely with hourly, weekly, monthly, or annual changes in the requirements of the area served. The minimum system load for a given period is termed the base load or the unity load-factor component. Maximum loads, resulting usually from temporary conditions, are called peak loads. Electric energy cannot feasibly be stored in large quantities; therefore the operation of the generating plants must be closely coordinated with fluctuations in the load.

Generating plants, often termed generating stations, contain apparatus that converts some form of energy to electric energy in bulk. Three significant types of generating plants are hydroelectric, fossil-fuel-electric, and nuclear electric.

The hydroelectric plant utilizes the potential energy released by the weight of water falling through a vertical distance called head. The plant consists basically of a dam to store the water in a forebay and create part or all of the head, a penstock to deliver the failing water to the turbine, a hydraulic turbine to convert the hydraulic energy released to mechanical energy, an alternating-current generator (alternator) to convert the mechanical energy to electric energy, and all accessory equipment necessary to control the power flow, voltage, and frequency, and to afford the protection required.

Pumped storage hydroelectric plants are being used increasingly. Under suitable geographical and geological conditions, electric energy can, in effect, be stored by pumping water from a low to a higher elevation and subsequently releasing this water to the lower elevation through hydraulic turbines. These turbines and their associated generators are reversible. The generators, operating in reverse direction as motors, drive their turbines as pumps to elevate the water. When this water is released through the turbines, electric power is produced by the generators. *See* HYDRAULIC TURBINE; WATERPOWER.

The fossil-fuel-electric plant utilizes the energy of combustion from coal, oil, or natural gas. A typical large plant consists of fuel processing and handling facilities, a combustion furnace and boiler to produce and superheat the steam, a steam turbine, an alternator, and the accessory equipment required for plant protection and for control of voltage, frequency, and power flow. A steam plant can frequently be built near a convenient load center, provided an adequate supply of cooling water and fuel is available, and is usually adaptable to either base loading or intermediate or peak loading. *See* STEAM TURBINE.

In the nuclear electric plant one or more of the nuclear fuels are utilized in a suitable type of nuclear reactor, which takes the place of the combustion furnace in the typical steam electric plant. The heat exchangers and boilers (if not combined in the reactor), the turbines, and alternating-current generators,complete with controls, accessories, and auxiliaries, make up the atomic electric plant. *See* NUCLEAR REACTOR.

The size or capacity of electric utility generating units varies widely, depending upon type of unit; duty required, that is base-, intermediate-, or peak-load service; and system size and degree of interconnection with neighboring systems. Base-load nuclear or coal-fired units may be as large as 1200 MW each, or more. Intermediate-duty generators, usually coal-, oil-, or gas-fueled steam units, are typically of 200 to 600 MW capacity each. Peaking units, combustion turbines or hydro, range from several tens of megawatts for the former to hundreds of megawatts for the latter. Hydro units, in both base-load and intermediate service, range in size up to 700 MW.

The total installed generating capacity of a system is typically 20 to 30% greater than the annual predicted peak load in order to provide reserves for maintenance and contingencies.

Voltage regulation is defined as change in voltage for specific change in load (usually from full load to no load) expressed as percentage of normal rated voltage. The voltage of an electric generator varies with the load and power factor; consequently, some form of regulating equipment is required to maintain a reasonably constant and predetermined potential at the distribution stations or load centers. Since the inherent regulation of most alternating-current generators is rather poor (that is, high percentagewise), it is necessary to provide automatic voltage control. The rotating or magnetic amplifiers and voltage-sensitive circuits of the automatic regulators, together with the exciters, are all specially designed to respond quickly to changes in the alternator voltage and to make the necessary changes in the main exciter output, thus providing the required adjustments in voltage. Electronic voltage control has been adapted to some generator and synchronous condenser installations. *See* VOLTAGE REGULATOR.

Computer-assisted (or online-controlled) load and frequency control and economic dispatch systems of generation supervision are being widely adopted, particularly for the larger new plants. Strong system interconnections greatly improve bulk power supply reliability but require special automatic controls to ensure adequate generation and transmission stability. Among the refinements found necessary in large, long-distance interconnections are special feedback controls applied to generator high-speed excitation and voltage regulator systems. *See* ALTERNATING-CURRENT GENERATOR; ELECTRIC POWER SYSTEMS. [E.C.S.]

Electric power measurement The measurement of the time rate at which work is done or energy is dissipated in an electric system. The work done in moving an electric charge is proportional to the charge and the voltage drop through which it moves. Charge per unit time defines electric current; electric power p is therefore defined as the product of the current i in a circuit and the voltage e across its terminals at a given instant. Expressed symbolically, $p = ei$.

A second important definition of power follows directly from Ohm's law. $p = i^2R$, where R is the resistance of the circuit.

The practical unit of electric power is the watt. The watt represents a rate of expending energy, and thus it is related to all other units of power; for example, in mechanics 1 watt = 1 joule/s and 746 watts = 1 horsepower. Commonly used small units are the milliwatt (0.001 watt) and the microwatt (0.000001 watt). Large units are the kilowatt (1000 watts) and the megawatt (1,000,000 watts).

Power measurements must cover the frequency spectrum from direct current through the conventional power frequencies, the audio and the lower radio frequencies, to the highest frequencies (up to 25,000 gigahertz). In general, different techniques are required in each frequency range. *See* ELECTRICAL MEASUREMENTS.

DC and ac power frequencies. In the measurement of power in a dc circuit not subject to rapid fluctuations, there is usually no difficulty in making simultaneous observations of the true values of voltage and current using common types of dc voltmeters and ammeters. The product of these observations then gives a sufficiently accurate measure of power in the given circuit, except that, if great accuracy is required, allowance must be made for the power used by the instruments.

In circuits with rapidly varying direct currents, pulsating rectified current, or, in general, alternating currents, the continuous averaging over short periods of time and the automatic multiplication of current and voltage values is accomplished by the wattmeter. *See* WATTMETER.

In ac circuits with steady effective values of voltage and current, the voltmeter-ammeter method may be used as in the dc case, except that, of course, ac meters are used, and a phase meter is also required to measure phase angle unless current and voltage are in phase. Because ac ammeters and voltmeters actually measure root-mean-square, or effective, values, these lead directly to values of average power.

Summation of power in the separate phases of a polyphase circuit is accomplished by combinations of single-phase wattmeters, or wattmeter elements, disposed according to the general rule, called Blondel's theorem, as follows: If energy is supplied to any system of conductors through N wires, the total power in the system is given by the algebraic sum of N wattmeters, so arranged that each of the N wires contains one current coil, the corresponding potential coil being connected between that wire and some point on the system which is common to all the potential circuits. If this common point is on one of the N wires and coincides with the point of attachment of the potential lead to that wire, the measurement may be effected by the use of $N - 1$ wattmeters.

Audio and radio frequencies. At frequencies above those used in the ordinary power-distribution systems, dynamometer-type wattmeters become inaccurate. For measurements of transmitted power at audio and the lower radio frequencies, no generally satisfactory substitute has been developed, and so measurements are confined to determinations of power dissipated in a load, or available from a source, and are deduced from measurements of impedance and current or voltage.

Power-output meters combine resistive loads, which can be adjusted to various known values, and voltmeters calibrated to indicate the power dissipated therein. They are used at audio frequencies to determine the maximum power output that can be obtained from a source, or to measure output power in studies of harmonic distortion, intermodulation, overload, frequency characteristic, and so forth.

At radio frequencies the principle of impedance matching is frequently used to determine the power input to a receiver, with a standard-signal generator serving as source.

The power dissipated in a resistive load can be determined from the heat developed therein, and several ingenious schemes for measuring this heat have been derived. A widely used technique depends upon dissipating the power in a resistive element of relatively high temperature coefficient of resistivity. The power is deduced from the change in resistance by one of several methods. Measurements of this general nature are known as bolometric methods. *See* BOLOMETER.

Microwave frequencies. Measurement of dissipated power at microwave frequencies can be carried out by methods similar to those used at lower frequencies. Because voltage and impedance become increasingly difficult to define precisely as the frequency is increased, however, calorimetric methods generally, and bolometric methods specifically, are almost universally used. [D.B.S.]

Electric power substation An assembly of equipment in an electric power system through which electrical energy is passed for transmission, distribution, interconnection, transformation, conversion, or switching. *See* ELECTRIC POWER SYSTEMS.

Specifically, substations are used for some or all of the following purposes: connection of generators, transmission or distribution lines, and loads to each other; transformation of power from one voltage level to another; interconnection of alternate sources of power; switching for alternate connections and isolation of failed or overloaded lines and equipment; controlling system voltage and power flow; reactive power compensation; suppression of overvoltage; and detection of faults, monitoring, recording of information, power measurements, and remote communications. Minor distribution or transmission equipment installation is not referred to as a substation.

Substations are referred to by the main duty they perform. Broadly speaking, they are classified as: transmission substations, which are associated with high voltage levels; and distribution substations, associated with low voltage levels. *See* ELECTRIC DISTRIBUTION SYSTEM.

Substations are also referred to in a variety of other ways:

1. Transformer substations are substations whose equipment includes transformers.

2. Switching substations are substations whose equipment is mainly for various connections and interconnections, and does not include transformers.

3. Customer substations are usually distribution substations on the premises of a larger customer, such as a shopping center, large office or commercial building, or industrial plant.

4. Converter stations are complex substations required for high-voltage direct-current (HVDC) transmission or interconnection of two ac systems which, for a variety of reasons, cannot be connected by an ac connection. The main function of converter stations is the conversion of power from ac to dc and vice versa. The main equipment includes converter valves usually located inside a large hall, transformers, filters, reactors, and capacitors.

5. Most substations are installed as air-insulated substations, implying that the bus-bars and equipment terminations are generally open to the air, and utilize insulation properties of ambient air for insulation to ground. Modern substations in urban areas are esthetically designed with low profiles and often within walls, or even indoors.

6. Metal-clad substations are also air-insulated, but for low voltage levels; they are housed in metal cabinets and may be indoors or outdoors.

7. Acquiring a substation site in an urban area is very difficult because land is either unavailable or very expensive. Therefore, there has been a trend toward increasing use of gas-insulated substations, which occupy only 5–20% of the space occupied by the air-insulated substations. In gas-insulated substations, all live equipment and bus-bars are housed in grounded metal enclosures, which are sealed and filled with sulfur hexafluoride (SF_6) gas, which has excellent insulation properties.

8. For emergency replacement or maintenance of substation transformers, mobile substations are used by some utilities.

An appropriate switching arrangement for "connections" of generators, transformers, lines, and other major equipment is basic to any substation design. There are seven switching arrangements commonly used: single bus; double bus, single breaker; double bus, double breaker; main and transfer bus; ring bus; breaker-and-a-half; and breaker-and-a-third. Each breaker is usually accompanied by two disconnect switches, one on each side, for maintenance purposes. Selecting the switching arrangement involves considerations of cost, reliability, maintenance, and flexibility for expansion.

A substation includes a variety of equipment. The principal items are transformers, circuit breakers, disconnect switches, bus-bars, shunt reactors, shunt capacitors, current and potential transformers, and control and protection equipment. *See* Bus-bar; Circuit breaker; Electric protective devices; Electric switch; Relay; Transformer; Voltage regulator.

Good substation grounding is very important for effective relaying and insulation of equipment; but the safety of the personnel is the governing criterion in the design of substation grounding. It usually consists of a bare wire grid, laid in the ground; all equipment grounding points, tanks, support structures, fences, shielding wires and poles, and so forth, are securely connected to it. The grounding resistance is reduced enough that a fault from high voltage to ground does not create such high potential gradients on the ground, and from the structures to ground, to present a safety hazard. Good overhead shielding is also essential for outdoor substations, so as to virtually eliminate the possibility of lightning directly striking the equipment. Shielding is provided by overhead ground wires stretched across the substation or tall grounded poles. *See* Grounding; Lightning and surge protection. [N.G.H.]

Electric power systems A complex assemblage of equipment and circuits for generating, transmitting, transforming, and distributing electrical energy.

Electricity in the large quantities required to supply electric power systems is produced in generating stations, commonly called power plants. Such generating stations, however, should be considered as conversion facilities in which the heat energy of fuel (coal, oil, gas, or uranium) or the hydraulic energy of falling water is converted to electricity. *See* Electric power generation; Hydraulic turbine; Nuclear reactor; Power plant; Steam turbine.

The transmission system carries electric power efficiently and in large amounts from generating stations to consumption areas. Such transmission is also used to interconnect adjacent power systems for mutual assistance in case of emergency and to gain for the interconnected power systems the economies possible in regional operation.

A relatively new approach to high-voltage long-distance transmission is high-voltage direct current (HVDC), which offers the advantages of less costly lines, lower transmission losses, and insensitivity to many system problems that restrict alternating-current systems. Its greatest disadvantage is the need for costly equipment for converting the sending-end power to direct current, and for converting the receiving-end direct-current power to alternating current for distribution to consumers. *See* Transmission lines.

As systems grow and the number and size of generating units increase, and as transmission networks expand, higher levels of bulk-power-system reliability are attained through properly coordinated interconnections among separate systems. Most of the electric utilities in the contiguous United States and a large part of Canada now operate as members of power pools, and these pools (except one in Texas) in turn are interconnected into one gigantic power grid known as the North American Power Systems Interconnection. The operation of this interconnection, in turn, is coordinated by the North American Power Systems Interconnection Committee (NAPSIC). Each individual utility in such pools operates independently, but has contractual arrangements with other members in respect to generation additions and scheduling of operation. Their participation in a power pool affords a higher level of service reliability and important economic advantages.

The majority of large electric systems have computerized energy centers whose functions are to control operation of the system so as to optimize economy and security. In some large interconnected networks of individual utility companies, such as those in the Pennsylvania–New Jersey–Maryland power pool, regional control centers have been established to perform the same function for a group of companies. New generation centers now use digital computers, or hybrid analog-digital units, rather than the analog machines previously used, and redundant systems are required to ensure security.

Power delivered by transmission circuits must be stepped down in facilities called substations to voltages more suitable for use in industrial and residential areas. *See* Electric power substation.

That part of the electric power system that takes power from a bulk-power substation to customers' switches, commonly about 35% of the total plant investment, is called distribution. *See* Electric distribution system. [W.C.H.]

Electric power transmission The transport of generator-produced electric energy to loads. An electric power transmission system interconnects generators and loads and generally provides multiple paths among them. Multiple paths increase system reliability because the failure of one line does not cause a system failure. Most transmission lines operate with three-phase alternating current (ac). The standard frequency in North America is 60 Hz; in Europe, 50 Hz. The three-phase system has three sets of phase conductors. Long-distance energy transmission occasionally uses high-voltage direct-current (dc) lines. *See* Alternating current; Direct current; Direct-current transmission.

The electric power system can be divided into the distribution, subtransmission, and transmission systems. With operating voltages less than 34.5 kV, the distribution system carries energy from the local substation to individual households, using both overhead and underground lines. With operating voltages of 69–138 kV, the subtransmission system distributes energy within an entire district and regularly uses overhead lines. With operating voltage exceeding 230 kV, the transmission system interconnects generating stations and large substations located close to load centers by using overhead lines. *See* Transmission lines.

Overhead alternating-current transmission. Overhead transmission lines distribute the majority of the electric energy in the system. A typical high-voltage line has three phase conductors to carry the current and transport the energy, and two grounded shield conductors to protect the line from direct lightning strikes. The usually bare conductors are insulated from the supporting towers by insulators attached to grounded towers or poles. Lower-voltage lines use post insulators, while the high-voltage lines are built with insulator chains or long-rod composite insulators. The normal distance between the supporting towers is a few hundred feet.

Transmission lines use ACSR (aluminum cable, steel reinforced) and ACAR (aluminum cable, alloy reinforced) conductors. In an ACSR conductor, a stranded steel core carries the mechanical load, and layers of stranded aluminum surrounding the core carry the current. An ACAR conductor is a stranded cable made of an aluminum alloy with low resistance and high mechanical strength. ACSR conductors are usually used for high-voltage lines, and ACAR conductors for subtransmission and distribution lines. Ultrahigh-voltage (UHV) and extrahigh-voltage (EHV) lines use bundle conductors. Each phase of the line is built with two, three, or four conductors connected in parallel and separated by about 1.5 ft (0.5 m). Bundle conductors reduce corona discharge. *See* CONDUCTOR (ELECTRICITY).

Transmission lines are subject to environmental adversities, including wide variations of temperature, high winds, and ice and snow deposits. Typically designed to withstand environmental stresses occurring once every 50–100 years, lines are intended to operate safely in adverse conditions.

Variable weather affects line operation. Extreme weather reduces corona inception voltage, leading to an increase in audible noise, radio noise, and telephone interference. Load variation requires regulation of line voltage. A short circuit generates large currents, overheating conductors and producing permanent damage.

The power that a line can transport is limited by the line's electrical parameters. Voltage drop is the most important factor for distribution lines; where the line is supplied from only one end, the permitted voltage drop is about 5%.

Conductor temperature must be lower than the temperature which causes permanent elongation. A typical maximum steady-state value for ACSR is 212°F (100°C), but in an emergency temperatures 10–20% higher are allowed for a short period of time (10 min to 1 h).

Corona discharge is generated when the electric field at the surface of the conductor becomes larger than the breakdown strength of the air. The oscillatory nature of the discharge generates high-frequency, short-duration current pulses, the source of corona-generated radio and television interference. Surface irregularities such as water droplets cause local field concentration, enhancing corona generation. Thus, during bad weather, corona discharge is more intense and losses are much greater. Corona discharge also generates audible noise with two components: a broad-band, high-frequency component, which produces crackling and hissing, and a 120-Hz pure tone. *See* CORONA DISCHARGE; ELECTRICAL INTERFERENCE.

Transmission-line conductors are surrounded by an electric field which decreases as distance from the line increases, and depends on line voltage and geometry. At ground level, this field induces current and voltage in grounded bodies, causes corona in grounded objects, and can induce fuel ignition. Utilities limit the electric field at the perimeter of right-of-ways to about 1000 V/m. An ac magnetic field around the transmission line also decreases with distance from the line. *See* ELECTRIC FIELD; MAGNETIC FIELD.

Lightning strikes produce high voltages and traveling waves on transmission lines, causing insulator flashovers and interruption of operation. Steel grounded shield conductors at the tops of the towers significantly reduce, but do not eliminate, the probability of direct lightning strikes to phase conductors. *See* LIGHTNING AND SURGE PROTECTION.

The operation of circuit breakers causes switching surges that can result in interruption of inductive current, energization of lines with trapped charges, and single-phase ground fault. Modern circuit breakers, operating in two steps, reduce switching surges to 1.5–2 times the 60-Hz voltage. *See* CIRCUIT BREAKER.

Line current induces a disturbing voltage in telephone lines running parallel to transmission lines. Because the induced voltage depends on the mutual inductance between the two lines, disturbance can be reduced by increasing the distance between the lines and shielding the telephone lines. *See* ELECTRICAL SHIELDING; INDUCTIVE COORDINATION.

Underground power transmission. Most cities use underground cables to distribute electrical energy. These cables virtually eliminate negative environmental effects and reduce electrocution hazards. However, they entail significantly higher construction costs.

Underground cables are divided into two categories: distribution cables (less than 69 kV) and high-voltage power-transmission cables (69–500 kV).

Extruded solid dielectric cables dominate in the 15–33-kV urban distribution system. In a typical arrangement (illus. *a*), the stranded copper or aluminum conductor is shielded by a semiconductor layer, which reduces the electric stress on the conductor's surface. Oil-impregnated paper-insulated distribution cables are used for higher voltages and in older installations (illus. *b*).

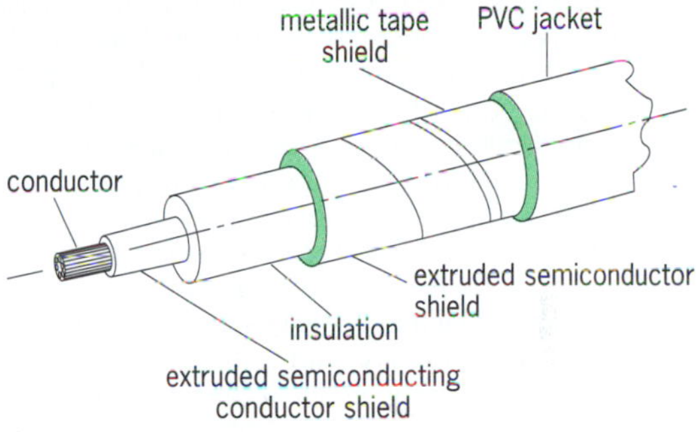

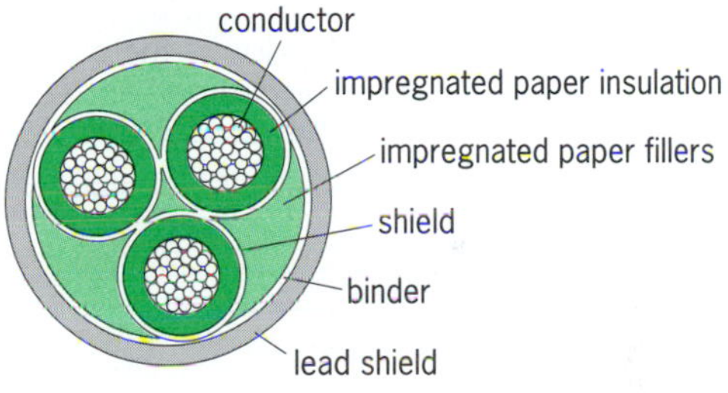

Underground distribution cables. (*a*) Extruded solid dielectric cable. (*b*) Three-phase oil-impregnated paper-insulated cable. (*After R. Bartnikas Srivastava, Elements of Cable Engineering, Sandford Educational Press, 1980*)

Cable temperatures vary with load changes, and cyclic thermal expansion and contraction may produce voids in the cable. High voltage initiates corona in the voids, gradually destroying cable insulation. Low-pressure oil-filled cable construction reduces void formation. A single-phase concentric cable has a hollow conductor with a central oil channel. Three-phase cables have three oil channels located in the filler.

Submarine cables. High-voltage cables are frequently used for crossing large bodies of water. Water provides natural cooling, and pressure reduces the possibility of void formation. A typical submarine cable has cross-linked polyethylene insulation, and corrosion-resistant aluminum alloy wire armoring that provides tensile strength and permits installation in deep water. *See* ELECTRIC POWER SYSTEMS. [G.G.K.]

Electric protective devices Equipment applied to electric power systems to detect abnormal and intolerable conditions and to initiate appropriate corrective action. Disturbances in the normal operation of electric power systems may be caused by natural phenomena, such as lightning, wind, or snow; by accidental means traceable to reckless drivers, inadvertent acts by plant maintenance personnel, or other acts of human beings; or by conditions produced in the system itself, such as switching surges, load swings, or equipment failure.

Protective devices must therefore be installed on a power system to ensure continuity of electrical service, to limit injury to personnel, and to limit damage to equipment when abnormal situations develop.

Protective devices, like any type of insurance, are applied commensurately with the degree of protection desired. For this reason, application of protective devices varies widely.

For the purpose of applying protection, the electric power system is divided into five major protection zones: generators; transformers; buses; transmission and distribution lines; and motors. Each block represents a set of protective relays and associated equipment selected to initiate correction or isolation of that area for all anticipated intolerable conditions or trouble. The detection is done by protective relays with a circuit breaker used to physically disconnect the equipment. For other areas of protection *see* GROUNDING; LIGHTNING AND SURGE PROTECTION.

Protective relays. These are compact analog or digital networks connected throughout the system to detect intolerable conditions within their assigned area or zone. They operate on voltage, current, current direction, power factor, power, impedance, temperature, and so forth, as well as combinations of these. System faults for which the relays respond are generally short circuits between the phase conductors, or between the phases and grounds. Some relays operate on unbalances between the phases, such as an open or reversed phase.

The most fundamental and widely used protection technique is the differential principle. The current flowing into the equipment to be protected is compared with the current flowing out. For normal and permissible operation, currents all sum to essentially zero. However, for internal trouble, they add up to flow through the relay.

While the application and protection principles are the same, the relay units may be either electromechanical or solid-state (also known as static) type. The plunger type (Fig. 1) is composed of a coil, plunger, and set of contacts. When current I flows in the coil, a force is produced that causes the plunger to move and close the relay contacts. The electromagnetic induction-disk relay (Fig. 2) responds to alternating current only. The main coil is connected to an external source. When current flows in the main coil, transformer action induces current in the secondary circuit connected to the upper poles. Fluxes produced by the currents flowing in the upper pole circuit induce eddy currents in the rotor disk. Interaction between rotor eddy currents and the flux from the lower pole produces torque on the rotor, causing it to move and thus closing the contacts. *See* RELAY.

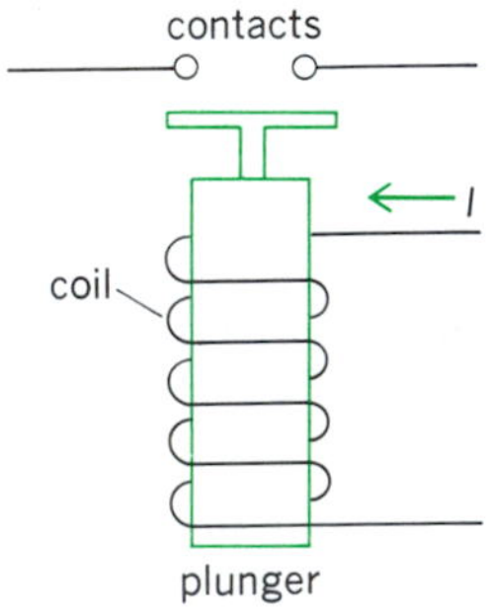

Fig. 1. The plunger relay, an electromechanical relay, operates on the principle of electromagnetic attraction.

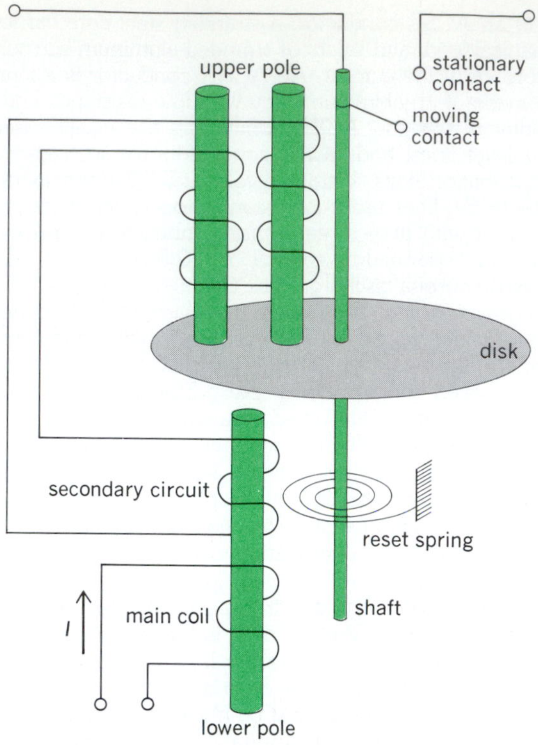

Fig. 2. An induction type of electromagnetic relay.

Overcurrent protection. This must be provided on all systems to prevent abnormally high currents from overheating and causing mechanical stress on equipment. Overcurrent in a power system usually indicates that current is being diverted from its normal path by a short circuit. In low-voltage, distribution-type circuits, such as those found in homes, adequate overcurrent protection can be provided by fuses that melt when current exceeds a predetermined value. *See* FUSE (ELECTRICITY).

Small thermal-type circuit breakers also provide overcurrent protection for this class of circuit. As the size of circuits and systems increases, the problems associated with interruption of large fault currents dictate the use of power circuit breakers. *See* CIRCUIT BREAKER.

Overvoltage protection. Lightning near power lines can cause very-short-time overvoltages in the system and possible breakdown of the insulation. Protection for these surges consists of lightning arresters connected between the lines and ground. Normally the insulation through these arresters prevents current flow, but they momentarily pass current during the high-voltage transient to limit overvoltage. Overvoltage protection is seldom applied elsewhere except at the generators, where it is part of the voltage regulator and control system.

Undervoltage protection. This must be provided on circuits supplying power to motor loads. Low-voltage conditions cause motors to draw excessive currents, which can damage the motors. If a low-voltage condition develops while the motor is running, the relay senses this condition and removes the motor from service.

Underfrequency protection. A loss or deficiency in the generation supply, the transmission lines, or other components of the system, resulting primarily from faults, can leave the system with an excess of load. Solid-state and digital-type underfrequency relays are connected at various points in the system to detect the resulting decline in the normal system frequency. They operate to disconnect loads or to separate the system into areas so that the available generation equals the load until a balance is reestablished.

Reverse-current protection. This is provided when a change in the normal direction of current indicates an abnormal condition in the system. In an ac circuit, reverse current implies a phase shift of the current of nearly 180° from normal. This is actually a change in direction of power flow and can be detected by ac directional relays.

Phase unbalance protection. This protection is used on feeders supplying motors where there is a possibility of one phase opening as a result of a fuse failure or a connector failure. One type of relay compares the current in one phase against the currents in the other phases. Another type monitors the three-phase bus voltages for unbalance.

Reverse-phase-rotation protection. Where direction of rotation is important, electric motors must be protected against phase reversal. A reverse-phase-rotation relay is applied to sense the phase rotation. This relay is a miniature three-phase motor with the same desired direction of rotation as the motor it is protecting. If the direction of rotation is correct, the relay will let the motor start. If incorrect, the sensing relay will prevent the motor starter from operating.

Thermal protection. Motors and generators are particularly subject to overheating due to overloading and mechanical friction. Temperature-sensitive elements, located inside the machine, form part of a bridge circuit used to supply current to a relay. When a predetermined temperature is reached, the relay operates, initiating opening of a circuit breaker or sounding of an alarm. [J.L.Bl.]

Electric rotating machinery

Any form of apparatus, having a rotating member, which generates, converts, transforms, or modifies electric power. The most common forms are motors, generators, synchronous condensers, synchronous converters, rotating amplifiers, phase modifiers, and combinations of these in one machine.

Most rotating machines consist of a stationary member, called the stator, and a rotating member, called the rotor. The rotor may be supported in bearings at both ends, or it may be supported at one or both ends by the shaft of another machine. *See* GENERATOR; MOTOR.

To function properly, rotating machines must have a magnetic circuit, usually involving both rotor and stator, and one or more insulated electrical circuits which interlink the magnetic circuit. To afford a low-reluctance magnetic path, the rotor and stator are separated only by a small clearance, called the air gap.

Rotating machinery must be ventilated to avoid overheating from internal losses. The principal cooling medium, usually air or hydrogen, is circulated by fans or blowers mounted on the rotor or separately driven. With conventional cooling, the cooling medium is blown over exposed surfaces of the insulated windings and core. In conductor cooling, the cooling medium flows in ducts within the major insulation wall. In large machines the superiority of conductor cooling is essential.

In all rotating machines, losses occur. Among them are I^2R losses, called copper losses, in the windings, connections, and brushes; stray load losses in windings, solid metal structures, and frame; core loss in the magnetic material and structural parts; windage and friction loss; and exciter and rheostat losses. *See* CORE LOSS. [L.T.R.]

Electric spark

A transient form of gaseous conduction. This type of discharge is difficult to define, and no universally accepted definition exists. It can perhaps best be thought of as the transition between two more or less stable forms of gaseous conduction. For example, the transitional breakdown which occurs in the transition from a glow to an arc discharge may be thought of as a spark. *See* ELECTRICAL CONDUCTION IN GASES.

Electric sparks play an important part in many physical effects. Usually these are harmful and undesirable effects, ranging from the gradual destruction of contacts in a conventional electrical switch to the large-scale havoc resulting from lightning discharges. Sometimes, however, the spark may be very useful. Examples are its function in the ignition system of an automobile, its use as an intense short-duration illumination source in high-speed photography, and its use as a source of excitation in spectroscopy. In the second case the spark may actually perform the function of the camera shutter, because its extinction renders the camera insensitive. *See* SPECTROSCOPY; STROBOSCOPIC PHOTOGRAPHY. [G.H.M.]

Electric susceptibility

A dimensionless parameter measuring the ease of polarization of a dielectric. The susceptibility χ is equal to the ratio of polarization $\mathbf{P}$ to the product of electric field strength $\mathbf{E}$ and vacuum permittivity ϵ_0, as in Eq. (1)

$$\chi = \mathbf{P}/\epsilon_0\mathbf{E} \qquad (1)$$

[$\epsilon_0 = 1$ in cgs electrostatic units, and 8.854×10^{-12} farad/m in mks units].

The electric susceptibility is related to the relative permittivity ϵ_r by Eq. (2), where γ is a geometrical factor equal to 4π or 1 in

$$\chi = (\epsilon_r - 1)/\gamma \qquad (2)$$

the cgs or mks systems, respectively. This ambiguity in the definition of χ is unfortunate and must be considered in evaluating published data. *See* DIELECTRIC CONSTANT; ELECTRICAL UNITS AND STANDARDS; POLARIZATION OF DIELECTRICS. [R.D.W.]

Electric switch

A device that makes, breaks, or changes the course of an electric circuit. Basically, an electric switch consists of two or more contacts mounted on an insulating structure and arranged so that they can be moved into and out of contact with each other by a suitable operating mechanism.

The term switch is usually used to denote only those devices intended to function when the circuit is energized or deenergized under normal manual operating conditions; as contrasted with circuit breakers, which have as one of their primary functions the interruption of short-circuit currents. Although there are hundreds of types of electric switches, their application can be broadly classified into two major categories: power and signal.

In power applications, switches function to energize or deenergize an electric load. On the low end of the power scale, wall switches are used in homes and offices for turning lights on and off; dial and push-button switches control power to electric ranges, washing machines, and dishwashers. On the high end of the scale are load-break switches and disconnecting switches in power systems at the highest voltages (several hundred thousand volts).

For power applications, when closed, switches are required to carry a certain amount of continuous current without overheating, and in the open position they must provide enough insulation to isolate the circuit electrically.

Load-break switches are required also to have the capability of interrupting the load current. Although this requirement is easily met in low-voltage and low-current applications, for high-voltage and high-current circuits, arc interrupters, similar to those used in circuit breakers are needed. In medium-voltage applications the most popular interrupter is the air magnetic type, in which the arc is driven into an arc chute by the magnetic field produced by the load current in a blowout coil. *See* BLOWOUT COIL; CIRCUIT BREAKER.

Some load-break switches may also be required to have the capability of holding the contacts in the closed position during short-circuit conditions so that the contacts will not be blown open by electromagnetic forces when the circuit breaker in the system interrupts the short-circuit current.

For signal applications, switches are used to detect a specified situation that calls for some predetermined action in the electrical circuit. For example, thermostats detect temperature; when a certain limit is reached, contacts in the thermostat energize or deenergize another electrical switching device to control power flow.

Switches for signaling purposes are often required to have long life, high speed, and high reliability. Contaminants and dust must be prevented from interfering with the operation of the switch. For this purpose, switches are usually enclosed and are sometimes hermetically sealed.

Switches frequently are composed of many single circuit elements, known as poles, all operated simultaneously or in a predetermined sequence by the same mechanism. Switches are often typed by the number of poles and referred to as single-pole or double-pole switches, and so on. It is also common to express the number of possible switch positions per pole, such as a single-throw or double-throw switch. [T.H.L.]

Electric transient A temporary component of current and voltage in an electric circuit which has been disturbed. In ordinary circuit problems, a stabilized condition of the circuit is assumed and steady-state values of current and voltage are sufficient. However, it often becomes important to know what occurs during the transition period following a circuit disturbance until the steady-state condition is reached. Transients occur only in circuits containing inductance or capacitance. In general, transients accompany any change in the amount or form of energy stored in the circuit.

The study of transient phenomena is very broad. The mathematical requirements become severe and go far beyond the borders of all known mathematics. Transient analysis often requires the use of calculating machines, models, and tests. Fourier and Laplace transforms have proven indispensable in the modern treatment of transients and these disciplines need be mastered by anyone going far in the study of transients. The analysis of lumped-parameter circuits is comparatively easy. An electric circuit or system under steady-state conditions of constant, or cyclic, applied voltages or currents is in a state of equilibrium. However, the circuit conditions of voltage, current, or frequency may change or be disturbed. Also, circuit elements may be switched in or out of the circuit. Any change of circuit condition or circuit elements causes a transient readjustment of voltages and currents from the initial state of equilibrium to the final state of equilibrium. In a sense the transient may be regarded as superimposed on the final steady state, so that Eq. (1) applies.

$$\begin{pmatrix}\text{Instantaneous}\\ \text{condition}\end{pmatrix} = \begin{pmatrix}\text{final}\\ \text{condition}\end{pmatrix} + \begin{pmatrix}\text{transient}\\ \text{terms}\end{pmatrix} \qquad (1)$$

Furthermore, since the instantaneous condition at the first instant of disturbance (time zero) must be the initial condition, it may be described by Eq. (2).

$$\begin{pmatrix}\text{Initial}\\ \text{condition}\end{pmatrix} = \begin{pmatrix}\text{final}\\ \text{condition}\end{pmatrix} + \begin{pmatrix}\text{transient terms}\\ \text{at time zero}\end{pmatrix} \qquad (2)$$

In many practical cases transients are one of three types:

1. Single-energy transients, in which only one form of energy storage (either electromagnetic or electrostatic) is present; the transient exhibits simple exponential decay from the initial to the final conditions.
2. Double-energy transients, in which both forms of energy storage are present; the transient is either aperiodic or a damped sinusoid.
3. Combination of 1 and 2. [L.V.B.]

Electric uninterruptible power system A system that provides protection against commercial power failure and variations in voltage and frequency. Uninterruptible power systems (UPS) have a wide variety of applications where unpredictable changes in commercial power will adversely affect equipment. This equipment may include computer installations, telephone exchanges, communications networks, motor and sequencing controls, electronic cash registers, hospital intensive care units, and a host of others. The uninterruptible power system may be used on-line between the commercial power and the sensitive load to provide transient free well-regulated power, or off-line and switched in only when commercial power fails.

There are three basic types of uninterruptible power system. These are, in order of complexity, the rotary power source, the standby power source, and the solid-state uninterruptible power system.

The rotary power source consists of a battery-driven dc motor that is mechanically connected to an ac generator. The battery is kept in a charged state by a battery charger that is connected to the commercial power line. In the event of a commercial power failure, the battery powers the dc motor which mechanically drives the ac generator. The sensitive load draws its power from the ac generator and operates through the outage.

The standby power source consists of a battery connected to a dc-to-ac static inverter. The inverter provides ac power for the sensitive load through a switch. A battery charger, once again, keeps the battery on full charge. Normally, the load operates directly from the commercial power line. In the event of commercial power failure, the switch transfers the sensitive load to the output of the inverter.

The solid-state uninterruptible power system has a general configuration much like that of the standby power system with one important exception. The sensitive load operates continually from the output of the static inverter. This means that all variations on the commercial power lines are cleaned and regulated through the output of the uninterruptible power system. A commercial power line, known as a bypass, is provided around the uninterruptible power system through a switch. Should the uninterruptible power system fail at some point, the commercial power is automatically transferred to the sensitive load through the switch. This scheme is known as an on-line automatic reverse-transfer uninterruptible power system.

An uninterruptible power system consists of four major subsystems: a method to put energy into a storage system, a battery charger; an energy storage system, the battery; a system to convert the stored energy into a usable form, the static inverter; and a circuit that electrically connects the sensitive load to either the output of the uninterruptible power system or to the commercial power line, the transfer switch. The position of the transfer switch is controlled by a monitor circuit. Generally the switch in an uninterruptible power system is a high-speed solid-state device that can transfer the load from one ac source to another with little or no break in power. *See* ELECTRIC POWER SYSTEMS; ELECTRIC SWITCH. [J.Su.]

Electrical breakdown A large, usually abrupt rise in electric current in the presence of a small increase in electric voltage. Breakdown may be intentional and controlled or it may be accidental. Lightning is the most familiar example of breakdown.

In a gas, such as the atmosphere, the potential gradient may become high enough to accelerate the naturally present ions to velocities that cause further ionization upon collision with atoms. If the region of ionization does not extend between oppositely charged electrodes, the process is corona discharge. If the region of ionization bridges the gap between electrodes, thereby breaking down the insulation provided by the gas, the process is ionization discharge. When controlled by the ballast of a fluorescent lamp, for example, the process converts electric power to light. In a gas tube the process provides controlled rectification. *See* GAS TUBE.

In a solid, such as an insulator, when the electric field gradient exceeds 10^6 volts/cm, valence bonds between atoms are ruptured and current flows. Such a disruptive current heats the solid abruptly. In a semiconductor if the applied backward or reverse potential across a junction reaches a critical level, current increases rapidly with further rise in voltage. This avalanche characteristic is used for voltage regulation in the Zener diode. In a transistor the breakdown sets limits to the maximum instantaneous voltage that can safely be applied between collector and emitter. *See* BREAKDOWN POTENTIAL; ELECTRICAL INSULATION; TRANSISTOR; ZENER DIODE. [F.H.R.]

Electrical codes Systematic bodies of rules governing the practical application, installation, and interconnection of electrically operated equipment, devices, and electrical wiring systems.

The basic code used throughout the United States is the National Electrical Code, prepared under the direction of the National Fire Protection Association (NFPA). It is approved by the American National Standards Institute. The National Electrical Code is purely advisory as far as the National Fire Protection Association is concerned, but it is very widely used for legal regulatory purposes, The code is administered by various local inspection agencies, whose decisions govern its application to individual installations. The National Electrical Code is incorporated bodily or by reference in many municipal building ordinances, often with additional provisions or restrictions applicable in the particular locality. Some large cities have independent electrical codes; however, the actual provisions in most such codes tend to be basically similar to the National Electrical Code.

Compliance with the provisions of the code can effectively minimize fire and accident hazards in any electrical design. It sets forth requirements that constitute a minimum standard for the framework of electrical design. As stated in its introduction, the code is concerned with the "practical safeguarding of persons and of buildings and their contents from hazards arising from the use of electricity for light, heat, power, radio, signaling and for other purposes." The National Electrical Code is recognized as a legal criterion of safe electrical design and installation. It is used in court litigation and by insurance companies as a basis for insuring buildings.

In addition to the National Electrical Code itself, other standards and recommended practices are made available by the National Fire Protection Association. These cover such special subjects as hospital operating rooms, municipal fire alarm systems, garages, aircraft hangars, and other equipment with great potential hazards due to improper design.

The National Electrical Safety Code (to be distinguished from the National Electrical Code) is published by the Institute of Electrical and Electronic Engineers, Inc. This code applies to the outdoor circuits of electric utility companies and to similar systems or equipment on commercial and industrial premises.

Standards on the construction and assembly of many types of electrical equipment, materials, and appliances are set forth in literature issued by the Underwriters' Laboratories, Inc. The Underwriters' Laboratories examines, tests, and determines the suitability of materials and equipment to be used according to code regulations. *See* WIRING. [J.F.McP.]

Electrical communications The science and technology by which information is collected from an originating source, transformed into electric currents or fields, transmitted over electrical networks or through space to another point, and reconverted into a form suitable for interpretation by a receiving entity.

Information to be transmitted comes from many sources. Some sources produce essentially continuously varying signals. Examples are speech, music, the output of TV cameras, and sensors of temperature, pressure, and the like. Others produce signals whose distinguishing characteristic is that they are on or off. Examples are the hand telegraph key, many operations of computers, above-or-below sensors of temperature, pressure, and so on.

For transmission over an electrical communication channel, a transducer is used to transform the original physical manifestation into an analogous electrical signal—a speech sound in air by a telephone transmitter, variations in light intensity in a scanned picture by a television camera, temperature observations by the opening and closing of the switch of a thermostat. *See* TRANSDUCER.

Transmission media. Media commonly used for electrical communication can be divided into two classes: (1) Those carrying electrical currents, such as a pair of wires or coaxial cable. (2) Those in which a transducer must be used to change electrical currents into a form suited to the medium. Examples are electromagnetic fields in space, such as radio, for which the antennas are the transducers; electromagnetic fields in guiding structures, such as a waveguide; pressure waves in water (sonar); light waves in space or optical fibers. *See* ANTENNA (ELECTROMAGNETISM); COAXIAL CABLE; COMMUNICATION CABLES; ELECTROMAGNETIC WAVE TRANSMISSION; MICROWAVE TRANSMISSION LINES; OPTICAL COMMUNICATIONS; OPTICAL FIBERS; SONAR; TRANSMISSION LINES; WAVEGUIDE.

Network configurations. Networks may be configured in several ways. Perhaps the most common is a channel between point A and point B. These may, as in the telephone network, be switched on a call or message basis among the very large number of points available. Another configuration is a treelike structure in which the channel from one source is divided and subdivided to connect to many receivers. A cable television (CATV) system is an example of this type of configuration. The third configuration is typified by radio and TV broadcasting: a single source sends out radio waves, and any number of receivers in the area can pick them up on a one-way basis. *See* CLOSED-CIRCUIT TELEVISION; PRIVATE BRANCH EXCHANGE; SWITCHING SYSTEMS (COMMUNICATIONS).

Signal amplification. Whatever the transmission medium, the signal becomes weaker with distance, dropping nearer to the inevitable noise in the medium, thus reducing the signal-to-noise (S/N) ratio, and so limiting the information-carrying capacity of the channel. It becomes desirable to place one or more amplifiers at periodic intervals along the channel to raise the signal power and hence maintain the desired S/N at the receiving end. *See* AMPLIFIER.

Frequency division multiplex. Radio broadcasting is the best-known example of many channels sharing a common medium, a technique known as frequency division multiplex. Each transmitter is assigned a position in the frequency spectrum and is permitted to occupy a specified bandwidth. The receivers include selective networks or filters, which, when tuned to a selected channel, accept that channel and reject the others. The process of shifting the input signal to the assigned frequency slot is called modulation, and the inverse process at the receiving end is called demodulation. The same processes are applied to the sharing of a common wire or coaxial cable medium. *See* ELECTRIC FILTER; MODULATION; RADIO BROADCASTING.

Time division multiplex. There is another way for a multiplicity of channels to share a common medium, called time division multiplex. Each channel is assigned a short time slot during which it occupies the whole available bandwidth, with the channels occupying time slots in turn. When all channels have had their turn, a frame has been completed, and the roll call is repeated. Obviously, there has to be an identifying symbol at the beginning of each frame, so that the channel time slots can be identified.

The processing of continuously varying input sources is much more complicated than in frequency-division AM or FM systems. It involves sampling, quantizing, coding, and organizing into words and frames, with framing symbols and precise timing. When only vacuum tubes were available, the economics of the resulting tradeoffs were completely unfavorable. With the advent of the transistor, and the subsequent flowering of the solid-state art into integrated circuits, the whole economic picture changed dramatically. Complicated processing could be accomplished by tiny, inexpensive chips. The extra complexity in many cases was well justified by savings in the cable system and its amplifiers.

At the same time, there was a rapidly growing demand for communication of essentially on-off signals, as between computers and machines. The time division multiplex is very well suited to these signals, with little processing. For these reasons, there has been rapidly expanding use of time division systems. *See* ELECTRONICS; INTEGRATED CIRCUITS; PULSE MODULATION; TRANSISTOR.

For a discussion of major forms of electrical communication systems *see* RADAR; RADIO; SOUND-REPRODUCING SYSTEMS; TELEGRAPHY; TELEMETERING; TELEVISION. [A.C.D.]

Electrical conduction in gases The process by means of which a net charge is transported through a gaseous medium. It encompasses a variety of effects and modes of conduction, ranging from the Townsend discharge at one extreme to the arc discharge at the other. The current in these two cases ranges from a fraction of 1 microampere in the first to thousands of amperes in the second. It covers a pressure range from less than 10^{-4} atm (10 pascals) to greater than 1 atm (100 kilopascals). *See* ARC DISCHARGE; TOWNSEND DISCHARGE.

In general, the feature which distinguishes gaseous conduction from conduction in a solid or liquid is the active part which the medium plays in the process. Not only does the gas permit the drift of free charges from one electrode to the other, but the gas itself may be ionized to produce other charges which can interact with the electrodes to liberate additional charges. Quite apparently, the current voltage characteristic may be nonlinear and multivalued. *See* ELECTROLYTIC CONDUCTANCE; SEMICONDUCTOR.

The applications of the effects encountered in this area are of significant commercial and scientific value. A few commercial applications are thyratrons, gaseous rectifiers, ignitrons, glow tubes, and gas-filled phototubes. These tubes are used in power supplies, control circuits, pulse production, voltage regulators, and heavy-duty applications such as welders. In addition, there are gaseous conduction devices widely used in research problems. Some of these are ion sources for mass spectrometers and nuclear accelerators, ionization vacuum gages, radiation detection and measurement instruments, and thermonuclear devices for the production of power. [G.H.M.]

Electrical connector A device that joins electric conductors mechanically and electrically to other conductors and to the terminals of apparatus and equipment. The term covers a wide range of devices designed, for example, to connect small conductors employed in communication circuits, or at the other extreme, large cables and bus-bars.

Electrical connectors are applied to conductors in a variety of ways. Soldered connectors have a tube or hole of approximately the same diameter as the conductor. The conductor and connector are heated, the conductor inserted, and solder flowed into the joint until it is filled. Solderless connectors are applied by clamping the conductor or conductors in a bolted assembly or by staking or crimping under great mechanical force.

Typical connector types are in-line splice couplers, T-tap connectors, terminal lugs, and stud connectors. Couplers join conductors end to end. T-tap connectors join a through conductor to another conductor at right angles to it (illustration *a*). Terminal lugs join the conductor to a drilled tongue for bolting to the terminals of equipment (illustration *b*). Stud connectors join the conductor to equipment studs; the stud clamp is threaded or smooth to match the stud.

Split-bolt connectors are a compact construction widely used for splices and taps in building wiring. The bolt-shape casting has a wide and deep slot lengthwise. The conductors are inserted in the slot and the nut is drawn up, clamping the conductors together inside the bolt (illustration *c*).

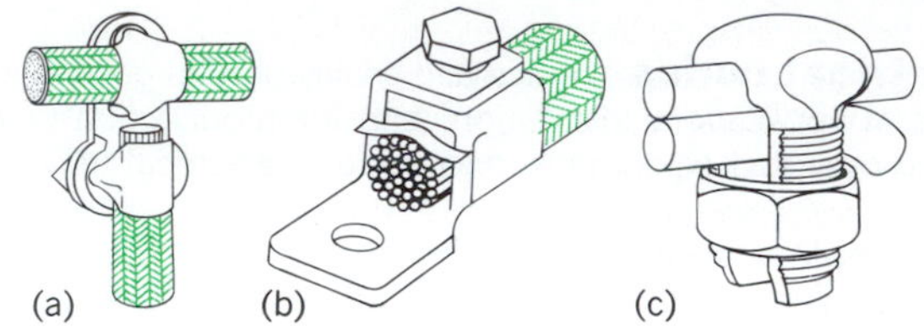

Types of connectors. (*a*) T-tap connector. (*b*) Terminal lug. (*c*) Split-bolt connector.

Expansion connectors or flexible connectors allow some limited motion between the connected conductors. The clamp portions of the connector are joined by short lengths of flexible copper braid and may also be held in alignment by a telescoping guide.

Separable types consist of matched plugs and receptacles, which may be readily separated to disconnect a conductor or group of conductors from the circuit or system. Separable connectors are commonly used for the connection of portable appliances and equipment to an electric wiring system.

Locking types are designed so that, when coupled, they may not be separated by a strain or pull on the cord or cable. In a typical construction the plug is inserted and twisted through a small arc, locking it securely in place.

Plug receptacles, sometimes called convenience outlets, are a type of wiring device distributed throughout buildings for the attachment of cord plug caps from portable lamps and appliances. In residences at least one such outlet must be provided for every 12 linear feet (3.66 m) or major fraction of wall perimeter. Grounding receptacles have an additional contact that accepts the third round or U-shaped prong of a grounding attachment plug. *See* WIRING. [J.F.McP.]

Electrical degree A unit equal to 1/360 of a complete cycle of electric current or voltage. In an electric machine it is 1/360 of the angle subtended at the axis by two consecutive field poles of like polarity, since the voltage wave generated in a conductor completes one cycle when it traverses one pair of poles. The term mechanical degree is used to designate the space angle between two positions about the axis of the machine. The number of electrical degrees between two positions about the axis equals the number of mechanical degrees multiplied by the number of pairs of poles on the machine. *See* ALTERNATING CURRENT. [A.R.E.]

Electrical engineering A branch of engineering dealing primarily with electricity and magnetism and devoted to utilization of the forces of nature and materials for the benefit of humans. Electrical engineering encompasses many phases of other engineering sciences and the physical sciences; it includes research, invention, development, design, application, and education. Many phases of electrical engineering are based on applications of higher mathematics. *See* ENGINEERING.

The need for better communications between electrical engineers led to the establishment of the American Institute of Electrical Engineers (AIEE) in 1884. In 1913 the Institute of Radio Engineers (IRE) was founded. In later years these two organizations were merged into a single organization which is now known as the Institute of Electrical and Electronics Engineers (IEEE).

Electrical engineers apply their abilities in other engineering fields that are not strictly electrical. There is hardly a field of technology to which electrical engineering has not made a contribution. For important related engineering fields *see* COMPUTER; CONTROL SYSTEMS; INSTRUMENTATION; SYSTEMS ENGINEERING.

[A.G.C.]

Electrical impedance The total opposition that a circuit presents to an alternating current. Impedance, measured in ohms, may include resistance, inductive reactance, and capacitive reactance. *See* REACTANCE.

Impedance may also be defined as the ratio of the rms voltage to the rms current, $Z = E/I$. This is a form of Ohm's law for ac circuits.

[B.L.R.]

Electrical instability A persistent condition of unwanted self-oscillation in an amplifier or other electrical circuit. Instability is usually caused by excessive positive feedback from the output to the input of an active network. If, in an audio-frequency amplifier, instability is at a low audible frequency, the output will contain a putt-putt sound, from which such instability is termed motorboating. The instability may also be at a high audible frequency, or it may be at frequencies outside the audible range. Although such oscillations may not be heard directly, they produce distortion by driving the amplifier beyond its linear range of operation. *See* AMPLIFIER.

[F.H.R.]

Electrical insulation A nonconducting material that provides electric isolation of two parts at different voltages. To accomplish this, an insulator must meet two primary requirements: it must have an electrical resistivity and a dielectric strength sufficiently high for the given application. The secondary requirements relate to thermal and mechanical properties. Occasionally, tertiary requirements relating to dielectric loss and dielectric constant must also be observed. A complementary requirement is that the required properties not deteriorate in a given environment and desired lifetime. *See* CONDUCTOR (ELECTRICITY).

Electric insulation is generally a vital factor in both the technical and economic feasibility of complex power and electronic systems. The generation and transmission of electric power depend critically upon the performance of electric insulation, and now plays an even more crucial role because of the energy shortage.

Requirements. The important requirements for good insulation are as follows.

The basic difference between a conductor and a dielectric is that free charge has high mobility on and in a conductor, whereas free charge has little or no mobility on or in a dielectric. Dielectric strength is a measure of the electric stress required to abruptly move substantial charge on or through a dielectric. It deteriorates with the ingress of water and with elevated temperature. For high-voltage (on the order of kilovolts) applications, dielectric strength is the most important single property of the insulation. *See* DIELECTRIC MATERIALS.

Resistivity is a measure of how much current will be drained away from the conductor through the bulk or along the surface of the dielectric. An insulator with resistivity equal to or greater than 10^{13} ohm-cm may be considered good. *See* ELECTRICAL RESISTIVITY.

When a dielectric is subjected to an alternating field, a time-varying polarization of the atoms and molecules in the dielectric is produced. The alternation of both the permanent and induced polarization in the dielectric results in power dissipation within the dielectric of which the power factor is a measure. This dielectric power loss is proportional to the product of the dielectric constant and the square of the electric field in the dielectric. Although it may be a loss that is small relative to other losses in most ambient-temperature applications, and even though it generally decreases at low temperatures, it is a relatively important loss for dielectrics to be used at cryogenic temperatures.

The dielectric constant, also known as the relative permittivity or specific inductive capacity, is a measure of the ability of the dielectric to become polarized, taken as the ratio of the charge required to bring the system to the same voltage level relative to the charge required if the dielectric were vacuum. It is thus a pure number, but is in fact not a constant, and may vary with temperature, frequency, and electric-field intensity.

In addition to the problem of intensification of the electric field in regions of relatively lower dielectric constant, a low-dielectric-constant insulation is desirable for two more compelling reasons. In ac transmission cables, the lower the dielectric constant, the more the current and the voltage will be in phase. This means that more usable power will be delivered, without the need for reactive compensation. Furthermore, in reducing the charging current (which is proportional to the dielectric constant), concomitant power losses (related to the square of the charging current) are also reduced. A high dielectric constant is desirable in capacitors, since the capacitance is proportional to it. *See* CAPACITANCE; DIELECTRIC CONSTANT.

Properties. All insulators may be classified as either solid or fluid. Solid insulation is further divided into flexible and rigid types.

Solid insulation. Flexible hydrocarbon insulation is generally either thermoplastic or thermosetting. Thermosets are initially soft, and can be extruded by using only pressure. Following heat treatment, when they return to ambient temperature, they are tougher and harder. After thermosetting, nonrubber thermosets are harder, stronger, and have more dimensional stability than the thermoplastics. Thermoplastics are softened by heating, and when cool become hard again. They are heat-extruded.

Cellulose paper insulation is neither thermoplastic nor thermosetting. It is widely used in cables and rotating machinery in multilayers and impregnated with oil. It has a relatively high dielectric loss that hardly decreases with decreasing temperature, which rules it out for cryogenic applications. Because of its high dielectric strength, the high loss has not been a deterrent to its use in conventional ambient-temperature applications. However, the high dielectric strength deteriorates quickly if moisture permeates the paper.

Rigid insulation includes glass, mica, epoxies, ceramoplastics, porcelain, alumina, and other ceramics. Rather than being used to insulate wires and cables, except for mica, these materials are used in equipment terminations (potheads) and as support insulators (in tension or compression) for overhead lines whose primary dielectric is air. These rigid structures must be shock-resistant, be relatively water-impervious, and be able to endure corona discharges over their surfaces.

Fluid insulation. Liquids, gases, and vacuum fall in the category of fluid insulation. For all of these, the electrical structure must be such as to contain the fluid in the regions of high electric stress.

The main types of insulating liquids are the mineral oils, silicones, chlorinated hydrocarbons, and the fluorocarbons with dielectric strengths on the order of megavolts per centimeter. Many other liquids also have good dielectric strength, such as carbon tetrachloride, toluene, hexane, benzene, chlorobenzene, alcohol, and even deionized water.

Most gases have a dielectric constant of about 1, and low dielectric loss. Air is used as a dielectric in a wide variety of applications, ranging from electronics to high-voltage (765-kV) and high-power (2000-MW) electric transmission lines. Dry air is a reasonably good insulator. However, its dielectric strength decreases with increasing gap.

Vacuum (that is, pressures of less than 10^{-5} torr or 10^{-3} pascal) has one of the highest dielectric strengths in the gap ranging 0.1 to 1 mm. However, as the gap increases, its dielectric strength decreases rapidly. A perfect vacuum might be expected to be a perfect insulator, since there would be no charge carriers present to contribute to electrical conductance. That this is not so in practice arises because of the effects of a high electric field or high voltage at the surface of electrodes in vacuum, rather than because a perfect vacuum is far from being realized in the laboratory. The dielectric properties of vacuum can degenerate rapidly because vacuum offers no resistance to the motion of charge carriers, once they are introduced into the vacuum region. [M.R.]

Electrical interference Any undesired electrical energy that tends to interfere with the reception of desired signals. It may be a signal generated by nearby electric motors, automotive ignition systems, radio transmitters, improperly operating television receivers, or a wide variety of other improperly operating electric devices. The interference signals either may be radiated through space as unguided electromagnetic waves or may travel along power lines as guided radiation. Interference may also be caused by atmospheric phenomena such as lightning. Such natural disturbances are sometimes termed sferics. Interfering signals are sometimes generated intentionally by powerful transmitters for the purpose of jamming communication and radar systems. *See* ELECTRICAL NOISE.

Interference from electrical devices is best suppressed at its source, by such means as minimizing sparking, connecting capacitors across sparking contacts, shielding the equipment with metal enclosures, and using choke coils or filters that prevent interfering signals from entering power lines. *See* ELECTRIC FILTER; ELECTRICAL SHIELDING; GROUNDING. [J.Mar.]

Electrical loading The addition of inductance to a transmission line in order to improve its transmission characteristics over the required frequency band. Loading coils are often inserted in telephone lines, at spacings as close as 1 mi (1.6 km), to counteract the capacitance of the line and thus make the line impedance more closely equivalent to a pure resistance. Similar coils are used between sections of lines used for carrier transmission. *See* TRANSMISSION LINES.

With coaxial cables and waveguides, loading is placed at the end of the line to absorb all power reaching the end, thereby achieving a nonreflecting termination. *See* WAVEGUIDE.

When an antenna is too short to give resonance at the desired frequency, a coil can be inserted in series with the antenna to give the required amount of loading needed for resonance. As an example, practically all auto radios have a loading coil in the antenna circuit, because the usual whip antenna is much too short for broadcast-band frequencies. *See* ANTENNA (ELECTROMAGNETISM). [J.Mar.]

Electrical measurements The measurement of any one of the many quantities by which the behavior of electricity is characterized. The knowledge of the quantitative behavior of electricity is essential to scientific and technical progress. Electrical measurements play a major role in industry, communications, and even in such unrelated fields as medicine.

Many electrical measurements can be made with direct-indicating instruments merely by connecting the instrument properly in the circuit. Thus a voltmeter provides a pointer which moves over a scale calibrated in volts, and an ammeter in the same way presents a reading of current in amperes. Other direct-reading instruments are wattmeters, frequency meters, power-factor or phase-angle meters, and ohmmeters. Many electrical quantities are measured both as instantaneous values and as values integrated over time. Some electrical measurements must be made with various specialized devices or systems requiring adjustment or balancing to obtain the measured value. Typical of these are potentiometers and bridges in many standard and specialized forms.

Because of differences in instruments and techniques, it is convenient to divide measurements into direct-current (dc) and alternating-current (ac) classes.

In dc circuits the measurement of voltage and current often suffices to define the operation of the circuit. The product of the two represents power. In the commercial sale of dc electricity the measurement of energy must be made with a dc watt-hour meter. Occasional use is made of a dc ampere-hour meter in battery-charging installations.

To measure high values of current, shunts are used to bypass all but a small fraction of the current around the measuring instrument. A newer technique employs a form of saturable reactor energized by alternating current to measure large direct currents. *See* CURRENT MEASUREMENT; ELECTRIC ENERGY MEASUREMENT; ELECTRIC POWER MEASUREMENT; VOLTAGE MEASUREMENT.

Alternating-current circuits involve more variables and hence more measurements than dc circuits. The most common measurements are voltage, current, and power; the last requires a wattmeter, as ac power cannot always be calculated directly from voltage and current. Also measured are frequency and power factor (or phase angle) and sometimes waveform or harmonic content. Energy is measured by means of the ac induction watt-hour meter. Direct-current instruments do not respond to ac quantities, but some may be adapted by the addition of rectifiers to convert alternating current to direct. The thermocouple is another form of convertor by which a dc instrument may be made to read ac quantities.

If alternating voltages and currents above the normal ranges of self-contained instruments are to be measured, instrument transformers may be used to extend the ranges of those instruments. In the study of ac waveform a qualitative evaluation may be made with an oscillograph or a cathode-ray oscilloscope. Quantitative measurement of harmonic content requires the use of a harmonic analyzer. *See* HARMONIC ANALYZER; INSTRUMENT TRANSFORMER.

In a laboratory, emphasis is normally placed on accuracy and on completeness of facilities to deal with all types of measurements. There is relatively little limitation on the size and complexity of equipment used.

The term field measurements is used to designate all measurements made outside a laboratory as in generating stations and substations, service shops, factory testing areas, ships, and aircraft. For these uses equipment is chosen to perform only specialized services. Accuracy well below that of laboratory measurements is usually permissible. Convenience, compactness, and often portability are prime considerations in choosing equipment.

All electric instruments draw some power from the circuits to which they are connected, and ac instruments generally take more power than dc instruments. This circuit loading may alter appreciably the quantity being measured. The magnitude of this error can usually be evaluated and minimized by proper choice of instruments. [I.F.K./E.C.St.]

Electrical model A mathematical description or electrical equivalent circuit that represents the behavior of a device or system. Models for complex systems are often represented by networks of models for simpler electrical devices such as resistors, capacitors, transistors, and transformers. By using analogies between current or voltage and other physical parameters, equivalent circuits can also be used to analyze thermal, mechanical, magnetic, and acoustic systems. A recurring issue in development and in application of models is the use of simplifying assumptions to allow a compromise between accuracy and complexity. A hierarchy of models often exists, ranging from highly accurate but complex physics-based nonlinear time-domain computer models to linear equivalent-circuit models suitable for hand calculation. The best model for a particular application is the simplest one that predicts the relevant behavior with acceptable accuracy. *See* EQUIVALENT CIRCUIT.

Electrical models can be divided into two categories: nonlinear, large-signal and linear, small-signal models. Each of these can be further divided into time-invariant and time- or frequency-dependent categories.

Large-signal models. These are usually derived by applying physical laws to generalized devices. This leads to a system of differential equations whose solution comprises the model. Such physically derived models are often highly complex, requiring lengthy computer solutions. An alternative approach is empirical modeling, in which a simple mathematical form is assumed for the model, with key physical parameters being determined by measurements on representative devices. *See* DIFFERENTIAL EQUATION.

There are two main application areas for large-signal models. One is to find the response of a system to time-varying inputs. The other is to determine the voltages and currents of a circuit with zero input variations, as a preliminary to small-signal analysis.

Small-signal modeling. In many applications, current and voltage variations are small enough that the device's characteristics are approximately independent of signal size. This allows the device to be represented by a simplified linear equivalent circuit. To find the values of the equivalent circuit's elements, the input signal is replaced with its average value, and all currents and voltages are found by using the large-signal model. The results represent the quiescent operating point or Q point for the circuit. All currents and voltages can be approximated by series expansions around the Q point.

Linearity has several benefits. One is the applicability of systematic solution methods for linear circuits that simplify both manual and computer solutions. Linearity also allows superposition: a linear system's response to the sum of two inputs equals the sum of its responses to the inputs applied separately. An important consequence of superposition is that the system is completely specified by its response to sinusoids of arbitrary frequency. This frequency response determines the output spectrum for arbitrary inputs. Linear circuits can be analyzed entirely in the frequency domain. *See* CIRCUIT (ELECTRONICS); GAIN; LINEARITY; RESPONSE; SUPERPOSITION THEOREM (ELECTRIC NETWORKS). [R.M.Fo.]

Electrical noise

Interfering and unwanted currents or voltages in an electrical device or system. Electrical noise, usually simply called noise, has an important effect on any electrical system which is used to gather, transmit, process, or present information. In such systems as telephone, radio, television, radar, radio navigation, telemetering, electronic control, or electronic computing, the desired signals carrying intelligence may be masked or distorted by noise.

Noise may originate either externally to the device in which it appears, as atmospheric static, or internally, as thermal noise in a resistor. It may result either from natural phenomena, as do both types of noise just mentioned, or from interference from devices such as nearby electric motors or generators.

Interference from electric devices can usually be nearly eliminated by good engineering design and proper location of equipment. Noise due to natural phenomena often cannot be reduced below certain fixed levels, and good engineering design can ensure only that the equipment will function as effectively as possible in the face of this irreducible noise. For example, a radio receiver cannot operate on received signals which are very weak (compared to some level determined by the thermal noise in the receiver), no matter how much amplification is used in the receiver, because the noise is amplified along with the signal.

Noise may conveniently be classified as either random noise or nonrandom noise. Random noise is defined as noise which is not predictable, although it may exhibit statistical regularities. Thermal noise is the random voltage which appears at the terminals of a resistor or any component with internal resistance because of the random motion of thermally excited electrons in the resistor. Contact noise, the noise in carbon resistors and carbon microphones, for example, is caused by randomly varying fluctuations in resistance.

Random noise also originates in transistors and other semiconductor devices. There are various mechanisms at work, and the terminology is not completely standard. Thermal noise, as above, is the noise caused by the random motion of thermally excited electrons. Shot noise, or generation-recombination noise, is produced by fluctuations in free carrier densities when an electric field is applied. Excess noise or modulation noise is caused by slow fluctuations in conductivity. Noise also originates in photoconductors. Most mechanisms of random noise generation share the principle that observable gross currents and voltages are the result of many random actions at microscopic level. *See* SEMICONDUCTOR.

Any electric device which must receive electromagnetic radiation will pick up radiated random noise as well as signals. Electrical disturbances in the atmosphere cause noise which is very irregular in character, often appearing in sharp bursts. Aside from atmospheric disturbances, an antenna receives a steady background of noise which is of thermal origin, thermal radiation from the gases of the atmosphere, and thermal radiation from heavenly bodies and systems. This last is sometimes called interstellar noise. The Sun radiates noise at all times, but during sunspot activity the intensity of its noise radiation is greatly increased. *See* ATMOSPHERIC ELECTRICITY; RADIO ASTRONOMY; STATIC.

Nonrandom noise is usually the result of radiation from other electric equipment, unwanted coupling with other systems, or spurious oscillations within an electrical circuit. *See* CROSSTALK; ELECTRICAL INTERFERENCE. [W.L.R.]

Electrical noise generator

A device which produces (usually random) electrical noise for use in electrical measurements. Electrical noise generators are commonly used in measuring the noise figure of a radio receiver or other amplifier. They are also used in other tests of the response of an electrical system to random noise, and in measurements of noise intensity. *See* ELECTRICAL NOISE.

Some standard types of noise generator are: hot-wire, diode, gas-discharge tube, and klystron. A hot-wire noise generator is commonly the filament of a lamp heated by a direct current. The filament is connected across the terminals where the noise is to be introduced, for example, the antenna terminals of a radio receiver. A reflex klystron, with reflector grid connected to the cavity to prevent oscillation, generates noise because of shot effect in the cathode current. *See* KLYSTRON; SCHOTTKY EFFECT. [W.L.R.]

Electrical resistance

That property of an electrically conductive material that causes a portion of the energy of an electric current flowing in a circuit to be converted into heat. In 1774 A. Henley showed that current I flowing in a wire produced heat, but it was not until 1840 that J. P. Joule determined that the rate of conversion of electrical energy into heat in a conductor, that is, power dissipation H/t, could be expressed by the relation given in notation (1).

$$H/t \propto I^2R \tag{1}$$

The day-today determination of resistance R by measuring the rate of heat dissipation is not practical. However, this rate of energy conversion is also VI, where V is the voltage drop across the element in question and I the current through the element, as in Eq. (2), from which the more conventional rela-

$$H/t \propto I^2R = VI \qquad (2)$$

tionship implied by Ohm's law, Eq. (3), is apparent.

$$R = V/I \qquad (3)$$

See ELECTRICAL RESISTIVITY; OHM'S LAW; RESISTANCE MEASUREMENT. [C.E.A.]

Electrical resistivity The electrical resistance offered by a homogeneous unit cube of material to the flow of a direct current of uniform density between opposite faces of the cube. Also called specific resistance, it is an intrinsic, bulk (not thin-film) property of a material. Resistivity is usually determined by calculation from the measurement of electrical resistance of samples having a known length and uniform cross section according to the following equation, where ρ is the resistivity, R the measured resistance, A the cross-sectional area, and l the length. In the mks system (SI), the unit of resistivity is the ohm-meter. Therefore, in the equation below, resistance is ex-

$$\rho = RA/l$$

pressed in ohms, and the sample dimensions in meters.

The room-temperature resistivity of pure metals extends from approximately 1.5×10^{-8} ohm-meter for silver, the best conductor, to 135×10^{-8} ohm-meter for manganese, the poorest pure metallic conductor. Most metallic alloys also fall within the same range. Insulators have resistivities within the approximate range of 10^8 to 10^{16} ohm-meters. The resistivity of semiconductor materials, such as silicon and germanium, depends not only on the basic material but to a considerable extent on the type and amount of impurities in the base material. Large variations result from small changes in composition, particularly at very low concentrations of impurities. Values typically range from 10^{-4} to 10^5 ohm-meters. *See* ELECTRICAL INSULATION; ELECTRICAL RESISTANCE; SEMICONDUCTOR.

The temperature coefficients (changes with temperature) of resistivity of pure metallic conductors are positive. Resistivity increases by about 0.4%/K at room temperature and is nearly proportional to the absolute temperature over wide temperature ranges. As the temperature is decreased toward absolute zero, resistivity decreases to a very low residual value for some metals. The resistivity of other metals abruptly changes to zero at some temperature above absolute zero, and they become superconductors.

Metals, and some semiconductors in particular, exhibit a change in resistivity when placed in a magnetic field. Theoretical relations to explain the observed phenomena have not been well developed. [C.E.A.]

Electrical shielding A means of avoiding pickup of undesired signals or noise, suppressing radiation of undesired signals, and confining wanted signals to desired paths or regions. These shielding objectives cannot be realized without some degree of modification of the electric and magnetic fields involved, although the effects of these modifications may not seriously interfere with wanted objectives.

A shield is a sheet, tube, screen, grid, or other object, usually of conducting material. Sometimes a shield is magnetic, or both magnetic and conducting, or even laminated. Fields may be largely confined to, or suppressed from, a specified region by shields. The ratio of the unwanted signal in a communication circuit when the source of shielding is present to the same signal when the shielding is absent is called the shield factor.

A change in an electric field gives rise to magnetic effects; similarly, a change in a magnetic field gives rise to electric effects. As a result, electromagnetic shielding is more commonly encountered than either electrostatic shielding or magnetostatic shielding.

Virtually complete shielding from external electrostatic fields is obtained by totally enclosing the space to be shielded with a highly conducting surface, usually grounded.

Substantially complete shielding from an external and virtually constant magnetic field is obtained by shunting the flux around the space to be shielded through a low-reluctance path. Ideally, the shield should present a complete magnetic path to the flux; that is, the space to be shielded should be surrounded by a ferromagnetic material, preferably of high permeability.

The magnetostatic shielding such as that described above is likewise effective when the magnetic field is changing with time. A different yet highly effective method of realizing electromagnetic shielding is by means of an electrostatic shield.

The cable sheath, which encloses many communication lines within a telephone cable, constitutes one of the most striking examples of electromagnetic shielding. *See* ELECTRICAL NOISE. [H.S.Bl.]

Electrical units and standards The standard in terms of which electrical quantities are evaluated, the quantities so adopted being known as units. The electrical units in practical use today, and also in extensive theoretical use, were designated by the Eleventh General Conference of Weights and Measures in 1960 as members of the International System of Units (Système International d'Unités, abbreviated SI in all languages). This action by the General Conference was the culmination of an effort initiated by A. Giorgi at the beginning of this century to bring the practical electrical units into a coherent system with appropriate mechanical units of the metric system.

Units. To accomplish the above objective, the base units for mechanical quantities were arbitrarily selected: the meter for the unit of length, the kilogram for the unit of mass, and the second for the unit of time. Units for other mechanical quantities are derived from these units in accordance with physical laws and concepts such as the unit of speed, the meter per second, and the unit for acceleration, the meter per second per second.

This system was originally called the mks system to distinguish it from the cgs system (based on the centimeter, gram, and second).

Two cgs systems of electric and magnetic units have been in use in scientific circles for a long time but both are rapidly giving way to the International System. They are the electrostatic system of units (esu) and the electromagnetic system of units (emu).

The electrostatic system defines a unit charge as that charge which exerts 1 cgs unit of force (1 dyne, which is equivalent to 10^{-5} newton) on another unit charge when separated from it by a distance of 1 centimeter in a vacuum. All other units of the system are derived from this definition by assigning unit coefficients in equations relating electric and magnetic quantities to each other. The units so derived are often referred to in terms of the SI units with the prefix "stat," for example, statvolts, statohms, and statamperes.

The electromagnetic system defines a unit magnetic pole (a highly fictitious concept) as that pole which exerts 1 cgs unit of force on another unit pole when separated from it by a distance of 1 centimeter in a vacuum. All other units of the system are derived from this definition in accord with the principles set forth above for the electrostatic system. Units so derived are often referred to in terms of the SI units with the prefix "ab," for example, abvolts, abohms, and abamperes. Special names are given to some magnetic units of the emu system such as the maxwell and the gauss, which correspond, respectively, to the weber and the tesla of SI, although differing from them in magnitude.

The magnitudes of corresponding units of the electrostatic and electromagnetic systems differ from each other by a factor theoretically equal to the speed of light, c (3×10^{10} centimeters per second, approximately), or its square.

The esu system is found convenient for handling purely electrostatic problems. A combination of the two systems in which electrostatic quantities are expressed in esu and magnetic and electromagnetic quantities in emu, with appropriate use of the conversion constant c between the two systems, is called the Gaussian system. All of these systems are rapidly giving way to use of SI units in treatment of electric, magnetic, and electromagnetic phenomena.

Ampere. The constant current which, if maintained in two straight parallel conductors of infinite length, of negligible circular sections, and placed 1 meter apart in a vacuum, would produce between these conductors a force equal to 2×10^{-7} mks unit of force per meter of length. The ampere (A) is the unit of electric current in the International System, and is one of the seven SI base units. See ELECTRIC CURRENT.

Volt. The difference of electric potential between two points of a conducting wire carrying a constant current of 1 ampere, when the power dissipated between these points is equal to 1 watt. The volt (V) is the unit of electric potential in the International System.

Ohm. The electric resistance between two points of a conductor when a constant difference of potential of 1 volt, applied between these two points, produces in the conductor a current of 1 ampere, the conductor not being the seat of any electromotive force. The ohm (Ω) is the unit of electric resistance in the International System. See ELECTRICAL RESISTANCE; ELECTROMOTIVE FORCE (CELLS).

Coulomb. The quantity of electricity transported in 1 second by a current of 1 ampere. The coulomb (C) is the unit of electric charge in the International System. See ELECTRIC CHARGE.

Farad. The capacitance of a capacitor between the plates of which there appears a difference of potential of 1 volt when it is charged by a quantity of electricity equal to 1 coulomb. The farad (F) is the unit of capacitance in the International System. See CAPACITANCE.

Henry. The inductance of a closed circuit in which an electromotive force of 1 volt is produced when the electric current in the circuit varies uniformly at a rate of 1 ampere per second. The henry (H) is the unit of inductance in the International System.

Weber. The magnetic flux which, linking a circuit of 1 turn, produces in it an electromotive force of 1 volt as it is reduced to zero at a uniform rate in 1 second. The weber (Wb) is the unit of magnetic flux in the International System. See MAGNETIC FLUX; UNITS OF MEASUREMENT.

Standards. Electrical standards are the physical embodiments by means of which the electrical units are realized and maintained. The only standards generally maintained in national standards laboratories are those for the volt, the ohm, the farad, and sometimes the henry, since very stable standards for these quantities can be produced. The other electrical quantities are determined from suitable combinations of these.

[A.G.McN.]

Electricity Electricity comprises those physical phenomena involving electric charges and their effects when at rest and when in motion. Electricity is manifested as a force of attraction, independent of gravitational and short-range nuclear attraction, when two oppositely charged bodies are brought close to one another. It is now known that the elementary (nondivisible) electric charges are possessed by electrons and protons. The charge of the electron is equal in magnitude to that of the proton, but is electrically opposite. The electron's charge is arbitrarily termed negative, and that of the proton, positive. Magnetism, those physical phenomena involving magnetic fields and their effects upon materials, manifests itself in the presence of moving electric charge. For this reason, magnetism was originally considered to be a part of electricity. See ELECTRIC CHARGE; MAGNETISM.

The sources of electricity in modern technology depend strongly on the application for which they are intended.

The principal use of static electricity today is in the production of high electric fields. Such fields are used in industry for testing the ability of components such as insulators and condensers to withstand high voltages, and as accelerating fields for charged-particle accelerators. The principal source of such fields today is the Van de Graaff generator. See PARTICLE ACCELERATOR; VAN DE GRAAFF GENERATOR.

The major use of electricity today arises in devices using electric currents alternating at low or zero frequency. The use of alternating current allows power transmission over long distances at very high voltages with a resulting low percentage power loss followed by highly efficient conversion to lower voltages for the consumer through the use of transformers. Large amounts of zero-frequency current, that is, direct current, are used in the electrodeposition of metals, both in plating and in metal production, for example, in the reduction of aluminum ore. To avoid power transmission difficulties, such facilities are frequently located near sources of abundant power. See ALTERNATING CURRENT; DIRECT CURRENT; ELECTRIC CURRENT; ELECTRIC POWER SYSTEMS.

The principal sources of low-frequency electricity are rotary generators whose operation is based on the Faraday induction principle. The force to drive such generators derives from the flow of water or the expansion of gases, as in steam and internal combustion engines. Other sources, such as fusion, solar, and oceanic, still do not have significant application.

Many high-frequency devices, such as communications equipment, television, and radar, involve the consumption of only moderate amounts of power, generally derived from low-frequency sources. If the power requirements are moderate and portability is needed, the use of ordinary chemical batteries is possible. See BATTERY. [W.Ar.]

Electroacoustics That part of electronics concerned with the conversion of acoustical energy into electrical energy or vice versa. The means for conversion from acoustical to electrical systems and vice versa are termed electroacoustic transducers or electroacoustic systems. Electroacoustic systems are used in the telephone, phonograph, radio, sound motion picture and magnetic tape reproducers, and sound-reinforcing systems for converting acoustical waves at the input into corresponding electrical waves, and for converting electrical waves at the output into corresponding sound waves. [H.F.O.]

Electrocardiography The study of the electrical manifestations of the heart beat as obtained from the surface of the body. The electrocardiogram (ECG, or from German EKG) is the record obtained, and the electrocardiograph is the recording instrument. Clinical electrocardiography is that part of the science which deals with the diagnostic interpretation of records obtained in humans based in part on empirical correlations, in part on fundamental physiologic and physical principles.

Electrocardiographic "leads" are records of differences in potential occurring at two specific electrode positions. Standard bipolar limb leads refer to measurements of potential differences between two extremities when each is connected to one of the input terminals of the recorder. Unipolar leads are obtained when one electrode (exploring electrode) is coupled with an indifferent electrode, whose potential is close to the mean potential of the body during the entire cardiac cycle. When the exploring electrode of a unipolar lead is placed close to the heart over the anterior chest, certain portions of the recorded curve are preferentially influenced by electrical events subjacent to the electrode position. This is an observation of considerable clinical usefulness. Other leads have been designed in an attempt to approach an undistorted three-dimensional recording system.

The clinical analysis of electrocardiograms is concerned with alterations in the form of the complexes induced by abnormalities in conduction pathways, rate of cellular changes, myocardial electrolyte imbalance, cardiac chamber enlargement, and pathologic changes secondary to alterations in myocardial blood supply. Interpretations of changes in cycle sequence, by itself a cause of variations in shape of the complexes, have laid the basis for a precise study on the nature of the irregular heart. *See* HEART DISORDERS. [H.H.He.]

Electrochemical equivalent The mass of a substance, according to Faraday's law, produced or consumed by electrolysis with 100% current efficiency during the flow of a quantity of electricity equal to 1 faraday or 96,487 coulombs (1 coulomb corresponds to a current of 1 ampere during 1 second). Electrochemical equivalents are essential in the calculation of the current efficiency of an electrode process.

The electrochemical equivalent of a substance is equal to the gram-atomic or gram-molecular mass of this substance divided by the number of electrons involved in the electrode reaction. For example, the electrochemical equivalent of zinc, for which two electrons are required in order to deposit one atom, is Zn/2 or 65.37/2 g. Thus, the faraday is equal to the product of the charge of the electron times the number of electrons (the Avogadro number) required to react with 1 atom- or molecule-equivalent of substance. *See* COULOMETER; ELECTROLYSIS.

[P.De.]

Electrochemical process The principles of electrochemistry may be adapted for use in the preparation of commercially important quantities of certain substances, both inorganic and organic in nature. *See* ELECTROCHEMISTRY.

Inorganic processes. Inorganic chemical processes can be classified as electrolytic, electrothermic, and miscellaneous processes including electric discharge through gases and separation by electrical means. In electrolytic processes, chemical and electrical energy are interchanged. Current passed through an electrolytic cell causes chemical reactions at the electrodes. Voltaic cells convert chemicals into electricity. Electrothermic processes use electricity to attain the necessary temperature for reaction. For related information *see* ELECTROCHEMISTRY; ELECTROLYSIS; ELECTROLYTIC CONDUCTANCE; ELECTROMOTIVE FORCE (CELLS).

Electrolysis in aqueous solutions. The electrolysis of water to form hydrogen and oxygen, according to the reaction $2H_2O \rightarrow 2H_2 + O_2$, may be considered as the simplest process for aqueous electrolytes. It does not compete with hydrogen from propane or from natural gas and with oxygen from liquid air, except in small installations. While simplicity, high hydrogen purity requirement, and lower capital cost (in small plants) have justified electrolytic plants, severely rising energy costs have limited such applications. Heavy water, or deuterium oxide, used in moderating nuclear reactors is also a by-product of the electrolysis of water. *See* DEUTERIUM; HEAVY WATER; HYDROGEN; OXYGEN.

Metallurgical applications. Protective or decorative coatings on a base metal such as steel are obtained by electroplating. Plating may also be used to replace worn metal or to provide a wear-resistant surface. Electrogalvanizing is preferred over hot dipping for applying zinc to steel. Tin plate for containers is electrolytic. *See* ELECTROPLATING OF METALS; METAL COATINGS.

Electroforming is a method of forming or reproducing articles by electrodeposition. In contrast to electroplating, the product is removed from the base surface or mold. Electrodeposition of metal powders is used to produce particles in the 1- to 1000-micrometer range for use in powder metallurgy and metallic pigments. Electrolytic polishing of metals is accomplished by making the article anodic in an electrolyte of mixed acids. Electrolytic machining of metals is accomplished by making the metal part anodic in a suitable electrolyte. Electrorefining is a process for purifying metals and recovering their impurities, which at times are more valuable than the original metal. Electrowinning, sometimes termed aqueous electrometallurgy, involves processing of metallic ores by leaching solutions to obtain metal-containing electrolytes which can be processed with insoluble anodes and metal cathodes. *See* ELECTROMETALLURGY; ELECTROPOLISHING.

Alkali-chlorine processes. Electrolysis of alkali halides is the basis of the alkali-chlorine and chlorate industries. Chlorine, Cl_2, and caustic soda, NaOH (or caustic potash, KOH), are made by electrolysis of brine, a solution of sodium chloride, NaCl, in water. Hydrochloric acid electrolysis is of interest for recovery of chlorine from HCl resulting as a by-product from organic chlorinations. *See* CHLORINE.

Oxidations and reductions. These reactions occur in all cells, but in a narrower sense oxidation reactions are those in which oxygen or chlorine at the anode oxidizes some material to form a new compound; reduction reactions are those in which hydrogen, liberated at the cathode, reduces a material to a new product. There are no commercial applications of inorganic electrochemical reductions by this narrow definition.

Ion-permeable membrane cells. These utilize diaphragms made of ion-exchange resins. Cation-permeable membranes permit cations to pass through but not anions, whereas the reverse holds for anion-permeable membranes. Purification of sea water is the most important application. Salt has been recovered from sea water which has been concentrated in this way. *See* ION-SELECTIVE MEMBRANES AND ELECTRODES.

Fused-salt electrolysis. Aluminum, barium, beryllium, cerium and misch metal, fluorine, lithium, magnesium, sodium, molybdenum, thorium, titanium, uranium, and zirconium are obtained by electrolysis of fused salts, because water interferes with the desired reaction. Raw materials must all be purified before addition to fused-salt cells, because purification of the electrolyte is not economical as in aqueous electrolytes. Metallizing is a process of depositing a metal as an alloy on a substrate from a fused complex metal salt.

Electrothermics. The manufacture of many products requires temperatures higher than can be obtained by combustion methods. Electric heat can usually be developed at, or close to, the point where it is required, so that it is relatively quick. It permits easy control of the atmosphere for oxidizing, reducing, or neutral conditions.

Products of the electric furnace include iron and steel; ferroalloys; nonferrous metals and alloys; the exotic metals titanium, zirconium, hafnium, thorium, and uranium; and nonmetallic products such as calcium carbide, calcium cyanamide, sodium cyanide, silicon carbide, boron carbide, and graphite. *See* ELECTRIC FURNACE; PYROMETALLURGY.

Zone refining of metals for the electronics industry, such as silicon for diodes and transistors, is accomplished by induction melting of the metal in a narrow zone and slow movement of the molten zone in the metal ingot from one end to the other in an evacuated or inert gas–filled enclosure. Impurities move toward the end of the ingot. The operation is repeated until the desired purity is obtained. *See* ZONE REFINING.

Electrodialysis. This is the separation of low-molecular-weight electrolytes from aqueous solutions by migration of the electrolyte through semipermeable membranes in an electric field. It is used on an industrial scale for deashing starch hydrolyzates and whey, and in many municipalities for producing potable water from saline water. Its uses also include the concentration of liquid foods such as dairy products and citrus juices, the recovery of sulfite pulp waste and pickling acid, and the isolation of proteins. *See* COLLOID; DIALYSIS; SALINE WATER RECLAMATION.

Electrophoretic deposition. This is the deposition of a nonconductive material in a finely divided state from a suspension in an inert medium. Electrophoresis is the migration of colloidal particles, which acquire positive or negative charges in an electric field. The process is useful in electropainting; for instance, electropainting of automobile bodies and other objects has now been adopted on a large scale. Rubber latex is an example of a negatively charged colloid which can be plated on an anode. Electronic components can be coated with inorganic salts, oxides, and ceramics suspended in organic media. *See* ELECTROPHORESIS.

Electroendosmosis. This is the movement of a liquid with respect to an immobilized colloid in an electric field. The process is used in the dehydration of peat, dye pastes, and clay. It is also used commercially for dewatering soils in mining, road building construction, and other civil engineering works.

Electrostatic technique. The deposition of charged particles from suspension in gases has many useful applications. The Cottrell electrostatic precipitator removes dusts and mists from gases. *See* DUST AND MIST COLLECTION; ELECTROSTATIC PRECIPITATOR.

Spray painting with a high voltage between the spray gun and the work is particularly effective in providing an even coating with an economical use of paint on irregular and open surfaces, such as a screen.

In xerography a sheet of plain paper is electrically sensitized in those areas corresponding to an original so that colored resin particles carrying an opposite charge are attracted and retained only on the sensitized areas, thus producing a visible image corresponding to the original.

Abrasive paper and cloth are coated with an adhesive and abrasive powders attracted to the base material in an electrostatic field. Pile fabrics can be produced in a similar manner, with the short fibers oriented by the electric field.

Organic processes. Organic electrochemistry was once regarded as a tantalizing area with many important laboratory achievements but few successes in commercial practice. This situation is changing, however, in that electroorganic processes are likely to prove commercially advantageous if they can fulfill either of two conditions: (1) performance under conditions of voltage corresponding thermodynamically to the conversion of an organic group to a reduced or oxidized group, with the cell products relatively easy to isolate and purify; (2) performance of a highly selective, specific technique to make an addition at a double bond, or to split a particular bond (for example, between carbon atoms 17 and 18 of a complex molecule having 25 carbon atoms).

Selectivity and specificity are highly important in electroorganic processes for the manufacture of complicated molecules of vitamins and hormones—as well as for the medicinal products whose action on pathogenic organisms is a function of their spatial arrangement, steric forms, and resonance.

The electrolytic approach can also be competitive for some low-cost, tonnage products. Here continuous processing is important, and only a single phase should be present, that is, a solution rather than an emulsion, dispersion, or mechanical mixture. Only for fairly valuable products is it practical to find a conducting solvent and then to engineer around it.

The electrolytic oxidation and reduction of organic compounds differ from the corresponding and more familiar inorganic reactions only in that organic reactions tend to be more complex and have low yields. The electrochemical principles are precisely those of inorganic reactions, while the procedures for handling the chemicals are precisely those of organic chemistry. [C.L.M.]

Electrochemical series A series in which the metals are listed in the order of their chemical reactivity, the most active at the top and the less reactive or more "noble" metals at the bottom. In a broader sense such an activity series need not be limited to the metals but may be carried on through the electronegative (nonmetallic) elements as well. See the table for a list of common elements.

Electrochemical series of the elements*

Lithium	Li	Aluminum	Al	Molybdenum	Mo
Potassium	K	Titanium	Ti	Tin	Sn
Rubidium	Rb	Zirconium	Zr	Lead	Pb
Cesium	Cs	Manganese	Mn	Germanium	Ge
Radium	Ra	Vanadium	V	Tungsten	W
Barium	Ba	Niobium	Nb	**Hydrogen**	**H**
Strontium	Sr	Boron	B		
Calcium	Ca	Silicon	Si	Copper	Cu
Sodium	Na	Tantalum	Ta	Mercury	Hg
Lanthanum	La	Zinc	Zn	Silver	Ag
Cerium	Ce	Chromium	Cr	Gold	Au
Magnesium	Mg	Gallium	Ga	Rhodium	Rh
Scandium	Sc	Iron	Fe	Platinum	Pt
Plutonium	Pu	Cadmium	Cd	Palladium	Pd
Thorium	Th	Indium	In	Bromine	Br
Beryllium	Be	Thallium	Tl	Chlorine	Cl
Uranium	U	Cobalt	Co	Oxygen	O
Hafnium	Hf	Nickel	Ni	Fluorine	F

*According to standard oxidation potentials E^0 at 25°C (77°F).

The electrochemical series as it applies to metals was first established by laboratory experiments in which the purpose was to determine which metals would displace others from solutions of their salts. By exhaustive experiments it becomes possible to draw up a complete list in the order of chemical activity, in which the metals at the top of the list are those which are found to give up their electrons most readily (that is, are the most electropositive elements). Such a list is shown in the table, where lithium exhibits the most reactivity as a metal.

To obtain an accurate and reproducible activity series, it is best to use the electrode potential, or oxidation-reduction potential, which is defined as the voltage developed by a sample of pure metal immersed in a solution of one of its salts (at unit activity and at 25°C or 77°F) versus a hydrogen electrode immersed in hydrochloric or sulfuric acid of equivalent concentration. *See* ELECTROCHEMISTRY; ELECTRODE POTENTIAL; ELECTRONEGATIVITY; OXIDATION-REDUCTION. [E.G.Ro.]

Electrochemical techniques Experimental methods developed to study the physical and chemical phenomena associated with electron transfer at the interface of an electrode and solution. The objective is to obtain either analytical or fundamental information regarding electroactive species in solution. Fundamental electrode characteristics may be investigated also.

The physical and chemical phenomena important in electrode processes generally occur very close to the electrode surface. Mass transfer of species involved in an electrode process to and from the bulk of solution is one important aspect. Inclusion of a large excess of inert electrolyte in most electrochemical systems eliminates electrical migration as an important means of mass transfer for electroactive species, and only convection and diffusion are considered.

Important chemical aspects of electrode processes include the oxidation or reduction occurring as a result of electron transfer, and coupled chemical reactions. Coupled reactions are initiated by production or depletion of the primary products or reactants at the electrode surface.

The primary experimental variables involved in electrochemical techniques are the potential, the current, and the time. Either the potential or current at the working electrode is controlled and the other observed as a function of time. The many ways in which either may be controlled give rise to the wide variety of controlled-potential or controlled-current techniques. *See* ELECTROCHEMICAL PROCESS; ELECTRODE; ELECTROLYSIS. [S.P.Pe.]

Electrochemistry The science dealing with the chemical changes accompanying the passage of an electric current, or the reverse process in which a chemical reaction is used as the source of energy to produce an electric current, as in a battery. Ionic conduction in electrolytes (liquid solutions, molten salts, and certain ionically conductive solids) is a phase of electrochemistry. Conduction in metals, semiconductors, and gases is generally considered a portion of physics. Other aspects of electrochemistry are described below. *See* ELECTROLYTIC CONDUCTANCE.

Galvanic cells. These are better known as electric batteries. Many chemical reactions can be arranged to produce electrical energy by physically separating the reaction into two half-reactions, one supplying electrons to an electrode forming the negative terminal of the cell, and the other removing the electrons from the positive terminal. *See* BATTERY; FUEL CELL; STORAGE BATTERY.

Electrodeposition. The most important type of chemical reaction brought about by the passage of electric current is the deposition of a metal at a cathode from a solution of its ions. Electroforming is a variety of electrodeposition in which an article to be reproduced is rendered conductive by spraying a thin metallic coating, then electroplated with a metallic deposit that is stripped from its substrate and filled with backing to reproduce the original article. Electrowinning is used for the commercial production of active metals, such as aluminum, magnesium, and sodium, from molten salts and others, such as copper, manganese, and antimony, from aqueous solution. Electrorefining is commonly used to purify metals such as silver, lead, and copper. The impure metal is used as the anode, and purified metal is deposited at the cathode. *See* ELECTROPLATING OF METALS.

Electrolytic processes. Many electrode reactions other than metal deposition are of commercial or scientific use. Electrolysis of brine to yield chlorine at the anode, hydrogen at the cathode, and sodium hydroxide in the electrolyte is an important industrial process. Many organic compounds can be prepared electrolytically. *See* ELECTROLYSIS.

Electroanalytical chemistry. Many electrochemical measurements are useful for analytical purposes. Electrodes that are commonly used for analytical purposes through measurement of their potentials include the glass electrode for pH measurements, and ion-selective electrodes for certain ions, such as sodium or potassium ion (special glass compositions), calcium ion (liquid membrane), and fluoride ion (doped lanthanum fluoride single crystals). Polarography involves the use of a dropping mercury electrode as one electrode of an electrolytic cell. Qualitative analysis is carried out by measurement of characteristic potentials (half-wave potentials) for electrode processes, and quantitative analysis by measurement of diffusion-controlled currents. Coulometry involves the application of Faraday's law for analytical purposes.

Other analytical applications of electrochemistry include chronopotentiometry (measurement of potential-time transients under constant current conditions), and linear sweep and cyclic voltammetry (measurement of currents with linear voltage scan). Several titration methods involve electrochemical measurements, for example, conductometric, potentiometric, and amperometric titrations. *See* ELECTRODEPOSITION ANALYSIS; ION-SELECTIVE MEMBRANES AND ELECTRONS; POLAROGRAPHIC ANALYSIS; TITRATION.

Miscellaneous phenomena. Electrochemical transport of ions through synthetic or natural membranes is important for processes, such as desalination of water and electrodialysis. In biological systems, the transmittal of nerve impulses and the generation of electrical signals, such as brain waves, are basically of electrochemical origin. A set of related phenomena can be grouped together under electrokinetic behavior, including the motion of colloidal particles in an electric field (electrophoresis), the motion of the liquid phase relative to a stationary solid under the influence of a potential gradient (electroosmosis), and the inverse generation of a potential gradient caused by a flowing liquid (streaming potential). Alternating-current phenomena, such as dielectric behavior, double-layer charging, and faradaic rectification, may also be included in a general definition of electrochemistry. Corrosion and passivation of metals are electrochemical in nature. [H.A.L.]

Electrochromic displays Devices that employ a reversible electrochemical reaction to cause a change in color of segments patterned to form alphanumeric characters.

Electrochromic displays are passive devices that only modulate ambient light, in contrast to a light-emitting diode. Hence, they operate at low voltages, and have a low enough energy requirement that a watch-size display can be operated for about 1 year from a small commercial battery. In the off state, the segments are typically colorless; in the on state, they are brightly colored, for example, blue or purple. *See* LIGHT-EMITTING DIODE.

The structure of an electrochromic display package is shown in the illustration. It consists of a top glass piece with the transparent electrodes, typically vacuum-deposited indium and tin oxides, on the inside surface. The lead from each segment continues to the edge of the glass. An insulating layer of silicon dioxide is deposited onto the leads to prevent their coloration. A counter electrode is deposited on the bottom piece. The two pieces are held together with a solder-glass, or epoxy, seal. To operate the device, a dc potential of 1–1.5 V is applied between the segments on the top plate and the counter electrode. Coloration and bleaching times are in the 200–500-millisecond range.

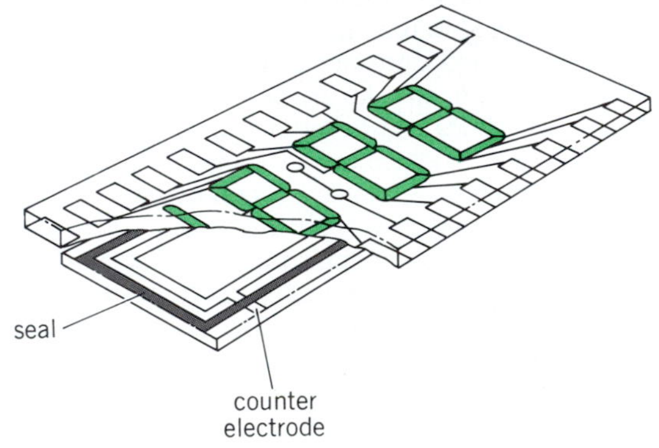

Structure of an electrochromic display.

Electrochromic displays can be categorized by the type of reaction that occurs when the potential is applied. Viologen electrochromic displays operate in an electroplating mode; with tungsten trioxide displays, the electrode material itself undergoes a color change. Electrochromic displays would be most efficiently utilized in applications requiring the display of large-format and slowly changing information, such as in clocks or message boards. *See* ELECTROCHEMICAL PROCESS; ELECTRONIC DISPLAY; ELECTROOPTICS. [G.G.B.]

Electroconvulsive shock The technique is that of eliciting convulsions by applying an electric current through the brain of a human patient or experimental animal for a brief period. Low current produces clonic convulsions, while high current produces tonic convulsions followed by clonic convulsions. Tonic convulsions cause loss of consciousness for a brief period. In the early 1930s drug-induced convulsions were used as treatment for mental disorders. In 1938 electroconvulsive shock treatments were introduced as a more convenient and reliable procedure. Electroconvulsive shock subsequently replaced convulsive drugs and is still used in some circumstances as a treatment for severe mental depression. The bases of the effectiveness of such clinical treatment are not known. [J.L.McG.]

Electrode An electrical conductor through which an electric current enters or leaves a conducting medium, whether it be an electrolytic solution, solid, molten mass, gas, or vacuum. For electrolytic solutions, many solids, and molten masses, an electrode is an electric conductor at the surface of which a change occurs from conduction by electrons to conduction by ions. For gases and vacuum, the electrodes merely serve to conduct electricity to and from the medium. *See* ELECTRODE POTENTIAL; ELECTRODEPOSITION ANALYSIS; ELECTROLYSIS; ELECTROMOTIVE FORCE (CELLS). [W.J.H.]

Electrode potential The potential which a metal or gas electrode takes up relative to a solution of ions,

The various electrodes encountered in electrochemical work may be grouped into seven types: (1) metal–metal ion, (2) amalgam, (3) nonmetal nongas, (4) gas, (5) metal–insoluble salt, (6) metal-insoluble oxide, and (7) oxidation-reduction. Any of these electrodes may be combined with any other to give a cell, the electromotive force (emf) of which is equal to the algebraic sum of the potentials of the two electrodes.

In order to measure the potential of any electrode, it is necessary, in principle, to combine the electrode with a reference electrode such as the hydrogen electrode, which has an arbitrarily assigned potential, and to measure the total voltage across the two half-cells. The potential of the reference electrode is then subtracted to give the required electrode potential. For various reasons, such as the difficulty in setting up a hydrogen gas electrode, several subsidiary reference electrodes, whose potentials are known on the hydrogen scale, have been devised. The most common of these is the calomel electrode, which consists of mercury in contact with a solution of potassium chloride saturated with mercurous chloride. *See* CALOMEL ELECTRODE.

A simple voltmeter cannot be used alone for measuring the emf of a small cell because the operation of the voltmeter draws some current, which causes chemical changes at the electrodes and produces a different voltage. To avoid these difficulties, the emf is measured with a potentiometer, that is, by balancing against a cell a known voltage under conditions such that practically no current flows.

The standard cell that is widely employed for emf measurements is some form of the Weston cell. It is highly reproducible, its emf remains constant over long periods of time, and it has a small temperature coefficient. *See* ELECTROCHEMISTRY; ELECTROMOTIVE FORCE (CELLS); ELECTRONEGATIVITY; OXIDATION-REDUCTION. [R.Gl.]

Electrodeposition analysis An electroanalytical chemistry technique in which the product of the electrode reaction is insoluble and is either deposited upon or dissolved into the electrode. This method usually results in the complete depletion in solution of the deposited substance. Electrodeposition is applied most commonly to the separation of metals from solution and to their subsequent quantitative determination. However, it may be used simply as a separation technique. *See* ELECTROLYSIS.

The quantities of metals deposited may be determined by weighing a suitable electrode before and after deposition (electrogravimetry). Among the elements that may be determined by deposition are cadmium, nickel, zinc, copper, cobalt, antimony, tin, silver, and gold. At the anode the oxides of lead, PbO_2, and manganese, MnO_2, may be deposited. Halides may be deposited on a silver anode as the respective silver halides. *See* COULOMETRIC ANALYSIS. [G.W.O'D.]

Electrodermal response A transient change in certain electrical properties of the skin, associated with the sweat gland activity and elicited by any stimulus that evokes an arousal or orienting response. Originally termed the psychogalvanic reflex, this phenomenon became known as the galvanic skin response. Phasic changes in skin conductance and potential (voltage) are designated as SCRs and SPRs, respectively, and electrodermal response has replaced galvanic skin response as the collective term.

The skin of a relaxed person has a low electrical conductance (high resistance), and the skin surface is some 40 mV negative with respect to interior tissues. Sweat gland activity changes these electrical properties by increasing skin conductance and by changing the balance of positive and negative ions in the secreted fluid.

Tonic skin conductance varies with psychological arousal, rising sharply when the subject awakens and rising further with activity, mental effort, or especially stress. Phasic skin conductance responses are wavelike increases in skin conductance that begin 1–2 seconds after stimulus onset and peak within about 5 seconds. The amplitude of the skin conductance response varies with the subjective impact of the eliciting stimulus, which in turn varies with the intensity of the stimulus, its novelty or unexpectedness for the subject, and its meaning or signal value. Aroused subjects display spontaneous skin conductance responses, generated apparently by mental events or other internal stimuli; their frequency, like the tonic skin conductance level, increases with the level of arousal.

Electrodermal responses are measured in studies of emotion and stress, conditioning, habituation, and cognitive processing; that is, when it is desired to assess the differential or changing impact of a series of stimuli. *See* LIE DETECTOR. [D.T.L.]

Electrodiagnosis Diagnosis using electrical equipment. Three major kinds of electrodiagnostic procedures are used clinically. Electrical activity of the brain is observed and measured by electroencephalography, of the heart by electrocardiography, and of skeletal muscle by electromyography. All three are also used extensively in basic physiological, psychophysiological, and pharmacological research, and each is based on the principle that living cells have an electrical potential difference between their interior and external surfaces. Electrodiagnosis utilizes equipment that amplifies and records changes in potential, especially in those cells that are highly specialized electrical conductors (heart and skeletal muscle or nerve cells) and whose principal function is related thereto. *See* ELECTROCARDIOGRAPHY; ELECTROENCEPHALOGRAPHY. [N.K.M.]

Electrodialysis Dialysis rates of ionic low-molecular-weight solutes may be increased greatly by applying an electric field to achieve an electrophoretic movement of ions through the dialytic membrane. This combined process of dialysis and electrophoretic transport of solutes through a membrane is known as electrodialysis. *See* ELECTROPHORESIS.

Ions do not migrate readily through membranes that carry in their pores charges of the same sign as the ion; ions of charge opposite to that of the membrane are not prevented from passing through. Such a selective permeability to ions by charged membranes has been used to improve the efficiency of electrodialysis. Membranes of ion-exchange resins are commercially available in cationic and anionic forms. *See* ION SELECTIVE MEMBRANES AND ELECTRODES.

For desalting small quantities of solutions for chromatography and radioisotope tracer studies, micro- and semimicroelectrodialyzers have proved valuable. On a larger scale, electrodialysis combined with ion-exchange membranes has been used to remove salt from sea water. *See* COLLOID; SALINE WATER RECLAMATION. [Q.V.W.]

Electrodynamic ammeter An ammeter in which a movable coil rotates in the magnetic field produced by a fixed coil. The instrument responds to alternating as well as direct current, and is thus a transfer instrument. Such instruments are precise and may be calibrated accurately and used for secondary standards, but the magnetic field of the fixed-coil system is relatively weak, 60 gauss (6 milliteslas) being a common value. The controlling forces are thus much lower than in other types, and the electrodynamic ammeter is mainly used in the laboratory for calibrating the more rugged types. *See* AMMETER; ELECTRODYNAMIC INSTRUMENT. [J.H.M./J.Be.]

Electrodynamic instrument A term used in preference to the older term electrodynamometer to refer to an instrument in which the electromagnetic reaction between two coils or sets of coils, each carrying electric current, can be measured. Because the reaction is proportional to the product of the current in the fixed and movable coils, the arrangement has found wide usage. It can be used to indicate ac or dc current, voltage, or power; with a crossed-coil movement it can be used for power-factor, phase-angle, frequency, and capacity measurements. For the crossed-coil instrument *see* PHASE METER.

A modern version of an electrodynamic instrument mechanism is shown in the illustration. Note the pair of fixed coils and the movable coil on a vertical staff with spiral countertorque springs for both control and current-carrying connections to the moving coil.

Electrodynamic instruments respond equally well to direct currents or to low-frequency alternating currents and therefore are used as transfer instruments to relate dc and ac currents. Transfer instruments are calibrated on direct current, using a standard cell and a standard ohm for reference, and are then used to measure correctly the root-mean-square value of an alternating current.

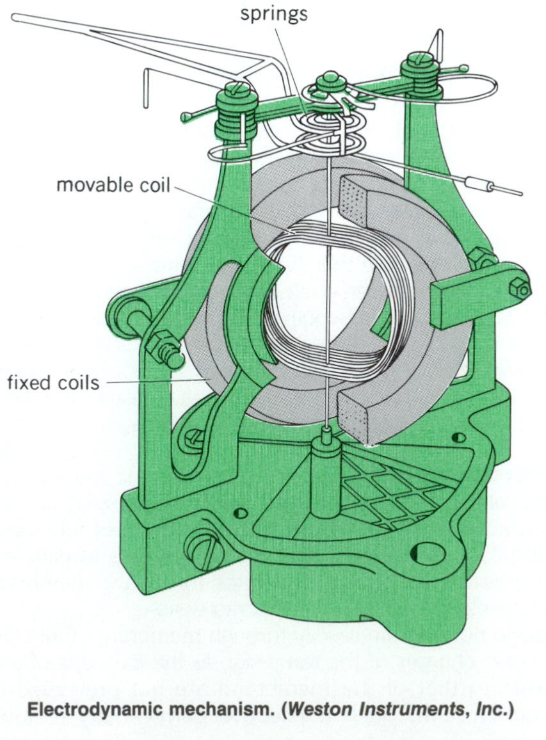

Electrodynamic mechanism. (*Weston Instruments, Inc.*)

The pointer deflection of the electrodynamic instrument is a function of the product of two currents; thus the instrument can be connected as a wattmeter. The main current flows through the fixed coils, and a small current proportional to the voltage flows through the movable coil. On alternating current where the current may lag the voltage, the instrument responds only to the in-phase product of the two currents and hence indicates in terms of true watts, irrespective of power factor. *See* WATTMETER. [J.H.M./J.Be.]

Electrodynamics The study of the relations between electrical, magnetic, and mechanical phenomena. This includes considerations of the magnetic fields produced by currents, the electromotive forces induced by changing magnetic fields, the forces on currents in magnetic fields, the propagation of electromagnetic waves, and the behavior of charged particles in electric and magnetic fields. Classical electrodynamics deals with fields and charged particles in the manner first systematically described by J. C. Maxwell, whereas quantum electrodynamics applies the principles of quantum mechanics to electrical and magnetic phenomena. Relativistic electrodynamics is concerned with the behavior of charged particles and fields when the velocities of the particles approach that of light. Cosmic electrodynamics is concerned with electromagnetic phenomena occurring on celestial bodies and in space. *See* COSMIC ELECTRODYNAMICS; ELECTROMAGNETISM; QUANTUM ELECTRODYNAMICS; RELATIVISTIC ELECTRODYNAMICS. [J.W.St.]

Electroencephalography The biomedical technology and science of recording the minute electric currents produced by the brains of humans and animals. Electroencephalography (EEG) has important clinical significance for the diagnosis of brain disease. The interpretation of EEG records has become a clinical specialty for neurological diagnosis.

The recording machine, the electroencephalograph, usually produces a 16-channel ink-written record of brain waves, the electroencephalogram. It is interpreted by an electroencephalographer. The placement of about 20 equally spaced electrodes pasted to the surface of the scalp is in accordance with the standard positions adopted by the International Federation of EEG, and is called the 10/20 system. About 10 patterns or montages of combinations of electrode pairs are selected for transforming the spatial location from the scalp to the channels which are traced on the EEG pen writer.

Electrical voltage is transduced from the scalp by differential input amplifiers and amplified about a million times in order to drive the pens for the paper record. The recording usually takes 30–60 min during a relaxed waking state, and also during sleep when possible. Often, activating procedures are used, such as a flickering light stimulator and hyperventilation or overbreathing for about 3 min.

EEG waves are defined by form and frequency. Various frequencies are given Greek letter designations. Alpha rhythm is defined as 8–12-Hz sinusoidal rhythmical waves. Alpha waves are normally present during the waking and relaxed state and enhanced by closing the eyes. They are suppressed or desynchronized when the eyes are open, or when the individual is emotionally aroused or doing mental work. They are of highest amplitude in the posterior regions of the brain. The alpha rhythm develops with age, reaching maturity by about 12 years, stabilizes, and then declines in frequency and amplitude in old age (over 65).

Beta rhythms are faster, low-voltage sinuosidal waves, usually about 14–30 Hz. They are more prominent in the frontal areas. They are prevalent during sedation with phenobarbital or with the use of tranquilizers and some sedative drugs.

Slower rhythms are theta and delta waves. Theta waves of 4–7 Hz usually replace the alpha rhythm during drowsiness and light sleep. Delta waves of 0.5–4 Hz are present during deep steep in normal people of all ages, and they are the primary waves present in the records of normal infants. Delta waves are almost always pathological in the waking records of adults.

Sleep stages are objectively defined by the use of EEG records. Theta waves are present in drowsiness and light sleep (stage I). Light sleep (stage II) is characterized by negative sharp waves at the vertex (top of head), called sleep humps, and spindle bursts of 13–15-Hz waves in the central regions of the head (over the Rolandic fissures). These are intermixed with theta waves, and seen only during light sleep, sometimes called the hump-and-spindle stage of sleep. Low-voltage, desynchronized delta waves are present during deep sleep (stage III), and higher-amplitude, slower, and more synchronized delta waves are seen in the deeper sleep stage (stage IV). Stage V sleep is called the dream stage or rapid-eye-movement (REM) state. That is when dreams take place, accompanied by rapid horizontal eye movements, as if the sleeper were scanning a real scene. The EEG pattern is more like waking, with alpha rhythm present. The REM state is the one in which it is hardest to wake up the subject, so that it is sometimes called paradoxical sleep. *See* SLEEP AND DREAMING.

The EEG reveals functional abnormalities of the brain, whether caused by localized structural lesions, essential paroxysmal states such as epilepsy, or toxic and abnormal metabolic conditions. The three major classes of abnormalities are asymmetries between the hemispheres, slow rhythms, and very sharp waves or spikes. Slow waves represent a depression of cerebral cortical activity or injury in the projection pathways beneath the recording electrodes. Sharp waves or spikes often indicate a hyperexcitable or irritable state of the cortex. During a full epileptic seizure attack, spikes become repetitive and synchronized over the whole surface of the brain. *See* EPILEPSY.

The EEG is frequently used for the evaluation of comatose states. The record is slowed in all areas in coma, with delta waves predominating. If the EEG becomes isoelectric or flat for several hours, brain function is not recoverable and the coma may be considered terminal. "Brain death" is determined by a flat EEG, recorded at the highest gain with widely spaced electrode positions and the absence of cerebral reflexes and spontaneous respiration. *See* DEATH. [J.Co.]

Electrokinetic phenomena Phenomena associated with the movement of charged particles through a continuous medium or with the movement of a continuous medium over a charged surface. The four principal electrokinetic phenomena are electrophoresis, electroosmosis, streaming potential, and sedimentation potential, or Dorn effect. These phenomena are related to one another through the zeta potential ζ of the electrical double layer which exists in the neighborhood of the charged surface. *See* ELECTROPHORESIS; STREAMING POTENTIAL.

[Q.V.W.]

Electroless plating A chemical reduction process which, once initiated, is autocatalytic. The process is similar to electroplating except that no outside current is needed. The metal ions are reduced by chemical agents in the plating solutions, and deposit on the substrate. An advantage of electroless plating with current is the more uniform thickness of the surface coating.

Electroless plating is used for coating nonmetallic parts. Decorative electroless plates are usually further coated with electrodeposited nickel and chromium. There are also applications for electroless deposits on metallic substrates, especially when irregularly shaped objects require a uniform coating. Electroless copper is used extensively for printed circuits, which are produced either by coating the nonmetallic substrate with a very thin layer of electroless copper and electroplating to the desired thickness or by using the electroless process only. Electroless iron and cobalt have limited uses. Electroless gold is used for microcircuits and connections to solid-state components. Deeply recessed areas which are difficult to plate can be coated by the electroless process. *See* ELECTROPLATING OF METALS. [R.W.]

Electroluminescence A luminescence resulting from the application of an electric field to a material, usually solid. Several different types of electroluminescence can be distinguished. For a discussion of the basic mechanism involved *see* LUMINESCENCE.

In 1936, G. Destriau in France observed that when suitably prepared zinc sulfide phosphor powders, activated with small additions of copper, are suspended in an insulator and an intense alternating electric field is applied with capacitorlike electrodes, light emission results (the Destriau effect). The effect is the basis for modern electroluminescent lamps. The insulator may be either liquid (oil) or solid (glass or plastic). Thin films of tin oxide or metal are used as transparent electrodes.

Maximum efficiency is not obtained for the same operating conditions as maximum brightness, and efficiency is normally low compared with other light sources. *See* LIGHT PANEL.

Injection electroluminescence results when a semiconductor *pn* junction or a point contact is biased in the forward direction. It is the result of radiative recombination of injected minority carriers with majority carriers across the energy band gap of the material. Such emission has been observed in a large number of semiconductors. In some cases the efficiency is high, approaching one emitted photon for each carrier passing through the junction. In suitable structures the excitation intensity can become so high that stimulated rather than spontaneous emission predominates, and laser action results, with spectral narrowing and coherent emission. *See* JUNCTION DIODE; LASER; SEMICONDUCTOR; SEMICONDUCTOR DIODE. [C.C.K.; J.H.S.]

Electrolysis A method by which reactions are carried out in solutions of electrolytes or in molten salts by use of electricity. As shown in the illustration, the electrodes of an electrolytic

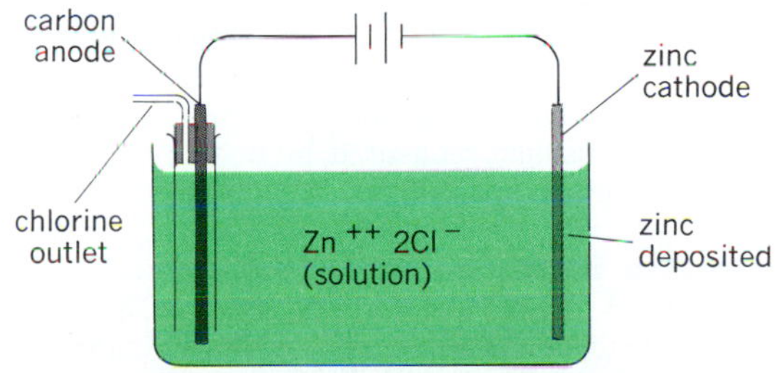

Electrolysis of zinc chloride solution.

cell are immersed in an electrolyte solution or in molten salts and are connected to a direct-current power supply. One or several reactions occur at each electrode when current flows through the cell. Reduction, a reaction in which electrons are consumed, occurs at the electrode called the cathode; oxidation occurs at the anode. For instance, sodium is produced at the cathode by reduction, and chlorine at the anode by oxidation in the electrolysis of molten sodium chloride.

Applications are important and varied: industrial production of chemicals, metallurgical extraction of metals, electroplating of metals, metal finishing, and production of electricity in batteries. Metallic corrosion often involves electrolytic processes. For application to analytical chemistry *see* ELECTRODEPOSITION

ANALYSIS; POLAROGRAPHIC ANALYSIS. *See also* ELECTROCHEMISTRY; ELECTROMETALLURGY; ELECTROPLATING OF METALS. [P.De.]

Electrolyte A chemical compound which when fused or dissolved in certain solvents, usually water, will conduct an electric current. The passage of the current is always accompanied by decomposition of the electrolyte, called electrolysis, which takes place at the electrodes. All electrolytes in the fused state or in solution give rise to ions which conduct the electric current. All acids, bases, and salts are electrolytes. *See* ELECTRIC CURRENT; ELECTRODE; ELECTROLYSIS; ION; SOLVENT.

Electrolytes are divided into strong and weak electrolytes. Strong electrolytes usually contain a stable ionic bond and are wholly ionized in solution and usually in the crystalline state. Weak electrolytes are only partially ionized in solution. Metallic hydroxides and salts are usually strong electrolytes, for example, potassium hydroxide and sodium chloride. A weak electrolyte contains a covalent bond which on dissolving in a solvent such as water may be transformed. into an ionic bond. A solution of a weak electrolyte contains both the ionic and the covalent forms in equilibrium; for example, acetic acid in water consists of a mixture of undissociated molecules, CH_3COOH, and of the ions CH_3COO^- and H^+. *See* CHEMICAL BONDING; IONIC EQUILIBRIUM. [T.C.W.]

Electrolytic conductance The transport of electric charges, under electric potential differences, by particles of atomic or larger size. This phenomenon is distinguished from metallic conductance, which is due to the movement of electrons. The charged particles that carry the electricity are called ions.

Positively charged ions are termed cations; the sodium ion, Na^+, is an example. The negatively charged chloride ion, Cl^-, is typical of anions. The negative charges are identical with those of electrons or integral multiples thereof. The unit positive charges have the same magnitude as those of electrons but are of opposite sign. Colloidal particles, which may have relatively large weights, may be ions, and may carry many positive or negative charges. Electrolytic conductors may be solids, liquids, or gases. Semiconductors have properties that are intermediate between the metallic and electrolytic types. *See* ELECTROCHEMISTRY; ELECTROMOTIVE FORCE (CELLS).

Conductances are usually reported as specific conductances κ, which are the reciprocals of the resistances of cubes of the materials, 1 cm (0.39 in.) in each dimension, placed between electrodes 1 cm square, on opposite sides. These units are sometimes called mhos, that is, ohm spelled backward. Conductances of solutions are usually measured by a method in which a Wheatstone bridge is employed. *See* WHEATSTONE BRIDGE. [H.A.L.]

Electrolytic tank A special type of computing machine based on the fact that in ideal fluid flow the velocity potential φ, and in planar flow the stream function ψ, satisfy Laplace's equations. Either the velocity potential or stream function or both can be related to the voltage through constants m and n called the scale factor as in the following equations:

$$\text{Analogy A: } \varphi = mV$$
$$\text{Analogy B: } \psi = nV$$

The fluid velocity components are proportional to the respective voltage gradients. Every ideal fluid-flow problem, therefore, has an electrical counterpart; for this reason the electrolytic tank is also known as an electrical tank analogy or potential flow analyzer. Its solution includes construction of a scaled electrical-flow model, its installation in an electrical tank with proper simulation of the physical boundary conditions both on the model contour and on the field boundaries, measurement of the electrical variables as required, and finally the translation of the measurements by numerical computation into meaningful fluid-flow terms. *See* FLUID FLOW; HYDRODYNAMICS; MAXWELL'S EQUATIONS. [W.B.Br.]

Electromagnet A soft-iron core that is magnetized by passing a current through a coil of wire wound on the core. Electromagnets are used to lift heavy masses of magnetic material and to attract movable magnetic parts of electric devices, such as solenoids, relays, and clutches.

The difference between cores of an electromagnet and a permanent magnet is in the retentivity of the material used. Permanent magnets, initially magnetized by placing them in a coil through which current is passed, are made of retentive (magnetically "hard") materials which maintain the magnetic properties for a long period of time after being removed from the coil. Electromagnets are meant to be devices in which the magnetism in the cores can be turned on or off. Therefore, the core material is nonretentive (magnetically "soft") material which maintains the magnetic properties only while current flows in the coil. All magnetic materials have some retentivity, called residual magnetism; the difference is one of degree. *See* MAGNETIZATION.

In an engineering sense the word electromagnet does not refer to the electromagnetic forces incidentally set up in all devices in which an electric current exists, but only to those devices in which the current is primarily designed to produce this force, as in solenoids, relay coils, electromagnetic brakes and clutches, and in tractive and lifting or holding magnets and magnetic chucks.

Electromagnets may be divided into two classes: traction magnets, in which the pull is to be exerted over a distance and work is done by reducing the air gap; and lifting or holding magnets, in which the material is initially placed in contact with the magnet. Examples of the latter type are magnetic chucks and circular lifting magnets. For examples of the first type *see* BRAKE; CLUTCH; RELAY; SOLENOID (ELECTRICITY). [J.Mei.]

Electromagnetic brake In commonly encountered electromagnetic brakes, the actuating force is applied by an electric current through a solenoid (see illustration). Direct current gives

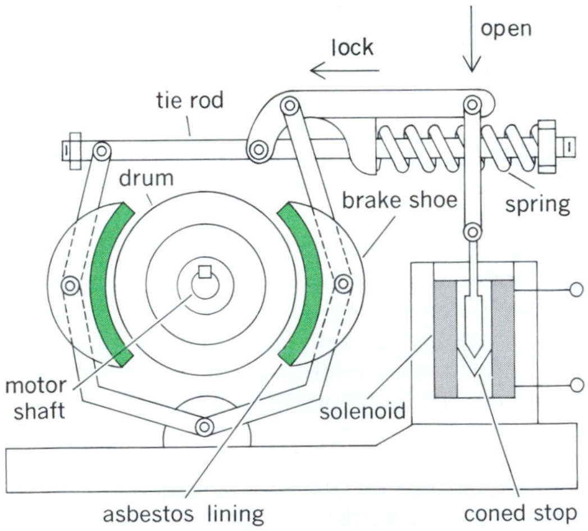

Elements of an electromagnetic brake.

greater braking force than alternating current, and is therefore used almost exclusively; ac is rectified if used. Usually the electromagnetic force releases the brake against a compression spring, thus providing braking action if the power fails and overspeed protection when the brake is used with dc series motors.

The plunger and its stop are cone-shaped to provide a higher force than without a stop, but without an abrupt increase in force as the gap closes. Because the solenoid force decreases somewhat when the brake is applied and the gap is thereby increased, the retraction current is slightly higher than the current at which the spring overcomes the solenoid and applies the brake. This operation is stable.

The solenoid is wound with high impedance for shunt operation across a power line. Shunt operation is common where loads and motor speeds vary widely or where a compound motor is used. Where a series motor is used for high starting torque, the solenoid is wound with low impedance for series operation in the circuit of the motor it protects. In this way the brake will be applied if the motor starts to run away under light load. *See* BRAKE; SOLENOID (ELECTRICITY). [H.P.H.]

Electromagnetic compatibility The capability of electronic equipment or systems to be operated in the intended electromagnetic environment at design levels of efficiency. The electromagnetic environment existing at the input or antenna terminal of a receiver is a mixture of signals from many sources. The mixture consists of both a desired signal and many undesired signals to which the receiver may be sensitive. These undesired or interference signals may come from equipment intended to radiate for the purpose of communications or from sources not intended to radiate electromagnetic energy but designed to transfer information. In addition to mutual interference between systems, the possibility of extraneous energy generated within a system interfering with its own operation exists. The subject of electromagnetic compatibility deals with reducing both the source of interference and its effect on the reception of signals. [S.H.C.]

Electromagnetic field A changing magnetic field always produces an electric field, and conversely, a changing electric field always produces a magnetic field. This interaction of electric and magnetic forces gives rise to a condition in space known as an electromagnetic field. The characteristics of an electromagnetic field are expressed mathematically by Maxwell's equation. *See* ELECTRIC FIELD; ELECTROMAGNETIC RADIATION; ELECTROMAGNETIC WAVE; MAGNETIC FIELD; MAXWELL'S EQUATIONS. [J.Be.]

Electromagnetic flowmeter A volume flow rate meter in which a magnetic field is applied across a metering tube carrying a flowing liquid, and the voltage generated between two perpendicularly located electrodes contacting the liquid is measured by a secondary or transmitter instrument. Since this voltage is strictly proportional to the average velocity, a linear volume flow signal results. Practical flow measurements can be made on any liquid that is at all conductive. The linear scale allows accurate flow measurements over a greater range than is possible in differential-pressure meters. The lack of obstruction to fluid flow makes it possible to measure thick slurries and gummy liquids. *See* ELECTROMAGNETIC INDUCTION; FLOW MEASUREMENT; VOLUME FLOW RATE METER. [M.Br.; L.P.E.]

Electromagnetic induction The production of an electromotive force either by motion of a conductor through a magnetic field in such a manner as to cut across the magnetic flux or by a change in the magnetic flux that threads a conductor. *See* ELECTROMOTIVE FORCE (EMF); FARADAY'S LAW OF INDUCTION; MAGNETIC FLUX.

If the flux threading a coil is produced by a current in the coil, any change in that current will cause a change in flux, and thus there will be an induced emf while the current is changing. This process is called self-induction. The emf of self-induction is proportional to the rate of change of current.

The process by which an emf is induced in one circuit by a change of current in a neighboring circuit is called mutual induction. Flux produced by a current in a circuit A threads or links circuit B. When there is a change of current in circuit A, there is a change in the flux linking coil B, and an emf is induced in circuit B while the change is taking place. Transformers operate on the principle of mutual induction. *See* TRANSFORMER.

The phenomenon of electromagnetic induction has a great many important applications in modern technology. For example, *see* COUPLED CIRCUITS; GENERATOR; INDUCTION HEATING; MICROPHONE; MOTOR; SERVOMECHANISM. [K.V.M.]

Electromagnetic propulsion Motive power for flight vehicles produced by high-speed discharge of a plasma fluid. Together with electrostatic (ion) propulsion, electromagnetic propulsion collectively designates several mechanisms capable of attaining specific impulses, exceeding by a considerable margin those of thermal propulsion devices. To heat fluids beyond the level of chemical reactions, a separate power source must be provided. Much more power can be transferred to a body of matter by electrical means than by heating, and material limitations impose a principal barrier to the rate of power transfer (propellant heating) in thermal propulsion systems. By increasing the power transfer enough to ionize the propellant material, it is possible to apply electric and electromagnetic driving forces, thereby circumventing the thermal barrier. However, different problems are encountered in the resulting electromagnetic and electrostatic drives. *See* ION PROPULSION; SPACECRAFT PROPULSION.

Electromagnetic propulsion uses highly ionized gases (plasma), whose behavior is determined by interaction between electric currents in the plasma and magnetic fields from the vehicle. The discharged plasma is electrically neutral. Discharge density is, therefore, not limited by electrostatic forces (space charge) present in an ion beam. Consequently, electromagnetic propulsion devices offer promise of higher thrust per unit discharge area. *See* MAGNETOHYDRODYNAMICS.

Electromagnetic propulsion is adaptable to a wide range of specific impulses. The optimization of a propulsion system within the framework of an overall vehicle system and a space mission calls for a compromise between acceleration and payload capability that best serves the particular purpose.

The accelerator provides direction and speed to the plasma flow and thus represents the thrust-producing mechanism. Electromagnetic propulsion devices can be divided, on the basis of accelerating mechanism, into steady-flow systems and pulsed systems. Of the presently recognized electromagnetic drives, the magnetoplasmadynamic arcjet shows the greatest potential because it offers the widest range of high specific impulses, highest thrust density, and longest operating life. [K.A.E.]

Electromagnetic pulse (EMP) A transient electromagnetic signal produced by a nuclear explosion in or above the Earth's atmosphere. Though not considered dangerous to people, the EMP is a potential threat to many electronic systems.

In a typical nuclear detonation, the nuclear reactions release tremendous amounts of energy as initial nuclear radiation in the form of neutrons and high-energy electromagnetic radiation, called gamma rays. As the prompt gammas move away from a high-altitude nuclear detonation (see illustration), those gamma rays moving toward the Earth penetrate a more dense region of the atmosphere called the source or deposition region. In this region the highly energetic gamma rays interact with the air molecules to form 1-MeV Compton electrons and less energetic gamma rays, which then proceed in the same general direction as the original gamma rays. The Compton electrons then spiral about the geomagnetic lines as they slow down. *See* COMPTON EFFECT; GAMMA RAYS; NUCLEAR REACTION; SYNCHROTRON RADIATION.

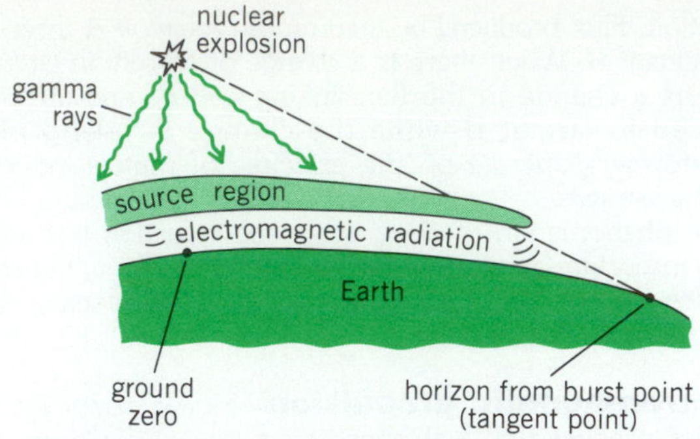

Schematic representation of the EMP in a high-altitude burst. The extent of the source region varies with the altitude and the yield of the explosion. (*After S. Glasstone and P. J. Dolan, eds., The Effects of Nuclear Weapons, U.S. Department of Defense and the Energy Research and Development Administration, 3d ed., 1977*)

If the observer is directly below the detonation (ground zero), the polarization of both electric and magnetic fields is predominantly horizontal; if the observer is at the horizon, the fields can have both horizontal and vertical components. Should the nuclear detonation occur closer to the Earth, the EMP generation process becomes far more complex and the electric and magnetic fields become very complicated. *See* POLARIZATION OF WAVES.

It is possible for initial nuclear radiation to directly interact with systems, causing EMP signals internal to structures. This phenomenon has been called internal or system-generated EMP and is potentially a serious problem for satellites and electronics in metallic enclosures.

The coupling of these rather wide-band (10^4 to 10^8 Hz) signals to systems of different topologies can be significant. For example, it is not unusual to predict voltage and current levels of hundreds of thousands of volts and a few thousand amperes coupled by high-altitude EMP onto extended systems. An estimate of about 1 joule (0.7 ft-lbf) of EMP coupled energy is considered reasonable for many systems. Even if the coupling onto circuits is inefficient, as little as 10^{-13} J can upset some semiconductor devices and 10^{-6} J can cause damage. The potential for such upset and damage in critical electronic systems has led to the development of semiconductor devices which are "hardened" or protected against EMP. This development is particularly advanced in communications systems whose disruption by EMP is considered an important civil and military vulnerability. *See* ELECTROMAGNETIC RADIATION; NUCLEAR EXPLOSION.

[R.A.Pf.]

Electromagnetic pumps Pumps that operate on the principle that a force is exerted upon a conductor (the fluid) carrying current in a magnetic field. The high electrical conductivity of liquid metals (used as heat-transfer media in some nuclear reactors) makes it possible to pump them by electromagnetic means.

For use in nuclear reactors, where a minimal amount of maintenance is desirable, electromagnetic pumps are often preferable to conventional mechanical pumps because they have no moving parts, bearings, or seals. Various methods are employed to cause current to flow in the liquid metal. See Nuclear reactor; Pump. [L.J.K.]

Electromagnetic radiation Energy transmitted through space or through a material medium in the form of electromagnetic waves. The term can also refer to the emission and propagation of such energy. Whenever an electric charge oscillates or is accelerated, a disturbance characterized by the existence of electric and magnetic fields propagates outward from it. This disturbance is called an electromagnetic wave. The frequency range of such waves is tremendous, as is shown by the electromagnetic spectrum in the table. The sources given are typical, but not mutually exclusive.

In theory, any electromagnetic radiation can be detected by its heating effect. This method has actually been used over the range from x-rays to radio. Ionization effects measured by cloud chambers, photographic emulsions, ionization chambers, and Geiger counters have been used in the γ- and x-ray regions. Direct photography can be used from the γ-ray to the infrared region.

Fluorescence is effective in the x-ray and ultraviolet ranges. Bolometers, thermocouples, and other heat-measuring devices are used chiefly in the infrared and microwave regions. Crystal detectors, vacuum tubes, and transistors cover the microwave and radio frequency ranges. *See* ANTENNA (ELECTROMAGNETISM); DIFFRACTION; ELECTROMAGNETIC WAVE; GAMMA RAYS; HEAT RADIATION; INFRARED RADIATION; INTERFERENCE OF WAVES; LIGHT; MAXWELL'S EQUATIONS; MICROWAVE; POLARIZATION OF WAVES; RADIATION; RADIO-WAVE PROPAGATION; REFLECTION OF ELECTROMAGNETIC RADIATION; REFRACTION OF WAVES; SCATTERING OF ELECTROMAGNETIC RADIATION; TRANSMISSION LINES; ULTRAVIOLET RADIATION; WAVE MOTION; X-RAYS.

[W.R.Sm.]

Electromagnetic wave A disturbance, produced by the acceleration or oscillation of an electric charge, which has the characteristic time and spatial relations associated with progressive wave motion. A system of electric and magnetic fields moves outward from a region where electric charges are accelerated, such as an oscillating circuit or the target of an x-ray tube. The wide wavelength range over which such waves are observed is shown by the electromagnetic spectrum. The term electric wave, or Hertzian wave, is often applied to electromagnetic waves in the radar and radio range. Electromagnetic waves may be confined in tubes, such as wave guides, or guided by transmission lines. They were predicted by J. C. Maxwell in 1864 and verified experimentally by H. Hertz in 1884. *See* ELECTROMAGNETIC RADIATION; ELECTROMAGNETIC WAVE TRANSMISSION.

[W.R.Sm.]

Electromagnetic wave transmission The transmission of electrical energy by wires, the broadcasting of radio signals, and the phenomenon of visible light are all examples of the propagation of electromagnetic energy. Electromagnetic energy travels in the form of a wave. Its speed of travel is approximately 3×10^8 m/s (186,000 mi/s) in a vacuum and is somewhat slower than this in liquid and solid insulators. An electromagnetic wave does not penetrate far into an electrical conductor, and a wave that is incident on the surface of a good conductor is largely reflected. *See* ELECTROMAGNETIC WAVE.

Electromagnetic waves originate from accelerated electric charges. For example, a radio wave originates from the oscillatory acceleration of electrons in the transmitting antenna. The light that is produced within a laser originates when electrons fall from a higher energy level to a lower one.

The waves emitted from a source are oscillatory and are described in terms of frequency of oscillation. The method of generating an electromagnetic wave depends on the frequency used, as do the techniques of transmitting the energy to another location and utilizing it when it has been received. Communication of information to a distant point is generally accomplished through the use of electromagnetic energy as a carrier.

The illustration shows the configuration of the electric and magnetic fields about a short vertical antenna in which flows a sinusoidal current. The picture applies either to an antenna in free space (in which case the illustration shows only the upper half of the fields), or to an antenna projecting above the surface of a highly conducting plane surface. In the latter case the conducting plane represents to a first approximation the surface of

Electromagnetic spectrum

Frequency, Hz	Wavelength, m	Nomenclature	Typical source
10^{23}	3×10^{-15}	Cosmic photons	Astronomical
10^{22}	3×10^{-14}	γ-rays	Radioactive nuclei
10^{21}	3×10^{-13}	γ-rays, x-rays	
10^{20}	3×10^{-12}	x-rays	Atomic inner shell
		Positron-electron annihilation	
10^{19}	3×10^{-11}	Soft x-rays	Electron impact on a solid
10^{18}	3×10^{-10}	Ultraviolet, x-rays	Atoms in sparks
10^{17}	3×10^{-9}	Ultraviolet	Atoms in sparks and arcs
10^{16}	3×10^{-8}	Ultraviolet	Atoms in sparks and arcs
10^{15}	3×10^{-7}	Visible spectrum	Atoms, hot bodies, molecules
10^{14}	3×10^{-6}	Infrared	Hot bodies, molecules
10^{13}	3×10^{-5}	Infrared	Hot bodies, molecules
10^{12}	3×10^{-4}	Far-infrared	Hot bodies, molecules
10^{11}	3×10^{-3}	Microwaves	Electronic devices
10^{10}	3×10^{-2}	Microwaves, radar	Electronic devices
10^{9}	3×10^{-1}	Radar	Electronic devices
		Interstellar hydrogen	
10^{8}	3	Television, FM radio	Electronic devices
10^{7}	30	Short-wave radio	Electronic devices
10^{6}	300	AM radio	Electronic devices
10^{5}	3000	Long-wave radio	Electronic devices
10^{4}	3×10^{4}	Induction heating	Electronic devices
10^{3}	3×10^{5}		Electronic devices
100	3×10^{6}	Power	Rotating machinery
10	3×10^{7}	Power	Rotating machinery
1	3×10^{8}		Commutated direct current
0	Infinity	Direct current	Batteries

the Earth. The fields have symmetry about the axis through the antenna. For pictorial simplicity only selected portions of the fields are shown in this illustration. The magnetic field is circular about the antenna, is perpendicular at every point to the direction of the electric field, and is proportional in intensity to the magnitude of the electric field, as in a plane wave. All parts of the wave travel radially outward from the antenna with the velocity equal to that of a plane wave in the same medium.

Often it is desired to concentrate the radiated energy into a narrow beam. This can be done either by the addition of more antenna elements or by placing a large reflector, generally parabolic in shape, behind the antenna. The production of a narrow beam requires an antenna array, or alternatively a reflector, that is large in width and height compared with a wavelength. The very narrow and concentrated beam that can be achieved by a laser is made possible by the extremely short wavelength of the radiation as compared with the cross-sectional dimensions of the radiating system.

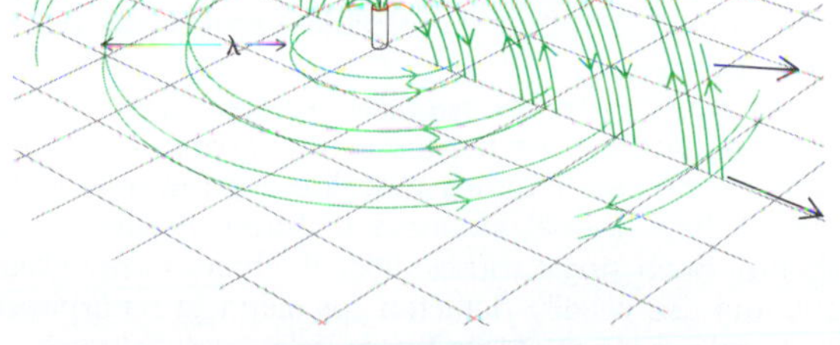

Configuration of electric and magnetic fields about a short vertical antenna. *E* = electric field intensity; *H* = magnetic field intensity; μ = absolute permeability of the medium; ε = permittivity of the medium; γ = wavelength.

The ground is a reasonably good, but not perfect, conductor; hence, the actual propagation over the surface of the Earth will show a more rapid decrease of field strength than that for a perfect conductor. Irregularities and obstructions may interfere. In long-range transmission the spherical shape of the Earth is important. Inhomogeneities in the atmosphere refract the wave somewhat. For long-range transmission, the ionized region high in the atmosphere known as the Kennelly-Heaviside layer, or ionosphere, can act as a reflector. *See* RADIO-WAVE PROPAGATION.

When an electromagnetic wave is introduced into the interior of a hollow metallic pipe of suitably large cross-sectional dimensions, the energy is guided along the interior of the pipe with comparatively little loss. The most common cross-sectional shapes are the rectangle and the circle. The cross-sectional dimensions of the tube must be greater than a certain fraction of the wavelength; otherwise the wave will not propagate in the tube. For this reason hollow waveguides are commonly used only at wavelengths of 10 cm or less (frequencies of 3000 MHz or higher). A dielectric rod can also be used as a waveguide. Such a rod, if of insufficient cross-sectional dimensions, can contain the electromagnetic wave by the phenomenon of total reflection at the surface. *See* WAVEGUIDE.

Electromagnetic energy can be propagated in a simple mode along two parallel conductors. Such a waveguiding system is termed a transmission line. Three common forms are the coaxial cable, two-wire line, and parallel strip line. As the wave propagates along the line, it is accompanied by currents which flow longitudinally in the conductors. These currents can be regarded as satisfying the boundary condition for the tangential field at the surface of the conductor. The conductors have a finite conductivity, and so these currents cause a transformation of electrical energy into heat. The energy lost comes from the stored energy of the wave, and so the wave, as it progresses, diminishes in amplitude. The conductors are necessarily supported by insulators which are imperfect and cause additional attenuation of the wave. *See* COAXIAL CABLE; TRANSMISSION LINES. [W.C.Jo.]

Electromagnetism The branch of science dealing with the observations and laws relating electricity to magnetism. Electromagnetism is based upon the fundamental observations that a moving electric charge produces a magnetic field and that a charge moving in a magnetic field will experience a force. The magnetic field produced by a current is related to the current, the shape of the conductor, and the magnetic properties of the medium around it by Ampère's law. The magnetic field at any point is described in terms of the force that it exerts upon a moving charge at that point. The electrical and magnetic units are defined in terms of the ampere, which in turn is defined from the force of one current upon another. The association of electricity and magnetism is also shown by electromagnetic induction, in which a changing magnetic field sets up an electric field within a conductor and causes the charges to move in the conductor. *See* AMPÈRE'S LAW; EDDY CURRENT; ELECTRICITY; ELECTROMAGNET; ELECTROMAGNETIC INDUCTION; FARADAY'S LAW OF INDUCTION; HALL EFFECT; INDUCTANCE; LENZ'S LAW; MAGNETIC CIRCUITS; MAGNETIC FIELD; MAGNETIC FLUX; MAGNETIC INDUCTION; MAGNETISM; MAGNETOMOTIVE FORCE; MAXWELL'S EQUATIONS; RELUCTANCE. [K.V.M.]

Electrometallurgy The branch of process metallurgy dealing with the use of electricity for smelting or refining of metals. The electrochemical effect of an electric current brings about the reduction of metallic compounds, and thereby the extraction of metals from their ores (electrowinning) or the pu-rification of the metals (electrorefining). *See* ELECTROREFINING; ELECTROWINNING.

In other metallurgical processes, electrically produced heat is utilized in smelting, refining, or alloy manufacturing. For a discussion of electrothermics, that is, the theory and applications of electric heating to metallurgy, *see* ELECTRIC FURNACE; ELECTRIC HEATING; ELECTROCHEMICAL PROCESS; STEEL MANUFACTURE. [P.Du.]

Electrometer A highly sensitive instrument used to measure a voltage, without drawing appreciable current, by measuring the electrostatic force exerted between two bodies that are charged with the voltage. In one form two suspended parallel strips of gold leaf are spread out at an angle proportional to the voltage to which they are charged; the amount of movement is measured with a microscope having a calibrated scale. In a string electrometer the gold strips are replaced by lightly stretched metalliized quartz fibers. Electrometers involving mechanical movements have largely been replaced by electronic devices, which are essentially voltage-measuring amplifiers having such a high input resistance (usually above 10^{10} ohms) that they draw practically no current. *See* ELECTROSCOPE; VOLTAGE MEASUREMENT; VOLTMETER. [J.Mar.]

Electromotive force (cells) When two dissimilar electrodes are connected through an external conducting circuit, a difference in electric potential exists between them. Although this difference in potential is sometimes called the potential difference of the electrode couple, it is customary to say that the galvanic cell composed of the two dissimilar electrodes exhibits an electromotive (driving) force. This electromotive force (emf) is the resultant of the relative potential forces of the two dissimilar electrodes at which electrochemical reactions occur during cell operation. The two dissimilar electrodes need not be of unlike metals; for example, the metals may both be copper, but with the two coppers immersed in solutions of different concentration or composition. Likewise, both electrodes may be of the same gas, but with the pressure of the gas different at the two electrodes. *See* ELECTRODE POTENTIAL.

Galvanic cells are of two general types, reversible and irreversible. If the chemical reactions at the electrodes can be exactly reversed by reversals in the direction of the current flow at the electrodes, the cell is said to be reversible; if the chemical reactions cannot be reversed, or if entirely different reactions occur on current reversal, the cell is then of the irreversible type. An example of a reversible cell is the familiar lead-acid storage cell (or battery), widely used for a variety of purposes, including starting, ignition, and lighting for automobiles. A cell prepared by immersing zinc and platinum electrodes in perchloric acid results in an irreversible cell, since the electrode reactions at each electrode for the charge differ from those obtained for the discharge. *See* STORAGE BATTERY.

In measuring the emf of a cell, a reference cell of known emf must be available to effect a comparison. This reference cell should have an emf that is known in terms of physical laws and units, and not one chosen arbitrarily. The standard cells for this purpose are the cadmium amalgam standard cells of the saturated type proposed by Edward Weston in 1892. This type of cell is the most reversible galvanic cell known and retains a constant emf to within a few microvolts for many decades. This cell consists of a cadmium amalgam anode (negative element), a mercury-mercurous sulfate cathode (positive element), and a saturated solution of cadmium sulfate containing crystals of $CdSO_4 \cdot {}^8\!/_3H_2O$. *See* BATTERY; CALOMEL ELECTRODE; ELECTROCHEMISTRY; POTENTIOMETER (VOLTAGE METER). [W.J.H.]

Electromotive force (emf) The electromotive force, represented by the symbol ε around a closed path in an electric field is the work per unit charge required to carry a small positive charge around the path. It may also be defined as the line integral of the electric intensity around a closed path in the field. The abbreviation emf is preferred to the full expression since emf, also called electromotance, is not really a force. The term emf is applied to sources of electric energy such as batteries, generators, and inductors in which current is changing.

The magnitude of the emf of a source is defined as the electrical energy converted inside the source to some other form of energy (exclusive of electrical energy converted irreversibly into heat), or the amount of some other form of energy converted in the source into electrical energy, when a unit charge flows around the circuit containing the source. For a discussion of motional emf *see* ELECTROMAGNETIC INDUCTION. For information on back, or counter, emf *see* DIRECT-CURRENT MOTOR. *See also* ELECTROMOTIVE FORCE (CELLS); MAGNETOMOTIVE FORCE. [R.P.Wi.]

Electromyography The detection and recording of electrical activity generated by muscle fibers. The basic elements of motor control in the body are the motor units which comprise motor neurons in the brainstem or spinal cord, their axons, and from ten to several hundred muscle fibers supplied by each motor neuron. Motor units vary in the size and properties of their motoneurons, the sizes and conduction velocities of their axons, the morphology of their nerve muscle junctions, and the structure and physiological properties of the muscle fibers supplied by each motor neuron.

Impulses originating in single motoneurons in response to various command signals from the central nervous system conduct to the periphery of the unit, normally causing all the muscle fibers in the unit to discharge. The electrical activity generated by the more or less synchronous discharges of all the muscle fibers in the unit may be detected by recording electrodes on the skin surface or by needles inserted into the muscle. Such potentials reflect the electrical activity generated by the whole motor unit.

Diseases affecting motor neurons are sometimes accompanied by spontaneous discharges of the axons. Additionally, degeneration of motor axons may leave some muscle fibers deprived of their normal innervation, some of which spontaneously fire. Such single muscle-fiber discharges are called fibrillations and are readily detected for diagnostic purposes by needle electrodes inserted into the muscle.

Electromyography may also be used to study primary muscle diseases such as the muscular dystrophies, and a wide variety of other metabolic inflammatory and congenital myopathies affecting the muscle fibers rather than motor neurons or their axons. *See* BIOPOTENTIALS AND IONIC CURRENTS; ELECTRODIAGNOSIS. [W.F.Br.]

Electron An elementary particle which is the negatively charged constituent of ordinary matter. The electron is the lightest known particle which possesses an electric charge. Its rest mass is $m_e \cong 9.1 \times 10^{-28}$ g, about $1/1836$ of the mass of the proton or neutron, which are, respectively, the positively charged and neutral constituents of ordinary matter. Discovered in 1895 by J. J. Thomson in the form of cathode rays, the electron was the first elementary particle to be identified. *See* ELECTRIC CHARGE; ELEMENTARY PARTICLE; NUCLEAR STRUCTURE.

The charge of the electron is $-e \cong -4.8 \times 10^{-10}$ esu = -1.6×10^{-19} coulomb. The sign of the electron's charge is negative by convention, and that of the equally charged proton is positive. This is a somewhat unfortunate convention, because the flow of electrons in a conductor is thus opposite to the conventional direction of the current.

Electrons are emitted in radioactivity (as beta rays) and in many other decay processes; for instance, the ultimate decay products of all mesons are electrons, neutrinos, and photons, the meson's charge being carried away by the electrons. The electron itself is completely stable. Electrons contribute the bulk to ordinary matter; the volume of an atom is nearly all occupied by the cloud of electrons surrounding the nucleus, which occupies only about 10^{-13} of the atom's volume. The chemical properties of ordinary matter are determined by the electron cloud. *See* BETA PARTICLES; MESON; RADIOACTIVITY.

The electron obeys the Fermi-Dirac statistics, and for this reason is often called a fermion. One of the primary attributes of matter, impenetrability, results from the fact that the electron, being a fermion, obeys the Pauli exclusion principle; the world would be completely different if the lightest charged particle were a boson, that is, a particle that obeys Bose-Einstein statistics. *See* BOSE-EINSTEIN STATISTICS; EXCLUSION PRINCIPLE; FERMI-DIRAC STATISTICS; POSITRON. [C.J.G.]

Magnetic moment. The electron has magnetic properties by virtue of (1) its orbital motion about the nucleus of its parent atom and (2) its rotation about its own axis. The magnetic properties are best described through the magnetic dipole moment associated with 1 and 2. The classical analog of the orbital magnetic dipole moment is the dipole moment of a small current-carrying circuit. The electron spin magnetic dipole moment may be thought of as arising from the circulation of charge, that is, a current, about the electron axis; but a classical analog to this moment has much less meaning than that to the orbital magnetic dipole moment. The magnetic moments of the electrons in the atoms that make up a solid give rise to the bulk magnetism of the solid.

Spin. That property of an electron which gives rise to its angular momentum about an axis within the electron. Spin is one of the permanent and basic properties of the electron. Both the spin and the associated magnetic dipole moment of the electron were postulated by G. E. Uhlenbeck and S. Goudsmit in 1925 as necessary to allow the interpretation of many observed effects, among them the so-called anomalous Zeeman effect, the existence of doublets (pairs of closely spaced lines) in the spectra of the alkali atoms, and certain features of x-ray spectra. *See* SPIN (QUANTUM MECHANICS).

The spin quantum number is s, where s is always $1/2$. This means that the component of spin angular momentum along a preferred direction, such as the direction of a magnetic field, is $\pm 1/2\hbar$, where $\hbar$ is Planck's constant h divided by 2π. The spin angular momentum of the electron is not to be confused with the orbital angular momentum of the electron associated with its motion about the nucleus. In the latter case the maximum component of angular momentum along a preferred direction is $l\hbar$, where l is the angular momentum quantum number and may be any positive integer or zero. *See* QUANTUM NUMBERS.

The electron has a magnetic dipole moment by virtue of its spin. The approximate value of the dipole moment is the Bohr magneton μ_0 which is equal to $eh/4\pi mc = 9.27 \times 10^{-21}$ erg/oersted, where e is the electron charge measured in electrostatic units, m is the mass of the electron, and c is the velocity of light. (In SI units, $\mu_0 = 9.27 \times 10^{-24}$ joule/tesla.) The orbital motion of the electron also gives rise to a magnetic dipole moment μ_l, that is equal to μ_0 when $l = 1$. [Ar.R.]

Electron accelerators Electrical devices which accelerate electrons to high energies. Almost all electron linear accelerators are of the traveling-wave type. The wave is produced in a loaded waveguide, which is a waveguide excited in a TM mode (having a longitudinal electric field component) and loaded by disks or other means to produce the correct phase velocity to match the particle. An unloaded or uniform waveguide is not suitable since the phase velocity in such a guide exceeds the velocity of light. *See* PHASE VELOCITY; WAVEGUIDE.

The energy possibilities of the electron linear accelerator are not limited by fundamental considerations since electrons accelerated in a straight line do not lose energy by radiation by an appreciable amount. The performance of the accelerator is closely tied to the available power sources. Some of the earlier machines were powered by magnetron oscillators; newer machines are powered by klystron amplifiers driven from a common master oscillator. The largest electron linac is the 2-mi-long (3.2-km) 20-GeV machine located at Stanford University. *See* KLYSTRON; MAGNETRON; MICROWAVE TUBE.

The superconducting electron linac can operate continuously with dissipations of about a watt per meter. However, there have been severe metallurgical and electron loading problems with this device, and the initial hopes of very high accelerating gradients have not been realized.

Proton accelerators. Existing proton accelerators are of the standing-wave type since protons, unlike electrons, do not reach relativistic velocities at injection voltages and traveling-wave structures are not feasible. The earliest and still popular design is the Alvarez structure. The drift tubes, supported by transverse stems, are metallic cylinders and no rf field is present inside. The distance between tube centers is called a cell and is one guide wavelength long.

The trend has been toward much higher beam currents, more precise energy, and much higher energies. These have been achieved largely by advances in structure design, in particular by the use of multiple resonances or cells of double periodicity which greatly relax the mechanical and beam-loading tolerances and allow efficient acceleration and precise control of the fields together with good power transfer along the structure. In modern designs the Alvarez structure with resonant posts or stems is preferred up to about 200 MeV. Above 200 MeV, coupled-cavity structures of great precision and efficiency have been developed. The 200–800 MeV section of the proton accelerator at the Los Alamos Meson Physics Facility (LAMPF) uses these structures.

Heavy-ion accelerators. Heavy-ion linacs are similar to low-energy proton accelerators, with differences dictated by the relatively low charge-to-mass ratios available from ion sources. For a given electric field and structure length, the ion velocities are much lower than for protons. After passing through one or more cavities and reaching velocities of about 4% of c (energies of 1 MeV per atomic mass unit, or amu), the ions are passed through a stripper—a very thin foil, gas-filled tube, or gas jet—and electrons are stripped from the ions; typically the ion charge is doubled, and about half the ions are in a single-charge state. Subsequent acceleration in later cavities is then twice as efficient. The final energy of the accelerators is in the range 7 to 15 MeV/amu (v/c of 0.12 to 0.18) and is chosen to allow nuclear reactions between any of the accelerated ions and the heaviest target nuclei.

Considerable development work has been undertaken to perfect a practical superconducting structure for heavy ions. Various problems remain to be solved with all of the proposed designs, but the principle is very attractive, particularly because of the very large power consumption of room-temperature heavy-ion linacs. *See* PARTICLE ACCELERATOR. [R.Ber.]

Electron affinity The amount of energy release when an electron at rest is captured by a species M, producing the negative ion M^-. The electron affinity of a species M can also be thought of as the ionization potential of the negative ion M^-. Stated in terms of a chemical equation, the electron affinity of a species M is equal to the exothermocity of the reaction $e + M \rightarrow M^-$, where the negative ion M^- is left in its lowest electronic, vibrational, and rotational state. *See* IONIZATION POTENTIAL.

If the electron affinity of M is negative, the M^- ion is unstable with respect to decomposition into $M + e$. Most atoms have positive electron affinities, even though there is no net Coulomb attraction between the electron and the atom until the electron is close enough to be "a part of the atom." The simple rules of chemical valency provide a qualitative guide to the magnitude of electron affinities. Thus the noble gases, which have a filled outer electronic shell and are chemically inert, are not capable of binding an additional electron to form a negative ion. The largest electron affinities are possessed by the halogens, atoms which require only one additional electron to fill the valence shell. *See* VALENCE.

The major exception to this concept is that multiply charged negative ions—for example, O^{2-}, one of many multiply charged negative ions which are stable in solution—are not stable in the gas phase. The ability to place more than one additional electron in the valence shell of a neutral atom or molecule appears to come from the medium; the solvent shell surrounding the ion in liquid solutions and the amorphous or crystalline region surrounding the ion in solids. [W.C.L.]

Electron beam welding A process wherein coalescence is produced by heat obtained from a concentrated beam composed primarily of high-velocity electrons impinging upon the surfaces to be joined. Electron beam welding equipment, which utilizes a heat source of electrons accelerated by an electric field to extremely high speeds and focused to a sharp beam by electrostatic or electromagnetic fields, is used for welding a wide range of metals in thicknesses ranging from foils to extremely thick sections. A weld of very high purity may be obtained by this technique. Of all welding techniques, electron beam welding seems most suited for use in outer space, where it has been successfully demonstrated in the United States and Russia. [M.M.S.]

Electron capture The process in which an atom or ion passing through a material medium either loses or gains one or more orbital electrons. In the passage of charged particles (defined here as nuclei having more or less than Z atomic electrons, where Z is the atomic number) through matter, the capture (and loss) of electrons is an important process in the slowing down of the particles and therefore has a strong influence on their range. Thus a neutral hydrogen atom loses only about half as much energy per centimeter as the positively charged proton in passing through matter consisting of light elements.

For the ordinary charged particles (alpha particles and protons) the capture process is important only at low energies, when the particle velocity is of the order of electron velocities in the stopping material, and thus is important at the end of the range. For fission fragments, however, which initially have a large excess of positive charges, electron capture occurs immediately and continues throughout the slowing-down process. This fact causes the energy-loss mechanisms at the latter part of the range to be different for fission fragments and protons or α-particles. *See* NUCLEAR FISSION.

The nuclear capture of electrons (*K* capture) occurs by a process quite different from atomic capture and is in fact a consequence of the general beta interaction. This general interaction includes β^- decay (the oldest known beta transformation and hence the name), β^+ decay (or positron decay), and *K* capture, the latter so called because the electron captured by the nucleus is taken from the *K* shell (the shell nearest the nucleus) of atomic electrons. A second-order process, called *L* capture, can also occur, in which (to speak pictorially and thus somewhat imprecisely) an *s* electron (from the *K* shell) is captured with the simultaneous transition of a *p* electron (from the *L* shell) to the *K* shell with the emission of gamma radiation. *See* RADIOACTIVITY. [McA.H.H.]

Electron configuration The orbital arrangement of an atom's electrons. Negatively charged electrons are attracted to a positively charged nucleus to form an atom or ion. Although such bound electrons exhibit a high degree of quantum-mechanical wavelike behavior, there still remain particle aspects to their motion. Bound electrons occupy orbitals that are somewhat concentrated in spatial shells lying at different distances from the nucleus. As the set of electron energies allowed by quantum mechanics is discrete, so is the set of mean shell radii. Both these quantized physical quantities are primarily specified by integral values of the principal, or total, quantum number n. The full electron configuration of an atom is correlated with a set of values for all the quantum numbers of each and every electron. In addition to n, another important quantum number is l, an integer representing the orbital angular momentum of an electron in units of $h/2\pi$, where h is Planck's constant. The values 1, 2, 3, 4, 5, 6, 7 for n and 0, 1, 2, 3 for l together suffice to describe the electron configurations of all known normal atoms and ions, that is, those that have their lowest possible values of total electronic energy. The first seven shells are also given the letter designations *K*, *L*, *M*, *N*, *O*, *P*, and *Q* respectively. Electrons with l equal to 0, 1, 2, and 3 are designated *s*, *p*, *d*, and *f*, respectively. *See* QUANTUM MECHANICS; QUANTUM NUMBERS.

In any configuration the number of equivalent electrons (same n and l) is indicated by an integral exponent (not a quantum number) attached to the letters *s*, *p*, *d*, and *f*. According to the Pauli exclusion principle, the maximum is s^2, p^6, d^{10}, and f^{14}. *See* EXCLUSION PRINCIPLE.

An electron configuration is categorized as having even or odd parity, according to whether the sum of *p* and *f* electrons is even or odd. Strong spectral lines result only from transitions between configurations of unlike parity. *See* PARITY (QUANTUM MECHANICS).

Insofar as they are known from spectroscopic investigations, the electron configurations characteristic of the normal or ground states of the first 103 chemical elements are shown in the following table:

Distribution of electrons in the atoms

Element and atomic number		K	L		M			N				O				Ground term	Ionization potential, eV
		1,0 1s	2,0 2s	2,1 2p	3,0 3s	3,1 3p	3,2 3d	4,0 4s	4,1 4p	4,2 4d	4,3 4f	5,0 5s	5,1 5p	5,2 5d	5,3 5f		
H	1	1	—	—	—	—	—	—	—	—	—	—	—	—	—	$^2S_{1/2}$	13.5981
He	2	2	—	—	—	—	—	—	—	—	—	—	—	—	—	$^1S_0$	24.5868
Li	3	2	1	—	—	—	—	—	—	—	—	—	—	—	—	$^2S_{1/2}$	5.3916
Be	4	2	2	—	—	—	—	—	—	—	—	—	—	—	—	$^1S_0$	9.322
B	5	2	2	1	—	—	—	—	—	—	—	—	—	—	—	$^2P^\circ_{1/2}$	8.298
C	6	2	2	2	—	—	—	—	—	—	—	—	—	—	—	$^3P_0$	11.260
N	7	2	2	3	—	—	—	—	—	—	—	—	—	—	—	$^4S^\circ_{3/2}$	14.534
O	8	2	2	4	—	—	—	—	—	—	—	—	—	—	—	$^3P_2$	13.618
F	9	2	2	5	—	—	—	—	—	—	—	—	—	—	—	$^2P^\circ_{3/2}$	17.422
Ne	10	2	2	6	—	—	—	—	—	—	—	—	—	—	—	$^1S_0$	21.564
Na	11				1	—	—	—	—	—	—	—	—	—	—	$^2S_{1/2}$	5.139
Mg	12				2	—	—	—	—	—	—	—	—	—	—	$^1S_0$	7.646
Al	13				2	1	—	—	—	—	—	—	—	—	—	$^2P^\circ_{1/2}$	5.986
Si	14	Neon configuration			2	2	—	—	—	—	—	—	—	—	—	$^3P_0$	8.151
P	15				2	3	—	—	—	—	—	—	—	—	—	$^4S^\circ_{3/2}$	10.486
S	16				2	4	—	—	—	—	—	—	—	—	—	$^3P_2$	10.360
Cl	17				2	5	—	—	—	—	—	—	—	—	—	$^2P^\circ_{3/2}$	12.967
Ar	18				2	6	—	—	—	—	—	—	—	—	—	$^1S_0$	15.759
K	19						—	1	—	—	—	—	—	—	—	$^2S_{1/2}$	4.341
Ca	20						—	2	—	—	—	—	—	—	—	$^1S_0$	6.113
Sc	21						1	2	—	—	—	—	—	—	—	$^2D_{3/2}$	6.54
Ti	22						2	2	—	—	—	—	—	—	—	$^3F_2$	6.82
V	23						3	2	—	—	—	—	—	—	—	$^4F_{3/2}$	6.74
Cr	24						5	1	—	—	—	—	—	—	—	$^7S_3$	6.765
Mn	25						5	2	—	—	—	—	—	—	—	$^6S_{5/2}$	7.432
Fe	26						6	2	—	—	—	—	—	—	—	$^5D_4$	7.870
Co	27	Argon configuration					7	2	—	—	—	—	—	—	—	$^4F_{9/2}$	7.86
Ni	28						8	2	—	—	—	—	—	—	—	$^3F_4$	7.635
Cu	29						10	1	—	—	—	—	—	—	—	$^2S_{1/2}$	7.726
Zn	30						10	2	—	—	—	—	—	—	—	$^1S_0$	9.394
Ga	31						10	2	1	—	—	—	—	—	—	$^2P^\circ_{1/2}$	5.999
Ge	32						10	2	2	—	—	—	—	—	—	$^3P_0$	7.899
As	33						10	2	3	—	—	—	—	—	—	$^4S^\circ_{3/2}$	9.81
Se	34						10	2	4	—	—	—	—	—	—	$^3P_2$	9.752
Br	35						10	2	5	—	—	—	—	—	—	$^2P^\circ_{3/2}$	11.814
Kr	36						10	2	6	—	—	—	—	—	—	$^1S_0$	13.999
Rb	37									—	—	1	—	—	—	$^2S_{1/2}$	4.177
Sr	38									—	—	2	—	—	—	$^1S_0$	5.693
Y	39									1	—	2	—	—	—	$^2D_{3/2}$	6.38
Zr	40									2	—	2	—	—	—	$^3F_2$	6.84
Nb	41	Krypton configuration								4	—	1	—	—	—	$^6D_{1/2}$	6.88
Mo	42									5	—	1	—	—	—	$^7S_3$	7.10
Tc	43									5	—	2	—	—	—	$^6S_{5/2}$	7.28
Ru	44									7	—	1	—	—	—	$^5F_5$	7.366
Rh	45									8	—	1	—	—	—	$^4F_{9/2}$	7.46
Pd	46									10	—	—	—	—	—	$^1S_0$	8.33

(*continued on next page*)

In the next-to-last column of the table, the spectral term of the energy level with lowest total electronic energy is shown. The main part of the term symbol is a capital letter, *S*, *P*, *D*, *F*, and so on, that represents the total electronic orbital angular momentum. Attached to this is a superior prefix, 1, 2, 3, 4, and so on, that indicates the multiplicity, and an anterior suffix, 0, ½, 1, 3⁄2, 2, 5⁄2, and so on, that shows the total angular momentum, or *J* value, of the atom in the given state. A sign ° above the *J* value signifies that the spectral term and electron configuration have odd parity.

The last column of the table presents the first ionization potential of the atom, the energy required to remove from an atom its least firmly bound electron and transform a neutral atom into a singly charged ion. *See* ATOMIC STRUCTURE AND SPECTRA; IONIZATION POTENTIAL. [J.E.B.]

Distribution of electrons in the atoms (cont.)

Element and atomic number		Configuration of inner shells	N	O				P			Q	Ground term	Ionization potential, eV
			4,3 4f	5,0 5s	5,1 5p	5,2 5d	5,3 5f	6,0 6s	6,1 6p	6,2 6d	7,0 7s		
Ag	47	Palladium configuration	—	1	—	—	—	—	—	—	—	$^2S_{1/2}$	7.576
Cd	48		—	2	—	—	—	—	—	—	—	$^1S_0$	8.993
In	49		—	2	1	—	—	—	—	—	—	$^2P^\circ_{1/2}$	5.786
Sn	50		—	2	2	—	—	—	—	—	—	$^3P_0$	7.344
Sb	51		—	2	3	—	—	—	—	—	—	$^4S^\circ_{3/2}$	8.641
Te	52		—	2	4	—	—	—	—	—	—	$^3P_2$	9.01
I	53		—	2	5	—	—	—	—	—	—	$^2P^\circ_{3/2}$	10.457
Xe	54		—	2	6	—	—	—	—	—	—	$^1S_0$	12.130
Cs	55	The shells 1s to 4d contain 46 electrons	—	The shells 5s to 5p contain 8 electrons		—	—	1	—	—	—	$^2S_{1/2}$	3.894
Ba	56		—			—	—	2	—	—	—	$^1S_0$	5.211
La	57		—			1	—	2	—	—	—	$^2D_{3/2}$	5.5770
Ce	58		1			1	—	2	—	—	—	$^1G^\circ_4$	5.466
Pr	59		3			—	—	2	—	—	—	$^4I^\circ_{9/2}$	5.422
Nd	60		4			—	—	2	—	—	—	$^5I_4$	5.489
Pm	61		5			—	—	2	—	—	—	$^6H^\circ_{5/2}$	5.554
Sm	62		6			—	—	2	—	—	—	$^7F_0$	5.631
Eu	63		7			—	—	2	—	—	—	$^8S^\circ_{7/2}$	5.666
Gd	64		7			1	—	2	—	—	—	$^9D^\circ_2$	6.141
Tb	65		9			—	—	2	—	—	—	$^6H^\circ_{15/2}$	5.852
Dy	66		10			—	—	2	—	—	—	$^5I_8$	5.927
Ho	67		11			—	—	2	—	—	—	$^4I^\circ_{15/2}$	6.018
Er	68		12			—	—	2	—	—	—	$^3H_6$	6.101
Tm	69		13			—	—	2	—	—	—	$^2F^\circ_{7/2}$	6.184
Yb	70		14			—	—	2	—	—	—	$^1S_0$	6.254
Lu	71		14			1	—	2	—	—	—	$^2D_{3/2}$	5.426
Hf	72	The shells 1s to 5p contain 68 electrons				2	—	2	—	—	—	$^3F_2$	6.865
Ta	73					3	—	2	—	—	—	$^4F_{3/2}$	7.88
W	74					4	—	2	—	—	—	$^5D_0$	7.98
Re	75					5	—	2	—	—	—	$^6S_{5/2}$	7.87
Os	76					6	—	2	—	—	—	$^5D_4$	8.5
Ir	77					7	—	2	—	—	—	$^4F_{9/2}$	9.1
Pt	78					9	—	1	—	—	—	$^3D_3$	9.0
Au	79	The shells 1s to 5d contain 78 electrons					—	1	—	—	—	$^2S_{1/2}$	9.22
Hg	80						—	2	—	—	—	$^1S_0$	10.43
Tl	81						—	2	1	—	—	$^2P^\circ_{1/2}$	6.108
Pb	82						—	2	2	—	—	$^3P_0$	7.417
Bi	83						—	2	3	—	—	$^4S^\circ_{3/2}$	7.289
Po	84						—	2	4	—	—	$^3P_2$	8.43
At	85						—	2	5	—	—	$^2P^\circ_{3/2}$	
Rn	86						—	2	6	—	—	$^1S_0$	10.749
Fr	87						—	2	6	—	1	$^2S_{1/2}$	
Ra	88						—	2	6	—	2	$^1S_0$	5.278
Ac	89						—	2	6	1	2	$^2D_{3/2}$	5.17
Th	90						—	2	6	2	2	$^3F_2$	6.08
Pa	91						2	2	6	1	2	$^4K_{11/2}$	5.89
U	92						3	2	6	1	2	$^5L_6$	6.05
Np	93						4	2	6	1	2	$^6L_{11/2}$	6.19
Pu	94						6	2	6	—	2	$^7F_0$	6.06
Am	95						7	2	6	—	2	$^8S^\circ_{7/2}$	5.993
Cm	96						7	2	6	1	2	$^9D^\circ_2$	6.02
Bk	97						9	2	6	0	2	$^6He^\circ_{5/2}$	6.23
Cf	98						10	2	6	0	2	$^5I_8$	6.30
Es	99						11	2	6	0	2	$^4I^\circ_{15/2}$	6.42
Fm	100						12	2	6	0	2	$^3H_6$	6.50
Md	101						13	2	6	0	2	$^2F^\circ_{7/2}$	6.58
No	102						14	2	6	0	2	$^1S_0$	6.65
Lw	103						(14)	2	6	(1)	(2)		

Electron diffraction The phenomenon associated with interference processes that occur when electrons are scattered by atoms to form diffraction patterns. The wave character of electrons is shown most strikingly, and doubtless most conclusively, by the phenomena of interference. For this reason, the diffraction of electrons presents the most obvious confirmation of quantum mechanics. Because of the dependence of the diffraction pattern on the distances between the atoms, electron diffraction is also an important tool for the study of the structure of crystals and of free molecules, analogous to the use of x-rays for these purposes. *See* X-RAY CRYSTALLOGRAPHY; X-RAY DIFFRACTION.

According to energy $E = eV$, two major techniques of structure analysis are distinguished: low-energy electron diffraction (LEED) [$E \simeq$ 5–500 eV] and high-energy electron diffraction (HEED) [$E \simeq$ 5–500 keV]; medium-energy electron diffraction (MEED) [$E \simeq$ 500 eV–5 keV] is of little importance. Unlike neutrons and x-rays, electrons penetrate matter only for a very short distance before they lose energy (by inelastic scattering) or are scattered elastically (diffracted). Due to the energy loss, the wavelength of an electron changes so that it can no longer interfere with the other incident electrons. This loss of coherence occurs in condensed matter typically after mean free paths for inelastic scattering ranging from several tenths of nanometers at low energies ($E \simeq$ 50 eV) to several tens of nanometers at high energies, which determines the application range of LEED and HEED. *See* COHESION (PHYSICS); DE BROGLIE WAVELENGTH; DIFFRACTION; INTERFERENCE OF WAVES; MEAN FREE PATH; QUANTUM MECHANICS.

Low-energy electron diffraction. LEED is used mainly for the study of the structure of single-crystal surfaces and of processes on such surfaces that are associated with changes in the lateral periodicity of the surface. A monochromatic, nearly parallel electron beam of 10^{-4} to 10^{-3} m in diameter strikes the surface, usually at normal incidence. The elastically backscattered electrons are separated from all other electrons by a retarding field and detected with a suitable movable collector or, more frequently—after acceleration to about 5 keV energy—on a hemispherical fluorescent screen with the crystal in its center. Typical LEED patterns obtained in the second detection mode are shown in the illustration. The numbers are indices (h,k) labeling diffraction spots of the surfaces.

An important contribution of LEED is to the understanding of chemisorption, which precedes corrosion and, in many cases, epitaxy. Here, not only the structure of many adsorption systems—mainly of gases such as H_2, O_2, N_2, and CO on metals such as tungsten, nickel, copper, and platinum, or metals on other metals and semiconductors—has been studied, but also the kinetics of the adsorption and desorption process as well as changes in the adsorption layer upon heating. The combination of LEED with Auger electron spectroscopy and with work-function measurements has proven particularly powerful in these studies, because such methods give the coverage and information on the location of the adsorbed atoms normal to the surface. Combining LEED with other complementary techniques such as ion scattering spectroscopy, electron energy loss spectroscopy, or photoelectron spectroscopy has become increasingly popular and can enable the elimination of ambiguities in the interpretation of many LEED results. *See* ADSORPTION; CORROSION; ELECTRON SPECTROSCOPY; SURFACE PHYSICS.

High-energy electron diffraction. HEED is used mainly for the study of the structure of thin foils, films, and small particles (thickness or diameter of 10^{-9} to 10^{-6} m), of molecules, and also of the surfaces of crystalline materials. A monochromatic, usually nearly parallel, electron beam with a diameter of 10^{-3} to 10^{-8} m is incident on the target. The forward-scattered electrons (backscattering is negligible) are detected by means of a fluorescent screen, a photoplate, or some other current-sensitive detector, usually without the inelastically scattered electrons being eliminated.

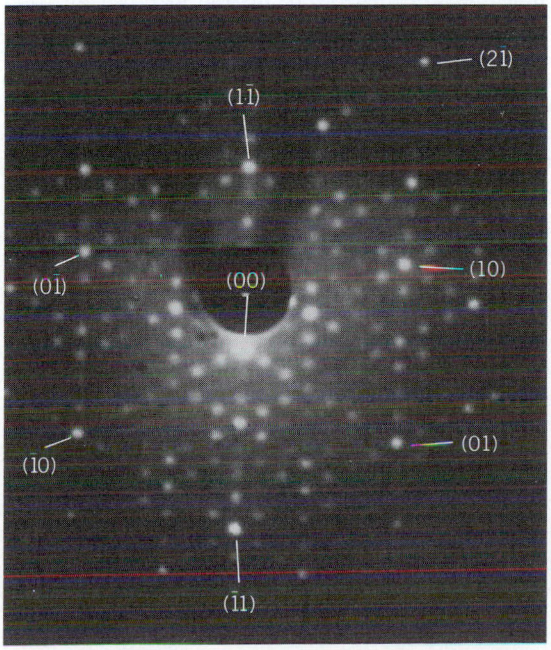

LEED pattern from clean silicon single-crystal (111) surface (7 × 7 structure), *E* = 85 eV.

The availability of ultra-high vacuum instruments has made it possible to study the structure of surfaces under clean conditions by reflection HEED (RHEED). Many experiments have confirmed the theoretical expectation that RHEED has a sensitivity comparable to that of LEED. Therefore, similar to LEED, RHEED can be used for the determination of the lateral arrangement of the atoms in the topmost layers of the surface, including the structure of adsorbed layers.

Like LEED, RHEED gives little information on the chemical nature of the atoms that produce the diffraction pattern. Therefore, RHEED has been combined with analytical tools, such as Auger electron spectroscopy or x-ray emission spectroscopy, in the same system. A second limitation inherent to both RHEED and LEED is the difficulty of determining the atomic positions normal to the surface. This is a consequence of multiple scattering effects and of absorption.

In scanning HEED (SHEED) the diffracted electrons are not recorded on photographic film but are directly measured electronically with sensitive detectors. By moving the detector across the diffraction pattern or by deflecting the diffracted electrons across a stationary detector (scanning), the intensity distribution in the diffraction pattern can be displayed quantitatively on an XY recorder. The main application of SHEED is in the study of processes which are accompanied by changes of the intensity distribution, such as the growth of thin films and annealing and corrosion processes.

The technological importance of thin film and interface devices has led to an upsurge of thin film growth studies by conventional transmission HEED (THEED), usually combined with transmission microscopy. Information obtained this way has been mainly on the orientation of the crystallites composing the film. [E.B.]

Gases and liquids. Electron diffraction in gases and liquids is similar in principle to that in solids; the differences arise from the lack in gases and liquids of any highly regular arrangement of the component atoms. In gases the low density makes it possible to study diffraction by individual atoms and molecules. The results obtained from monatomic gases represent the density of electronic charge in the atom as a function of the distance from the nucleus. The results from gaseous polyatomic molecules represent the equilibrium distances between the atomic nuclei and the average amplitudes of vibration associated with these distances. Liquids have been studied much less thoroughly, both in theory and in practice, than have gases. *See* SCATTERING EXPERIMENTS (ATOMS AND MOLECULES); SCATTERING EXPERIMENTS (NUCLEI). [L.O.B.]

Electron emission The liberation of electrons from a substance into vacuum. Since all substances are built up of atoms and since all atoms contain electrons, any substance may emit electrons; usually, however, the term refers to emission of electrons from the surface of a solid.

The process of electron emission is analogous to that of ionization of a free atom, in which the latter parts with one or more electrons. The energy of the electrons in an atom is lower than that of an electron at rest in vacuum; consequently, in order to ionize an atom, energy must be supplied to the electrons in some way or other. By the same token, a substance does not emit electrons spontaneously, but only if some of the electrons have energies equal to, or larger than, that of an electron at rest in vacuum. This may be achieved by various means, such as by heating, irradiation with light (photoemission), bombardment with charged particles (secondary emission), or use of a strong electric field (field, or cold, emission). *See* FIELD EMISSION; PHOTOEMISSION; SECONDARY EMISSION; THERMIONIC EMISSION. [A.J.D.]

Electron-hole recombination The process in which an electron, which has been excited from the valence band to the conduction band of a semiconductor, falls back into an empty

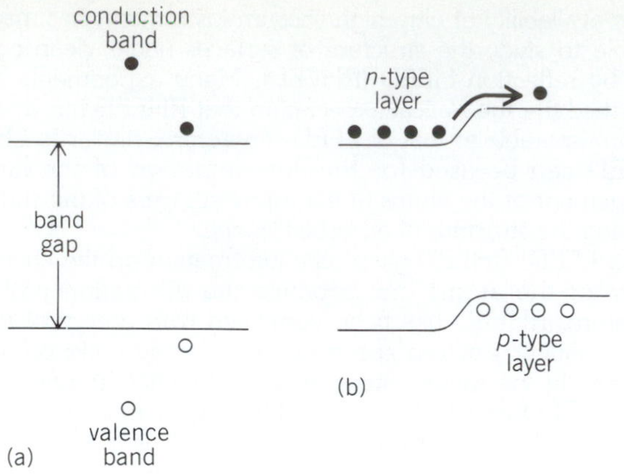

Recombination of electrons and holes generated by (a) optical absorption and (b) a forward-biased *pn* junction.

state in the valence band, which is known as a hole. *See* BAND THEORY OF SOLIDS.

Light with photon energies greater than the band gap can be absorbed by the crystal, exciting electrons from the filled valence band to the empty conduction band (see illustration *a*). The state in which an electron is removed from the filled valence band is known as a hole. It is analogous to a bubble in a liquid. The hole can be thought of as being mobile and having positive charge. The excited electrons and holes rapidly lose energy (in about 10^{-12} s) by the excitation of lattice phonons (vibrational quanta). The excited electrons fall to near the bottom of the conduction band, and the holes rise to near the top of the valence band, and then on a much longer time scale (of 10^{-9} to 10^{-6} s) the electron drops across the energy gap into the empty state represented by the hole. This is known as electron-hole recombination. An energy approximately equal to the band gap is released in the process. Electron-hole recombination is radiative if the released energy is light and nonradiative if it is heat. *See* PHONON.

Electron-hole recombination requires an excited semiconductor in which both electrons and holes occupy the same volume of the crystal. This state can be produced by purely electrical means by forward-biasing a *pn* junction. The current passing through a *pn* diode in electrons per second equals the rate of electron-hole recombination (see illustration *b*). A major application of this phenomenon is the light-emitting diode. *See* LIGHT-EMITTING DIODE; LUMINESCENCE; SEMICONDUCTOR DIODE.

Efficient radiative recombination between free electrons and holes takes place only in direct-bandgap semiconductors. During an optical transition, momentum is conserved, and since the photon carries away negligible momentum, transitions take place only between conduction-band and valence-band states having the same momentum. This is easily satisfied in direct-bandgap semiconductors, because electrons and holes collect at the conduction band at minimum and the valence band at maximum, and both extrema have the same momentum. However, for indirect-bandgap semiconductors, the conduction-band minimum and valence-band maximum have very different momenta, and consequently optical transitions between free electrons and holes are forbidden. Radiative electron-hole recombination is possible in indirect-band-gap semiconductors when the transition is assisted by lattice phonons and impurities. *See* CRYSTAL.

Apart from its application in light-emitting diodes and laser operation, radiative recombination, especially at low temperatures (approximately 2 K or −456°F), has been a very important tool for studying the interaction of electrons and holes in semiconductor crystals. *See* EXCITON.

Competing with radiative recombination are the nonradiative recombination processes of multiphonon emission and Auger recombination. It is suspected that nonradiative recombination by multiphonon emission drives the movement of atoms at room temperature that are responsible for device degradation phenomena such as the climb of dislocations found in GaAs light-emitting diodes and lasers. Auger recombination has been shown to limit the performance of long-wavelength (1.3–1.6 micrometer) lasers and light-emitting diodes used in optical communication systems. *See* AUGER EFFECT; DIFFUSION IN SOLIDS; LASER; OPTICAL COMMUNICATIONS; SEMICONDUCTOR.

[C.H.He.]

Electron lens An electric or magnetic field, or a combination thereof, which acts upon an electron beam in a manner analogous to that in which an optical lens acts upon a light beam. Electron lenses find application for the formation of sharply focused electron beams, as in cathode-ray tubes, and for the formation of electron images, as in infrared converter tubes, various types of television camera tubes, and electron microscopes.

Any electric or magnetic field which is symmetrical about an axis is capable of forming either a real or a virtual electron image of an object on the axis which either emits electrons or transmits electrons from another electron source. Hence, an axially symmetric electric or magnetic field is analogous to a spherical optical lens.

The lens action of an electric and magnetic field of appropriate symmetry can be derived from the fact that it is possible to define an index of refraction for electron paths in such fields. This index depends on the field distribution and the velocity and direction of the electrons.

Electron lenses differ from optical lenses both in the fact that the index of refraction is continuously variable within them and that it covers an enormous range. Furthermore, in the presence of a magnetic field, the index of refraction depends both on the position of the electron in space and on its direction of motion. It is not possible to shape electron lenses arbitrarily. *See* ELECTROSTATIC LENS; MAGNETIC LENS. [E.G.R.]

Electron microscope A device for forming greatly magnified images of objects by means of electrons. Electron microscopes serve primarily two purposes: (1) the visual examination of structures too fine to be resolved with ordinary, or light, microscopes, and (2) the study of surfaces that emit electrons. The first function made transmission electron microscopes essential research tools in biology, chemistry, and metallurgy. Beginning in the 1960s the scanning electron microscope came to play an increasingly important role in the study of the surfaces of solid objects at more moderate magnifications. Various emission electron microscopes serve more specialized research purposes. *See* EMISSION ELECTRON MICROSCOPE; SCANNING ELECTRON MICROSCOPE.

A transmission electron microscope consists in its simplest form of a source supplying a beam of electrons of uniform velocity, a condenser lens for concentrating the electrons on the specimen, a specimen stage for displacing the specimen which transmits the electron beam, an objective lens, a projector lens, and a fluorescent screen on which the final image is observed. For permanent record of the image, the fluorescent screen is replaced by a photographic plate or film.

Electrons are strongly scattered by all forms of matter including air. Hence the entire instrument must be evacuated to about 10^{-4} mmHg (10^{-7} atm). Furthermore, the lenses cannot be material in nature. Instead, they are electric or magnetic fields, symmetrical about the axis of the instrument, that have the property of bending the electron paths toward the axis, just as converging glass lenses bend light rays toward their axis. Lens strength is varied by varying the current. Most electron microscopes employ magnetic lenses of this type. These have yielded the highest resolution and magnification attained. However, good results have also been obtained with electron

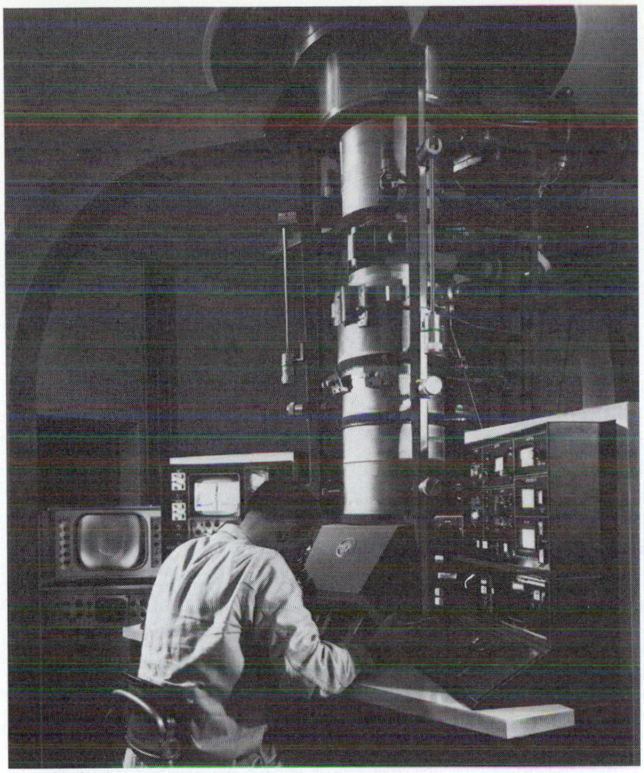

Megavolt electron microscope at U.S. Steel research laboratory at Monroeville, Pennsylvania. (*Radio Corporation of America*)

microscopes employing unipotential electrostatic lenses and magnetic lenses excited by permanent magnets. *See* ELECTRON LENS; ELECTROSTATIC LENS; MAGNETIC LENS.

A microscope can, at best, permit the discrimination of two point objects greater than $0.6\lambda/\sin\theta$ apart. Here λ is the wavelength of the illuminating radiation, and θ is the aperture angle of the cone of radiation that participates in forming the image. For green light $\lambda = 500$ nanometers. Even for ultraviolet radiation in an immersion medium of refractive index 1.5, λ is no less than 170 nm. Since light and ultraviolet microscope objectives can be designed to utilize practically all the radiation passing through the specimen, $\sin\theta \cong 1$ and the least resolvable distance for the ultraviolet microscope is about 100 nm. For 50- to 100-kilovolt electrons, such as are commonly employed in electron microscopes, the wavelength range is 0.0053–0.0037 nm. Hence, even though a cone of radiation with an aperture angle less than 0.01 radian contributes to an image of optimum sharpness, object separations smaller than 0.3 nm have been resolved with the electron microscope. Thus the electron microscope has several hundred times the resolving power of the light microscope. Similarly, whereas the maximum useful magnification of the light microscope is about 2000, that of the electron microscope may approach 1,000,000. The maximum useful magnification is the least magnification of the image that reveals to the observer all the specimen detail that the microscope is capable of conveying.

Electrons are commonly emitted from the tip of a fine tungsten-wire hairpin filament or, to further reduce the size of the effective electron source, from a sharply pointed segment of wire welded to the filament tip. Image contrasts are formed by the scattering of electrons out of the narrow cone that contributes to the formation of the image; denser or thicker portions of the specimen scatter more electrons and hence appear darker in the image.

In addition to the standard transmission microscopes operating at 50–100 kV, a number of very-high-voltage instruments have been constructed. A microscope at the Laboratory for Electron Optics at Toulouse, France, is designed for operation up to 1500 kV (see illustration), The advantage of high-voltage electron microscopy does not lie in greater resolving power but in increased penetration, which is particularly valuable in the direct study of metal sections prepared with a microtome.

[E.G.R.]

Electron-nuclear double resonance A technique in which the magnetic resonance of a nucleus is detected by observing that of a nearby electron. The magnetic coupling between the nuclear and electronic magnetic moments gives rise to a back reaction on the electron resonance when the nuclei are brought into resonance. The sample under study is placed in a conventional electron resonance apparatus. In addition, an oscillator drives a coil to produce alternating magnetic fields at the sample under study, the frequency of alternation being in an appropriate range to produce nuclear transitions. Typically, the static magnetic field is adjusted to produce electron resonance. With the static magnetic field and the frequency of the electron resonance apparatus held fixed, the frequency of the nuclear resonance oscillator is then swept. As it passes through the resonant frequency of any nucleus that is coupled to the electron, a change in the electron absorption occurs. *See* MAGNETIC RESONANCE. [C.P.S.]

Electron optics The branch of physics concerned with the motion of free electrons under the influence of electric and magnetic fields. The term electron optics is derived from the fact that the laws governing electron paths in such fields are formally identical with those governing light rays in media of varying refractive index. Both may be derived from Fermat's law. This law states that the actual light ray or electron path passing through two prescribed points A and B is that which makes the integral

$$\int_A^B n\, ds$$

carried out over it a minimum. The refractive index n is a function of position and, in the presence of a magnetic field, of the direction of the electron path. A similar dependence of the refractive index on the direction of a light ray is encountered in crystal optics. *See* CRYSTAL OPTICS.

The study of electron paths, analogous to the study of light rays, is more properly called geometrical electron optics. The electron paths may be regarded as normals to electron waves, whose amplitude determines the statistical density of electrons, just as the amplitude of a light wave determines the density of light quanta, or photons. The study of the wave motion associated with electrons is called electron wave optics. It describes diffraction and interference effects between electron beams which are in every way similar to the diffraction and interference effects observed with light and x-rays.

Electron optics finds application in the formation of electron beams, as in cathode-ray tubes and television camera tubes; in the deflection of such beams by electric and magnetic fields; and in the formation of electron images, as in electron microscopes and image tubes. *See* CATHODE-RAY TUBE; ELECTRON DIFFRACTION; ELECTRON LENS; ELECTRON MICROSCOPE; ELECTROSTATIC LENS; MAGNETIC LENS. [E.G.R.]

Electron paramagnetic resonance (EPR) spectroscopy The study of magnetic resonance spectra of materials which show paramagnetism because of the magnetic moment of unpaired electrons. EPR spectra are usually presented as plots of the absorption or dispersion of the energy of an oscillating magnetic field of fixed radio frequency versus the intensity of an applied static magnetic field. *See* MAGNETIC RESONANCE; PARAMAGNETISM; SPIN (QUANTUM MECHANICS).

EPR spectroscopy has been used for detection and identification of paramagnetic materials, for determinations of electronic structure, for studies of interactions between molecules, and for measurements of nuclear spins and moments. Among the wide variety of paramagnetic substances to which EPR spectroscopy has been applied are free radicals (including free atoms), impurity centers, and compounds of the transition elements, rare earths, and actinides. EPR spectra have been obtained from gases, liquids, and solids.

Spectra characteristic of individual paramagnetic molecules, uncomplicated by magnetic interactions with neighboring paramagnetic molecules, may be obtained only from dilute solutions in diamagnetic solvents. The required degree of dilution depends on the nature of the magnetic molecules. In some cases spin-lattice relaxation (exchange of magnetic energy with thermal motions of the environment) obscures the spectra. The effects are especially pronounced in inorganic magnetic ions and frequently require the use of low temperatures.

Greatly enhanced resolution may be obtained in many cases through simultaneous irradiation with several frequencies. The most useful of these methods is electron nuclear double resonance (ENDOR). The material is simultaneously irradiated at one of its EPR resonant frequencies and by a second oscillatory field whose frequency is swept over the range of nuclear frequencies. The intensity of the EPR response is measured as a function of the second frequency. Changes in EPR intensity accompany passage through nuclear resonance. Resonances as narrow as 10 kHz are observed. *See* NUCLEAR MAGNETIC RESONANCE (NMR). [S.I.W.]

Electron-positron pair production A process in which a negative electron (negatron) and a positive electron (positron) are simultaneously created in the vicinity of a nucleus or an elementary particle. In external pair production, an electromagnetic wave (photon) is absorbed and creates an electron pair. The absorption of high-energy γ-rays is due mainly to this effect. Internal pair production is not associated with observable electromagnetic radiation and may occur when an excited nucleus releases some of its internal energy. *See* ELECTRON; POSITRON.

Pair production is of considerable theoretical interest, not only as an example of the materialization of energy, but also as a striking confirmation of the relativistic quantum theory proposed by P. A. M. Dirac. This theory has made possible quantitative predictions of production probability, differential electron distribution, and kinetic energy partition. The results are in satisfactory agreement with experimental findings. *See* RELATIVISTIC QUANTUM THEORY. [G.B.]

Electron-probe microanalysis A method used for determining the elemental composition of materials, based on the x-rays emitted by different elements when bombarded with high-energy electrons. It is a micro method that can detect x-ray photons emitted by the atoms within a small volume excited by an electron beam focused to 10 nanometers diameter or less. In biology electron-probe microanalysis can be used to determine the composition of cell organelles without isolating them and therefore altering the distribution of diffusible elements.

High-energy electrons can ionize atoms, ejecting an inner-shell electron. To fill the resultant vacancy, an outer-shell electron falls into the ionized shell; the atom remains in the higher-energy excited state. The emission of a characteristic x-ray by the ionized atom is one of the mechanisms for releasing its excess energy. Another mechanism is the emission of Auger electrons; the probability of ejecting an x-ray instead of an Auger electron is the fluorescence yield. In the case of electron-probe microanalysis the source of the exciting electrons is the electron gun, which, in modern electron microscopes equipped with field-emission guns, can produce a focused beam narrower than 1 nm. The x-rays emitted as the result of atomic ionization are called characteristic x-rays, because their energy is characteristic of the core shell of the ionized element and of the shell from which the electron relaxed into the vacancy. *See* AUGER EFFECT; SURFACE PHYSICS; X-RAY FLUORESCENCE ANALYSIS. [A.P.S.]

Electron ring accelerator A device for imparting to protons or other positive ions kinetic energies of several to several hundred MeV per nucleon by the electron drag force of stable ringlike configuration of electrons that is accelerated to the requisite velocity.

Conventional particle accelerators are subject to limitations in the strengths of the fields that can feasibly be employed to guide and accelerate the desired particles, and accordingly may be massive and expensive. This situation has motivated the development of techniques for accelerating a small group of particles by means of their interaction with another group of charges or with some form of plasma wave. This "collective acceleration" employs fields developed by charged particles within the accelerator itself. Such fields can be quite large and are not subject to the technological limitations that restrict externally applied fields.

A direct and attractive method of collective acceleration is employed in the electron ring accelerator. The fields of significance in this accelerator are produced by a slender ring of electrons circulating in a magnetic field. The electron ring is formed by injecting an electron beam of several hundred amperes into a magnetic field that is rapidly deformed to trap one or more injected turns into closely spaced circular orbits and that then is further pulsed to shrink the ring dimensions to suitable size. The pulsed magnetic field serves both to increase the energy of the circulating electrons (through the transformer action of induced electric fields) and to reduce the major and minor dimensions of the ring (compression). In alternative experimental devices these rings are formed by injecting a hollow relativistic electron beam into a cusp-shaped magnetic field, wherein the forward motion of the electrons is converted into circulatory motion. *See* PARTICLE ACCELERATOR. [U.S.]

Electron spectroscopy A form of spectroscopy which deals with the emission and recording of the electrons which constitute matter—solids, liquids, or gases. The usual form of spectroscopy concerns the emission or absorption of photons (x-rays, ultraviolet rays, visible or microwave wavelengths, and so on). Electron spectra can be excited by x-rays, which is the basis for electron spectroscopy for chemical analysis (ESCA), or by ultraviolet photons, or by ions (electrons). By means of ESCA, complete sets of photoelectron lines can be excited from the internal (core) levels as well as from the external (valence) region. Also, complete sequences of the Auger electron lines are automatically obtained in this mode. *See* AUGER EFFECT.

The electron lines in an ESCA spectrum are extremely sharp and well suited for precision measurements. With a high-resolving ESCA spectrometer which has a magnetic or electrostatic focusing dispersive system, the electron lines have widths which are set by the limit caused by the uncertainty principle (the "inherent" widths of atomic levels). With a suitable choice of radiation, electron spectroscopy reproduces directly the electronic level structure from the innermost shells (core electrons) to the atomic surface (valence or conduction band). Furthermore, all elements from hydrogen to the heaviest ones can be studied even if the element occurs together with several other elements and even if the element represents only a small part of the chemical compound. *See* LINE SPECTRUM.

When applied to solid materials, ESCA is a typical surface spectroscopy with applications to problems such as chemical surface reactions, for example, corrosion or heterogeneous catalysis. ESCA also reproduces bulk matter properties such as valence electron band structures. Electron spectroscopy can supply a detailed knowledge of the valence orbital structure for all molecules which can be brought into gaseous form with pressures of 10^{-5} torr (10^{-3} pascal) or more. Under certain conditions, liquids and solutions of various compositions can be studied by ESCA techniques.

A unique feature of ESCA is that, if the exact position of the electron lines characteristic of the various elements in the molecule is measured, the area of inspection can be moved from one atomic species to another in the molecular structure. If the structure of the molecule is known, the charge distribution can be estimated in a simple way by using, for example, the electronegativity concept and assuming certain resonance structures. More sophisticated quantum-chemical treatments can also be applied. Conversely, if, by means of ESCA, the approximate charge distribution is known, conclusions concerning the structure of the molecule can be drawn. *See* ATOMIC STRUCTURE AND SPECTRA; ELECTRON CONFIGURATION; ELECTRONEGATIVITY; MOLECULAR ORBITAL THEORY; SPECTROSCOPY. [K.S.]

Electron-transfer reaction A reaction in which one electron is transferred from one molecule or ion to another molecule or ion. Electron-transfer reactions are ubiquitous in nature. Some are deceptively simple [for example, reaction (1),

$$^{*}Fe(H_2O)_6^{3+} + Fe(H_2O)_6^{2+} \rightarrow {}^{*}Fe(H_2O)_6^{2+} + Fe(H_2O)_6^{3+} \quad (1)$$

where the asterisk is used to identify a specific isotope]; others look very complicated (for example, the long-range electron transfers found in biology). The widespread occurrence of electron-transfer reactions has stimulated much theoretical and experimental work. *See* CHEMIOSMOSIS; PHOTOSYNTHESIS.

The simplest reactions in solution chemistry are electron self-exchange reactions (2), in which the reactants and products

$$^{*}A_{ox} + A_{red} \rightarrow {}^{*}A_{red} + A_{ox} \quad (2)$$

are the same (the asterisk is used to identify a specific isotope). The only way to determine chemically that a reaction has taken place is to introduce an isotopic label. There is no change in the free energy ($\Delta G^\circ = 0$) for this type of reaction.

Much more common are cross reactions (3), where A_{ox} is the

$$A_{ox} + B_{red} \rightarrow A_{red} + B_{ox} \quad (3)$$

oxidized reactant, B_{red} is the reduced reactant, A_{red} is the reduced product, and B_{ox} is the oxidized product. For these reactions, $\Delta G^\circ \neq 0$. *See* GIBBS FUNCTION.

Both types of electron-transfer reactions (self-exchange and cross reactions) can be classified broadly as inner sphere or outer sphere. In an inner-sphere reaction, a ligand is shared between the oxidant and reductant in the transition state. An outer-sphere reaction, on the other hand, is one in which the inner coordination shells of both the oxidant and reductant remain intact in the transition state. There is no bond breaking or bond making, and no shared ligands between redox centers. Long-range electron transfers in biology are all of the outer-sphere type. *See* OXIDATION-REDUCTION.

Electron-transfer theory. The simplest electron transfer occurs in an outer-sphere reaction. The changes in oxidation states of the donor and acceptor centers result in a change in their equilibrium nuclear configurations. This process involves geometrical changes, the magnitudes of which vary from system to system. In addition, changes in the interactions of the donor and acceptor with the surrounding solvent molecules will occur. The Franck-Condon principle governs the coupling of the electron transfer to these changes in nuclear geometry: during an electronic transition, the electronic motion is so rapid that the nuclei (including metal ligands and solvent molecules) do not have time to move. Hence, electron transfer occurs at a fixed nuclear configuration. In a self-exchange reaction, the energies of the donor and acceptor orbitals (hence, the bond lengths and bond angles of the donor and acceptor) must be the same before efficient electron transfer can take place. *See* FRANCK-CONDON PRINCIPLE.

Long-range electron transfer. The rate of long-range electron transfer between an electron donor (B_{red}) and an electron acceptor (A_{ox}) depends on both the electronic coupling between A_{ox} and B_{red} (which is a function of the intersite $A_{ox}//B_{red}$ distance d, the nature of the intervening medium, and the relative $A_{ox}//B_{red}$ orientation, where // represents the protein medium that separates the donor and the acceptor) and an activation energy term. A standard theoretical rate equation (4) expresses k_{et} in terms of these factors; here, ν is a

$$k_{et} = \nu\,[\exp(-ßd)]\{\exp[-(\lambda + \Delta G^\circ)^2/4\lambda RT]\} \quad (4)$$

frequency factor, ß is a medium- and orientation-dependent quantity, d is the intersite distance, λ is the reorganization energy, ΔG° is the reaction free energy of the electron-transfer process, R is the universal gas constant, and T is the absolute temperature of the system. Experiments in several laboratories have been designed to estimate the values of λ and ß in modified metalloproteins, rigid organic molecules, and protein–protein complexes. [B.E.Bo.; W.R.El.; H.B.Gr.; T.J.Me.]

Electron tube A device in which conduction takes place by the movement of electrons or ions between electrodes through a vacuum or ionized gas within a gas-tight envelope. Electron tubes include all partially evacuated tubes whose electrical characteristics are derived from the flow of electrons through the tube. Two subclasses of electron tubes are vacuum tubes and gas-filled tubes. Vacuum tubes are primarily used in applications where low noise and high frequency are involved. In contrast, gas tubes are used for high-current, low-frequency applications. *See* GAS TUBE; VACUUM TUBE.

Although semiconductor devices have largely replaced vacuum tubes in many areas of application, tubes are still essential in some areas. Thus, cathode-ray tubes, a specialized form of vacuum tube, are at present indispensable to television receivers, oscilloscopes, and radar display units; camera tubes are used in television cameras; high-power tubes are used in broadcasting transmitters; and microwave tubes are used in radar, telephony, space communication and control, scientific research, and high-frequency ovens. *See* CATHODE-RAY TUBE; MICROWAVE TUBE; NUMERICAL INDICATOR TUBE; PHOTOTUBE; SEMICONDUCTOR; X-RAY TUBE. [H.J.R.]

Electron wake The region behind a fast-moving ion in a metallic solid within which most of the effects of the ion on

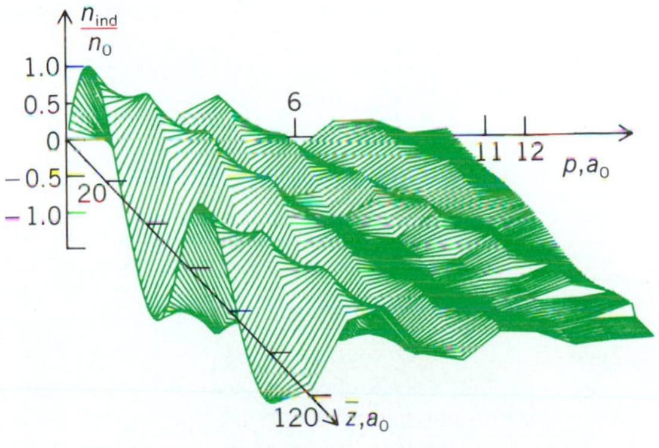

Electron wake of a completely stripped sulfur (S^{+16}) ion with an energy of 0.93 MeV per nucleon in aluminum. *(After W. Schafer et al., Mach-shock electron distributions from solids, Z. Phys., B36:319–322, 1980)*

the conduction electrons in the solid are confined. The ion moving with velocity v, much greater than the highest electron velocity in the solid, induces a polarization charge density due to excitation of collective plasma waves. This creates characteristic oscillatory fluctuations in the electron density, which is alternately enhanced and depleted relative to the equilibrium density (see illustration). The various waves resemble Mach shock waves.

When a molecular ion beam enters the solid target, its electrons are stripped off and the ions fly apart because of their mutual Coulomb repulsion, The leading fast-moving ion creates the electron wake while the slow-trailing ion is used as a probe. In the double-foil experiment, the ions are made to pass through a vacuum before entering the second foil. The trailing ion probes the extended wake region in the second foil. The ratio of intensities of the fast- to slow-moving ions transmitted at a scattering angle of 0°, when plotted against the separations of the foils, exhibits the signature of the characteristic oscillatory nature of the wake. *See* COULOMB EXPLOSION; FREE-ELECTRON THEORY OF METALS; PLASMA PHYSICS; POLARIZATION OF DIELECTRICS. [W.Gr.]

Electronegativity According to L. Pauling, "the power of an atom in a molecule to attract electrons to itself." Quantitative definitions and scales of electronegativity have been based not on electron distribution itself but on properties which were assumed to reflect electronegativity.

The electronegativity of an element depends upon its valence state and thus is not an invariant atomic property. As an example, the electron-withdrawing ability of an sp^n hybrid orbital centered on carbon and directed toward hydrogen increases as the percentage of *s* character in the orbital increases in the series ethane < ethylene < acetylene. Thus, according to this concept of orbital electronegativity, each element exhibits a range of electronegativity values.

The original scale, proposed by Pauling in 1932, is based upon the difference between the energy of the A-B bond in the compound AB_n and the mean of the energies of the homopolar bonds A-A and B-B (see table). R. S. Mulliken proposed that the electronegativity of an element is given by the average of the valence-state ionization potential and electron affinity. The Mulliken approach is consistent with Pauling's original definition and gives orbital electronegativities, not invariant atomic electronegativities. Electronegativity was defined by A. L. Allred and E. G. Rochow as the force of attraction between a nucleus and an electron from a bonded atom. A quantum-defect electronegativity scale has been developed from potentials based on atomic spectral data, and a nonempirical scale has been calculated by an ab initio method using floating gaussian orbitals.

Average electronegativities from thermochemical data

Element	Value	Element	Value
H	2.20	Al	1.61
Li	0.98	Ga	1.81
Na	0.93	In	1.78
K	0.82	Tl	2.04
Rb	0.82	C	2.55
Cs	0.79	Si	1.90
Be	1.57	Ge	2.01
Mg	1.31	Sn	1.96
Ca	1.00	Pb	2.33
Sr	0.95	N	3.04
Ba	0.89	P	2.19
Sc	1.36	As	2.18
Ti	1.54	Sb	2.05
V	1.63	Bi	2.02
Cr	1.66	O	3.44
Mn	1.55	S	2.58
Fe	1.83	Se	2.55
Co	1.88	F	3.98
Ni	1.91	Cl	3.16
Cu	1.90	Br	2.96
Zn	1.65	I	2.66
B	2.04		

Other methods for calculating electronegativities utilize such observables as bond-stretching force constants, electrostatic potentials, spectra, and covalent radii. The measurement of electronegativities involves observations of properties dependent upon electron distribution. Close agreement of electronegativity values obtained from measurements of several diverse properties lends confidence and utility to the concept. [A.L.A.]

Electronic air cleaner A type of air filter limited primarily to applications requiring high air-cleaning efficiency. These devices operate on the principle of passing the airstream through an ionization field where a 12,000-V potential imposes a positive charge on all airborne particles. The ionized particles are then passed between aluminum plates, alternately grounded and connected to a 6000-V source, and are precipitated onto the grounded plates. *See* AIR FILTER. [M.A.B.]

Electronic display An electronic component used to convert electric signals into visual imagery in real time suitable for direct interpretation by a human operator. It serves as the visual interface between human and machine. The visual imagery is processed, composed, and optimized for easy interpretation and minimum reading error. The electronic display is dynamic in that it presents information within a fraction of a second from the time received and continuously holds that information, using refresh or memory techniques, until new information is received. The image is created by electronically making a pattern from a visual contrast in brightness on the display surface without the aid of mechanical or moving parts.

Electronic display is widely used for presentation of graphs, symbols and alphanumerics, and video pictures. Electronic displays are rapidly replacing traditional mechanical and hardcopy (paper) means for presenting information. This is due to the increased use of computers and microprocessors, both of which employ low-cost large-scale integration (LSI) electronics, and digital mass memories. The success of the hand-held calculator is directly attributable to the availability of low-cost LSI electronics and low-cost electronic numeric displays. *See* CALCULATORS; COMPUTER; COMPUTER GRAPHICS; INTEGRATED CIRCUITS; MICROCOMPUTER; MICROPROCESSOR.

Electronic transducers and four-digit (or more) flat-panel displays are replacing the galvanometer movement, thermometer scale, barometer movement, and other forms of scientific instrumentation. Large signs, arrival and departure announcements, and scoreboards use electronic means to portray changing messages and data. One of the major electronic displays applications is in home color television. The computer terminal using a cathode-ray tube is one of the most important industrial applications of electronic displays. The standard computer terminal displays 25 lines of 80 characters, for a total of 2000 characters, which correspond to a typed page.

Electronic displays can be categorized into four classifications, as shown in the table. Each classification is defined by natural technical boundaries and cost considerations. The categorization is useful in visualizing the extent to which electronic displays are used.

The primary applications of the cathode-ray tube are in home entertainment television, scientific and electrical engineering oscilloscopes, radar display, and alphanumeric and graphic electronic displays. *See* CATHODE-RAY TUBE; OSCILLOSCOPE; RADAR; TELEVISION.

Because of the depth dimension of the cathode-ray tube, there has been a concentrated effort to invent a flat-panel display. A primary motivating factor has been to achieve a flat

Electronic display spectrum of applications

Classification	Characteristics	Applications	Electronic technologies
Pseudoanalog	Dedicated arrangement of discrete pixels used to present analog or qualitative information	Meterlike presentations, go/no-go messages, legends and alerts, analoglike (watch) dial	Gas discharge, light-emitting diodes, liquid crystal, incandescent lamps
Alphanumeric	Dedicated alphanumeric pixel font of normally less than 480 characters; most common are 4- and 8-character numeric displays	Digital watches, calculators, digital multimeters, message terminals, games	Liquid crystal, light-emitting diodes, vacuum fluorescent, gas discharge, incandescent lamps
Vectorgraphic	Large orthogonal uniform array of pixels which are addressable at medium to high speeds; normally, monochromatic with no gray scale; may have memory; normally, over 480 characters or simple graphics	Computer terminals, TWX terminals, arrivals and departures, scheduling terminals, weather radar, air-traffic control, games	Cathode-ray tube, plasma panels, gas discharge, vacuum fluorescent, electroluminescence, other technologies in advanced development
Video	Large orthogonal array of pixels which are addressed at video rates (30 frames/s); monochromatic with gray scale or full color; standardized raster scan addressing interface, arrays of pixels approximately 512 rows by 512 columns	Entertainment television, graphic arts, earth resources, video repeater, medical electronics, aircraft flight instruments, computer terminals, command and control, games	Cathode-ray tube, other technologies in advanced development

television receiver which could be hung on a wall. The electrical phenomena most extensively developed for flat-panel displays are gas discharge (plasma), electroluminescence, light-emitting diode, cathodoluminescence, and liquid crystallinity. The cost of flat-panel displays is higher than cathode-ray tubes on a per-character basis for higher information content displays such as the 2000-character computer terminal. The flat-panel technologies are utilized extensively in portable displays and numeric displays using several hundred characters and less. *See* CATHODOLUMINESCENCE; ELECTROLUMINESCENCE; LIGHT-EMITTING DIODE; LIQUID CRYSTALS.

Color can be created on a cathode-ray tube equipped with a shadow mask duplicating quite closely all colors that occur in nature. This is done in the cathode-ray tube by using three different electron guns and three phosphors in a triad of red, green, and blue on the screen at each pixel. The shadow mask is a metal screen with a hole for each pixel. It is located in the path of the beam between the deflection area and phosphor screen. Electron beams from each of three guns are constrained by each shadow mask hole to hit each respective phosphor dot.

Penetration phosphors are also used to create color on cathode-ray tube displays to eliminate the need for the shadow mask and extra guns. However, the color is limited and the brightness is low. Normally, two phosphors are placed on the screen in two layers or in microspheres of two layers. The gun and cathode-ray-tube anode are operated in two energy states to produce either a high-energy or a low-energy electron beam switchable in time.

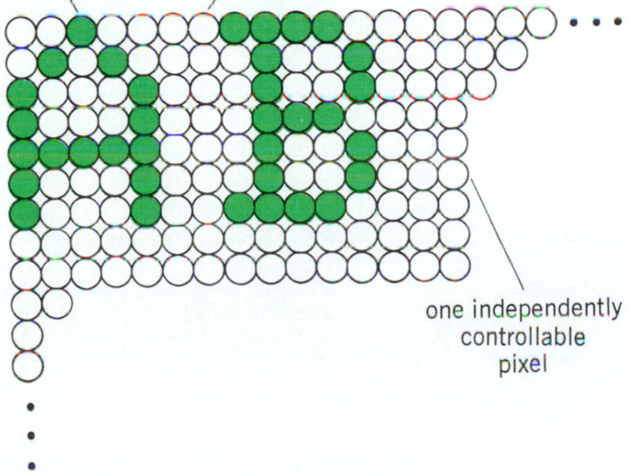

Fig. 1. Pixel array used for creating electronic display images.

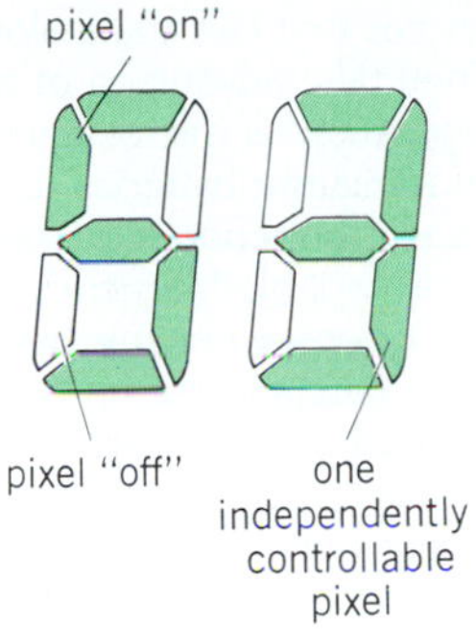

Fig. 2. Seven-bar numeric font.

Monochromatic as well as full color is readily produced by flat-panel display technologies. Full color is produced by using a three-color triad of red, green, and blue for each pixel. Full color is very important for entertainment television displays. Most industrial electronic displays do not need or use full color. The most efficient monochrome from each display technology is normally used; for example, red for light-emitting diode, orange for gas discharge, and orange-yellow for electroluminescence. Limited color displays are sometimes used in special industrial applications such as aircraft weather radar and artificial horizons, computer-aided multilayer circuit design, earth resources studies, air-route traffic control, and medical electronic displays. In all of these applications, the display instrument is usually a cathode-ray tube using the shadow-mask color technique.

The essence of electronic displays is based upon the ability to turn on and off individual picture elements (pixels; Fig. 1). The pixel is the smallest controllable element of the display. A typical high-information-content display will have a quarter million pixels in an orthogonal array, each under individual control by the electronics. The pixel resolution is normally just at or below the resolving power of the eye. Thus a good-quality picture can be created from a pattern of activated pixels. The pixel concept for electronic displays has evolved from the modern flat-panel display technologies and digital electronics.

With flat-panel displays and cathode-ray-tube digital-raster displays, alphanumeric character fonts are created by turning on the appropriate pixels in an array. One standard size is a 5×7 array with two pixels between characters and two pixels between rows, as shown in Fig. 1. All the letters and numbers can be created on this common array format. A very efficient and elegant array has evolved for portraying numeric characters only, called the seven-bar font (Fig. 2). Each bar is a pixel by definition. A similar 14-bar font is used for alphanumeric characters. [L.E.T.]

Electronic listening devices Devices which are used to capture the sound waves of conversation originating in an ostensibly private setting in a form, usually as a magnetic tape recording, which can be used against the target by adverse interests.

There are two kinds of electronic listening devices. One takes advantage of equipment already present on the target's premises, such as a telephone, radio, phonograph, television set, public-address loudspeaker, or tape recorder, to act as a microphone, transmitter, or power supply. The other does not. In the former case, the target's equipment is said to have been compromised.

Compromise of the target's own equipment takes advantage of the fact that any loudspeaker is capable of functioning just as well as a microphone, that convenient sources of dc power are available within the equipment, or that the equipment is connected to power or signal lines that can transmit the intercepted conversation to some place where recording can conveniently be accomplished. The equipment most frequently compromised is the telephone handset.

Eavesdropping devices that can stand alone are known commonly as "bugs." They take advantage of many developments of modern technology, such as microcircuits, miniature ceramic microphones, and miniature batteries. Electronically a bug is often just a two-stage frequency-modulated transmitter: an audio amplifier and a variable-frequency radio-frequency (rf) oscillator. Bugs may operate on any frequency from 20 to 1000 MHz, but usually they snuggle up beside a powerful local FM or vhf television station. *See* Amplifier; Oscillator; Radio transmitter.

A popular hybrid between a compromise device and a bug is the telephone drop-in. In this design, an FM transmitter is made in the form of a telephone microphone. The eavesdropper can casually unscrew the mouthpiece of his target's telephone handset and substitute the drop-in for the original microphone. The range of this device is about 240 ft (75 m). It has the added advantage of drawing its dc power from the telephone company central battery. [J.M.C.]

Electronic mail Any of various systems that transmit some form of electronic (usually digital) representation of a page or message from one location to another, The two locations may be as close as two adjacent terminals or as far as halfway around the world. Some forms of transmission capability have existed for a long time, but the term electronic mail appeared with the computer and integrated-circuit technology explosion. The emergence of inexpensive, high-speed scanners (which convert hard copy to an electronic representation) and error-free compression (a method of reducing the transmission time required) may be the primary catalysts that created the term electronic mail. The widespread usage of this term, however, probably results from the ability of large numbers of users to create electronic messages at readily available computer terminals. Because the term electronic mail is so commonly used in the field of communications, it has acquired a variety of meanings and interpretations, with regard to both the role of the originator and that of the recipient. *See* Integrated circuits.

The small-business-computer industry uses the term electronic mail to describe the ability to receive on one terminal a textual message that originated on another terminal. The output is normally a pictorial display on a computer monitor, which can be printed if the receiving terminal is so equipped. *See* Computer; Microcomputer.

In 1980 the Consulting Committee for International Telephone and Telegraphy (CCITT) agreed to a standard for facsimile machines which could transmit a page in less than 1 min. The popularity of this form of electronic mail is due to the high quality of the output copy. *See* Facsimile.

Some services which have been available for a long time, such as telegrams, Telex, and TWX, have been undergoing a major modernization with computer-controlled switching and advanced designed terminal equipment, which may justify using a modern name such as electronic mail. *See* Telegraphy.

The modern office automation systems industry uses the term electronic mail in conjunction with the ability of their equipment to transfer correspondence. The correspondence is generated on one terminal, stored in a central file, and retrieved by another compatible terminal with access to that central file. *See* Word processing.

Finally, the term electronic mail is applied to specially designed services that transfer a message page from one geographic location to another by electronic communication methods. *See* Electrical communications. [V.P.B.]

Electronic navigation systems Systems deriving navigationally useful parameters through the application of the electronic sciences. The older term for these systems was "radio aids to navigation." However, several of the modern devices, although employing electronics, do not employ radio waves. Therefore, the term electronic navigation systems is employed to cover all of the devices which use electronics. In electronic navigation systems there is a blending of the sciences of electronics and navigation. Therefore, the subject may be discussed from both points of view.

Electronic classifications. From the electronic point of view, it is possible to classify all systems as classical or self-contained. Classical systems employ at least one radio transmitter and one radio receiver. The transmitter emits energy which travels over one or more paths to the receiver. The navigational parameter is derived by a measurement made on the delay incurred in the transmission. All classical systems are based on the assumption that wave propagation is rectilinear and that the velocity of propagation is constant. The table shows the total contribution of electronics to navigation systems.

The self-contained systems consist of devices which make observations on certain natural phenomena and computers which derive navigational parameters from the observations.

Electronic navigation systems

Classical	
Single-path (circular LOP)	
One-way path	
Dectra (distance)	Navarho (distance)
Global Positioning System	
Round-trip path	
Benito	Oboe
Condar	Radar (distance)
DME distance-measuring equipment)	Rebecca
	Shoran
Global Positioning System	Tacan (distance)
JTID (distance)	Vortac (distance)
Multiple paths (hyperbolic)	
Decca	Omega
Dectra (bearing)	Radux
Gee	Raydist
Loran	
Radial	
Adcock direction finder	Navaglobe
Consol	Radar bearing
Four course range	Tacan (bearing)
Glide slope	VOR (vhf omnidirectional radio range)
JTID (bearing)	Vortac (bearing)
Loop direction finder	Wullenberger direction finder
Localizer	
Self-contained	
Astro	Inertial
Doppler	Pressure

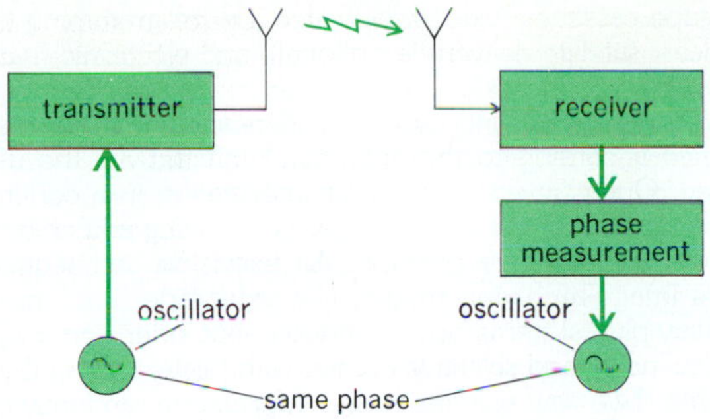

Fig. 1. Single-path system.

Classical systems have often been classified by the various types of position lines which they produce, but actually they are of only two fundamental types. These are the single-path and the multiple-path systems. Two main variations of each of these types exist. The single-path systems measure absolute transmission time and circular lines of position (Fig. 1). The multiple-path systems measure differences in, or otherwise compare, transmission times (Fig. 2). Since the locus of all points having a constant difference from two other points is a hyperbola, these systems should produce hyperbolic lines of positions. Actually, many of these multiple-path systems are so instrumented that they can only determine when the difference in transmission time is zero (that is, the times are equal). With these systems the line of position is a hyperbola of zero curvature, that is, a straight line. The simplified multiple-path systems are known as radial systems. These radial systems were the first type developed and included direction finders, radio ranges, and the directional portion of radar. The true hyperbolic systems must use transmissions which are diverse in either time or frequency. *See* HYPERBOLIC NAVIGATION SYSTEM.

There are four types of so-called self-contained navigational systems. The term is somewhat misleading, because these systems actually depend on observations made on external phenomena as the basis of their measurements. From these observations, course and speed are determined and used to compute an advanced position from a known position. The classification of these systems is based on the phenomena on which the observation is made. Self-contained systems include astro devices, which make observations on heavenly bodies; pressure devices, which observe the atmospheric pressure system; Doppler devices, which observe changes in radio frequency due to the Doppler effect; and inertial devices, which measure forces in inertial space.

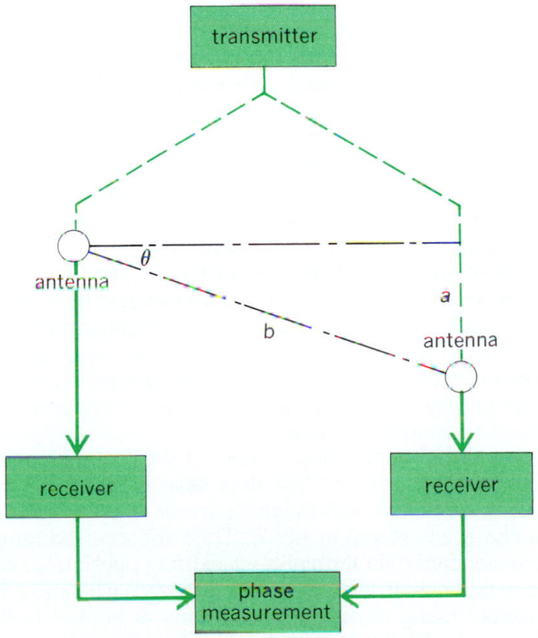

Fig. 2. Multipath system. Transmission paths (broken lines) are assumed parallel for long distances. *a* = difference in path lengths; *b* = base line; phase difference in electrical degrees equals θ or 2θ.

Operational classifications. Electronic navigation systems must also be discussed from the standpoint of navigation. Air operations are considered to take place in four zones: the en route long-distance zone, the en route short-distance zone, the approach and landing zone, and the airport zone. Electronic navigation devices serve three purposes in marine operations: anticollision, position fixing, and harbor control. *See* AIR NAVIGATION; AIR-TRAFFIC CONTROL; INSTRUMENT LANDING SYSTEM (ILS); MARINE NAVIGATION; UNDERWATER NAVIGATION. [P.C.S.]

Electronic packaging The technology relating to the establishment of electrical interconnections and appropriate housing for electrical circuitry. Electronic packages provide four major functions: interconnection of electrical signals, mechanical protection of circuits, distribution of electrical energy (that is, power) for circuit function, and dissipation of heat generated by circuit function.

Printed circuitry. As solid-state transistors started to replace vacuum-tube technology, it became possible for electronic components, such as resistors, capacitors, and diodes, to be mounted directly by their leads into printed circuit boards or cards, thus establishing a fundamental building block or level of packaging that is still in use. *See* PRINTED CIRCUIT.

Packaging hierarchy. Complex electronic functions often require more individual components than can be interconnected on a single printed circuit card. Multilayer card capability was accompanied by development of three-dimensional packaging of daughter cards onto multilayer mother boards.

Integrated circuitry allows many of the discrete circuit elements such as resistors and diodes to be embedded into individual, relatively small components known as integrated circuit chips or dies. In spite of incredible circuit integration, however, more than one packaging level is typically required, in part because of the technology of integrated circuits itself.

Integrated circuit chips are quite fragile, with extremely small terminals. First-level packaging achieves the major functions of mechanically protecting, cooling, and providing capability for electrical connections to the delicate integrated circuit. At least one additional packaging level, such as a printed circuit card, is utilized, as some components (high-power resistors, mechanical switches, capacitors) are not readily integrated onto a chip. For very complex applications, such as mainframe computers, a hierarchy of multiple packaging levels is required. *See* INTEGRATED CIRCUITS.

First-level packaging. Chip-to-package interconnection is typically achieved by one of three techniques: wire bond, tape-automated bond, and solder-ball flip chip. The wire bond is the most widely used first-level interconnection; it employs ultrasonic energy to weld very fine wires mechanically from metallized terminal pads along the periphery of the integrated circuit chip to corresponding bonding pads on the surface of the substrate. In the tape-automated bond, photolithographically defined gold-plated copper leads are formed on a polyimide carrier that is usually handled like 35-mm photographic roll film, with perforated edges to reel the film or tape. The solder-ball flip-chip technique involves the formation of solder bump contacts on the terminals of the integrated chips and reflowing the solder with the chip flipped in such a way that the bump contacts touch and wet to matching pads on the substrate.

Substrates for first-level packages are quite varied. A major

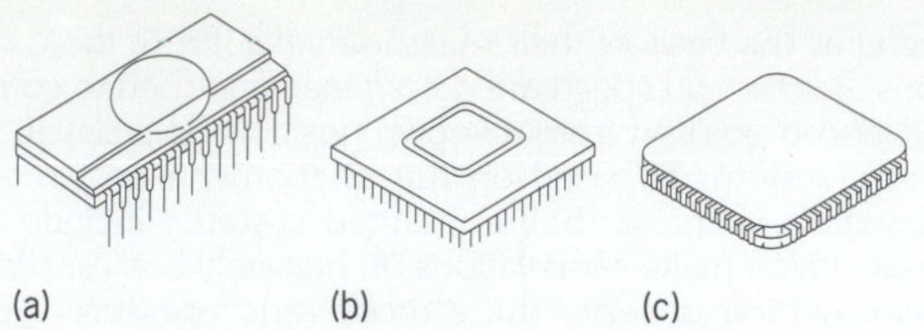

Typical first-level packages. (*a*) Dual in-line package. (*b*) Pin grid array. (*c*) Leadless chip carrier.

classification is whether the package supports a single integrated circuit chip (single-chip module) or more than one chip (multichip module). The former is by far the most common. Substrate insulator materials for multichip and single-chip modules are selected from one of two broad groups of materials, organics and ceramics.

Means for interconnecting to the second-level package often dictate the general form of the first-level package (see illus.). *See* ELECTRONICS. [P.J.Br.]

Electronic power supply A source of electric energy employed to furnish the tubes and semiconductor devices of an electronic circuit with the proper electric voltages and currents for their operation. The more common sources of energy are chemical batteries and alternating-current mains or lines. Batteries are useful as portable sources but are expensive and have small capacities. Alternating-current mains are not portable but are relatively inexpensive and have a large capacity. *See* ALTERNATING CURRENT; BATTERY.

One common method of classifying power supplies is by their use in electronic circuits. Most vacuum and gas tubes require a source of energy to energize their filaments or heaters; this type is known as a filament or heater power supply. Tubes require plate or anode voltages, and transistors need voltages for their collectors and emitters. These voltages are from a direct-current power supply which may provide about 5 to 40 volts for circuits involving transistors and about 50 to 400 volts for circuits using tubes. *See* GAS TUBE; SEMICONDUCTOR; TRANSISTOR; VACUUM TUBE.

If the source is one of alternating voltage, a transformer is usually used to raise or lower the voltage to the required level. The alternating current usually must be changed to a direct current; this is accomplished by a rectifier. A rectifier allows current to flow mostly in one direction, and pulsating current results. This pulsating direct current is not suitable for most purposes and must be smoothed by a power-supply filter or voltage regulator. *See* RECTIFIER; TRANSFORMER.

A power-supply filter stores energy when the current is high and gives it up when the current is lower. The net result is to smooth out the variations in the current. The voltage regulator performs a similar function, but its operation is quite different. The Zener diode type of voltage regulator has a voltage which is nearly constant as the current is varied over a considerable range of values. The gas-tube type of voltage regulator has a characteristic constant voltage over a large range of current values. Vacuum-tube and transistor regulators usually operate as variable resistances; the resistance decreases when the load is heavy and increases when it is lighter. Voltage regulators deliver an almost constant output voltage in spite of variation in load or in input ac supply voltage. *See* VOLTAGE REGULATOR; ZENER DIODE. [D.L.W.]

Electronic publishing A technology encompassing a variety of activities which contain or convey information with a high editorial and value-added content in a form other than print. Included in the list of presently practiced electronic publishing activities are: on-line data bases, videotext, teletext, videotape cassettes, videodisks, cable TV programming (direct broadcast satellite delivery is optional), and electronic mail and messaging.

For convenience and ease of understanding, these may be classified according to the amount of interactivity the user is allowed. One-way systems deliver information in a continuous manner and allow the user only passive viewing and crude control over sequence or position. All television and sequential-access image-archiving media, like videotape, are one-way. One-way-plus systems are continuous but offer some option for sequencing and selective access, with teletext being the best example. Two-way systems access information randomly under direct user control.

Publishing environment. The immense growth of interest in electronic publishing has resulted primarily from technological advances affecting the areas of information distribution, display, and storage, combined with a dramatic decrease in the cost, and increase in the speed and power, of telecommunications.

In the area of communications, satellites have already begun to revolutionize commercial long-distance data transmissions. In the late 1970s and early 1980s the United States witnessed the rise of a new industry, dedicated to creation of nationwide data communication networks using packet switching technology, such as GTE's Telenet and Tymeshare's Tymenet networks. These networks are easy and inexpensive to use when compared to long-distance telephone lines. For most users, access to the network is through a local telephone call on a regular telephone. The networks are designed to allow data communications and computing devices manufactured by many different companies to connect with one another intelligently, that is, without the users having to concern themselves with the compatibility of individual devices. Due to their low cost, ease of use, and effect on the device compatibility issue, packet switching networks have created a strong and stable data telecommunications industry. *See* DATA COMMUNICATIONS; SWITCHING CIRCUIT.

Since packet switching networks depend for most of their operation on satellites which are in a geostationary orbit 22,500 mi (36,000 km) above the Earth, surface distances are relatively immaterial to the cost of telecommunications. Other resources used are equivalent to those used in a local telephone call.

In this environment, electronic publishers can reach a national market of potential subscribers without the costs of logistical problems associated with the nationwide distribution of printed material. Since electronic publishing is based on a pay-per-use concept and the data base of text or images is continuously available, there are dramatically different economics from those associated with traditional publishing, where manufacturing and distribution of books, magazines, or newspapers represent by far the majority of costs.

Crucial to the economical production of new information packages is the ability to manipulate images and type in totally digital formats. This was available only experimentally and at great expense until mid-1981. Moreover, software and engineering expertise for type- and image-processing systems, most commonly referred to as pagination or image-assembly devices, are maturing at a rapid rate.

On-line data bases. Most mature of the electronic publishing activities is that of on-line data base publishing. On-line data bases utilize standard magnetic media and are composed of text characters stored in ASCII. They are accessed remotely by using standard data terminals via existing public telecommunications facilities at low to moderate speed, with most hard-copy output being done on line printers at central facilities. Direct output of full-text data from electronic text-processing systems to on-line data bases is already a reality in highly automated newspaper or magazine publishing environments.

Inability to incorporate graphic material has been a major

limitation and block to greater acceptance of on-line data bases. In 1982 the first attempt was made to use videodisk, attached locally to the access terminal, as a way of integrating image material into the on-line data base presentation.

Videotex. Videotex is the generic term for electronic home information delivery systems. Within this broad term there are two specific approaches: teletext, and viewdata or videotext. Teletext is a one-way communications medium. Images, each constituting a single frame of TV data in a special, compressed format, are transmitted in a continuous sequence. Users indicate which frame they would like to see by interaction with the decoding unit in their local TV set. Videotext, or viewdata, is a potentially more powerful service since it involves users' ability to directly access a central data base interactively from their local TV.

Videodisk. Laser videodisk is the most exciting of the new visual media. It has rich potential since it employs a random-access technique. Available videodisks store single-frame television images in the form of analog FM-encoded TV signals. These are decoded and played back by using a low-power laser. Future optical disks will be memory systems rather than frame storage devices. However, because of the enormous quantities of digital data they will be capable of storing in binary form, the creation of all-digital image archives will be their prime application. *See* CONSUMER ELECTRONICS; VIDEODISK RECORDING. [D.H.G.]

Electronic switch An electronic device to which two input waveforms can be applied and which delivers at a pair of

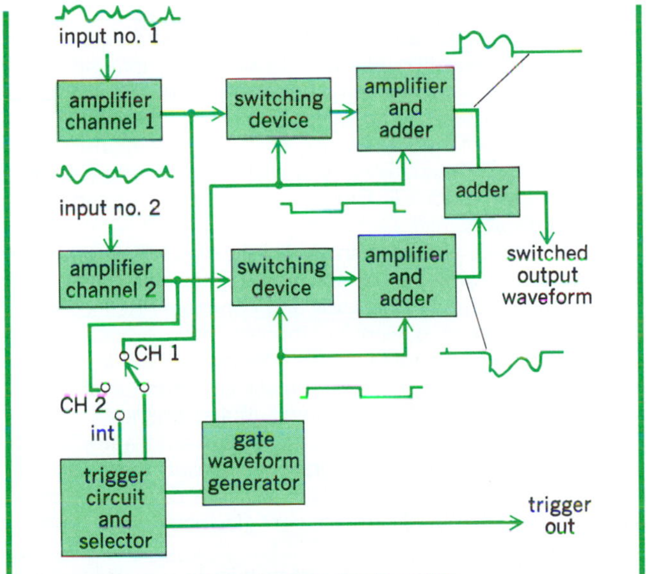

Elements of electronic switch.

output terminals a signal that is alternately a replica of each of the input signals (see illustration). The transmission gate performs the basic switching function of the electronic switch, and sometimes such a gate circuit is itself defined as an electronic switch. However, the electronic switch is usually considered to be a separate instrument to which periodically recurrent input signals, of arbitrary waveform but with synchronously related periods, are connected. The output is switched between the two waveforms at a rate that is synchronous with the period of the input waveforms. One of the input signals often provides synchronizing information to supply periodic trigger pulses, which are then applied to the gate signal generators within the switch. *See* GATE CIRCUIT.

A frequent use of the electronic switch is to provide means for displaying two time-related signals on the screen of a cathode-ray oscilloscope without requiring two independent deflection systems within the cathode-ray tube. *See* SWITCHING CIRCUIT. [G.M.G.]

Electronic test equipment Electronic apparatus used for the purpose of calibration, measurement, or evaluation.

Electrical instruments has evolved from simple mechanical devices for the measurement of electrical properties into complex computer-controlled systems. The most significant of the early electromechanical instruments is the moving-coil meter. Another early instrument is the bridge circuit. *See* AMMETER; BRIDGE CIRCUIT; GALVANOMETER.

An instrument ideally should provide consistent measurements for years under various environmental conditions. Performance depends upon design and manufacture. Many sources of possible error are eliminated by the design of the bridge circuit. Thus a change in battery voltage affects both branches of the circuit equally and does not change bridge balance conditions.

Even perfectly accurate instruments may make less than perfect measurements. The mere act of attaching test equipment alters performance slightly. Thus a very small amount of current is drawn from a battery by a voltmeter, which causes a correspondingly small drop in the battery voltage. *See* CALIBRATION; INSTRUMENTATION; PHYSICAL MEASUREMENT.

Electronic instruments. Early test instruments had limited ability to avoid such loading problems, because they required significant amounts of power. Improvement came in the first decade of the twentieth century with the development of the vacuum tube. The vacuum tube was followed by the transistor, which offered great improvements in performance, economy, and reliability. These electronic devices provide control of large output signals by small input signals. A moving-coil meter placed in the output circuit of an electronic amplifier greatly increases its sensitivity from the amplifier input and diminishes its potential for loading error. *See* TRANSISTOR; VACUUM TUBE.

New techniques, such as sampling, added to the remarkable development of semiconductors, have made measurements practical at higher and higher frequencies. These advances were necessary for faster computing and improved communication. Electronics has also led to the development of digital techniques, which are providing measurements with unparalleled precision and allow the digital display of the results. *See* ANALOG-TO-DIGITAL CONVERTER; DIGITAL COUNTER; VOLTMETER.

Embedded-computer instruments. The semiconductor industry has provided high-level integrated circuits, including computers, for instruments, allowing error-compensating and noise-canceling techniques. The accuracy of measurement may be improved by self-calibration. A self-calibrating instrument periodically compares a measurement to an internal standard, thereby reducing instrument drift. Automated averaging also improves accuracy of measurement. Another benefit of embedded computers is in the derivation of desired quantities from measurements. *See* EMBEDDED SYSTEMS; INTEGRATED CIRCUITS; MICROPROCESSOR.

Test equipment also includes signal generators, even though they do not directly make measurements. Signals of unusual waveshape may be generated by digital techniques involving high-speed memory and digital-to-analog conversion. *See* SIGNAL GENERATOR.

Computers also permit automation of test equipment. Long sequences of settings and readings can occur rapidly and without error. [J.K.S.]

Electronic warfare Use of techniques, devices, and equipment by an adversary to deny or counteract the enemy's use of radar, communications, guidance, or other radio-wave devices. Because of the use of optical and infrared techniques for communications, guidance, detection, and control, electronic warfare techniques are sometimes called electromagnetic, rather than electronic, to convey more adequately the idea that countermeasures are not confined to the portion of the spectrum where electronic techniques alone are applicable but may be used throughout the electromagnetic spectrum.

Traditionally, electronic warfare equipment and techniques are categorized as active and passive, depending on whether or not they radiate their own energy. The passive category includes reconnaissance or surveillance equipment that detects and analyzes electromagnetic radiation from radar and communications transmitters in a potential enemy's aircraft, missiles, ships, satellites, and ground installations. The reconnaissance devices may be used to identify and map the location of emitters without in any way altering the nature of the signals they receive. Other types of passive electronic warfare equipment do seek to enhance or change the nature of the energy reflected back to enemy radars, but they do not generate their own energy.

Active electronic warfare equipment generates energy, either in the form of noise to confuse an enemy's electromagnetic sensors or by generating false or time-delayed signals to deceive radio or radar equipment and their operators. *See* ELECTRICAL NOISE.

Passive systems. Reconnaissance or surveillance electronic warfare systems are carried by Earth-orbiting satellites, aircraft, ships, uncrewed (drone) aircraft, and automotive vehicles. Some are located on the ground. A few systems are small enough to be carried by a person. Reconnaissance systems are interchangeably called ferret or electronic intelligence (elint) systems. They consist of sensitive receivers electromechanically or electronically tuned over desired portions of the spectrum in search of transmissions of interest. Bearing to an intercepted signal can be determined by direction-finding techniques. Once secured, the signals can be displayed for analysis by an operator or stored on tape or other storage media for subsequent analysis.

Warning-receiver systems, which came into widespread use on United States tactical and transport aircraft during the Vietnamese war, are a more limited form of the elint system. Unlike the latter, which is intended to search for signals over a broad range of the spectrum, the warning receiver is programmed to alert a pilot when the aircraft is being illuminated by a specific radar signal above predetermined power thresholds anticipated by elint systems. When the pilot has been alerted, the aircraft can be maneuvered to evade the threat or initiate counteraction with onboard electronic warfare capability.

One of the oldest passive electronic warfare techniques, made famous by a Royal Air force raid on Hamburg in 1943, is the use of chaff. These are metallic strips cut to lengths resonant at the defense radar frequency so that they return spurious radar echoes to enemy radar. Chaff can confuse an enemy by generating false targets, or noise, forcing the enemy to take time to analyze the returns and sort real from false targets. Chaff can screen or mask aircraft or higher-speed ships so that the enemy is unable to determine their presence, or chaff can help an aircraft break track once it is alerted by its warning receiver that it is being tracked by radar.

Other passive electronic warfare techniques include the use of special radar-absorbing materials, such as ceramics or ferrites, which reduce reflection coefficients so that the amount of radar energy returned to the illuminating radar is reduced; the special shaping of bodies, specifically in missile reentry systems, that reduces the vehicle's radar cross section; the use of corner reflectors, or Luneberg lenses, which concentrate the energy they reflect back to the radar.

Active systems. The many active electronic warfare techniques can be classified broadly either as noise or deception jamming. The former is the oldest, simplest, and most straightforward, but requires higher average power levels and is more expensive. Deception jamming is the more artful and sophisticated technique, operating on the characteristics of the pulse train generated by threat radars.

Deception-jamming techniques are predicated on the idea of operating on pulses received from the enemy so that the signal reradiated from the target deceives the enemy radar or its operators. For instance, the deception set may receive an enemy radar pulse, circulate it through a delay line, amplify it, and reradiate it back toward the enemy. Because the enemy determines the position of the target by the round-trip transit time of the radar energy, the radar decision circuitry will conclude that the target is at a greater distance than it actually is because of the deceptive pulse delay inserted in that round-trip period by the active set. Similarly, the deception set may operate on the radar pulse train, returning many pulses instead of one, in an effort to deceive the enemy into believing there are many targets spaced at different positions. [B.Mi.]

Electronics The branch of science and technology relating to the conduction and control of electricity flowing through semiconducting materials or through vacuum or gases. Electronics is concerned with the study and applications of the motions of charge carriers (electrons, holes, and ions) under the influence of externally applied voltage or current, or in relation to the incidence or production of radiant energy. While electronics is properly a part of electrical engineering, the latter term is often reserved for applications involving power generation, distribution, and use at low frequencies, for example, in utility systems and industry. Since the 1960s, the dominant segment of electronics has been that known as solid-state, which involves transistors and other semiconductor devices and assemblies.

The electronics industry is concerned with the design, manufacture, and application of transistors, diodes, integrated circuits, and other semiconductor devices, and to a much lesser extent, electron tubes. The major uses of electron tubes, not taken over by solid-state devices, are the generation of large amounts of radio-frequency power and the perception and display of images, as in television camera and picture tubes and computer displays.

These devices are used in a rapidly expanding range of applications making use of so-called digital electronics: that basic to the electronic calculator and computer. Reductions in the size and cost of powerful computers have produced, in a single large-scale integrated circuit, devices used in the automotive field to control fuel use, in "intelligent" control of household appliances, satellite communications, financial transactions, industrial instrumentation, control and automation, medical devices and implants, and the aerospace industry and numerous aspects of national defense. To a degree not envisaged as late as 1970, electronics pervades nearly every avenue of human endeavor.

The manifold applications of electronic devices depend fundamentally on the physical fact that the force exerted on an electron or a hole by the voltage applied to such devices is very great in proportion to the mass of the charge carrier. Consequently the electric charge is very rapidly accelerated and can achieve its maximum velocity in a billionth of a second or less. This rapid reaction permits the device to amplify and generate currents alternating at frequencies of billions of cycles per second (gigahertz) or to respond to stimuli in less than a billionth of a second (nanosecond).

While these numbers apply primarily to low-power devices, lower-frequency semiconductor devices, such as the silicon controlled rectifier (SCR), can control as much as 1000 MW of power at commercial frequencies. This requires the use of large arrays of silicon controlled rectifiers. These devices are also finding use in the transformation and control of power in electric locomotives and in similar heavy-current uses. These have given rise to the term power electronics. *See* SEMICONDUCTOR RECTIFIER. [D.G.F.]

Electronvolt A unit of energy used for convenience in atomic systems. Specifically, it is the change in energy of an electron, or of any particle having a charge numerically equal to that of an electron, when it is moved through a difference of potential of 1 mks volt. Its value (in mks units) is obtained from the equation $W = qV$, where W is energy in joules, q the charge in coulombs, and V the potential difference in volts. For a potential difference of 1 volt and the electronic charge of 1.601×10^{-19} coulomb, the electronvolt is 1.601×10^{-19} joule. *See* ELECTRON; IONIZATION POTENTIAL. [G.H.M.]

Electrooptics The branch of physics which deals with the influence of an electric field on the optical properties of matter, especially in its crystalline form. These properties include transmission, emission, and absorption of light.

An electric field applied to a transparent crystal can change its refractive indexes and, therefore, alter the state of polarization of light propagating through it. When the refractive-index changes are directly proportional to the applied field, the phenomenon is termed the Pockels effect. When they are proportional to the square of the applied field, it is called the Kerr effect. *See* KERR EFFECT; POLARIZED LIGHT; REFRACTION OF WAVES.

The Pockels effect is used in a light modulator called the Pockels cell. This device (see illustration) consists of a crystal C

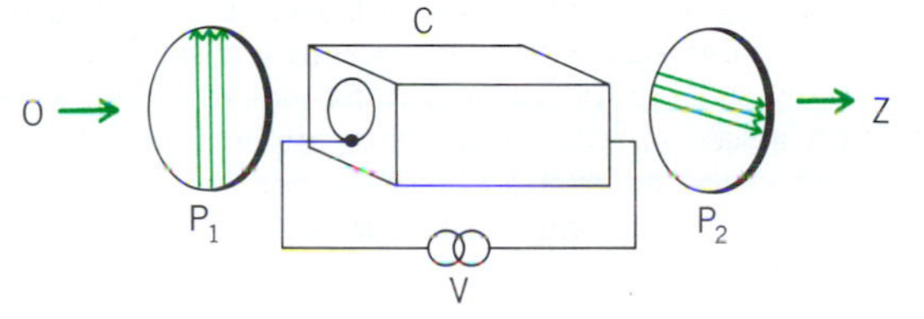

Pockels cell light modulator. The arrows represent a light beam.

(usually potassium dihydrogen phosphate, or KDP) placed between two polarizers P_1 and P_2 whose axes are crossed. Ring electrodes bonded to two crystal faces allow an electronic driver V to apply an electric field parallel to the axis OZ along which a light beam (for example, a laser beam) is made to propagate. Pockels cells can be switched on and off in well under 1 nanosecond. *See* LASER; OPTICAL MODULATORS.

The linearity and high-speed response of the Pockels effect within an electrooptic crystal make possible a unique optical technique for measuring the amplitude of repetitive high-frequency (greater than 1 GHz) electric signals that cannot be measured by conventional means. The technique, known as electrooptic sampling, employs a special traveling-wave Pockels cell between crossed polarizers. It is used to analyze ultrafast electric signals such as those generated by high-speed transistors and optical detectors. [M.A.D.; J.A.V.]

A self-electrooptic-effect device (SEED) is a combination of a quantum-well electrooptic modulator with a photodetector which, when light shines on it, changes the voltage on the modulator. Although the device relies internally on an electrooptic effect, the output from the modulator is controlled by the light shining on the photodetector, giving an optically controlled device with an optical output. Most of these devices rely on the quantum-confined Stark effect in semiconductor quantum-well heterostructures as the electrooptic mechanism and utilize the changes in optical absorption resulting from this mechanism. *See* SEMICONDUCTOR HETEROSTRUCTURES; STARK EFFECT. [D.A.B.M.]

Electrophilic and nucleophilic reagents Electrophilic reagents are chemical species which, in the course of chemical reactions, acquire electrons, or a share in electrons, from other molecules or ions. Although this definition embraces all oxidizing agents and all Lewis acids, electrophilic reagents are ordinarily thought of as cationic species, such as H^+, NO_2^+, Br^+, or SO_3 (or carriers of these species such as HCl, CH_3COONO_2, or Br_2), which can form stable covalent bonds with carbon atoms. Electrophilic reagents frequently are positively charged ions (cations). *See* ACID AND BASE.

Nucleophilic reagents are the opposite of electrophilic reagents. Nucleophilic reagents give up electrons, or a share in electrons, to other molecules or ions in the course of chemical reactions. Nucleophilic reagents frequently are negatively charged ions (anions). Typical nucleophilic reagents are hydroxide ion (OH^-), halide ions (F^-, Cl^-, Br^-, and I^-), cyanide ion (CN^-), ammonia (NH_3), amines, alkoxide ions (such as CH_3O^-), and mercaptide ions (such as $C_6H_5S^-$). *See* SUBSTITUTION REACTION. [J.F.B.]

Electrophoresis The migration of electrically charged particles in solution or suspension in the presence of an applied electric field. Each particle moves toward the electrode of opposite electrical polarity. For a given set of solution conditions, the velocity with which a particle moves divided by the magnitude of the electric field is a characteristic number called the electrophoretic mobility. The electrophoretic mobility is directly proportional to the magnitude of the charge on the particle, and is inversely proportional to the size of the particle. An electrophoresis experiment may be either analytical, in which case the objective is to measure the magnitude of the electrophoretic mobility, or preparative, in which case the objective is to separate various species which differ in their electrophoretic mobilities under the experimental solution conditions.

The original apparatus known as the Tiselius cell detects electrophoretic motion by the moving-boundary method, in which a boundary is created between the solution of particles to be examined and a sample of pure solvent (see illustration). As the particles migrate in an electric field, the boundary between solution and solvent can be observed to move, and if there are a number of species in the solution with different electrophoretic mobilities, a series of boundaries of various shapes and magnitudes can be detected. The moving-boundary method was used for three decades to separate complex mixtures of charged macromolecules in solution and to study the physical characteristics of solutions of proteins and other macromolecules of biological and industrial importance.

The resolving power of electrophoresis has been greatly

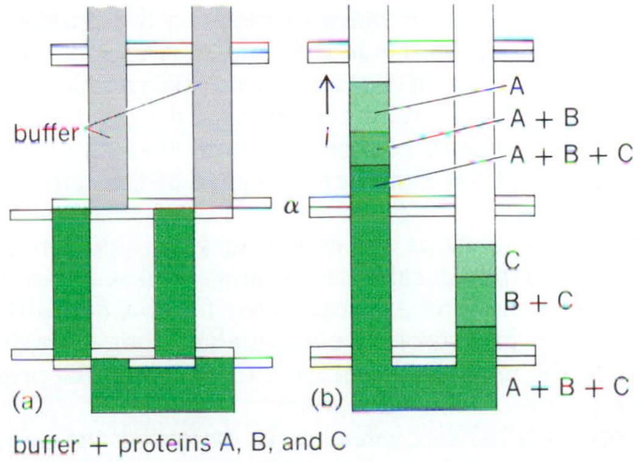

Diagrammatic representation of Tiselius electrophoresis cell, showing *(a)* formation and *(b)* motion of electrophoretic boundaries of a mixture of proteins.

improved by the introduction of the use of gel supporting media. The gel matrix prevents thermal convection caused by the heat which results from the passage of electric current through the sample, which reduces greatly the mixing of the various parts of the sample, and therefore allows for more stable separation. This method is now the most common technique for the determination of molecular weights of proteins in biochemical studies.

The methodology of electrophoresis may be modified considerably when the particles undergoing analysis are of sufficient size to be viewed either with the naked eye or with the assistance of an optical microscope. This general area of particle electrophoresis has its most important applications in the analysis of the surface charge of living cells and in the study of various types of particles used in industrial coating processes.

Application of the optical laser to electrophoretic detection has resulted in the development of a technique which can be used for analytical electrophoresis experiments on particles of all sizes. The basic principle is that the highly monochromatic laser light impinges upon the particles, and is scattered from the particles in all directions. Observation of the laser light which has been scattered from a moving particle permits detection of a slight shift in the frequency of the light as a result of the motion of the particle. This is the Doppler effect. The application of the laser Doppler principle to electrophoresis experiments, often called electrophoretic light scattering (ELS), has become an important method for the rapid determination of electrophoretic velocities. Electrophoretic light scattering has been used for the study of many types of living cells, cell organelles, viruses, proteins, nucleic acids, and synthetic polymers. *See* DOPPLER EFFECT; ELECTROLYTIC CONDUCTANCE; ISOELECTRIC FOCUSING; ISOTACHOPHORESIS. [B.R.W.]

Electroplating of metals The process of electrodepositing metallic coatings to alter the existing surface properties or the dimensions of an object. Electroplated coatings are applied for decorative purposes, to improve resistance to corrosion or abrasion, or to impart desirable electrical or magnetic properties. Plating is also used to increase the dimensions of worn or undersized parts. An example of a decorative coating is that of nickel and chromium on automobile bumpers. However, in this application, corrosion and abrasion resistance are also important. An example of electrodeposition used primarily for corrosion protection is zinc plating on such steel articles as nuts, bolts, and fasteners. Since zinc is more readily attacked by most atmospheric corrosive agents, it provides galvanic or sacrificial protection for steel. An electrolytic cell is formed in which zinc, the less noble metal, is the anode, and steel, the more noble one, the cathode. The anode corrodes, and the cathode is protected. *See* ELECTROCHEMISTRY.

The electroplating process consists essentially of connecting the parts to be plated to the negative terminal of a direct-current source and another piece of metal to the positive pole, and immersing both in a solution containing ions of the metal to be deposited. The part connected to the negative terminal becomes the cathode, and the other piece the anode. In general, the anode is a piece of the same metal that is to be plated. Metal dissolves at the anode and is plated at the cathode. *See* ELECTROLYSIS.

Most plating solutions are of the aqueous type. There is a limited use of fused salts or organic liquids as solvents. Nonaqueous solutions are employed for the deposition of metal with lower hydrogen overvoltages; that is, hydrogen rather than the metal is reduced at the cathode in the presence of water.

In order for adherent coatings to be deposited, the surface to be plated must be clean, that is, free from all foreign substances such as oils and greases, as well as oxides or sulfides. The two essential steps are cleaning and pickling.

Three principal cleaning methods are employed to remove grease and attached solids. (1) In solvent cleaning, the articles undergo vapor degreasing, in which a solvent such as tri- or tetrachloroethylene is boiled in a closed system, and its vapors are condensed on the metal surfaces. (2) In emulsion cleaning, the metal parts are immersed in a warm mixture of kerosine, a wetting agent, and an alkaline solution. (3) In electrolytic cleaning, the articles are immersed in an alkaline solution, and a direct current is passed between them and the other electrode, which is usually steel. Ultrasonic cleaning is also used extensively, especially for blind holes or gears packed with soils. Ultrasonic waves introduced into a cleaning solution facilitate and accelerate the detachment of solid particles embedded in crevices and small holes.

In the pickling process, oxides are removed from the surface of the basis metal. For steel, warm, dilute sulfuric acid is used in large-scale operations because it is inexpensive; but room-temperature, dilute hydrochloric acid is also used for pickling because it is fast-acting. Hydrogen embrittlement may be caused by the diffusion of hydrogen in steel (especially high-carbon steel) during pickling and also in certain plating operations. *See* EMBRITTLEMENT.

Electropolishing is used when a thin, bright deposit is to be produced. In this case, the substrate surface contours are essentially copied by the deposit. The substrate surface therefore must have a bright finish which can be attained by electropolishing. *See* ELECTROPOLISHING.

Special processes, such as electroforming and anodizing, are required for certain applications. Electroforming is a special type of plating in which thick deposits are subsequently removed from the substrate, which acts as a mold. The process is particularly suitable for forming parts which require intricate designs on inside surfaces. Important applications of electroforming are in the production of phonograph record masters, printing plates, and some musical instruments and fountain pen caps as well in waveguides. In anodizing, a process related to plating, an oxide is deposited on a metal which is the anode in a suitable solution. The process is primarily used with aluminum, but it can be applied to beryllium, magnesium, tantalum, and titanium. Colored coatings can be produced by the incorporation of dyes. *See* METAL COATINGS. [R.W.]

Electropolishing A method of polishing metal surfaces by applying an electric current through an electrolytic bath in a process that is the reverse of plating. The metal to be polished is made the anode in an electric circuit. Anodic dissolution of protuberant burrs and sharp edges occurs at a faster rate than over the flat surfaces and crevices, possibly because of locally higher current densities. The result produces an exceedingly flat, smooth, brilliant surface.

Electropolishing is used for many purposes. The brilliance of the polished surface makes an attractive finish. Because the polished surface has the same structural properties as the base metal, it serves as an excellent surface for plating. Electropolishing avoids causing differential surface stresses, one of the requirements for the formation of galvanic cells which cause corrosion. Because no mechanical rubbing is involved, work hardening is avoided. Contaminants, which often are associated with the use of abrasives and polishing compounds, are also avoided. The surface is left clean and may require little or no preparation for subsequent treatment or use. Electropolishing also minimizes loss of high-temperature creep-rupture strength. *See* ELECTROPLATING OF METALS. [W.W.Sn.]

Electrorefining A purification process in which an impure metal anode is dissolved electrochemically in a solution of a salt of the metal being refined; the pure metal is recovered by electrodeposition at the cathode.

Electrorefining is the most economical method for securing

the high purity required for many uses of nonferrous metals. Since the same electrochemical reaction proceeds in opposite directions at the anode and the cathode, the overall chemical change is a small change in the activity of the metal at the two electrodes, and the reversible cell voltage is practically zero. The anodic and cathodic overvoltages, required for an economical reaction rate, are small, and the operating cell voltage consists mainly of the ohmic voltage drop through the electrolyte and at the electrode contacts. The power consumption is moderate, and electrorefining is less sensitive to the cost of electric power than are other electrometallurgical processes.

Electrochemical refining is a more efficient purification process than other chemical methods are because it is very selective. In particular, for metals such as copper, silver, gold, and lead, the operating electrode potential is close to the reversible potential, and a sharp separation is accomplished, both at the anode and at the cathode. Precious metals do not dissolve at the potential of the anode, and they remain as metals in the slimes, along with insoluble selenium and tellurium compounds and some metallic copper particles that fall from the anode. Metals less noble than copper do not deposit at the cathode. In particular, nickel and arsenic accumulate in the electrolyte, and their concentration is controlled by partial withdrawal and purification of the solution. *See* ELECTROCHEMICAL PROCESS. [P.Du.]

Electroscope An instrument for detecting the presence and sign of an electric charge. It is the simplest type of ionization chamber. *See* IONIZATION CHAMBER.

The illustration shows a common type of simple gold-leaf electroscope. Gold leaf (L) is used because it is an extremely thin conducting foil which has low mass per unit area and is very flexible. Hence, it responds quickly and vigorously to small electrostatic forces. In the illustration H serves as a grounded electrostatic shield, as well as a shield against air currents. The hard-rubber rod R (illustration *a*) with its negative charge has

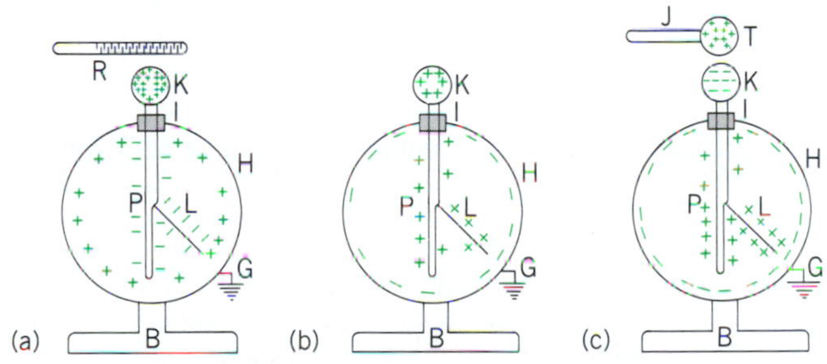

Electroscope. (P is a metal support terminating in knob K; B is the base; I is an insulator; and H is a cylindrical metal housing with flat ends and windows.) (*a*) Being charged by induction by negative charge on hard-rubber rod R. (*b*) Positive charge left on its leaf after induction process is complete. (c) Testing the sign of an unknown charge on test ball T.

set up the charge distribution by the process of electrostatic induction. The response shown is a test for the fact that R has a charge. *See* ELECTROSTATICS.

To leave the electroscope with a net charge, a grounded conductor is touched to K so that the surplus electrons on P and L go off to ground, leaving the bound positive charge on K. The ground connection is then broken and R is removed. At this stage (illustration *b*) the electroscope is said to have a positive charge because there is a positive charge on its leaf system.

If an electroscope has a charge of known sign, as in illustration *b*, it can be used to test the sign of an unknown charge, as in illustration *c*, where the metal test ball (T), with its insulating handle (J), has the unknown charge. In the situation pictured, L moves farther away from P as T is brought slowly up toward K, showing that T has a positive charge. If T had a negative charge, L would move toward P, as T slowly approaches K. The converse situation, if the leaf system in illustration *c* had a negative charge initially, can be readily visualized.

Although electroscopes have been built with a wide variety of geometries, the principle of operation is essentially the same for all. If an electroscope has a scale, permitting quantitative measurements, it is called an electrometer or electrostatic voltmeter. For information on electrometers *see* VOLTMETER. [R.P.Wi.]

Electrostatic accelerator A device that accelerates atomic or subatomic particles to high energies by means of a high voltage potential that is maintained through some electrical or mechanical means of transporting charge from ground to the high voltage potential. Modern machines of larger size all use mechanical charge transport, but electrical transport systems are widely used in small Cockcroft-Walton or dynamitron-type electrostatic accelerator units. The most commonly used electrostatic generator was invented in the 1930s by R. J. Van de Graaff and utilizes a high-speed rubberized-fabric belt to transport the charge.

The Pelletron electrostatic accelerator utilizes a chain-charging system instead of the rubberized-fabric belt. A large Pelletron accelerator at the Oak Ridge National Laboratory is designed to operate at 25 MV. *See* COCKCROFT-WALTON ACCELERATOR; DYNAMITRON ACCELERATOR; PELLETRON ACCELERATOR; RESONANCE TRANSFORMER; VAN DE GRAAFF GENERATOR.

All particle accelerations are carried out inside an evacuated tube so that the accelerated particles do not collide with air molecules or atoms and may follow trajectories characterized specifically by the electric fields utilized for the acceleration. Usually the evacuated tube is provided with a series of electrodes arranged to have gradually increasing potentials from ground to the maximum high-voltage potential. In this way, the high voltage is distributed uniformly along the acceleration tube, and the acceleration process is thereby simplified and better control permitted.

Most electrostatic accelerators are housed inside a large high-pressure vessel that is filled with very dry insulating gas. This high-pressure insulating gas allows the machine to be housed in a much smaller space than would be possible in air at atmospheric pressure.

Large electrostatic accelerators are constructed in the form of tandem electrostatic accelerators. This device gets its name from the fact that it is effectively two electrostatic accelerators back to back, arranged so that the high voltage potential is first used to accelerate negative ions from ground potential up to the high positive potential. The charge of the particle is then switched from negative to positive, and the positive ions are additionally accelerated from the high voltage potential back to ground, continuing in the same direction as the original acceleration. A third stage of acceleration can be accomplished by first accelerating negative particles from high voltage to ground and then injecting them into a conventional tandem. The charge-changing in the high-voltage terminal is accomplished by either a gas or foil stripper.

The folded tandem design has also been developed for large electrostatic accelerators. They are constructed like a large single-stage electrostatic accelerator; however, the column which supports the high-voltage terminal contains two acceleration tubes—one for the acceleration of negative ions from ground potential to the high-voltage terminal, and the other for acceleration of positive ions from the high-voltage terminal down to ground potential.

The larger electrostatic accelerators are used primarily for basic research in the study of nuclear structure and heavy-ion reactions, including fusion and fission processes. The smaller machines are used throughout industry for many applications, including neutron and x-ray production for diagnostic measurements, polymerization and other treatments of plastics and various materials, ion implantation, thin-film analysis, and many other related applications. *See* ION IMPLANTATION; PARTICLE ACCELERATOR. [H.E.W.]

Electrostatic lens An electrostatic field with axial or plane symmetry which acts upon beams of charged particles of uniform velocity as glass lenses act on light beams. The action of electrostatic fields with axial symmetry is analogous to that of spherical glass lenses, whereas the action of electrostatic fields with plane symmetry is analogous to that of cylindrical glass lenses. Plane symmetry as used here signifies that the electrostatic potential is constant along any normal to a family of parallel planes.

The action of an electrostatic lens on the paths of charged particles passing through it is most readily visualized with the aid of an equipotential plot of the fields in a plane of symmetry of the lens. The equipotential lines in the plot indicate the intersection with the plane of the drawing of surfaces on which the electrostatic potential is a constant. The paths of charged particles in the electrostatic field are bent toward the normals of the equipotentials as the particles are accelerated, and away from the normals as the particles are decelerated.

Axially symmetric lenses are generally formed at or between circular apertures and cylinders maintained at suitable potentials. For any of these it is possible to define focal points, principal planes, and focal lengths in the same manner as for light lenses and to determine with their aid image magnification for any object position.

Lenses of plane symmetry, analogous to cylindrical glass lenses, are formed between parallel planes and at slits, replacing the circular cylinders and apertures of lenses with axial symmetry. [E.G.R.]

Electrostatic precipitator A device used to remove liquid droplets or solid particles from a gas in which they are suspended. The process depends on two steps. In the first step the suspension passes through an electric discharge (corona discharge) area where ionization of the gas occurs. The ions produced collide with the suspended particles and confer on them an electric charge. The charged particles drift toward an electrode of opposite sign and are deposited on the electrode where their electric charge is neutralized. The phenomenon would be more correctly designated as electrodeposition from the gas phase.

The use of electrostatic precipitators has become common in numerous industrial applications. Among the advantages of the electrostatic precipitator are its ability to handle large volumes of gas, at elevated temperatures if necessary, with a reasonably small pressure drop, and the removal of particles in the micrometer range. Some of the usual applications are: (1) removal of dirt from flue gases in steam plants; (2) cleaning of air to remove fungi and bacteria in establishments producing antibiotics and other drugs, and in operating rooms; (3) cleaning of air in ventilation and air conditioning systems; (4) removal of oil mists in machine shops and acid mists in chemical process plants; (5) cleaning of blast furnace gases; (6) recovery of valuable materials such as oxides of copper, lead, and tin; and (7) separation of rutile from zirconium sand. [G.S.M.; W.O.M.]

Electrostatics The study of electric charges at rest under conditions where the positive and negative charges are separated from each other. The term static electricity is often used to refer to electric charges at rest.

Electrification by friction or rubbing occurs when one substance is rubbed with another and a surplus of electrons, each with its negative charge, is rubbed onto one of the bodies, leaving a deficiency of electrons, that is, a positive charge, on the other body. The most commonly used materials for electrification by friction are hard rubber and fur. The hard rubber has a negative charge after being rubbed with cat fur. Only good insulators show a net charge after rubbing, because electrical conduction permits neutralization of the charge very rapidly in other materials. *See* ELECTRIC CHARGE; ELECTRICAL INSULATION.

Electrostatic induction affords another way of charging a body where a charged, hard-rubber rod is responsible for the charging process. Since like charges repel each other and unlike charges attract each other, the negative charge on the rod repels some electrons to the opposite side of a metal sphere. This leaves an equal positive charge on the part of the sphere near the rod. If a grounded lead (for example, the experimenter's finger) is touched to the metal sphere, the unbound electrons flow off to ground leaving the positive charge bound by the presence of the charge on the rod. Then if the ground lead is removed and next the rod is removed, the metal sphere is left with a net positive charge which, in the end, is uniformly distributed over the surface of the sphere. When the charge on a conductor is static, that is, after it has completed its redistribution, it is in any case all on the surface of the conductor.

Electrostatic generators, sometimes called static machines, depend on electrification by friction and by induction for their operation. The simplest electrostatic generator is the electrophorus shown in the illustration. The hard-rubber plate R

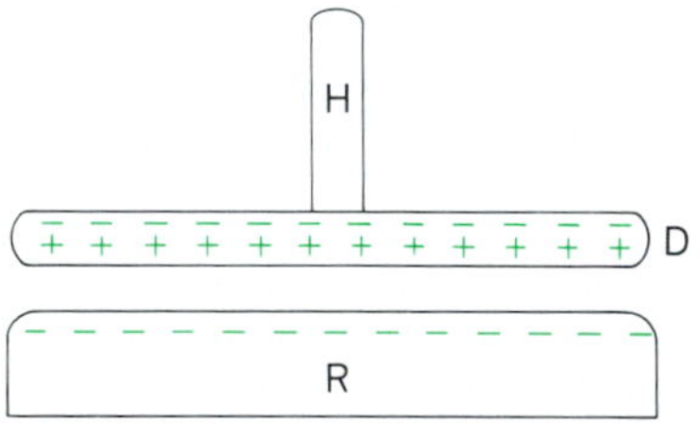

Electrophorus. When the metal plate D with insulating handle H is placed on the rubber plate R, charge is induced as shown.

has been given a negative charge by rubbing it with cat fur. The metal plate D has an insulating handle H. When D is set on R, it touches it at only a few high points because R is microscopically rough, even though it appears polished to the eye. A positive charge is induced on the lower surface of D, and an equal, negative charge appears on the upper surface of D in the induction process. Now if a grounded conductor is touched to D, the electrons go off to ground and the ground connection is then removed. Next, by using the insulating handle H, D is lifted from the rubber plate, and there is a net positive charge on D which can be shared with a receiver and used for experimentation. Since no charge is removed from R while charging by induction, D can be recharged from R as many times as the above process is repeated, and a large charge can be built up on the receiver. A rotating machine which would carry out the repetition of the induction process might be built from the electrophorous. The Wimshurst machine is an electrostatic generator based on this method of operation. For information on the important electrostatic devices used in nuclear physics *see* COCKCROFT-WALTON ACCELERATOR; IMPULSE GENERATOR; PARTICLE ACCELERATOR; RESONANCE TRANSFORMER; VAN DE GRAAFF GENERATOR. [R.P.Wi.]

Electrostriction A form of elastic deformation of a dielectric induced by an electric field; specifically, the term applies to those components of strain which are independent of reversal of the field direction. Electrostriction is a property of all dielectrics and is thus distinguished from the converse piezoelectric effect, a field-induced strain which changes sign upon field reversal and which occurs only in piezoelectric materials. *See* DIELECTRIC MATERIALS; PIEZOELECTRICITY.

The electrostrictive effect in certain ceramics is employed for commercial purposes in electromechanical transducers for sonic and ultrasonic applications. *See* MICROPHONE; UNDERWATER TRANSDUCER. [R.D.W.]

Electrotherapy Medical treatment which uses electrical techniques to change structure or function of the body or a body part. In some methods the electricity is clearly not the major active agent; in others the mechanism of action is unknown. Therefore the classification is of questionable significance except as it indicates technique: (1) High-frequency currents are used within physiological limits as in diathermy, or beyond such limits as in the destruction or removal of tissue by so-called electrosurgery. Another application is in closing small bleeding vessels in surgery, electrocauterization. (2) Low-frequency currents are used for passive exercise of paralyzed muscles. (3) Iontophoresis depends upon the movement of ions in a voltage gradient. If two electrodes are in contact with tissue (such as skin and mucous membranes), then positively charged ions can be driven into the tissue by a positive electrode, and negatively charged ions by a negative electrode. (4) Electroshock therapy is a method of producing violent excitation of the central nervous system which seems beneficial in certain types of psychopathology. *See* ELECTROCONVULSIVE SHOCK. [J.M.To./N.K.M.]

Electrothermal propulsion Vehicular propulsion that involves electrical heating to raise the energy level of the propellant. In contrast, chemical rockets use the chemical energy of one or more propellants to heat and accelerate the decomposition products (monopropellants) or combustion products (bipropellants) for thrusting purposes. In both instances, the high-energy propellant gases are exhausted through a nozzle where they are accelerated to a high velocity, and thrust is produced by reaction. By decoupling the heating or energy addition process from the restraints of propellant chemistry considerations, electrothermal devices can be operated on a wide variety of materials, many of which would not otherwise be considered to be propellants. Water and space station liquid-waste streams are two examples of such propellants being considered for electrothermal propulsion purposes. *See* ROCKET PROPULSION.

Practical electrothermal thrusters come in two forms, resistojets and arcjets. In resistojets, which are now flying in a station-keeping role on many communications satellites, the electrical energy is first deposited in a heater or resistive element and then transferred to the propellant. The need to first heat a material limits the maximum operating temperature and the maximum enthalpy of the propellant. As a consequence, the essential simplicity of the device is balanced by well-defined limitations on exhaust velocity or specific impulse. Arcjets circumvent this limitation by using the propellant as the heater element. An electric arc discharge passes directly through the propellant. High temperatures and specific-impulse values can be achieved but only at the price of design complexity. *See* ARC DISCHARGE; SPECIFIC IMPULSE.

One further and fundamental distinction is that electrical thrusters are power-limited whereas chemical thrusters are energy-limited. By definition, electrical propulsion systems must have an associated power supply for operation. Solar panels or nuclear power supplies can supply well-defined power to the thruster for essentially unlimited time. Consequently, although the power level is well constrained, the total energy available is virtually boundless. Chemical systems are exactly opposite. In this case, the total energy available for propulsion is well defined by the propellant volume. However, the rate at which this propellant is used, the rate of energy usage per unit time, or the power can be exceedingly high. Chemical thrusters are ideal for escaping the Earth's gravity well; electrical thrusters are ideal for moving payloads in low-acceleration conditions removed from gravity wells, that is, for the more ambitious missions far removed from low Earth orbit. *See* SPACECRAFT PROPULSION. [G.W.Bu.]

Electroweak interaction At its most fundamental level, the interaction between quarks and leptons due to the exchange of photons, $W^{\pm}$ mesons, and Z^0 mesons. Since protons, neutrons, and other baryons are made of quarks, nuclei are made of protons and neutrons, and atoms are made of nuclei and electrons (the lightest charged leptons), the electroweak interaction appears as forces and transitions among nuclei and atoms. *See* INTERMEDIATE VECTOR BOSON; LEPTON; PHOTON; QUARKS.

The electroweak interaction is a partial unification, in a sense to be explained later, of the electromagnetic and weak interactions. The prototype electromagnetic interaction is the attraction of unlike and the repulsion of like electric charges; however, the electromagnetic interaction is itself a unification of electric and magnetic interactions so that magnetic forces are also part of this interaction. The prototype weak interaction is the beta decay of the neutron, $n \rightarrow pe^-\bar{\nu}_e$, where n, p, e^-, and ν_e are the neutron, proton, electron, and electron neutrino, respectively. (A bar over the symbol for a particle denotes the corresponding antiparticle.) In nuclei, the weak interaction produces nuclear beta decay, which involves a transmutation of elements. *See* ELECTRIC FIELD; MAGNETIC FIELD; MAXWELL'S EQUATIONS; RADIOACTIVITY; WEAK NUCLEAR INTERACTIONS.

Strength, range, and symmetry. The electroweak interaction is one of the fundamental interactions of nature. The others are the strong interaction and gravity. These interactions differ in their strength, range, and symmetry properties. For the electromagnetic interaction, the strength is $\alpha = e^2/(4\pi\epsilon_0\hbar qc) = 1/137$, where α is the fine structure constant, $\hbar$ is Planck's constant divided by 2π, ϵ_0 is the permittivity of free space, and c is the speed of light; the range is infinite; and electromagnetism is invariant under space inversion (parity, P) and particle-antiparticle interchange (charge conjugation, C). For the weak interaction, the strength is $G_F M_p^2 c/\hbar^3 = 1.02 \times 10^{-5}$, where G_F is the Fermi constant and M_p is the mass of the proton; the range (for charge-changing interactions) is $\hbar/(M_W c) = 9.42 \times 10^{-18}$ m, where M_W is the mass of the W meson; and the weak interaction violates both P and C maximally, violates the combined symmetry CP by about 0.2%, and obeys CPT (T is time reversal), which is a general consequence of local relativistic quantum field theory. *See* CPT THEOREM; FUNDAMENTAL CONSTANTS; FUNDAMENTAL INTERACTIONS; PARITY (QUANTUM MECHANICS); STANDARD MODEL; SYMMETRY LAWS (PHYSICS).

Gauge theory. The electroweak interaction is a Yang-Mills (or gauge) theory based on the product group $SU(2)_L \times U(1)_Y$, with two independent coupling constants, one for each group, denoted g and g'. As discussed below, the $SU(2)_L$ group gives an interaction between the gauge bosons and the left-handed projections of the fermions; the subscript L stands for left-handed. The $U(1)_Y$ group gives an interaction proportional to the hypercharge Y. Because it has two groups and two coupling constants, the theory does not fully unify the weak and electromagnetic interactions; however, they are partially unified because of the mixing of the neutral fields associated with generators in each of the groups. The gauge principle applied to this group produces four gauge particles which are spin-1 mesons. Three of the mesons are associated with the generators of $SU(2)_L$ and one with the generator of $U(1)_Y$. *See* GAUGE THEORY; GROUP THEORY; UNITARY SYMMETRY.

Higgs-Kibble mechanism. When the gauge group is unbroken, the gauge fields are massless and give rise to a long-range force. The short range of the observed weak interactions can be accommodated by breaking the gauge symmetry spontaneously, using an idea proposed by P. W. Higgs, T. W. B. Kibble, and others. *See* HIGGS BOSON; SYMMETRY BREAKING.

Outlook for the theory. The electroweak theory agrees with experiment over a great range of energies, from low-energy atomic parity violation experiments to high-energy deep inelastic scattering experiments. The theory also accounts for a vast amount of data with relatively few parameters. However, the electroweak theory does not predict the mass of the Higgs meson, which so far has eluded experimental detection. It has been conjectured that the role of the Higgs meson in breaking the symmetry may be accomplished in some other way and that the Higgs meson may not exist. Thus, despite the great successes of the electroweak theory, the unsatisfactory properties of the Higgs sector, the use of two gauge groups with two coupling constants, and the lack of understanding of the fermion masses and mixings give impetus to the search for a more comprehensive and more predictive theory. *See* ELEMENTARY PARTICLE. [O.W.G.]

Electrowinning

Electrowinning An electrometallurgical process involving the recovery of a metal, usually from its ore, by dissolving a metallic compound in a suitable electrolyte and reducing it electrochemically through passage of a direct electric current. Metals recovered by the electrowinning process include zinc, manganese, copper, aluminum, alkali metals, alkaline-earth metals, rare-earth metals.

Following beneficiation and, sometimes, chemical pretreatment, the metal-bearing constituent of the ore is dissolved in an aqueous solution or in a molten salt. The electrolysis of the purified electrolyte with direct current yields the reduced metal at the cathode (negative electrode of an electrolytic cell), and the nonmetal is the oxidation product at the anode (positive electrode). Since the process is very selective, the purity of the electrowon metal is high, and no further refining is needed. [P.Du.]

Element 110

Element 110 A transuranium element, atomic number 110. Transuranium elements have atomic numbers greater than 92 (that of uranium) and are created artificially. Element 110 was first reported in 1994. Just a few atoms of it were created in a single experiment; they undergo radioactive decay within less than a millisecond after their birth. *See* ATOMIC NUMBER; TRANSURANIUM ELEMENTS.

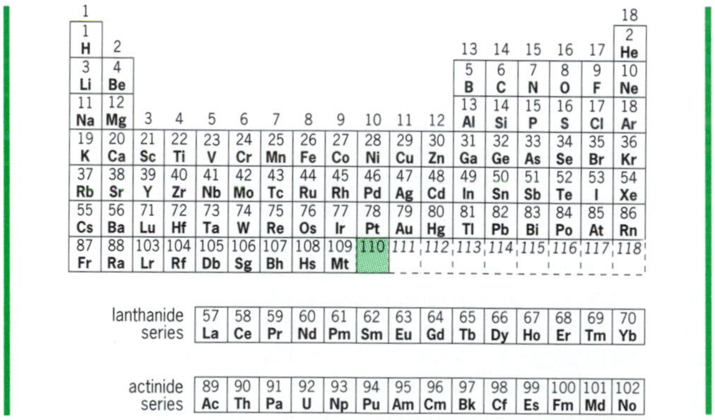
1 18
1 H 2 He
2 13 14 15 16 17
3 Li 4 Be 5 B 6 C 7 N 8 O 9 F 10 Ne
11 Na 12 Mg 3 4 5 6 7 8 9 10 11 12 13 Al 14 Si 15 P 16 S 17 Cl 18 Ar
19 K 20 Ca 21 Sc 22 Ti 23 V 24 Cr 25 Mn 26 Fe 27 Co 28 Ni 29 Cu 30 Zn 31 Ga 32 Ge 33 As 34 Se 35 Br 36 Kr
37 Rb 38 Sr 39 Y 40 Zr 41 Nb 42 Mo 43 Tc 44 Ru 45 Rh 46 Pd 47 Ag 48 Cd 49 In 50 Sn 51 Sb 52 Te 53 I 54 Xe
55 Cs 56 Ba 71 Lu 72 Hf 73 Ta 74 W 75 Re 76 Os 77 Ir 78 Pt 79 Au 80 Hg 81 Tl 82 Pb 83 Bi 84 Po 85 At 86 Rn
87 Fr 88 Ra 103 Lr 104 Rf 105 Db 106 Sg 107 Bh 108 Hs 109 Mt 110 111 112 113 114 115 116 117 118
lanthanide series 57 La 58 Ce 59 Pr 60 Nd 61 Pm 62 Sm 63 Eu 64 Gd 65 Tb 66 Dy 67 Ho 68 Er 69 Tm 70 Yb
actinide series 89 Ac 90 Th 91 Pa 92 U 93 Np 94 Pu 95 Am 96 Cm 97 Bk 98 Cf 99 Es 100 Fm 101 Md 102 No

Element 110 was created by using a heavy-ion linear accelerator and a means of detecting the alpha-particle-decay patterns that are characteristic for radioactive elements. The first atoms of element 110 with an atomic mass of 269 were created by bombarding atoms of lead-208 with atoms of nickel-62 in the accelerator. *See* ALPHA PARTICLES; PARTICLE ACCELERATOR; RADIOACTIVITY.

The chemical and physical properties of element 110 cannot be investigated because it has such a fleeting half-life. However, in the periodic table it falls in group 10, and is considered a metal. *See* PERIODIC TABLE. [B.Ri.]

Elementary particle

Elementary particle A particle which, in the present state of knowledge, cannot be described as compound. The elementary particles are the fundamental constituents of all matter. Ordinary matter and radiation are composed of only four kinds of elementary particles, namely protons, neutrons (bound in stable nuclei), electrons, and photons. (Plus, strictly speaking, gravitons and neutrinos, but these interact so weakly as to be negligible.) This is because other known elementary particles are unstable; they decay spontaneously into two or more other particles in an average time, called their lifetime or mean life, which ranges from 2×10^{-6} s (muon) to less than 10^{-23} s (hadron resonances). Free neutrons are also unstable. On the other hand, although free positrons and antiprotons are stable, in the presence of ordinary matter they are rapidly destroyed by annihilation with electrons or nucleons, respectively. *See* ANTIPROTON; POSITRON; RADIOACTIVITY.

The unstable elementary particles must be studied within a short time of their creation, which occurs in the collision of a fast (high-energy) particle with another particle. In this process some of the kinetic energy of the colliding particles is transformed into the rest mass of the created particles, in accordance with Einstein's mass-energy relation, $E = mc^2$; in the decay of particles, the reverse energy transformation takes place. Such fast particles exist in nature, namely the cosmic rays, but their flux is small; thus most elementary-particle research is based on high-energy particle accelerators. *See* NUCLEAR REACTION; PARTICLE ACCELERATOR; PARTICLE DETECTOR; REST MASS.

A particle, by its quantum mechanical definition, has a definite mass and spin. These quantities are eigenvalues of symmetry operators generating translations and rotations of space-time. A particle is also characterized by other quantum numbers, such as parity and lepton number, which are eigenvalues of further symmetry operators that commute with mass and spin. *See* EIGENVALUE (QUANTUM MECHANICS); QUANTUM MECHANICS; QUANTUM NUMBERS; SYMMETRY LAWS (PHYSICS).

All the particles of one kind are identical, and can be described as quanta of fields; for each kind of particle there is an associated field. Each particle is either a boson or a fermion, according to whether its spin is an integral or a half-odd-integral multiple of $\hbar$, Planck's constant divided by 2π. *See* QUANTUM FIELD THEORY; QUANTUM STATISTICS; SPIN (QUANTUM MECHANICS).

To each kind of particle there corresponds an antiparticle, or charge-conjugate particle; if self-conjugate, the particle and antiparticle are identical. The elementary particles can be grouped conveniently into three classes:

1. Classons, massless bosons which are quanta of the two classical fields, gravitational and electromagnetic.
2. Leptons, fermions which have interactions with the classical fields and also weak interactions (beta interactions). The leptons are the electron, the muon, formerly called the mu meson, the tau, and their associated neutrinos.
3. Hadrons, or strongly interacting particles, which in addition have strong interactions. Boson hadrons are called mesons. Fermion hadrons are called baryons and carry one unit of an exactly conserved quantity, the baryon number. Baryons include nucleons (proton and neutron), quasistable

hyperons, and numerous resonances. For detailed information on the various particles *see* BARYON; ELECTRON; GRAVITON; HADRON; HYPERON; J PARTICLE; LEPTON; MESON; NEUTRINO; NEUTRON; NUCLEON; PROTON; STRANGE PARTICLES; UPSILON PARTICLES.

The conventional restriction of the term elementary particle to the types of particles named above excludes electromagnetically bound particles (such as atoms, molecules, and macroscopic particles) and hadrons with baryon number greater than unity (atomic nuclei). The former can be described very accurately by means of quantum electrodynamics; the latter can be described, though less accurately, as bound or resonant states of baryons. The success in describing them as compound particles is due to the smallness of their binding energy, which is at most a few percent of the mass of their constituent particles. The weakness of the binding between baryons is attributed to a "repulsive core" in their interaction. This is lacking in other combinations of hadrons (baryon-meson, meson-meson), so that their binding is much stronger. Consequently, the latter particles, hadrons with baryon number $B = -1,0,1$, cannot easily be described as compounds, and so are termed elementary.

Hadrons can be divided into the quasistable or (semistable) and the unstable. The quasistable hadrons are simply those that are too light to decay into other hadrons by way of the strong interactions, such decays being restricted by the requirement that isotopic spin, **I**, strangeness, *s*, and charm, *c*, be conserved. The unstable hadrons are also called elementary-particle resonances or excited states. Their lifetimes, of the order of 10^{-23} s, are much too short to be observed directly. Instead they appear, through the uncertainty principle, as spreads in the masses of the particles—that is, in their widths—just as in the case of nuclear resonances. *See* UNCERTAINTY PRINCIPLE.

A characteristic of the hadrons is that they are grouped into isotopic-spin (*i*-spin) multiplets (for example, n,p; π^-,π^0,π^+); the masses of the particles in each multiplet differ by only a few megaelectronvolts. The hadrons exhibit a further grouping into supermultiplets of *i*-spin multiplets, the masses of the multiplets in a supermultiplet differing by a few hundred megaelectronvolts. These supermultiplets appear to be the consequence of a symmetry that is less exact than *i*-spin symmetry. This symmetry, termed unitary symmetry, is the symmetry with respect to a group of transformations SU_3.

Elementary particles are classified by the values of certain conserved quantities, or quantum numbers, which they carry. The spin **J** and parity *P* correspond to symmetries of ordinary space. Quantum numbers corresponding to an internal symmetry space are charge *Q*, charge parity *C*, lepton number, baryon number, and, for hadrons, the not exactly conserved quantities *i*-spin **I**, strangeness *s*, charm *c*, *G* (if baryon number $B = 0$ and $s = c = 0$), and unitary spin, that is, the SU_3 supermultiplet to which the particle belongs, such as octet or deket.

There are more than 75 known hadron multiplets; the actual number may well be infinite. Attempts have been made to construct a theory which will explain the existence and properties of these particles in an economical way. Broadly speaking, two kinds of such theories have been proposed, constituent and self-consistent. In constituent theories, hadrons are supposed to be composed of truly elementary particles, analogous to the way in which atomic nuclei are composed of nucleons. On the other hand, in self-consistent theories, the properties of hadrons are supposed to be determined by the requirement that their scattering amplitudes contain resonances, that is, hadrons, with the same properties.

In the most popular constituent theories the constitutents are called quarks. By assuming that there are three equivalent quarks, all states made from them will have SU_3 symmetry. If it is further assumed that mesons consist of a quark and an antiquark, $q\bar{q}$, and baryons consist of three quarks, qqq (each q in these formulas stands for any one of the three quarks), then mesons will comprise only the multiplets 1 and 8, and baryons will comprise only 1, 8, and 10, in agreement with the data. For the baryons to be fermions, the quarks must be so too; it is simplest to assume that they have spin 1/2. A curious consequence of the construction of baryons as qqq is that each quark must carry one-third of a unit of baryon number and a charge which is a multiple of one-third of the electronic charge *e*.

Despite extensive searches, no observation of a free quark has been independently confirmed. However, a theoretical structure, quantum chromodynamics, has been developed to understand and predict the behavior of hadrons, based on the interactions of their constituent quarks. It has been conjectured that one consequence of this theory is quark confinement, whereby it is impossible for two quarks to separate freely to large distances, so that free quarks cannot be isolated. *See* QUANTUM CHROMODYNAMICS; QUARKS.

In 1974 the first of a new class of mesons was discovered independently at Brookhaven National Laboratory and the Stanford Linear Accelerator Laboratory, and was named *J* and ψ, respectively, by the discoverers. The J/ψ is distinct from the previously known mesons due to the combination of its large mass and small decay rate, which together means that it is strongly forbidden to decay hadronically. The quark model can describe the J/ψ if extended by the supposition that there is a fourth type or flavor of quark, the so-called charmed quark, *c*. The *c* quark must be supposed much more massive than the other three in order to explain the large mass of the J/ψ which is described as the ground state of $c\bar{c}$. The system $c\bar{c}$ is called charmonium (in analogy to positronium, $e\bar{e}$). Since it is a $c\bar{c}$ state the J/ψ has no net charm. The lightest charmed mesons are the D^+ and D^0), respectively $c\bar{d}$ and $c\bar{u}$ states, and their antiparticles, first observed in 1976. *See* CHARM.

In 1977 a new meson resonance was discovered at an energy of about 9.4 GeV in a study of the production of $\mu^+\mu^0$ pairs in proton-nucleus collisions. Subsequent studies of these reactions and of electron-positron annihilation showed that the new particle, the Υ (upsilon), is evidence of a fifth type or flavor of quark, called *b* (for bottom or beauty). [C.J.G.]

Elements An element is a substance made up of atoms with the same atomic number. Some common elements are oxygen, hydrogen, iron, copper, gold, silver, nitrogen, chlorine, and uranium. Approximately 75% of the elements are metals and the others are nonmetals. Most of the elements are solids at room temperature, two of them (mercury and bromine) are liquids, and the rest are gases. A few of the elements are found in nature in the free (uncombined) state. Some of these are oxygen, nitrogen, the noble gases (helium, neon, argon, krypton, xenon, and radon), sulfur, copper, silver, and gold. Most of the elements in nature are combined with other elements in the form of compounds.

The elements are classified in families or groups in the periodic table. Elements are also frequently classified as metals and nonmetals. A metallic element is one whose atoms form positive ions in solution, and a nonmetallic element is one whose atoms form negative ions in solution. *See* PERIODIC TABLE.

Atoms of a given element have the same atomic number, but may not all have the same atomic weight (see table). Atoms with identical atomic numbers but different atomic weights are called isotopes. All the elements have isotopes, although in certain cases only synthetic isotopes are known. Many of the isotopes of the different elements are unstable, or radioactive, and hence disintegrate to form stable atoms either of that element or of some other element. *See* ATOMIC WEIGHT; ISOTOPE; RADIOACTIVITY.

The origin of the chemical elements is believed to be the result of the synthesis by fusion processes at very high temperatures (in the order of 180,000,000°F or 100,000,000°C and higher) of the simple nuclear particles (protons and neutrons) first to heavier atomic nuclei such as those of helium and then on to the heavier and more complex nuclei of the light elements (lithium, boron, beryllium, and so on). The helium atoms bombard the atoms of light elements and produce neutrons. The neutrons are captured by the nuclei of elements and produce heavier elements. These two processes—fusion of protons and neutron capture—are the main processes forming the chemical elements. *See* NUCLEOSYNTHESIS.

A number of elements, found in only very slight traces or not at all in nature, have been synthesized. Those elements are technetium, promethium, astatine, francium, and all the elements with atomic numbers above 92. *See* CHEMISTRY; TRANSURANIUM ELEMENTS. [A.B.G.]

Cosmic abundance. The abundance of the elements in surface rocks of the Earth, in the Earth as a whole, in meteorites, in the solar system, in galaxies, or in the total universe corresponds to the average relative amounts of the chemical elements present, or, in other words, to the average chemical composition of the respective object. Element abundances are given in numbers of atoms of one element relative to a certain number of atoms of a reference element. Silicon is commonly taken as the reference element in the study of the composition of the Earth and the meteorites, and the data are given in atoms per 10^6 atoms of silicon. The results of astronomical determinations of the composition of the Sun and of the stars are often expressed in atoms per 10^{10} atoms of hydrogen. Ordinary chemical analyses, including advanced techniques for trace element studies (such as neutron activation or isotope dilution), are used for determination of the composition of rocks and meteorites. The composition of the Sun and of stars can be derived by quantitative spectral analysis. On the surface of the Earth, the most abundant elements are oxygen, silicon, magnesium, calcium, aluminum, and iron. In the universe as a whole, hydrogen and helium constitute more than 95% of the total matter. *See* ACTIVATION ANALYSIS; ASTRONOMICAL SPECTROSCOPY; ISOTOPE DILUTION TECHNIQUES.

The isotopic composition of the elements is practically the

Table of atomic weights, 1973*

Name	Symbol	Atomic number	Atomic weight
Actinum	Ac	89	(227)
Aluminum	Al	13	26.98154
Americium	Am	95	(243)
Antimony	Sb	51	121.75
Argon	Ar	18	39.948
Arsenic	As	33	74.9216
Astatine	At	85	(210)
Barium	Ba	56	137.33
Berkelium	Bk	97	(247)
Beryllium	Be	4	9.01218
Bismuth	Bi	83	208.9804
Boron	B	5	10.81
Bromine	Br	35	79.904
Cadmium	Cd	48	112.41
Calcium	Ca	20	40.08
Californium	Cf	98	(251)
Carbon	C	6	12.011
Cerium	Ce	58	140.12
Cesium	Cs	55	132.9054
Chlorine	Cl	17	35.453
Chromium	Cr	24	51.996
Cobalt	Co	27	58.9332
Copper	Cu	29	63.546
Curium	Cm	96	(247)
Dysprosium	Dy	66	162.50
Einsteinium	Es	99	(254)
Erbium	Er	68	167.26
Europium	Eu	63	151.96
Fermium	Fm	100	(257)
Fluorine	F	9	18.998403
Francium	Fr	87	(223)
Gadolinium	Gd	64	157.25
Gallium	Ga	31	69.72
Germanium	Ge	32	72.59
Gold	Au	79	196.9665
Hafnium	Hf	72	178.49
Helium	He	2	4.00260
Holmium	Ho	67	164.9304
Hydrogen	H	1	1.0079
Indium	In	49	114.82
Iodine	I	53	126.9045
Iridium	Ir	77	192.22
Iron	Fe	26	55.847
Krypton	Kr	36	83.80
Lanthanum	La	57	138.9055
Lawrencium	Lr	103	(260)
Lead	Pb	82	207.2
Lithium	Li	3	6.941
Lutetium	Lu	71	174.97
Magnesium	Mg	12	24.305
Manganese	Mn	25	54.9380
Mendelevium	Md	101	(258)
Mercury	Hg	80	200.59
Molybdenum	Mo	42	95.94
Neodymium	Nd	60	144.24
Neon	Ne	10	20.179
Neptunium	Np	93	237.0482
Nickel	Ni	28	58.70
Niobium	Nb	41	92.9064
Nitrogen	N	7	14.0067
Nobelium	No	102	(255)
Osmium	Os	76	190.2
Oxygen	O	8	15.9994
Palladium	Pd	46	106.4
Phosphorus	P	15	30.97376
Platinum	Pt	78	195.09
Plutonium	Pu	94	(244)
Polonium	Po	84	(209)
Potassium	K	19	39.0983
Praseodymium	Pr	59	140.9077
Promethium	Pm	61	(145)
Protactinium	Pa	91	231.0359
Radium	Ra	88	226.0254
Radon	Rn	86	(222)
Rhenium	Re	75	186.207
Rhodium	Rh	45	102.9055
Rubidium	Rb	37	85.4678
Ruthenium	Ru	44	101.07
Samarium	Sm	62	150.4
Scandium	Sc	21	44.9559
Selenium	Se	34	78.96
Silicon	Si	14	28.0855
Silver	Ag	47	107.868
Sodium	Na	11	22.98977
Strontium	Sr	38	87.62
Sulfur	S	16	32.06
Tantalum	Ta	73	180.9479
Technetium	Tc	43	(97)
Tellurium	Te	52	127.60
Terbium	Tb	65	158.9254
Thallium	Tl	81	204.37
Thorium	Th	90	232.0381
Thulium	Tm	69	168.9342
Tin	Sn	50	118.69
Titanium	Ti	22	47.90
Tungsten†	W	74	183.85
Uranium	U	92	238.029
Vanadium	V	23	50.9414
Xenon	Xe	54	131.30
Ytterbium	Yb	70	173.04
Yttrium	Y	39	88.9059
Zinc	Zn	30	65.38
Zirconium	Zr	40	91.22

SOURCE: Inorganic Chemistry Division, Commission on Atomic Weights, International Union of Pure and Applied Chemistry. Incorporates changes approved at 28th IUPAC Conference, Madrid, 1975. Published in *Pure Appl. Chem.*, 37(4):591–603, 1974.

*Scaled to the relative atomic mass. A_r (^{12}C) = 12.

†Also known as wolfram.

same in all terrestrial material and in meteorites. From the isotopic composition of an element and its cosmic abundance, the nuclear abundances of its isotopes can be calculated. *See* ISOTOPE; NUCLEAR STRUCTURE.

Nuclear abundance values show a clear correlation with certain nuclear properties, and can be assumed to represent in good approximation the original yield distribution of the thermonuclear processes that led to the formation of the elements. The empirical abundance values can therefore serve as the basis for theoretical considerations about the origin of matter and of the universe, and have led to the following conclusion: No simple, single mechanism exists by which the elements in their observed isotopic composition can have formed. The matter of the cosmos appears to be a mixture of material that formed under different conditions by different types of nuclear processes. *See* NUCLEOSYNTHESIS. [H.E.S.]

Geochemical distribution. The distribution of chemical elements in major zones (crust, mantle, and core) of the Earth is dependent upon the early history and subsequent evolution of both the Earth and solar system. Since these events occurred long ago and there is no direct evidence for what actually happened, considerable speculation is necessarily involved in explaining the postulated present-day distribution of elements among the major zones of the Earth.

The proto-solar system, before it evolved into the Sun and orbiting planets, was probably a rotating, swirling, lens-shaped cloud of gas, dust, and other matter. The interior of this cloud contracted and heated primarily through gravitational attraction, until finally temperatures and pressures were high enough to initiate nuclear reactions, giving off light and heat. Matter in swirling vortices within peripheral regions of the cloud eventually coalesced and formed individual planets. Portions of lighter, more volatile elements (such as N, C, O, and H) escaped outward from the hotter interior of the system and were preferentially enriched in the less dense, larger outer planets (Jupiter, Saturn, Uranus, and Neptune). Heavier, less volatile elements (such as Ca, Na, Mg, Al, Si, K, Fe, Ni, and S) tended to remain closer to the center of the system, eventually becoming preferentially enriched in the more dense, smaller inner planets (Mercury, Venus, Earth, and Mars). *See* SOLAR SYSTEM.

Accretion of the Earth is believed to have taken place from a cloud whose composition was quite similar to that of the type of stony meteorites known as chondrites. The protoearth was probably a homogeneous, unzoned spheroid of approximately chondritic composition. *See* METEORITE.

Assuming the hypothesis of an initially unzoned Earth, and the chondritic Earth model hypothesis, the Ni-Fe alloy eventually formed the core, and the remaining phases formed the mantle. At a very early stage in its history ($4–5 \times 10^9$ years ago), the Earth probably was mostly in solid form. Most earth scientists now believe that subsequent heating, due to adiabatic contraction and radioactive decay, led to a widespread melting event. The lowest-melting component, the Ni-Fe alloy, is believed to have melted initially; because of its much greater density, this alloy would have settled and formed the core. This event has been referred to as the iron catastrophe. Continued melting would have created three immiscible liquids: silicate, sulfide, and alloy. The remaining silicates, sulfides, and other compounds then would have formed the surrounding mantle.

New oceanic crust, composed primarily of basaltic rocks, is thought to be forming at the mid-oceanic ridges (spreading centers) by means of partial melting of the underlying mantle. Relative to the mantle, the basaltic crust is comparatively enriched in Si, Al, Ca, Na, K, and large ionic lithophile elements, but is depleted in Mg, Fe, and certain transition metals (group VIII in particular). The process of partial melting of upper mantle material and ascent of magma, eventually forming new crust, may be the dominant mechanism for concentrating the enriched-in-crust elements in the overlying crust at the expense of the mantle.

Partial melting also goes on within the continental crust, leading to the formation and ascent of magmas comparatively rich in enriched-in-crust elements and poor in enriched-in-mantle elements relative to the rocks from which the magmas derive. These magmas tend to move upward through time, eventually solidifying and forming a zoned continental crust, with the upper zone (sial) having an overall granitic composition and the lower zone (sima) an unknown composition, probably similar to that of basalt. The upper granitic crust is even more abundant in enriched-in-crust elements. Further modifications of the upper continental crust can occur through processes such as weathering sedimentation, metamorphism, and igneous differentiation. *See* GEOCHEMISTRY; PLATE TECTONICS. [P.G.R.]

Elephant The common name for two living species of mammals in the family Elephantidae, one of several families included in the order Proboscidea. The remaining families contain extinct animals, such as the mammoth. One of the living species (*Loxodonta africana*) is indigenous to Africa, and the other (*Elephas maximum*) ranges throughout Southeast Asia.

These animals are terrestrial and entirely herbivorous. The nostrils and upper lip are elongated into a proboscis, the trunk, which is a powerful and sensitive organ specialized to form a prehensile, food-gathering structure. The upper incisors protrude from the mouth on either side of the trunk as tusks, which continually grow throughout the life of the animal. The hard, thick skin of the elephant is sparsely covered with hair, and serves as an insulator. The eyes are tiny but vision is keen. The tail is short; large columnar legs support the massive body.

Elephants live and travel in herds which were originally composed of several hundred individuals, but the usual number is now around 20 animals, with a mature bull, a number of cows and calves, and some younger bulls. Elephants are at ease in water, and they bathe and roll in the mud as a protective measure since, despite the thickness of the skin, they are sensitive to intense sun and insects.

Bulls reach sexual maturity at the age of 15, while cows mature earlier. The gestation period averages 20–22 months with a single calf being born. Twins are rare. The newborn, which may be 3 ft (0.9 m) tall and weigh 200 lb (90 kg), grows rapidly. *See* MAMMALIA; PROBOSCIDEA. [C.B.C.]

Eleutherozoa One of the two subphyla into which the phylum Echinodermata had been customarily divided. The Eleutherozoa are now best considered as comprising at least two distinct subphyla: (1) the Echinozoa, spherical-bodied forms with meridional symmetry; and (2) the Asterozoa, star-shaped forms with radially divergent axes of symmetry. *See* ECHINODERMATA; ECHINOZOA. [H.B.F.]

Elevating machines Materials-handling machines that lift and lower a load along a fixed vertical path of travel with intermittent motion. In contrast to hoisting machines, elevating machines support their loads instead of carrying them suspended, and the path they travel is both fixed and vertical. They differ from vertical conveyors in operating with intermittent rather than continuous motion. Industrial lifts, stackers, and freight elevators are the principal classes of elevating machines.

A wide range of mechanically, hydraulically, and electrically powered machines are classified as industrial lifts (Fig. 1). They are adapted to such diverse operations as die handling and feeding sheets, bar stock, or lumber. In some locations with differences in floor level between adjacent buildings, lifts take the form of broad platforms to serve as floor levelers to obviate the need for ramps. They are also used to raise and lower loads between the ground and the beds of carriers when no loading platform exists. Lifting tail gates attached to the rear of trucks are similarly used for loading or unloading merchandise on sidewalks or roads and at points where the lack of a raised dock would make loading or unloading difficult.

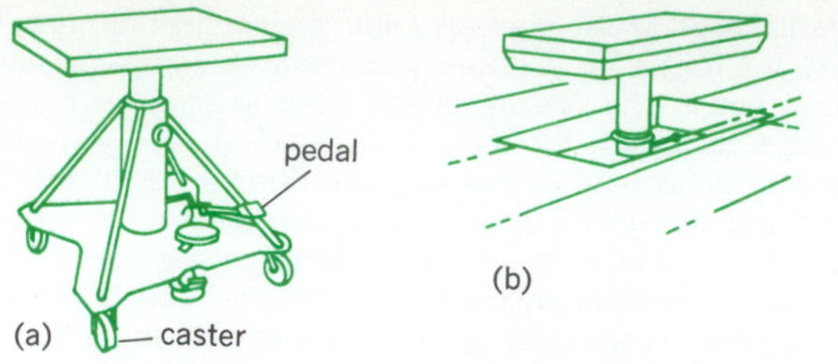

Fig. 1. Examples of industrial lifts. (*a*) Hydraulic elevating work table. (*b*) Hydraulic lift floor leveler.

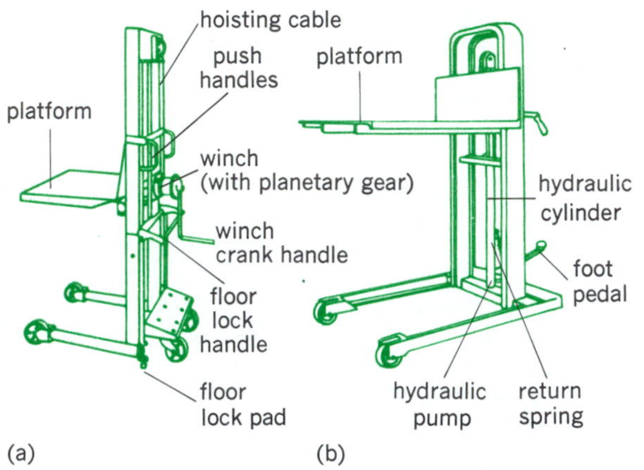

Fig. 2. Two types of electric and hydraulic stackers. (*a*) Hand type. (*b*) Hydraulic foot type.

Stackers are tiering machines and portable elevators used for stacking merchandise with basically portable vertical frames that support and guide the carriage, to which is attached a platform, pair of forks, or other suitable lifting device (Fig. 2). Horizontal movement is effected by casters on the bottom of the vertical frame, and can be accomplished manually, or mechanically, by using the same power source as the lifting mechanism. These casters are usually provided with floor locks bolted in position during the elevating or lowering operation. Used in conjunction with cranes, stackers are widely applied to the handling of materials on storage racks and die racks.

Examples of industrial elevators range from those set up temporarily on construction jobs for moving materials and personnel between floors to permanent installations for mechanized handling in factories and warehouses. Dumbwaiters are a type of industrial elevator; they carry parts, small tools, samples, and similar small objects between buildings, but are not permitted to carry people. The most common and economical elevator employs electric motors, cables, pulleys, and counterweights. *See* MATERIALS-HANDLING EQUIPMENT. [A.M.P.]

Elevator A platform or enclosure that is raised and lowered in a vertical hoistway to transport freight or people. The term elevator can also encompass all of the hoisting equipment, motor, cables, and accessories. *See* ELEVATING MACHINES.

The closed passenger car of a modern elevator rests inside a steel frame. The car and the car frame ride up and down on steel rails in an elevator shaft or hoistway. Guide shoes or rollers on the frame keep the car in place on the rails. Most elevators also have a heavy weight, called a counterweight, attached to the other end of the steel hoisting ropes that pass over the driving machine pulley. The counterweight offsets much of the weight of the car and passengers, thereby reducing power requirements.

The typical elevator control system is made up of a speed-sensing device known as a governor, a clamping device (safety) mounted under each end of the car frame that grips the guide rail when tripped, a tension sheave (pulley) in the pit, and a steel rope. *See* GOVERNOR.

Control devices are also built into the door and its control circuit. When the doors open, control circuits prevent the car from moving away from the landing, but permit releveling if the car moves as passengers enter or leave the elevator (load changes).

In modern elevators, microprocessor computer systems control elevator position, direction of travel, speed, door operation, passenger waiting time, flight time, energy consumption, and system diagnostics. Modern elevators include Braille buttons and voice announcements of the floors to help the sight-impaired. [C.DiT.]

Elevator (aircraft) The hinged rear portion of the longitudinal stabilizing surface or tail plane of an aircraft used to obtain longitudinal- or pitch-control moments. The angular setting of the elevator is controlled by the human or automatic pilot through the flight-control system. A typical elevator control surface is shown in the illustration. *See* FLIGHT CONTROLS.

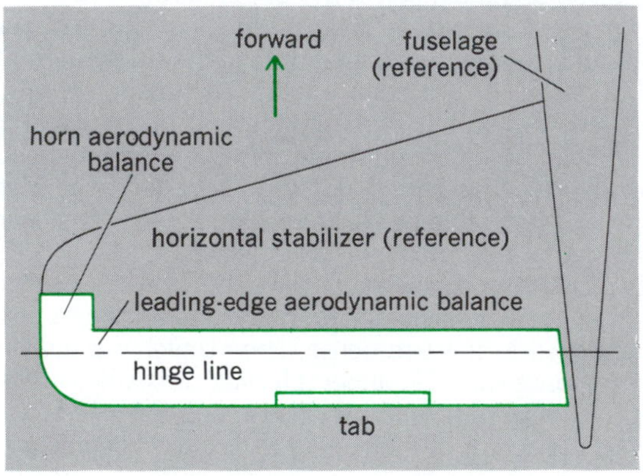

Elevator control surface (left-hand side).

The elevator is used to perform pitching maneuvers, or maneuvers in which the aircraft's plane of symmetry is not disturbed. These maneuvers include airspeed adjustments and acceleration normal to the flight path (pull-ups or push-downs). The elevator also serves to adjust the aircraft's attitude with respect to the ground for takeoff and landing. [M.J.A.]

Elevon The hinged rear portion of an aircraft wing, moved in the same direction on each side of the aircraft to obtain longitudinal control and differentially to obtain lateral control. The word elevon is a combination of the words elevator and aileron, to denote that elevons combine the functions of aircraft elevators and ailerons. *See* AILERON; ELEVATOR (AIRCRAFT).

Elevons are used exclusively on tailless aircraft, or specifically those without horizontal stabilizers. Aircraft in this category include the modern delta wing fighter aircraft. In addition, elevons are also required on lifting-body versions of crewed space vehicles. [M.J.A.]

Ellipse A member of the class of curves that are intersections of a plane with a cone of revolution. The ellipse is obtained when the plane cuts all the elements of one nappe, and does not go through the apex. In the illustration, denote the distance between two points F, F' of a plane by $2c$, $c > 0$, and let $2a$ be a constant, with $a > c$. The ellipse with foci F and F' and major axis $2a$ is the locus of points P of the plane such that $PF + PF' = 2a$, where PF denotes the distance of P and F. This suggests the following construction of an ellipse.

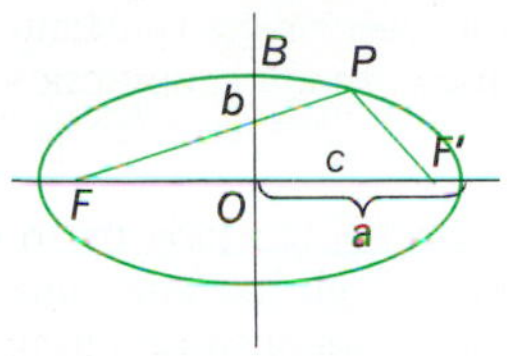

An ellipse, as described in the text.

Put pins at F and F', and slip over them a loop of thread of length $2a + 2c$, pulling the thread taut with a pencil. If the pencil is moved, keeping the thread taut, its point traces an ellipse. *See* CONIC SECTION.

The midpoint of F, F' is the center O of the ellipse, and the chord through O perpendicular to the major axis is the minor axis, whose length is denoted by $2b$. If B is a point in which the minor axis intersects the ellipse, then $BF = BF' = a$, and so $c^2 = a^2 - b^2$. The ratio $c/a = \epsilon < 1$ is the eccentricity of the ellipse. *See* ANALYTIC GEOMETRY. [L.M.Bl.]

Ellipsoid and spheroid An ellipsoid is a quadric surface of finite extent representable in analytic geometry by an equation of the second degree in three variables (such as x, y, z). Ellipsoids include spheres, spheroids obtained by stretching or shrinking a sphere in one direction, and surfaces obtained by stretching a sphere in two mutually perpendicular directions. Every plane section of an ellipsoid is either an ellipse or a single point (provided that a circle is considered to be a special case of an ellipse). The centers of a set of parallel plane sections of an ellipsoid lie on a line segment called a diameter of the ellipsoid, whose midpoint is the center of the ellipsoid. Chords of the ellipsoid through the centers are diameters and are bisected by the center. *See* ANALYTIC GEOMETRY; SURFACE AND SOLID OF REVOLUTION. [J.S.F.]

Ellipsometry A technique for determining the properties of a material from the characteristics of light reflected from its surface. The materials studied include semiconductors, liquids, and metals.

When the electromagnetic wave is reflected from the surface of a material, the amplitude of the reflected wave depends upon the properties of the material, the angle of incidence, and the polarization of the wave. An *s* wave has its electric vector parallel to the surface, while a *p* wave has its electric vector in the plane of incidence (the plane formed by the incident and reflected waves). The reflection amplitudes are written as r_s, and r_p. These are the ratios of the reflected- to incident-wave amplitudes and include a phase shift in the wave.

Ellipsometry is a technique which measures $r = r_p/r_s$. This can be done by shining light which is linearly polarized at 45° to both the *s* and *p* directions so that the *s* and *p* incident amplitudes are equal and in phase. The electric vector of the reflected light in general traces an ellipse, where the orientation and the lengths of the major and minor axes depend on the magnitudes of r_p and r_s, and on the relative phase difference between the two waves. *See* POLARIZED LIGHT.

The two chief applications of ellipsometry are the study of surface properties and the area of spectroscopic ellipsometry. When the media on each side of the surface are nonabsorbing, and the dielectric constant undergoes an abrupt step at the surface, then both r_p and r_s are real. As a result, linearly polarized incident light gives linearly polarized reflected light. In practice, however, no interface is abrupt, and there is always a transition region in the variation from medium 1 to medium 2. Then both r_p and r_s are altered by an amount proportional to d/λ, where d is the transition region thickness and λ is the wavelength of the light. This effect has been used to study surfaces of glass, interfaces between two liquids, interfaces between a liquid and its vapor, and layers of organic molecules floating on water. When the medium reflecting the wave is absorbing, r is complex even for an abrupt boundary. A common use for ellipsometry is to study the electronic excitations of materials from measurements of r as a function of incident wavelength. This is called spectroscopic ellipsometry. *See* MONOMOLECULAR FILM; SURFACE PHYSICS. [D.Be.]

Elliptic function and integral In a certain sense, elliptic integrals are the simplest integrals not expressible in terms of elementary functions; elliptic functions arise as the inverse functions of certain elliptic integrals.

Let R be a rational function of x and y, and set $I = \int R(x,y)dx$. I can be expressed in terms of elementary functions if y^2 is a polynomial of degree 2 or less in x. If y^2 is a polynomial of degree 3 or 4 in x, I cannot in general be expressed in terms of elementary functions and is called an elliptic integral. The standard elliptic functions are analogous to trigonometric functions. Trigonometric functions may be defined as the inverse functions of certain integrals of the form I; they satisfy differential equations, are periodic functions, and may alternatively be obtained as the "simplest" periodic functions. The standard elliptic functions are the inverse functions of certain elliptic integrals; they satisfy differential equations of order 1 and degree 2, are doubly periodic functions, and may alternatively be obtained as the "simplest" doubly periodic functions.

In geometry elliptic functions or integrals arise in determining the length of an arc of an ellipse, hyperbola, or lemniscate, the surface of an ellipsoid, geodesics on quadrics of revolution, parametric representations of plane cubic curves or, more generally, curves of genus 1, conformal representation problems, and other problems. There are applications to differential equations (Lamé's equation, diffusion equation, and others) in analysis, and there are applications in a great variety of problems in number theory. In the physical sciences elliptic functions or integrals appear in potential theory both through conformal representations and in the potential of an ellipsoid, in the theory of elastica, the pendulum, in rigid body motion, in Green's functions in heat conduction and diffusion theory, and many other problems. [A.Er.]

Elm Any species of *Ulmus*, a genus of hardwood trees in the Northern Hemisphere, with simple, serrate, deciduous leaves. The American or white elm (*U. americana*) is the most important species. It ranges from the eastern half of the United States westward as far as the base of the northern Rockies and southward through central Texas to the Gulf of Mexico. The tree is also found in southern Canada.

The tree was very popular as a shade tree, perhaps better adapted as a street tree than any other species because the upper branches spread, joining with elms across the street to form an arch. Once abundant, the elm has been severely attacked by the lethal Dutch elm disease imported from Europe. The future of the species is uncertain. [A.H.G./K.P.D.]

Elopiformes A primitive order of soft-rayed, teleost fishes, including the tarpons, bonefishes, and their relatives, that was formerly included in the Clupeiformes. The elopiforms have a single dorsal fin composed of soft rays only (see illustration).

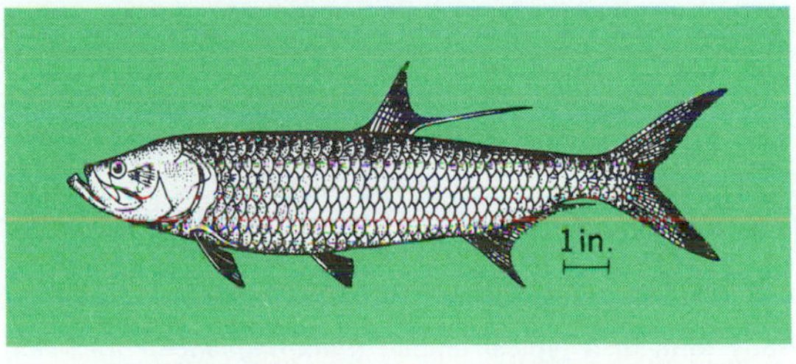

Tarpon (*Megalops atlantica*). 1 in. = 2.54 cm. (*After G. B. Goode, Fishery Industries of the United States, Sect. 1, 1884*)

The most important characteristic of the order involves the occurrence in early development of a leptocephalous larval stage. This larva is translucent, ribbonlike, and strongly toothed; eventually it passes through a marked transformation to adult form. This developmental pattern characterizes only two other orders of teleosts, the Notacanthiformes (spiny eels and halosaurs) and the Anguilliformes (true eels), groups which are therefore believed to be descendants of elopiforms. *See* ANGUILLIFORMES; CLUPEIFORMES; NOTACANTHIFORMES.

The elopiforms have a well-represented paleontological history since the Jurassic, on all continents. They were especially abundant in the Cretaceous. Extant forms are classified in 2 families, 4 genera, and 12 species. Most inhabit tropical to temperate shore waters of all oceans, but tarpons invade rivers, sometimes for long distances. *See* ACTINOPTERYGII; TARPON; TELEOSTEI. [R.M.B.]

Embedded systems Computer systems that cannot be programmed by the user because they are preprogrammed for a specific task and are buried within the equipment they serve. The term derives from the military, where computer systems are generally activated by the flip of a toggle switch or the push of a button.

The emergence of extremely small microcomputers on silicon chips about ¼ in. (6 mm) square made the concept of embedded systems for the military all the more desirable. With very inexpensive microcomputer chips, not one but dozens of computers can be squeezed into a small piece of equipment the size of a lunch box, each preprogrammed to do a specific job. By doing only its own task, each microcomputer can devote its entire computational and analytical powers to that task, making possible extremely powerful military systems. *See* INTEGRATED CIRCUITS; MICROCOMPUTER; MICROPROCESSOR.

For its software needs, the Deparment of Defense adopted Ada as the single high-level software programming language for all three branches of the service to simplify procurements, the interchange of software programs among the services' branches, and training. *See* PROGRAMMING LANGUAGES. [R.Al.]

Embioptera A peculiar order of silk-spinning, orthopteroid insects related to termites, commonly called the embiids or web spinners. This order comprises about 1000 species which are chiefly tropical in distribution. The body is linear and supple (see illustration). The legs are short with three-segmented tarsi. The forelegs are adapted for spinning silk, and the hindlegs for reverse locomotion. Metamorphosis is incomplete. The females are neoteinic and wingless (apterous). Males are usually winged (alate), but in certain genera and species they are apterous. The wings are subequal and elongate. The wings are flexible when in repose and folded over the back, but are stiffened when extended for flight by the blood pressure in the saclike veins. Flight is a poorly directed, whirling flutter.

The silk galleries, which constitute a safe shelter for all embiid activities except adult dispersal, radiate on or in the food supply which also constitutes the habitat and consists of bark, lichens, moss, dead leaves, or grass. Many individuals, usually the brood of one female, may occupy one gallery system. *See* INSECTA; ORTHOPTERA. [E.S.R.]

Body form of a typical embiid (*Pararhagadochir trachelia*). (From E. S. Ross, *Insects Close Up*, University of California Press, 1953)

Embolism The sudden blocking of an artery or vein by a clot or other substance which has been brought to its place by the blood current. The material carried in the circulation in this process is an embolus. Emboli may be composed of thrombi, fat, air, tumor cells, masses of bacteria or parasites, bone marrow, amniotic fluid, or atheromatous material from the vessel wall. With complete obstruction of a vessel by an embolus an infarct may result. *See* EMBOLUS; INFARCTION. [R.A.V.]

Embolus An abnormal mass or particles carried in the bloodstream to a site too small to allow passage so that blood obstruction usually follows. A majority of all emboli originate as fragments of thrombi which have broken off the fixed clot. Clusters of tissue cells, bacteria, or parasites, and bubbles of gas, air, or foreign materials may also become emboli. *See* EMBOLISM.

Venous emboli, 95% of which originate in the leg veins, may pass through the heart to lodge in the lungs. These frequently fatal pulmonary emboli originate in patients with poor circulation produced by age, lack of activity, pregnancy, or varicosities, or following operations or other conditions in which the venous blood is slowed down or immobilized so that an embolus forms. *See* CIRCULATION DISORDERS.

Arterial emboli commonly arise from intracardiac or aortic thrombi. In this type lodgement occurs in the brain, legs, spleen, and kidneys, and damage may be severe or fatal.

Air embolism may be a complication of a surgical procedure, particularly those about the neck. It can also result from rapid decompression, as in caisson disease, with the formation of bubbles of nitrogen in the blood. Infected emboli can form new foci of infection at their sites of lodgement. *See* DECOMPRESSION ILLNESS.

Trauma to fatty tissues or the fracture of bones containing fatty marrow may release small fat globules. Since most of the droplets can pass singly through various circulatory beds, any effects are usually the result of accumulation, which occurs most often in the lungs and in the brain with variable results.

Bacteria may, in certain cases, form septic emboli which are characteristic of some infections. One common method of tumor spread is by means of blood-borne, malignant emboli which have become dislodged from the parent neoplasm, *See* ONCOLOGY. [E.G.St./N.K.M.]

Embrithopoda An order established for the unique mammal *Arsinoitherium*, which has been found only in early Oligocene deposits in northern Egypt. This animal was of rhinoceros size with a large body and short pillarlike legs (see illustration). There were two huge, scimitarlike horns over the nose

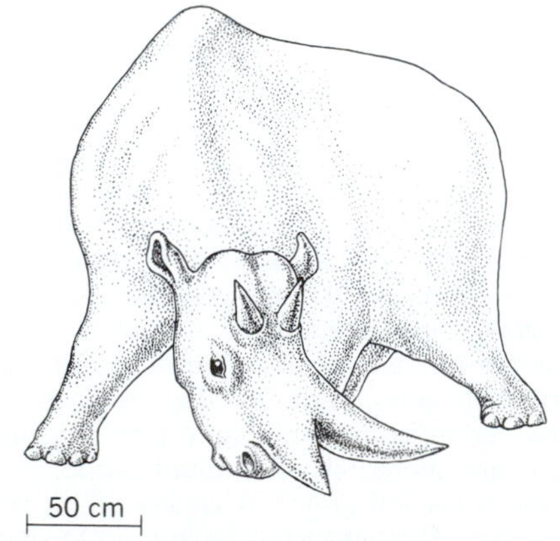

***Arsinoitherium*, the early Oligocene embrithopod from Egypt. (After C. R. Knight)**

and two much smaller, peglike horns over the eyes. The exact relationship of this exotic order to other orders is not clear, but similarities to both the uintatheres, or Dinocerata, and Hyracoidea have been interpreted by some authorities as evidence of kinship. *See* DINOCERATA; MAMMALIA. [M.C.McK.]

Embrittlement A general set of phenomena whereby materials suffer a marked decrease in their ability to deform (loss of ductility) or in their ability to absorb energy during fracture (loss of toughness), with little change in other mechanical properties, such as strength and hardness. Embrittlement can be induced by a variety of external or internal factors, for example, (1) a decreasing or an increasing temperature; (2) changes in the internal structure of the material, namely, changes in crystallite (grain) size, or in the presence and distribution of alloying elements and second-phase particles; (3) the introduction of an environment which is often, but not necessarily, corrosive in nature; (4) an increasing rate of application of load or extension; and (5) the presence of surface notches.

Low-temperature embrittlement results from a competition between deformation and brittle fracture, with the latter becoming preferred at a critical temperature. For a material to be useful structurally, it is desirable that this critical temperature be below the minimum anticipated service temperature; in most cases, this is room temperature. At high temperatures, internal structural changes that lead to intergranular embrittlement can occur. Embrittlement usually occurs in the creep temperature range, a temperature at which deformation can occur under very low stresses; and the two processes are believed to be connected.

In many metals, particularly structural steels, annealing or heat treating in certain temperature ranges sensitizes the grain boundaries in such a way that intergranular embrittlement subsequently occurs during service. To reduce the brittleness, the steel undergoes an annealing treatment called tempering, which, while decreasing the strength, usually increases the toughness. The exception to this trade-off occurs when the steel is tempered at 1000°F (538°C). This can lead to a mode of intergranular fracture called temper embrittlement; such a process has led to catastrophic failures in turbines, rotors, and other high-strength steel parts. In other metals, there are less specific but similar types of embrittlement resulting from critical heat treatments. *See* HEAT TREATMENT (METALLURGY); TEMPERING.

Metals can fracture catastrophically when exposed to a variety of environments. These environments can range from liquid metals to aqueous and nonaqueous solutions to gases such as hydrogen.

If a thin film of a liquid metal is placed on the oxide-free surface of a solid metal, the tensile properties of the solid metal will not be affected, but the fracture behavior can be markedly different from that observed in air. Although many different liquid metals are capable of inducing embrittlement in a variety of solid metals, some of the more common couples, many of which have important engineering and design consequences, are mercury embrittlement of brass, lead embrittlement of steel, and gallium embrittlement of aluminum.

Stress corrosion cracking can occur when a metal is stressed and simultaneously exposed to an environment which may be, but is not necessarily, corrosive in nature. Both stress and environment are required; if only one of these elements is present, the metal usually displays no embrittlement. *See* CORROSION.

Hydrogen embrittlement is a form of embrittlement often considered to be a type of stress corrosion cracking. Hydrogen atoms can enter a metal, causing severe embrittlement, again with little effect on other mechanical properties. This phenomenon was originally observed, and is most critical in, steels, but it is not documented to occur in titanium and nickel alloys, and may lead to cracking in other alloy systems as well.

Factors such as notches and the rate of application of stress can modify the response of a material to a specific type of embrittlement. In general, notches or surface flaws always enhance embrittlement, both by acting as a stress raiser and by providing a preexisting crack. [I.M.B.]

Embryobionta One of the two plant subkingdoms, the other being the Thallobionta. The Embryobionta are here considered to include eight divisions, the Rhyniophyta, Bryophyta, Psilotophyta, Lycopodiophyta, Equisetophyta, Polypodiophyta, Pinophyta, and Magnoliophyta. The Rhyniophyta are represented only by Paleozoic fossils, but the other seven divisions have both modern and fossil representatives. See the separate articles on each division.

The Embryobionta differ from the green algae (Chlorophyta) and from most Thallobionta in that the normal life cycle of the Embryobionta shows a well-marked alternation of generations in which the sporophyte (spore-producing, typically diploid) generation always begins its development as a parasite on the gametophyte (gamete-producing, typically haploid) generation. The young sporophyte is called an embryo.

The more primitive divisions of Embryobionta have the gametes produced in multicellular sex organs (archegonia and antheridia), in contrast to the unicellular oogonia and antheridia of the Thallobionta in general. In the more advanced divisions (Pinophyta and Magnoliophyta) of Embryobionta, the antheridia and archegonia are highly modified or entirely suppressed, in conformity with the general reduction of the gametophyte generation.

All divisions of Embryobionta except the Bryophyta have specialized conducting tissues (xylem and phloem) in the sporophyte. With the exception of most bryophytes, they also commonly have a characteristic stomatal apparatus which controls the opening and closing of numerous tiny pores (stomates) in the leaves and stems in response to environmental conditions. These specializations, together with the progressive reduction of the gametophyte in the more advanced divisions, reflect the progressive evolutionary adaptation of the Embryobionta to life on dry land instead of in the ancestral water. The Embryobionta are therefore often called the land plants, in spite of the fact that many of them, such as the water lilies, have returned to an aquatic habitat. The seven divisions of Embryobionta which characteristically have xylem and phloem have sometimes been treated as a single comprehensive division under the name Tracheophyta. *See* PLANT KINGDOM; THALLOBIONTA. [A.Cr.]

Embryogenesis The formation of an embryo from the egg, showing the basic pattern of organization of the animal, with the rudiments of most organ systems at least outlined. The axial organization of the embryo (anteroposterior and dorsoventral axes, giving the embryo its bilateral symmetry) can often be traced back to the egg before or after fertilization. *See* DEVELOPMENTAL BIOLOGY; MEROGONY. [G.Fa.]

Embryology The study of the development of the organism from the zygote, or fertilized egg. It confines itself mainly to the study of the development of many-celled organisms, since reproduction in one-celled organisms is usually carried out by means other than through the formation of zygotes. Embryology classifies the processes of development into a number of overlapping phases. These are gametogenesis, fertilization, cleavage, and gastrulation. *See* BLASTULATION; CLEAVAGE (EMBRYOLOGY); GAMETOGENESIS; GASTRULATION; REPRODUCTION (ANIMAL).

Prior to gastrulation, the cells of any particular embryo may differ from each other in their size, inclusions, pigments, and so forth. However, they are still essentially similar in that they are undifferentiated; that is, they have not yet attained the structural and functional specialization that marks their adult state. The beginning manifestations of their differentiation become visibly apparent usually after gastrulation. The process of gastrulation therefore has received great attention from embryologists who are interested in investigating the factors that influence cellular differentiation.

Embryologists study not only the differentiation of individual cells but also changes in the size and shape of aggregates of cells constituting tissues, organs, and the whole organism. The sum total of these processes is called morphogenesis. Morphogenesis, or the development of form, occurs concomitantly with all the other phases of development and includes them all since it deals with the differentiation and growth of the whole and its parts. *See* ANIMAL MORPHOGENESIS; EMBRYONIC DIFFERENTIATION.

The constitution and organization of eggs, and their specific patterns and procedures of development, differ so markedly in different species that embryology has not succeeded in formulating any satisfactory general theory of development. The main generalizing concept to have emerged is the principle of progressive differentiation, which postulates that the events of each phase of development are intimately related to events of the phases immediately preceding and following. *See* DEVELOPMENTAL BIOLOGY. [J.M.O.]

Embryonated egg culture Embryonated eggs are among the most useful and available forms of living animal tissue for the isolation and identification of animal viruses, for titrating viruses, and for quantity cultivation in the production of viral vaccines. The embryo proper, chorioallantoic membrane, yolk sac, allantoic sac, or amniotic sac may be inoculated in hen eggs of various ages, so that a wide choice of types of tissue is available to fit the characteristics of the virus under study or for special studies. The chorioallantoic membrane is frequently used; in some infections, such as smallpox, vaccinia, and herpes simplex, characteristic lesions are produced which in some cases may resemble those in the natural host. *See* ANIMAL VIRUS. [J.L.Me.]

Embryonic differentiation The process by which specialized and diversified structures arise during development of the embryo. The process involves (1) an increase in the number of cell types, and (2) an increase in morphological heterogeneity through the arrangement of cells into increasingly complex structural patterns in the form of tissues and organs. *See* HISTOGENESIS.

Differentiation begins in most organisms with fertilization of an egg with a sperm, after which the relatively large egg divides into many smaller cells called blastomeres. The blastomeres receive unequal portions of the cytoplasmic materials of the egg and are therefore initially somewhat different from each other. At the end of cleavage, the blastomeres are organized into a blastula, commonly either a hollow ball of cells or a flattened two-layered disk of cells. The cells of the blastula lie in different relative positions from those that will be occupied by their descendants in the adult organism. By a process known as gastrulation, they move to their approximate final positions and are arranged into three basic layers, called germ layers. However, only two layers form in the simpler multicellular organisms. The outer layer is the ectoderm, from which arise the nervous system and the epidermal layer of the skin. The innermost germ layer, the endoderm, forms the epithelial lining of the digestive tract and contributes the essential tissue of associated organs. In all but the most primitive animals a third germ layer, the mesoderm, is formed by cells which come to lie in the area between the other two layers. In higher animals the mesoderm gives rise to most of the cells of the organism, such as those found in the muscles, skeleton, blood, connective tissue, kidneys, gonads, and certain other organs. The molding of groups of embryonic cells into such diverse tissues and organs proceeds through a variety of morphogenetic processes, such as migration, aggregation, dispersion, delamination, folding, and differential local growth of cells. *See* BLASTULATION; GASTRULATION; GERM LAYERS.

The mechanisms by which the course of cellular differentiation is realized are not precisely known. The factors involved may, however, be divided into two classes: (1) intrinsic, those operating within the cell, and (2) extrinsic, those brought to bear upon the cell from outside. The relative importance of these factors varies considerably from one cell strain to another and also within the same cell at different stages in its development.

The fertilized egg begins development with a rich endowment, consisting of a nucleus with a set of paternal and maternal chromosomes together with a complexly organized cytoplasm. The emergence of new cell characteristics may be attributed to an oscillating interaction between the intrinsic gene makeup of the cell and the surrounding cytoplasm. The dynamic imbalance existing between these interacting components drives the cell along its path of differentiation. In certain kinds of invertebrate embryos, interactions within each separate cell seem sufficient for guiding differentiation to its terminal state. Such embryos exhibit mosaic development. By contrast, in the embryos of vertebrates and certain invertebrates such as echinoderms, influences from adjacent cells are an essential part of the differentiation process. These embryos show regulative development. *See* CELL LINEAGE; EMBRYONIC INDUCTION. [C.L.Ma.]

Embryonic induction Each tissue of multicellular animals is formed during embryogenesis by a specific cell group located at a definite site in the early embryo. In many cases such a cell group is able to develop into the corresponding tissue only under a specific influence of a neighboring cell group. Such an influence is called embryonic induction. Thus in embryonic induction two cell groups interact: the reacting system which develops into the tissue in question, and is influenced by the other cell group in its developmental pathway; and the acting system which contributes to formation of the tissue in question by controlling the mode of development of the former group, but not by providing the induced tissue with cellular material. It should be emphasized, however, that in the absence of the acting system the reacting system can proceed in development and form a specific tissue which is different from that produced in the presence of the acting system. Thus embryonic induction consists in a shift in the developmental pathway of the reacting system, rather than in initiating development in the same system. The acting system is often called the inducer or inductor. These two expressions, however, are also used to denote chemical or physical factors which affect the developmental pathway of the reacting system.

The ability of the reacting system to respond to the inductive stimulus, often called competence, starts to appear at a developmental stage and disappears in many cases at a later stage. These critical stages are different in different types of embryonic induction. *See* DEVELOPMENTAL BIOLOGY. [T.Y.]

Emerald The medium- to dark-green gem variety of the mineral beryl, crystallizing in the hexagonal system. A flawless emerald with good color is one of the most sought after and highly prized of all precious gems. Emerald is restricted in its occurrence, and only infrequently are exceptional stones found; most emeralds are flawed and cloudy, and few stones command high prices.

In contradistinction to beryl and its other gem varieties, emeralds have only been found in mica schists or metasomalized limestones. The most outstanding occurrences include the Muzo and El Chivor mines in Colombia. Noteworthy occurrences in mica schists include Tokovoja in the Ural Mountains, where emerald occurs with the beryllium minerals chrysoberyl (and its gem variety alexandrite) and phenakite; Habachtal,

Austria; Transvaal, South Africa; and Kaliguman, India. The ultimate source of an emerald can often be assessed by a study of its inclusions. *See* BERYL; GEM. [P.B.M.]

Emergency medicine The medical specialty that comprises the immediate decision making and action necessary to prevent death or further disability under emergency conditions. It is based primarily in hospital emergency departments, but with extensive responsibilities for supervising emergency medical systems outside the hospital (paramedics).

Prehospital care is professional emergency care delivered before the patient reaches the hospital. Care is provided by paramedics and emergency medical technicians. Paramedics are technicians trained in techniques such as electric shock for defibrillation of the heart, insertion of a tube into the trachea to assist respiration, and the use of intravenous drugs normally administered by physicians. They operate under the supervision of an emergency physician, usually by radio communication. Emergency medical technicians (EMTs) are trained in first aid, give no medication, seldom communicate with a doctor, and often receive very little medical supervision.

Trauma is physical injury, whether minor or major; it does not include medical emergencies due to other causes. In major trauma (usually caused by vehicular or industrial accidents, stabbing or gunshot wounds, or falls) the best care requires rapid transport to the hospital. Proper treatment of trauma requires prompt assessment by qualified paramedics, who may have to immobilize the neck to prevent spinal cord injury, protect the breathing passages, or administer intravenous fluid to replace blood lost by hemorrhage. Further evaluation and treatment by a qualified emergency physician must take place immediately upon arrival at the emergency department or trauma center. Trauma care has also been revolutionized by the advent of computerized tomographic x-ray. *See* COMPUTERIZED TOMOGRAPHY.

Burns are a very serious injury because the protective barrier of the skin is destroyed, leading to massive loss of body fluids, development of shock, and invasion by bacteria. Patients with major burns are best treated in specialized burn centers and should be taken there promptly. Burn mortality has dropped significantly because of prompt intravenous administration of very large volumes of fluid to replace body fluid losses, and the use of antibiotics. *See* BURN.

Cardiac arrest occurs when the heart suddenly goes into an abnormal rhythm so that it does not pump blood; it is the chief cause of death in heart attack. Cardiac arrest tends to occur in the first few hours of heart attack, often before the patient has even decided to go to the hospital, and sometimes is not preceded by any symptoms. If treated by defibrillation (electric shock to the heart) within the first 6 min, it is reversible. Otherwise, death is certain. Therefore, to save lives, emergency medical service programs must deliver paramedics to the patient's side within 6 min of the arrest.

Innumerable poisonings occur every year in the United States, involving thousands of products. Poison centers have been established in every state, where trained physicians and pharmacologists equipped with electronic and on-line references provide emergency telephone consultation. Poison centers have greatly improved the quality of care in poisoning cases. [M.Ca.]

Emery A natural mixture of corundum with magnetite or with hematite and spinel. Because the mixture is very intimate and appears to be quite homogeneous, it was considered to be a single mineral species until the middle of the 19th century. The aggregate has a gray-to-black color and is extremely tough and difficult to break. The specific gravity varies from 3.7 to 4.3, depending upon the relative amounts of the constituent minerals. The hardness is about 8 (Mohs scale), less than that of pure corundum which is 9, and is more dependent upon the physical state of aggregation than on the percentage of corundum. *See* CORUNDUM.

Although synthetic abrasives have replaced emery in many of its earlier uses, it is still used as an abrasive and polishing material by lapidaries and in the manufacture of lenses, prisms, and other optical equipment. Emery wheels, emery paper, and emery cloth are used not only by lapidaries but also by machinists in the grinding and polishing of steel. *See* ABRASIVE; GRINDING; MAGNETITE. [C.S.Hu.]

Emission electron microscope An electron microscope in which electrons emitted from a metal surface are projected on a fluorescent screen, with or without focusing. These may take several forms (see illustration). In the immersion electron microscope the specimen is a flat conducting surface which may be heated, illuminated, or bombarded by high-velocity electrons or ions so as to emit low-velocity thermionic, photo-, or secondary electrons. These are accelerated to a high velocity in an immersion objective or cathode lens and imaged as in a transmission electron microscope. *See* ELECTRON MICROSCOPE.

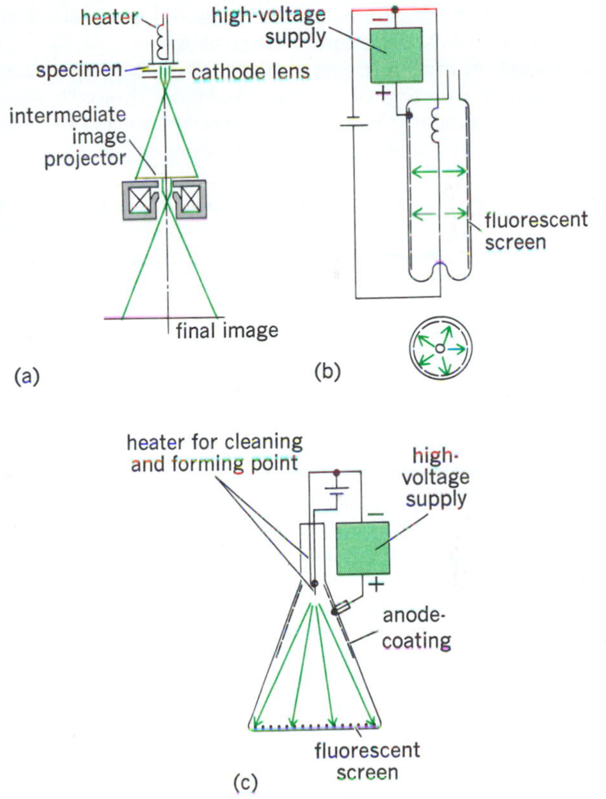

Emission electron microscopes. (*a*) Immersion electron microscope. (*b*) Cylindrical thermionic-emission microscope. (*c*) Field-emission microscope.

Other emission microscopes employ simple projection without lenses. One example is the cylindrical microscope used for examining the thermionic emission from a thin wire stretched along the axis of a cylindrical tube whose walls at high positive potential are coated with a fluorescent substance. Another is the field-emission microscope, in which field electron currents are drawn from an extremely fine, rounded point on the end of a wire of a metal such as tungsten. The electrons drawn out of the point follow nearly straight lines toward a positive-potential screen. The display on the screen shows variations in the emission characteristics (as determined by the work function and surface nonuniformities) over the surface of the point. *See* Field-emission microscopy; Thermionic emission. [E.G.R.]

Emission spectrochemical analysis A technique conducted by monitoring and measuring the spectrum of light caused to be emitted by the material to be analyzed. In general, there are many ways in which to conduct an emission spectrometric measurement; the differences among approaches result mainly from the choice of location within the electromagnetic spectrum at which to observe emitted radiation. However, emission spectrochemical analysis traditionally refers to those analytical determinations based on radiation in the visible through vacuum ultraviolet region of the electromagnetic spectrum (wavelengths about 800 to 100 nanometers). The technique is used principally to detect (qualitative analysis) and determine (quantitative analysis) metals and some nonmetals. Under optimum conditions, as little as 10^{-10} gram of an element per gram of sample can be determined.

The steps in emission spectrochemical analysis are: vaporization and atomization of sample; excitation of atomic vapor; resolution of emitted radiation; and observation and measurement of resolved radiation. *See* Spectroscopy. [A.T.Z.]

Emissivity The ratio of the radiation intensity of a nonblackbody to the radiation intensity of a blackbody. This ratio, which is usually designated by the Greek letter ϵ, is always less than or just equal to one. The emissivity characterizes the radiation or absorption quality of nonblackbodies. Published values are readily available for most substances. Emissivities vary with temperature and also vary throughout the spectrum. For an extended discussion of blackbody radiation and related information *see* Heat radiation. [H.G.S.; P.J.W.]

Emitter follower A circuit that utilizes a common-collector transistor which provides less than unity voltage gain but high input resistance and low output resistance. This circuit is used extensively to provide isolation or impedance matching between two electronic circuits.

The most common use for the emitter follower is as a device which performs the function of impedance transformation over a wide range of frequencies with voltage gain close to unity. In addition, the emitter follower increases the power level of the signal. [C.C.H.]

Emotion An umbrella concept in the common language, typically defined by instantiation by reference to a variety of mental and behavioral states. These range from lust to a sense of liking, from joy to hostile aggression, and from esthetic appreciation to disgust. Emotions are usually considered to be accompanied by some degree of internal, frequently visceral, excitement, as well as strong evaluative components. Emotions are also often described as irrational, that is, not subject to deliberative cogitation, and as interfering with normal thought processes.

These latter qualities are often exacerbated in the emotional behavior and expression seen in clinical cases. The expression of strong emotions is typically considered to be symptomatic of some underlying conflict, and even the positive emotions are used as indices of unusually strong attachments and atypical earlier experiences. Sigmund Freud introduced the concept of repression to describe a defense mechanism against the occurrence of strong emotional experiences. From the psychoanalytic point of view, what is repressed is not the emotion itself, since the very concept of emotion implies conscious experience, but rather the memory of an event which, if it became conscious, would lead to strong conflicts and emotional consequences. Many other defense mechanisms, such as rationalization and compulsive or obsessive neurotic symptoms, are also seen as serving the purpose of avoiding conscious conflict and emotional sequelae. *See* Psychoanalysis. [G.M.]

Emphysema A disorder of pulmonary inflation characterized by enlargement and destruction of the air spaces. The key element in this definition is the word destruction for it implies the irreversible loss of a given area of the pulmonary parenchyma. Certain variants of this condition do not necessarily imply irreparable destruction of pulmonary tissue but rather overdistention of air spaces, and consequently are not properly classified as emphysema.

Generalized emphysema probably has many causes; most share chronic bronchiolitis as a factor. Narrowing at this level would cause retention of air, leading to dilatation and rupture of alveolar septa. Increasing attention is being given to heavy cigarette smoking and air pollution as contributing factors. Given the dilatation of the air spaces, the total air space in the lungs is increased. However, the lungs cannot be properly emptied and are functionally impaired.

Emphysema, if widespread, will cause very serious limitation in physical activity. Many cases, however, are compatible with long survival. Complications of severe emphysema include right heart failure (cor pulmonale), respiratory acidosis, and rupture of bullae with development of pneumothorax.

The important variants of emphysema are as follows. Centrilobular emphysema affects predominantly respiratory bronchioles without involvement of the more peripheral elements. In diffuse vesicular emphysema, the most common form, all elements of the respiratory unit (respiratory bronchiole, alveolar ducts, alveolar sacs, and alveoli) are dilated. Senile emphysema was formerly applied to barrel-chested elderly people; however, functional impairment is, in most cases, inconspicuous. This condition is also known as aging lung, *See* Bronchial disorders. [V.E.G.]

Empirical method The empirical method is generally characterized by the collection of a large amount of data before much speculation as to their significance, or without much idea of what to expect, and is to be contrasted with more theoretical methods in which the collection of empirical data is guided largely by preliminary theoretical exploration of what to expect. The empirical method is necessary in entering hitherto completely unexplored fields and becomes increasingly less purely empirical as the acquired mastery of the field increases. Successful use of an exclusively empirical method demands a high degree of intuitive ability in the practitioner.

[P.W.Br./G.Ho.]

Emulsion A dispersion of one liquid in a second immiscible liquid. Since the majority of emulsions contain water as one of the phases, it is customary to classify emulsions into two types: the oil-in-water (O/W) type consisting of droplets of oil dispersed in water, and the water-in-oil (W/O) type in which the phases are reversed. The continuous liquid is referred to as the dispersion medium, and the liquid which is in the form of droplets is called the disperse phase.

A stable emulsion consisting of two pure liquids cannot be prepared; to achieve stability, a third component, an emulsifying agent, must be present. Generally, the introduction of an emulsifying agent will lower the interfacial tension of the two phases. A large number of emulsifying agents are known; they can be classified broadly into several groups. The largest group is that of the soaps, detergents, and other compounds whose basic structure is a paraffin chain terminating in a polar group. Some solid powders can act as emulsifiers by being wetted more by one phase than by the other. Whichever phase shows the greater wetting power will become the dispersion medium. Many naturally occurring emulsions, such as milk or rubber latex, are stabilized by proteins. Egg yolk proteins stabilize mayonnaise and salad dressing. Certain hydrophilic colloids such as gum arabic or gelatin also stabilize water-in-oil emulsions by a similar mode of action.

Emulsions may be prepared readily by shaking together the two liquids or by adding one phase drop by drop to the other phase with some form of agitation, such as irradiation by ultrasonic waves of high intensity. In industry, emulsification is accomplished by means of emulsifying machines.

The breaking of emulsions is necessary in many industrial operations, for example, in the separation of water-in-oil emulsions in the petroleum industry and in product recovery from emulsions produced by the steam distillation of organic liquids. Emulsions may be broken by (1) addition of multivalent ions of charge opposite to the emulsion droplet, (2) chemical action (addition of acids to emulsions stabilized by soaps), (3) freezing, (4) heating, (5) aging, (6) centrifuging, (7) application of high-potential alternating electric fields, and (8) treatment with ultrasonic waves of low intensity. *See* COLLOID; DETERGENT; SOAP.

[G.S.M.; W.O.M.]

Enamel A type of paint usually formulated by the suspension of pigments in resin and having excellent flow and leveling properties. Early enamels cured to a hard, smooth film with high gloss that simulated porcelain enamel. Modern formulation techniques now make available semigloss enamel finishes comparable in sheen and texture to satin fabric and flat finishes, as well as the high-gloss materials, all of which level out free from brush marks and other surface irregularities. *See* PAINT.

[C.R.Ma.; C.W.Si.]

Enantiomer One of an isomeric pair of chemical compounds whose molecules are nonsuperimposable mirror images. One molecular configuration of such dissymmetric substances is capable of rotating plane-polarized light to the right, dextro or (+) form, while the mirror image rotates the light equally to the left, levo or (−) form. Each member of such a pair (optical isomers) possesses identical chemical and physical properties except for interaction with other dissymmetric systems, that is, other optically active substances or plane-polarized or circularly polarized light. *See* OPTICAL ACTIVITY. [W.R.V.]

Enargite A mineral having composition Cu_3AsS_4. In some places enargite is a valuable ore of copper. The mineral has perfect prismatic cleavage, metallic luster, and grayish-black color. The hardness is 3 on Mohs scale, and the specific gravity is 4.44. Enargite is one of the rarer copper ore minerals and it has been mined in Yugoslavia, Peru, the Philippines, and the United States at Butte, Montana, and Bingham Canyon, Utah. Probably the largest deposit is at Chuquicamata, Chile. [C.S.Hu.]

Encalyptales An order of the true mosses (subclass Bryidae) that grow in dull, dark tufts on soil or soil-covered rock, generally in calcareous areas. The Encalyptales consist of a single family and two genera, the better known being *Encalypta*, the extinguisher moss, so called because of its long calyptra of candlesnuffer form.

Encalyptales are characterized by broad, papillose leaves and erect capsules covered by very long calyptrae. The stems are erect and simple or forked with folded, incurved leaves. Leaves are broadly acute to rounded and often abruptly short-pointed to hair-pointed with a strong midrib ending at or beyond the leaf tip. The sporophytes are terminal, with elongate setae, and erect, cylindric, and ribbed capsules. The operculum is long-rostrate, and the peristome variable in structure. The spores are often large and quite various in sculpturing. *See* BRYIDAE; BRYOPHYTA; BRYOPSIDA; POTTIALES. [H.Cr.]

Endangered species A plant or animal species that is in immediate danger of extinction throughout all or part of its range. A species may also be classified as threatened if it is likely to become endangered in the near future. These technical definitions are used primarily in the United States. Elsewhere, species are often considered endangered if factors that have caused their decline, such as habitat destruction, continue, or if the species are considered vulnerable and likely to become endangered if no action is taken to halt their decline. *See* EXTINCTION (BIOLOGY).

The large-scale clearing of forests, draining of swamps, cultivation of soils, and general alteration of the environment result in fragmentation of species distributions, reduction of population sizes, and isolation of species segments. These isolated populations may diverge in relation to the rest of the species, but are highly vulnerable to local extinction events because of reduced genetic variability and reduced ability to adapt to environmental change. As habitats become fragmented and reduced in size, a rapid reduction in species occurs. *See* ECOSYSTEM.

An increasing number of species are becoming endangered because of the effects of pollution. The devastating effect of the pesticide dichlorodiphenyltrichloroethane (DDT) on bird egg shells is believed to be responsible for the rapid decline of many birds of prey, such as the peregrine falcon. This endangered species, along with others formerly considered to be endangered such as the bald eagle, has made significant recovery with the removal of DDT from the market and the subsequent protection afforded by the Endangered Species Act.

A wide variety of aquatic organisms are negatively affected by pollution and water management programs such as flood control. Entire groups of fishes such as sturgeons are now endangered or extinct primarily because of pollution of their habitats. Desert river systems, such as the Rio Grande in the southwestern United States, have lost a large proportion of their native fish faunas and invertebrate species.

Populations of flora and fauna are now affected on a global scale because of widespread pollution in the form of acid rain, increased ultraviolet radiation from ozone depletion, and global warming. These types of pollution are especially harmful to larval vertebrates and invertebrates and create effects across a wide geographic area. The global reduction in the number of frog species is believed to be due to the thinning of the ozone layer and the resulting increase in ultraviolet radiation, which is harmful to larval frogs. *See* ACID RAIN; AIR POLLUTION; ATMOSPHERIC OZONE; GREENHOUSE EFFECT. [T.L.Y.]

Endocardial fibroelastosis A cardiomyopathy characterized by marked fibrous and elastic thickening of the endocardium (the lining of the heart wall) which extends into the immediately subjacent myocardium. It is an important cause of cardiac death in infants between the age of 1 and 2 years. Some are associated with congenital malformations of the heart or a major blood vessel. However, in most instances it occurs independently of other heart lesions. The etiology and pathogenesis of this lesion is completely unknown. As a result of the thickening of the endocardium, the efficiency of the contraction of the heart is markedly diminished, and congestive heart failure ensues. *See* HEART (VERTEBRATE); HEART DISORDERS.

[N.K.M.]

Endocarditis An inflammation of the lining of the heart or endocardium, which also includes the tissues which form the heart valves. The two main forms of endocarditis are bacterial and nonbacterial. In the former there is active infection on the surface of a valve. This is usually fatal within a period of a few months unless successfully treated. The most common cause of nonbacterial endocarditis is rheumatic fever, which is sometimes characterized by a slowly progressive scarring of a heart valve. After some years this may cause serious mechanical impairment of cardiac function, due either to narrowing (stenosis) or incompetence (regurgitation) at the valve orifice. *See* RHEUMATIC FEVER. [P.B.B.]

Endocrine system (invertebrate) The chemical integrating system in animals that lack a vertebral (spinal) column. An endocrine system consists of those glandular cells, tissues, and organs whose products (hormones) supplement the rapid, short-term coordinating functions of the nervous system. Almost all of the information about invertebrates pertains to the more highly evolved groups that will be discussed below, the annelids, echinoderms, mollusks, and most particularly two classes of arthropods, the insects and crustaceans. Several of the hormones in invertebrates are neurohormones, that is, they are produced by nerve cells. *See* NEUROSECRETION.

Insects. Increase in linear dimensions of an insect can only occur at periodic intervals when the restricting exoskeleton is shed during a process known as molting. Once an insect becomes an adult, it ceases to molt. The orderly sequence of molts that leads from the newly hatched insect to the adult is controlled by three hormones. The brain produces a neurohormone which stimulates a pair of glands in the prothorax, the prothoracic glands, causing release of the molting hormone, ecdysone. A third hormone, the juvenile hormone, produced by a pair of glands near the brain, functions during the juvenile molts to suppress the differentiation of adult tissues. Juvenile hormone permits growth but prevents maturation. *See* ECDYSONE.

Two neurohormones with antagonistic actions are involved in regulating the water content of insects. One, the diuretic hormone, promotes water loss by increasing the volume of fluid secreted into the Malpighian tubules, the excretory organs. The second, the antidiuretic hormone, acts to conserve water by causing the wall of the rectum to increase the volume of water resorbed from its lumen while lowering the excretion rate from the Malpighian tubules.

Bursicon, a protein neurohormone, is responsible for the tanning and hardening of the newly formed cuticle. During the development of some insects, a period of arrested development occurs, termed the diapause. The mechanisms controlling the onset and the termination of diapause are largely unknown. However, in some insects there is evidence for a hormone, proctodone, that reinitiates development. A very few species of insects, most notably the stick insect (*Carausius morosus*), have the ability to change color. This insect becomes darker at night and lighter by day as a result of the rearrangement of pigment granules within the epidermal cells. Darkening is due to a hormone produced in the brain.

Crustaceans. Higher crustaceans have a structure, the sinus gland, which in most stalk-eyed species lies in the eyestalk and is the storage and release site of a molt-inhibiting hormone. The Y-organs, a pair of structures found in the anterior portion of the body near the excretory organs, are the source of the crustacean molting hormone, crustecdysone. Chemically, crustecdysone is very similar to the ecdysone from insects; both substances can cause molting in crustaceans and insects. The sinus gland is a neuroendocrine structure, but the Y-organs are nonneural.

Crustacean pigment cells (chromatophores) are under hormonal control, as in insects. The active substances are released from the sinus glands and the postcommissural organs, which lie near the esophagus. Substances have been found that cause dispersion of the pigment within the chromatophore, as well as substances with the opposite action. *See* CHROMATOPHORE.

The compound eye of crustaceans has three retinal pigments. The movements of these pigments with illumination level control the amount of light impinging on the photosensitive cells. A substance has been found that causes the migration of these pigments toward the light-adapted positions and another substance that causes migration toward dark-adapted positions.

The pericardial organs, which are found near the heart, are neuroendocrine organs that cause an increase in the amplitude of the heart beat. The sinus gland contains the hyperglycemic hormone, which causes a rise in blood glucose.

Annelids. Strong evidence for hormones in annelids has been obtained from studies of the reproductive system. One substance that has been found in some marine annelids inhibits maturation of the gametes. This substance was thought to have been produced by the brain, but studies show that another structure, the infracerebral gland, which lies ventral to the brain, may be involved.

Echinoderms. The radial nerves of starfishes contain two substances that are required for the maintenance of a normal reproductive cycle. One, the shedding substance, induces spawning. The second, shedhibin, inhibits the former.

Mollusks. The best-established endocrine organs in mollusks, the optic glands, occur in the octopus and squid. They are a pair of small structures, found near the brain, that produce a substance which causes gonadal maturation. The optic glands in turn are regulated by inhibitory nerves from the brain. There is also some evidence that a portion of the nervous system (the pleural ganglia) may secrete a hormone that affects water balance. Removal of the pleural ganglia from a freshwater snail results in swelling of the animal due to the influx of water. [M.F.]

Endocrine system (vertebrate) A system that consists of localized groups of specialized cells, tissues, and discrete ductless endocrine organs that synthesize and secrete into the internal environment specific organic substances (hormones) that regulate other kinds of cells in the body. In performing this role, the endocrine system shares with the nervous system the responsibility for adjusting body structure, function, and behavior to the constantly changing external environment and for maintaining a constant and appropriate internal physiological state. *See* NERVOUS SYSTEM (VERTEBRATE).

Endocrine tissues. Many endocrine glands form discrete organs. Among these are the pituitary, gonads, thyroid, parathyroids, adrenals, pineal body, ultimobranchial bodies, and thymus. Some endocrine cells are arranged more or less diffusely in restricted regions, often embedded in other organs. Most intensively studied of the diffuse tissues are islands of endocrine cells embedded in the pancreas, in the lining of the stomach, intestine, and uterus, and in the placenta. Additional endocrine tissue is embedded in the liver and kidneys, and groups of atypical neurons (neurosecretory neurons) in the brain and, in fish, in the spinal cord (urophysis) secrete important endocrine substances. Associated with all endocrine tissues are rich networks of capillaries from which endocrine cells receive raw materials and chemical instructions, and into which they secrete their products, hormonal and otherwise. *See* ADRENAL GLAND; OVARY; PANCREAS; PARATHYROID GLAND; PITUITARY GLAND; TESTIS; THYMUS GLAND; THYROID GLAND.

Targets. Vertebrate hormones are organic molecules of various sizes, ranging from amines and derivatives of fatty acids to small polypeptides, huge protein molecules, and steroids. Hormones are usually transported by the bloodstream, and thus bathe every living cell of the body. Nevertheless, not all cells are competent to respond. For example, thyroid cells respond to thyroid-stimulating hormone from the pituitary but not to antidiuretic hormone from the same source. Kidney cells, on the other hand, respond to antidiuretic hormone. Thus the thyroid, but not the kidney, is a "target" for thyrotropic hormone. A few hormones—general metabolic ones that influence basic cellular metabolism such as utilization of energy—may have as targets every cell of the body. *See* HORMONE.

In order for a cell to be a target for a hormone, the cell must contain "receptors" for that specific hormone. Receptors are molecules of specific configuration that are capable of forming a bond with a specific hormone molecule. The resulting combination, hormone plus receptor, triggers a chain reaction that results in the characteristic response of the target cell.

Hormones are not always circulated by the bloodstream. When targets are close enough, the hormone may reach the target via microscopic intercellular spaces. For instance, male sex hormones produced by endocrine cells in a rat's testis diffuse to adjacent seminiferous tubules, where they stimulate formation of sperm.

Roles. The endocrine and nervous systems of vertebrates share responsibility for integrating body activities. Rapid, but quickly reversible, changes in the external environment are compensated for partly through the mediation of the nervous system, which may evoke immediate but short-term contraction of muscles that drive an animal deeper into a friendly environment or cause shivering in cold, temporary alteration of heartbeat, or a spurtlike release of glucose into the bloodstream. More gradual environmental changes, such as seasonal fluctuations in day lengths, temperatures, food supply, moisture, or salinity, evoke more gradual and sustained responses; these are mediated ultimately by the endocrine system. Endocrine responses prepare the animal for the changing seasons, and since reproduction is usually seasonal, they regulate reproduction. Hormones maintain homeostasis, and regulate the long-term availability of energy-containing compounds and utilization of energy by the cells. Hormones in developing vertebrates regulate such processes as cell division, growth (by regulating protein synthesis), differentiation, and metamorphosis. It might be said that the nervous system copes with the environment on an immediate basis, whereas the endocrine system copes with it on a sustained basis. *See* HOMEOSTASIS.

Endocrine rhythms. The concentration of most hormones in the bloodstream of vertebrates rises and falls rhythmically. Daily rhythms in serum levels of these hormones result in daily fluctuations in metabolism and behavior. As a result, human beings traveling by air through several time zones find their body rhythms affected, and animals placed in an artificial environment in which night and day are reversed continue to show daytime metabolism and behavior in the dark, and vice versa, until the hormone rhythms have been modified by the new lighting schedule. Annual rhythms also occur, and a hormone that peaks in the bloodstream just before dawn in summer may peak at the end of the daylight period in winter. Since different hormones peak at different times, ratios of one hormone to another vary daily and annually. Daily and annual fluctuations in adrenal steroid hormones, in particular, appear to play a prominent role in the temporal organization of metabolism and behavior in vertebrates. Birds that normally migrate northward in spring can be caused to migrate southward at that time from the same starting point if the ratio of corticosterone (an adrenal steroid) to prolactin is experimentally altered so as to mimic that of a nonexperimental bird about to migrate south. [G.C.K.]

Endocrinology The study of the glands of internal secretion, the endocrine glands, and the hormones which they synthesize and secrete. These glands are ductless; the hormones are secreted directly into the blood to be carried to the target tissue or organ. The hormones, or chemical messengers, are highly specific and their action may be selective or generalized. *See* ENDOCRINE SYSTEM (VERTEBRATE); HORMONE. [S.P.P.]

Endocytosis The process of uptake or internalization of particles, macromolecules, and fluid droplets by living cells; the first observation of this process was of the uptake of fluid droplets into macrophages by W. H. Lewis in 1930, an event he termed pinocytosis. The term endocytosis includes very distinct mechanisms which cells use for very specific purposes, and can be divided into categories according to the morphologic processes that are involved, or into functional categories.

Cells have large protrusions on their surfaces, called pseudopods or ruffles, which are thin extensions of the cytoplasm into the surrounding medium. Within a few minutes a ruffle can be extended from the cell, fall back onto the adjacent cell surface, and the membranes of the ruffle and cell can fuse at the edges, trapping a bubble of fluid, which is then, at least functionally, inside the cell. The process is called macropinocytosis.

Another endocytosis mechanism also involves the internalization of fluid from the outside of the cell. However, this process is mediated by indentations in the surface membrane (caveolae) which they pinch off from the plasma membrane as very tiny internal vesicles called micropinosomes.

The third, and most unusual, morphologic structure is a very specialized area of the surface membrane that is coated on the inner cytoplasmic surface by a layer of spikes called a bristle coat. By some mechanism, the membrane of the coated pit enters the cell and becomes an intracellular vesicle.

The internalization of soluble materials and fluid from the outside of the cell is called fluid-phase or bulk-phase endocytosis. It is a process by which materials from the outside of the cell neither bind selectively to the cell surface nor are concentrated before entering the cell. Macromolecules, particularly proteins, which bind selectively to the cell surface, enter the cell by adsorptive endocytosis; that is, they bind to something on the cell surface which they recognize, and are taken into the cell. The third form of functional endocytosis is a very specialized type, concentrative adsorptive endocytosis.

Almost all animal cells have specific molecules on their surface which bind other molecules in the medium they wish to take in. These surface-binding molecules are called receptors, and for many physiologically important systems, the receptors are surface proteins that are part of the cell membrane.

While not all of the functions of endocytosis are clear, some guesses can be made about its significance. Fluid-phase pinocytosis brings about the release of nutrients in the cell by lysosomal degradation of materials in the medium. A more likely major role for this pinocytosis, however, is to recycle, or cleanse, the surface membrane of the cell to ensure a viable and functioning barrier between the inside of the cell and its environment. Receptor-mediated endocytosis may either deliver specifically valuable surface-bound materials into the interior or selectively remove receptors with their ligands from the surface. In some cases, such as some polypeptide hormones, the cell responds to a hormone only when it is bound to the surface on the outside of the cell. Endocytosis could therefore be a way of regulating the response to a hormone on the outside. *See* PHAGOCYTOSIS. [M.C.W.]

Endodermis The single layer of plant cells that is located between the cortex and the vascular (xylem and phloem) tissues. It has its most obvious development in roots and subaerial stems. The endodermis has many apparent functions: absorption of water, selection of solutes and ions, and production of oils, antibiotic phenols, and acetylenic acids.

The endodermis has been found to have extra sets of chromosomes as compared with cortical and other cells in the plant. In some plants the chromosome numbers may be so high in the endodermis that four sets of chromosomes may occur in each endodermal cell. The larger amount of nuclear material and nucleic acid in the cells of the endodermis may in part account for the great capacity of endodermal cells to produce large amounts of chemical substances, such as acetylenic oils, high in caloric energy. *See* CORTEX (PLANT); HYPODERMIS. [D.S.V.F.]

Endoplasmic reticulum A finely divided vacuolar system of the cytoplasm which, in differentiated cells, assumes a variety of forms and specialized functions.

The endoplasmic reticulum can be pictured as a complex of vesicles and connecting tubules that extends to almost all parts of the cytoplasm. In some cells, especially those most active in the manufacture of proteins, the predominant vesicles are thin (75 nanometers) and lamellar, like pancakes. These vesicles, or cisternae, are often arranged parallel and in stacks of several units. In other cells the system is constructed of tubules (75–100 nm in diameter) which interconnect to form a complex three-dimensional lattice.

One of the most constant and remarkable features of the endoplasmic reticulum is its continuity with the nuclear envelope; the two are properly regarded as parts of a single system. This means that part of the reticulum, represented by the envelope, is in intimate contact with the genetic material of the cell. *See* CHROMOSOME.

Among several associations with other components of the cell, none has greater significance than that with ribosomes, the sites of protein synthesis (see illustration). As a result of this association the polypeptides assembled in the ribosomes are simultaneously pushed through the membranes and into the cavities of the endoplasmic reticulum where they undergo the secondary and tertiary foldings that give the protein molecule its mature functional form. *See* RIBOSOMES.

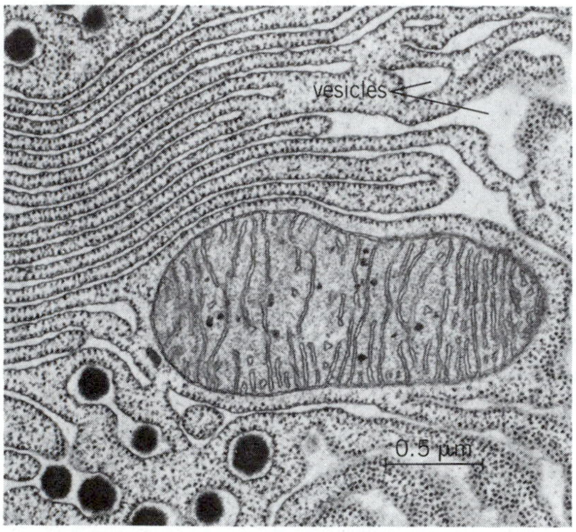

Electron photomicrograph of the cytoplasm of an exocrine cell of the bat pancreas. Vesicles are limited by membranes of the endoplasmic reticulum, seen in lamellar formation in the upper left. Numerous ribosomes (small dense granules) are seen on the surfaces of the membranes. Large dense granules inside vesicles are condensations of protein synthesized by ribosomes. Note mitochondrion in center of micrograph.

The ribosome-rich, or rough, form of the reticulum plays a role in protein synthesis and sequestration. Some fragments of the endoplasmic reticulum from muscle fibers have an unusual capacity to concentrate calcium ion, and this has helped define a role of the endoplasmic reticulum in muscle contraction and relaxation. The ribosome-free, or smooth, form of the endoplasmic reticulum in liver cells has the capacity to metabolize and detoxify a wide range of drugs and toxic substances. This points to the extraordinary fact that the membranes of special parts of the endoplasmic reticulum (usually free of ribosomes) contain enzymes capable of performing highly specialized functions. [K.R.P.]

Endopterygota A division of the insects which comprises part of the subclass Pterygota. The group includes those orders of insects which undergo a complete or holometabolous metamorphosis during their life cycle; that is, they undergo four stages of development during metamorphosis: the egg, larva, pupa, and imago or adult. [B.E.R.]

Endorphins Peptides manufactured within the central nervous system which share with plant alkaloids derived from the opium poppy the ability to alleviate pain. Endorphins are relatively short chains of amino acids which function as neurohumoral transmitters or as neuromodulators in a system descending from the brain through the spinal cord which antagonizes painful impulses coming from the periphery toward the brain. The term endorphin (from elision of *endo*genously manufactured *morphine*like substance) is now used generically to describe a family of endogenous opiates which include leucine enkephalin; methionine enkephalin; alpha, beta, gamma, and delta endorphin; and a growing number of synthetic compounds. Narcotic analgesic drugs are thought to mimic the actions of one or another of the endorphins by occupying stereospecific opiate receptors in the brain, spinal cord, and gut normally occupied by endorphins. *See* ALKALOID; ANALGESIC; NARCOTIC; OPIATES; PAIN. [L.M.H.]

Endospores (bacteria) Resistant and metabolically dormant bodies produced by the gram-positive rods of *Bacillus* (aerobic or facultatively aerobic) and *Clostridia* (strictly anaerobic), by the coccus *Sporosarcina*, and by certain other bacteria. Sporeforming bacteria are found mainly in the soil and water and also in the intestines of human beings and animals. Some sporeformers are pathogens. Endospores seem to be able to survive indefinitely. Spores kept for more than 50 years have shown little loss of their capacity to germinate and propagate by cell division. The unique property of bacterial spores is their extreme resistance to heat, radiation from ultraviolet light and x-rays, organic solvents, chemicals, and desiccation.

The endospore appears as a light-refractile body inside another cell (sporangium). Each sporangium produces one endospore with a characteristic size, shape, and position within the cell. The mature spore has a complex structure which contains a number of layers. The outermost envelope, surrounding the spore, is called the exosporium and is a thin, membranous covering. Beneath the exosporium lies the spore coat, which is composed of several layers, largely of a protein nature; each is about 2–2.5 nanometers thick. Beneath these is a thin membrane which separates the spore coat from an area of low electron density called the cortex, primarily a modified peptidoglycan structure. It occupies approximately half the volume of the spore. A wall and a thin membrane separate the cortex from the cytoplasm of the dormant spore.

The three processes involved in the conversion of the spore into a vegetative cell are (1) activation (usually by heat or aging), which conditions the spore to germinate in a suitable environment; (2) germination, an irreversible process which results in the loss of the typical characteristics of a dormant spore; and (3) outgrowth, in which new classes of proteins and structures are synthesized so that the spore is converted into a new vegetative cell. [H.O.H.; K.Hu.; C.O.]

Endotoxin A toxic product that is made by certain bacterial species and is an integral part of the cell wall. Endotoxins are large molecules, consisting of carbohydrate and lipid, referred to as lipopolysaccharides. The lipid part, called lipid A, is largely responsible for the biological effects. The polysaccharides consist of two parts, the core and the side chains. The latter differ in chemical composition from species to species and from type to type and endow these bacteria with their antigenic specificity. The side chains are attached to the core. When smooth strains mutate to rough cells, the side chains are lost and the complexity of the core decreases with increasing roughness.

Lipopolysaccharides are produced by gram-negative bacteria, both cocci and bacilli and both aerobes and anaerobes. They are synthesized by pathogenic and nonpathogenic bacteria.

There is no bacterial product that induces so many alterations in susceptible hosts as do endotoxins. Some animal species are highly susceptible and others are not. Endotoxin-induced reactions include fever, shock, alteration of metabolism, changes in the blood-forming tissues, hemorrhagic lesions, necrosis of certain tumors, activation of complement, either enhancement or suppression of the immune response, and increased resistance to infection. Many of these endotoxic reactions are due to the effects of host factors, the so-called mediators. *See* MEDICAL BACTERIOLOGY; VIRULENCE. [E.Ne.]

Energy The ability of one system to do work on another system. There are many kinds of energy: chemical energy from fossil fuels, electrical energy distributed by a utility company, radiant energy from the Sun, and nuclear energy from a reactor. The units of energy include ergs, joules, foot-pounds, and foot-poundals. Work and heat have the same units as energy, but are entirely different physical concepts. *See* HEAT; WORK.

Any particle or system of particles subject to conservative forces has two kinds of energy, potential energy and kinetic energy. Potential energy is the energy due to position or configuration, and kinetic energy is the energy due to motion. *See* KINETIC ENERGY; POTENTIAL ENERGY.

Energy is conserved for all isolated mechanical systems. This is because if a system A is isolated, there is no other system B that it can give any energy to, and its total energy must remain constant. This system A can convert kinetic energy to potential energy, and it can convert one form of potential energy to another, but the total energy must remain the same. The meaning of conserved total energy is that the system has the same value of total energy at all times. *See* CONSERVATION OF ENERGY.

In 1905 A. Einstein showed that at high velocities near the speed of light important modifications must be made in physical concepts. One particularly radical idea which he advanced was that space and time are not independent, but rather are two aspects of the same object, a space-time manifold. This necessitated a reexamination of the concept of energy and led to the conclusion that the inertia, or mass m, depends upon its energy through the mass-energy relation shown below, where

$$E = mc^2$$

c is the speed of light in vacuum. Furthermore, energy and momentum conservation become joined in a single four-momentum conservation law in special relativity. *See* INTERNAL ENERGY; RELATIVITY. [B.DeF.]

Energy conversion The process of changing energy from one form to another. There are many conversion processes that appear as routine phenomena in nature, such as the evaporation of water by solar energy or the storage of solar energy in fossil fuels. In the world of technology the term is more generally applied to operations of human origin in which the energy is made more usable; for instance, the burning of coal in power plants to convert chemical energy into electricity, the burning of gasoline in automobile engines to convert chemical energy into propulsive energy of a moving vehicle, or the burning of a propellant for ion rockets and plasma jets to provide thrust.

There are well-established principles in science which define the conditions and limits under which energy conversions can be effected, for example, the law of the conservation of energy, the second law of thermodynamics, the Bernoulli principle, and the Gibbs free-energy relation. Recognizable forms of energy which allow varying degrees of conversion include chemical, atomic, electrical, mechanical, light, potential, pressure, kinetic, and heat energy. In some conversion operations the transformation of energy from one form to another, more desirable form may approach 100% efficiency, whereas with others even a "perfect" device or system may have a theoretical limiting efficiency far below 100%. *See* ENERGY SOURCES. [T.Ba.]

Energy level (quantum mechanics) Self-contained physical systems of molecular size or smaller are found to have states of internal motion or excitation in which their energies take on definite stationary values. The energies of these states are called energy levels.

The existence of stationary states was first clearly seen in the study of atomic spectra. In 1900 Johannes Rydberg remarked that the frequencies of many observed spectral lines could be expressed in terms of the differences between pairs of numbers in a certain set called spectral terms; in 1908 Walter Ritz proposed this as a fundamental physical law, which later became known as the combination principle. Five years later Niels Bohr explained the combination principle in dynamical terms: Atoms exist in stationary states; when an atom passes from one stationary state to another of lower energy, it delivers the extra energy to a single quantum of light. Since according to Max Planck the energy of a quantum is proportional to its frequency, it follows that the spectral terms, written in appropriate units, are the energy values of the stationary states and that Ritz's combination principle merely expresses the conservation of energy. *See* ATOMIC STRUCTURE AND SPECTRA.

For nuclear energy levels, the continuous spectrum begins rather low, that is, typical level spacings are a few MeV (a million times greater than those of atomic levels), and a few more MeV disrupt a nucleus. Further, there are several ways in which a nucleus in an excited energy level can lose energy in addition to the radiation of a photon: It can lose an electron, a positron, or more rarely one or two protons or neutrons, a deuteron, or an alpha particle. The emission of anything but a photon changes the identity of the nucleus.

In reactions involving the so-called elementary particles, the same considerations enter, but often in simpler form because the internal structure of these particles, whatever it may be, is presumably simpler than those of all but the simplest nuclei. The situation with particles is, however, complicated by the existence of transitions and selection rules that have no counterpart among nuclei. *See* ELEMENTARY PARTICLE. [D.P.]

Energy metabolism Living cells all carry on sequences of chemical reactions, the overall energetics of which are such as to label the processes as spontaneous, that is, exergonic, in terms of the change in free chemical energy. A few of the chemical intermediates in such sequences are sufficiently reactive as they are so formed that inorganic phosphate may in turn be "activated" chemically and thus be able to form substituted phosphoric acid anhydrides, such as adenosinetriphosphate (ATP) or its energetic equivalents. The ATP becomes a reactant in numerous energy-requiring reactions, both relatively simple (fatty acid activation) and relatively complex (muscle contraction). *See* ADENOSINETRIPHOSPHATE (ATP).

There is a remarkable unity and also a remarkable simplicity in the patterns of energy transformation in living cells. A surprisingly small number of chemical entities and appropriate catalysts (enzymes) and reaction sequences are found in the amazingly diverse forms of life. Living cells are chemical engines whose function is to synthesize ATP (or equivalent) accompanying the overall energy-yielding degradation of nutrients. The estimated thermodynamic efficiency is 60–70%, which is very high. *See* CELL (BIOLOGY); ENZYME; KREBS CYCLE; METABOLISM. [C.S.V.]

Energy sources Sources from which energy can be obtained to provide heat, light, and power. Sources of energy have evolved from human and animal power to fossil fuels, uranium, water power, wind, and the Sun.

The principal fossil fuels are coal, lignite, peat, petroleum, and natural gas; other potential sources of fossil fuels include oil shale and tar sands. As fossil fuels become depleted, nonfuel sources and fission and fusion sources will become of greater importance since they are renewable. Nuclear power is based on the fission of uranium, thorium, and plutonium, and the fusion power is based on the forcing together of the nuclei of two light atoms such as deuterium, tritium, or helium-3. *See* Coal; Natural gas; Nuclear power; Oil sand; Oil shale; Petroleum.

Nonfuel sources of energy include wastes, water, wind, geothermal deposits, biomass, and solar heat. *See* Biomass; Geothermal power; Solar energy, Waterpower; Wind power.

Fuels which do not exist in nature are known as synthetic fuels. They are synthesized or manufactured from varieties of fossil fuels which cannot be used conveniently in their original forms. Substitute natural gas (SNG) is manufactured from coal, peat, or oil shale. Synthetic liquid fuels can be produced from coal, oil shale, or tar sands. Both gaseous and liquid fuels can be synthesized from renewable resources, collectively called biomass. These carbon sources are trees, grasses, algae, plants, and organic waste. Production of synthetic fuels, particularly from renewable resources, increases the scope of available energy sources.

Energy management includes not only the procurement of fuels on the most economical basis, but the conservation of energy by every conceivable means. Whether this is done by squeezing out every Btu through heat exchangers, or by room-temperature processes instead of high-temperature processes, or by greater insulation to retain heat which has been generated, each has a role to play in requiring less energy to produce the same amount of goods and materials. [G.C.G.]

Energy storage The general method and specific techniques for storing energy derived from some primary source in a form convenient for use at a later time when a specific energy demand is to be met, often in a different location.

In the past, energy storage on a large scale had been limited to storage of fuels. For example, large amounts of natural gas and petroleum are routinely stored. On a smaller scale, electric energy is stored in batteries that power automobile starters and a great variety of portable appliances. In the future, energy storage in many forms is expected to play an increasingly important role in shifting patterns of energy consumption away from scarce to more abundant and renewable primary resources.

An example of growing importance is the storage of electric energy generated at night by coal or nuclear power plants to meet peak electric loads during daytime periods. This is achieved by pumped hydroelectric storage, that is, pumping water from a lower to a higher reservoir at night and reversing this process during the day, with the pump then being used as a turbine and the motor as a generator.

Off-peak electric energy can also be converted into mechanical energy by pumping air into a suitable cavern where it is stored at pressures up to 80 atm (8 megapascals). Turbines and generators can then be driven by the air when it is heated and expanded.

The development of advanced batteries (such as nickel-zinc, nickel-iron, zinc-chloride, and sodium-sulfur) with characteristics superior to those of the familiar lead-acid battery could result in use of battery energy storage on a large scale. For example, batteries lasting 2000 or more cycles could be used in installations of several-hundred-thousand–kilowatt-hour capacity in various locations on the electric power grid, as an almost universally applicable method of utility energy storage. Batteries combining these characteristics with energy densities (storage capacity per unit weight and volume) well above those of lead-acid batteries could provide electric vehicles with greater range. *See* Battery; Storage battery.

Ceramic brick "storage heaters" that store off-peak electricity in the form of heat have gained wide acceptance for heating buildings in Europe, and the barriers to their increased use in the United States are more institutional and economic than technological.

Solar hot-water storage is technically simple and commercially available. However, the use of solar energy for space heating requires relatively large storage systems, with water or rock beds as storage media, and difficulties can arise in integrating this storage with existing buildings while keeping costs within acceptable limits. *See* Solar energy; Solar heating and cooling.

Heat or electricity may be stored by using these energy forms to force certain chemical reactions to occur. Such reactions are chosen so that they can be reversed readily with release of energy; in some cases the products can be transported from the point of generation to that of consumption. For example, reactions which produce hydrogen could become attractive since hydrogen could be stored for extensive periods of time and then conveniently used in either combustion devices or in fuel cells. *See* Fuel cell.

Electrical energy can be stored directly in the form of large direct currents used to create fields surrounding the superconducting windings of electromagnets. In principle such devices appear attractive because their storage efficiency is high. However, the need for maintaining the system at temperatures approaching absolute zero and, particularly, the need to physically restrain the coils of the magnet when energized require expensive auxiliary equipment (insulation, vacuum vessels, and structural supports). *See* Superconducting devices.

Storage of kinetic energy in rotating mechanical systems such as flywheels is attractive where very rapid absorption and release of the stored energy is critical. However, research indicates that even advanced designs and materials are likely to be too expensive for utility energy storage on a significant scale, and applications will probably remain limited to systems where high power capacity and short charging cycles are the prime consideration. [F.K.; T.R.S.]

Engine A machine designed for the conversion of energy into useful mechanical motion. The principal characteristic of an engine is its capacity to deliver appreciable mechanical power. By usage an engine is usually a machine that burns or otherwise consumes a fuel, as differentiated from an electric machine that produces mechanical power without altering the composition of matter. Similarly, a spring-driven mechanism is said to be powered by a spring motor; a flywheel acts as an inertia motor. By this definition a hydraulic turbine is not an engine, although it competes with the engine as a prime source of mechanical power. *See* Hydraulic turbine; Motor; Prime mover.

Traditionally, engines are classed as external or internal combustion. External combustion engines consume their fuel or other energy source in a separate furnace or reactor. A further basis of classification concerns the working fluid. If the working fluid is recirculated, the engine operates on a closed cycle. If the working fluid is discharged after one pass through boiler and engine, the engine operates on an open cycle. The commonest types of engine use atmospheric air in open cycles both as the principal constituent of their working fluids and as oxidizer for their fuels. *See* Diesel engine; Gas turbine; Internal combustion engine; Nuclear reactor; Rotary engine; Steam-generating furnace; Stirling engine; Turbine propulsion. [F.H.R.]

Engine cooling Cylinder gas temperatures in internal combustion engines may reach 4500°F (2500°C). This is well above the melting point of the engine parts in contact with the gases, so that it is necessary to control the temperature of the parts or they will become too weak to carry the gas pressure stresses.

If the outer cylinder surface is placed in contact with a cool fluid such as air or water and there is sufficient contact area to cause a rapid heat flow, the resulting temperature drop produced by the heat flow in the inside boundary layer keeps the cylinder wall temperature much closer to the coolant temperature than to the combustion gas temperature.

If the coolant is water, it is usually circulated by a pump through jackets surrounding the cylinders and cylinder heads. The water is circulated fast enough to limit the water's temperature rise through the engine to about 15°F (8°C). The warmed coolant is then piped to an air-cooled heat exchanger called a radiator. To prevent freezing, the water coolant is often mixed with alcohol or ethylene-glycol. *See* ANTIFREEZE MIXTURE.

Engines are often cooled directly by a stream of air without the interposition of a liquid medium. The heat-transfer coefficient between the cylinder and airstream is much less than with a liquid coolant, so that the cylinder temperatures must be much greater than the air temperature to transfer to the cooling air the heat flowing from the cylinder gases. To remedy this situation and to reduce the cylinder wall temperature, the outside area of the cylinder, which is in contact with the cooling air, is increased by finning. *See* HEAT EXCHANGER; HEAT TRANSFER; INTERNAL COMBUSTION ENGINE. [A.R.R.]

Engine lubrication Continuous supply of a viscous film between moving surfaces in an engine during its operation. Relative motion between unlubricated surfaces of engine parts would result in excessive wear, overheating, and seizure, with destruction of the contacting surfaces. The energy loss associated with the wear and overheating would be evident in reduced engine power and efficiency. The purpose of lubrication is to minimize these effects by interposing a viscous fluid called the lubricant between the parts.

Engines are generally lubricated with petroleum base oils containing chemical additives to improve their natural properties. The absolute viscosity is a measure of the force required to move one layer of the oil film over another. With low viscosity a protecting oil film between the parts is not formed. With high viscosity, the flow of oil through the engine is retarded. The viscosity of oil tends to decrease as its temperature is raised. It is therefore difficult to maintain the proper viscosity as the engine warms up, or as other operating conditions change. *See* LUBRICANT.

The function of the lubrication system is to supply clean oil cooled to the proper viscosity to critical points in the engine. In all but the most primitive lubrication systems, the oil, which has been heated by friction and by heat transfer from the hot engine parts, is collected, cooled, and pumped back into the engine. *See* INTERNAL COMBUSTION ENGINE. [A.R.R.]

Engine manifold The branch pipe arrangement which connects the valve ports of multicylinder engines to a single carburetor or to a muffler. The manifold may have considerable effect on engine performance.

The intermittent nature of the gas flow through each cylinder may develop resonances (similar to the vibrations in organ pipes) in the airflow at certain speeds. These may increase the volumetric efficiency and thus the power at certain engine speeds at the expense of loss at other speeds, depending on manifold dimensions.

Another very important influence which intake manifolds may have on engine performance results from the fact that it is here that most of the heat is supplied to vaporize the fuel droplets metered to the airstream as it passes through the carburetor. The resulting temperature of the fuel mixture should be held within a comparatively narrow range for best engine performance with a fuel of a given volatility. If the temperature is not high enough, the unvaporized fuel in the manifold makes engine response to the throttle during acceleration sluggish and probably unsatisfactory. If the temperature is unnecessarily high, it causes loss of power, due to the reduced weight of the mixture supplied to the cylinders and the increased tendency of the engine to develop vapor lock from fuel boiling in the metering system; moreover, fuel knock can develop. *See* CARBURETOR; INTERNAL COMBUSTION ENGINE. [N.MacC.]

Engineering Most simply, the art of directing the great sources of power in nature for the use and the convenience of people. In its modern form engineering involves people, money, materials, machines, and energy. It is differentiated from science because it is primarily concerned with how to direct to useful and economical ends the natural phenomena which scientists discover and formulate into acceptable theories. Engineering therefore requires above all the creative imagination to innovate useful applications of natural phenomena. It seeks newer, cheaper, better means of using natural sources of energy and materials.

The typical modern engineer goes through several phases of career activity. Formal education must be broad and deep in the sciences and humanities. Then comes an increasing degree of specialization in the intricacies of a particular discipline, also involving continued postscholastic education. Normal promotion thus brings interdisciplinary activity as the engineer supervises a variety of specialists. Finally, the engineer enters into the management function, weaving people, money, materials, machines, and energy sources into completed processes for the use of society.

For specific articles on various engineering disciplines *see* CHEMICAL ENGINEERING; CIVIL ENGINEERING; ELECTRICAL ENGINEERING; INDUSTRIAL ENGINEERING; MANUFACTURING ENGINEERING; MARINE ENGINEERING; MECHANICAL ENGINEERING; METHODS ENGINEERING; MINING; NUCLEAR ENGINEERING. [J.W.B.]

Engineering design Engineering is concerned with the creation of systems, devices, and processes useful to, and sought by, society. The process by which these goals are achieved is engineering design.

The process can be characterized as a sequence of events as suggested in Fig. 1. The process may be said to commence upon the recognition of, or the expression of, the need to satisfy some human want or desire, the "goal," which might range from the detection and destruction of incoming ballistic missiles to a minor kitchen appliance or fastener.

Concept formulation. The first obligation of the engineer is to develop more detailed quantitative information which

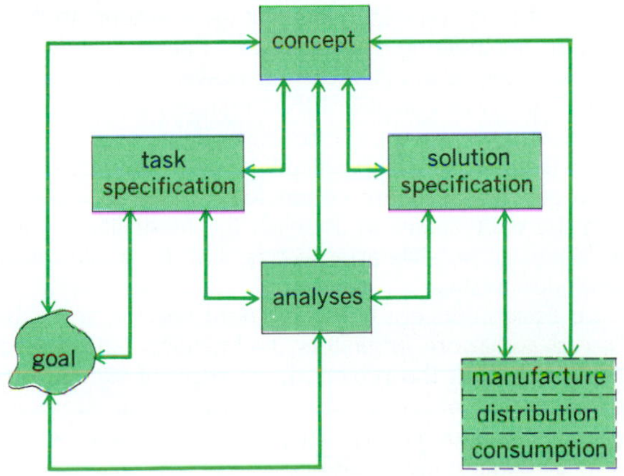

Fig. 1. The engineering design process.

defines the task to be accomplished in order to satisfy the goal, labeled on Fig. 1 as task specification. At this juncture the scope of the problem is defined, and the need for pertinent information is established. Generation of ideas for possible solutions to the problem is the creative stage, called the concept formulation. When great strides in engineering are made, this represents ingenious, innovative, inventive activity; but even in more pedestrian situations where rational and orderly approaches are possible, the conceptual stage is always present. The concept does not represent a solution, but only an idea for a solution. It can only be described in broad, qualitative, frequently graphical terms. Concepts for possible solutions to engineering challenges arise initially as mental images which are recorded first as sketches or notes and then successively tested, refined, organized, and ultimately documented by using standardized formats.

Concepts are accompanied and followed by, sometimes preceded by, acts of evaluation, judgment, and decision. It is in fact this testing of ideas against physical, economic, and functional reality that epitomizes engineering's bridge between the art of innovation and science. The process of analysis is sometimes intuitive and qualitative, but it is often mathematical, quantitative, careful, and precise.

Production considerations can have a profound influence on the design process, especially when high-volume manufacture is anticipated. Evolutionary products manufactured in large numbers, such as the automobile, are tailored to conform with existing production equipment and techniques such as assembly procedures, interchangeability, scheduling, and quality control. New techniques such as those associated with space exploration, where volume production is not a central concern, factor into the engineering design process in a very different fashion.

Similarly, the design process must anticipate and integrate provisions for distribution, maintenance, and ultimate replacement of products. Well-conceived and executed engineering design will encompass the entire product cycle from definition and conception through realization and demise and will give due consideration to all aspects.

Hierarchy of design. An adequate description of the engineering design process must have both general validity and applicability to a wide variety of engineering situations: tasks simple or complex, small- or large-scale, short-range or far-reaching. That is to say, there is a hierarchy of engineering design situations.

Systems engineering occupies one end of the spectrum. The typical goal is very broad, general, and ambitious, and the concepts are concerned with the interrelationships of a variety of subsystems or components which, taken together, make up the system to accomplish the desired goal. *See* SYSTEMS ENGINEERING.

Time–worker-power resource dynamics. Another dimension of the dynamics of the engineering design process is the elapse of time and expenditure of worker-hours in the evolution of an engineering design project. Figure 2 plots time as the abscissa and resources (worker-power or dollars) as the ordinate. The various stages of the engineering design process are set out in time sequence from left to right.

Goal refinement, task specification, and first-order concept and analyses iterations are conducted by one to a few engineers in the early stages to establish the feasibility of the idea and to block out possible approaches. This is usually called the advanced design stage.

As the design concept becomes more specific and substantive, more and more engineers, technicians, and draftsmen become involved in the project. In projects of significant size, the problem of coordinating and integrating the efforts of the many participants of different talents and skills becomes itself a major consideration. *See* PERT.

Use of the computer in design. The use of the computer, both analog and digital, in the engineering design process has

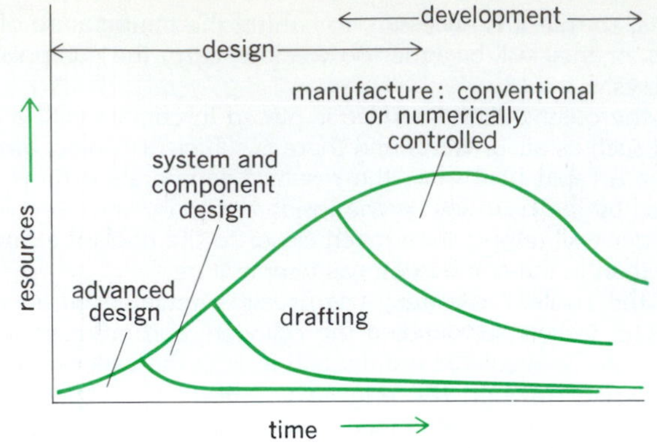

Fig. 2. Elapse of time and resources in an engineering design project, showing various stages in sequence.

increased. Where economically justified, the overall engineering design process for a product is mechanized via computer programming. *See* COMPUTER; COMPUTER-AIDED DESIGN AND MANUFACTURING; COMPUTER STORAGE TECHNOLOGY; MULTIACCESS COMPUTER.

The speed, memory, and accuracy of the computer to iteratively calculate, store, sort, collate, and tabulate have greatly enhanced its use in design and encouraged the study, on their own merits, of the processes and subprocesses exercised in the design process. These include optimization or sensitivity analysis, reliability analysis, and simulation as well as design theory. *See* DIGITAL COMPUTER.

Optimization analyses, given a model of the design and using linear and nonlinear programming, determines the best values of the parameters consistent with stated criteria and further studies the effects of variations in the values of the parameters.

Reliability is a special case of optimization where the emphasis is to choose or evaluate a system so as to maximize its probability of successful operation, for example, the reliability of electronics. *See* CIRCUIT (ELECTRONICS); RELIABILITY (ENGINEERING).

Simulation, as of dynamic systems, is mathematical modeling to study the response of a design to various inputs and disturbances. The analog computer has been widely used for simulation through its physical modeling of the mathematic analytical relationships of the proposed design. *See* ANALOG COMPUTER; SIMULATION.

Decision theory deals with the general question of how to choose between a great number of alternatives according to established criteria. It proposes models of the decision process as well as defining techniques, that is, programs or algorithms, of calculation by which to make choices. *See* DECISION THEORY.

Graphical input/output. In many fields of design—notably architecture; design of airplanes, automobiles, and ships; and almost all mechanical design—the designer works largely in visual terms. A way has been found to set up the computer to interpret drawings. Moreover, the computer has become an active partner in the act of drawing, so that it can provide a certain superskill in preparing the drawing once the human operator has made intentions clear. *See* COMPUTER GRAPHICS; DATA-PROCESSING SYSTEMS. [R.W.M.]

Engineering drawing

Engineering drawing A graphical language used by engineers and other technical personnel associated with the engineering profession. The purpose of engineering drawing is to convey graphically the ideas and information necessary for the construction or analysis of machines, structures, or systems. *See* COMPUTER GRAPHICS; DRAFTING; GRAPHIC METHODS; SCHEMATIC DRAWING.

The basis for much engineering drawing is orthographic representation (projection). Objects are depicted by front, top, side, auxiliary, or oblique views, or combinations of these. The complexity of an object determines the number of views shown. At times, pictorial views are also shown. *See* DESCRIPTIVE GEOMETRY; PICTORIAL DRAWING.

Engineering drawings often include such features as various types of lines, dimensions, lettered notes, sectional views, and symbols. They may be in the form of carefully planned and checked mechanical drawings, or they may be freehand sketches. Usually a sketch precedes the mechanical drawing. Final drawings are usually made on tracing paper, cloth, or Mylar film, so that many copies can be made quickly and cheaply by such processes as blueprinting, ammonia-developed (diazo) printing, or lithography. *See* PHOTOCOPYING PROCESSES.

[C.J.B.]

Engineering geology The application of education and experience in geology and other geosciences to solve geological problems posed by civil engineering works. The branches of the geosciences most applicable are surficial geology, petrofabrics, rock and soil mechanics, geohydrology, and geophysics, particularly exploration geophysics and earthquake seismology.

The terms engineering geology and environmental geology often seem to be used interchangeably. Specifically, environmental geology is the application of engineering geology in the solution of urban problems; in the prediction and mitigation of natural hazards such as earthquakes, landslides, and subsidence; and in solving problems inherent in disposal of dangerous wastes and in reclaiming mined lands.

Another relevant term is geotechnics, the combination of pertinent geoscience elements with civil engineering elements to formulate the civil engineering system that has the optimal interaction with the natural environment.

Engineering properties of rock. The civil engineer and the engineering geologist consider most hard and compact natural materials of the earth crust as rock, and their derivatives, formed mostly by weathering processes, as soil. A number of useful soil classification systems exist. Because of the lack of a rock classification system suitable for civil engineering purposes, most engineering geology reports use generic classification systems modified by appropriate rock-property adjectives. *See* ROCK; ROCK MECHANICS; SOIL MECHANICS.

The properties of a rock element can be determined by tests on cores obtained from boreholes. The rock properties most useful to the engineering geologist are compressive and triaxial shear strengths, permeability, Young's modulus of elasticity, erodability under water action, and density (in pounds per cubic foot, or pcf; or in kilograms per cubic meter). *See* YOUNG'S MODULUS.

The compressive (crushing) strength of rock generally is defined as the amount of stress required to fracture a sample unconfined on the sides and loaded on the ends. The wide variety of classification systems used for rock results in a broad range of compressive strengths for rocks having the same geologic name. The value of compressive strength to be used in an engineering design must be related to the direction of the structure's load and the orientation of the bedding, discontinuities, and structural weaknesses in the foundation rock.

Shearing stresses tend to separate portions of the rock (or soil) mass. Faults and folds are examples of shear failures in nature. In engineering structures, every compression is accompanied by shear stresses. The application of loads over long periods of time on most rocks will cause them to creep or even to flow like a dense fluid (plastic flow). *See* CREEP (MATERIALS); STRUCTURAL GEOLOGY.

Ambient stress in a rock system is actually potential energy, probably created by ancient natural forces, recent seismic activity, or nearby human-caused disturbances. Ambient (residual, stored, or primary) stress may remain in rock long after the disturbance is removed. An excavation, such as a tunnel or quarry, will relieve the ambient stress by providing room for displacement of the rock, and thus the potential energy is converted to kinetic energy. In tunnels and quarries, the release of this energy can cause spalling, the slow outward separation of rock slabs from the rock massif; when this movement is rapid or explosive, a rock burst occurs. *See* ROCK BURST; UNDERGROUND MINING.

One of two fundamental principles generally is used to predict the possible rock load on a tunnel roof, steel, or timber supports, or a concrete or steel lining: (1) The weight of the burden (the rock and soil mass between the roof and the ground surface) and its shear strength control the load, and therefore the resultant stresses are depth-dependent, or (2) the shear strength of the rock system and the ambient stresses control the stress distribution, so the resultant loading is only indirectly dependent upon depth. *See* TUNNEL.

Rock as a construction material is used in the form of dimension, crushed, or broken stone. Broken stone is placed as riprap on slopes of earth dams, canals, and river banks to protect them against water action. Also, it is used as the core and armor stone for breakwater structures. Dimension stone (granite, limestone, sandstone, and some basalts) is quarried and sawed into blocks of a shape and size suitable for facing buildings or for interior decorative panels. Crushed stone (primarily limestone but also some basalt, granite, sandstone, and quartzite) is used as aggregate in concrete and in bituminous surfaces for highways, as a base course or embankment material for highways, and for railroad ballast (to support the ties). *See* PETROGRAPHY; STONE AND STONE PRODUCTS.

Geotechnics. Glacial and alluvial deposits contain heterogeneous mixtures of pervious (sand and gravel) and impervious (clay, silt, and rock flour) soil materials. The pervious materials can be used for highway subgrade, concrete aggregate, and filters and pervious zones in earth embankments. Dam reservoirs may be endangered by the presence of stratified or lenticular bodies of pervious materials or ancient buried river channels filled with pervious material, Deep alluvial deposits in or close to river deltas may contain very soft materials such as organic silt or mud. *See* DELTA; FLOODPLAIN.

Concrete or earth dams can be built safely on sand foundations if the latter receive special treatment. One requirement is to minimize seepage losses by the construction of cutoff walls (of concrete, compacted clay, or interlocking-steel-sheet piling) or by use of mixed-in-place piles 3 ft (0.9 m) or more in diameter.

Loess, a relatively low-density (± 0.044 ton/ft^3 or 1.4 metric tons/m^3) soil composed primarily of silt grains cemented by clay or calcium carbonate, has a vertical permeability considerably greater than the horizontal. When a loaded loess deposit is wetted, it rapidly consolidates, and the overlying structure settles. Eolian sand deposits present the problem of stabilization for the continually moving sand. This can be done by planting such vegetation as heather or young pine or by treating it with crude oil. *See* EOLIAN LANDFORMS; LOESS.

Excessive settlement will occur in structures founded on muskeg terrain. Embankments can be stabilized by good drainage, the avoidance of cuts, and the removal of the organic soil and replacement by sand and gravel. Structures imposing concentrated loads are supported by piling driven through the soft layers into layers with sufficient bearing power. *See* MUSKEG; TUNDRA.

Residual soils are derived from the in-place deterioration of the underlying bedrock. There are regions (such as California, Australia, and Brazil) where disintegrated granite (locally termed "DG") may be hundreds of feet thick; although it may be competent to support moderate loads, it is unstable in open excavations and is pervious. Laterite (a red clayey soil) derived from the in-place disintegration of limestone in tropical to semitropical climates is another critical residual soil. It is unstable in open cuts on moderately steep slopes, is compressible under load, and when wet produces a slick surface that is unsatisfactory for vehicular traffic.

Clays supporting structures may consolidate slowly over a long period of time and cause structural damage. When clay containing montmorillonite is constantly dried and rewetted by climatic or drainage processes, it alternately contracts and expands. During the drying cycle, extensive networks of fissures are formed that facilitate the rapid introduction of water during a rainfall. This cyclic volume change of the clay can produce uplift forces on structures placed upon the clay or compressive and uplift forces on walls of structures placed within the clay. A thixotropic or "quick" clay has a unique lattice structure that causes the clay to become fluid when subjected to vibratory forces. Various techniques are used to improve the foundation characteristics of critical types of clay: (1) electroosmosis that uses electricity to force redistribution of water molecules and subsequent hardening of the clay around the anodes inserted in the foundation; (2) provision of adequate space beneath a foundation slab or beam so the clay can expand upward and not lift the structure; (3) belling, or increasing in size, of the diameter of the lower end of concrete piling so the pile will withstand uplift forces imposed by clay layers around the upper part of the pile; (4) treatment of the pile surface with a frictionless coating (such as Teflon or a loose wrapping of asphalt-impregnated paper) so the upward-moving clay cannot adhere to the pile; (5) sufficient drainage around the structure to prevent moisture from contacting the clay; and (6) replacement of the clay by a satisfactory foundation material. Where none of these solutions is feasible, the structure then must be relocated to a satisfactory site or designed so it can withstand uplift or compressive forces without expensive damage. *See* CLAY.

Geotechnical investigations for engineering projects may include preliminary studies, preconstruction or design investigations, consultation during construction, and the maintenance of the completed structure.

Preliminary studies are made to select the best location for a project and to aid in formulating the preliminary designs for the structures. The first step in the study is a search for pertinent published material in libraries, state and Federal agencies, private companies, and university theses. Sources of geologic information are listed in the *Directory of Geological Material in North America* by J. A. Howell and A. I. Levorsen (NAS-NRC Publ. 556, 1957). Air photos and other remote sensing techniques such as pulsed or side-looking radar or false color can be used to supplement map information (or may be the only surficial information readily available). Airborne geophysical techniques using magnetometers or gravimeters also may be useful to delineate surface and subsurface geologic conditions. *See* AERIAL PHOTOGRAPH; REMOTE SENSING; TOPOGRAPHIC SURVEYING AND MAPPING.

Field reconnaissance may include the collection of rock and soil specimens; inspection of road cuts and other excavations; inspection of the condition of nearby engineering structures such as bridges, pavements, and buildings; and location of sources of construction material. Aerial reconnaissance is essential at this stage and can be performed best in helicopters and second-best in slow-flying small planes.

Surface and subsurface investigations are required prior to design and construction. Surface studies include the preparation of a detailed map of surficial geology, hydrologic features, and well-defined landforms. For dam projects, a small-scale geologic map (for example, 1:5000) is made of the reservoir area and any adjacent areas that may be directly influenced by the project; in addition, a large-scale geologic map (for example, 1:500) is required of the various specific sites of the main structures (the dam, spillway, power plant, tunnels, and so on).

Subsurface investigations may include test pits, trenches, short tunnels (drifts or adits), and the drilling of vertical, horizontal, or oblique (angle) boreholes. The geology disclosed by subsurface investigations is "logged" on appropriate forms. *See* WELL LOGGING.

Geotechnical supervision is desirable during construction in or on earth media. The engineering geologist must give advice and keep a record of all geotechnical difficulties encountered during the construction and of all geological features disclosed by excavations. The engineering geologist also is considered an important member of the team assigned to the task of assessing the safety of existing dams as now required by Federal legislation. An important consideration for the engineering geologist is the possibility of a contractor making legal claims for damages, purportedly because of unforeseen geologic conditions (generally referred to legally as charged conditions) encountered during construction. Legal support for such claims can be diminished if the engineering geologist supplies accurate and detailed geologic information in the specifications and drawings used for bidding purposes.

Special geotechnical problems. In arctic zones, structures built on permafrosted soils may be heaved or may cause thawing and subsequent disastrous settlement. The growth of permafrost upward into earth dams seriously affects their stability and permeability characteristics. Sedimentation studies are required for the design of efficient harbors or reservoirs because soil carried by moving water will settle and block or fill these structures. In areas with known earthquake activity, aseismic (earthquakeproof) design requires knowledge of the intensity and magnitude of earthquake forces. The prevention and rehabilitation of slides (landslides) in steep natural slopes and in excavations are important considerations in many construction projects and are particularly important in planning reservoirs.

Geohydrologic problems can arise in the foundation material under a structure from pore water locked into the interstices or pores of the soil or rock, free water that is moving through openings in the earth media, or included water that is a constituent or chemically bound part of the soil or rock. When the structure load compresses the foundation material, the resulting compressive forces on the pore water can produce undesirable uplift pressures on the base of the structure. Included or pore water generally is determined by laboratory tests on cores; these are shipped from the borehole to the laboratory in relatively impervious containers that resist loss of moisture from the core.

Civilian and military structures may be designed to minimize the effects of nuclear explosions. The most effective protection is to place the facility in a hardened underground excavation. A hardened facility, including the excavation and its contents, is able to withstand the effects generated by a specified size of nuclear weapon such as displacement, acceleration, and particle velocity that occurs in the earth media and the adjacent structure. Desirable depths and configurations for hardened facilities are highly dependent upon the shock-wave characteristics of the surrounding earth media, for example, the type of rock, discontinuities in the rock system, free water, and geologic structure. *See* NUCLEAR EXPLOSION.

Projected uses of nuclear energy for the efficient construction of civil engineering include rapid excavation, increasing production of natural gas by opening fractures in the reservoir rock, expediting production of low-grade copper ore by causing extensive fracturing and possible concentration of the ore, and by the underground "cracking" of oil shale. The production of electrical energy by nuclear fission would require engineering geology inputs during the planning and design of the power plant; for example, a major question to be answered is the presence or absence of faults and an estimate of when the last movement on the fault occurred.

Another geotechnical problem occurs in the use of nuclear energy for generation of power or radioisotopes: safe disposal of the radioactive waste products. This is a problem that is under serious study. *See* NUCLEAR POWER; NUCLEAR REACTOR. [W.R.J.]

Enopla A class of the phylum Rhynchocoela which is divided into the orders Hoplonemertini (free-living) and Bdellonemertini or Bdellomorpha (symbiotic). In both orders the mouth is anterior to the brain, the nervous system is inside the body wall musculature, and the vascular system is well developed. *See* ANOPLA; BDELLONEMERTINI; HOPLONEMERTINI; RHYNCHOCOELA. [J.B.J.]

Enoplida An order of nematodes having the cuticle of the cephalic region doubled or formed into a cap or helmet. The helmet results from the infolding of the cuticle over the extreme anterior end of the esophagus. The stoma may possess armature in the form of teeth or movable jaws. Even though the stoma is surrounded by esophageal tissue, the anterior portion of the movable jaws is of cheilostomal origin. The well-developed esophagus is nearly cylindrical, and the esophagointestinal valve is prominent. The five esophageal glands are located in the posterior region of the esophagus; however, the orifices for these glands are located anterior to the nerve ring. The dorsal and two anterior subventral glands open at the base of the stoma or through the stomatal teeth. The amphid apertures may be transverse slits or elongate ovals. The cephalic sensilla are most often in two whorls: an anterior circumoral whorl of 6 and a posterior whorl of 10 (6 + 4); in some taxa this whorl is clearly separated into the ancestral state of a circlet of 6 and one of 4. A single medioventral excretory cell is usually present; the orifice is generally at the level of or anterior to the nerve ring. Medioventral supplementary organs may be present on males. The male spicules are paired and accompanied by a gubernaculum. Caudal glands and a cuticular spinneret are found in males and females.

There are two enoplid superfamilies, Enoploidea and Oxystominoidea. The Enoploidea are free-living marine nematodes comprising small to very large species in five well-defined families: Leptosomatidae, Lauratonematidae, Phanodermatidae, Thoracostomopsidae, and Enoplidae. While the feeding habits are largely unknown, it has been postulated that the group is predaceous or omnivorous. The group is almost exclusively marine in habitat, but a few species have been reported from brackish water or soils. Oxystominoidea contain species that maintain the most ancestral of characters known in the phylum. They are separated from other Enoplida by not having the lip region (=cephalic region) set off by a groove. *See* NEMATA. [A.R.M.]

Enstatite The name given to the magnesian end member ($MgSiO_3$) of the orthorhombic pyroxene solid-solution series. The mineral is usually yellowish-gray, becoming greenish with a little iron present, and transparent in thin sections.

Enstatite is an important mineral in the upper mantle of the Earth, and coexists with clinopyroxene, garnet, olivine, and plagioclase. Enstatite, clinopyroxene, and garnet rocks called eherzolites are common in certain alpine mountain terrains and as nodules in some diamond-bearing rocks called kimberlites. *See* PYROXENE. [G.W.DeV.]

Enterobacter A genus of the bacteria, formerly called *Aerobacter*. Its members are motile, ferment lactose, form gas from glucose, decarboxylate ornithine, give a positive Voges-Proskauer reaction and a variable urease reaction, and utilize citrate, but do not produce indole or hydrogen sulfide.

The two recognized species, *E. cloacae* and *E. aerogenes*, are widely distributed in nature (soil, water sources, sewage) and are also found in human and animal feces. Human infections often follow primary infection with other pathogens (*Escherichia coli*, gram-positive cocci) or instrumentation, for example, of the urinary tract. Septicemia is rare. A majority of strains are hospital-acquired and resistant to the cephalosporins and to ampicillin.

A new species, *E. agglomerans*, has been proposed for strains formerly assigned to the genus *Erwinia*, species *herbicola*, *lathyri*, *ananas*, *milletiae*, *uredovora*, and *cassavae*. They have been observed as human pathogens in a variety of conditions as primary invaders, for example, after infusion of contaminated solutions. [A.W.C.V.G.]

Enterobacteriaceae A family of bacteria consisting of straight, rod-shaped gram-negative, non-spore-forming cells, motile with peritrichous flagella or nonmotile. All members are facultative chemosynthetic heterotrophs that obtain energy by the oxidation of a variety of simple organic compounds or by the fermentation of sugars and, in some cases, organic acids and polyalcohols. Most species grow in synthetic media with a single carbon source and an inorganic nitrogen source, although some species require specific amino acids or water soluble vitamins. The family contains many species of direct significance to human affairs, including important human and plant pathogens.

A stable classification of the group has yet to be achieved, but as now constituted in *Bergey's Manual* (eighth edition), the family is divided into five fairly distinct groups, which, for purposes of reference, were given tribe designations. The separations are based on general and specific metabolic characteristics and on DNA composition. Antigenic analysis, initially undertaken for epidemiological reasons, has shown the existence of about 2000 different *Salmonella*, with respect to the combinations of somatic and flagellar antigens present, and these differences have served as the basis of various schemes of classification. Although they have not been investigated as extensively, it appears that *Escherichia* are equally diverse antigenically, and *Shigella* somewhat less so. The presence of antigens common to species of different genera of the family has been noted frequently and has been interpreted as indicating a common phylogeny for the group.

The tribe Escherichieae is composed of species whose normal habitat is primarily the intestinal tract of warm-blooded animals including humans. Included are the *Escherichia* and *Citrobacter*, which are a characteristic part of the normal intestinal flora of humans. The *Salmonella* species are pathogens of mammals and birds, and the *Shigella* species are pathogens of primates. *See* ESCHERICHIA; SALMONELLA.

Members of the Klebsielleae differ from the Escherichieae in producing neutral solvents during sugar fermentation rather than the mixed acids characteristic of the Escherichieae. Although *Enterobacter*, *Hafnia*, and some strains of *Klebsiella pneumoniae* are normal to the intestinal tract, Klebsielleae are also respiratory and genitourinary tract parasites or pathogens. Other Klebsielleae, including the *Serratia*, are widely distributed, free-living forms common in soil, water, grain, and so on.

The tribe Proteeae, genus *Proteus*, includes forms common in feces and putrefying materials which may be involved in intestinal and urinary tract infections in humans. Characteristically, the representatives are motile, ferment glucose but not lactose with gas production, and produce urease.

The tribe Yersinieae, genus *Yersinia*, results from a subdivision of the previously accepted genus *Pasteurella*. Its most prominent species, *Yersinia pestis*, is the causative agent of plague in humans and rodents.

The tribe Erwinieae, with the single genus *Erwinia*, comprises a group of plant pathogens that causes a variety of galls, wilts, and soft rots, and includes the causative agent of fire blight of pear, the first bacterial plant pathogen to be described. The production of pectinases is characteristic of the soft rot group. [S.C.R.]

Enteropneusta A class of the Hemichordata, known as the acorn worms or tongue worms. They are free-living solitary animals without an exoskeleton. Numerous gill slits, a straight gut, and marked distinction between proboscis, collar, and trunk are characteristic of this class (see illustration). Often, a free-swimming larva of the tornaria type occurs. Many of the dozen or so genera are widespread. Some species are locally abundant, but in general the group is uncommon. The distribution suggests that this is a group of great antiquity. Well-known genera include *Balanoglossus* and *Saccoglossus*.

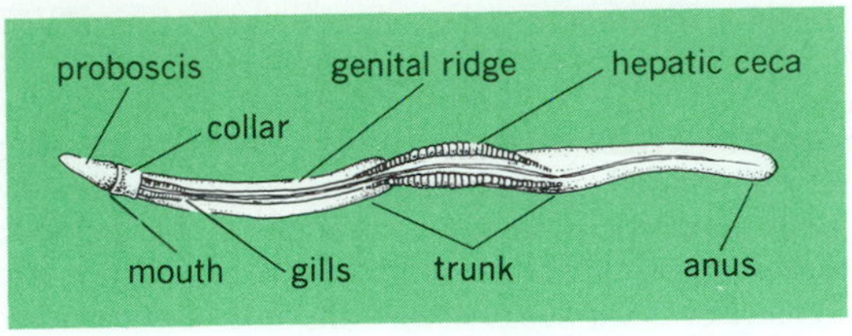

***Saccoglossus*, the tongue worm or acorn worm, dorsal view. (After T. I. Storer and R. L. Usinger, *General Zoology*, 4th ed., McGraw-Hill, 1965)**

These excessively fragile, long, wormlike animals are usually burrowers in sand or mud. They feed upon the organic matter in the substratum which is passed through the gut. Some species throw up conspicuous castings. Many are described as smelling like iodoform; one is recorded as swimming; another is luminous. *See* HEMICHORDATA. [T.H.B.]

Enterovirus One of the two subgroups of human picornaviruses. Enteroviruses include the polioviruses, the coxsackieviruses, and the echoviruses. They are small (17–28 nanometers in diameter), contain ribonucleic acid (RNA), and are resistant to ether. The enteroviruses multiply chiefly in the alimentary tract. The polioviruses, most echoviruses, and a number of the coxsackieviruses can be grown in cell cultures of monkey origin, as well as in human cells.

Enteroviruses are widespread during summer and fall in temperate climates, but may circulate throughout the year in tropical areas. The majority of enterovirus infections are benign and inapparent. However, when these viruses invade tissues other than the enteric tract, serious diseases may result, as when poliovirus invades the spinal cord or when some of the coxsackievirus types invade the heart muscle. *See* ANIMAL VIRUS; COXSACKIEVIRUS; ECHOVIRUS; PICORNAVIRIDAE; POLIOMYELITIS; RHINOVIRUS. [J.L.Me.]

Enthalpy For any system, that is, the volume of substance under discussion, enthalpy is the sum of the internal energy of the system plus the system's volume multiplied by the pressure exerted by the system on its surroundings. The sum is given the special symbol H primarily as a matter of convenience because this sum appears repeatedly in thermodynamic discussion. Previously, enthalpy was referred to as total heat or heat content, but these terms are misleading and should be avoided. Enthalpy is, from the viewpoint of mathematics, a point function, as contrasted with heat and work, which are path functions. Point functions depend only on the initial and final states of the system undergoing a change; they are independent of the paths or character of the change. For change in enthalpy with pressure or temperature *see* THERMODYNAMIC PRINCIPLES. *See also* ENTROPY; THERMODYNAMIC PROCESSES. [H.C.W.; W.A.S.]

Entodiniomorphida An order of the Spirotrichia. These are strikingly different-looking ciliates, covered with a smooth, firm pellicle. They are devoid of external ciliature except for the adoral zone of membranelles and, occasionally, one or two other tufts or zones of other specialized cilia (see illustration). Internal organization of the body is very specialized and complex. These organisms are considered to be highly evolved. Entodiniomorphids occur exclusively as endocommensals of herbivorous mammals, either in the rumen and reticulum of ruminants or in the colon of certain higher mammals. *See* CILIOPHORA; PROTOZOA; SPIROTRICHIA. [J.O.C.]

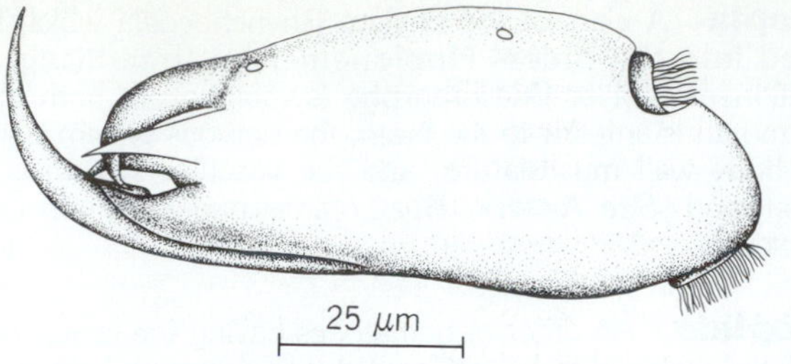

***Ophryoscolex*, an entodinomorphid.**

Entomology The broad field of the biological sciences which treats all aspects of the study of insects. Because of the large number of insect species and their economic significance, many specialized areas of entomology have evolved. Numerous insects are of benefit to humans. They are necessary for flower pollination upon which much of the fruit crop depends. Many plants are attacked by fungal, viral, and bacterial diseases that are transmitted by insects. Insect parasites have been studied as a method of controlling destructive insects. Such studies form the basis of economic entomology. The specialized field of medical entomology is concerned with insects that are vectors for numerous diseases and parasitic infestations of humans and domestic animals. *See* ENTOMOLOGY, ECONOMIC; EPIDEMIOLOGY; INSECTA; PLANT PATHOLOGY. [C.B.C.]

Entomology, economic The study of insects that have a direct influence on humanity. Though this includes beneficial as well as harmful species, most attention is devoted to the latter and how they become pests and are controlled. The emphasis on managing harmful insects reflects the immediacy and seriousness of pest problems, particularly the destruction of food and the transmission of disease. These are highly visible problems, whereas the benefits gained from useful insects are not so clearly understood, nor so well documented economically.

Central to the definition of a pest is determination of the economic threshold. Any insect population, when introduced into a favorable environment, increases numbers until reaching an environmental carrying capacity. In pest insects, there exists a density above which the insect population interferes with human health, comfort, convenience, or profits. When this economic threshold is reached, a decision must be made to utilize some control measure to prevent further increase in numbers. Often, the presence of even a single insect is sufficient to warrant control measures, for instance, when that insect is a flea harboring the plague bacillus, or a mosquito capable of transmitting malaria. Also, consumer expectations in most markets are for insect-free produce, so that the economic threshold can be very low on items that people eat. However, economic thresholds may be higher for insects that damage only the inedible portions of crop plants such as the leaves of beans, tomatoes, and apple trees. In any case, knowledge of the amount of injury which is due to different densities of insects is an important prerequisite for efficient management.

Pest management. Management of insect pests begins with prevention. Many of the United States' most noxious insects have been imported from overseas: most domestic cockroaches, the gypsy moth, Japanese beetle, corn borer, housefly, cabbageworm, and codling moth are just a few. Some North American insects have spread elsewhere too—the Colorado potato beetle to Europe and the fall webworm to Japan. To stem the flow of insect invasions, the federal government's Animal and Plant Health Inspection Service maintains inspection facilities for the examination of all incoming shipments of plant or other material that may harbor pests.

Once a pest is established, its spread can sometimes be slowed by an efficient system of local quarantines, early detection, and local eradication. A widely used eradication method is the application of synthetic chemical insecticides. Insecticides were once regarded as a panacea for pest problems, but the development of resistant strains of major insect pests, together with the rising cost of materials and application, has led to recognition that insecticides are more efficiently utilized in a program integrating them with other techniques in a framework of total crop management. *See* INSECTICIDE.

Another method is biological control, in which insects have had their numbers checked by natural enemies. Economic entomologists have effectively reduced densities of several pests by releasing parasites or predators. Natural enemies that are mobile and relatively restricted in diet do the best job of biological control. *See* INSECT CONTROL, BIOLOGICAL.

Crop rotation is a standard agronomic practice that often reduces damage due to insects. Rotation of alfalfa or soybeans with corn reduces populations of corn rootworms, wireworms, and white grubs. The physical disruption of autumnal plowing and disking destroys many insects that could overwinter in stubble or on the soil surface.

The cleanup of breeding and gathering sites is useful, especially in management of medically important insects, many of which have evolved resistance to the commoner insecticides. In general, the most successful programs of insect pest management rely on integrated control or the use of several methods in concert to control a complex of pests.

Beneficial insects. It has been estimated that the dollar value gained from a single insect, the honeybee, equals the loss from damage plus cost of control for all pests combined. Honeybees are managed for their honey and beeswax, but their most valued service is pollination of crop plants. Bees of many species are the chief pollinators, though wasps, flies, moths, butterflies, and beetles pollinate as well. *See* POLLINATION.

Silk is produced by larvae of the silkworm, an insect so thoroughly domesticated that it cannot climb its food plant, mulberry, with its degenerated legs. The silkworm apparently no longer survives in the wild. Many uses of silk have been taken over by less expensive synthetic materials.

Other insects may be equally beneficial, but their value is not so easily calculated. Foremost among these are predatory insects of several orders. These predators may prevent other insects from ever reaching an economic threshold and thus becoming pests. Innumerable insect species are scavengers, quietly but efficiently breaking down the remains of dead plants and animals. A lack of scavenging insects would, however, result in a great increase of decomposing organic material lying about. Plant-eating insects have been set to beneficial use when their diets consist mainly of unwanted weeds.

Certain rare and showy butterflies and beetles are sought after so that they have considerable economic worth. Conservation of rare and endangered insects incurs some expense as well. Habitat management to conserve rare insects is a valid and growing concern of economic entomologists.

Finally, insects have rendered invaluable service to science, and thus to humanity, as easily reared experimental animals for investigation of basic principles of genetics, biochemistry, development, and behavior. *See* INSECTA. [D.J.Hor.]

Entoprocta A phylum of sessile, aquatic, often colonial invertebrates having a looped gut with both mouth and anus situated inside a circlet of tentacles, a pseudocoelomate body cavity, and no hardened skeleton. The phylum is divided into two orders, Loxosomatida (or Solitaria), with about 100 species, and Pedicellinida (or Coloniales), with about 30. No fossils are known.

Entoprocts are mainly marine, and may be encountered as small colonies on seaweeds and bryozoans near low tidemark. The colonies consist of threadlike, creeping stolons and erect zooids arising at intervals from them (see illustration).

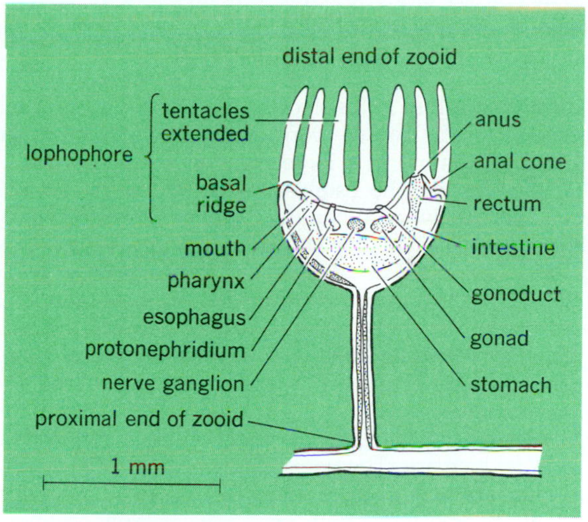

Morphological features of an entoproct zooid, in vertical cross section.

Noncolonial forms have similar zooids but lack the stolons. Each zooid is goblet-shaped, consisting of a basal stalk supporting a rounded or pyriform viscera-containing dilatation, the calyx.

The entoprocts feed on particles filtered from the surrounding water. Long lateral cilia drive water inward between the tentacles, while shorter cilia on the inner face convey particles trapped in mucus down the tentacles. These frontal tracts join an annular ciliated groove, which dips into the mouth.

Entoprocts reproduce asexually and sexually. Dispersal is by means of sexually produced larvae. Zooids may be unisexual or hermaphroditic. Gonads are paired, with the ducts uniting before opening into the coelom. Fertilization is internal. [J.S.R.]

Entropy A function first introduced in classical thermodynamics to provide a quantitative basis for the common observation that naturally occurring processes have a particular direction. Subsequently, in statistical thermodynamics, entropy was shown to be a measure of the number of microstates a system could assume. Finally, in communication theory, entropy is a measure of information. Each of these aspects will be considered in turn. Before the entropy function is introduced, it is necessary to discuss reversible processes.

Reversible processes. Any system under constant external conditions is observed to change in such a way as to approach a particularly simple final state called an equilibrium state. For example, two bodies initially at different temperatures are connected by a metal wire. Heat flows from the hot to the cold body until the temperatures of both bodies are the same. It is common experience that the reverse processes never occur if the systems are left to themselves; that is, heat is never observed to flow from the cold to the hot body. Max Planck classified all elementary processes into three categories: natural, unnatural, and reversible. Natural processes do occur, and proceed in a direction toward equilibrium. Unnatural processes move away from equilibrium and never occur. A reversible process is an idealized natural process that passes through a continuous sequence of equilibrium states.

Entropy function. The state function entropy S puts the foregoing discussion on a quantitative basis. Entropy is related to q, the heat flowing into the system from its surroundings, and to T, the absolute temperature of the system. The important properties for this discussion are:

1. $dS > q/T$ for a natural change.
   $dS = q/T$ for a reversible change.
2. The entropy of the system S is made up of the sum of all the parts of the system so that $S = S_1 + S_2 + S_3 \cdots$. *See* HEAT; TEMPERATURE; THERMODYNAMIC PRINCIPLES.

Nonconservation. In his study of the first law of thermodynamics, J. P. Joule caused work to be expended by rubbing metal blocks together in a large mass of water. By this and similar experiments, he established numerical relationships between heat and work. When the experiment was completed, the apparatus remained unchanged except for a slight increase in the water temperature. Work (W) had been converted into heat (Q) with 100% efficiency. Provided the process was carried out slowly, the temperature difference between the blocks and the water would be small, and heat transfer could be considered a reversible process. The entropy increase of the water at its temperature T is $\Delta S = Q/T = W/T$. Since everything but the water is unchanged, this equation also represents the total entropy increase. The entropy has been created from the work input, and this process could be continued indefinitely, creating more and more entropy. Unlike energy, entropy is not conserved. *See* CONSERVATION OF ENERGY; THERMODYNAMIC CYCLE; THERMODYNAMIC PROCESSES.

Degradation of energy. Energy is never destroyed. But in the Joule friction experiment and in heat transfer between bodies, as in any natural process, something is lost. In the Joule experiment, the energy expended in work now resides in the water bath. But if this energy is reused, less useful work is obtained than was originally put in. The original energy input has been degraded to a less useful form. The energy transferred from a high-temperature body to a lower-temperature body is also in a less useful form. If another system is used to restore this degraded energy to its original form, it is found that the restoring system has degraded the energy even more than the original system had. Thus, every process occurring in the world results in an overall increase in entropy and a corresponding degradation in energy. [W.F.J.]

Measure of information. The probability characteristic of entropy leads to its use in communication theory as a measure of information. The absence of information about a situation is equivalent to an uncertainty associated with the nature of the situation. This uncertainty is the entropy of the information about the particular situation. *See* INFORMATION THEORY. [F.H.R.]

Enumeration of arrangements

A branch of combinatorial theory which attempts to determine how many arrangements of a given type exist, without regard to their properties. In enumeration of arrangements it is always essential to know which arrangements are considered the same and which different. When k objects are to be selected from a set S and different orderings on the selected objects are to be counted separately, the arrangements are called permutations or relabelings of the set of k objects. But if order is to be disregarded, so that attention is focused only on the subset consisting of the k selected elements, the choice is termed a combination. If elements of S are allowed to be selected more than once, then the arrangements are called permutations with repetition or combinations with repetition. If S has n different elements, Eqs. (1)–(4) enumerate these four basic cases. The number of k-permutations is shown in Eq. (1). In particular, $P(n,n) =$

$$P(n,k) = n(n-1)(n-2)\cdots(n-k+1) = \frac{n!}{(n-k)!} \qquad (1)$$

$n! = n(n-1)\cdots(2)(1)$ is the number of ways to order S. The number of k-combinations (or k-element subsets of S) is shown in Eq. (2). When repetition is allowed, Eq. (3) gives the

$$C(n,k) = \binom{n}{k} = \frac{P(n,k)}{k!} = C(n,n-k) \qquad (2)$$

$$\overline{P}(n,k) = n^k \qquad (3)$$

number of permutations with repetition (or k-tuples) and Eq. (4) gives the number of combinations with repetition. *See* CALCULUS.

$$\overline{C}(n,k) = \binom{n+k-1}{k} \qquad (4)$$

The symbol $\binom{n}{k}$ in Eq. (2) is sometimes called a binomial coefficient, because it is the coefficient of x^k in the expansion of $(1+x)^n$, given in Eq. (5). When a finite or infinite sequence

$$C(n,x) = (1+x)^n = \sum_{k=0}^{n} \binom{n}{k} x^k \qquad (5)$$

$(a_0, a_1, \ldots)$ is exhibited in this way as the coefficients of a polynomial or series $p(x) = \sum a_i x^i$, then $p(x)$ is called a generating function for the sequence. The general method of generating functions was developed by P. S. Laplace in 1812 but had already been used extensively in the 18th century by L. Euler. For example, substituting $x = -1$ in Eq. (5) shows that there are exactly as many even subsets of a set as odd subsets (and thus the probability of an even number of heads in n flips of a coin is 1/2). *See* BINOMIAL THEOREM; SERIES.

Occasionally, only a rough estimate of the number of arrangements of a certain kind is desired (for example, to estimate how long a computer will take to list all such arrangements), and then an asymptotic enumeration formula is sought. Thus, $n!$, which is the number of (complete) permutations $P(n,n)$, obeys for large n Stirling's formula (6), where

$$n! \sim \sqrt{2\pi n}\left(\frac{n}{e}\right)^n \qquad (6)$$

$e = 2.71828\cdots$ [here $f(n) \sim g(n)$ means that as n grows without bound the ratio $f(n)/g(n)$ approaches one]. *See* CATALAN NUMBERS; COMBINATORIAL THEORY; E (MATHEMATICS); FIBONACCI NUMBERS; INTEGER PARTITION; MAGIC SQUARE; MÖBIUS INVERSION. [T.Br.]

Environment

Ecologically, the environment is the sum of all external conditions and influences affecting the life and development of organisms. Various ecological principles and concepts have been developed in regard to the environment. Two main aspects of the environment are usually considered, the abiotic and the biotic. These divisions are artificial in the sense that neither can be separated when organisms are studied. All environmental aspects and their influences on living organisms must be considered together.

Abiotic environment. The physical or abiotic environment includes all those physical and nonliving chemical aspects which exert an influence on living organisms. Among these factors are soils, water, and the atmosphere, as well as energy from various sources (gravity, pressure, sound, and so on).

Various forms of energy exert an influence on, or modify, the environment in which organisms are found. With minor exceptions, the ecologist is concerned with radiant energy received directly from the Sun, infrared radiation or heat, visible radiation, and ionizing radiation. These energies, coming from the same source, are subject to the same modifying factors, the primary one being daily and seasonal cycles which in turn are modified by latitude, altitude, slope, exposure, atmospheric conditions, and the color, texture, and cover of the ground surface.

The intensity aspect of heat energy is temperature, and, with moisture and light, it is one of the most familiar of environmental conditions. With the exception of birds and mammals, the body temperature of plant and animal life is determined primarily by the external environment. *See* HIBERNATION; HOMEOSTASIS; HYPOTHERMIA.

The source of light may be the Sun, Moon, stars, or luminescence, but usually only the first two are of sufficient intensity to influence life. In addition to the well-known uses of light for photosynthesis in plants and vision in more advanced animals, it also affects growth, development, and survival, and through many types of photoperiodicities, affects the behavior, life cycle, and distribution of many plants and animals.

Ultraviolet, nuclear, and cosmic radiations have sufficient energy to damage protoplasm on relatively short exposures. The atmosphere is opaque to all ultraviolet light shorter than 290 nanometers, so that little or no injury results at moderate altitudes. Nuclear radiation, composed of alpha and beta particles and gamma rays, comes from radioactive substances naturally occurring in the crust of the Earth, and from living and nonliving materials which have these incorporated in them. The intensity of this background radiation varies on an average from 0.1 to 0.5 roentgen unit (r) per year, although many areas give much higher readings. Cosmic rays come from space and vary appreciably with altitude. Some geneticists estimate that 10% of the mutations in geological time are attributable to these ionizing radiations. Very little is yet known about their direct effects at natural levels on organisms. *See* COSMIC BACKGROUND RADIATION; ENVIRONMENTAL RADIOACTIVITY; RADIATION BIOLOGY.

Water and air are the fundamental media in which life exists. As such, they provide the basis for the division of the world into two major environments, aquatic and terrestrial. Water not only covers 70% of the Earth's surface, but also provides for the existence of life throughout its depths. As the chief components of protoplasm, water and air have many common physicochemical characteristics. *See* HYDROSPHERE.

The atmosphere is composed of 78% nitrogen, 21% oxygen, 0.03% carbon dioxide, several other gases in very small quantities, and varying amounts of water vapor. In addition to the physical characteristics of the atmosphere as a medium, its oxygen and carbon dioxide chemically affect all forms of life through photosynthesis and respiration, that is, through their reciprocal relationship of synthesis and decomposition. *See* ATMOSPHERE; PHOTOSYNTHESIS; RESPIRATION.

The substratum is any solid on whose surface an organism rests or moves, or within which it lives, thus providing anchorage, shelter, and food. Aquatic substrata are rock, sand, and mud, while terrestrial substrata are rock, sand, and soil, soil being the most complex.

Biotic environment. The biotic environment consists of living organisms, which both interact with each other and are inseparably interrelated with their abiotic environment.

Within a population, interactions between organisms include such aspects as density, birth rate, death rate, age distribution, dispersion, and growth forms. At the interspecies population level interactions include competition, predation, parasitism, commensalism, cooperation, and mutualism. At the community levels, interactions include dominance, where one or several species control the habitat. In succession, there is an orderly process of community change. Rhythmic change in community activity is termed periodicity. *See* ECOLOGICAL INTERACTIONS.

Interactions with abiotic environment include (1) the effects of organisms on the microenvironment through temperature, water, wind, and light, (2) their modification of the substrate, as through soil building, and (3) their modification of the medium, as in aquatic habitats. *See* BIOSPHERE; ECOSYSTEM. [R.B.P.]

Environmental engineering The discipline which evaluates the effects of humans on the environment and develops controls to minimize environmental degradation.

In the 1960s the United States became acutely aware of the deterioration of its air, water, and land. The roots of the problem lie in the rapid growth of the national population and the industrial development of natural resources which has given Americans the highest standard of living in the world. Since 1964 there has been enactment of national, state, and local legislation directed toward the preservation of these resources.

It is the feeling of industry and governmental agencies that the ultimate goals should be the design of processes and systems which need minimal treatment for pollution control and the ultimate recycling of all wastes for reuse. This philosophy is both logical and necessary in a society that is rapidly depleting its nonrenewable natural resources.

Governmental policy. The federal government enacts environmental air, water, and land-use laws which in most cases require the individual states to develop programs and enforcement policies to ensure that the goals are fulfilled. Much environmental legislation was enacted in the 1970s by both the state and federal governments.

The significant federal legislation has included the Wilderness Act of 1964; the Air Quality Act of 1967, followed by the Clean Air Act of 1970 and the very significant Clean Air Act Amendments of 1977; the Water Quality Act of 1965, replaced by the Water Pollution Control Act of 1972 with amendments to the act in 1977; the National Environmental Policy Act of 1969; Executive Order No. 11574 of December 1970, restating the Refuse Act of the Rivers and Harbors Act of 1899; the Noise Control Act and the Coastal Zone Management Act, both of 1972; the Resource Conservation and Recovery Act of 1976; the Surface Mining Control and Reclamation Act and the Toxic Substances Control Act, both of 1977. The sweeping federal legislation program for environmental protection has also led to the establishment of complementary legislation, regulations, and requirements at the state level. All states now have extensive programs for environmental control and protection.

The National Environmental Policy Act is an example of the all-inclusiveness of government regulation. This act makes it mandatory that an in-depth study of environmental impacts be made in connection with any new industrial or government activity that may involve the federal government, directly or indirectly. Similarly, state environmental policy acts, patterned after this federal act, often require more detailed studies which must withstand more vigorous scrutiny.

In December 1970, by presidential order, the Environmental Protection Agency (EPA) was formed. This agency embodies under one administrator the responsibility for setting standards and a compliance timetable for air and water qualities improvement. This agency also administers the Noise Control Act, the Resource Conservation and Recovery Act, and the Toxic Substances Control Act. In addition, the agency administers grants to state and local governments for construction of wastewater treatment facilities and for air and water pollution control.

Industrial policy. Through its interdisciplinary environmental teams, industry is directing large amounts of capital and technological resources both to define and resolve environmental challenges. Each air and water problem has its own unique approach and solution. Restrictive standards necessitate high retention efficiencies for all control equipment. Off-the-shelf items, which were applicable in the past, no longer suffice. Controls must now be specifically tailored to each installation. Liquid wastes can generally be treated by chemical or physical means, or by a combination of the two, for removal of contaminants with the expectation that the majority of the liquid can be recycled. Air or gaseous contaminants can be removed by scrubbing, filtration, absorption, or adsorption and the clean gas discharged into the atmosphere. The removed contaminants, either dry or in solution, must be handled wisely, or a new water- or air-pollution problem may result.

Industries that extract natural resources from the earth, and in so doing disturb the surface, are being called on to reclaim and restore the land to a condition and contour that is equal to or better than the original state. The Surface Mining Control and Reclamation Act, which now covers only the mining of coal, requires the states to have an EPA-approved program for controlling surface mining operations and the reclaiming of abandoned mined lands. Most states already have reclamation laws which cover all types of mining activities; most require an approved restoration plan and bonding to assure that restoration is accomplished. *See* AIR POLLUTION; ENVIRONMENTAL PROTECTION; SURFACE MINING; WATER POLLUTION. [L.N.B.; J.C.Sp.; W.H.U.]

Environmental protection The system of procedures which limit the impairment of the quality of water humans use, of the air they breathe, and of the land that sustains them. It includes the means to control the physical energies of ionizing radiations, nonionizing radiations, sound, air pressure changes, and heat and cold. Human activities produce wastes that are vapors or gases, solids, liquids, or energy states. Humans seek to disperse these to the open environment of water, air, or land. The receptors are all forms of life on Earth, with people the primary concern.

Objectives. Environmental protection has three objectives. The first is to protect people from physiological damage from pathogenic organisms, from toxic chemicals, and from excesses of physical energies. The second is to spare humans annoyance, irritation, and discomfort from offensive conditions in water, in air, and on the land. The third objective is to safeguard the balances in the Earth's ecosystems and to conserve natural resources. Many people strongly advocate that this should be the primary goal of environmental protection. Fortunately, the three objectives are not incompatible, although conflicts arise. The drainage of a swamp which is a breeding place for anopheline mosquito vectors of malaria obviously changes the ecosystem that has existed there. Thus there are differences of opinions on which environmental actions should be given priority when the three objectives are not compatible.

Pollution. When humans' waste-loads on the water, air, and land overwhelm the natural process of assimilation of such wastes, pollution occurs. The condition of pollution may jeopardize one or more of the three objectives of environmental protection. Large urban-industrial areas with the massive use of individual internal combustion engines for transportation have both chronic and acute conditions of air pollution. The airborne waste loads exceed the capacity of horizontal and vertical air movement to disperse the materials. Additionally, conditions of inversion, stagnation, and ultraviolet radiation produce reactants from the primary pollutants. On land and in water, not all wastes are usable as a food for the natural biota; these are labeled nonbiodegradable. The chlorinated hydrocarbon pesticides are high on the list of persistent contaminants which change slowly in the open environment.

Control of pollutants. To keep the balance, the alternatives are: eliminate the source; eliminate the waste; treat the waste to reduce the deleterious load on the open environment; or augment the environmental capacity to assimilate the waste. All of these are applied in one way or another to manage liquid, solid, and airborne wastes.

Abatement action is taken to prevent injury to humans and animal and plant life, to protect property and resource values, and to limit conditions that are offensive to people apart from physiological damage. The mainstay of wastewater treatment is biooxidation. Trickling filters, activated sludge, and stabilization ponds depend on the same natural process as a well-aerated stream. That process is the biological feeding, primarily by bacteria, on the carbohydrates, proteins, and fats in the wastewater in the presence of ample oxygen. The settled solids, or sludge, are separated and made the food of anaerobic organisms in sludge digestors. Sewage sludge is an example of removing a waste from a mainstream and then having to deal with it further. The same holds for removing particulates from an airstream or from flue gas. The captured material requires further handling. There are always possibilities of recovery of the isolated material for economically useful purposes. *See* WATER TREATMENT.

Environmental Protection Agency. The initiatives of environmental protection in the United States have passed to the federal government. The National Environmental Protection Act (NEPA) consolidated all federal activities on air and water pollution, solid wastes, pesticides, noise, and environmental radiation in a new organization, the Environmental Protection Agency (EPA). Many states followed suit by organizing all or most of the environmental protection work in an independent or autonomous unit. States follow the EPA leads as surrogates to keep out direct federal intervention and to qualify for various forms of monetary subsidies. When the subsidies go to local projects, the state agency is the control gate. An extensive and sometimes intensive bureaucratic process has evolved to move the authorities and mandates of Congress to EPA, sometimes directly and sometimes through the states to the point where the problem can be defined and solved.

EPA regulations, standards, and requirements are nationwide. Much has been accomplished. Environmental protection in governmental agencies is firmly institutionalized. It has greater strength and support than ever. Pollutant sources are being controlled. Many streams and air sheds are showing improvement. Ocean dumping has been greatly reduced. Solid wastes are being managed much better than in the past. Air loading of particulates, hydrocarbons, and carbon monoxide on a national scale has been reduced. State agencies for environmental protection have been strengthened.

Low-level intakes through a lifetime. The matter of low-level intakes of known or suspected toxicants through a life of 65 to 75 years cannot be resolved by present toxicological and epidemiological information or methods. Decisions on the use and control of such substances cannot be made solely on existing scientific evidence. These issues are politically and socially sensitive and highly emotional when laboratory animal tests indicate the possibility of the substances being carcinogenic, mutagenic, or teratogenic. *See* ENVIRONMENTAL ENGINEERING.

[E.T.C.]

Environmental radioactivity Radioactivity that originates from natural and anthropogenic sources, including radioactive materials in food, housing, and air; radioactive materials used in medicine; nuclear weapon tests in the open atmosphere; and radioactive materials used in industry and power generation.

Natural radioactivity, which is by far the largest component to which humans are exposed, is of both terrestrial and extraterrestrial (cosmic) origin. About 340 nuclides are known in nature, of which 70 are radioactive and are found mainly among the heavy elements. Three nuclides which are responsible for most of the terrestrial component are potassium-40, uranium-238, and thorium-232.

The average person in the United States receives 80–180 mrem/year (0.8–1.8 millisieverts/year) from natural sources of ionizing radiation. Most of this dose originates from radioactive materials in the Earth's crust. The external dose due to cosmic rays is an average of about 28 mrem/year (0.28 mSv/year), a value that increases with altitude due to reduced shielding of cosmic radiation by the atmosphere.

A number of radionuclides are produced in nature by the interaction of cosmic rays with atmospheric gases, and these add slightly to the exposure received from natural sources. Carbon-14 is the most important of these nuclides which, like potassium-40, is present in the body to the extent of about 0. 1 microcurie (3.7×10^3 becquerel), but which delivers a dose of only about 1 mrem/year (10 microsieverts/year) because of the unusual softness, that is, low energy, of its beta emission.

There are wide deviations from the average doses. Thus, at one extreme, miners working underground in the presence of radioactive ore can be exposed to such high levels of atmospheric radon that they develop lung cancer. There are also geographical areas where the levels of natural radioactivity are unusually high. Six types of anomalies that can be important from the point of view of population exposure are: monazite sands and other placers, alkaline intrusives and granites of the Conway type in New Hampshire, bauxites and intensely weathered soils, uraniferous phosphate rock (and soils), groundwaters enriched in radium and radon, and black shales and related organic accumulations. The natural radioactive environment can also be altered by human activities, such as building construction, combustion of fossil fuels, aircraft travel, medical procedures, and nuclear weapons testing.

Although various national and international regulatory organizations have proposed guidelines that limit the per capita dose received by individuals in the general population to 170 mrem/year (1.7 mSv/year), it has become evident that nuclear power plants can be routinely operated so that the general population will not be exposed to more than 1% of this limit. *See* NUCLEAR REACTOR; RADIOACTIVE WASTE MANAGEMENT; RADIOACTIVITY. [M.E.]

Environmental test A laboratory test conducted to determine the functional performance of a component or system under conditions that simulate the real environment in which the component or system is expected to operate. The performance of all components and systems is influenced to varying degrees by the environment. Moral, economic, and design considerations often preclude testing for environmental effects by actual operation under real conditions.

Temperature testing in the laboratory is conducted in a sealed chamber which can be cooled or heated. The test chamber must be capable not only of maintaining a desired temperature for extended periods of time, but also of varying the temperature of the device under test. The test chamber should be designed so that the component can actually operate during temperature testing.

Many components and systems must perform satisfactorily during and after exposure to wide ranges of humidity, when subject to fresh- and salt-water spray, or complete immersion, and when blasted by windborne sand. Humidity testing is usually done in conjunction with temperature testing. In the case of spray, immersion, and sand testing, specific test conditions must be determined.

Components may be expected to operate at low pressures corresponding to high altitudes or outer space, or at high pressures characteristic of great depths beneath the surface of the sea or perhaps in high-pressure transients. When such requirements are posed, a satisfactory environmental test procedure must provide laboratory facilities for duplicating, insofar as is possible, the actual pressure levels and pressure gradients to which the components will be subjected.

Many components and systems, as for example, the control components in a missile rocket-engine orienting system, are subject in actual operation to high accelerations, vibrations, and shocks. To provide some assurance of the satisfactory performance of such a component in actual flight, it is essential to duplicate in the laboratory as many as possible of these several environmental factors and to explore the performance of the component while it is subjected to one or more of these disturbing influences. Acceleration testing is conducted on a centrifuge or a sled. Acceleration testing in pure translation is accomplished by means of rocket-accelerated sleds on which the components to be tested are mounted. Shock testing consists of exposing the component to high rates of change of deceleration or acceleration, by either dropping the component onto a resilient or absorbent surface such as rubber, cork, or sand, or subjecting it to a pressure pulse from a compressed gas.

Vibration tests measure the ability of a component to perform satisfactorily while subjected to vibration by mounting it on a structure whose frequency and amplitude of vibration can be controlled. *See* SYSTEMS ENGINEERING. [R.W.M.]

Environmental toxicology A broad field of study encompassing the production, fate, and effects of natural and synthetic pollutants in the environment. The breadth of this field depends on the definition of environment. It can be defined as narrowly as the home and workplace or as broadly as the entire Earth and its biosphere. Environmental toxicology is truly an interdisciplinary science. The effects of a pollutant on the environment depend on the amount released (the dose) and its chemical and physical properties. Pollutants can be grouped according to their origin and effects.

Pollution from nutrients is generally a problem of aquatic systems. Carbon, nitrogen, and phosphorus are essential nutrients and, when present in excess, can result in an overstimulation of microbial and plant growth. Nutrients enter the environment in runoff from fertilized agricultural areas, in effluents from municipal and industrial wastewater treatment facilities, and in dead and decaying plant material. *See* EUTROPHICATION.

Pathogenic bacteria and protozoa can be a major source of pollution in areas that receive untreated sewage, items from ocean dumping, and improperly discarded hospital waste. Toxic metabolites of fungal origin (mycotoxins) are also potential pollutants. *See* AFLATOXIN.

Forest fires, volcanic eruptions, and dust storms can be major sources of suspended materials. These materials can also originate in runoff from agricultural areas, construction and mining sites, and roads and other paved areas. Truck and automobile exhaust and industrial discharge to the atmosphere are also sources of suspended solids.

Metabolic processes and natural combustion and thermal activity (such as forest fires and volcanoes) can release large amounts of gaseous by-products to the atmosphere. However, natural inputs are minor compared to atmospheric pollutants due to human activity. Although most anthropogenic air pollution is produced by the various forms of transportation, emissions from stationary sources of fuel combustion (for example, factories and power plants) are responsible for the greatest amount of hazardous materials released. *See* ACID RAIN; ATMOSPHERIC OZONE; GREENHOUSE EFFECT.

Metals, when present at concentrations above the level of homeostatic regulation, can be toxic. In addition, there are metals that are chemically similar to, but higher in molecular weight than, the essential metals (heavy metals).

Organic solvents are used widely and in large amounts in industries, laboratories, and homes. They are released to the atmosphere as vapor and can pose a significant inhalation hazard. *See* WATER POLLUTION.

The pesticides represent an important group of materials that can enter the environment as pollutants. They are highly toxic, and many nontarget organisms can suffer harmful effects if misuse or unintended release occurs. *See* PESTICIDE.

Coal and pretroleum-derived materials and by-products are major environmental pollutants. The world's economy is highly dependent on fossil fuels in energy production, industry, and transportation. Widespread use has led to enormous releases to the environment of distillate fuels, crude oils, runoff from coal piles, exhaust from internal combustion engines, emissions from coal-fired power plants, industrial emissions, and emissions from municipal incinerators. *See* MUTAGENS AND CARCINOGENS.

Polychlorinated biphenyls (PCBs) are produced by the chlorination of biphenyl, giving rise to mixtures of up to 210 possible products. They have been used worldwide in electrical equipment, vacuum pumps, hydraulic fluids, heat-transfer systems, lubricants, and inks. The related polybrominated biphenyls (PBBs) have been used as fire retardants. Chlorinated dibenzo-*p*-dioxins and dibenzofurans have been identified as potential contaminants in the herbicide 2,4,5-T. They can be formed during the incineration of municipal wastes, polychlorinated biphenyls, or plant materials treated with chlorophenols. *See* ENVIRONMENTAL ENGINEERING; TOXICOLOGY.

[J.T.O.]

Enzyme A catalytic protein produced by living cells. The chemical reactions involved in the digestion of foods, the biosynthesis of macromolecules, the controlled release and utilization of chemical energy, and other processes characteristic of life are all catalyzed by enzymes. In the absence of enzymes, these reactions would not take place at a significant rate. Several hundred different reactions can proceed simultaneously within a living cell, and the cell contains a comparable number of individual enzymes, each of which controls the rate of one or more of these reactions. Some representative enzymes, their sources, and reaction specificities are shown in the table.

Chemical nature. All enzymes are proteins. Their molecular weights range from about 10,000 to more than 1,000,000. An enzyme molecule may contain one or more polypetide chains. The sequence of amino acids within the polypeptide chains is characteristic for each enzyme and is believed to determine the unique three-dimensional conformation in which the chains are folded. This conformation, which is necessary for the activity of the enzyme, is stabilized by interactions of amino acids in different parts of the peptide chains with each other and with the surrounding medium. These interactions are relatively weak and may be disrupted readily by high temperatures, acid or alkaline conditions, or changes in the polarity of the medium. Such changes lead to an unfolding of the peptide chains (denaturation) and a concomitant loss of enzymatic activity, solubility, and other properties characteristic of the native enzyme. Enzyme denaturation is sometimes reversible. *See* PROTEIN.

Many enzymes contain an additional, nonprotein component, termed a coenzyme or prosthetic group. This may be an organic molecule, often a vitamin derivative, or a metal ion. The coenzyme, in most instances, participates directly in the catalytic reaction. Some enzymes have coenzymes that are tightly bound to the protein and difficult to remove, while others have coenzymes that dissociate readily. When the protein moiety (the apoenzyme) and the coenzyme are separated from each other, neither possesses the catalytic properties of the original conjugated protein (the holoenzyme). *See* COENZYME.

Classification and nomenclature. Enzymes are usually classified and named according to the reaction they catalyze.

Oxidoreductases are enzymes which catalyze reactions involving electron transfer, and play an important role in cellular respiration and energy production. Some of them participate in the process of oxidative phosphorylation, whereby the energy released by the oxidation of carbohydrates and fats is utilized for the synthesis of adenosinetriphosphate (ATP) and thus made directly available for energy-requiring reactions. *See* ENERGY METABOLISM.

Transferases catalyze the transfer of a particular chemical group from one substance to another. Thus, transaminases transfer amino groups, transmethylases transfer methyl groups, and so on. An important subclass of this group are the kinases, which catalyze the phosphorylation of their substrates by transferring a phosphate group, usually from ATP, thereby activating an otherwise metabolically inert compound for further transformations.

Hydrolases catalyze the hydrolysis of proteins (proteinases and peptidases), of nucleic acids (nucleases), of starch (amylases), of fats (lipases), of phosphate esters (phosphatases), and of other substances. Many hydrolases are secreted by the stomach, pancreas, and intestines and are responsible for the digestion of foods. Others participate in more specialized cellular functions. For example, cholinesterase, which catalyzes the hydrolysis of acetylcholine, plays an important role in the transmission of nervous impulses.

Lyases catalyze the nonhydrolytic cleavage of their substrate with the formation of a double bond. Examples are decarboxylases, which remove carboxyl groups as carbon dioxide, and dehydrases, which remove a molecule of water. The reverse reactions are catalyzed by the same enzymes, but are difficult to demonstrate in some cases because of an unfavorable equilibrium.

Isomerases are enzymes which catalyze the interconversion of isomeric compounds. For example, triose phosphate isomerase catalyzes the reaction shown below.

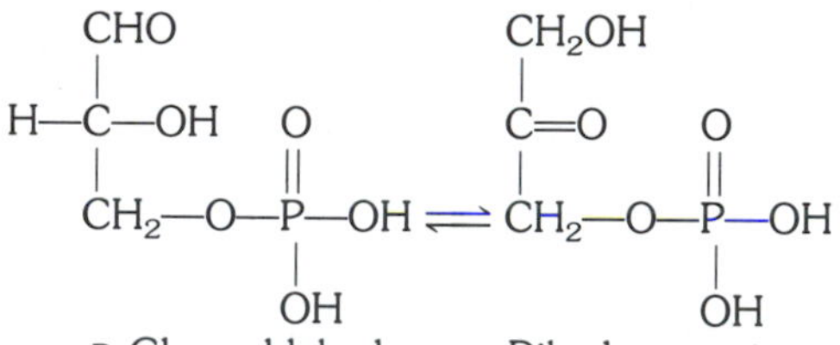

D-Glyceraldehyde-3-phosphate ⇌ Dihydroxyacetone phosphate

Ligases or synthetases are enzymes that catalyze endergonic syntheses coupled with the exergonic hydrolysis of ATP. They

Some representative enzymes, their sources, and reaction specificities

Enzyme	Some sources	Reaction catalyzed
Pepsin	Gastric juice	Hydrolysis of proteins to peptides and amino acids
Urease	Jack bean, bacteria	Hydrolysis of urea to ammonia and carbon dioxide
Amylase	Saliva, pancreatic juice	Hydrolysis of starch to maltose
Phosphorylase	Muscle, liver, plants	Reversible phosphorolysis of starch or glycogen to glucose-1-phosphate
Transaminases	Many animal and plant tissues	Transfer of an amino group from an amino acid to a keto acid
Phosphohexose isomerase	Muscle, yeast	Interconversion of glucose-6-phosphate and fructose-6-phosphate
Pyruvic carboxylase	Yeast, bacteria, plants	Decarboxylation of pyruvate to acetaldehyde and carbon dioxide
Catalase	Erythrocytes, liver	Decomposition of hydrogen peroxide to oxygen and water
Alcohol dehydrogenase	Liver	Oxidation of ethanol to acetaldehyde
Xanthine oxidase	Milk, liver	Oxidation of xanthine and hypoxanthine to uric acid

allow the chemical energy stored in ATP to be utilized for driving reactions uphill.

Active site. Since enzymes are large molecules and their substrates are often of low molecular weight, it appears probable that only a small portion of the enzyme protein, the active site, comes into contact with the substrate and is directly involved in catalyzing the reaction. The active site consists of a few amino acid residues. The active site also includes any coenzyme which may be required for activity. The function of the remainder of the protein molecule may be primarily to hold the components of the active site in the proper relative position and orientation. Part of the active site is involved in binding the substrate, while another part is responsible for the making or breaking of chemical bonds.

Specificity. The high degree of specificity exhibited by enzymes is one of their most characteristic properties. The majority of enzymes catalyze only one type of reaction and act on only one compound or on a group of closely related compounds. There must exist between an enzyme and its substrate a close fit, or complementarity. This relationship has been compared to that between a lock and key. In many cases, a small structural change, even in a part of the molecule remote from that altered by the enzymatic reaction, abolishes the ability of a compound to serve as a substrate. On the other hand, some enzymes exhibit a less restricted specificity and act on a number of different compounds that possess a particular chemical group. This is termed group specificity.

A remarkable property of many enzymes is their high degree of stereospecificity, that is, their ability to discriminate between asymmetric molecules of the right-handed and left-handed configurations. An example of a stereospecific, as well as a group-specific, enzyme is L-amino acid oxidase. *See* ALLOSTERIC ENZYME; ENZYME INHIBITION; ISOZYME. [D.W.]

Enzyme inhibition The prevention of an enzymic process as a result of the interaction of some substance with an enzyme so as to decrease the rate of the enzymic reaction. The substance causing such an effect is termed an inhibitor. Enzyme inhibitors are important as chemotherapeutic agents, as regulators in normal control of enzymic processes in living organisms, and as useful agents in the study of biochemistry. *See* ANTIBIOTIC; ANTIMETABOLITE; CHEMOTHERAPY; ENZYME.

Inhibitors have been classified as competitive, noncompetitive, and uncompetitive. The effect of a competitive inhibitor is to bind only free enzyme. This can be reversed by sufficiently increased substrate concentrations, so that essentially all of the enzyme is bound into an enzyme-substrate complex. Since both noncompetitive and uncompetitive inhibitors interact with the enzyme-substrate complex, their effects are not nullified by increased concentrations of substrate. An uncompetitive inhibitor exerts less effect (as percent of control) at low than at high substrate concentrations, since less of the enzyme is in the form of the enzyme-substrate complex, with which it interacts. A noncompetitive inhibitor, which reacts with both free enzyme and the enzyme-substrate complex, exerts comparable effects at all substrate concentrations. [W.Sh.]

Eocanthocephala An order of the Acanthocephala characterized by the presence of a small number of giant subcuticular nuclei which are similar to the embryonic nuclei. Body spines may or may not be present and the chief lacunar vessels are dorsal and lateral. Proboscis hooks are few in number and arranged in circles. Eggs are ellipsoidal and thin shelled. These worms are parasitic in cold-blooded vertebrates (turtles, fish). The cystacanth occurs in crustaceans. *See* ACANTHOCEPHALA. [D.V.Mo.]

Eocene The next to the oldest of the five major worldwide divisions (epochs) of the Tertiary Period (Cenozoic Era); the epoch of geologic time extending from the end of the Paleocene Epoch to the beginning of the Oligocene Epoch; the middle epoch of the older Tertiary (Paleogene or Nummulitic). *See* CENOZOIC; TERTIARY.

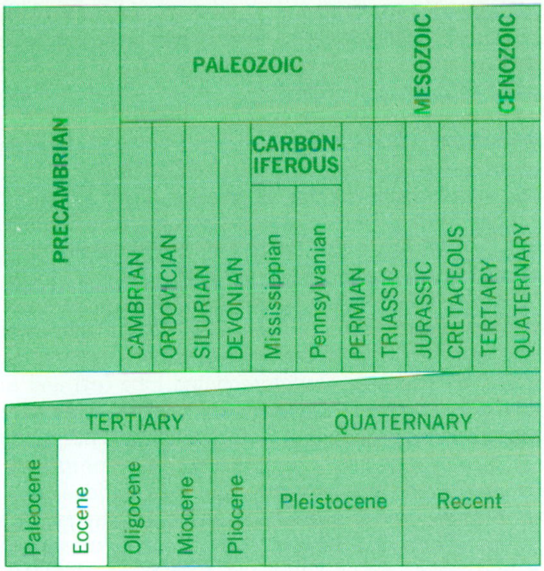

Eocene strata include all the common sedimentary types, varying from marine through estuarine to terrestrial in origin. Igneous activity was still not as widespread as in the later Tertiary, though it was notable in some areas (Oregon, Washington, British Columbia).

Eocene marine waters expanded far beyond those of the Paleocene. Generally, the early Eocene seas transgressed over land areas of low relief. Mid-Eocene diastrophic disturbances modified somewhat the previous low-relief surfaces; breaks in the depositional record do not indicate intensive folding, but in many areas diastrophism is reflected by a coarsening in texture of the sedimentary deposits. A widespread threefold upper Eocene sequence produced strata of shallow-water origin below and above, and of deeper water between.

Eocene life was essentially an amplification of that of the Paleocene. Birds diversified and the modern families appeared. True bony fish and marine siphonate gastropods flourished. The pelecypod *Venericardia planicosta* characterized the world's last tropicopolitan faunas of bottom-living shellfish. Echinoids thrived. Nummulites multiplied and diversified, and the worldwide discocycline foraminifers made their last stand. Microscopic floating life (siliceous diatoms, calcareous coccoliths, and foraminifers) flowered in the widespread open oceans.

On land nipa palms grew where London is today, and the flora there was much as in Japan and Australia. On the North American Pacific Coast palms reached Puget Sound and southern Alaska. On the shrunken isolated Eocene land masses, large flightless birds developed locally. Bats, flying lemurs, creodont carnivores, perissodactyls, artiodactyls, notoungulates (mostly South American), and edentates reflected the diversification of primitive placental mammals; uintatheres, rhinoceroslike titanotheres, and other huge grotesque browsers eventually appeared alongside the last multituberculate mammals, and Eohippus (*Hyracotherium*), the multitoed, little "dawn horse," was already present. Toward the close of the Eocene, the aquatic ancestors of the proboscidians developed in Egypt, and mammals began their trek into the free-swimming life of the ocean: Zeuglodonts (the first whales) and the first sea cows appeared.

Australia, already separated from Paleocene southeastern Asia, was further isolated by the expansion of the warm-water Eocene seas over much of the East Indies, laying the foundations for the most sharply drawn of all modern terrestrial life zones, Wallace's Line. Eocene marine expansions in Middle America cut off South America from North, isolating its fauna until toward the end of the Tertiary. *See* PALEOBOTANY; PALEONTOLOGY. [R.M.Kl.]

Eocrinoidea A class of extinct Pelmatozoa, in the subphylum Crinozoa, that had biserial brachioles like those of cystoids, combined with a theca like that of crinoids. Their affinities have been sought among both cystoids and crinoids. Eocrinoidea appeared in the Lower Cambrian until the Middle Ordovician. *See* CRINOIDEA; CRINOZOA; ECHINODERMATA; PELMA-TOZOA. [H.B.F.]

Eognathostomata A superorder of the subclass Euechinoidea comprising primitive irregular sea urchins (Echinoidea) that have simple, undifferentiated tube feet and tuberculation and that retain a high degree of radial symmetry. They have a fully formed functioning lantern when juvenile, and they retain it throughout life. Eognathostomata comprises two orders, Pygasteroida and Holectypoida. Pygasteridae are probably paraphyletic, being ancestral to all other irregular enchinoid groups. They first appear in the Lower Jurassic. Pygasteroids and holectypoids flourished during the Mesozoic but became extinct in the Late Cretaceous, during the Campanian and Maastrichtian stages, respectively. The Echinoneioids first appeared in the Late Jurassic and, although never common, survive to the present day represented by two genera. *See* ECHINODERMATA; EUECHINOIDEA; SEA URCHIN. [A.Sm.]

Eolian landforms Topographic features generated by the wind. The most commonly seen eolian landforms are sand dunes created by transportation and accumulation of windblown sand. Blankets of wind-deposited loess, consisting of fine-grained silt, are less obvious than dunes, but cover extensive areas in some part of the world.

Where abundant loose sand is available for the wind to carry, sand dunes develop. As soon as enough sand accumulates in one place, it interferes with the movement of air and a wind shadow is produced which contributes to the shaping of the pile of sand. Dunes advance downwind by erosion of sand on the windward side and redeposition on the slip face. Dunes may have a variety of shapes, depending on wind conditions, vegetation, and sand supply. The fine silt and clay winnowed out from coarser sand is often blown longer distances before coming to rest as a blanket of loess mantling the preexisting topography. Thick deposits of loess are most often found in regions downwind from glacial outwash plains or alluvial valleys. *See* DUNE; LOESS; SAND. [D.J.E.]

Eosuchia The oldest, most primitive, and only extinct order of lepidosaurian reptiles; they are clearly ancestral of the other two orders (Rhynchocephalia and Squamata). They have a typical diapsid skull, with at most a partial reduction of the lower arcade. Tooth implantation is subthecodont.

Of the four suborders, the oldest is the Younginiformes, including forms such as *Youngina* (see illustration). These, ranging from the Middle Permian to the Lower Triassic and found mainly in South Africa, were small unspecialized lizardlike reptiles. The Prolacertiformes, also terrestrial, range throughout the Triassic. They include forms in which a reduction of the lower temporal arcade and other features indicate a transition toward the Squamata (lizards). The Thalattosauria include a few marine reptiles from the Middle Triassic. The Choristodera include only one genus, *Champsosaurus*, a much later reptile (Upper Cretaceous to Lower Eocene), which had amphibious habits and seems to have been an eosuchian offshoot. *See* LEPIDOSAURIA; REPTILIA; RHYNCHOCEPHALIA; SQUAMATA. [A.J.C.]

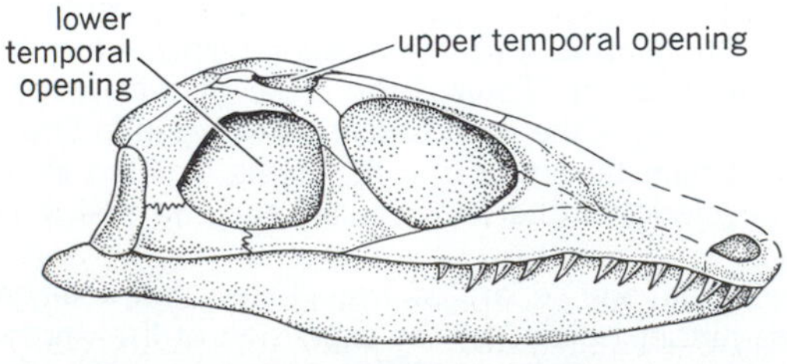

Lateral view of *Youngina* skull, Upper Permian to Lower Triassic. (*After A. S. Romer, Vertebrate Paleontology, 3d ed., University of Chicago Press, 1966*)

Ephedrales An order of the class Ginkgoopsida having about 35 species in the genus *Ephedra*. These plants are mostly of arid regions. They include freely branched, low shrubs with reduced, scalelike leaves and green photosynthetic twigs. The species are dioecious (seldom gynodioecious). The genus is known from Asia to Europe, in northern Africa, in the United States and Mexico, and from Bolivia to Patagonia. It is not known in the fossil record. *See* GINKGOOPSIDA; PINOPHYTA; PLANT KINGDOM. [T.A.Z.]

Ephemeris A table of data, especially astronomical data, that depend on the time, usually arranged with values of the time in the left-hand column. A lunar ephemeris, for example, may give the right ascension and declination of the Moon for every hour of a particular year. *The American Ephemeris and Nautical Almanac* is an annual volume published by the Nautical Almanac Office, U.S. Naval Observatory, with ephemerides of the Sun, Moon, planets, and satellites, and other astronomical data. *See* NAUTICAL ALMANAC. [G.M.C.]

Ephemeroptera An order of insects commonly known as mayflies. They are aquatic and live in clean, fresh waters during their immature, or nymphal, lives. Nymphs are adapted to aquatic environments ranging from ponds to mountain streams. Most mayfly nymphs are vegetarians, and most species are present throughout the year as nymphs. As immature insects, they constitute a year-round basic food supply for carnivores of their communities, especially for fishes, and trout in particular. Nymphs and adults are models for artificial bait flies (see illustration).

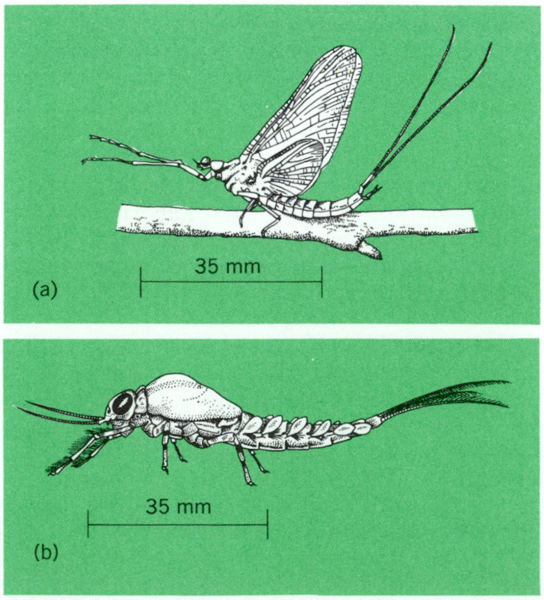

Ephemeroptera. (*a*) Mayfly, adult male, *Hexagenia* sp., in usual resting position. (*b*) Mayfly nymph, *Isonychia* sp. (*After A. H. Morgan, Field Book of Ponds and Streams, copyright 1930 by G. Putnam's Sons*)

Mayfly nymphs are distinguished from all other aquatic insects by the paired tracheal gills on the back of each of the first seven abdominal segments. The body ends in two or three finely segmented tailpieces. The mouthparts are highly interesting examples of adaptations for scraping, cutting, and crushing the various plant cells, which are the chief food.

As subadults, mayflies have the general adult form and useless mouthparts. They are clothed by a thin, furry, grayish skin which accounts for the common name, duns. The adult has transparent wings, shining body, and useless mouthparts; the males generally have extremely large eyes, which may be divided, and long front legs. This stage is brief, lasting from a few hours to several days. In a few mayflies, the legs of the female, and those of the male, except for the front ones, are nonfunctional and the brief adult life is spent entirely in flight. The quivering of the tailpieces during the mating flight gave them the name spinners. Thousands join in the twilight mating swarms, rhythmic dancing flights unequaled by other insects. *See* INSECTA. [A.H.M./L.Ber.]

Epicaridea A suborder of the Isopoda which are parasitic on various crustaceans, mainly marine forms. The females are sometimes modified so strongly as to leave almost no indication of their isopod nature. The female pierces the host's skin with the aid of styliform mandibles to suck blood. The dwarf male attaches to the female, from which it takes nourishment, and retains its isopod structure.

The suborder is divided into two tribes, the Cryptoniscina and Bopyrina. The Cryptoniscina live on Entomostraca and are protandrous hermaphrodites. The cryptoniscium larva attaches to a suitable female, develops testes without any change in its external appearance, and acts as a male. After the death of the host, it metamorphoses into a female, ultimately undergoing anatomical degradation that leaves it a mere bulky sac distended with developing embryos.

The Bopyrina are dioecious, but it has been shown experimentally in some forms that the cryptoniscium is ambipotent, and a presumptive young male removed from the female develops into either a female or intersex. The tribe includes three families: Bopyridae, Entoniscidae, and Dajidae. *See* ISOPODA; PARASITIC CASTRATION. [S.M.S.]

Epicycloid A curve traced by a point *P* on a circle with radius *r* and center *O'* that rolls on the convex side of a fixed circle with radius *R* and center *O* (see illustration). The term is also occasionally applied to the curve generated by a point on the prolongation of the radius of a circle as the circle rolls on a straight line. It is a cardioid when the rolling circle has the same radius as the fixed circle. *See* CARDIOID. [L.M.Bl.]

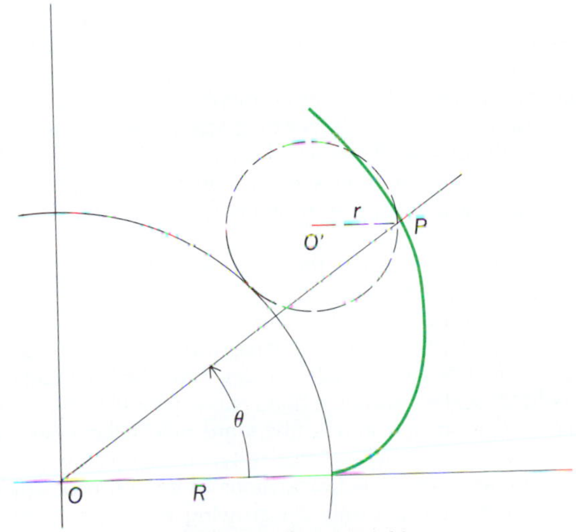

Diagram of an epicycloid.

Epidemic Any sudden increase in the incidence rate of a disease. The rate must be in excess of the normal, but the actual value required to justify the name epidemic is not fixed. The usual practice is to compare the current incidence of the disease with the level in the same population, at the same time of the year, for several recent nonepidemic years. The judgment which is then made is influenced by various administrative considerations. A single case of a dreaded disease, such as smallpox, arising in an area from which it has been long absent, may be regarded as the potential beginning of an epidemic; whereas in the case of diseases calling for less community action, such as mumps or colds, there may be a considerable increase in the incidence rate without any reference being made to an epidemic.

A worldwide epidemic is given the name of pandemic. An example is the Black Death of the Middle Ages, a pandemic of plague, which in London alone killed 70,000 persons in a population of 500,000. A more recent example is the pandemic of influenza which occurred in 1918. In the United States there were 20,000,000 cases and 500,000 deaths. *See* INFLUENZA; PLAGUE.

At the other end of the scale from the pandemic is the endemic state, which may be defined as the occurrence of cases of disease at a rate less than the epidemic level. The sporadic cases of the disease that occur in the endemic period provide a reservoir of infection. If conditions become suitable for more rapid transmission of the disease, an epidemic may break out. *See* EPIDEMIOLOGY. [C.W.]

Epidemic neuromyasthenia A disease sometimes confused with mild poliomyelitis, as well as with other diseases in which there appears to be involvement of the nervous system. Various names (Iceland disease, Akureyri disease, benign myalgic encephalomyelitis, epidemic vegetative neuritis, acute infective encephalomyelitis) have been applied to the illness in different outbreaks reported first in Iceland, then in Europe and the United States, during 1950–1960.

The main features of the illness are fatigue, headache, intense muscle pain, slight and transient paresis, emotional and mental disturbances, and objective evidence of diffuse involvement of the central nervous system. Sensory disturbance (paresthesia) may occur. The disease often takes a course which is unaccountably prolonged and debilitating, sometimes with recrudescences of the acute symptoms over a period of months following onset. No etiologic agent has been isolated, although viruses are believed to play a role. [J.L.Me.]

Epidemic typhus fever An acute, febrile human disease caused by *Rickettsia prowazeki*, and transmitted by the human louse, *Pediculus humanus*. It is the most universally known and longest recognized of the rickettsioses because of its past ravages during wars and famine. *See* MURINE TYPHUS FEVER; RICKETTSIALES; RICKETTSIOSES.

In humans, some 10–14 days after exposure, onset is usually abrupt with headache, malaise, and generalized aches and pains. Fluctuating but sustained fever between 100–105°F (37.8–40.6°C) lasts over 2 weeks. A generalized macular rash occurs about the fourth or fifth day, accompanying marked prostration. The second to third weeks of illness are critical, the patients frequently becoming stuporous and disoriented. Fatality rates vary but average about 20% and have reached as high as 70% in severe epidemics. Recovery is surprisingly rapid and sustained aftereffects are rare for so severe an illness accompanied by lesions in the brain. Immunity is generally lasting, but recrudescences many years afterward in the absence of lice are know to occur as Brill's disease. [C.B.P.]

Epidemic viral gastroenteritis A clinical syndrome characterized by a watery diarrhea, anorexia, nausea, vomiting, and abdominal cramps, with little or no fever. Probably more than one virus is responsible. This disease is also known as acute infectious gastroenteritis, winter vomiting disease, and febrile nonbacterial gastroenteritis. It is a common syndrome of wide geographic distribution, highly communicable, occurring frequently in epidemic form, particularly in institutions, although sporadic cases also occur. Infants, as well as adults of all ages, are affected. The disease is of short duration, rarely lasting longer than 2 or 3 days, and the recovery, like the onset, is sudden. Immunity is not lasting. The disease appears to be distinct from epidemic diarrhea of the newborn, some outbreaks of which have been associated with the echoviruses. *See* INFANT DIARRHEA. [J.L.Me.]

Epidemiology The study of the mass aspects of disease. The word epidemic literally means "upon the people," and was coined to describe the way an infectious disease, in spreading through a group of people, gives the impression of an affliction that has been placed upon the community as a whole.

In recent years diseases other than infectious diseases have been included in the subject matter of epidemiology. It is quite natural, therefore, that these massive noninfectious diseases should be studied as group phenomena, and that one should speak of the epidemiology of accidents, cancer, heart disease, or mental illness. For both infectious and noninfectious diseases there are similar general problems such as identifying cases in the population at risk, and finding which circumstances lead to an increase and which to a decrease in the disease. In spite of the differences in detail arising in a study of Asian influenza and a study of suicide in a particular population, both are fundamentally investigations of disease rates and a comparison of the way these rates vary from one subgroup to another.

The mortality rate for a given year is the number of deaths occurring in the year per 1000 total midyear population. However, there are certain diseases which cannot be studied by a means of mortality rates of any kind. Mental illness, arthritis, and the common cold are examples of conditions the epidemiology of which is reflected inadequately by death rates. A more useful approach is to consider morbidity rates which give the number of cases of disease per 100,000 population, either as cases existing at a particular point in time or as cases occurring in a particular period of time.

The incidence rate is concerned with the cases of disease that develop or, more strictly, that come under diagnosis in a given period of time. The incidence rate per 100,000 population for a given year is equal to the number of new cases occurring during the year times 100,000, divided by the midyear population.

The prevalence ratio is concerned with the cases of disease that exist at a particular point in time. The concept is clearest when the point of time is thought of as being instantaneous; for example, one might ask how many cases of tuberculosis per 100,000 population existed at a specified moment, and this would then be a prevalence ratio for the disease. Prevalence ratios are also computed, especially for chronic diseases, over periods as long as a year.

In addition to studying the presence or absence of a disease, the epidemiologist often uses some measure of the severity of the disease. One index of severity is the case fatality ratio, which is defined as the number of deaths expressed as a percentage of the number of cases. *See* DISEASE; EPIDEMIC. [C.W.]

Epidermal ridges Minute corrugations of the skin. They compose a sculpturing, termed dermatoglyphics, which characterizes the palmar and plantar surfaces of all primates. These areas lack hair and sebaceous (oil) glands, but sweat glands are numerous. In certain kinds of monkeys, a portion of the undersurface of the tail bears similarly specialized skin.

The ridges over the human hand as a whole, in young adult males, average 0.48 mm in breadth. They are slightly narrower in females, 0.43 mm. Ridges serve two functions: (1) They increase security of contact with objects, in the manner of the milling of a tool handle. Ducts of sweat glands open on the summits of ridges, and moistening of the skin augments the security of contact. (2) They enhance the sense of touch. In passing the fingers or palm over an object for judging its texture, the slight displacement of ridges heightens stimulation of the underlying nerve endings.

The characteristics of individual ridges and their collective configurations are not ordinarily studied by direct inspection of the skin. Instead, they are examined in prints, usually impressions of the inked surface, made for the purpose of record, or natural imprints made by chance contact. *See* FINGERPRINT. [H.Cu.]

Epidermis (plant) The outermost layer (occasionally several layers) of cells on the primary plant body. Its structure is variable; this article singles out five structural components of the tissue: (1) cuticle; (2) stomatal apparatus (including guard cells and subsidiary cells); (3) bulliform (motor) cells; (4) trichomes; and (5) root hairs.

Leaves, herbaceous stems, and floral organs usually retain the epidermis through life. Most woody stems retain it for one to many years, after which it is replaced. In roots it is usually short-lived. *See* LEAF; PERIDERM.

Cutin is a mixture of fatty substances characteristically found in epidermal cells. It impregnates the outer cell walls and occurs as a continuous layer (cuticle) on the outer surface. The cuticle covers the surfaces of young stems, leaves, floral organs, and even apical meristems. Waxes appear as a deposit on the outside of the cuticle in many plants; the bloom on purple grapes and plums is an example. Most often the waxes are present in small quantity, but the leaves of some plants may be almost white with wax (*Echeveria subrigida*). The waxes of a few species are of great commercial value in the manufacture of polishes for floors, furniture, automobiles, and shoes. Other substances, such as gums, resins, and salts, usually in crystalline form, may be deposited on the outside of the cuticle.

The apertures in the epidermis which are surrounded by two specialized cells, the guard cells, are known as stomata. The singular form, stoma, is derived from the Greek word for mouth. However, some authorities prefer to include both aperture and guard cells within the concept of stoma. The apertures of stomata are contiguous with the intercellular space system of underlying tissues and thus permit gas exchange between internal cells and the external environment. The opening and closing of the stomatal aperture is caused by relative changes in turgor between the guard cells and surrounding epidermal cells.

Bulliform (motor) cells are large, highly vacuolated cells that occur on the leaves of many monocotyledons but are probably best known in grasses. They are thought to play a role in the unfolding of developing leaves and in the rolling and unrolling of mature leaves in response to alternating wet and dry periods.

Appendages derived from the protoderm are known as trichomes; the simplest are protrusions from single epidermal cells. Included in the concept, however, are such diverse structures as uniseriate hairs, multiseriate hairs (*Begonia*, *Saxifraga*), anchor hairs, stellate hairs, branched (candelabra) hairs, peltate scales, stinging hairs, and glandular hairs (see illustration). Cotton and kapok fibers are unicellular epidermal hairs.

Root hairs are thin-walled extensions of certain root epidermal cells. They develop only on growing root tips and may

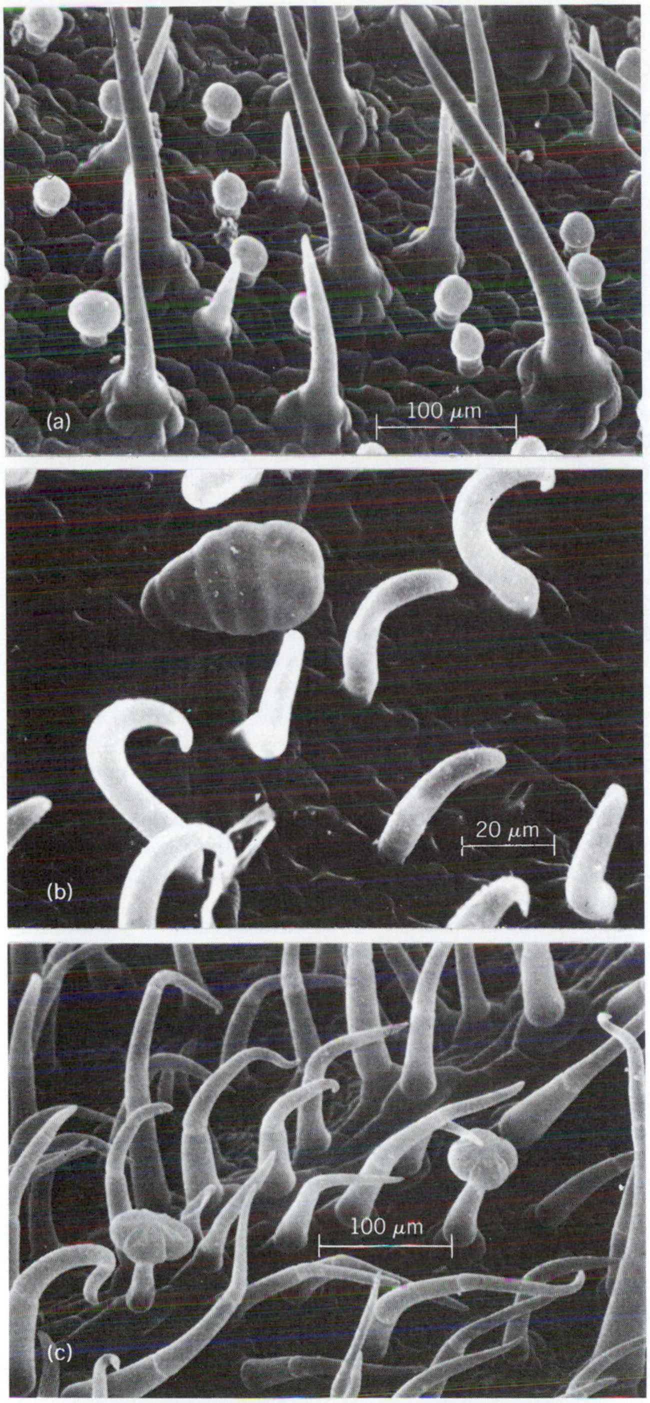

Trichomes. (*a*) Unicellular and glandular (colleters) hairs of the geranium (*Pelargonium*). (*b*) Unicellular-hooked and uniseriate, club-shaped hairs of the bean (*Phaseolus*). (*c*) Uniseriate and glandular hairs of the tomato (*Lycopersicon*).

arise from any epidermal cell, or from specialized cells known as trichoblasts. The life of a given root hair is usually numbered in days. *See* ROOT (BOTANY); SECRETORY STRUCTURES (PLANT). [N.H.B.]

Epidiorite A rock name applied to an altered diorite or gabbro. Gabbroic rocks exposed to mechanical pressure and hydrothermal action easily alter into green rocks. Many greenstones are in reality epidiorites. The term epidiorite has been generally used to cover rocks of this category. Minor recrystallized constituents may be epidote, garnet, sphene, and granular quartz. Epidiorites have a widespread distribution among low-metamorphic schists and phyllites of mountain ranges in Scotland, the Alps, the Appalachians, and the Rocky Mountains. *See* METAMORPHIC ROCKS. [T.F.W.B.]

Epidosite A rare metamorphic rock composed of epidote and quartz. It may form by normal metamorphism of calcareous grits, but most epidosites are of metasomatic-hydrothermal origin. The truly metasomatic epidosites are found in connection with limestones that have reacted with adjacent silicate rocks under the conditions of the epidote-amphibolite facies. Or they form bands, veins, streaks, or nodules in amphibolites, hornblende schists, and masses of epidiorite. Epidosite from such occurrences is sometimes cut and polished and used for ornamental purposes. *See* EPIDOTE; METAMORPHIC ROCKS; METASOMATISM. [T.F.W.B.]

Epidote A group of minerals having monoclinic symmetry and the ideal composition $Ca_2Fe^{3+}Al_2O(OH)[Si_2O_7][SiO_4]$. The common rock-forming epidote group minerals include zoisite (gemstone variety, tanzanite), clinozoisite, epidote, and piemontite (manganiferous epidote).

Epidote crystals are usually a distinctive yellowish-green (pistachio), but it may be gray, brown, or blackish-green. Hardness is 6–7 on Mohs scale and specific gravity is 3.2–3.5. In petrographic thin section, epidote may have yellow-green pleochroism with spectacular color-zoning common.

Epidote is a common rock-forming mineral in regionally metamorphosed mafic rocks from low to medium-grade terranes. It is not characteristic of very-low-grade conditions, and in rocks of appropriate bulk compositions it replaces early-formed pumpellyite. Epidote usually occurs with quartz, sodium-rich plagioclase, chlorite, muscovite, and calcium-rich amphibole. It is a common constituent of igneous rocks, crystallizing as a late-stage mineral in cavities or as an alteration of primary calcium-aluminum silicates such as plagioclase feldspar. *See* SILICATE MINERALS. [G.H.My.]

Epilepsy A disorder of the central nervous system in which there is a chronic recurrence of seizures. A seizure (ictus) is an event in which there is a sudden alteration in function of nerve cells, most commonly involving an excessive electrical activity of the cells. This sudden change in nerve cell function is usually relatively brief, lasting seconds to minutes. Soon after a seizure, the brain may function quite normally. The manifestation of a seizure varies depending on which area of the brain is involved.

The following are examples of the four common seizure types.

1. Focal motor epilepsy is a type of seizure which is manifested by uncontrolled rhythmic jerking of the face, arm, or leg, caused by excessive abnormal discharges of nerve cells within the area of the brain which under usual circumstances controls movement in that part of the body.

2. Temporal lobe seizures may be manifested by a myriad of symptoms depending upon which part of the lobe is involved. Psychomotor seizures, the most common type, are characterized by an alteration in the state of consciousness, and performance of repetitive, patterned, non-goal-directed activity. Usually, the person is aware of having had the seizure, but may not remember what happened during it.

3. Grand mal seizures, also referred to as generalized tonic-clonic convulsive seizures, occur if the abnormal discharges involve the entire brain all at once. In this condition there are forceful, generalized, symmetrical musculature contractions accompanied by the loss of consciousness. A seizure comes to an end abruptly or gradually, and depending on the pattern and intensity of the attack, the person may return immediately

to normal preictal conditions, or the person may continue to show various physical or mental deficits or both for some time. Amnesia is rather common, and after major seizures the subject often falls into a sleeplike state or goes through a period of confusion.

The relatively rare but serious life-threatening condition in which one seizure occurs after another without recovery between them and lasts for hours is called status epilepticus.

4. Generalized nonconvulsive seizures, also called absences or petit mal seizures, consists of brief periods of loss of consciousness and immediate recovery. The subjects are often unaware of having had the episode.

Epilepsy is not a disease in itself. It is a symptom of an underlying disease process. That disease process may be metabolic, such as uremia, decreased brain oxygen, or low calcium levels; or structural brain damage, such as from head trauma at birth, brain injuries, brain tumors, strokes, or congenital malformations, or previous encephalitis or meningitis. In some types of epilepsy there is apparently also a genetic component.

The ideal treatment of epilepsy is removal of the cause, such as excision of a tumor, or correction of a metabolic disorder, such as uremia or hypocalcemia. In many instances, however, the cause cannot be established or may not be amenable to direct treatment. When the cause cannot be removed, the symptoms (seizures) are treated. The following drugs are commonly used to control seizures. Phenytoin, carbamazepine, phenobarbital, or primidone are used for individuals with partial (focal) or generalized convulsive seizures. Ethosuximide, dipropylacetic acid, or trimethadione may be beneficial in generalized, nonconvulsive seizures (absences, petit mal). In individuals whose seizures are more difficult to control, combinations of these drugs might be used, or other agents may be added.

The occurrence of epilepsy can be decreased by measures to reduce head injury and birth defects. There are some indications that epilepsy can also be reduced by prophylactic administration of antiepileptic drugs in conditions which, if untreated, may lead to development of chronically recurring seizures. These include febrile convulsions of childhood, traumatic head injuries, and postoperative conditions of patients who have had brain surgery for other reasons. *See* CONVULSIVE DISORDERS.

[L.M.O.]

Epinephrine A hormone which is the predominant secretion from the adrenal medulla; also known as adrenalin, it has the structure shown. Epinephrine is a sympathomimetic sub-

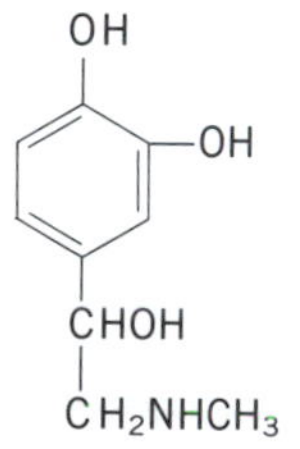

stance; that is, it acts on tissue supplied by sympathetic nerves, and generally the effects of its action are the same as those of other nerve stimuli. Conversely, the stimulation of the splanchnic or visceral nerves will cause the rapid release of the hormone from the medullary cells of the adrenal gland. Thus, epinephrine plays an important role in preparing the organism to meet conditions of physiologic emergency.

When injected intravenously, epinephrine causes an immediate and pronounced elevation in blood pressure, which is due to the coincident stimulation of the action of the heart and the constriction of peripheral blood vessels. The chief metabolic changes following the injection of epinephrine are a rise in the basal metabolic rate and an increase of blood sugar. These effects of epinephrine are transitory. *See* ADRENAL GLAND; CARBOHYDRATE METABOLISM.

[C.H.L.]

Epithelium One of the four primary tissues of the body, which constitutes the epidermis and the lining of respiratory, digestive, and genitourinary passages. The major characteristic of epithelium is that the cells are close together, separated by a very small amount of intercellular substance. Epithelium may be derived from any of the three primary germ layers of the very early embryo—ectoderm, entoderm, or mesoderm. With very few exceptions, epithelium is free of blood vessels.

The functions of epithelium are varied and include (1) protective function, by completely covering the external surface (including the gastrointestinal surface—and the surface of the whole pulmonary tree including the alveoli); (2) secretory function, by secreting fluids and chemical substances necessary for digestion, lubrication, protection, excretion of waste products, reproduction, and the regulation of metabolic processes of the body; (3) absorptive function, by absorbing nutritive substances and preserving water and salts of the body; (4) sensory function, by constituting important parts of sense organs, especially of smell and taste; and (5) lubricating function, by lining all the internal cavities of the body, including the peritoneum, pleura, pericardium, and the tunica vaginalis of the testis.

The forces which hold the epithelial cells together are not satisfactorily understood. The intercellular substance between the cells, also called cement substance, is undoubtedly important. The interdigitation of adjacent cell surfaces and the occurrence of intercellular bridges in certain cells may also be important in holding the cells together. Finally, in certain cells local modifications of contiguous surfaces and the intervening intercellular substances, which together form the terminal bars, may be effective in the same way.

The outstanding property of the arrangement of most of the epithelium of the body is the economy of space achieved in the face of a broad exposure of the cell surfaces. The efficiency is achieved by the presence of numerous folds, which may be gross or microscopic and temporary or permanent. A part must also be attributed to the surface specialization of the

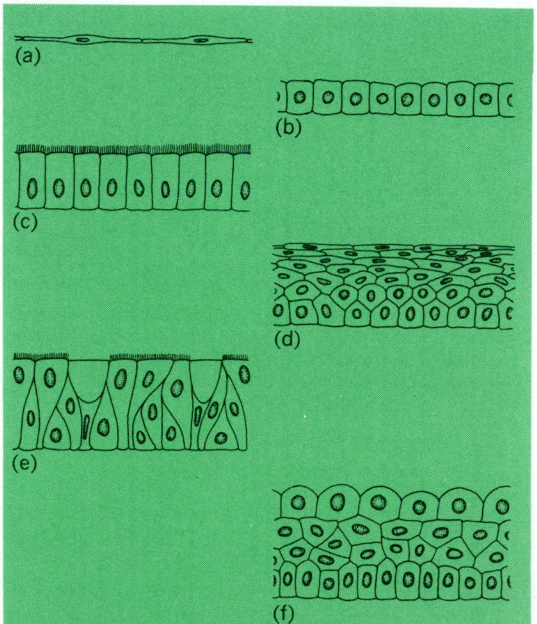

Cellular arrangements in epithelial tissues. (*a*) Squamous. (*b*) Cuboidal. (*c*) Columnar. (*d*) Stratified squamous. (*e*) Pseudostratified. (*f*) Transitional.

epithelial cells themselves, such as their minute, fingerlike processes. Another specialization of the surface or epithelial cell is the occurrence of motile cilia. *See* CILIA AND FLAGELLA.

Classification of epithelia is based on morphology, that is, on the shape of the cells and their arrangement (see illustration):

I. Single-layered.
  A. Squamous (mesothelium, descending loop of Henle in the kidney)—thin, flat.
  B. Cuboidal (duct, thyroid, choroid plexus)—cubelike.
  C. Columnar (intestine), sometimes ciliated (Fallopian tube, or oviduct)—tall.
II. Multiple-layered or stratified.
  A. Squamous (skin, esophagus, vagina)—superficial cells thin and flat, deeper cells cuboidal and columnar.
  B. Columnar (pharynx, large ducts of salivary glands), sometimes ciliated (larynx)—two or more layers of tall cells.
III. Pseudostratified (male urethra), sometimes ciliated (respiratory passages)—all cells reach to basement membrane but some extend toward the surface only part of the way, while others reach the surface.
IV. Transitional (urinary bladder)—like stratified squamous in the fully distended bladder; in the empty bladder, superficial cells rounded, almost spherical.

An important property of epithelium is the ability of its cells to glide over surfaces. This allows replacement of dead cells to take place in the normal state, while presenting a closed surface to the external environment; replacement is especially important in wound repair. Gliding ability is also manifested normally in the movement of cells which slide over each other in transitional epithelium, for example, when the urinary bladder is being distended or contracted. *See* GLAND. [I.G.]

Epoxidation The conversion of olefins (or other substances containing carbon-carbon double bonds) into epoxy (oxirane) compounds. Direct oxidation with oxygen to oxiranes has been used industrially only with ethylene, where oxidation over silver yields ethylene oxide [reaction (1)].

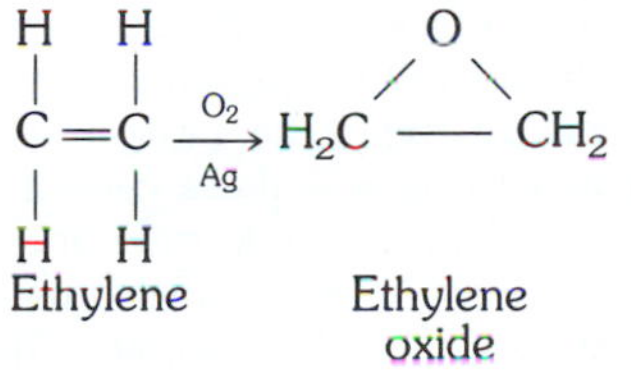

(1)

Epoxidation with peroxy acids is the most generally applied method [reaction (2)]. Other epoxidation procedures involve

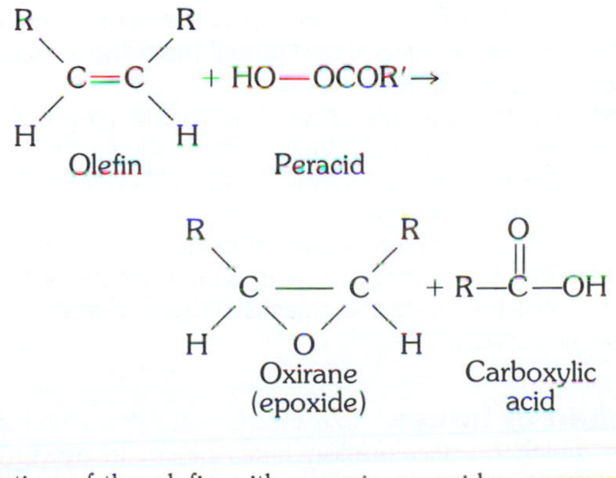

(2)

oxidation of the olefin with organic peroxides, permanganates, chromates, or dehydrochlorination of chlorohydrins with caustic alkalies. *See* ALKENE; SUBSTITUTION REACTION. [F.W.]

Epoxy resin A polyether resin formed by the polymerization of bisphenol A and epichlorohydrin. Epoxy resins are used as coatings, adhesives, castings, and foams. Laminates of epoxy resin and glass cloth have been used to make pipes, to repair damaged automobile bodies, and to make small-boat hulls. *See* POLYETHER RESINS. [M.Ye.]

Epsomite A mineral with the chemical composition $MgSO_4 \cdot 7H_2O$. Epsomite, or epsom salt, occurs in clear, needlelike, orthorhombic crystals. More commonly it is massive or fibrous, although crystals from salt lakes on Kruger Mountain near Orville, Washington, are reported to be several feet long. Fracture is conchoidal. Luster varies from vitreous to silky. Hardness is 2–2.5 on Mohs scale and specific gravity is 1.68. The mineral has a salty bitter taste and is soluble in water. Epsomite is found as a capillary coating in limestone caves and in coal or metal mine galleries. It is also found associated with gypsum and in thin layers in salt deposits of oceanic origin or from salt lakes. [E.C.T.C.]

Epstein-Barr virus An antigenically distinct member of the herpesvirus group of viruses, whose genome is DNA. EB virus is the cause of one benign disease (infectious mononucleosis), and is associated with certain types of cancer; however, the great majority of EB virus infections are clinically inapparent. The virus was detected initially by electron microscopy in a small proportion of cells in continuous lymphoblastoid cell lines derived from Burkitt's lymphoma (but particles have not been seen in cells of the tumor itself). The virus also has been detected in cell lines derived from nasopharyngeal carcinomas, a type of cancer found with high frequency in persons from southern China. The virus is found in peripheral blood leukocytes from normal individuals and from patients with infectious mononucleosis. *See* INFECTIOUS MONONUCLEOSIS; LYMPHOMA.

If EB virus is indeed confirmed as having a role in the development of human malignancies, then one major question to be resolved is how a virus so ubiquitous can be involved in so wide a variety of responses. However, it should be recalled that many virus infections (for example, polio virus, hepatitis viruses, certain of the arboviral encephalitides) have a wide spectrum of outcomes, ranging from inapparent infection to severe syndromes. *See* ANIMAL VIRUS; ONCOLOGY. [J.L.Me.]

Equalizer A class of electronic filters designed to augment or adjust electronic or acoustic systems. Equalizers can be fixed or adjustable.

Equalizers commonly modify the frequency response of the signal passing through them; that is, they modify the amplitude versus frequency characteristics. There are also fixed equalizers that modify the phase response of the transmitted signals without disturbing the frequency content. These are referred to as all-pass, phase-shifting, or time-delay equalizers. *See* ALL-PASS CIRCUIT.

Equalizers find use primarily in the fields of instrumentation and sound reproduction. Within the field of instrumentation, vibrational analyses use equalizers to reduce unwanted noise and enhance critical frequencies. Sonar uses equalizers to create a narrow listening window that rejects all frequencies except those of interest. Ultrasonic applications use equalizers to reduce interference from and to other instrumentation. Noise-control studies use equalizers to shape noise measurements to compensate for the nonflat frequency response of the ear. *See* ACOUSTIC NOISE; INSTRUMENTATION; SONAR; ULTRASONICS; VIBRATION.

Equalizers used in sound recording and sound reproduction assume many forms, all of which classify as frequency-modifying devices. Fixed equalizers are used in tape recorders, phonograph preamplifier stages, compact disk (CD) players, and tuners. All these applications provide the reverse frequency response from that used to record or broadcast the audio-frequency signals. Frequency shaping is commonly used to enhance the noise performance and capture the dynamic range encountered in these applications. *See* COMPACT DISK; DISK RECORDING; MAGNETIC RECORDING; SOUND RECORDING; SOUND-REPRODUCING SYSTEMS; TUNING.

Adjustable equalizers provide users of sound reproduction equipment with the means to change the frequency response of program material. In the strictest sense, these are not equalizers, since the intent is to change the response, not to equalize it to the original. Consumer products feature equalizers with rotary and slide controls. Most simple high-fidelity tone control circuits are rotary controlled equalizers. More expensive models feature control over 7, 10, or more bands and use slide controls. These units are called graphic equalizers, since the position of each slide control graphs the affected frequency response. In professional sound reproduction, graphic equalizers feature as many as 31 separate controls for each audio channel. The most complex equalizers afford increased control over where (variable center frequency) the boost or cut amplitude change takes place, and how wide (variable bandwidth) an area is affected. Since center frequency, bandwidth, and amplitude define the parameters for band-pass filters, these units are called parametric equalizers. [D.A.B.]

Equation of continuity An equation of continuity appears in many branches of physics. In dynamic field theory, it is essentially a statement that charge is conserved or that the rate of increase of charge in any region equals the current $\mathbf{i}$ flowing into that region. If v is the volume enclosed by the surface S, this statement may be expressed in integral form, as given in Eq. (1), where ρ is the charge density and $\mathbf{n}$ is a unit

$$\int_S \mathbf{i}\cdot\mathbf{n}\,dS = \int_v \nabla\cdot\mathbf{i}\,dv = -\frac{\partial}{\partial t}\int_v \rho\,dv \qquad (1)$$

vector normal to S. The second integral comes from the first by Gauss' divergence theorem. The volume in Eq. (1) is arbitrary, so the integrands in the last two integrals are equal. This leads to the differential form in Eq. (2), where i_x, i_y, and i_z are

$$\nabla\cdot i = \frac{\partial i_x}{\partial x} + \frac{\partial i_y}{\partial y} + \frac{\partial i_z}{\partial z} = -\frac{\partial\rho}{\partial t} \qquad (2)$$

the rectangular components of $\mathbf{i}$. Maxwell's equations satisfy Eq. (1). *See* GAUSS' THEOREM; MAXWELL'S EQUATIONS. [W.R.Sm.]

Equation of time The quantity added to mean solar time to obtain apparent solar time. Apparent solar time equals the hour angle of the true Sun (angle west of the meridian) plus 12 h. The Sun moves apparently with respect to the stars, in the ecliptic inclined 23½° to the Equator, and not at a uniform angular rate because of the Earth's elliptic orbit. In consequence, apparent solar time has annual variations. *See* TIME.

The introduction of fairly accurate clocks and watches about 1800 made it necessary to introduce the term mean solar time, defined as 12 h plus the hour angle of the mean sun. The mean sun is a fictitious body that moves in the celestial equator at the same average speed as the true Sun and has the same average position. [W.M.]

Equations, theory of The branch of mathematics concerned with finding facts concerning the roots of algebraic equations and finding methods for obtaining them. The most important type of algebraic equation is the polynomial equation in one unknown which is an expression of the form $f(x) = a_n x^n + a_{n-1} x^{n-1} + \cdots + a_1 x + a_0 = 0$, where x is called the unknown, or variable; n is a positive whole number; and the a_i, with $i = 0, 1, \ldots, n$, are constants, or fixed numbers, called coefficients of the equation. The left member of the equation is called a polynomial in one variable of degree n. A root of such an equation is a number which, when substituted for the variable x, makes the left member zero. For example, 3 is a root of the equation $x^3 + 2x^2 - 13x - 6 = 0$. In addition, systems of equations in one or more variables are considered, and here the problem is to find values for the variables which simultaneously satisfy each equation of the system. *See* LINEAR SYSTEMS OF EQUATIONS; POLYNOMIAL SYSTEMS OF EQUATIONS.

The topics covered in a systematic study of the theory of equations can be placed in the following principal subdivisions: properties of a polynomial which do not depend on the particular number system containing the coefficients of the polynomial; factorization of polynomials; equations with coefficients which are rational, real, or complex numbers; determination of bounds for real roots, and systematic methods for approximating real roots of equations; the solution of quadratic, cubic, and quartic equations by radicals.

The factorization of a polynomial $f(x)$ depends on the particular number field F under consideration. For example, the polynomial $x^5 - \frac{1}{2}x^4 - x^3 + \frac{1}{2}x^2 - 2x + 1$, having coefficients which are rational numbers, factors as

$$(x^2 + 1)(x^2 - 2)(x - \tfrac{1}{2})$$

over the rational numbers, as

$$(x^2 + 1)(x - \sqrt{2})(x + \sqrt{2})(x - \tfrac{1}{2})$$

over the real numbers, and as

$$(x + i)(x - i)(x - \sqrt{2})(x + \sqrt{2})(x - \tfrac{1}{2})$$

over the complex numbers. A polynomial $f(x)$ with coefficients in F is irreducible over F if it cannot be expressed as a product of polynomials of lower degree. Every polynomial can be expressed in essentially one way as a product of irreducible factors, although there is no general algorithm which enables one to obtain this expression.

The fundamental theorem of algebra states that a polynomial equation with complex coefficients has a complex root. From this it follows immediately that a polynomial of degree n with complex coefficients factors into n linear factors over the complex numbers.

Polynomial equations of degree 2, 3, and 4 are solvable by radicals. This means that there are formulas which give the roots in terms of the coefficients of the equation and that these formulas involve only the rational operations and the operation of extraction of roots. By use of the Galois theory of equations, it can be proved that for $n > 4$ there cannot exist a formula involving only rational operations and root extractions for expressing the roots of every polynomial equation of degree n in terms of the coefficients. *See* DETERMINANT; MATRIX THEORY. [R.A.Be.]

Equator The great circle around the Earth, equally distant from the North and South poles, which divides the Earth into Northern and Southern hemispheres. It is the greatest circumference of the Earth because of centrifugal force from rotation, and resultant flattening of the polar areas.

The Earth's rotational axis is vertical to the plane of the Equator, and because the inclination of the axis is 66½° from the plane of the ecliptic, the plane of the Equator is always inclined 23½° from the ecliptic.

The celestial equator in astronomy is equally distant from the celestial poles and is the great circle in which the plane of the terrestrial Equator intersects the celestial sphere. *See* ASTRONOMY; MATHEMATICAL GEOGRAPHY. [V.H.E.]

Equilibrium of forces In a mechanical system the condition under which no acceleration takes place. In newtonian mechanics the law of motion states that a particle of mass m acted on by resultant force $\mathbf{F}$ has acceleration $\mathbf{a}$ in accordance with the equation $\mathbf{F} = km\mathbf{a}$. Therein, k is a positive constant whose value depends upon the units in which **F**, m, and **a** are measured.

The action-reaction law states that when one particle exerts force on another, the other particle exerts on the one a collinear force equal in magnitude but oppositely directed.

A particle at rest or in uniform (unaccelerated) motion in a newtonian reference frame is in equilibrium. With few exceptions, a frame of reference fixed with respect to Earth is considered to be newtonian and the equation $\mathbf{F} = km\mathbf{a}$ applied. Accordingly, when a particle is in equilibrium on Earth, $\mathbf{a} = 0$ and thus $\mathbf{F} = 0$. Also, when $\mathbf{F} = 0$, $\mathbf{a} = 0$ because k and m are not zero. These considerations lead to a theorem and its corollary: The resultant of forces exerted on a particle in equilibrium is zero; and if the resultant of forces exerted on a particle is zero, the particle is in equilibrium.

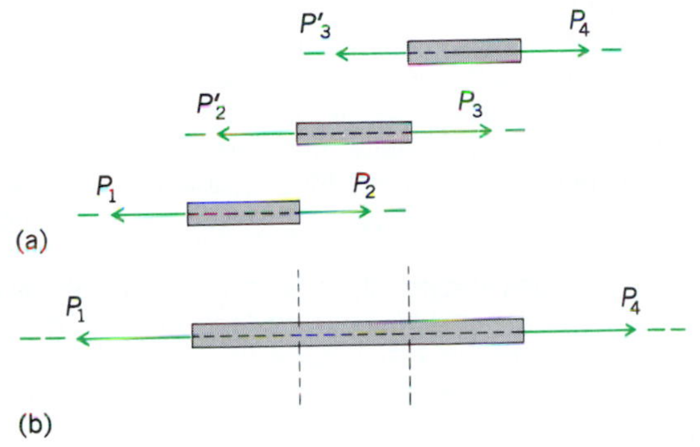

Examples of equilibrium forces. (a) Action forces and reaction forces acting on particles of a body. (b) Balanced external forces on a body.

A body acted upon by force is in equilibrium when its constituent particles are in equilibrium. The forces exerted on its particles (and therefore on the body) are either internal or external to the body. An internal force is one exerted by one particle on another in the same body. An external force is one exerted on a particle or the body by a particle not of the body.

Being particle actions, internal forces obey the action-reaction law. They occur in pairs whose individual and combined resultants are zero. Hence a further theorem is that the resultant of all forces internal to a body is zero.

The resultant of forces on each particle and therefore on all particles of a body in equilibrium is zero. Because the forces are either internal or external and the resultant of internal forces is zero, it follows as a theorem that the resultant of all external forces acting upon a body in equilibrium is zero.

To determine the forces which act upon stated matter at rest on Earth is the usual problem of equilibrium. The problem is statically determinate if solution is possible by employing the equations of equilibrium only. These equations relate all external forces; hence accurate solution requires that external forces be recognized and described according to their agencies of application. To summarize this information the free-body diagram is conveniently employed (see illustration). A free-body diagram is the sketch of a body of interest with all external forces acting thereon symbolized by arrows drawn on the sketch. *See* DYNAMICS; FORCE; KINETICS (CLASSICAL MECHANICS); STATICS. [N.S.F.]

Equine encephalitis A mosquito-borne viral infection of horses and mules. Two types are known in the United States. Western equine encephalitis (WEE) occurs chiefly west of the Mississippi River, and Eastern equine encephalitis (EEE) in the eastern and southern parts of the country. Both viruses have been isolated in Canada, as well as in South America. The chief vector of WEE is a culicine mosquito, *Culex tarsalis*. The vector of EEE is not known.

The virus, a member of arbovirus group A, is believed to be maintained in nature by a cycle in birds and mosquitoes. If a human happens to be bitten by an infected mosquito, an encephalitic disease may be produced. *See* ARBOVIRAL ENCEPHALITIDES.

Venezuelan equine encephalitis (VEE) is distributed in northern South America, the Amazon valley, Trinidad, and Panama. It is a disease chiefly of equine animals, but it may be transmitted to humans, in whom it usually produces a mild, febrile disease. [J.L.Me.]

Equine infectious anemia An infectious disease of members of the horse family caused by a persistent lentivirus infection and characterized by anemia, recurring fever, weight loss, and edema. Inapparent infections predominate and can be found in most areas of the world.

Equine infectious anemia virus is classified in the lentivirus subfamily of retroviruses and is closely related to the human immunodeficiency virus (HIV). The virus preferentially replicates in tissue macrophages and causes a persistent infection in horses. It can be found in the blood of an infected house throughout its lifetime. The mechanical transmission of blood between horses by blood-feeding insects is thought to be the most important method of natural transmission of the virus. Historically, transmission of the virus has been induced inadvertently by the veterinarian, contributing to its rapid dissemination, although this has been greatly reduced by the promulgation of sanitary precautions.

Responses of members of the horse family to infection with equine infectious anemia virus appear to be determined by the virus strain, in part related to the genetic composition of the virus, and the species of equine. Clinical signs of equine infectious anemia are usually observed shortly after exposure to the virus (incubation periods of 10–45 days are common) and consist of fever, lethargy, loss of appetite, mild anemia, and hemorrhage. Horses that survive this acute episode test positive for the antibody and may develop signs of chronic equine infectious anemia, which include recurring fever with more severe anemia, weight loss, edema, and hemorrhage. These recurring clinical bouts appear to be caused by the emergence and replication of mutant strains of equine infectious anemia virus which escape immunosurveillance by the host. *See* IMMUNITY.

This disease is controlled in most areas of the world by serology tests of horses for the presence of antibodies to the major core protein of the virus. Once the horse tests positive, segregation or quarantine methods can prevent transmission of the virus. The horse should be identified with a visible and permanent mark. The application of these techniques resulted in about a tenfold decrease in the reported rate of positive tests in the United States between 1975 and 1995. A modified live virus has been administered to horses to control clinical equine infectious anemia in China; its ability to engender immune responses that protect against infection with field strains is not known. *See* RETROVIRUS; VIRUS INFECTION, LATENT, PERSISTENT, SLOW. [C.I.]

Equinox The date of the year and the point on the celestial sphere at which the Sun's rays at noon are 90° above the horizon at the Equator, or at an angle of 90° with the Earth's axis, and neither North nor South Pole is inclined toward the Sun. This phenomenon occurs on two days of the year, approximately March 21 and September 23. In the Northern Hemisphere the event in March is referred to as the vernal equinox, and that in September as the autumnal equinox. *See* ECLIPTIC.

With the Sun's rays at 90° to the Earth's axis the rays illuminate half the globe. Therefore the Sun appears at all places on the Earth and gives 12 h of sunlight and 12 h of darkness, not considering twilight. In other words, at all places the Sun at the equinox appears in the east at 6 A.M., local Sun time, and sets at 6 P.M., in the west; thus the term equinox, meaning equal night. *See* ASTRONOMICAL COORDINATE SYSTEMS; PRECESSION OF EQUINOXES. [V.H.E.]

Equisetales An order of the class Equisetopsida of the division Equisetophyta, subkingdom Embryobionta. The Equisetales, commonly known as horsetails, is represented by a single living genus, *Equisetum*, with about 25 species found both in moist and dry habitats. These plants grow throughout the world, except in Australia and New Zealand. The plants range from herbaceous to shrubby.

The plant body is commonly composed of perennial underground stems (rhizomes) with various types of aerial stems (see illustration) in different species. Some of these are perennial, others annual; some are unbranched and reproductive, others much branched and vegetative. The bushy structure of the latter is suggestive of the common name, horsetail. *See* EQUISETOPHYTA; EQUISETOPSIDA. [P.A.V.]

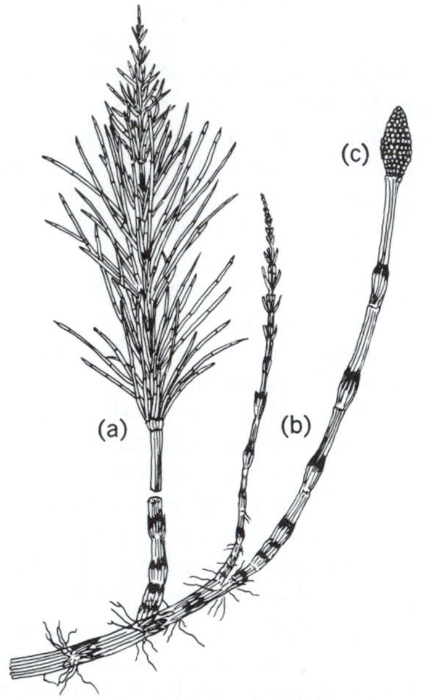

***Equisetum arvense*. (*a*) Sterile shoot. (*b*) Fertile shoot growing from an underground rootstock. (*c*) Cone. (*After E. W. Sinnott and K. S. Wilson, Botany: Principles and Problems, 6th ed., McGraw-Hill, 1963*)**

Equisetophyta One of the eight divisions of the subkingdom Embryobionta. The division (formerly called Equisetineae) is divided into three classes: Hyeniopsida, known only from fossils, Sphenophyllopsida, and Equisetopsida. Because of deposits of silicon in the epidermis, the plants were formerly used for cleaning pots and pans and were called scouring rushes. This division had its beginnings in the Devonian, its peak in the Carboniferous, and exists today only as a remnant group. Its fossil members contributed greatly to the Carboniferous coal seams. *See* HYENIALES; SPHENOPHYLLALES. [P.A.V.]

Equivalent circuit A representation of an actual electric circuit or device by a simplified circuit whose behavior is identical to that of the actual circuit or device over a stated range of operating conditions. These conditions depend on the elements contained in the actual circuit and may include such variables as frequency, temperature, and pressure, in addition to voltage and current. Equivalent circuits are often used by engineers to simplify circuit analysis since they show the relation between the variables more clearly than the actual circuit. [R.L.R.]

Equivalent weight The number of parts by weight of an element or compound which will combine with or replace, directly or indirectly, 1.008 parts by weight of hydrogen, 8.00 parts of oxygen, or the equivalent weight of any other element or compound. For all elements, the atomic weight is equal to the equivalent weight times a small whole number, called the valence of the element. An element can have more than one valence and therefore more than one equivalent weight. *See* ATOMIC WEIGHT; ELEMENTS; VALENCE.

The concept of equivalent weight, together with that of gram-equivalent weight, tends to have been abandoned, and relations are expressed in terms of balanced stoichiometric chemical equations and relative numbers of moles reacting. *See* ELECTROCHEMICAL EQUIVALENT; MOLE (CHEMISTRY); STOICHIOMETRY. [T.C.W.]

Erbium A chemical element, Er, atomic number 68, atomic weight 167.26, belonging to the rare-earth group. The naturally occurring element is made up of the six stable isotopes. The rose-pink oxide, Er_2O_3, dissolves in mineral acids to give rose-colored solutions. The salts are paramagnetic and the ions are trivalent. At low temperatures the metal is antiferromagnetic and at still lower temperatures becomes strongly ferromagnetic. For properties of the metal *see* RARE-EARTH ELEMENTS. [F.H.Sp.]

1	2	3	4	5	6	7	8	9	10	11	12	13	14	15	16	17	18
1 **H**																	2 **He**
3 **Li**	4 **Be**											5 **B**	6 **C**	7 **N**	8 **O**	9 **F**	10 **Ne**
11 **Na**	12 **Mg**											13 **Al**	14 **Si**	15 **P**	16 **S**	17 **Cl**	18 **Ar**
19 **K**	20 **Ca**	21 **Sc**	22 **Ti**	23 **V**	24 **Cr**	25 **Mn**	26 **Fe**	27 **Co**	28 **Ni**	29 **Cu**	30 **Zn**	31 **Ga**	32 **Ge**	33 **As**	34 **Se**	35 **Br**	36 **Kr**
37 **Rb**	38 **Sr**	39 **Y**	40 **Zr**	41 **Nb**	42 **Mo**	43 **Tc**	44 **Ru**	45 **Rh**	46 **Pd**	47 **Ag**	48 **Cd**	49 **In**	50 **Sn**	51 **Sb**	52 **Te**	53 **I**	54 **Xe**
55 **Cs**	56 **Ba**	71 **Lu**	72 **Hf**	73 **Ta**	74 **W**	75 **Re**	76 **Os**	77 **Ir**	78 **Pt**	79 **Au**	80 **Hg**	81 **Tl**	82 **Pb**	83 **Bi**	84 **Po**	85 **At**	86 **Rn**
87 **Fr**	88 **Ra**	103 **Lr**	104 **Rf**	105 **Db**	106 **Sg**	107 **Bh**	108 **Hs**	109 **Mt**	110	*111*	*112*	*113*	*114*	*115*	*116*	*117*	*118*

lanthanide series	57 **La**	58 **Ce**	59 **Pr**	60 **Nd**	61 **Pm**	62 **Sm**	63 **Eu**	64 **Gd**	65 **Tb**	66 **Dy**	67 **Ho**	68 **Er**	69 **Tm**	70 **Yb**
actinide series	89 **Ac**	90 **Th**	91 **Pa**	92 **U**	93 **Np**	94 **Pu**	95 **Am**	96 **Cm**	97 **Bk**	98 **Cf**	99 **Es**	100 **Fm**	101 **Md**	102 **No**

Ergosterol A steroid that belongs to the class of unsaponifiable lipids. It was first isolated from ergot bodies. Ergosterol is a white crystalline compound, insoluble in water and soluble in organic solvents, with the structural formula shown.

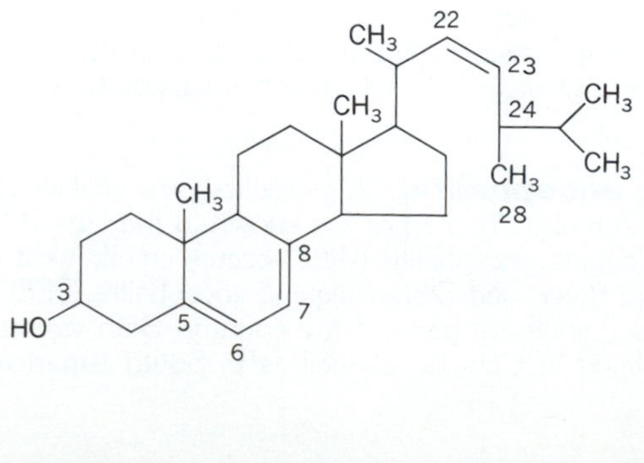

Ergosterol differs from cholesterol by having three positions of unsaturation, at carbon atoms 5-6, 7-8, and 22-23, and containing a methyl group (carbon atom 28) substituted for a hydrogen atom at carbon atom 24. *See* CHOLESTEROL.

Ultraviolet irradiation of ergosterol leads to the formation of calciferol (vitamin D_2). Under controlled conditions, a more than 50% yield of vitamin D_2 can be obtained. The process is used commercially. *See* STEROID; VITAMIN D. [W.J.V.W.]

Ergot The seedlike body of fungi (molds) of the genus *Claviceps*. Ingestion of these long, hard, purplish-black structures which contain poisonous alkaloids may lead to convulsions, abortion, hallucinations, or death. There are 32 recognized species of *Claviceps*, most of which infect members of the grass family. Only three species are parasitic on the rushes and sedges.

The use of ergot in medicine has been reported in the past, but only in the 20th century did it become possible to produce stable ergot preparations (ergotoxine and ergotamine). However, because of side effects, researchers continued to look for other compounds and in 1935, a new water-soluble ergot alkaloid was reported. It is called ergonovine in the United States, ergobasine in Switzerland, and ergometrine in other countries. Ergonovine is a small lysergic acid derivative. Ergotamine is a large, tetracyclic peptide linked to lysergic acid. Ergotoxine was found to be a mixture of similar large alkaloids (ergocornine, ergokryptine, and ergocristine). Many other important lysergic acid derivatives were produced in the decade that followed. Of these derivatives, LSD-25 (*d*-lysergic acid diethylamide) is undoubtedly the most famous.

Many uses for ergot alkaloids have been found. Ergotamine and dihydroergotamine are used to treat migraines and methysergine is used in migraine prophylaxis. Dihydroergotoxine is prescribed for hypertension, high blood pressure, cerebral sclerosis, and peripheral vascular disorders. Ergocornine and the less toxic agroclavine have been reported as unusual experimental birth-control agents. *See* ERGOTISM. [R.L.Mo.]

Ergotism A complex disease of humans and certain domestic animals caused by ingestion of grains and cereals infested with ergot. There are three types of ergotism (gangrenous, convulsive, and hallucinogenic). Their symptoms often overlap; the hallucinogenic form is usually observed in combination with one of the other two. High mortality rates have been associated with severe epidemics involving any of the three forms of ergotism.

The onset of gangrenous ergotism is generally characterized by lassitude, nausea, and pains in the limbs, Then, alternating sensations of intense heat and cold occur. As bodily extremities become numb, livid watery vesicles may appear on the affected parts (usually arms and legs). Finally, the diseased area turns black, dry, and becomes mummified. During large epidemics it was not uncommon for the whole rotted limb to spontaneously break off at the joint.

In convulsive ergotism, various parts of the body become grossly deformed as a result of clonic or tonic convulsions, or both. Generalized neurological stimulation causes epileptiform seizures, whereas specific stimulation might involve ravenous hunger, violent retching, tongue biting, or unusual breathing patterns. This form involves a longer recovery period and often results in permanent nerve damage and subsequent sensitization.

The hallucinogenic form often includes symptoms of one of the other types. In its more pure form, it is referred to as choreomania, St. Vitus's dance, or St. John's dance. Vivid hallucinations are accompanied by psychic intoxication reminiscent of the effects of many of the modern psychedelic drugs (nervousness, physical and mental excitement, insomnia, and disorientation).

The success with which the disease is controlled in humans has been brought about by (1) agricultural inspection, (2) use of wheat, potatoes, and maize instead of rye, (3) limited control of ergot, (4) reserves of sound grain, and (5) forecasting severe ergot years. The most recent and best-recorded epidemic was in southern France in 1951 when an unscrupulous miller used moldy grain to make flour. *See* ERGOT. [R.L.Mo.]

Ericales An order of flowering plants, division Magnoliophyta (Angiospermae), in the subclass Dilleniidae of the class Magnoliopsida (dicotyledons). The order consists of seven families and a little more than 3000 species. The Ericaceae, with more than 2500 species, are by far the largest family, followed by the Epacridaceae, with about 400. The Ericales are sympetalous (having the petals joined by their margins, at least at the base) or less often polypetalous (having the petals separate from each other) plants. The stamens are typically twice as many as the petals and usually attached directly to the receptacle. Generally, there are many ovules in each locule of the ovary. The anthers very often have prominent appendages (like horns or tails) and open by terminal pores.

A large proportion of the species are strongly mycorrhizal, and some of them are so dependent on their fungal symbiont that they do not even produce chlorophyll. Heath (*Erica*), heather (*Calluna*), blueberries (*Vaccinium*), rhododendron, wintergreen (*Gaultheria procumbens*), and Indian pipe (*Monotropa uniflora*) are familiar members of the Ericales. *See* BLUEBERRY; DILLENIIDAE; MAGNOLIOPSIDA. [A.Cr.]

Eriocaulales An order of flowering plants, division Magnoliophyta (Angiospermae), subclass Commelinidae of the class Liliopsida (monocotyledons). The order consists of the single family Eriocaulaceae, with about 1200 species. The Eriocaulales are Commelinidae with a reduced (or no) perianth and with unisexual flowers aggregated into a dense, involucrate head that is elevated above the clustered, basal leaves on a long peduncle. Although the individual flowers are small and inconspicuous, the heads are more or less showy and pollination is usually by insects, in spite of the absence of nectar and nectaries. The order is of negligible economic importance. *See* COMMELINIDAE; LILIOPSIDA. [A.Cr.]

Erosion The loosening and transportation of rock debris at the Earth's surface. Erosion is one of the most important geologic processes at the surface and operates to move earth materials to lower levels. Agents of erosion include moving surface, ground, and ocean waters; ice; wind; organisms; temperature changes; and gravity. They work by moving earth materials in two ways: physically, that is, without change in chemical composition of the earth materials; and chemically, in which the rock materials have been dissolved or decomposed. Counteracting the effects of erosion are geologic processes which raise the Earth's surface, such as volcanism (volcanoes and intrusions) and diastrophism (folding, faulting, plate tectonics). *See* WEATHERING PROCESSES.

Within a given region, distinctive landforms, enjoyed as scenery, are produced by stream erosion, mountain glaciation, continental glaciation, wind (desert) erosion, landslides and mass wastage, and groundwater solution (karstic) features. *See* GLACIOLOGY; MASS WASTING; STREAM TRANSPORT AND DEPOSITION.

Not only does erosion affect human beings, but the human race in turn is an almost violent agent of erosion, with its practices of agriculture, overgrazing, lumbering, and urban development on continental land slopes and coastal topographic features. *See* SOIL CONSERVATION. [W.D.K.]

Errantia A group of 34 families of Polychaeta in which the anterior, or cephalic, region is more or less fully exposed and the body is often long, linear to short, and depressed. The segments of these worms are similar or change gradually. The pharynx is often heavily muscularized and eversible, and its inner walls are fortified with calcified or chitinized plates or jaws. Some families are benthic throughout their life, others are entirely pelagic, and some have pelagic larval and reproductive stages and benthic trophic development. Errantia occur in all seas, at all depths, and in inland seas or lakes. They range in length from a few tenths of an inch (few millimeters) to 6 ft (1.8 m). *See* POLYCHAETA.

There are 6 families of scale bearers and allies: Aphroditidae (9 genera and about 85 species) include the sea mouse (*Aphrodita*) and the giant cold-water *Laetmonice*. Polynoidae (75 genera, 600 species) are in all seas and at all depths. Most are benthic, some are pelagic (*Drieschia*, *Nectochaeta*), and others are commensal (*Arctonoe*, *Hesperonoe*, *Polynoe*); some are known only from abyssal depths (*Macellicephala*). Common genera are *Harmothoe*, *Lepidonotus*, and *Halosydna*. Polyodontidae (8 or 9 genera and about 35 species) are tubicolous and often large-bodied. Some secrete fibers to construct tubes. They are worldwide, occurring chiefly in tropical latitudes. Typical genera are *Eupanthalis*, *Panthalis*, and *Polyodontes*. Sigalionidae (10 genera and about 130 species) are worldwide and occur in all latitudes, from littoral to abyssal depths. The best-known genera are *Leanira*, *Psammolyce*, *Sigalion*, *Sthenelais*, and *Thalenessa*. The other 5 families are small: Peisidicidae (1 genus, 3 species); Pareulepidae (1 genus, 8 species); Pisionidae (5 genera, 12 species); Chrysopetalidae (4 genera, 18 species); and Palmyridae (1 genus, 2 species).

There are 3 Amphinomorpha families: Amphinomidae (18 genera and about 120 species) comprise the stinging or fire worms; a burning sensation results from handling them. Common genera are the large, brilliantly colored *Amphinome* and *Hermodice*—the "scorpions of the sea." Euphrosinidae (Fig. 1; 2 genera, 45 species) and Spintheridae (1 genus, 8 species) are short and depressed and are associated with sponges and other colonial animals.

The leaf-bearing and pelagic groups are composed of 8 families. The Phyllodocidae (30 genera, more than 240 species) are often brilliantly iridescent and are highly motile, with arresting colors and ornamentation. They occur in all seas and at all depths. Common genera are *Anaitides*, *Eulalia*, *Eteone*, and *Phyllodoce*. The pelagic families allied to the Phyllodocidae are Alciopidae (10 genera, 41 species); Lopadorrhynchidae (2 genera, 5 species); Iospilidae (4 genera, 7 species); Pontodoridae (single genus and species); and Typhloscolecidae (3 genera, 14 species). The benthic families are Lacydonidae (3 genera, 6 species), and Tomopteridae, or glass worms (2 genera and about 40 species).

The Hesionidae (22 genera, more than 82 species), and the Pilargidae (6 genera and about 25 species) differ from other depressed; a few are long and linear. Some are free-living; others commensal. The best-known genera are *Hesionie*, *Leocrates*, *Oxydromus*, and *Podarke* in the Hesionidae and *Ancistrosyllis* and *Pilargis* in the Pilargidae.

Syllidae comprise the second largest of polychaete families; they have 55–60 genera and more than 400 species. Most occur along intertidal rocky shores among epiphytic plants or animals or in encrusting sponges. They are chiefly foraging predators or scavengers, identified by their long, linear, translucent bodies with articulated cirri (Fig. 2).

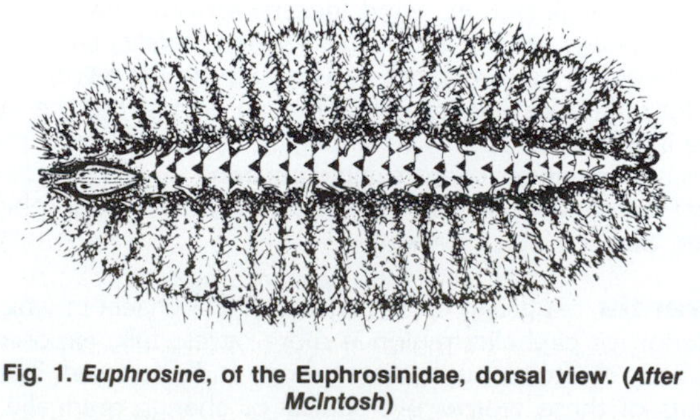

Fig. 1. ***Euphrosine*, of the Euphrosinidae, dorsal view. (*After* McIntosh)**

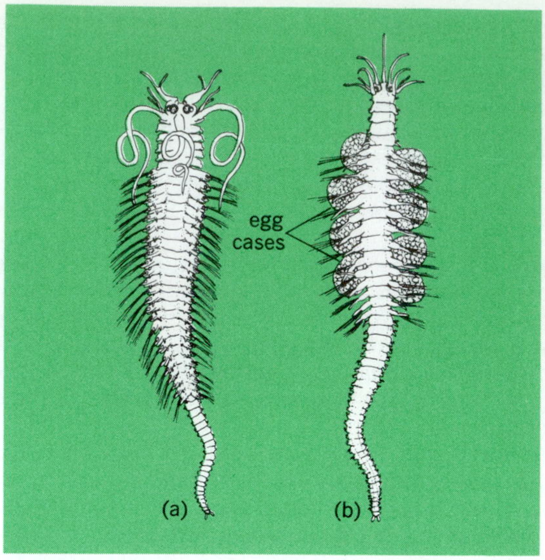

Fig. 2. ***Procerastea* of the Syllidae (Autolytinae). (*a*) Epitokous male, dorsal view. (*b*) Epitokous female with attached egg cases, dorsal view.**

Nereidae is known for about 27 genera and more than 360 species. Most are marine, occurring in worldwide seas and at all depths; a few are euryhaline, entering tidal streams or inland seas, or occur in thermal springs and high mountain lakes. Common genera are *Nereis*, *Neanthes*, *Ceratocephale*, *Platynereis*, and *Perinereis*.

Nephtyidae (4 genera and about 85 species) are sometimes called sandworms for their occurrence in intertidal sands, where their highly opalescent colors and rapid movements render them almost invisible. They occur in worldwide seas at all depths. Common genera are *Nephtys* and *Aglaophamus*.

Sphaerodoridae (2 genera and about 18 species) are small, short- to long-bodied, and usually papillated; usually their occurrence is rare.

Glyceridae (3 genera and about 66 species) are characterized by the enormous eversible proboscis used in food capture and forward progression. Closely allied, the family Goniadidae (7 genera and more than 60 species) occurs in all seas and at all depths. *Glycinde*, *Goniada*, *Goniadides*, and *Ophioglycera* are worldwide in occurrence.

The superfamily Eunicea comprises 6 families. Onuphidae (10 genera and about 100 species) are tubicolous, grazing, herbivorous, scavenging, are worldwide in occurrence, and exist in intertidal to abyssal depths. Eunicidae (7 genera and about 225 species) are mainly from tropical seas; included are the coral-eating *Lysidice*, and a bait worm, *Marphysa*. Lumbrineridae (3 genera and about 130 species) are known in all seas. Arabellidae (9 genera and about 66 species) chiefly include free-living and parasitic species. Lysaretidae (3 genera, 8 species) include large fish-bait worms (*Lysarete* and *Halla*) in tropical regions where they occur. *Iphitime* is an ectoparasite on large crabs. Dorvilleidae (3 genera and about 40 species) are small to minute and are chiefly intertidal in epiphytic growth.

Two parasitic families allied to the Eunicea are the Histriobdellidae, ectoparasites of crayfishes (2 genera, 5 species), and the Ichthyotomidae, parasites of fishes (single genus and species). *See* ANNELIDA. [O.H.]

Erysipelothrix A genus of bacteria comprising the single species *Erysipelothrix rhusiopathiae*, an infectious organism that primarily afflicts animals. Infections in humans are acquired by contact with products from infected animals.

Microscopic examination of *E. rhusiopathiae* in infected tissue, blood, or artificial culture media reveals small, slender, gram-positive, rod-shaped organisms with rounded ends, measuring 0.2–0.4 micrometer wide by 0.8–2 micrometers long. Some cells may be slightly curved. They may occur singly, in pairs, and in short chains. As cultures age, many cells develop long filaments.

Soil and water contaminated with urine and feces excreted by infected animals (including birds) serve as sources of infection for healthy animals, which ingest the organisms while feeding. Organisms deposited in soil are capable of surviving 2–5 years, thus providing exposure and opportunity for subsequent disease in domestic and wild animals and birds. *See* MEDICAL BACTERIOLOGY. [H.J.W.]

Erysiphales An order of ascomycetous fungi which are obligate parasites of seed plants. In the Erysiphales the ascocarp consists of a tough, dark rind enclosing one ascus or several asci arising at the same level (see illustration). Asci and ascospores are liberated by rupture of the ascocarp wall. In most Erysiphales the filamentous mycelium is superficial except for absorptive branches (haustoria) which penetrate host cells.

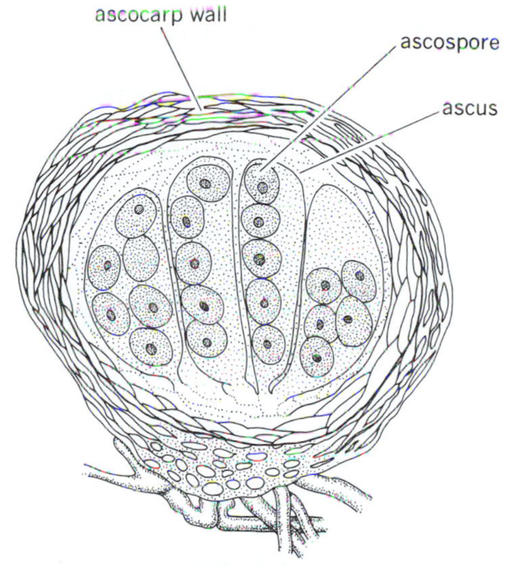

Longitudinal section showing several asci in ascocarp of *Erysiphe aggregata*.

About 70 genera and 1000 species are divided into two families: the Erysiphaceae (powdery mildews), with light-colored mycelium and conidia, and the Meliolaceae (sooty molds), with dark mycelium and conidia.

Erysiphaceae are common in temperate regions. Important powdery mildews include those of gape (*Uncinula necator*), apple (*Podosphaera leucotricha*), hops (*Sphaerotheca humuli*), roses (*S. pannosa* var. *rosae*), beans (*Erysiphe polygoni*), and grains (*E.* graminis). *See* ASCOMYCOTINA. [R.M.P.]

Erythrocytosis An increase in the number of circulating red blood cells (erythrocytes). This condition is also known as polycythemia or erythremia. Since the erythrocyte is largely responsible for the viscosity of whole blood, serious circulatory complications are frequently associated with increases that are significantly above average normal values.

There are several forms of polycythemia. Absolute erythrocytosis is distinguished by an increase in the total mass of red corpuscles. It may be classified as primary (polycythemia rubra vera), of unknown cause, and secondary, which develops from a variety of factors. Some cases represent a physiological response to conditions of hypoxia; that is, defective oxygen saturation of arterial blood stimulates red blood cell production. Other cases are associated with renal disease (hydronephrosis and polycystic kidney) and tumors (renal and hepatic carcinomas, cerebellar aggregations of blood vessels, and adrenal cortical tumors related to aldosteronism and Cushing's disease).

Relative erythrocytosis occurs when the concentration of red blood cells in the circulating blood increases through loss of blood plasma. This is the consequence of dehydration, due to reduced fluid intake or loss of body fluid or stress. [M.Wu.]

Erythromycin A relatively stable crystalline antibiotic produced by cultures of *Streptomyces erythreus*, and of primary importance in the treatment of gram-positive bacterial infections. It is especially useful in controlling diseases which have become resistant to other antibiotics, such as penicillin.

Erythromycin has been used successfully in the treatment of staphylococcal bacteremia, streptococcal bacteremia, and pneumococcal septicemia. The antibiotic has controlled infections of the nose, throat, tracheobronchial tree, the skin and soft tissue, and certain types of skeletal infections. It is useful for treating gonorrhea or syphilis in patients who are allergic to penicillin.

Erythromycin is given intravenously or, more usually, orally. The antibiotic is readily absorbed and diffuses throughout most of the body tissues. It tends to concentrate in the liver and is excreted in biologically active form in the bile, urine, and feces. *See* ANTIBIOTIC. [F.R.H.]

Escalator A continuously moving stairway and handrail. The escalator (see illustration) transports a continuous stream of passengers from floor to floor. Usually speed is 90 ft/min (0.46 m/s); slope is standardized at 30°. Steps ride on resilient rollers running on tracks; endless roller chains propel the steps, the chains being driven by sprockets and a worm drive. An electric motor at the top of the escalator provides the motive force. The direction of travel can be reversed in accordance with the flow of traffic. [F.H.R.]

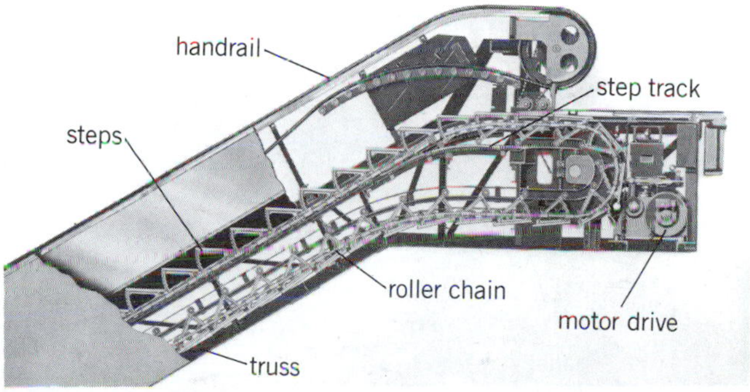

Cutaway view of escalator. (*Otis Elevator Co.*)

Escape velocity Minimum speed away from a parent body that a particle must acquire to escape permanently from the gravitational attraction of the parent. Escape velocity is also termed parabolic velocity. *See* ORBITAL MOTION.

Earth retains an atmosphere because the escape velocity is considerably higher than the mean velocity of the gas molecules in its atmosphere. For a space ship to escape from Earth and travel to another planet or orbit about the Sun, it must reach escape velocity. This velocity can be calculated by equating the kinetic energy of the moving body to the work necessary to overcome the gravity at the surface of the parent. For Earth, $v_{escape} = 11.2 \times 10^3$ m/s. *See* SATELLITE (SPACECRAFT). [R.L.Du.]

Escapement A mechanism in which a toothed wheel engages alternate pallets attached to an oscillating member. The escapement is found principally in timepieces but may be employed wherever oscillating motion is required.

In a mechanical clock or watch, the escapement intervenes between the energy source (spring or elevated weight) and the regulating device (pendulum or balance wheel). It is acted upon by both. The escape wheel is mounted on the same shaft as the last wheel of the gear train, and impulses are delivered from the escape wheel to operate the regulating device. The regulating device, which has a natural period of oscillation, determines the rate at which it will receive these impulses, and thus regulates the rate of going of the timepiece.

The anchor recoil escapement (Fig. 1) is used with a pendulum and takes its name from the shape of its oscillating member and its action. In the position shown, pallet *B* has just received an impulse from the escape wheel, the impulse swinging the pendulum to the left. When pallet *B* has cleared the tooth, allowing the wheel to escape, pallet *C* will be in position to arrest the wheel. Recoil, or momentary reversal of the escape wheel, occurs just after it is arrested because the pendulum has not quite completed its swing. The wheel tooth in contact with pallet *C* will then give the oscillating parts an impulse in the opposite direction.

Modern watches generally employ a detached-lever escapement (Fig. 2), which has banking pins *B* to limit the oscillation of the anchor and its lever. An escapement is termed detached when the regulating device, in this case the balance wheel, is given an impulse during only a small part of its operating cycle. When the fork reaches the end of its swing, it is lightly locked by a wheel tooth and remains stationary until the returning impulse pin *E* causes sufficient recoil of the escape wheel to release the pallet. *See* CHRONOMETER; CLOCK.

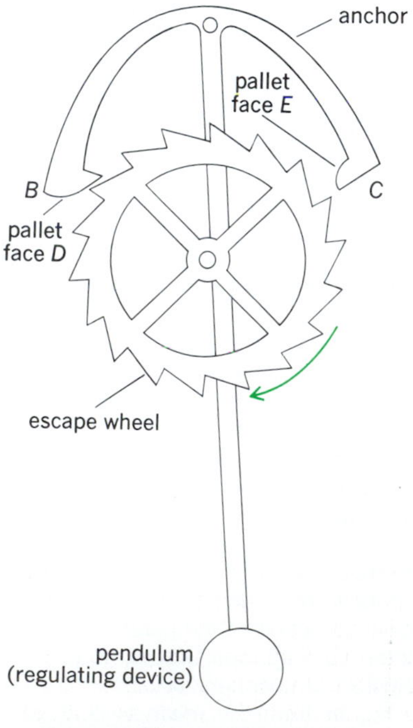

Fig. 1. Anchor recoil escapement.

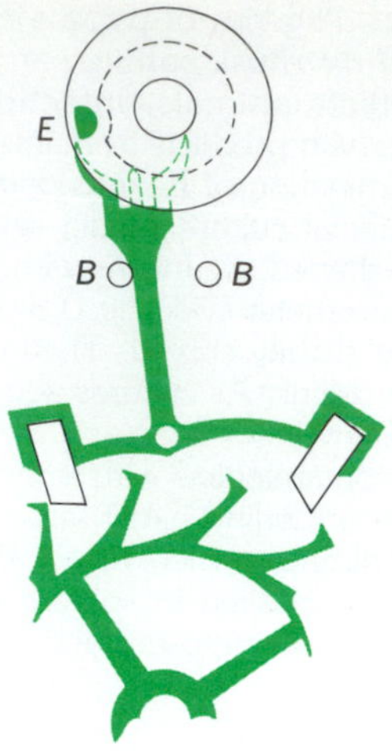

Fig. 2. Detached-lever escapement, often used in modern watches. (*After F. J. Britten, Britten's Old Clocks and Watches and Their Makers, 7th ed., Dutton, 1956*)

Some modern watches employ a light ratchet powered by a vibrating reed at 360 Hz. *See* WATCH. [D.P.Ad.]

Escarpment A long line of cliffs or steep slopes that break the general continuity of the land by separating it into two level or sloping surfaces. Some very high escarpments, or scarps, may form by vertical movement along faults. Often a whole block of land may be forced upward while the adjacent block is downfaulted. *See* FAULT AND FAULT STRUCTURES.

Other types of escarpments form by differential weathering and erosion of contrasted rock types. Less resistant rocks, such as clay or shale, are often eroded from beneath resistant cap rocks, such as sandstone and limestone. With support removed from below, the cap rock fails and the escarpment retreats. Escarpments are often very prominent in arid regions, where hardened weathering products may form extensive cap rocks known as duricrusts.

Some of the largest known escarpments occur on the planet Mars, where erosion has presumably been much slower than on the Earth in reducing primary structural relief. *See* MARS. [V.R.B.]

Escherichia A genus of the bacterial family Enterobacteriaceae, with only one species, *Escherichia coli*. Most strains ferment lactose, form gas from glucose, and are motile; they do not produce hydrogen sulfide. Indole is produced, urea is not split, citrate is not utilized, and the Voges-Proskauer reaction is negative. The term "coliform bacteria" generally refers to the group of genera made up of *Escherichia*, *Klebsiella*, *Enterobacter*, *Serratia*, and *Citrobacter*.

The bacterium is regularly present in the large intestine (and stool) of humans and warm-blooded animals. Its presence in drinking water is considered an indication of fecal contamination. *Escherichia coli* is the most commonly isolated gram-negative rod in the clinical laboratory, and the most common agent of urinary tract infection in humans. It is one of the most studied microorganisms and is extensively used in research. Genetically altered *E. coli* are used for the production of insulin and interferon. *See* ENTEROBACTERIACEAE. [A.W.C.V.G.]

Esker A sinuous ridge composed predominantly of sand and gravel deposited by glacial meltwater. Eskers vary in degree of continuity, and range in size from a few meters (1 m = 3.3 ft) to tens of meters high and from a few meters to a hundred or more kilometers long. They have steep ice-contact slopes and were deposited in channels confined by ice. Most eskers generally parallel the direction of ice flow, and while most follow valleys and have a normal down-drainage slope, some trend up a regional or local slope. [W.H.J.]

Esophagus A section of the alimentary canal that is interposed between the pharynx and the stomach. In humans it is a tube running the full length of the neck and the thorax, held in its position ventral to the vertebral centra by a tunica adventitia of loose connective tissue. It has an inner lining of folded mucous membrane with an exceptionally thick lamina propria, a submucosa of elastic and collagenous connective tissue, and two layers of muscle. It is supplied with autonomic nerve fibers.

Although normally collapsed, the human esophagus is capable of considerable distension during the rapid passage of swallowed material, under which condition the folds of mucous membrane and lamina propria are temporarily smoothed out. *See* DIGESTIVE SYSTEM. [W.W.B.]

Pathology of the esophagus can be of congenital, inflammatory, vascular, neoplastic, and mechanical origin. The two major symptoms of esophageal disorders are dysphagia and hematemesis, or vomiting of blood.

Congenital defects are infrequent, but stenosis (a narrowing) and atresia (a failure of development) are seen with some regularity.

Esophagitis, or inflammation, most commonly follows the accidental swallowing of corrosive agents by children. Other causes are infections, persistent vomiting, and reflux of stomach acid into the esophagus.

Carcinoma of the esophagus is the most important of the neoplastic processes, and it accounts for more than 5% of all malignancies other than cutaneous neoplasms in men. It occurs most often in men over 50 years of age and in any case of dysphagia, or difficulty in swallowing, carcinoma must be excluded before another diagnosis is made.

Mechanical lesions include diverticula which may be of two main types. Traction diverticula are outpouchings of the esophageal wall caused by adhesions following previous inflammatory disease of the adjacent lymph nodes or mediastinal tissue, usually following tuberculosis. Pulsion diverticula are small to large pouches which extrude through surrounding tissues, such as the pharyngeal muscles or the diaphragm. E.G.St./N.K.M.]

Essential oils Odoriferous, oily products of plant origin which are distillable. The essential oils occur in the leaves, twigs, blossoms, fruit, trunk, and roots of plants. Relatively few of the great number of species of plants have been used to produce the oils of commerce. These plants are widespread and occur even in remote geographical regions. The tropic zone produces many of them.

The principal constituents of essential oils are the terpenes. Benzenoid and aliphatic compounds may also be present. Most of the constituents are hydrocarbons and oxygenated derivatives of hydrocarbons. A few contain nitrogen and sulfur. For example, oil of mustard contains organic isothiocyanates; garlic and onion oils contain organic sulfides. Indole, skatole, and anthranilates sometimes occur in small amounts. Most essential oils are exceedingly complex mixtures of terpene and nonterpene ingredients. For example, 75 identifiable compounds have been isolated from camphor oil. *See* TERPENE.

No essential oil has been successfully reconstructed. Synthetic methyl salicylate can be differentiated by odor and taste from oil of sweet birch even though the natural oil contains 98% methyl salicylate. However, synthetic methods have made possible the commercial production of important constituents of essential oils, such as linalool, linalyl acetate, nerolidol, and citral. [W.Mos.]

Ester The product of a condensation reaction (esterification) in which a molecule of an acid unites with a molecule of alcohol with elimination of a molecule of water as shown in the following reaction. *See* ESTERIFICATION.

$$\underset{\text{Acid}}{\mathrm{R\overset{\overset{\large O}{\|}}{C}OH}} + \underset{\text{Alcohol}}{\mathrm{HOR'}} \rightarrow \underset{\text{Ester}}{\mathrm{R\overset{\overset{\large O}{\|}}{C}OR'}} + H_2O$$

At one time it was thought that esterification was analogous to neutralization, and esters are still named as though they are "alkyl salts" of carboxylic acids.

Esters are generally insoluble in water and have boiling points slightly higher than hydrocarbons of similar molecular weight.

Ethyl and butyl acetates are volatile industrial solvents, used particularly in the formulation of lacquers. Higher-boiling esters such as butyl phthalate are used as softening agents (plasticizers) in the compounding of plastics. The natural waxes of biological origin are largely simple esters. For example, a principal component of beeswax is myricyl palmitate. *See* SOLVENT; WAX, ANIMAL AND VEGETABLE.

Esters of cellulose (cellulose triacetate) are used in photographic film, as a textile fiber (acetate rayon), and several have become important as thermoplastic materials. Cellulose nitrate, called celluloid pyroxylin, forms celluloid, dynamite cotton, and gun cotton. Cordite and ballistite are made from gun cotton. Dimethyl and diethyl sulfates (esters of sulfuric acid) are excellent agents for alkylating organic molecules that contain labile hydrogen atoms, for example, starch and cellulose.

Esters of unsaturated acids, for example, acrylic or methacrylic acid, are reactive and polymerize rapidly, yielding resins; thus, methyl methacrylate yields a polymethyl methacrylate resin (Lucite). Analogously, esters of unsaturated alcohols are reactive and readily react with themselves; thus, vinyl acetate polymerizes to polyvinyl acetate.

Many low-molecular-weight esters have characteristic, fruitlike odors: banana (isoamyl acetate), rum (isobutyl propionate), and pineapple (butyl butyrate). These esters are used to some extent in compounding synthetic flavors and perfumes. *See* ALCOHOL; CARBOXYLIC ACID; DRYING OIL; FAT AND OIL; POLYESTER RESINS. [P.E.F.]

Esterification In the broadest sense, any reaction in which at least one of the products is an ester. There are many routes to the formation of esters. Some of the more important reactions for preparing esters take place between the following pairs of compounds: (1) an acid and an alcohol, (2) an acid anhydride and an alcohol, (3) an acid chloride and an alcohol, (4) an acid and an unsaturated hydrocarbon such as an olefin or an acetylene, (5) an ester and an alcohol, (6) an ester and an acid, and (7) two different esters. *See* TRANSESTERIFICATION.

Esterification reactions are generally reversible and accompanied by relatively small heat effects of the order of a few kilocalories per mole of ester. Although the reactions generally take place in a single liquid phase in the presence of a catalyst, a limited number of esters have been prepared by passing the reactant vapors over a solid catalyst.

In a typical industrial procedure for the preparation of ethyl acetate, a mixture of acetic acid, excess ethanol, and sulfuric acid is passed into an esterifying column heated to reflux. A ternary azeotrope containing 70% ethanol, 20% ester, and 10% water separates into layers, one of which contains 85% ethyl acetate. The ester may be purified by fractional distillation, and the recovered starting materials are recycled. Aspirin, the world's most used analgesic, is prepared by the reaction of salicylic acid with acetic anhydride below 190°F (90°C), and is purified by recrystallization. *See* ASPIRIN; AZEOTROPIC MIXTURE; ESTER. [P.E.F.]

Estimation theory A branch of probability and statistics concerned with deriving information about properties of random variables, stochastic processes, and systems based on observed samples. Some of the important applications of estimation theory are found in control and communication systems, where it is used to estimate the unknown states and parameters of the system.

The estimation problem for dynamic systems may be divided into two parts: parameter estimation and state estimation. The basic difference between a parameter and the state is that the former either does not change at all or changes slowly in time, whereas the latter continuously evolves in time. For example, the state of a satellite is a six-dimensional vector consisting of three position variables and three velocity variables along the axes of an orthogonal coordinate system. The parameters of the satellite are its mass, inertia, and so on. In many control and communication problems, some of the system parameters are not known with desired accuracy. The problem of estimating these parameters from observed data is called parameter identification, though it is basically a problem of estimation. The more general problem of developing a mathematical model of the system from observed data is called system identification. On the other hand, the problem of state estimation is described by names such as signal processing, filtering, and smoothing. The problem belongs to the theory of stochastic processes and is also commonly known as time series analysis. *See* STOCHASTIC PROCESS.

Three basic approaches used for estimation are least-squares, maximum-likelihood, and Bayesian. An estimator is defined as a function of the observations possessing certain desirable properties such as unbiasedness, consistency, and minimum variance. A Kalman filter provides estimates that are optimal in the least-squares, maximum-likelihood and Bayesian sense for a Gauss-Markov model. (A stochastic process is Markov if, given its present state, its future is independent of its past.)

Since their introduction in 1960, Kalman filters and their extensions have found numerous applications. Initially, these filters were developed for space applications such as satellite orbit determination, inertial navigation, Apollo lunar landing module guidance, and so on. The applications to power systems and industrial processes were developed shortly thereafter. Kalman filters have been used for forecasting, water quality prediction, hurricane tracking, aircraft landing systems, and stochastic control. *See* CONTROL SYSTEMS; FLIGHT CONTROLS; GUIDANCE SYSTEMS; PROCESS CONTROL; STATISTICS; SYSTEMS ENGINEERING. [R.K.M.]

Estrogen A substance that maintains the secondary sex characters and organs, such as mammary glands, uterus, vagina, and fallopian tubes, of mammalian females. Naturally occurring substances with this activity are steroid hormones. The principal estrogenic hormone substances are 17(ß)-estradiol (with the structure shown), estrone, and estriol. They are

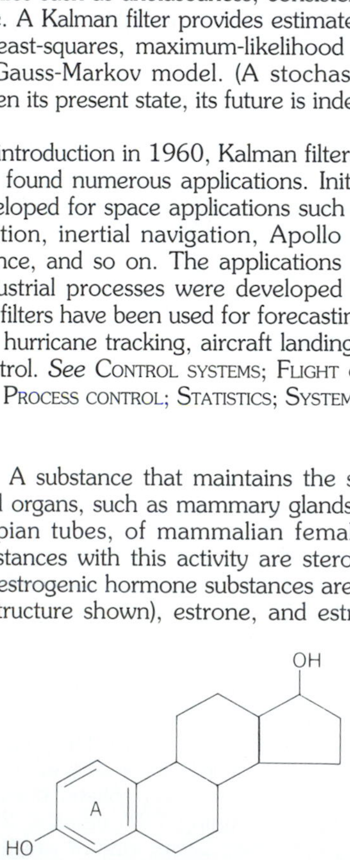

17(*β*)-Estradiol

produced and secreted directly into the bloodstream by the ovary, testis, adrenal, and placenta of pregnancy. Two other naturally occurring estrogenic hormones, equilin and equilenin, have been obtained only from the urine of pregnant mares and are apparently peculiar to that species. Stilbestrol, a synthetic compound with considerable estrogenic activity, has been used extensively in medical practice. *See* HORMONE; STEROID. [R.I.D.]

Estrus The period in mammals during which the female ovulates and is receptive to mating. It is commonly referred to as rut or heat. From one estrus period to the next there occurs a series of changes, particularly in the ovary, uterus, and vagina, termed the estrous cycle. With reference to the ovary, the cycle can be divided into a follicular phase, during which the Graafian follicles are ripening, and a luteal phase, during which the corpora lutea develop in the ovulated follicles. During these two phases mainly estrogen and progesterone, respectively, are secreted, and these hormones control the uterine and vaginal changes. The beginning of the follicular phase is termed proestrus and the luteal phase metestrus. Following the latter, there is a period of relatively little change, termed diestrus. In species in which the latter is prolonged, it is termed anestrus. *See* ESTROGEN; MENSTRUATION; OVUM; PROGESTERONE; REPRODUCTION (ANIMAL). [A.T./H.L.H.]

Estuarine oceanography The study of the physical, chemical, biological, and geological characteristics of estuaries. An estuary is a semienclosed coastal body of water which has a free connection with the sea and within which the sea water is measurably diluted by fresh water derived from land drainage. Many characteristic features of estuaries extend into the coastal areas beyond their mouths, and because the techniques of measurement and analysis are similar, the field of estuarine oceanography is often considered to include the study of some coastal waters which are not strictly, by the above definition, estuaries. Also, semienclosed bays and lagoons exist in which evaporation is equal to or exceeds freshwater inflow, so that the salt content is either equal to that of the sea or exceeds it. Hypersaline lagoons have been termed negative estuaries, whereas those with precipitation and river inflow equaling evaporation have been called neutral estuaries. Positive estuaries, in which river inflow and precipitation exceed evaporation, form the majority, however.

Within estuaries, the river discharge interacts with the sea water, and river water and sea water are mixed by the action of tidal motion, by wind stress on the surface, and by the river discharge forcing its way toward the sea. There is a small difference in salinity between river water and sea water, but it is sufficient to cause horizontal pressure gradients within the water which affect the way it flows. Salinity is consequently a good indicator of estuarine mixing and the patterns of water circulation.

Estuarine ecological environments are complex and highly variable when compared with other marine environments. They are richly productive, however. Because of the variability, fewer species can exist as permanent residents in this environment than in some other marine environments, and many of these species are shellfish that can easily tolerate short periods of extreme conditions. Motile species can escape the extremes. A number of commercially important marine torms are indigenous to the estuary, and the environment serves as a spawning or nursery ground for many other species. *See* MARINE ECOLOGY.

The patterns of sediment distribution and movement depend on the type of estuary and on the estuarine topography. The type of sediment brought into the estuary by the rivers, by erosion of the banks, and from the sea is also important; and the relative importance of each of these sources may change along the estuary. Fine-grained material will move in suspension and will follow the residual water flow, although there may be deposition and re-erosion during times of locally low velocities. The coarser-grained material will travel along the bed and will be affected most by high velocities and, consequently, in estuarine areas, will normally tend to move in the direction of the maximum current. [K.R.D.]

Ethane A member of the alkane or paraffin series of hydrocarbons, formula CH_3CH_3. It occurs in small quantities in natural gas. It is a colorless, odorless, normally gaseous hydrocarbon having a freezing point of $-183.3°C$ ($-297.9°F$) and a boiling point of $-88.6°C$ ($-127.5°F$). Pyrolysis of the ethane and propane portion of natural gas is used as an industrial method for production of ethylene. *See* ALKANE. [L.S.]

Ether One of a class of organic compounds characterized by the structural feature of an oxygen atom linking two hydrocarbon groups, R—O—R′. Ethers are used widely as solvents, both in chemical manufacture and in the research laboratory.

The hydrocarbon radicals R and R′ may be identical (simple ether) or different (mixed ether). They may be aromatic or aliphatic, and the names of the ethers correspond to the hydrocarbon groups present. Thus, $CH_3—O—CH_3$ is methyl ether, rarely dimethyl ether, and $C_6H_5—O—CH_3$ is phenyl methyl ether.

Simple ethers may be considered to be the anhydrides of alcohols and are manufactured from alcohols by catalytic dehydration, as in reaction (1), or from olefins by controlled catalytic hydration, as in reaction (2).

$$2ROH \rightarrow ROR + H_2O \qquad (1)$$

$$2CH_3CH{=}CH_2 + HOH \rightarrow (CH_3)_2\,CH—O—CH(CH_3)_2 \qquad (2)$$

Ethers are less soluble in water than are the corresponding alcohols, but are miscible with most organic solvents. Low-molecular-weight ethers have a lower boiling point than the corresponding alcohols, but for those ethers containing radicals larger than butyl, the reverse is true. Inertness at moderate temperatures, an outstanding chemical characteristic of the saturated alkyl ethers, leads to their wide use as reaction media.

The best known of the ethers is ethyl ether, sometimes called diethyl ether or simply ether, $CH_3CH_2OCH_2CH_3$. It is used in industry as a solvent and in medicine as an anesthetic. When ethyl ether is used as a solvent, its high volatility can cause loss. However this volatility is advantageous in that the ether can be readily removed from the concentrated or crystallized product. The toxicity to humans is low, and recovery from overexposure is rapid and complete. It readily forms explosive mixtures with air, and on standing in containers which have been opened, it forms dangerous peroxides.

Several cyclic ethers (epoxides) are of special importance and interest. The simplest of these is ethylene oxide or oxirane (I), made industrially by the oxidation of ethylene with air over a silver catalyst. Dioxane or 1,4-dioxane (II) is prepared by the catalytic dimerization of ethylene oxide. It is used extensively as an industrial solvent. Furan (III), made by the decarbonylation of furfural, is the most important ether obtained from an agricultural source. Most of it is hydrogenated to form the useful solvent tetrahydrofuran (IV). *See* ETHYLENE OXIDE; FURAN.

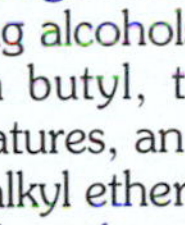
(I)

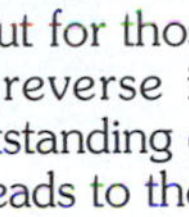
(II)

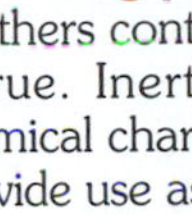
(III)

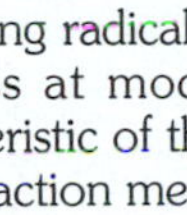
(IV)

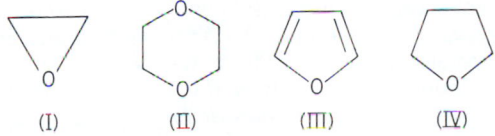

Certain large-ring polyethers, the crown ethers, are able to increase the solubility of alkali metal salts in nonpolar organic solvents. Specific metal complexes are formed by these crown ethers with alkali metal cations, the specificity for a given cation depending upon the hole in the middle of the crown ether structure. *See* CROWN ETHERS; HETEROCYCLIC COMPOUNDS. [P.E.F.]

Ether hypothesis James Clerk Maxwell and his contemporaries in the 19th century found it inconceivable that a wave motion should propagate in empty space. They therefore postulated a medium, which they called the ether, that filled all space and transmitted electromagnetic vibrations.

Direct experimental attempts to establish the existence of an absolute ether frame of reference, in which Maxwell's equations hold and light has the velocity c, have failed. The best known of these is the Michelson-Morley experiment, in which an attempt was made to measure the velocity of the Earth relative to the ether. At present, there is no evidence whatever that the ether exists. *See* LIGHT; MAXWELL'S EQUATIONS; RELATIVITY. [W.R.Sm.]

Ethology The study of animal behavior. Ethology differs from psychology in that the interest is in behavior that occurs in a natural context. Although laboratory experiments are often done, the ethologist focuses on the adaptive value of behavior.

Some patterns of coordination show little variability while other movements are graded and more sensitive to changes in the environment. All degrees of intermediacy fall between these poles. Relatively form-constant behavior patterns that are easily recognized have been called modal action patterns. The best known are usually concerned with such social actions as courtship, fighting, or parental behavior. But they also serve nonsignal activities, such as the grooming movements of crickets or mice and the specialized feeding behavior of a variety of animals.

In animal behavior the same stimulus is responded to in differing ways at different times; there is no simple one-to-one relationship. The analysis of the factors leading to this variation in responsiveness is termed the study of motivation.

Stimulus filtering refers to apparent awareness of only a few of the great number of stimuli bombarding the animal. It may involve any sensory modality, such as the visual, chemical, auditory, or mechanical.

Among the classic works in ethology are Konrad Lorenz's experiences with young geese which adopted him as their mother. Lorenz characterized imprinting as follows. (1) There is a very early and brief sensitive period. (2) The object encountered in that time is the one that is followed, then and later. (3) This "mother" will become the object of sexual responses in the mature animal; thus the object of response is learned long before the performance of that response. (4) The learning is irreversible in that the animal will respond to the same object throughout life. (5) The response generalizes to all members of the species. While others before Lorenz had observed similar phenomena, none had dealt with it in a significant conceptual framework.

This concept has been hotly debated, in part because so many and widely different phenomena have been called imprinting; it is doubtful that these share the same causal mechanisms. Also a disputed issue is whether imprinting differs from more traditionally defined learning. Most research in this area is directed toward the sensitive period and what causes it to end. Lack of social contact is one way of prolonging it. While much argued, the development of fear responses, perhaps by becoming familiar with one's general environment, seems important in ending the sensitive period. [G.W.B.]

Ethyl alcohol Probably the best known of the alcohols, ethyl alcohol (C_2H_5OH) is also called alcohol, ethanol, grain alcohol, industrial alcohol, fermentation alcohol, cologne spirits, ethyl hydroxide, and methylcarbinol. Pure ethyl alcohol is a colorless, volatile liquid which is flammable and toxic and has a pungent taste. It boils at 78.4°C (173°F) and melts at $-112.3°C$ ($-170.1°F$), has a specific gravity of 0.7851 at 20°C (68°F), and is soluble in water and most organic liquids. It is one of the most important industrial organic chemicals. Billions of pounds of it are produced annually. Ethyl alcohol is produced by chemical synthesis and by fermentation or biosynthetic processes.

Ethyl alcohol is used as a solvent, extractant, antifreeze, and intermediate in the synthesis of innumerable organic chemicals. It is also an essential ingredient of alcoholic beverages.

Various grades of ethyl alcohol are produced, depending on their intended use. U.S. Pharmaceutical grade is 95% ethyl alcohol by volume. National Formulary grade is 99+% ethyl alcohol by weight; it is also called absolute, or anhydrous, alcohol. Denatured alcohol contains a small amount of a malodorous or obnoxious material to prevent the use of this nontaxed grade of ethyl alcohol for beverage purposes. Ethyl alcohol which is employed in the manufacture of beverages, medicine, and flavoring is taxed and all ethyl alcohol production is closely supervised by the government.

The concentration of ethyl alcohol is often expressed as proof, which is simply twice the volume percent of ethyl alcohol. Thus, a 100-proof whiskey contains 50% by volume of ethyl alcohol. *See* ALCOHOL; FERMENTATION. [J.W.L.]

Ethylene A colorless gas, formula $H_2C{=}CH_2$, with a boiling point of $-103.8°C$ ($-155°F$) and a melting point of $-169.4°C$ ($-273°F$). Ethylene is the most important synthetic organic chemical in terms of volume, sales value, and number of derivatives. Polyethylene, the most important derivative of ethylene, is produced by both high- and low-pressure processes to make high- and low-density, high-molecular-weight thermoplastic polymers. Other important derivatives are ethylene chlorohydrin, ethylene dichloride, vinyl chloride, ethanol, vinyl acetate, and acetaldehyde. *See* ALKENE; POLYMER. [R.K.Ba.]

Ethylene glycol A colorless, nearly odorless, sweet-tasting, hygroscopic liquid, formula $HOCH_2CH_2OH$. It is relatively nonvolatile and viscous and is the simplest member of the glycol family. Ethylene glycol freezes at $-13°C$ (8.6°F), boils at 197.6°C (387.7°F), and is completely soluble in water, common alcohols, and phenol. Low molecular weight, low volatility, water solubility, and low solvent action on automobile finishes make ethylene glycol ideal as a radiator antifreeze and coolant. Other uses for this commodity chemical are in polyester resins, explosives, brake and shock-absorber fluids, and alkyl-type resins. *See* ANTIFREEZE MIXTURE; GLYCOL; POLYESTER RESIN.

Dibasic acids or anhydrides react with ethylene glycol to form polyester condensation polymers. Reaction with terephthalic acid or its esters produces polyester resins which can be spun to fibers that find wide use in clothing and general fabrics applications. The polymer also has important film applications. *See* ETHYLENE OXIDE; POLYETHYLENE GLYCOL. [R.K.Ba.]

Ethylene oxide The simplest cyclic ether or epoxide, with the formula C_2H_4O, and the structure shown. It is also called

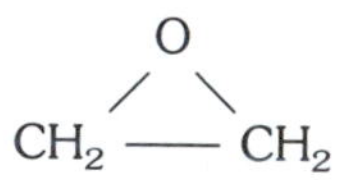

epoxyethane and oxirane. Commercial processes use either air or oxygen to oxidize ethylene to ethylene oxide. *See* ETHER; ETHYLENE; HETEROCYCLIC COMPOUNDS.

Ethylene oxide is a colorless gas boiling at 10.4°C (50.7°F) and melting at $-112°C$ ($-169.6°F$). Its vapors are flammable and explosive, and it is considered a relatively toxic liquid and gas. It is miscible in all proportions with water, alcohols, ethers, and other organic solvents.

About 50% of the ethylene oxide produced is converted to ethylene glycol; about 30% is used in the manufacture of nonionic surfactants, glycol ethers, and ethanolamines. Gaseous ethylene oxide in CO_2 or difluoromethane is used as a fumigant and a sterilizing agent for medical equipment. *See* ETHYLENE GLYCOL. [R.K.Ba.]

Ethylenediaminetetraacetic acid A chelating agent for metallic ions, abbreviated EDTA. Tetrasodium EDTA is the most common form in commerce, but other metallic chelates are marketed, for example, iron, zinc, and calcium. Tetrasodium EDTA is a white solid, very soluble in water and forming a basic solution. Prepared from ethylenediamine, formaldehyde, and sodium cyanide in basic solution, or from ethylenediamine and sodium chloroacetate, EDTA is a strong complexing and chelating agent. It reacts with many metallic ions to form soluble chelates, which are used in a variety of applications. *See* CHELATION. [F.W.]

Eubrya A subclass of the Bryopsida (Musci) which is composed of many orders of living mosses. There are approximately 80 families, 650 genera, and 14,000 species. The leafy gametophytes arise from buds on the protonemata, which are nearly always filamentous or branched green threads and are attached to the substratum by rhizoids.

In a cross section of the stem an epidermis, a cortical area, and a central cylinder are commonly distinct (Fig. 1). The leaves are unistratose in the majority of species, bistratose in some, and may be of more than two layers in other species. A costa or midrib more than one cell in thickness is present in many species.

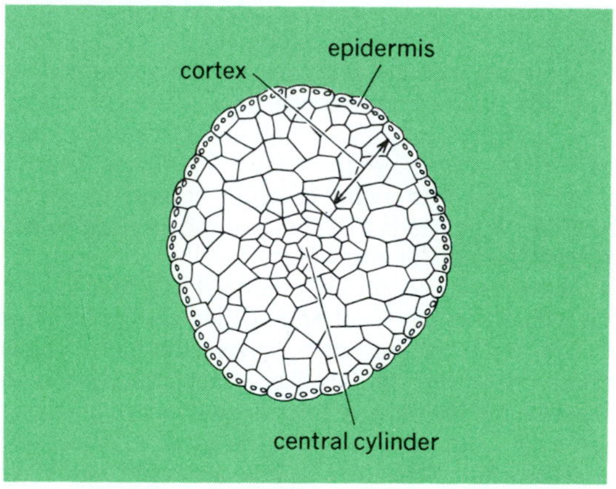

Fig. 1. Cross section of stem of gametophyte of *Funaria*. (*After G. M. Smith, Cryptogamic Botany, vol. 2, 2d ed., McGraw-Hill, 1955*)

The sporophyte consists of the foot, seta, and capsule. The setae may be very short to much elongated, but in the Eubrya there is never a pseudopodium to elevate the capsule.

The columella (absent in *Archidium*) and spores develop from the endothecium. The columella penetrates the spore-producing layer, the archesporium. The sporogenous tissue, therefore, is not in the form of a dome over the upper portion of the columella, as in the Sphagnobrya and Andreaeobrya. An air cavity, not present in the Andreaeobrya and Sphagnobrya, separates the spore sac from the wall of the capsule.

In the Eubrya the mouth of the urn frequently bears an annulus and one or more rows or rings of processes, resembling a fringe, the peristome. When the spores are ripe and the operculum is shed or lost, the teeth of the peristome, which are hygroscopic, bend into the cavity of the urn (Fig. 2). As the teeth dry, they straighten, lifting the spores to disseminate them in the wind. If the peristome is single, there is one row of teeth. If the peristome is double, it consists of an outer ring of teeth, termed the outer peristome, and an inner ring of segments. Often one or more cilia occur between the segments.

For convenience in classification, the subclass Eubrya is

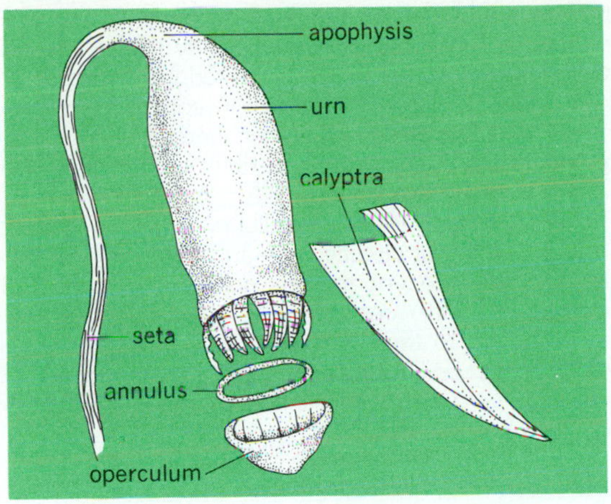

Fig. 2. Sporangium of a moss. (*After W. W. Robbins et al., Botany, 3d ed., Wiley, 1964*)

often divided into plants acrocarpous (sporophyte at end of leafy gametophyte or terminal) and pleurocarpous (sporophyte in axils of leaves along the side of the stem or lateral).

The habitats of the plants in the Eubrya may be dry, moist, or wet; and the plants can exist in the tropics, temperate zones, or in the arctic regions. Some mosses grow in water but the majority live on rock, bark, wood, leaves, and soil, most frequently in shaded areas. *See* BRYOPSIDA. [W.H.W.]

Eubryales An order of mosses. The plants are acrocarpous, vary from small to robust (0.2–3 in. or 0.5–8 cm), and generally grow in tufts. The costate leaves, of various shapes, are arranged on the stem in several rows. The terminal sporophyte consists of an elongated seta and a cernuous, horizontal, or pendant capsule. The peristome is commonly double. The outer row of processes consists of 16 teeth, the inner of 16 segments which are carinate or carinately split, and cilia are often present. *See* BRYOPSIDA. [W.H.W.]

Eucalyptus A large genus of evergreen trees, with about 600 species, belonging to the myrtle family (Myrtaceae), and occurring in Australia, New Guinea, and the Sunda Islands. It has been widely planted as an ornamental and for wood production. Some species are valuable for fuel and lumber. The leaves of blue gum (*E. globulus*) contain an essential oil used in medicines and perfumes. *See* MYRTALES. [P.D.St./K.P.D.]

Eucarida A large and economically important superorder of higher Crustacea (Malacostraca), including shrimps, lobsters, hermit crabs, and crabs. They have the following characteristics: The shell is fused dorsally with all the thoracic segments; and the eyes are on movable stalks. Eggs are usually attached to the abdominal appendages. The liver is much ramified; the heart is short and thoracic. Development occurs as a rule with metamorphosis and, in primitive forms, a free-swimming nauplius stage. Eucarida comprises two orders, the Euphausiacea and Decapoda. *See* DECAPODA (CRUSTACEA); EUPHAUSIACEA; MALACOSTRACA. [I.Gor.]

Euclasterida An order of Asteroidea (starfishes) in which there are many deciduous arms clearly differentiated from a small discoidal body. Euclasterids normally inhabit deep water and bear a superficial resemblance to ophiuroids, but they have typically open asteroid ambulacra (these are closed in ophiuroids). They range in time from the lower Oligocene to the present. *See* ASTEROIDEA; ECHINODERMATA. [A.C.C.]

Euclidean geometry The chief subject matter of the monumental 13-volume work called *The Elements*, written about 300 B.C. by the Greek mathematician Euclid, who taught and founded a school of geometry at Alexandria. One of the milestones in the history of scientific thought, these books of Euclid still occupy an important position in mathematical instruction today.

Geometry, as developed in Euclid, was a systematic body of mathematical knowledge, built by deductive reasoning upon a foundation of three main pillars: (1) definitions of such things as points, lines, planes, angles, circles, and triangles; (2) the assumption of certain geometrical postulates regarded as true but perhaps not self-evident; and (3) the assumption of certain axioms or common notions which were taken to be self-evident truths. The body of Euclid's great work consists of a set of propositions or theorems, each derived systematically and logically from the definitions, axioms, and postulates of his foundation and from theorems already proved.

Five postulates may be paraphrased as:

1. A unique straight line can be drawn from any point to any other point.
2. A finite straight line can be extended continuously in either direction in a straight line.
3. A circle can be described with any given center and radius.
4. All right angles are equal.
5. If a straight line falling on two straight lines makes interior angles on the same side with a sum less than two light angles, the two straight lines, if produced indefinitely, meet on that side on which the angle sum is less than two right angles.

This fifth postulate is known as the parallel postulate and is essentially equivalent to the statement that "a unique line parallel to a given line can be constructed through any point not on the line." For 2000 years many unsuccessful attempts were made to prove that this postulate was a consequence of Euclid's other postulates, definitions, and axioms. Only as recently as the 19th century did N. I. Lobachevski (1793–1856), J. Bolyai (1802–1860), and G. F. B. Riemann (1826–1866) show that the parallel postulate was independent of the others, by constructing so-called noneuclidean geometries in which the fifth postulate is not valid.

Certain axioms were stated by Euclid and treated as self-evident truths.

1. Things which are equal to the same thing are equal to each other.
2. If equals be added to equals, the wholes are equal.
3. If equals be subtracted from equals, the remainders are equal.
4. Things which coincide with one another are equal to one another.

A fifth axiom, ascribed by Proclus to Euclid, is believed to have been added by later writers.

5. The whole is greater than the part.

This has also been replaced by the statement: The whole is equal to the sum of its parts.

The first six books of Euclid include most of the subject matter commonly taught in high school geometry. Book 1, after listing a large number of definitions and the postulates and axioms described above, contains 48 propositions, including many properties relating to perpendicular and parallel lines, angles, and congruent triangles, and the book culminates with the important Pythagorean theorem.

Book 2 contains 14 propositions on geometrical algebra, establishing relationships between the areas of certain related squares and rectangles. Book 3 deals with circles, chords, inscribed angles, and other figures related to the circle; book 4 with inscribed and circumscribed circles and polygons; book 5 with ratios; and book 6 with areas and similar triangles. Books 7, 8, 9, and 10 are concerned with number theory, square and cube roots, and incommensurable magnitudes; books 11 and 12 are concerned with solid geometry and mensuration; and book 13 is concerned with extreme and mean ratio and the five regular solids. *See* DIFFERENTIAL GEOMETRY; POLYHEDRON; PROJECTIVE GEOMETRY; SOLID (GEOMETRY). [J.S.F.]

Eucoccida The most important order of the protozoan subclass Coccidia. In this group there is alternation of asexual and sexual phases of the life cycle, and the parasitic stages occur within the cells of vertebrates or invertebrates. There are three suborders.

In the suborder Adeleina, the macrogamete and microgametocyte are joined during development (syzygy) and the microgametocytes produce only a few microgametes. Members of this suborder are parasites of invertebrates and lower vertebrates; a few even occur in higher vertebrates.

In the suborder Eimeriina, there is no syzygy and the microgametocytes produce a large number of microgametes. This group contains several hundred species of coccidia, most of which occur in the intestinal cells of vertebrates.

In the suborder Haemosporina, there is no syzygy and the microgametocytes produce a moderate number of microgametes. Asexual development takes place in the blood cells of vertebrates, and sexual development in blood-sucking insects or mites. This suborder contains *Plasmodium*, species of which cause malaria in humans, lower primates, birds, and other vertebrates. *See* COCCIDIA; MALARIA; PROTOZOA. [N.D.L.]

Eucommiales An order of flowering plants, division Magnoliophyta (Angiospermae), in the subclass Hamamelidae of the class Magnoliopsida (dicotyledons). The order consists of a single family, genus, and species, *Eucommia ulmoides*, of China. It is a simpleleaved tree with reduced, solitary flowers that lack a perianth, and leaves without stipules, the paired basal appendages found on many leaves. *See* HAMAMELIDAE; MAGNOLIOPSIDA. [A.Cr.]

Euechinoidea A subclass of Echinoidea (sea urchins, sand dollars, and heart urchins) in which the test plates are consistently arranged in double rows or columns. Five double rows bear the tube-feet (the ambulacral rows), and these alternate with five which do not (the interambulacral rows). *See* ATELOSTOMATA; DIADEMATACEA; ECHINACEA; ECHINODERMATA; ECHINOIDEA; GNATHOSTOMATA (ECHINODERMATA). [A.C.C.]

Eugenics The study of factors that influence the hereditary qualities of future generations. It may be thought of as both a science and a social movement. Eugenics proposes to improve humanity's future by increasing the number of children produced by persons who are, by some definition, superior and by reducing the number produced by persons who are physically or mentally deficient. Attempts to encourage larger families from superior parents are called positive eugenics, attempts to reduce the number of children from defective parents negative eugenics.

There have been a number of proposals regarding negative eugenics in the United States. Several of the states have laws providing for sterilization of persons with severe mental or physical defects. Segregation of persons in public institutions also has the effect of reducing the number of births. Finally there is the advocacy of greater use of contraceptives by persons likely to transmit hereditary weaknesses. Positive eugenics has consisted largely of exhortations for the well endowed to have more children.

Such eugenic proposals have not had complete acceptance. There are objections on religious grounds. There are those who think that eugenic laws, particularly if they involve compulsory sterilization, are an unwarranted infringement on basic human rights. A different reservation comes from those who point out that knowledge of human heredity is not very complete and that prediction of the attributes of children from their parents is not very exact.

The spectacular success of cellular and molecular genetics and the possibility of directly changing, replacing, or otherwise influencing the deoxyribonucleic acid (DNA) in bacteria and viruses have led to considerable speculation about what humans can do to their heredity if and when the techniques known for microbes become available. However, such possibilities are still very much in the future, nobody knows how far away. *See* HUMAN GENETICS. [J.F.Cr.]

Euglenida An order of the class Phytamastigophorea. This order of protozoans, also known as Euglenoidina, includes the largest green noncolonial flagellates, *Euglena ehrenbergii*, which are 400 micrometers long. Many of the colorless members are also large. They have one or two equal or subequal flagella. There are relatively few genera of Euglenida. *See* PHYTAMASTIGOPHOREA.

Euglenids generally show an anteroposterior elongation; however, *Trachelomonas volvocina* is practically spherical, and *Euglena acus* is extremely needlelike. Many are enclosed in a thick pellicle which is sculptured into keels. The pellicle is sometimes ornamented with striae or small wartlike protuberances. Colony formation is restricted to *Colacium*, in which a flagellated euglenoid cell usually attaches to a copepod by its anterior end and develops into an arboroid structure of several nonflagellated cells. Tests or shells are common.

Euglenids have two distinguishing features. They synthesize a starchlike polysaccharide, paramylum; and they have a gullet-reservoir system.

In habitat the Euglenida are widespread. Green members occur mostly in fresh water and frequently in such numbers as to form blooms. Unlike the Phytomonadida, the Euglenida appear in warmer waters but do not seem to occur in hot springs. They abound in citrus-waste lagoons and in later stages of sewage-treatment waters, an indication of their partially saprozoic nutrition. *See* MASTIGOPHORA; PROTOZOA. [J.B.L.]

Euglenophyceae A class coextensive with the division Euglenophycota, comprising unicellular colorless or photosynthetic flagellates with very distinctive cytological characters. In protozoological classification, these organisms constitute an order, Euglenida, of the class Phytomastigophora. Although photosynthetic euglenoids are like Chlorophycota in containing chlorophyll *a* and *b*, in most other respects they are so different from those algae as to suggest an independent phylogenetic origin of their pigments. About 1000 species have been described and classified into about 40 genera and 6 orders. *See* EUGLENIDA.

Most euglenoids are free-swimming and have two flagella, one of which may be nonemergent, arising from an anterior invagination known as a reservoir. Photosynthetic euglenoids contain one to many grass-green chloroplasts, which vary from minute disks to expanded plates or ribbons. Colorless euglenoids depend on osmotrophy or phagotrophy for nutrient assimilation.

Euglenoids are found most commonly in fresh water rich in organic matter, but they also occur in marine or brackish habitats, on mud or sand, and in ice or snow. A few species prefer very acidic water. [P.C.Si.; R.L.Moe]

Eugregarinida An order of the subclass Gregarinia. These protozoans are common parasites of invertebrates such as the arthropods and annelids. Only sexual reproduction occurs in the life history of these animals; schizogony is lacking. This order is divided into the suborders Cephalina and Acephalina. In North America grasshoppers of the genus *Melanoplus* are commonly the host for the cephaline *Gregatina rigida*, inhabiting the lumen of the gut. *See* GREGARINIA; INVERTEBRATE PATHOLOGY; PROTOZOA. [E.R.B./N.D.L.]

Eukaryotae The vast array of living and fossil organisms comprising all taxonomic groups above the primitive unicellular prokaryotic level typified by the bacteria. The Eukaryotae thus include all plants and animals, the unicellular protists, and the fungi.

All organisms in this group possess an organized nucleus (or nuclei) in their cells, with a surrounding nuclear envelope and paired deoxyribonucleic acid–containing chromosomes; and elaborate cytoplasmic organelles, such as mitochondria, Golgi bodies, lysosomes, peroxisomes, endoplasmic reticulum, microfilaments and microtubules in various arrays, and, in photosynthetic forms, plastids or chloroplasts. Characteristically, centrioles, cilia, or flagelia are also present, the latter locomotory organelles composed mainly of tubulin with microtubules exhibiting a universal 9 + 2 pattern. *See* CELL (BIOLOGY); CILIA AND FLAGELLA; PROKARYOTAE.

Organization at the organismal level ranges from solitary unicellular to colonial unicellular, mycelial, syncytial (coenocytic), and truly multicellular with extensive tissue differentiation. Modes of nutrition run the gamut: absorptive, ingestive, photoautotrophic, plus combinations of these three major kinds. Life cycles vary tremendously, with haploidy more characteristic of "lower" groups and diploidy of the "higher" taxa. Reproduction, similarly, includes both asexual and sexual methods. In the "higher" multicellular forms, true embryos develop from the diploid zygote stage, which has resulted from fusion of sperm and egg cells.

Aerobic metabolism is commonly exhibited, especially by aquatic and terrestrial forms; anaerobic mechanisms exist, however, for numerous species found in poorly oxygenated habitats, including various sites within bodies of host organisms.

Size of species range from 1 micrometer (certain protozoa and algae) to many meters (whales, trees), Habitats cover all possible ecological niches: aquatic, terrestrial, and aerial, for free-living forms (many of which are motile, with the major exception of the trophic stage of various plants); and in or on all kinds of hosts, for symbiotic or parasitic forms (internal habitats, including cells, tissues, organs, or various body cavities). Dormant stages include cysts, spores, and seeds; these are often involved in dispersion or propagation of the species. *See* ANIMAL KINGDOM; FUNGI; PLANT KINGDOM; PROTOZOA. [J.O.C.]

Euler angles Three angular parameters that specify the orientation of a body with respect to reference axes. They are used for describing rotating systems such as gyroscopes, tops, molecules, and nonspherical nuclei. They are not symmetrical in the three angles but are simpler to use than other rotational parameters.

Unfortunately, different definitions of Euler's angles are used, and therefore it is confusing to compare equations in different references. The definition given here is the majority convention according to H. Margenau and G. Murphy.

Let *OXYZ* be a right-handed cartesian (right-angled) set of fixed coordinate axes and *Oxyz* a set attached to the rotating body (see illustration).

The orientation of *Oxyz* can be produced by three successive rotations about the fixed axes starting with *Oxyz* parallel to *OXYZ*. Rotate through (1) the angle ψ counterclockwise about *OZ*, (2) the angle θ counterclockwise about *OX*, and (3) the angle ϕ counterclockwise about *OZ* again. The line of intersection *OK* of the *xy* and *XY* planes is called the line of nodes.

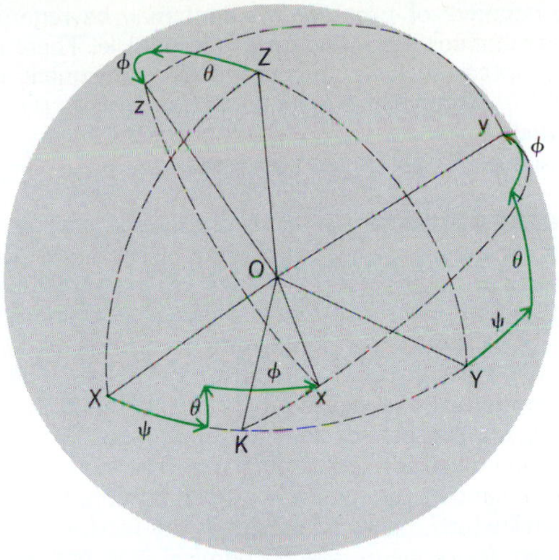

Euler angles. The successive movements of the axes on a unit sphere described in the text are shown by arrows. The complete rotation may also be obtained by a different sequence of rotations, namely, first through ϕ about *OZ*, then through θ about the displaced x axis (which is *OK*), then through ψ about *OZ*.

Denote a rotation about *OZ*, for example, by *Z* (angle). Then the complete rotation is, symbolically, given by the equation below where the rightmost operation is done first.

$$R(\psi,\theta,\phi) = Z(\phi)X(\theta)Z(\psi)$$

[B.G.]

Euler's equations of motion A set of three differential equations expressing relations between the force moments, angular velocities, and angular accelerations of a rotating rigid body. They are equations of motion in the usual dynamical sense, having the form of Eqs. (1)–(3).

$$I_1(d\omega_1/dt) + (I_3 - I_2)\omega_2\omega_3 = M_1 \quad (1)$$

$$I_2(d\omega_2/dt) + (I_1 - I_3)\omega_3\omega_1 = M_2 \quad (2)$$

$$I_3(d\omega_3/dt) + (I_2 - I_1)\omega_1\omega_2 = M_3 \quad (3)$$

The formulation employs as coordinate axes the three principal axes of rotational inertia of the body that can rotate about a body-fixed point, which is the center of mass if constraints are absent. These reference axes, which form a righthand set, are indicated by subscripts 1, 2, and 3 in the equations, where I_1, I_2, and I_3 represent the principal moments of inertia; ω_1, ω_2, and ω_3 the angular velocities about the axes; M_1, M_2, and M_3 the corresponding force moments; and t the time. *See* RIGID-BODY DYNAMICS. [R.A.Fl.]

Euler's momentum theorem A principle of fluid mechanics which states that momentum of the particles in a moving frictionless, or inviscid, fluid is conserved. This theorem is expressed by Euler's hydrodynamical equations, a set of nonlinear partial differential equations. When viscous shear forces are included, the momentum equations are called the Navier-Stokes equations. The eulerian method of viewing fluid motion is to consider it as a velocity-pressure field; that is, the velocity of the fluid particles is considered as a vector function of position in space and time, and the pressure as a scalar function of position and time. Euler's equations are essentially a restatement of Newton's single-particle law, force equals mass times acceleration, adapted to the many-particle continuum concept of fluid motion.

The negative of pressure gradient may be regarded as a force per unit volume acting on a fluid particle. There may also be body forces such as gravity and electromagnetic forces (if the fluid is electrically conducting). *See* BERNOULLI'S THEOREM; GAS DYNAMICS; KELVIN'S CIRCULATION THEOREM; LAPLACE'S IRROTATIONAL MOTION; NAVIER-STOKES EQUATIONS. [A.E.Br.]

Eumalacostraca A series of the class Crustacea which contains the superorders Syncarida, Peracarida, Eucarida, Hoplocarida, and Pancarida. These higher crustaceans are shrimplike in appearance and have several invariable characteristics. The thorax has eight segments and the abdomen has six segments in addition to the telson or tail fan. The last pair of pleopods has a specialized form and function. In many orders of Eumalacostraca the first or second pleopods in the male are modified for sexual purposes. The stomach is divided into two chambers, the anterior cardiac portion, containing the gastric mill which mechanically grinds the food, and the posterior pyloric portion, in which chemical digestion is initiated. Characteristic larval stages are the nauplius, contained within the egg, and the mysis and zoaea, which are free-swimming. These organisms are free-living or parasitic. The parasites show a great adaptability to their hosts which are fish or other crustaceans. The majority of the species are marine and many, except the Hoplocarida, show adaptive tendencies to brackish and fresh-water environments. The oviducts open on the sixth thoracomere while the sperm ducts open on the eighth. *See* EUCARIDA; HOPLOCARIDA; PERACARIDA; SYNCARIDA. [H.J.]

Eumycetozoida An order of Mycetozoia. These are slime molds which form a plasmodium, a multinucleate stage that may in some species measure a foot or more across. The plasmodium, typically found on decaying plant material, is a migratory phagotroph preying on microorganisms. A mature plasmodium is a sheet of protoplasm containing a network of channels through which rapid endoplasmic streaming occurs.

Resistant cysts (spores), possibly haploid, are produced in sporangia. Liberated spores hatch into myxoflagellates or myxamebae (which may change into flagellates). Either type may represent gametes, and mating types have been reported in certain species. [R.P.H.]

Eumycota True fungi; a division of heterotropic organisms with absorptive nutrition, capable of utilizing insoluble food outside the cell by secretion of digestive enzymes. Storage products are chiefly glycogen or lipid. Fungi have a well-defined cell wall formed principally of chitin; however, in some taxa it is made of cellulose, or chitin and cellulose. Typically, the fungus thallus consists of microscopic, branched, threadlike hyphae (collectively called the mycelium) which produce radiating, macroscopic colonies on a substrate or host. The filamentous hypha is divided by cross walls (septa) into a single row of cells; it grows apically. Some species are coenocytic (without cross walls), and others are unicellular.

Reproductive bodies are highly variable in morphology and size. Sexual reproduction is known in many species. Of cosmopolitan distribution, Eumycota are found in practically every type of habitat (soil, air, fresh or brackish water, ocean, and so forth). Most are strictly aerobic, although some are facultative anaerobes. They occur as parasites of animals and especially of plants, or as symbionts, but the majority are saprobes. *See* FUNGI. [D.J.S.B.]

Euphausiacea A well-defined marine order of the class crustacea. All are shrimplike in appearance. The order contains two families, one with a single species, the other with 84 species divided among 10 genera. Euphausiids are predominantly pelagic organisms, but a few species are neritic. Euphausiids form a significant proportion of the total planktonic biomass.

Euphausiids, with the exception of *Bentheuphausia amblyops*, possess photophores that emit a brilliant blue-green light. The eyes are compound and contain three pigments: a carotenoid (astaxanthin), a melanoid, and a photolabile substance that is probably a visual pigment. Respiration is by means of foliose, digitiform gills located at the bases of the second to eighth thoracic appendages.

The blood is a pale, leukocyte-bearing fluid with hemocyanin as the respiratory pigment. The heart is compact and has two pairs of ostia.

All species are strictly marine and do not occur in freshwater or brackish environments. Species are found in all oceans and seas, with the exception of the brackish Baltic and Black seas. Known as krill to whalers, they constitute the diet of many whales, particularly the whalebone whales. The euphausiids can be processed to yield large quantities of protein. Krill pastes and meal can be manufactured to feed domestic animals or used in therapeutic diets. Krill sausages, krill-stuffed eggs, and shrimp butter are already marketable products in Japan, Russia, and Germany. *See* CRUSTACEA; DECA-PODA (CRUSTACEA); MALACOSTRACA. [J.Mau.]

Euphorbiales An order of flowering plants, division Magnoliophyta (Angiospermae), in the subclass Rosidae of the class Magnoliopsida (dicotyledons). The order consists of the large family Euphorbiaceae (about 7500 species) and four small satellite families which have fewer than 150 species among them. The Euphorbiales are a group in which the flowers have become unisexual and then undergone further reduction accompanied by aggregation.

The Christmas poinsettia (*Euphorbia pulcherrima*) and the para rubber tree (*Hevea brasiliensis*) are well-known members of the Euphorbiaceae. Many African euphorbiads are spiny stem-succulents which resemble cacti in habit. Aside from the pronounced floral differences, the euphorbiads have a milky juice, which the cacti do not. *See* MAGNOLIOPSIDA; ROSIDAE. [A.Cr.]

Europe Although long called a continent, in many physical ways Europe is but a great western peninsula of the Eurasian landmass. Its eastern limits are arbitrary and are conventionally drawn along the water divide of the Ural Mountains, the Ural River, the Caspian Sea, and the Caucasus watershed to the Black Sea. On all other sides Europe is surrounded by salt water. Of the oceanic islands of Franz Josef Land, Spitsbergen (Svalbard), Iceland, and the Azores, only Iceland is regarded as an integral part of Europe; thus the northwestern boundary is drawn along the Danish Strait.

Europe is not only peninsular but has a large ratio of shoreline to land area reflecting a notable interfingering of land and sea. Excluding Iceland, the maximum north-south distance is (3529 mi) (5680 km); and the greatest east-west extent is 2398 mi (3860 km). Of Europe's area of 3,881,000 mi^2 (10,050,000 km^2) 73% is mainland, 19% peninsulas, and 8% islands. Also, 51% of the land is less than 155 mi (250 km) from shores and another 23% lies closer than 310 mi (500 km). This situation is caused by the inland seas that enter, like arms of the ocean, deep into the northern and southern regions of Europe, which thus becomes a peninsula of peninsulas. The most notable of these branching arms of salt water are the White Sea, the North Sea, the Baltic Sea with the Gulf of Bothnia, the English Channel (La Manche), the Mediterranean Sea with its secondary branches, and finally, the Black Sea. Even the Caspian Sea, presently the largest salt-water lake of the world, formed part of the southern seas before the folding of the Caucasus. The penetration of the landmass by these seas brings marine influences deep into the continent and provides Europe with a balanced climate favorable for human evolution and settlement.

Europe has a unique diversity of land forms and natural resources. The relief, as varied as that of other continents, has an average elevation of 980 ft (300 m) as compared with North America's 1440 ft (440 m). The shape and the overall physiographic aspect of the great peninsula are controlled by geologic structure which delimits the major regional units.

Climate is determined by a number of factors. Probably the most important are a favorable location between 35° and 71°N latitudes on the western or more maritime side of the world's largest continental mass; the west-to-east trend (rather than north-south) of the lofty southern ranges and the Central Lowlands, as well as of the inland seas, which permit the prevailing westerly winds of these latitudes to carry marine influences deep into the continent; the beneficial influence of the North Atlantic Drift, which makes possible ice-free coasts far within the Arctic Circle; and the low elevation of the northwestern mountain ranges and the Urals, which allows the free shifting of air masses over their crests.

The intricate relief and the climates of Europe are well reflected in the drainage system. Extensive drainage basins with large slow-flowing rivers are developed only in the Central Lowlands, especially in the eastern part. Streams with the greatest discharge empty into the Black Sea and the North Sea, although Europe's longest river, the Volga, feeds the Caspian Sea. Second in dimension is the Danube, which crosses the Carpathian Basin and cuts its way twice through mountain ranges at the Gate of Bratislava and at the Iron Gate. The Rhine and Rhone are the two major Alpine rivers with headwater sources close to each other but feeding the North Sea and the Western Mediterranean Basin, respectively. Abundant precipitation throughout the year, as well as the permeable soils and the dense vegetation which temporarily store the water, provides the streams of Europe north of the Southern Highlands with ample water throughout the seasons. The combined effects of poor vegetation, rocky and desolate limestone karstlands, and slight annual precipitation result in intermittent flow of the rivers along the Mediterranean coast, especially on the eastern side of peninsulas. Only the Alpine rivers carry enough water, and if it were not for the Danube and Rhone, both originating in regions north of the Alps, the only major river of the Mediterranean basin would be the Po. *See* ATLANTIC OCEAN; BALTIC SEA; BLACK SEA; CONTINENT; MEDITERRANEAN SEA. [G.T.]

Europium A chemical element, Eu, atomic number 63, atomic weight 151.96, a member of the rare-earth group. The stable isotopes, ^{151}Eu and ^{153}Eu, make up the naturally occurring element. The metal is the second most volatile of the rare earths and has a considerable vapor pressure at its melting point. It is very soft, is rapidly attacked by air, and really belongs more to the calcium-strontium-barium series than to the rare-earth series.

The element is attractive to the atomic industry, since the elements can be used in control rods and as nuclear poisons. These poisons are materials added to a nuclear reactor to balance the excess reactivity at start-up, and are so chosen that the poisons burn out at the same rate as the excess activity decreases. The television industry uses considerable quantities of phosphors, such as europium-activated yttrium orthovanadates, and other europium-activated yttrium phosphors have been patented. These phosphors give a brilliant red color and are used in the manufacture of television screens. *See* RARE-EARTH ELEMENTS. [F.H.Sp.]

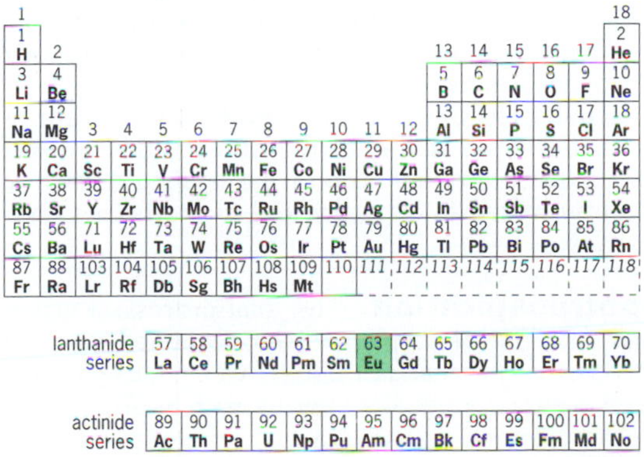

Eurotiales An order of ascomycetous fungi bearing ascospores in globose or broadly oval, evanescent asci without a pore. The asci are arranged at various levels, typically within a closed fruiting body or cleistothecium. The order was previously known as Aspergillales, and Plectascales. The order is divided into five families: Gymnoascaceae, Eurotiaceae, Onygenaceae, Trichocomaceae, and Elaphomycetaceae. *See* ASCOMY-COTINA; FUNGI; MEDICAL MYCOLOGY. [C.J.A.]

Euryapsida A subclass of the reptiles, extinct since the end of the Mesozoic Era. In the Euryapsida each side of the skull has only an upper temporal opening, bounded laterally by postorbital and squamosal bones (see illustration); the lower part of the cheek is never truly fenestrated.

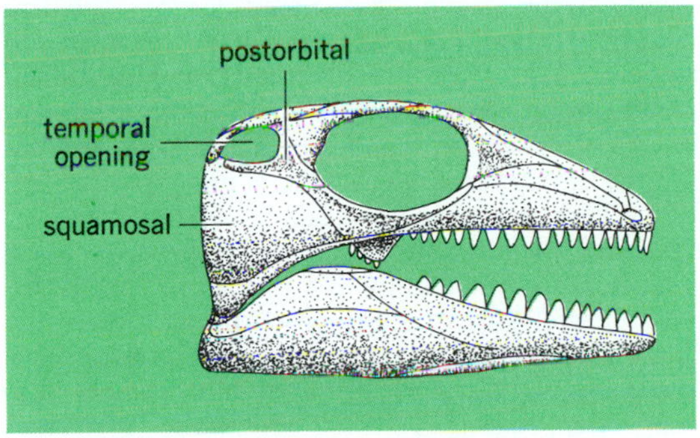

Lateral view of *Araeoscelis* skull showing temporal opening. (After *A. S. Romer, Vertebrate Paleontology, 3d ed., University of Chicago Press, 1966*)

In the classification adopted here, the Euryapsida comprise three orders: the Permian and Triassic Araeoscelidia, little-known forms, generally small and essentially terrestrial; the Placodontia of the Triassic which were mollusk-eating aquatic reptiles, usually with enormous flat crushing teeth; and the Sauropterygia, the small to medium-sized amphibious nothosaurs of the Triassic and the marine plesiosaurs which range throughout the Mesozoic. *See* ARAEOSCELIDIA; PLACODONTIA; REPTILIA; SAUROPTERYGIA. [A.J.C.]

Eurypterida Primitive aquatic arthropods of the subphylum Chelicerata and class Merostomata. The Eurypterida were dominant from the Ordovician to the end of the Paleozoic. Like other Merostomata, their elongate-lanceolate bodies were encased in a chitinous exoskeleton comprising a cephalothorax (prosoma) and an abdomen (opisthosoma). The Eurypterida, presumably descended from aglaspid ancestors, apparently left no direct descendants. *See* MEROSTOMATA. [G.O.R.]

Eustigmatophyceae A small class of nonmotile, photosynthetic, unicellular algae, segregated from the class Xanthophyceae (Tribophyceae) on the basis of cytological, ultrastructural, and biochemical features of the vegetative cell and zoospore. Because of their lack of chlorophyll *b*, these organisms may be placed in the division Chromophycota, but their lack of chlorophyll *c*, in contrast to other chromophytes, supports recognition at a higher level (such as the division Eustigmatophyta). Only a dozen species in three genera are known.

These live chiefly in fresh water, but also in marine habitats and in soil. *See* ALGAE; XANTHOPHYCEAE. [P.C.Si.; R.L.Moe]

Eutardigrada An order of tardigrades, lacking a cirrus lateralis, a sensory cephalic appendage, and a clava, or club-shaped appendage. Pharyngeal pockets are strengthened by separated rods or macroplacoids or are without thickenings. Claws are of different size, arranged in two pairs in which a larger and smaller claw are united. In *Milnesium* the claws are separated. *Haplomacrobiotus* has two simple claws. *See* HETEROTARDIGRADA; TARDIGRADA. [E.M./Ev.M.]

Eutectics The microstructures that result when a solution of metal of eutectic composition solidifies. The eutectic reaction must be distinguished from eutectic microstructures. The eutectic reaction is a reversible transformation of a liquid solution to two or more solids, under constant pressure conditions, at a constant temperature denoted as the eutectic temperature. Microstructures which are wholly eutectic in nature can occur only for a single, fixed composition in each alloy system demonstrating the reaction.

Although technologically important alloy systems, particularly the Fe-C system (all cast irons used commercially pass through a eutectic reaction during solidification), exhibit at least one eutectic reaction, there has been little exploitation of wholly eutectic microstructures for structural purposes. Some eutectic fusible alloys are used as solders, as heat-transfer media, for punch and die mold and pattern applications, and as safety plugs. A silver-copper eutectic alloy is also used for high-temperature soldering applications. *See* SOLDERING.

Directional solidification of eutectic alloys so as to create a microstructure well aligned parallel to the growth direction produces high-strength, multiphase composite materials with excellent mechanical properties. Among the major advantages of these alloys are extraordinary thermal stability of unstressed microstructures, retention of high strength to very close to the eutectic temperature of the respective alloys, and the ability to optimize strength by appropriate alloying additions to induce either solid-solution strengthening or intraphase precipitation of additional phases.

The most likely future applications for aligned eutectics are as gas turbine engine materials (turbine blades or stator vanes) or in nonstructural applications such as superconducting devices in which directionality of physical properties is important. *See* ALLOY; METAL; SOLID SOLUTION. [N.S.S.]

Eutheria An infraclass of therian mammals including all living mammals except the monotremes and marsupials. Eutherians, often called placental mammals, are distinguished from their nearest relatives, the marsupials, by numerous characters. The braincase is large, and the angular process of the mandible is not inflected. The young are born in a relatively advanced state of development, and are never sheltered in a pouch after birth. Living eutherians are divided into 16 orders; at least 10 additional orders are extinct and known only as fossils. The largest orders are the Rodentia, Chiroptera, Carnivora, and Primates. *See* MAMMALIA; METATHERIA; THERIA. [D.D.D./F.S.S.]

Eutrophication The deterioration of the esthetic and life-supporting qualities of lakes and estuaries, caused by excessive fertilization from effluents high in phosphorus, nitrogen, and organic growth substances. Algae and aquatic plants become excessive, and, when they decompose, a sequence of objectionable features arise. Water supplies drawn from such lakes must be filtered and treated. Diversion of sewage, better utilization of manure, erosion control, improved sewage treatment and harvesting of the surplus aquatic crops alleviate the symptoms. Prompt public action is essential. *See* WATER CONSERVATION; WATER POLLUTION. [A.D.H.]

Evaporation The process by which a substance in the liquid state is converted into the vapor state. The molecules of substances in a condensed state are held to one another by strong forces of attraction, which are balanced by equally strong repulsive forces. Tending to overcome the potential energy of attraction is the escaping tendency of molecules, which arises from their kinetic energy. The kinetic energy, and therefore the escaping tendency of molecules, is a function of temperature.

The following factors affect the rate of evaporation of a liquid: the rate at which heat is supplied to the liquid to furnish the latent heat of vaporization; the rate at which the liquid is stirred to bring to the surface molecules having sufficient kinetic energy to escape; and the rate at which the vapor above the liquid is changed to provide the best conditions for escape of molecules from the surface of the liquid. *See* EVAPORATOR; EVAPOTRANSPIRATION; HUMIDIFICATION; LIQUID; VAPOR PRESSURE. [N.H.N.]

Evaporator A device used to vaporize part or all of the solvent from a solution. The valuable product is usually either a solid or a concentrated solution of the solute. If a solid, the heat required for evaporation of the solvent must have been supplied to a suspension of the solid in the solution, otherwise the device would be classed as a drier. The vaporized solvent may be made up of several volatile components, but if any separation of these components is effected, the device is properly classed as a still or distillation column. When the valuable product is the vaporized solvent, an evaporator is sometimes mislabeled a still, such as water still, and sometimes is properly labeled, such as boiler-feedwater evaporator. In the great majority of evaporator installations, water is the solvent that is removed. *See* DISTILLATION.

Evaporators are used primarily in the chemical industry. For example, common salt is made by boiling a saturated brine in an evaporator. The salt precipitates as a solid in suspension in the brine. This slurry is pumped continuously to a filter, from which the solids are recovered and the liquid portion returned for further evaporation. Evaporators are widely used in the food industry, usually as a means of reducing volume to permit easier storage and shipment. Evaporators are also the most commonly used means of producing potable water from sea water or other contaminated sources. *See* SALINE WATER RECLAMATION.

The vaporization of solvent requires large amounts of heat. Provisions for transferring this heat to the solution constitute the largest element of evaporator cost and the principal means of distinguishing between types of evaporators. Practically all evaporators fall into one of the following categories:

1. Submerged-combustion evaporators: those heated by a flame that burns below the liquid surface, and in which the hot combustion gases are bubbled through the liquid.
2. Direct-fired evaporators: those in which the flame and combustion gases are separated from the boiling liquid by a metal wall, or heating surface.
3. Stem-heated evaporators: those in which steam or other condensable vapor is the source of heat, and in which the steam condenses on one side of the heating surface and the heat is transmitted through the wall to the boiling liquid. *See* EVAPORATION. [F.C.S.]

Evapotranspiration The total process of water vapor transfer into the atmosphere from vegetated land surfaces. evaporation is the change of liquid water into gaseous water vapor; it occurs from bodies of water such as lakes and streams or from other wet surfaces such as wet soil surfaces. Transpiration is the process whereby water is absorbed by plant roots, transported through the plant, and evaporated from plant surfaces, especially from leaves into the air above. The energy required to change liquid water into water vapor comes from sources traceable to the Sun. Energy for evapotranspiration may come from the transfer of heat from warm air to a cooler plant or soil surface, but the major source of energy is generally net radiation. Net radiation is the difference between the incoming and the outgoing shortwave and longwave radiation streams.

In general, the amount of evapotranspiration will be less than or equal to the amount of precipitation that falls in a given region. Exceptions arise when precipitation is supplemented with irrigation, or in instances when plant roots may extract water from underground water tables located within a meter or two of the soil surface. [B.L.B.]

Evergreen plants Plants that retain their green foliage throughout the year. Popularly, needle-leaved trees (pine, fir, juniper, spruce) and certain broad-leaved shrubs (rhododendron, laurel) are called evergreens. In warm regions many broad-leaved trees (magnolia, live oak) are evergreen, and in the tropics most trees are evergreen and nearly all have broad leaves. [N.A.]

Excavator Any of a variety of bucket-equipped construction machines used for digging earth and rock. The most common excavator is basically a modified crane. When modified for digging, the unit made up of a crane and its digging attachment is commonly called by the name of the attachment. Thus a crane carrying a dragline bucket is called a dragline.

There are four principal modifications of cranes for use as an excavator: dragline, clamshell, power shovel, and backhoe. The dragline uses a long latticed crane boom, two working cables, and a bucket designed for horizontal digging. The clamshell uses a crane boom and a clamshell bucket. Digging with a clamshell is not a power operation; the clamshell is filled only by its own weight and the bite of its jaws as they are closed. The power shovel uses a heavy but relatively short boom and a dipper stick to which is attached a hinged-bottom digging bucket or dipper. The backhoe is a short-boomed machine with an inverted bucket at the end of its stick.

The hydraulic backhoe is a variation in which the operations are activated by hydraulic pistons rather than cables. The front-end loader consists of a continuous-track or wheeled tractor with a wide hydraulically operated bucket mounted on jointed arms at its front end. The trencher, designed especially for excavating trenches, is a crawler tractor mounting either a movable wheel or a continuous chain, on which there are numerous buckets. *See* BULK-HANDLING MACHINES; CONSTRUCTION EQUIPMENT. [E.M.Y.]

Exchange interaction A quantum-mechanical phenomenon that gives the energy of two elementary particles. Exchange effects arise for all kinds of elementary particles, but these effects were first introduced into physics in consideration of atomic structure and the energy of the electrons in an atom. In this context, they arise as a consequence of two facts: electrons are indistinguishable, and they obey the Pauli exclusion principle. *See* ELEMENTARY PARTICLE.

Indistinguishability demands that the wave function describing two electrons either is unchanged or changes sign when the labels of the two electrons are exchanged (that is, it is either symmetric or antisymmetric to this exchange). In the case of electrons, the wave function is antisymmetric; this is a more precise statement of the Pauli exclusion principle. *See* ELECTRON; EXCLUSION PRINCIPLE.

When the spins of the two electrons are parallel the spatial part of the wave function is antisymmetric under exchange, and when they are opposed it is symmetric. The distribution in space of the two electrons is different in these two states, and so their mutual electrostatic energy is different. This difference in the electrostatic energy, called the exchange energy, appears as an interaction between the two electrons which depends on their relative orientation (although this dependence is incidental).

Exchange is the mechanism by which the electron spins in many magnetic materials are lined up parallel or opposed (for example, in α-iron and the ferrites). It is also the mechanism whereby the electron spins are parallel in the first excited state of the helium atom, and the spins of the two electrons are opposed in the carbon-carbon covalent bond. *See* ATOMIC STRUCTURE AND SPECTRA; CHEMICAL BONDING; ELECTRON CONFIGURATION; FERROMAGNETISM; QUANTUM MECHANICS; SPIN (QUANTUM MECHANICS). [J.M.D.]

Excitation Application of energy to one portion of a system or apparatus in a manner that enables another portion to carry out a specialized function. Excitation establishes a condition essential to that function.

Excitation of a system or apparatus serves either to permit a transfer of energy or to control a flow of energy elsewhere in the system. Excitation energy may differ from the output energy in source, form, level, or location. Excitation produces a primary effect that is linked through an intermediate physical phenomenon to a dependent secondary effect. [W.W.Sn.]

Excitation potential The difference in potential between an excited atomic or molecular state and the ground state. The term is most generally used in connection with electron excitation, but it can be applied to excited molecular vibrational and rotational states.

A closely related term is excitation energy. If the unit of potential is taken as the volt and the unit of energy as the electron volt, then the two are numerically equal. According to the Bohr theory, there is a relationship between the wavelength of the photon associated with the transition and the excitation energies of the two states. Thus the basic equation for the emission or absorption of energy is as shown below, where h

$$\frac{hc}{\lambda} = E_i - E_f$$

is Planck's constant, c the velocity of light, λ the wavelength of the photon, and E_f and E_i the energies of the final and initial states, respectively. *See* EXCITED STATE; GROUND STATE; IONIZATION POTENTIAL. [G.H.M.]

Excited state In quantum mechanics, a stationary state of higher energy than the lowest stationary state or ground state of a particle or a system of particles. Customarily, only bound stationary states, which generally are at most denumerably infinite in number, are spoken of as excited, although the formal quantum theory often treats the noncountable unbound stationary states on an equal footing with the bound states. *See* GROUND STATE; METASTABLE STATE; STATIONARY STATE. [E.G.]

Exciton In a nonmetallic crystal, a quantum of electronic excitation which transports energy but not charge. There are two customary ways of picturing an exciton, termed the Frenkel model and the Wannier model.

The Frenkel model applies to a solid consisting of weakly interacting atoms or molecules, a prime example being organic crystals such as anthracene and naphthalene. The excitons are very much like the excited electronic states of the isolated molecules which make up the crystal. Although largely localized on a single molecule, an exciton can hop readily from one molecule to another in the crystal.

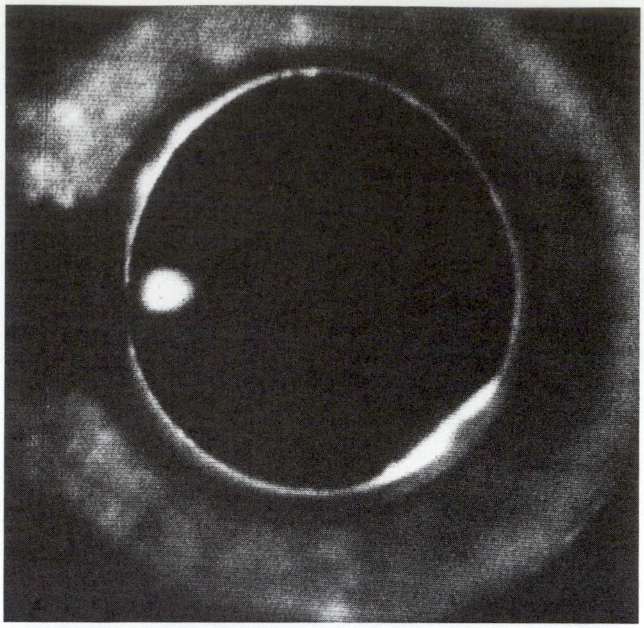

An electron-hole droplet (light spot) consisting of condensed excitons in a circular sample of germanium. (*From J. P. Wolfe et al., Photograph of an electron-hole drop in germanium, Phys. Rev. Lett., 34:1292–1293, 1975*)

The Wannier model begins at the opposite extreme, with an electron in the conduction band and a hole in the valence band which are bound weakly to each other by the Coulomb force. This exciton overlaps many atoms in the crystal and does not resemble the excitation of any constituent atom. The dielectric screening of the crystal serves to weaken the attraction between electron and hole, and thus to increase their average separation. The Wannier model is therefore most appropriate for materials with a high dielectric constant—semiconductors such as germanium and silicon, in particular. *See* BAND THEORY OF SOLIDS; DIELECTRIC CONSTANT; SEMICONDUCTOR.

An exciton is not stable against electron-hole recombination and can thus exist for only a brief interval before it decays by emitting a photon (recombination luminescence) or by giving up its energy to the crystal as heat or as stored energy. In most materials, the lifetime is long enough that the exciton becomes trapped before it decays. *See* ELECTROLUMINESCENCE; LUMINESCENCE; TRAPS IN SOLIDS.

The production of a high concentration of excitons in Ge and Si crystals at very low temperatures leads to the phenomenon of exciton condensation, which is similar in many respects to the condensation of water molecules in air to form fog droplets. The condensed phase is referred to as an electron-hole droplet (see illustration) since the excitons lose their individual character, and the droplet becomes an electrically conducting plasma. [M.N.K.]

Exclusion principle No two electrons may simultaneously occupy the same quantum state. This principle, often called the Pauli principle, was first formulated by Wolfgang Pauli in 1925 and, for time-independent quantum states, it means that no two electrons may be described by state functions which are characterized by exactly the same quantum numbers. In addition to electrons, all known particles having half-integer intrinsic angular momentum, or spin, obey the exclusion principle. It plays a central role in the understanding of many diverse phenomena, including the periodic table of the elements and their chemical activities, the electron contribution to the specific heat of metals, the shell structure in the atomic nucleus analogous to that of electrons in atoms, and certain symmetries in the scattering of identical particles. *See* ANGULAR MOMENTUM; QUANTUM NUMBERS; SPIN (QUANTUM MECHANICS).

Using the fact that a system will try to occupy the state of lowest possible energy, the electron configuration of atoms may be understood by simply filling the single-particle energy levels according to the Pauli principle. This is the basis of Niels Bohr's explanation of the periodic table. *See* ATOMIC STRUCTURE AND SPECTRA; ELECTRON CONFIGURATION. [S.A.Wi.]

Exobiology The study of the origin and distribution of life in the universe. With time, exobiology has grown to include the study of the origin, evolution, and distribution not only of life, but of the chemicals necessary for life, both on Earth and throughout the universe. It strives to determine the relationship of life to the evolution of planets. Planetary systems are now believed to be a companion event in the formation of stars, and interstellar gas clouds have been found to contain biologically important molecules made of carbon, hydrogen, and other elements. Thus there is reason to believe that life could be widespread in the universe.

In 1976 two Viking robot spacecraft landed on Mars. One instrument showed that the Mars soil contained no organic matter. It was concluded that the intense ultraviolet radiation from the Sun dissociates organic compounds exposed to it. A second instrument to detect biological reactions in the Martian soil indicated no life. Mars is still an enigma, since final conclusions cannot be drawn from such limited testing. *See* MARS.

It is quite possible that the outer planets (Jupiter, Saturn, Uranus, and Neptune) and some of their satellites may be veritable laboratories of chemical evolution, since they have atmospheres which appear to be very similar to that of the primitive Earth. Comets, meteorites, and asteroids, like the outer planets, may also contain evidence of chemical evolution. Organic geochemical analyses of carbonaceous chrondrites has revealed the presence of amino acids and other organic molecules of biological interest. Organic compounds have even been detected in the interstellar medium with powerful radio telescopes. The apparent ubiquity of abiotically produced organic molecules supports the universality of chemical evolution and suggests that the processes which led to the origin of life on Earth may be quite commonplace throughout the universe.

It is now generally believed that the origin of life on Earth was closely related to the origin of the solar system. It is thought that soon after Earth formed, its primitive atmosphere, which contained no oxygen and was composed mainly of reduced gases like hydrogen, methane, ammonia, and water vapor, was activated by energy sources much as lightning and ultraviolet radiation from the Sun. The gas molecules broke apart and recombined repeatedly, forming newer and increasingly complex molecules, advancing through prebiotic synthesis of organic matter to a point where the assembled molecules exhibited properties of living things: respiration, metabolism, reproduction, and the transfer of genetic information.

The accumulation of information supporting other life in the universe has stimulated a search for life beyond the solar system. Terrestrial radio telescopes listen for signals from different stars. Intelligently contrived signals would constitute proof of the existence of life elsewhere in the universe. [T.W.H.]

Exon In split genes, a portion that is included in the ribonucleic acid (RNA) transcript of a gene and survives processing of the RNA in the cell nucleus to become part of a spliced messenger RNA (mRNA) or structural RNA in the cell cytoplasm. Split genes are those in which regions that are represented in mature mRNAs or structural RNAs (exons) are separated by regions that are transcribed along with exons in the primary RNA products of genes, but are removed from within the primary RNA molecule during RNA processing steps (introns). *See* INTRON; RIBONUCLEIC ACID (RNA).

Exons comprise three distinct regions of a protein-coding gene. The first is a portion that is not translated into protein, but contains the signal for the beginning of RNA synthesis, and sequences that direct the mRNA to ribosomes for protein synthesis. The second is a set of exons containing information that is translated into the amino acid sequence of a protein. The third region of a gene that becomes part of an mRNA is an untranslated end portion that contains signals for transcription termination and for the addition of a polyadenylate tract at the end of a transcript.

The mechanism by which the exons are joined in mRNA copies of genes is called RNA splicing, and it is part of the maturation of mRNA from primary transcripts of genes. The process is not fully understood, but it involves short nucleotide sequences that are universally at the junctions of exons and introns, and participate in the precise removal of intron segments from RNA. It is almost certain that the removal of introns and exon splicing are catalyzed in part by small nuclear RNA-protein particles that serve at least to properly align exons for intron excision and subsequent splicing. *See* GENE; GENETIC CODE; RIBOSOMES. [P.M.M.R.]

Exopterygota One of the two major divisions of the subclass Pterygota, the winged Insects, including the insects often referred to as the Hemimetabola. The developing wings appear as external pads during the early stages of the insect. The immature forms, or nymphs, usually live in the same environment as the adults and feed on the same type of food. *See* ENDOPTERYGOTA; INSECTA; PTERYGOTA. [F.M.C.]

Exotic nuclei Nuclei with ratios of neutron number N to proton number Z much larger or much smaller than those of nuclei found in nature. Studies of nuclear matter under extreme conditions, in which the nuclei are quite different in some way from those found in nature, are at the forefront of nuclear research. Such extreme conditions include nuclei at high temperature and at high density (several times normal nuclear density), as well as those with larger or smaller N/Z ratios. The N/Z ratio depends on the nature of the attractive nuclear force that binds the protons and neutrons in the nucleus and its competition and complex interplay with the disruptive Coulomb or electrical force that pushes the positively charged protons apart.

A chart of the nuclides is shown in the illustration. The squares are the stable nuclei ($Z \leq 83$) and the very long-lived nuclei (with half-lives of the order of 10^9 years or more) found in nature. The first jagged lines to either side are the limits of the presently observed nuclei; very little is known about those at the edges. The light, stable nuclei (such as ^{4}He) have $Z = N$, reflecting the preference of the nuclear forces for $N = Z$ symmetry, but as Z increases, the strength of the Coulomb force demands more neutrons than protons to make a particular element stable, and $N \approx 1.6\ Z$ for the heaviest long-lived nuclei (such as ^{238}U). For $Z > 92$, the Coulomb force causes most nuclei to spontaneously fission. *See* NUCLEAR FISSION.

Stable nuclei lie in the so-called valley of beta stability. As the N/Z ratio decreases (proton-rich nuclei) or increases (neutron-rich nuclei) compared to that of the stable isotopes, there is, respectively, energy for a proton or neutron in the nucleus to undergo beta (β^+, β^-) decay to move the nucleus back toward stability. Most knowledge of nuclear structure and decay has been gained from nuclei in or near the valley of beta stability. *See* RADIOACTIVITY.

The spherical shell model was developed to explain the so-called spherical magic numbers for protons and neutrons, which give nuclei with these numbers a very stable structure and spherical shape. Spherical magic Z and N of 2, 8, 20, 28, 50, 82, and 126, and the weaker magic Z and N at 40, are

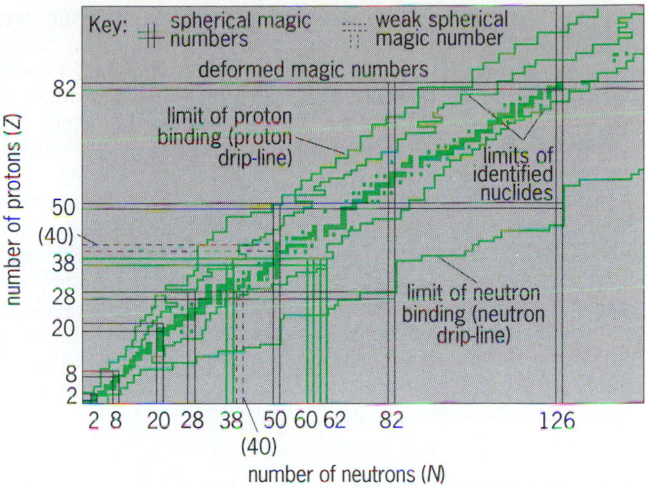

Chart of the nuclides. The squares are the stable and very long-lived nuclei found in nature.

shown in the illustration as horizontal and vertical lines. The nuclear shell model resembles the atomic shell model, where the noble gases (helium, neon, argon, and so forth) have filled shells and then there is a gap in the binding energy to the next electron shell (or orbit). A major question concerns what happens to these spherical magic proton and neutron gaps (orbits) in exotic nuclei. *See* NUCLEAR STRUCTURE.

Another major question concerns the decay modes whereby a nucleus rids itself of excess energy and returns to the stable forms of nuclear matter. As N/Z decreases or increases from the stable values, a point is reached for a given Z where, if one more neutron is pulled out, one proton becomes unbound (an isotope of that element cannot exist with that number of neutrons), or, if one more neutron is added, that neutron is not bound to that nucleus. These limits define, respectively, the proton and neutron drip-lines (the jagged lines furthest from the valley of stability in the illustration).

Answers to the above questions give significant insights into the structure and decay modes of nuclear matter that make it possible to test and extend theoretical models of the nucleus and the understanding of the nature of both the strong and weak nuclear forces. These insights could not be gained by studies of nuclei in and near the valley of stability.

These insights include the discovery of nuclear shape coexistence. Competing bands of levels occur in one nucleus, which overlap in energy but are quite separate in their decays because they are built on quite different coexisting nuclear shapes. Shape coexistence is now known to be important in many nuclei throughout the periodic table.

A significant advance was made in the nuclear shell model with the discovery of new magic numbers associated with shell gaps in the energies of the proton and neutron orbitals. These new numbers may be called deformed magic numbers because they stabilize a nucleus in a deformed shape, just as the spherical magic numbers give stability to a spherical shape. The deformed magic numbers (shell gaps) identified so far include N and Z of 38 and N of 60 and 62.

Exotic nuclei exhibit decay modes not seen near stability, such as proton radioactivity and beta-delayed particle emission. (After beta decay, the highly excited nucleus can emit one or more particles, such as one or two protons, an alpha particle, or one or two neutrons.) [J.H.H.]

Exotic viral diseases Viral diseases that occur only rarely in human populations of developed countries. However, many of these diseases cause significant human morbidity and mortality in underdeveloped areas, and have the proven capacity to be transported to population centers in developed countries and to cause explosive outbreaks or epidemics. Most of the exotic viruses are zoonotic, that is, they are transmitted to humans from an ongoing life cycle in animals or arthropods; the exception is smallpox.

Important diseases caused by exotic viruses include yellow fever, Venezuelan equine encephalitis, Rift Valley fever, tick-borne encephalitis, Crimean hemorrhagic fever, rabies, Lassa fever, hemorrhagic fever, and Marburg and Ebola hemorrhagic fevers. Control of these diseases is often very difficult because of the lack of detailed knowledge about the natural history of the viruses in their natural animal hosts, and because of the difficulty of controlling natural populations of alternative hosts such as insects or rodents. *See* ANIMAL VIRUS. [F.A.M.]

Expert systems Methods and techniques for constructing human-machine systems with specialized problem-solving expertise. The pursuit of this area of artificial intelligence research has emphasized the knowledge that underlies human expertise and has simultaneously decreased the apparent significance of domain-independent problem-solving theory. In fact, new principles, tools, and techniques have emerged that form the basis of knowledge engineering.

Expertise consists of knowledge about a particular domain, understanding of domain problems, and skill at solving some of these problems. Knowledge in any specialty is of two types, public and private. Public knowledge includes the published definitions, facts, and theories which are contained in textbooks and references in the domain of study. But expertise usually requires more than just public knowledge. Human experts generally possess private knowledge which has not found its way into the published literature. This private knowledge consists largely of rules of thumb or heuristics. Heuristics enable the human expert to make educated guesses when necessary, to recognize promising approaches to problems, and to deal effectively with erroneous or incomplete data. The elucidation and reproduction of such knowledge are the central problems of expert systems.

Researchers in this field suggest several reasons for their emphasis on knowledge-based methods rather than formal representations and associated analytic methods. First, most of the difficult and interesting problems do not have tractable algorithmic solutions. This is reflected in the fact that many important tasks, such as planning, legal reasoning, medical diagnosis, geological exploration, and military situation analysis, originate in complex social or physical contexts, and generally resist precise description and rigorous analysis. *See* ALGORITHM.

The second reason for emphasizing knowledge is pragmatic: human experts achieve outstanding performance because they are knowledgeable. If computer programs embody and use this knowledge, they too attain high levels of performance. This has been proved repeatedly in the short history of expert systems. Systems have attained expert levels in several tasks: mineral prospecting, computer configuration, chemical structure elucidation, symbolic mathematics, chess, medical diagnosis and therapy, and electronics analysis.

The third motivation for focusing on knowledge is the recognition of its intrinsic value. Traditionally, the transmission of knowledge from human expert to trainee has required education and internship periods ranging from 3 to 20 years. By extracting knowledge from humans and transferring it to computable forms, the costs of knowledge reproduction and exploitation can be greatly reduced. At the same time, the process of knowledge refinement can be accelerated by making the previously private knowledge available for public test and evaluation.

Most of the knowledge-engineering applications fall into a few distinct types, summarized in the table.

Generic categories of knowledge engineering applications

Category	Problem addressed
Interpretations	Inferring situation descriptions from sensor data
Prediction	Inferring likely consequences of given situations
Diagnosis	Inferring system malfunctions from observables
Design	Configuring objects under constraints
Planning	Designing actions
Monitoring	Comparing observations to plan vulnerabilities
Debugging	Prescribing remedies for malfunctions
Repair	Executing a plan to administer a prescribed remedy
Instruction	Diagnosing, debugging, and repairing students' knowledge weaknesses
Control	Interpreting, predicting, repairing, and monitoring system behaviors

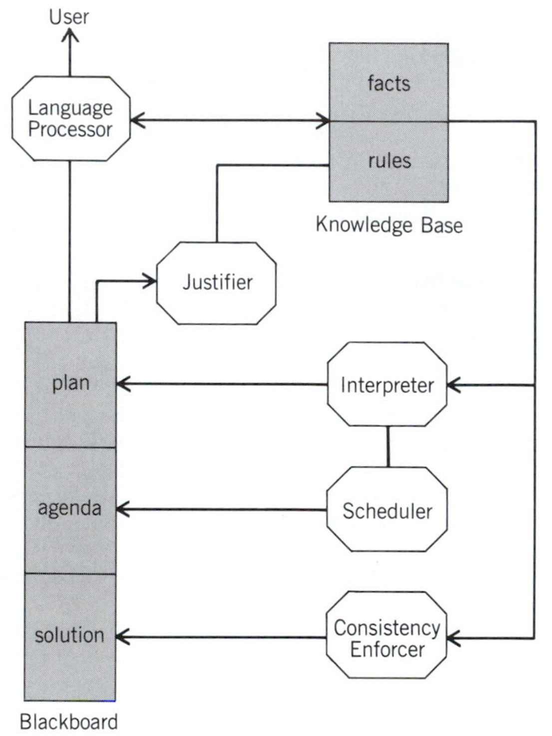

Anatomy of an ideal expert system.

The illustration shows an idealized representation of an expert system. No existing expert system contains all components shown, but one or more components occur in every system. The ideal expert system contains: a language processor for problem-oriented communications between the user and the expert system; a blackboard for recording intermediate results; a knowledge base comprising facts plus heuristic planning and problem-solving rules; an interpreter that applies these rules; a scheduler to control the order of rule processing; a consistency enforcer that adjusts previous conclusions when new data or knowledge alter their bases of support; and a justifier that rationalizes and explains the system's behavior. *See* ARTIFICIAL INTELLIGENCE; DATA-PROCESSING SYSTEMS. [F.H.-Ro.]

Explosive forming The shaping or modifying of metals by means of explosions. The explosives may be of either the detonating or deflagrating type. Explosive gas mixtures or stored gas at high pressure may also provide the motive power.

Cold welds can be made between dissimilar metals by driving the two parts together under explosive impact. In other applications of explosive-forming methods, powders are pressed into solid billets. In a different application, high explosives are used to cut large blocks of metal and even to split thin sheets into two layers of exactly one-half the original thickness. Explosives can also be employed to extrude metal shapes and to punch hard metals with the aid of dies. Shapes produced explosively are very exact and free from the fine cracks that sometimes result when pressure is applied slowly. *See* METAL FORMING. [W.E.Go.]

Exponent In mathematics a symbol or number written to the right of and above another symbol or number to denote how many times the latter is to be multiplied by itself. For example, $7^3 = 7 \cdot 7 \cdot 7$, and $a^3 = a \cdot a \cdot a$. By use of this convenient mathematical device, the number 420,000,000,000,000 can be expressed unmistakably and more compactly as 4.2×10^{14}, which is read as, "four and two-tenths times ten multiplied by itself fourteen times" or, more properly, "four and two-tenths times ten to the fourteenth power." This abbreviated notation is particularly valuable in expressing the extremely large or extremely small numbers encountered in modern scientific work. For example, 0.000000143 may be expressed as

$$\frac{143}{1{,}000{,}000{,}000} \quad \text{or} \quad \frac{143}{10^9} \quad \text{or} \quad 1.43 \times 10^{-7}$$

Operations with exponents are governed by the following rules:

1. $x^a x^b = x^{a+b}$. Example, $x^2x^3 = xx \cdot xxx = x^5$.
2. $(x^a)^b = x^{ab}$. Example, $(x^2)^3 = (x^2)(x^2)(x^2) = x^6$.
3. $(xy)^a = x^a y^a$. Example, $(xy)^3\ xyxyxy = xxxyyy = x^3y^3$.
4. If $y \neq 0$, $\left(\frac{x}{y}\right)^a = \frac{x^a}{y^a}$. Example, $\left(\frac{x}{y}\right)^3 = \frac{x}{y} \cdot \frac{x}{y} \cdot \frac{x}{y} = \frac{x^3}{y^3}$.
5. If $x \neq 0$, $\frac{x^a}{x^b} = x^{a-b}$. Example, $\frac{x^5}{x^2} = \frac{xxxxx}{xx} = xxx = x^3$.

See ALGEBRA. [A.N.L./S.Bo.]

Exposure meter An indicating instrument used in photography to determine lens aperture and shutter speed. An exposure meter may be used either in the darkroom to determine approximate printing time for a contact print or an enlargement or, more usually, with a camera to determine exposure of film.

Exposure meters are of two basic types, photovoltaic and photoconductive. In photovoltaic meters, a selenium cell converts photons to electrons, producing a current directly proportional to received light. A sensitive microammeter indicates this current to the user.

In the photoconductive meter, a cadmium sulfide cell changes conductivity in proportion to the received light. A battery (usually a mercury cell) supplies power through the cadmium sulfide cell to the meter movement. Because the received light serves to control the power from the battery, this type of instrument is about three orders of magnitude more power-sensitive than the photovoltaic type.

Used at the camera, either type of meter serves as a brightness meter to measure reflected light from the subject. Used at the subject, with a diffuser head over the exposure meter lens, the instrument serves as an illumination meter to indicate incident light. *See* PHOTOGRAPHY; PHOTOMETRY. [F.H.R.]

Extended x-ray absorption fine structure (EXAFS) Oscillations in the total x-ray absorption cross section, observable from several tens to several hundreds of electronvolts above an inner-shell ionization threshold, produced by scattering of the ionized electron off neighboring atoms. This backscattered wave interferes with the direct wave, leading to a maximum if the interference is constructive and a minimum if it is destructive. A single equation, the EXAFS equation, relates the amplitudes and phases of these maxima and minima to the number, type, and positions of the scatterers. The principal contribution to the interference comes from those atoms nearest the one being ionized; thus the oscillations are a reflection of the short-range order in the system. *See* ABSORPTION; INTERFERENCE OF WAVES; IONIZATION; X-RAYS.

With the development of electron synchrotrons and storage rings as high flux sources of tunable x-radiation, EXAFS has become a powerful analytic tool. It is element-specific and can be observed in any system which contains more than one atom and in any physical state. The principal result of the experiment, information on the number and average distances of the scatterers from the newly formed ion, may represent, for example, the bonding of a metal atom in a biological molecule, the coordination of a metal catalyst on a substrate, or the nature of the adhesion of a molecule to a surface. *See* CHEMICAL BONDING; CHEMICAL STRUCTURES; SPECTROSCOPY; SURFACE PHYSICS; SYNCHROTRON RADIATION. [C.D.C.]

Exterior ballistics The science dealing with the motions of projectiles not under propulsive power, such as a shell after it has left the cannon or a rocket after burning has ceased. The motion of a body projected at an angle to the Earth's surface is perturbed by several forces: drag, cross wind forces, and the Coriolis effect. Their magnitude depends upon the shape and mass of the body, the density of the air, the wind currents, and the length of the path along the trajectory. *See* TRAJECTORY.

The trajectory of a projectile under the influence of the Earth's gravitational field and in vacuum would be an arc of an ellipse with the center of the Earth as one of the foci. This arc can be closely approximated by a parabola if the motion does not carry the projectile so high that the gravitational force differs significantly from the force at the Earth's surface.

As a projectile moves through a fluid such as air, the fluid particles are accelerated as it collides with them, and this results in a force, called drag, which robs the projectile of some of its energy. The drag is a function of the density of the fluid, the area of the moving body, and the velocity of the body squared.

For projectiles, such as rockets, with high velocities and high quadrant angles of departure, the variations of the drag coefficient and of atmospheric density along the trajectory are more complex. Solutions for these trajectories involve the method of approximate numerical integration, long used in astronomy. Computers with large data-handling and storage capacity have been used extensively to perform the necessary calculations. *See* NUMERICAL ANALYSIS.

When the range of the projectile becomes great, the ballistics analysis must be modified to accommodate the fact that the force of gravity varies with height, is directed toward the center of the Earth, and therefore changes direction as the projectile moves forward. Also, the impact point is depressed because of the Earth's curvature, and the range therefore is extended. These effects are customarily taken care of by the method of differential corrections.

Once the projectile is clear of, say, the gun muzzle or the launching stand, it is a separate entity free of the Earth's rotation. One component of its initial velocity, however, is the velocity of the surface of the Earth at the point of firing. For this reason, when the projectile arrives at the impact point, it will miss by an amount dependent on the azimuth of the trajectory, the range, and the height of the trajectory. This is called the Coriolis effect. For intermediate trajectories, such as those with pronounced Earth curvature, the Coriolis effect is handled by means of differential corrections. For very-long-range missiles, the approach of celestial mechanics with nonrotating Earth-centered coordinates is used. *See* CORIOLIS ACCELERATION.

In order to stabilize a projectile in flight, it is frequently spun about its longitudinal axis. The spin of a shell is imparted by rifling the gun barrel, that of a rocket by canting the exhaust nozzles or the fins. While spin does tend to stabilize the orientation, in practice the spin axis will precess so that the nose of the projectile follows a spiral path. This causes troublesome complexities, but the equation of motion of a spinning projectile can be integrated by high-speed computing machines. [J.P.H.]

Extinction (biology) The death and disappearance of a species. The fossil record shows that extinctions have been frequent in the history of life. Mass extinctions refer to the loss of a large number of species in a relatively short period of time. Episodes of mass extinction occur at times of rapid global environmental change; five such events are known from the fossil record of the past 600 million years. Human activity is causing extinctions on a scale comparable to the mass extinctions in the fossil record.

Record. An extinction may be of two types; phyletic or terminal. Phyletic extinction occurs when one species evolves into another with time; in this case, the ancestral species can be called extinct. However, because the evolutionary lineage has continued, such extinctions are really pseudoextinctions. In contrast, terminal extinction marks the end of an evolutionary lineage, termination of a species without any descendants. Most extinctions recorded in the fossil record and those occurring today are terminal. It has been estimated that 99% of all species that have ever lived are now extinct. *See* ORGANIC EVOLUTION.

The fossil record is best known for marine organisms. The mass extinctions of the marine fossil record occurred during the Late Ordovician, Late Devonian, Late Permian, Late Triassic, and Late Cretaceous. These mass extinctions affected a variety of organisms in many different ecological settings. Terrestrial and marine mass extinctions seem to occur at about the same time. The Late Permian, Late Triassic, and Late Cretaceous are also times of extinction for terrestrial vertebrates; the most dramatic extinction of terrestrial vertebrates took place at the end of the Cretaceous, when the last dinosaurs died off. *See* DINOSAUR.

The best record of terrestrial vertebrate extinction is that of the Pleistocene. Late Pleistocene extinctions in North America are especially well known—33 genera of mammals vanished during the last 100,000 years. These extinctions were concentrated among the large mammals—those over 100 lb (44 kg) in weight—and most occurred during a short time interval approximately 11,000 years ago.

Causes. Ever since the work of Georges Cuvier, the French naturalist who demonstrated the reality of extinction, explanations have fascinated both scientists and the general public. Cuvier invoked sudden catastrophic events, whereas his contemporaries favored more gradual processes. These two themes, catastrophism and gradualism, are still debated.

In 1980 high concentrations of iridium were reported precisely at the Cretaceous-Tertiary boundary. Iridium is rare in most rocks but more abundant in meteorites. It was proposed, therefore, that an asteroid struck the Earth 65 million years ago. The impact darkened the atmosphere with dust, caused a catastrophic short-term cooling of the climate, and thus led to the extinction of dinosaurs and many other Cretaceous species. The iridium-rich layer at the boundary marks this terminal Cretaceous event.

Astronomical theories have been put forward to explain the Late Cretacous extinctions as well as the 26-million-year periodicity. In one theory, the Sun has a distant companion star that would pass in orbit near the solar system's cloud of comets every 26 million years. This might perturb many comets, sending a few into the Earth. A comet would produce the same effects as an asteroid.

Other explanations for mass extinctions include lowered sea level, climatic cooling, and changes in oceanic circulation. Biotic processes such as disease, predation, and competition may also cause the extinction of species but are difficult to prove from the fossil record because they leave little evidence. Biotic factors usually affect only one or a few interdependent species. Predation and competition are important causes of more recent extinctions, which continue today. Human activities such as hunting and fishing (predation), habitat alteration (competition for space), and pollution have probably destroyed thousands of species. These activities, together with continued tropical deforestation and resulting changes in climate, are likely to cause extinctions that will be comparable to the mass extinctions seen in the fossil record. *See* FOSSIL. [K.W.F.]

Extraction A method of separating the constituents of a mixture utilizing preferential solubility of one or more components in a second phase. Commonly, this added second phase is a liquid, while the mixture to be separated may be either solid or liquid. If the starting mixture is a liquid, then the added solvent must be immiscible or only partially miscible with the original and of such a nature that the components to be separated have different relative solubilities in the two liquid phases.

Solvent extraction processes can be divided into two broad categories according to the origins of the differential solubility. On the one hand, it arises from purely physical differences between the two solutes, such as polarity, while in other cases it can be traced to definite chemical interaction between solute and solvent. Categories of major importance for the latter cases are ion-association systems and chelate compounds.

Liquid/solid extraction may be considered as the dissolving of one or more components in a solid matrix by simple solution, or by the formation of a soluble form by chemical reaction. The largest use of liquid/solid extraction is in the extractive metallurgical, vegetable oil, and sugar industries. The field may be subdivided into the following categories: leaching, washing extraction, and diffusional extraction. Leaching involves the contacting of a liquid and a solid (usually an ore) and the imposing of a chemical reaction upon one or more substances in the solid matrix so as to render them soluble. In washing extraction the solid is crushed to break the cell walls, permitting the valuable soluble product to be washed from the matrix. In diffusional extraction the soluble product diffuses across the denatured cell walls (no crushing involved) and is washed out of the solid.

Liquid/liquid extraction separates the components of a homogeneous liquid mixture on the basis of differing solubility in another liquid phase. Because it depends on differences in chemical potential, liquid/liquid extraction is more sensitive to chemical type than to molecular size. This makes it complementary to distillation as a separation technique. One of the first large-scale uses was in the petroleum industry for the separation of aromatic from aliphatic compounds. Liquid/liquid extraction also has found application for many years in the coal tar industry. On a smaller scale, extraction is a key process in the pharmaceutical industry for recovery of antibiotics from fermentation broths, in the recovery and separation of vitamins, and for the production of alkaloids from natural products. *See* CHEMICAL SEPARATION TECHNIQUES. [B.M.S.]

Extrapolation A process in mathematics used to find the value of a function outside its tabulated values. This is done as in interpolation by assuming that over a small range of x the function may be closely approximated by a polynomial or some other readily computed function. Any of the interpolation formulas can be used, therefore, and the desired value of x substituted in them. *See* INTERPOLATION. [K.S.K.]

Extrasensory perception (ESP) The phenomenon of awareness or reaction to external events in the absence of any known physical energy conveying information about those events to the person perceiving or reacting to them.

Although a wide variety of unusual events occur which may represent forms of ESP, only telepathy, clairvoyance, and precognition have been extensively studied under laboratory conditions. Mental telepathy refers to the case in which one person is aware of or thinking about something and another, sensorially separated person shows knowledge of the first person's thoughts. Clairvoyance refers to direct knowledge of physical-world events that are sensorially shielded from the knower when, at the time the knowledge is obtained, no one else has knowledge (no mental representations) of those events. Precognition refers to knowledge of future events which is not predictable by logical inference from known facts.

There seems to be a paranormal motor analog as well as sensory analog, namely psychokinesis (PK), or the direct effect of the mind on matter. The three kinds of ESP and psychokinesis are collectively referred to as psi. Two major problems hinder advances in understanding psi phenomena. First, since psi phenomena are paraconceptual, some scientists prefer to believe that all the evidence for them must be due to error and fraud, and psi is not a proper subject for scientific study. As a result, parapsychological research has been an extremely small-scale activity. The other major problem in understanding psi is the illusiveness of the phenomena. While experiments in general replicate in a statistical sense, no one knows how to produce psi phenomena on demand in any particular experiment. [C.T.T.]

Extraterrestrial intelligence The potential existence beyond the Earth of other advanced civilizations with a technology at least as developed as that on Earth. The idea that life, especially life with intelligence, might exist in other parts of the universe is a very old one. Early ideas were based on an intuitive belief in the enormity of the universe and in what is now called the mediocrity principle, namely that there is nothing special about the Sun, the Earth, and the human race.

Present ideas are also based on the mediocrity principle supported by the universality of the laws of physics and chemistry, and by the enormity of the universe. The Sun is one of 2×10^{11} stars of the Galaxy (the Milky Way), and there are about 10^{11} galaxies in the visible universe. The chemical evolution, that is, the natural formation of complex organic compounds, that led to the origin of life on Earth is quite common in the universe. Life on Earth started at least 3.5×10^9 years ago, that is, soon after the formation of the oceans, indicating a rather straightforward natural process. Through mutations and Darwinian selection, evolution advanced slowly from primitive unicellular microorganisms to advanced multicellular organisms with intelligence. Intelligence, which is favored by evolution because it has a high survival value, evolved into a technological society.

The search for extraterrestrial intelligence was initiated only after the development of radio astronomy and large radio telescopes, because radio waves seem to be the most efficient means of communication over interstellar distances. At least 45 radio searches have been carried out, accumulating more than 10,000 h of observations. The results continue to be negative, but selecting the proper frequency, bandwidth, polarization, target, and so forth, is an extremely complex problem. *See* RADIO ASTRONOMY; RADIO TELESCOPE.

In the early 1060s, F. Drake developed an equation to estimate the number N of advanced technological civilizations currently active in the Galaxy, based on the scientific knowledge presently available. The Drake equation, $N = R \times P \times L$, gives N in terms of the rate R at which new stars are born in the Galaxy, the probability P (actually a product of probability factors) that any one of these stars will possess the necessary conditions (luminosity, planets at the appropriate distances, and so forth) for life to originate and to slowly evolve to a technological civilization, and the average longevity L of such civilizations. The values advocated by the proponents of the Drake equation yield a value of N of approximately 200,000 stellar civilizations. However, the uncertainties in P and L are very large and some scientists think that the human race is probably the only advanced civilization in the Galaxy; that is, N equals 1.

In the mid-1970s, the possibility of large human colonies in space began to be seriously considered, and the idea that such self-sustained habitats could undertake multigeneration trips of several centuries to other stars began to gain acceptance. The possibility of galactic colonization, in combination with the natural tendency of life to expand into all available space and with the innate desire of intelligent life to explore all unknown territories, makes many scientists think that such colonization is not only possible but an almost inevitable consequence of the evolution of intelligence and technology. This concept, however, leads to two extreme alternatives: either the Galaxy has already been colonized, in which case N must be very large; or it has not been colonized, because no one was there to do so, in which case N must be very small. Both alternatives are contrary to the results of the Drake equation, which negates galactic colonization and predicts an intermediate value of N. All three alternatives, however, contain serious contradictions which are not easy to reconcile. [M.D.Pa.]

Extrusion The forcing of solid metal through a suitably shaped orifice under compressive forces. Extrusion is somewhat analogous to squeezing toothpaste through a tube, although some cold extrusion processes more nearly resemble forging, which also deforms metals by application of compressive forces. Most metals can be extruded, although the process may not be economically feasible for high-strength alloys.

The most widely used method for producing extruded shapes is the direct, hot extrusion process. In this process, a heated billet of metal is placed in a cylindrical chamber and then compressed by a hydraulically operated ram (see illustration). The opposite end of the cylinder contains a die having an orifice of the desired shape; as this die opening is the path of least resistance for the billet under pressure, the metal, in effect, squirts out of the opening as a continuous bar having the same cross-sectional shape as the die opening. By using two sets of dies, stepped extrusions can be made.

The extrusion of cold metal is variously termed cold pressing, cold forging, cold extrusion forging, extrusion pressing, and impact extrusion. The term cold extrusion has become popular in the steel fabrication industry, while impact extrusion is more widely used in the nonferrous field.

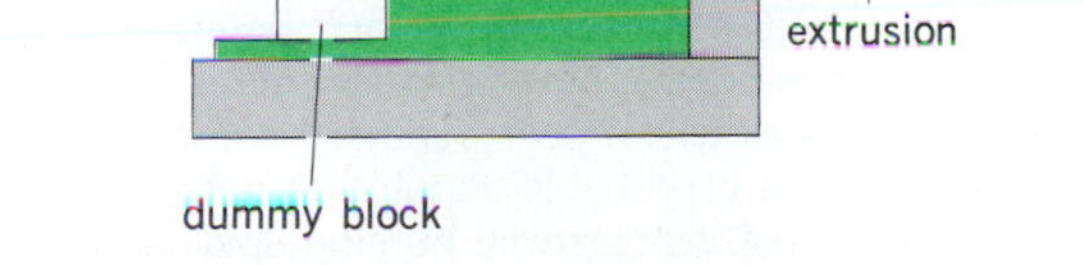

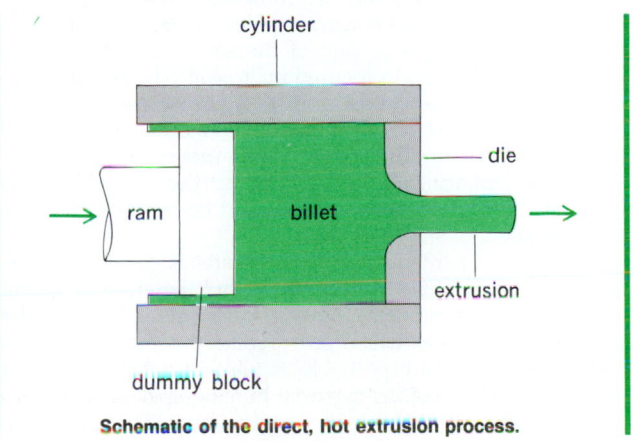

Schematic of the direct, hot extrusion process.

The original process (identified as impact extrusion) consists of a punch (generally moving at high velocity) striking a blank (or slug) of the metal to be extruded, which has been placed in the cavity of a die. Clearance is left between the punch and die walls; as the punch comes in contact with the blank, the metal has nowhere to go except through the annular opening between punch and die. The punch moves a distance that is controlled by a press setting. This distance determines the base thickness of the finished part. The process is particularly adaptable to the production of thin-walled, tubular-shaped parts having thick bottoms, such as toothpaste tubes.

Advantages of cold extrusion are higher strength because of severe strain-hardening, good finish and dimensional accuracy, and economy due to fewer operations and minimum of machining required. *See* METAL FORMING. [R.L.Fr.]

Eye (invertebrate) An eye may be defined as an aggregation of photoreceptor cells. It is one of the most frequently encountered sense organs of animals, and is possessed by some species of virtually every major animal phylum. However, the complexity of eyes varies greatly and this sense organ undoubtedly evolved independently a number of times within the animal kingdom.

Structure and function. The simplest invertebrate eyes are clusters of photoreceptor cells located on the surface of the body. Pigment cells are usually interspersed among the photoreceptors, giving the eye a red or black color. Simple eyes of this type, called pigment spot ocelli, are found in such invertebrates as jellyfish, flatworms, and sea stars.

The more common type of invertebrate eye probably arose from such patches of photoreceptor cells by an in-sinking of the sensory epithelium to form a cup. The cup, which represents the retina, may have remained as a pit open to the surface, may have sunk beneath the surface, or may have become closed in conjunction with the evolution of a cornea and lens.

Simple light detection is still the sole function of the great majority of invertebrate eyes, even those with lens and cornea. Cornea and lens merely provide the eye with a protective covering and concentrate the light on the retinal surface. Image formation has evolved as an additional function of the eyes of some invertebrates. The number of photoreceptor cells composing the retinal surface is of primary importance, since each photoreceptor cell or group of cells acts as the transmitter for one point of light. The great majority of invertebrate eyes cannot form an image because they do not possess a sufficient number of photoreceptor cells. The number of photoreceptor cells might be sufficient to detect movement of an object, but is inadequate to provide information about the object's form. *See* PHOTORECEPTION.

The focusing mechanisms of invertebrate eyes vary considerably. The focus of arthropod eyes tends to be fixed, that is, the distance between the light-bending structure—the cornea-lens—and the retina cannot be changed. Thus objects are in focus only a certain distance from the eye. The oceanic family of swimming polychaete worms, Alciopidae, have the most highly developed eyes of any of the segmented worms. The eyes of these worms are focused hydrostatically. A bulb to one side of the eye injects fluid into the space between the retina and the lens, forcing the lens outward. Another mechanism is employed in octopods whereby lens movement is brought about by a ciliary muscle attached to the lens. However, octopods tend to be rather nearsighted because of the somewhat flattened, or shortened, eye.

Visual object perception in invertebrates is correlated with a variety of behavioral adaptations. In fact, behavior frequently indicates that the eye is capable of some degree of image formation. Invertebrates which are predatory commonly possess complex eyes that are probably capable of detecting and recognizing objects, such as cursorial hunting spiders, squids and octopods, certain insects, some crabs and other crustaceans; those with highly directional flight, such as bees, wasps, flies, and some butterflies; those displaying courtship behavior, such as jumping spiders and fiddler crabs; or those whose orientation is by celestial clues, such as bees and some spiders.

Compound eyes. The compound eye of crustaceans, insects, centipedes, and horseshoe crabs has a different construction from that of other invertebrates. The term compound is somewhat misleading, because the eye functions basically like that of other eyes and is not compound in the sense of representing an aggregate of eyes. Each light-transmitting photoreceptor unit has a refractive body instead of a single cornea-lens structure for the entire eye.

The structural unit of the compound eye is called ommatidium (see illustration). The outer end of the ommatidium is composed of a cornea, which appears on the surface of the eye as a facet. Beneath the cornea is an elongated, tapered crystalline cone; the cornea and cone together function as a lens. The receptor element at the inner end of the ommatidium is called the retinula and is composed of a central translucent cylinder (rhabdome), around which are located usually seven photoreceptor cells plus one eccentric cell. The photoreceptor element of each ommatidium functions as a unit and can respond only to one point of light. Thus image formation is dependent upon the number of photoreceptor units present. Pigment granules surround the ommatidium proximally and distally, forming a light screen that separates one ommatidium from another. The pigment granules migrate, depending upon the amount of light.

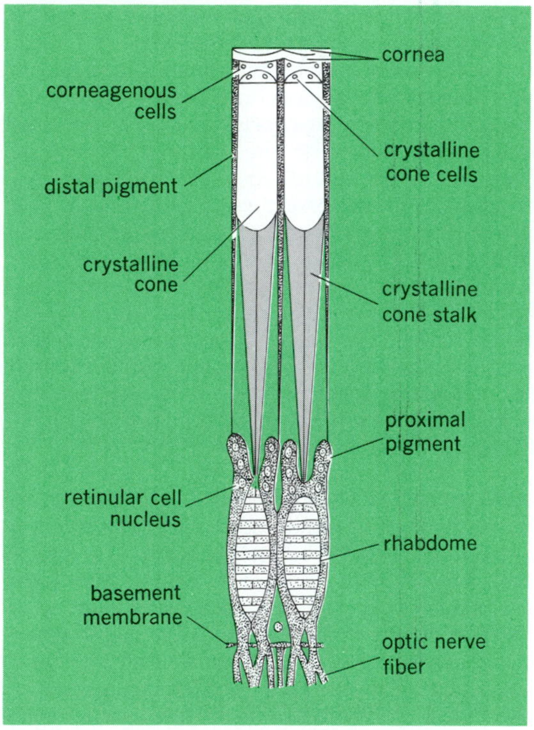

Two ommatidia from compound eye of crayfish *Astacus*. (*After R. D. Barnes, Invertebrate Zoology, 2d ed., W. B. Saunders Co., 1968*)

The advantage of the compound eye may well be its ability to detect movement. A slight shift in a point of light results in a corresponding shift in the ommatidia being stimulated; this feature is valuable in emphasizing movement since several ommatidia will respond to the same point of light. Another advantage of the compound eye of many arthropods is that the total corneal surface is greatly convex, resulting in a wide visual field. For example, the compound eye at the end of the eyestalk of a crab may cover an arc of 180°.

Nocturnal vision. The eyes of many nocturnal invertebrates are adapted for functioning in reduced light. Wolf spiders, hunters that run over the ground at night, illustrate eyes of this type. Of the eight eyes, six are indirect and are supplied with a layer of reflecting pigment which reflects light onto the photoreceptors. Like those of a cat, the eyes of these spiders appear to glow in the dark when facing a beam of light; the eyes of specimens the size of a quarter can be seen from a distance of 50 ft (15 m). [R.D.B.]

Eye (vertebrate) A sense organ that acts as a photoreceptor capable of image formation. The eye of vertebrates is constructed along a basic anatomical pattern which, in the diversification of animals, has undergone a variety of structural and functional modifications associated with different ecologies and modes of living. Often compared with a camera, the vertebrate eye is conveniently described in terms of its wall, cavities, and lens (see illustration).

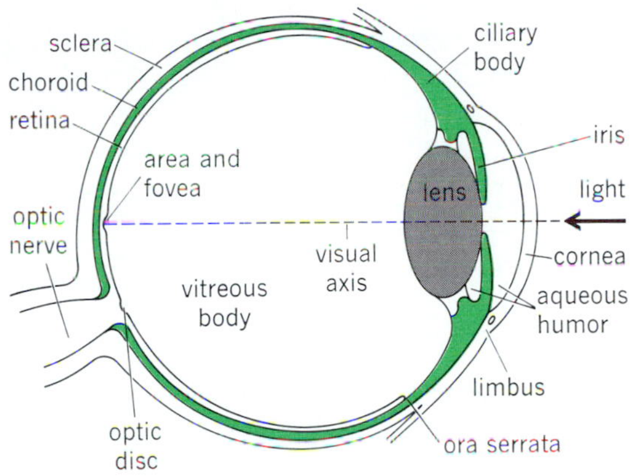

Horizontal section through human eye.

The wall. The wall of the eye consists of three distinct layers or tunics which, from outward to inward, are termed the fibrous, vascular, and sensory tunics.

Fibrous tunic. This continuous, outermost fibrous tunic comprises a transparent anterior portion, the cornea, and a tough posterior portion, the sclera. In the human, the cornea represents about one-sixth of the fibrous tunic, the sclera five-sixths.

The vertebrate cornea exhibits very few modifications in structure regardless of environmental influences. Its major constituent is connective tissue (both cells and fibers), regularly arranged and bordered on both anterior and posterior surfaces by an epithelium. The anterior epithelium is stratified, ectodermal in origin, and continuous with the (conjunctival) epithelium lining the eyelids. The transparency of the cornea is attributed to the geometric organization of its connective tissue elements, its constant state of deturgescence, and its chemical composition. It is the first ocular component traversed by the incoming light.

The sclera, a tough connective tissue tunic, provides support for the eye and serves for the attachment (insertions) of the muscles that move it.

The limbus is located at the angle of the anterior chamber. This small, circular transitional zone between the cornea and the sclera houses the major route for the discharge of aqueous humor from the anterior chamber.

Vascular tunic. The vascular tunic or uvea makes up the middle layer of the wall of the eye. It does not form a continuous layer around the eye but is deficient anteriorly, where the opening is termed the pupil. Beginning at the pupil, three continuous components of the uvea can easily be recognized: the iris, ciliary body, and choroid.

The iris is a spongy, circular diaphragm of loose, pigmented connective tissue separating the anterior and posterior chambers and housing a hole, the pupil, in its center. When heavily pigmented, the human iris appears brown; when lightly pigmented, blue.

The ciliary body is continuous with the root of the iris. The posterior epithelium of the iris continues along the internal surface of the ciliary body as a double layer of cells (ciliary part of the retina) which assumes many folds for the attachment of the suspensory ligament of the lens. This ligament holds the lens in position and shape, and marks the posterior boundary of the posterior chamber. The inner layer of the ciliary epithelium contains no pigment. It produces aqueous humor which flows into the posterior chamber and thence into the anterior chamber (via the pupil). The continual production and removal of this fluid maintain the intraocular pressure of the eye (which is increased in glaucoma).

The choroid is the most posterior portion of the uvea. It is directly continuous with the subepithelial portion of the ciliary body and consists primarily of blood vessels embedded within deeply pigmented connective tissue.

Sensory tunic. The retina is the sensory tunic of the eye. It has the form of a cup closely applied to the inner portion of the choroid, and, internally, it is slightly adherent to the semisolid vitreous body. The vertebrate retina contains the light-sensitive receptors (visual cells) and a complex of well-organized impulse-carrying nerve cells (neurons), all arranged into discrete layers.

Pigmented epithelium forms an important barrier between the light-sensitive receptors (visual cells) and their blood supply, the choroid. As in the choroid, the pigmentation serves to absorb light and prevent its reflection.

The rods and cones of vertebrates generally occur as single units, but combinations of each type are frequently encountered in several vertebrate classes. Cones appear to be adapted for photopic, or daylight, vision, based on correlations with the visual habits of the animals involved. Rods, which predominate in nocturnal vertebrates, are adapted for scotopic, or night, vision. Except for their external process, the structure of these cells does not reflect these differences. *See* PHOTORECEPTION; VISION.

An important adaptation for improving visual detail in vertebrates is the formation of circumscribed thickenings of the retina resulting from localized increases in the number of visual cells and the other retinal neurons associated synaptically with them. Such thickenings, termed areas of acute vision, appear in some members of all vertebrate classes and reach their greatest development in birds, in which one to three distinct areas may be found in the same retina. Only a single area occurs in the human; it is colored yellow and is called the macula. The macula is situated in the center of the fundus and contains only cones.

The cavities. Three cavities or chambers are present within the vertebrate eye: anterior, posterior, and vitreous. The anterior and posterior chambers are continuous with one another at the pupil and are filled with the aqueous humor. The eye is normally maintained in a distended state by the (intraocular) pressure created by this fluid. The vitreous cavity, on the other hand, is filled with a semisolid material, the vitreous body, which is fixed in amount and relatively permanent. Its consistency is not uniform in all vertebrates, however.

The lens. The lens is a transparent body, supported by thin suspensory fibers and by the vitreous body behind and by the iris in front. It is completely cellular, the anterior cells forming a thin epithelium, and the posterior cells, much elongated, forming the so-called lens fibers. The entire lens is surrounded by an elastic capsule which serves for the attachment of the ciliary zonule. In all vertebrates the lens functions in accommodation, either by moving backward and forward or by changing its shape. An opacity of the lens is termed a cataract.

[D.B.M.; R.O'Ra.]

Eyepiece A lens or optical system which offers to the eye the image originating from another system (the objective), at a suitable viewing distance. The image can be virtual. *See* Optical image.

In modern instruments, most eyepieces (also called oculars) are not independently corrected for all errors. They are designed to balance out certain residual aberrations of the objective or (as in the microscope) of a group of objectives, for instance, chromatic difference of magnification. *See* Aberration (optics).

The Ramsden eyepiece consists of two planoconvex lenses, the field lens and the eye lens, with their plane sides out. Both of these lenses have the same power and focal length; their separation is equal to their common focal length. *See* Geometrical optics; Lens (optics).

The Huygens eyepiece also consists of two planoconvex lenses, but the plane sides of both lenses face the eye, The focal length of the field lens is in general three times that of the eye lens, and the separation is twice the focal length of the eye lens. [M.J.H.]

F

Fabales An order of flowering plants, division Magnoliophyta (Angiospermae), in the subclass Rosidae of the class Magnoliopsida (dicotyledons). The order consists of three closely related families (Mimosaceae, Caesalpiniaceae, Fabaceae) collectively called the legumes. Members of the order typically have stipulate, compound leaves, 10–many stamens which are often united by the filaments, and a single capel which gives rise to a dry fruit (legume) that opens at maturity by splitting along two sutures, releasing the nonendospermous seeds. Many, or perhaps most, members of the order harbor symbiotic nitrogen-fixing bacteria in the roots. *See* ALFALFA; BEAN; CLOVER; COWPEA; LESPEDEZA; LOCUST (FORESTRY); MAGNOLIOPSIDA; PEA; PEANUT; SOYBEAN; VETCH. [T.M.Ba.]

Facies (geology) Any observable attribute of rocks, such as overall appearance, composition, or conditions of formation, and changes that may occur in these attributes over a geographic area. The term facies is widely used in connection with sedimentary rock bodies, but is not restricted to them. In general, facies are not defined for sedimentary rocks by features produced during weathering, metamorphism, or structural disturbance. In metamorphic rocks specifically, however, facies may be identified by the presence of minerals that denote degrees of metamorphic change.

Sedimentary facies. The term sedimentary facies is applied to bodies of sedimentary rock on the basis of descriptive or interpretive characteristics. Descriptive facies are based on lithologic features such as composition, grain size, bedding characteristics, and sedimentary structures (lithofacies) or on biological (fossil) components (biofacies), or on both. Individual lithofacies or biofacies may be single beds a few millimeters thick or a succession of beds tens to hundreds of meters thick. For example, a river deposit may consist of decimeters-thick beds of a conglomerate lithofacies interbedded with a cross-bedded sandstone lithofacies. The fill of certain major Paleozoic basins may be divided into units hundreds of meters thick comprising a shelly facies, containing such fossils as brachiopods and trilobites, and a graptolitic facies. *See* SEDIMENTARY ROCKS.

The term facies can be used also in an interpretive sense for groups of rocks that are thought to have been formed under similar conditions. This usage may emphasize specific depositional processes, such as a turbidite facies, or a particular depositional environment such as a shelf carbonate facies, encompassing a range of depositional processes. Finally, widespread use is made of two facies terms which are defined primarily on the basis of tectonic control. The flysch facies comprises marine sediments, typically turbidites and other mass-flow deposits, formed on continental margins under plate tectonic conditions of convergence, prior to the development of major orogenic uplifts. The molasse facies consists of nonmarine and shallow marine sediments formed within and flanking fold belts during and following their elevation into mountain ranges, under conditions of plate subduction. *See* TURBIDITE.

Groups of facies (usually lithofacies) that are commonly found together in the sedimentary record are known as facies assemblages or facies associations. These groupings provide the basis for defining broader, interpretive facies for the purpose of paleogeographic reconstruction. *See* PALEOGEOGRAPHY. [A.D.M.]

Metamorphic facies. A metamorphic facies refers to a set of mineral assemblages produced from primary igneous or sedimentary parent rocks in response to changes in temperature and pressure during burial in the Earth's crust. *See* METAMORPHIC ROCKS.

The facies concept of metamorphism, with its reliance on stable mineral assemblages as indicators of metamorphic conditions, complements and refines the so-called index mineral approach to metamorphic petrogenesis, whereby individual minerals such as chlorite, kyanite, and lawsonite are used to signify relative metamorphic grade. *See* METAMORPHISM.

Metamorphic facies are often correlated with formation in a particular geological environment. For example, the blueschist facies is usually interpreted as having formed in the subduction zones of converging lithospheric plates and the eclogite facies rocks near the base of the continental crust. *See* BLUESCHIST METAMORPHISM; PETROLOGY; PLATE TECTONICS. [J.J.F.]

Facsimile Transmission of a fixed image by wire or radio. The image or subject copy is usually in the form of a photograph, handwriting, map, or drawing. Facsimile equipment in the United States is used for telegram pickup and delivery and for business document transmission. Almost all newspaper photographs of events outside the local city in which the newspaper is published are sent by facsimile. Complete newspapers are printed from facsimile transmissions originating in distant cities. Railroads employ high-speed facsimile for transmission of waybills. Cloud-cover photographs and other types of graphic data are recorded by facsimile from signals sent by orbiting and stationary satellites. X-ray photographs have been sent from hospitals to radiological interpretation centers. Another use that is becoming increasingly important is information storage and retrieval. Facsimile will provide rapid, remote access to documents stored on microfilm at central locations such as libraries.

In the black-and-white scanning and transmitting process the subject copy is divided into dots, or elemental areas. Each elemental area ($1/100 \times 1/100$ in. or 0.25×0.25 mm in a typical scanning raster) is transmitted as an electrical signal that represents its blackness. In an amplitude-modulated system, elemental areas are transmitted at a constant rate, usually 1800–4000 per second, when transmission is over a voice-frequency circuit. Newspaper pages are transmitted at rates of up to 1,500,000 elemental areas per second over microwave circuits.

Some circuits cannot retain the true amplitude relationship of the signal in transmission. For example, high-frequency radio circuits are subject to considerable fading. In such cases amplitude modulation is unsatisfactory. Frequency-shift modulation is used so that the received copy will not be degraded by fading or level changes.

The received signals are recorded on a record sheet by any one of many processes. In a direct-recording facsimile system, an image is visible while being recorded. In photofacsimile systems using photographic film or paper, processing is necessary before the image can be viewed.

For transmission of color photographs, three separate facsimile transmissions are made from the original color print. Color-separation filters in the optical system of the facsimile transmitter produce black-and-white color-separation negatives at the facsimile recorder. The technique is similar to that used for color-separation negatives made by photography.

One type of digital facsimile system sends only black or white picture information; no gray scale shades are reproduced. Because changes from black to white can occur only at discrete points along a recording line, vertical lines have a more ragged appearance than in comparable analog facsimile systems. However, noise on long circuits is less apt to degrade the received picture. Another advantage is secrecy. For example, digital signals may be connected to military encrypting equipment for sending secret information. [K.R.McC.]

Factor analysis A method of quantitative multivariate analysis with the goal of representing the interrelationships among a set of continuously measured variables (usually represented by their intercorrelations) by a number of underlying, linearly independent reference variables called factors. Although the term factor analysis has come to represent a family of analysis methods, the two most commonly used approaches are the full component model, in which the entire variance of the variables (represented by unities inserted in the principal diagonal of the correlation matrix) is analyzed, and the common factor model, in which the proportion of the variance that is accounted for by the common factors (represented by communality estimates inserted in the principal diagonal) is analyzed.

The method was developed in England around the turn of the century and was first applied to the study of the structure of intellectual abilities. Since then it has been used in many disciplines, from agriculture to zoology, in which the underlying structure of multiple variables and their representation in that structure are of interest. Another application of factor analysis is to represent parsimoniously the variables in the set on which the observations are made by a smaller number of underlying reference variables or factors. *See* STATISTICS. [B.Fr.]

Fagales An order of flowering plants, division Magnoliophyta (Angiospermae), in the subclass Hamamelidae of the class Magnoliopsida (dicotyledons). The order consists of three families, the Fagaceae, Betulaceae, and Balanopaceae. The Fagales are simple-leaved, woody plants with much reduced, mostly unisexual flowers, the pistillate ones with one or two anatropous ovules in each of the two or three locules of the ovary, the staminate ones usually grouped into catkins. Birch (*Betula*), beech (*Fagus*), and oak (*Quercus*) are well-known members of the Fagales. *See* BEECH; BIRCH; HAMAMELIDAE; MAGNOLIOPSIDA. [A.Cr.]

Falconiformes The order containing the diurnal birds of prey. Five living families are usually recognized, as well as two extinct families, the Neocathartidae (Upper Eocene) and Teratornithidae (Pleistocene). The largest and most diverse family is the Accipitridae, including rapacious birds such as the hawks, eagles, and kites, adapted for capturing living prey ranging from ungulates to small invertebrates. It also includes a subfamily (Aegypiinae) of carrion-feeders, the Old World vultures. Usually placed in this subfamily is the only primarily vegetarian member of the order, the palm-nut vulture (*Gypohierax angolensis*) of Africa.

Similar to the Aegypiinae in food habits and superficially similar in appearance are the New World vultures (Cathartidae), among which is the Andean condor, one of the largest of extant flying birds, attaining a wingspread of 12 ft (3.6 m). The Cathartidae differ from other Falconiformes in many anatomical and behavioral characters.

The falcons (Falconidae) constitute another anatomically distinct family composed principally of long-winged rapacious birds. One New World subfamily, the caracaras (Caracarinae), has become secondarily adapted to carrion-feeding.

Two somewhat specialized species of this order are each given family rank. The widely distributed osprey or fish hawk (Pandionidae) has a reversible hind toe, especially well-developed claws, and spicules on the scales of its feet, all assisting in the capture of slippery prey. The long-legged, primarily terrestrial secretary bird (Sagittariidae) of Africa, named for a fancied resemblance of its nuchal plumes to quill pens stuck behind an ear, is famous for its ability to prey on poisonous snakes, although it also feeds on many other vertebrates. *See* AVES; BIRDS OF PREY. [K.C.P.]

Fall line The zone or boundary between resistant rocks of older land and weaker strata of plains. Here the easily eroded strata of the plains lap onto the more resistant rocks of the old land. Streams that cross this boundary have steeper gradients along the width of the zone and make sudden descents over falls and rapids. In geologic and geographic literature, a line connecting such falls on several main rivers is referred to as a fall line. [S.Ju.]

Fallopian tube The upper part of the female oviduct present in the humans and other higher vertebrates. The Fallopian tube extends from the ovary to the uterus and transports ova from the ovary to the cavity of the uterus. Each tube is about 5 in. (12.5 cm) long; one lies on either side of the uterus and is attached at the upper portion. Each curves outward to end in a hoodlike opening, the infundibulum, with many fingerlike projections; the cavity of the Fallopian tube is continuous with the cavity of the coelom. The ovaries lie below and inside the tubal curve.

The ovum remains viable in the oviduct for about 1–3 days only. If fertilization occurs, the ovum moves into the cavity of the uterus and then implants on its wall. If fertilization fails to occur, the ovum degenerates in the uterus. Occasionally, a fertilized ovum fails to enter the uterus, or may be freed into the abdominal cavity, so that an ectopic pregnancy results if the ovum finds a site for implantation. *See* PREGNANCY DISORDERS; REPRODUCTIVE SYSTEM. [W.B.]

Fan A fan moves gases by producing a low compression ratio, as in ventilation and pneumatic conveying of materials. The increase in density of the gas in passing through a fan is generally negligible; the pressure increase or head is usually measured in inches of water.

Blowers are fans that operate where the resistance to gas flow is predominantly downstream of the fan. Exhausters are fans that operate where the flow resistance is mainly upstream of the fan.

Fans are further classified as centrifugal or axial (see illustration). The housing provides an inlet and an outlet and confines the flow to the region swept out by the rotating impeller. The impeller imparts velocity to the gas, and this velocity changes to a pressure differential under the influence of the housing and ducts connected to inlet and outlet. *See* DUCTED FLOW. [T.Ba.]

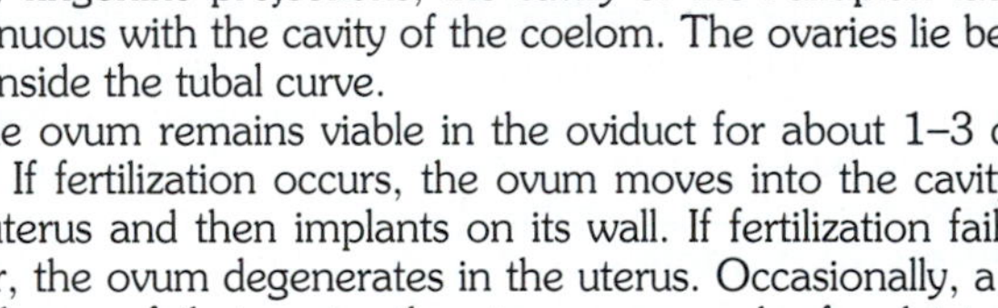

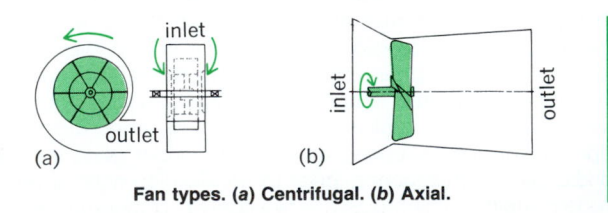

Fan types. (*a*) Centrifugal. (*b*) Axial.

Faraday effect Rotation of the plane of polarization of a beam of linearly polarized light when the light passes through matter in the direction of the lines of force of an applied magnetic field. Discovered by M. Faraday in 1846, the effect is often called magnetic rotation. *See* MAGNETOOPTICS.

The Faraday effect is particularly simple in substances having sharp absorption lines, that is, in gases and in certain crystals, particularly at low temperatures. Here the effect can be fully explained from the fundamental properties of the atoms and molecules involved. In other substances the situation may be more complex, but the same principles furnish the explanation.

Rotation of the plane of polarization occurs when there is a difference between the indices of refraction n^+ for right-handed polarized light and n^- for left-handed polarized light. Most substances do not show such a difference without a magnetic field, except optically active substances such as crystalline quartz or a sugar solution. It should be noted that the index of refraction in the vicinity of an absorption line changes with the frequency. *See* ABSORPTION; POLARIZED LIGHT. [G.H.Di.; W.W.W.]

Faraday's law of induction A statement relating an induced electromotive force (emf) to the change in magnetic flux that produces it. For any flux change that takes place in a circuit, Faraday's law states that the magnitude of the emf ξ induced in the circuit is proportional to the rate of change of flux as in the expression below.

$$\xi \propto \frac{d\Phi}{dt}$$

The time rate of change of flux in this expression may refer to any kind of flux change that takes place. If the change is motion of a conductor through a field, $d\Phi/dt$ refers to the rate of cutting flux. If the change is an increase or decrease in flux linking a coil, $d\Phi/dt$ refers to the rate of such change. It may refer to a motion or to a change that involves no motion. *See* ELECTROMAGNETIC INDUCTION; MAGNETIC FLUX. [K.V.M.]

Fascioliasis The accidental infection of humans with the cattle liver fluke (*Fasciola hepatica*). The leaflike fluke inhabits the bile ducts; infected animals pass the eggs in feces; the eggs embryonate in water. Humans are exposed by ingesting aquatic vegetables such as watercress. Incidence is generally low, but serious outbreaks have been recorded in humans from various parts of the world. Symptoms are usually chronic and directly referable to biliary malfunction, which may become confused with various liver diseases. *See* TREMATODA. [J.F.M.]

Fasciolopsiasis The presence of the trematode *Fasciolopsis buski* in the small intestine. The parasite is native to eastern Asia and the southwestern Pacific; cases have been reported from Ecuador. Hogs act as natural reservoirs. Humans become parasitized by using their teeth to peel various aquatic raw vegetables on which the parasite encysts after its development in a molluscan host. The fluke lives attached to the intestinal mucosa by its two powerful suckers, producing inflammation and ulceration. A light infection is characterized by toxic diarrhea. A heavy one produces toxic symptoms leading to edema and heart failure. *See* TREMATODA. [J.F.M.]

Fat and oil Naturally occurring esters of glycerol and fatty acids that have commercial uses. Since fats and oils are triesters, they are commonly called triglycerides or simply glycerides. A glyceride may be designated a fat or oil, depending on its melting point. A fat is solid and an oil is liquid at room temperature. Some liquid waxes are incorrectly referred to as oils. *See* ESTER; TRIGLYCERIDE; WAX, ANIMAL AND VEGETABLE.

The structure of triglycerides is shown below, where R_1, R_2, and R_3 represent the alkyl chain of the fatty acid.

$$\begin{array}{c} H_2COC(O)R_1 \\ | \\ HCOC(O)R_2 \\ | \\ H_2COC(O)R_3 \end{array}$$

The physical and chemical properties of fats and oils are determined to a large extent by the types of fatty acids in the glyceride. It is possible for all the acids to be identical, but this is rare. Usually there are two or even three different acids esterified to each glycerol molecule.

In all commercially important glycerides, the fatty acids are straight-chain, and nearly all contain an even number of carbon atoms. Most fats and oils are based on C_{16} and C_{18} acids with zero to three ethylenic bonds. There are exceptions, such as coconut oil, which is rich in shorter-chain acids, and some marine oils, which contain acids with as many as 22 or more carbons and six or more ethylenic linkages.

The majority of fats and oils come from only a few sources. Plant sources are nuts or seeds, and nearly all terrestrial animal fats are from adipose tissue. Marine oils come principally from the whole body, although a small amount comes from trimmings. Plant fats and oils are obtained by crushing and solvent extraction. Animal and marine oils are nearly all recovered by rendering. This is a process of heating fatty tissue with steam or hot water to melt and free the glyceride, followed by separating the oil or fat from the aqueous layer. *See* SOLVENT EXTRACTION.

There are some uses for fats and oils in their native state, but ordinarily they are converted to more valuable products. The most important changes are hydrolysis and hydrogenation. Hydrolysis is commonly called splitting. The purpose is to hydrolyze the ester into its constituent glycerol and fatty acids, which are valuable intermediates for many compounds. Hydrogenation is the catalytic addition of hydrogen to ethylenic bonds, and is applicable to both acids and glycerides. The purpose of hydrogenation is usually to raise the melting point or to increase the resistance to oxidation. *See* HYDROGENATION; HYDROLYTIC PROCESSES. [H.M.H.]

Fat and oil (food) One of the three major classes of basic food substances, the others being protein and carbohydrate. Fats and oils are a source of energy. They also aid in making both natural and prepared foods more palatable by improving the texture and providing a more desirable flavor. *See* FOOD.

Fats are grouped according to source. Animal fats are rendered from the fatty tissues of hogs, cattle, sheep, and poultry. Butter is obtained from milk. Vegetable oils are pressed or extracted from various plant seeds, primarily from soybean, cottonseed, corn (germ), peanut, sunflower, safflower, olive, rapeseed, sesame, coconut, oil palm (pulp and kernel separately), and cocoa beans. Marine oils, which are not consumed in the United States, are obtained mostly from herring, sardine, and pilchard.

Fats and oils are important in the diet. They are the most concentrated form of food energy, contributing about 9 cal/g (38 joules/g), as compared to about 4 cal/g (17 joules/g) for carbohydrates and proteins. Fats make a meal more satisfying by creating a feeling of fullness, and also delay the onset of hunger. Contrary to popular belief, fats are highly digestible, with 94–98% of the ingested fat being absorbed from the intestinal tract.

The polyunsaturated fatty acids, primarily linoleic and arachidonic, are essential nutrients; that is, they are not synthesized by the body but are required for tissue development. Absence of these fatty acids from the diet results in an essential fatty acid syndrome and in a specific form of eczema in infants.

Vegetable oils are an excellent source of linoleic acid, while meat fats provide arachidonic acid in small but significant amounts. Fats and oils are carriers of the oil-soluble vitamins A and D, and are the main source of vitamin E. They also have a sparing action on some of the B complex vitamins. *See* NUTRITION.

Several forms of deterioration may occur in fats and oils. Flavor and odor may develop after deodorization of a product to complete blandness. The flavor is generally characteristic of the oil source and is therefore usually acceptable. However, soybean oil can develop disagreeable flavors described as beany, grassy, painty, fishy, or like watermelon rind. Beef fat can become tallowy, which is also objectionable. Reversion is apparently caused by changes in substances which have been oxidized prior to, but not removed by, deodorization.

Oxidative rancidity is a serious flavor defect and highly objectionable. It starts with the formation of hydroperoxides which then decompose to form aldehydes which have a pungent, disagreeable flavor and odor. Retardation of oxidation is brought about by using opaque, airtight containers, or nitrogen blanketing if clear glass bottles are used. Antioxidants are required in meat fats, since lard, tallow, and so on contain no natural antioxidant material. Vegetable oils contain tocopherols. Additional antioxidant, with the exception of tertiary butyl hydroquinone (TBHQ), has little benefit for these oils.

Hydrolytic rancidity results from the liberation of free fatty acids by the reaction of fats and oils with water. While most fats show no detectable off flavors, coconut and other lauric acid oils develop a soapy flavor, and butter develops the strong characteristic odor of butyric acid. Packaged coconut-oil products and lauric-type hard butters sometimes contain added lecithin, which acts as a moisture scavenger, thereby retarding hydrolytic rancidity development.

Fats and oils used in deep fat frying can break down under adverse conditions, especially where frying is intermittent or the fryer capacity is not fully used. Deterioration ultimately results in the oil becoming very dark in color, viscous, foul-odored, and foaming badly during frying. The oil becomes oxidized and then polymerized, requiring that it be discarded, since it imparts strong off flavors to the fried food.

Crystal structure transformation of packaged shortening results in formation of a grainy, soft product, which may also lose incorporated gas and take on the appearance of petroleum jelly. Similar changes in crystal structure can cause bloom in chocolate coatings, a defect which gives a white haze or even open grain on an originally smooth, glossy surface. Chocolate bloom can be inhibited by the addition of lecithin or polysorbates or both. *See* LIPID. [T.J.W.]

Fault analysis The detection and diagnosis of malfunctions in technical systems. Such systems include production equipment, transportation vehicles, and household appliances. While the need to detect and diagnose malfunctions is as old as the construction of such systems, advanced fault detection has been made possible only by the proliferation of the computer. Fault detection and diagnosis actually means a scheme in which a computer monitors the technical equipment to signal any malfunction and determine the components responsible for it. The detection and diagnosis of the fault may be followed by automatic actions enabling the fault to be corrected, such that the system may operate successfully even under the particular faulty condition.

Fault detection and diagnosis applies to both the basic technical equipment and the actuators and sensors attached to it. Actuator and sensor fault detection is very important because these devices are quite prone to faults.

The on-line or real-time detection and diagnosis of faults means that the equipment is constantly monitored during its regular operation by a permanently connected computer, and any discrepancy is signaled almost immediately. On-line monitoring is very important for early detection of any component malfunction, before it can lead to more substantial equipment failure. In contrast, off-line diagnosis involves the monitoring of the system by a special, temporarily attached device, under special conditions (for example, car diagnostics at a service station).

The diagnostic activity may be broken down into several logical stages. Fault detection is the indication of something going wrong in the system. Fault isolation is the determination of the fault location (the component which malfunctions), while fault identification is the estimation of its size.

Fault detection and isolation can never be performed with absolute certainty, because of circumstances such as noise, disturbances, and model errors. There is always a trade-off between false alarms and missed detections, the proper balance depending on the particular application. [J.J.G.]

Fault and fault structures Products of fracturing and differential movements along fractures in continental and oceanic crustal rocks. Faults range in length and magnitude of displacement from small structures visible in hand specimens, displaying offsets of a centimeter or less, to long, continuous crustal breaks, extending hundreds of kilometers in length and accommodating displacements of tens or hundreds of kilometers. Faults exist in deformed rocks at the microscopic scale, but these are generally ignored or go unrecognized in most geological studies. Alternatively, where microfaults systematically pervade rock bodies as sets of very closely spaced subparallel, planar fractures, they are recognized and interpreted as a type of cleavage which permitted flow of the rock body. Fractures along which there is no visible displacement are known as joints. Large fractures which have accommodated major dilational opening (a meter or more) perpendicular to the fracture surfaces are known as fissures. Formation of fissures is restricted to near-surface conditions, for example, in areas of crustal stretching of subsidence. *See* JOINT (GEOLOGY).

In addition to describing the physical and geometric nature of faults and interpreting time of formation, it has been found to be especially important to determine the orientations of minor fault structures (such as striae and drag folds) which record the sense of relative movement. Evaluating the movement of faulting can be difficult, for the apparent relative movement (separation) of fault blocks as seen in map or out-

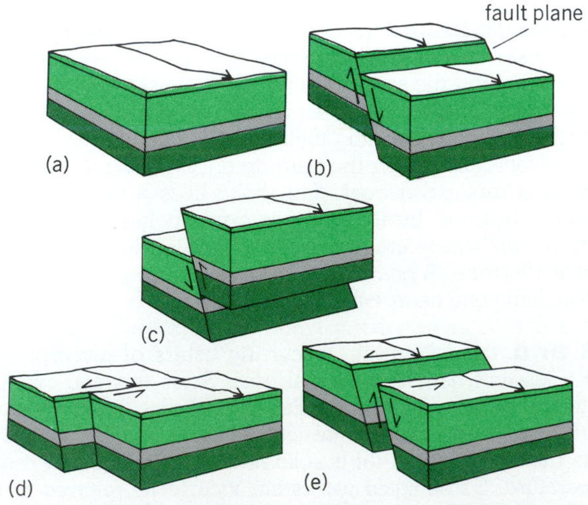

Slip on faults. (*a*) Block before faulting; (*b*) normal-slip; (*c*) reverse-slip; (*d*) strike-slip; (e) oblique-slip. (*After F. Press and R. Siever, Earth, 2d ed., 1978*)

crop may bear little or no relation to the actual relative movement (slip). The slip of the fault is the actual relative movement between two points or two markers in the rock that were coincident before faulting (see illustration). Strike-slip faults have resulted in horizontal movements between adjacent blocks; dip-slip faults are marked by translations directly up or down the dip of the fault surface; in oblique-slip faults the path of actual relative movement is inclined somewhere between horizontal and dip slip.

Recognizing even the simplest translational fault movements in nature is often enormously difficult because of complicated and deceptive patterns created by the interference of structure and topography, and by the absence of specific fault structures which define the slip path. While mapping, the geologist mainly documents apparent relative movement (separation) along a fault, based on what is observed in plan-view or cross-sectional exposures. *See* TRANSFORM FAULT. [G.H.D.]

Fault-tolerant systems Systems, predominantly computing and computer-based systems, which tolerate undesired changes in their internal structure or external environment. Such changes, generally referred to as faults, may occur at various times during the evolution of a system, beginning with its specification and proceeding through its utilization. Faults that occur during specification, design, implementation, or modification are called design faults; those occurring during utilization are referred to as operational faults. The use of fault tolerance techniques is based on the premise that a complex system, no matter how carefully designed and validated, is likely to contain residual design faults and to encounter unpreventable operational faults.

Generally, fault tolerance techniques attempt to prevent lower-level errors (caused by faults) from propagating into system failures. By using various types of structural and informational redundancy, such techniques either mask a fault (no errors are propagated to the faulty subsystem's output) or detect a fault (via an error) and then effect a recovery process which, if successful, prevents a system failure. In the case of a permanent internal fault, the recovery process usually includes some form of structural reconfiguration (for example, replacement of a faulty subsystem with a spare or use of an alternate program) which prevents the fault from causing further errors. Typically, a fault-tolerant system design will incorporate a mix of fault tolerance techniques which complement the techniques used for fault prevention. *See* SOFTWARE ENGINEERING. [J.F.Me.]

Feather A specialized keratinous outgrowth of the skin, which is a unique characteristic of birds. Feathers are highly complex structures that provide insulation, protection against mechanical damage, protective coloration, and also function significantly in behavior. One special functional role is in flight, where feathers provide propulsive surfaces and a body surface aerodynamically suitable for flight. Feathers are used in maintenance of balance and occasionally in the capture of prey and various specialized displays.

A representative definitive feather contains a single long central axis which supports a row of small branchlike structures along each side (barbs). Barbs form the vane, or web, of the feather. Individual barbs branch off at variable angles and point toward the outer tip of the feather. The barbules are small branches from the barbs. They lie in the same plane as the barbs and arise in rows from their anterior and posterior surfaces. The anterior barbules have a flattened base and a series of small hooklike projections which attach to the proximal ridge of the posterior barbules of the next barb, forming an interlocking structure characterized by its great strength and light weight. All feather types consist basically of these structural elements.

Most of the superficial feathers are contour feathers (pennae). These include the large flight feathers (remiges) of the

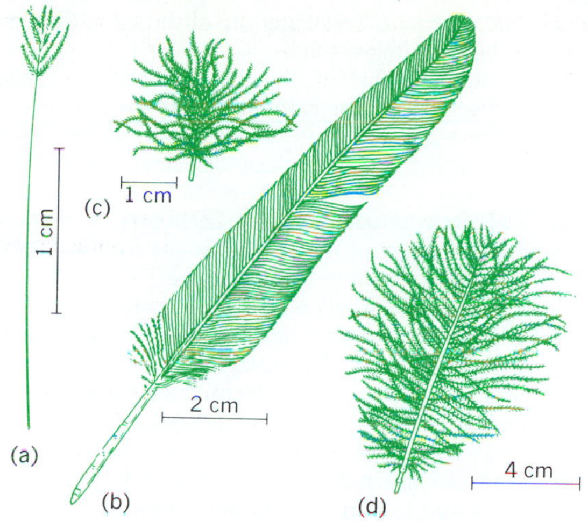

Types of feathers, (*a*) Filoplume. (*b*) Vane or contour. (*c*) Down. (*d*) Semiplume. (*After J. C. Welty, The Life of Birds, Saunders, 1962*)

wing and the long tail feathers (rectrices). Other common feather types include the down feathers (plumulae), intermediate types (semiplumes), and filoplumes (see illustration).

Feathers normally undergo attrition because of the physical abuse attendant to the normal activity of birds. In most species, feathers are replaced completely at least annually, and many of the feathers are replaced more frequently. The sequence of feather molt is surprisingly orderly. Penguins, which shed large patches of feathers in an irregular pattern, are an exception. In most species the power of flight is retained during molt. The molt, that is, the normal shedding of feathers and their replacement by a new generation of feathers, is a single growth process which is actively concerned only with the production of the new generation of feathers. The old feathers are pushed out of the follicles passively.

A major physiological role of feathers is to provide insulation. This is accomplished by regulating the configuration of feather and skin in such a way that differing amounts of air are trapped in the dead space so formed. A second mechanism for control of heat dissipation is the balance of the exposure of feathered and unfeathered body parts.

Feathers act as a protective boundary in their role of providing waterproofing. Water repellency is a structural feature of feathers and is the result of precise geometric relationships between the diameter and spacing of barbs and barbules. Preening appears to be more important in the maintenance of this structure than it is for the application of oils or any other natural product, as was once thought.

A third function of the surface configuration and overall pattern of feathers is in the area of behavioral adaptations. These may be of two types. First is concealment, when the bird is cryptically marked to match its background and escape detection. The second type consists of various types of advertisement. *See* PROTECTIVE COLORATION. [A.H.B.]

Feces The waste material eliminated by the gastrointestinal tract. The components and form of feces vary with diet, metabolic activities, and the state of health; illness, changes in diet or water intake and excretion, emotional stimuli, and many drugs and chemicals may alter the feces both qualitatively and quantitatively. *See* CONSTIPATION; DIARRHEA.

In general, the feces contain undigested food materials and excretions and secretions of the digestive system, including salts, pigments, bile products, and cellular debris. Food passes from the small intestine to the lower bowel (large intestine) in a semiliquid state; nutrients and water are absorbed in the bowel, and the feces become more solid.

Feces constitute important objective evidence in disease. Defecation patterns together with laboratory analyses of fecal components are helpful, though not definitive, in the differential diagnosis of digestive system disorders. *See* BILIRUBIN. [S.P.P.]

Federal Telecommunications System A private telephone network throughout the United States for the exclusive use of all Federal government offices. The Federal Telecommunications System (FTS) consists of leased private-line circuits arranged for automatic switching at centers located on telephone company premises. This unique system can interconnect agency telephones by using techniques similar to those employed in the Bell System Direct Distance Dialing (DDD) network. In addition, there are telephones that can have access to the network but are not directly connected. The network has been engineered to permit many simultaneous messages to be handled with ease. The network can be rapidly augmented when needed for emergencies such as floods and storms. The network stretches from coast to coast and covers approximately 11,000,000 route miles (18,000,000 route kilometers). In addition to serving the continental United States, FTS is also installed in Alaska, Puerto Rico, and Hawaii. The system is used to handle normal administrative telephone calls and data transmission. *See* TELEPHONE; TELEPHONE SERVICE; TELEPHONE SIGNALING; TELEPHONE SYSTEMS CONSTRUCTION. [R.M.S.]

Feedback circuit A circuit that returns a portion of the output signal of an electronic circuit or control system to the input of the circuit or system. When the signal returned (the feedback signal) is at the same phase as the input signal, the feedback is called positive or regenerative. When the feedback signal is of opposite phase to that of the input signal, the feedback is negative or degenerative.

The use of negative feedback in electronic circuits and automatic control systems produces changes in the characteristics of the system which improve the performance of the system. In electronic circuits, feedback is employed either to alter the shape of the frequency-response characteristics of an amplifier circuit and thereby produce more uniform amplification over a range of frequencies, or to produce conditions for oscillation in an oscillator circuit. It is also used because it stabilizes the gain of the system against changes in temperature, component replacement, and so on. Negative feedback also reduces nonlinear distortion. In automatic control systems, feedback is used to compare the actual output of a system with a desired output, the difference being used as the input signal to a controller. *See* AMPLIFIER; NEGATIVE-RESISTANCE CIRCUITS; SERVOMECHANISM. [H.F.K.]

Feldspar A group of silicate minerals that constitute the most abundant group of minerals and are also important economic minerals, particularly in the ceramic and glass industries which utilize most of the output as raw materials. Feldspars are also used as gemstones if distinguished by an attractive luster or color. *See* SILICATE MINERALS.

The feldspars are silicates of aluminum, Al, with the metals potassium, K, sodium, Na, and calcium, Ca, and, rarely, barium, Ba. They occur as components of all kinds of rocks. Feldspars have no color of their own but are frequently colored (yellow, brown, reddish, or dirty green to black) by impurities. The cause of the beautiful bright green color of amazonite is unknown. The hardness is 6 on Mohs scale. Feldspars are brittle but usually show excellent cleavage.

Alkali feldspars is the name for feldspars composed of orthoclase (Or) and albite (Ab) in any ratio. Usually some anorthite (An) is present as a minor constituent. There are many names for the different varieties of alkali feldspars, most of which include the following: sanidine, microcline, orthoclase, albite, analbite, monalbite, perthite, antiperthite, and anorthoclase.

Plagioclase feldspars form at high temperature a complete series of solid solutions, ranging from $NaAlSi_3O_8$(Ab) to $CaAl_2Si_2O_8$(An). The plagioclase series is arbitrarily subdivided and named according to increasing An content as follows:

$Ab_{100}An_0$–$Ab_{90}An_{10}$ (albite); $Ab_{90}An_{10}$–$Ab_{70}An_{30}$ (oligoclase); $Ab_{70}An_{30}$–$Ab_{50}An_{50}$ (andesine); $Ab_{50}An_{50}$–$Ab_{30}An_{70}$ (labradorite); $Ab_{30}An_{70}$–$Ab_{10}An_{90}$ (bytownite); and $Ab_{10}An_{90}$–Ab_0An_{100} (anorthite). Because the content of silicic acid (in its anhydrous form SiO_2) is relatively higher in $NaAlSi_3O_8$ than in $CaAl_2Si_2O_8$, the Ab-rich members albite-oligoclase are frequently called acid plagioclases and the An-rich members bytownite-anorthite basic plagioclases. *See* ALBITE; ANDESINE; ANORTHOSITE; BYTOWNITE; LABRADORITE; OLIGOCLASE. [F.H.L.]

Feldspathoid A member of the feldspathoid group of minerals. Members of this group are characterized by the following related features: (1) All are aluminosilicates with one or more of the large alkali ions (for example, sodium, potassium) or alkaline-earth ions (for example, calcium, barium). (2) The proportion of aluminum relative to silicon, both of which are tetrahedrally coordinated by oxygen, is high. (3) Although the crystal structures of many members are different, they are all classed as tektosilicates. (4) They occur principally in igneous rocks, but only in silica-poor rocks, and do not coexist with quartz (SiO_2). Feldspathoids react with silica to yield feldspars, which also are alkali–alkaline-earth aluminosilicates. Feldspathoids commonly occur with feldspars. *See* FELDSPAR; SILICATE MINERALS.

The principal species of this group are the following:

Nepheline	$KNa_3[AlSiO_4]_4$
Leucite	$K[AlSi_2O_6]$
Cancrinite	$Na_6Ca[CO_3 \mid (AlSiO_4)_6] \cdot 2H_2O$
Sodalite	$Na_8[Cl_2 \mid (AlSiO_4)_6]$
Nosean	$Na_8[SO_4 \mid (AlSiO_4)_6]$
Haüyne	$(Na,Ca)_{8-4}[(SO_4)_{2-1} \mid (AlSiO_4)_6]$
Lazurite	$(Na,Ca)_8[(SO_4,S,Cl)_2 \mid (AlSiO_4)_6]$

The last four species (sodalite group) are isostructural, and extensive solid solution occurs between end members; but members of the sodalite group, cancrinite, leucite, and nepheline have different crystal structures. *See* CANCRINITE; LAZURITE; LEUCITE; LEUCITE ROCK; NEPHELINE SYENITE; NEPHELINITE; SODALITE. [D.R.P.]

Feline infectious peritonitis A fatal disease of both domestic and exotic cats which is caused by infection with feline infectious peritonitis virus, a member of the Coronaviridae family. There are multiple strains of the virus which vary in virulence. Feline infectious peritonitis virus is closely related morphologically, genetically, and antigenically to other members of the Coronaviridae. These members include feline enteric coronaviruses, which infect cats but generally induce very mild or inapparent gastroenteritis, as well as transmissible gastroenteritis virus of pigs, and canine coronavirus. These viruses may be host-range mutants of a single prototypical virus, given their ability for cross-species infection among cats, pigs, and dogs.

Feline coronaviruses are contracted via ingestion or inhalation during direct cat-to-cat contact. Feline infectious peritonitis virus was traditionally considered, like other enveloped viruses, to be very labile once outside a cat's body. However, the virus may be able to survive for as long as 7 weeks if dried onto a surface, although it is readily inactivated by most disinfectants.

Feline infectious peritonitis is commonly seen in cats between 6 months and 2 years of age. There are two clinical forms of feline infectious peritonitis, the wet (effusive) form and the dry (granulomatous) form. Infected cats show fever, lack of appetite, weight loss, and lethargy. Cats with the wet form of disease accumulate fluid in the abdomen or chest, or both, and may exhibit abdominal enlargement and difficulty in breathing. The signs associated with the dry form depend upon the particular organ affected, and thus may mimic disease of the liver, pancreas, kidneys, central nervous system, and eyes.

As with most viral infections, there is no specific antiviral drug of proven efficacy in the treatment of feline infectious peritonitis. Clinical management still rests upon palliative treatment of the specific signs exhibited by each cat, and upon antibiotics, when indicated, to reduce secondary bacterial infections. The most important therapeutic approach involves the administration of immunosuppressive doses of corticosteroids to reduce the cat's immune response to the virus and thus reduce the level of deleterious antibody produced. *See* ANIMAL VIRUS; VIRUS. [C.W.O.]

Feline leukemia A type of cancer caused by the feline leukemia virus, a retrovirus which affects only a small percentage of freely roaming or domestic cats. The feline leukemia virus is genetically and morphologically similar to murine leukemia virus, from which it presumably evolved several million years ago.

About 1–5% of healthy-appearing wild or freely roaming domestic cats have lifelong (persistent) infections. These carrier cats shed the virus in urine, feces, and saliva. The principal route of infection is oral. Infections occurring in nature are usually inapparent or mild, and 95% of such cats recover without any signs of illness. Mortality due to a feline leukemia virus infection occurs mainly among persistently infected cats and at a rate of around 50% per year.

There is no treatment that eliminates the virus. Supportive or symptomatic treatment may prolong life for weeks or months, depending on the particular disease manifestation. All cats should be tested for the presence of the virus prior to putting them in contact with feline leukemia virus–free animals. Healthy-appearing or ill infected cats should not be in intimate contact with noninfected cats, even if the latter have been vaccinated. Feline leukemia virus vaccines are available and should be administered annually, although they should not be considered a substitute for testing, elimination, and quarantine procedures. *See* LEUKEMIA; RETROVIRUS. [N.C.P.]

Feline panleukopenia An acute viral infection of cats, also called feline viral enteritis and (erroneously) feline distemper. The virus infects all members of the cat family (Felidae) as well as some mink, ferrets, and skunks (Mustelidae); raccoons and coatimundi (Procyonidae); and the binturong (Viverridae). Panleukopenia is the most important infectious disease of cats. This disease occurs worldwide, and nearly all cats are exposed by their first year because the virus is stable and ubiquitous; the disease is rarely seen in older cats. Without treatment, this disease is often fatal.

Feline panleukopenia virus is classified as a parvovirus, and is one of the smallest known viruses. It is antigenically identical to the mink enteritis virus, and only minor antigenic differences exist between feline panleukopenia virus and canine parvovirus. It is believed that canine parvovirus originated as a mutation from feline panleukopenia virus.

The disease is severe and life threatening in 20–50% of cases. The cat is depressed and may refuse food or water; vomiting and diarrhea are common, resulting in severe dehydration. The cat may have a fever or a subnormal temperature. A low white blood cell count confirms the diagnosis as panleukopenia. Diagnosis can be made by autopsy and evidence of the destruction of the intestinal crypts and villus shortening.

Highly effective and safe vaccines are available for the prevention of panleukopenia. Premises contaminated by feline panleukopenia virus are extremely difficult to disinfect; chlorine bleach, formaldehyde, or a certain quaternary ammonium disinfectant will destroy the virus. A cat should be successfully immunized before being introduced to premises where a panleukopenia-infected cat previously lived. *See* ANIMAL VIRUS. [J.H.Car.]

Felsite An igneous rock with a felsitic or aphanitic (not visibly crystalline) texture, composed largely of light-colored (felsic) minerals (quartz and feldspar). The constituents may consist of glass, very fine crystalline material, or both. *See* APHANITE; IGNEOUS ROCKS. [C.A.C.]

Fennel The culinary spice and the plant *Foeniculum vulgare* (Umbelliferae), a tall perennial herb native to the Mediterranean region. It now occurs in all parts of the world. The fruits are used in cookery, confectionary, and for flavoring beverages. In modern French and Italian cooking, fennel is indispensable. Oil of fennel is used in medicine, soaps, and perfumes. *See* APIALES. [P.D.St.; E.L.C.]

Fermentation Decomposition of foodstuffs generally accompanied by the evolution of gas. The best-known example is alcoholic fermentation, in which sugar is converted into alcohol and carbon dioxide. During fermentation organic matter is decomposed in the absence of air (oxygen); hence, there is always an accumulation of reduction products, or incomplete oxidation products. Some of these products (for example, alcohol and lactic acid) are of importance to humans, and fermentation has therefore been used for their manufacture on an industrial scale. There are also many microbiological processes that go on in the presence of air while yielding incomplete oxidation products. Good examples are the formation of acetic acid (vinegar) from alcohol by vinegar bacteria, and of citric acid from sugar by certain molds (for example, *Aspergillus niger*). These microbial processes, too, have gained industrial importance, and are often referred to as fermentations, even though they do not conform to Pasteur's concept of fermentation as a decomposition in the absence of air. *See* INDUSTRIAL MICROBIOLOGY. [C.B.V.N.]

Fermi-Dirac statistics The statistical description of particles or systems of particles that satisfy the Pauli exclusion principle. This description was first given by E. Fermi, who applied the Pauli exclusion principle to the translational energy levels of a system of electrons. It was later shown by P. A. M. Dirac that this form of statistics is also obtained when the total wave function of the system is antisymmetrical. *See* EXCLUSION PRINCIPLE.

Such a system is described by a set of occupation numbers $\{n_i\}$ which specify the number of particles in energy levels ϵ_i. It is important to keep in mind that ϵ_i represents a finite range of energies, which in general contains a number, say g_i, of nondegenerate quantum states. In the Fermi statistics, at most one particle is allowed in a nondegenerate state. (If spin is taken into account, two particles may be contained in such a state.) This is simply a restatement of the Pauli exclusion principle, and means that $n_i \leqq g_i$. The probability of having a set $\{n_i\}$ distributed over the levels ϵ_i, which contain g_i nondegenerate levels, is described by Eq. (1), which gives just the number

$$W = \prod_i \frac{g_i!}{(g_i - n_i)!\, n_i!} \qquad (1)$$

of ways that n_i can be picked out of g_i, which is intuitively what one expects for such a probability. The equilibrium state which actually exists is the set of n's that makes W a maximum, under the auxiliary conditions given in Eqs. (2*a*) and (2*b*).These conditions express the fact that the total engergy E

$$\sum n_i = N \qquad (2a)$$

$$\sum n_i \epsilon_i = E \qquad (2b)$$

and the total number of particles N are given. Equation (3) holds for this most probable distribution. Here A and β are

$$n_i = \frac{g_i}{\frac{1}{A}\,\epsilon^{\beta\epsilon_i} + 1} \qquad (3)$$

parameters, to be determined from Eq. (3); in fact, $\beta = 1/kT$, where k is Boltzmann's constant and T is the absolute temperature. When the 1 in the denominator may be neglected, Eq. (3) goes over into the Boltzmann distribution.

Classical conditions pertain when the volume per particle is much larger than the volume associated with the de Broglie wavelength λ of a particle. For electrons in a metal at 300 K, the ratio of the volume per particle to λ^3 has the value 10^{-4}, showing that classical statistics fail altogether. When the classical distribution fails, a degenerate Fermi distribution results. A somewhat lengthy calculation yields the result that in this case the contribution of the electrons to the specific heat is negligible. This resolves an old paradox, for, according to the classical equipartition law, the electronic specific heat C should be $(3/2)Nk$, whereas in reality it is very small. See BOSE-EINSTEIN STATISTICS; KINETIC THEORY OF MATTER; QUANTUM STATISTICS; STATISTICAL MECHANICS. [M.Dr.]

Fermi surface A surface of constant energy in a space defined by the components of the wave vectors of a system of half-integral spin particles. This is not a real surface, but is a geometrical description of the dynamical behavior of conduction electrons in solids. Its name comes from the fact that half-integral spin particles obey Fermi-Dirac statistics and at the zero of temperature T fill energy levels up to a maximum energy, the Fermi energy, and no energy levels are occupied above this energy. The wave vectors of electrons having the Fermi energy at $T = 0$ define the Fermi surface. See FERMI-DIRAC STATISTICS.

Fermi surfaces are used in the discussion of the electrical conduction properties of solids. As a first approximation to consider, assume that the electrons responsible for electrical conduction in a solid constitute a free-electron gas. In this approximation the energy of the conduction electrons is entirely kinetic, with vanishingly small potential interactions with either the atomic electrons or other conduction electrons. The only interaction is the fact that they obey Fermi-Dirac statistics. Thus, the energy of electrons at the Fermi energy is given by the equation below, where k_F is the wave vector of electrons of

$$E_F = \frac{\hbar^2 k_F^{\,2}}{2m}$$

this energy and m is the electronic mass. At $T = 0$ all states with k less than k_F are filled, and those with k greater than k_F are empty. This situation can be represented geometrically by a sphere of radius k_F. All states inside the sphere are filled, and states outside are empty. See FREE-ELECTRON THEORY OF METALS.

Because of the Pauli exclusion principle, only a small number of electrons having energies near the Fermi energy can contribute to conduction processes, and this makes the details of the properties of the Fermi surface of prime importance to understanding electrical conduction in solids. Two phenomena not included in this free-electron approximation cause real metals to have Fermi surfaces which are geometrical shapes other than spheres: conduction electrons experience potential interactions with the atoms in the crystal, and they experience Bragg reflections at the boundaries of the Brillouin zone. See BAND THEORY OF SOLIDS; BRILLOUIN ZONE.

Other techniques for experimentally determining the dimensions and other detailed properties of Fermi surfaces of metals which involve the application of a magnetic field include the de Haas–van Alphen effect, the radio-frequency size effect, cyclotron resonance, acoustical geometrical resonance, and magnetoresistance measurements. See DE HAAS–VAN ALPHEN EFFECT; MAGNETORESISTANCE; ULTRASONICS. [R.G.G.]

Fermium A chemical element, Fm, atomic number 100, the eleventh element in the actinide series. Fermium does not occur in nature; its discovery and production have been accomplished by artificial nuclear transmutation of lighter elements. Radioactive isotopes of mass number 244–259 have been discovered. The total weight of fermium which has been synthesized is much less than one-millionth of a gram. See ACTINIDE ELEMENTS; RADIOACTIVITY.

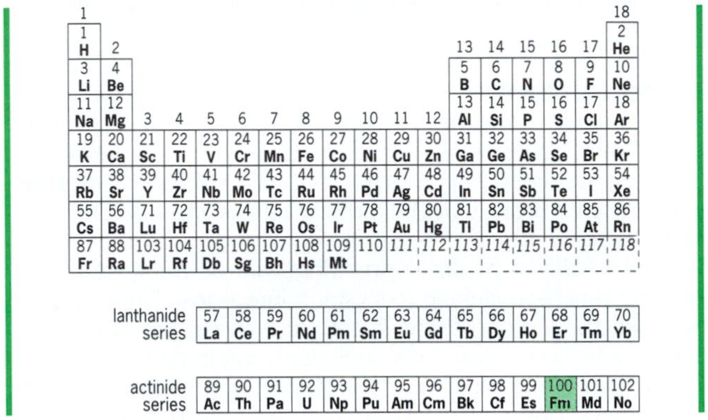

Spontaneous fission is the major mode of decay for ^{244}Fm, ^{256}Fm, and ^{258}Fm. The longest-lived isotope is ^{257}Fm, which has a half-life of about 100 days. Fermium-258 decays by spontaneous fission with a half-life of 0.38 millisecond. This suggests the existence of an abnormality at this point in the nuclear periodic table. See NUCLEAR CHEMISTRY; NUCLEAR REACTION; TRANSURANIUM ELEMENTS. [G.T.S.]

Ferret The name for the largest member of the weasel family, Mustelidae. This carnivorous animal, also known as the black-footed ferret (*Mustela nigripes*), is an inhabitant of the western states in the Rocky Mountain area, where it is referred to as the prairie dog ferret or prairie dog hunter. Though once abundant, ferrets are now very rare; they suffered from the poisoning campaigns against prairie dogs and larger carnivores.

Two litters are born each year after a gestation period of about 60 days, with 5–10 young in each litter. The maximum life is 13 years. See CARNIVORA; WEASEL. [C.B.C.]

Ferrimagnetism A specific type of ordering in a system of magnetic moments or the magnetic behavior resulting from such order. In some magnetic materials the magnetic ions in a crystal unit cell may differ in their magnetic properties. This is clearly so when some of the ions are of different species. It is also true for similar ions occupying crystallographically inequivalent sites. Such ions differ in their interactions with other ions, because the dominant exchange interaction is mediated by the neighboring nonmagnetic ions. They also experience different crystal electric fields, and these affect the magnetic anisotropy of the ion. A collection of all the magnetic sites in a crystal with identical behavior is referred to as a magnetic sublattice. A material is said to exhibit ferrimagnetic order when, first, all moments on a given sublattice point in a single direction and, second, the resultant moments of the sublattices lie parallel or antiparallel to one another. The notion of such an order is due to L. Néel, who showed in 1948 that its existence would explain many of the properties of the magnetic ferrites. See FERRITE; FERROMAGNETISM.

In general, there is a net moment, the algebraic sum of the sublattice moments, just as for a normal ferromagnet. However, its variation with temperature rarely exhibits the very simple behavior of the normal ferromagnet. For example, in some materials, as the temperature is raised over a certain range, the magnetization may first decrease to zero and then increase again. Ferrimagnets can be expected, in their bulk properties, measured statically or at low frequencies, to resemble ferromagnets with unusual temperature characteristics. *See* CURIE TEMPERATURE. [L.R.W.]

Ferrite Any of the class of magnetic oxides. Typically the ferrites have a crystal structure which has more than one type of site for the cations. Usually the magnetic moments of the metal ions on sites of one type are parallel to each other, and antiparallel to the moments on at least one site of another type. Thus ferrites exhibit ferrimagnetism. *See* FERRIMAGNETISM; MAGNETIC MATERIALS.

There are three important classes of commercial ferrites. One class has the spinel structure, with the general formula $M^{2+}Fe_2^{3+}O_4$, where M^{2+} is a divalent metal ion. So-called linear ferrites used in inductors and transformers are made of Mn and Zn (for frequencies up to 1 MHz) and Ni and Zn (for frequencies greater than 1 MHz). MgMn ferrites are used in microwave devices such as isolators and circulators. Until the late 1970s, ferrites with square loop shapes held a dominant position as computer memory-core elements, but these gave way to semiconductors. *See* COMPUTER STORAGE TECHNOLOGY.

The second class of commercially important ferrites have the garnet structure, with the formula $M_3^{3+}Fe_5^{3+}O_{12}$, where M^{3+} is a rare-earth or yttrium ion. Yttrium-based garnets are used in microwave devices. Thin monocrystalline films of complex garnets have been developed for bubble domain memory devices. *See* MAGNETIC BUBBLE MEMORY.

The third class of ferrites has a hexagonal structure, of the $M^{2+}Fe_{12}^{3+}O_{19}$ magnetoplumbite type, where M^{2+} is usually Ba, Sr, or Pb. Because of their large magnetocrystalline anisotropy, the hexagonal ferrites develop high coercivity and are an important member of the permanent magnet family.

Another magnetic oxide, γ-Fe_2O_3, also has the spinel structure, but has no divalent cations. It is the most commonly used material in the preparation of magnetic recording tapes.

The largest usage of ferrite measured in terms of material weight is in the nonlinear *B/H* range, and is found in the form of deflecting yokes and flyback transformers for television receivers. The cores for these devices must have high saturation induction B_s along with high maximum permeability μ_m at the knee of the *B/H* curve to frequencies as high as 100 kHz, the effective flyback frequency used in scanning a television tube. Again, MnZn and NiZn ferrites dominate the use in these devices. *See* TELEVISION RECEIVER. [G.Y.C.]

Ferroalloy A member of an important group of metallic raw materials required for the steel industry. Ferroalloys are the principal source of such additions as silicon and manganese which are required for even the simplest plain-carbon steels; and chromium, vanadium, tungsten, titanium, and molybdenum, which are used in both low- and high-alloy steels. Ferroalloys are unique in that they are brittle and otherwise unsuited for any service application, but they are important as the most economical source of these elements for use in the manufacture of the engineering alloys. These same elements can also be obtained, at much greater cost in most cases, as essentially pure metals. The ferroalloys contain significant amounts of iron and usually have a lower melting range than the pure metals and are therefore dissolved by the molten steel more readily than the pure metal. In other cases, the other elements in the ferroalloy serve to protect the critical element against oxidation during solution and thereby give higher recoveries. Ferroalloys are used both as deoxidizers and as a specified addition to give particular properties to the steel. *See* STEEL. [G.D.]

Ferrocene Dicyclopentadienyl iron, $(C_5H_5)_2Fe$, an orange crystalline solid with a melting point of 174° (345°F). The compound sublimes at 100°C (212°F), is diamagnetic, and has a dipole moment of zero. The structure of ferrocene, in which an atom of iron is sandwiched between two parallel cyclopentadiene rings, is shown below.

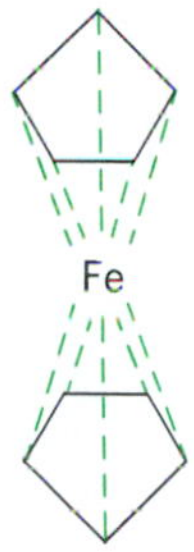

The most outstanding feature of the chemistry of ferrocene is its aromatic character. The aromaticity of ferrocene is evidenced by the manner in which it behaves toward electrophilic reagents. The molecule undergoes typical Friedel-Crafts acylation to yield either the mono- or diacetylated product. Ferrocene also undergoes other electrophilic substitution reactions typical of a reactive aromatic compound. *See* AROMATIC; METALLOCENES.
[F.D.P.]

Ferroelectrics Crystalline substances which have a permanent spontaneous electric polarization (electric dipole moment per cubic centimeter) that can be reversed by an electric field. In a sense, ferroelectrics are the electrical analog of the ferromagnets, hence the name. The spontaneous polarization is the so-called order parameter of the ferroelectric state. The names Seignette-electrics or Rochelle-electrics, which are also widely used, are derived from the name of the first substance found to have this property, Seignette salt or Rochelle salt. *See* FERROMAGNETISM.

From a practical standpoint ferroelectrics can be divided into two classes. In ferroelectrics of the first class, spontaneous polarization can occur only along one crystal axis; that is, the ferroelectric axis is already a unique axis when the material is in the paraelectric phase. Typical representatives of this class are Rochelle salt, monobasic potassium phosphate, ammonium sulfate, guanidine aluminum sulfate hexahydrate, glycine sulfate, colemanite, and thiourea.

In ferroelectrics of the second class, spontaneous polarization can occur along several axes that are equivalent in the paraelectric phase. The following substances belong to this class: barium(IV) titanate-type (or perovskite-type) ferroelectrics; cadmium niobate; lead niobate; certain alums, such as methyl ammonium alum; and ammonium cadmium sulfate.

From a scientific standpoint, one can distinguish proper ferroelectrics and improper ferroelectrics. In proper ferroelectrics, the structure change at the Curie temperature can be considered a consequence of the spontaneous polarization. In improper ferroelectrics, the spontaneous polarization can be considered a by-product of another structural phase transition. Examples of such systems are gadolinium molybdate and boracites.

The spontaneous polarization can occur in at least two equivalent crystal directions; thus, a ferroelectric crystal consists in general of regions of homogeneous polarization that differ only in the direction of polarization. These regions are called ferroelectric domains. Ferroelectrics of the first class consist of domains with parallel and antiparallel polarization, whereas ferroelectrics of the second class can assume much more complicated domain configurations. The region between two adjacent domains is called a domain wall. Within this wall, the spontaneous polarization changes its direction.

As a rule, the dielectric constant ϵ measured along a ferroelectric axis increases in the paraelectric phase when the Curie temperature is approached. In many ferroelectrics, this increase can be approximated by the Curie-Weiss law. *See* Curie-Weiss law; Dielectric constant.

Ferroelectrics can be divided into two groups according to their piezoelectric behavior. The ferroelectrics in the first group are already piezoelectric in the unpolarized phase. Those piezoelectric moduli which relate stresses to polarization along the ferroelectric axis have essentially the same temperature dependence as the dielectric constant along this axis, and hence become very large near the Curie point. The spontaneous polarization gives rise to a large spontaneous piezoelectric strain which is proportional to the spontaneous polarization.

The ferroelectrics in the second group are not piezoelectric when they are in the paraelectric phase. However, the spontaneous polarization lowers the symmetry so that they become piezoelectric in the polarized phase. This piezoelectric activity is often hidden because the piezoelectric effects of the various domains can cancel. However, strong piezoelectric activity of a macroscopic crystal or even of a polycrystalline sample occurs when the domains have been aligned by an electric field. The spontaneous strain is proportional to the square of the spontaneous polarization. *See* Piezoelectricity.

Antiferroelectric crystals are characterized by a phase transition from a state of lower symmetry (generally low-temperature phase) to a state of higher symmetry (generally high-temperature phase). The low-symmetry state can be regarded as a slightly distorted high-symmetry state. It has no permanent electric polarization, in contrast to ferroelectric crystals. The crystal lattice can be regarded as consisting of two interpenetrating sublattices with equal but opposite electric polarization. This state is referred to as the antipolarized state. In a certain sense, an antiferroelectric crystal is the electrical analog of an antiferromagnetic crystal.

The piezoelectric effect of ferroelectrics (and certain antiferroelectrics) finds numerous applications in electromechanical transducers. The large electrooptical effect (birefringence induced by an electric field) is used in light modulators. In certain ferroelectrics, light can induce changes of the refractive indices. These substances can be used for optical information storage and in real-time optical processors. The temperature dependence of the spontaneous polarization corresponds to a strong pyroelectric effect which can be exploited in thermal and infrared sensors. [W.K.]

Ferromagnetism A property exhibited by certain metals, alloys, and compounds of the transition (iron group), rare-earth, and actinide elements in which, below a certain temperature called the Curie temperature, the atomic magnetic moments tend to line up in a common direction. Ferromagnetism is characterized by the strong attraction of one magnetized body for another.

Atomic magnetic moments arise when the electrons of an atom possess a net magnetic moment as a result of their angular momentum. The combined effect of the atomic magnetic moments gives rise to a relatively large magnetization, or magnetic moment per unit volume, for a given applied field. Above the Curie temperature, a ferromagnetic substance behaves as if it were paramagnetic: Its susceptibility approaches the Curie-Weiss law. The Curie temperature marks a transition between order and disorder of the alignment of the atomic magnetic moments. Some materials exhibit a special form of ferromagnetism below the Curie temperature called ferrimagnetism. *See* Curie temperature; Curie-Weiss law; Ferrimagnetism; Magnetic susceptibility; Paramagnetism; Spin (quantum mechanics).

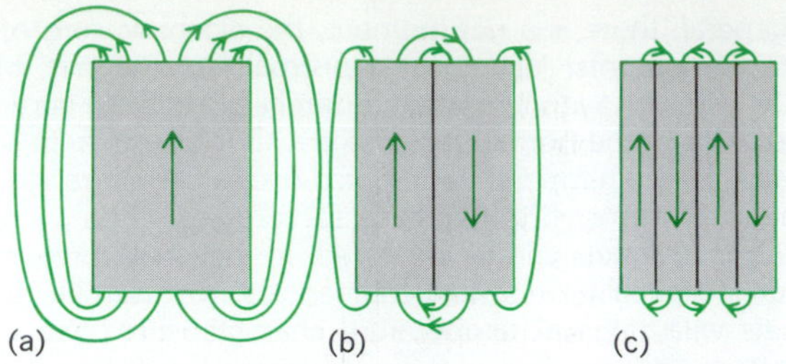

Lowering of magnetic field energy by domains. (*a*) Lines of force for a single domain. (*b*) Shortening of lines of force by division into two domains. (*c*) Reduction of field energy by further subdivision.

The characteristic property of a ferromagnet is that, below the Curie temperature, it can possess a spontaneous magnetization in the absence of an applied magnetic field. Upon application of a weak magnetic field, the magnetization increases rapidly to a high value called the saturation magnetization, which is in general a function of temperature. *See* Magnetization.

Small regions of spontaneous magnetization, formed at temperatures below the Curie point, are known as domains. As shown in the illustration, domains originate in order to lower the magnetic energy. In illustration *b* it is shown that two domains will reduce the extent of the external magnetic field, since the magnetic lines of force are shortened. On further subdivision, as in illustration *c*, this field is still further reduced.

An alternate way to describe the energy reduction is to note that the interior demagnetizing fields, coming from surface poles, are much smaller in the long, thin domains of illustration *c* than in the "fat" domain of illustration *a*.

The question arises as to how long this subdivision process continues. With each subdivision there is a decrease in field energy, but there is also an increase in Heisenberg exchange energy, since more and more magnetic moments are aligning antiparallel. Finally a state is reached in which further subdivision would cause a greater increase in exchange energy than it would cause decrease in field energy, and the ferromagnet will assume this state of minimum total energy.

Materials easily magnetized and demagnetized are called "soft"; these are used in alternating-current machinery. The ideal cheap soft material would be an iron alloy fabricated by some inexpensive technique which results in all crystal grains being oriented in the same or nearly the same direction. Various complicated rolling and annealing methods have been discovered in the continued search for better grain-oriented or "cube-textured" steels. *See* Magnetic materials.

Materials which neither magnetize nor demagnetize easily are called "hard"; these are used in permanent magnets. The fine-particle precipitates have dominated the market. These alloys consist basically of small and elongated magnetic particles embedded in a nonmagnetic matrix. [E.A.; F.Ke.]

Ferry A vessel having provision for passengers or vehicles, or both, operating on a short run with a frequent schedule between two points over the most direct water route and offering a public service of a type normally attributed to bridge or tunnel. With a few notable exceptions, passengers on ferry vessels are considered to be day passengers, meaning that seating is provided similar to a commercial airliner, bus, or passenger train without staterooms or overnight sleeping accommodations. Vehicles may include railroad rolling stock in addition to passenger automobiles, recreation vehicles, commercial trucks and trailers.

In the strict sense, ferry vessels are normally confined to crossing rivers, canals, lakes, bays, and sounds. However, with the expansion of ferry systems worldwide, the specific terms ocean ferry and coastwise ferry have been added to ferry vessel designations. Ocean ferries include all ferry vessels navigating any ocean or the Gulf of Mexico, the Caribbean Sea, or the Gulf of Alaska more than 20 nautical miles (37 km) offshore. Coastwise ferries include all ferry vessels normally navigating these same bodies of water 20 nmi or less offshore. Ocean and coastwise ferries are generally equipped with staterooms for overnight passenger accommodations.

In many cases, ferry vessels travel to foreign ports and therefore carry the additional classification of short international voyage. This means an international voyage in the course of which a vessel is not more than 200 nmi (370 km) from a port or place in which the passengers and crew could be placed in safety, and which does not exceed 600 nmi (1111 km) in length between the last port of call in the country in which the voyage begins and the final port of destination.

On short water routes where ferries shuttle back and forth across a river, canal, or other inland waterway, the vessels are generally designed to be double-ended. This term describes a vessel with a propeller and rudder at each end and a hull form which is identical at each end.

On longer water routes, where speed, weather conditions, or certain other operational features demand, a more normal hull form is adopted with the propellers and rudders at the stern. These vessels proceed in one direction and are designated as being single-ended. They must maneuver at the terminals in order to properly load and discharge their vehicles. [P.F.S.]

Fertilizer Materials added to the soil, or applied directly to crop foliage, to supply elements needed for plant nutrition. These materials may be in the form of solids, semisolids, slurry suspensions, pure liquids, aqueous solutions, or gases.

The chemical elements nitrogen, phosphorus, and potassium are the macronutrients, or primary fertilizer elements, which are required in greatest quantity. Sulfur, calcium, and magnesium, called secondary elements, are also necessary to the health and growth of vegetation, but they are required in lesser amounts compared to the macronutrients. The other elements of agronomic importance, called micronutrients and provided for plant ingestion in small (or trace) amounts, include boron, cobalt, copper, iron, manganese, molybdenum, and zinc. All these fertilizer elements, along with other chemical elements, occur naturally in agricultural soils in varying concentrations and mineral compositions which may or may not be in forms readily accessible to root systems of plants.The addition of fertilizer to soils used for the production of commercial crops is necessary to correct natural deficiencies and to replace the components absorbed by the crops in their growth.

Crop requirements of fertilizer components could be satisfied by the spreading of individual materials for each element deficient in the soil. However, economy favors the single application of a balanced mixture that satisfies all nutritional needs of a crop. Many commercial fertilizers therefore contain more than one of the primary fertilizer elements.

The compositions of fertilizer mixtures, in terms of the primary fertilizer elements, are identified by an N-P-K code: N denotes elemental nitrogen; P denotes the anhydride of phosphoric acid (P_2O_5); K denotes the oxide of potassium (K_2O). All are expressed numerically in percentage composition, or units of 20 lb each per short ton (10 kg per metric ton) of finished fertilizer as packaged. Formula 8-32-16 thus contains a mixture aggregating 8 wt % N in some form of nitrogen compounds, 32 wt % P_2O_5 in some form of phosphates, and 16 wt % K_2O in some form of potassium compounds, to give a product with a total of 56 fertilizer units. The commercial N-P-K formulas are generally in whole numbers. None of the N-P-K formulas totals 100% plant nutrients because the formulas indicate only the nutrient portions of the primary-element compounds and do not account for any other materials present.

Other unidentified materials present in the fertilizer products include free water and water content of the hydrated primary components, uncredited cation and anion portions of salts (sodium, silica, chloride, sulfate, and so on), impurities remaining in the chemical ingredients as processed, and deliberately added inert materials (called ballast), to make whole-number formula compositions.

Growing plants can assimilate only fertilizer elements in the combined state of inorganic compounds that are amenable to osmotic absorption. Many modern fertilizer materials consist of compounds that are immediately usable by the crops to which they are applied. Others are quickly converted within the soil to forms that can be assimilated. Some fertilizer chemicals are specifically designed to dissolve slowly or to delay reaction within the soil and therefore prolong the release of easily absorbed compounds to provide sustained feeding over the growth cycle of the plants. The relatively few plants for production of large quantities of these basic materials are usually located near the mining operations. In the case of nitrogen fertilizer products, the large production complexes are located near natural gas sources, which may be wells or major natural gas pipelines.

Aqueous solutions of urea, ammonia, and ammonium nitrate (UAN solutions) are used directly by the farmers as well as in the preparation of granular N-P-K products by mixing with other materials, such as normal superphosphate and triple superphosphate. UAN solutions are also spread directly by field application or used to prepare complete N-P-K fertilizer solutions or suspensions. Suspension fertilizers consist of aqueous slurries of fine crystals in saturated solutions that are stabilized by small amounts of gelling materials, such as attapulgite clay. Suspensions can be maintained in uniform composition during spreading on the fields, and give better dispersion than granular material. *See* FERTILIZING. [A.Lo.]

Fertilizing Addition of elements or other materials to the soil to increase or maintain plant yields. Fertilizers may be organic or inorganic. Organic fertilizers are usually manures and waste materials which in addition to providing small amounts of growth elements also serve as conditioners for the soil. Commercial fertilizers are most often inorganic. *See* FERTILIZER.

Methods of applying fertilizers vary widely and depend on such factors as kind of crop and stage of growth, application rates, physical and chemical properties of the fertilizer, and soil type. Two basic application methods are used, bulk spreading and precision placement. Time and labor are saved by the practice of bulk spreading, in which the fertilizer is broadcast over the entire area by using large machines which cover many acres in a short time. Precision placement, in which the fertilizer is applied in one or more bands in a definite relationship to the seed or plants, requires more equipment and time, but usually smaller amounts of fertilizer are needed to produce a given yield increase.

For some deep-rooted plants, subsoil fertilization to depths of 12–20 in. (30–50 cm) is advantageous. This is usually a separate operation from planting, and uses a modified subsoil plow followed by equipment to bed soil over the plow furrow, thereby eliminating rough soil conditions unfavorable for good seed germination. Top-dressings are usually applied by broadcasting over the soil surface for closely spaced crops such as small grains.

Since solid fertilizers range from dense heavy materials to light powders and liquid fertilizers range from high pressure to zero pressure, a variety of equipment is required for accurate metering and placement. In addition, application rates may be as low as 50 lb/acre (56 kg/hectare) or as high as 6 tons/acre (13.3 metric tons/hectare). Large bulk spreaders usually use drag chains or augers to force the material through a gate or opening whose size is varied to regulate the amount passing through and falling on the spreader.

Liquid fertilizer of the high-pressure type (for example, anhydrous ammonia) is usually regulated by valves or positive displacement pumps. The size of the orifice may be controlled manually or automatically by pressure-regulating valves. Low-pressure solutions may be metered by gravity flow through orifices, but greater accuracy is obtained by using compressed air or other gases to maintain a constant pressure in the tank. This method eliminates the effect of temperature and volume changes. Nonpressure solutions may be metered by gravity or by gear, roller, piston, centrifugal, or hose pumps. The accuracy of the gravity type can be improved by the use of a constant head device by which all air is introduced into the tank at the bottom.

Nonvolatile fertilizer solutions are often pumped into the supply lines of irrigation systems to allow simultaneous fertilization and irrigation. With the exception of bulk spreaders and other broadcasters, most fertilizer application devices are built as attachments which can be mounted in conjunction with planters, cultivators, and herbicide applicators. Often the tanks, pumps, and controls used for liquid fertilizers are also used for applying other chemicals such as insecticides. [J.G.F.]

Fescue A group of grasses, of which the most important species in agricultural use is tall fescue (*Festuca arundinacea*). Other important fescues are Meadow fescue (*F. elatior*), red fescue (*F. rubra*), Idaho fescue (*F. idahoensis*), sheep fescue (*F. ovina*), and hard fescue (*F. diriuscula*). [H.B.S.]

Fetal alcohol syndrome A spectrum of changes in the offspring of women who consume alcoholic beverages during pregnancy, ranging from severe growth deficiency, mental retardation, and abnormal facial features to mild mental changes. Physical abnormalities include limitation of joint motion, increased congenital heart malformation rate, short palpebral fissures, and maxillary hypoplasia with relative prognathism.

The frequency of adverse outcome of pregnancies of chronic alcoholic women is very high. There is an eightfold increase in perinatal mortality, and the frequency of the fetal alcohol syndrome in the surviving children is high. Almost half of the offspring have IQs below 80, and a smaller percentage have the dysmorphic features noted above.

Medical advice regarding the consumption of alcoholic beverages during pregnancy involves three principles: heavy drinking is clearly deleterious to the fetus; abstinence from alcoholic beverages immediately preceding and during pregnancy is the only assured safe course of action presently known; and therapeutic attention should also be directed to the problems leading to the alcoholism so that the psychological environment of the offspring may be improved. *See* ALCOHOLISM; BEHAVIORAL TOXICOLOGY. [N.K.M.]

Fetal membrane One of the membranous structures which surround the embryo during its developmental period. Since such membranes are external to the embryo proper, they are called extraembryonic membranes. They function in the embryo's protection, nutrition, respiration, and excretion.

There are four fetal membranes—the amnion, chorion, yolk sac, and allantois. In the course of development, the chorion becomes the outermost, and the amnion the innermost, membrane surrounding the developing embryo. As the allantois increases in size, it expands and becomes closely associated, if not fused, with the chorion. The two membranes together are known as the chorioallantoic membrane.

The amniotic cavity within which the embryo is enclosed becomes filled with an aqueous fluid which gives osmotic and physical protection to the embryo during the remainder of its fetal existence. Smooth muscle fibers in the amnion spontaneously contract and gently rock the embryo before it develops the capacity for spontaneous movement.

As the stored nutrients of the yolk are depleted during development, the yolk sac gradually decreases in size and is eventually incorporated into the midgut of the embryo. The yolk sac in the nonyolky eggs of placental mammals is vestigial. It has evolutionary but essentially no functional significance.

At the time of birth or hatching, the embryo becomes completely separated from the amnion and chorion and from the major portion of the allantois. The proximal portion of the latter remains within the embryo, however, as the urinary bladder. *See* ALLANTOIS; AMNION; CHORION; YOLK SAC. [A.R.B.]

Fever An elevation in the central body temperature of warm-blooded animals caused by abnormal functioning of the thermoregulatory mechanisms. Fever accompanies a wide variety of disease states, both infectious and noninfectious, and in the great majority of instances is due to an abnormality in the regulation of body temperature by the central nervous system. *See* THERMOREGULATION.

Experimental studies on the cause of fever suggest that leukocytes, as well as the fixed macrophages can be activated by various stimuli to produce a fever-inducing substance—endogenous pyrogen (EP). Endogenous pyrogen has been characterized as a protein of relatively low molecular weight (13,000) with an essential lipid moiety.

The mechanisms responsible for elevating body temperature include: reduction in heat loss by constriction of peripheral vessels whose tone is under control of the sympathetic nervous system; inhibition of panting and sweating, the latter by way of the cholinergic nerves; and increased heat production by means of shivering in voluntary muscles innervated by somatic motor nerves.

There is no clear evidence that elevated body temperature evoked by most infections is directly injurious to microbial invaders. As fever regularly accompanies inflammation, however, increased body temperature may well accelerate certain biochemical reactions of use to the host in combating infection. *See* HOMEOSTASIS; INFLAMMATION. [E.At.]

Feynman diagram A pictorial representation of elementary particles and their interactions. Feynman diagrams show paths of particles in space and time as lines, and interactions between particles as points where the lines meet.

The illustration shows Feynman diagrams for electron-electron scattering. In each diagram, the straight lines represent space-time trajectories of noninteracting electrons, and the wavy lines represent photons, particles that transmit the electromagnetic interaction. External lines at the bottom of each diagram represent incoming particles (before the interactions), and lines at the top, outgoing particles (after the interactions). Interactions between photons and electrons occur at the vertices where photon lines meet electron lines. *See* ELECTRON; PHOTON.

Each Feynman diagram corresponds to the probability amplitude for the process depicted in the diagram. The set of all distinct Feynman diagrams with the same incoming and outgoing lines corresponds to the perturbation expansion of a matrix element of the scattering matrix in field theory. This correspondence can be used to formulate the rules for writing the amplitude associated with a particular diagram. The perturbation expansion and the associated Feynman diagrams are

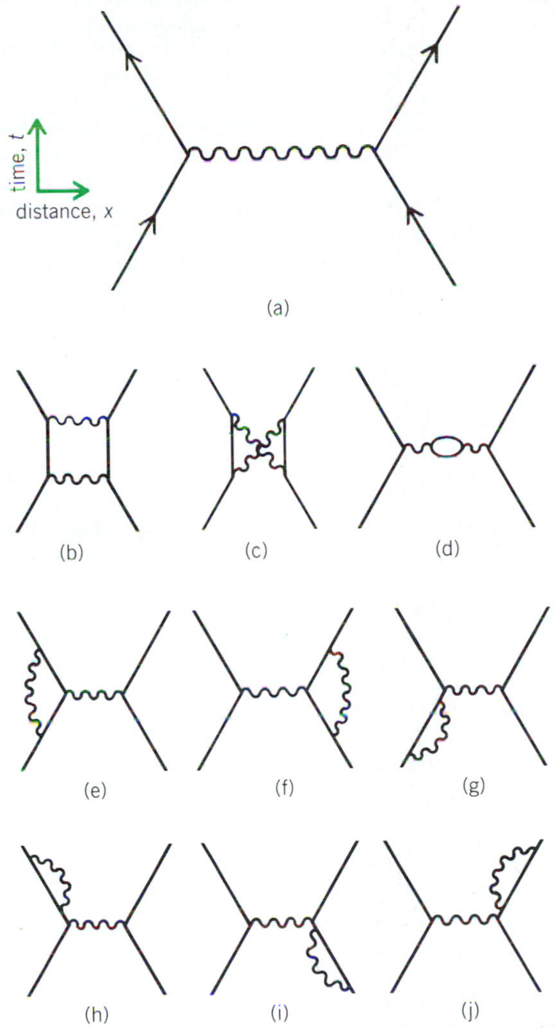

Feynman diagrams for electron-electron (Møller) scattering: (*a*) second-order diagram (two-vertices); (*b–j*) fourth-order diagrams.

useful to the extent that the strength of the interaction is small, so that the lowest-order terms, or diagrams with the fewest vertices, give the main contribution to the matrix element. *See* PERTURBATION (QUANTUM MECHANICS); SCATTERING MATRIX.

Since their introduction in quantum electrodynamics, Feynman diagrams have been widely applied in other field theories. They are employed in studies of electroweak interactions, certain situations in quantum chromodynamics, and in many-body theory in atomic, nuclear, plasma, and condensed matter physics. *See* ELEMENTARY PARTICLE; FUNDAMENTAL INTERACTIONS; QUANTUM CHROMODYNAMICS; QUANTUM ELECTRODYNAMICS; QUANTUM FIELD THEORY; WEAK NUCLEAR INTERACTIONS; WEINBERG-SALAM MODEL. [P.M.]

Feynman integral A technique, also called the sum over histories, which is basic to understanding and analyzing the dynamics of quantum systems. It is named after fundamental work of Richard Feynman. The crucial formula gives the quantum probability density for transition from a point q_0 to a point q_1 in time t as the expression below, where S(path)

$$\int \exp\,[iS(\text{path})/\hbar]\; d(\text{path})$$

is the classical mechanical action of a trial path, and $\hbar$ is the rationalized Planck's constant. The integral is a formal one over the infinite-dimensional space of all paths which go from q_0 to q_1 in time t. Feynman defines it by a limiting procedure using approximation by piecewise linear paths.

Feynman integral ideas are especially important in quantum field theory, where they not only are a useful device in analyzing perturbation series but are also one of the few nonperturbative tools available. *See* QUANTUM FIELD THEORY.

An especially attractive element of the Feynman integral formulation of quantum dynamics is the classical limit, $\hbar \to 0$. Formal application of the method of stationary phase to the above expression says that the significant paths for small $\hbar$ will be the paths of stationary action. One thereby recovers classical mechanics in the hamiltonian stationary action formulation. *See* LEAST-ACTION PRINCIPLE. [B.Si.]

Fiber bundle A type of structure arising frequently in various branches of mathematics, especially differential geometry, Lie groups, and algebraic geometry. A fiber bundle is a principal tool for the applications of algebraic topology to these subjects.

A fiber bundle is a topological space B which is decomposed in a smooth fashion into a family of closed disjoint subsets, called fibers, each of which is homeomorphic to a fixed space Y. The base space X of the bundle is the space whose points are the fibers. The projection $p:B \to X$ is the function assigning to each point the fiber which contains it, and X is topologized so that p is continuous. The requirement of smoothness reads as follows: For each $x \in X$, there is a neighborhood V of x, and a homeomorphism $\phi: V \times Y \to p^{-1}(V)$ satisfying $p\phi(x',y) = x'$ for all $x' \in V$ and $y \in Y$. *See* TOPOLOGY.

The product space $X \times Y$ and its projection into X satisfy these conditions with each $V = X$. The Möbius band (see illustration) is one of the simplest examples of a bundle which is not a product. The fibers are the line segments perpendicular to its center line, and its base space is a circle. The corresponding product space is a cylindrical surface. Because of such examples, fiber bundles are sometimes called twisted products. [N.E.S.]

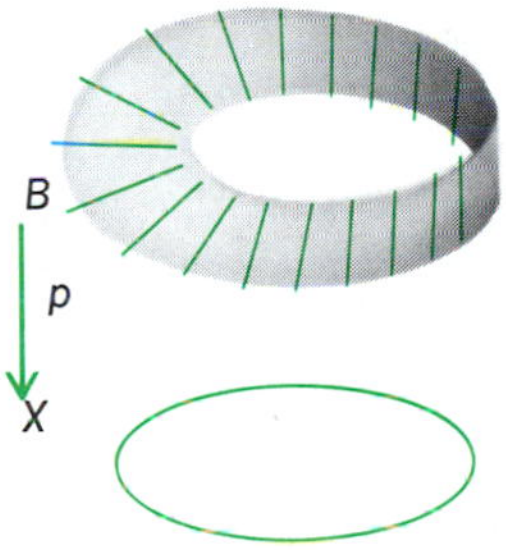

Projection of topological space *B* (a Möbius band) into its base space *X* (a circle).

Fiber-optic circuit The path of information travel, usually from one electrical system to another, in which light acts as the information carrier and is propagated by total internal reflection through a transparent optical waveguide. An electrooptic modulator and an optoelectric demodulator are required to convert the electrical signals into light and back again at the transmit and receive ends of the link, respectively.

A fiber-optic circuit, or link, is used for data transmission when a shielded twisted pair or a coaxial cable fails to meet one or more required performance criteria of the system designer. Depending upon fiber type, the distance-bandwidth product of a fiber is tens to thousands of times larger than that of electrical transmission. An optical communication fiber is a nearly perfect waveguide for light, meaning that little or no energy escapes through radiation. Thus, the data traveling in the fiber are secure from eavesdropping, as well as being harmless in or around equipment sensitive to electromagnetic interference. Telecommunication fiber also has a very small diameter, 5–10 micrometers (0.0002–0.0005 in.), which allows telecommunication cables to be fabricated with a much higher packing density. Also, the most common materials used to make the fibers, silica and plastic, are less dense than copper, making the cable lighter. Lastly, since the fiber is a dielectric it can be used in volatile or sensitive environments that require electrical isolation. *See* ELECTRICAL INTERFERENCE; OPTICAL COMMUNICATIONS; OPTICAL FIBERS.

The transmitter generally consists of a silicon integrated circuit that converts input voltage levels from a personal computer or a mainframe into current pulses. These, in turn, drive a light-emitting diode (LED). *See* INTEGRATED CIRCUITS; LIGHT-EMITTING DIODE.

Light from the fiber is focused onto a reverse-biased *pn*-junction photodiode that generates an electron-hole pair for each photon impinging on or near its active area. Another circuit, usually a silicon integrated circuit, amplifies this electron-hole current and converts it into voltage levels suitable for interfacing with the computer at the receiving end. *See* PHOTODIODE; PHOTOVOLTAIC EFFECT. [K.W.Li.]

Fiber-optic sensor A sensor that uses thin optical fibers to carry light to and from a location to be probed. In performing the sensing, light can be lost from the fibers or modified in velocity by the action of the phenomena on the fiber. Fiber-optic sensors are ideal for probing in remote or hostile locations, where miniature sensors are required such as in the body, or where extreme sensitivity is required. Two classes of fiber sensors have evolved: intensity sensors, in which the amplitude of light in the fiber is changed during sensing, and interferometric sensors, in which the velocity of light or its phase is modified during sensing. The latter class has proved to be extremely sensitive; intensity sensors are used where moderate performance is acceptable and lower cost is important. Depending on their design, intensity sensors can respond to pressure, temperature, liquid level, position, flow, smoke, displacement, electric and magnetic fields, chemical composition, and numerous other conditions.

Optical interferometry is one of the most sensitive means of detecting displacements as small as 10^{-13} m. Interferometric fiber sensors apply this technology to sense many physical phenomena. A fiber gyro based on the Sagnac effect is formed by making a fiber loop which, when rotated, causes the light traveling in both directions in the loop to experience different velocities with or against the rotation. A second type of interferometric sensor is constructed by using the Mach-Zehnder interferometer. Sensitivities equal to or surpassing the best conventional technologies have been achieved in these sensors. *See* GYROSCOPE; INTERFEROMETRY. [T.G.G.]

Fiber-optics imaging The use of fiber optics in image transmission, based on the ability of a precisely aligned bundle of optical fibers to transmit an image from one end of the bundle to the other. Each fiber transmits one element of the image so that the complete image is made up of a matrix of dots (pixels) which blend into a recognizable image like a halftone picture on a printed page.

Light is transmitted through individual fibers by means of total internal reflection from the fiber walls. For efficient transmission, each fiber has a highly transparent core (usually glass) coated with a layer (cladding) of lower refractive index (also usually glass). This cladding prevents light from leaking into neighboring fibers (crosstalk) and protects the core from contamination and wear. Fiber bundles may be either flexible or rigid. *See* CROSSTALK; OPTICAL MATERIALS; REFRACTION OF WAVES.

Flexible image bundles are used in a wide variety of industrial and medical fiberscopes or endoscopes to transmit the image from an objective lens at the distal end to an eyepiece in the control handle. Industrial fiberscopes ranging in length from 3 to 8 ft (1 to 2.5 m) are used in aircraft engine inspection, for example, examination of turbine blades. Longer industrial fiberscopes are used for pipe and weld inspection, including those in nuclear power plants.

Medical endoscopes range in size from less than 0.08 in. (2 mm) for viewing inside the arteries (cardioscope) to 0.6 in. (15 mm) for examining the colon (colonoscope). The latter has an operating channel large enough to perform surgery via remotely controlled forceps or electrosurgical snares. Intermediate-sized endoscopes are used to view the bronchi (bronchoscope), the kidneys (nephroscope), the bladder (cystoscope), the throat (laryngoscope), and the stomach (gastroscope).

Fused-fiber boules are the starting point for various image-transmitting components, including faceplates (windows) for image-intensifier tubes and cathode-ray tubes, image inverters (twisters), magnifiers, and rigid conduits. Image intensifier tubes are used in military night-vision devices, low-light television cameras, and astronomical telescopes. In a cathode-ray tube the fiber-optic faceplate is used to transmit the image on the phosphor screen to the exit face for printing directly onto photographic material. *See* CATHODE-RAY TUBE; LIGHT AMPLIFIER; OPTICAL FIBERS. [W.P.S.]

Fibonacci numbers A series of numbers defined by the elementary recursion $f_{n+2} = f_n + f_{n+1}$; $f_0 = 0$, $f_1 = 1$. Each f_n counts, for example, the number of sequences of $n - 2$ zeros and ones with no consecutive zeros. The generating function $f(x)$ of the Fibonacci numbers satisfies the identity $(1 - x - x^2)f(x) = x$, from which the closed form $f_n = (\tau^n - (1 - \tau)^n)/\sqrt{5}$, where $\tau = (1 + \sqrt{5})/2$, may readily be deduced. *See* COMBINATORIAL THEORY; ENUMERATION OF ARRANGEMENTS. [T.Br.]

Fibrinogen A plasma protein synthesized by the parenchymal cells of the liver. Fibrinogen (factor I) is the precursor of fibrin, the insoluble fibrous protein that polymerizes to form the clot during blood coagulation. Congenital or acquired deficiencies of circulating fibrinogen may be associated with severe symptoms of a bleeding tendency. *See* HEMOPHILIA; HEMORRHAGE; PLASMIN. [O.D.R.]

Fibrous protein One of the two main classes of protein molecules, the other being termed globular proteins. Examples of fibrous proteins are keratin in hair, wool, horn, and feathers; myosin in muscle; fibrinogen, a plasma protein that forms fibrin clots; collagen in tendon, cartilage, and fish scales; fibroin in silk; and elastin, the structural element of elastic fibers. These proteins are predominantly arranged as long fibers along a single axis. Alignment between fibers is accomplished by hydrogen bonding such that the polypeptide backbone of the protein assumes a "pleated" conformation.

In contrast to globular proteins, which do not usually contain covalent bonds between molecules, the fibrous proteins can form a network through a small number of covalent bonds between each fiber. Thus, these tightly linked molecules are generally insoluble in aqueous solvents and are highly resistant to digestion by proteolytic enzymes. *See* COLLAGEN; GELATIN; PROTEIN. [J.M.M.]

Fidelity The degree to which the output of a system accurately reproduces the essential characteristics of its input signal. Thus, high fidelity in a sound system means that the reproduced sound is virtually indistinguishable from that picked up by the microphones in the recording or broadcasting studio. Similarly, a television system has a high fidelity when the picture seen on the screen of a receiver corresponds in essential respects to that picked up by the television camera. Fidelity is achieved by designing each part of a system to have minimum distortion, so that the waveform of the signal is unchanged as it travels through the system. *See* DISTORTION (ELECTRONIC CIRCUITS); SOUND-REPRODUCING SYSTEMS. [J.Mar.]

Field emission The emission of electrons from a metal or semiconductor into vacuum (or a dielectric) under the influence of a strong electric field. In field emission, electrons tunnel through a potential barrier, rather than escaping over it as in thermionic or photoemission. The effect is purely quantum-mechanical, with no classical analog. It occurs because the wave function of an electron does not vanish at the classical turning point, but decays exponentially into the barrier (where the electron's total energy is less than the potential energy). Thus there is a finite probability that the electron will be found on the outside of the barrier. *See* PHOTOEMISSION; QUANTUM MECHANICS; THERMIONIC EMISSION.

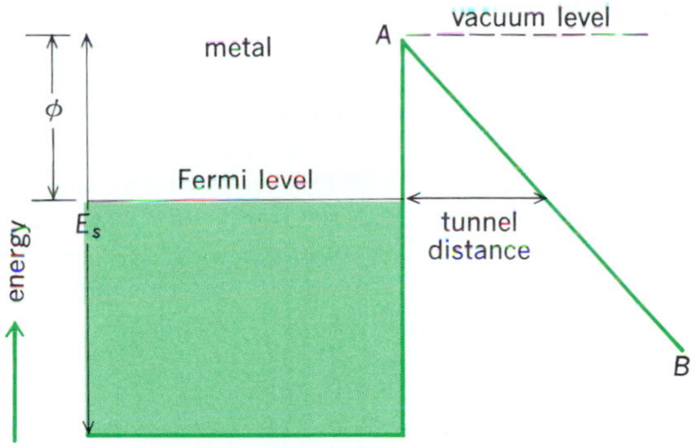

Diagram of the energy-level scheme for field emission from a metal at absolute zero temperature.

For a metal at low temperature, the process can be understood in terms of the illustration. The metal can be considered a potential box, filled with electrons to the Fermi level, which lies below the vacuum level by several electronvolts. The distance from Fermi to vacuum level is called the work function, ϕ. The vacuum level represents the potential energy of an electron at rest outside the metal, in the absence of an external field. In the presence of a strong field, the potential outside the metal will be deformed along the line *AB*, so that a triangular barrier is formed, through which electrons can tunnel. Most of the emission will occur from the vicinity of the Fermi level where the barrier is thinnest. *See* ELECTRON EMISSION; FIELD-EMISSION MICROSCOPY. [R.Gom.]

Field-emission microscopy A technique that uses field emission of electrons or positive ions to produce a magnified image of the emitter surface on a fluorescent screen. The field-emission (or field-ion) microscope is the most powerful microscope known. The tip of the needle-shaped emitter shown in Fig. 1 is projected directly without lenses, with a magnification up to 1,000,000 diameters and a resolution, with electrons as the emitted particles, of 2 nanometers. With this microscope, adsorption of various gases has been studied in the temperature range from 4 to above 2000 K (− 452 to above 3140°F). *See* FIELD EMISSION.

The condensation and surface migration of evaporated materials and surface reactions important in catalysis have been investigated. Medium-sized individual molecules such as phthalocyanin have been made visible.

Field-ion microscope. Rather than by electrons, the emitter tip surface can be imaged by positive ions. In the field-ion

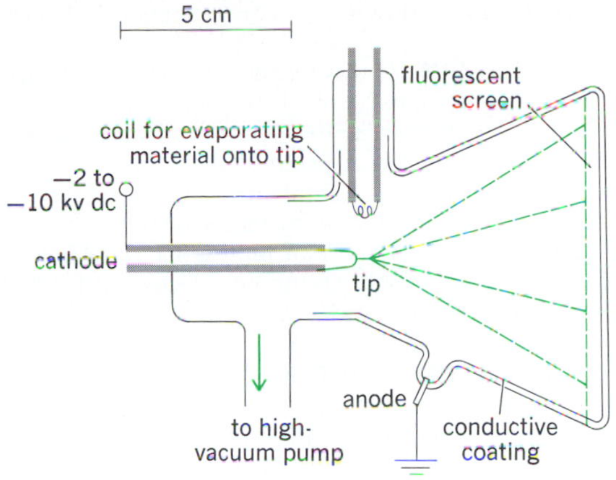

Fig. 1. Electron-operated field-emission microscope.

microscope (FIM), the tip is kept at a high positive potential, while the microscope vessel is backfilled with about 1 millitorr (0.1 pascal) of helium, neon, or hydrogen. Field ionization occurs when the electron tunnels out of an image gas molecule in the region of exceedingly high field strength above each protruding surface atom. When the tip is cooled by heat conduction through its leads, preferably by using liquid hydrogen or liquid helium, the reduced random velocity of the ions, combined with their short de Broglie wavelength, assures a resolution approaching 0.2 nm. Thus the atoms constituting the specimen surface can be seen individually, a feat not possible with any other type of microscope presently known. The FIM is used as a research tool for the study of crystal lattice defects

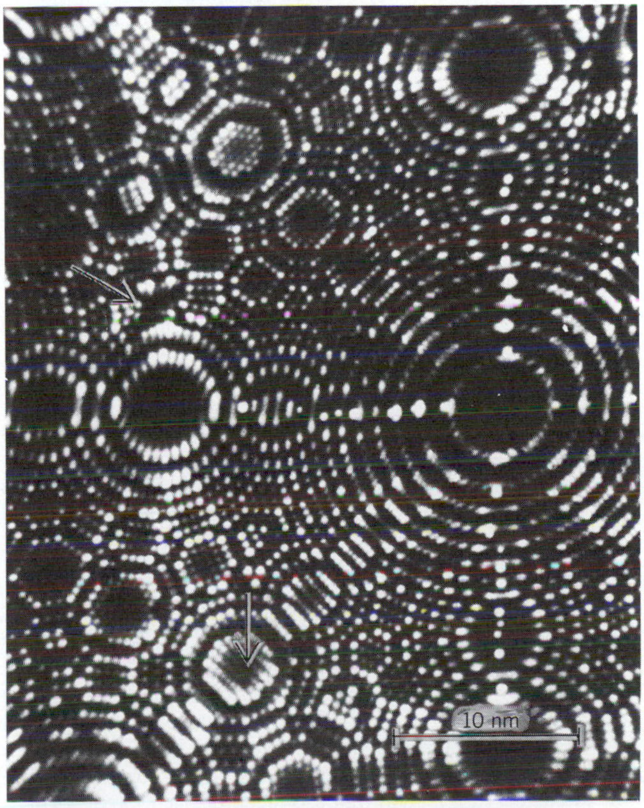

Fig. 2. Field-ion micrograph of the atomic structure of an iridium crystal surface, with a vacancy and a dislocation shown by arrows, and many fully resolved net planes of atoms.

in metals. Figure 2 shows the atomic structure of an iridium crystal surface.

Atom-probe FIM. A modification is the atom-probe FIM. The tip of this FIM can be tilted to place an interesting image spot onto a small hole in the screen. When a high-voltage pulse has ripped off the selected atom, it travels through the probe hole into a time-of-flight mass spectrometer of single particle sensitivity, where it is chemically identified by its atomic mass. Thus the atom-probe FIM represents the ultimate in microanalysis, requiring only one atom selected at the discretion of the operator. *See* CRYSTAL. [E.W.M.]

Field theory (mathematics) In algebra, the term field is used to designate an algebraic system or structure containing at least two elements and having two binary rules of composition: addition and multiplication (that is, if a and b are any two elements of the field, then $a + b$ and ab are defined and are elements of the field). The structure rules are as follows: The elements form an abelian (commutative) group under addition with the additive identity denoted by 0; that is, $a + 0 = a$ for all elements a. The set of nonzero elements (and there are some since the field has at least two elements) form an abelian group under multiplication with the multiplicative identity denoted by 1. It follows that all nonzero elements have a multiplicative inverse or reciprocal. The two rules of composition are related by the distributive law: $(a + b)c = ac + bc$ for all elements a, b, c. It follows from the distributive law that $a \cdot 0 = 0$ for all elements a, since $1 \cdot a = (1 + 0)a = 1 \cdot a + 0 \cdot a$, and consequently $0 = 0 \cdot a$. *See* GROUP THEORY.

The most familiar example of a field is the set **Q** of all rational numbers ($0, 1, 2, \frac{1}{2}, -1, -2, -\frac{1}{2}, 3, \frac{1}{3}, \frac{2}{3}, \ldots$) with the usual rules for equality, addition, and multiplication. Other examples are the set **R** of all real numbers and the set **C** of all complex numbers (both with the usual rules of addition and multiplication). The set of rational functions (formal quotients of two polynomials with rational coefficients) also constitute a field under the usual rules for equality, addition, and multiplication. *See* ALGEBRA; COMPLEX NUMBERS; REAL VARIABLE.

There are fields which contain only finitely many elements. The simplest example is obtained by considering the residue class ring of the integers modulo a prime p. This field, which may be denoted by $\mathbf{F}_p$, has as its elements the p distinct arithmetic progressions or residue classes modulo the prime p, namely $I_0, I_1, \ldots, I_p - 1$ where I_j consists of the integers j, $j \pm p$, $j \pm 2p$, ... The rules of composition are given by Eqs. (1) and (2).

$$I_j + I_k = I_s \quad \text{if } j + k \text{ lies in } I_s \tag{1}$$

$$I_j I_k = I_t \quad \text{if } jk \text{ lies in } I_t \tag{2}$$

If F is a field, and a subset A of F is a field under the rules of composition of F, then A is said to be a subfield of F, and F is said to be an extension of A. Thus the field of rational numbers **Q** is a subfield of the field of real numbers **R**, and **R** is an extension of **Q**. If a field F is an extension of the field A, then F is a vector space over A. If, as a vector space, F is of finite dimension n over A, then F is said to be an extension of A of degree n. *See* LINEAR ALGEBRA.

If a field F is such that every polynomial equation over F of positive degree has a solution in F, then F is said to be algebraically closed. The field of all complex numbers is algebraically closed; but this is a theorem in analysis. The field of rational numbers, the fields of residue classes modulo a prime p, and the field of real numbers are not algebraically closed. Given any field F, there exists a unique (to within isomorphism) minimal extension $\bar{F}$ of F such that $\bar{F}$ is algebraically closed. [D.J.L.]

Fièvre boutonneuse A generally mild febrile disease of humans due to *Rickettsia conori*. This condition is also called boutonneuse or Marseille fever. While the disease resembles Rocky Mountain spotted fever in some respects (for example, the rash), it differs in exhibiting a characteristic tache noire, or primary ulcer, at site of tick attachment, and swollen lymph glands which drain the area; furthermore, it does not exhibit localities of high virulence. The generalized rash has a similar maculopapular appearance and includes the palms and soles, but the case fatality rate is less than 3%.

The disease occurs in parts of the Old World, particularly the Mediterranean littoral, the Crimea, and parts of India. The primary cycle in most areas involves domestic dogs and the brown dog tick (*Rhipicephalus sanguineus*), which invades premises inhabited by their hosts. *See* RICKETTSIOSES; ROCKY MOUNTAIN SPOTTED FEVER. [C.B.P.]

Fig A deciduous tree, *Ficus carica*, with a milky juice growing in subtropical regions of Asia, Africa, Europe, and the United States. The fruit consists of a fleshy hollow receptacle lined with pistillate, or female, flowers and with a small opening at the end. Most edible types and varieties develop fruit without pollination. In the United States figs are grown primarily in California with minor production in Arizona, Texas, and Florida. [C.A.Sch.]

Filariasis A disease caused by Filarioidea in humans or lower animals. More loosely, the term is used to indicate mere infection by such organisms. In human medicine, filariasis commonly refers to the disease caused by, or infection with, one of the mosquito-borne, elephantoid-producing Filarioidea—most frequently *Wuchereria bancrofti*, less frequently *Brugia malayi*, and more recently including *B. timori*.

Wuchereria bancrofti has worldwide distribution in the tropics and extends into both north and south temperate zones. However, the very high infection rates and the high prevalence of the resulting grotesque elephantiasis are localized at sea-level areas with year-round uncontrolled high mosquito densities in the tropics. Introduction into the temperate zones commonly results in a few secondary cases before the infection dies out.

In most areas of the world, the microfilariae of *W. bancrofti* have a conspicuous nocturnal periodicity. There are 50 to 150 times as many microfilariae in the midnight blood as in noon blood. Often none may be found in midday blood. This periodic variety of *W. bancrofti* is transmitted by domestic, anthropophilous, night-biting mosquitos, which readily enter houses to feed. *Culex fatigans* (*C. quinquefasciatus*) and some of the important malaria vectors such as *Anopheles darlingi*, *A. funestus*, and *A. farauti* are important vectors. *See* FILARIOIDEA. [G.F.O.]

Filbert A nut from any plant in the genus *Corylus* (Betulaceae); also called hazelnut. About 15 species are recognized, distributed widely over the north temperate zone and ranging in size from shrubs to tall trees. Filberts in commerce are derived mostly from the European species *C. avellana* and *C. maxima*. Hybrids of these with the American species, *C. americana*, are grown sparingly in the eastern United States.

Filberts are grown widely in Europe and the Mediterranean basin, where much of the crop is consumed locally. Imports to the United States come mostly from Turkey. In the United States filberts are grown in Oregon and Washington. Filberts are sold in shell, or the kernels are roasted and used in confectionery and baked goods. [L.H.MacD.]

Filler A coating material used to smooth and fill porous wood surfaces before finishing coats are applied. Fillers normally are colorless materials that harden rapidly with minimum shrinkage. A typical formulation includes silica or another hard pigment dispersed in fast-drying varnish.

Materials specially formulated to fill crevices before surfaces such as wallboard are painted are sometimes referred to as fillers, as are surfacers used to smooth dents or irregularities in metals before finishing. The term is sometimes used to describe transparent extender pigments in paint formulations. *See* PAINT; SURFACE COATING; WOOD FINISHING. [C.R.Ma.; C.W.Si.]

Film (chemistry) A material in which one spatial dimension, thickness, is much smaller than the other two. Films can be conveniently classified as those that support themselves and those that exist only as layers on top of a supporting substrate. The latter are known as thin films and have their own specialized science and technology.

Thin films, from one to several hundred molecular layers, are generally defined as those that lie on a substrate, either liquid or solid. Monomolecular films on the surface of water can be made by adding appropriate materials in extremely small quantity to the surface. These materials are inappreciably soluble in water, rendered so by a large hydrocarbon or fluorocarbon functional group on the molecule or by other means; they are, however, able to dissolve in the surface of water by virtue of a hydrogen-bonding group or groups on the molecule.

Thin films can also be deposited directly onto solid substrates by evaporation of material in a vacuum, by sputtering, by chemical reaction (chemical vapor deposition), by ion plating, or by electroplating. *See* DIAMOND; ELECTROPLATING OF METALS; MONOMOLECULAR FILM; SPUTTERING; VACUUM METALLURGY; VAPOR DEPOSITION.

Thin films play an extensive role in both traditional and emerging technologies. Examples are foams and emulsions, the active layers in semiconductors, the luminescent and protective layers in electroluminescent thin-film displays, and imaging and photoelectric devices. *See* ELECTROLUMINESCENCE; ELECTRONIC DISPLAY; EMULSION; FOAM; SOLAR CELL.

Self-supported films have a nominal thickness not larger than 250 micrometers. Films of greater thickness are classified as sheets or foils. Self-supported films are commonly composed of organic polymers, either thermoplastic resins or cellulose-based materials. *See* POLYMER. [S.Ro.]

Film badge A device worn for the purpose of indicating the absorbed dose of ionizing radiation received by the wearer. A film badge (or badge meter, as it is commonly called) usually is made of metal, plastic, or paper and is loaded with one or more film packets (ordinarily standard dental film packets). It is attached to the clothing by means of a clip or pin and is frequently combined with an identification badge. In some cases, it is made into a film ring or a film bracelet to indicate specifically the absorbed dose received by the hands or forearms. The simplest type of film badge consists of a small paper envelope containing a dental film, one-half of which is surrounded by a thin lead foil.

The badge must contain one or more filters (usually made of cadmium, lead, silver, aluminum, or copper) so that a comparison can be made of the relative blackening of the developed film from behind the various filters. This comparison reveals the extent of exposure to various types of radiation. *See* DOSIMETER; MONITORING OF IONIZING RADIATION. [K.Z.M.]

Filosia A subclass of Rhizopodea characterized by slender filopodia which rarely anastomose. The two orders are Aconchulinida and Gromiida. *See* ACONCHULINIDA; GROMIIDA; PROTOZOA; RHIZOPODEA. [R.P.H.]

Filtration The separation of solid particles from a fluid-solids suspension of which they are a part by passage of most of the fluid through a septum or membrane that retains most of the solids on or within itself. The septum is called a filter medium, and the equipment assembly that holds the medium and provides space for the accumulated solids is called a filter. The fluid may be a gas or a liquid. The solid particles may be coarse or very fine, and their concentration in the suspension may be extremely low (a few parts per million) or quite high (>50%).

The object of filtration may be to purify the fluid by clarification or to recover clean, fluid-free particles, or both. In most filtrations the solids–fluid separation is not perfect. In general, the closer the approach to perfection, the more costly the filtration; thus the operator of the process cannot justify a more thorough separation than is required.

Gas filtration involves removal of solids (called dust) from a gas-solids mixture because: (1) the dust is a contaminant rendering the gas unsafe or unfit for its intended use; (2) the dust particles will ultimately separate themselves from the suspension and create a nuisance; or (3) the solids are themselves a valuable product that in the course of its manufacture has been mixed with the gas.

Three kinds of gas filters are in common use. Granular-bed separators consist of beds of sand, carbon, or other particles which will trap the solids in a gas suspension that is passed through the bed. Bag fitters are bags of woven fabric, felt, or paper through which the gas is forced; the solids are deposited on the wall of the bag. Air filters are light webs of fibers, often coated with a viscous liquid, through which air containing a low concentration of dust can be passed to cause entrapment of the dust particles. *See* AIR FILTER; DUST AND MIST COLLECTION.

Liquid filtration is used for liquid-solids separations in the manufacture of chemicals, polymer products, medicinals, beverages, and foods; in mineral processing; in water purification; in sewage disposal; in the chemistry laboratory; and in the operation of machines such as internal combustion engines.

Liquid filters are of two major classes, cake filters and clarifying filters. The former are so called because they separate slurries carrying relatively large amounts of solids. They build up on the filter medium as a visible, removable cake which normally is discharged "dry" (that is, as a moist mass), frequently after being washed in the filter. It is on the surface of this cake that filtration takes place after the first layer is formed on the medium. The feed to cake filters normally contains at least 1% solids. Clarifying filters, on the other hand, normally receive suspensions containing less than 0.1% solids, which they remove by entrapment on or within the filter medium without any visible formation of cake. The solids are normally discharged by backwash or by being discarded with the medium when it is replaced. *See* CLARIFICATION. [S.A.M.]

Fine structure (spectral lines) The closely spaced groups of lines observed in the spectra of the lightest elements, notably hydrogen and helium. The components of any one such group are characterized by identical values of the principal quantum number n, but different values of the azimuthal quantum number l and the angular momentum quantum number j.

In atoms having several electrons, this fine structure becomes the multiplet structure resulting from spin-orbit coupling. This gives splittings of the terms and the spectral lines that are "fine" for the lightest elements but that are very large, of the order of an electronvolt, for the heavy elements. [F.A.J./W.W.W.]

Finger Lakes Long, narrow, comparatively straight lakes developed in the north border of the Appalachian Plateau in western New York. These lakes resemble fiords; the bottoms of some are below sea level. The lakes occupy preglacial or interglacial valleys that were deepened by glacial erosion. Some tributary valleys hang above the lakes, creating falls. *See* FIORD. [F.T.T./R.F.Fl.]

Fingerprint Distinctive ridges that appear on the bulbs of the inside of the end joints of the fingers and thumbs. Fingerprints are an infallible means of identification. In addition to their value in criminal matters, fingerprints can ensure personal identification for humanitarian reasons, such as in cases of amnesia, missing persons, or unknown deceased. Fingerprints are invaluable in effecting identifications in tragedies such as fire, flood, and vehicle crashes. The vast majority of fingerprints maintained in the Identification Division of the Federal Bureau of Investigation of the United States, the largest repository of fingerprints in the world, are civil records.

Fingerprints fall into three general types of patterns (see illustration). Each group bears the same general characteristics or family resemblance. The three general pattern types may be further divided into subgroups by means of smaller differences existing between the patterns in the same general group. The arch group includes the plain arch and the tented arch. The loop group includes the radial and ulnar loops. The whorl group includes four types of whorls, the plain whorl, central pocket loop, double loop, and accidental whorl.

The pattern area is that part of a fingerprint in which appear the cores, deltas, and ridges that are used for classification.

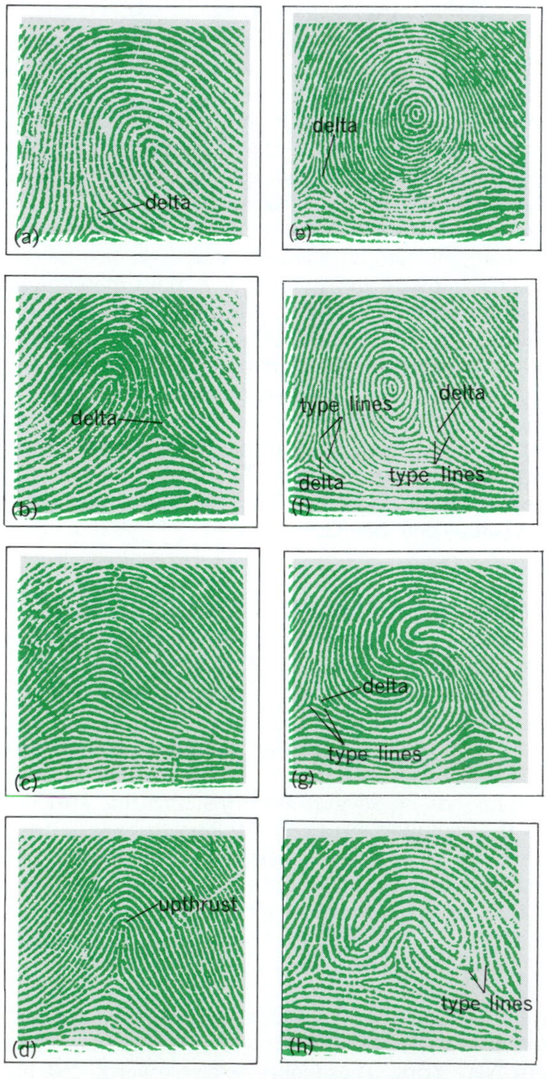

Types of fingerprint patterns. (*a–b*) Loop patterns: (*a*) ulnar loop; (*b*) radial loop. (c–*d*) Arch patterns: (c) plain arch; (*d*) tented arch. (*e–h*) Whorl patterns: (*e*) plain whorl; (*f*) central pocket loop; (*g*) double loop; (*h*) accidental whorl. (*Federal Bureau of Investigation*)

The pattern areas of loops and whorls are enclosed by type lines, which may be defined as the two innermost ridges which start parallel, diverge, and surround or tend to surround the pattern area.

Within the pattern areas of loops and whorls are the focal points which are used in the detailed classification. These points are called the delta and the core. The delta is that point on a ridge at or in front of and nearest the center of the divergence of the type lines. A core may be defined as that point on a ridge which is located in the approximate center of the finger impression.

The distinctive ridges of the palms of the hands and the soles of the feet closely resemble those of the fingerprint and can be identified in the same way. Footprints of persons having no fingers are taken for record purposes. Fingerprint, palm print, and sole print identifications have all been accepted as evidence by the courts. *See* EPIDERMAL RIDGES.
[J.E.Ho.]

Finite element method A numerical analysis technique for obtaining approximate solutions to many types of engineering problems. The need for numerical methods arises from the fact that for most practical engineering problems analytical solutions do not exist. While the governing equations and boundary conditions can usually be written for these problems, difficulties introduced by either irregular geometry or other discontinuities render the problems intractable analytically. To obtain a solution, the engineer must make simplifying assumptions, reducing the problem to one that can be solved, or a numerical procedure must be used. In an analytic solution, the unknown quantity is given by a mathematical function valid at an infinite number of locations in the body, while numerical methods provide approximate values of the unknown quantity only at discrete points in the body. In the finite element method, the region of interest is divided up into numerous connected subregions or elements within which approximate functions are used to represent the unknown quantity.

The physical concept on which the finite element method is based has its origins in the theory of structures. The idea of

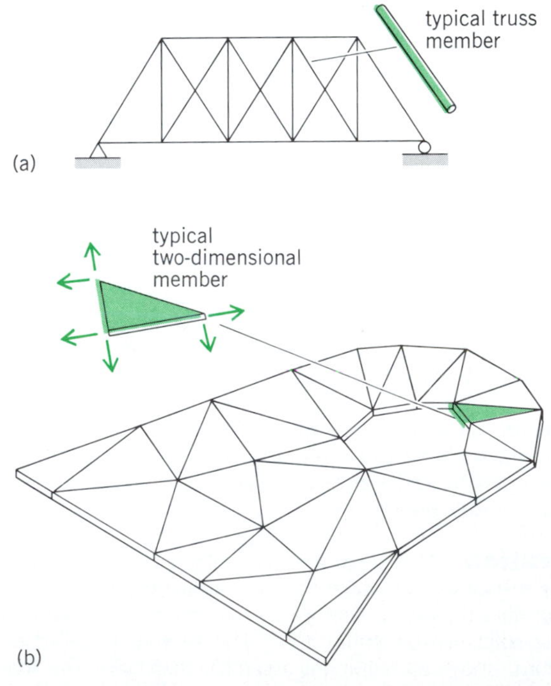

Structures modeled by fitting together structural elements: (*a*) truss structure; (*b*) two-dimensional planar structure.

building up a structure by fitting together a number of structural elements (see illustration) was used in the early truss and framework analysis approaches employed in the design of bridges and buildings in the early 1900s. By knowing the characteristics of individual structural elements and combining them, the governing equations for the entire structure are obtained. The limitation on the number of simultaneous algebraic equations that could be solved posed a severe restriction to the analyses of that period. The introduction of the digital computer has made possible the solution of the large-order systems of equations.

The finite element method is one of the most powerful approaches for the approximate solution of a wide range of problems in mathematical physics. As such, its application has extended beyond structural analysis to soil mechanics, heat transfer, fluid flow, magnetic field calculations, and other areas. [T.B.]

Finite mathematics Those parts of mathematics which deal with finite sets. More generally, finite mathematics is taken to include those fields which make no use of the concept of limit; thus linear programming, which finds maxima and minima through the four arithmetic operations, is included, whereas classical optimization, which uses differential calculus, is excluded. Other subjects generally considered in the finite mathematics include finite set theory; logic; the study of finite groups, rings, and fields; combinatorial theory; linear algebra; probability; game theory; dynamic programming; and networks. *See* DYNAMIC PROGRAMMING; FIELD THEORY (MATHEMATICS); GAME THEORY; GROUP THEORY; LINEAR ALGEBRA; LINEAR PROGRAMMING; LOGIC; NETWORK (MATHEMATICS); OPTIMIZATION; PROBABILITY; RING THEORY; SET THEORY. [G.Ow.]

Fiord A segment of a troughlike glaciated valley partly filled by an arm of the sea. It differs from other glaciated valleys only in the fact of submergence. The floors of many fiords are elongate basins excavated in bedrock, and in consequence are shallower at the fiord mouths than in the inland direction. The seaward rims of such basins represent lessening of glacial erosion at the coastline, where the former glacier ceased to be confined by valley walls and could spread laterally. Some rims are heightened by glacial drift deposited upon them in the form of an end moraine.

Fiords occur conspicuously in British Columbia and southern Alaska, Greenland, Arctic islands, Norway, Chile, New Zealand, and Antarctica—all of which are areas of rock resistant to erosion, with deep valleys, and with strong former glaciation. [R.F.Fl.]

Fir Any tree of the genus *Abies*, of the pine family, characterized by erect cones, by the absence of resin canals in the wood but with many in the bark, and by flattened needlelike leaves which lack definite stalks. The leaves usually have two white lines on the underside and leave a circular scar when they fall.

The native fir of the northeastern United States and adjacent Canada is *A. balsamea*. Its principal uses are for paper pulp, lumber, boxes and crates, and as a source of the liquid resin called Canada balsam. In the eastern United States the fir is commonly used as a Christmas tree.

The Fraser fir (*A. fraseri*) is a similar species found in the southern Appalachians.

Several species of *Abies* grow in the Rocky Mountains region and westward to the Pacific Coast. The most important commercially is the white fir (*A. concolor*), also known as silver fir. Other western species of commercial importance are the subalpine fir (*A. lasiocarpa*), grand fir (*A. grandis*), Pacific silver fir (*A. amabilis*), California red fir (*A. magnifica*), and noble fir (*A. procera*). *See* PINALES. [A.H.G./K.P.D.]

Fire A rapid but persistent chemical reaction accompanied by the emission of light and heat. The reaction is self-sustaining, unless extinguished, to the extent that it continues until the fuel concentration falls below a minimum value. Most commonly, it results from a rapid exothermic combination with oxygen by some combustible material. Flame, the visible manifestation of fire, results from a heating to incandescence of minute particulate matter composed principally of incompletely burned fuel. *See* COMBUSTION; FLAME. [F.J.J.]

Fire assaying Analysis of a metal-bearing material by heating a sample with a suitable flux and measuring the content of resultant metal beads by weighing or some instrumental technique.

Fire assaying is used almost exclusively for measuring the content of silver, gold, and platinum metals in ores, alloys, plating solutions, and reclaimed materials, such as sweepings, commonly called sweeps. Reasons for the continued use of the fire-assay method include complete decomposition of large refractory representative samples, high sensitivity, high selectivity, and ready adaptability to routine work. These reasons become apparent on considering the essentials of the method. [S.K.]

Fire detector A temperature-sensing device designed to sound an alarm, to turn on a sprinkler system, or to activate some other fire preventive measure at the first signs of fire. Sprinkler systems are usually equipped with fusible metal links, and these are perhaps the most common fire detector devices. They serve a double function, because they sense the rise in temperature due to the fire and put the defense measures into motion. More elaborate fire alarm systems use detectors similar to thermostats. They are often actuated electrically.

Smoke detectors are also common. These detectors are actuated when the smoke interrupts a light beam that shines on a photoelectric cell.

Photosensing devices respond to a stimulus in the infrared part of the spectrum and can thus detect the rise in temperature at the source of the fire itself. Because of the rapid response of such detectors, extinguishing apparatus can be actuated in a fraction of a second. Therefore, it is possible to stop the development of an explosion in the fuel tank of an aircraft, for example, by placing the detector and the extinguishing equipment inside the tank itself. *See* ALARM SYSTEMS; AUTOMATIC SPRINKLER SYSTEM; FIRE EXTINGUISHER. [W.E.Go.]

Fire extinguisher Fire may be extinguished by the following methods: (1) cooling the burning materials; (2) blanketing the fire with inert gas that chokes it for lack of oxygen; (3) introducing materials that inhibit combustion; and (4) covering the burning matter with a blanket or a layer of solid particles that prevent access of air. Fire extinguishers operate on one or a combination of these principles.

Water is the most effective cooling agent used in fire extinguishing. The generation of steam also drives away the air and forms a blanket, but being less dense than air, it is rapidly displaced. Wetting agents and foaming agents increase the effectiveness of water. In the small, first-aid, water fire extinguishers, a propellant must be provided. Usually this is carbon dioxide, which is either generated when needed or stored in a cartridge. Water should not be used on oil fires or on electrical fires.

Automatic water sprinkler systems are a common form of fire protection in industrial plants and large buildings. The installation of these systems is the greatest single factor to be credited for the sharply reduced incidence of disastrous fires in recent years. See AUTOMATIC SPRINKLER SYSTEM.

Carbon dioxide is a safe and effective extinguisher for all confined fires. It acts as an inert blanket, and because it is heavier than air, it will exclude oxygen very efficiently from a fire on the floor of a building or in a vat or similar vessel. It is not effective in an elevated location or outdoors where the wind can blow the gas away.

A dry powder, consisting principally of sodium bicarbonate, may also be used as a fire extinguisher. The powder must have the correct particle size and contain materials that prevent it from caking. The action of the powder is threefold: to generate carbon dioxide; to cool the burning material; and to provide a shielding to prevent access of air. Dry chemical is useful for small fires, and especially electrical fires.

Carbon tetrachloride, CCl_4, has had a long history as a fire-extinguishing agent. As it is customarily used, in small quantities, the principal action is to supply a heavy blanket of vapor over the fire. In addition, carbon tetrachloride, in common with all the halogenated compounds, has a definite chemical inhibiting effect on combustion. Other halogenated hydrocarbons that have been used are chlorobromomethane, and several of the fluorinated hydrocarbons known as freons. The principal difficulty, however, with all the halogenated hydrocarbons, and with CCl_4, in particular, is toxicity.

Other extinguishing methods should be mentioned that require no special equipment. For a household fire, especially involving clothing on a person, a blanket or a rug provides an effective means to smother the fire. Small fires around a kitchen stove may be snuffed out with salt or, better still, with bicarbonate of soda. A bucket of sand, strategically located, is also useful against domestic fire hazards. [W.E.Go.]

Fire-tube boiler A steam boiler in which hot gaseous products of combustion pass through tubes surrounded by boiler water. The water and steam in fire-tube boilers are contained within a large-diameter drum or shell, and such units often are referred to as shell-type boilers. Heat from the products of combustion is transferred to the boiler water by tubes or flues of relatively small diameter (approximately 3–4 in. or 7.5–10 cm) through which the hot gases flow. The tubes are connected to tube sheets at each end of the cylindrical shell and serve as structural reinforcements to support the flat tube sheets against the force of the internal water and steam pressure. Braces or tension rods also are used in those areas of the tube sheets not penetrated by the tubes.

Fire-tube boilers may be designed for vertical, inclined, or horizontal positions. One of the most generally used types is the horizontal-return-tube (HRT) boiler (see illustration). In the HRT boiler, part of the heat from the combustion gases is transferred directly to the lower portion of the shell. The gases then make a return pass through the horizontal tubes or flues before being passed into the stack. See STEAM-GENERATING UNIT. [G.W.K.]

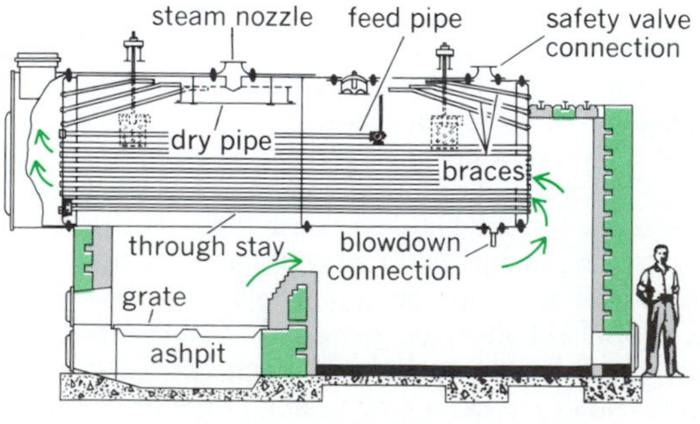

Horizontal-return-tube boiler.

Fischer-Tropsch process The synthesis of hydrocarbons and, to a lesser extent, of aliphatic oxygenated compounds by the catalytic hydrogenation of carbon monoxide. The synthesis was discovered in 1923 by F. Fischer and H. Tropsch at the Kaiser Wilhelm Institute for Coal Research in Mülheim, Germany. The reaction is highly exothermic, and the reactor must be designed for adequate heat removal to control the temperature and avoid catalyst deterioration and carbon formation. The sulfur content of the synthesis gas must be extremely low to avoid poisoning the catalyst. See COAL GASIFICATION. [J.H.Fi.]

Fisher A carnivorous mammal that is a member of the family Mustelidae, which also includes the minks, ferrets, martens, weasels, badgers, otters, skunks, and wolverine. The American fisher (*Martes pennanti*) occurs in forested areas of North America. The fur, which is valuable, is variable in color, but always dark, and does not change color seasonally. The fisher has a long, lithe powerful body with short legs and a long muzzle. These animals breed once a year, and the two to four young are born in the nest high above the ground. This species is now nearly extinct in the eastern United States. See BADGER; CARNIVORA; MARTEN; OTTER; SKUNK; WEASEL; WOLVERINE. [C.B.C.]

Fishery The taking or propagation of fishes or other aquatic life in inland or oceanic waters. A fishery may be operated for subsistence, for pleasure (sport fishery), or for profit (commercial fishery). The "fishes" include: fin fishes; invertebrate shellfishes such as the mollusks (including clams, mussels, and oysters) and the shrimps and their relatives, the marine lobsters and the fresh-water crayfishes; aquatic plants; sponges; coral; sea cucumbers; amphibians, chiefly frogs; reptiles such as turtles, alligatrs, and crocodiles; and mammals, including whales, seals, and walruses.

Fisheries are important for the production of food and of raw materials for industry, and for recreation. They occur in all the many kinds of waters that together cover nearly three-fourths of the Earth's surface. In the international accounting system of world fish catch maintained by the UN Food and Agriculture Organization (FAO), there are three major fisheries—marine (salt-water), fresh-water, and diadromous (involving fishes that ascend fresh-water streams to spawn, like salmon, or descend from streams to reproduce in the oceans, like certain eels). See AQUACULTURE; MARINE FISHERIES. [K.F.L.]

Fissidentales An order of the Bryopsida. The plants appear flattened because the two rows of leaves are arranged on the stem in one plane. They are described as being distichous or two-ranked. Most of the species are terrestrial, but a few are aquatic. See BRYOPSIDA. [W.H.W.]

Fission track dating A method of dating geological and archeological specimens by counting the radiation-damage tracks produced by spontaneous fission of uranium impurities in minerals and glasses. During fission two fragments of the uranium nucleus fly apart with high energy, traveling a total distance of about 25 micrometers (0.001 in.) and creating a single, narrow but continuous, submicroscopic trail of altered material, where atoms have been ejected from their normal positions. Such a trail, or track, can be revealed by using a chemical reagent to dissolve the altered material, and the trail can then be seen in an ordinary microscope. The holes produced in this way can be enlarged by continued chemical attack until they are visible to the unaided eye.

Track dating is possible because most natural materials contain some uranium in trace amounts and because the most abundant isotope of uranium, ^{238}U, fissions spontaneously. Over the lifetime of a rock substantial numbers of fissions occur; their tracks are stored and thus leave a record of the time elapsed since track preservation began. The number of tracks produced in a given volume of material depends on the uranium content as well as the age, so that it is necessary to measure the uranium content before an age can be determined.

One feature unique to this dating technique is the time span to which it is applicable. It ranges from less than 100 years for certain synthetic, decorative glasses to approximately 4,500,000,000 years, the age of the solar system. A second useful feature is that measurements can sometimes be made on extremely minute specimens, such as chips of meteoritic minerals or fragments of glass from the ocean bottom. A third useful feature is that each mineral dates the last cooling through the temperature below which tracks are retained permanently. Since this temperature is different for each mineral, it is possible to measure the cooling rate of a rock by dating several minerals—each with a different track-retention temperature. *See* AMINO ACID DATING; GEOLOGICAL TIME SCALE; ROCK AGE DETERMINATION. [R.L.F.; P.B.P.; R.M.W.]

FitzGerald-Lorentz contraction The contraction of a moving body in the direction of its motion. In 1892 G. F. FitzGerald and H. A. Lorentz proposed independently that the failure of the Michelson-Morley experiment to detect an absolute motion of the Earth in space arose from a physical contraction of the interferometer in the direction of the Earth's motion. According to this hypothesis, as formulated more exactly by Albert Einstein in the special theory of relativity, a body in motion with speed v is contracted by a factor $\sqrt{1 - v^2/c^2}$ in the direction of motion, where c is the speed of light. *See* LIGHT; RELATIVITY. [E.L.Hi.]

Flabellifera The largest and morphologically most generalized suborder of isopod crustaceans. The biramous uropods are attached to the sides of the abdomen and may with the last abdominal segment form a caudal fan (see illustration). More than 1400 species and over 10 families are known.

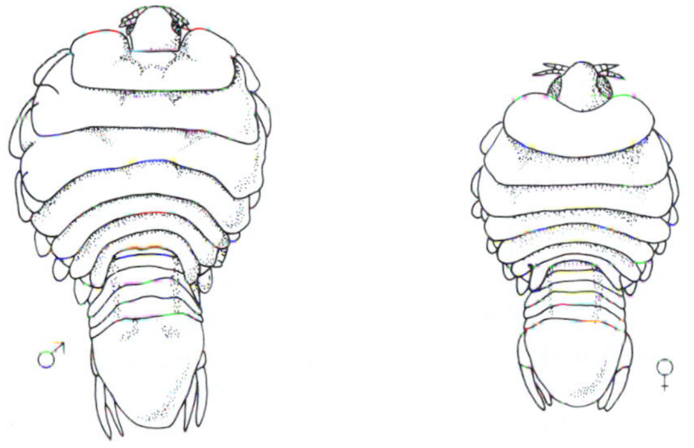

***Ichthyoxenus jellinghausii* Herklots, male and female. (*Proceedings of U.S. National Museum*)**

Members of the family Cirolanidae are actively swimming predators and scavengers with biting mouthparts. They feed on dead and living animals and are notorious for their attacks on the bait of line fishermen and on fishes caught in nets. The legs of bathers are also sometimes bitten.

Members of the families Corallanidae and Excorallanidae are sometimes found parasitizing fishes, but are often free-living. The families Aegidae and Cymothoidae include fish parasites; the Aegidae feed by sucking blood. The Cymothoidae are more highly adapted for a parasitic life than other Flabellifera.

The gribbles, Limnoriidae, are well known for the damage they cause to marine structures of wood. In the family Sphaeromatidae, some species of *Sphaeroma* are economically important in tropical seas because of their habits of boring into wood, clay, and rock. Isopods of the family Serolidae are greatly flattened forms which live partially buried on sandy bottoms. *See* ISOPODA. [T.E.B.]

Flame A reaction front (or wave) in a gaseous medium into which the reactants flow and out of which the products flow. Solid or liquid particles may also be carried in by the gas and provide fuel for the flame, and sometimes the product gases contain solid particles, such as soot. Flames produce heat and, usually, light.

A seeming paradox is worthy of note, however. A faint, luminous phenomenon, known as a cool flame, is observed when, for example, a mixture of ether vapor and oxygen is slowly heated. Such flames are not really cool; they do produce some heat. But unlike ordinary flames that advance from layer to layer in the gas mainly by heat conduction, cool flames proceed by diffusion of reactive molecules, or free radicals, which initiate chemical processes as they go. The light from cool flames is a chemiluminescence that originates from excited molecules created by the reaction. *See* CHEMILUMINESCENCE.

Combustion flames, in which oxygen or some other oxidizing substance combines with a fuel, are more common. But decomposition flames, in which a molecule such as ozone, O_3, breaks down into simple molecules, in this case oxygen, O_2, are also known. Not all combustion or decomposition reactions, however, produce a flame, even though they may evolve heat. For example, charcoal undergoes surface combustion and, although it glows, there is no flame. *See* COMBUSTION. [W.E.Go.]

Flame arc lamp An arc lamp that radiates light from the arc instead of from the electrode. The carbon electrode is impregnated with chemicals, which are more volatile than the carbon and, when driven into the arc, become luminous. The chemicals commonly used are calcium, barium, titanium, and strontium, which produce their characteristic color of light as chemical flames and radiate efficiently in their specific wavelengths in the visible spectrum. Some flames use several chemicals, some for light and others for high arc temperature and steady arc operation. Other chemicals may be introduced to produce radiation outside the visible spectrum for specific commercial or therapeutic purposes. *See* ARC LAMP. [J.O.K.]

Flame photometry A branch of spectrochemical analysis in which samples in solution are excited to luminescence by introduction into a flame. Flame photometry is particularly useful in determining small amounts of lithium, sodium, and potassium in solution. Only small samples are required, and an analysis usually can be carried out in relatively short time.

Most emission flame photometry in the United States is done with the Beckman burner, which aspirates solution directly into a hydrogen-oxygen or acetylene-oxygen flame. Since flame temperatures are comparatively low, only the more volatile compounds are vaporized from the residue. Also, only the less stable molecules are dissociated, and only the more easily excited spectral lines and bands are emitted. Flame line-emission spectra are thus relatively simple, although molecular band spectra caused by the fuel or metallic oxides sometimes cause spectral interference. The line or band of interest is isolated with a monochromator, and its intensity measured photoelectrically. The size and type of monochromator required depend on the spectral region studied and the complexity of the spectrum.

Emission flame photometry is more accurate, precise, and sensitive than wet chemical methods for the determination of small amounts of alkali metals, and it rivals wet methods in the determination of some other elements. *See* ATOMIC STRUCTURE AND SPECTRA; SPECTROCHEMICAL ANALYSIS.

Atomic absorption analysis is a method of determining the concentration of a metal, usually in solution. The solution is converted to a vapor in which at least some of the metal is in the form of individual neutral atoms. The absorbance of this vapor is measured at a resonant absorption wavelength of the metal in question. This absorbance is then compared with that given by standard solutions. For some elements (for example, zinc, cadmium, antimony, lead, bismuth, mercury, and tellurium) detection limits are much better than those achieved by emission flame photometry; for most elements the two methods are comparable in sensitivity. The atomic absorption method has fewer spectral line interference problems than does emission flame photometry. [C.F.]

Flameproofing The process of treating materials so that they will not support combustion. Although cellulosic materials such as paper, fiberboard, textiles, and wood products cannot be treated so that they will not be destroyed by long exposure to fire, they can be treated to retard the spreading of fire and to be self-extinguishing after the igniting condition has been removed.

Numerous methods have been proposed for flameproofing cellulosic products. One of the simplest and most commonly used for paper and wood products is impregnation with various soluble salts, such as ammonium sulfate, ammonium phosphate, ammonium sulfamate, borax, and boric acid. Special formulations are often used to minimize the effects of these treatments on the color, softness, strength, permanence, or other qualities of the paper. For some applications, these treatments are not suitable because the salts remain soluble and leach out easily on exposure to water. A limited degree of resistance to leaching can be achieved by the addition of latex, lacquers, or waterproofing agents. In some cases the flameproofing agent can be given some resistance to leaching by causing it to react with the cellulose fiber (for example, urea and ammonium phosphate).

Leach-resistant flameproofing may also be obtained by incorporating insoluble retardants in the paper during manufacture, by application of insoluble materials in the form of emulsions, dispersions, or organic solutions, or by precipitation on, or reaction with, the fibers in multiple-bath treatments. The materials involved are of the same general types as those used for flameproofing textiles and include metallic oxides or sulfides and halogenated organic compounds. *See* COMBUSTION; TEXTILE CHEMISTRY. [T.A.H.]

Flash-photolysis resonance fluorescence A technique of fluorometric analysis for measuring absolute rate constants for gas-phase reactions involving free radicals (principally atoms and diatomic radicals). The method combines pulsed (about 10^{-5} s duration) photolytic production of the radical with real-time resonance fluorescence monitoring of its subsequent temporal behavior. This technique revolutionized gas-phase chemical kinetics research in the 1970s.

The typical experimental apparatus includes a vacuum reaction chamber, a pulsed photolysis source (flash lamp), an analytical light source (reasonance lamp), and a detection system. In a typical experiment a gas mixture consisting of a photolytic radical source, molecular reactant, and inert diluent are flash-photolyzed in the cell under preselected conditions of light intensity and wavelength distribution. In this way the desired uniform concentration of transient reactive species is produced across the central region of the cell. The analytical light source probes an intersection of this central region continuously, thereby generating a fluorescence emission from the radical species.

The application of resonance fluorescence analysis for real-time kinetics investigations is limited only by the fluorescing ability of the free radical and the availability of a suitable resonance light source. This versatility has led to the adaptation of this measurement technique to other kinetics experiments utilizing entirely different production and analysis methodologies. *See* CHEMICAL DYNAMICS; FLUORESCENCE; OPTICAL METHODS OF CHEMICAL ANALYSIS. [M.J.Ku.]

Flash point The temperature at which a flash of flame is first observed when a small lighted flame is passed over a sample of a flammable substance in a cup. The flash point is an important test for fire and safety regulations. So also is the fire point, or temperature at which sustained burning first occurs. These tests are indicative of the lowest-boiling components in the oil and are applied to solvents, distillate fuels, lubricating oils, and many other petroleum products. [M.Sou.]

Flash welding A form of resistance welding that is used for mass production. The welding circuit consists of a low-voltage, high-current energy source (usually a welding transformer) and two clamping electrodes, one stationary and one movable.

The two pieces of metal to be welded are clamped tightly in the electrodes, and one is moved toward the other until they meet, making light contact. Energizing the transformer causes a high-density current to flow through small areas that are in contact with each other. Flashing starts, and the movable workpiece must accelerate at the proper rate to maintain an increasing flashing action. After a proper heat gradient has been established on the two edges to be welded, an upset force is suddenly applied to complete the weld. This upset force extrudes slag, oxides, molten metal, and some of the softer plastic metal, making a weld in the colder zone of the heated metal. *See* RESISTANCE WELDING. [E.J.L.]

Flat-bed chromatography A common form of liquid chromatography that takes place in flat chromatographic beds. Two methods which belong in this category are paper chromatography and thin-layer chromatography.

In paper chromatography, a strip of filter paper that has been equilibrated in a moist atmosphere is spotted near the bottom with a sample (mixture) and placed inside the developing chamber. The water in the moist atmosphere that has been adsorbed by paper is most commonly used as the stationary phase (the paper is a support). Papers can also be impregnated by other liquids. The mobile phase, usually an organic solvent or a mixture of solvents, ascends the paper strip by means of capillary forces, and the components of the sample mixture are thus separated by the partition effect.

This stationary phase in thin-layer chromatography is a suspension which forms a layer (approximately 250 micrometers thick) on a metal or glass plate. It is most frequently an adsorbent (with a particle size of several micrometers) suspended in a suitable solvent, uniformly spread on a plate, and dried. The mobile phase is a liquid that ascends the plate by capillary action. The development of chromatograms is similar to that in paper chromatography in its simplicity. But thin-layer chromatography is superior in terms or rapidity, the variety of stationary phases that can be used, and detection methods. *See* CHROMATOGRAPHY; LIQUID CHROMATOGRAPHY. [M.V.N.]

Flatfish A number of asymmetrical fish which make up the order Pleuronectiformes. The body of these fish is laterally compressed, and both eyes are on the same side of the head.

The larva is symmetrical, but after further development and assumption of its bottom-dwelling habits, the young turns to one side or the other and there is a movement or migration of the lower eye to the upper surface of the head. In addition, the body surface of the fish that is in contact with the substratum is unpigmented; many species are able to modify the pattern and intensity of their pigmented skin to match their variable substrate. These are mainly marine species that live on the continental shelf. The more common flatfish are economically important species which include the turbot, halibut, plaice, flounder, and sole. *See* PLEURONECTIFORMES. [C.B.C.]

Flavobacterium A genus of uncertain affiliation. Its members include gram-negative, nonsporeforming rod-shaped cells which do not show motility in a hanging drop. Flavobacteria are strictly aerobic. The genus is taxonomically uncertain since many of its dozen or so species are similar to the genus *Cytophaga*, including species isolated from human pathological materials. *Flavobacterium aquatile* is the type species. Flavobacteria are widely distributed in soil and water and are commonly found in vegetables, dairy products, meat, poultry, and hospital environments. Some species of flavobacteria are quite resistant to antimicrobial agents, showing susceptibility only to novobiocin, clindamycin, and trimethoprim-sulfamethoxazole. Other species are susceptible to the penicillins, the cephalosporins, tetracycline, chloramphenicol, and trimethoprim-sulfamethoxazole. *See* ANTIBIOTIC; MEDICAL BACTERIOLOGY. [G.L.Gi.]

Flavonoid Any of a series of widely distributed plant constituents related in one way or another to the aromatic heterocyclic skeleton of 2-phenylbenzopyran, or flavan. The three most important classes are the anthocyanins, the flavones, and the flavon-3-ols, or flavonols. Other classes are the chalcones and aurones (yellow flower pigments); isoflavones (some are weak estrogens); flavanones (such as naringin, the bitter principle of the lemon); and leucoanthocyanidins (condensed tannins). The various flavonoid classes are biogenetically related, all being derived from a common precursor. *See* ANTHOCYANIN.

Flavonoids have no single overriding function in plants. Many are highly colored and serve to attract insects and animals to the flowers and fruits for the purpose of pollination and seed dispersal. Others have been implicated in growth regulatory processes or as "phytoalexins," (agents for protecting plants from fungal attack). Others present in leaves, because of their taste properties, protect plants from being eaten by insects or wild animals.

Flavonoids are found in all groups of plants except algae, bacteria, and fungi. Their occurrence in insects, in the wings of a number of butterflies, is due to their ingestion in the food and a subsequent failure of the caterpillar to metabolize them. All animals above insects break down the dietary flavonoids into small fragments, which are then excreted in the urine.

In contrast to the alkaloids, flavonoids are rather inactive pharmacologically; the quercetin glycoside rutin is the only one that is used much in medicine. It has a protective function in preventing capillary fragility. An economically significant property of flavonoids is their contribution to taste and flavor in foods. Some compounds are bitter, others very sweet, and yet others astringent. Bitterness is largely confined to certain flavanone glycosides of citrus fruits.

From the point of view of the food industry, the most important classes of flavonoids are the catechins and flavan-3,4-diols, or leucoanthocyanidins. These compounds are widely distributed in drinks such as cocoa, tea, beer, and cider and in many fruits and vegetables. They are responsible for astringency, a necessary counterbalance in certain foods to insipid sweetness. Leucoanthocyanidins also have the property of tanning protein, and most of the condensed tannins used in the leather industry are flavonoid polymers. *See* TANNIN. [J.B.Ha.]

Flavor A generic term referring to quarks, the most elementary constituents of matter. The evidence is conclusive that all matter is constituted of quarks of at least five different types of flavors, with a sixth predicted. They are labeled *u*, *d*, *s*, *c*, *b*, and *t*. The last, the *t* quark, is predicted from symmetry considerations. Corresponding mnemonics are up, down, strange, charmed, bottom, and top. The *u*, *c*, and *t* flavors each carry a positive electric charge equal in magnitude to two-thirds that of the electron, and the *d*, *s*, and *b* flavors have a negative charge one-third that of the electron. *See* ELEMENTARY PARTICLE; QUARKS. [V.F.]

Flax The flax plant (*Linum usitatissimum*) is the source of two products: flaxseed for linseed oil and fiber for linen products. Plants with two distinct types of growth are used for seed and fiber production.

Flax for fiber requires fertile, well-drained, and well-prepared soil and a cool, humid climate. The best-known use for flax fiber is in the manufacture of fine linen fabrics; other uses are linen thread, linen twine, toweling, and canvas. Seed from fiber flax is used for replanting and for oil. The major producer of flax fiber is Russia, but the world's best fiber comes from Belgium and adjoining countries. Most Irish linen is manufactured from fiber produced in Belgium. *See* NATURAL FIBER; TEXTILE.

Seedflax is the source of linseed oil. The principal uses of linseed oil are in the manufacture of protective coatings—paint, varnish, and lacquer. It is also used in linoleum, printer's ink, and patent and imitation leathers, and as core oil for making sand forms in metal casting. The linseed cake remaining after the oil is extracted is used as a high-protein livestock feed. Large quantities of clean (weed-free) straw from seedflax are used in the production of a high-grade paper. This product is used to make cigarette and other fine papers. The world's major producers of flaxseed are Argentina, Canada, India, Russia, and the United States. [E.G.N; J.O.Cu.]

Flea A member of the insect order Siphonaptera; a flattened, wingless ectoparasite. Most fleas are less than 1/5 in. (5 mm) long. They are intermittent parasites, readily changing hosts and able to live up to 4 months without feeding. Adults have piercing-sucking mouthparts with which they take blood from mammals and birds. Most fleas parasitize mammals, and while human, cat, dog, and rodent fleas each have their preferred host, they opportunistically may bite any species of mammal encountered. This can cause irritation to pet lovers, and a serious consequence is transmission of plague from rodents to humans. *See* PLAGUE; SIPHONAPTERA. [D.J.Hor.]

Fletcher-Munson contours A set of curved lines showing how the intensities of tones of different frequency should be adjusted so that all appear to be equally loud to the listener. This important characteristic of hearing is shown in the illustration in the form adopted as an international standard in 1961.

Equal loudness contours are obtained from tests in which a group of observers listen to electronically generated pure tones of different pitch coming from a loudspeaker. The tones are heard in pairs, one after the other, and volume controls are adjusted until the two tones of different frequency appear to be equally loud. Then the tones are measured with a sound level meter, and it is generally found that the sound pressures are not the same but differ systematically as the frequencies of the tones are changed. It follows that the "sound level," measured

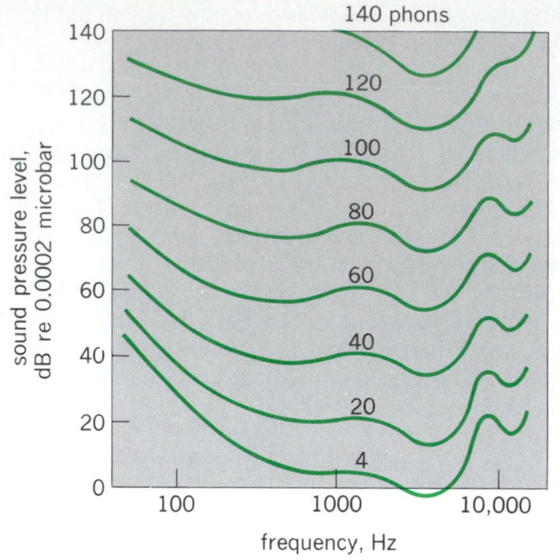

Equal loudness level contours.

with a meter, is not the same as the "loudness level," where the ear is used as the measuring device. The differences can be obtained from the equal loudness contours shown in the illustration. All tones represented by different points on the same contour line sound equally loud to the average listener. *See* HEARING (HUMAN); LOUDNESS. [W.A.M.]

Flexibilia An extinct subclass of stalked or creeping Crinoidea which includes some 50 Paleozoic genera, ranging from Ordovician to Permian times. A flexible tegmen was present with open ambulacral grooves. One of the five oral plates served as a madreporite. The anus was at the tip of a short siphon. The uniserial arms branched freely, arching inward to form a globular crown (see illustration). The cylindrical stem

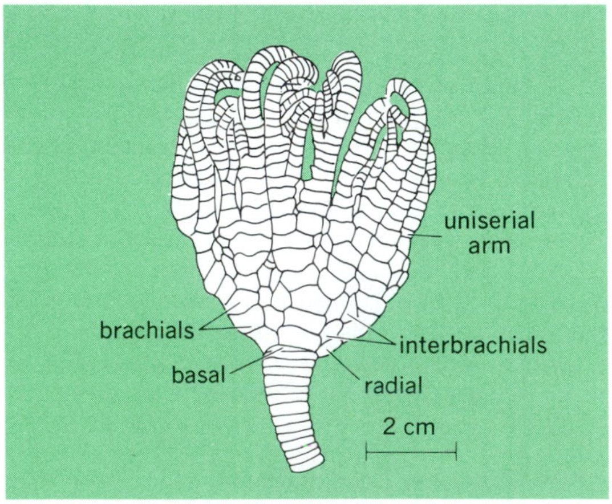

***Talanterocrinus* sp.**

lacked cirri and was sometimes discarded to produce a creeping adult. *See* CRINOIDEA; ECHINODERMATA. [H.B.F.]

Flexible manufacturing system A factory or part of a factory made up of programmable machines and devices that can communicate with one another. Materials such as parts, pallets, and tools are transported automatically within the flexible manufacturing system and sometimes to and from it. Some form of computer-based, unified control allows complete or partial automatic operation. Flexible manufacturing systems are part of a larger computer-based technology, termed computer-integrated manufacturing (CIM), which encompasses more than the movement and processing of parts on the factory floor. *See* AUTOMATION; COMPUTER-AIDED DESIGN AND MANUFACTURING; COMPUTER-INTEGRATED MANUFACTURING.

The programmable machines and devices are numerically controlled machine tools, robots, measuring machines, and materials-handling equipment. Each programmable machine or device typically has its own controller which is, in effect, a dedicated digital computer; programs must be written for these controllers, usually in special-purpose languages designed to handle the geometry and machining. Increasingly, numerically controlled machines are being programmed by graphical presentations on computer screens, that is, graphical computer interfaces. This allows the programmer to follow the machining operation and specify desired operations without the need for statements in a programming language. Robots have usually been programmed by so-called teaching, where the robot is physically led through a sequence of movements and operations; the robot remembers them and carries them out when requested. *See* COMPUTER GRAPHICS; DIGITAL COMPUTER; INTELLIGENT MACHINE; MATERIALS-HANDLING EQUIPMENT; PROGRAMMING LANGUAGES; ROBOTICS; TOOLING.

The programmable machines and devices communicate with one another via an electronic connection between their controllers. Increasingly, this connection is by means of local-area networks, that is, communication networks that facilitate high-speed, reliable communication throughout the entire factory. *See* LOCAL-AREA NETWORK.

The automatic material transport system is usually a guided, computer-controlled vehicle system. The vehicles are usually confined to a fixed network of paths, but typically any vehicle can be made to go from any point in the network to any other point. The network is different from a classical assembly line in that it is more complex and the flow through it is not in one direction.

Commands and orders to the flexible manufacturing system are sent to its computer-based, unified control. The control, in turn, issues orders for the transport of various kinds of material, the transfer of needed programs, the starting and stopping of programs, the scheduling of these activities, and other activities.

Flexible manufacturing systems are flexible in the sense that their device controllers and central control computer can be reprogrammed to make new parts or old parts in new ways. They can also often make a number of different types of parts at the same time. However, this flexibility is limited to a certain family of parts, for example, axles. A general goal for designers is to increase flexibility, and advanced flexible manufacturing systems are more flexible than the earlier ones. [A.W.N.]

Flight Traditionally, flight has dealt with motion in or through the air, but flight now includes motion at distances so far from the Earth's surface that the air has negligible effect. Natural animal flight is necessarily restricted to the denser portion of the Earth's atmosphere. Artificial flight can either use the air or be independent of it. *See* AERONAUTICS; ASTRONAUTICS.

The ballistic flight of a bullet or missile is a consequence of energy imparted to it during its ejection from a gun barrel or during the burning period of its engine. At the opposite extreme is the soaring flight of a glider or a hawk that depends on the energy from local rising air currents. True natural flight, as for birds, bats, and insects, and true artificial flight, as for humans in aircraft, require the continuous expenditure of energy to achieve fully controlled progress through the air. Historically, insects flew fast, followed by reptile groups that are now extinct, and then birds, bats, and finally humans.

The greater the size of an animal, the greater is the problem of remaining airborne, because, although the weight increases as the cube of the linear dimensions, the surface area increases only as the square. The practical limit seems to have been reached in the largest birds.

In all flying animals both lift and thrust are provided by the predominantly up-and-down movement of a pair of laterally extended planes. The movements of animal wings in level flight usually show a forward component on the downstroke and a backward component on the upstroke, either relative to the body axis or to the vertical or both. These wing movements have the effect of giving a downward component to the main direction of the slipstream and hence contribute lift. In hovering flight the movements are usually tremendously exaggerated. [B.Hoc.]

Flight controls The devices and systems which govern the attitude of an aircraft and, as a result, the flight path followed by the aircraft. Flight controls are classified as primary flight controls, auxiliary flight controls, and automatic controls. In the case of many conventional airplanes, the primary flight controls utilize hinged, trailing-edge surfaces called elevators for pitch, ailerons for roll, and the rudder for yaw. These hinged surfaces are operated by the human pilot in the cockpit or by an automatic pilot. In the case of vertically rising aircraft, a lift control is provided. *See* AILERON; AIRCRAFT RUDDER; ELEVATOR (AIRCRAFT); ELEVON.

Auxiliary flight controls may include trimming devices for the primary flight controls, as well as landing flaps, leading-edge flaps or slats, an adjustable stabilizer, a wing with adjustable sweep, and dive brakes or speed brakes.

Controls to govern the engine power and speed, while not usually classified as flight controls, are equally important in the overall control of the aircraft.

Automatic controls include systems which supplement or replace the human pilot as a means of controlling the attitude or path of the aircraft. Such systems include automatic pilots, stability augmentation systems, automatic landing systems, and active controls. Active controls encompass automatic systems which result in performance improvement of the aircraft by allowing reductions in structural weight or aerodynamic drag, while maintaining the desired integrity of the structure and stability of flight.

The control system incorporates a set of cockpit controls which enable the pilot to operate the control surfaces. Because of the approximately fixed size and strength of the human pilot and the need to standardize the control procedures for airplanes, the primary controls are similar in most types of airplanes. The cockpit controls incorporate a control stick (sometimes called the joy stick) which operates the elevators and ailerons, and pedals which operate the rudder. Sometimes a column/wheel arrangement is used to operate the elevators and ailerons, respectively. The cockpit controls for auxiliary control devices are not as completely standardized as those for the primary controls.

Control systems with varying degrees of complexity are required, depending on the size, speed, and mission of an airplane. In relatively small or low-speed airplanes, the cockpit controls may be connected directly to the control surfaces by cables or pushrods, so that the forces exerted by the pilot are transmitted directly to the control surfaces. In large or high-speed airplanes, the forces exerted by the pilot may be inadequate to move the controls. In these cases, either an aerodynamic activator called a servotab or spring tab may be employed, or a hydraulic activator may be used. In some airplanes, particularly those with swept wings and those which fly at high altitudes, the provision of adequate static stability and damping of oscillations by means of the inherent aerodynamic design of the airplane becomes difficult. In these cases, stability augmentation systems are used. These systems utilize sensors such as accelerometers and gyroscopes to sense the motion of the airplane. These sensors generate electrical signals which are amplified and used to operate the hydraulic actuators of the primary control surfaces to provide the desired stability or damping.

The weight and complication of mechanical control linkages and the extensive reliance on electrical signals in automatic controls have led to the development of control systems in which the control inputs from the pilot, as well as those from the stability augmentation sensors, are transmitted to the primary control actuators by electrical signals. Systems of this type are called electrical signaling systems or fly-by-wire systems. The electrical signals are readily compatible with computers, either analog or digital, which can perform the functions of combining the signals from the pilot and the sensors, as well as introducing additional stabilization, navigation, and guidance functions. Because of the need for high reliability, these systems employ the principle of redundancy; that is, the components are installed in multiple configurations, and provision is made for detecting a failed component and removing its influence from the system.

The availability of reliable, full-authority electronic control systems makes possible the use of closed-loop control systems to improve the performance of the airplane. The applications most frequently considered in order to take advantage of this capability are reduced static stability, maneuver load control, gust load alleviation, and structural mode damping.

The flight of an aircraft may be controlled automatically by providing the necessary signals for navigation as inputs to the control system. In practice, automatic pilots are used to relieve the human pilot of routine flying for long periods, and automatic control systems are used to make precision landings and takeoffs under conditions of reduced visibility. [W.H.P.]

Flight dynamics The study of the motion of an aircraft or missile. Flight dynamics is generally concerned with transient or short-term effects, which have to do with the stability and control of the vehicle, rather than with calculating such performance as ultimate range, attitude, or velocity of the vehicle. Sometimes, however, the two are treated together, particularly in the case of aircraft. *See* AEROELASTICITY; FLIGHT CONTROLS. [J.R.Se.]

Flight science The sum total of all knowledge that enables humans to accomplish flight. Flight science is compounded of both science and engineering. It is concerned with airplanes, missiles, and crewed and uncrewed space vehicles.

The scope of flight science is illustrated by some of the diverse fields which are included within it, such as electronics, aerodynamics, propulsion engineering, structural engineering, nuclear engineering, metallurgy, chemistry, space medicine, certain parts of astronomy, mathematics, classical and modern physics, and other branches of engineering, such as civil engineering for the planning and construction of airports. These constitute the major branches of science and technology needed to solve modern flight problems. [J.R.Se.]

Flocculation The formation of larger particles of a solid phase dispersed in a solution by the gathering together of smaller particles. The process whereby initial aggregates (having dimensions of a few unit cells) in a solution develop spontaneously into particles of a new stable phase is known as nucleation. When these particles grow to size sufficient to scatter visible light, they are known as colloids. By changing the ionic environment in which colloidal particles exist, the colloidal particles can be made to undergo further aggregation, or flocculation. *See* COLLOID; NUCLEATION; PRECIPITATION (CHEMISTRY). [L.Go./R.W.Mu.]

Floodplain The relatively broad and smooth valley floor that is constructed by an active river and periodically covered with floodwater from that river during intervals of overbank flow. Engineers consider the floodplain to be any part of the valley floor subject to occasional floods that threaten life and property. Various channel improvements or impoundments may be used to restrict the natural process of overbank flow. Geomorphologist consider the floodplain to be a surface that develops by the active erosional and depositional processes of a river. Floodplains are underlain by a variety of sediments, reflecting the fluvial history of the valley.

Most floodplains consist of the following types of deposits: colluvium—slope wash and mass-wasting products from the valley sides, as is common in small, narrow floodplains; channel lag—coarse debris marking the bottoms of former channels; lateral accretion deposits—sand and gravel deposited as the meandering river migrates laterally; vertical accretion deposits—clay and silt deposited by overbank flooding of the river; crevasse-splay deposits—relatively coarse sediment carried through breaks in the natural river levees and deposited in areas that usually receive overbank deposition; and channel-fill deposits—fills of former river channels. Channel fills may be coarse for sandy rivers. The noncohesive character of coarse sediments allows these rivers to easily erode laterally. *See* STREAM TRANSPORT AND DEPOSITION. [V.R.B.]

Floor construction The selection of the type of construction for the floors of a building should be based on the building's architectural and structural requirements and on cost.

A medium-priced home might contain a wooden floor supported by wooden joists. A floor system frequently used in small commercial buildings, apartment buildings, and higher-priced homes consists of a solid concrete slab supported on steel or concrete beams or open-web steel joists. Open-web joists are available for clear spans up to 120 ft (36 m). A thin, lightly reinforced concrete slab cast in one piece with its supporting reinforced concrete joists is particularly economical for short spans. Light-gage steel cellular floor decking is often used for steel-framed buildings. [C.N.G.]

Floriculture The segment of horticulture concerned with commercial production, marketing, and sale of bedding plants, cut flowers, potted flowering plants, foliage plants, flower arrangements, and noncommercial home gardening.

Commercial crops are grown either in the field or under protected cultivation, such as in glass or plastic structures. Field production is confined to warm climates or to summer months in colder areas. Typical field crops are gladiolus, peonies, stock, gypsophila, asters, and chrysanthemums. Greenhouse production is not as confined by climate or season, but most greenhouses are located in areas that have advantages such as high light intensity, cool night temperatures, or ready access to market. Jet air transportation resulted in major changes in international crop production.

Pronounced improvements in cultivars have been realized because of excellent breeding programs conducted by commercial propagators and by some horticulture departments. Modern cultivars have traits such as more attractive flower colors and forms, longer-lasting flowers, better growth habit, increased resistance to insects and disease organisms, or ability to grow and flower at cooler night temperatures. *See* BREEDING (PLANT). [R.A.L.]

Flotation A process used to separate particulate solids, which have been suspended in a fluid, by selectively attaching the particles to be removed to a light fluid and allowing this mineralized fluid aggregation to rise to where it can be removed. The principal use of the process is to separate valuable minerals from waste rock, or gangue, in which case the ground ore is suspended in water and, after chemical treatment, subjected to bubbles of air. The minerals which are to be floated attach to the air bubbles, rise through the suspension, and are removed with the froth which forms on top of the pulp. Although most materials subjected to flotation are minerals, applications to chemical and biological materials have been reported. [R.L.At.]

Flow measurement The determination of the quantity of a fluid, either a liquid, vapor, or gas, that passes through a pipe, duct, or open channel. Flow may be expressed as a rate of volumetric flow (such as liters per second, gallons per minute, cubic meters per second, cubic feet per minute), mass rate of flow (such as kilograms per second, pounds per hour), or in terms of a total volume or mass flow (integrated rate of flow for a given period of time).

Flow measurement, though centuries old, has become a science in the industrial age. This is because of the need for controlled process flows, stricter accounting methods, and more efficient operations, and because of the realization that most heating, cooling, and materials transport in the process industries is in the form of fluids, the flow rates of which are simple and convenient to control with valve or variable speed pumps. *See* PROCESS CONTROL.

Measurement is accomplished by a variety of means, depending upon the quantities, flow rates, and types of fluids involved. Many industrial process flow measurements consist of a combination of two devices: a primary device that is placed in intimate contact with the fluid and generates a signal, and a secondary device that translates this signal into a motion or a secondary signal for indicating, recording, controlling, or totalizing the flow. Other devices indicate or totalize the flow directly through the interaction of the flowing fluid and the measuring device that is placed directly or indirectly in contact with the fluid stream.

For discussion of various types of fluid meters *see* ELECTROMAGNETIC FLOWMETER; FLUIDIC-FLOW MEASUREMENT; LASER DOPPLER VELOCIMETER; MASS FLOW RATE METER; METERING ORIFICE; NUCLEAR MAGNETIC RESONANCE FLOWMETER; PITOT TUBE; POSITIVE-DISPLACEMENT FLOWMETER; QUANTITY FLOWMETER; ROTATING IMPELLER FLOWMETER; TARGET FLOWMETER; THERMAL FLOW MEASUREMENT DEVICES; ULTRASONIC FLOW MEASUREMENT; VARIABLE-AREA FLOWMETER; VENTURI TUBE; VOLUME FLOW RATE METER; VORTEX FLOWMETER. [M.Br.; L.P.E.]

Flower A higher plant's sexual apparatus in the aggregate, including the parts that produce sex cells and closely associated attractive and protective parts (Fig. 1). "Flower" as used in this article will be limited, as is usual, to the angiosperms, plants with enclosed seeds and the unique reproductive process called double fertilization. In its most familiar form a flower is made up of four kinds of units arranged concentrically. The green sepals (collectively termed the calyx) are outermost, showy petals (the corolla) next, then the pollen-bearing units (stamens, androecium), and finally the centrally placed seedbearing units (carpels, gynoecium). This is the "complete" flower of early botanists, but it is only one of an almost overwhelming array of floral forms. One or more kinds of units may be lacking or hard to recognize depending on the species, and evolutionary modification has been so great in some groups of angiosperms that a flower cluster (inflorescence) can took like a single flower.

Flora diversity. Most botanical terms are descriptive, and a botanist must have a large store of them to impart the multiformity of flowers. The examples that follow are only a smattering. An extra series of appendages alternating with the sepals, as in purple loosestrife, is an epicalyx. A petal with a broad distal region and a narrow proximal region is said to have a blade and claw: the crape myrtle has such petals. The term perianth,

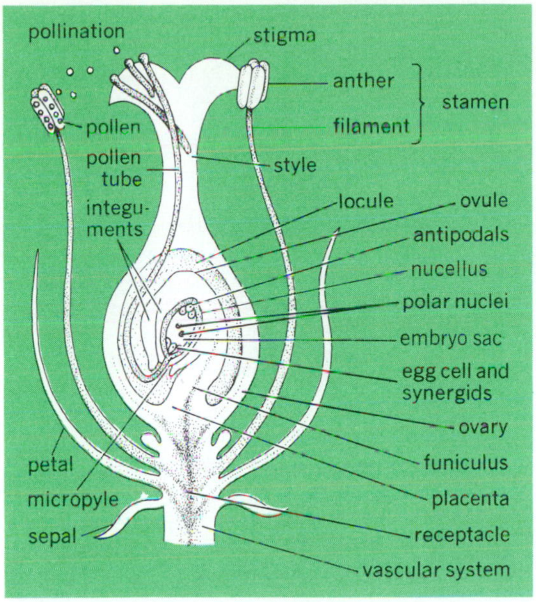

Fig. 1. Flower structure, median longitudinal section.

which embraces calyx and corolla and avoids the need to distinguish between them, is especially useful for a flower like the tulip, where the perianth parts are in two series but are alike in size, shape, and color. The members of such an undifferentiated perianth are tepals. When the perianth has only one series of parts, however, they are customarily called sepals even if they are petallike, as in the windflower.

A stamen commonly consists of a slender filament topped by a four-lobed anther, each lobe housing a pollen sac. In some plants one or more of the androecial parts are sterile rudiments called staminodes: a foxglove flower has four fertile stamens and a staminode. Carpellode is the corresponding term for an imperfectly formed gynoecial unit.

A gynoecium is apocarpous if the carpels are separate (magnolia, blackberry) and syncarpous if they are connate (tulip, poppy). Or the gynoecium may regularly consist of only one carpel (bean, cherry). A solitary carpel or a syncarpous gynoecium can often be divided into three regions: a terminal, pollen-receptive stigma; a swollen basal ovary enclosing the undeveloped seeds (ovules); and a constricted, elongate style between the two. The gynoecium can be apocarpous above and syncarpous below; that is, there can be separate styles and stigmas on one ovary (wood sorrel).

Every flower cited so far has a superior ovary: perianth and androecium diverge beneath it (hypogyny). If perianth and androecium diverge from the ovary's summit, the ovary is inferior and the flower is epigynous (apple, banana, pumpkin). A flower is perigynous if the ovary is superior within a cup and the other floral parts diverge from the cup's rim (cherry). A syncarpous ovary is unilocular if it has only one seed chamber, plurilocular if septa divide it into more than one. The ovules of a plurilocular ovary are usually attached to the angles where the septa meet; this is axile placentation, a placenta being a region of ovular attachment. There are other ways in which the ovules can be attached—apically, basally, parietally, or on a free-standing central placenta—each characteristic of certain plant groups (Fig. 2).

The term bract can be applied to any leaflike part associated with one or more flowers but not part of a flower. Floral bracts are frequently small, even scalelike, but the flowering dogwood has four big petallike bracts below each flower cluster. The broad end of a flower stalk where the floral parts are attached is the receptacle. The same term is used, rather inconsistently, for the broad base bearing the many individual flowers (florets) that make up a composite flower like a dandelion or a sunflower.

Sexuality. A plant species is diclinous if its stamens and carpels are in separate flowers. A diclinous species is monoecious if each plant bears staminate and carpellate (pistillate) flowers, dioecious if the staminate and carpellate flowers are on different plants. The corn plant, with staminate inflorescences (tassels) on top and carpellate inflorescences (ears) along the stalk, is monoecious. Hemp is a well-known dioecious plant.

Nectaries. Flowers pollinated by insects or other animals commonly have one or more nectaries, regions that secrete a sugar solution. A nectary can be nothing more than a layer of tissue lining part of a floral tube or cup (cherry), or it can be as conspicuous as the secretory spur of a nasturtium or a larkspur. It can be a cushionlike outgrowth at the base of a superior ovary (orange blossom) or atop an inferior ovary (parsley family). Gladiolus and a number of other monocotyledons have septal nectaries, deep secretory crevices where the carpels come together. Substances that give off floral odors—essential oils for the most part—ordinarily originate close to the nectar-producing region but are not coincident with it. Production by the epidermis of perianth parts is most common, but in some species the odor emanates from a more restricted region and may even come from a special flap or brush. Most insect-pollinated plants have visual cues, some of them outside the human spectral range, as well as odor to bring the pollinators to the flowers and guide them to the nectar. *See* SECRETORY STRUCTURES (PLANT).

Inflorescence. Inflorescence structure, the way the flowers are clustered or arranged on a flowering branch, is almost as diverse as floral structure. To appreciate this, one need only contrast the drooping inflorescences (catkins) of a birch tree with the coiled flowers of a forget-me-not or with the solitary flower of a tulip. In some cases one kind of inflorescence characterizes a whole plant family. Queen Anne's lace and other members of the parsley family (Umbelliferae) have umbrellalike inflorescences with the flower stalks radiating from almost the same point in a cluster. The stalkless flowers (florets) of the grass family are grouped into clusters called spikelets, and these in turn are variously arranged in different grasses.

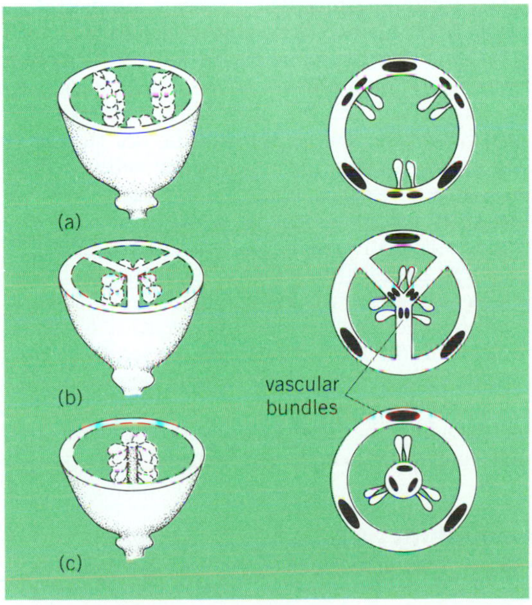

Fig. 2. Placentation: (a) parietal, (b) axile, and (c) free central. (After P. H. Raven, R. F. Evert, and H. Curtis, *Biology of Plants*, Worth Publishers, 1976)

Flowers of the arum family (calla lily, jack-in-the-pulpit), also stalkless, are crowded on a thick, fleshy, elongate axis. In the composite family, florets are joined in a tight head at the end of the axis; the heads of some composites contain two kinds, centrally placed florets with small tubular corollas and peripheral ray florets with showy, strap-shaped corollas (the "petals" one plucks from a daisy). *See* INFLORESCENCE.

Anatomy. Some of the general anatomical features of leaves can be found in the floral appendages. A cuticle-covered epidermis overlies a core of parenchyma cells in which there are branching vascular bundles (solitary bundles in most stamens). Sepal parenchyma and petal parenchyma are often spongy, but palisade parenchyma occurs only rarely in flowers and then only in sepals. As in other parts of the plant, color comes mostly from plastids in the cytoplasm and from flavonoids in the cell sap. Cells of the petal epidermis may have folded side walls that interlock so as to strengthen the tissue. In some species the outer walls of the epidermis are raised as papillae; apparently, this is part of the means of attracting pollinators, for the papillae are light reflectors.

Stamen. As a stamen develops, periclinal divisions in the second cell layer of each of its four lobes start a sequence that will end with the shedding of pollen. The first division makes two cell layers. The outer daughter cells give rise to the wall of the pollen sac, and the inner ones are destined to become pollen after further divisions. When mature, a pollen sac typically has a prominent cell layer just below a less distinctive epidermis. The inner wall and the side walls of an endothecium cell carry marked thickenings, but the outer wall does not. Splitting of the ripe anther is due partly to the way in which these differentially thickened walls react to drying and shrinking and partly to the smaller size of the cells along the line of splitting. *See* POLLEN.

Carpel. Like other floral parts, a carpel is made up of epidermis, parenchyma, and vascular tissue. In addition, a carpel commonly has a special tissue system on which pollen germinates and through which, or along which, pollen tubes are transmitted to the ovules. Most angiosperms have solid styles, and the transmitting tissue is a column of elongate cells whose softened walls are the medium for tubal growth. The epidermis at the stigmatic end of a carpel usually changes to a dense covering of papillae or hairs; the hairs can be unicellular or pluricellular, branched or unbranched. In taxa with hollow styles, the transmitting tissue is a modified epidermis running down the stylar canal. There are two kinds of receptive surfaces, and they are distributed among the monocotyledons and the dicotyledons with taxonomic regularity. One kind has a fluid medium for germinating the pollen, and the other has a dry proteinaceous layer over the cuticle. The proteins of the dry stigmas have a role in the incompatibility reactions that encourage outbreeding.

Ovule. Ovule development usually takes place as the gynoecium forms, but it may be retarded when there is a long interval between pollination and fertilization (oaks, orchids). A typical ovule has a stalk (funiculus), a central bulbous body (nucellus), and one or two integuments (precursors of seed coats), which cover the nucellus except for a terminal pore (micropyle). Orientation of the ovule varies from group to group. It can be erect on its stalk or bent one way or another to differing degrees. There are also taxonomic differences in the extent to which the ovule is vascularized by branches from the gynoecial vascular system. *See* FRUIT; REPRODUCTION (PLANT).

[R.H.E.]

Fluid coupling A device for transmitting rotation between shafts by means of the acceleration and deceleration of a hydraulic fluid. Structurally, a fluid coupling consists of an impeller on the input or driving shaft and a runner on the output or driven shaft. The two contain the fluid (see illustration). Impeller and runner are bladed rotors, the impeller acting as a pump and the runner reacting as a turbine. Basically, the impeller accelerates the fluid from near its axis, at which the tangential component of absolute velocity is low, to near its periphery, at which the tangential component of absolute velocity is high. This increase in velocity represents an increase in kinetic energy. The fluid mass emerges at high velocity from the impeller, impinges on the runner blades, gives up its energy, and leaves the runner at low velocity. *See* HYDRAULICS.

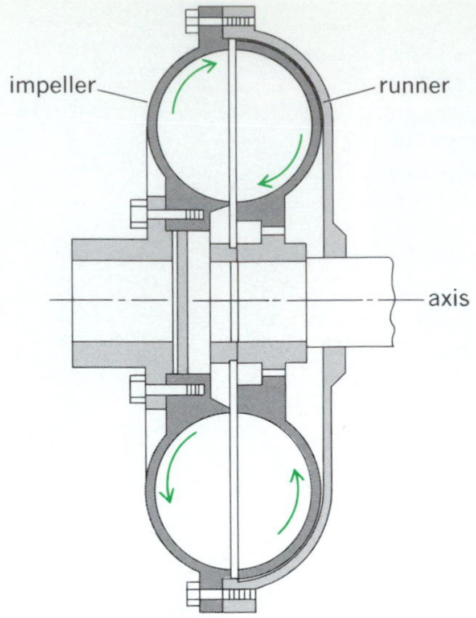

Basic fluid coupling.

[H.J.Wir.]

Fluid dynamics The science of fluids in motion. Fluid dynamics attempts to describe the motion of a fluid as it is displaced and deformed by the action of moving or fixed boundaries. Fluid dynamics may be divided into two parts, hydrodynamics and aerodynamics. *See* FLUID MECHANICS; GAS DYNAMICS; HYDRODYNAMICS.

[V.L.S.]

Fluid flow Motion of a fluid as a continuum. Fluids flow whereas solids move as bodies. In flow, the individual particles of a substance move relative to each other as well as to their surroundings. A fluid is a substance that flows under the slightest stress. Thus, the pressure of a gas is sufficient to cause it to flow throughout a container to which it is admitted. The weight of a liquid is sufficient to cause it to flow in a container but without significant change in volume.

A fluid may be liquid, vapor, or gas. A liquid will fill the container which holds it, but it may have a free surface, that is, a surface from which all pressure is removed except that of its own vapor. All liquids are relatively incompressible. *See* FLUIDS.

A vapor is a gas whose temperature and pressure are such that it is very near the liquid phase. Thus, steam is considered to be a vapor because its state is not far from that of water.

A gas may be defined as a highly superheated vapor; that is, its state is far removed from the liquid phase. Thus, air is considered to be a gas because its state is normally very far from that of liquid air. A gas is very compressible, and when all external pressure is removed, it tends to expand indefinitely. A gas is therefore in equilibrium only when it is completely enclosed.

The volume of a liquid is altered only slightly by changes in either pressure or temperature unless the temperature change is considerable or unless the initial temperature is near the critical temperature, but the volume of a gas or a vapor is greatly affected by changes in either pressure or temperature. *See* GAS; LIQUID.

When there is no friction between adjacent moving particles, flow is termed ideal; that is, viscosity is zero. In ideal flow, internal forces at any section are always normal to the section. Forces are purely pressure forces. Such flow is approached but never achieved in reality.

In a real fluid, tangential or shearing forces always come into being whenever motion takes place, thus giving rise to fluid friction, because these forces oppose the sliding of one particle past another. These friction forces are due to a property called viscosity. *See* VISCOSITY.

All liquids are relatively incompressible, and for most purposes water in particular can be treated as incompressible; yet it is 10 times as compressible as steel. The passage through water of a sound wave, which is really a pressure wave, is a result of the water's compressibility.

In dealing with the flow of air or other gases when the change in pressure is small, so that the change in density is negligible, even gases can be treated as incompressible. For an airplane flying at speeds of less than 250 mi/h (110 m/s) the air can be considered to be of constant density. However, as the speed approaches that of sound in air, which is of the order of 700 mi/h (310 m/s) the pressure of the air adjacent to the body becomes materially different from that at some distance away, and air must then be considered to be compressible. *See* COMPRESSIBLE FLOW. For various types of fluid flow *see also* BOUNDARY-LAYER FLOW; IRROTATIONAL FLOW; ISENTROPIC FLOW; LAMINAR FLOW; NON-NEWTONIAN FLUID; ROTATIONAL FLOW; STEADY FLOW; TURBULENT FLOW; UNIFORM FLOW; UNSTEADY FLOW; WAKE FLOW. [R.L.D.]

Fluid mechanics The science concerned with fluids, either at rest or in motion. It deals with pressures, velocities, and accelerations in the fluid, including fluid deformation and compression or expansion. Fluid mechanics may be divided into two branches, fluid statics and fluid dynamics; the first deals with pressure intensities and forces exerted by a fluid at rest, and the second with forces exerted on fluids and their resulting motions. *See* FLUID DYNAMICS; FLUID STATICS.

The laws of fluid mechanics control a great portion of natural phenomena. The flight of an insect or bird, the motion of a fish through water, the relative movement of air masses as in frontal weather systems, and the eruption of a volcano are examples of flow that follow the laws of fluid mechanics. The science of fluid mechanics is involved in many phases of aeronautical, chemical, civil, and mechanical engineering. It requires the combination of theoretical analysis and orderly experimentation. *See* FLUID FLOW. [V.L.S.]

Fluid statics The determination of pressure intensities and forces exerted by liquids and bases at rest. Hydrostatics, although implying the statics of water alone, applies to liquids in general. By definition, a fluid at rest cannot sustain a shear stress; therefore, a fluid force exerted on an element of boundary area must act normal to the area. For example, consider the small free body of fluid shown in the illustration. Equilibrium requires that $dp = -\gamma dz$, in which p is the absolute pressure, γ is the specific weight (weight of a unit volume of fluid), and z is the elevation, measured vertically upward. For liquids, γ is substantially constant, and the equation shows that pressure decreases linearly as the elevation increases. *See* HYDROSTATICS.

With gases, the variation of γ with pressure or elevation must be known in order to integrate the equation. For an isothermal gas ($p/\gamma = p_0\gamma_0$ = constant), the pressure variation with elevation is given by the equation below, in which p_0 and

$$p = p_0 \exp[-(z - z_0)/(p_0/\gamma_0)]$$

γ_0 are the values of p and γ at elevation z_0.

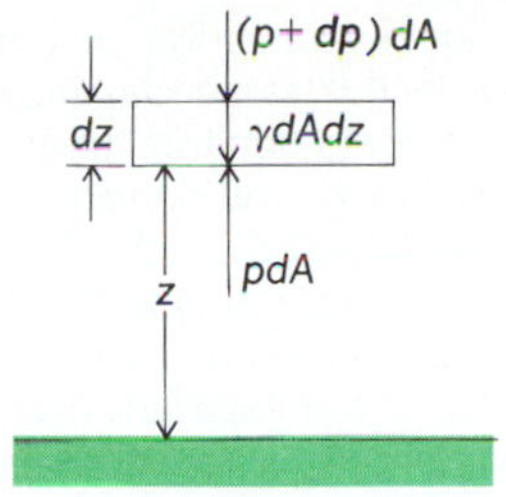

Free-body diagram for vertical forces acting on fluid element.

Fluid forces on plane surfaces may be determined by integration of $p\,dA$ over the surface, with p the pressure intensity and dA an element of the surface area. For liquids, the magnitude of the force is the product of the area and the pressure at the centroid of the area. *See* ARCHIMEDES' PRINCIPLE; BUOYANCY; FLUID MECHANICS. [V.L.S.]

Fluidic-flow measurement Use of fluidic phenomena to measure the rate of fluid flow. Two types of flowmeters based on such phenomena have been developed: the fluidic oscillator meter and the fluidic flow sensor. *See* FLUIDICS.

The fluidic-oscillator meter works on the principle of the Coanda effect, the tendency of fluid coming out of a jet to follow a wall contour. The fluid entering the meter will attach to one of two opposing diverging side walls. A small portion of the stream is split off and channeled back through the feedback passage on that side to force the incoming stream to attach to the other side wall. The frequency of the oscillation is directly proportional to the volume flow through the meter. A sensor detects the oscillations and transmits the signal. This principal is used in both liquid and gas flowmeters.

The fluidic-flow sensor (also known as a deflected-jet fluidic flowmeter) is used in gas flows, particularly in dirty environments. It consists of a jet of air or other selected gas directed back from the outer nozzle onto two adjacent small openings. The flow of the gas being measured will deflect the jet to change the relative pressure on the two ports and thereby give a signal corresponding to the gas velocity. *See* FLOW MEASUREMENT. [M.Br.; L.P.E.]

Fluidics A body of techniques which use flowing fluids as a signal-bearing medium and fluid dynamics phenomena as a basis for signal switching and amplification. Most work and interest in the field dates between 1960 and 1968. The work has natural analogies to electronics, electronic devices such as transistors, and even electronic circuit board technology. *See* AMPLIFIER; ELECTRONICS; LOGIC CIRCUITS; TRANSISTOR.

Fluidic devices are capable of switching speeds in the kilohertz range. They function with a variety of fluids and can be included directly in hydraulic designs without transducers. In principle, fluidics is appropriate to applications where electronics break down, as in high electronic interference or nuclear radiation. The relative disadvantages of the devices are their potential to clog and the increasing availability of very-high-speed inexpensive electronic computation. For these latter reasons, the early interest has diminished. *See* FLUID DYNAMICS; HYDRAULICS. [E.H.Br.]

Fluidization The processing technique employing a suspension or fluidization of small solid particles in a vertically rising stream of fluid—usually gas—so that fluid and solid come into intimate contact. This is a tool with many applications in the petroleum and chemical process industries. Suspensions of solid particles by vertically rising liquid streams are of lesser interest in modern processing, but have been shown to be of use, particularly in liquid contacting of ion-exchange resins. However, they come in this same classification and their use involves techniques of liquid settling, both free and hindered (sedimentation), classification, and density flotation. *See* ION EXCHANGE; MECHANICAL CLASSIFICATION.

The interrelations of hydromechanics, heat transfer, and mass transfer in the gas-fluidized bed involve a very large number of factors. Because of the excellent contacting under these conditions, numerous chemical reactions are also possible—either between solid and gas, two fluidized solids with each other or with the gas, or most important, one or more gases in a mixture with the solid as a catalyst. In the usual case, the practical applications in plants have far outrun the exact understanding of the physical, and often chemical, interplay of variables within the minute ranges of each of the small particles and the surrounding gas phase.

With such excellent opportunities for heat and mass transfer to or from solids and fluids, fluidization has become a major tool in such fields as drying, roasting, and other processes involving chemical decomposition of solid particles by heat. An important application has been in the catalysis of gas reactions, wherein the excellent opportunity of heat transfer and mass transfer between the catalytic surface and the gas stream gives performance unequaled by any other system. *See* CATALYSIS; CRACKING; FLUIDIZED-BED COMBUSTION; GAS ABSORPTION OPERATIONS; HEAT TRANSFER; MASS-TRANSFER OPERATION; UNIT OPERATIONS.

[D.F.O.]

Fluidized-bed combustion A method of burning fuel in which the fuel is continually fed into a bed of reactive or inert material while a flow of air passes up through the bed, causing it to act like a turbulent fluid. Fluidized beds have long been used for the combustion of low-quality, difficult fuels and have become a rapidly developing technology for the clean burning of coal. *See* FLUIDIZATION.

A fluidized-bed combustor is a furnace chamber whose floor is slotted, perforated, or fitted with nozzles. Air is forced through the floor and upward through the chamber. The chamber is partially filled with particles of either reactive or inert material, which will fluidize at an appropriate air flow rate. When fluidization takes place, the bed of material expands (bulk density decreases) and exhibits the properties of a liquid. As air velocity increases, the particles mix more violently, and the surface of the bed takes on the appearance of a boiling liquid. If air velocity were increased further, the bed material would be blown away.

Once the bed is fluidized, its temperature can be increased with ignitors until a combustible material can be injected to burn within the bed. Proper selection of air velocity, operating temperature, and bed material will cause the bed to act as a chemical reactor. The three broad areas of application of fluidized-bed combustion are incineration, gasification, and steam generation. *See* COAL GASIFICATION; COMBUSTION; GAS TURBINE; STEAM-GENERATING UNIT.

[M.Po.]

Fluids Substances with no reference configuration of permanent significance. Aggregates of matter in which the molecules are able to flow past each other without limit and without fracture planes forming are usually classified as fluids. The subdivisions of fluids known as gases, vapors, and liquids, each of which exhibits successively closer association of the molecules and is distinguished by different thermodynamic and mechanical properties, compose a group of easily distinguishable fluids, some having mainly newtonian and the rest mainly non-newtonian flow properties.

There are some substances, the semisolids, which appear to be able to flow without fracture and yet which also have some of the attributes of solids such as the ability to form freestanding figures. Butter in a temperate environment is an excellent example of a material with this dual nature.

Similarly, some fluids possess elastic properties, the so-called viscoelastic fluids. Such substances are not easy to classify in one or another category, and their properties are not well understood in depth. *See* GAS; LIQUID; NON-NEWTONIAN FLUID.

[J.Har.; W.L.Wi.]

Fluke Any of more than 40,000 species of parasitic flatworms that form the class Trematoda in the phylum Platyhelminthes. The class can be separated into two sharply demarcated subclasses: Monogenea, the majority of which are ectoparasites of cold-blooded vertebrates; and Digenea, the largest group of flatworms, endoparasites of vertebrates and invertebrates. *See* DIGENEA; MONOGENEA; TREMATODA. [C.B.C.]

Fluoborate In the broadest sense, the term fluoborate refers to a group of compounds related to the borates in which one or more oxygens have been replaced by fluorines. In a strict sense, fluoborate refers to the BF_4^- ion which is derived from tetrafluoroboric acid, HBF_4. The fluoborates are used as fluxes and in electroplating. *See* BORATE; FLUORINE. [E.E.W.]

Fluorescence Fluorescence is generally defined as a luminescence emission that is caused by the flow of some form of energy into the emitting body, this emission ceasing abruptly when the exciting energy is shut off. In attempts to make this definition more meaningful it is often stated, somewhat arbitrarily, that the decay time, or afterglow, of the emission must be of the order of the natural lifetime for allowed radiative transitions in an atom or a molecule, which is about 10^{-8} s for transitions involving visible light. Perhaps a better distinction between fluorescence and its counterpart, phosphorescence, rests not on the magnitude of the decay time per se, but on the criterion that the fluorescence decay is temperature-independent.

In the literature of organic luminescence, the term fluorescence is used exclusively to denote a luminescence which occurs when a molecule makes an allowed optical transition. Luminescence with a longer exponential decay time, corresponding to an optically forbidden transition, is called phosphorescence, and it has a different special distribution from the fluorescence. *See* PHOSPHORESCENCE.

The decay time of fluorescent materials varies widely, from the order of 5×10^{-9} s for many organic crystalline materials up to 2 s for the europium-activated strontium silicate phosphor. Fluorescent materials with decay times between 10^{-9} and 10^{-7} s are used to detect and measure high-energy radiations, such as x-rays and gamma rays, and high-energy particles such as alpha particles, beta particles, and neutrons. These agents produce light flashes (scintillations) in certain crystalline solids, in solutions of many polynuclear aromatic hydrocarbons, or in plastics impregnated with these hydrocarbons. The so-called fluorescent lamps employ the luminescence of gases and solids in combination to produce visible light. *See* ABSORPTION; FLUORESCENT LAMP; LUMINESCENCE. [J.H.S.; C.C.K.]

Fluorescence microscope A variation of the compound laboratory light microscope which is arranged to transmit ultraviolet, violet, and sometimes blue radiations to a specimen. The specimen then fluoresces, that is, appears to be self-luminous and often colored. The phenomenon of fluorescence is thought to involve an electronic rearrangement in the irradiated substance. *See* FLUORESCENCE; OPTICAL MICROSCOPE.

Materials with absorption-spectra maxima below 320 nanometers require a quartz condenser in place of the usual glass condenser of the microscope. The microscope should have an aluminized front-surface mirror because silver is a less efficient reflector of ultraviolet.

The fluorescent image looks bright and has good contrast, although the amount of light is small. With weak fluorescence a monocular microscope must be used and the work is done in a darkened room. With more intense fluorescence a binocular microscope should be coated to decrease loss of light.

The absorbing filter is usually in the microscope between the objective and the observer's eye to remove any other exciting radiation not absorbed by the specimen. This may be a nearly colorless filter absorbing only ultraviolet when the specimen is to be observed in full color, or it may be of a color complementary to the radiation used. For example, with materials absorbing blue and ultraviolet radiation a yellow filter absorbs the blue radiation beyond the specimen and passes the yellow fluorescence of the specimen to the eye. The cross-filter combination is chosen to give the best visibility for the specimen. [O.W.R.]

Fluorescence spectroscopy A technique of fluorometric analysis in which the emitted fluorescent light is passed through a monochromator so that the fluorescence emission spectrum can be recorded. It is possible to measure several fluorescing compounds in the same solution, provided that they have sufficiently different fluorescence emission spectra. In obtaining this selectivity, some sensitivity is lost since some fluorescent light is lost in the monochromator, and only a small portion of the total fluorescent energy emitted is measured at any one wavelength. *See* FLUORESCENCE; OPTICAL METHODS OF CHEMICAL ANALYSIS. [R.F.G.; J.N.L.]

Fluorescent lamp A lamp which produces light largely by conversion of ultraviolet energy from a low-pressure mercury arc to visible light. Chemicals that absorb radiant energy of a given wavelength and reradiate at longer wavelengths are called phosphors. The phosphors produce most of the light provided by fluorescent lamps.

The fluorescent lamp consists of a glass tube containing two electrodes, a coating of powdered phosphor, and small amounts of mercury. The glass tube seals the inner parts of the lamps from the atmosphere. The electrodes provide a source of free electrons to initiate the arc and are connected to the external circuit through the ends of the lamp. The phosphor is a chemical or mixture of chemicals that converts shortwave ultraviolet energy into light. The mercury, when vaporized in the arc, produces the ultraviolet radiation that causes fluorescence. Argon gas, introduced in small quantities, provides the ions that facilitate starting of the lamp (see illustration). *See* FLUORESCENCE.

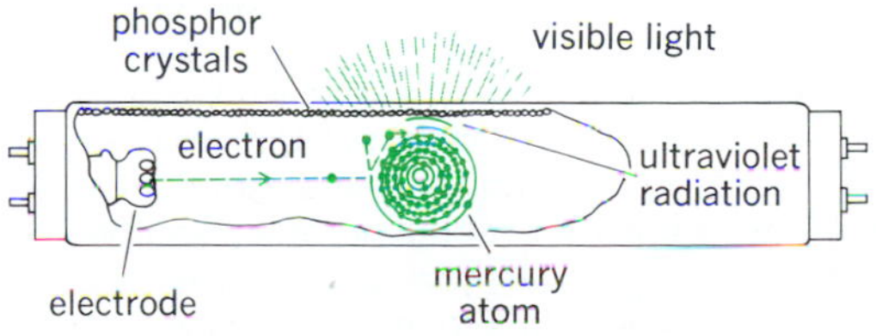

Parts of a typical fluorescent lamp. (*Westinghouse Lighting Handbook*)

Fluorescent lamps are usually operated on ac circuits with a frequency of 60 Hz. However, higher frequencies permit higher-efficiency lamp operation with simpler ballasts of lower power dissipation per watt of lamp. Consequently, systems have been developed for the operation of fluorescent lamps at frequencies from 360 to 3000 Hz. These systems employ various types of frequency converters to obtain the high-frequency power.

Fluorescent lamps provide light at several times the efficiency of incandescent lamps, the exact ratio depending on the fluorescent lamp color. Lamp color is determined by the selection of chemicals used in the phosphors. Several types of essentially white fluorescent lamps are available commercially, as well as a range of tinted and saturated colors. [A.M.]

Fluoride A compound derived from hydrofluoric acid, HF, in which the fluorine atom is in the 1− oxidation state. The small size of the fluoride ion means that the extra electron is held tightly; it is difficult to convert the fluoride ion to free fluorine. The small size also means that there is less tendency to form covalent bonds in simple salts such as mercuric fluoride, HgF_2, which has an ionic lattice, whereas its chlorine analog, mercuric chloride, $HgCl_2$, has a molecular crystal. In direct contrast to this, many fluorides are known in which the central atom is in a higher oxidation state than that observed for other halides and these compounds are often covalent, for example, vanadium pentafluoride, VF_5, chromium tetrafluoride, CrF_4, bismuth pentafluoride, BiF_5, and sulfur hexafluoride, SF_6. *See* FLUOBORATE; FLUOROCARBON; FLUOSILICATE; HALOGENATED HYDROCARBON. [E.E.W.]

Fluorine A chemical element, symbol F, atomic number 9, the member of the halogen family that has the lowest atomic number and atomic weight. Although only the isotope with atomic weight 19 is stable, the other, radioactive isotopes between atomic weight 17 and 22 have been artificially prepared. Fluorine is the most electronegative element, and by a substantial margin the most chemically energetic of the nonmetallic elements.

Properties. The element fluorine is a pale yellow gas at ordinary temperatures. The odor of the element is somewhat in doubt. Some physical properties are listed in the table. The reactivity of the element is so great that it will react readily at ordinary temperatures with many other elementary substances, such as sulfur, iodine, phosphorus, bromine, and most metals. Since the products of the reactions with the nonmetals are in the liquid or gaseous state, the reactions continue to the complete consumption of the fluorine, frequently with the evolution of considerable heat and light. Reactions with the metals usually form a protective metallic fluoride which blocks further reaction, unless the temperature is raised. Aluminum, nickel, magnesium, and copper form such protective fluoride coatings.

Fluorine reacts with considerable violence with most hydrogen-containing compounds, such as water, ammonia, and all organic chemical substances whether liquids, solids, or gases. The reaction of fluorine with water is very complex, yielding mainly hydrogen fluoride and oxygen with less amounts of hydrogen peroxide, oxygen difluoride, and ozone. Fluorine displaces other nonmetallic elements from their compounds, even those nearest fluorine in chemical activity. It displaces chlorine

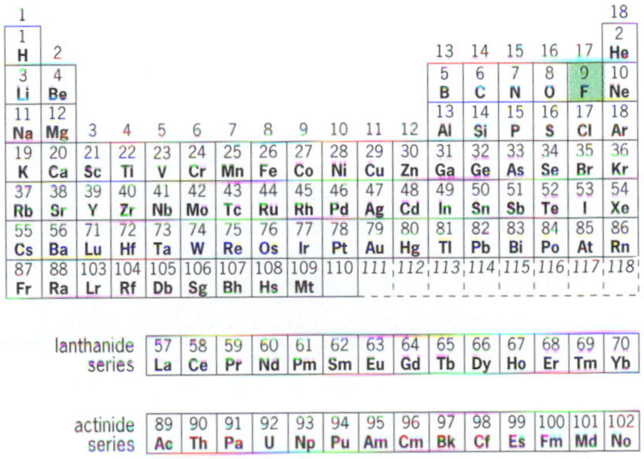

Physical properties of fluorine

Property	Value
Atomic weight	18.998403
Boiling point, °C	−188.13
Freezing point, °C	−219.61
Critical temperature, °C	−129.2
Critical pressure, atm*	55
Density of liquid at b.p., g/ml	1.505
Density of gas at 0°C + 1 atm*, g/liter	1.696
Dissociation energy, kcal/mol	36.8
Heat of vaporization, cal/mol	1510
Heat of fusion, cal/mol	121.98
Transition temperature (solid), °C	−227.61

*1 atm = 101.325 kilopascals.

from sodium chloride, and oxygen from silica, glass,, and some ceramic materials. In the absence of hydrofluoric acid, however, fluorine does not significantly etch quartz or glass even after several hours at temperatures as high as 390°F (200°C).

Fluorine is a very toxic and reactive element. Many of its compounds, especially inorganic, are also toxic and can cause severe and deep burns. Care must be taken to prevent liquids or vapors from coming in contact with the skin or eyes.

Natural occurrence. At an estimated 0.065% of the Earth's crust, fluorine is roughly as plentiful as carbon, nitrogen, or chlorine, and much more plentiful than copper or lead, though much less abundant than iron, aluminum, or magnesium. Compounds whose molecules contain atoms of fluorine are widely distributed in nature. Many minerals contain small amounts of the element, and it is found in both sedimentary and igneous rocks.

Uses. Fluorine-containing compounds are used to increase the fluidity of melts and slags in the glass and ceramic industries. Fluorspar (calcium fluoride) is introduced into the blast furnace to reduce the viscosity of the slag in the metallurgy of iron. Cryolite, Na_2AlF_6, is used to form the electrolyte in the metallurgy of aluminum. Aluminum oxide is dissolved in this electrolyte, and the metal is reduced electrically from the melt. The use of halocarbons containing fluorine as refrigerants was patented in 1930, and these volatile and stable compounds found a market in aerosol propellants as well as in refrigeration and air-conditioning systems. However, use of fluorocarbons as propellants has declined sharply because of concern over their possible damage to the ozone layer of the atmosphere. A use for fluorine that became prominent during World War II is in the enrichment of the fissionable isotope ^{235}U; the most important process employed uranium hexafluoride. This stable, volatile compound was by far the most suitable material for isotope separation by gaseous diffusion.

While consumers are mostly unaware of the fluorine compounds used in industry, some compounds have become familiar to the general public through minor but important uses, such as additive to toothpaste and nonsticking fluoropolymer surfaces on frying pans and razor blades (for example, Teflon).

Compounds. In all fluorine compounds the high electronegativity of this element suggests that the fluorine atom has an excess of negative charge. It is convenient, however, to divide the inorganic binary fluorides into saltlike (ionic lattice) nonvolatile metallic fluorides and volatile fluorides, mostly of the nonmetals. Some metal hexafluorides and the noble-gas fluorides show volatility that is frequently associated with a molecular compound. Volatility is often associated with a high oxidation number for the positive element.

The metals characteristically form nonvolatile ionic fluorides where electron transfer is substantial and the crystal lattice is determined by ionic size and the predictable electrostatic interactions. When the coordination number and valence are the same, for example, BF_3, SiF_4, and WF_6, the binding between metal and fluoride is not unusual, but the resulting compounds are very volatile, and the solids show molecular lattices rather than ionic lattice structures. For higher oxidation numbers, simple ionic lattices are less common and, while the bond between the central atom and fluorine usually still involves transfer of some charge to the fluorine, molecular structures are identifiable in the condensed phases.

In addition to the binary fluorides, a very large number of complex fluorides have been isolated, often with a fluoroanion containing a central atom of high oxidation number. The binary saltlike fluorides show a great tendency to combine with other binary fluorides to form a large number of complex or double salts.

The fluorine-containing compounds of carbon can be divided into fluorine-containing hydrocarbons and hydrocarbon derivatives (organic fluorine compounds) and the fluorocarbons and their derivatives. The fluorine atom attached to the aromatic ring, as in fluorobenzene, is quite unreactive. In addition, it reduces the reactivity of the molecule as a whole. Dyes, for example, that contain fluorine attached to the aromatic ring are more resistant to oxidation and are more light-fast than dyes that do not contain fluorine. Most aliphatic compounds, such as the alkyl fluorides, are unstable and lose hydrogen fluoride readily. These compounds are difficult to make and to keep and are not likely to become very important. *See* FLUOROCARBON; HALOGEN ELEMENTS. [I.S.]

Fluorite A mineral of composition CaF_2. It is the most abundant fluorine-bearing mineral, and occurs as cubes or compact masses and more rarely as octahedra with complex modifications. Fluorite has a perfect octahedral cleavage, hardness 4 (Mohs scale), and specific gravity 3.18. The color is extremely variable, the most common being green and purple; but fluorite may also be colorless, white, yellow, blue, or brown. Colors may result from the presence of impurity ions. Fluorite frequently emits a blue-to-green fluorescence under ultraviolet radiation, especially if rare-earth or hydrocarbon material is present. Some fluorites are thermoluminescent; that is, they emit light when heated.

Fluorite occurs as a typical hydrothermal vein mineral with quartz, barite, calcite, sphalerite, and galena. Crystals of great beauty from Cumberland, England, and Rosiclare, Illinois, are highly prized by mineral fanciers. It also occurs as a metasomatic replacement mineral in limestones and marbles. [P.B.M.]

Fluorocarbon An organic compound containing fluorine directly bonded to carbon. Simple fluorocarbons are made by the liquid-phase reaction of carbon tetrachloride with hydrogen fluoride catalyzed by antimony trifluoride. CCl_3F is employed mainly as an aerosol propellant, and CCl_2F_2 is used as a refrigerant. They are low-boiling, odorless, nonflammable gases. Higher fluorocarbons are nontoxic. For example, tetrafluoroethylene, CF_2CF_2, Undergoes polymerization in the presence of free-radical initiators to form the highly inert polytetrafluoroethylene.

Consumption of the fluorocarbon aerosol propellants has been declining because of concern over their possible effect on the stratospheric ozone layer, and there is a government prohibition for certain applications, such as deodorants, hairsprays, perfumes, polishes, some insecticides, and food products. The largest use for these products is as a refrigerant. Some fluorocarbons are used as blowing agents in the manufacture of both flexible and rigid polyurethane foams and in polystyrene and polyethylene foams. Almost all the remaining fluorocarbons are used as grease-removing solvents, especially in electronic and aerospace industries, and for the production of fluoropolymers. *See* FREON; HALOGENATED HYDROCARBON; POLYFLUOROOLEFIN RESINS. [F.W.]

Fluosilicate The SiF_6^{2-} ion which is derived from fluosilicic acid, H_2SiF_6. The fluosilicates are prepared from solutions of fluosilicic acid. The salts of barium, potassium, rubidium, cesium, and sodium are only sparingly soluble, whereas many of the heavy-metal salts are quite soluble.

Sodium fluosilicate, Na_2SiF_6, is used in the fluoridation of water supplies and in laundry scours, enamel frits, insecticides, and wood preservatives. *See* FLUORINE. [E.E.W.]

Flutter (aeronautics) An aeroelastic self-excited vibration with a sustained or divergent amplitude, which occurs when a structure is placed in a flow of sufficiently high velocity. Flutter is an instability that can be extremely violent. At low speeds, in the presence of an airstream, the vibration modes of an aircraft are stable; that is, if the aircraft is disturbed, the ensuing motion will be damped. At higher speeds, the effect of the airstream is to couple two or more vibration modes such that the vibrating structure will extract energy from the airstream. The coupled vibration modes will remain stable as long as the extracted energy is dissipated by the internal damping or friction of the structure. However, a critical speed is reached when the extracted energy equals the amount of energy that the structure is capable of dissipating, and a neutrally stable vibration will persist. This is called the flutter speed. At a higher speed, the vibration amplitude will diverge, and a structural failure will result. *See* AEROELASTICITY.

Aircraft manufacturers now have engineering departments whose primary responsibility is fluffer safety. Modern flutter analyses involve extensive computations, requiring the use of large-capacity, high-speed digital computers. Flutter engineers contribute to the design by recommending stiffness levels for the structural components and control surface actuation systems and weight distributions on the lifting surfaces, so that the aircraft vibration characteristics will not lead to flutter within the design speeds and altitudes. *See* AIRFRAME; WING. [W.P.R.]

Flutter and wow All sound recording and reproducing processes employ mechanical systems for the translation of electric waves or sound waves into and from the permanent record. If the speeds of either or both the recording mechanism and reproducing mechanism vary at some rate, then a constant frequency impressed upon the recorder will not be reproduced as a constant frequency by the reproducer. The frequency of the tone will vary above and below the normal pitch depending upon the extent of the speed variation, and in frequency at the rate of the speed variation. The manifestation of this phenomenon is termed flutter. The term wow is used to designate this phenomenon for lower rates of flutter, such as those encountered in phonograph recording.

Flutter and wow are produced in disk recording systems by variations in the rotational speed of the recording and reproducing turntables and by a nonsymmetrically located center hole in the disk record. Flutter and wow are produced in tape and film reproducers by a nonuniform rotational speed of the capstan in the recording and reproducing machines, by nonuniform tension in the take-up and pay-off reel systems, by the variety of idlers, and by a mechanical distortion of the tape in the recording or reproducing process. *See* DISK RECORDING; MAGNETIC RECORDING. [H.F.O.]

Fluvial sediments Deposits formed on land from the sediment load carried by rivers. Many such deposits are ephemeral and are reeroded as the river continues to degrade its bed. Geologically more significant deposits are those preserved in areas of net aggradation (vertical sediment accretion), where the sedimentary basin subsides at a rate faster than downward erosion.

Fluvial sediments range in grain size from boulder conglomerates to the finest mud and contain a variety of internal structures. The nature of any deposit depends on the morphology of the river which forms it, the river morphology itself being dependent on climate, discharge (amount and variability), sediment load, and slope. *See* DELTA; FLOODPLAIN; STREAM TRANSPORT AND DEPOSITION.

Many deposits of nonrenewable fuels and minerals occur in fluvial sediments. Placer gold, uranium, and diamond deposits (in the Yukon and South Africa, for example) consist of detrital fragments transported and concentrated on riverbeds. Porous gravel and sand units make excellent aquifers and may be conduits and traps for migrating fluids such as oil and gas (as in the Gulf Coast and Alberta Province) or metallic mineralizing fluids, including copper, tin, and uranium (as in Texas and the Rocky Mountain region). Coal is a floodplain deposit which typically occurs in coastal river and delta systems. [A.D.M.]

Fluxional compounds Molecules that undergo rapid intramolecular rearrangements among equivalent structures in which the component atoms are interchanged. The rearrangement process is usually detected by nuclear magnetic resonance (NMR) spectroscopy. With sufficiently rapid rates, a single resonance is observed in the NMR spectrum for a molecule that might be expected to have several nonequivalent nuclei on the basis of its instantaneous structure.

Within organic chemistry, degenerate Cope rearrangements represented some of the first examples of interconversions between equivalent structures, but these were relatively slow. The rate of this rearrangement is rapid in more complex molecules. The epitome of degeneracy is reached in bullvalene, which has more than 1,200,000 equivalent structures and rapidly interconverts among them.

Fluxional molecules are frequently encountered in organometallic chemistry, and rapid rearrangements which involve migrations about unsaturated organic rings are commonly observed. The best known (called ring-whizzers) are cyclopentadienyl and cyclooctatetraene complexes of iron.

Inorganic structures also exhibit fluxional phenomena, and five-coordinate complexes provide the greatest number of well-known examples, one being phosphorus pentafluoride (PF_5).

The rearrangement of PF_5 involves interconversion of the trigonal bipyramidal molecule to a square pyramidal configuration and back. If two such nonequivalent structures are present in observable concentrations but interconvert rapidly to cause averaging in the NMR experiments, they are said to be stereochemically nonrigid. This term is generally taken to embrace all compounds that undergo rapid reversible intramolecular rearrangements. Thus, fluxional compounds are a subset of nonrigid compounds with equivalent structures. Nonequivalent structures, that is, tautomers, might be stereochemically nonrigid if they rearranged rapidly, but would not be considered fluxional.

Some workers prefer to reserve the term fluxional for molecules in which bonds are broken and reformed in the rearrangement process. Hence, of the examples above, only bullvalene and the iron complexes would be termed fluxional, whereas all would be considered stereochemically nonrigid. *See* NUCLEAR MAGNETIC RESONANCE (NMR); RESONANCE (MOLECULAR STRUCTURE); TAUTOMERISM. [J.W.F.]

Fluxmeter An instrument designed to measure magnetic flux. The fluxmeter is essentially a moving-coil ballistic galvanometer with a long period and restoring torque reduced to minimum. The low restoring torque (ideally zero) may be obtained by substituting a silk fiber for the usual metallic strip and making connection to the moving coil by light metallic spirals designed to exert negligible control of the motion. *See* GALVANOMETER; MAGNETIC FLUX.

A search coil is connected to the fluxmeter, as shown in the illustration, and the change in the flux that produces the induced emf occurs within the search coil. This change may be

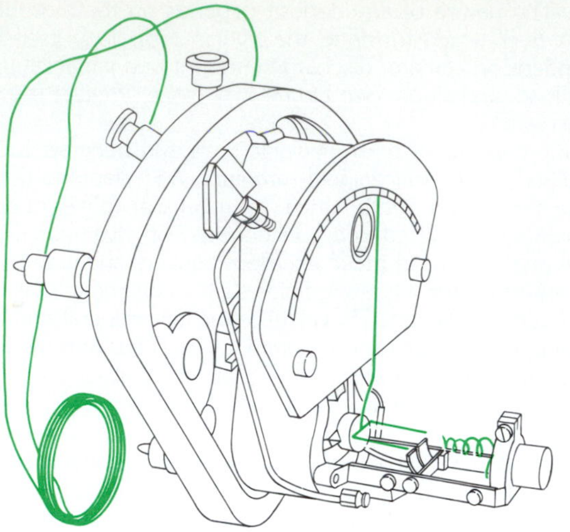

Fluxmeter, with search coil connected. (*After F. A. Laws, Electrical Measurements, 2d ed., McGraw-Hill, 1938*)

caused by a change in the flux that links a stationary coil or by a motion of the search coil in the field so that there is a change in the flux linkage of the coil. *See* SEARCH COIL. [K.V.M.]

Fly A member of the insect order Diptera. Insects of other orders are popularly called flies; mayflies, stoneflies, dragonflies, dobsonflies, and caddis flies all have four wings and therefore are not true flies. Mosquitoes, gnats, and midges all have two wings and are therefore also true flies, order Diptera. Hindwings of all Diptera are greatly reduced balancing organs called halteres.

Flies are a numerous and diverse lot, with over 85,000 described species worldwide. North America has 16,000 species belonging to 107 families. Flies are old, their fossils dating from the Triassic, over 200,000,000 years ago.

The flies of greatest importance to humanity are those that suck blood from people or domestic animals. Females of many mosquitoes, deerflies, horseflies, blackflies, and gnats require a meal of blood before producing eggs. When biting, they may introduce pathogenic microorganisms. Mosquitoes may transmit malaria, yellow fever, viral encephalitis, or parasitic roundworms. Tropical blackflies transmit onchocerciasis (river blindness), and sandflies transmit leishmaniasis, a debilitating protozoan infection. Tsetse flies, in which both sexes bite, transmit African sleeping sickness.

Some flies that do not bite may become a nuisance because of their sheer numbers and association with human habitations. The ubiquitous housefly is bothersome and, in unsanitary situations, may contaminate food with the pathogens of hepatitis, polio, cholera, typhoid, or tuberculosis.

Many thousands of flies are predatory, and they doubtless help to suppress populations of insect pests. Especially important are larvae of the family Syrphidae that eat up to 50 aphids per day. Scavenging Diptera are quite important in aiding the quick breakdown of dead animals and plants. *See* DIPTERA; ENTOMOLOGY, ECONOMIC. [D.J.Hor.]

Flyingfish About 65 species of marine fishes which form the family Exocoetidae in the order Atheriniformes. The enlarged pectoral fins are used for gliding; the fish are incapable of true flight. None of the species is economically important.

Flyingfish travel in large schools in the open seas. These fish attain a considerable speed under water, then suddenly rise to the surface and pass into the air, propelled by the large forked caudal fin. The pectoral fins, which are pressed to the body when in water, are spread out to form a gliding surface. Some species may glide as far as 400 yd (365 m), but always near the surface of the water. *See* ATHERINIFORMES. [C.B.C.]

Flywheel A rotating mass used to maintain the speed of a machine between given limits while the machine releases or receives energy at a varying rate. A flywheel is an energy storage device. It stores energy as its speed increases, and gives up energy as the speed decreases. The specifications of the machine usually determine the allowable range of speed and the required energy interchange.

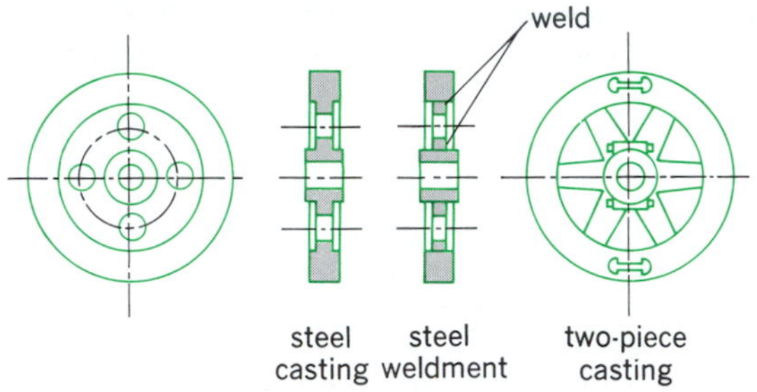

Typical flywheel structures.

The difficulty of casting stress-free spoked flywheels leads the modern designer to use solid web castings or welded structural steel assemblies. For large, slow-turning flywheels on heavy-duty diesel engines or large mechanical presses, cast-spoked flywheels of two-piece design are standard (see illustration). *See* ENERGY STORAGE. [L.S.L.]

Foam A material made up of gas bubbles separated from one another by films of liquid. The bubbles are spherical when the liquid films separating them are thick (approximately 0.01 mm). Pure liquids do not foam; that is to say, they cannot produce liquid films of any permanence. Relatively permanent films are created only when a substance is present that is adsorbed at the surface of the liquid. Substances capable of being so adsorbed may be in true solution in the liquid or may be particles of a finely divided solid, which, because of poor wetting by the liquid, remain at the surface. In both cases, surface layers of the added substance are produced. The reluctance of the adsorbed substance to enter the bulk of the liquid preserves the surface and, hence, the thermodynamic stability of the foam. *See* SURFACTANT.

Although thermodynamically stable, a foam is mechanically fragile. Offsetting this fragility to some extent are mechanisms that provide the liquid films with resiliency and plasticity.

Although foams of exceptional stability are desired in some commercial applications, foam is a nuisance in many situations. A common recourse is the addition of chemical antifoams, which are usually insoluble liquids of very low surface tension. When a droplet of such a liquid is sprayed onto the foam or is carried into it by mechanical agitation, it spreads spontaneously and rapidly at the surface of the film, virtually sweeping the film away as it does so. *See* ADSORPTION; INTERFACE OF PHASES; SURFACE TENSION. [S.Ro.]

Focal length A measure of the collecting or diverging power of a lens or an optical system. Focal length, usually designated f' in formulas, is measured by the distance of the focal point (the point where the image of a parallel entering bundle of light rays is formed) from the lens, or more exactly by the distance from the principal point to the focal point. *See* GEOMETRICAL OPTICS.

The power of a lens system is equal to n'/f', where n' is the refractive index in the image space (n' is usually equal to unity). A lens of zero power is said to be afocal. Telescopes are afocal lens systems. *See* DIOPTER; LENS (OPTICS); TELESCOPE. [M.J.H.]

Fog Water droplets or, rarely, ice crystals suspended in the air so as to reduce visibility appreciably; a cloud resting on the ground. Fog is most common near seacoasts or large lakes, on mountains, in small valleys of nonarid regions, and over oceans where the water is relatively cold, as in summer off Newfoundland or California. On land, fog is most frequent in the early morning.

Fog results either when air is cooled to its dew point and below or from an increase of moisture through evaporation from water that is warmer than the air. Causes of cooling are the especially rapid nighttime radiation that develops with clear skies; advection (air moving over a colder surface); and expansional (adiabatic) cooling, as when air moves up a mountain slope. For radiation to cause fog, winds must be light or calm; otherwise, mixing prevents fog or produces low clouds.

Before fog can form, there must be sufficient condensation nuclei in the air. This condition is usually met, but ice nuclei are often lacking in sufficient quantity so that many fogs below freezing are water fogs. Fog with snow cover, however, is common only near freezing, and again, at temperatures below about −40°C (−40°F). *See* DEW; DEW POINT; HUMIDITY; PRECIPITATION (METEOROLOGY); SMOG. [J.R.F.]

Folic acid A yellow vitamin, slightly soluble in water, which is usually found in conjugates containing varying numbers of glutamic acid residues. It is also known as pteroylglutamic acid (PGA), and has the structural formula shown.

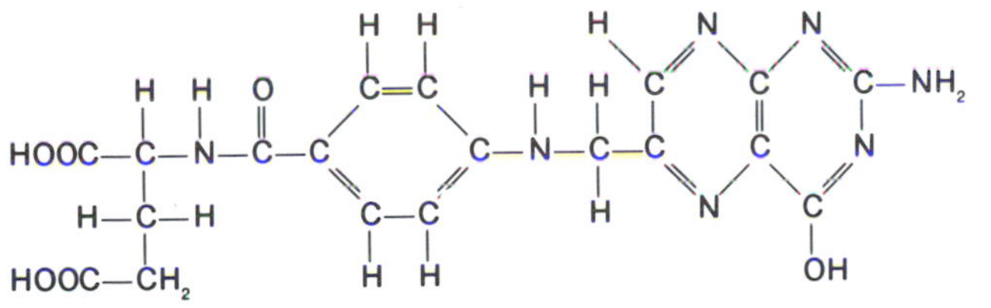

Folic acid is so widespread in nature and intestinal synthesis is so great that a folic acid deficiency in humans because of low dietary intake is probably not very common. Deficiencies of other nutrients (particularly iron, ascorbic acid, or vitamin B_{12}) may lead to a number of clinical conditions in which folic acid deficiency is involved. These include various nutritional macrocytic anemias, sprue, idiopathic steatorrhea, and pernicious anemia. *See* VITAMIN. [S.N.G.]

Fontéchevade man A fossil man, known from a skullcap of an old adult and a brow fragment of an adolescent excavated in 1947 by Germaine Henri-Martin in the Fontéchevade cave, Charente, France. The fossils, as reconstructed by Henri Vallois, appeared similar to modern humans who lack large browridges. Studies have shown the fossils to be similar to other Last Interglacial human fossils and possibly ancestral to the more recent Neandertals. *See* FOSSIL HUMAN; NEANDERTALS. [E.T.]

Food The material that enables humans to grow and to maintain and reproduce themselves. Food is composed of chemicals (natural or synthetic) in various combinations and is as essential to life as water and oxygen, which are also important constituents of foods. *See* NUTRITION.

Staple foods classified by origin

Plant origin	Animal origin
Cereals: wheat, corn (maize), rye, barley, oats, rice	Meats: beef, pork, mutton, poultry
Sugars: sugarcane, sugarbeets	Eggs
Vegetables: potatoes, sweet potatoes, cassava, peas, beans, onions, tomatoes, soybeans	Dairy products
Fruits: banana, citrus, plantain, apples, berries	Fish and marine products
Nuts: groundnuts (peanuts), coconuts, various seeds	
Microbes: molds, mushrooms, yeast	

Food scientists, technologists, and engineers are concerned with the many-faceted problems of transfer of food from field to consumer in the most wholesome and palatable condition possible. Their concern involves such problems as freshness, appearance, stability, avoidance of contamination, and prevention of spoilage, as well as development of new products from food components. In particular, they are responsible for the maintenance or enhancement of the palatability and nutritional value of the product through various processing procedures. *See* FOOD ENGINEERING; FOOD MANUFACTURING.

In any dietary regimen minimal amounts of three types of materials must be provided: energy-producing compounds, structure-producing compounds, and compounds necessary to energy-exchanging reactions. These three types include proteins, carbohydrates, fats, vitamins, and mineral elements, which constitute the classes of chemical compounds used to describe the composition of any particular food. Water may be considered essential for their utilization. *See* CARBOHYDRATE; FAT AND OIL (FOOD); PROTEIN; VITAMIN.

The staple foods form the basis of the dietaries of civilized humans and may be classified by origin (see table). The first step in the natural synthesizing of chemical elements into food molecules takes place in plants. The plant concentrates the foods so produced in seeds and nuts (less so in fruits) and stores them in leaves and stems. These basic plant processes give rise to the wide variety of cereals, vegetables, and other plant foods.

Foods of animal origin are composed predominantly of protein, are low in carbohydrate content, and contain varying amounts of fat. Of the many animal foods, milk most nearly approaches the perfect food, containing proportionately significant amounts of all nutrients except iron. Milk by-products are cheese and butter. [J.H.N./B.E.P.]

Food additives Substances added to foods during processing to retain or improve desirable characteristics or quality. Use of food additives has extended beyond mere preservation to include all stages of production, processing, and distribution of food. Additives may enhance flavor, improve texture, heighten color, better the appearance, or retain or increase nutritive value; facilitate mass production of products on uniform quality and properties; prevent microbial spoilage or oxidative changes (especially during transportation and distribution); increase effectiveness of sanitary packaging and protection of foodstuffs for retail sale; or add to ease, rapidity, and uniformity of preparation in the home. Some additives serve more than one function when present in certain foods. However, they may be classified according to their intended purpose as described below.

Flavoring agents. Flavoring agents make up the largest and most important group of food additives. Many of the most familiar of these are the natural spices and essential oils, and extracts, such as clove, ginger, citrus oils, and vanilla. Equally important, though less well known, are such synthetic aromatic chemicals as aldehydes, esters, and alcohols, which may be used to impart to foods such flavors as strawberry, cherry, pineapple, walnut, wintergreen, and many others. Other compounds, of which monosodium glutamate and sodium inosinate are examples, serve as flavor enhancers in meat products.

Nutrients. Minerals and vitamins are added to foods to restore nutritional values lost in processing or to supplement natural content of nutrients. Iodine in the form of iodized table salt was the first nutritional supplement to be added to a food. Vitamin D added to fluid milk has made rickets rare in the United States. Flour and bread are enriched with iron, thiamine, riboflavin, and niacin. Calcium as carbonate or phosphate is added to special dietary foods. Sodium fluoride and silicon fluoride are introduced into water supplies deficient in fluorine in order to improve dentition.

Preservatives. During storage and distribution, undesirable changes occur in flavor, color, texture, and appetite appeal. The food producer uses various preservatives to delay these changes. *See* FOOD PRESERVATION.

Antioxidants. Oxygen in the air may cause off-colors or undesirable flavors; for example, browning of apple or peach slices or rancidity in lard. Antioxidants are used to retard such effects. Dipping of fresh fruit in ascorbic acid solutions (vitamin C) prevents enzymatic browning. Sodium ascorbate retards oxidation of cooked, cured, and comminuted meat products. Off-flavor development in lard, vegetable oils, hydrogenated shortening, crackers, biscuits, breakfast cereals, dry cake mixes, and soup mixes due to oxidation is safely retarded by addition of butylated hydroxyanisole, butylated hydroxytoluene, lecithin, and tocopherols. Sulfur dioxide and sulfites are useful antioxidants for wine and beer, sugar syrups, and cut, peeled, or dried fruits and vegetables.

Antimicrobials. Molds, bacteria, and yeasts produce another kind of food spoilage, but good sanitary practices in the modern food-processing plant, together with additives effective in controlling such spoilage organisms, keep losses to a minimum. Common table salt (sodium chloride) has been used for centuries to keep meat and fish from spoiling. High concentration of sugars helps preserve fruit jams and jellies, as well as canned and frozen fruits, against microbial spoilage. Chlorine and hypochlorites control microorganisms in public water supplies. Sorbic acid and sorbates inhibit mold in cheese wraps, pie fillings, and fruit syrups. Benzoic acid and sodium benzoate have long been used as safe preservatives in pickles, catsup, relishes, soft drinks, jellies, jams, and preserves. Bread no longer becomes moldy or ropy quickly in hot weather because the baker adds calcium propionate or sodium diacetate to the dough. *See* ANTIOXIDANT.

Emulsifiers. Desirable consistency in many foods is difficult to achieve or maintain unless an emulsifying agent is added. These surfactants facilitate and stabilize oil-in-water (or water-in-oil) dispersions. The most important emulsifiers are the mono- and diglycerides of various fatty acids and their derivatives.

Thickeners. The appetizing mouth feel of many foods, as well as their flavor appeal, is traceable to smooth, uniform texture derived from added stabilizers or thickeners. Such agents impart body or viscosity to fluids or emulsions. They include glycerol, sorbitol, gelatin, pectin, vegetable gums (alginates, carrageenan, and karaya), and cellulose gums (carboxymethylcellulose). They are used in processed cheese and salad dressings; in ice cream and frozen desserts they help prevent formation of large, coarse ice crystals.

Coloring agents. Coloring agents are often employed to improve food appearance. Coloring agents may be of natural origin, such as turmeric, annatto, caramel, carmine, and carotene; or they may be one of the synthetic certified food colors. Principal uses are in soft drinks, confectionery, icings, jellies, baked goods, fillings, cake mixes, cereals, popcorn, packed fruits, cheese, butter, margarine, and meat products.

Bleaching agents. The yellow color of freshly milled flour gradually disappears on aging as a result of oxidation. This maturing makes flour satisfactory for baking. Bleaching with chemical agents accelerates this process and produces uniformity in color. For this purpose oxides of nitrogen, chlorine, chlorine dioxide, nitrosyl chloride, or ammonium persulfate are employed.

Anticaking agents. Some powdered foods tend to lump or cake on storage. Anticaking agents or desiccants prevent this occurrence in table salt, confectioner's sugar, malted milk powder, onion or garlic salt, and meat-curing mixes. Among the chemicals used as anticaking agents are calcium phosphate, magnesium carbonate, calcium or magnesium silicate, silica gel, talc, and starch.

Clarifying agents. Fruit juices, vinegar, wine, beer, and soft drinks frequently exhibit turbidity, damaging the appearance of the liquids. A variety of compounds are used as clarifying agents to improve quality. These include tannin, gelatin, albumin, methyl cellulose, pectinases, and proteinases (papain, bromelin, and fungal enzyme preparations).

Artificial sweeteners. Nonnutritive sweeteners, such as saccharin, impart sweetness to foods without adding calories. Their sole use is in low-calorie or diabetic foods, such as beverages, canned fruits, and desserts.

Foam regulators. Food processors use foam regulators either to stabilize foams or to depress them. Dextrins, peptones, and cellulose gums are added to pressure-packed toppings so that they will be propelled from the containers fully foamed. They ensure that beer and root beer will produce a head when dispensed.

Aerating agents. Carbonated beverages depend on carbon dioxide, added under pressure, for their flavor appeal. Cream is dispensed from pressurized containers charged with nitrous oxide or food-grade Freon.

Buffering agents. With many processed foods, quality depends on adjusting and stabilizing pH (hydrogen ion concentration) to an optimum value. This adjustment is accomplished by means of buffering agents. Sodium salts of acetic, citric, phosphoric, and pyrophosphoric acids are employed for this purpose. They are used in confectionery, jellies, jams, preserves, soft drinks, prepared cereals, meat cures, and evaporated milk. Lactic, citric, and acetic acids serve to adjust pH to the desired end point in processed cheese, pickles, and brines, canned vegetables, and some kinds of confectionery.

Enzymes. In various food processing operations, enzymes also serve as valuable additives. In practice, enzymes are permitted to react only to an established optimum condition and are then inactivated by heat. Their function is to effect hydrolysis of such food components as proteins, carbohydrates, and hemicelluloses. *See* ENZYME.

Other additives. Among additives of lesser importance are clouding agents used to impart turbidity to beverages; coating agents for glazing oranges and chocolate confectionery and for facilitating removal of baked goods from pans or confectionery from slabs; hydrolytic agents to effect cleavage of proteins, carbohydrates, and cellulosic materials; and growth stimulators, such as antibiotics or phenylarsonic acid to promote rapid growth of poultry and livestock. *See* FOOD ENGINEERING. [V.O.W.]

Food engineering The technical discipline involved in food manufacturing and refined foods processing. It encompasses the practical application of food science in the efficient industrial production, packaging, storing, and physical distribution of nutritious and convenient foods that are uniform in quality, palatable, and safe. Controlled biological, chemical, and physical processes and the planning, design, construction, and operation of food factories and processes are usually involved. *See* FOOD MANUFACTURING.

Food engineering is the food industry equivalent of chemical engineering. Food science in industry converts agricultural materials into products that are marketable because they meet a consumer need and can be profitably sold at reasonable prices by virtue of being economically produced, packaged, and distributed.

Food engineering is a vital link, therefore, between farms and food stores. Without it, food would be available only at farms, in forms produced by nature, and only in season.

Many natural foods must be preserved in order to be stored and shipped long distances to population centers. To provide a varied, nutritious diet, many raw food materials must be refined, and others combined with various ingredients and processed to produce food in new forms, such as bread, cheese, ice cream, frankfurters, soft drinks, cake and dessert mixes, candy, many breakfast cereals, ketchup, and salad dressings.

Because food engineering is applied in food manufacturing and refined foods processing, it requires a knowledge of unit operations and processes such as cleaning, separating, mixing, forming, heat transfer, moisture removal, fermenting, curing, packaging, and materials handling. This requirement to a large extent differentiates food engineering from food science. Yet these operations involve applied food science, just as chemical engineering encompasses chemistry. That is why the food engineer must have a working knowledge of food chemistry, bacteriology, and industrial microbiology, as well as of physics, mathematics, and basic engineering disciplines. *See* FOOD MICROBIOLOGY. [F.K.L.]

Food fermentation Production of food with the aid of microorganisms, which may be yeasts, molds, or bacteria. Fermented commodities include cereals such as wheat, rice, sorghum, corn; legumes, peanuts, soybeans, pulses; red meat, sausages, pork; milk; fish, shellfish; and plant juices. There are countless varieties of food fermentations in the world, sometimes known under different names in different countries and varying greatly in complexity. *See* FERMENTATION.

Among the molds used in food fermentations are *Aspergillus, Rhizopus, Penicillium, Neurospora, Actinomucor, Mucor, Amylomyces,* and *Monascus*. The yeasts include species of *Saccharomycopsis, Zygosaccharomyces,* and *Candida*. The bacteria are *Bacillus* and lactic acid bacteria, including *Lactobacillus, Streptococcus, Pediococcus*, and *Leuconostoc*.

The inoculum or fermentation starter may be of four types. The substrate may be moistened, heat-sterilized, and inoculated with a single organism. In a second type, more than one strain of a single species is used. A third type of inoculum contains more than one species of microorganism. Finally, some substrates contain a complex inoculum in which many different microorganisms of unknown identity are present. [C.W.H.]

Food manufacturing The commercial production and packaging of food products that are fabricated by processing or combining various ingredients or both. Manufactured foods are basically different from any found in nature, but they are fabricated principally from natural ingredients, and often include small quantities of chemical food additives to improve nutritive value, taste, color, shelf life, and convenience in use. A limited quantity of simulated meatlike foods has been fabricated from soya protein and other ingredients. Development is being done on the production of intermediate-moisture foods for human consumption, using a process similar to the one for making intermediate-moisture pet foods.

Mechanical manipulation, chemical treatment, biologic processes, heat treatment, or freezing may be involved. For example, bread is manufactured by combining flour, shortening, yeast, and other minor ingredients to make dough. Kneading and fermentation of the dough change the characteristics and develop flavor. The dough is formed into loaf-sized pieces, put into bake pans, proofed, and baked to form bread, which is sliced and wrapped.

Some foods, however, are prepared for consumption by refining processes which do not essentially change the nature of the material. Milk, for example, is clarified by centrifuging and pasteurized by heat treatment to kill pathogenic organisms. This treatment makes it safe for human consumption, but does not change it basically.

Food manufacturing and processing involve many unit operations, such as separating, mixing, forming, application of heat or refrigeration, packaging, materials handling, and process control. These are carried out in plants ranging in size from small retail bakeries to a processing plant covering 35 acres (141,600 m^2).

Originally an enlarged version of a kitchen, the typical food factory has been developed through engineering into a highly efficient, technical industrial plant. Many processes have been converted from batch to continuous methods and put under automatic control. Materials and finished products are handled with high efficiency and little labor, with the aid of bulk pneumatic conveyors, pumps, mechanical conveyors, and lift trucks. Modern food factories are built with sanitary floors and walls of tile and brick, and much of the equipment is made of stainless steel. Most of the newer plants are in suburban areas on attractively landscaped plots.

Food factories are inspected by Federal health authorities if they do interstate business, and by state and municipal authorities when they do business locally. Health agencies also require that additives and packaging materials used in food manufacturing be approved as safe before they may be employed.

Manufacturers usually sell their products to retail stores through wholesalers, though some sales are made direct to large chain-store organizations. Sale of processed and manufactured foods to restaurants and caterers is a large business; and so is sale of convenience products via vending machines.

Major branches of the food manufacturing and refining industry are those producing bakery products; flour and cereal products; milk and other dairy products; canned, frozen, and dehydrated fruits and vegetables; fruit and vegetable juices; preserves; meat; poultry; fish; soft drinks; beer and ale; wine; confections; coffee; tea; spices and condiments; sweeteners; syrups; baby foods; dessert mixes; roasted nuts and nut butters; oleomargarine; shortenings and edible oils; and mayonnaise and salad dressings. *See* FOOD; FOOD ENGINEERING; FOOD PRESERVATION; FOOD SCIENCE; FOOD SPOILAGE. [F.K.L.]

Food microbiology The branch of biology that deals with the microorganisms involved in the spoilage, contamination, and preservation of food. Food microbiologists must know how to detect the disease-producing organisms, how to prevent their entrance into food, and how to destroy them if they are present.

Another aspect of food microbiology concerns food preservation. Some foods are preserved and desirably flavored by fermentation processes; therefore, the specialist needs to be familiar with the organisms involved as well as their physiology and behavior in fermentation. *See* BOTULISM; FOOD PRESERVATION; FOOD POISONING; FOOD SPOILAGE. [H.O.H.]

Food packaging A total system which assures separation of the contained product from the environment and protection against external chemical, physical, and biological influences. As a system it provides protection, containment, communication, and unitization while affording sanitation, retention of product quality, and consumer convenience.

Food packaging might be categorized by several schemes, such as by materials (metal, glass, paper, plastic, and composites); package form (can, bottle, pouch, bag, wrap, carton, and case); packaging process (canning, bottling, pouching, bagging, overwrapping, collating, filling, closing, labeling, coding, and casing); food product; or packaging level (primary, secondary, tertiary, and quaternary). Each method of classification has its merits and disadvantages, and none is without considerable overlap into other categories.

Labeling. The Truth in Labeling Act instituted in the mid-1960s stipulated that all food packages bear a uniform and accurate representation of the weight, volume, or count of the package contents. The law further provided for more accurate graphic depiction of the nature of the contents. Accompanying the mandate were provisions encouraging food packagers to voluntarily reduce the number and variety of sizes made available to consumers.

Indirect additives. The 1958 Delaney Amendment to the 1938 Food, Drug and Cosmetic Act stated that any chemical additive capable of causing cancer was specifically prohibited from food products. Chemical additives are defined as any compounds which enter the product intentionally or unintentionally, or are entered directly or indirectly. Since many packaging materials are in intimate contact with the foods contained, components of the materials may be extracted into the food. These components or indirect additives may be compounds having a functional value in food packaging or may be compounds which are present incidentally.

Degradable packaging. The environmental issue of litter has stimulated many studies in the development of biodegradable packaging, that is, packaging that would disintegrate to invisible components as a result of exposure to natural elements. Biodegradable packaging would be counter to the need for stable waste required for sanitary landfill. For effective use of solid-waste dumps, it is necessary that the waste be resistant to the elements so that the recovered land does not collapse due to biodegradation. Further, biodegradability contradicts the concept of the protective value of packaging.

Yet another direction of research has been to find better means of ultimate useful degradation of packaging, as in incineration. Much of the packaging which is functionally effective does not burn well when it is being destroyed or being used to provide thermal energy to, say, generate electrical power. Among the problems encountered are that glass and metal must be separated out as nonburnable, and that some plastics such as PVC might possibly emit quantities of corrosive HCl.

Universal Product Code (UPC). In a long-sought answer to the problem of improving productivity in retail food store operations, Universal Product Code was created to appear on all food packages. This code, designating the manufacturer, brand, product, and package size, was designed to be read by electronic reading devices, thus eliminating the need for package price marking and for manual checkout procedures in retail stores. These electronic devices scan the code and transmit their reading to a computer memory which records the item and price and totals the purchase automatically.

Packaging materials. A broad spectrum of materials is employed for packaging, depending upon the ultimate package configuration and upon economics. Immediately containing the food product is the primary package, which is no longer a simple material. In a continuing effort to provide the most cost-effective functional package, developments have been directed to combining materials to provide an optimum synergy of the desired attributes of the basic materials. For example, in a glass package, the glass would contribute impermeability, rigidity, inertness in contact with the food, and vertical stacking strength in distribution. In the same glass package, drawn aluminum with a plastic core would provide the closure, paper and adhesive the label, and a thin plastic coating protection to the glass surface to minimize the potential hazard from breakage.

Primary packages might be generally classified as cans, bottles and jars, bags and pouches, wraps and cartons. Secondary packaging is usually composed of paperboard cartons or wraps, but considerable quantities of plastic unitizers are being increasingly employed. Tertiary packaging is usually the corrugated shipping container, but can also be paperboard or plastic wrap. [A.L.B.]

Nutrition labeling. This is a standardized way to tell the consumer through labeling the chief nutritional characteristics of foods. The nutritional properties stated are based on a serving of the food. The size of the serving in common household units is stated on the label and also, for packaged foods, the number of servings in the package. The label then states the number of calories and the number of grams each of protein, carbohydrate, and fat in the serving of food. If so desired, the purveyor may further qualify the declaration of fat by saying how much of it is saturated and how much is polyunsaturated (only for significant sources of fat). The amount of cholesterol present may also be stated if the food is a significant source.

The next section of the label sets forth the content of protein (restated in different units) and micronutrients per serving. Each nutritionally labeled product must declare the amount of vitamins A and C, thiamine, riboflavin, niacin, calcium, and iron. Optionally, it may also declare vitamins D, E, B_6, and B_{12}, folic acid, phosphorus, iodine, magnesium, zinc, copper, biotin, or pantothenic acid. In this section, the nutrient levels are declared as percentages of the United States Recommended Daily Allowances (RDA). These United States RDAs are based on but not quite identical to the Recommended Daily Dietary Allowances of the Food and Nutrition Board of the National Research Council (NRC). *See* NUTRITION. [V.O.W.]

Food poisoning

Food poisoning The two types of bacterial food poisoning are infection of the host by organisms carried by food; and intoxication by poisonous substances produced by bacteria in food before it is consumed.

Salmonella food poisoning is the most common food-borne infection. Large numbers of these bacteria are necessary to establish human infection; the infection is usually acute, but mortality is very low. Food becomes contaminated by contact with animal or human carriers. *See* SALMONELLA.

Food-borne intoxication is most frequently caused by ingestion of staphylococcus enterotoxin. Certain strains of *Staphylococcus pyogenes* are capable of producing this heat-stable toxin as an excretion, or exotoxin, during rapid growth in food products. The gastrointestinal symptoms are severe, but recovery is complete in 3–4 days. *See* STAPHYLOCOCCUS.

Botulism is a rare intoxication usually associated with foods of low acid content canned under insufficiently rigorous conditions. The heat-labile toxin formed during growth of *Clostridium botulinum* acts upon the central nervous system. Mortality averages about 60%. *See* BOTULISM.

Ptomaine poisoning is an archaic and erroneous term once used as a synonym for food poisoning. Ptomaines are diamines found in putrefying material. They are in part responsible for foul odors but their toxicity is low. They have long ceased to be regarded as important in food poisoning. [J.H.Si.]

Food preservation

Food preservation The branch of food science and technology that deals with the practical control of factors capable of adversely affecting the safety, nutritive value, appearance, texture, flavor, and keeping qualities of raw and processed foods. Since thousands of food products differing in physical, chemical, and biological properties can undergo deterioration from such diverse causes as microorganisms, natural food enzymes, insects and rodents, industrial contaminants, heat, cold, light, oxygen, moisture, dryness, and storage time, food preservation methods differ widely and are optimized for specific products. *See* FOOD SCIENCE.

Food preservation methods involve the use of heat, refrigeration, freezing, concentration, dehydration, irradiation, pH control, chemical preservatives, and packaging applied to produce various degrees of preservation in accordance with the differing use patterns and shelf-life needs of unique products.

Thermal processes to preserve foods vary in intensity. True sterility to ensure total destruction of the most heat-resistant bacterial spores in nonacidic foods may require a treatment of at least 250°F (121°C) of wet heat for at least 15 min to be delivered throughout the entire food mass. The term commercial sterility refers to a less severe condition that still assures destruction of all pathogenic organisms, as well as organisms that, if present, could grow in the product and produce spoilage under normal conditions of handling and storage. *See* STERILIZATION.

Many foods are subjected to still less severe heating by methods that produce pasteurization to assure destruction of pathogens and extend product shelf life. *See* PASTEURIZATION.

The slowing of biological and chemical activity with decreasing temperature is the principle behind cooling (refrigeration) and freezing preservation. In addition, when water is converted to ice, free water required for its solvent properties by all living systems is removed. Commercial freezing methods utilize refrigerated still air; high-velocity air, which is faster and more efficient; and high-velocity air made to suspend particulate foods, such as peas, as in a fluidized-bed fast freezer. Indirect-contact freezing utilizes hollow flat plates chilled with an internally circulated refrigerant to freeze solid foods, or with refrigerated tubular heat exchangers that rapidly slush-freeze liquids. Immersion freezing involves direct contact of the food or its container with refrigerants approved for food or a fast-freezing cryogenic liquid. *See* COLD STORAGE; CRYOGENIC ENGINEERING.

When sufficient water is removed from foods, microorganisms will not grow, and many enzymatic and nonenzymatic reactions will cease or be markedly slowed. Concentration preservation can be achieved by physically removing water, as by boiling or with lower-temperature vacuum evaporation, or by binding water through the addition of sugar, salt, or other solutes.

Foods preserved by dehydration contain considerably lower water activity and less total water than concentrated foods. Most dehydration methods utilize heat to vaporize and remove water. The heat and oxygen sensitivity of many foods necessitates vacuum dehydration for high quality. Under vacuum, water can be removed at reduced temperature, and oxidative changes are minimized. In freeze-drying, foods are frozen quickly and placed in a chamber under high vacuum. A food's structure remains rigid as it goes directly from the frozen state to dryness. *See* SUBLIMATION.

Food irradiation remains highly controversial, partly because of fears that the safety of products and processes cannot be adequately regulated. The natural acids of certain fruits and vegetables, acid added as a chemical, and acid produced by fermentation can inhibit or partially inhibit several pathogenic and spoilage organisms. The pH of acidic foods, however, is rarely sufficiently low to assure long-term preservation from acid alone. Many acidic and fermented foods further depend upon prior pasteurization of their ingredients, the addition of salt and other chemicals, and refrigeration. *See* PH.

The U.S. Food and Drug Administration and comparable agencies in various countries vigorously regulate the chemicals that may be added to foods as well as the conditions of their use. Chemical preservatives and similar substances include antimicrobials, enzyme inhibitors, and antioxidants. There is much pressure to remove chemicals from the food supply, especially where their effects can be achieved by other means.

Packaging protects foods from contamination, moisture gain or loss, flavor loss and odor pickup, the adverse effects of light, physical damage, and intentional tampering. Ultimately, a food product's quality and storage life are determined largely by its package. *See* FOOD ENGINEERING; FOOD MANUFACTURING; FOOD MICROBIOLOGY. [N.N.P.]

Food science The applied science which deals with the chemical, biochemical, physical, physicochemical, and biological properties of foods. Chemical properties of foods include: (1) composition, that is, constituents, such as proteins, fats, and carbohydrates, which make up food; (2) chemical reactions which occur when foods are processed (canned, frozen, dried, irradiated, and so on), packaged, or stored; and (3) interactions of food constituents with chemical additives, such as bleaching agents and antioxidants.

Biochemical properties of foods are related to postharvest (plant) or postmortem (animal) enzymatic activity, and to the presence of physiologically active constituents, such as vitamins and other nutrients which are essential to humans or animals. Physical properties include color, specific gravity, refractive index, viscosity, texture, and various thermal constants. Physicochemical properties relate to the behavior of colloids, solutions, crystals, and so on, occurring in foods. Biological properties pertain to the activity of macroorganisms, such as insects or parasites, and microorganisms, including bacteria, molds, yeasts, and viruses, found in foods (which are involved in spoilage and fermentation), as well as the effects of natural or induced toxicants.

Food science provides the basic knowledge for food technology, which is the application of science and engineering to the research, production, processing, packaging, distribution, preparation, and utilization of foods. [G.E.Liv.; M.So.]

Food spoilage Foods undergo many different types of spoilage, resulting in abnormal colors, flavors, odors, or in other changes, such as consistency. Spoilage cannot be equated to danger to the consumer; indeed, in most instances, foods which contain poison of either chemical or biological origin show no external evidence of spoilage.

Chemical changes are responsible for spoilage. These changes are brought about through enzyme action. These enzymes may have their origin in the food material or may be elaborated by microorganisms such as yeasts, molds, and bacteria which contaminate the product. In general, enzymes derived from foods themselves are of secondary importance to those elaborated by microorganisms; however, when foods are held under conditions which preclude microbial action, the natural enzymes in the food materials may be the primary reason for spoilage. The fat-splitting enzymes in frozen foods cause this type of spoilage.

Abnormal or unusual spoilages of food products occur periodically. Such spoilage may be seasonal, regional, or entirely haphazard in appearance. In extreme cases a food plant may have to cease manufacture of a product due to contamination with undesirable microorganisms responsible for peculiar spoilage problems. Examples of such unusual spoilage include green discoloration of milk; gas pockets in, and in extreme cases the explosion of, cheese. In most cases, a particular organism or group is responsible for each of these conditions. *See* FOOD ENGINEERING; FOOD MICROBIOLOGY; FOOD PRESERVATION. [J.H.Si.]

Food web A diagram of the trophic (feeding) interrelatioships within a community or ecosystem. The basic data constitute a catalog of the known diet of each community member. Each node of the diagram, representing a single species or an aggregation of ecologically similar species, is connected to at least one other node (see illustration). The community web therefore represents the connection, or the coalescing, of a number of individual food chains (diatom → zooplankton → fish → fish; or, rosebush → aphid → warbler → hawk). [The arrows point in the direction of the species or ecological unit doing the consuming.]

The base of almost all food webs, the first trophic level, is autotrophic, usually comprising photosynthetically active plants. Two different kinds of webs derive from these primary producers. In grazing food webs, each additional trophic level

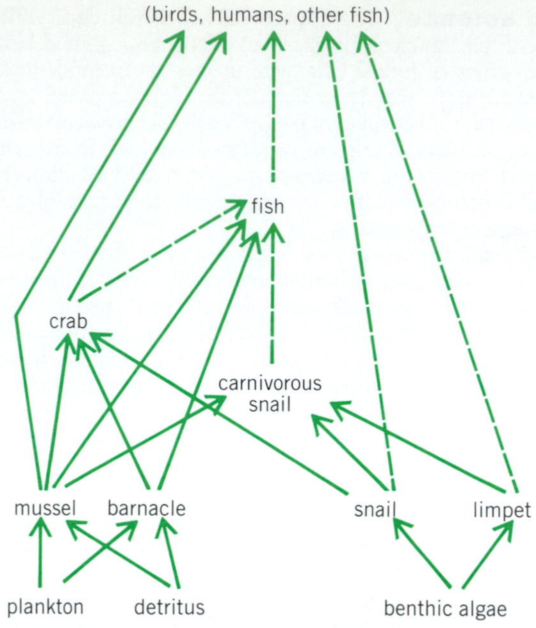

A typical food web characteristic of the protected rocky shore community of southern New England. Solid lines suggest major trophic interactions, broken lines minor ones. Such webs provide useful but incomplete biological descriptions, since not all prey are listed, and certain categories of consumers (birds, snails) and prey (detritus, benthic algae) can include many individual species. (*After D. C. Edwards et al., Mobile predators and the structure of marine intertidal communities, Ecology, 63(4):1175–1180, 1982*)

is, on the average, one link further removed from the first level: producers → herbivores → primary carnivores → secondary carnivores. Usually there are no more than five levels, or four links, in such communities. Omnivores are represented by connections to two or more lower levels. In detrital food webs, the majority of the photosynthetically fixed energy goes directly to saprophages or decomposers, organisms feeding on dead or dying organisms or their products. Their activities bear important implications for ecosystem functioning, especially nutrient cycling. All natural communities are characterized by the presence of both consumer and detrital food webs.

Food web diagrams can indicate the presence of symbioses (mutually beneficial interactions), the trophic position of cannibalism (where a species eats a conspecific), or the existence of compartments or subwebs, groups of species interacting more frequently or strongly with themselves than with other species within the same community. *See* ECOLOGICAL COMMUNITY; ECOLOGICAL INTERACTIONS; ECOSYSEM. [R.Pa.]

Foot (anatomy) The terminal portion of the lower, or inferior, extremity. Its bony framework consists of 7 tarsal or ankle bones, 5 metatarsal or foot bones, and 14 phalanges or toe bones. The prominent heel is a projection of the calcaneus, a tarsal bone (see illustration). A strong but elastic system of ligaments holds the bones in position, producing the arched structure characteristic of the human foot, and strengthens the joints of the foot. Ankle, foot, and toe movements are produced by the action of three groups of leg muscles and six groups of foot muscles. [T.S.P.]

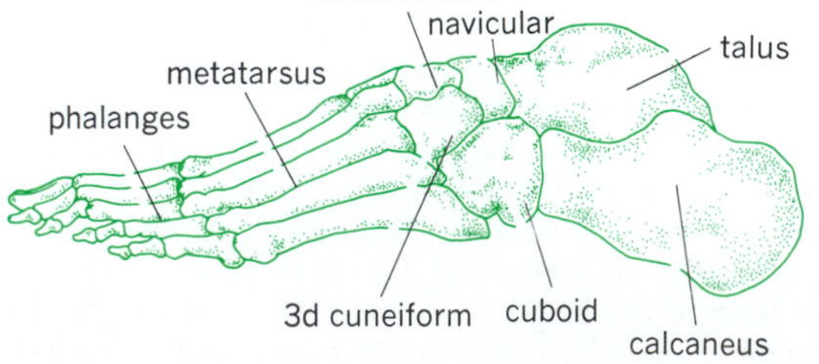

The skeleton of the human left foot, shown in a lateral view. (*After W. H. Lewis, ed., Gray's Anatomy of the Human Body, 22d ed., Lea and Febiger, 1930*)

Foot-and-mouth disease A highly contagious virus disease of economic importance, infecting cattle, pigs, sheep, and goats. It is characterized by fever, salivation, and formation of vesicles in the mouth and pharynx and on the feet. It may be transmitted to humans.

Attenuated-virus vaccines have been developed and used in the field with success. In serious epizootics all susceptible animals in the area are slaughtered, and strict quarantine is observed. To prevent human infection, boiling or pasteurizing of farm products under suspicion should be adequate. *See* ANIMAL VIRUS; EPIDEMIOLOGY. [J.L.Me.]

Foot disorders Musculoskeletal, neurological, or dermatologic abnormalities of the foot. They may be of developmental or acquired origin.

Clubfoot, also known as talipes, is a congenital deformity of the foot characterized by a club-shaped appearance, and occurs in approximately 1 in 1000 live births. Clubfoot takes a number of forms, the most common being talipes equinovarus, an extreme turning down and under of the foot.

During normal intrauterine development, and normally continuing after birth, the position of the foot changes gradually from one of inversion such that the soles of both feet face each other to one of alignment with the leg. In the flatfoot that positional change is not always complete, so that the foot undergoes a series of complex compensations to reach the ground that is called pronation. Twisting of the ankle and obliteration of the normal arch of the foot result. In addition, the joints of the foot tend to lose their functional articular contacts, and the resulting hypermobility at the joint interfaces predisposes the foot to arthritis. Bunions and hammertoe deformities, with painful stretching of the soft tissues of the sole, are additional complications that result from hypermobility.

Heel spurs, or calcaneal exostoses, are bony growths produced by excessive musculoskeletal tension at the heel. When the pull by ligaments and connective tissue becomes excessive, as is seen with flatfoot deformities and tight or shortened heel cords, the outer covering of the heel bone known as the periosteum enlarges. The bone growth is known as an exostosis or spur.

The nerves of the foot are subject to irritation from pressure originating from the structures of the foot and from external forces. In cases of Morton's neuroma, the third intermetatarsal nerve thickens over a period of years, and a small benign fusiform tumor eventually forms in the space between the third and fourth toes. The posterior tibial nerve is located at the inside of the ankle behind the ankle bone, and it travels to the bottom of the foot through a channel known as the tarsal canal. In tarsal tunnel syndrome, the nerve becomes compressed and damaged within the tarsal canal.

The skin of the foot undergoes stress from pressure and friction. As a response, it protects itself with defensive modifications such as corns and calluses that themselves cause pain.

Warts are tumors of the skin induced by the human papilloma virus. Second only to corns and calluses in prevalence, warts are benign skin lesions. Transmission of warts from person to person may occur, but individual immunity plays an important role in susceptibility to these infectious lesions.

Ingrown toenail, also called onychocryptosis, is caused by the penetration of the free portion, or edge, of the nail into the surrounding soft tissue. The cause may be related to congenital variations in the shape of the nail itself or the nail fold (the skin adjacent to the edges of the nail), biomechanical abnormalities in foot structure or function, nail disease, or ill-fitting shoegear. [H.Lem.]

Forage crops Grasses and legumes that make up grasslands and are used as forages for livestock. The grasslands represent an ancient renewable natural resource. They form 25% of the world's vegetation and occupy the largest area of any single plant type. They benefit humanity indirectly by providing food for both wild animals and domesticated livestock, some of which are ruminants that, because their digestive systems contain microorganisms, are able to digest fibrous forage material. Thus, the prime value of grassland areas lies in the meat, milk, or work produced by the livestock that graze on them. *See* GRASSLAND ECOSYSTEM; LEGUME.

Grasslands have other attributes as well. In order to withstand the tug and pull of the grazing animal, a forage plant must have an extensive root system, and this makes a contribution to soil fertility. When the plant is grazed or cut, the photosynthetic area that remains is not large enough to provide sugars to maintain root respiration, and so part of the root system dies and adds to the organic material in the soil and to soil structure. Such is the basis of many highly productive crop rotations that maintain soil fertility without expensive fertilizer applications. Most forage plants are long-lived perennials that can be defoliated repeatedly during a growing season.

Forages also protect the soil from erosion by both wind and water. In fact, where row crop farming has led to soil deterioration, as it did when dust bowl conditions prevailed in North America, forages are used to restore and stabilize the land. *See* EROSION. [P.D.W.]

Foraminiferida An order of Granuloreticulosia in the class Rhizopodea. Foraminiferans are dominantly marine protozoans, with a secreted or agglutinated shell, or test, enclosing the continually changing ameboid body (see illustration) that characterizes this and other orders of the superclass Sarcodina. Their unique combination of long geologic history, ubiquitous geographic distribution, and exceptional diversity of test composition, form, and structure make the foraminiferans the most useful of all marine fossils for stratigraphic correlation, geologic age dating of sediments, and paleoecologic interpretation. Their tests accumulated in great numbers and are recoverable from small quantities of sediment, rock outcroppings, well cores or cuttings, or ocean dredging and submarine coring.

Pseudopodia from the ameboid body form a delicate anastomosing network of 2–10 times the test diameter. The pseudopodial net may arise from the apertural region of the test alone in those with tectinous, porcelaneous, and agglutinated walls, or may radiate in all directions through many tiny perforations of the hyaline test wall. The pseudopodia variously serve in capture, ingestion, and digestion of food, in test and temporary cyst construction, for anchorage, and for locomotion. The characteristic granular streaming of the continuously moving cytoplasm differentiates the Foraminiferida from the other Sarcodina orders.

The foraminiferan constructs its own test. Growth of the individual may cease after test construction, or the test may enlarge by continued growth in one or more directions (tubular or branching tests) or by periodic formation of separate but always interconnected chambers. Growth may be either continuous or periodic and may be in a straight line, a planispiral or trochospiral coil, a cycle, or a zigzag. The latter (zigzag) is expressed by chambered forms in biseriality. Variations and combinations of these growth patterns occur repeatedly in different lineages (isomorphism).

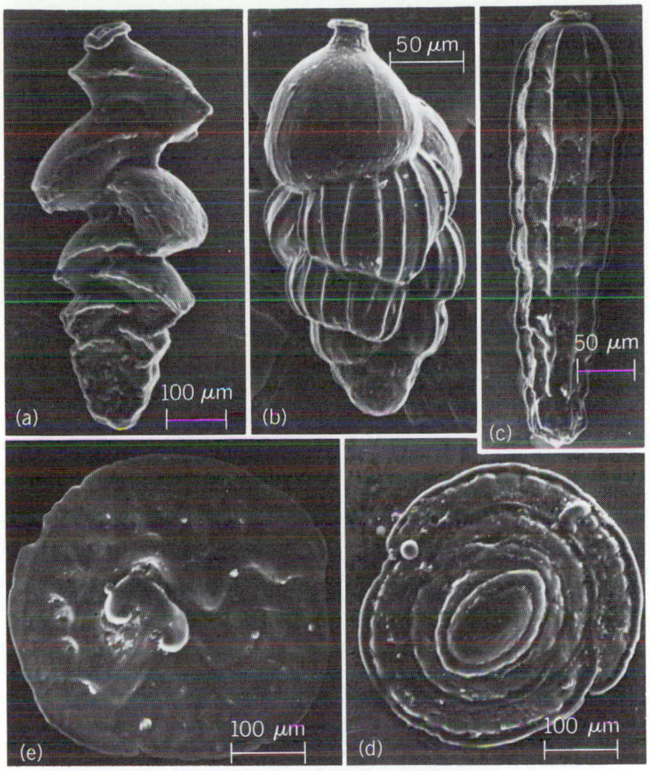

Scanning electron micrographs of foraminiferans of suborder Rotaliina. Superfamily Buliminacea: (*a*) *Eouvigerina*, elongate test. (*b*) *Uvigerina*, elongate triserial test. (*c*) *Siphogenerinoides*, reduced early biserial stage. Superfamily Spirillinacea; *Patellina*: (*d*) spiral view of low conical test; (*e*) umbilical view. (*R. B. MacAdam, Chevron Oil Field Research Co.*)

As is true of most protists with skeletons or tests, systematic differentiation and classification of foraminiferans is based on test composition, microstructure, and gross morphology. Information currently available concerning cytoplasmic characters, life cycles, and so on has shown good agreement with this classification, although the function and origin of many shell characters believed to be of systematic importance (canal systems, pores, septal doubling, and apertural tooth plates) are yet undetermined. There are 5 suborders with 19 superfamilies that comprise 102 families and nearly 1400 genera.

Foraminiferans have shown a remarkable diversification and rapid evolutionary development over their 500,000,000-year known history. Of the 34,000 living and fossil species, possibly 4000 are living. Some with uncomplicated tests have been reported to have a long geologic range, but most with diagnostic shell structure and morphology had relatively short histories, each successively replaced by others. The oldest-known and presumably primitive tests are the morphologically simple tectinous and agglutinated one of the Cambrian and Ordovician, of globular, tubular, branching, or irregular form.

Most foraminiferans are benthonic, living upon the sea floor, within the upper few centimeters of ooze, or upon benthic algae or other organisms. They occur from the intertidal zone to oceanic depths, in brackish, normal marine, or hypersaline waters, and from the tropics to the poles. Some modern Lagynacea live in fresh water, but none are known as fossils. Assemblages vary widely in response to local conditions, with the greatest diversity occurring in warm, shallow water. A smaller number, the Globigerinacea, are planktonic, living at various depths in the water column from the surface to the bottom, being most numerous between 20 and 100 ft (6 and 30 m). Vertical migration may be diurnal and may occur during ontogenetic development. The preferred depth range of a species may vary geographically in response to temperature differences or to changes in water density. *See* GRANULORETICULOSIA; RHIZOPODEA; SARCODINA. [H.T.]

Force Force may be briefly described as that influence on a body which causes it to accelerate. In this way, force is defined through Newton's second law of motion.

This law states in part that the acceleration of a body is proportional to the resultant force exerted on the body and is inversely proportional to the mass of the body. An alternative procedure is to try to formulate a definition in terms of a standard force, for example, that necessary to stretch a particular spring a certain amount, or the gravitational attraction which the Earth exerts on a standard object. Even so, Newton's second law inextricably links mass and force. *See* ACCELERATION; MASS.

One may choose either the absolute or the gravitational approach in selecting a standard particle or object. In the so-called absolute systems of units, it is said that the standard object has a mass of one unit. Then the second law of Newton defines unit force as that force which gives unit acceleration to the unit mass. Any other mass may in principle be compared with the standard mass (m) by subjecting it to unit force and measuring the acceleration ($\mathbf{a}$), with which it varies inversely. By suitable appeal to experiment, it is possible to conclude that masses are scalar quantities and that forces are vector quantities which may be superimposed or resolved by the rules of vector addition and resolution.

In the absolute scheme, then, the equation below is written

$$\mathbf{F} = m\mathbf{a}$$

for nonrelativistic mechanics; here boldface type denotes vector quantities. This statement of the second law of Newton is in fact the definition of force. In the absolute system, mass is taken as a fundamental quantity and force is a derived unit of dimensions MLT^{-2} (M = mass, L = length, T = time).

The gravitational system of units uses the attraction of the Earth for the standard object as the standard force. Newton's second law still couples force and mass, but since force is here taken as the fundamental quantity, mass becomes the derived factor of proportionality between force and the acceleration it produces. In particular, the standard force (the Earth's attraction for the standard object) produces in free fall what one measures as the gravitational acceleration, a vector quantity proportional to the standard force (weight) for any object. It follows from the use of Newton's second law as a defining relation that the mass of that object is $m = w/g$, with g the magnitude of the gravitational acceleration and w the magnitude of the weight. The derived quantity mass has dimensions $FT^2 L^{-1}$. *See* FREE FALL. [G.E.P.]

Force fit A means for holding mating mechanical parts in fixed position relative to each other. In a force fit of cylindrical parts, the inner member has a greater diameter than the hole of the outer member; that is, the metals of the two parts interfere. In a true force fit, the parts are highly stressed, the interference amounting to 0.002 or 0.003 in. (0.05 or 0.07 mm) for parts with a basic diameter of 1 in. *See* ALLOWANCE; SHRINK FIT. [P.H.B.]

Forced oscillation An oscillation produced in a simple oscillator or equivalent mechanical system by an external periodic driving force. Forced oscillations are to be contrasted with free oscillations which occur when the system is displaced from equilibrium and released. *See* HARMONIC MOTION; MECHANICAL VIBRATION; OSCILLATION; VIBRATION.

Application of the driving force results at first in two simultaneous oscillations, one at the frequency of the free oscillations, and one at the frequency of the impressed driving force. The former, the so-called transient, eventually decays, whereas the latter builds up to the steady-state value. Small damping is associated with slow decay of the transient and slow establishment of the steady state, whereas large damping is associated with fast decay. *See* DAMPING. [M.Gr.]

Forcipulatida An order of Asteroidea characterized by straight or crossed pedicellariae or both. These may be carried on stalks, and in some cases are arranged in rosettes around the bases of the spines. The spines usually occur singly, and are ungrouped. The body is stellate in outline, but is not necessarily restricted to five arms or radial axes.

The order contains four families, whose genera are numerous and widely distributed. The Brisingidae and Zoroasteridae are characteristically deep-water families, and the former includes species with as many as 44 arms. The Asteridae is the largest family, and includes many well-known predatory starfishes, some of which are economically significant because of the damage they inflict upon oyster and clam beds. The fourth family is the Heliasteridae. *See* ASTEROIDEA; ECHINODERMATA. [A.C.C.]

Forensic chemistry The application of chemistry to the study of physical materials or theoretical problems, the results of which may be entered into court as technical evidence. Boundaries are not sharply defined for forensic chemistry, and it includes topics that are not entirely chemical in nature.

Some of the items most often encountered in crime laboratories, and the information sought in regard to them, are: (1) body fluids and viscera to be analyzed for poisons, drugs, or alcohol, quantitation of which may assist in determining the dosage taken or the person's behavior prior to death; (2) licit and illicit pills, vegetable matter, and pipe residues for the presence of controlled substances; (3) blood, saliva, and seminal stains, usually in dried form, to be checked for species, type, and genetic data; (4) hairs, to determine if animal or human; if human, the race, body area of origin, and general characteristics; (5) fibers, to determine type (animal, vegetable, mineral, or synthetic), composition, dyes used, and processing marks; (6) liquor, for alcoholic proof, trace alcohols, sugars, colorants, and other signs of adulteration; (7) paint, glass, plastics, and metals, usually in millimeter-sized chips, to classify and compare to known materials; (8) inks on documents, to determine type, dye content, or possible age; also chemical obliterations and restoration of chaffed papers; (9) swabs from the hands of suspects, to be checked for the presence of gunshot residue; (10) debris from a fire or explosion scene, for the remains of the accelerant or explosive used. *See* ANALYTICAL CHEMISTRY; FORENSIC MEDICINE; FORENSIC TOXICOLOGY. [M.J.C.]

Forensic medicine The use of medical evidence or medical opinion for purposes of settling disputes in civil matters or providing medical evidence in issues of criminal action. All branches of medicine may provide such information. Medical specialists who frequently contribute evidence in legal disputes are from the disciplines of forensic pathology, forensic psychiatry, forensic dentistry, orthopedic surgery, general surgery, and general practice. The expert medical witness is called upon to state an opinion of the medical facts at issue. In industrial compensation cases, medical evidence is used to assess the nature and extent of injury or poor health resulting from employment.

All state and local governments in the United States establish by constitution or statute an office or agency to investigate deaths which are of public interest. Most states have a medical examiner, that is, a medical expert to investigate violent and sudden unexpected death. The physician has the responsibility of ruling the violent death an accident, a suicide, a homicide, or the result of violence of undetermined origin. About 25% of all deaths require some level of investigation. There may be only an investigation into the decedent's medical history and the circumstances surrounding death. Other deaths may require a complete investigation, including autopsy.

Special problems involved in forensic deaths but rarely encountered in other branches of medicine include identification of decomposed bodies, parts of bodies, and skeletal remains. Other forensic disciplines intersect with forensic pathology particularly in the area of identification, where forensic anthropology and forensic dentistry provide the means. Anthropological measurements are used to determine stature, age, race, and sex of skeletal remains. Dental records can provide the means to establish the identity of a badly burned body. In assessing the nature of violence, the forensic pathologist must be skilled in the examination of microscopic tissue to determine the age of injury. The pathologist also must approximate the time of death by assessing the general condition of the body. Other tests are used to establish evidence of sexual battery; blood group determination of stains present on clothing or the body can be most helpful in identifying a murder suspect. Recovery of bullets from tissue can provide ballistic evidence which conclusively establishes that a bullet was fired from a particular weapon. *See* ANTHROPOMETRY; BLOOD GROUPS; DEATH; FORENSIC CHEMISTRY; FORENSIC TOXICOLOGY; PUBLIC HEALTH.

The majority of deaths which are investigated by the local forensic authority are natural in character. Such deaths come under his or her purview when death occurs suddenly and unexpectedly in people who are in apparent good health.

[D.T.R.]

Forensic toxicology An interdisciplinary science dealing with the analysis and interpretation of drugs and chemicals in biological samples for medical-legal purposes. There are three major case load areas in most forensic toxicology laboratories: drug abuse—cases resulting from the illegal use of drugs; police cases—toxicological aspects of criminal investigations; and postmortem cases—analytical studies in support of the medical examiner to determine the cause of death.

In the United States poisons are responsible for approximately 10,000 deaths annually (3% of all accidental deaths and 25% of all suicides). The most commonly encountered drugs and chemicals include ethyl alcohol, barbiturates, carbon monoxide, morphine (most often from heroin use), propoxyphene, and benzodiazepines (diazepam, chlordiazepoxide, and so on). The role of ethyl alcohol in death is rarely due to its direct toxic effect but to its indirect role in accidents. Similarly hallucinogens rarely cause death by their direct toxicity, but are indirectly responsible for accidents, suicides, and so forth. Carbon monoxide, barbiturates, and analgesics exert a direct toxic effect (asphyxia, central nervous system depression, and respiratory depression). *See* FORENSIC MEDICINE; TOXICOLOGY. [J.Bid.]

Forest and forestry A forest is a community of trees, other plants, and animals which live in and thrive on the forest environment. Trees are the dominant form of vegetation. The forest community or ecosystem also includes shrubs, herbs, mosses, fungi, insects, reptiles, birds, and mammals. All these organisms live on or in the soil, water, and air of the forest. Each is a part of the community, and each reacts with all the other parts, but all require the warmth and energy provided by the Sun to survive.

The natural affiliation of some trees for specific environmental conditions in the forests of the United States has resulted in hundreds of groupings of trees known as forest types and subtypes. These can be classified into six major forests or natural forest regions. Three are east of the Great Plains: the Northern Forest, the Central Hardwood Forest, and the Southern Forest. The Rocky Mountain, Pacific Coast, and Alaskan Forests are west of the Prairie states.

To facilitate forest management, foresters have arbitrarily classified forest trees into several groups. One classification, based on the size of the individual trees, refers to them as seedlings, saplings, poles, standards, or veterans. A forest containing trees ranging from small seedlings to poles to large veterans is said to be all-aged. Forests with trees essentially the same size and age are called even-aged. Foresters talk of a pure forest if it is composed mainly of one species, and of a mixed forest if it contains several species. Botanists and foresters refer to tolerant trees as those which can grow in the shade of other trees. Species which cannot survive in shade are said to be intolerant.

The trees in a forest compete with each other for soil, water, and light on any given site. Some tree species may grow better on hills facing north, while others do better on slopes facing south. This forest competition may result in one combination of trees being more successful than another for a given type of environment. As new species invade a site, others may decline in importance and bring about a forest community change called forest succession.

Forestry. In 1971 the Society of American Foresters defined forestry as "the science, the art and practice of managing and using for human benefit the natural resources that occur on and in association with forest lands." Other definitions might note specifically that the basic "natural resources" were wood, forage, water, wildlife, and recreation. The major divisions of forestry which deal specifically with biological fields are forest soils, dendrology, silvics, and silviculture. Forest pathology, forest zoology, and forest entomology also fall into this category. The part that forest soils play in providing growth and support of forest trees is essential. Soil composition, physical and chemical character, and the behavior of soils under different forest conditions must be studied and understood. Dendrology covers the description, classification, and identification of the numerous species of trees with which foresters must work. Silvics deals with the manner in which environmental factors such as climate, slope, soil, fire, and biological conditions affect forest sites. Silviculture deals with the technical problems of establishing new forests, naturally or artificially, making thinnings or cuttings to stimulate growth on the remaining trees, and making harvest cuts of forest stands in such a way as to assure their perpetuation. *See* DENDROLOGY; SILVICULTURE.

Forest management. The fact that forest crops of timber take from 20 to 100 years to mature makes it essential that planning and decision making should be based on the best facts and forecasts available. Planning for use of resources other than timber also requires careful long-term planning and integration with all the environmental factors of the forest. Forest management refers to the practical application of scientific, economic, and social principles affecting the administration and working of a forest estate for specified objectives. Forest management includes mensuration (forest measurement), regulation, and preparation of working plans. The working plans also involve forest finance, forest economics, and forest policy, and should meet the requirements of an environmental impact statement. This statement, which is now a prerequisite of all management plans of national forests, must show the probable impact that the management plan will have on the wildlife, watershed, and recreational resources. The statement may even include integration with range (grazing of livestock) management resources and, in some cases, with municipal and utility plans and resources. Management plans and policies for forests or public lands may have to submit their plans and environmental impact statements to a series of public hearings and amendments before they are finally adopted.

Utilization. Forests cannot be used or timber harvested without a system of roads, bridges, culverts, and some administrative structures. Their planning and construction and the planning and conduct of logging operations make up a body of knowledge called forest engineering. In terms of volume and dollar value, the greatest direct product of most forests is wood. Study of its variations of structure, physical and mechanical properties, and behavior of wood under varying conditions is labeled wood technology. Utilization also includes the manufacture, transportation, and marketing of forest products, and it most certainly includes the utilization of all wood (and bark) materials left over after logging operations in the woods and manufacturing at the mills. The use of leftovers is resulting in a minimum of waste in the woods and provides a major source of increased yields of products and fuel for the forest. *See* LOGGING (FORESTRY); WOOD PRODUCTS. [W.S.Br.]

Ecosystem analysis. The term "ecosystem" is the most concise formulation of the concept of an interacting system comprising living organisms together with their site. The goal of the study of whole ecosystems is to understand their processes and functions. Such an understanding is urgent because of the need to predict the long-term as well as the short-term consequences of humans' actions in ecosystems. Greatest emphasis has been placed on photosynthesis and respiration, on cycling of water and nutrients, and on the relationship of these to forest productivity.

Studies of nutrient cycling in hardwood forests of the northeastern United States show the enormous amount of nutrients available for uptake and regeneration of a new forest after cutting of the mature forest. Conventional forest cuttings pose little threat to site productivity, but short-rotation cropping and whole-tree harvesting can lead to nutrient deficiencies. This kind of integrated study will increasingly provide the basis for management of terrestrial and adjacent aquatic ecosystems. *See* ECOSYSTEM.

Forest succession. Since the developing forest changes the forest site, it follows that the changed site may itself be more favorable to a new group of tree species other than those currently occupying it. As a result, successive forest communities occupy the site in the absence of fire, logging, windstorm, or other disturbance. Thus, in the northeastern United States early successional or pioneer communities characterized by aspens, cherries, birches, and pines occupy open sites. These in time are replaced on fertile sites by midsuccessional communities containing oaks, ash, and other species until, in several hundred years, the late successional or climax community is attained. This climax contains species, such as hemlock, sugar maple, and beech, that are capable of reproducing under their own cover in the absence of disturbance. On dry, sandy sites, especially, periodic fires enable pines to succeed themselves over long time periods, and a fire climax prevails. Thus, species composition of late successional communities is dependent on prevailing site factors; several climax types are typically recognized in each forest region.

Effects of fire on forest stands. A light surface fire burns grass and litter of this stand in central Idaho. (*U.S. Forest Service*)

Throughout the world, fire greatly influences forest succession, setting it back or maintaining fire species as climax types. In the southeastern United States, pine types predominate and constitute the most important commercial forests. Pines owe their existence to periodic fires; without fire they would be replaced successionally on many sites by mixed hardwood forests. Similarly, in western North America, the extensive pure stands of Douglas-fir owe their existence to fire and periodic felling by wind. They would be replaced by less valuable, mixed coniferous types if these disturbances were eliminated. Also in western North America the esthetically pleasing and commercially important parklike stands of ponderosa pine are perpetuated and shaped by fire (see illustration). In present-day silviculture humans increasingly use controlled fire together with cutting, chemicals, and mechanical treatments to maintain the forest on a given site at a stage of forest succession most valuable to humans. *See* ECOLOGY. [B.V.B.]

Forest engineering Application of engineering principles to the solution of forestry problems, such as those dealing with harvesting, forest transportation, materials handling, and mechanical silviculture, with regard to long-range environmental and economic effects. The work that forest engineers perform varies widely throughout the United States. In the Northeast and Southeast, tasks include mechanization of harvesting, site preparation, planting, and product handling. In the West, planning, design, and construction of road systems are major operational and environmental challenges. The skills possessed by forest engineers in the eastern United States parallel most closely those of the mechanical or agricultural engineer. In the West, the skills are more closely aligned with civil engineering.

Forest engineering encompasses many important and diverse subjects. Logging is one of the most important aspects. It is very critical that the logging engineer not only log at the lowest cost possible but also consider other factors, such as safety, water and soil protection, ergonomics, esthetics, machine design and development, and systems development. *See* LOGGING (FORESTRY).

Protecting the forest environment is increasingly important to the forest engineer. The "cut and get out" attitude toward forest properties is a thing of the past. Today's engineer has an obligation to preserve forests, whether they are publicly or privately owned. Sometimes, this creates a conflict among interested parties in various phases of forest management. Environmental issues include water quality, soil disturbance, preservation of scenic values and recreation and wilderness areas, presence of wildlife, controversial logging methods like clearcutting, smoke due to slash burning, and forest pest control. *See* FOREST AND FORESTRY. [J.E.O'L.]

Forest fire The term wildfire refers to all uncontrolled fires that burn surface vegetation (grass, weeds, grainfields, brush, chaparral, tundra, and forest and woodland); often these fires also involve structures. In addition to the wildfires, several million acres of forest land are intentionally burned each year under controlled conditions to accomplish some silvicultural or other land-use objective or for hazard reduction.

Most wildfires are caused by human beings, directly or indirectly. In the United States less than 10% of all such fires are caused by lightning, the only truly natural cause. In the West (the 17 Pacific and Rocky Mountain states) lightning is the primary cause, with smoking (cigarettes, matches, and such) the second most frequent. Combined they account for 50 to 75% of all wildfires. In the 13 southern states (Virginia to Texas) the primary cause is incendiary. This combined with smoking and debris burning make up 75% of the causes. The 20 eastern states have smoking and debris burning as causing close to 50% of all wildfires. Miscellaneous causes of wildfires are next in importance in most regions. The other causes of wildfires are machine use and campfires. Machine use includes railroads, logging, sawmills, and other operations using equipment.

The manner in which fuel ignites, flame develops, and fire spreads and exhibits other phenomena constitutes the field of fire behavior. Factors determining forest fire behavior may be considered under four headings: attributes of the fuel, the atmosphere, topography, and ignition. A forest fire may burn primarily in the crowns of trees and shrubs—a crown fire; primarily in the surface litter and loose debris of the forest floor and small vegetation—a surface fire; or in the organic material beneath the surface litter—a ground fire. The most common type is a surface fire.

The U.S. Forest Service has developed a National Fire Danger Rating System (NFDRS) to provide fire-control personnel with numerical ratings to help them with the tasks of fire-control planning and the suppression of specific fires. The system includes three basic indexes: an occurrence index, a burning index, and a fire load index. Each of these is related to a specific part of the fire-control job. These indexes are used by dispatchers in making decisions on setting up firefighting forces, lookout systems, and so forth. [W.S.Br.]

Forest genetics and breeding Forest tree improvement has greatest application to artificially established stands. The genetic gain that can be realized from manipulating natural stands is usually so small and uncertain as to make the effort impractical.

Natural variation occurs among species of trees; within a species variation occurs among provenances (sources), stands, and trees in stands. The first step in a tree improvement program is to establish tests to determine the species and provenances best suited for local conditions. This evaluation is especially critical for establishment of plantations outside the range of the species. The variations among trees of a species are usually greater within a stand than among stands and provenances.

Forest tree improvement is the combined application of forest genetics and silvicultural practice. The trees can be genetically manipulated because of their inherent similarities and differences. The different genotypes produced from cross-fertilization among selected trees are likewise amenable to cultural practices, such as cultivation, spacing, and nutrient and water supplementation. Tree improvement programs are most often initiated by selecting superior phenotypes from a base population which could be a natural stand or a plantation. The phenotype reflects the superior physical appearance of the tree, for example, its total height, straightness, and freedom from pests. *See* BREEDING (PLANT); SILVICULTURE.

Seed orchards are most often established by using grafts in which twigs (scions) from the upper crown of the phenotypically superior trees are grafted on 2-ft-tall (0.6-m) seedlings of the same species. Results from genetic tests which are established separately from the seed orchard are used to remove grafts of the poorest clones (genotypes) from the orchard, leaving only the best clones to cross-pollinate for commercial seed production. The alternative to grafted seed orchards are seedling seed orchards in which the genetic test becomes the orchard following removal of the poorest families and the poorest trees of the best families. The best trees of the best families are left to cross-pollinate for commercial seed production as is done for clonal or grafted seed orchards. *See* POLLINATION.

The genetic tests established to determine the genetic worth of the parent trees also serve as the base population of selection for the second generation, and their progeny are the base population of the third generation, and so on. This procedure can cause a rapid constriction in the genetic base if precautions are ignored. One solution to this potential problem is the use of a wide initial genetic base and then the development of breeding units. Various levels of inbreeding (crosses among relatives) are tolerated within the breeding units, but only nonrelated trees make up the advance-generation seed orchards, assuring outcrossing (nonrelated breeding) for the production of seeds and seedlings for the operational plantations. Hybridization programs are applied to species having minimal genetic variation in which selection programs are ineffective. They also are used for adverse environments in which favorable genes for traits, such as tree form and disease resistance, of the intolerant species are combined with the tolerant species. Optimum genetic gains will be realized only by combining the improved genetic stock with good silvicultural practices, which include matching species to site, intensive site preparation, competition control, nutrient supplementation, and thinning. *See* FOREST AND FORESTRY. [R.C.K.]

Forest pest control The techniques and methods used in protecting forests from pests and diseases. In the natural, unmanaged forest, pests are most often of little significance because there are usually many kinds of trees of different ages, which discourages the buildup of pest populations. In many of today's forested areas, however, this natural balance has been changed, and trees of more or less continuous age and species composition are grown in large areas. In such a management procedure, known as monoculture, pest problems take on increased significance.

As trees go through their natural developmental process from seed to seedling to mature and finally overmaturity, each stage is subject to attack by a different series of pests. A variety of insects and fungi attack seeds on the trees. Birds, squirrels, and other rodents also cause major losses to seed under natural field conditions. Seedlings grown in a forest tree nursery are subjected to a number of nematodes, insects, and fungi that attack roots, stems, and foliage. Under these conditions, pests that would not be a problem under field conditions can cause serious losses. In the field, seedlings become susceptible to a wider variety of pests. The most severe problem at this stage, however, is weed competition and drought.

As the seedling becomes a sapling, some of the seedling pests persist and new pests, including sawflies, foliage diseases, canker diseases, borers, and root rots begin to play a major role. The poletimber stage provides the timber for forest products, but some of the more destructive pests are found in this stage, including bark beefies, borers, rust fungi, root rots, cankers, and decay. Trees of sawtimber size are the largest and most valuable, but experience the greatest losses from pests. Vascular wilts, bark beetles, root rots, hardwood borers, defoliators, decay, dwarf mistletoes, and a variety of other pests cause serious damage at this stage.

In some cases, a complex decline is caused by a combination of age, defoliating agents, root diseases, borers, and environmental stresses all working together. Sometimes, either through lack of continuous management or special use, trees are permitted to become overmature. When this happens, decay, borers, bark beetles, root rots, and other pests work overtime to reduce the number and quality of remaining trees.

However, these stands may still be valuable for use other than as lumber.

Pest problems are often interrelated. Pests frequently aggravate one another's impact due to the compounding stress factors. One option is integrated pest management, a strategy that manipulates forest pests to achieve resource management objectives. It is the planned and systematic use of all necessary and appropriate cultural, biological, chemical, and mechanical means to prevent or reduce pest-caused damage and losses to levels that are economically, environmentally, and esthetically acceptable. Key elements in this process include systematic detection and evaluation of forest pests and hazard rating of stands. *See* PLANT PATHOLOGY. [R.L.An.]

Forestry, urban The management of tree resources in and around cities and towns. A forest is a community of woody plants, predominantly trees, and related organisms. By this definition, many cities and towns are forests, and in many ways are similar to rural forests. There are many differences, however, resulting primarily from intense interaction of vegetation and people.

There is a diversity of growing space for woody plants in urban areas. Growing sites may be as small as planting pots or holes carved in sidewalks in the central city, or as large as parks, green corridors, median strips, and vacant lots in commercial areas. In residential sections, yards of homes and open spaces along streets dominate the landscape. At the city's edge, trees of the urban forest blend in with rural woodlands. While not visibly distinguishable, these woodlands are different in that they are owned largely by urban people and have potential for development into urban properties.

In rural forests, plant communities develop as a result of existing site conditions. Soil nutrients, light, and climate play key roles in development of forest types. These forests, left alone by humans, will develop through a series of predictable successional changes. Development of urban forests is less predictable, however, because of human influence: growing sites have been altered; different, often exotic, species are planted; and trees are intended to serve various purposes. *See* ECOLOGICAL SUCCESSION.

The makeup of urban forest differs among central cities, suburbs, and urban fringes. Urban soils are often infertile or have a poor structure, following disruption and compaction by construction equipment. The resulting loss of nutrients, structure, and air space creates difficult conditions for tree growth. At the urban fringe, soils develop naturally and retain most of the elements required for tree growth.

Management of urban forests requires an understanding of both forest resources and people. Planning during early stages of development offers the greatest opportunity for considering soils and other urban forest features in a community. Urban foresters have often been able to influence the planning process and incorporate natural features into new developments. *See* FOREST AND FORESTRY. [G.A.Mo.; G.Gr.]

Forging The plastic deformation of metals, usually at elevated temperatures, into desired shapes by compressive forces exerted through a die. Forging processes are usually classified either by the type of equipment used or by the geometry of the end product. The simplest forging operation is upsetting, which is carried out by compressing the metal between two flat parallel platens. From this simple operation, the process can be developed into more complicated geometries with the use of dies. A number of variables are involved in forging; among major ones are properties of the workpiece and die materials, temperature, friction, speed of deformation, die geometry, and dimensions of the workpiece.

In practice, forgeability is related to the material's strength, ductility, and friction. In terms of factors such as ductility, strength, temperature, friction, and quality of forging, various engineering materials can be listed as follows in order of decreasing forgeability: aluminum alloys, magnesium alloys, copper alloys, carbon and low-alloy steels, stainless steels, titanium alloys, iron-base superalloys, cobalt-base superalloys, columbium alloys, tantalum alloys, molybdenum alloys, nickel-base superalloys, tungsten alloys, and beryllium. *See* METAL.

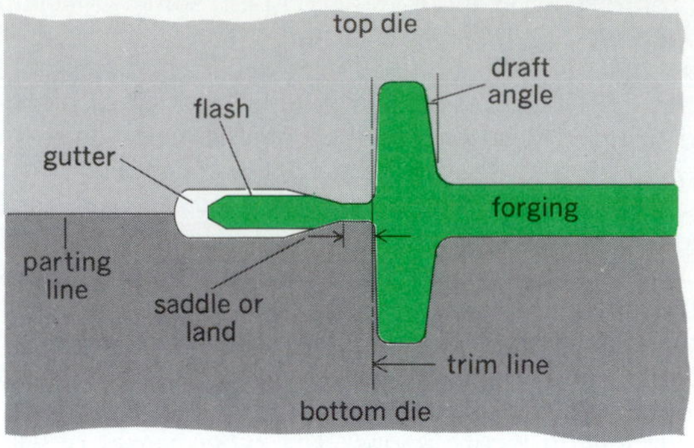

Closed-die forging terminology.

Some of the terminology in forging is shown in the illustration. Draft angles facilitate the removal of the forging from the die cavity. The purpose of the saddle or land in the flash gap is to offer resistance to the lateral flow of the material so that die filling is encouraged. Die filling increases as the ratio of land width to thickness increases up to about 5; larger ratios do not increase filling substantially and are undesirable due to increased forging loads and excessive die wear. The purpose of the gutter is to store excess metal. The flash is removed either by cold or hot trimming or by machining.

A number of methods produce the necessary force and die movement for forging. Two basic categories are open-die and closed-die forging. Drop hammers supply the energy through the impact of a failing weight to which the upper die is attached. Another type of forging equipment is the mechanical press. For large forgings the hydraulic press is the only equipment with sufficient force. However, the speed for such presses is about one-hundredth that of hammers. *See* METAL FORMING; SWAGING. [S.Ka.]

Formaldehyde The simplest aldehyde, formula HCH═O. Because of its extreme reactivity, even with itself, it cannot be readily isolated or handled in the pure state. Therefore, it is produced and marketed as an aqueous solution (usually 37–50% formaldehyde by weight), sometimes known as Formalin. It is also sold as the solid hydrated polymer known as paraformaldehyde or paraform.

Formaldehyde is used principally to produce synthetic resins and adhesives by reaction with phenols, urea, and melamine. Other uses are in the manufacture of textiles, dyes, drugs, paper, leather, photographic materials, embalming agents, disinfectants, and insecticides. *See* ALDEHYDE; PHENOLIC RESIN. [L.M.]

Formic acid A colorless, pungent liquid having a melting point of 8.4°C (47°F) and the formula HCOOH. Formic acid is miscible with water, alcohol, ether, and glycerol. It is toxic, and causes painful skin wounds. It is a constituent of the stings of ants, stinging caterpillars, and stinging nettles.

Formic acid is stronger than acetic acid. With carbonates, oxides, and hydroxides of alkali or alkaline-earth metals, it gives formates (salts); with alcohols it forms esters (formates). *See* ESTER.

It is used as a reducing agent for metallic ions and dyes, in dehairing and tanning, as a latex coagulant, in the preparation of a large number of esters, as a source of allyl alcohol, and in the preparation of oxalic acid. [E.B.R.]

Fossil The remains and traces, such as skeletons, animal tissues, footprints, and trails, of organisms preserved from the geologic past. The term fossil was originally used for a wide range of curios, both organic and mineral, found in rocks; but it came to be restricted to organic remains about the close of the 17th century. The term fossil is still used as an adjective and, in a figurative sense, to indicate that nonorganic structures in the rocks, such as fossil mudcracks, rain imprints, ripple marks, and even fuels, were formed in the geologic past.

Several different modes of preservation are known. Generally only the hard parts, such as bones and shells, have been preserved, but in some instances even the soft tissues are preserved almost intact. Some organic objects are turned to stone. This petrifaction usually involves only the hard parts such as bones, shells, and wood. *See* EDIACARAN FAUNA; PETRIFACTION.

Mold, casts, and imprints form another category of fossils. They are formed when the sediment has solidified about an organic object and the latter is subsequently dissolved, leaving a hole in the rock. This is a mold. Deposition of mineral matter from underground solutions may fill the hole, producing a cast. Molds of thin objects such as leaves are commonly called imprints.

Tracks, trails, and burrows are traces that constitute another category of fossils. Although involving no remnant of the organism, they are structures made by a living animal and, as such, may give evidence of its size, shape, and habits.

Coprolites are petrified excrement. Undigested particles of food preserved in such fossils may indicate the feeding habits of animals long since extinct. *See* PALEOBOTANY; PALEONTOLOGY. [C.O.D.]

Fossil fuel Carbon-containing materials that are burned with air, or with oxygen derived from the air to supply heat for any purpose.

Fuels are characterized (1) by their physical form (solid, liquid, or gas) at normal temperatures; (2) by their heating value, that is, by the amount of heat given off when a unit weight or volume of the material is burned under standard conditions; and (3) by their combustion characteristics, which cover such points as ease of ignition, rate of combustion, flame temperature, and flame luminosity.

Fuels may be used to supply heat directly, as in a furnace. They may also be used to supply heat to perform some other function, such as to produce steam in a boiler, the steam then being used to produce power in a turbine or engine. The heat may be used to raise the temperature and pressure of the products of the combustion to provide a driving force, as in the internal combustion engine or gas turbine.

Solid fuels comprise primarily the various ranks of coal (anthracite, bituminous, subbituminous, and lignite) and the coke or char derived from them. The important factors governing choice of coal for a given use are cost per unit of heating value, size range, moisture and ash content, and amount of smoke evolved. *See* COAL; COKE.

Liquid fuels are derived almost entirely from petroleum. In general, the liquid fuels suitable for internal combustion engines command a premium, and the fuels designed for external combustion are derived from those portions of the petroleum that have the least potential for gasoline and diesel fuel manufacture. Factors in choice of liquid fuels are heating value, fluidity, boiling range, impurities such as sulfur, and freedom from water and sediment. *See* DIESEL FUEL; FUEL OIL; GASOLINE; OIL SAND; OIL SHALE; PETROLEUM.

Of the gaseous fuels, the most important are those derived from natural gas or petroleum, chiefly methane. Other gaseous fuels are those that are by-products of some manufacturing process and therefore available at a cost that makes them competitive with natural gas. These include coke-oven gas, blast-furnace gas, and producer gas. *See* ENERGY SOURCES; FUEL GAS; NATURAL GAS. [H.R.B.]

Fossil human All prehistoric skeletal remains of humans which are archeologically earlier than Neolithic (necessarily an imprecise limit), regardless of degree of mineralization or fossilization of bone, and regardless of whether the remains may be classed as *Homo sapiens sapiens*, modern humans.

Discoveries began in the early years of the 19th century, although their meaning and antiquity were not recognized before the finding of Neandertal man in 1856. Remains have come principally from Europe and adjacent portions of Asia, from Africa, from China, and from Java. Because of the rather late entry of humans into the New World, American Indian remains are all of relatively recent origin and recognizable as *H. s. sapiens*.

Prehuman ancestry. The human is a catarrhine primate, part of a group including Old World monkeys, apes, and various extinct forms. The oldest certain representatives of the Catarrhini are fossils from the Fayum beds of northern Egypt, dated around 30 million years (m.y.) old. The best-known of these is *Propliopithecus* (= *Aegyptopithecus*) *zeuxis*, a possible ancestor for apes and humans (see illustration). This animal was the size of a cat, with a monkeylike body, apelike teeth, and a distinctive skull: a mosaic of evolutionary features. Between about 23 and 15 m.y. ago, in the early Miocene, several species of *Dryopithecus* (= *Proconsul*) in East Africa represent relatives of the ancestors of humans. Evidence from both comparative morphology and molecular studies of proteins shows that humans' closest living relatives are the African apes: the chimpanzee and the gorilla. *See* AEGYPTOPITHECUS; MONKEY; PRIMATES.

The first evidence of a distinctive line of human evolution is provided by a group of animals known colloquially as ground apes. These comprise the genera *Sivapithecus*, *Ramapithecus* (= *Kenyapithecus*), and *Gigantopithecus*, known by about seven species from East Africa, eastern Europe, Turkey, Pakistan, India, and China in the middle and late Miocene (about 15–8 m.y. ago), contemporary with the younger species of *Dryopithecus* from western Europe. *See* KENYAPITHECUS; RAMAPITHECUS.

Pliocene Homininae. *Australopithecus*, the first truly humanlike beings, appear in the fossil record in quantity some 4 m.y. ago, during the Pliocene. One partial jaw with a single tooth has been found that may be as old as 6 m.y., and a few other fragments are a bit younger, but the majority of more complete finds date between 1 and 4 m.y. It now appears that between 6 and 4 m.y. ago (when few fossils are known), a generalized type of *Australopithecus* inhabited eastern (and perhaps southern) Africa. Recently interpreted finds from Hadar and Laetoli, between 4 and 3 m.y. in age, document such a creature, which has been named *A. afarensis*. Most exciting was the find in 1974 of a partial skeleton, identified as a female by its pelvic bones and named "Lucy."

There is some disagreement as to the place of *A. afarensis* in human evolution, whether a possible ancestor of all later forms or only related to one of several lineages. By 2.5 to 3 m.y. ago, the two later varieties of *Australopithecus* had diverged. *Australopithecus africanus*, the so-called gracile form, probably was little changed from its putative ancestor; it is known especially in South Africa, and less certainly in East Africa, between 3 and 2 m.y. ago. The second form, known as *A. robustus* (or also *A. boisei*, and sometimes given generic or subgeneric rank as *Paranthropus*), lived in East and South Africa between roughly 2.25 and 1.25 m.y. ago. At some sites, bones of the two forms may occur together. Specimens

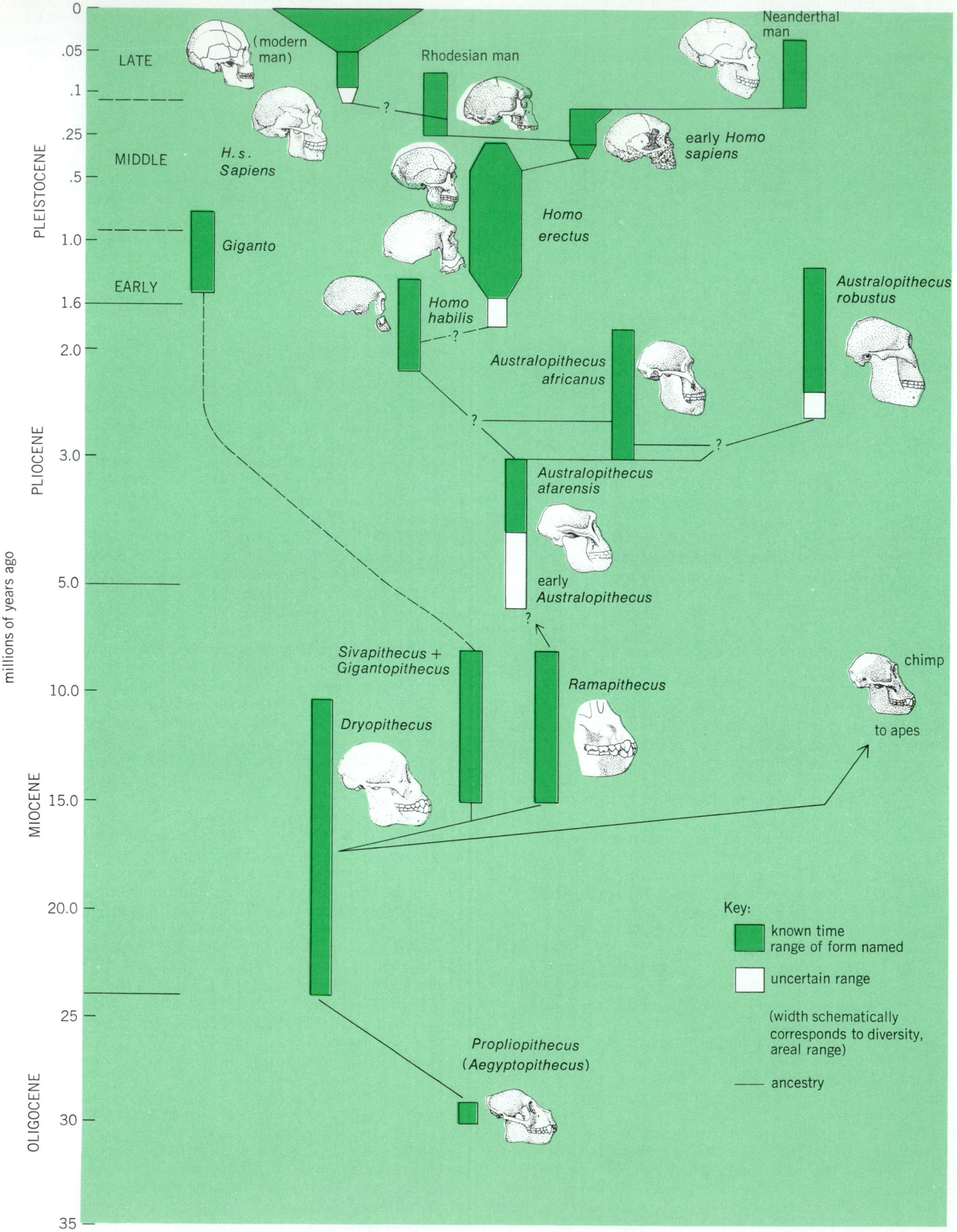

Human phylogeny from the Oligocene to the present time, showing the skulls of the major known fossil relatives and possible ancestors of modern humans.

of *A.* (*P.*) *robustus* may be distinguished, however, by their larger size and their craniodental specializations. *See* AUSTRALOPITHECUS.

Early Pleistocene. The only other genus of the subfamily Homininae (or family Hominidae of many authors) is *Homo*, true humans, into which all later forms (including ourselves) are placed. The identification of the earliest specimens of *Homo* is a subject of debate among paleoanthropologists. In the late 1970s the scientific pendulum had swung back to an idea proposed on less secure grounds by L. S. B. Leakey and colleagues in 1964. They named a species *Homo habilis*, based on several finds from Olduvai. Especially significant was the discovery of the 1.8-m.y.-old remains of a juvenile's lower jaw, with teeth much like those of *A. africanus*, and its partial skull, with an estimated cranial capacity of about 650 to 700 cm^3. Although most experts did not consider this find to be evidence of a new species, and showed that other specimens involved here might really be a later type of human, newer discoveries suggested that there was indeed an early form of *Homo* in East Africa, and possibly elsewhere, between 1.5 m.y. and more than 2 m.y. years ago.

Homo erectus. While *H. habilis* is apparently a short-lived and rare East African species, its descendant, *Homo erectus*, is common, widespread, and long-surviving. The first fossils were found in Java in 1893 and termed *Pithecanthropus erectus*. Later finds in China and across Africa were each given distinctive generic and specific names, but are now all considered local variants or subspecies of the single species *H. erectus*. *See* JAVA MAN; PEKING MAN.

The earliest known representatives are once again from East Africa, dating from 1.5 to 1 m.y. ago. *Homo erectus* probably spread from Africa into Eurasia more than 1 m.y. ago, perhaps the first type of human to do so in large numbers. Fossils in Asia may date from over 1 m.y. through nearly 250,000 years ago. New finds in Africa document *H. erectus* as young as about 400,000 years ago. In Europe, however, there is definite evidence of human-made tools between 1 m.y. and 500,000 years ago, but no human fossils are older than about 350,000 years. The earliest of these, in central and southeastern Europe, already seem to be more advanced in some ways than Afro-Asian *H. erectus*.

Premodern Homo sapiens. It is possible that the increased rigor of the glacial climate in Europe at this time was the impetus leading to the evolution of *H. sapiens*, modern-type humans. It is about equally likely that *H. sapiens* may have evolved elsewhere, but between about 500,000 and 350,000 years ago, the only known human fossils are those from Europe and those of typical *H. erectus* in China and North Africa. *See* HEIDELBERG MAN.

For the next 200,000 years there is still a dearth of evidence, but it may be suggested that local varieties of *H. sapiens* developed in Africa, Europe, and perhaps northern Asia. The best-known of these varieties is Neandertal man, mostly from Europe and the Near East, who was formally termed *Homo sapiens neanderthalensis*. *See* FONTÉCHEVADE MAN; NEANDERTALS.

Spread of modern humans. There is virtually no evidence of the area of origin and early history of anatomically modern *H. s. sapiens*, characterized by a small upright face, small teeth and brow ridges, chin, and high rounded skull. There are some early (about 100,000 year old) fossils from East and especially South Africa which suggest this region as a possible source. By about 35,000 years ago, modern *H. s. sapiens* was the sole form of human to be found anywhere. One reason for the success of *H. s. sapiens* may have been their greater toolmaking efficiency. In many parts of the world, they also engaged in artistic pursuits, including carving small animal statues and perhaps calendars as well as painting on the walls of deep caves. [E.D.]

Foucault pendulum A pendulum or swinging weight, supported by a long wire, by which J. B. L. Foucault demonstrated in 1851 the rotation of Earth on its axis. Foucault used a 62-lb (28-kg) iron ball suspended on about a 200-ft (60-m) wire in the Pantheon in Paris. The upper support of the wire restrains the wire only in the vertical direction. The bob is set swinging along a meridian in pure translation (no lateral or circular motion). In the Northern Hemisphere the plane of swing appears to turn clockwise; in the Southern Hemisphere it appears to turn counterclockwise, the rate being 15 degrees times the sine of the local latitude per sidereal hour. Thus, at the Equator the plane of swing is carried around by Earth and the pendulum shows no apparent rotation; at either pole the plane of swing remains fixed in space while Earth completes one rotation each sidereal day. *See* DAY; INERTIAL GUIDANCE SYSTEM; PENDULUM; SCHULER PENDULUM. [F.H.R.]

Foundations Structures or other constructed works are supported on the earth by foundations. The word "foundation" may mean the earth itself, something placed in or on the earth to provide support, or a combination of the earth and the elements placed on it. The foundation for a multistory office building could be a combination of concrete footings and the soil or rock on which the footings are supported. The foundation for an earth-fill dam would be the natural soil or rock on which the dam is placed. Concrete footings or piles and pile caps are often referred to as foundations without including the soil or rock on which or in which they are placed. The installed elements and the natural soil or rock of the earth form a foundation system; the soil and rock provide the ultimate support of the system. Foundations that are installed may be either soil-bearing or rock-bearing. The reactions of the soil or rock to the imposed loads generally determine how well the foundation system functions. In designing the installed portions, the designer must determine the safe pressure which can be used on the soil or rock and the amount of total settlement and differential settlement which the structure can withstand.

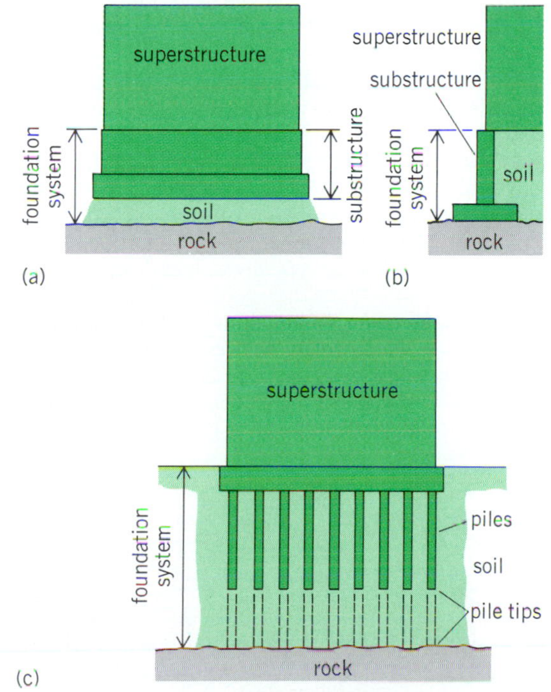

Examples of foundation systems. (*a*) Structure supported on a foundation bearing on soil. (*b*) Structure supported on a foundation bearing on rock. (c) Structure supported by a pile foundation.

The installed parts of the foundation system may be footings, mat foundations, slab foundations, and caissons or piles, all of which are used to transfer load from a superstructure into the earth. These parts, which transmit load from the superstructure to the earth, are called the substructure (see illustration).

Footings or spread foundations are used to spread the loads from columns or walls to the underlying soil or rock. Normally, footings are constructed of reinforced concrete; however, under some circumstances they may be constructed of plain concrete or masonry. When each footing supports only one column, it is square. Footings supporting two columns are called combined footings and may be either rectangular or trapezoidal. Cantilever footings are used to carry loads from two columns, with one column and one end of the footing placed against a building line or exterior wall. Footings supporting walls are continuous footings.

Mat or raft foundations are large, thick, and usually heavily reinforced concrete mats which transfer loads from a number of columns or columns and walls to the underlying soil or rock. Mats are also combined footings, but are much larger than a footing supporting two columns. They are continuous footings and are designed to transfer a relatively uniform pressure to the underlying soil or rock.

Slab foundations are used for light structures wherein the columns and walls are supported directly on the floor slab. The floor slab is thickened and more heavily reinforced at the places where the column and wall loads are imposed. *See* CAISSON FOUNDATION; PILE FOUNDATION; RETAINING WALL. [G.M.R]

Four-bar linkage A basic linkage mechanism used in machinery and mechanical equipment. The term has been applied to three types of linkages: plane, spherical, and skew.

The plane four-bar linkage (Fig. 1) consists of four pin-connected links forming a closed loop, in which all pin axes are parallel. The spherical four-bar linkage consists of four pin-connected links forming a closed loop, in which all pin axes intersect at one point. The skew four-bar linkage (Fig. 2) consists of four jointed links forming a closed loop, in which crank 2 and link 4 are pin-connected to ground 1 and the axes of the pins are generally nonparallel and nonintersecting; coupler 3 is connected to crank 2 and link 4 by ball joints.

Four-bar linkages are most frequently used to convert a uniform continuous rotation (the motion of crank 2) into a nonuniform rotation or oscillation (the motion of link 4). In instrument applications the primary function of the linkage is the conversion of motion, while in power applications both motion conversion and power transmission are fundamental.

Each of the above linkages can be proportioned for three types of motion, or linkage types: crank-and-rocker, drag, and double-rocker.

Crank-and-rocker linkages have a motion in which the crank (link 2) is capable of unlimited rotation, while the output link (link 4) oscillates or rocks through a fraction of one turn (usually less than 90°). This is the most common form of the plane and the skew four-bar linkage, and is used in machinery and appliances of all types.

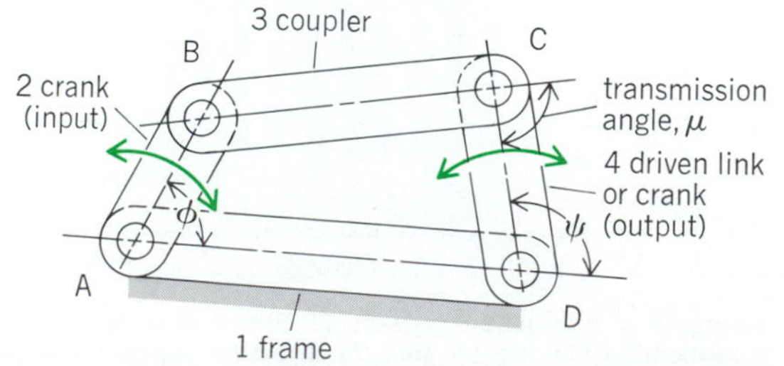

Fig. 1. Plane four-bar linkage with joints at *A*, *B*, *C*, and *D*. ϕ, ψ, and μ are angles defining orientations of joints.

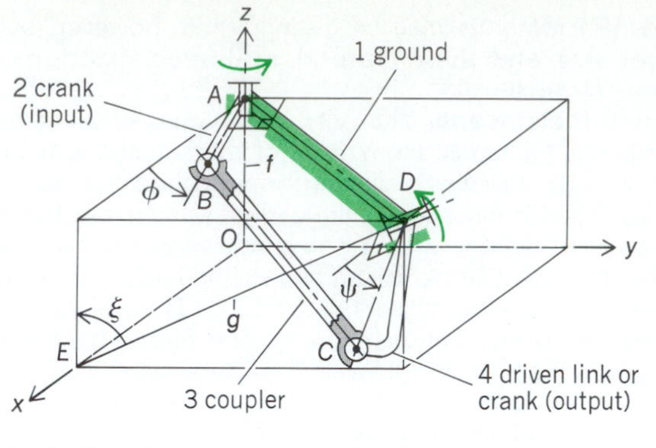

Fig. 2. Skew four-bar linkage with joints at *A*, *B*, *C*, and *D*. *OA* = *f*; *ED* = *g*; *OE* = common perpendicular between axes of pin joints at *A* and *D*; ϕ, ψ, and ξ are angles defining orientations of joints.

In drag linkages the motions of cranks 2 and 4 are both capable of unlimited rotations. The plane drag linkage has been used for quick-return motions. The most common drag linkage is the spherical drag linkage. One such linkage is the Hooke-type universal joint, or hooke joint. *See* UNIVERSAL JOINT.

In double-rocker linkages, neither crank 2 nor 4 is capable of complete rotations. Such motions occur in hand tools and mechanical equipment in which only limited rotations are required. *See* LINKAGE (MECHANISM); SLIDER CRANK MECHANISM; STRAIGHT-LINE MECHANISM. [F.F.]

Fourier series A mathematical tool for the study of periodic phenomena, indispensable in the theory of wave motion (for example, light waves and sound waves), oscillatory mechanical systems (such as vibrating strings), and celestial orbits. Fourier series and related topics have important applications to other branches of mathematics, of which the theories of probability and partial differential equations are worthy of special mention. Finally, the disciplines motivated by the subject itself enjoy a prominent position in pure mathematical research.

A function f of a real variable is said to be periodic with period T if $f(t + T) = f(t)$ for each t. The simplest examples of functions with a given period T are pure harmonics, that is, functions of the form $f(t) = a_n \cos n\omega t + b_n \sin n\omega t$, where $\omega = 2\pi T^{-1}$ is the fundamental frequency and a_n, b_n are constants. The basic idea for the applications of Fourier series is that any function f of period T which satisfies rather mild restrictions can be represented as a superposition of pure harmonics, as in relation (1), or, in a more convenient form, using complex exponentials, as relation (2).

$$f(t) \sim \sum_{n=0}^{\infty} (a_n \cos n\omega t + b_n \sin n\omega t) \quad (1)$$

$$f(t) \sim \sum_{\infty}^{-\infty} c_n e^{in\omega t} \quad (2)$$

If term-by-term integration of relation (2) is assumed to be legitimate, an easy calculation shows that Eq. (3) can be formed.

$$c_n = T^{-1} \int_{-T/2}^{T/2} f(t) e^{-in\omega t} dt \quad (3)$$

(The interval of integration could be any interval of length T). This motivates the formal definition of Fourier series. Suppose that f is a function of period T such that expression (4) exists and is finite. The coefficients $\{c_n\}$ defined by Eq. (3)

$$\int_{-T/2}^{T/2} |f(t)|\, dt \quad (4)$$

are the Fourier coefficients of f, and the series in relation (2), the Fourier series expansion of f. The coefficients uniquely determine the function; that is, if $c_n = 0$ for each n, then f is essentially the zero function. In addition, many formal operations in which the series is substituted term by term for the function can be justified. *See* SERIES. [C.S.He.]

Fractals Geometrical objects that are self-similar under a change of scale, for example, magnification. The concept is helpful in many disciplines to allow order to be perceived in apparent disorder. For instance, in the case of a river and its tributaries, every tributary has its own tributaries so that it has the same structure organization as the entire river except that it covers a smaller area. The branching of trees and their roots as well as that of blood vessels, nerves, and bronchioles in the human body follows the same pattern. Other examples include a landscape with peaks and valleys of all sizes, a coastline with its multitude of inlets and peninsulas, the mass distribution within a galaxy, the distribution of galaxies in the universe, and the structure of vortices in a turbulent flow. The rise and fall of economic indices has a self-similar structure when plotted as a function of time. *See* GALAXY, EXTERNAL; TURBULENT FLOW; UNIVERSE.

The triadic Koch curve, shown in the illustration, is a good example of how a fractal may be constructed. The procedure begins with a straight segment. This segment is divided into three equal parts, and the (single) central piece is replaced by two similar pieces (illus. *a*). The same procedure is now applied to each of the four new segments (illus. *b*), and this is repeated an infinite number of times. The curve is self-similar, because a magnification by 3 of any portion will look the same as the original curve.

Fractals came into natural sciences when it was recognized that natural objects are random versions of mathematical fractals. They are self-similar in a statistical sense; that is, given a sufficiently large number of samples, a suitable magnification of a part of one sample can be matched closely with some member of the ensemble. Unlike the Koch curve which must be magnified by an integral power of 3 to achieve self-similarity, natural fractal objects are usually self-similar under arbitrary magnification.

Physicists have used the concept of fractals to study the properties of amorphous solids and rough interfaces and the dynamics of turbulence. It has also been found useful in physiology to analyze the heart rhythm and to model blood circulation, and in ecology to understand population dynamics. In computer graphics it has been shown that the vast amount of information contained in a natural scene can be compressed very effectively by identifying the basic set of fractals therein together with their rules of construction. When the fractals are reconstructed, a close approximation of the original scene is reproduced. *See* AMORPHOUS SOLID; CARDIOVASCULAR SYSTEM; COMPUTER GRAPHICS. [S.H.L.]

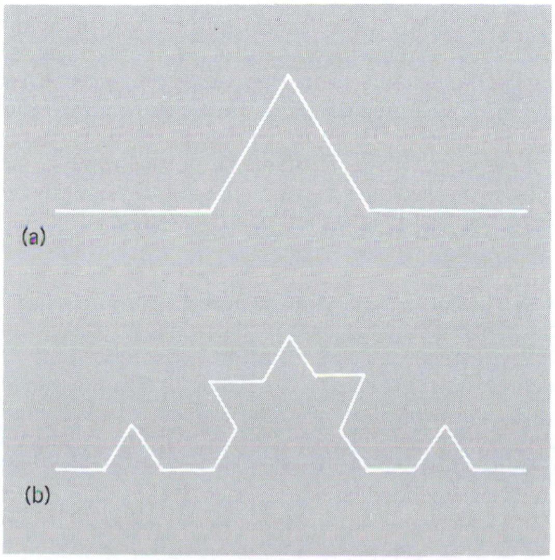

Koch curve, (*a*) first and (*b*) second stages.

Fractionating column An apparatus used widely for countercurrent contacting of vapor and liquid to effect separations by distillation or absorption. In general, the apparatus consists of a cylindrical vessel with internals designed to obtain multiple contacting of ascending vapor and descending liquid, together with means for introducing or generating liquid at the top and vapor at the bottom (see illustration).

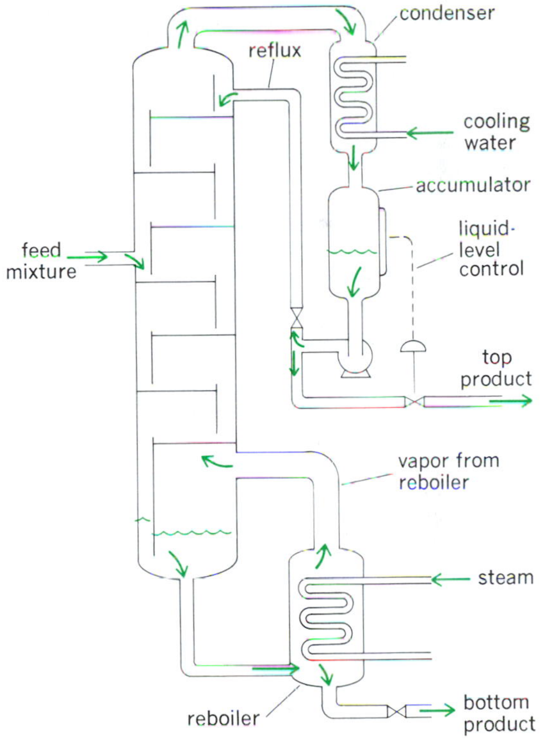

Elements of a fractionating column.

Fractionating columns used in industrial plants range in diameter from a few inches to 40 ft (12 m) and in height from 10 to 200 ft (3 to 60 m). They operate at pressures as low as a few millimeters of mercury and as high as 3000 psi (21 megapascals); at temperatures from −300 to 700°F (−180 to 370°F). They are made of steel and other metals, of ceramics and glass, and even of such materials as bonded carbon and plastics. *See* DISTILLATION; GAS ABSORPTION OPERATIONS. [M.Sou.]

Frame of reference A base to which to refer physical events. A physical event occurs at a point in space and at an instant of time. Each reference frame must have an observer to record events, as well as a coordinate system for the purpose of assigning locations to each event. The latter is usually a three-dimensional space coordinate system and a set of standardized clocks to give the local time of each event. For a discussion of the geometrical properties of space-time coordinate systems *see* SPACE-TIME. *See also* RELATIVITY.

In the ordinary range of experience, where light signals, for all practical purposes, propagate instantaneously, the time of an event is quite distinct from its space coordinates, since a single clock suffices for all observers, regardless of their state of relative motion. The set of reference frames which have a common clock or time is called newtonian, since Isaac Newton regarded time as having invariable significance for all observers.

For discussion of other types of reference frames *see* ACCELERATED REFERENCE FRAME; ASTRONOMICAL REFERENCE FRAME; INERTIAL REFERENCE FRAME; ROTATING REFERENCE FRAME. [B.G.]

Francisella A genus of very small, coccoid to ellipsoidal, pleomorphic, nonmotile and nonsporulating, gram-negative, rod-shaped bacteria. Fastidious and strictly aerobic, it grows at 98.6°F (37°C) within several days, but only in enriched media such as coagulated egg yolk or glucose-cysteine-blood-agar. The organisms occur in natural waters of the Northern Hemisphere, and can be parasitic and pathogenic in birds, anthropods, and mammals, including humans. *Francisella tularensis* has been found in many wild animals throughout North America, continental Europe, and Asia. Tularemia in humans is acquired via transmission by blood-sucking arthropods or by contact with infected animals, most frequently by hunters and butchers. *See* ANTIBIOTIC; MEDICAL BACTERIOLOGY; TULAREMIA. [W.Ma.]

Francium A chemical element, Fr, atomic number 87, an alkali metal element falling below cesium in group I of the periodic table. Distinguished by nuclear instability, francium exists only in short-lived radioactive forms, the most durable of which has a half-life of 21 min. The chief isotope of francium is actinium-K, an isotope of mass 223, which arises from the radioactive decay of the element actinium. From the properties of the known isotopes, it is reasonably certain that no long-lived form of element 87 will ever be found in nature or synthesized artificially.

1	2	3	4	5	6	7	8	9	10	11	12	13	14	15	16	17	18
1 H																	2 He
3 Li	4 Be											5 B	6 C	7 N	8 O	9 F	10 Ne
11 Na	12 Mg											13 Al	14 Si	15 P	16 S	17 Cl	18 Ar
19 K	20 Ca	21 Sc	22 Ti	23 V	24 Cr	25 Mn	26 Fe	27 Co	28 Ni	29 Cu	30 Zn	31 Ga	32 Ge	33 As	34 Se	35 Br	36 Kr
37 Rb	38 Sr	39 Y	40 Zr	41 Nb	42 Mo	43 Tc	44 Ru	45 Rh	46 Pd	47 Ag	48 Cd	49 In	50 Sn	51 Sb	52 Te	53 I	54 Xe
55 Cs	56 Ba	71 Lu	72 Hf	73 Ta	74 W	75 Re	76 Os	77 Ir	78 Pt	79 Au	80 Hg	81 Tl	82 Pb	83 Bi	84 Po	85 At	86 Rn
87 Fr	88 Ra	103 Lr	104 Rf	105 Db	106 Sg	107 Bh	108 Hs	109 Mt	110	*111*	*112*	*113*	*114*	*115*	*116*	*117*	*118*

lanthanide series	57 La	58 Ce	59 Pr	60 Nd	61 Pm	62 Sm	63 Eu	64 Gd	65 Tb	66 Dy	67 Ho	68 Er	69 Tm	70 Yb
actinide series	89 Ac	90 Th	91 Pa	92 U	93 Np	94 Pu	95 Am	96 Cm	97 Bk	98 Cf	99 Es	100 Fm	101 Md	102 No

The chemical properties of francium can be studied only on the tracer scale. The element has all the properties expected of the heaviest alkali element. With few exceptions, all the salts of francium are water-soluble. *See* ACTINIUM; ALKALI METALS. [E.K.H.]

Franck-Condon principle In any molecular system the transition from one energy state to another occurs so near to instantaneously that the nuclei of the atoms involved can be considered to be stationary during the transition. This principle, proposed by J. Franck in 1925 and developed on the basis of quantum mechanics by E. U. Condon in 1928, is important in discussing systems where more than one atom is involved. It is therefore valuable in problems of molecular spectroscopy and in the interpretation of the optical properties of liquids and solids. *See* LUMINESCENCE; MOLECULAR STRUCTURE AND SPECTRA. [C.C.K.; J.H.S.]

Franklinite A natural member of the spinel structure type, with composition $Zn^{2+}Fe_2^{3+}O_4$. The habit is octahedral, often modified by the cube and dodecahedron, but the mineral usually occurs as bands of isolated rounded grains, blebs, or compact masses. It is black with a metallic luster and red internal reflections; its hardness is 6 (Mohs scale); specific gravity is 5.3; and it is weakly magnetic.

Franklinite is confined in its occurrence to the unique ore bodies at Franklin and Sterling Hill (Ogdensburg), Sussex County, New Jersey. Franklinite is a major ore mineral and is still mined at Sterling Hill for spiegeleisen and zinc. *See* WILLEMITE; ZINC; ZINCITE. [P.B.M.]

Fraunhofer lines Dark absorption features in the solar spectrum. J. von Fraunhofer first studied them in 1814. They occur from the ultraviolet at about 180 nanometers to the infrared at 20 micrometers. Each line represents the net absorption of light by a particular atom or molecule. Most lines form in the Sun's atmosphere, although the Earth's telluric spectrum contributes lines of molecular oxygen (O_2), carbon monoxide (CO), and other molecules. Some lines, such as Fraunhofer's C line in the red (hydrogen-alpha), can be seen with a pocket spectroscope. Powerful research instruments reveal millions of lines, most of which are weak and blended together in an almost inextricable tangle.

A spectrum line is caused by the absorption of photons of light that excite the atom from a lower to a higher energy level. Spontaneous decay back to the atom's lower level then follows, accompanied by the isotropic emission of light at the wavelength of the line. The result is a loss of light in the Sun-Earth direction.

The study of the Fraunhofer spectrum is the principal means of learning about physical conditions in the solar atmosphere. On the resolved solar disk, variations in line strength from point to point convey information about temperature, Doppler shifts of the lines reveal gas motions, and line splitting from the Zeeman effect maps magnetic fields. Because each line represents a chemical element, the composition of the solar atmosphere can be deduced. *See* ASTRONOMICAL SPECTROSCOPY; DOPPLER EFFECT; SOLAR MAGNETIC FIELD; SUN; SUPERGRANULATION; ZEEMAN EFFECT. [W.C.Li.]

Free-electron laser A device in which laser action involves passage of electrons through a spatially varying magnetic field which causes the electron beam to "wiggle" and hence to radiate. Free-electron lasers are of interest because of their potential for efficiently producing very high-power radiation, tunable from the millimeter to the x-ray region. However, operation has been demonstrated only at 3.4 nanometers and at very low-power levels. The large Doppler upshift due to relativistic electron velocities can be adjusted, resulting in tunable emission at optical frequencies. *See* DOPPLER EFFECT; LASER. [S.F.J.; A.L.S.]

Free-electron theory of metals A model of a metal in which the free electrons, that is, those giving rise to the conductivity, are regarded as moving in a potential—the sum of the average potential due to the metal ions in the lattices and to all the remaining free electrons—which is approximated as constant everywhere inside the metal. The free electrons are thus assumed to move independently of one another throughout the space bounded by the surfaces of the metal. Interaction of the free electrons with the ion cores is assumed to be negligible. The free electrons are taken to be identical to the valence electrons of the free metal atoms; thus, alkali metals contain one free electron per atom, and aluminum contains three. The free electron theory should actually be termed the quasi-free-electron theory, to distinguish between electrons in a metal which behave nearly as do free electrons, and truly free electrons, such as those in a vacuum tube.

A number of important physical properties of some metals, in particular the simple monovalent metals, are explained satisfactorily in terms of the quasi-free-electron model. Among these are the electronic specific heat, the magnetic susceptibility, and the electrical and thermal conductivity. To understand the behavior of the more complex metals and of semiconductors, it became necessary to introduce the more complicated band theory of solids. *See* BAND THEORY OF SOLIDS. [F.J.B.]

Free energy A term in thermodynamics which in different treatments may designate either of two functions defined in terms of the internal energy E or enthalpy H, and the temperature-entropy product TS.

The function $(E - TS)$ is the Helmholtz free energy and is the function ordinarily meant by free energy in European references. The Gibbs free energy is the function $(H - TS)$. For the Lewis and Randall school of American chemical thermodynamics, this is the function meant by the free energy F. To avoid confusion with the symbol F as applied elsewhere to the Helmholtz free energy, the symbol G has also been used. Another development was the introduction of the name free enthalpy, with symbol G, for the Gibbs function. *See* WORK FUNCTION (THERMODYNAMICS).

For a closed system (no transfer of matter across its boundaries), the work which can be done in a reversible isothermal process is given by the series shown in Eq. (1). For these conditions, $T\Delta S$ represents the heat given up to the surroundings.

$$W_{rev} = -\Delta A = -\Delta(E - TS) = -(\Delta E - T\Delta S) \quad (1)$$

Should the process be exothermal, $T\Delta S < 0$, then actual work done on the surroundings is less than the decrease in the internal energy of the system. The quantity $(\Delta E - T\Delta S)$ can then be thought of as a change in free energy, that is, as that part of the internal energy change which can be converted into work under the specified conditions. This then is the origin of the name free energy. Such an interpretation of thermodynamic quantities can be misleading, however; for the case in which $T\Delta S$ is positive, Eq. (1) shows that the decrease in "free" energy is greater than the decrease in internal energy. *See* CHEMICAL THERMODYNAMICS.

For constant temperature and pressure in a reversible process the decrease in the Gibbs function G for the system again corresponds to a free-energy change in the above sense, since it is equal to the work which can be done by the closed system other than that associated with its change in volume ΔV under the given constant pressure P. The relations shown in Eq. (2) can be formed since $\Delta H = \Delta E + P\Delta V$.

$$\Delta G = -(\Delta H - T\Delta S) = W_{net} = W_{rev} - P\Delta V \quad (2)$$

Each of these free-energy functions is an extensive property of the state of the thermodynamic system. For a specified change in state, both ΔA and ΔG are independent of the path by which the change is accomplished. Only changes in these functions can be measured, not values for a single state.

The thermodynamic criteria for reversibility, irreversibility, and equilibrium for processes in closed systems at constant temperature and pressure are expressed naturally in terms of the function G. For any infinitesimal process at constant temperature and pressure, $-dG \geqq \delta w_{net}$. If δw_{net} is never negative, that is, if the surroundings do no net work on the system, then the change dG must be negative or zero. For a reversible differential process, $-dG > \delta w_{net}$; for an irreversible process, $-dG > \delta w_{net}$. The free energy G thus decreases to a minimum value characteristic of the equilibrium state at the given temperature and pressure. At equilibrium, $dG = 0$ for any differential process taking place, for example, an infinitesimal change in the degree of completion of a chemical reaction. A parallel role is played by the work function A for conditions of constant temperature and volume. Because temperature and pressure constitute more convenient working variables than temperature and volume, it is the Gibbs free energy which is the more commonly used in thermodynamics. *See* ENTROPY; THERMODYNAMIC PRINCIPLES. [P.J.B.]

Free fall The accelerated motion toward the center of the Earth of a body acted on by the Earth's gravitational attraction and by no other force. If a body falls freely from rest near the surface of the Earth, it gains a velocity of approximately 9.8 m/s every second. Thus, the acceleration of gravity g equals 9.8 m/s^2 or 32.16 ft/s^2. This acceleration is independent of the mass or nature of the falling body. For short distances of free fall, the value of g may be considered constant. After t seconds the velocity v_t of a body failing from rest near the Earth is given by Eq. (1).

$$v_t = gt \quad (1)$$

If a falling body has an initial constant velocity in any direction, it retains that velocity if no other forces are present. If other forces are present, they may change the observed direction and rate of fall of the body, but they do not change the Earth's gravitational pull; therefore a body may still be thought of as freely "failing" even though the resultant observed motion is upward.

For a body failing a very large distance from the Earth, the acceleration of gravity can no longer be considered constant. According to Newton's law of gravitation, the force between any two bodies varies inversely with the square of the distance between them; therefore with increasing distance between any body and the Earth, the acceleration of the body toward the Earth decreases rapidly. The final velocity v_f, attained when a body falls freely from an infinite distance to the surface of the Earth, is given by Eq. (2), where R is the radius of the Earth,

$$v_f = \sqrt{2gR} \quad (2)$$

which gives a numerical value of 11.3 km/s or 7 mi/s. This is consequently the "escape velocity," the initial upward velocity for a rising body to completely overcome the Earth's attraction.

Because of the independent action of the forces involved, a ball thrown horizontally or a projectile fired horizontally with velocity v will be accelerated downward at the same rate as a body falling from rest, regardless of the horizontal motion.

At a sufficiently large horizontal velocity, a projectile would fall from the horizontal only at the same rate that the surface of the Earth curves away beneath it. The projectile would thus remain at the same elevation above the Earth and in effect become an earth satellite. *See* BALLISTICS; GRAVITATION. [R.D.Ru.]

Free radical Any molecule or atom which possesses one unpaired electron. There are some molecules which contain more than one unpaired electron (for example, oxygen); they normally are not considered as free radicals. Free radicals can be chemically very reactive (for example, the methyl radical) or they can be very stable entities (for example, nitric oxide).

Free radicals can be grouped into three major classes: atoms (for example, H, F, and Cl), inorganic radicals (for example, OH, CN, NO_2, and ClO_3), and organic radicals (for example, CH_3, CH_3CH_2, and $C_6H_6^{\cdot -}$). Such radicals are of great importance since they often appear as intermediates in thermal and photochemical reactions. Radicals are also known to initiate and propagate polymerization and combustion reactions.

In general, free radicals are formed by the rupture of a bond in a stable molecule with the production of two fragments, each with an unpaired electron. The resulting free radicals may participate in further reactions or may combine to reform the original compound.

There are many ways in which radicals can be generated—among these are thermal decomposition, electric discharge photochemical reactions, electrolysis at an electrode such as mercury or platinum, rapid mixing of two reactants, and gamma- or x-ray irradiation. [J.R.Bo.]

Freon One of a group of polyhalogenated derivatives of methane and ethane containing fluorine and, in most cases, chlorine or bromine. These include such compounds as trichlorofluoromethane and dichlorodifluoromethane, also known as freon-11 and freon-12, respectively. The freons possess such noteworthy characteristics as nonflammability, excellent chemical and thermal stability, and low toxicity. Additional properties include high density, low boiling point, low viscosity, and low surface tension.

This unique combination of properties has made the freons particularly suitable for use as refrigerants. In addition to their use as refrigerants, the freons serve as propellants in aerosol products, as solvents, and as intermediates in the synthesis of other fluorine compounds. Further application has been in the field of polymers and plastics, where several of the freons are important intermediates. *See* FLUOROCARBON; HALOGENATED HYDROCARBON; REFRIGERATION. [E.C.; M.G.G.]

Frequency (wave motion) The number of times which sound pressure, electrical intensity, or other quantities specifying a wave vary from their equilibrium value through a complete cycle in unit time. The most common unit of frequency is the hertz (Hz), which is equal to 1 cycle per second. In one cycle there is a positive variation from equilibrium, a return to equilibrium, then a negative variation, and return to equilibrium. This relationship is often described in terms of the sine wave, and the frequency referred to is that of an equivalent sine-wave variation in the parameter under discussion. *See* ANGULAR FREQUENCY; FREQUENCY MEASUREMENT; SINE WAVE; WAVE MOTION. [W.J.G.]

Frequency bridge A bridge network that is used to measure frequency. An ac bridge is generally made up of a four-arm network, where an ac source is applied to a pair of opposite terminals and a current-detecting device is connected to the remaining two terminals. With the bridge network energized, the current in the detecting device can be made zero by adjusting suitable values of resistance, capacitance, or inductance in the four bridge arms, a process known as balancing.

Bridge networks in which the balancing action depends upon the supply frequency may be used for frequency measurement. The choice of frequency bridge is dependent on the frequency range, the available apparatus, and the ease with which the bridge can be set up and balanced. *See* BRIDGE CIRCUIT; FREQUENCY MEASUREMENT. [E.B.C.]

Frequency counter An electronic device capable of counting the number of cycles in an electrical signal during a preselected time interval. The modern high-speed electronic counter (*see* illustration) is a useful tool in the measurement of frequency when an accurate time base is available. It provides a digital counting or scaling device for registering the total number of events occurring during a given time interval.

The principal requirements are for signals of adequate level, free from interference and noise. With these conditions satisfied it is usually possible to obtain a precision of reading well beyond the stability of the overall system if a sufficiently long counting time is used. An extension of the frequency-measurement range is made possible by extending the range of the converter or heterodyning system. It is also feasible to add a transfer oscillator with harmonics available up into the microwave region. *See* ELECTRICAL MEASUREMENTS; FREQUENCY MEASUREMENT. [F.D.L.]

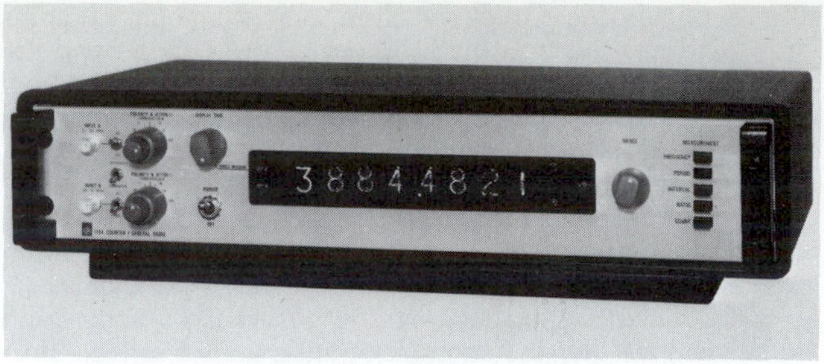

Electronic digital counter incorporating multiple functions which are selectable from panel, and using integrated circuits. (*General Radio Co.*)

Frequency divider An electronic circuit that produces an output signal at a frequency that is an integral submultiple of the frequency of the input signal.

Several information-processing and transmitting techniques require frequency division. In television, for example, it is essential to maintain a precise relationship between the horizontal-scanning frequency and the vertical-scanning frequency. Frequency division can be conveniently accomplished in two ways, digital division and division by triggering a subharmonic.

Many circuits are available to count pulses. A bistable or flip-flop circuit produces one output pulse for every two input pulses. By cascading successive flip-flops, any desired degree of division can be obtained. This is the method of digital division.

Any circuit which has a characteristic resonance responds to certain types of input energy by ringing, that is, by going through one or more cycles of electrical activity caused by the nature of the circuit rather than by the nature of the input. This characteristic can be used to accomplish frequency division, provided the input frequency does not vary over any extensive frequency range. This is the method for triggering a subharmonic. [W.W.Sn.]

Frequency measurement Measurement of the frequency of a periodic quantity, defined as the number of times a cyclic phenomenon occurs per unit of time. The second is the universally used unit of time. Conversely, time may be measured by observing the number of cycles occurring at constant frequency. The ordinary pendulum and household electric clocks are familiar examples of such time-measuring devices.

The only primary frequency standards acceptable for use in national standards laboratories for frequency reference are atomic standards of the cesium-beam type. Cesium-beam atomic clocks have superseded other types of primary frequency standards as a consequence of the adoption of the atomic second as the unit of time. *See* ATOMIC CLOCK; ATOMIC TIME.

The frequency of a secondary standard is determined by comparison with a primary standard or by comparison with another secondary standard originally compared with a primary standard. Secondary standards include (1) quartz-crystal frequency standards, with or without clock indicators, and (2) gas-cell atomic standards such as rubidium-cell standards, and (3) ammonia masers. Rubidium-cell standards are relatively rugged and can be made to exhibit low drift rates, thus enabling relatively long intervals between calibrations. Quartz-crystal frequency standards are the least expensive initially and may provide the most satisfactory working standards if calibration means are available. *See* QUARTZ CLOCK. [F.D.L.]

Audio-frequency. A fairly wide variety of audio-frequency meters has been developed and is commercially available.

Broadly speaking they can be grouped into two classes, the resonant and ratiometer types. The resonant type may be further subdivided into instruments employing resonant reeds and those having electrically resonant circuits. The latter, as well as the ratiometer type, are classified as deflection-type instruments embodying moving systems with pointers and scales. *See* FREQUENCY BRIDGE; FREQUENCY COUNTER; MOVING-COIL FREQUENCY METER; MOVING-IRON FREQUENCY METER; REED-TYPE FREQUENCY METER.

The most widely used of the simpler methods is aural comparison. An unknown frequency can be determined by direct comparison with a known adjustable standard frequency by using the ear. Typically, headphones are connected directly to the two audio sources. The unknown frequency is fed into one earpiece while the standard frequency supplied by a resistance-tuned or a beat-frequency oscillator is fed into another earpiece. A beat note between the two frequencies can be detected by the ear when the frequencies are not exactly equal. A meter or an electric-eye indicator can take the place of the ear as a detector.

Cathode-ray oscilloscopes provide a most convenient means of comparing two frequencies. The usual method consists of applying a voltage of unknown frequency to one pair of deflecting plates and a voltage of known frequency to the other. The resulting patterns, known as Lissajous figures, are a function of the difference in amplitude, frequency, phase, and the waveforms of the two applied voltages and their frequency ratio. Knowing one of the frequencies and determining the difference or the ratio of the frequencies, it is possible to calculate the second frequency. *See* LISSAJOUS FIGURES. [E.B.C.]

Radio-frequency. Precise measurement of frequencies above the audible range may be performed by various techniques. Basic measuring systems may consist of (1) a calibrated oscillator with some means of comparison with the unknown frequency, such as an oscilloscope or a heterodyne detector unit, (2) a digital counting or scaling device which registers the total number of events occurring during a given time interval, or (3) an electronic circuit for producing a direct current proportional to the frequency of its input signal which then may be indicated by a dc meter. In practice, each of these basic systems has an upper frequency limit for practical operation.

In order to extend the range of frequency measurement above the practical limits mentioned above, it is customary to generate a fixed standard frequency and to select a harmonic of it near the unknown frequency, after which the unknown frequency may be reduced to a lower value by subtraction of the standard-frequency harmonic in a mixer or heterodyne detector. This lower beat frequency may then be measured by application of the basic methods outlined above, which thus serve to interpolate between the known standard frequencies.

[F.D.L.]

Frequency modulation A special kind of angle modulation in which the instantaneous frequency of a sine-wave carrier is varied by an amount proportional to the magnitude of the modulating wave. *See* ANGLE MODULATION; PHASE MODULATION.

Either amplitude modulation (AM) or frequency modulation (FM) offers a solution to the important problem of how to impress the message wave to be communicated upon a high-frequency oscillation. However, FM offers important advantages in exchange for extra bandwidth occupancy. Also, FM with negative feedback minimizes noise problems and receiver distortion. *See* MODULATION.

Frequency modulation is defined in terms of a generalized concept known as instantaneous frequency which is directly proportional to the time rate of change of the angle of a sine function, the argument of which is a function of time. When the argument is expressed in radians and the time in seconds, the instantaneous frequency in hertz is the time rate of change of the angle divided by 2π.

In frequency modulation the instantaneous frequency is linearly proportional to the magnitude of the modulating wave. Louder tones with AM mean greater changes in amplitude. Louder tones with FM mean greater changes in frequency.

In radio broadcasting, provided the frequency deviation (peak difference between instantaneous and carrier frequencies) is large and provided multipath transmission effects are small, FM is capable of high-fidelity reception combined with the advantages of reduced noise, less interference between stations, and less transmitter power to cover a given area. Constant average power and constant peak power that is only twice the average power are two factors that permit a ready realization of a simple high-efficiency transmitter, simplify problems of automatic volume control, and allow amplifiers and other devices to operate closer to their maximum power capability without the penalties normally associated with nonlinearities.

Practically, when estimating approximate bandwidth occupancy, a rule of thumb states that angle modulation requires the band traversed by the instantaneous frequency plus the bandwidth of the modulating wave added at both top and bottom. For some purposes an even wider band may be required.

Moreover, unambiguous representation and recovery of the wanted message by angle-modulation techniques also require that the unmodulated carrier frequency comfortably exceed the sum of the frequency deviation in the down direction plus the bandwidth of the modulating wave. In other words, in FM the carrier frequency must be high compared to the maximum frequency deviation.

For certain types of noise disturbance characterized by a noise spectrum that is uncorrelated and independent of frequency, the ratio of average signal power to average noise power in the output of the FM receiver will be proportional to the square of the peak-frequency deviation. Therefore under certain important conditions, the signal-to-noise ratio of an angle-modulation system improves 6 decibels for each 2:1 increase in bandwidth occupancy.

However, the noise advantage of FM cannot be increased indefinitely. As the bandwidth occupancy is continually increased to accommodate an increased frequency deviation, more noise reaches the FM detector. Presently, the assumption that the noise is less than the so-called improvement threshold is violated, whereupon the noise advantage of FM is quickly lost. FM with negative feedback acts differently in that the improvement threshold is minimized and held constant, independent of the bandwidth occupied by the incoming FM signal. FM with negative feedback is accompanied by a decrease in distortion originating within the FM receiver and by an increased tolerance to noise failing within the frequency band occupied by the incoming FM signal. These two important advantages are not possessed by nonfeedback receivers. Substantial benefits are realized only when the amount of negative feedback is large. In common with nonfeedback FM systems, any large reduction in noise must be paid for by a corresponding increase in the bandwidth occupied by the transmitted FM signal.

Many schemes for production and detection are possible and nearly all use spectrum translation. For the production of FM most schemes resort to spectrum multiplication. Spectrum translation of an angle-modulated wave is accomplished by single-sideband modulation. The translated spectrum, with or without inversion, is centered about a new carrier. Otherwise its significant properties are unchanged. Spectrum multiplication implies angle multiplication. By generating the xth harmonic of an angle-modulated wave, the angle is multiplied by x. If the required multiplication is too great, then after a convenient number of multiplications the resulting spectrum may be translated downward and multiplication resumed. *See* FREQUENCY-MODULATION DETECTOR; FREQUENCY-MODULATION RADIO.

[H.S.Bl.]

Frequency-modulation detector Detection or demodulation of a frequency-modulated (FM) wave. FM detectors operate in several ways. In one class of detector, known as a discriminator, the frequency modulation is first converted to amplitude modulation, which is then detected by an amplitude-modulation detector. Another type of FM detector employs a phased-locked oscillator to recover the modulation. A still different type converts the frequency modulation to pulse-rate modulation, which can be converted to the desired signal by use of an integrating circuit. *See* AMPLITUDE-MODULATION DETECTOR. [C.L.A.]

Frequency-modulation radio Radio transmission accomplished by symmetrical variation of the carrier frequency by an analog input signal. The amount of swing from center frequency is dependent upon the peak value of the modulating voltage, as well as upon its frequency. The frequency of the modulation signal governs the rate at which the changes in carrier frequency occur. Frequency-modulation (FM) sidebands are formed during the modulation process and are separated from one another by an amount equal to the audio frequency. The amplitude of the sidebands diminishes progressively as the sidebands occur farther and farther from the center frequency, and the number of significant sidebands depends on the amplitude of the modulating signal. The deviation ratio of the carrier-frequency variation to the highest signal frequency transmitted may be any selected value, from fractional to large values.

Because there is no amplitude change in the output of an FM transmitter, whatever the deviation ratio, this mode is an almost perfect cure for amplitude-related interference problems that plague radio-frequency reception. Another property of FM receivers is the relative freedom from interference between distant and local stations using the same channel; only the strongest signal is received, even if the wanted signal is only 3–6 dB stronger than the interfering one. This characteristic is known as the capture effect. In contrast, AM radio signals differing in strength by 35 dB result in noticeable interference.

Since most FM receivers are equipped with a muting circuit (squelch) that silences the audio channel when no signal is present, an irritating hiss noise is not emitted from the loudspeaker when the frequency (or channel) being monitored is not in use.

Frequency modulation is used mainly for transmissions above 25 megahertz (MHz). Typical uses are in broadcasting, television sound, mobile radio telephony, radio paging systems, space telemetry, intercity microwave relaying of all classes of public traffic including voice channels, teleprinting, facsimile, broadcast network programs, and television and computer data, and intercontinental telecommunications via satellite. Frequency modulation is used for both analog and digital communications, and phase modulation as well as frequency modulation is employed.

FM broadcasting. The frequency band 88–108 MHz is allocated to FM broadcasting in a large part of the world by international agreement. For a channel spacing of 200 kHz, there are 100 allocatable channels for transmission of an audio range of 50–15,000 Hz, with a frequency-deviation ratio of 5. This means that there are five significant sidebands above and below the carrier, the carrier is deviated a maximum of ±75 kHz, and the emitted spectrum is twice this value. Because of the relatively small signal power in the modulating frequencies above 4 kHz, the received signal-to-noise ratio is improved substantially by preemphasis of the audio signal in transmission, necessitating complementary deemphasis in the receiver to restore natural program balance. In fact, preemphasis produces a sort of hybrid form of modulation, being pure frequency modulation at the lowest audio frequencies, and gradually changing to phase modulation at the highest. *See* FREQUENCY MODULATION.

FM mobile transmission. Millions of land, maritime, and aeronautical mobile FM transceivers are employed by police, firemen, public safety agencies, industrial and commercial enterprises, private citizens, and radio amateurs who desire the benefits of enroute telephony. The intensity of such usage has grown exponentially, mainly because of the availability of reliable, small, low-power-consumption solid-state equipments that are economical, and also because of the public realization of the benefits of having such communications.

Radio relaying. Frequency modulation is used for microwave radio relaying over land, over water, and to great distances using satellites, sometimes carrying thousands of simultaneous telephone conversations or several television channels.

The advent of requirements for short-haul services, local distribution networks within cities, television relay, and a wide variety of optical communication services including high-speed computer communications, electronic mail, data transmission, and other services, where it may be cost-effective to avoid the local telephone loop, presents another application for FM radio relaying.

Telegraphy. Telegraphy, including teleprinting and binary digital data transmission, is based on shifting the carrier frequency or its phase between two limiting values, one of which represents a mark signal and the other a space signal. This frequency shift (or phase shift) is a form of FM signaling used over a wire, cable, or radio. *See* TELEGRAPHY.

Facsimile. Black-and-white images (line drawings and typed copy) can be transmitted by employing the principles used in FM telegraphy; one limit frequency corresponds to black, and the other to white, on the image to be transmitted. A continuous gray scale can be transmitted and recorded if, instead of just two frequencies, a continuous frequency shift is employed between some low frequency (say, 1500 Hz) and some higher frequency (say, 2700 Hz), the exact frequency at any instant being proportional to the gray level of the image. *See* FACSIMILE.

Telemetry. Frequency modulation is the preferred method for transmission of information or data from a remote or inaccessible location such as a rocket vehicle in flight. Each condition to be remotely observed actuates one subchannel, which, when multiplexed with other channels reporting other status conditions, modulates the radio carrier by frequency modulation. *See* TELEMETERING. [J.S.B]

Frequency modulator An electronic circuit or device producing frequency modulation. The frequency modulator changes the frequency of an oscillator in accordance with the amplitude of a modulating signal. If the modulation is linear, the frequency change is proportional to the modulating voltage.

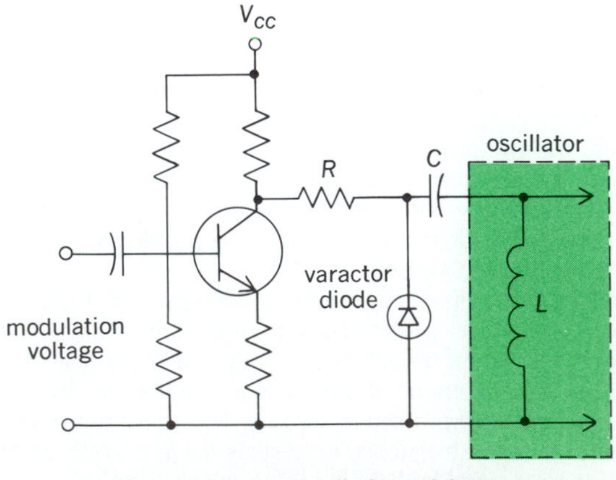

Elementary varactor-diode modulator.

High-frequency oscillators usually employ *LC* tuned circuits to establish the frequency of oscillation, and this frequency can be controlled by changing the effective capacitance or inductance in accordance with the modulating signal. Practical circuits usually employ either a reactance tube or a varactor diode to change the oscillator frequency in accordance with a modulating voltage.

The varactor diode is used commonly as a frequency modulator in modern equipment. A typical varactor modulator is shown in the illustration. In this circuit the collector voltage of the modulating amplifier controls the reverse bias across the diode. But since the junction capacitance of an abrupt junction diode is inversely proportional to the square root of the junction voltage, the oscillator tuning capacitance, which consists of capacitor *C* and the diode in series, is therefore controlled by the modulation signal. The resistance of *R* must be small in comparison with the reactance of either capacitor *C* or the diode at all modulating frequencies. Otherwise, the high modulating frequencies will be attenuated. Also, the resistance of *R* must be large in comparison with either the reactance of *C* or the reactance of the diode at the oscillator frequency. Otherwise the *Q* of the tuned circuit will be degraded. *See* FREQUENCY MODULATION; PHASE MODULATION. [C.L.A.]

Frequency multiplier An electronic circuit that produces an output frequency which is an integral multiple of the input frequency. There are two basic types of frequency multipliers. The first type is a nonlinear amplifier which generates harmonics in its output current and a tuned load that resonates at one of these harmonics. The second type uses the nonlinear capacitance of a junction (semiconductor) diode to couple energy from the input circuit, which is tuned to the fundamental, to the output circuit, which is tuned to the desired harmonic.

A highly efficient doubler can be devised by using two amplifiers, such as transistors, with their inputs driven with opposite polarity, obtained from opposite ends of a center-tapped coil, and their outputs connected in parallel. *See* AMPLIFIER; SEMICONDUCTOR. [C.L.A.]

Frequency-response equalization The process of obtaining a desired overall frequency-response characteristic in an audio-frequency circuit by introducing corrective electrical networks of various types, termed equalizers, into the circuit.

Equalizers are used both in communications networks and in systems for the recording and reproducing of photographic film, magnetic tape, and disk phonograph records. These equalizers are in the form of electrical networks of resistance, inductance, and capacitance. *See* AMPLIFIER; DISK RECORDING. [H.F.O.]

Fresh-water ecosystem An ecosystem is the functional unit of ecology, consisting of living organisms and the nonliving environment interacting upon each other to produce an exchange of energy and materials between the living and nonliving components. A lake or pond in its entirety is such an ecosystem. A stream is also an ecosystem, although less well differentiated. Limnology is the comprehensive study of all the components of inland aquatic ecosysterms and their interrelationships. *See* ECOSYSTEM; LIMNOLOGY.

The energy of fresh-water ecosystems is derived mainly from photosynthesis accomplished by the algae suspended in the water and by the higher plants and algae growing on or in the bottom. A variable proportion of the total energy available is derived from allochthonous organic matter, such as leaves and pollen, produced by terrestrial communities. The materials cycled within an ecosystem are derived ultimately from the weathering of rocks and the leaching of soil in the watershed, and to a lesser extent from the air.

Fresh-water habitats are conveniently divided into a lenitic or basin series, such as lakes, reservoirs, ponds, and bogs, and a lotic or channel series, such as rivers, streams, brooks, springs, and groundwater. The lotic series is distinguished by a continual flow of water in one direction. [D.G.Fre.]

Friction Resistance to sliding, a property of the interface between two solid bodies in contact. Many everyday activities like walking or gripping objects are carried out through friction, and most people have experienced the problems that arise when there is too little friction and conditions are slippery. However, friction is a serious nuisance in devices that move continuously, like electric motors or railroad trains, since it constitutes a dissipation of energy, and a considerable proportion of all the energy generated by humans is wasted in this way. Most of this energy loss appears as heat, while a small proportion induces loss of material from the sliding surfaces, and this eventually leads to further waste, namely, to the wearing out of the whole mechanism. *See* WEAR; WORK.

In stationary systems, friction manifests itself as a force equal and opposite to the shear force applied to the interface. Thus, as in the illustration, if a small force *S* is applied, a friction force *P* will be generated, equal and opposite to *S*, so that the surfaces remain at rest. *P* can take on any magnitude up to a limiting value *F*, and can therefore prevent sliding whenever *S* is less than *F*. If the shear force *S* exceeds *F*, slipping occurs. During sliding, the friction force remains approximately equal to *F* and always acts in a direction opposing the relative motion. The friction force is proportional to the normal force *L*, and the constant of proportionality is defined as the friction coefficient *f*. This is expressed by the equation $F = fL$.

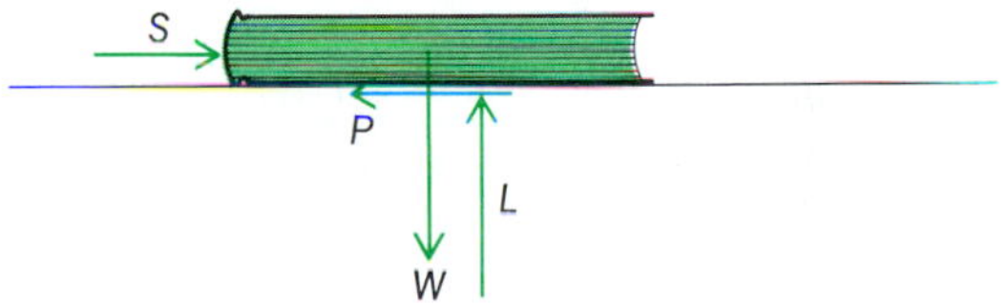

The forces acting on a book resting on a flat surface when a sheer force *S* is applied. The friction from *P* is equal to *S* (up to a limiting value *F*), while *L*, the normal force, is equal to the weight *W* of the book.

In prehistoric and early historic times, humans' main interest in friction was to reduce the friction coefficient, to reduce the labor involved in dragging heavy objects. This led to the invention of lubricants, the first of which were animal fats and vegetable oils. A great breakthrough was the use of rolling action, first in the form of rolling logs and then in the form of wheels, to take advantage of the lower friction coefficients of rolling systems. *See* LUBRICANT.

In modern engineering practice available materials and lubricants reduce friction to acceptable values. In special circumstances when energy is critical, determined efforts to minimize friction are undertaken. Friction problems of practical importance are those of getting constant friction in brakes and clutches, so that jerky motion is avoided, and avoiding low friction in special circumstances, such as when driving a car on ice or on a very wet road. Also, there is considerable interest in developing new bearing materials and new lubricants that will produce low friction even at high interfacial temperatures and maintain these properties for long periods of times, thus reducing maintenance expenses. Perhaps the most persistent problem is that of avoiding frictional oscillations, a constant cause of noise pollution of the environment. [E.R.]

Friction brake A type of brake in which a fixed surface contacts a moving part that is to be slowed or stopped. Friction brakes are classified according to the kind of friction element employed and the means of applying the friction forces. *See* FRICTION.

The simplest form of brake, the single-block, consists of a short block fitted to the contour of a wheel or drum and pressed up against the surface by means of a lever on a fulcrum, as widely used on railroad cars. The block may have the contour lined with friction-brake material, which gives long wear and a high coefficient of friction. The fulcrum may be located with respect to the lever in a manner to aid or retard the braking torque of the block. The lever may be operated manually or by a remotely controlled force (illustration *a*).

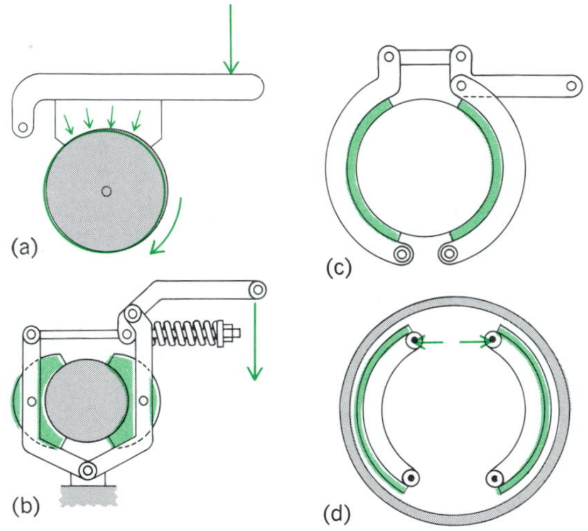

Friction brakes. (*a*) Single-block brake. Block fixed to operating lever; force in direction of arrow applies brake. (*b*) Double-block brake. Blocks pivoted on their levers; force in direction of arrow releases brake. (c) External-shoe brake. Shoes are lined with frictional material. (*d*) Internal-shoe brake with lining.

Two single-block brakes in symmetrical opposition, where the operating force on the end of one lever is the reaction of the other, make up a double-block brake (illustration *b*). External thrust loads are balanced on the rim of the rotating wheel.

An external-shoe brake operates in the same manner as the block brake, and the designation indicates the application of externally contracting elements. In this brake the shoes are appreciably longer, extending over a greater portion of the drum (illustration *c*). It is used on elevator installations for locking the hoisting sheave by means of a heavy spring when the electric current is off and the elevator is at rest.

An internal shoe has several advantages over an external shoe. Because it works on the inner surface of the drum, it is protected from water and grit (illustration *d*). It may be designed in a more compact package, is easily activated, and is effective for drives with rotations in both directions. The internal shoe meets the requirement of the automobile. Early mechanical brakes had an elliptical cam; modern brakes have hydraulic piston-actuated shoes.

Hoists, excavating machinery, and hydraulic clutch-controlled transmissions have band brakes. They operate on the same principle as flat belts on pulleys. In the simplest band brake, one end of the belt is fastened near the drum surface and the other end is then pulled over the drum in the direction of rotation so that a lever on a fulcrum may apply a tension to the belt.

Disk brakes have long been used on hoisting and similar apparatus. Because more energy is absorbed in prolonged braking than in a clutch startup, additional heat dissipation must be provided in equivalent disk brakes. A significant application of disk brakes is on the wheels of aircraft. A familiar application is the bicycle coaster brake.

The caliper disk brake consists of a revolving disk which can be gripped between two brake pads. A brake on a bicycle is an example. The automotive caliper disk brake is hydraulically operated, and the pads cover between one-sixth and one-ninth of the swept area of the disk. It is used on the front axle of a car. *See* AUTOMOTIVE BRAKE. [H.P.H.]

Friction factor The name given to the proportionality factor in the equation below relating the friction or head loss en-

$$E_f = 4\tau_w(L/\rho D) = af(L/D)(u^2/2g_c)$$

countered when a fluid flows in a pipe to the kinetic energy of the flowing fluid and the length and diameter of the pipe. Here E_f is the energy converted into heat as the result of friction [(ft-lb force)/(lb fluid flowing) or J/kg of fluid flowing], τ_w is the shear stress at the wall (lb force/ft^2 or N/m^2), ρ is the fluid density (lb mass/ft^3 or kg/m^3), f is a dimensionless friction factor, L is pipe length (ft or m), D is pipe diameter (ft or m), u is average fluid velocity in the pipe (ft/s or m/s), g_c is the conversion factor [32.17 (lb mass/lb force)(ft/s^2) or 1 (kg/N)(m/s^2)], and a is an arbitrary dimensionless constant. *See* FLUID FLOW. [C.E.La.]

Friedel-Crafts reaction A substitution reaction, catalyzed by aluminum chloride, in which an alkyl (R—) group or an acyl (RCO—) group replaces a hydrogen atom of an aromatic nucleus. This general reaction is the most important member of a larger group of aromatic substitution reactions known to be catalyzed by conventional or Lewis acids.

In the classical alkylation reaction, an alkyl halide (RX) serves as the alkylating agent. Alkenes may be substituted for alkyl halides. For acylation of aromatic hydrocarbons, acyl halides have proved most valuable although acid anhydrides have also been used. *See* AROMATIC HYDROCARBON; SUBSTITUTION REACTION. [C.K.B.]

Fringe (optics) One of the light or dark bands produced by interference or diffraction of light. Distances between fringes are usually very small, because of the short wavelength of light. Fringes are clearer and more numerous when produced with light of a single color.

Diffraction fringes are formed when light from a point source, or from a narrow slit, passes by an opaque object of any shape. Interference fringes are obtained by bringing together two or more beams of light that have originated from a common source. This is usually accomplished by means of an apparatus especially designed for the purpose called an interferometer, although interference fringes may also be seen in nature. Examples are the colors in a soap film and in an oil film on water. *See* DIFFRACTION; INTERFERENCE OF WAVES; INTERFEROMETRY; RESOLVING POWER (OPTICS). [F.A.J./W.W.W.]

Frit A ground glass used in making glazes and enamels. It is formed by melting the ingredients and rapidly cooling (quenching) them, for example, by pouring into water or onto water-cooled steel rolls.

The main reason for fritting is to convert water-soluble raw materials into an insoluble glass, making it easier to keep them uniformly distributed in the glaze or enamel during processing. Other advantages of fritting are the conversion of poisonous materials (for example, lead compounds) into less dangerous forms, the more even distribution of color, and the reduction in loss of volatile materials. *See* GLAZING; METAL COATINGS. [J.F.McM.]

Frog The name for a number of species of animals in the order Anura of the class Amphibia. These animals lack a tail in the adult stage, the hindlegs are adapted for jumping, and the skin is moist and smooth with no scales. The eyes are large and have a functional third eyelid, or nictitating membrane. External ears are absent, but there is a tympanum to receive vibratory stimuli. The mobile tongue, attached anteriorly to the floor of the mouth and free and notched at the posterior, can be projected at the prey. Insects adhere to the tongue, which is coated with sticky secretions from tongue glands.

The larval form, the tadpole has a tail and external gills at first, but these structures soon disappear as the gills are replaced by internal gills in the pharyngeal region and the tail is resorbed. Metamorphosis of the larva is complete soon after the appearance of the limbs.

Frogs are poikilothermous animals; that is, body temperature is dependent upon the environment. There are larger numbers of individuals and species in moist areas of temperate and tropical regions. *See* ANURA. [C.B.C.]

Front An elongated, sloping zone in the troposphere, within which changes of temperature and wind velocity are large compared to changes outside the zone. Thus the passage of a front at a fixed location is marked by rather sudden changes in temperature and wind and also by rapid variations in other weather elements such as moisture and sky condition.

In its idealized sense, a front can be regarded as a sloping surface of discontinuity separating air masses of different density or temperature. In practice, the temperature change from warm to cold air occurs mainly within a zone of finite width, called a transition or frontal zone. The three-dimensional structure of the frontal zone is shown in the illustration. In typical cases, the zone is about 1 km (0.6 mi) in depth and 100–200 km (60–120 mi) in width, with a slope of approximately 1/100. The cold air lies beneath the warm in the form of a shallow wedge. Temperature contrasts generally are strongest at or near the earth's surface, the frontal zone usually being narrowest near the ground and becoming wider and more diffuse with height.

The surface separating the frontal zone from the adjacent warm air mass is referred to as the frontal surface, and it is the line of intersection of this surface with a second surface, usually horizontal or vertical, that strictly speaking constitutes the front. According to this more precise definition, the front represents a discontinuity in temperature gradient rather than in temperature itself. The boundary on the cold air side is often ill-defined, especially near the earth's surface, and for this reason is not represented in routine analysis of weather maps. *See* WEATHER MAP.

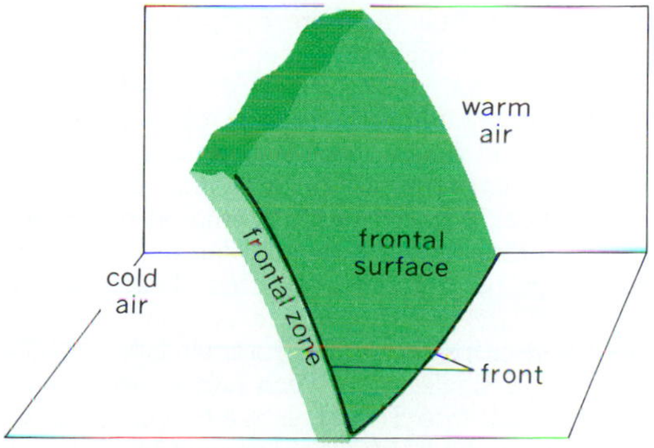

Schematic diagram of the frontal zone, angle with Earth's surface much exaggerated.

The wind gradient, or shear, like the temperature gradient, is larger within the frontal zone than on either size of it. In well-developed fronts the shift in wind direction often is concentrated along the frontal surface, while a more gradual change in wind speed may occur throughout the frontal zone. An upper-level jet stream normally is situated above the frontal zone. *See* JET STREAM. [S.E.M.]

Frost A covering of ice in one of several forms produced by the freezing of supercooled water droplets on objects colder than 32°F (0°C). The partial or complete killing of vegetation, by freezing or by temperatures somewhat above freezing for certain sensitive plants, also is called frost. Air temperatures below 32°F (0°C) sometimes are reported as "degrees of frost"; thus, 10°F (−12°C) is 22 degrees of frost (this usage is confined to the Fahrenheit scale and is not applied to Celsius temperatures).

Frost forms in exactly the same manner as dew except that the individual droplets that condense in the air a fraction of an inch from a subfreezing object are themselves supercooled, that is, colder than 32°F (0°C). When the droplets touch the cold object, they freeze immediately into individual crystals. When additional droplets freeze as soon as the previous ones are frozen, and hence are still close to the melting point because all the heat of fusion has not been dissipated, amorphous frost or rime results.

At more rapid rates of condensation, the drops form a thin layer of liquid before freezing, and glaze or glazed frost ("window ice" on house windows, "clear ice" on aircraft) generally follows. Glaze formation on plants, buildings and other structures, and especially on wires sometimes is called an ice storm, a silver frost storm, or thaw.

At slower deposition rates, such that each crystal cools well below the melting point before the next joins it, true crystalline or hoar frosts form. These include fernlike assemblages on snow surfaces, called surface hoar; similar feathery plumes in cold buildings, caves, and crevasses, called depth hoar; and the common window frost or ice flowers on house windows.

Killing frosts or freezes damage or kill vegetation depending on their duration and their intensity, that is, how far the plant temperatures go below 32°F (0°C). Such conditions result from advection of much colder air, which then cools the plants, as in the infamous cold waves of the north-central United States; or from radiational cooling of the plants themselves, by long-wave radiation to clear skies at night. In either case, the extent to which plant fluids freeze determines the severity of the frost. *See* AIR TEMPERATURE; DEW; DEW POINT. [A.Cou.]

Frost action The mechanical weathering process caused by alternate or repeated cycles of freezing and thawing of water in pores, cracks, and other openings, usually at the surface, and the resulting effects of frost action on materials and structures. The term is loosely used to indicate any action in which frost is involved.

Frost action is initiated during growth of ice in the ground from which may result either hydrostatic pressure or directed pressure by individual ice crystals, usually growing in the direction of heat loss. The ice grows slowly as water migrates molecule by molecule through "suction" or capillary action to the freezing zone. Plant roots can be broken, soils stiffed and sorted, and many small morphologic features may grow overnight or within a few years. In fine-grained unconsolidated materials, ice much in excess of the volume of pores is commonplace. Under these conditions thawing yields suspensions of soil or high pore-water pressures that promote loss of strength and subsequent slump and flow. Under such conditions the common potholes in roads are initiated. *See* EROSION; MASS WASTING; WEATHERING PROCESSES. [R.F.B.]

Froude number In fluid mechanics, the ratio v^2/gd of the inertia force v^2 to the gravitational force gd, where v is velocity, g is gravitational acceleration, and d is a characteristic length. Froude number is significant in the design of a model of any system in which the effect of gravity is important in controlling the velocities or the flow pattern. For example, it is used in the evaluation of the drag or the slipstream velocity of a ship producing surface waves. *See* DYNAMIC SIMILARITY; MODEL THEORY. [G.Mu.]

Fructose A sugar that is the commonest of ketoses and the sweetest of the sugars. It is also known as D-fructose, D-fructopyranose, and levulose fruit sugar. It is found in free state, usually accompanied by D-glucose and sucrose in fruit juices, honey, and nectar of plant glands. D-Fructose is the principal sugar in seminal fluid. *See* CARBOHYDRATE.

Fructose is readily utilized by diabetic animals. In persons with diabetes mellitus or parenchymal hepatic disease, the impairment of fructose tolerance is relatively small and not at all comparable to the diminution in their tolerance to glucose. *See* MONOSACCHARIDE. [W.Z.H.]

Fruit A matured carpel or group of carpels (the basic units of the gynoecium or female part of the flower) with or without seeds, and with or without other floral or shoot parts (accessory structures) united to the carpel or carpels. Carpology is the study of the morphology and anatomy of fruits. The ovary develops into a fruit after fertilization and usually contains one or more seeds, which have developed from the fertilized ovules. Parthenocarpic fruits usually lack seeds. Fruitlets are the small fruits or subunits of aggregate or multiple fruits. Flowers, carpels, ovaries, and fruits are, by definition, restricted to the flowering plants (angiosperms), although fruitlike structures may enclose seeds in certain other groups of seed plants. The fruit is of ecological significance because of seed dispersal. *See* SEED.

Morphology. A fruit develops from one or more carpels. Usually only part of the gynoecium, the ovary, develops into a fruit; the style and stigma wither. Accessory (extracarpellary or noncarpellary) structures may be closely associated with the carpel or carpels and display various degrees of adnation (fusion) to them, thus becoming part of the fruit. Such accessory parts include sepals (as in the mulberry), the bases of sepals, petals, and stamens united into a floral tube (apple, banana, pear, and other species with inferior ovaries), the receptacle (strawberry), the pedicel and receptacle (cashew), the peduncle (fleshy part of the fig), the involucre composed of bracts and bracteoles (walnut and pineapple), and the inflorescence axis (pineapple). *See* FLOWER.

A fruit derived from only carpellary structures is called a true fruit, or, because it develops from a superior ovary (one inserted above the other floral parts), a superior fruit (corn, date, grape, plum, and tomato). Fruits with accessory structures are called accessory (or inaptly, false or spurious) fruits (pseudocarps), or, because of their frequent derivation from inferior ovaries (inserted below the other floral parts), inferior fruits (banana, pear, squash, and walnut).

Fruits can be characterized by the number of ovaries and flowers forming the fruit. A simple fruit is derived form one ovary, an aggregate fruit from several ovaries of one flower (magnolia, rose, and strawberry). A multiple (collective) fruit is derived from the ovaries and accessory structures of several flowers consolidated into one mass (fig, pandan, pineapple, and sweet gum).

The fruit wall at maturity may be fleshy or, more commonly, dry. Fleshy fruits range from soft and juicy to hard and tough. Dry fruits may be dehiscent, opening to release seeds, or indehiscent, remaining closed and containing usually one seed per fruit. Fleshy fruits are rarely dehiscent.

The pericarp is the fruit wall developed from the ovary. In true fruits, the fruit wall and pericarp are synonymous, but in accessory fruits the fruit wall includes the pericarp plus one or more accessory tissues of various derivation. Besides the fruit wall, a fruit contains one or more seed-bearing regions (placentae) and often partitions (septa).

Anatomy. Anatomically or histologically, a fruit consists of dermal, ground (fundamental), and vascular systems and, if present, one or more seeds. After fertilization the ovary and sometimes accessory parts develop into the fruit; parthenocarpy is fruit production without fertilization. The fruit generally increases in size and undergoes various anatomical changes that usually relate to its manner of dehiscence, its mode of dispersal, or protection of its seeds. The economically important, mainly fleshy fruits have received the most histological and developmental study.

Size increase of fruits is hormonally controlled and results from cell division and especially from cell enlargement. Cell number, volume, and weight thus control fruit weight. Cell division generally is more pronounced before anthesis (full bloom); cell enlargement is more pronounced after.

Functional aspects. Large fruits generally require additional anatomical modifications for nutrition or support or both. The extra phloem in fruit vascular bundles and the often increased amount of vascular tissue in the fruit wall and septa supply nutrients to the developing seeds and, especially in fleshy fruits, to the developing walls. Large, especially fleshy fruits (apple, gourd, and kiwi) usually contain proportionally more vascular tissue than small fruits. Vascular tissue also serves for support and in lightweight fruits may be the chief means of support.

Crystals, tannins, and oils commonly occur in fruits and may protect against pathogens and predators. The astringency of tannins, for example, may be repellent to organisms. With fruit maturation, tannin content ordinarily decreases, so the tannin repellency operative in early stages is superseded in fleshy fruits by features (tenderness, succulence, sweetness through odor and increased sugar content, and so on) attractive to animal dispersal agents. Many fruits are dispersed by hairs, hooks, barbs, spines, and sticky mucilage adhering the fruit to the surface of the dispersal agent. Lightweight fruits with many air spaces or with wings or plumes may be dispersed by wind or water. Gravity is always a factor in dispersal of fruits and seeds. [R.S.]

Fruit, tree Tree fruits include temperate, subtropical and tropical zone species. Most temperate zone fruits are grown principally in regions protected from prolonged summer heat and severe winter cold (above −10 to −15°F or −23 to −26°C). The principal deciduous tree fruits grown in the United States are apple, peach, pear, plum, apricot, sweet cherry, tart cherry, and nectarine. Tree nuts, such as almond, pecan, walnut, and filbert, are sometimes classified as deciduous tree fruit crops.

Most subtropical fruit trees are evergreen. They will withstand temperatures somewhat below freezing during their dormant or semidormant season, but not the extreme temperatures tolerated by the temperate zone crops. Major subtropical fruits in the United States are the citrus group (for example, orange, grapefruit, lemon), olive, avocado, fig, and others of lesser importance. The Japanese persimmon might be considered as a borderline case between the temperate and subtropical zone groups, and avocado is often classed as a tropical fruit.

The tree fruit crops are grown commercially in orchards or groves, usually in single rows which permit necessary cultural operations for each tree. Nearly all tree fruit crops have similar requirements of training, pruning, spraying to control diseases and insects, and cultivation or chemical control of weeds. *See* separate articles on the common named fruits. [R.P.L.]

Fucales A large and diverse order of conspicuous brown algae (Phaeophyceae) including such well-known seaweeds as *Fucus* and *Sargassum*. Definitive features include apical growth, production of sex organs in cavities, and a life history with only one somatic phase, which is diploid. The mature thallus consists typically of a holdfast from which arise several axes with dichotomous or radial primary branching. Sexual reproduction is oogamous, with oogonia and antheridia produced in flask-shaped cavities called conceptacles.

In the family Fucaceae (rockweeds), the thallus typically is composed of a discoid or conical holdfast with a cluster of erect fronds that are branched dichotomously in one plane. *Fucus* is common throughout the North Atlantic and North Pacific. *Pelvetia* is also common in the Northern Hemisphere, but is not found on the American side of the Atlantic. *Ascophyllum* occurs abundantly in the North Atlantic, both in exposed sites and in estuaries far from the open sea. In subtropical and tropical regions of all oceans, the family Sargassaceae is common, dominating the biomass of benthic seaweeds. The holdfast is discoid, conical, hapteroid (with rootlike outgrowths), or rhizomatous. *Sargassum* is by far the largest and most important genus in the Sargassaceae. Hundreds of species and varieties have been described.

Rockweeds are well known to coastal peoples of the temperate regions of the Northern Hemisphere. Traditionally, the seaweed was either plowed into the soil as a green manure or burned to an ash that was spread on the fields. In modern chemical industry, *Ascophyllum* is an important source of alginate, especially in Norway. *See* ALGAE; ALGINATE; PHAEOPHYCEAE. [P.C.Si.; R.L.Moe]

L-Fucose A methyl pentose (also known as 6-deoxy-L-galactose, L-galactomethylose, L-rhodeose, and L-fucopyranose) present in some algae (seaweed), especially in *Laminaria digitata*, and in a number of gums; it has been identified in the polysaccharides of blood group–specific substances from animal and human sources. L-Fucose has also been found to be a constituent of certain bacterial polysaccharides. *See* BLOOD GROUPS; MONOSACCHARIDE. [W.Z.H.]

Fuel cell An electric cell that converts the chemical energy of a fuel directly into electric energy in a continuous process. The efficiency of this conversion can be made much greater than that obtainable by thermal-power conversion. In the latter the chemical reaction is made to produce heat by combustion. The heat is then transformed partially into mechanical energy by a heat engine, which drives a generator to produce electric energy. Further loss is involved if the direct current generated is converted into alternating current.

Although, in principle, the nature of the reactants is not limited, the fuel-cell reaction almost always involves the combination of hydrogen with oxygen. If the reaction is harnessed in a galvanic cell working at 100% efficiency, a cell voltage of 1.23 volts results. Fuel cells are of 200–500 watts capacity and 50–100 mA/cm^2 current density. Larger prototypes have been produced.

In the present stage of development, it is difficult to make a classification of the fuel-cell types. The most popular and successful type remains the classical H_2-O_2 fuel cell of the direct or indirect type. In the direct type, hydrogen and oxygen are used as such, the fuel being produced in independent installations. The indirect type employs a hydrogen-generating unit which can use as raw material a wide variety of fuel. *See* DRY CELL; ELECTRODE POTENTIAL; ELECTROMOTIVE FORCE (EMF). [J.D.; L.R.]

Fuel gas A fuel in the gaseous state whose potential heat energy can be readily transmitted and distributed through pipes from the point of origin directly to the place of consumption. The types of fuel gases are natural gas, LP gas, refinery gas, coke oven gas, and blast-furnace gas. The last two are used in steel mill complexes.

Most fuel gases are composed in whole or in part of the combustibles hydrogen, carbon monoxide, methane, ethane, propane, butane, and oil vapors and, in some instances, of mixtures containing the inerts nitrogen, carbon dioxide, and water vapor. *See* COAL GASIFICATION; LIQUEFIED NATURAL GAS (LNG); LIQUEFIED PETROLEUM GAS (LPG); NATURAL GAS. [J.Hu.]

Fuel injection Implicitly, the delivery of fuel to an internal combustion engine by pressure from a mechanical pump. It must be used in diesel engines because of the high pressures necessary to deliver fuel against the highly compressed air in the engine cylinders at the end of the compression stroke, as required for combustion in the diesel cycle. In addition, considerable pressure is required to break up the fuel oil, which has low volatility and is often viscous, into the fine-spray pattern required for good combustion. *See* DIESEL ENGINE; INTERNAL COMBUSTION ENGINE.

Fuel injection has been used also for spark ignition engines with volatile fuels, notably racing automobile engines and reciprocating piston airplane engines. The advantages over the almost universally used aspirated carburetors on these engines are the additional power which may be obtained from an engine of a given size, elimination of the warm-up time, and freedom from stalling due to ice formation on the throttle valves. *See* CARBURETOR. [N.MacC.]

Fuel oil Any of the petroleum products which are less volatile than gasoline and are burned in furnaces, boilers, or other types of heaters. The two primary classes of fuel oils are distillate and residual. Distillate fuel oils are composed entirely of material which has been vaporized in a refinery distillation tower. Consequently, they are clean, free of sediment, relatively low in viscosity, and free of inorganic ash. Residual fuel oils contain fractions which cannot be vaporized by heating. These fractions are black and viscous and include any inorganic ash components which are in the crude. In some cases, whole crude is used as a residual fuel.

Distillate fuel oils are used primarily in applications where ease of handling and cleanliness of combustion are more important than fuel price. The most important use is for home heating. They are also used in certain industrial applications where low sulfur or freedom from ash is important. Increasing amounts of distillate fuel oils have been burned in gas turbines used for electricity generation. *See* GAS TURBINE.

Residual fuel oils are used where fuel cost is an important enough economic factor to justify additional investment to overcome the handling problems they pose. They are particularly attractive where large volumes of fuel are used, as in electric power generation, industrial steam generation, process heating, and steamship operation. *See* ELECTRIC POWER GENERATION; OIL FURNACE. [C.W.S.]

Fuel pump A pump for drawing fuel from a storage tank and forcing it to an engine or furnace. Choice of pump type for fuel depends to a great extent on the volatility of the liquid to be pumped.

In a gasoline engine the fuel is highly volatile at ambient temperature; therefore a sealed diaphragm positive-displacement pump is used (see illustration). With such a pump the fuel line is completely sealed from tank to carburetor to prevent escape of fuel and to enable the pump to purge the line of vapor in event of a "vapor lock" caused by fuel vaporizing due to abnormally high ambient temperature. *See* FUEL SYSTEM.

In a diesel engine, where fuel is injected at high pressure through a nozzle into the combustion cylinder, a piston serves as its own inlet valve and as the compression member of the fuel pump. *See* DIESEL ENGINE; FUEL INJECTION.

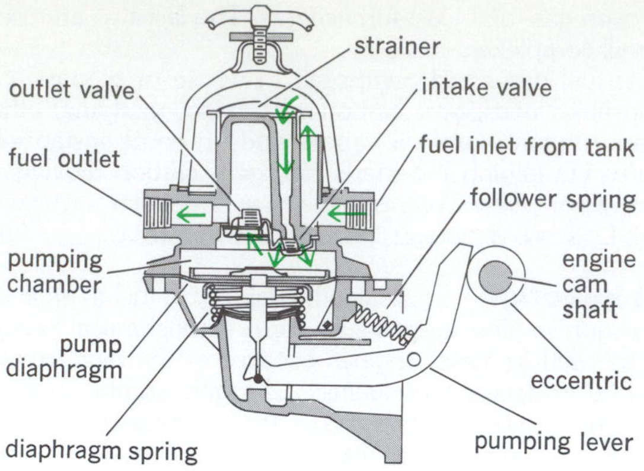

Typical automobile fuel supply pump.

Fuel pumps must operate before the engine or furnace that they serve can function; consequently an auxiliary or starter motor is usually an essential adjunct to a fuel pump. [F.H.R.]

Fuel system A system which stores fuel for present use and delivers it as needed to an engine. The commonly used components for automobiles and stationary gasoline engines are fuel tank, remote reading fuel gage, filter, fuel pump, and carburetor. *See* CARBURETOR; FUEL PUMP.

The presence of multiple engines and multiple tanks complicates the aircraft fuel system. Components of a typical aircraft fuel system include one main and two auxiliary tanks with their gages; booster, transfer, and engine-driven pumps; various selector valves; and a fuel jettisoning or defuel valve and connection, which is typical also of what would be needed for either single-point ground or flight refueling (see illustration). The arrangement

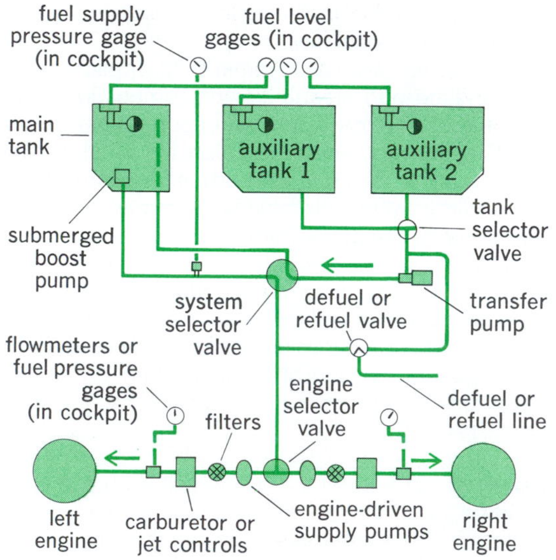

Diagram of a typical aircraft fuel system.

is usually such that all the fuel supply will pass to the engines by way of the main tank, which is refilled as necessary from the auxiliary tanks. In case of emergency, the system selector valve may connect the auxiliary tanks to the engines directly.
[F.C.M./J.A.B.]

Fugacity A function introduced by G. N. Lewis to facilitate the application of thermodynamics to real systems. Thus, when fugacities are substituted for partial pressures in the mass action equilibrium constant expression, which applies strictly only to the ideal case, a true equilibrium constant results for real systems as well.

The fugacity f_i of a constituent i of a thermodynamic system is defined by the following equation (where μ_i is the chemical

$$\mu_i = \mu_i^* + RT \ln f_i$$

potential and μ_i^* is a function of temperature only), in combination with the requirement that the fugacity approach the partial pressure as the total pressure of the gas phase approaches zero. At a given temperature, this is possible only for a particular value for μ_i^*, which may be shown to correspond to the chemical potential the constituent would have as the pure gas in the ideal gas state at 1 atm pressure. This definition makes the fugacity identical to the partial pressure in the ideal gas case. For real gases, the ratio of fugacity to partial pressure, called the fugacity coefficient, will be close to unity for moderate temperatures and pressures. At low temperatures and appropriate pressures, it may be as small as 0.2 or less, whereas at high pressures at any temperature it can become very large. *See* ACTIVITY (THERMODYNAMICS); CHEMICAL EQUILIBRIUM; CHEMICAL THERMODYNAMICS; GAS. [P.J.B.]

Fullerene A hollow, pure carbon molecule in which the atoms lie at the vertices of a polyhedron with 12 pentagonal faces and any number (other than one) of hexagonal faces. The molecule was named after R. Buckminster Fuller, the inventor of geodesic domes, which conform to the same underlying structural formula.

Buckminsterfullerene (C_{60} or fullerene-60; Fig. 1) is the archetypal member of the fullerenes. Other stable members of the fullerene family have similar structures (Fig. 2). The fullerenes can be considered, after graphite and diamond, to be the third well-defined allotrope of carbon. Macroscopic amounts of various fullerenes were first isolated in 1990, and since that time it has been discovered that members of this class of spheroidal organic molecules have numerous novel physical and chemical properties. The fullerenes promise to have synthetic, pharmaceutical, and industrial applications. Derivatives have been found to exhibit fascinating electrical and magnetic behavior, in particular superconductivity and ferromagnetism. *See* CARBON; DIAMOND; GRAPHITE.

Structures and properties. In the fullerene molecule an even number of carbon atoms are arrayed over the surface of a closed hollow cage. Each atom is trigonally linked to its three

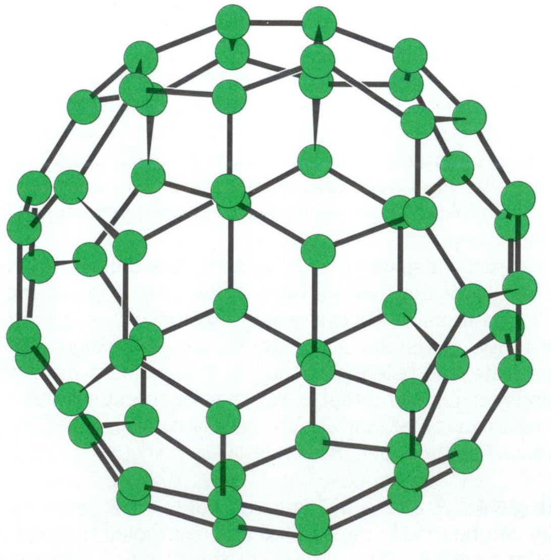

Fig. 1. Structure of fullerene-60 (C_{60}).

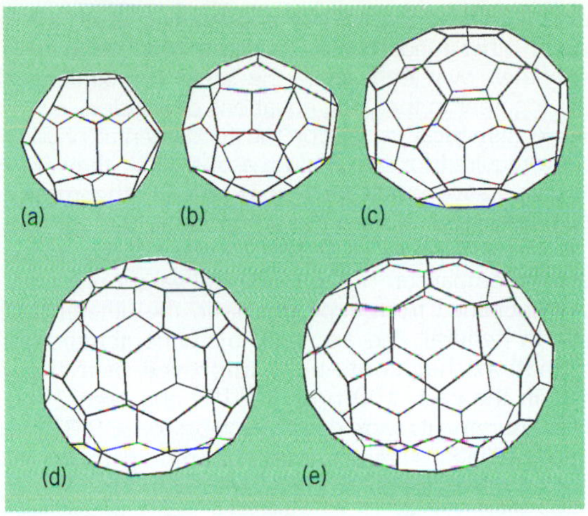

Fig. 2. Some of the more stable members of the fullerene family. (*a*) C_{28}. (*b*) C_{32}. (*c*) C_{50}. (*d*) C_{60}. (*e*) C_{70}.

near neighbors by bonds that delineate a polyhedral network, consisting of 12 pentagons and *n* hexagons. (Such structures conform to Euler's theorem for polyhedrons in that *n* may be any number other than one including zero.) All 60 atoms in fullerene-60 are equivalent and lie on the surface of a sphere distributed with the symmetry of a truncated icosahedron. The 12 pentagons are isolated and interspersed symmetrically among 20 linked hexagons.

In benzene solution, fullerene-60 is magenta and fullerene-70 red. Fullerene-60 forms translucent magenta face-centered cubic (fcc) crystals that sublime. The ionization energy is 7.61 eV and the electron affinity is 2.6–2.8 eV. The strongest absorption bands lie at 213, 257, and 329 nanometers. Studies with nuclear magnetic resonance spectroscopy yield a chemical shift of 142.7 parts per million; this result is commensurate with an aromatic system.

Solid C_{60} exhibits interesting dynamic behavior in that at room temperature the individual round molecules in the face-centered cubic crystals are rotating isotropically (that is, freely) at around 10^8 Hz. At around 260 K (8.3°F) there is a phase transition to a simple cubic (sc) lattice accompanied by an abrupt lattice contraction. Rotation is no longer free, and the individual molecules make rotational jumps between two favored (relative) orientational configurations—in the lower-energy one a double bond lies over a pentagon, and in the other it lies over a hexagon. At 90 K (−300°F) the individual molecules stop rotating altogether, freezing into an orientationally disordered crystal involving a mix of the two configurations. *See* CRYSTAL.

Chemistry and formation. Fullerene-60 behaves as a soft electrophile, a molecule that readily accepts electrons during a primary reaction step. It can readily accept three electrons and perhaps even more. The molecule can be multiply hydrogenated, methylated, ammonated, and fluorinated. It forms exohedral complexes in which an atom (or group) is attached to the outside of the cage, as well as endohedral complexes in which an atom is trapped inside the cage.

The C_{60} molecule behaves as though it has only a single resonance form—one in which the 30 double bonds are localized in the bonds that interconnect the pentagons. This is a key factor, as addition to these double bonds is the most important reaction as far as the application of C_{60} in synthesis is concerned.

On exposure of C_{60} to certain alkali and alkaline earth metals, exohedrally doped crystalline materials are produced that exhibit superconductivity at relatively high temperatures (10 to 33 K or −440 to −400°F). The C_{60} molecule has a triply degenerate lowest unoccupied molecular orbital (LUMO), which in the superconducting materials is half filled, containing three electrons. Other ionic phases, such as M_nC_{60} (n = 1, 2, 4, 6, where M is the intercalated metal atom), exist but are not superconducting—they appear to be metallic or semiconductor/insulators. *See* MOLECULAR ORBITAL THEORY; SUPERCONDUCTIVITY.

Perhaps the most important aspect of the fullerene discovery is that the molecule forms spontaneously. This fact has important implications for understanding the way in which extended carbon materials form, and in particular the mechanism of graphite growth and the synthesis of large polycyclic aromatic molecules. It has become clear that as far as pure carbon aggregates of around 60–1000 atoms are concerned, the most stable species are closed-cage fullerenes. [J.Hare; H.W.Kr.]

Fuller's earth Any natural earthy material (such as clay materials) which decolorizes mineral or vegetable oils to a sufficient extent to be of economic importance. It has no mineralogic significance. The clay minerals present in fuller's earth may include montmorillonite, attapulgite, and kaolinite. However, some materials containing these clay minerals often have very low decolorizing power. *See* CLAY MINERALS.

Fuller's earth deposits are distributed throughout the world, but the largest deposits have been found in the United States, England, and Japan. In the United States, principal deposits are found in Georgia, Florida, Illinois, Texas, Nevada, and California. [F.M.W.; R.E.Gr.]

Fulminate A compound containing the —ONC group and derived from fulminic acid, HONC. Fulminates are isomeric with cyanates; that is, cyanates have the same atoms but in different arrangement, —OCN.

The fulminates have been commercially important because of the use of mercury fulminate, $Hg(ONC)_2$, in priming compositions and as an initial detonating agent; lead azide is replacing mercury fulminate as a detonating agent. *See* CYANATE. [E.E.W.]

Fumigant A pesticidal chemical or chemical formulation that functions in a gaseous state. Chemical formulations are designed to increase toxicity, reduce flammability, give off warning odors, and provide for sorption at different rates.

Physical types of fumigants include gases, liquids, and solids. There are several chemical types of fumigants. These include: halogenated hydrocarbons, such as carbon tetrachloride and ethylene dibromide; sulfur-containing compounds, such as carbon disulfide and sulfur dioxide; cyanides, such as hydrogen cyanide and calcium cyanide; and others, such as phosphine and ethylene dioxide.

Fumigants are used in space fumigations to disinfest food-processing plants, warehouses, grain elevators, boxcars, shipholds, stores, and households, and in spot fumigations within those structures. They are used in atmospheric vaults and vacuum chambers and are applied extensively to stacked bags of grain or stored foods under polyethylene sheets, to trees under tents to control scale insects, and to areas of land to destroy weeds, soil-infesting insects, and nematodes. [D.A.W.]

Funariales An order of mosses. The plants are usually annual, occasionally biennial, and terrestrial, growing in clusters or loose tufts. The stems are erect, short, simple, or sparingly branched. The leaves are soft, broad, and costate. The calyptra varies but is often inflated, usually cucullate and rostrate. The sporophyte is acrocarpous, with a rudimentary to elongated seta. The capsule is erect and symmetrical or cernuous and unsymmetrical, often with a distinct and narrower neck. *See* BRYOPSIDA. [W.H.W.]

Function generator An electronic instrument which generates periodic voltage or current waveforms that duplicate various types of well-defined mathematical functions. The simplest function generator usually generates a combination of square waves, triangular waves, and sine waves.

One electronic circuit approach to the design of a simple function generator is to begin with a bistable multivibrator or "flip-flop" controlled in time by a succession of clock pulses which generates the square wave. The triangular waveform is obtained by integrating the square wave through the use of the operational amplifier integrator. The sine wave is obtained by applying the triangular wave to a shaping circuit consisting of a combination of resistors and diodes. *See* AMPLIFIER; MULTIVIBRATOR; WAVE-SHAPING CIRCUITS.

Alternatively the sine wave may be generated by a sinusoidal oscillator. From this output, the square wave may be obtained by amplication, limiting, and clipping of the sine wave. Then the triangular wave may be obtained using an integrator as before. *See* CLIPPING CIRCUIT; LIMITER CIRCUIT.

A combination of counters, programmed read-only memories (PROMS), and a digital-to-analog converter can be used as a function generator, generating almost any function desired to almost any degree of accuracy. *See* COMPUTER STORAGE TECHNOLOGY; DIGITAL-TO-ANALOG CONVERTER. [G.M.G.]

Fundamental constants That group of physical constants which play a fundamental role in the basic theories of physics. These constants include the speed of light in vacuum, c; the magnitude of the charge on the electron, e, which is the fundamental unit of electric charge; the mass of the electron, m_e; Planck's constant, h; and the fine-structure constant, α. These five quantities typify the different origins of the fundamental constants: c and h are examples of quantities which appear naturally in the mathematical formulation of certain physical theories—Einstein's theories of relativity, and quantum theory, respectively; e and m_e are examples of quantities which characterize the elementary particles of which all matter is constituted; and α, the fundamental constant of quantum electrodynamics (QED), is an example of quantities which are combinations of other fundamental constants, but are actually constants in their own right since the same combination always appears together in the basic equations of physics.

Reliable numerical values for the fundamental physical constants are required for two main reasons. First, they are necessary if quantitative predictions from physical theory are to be obtained. Second, and even more important, the self-consistency of the basic theories of physics can be critically tested by a careful intercomparison of the numerical values of fundamental constants obtained from experiments in the different fields of physics. In general, the accuracy of fundamental constants determinations has continually improved over the years. Whereas in the past, 100 ppm (0.01%) and even 1000 ppm (0.1%) measurements were commonplace, today 0.01 ppm and better determinations are not unusual (ppm = parts per million).

Complex relationships can exist among groups of constants and conversion factors, and a particular constant may be determined either directly by measurement or indirectly by an appropriate combination of other directly measured constants. If the direct and indirect values have comparable accuracy, then both must be taken into account in order to arrive at a best value for that quantity. (By best value is meant that value believed to be closest to the true but unknown value.) Generally, each of the several routes which can be followed to a particular constant, both direct and indirect, will give a slightly different numerical value. Such a situation may be satisfactorily handled by the mathematical method known as least-squares. This technique provides a self-consistent procedure for calculating best "compromise" values of the constants from all of the available data. It automatically takes into account all possible routes and determines a single final value for each constant being calculated. It does this by weighting the different routes according to their relative uncertainties. The appropriate weights follow from the uncertainties assigned the individual measurements constituting the original set of data.

The 1986 least-squares adjustment, carried out under the auspices of the CODATA Task Group on Fundamental Con-

Recommended values (1986) of selected fundamental physical constants

Quality	Symbol	Numerical value*	Units†	Relative uncertainty, ppm
Speed of light in vacuum	c	299792458	m/s	(defined)
Constant of gravitation	G	6.67259(85)	10^{-11} m^3/(kg · s^2)	128
Planck constant	h	6.6260755(40)	10^{-34} J · s	0.60
Elementary charge	e	1.60217733(49)	10^{-19} C	0.30
Magnetic flux quantum, $h/(2e)$	Φ_0	2.06783461(61)	10^{-15} Wb	0.30
Fine-structure constant,	α	7.29735308(33)	10^{-3}	0.045
$\mu_0 ce^2/(2h)$	α^{-1}	137.0359895(61)		0.045
Electron mass	m_e	9.1093897(54)	10^{-31} kg	0.59
Proton mass	m_p	1.6726231(10)	10^{-27} kg	0.59
Neutron mass	m_n	1.6749286(10)	10^{-27} kg	0.59
Proton-electron mass ratio	m_p/m_e	1836.152701(37)		0.020
Rydberg constant, $m_e c\alpha^2/(2h)$	R_∞	1.0973731534(13)	m^{-1}	0.0012
Bohr radius, $\alpha/(4\pi R_\infty)$	a_0	5.29177249(24)	10^{-11} m	0.045
Compton wavelength of the electron, $h/(m_e c) = \alpha^2/(2R_\infty)$	λ_c	2.42631058(22)	10^{-12} m	0.089
Classical electron radius, $\mu_0 e^2/(4\pi m_e) = \alpha^3/(4\pi R_\infty)$	r_e	2.81794092(38)	10^{-15} m	0.13
Bohr magneton, $eh/(4\pi m_e)$	μ_B	9.2740154(31)	10^{-24} J/T	0.34
Electron magnetic moment in Bohr magnetons	μ_e/μ_B	1.001159652193(10)		10^{-5}
Nuclear magneton, $eh/(4\pi m_p)$	μ_N	5.0507866(17)	10^{-27} J/T	0.34
Proton magnetic moment in nuclear magnetons	μ_p/μ_N	2.792847386(63)		0.023
Boltzmann constant	k	1.380658(12)	10^{-23} J/K	8.5
Avogadro constant	N_A	6.0221367(36)	10^{-23}/mol	0.59
Fataday constant, $N_A e$	F	96485.309(29)	C/mol	0.30
Molar gas constant, $N_A k$	R	8.314510(70)	J/(mol-K)	8.4

*The digits in parentheses represent the one-standard-deviation uncertainties in the last digits of the quoted value.
†C = coulomb, J = joule, kg = kilogram, m = meter, mol = mole, s = second, T = tesla, Wb = weber.

stants, succeeded a CODATA adjustment in 1973 by E. R. Cohen and B. N. Taylor; CODATA, the Committee on Data for Science and Technology, is an interdisciplinary committee of the International Council of Scientific Unions. Recommended values are shown in the table. [E.R.Co.]

Fundamental frequency The lowest frequency at which a system vibrates freely. The mode of vibration having this frequency is called the fundamental mode. For a complex wave consisting of a sum of sinusoidal components, that component having the lowest frequency is the fundamental and its frequency is the fundamental frequency. *See* HARMONIC (PERIODIC PHENOMENA); MODE OF VIBRATION; PARTIAL TONE. [R.W.Y.]

Fundamental interactions Fundamental forces that act between the elementary particles of which all matter is assumed to be composed. At present, four fundamental interactions are distinguished. The gravitational interaction manifests itself as a long-range force of attraction between all elementary particles. The electromagnetic interaction is responsible for the long-range force of repulsion of like and attraction of unlike electric charges. At comparable distances, the ratio of gravitational to electromagnetic interactions (as determined by the strength of respective forces between an electron and a proton) is approximately 4×10^{-37}. *See* GRAVITATION; MAXWELL'S EQUATIONS; QUANTUM ELECTRODYNAMICS.

The third fundamental interaction is the weak nuclear interaction, whose characteristic strength for low-energy phenomena is a thousand times weaker than electromagnetic. Weak interactions are short-range, the range of the force being much smaller than 10^{-15} cm. The strong nuclear interaction between protons and neutrons is also short-range, though the range is approximately 10^{-13} cm rather than $\ll 10^{-15}$ cm. As the name implies, within this range of distances the strong force overshadows all other forces between these particles. *See* QUANTUM CHROMODYNAMICS; STRONG NUCLEAR INTERACTIONS; WEAK NUCLEAR INTERACTIONS.

Ever since the discovery and clear classification of these four interactions, particle physicists have attempted to unify these interactions as aspects of one basic interaction between all matter.

Unification of weak and electromagnetic interactions was suggested by S. Glashow and by Abdus Salam and J. C. Ward. This followed a parallel between these two interactions, pointed out by J. S. Schwinger, providing one assumed that the then known classes of weak interactions were mediated by exchanges of W^+ and W^-, quanta of weak interactions, just as electromagnetism is mediated by the exchange of quanta of light, the photons. The empirical properties of weak interaction phenomena had revealed that if $W\pm$ existed, these quanta of the weak interaction must carry an intrinsic spin of magnitude $\hbar$ (Planck's constant divided by 2π), like the photon. *See* QUANTUM FIELD THEORY.

One important consequence of this unification is the prediction of the existence of weak interactions exhibiting neutral-current phenomena, in which the nature of the interacting particles is not changed during the interaction, or, equivalently, the existence of a new heavy intermediate weak-quantum (called Z^0) which, unlike the W^+ and W^- (which carry the same electric charges as the proton and the electron), would be electrically neutral. Since 1973, the verification of neutral current phenomena has given decisive support to the notion of a basic unity between weak and electromagnetic interactions into a single so-called electroweak interaction.

The gauge unification of weak and electromagnetic interactions, which started with the observation that the mediating quanta ($W\pm$,Z^0,γ) for these interactions possess intrinsic spin $\hbar$, can be carried further to include strong nuclear interactions as well, if strong interactions are also mediated through quanta (gluons) carrying spin $\hbar$. Indirect experimental evidence exists for the existence of gluons, and it should be possible to verify if their spin is indeed $\hbar$. If this is so, a complete gauge unification of all three forces, electromagnetic, weak-nuclear, and strong-nuclear, into a single electronuclear force seems plausible. Such a unification necessarily means that the distinction between quarks on the one hand and neutrinos, electrons, and muons on the other must disappear and that all interactions (weak, electromagnetic, and strong) are facets of a universal force, with primitive strength of order of α, this universality manifesting itself for experiments carried out at sufficiently high energies. Another consequence of this disappearance of distinction between quarks and neutrinos and electrons appears to be the possibility of proton decays (with a lifetime of approximately 10^{31} years).

It is possible that the gravitational interaction will also be unified with the other three interactions by extensions of gauge ideas to the so-called supersymmetry between particles and quanta. *See* ELEMENTARY PARTICLE. [A.S.]

Fundamental stars The term applied to a relatively small number of the brighter stars, distributed as uniformly as possible over the entire sky. The fundamental stars are the celestial analog of the so-called first-order triangulation points on the Earth. Thus the coordinates of the fundamental stars are first measured, using elaborate special techniques designed to measure large angular distances; the coordinates of the remaining stars are then inferred by other, less elaborate techniques. *See* ASTROMETRY; ASTRONOMICAL COORDINATE SYSTEMS.

Since the coordinates of the stars are continuously changing as a result of the motion of the Earth and, to a lesser degree, the motions of the stars themselves, it is equally necessary to measure the changes as to measure the coordinates at a definite specified time. [G.M.C.]

Fungal biotechnology The application of scientific and engineering principles to fungal metabolism for the purpose of obtaining a useful product or benefit. This is possible because fungi, including the yeasts, produce enzymes that catalyze a wide variety of biochemical activities. *See* ENZYME.

Fungal biotechnology is usually thought of in terms of the fermentation industry. The products of fungal fermentations are actually fungal metabolites. Primary metabolites, produced by fungi for their own growth and metabolism, include organic acids, alcohols, polysaccharides, amino acids, and vitamins. Secondary metabolites, not essential for fungal growth, are produced during the stationary or nongrowth phase; examples are antibiotics and gibberellins. Both primary and secondary metabolites result from the action of fungal enzymes, and in some fermentations the enzyme itself is the desired product.

Because they are capable of secreting large quantities of certain proteins in liquid culture, fungi have proven to be useful cloning hosts for the production of recombinant proteins of fungal and human origin. Previously rare proteins, such as human growth hormone and the interferons, have thus become available in greater quantities. Fungi commonly used as cloning hosts are *Saccharomyces cerevisiae, Aspergillus awamori, A. nidulans, A. niger, A. oryzae,* and *Pichia pastoris. Aspergillus nidulans* has been designed to produce many human therapeutic proteins, including growth factors and protein hormones. The yeasts *S. cerevisiae* and *P. pastoris* have been utilized for the expression of proteins, such as human interferon and serum albumin. Yeast chromosomes are also being employed in the mapping of the human genome.

In addition to the important role of fungi in the fermentation industry, applications for their metabolic activities are found in other industries, agriculture, environmental remediation, food processing, and medicine. A major characteristic of fungi is their ability to enzymatically degrade simple and complex polymers. Their ability to degrade organic compounds makes them good candidates for the degradation of organic pollutants found in waste effluents, sludges, sediments, and landfills. As an alternative to chemical pesticides, fungi may be used as biocontrol agents to reduce the crop damage from insects, nematodes, weeds, and plant pathogenic microorganisms. Another important characteristic of fungi is their ability to produce a number of compounds that have antibacterial or antifungal activity. Because of its life-saving properties, the most important fungal metabolite is the antibiotic penicillin obtained from *Penicillum chrysogenum*. *See* ANTIBIOTIC; FOOD FERMENTATION; FOOD MICROBIOLOGY; FUNGI; GIBBERELLIN; PENICILLIN; YEAST. [J.M.Bi.]

Fungal ecology The subdiscipline in mycology and ecology that examines fungal community composition and structure; responses, activities, and interactions of single species; and the functions of fungi in ecosystems.

Fungi occur in many different habitat types, such as leaf surfaces, root surfaces and internal tissues, leafy debris and woody debris in successional stages in decomposition, and bark and soil—a milieu of microhabitats very different from those above ground. Fungi also are found in the internal tissues of leaves and stems, in fresh and marine waters, and on or associated with other fungi, lichens, bacteria, algae, dung, and rumens. Fungi are the principal pathogens of vascular plants, however, and in native and field sites they inhabit probably all plants as mutualists in roots (mycorrhizae), stems, or leaves. They are the essential, mutualistic partners of algae or cyanobacteria in all lichens. Many fungal associations with nematodes, mites, protozoa, and insects occur. Trichomycetes inhabit the digestive tracts of arthropods. Laboulbeniales grow externally on insects (approximately 12 orders of ants, flies, and beetles). Ascosphaeriales live in the nests and bodies of bees, and also are consumed by mites. Mutualistic fungi inhabit the guts of wood-boring insects.

Fungi possess unique features that affect their capacity to adapt and to function in ecosystems:

1. Fungi are composed of a vegetative body (hyphae or single cells) capable of rapid expansion and made up of essentially autonomous cells that are in direct contact with the substrate. The cells secrete enzymes, digest externally, and absorb the degradation products.

2. Filamentous fungi are able to mechanically penetrate the substrate.

3. Structural and biochemical-physiological features facilitate absorption and accumulation of nitrogen, phosphorus, potassium, and other mineral elements, including toxins (such as arsenic, selenium, and mercury).

4. Many species of fungi have a capacity to shift their mode of nutrition. The principal modes of nutrition in fungi are saprotrophy, the utilization of dead organic matter; and biotrophy, which is characteristic in parasitic, predaceous, and mutualistic fungi, including mycorrhizal and lichen-forming fungi.

5. Fungi have the capacity for indeterminate growth, longevity, resilience, and asexual reproduction. *See* FUNGAL ECOLOGY; MYCORRHIZAE.

Fungi have two primary functions in ecosystems: decomposition of organic materials, which invariably is accompanied by mineralization or immobilization of all other elemental constituents; and facilitation of nutrient cycling. Nutrient cycling, which includes mineralization and immobilization, is highly complex in that foods and mineral nutrients can be released directly into the general soil milieu but probably more often move from the fungal decomposers through food webs of many kinds. *See* FOOD WEB. [M.Ch.]

Fungal virus Any of the defective and largely functionless replicative elements that permeate fungi. These viruses are endogenous to the host cell not only during their replication but also during their transmission. With few exceptions, fungal viruses are nonpathogenic as to phenotype, and their transmission is intimately associated with either fungal spore dispersal (serial transmission) or fusion of compatible fungal cells (lateral transmission). The transmission of these viruses is a cell-mediated phenomenon and does not require infection by free particles across the host cell surface. Thus, these viruses are heritable rather than infectious in nature.

The physiochemical characteristics of these particles are clearly viral in nature. Most fungal viruses are spherical to hexagonal isometric particles (25–50 nanometers in diameter) with double-stranded ribonucleic acid (RNA) genomes and protein coats (capsids). However, the group shows considerable heterogeneity for genome size and for the distribution of double-stranded RNA segments among particles, and for antigenicity, particle morphology, and mode of replication.

These viruses have simple, single-shelled capsids, and the genome for a given virus may be segmented and dispersed among a population of particles rather than contained in a single particle. Such features are consistent with loss of functions related to infectively, as neither an outer capsid shell for cell-surface reception nor a complete genome per given particle for unit infectivity would be required with cytoplasmic or cell-mediated transmission of particles.

The viruses and plasmids that permeate fungi are unlike the viruses and plasmids of other life forms because these elements are heritable rather than infectious, and with few exceptions they are nonpathogenic. Their physiochemical properties, especially nucleotide sequence information, reveals their homology to a variety of infectious agents that are represented in other life forms. *See* FUNGI; PLANT PATHOLOGY; VIRUS; VIRUS CLASSIFICATION. [P.A.L.]

Fungi Nucleated, usually filamentous, spore-bearing organisms devoid of chlorophyll; typically reproducing both sexually and asexually; living as parasites in plants, animals, or other fungi, or as saprobes on plant or animal remains, in aquatic, marine, terrestrial, or subaerial habitats. Yeasts, mildews, rusts, mushrooms, and truffles are examples of fungi.

In one system of classification, the fungi constitute the division of Eumycetes of the plant kingdom. The division is subdivided into eight classes: Chytridiomycetes, Hyphochrytridiomycetes, Plasmodiophoromycetes, Oomycetes, Zygomycetes, Trichomycetes, Ascomycetes, and Basidiomycetes, and two form classes, Deuteromycetes (Fungi Imperfecti) and Lichenes. The first six classes are grouped under the general term Phycomycetes. Some mycologists also include the class Myxomycetes (Mycetozoa, Myxomycophyta) in the fungi.

Another system recognizes the division of Eumycota with 16 classes in five subdivisions: Mastigomycotina (Chytridiomycetes, Hyphochytridiomycetes); Zygomycotina (Zygomycetes, Trichomycetes); Ascomycotina (Hemiascomycetes, Loculoascomycetes, Plectomycetes, Pyrenomycetes, Discomycetes); Basidiomycotina (Teliomycetes, Phragmobasiomycetes, Hymenomycetes, Gasteromycetes); and Deuteromycotina (Hyphomycetes, Coelomycetes). [C.J.A.]

The characteristics of this group are described for the four classes of the older system of classification.

Phycomycetes, class I, are characterized by mycelia of nonseptate hyphae. Examples are water molds, downy mildews, and black bread molds.

Ascomycetes, class II, are characterized by ascospores, the final product of the sexual process. Examples are yeasts, powdery mildews, some blue and green molds, pink bread mold, cup fungi, morels, and truffles.

Basidiomycetes, class III, are characterized by basidiospores, the final product of the sexual process. Examples are rusts, smuts, bracket fungi, jelly fungi, mushrooms, puffballs, earth stars, and stink horns.

Fungi Imperfecti, class IV, are a group of vegetative molds, predominantly allied with class II, which lack sexual reproduction. Examples are most blue and green molds, some yeasts, and most human pathogenic fungi.

The mycelium, generally the vegetative body of fungi, is extremely variable. Most mycelia are white, but a wide variety of pigments can be synthesized by specific forms and may be secreted into the medium or deposited in cell walls and protoplasm. Mycelial consistency varies from loose, soft wefts of hyphae to compact, hardened masses that resemble leather. Each cell is usually able to regenerate the entire mycelium, and vegetative propagation commonly results from mechanical fragmentation of the mycelium.

Asexual reproduction, propagation by specialized elements that originate without sexual fusion, occurs in most species and is extremely diverse. The most common and important means of asexual reproduction are unicellular or multicellular spores of various types that swim, fall, blow, or are forcibly discharged from the parent mycelium.

Sexual reproduction occurs in a majority of species of all classes except the Fungi Imperfecti. Juxtaposition and fusion of compatible sexual cells are achieved by four distinct sexual mechanisms. [J.R.Ra.]

Fungicide Synthetic or biosynthetic compounds used to control fungal diseases in plants. Chemical compounds are used to control plant diseases caused by fungi. Fungicides now used include both inorganic and organic compounds. Agricultural fungicides must have certain properties and conform to very strict regulations, and of thousands of compounds tested, few have reached the farmer's fields.

The manner in which these compounds are applied, that is, as wettable powders, dusts, or emulsions, is often essential to the success of agricultural fungicides. Raw fungicides must be pulverized to uniform particles of the most effective size, mixed with wetting agents, or dissolved in solvents. These carriers or diluents must not degrade the fungicides or injure the plants.

Foliage fungicides are applied to aboveground parts of plants, usually to prevent disease rather than cure it. Because they are intended to form a protective coating on the plant surface that kills fungus spores before infection occurs, foliage fungicides must adhere to foliage despite weathering. Protective fungicides must be insoluble in water in order to remain on foliage. Certain foliage fungicides, however, are water-soluble. These materials destroy the fungus in disease spots after infection. Fungicides of this type are called eradicant or contact fungicides.

Seeds and seedlings are protected against fungi in the soil by treating the seeds and the soil with fungicides. Seed-treating materials must be safe for seeds and must resist degradation by soil and soil microorganisms. Some soil fungicides are safe to use on living plants.

Systemic fungicides are compounds that permeate plants to protect new growth or to eliminate infections that have already occurred.

Formic acid, acetic acid, and propionic acid up through pelargonic acid and capric acid (the C_1–C_{10} volatile fatty acids) possess significant fungicidal activity. Many of them are present in natural foodstuffs that are resistant to fungal attack. These lower volatile fatty acids are used extensively to prevent fungal growth and so preserve the nutritious quality of cattle fodders. Since these volatile fatty acids stop fungal growth, they prevent mycotoxin generation and lessen the risk of cancer from exposure to mycotoxins. [J.G.H.; S.Ri.]

Furan One of a group of organic heterocyclic compounds containing a diunsaturated ring of four carbon atoms and one oxygen atom. Furan (I) is a typical member of the group. Furfural (II) and some of its close relatives, such as furfuryl alco-

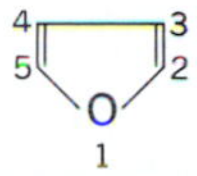

(I)

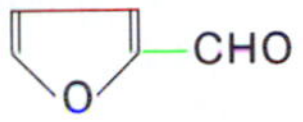

(II)

hol, tetrahydrofurfuryl alcohol, and tetrahydrofuran, are important chemicals of commerce. *See* FURFURAL; HETEROCYCLIC COMPOUNDS; NITROFURAN; PYRAN.

Furan (I) is a colorless, volatile liquid, bp 31.4°C (88.5°F), which is stable to alkali but not to mineral acid. Its water solubility is approximately 1% at room temperature. On exposure to air, furan decomposes very slowly by autoxidation. Substituted furans, particularly negatively substituted furans, are much less sensitive. The furan system is aromatic. Nitration, halogenation, acylation, mercuration, and sulfonation reactions occur with relative ease. [W.J.Ge.]

Furfural When pure, a colorless liquid aldehyde boiling at 161.7°C (323.1°F); also called furfuraldehyde, fural, 2-furaldehyde, or 2-furancarboxaldehyde. It is obtained by the digestion of corn cobs or oat hulls with dilute mineral acid. A wide variety of agricultural by-products also yield furfural if treated in a similar manner; thus, the potential supply of this highly reactive aldehyde is virtually limitless.

Furfural is used as a general synthetic intermediate in the preparation of chemicals, many of which compete with the same chemicals derived from coal and petroleum. Large amounts are also used in the preparation of molding resins and other polymers of value to the plastics industry. Its unusual solvent properties make it useful in the refining of vegetable and lubricating oils and in extracting certain components, such as butadiene, from cracked refinery gases. Other uses to which furfural has been put are as an insecticide, herbicide, fungicide, and embalming fluid. *See* ALDEHYDE; FURAN. [A.E.B.]

Furnace An apparatus in which heat is liberated and transferred directly or indirectly to a solid or fluid mass for the purpose of effecting a physical or chemical change. The source of heat is the energy released in the oxidation of fossil fuel (commonly known as combustion) or the flow of electric current through adjacent semiconductors or through the mass to be heated. According to the source of heat and method of its application, there are four categories of furnaces: combustion, electric, nuclear, and solar, in the order of their present commercial or industrial importance. *See* ELECTRIC FURNACE; STEAM-GENERATING FURNACE; TUBE-STILL HEATER.

Furnaces employing combustion vary widely in construction, depending upon the application of the heat released, whether direct or indirect. Direct heat transfer is used, in regenerative refractory-type heaters, in flow systems in which reactants are injected into the combustion gases. Indirect heat transfer is employed in heaters in which the mass to be heated is kept separate from the combustion gases and made to flow in tubes which absorb and transmit the heat to the fluid to be heated. *See* COMBUSTION.

Furnaces developed for indirect heat transfer can be divided into two classes. One class of heaters is used solely for general utility purposes, such as all types of boilers; and the second class is applied in the petroleum and chemical industries as an essential unit operation in refining or processing plants. *See* BOILER. [H.C.S./H.L.B.]

Fuse (electricity) An expendable device for opening an electric circuit when the current therein becomes excessive. An electric fuse consists principally of a section of conductor, known as a fusible element, of such properties and proportions that excessive current melts it and thereby severs the circuit. Fuses are used in nearly all types of electric circuits to protect circuit conductors and apparatus from damage that could result from sustained excessive current.

The most familiar type of fuse is the screw-plug fuse used in domestic electric systems. It consists of a shell, similar to the base of an incandescent lamp, a fusible element connected between the screw shell and the center terminal, and a transparent protective cover.

Another widely used type is the cartridge fuse, in which the fusible element is connected between metal ferrules at either end of an insulating tube. In an expendable cartridge fuse, this tube is usually filled with an insulating material, such as chalk, sand, or a suitable liquid, which cools and quenches the arc that forms when the fusible element melts. Some fuses, known as renewable fuses, are so constructed that a new fusible element can be inserted to replace one which has been melted. In some types the cartridge is unfilled and open at one end to allow the arc gases to escape to the atmosphere. These are known as expulsion fuses.

In large power systems available short-circuit current can be enormous. In such systems, it is often desirable to have a fuse that can melt and interrupt rapidly. These fuses are known as current-limiting fuses. To accomplish the current-limiting effect, the fuse must be able to melt rapidly at high currents and to form an arc that the system voltage cannot sustain. *See* ARC DISCHARGE; CIRCUIT BREAKER; ELECTRIC PROTECTIVE DEVICES. [T.H.L.]

Fuse (explosive) A conduit that leads fire from one place to another. The firecracker fuse is a familiar example; it consists of a thin train of black powder wrapped in a tube of tissue paper. The miner's safety fuse is similar, the tissue paper being replaced by a woven and waxed fabric.

The burning rate of a black-powder fuse varies widely and depends on the grain size and physical distribution of the powder and especially on the cord construction.

In one type of fuse used with high explosives in mining and quarrying, a woven fabric tube, often reinforced with wrappings of fiber or metal, and impregnated with asphalt and wax, is filled with a core of high explosive, such as pentaerythritol tetranitrate (PETN).

The word fuse should not be confused with fuze (sometimes also spelled fuse), which applies to mechanisms used to fire a shell at the moment it strikes a target, for example, or a hand grenade at a certain interval after it is thrown. *See* DETONATOR; PYROTECHNICS. [W.E.Go.]

Fuselage The component of an aircraft that provides the payload containment and the structural connection for the wing and the empennage (tail assembly). The fuselage and the wing are major structural components of an aircraft. The fuselage is the mounting structure for the horizontal and tail surfaces that provides stability as well as the means of introducing pitch and yaw control to the aircraft. For some aircraft like fighter and private aircraft, the fuselage houses the engine or engines. The nose or tail gear and the main landing gear are often attached to the fuselage structure.

The history of the construction of aircraft fuselages has evolved through the early wood truss structural arrangements to the current metal semi-monocoque shell structures. A majority of aircraft fuselages are fabricated from aluminum alloys and are produced by a process of automatic machining of the skins and stringers (see illustration), with much of the assembly being done by automatic drilling, countersinking, and fastener installation. In some areas, adhesive bonding is used as a means of attaching doublers to reinforce skin panels. In many of the high-performance aircraft, such as fighters and bombers, extensive use is made of titanium and high-strength steel. *See* AIRFRAME. [J.E.McC.]

Boeing 747 fuselage with stringer-stiffened skin supported by frames. (*Boeing Co.*)

Fusulinacea An extinct superfamily of marine organisms in the phylum Protozoa, order Foraminiferida. Fusulinaceans first appeared late in the Mississippian Period, and in the succeeding 100,000,000 years evolved into more than 125 genera and 6000 species before becoming extinct near the close of the Permian Period. Among protozoans fusulinaceans became giants, many reaching 0.4 in. (1 cm) in length and some even attaining 4 in. (10 cm) in length. The group is characterized by rapid evolution of distinctive morphological features, which have enabled paleontologists to establish a detailed fusulinacean phylogeny and biostratigraphic zonation.

In general, the calcareous shell of fusulinaceans is constructed on a simple plan. The initial chamber, the proloculus, is a small calcareous sphere with an opening on one side. As the individual grew and the volume of protoplasm increased, chambers were added successively in a planispiral coil around the proloculus. In most genera these chambers overlap the ends of the previous chambers and become elongate, parallel to the axis of coiling. The spiral wall (spirotheca) is composed of several layers. The initial wall is formed of a thin, dark organic-rich layer (tectum), beneath which is a thicker translucent layer (diaphanotheca) formed of calcium carbonate. *See* FORAMINIFERIDA. [C.A.R.]

Fuzzy sets and systems A fuzzy set is a generalized set to which objects can belong with various degrees (grades) of memberships over the interval [0,1]. Fuzzy systems are processes that are too complex to be modeled by using conventional mathematical methods. In general, fuzziness describes objects or processes that are not amenable to precise definition or precise measurement. Thus, fuzzy processes can be defined as processes that are vaguely defined and have some uncertainty in their description. The data arising from fuzzy systems are, in general, soft, with no precise boundaries. Examples of such systems are large-scale engineering complex systems, social systems, economic systems, management systems, medical diagnostic processes, and human perception. *See* SET THEORY.

The mathematics of fuzzy set theory was originated by L. A. Zadeh in 1965. It deals with the uncertainty and fuzziness arising from interrelated humanistic types of phenomena such as subjectivity, thinking, reasoning, cognition, and perception. This type of uncertainty is characterized by structures that lack sharp (well-defined) boundaries. This approach provides a way to translate a linguistic model of the human thinking process into a mathematical framework for developing the computer algorithms for computerized decision-making processes. The theory has grown very rapidly. Many fuzzy algorithms have been developed for application to process control, medical diagnosis, management sciences, engineering design, and many other decision-making processes where soft data are generated. Thus, fuzzy mathematics provides a modeling link between the human reasoning process, which is vague, and computers, which accept only precise data.

For example, in the design of many engineering systems, process information is not available both because it is difficult to understand precisely the complexity of the phenomena and because human reasoning is inexact and is based upon subjective perception. However, by virtue of knowledge and experience, which is inexact, it is possible to build increasingly good systems. In fact, fuzziness in thinking and reasoning processes is an asset since it makes it possible to convey a large amount of information with a very few words. However, in order to emulate this experience and these reasoning processes on a computer, for example, for intelligent robotics applications and medical diagnosis, a mathematical precision must be given to the vagueness of the information so that a computer can accept it. This is done by using the theory of fuzzy sets. Probability theory deals with the uncertainty or randomness that arises in mechanistic systems, whereas fuzzy set theory has been created to deal with the uncertainty that arises in human cognitive processes. *See* PROBABILITY.

The premise of fuzzy set theory is that the key elements in human reasoning processes are not numbers but labels of fuzzy sets. The degree of membership is specified by a number between 1 (full membership) and 0 (full nonmembership). An ordinary set is a special case of a fuzzy set, where the degree of membership is either 0 or 1. By virtue of fuzzy sets, human concepts like small, big, rich, old, very old, and beautiful can be translated into a form usable by computers. [M.M.G.]

G

Gabbro A rock made up of centimeter-sized crystals mostly of calcium-rich plagioclase (35–65 vol%) and pyroxene in subequal amounts. The terms gabbro, diabase (or dolerite), and basalt refer to mineralogically identical rocks which differ in grain size. The textures of most gabbros are indicative of solidification from a melt. Gabbro occurs widely, but is not as common as basalt (its extrusive equivalent). *See* BASALT.

Gabbro is primarily an intrusive igneous rock. It forms irregular bodies as large as about 39 mi^2 (100 km^2) in area, as well as dikes, sills, and laccoliths. The contacts of bodies of gabbro with surrounding rocks commonly are sharp, and the gabbro generally cuts across trends in older adjacent rock. Many gabbros are massive, but many have either a foliation or a layering or both. *See* IGNEOUS ROCKS.

Gabbros occur in all major tectonic environments (oceanic ridge, continental margin, oceanic island, and continental interior). Many continental interior occurrences are in fossil environments of continental margins and, possibly, oceanic ridges. On the Moon, as on Earth, gabbros are less common than are basalts. The lunar gabbros have textures similar to those of terrestrial gabbros. [A.T.A.]

Gadiformes A well-defined order of actinopterygian fishes, also known as the Anacanthini, which includes the codfishes and grenadiers, or rattails. Structurally the group is more or less intermediate between typical soft-rayed and spiny-rayed fishes. As in the former, there are no fin spines, the scales are cycloid, and the pelvic fin often has many rays. As in typical perciforms, however, the swim bladder has no duct, the orbitosphenoid and mesocoracoid are absent, and the upper jaw is bordered only by the protractile premaxillae. The pelvic fins, if present, are jugular in position but are attached to the cleithra only by ligaments. *See* CODFISH; PERCIFORMES.

Gadiform fishes are known from the Paleocene. Recent forms are classified into 8 families, about 185 genera, and nearly 730 species. The best-known family, the Gadidae, comprises 60 species that live in northern seas, where cod, haddock, pollock, and hake form the basis for extensive commercial fisheries. *See* ACTINOPTERYGII. [R.M.B.]

Gadolinium A metallic chemical element, Gd, atomic number 64 and atomic weight 157.25, belonging to the rare-earth group. The naturally occurring element is composed of eight isotopes. It is named in honor of the Swedish scientist J. Gadolin. The oxide, Gd_2O_3, in powdered form is white, and solutions of the salt are colorless. Gadolinium metal is paramagnetic and becomes strongly ferromagnetic below room temperatures. The Curie point, where this transition occurs, is about 16 K. *See* RARE-EARTH ELEMENTS. [F.H.S.]

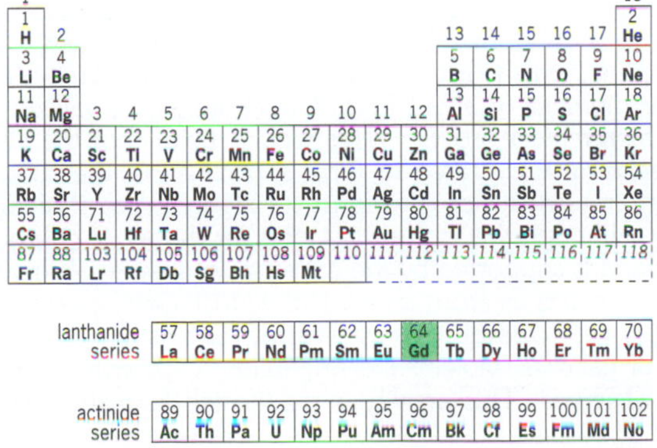

Gage A device for determining the relative size or shape of objects. The function of gages is to determine whether parts are within or outside of the specified tolerances, which are expressed in a linear unit of measurement. Gages are the most widely used production tools for controlling linear dimensions during manufacture and for assuring interchangeability of finished parts. A gage may be an indicating type that measures the amount of deviation from a mean or basic dimension, or it may be a fixed type that simply accepts parts within tolerance and rejects parts outside tolerance. *See* GAGE BLOCK; INDICATING GAGE; PLUG GAGE; RECEIVING GAGE; RING GAGES; SCREW-THREAD GAGE; SNAP GAGE; TAPER PIPE THREAD CAGE; TOLERANCE. [R.A.Bo.]

Gage block The standard of lineal measurement for most manufacturing processes is the gage block (see illustration). Gage blocks are used for setting gages, for setting machines, and for comparative measurements. *See* COMPARATOR.

Gage blocks are made of chrome steel and may be chrome-plated or have carbide wear faces. Individual blocks have two surfaces which are flat and parallel. The parallel distance be-

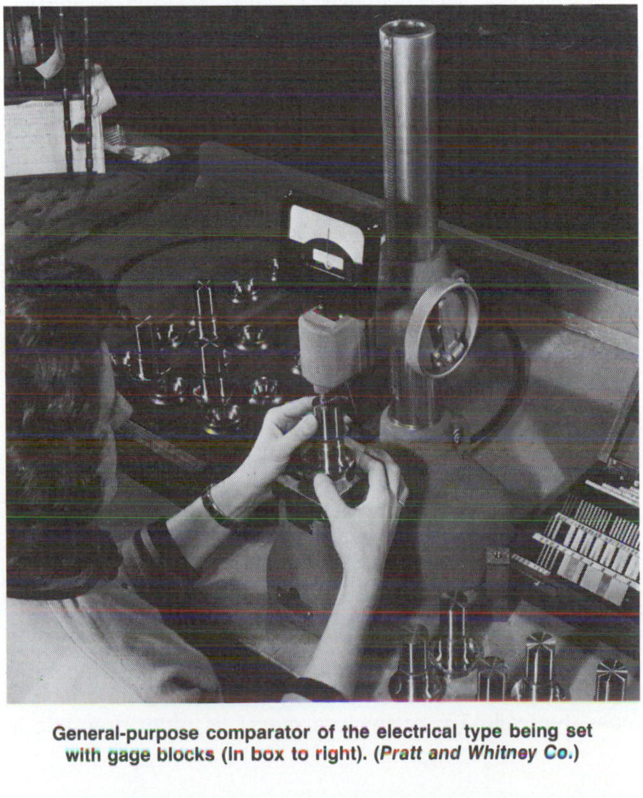

General-purpose comparator of the electrical type being set with gage blocks (in box to right). (*Pratt and Whitney Co.*)

tween the surfaces is the size marked on the block to a guaranteed accuracy of 0.000002, 0.000005, or 0.000008 in. (0.05, 0.13, or 0.20 micrometer) at an ambient temperature of 68°F (20°C). Blocks are sold in sets so that combinations of individual blocks, when wrung firmly together, will produce a new end standard equal to the sum of the sizes of the individual blocks.

Gage blocks are measured by interferometry. Their actual sizes are determined by comparison with the wavelength of red light in the color spectrum, the ultimate standard of lineal measurement being the wavelength of light. Gage blocks translate this basic lineal standard into a practical form for shop use. *See* OPTICAL FLAT. [R.A.Bo.]

Gain An increase in signal power or voltage produced by an amplifier in transmitting a signal from one point to another. The amount of gain is usually expressed in decibels above a reference level. *See* AMPLIFIER.

Antenna gain is a measure of the effectiveness of a directional antenna as compared to a nondirectional antenna. *See* ANTENNA (ELECTROMAGNETISM). [J.Mar.]

Galactose A monosaccharide and a constituent of oligosaccharides, notably lactose, melibiose, raffinose, and stachyose. It is also known as D-galactose and cerebrose (see illustration). Agar, gum arabic, mesquite gum, larch arabo galactan, and a variety of other gums and mucilages contain D-galactose. *See* AGAR; GUM; MONOSACCHARIDE.

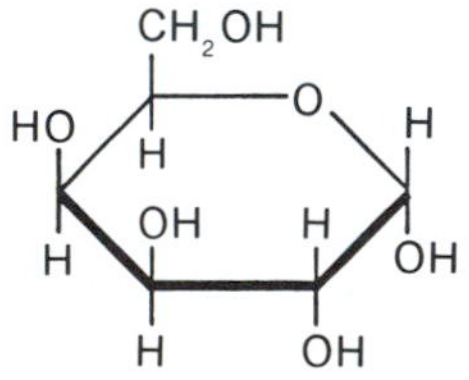

Structural formula for α-D-galactose.

L-Galactose (enantiomorph of D-galactose) occurs in several polysaccharides, including agar, flaxseed mucilage, snail galactogen, and chagual gum. Since D-galactose is usually also present, hydrolysis of these polysaccharides produces DL-galactose. *See* CARBOHYDRATE; POLYSACCHARIDE. [W.Z.H.]

Galaxy, external One of the large, self-gravitating aggregates of stars, gas, and dust which contain nearly all of the observed matter in the universe. Typical large galaxies have symmetric and regular forms, are about 50,000 light-years (3×10^{17} mi or 5×10^{17} km) in diameter, and are roughly 5×10^{10} times more luminous than the Sun. The stars and other material within a galaxy move through it, often in regular rotation, with periods of a few hundred million years.

The hundreds of billions of stars making up a galaxy are not generally individually observable with current telescope technology because they are too faint and distant. Only the brightest stars in the nearest galaxies can be observed directly with large telescopes. Two general types of stellar populations are distinguished: One type (population I) is characterized by the presence of young stars and by ongoing star formation. It is usually associated with the presence of gas. The second type (population II) shows an absence of gas and young stars as well as other indications that star formation ceased long ago. The Sun is a population I star. *See* STAR.

Galaxies contain gas (mostly un-ionized hydrogen) in amounts varying from essentially zero up to a considerable fraction of their total mass. Dust in galaxies, although small in mass (typically 1% of the gas mass), is often dramatic in appearance because it efficiently obscures the starlight. *See* INTERSTELLAR MATTER.

The "Whirlpool" galaxy (NGC 5194), type Sc, and a companion irregular satellite (NGC 5195).

Galaxies generally display strikingly regular forms. The most common form is a disk with a central bulge. Its appearance is characterized by radially decreasing brightness with a superposed spiral or bar pattern, or both (see illustration). The central bulge may vary in size from hundreds to many thousands of light-years and may be spherical or barlike in shape. Such galaxies are classified as spirals (S) or barred spirals (SB) and subclassified a, b, or c (for example, Sa, SBc) to distinguish increasingly open spiral structure and small bulge size. *See* MILKY WAY GALAXY.

Another common type of galaxy is a featureless ellipsoid with radially decreasing brightness. These galaxies are classified as ellipticals. They may vary in size from thousands to several hundred thousand light-years. They are most commonly found in clusters of galaxies and rarely contain gas or dust. The brightest galaxies are usually ellipticals.

Although galaxies are scattered through space in all directions for as far as they can be observed, their distribution is not uniform or random. Most galaxies are found in associations containing from two to hundreds of individual galaxies (with a median number of about five). The clustering of galaxies may be a hierarchical process. This would mean that the clusters themselves are clustered to form superclusters, the superclusters are in turn clustered to form yet higher order clusters, and so on. Although this is the most popular theory of galaxy clustering and is supported by the existence of several observed superclusters, it is far from an established fact. *See* SUPERCLUSTERS.

In the very central regions (sizes at least as small as a light-year) of galaxies, violent and apparently explosive behavior is often observed. This activity is manifested in many ways, including the high-velocity outflow of gas, strong nonthermal radio emission (implying relativistic particles and magnetic fields), intense and often polarized and highly variable radiation at infrared, optical, ultraviolet, and x-ray wavelengths, and ejection of jets of relativistic material. In the most extreme cases (called Seyfert and N galaxies) the energy in the nuclear activity surpasses that in the rest of the galaxy combined. The ultimate source of this activity, whether in some extreme stellar population or in some previously unobserved type of object (for example, black holes) remains completely unknown. There are many astronomers who suspect that active galactic nuclei are related to the even more extreme, but in many ways similar, phenomenon of quasistellar objects. *See* BLACKHOLE; QUASAR.

The nearest galaxies to the Milky Way Galaxy are the Large and Small Magellanic Clouds, two small irregulars lying about 75,000 light-years away. The nearest bright galaxy is the Andromeda Nebula, M31, visible to the naked eye in the constellation of the same name. It is at a distance of almost 2×10^6 light-years. These galaxies and a handful of fainter ones (all within a few million light-years) make up the Local Group, of which the Milky Way Galaxy is a member. *See* ANDROMEDA GALAXY; MAGELLANIC CLOUDS. [E.L.T.; J.B.T.]

Galena A mineral with composition PbS (lead sulfide) and belonging to the rock salt (NaCl) structure type. Galena usually occurs as cubes, sometimes modified by the octahedral form, with perfect cubic cleavage, brilliant metallic luster, color lead gray, specific gravity 7.5, and hardness $2\frac{1}{2}$ on Mohs scale.

Galena is widely distributed and constitutes by far the most important ore for lead. Silver, antimony, arsenic, copper, and zinc minerals often occur in intimate association with galena; consequently, galena ores mined for lead also include many valuable by-products. Important localities include Broken Hill, Australia; the tristate district of Missouri, Kansas, and Oklahoma; and numerous occurrences in Colorado, Montana, and Idaho. [P.B.M.]

Galilean transformations The family of mathematical transformations used in newtonian mechanics to relate the space and time variables of uniformly moving (inertial) reference systems. In the simple case of two similarly oriented cartesian reference frames, moving along their common (x,x') axis, the transformation equations can be put in the form of the following equations, where x, y, z and x', y', z' are the

$$x' = x - vt \qquad y' = y \qquad z' = z \qquad t' = t$$

space coordinates of a given particle, and v is the speed of one system relative to the other. *See* FRAME OF REFERENCE. [E.L.Hi.]

Gallbladder A hollow muscular organ, present in humans and most vertebrates, which receives dilute bile from the liver, concentrates it, and discharges it into the duodenum. Although not a vital organ, it is of great importance in humans because it stores bile, regulates binary tract pressures, and, when diseased, enhances precipitation of various constituents of the bile as gallstones. *See* CHOLECYSTITIS; GALLSTONES.

The system of bile ducts lying outside the liver is known as the extrahepatic biliary tract. In humans (see illustration) right and left hepatic ducts empty into the common hepatic duct, which continues to the duodenum as the common bile duct, or ductus choledochus. The gallbladder and cystic duct thus appear to be accessory organs and therefore are removable. However, they are converted into main-line structures by the presence of a sphincter at the choledochoduodenal junction. Tonic contraction of this sphincter between meals forces the bile to back up into the gallbladder. *See* LIVER.

In humans, evacuation of the gallbladder is accomplished by a trigger mechanism which is set off by the presence of fatty foods, meat, and hydragogue cathartics in the duodenum and

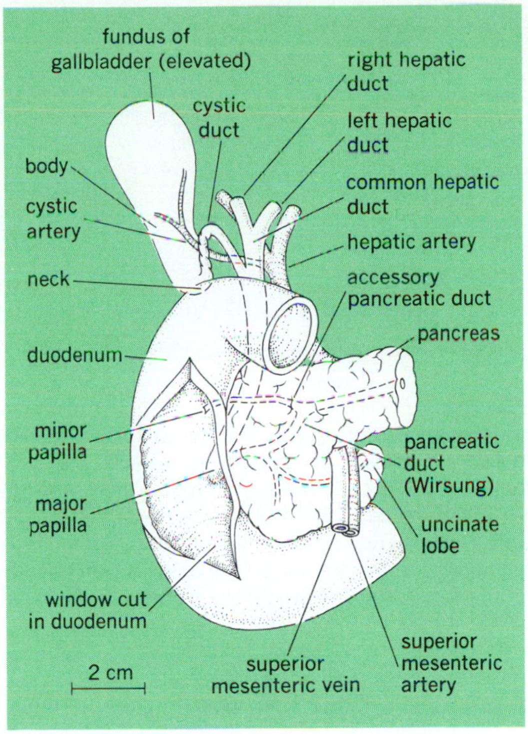

Extrahepatic biliary tract in humans.

upper jejunum. Absorption of these substances by the mucous membrane results in the release of cholecystokinin, a hormone which rapidly circulates in the bloodstream and simultaneously produces contraction of the gallbladder and relaxation of the sphincter of Oddi. The most effective food is egg yolk which contains certain *l*-amino acids. [E.A.B.]

Galliformes The order of birds of great economic significance because it includes important domestic and game birds. Seven families of living Galliformes are usually recognized: Megapodiidae (mound birds and brush turkeys), Cracidae (chachalacas, guans, and curassows), Tetraonidae (grouse), Phasianidae (pheasants, partridge, and quail), Numididae (guinea fowl), Meleagrididae (turkeys), and Opisthocomidae (hoatzins). *See* AVES; BIRDS, UPLAND GAME. [K.C.P.]

Gallium A chemical element, Ga, atomic number 31 and atomic weight 69.72, Gallium was discovered by Lecoq de Boisbaudran in France in 1875. Gallium has a great temperature range in the liquid state, prompting its use to some extent in high-temperature thermometers and manometers. In an

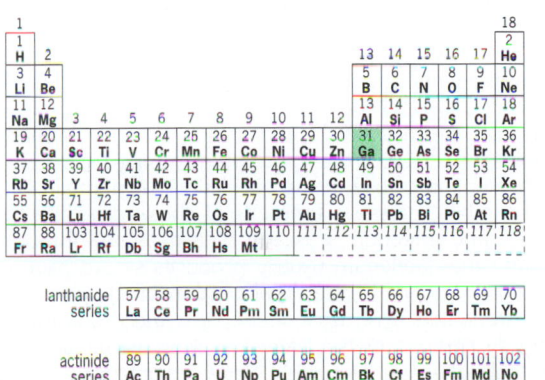

1	2	3	4	5	6	7	8	9	10	11	12	13	14	15	16	17	18
1 H																	2 He
3 Li	4 Be											5 B	6 C	7 N	8 O	9 F	10 Ne
11 Na	12 Mg											13 Al	14 Si	15 P	16 S	17 Cl	18 Ar
19 K	20 Ca	21 Sc	22 Ti	23 V	24 Cr	25 Mn	26 Fe	27 Co	28 Ni	29 Cu	30 Zn	31 Ga	32 Ge	33 As	34 Se	35 Br	36 Kr
37 Rb	38 Sr	39 Y	40 Zr	41 Nb	42 Mo	43 Tc	44 Ru	45 Rh	46 Pd	47 Ag	48 Cd	49 In	50 Sn	51 Sb	52 Te	53 I	54 Xe
55 Cs	56 Ba	71 Lu	72 Hf	73 Ta	74 W	75 Re	76 Os	77 Ir	78 Pt	79 Au	80 Hg	81 Tl	82 Pb	83 Bi	84 Po	85 At	86 Rn
87 Fr	88 Ra	103 Lr	104 Rf	105 Db	106 Sg	107 Bh	108 Hs	109 Mt	110	*111*	*112*	*113*	*114*	*115*	*116*	*117*	*118*

lanthanide series	57 La	58 Ce	59 Pr	60 Nd	61 Pm	62 Sm	63 Eu	64 Gd	65 Tb	66 Dy	67 Ho	68 Er	69 Tm	70 Yb
actinide series	89 Ac	90 Th	91 Pa	92 U	93 Np	94 Pu	95 Am	96 Cm	97 Bk	98 Cf	99 Es	100 Fm	101 Md	102 No

Selected physical properties of gallium	
Property	Value
Melting point	29.78°C (85.60°F)
Boiling point	2237°C (4059°F)
Hardness, Mohs scale	1.5
Density	
Solid	
at 20°C (68°F)	5.907 g/cm^3
at 29.67°C (85.35°F)	5.9037 g/cm^3
Liquid	
at −35.6°C (−32.1°F)	6.153 ± 0.008 g/cm^3
at −16.3°C (−2.66°F)	6.136 g/cm^3
at 29.8°C (85.65°F)	6.0948 g/cm^3
at 1100°C (2012°F)	5.445 g/cm^3

alloy with silver and tin, gallium helps to form a suitable replacement for amalgam in dental fillings. Gallium is also used to solder nonmetallic materials, including gems, to metals. Gallium arsenide can be used in a system for translating mechanical motion into electrical impulses. Synthetic superconducting articles are prepared by making a porous matrix of vanadium or tantalum and impregnating it with gallium hydride. Gallium has given excellent results in a semiconductor for use in rectifiers, transistors, photoconductors, light sources, laser or maser diodes, and refrigerating devices.

Solid gallium looks bluish gray when exposed to the atmosphere. Liquid gallium is silver white with a bright mirror surface. The freezing point of gallium is lower than that of any metal except mercury (−39°C or −38°F) and cesium (28.5°C or 83.3°F). Some properties of gallium are given in the table.

Gallium is chemically similar to aluminum. It is amphoteric but slightly more acid than aluminum. The normal valence of gallium is 3+, and it forms hydroxides, oxides, and salts. Gallium burns in air when heated to 500°C (930°F). It reacts vigorously with boiling water, but only slightly in water at room temperature. The salts of gallium are colorless; they are prepared directly from the metal, since purification of the metal is simpler than purification of the salts.

Gallium forms low-melting eutectic alloys with several metals, and intermetallic compounds with many others. All aluminum contains small amounts of gallium as a harmless impurity, but intergranular penetration by larger amounts of gallium at 30°C (86°F) causes catastrophic failure on loading. [W.D.Wi.]

Gallstones Round, oval, or faceted concretions formed within the gallbladder from the salts and pigment of bile. Such stones may also be formed in any of the bile ducts within or outside of the liver, but the incidence there is low compared to the number originating in the gallbladder. The occurrence of gallstones increases with age, and 20–30% of all elderly adults have one or more such stones. They are particularly common in women, blacks, and individuals with diabetes. Usually about the size of a pea or a marble, gallstones may be extremely tiny, or so large that a single stone completely fills the gallbladder. They may be composed solely of calcium, cholesterol, or bilirubin, but usually represent combinations or mixtures of these three ingredients precipitated from supersaturated bile. *See* CHOLESTEROL; GALLBLADDER.

Although the mechanism and reason for their formation is not clearly understood, the major predisposing factors are prolonged retention of bile in the gallbladder, abnormal composition of the bile (excessive amounts of cholesterol, bilirubin, or calcium), and infection. Passage of a gallstone through the ducts into the duodenum usually produces severe pain, called biliary colic. Lodged in the distal portion of the common bile duct, a stone may obstruct that structure and produce jaundice and finally cirrhosis of the liver. The most frequent complication of gallstones is cholecystitis. *See* CHOLECYSTITIS. [M.R.H.]

Galvanizing The generic term for any of several techniques for applying thin coatings of zinc to iron or steel stock or finished products to protect the ferrous base metal from corrosion; more specifically, the hot dipping that is widely practiced with mild steel sheet and corrugated sheets. During dipping, molten zinc reacts with the steel to form a brittle zinc-iron alloy. For marine use, magnesium is added.

An electrolytic process (also called cold galvanizing or electrogalvanizing) is also used for wire, as well as for applications requiring deep drawing. An alloy layer does not form, hence the smooth electroplated coating does not flake in the drawing die. *See* METAL COATINGS. [F.H.R.]

Galvanomagnetic effects Electrical and thermal phenomena occurring when a current-carrying conductor or semiconductor is placed in a magnetic field. The galvanomagnetic effects are closely related to the thermomagnetic effects. *See* THERMOMAGNETIC EFFECTS.

Let the electric current density j be transverse to the magnetic field H_z, for example, along x. Then the following transverse-transverse effects are observed: (1) Hall effect, an electric field along y. (2) Ettingshausen effect, a temperature gradient along y. Also the following transverse-longitudinal effects are observed: (3) Transverse magnetoresistance, an electrical potential change along x. (4) Nernst effect, a temperature gradient along x. *See* HALL EFFECT; MAGNETORESISTANCE.

Let the electric current density j be along H. Then, the most important effect is longitudinal magnetoresistance, or an electrical potential change along H. [E.A.; F.Ke.]

Galvanometer A device for indicating very small electric currents. Although the deflection of a galvanometer results from current in the moving coil, the voltage in a closed circuit producing this current is frequently the quantity of interest to the

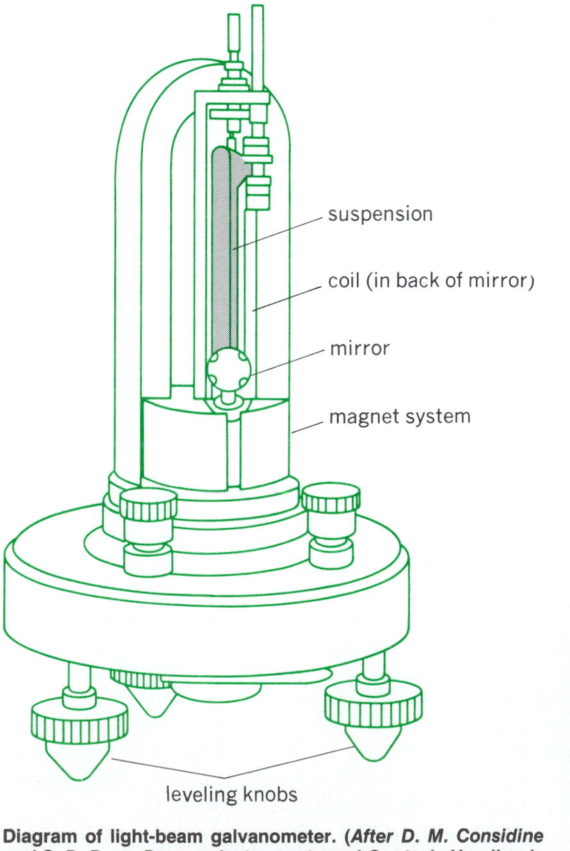

Diagram of light-beam galvanometer. (*After D. M. Considine and S. D. Ross, Process Instruments and Controls Handbook, 2d ed., McGraw-Hill, 1974*)

user. Galvanometers may also be used ballistically to integrate a transient current, as from the discharge of a capacitor, or a transient voltage, as produced when a coil moves relative to a magnetic field. *See* CURRENT MEASUREMENT; VOLTAGE MEASUREMENT.

While great sensitivity is desirable, reasonably rapid and controlled response is also required so that readings may be taken promptly and accurately. The degree to which the movable system is damped is thus of great importance.

The moving systems of many galvanometers carry a small mirror which deflects a light beam to give an indication. This serves as a means for amplifying small motions, and numerous variations have been devised to obtain optimum results in limited space.

The d'Arsonval galvanometer is the most common type, and is widely used. Its indicating system consists of a light coil of wire suspended from a copper or gold ribbon a few thousandths of an inch wide and less than 0.001 in. (0.025 mm) thick. This coil, free to rotate between the poles of a permanent magnet, carries a small mirror which serves as an optical pointer and indicates the coil position by reflecting a light beam onto a fixed scale (see illustration). Current is conducted to and from the coil by the suspension ribbons. The torque which deflects the indicating element is produced by the reaction of the coil current with the magnetic field in which it is suspended. *See* AMMETER; ELECTRICAL MEASUREMENTS; VOLTMETER. [F.K.H.]

Game theory The theory of games of strategy can briefly be characterized as the applicaiton of mathematical analysis to abstract models of conflict situations. The first such models analyzed by the theory were parlor games such as chess, poker, and bridge. Since then, models arising from the behavioral sciences such as economics, sociology, and political science have been analyzed. Game theory is used in or closely connected to other areas such as linear programming, statistical decisions, management science, operations research, and military planning. In certain areas, the language and concepts of the theory are sometimes used even though the corresponding mathematics is not. *See* LINEAR PROGRAMMING; OPERATIONS RESEARCH.

Games in extensive form. The players of a game are called persons, and such a person may actually consist of one or more people (for instance, in bridge the pairs of partners, east-west and north-south, each make up a player in the game). Chance moves occur when hands are dealt from a shuffled pack, dice are rolled, or pointers are spun. One says that all chance moves are allotted to the chance player—a fiction that is useful in abstracting properties of games.

When specified by a list of rules, a game is said to be in extensive form. For mathematical purposes, it is convenient to have games in normalized form, and for that, the idea of a pure strategy is needed.

A pure strategy for a player (not the chance player) is a complete list of choices of legal moves that he or she will make for every possible situation that can occur during the game. This is a much more complete list of decisions than that commonly called a strategy. The number of pure strategies in a game can be astronomical even for childish games such as tic-tac-toe. Because of the enormous number of pure strategies, the actual applications of game theory even to parlor games have been severely limited by computational difficulties. Simplified versions of the games have been developed for which computations have been completely carried out.

Games in normalized form. After players have chosen pure strategies in a game, they need not physically play the game. Instead they could hand their strategies to a neutral person, or umpire, who could then carry out their instructions and make the moves they would have made. This intuitively obvious idea leads naturally to the normalized form of the game.

Assume for the moment there are no chance moves in the game, that is, that there are n real players but no chance player. Denote by $s_1, s_2, \ldots, s_n$ specific pure strategies for players 1, 2,…, n, respectively. Given these, the game must be played in exactly one way and a unique outcome will result. Let $P_i(s_1,s_2,\ldots,s_n)$ be the monetary outcome to player i for this play of the game.

Before the effect of the chance player can be introduced, the important concept of mathematical expectation must be explained. Suppose that $O_1, O_2, \ldots, O_k$ are mutually exclusive monetary outcomes of a chance event, and suppose further that they happen with probabilities, $p_1, p_2, \ldots, p_k$ where $p_i > 0$ and $p_1 + p_2 + \ldots + p_k = 1$. Then the mathematical expectation E of the chance event is defined to be the sum $E = p_1O_1 + p_2O_2 + \ldots + p_kO_k$. *See* PROBABILITY.

If there are chance moves in the game, a set of pure strategies, one for each player, will not determine a unique outcome of the game but merely a set of possible outcomes. These outcomes will be mutually exclusive and have probabilities depending on the chance moves associated with their occurrence. Hence in this case one can let $P_i(s_1,s_2,\ldots,s_n)$ be the expected payoff to player i for each $i = 1,2,\ldots, n$.

Now the normalized form of a game is defined as the list of all expected payoffs to each player for every possible combination of pure strategies. In the case of two-person games it is most convenient to list these in tables called matrices. *See* MATRIX THEORY.

Classification of games. A game is called zero-sum if, for every possible n-tuple of pure strategies $s_1,\ldots, s_n$, the sum of the payoffs to all players is zero, that is, Eq. (1) holds. If this

$$P_1(s_1,\ldots,s_n) + P_2(s_1,\ldots,s_n) + \cdots + P_n(s_1,\ldots,s_n) = 0 \qquad (1)$$

sum is not zero for some n-tuple of pure strategies, the game is called nonzero-sum.

A game in extensive form is said to have perfect information if every player knows and remembers each move of each of her or his opponents as they are made. A game is said to have perfect recall if each player knows and remembers everything he or she did (but not necessarily what the opponent did). Bridge does not have perfect recall, because the personality of a player (= team) is divided between two actual persons and, for instance, north does not see what is in south's hand (except when south is dummy) even though they are members of the same team.

Games can then be classified according to the number of players they have, whether or not they are zero-sum, and whether or not they have perfect information or perfect recall.

Nonzero-sum games. By far the most satisfactory part of the theory of games consists of the zero-sum two-person cases, that is, in matrix games. Applications of the theory to such areas as economics, sociology, and political science almost invariably lead to many-person nonzero-sum games. Although no universally accepted theory has been developed to cover these games, many interesting and useful attempts have been made to deal with them.

When more than two persons are involved in a conflict situation, the important feature of the game becomes the coalition structure of the game. A coalition is a group of players who band together and, in effect, act as a new player in the game. There are two extremes to be considered. One is the noncooperative game in which such coalitions are banned by some means. Equilibrium-point solutions, discussed below, provide reasonably satisfactory solutions to such games. The other extreme is that in which all the players join together in a coalition to maximize jointly their total payoff. A game in which coalitions are permitted is called a cooperative game.

In the noncooperative game, each player is solely interested in his or her own payoff. By an equilibrium point in such a game is meant a set of mixed strategies $s_1,\ldots, s_n$ such that Eq. (2)

$$P_i(s_1,\ldots, s_i,\ldots, s_n) = P_i(s_i,\ldots,s'_n,\ldots,s_n) \qquad (2)$$

holds for each $i = 1,\ldots, n$ for all strategies s_i, of player i. What this means is that no player can, by changing strategy and assuming that the other strategies stay fixed, improve the payoff. By a theorem of J. Nash, every game has at least one equilibrium-point solution (commonly there are several).

Simple games. An important class of n-person games for application to political behavior are the so-called simple games. Each coalition in such a game can be either winning, losing, or blocking. For instance, a winning coalition may be a set of voters who can elect their candidate, or a group of lawmakers who can pass their bill. The players not in a given winning coalition form a losing coalition. Finally, a coalition is blocking if neither it nor the players not in it can enforce their wishes.

Continuous games. If, in the normalized form of the game, each player is permitted to have a continuous range of pure strategies and the payoff function is permitted to be a function of the two real variables that range over each player's strategies, the result is a continuous game.

Game-playing machines. One of the first applications of large electronic computers to numerical problems was in solving large matrix games. Several methods have been devised for finding such solutions. One such method is the simplex method. So-called decomposition methods can extend these methods to certain problems having thousands or even millions of variables and constraints. The principal application of the method is to solve linear programming problems, which can be shown to be equivalent to matrix games.

Computers have also been programmed to play board games such as checkers and chess. Strictly speaking they do not use the theory of games at all at present, but instead use some of the game sense of the people who devised the codes. *See* ARTIFICIAL INTELLIGENCE; DECISION THEORY; INFORMATION THEORY; STOCHASTIC PROCESS. [G.L.T.]

Gametogenesis The meiotic process of formation of the gametes, either eggs in the female or sperm in the males, which is common to all animals. This process involves changes in the chromosome number from the diploid ($2N$) number in somatic cells to the haploid (N) number in the gametes. In the process, the chromosomes (deoxyribonucleic acid, or DNA) are duplicated to form four chromatids (tetrads) with DNA quantity equal to $4N$.

Meiotic divisions in eggs generally occur in association with ovulation and are completed with fertilization. Meiotic divisions are fundamentally the same in plants, with the formation of asexual spores called tetraspores which produce the gametophyte with mitotic divisions. *See* MEIOSIS; OOGENESIS, SPERMATOGENESIS. [G.W.H.]

Gamma function A particular mathematical function that can be used to express many definite integrals. There are, however, no significant applications where the gamma function by itself constitutes the essence of the solution. Instead it occurs usually in connection with other functions, such as Bessel functions and hypergeometric functions.

A special case of the gamma function is the factorial $n! = 1 \cdot 2 \cdot 3 \cdots \cdot n$ (for example, $1! = 1$, $2! = 2$, $3! = 6$, $4! = 24$). It is defined only for integral positive values of n. The factorial occurs, for instance, in the expansion

$$\exp z = 1 + z/1! + z^2/2! + z^3/3! + \cdots$$

The binomial coefficient (N/n) can be expressed in terms of factorials as $N!/[n!(N - n)!]$. Many occurrences of the factorial are found in combinatorial theory (for instance, $n!$ is the number of permutations of n different elements), in probability theory, and in the applications of this theory to statistical mechanics. For large values of n, the factorial can be easily, although only approximately, computed with Stirling's formula,

$$n! \approx (n/e)^n(2\pi n)^{1/2}$$

See BESSEL FUNCTIONS. [J.Meix.]

Gamma-ray astronomy The study of gamma rays of cosmic origin. The field may be divided into the study of high-energy gamma rays and that of soft gamma-ray bursts. The gamma-ray bursts have a typical energy 100 times lower, and do not seem at the present time to be related to the high-energy gamma rays. *See* GAMMA-RAY BURSTS; GAMMA RAYS.

Because, in the high-energy gamma-ray range, the photons (simply called gamma rays in the following) are completely absorbed by the atmosphere, it is necessary to place the gamma-ray detectors aboard high-altitude balloons, or better still, artificial satellites. The heart of a gamma-ray telescope is a spark chamber, similar to chambers used with particle accelerators, but much smaller. The incoming gamma rays are absorbed in the plates and converted into electron-positron pairs. Since the gamma rays are invisible, it is the analysis of the electron and positron tracks, visualized by sparks, that gives the directional and energetic information about the detected photons. The location of the sparks is obtained through a digitized wire array. *See* GAMMA-RAY DETECTORS; SPARK CHAMBER.

The measurement of the energy of the incoming gamma rays is not very precise. At present, it involves an uncertainty of about 50–100%. Astronomical limitations of gamma-ray telescopes result from the weak intrinsic fluxes of gamma rays and from background problems.

The gamma-ray image of the Milky Way appears as a bright lane of diffuse emission, much like the optical image, although thinner (2° wide). It spans roughly the same angular extent, from −60° to +60° in galactic longitude. Two bright spots, in the constellations Cygnus and Vela, are also visible. There is a fainter emission in other parts of the galactic disk.

The gamma-ray sky is almost devoid of "stars," except along the galactic plane, where there is a strong concentration of gamma-ray sources (that is, "stars" of 1° radius). This is remarkable, as it is very different from what is seen at other wavelengths, even in x-rays. [T.Mo.]

Gamma-ray bursts Brief, intense bursts of cosmic gamma rays which are not obviously associated with any known class of astronomical objects. They were discovered in 1967 by the *Vela* satellites, developed to monitor the Nuclear Test Ban Treaty, and their discovery was announced in 1973.

The soft gamma-ray bursts typically last 1–10 s; however, some are as short as 100–300 milliseconds, and the longest burst observed so far had a duration of about 80 s. Most of the bursts exhibit multiple (nonperiodic) pulses with significant structure occurring on time scales of 100 ms. Some even show variations on a time scale as short as the 16-ms resolution of the *Vela* detectors. Such short time scales imply a source size less than 10^9 cm (10^7 m) or a highly relativistic emission process.

The duration and the fast-time-scale structure of the gamma-ray bursts lie in the range that is characteristic of the dynamical time scales (10^{-4} to 1 s) associated with degenerate dwarfs, neutron stars, and black holes. There are believed to be a large number of such compact objects within the 100-parsec distance that is characteristic if the gamma-ray-burst sources lie within the Galaxy. These considerations have led theorists to propose models involving such compact objects; several of these models invoke neutron star starquakes. *See* BLACK HOLE; GRAVITATIONAL COLLAPSE; WHITE DWARF STAR. [D.Q.L.]

Gamma-ray detectors Devices for detecting gamma (γ) rays. Most detectors are capable of detecting and registering the passage of individual γ-rays, or photons. A detector which simultaneously determines the energy of a γ-ray may be properly termed a spectrometer. The term "detector" is frequently used as a generic name for devices that are also spectrometers. *See* GAMMA RAYS.

The two most popular devices in this category are the (thallium-activated) sodium-iodide [NaI(Tl)] detector and the (lithium-

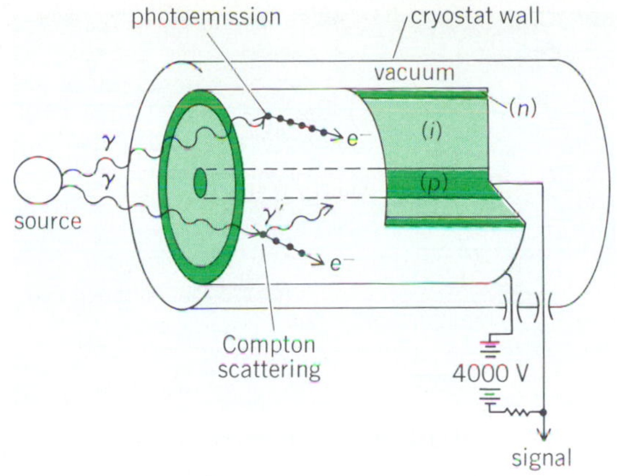

Schematic diagram of a coaxial Ge(Li) detector.

drifted) germanium [Ge(Li)] detector (see illustration). Since γ-rays have no intrinsic charge or mass, both detectors depend on the fact that the three processes whereby γ-rays interact with matter (photoemission, Compton effect, and pair production) result in the production of energetic electrons whose total energy is exactly proportional to the original γ-ray energy. The electrons, in turn, lose energy by the ionization of the material (Ge or NaI), producing thereby a number of ion pairs exactly proportional to the electron energy. In both the NaI(Tl) and Ge(Li) detectors, the ionization energy is ultimately converted to a voltage-current pulse proportional to the total ionization produced by the γ-ray interaction. The output pulses from the detector (after amplification) are then analyzed by a pulse-height analyzer to provide a quantitative measure of the pulse-height spectrum. *See* ELECTRON-POSITRON PAIR PRODUCTION; PHOTOEMISSION. [J.W.Ol.]

Gamma rays Electromagnetic radiation emitted from excited atomic nuclei as an integral part of the process whereby the nucleus rearranges itself into a state of lower excitation (that is, energy content). *See* NUCLEAR STRUCTURE; RADIOACTIVITY.

The gamma (ray) is an electromagnetic radiation pulse—a photon—of very short wavelength. The electric (**E**) and magnetic (**H**) fields associated with the individual radiations oscillate in planes mutually perpendicular to each other and also to the direction of propagation with a frequency υ which characterizes the energy of the radiation. The **E** and **H** fields exhibit various specified phase-and-amplitude relations, which define the character of the radiation as either electric or magnetic. Gamma rays range in energy from a few keV to 100 MeV, although most radiations are in the range 50–6000 keV. As such, they lie at the very upper high-frequency end of the family of electromagnetic radiations, which include also light rays and x-rays. *See* ELECTROMAGNETIC RADIATION; MULTIPOLE RADIATION; PHOTON.

The dual nature of γ-rays is well understood in terms of the wavelike and particlelike behavior of the radiations. For a γ-ray of intrinsic frequency υ, the wavelength is $\lambda = c/\upsilon$, where c is the velocity of light; energy is $E = h\upsilon$, where h is Planck's constant. *See* LIGHT; QUANTUM MECHANICS; RELATIVITY.

Various nuclear species exhibit distinctly different nuclear configurations; the excited states, and thus the γ-rays which they produce, are also different. Precise measurements of the γ-ray energies resulting from nuclear decays may therefore be used to identify the γ-emitting nucleus. This has ramifications for nuclear research and also for a wide variety of more practical applications. One of the most useful studies of the nucleus involves the bombardment of target nuclei by energetic nuclear projectiles to form final nuclei in various excited states. Measurements of the decay γ-rays are routinely used to identify the various final nuclei according to their characteristic γ-rays.

In practical applications, the presence of γ-rays is used to detect the location or presence of radioactive atoms which have been deliberately introduced into the sample. In irradiation studies, for example, the sample is activated by placing it in the neutron flux from a reactor. The resultant γ-rays are identified according to isotope, and thus the composition of the original sample can be inferred. Such studies have been used to identify trace elements found as impurities in industrial production, or in ecological studies of the environment, such as minute quantities of tin or arsenic in plant and animal tissue. *See* ACTIVATION ANALYSIS.

In tracer studies, a small quantity of radioactive atoms is introduced into fluid systems (such as the human blood stream), and the flow rate and diffusion can be mapped out by following the radioactivity. Local concentrations, as in tumors, can also be determined. *See* RADIOACTIVE TRACER.

For the three types of interaction with matter which together are responsible for the observable absorption of γ-rays, namely, Compton scattering, the photoelectric effect, and pair production, *see* COMPTON EFFECT; ELECTRON-POSITRON PAIR PRODUCTION; PHOTOEMISSION. [J.W.Ol.]

Gammaridea A suborder of Amphipoda. These crustaceans are commonly known as scuds in aquatic environments, and as sandhoppers on beaches. The suborder comprises nearly 4750 species in 880 genera and 91 families. The majority of the species are marine. The terrestrial species are confined mainly to beaches in high latitudes but in the IndoPacific insular region they penetrate far inland and to an altitude of 9900 ft (3000 m) in moist environments.

Gammaridea are usually compressed laterally and are poor walkers except for the families Corophiidae and Cheluridae, which have depressed bodies (see illustration). The head and segments are free, lacking a carapace. The head bears two pairs of antennae and six kinds of mouthparts, some of them paired. The mandibles usually bear cutting and trituration surfaces for chewing, but a few species lack these devices or have the mouthparts modified for sucking the tissues of animals such as compound ascidians.

Gammaridea are largely scavengers, feeding on organic debris or detritus which falls to the ocean bottom. Species in the families Ampeliscidae, Photidae, and Corophiidae build nest-

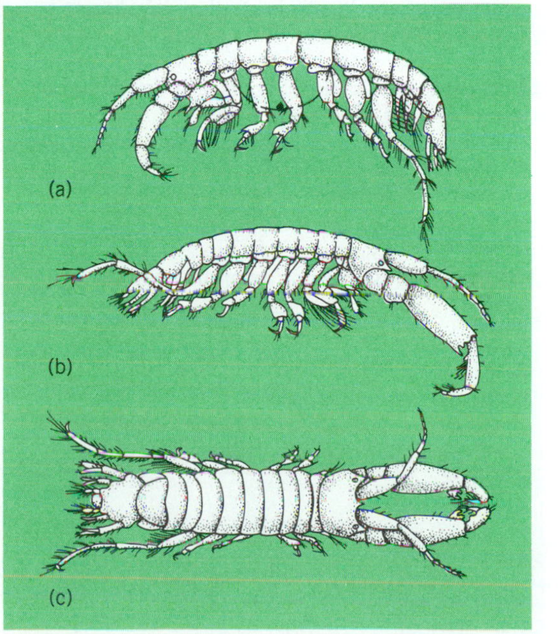

Corophium crassicorne **(Corophiidae). (*a*) Female. (*b*) Male. (*c*) Male, dorsal view. (*After G. O. Sars, An Account of the Crustacea of Norway, vol. 1, 1895*)**

ing tubes attached to solid intertidal objects or lying on the sea bottom. The tubes are spun either from secreting glands on the first two pairs of pereiopods or from cuticular glands on the body. These animals use their well-developed antennae to strain food particles. Semiparasitic and commensal species with sucking or lapping mouthparts are known in the families Stenothoidae, Leucothoidae, and Dexaminidae. *See* AMPHIPODA. [J.L.B.]

Ganglion A group of nerve cell bodies, usually located outside the brain and spinal cord. A ganglion located inside the central nervous system is called a nucleus.

The dorsal root ganglia are rounded clusters of cell bodies and fibers, surrounded by a connective tissue covering, located on the dorsal, or sensory, root of each spinal nerve. Other ganglia are given specific names which indicate their function or location, such as acoustic, cardiac, carotid, jugular, celiac, and sympathetic ganglia. Sympathetic ganglia, lying on either side of the vertebral column, unite by fiber strands to form a sympathetic chain. *See* SYMPATHETIC NERVOUS SYSTEM.

The term ganglion may be applied to a tumorlike, often cystic growth found on tendons, joints, and other connective tissues, but this usage is rare. *See* BRAIN; SPINAL CORD. [W.B.]

Gangrene A form of tissue death, or necrosis, usually occurring in an extremity and due to insufficient blood supply.

If no bacterial contamination is present, the part becomes dry, greenish-yellow, and finally turns brown or black. This is known as mummification. A sharp inflammatory border marks the edge of the adjacent viable tissue. This dry gangrene is seen most often in small portions of the extremities, such as the fingers and toes. Senile gangrene is the form caused by deterioration of blood supply in the elderly, usually as the result of progressive arteriosclerosis. Similar types are often present in diabetes, Reynaud's disease, and Buerger's disease (thromboangiitis obliterans). *See* ARTERIOSCLEROSIS; BUERGER'S DISEASE; DIABETES.

When bacterial infection intervenes, putrefaction ensues, thus producing the moist or wet type of gangrene. Occasionally the invading microorganisms are gas-producing species of the genus *Clostridium*, and rapidly progressive gas gangrene is the result. Gas gangrene is characterized by extensive edema with gas production and discoloration of the tissue, and often accompanied by a putrefactive odor. Moist gangrene may occur anywhere in the body that the blood supply is blocked and bacterial contamination occurs. [E.G.St./N.K.M.]

Gantt chart A visual control device developed by Henry L. Gantt and used in production planning and control. In concept, the Gantt chart is a bar chart of several interrelated activities with time on the horizontal axis. The chart specifies the work required by a project (comprising a series of tasks) and the order in which the tasks must be done.

There are two basic types of Gantt charts: the planning chart and the progress chart. Preparation for the chart begins

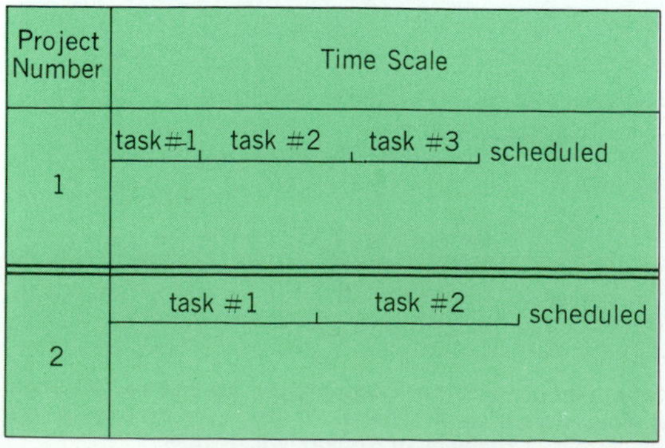

Fig. 1. Gantt planning chart.

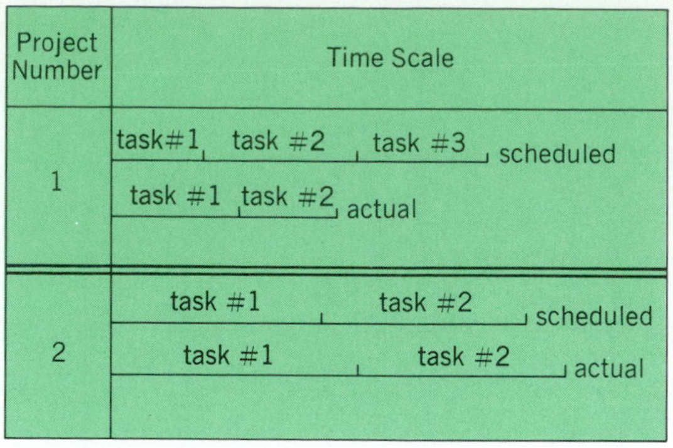

Fig. 2. Gantt progress chart.

by listing the tasks that must be done to complete the project. A technological sequence of the activities for each task is required. After this step is completed, time estimates for each activity must be obtained. Next the Gantt planning chart (Fig. 1) is prepared, making a definite plan for each task necessary, This is one of the advantages of Gantt charts. It forces a thinking-through of the things that will be encountered and for which provisions must be made. As work progresses, a bar is drawn on the chart with a length proportional to the amount of work done during each time interval. Work done is initially plotted on the time interval during which it is accomplished. Later a cumulative bar may be added to summarize the status of work for an extended period. This second type of Gantt chart is known as the progress chart (Fig. 2). *See* CRITICAL PATH METHOD (CPM); INDUSTRIAL COST CONTROL; PERT. [N.M.F.]

Gar The name for about seven species of bony ganoid fishes (sometimes called garpikes) of the genus *Lepisosteus*. All have a slim form and an elongate snout (see illustration) and are

Long-nosed gar.

covered with thick, close-set scales. These fish are a small remnant of a formerly large group of otherwise fossil fishes, usually regarded by paleontologists as constituting the order Semionotiformes. *See* ACTINOPTERYGII; SEMIONOTIFORMES. [C.B.C.]

Garlic A hardy perennial, *Allium sativum*, of Asiatic origin and belonging to the plant order Liliales, garlic is grown for its pungent bulbs, segments of which are used primarily for seasoning. Popular varieties are Italian, Tahiti, and Creole or Mexican. Garlic salt is made from dehydrated cloves. California is the most important producing state; smaller acreages are planted to garlic in Louisiana and Texas. *See* LILIALES. [H.J.C.]

Garnet A hard, dense silicate mineral which occurs as crystals of cubic symmetry in a wide range of geologic environments. The general chemical formula of the silicate garnet group is $A_3B_2(SiO_4)_3$ where, in natural occurrences, the A cations are dominantly Fe^2, Mn^2, Mg, and Ca, and the B cations are Al, Fe^3, and Cr^3.

The garnet mineral group is generally divided into a number of individual species on the basis of chemical composition. The more common of these species are pyrope, almandine, spessartine, grossular, andradite, and uvarovite.

Garnets are substantially denser than most chemically analogous silicates, with specific gravities ranging between 3.58

(pyrope) and 4.32 (almandine). They also have high refractive indices (1.71–1.89) and hardness, on Mohs scale, of 6½ to 7½. The relative hardness, coupled with the absence of cleavage, has led to the use of garnet as an abrasive. The color of garnet is primarily controlled by its chemical composition. Uvarovite is emerald green; gem varieties of garnet are generally clear, deep red pyrope. *See* GEM; SILICATE MINERALS.

Garnets are widespread in their occurrence, particularly in rocks which formed at high temperatures and pressures. Because of the large, readily identifiable crystals which form, the first appearance of garnet is commonly used by geologists as an index of the intensity, or grade, of metamorphism. Garnets are strongly resistant to weathering and alteration and are hence widespread constituents of sands and sediments in areas of garnetiferous primary rocks. *See* METAMORPHIC ROCKS. [B.J.Wo.]

Garnierite A mineral used as a minor ore of nickel. The hardness is 2–3 on Mohs scale, and the specific gravity is 2.2–2.8. The luster is earthy and the color apple green to white. The composition may be expressed by the formula $(Ni,Mg)SiO_3 \cdot nH_2O$, but varies greatly in the relative amounts of nickel and magnesium and in the amount of water. Garnierite has been mined as an ore of nickel in New Caledonia, the Soviet Union, South Africa, and Madagascar. In the United States it is found at Riddle, Oregon, and Webster, North Carolina. *See* NICKEL. [C.S.Hu.]

Garvey-Kelson mass relations A set of equations relating the masses of atomic nuclei with slightly different numbers of neutrons and protons. These relationships are derived from general physical arguments and are observed to be well obeyed. The simplest form that the Garvey-Kelson mass relations take is given by the equation below, where N_0 and Z_0

$$M(N_0, Z_0 - 1) - M(N_0 - 1, Z_0) + M(N_0 - 1, Z + 1) - M(N_0 + 1, Z_0 - 1) + M(N_0 + 1, Z) - M(N_0, Z_0 + 1) = 0$$

can take any value provided that the absolute value of $N_0 - Z_0$ is equal to or greater than 2.

Starting with relatively few known masses, the equation can be applied over and over again to fill in a complete mass table. Thus, the particle stability of such unusual species as ^{8}He, ^{11}Li, and ^{14}B has been predicted and subsequently observed in laboratory experiments. Further, the stability of species as neutron-rich as ^{70}Ca are also predicted. While it is unlikely that such species can be found in the laboratory, their potential existence in very neutron-rich stellar environments has significant consequences for nucleosynthesis of the elements. *See* NUCLEAR STRUCTURE; NUCLEOSYNTHESIS. [G.T.G.]

Gas A phase of matter characterized by relatively low density, high fluidity, and lack of rigidity. A gas expands readily to fill any containing vessel. Usually a small change of pressure or temperature produces a large change in the volume of the gas. The equation of state describes the relation between the pressure, volume, and temperature of the gas. In contrast to a crystal, the molecules in a gas have no long-range order.

At sufficiently high temperatures and sufficiently low pressures, all substances obey the ideal-gas, or perfect-gas, equation of state below, where p is the pressure, T is the absolute

$$p\bar{V} = RT$$

temperature, $\bar{V}$ is the molar volume, and R is the gas constant. Absolute temperature T is expressed on the Kelvin scale. The gas constant is 8.314 joules/(mole K). The molar volume is the molecular weight divided by the gas density.

At lower temperatures and higher pressures, the equation of state of a real gas deviates from that of a perfect gas. Various empirical relations have been proposed to explain the behavior of real gases. *See* GAS DYNAMICS. [C.F.C.; J.O.H.]

Gas absorption operations The separation of solute gases from gaseous mixtures of noncondensables by transfer into a liquid solvent. This recovery is achieved by contacting the gas stream with a liquid that offers specific or selective solubility for the solute gas or gases to be recovered. The operation of absorption is applied in industry to purify process streams or recover valuable components of the stream. It is used extensively to remove toxic or noxious components (pollutants) from effluent gas streams. *See* ABSORPTION.

The absorption process requires the following steps: (1) diffusion of the solute gas molecules through the host gas to the liquid boundary layer based on a concentration gradient, (2) solvation of the solute gas in the host liquid based on gas-liquid solubility, and (3) diffusion of the solute gas based on concentration gradient, thus depleting the liquid boundary layer and permitting further solvation. The removal of the solute gas from the boundary layer is often accomplished by adding neutralizing agents to the host liquid to change the molecular form of the solute gas. This process is called absorption accompanied by chemical reaction. *See* DISTILLATION; MASS-TRANSFER OPERATION. [A.J.T.]

Gas adsorption operations Processes for separation of gases based on the adsorption effect. When a pure gas or a gas mixture is contacted with a solid surface, some of the gas molecules are concentrated at the surface due to gas–solid attractive forces, in a phenomenon known as adsorption. The gas is called the adsorbate and the solid is called the adsorbent. *See* ADSORPTION.

Adsorption can be either physical or chemical. Physisorption resembles the condensation of gases to liquids, and it may be mono- or multilayered on the surface. Chemisorption is characterized by the formation of a chemical bond between the adsorbate and the adsorbent.

If one component of a gas mixture is strongly adsorbed relative to the others, a surface phase rich in the strongly adsorbed species is created. This effect forms the basis of separation of gas mixtures by gas adsorption operations. Gas adsorption is an important unit operation for the chemical and petrochemical industries; it is applied to solve many different kinds of gas separation and purification problems of practical importance. By far the most frequent industrial applications of gas adsorption are the drying of gases, solvent vapor recovery, and removal of impurities or pollutants. *See* UNIT OPERATIONS.

The extent of adsorption is measured in terms of the surface excess (n^e), which is the actual amount adsorbed (n) minus the amount of gas that would be in the adsorbed phase if the solid exerted no intermolecular forces. The n^e is approximately equal to n when the gas pressure (P) is low. Both n and n^e are expressed in specific terms, that is, in number of moles of gas per unit mass of adsorbent. The relationship between n and P for a pure gas is called the adsorption isotherm. Its shape depends on the nature of the gas and the solid, as well as on the temperature. Practical considerations require that the adsorbents have a high surface area so that a large amount of gas can be adsorbed in a small volume of the solid. This is achieved by using porous adsorbents which provide a large surface area on the internal pore walls.

The actual physisorption process is generally very rapid. However, the measured rate of uptake of a gas by a porous adsorbent particle may be limited by its rate of transfer to the adsorption sites. Slow mass transfer may be caused by: the gas-film resistance outside the particle; an anisotropic skin of binding material at the surface of the particle; and internal macro- and micropore diffusional resistances. The rate of uptake may also be limited by the rate of dissipation of heat liberated during the sorption process.

Microporous adsorbents like zeolites, activated carbons, silica gels, and aluminas are commonly employed in industrial gas separations. These solids exhibit a wide spectrum of pore structures, surface polarity, and chemistry which makes them specifically selective for separation of many different gas mixtures. *See* ACTIVATED CARBON; MOLECULAR SIEVE; ZEOLITE [A.L.M.; S.Sir.]

Gas chromatography A method for the separation and analysis of complex mixtures of volatile organic and inorganic compounds. Most compounds with boiling points less than about 250°C (480°F) can be readily analyzed by this technique. A complex mixture is separated into its components by eluting the components from a heated column packed with sorbent by means of a moving-gas phase. *See* CHROMATOGRAPHY.

Gas chromatography may be classified into two major divisions: gas-liquid chromatography, where the sorbent is a nonvolatile liquid called the stationary-liquid phase, coated as a thin layer on an inert, granular solid support, and gas-solid chromatography, where the sorbent is a granular solid of large surface area. The moving-gas phase, called the carrier gas, is an inert gas such as nitrogen or helium which flows through the chromatographic column packed with the sorbent. The solute partitions, or divides, itself between the moving-gas phase and the sorbent and moves through the column at a rate dependent upon its partition coefficient, or solubility, in the liquid phase (gas-liquid chromatography) or upon its adsorption coefficient on the packing (gas-solid chromatography) and the carrier-gas flow rate.

The apparatus used in gas chromatography consists of four basic components: a carrier-gas supply and flow controller, a sample inlet system providing a means for introduction of the sample, the chromatographic column and associated column oven, and the detector system.

Qualitative and quantitative information is obtained from analyzing the peaks appearing on a chromatogram. Combination of gas chromatography with mass spectrometry provides the ultimate in qualitative information and has been used extensively in research. *See* MASS SPECTROMETRY; QUALITATIVE CHEMICAL ANALYSIS. [R.S.J.]

Gas constant Boyle's law and Charles' law may be combined into a single expression, pV/T = a constant, showing how the volume V of a given mass of gas depends upon its temperature T and pressure p. If the mass of gas chosen is 1 gram-molecule, then the constant, known as the gas constant, is written as R. Hence, $pV = RT$ for 1 g-mole. *See* BOYLE'S LAW; CHARLES' LAW.

The numerical value of the gas constant R is obtained by dividing the molar or gram-molecular volume of a perfect gas at the melting point of ice and a pressure of 760 mm of mercury (101.325 kPa; the standard atmosphere) by the absolute temperature at the ice point, $R = V_0/T_0$. The value for R is given below.

$$R = 8.314 \text{ joules/(mole K)}$$

See GAS. [T.C.W.]

Gas discharge A system made up of a gas, electrodes, and an enclosing wall in which an electric current is carried by charged particles in response to an electric field, the gradient of the electric potential, or the voltage between two surfaces. The gas discharge is manifested in a variety of modes (including Townsend, glow, arc, and corona discharges) depending on parameters such as the gas composition and density, the external circuit or source of the voltage, electrode geometry, and electrode material. For discussion of these modes *see* ARC DISCHARGE; CORONA DISCHARGE; ELECTRICAL BREAKDOWN; GLOW DISCHARGE; TOWNSEND DISCHARGE.

Gas discharges are useful both as tools to study the physics existing under various conditions and in technological applications such as in the lighting industry and in electrically excited gas lasers. New applications in gas insulation, in high-power electrical switching, and in materials reclamation and processing will assure a continuing effort to better understand all aspects of gas discharges. *See* GAS TUBE; LASER; VAPOR LAMP.

Electrons, rather than ions, are the main current carriers in gas discharges because their mass is smaller and their mobility is correspondingly much higher than that of ions. Electrons are produced by ionization of the gas itself, or they originate at the electrodes present in the system. Gas ionization can be accomplished in several ways, including electron impact ionization, photoionization, and associative ionization. Bombardment by photons, energetic ions or electrons, and excited neutral particles can cause secondary emission from the electrodes. A high-energy-per-unit electrode surface area can induce thermionic or field emission of electrons. Each of these means of producing electrons leads to a different response of the gas discharge as a circuit element. *See* ELECTRICAL CONDUCTION IN GASES; ELECTRON EMISSION; FIELD EMISSION; IONIZATION; PHOTOEMISSION; SECONDARY EMISSION; THERMIONIC EMISSION. [L.C.P.]

Gas-discharge laser A device which makes use of nonequilibrium processes in a gas discharge to achieve the population inversion required for laser action. At moderately low pressures (on the order of 1 torr or 10^2 pascals) and fairly high currents, the population of energy levels is far from an equilibrium distribution. Some levels are populated especially rapidly by the fast electrons in the discharge. Other levels empty particularly slowly and so accumulate large numbers of excited atoms. Thus laser action can occur at many wavelengths in any of a large number of gases under suitable discharge conditions. For some gases, a continuous discharge, with the use of either direct or radio-frequency current, gives continuous laser action. Wavelengths generated span the ultraviolet and visible regions and extend out beyond 700 micrometers in the infrared. They thus provide the first intense sources of radiation in much of the far-infrared region of the spectrum.

The earliest, and still most widely used, gas-discharge laser utilizes a mixture of helium and neon. Various infrared and visible wavelengths can be generated, but most commonly they produce red light at a 632.8-nanometer wavelength. Argon and krypton ion gas-discharge lasers provide a number of visible and near-ultraviolet wavelengths. Many molecular gases, such as hydrogen cyanide, carbon monoxide, and carbon dioxide, can provide infrared laser action. *See* LASER; PULSED GAS LASER. [S.F.J.; A.L.S.]

Gas-dynamic laser A device in which laser action is achieved by the sudden expansion of a hot molecular gas through a nozzle. The gas cools quickly, but different excited states lose energy at different rates. It can happen that, just after cooling, some particular upper state has more molecules than some lower one, so that amplification by stimulated emission can occur. Very high continuous power outputs have been generated from carbon dioxide in large gas-dynamic lasers. *See* LASER. [S.F.J.; A.L.S.]

Gas dynamics The study of the motion of gases which takes into account thermal effects generated by the motion. Gas dynamics combines fluid mechanics and thermodynamics and differs from gas statics in that there is motion and from gas kinematics in that the forces exerted on or by the gas are considered. *See* GAS KINEMATICS.

Several other names are used to define the subject. The most important ones are aerothermodynamics, aerothermochemistry, fluid dynamics, compressible fluid flow, and supersonic aerodynamics. This terminology reflects the fact that in each particular case different aspects of gas dynamics are emphasized. Magnetogasdynamics, which includes effects due to magnetism and electricity, has applications in rocket reentry, propulsion, and astrophysics. *See* AEROTHERMODYNAMICS; COMPRESSIBLE FLOW; FLUID DYNAMICS.

Gas dynamics deals with the motion of a continuous medium and is not concerned with the behavior of individual atoms or molecules which constitute the gas. At low pressures, however, such as may be encountered at very high altitudes, the particle mean free path is very large and continuum considerations may not be applicable.

The fundamental conservation principles of mechanics and thermodynamics constitute the theoretical basis of gas dynamics. The conservation laws can be derived in lagrangian form with respect to a specific mass of flowing gas or in eulerian form with respect to the rate at which gas enters and leaves a fixed control volume in space. The eulerian conservation laws are expressed by (1) the continuity equation (conservation of mass); (2) the Navier-Stokes equations (conservation of momentum); and (3) the energy equation (conservation of energy). *See* Conservation of energy; Equation of continuity; Fluid flow; Navier-Stokes equations. [J.Men.; A.B.C.]

Gas field and gas well Petroleum gas, one form of naturally occurring hydrocarbons of petroleum, is produced from wells that penetrate subterranean petroleum reservoirs of several kinds. Oil and gas production are commonly intimately related, and about one-third of gross gas production is reported as derived from wells classed as oil wells. If gas is produced without oil, production is generally simplified, in part at least because the gas flows naturally without lifting, and also because of fewer complications in reservoir problems. As for all petroleum hydrocarbons, the term field designates an area underlain with little interruption by one or more reservoirs of commercially valuable gas. *See* Natural gas; Oil and gas field exploitation; Oil and gas well drilling; Petroleum; Petroleum geology; Petroleum reservoir engineering. [C.V.C.]

Gas furnace An enclosure in which a gaseous fuel is burned. Domestic heating systems may have gas furnaces. Some industrial power plants are fired with gases that remain as a by-product of other plant processes. Utility power stations may use gas as an alternate fuel to oil or coal, depending on relative cost and availability. Some heating processes are carried out in gas-fired furnaces. Among the gaseous fuels are natural gas, producer gas from coal, blast furnace gas, and liquefied petroleum gases such as propane and butane. *See* Fuel gas; Furnace; Steam-generating furnace. [R.M.H.]

Gas kinematics The motion of a gas considered by itself, without regard for the cause of the motion. Various flow phenomena arise in widely different applications, yet the phenomena dealt with independently of their causes can be classed into relatively few fundamental patterns. Gas kinematics then deals with these abstracted flows. *See* Gas dynamics; Sink flow; Source flow. [F.H.R.]

Gas kinetics The effects of gases due to motion. The method of science is to isolate one phenomenon at a time for study, holding all else constant. However, in actual gas flows, motion is accompanied by other phenomena, so that for practical purposes a more comprehensive view is necessary. *See* Aerodynamic force; Bernoulli's theorem; Buoyancy; Compressible flow; Euler's momentum theorem; Gas dynamics; Viscosity. [F.H.R.]

Gas mechanics The action of forces on gases. One of the simplest forces on a gas is illustrated by the gravitational attraction of the Earth for its atmosphere. From this and other observations, the weight of an ideal gas is found to be proportional to its molecular weight. Another mechanical property of a gas is its ability to transmit pressure in any direction. In this respect a gas and a liquid behave similarly. *See* Avogadro's law; Fluid mechanics; Pascal's law. [F.H.R.]

Gas thermometry A method of measuring temperatures with a gas as the thermometric fluid. It is the most accurate means available for determining thermodynamic temperatures over the range from 2.5 to 933 K. The unit of thermodynamic temperature is one of the base units of the International System (SI), and is defined by assigning the value of 273.16 K to the temperature of the triple point of water. Any other temperature to be determined by gas thermometry must be measured relative to the temperature of that triple point.

Of the several types (constant-volume, constant-pressure, and constant-bulb-temperature), constant-volume gas thermometry is potentially the most accurate, both because the equipment is relatively simple and because highly accurate absolute volume measurements are unnecessary. Accurate results are painstakingly difficult to obtain by any type of gas thermometry, however, so that reliable determinations of thermodynamic temperatures require special projects of substantial duration. The results are disseminated by recording them on the International Practical Temperature Scale—the international laboratory scale capable of precise and relatively easy realization. It is revised periodically to bring it into as close conformity with thermodynamic values as is known at a given time. *See* Low-temperature thermometry; Physical measurement; Temperature measurement. [L.A.G.]

Gas tube An electron tube that contains gas or vapor at low pressure in which an electrical discharge takes place. Gas tubes are of two general types: cold-cathode tubes, in which the phenomenon known as glow discharge serves to maintain a conducting path between the electrodes, and hot-cathode tubes, in which an arc discharge conducts the current. The cold-cathode glow tubes are characterized by a relatively high voltage drop and a low current, while the hot-cathode arc tubes are characterized by a low voltage drop and relatively high current. *See* Electron tube.

Cold-cathode gas tubes, or glow tubes, require no cathode heater power and are therefore always ready for instant service. Glow tubes are commonly constructed as two-element (diode) or three-element (triode) tubes.

Three representative types of hot-cathode gas tubes may be distinguished: the Tungar (sometimes called a Rectigon), the phanotron, and the thyratron. The first two are simple rectifier tubes while the third is a control tube having one or two grids between cathode and anode. Both phanotrons and thyratrons may be built with glass, ceramic, or metal envelopes [J.D.Co.]

Gas turbine A heat engine that converts some of the energy of fuel into work by using gas as the working medium and that commonly delivers its mechanical output through a rotating shaft. In the usual gas turbine, the sequence of thermodynamic processes consists basically of compression, addition of heat in a combustor, and expansion through a turbine. The flow of gas during these thermodynamic changes is continuous in the basic, simple, open-cycle arrangement.

The various gas-turbine cycle arrangements can be operated as open, closed, or semiclosed types. In the open-cycle gas turbine, there is no recirculation of working medium within the structural confines of the power plant, the inlet and exhaust being open to the atmosphere. This cycle offers the advantage of a simple control and sealing system. It also can be designed for high power-to-weight ratios (aircraft units) and for operation without cooling water. Most gas turbine plants are of this type.

In the closed-cycle gas turbine, essentially all the working medium (except for seal leakage, bleed loss, and any addition or extraction of working medium for control purposes) is continuously recycled. Heat from a source such as fossil fuel (or, possibly, nuclear reaction) is transferred through the walls of a closed heater to the cycle. The closed cycle can be charged with gases other than air such as helium, carbon dioxide, or nitrogen. This is a particular advantage with a nuclear heat source. Other advantages of the closed cycle are (1) clean working fluid; (2) control of the pressure and composition of the working fluid; (3) high absolute pressure and density of the working fluid; and (4) constant efficiency over wide load range.

In the semiclosed cycle gas turbine, a portion of the working fluid is recirculated. This type requires a precooler for the recirculated gas, and a charging compressor to provide the necessary air for combustion. The semiclosed cycle can operate at high densities. The major disadvantages of this cycle are the corrosion and fouling which occur with the recirculation of the products of combustion, particularly when the fuels used have high sulfur or ash content.

To achieve overall performance, each process is carried out in the engine by a specialized component. Air for the combustion chamber is forced into the engine by a compressor. Fuel is mixed with the compressed air and burned in combustors. The heat energy thus released is converted by the turbine proper into rotary energy. To improve efficiency, heat exchangers can be added on the gas turbine exhaust to recover heat energy and to return it to the working medium after compression and prior to its combustion.

In the open-cycle plant, products of combustion come in direct contact with the turbine blading and heat-exchanger surfaces. This requires a fuel in which the products of combustion are relatively free of corrosive ash and of residual solids that could erode or deposit on the engine surfaces. Natural gas, refinery gas, blast-furnace gas, and distillate oil have proved to be ideal fuels for open-cycle gas turbines. In closed-cycle turbines, gaseous fuel, distillate oils, and coal are satisfactory.

Fuels for semiclosed-cycle turbines must be selected to minimize both erosion and corrosion. These requirements limit the use of the semiclosed plant to fuels of low ash and sulfur content. *See* TURBINE. [T.J.P./J.A.B.]

Gas welding A fusion welding process which uses a flame provided by the combustion of a fuel gas. The gas flame is a less concentrated and lower-temperature source of heat than an arc. For this reason, a welding torch is often used for brazing, for welding thin material, and for applications requiring low gradients of temperature to avoid cracking, as in welding cast iron. The oxyacetylene flame, using a mixture of equal volumes of oxygen and acetylene, C_2H_2, has a higher temperature than any other commercially available fuel gas combination. High flame temperature is essential for the rapid localized heating required in welding. *See* TORCH.

Flux is not used in welding steel but is necessary for gas-welding nonferrous metals. The use of other fuel gases, such as hydrogen, is used in the welding of metals of low melting point, such as lead, aluminum, and magnesium. [M.M.S.]

Gasket Deformable material used to make a pressure-tight joint between stationary parts, such as cylinder head and cylinder, that may require occasional separation. Gaskets are known as static seals, as compared with packing or dynamic seals. In packings the parts are frequently in motion, as in piston rods and valve stems.

Gaskets are made of sheet materials such as natural or synthetic rubber, cork, vegetable fiber such as paper, asbestos and plastic pastes, or of soft metallic materials such as lead and copper. Rubber in the form of O-rings is used for light pressure. *See* PRESSURE SEAL. [P.H.B.]

Gasoline A mixture of hydrocarbons used to fuel the spark-ignited internal combustion engine. This mixture consists of more than 100 different hydrocarbons. The lightest, that is, lowest-boiling, is isobutane, while the heaviest hydrocarbons consist of a variety of substituted naphthalenes.

The volatility of gasoline is controlled to provide a balance between a lack of vehicle performance due to insufficient vaporization and poor performance due to an excess of vapor. In order for the spark-ignited internal combustion engine to function, there must be a gaseous mixture of fuel vapor and air that sustains combustion when ignited by the spark discharge. That is, in cold weather there must be enough low-boiling hydrocarbons to form a flammable mixture in the cold cylinder. Shortly after the engine is running, it is expected to produce power and propel the vehicle. As power is required, greater quantities of vapor (up through the middle of the boiling range) are required. However, once the engine is fully warmed up, this same gasoline which was volatile enough to start the cold engine must not generate so much vapor as to create another set of operating problems. *See* INTERNAL COMBUSTION ENGINE.

A gasoline's octane number is an indication of the fuel's ability to prevent the occurrence of spark knock in an engine. Spark knock takes place when chain-branching reactions lead to an increasingly rapid buildup of hydroperoxide radicals in the region ahead of the advancing flame front. Tetraethyllead is an effective antiknock agent, because its decomposition gives rise to a nonstoichiometric lead oxide which acts as a radical trap. *See* ANTIKNOCK AGENT; OCTANE; OCTANE NUMBER; TETRAETHYLLEAD.

A host of properties exhibited by gasolines results from the use of additives. Gasolines are colored with oil-soluble dyes to differentiate between the various grades marketed. Oxidation inhibitors make gasolines less susceptible to oxidation; this allows the gasoline to be stored for many weeks without excessive "gum" formation. The "gums," now only partially soluble in gasoline, can deposit on critical parts of the carburetor and cause various parts to stick and thus fail to operate properly. Numerous organic compounds with an affinity for polar surfaces are used as rust inhibitors.

Most of the desirable features of gasolines require the deliberate application of technology during the refining and marketing of gasoline. The particular balance of properties arrived at will not be the same in all instances. Technical capability and economic considerations both play a part in determining which properties are imparted and emphasized. [H.F.Sh.]

Gasteromycetes A class of fungi in the subdivision Basidiomycotina in which basidiospores are produced in a mass (gleba) and enclosed within a membrane called the peridium. Such enclosed fruit bodies evolved many times; hence the class is heterogeneous. Most Gasteromycetes have lost the ability to discharge spores ballistically off basidia directly into air. Since internal sporulation prevents normal spore drop into air currents, these fungi have developed other dispersal mechanisms. Puffballs (Lycoperdales) typically form powdery spores in a perforated peridium that acts like a bellows when hit by rain. Earthstars (*Geastrum*) disperse similarly, but the peridia are elevated on arched rays. *Bovista* scatters spores while tumbling in the wind; *Calvatia* cracks open in place. In the gleba, cottony threads called capillitia slow the rate of dispersal. Stalked puffballs (Tulostomatales and Podaxales) characteristically form underground, emerging on stalks after spore maturation; they are adapted to desert life by their protected hypogeous maturation. False puffballs (Sclerodermatales) typically have thicker peridial walls than true puffballs.

The highly evolved Nidulariales have funnel-shaped structures that hold seedlike, splash-dispersed, sporulating disks called peridioles. These fruit bodies resemble miniature bird nests. The related *Sphaerobolus* violently everts an inner membrane, tossing a sticky glebal ball several centimeters. Stinkhorns (Phallales) mature their sticky, foul-smelling glebas in a hypogeous egg from which erupt showy phallic, floral or cagelike receptacles designed to attract insects. Flies or beetles ingest the spores and disperse them. Hymenogastrales (and possibly Melanogastrales and Gauteriales) are dispersed as spores in feces after ingestion by animals lured to these false truffles by characteristic odors.

Gasteromycetes are either saprophytic or mycorrhizal. Historically, mature dry spores of puffballs were used to clot wounds. Young puffballs are edible, but false puffballs are poisonous. Because of their phallic shape, stinkhorns have been associated with witchcraft. *See* BASIDIOMYCOTINA; EUMYCOTA; FUNGI. [S.A.R.]

Gasterosteiformes An order of actinopterygian fishes which includes the groups Solenichthyes, Thoracostei, Hemibranchii, Lophobranchii, and Syngnathiformes of other classifications. Members of this order, commonly known as sticklebacks, tube-snouts, snipefishes, pipefishes, and seahorses, may be classified in 3 suborders, 7–9 families, 50–55 genera, and 200 or more species. The group dates from the Eocene.

Common characters include a swim bladder without a duct; pelvic fin, if present, with or without a spine and abdominal to subthoracic in position; pelvic girdle not in contact with the cleithrum; and articulation of the quadrate with the lower jaw in advance of the orbit. Fin spines are present or absent. The snout is commonly produced, the mouth opening at the end of a long tube (see illustration).

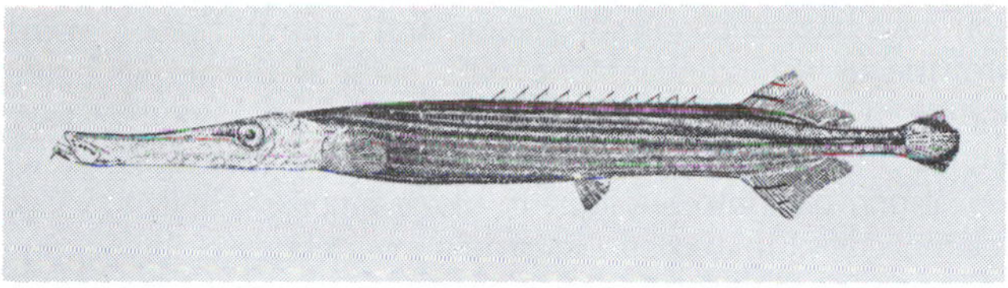

Trumpetfish (*Aulostomus maculatus*). (*After D. S. Jordan and B. W. Evermann, The Fishes of North and Middle America, U.S. Nat. Mus. Bull. no. 47, 1900*)

Most species are inhabitants of warm seas, but some sticklebacks live in fresh waters of northern continents, and a few tropical pipefishes enter or reside permanently in rivers. *See* Actinopterygii; Sea horse; Teleostei. [R.M.B.]

Gastornithiformes A group of extinct flightless birds known only from upper Paleocene and lower to middle Eocean deposits in North America and Europe. *Diatryma* was a huge bird (up to 6.5 ft or 2 m tall) with powerful legs, very small wings, and a large head and beak (see illustration). It lived during the period just after the extinction of dinosaurs and just before the major radiation of very large mammals. The apparently powerful nature of the locomotory and feeding apparatus of *Diatryma* suggests that it was an active predator. In fact, *Diatryma* may have been one of the main enemies of the terrestrial mammals of its era. Among living birds, the Gastornithiformes are probably distantly related to the Gruiformes, which include the cranes, limpkins, trumpeters, and seriemas. *See* Aves; Gruiformes; Neornithes. [D.W.S.]

Diatryma steini (reconstruction). (*American Museum of Natural History*)

Gastrolith Any of the pebbles swallowed by animals and retained for a time in the gizzard or stomach, where they serve to grind up the food and in so doing become rounded and highly polished. Birds generally use such pebbles, as do some living reptiles, notably the crocodile and certain lizards. Some of the Mesozoic reptiles also used gastroliths. [C.O.D.]

Gastropoda The largest and most varied class in the phylum Mollusca, possibly numbering over 74,000 species and commonly known as snails.

General characteristics. The shell is in one piece which, in the majority of forms, grows along a turbinate (equiangular) spiral (see illustration), but which is modified into an open cone in various limpets or is secondarily lost in various slugs.

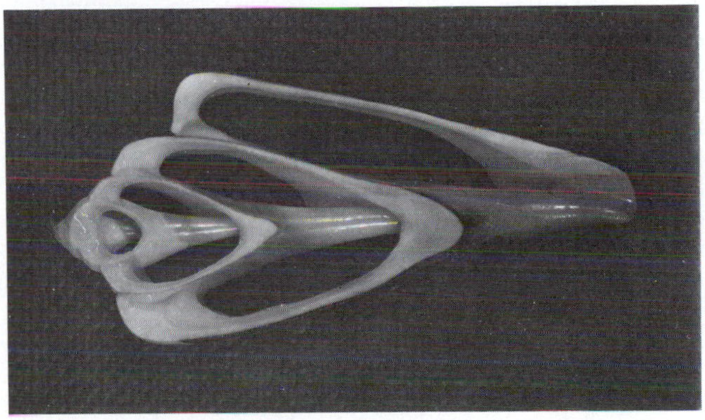

Longitudinal section ground through the shell of a specimen of *Conus spurius* to reveal the central columella and spiral of whorls expanding to the aperture. (*From W. D. Russell-Hunter, A Life of Invertebrates, Macmillan, 1979*)

All gastropods, at some time in their phylogeny and at some stage in their development, have undergone torsion. The process does not occur in any other mollusks. It implies that the visceral mass and the mantle shell covering it have become twisted through 180° in relation to the head and foot. As a result of torsion, all internal organs are twisted into a loop. Similarly in gastropods, the mantle cavity (the semi-internal space enclosed by the pallium or mantle) containing the characteristic molluscan gills (ctenidia) has become anterior and placed immediately above and behind the head. The most primitive gastropods retain a pair of aspidobranch (bipectinate or featherlike) gills, each with alternating ctenidial leaflets on either side of a ctenidial axis in which run afferent and efferent blood vessels. Lateral cilia on the faces of the leaflets create a respiratory water current (toward the midline and anteriorly) in the direction opposite to the flow of blood through the gills, to create the physiological efficiency of a countercurrent exchange system.

As also occurs in other mollusks, the ctenidia form a gill curtain which functionally divides the mantle cavity into an inhalant part (lateral and below) with characteristic osphradia as water-testing sense organs, and an exhalant part into which the anus and renogenital ducts discharge.

Classification and diversity. The usual systematic arrangement of the class Gastropoda involves three somewhat unequal subclasses. The first, the largest and most diverse, is the subclass Prosobranchia, which is made up largely of marine snails all retaining internal evidence of torsion. Prosobranchs are divided into four orders: Archaeogastropoda, Neritacea, Mesogastropoda and Neogastropoda (or Stenoglossa). The other two subclasses (Opisthobranchia and Pulmonata) are each considerably more uniform than the subclass Prosobranchia and, in both, the effects of torsion are reduced or obscured by secondary processes of development and growth. *See* OPISTHOBRANCHIA; PROSOBRANCHIA; PULMONATA.

More than half of all molluscan species are gastropods, and they encompass a range from marine zygobranchs, which can be numbered among the most primitive of all living mollusks, to the highly evolved terrestrial air-breathing slugs and snails. Pulmonates and certain mesogastropod families are the only successful molluscan colonizers of land and fresh waters.

[W.D.R.-H.]

Fossils. Fossil gastropods have a long geologic history, being common throughout the Paleozoic and increasingly abundant in the Mesozoic and Cenozoic. All three subclasses are known in the fossil record; many superfamilies, particularly prosobranchs, are extinct. Average duration of a genus has been estimated to range from 30,000,000 to 90,000,000 years.

Marine gastropods are important stratigraphic indicators in Cenozoic strata and locally are abundant in Cretaceous rocks. They are less common and less useful in the Jurassic and Triassic. Although individual genera have stratigraphic utility within the Paleozoic, it is only in the Ordovician that they are significant for correlation.

In North America, late Cenozoic nonmarine gastropods are becoming increasingly important for correlation and ecological interpretation of terrestrial deposits. Marine Cenozoic and later Mesozoic gastropods are interpreted as having lived under essentially the same ecologic conditions as related Recent gastropods. The possible ecology of Paleozoic and earlier Mesozoic gastropods has not been thoroughly investigated. In the Ordovician and Silurian, gastropods are most common in limestones. From the Silurian onward, marine forms are more common in shales and some sandstones. Most occurrences of abundant fossil gastropods probably can be interpreted as indicating shallow water conditions. *See* MOLLUSCA. [E.L.Y.]

Gastrotricha

A phylum of minute metazoan animals (formerly placed in the aschelminth group) numbering 500 described species worldwide. Some 300 species have been reported from the marine habitat, with new ones being described every year.

Gatrotrichs comprise two orders, the Macrodasyida and the Chaetonotida. The term gastrotricha refers to the ventral locomotor cilia by which the animals glide gracefully over the substratum or through its interstices; unlike many other ciliated animals, they cannot move in reverse. Gastrotrichs have a complete digestive tract, with a sucking pharynx, a simple intestine with a wall only a single cell thick, and an anus. They appear to be selective feeders on bacteria, very small protozoa, and yeasts. Most have protonephridia, accounting in part for their broad salinity tolerances.

Gastrotrichs appear to be regionally cosmopolitan, with 20–30% having broad distributions within continents, and 10–15% between continents; endemism probably does not exceed 20%.

The phylum Gastrotricha is the most primitive in the aschelminth group of phyla. Gastrotrichs and nematodes probably share a common ancestor, which in turn was descended from a stock that included gnathostomulids and turbellarianoid animals. *See* CHAETONOTIDA; GNATHOSTOMULIDA; MACRODASYIDA; NEMATA; TURBELLARIA. [W.D.Hu.]

Gastrulation

The formation of the primordial gut, the archenteron, or digestive cavity of an early animal embryo. More generally, and originally, the term gastrulation referred to the process by which the gastrula stage of the embryo is formed. Thus to 19th-century embryologists, gastrulation was the process by which the single-layered blastula, a hollow ball of cells, is converted into the double-layered gastrula. The term has now come to have a still more general meaning, namely, the process by which the three germ layers, or primordial tissues of the embryo, are brought into the positions and relations characteristic of the late gastrula stage, with ectoderm (outer skin), mesoderm (middle skin), and endoderm (inner skin) from the outside to the inside. The terms epiblast, mesoblast, and hypoblast are also used to denote ectoderm, mesoderm, endoderm, respectively. *See* BLASTULATION.

Two general but not mutually exclusive methods of gastrulation have been recognized: epiboly and emboly. Epiboly is the growing or extending of one part, such as the upper hemisphere of a spherical blastula, over and around another part, such as the lower hemisphere. Emboly is the pushing or growing of one part into another. In many embryos, both types of cell movement may occur; in certain invertebrate embryos, one type may predominate almost to the exclusion of the other. Generally speaking, epiboly tends to be the major, but not the only, method of gastrulation in forms with large, yolky eggs. *See* GERM LAYERS. [N.T.S.]

Gate circuit

An electronic circuit having one output and one or more inputs, in which the output is a function of the inputs in a prescribed manner and in a controllable time sequence. Gate circuits may be classified as transmission gates or as switching gates, of which the logic gate is the most notable example.

The transmission gate, sometimes called a sampling gate, delivers an output waveform that is a replica of a selected input during a specific time interval. The control signal, which determines the time interval, is called the gating signal or waveform; the gating generator is often a multivibrator. An operational amplifier with a separate strobe input which internally disconnects the input from the output sometimes functions as a transmission gate. The term analog switch is sometimes used synonymously with transmission gate. *See* AMPLIFIER.

In the switching, or logical, gate an output having a constant amplitude is registered during a time interval if a particular combination of input signals exists. The output indicates only the existence of a particular combination of similar input signals. Examples are the OR, AND, NOR, and NAND circuits, which are basic building blocks of digital computers. *See* LOGIC CIRCUITS. [G.M.G.]

Gauge theory

In fundamental physics, theories conforming to the gauge principle of interaction between constituents of matter. It is the basis of all currently known interactions in physics.

H. Weyl proposed in 1918 a theory of electromagnetism which he called gauge theory, because it has a gauge invariance. The idea was that electromagnetism causes a space-time-dependent scale change, so that if a particle is moved from the point x^μ to $x^\mu + dx^\mu$, its scale would change by $1 + S_\mu dx^\mu$, where S_μ is a space-time-dependent vector function. Weyl tried to identify S_μ with a numerical constant times the electromagnetic four-vector potential A_μ. A. Einstein raised an objection to Weyl's idea that made it appear that it could not be valid. *See* ELECTROMAGNETISM; RELATIVITY; SPACE-TIME.

In 1927 V. Fock observed that in quantum mechanics the expression for a charged particle in classical electrodynamics,

$p_\mu - eA_\mu$, should be replaced as in Eq. (1). F. London then

$$p_\mu - eA_\mu \rightarrow -i\hbar\partial_\mu - eA_\mu = -i\hbar[\partial_\mu - ie\hbar^{-1}A_\mu] \qquad (1)$$

pointed out that the expression in the square brackets would be the same as Weyl's $[\partial_\mu + S_\mu]$ if one makes the identification in Eq. (2). This is similar to Weyl's identification, but with

$$S_\mu = -ie\hbar^{-1}A_\mu \qquad (2)$$

the insertion of the important factor $i = \sqrt{-1}$. That, of course, implies that the scale change $1 + S_\mu dx^\mu$ becomes Eq. (3), which is a phase change, not a scale change. Thus was

$$1 - ie\hbar^{-1}A_\mu dx^\mu = \exp[-ie\hbar^{-1}A_\mu dx^\mu] \qquad (3)$$

born the concept of electromagnetism as a gauge theory in quantum mechanics. With the scale change replaced by a phase change, the Einstein objection is rendered inoperative. *See* QUANTUM MECHANICS.

In 1954, C. N. Yang and R. L. Mills generalized the electromagnetic gauge principle. In their theory, the phase [Eq. (3)] is replaced by a generalized phase, a concept developed by S. Lie, E. Cartan, and others. Lie explored the theory of Lie groups, the elements of which can be taken as generalized phases. Lie groups can be abelian or nonabelian. With a nonabelian Lie group, the resultant gauge theory is called a nonabelian gauge theory, and has a particularly intricate and beautiful structure. For abelian gauge theories, such as electromagnetism, the field itself is not charged. But for a nonabelian theory, the field is itself charged with the same conserved quantity that generates the field. In other words, a nonabelian gauge field generates itself. *See* GROUP THEORY; LIE GROUP.

Theoretical and experimental developments since 1970 have led to the now universally accepted notion that all fundamental interactions of nature (including the weak interactions, described by the Weinberg-Salam model; the strong interactions, described by quantum chromodynamics; and gravity) are gauge theories. Since gauge theories are based on the principle of local phase symmetry, the current notion about the origin of interactions is summarized by the statement: symmetry dictates interactions. *See* ELEMENTARY PARTICLE; FUNDAMENTAL INTERACTIONS; GRAVITATION; QUANTUM CHROMODYNAMICS; SYMMETRY LAWS (PHYSICS); WEINBERG-SALAM MODEL. [C.N.Y.]

Gauss' theorem The assertion, under certain light restrictions, that the volume integral through a volume V of the divergence of vector function $\mathfrak{F}(x,y,z)$ is equal to the surface integral of the exterior normal component of $\mathfrak{F}$ over the boundary surface S of V, or in symbols $\iiint \nabla \cdot \mathfrak{F}\, dV = \iint \mathfrak{F} \cdot \mathbf{v}\, dS$, with $\mathbf{v}$ the unit exterior normal to S. This theorem is also known as the divergence theorem and as Green's theorem.

The divergence theorem plays an important role in a variety of subjects such as electricity and magnetism, mechanics of continuous media (including fluid dynamics), heat flow, partial differential equations, and potential theory. *See* CALCULUS OF VECTORS; GREEN'S THEOREM. [H.V.C.]

Gavial The name of two species of reptiles which form the family Gavialidae in the order Crocodilia. The Indian gavial (*Gavialis gangeticus*) is confined to the Ganges River and its tributaries, while the Malayan gavial (*Tomistoma schlegeli*) is found in Borneo and Sumatra.

The gavial is distinguished from other members of the order by its extremely long, slender snout (see illustration). The tip of the snout is enlarged, and fleshy elevations surround the openings of the nostrils. The females lay about 30–40 eggs in nests which they have prepared on the riverbanks. The newly hatched young are about 1 ft (30 cm) long and quite active. *See* ALLIGATOR; CROCODILE; CROCODYLIA; REPTILIA. [C.B.C.]

The Indian gavial, which may attain a length of 30 ft (9 m).

Gaviiformes The order of birds containing the loons, which are known as divers in Europe. There is but a single family, Gaviidae, and one living genus, *Gavia*, with four species, all of which nest in cooler parts of the Northern Hemisphere. The best-known and most southerly species is the common loon, or great northern diver (*G. immer*), whose cry is legendary.

Loons are goose-sized, fish-eating, diving birds that are highly adapted for aquatic life. Their legs are placed so far back on the body that they can neither walk well nor gain the air from dry land. The nest is usually placed at the water's edge, where the incubating bird can slip off and swim away quietly if disturbed. The feet of loons are webbed, the tarsi compressed and bladelike, and the bill heavy and pointed. Although their wings are exceptionally small in relation to body size, loons regularly perform long migrations. *See* AVES. [K.C.P.]

Gear A machine element used to transmit motion between rotating shafts when the center distance of the shafts is not too large. Toothed gears provide a positive drive, maintaining exact velocity ratios between driving and driven shafts, a factor that may be lacking in the case of friction gearing which is subject to slippage. *See* BELT DRIVE; CHAIN DRIVE; ROLLING CONTACT.

The application of gears for power transmission between shafts falls into three general categories: those with parallel shafts, those for shafts with intersecting axes, and those whose shafts are neither parallel nor intersecting but skew. *See* BEVEL GEARS; GEAR TRAIN; SPUR GEARS; WORM GEARS. [J.R.Z.]

Gear drive Transmission of motion or torque from one shaft to another by means of direct contact between toothed wheels. The active parts of the gear teeth are usually made in the form of involutes and transmit the force smoothly from a tooth on one gear to a tooth on another gear. *See* GEAR; GEAR TRAIN. [R.M.Ph.]

Gear train A combination of two or more gears used to transmit motion between two rotating shafts or between a shaft and a slide. In theory two gears can provide any speed ratio in connecting shafts at any center distance, but it is often not practical to use only two gears. If the ratio is large or if the center distance is relatively great, the larger of the two gears may be excessively large. Moreover, an additional gear may be necessary simply to give the proper direction to the output gear. Belt, rope, and chain drives are frequently used in conjunction with gear trains. *See* BELT DRIVE; CHAIN DRIVE; GEAR; PLANETARY GEAR TRAIN.

The most important distinction in classifying gear trains is that between ordinary and epicyclic gear trains. In ordinary trains

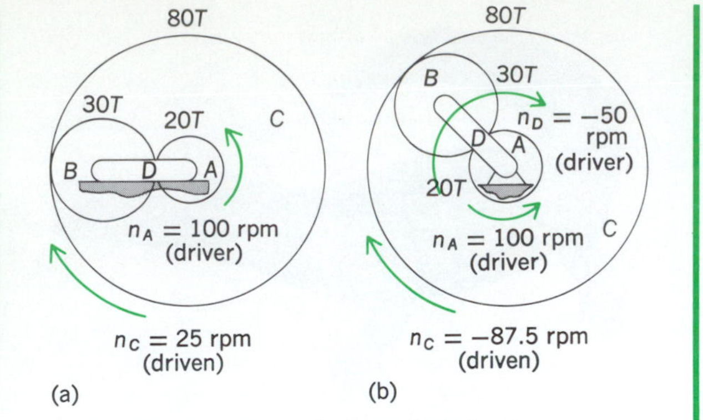

Gear trains. (a) Ordinary. (b) Epicyclic.

(illustration *a*), all axes remain stationary relative to the frame. But in epicyclic trains (illustration *b*), at least one axis moves relative to the frame. In illustration *b*, gear *B*, whose axis is in motion, is called a planet. The gears *A* and *C* are sun gears.

A simple gear train is one in which each gear is fastened to a separate shaft, as in illustration *a*. If at least one shaft has two or more gears fastened to it, the train is said to be compound. The train is a reverted gear train when the input and output shafts are in line. If the input shaft does not line up with the output shaft, the train is said to be nonreverted.

If a machine must be operated at any one of several output speeds, a multiple-speed gearbox, or transmission, may be used as a component part. Machine tools and motor vehicles are familiar instances of the need for transmissions. The speed of the output shaft of a transmission can be varied by sliding gears in and out of contact or by connecting gears in continual mesh to shafts by means of clutches. *See* AUTOMOTIVE TRANSMISSION. [J.R.Z.]

Gecko The name for about 300 species of reptiles that form the family Gekkonidae in the order Squamata. They are small lizards, primarily arboreal and nocturnal, which occur in the warm regions of the world. The body is flattened; most species have five digits, while some have only four; the toes often have adhesive pads to assist the animal when climbing on smooth surfaces. Geckos feed on small animals, especially insects, and all species have a long sensitive tongue to aid in capturing their prey. Most species of gecko are oviparous.

The largest, most aggressive species is the orange-spotted Tokay (*Gekko gecko*), which is indigenous to Southeast Asia. Another species is the flying gecko (*Ptychozoon homalocephalum*), which is essentially arboreal and is well adapted to leaping and gliding because it has folds of skin on either side of the body that can be opened out to form a planing surface. The banded gecko (*Coelonyx variegatus*) of the southwestern United States is one of the few species with movable eyelids, and with claws instead of pads on the toes. *See* SQUAMATA. [C.B.C.]

Geiger-Müller counter A detector of ionizing radiation. When a fast-moving charged particle traverses a Geiger-Müller counter, an electrical impulse is produced. These impulses can readily be counted by electronic circuits. Geiger-Müller (GM) counters, usually referred to simply as Geiger counters, are widely used to indicate the presence and intensity of nuclear radiations.

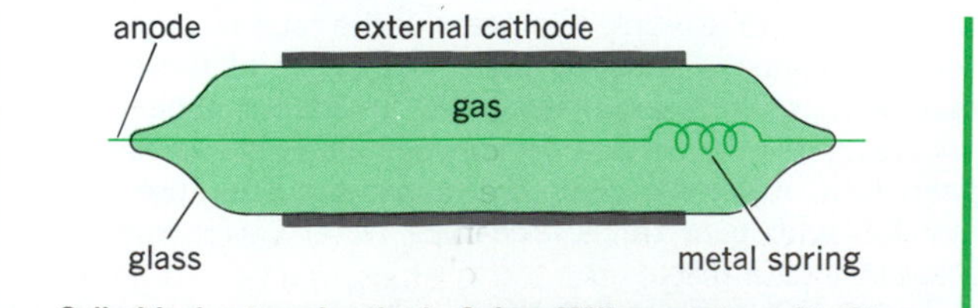

Cylindrical external-cathode Geiger-Müller counter, with thin soda glass and central wire of 0.003-in.-diameter (0.008-cm) tungsten. Metal spring keeps central wire taut.

A GM counter consists of a gas between two electrodes (see illustration). One electrode, usually cylindrical and hollow, is the cathode. The other electrode, a fine "wire stretched along the axis of the cylinder, is the anode. A potential of about 1000 volts is placed on the wire. When an atom of the gas between the two electrodes is ionized by collision with a charged particle passing through the gas, the electron produced in the collision is drawn toward the central wire. The electron then collides with the atoms of the gas. Near the central wire the electric field is very intense, and the electron may acquire enough energy between two collisions to allow it to ionize another atom. A second electron is then set free, and by successive collisions, an avalanche of electrons is produced which is then collected as charge on the central wire. This charge produces an electrical impulse which in typical cases may be 50 volts. *See* IONIZATION CHAMBER; PARTICLE DETECTOR. [W.B.Fr.]

Gel A two-phase colloidal system consisting of a solid and a liquid. Gels behave as elastic solids and retain their characteristic shape, whereas sols (colloidal dispersions) possess the shape of the container. Commonly, gels have a low solid content. Gels include jellies or transparent elastic gels rich in liquid, and gelatinous precipitates which are believed to consist of minute particles of jelly. Gels or jellies which have dried until apparently solid may be called zerogels, and sometimes will swell or redisperse to form a sol when treated with a suitable solvent. *See* COLLOID; GELATIN; RHEOLOGY. [W.O.M.]

Gel chromatography A form of liquid chromatography in which separation is based on a selective process of penetration of molecules of different sizes and shapes through a porous gel medium. In this process, different molecules are "sized" and thus separated from each other. The largest molecules in the mixture are not allowed to penetrate the porous structure at all; the medium-size molecules can penetrate only some pores; and the small molecules can diffuse rather freely inside the medium. Consequently, if the porous material is contained in a column, mixtures of components with differing molecular weights can be effectively resolved. This molecular sieve effect is extremely useful in many important separations. Gel chromatography has resulted in tremendous progress in the chemistry of biomacromolecules, *See* CHROMATOGRAPHY; LIQUID CHROMATOGRAPHY; MOLECULAR SIEVE. [M.V.N.]

Gel permeation chromatography A separation technique involving passage of a liquid moving phase through a column containing the stationary phase which consists of a porous material. Gel permeation chromatography (GPC) affords a rapid method for separating high-molecular-weight materials, with separation based on differences in molecular size. It is of particular importance for research in biological systems and synthetic polymers.

The stationary phase is a porous inorganic solid or cross-linked gel containing a range of pore sizes. The mobile phase is a liquid, usually water or a buffered solution for biological systems. Organic solvents are used for synthetic polymers. The mobile phase is circulated through the column either by pumping or by gravity flow. The sample to be separated is introduced at the head of the column. As it progresses through the column, small molecules can enter all pores larger than the molecule, while larger molecules can fit into a smaller number of pores, again only those larger than the molecule. Thus the larger the molecules, the smaller the amount of pore volume available into which the molecules can enter. The sample emerges from the end of the column in inverse order of molecular size, with the larger molecules eluting first followed by the smaller ones. In order to determine the amount of sample emerging, a detector is used which produces a signal proportional to sample concentration. The volume of the mobile phase is also recorded to provide a means of characterizing the molecular size of the emerging molecule. See CHROMATOGRAPHY. [A.R.C.; J.F.J.]

Gelatin A protein extracted after partial hydrolysis of collagenous raw material from the skin, white connective tissue, and bone of animals. It is a linear polymer of amino acids, most often with repeating glycine-proline-proline and glycine-proline-hydroxyproline sequences in the polypeptide linkages. See COLLAGEN; PROTEIN.

The unique characteristics of gelatin are: reversible sol-to-gel formation, amphoteric properties, swelling in cold water, film-forming properties, viscosity-modifying properties, and protective colloid properties. Gelatin contains 26.4–30.5% glycine, 14.8–18% proline, 13.3–14.5% hydroxyproline, 11.1–11.7% glutamic acid, 8.6–11.3% alanine, and in decreasing order arginine, aspartic acid, lysine, serine, leucine, valine, phenylalanine, threonine, isoleucine, hydroxylysine, histidine, methionine, and tyrosine. Absence of only two essential amino acids—tryptophan and methionine—makes gelatin a good dietary food supplement.

The principal uses of gelatin are in foods, pharmaceuticals, and photographic industries. Other uses of industrial gelatin are in the field of microencapsulation, health and cosmetics, and plastics. [F.V.]

Gem A mineral or other material that has sufficient beauty for use as personal adornment and has the durability to make this feasible. With the exception of a few materials of organic origin, such as pearl, amber, coral, and jet, and inorganic substances of variable composition, such as natural glass, gems are lovely varieties of minerals.

Natural gems. Each distinct mineral is called a species by the gemologist. Two stones that have the same essential composition and crystal structure but that differ in color are considered varieties of the same species. Thus ruby and sapphire are distinct varieties of the mineral species corundum, and emerald and aquamarine are varieties of beryl. See MINERAL; MINERALOGY.

Most gemstones are crystalline (that is, they have a definite atomic structure) and have characteristic properties, most of which are related directly to either beauty or durability. Each mineral has a characteristic hardness (resistance to being scratched) and toughness (resistance to cleavage and fracture). With few exceptions, the most important gemstones are those at the top of the Mohs hardness scale; for example, diamond is 10, ruby and sapphire are 9, chrysoberyl is 8½, and topaz, beryl (emerald and aquamarine), and spinel are 8.

Optical properties are particularly important to the beauty of the various gem materials. The important optical properties include color; dispersion (or "fire"); refractive index (relating the breaking up of white light into colors—a rough measure of brilliancy); and pleochroism (the property of some doubly refractive materials of absorbing light unequally in the different directions of transmission, resulting in color differences). Gemstones usually are cherished for their color, brilliancy, fire, or one of the several optical phenomena, such as asterism (the star effect caused by certain reflections of light); chatoyancy, or a cat's-eye effect; play of color, such as displayed by an opal; and adularescence (the billowy light effect seen in adularia or moonstone varieties of orthoclase feldspar).

Gemstones are commonly designated as precious or semiprecious. This is a somewhat meaningless practice, however, and often misleading, since many of the so-called precious gem varieties are inexpensive and many of the more attractive varieties of the semiprecious stones are exceedingly expensive and valuable. For example, a piece of fine-quality jadeite may be valued at approximately 100 times the price per carat of a low-quality star ruby. Fine black opals, chrysoberyl cat's-eyes, and alexandrites are often much more expensive than many sapphires of certain colors. See PRECIOUS STONES.

Hardness, specific gravity, and refractive indices of gem materials

Gem material	Hardness	Specific gravity	Refractive index
Amber	2–2½	1.05	1.54
Beryl	7½–8	2.67–2.85	1.57–1.58
Chrysoberyl	8½	3.73	1.746–1.755
Corundum	9	4.0	1.76–1.77
Diamond	10	3.52	2.42
Feldspar	6–6½	2.55–2.75	1.5–1.57
Garnet			
Almandite	7½	4.05	1.79
Pyrope	7–7½	3.78	1.745
Rhodolite	7–7½	3.84	1.76
Andradite	6½–7	3.84	1.875
Grossularite	7	3.61	1.735
Spessartite	7–7½	4.15	1.80
Hematite	5½–6½	5.20	
Jade			
Jadeite	6½–7	3.34	1.66–1.68
Nephrite	6–6½	2.95	1.61–1.63
Lapis lazuli	5–6	2.4–3.05	1.50
Malachite	3½–4	3.34–3.95	1.66–1.91
Opal	5–6½	2.15	1.45
Pearl	4	2.7	
Peridot	6½–7	3.34	1.654–1.690
Quartz			
Crystalline	7	2.65	1.54–1.55
Chalcedonic	6½–7	2.60	1.535–1.539
Spinel	8	3.60	1.72
Spodumene	6–7	3.18	1.66–1.676
Topaz	8	3.53	1.61–1.62
Tourmaline	7–7½	3.06	1.624–1.644
Turquois	5–6	2.76	1.61–1.65
Zircon			
Blue and colorless	7½	4.7	1.92–1.98
Green	6	4.0	1.81

More than 100 natural materials have been fashioned at one time or another for ornamental purposes. Of these, however, only a relatively small number are likely to be encountered in jewelry articles. See GEM CUTTING.

The table lists the important gem minerals and the properties most useful in identification. For further information on the individual species or groups see AMBER; AMETHYST; AZURITE; BERYL; CAMEO; CHRYSOBERYL; CORUNDUM; DIAMOND; EMERALD; FELDSPAR; GARNET; INTAGLIO (GEMOLOGY); JADE; JET (GEMOLOGY); LABRADORITE; LAZURITE; MALACHITE; MICROCLINE; OLIVINE; ONYX; OPAL; ORTHOCLASE; PEARL; QUARTZ; RUBY; SAPPHIRE; SPINEL; SPODUMENE; TOPAZ; TOURMALINE; TURQUOISE; ZIRCON.

Manufactured gems. A mineral or other material that has sufficient beauty and durability for use as a personal adornment can be manufactured. The term "manufactured," as used here, does not include such processes as shaping, faceting, and polishing, but only the processes that affect the material from which the finished gem is produced. These processes are (1) those that change the mineral in some fundamental characteristic, such as color, called a treated gem; (2) those by which a material is made that is identical with the naturally occurring mineral, called a synthetic gem; and (3) those that produce a simulated material with the appearance but not both the composition and structure of the natural gem, called an imitation gem.

Treated gems. There are four basic methods of treatment: (1) dyeing and staining, (2) plastic or other impregnation, (3) heat treatment, and (4) radiation. When the process to which a gem material is subjected changes its structure or adds something, such as a dye or a plastic binder, an effect on value takes place. When such changes are made, the nature of the alteration must be disclosed. An example of the second category of treatment, wherein there is no obvious effect on value, is gentle heating of amethyst to even its color, or stronger heating to change it to yellow or brown citrine. The change is permanent and nothing but temporary heat has been added. Many colored stones, including most green tourmaline, aquamarine, and colorless and flame-colored zircon and all pink topaz and blue zircon have been heated to improve their color.

Synthetic gems. The U.S. Federal Trade Commission has restricted the term synthetic gems to manufactured materials that have the same chemical, physical, and optical properties as their naturally occurring counterparts. Many gem materials, including diamond, have been synthesized, but in such small crystals or poor quality that they are unsatisfactory as gemstones. Some attempts to make gemstones have resulted in producing substances hitherto not known, many of which are of great importance industrially. Others have resulted in significant improvements in existing processes.

Imitation gems. Since prehistoric times glass has been the most widely used gem imitation. Since World War II colored plastics have replaced glass to a great extent in the least expensive costume jewelry.

Identification. Materials made by a flame-fusion process almost always contain gas bubbles, which are usually spherical or nearly so. Those with medium to dark tones of color often show color banding, or striae, with a curvature corresponding to that of the top of the boule. Natural gem materials are characterized by angular inclusions and straight color bands, if any are present.

Flux-fusion synthetic emeralds have distinctly lower refractive indices and specific gravities than natural emeralds and are characterized by wisplike or veillike flux inclusions. They show a red fluorescence under ultraviolet light, whereas most natural emeralds are inert.

Hydrothermally made synthetic emeralds have properties similar to many natural emeralds, but their inclusions differ and they are characterized by a very strong red fluorescence under ultraviolet.

The cheaper forms of glass are cast in molds and, under a hand lens, show rounded edges at the intersections of facets; the facets are often concave. The better grades, known as cut glass, have been cut and polished after first being molded approximately into the desired form. Cut glass has facets that intersect in sharp edges. Both types may contain gas bubbles or have a roiled appearance in the heart or have both, in contrast to most of the colored stones they imitate. [R.T.L.]

Gem cutting The polishing of rough gem materials into faceted or rounded forms for use in jewelry is called lapidary or gem cutting. The term lapidary is limited in application to the cutting of colored stones. One who fashions colored stones is called a lapidary, lapidarist, or lapidist.

Commercial colored-stone and diamond cutting are separate and distinct fields. Mechanically, the problems of colored-stone cutting are made simpler by the fact that all colored stones are soft enough to be shaped readily using silicon carbide or alumina powder as an abrasive. In contrast to diamond, they may be sawed or polished in any direction. A commercial gem cutter determines by eye the angles to which facets should be cut in relation to the girdle plane or the top facet and creates by eye the perfect symmetry one sees when viewing the finished stone from above.

The first step in the usual cutting operation is to saw the crystal or piece of massive gem material to obtain the size and general shape desired. If the material is relatively inexpensive, it may be ground to shape rather than reduced by cutting. Next, if the stone is soft enough, it is usually ground into the approximate final shape against a silicon carbide wheel. Facet grinding is usually accomplished on copper, iron, or lead laps, using diamond or silicon carbide powder for ruby and sapphire or their synthetic counterparts, and silicon carbide powder for the softer materials. Polishing is accomplished on tin, wood, or plastic laps, with the wooden laps usually covered with leather or cloth.

Innumerable cutting styles are used to fashion gemstones. The two basic types are those that employ curved surfaces, which are called cabochons, and those with flat surfaces, which are said to be faceted.

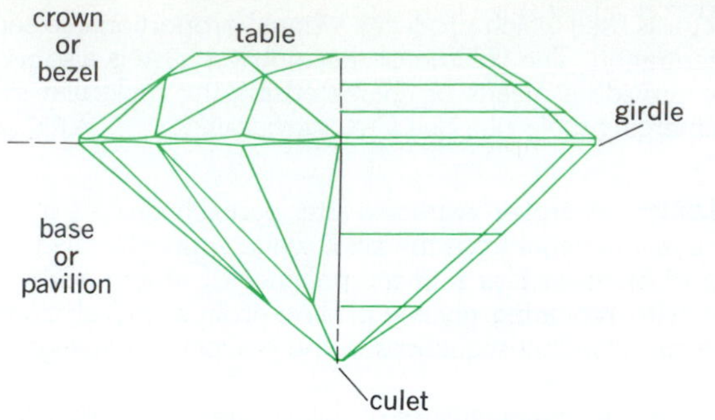

Arrangement of faceted cuts, showing brilliant cut (left side) and step cut (right side).

The two basic arrangements of facets are known as the brilliant and the step cut, from which most cutting styles derive. Both are characterized by a large facet (table) topping the crown facets, which slope to the periphery of the stone (girdle), and by a base portion, which slopes from the periphery to a tiny facet (culet) at the lowest point (see illustration). The two styles differ in that the facets on step cuts are parallel to the table and the girdle, so that they are all trapeze-shaped, whereas brilliant-cut facets, other than the table and culet, are triangular or kite-shaped. *See* GEM; GEM MOUNTING. [R.T.L.]

Gem mounting Gemstones are usually set in rings or other jewelry pieces made of one of the precious metals. Those metals most widely used for jewelry purposes are yellow gold, white gold, platinum, and palladium. When a stone is set in a ring, the portion in which the stone is placed is called the setting (see illustration). The stone may be held by small beads of metal, which are pushed over the edges by the stonesetter; by prongs, which are also bent over the edges of the stone to hold it securely; or by a rim of metal, which the stonesetter turns over the edges all the way around the stone. Tiffany settings are those in which naked prongs hold the stone in place high above the finger. *See* GEM; GEM CUTTING. [R.T.L.]

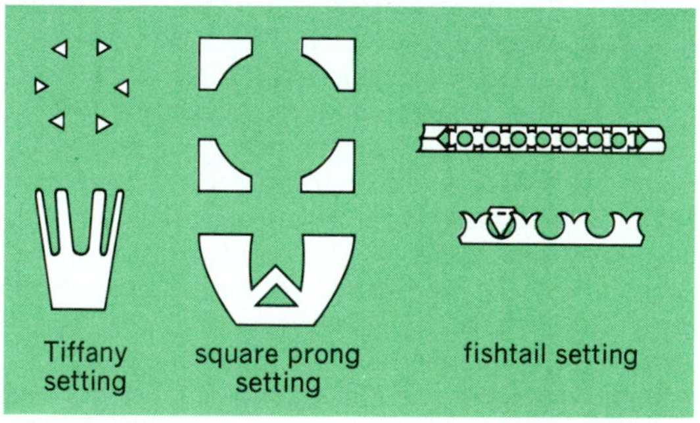

Gemstone settings. (*Gemological Institute of America*)

Geminga A relatively nearby neutron star that emits pulsed x-rays and gamma rays, has no detectable radio signal, and has also been observed at optical wavelengths. Its proximity provides unique opportunities for studying the properties of neutron stars.

Owing to their large, rapidly spinning magnetic fields, neutron stars emit significant nonthermal luminosity, that is, radiation produced by the interaction of accelerated particles and electromagnetic fields, or by the spinning fields themselves. This radiation was first detected at radio frequencies with the discovery of pulsars in 1968. Subsequently, notably with the launch of the *Compton Gamma Ray Observatory* in 1991, came the observation of gamma-ray pulsars. These are radio pulsars that emit a large flux of high-energy (100-MeV) gamma-ray photons. At the same time, a gamma-ray source was discovered that had all the markings of being connected with a radio pulsar but had no such object in its positional error box in the sky. *See* GAMMA-RAY ASTRONOMY; RADIO ASTRONOMY; SATELLITE ASTONOMY.

The object, named Geminga, is indeed a rotating neutron star, differing from ordinary pulsars only because it shows no detectable radio signal. It also emits soft x-rays, a fact which played a key role in the identification process, because the pulsating nature of the emission from Geminga was first observed in this domain. *See* X-RAY ASTRONOMY.

To observe Geminga at optical wavelengths has proven extremely difficult. Its faint glow is captured only by the largest ground-based facilities and by the Hubble Space Telescope. But the key to understanding the object, and indeed possibly the whole population of gamma-ray pulsars, is in these observations. The optical data have shown that the object has a large proper motion, implying a small distance for Geminga, not more than a few hundred light-years. This provides a measure of its overall emitted energy, which can be readily compared to that lost by rotation. The evidence is that Geminga, and most gamma-ray pulsars, is transforming nearly all its mechanical energy into very high energy photons, a finding which suggests a new approach to the understanding of neutron stars. *See* NEUTRON STAR; PULSAR. [G.F.Bi.]

Gemini The Twins, in astronomy, is a winter zodiacal constellation. Gemini is the third sign of the zodiac. It is conspicuous, containing first-, second-, and third-magnitude stars. These stars, α, β, γ, and μ, form a rough quadrilateral figure (see illustration).

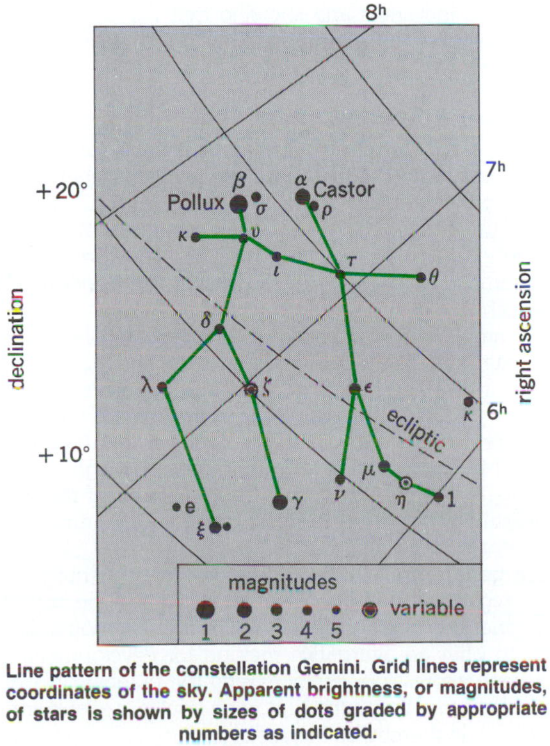

Line pattern of the constellation Gemini. Grid lines represent coordinates of the sky. Apparent brightness, or magnitudes, of stars is shown by sizes of dots graded by appropriate numbers as indicated.

The constellation is pictured as the figures of the twin heroes Castor and Pollux, with the two brightest stars of the same names representing the heroes' heads. Pollux, slightly brighter than Castor, is a navigational star. The Sun is in this constellation at the time of the summer solstice. *See* CONSTELLATION. [C.-S.Y.]

Gemology The science of those minerals and other materials which possess sufficient beauty and durability to make them desirable as gemstones. It is concerned with the identification, grading, evaluation, fashioning, and other aspects of gemstones. *See* GEM. [R.T.L.]

Gene The basic unit in inheritance. There is no general agreement as to the exact usage of the term, since several criteria that have been used for its definition have been shown not to be equivalent. The nature of this difficulty can be indicated most easily after a description of the earlier position. The facts of mendelian inheritance indicate the presence of discrete hereditary units that replicate at each cell division, producing remarkably exact copies of themselves, and that in some highly specific way determine the characteristics of the individuals that bear them. The evidence also shows that each of these units may at times mutate, to give a new equally stable unit, which has more or less similar but not identical effects on the characters of its bearers. Each unit can be shown to occupy a specific locus in a chromosome, and the new units (alleles) to which it gives rise occupy the same locus. *See* ALLELE; MENDELISM.

These hereditary units are the genes, and the criteria for the recognition that certain genes are alleles have been that they (1) arise from one another by a single mutative step, (2) have similar effects on the characters of the organism, and (3) occupy the same locus in the chromosome. It has long been known that there were a few cases where these criteria did not give consistent results, but these were explained by special hypotheses in the individual cases. But, such cases have been found to be so numerous that they seem the rule rather than the exception. *See* GENE ACTION; MUTATION; RECOMBINATION (GENETICS).

Some authorities use the term gene to indicate the smallest unit of recombination. This is a logical procedure, but a somewhat inconvenient one operationally, since there is no method of determining when the limit of divisibility has been reached. Other authorities use the term to designate an area in a chromosome made up of subunits that are closely related in their action and that must be present in an unbroken unit to give their characteristic effect. This definition has the disadvantage of being rather indefinite, since intermediate conditions are known. *See* CHROMOSOME; DEOXYRIBONUCLEIC ACID (DNA); NUCLEIC ACID; OPERON; TRANSPOSONS. [A.H.St.]

In molecular terms a gene usually corresponds to a region on the linear DNA molecule (a cistron) that codes for a polypeptide. This polypeptide, built up from at most 20 kinds of amino acids, contains in itself all the information needed to fold up either alone or in association with other polypeptides into a functional protein (such as an enzyme or a structural complex). The linear order of the amino acids is specified by the nucleotide sequence of the gene. This information, encoded in the DNA, is first copied into a messenger RNA. DNA as well as RNA contains a linear arrangement of four different monomers (deoxynucleotides in DNA, ribonucleotides in RNA). The messenger RNA is then translated sequentially by the protein-synthesizing machinery of each cell (ribosomes, transfer RNA, enzymes, and so on). A group of three nucleotides constitutes a code word that corresponds to a given amino acid. Because there are four different nucleotides in the genetic textbook, there are 64 different code words to code for 20 amino acids and for initiation and stop signals. Genes carry information not only to specify a protein but to control precisely their expression, often in various ways. *See* GENETIC CODE. [R.Co.; W.Fi.]

Gene action The functioning of genes (hereditary units) in determining the structural and functional characteristics of an individual, that is, its phenotype. Gene action is studied by two somewhat different but complementary approaches: (1) the analysis of changes which occur in the phenotype when a gene mutates, or is changed in dosage, or in position relative to other genes; this is frequently called the study of phenogenetics; and (2) the more direct approach, which attempts to determine the actual means by which genes exert their control over metabolism. The more direct approach is best described as study of primary gene action, but includes study of the interaction of primary or secondary products of gene action. *See* GENE.

All genes, with the exception of those in ribonucleic acid (RNA) viruses, are constituted of deoxyribonucleic acid (DNA), and the primary action of the great majority of them is to initiate a series of events leading to the determination of the amino acid sequences of specific polypeptides. The base sequence of probably only one of the strands of the DNA double helix of the gene is transcribed in the presence of RNA polymerase or polymerases to an RNA molecule known as messenger RNA (mRNA), which leaves the site of synthesis on the gene and participates in protein synthesis through the participation of transfer RNA (tRNA), ribosomes, and the necessary enzymes, amino acids, and cofactors. Each gene determines a specific mRNA, which in turn determines by a translation of the code a specific polypeptide. This polypeptide becomes a functional protein by folding and in most cases aggregating into a multimer. Therefore, there must exist in each organism at least as many genes as there are different kinds of polypeptides. *See* DEOXYRIBONUCLEIC ACID (DNA); GENETIC CODE; PROTEIN; RIBONUCLEIC ACID (RNA); RIBOSOMES.

It should be recognized that transcription is in essence the primary action of the gene. It is this step which determines the specificity of a protein. Most changes in the DNA sequence will result in an altered protein.

In addition to coding for specific proteins, such as enzymes, which function in metabolism, genes also code for the RNA species such as tRNA, ribosomal RNA, and proteins involved as general tools in protein synthesis or in the control of gene action as discussed below.

The actual effects of gene action are recognized for the most part by noting the effects of gene mutation on the phenotype, but in complex multicellular organisms the final phenotypic effect observed superficially may be far removed from the initial action of the gene itself, for example, a change in shape of the ear or a change in eye color. However, through study of the lower organisms, such as bacteria and fungi, it has become evident that there is a close relationship between genes and proteins. This has also been borne out by studies of the higher organisms. It is apparent that some phenotypic changes produced by gene mutation are due to changes in the protein-synthesis machinery brought about by the genes controlling tRNA and ribosomal RNA synthesis.

Because of the ramifying effects of a change in a single protein, phenotypic change may be drastic, leading to lethality or changes in a number of characteristics of the organism (pleiotropy) rather than just in one. *See* LETHAL GENE; MUTATION.

Under certain conditions, an increase in the number of times a particular gene is present has a direct quantitative effect on one or more aspects of the phenotype. This effect is considered to be a manifestation of quantitative gene activity or gene dosage. It means that the gene does something, and that the higher the dose with which it is present, the greater the physiological or chemical end result. *See* POLYPLOIDY.

Some heterozygous combinations of mutant alleles do not produce the phenotype expected from the phenotypes of the homozygotes. This is defined as a manifestation of allelic interaction. It is in contrast to those situations in which one allele is dominant over the other, so that the phenotype of the heterozygote is very similar or identical to that of the homozygous dominant. Also, it is different from those situations in which the two alleles show an additive effect, and the phenotype of the heterozygote is intermediate between those expected from the homozygotes. *See* ALLELE; DOMINANCE.

The final phenotype of an organism is the resultant of the action of all the active genes in its cell or cells. These genes may act independently in producing their respective primary products, but the primary products, and the products of their activity, that is, enzymes and other macromolecules, interact at the level of extragenic metabolism to give the final phenotype. Thus, there is really no one gene determining the shape of an organ, or even the production of a certain pigment. These end products are determined by many genes acting together through their respective immediate and then succeeding interrelated products. This is true even though the mutation of only a single one of the genes may cause a characteristic unit change in the phenotype. The manifestation of these interactions is called gene interaction. This term does not necessarily imply that the genes themselves interact. In general, the term is applied to apparent interactions between genes.

The phenotype which develops in any organism is the resultant of two guiding influences, the genotype and the environment. Hence, the environment must be considered in any analysis of gene action, if it is desired to arrive at an understanding of how genes act toward the production of the phenotype.

The environment of genes is obviously a complex one. For convenience, two areas can be defined: (1) that immediately around the genes, the intracellular environment of the rest of the cell; and (2) the extracellular and extraorganismal environment. The intracellular environment can be changed by the mutation of other genes, which may then modify the action of a gene under study. Gene interaction is thus seen to be in part an aspect of the study of the internal environment of the cell. Extracellular environmental factors, such as light and heat, may also influence the action of genes greatly.

Changes in phenotype which occur against a constant genetic background are in reality responses to the environment by the extragenic part of the living system. The genes themselves are not changed, as can be readily demonstrated by changing the environment back to the original condition, or by breeding the individual and showing that the offspring inherit the original parental genotype. *See* DEVELOPMENTAL GENETICS; GENETICS. [R.P.W.]

Gene amplification The process by which a cell specifically increases the copy number of a particular gene to a greater extent than it increases the copy number of genes composing the remainder of the genome (all the genes which make up the genetic machinery of an organism). It is therefore distinguished from duplication, which is a precise doubling of the genome preparatory to cell division, and endoreduplication, which leads to endopolyploidy.

Gene amplification results from the repeated replication of the deoxyribonucleic acid (DNA) in a limited portion of the genome, in the absence of or to a much greater extent than replication of DNA composing the remainder of the genome. Thus is formed a cell in which the genes composing a limited portion of the genome are present in relatively high copy number, while the genes composing the remainder of the genome are present in approximately normal copy number. *See* DEOXYRIBONUCLEIC ACID (DNA).

Since gene amplification increases the copy number of a specific region of the genome without altering the copy number of genes composing the remainder of the genome, it would appear to offer an alternative method for developmental control of gene expression. By increasing the number of copies of a particular gene, the number of gene copies available for transcription could thereby be increased.

In a number of instances of gene amplification, the amplification phenomenon appears to be developmentally regulated, and the amplified copies of the gene are subsequently lost from the cell. Studies on cells in culture have demonstrated "amplification" of genes involved in resistance to specific drugs. *See* GENE; GENE ACTION. [M.D.C.]

General aviation All aviation activity not associated with either certificated air carriers or the military. Consequently, general aviation consists of the largest number of aircraft and pilots and accounts for the largest number of flying hours among all classifications of aviation within the United States.

A general aviation aircraft can range in size from the smallest pleasure craft designed and built by an individual to the largest air transport manufactured by a large airframe manufacturer. Of the active general aviation aircraft registered by the Federal Aviation Administration (FAA), about 50% are four-to-six-place single-engine airplanes that are capable of speeds below 230 mi/h (103 m/s). The fleet also includes multiengine aircraft, mostly powered by two wing-mounted intermittent-combustion power plants, jet aircraft, and turboprop vehicles. The remainder of the general aviation fleet, which numbers about 60,000, are gliders and other recreational-use aircraft and two-or-three-place training and sport aircraft. *See* GLIDER.

Business applications, usually involving the transportation of executive or technical personnel in aircraft that range from single-engine vehicles flown personally by a business person to corporate jets piloted by a professional crew, account for more flying hours than any other segment of general aviation. An area that has responded to the changing nature of certificated air carriers is the commercial sector of general aviation, specifically the commuter airline and air taxi operators.

Commuters are very similar to the certificated air carriers in intent, although many lack the sophistication of the major airlines in execution. Commuters are the only airlines serving many communities, and they provide many aspects of the service that once was offered by a certificated air carrier. Air taxis provide on-demand air transportation for hire in aircraft that cover the spectrum of general aviation vehicles. General aviation aircraft are used in a variety of other commercial applications, including law enforcement, aerial mapping, pipeline patrol, counting cattle, surveying farm lands, fighting fires, and providing air ambulance services.

About 25% of general aviation involves the use of small aircraft for personal transportation and pleasure flying. Some aviators use their own or a rental aircraft to conduct vacation travel, while others just enjoy the pleasure of being aloft on a sunny day. Not all activities in this segment are leisure-oriented, however, since some proficiency flying (which is required by the FAA for the maintenance of a pilot's license) falls into this category. *See* AVIATION. [J.W.O.]

Generator Any machine by which mechanical power is transformed into electric power. Generators fall into two main groups, alternating-current (ac) and direct-current (dc). They may be further classified by their source of mechanical power, called the prime mover. Generators are usually driven by steam turbines, hydraulic turbines, engines, gas turbines, or motors. Small generators are sometimes powered from windmills or through gears, belts, friction, or direct drive from parts of vehicles or other machines. *See* ALTERNATING-CURRENT GENERATOR; DIRECT-CURRENT GENERATOR; PRIME MOVER.

In practice, permanent-magnet fields are used only in small generators. Large generators, except induction generators, are equipped with dc field windings. The field coils are wound on the stators of most dc generators to permit mounting the armature coils and commutator on the rotor. On ac generators, the field coils are normally located on the rotors.

Generators driven by steam or gas turbines are sometimes called turbogenerators. Although in small sizes these may be gear-driven, and some may be dc generators, the term turbogenerator generally means an ac generator driven directly from the shaft of a steam turbine. *See* ELECTRIC POWER GENERATION; ELECTRIC ROTATING MACHINERY. [L.T.R.]

Genetic algorithms Search procedures based on the mechanics of natural selection and genetics. Such procedures are known also as evolution strategies, evolutionary programming, genetic programming, and evolutionary computation. Genetic algorithms are increasingly by solving difficult search, optimization, and machine-learning problems that have previously resisted automated solution. They can solve hard problems quickly and reliably, are easy to interface to existing simulations and models, are extensible, and are easy to hybridize.

Motivation. Just as natural selection and genetics have filled a variety of niches by creating genotypes (sets of chromosomes) that result in well-adapted phenotypes (or organisms), so too can genetic algorithms solve many artificial problems by creating strings (artificial chromosomes) that result in better solutions. Users ultimately turn to genetic algorithms for robustness, that is, for algorithms that are broadly applicable, relatively quick, and sufficiently reliable. This emphasis on robustness contrasts starkly with the philosophy of operations research, where new algorithms must be tailored to specific problems. The need to invent a new method for each new problem class is daunting, and users look for methods that can solve complex problems without this requirement. *See* GENETICS; OPERATIONS RESEARCH; ORGANIC EVOLUTION.

Mechanics. For concrete exposition, the discussion is limited to a simple genetic algorithm that processes a finite population of fixed-length, binary strings. A simple genetic algorithm consists of three operators: selection, crossover, and mutation.

Selection is the survival of the fittest within the genetic algorithm. The key notion is to give preference to better individuals. Of course, for selection to function, there must be some way of determining what is good. This evaluation can come from a formal objective function, or it can come from the subjective judgment of a human observer or critic.

If genetic algorithms were to do nothing but selection, the trajectory of populations could contain nothing but changing proportions of the strings in the original population. To do something more sensible, the algorithm needs to explore different structures. A primary exploration operator used in many genetic algorithms is crossover. Simple, one-point crossover proceeds in three steps: (1) two individuals are chosen from the population by using the selection operator, and these two structures are considered to be mated; (2) a cross site along the string length is chosen uniformly at random; and (3) position values are exchanged between the two strings following the cross site.

In a binary-coded genetic algorithm, mutation is the occasional (low probability) alteration of a bit position, and with other codes a variety of diversity-generating operators may be used. When used together with selection and crossover, mutation acts both as an insurance policy against losing needed diversity and as a hill-climbing algorithm. [D.E.Go.]

Genetic code The key to translation from the four-letter (nucleotides) alphabet of nucleic acid, in which genetic information is stored, to the 20-letter (amino acids) alphabet of protein, where the genetic information finds its primary expression. The nucleic acid language consists of a vocabulary of 64

first position (5′ end)	second position: U	C	A	G	third position (3′ end)
U	phe	ser	tyr	cys	U
U	phe	ser	tyr	cys	C
U	leu	ser	C.T.	C.T.	A
U	leu	ser	C.T.	try	G
C	leu	pro	his	arg	U
C	leu	pro	his	arg	C
C	leu	pro	gln	arg	A
C	leu	pro	gln	arg	G
A	ile	thr	asn	ser	U
A	ile	thr	asn	ser	C
A	ile	thr	lys	arg	A
A	C.I. met	thr	lys	arg	G
G	val	ala	asp	gly	U
G	val	ala	asp	gly	C
G	val	ala	glu	gly	A
G	C.I. val	ala	glu	gly	G

Allocation of the 64 codons, derived mainly from studies of *Escherichia coli*. C.T. indicates codons thought to be signals for polypeptide chain termination; the two codons marked C.I. are polypeptide chain initiators.

three-nucleotide sequences, or codons; a sequence of such codons (on average about 300) constructs the message for a polypeptide. Translation into the protein language, which consists of a vocabulary of 20 amino acids, takes place in the protein-synthesizing apparatus of the cell strictly following the rules given by the genetic code. *See* RIBOSOMES.

The genetic information of an organism is encoded and maintained in the linear sequence of deoxyribonucleotides in its deoxyribonucleic acid (DNA), although in some viruses it is in ribonucleotides in their ribonucleic acid (RNA). A major problem of biology has been to elucidate how the information content of a particular DNA molecule becomes expressed by the production of a particular type of protein molecule. One only of the two antiparallel strands of the DNA double helix contains genetic information, and this information is first transcribed from the four-letter alphabet of DNA into the slightly different four-letter alphabet of RNA. *See* DEOXYRIBONUCLEIC ACID (DNA); PROTEIN; RIBONUCLEIC ACID (RNA).

A codon in RNA is the basic unit which is translated into, and therefore corresponds to, one amino acid in protein. The coding ratio is 3 and the nucleic acid language is a vocabulary of 64 codons which construct different messages depending on their numbers and their linear sequence. The genetic code is completely deciphered for *Escherichia coli*. The full catalog of code words is listed in the illustration. Apparently this code is universal, or nearly so, for organisms ranging from viruses to mammals.

The genetic code is a nonoverlapping triplet code; that is, a nucleotide which is part of one codon is never also part of a preceding or a following codon. The code is highly degenerate in a semisystematic way; that is, more than one codon is usually assigned for a particular amino acid (the only exceptions are the amino acids tryptophan: UGG; and methionine: AUG) and where several codons are assigned for the same amino acid, almost invariably they are closely related, differing in only one nucleotide, which is usually the third (for example, valine: GUU, GUC, GUA, and GUG).

As there are only four nucleotides in messenger RNA (A = adenylic acid; U = uridylic acid; C = cytidylic acid; G = guanylic acid), there are $4^3 = 64$ possible codons, and of these, 61 are known to code for one amino acid or another. The three remaining codons apparently do not code for an amino acid but are signals for termination of the growing polypeptide chain. *See* GENE; GENE ACTION; GENETICS. [H.S.S.]

Genetic engineering The joining together of two DNA molecules in a test tube and the subsequent insertion of this recombinant molecule into a living system. It does not necessarily imply that the two molecules never join together in nature, since frequently genetic engineering is used to achieve quickly and in high yield a result which occurs only very rarely in nature. When applied to higher organisms, the term is usually taken to imply the introduction into an organism of a function, usually a gene or genes which either have not previously been detected in that organism or exist in it in a different form.

Plants. The transfer of genetic functions between plants by breeding methods is possible only between closely related species. The barrier to breeding distantly related species imposes limits on the extent to which desirable genes can be introduced into many economically important plants. It is therefore of importance to develop artificial methods which can bypass this breeding barrier. *See* BREEDING (PLANT).

The majority of genes that would be of interest to transfer between distantly related plant species do not give a selectable phenotype. They would, therefore, have to be coupled to a marker gene that would enable transformed plants or cells to be selected from a population comprising both transformed and untransformed individuals. In attempting to use auxotrophic markers (for instance, a requirement for an exogenously applied amino acid), considerable effort must be expended in producing mutant forms of the receptor plant species. Little success has been achieved in the production of such mutants.

An entirely different but parallel approach attempts to make use of natural systems which modify or interfere with the expression of plant genomes. These systems are potential natural vectors in that they are based on DNA molecules, which, as a result of natural processes, are transferred to and replicate in plant cells. They are naturally marked in that they cause plant diseases which in some cases do not directly kill the infected plant but only cause a loss of vigor.

The other natural plant modification system is the plant crown gall system. Crown galls are plant tumors produced in response to infection through a wound by the soil bacterium *Agrobacterium tumefaciens*. The "agent" or tumor-inducing principle within the bacterium is a group of genes carried on a tumor-inducing (Ti) plasmid. This relatively small ($90–160 \times 10^6$ daltons) circular DNA molecule codes for several distinct types of function. In *Agrobacterium*, there is a system which transfers a fairly clearly defined piece of DNA to plant cells, where it is integrated into the plant genome. The observation that bacterial transposons inserted into the T-DNA (T for "transforming") are carried by it into plant cells and integrated as part of the T-DNA shows that the system is capable of acting as a vector for foreign DNA. The fact that some tumors produce morphologically normal plants shows that tumorigenic functions can be suppressed, while other, nononcogenic T-DNA functions (such as nopaline synthesis) continue to be expressed, thus suggesting that the deleterious side effects of the system can be removed. *See* CROWN GALL.

Animals. In contrast, the genetic engineering of animal cells poses a different set of problems but, at the same time, carries a number of advantages. For mammalian species, genome modification has been shown to be possible at many levels.

Cell fusion techniques have been extensively used in human genetics for locating individual genes on specific chromosomes. Where mouse and human cells have been fused, it is usually the human chromosomes which are preferentially lost. When the final stage has been reached of a mouse cell containing one human chromosome, the cells can be screened for proteins characteristic of human cells. The genes for these can then be assigned to the particular chromosome contained in the mouse cell. *See* SOMATIC CELL GENETICS.

Several tumor viruses are capable of integrating their DNA, or complementary DNA, into the host cell genome. After infection, they transform their hosts to a tumor phenotype essentially by synthesizing a protein or proteins which interfere with the normal cell-division control processes. Such viruses could be used as vectors for the introduction of foreign genes into animal cells in rather the same way that the T-DNA of the *Agrobacterium* Ti plasmid can be used for the modification of plant cells. In the case of the animal tumor viruses, however, the extent of virus genome integration can be variable. *See* TUMOR VIRUSES.

There is one final system of animal cell modification which has considerable potential and involves the transformation of cells with pure DNA, which results in the integration of that DNA into the genomes of a subpopulation of integration-competent cells. However, the complete inability to regenerate whole animals from cultured cells severely limits the use of the system. This is in contrast to plant tissue cultures, where the ability to regenerate whole plants from undifferentiated cultured cells is a well-established phenomenon. *See* DEOXYRIBONUCLEIC ACID (DNA); GENE; GENETICS. [A.G.He.]

Genetic homeostasis The tendency of mendelian populations to maintain a constant genetic composition in the face of external pressure. When subjected to such pressures as artificial selection (usually for some quantitative trait) or temporary environmental changes, genetic homeostatic mechanisms tend to restore to equilibrium gene frequencies that may have shifted from mean optimal values. *See* MENDELISM; POPULATION GENETICS.

The mechanisms responsible for genetic homeostasis are varied. The two most commonly invoked, which are also responsible for maintenance of genetic polymorphisms, are frequency-dependent selection and heterozygote advantage. Under the first, the selective value of a gene depends on its frequency in the gene pool with a loss of selective advantage by an allele which was rare and has become common. Under the second, when either single alleles or blocks of linked genes are involved, the relative fitness of homozygotes is lower than that of heterozygotes. *See* ALLELE; POLYMORPHISM (GENETICS). [I.M.L.]

Genetic mapping The resolution of the genetic material of an organism into the linear sequence of the material's elements. In decreasing order of size these elements are: the chromosome, the genes, the codons, and, ultimately, the nucleotides of deoxyribonucleic acid, DNA (ribonucleic acid, RNA, in certain viruses). Complete mapping of the genetic material of an organism implies the complete description of its DNA sequences. This has been achieved in a few cases, with small viruses with sequences only a few thousand nucleotides long. Higher plants or animals have vastly longer sequences; in mammals, for example, some sequences are 10^9 nucleotides long. Thus genetic mapping must proceed by successive approximations. Once the chromosome set and a number of genes have been identified, mapping proceeds by establishing (1) the chromosome or linkage group to which each gene belongs; (2) the order of genes in a specific chromosome or linkage group; and (3) the distances between the genes. A further step is "fine" genetic analysis at the intragenic level, that is, the resolution of the nucleotides within a single gene or a small segment of DNA. *See* CHROMOSOME; DEOXYRIBONUCLEIC ACID (DNA); GENE; RIBONUCLEIC ACID (RNA). [G.Po.]

Genetics The science that is concerned with biological inheritance, that is, with the causes of the resemblances and differences among related individuals. Genetics occupies a central position in biology, for essentially the same principles apply to all animals and plants, and an understanding of inheritance is basic for the study of evolution and for the improvement of cultivated plants and domestic animals. It has also been found that genetics has much to contribute to the study of embryology, biochemistry, pathology, anthropology, and other subjects.

Genetics may also be defined as the science that deals with the nature and behavior of the genes, the fundamental hereditary units. From this point of view, evolution is seen as the study of changes in the gene composition of populations, whereas embryology is the study of the effects of the genes on the development of the organism. *See* BREEDING (ANIMAL); BREEDING (PLANT); CHROMOSOME; GENE ACTION; GENETIC CODE; HETEROSIS; HUMAN GENETICS; MENDELISM; MOLECULAR BIOLOGY. [A.H.St.]

Gentamicin A broad-spectrum antibiotic produced by a species of microorganism which belongs to the genus *Micromonospora*. The producing organism has been named *M. purpurea* for the purple color produced in the mycelium when grown in submerged culture.

Gentamicin is active in animals against a wide variety of gram-positive and gram-negative bacteria, mycobacteria, mycoplasma, and rickettsial organisms. Gentamicin is used clinically in a topical and parenteral form against susceptible organisms. The topical preparation has been shown to be effective for use on primary and secondary skin infections and skin ulcers and in the treatment and prophylaxis of infections due to burns. The injectable form is used for the treatment of acute and chronic urinary tract infections, pneumonias, and septicemias due to susceptible organisms, particularly *Pseudomonas* and *Proteus* strains. *See* ANTIBIOTIC. [M.J.W.]

Gentianales An order of flowering plants, division Magnoliophyta (Angiospermae), in the subclass Asteridae of the class Magnoliopsida (dicotyledons). The order consists of four families and about 5600 species. The Apocynaceae and Asclepiadaceae have about 2000 species each, the Gentianaceae have about 1100, and the Loganiaceae have about 500.

The Gentianales have well-developed internal phloem and opposite, simple, mostly entire leaves. The flowers are hypogynous, with a regular corolla, the lobes usually contorted in bud. The stamens form a single cycle, alternate with the corolla lobes. Many of the genera contain alkaloids or (in the Gentianaceae) glucosides. *See* ASTERIDAE; RAUWOLFIA. [A.Cr.]

Geobotanical indicators Vegetation whose characteristics provide information useful in interpreting geologic phenomena. Species assemblages and plant density are used to map strata of different chemical composition and reservoir capacity. Plants are also used in prospecting by analyzing plant tissue for metal content and by observing the species association, toxicity effects, and morphological changes in plants whose roots are in contact with ore.

Variations in plant associations are often useful in mapping the areal distribution of specific rock formations. A significant relationship between the vegetation and underlying rocks is commonly found in arid regions where the soils bear a chemical similarity to the rocks from which they have been derived. Plants are especially useful in geologic mapping wherever the dominant plant species differ on the formations being mapped.

Some plant species exhibit dependence upon groundwater in contrast to the ability of others to survive on occasional rainfall. Therefore plants are useful as indicators of the quality of

Useful indicator plants in mineral prospecting

Plant	Ore element	Area
Eriogonum ovalifolium v. *ovalifolium*	Cu	Montana
Gypsophila patrinii	Cu	Soviet Union
Polycarpea spyrostylis	Cu	Australia
Tephrosia sp.	Cu	Australia
Elscholtzia haichowensis	Cu	China
Haumaniastrum robertii	Cu	Katanga
Becium homblei	Cu	Katanga
Merceya ligulata	Cu	Sweden
Merceya latifolia	Cu	Sweden
Mielichhoferia macrocarpa	Cu	Sweden, Alaska
Armeria maritima	Cu	Wales
Armeria maritima (vulgaris)	Zn	Aachen
Armeria halleri	Zn	Pyrenees
Hutchinsia alpina	Zn	Pyrenees
Thlaspi alpestre v. *calaminare*	Zn	Aachen
Stellaria verna	Zn	Aachen
Viola lutea v. *calamineria*	Zn	Aachen
Silene cobalticola	Co	Katanga
Crotalaria cobalticola	Co	Katanga
Dicoma niccolifera	Ni	Zambia
Hybanthus caledonicus	Ni	New Caledonia
Astragalus bisulcatus	Se	U. S. (western)
Astragalus pattersoni	Se, U	U. S. (western)
Astragalus garbancillus	Se, U	Peru
Mechovia grandiflora	Mn	Katanga

water and depth of groundwater; the depth can be estimated from the known depth of root penetration of the species present.

The use of plant distribution, plant appearance, and growth anomalies in prospecting for ore deposits has been called geobotanical prospecting, in contrast to biogeochemical prospecting, in which common plant species are used as a sampling medium for chemical analysis. Plants have a third-dimensional advantage in prospecting because the roots may extend through many feet of surficial material for contact with mineralized water or rock. *See* GEOCHEMICAL PROSPECTING.

A plant species that grows more extensively near an ore body because it requires one or more of the ore elements or because an element on which this particular species depends is more available, is called an indicator. Some useful indicator plants are given in the table. A plant's distribution may be useful even though it is controlled by an associated pathfinder element rather than the most economically valuable element of the suite. [H.L.C.]

Geochemical prospecting The use of chemical properties of naturally occurring substances (including rocks, glacial debris, soils, stream sediments, waters, vegetation, and air) as aids in a search for economic deposits of metallic minerals or hydrocarbons. In exploration programs, geochemical techniques are generally integrated with geological and geophysical surveys. *See* GEOCHEMISTRY; GEOPHYSICAL EXPLORATION.

General principles. Mineral deposits represent anomalous concentrations of specific elements, usually within a relatively confined volume of the Earth's crust. Most mineral deposits include a central zone, or core, in which the valuable elements or minerals are concentrated, often in percentage quantities, to a degree sufficient to permit economic exploitation. The valuable elements surrounding this core generally decrease in concentration until they reach levels, measured in parts per million (ppm) or parts per billion (ppb), which appreciably exceed the normal background level of the enclosing rocks. These zones or halos afford means by which mineral deposits can be detected and traced; they are the geochemical anomalies being sought by all geochemical prospectors.

The zone surrounding the core deposit is known as a primary halo or anomaly, and it represents the distribution patterns of elements which formed as a result of primary dispersion. Primary dispersion halos vary greatly in size and shape as a result of the numerous physical and chemical variables that affect fluid movements in rocks. Some halos can be detected at distances of hundreds of meters from their related ore bodies; others are no more than a few centimeters in width.

Abnormal chemical concentrations in weathering products are known as secondary dispersion halos or anomalies and are more widespread. They are sometimes referred to as dispersion trains. The shape and extent of secondary dispersion trains depend on a host of factors, of which topography and groundwater movement are perhaps most important. Groundwaters frequently dissolve some of the constituents of mineralized bodies and may transport these for considerable distances before eventually emerging in springs or streams. Further dispersion may ensue in stream sediments when soil or weathering debris that has anomalous metal content becomes incorporated through erosion in stream sediment. Analysis of the fine sand and silt of stream sediment can be a particularly effective method for detection of mineralized bodies within the area drained by the stream.

Survey design. The degree of success of a geochemical survey in a mineral exploration program is often a reflection of the amount of care taken with initial planning and survey design. This phase of activity is often referred to as an orientation survey; its practical importance cannot be overstressed.

When a geochemical prospecting survey is contemplated, four basic considerations must be addressed: the nature of the mineral deposits being sought; the geochemical properties of the elements likely to be present in the target mineral deposit; geological factors likely to cause variations in geochemical background; and environmental, or landscape, factors likely to influence the geochemical expression of the target mineral deposit. Elucidation of these factors in an orientation survey will permit design of a geochemical prospecting survey that is most likely to prove effective under the prevailing conditions. *See* PROSPECTING.

Geochemical prospecting surveys fall into two broad categories, strategic or tactical, which may be further subdivided according to the material sampled. Strategic surveys imply coverage of a large area (generally several thousands of square kilometers) where the primary objective is to identify districts of enhanced mineral potential; tactical surveys comprise the more detailed follow-up to strategic reconnaissance. Typically the area covered by a tactical survey is divided into discrete areas of high mineral potential within the general anomalous district.

Soil and glacial till surveys have been used extensively in geochemical prospecting and have resulted in the discovery of a number of ore bodies. Generally, such surveys are of a detailed nature and are run over a closely spaced grid.

Biogeochemical surveys are of two types. One type utilizes the trace-element content of plants to outline dispersion halos, trains, and fans related to mineralization; the other uses specific plants or the deleterious effects of an excess of elements in soils on plants as indicators of mineralization. The latter type of survey is often referred to as a geobotanical survey. *See* GEOBOTANICAL INDICATORS.

Rock geochemical surveys are reconnaissance surveys carried out on a grid or on traverses of an area, with samples taken of all available rock outcrops or at some specific interval. One or several rock types may be selected for sampling and analyzed for various elements. Geochemical maps are compiled from the analyses, and contours of equal elemental values are drawn. These are then interpreted, often by using statistical methods. Under favorable conditions, mineralized zones or belts may be outlined in which more detailed work can be concentrated. If the survey is executed over a large expanse of territory, geochemical provinces may be outlined.

Isotopic surveys are applicable to elements which exist in two or more isotopic forms. They employ the ratios between isotopes such as ^{204}Pb, ^{206}Pb, ^{207}Pb, ^{208}Pb, or ^{32}S and ^{34}S to "fingerprint" or indicate certain types of mineral deposits which may share a common origin. Isotopic ratios may also be used to determine the ages of minerals or given rock types and may, thus, assist in elucidating questions of ore formation.

Geochemistry applied to hydrocarbon exploration differs from that in the search for metallic mineral deposits; the former chiefly involves detection and study of organic substances found during drilling; the latter, detection and study of inorganic substances at the surface. Once hydrocarbon accumulations have been discovered, their classification into geochemical families is important. The final stages of detailed exploration may involve complex multivariate computer-aided modeling of all available geological, geochemical, geophysical, and hydrological data—to determine the ultimate hydrocarbon potential of a given basin. [R.F.H.]

Geochemistry The study of the abundance of elements and nuclides in the Earth, the distribution of elements among the geochemical phases of the Earth, and the laws governing these abundance and distribution relationships. With the current emphasis on lunar materials, the definition should probably be expanded to include other parts of the cosmos accessible to exact observations. Classical geochemistry was directed largely at defining the abundance of the elements in rocks and minerals. Such data were essential to attempts to understand underlying nuclear causes that produced the observed abundance of the elements.

Geochemistry is subdivided into bio-, atmo-, hydro-, and lithogeochemistry. The distribution of the elements depends on the atomic structure of the elements (their chemistry), whereas the quantitative composition of the Earth depends on the nuclear structure of the elements.

Advances in instrumentation techniques, including mass spectrometry, neutron activation analysis, and ion microprobe analysis, have led to improvements in accuracy of the analyses. Studies on phase equilibria of simple petrologic systems conducted under high temperature and pressure have increased knowledge of chemical reactions in the Earth's crust and have provided new data on origin and emplacement of ore bodies within the Earth. Modern geochemical research places considerable stress on the quantitative determination of temperature, pressure, and time in geologic systems. Methods based on isotopic fractionation and radioactivity which utilize mass spectrometry have proved extremely useful. *See* ACTIVATION ANALYSIS; COSMOCHEMISTRY; GEOCHEMICAL PROSPECTING; GEOLOGIC THERMOMETRY; PALEOBIOCHEMISTRY; PETROLOGY. [D.Ay.; M.W.R.]

Geochronometry The study of the absolute age of the rocks of the Earth using methods based on the radioactive decay of such isotopes as ^{238}U, ^{235}U, ^{232}Th, ^{87}Rb, ^{40}K, and ^{14}C. Measurements using these decay schemes permit definition of the age of the Earth and meteorites (4.5×10^9 years); the oldest available rocks (3×10^9 years); significant events in the geologic time scale; the rate of evolution of organisms; the time and duration of major events in crustal evolution; the advance and retreat of continental glaciers; and the evolution of human culture. *See* FISSION TRACK DATING; GEOLOGICAL TIME SCALE; RADIOCARBON DATING; ROCK AGE DETERMINATION. [J.L.K.]

Geode A roughly spheroidal hollow body, lined on the inside with inward-projecting small crystals (see illustration). Geodes are found most frequently in limestone beds but may occur in some shales. Typically, a geode consists of a thin outer shell of dense chalcedonic silica and an inner shell of quartz crystals. Many geodes are filled with water; others, having been exposed for some time at the surface, are dry. Calcite or dolomite crystals line the interior of some geodes, and a host of other minerals are less commonly found. In some geodes there is an alternation of layers of silica and calcite, but almost all geodes show some banding suggestive of rhythmic precipitation. *See* CHALCEDONY. [R.Si.]

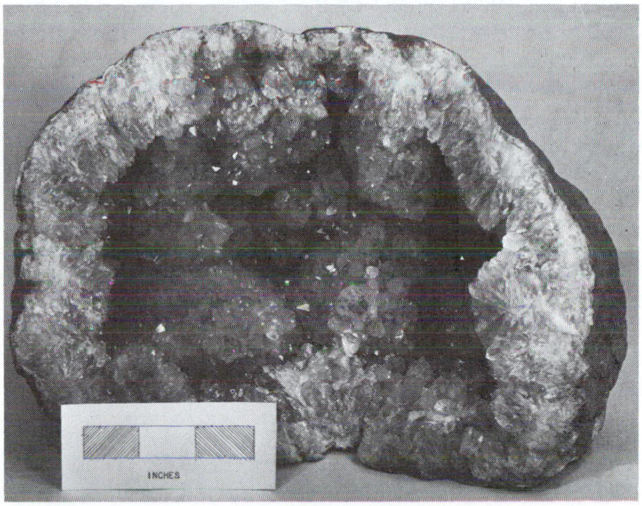

Geode, lined with quartz crystals, Keokuk, Iowa. (*Brooks Museum, University of Virginia*)

Geodesic dome A curved lattice grid dome that utilizes the equilateral triangle as the basis of its surface grid geometry.

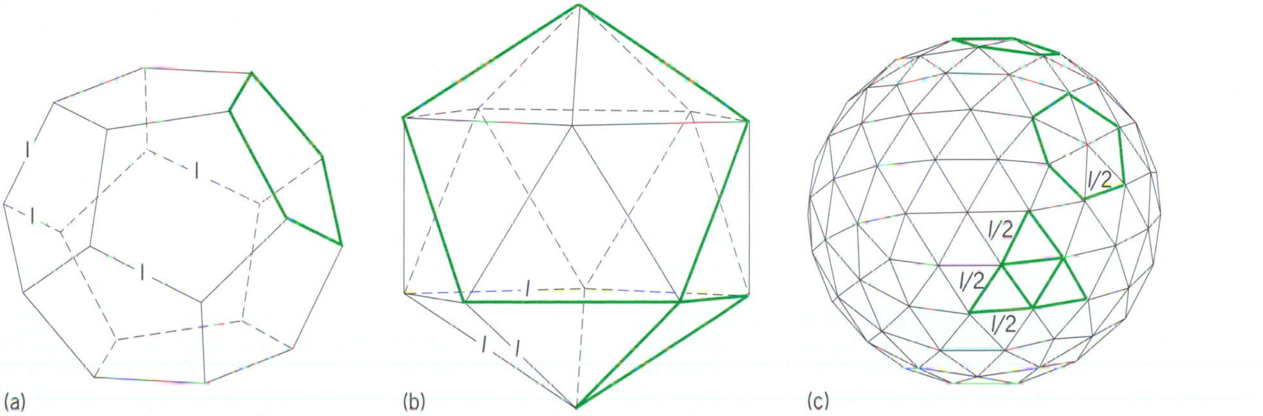

Geometry of geodesic domes. (*a*) Dodecahedron: a regular pentagon is typical of each face; every point is an apex because all apexes are on the sphere; each strut (*l*) is the same length. (*b*) Icosahedron: an apex is above the center of each polygon and on the surface of the sphere; the equilateral triangle typical of each face is highlighted; each strut is the same length. (*c*) Larger dome based on the icosahedron: subdivision is formed by connecting mid points of struts of equilateral triangles (each half strut is labeled *l*/2); the original pentagon is shown at the top, and a formed hexagon is also shown.

R. Buckminster Fuller, the inventor and champion of the geodesic dome, obtained a patent in 1954 that described a method of dividing a spherical surface into equilateral triangles. The two regular polyhedra that can be inscribed in a sphere are the dodecahedron (12 faces, each of which is a regular polygon; illus. *a*) and the more utilized icosahedron (20 faces, each of which is an equilateral triangle; illus. *b*). *See* POLYHEDRIN.

The geodesic dome has been used for everything from great exhibition spaces and halls to outdoor tent supports and jungle gyms. By utilizing the icosahedron as the basic building block of the geodesic dome, larger domes are possible with additional triangular subdivisions. This subdivision is known as the frequency. The first frequency is to interconnect the projected midpoints of the struts of each equilateral triangle of the icosahedron as they will project on the spherical surface. The result is four almost equilateral triangles where there was one before. The resulting lattice has similar but not exactly equilateral triangles if the grid is to remain on the spherical surface. This subdivision process can continue. The resulting grids have both triangular and hexagonal grids as a by-product within the basic geodesic dome geometry, with pentagons around the apex of the basic underlying icosahedron framework (illus. *c*). [I.P.L.]

Geodesy The science of surveying and mapping the Earth's surface. Geodesy supplies locations, distances, directions, elevations, and gravity information for the needs of civil engineering, boundary demarcations, navigation, and prospecting, as well as for other geosciences.

The many-faceted subject matter of geodesy includes the size and shape of the Earth, its external gravity field, horizontal and vertical control networks, and time-dependent phenomena. Included also are the utilization and adaptation of advanced technology (such as satellite geodesy and high-speed electronic computers) and the refinement of theories. Other sciences depend heavily upon geodesy, so that the scope of geodetic concerns embraces certain aspects of astronomy (astronomical positioning, orbit theory), geography (cartography, map projections), geophysics (gravity), geodynamics (crustal movements, polar motion), and oceanography (marine geodesy) and it even extends to the Moon (selenodesy) and planets. *See* CARTOGRAPHY; TERRESTRIAL COORDINATE SYSTEMS; TOPOGRAPHIC SURVEYING AND MAPPING. [I.K.F.]

Geoelectricity Electromagnetic phenomena and electric currents, mostly of natural origin, that are associated with the Earth. Geophysical methods utilize natural and artificial electric currents to explore the properties of the Earth's interior and to search for natural resources (for example, petroleum, water, and minerals). Geoelectricity is sometimes known as terrestrial electricity. All electric currents (natural or artificial, local or worldwide) in the solid Earth are characterized as Earth currents. The term telluric currents is reserved for the natural, worldwide electric currents whose origins are almost entirely outside the atmosphere. Geoelectromagnetism is a more comprehensive term than geoelectricity. Time variations of any magnetic field are associated with an electric field that induces electric currents in conducting media such as the Earth.

Magnetic fields, electric fields, and electric currents are the constituents of electromagnetism, and are related by Maxwell's equations. For instance, the illustration shows the time variations of the natural magnetic and electric fields simultaneously measured at one location at the surface of the Earth. These two traces are related to each other, not only by Maxwell's equations but also by the physical properties of the subsurface rocks in the vicinity of the measuring site. Either one of the two traces may be computed synthetically from the other if the properties of the subsurface rocks are known. Conversely, the two traces together can yield geologic information; this is a form of geophysical exploration or prospecting. Thus, the terms geoelectricity, geomagnetism, and geoelectromagnetism are essentially interchangeable, although each one may have a somewhat different emphasis. For example, the term geomagnetism is sometimes used for the study of the Earth's quasistationary main magnetic field. *See* GEOMAGNETISM; GEOPHYSICAL EXPLORATION. [S.H.Y.]

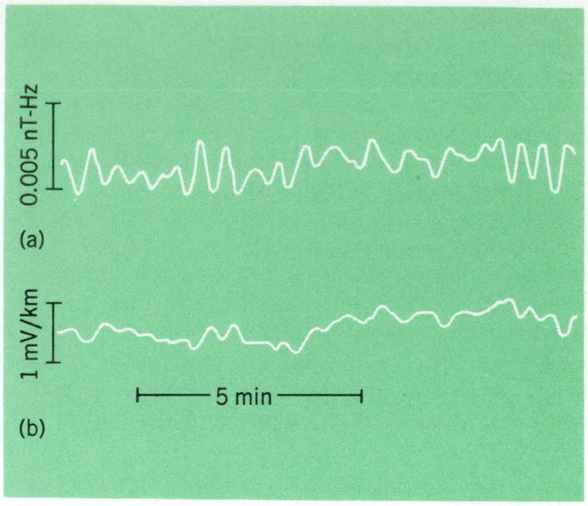

Time variations of the horizontal orthogonal components of the natural (*a*) magnetic and (*b*) electric fields, simultaneously measured at one site at the surface.

Geographic information systems Computer-based technologies for the storage, manipulation, and analysis of geographically referenced information. Attribute and spatial information is integrated in geographic information systems (GIS) through the notion of a data layer, which is realized in two basic data models: raster and vector. The major categories of applications comprise urban and environmental inventory and management, policy decision support, and planning; engineering and defense applications; and scientific analysis and modeling.

A geographic information system differs from other computerized information systems in two major respects. First, the information in this type of system is geographically referenced (geocoded). Second, a geographic information system has considerable capabilities for data analysis and scientific modeling, in addition to the usual data input, storage, retrieval, and output functions.

A geographic information system is composed of software, hardware, and data. The notion of data layer (or coverage) and overlay operation lies at the heart of most software designed for geographic information systems.

Two fundamental data models, the vector and raster models, embody the overlay idea in geographic information systems. In a vector geographic information system, the geometrical configuration of a coverage is stored in the form of points, arcs (line segments), and polygons, which constitute identifiable objects in the database. In a raster geographic information system, a layer is composed of an array of elementary cells of pixels, each holding an attribute value without explicit reference to the geographic feature of which the pixel is a part. *See* COMPUTER GRAPHICS; ELECTRONIC DISPLAY; IMAGE PROCESSING.

A data layer or coverage integrates two kinds of information: attribute and spatial (geographic). The functionality of a geographic information system consists of the ways in which that information may be captured, stored, manipulated, analyzed, and presented to the user. Spatial data capture (input) may be from primary sources such as remote sensing scanners, radars, or global positioning systems, or from scanning or digitizing images and maps derived from remote sensing. Output (whether as a display on a cathode-ray tube or as hard copy) is usually in map or graph form, accompanied by tables and reports linking spatial and attribute data. The critical data management and analysis functions fall into four categories: retrieval, classification, and measurement; overlay functions; neighborhood operations; and connectivity functions. *See* DATABASE MANAGEMENT SYSTEMS; DIGITAL COMPUTER; REMOTE SENSING. [H.Co.]

Geography The study of the Earth's landscapes and their change with time. Landscapes may include ever-changing natural physical or biological systems such as climate, landforms, or vegetation (physical geography), or they may include ever-changing human designs such as cities, road systems, or barn types (human geography). Geography is, therefore, a very broad science that deals with subject matter of a spatial nature that changes through time. As such, its boundaries often blur and overlap with other sciences. *See* PHYSICAL GEOGRAPHY.

Because the boundaries of geography blur and overlap with other sciences, geography is viewed by many as an interdisciplinary science. Thus geographers often find close allies in cognate sciences such as geology, economics, political science, and meteorology, to mention but a few. *See* BIOGEOGRAPHY; CLIMATOLOGY; GEOMORPHOLOGY; HYDROLOGY; MATHEMATICAL GEOGRAPHY; PEDOLOGY.

The word geography immediately seems to bring to mind maps and mapmaking. However, maps are simply useful tools of geographers, just as they are useful tools of, say, geologists and politicians. Other useful tools of geographers are air photos and other remotely sensed imagery, and statistics. In fact, one of the more recent important trends in geography is the increasing use of statistics and remotely sensed imagery (radar, photography, impulses) from high-flying aircraft and spacecraft. *See* AERIAL PHOTOGRAPH; CARTOGRAPHY; REMOTE SENSING. [D.L.J.]

Geologic thermometry The measurement or estimation of temperatures at which geologic processes take place. Methods used can be divided into two groups, nonisotopic and isotopic. Nonisotopic methods involve measurements of earth temperatures either directly by surface and near-surface features or indirectly from various properties of minerals and fossils. The isotopic methods involve the determination of distribution of isotopes of the lighter elements between pairs of compounds in equilibrium at various temperatures, and application of these data to problems of the temperature at which these compounds (commonly minerals) form in nature. [E.I.]

Geological time scale The assumed absolute time intervals occupied by the various geological periods. The evolution of plants and animals and the sequence of depositional and orogenic events provide the data of Earth history. In order to relate these events to absolute time and therefore establish a geological time scale, an independent means of dating rocks is required. Before the discovery of radioactivity, crude estimates of the length of geologic history were made based on the total thicknesses of sedimentary rock and assumed rates of erosion and sedimentation. These estimates varied by as much as a factor of 10.

The modern geologic time scale is based on many measurements of various rock types by the three quantitative isotopic chronometers (uranium-lead, U-Pb; rubidium-strontium, Rb-Sr; and potassium-argon, K-Ar). Unfortunately it is not yet possible to measure directly the absolute age of a sedimentary rock. A modern time scale is shown in the table. *See* ROCK AGE DETERMINATION. [J.L.K.]

Geological time scale

ERA / Period / Epoch	Age of beginning of period, 10^6 years	Period length, 10^6 years
CENOZOIC		
Quaternary		
Holocene	0.011	
Pleistocene	0.6	
Tertiary		65
Pliocene	12	
Miocene	20	
Oligocene	35	
Eocene	55	
Paleocene	65	
MESOZOIC		
Cretaceous		75
Upper (Base Santonian)	90	
Middle (Base Albian)	120	
Lower	140	
Jurassic		45
Upper (Base Callovian)	155	
Middle (Base Bajocian)	170	
Lower	185	
Triassic		45
Upper	200	
Middle	215	
Lower	230	
PALEOZOIC		
Permian		45
Upper (Base Ochoan)	245	
Middle (Base Guadalupian)	260	
Lower	275	
Pennsylvanian		35
Upper		
Middle		
Lower	310	
Mississippian		40
Upper		
Middle		
Lower	350	
Devonian		50
Upper	365	
Middle	385	
Lower	400	
Silurian		20
Upper		
Middle		
Lower	420	
Ordovician		70
Upper	440	
Middle	460	
Lower	490	
Cambrian		50+
Upper (Base Croixian)	510	
Middle	540	
Lower		
— — — ? — ? — — —		
PRECAMBRIAN		
Sinian sedimentation, U.S.S.R.		600–1200
Grenville orogeny		900–1200
Central and western United States basement		1300–1400
Rapakivi, Finland, and U.S.S.R.		1600
Sveco-Fennidic orogeny		1800–1900
Central Canadian Shield, Western Australia, Ukraine		2500–2800
Oldest reported basement, Kola Peninsula, U.S.S.R.		~3400
Probable age of the Earth		~4600

Geology The study or science of the Earth. Geologists are concerned primarily with rocks that make up the outer part of the Earth. Rocks are chemical systems adjusting to the physical environment in attempts to maintain thermodynamic equilibrium. Many rocks represent the consequences of processes occurring deep within the Earth, and modern geology is firmly rooted in the geophysics and geochemistry of the whole Earth. The study of geology involves the complex interrelationships between the solid Earth and its fluid envelopes of water and air, which are modified by the action of the living biosphere and modulated by energy received from and lost to space. Geology itself is an interdisciplinary subject, and it overlaps other disciplines. Study and mapping of surface forms and rocks are shared by geology and geodesy. The study of the Earth's waters as related to geologic processes is shared by hydrology, oceanography, and meteorology. Paleontology, the study of records left in rocks by animals and plants that lived in past ages, is an essential part of geology, though it involves many fundamental aspects of biologic science. *See* GEODESY; HYDROLOGY; METEOROLOGY; OCEANOGRAPHY.

The geological sciences have classically been subdivided into physical geology and historical geology, each with many branches. Physical geology is concerned with understanding the composition of the Earth and the physical changes occurring on it, which are derived from a study of the Earth as a whole, and of minerals and rocks, their structures and formations, and their processes of origin and alteration. Historical geology is based on two complementary disciplines, stratigraphy and paleontology. Paleontology is the study of fossilized plants and animals, with regard to their distribution in time. The stratigraphic record shows a succession of events through eons of time, with absolute dates supplied by geochemical studies of radiometric isotopes. Geophysical studies of the magnetic properties of rocks have become increasingly important in dating stratigraphic units. *See* PALEONTOLOGY; ROCK MAGNETISM; STRATIGRAPHY.

Mineralogy concerns the study of minerals, the basic units in the composition of most rocks. Although many hundreds of mineral species are known, comparatively few are important in kinds of rocks that are abundant in the visible part of the Earth crust. *See* MINERAL; MINERALOGY.

Petrology involves the study of rocks, their physical and chemical properties, and their modes of origin. Known rocks are divisible into three groups: igneous rocks, which have solidified from molten matter (magma); sedimentary rocks, made of fragments derived from preexisting rocks, of chemical precipitates, or of organic products; and metamorphic rocks derived from igneous or sedimentary rocks under conditions that brought about changes in mineral composition, texture, and internal structure. *See* IGNEOUS ROCKS; METAMORPHIC ROCKS; PETROGRAPHY; PETROLOGY; ROCK; SEDIMENTARY ROCKS.

An understanding of weathering, erosion, and sedimention is all important in geology. Weathering prepares the way for removal of rock materials and reshaping of land surfaces by several agents of erosion. The most obvious of these agents is running water. Terrestrial sediments, widespread and highly varied, are laid down chiefly through the agencies of mass wasting, running water, glacier ice, and wind. The deposits are in large part temporary, as the tendency is for them to shift seaward in continued erosion of the lands. Marine sediments, deposited beneath oceans, comprise material derived from the land, from shells and skeletons of marine animals and plants, from the ocean by chemical precipitation, and from space (particles of meteorites). *See* EROSION; MARINE SEDIMENTS; MASS WASTING; SEDIMENTOLOGY; WEATHERING PROCESSES.

A geologic map represents the lithology and, so far as possible, the geologic age of every important geologic unit in a given area. Each distinctive unit that can be shown effectively to the scale of the map is a geologic formation. A good topographic base map is essential for representing relations of bedrock to land surface forms. *See* AERIAL PHOTOGRAPH; ECONOMIC GEOLOGY; REMOTE SENSING; STRUCTURAL GEOLOGY; TOPOGRAPHIC SURVEYING AND MAPPING. [C.R.L.; P.J.Wy.]

Geomagnetism A term signifying both the magnetism of the Earth and the branch of science that deals with the Earth's magnetism. The term geomagnetism is now given some preference over the older term terrestrial magnetism.

The main geomagnetic field is specified at any point O by its vector magnetic intensity F. Its direction is that of the line $P'OP$ from the negative end P' to the positive end P of a magnetized needle $P'P$, perfectly balanced before it is magnetized, and freely pivoted about O, when in equilibrium. The positive pole P is the one that at most places on the Earth takes the more northerly position. Over most of the Northern Hemisphere, P is below O; the needle is said to dip below the horizontal by an angle I, called the magnetic inclination. Over about half the Earth, however, P will be above O; the inclination I (or magnetic dip) is then reckoned as negative. This is the case over most of the Southern Hemisphere. The value of I thus ranges from 90 to $-90°$. A point where $I = \pm 90°$ is called a magnetic pole of the Earth.

There are two main magnetic poles; their approximate positions are 76.1°N, 100°W and 65.8°S, 139°E. In a few places, near strongly magnetized mineral deposits, there may be local magnetic poles.

The distribution of I over the Earth's surface can be indicated on a globe, or on a map (with any kind of projection), by lines called isoclinic lines or isoclines, along each of which I has the same value. The isocline for which $I = 0$ (where the balanced magnetized needle rests horizontal) is called the magnetic or dip equator.

A compass needle is magnetized and pivoted at a slightly noncentral point so as to rest and move in the horizontal plane. The deviation of its direction from geographic (gg) north is called the magnetic declination D. This is reckoned positive to the east and negative to the west; alternatively, declinations can be specified by a positive (eastward) value ranging up to 360°.

Over the greater part of the Earth D is numerically less than 90°, but along any small circuit around the magnetic poles it takes all values. At the magnetic poles themselves the compass needle takes no definite direction. At the gg poles the needle takes a definite direction, but as the northward direction changes through a whole revolution along any small circuit around the pole, D likewise takes all values around the circuit.

The distribution of the declination over the Earth's surface can be indicated by lines along each of which D is constant. These are called isogonic lines, or isogones. The isogones for which $D = 0$ are called agonic lines. Along these lines, the compass points to true north.

The distribution of D over the Earth can also be indicated by magnetic meridians, which at each point P have the direction of the compass needle at P. These lines extend from the south magnetic pole to the north and have no complication (as do the isogones) at the gg poles. They indicate, more clearly than the isogones, the general distribution of the compass direction over the Earth. But the isogonic map is more convenient in enabling the compass direction at any point to be read or estimated without using an angle measurer. Hence it is the one used by navigators and travelers on sea and land and in the air. *See* MAGNETIC COMPASS. [S.C.]

Geometric phase A unifying mathematical concept that describes the relation between the history of internal states of a system and the system's resulting orientation in space. Under various aspects, this concept occurs in geometry, astronomy, classical mechanics, and quantum theory. In geometry it is known as holonomy. In quantum theory it is known as Berry's phase, after M. Berry, who isolated the concept (which was already known in special cases) and explained its wide-ranging significance.

A system is envisioned whose possible states can be visualized as points in a suitable abstract space. At the same time, the system has some position or orientation in another space. A history of internal states can be represented by a curve in the first space; and the effect of this history on the disposition of the system, by a curve in the second space. The mapping between these two curves is described by the geometric phase. Especially interesting is the case when a closed curve (cycle) in the first space maps onto an open curve in the second, for then there is no net change in internal state, yet the disposition of the system with respect to the outside world is altered.

The power of the geometric phase ideas is that they make it possible, in complex dynamical problems, to find some simple universal regularities without having to solve the complete equations. Significant uses of these ideas include demonstrations of the fractional electric charge and quantum statistics of the quasiparticles in the quantum Hall effect, and of the occurrence of anomalies in quantum field theory. *See* ANYONS; HALL EFFECT; QUANTUM FIELD THEORY. [F.Wil.]

Geometrical optics The geometry of light rays and their imagery through optical systems. The phenomena of diffraction due to the finite apertures of the lens systems are neglected in geometrical optics. *See* DIFFRACTION.

Light moves in straight lines through homogeneous media and changes its direction at the surface separating two such media, for instance, air and glass. An incident ray at the bounding surface is divided into two; one is reflected back into the first medium and the other penetrates the second medium after being bent or refracted. The incident, reflected, and refracted rays all lie in one plane containing the surface normal and form angles i, i_r, and i' with the surface normal such that Eqs. (1) and (2) hold where n and n' are the refractive indices

$$i_r = \pi - i \quad (1)$$

$$n \sin i = n' \sin i' \quad (2)$$

of the media separated by the refracting surface. In order to obtain a solution of Eq. (2), i and i' must be chosen so that they are in the same quadrant. *See* LIGHT; REFLECTION OF ELECTROMAGNETIC RADIATION; REFRACTION OF WAVES.

These formulas together with pure geometry make it possible to trace a ray through a system of lenses. The specific form of the ray-tracing formula should be adapted to the tools of the lens designer; that is, it will be different for a person using logarithm tables and for one using an electric desk machine or an electronic computer.

The aim of the optical designer is to see what happens to all the rays coming from every point of the object to be imaged. Moreover, the designer (1) wants to direct the rays so that all of them coming from a fixed object point are collected at a fixed image point (freedom from aberration); (2) wants all these image points to lie on a plane (freedom from field curvature); and (3) wants the image to be similar in shape to the object (freedom from distortion). Finally, correction should be achieved for light of different wavelengths (freedom from chromatic aberrations). *See* ABERRATION (OPTICS); CHROMATIC ABERRATION. [M.J.H.]

Geometry A branch of mathematics concerned with the properties of space, including points, lines, curves, planes and surfaces in space, and figures bounded by them. For discussion of various branches of geometry *see* ALGEBRAIC GEOMETRY; DIFFERENTIAL GEOMETRY; EUCLIDEAN GEOMETRY; PROJECTIVE GEOMETRY; RIEMANNIAN GEOMETRY. [J.S.F.]

Geomorphology The study of landforms, including the description, classification, origin, development, and history of planetary surface features. Emphasis is placed on the genetic interpretation of the erosional and depositional features of the Earth's surface. However, geomorphologists also study primary relief elements formed by movements of the Earth's crust, topography on the sea floor and on other planets, and applications of geomorphic information to problems in environmental engineering.

Geomorphologists analyze the landscape. Their purview includes the structural framework of landscape, weathering and soils, mass movement and hillslopes, fluvial features, eolian features, glacial and periglacial phenomena, coastlines, and karst landscapes. Processes and landforms are analyzed for their adjustment through time, especially the most recent portions of Earth history. Geomorphologists consider processes from the perspectives of pedology, soil mechanics, sedimentology, geochemistry, hydrology, fluid mechanics, remote sensing, and other sciences. The complexity of geomorphic processes has required this interdisciplinary approach, but it has also led to a theoretical vacuum in the science. At present many geomorphologists are organizing their studies through a form of systems analysis. The landscape is conceived of as a series of elements linked by flows of mass and energy. Process studies measure the inputs, outputs, and transfers for these systems. Systems analysis provides an organizational framework within which geomorphologists are developing models to predict selected phenomena. [V.R.B.]

Geophagia Soil ingestion by animals. Grazing animals such as sheep and cattle ingest varying amounts of soil when they graze herbage contaminated with it. Pastures become contaminated with soil when livestock walk across the herbage, particularly in wet conditions.

The amount of soil ingested by sheep and cattle is influenced by soil type, stock density, earthworm activity, management practices, and various seasonal factors. Soils that are well drained and have a strong structure do not break up so readily and contaminate the herbage as is the case for poorly drained, weak-structured soils. When the density of stock grazing in a given area of herbage is increased, the amount of treading is increased, while the herbage is grazed more closely. The overall effect is that more soil is transferred to the herbage and ingested.

Geophagia is subject to seasonal variations. The wetter and cooler conditions of autumn and winter result in muddier herbage and an increase in soil ingestion by grazing animals. During the spring and summer, the greater growth of the herbage and drier conditions result in cleaner herbage, and there is a marked decrease in intake of soil.

Soil can be a source of mineral nutrients. Since soils are higher than herbage in iron, manganese, zinc, copper, cobalt, selenium, and iodine, they may contribute to the mineral nutrition of the grazing animals. *See* AGRICULTURE; SOIL CHEMISTRY. [N.D.G.]

Geophysical exploration Making, processing, and interpreting measurements of the physical properties of the Earth with the objective of practical application of the findings. Most exploration geophysics is conducted to find commercial accumulations of oil, gas, or other minerals, but geophysical investigations are also employed with engineering objectives, in studies aimed at predicting the nature of the Earth for the foundations of roads, buildings, dams, tunnels, nuclear power plants, and other structures, and in the search for geothermal areas, water resources, archeological ruins, and so on.

Geophysical exploration is also called applied geophysics or geophysical prospecting. The physical properties and effects of subsurface rocks and minerals that can be measured at a distance include density, electrical conductivity, thermal conductivity, magnetism, radioactivity, elasticity, and other properties. Exploration geophysics is often divided into subsidiary fields according to the property being measured, such as magnetic, gravity, seismic, electrical, thermal, or radioactive properties.

Magnetic exploration. Rocks and ores containing magnetic minerals become magnetized by induction in the Earth's magnetic field so that their induced field adds to the Earth's field. Magnetic exploration involves mapping variations in the magnetic field to determine the location, size, and shape of such bodies. The magnetic susceptibility of sedimentary rock is generally orders of magnitude less than that of igneous or metamorphic rock. Consequently, the major magnetic anomalies observed in surveys of sedimentary basins usually result from the underlying basement rocks. Determining the depths of the tops of magnetic bodies is thus a way of estimating the thickness of the sediments. *See* GEOMAGNETISM; MAGNETOMETER; ROCK MAGNETISM.

Except for magnetite and a very few other minerals, mineral ores are only slightly magnetic. However, they are often associated with bodies such as dikes that have magnetic expression so that magnetic anomalies may be associated with minerals empirically. For example, placer gold is often concentrated in stream channels where magnetite is also concentrated.

Gravity exploration. Gravity exploration is based on the law of universal gravitation: the gravitational force between two bodies varies in direct proportion to the product of their masses and in inverse proportion to the square of the distance between them. Because the Earth's density varies from one location to another, the force of gravity varies from place to place. Gravity exploration is concerned with measuring these variations to deduce something about rock masses in the immediate vicinity. Gravity surveys are used more extensively for petroleum exploration than for metallic mineral prospecting. The size of ore bodies is generally small; therefore, the gravity effects are quite small and local despite the fact that there may be large density differences between the ore and its surroundings. *See* GRAVITY METER; PROSPECTING.

Seismic exploration. Seismic exploration is the predominant geophysical activity. Seismic waves are generated by one of several types of energy sources and detected by arrays of sensitive devices called geophones or hydrophones. The most common measurement made is of the travel times of seismic waves, although attention is being directed increasingly to the amplitude of seismic waves or changes in their frequency content or wave shape. *See* SEISMOLOGY.

Electrical and electromagnetic exploration. Variations in the conductivity or capacitance of rocks form the basis of a variety of electrical and electromagnetic exploration methods, which are used primarily in metallic mineral prospecting. Both natural and induced electrical currents are measured. Direct currents and low-frequency alternating currents are measured in ground surveys, and ground and airborne electromagnetic surveys involving the lower radio frequencies are made. *See* GEOELECTRICITY.

Radioactivity exploration. Natural radiation from the Earth, especially of gamma rays, is measured both in land surveys and airborne surveys. Natural types of radiation are usually absorbed by a few feet of soil cover, so that the observation is often of diffuse equilibrium radiation. The principal radioactive elements are uranium, thorium, and potassium; radioactive exploration has been used primarily in the search for uranium and other ores, such as columbium, which are often associated with them. The Geiger counter and scintillation counter are instruments generally used to detect and measure the radiation. *See* GEIGER-MÜLLER COUNTER; SCINTILLATION COUNTER.

Remote sensing. Measurements of natural and induced electromagnetic radiation made from high-flying aircraft and earth satellites are referred to collectively as remote sensing. This comprises both the observation of natural radiation in various spectral bands, including both visible and infrared radiation, such as by photography and measurements of the reflectivity of infrared and radar radiation. *See* REMOTE SENSING.

Well logging. A variety of types of geophysical measurements are made in boreholes, including self-potential, electrical conductivity, velocity of seismic waves, natural and induced radioactivity, and temperature variations. Borehole logging is used extensively in petroleum exploration to determine the characteristics of the rocks which the borehole has penetrated, and to a lesser extent in mineral exploration. *See* BOREHOLE LOGGING; WELL LOGGING. [R.E.Sh.]

Geophysics Those branches of earth sciences in which the principles and practices of physics are used to study the Earth. Geophysics is considered by some to be a branch of geology, by others to be of equal rank. It is distinguished from the other earth sciences largely by its use of instruments to make direct or indirect measurements of the parts of the Earth being studied, in contrast to the more direct observations which are typical of geology. *See* GEOLOGY.

Geophysics consists of several principal fields plus parallel and subsidiary divisions. These are commonly considered to include plutology, with geodesy, geothermometry, seismology, and tectonophysics as subdivisions; hydrospheric studies, hydrology (groundwater studies), oceanography, and glaciology; atmospheric studies, meteorology, and aeronomy; and several fields of geophysics which overlap one another, including geomagnetism and geoelectricity, geochronology, geocosmogony, and geophysical exploration and prospecting. Planetary sciences, the study of the planets and satellites aside from the Earth, are usually considered a branch of geophysics because the techniques used have been, until the first landings on each, entirely instrumental rather than directly observational. *See* GEOCHRONOMETRY; GEODESY; GEOELECTRICITY; GEOLOGIC THERMOMETRY; GEOMAGNETISM; GLACIOLOGY; HYDROLOGY; METEOROLOGY; OCEANOGRAPHY; SEISMOLOGY; TECTONOPHYSICS. [B.F.H.]

Geostrophic wind A hypothetical wind based upon the assumption that a perfect balance exists between the horizontal components of the Coriolis force and the horizontal pressure gradient force per unit mass, with the implication that viscous forces and accelerations are negligible. The geostrophic wind blows parallel to the isobars (lines of equal pressure) with lower pressure to the left of the direction of the wind in the Northern Hemisphere and to the right in the Southern Hemisphere. It represents a good approximation to the actual wind at elevations greater than about 3000 ft (900 m), except in instances of strongly curved flow and in the vicinity of the Equator.

The term thermal wind denotes the net change in the geostrophic wind over some specific vertical distance. This change arises because the rate of change of pressure in the vertical is different in two air columns of different air density, so that the horizontal component of the pressure gradient force per unit mass varies in the vertical. The thermal wind is directed approximately parallel to the isotherms of air temperature with cold air to the left and warm air to the right in the Northern Hemisphere, and vice versa in the Southern Hemisphere. Thus, for example, the increasing predominance of westerly winds aloft may be viewed as a consequence of the warmth of tropical latitudes and the coldness of polar regions. *See* CORIOLIS ACCELERATION; GRADIENT WIND; WIND; WIND STRESS. [F.S.]

Geosyncline A linear part of the crust of the Earth that sagged deeply through time, that is, a great trough hundreds of kilometers long and tens of kilometers wide that subsided as it received thousands of meters of sedimentary and volcanic

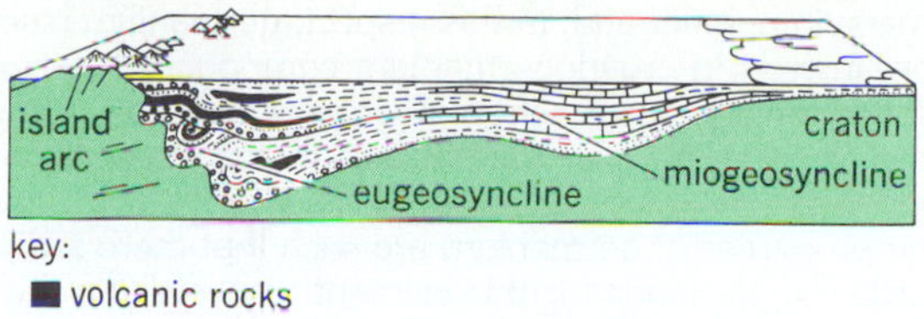

Diagrammatic section of Cordilleran geosyncline in southeastern Alaska and British Columbia at the close of Permian time (about 230,000,000 years ago). (*After A. J. Eardley, J. Geol., 55:319–342, 1947*)

rocks through millions of years. Thicknesses are roughly 10 times greater than synchronous strata in adjacent stable regions. Linear geosynclinal belts were great subsiding tectonic divisions of the crust that lay between more stable parts of continents (cratons) or the deep abyssal ocean basins of their time, or both. Geosynclines generally contain volcanic (eugeosynclinal) and nonvolcanic (miogeosynclinal) zones (see illustration). Although defined on the basis of thickness and types of rocks, the geosyncline has always been so closely linked with the origin of mountains that it has little meaning in any other context. *See* Cordilleran belt; Orogeny; Syncline. [R.H.Dot.]

Geotextile Any permeable textile used with foundations, soils, rock, earth, or other geotechnical material as an integral part of a manufactured project, structure, or system. These textile products are made of synthetic fibers or yarns, constructed into woven or nonwoven fabrics that weigh from 3 to 30 oz/yd^2 (100 to 1000 g/m^2). Geotextiles are more commonly known by other names, for example, filter fabrics, civil engineering fabrics, support membranes, and erosion control cloth. *See* Manufactured fiber.

Geotextiles perform three basic functions in earth structures: separation (the fabric provides a boundary that segregates materials); reinforcement (the fabric imparts tensile strength to the system, thereby increasing its structural stability); and filtration (the fabric retains soil particles while allowing water to pass through). Geotextiles can thus be adapted to numerous applications in earthwork construction. The major end-use categories are: stabilization (for roads, parking lots, embankments, and other structures built over soft ground); drainage (of subgrades, foundations, embankments, dams, or any earth structure requiring seepage control); erosion control (for shoreline, riverbanks, steep embankments, or other earth slopes to protect against the erosive force of moving water); and sedimentation control (for containment of sediment runoff from unvegetated earth slopes). *See* Dam; Engineering geology; Foundations; Pavement. [R.G.C.]

Geothermal power Thermal or electrical power produced from the thermal energy contained in the Earth (geothermal energy). Use of geothermal energy is based thermodynamically on the temperature difference between a mass of subsurface rock and water and a mass of water or air at the Earth's surface. This temperature difference allows production of thermal energy that can be either used directly or converted to mechanical or electrical energy.

There are several types of natural geothermal reservoirs. All the reservoirs developed to date for electrical energy are termed hydrothermal convection systems and are characterized by circulation of meteoric (surface) water to depth. The driving force of the convection systems is gravity, effective because of the density difference between cold, downward-moving, recharge water and heated, upward-moving, thermal water. Depending on the physical state of the pore fluid, there are two kinds of hydrothermal convection systems: liquid-dominated, in which all the pores and fractures are filled with liquid water that exists at temperatures well above boiling at atmospheric pressure, owing to the pressure of overlying water; and vapor-dominated, in which the larger pores and fractures are filled with steam. Liquid-dominated reservoirs produce either water or a mixture of water and steam, whereas vapor-dominated reservoirs produce only steam, in most cases superheated.

Natural geothermal reservoirs also occur as regional aquifers. In some rapidly subsiding young sedimentary basins, porous reservoir sandstones are compartmentalized by growth faults into individual reservoirs that can have fluid pressures exceeding that of a column of water and approaching that of the overlying rock. The pore water is prevented from escaping by the impermeable shale that surrounds the compartmented sandstone. The energy in these geopressured reservoirs consists not only of thermal energy, but also of an equal amount of energy from methane dissolved in the waters plus a small amount of mechanical energy due to the high fluid pressures. *See* Aquifer; Groundwater hydrology.

The most conspicuous use of geothermal energy is the generation of electricity. Hot water from a liquid-dominated reservoir is flashed partly to steam at the Earth's surface, and this steam is used to drive a conventional turbine-generator set. Equally important worldwide is the direct use of geothermal energy, often at reservoir temperatures less than 212°F (100°C). Geothermal energy is used directly in a number of ways: to heat buildings (individual houses, apartment complexes, and even whole communities); to cool buildings (using lithium bromide absorption units); to heat greenhouses and soil; and to provide hot or warm water for domestic use, for product processing (for example, the production of paper), for the culture of shellfish and fish, for swimming pools, and for therapeutic (healing) purposes. [L.J.P.M.]

Geraniales An order of flowering plants, division Magnoliophyta (Angiospermae), in the subclass Rosidae of the class Magnoliopsida (dicotyledons). The order consists of 5 families and about 200 species; several other families have often been included in other systems of classification. The Geraniales are herbs or soft shrubs with a superior ovary and generally either with compound or deeply cleft leaves (see illustration) or with strongly irregular flowers in which one of the sepals has a conspicuous free spur. Wood sorrel (*Oxalis*, family Oxalidaceae), geranium (*Geranium* and *Pelargonium*, family Geraniaceae), garden nasturtium (*Tropaeolum*, family Tropaeolaceae), and

A common eastern United States geranium (*Geranium maculatum*), which is characteristic of order Geraniales. (*Courtesy of A. W. Ambler, from National Audubon Society*)

touch-me-not (*Impatiens*, family Balsaminaceae) are familiar members of the Geraniales. *See* MAGNOLIOPSIDA; ROSIDAE. [A.Cr.]

Gerbil The common name for 106 species of rodents, known as the sand rats, composing the subfamily Gerbillinae in the family Muridae. They inhabit the more arid regions of Africa and Asia. All species have light, soft fur and hindlegs that are longer than the front ones, enabling them to hop kangaroo fashion, although most frequently they run like typical rodents. These animals are active nocturnally and remain in their burrows during the heat of the day. In captivity they are gregarious and make good pets. *See* RODENTIA. [C.B.C.]

Germ layers The primitive cell layers or first tissues, which appear early in the development of animals, and from which the embryo body and its auxiliary membranes, when present, are constructed. The germ layers are more or less distinct anatomically, but do not necessarily have sharp boundaries of demarcation. Germ layers are almost universal among animal embryos.

Three kinds of germ layers are recognizable: (1) the ectoderm or outer skin, (2) the endoderm or inner skin, and (3) the mesoderm or middle skin. The layers have thus been named in accordance with their definitive positions in the spherical type of gastrula such as that of the sea urchin or amphibian. The terms epiblast, mesoblast, and hypoblast are sometimes used as synonyms for ectoderm, mesoderm, and endoderm, respectively. The three primary germ layers are present as a basic structural plan in all Metazoa except the coelenterates and the Porifera, in which a distinct mesodermal layer is absent. *See* GASTRULATION. [N.T.S.]

Germanium A brittle, silvery-gray, metallic chemical element, Ge, atomic number 32, atomic weight 72.59, melting point 937.4°C (1719°F), and boiling point 2830°C (5130°F), with properties between silicon and tin. Germanium is distributed widely in the Earth's crust in an abundance of 6.7 parts per million (ppm). Germanium is found as the sulfide or is associated with sulfide ores of other elements, particularly those of copper, zinc, lead, tin, and antimony.

Germanium has a metallic appearance but exhibits the physical and chemical properties of a metal only under special conditions since it is located in the periodic table where the transition from nonmetal to metal occurs. At room temperature there is little indication of plastic flow and consequently it behaves like a brittle material.

As it exists in compounds, germanium is either divalent or tetravalent. The divalent compounds (oxide, sulfide, and all four halides) are easily reduced or oxidized. The tetravalent compounds are more stable. Organogermanium compounds are many in number and, in this respect, germanium resembles silicon. Interest in organogermanium compounds has centered around their biological action. Germanium in its derivatives appears to have a lower mammalian toxicity than tin or lead compounds.

The properties of germanium are such that there are several important applications for this element, especially in the semiconductor industry. The first solid-state device, the transistor, was made of germanium. Single-crystal germanium is used as a substrate for vapor-phase growth of GaAs and GaAsP thin films in some light-emitting diodes. Germanium lenses and filters are used in instruments operating in the infrared region of the spectrum. Mercury-doped and copper-doped germanium are used as infrared detectors; synthetic garnets with magnetic properties may have applications for high-power microwave devices and magnetic bubble memories; and germanium additives increase usable ampere-hours in storage batteries. [P.S.G.]

1	2	3	4	5	6	7	8	9	10	11	12	13	14	15	16	17	18
1 H																	2 He
3 Li	4 Be											5 B	6 C	7 N	8 O	9 F	10 Ne
11 Na	12 Mg											13 Al	14 Si	15 P	16 S	17 Cl	18 Ar
19 K	20 Ca	21 Sc	22 Ti	23 V	24 Cr	25 Mn	26 Fe	27 Co	28 Ni	29 Cu	30 Zn	31 Ga	32 Ge	33 As	34 Se	35 Br	36 Kr
37 Rb	38 Sr	39 Y	40 Zr	41 Nb	42 Mo	43 Tc	44 Ru	45 Rh	46 Pd	47 Ag	48 Cd	49 In	50 Sn	51 Sb	52 Te	53 I	54 Xe
55 Cs	56 Ba	71 Lu	72 Hf	73 Ta	74 W	75 Re	76 Os	77 Ir	78 Pt	79 Au	80 Hg	81 Tl	82 Pb	83 Bi	84 Po	85 At	86 Rn
87 Fr	88 Ra	103 Lr	104 Rf	105 Db	106 Sg	107 Bh	108 Hs	109 Mt	110	*111*	*112*	*113*	*114*	*115*	*116*	*117*	*118*

lanthanide series	57 La	58 Ce	59 Pr	60 Nd	61 Pm	62 Sm	63 Eu	64 Gd	65 Tb	66 Dy	67 Ho	68 Er	69 Tm	70 Yb
actinide series	89 Ac	90 Th	91 Pa	92 U	93 Np	94 Pu	95 Am	96 Cm	97 Bk	98 Cf	99 Es	100 Fm	101 Md	102 No

GERT A procedure for the formulation and evaluation of systems using a network approach. Problem solving with the GERT (graphical evaluation and review technique) procedure utilizes the following steps:

1. Convert a qualitative description of a system or problem to a generalized network similar to the critical path method—PERT type of network.
2. Collect the data necessary to describe the functions ascribed to the branches of a network.
3. Combine the branch functions (the network components) into an equivalent function or functions which describe the network.
4. Convert the equivalent function or functions into performance measures for studying the system or solving the problem for which the network was created. These might include either the average or variance of the time or cost to complete the network.
5. Make inferences based on the performance measures developed in step 4.

Both analytic and simulation approaches have been used to perform step 4 of the procedure. GERTE was developed to analytically evaluate network models of linear systems through an adaptation of signal flow-graph theory. For nonlinear systems, involving complex logic and queuing situations, Q-GERT was developed. In Q-GERT, a simulation of the network is performed in order to obtain statistical estimates of the performance measures of interest.

GERT networks have been designed, developed, and used to analyze the following situations: claims processing in an insurance company, production lines, quality control in manufacturing systems, assessment of job performance aids, burglary resistance of buildings, capacity of air terminal cargo facilities, judicial court system operation, equipment allocation in construction planning, refueling of military airlift forces, planning and control of marketing research, planning for contract negotiations, risk analysis in pipeline construction, effects of funding and administrative strategies on nuclear fusion power plant development, research and development planning, and system reliability. *See* CRITICAL PATH METHOD (CPM); DECISION THEORY; PERT; SIMULATION. [A.A.B.P.]

Gestation period That period in mammals from fertilization to birth. In mammals other than humans the gestation period may vary, since it depends on such factors as maternal care, metabolic rates, environmental conditions, and types of placentation (see table). In humans the gestation period lasts 270–295 days; for clinical purposes the period is divided into three trimesters.

Gestation periods of common mammals

Animal	Gestation, days	Animal	Gestation, days
Armadillo	150	Guinea pig	68–71
Bear		Horse	330–380
Black	210	Human	270–295
Polar	240	Kangaroo	40–45*
Cat	60	Lion	106
Chimpanzee	250	Mole	30
Cow	282	Mouse	20–21
Dog	58–65	Opossum	13*
Donkey	365–380	Rabbit	30–43
Elephant		Whale	334–365
African	641	Wolf	63
Indian	607–641	Zebra	300–345
Giraffe	450		

*Marsupial, with extended stay in maternal pouch.

For humans, loss of the embryo or young fetus prior to the stages of viability is termed abortion; later loss is termed premature delivery until the fetus weighs more than 88 oz (2500 g), the lower limit of normal weight of the newborn. *See* PREGNANCY; PREGNANCY DISORDERS. [S.P.P.]

Geyser A natural spring or fountain which discharges a column of water or steam into the air at more or less regular intervals. Perhaps the best-known area of geysers is in Yellowstone Park, Wyoming, where there are more than 100 active geysers and more than 3000 noneruptive hot springs.

The eruptive action of geysers is believed to result from the existence of very hot rock not far below the surface. The neck of the geyser is usually an irregularly shaped tube partly filled with water which has seeped in from the surrounding rock. Far down the pipe the water is at a temperature much above the boiling point at the surface, because of the pressure of the column of water above it. Its temperature is constantly increasing, because of the volcanic heat source below. Eventually the superheated water changes into steam, lifting the column of water out of the hole. [A.N.S./R.K.Li.]

Ghost image (optics) An undesired image appearing at the image plane of an optical system. Each surface of an optical system divides the incoming light into two parts: (1) the reflected light, which returns into the first medium, and (2) the refracted light. The reflected light is again divided into two parts when it in turn strikes another dividing surface. The light thus reflected twice forms an image which may be near the plane of the primary image. This may be a false image of the object or an out-of-focus image of a bright source of light in the field of the optical system. Thus a large number of undesired or ghost images may appear. *See* OPTICAL IMAGE; REFLECTION OF ELECTROMAGNETIC RADIATION; REFRACTION OF WAVES.

If the ghost images are far out of focus, they only diminish the contrast in the primary image, a condition known as flare. But if the ghost images are near the focal plane, they are very disturbing. This effect is especially noticeable if there is a bright light source in the field of the instrument, since the ghost image of the light source may have an even greater brightness than the image of the desired object. The coating of lenses with layers of fluorite and other materials has nearly eliminated ghost images from modern optical systems. [M.J.H.]

Giant nuclear resonances Systematic excitations of the atomic nucleus which occur with great strength in a concentrated energy region. When high-energy electromagnetic radiation (gamma radiation) impinges on a nucleus, it can be strongly absorbed into a number of high-lying resonances: one in which the nucleus is excited in a dipole mode of oscillation that is electric in nature (E1), a second dipole mode that is magnetic in nature (M1), and a third excitation that can be identified as an electric quadrupole oscillation (E2). When excited into these resonances, the nucleus can deexcite by emitting gamma radiation or, if energetically allowed, by emitting particles, especially protons and neutrons. If proton and neutron emissions are allowed, it is possible to study these resonances by the inverse process in which the proton or the neutron is captured in the giant resonance and gamma radiation is emitted. These giant resonances can also be excited by inelastic scattering of electrons and other particles such as protons, deuterons, ^{3}He, and α-particles. For inelastic excitation with particles, greater momentum can be imparted to the nucleus, and higher modes such as octupole oscillations can be excited. The particles can also excite the electric monopole vibration (E0), which corresponds to a "breathing" mode whose energy determines the incompressibility of a nucleus, a property that is basic to understanding the force between neutrons and protons. *See* GAMMA RAYS; MULTIPOLE RADIATION; NUCLEAR MOMENTS; NUCLEAR REACTION; NUCLEAR SPECTRA; NUCLEAR STRUCTURE. [S.S.H.]

Giant star An intermediate state in the evolution of a star in which it swells to enormous proportions before its death. During the longest and most stable phase of a star's life, the star, like the Sun, derives its energy from the thermonuclear fusion of hydrogen into helium deep in its dense, hot core. When the hydrogen fuel is exhausted, the core contracts and heats under the action of gravity, fresh hydrogen is ignited in a shell that surrounds the spent core, and the star becomes much more luminous, larger, and cooler at its surface. The lower surface temperature produces a redder color, hence the common term red giant. Stars like the Sun brighten by a factor of 100 and grow in radius by a factor of nearly 50.

There are actually two separate giant states. The first, described above, is terminated when the core temperature climbs so high that the helium ignites and fuses into carbon. This event stabilizes the star, but when this helium is exhausted, the earlier behavior is repeated. The star then swells to enormous proportions, perhaps two astronomical units (1.8×10^8 mi or 3×10^8 km), becoming even redder than before. It may pulsate and loses much or most of its mass through a strong wind. *See* HERTZSPRUNG-RUSSELL DIAGRAM; STELLAR EVOLUTION. [J.B.Ka.]

Giardiasis A disease caused by the protozoan parasite *Giardia lamblia*, characterized by chronic diarrhea that usually lasts 1 or more weeks. The diarrhea may be accompanied by one or more of the following: abdominal cramps, bloating, flatulence, fatigue, or weight loss. The stools are malodorous and have a pale greasy appearance. Infection without symptoms is also common.

Giardiasis occurs worldwide. In community epidemics caused by contaminated drinking water, as many as 50 to 70% of the residents have become infected. Outbreaks also occur among backpackers and campers who drink untreated stream water. Both human and animal (beaver) fecal contamination of stream water has been implicated as the source of *Giardia* cysts in waterborne outbreaks. *Giardia* species in dogs and possibly other animals are also considered infectious for humans. Epidemics resulting from person-to-person transmission occur in day-care centers for preschool-age children and institutions for the mentally retarded. Infants and toddlers in day-care centers are more commonly infected than older children who have been toilet-trained. *See* EPIDEMIOLOGY. [D.D.J.]

Gibberellin Any member of a family of compounds found in fungi and plants, some being plant hormones that usually have a broad spectrum of growth-promoting activity.

The gibberellins are cyclic diterpenes possessing 19 or 20 carbon atoms. Fifty-seven gibberellins have been identified, and the trivial system of nomenclature identifies them as GA_1–GA_{57} inclusively.

The gibberellins are of widespread occurrence, being found in bacteria, fungi, algae, mosses, ferns, gymnosperms, and angiosperms. Although they seem to be ubiquitous, the gibberellins are generally found in very small amounts. Exceptions to this rule include the fungus *Gibberella fujikuroi* and immature seeds of some plants, which are rich sources of gibberellins.

The gibberellins play an important role in the integration of growth and development of plants. It is only in the flowering plants, however, that their function as hormonal agents has been conclusively established. This function is inferred for the lower vascular plants and algae and bryophytes because application of gibberellin to these plants influences aspects of their growth and differentiation. *See* PLANT HORMONES. [R.L.J.]

Gibbon A medium to very large animal which lacks a tail and is the least anthropoid of the order Primates and smallest of the apes. The seven species are included in the single genus *Hylobates* and have a general distribution throughout Southeast Asia.

Gibbons live in family groups which normally include an adult male and female with the young. Like many animals, each family has a territory which it defends. The gibbons live almost entirely in the trees. They are vegetarians and usually eat fruits and nuts but will on occasion eat bird eggs and small vertebrates. The young are mature at the age of 6 years. *See* PRIMATES. [C.B.C.]

Gibbs function The Gibbs function, G, also known as Gibbs free energy or free enthalpy, is defined in the equation shown, where E is the internal energy, p is the pressure, v is

$$G = E + pv - TS$$

the volume, T is the absolute temperature and S is the entropy. The Gibbs function is most useful in analyzing systems held at constant temperature and pressure. Under these conditions, the change in the Gibbs function, ΔG, of a system is a measure of the maximum attainable work, not including the work of displacing the environment. Since chemical processes frequently occur at constant temperature and pressure, the Gibbs function is extensively used in chemical engineering for calculating phase equilibrium and reaction equilibrium. *See* CHEMICAL THERMODYNAMICS; FREE ENERGY. [W.F.J.]

Gila monster The name for one of two species of reptiles in the family Helodermatidae that occurs in the southwestern United States and Mexico. *Heloderma suspectum* is the Gila monster; the other, larger species, *H. horridum*, is known as the beaded lizard. The body of both species has a rounded appearance and is covered with many-colored beadlike tubercles that form patterns of orange, red, or yellow on a dark-brown or black background.

These are the only known poisonous lizards. They are carnivorous and feed on small mammals, birds, and eggs but can survive for months without food as they store fat in their tails. They are oviparous and lay white, tough-shelled eggs which are large in proportion to the size of the parent. *See* REPTILIA; SQUAMATA. [C.B.C.]

Gill A structure in many vertebrates (Acrania, elasmobranchs, Holocephali, cyclostomes, Dipnoi, bony fishes, and amphibians) involved in external respiration. The gills in fish are considered here because these animals have been studied more extensively than others. The gills of fish serve for the exchange of materials between the blood and water. There are two kinds of gills, external and internal (see illustration). They are permeable to certain substances of importance to the economy of the organism, notably oxygen, carbon dioxide, and the nitrogenous wastes, ammonia and urea. These are transferred through the gill by simple diffusion. In addition, the gills are somewhat permeable to water, a circumstance for which most fish must compensate by the active transport of certain ions to maintain their osmotic balance. A significant part of this latter function is also carried out by the gills. The gills cannot be considered by themselves alone, because their functioning depends as much on the systems which bathe their external surface with water and their internal surface with blood as it does on the gills themselves.

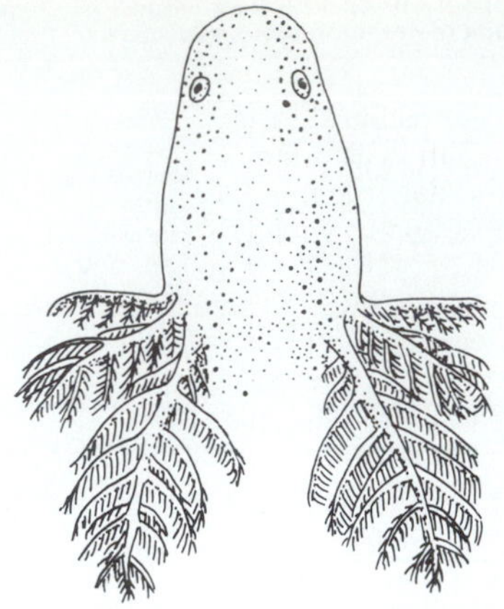

External gill in the adult urodele *Pseudobranchus striatus*.

The fundamental unit of the gill is the gill filament. The filaments are long, flattened, slightly crescent-shaped structures extending laterally from openings in the side of the pharynx. Filaments on the opposite sides of a given pharyngeal opening arch toward each other, at least at their lateral extremities, their tips meeting and thus enclosing a space between the filaments and the pharyngeal opening that may be termed the prebranchial chamber. External to the arched free portions of the filaments is a postbranchial chamber with an opening in the body wall.

The gill filaments bear two important types of structure. The first of these is the secondary lamella, which provides the actual surface for diffusion. The lamellae are almost microscopic, somewhat semicircular, leaflike extensions transverse on the flat faces of the filament. They consist of a loose web of supporting tissue covered by an epithelium of a single layer of cells so that there are extensive blood lacunae within them, separated from the water by the least barrier to diffusion. The other type of structure probably of great importance to the special functions of the gill is the so-called chloride-excreting cell. These are cells that are associated with the afferent blood supply. They are clustered about the bases of the lamellae and occur on the face of the filament itself.

There are comparative differences in each component of the branchial system. There are wide quantitative differences in the area of the gill surface and in the dimensions and the spacing of the lamellae. Such differences are largely associated with adaptive radiation. Active pelagic species tend to have large, finely divided surfaces, while more sluggish demersal species tend to have less lamellar surface and somewhat coarser channels. Those fish which have organs for breathing air, in addition to gills, have reduced gill surfaces. There are also wide quantitative differences in the oxygen capacity of the blood and in the nature of the loading curve of the hemoglobin, which also appear to be of adaptive significance. Active fish tend to have blood of higher oxygen capacity; the blood of those which live in well-aerated waters requires a relatively high partial pressure of oxygen for saturation. There are also major differences in the sensitivities of fish bloods with respect to the effect of the presence of carbon dioxide on the oxygen capacity. These, too, probably have their part in fitting the various species for the environments in which they are found. *See* RESPIRATORY SYSTEM. [F.E.J.F.]

Gilsonite An asphaltite that occurs in the Uinta Basin of northeastern Utah. Gilsonite is black and has high luster, coarsely conchoidal fracture, and a red-brown streak. The material is friable and in mining or crushing breaks to a fine, penetrating dust.

In the petroleum industry, the only major products from gilsonite are coke for use in the aluminum industry, gasoline for local consumption, and gas, valuable as a fuel. Gilsonite is used in the manufacture of paints, battery boxes, asphalt floor tiles, brake linings, printing inks, and electrical insulation. It is also used for waterproofing and insulating high-temperature piping. *See* ASPHALT AND ASPHALTITE. [I.A.B.]

Ginger An important spice or condiment; also the plant from which it is obtained, *Zingiber officinale*, of the ginger family (Zingiberaceae). The plant is a native of southeastern Asia. It is an erect perennial herb having thick, scaly, branched rhizomes which contain starch, gums, an oleoresin (gingerin) responsible for the pungent taste, and an essential oil which imparts the aroma. Ginger is used in medicine, in culinary preparations, and for flavoring beverages such as ginger ale and ginger beer. The plant is grown in China, Japan, Sierra Leone, Jamaica, Australia (Queensland), and Indonesia. *See* ZINGIBERALES. [P.D.St./E.L.C.]

Ginkgoales An order of nearly extinct gymnosperms (Pinophyta) having only one living species, the maidenhair tree (*Ginkgo biloba*). The fan-shaped leaves resemble those of the maidenhair fern. The leaves are dichotomously veined (fork-veined) and often more or less lobed (see illustration). The species is dioecious; that is, the male and female reproductive structures occur on separate trees. *Ginkgo* is widely planted on street borders, on home grounds, and in city parks. *See* PINOPHYTA. [A.Cr.]

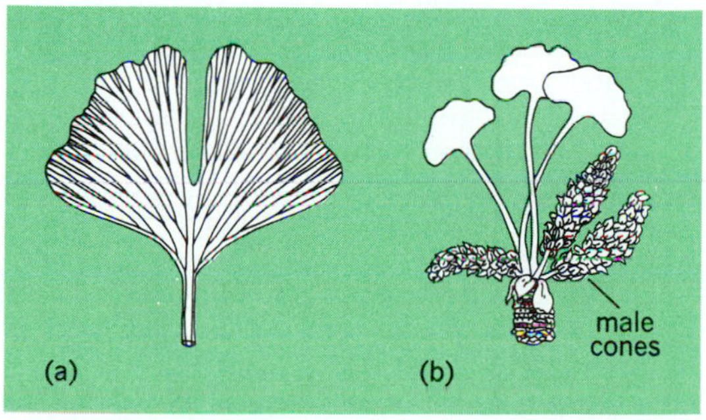

Essential features of *Ginkgo*. (*a*) Leaf showing dichotomous venation of blade. (*b*) Spur branch with mature male cones. (*After G. M. Smith et al., A textbook of General Botany, 5th ed., Macmillan, 1953*)

Ginkgoopsida A class of the division Pinophyta (Gymnospermae). It is known mostly from the fossil record, but has about 35 living species in the genera *Ephedra* of the order Ephedrales, and 1 species, *Ginkgo biloba*, in the order Ginkgoales. There are seven other orders known as fossils. The class is an ancient one with representatives in the early Carboniferous. Proliferation within the group occurred in the late Carboniferous to the Triassic. The order Ephedrales was traditionally referred to the "Gnetopsida," but for anatomical reasons is now tentatively assigned to the Ginkgoopsida. *See* EPHEDRALES; GINKGOALES; PINOPHYTA; PLANT KINGDOM. [T.A.Z.]

Giraffe Member of the family Giraffidae represented by a single species, *Giraffa camelopardalis*. The giraffe occurs in the savanna regions of tropical Africa. Giraffes are ruminants and belong to the mammalian order Artiodactyla (even-toed ungulates). The giraffe is the tallest of all mammals and may reach a height of 18 ft or 5.5 m. There are two prominent horns on the forehead which are bony outgrowths covered by skin, and there is a short mane along the back of the neck.

The giraffe lives in small herds with many females and usually one mature and several immature males. Gestation for the giraffe lasts about 15 months, and a single young is born. The young is about 6 ft (1.8 m) tall at the time of birth. *See* ARTIODACTYLA. [C.B.C.]

Glacial epoch An informal reference to a time during the history of the Earth when there were larger ice sheets (continental size) and mountain glaciers than today. The most recent glacial epoch, better known as the Pleistocene glacial epoch, and also by the older term Quaternary period, encompassed at least the last 3,000,000 years.

Many side effects resulted from the existence of these ice sheets and glaciers, including climate changes, sea-level rise and fall, depressions of the Earth's crust, and large-scale migrations of plants, animals, and humans as well as mass extinctions. Mountain landscapes were sculptured by glaciers, and erosional and depositional landforms were formed. Ocean temperatures were cold during glaciations and warm at times of interglacials. Early human evolution, development, and migrations resulted from the ever-changing climates closely related to glacier advances and retreats.

Glacial epochs seem to recur at intervals of 200,000,000 to 250,000,000 years. In overall occurrence, all the glacial epochs that have ever occurred occupy only 5 to 10% of all geologic time. During major glacial epochs, great ice sheets formed in the high latitudes and spread out to cover as much as 40% of the Earth's land surface. Accompanying drops in temperature during some glacial epochs may have been as much as 25°F (14°C) in the mid-latitudes. During a glacial epoch, major glaciations are short-lived, each lasting less than 10,000 years, with the interglacials persisting for only about 10,000 years, so that for most of an epoch, the ice sheets either grow or diminish in size. The Pleistocene glacial epoch was distinguished by seven or eight glacial advances within the last 700,000 years. Its last glaciation ended about 9000 years ago in Fennoscandia, and less than 8000 years ago in north-central Canada. *See* GLACIAL GEOLOGY; GLACIOLOGY; PLEISTOCENE. [S.E.Wh.]

Glacial geology The scientific study of the effects of glaciers on the broad land areas, on the oceans, and on climate, of their erosion and deposition, and of their modification of the Earth's surface in detail. Included in the realm of glacial geology is the history of glacial theory, consideration of the origin of glacial ages, extent and times of past glaciations, erosion and sculpturing of plains and mountains, deposition of ice-contact and meltwater sediments, and the consequences of glaciers on worldwide climate, and also on local climate around their edges. Quite distinct from glacial geology, however, is the separate, growing subscience of glaciology, the study of glaciers themselves. *See* GLACIOLOGY.

Features on the Earth's surface explained by former worldwide glaciation are numerous, embracing, for example, glacially eroded and molded valleys and mountains; ice-transported and deposited sediments and nonglacial sediments; abandoned stream channels with associated floodwater deposits; elevated silts and clays that collected around continental edges when sea level was higher; valleys eroded across and into continental shelves and slopes when sea level was much lower; communities of plants and animals similar to each other but separated by shallow seaways where land bridges once existed; fossil shells and microorganisms in deep-sea sediments reflecting colder or warmer water temperatures than today; vegetated sand dunes aligned to wind systems no longer operating; ancient shorelines and beach ridges ringing dry empty lake basins far inland; and orderly patterns of stones and fine sediments next to glacier margins in polar regions and high mountains. *See* CIRQUE; DRUMLIN; GLACIAL EPOCH; MORAINE. [S.E.Wh.]

Glaciology The study of existing or modern glaciers in their entirety, involving all related scientific disciplines. Glaciology is largely concerned with present glacial characteristics and processes, as opposed to studies in glacial geology which relate to the nature and effects of former glaciation. *See* GLACIAL GEOLOGY.

A glacier is a naturally accumulating mass of ice that moves in the process of discharging from head or center to its margins or a terminal dissipation zone. Glaciers are nourished in areas of snow accumulation that lie above the mean climatological or orographical snow line, which on the glacier surface is referred to as the névé line or firn line. The most active glaciers are generally found in regions receiving the heaviest snowfall, such as the maritime flanks of high coastal mountain ranges. *See* SNOW LINE; SNOWFIELD AND NÉVÉ.

Glaciers are composed of three substances: snow, firn, and ice. The main material of glaciers is bubbly glacier ice composed of myriads of interlocking crystals, hence it is a polycrystalline material containing air pockets and entrapped water bubbles. Below the late-summer glacier snow line (névé line) only bubbly glacier ice is exposed. Above the névé line, the other categories of snow and firn (and firn ice) exist to depths of a few to hundreds of feet. It is deeper in polar (colder) firn packs. (Firn is a consolidated granular transition of snow not yet changed to glacier ice.)

Glacier deformation is a composite of internal and external movement. The internal movement is dominated by a continuous plastic creep (flow deformation), and the external in some cases by a fracture type of discontinuous movement at the bed or near the glacier margins. As a general rule, flow deformation, hence the rate of movement, is greater in the upper portion of a glacier than in its basal section.

Research has suggested that glacier surges, expressed as sudden catastrophic advances or raising and lowering of the ice surface, are relatively common phenomena in some regions. Such abnormal surges are characterized by marked and seemingly anomalous increases in flow velocity and often by a rapid transfer of ice from the névé to the terminus. The upper glacier surface may sink or be substantially lowered with this volume loss, expressed as a comparable thickening in the lower valley sector. Such surges are believed to be kinematic in nature, in that the wave moves through the glacier at a substantially faster rate than the actual discharge of ice. [M.M.Mi.]

Gland A structure which produces a substance or substances essential and vital to the existence of the organism and species. Glands are classified according to (1) the nature of the product; (2) the structure; (3) the manner by which the secretion is delivered to the area of use; and (4) the manner of cell activity in forming secretion. A commonly used scheme for the classification of glands follows.

- I. Morphological criteria
  - A. Unicellular (mucous goblet cells)
  - B. Multicellular
    - 1. Sheets of gland cells (choroid plexus)
    - 2. Restricted nests of gland cells (urethral glands)
    - 3. Invaginations of varying degrees of complexity
      - *a.* Simple or branched tubular (intestinal and gastric glands)—no duct interposed between surface and glandular portion
      - *b.* Simple coiled (sweat gland)—duct interposed between glandular portion and surface
      - *c.* Simple, branched, acinous (sebaceous gland)—glandular portion spherical or ovoid, connected to surface by duct
      - *d.* Compound, tubular glands (gastric cardia, renal tubules)—branched ducts between surface and glandular portion
      - *e.* Compound tubular-acinous glands (pancreas, parotid gland)—branched ducts, terminating in secretory portion which may be tubular or acinar
- II. Mode of secretion
  - A. Exocrine—the secretion is passed directly or by ducts to the exterior surface (sweat glands) or to another surface which is continuous with the external surface (intestinal glands, liver, pancreas, submaxillary gland)
  - B. Endocrine—the secretion is passed into adjacent tissue or area and then into the bloodstream directly or by way of the lymphatics; these organs are usually circumscribed, highly vascularized, and usually have no connection to an external surface (adrenal, thyroid, parathyroid, islets of Langerhans, parts of the ovary and testis, anterior lobe of the hypophysis, intermediate lobe of the hypophysis, groups of nerve cells of the hypothalamus, and the neural portion of the hypophysis)
  - C. Mixed exocrine and endocrine glands (liver, testis, pancreas)
  - D. Cytocrine—passage of a secretion from one cell directly to another (melanin granules from melanocytes in the connective tissue of the skin to epithelial cells of the skin)
- III. Nature of secretion
  - A. Cytogenous (testis, perhaps spleen, lymph node, and bone marrow)—gland "secretes" cells
  - B. Acellular (intestinal glands, pancreas, parotid gland)—gland secretes noncellular product
- IV. Cytological changes of glandular portion during secretion
  - A. Merocrine (sweat glands, choroid plexus)—no loss of cytoplasm
  - B. Holocrine (sebaceous glands)—gland cells undergo dissolution and are entirely extruded, together with the secretory product
  - C. Apocrine (mammary gland, axillary sweat gland)—only part of the cytoplasm is extruded with the secretory product
- V. Chemical nature of the product
  - A. Mucous goblet cells (submaxillary glands, urethral glands)—the secretion contains mucin
  - B. Serous (parotid gland, pancreas)—the secretion does not contain mucin [O.E.N.]

Glanders An uncommon disease of equines known since the days of Aristotle and Hippocrates, which is occasionally transmitted to humans. Glanders still exists in parts of Europe, Asia, and Africa in horses, mules, donkeys, and humans. The causative organism is *Pseudomonas mallei*.

Glanders in equines occurs in two forms: acute or chronic respiratory infection, and farcy, a chronic infection of skin and lymphatics. Transmission occurs by food and water contaminated with nasal or skin discharge.

Glanders in humans is acquired by contact with discharge from infected equines or through accidental laboratory exposure. The portal of entry may be respiratory, alimentary, or abraded skin. The usual acute form of the disease produces chills, fever, and marked prostration. Subacute and chronic forms produce semigranulomatous pyemic nodules and abscesses in many organs. [W.R.Mi.]

Glass Materials made by cooling certain molten materials in such a manner that they do not crystallize but remain in an amorphous state, their viscosity increasing to such high values that, for all practical purposes, they are solid. Materials having this ability to cool without crystallizing are relatively rare, silica, SiO_2, being the most common example. Although glasses can be made without silica, most commercially important glasses are based on it. The most important properties are viscosity; strength; index of refraction; dispersion; light transmission (both total and as a function of wavelength); corrosion resistance; and electrical properties.

Chemically, most glasses are silicates. Silica by itself makes a good glass (fused silica), but its high melting point (1723°C or 3133°F) and its high viscosity in the liquid state make it difficult to melt and work. To lower the melting temperature of silica to a more convenient level, soda, Na_2O, is added in the form of sodium carbonate or nitrate, for example. This has the desired effect, but unfortunately the resulting glass has no chemical durability and is soluble even in water (water glass). To overcome this problem, lime, CaO, is added to the glass to form the basic soda-lime-silica glass composition which is used for the bulk of common glass articles, such as bottles and sheet (window) glass. Although these are the main ingredients, commercial glass contains other oxides (aluminum and magnesium oxides) and ingredients to help in oxidizing, fining, or decolorizing the glass batch.

Special kinds of glass have other oxides as major ingredients. For example, boron oxide is added to silicate glass to make a low-thermal-expansion glass for chemical glassware which must stand rapid temperature changes, for example, Pyrex glass. Also, lead oxide is used in optical glass because it gives a high index of refraction. [J.F.McM.]

Glass switch A glassy, solid-state device used to control the flow of electric current. Useful solid-state devices can be made from glassy as well as crystalline semiconductors. A glass is a special case of a noncrystalline class of materials, namely, amorphous solids. These do not exhibit long-range order, although they tend to have the same local structure (that is, short-range order) as the corresponding crystal. A glass is an amorphous solid that is formed by cooling rapidly from the liquid phase. *See* GLASS.

The first applications of glassy semiconductors were switches made from chalcogenide (that is, alloys containing tellurium, selenium, or sulfur) glasses. The two basic structures are known as the Ovonic Threshold Switch (OTS) and the Ovonic Memory Switch (OMS). They are active devices consisting simply of a thin film (about 1 micrometer thick) of glass between two metallic contacts. The device characteristics depend on the bulk properties of the semiconductor material rather than on the contacts. Consequently, the switches are symmetrical in that they respond identically to voltages and currents of both polarities.

Amorphous semiconductors, as opposed to crystalline semiconductors, can be doped or made insensitive to the effects of impurities, depending on the desired application. For switching devices, it is ordinarily preferable to use an impurity-independent structure to take advantage of lower costs and stable operation. Ovonic switches are also highly resistant to the effects of radiation.

Both the OTS and OMS show a rapid and reversible transition between a highly resistive and a conductive state effected by applied electric fields. The main difference between the two devices is that, after being brought from the highly resistive state to the conducting state, the OTS returns to its highly resistive state when the current falls below a holding current value. On the other hand, the OMS remains in the conducting state until a current pulse returns it to its highly resistive state. The OMS thereby remembers the last applied switching command, and it is from this property that the device receives its name.

Integrated arrays of OMSs can be used as electronically alterable read-only memories. They fill the gap in the computer memory spectrum between permanent read-only memories and volatile random-access memories. Photographic films have been developed in which the quality and amount of structural changes can be controlled by the amount of energy incident upon the film. Since these films do not contain silver, their cost does not vary with that of any precious metal. A transistor, using an OTS as the emitter, has been developed. This can be used as a threshold amplifier, as a threshold latching amplifier, or as the basis for a computer using ternary logic. Other promising application areas include ac control and microwave generation. *See* AMORPHOUS SOLID; SEMICONDUCTOR. [S.R.O.; D.A.]

Glass transition The transition that occurs when a liquid is cooled to an amorphous or glassy solid. This can occur only if the cooling rate is fast enough to prevent crystallization which would otherwise occur if time had been sufficient for the sample to reach true equilibrium at each temperature. Since the crystal is invariably the thermodynamically stable low-temperature phase, the glass transition corresponds to a transition from a high-temperature liquid into a nonequilibrium metastable low-temperature solid. *See* AMORPHOUS SOLID; CRYSTAL; VISCOSITY.

For many organic and polymeric systems, the difficulty of molecular packing and the steric hindrances are sufficient to prevent crystallization, and glass formation in these systems is relatively easy. In other systems, for example, metallic systems, rapid quench rates on the order of 10^6 K/s (2×10^6 °F/s) may be necessary to avoid crystallization, suggesting that any system can be quenched from the liquid state to an amorphous glassy state assuming that the system can be cooled rapidly enough. *See* GLASS; METALLIC GLASSES. [G.S.G.]

Glaucoma A disease in which an elevated intraocular pressure damages the eye. It is one of the most common and most serious disorders of the eye, exceeded in the United States only by cataracts as a cause of blindness.

Glaucoma is generally classified into four major groups: congenital or infantile glaucomas (occurring at birth or shortly after); primary open-angle glaucoma; primary angle-closure glaucoma (the acute, painful kind; can be subdivided into acute and subacute or chronic angle-closure glaucoma); and secondary glaucoma (associated with other ocular disease such as a swollen cataract, inflammations, trauma, or aftereffects of surgical procedures).

Infantile glaucoma results in many cases in a large eye because of the lack of the rigidity of the sclera, and this condition is known as buphthalmos. Angle-closure glaucoma occurs when there is a sudden increase in intraocular pressure due to a block of the anterior chamber angle in such a way that the root of the iris cuts off the aqueous outflow. A severe, sudden pressure rise causes pain and sudden visual loss. Such an eye will show a steamy or cloudy cornea, a semidilated and fixed pupil (that will not respond to a light shining onto it), and an elevated intraocular pressure. The chief threat of the most common kind of glaucoma, the so-called open-angle, symptomless glaucoma, is insidious visual impairment. The disease is bilateral, and there is progressive field loss, occurring mostly in the periphery, so that the person is unaware of losing vision until it is too late.

In most cases, blindness can be prevented if treatment is instituted early. The object of the treatment is to maintain the intraocular pressure at a normal level. In cases in which pressure cannot be controlled with medication, such as angle-closure glaucoma, surgery can be done to create an artificial outflow and lower the intraocular pressure. Congenital glaucoma, which occurs as a result of a defect in the development of the angle, must be treated surgically. [J.Hart.]

Glauconite The term glauconite as currently used has a two-fold meaning. It is used as both a mineralogic and morphologic term. The mineral glauconite is defined as an illite type of clay mineral. A fundamental characteristic of glauconite is that the unit cell is composed of a single silicate layer rather than the double layer of most other dioctahedral micas. *See* CLAY MINERALS; ILLITE.

Glauconite is known to occur in flakes and as pigmentary materials. When used in the morphological sense, the term glauconite often refers to small, green, spherical, earthy pellets. Some of these pelletal varieties are composed solely of the mineral described above, others are a mixed-layer association of this mineral and other three-layer structures.

Glauconite forms during marine diagenesis, in relatively shallow water, and at times of slow or negative deposition. Glauconite has been identified in both recent and ancient sediments. It is a major component in some "greensand" deposits and has been used commercially for the extraction of potassium from such sources. *See* AUTHIGENIC MINERALS; DIAGENESIS; MARINE SEDIMENTS. [F.M.W.; R.E.Gr.]

Glaucophane The name given to the monoclinic sodium amphiboles of the general composition $Na_2Mg_3Al_2Si_8O_{22}(OH)_2$. Like most sodium amphiboles, glaucophane is blue to black with marked pleochroism (color change on rotation in plane-polarized light) in thin sections from yellowish green to deep blue. The crystals are prismatic and sometimes fibrous. Glaucophane occurs in a restricted group of metamorphic schists that commonly exhibit an unusual mineral association with dense minerals, such as lawsonite, jadeite, pumpellyite, and garnet, and less dense minerals, such as epidote, mica, chlorite, and albite. *See* AMPHIBOLE. [G.W.DeV.]

Glazing The application of finely ground glass, or glass-forming materials, or a mixture of both, to a ceramic body and heating (firing) to a temperature where the material or materials melt, forming a coating of glass on the surface of the ware. Glazes are used to decorate the ware, to protect against moisture absorption, to give an easily cleaned sanitary surface, and to hide a poor body color.

Glazes are classified and described by the following characteristics: surface—glossy or matte; optical properties—transparent or opaque; method of preparation—fritted or raw; composition—such as lead, tin, or boron; maturing temperature; and color. Opaque glazes contain small crystals embedded in the glass, but special glazes in which a few crystals grow to recognizable size are called crystalline glazes. *See* CERAMICS; FRIT; GLASS. [J.F.McM.]

Glide-path indicator An aircraft landing instrument that provides the pilot with a set of vertical and horizontal cross pointers that indicate deviation from a radio-transmitted course to the threshold of the runway. A dual-frequency transmitter sends out one frequency to the right of the runway center line and a second frequency to the left of the runway center line. The reception of these signals in the aircraft biases the vertical needle to the left or right depending on the position of the aircraft relative to the transmitted 5° localizer path. Simultaneously, another dual-frequency transmitter causes a horizontal needle to indicate high or low as the aircraft descent path is compared to the transmitted 3° glide path. [J.W.A.]

Glider An unpowered flying device that attempts to copy the flight of soaring birds as accurately as possible. In October 1911, Orville Wright made a gliding flight of nearly 10 min duration, and demonstrated that gliders could stay up for long periods in rising air. This condition of flight, called slope soaring, was the basic method of soaring flight until about 1930. Thermal soaring, the next step, was accomplished by flying in areas of rising convection currents. By the use of thermal flight, the modern glider can fly almost anywhere in the world for extended time and distances over 800 km (500 mi) in one flight. Other methods of soaring make use of clouds and standing-wave phenomena in the atmosphere. High-performance gliders (sailplanes) may be launched by towing behind powered aircraft or by car towing, which is used to a lesser extent. Some gliders have been fitted with a small motor and propeller, which enables them to take off and climb to an altitude where rising air permits them to soar unpowered. *See* SAILPLANE.

Human-powered aircraft *Gossamer Albatross* in flight. (*Photograph by Don Monroe*)

Modern foot-launched hang gliders with aluminum tube frames have flown over 160 km (100 mi) in straight-line distance, have reached about 6000 m (20,000 ft) altitude, and have remained aloft more than 15 h. But it is not so much their performance that makes hang gliders popular, as their low cost, their convenience of folding into a small package for transport or storage, and the fact that no license is required for glider or pilot.

In 1977 a team of workers directed by Paul B. MacReady constructed a very large but very light hang glider capable of human-powered flight. This aircraft, the *Gossamer Condor*, was pedaled around a figure-eight course. MacReady and his team improved their design, and in 1979 made the first human-powered flight across the English Channel with the *Gossamer Albatross* (see illustration). [F.M.R.]

Globe (Earth) A sphere on the surface of which is a map of the world. The map may be drawn, engraved, or painted directly on the surface but is more commonly prepared as a series of gores, or segments in other designs, to be affixed to the globe ball (see illustration).

Globes are both artistically interesting and scientifically useful. Their principal value is in stimulating sound concepts of worldwide patterns and in rectifying errors induced by the limitations of flat maps. All flat maps distort the Earth's surface patterns, but carefully made globes constitute truer scale models of the Earth, with correct areas, shapes, and distances as well as continuity of surface. Globes have long been used as aids in navigation, in the teaching of earth sciences, and as room ornaments.

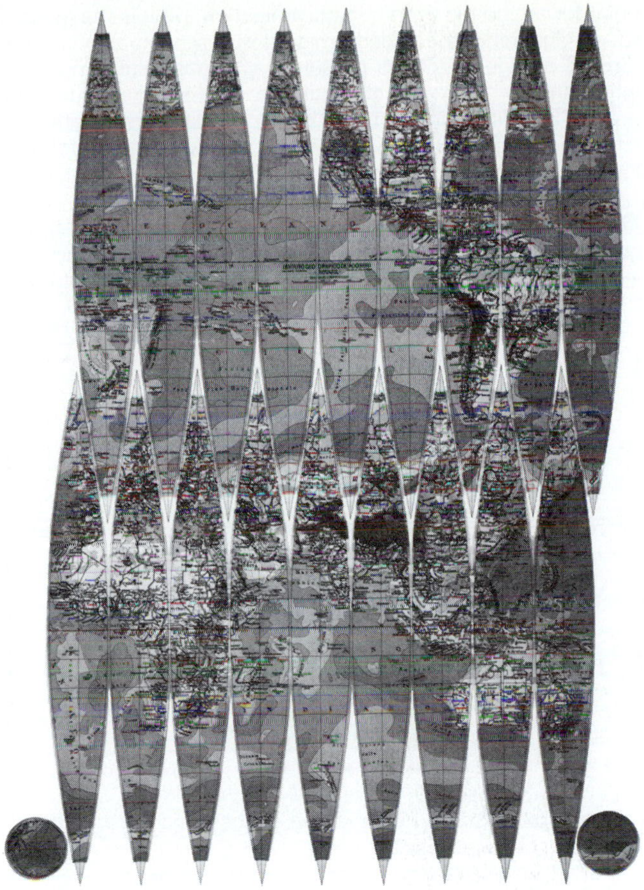

Globe gores from collections of Library of Congress. (*Istituto Geografico de Agostini, Novara, Italy*)

Many modern globes have special attachments to improve their utility. A meridian ring, extending from pole to pole, may be calibrated in degrees to measure latitude. The longitude of points directly beneath that ring will be indicated at the intersection of the ring with the equatorial scale. A horizon ring at right angles to the meridian ring may be calibrated in miles or in meters, degrees, and hours to expedite distance and time measurement. A hinged horizon ring may be lifted to serve as a meridian ring, or placed in an oblique position to show great circle routes and distances. [A.C.G.]

Globule A roundish blob of cosmic dust (that is, a roundish dark nebula) seen projected against the starlit sky. The diameter of a typical nearby globule is generally about 3–5 minutes of arc. Globules look like "holes in the heavens" and are made detectable by the fact that the cosmic dust inside is so thick that the stars on the far side of the globule cannot be seen.

Theoretical calculations show that globules should collapse into protostars in times of the order of a half million years. There will probably be considerable mass loss in the process of collapse. The result of the collapse will probably be a star (or binary pair of stars) with masses of 1 to a few solar masses. It appears that about one in every five to eight stars in the Milky Way Galaxy is born from a globule. *See* INTERSTELLAR MATTER; STELLAR EVOLUTION. [B.J.B.]

Globulin A general name for any member of a heterogeneous group of serum proteins precipitated by 50% saturated ammonium sulfate. *See* PROTEIN; SERUM.

The introduction of electrophoresis during the 1930s permitted subdivision of the globulins into alpha, beta, and gamma globulins on the basis of relative mobility at alkaline pH (8.6). However, each of these subgroups, though electrophoretically homogeneous, consists of a great variety of proteins with different biological properties and markedly different sizes and chemical properties other than net charge. Thus the α_2-globulins, for example, as defined by moving boundary or paper electrophoresis, contain proteins ranging in molecular weight from approximately 50,000 to approximately 1,000,000 (α_2-macroglobulin), each with differing functions. *See* ELECTROPHORESIS; IMMUNOGLOBULIN. [H.H.F.]

Glomerulonephritis A common kidney disorder affecting the whole nephron, including the surrounding interstitial tissue. There are many diseases with which the acute process may be associated, the most common being a streptococcal infection, usually of the pharynx. The signs of nephritis—hypertension, edema, decreased urine output, and red blood cells and protein in the urine—typically begin 10–14 days after the acute infection. This disease is common in children and complete healing without treatment is the general rule. In affected adults, nearly 40% develop some degree of chronic disease. The histologic lesion is characterized by an increased number of cells in swollen glomeruli.

The course of this disease averages 15.7 years from the time of diagnosis. Treatment is purely symptomatic entailing management of high blood pressure and restriction of salt intake to reduce the amount of solute that the kidneys must excrete. *See* KIDNEY; NEPHRITIS. [G.E.Str.]

Glow discharge A mode of electrical conduction in gases. Glow discharge commonly occurs under conditions of relatively low pressure and generally in the pressure range of 1–10 mm of mercury (10^2–10^3 pascals), The discharge typically gives off light, so that the region of the discharge appears to glow with considerable intensity. This glow is quite diffuse as contrasted to a higher-pressure discharge, such as a high-pressure arc. Typical currents may be of the order of tens or hundreds of milliamperes, whereas the potential drop may be of the order of 100 volts.

The most important application of the glow discharge is in the so-called voltage regulator or voltage reference tube. This device maintains a relatively constant difference of potential across itself as the current is varied over an appreciable range, and consequently is very useful in cases where a constant reference potential is required. *See* ELECTRIC SPARK; ELECTRICAL CONDUCTION IN GASES. [G.H.M.]

Glucagon The protein hormone secreted by the pancreas which is known to influence a wide variety of metabolic reactions. Glucagon, along with insulin and other hormones, plays a role in the complex and dynamic process of maintaining adequate supplies of sugar in the blood. Glucagon has often been called the hyperglycemic-glycogenolytic factor because it causes the breakdown of liver glycogen to sugar (a process known as glycogenolysis) and thereby increases the concentration of sugar in the bloodstream (a condition known as hyperglycemia). Glucagon may also be involved in the regulation of protein and fat metabolism, gastric acid secretion and gut motility, excretion of electrolytes (such as sodium, potassium and chloride) by the kidney, contractility of heart muscle, and release of insulin from the pancreas. Glucagon is used in human medicine chiefly in certain diabetic conditions when a dangerously low blood sugar must be rapidly raised. *See* CARBOHYDRATE METABOLISM; DIABETES; GLYCOGEN; HORMONE; INSULIN; PANCREAS. [W.W.Bro.]

Glucose A monosaccharide also known as D-glucose, D-glucopyranose, grape sugar, corn sugar, dextrose, and cerelose. The structure of glucose is shown in the illustration.

Glucose in free or combined form is not only the most common of the sugars but is probably the most abundant organic

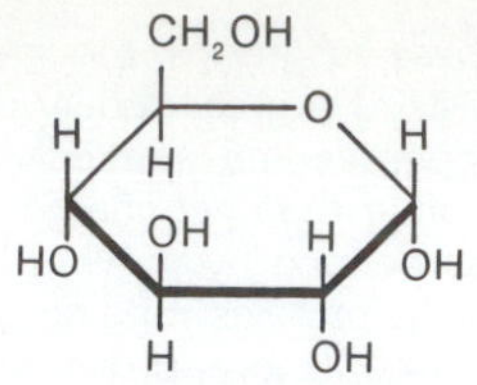

Structural formula for α-D-glucose.

compound in nature. It occurs in free state in practically all higher plants. It is found in considerable concentrations in grapes, figs, and other sweet fruits and in honey. In lesser concentrations, it occurs in the animal body fluids, for example, in blood and lymph. Urine of diabetic patents usually contains 3–5%.

Cellulose, starch, and glycogen are composed entirely of glucose units. Glucose is also a major constituent of many oligosaccharides, notably sucrose, and of many glycosides. It is produced commercially from cornstarch by hydrolysis with dilute mineral acid. The commercial glucose so obtained is used largely in the manufacture of confections and in the wine and canning industries. *See* CELLULOSE; GLYCOGEN; STARCH.

D-Glucose is the principal carbohydrate metabolite in animal nutrition; it is utilized by the tissues, and it is absorbed from the alimentary tract in greater amounts than any other monosaccharide. Glucose could serve satisfactorily in meeting at least 50% of the entire energy needs of humans and various animals.

Glucose enters the bloodstream by absorption from the small intestine. It is carried via the portal vein to the liver, where part is stored as glycogen, the remainder reentering the circulatory system. Another site of glycogen storage is muscle tissue.

Glucose is readily fermented by yeast, producing ethyl alcohol and carbon dioxide. It is also metabolized by many bacteria, resulting in the formation of various degradation products, such as hydrogen, acetic and butyric acids, butyl alcohol, acetone, and many others. *See* CARBOHYDRATE; MONOSACCHARIDE.

Gluons The hypothetical force particles which are believed to bind quarks into "elementary" particles. Although theoretical models in which the strong interactions of quarks are mediated by gluons have been successful in predicting, interpreting, and understanding many phenomena in particle physics, free gluons remain undetected in experiments (as do free quarks). According to prevailing opinion, an individual gluon cannot be isolated. *See* ELEMENTARY PARTICLE; FUNDAMENTAL INTERACTIONS; QUARKS. [C.Q.]

Glutamic acid An amino acid. Glutamic acid has many important functions, including the following: (1) It is the principal point by which ammonia enters organic compounds (reductive amination of (α-ketoglutarate), and the central distribution

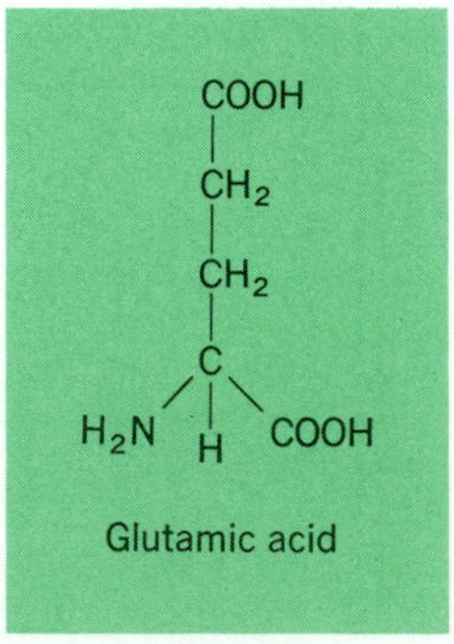

Glutamic acid

point for amino nitrogen by transamination. (2) It is the precursor of glutamine, the added amide group serving as a storage form of nitrogen and as a precursor of certain nitrogen atoms in purines, histidine, and glucosamine. (3) It is the precursor of proline and of arginine. (4) It is incorporated into both glutathione and folic acid. *See* AMINO ACIDS; ARGININE; FOLIC ACID; HISTIDINE; PROLINE. [E.A.Ad.]

Glutamine An amino acid. Glutamine with its amide group is an important storage form of nitrogen in plants and animals; it also serves as the precursor of certain ring nitrogen

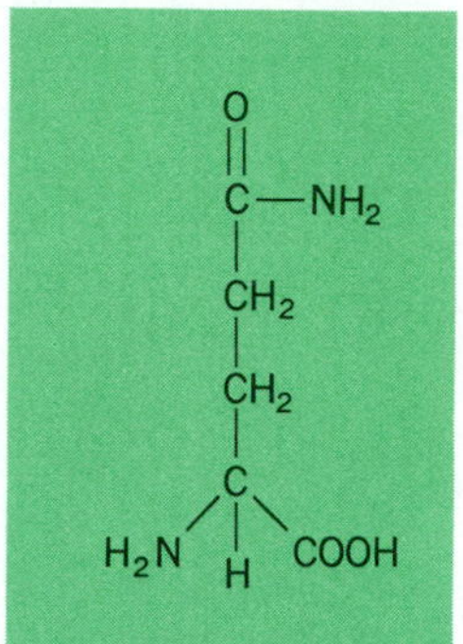

Glutamine

atoms in purines and histidine and of the amino group in glucosamine. *See* AMINO ACIDS. [E.A.Ad.]

Glycerol The simplest trihydric alcohol, with the formula $CH_2OHCHOHCH_2OH$. The name glycerol is preferred for the pure chemical, but the commercial product is usually called glycerin. It is widely distributed in nature in the form of its esters, called glycerides. The glycerides are the principal constituents of the class of natural products known as fats and oils.

When pure, glycerin is a colorless, odorless, viscous liquid with a sweet taste. It is completely soluble in water and alcohol but is only slightly soluble in many common solvents, such as ether, ethyl acetate, and dioxane. Glycerin is insoluble in hydrocarbons. It boils at 290°C (554°F) at atmospheric pressure and melts at 17.9°C. Its specific gravity is 1.262 at 25°C (77°F) referred to water at 25°C, and its molecular weight is 92.09. It has a very low mammalian toxicity.

Glycerin is used in nearly every industry. With dibasic acids, such as phthalic acid, it reacts to make the important class of products known as alkyd resins, which are widely used as coating and in paints. It is used in innumerable pharmaceutical and cosmetic preparations; it is an ingredient of many tinctures, elixirs, cough medicines, and anesthetics; and it is a basic medium for toothpaste. In foods, it is an important moistening agent for baked goods and is added to candies and icings to prevent crystallization. It is used as a solvent and carrier for extracts and flavoring agents and as a solvent for food colors. Many specialized lubrication problems have been solved by using glycerin or glycerin mixtures. Many millions of pounds are used each year to plasticize various materials.

Several grades of glycerin are marketed, including high gravity, dynamite, yellow distilled, USP (U.S. Pharmacopoeia), and CP (chemically pure). USP grade is water-white and suitable for use in foods, pharmaceuticals, and cosmetics, or for any purpose where the product is designed for human consumption. *See* ALCOHOL; FAT AND OIL; POLYOL. [P.H.C.]

Glycine An amino acid. Glycine has many important functions, including the following: (1) It is incorporated intact into

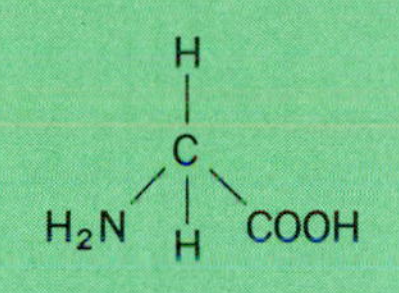
Glycine

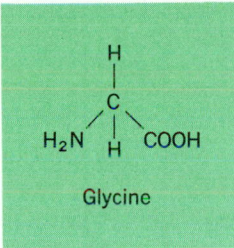

purines. (2) The α-carbon and nitrogen atoms are incorporated into the pyrrole rings of porphyrin. (3) It accepts an amidine group from arginine and a methyl group from methionine to form creatine. (4) It is a constituent of the tripeptide coenzyme glutathionine (γ-glutamylcysteinylglycine). (5) It can accept formaldehyde from hydroxymethyltetrahydrofolic acid to become serine. This reaction is reversible. *See* AMINO ACIDS. [E.A.Ad.]

Glycogen The primary reserve polysaccharide of the animal kingdom. It is found in the muscles and livers of all higher animals, as well as in the cells of lower animals. Because of its close relationship to starch, it is often called animal starch, although glycogen is found in some lower plants, fungi, yeast, and bacteria. *See* STARCH.

Glycogen is a nonreducing, white, amorphous polysaccharide which dissolves readily in cold water, forming an opalescent, colloidal solution. The molecular weight of glycogen is usually very high, and it varies with the source and the method of preparation; molecular weights of the order of $1\text{–}20 \times 10^6$ have been reported. Chemical studies show glycogen to possess a branched structure similar to the amylopectin starch fraction.

In its biochemical reactions, glycogen is similar to starch. It is attacked by the same plant amylases that attack starch, and like starch, it is degraded to maltose and dextrins. Both glycogen and starch are broken down by animal or plant phosphorylase enzyme in the presence of inorganic phosphate with the production of α-D-glucose-1-phosphate. *See* CARBOHYDRATE METABOLISM.

The metabolic formation of glycogen from glucose in the liver is frequently termed glycogenesis. In fasted animals, glycogen formation can be induced by the feeding, not only of materials that can be hydrolyzed to glucose and other monosaccharides, such as fructose, but also of various other materials. A number of L-amino acids, such as alanine, serine, and glutamic acid, upon deamination in the liver give rise to substances, such as pyruvic acid and α-ketoglutaric acid, that can be converted in the liver to glucose units which are subsequently converted to glycogen. Furthermore, substances such as glycerol derived from fats, dihydroxyacetone, or lactic acid can all be utilized for glycogen synthesis in the liver. Such noncarbohydrate precursors are termed glycogenic compounds. The process of glycogen formation from these precursors is known as glyconeogenesis. The term glycogenolysis is used to connote glycogen breakdown. *See* POLYSACCHARIDE. [W.Z.H.]

Glycol Any of a class of compounds characterized by having two hydroxyl (—OH) groups on separate carbons of an organic structure, usually linear and aliphatic. Products in this class are also called diols, dihydric alcohols, or in special cases, polyglycols. Glycols are generally water-soluble and hygroscopic, viscous liquids. Polyethylene glycols, that is, polyglycols derived by polymerization of ethylene oxide, are water-soluble at all molecular weights. Polypropylene glycols from propylene oxide become increasingly less water-soluble at molecular weights above 425.

Glycols undergo typical reactions of alcohols such as esterification, etherification, salt formation, and additions to activated double bonds. They also react with diacids or diesters to produce long-chain polyester polymers, an important family of products which find broad application in the manufacture of fibers, films, and molded plastic products. Reaction of glycols, especially polypropylene glycols, with diisocyanates produces another important family of products, polyurethanes, used in rigid and flexible foams and elastomers. *See* POLYESTER RESINS; POLYETHYLENE GLYCOL; POLYMER; POLYVINYL RESINS. [R.K.Ba.]

Glycolipid One of a class of complex lipids which contains carbohydrate residues. Cerebrosides are glycosides of ceramide (fatty acid amide of sphingosine). Galactose is the major constituent of animal cerebrosides. Glucocerebrosides are also found normally and accumulate in large amounts in certain pathological conditions. A glycoside of ceramide containing both galactose and glucose has been isolated from a human carcinoma.

Ganglioside, hematoside, and globoside are glycosides of ceramide containing several sugar residues. Ganglioside, which is found in the brain, contains glucose, galactose, galactosamine, and neuraminic acid as sugar residue. *See* GLYCOSIDE; LIPID; SPHINGOLIPID. [H.E.Ca.; R.H.G.]

Glycoprotein A compound in which carbohydrate (sugar) is covalently linked to protein. The carbohydrate may be in the form of monosaccharides, disaccharides, oligosaccharides, or polysaccharides, and is sometimes referred to as glycan. The sugar may be linked to sulfate or phosphate groups. In different glycoproteins, 100–200 glycan units may be present. Therefore, the carbohydrate content of these compounds varies markedly, from 1% (as in the collagens), to 60% (in certain mucins), to >99% (in glycogen). *See* COLLAGEN; GLYCOGEN.

Glycoproteins are ubiquitous in nature, although they are relatively rare in bacteria. They occur in cells, in both soluble and membrane-bound forms, as well as in the intercellular matrix and in extracellular fluids, and include numerous biologically active macromolecules. A number of glycoproteins are produced industrially by genetic engineering techniques for use as drugs; among them are erythropoietin, interferons, colony stimulating factors, and blood-clotting factors. *See* GENETIC ENGINEERING.

In most glycoproteins, the carbohydrate is linked to the polypeptide backbone by either N- or O-glycosidic bonds. A different kind of bond is found in glycoproteins that are anchored in cell membranes by a special carbohydrate-containing compound, glycosylphosphatidylinositol, which is attached to the C-terminal amino acid of the protein. A single glycoprotein may contain more than one type of carbohydrate-peptide linkage. N-linked units are typically found in plasma glycoproteins, in ovalbumin, in many enzymes (for example, the ribonucleases), and in immunoglobulins. O-linked units are found in mucins; collagens; and proteoglycans (typical constituents of connective tissues), including chondroitin sulfates, dermatan sulfate, and heparin. *See* ALBUMIN; CARBOHYDRATE; ENZYME; IMMUNOGLOBULIN; MONOSACCHARIDE; OLIGOSACCHARIDE; POLYSACCHARIDE; PROTEIN.

Within any organism, all molecules of a particular protein are identical. In contrast, a variety of structurally distinct carbohydrate units are found not only at different attachment sites of a glycoprotein but even at each single attachment site—a phenomenon known as microheterogeneity. For instance, ovalbumin contains one glycosylated amino acid, but over a dozen different oligosaccharides have been identified at that site, even in a preparation isolated from a single egg of a purebred hen. [N.Sh.]

Glycoside A member of a large class of carbohydrate derivatives characterized by the replacement of the hydrogen

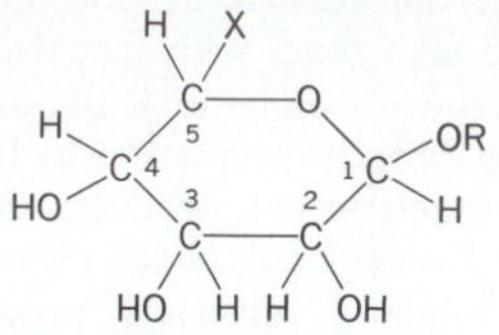

Glycoside structure. When R = H and X = —CH_2OH, it is a hexose sugar; when R = H and X = H, it is a pentose sugar; when R = alkyl or aryl, it is a glycoside.

atom of the hemiacetal hydroxyl group by an alkyl or aryl moiety, the aglycon. In nature, most glycosides are those formed from five-carbon (pentose) sugars or six-carbon (hexose) sugars occurring predominantly as pyranoside (six-membered) rings (see illustration).

Natural glycosides constitute a large and biologically important class of sugar derivatives. Many natural pigments and dyestuffs, poisons, drugs, and aromatic principles are glycosides. Heart stimulants, cerebrosides, and gangliosides are glycosides. It is believed that the hydrophilic sugar moiety in the glycosides serves to solubilize the hydrophobic aglycon. Frequently the biochemical interest or biological activity of glycosides resides solely, or largely, in the aglycon. *See* ANTHOCYANIN; ANTHRAQUINONE PIGMENTS; FLAVONOID; STEROID. [G.N.R.; R.L.Wh.; J.R.D.]

Gnathiidea A suborder of the Isopoda. These animals are characterized as having a much reduced second thoracomere which is incorporated with the cephalothorax. The antennules are short and each has a flagellum with four or five joints; the terminal three bear single sensory palps. The antennae are also short, with flagella having five to eight joints, but generally there are seven. Thoracomeres 3–7 have a normal appearance and bear pereiopods, while the eighth thoracomere is vestigial and lacks appendages (see illustration).

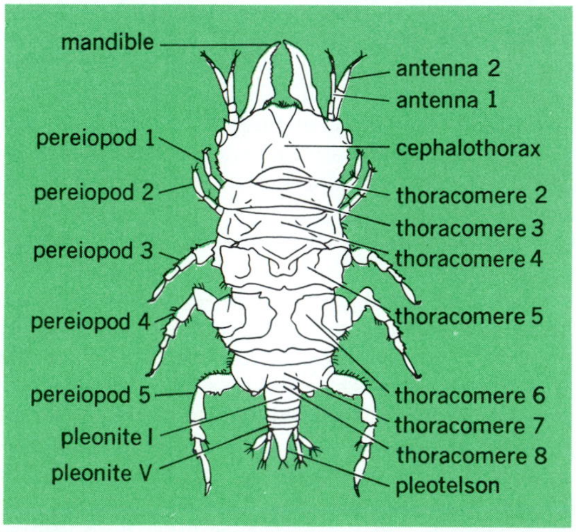

***Gnathia calva*, male specimen, dorsal view. (*After W. H. Tattersall, 1921*)**

The group comprises marine forms exclusively. They are found in all latitudes and at all depths. The larvae are found as parasites on fish. *See* ISOPODA. [T.M.]

Gnathostomata A group of the subphylum Vertebrata which possesses jaws, teeth, paired appendages, and girdles (secondarily lost in some), as well as other advanced features in contrast to the more primitive Agnatha, which lack jaws and paired appendages. The Gnathostomata are divided into two major subdivisions, the superclasses Pisces and Tetrapoda, which respectively constitute roughly the aquatic and the terrestrial vertebrates. *See* AGNATHA; PISCES (ZOOLOGY); TETRAPODA; VERTEBRATA. [W.J.B.]

Gnathostomulida Microscopic marine worms of uncertain systematic relationship, mainly characterized by cuticular structures in the pharynx and a monociliated skin epithelium. It is the most recently described phylum of the animal kingdom. The total number of species probably exceeds 1000.

Gnathostomulids are worm-shaped, cylindrical or slightly depressed, and semitransparent (or bright red), and sometimes have the external division of head and tail. The skin is a one-layered epithelium that is completely monociliated; that is, each of the polygonal epidermal cells bears only one cilium. The sensory system usually consists of 1–2 pairs of simple and 3–4 pairs of compound bristles (frontally and laterally), and a bundle of stiff cilia (dorsally on the head). The reproductive system consists of a dorsal ovarium and in most cases two caudolateral groups of testes follicles in the same specimen. Fertilization is internal.

The distribution of gnathostomulids is worldwide, the majority of localities being known from European coasts, some from the North American east coast, and some scattered over the western Pacific. [R.J.R.; W.E.S.]

Gneiss An old term in mining and geology that was first applied by the miners of the Erzgebirge, Austria, to designate the country rock adjacent to the ore veins. In its widest sense, gneiss is used as a structural rather than compositional term. Thus it applies to a great variety of rocks with a banded or coarsely foliated structure. A modern version includes in the definition of gneiss (primary gneiss) that it is of plutonic igneous origin with banding caused by flow movements in a crystallizing heterogeneous magma.

Typical gneisses are highly metamorphic, coarse-grained, irregularly banded or foliated rocks composed mainly of quartz and feldspar with some biotite (= granitic composition). As the grain becomes finer and the amount of mica increases, gneisses pass into pelitic schists. With fine grain and decrease of mica, or garnet substituting for mica, they become less foliated and pass into granulite. *See* GRANULITE; SCHIST.

Primary gneiss is a term used for rocks of metasomatic, migmatitic origin which are related to regional processes of granitization. Paragneiss shows a sedimentary parentage. Orthogneisses are primary igneous rocks in which a gneiss structure has been induced by essentially kinetic metamorphism. Ordinarily the term gneiss refers to a granitic composition; anything else should be specially noted. *See* METASOMATISM; MIGMATITE. [T.F.W.B.]

Gnetales An order of the class Cycadopsida having about 35 species, in the single genus *Gnetum*. All species are tropical, with more than half being native to tropical Asia (mainly Malaysia); species also occur in Africa and South America. Most of the species are lianas (climbing, woody vines), and a few are shrubs or small trees. The leaves are opposite, oval, entire-margined, and net-veined, appearing very similar vegetatively to the angiosperms. The species are dioecious, with the sporophylls being borne in whorls on the slender, linear axis. The mature cone is fleshy with a single seed. The order is not definitely known in the fossil record. *See* CYCADOPSIDA; PINOPHYTA; PLANT KINGDOM. [T.A.Z.]

Gneticae A subdivision of the division Pinophyta, represented by a single class, Gnetopsida, with three orders, Ephedrales, Gnetales, and Welwitschiales; each order is represented by a single genus: *Ephedra*, *Gnetum*, and *Welwitschia*, respectively. *See* EPHEDRALES; GNETALES; WELWITSCHIALES.

The genus *Ephedra* has about 35 species, mostly of arid regions such as the southwestern United States. The plants are freely branched, low shrubs with reduced, scalelike leaves and green, photosynthetic twigs.

The genus *Gnetum* also has about 35 species, all of which are tropical, with more than half of them native to the Malaysian region. Most of the species are lianas (climbing, woody vines); only a few are shrubs or small trees.

A single species, *mirabilis*, native to very dry deserts of southwestern Africa, is assigned to the genus *Welwitschia*. One of the most bizarre plants in the world, it has a very short, unbranched, woody main stem which is cushion-shaped or saucer-shaped and tapers quickly to a very long taproot. There are only two leaves and these persist throughout the life of the plant, perhaps 100 years. *See* PINOPHYTA. [A.Cr.]

Goat A number of wild and domestic species of the mammalian order Artiodactyla. They are swift, climbing animals with a strong odor, particularly in the male. Goats are closely related to sheep, but differ in having a narrower head, a bearded chin in the male, horns that are close at the base and sweep up and back and are ringed or twisted, a short upturned tail, and no facial gland. Since these animals are herbivores, the upper incisors and canines are lacking. Some species of wild goats are listed in the table.

Common name and geographic distribution of some wild goat species

Scientific name	Common name	Distribution
Capra hircus	Pasang, wild goat	Asia Minor
C. falconeri	Markhor	Himalayas
C. pyrenaica	Spanish ibex	S.W. Europe
C. ibex sibirica	Sakin, Asiatic ibex	Siberia
C. ibex nubiana	Beden, Nubian ibex	North Africa
C. ibex ibex	Steinbock, Alpine ibex	Alpine areas
C. caucasiaca	Tur	Caucasus

Domestic goats are hardy animals, resistant to disease, and produce two to three times the amount of milk that the cow does, relative to size. Goat's milk does not harbor the tubercular bacillus and provides excellent cheese and butter. The Alpine goat is the most important milk-producing breed. The Swiss or Alpine goat, usually divided into two breeds, the Saanen and Toggenburg, has been bred to local, native goats in many countries to improve milk production. The Nubian or Oriental goat has been imported into many countries also. It has long, drooping ears about the size of the head, and the horns are small or entirely absent. In South Africa, a Swiss-Nubian cross is referred to as the Boer goat. The Angora and Kashmir goats have been raised for wool. *See* ARTIODACTYLA. [C.B.C.]

Gobiesociformes An order of bony fishes, also known as the Xenopterygii, or clingfishes, equipped with a thoracic sucking disk which serves for attachment to the substrate. There are single dorsal and anal fins that lack fin spines (see illustration). The body is scaleless, and there are no ribs and no swim bladder. The order consists of a single family that is classified into 8 subfamilies, 33 genera, and nearly 100 Recent species. *See* ACTINOPTERYGII. [R.M.B.]

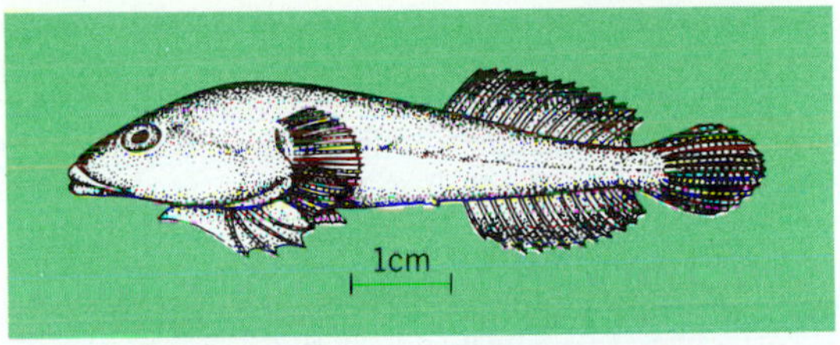

Northern clingfish (*Gobiesox maeandricus*). (After D. S. Jordan and B. W. Evermann, *The Fishes of North and Middle America*, U.S. Nat. Mus. Bull. no. 47, 1900)

Gödel's theorem The result, proved by K. Gödel in 1931, that any sufficiently advanced mathematical system must be incomplete in that there must always be a true sentence that is not provable in the system. Roughly speaking, Gödel showed how, for each such system, a sentence could be constructed that asserted its own nonprovability in the system.

Gödel considered mathematical systems that involve numbers, sets of numbers, and other purely mathematical entities. He demonstrated a still more remarkable result for these same systems, which include the most comprehensive mathematical systems known. He showed that if these systems are consistent, they cannot prove their own consistency. This is known as Gödel's second incompleteness theorem.

Unfortunately, there are widespread misconceptions about this result. For example, Gödel's second theorem is sometimes thought to imply the impossibility of knowing that these systems are consistent. In reality, the fact that a system cannot prove its own consistency does not constitute the slightest grounds for doubting its consistency. Indeed, if a system could in fact prove its own consistency, that of course would not be any guarantee that the system was consistent, since an inconsistent system can prove anything. The consistency of the systems considered by Gödel is known rather by the self-evident nature of the axioms and the obvious correctness of the rules of reasoning. Still, it is of interest that the systems, though obviously consistent, cannot prove their own consistency.

In the eighteenth century, G. Leibniz envisioned a universal calculating machine that could solve all mathematical problems. The impossibility of such a device has been conclusively demonstrated by further ramifications of Gödel's work developed by A. Church and A. Turing. This work shows that mathematics cannot be mechanized, and that creativity and ingenuity will always be required. This is perhaps the most important consequence of Gödel's work. Another way of stating this consequence is that human beings can never eliminate the necessity of using their own intelligence, regardless of how cleverly they try. *See* LOGIC; RECURSIVE FUNCTION. [R.M.Sm.]

Goethite A mineral of composition FeO · OH, crystallizing in the orthorhombic system. Crystals are rare, and the mineral is usually in reniform or stalactitic masses which have a radiating fibrous internal structure. The luster is adamantine to dull, and the color light to dark brown. The Mohs hardness is 5.0–5.5, and the density is 4.28 for crystals and 3.3–4.3 for massive material. Most of the common, yellow-brown, earthy ferric oxides known as limonite are mixtures composed largely of cryptocrystalline goethite.

Goethite is one of the most common minerals. It is the major constituent of the gossan at the surface of metalliferous deposits rich in iron-bearing sulfides, as at Bisbee, Arizona, and of laterites, as in Cuba. Well-formed crystals are found at Pribram, Bohemia, and Cornwall, England. It is an important iron ore in Alsace-Lorraine, in the Lake Superior hematite deposits, and in the southern Appalachians. *See* LIMONITE. [L.Gr.]

Goiter An enlargement of all or part of the thyroid gland, due to various causes, which may be accompanied by a hormonal dysfunction.

Colloid, endemic, or simple goiter results from an iodine deficiency. Endemic goiter is peculiar to many geographic areas in the world where iodine is lacking in food, soil, or water. However, there is often a marked difference in the incidence of the disease in endemic areas, thus indicating other etiological factors.

Diffuse primary hyperplasia, or exophthalmic goiter, results from an overall glandular enlargement accompanied almost always by excess thyroxine secretion. The exact cause of this disorder, also called Grave's disease, is unknown.

Nodular goiter, or multiple colloid goiter, is usually the result of repeated cyclic changes in a part of the gland subjected to other abnormal processes such as the two preceding conditions. Thyroid output may be decreased but more often is normal or increased.

Goiter does not necessarily mean thyroid dysfunction, since enlargement may be seen in hypothyroid, euthyroid, and hyperthyroid states. Conversely, dysfunction may occur in a gland that shows no visible change. *See* THYROID GLAND. [E.G.St./N.K.M.]

Gold A chemical element, Au, atomic number 79 and atomic weight 196.967, a deep yellow, soft, and very dense metal. Gold is classed as a heavy metal and as a noble metal; commercially, it is the most familiar of the precious metals. Copper, silver, and gold are in the same group of the periodic table of elements. The Latin name for gold, *aurum* (glowing dawn), is the source of the chemical symbol Au. There is only one stable isotope of gold, that of mass number 197.

Uses. Consumption of gold in jewelry accounts for about three-fourths of the world's production of gold. Industrial applications, especially electronic, consume another 10–15%. The remainder is divided among medical and dental uses, coinage, and bar stock for governmental and private holdings. Gold coins and most decorative gold objects are actually gold alloys, because the metal itself is too soft (2.5–3 on Mohs scale) to be useful with frequent handling. *See* GOLD ALLOYS.

Radioactive ^{198}Au is used in medical irradiation, in diagnosis, and in a number of industrial applications as a tracer. Another tracer use is in the study of movement of sediment on the ocean floor in and around harbors. The properties of gold toward radiant energy have led to development of efficient energy reflectors for infrared heaters and cookers and for focusing and retention of heat in industrial processes.

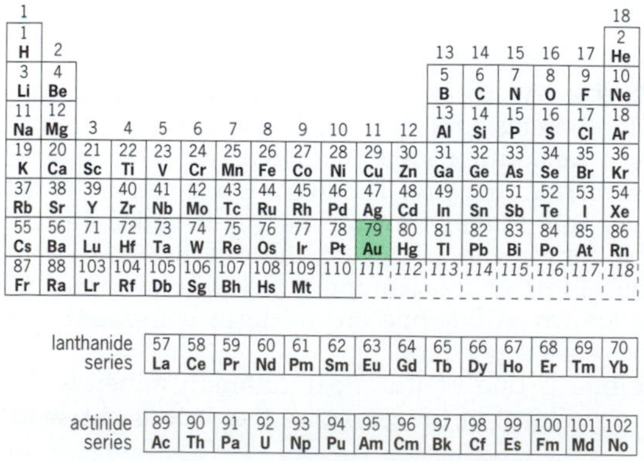

Occurrence. Gold occurs widely throughout the world, but usually very sparsely, so that it is quite a rare element. Sea water contains low concentrations of gold, on the order of 10 μg per ton (10 parts of gold per trillion parts of water). Somewhat higher concentrations accumulate on plankton or on the ocean bottom. At present, no economically feasible process is visualized for extracting gold from the sea. Native, or metallic, gold and various telluride minerals are the only forms of gold found on land. Native gold may occur in veins among rocks and ores of other metals, especially quartz or pyrite, or it may be scattered in sands and gravel (alluvial gold). *See* GOLD METALLURGY.

Properties. The density of gold is 19.3 times that of water at 20°C (68°F), so that 1 ft^3 of gold weighs about 1200 lb (1 m^3, about 19,000 kg). Masses of gold, like those of other precious metals, are measured on the troy scale, which counts 12 oz to the pound. Gold melts at 1064.43°C (1947.97°F) and boils at 2860°C (5180°F). It is somewhat volatile well below its boiling point. Gold is a good conductor of heat and electricity. It is the most malleable and ductile metal. It can easily be made into translucent sheets 0.0000039 in. (0.00001 mm) thick or drawn into wire weighing only 0.00005 oz/ft (0.5 mg/m). The quality of gold is expressed on the fineness scale as parts of pure gold per thousand parts of total metal, or on the karat scale as parts of pure gold per 24 parts of total metal. Gold readily dissolves in mercury to form amalgams. Gold is one of the least active metals chemically. It does not tarnish or burn in air. It is inert to strong alkaline solutions and to all pure acids except selenic acid.

Compounds. Gold may be either unipositive or tripositive in its compounds. So strong is the tendency for gold to form complexes that all the compounds of the 3+ oxidation state are complex. The compounds of the 1+ oxidation state are not very stable and tend to be oxidized to the 3+ state or reduced to metallic gold. All compounds of either oxidation state are easy to reduce to the metal.

In its complex compounds gold forms bonds most readily and stably with halogens and sulfur, less stably with oxygen and phosphorus, and only weakly with nitrogen. Bonds between gold and carbon are fairly stable, as in the cyanide complexes and a variety of organogold compounds. [W.E.C.]

Gold alloys Combinations of gold and other metals. Pure gold is soft. The addition of copper hardens the gold, and ultimately gold-copper alloys became standard for coinage. Gold coins in the United States contained 10% copper, the balance gold.

Pure gold is weak, having a tensile strength of less than 20,000 psi (138 megapascals) when annealed; however, by alloying with copper, sometimes in conjunction with silver or nickel, and often a little zinc, gold alloys with strengths of 60,000–100,000 psi (414–690 MPa) may be made. Addition of these metals changes the color of gold so that red, yellow, greenish, and white golds result. The proportion of gold in solid gold jewelry is designated in karats (k); pure gold is 24 k, 18 k is 18/24 or 75% pure gold, and 14 k is 14/24 or 58.3% pure gold.

Industrial uses of gold depend primarily upon the corrosion resistance and secondarily upon the strength that can be secured by alloying alone or by alloying and heat treating. Many alloys used in dentistry contain gold, silver, and copper, often with small amounts of platinum and palladium; these alloys can be heat-treated to develop strengths above 150,000 psi (1.0 gigapascal). The latter have good spring properties. Alloys of this type find many electrical uses as contacts, particularly where rubbing is involved. Gold electroplate, often thin, is employed on high-frequency conductors, such as those in radar equipment, because of the high electrical conductivity and tarnish resistance of gold. For the same reason gold is employed in the construction of many transistors, microcircuits, printed circuits, and integrated circuits. Most such devices are so small that the cost of gold is relatively unimportant.

Because gold does not oxidize when heated in air, appropriate gold compounds can be decomposed by heat to liberate the metal. Compounds of this type are used in the decoration of china and also for the production of printed electrical circuits on ceramics. These materials, known as liquid bright golds, are applied in the form of varnish, which is dried and then heated to redness, leaving a thin film of gold firmly attached to the underlying ceramic. *See* GOLD; GOLD METALLURGY. [C.R.M.; G.S.]

Gold metallurgy Extracting gold from ores, refining it, and preparing it for use. The principal gold-producing countries in order of productivity are South Africa, Russia, Canada, and the United States.

In placer deposits the gold particles are generally free, and may be concentrated by simple methods based on gravity (planning, rocking, and hydraulic mining). In ores mined from solid rock, now the chief source of gold, the metal is usually finely dispersed in quartz veins or lodes or disseminated in rock, and is usually associated with sulfide minerals. Such ores must be finely ground and subjected to chemical treatment for extraction of the gold.

Cyanidation is the predominant process for extracting gold from its ores. A dilute cyanide solution is added to the finely ground ore for dissolution of the gold and silver. After extraction of the gold, the solution is separated from the tailings by filtration or countercurrent decantation and then deaerated, and the gold is precipitated by the addition of zinc dust. The precipitate, containing gold as well as some silver and unreacted zinc, is smelted to eliminate the zinc. The resulting gold-silver alloy is refined by parting with nitric acid or hot concentrated sulfuric acid, by chlorination of the molten metal, or by electrolysis. *See* GOLD. [D.N.H.]

Goldhaber triangle The phase space triangle, or Goldhaber triangle, corresponds to the kinematically allowed boundary for a high-energy reaction leading to four or more particles. In a high-energy reaction between two particles a and b yielding four particles 1, 2, 3, and 4 in the final state ($a + b \rightarrow 1 + 2 + 3 + 4$), it is convenient to consider the reaction in terms of the production of two intermediate-state quasi-particle composites x and y, which then decay into two particles each, as in expression (1).

$$a + b \rightarrow x + y, \quad x \rightarrow 1 + 2, \quad y \rightarrow 3 + 4 \qquad (1)$$

Most high-energy interactions indeed proceed through such intermediate steps, in which, for specific values of the invariant masses $m_x = M_x^*$ and $m_y = M_y^*$, the quasi-particle composites may form resonances. However, the description in terms of the composites x and y, with the invariant masses m_x and m_y as variables, is valid irrespective of whether or not these composites form resonances.

The kinematical limits in this representation are particularly simple, namely, they form a right-angle isosceles triangle. A Goldhaber triangle plot corresponds to plotting a point (m_x, m_y) for each event occurring in the above high-energy reaction. Because of the kinematical constraints, these points must all lie inside the triangle.

If one considers the general reaction given in Eq. (1), then the length of each of the two equal sides of the triangle is Q, defined in Eq. (2). Here W is the total energy in the center of

$$Q = W - \sum_{i=1}^{4} m_i \qquad (2)$$

mass of particles a and b, and m_i, for $i = 1$ to 4, is the mass of the particles 1 to 4. Hence Q corresponds to the total kinetic energy available in the reaction.

In the triangle corresponding to the general reaction (see illustration), the vertical and horizontal bands indicate resonances at masses M_x^* and M_y^* with full width at half-maximum height or Γ_x and Γ_y, respectively. The bands shown of width 2Γ represent the regions usually chosen if the events corresponding to a given resonance are selected. [G.G.]

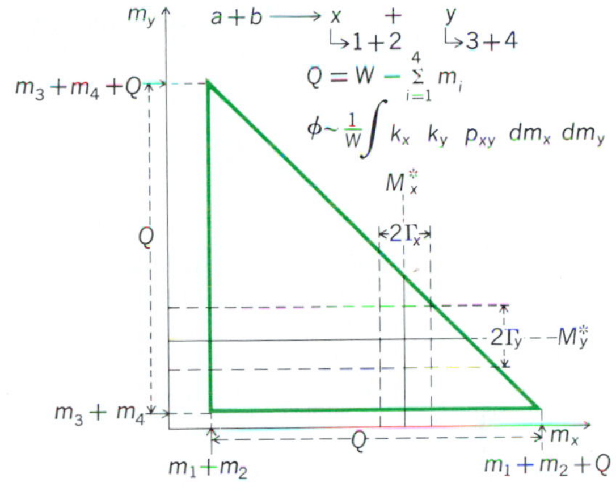

Definition of the kinematical boundary of the Goldhaber triangle for four particles.

Golgi apparatus A part of the endomembrane system of the cell, composed of regions of stacked cisternae (saccules), or flattened saccules, lacking membrane-attached ribosomes. A major role of the Golgi apparatus is to form vesicles whose limiting membranes can fuse with the external or plasma membrane of the cell. These vesicles function as vehicles for the transport of substances (membranes and secretory products) to the external cell surface. *See* CELL MEMBRANES.

The terms Golgi apparatus, Golgi component, Golgi zone, Golgi complex, Golgi body, and dictyosome have been used for the apparatus or its components. The term dictyosome denotes an individual stack of cisternae or saccules and the term Golgi apparatus usually denotes a system of dictyosomes where the dictyosomes are structurally and functionally associated (see illustration).

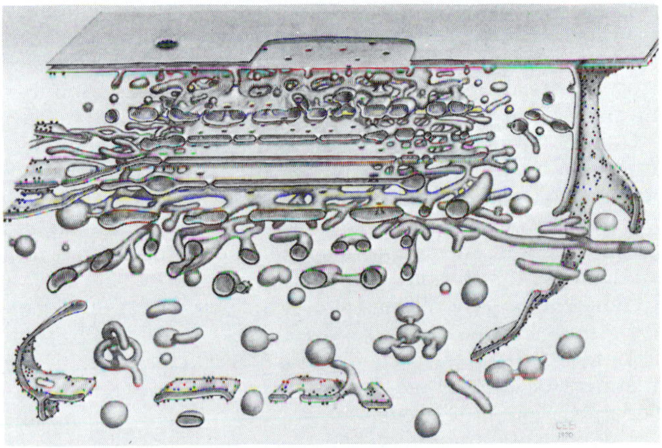

A diagrammatic interpretation of a portion of a dictyosome composed of six cisternae. Cisternal maturation is depicted from top to bottom. (*Courtesy of Prof. Charles Bracker*)

In secretory cells a major function of the Golgi apparatus is to compartmentalize products into membrane-bounded packets (the secretory vesicles), which may function as discrete cell components (for example, condensing vacuoles and primary lysosomes) before the vesicle contents are discharged at the cell surface. During discharge the vesicle membranes become part of the plasma membrane and, in certain rapidly growing or dividing cells, may be an important source of plasma membrane. Depending on cell type, secretory vesicles derived from the Golgi apparatus may contain enzymes, structural proteins, glycoproteins, lipoproteins, complex carbohydrates, polysaccharides, particles of high lipid content, or pigments. *See* LYSOSOME; VACUOLE.

Another important function of the Golgi apparatus is in the biosynthesis and delivery of membranes and associated informational receptor molecules to the surface of the cell.

[H.H.M.; D.J.M.]

Gonad One of the paired primary sex glands—the testes in the male and the ovaries in the female. The hormones secreted by these glands are responsible for the development of differential secondary sex characteristics at puberty, for example, distribution of hair, skeletal differences, voice, and mammary development, as well as the more obvious influence on the primary sex organs. The growth, development, and functions of female and male gonads are under the control of gonadotropic hormones produced by the anterior lobe of the pituitary. *See* OVARY; TESTIS. [C.H.L.]

Gonorrhea A common sexually transmitted disease caused by the bacterium *Neisseria gonorrhoeae*. In infection, *N. gonorrhoeae* colonizes the epithelium of the mucous membranes in the human genital tract, pharynx, or rectum. Local epithelial cell destruction is the usual outcome, but the infection may disseminate and result in arthritis or dermatitis and, rarely, in endocarditis or meningitis. Although asymptomatic infections occur, ectopic pregnancy and infertility are serious consequences of gonococcal pelvic inflammatory disease. Gonococcal infection during pregnancy is a risk factor for premature delivery and the transmission of the infection to the neonate in the form of ophthalmia neonatorum during passage through the birth canal or in the postpartum period.

Although gonorrhea continues to be the most widely reported communicable disease in the United States, the incidence of gonorrhea has been declining in both the United States and in European countries. In developing countries that report the incidence of gonorrhea, a comparable decline has not been found.

Transmission of *N. gonorrhoeae* requires direct contact between mucous membranes and is not transmitted by fomites or nonsexual contact. The risk of infection depends on anatomic site and the number of exposures. Not all infections produce symptoms; those individuals who are asymptomatic probably are the primary transmitters of the disease and thus are responsible for the spread of gonorrhea.

Control of gonorrhea depends on early diagnosis and effective treatment as well as identification of asymptomatic individuals in high-risk groups through large-scale screening programs. Prevention of gonorrhea relies on health education, the use of condoms, and antimicrobial therapy, because a practical vaccine has not yet been developed.

Although *N. gonorrhoeae* strains are susceptible to a number of antimicrobial agents, several strains are resistant to penicillin, tetracycline, or spectinomycin, so that ceftriaxone is the treatment of choice. *See* ANTIBIOTIC; SEXUALLY TRANSMITTED DISEASE. [S.A.La.]

Gonorynchiformes A small order of soft-rayed teleost fishes which at one time were included in the large order Clupeiformes (or Isospondyli). Gonorynchiforms are fusiform or moderately compressed fishes, varying from less than 2 in. (5 cm) to 5 ft (1.5 m) in length. There are single short dorsal and anal fins, and no adipose fin; the caudal fin is usually forked; and the pelvic fins are placed well back. The jaws are weak and toothless.

Modern gonorynchiforms are classified in 2 suborders, 4 families, 6 genera, and about 12 species. They have a wide fossil distribution that goes back to the Lower Cretaceous. *Gonorynchus* lives in marine shore waters of the Indo-Pacific area; the four genera of the Kneriidae and Phractolaemidae live in tropical African fresh waters; and the milkfish (*Chanos chanos*, family Chanidae) lives in marine and estuarine waters of the tropical Indo-Pacific. Milkfish are pond-cultured for food in Southeast Asia. *See* ACTINOPTERYGII; CYPRINIFORMES; TELEOSTEI.

[R.M.B.]

Gooseberry A small fruit represented by about six species of the genus *Ribes* of the plant order Rosales. The gooseberry is a thorny, spreading bush which produces red, yellow, or green berries. The most desirable hardier types in the United States are of American parentage, or are hybrids between American and European species. Commercial culture is limited to a few states, notably Oregon, Michigan, and Washington.

The fruit is very acid and only a few European varieties, when fully ripe, are suitable for eating fresh. The fruit may be canned or frozen for use in pies or as preserves. *See* FRUIT; ROSALES. [J.H.Cl.]

Gopher The name for the North American rodents of the family Geomyidae. There are 39 species in 8 genera; 11 species are found in the United States. The distinctive name pocket gopher is applied to these animals, since they have large, furlined cheek pouches which open outward on the side of the face. Two common species are the northern pocket gopher (*Thomomys talpoides*), a small burrowing species, most numerous in the western United States and Mexico, and the prairie pocket gopher (*Geomys bursarius*). *See* RODENTIA. [C.B.C.]

Gorgonacea An order of the coelenterate subclass Alcyonaria. The Gorgonacea are the horny corals which often form fanlike or featherlike colonies with branches spread rapidly or oppositely in one plane (see illustration). They attach to objects by somewhat enlarged bases or tufts of stolons. They are more are more widely distributed than the Alcyonacea and extend from the littoral zone to some great depth. *See* ALCYONARIA. [K.At.]

Gorgonacea colonies. (a) Skeleton of *Corallium knojoi*. (b) *Melitodes* sp. (preserved specimen). (c) *Anthoplexaura dimorpha* (preserved specimen).

Gorilla The largest of the living primates, of which a single species, *Gorilla gorilla*, is recognized. The gorilla is found in the forested areas of central-western Africa. There are two subspecies, the lowland gorilla, which is still found in fair numbers, and the mountain gorilla, which is comparatively rare.

The adult male may reach a height of 66 in. (1.68 m) and weigh more than 400 lb (180 kg). In addition to its impressive size, the gorilla has ferocious features, with eyes sunk deeply under heavy brows. However, it never kills to eat and exists on a fruit and vegetable diet.

The male gorilla lives in a family unit with his females and young. He is extremely protective and dangerous when accompanied by a female and will not hesitate to attack an intruder. At times some families band together to form a tribe of about 40 animals.

Gorillas become sexually mature at about the age of 12; since they seldom breed in captivity, little is known of their reproductive behavior. *See* MAMMALIA; PRIMATES. [C.B.C.]

Gout A hereditary disease due to abnormal purine metabolism. The disease is characterized by increased amounts of blood uric acid (hyperurilcemia), acute and chronic inflammatory arthritis, tophaceous deposits of uric acid crystals, and renal insufficiency. The increase of uric acid is thought to be caused by increased production or decreased excretion of uric acid from the kidney or both. *See* ARTHRITIS; PURINE.

Primary gout occurs most frequently in middle-aged males and is passed as a familiar or hereditary trait which for some unknown reason does not appear as often in females. Secondary gout refers to the disease when associated with some underlying disorder causing an elevation in uric acid production (for example, myeloproliferative disorders).

The uric acid becomes deposited as urates in soft tissues, especially around the joints, in the cartilages of the ear, along the shafts of long bones, in the kidney, and occasionally on the heart valves. Such deposits, when superficial, can be seen grossly as reddish, inflamed masses called tophi. *See* PROTEIN METABOLISM; URIC ACID. [R.Se.]

Governor A device used to control the speed of a prime mover. A governor protects the prime mover from overspeed and keeps the prime mover speed at or near the desired revolutions per minute. When a prime mover drives an alternator supplying electrical power at a given frequency, a governor must be used to hold the prime mover at a speed that will yield this frequency. An unloaded diesel engine will fly to pieces unless it is under governor control. *See* PRIME MOVER.

A governor regulates the speed of a prime mover by properly varying the flow of energy to or from it. In the case of gas and steam turbines and internal combustion engines, the fuel furnishes the energy to the prime mover. For such applications, the governor usually controls the speed of the unit by regulating the rate at which fuel, and hence energy, is furnished to the prime mover. The governor controls the fuel flow so that the speed of the prime mover remains constant regardless of load and other disturbances, or changes in accordance with such operating conditions as changes in speed setting.

The speed of a prime mover is usually measured by a ballhead that contains flyweights driven at a speed proportional to the speed of the prime mover. The force from the flyweights is balanced, at least in part, by the force of compression of a speeder spring (see illustration). The upper end of this spring is positioned according to the speed setting of the governor.

To increase the power output of a governor, a hydraulic amplifier is often employed. A governor that keeps the speed of a prime mover constant is said to be isochronous. In a simple isochronous governor, the ballhead senses the speed and strokes a pilot valve plunger that regulates the flow of fluid to a servomotor. The performance of the simple isochronous governor is often greatly improved by the introduction of a dashpot in the feedback path from the output to the ballhead. If there is little damping in the prime mover, instability often occurs when the simple isochronous governor is used, whereas this instability is removed when the dashpot is incorporated. Acceleration governors are sometimes used in place of governors with dashpots. In such governors a flywheel is employed instead of a dashpot. The prime mover drives the flywheel through a spring. [R.O.]

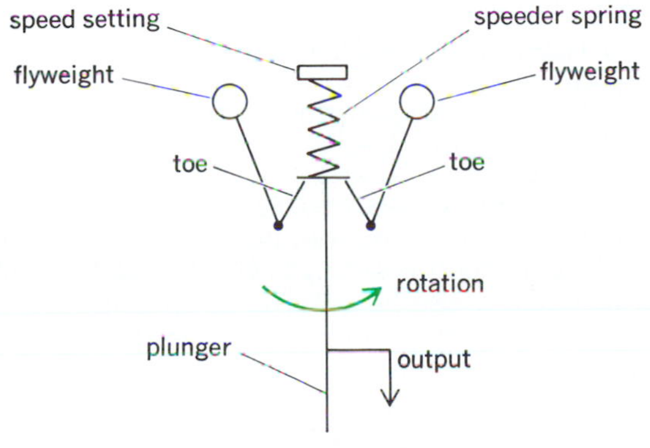

Ballhead governor.

Grab bucket A digging and materials-handling tool with hinged sides or leaves. It is particularly suited for digging deep holes or for raising material to high places. A grab bucket may have two, three, or four movable sides. A two-sided bucket is somewhat rectangular in shape and is called a clamshell. A three- or four-leafed bucket is ball-shaped and is called an orange peel. For hard digging, clamshells are equipped with hardened prongs or teeth. Orange peels are used mainly for cleaning sewer catch basins, vertical pipes, caissons, and other small vertical openings. [E.M.Y.]

Graben A block of the Earth's crust, generally with a length much greater than its width, that has been dropped relative to the blocks on either side (see illustration). The size of a graben may vary. The faults that separate a graben from the adjacent rocks are inclined from 50 to 70° toward the downthrown block and have displacements ranging from inches to thousands of feet. The direction of slip on these indicates that they are gravity faults. *See* FAULT AND FAULT STRUCTURES; HORST; RIFT VALLEY. [P.H.O.]

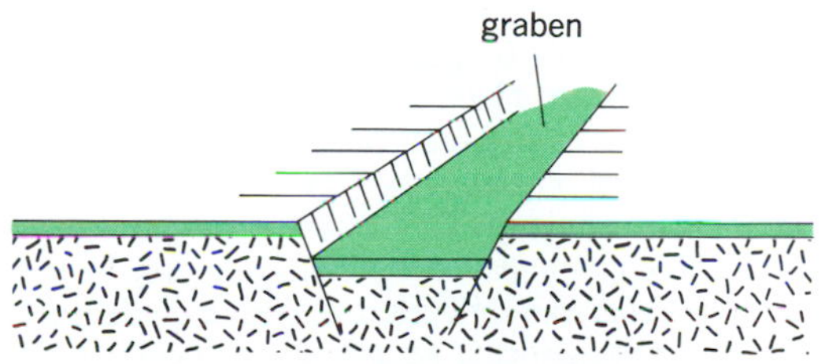

Diagram of simple graben. (*After A. K. Lobeck, Geomorphology, McGraw-Hill, 1939*)

Gradient of a scalar The result of the application of the distributive vector differential operator ∇, $\nabla = \mathbf{i}\,\partial/\partial x + \mathbf{j}\,\partial/\partial y + \mathbf{k}\,\partial/\partial z$, to a differentiable scalar function $S(x,y,z)$; thus $\nabla S = \mathbf{i}\,\partial S/\partial x + \mathbf{j}\,\partial S/\partial y + \mathbf{k}\,\partial S/\partial z$. The letters $\mathbf{i}$, $\mathbf{j}$, $\mathbf{k}$ are symbols for the base vectors associated with x, y, z. The gradient of S is also denoted by grad S, while the symbol ∇ is usually called del and less frequently nabla. *See* CALCULUS OF VECTORS. [H.V.C.]

Gradient wind A hypothetical wind based upon the assumption that the sum of the horizontal components of the Coriolis force and the atmospheric pressure gradient force per unit mass is equivalent to a wind acceleration which is normal to the direction of the wind itself (centripetal acceleration), with the implication that there are no viscous forces acting. The direction of the gradient wind is the same as that of the geostrophic wind. The gradient wind speed is less than the geostrophic speed when the air moves in a cyclonically curved path and greater when the air moves in an anticyclonically curved path. The gradient wind is a good approximation of the actual wind and is often superior to the geostrophic wind, particularly when the flow is strongly curved in the cyclonic sense. *See* CORIOLIS ACCELERATION; GEOSTROPHIC WIND. [F.S.]

Grain boundaries The surfaces separating individual crystals, or grains, in a crystalline solid. Many solids, especially metallic and ceramic materials, are characterized by a periodic internal structure. This structure is defined by arrangement of atoms (or ions) in a precise three-dimensional array to form a crystal. The extent of the arrangement is large with respect to the size of the atom or ion. Although solids may be a single crystal, it is much more common for solids to be composed of many crystals, which retain their characteristic structure but are misoriented with respect to neighboring crystals by a relative rotation of their principal axes. The boundary separating these individual grains is known as a grain boundary and is characterized by a two-dimensional structure which may be related to the structure of neighboring grains. Therefore, grain boundaries represent planar defects in an otherwise perfect crystalline material.

The microscopic structure of grain boundaries is intimately related to many macroscopic properties, such as electrical resistivity, corrosion resistance, tensile strength, and fracture toughness. Perhaps the effect of grain boundaries on mechanical properties is more important than on any other class of properties. In general, strength and ductility are controlled by the motion of dislocations under the influence of applied forces, which is responsible for plastic deformation. Grain boundaries provide an effective barrier to this motion so that strength increases and ductility decreases with decreasing grain size. Curiously, the role of grain boundaries is reversed at elevated temperatures, at which small grain size, that is, a large amount of grain boundary area, facilitates plastic deformation by mass transport of constituent atoms along these boundaries. This phenomenon is known as creep. Fine-grained polycrystalline materials sometimes exhibit extraordinarily large degrees of plasticity at elevated temperatures. This large degree of plasticity, termed superplasticity, actually is an accelerated form of creep. *See* CREEP (MATERIALS); SUPERPLASTICITY. [C.L.B.]

Grain crops Crop plants that belong to the grass family (Gramineae), generally grown for their edible starchy seeds. They also are referred to as cereal crops and include wheat, rice, maize (corn), barley, rye, oats, sorghum (jowar), and millet. The grain of all these is used directly for human food and also for livestock, especially maize, barley, oats, and sorghum.

An important attribute of these grain crops is the easy manner in which they can be stored. The grain often dries naturally before harvest to a safe moisture content (10–12%), or can easily be dried with modern equipment. Grain placed in adequate storage facilities can then be protected against insect infestations and maintained in sound condition for years. *See* CORN; MILLET; OATS; RICE; SORGHUM; WHEAT. [E.G.H.]

Gram-molecular weight The molecular weight of an element or compound expressed in grams (g), that is, the molecular weight on a scale on which the atomic weight of the ^{12}C isotope of carbon is taken as 12 exactly. This replaces the earlier scale on which the atomic weight of oxygen was taken as 16.00 g. In the International System of Units, gram-molecular weight is replaced by the mole.

The ratio of the gram-molecular weights of any two elements or compounds must be identical with the ratio of the absolute weights of their individual molecules. Therefore, the gram-molecular weights of all elements or compounds contain the same number of molecules. This number, called the Avogadro number, N, is 6.022×10^{23}. *See* AVOGADRO NUMBER; MOLE (CHEMISTRY); MOLECULAR WEIGHT; RELATIVE MOLECULAR MASS. [T.C.W.]

Gram's stain A differential stain used in bacteriology as one of the criteria in the identification and classification of bacteria. The Gram's stain reaction is based on differences among stained bacteria in their resistance to decolorization by neutral solvents such as alcohol or acetone. The fixed smear is stained with a slightly alkaline solution of a suitable basic dye, treated with a mordant and then with a neutral decolorizing agent, and usually counterstained with a dye solution of a contrasting color. Cells or structures that hold the color of the primary stain are called gram-positive, and those that take the color of the counterstain, gram-negative. [G.K.]

Grand unification theories Attempts to unify three fundamental interactions—strong, electromagnetic, and weak—with a postulate that the three forces, with the exception of gravity, can be unified into one at some very high energy. The basic idea is motivated by the incompleteness of the electroweak theory of S. Weinberg, A. Salam, and S. Glashow, which has been extremely successful in the energy region presently accessible with the use of accelerators, and by the observation that the coupling constant for strong nuclear forces becomes smaller as energy increases whereas the fine-structure constant ($\alpha = 1/137$) for electromagnetic interactions is expected to increase with energy. *See* GRAVITATION; STRONG NUCLEAR INTERACTIONS; WEAK NUCLEAR INTERACTIONS; WEINBERG-SALAM MODEL.

The simplest grand unification theory (GUT), proposed by H. Georgi and Glashow, is based on the assumption that the new symmetry that emerges when the three forces are unified is given by a special unitary group SU(5) of dimension 24. This symmetry is not observable in the low-energy region since it is badly broken. In this model, as in most GUTs, the coupling constants for the three interactions merge into one at an energy of about 10^{14} GeV. Quarks and leptons belong to the same multiplets, implying that distinctions between them disappear at the energy of 10^{14} GeV or above. In addition to the known 12 quanta of strong, electromagnetic, and weak interactions, there appear, in this model, 12 new quanta with the mass of 10^{14} GeV. These generate new but extremely weak interactions that violate baryon- and lepton-number conservation. The most spectacular prediction of GUTs is the instability of the proton, which is a consequence of baryon-number (and lepton-number) violation. *See* LEPTON; PROTON; QUARKS; SYMMETRY BREAKING; SYMMETRY LAWS (PHYSICS).

GUTs, in general, explain why the charge of the electron is precisely that of the proton with the opposite sign. Massive neutrinos are a distinct possibility in GUTS, and the smallness of their mass can also be understood. Neutrinos with the appropriate mass may explain the missing mass in galaxies and clusters and perhaps the formation of galaxies. *See* COSMOLOGY; NEUTRINO.

According to the scenario based on the GUTS, the universe underwent a phase transition when its temperature cooled to

10^{27} K, which corresponds to 10^{14} GeV in energy and to the first 10^{-35} s after the big bang. The phase transition caused an exponential expansion (10^{30}-fold in 10^{-32} s) of the universe, which explains why the observed 3 K microwave background radiation is uniform (the horizon problem), and why the universe behaves as if space is practically flat (the flatness problem). *See* BIG BANG THEORY; INFLATIONARY UNIVERSE COSMOLOGY; PHASE TRANSITIONS; UNIVERSE.

In spite of its theoretical triumph and spectacular predictions, the simple SU(5) model is practically untested by experiment and appears to be incomplete or even incorrect. No experimental evidence of proton decay has been established, and the problems which GUTs leave unsolved are numerous. *See* ELEMENTARY PARTICLE; FUNDAMENTAL INTERACTIONS; SUPERGRAVITY; SUPERSYMMETRY. [C.W.K.]

Granite A crystalline igneous rock that consists largely of alkali feldspar (typically perthitic microcline or orthoclase), quartz, and plagioclase (commonly calcic albite or oligoclase). Its average grain size is 0.04–1.0 in. (1–25 mm); finer-grained rocks of this composition include rhyolite and aplite, and coarser-grained ones are granite pegmatite. *See* APLITE; PEGMATITE; RHYOLITE.

The revised nomenclature of the International Union of Geological Sciences (IUGS) subcommission defines granite as containing 80–100% by volume quartz, alkali feldspar, and plagioclase in the proportions given in the illustration, and 20–0% accessory minerals. The three essential minerals must include 20–60% quartz, and alkali feldspar must constitute 65–90% of the total feldspar. The variety alkali feldspar granite is similar except that alkali feldspar constitutes 90–100% of its total feldspar. The term granitic rocks includes granodiorite and tonalite as well as granite, and as used by some geologists may include quartz syenite to quartz diorite (see illustration). *See* FELDSPAR; GRANODIORITE; HORNBLENDE; MUSCOVITE; TONALITE.

Granites may be divided into three major types: calc-alkaline, peraluminous, and alkaline. Calc-alkaline granites typically are biotite or biotite-hornblende granites, some contain augite, and sphene is a common accessory.

Peraluminous granites, also known as S-type granites, contain aluminum in excess of that contained in feldspars and biotite; thus muscovite is an accessory mineral. Other aluminous minerals such as andalusite, sillimanite, cordierite, or garnet also may be accessory. Subalkaline to alkaline granites are characterized by iron-rich mafic minerals and relatively sodic alkali feldspar. The subalkaline type typically contains ferruginous biotite or hornblende, or both, but varieties containing ferrohedenbergite or fayalite are not uncommon. Allanite and zircon are common accessories. The alkaline type contains the Na-Fe minerals aegirine or riebeckite-arfvedsonite, or all three, and may also contain ferruginous biotite or even astrophyllite, eudialyte, or other rare minerals. *See* GRANITIZATION; IGNEOUS ROCKS. [F.B.]

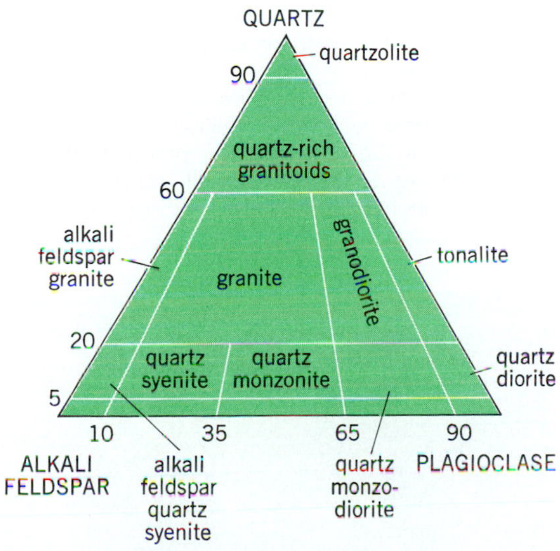

Classification of granitic rocks by the IUGS scheme. All proportions are by volume percentages.

Granitization The process whereby various types of rocks may be converted, essentially in the solid state, to granite or closely related material, such as granodiorite—a process which cannot be considered independently of metamorphism, metasomatism, migmatization, the partial fusion of metamorphic rocks, and the intrusion of granitic magma. The origin of granitic rocks has been the focus of controversy between petrologists (magmatists) who believed that the rocks crystallized for the most part from granitic magmas, and petrologists (transformists, granitizers) who believed that the rocks were formed by a metasomatic process of granitization, without the intervention of a silicate melt. The interpretation of migmatites has played a central role in the granite controversy. *See* MAGMA; METAMORPHIC ROCKS; METAMORPHISM; MIGMATITE.

Granitization is generally a large-scale change and is considered to have operated in orogenic zones essentially contemporaneously with metamorphism. Here stratified rocks, undergoing reconstitution and recrystallization, may suffer changes in bulk composition adequate to convert them to rocks of granitic character. Such stratified rocks have thus been granitized, and the granites formed thereby may be indistinguishable from those of magmatic origin. In fact, one type of granite may pass imperceptibly into the other. *See* GRANITE; GRANODIORITE.

Granitization on a small scale, more or less confined to contacts of magmatic bodies at high levels in the Earth's crust, might more appropriately be referred to as contact metasomatism or feldspathization. *See* CONTACT AUREOLE.

The change in bulk composition during granitization involves the introduction of certain constituents and removal of others. This substitution, in essentially solid rock and with no appreciable overall volume change, is known as metasomatism. Bodily movement of fluids (along fissures and openings) carrying requisite substances in solution may account for the compositional changes in the rocks. Where openings are too small to permit fluid flow, diffusion along intergrain surfaces may occur. For short distances diffusion through the crystal structure may be important. Extensive deformation accompanying metamorphism may maintain adequate channelways for the migrating material. *See* METASOMATISM. [P.J.Wy.]

Granodiorite A phaneritic (visibly crystalline) plutonic rock composed chiefly of sodic plagioclase (oligoclase or andesine), alkali feldspar (microcline or orthoclase, usually perthitic), quartz, and subordinate dark-colored (mafic) minerals (biotite, amphibole, or pyroxene). Granodiorite is intermediate between granite and quartz diorite (tonalite). For convenience granite and granodiorite are commonly grouped and referred to as granite. *See* GRANITE; IGNEOUS ROCKS. [C.A.C.]

Granulite An important class of metamorphic rocks exposed at the surface of the Earth's crust, and inferred to make up a large portion of the deeper crust. Granulites are known to have formed at higher temperatures, and in many cases, higher pressures, than most other crustal rock assemblages. Thus, they are believed to have formed at considerable depths in the crust. *See* METAMORPHIC ROCKS.

Granulites may be of many different bulk compositions, inherited from precursor sedimentary, igneous, or lower-grade metamorphic rocks. The high temperatures of crystallization have resulted in very low water content, reflected in nearly anhydrous mineralogy. Characteristic minerals of granulite metabasalts are plagioclase, orthopyroxene, clinopyroxene, hornblende, and garnet. These minerals are also characteristic of granulites of intermediate to granitic composition, together with progressively greater amounts of quartz and potassium feldspar. The association of potassium feldspar with orthopyroxene is definitive for charnockite, a granulite of approximately granitic composition characteristic of ancient high-grade terrains. *See* BASALT; GRANITE.

Granulites characteristically contain CO_2-rich fluid inclusions in the mineral grains, in contrast to the more aqueous fluid inclusions of other kinds of rock. This has suggested that action of volatiles low in H_2O and rich in CO_2, probably of subcrustal origin, were important in crustal metamorphism early in the Earth's history. *See* METAMORPHISM. [R.C.N.]

Granuloma inguinale A mildly infectious, chronic, granulomatous disease principally affecting skin and subcutaneous tissues of the genital and rectal areas. The causative organism is *Calymmatobacterium granulomatis* (*Donovania granulomatis*). Although rare in the United States, the disease is very common in New Guinea, the Caribbean, and other tropical and subtropical areas.

Although the method of transmission is controversial, there is a definite correlation with sexual activity and a frequent association with homosexual behavior. Tetracyclines are the drugs of choice, with streptomycin an effective alternative. [D.S.K.]

Granuloreticulosia A subclass of the Rhizopodea. These Protozoa produce reticulopodia which often fuse into networks greatly exceeding the area of the body, as in various Foraminiferida. Such nets are effective food traps and may also be important in construction of tests and cyst walls. A characteristic feature of reticulopodia is the bidirectional flow of cytoplasm. Granuloreticulosia include three orders: Athalamida, Foraminiferida, and Xenophyophorida. *See* ATHALAMIDA; FORAMINIFERIDA; RHIZOPODEA; XENOPHYOPHORIDA. [R.P.H.]

Grape The two genera of grapes are *Vitis* and *Muscadinia*. *Vitis vinifera* has intermittent forked tendrils, bark that sheds, and elongated clusters with berries that adhere to the pedicels at maturity. This species also has thin, smooth, shiny leaves with three, five, or seven lobes. Berries may be round or oval and have edible skins that adhere to the flesh. In the American species skins slip from the pulp. Many American species have a characteristic musky or foxy odor and taste. *Muscadinia* can be easily distinguished from *Vitis* by bark that does not shed and simple tendrils that do not fork.

Viticulture is the science of grape production. In a broad sense, viticulture includes studies of grape varieties; methods of culture such as trellising, pruning, and training; insect and disease control; propagation; and raisin production.

In the United States, *V. vinifera* is grown on the west coast, and most of the grapes cultivated east of the Rocky Mountains have been derived from American native species such as *V. labrusca* and *V. aestivalis*, or from crosses between them and *V. vinifera*. There is also a native Caribbean species and several Asiatic species. There are three main species of *Muscadinia* that are found mostly in the southeast portion of the United States.

Table grapes are utilized for food and decorative purposes. Some of the leading table grapes in California are Emperor, Tokay, Thompson Seedless, Cardinal, and Perlette. Some of the principal commercial American varieties are Concord, Catawba, Delaware, and Niagara. Some of the important varieties of *M. rotundifolia*, the Muscadine grape, are Scuppernong, Thomas, and Hunt.

Important wine grapes in California include Cabernet Sauvignon, Carignane, Chardonnay, Grenache, French Colombard, and Zinfandel. Many of the North American and *rotundifolia* species that are used for eating purposes are also used for wine. [R.J.W.]

Grapefruit A citrus fruit, *Citrus paradisi*, including both seedy and seedless varieties, with pink, white, or red flesh. The trees are vigorous and evergreen, with well-rounded tops. *See* EVERGREEN PLANTS.

Of the world supply of grapefruit, almost 80% is produced in the United States. Approximately one-half of this crop is marketed as fresh fruit; the rest is processed as canned juice and segments. Other countries producing significant amounts of grapefruit are Israel, Argentina, and the Republic of South Africa. [F.E.G.]

Graph theory A branch of mathematics that belongs parly to combinatorial analysis and partly to topology. Its applications occur (sometimes under other names) in electrical network theory, operations research, organic chemistry, theoretical physics, and statistical mechanics, and in sociological and behavioral research. Both in pure mathematical inquiry and in applications, a graph is customarily depicted as a topological configuration of points and lines, but usually is studied with combinatorial methods. *See* COMBINATORIAL THEORY; TOPOLOGY.

A graph consists of a set of points, a set of lines, and an incidence relation that designates the end points of each line. In many applications no line starts and ends at the same point. (Such a line would be called a loop.) Also, no two lines have the same pair of end points. A graph whose lines satisfy these conditions is called simplicial. The valence of a point is the number of lines incident on it, calculated so that a loop is twice incident on its only end point. Two graphs are isomorphic if there is one-to-one correspondence from the point set and line set of one onto the point set and line set, respectively, of the other that preserves the incidence relation. An automorphism of a graph is an isomorphism of a graph with itself. Two graphs are homeomorphic if, after smoothing over all points of valence 2, the resulting graphs are isomorphic. *See* GROUP THEORY.

Drawing a graph on a surface decomposes the surface into regions. One colors the regions so that no two adjacent regions have the same color, rather like a political map of the world. It is a remarkable fact that for a given surface, there is a single number of colors that will always be enough no matter how many regions occur in a decomposition of the surface. The smallest such number is called the chromatic number of that surface. It is easy to draw a plane map that requires four colors. In 1976 K. Appel and W. Haken settled a problem dating back to about 1850, by showing that four colors are always enough for plane maps.

A graph is planar if it can be drawn in the plane so that none of its lines cross each other. Neither of the two graphs in the illustration can be drawn in the plane. K. Kuratowski proved in 1930 that a graph is planar if and only if it contains no subgraph homeomorphic to either of those two graphs.

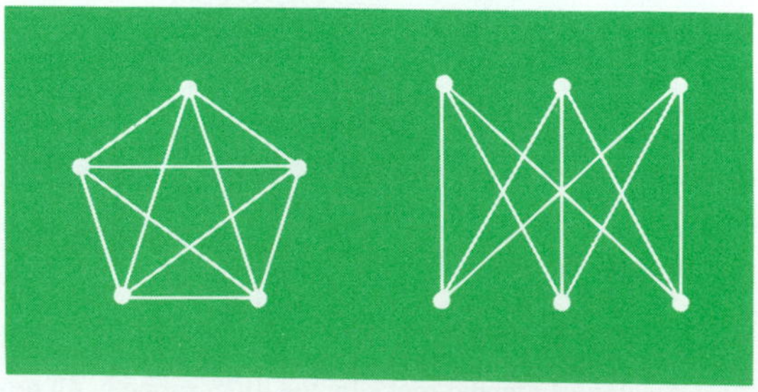

Prototypes of all nonplanar graphs.

In a directed graph, or digraph, each line *ab* is directed from one end point *a* to the other end point *b*. There is at most one line from *a* to *b*.

An oriented graph is obtained from an ordinary graph by assigning a unique direction to every line. If there is one line between every pair of points and no loops, an ordinary graph is called complete. An oriented complete graph is called a tournament. [J.L.Gro.]

Graphic methods Pencil and paper methods that employ the geometry of a plane to express mathematical relationships and to carry out mathematical operations in analog form. Some of the specific topics that arise under this broad definition are the plotting of experimental data, curve fitting and development of empirical formulas, graphical interpolation and extrapolation, nomographs and alignment charts, graphical differentiation and integration, and graphical statics.

Since graphical methods are often rapid and easily understood, they find widespread application in engineering. Graphical techniques are also useful in statistical studies to determine the correlation between two quantities, to obtain a frequency distribution from experimental data, to smooth data having an appreciable statistical fluctuation, and to represent the conclusions of a statistical study. *See* COORDINATE SYSTEMS; CURVE FITTING; EXTRAPOLATION; INTERPOLATION; NOMOGRAPH; NUMERICAL ANALYSIS; STATICS; STATISTICS. [K.S.K.]

Graphic recording instruments Instruments that make a graphic record of one or more quantities as a function of another variable, usually time. Recording instruments may have any type of sensing device, such as pressure, temperature, voltage, or weight. The sole identifying feature is that they make a graphic record of the quantity being measured. An instrument with a recording device often uses the suffix graph, such as a barograph or oscillograph, in place of the suffix used with a similar nonrecording instrument, such as a barometer or oscilloscope.

Recording instruments are classified according to their principle of operation. They are either direct-acting, in which case the marking device is mechanically connected to and directly operated by the primary detector, or indirect-acting, in which case the measurement energy of the primary detector is increased through some intermediate means to actuate the marking device. These intermediate means are usually mechanical, electrical, electronic, or photoelectric.

In addition to the primary classification, recording instruments are classified by their exhibiting means, recording means, number of marking devices, and marking means. The exhibiting means are either circular charts or strip charts. The recording means may be continuous, intermittent (marking device retracted from chart between measurements), or sequential (more than one variable is intermittently recorded). The instrument may have one or more marking devices. The marking means may be an inking pen on a paper chart, an inked typed impression on a paper chart, a heated stylus on coated paper, a mechanical stylus on coated or chemically treated paper, an electric stylus on current-sensitive paper, or a light beam on film or light-sensitive paper which can be continuously developed during use. [A.W.J.]

Graphite A low-pressure polymorph of carbon (the common high-pressure polymorph being diamond). Graphite is metallic in appearance and very soft. Crystals of graphite are infrequently encountered since the mineral usually occurs as earthy, foliated, or columnar aggregates often mixed with iron oxide, quartz, and other minerals. *See* CARBON.

The sheetlike character of the graphite atomic arrangement results in distinctive physical properties. The mineral is very soft, with hardness 1½; it soils the fingers and leaves a black streak on paper, hence its use in pencils. The specific gravity is 2.23, often less because of the presence of pore spaces and impurities. The color is black in earthy material to steel-gray in plates, and thin flakes are deep blue in transmitted light. Graphite is a conductor of electricity.

The major sources of graphite are in gneisses and schists, where the mineral occurs in foliated masses mixed with quartz, mica, and so on. Noteworthy localities include the Adirondack region of New York, Korea, and Ceylon. In Sonora, Mexico, graphite occurs as a product of metamorphosed coal beds. Graphite is also observed in meteorites. [P.B.M.]

Synthetic graphite. Commercially produced synthetic graphite is a mixture of crystalline graphite and cross-linking intercrystalline carbon. Its physical properties are the result of contributions from both sources. Thus, among engineering materials, synthetic graphite is unusual because a wide variation in measurable properties can occur without significant change in chemical composition.

At room temperature the thermal conductivity of synthetic graphite is comparable to that of aluminum or brass. An unusual property of graphite is its increased strength at high temperature. Graphite is resistant to thermal shock because of its high thermal conductivity and low elastic modulus. It is one of the most inert materials with respect to chemical reaction with other elements and compounds. It is subject to oxidation, and reaction with and solution in some metals.

Graphite has many uses in the electrical, chemical, metallurgical, nuclear, and rocket fields: electrodes in electric furnaces producing carbon steel, alloy steel, and ferroalloys; anodes for the electrolytic production of chlorine, chlorates, magnesium, and sodium; motor and generator brushes; sleeve-type bearings and seal rings; rocket motor nozzles; missile nose cones; metallurgical molds and crucibles; linings for chemical reaction vessels; and, in a resin-impregnated impervious form, for heat exchangers, pumps, pipings, valves, and other process equipment.

Graphite (carbon) fibers. Carbon fibers are filamentary forms of carbon, with a fiber diameter normally in the 6–10-micrometer range. The product is offered in the form of yarns or tows containing from 1000 to 500,000 filaments per strand. The fibers offer a unique combination of properties. They are flexible, lightweight, thermally and to a large extent chemically inert, and are good thermal and electrical conductors. In their high-performance varieties, carbon fibers are very strong and can be extremely stiff.

The principal use of high-performance carbon fibers is as the reinforcing component in structural epoxy matrix composites. Due to initially high cost, the original applications were almost exclusively for lightweight, high-stiffness, and high-strength composites for the aerospace industry. The second major usage of high-performance carbon fibers is in sporting goods, such as golf club shafts, tennis rackets, fishing rods, and sailboat structures. The major matrix material for both applications is epoxy. *See* COMPOSITE MATERIAL. [H.F.V.]

Graptolithina A class of extinct marine colonial animals believed to be related to the Hemichordata. They have an organic skeleton which appears to be made of collagen fibrils. The skeleton forms a series of tubes varying in diameter from 0.0012 to 0.8 in. (30 micrometers to 2 millimeters) whose walls are built of a series of half rings usually strengthened by supplementary cortical material, mainly on the outside. This structure is similar to that seen in the extant *Rhabdopleura* and its allies. The colonies range in size from a few millimeters diameter to long, slender ones which may be more than a meter long but less than a centimeter wide. Graptolites are found in Paleozoic rocks of all continents, but are most abundant in dark shales of the Ordovician and Silurian which often contain few other fossils.

The class Graptolithina is divided into several orders, of which the Dendroidea and Graptoloidea are the most important. The other orders and some doubtfully associated groups are known from relatively few occurrences. [I.St.]

Grass Grasses provide about 53% of the total feed units (forages and grains) consumed by all domestic livestock, in the form of pastures, native range grazing, and harvested forage (hay, green feed, and grass silage). Grasses fill other important roles in the national economy: in conservation of soils, in development and conservation of water resources, in upstream flood control, in wildlife and game management, and in a wide range of outdoor recreation. *See* GRASSLAND ECOSYSTEM.

Grass stems have solid joints (nodes) and leaves arranged in two rows, with one leaf at each joint (see illustration). The leaves consist of the sheath, which fits around the stem like a split tube, and the blade, which is commonly long and narrow. Seed heads are made up of minute flowers on tiny branchlets, often several crowded together, but always two-ranked like the leaves. The flowers are generally wind-pollinated. The seeds are enclosed between two bracts, or glumes, which remain on the seed when ripe. *See* FLOWER.

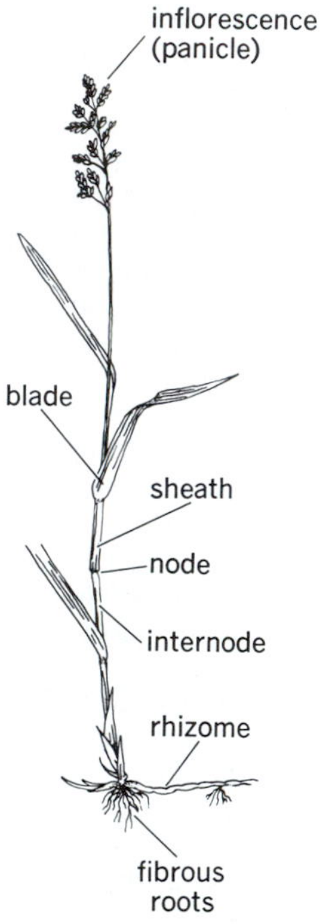

Typical grass plant. (*After P. D. Strausbaugh and E. L. Core, Flora of West Virginia, West Va. Univ. Bull., ser. 52, no. 12–2, pt. 1, p. 67, 1952*)

Grass species may be annual or perennial. Some perennial species are perpetuated by creeping stems as well as by seed. Stems that creep underground are termed rhizomes, or rootstocks, whereas surface-creeping stems are called runners, or stolons. Species having creeping stems produce a well-knit sod. These generally tolerate continuous close grazing and are utilized primarily for pastures. Species that do not creep vegetatively tend to grow erect. They make an open or loose sod, and are best suited for hay or silage. In subhumid to semiarid regions, range species are predominantly perennial and are perpetuated mostly by seed. In other areas both annuals and perennials commonly occur. All grasses have a fibrous type of root system that permeates the soil extensively and is effective in preventing soil erosion and restoring soil humus content. Grasses are worldwide in distribution; adapted species are available for nearly all conditions of soil and climate. See separate articles for grasses under their common names. [H.B.S.]

Grassland ecosystem Grassland vegetation comprises herbaceous (not woody) plant species; graminoids (grasses and grasslike plants) usually dominate. Natural grasslands once occupied at least one-third of the land surface of the Earth, occurring in regions which were too arid for forest, but where the climate was not so adverse as to prevent the development of a closed perennial herbaceous cover that is lacking in desert. In addition, grassland with scattered trees (savannas) occurs in transition zones between grassland and forest, particularly in the tropics and subtropics, and in semidesert regions shrubs are scattered throughout the grassland. Humans have converted some forest areas into seminatural grasslands, and they have modified many natural grasslands by tillage and by grazing domesticated livestock. The most extensive areas of natural open grassland occur in the temperate zone (south-central North America, north-central Asia, eastern Europe, southeastern South America, and southern Africa). Savannas are widespread in the tropical regions of Africa, South America, and Australia; shrubby grasslands lie adjacent to deserts.

While there are a large number of plant species in the grassland ecosystems of the world, usually very few grasses dominate in any one location: commonly 50–200 species of flowering plants occur in the same grassland, but only a few of these are considered dominants. Grassland complexity (in terms of numbers of plant species) tends to increase with increasing favorableness of habitat and to decrease with increasing aridity.

The treelessness of natural grassland is often the result of aridity. Relatively low precipitation and high evaporation restrict the supply of soil moisture. Most trees cannot compete effectively with grasses where upper soil layers are intermittently moist, but deeper layers are continuously dry. Other factors that have been cited to account for the absence of trees in various grassland areas are impermeable soil, excessive salinity, and poor drainage.

Natural grasslands have developed in areas where fires are characteristic, since herbaceous species are much better adapted than trees and shrubs to withstand the effects of fire. This is because their perennating buds are located near the soil surface where they are less exposed, and because they have no perennial stems.

The two principal purposes for which natural grasslands are used are crop production and domesticated livestock grazing. Only a small fraction of the world's grasslands have been brought under tillage because of constraints imposed by climate, aridity, soil conditions (including stoniness, salinity, and erodibility), and topography. Most of the remainder are in some form of ranching enterprise. *See* ECOSYSTEM; GRASS.

[R.T.C.; G.V.D.]

Gravel A loose, or unconsolidated, deposit of rounded pebbles, cobbles, or boulders, whose size is greater than 0.08 in. (2 mm) in diameter. Gravels may consist only of pebbles, cobbles, and boulders, with large voids between the particles; these are the openwork gravels. Commonly, however, gravels have sand filling the interstices between the pebbles. The sand filling the pores is referred to as the matrix and commonly makes up about 30% of the rock. [R.Si.]

Gravimetric analysis That branch of quantitative chemical analysis in which a desired constituent is converted (usually by precipitation) to a pure compound or element of definite, known composition, and is weighed. In a few cases, a compound or element is formed which does not contain the constituent but bears a definite mathematical relationship to it. In either case, the amount of desired constituent can be determined from the weight and composition of the precipitate.

At least two weighings are required for each analysis—the original sample, and the dried or ignited residue. From these weights, the percentage or proportion of the desired constituent may be calculated from the equation below.

$$\%A = \frac{\text{wt of residue} \times \text{factor} \times 100}{\text{wt of sample}}$$

The factor is determined from a knowledge of the chemical relationships between the weight of substance A contained in, or equivalent to, a fixed weight of residue of known composition. *See* ANALYTICAL CHEMISTRY; ELECTRODEPOSITION ANALYSIS; QUANTITATIVE CHEMICAL ANALYSIS; STOICHIOMETRY. [S.G.S.]

Gravitation The mutual attraction between all masses and particles of matter in the universe. In a sense this is one of the best-known physical phenomena. During the 18th and 19th centuries gravitational astronomy, based on I. Newton's laws, attracted many of the leading mathematicians and was brought to such a pitch that it seemed that only extra numerical refinements would be needed in order to account in detail for the motions of all celestial bodies. In the present century, however, A. Einstein shattered this complacency, and the subject is currently in a healthy state of flux.

Newton's law of gravitation. Newton's law of universal gravitation states that every two particles of matter in the universe attract each other with a force that acts in the line joining them, the intensity of which varies as the product of their masses and inversely as the square of the distance between them. Put into symbols, the gravitational force F exerted between two particles with masses m_1, and m_2, separated by a distance d is given by the equation below, where G is called the constant of gravitation.

$$F = Gm_1m_2/d^2$$

Gravitational constant. In 1774 N. Maskelyne determined G by measuring the deflection of the vertical by the attraction of a mountain. This method is much inferior to the laboratory method in which the gravitational force between known masses is measured. In the torsion balance two small spheres, each of mass m, are connected by a light rod, suspended in the middle by a thin wire. The deflection caused by bringing two large spheres each of mass M near the small ones on opposite sides of the rod is measured, and the force is evaluated by observing the period of oscillation of the rod under the influence of the torsion of the wire (see illustration). This is known as the Cavendish experiment, in honor of H. Cavendish who achieved the first reliable results by this method in 1797–1798. Modern determinations yield the results: constant of gravitation $G = 6.67 \times 10^{-11}$ mks units; mass of Earth = 5.98×10^{27} g. The uncertainty of these results is probably about one-half unit of the last place given.

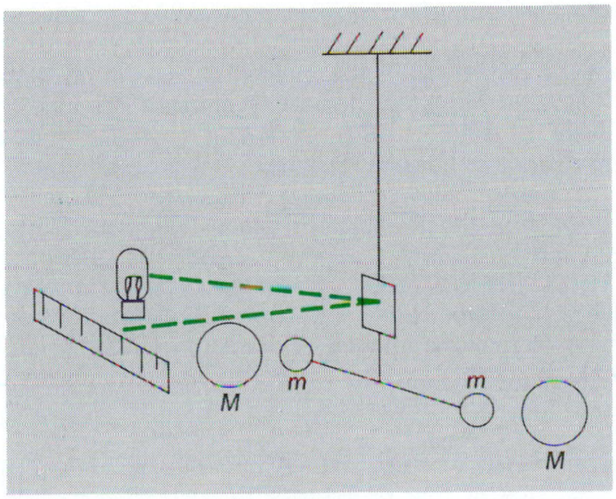

Torsion balance.

In newtonian gravitation G is an absolute constant, independent of time, place, and the chemical composition of the masses involved. Partial confirmation of this was provided before Newton's time by the experiment attributed to Galileo when different weights released simultaneously from the top of the Tower of Pisa reached the ground at the same time. Newton found further confirmation, experimenting with pendulums made out of different materials. Early in this century, R. Eötvös found that different materials fall with the same acceleration to within 1 part in 10^7. The accuracy of this figure was extended by R. H. Dicke to 1 part in 10^{11}, using aluminum and gold, and by V. B. Braginskii and V. 1. Panov to 0.9×10^{-12} with a confidence of 95%, using aluminum and platinum.

Mass and weight. A cosmology with changing physical "constants" was first proposed in 1937 by P. A. M. Dirac. Field theories applying this principle have since been proposed by P. Jordan and D. W. Sciama and by C. Brans and Dicke. In these theories G is diminishing; for instance Brans and Dicke suggest a change of about 2×10^{-11} per year. This would have profound effects on phenomena ranging from the evolution of the universe to the evolution of the Earth.

In the equations of motion of newtonian mechanics, the mass of a body appears as inertial mass as a factor of the acceleration, and as gravitational mass in the expression of the gravitational force. The equality of these masses is confirmed by the Eötvös experiment. It justifies the assumption that the motion of a particle in a gravitational field does not depend on its physical composition. In Newton's theory the equality can be said to be a coincidence, but not in Einstein's theory, where inertia and gravitation are unified.

While mass in newtonian mechanics is an intrinsic property of a body, its weight depends on certain forces acting on it. For example, the weight of a body on the Earth depends on the gravitational attraction of the Earth on the body and also on the centrifugal forces due to the Earth's rotation. The body would have lower weight on the Moon, even though its mass would remain the same. *See* CENTRIFUGAL FORCE.

Gravity. This should not be confused with the term gravitation. Gravity is the older term, meaning the quality of having weight, and so came to be applied to the tendency of downward motion on the Earth. Gravity or the force of gravity are today used to describe the intensity of gravitational forces, usually on the surface of the Earth or another celestial body. So gravitation refers to a universal phenomenon, while gravity refers to its local manifestation. *See* GRAVITY.

Accuracy of newtonian gravitation. A discrepancy in newtonian gravitation was discovered by U. J. J. Leverrier in the orbit of Mercury. Because of the action of the other planets, the perihelion of Mercury's orbit advances. But allowance for all known gravitational effects still left an observed motion of about 43 seconds of arc per century unaccounted for by Newton's theory. Attempts to account for this by adding an unknown planet or by drag with an interplanetary medium were unsatisfactory, and a very small change was suggested in the exponent of the inverse square of force. This particular discordance was accounted for by Einstein's general theory of relativity in 1916, but the final word on the subject has yet to be said. *See* RELATIVITY.

Gravitational lens. Light is deflected when it passes through a gravitational field, and an analogy can be made to the refraction of light passing through a lens. It has been suggested that a galaxy situated between an observer and a more distant source might have a focusing effect, and that this might account for some of the observed properties of quasistellar objects. The multiple images of one quasar (QSO 0957 + 567 A,B) are almost certainly caused by the light from a single body passing through a gravitational lens. *See* GRAVITATIONAL LENS; QUASAR.

Relativistic theories. In spite of its success and the absence of a reasonable alternative, Newton's theory was heavily criticized, not least with regard to its requirement of "action at a distance" (that is, through a vacuum). Newton himself considered this to be "an absurdity," and he recognized the weaknesses in postulating in his system of mechanics the existence of preferred reference systems (that is, inertial reference systems) and an absolute time.

Einstein showed in his special theory of relativity that these postulates were physically unacceptable. In his general theory of relativity he incorporated gravitational phenomena in such a way that there was no longer any preferred reference system. He treated the phenomenon of gravitation as a consequence of the geometric properties of space-time, this geometry being affected by the presence of gravitating matter. The acceleration of a body is determined by the local geometry, and the newtonian concepts of gravitational force and action at a distance are abandoned. Mathematically the theory is far more complicated than Newton's. Instead of the single potential described above, Einstein worked with 10 quantities that are members of a tensor.

An important step in Einstein's reasoning is his "principle of equivalence": that a uniformly accelerated reference system imitates completely the behavior of a uniform gravitational field. This principle requires that all bodies fall in a gravitational field with precisely the same acceleration and this is confirmed by the Eötvös experiment mentioned earlier. Also, if matter and antimatter were to repel one another, it would be a violation of the principle. Einstein's theory requires that experiments should have the same results irrespective of the location or time. This has been said to amount to the "strong" principle of equivalence. *See* FREE FALL.

Gravitational waves. The existence of gravitational waves, or gravitational "radiation," was predicted by Einstein shortly after he formulated his general theory of relativity. They are now a feature of any relativity theory. Gravitational waves are "ripples in the curvature of space-time." In other words, they are propagating gravitational fields, or propagating patterns of strain, traveling at the speed of light. They carry energy and can exert forces on matter in their path, producing, for instance, very small vibrations in elastic bodies. The gravitational wave is produced by change in the distribution of some matter. It is not produced by a rotating sphere, but would result from a rotating body not having symmetry about its axis of rotation: a pulsar, perhaps, In spite of the relatively weak interaction between gravitational radiation and matter, the measurement of this radiation is now technically possible and may already have been achieved. [J.M.A.D.]

Gravitational collapse The rapid implosion of a star or other astronomical body from its initial, normal size to a size hundreds or thousands of times smaller. Astrophysicists have devoted considerable effort to understanding the mechanisms behind such explosive astrophysical phenomena as supernovae, quasars, the origin of strong radio sources, and explosions in the nuclei of galaxies. Because the explosive energy in these objects may be a sizable fraction of the rest mass, and because thermonuclear reactions, such as occur in the Sun and in hydrogen bombs, can convert no more than 0.9% of the rest mass to energy, a search has been made for other, more powerful mechanisms of energy release. *See* NUCLEAR FUSION; PLASMA PHYSICS; QUASAR; SUPERNOVA; THERMONUCLEAR REACTION.

The most likely alternative mechanism seems to be gravitational collapse. If a star or other body collapses to a small fraction of its original size and then releases the final kinetic energy of collapse as radiation and high-energy particles, a sizable fraction of its original mass is converted to energy. There is no fully compelling reason to believe that gravitational collapse is important in quasars, in the origin of strong radio sources, or in galactic explosions. However, theoretical and observational studies provide convincing evidence that collapse is important in supernovae.

The effectiveness of gravitational collapse as a mechanism for converting mass to explosive energy is severely limited by general relativistic phenomena. J. Robert Oppenheimer and his students showed in 1939 that any spherical star which collapses to a radius smaller than its Schwarzschild radius, given by the equation below, can no longer release its kinetic energy

$$R_{Sch} = 3.0(M/M_{\odot})\ \text{km}$$

of collapse in any form to the outside universe; it must continue to collapse until gravitation forces have compressed it to infinite density (general relativistic singularity). The work of Oppenheimer and his students and of S. Chandrasekhar also revealed that any star exceeding about 1.5 solar masses (the critical value is uncertain by a factor 2) must eventually reduce its mass below this value—for example, by a supernova explosion—or face inevitable collapse past its Schwarzschild radius and into the singularity. A set of theorems due to R. Penrose and S. Hawking generalize to situations without spherical symmetry the concept of being inside the Schwarzschild radius, and the consequent inevitability of continued collapse to a singularity.

Evidence has suggested strongly that a black hole, produced millions of years ago by stellar collapse, is now in orbit around a massive, bright star in the constellation Cygnus. The hole seems to be pulling gas off its companion. As the gas falls inward toward the hole's Schwarzschild radius, it emits x-rays which have been detected by satellites above the Earth's atmosphere. The name of this possible black hole is Cygnus X-1. *See* BLACK HOLE. [K.S.T.]

Gravitational lens A massive body producing distorted, magnified, or multiple images of more distant objects when its gravitational fields bend the paths of light rays. Lenses have been observed when the light from very distant quasars is affected by intervening galaxies and clusters of galaxies, producing several different images of the same quasar. Albert Einstein predicted the occurrence of this phenomenon in 1936, but the discovery of real gravitational lenses did not occur until 1979. Gravitational lenses, in addition to being intrinsically interesting, can reveal the intrinsic properties of galaxies, active galaxies, and quasars, and provide information on the overall geometry and scale of the universe. *See* GRAVITATION.

Gravitational lenses can make distant quasars appear brighter than they would otherwise be, since their focusing action beams more light in the direction of the Earth. This effect could cloud the interpretation of the quasar statistics, which currently argues for a "quasar era" early in cosmic evolution.

When the brightness of the multiple images in gravitational lenses has been monitored for a long time, lenses may reveal the cosmos in some more fundamental ways. If the energy-emitting regions of quasars are small (as is believed), then stars in the halos of galaxies could also act as lenses, changing the intensity of the light from quasar images when the stars pass between the Earth and the quasar. The lens phenomenon could be used to detect the existence of these stars, even if they were too faint to be seen optically. Variations in the brightness of multiple images of the same quasar could also determine the distance to the lensing galaxy and thus the scale of the universe. *See* COSMOLOGY. [H.L.Sh.]

Gravitational radiation A type of wave generated by accelerated masses that propagates through vacuum with the speed of light. The existence of gravitational radiation is an inescapable consequence of A. Einstein's general theory of relativity.

Properties. The general theory of relativity posits that matter (and energy) introduces curvature into four-dimensional spacetime and that matter moves in response to this curvature. The degree of curvature generated by a distribution of matter can be calculated by using the Einstein field equations, which are analogous to the Maxwell field equations of electromagnetism. The Einstein equations admit solutions corresponding to weak, transverse waves that propagate in vacuum with the speed of light just like the electromagnetic waves predicted by Maxwell's equations and demonstrated experimentally by H. Hertz in 1884. However, while electromagnetic waves are fluctuations in the electric field vector which can be measured by the acceleration of a single charged particle, gravitational waves are fluctuations in the tidal gravitational force which is measured by a tensor and can be detected by the relative acceleration induced between two free test masses (see illus.). *See* ELECTROMAGNETIC RADIATION; MAXWELL'S EQUATIONS.

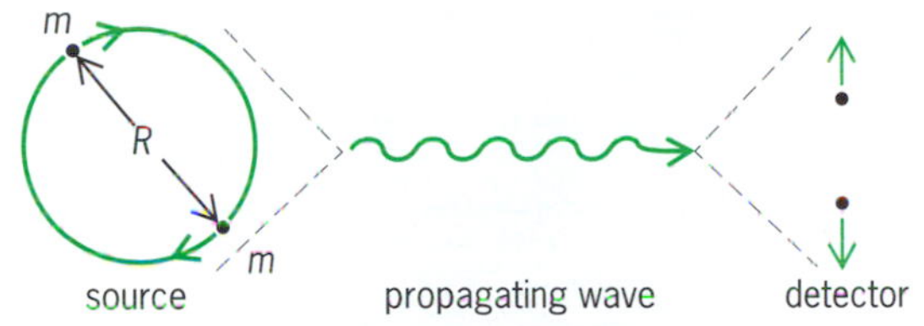

Generation and detection of gravitational waves. A powerful source of these waves, such as a neutron star binary with a circular orbit, radiates with a fundamental period equal to half the orbital period. The presence of the wave can be demonstrated conceptually by measuring the relative acceleration between two free masses.

Sources. In 1974, R. Hulse and J. Taylor discovered the first binary pulsar, now known as PSR 1913+16. This is believed to comprise two neutron stars, one of which spins with a period of 59 milliseconds and emits regular radio pulses with an equal period. Radio astronomers on Earth observe a pulse period that varies slightly as the pulsar source traces an elliptical orbit around its companion due to the Doppler effect. This system emits gravitational waves with a fundamental period equal to the orbital period of 7.8 h, and these waves carry energy away from the binary, causing it to become more tightly bound. The orbital period decreases as the two stars approach one another. The fractional change in the measured orbital period agrees with the relativistic prediction to approximately 0.5%, a striking verification of general relativity. *See* DOPPLER EFFECT; NEUTRON STAR; PULSAR.

When such binary stars shrink so that their orbital periods are only a few milliseconds, the rate of radiation increases dramatically. The waves emitted by such coalescing stars may be strong enough to measure at Earth even when the stars reside in quite distant galaxies. Other proposed sources of detectable gravitational radiation include supernovae, binary black holes in active galactic nuclei, and cosmic strings. *See* BLACK HOLE; COSMIC STRING; SUPERNOVA. [R.D.Bl.]

Detectors. The direct detection of gravitational waves has not yet been achieved, though there is an active research program directed toward the construction of appropriate ultrasensitive detectors. The two classes of detectors are resonant-mass and free-mass antennas. Both types comprise a mechanical element and a transducer that converts mechanical motion to an electronic signal. *See* TRANSDUCER.

The resonant-mass antennas are commonly cylindrical bars of a few tons mass, although a mass in the shape of a icosahedron (a body with 60 faces, like a soccer ball) would be superior to a cylinder. The material most commonly used is aluminum. To reduce thermal noise, the bar is brought to cryogenic temperatures in a vacuum vessel cooled by liquid helium and is isolated against vibrations from the terrestrial environment by a series of mechanical filters. The lowest-frequency longitudinal mode of the antenna, in which the two end faces of the cylinder are displaced in opposition, is most strongly excited by a passing gravitational wave. An electromechanical transducer is attached to one end face of the bar to detect the bar's vibration.

The free-mass antennas are composed of almost inertial masses which are actually very low frequency pendulums. A common design is the arrangement of three such masses at the vertices of a right triangle with equal legs lying on the surface of the Earth. The passage of a gravitational wave in a direction perpendicular to the plane of the free-mass antenna lengthens one leg of the triangle relative to the perpendicular leg. This change in length can be detected by a laser interferometer which is composed of mirrors mounted on the masses and a high-power visible laser light source. Each mass is rigorously isolated from vibrations in the environment, and the entire system is placed in a high-vacuum chamber to eliminate light scattering by gas and dust particles. *See* INTERFEROMETRY. [M.F.B.; D.H.D.]

Gravitational redshift A shift toward longer wavelengths of spectral lines emitted by atoms in strong gravitational fields. One of three famous predictions of the general theory of relativity, this shift results from the slowing down of all periodic processes in a gravitational field. The amount of the shift is proportional to the difference in gravitational potential between the source and the receiver. For starlight received at the Earth the shift is proportional to the mass of the star divided by its radius. In the solar spectrum the shift amounts to about 0.001 nanometer at a wavelength of 500 nanometers. In the spectra of white dwarfs, whose ratio of mass to radius is about 30 times that of the Sun, the shift is about 0.03 nanometer, which can easily be measured if it can be separated from the Doppler effect. *See* WHITE DWARF STAR. [D.La.]

Graviton A theoretically deduced particle postulated as the quantum of the gravitational field. According to Einstein's theory of general relativity, accelerated masses (or other distributions of energy) should emit gravitational waves, just as accelerated charges emit electromagnetic waves. And according to quantum field theory, such a radiation field should be quantized; that is, its energy should appear in discrete quanta, called gravitons, just as the energy of light appears in discrete quanta, namely photons. *See* ELEMENTARY PARTICLE; GRAVITATION; QUANTUM FIELD THEORY; RELATIVITY. [C.J.G.]

Gravity The gravitational attraction at the surface of a planet or other celestial body. The quantity g is often referred to simply as "gravity" or "the force of gravity" of Earth, both of which are incorrect. The force of gravity means the force with which a celestial body attracts an object, that is, the weight of the object. The letter g represents the acceleration caused by the gravitational force and, of course, has the dimensions of acceleration. *See* GRAVITATION. [D.Br./G.M.C.]

Gravity meter An instrument that measures the acceleration due to gravity; it is also known as a gravimeter. Different types of instruments are used, depending on whether the information required is an absolute value, a gravity difference between two or more locations, or long-term variations at a single point. Gravity acceleration is usually measured in gals (1 Gal = 1 cm/s^2), which is the acceleration produced by a force of 1 dyne acting on a mass of 1 gram. For small differences, it is useful to use a smaller unit, the milligal (mGal), which equals 10^{-3} Gal. The acceleration of gravity at the Earth's surface (g) is approximately 980 Gals.

Absolute gravity measurements are made by direct measurement of the acceleration of bodies failing in a vacuum. Relative gravity measurements between stations a few meters to a few thousand kilometers apart are made with four types of apparatus: spring, vibrating string, force-balance, and pendulum. Each type has certain advantages, depending on the mode of transport on land, sea, or air or on the ability to fit within a narrow-diameter drill hole. *See* ACCELEROMETER.

Absolute gravity systems utilize failing reflectors in high vacuums. The rate of fall is determined by an optical interferometer with a laser for a light source. Multiple drops are made, usually at least 50, over a vertical distance of about 1.5 ft (0.5 m). These systems provide absolute gravity values to an accuracy of about 0.03 mGal (standard error).

In principle, a spring-type gravity meter may be visualized as a simple mass m on a spring. If the meter is taken to another location where gravity is greater by an amount Δg, the spring will be stretched by some distance s, and this distance is proportional to the difference in gravity, Eq. (1) or (2), where k is the sensitivity of the spring.

$$\Delta s = k\Delta g \qquad (1)$$

$$\Delta g = \Delta s/k \qquad (2)$$

The natural frequency of a vibrating string depends on its tension and the mass per unit length of the string. If the tension is provided by a suspended mass, the frequency of the string changes with variations in gravity, and the specific change in frequency can be monitored with a high precision.

Changes in the gravitational pull on a ferromagnetic mass can be balanced by electromagnetic forces, and these forces can be measured very accurately by measuring electric current. This principle is employed in the Bell gravity meter used on surface ships at sea, and in the cryogenic gravity meter. The latter consists of an iron ball held in place by a magnetic field. This field is produced by a superconducting electromagnet kept within 1°C of absolute zero (−273°C). The stability of the magnetic field at this low temperature makes it possible for the cryogenic meter to measure changes in the Earth's gravity field as small as 1 microGal (10^{-6} Gal). *See* GEOPHYSICAL EXPLORATION; GRAVITY. [H.W.O.]

Graybody An energy radiator which has a blackbody energy distribution, reduced by a constant factor, throughout the radiation spectrum or within a certain wavelength interval. The designation "gray" has no relation to the visual appearance of a body but only to its similarity in energy distribution to a blackbody. Most metals, for example, have a constant emissivity within the visible region of the spectrum and thus are graybodies in that region. *See* BLACKBODY; HEAT RADIATION. [H.G.S.; P.J.W.]

Graywacke An argillaceous sandstone ("dirty" sandstone) that is characterized by an abundance of unstable mineral and rock fragments and a fine-grained clay matrix binding the larger sand-size detrital fragments. One of the most important characteristics is the clay matrix, which constitutes a minimum of 15% of the rock and may, in extreme cases, make up almost half of the bulk.

The unstable mineral fragments include feldspar, augite, hornblende, serpentine, biotite, chlorite, and magnetite. Rock fragments include many varieties of low- and high-grade metamorphic rocks (phyllite, slate, quartzite, and granulite), as well as aphanitic (microscopically crystalline) volcanic rocks, typically mafic in composition. The clay matrix is composed of a mixture of micaceous miners.

The sorting of graywackes is poor. Thus the size distribution curve is skewed toward the fine sizes. Even when the matrix is not considered, the larger detrital grains are poorly sorted. The grains tend to be sharp and angular and may show little or no evidence of abrasion. *See* SANDSTONE; SEDIMENTARY ROCKS; SUBGRAYWACKE. [R.Si.]

Great circle, terrestrial A circle or near-circle described on the Earth's surface by a plane passing through the center of the Earth (see illustration). The most important such circle is the Equator, cut by the plane which bisects, and is perpendicular to, the Earth's axis. Planes through the poles cut the Earth along meridians. The described forms are not quite circular, being slightly flattened ellipsoids. All other great circles have shapes between those of the Equator and the meridians. Even the equatorial circle and the ellipsoids have minor irregularities. *See* GEODESY; MATHEMATICAL GEOGRAPHY.

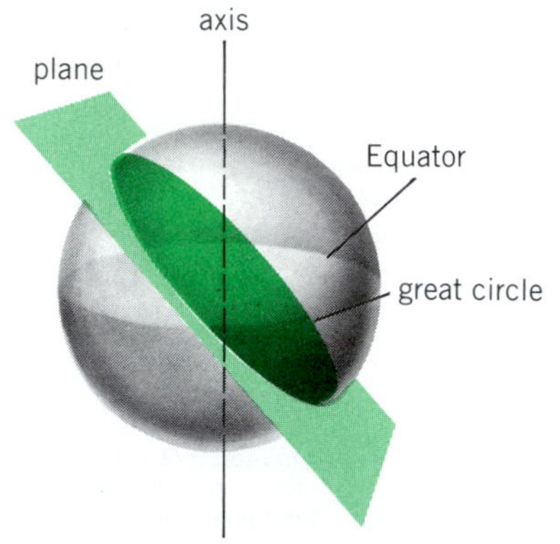

Diagram of a great circle described by a plane through the center of the Earth.

Segments of great circles extending from beginning to terminus of proposed global travel make the shortest routes on the Earth's surface. Ships sailing for short distances, however, follow compass directions, also called rhumb lines, or loxodromes. Except where they coincide with meridians or the Equator, loxodromes are not segments of great circles. In transoceanic voyages, distance can be shortened by following great-circle routes called orthodromes. These are laid out on gnomonic charts which show every orthodrome as a straight line. Ships, however, navigate by compass; thus orthodromes are transferred to a Mercator chart on which compass directions appear as straight lines; steering is actually by a series of short segments of compass directions. The longer the distance, the greater the deviation between loxodromes and orthodromes. *See* CARTOGRAPHY; MAP PROJECTIONS; NAVIGATION.

Great circles are used in various technical fields. Radio waves propagate along great circles from transmitting to receiving stations. This is utilized in loran and similar navigational systems. Great-circle directions are important in seismology, ballistics, intercontinental missiles, satellites, and like technical aspects of travel and transmission. [E.Ra./W.C.]

Greenhouse effect The Earth's atmosphere acts as the glass walls and roof of a greenhouse in trapping heat from the Sun. Like the greenhouse, it is largely transparent to solar radiation, but it strongly absorbs the longer-wavelength radiation from the ground. Much of this long-wave radiation is reemitted downward to the ground, with the paradoxical result that the Earth's surface receives more radiation than it would if the atmosphere were not between it and the Sun.

The absorption of long-wave (infrared) radiation is effected by small amounts of water vapor, carbon dioxide, and ozone in the air and by clouds. Clouds are almost completely opaque to infrared radiation, unless they are extremely thin. The appearance even of cirrus clouds after a period of clear sky at night is enough to cause the surface air temperature to increase rapidly by several degrees because of radiation from the cloud. *See* OZONE; TERRESTRIAL RADIATION. [L.D.K.]

Greenockite A mineral having composition CdS (cadmium sulfide). Greenockite usually occurs as earthy coatings with resinous luster and yellow-to-orange color. There is good prismatic cleavage; the hardness is 3 (Mohs scale) and specific gravity is 4.9. Greenockite and wurtzite, ZnS, are isostructural, and a complete solid-solution series exists between the two minerals. Although greenockite is the most common cadmium mineral, no deposits of it are sufficiently large to warrant mining it solely as a source of cadmium. *See* CADMIUM. [C.S.Hu.]

Green's function A solution of a partial differential equation for the case of a point source of unit strength within the region under examination. The Green's function is an important mathematical tool that has application in many areas of theoretical physics including mechanics, electromagnetism, acoustics, solid-state physics, thermal physics, and the theory of elementary particles. The underlying physics in each of these areas is generally described by some linear partial differential equation which relates the physical variable of interest (electrostatic potential or pressure amplitude in a sound wave, for example) to a source function. For present purposes the source may be regarded as an independent entity, although in some applications (for example, particle physics) this view masks an inherent nonlinearity. The source may be physically located within the region of interest, it may be simulated by certain boundary conditions on the surface of that region, or it may consist of both possibilities. A Green's function is a solution to the relevant partial differential equation for the particular case of a point source of unit strength in the interior of the region and some designated boundary condition on the surface of the region. Solutions to the partial differential equation for a general source function and appropriate boundary condition can then be written in terms of certain volume and surface integrals of the Green's function. *See* DIFFERENTIAL EQUATION. [P.Sh.]

Green's theorem A term used variously in mathematical literature to denote either (1) the Gauss divergence theorem

$$\iiint \nabla \cdot \mathfrak{F}\, dV = \iint \mathfrak{F} \cdot v\, dS$$

or some one of several forms or immediate consequences of this theorem, or (2) the plane case of Stokes' theorem. *See* CALCULUS OF VECTORS; GAUSS' THEOREM; STOKES' THEOREM. [H.V.C.]

Gregarinia A subclass of the class Telosporea. These protozoans occur principally as extracellular parasites in the digestive tracts and body cavities of invertebrates. There are three orders: the Archigregarinida, whose life cycle embraces both sexual and asexual phases; the Eugregarinida, which increases only by sporogony; and the Neogregarinida, whose life cycle involves schizogony and gamont formation. *See* ARCHIGREGARINIDA; EUGREGARINIDA; NEOGREGARINIDA; TELOSPOREA. [E.R.B./N.D.L.]

Greisen A pneumatolytically altered granite consisting of mainly quartz and a light-green (white) mica; the feldspars and biotites of the original granite have been replaced more or less completely by muscovite, lithium mica, kaolinite, and tourmaline. Other substances usually introduced during alteration are topaz, apatite, fluorite, and ores of iron, and in some places ores of tungsten and tin. *See* GRANITE; PNEUMATOLYSIS.

Greisen occurs in bands and veins intersecting granite. The bands are rarely more than 2 ft (0.6 m) thick with indefinite margins that grade into unaltered granite. They were evidently formed by flow of vapors and gases through fissures. [T.F.W.B.]

Grignard reaction A reaction between an alkyl or aryl halide and magnesium metal in a suitable solvent, usually absolute ether. The organomagnesium halides produced by this reaction are known as Grignard reagents and are useful in many chemical syntheses. The structure of a Gignard reagent is usually written RMgX, where X represents a halogen.

The scope of the Grignard reaction is extremely broad, and Grignard reagents have been prepared from many kinds of alkyl and aryl halides. In general, alkyl chlorides, bromides, and iodides and aryl bromides and iodides react readily. A few halides, such as aryl chlorides, react very sluggishly. *See* ORGANOMETALLIC COMPOUND. [P.E.F.]

Grimmiales An order of mosses. In size, the plants vary from slender to somewhat robust. They commonly grow in dense tufts or cushions, generally upon rock. These mosses usually range in height from about ⅕ in. (5 mm) to 3 in. (8 cm) in the erect species and in length from 1½ to 12 in. (4 to 30 cm) in the procumbent species. The leaves are very hygroscopic and are arranged on the stem in many rows, crowded, costate, usually lanceolate to ovate-lanceolate, but occasionally linear-subulate. The single peristome consists of 16 teeth, which may be entire, cribrose, or irregularly 2–3-cleft at the apex. *See* BRYOPSIDA. [W.H.W.]

Grinding The process of removing material from workpieces by the cutting action of a solid, rotating, grinding wheel composed of abrasive grains or grit plus a bonding material. The abrasive grains of the wheel perform a multitude of minute machining cuts on the workpiece. Although grinding is sometimes used as the complete machining operation on surfaces, it is generally considered a finishing process used to obtain a fine surface finish and extremely accurate dimensional tolerances. *See* ABRASIVE.

The grinding process is used in machining a wide range of metals in addition to such materials as cemented carbide, marble, stone, and certain ceramics. Because commercial grinding abrasives are many times harder than the metals to be machined, the grinding process may be used on metals too hard to machine otherwise. [A.H.T.]

Grinding mill A machine that reduces the size of particles of raw material fed into it. The size reduction may be to facilitate removal of valuable constituents from an ore or to prepare the material for industrial use, as in preparing clay for pottery making or coal for furnace firing. Coarse material is first crushed.

Grinding mills are of three principal types, as shown in the illustration. In ring-roller pulverizers, the material is fed past spring-loaded rollers. The rolling surfaces apply a slow large force to the material as the bowl or other container revolves. The fine particles may be swept by an air stream up out of the mill. In tumbling mills the material is fed into a shell or drum that rotates about its horizontal axis. The attrition or abrasion between particles grinds the material. The grinding bodies may be flint pebbles, steel balls, metal rods lying parallel to the axis of the drum, or simply larger pieces of the material itself. In hammer mills, driven swinging hammers reduce the material

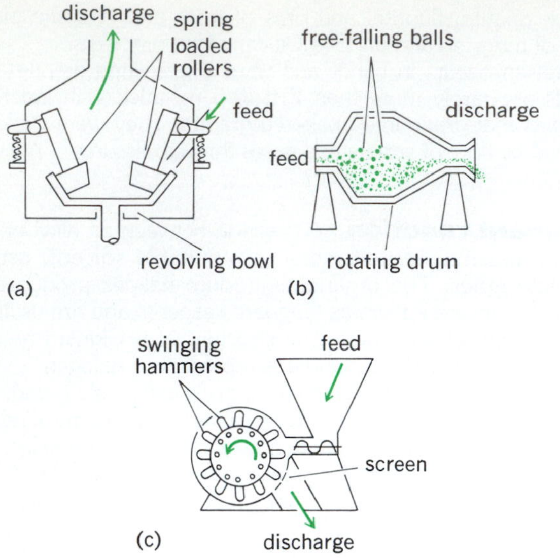

Basic grinding mills. (*a*) Ring-roller mill. (*b*) Tumbling mill. (*c*) Hammer mill.

by sudden impacts. *See* BALL-AND-RACE-TYPE PULVERIZER; CRUSHING AND PULVERIZING; PEBBLE MILL; TUMBLING MILL. [R.M.H.]

Griseofulvin An antifungal antibiotic effective in the treatment of ringworm and fungus infections of the skin, hair, and nails. Griseofulvin is a colorless, neutral, thermostable substance which is only very slightly soluble in water. The antibiotic is a metabolic product of various species of *Penicillium*.

Griseofulvin is indicated in the treatment of infections caused by dermatophytic fungi of the hands, feet, fingernails, toenails, glabrous skin, beard, and scalp. Onychomycosis and the following tineas respond to oral therapy: capitis, barbae, cruris, corporis, pedis, tonsurans, and rubrum. *See* ANTIBIOTIC. [M.J.W.]

Groin (engineering) A structure oriented perpendicular to a shoreline and extending into the sea for the purpose of retarding erosion or trapping littoral drift, thus establishing, stabilizing, or widening a beach.

A single structure may suffice, but a series of structures uniformly spaced are usually more efficient. Groins may be permeable or impermeable, high or low, fixed or adjustable, and may be constructed of timber, steel, stone, concrete, and other materials. *See* COASTAL ENGINEERING. [E.J.Q.]

Gromiida An order of the subclass Filosia. The test of these protozoa is mostly chitinous in some species, rather thin and a bit flexible in others. Siliceous particles of endogenous origin, or sometimes minute platelets, may be embedded in the chitinous layer (see illustration); some species typically reinforce the test with sand grains or other extraneous particles. A few species are marine, the rest fresh-water types. Little is known about most Gromiida. *See* FILOSIA. [R.P.H.]

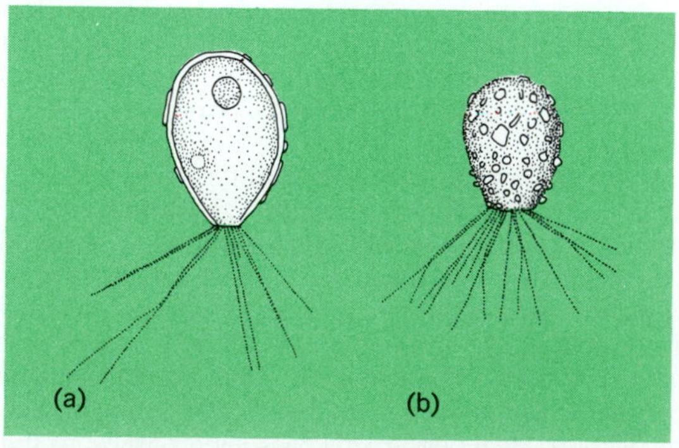

Representative Gromiida. (*a*) *Plagiophrys parvipunctata*. (*b*) *Pseudodifflugia fulva*. (*After R. P. Hall, Protozoology, Prentice-Hall, 1953*)

Ground proximity warning system An airborne computer that provides a pilot with visual and audible warning when it senses that the aircraft is in danger of inadvertent contact with the ground. The unit is typically 3 × 7 × 15 in. (8 × 18 × 38 cm) in size and weighs about 8 lb (4 kg). It receives inputs from the barometric altimeter, radio altimeter, and instrument landing system (ILS) glide-slope signal. Also, so as not to alarm the pilot unduly when the aircraft is making an intentional landing, it receives inputs from the landing gear (up or down) and the flaps (up or down). *See* AIR NAVIGATION; ALTIMETER; INSTRUMENT LANDING SYSTEM (ILS).

The system is not a cure-all for the aircraft-to-ground collison warning problem. Its sensors do not see the terrain ahead of the aircraft and therefore do not warn of impending collision with steeply rising terrain. It does not take care of the situation where a pilot knowingly flies below the ILS glide slope in order to touch down at the earliest possible point on the runway. Overall, however, the consensus is that the system has proved its usefulness. [S.H.D.]

Ground state In quantum mechanics, the stationary state of lowest energy of a particle or a system of particles. The ground state may be bound or unbound; when bound, its energy generally is a finite amount less than the energy of the next higher or first excited state. In the typical circumstance that the potential energy is zero at infinite separation, the magnitude of the negative ground-state energy is the binding energy, that is, the energy required to separate all the particles infinitely. *See* ENERGY LEVEL (QUANTUM MECHANICS); EXCITED STATE; NUCLEAR BINDING ENERGY; STATIONARY STATE. [E.G.]

Grounding Intentional electrical connections to a reference conducting plane, which may be earth (hence the term ground), but which more generally consists of a specific array of interconnected electrical conductors, referred to as the grounding conductor. The symbol which denotes a connection to the grounding conductor is three parallel horizontal lines, each of the lower two being shorter than the one above it (Fig. 1). The electric system of an airplane or ship observes specific grounding practices with prescribed points of grounding, but

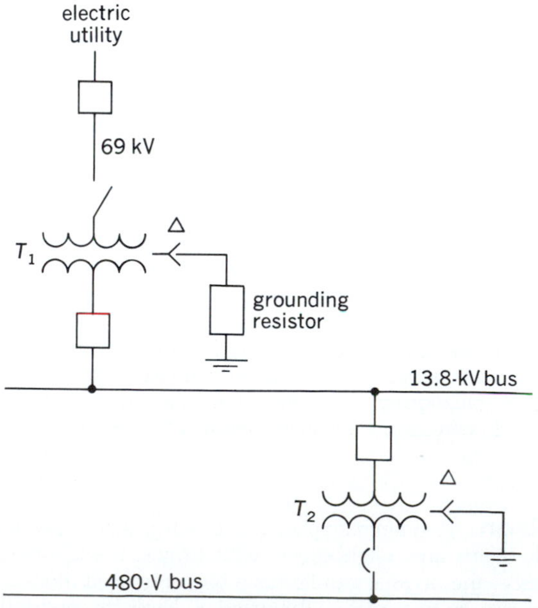

Fig. 1. Each conductively isolated portion of a distribution system requires its ground.

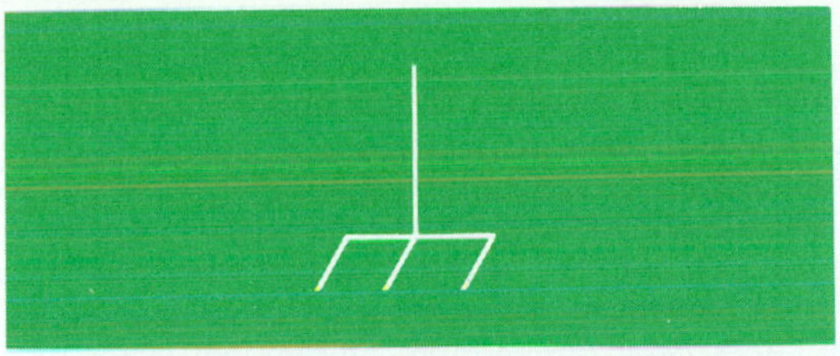

Fig. 2. Symbol to denote connection to a reference ground that is independent of earth.

no connection to earth is involved. A connection to such a reference grounding conductor which is independent of earth is denoted by use of the symbol shown in Fig. 2.

The subject of grounding may be conveniently divided into two categories: system grounding and equipment grounding. System grounding relates to a grounding connection from the electric power system conductors for the purpose of securing superior performance qualities in the electrical system. Equipment grounding relates to a grounding connection from the various electric-machine frames, equipment housings, metal raceways containing energized electrical conductors, and closely adjacent conducting structures judged to be vulnerable to contact by an energized conductor. The purpose of such equipment grounding is to avoid environmental hazards such as electric shock to area occupants, fire ignition hazard to the building or contents, and sparking or arcing between building interior metallic members which may be in loose contact with one another. The design of outdoor open-type installations presents special problems.

Installations in which earth is used as a reference ground plane present special problems. To design an earth "floor surface" for an outdoor open-type substation which will be free of dangerous electric shock voltage exposure to persons around the station is a difficult task. [R.H.K.]

Groundwater hydrology The occurrence, circulation, distribution, and properties of any liquid water residing beneath the surface of the earth. Generally groundwater is that fraction of precipitation which infiltrates the land surface and subsequently moves, in response to various hydrodynamic forces, to reappear once again as seeps or in a more obvious fashion as springs. Most of groundwater discharge is not evident because it occurs through the bottoms of surface water bodies. *See* SPRING (HYDROLOGY).

Groundwater can be found, at least in theory, in any geological horizon containing interconnected pore space. Thus a groundwater reservoir (an analogy to an oil reservoir) can be a classical porous medium, such as sand or sandstone; a fractured, relatively impermeable rock, such as granite; or a cavernous geologic horizon, such as certain limestone beds. Groundwater reservoirs which readily yield water to wells are known as aquifers; in contrast, aquitards are formations which do not normally provide adequate water supplies, and aquicludes are considered, for all practical purposes, to be impermeable. These terms are, of course, subjective descriptions; the flow of water which constitutes an economically viable supply depends upon the intended use and the availability of alternative sources. *See* AQUIFER.

To effectively utilize groundwater as a natural resource, it is necessary to be able to forecast the impact of exploitation on water availability. When groundwater is used for water supply, a concern is the potential energy in the aquifer as reflected in the water levels in the producing well or neighboring wells. When a groundwater reservoir which does not readily transmit water is tapped, the energy loss associated with flow to the well can be such that the well must be drilled to prohibitively great depths to provide adequate supplies. On the other hand, in a formation able to transmit fluid easily, water levels may drop because the reservoir is being depleted of water. This is generally encountered in reservoirs of limited areal extent or those in which natural infiltration has been reduced either naturally or through human activities.

Problems involving groundwater quantity were once the primary concern of hydrologists; interest is now focused on groundwater quality. Groundwater contamination is a serious problem, particularly in the highly urbanized areas of the United States. *See* HYDROLOGY; WATER POLLUTION. [G.F.P.]

Group theory Any set of elements which is equipped with an operation (called multiplication) that is associative, has an identity element, and has an inverse element for each element of the set, is called a group. [These requirements are stated explicitly in Eqs. (1), (2) and (3) below.] Group theory is the branch of mathematics devoted to the properties of groups. Group theory has applications in the theories of relativity, quantum mechanics, and crystallography, and also in some branches of algebra and in analytic function theory.

The group operation is supposed to give for every pair of elements, for example, g and h, in the group, another element gh (called their product). Multiplication is supposed to satisfy the following requirements.

1. If g_1, g_2, and g_3 are any elements of the group, then Eq. (1) is satisfied; that is, in forming a product of three elements one

$$(g_1g_2)g_3 = g_1(g_2g_3) \qquad (1)$$

obtains the same result by first multiplying g_1 and g_2 and then multiplying the result and g_3, as first mutiplying g_2 and g_3 and then multiplying g_1 and the result. This is called the associative law.

2. There is in the group an element e (called the identity) with the property that Eq. (2) is satisfied for every element g of the group.

$$eg = ge = g \qquad (2)$$

3. For each element g of the group there is an element g^{-1} (called the inverse of g) which satisfies Eq. (3).

$$g^{-1}g = gg^{-1} = e \qquad (3)$$

An example of a group is the set of positive real numbers equipped with the operation of ordinary multiplication. The identity element is then the number 1, and the inverse of an element is its reciprocal. Another example of a group is the set of all real numbers equipped with the operation of ordinary addition. The identity element is then the number 0, and the inverse of an element is its negative.

Both these examples are commutative groups (also called abelian groups after the mathematician Niels Abel) because their multiplication law satisfies Eq. (4) for every g and h in the

$$gh = hg \qquad (4)$$

group. An example of a noncommutative group (also called nonabelian) is the group of rotations of a three-dimensional rigid body around a point.

These three examples are all of groups with an infinite number of elements, that is, infinite groups. There are also finite groups. A simple example is the group with two elements, the identity e and another e_1 satisfying $e_1e_1 = e$.

For infinite groups, it often occurs that a group has a natural geometry. For example, in the group of positive real numbers described above one has the geometry of the real numbers. This geometry is compatible with the group operations in the sense that the product gh is a continuous function of g and h, and the inverse g^{-1} is a continuous function of g. Generalizing this scheme to arbitrary groups, one arrives at the notion of a topological group (or less frequently a continuous group) which is a set of elements equipped not only with a group operation but also with a topology, the two notions being required to be compatible in the above sense. *See* TOPOLOGY.

There is a particular class of topological groups which is of great importance. These are the so-called Lie groups (named after the mathematician Sophus Lie). Most of the topological groups occurring in applications are Lie groups. *See* LIE GROUP.

The principal applications of group theory outside mathematics itself have to do with the classification and exploitation of symmetries in physical systems. The applications of group theory within mathematics itself are numerous and important. The part of mathematics in which the notion of group was first clearly isolated was the theory of algebraic equations. [A.S.W.]

Group velocity The velocity of propagation of a group of waves forming a wave packet; also, the velocity of energy flow in a traveling wave or wave packet. The pure sine waves used to define phase velocity υ_p do not ever really exist, for they would require infinite extent. What do exist are groups of waves, wave packets, which are combined disturbances of a group of sine waves having a range of frequencies and wavelengths. Good approximations to pure sine waves exist, provided the extent of the media is very large in comparison with the wavelength of the sine wave. In nondispersive media, pure sine waves of different frequencies all travel at the same speed υ_p, and any wave packet retains its shape as it propagates. In this case, the group velocity υ_g is the same as υ_p. But if there is dispersion, the wave packet changes shape as it moves, because each different frequency which makes up the packet moves with a different phase velocity. If υ_p is frequency-dependent, then υ_g is not equal to υ_p. *See* PHASE VELOCITY; SINE WAVE; WAVE MOTION; WAVE PACKET. [S.A.Wi.]

Grout A mixture of cement and water called neat cement grout, or a mixture of cement, water, and fine aggregate that hardens with hydration of the cement to form a solid. The latter mixture is also called mortar. Grout may be pumped through drilled holes under pressure to fill voids or cracks in foundations for structures in order to improve load-bearing properties or watertightness of the material. Grout serves as a filling for joints in dams and for cable ducts in posttensioned concrete structures. It may also be used to solidify gravel or stone fill, as in Prepakt concrete. *See* CONCRETE. [I.L.T.]

Growing season Generally, the period of the year when climatic conditions are favorable for plant growth, common to a place or an area. In middle latitudes, with sufficient moisture, the growing season generally falls between the date of the last freeze in spring and the time of first freeze of autumn. This would apply to most annual plants and crops and to many perennials and biennials. However, some more hardy plants extend their growing season beyond these limits and continue some growth until the soil freezes; but some less hardy plants require minimum temperatures of 40°F (4°C), as in the tropics or subtropics. *See* AGRICULTURAL METEOROLOGY. [M.L.B.]

Growth factor Any of a group of biologically active polypeptides which function as hormonelike regulatory signals, controlling the growth and differentiation of responsive cells. Indeed, the distinction between growth factors and hormones is frequently arbitrary and stems more from the manner of their discovery than from a clear difference in function. *See* CELL DIFFERENTIATION; HORMONE.

The sequence of amino acids has been determined for several growth-factor polypeptides. This information permits a number of growth factors to be placed into families, members of which have related amino acid sequences, suggesting that they evolved from a single ancestral protein. The insulin family comprises somatemedins A and C, insulin, insulinlike growth factor (IGF), and multiplication-stimulating factor (MSF). A second family consists of sarcoma growth factor (SGF), transforming growth factors (TGFs), and epidermal growth factor (EGF). In addition, there are growth factors, such as nerve growth factor (NGF), fibroblast growth factor (FGF), and platelet-derived growth factor (PDGF), for which structural homologs have not been identified. *See* INSULIN; PROTEIN.

The stimulation of cell proliferation by several growth factors is similar in some ways to the rapid cell proliferation characteristic of tumor cells. Furthermore, the growth factor receptors are similar to the tumor-causing proteins produced by several RNA tumor viruses. It has been demonstrated that platelet-derived growth factor is virtually identical to the tumor-causing protein of the RNA tumor virus, simian sarcoma virus. Some forms of cancer involve improper function of growth factors. *See* CANCER (MEDICINE); ONCOLOGY; TUMOR VIRUSES. [M.Bo.]

Gruiformes A rather heterogeneous order of birds, especially well represented in the fossil record; of some 22 families, 10 are known only from fossils, and many fossil genera and species are allocated to living families. The order as a whole is cosmopolitan, but among the living families are several of small size and limited distribution, notably the monotypic family Rhynochetidae, containing only the peculiar kagu of New Caledonia. The largest and most widely distributed family is the Rallidae, comprising rails, gallinules, and coots. Largest living members of the order are the cranes (Gruidae), which are virtually cosmopolitan except in the Americas south of Mexico. *See* AVES. [K.C.P.]

Guava A plant, *Psidium guajava*, of tropical America. It is a shrub or low tree which belongs to the myrtle family (Myrtaceae). The fruit is a berry, yellow when ripe, and quite variable in size depending on variety and growing conditions. The guava is quite aromatic, sweet, and juicy. It is used mostly for jellies and preserves, but also as a fresh fruit. *See* FRUIT; MYRTALES. [P.D.St./E.L.C.]

Guayule A desert plant, *Parthenium argentatum*, of the composite family (Compositae), which produces rubber. It is a native perennial shrub growing in the Chihuahuan Desert of north-central Mexico and southwestern Texas. The plant is bushy with dense branches, thick clusters of silverlike leaves, a strong taproot, and a thick crown.

During the early decades of this century, guayule was cultivated as an alternative source for rubber. With the development of synthetic rubber and the end of World War 11, economic and political incentives for guayule production disappeared, and all projects were terminated by 1959. Later, in 1976, research was reinitiated as the rising price of oil and the accompanying political difficulties with oil-producing countries rekindled interest in guayule as a source of natural rubber. *See* CAMPANULALES; RUBBER. [D.D.Ru.; L.J.Ca.]

Guidance systems Apparatus for guiding the path of a vehicle, often remotely and automatically. Guidance systems are necessary in missiles and spacecraft of all kinds, and in drone (pilotless) aircraft. They are also employed to assist the human navigator in piloted aircraft, ships, and submarines.

In common usage the term guidance implies supervision of the navigation of a vehicle on its path from present location to desired future location. The companion term, flight control, is used to connote control of the vehicle, such as in attitude, heading, and speed. The equipment that performs the flight-control function is called an automatic pilot, or autopilot, and is an essential component of every guidance system. *See* AUTOPILOT.

In addition to the autopilot, a guidance system consists of a command computer—for accepting information and commands and, from them, determining the desired flight path—

and sensors for measuring the actual flight path of the vehicle. The command computer determines the path to be flown and instructs the autopilot accordingly; actual flight path is measured by the sensors to be sure the guidance commands are being carried out; and errors are corrected by the usual feedback technique, that is, by making a corrective change proportional to error. *See* CONTROL SYSTEMS.

A revolutionary trend in the development of guidance systems of all kinds has been the increasing proportion of the guidance computation burden performed on board the vehicle. This trend was made possible by the advent of small digital computers of truly large capacity and by the spectacular rate at which the technology of microcircuitry is shrinking their size and power requirements and increasing their reliability. *See* INTEGRATED CIRCUITS.

Drone aircraft are often guided by ground radar tracking and radio command to the autopilot. Piloted aircraft are sometimes similarly guided by ground equipment for special purposes, such as ground-controlled intercept (GCI) of another aircraft or ground-aided approach to an airport. Except for such special purposes, however, piloted aircraft are navigated from on board with numerous internal and external aids. *See* ELECTRONIC NAVIGATION SYSTEMS; INERTIAL GUIDANCE SYSTEM; RADAR; STAR TRACKER.

Guidance of a missile to intercept a moving target such as an aircraft can be accomplished by either remote (launcher-based) command or by an on-board homing system. Alternatively, the desired missile course may simply be preset as a function of time in an internal computer, as is common with torpedoes.

A command guidance system has the advantage that the missile carries only a radar receiver and an autopilot; the computer, being on the ground, may be as sophisticated as desired. However, the range of the system is limited, as is also the accuracy with which it can resolve geometric angles involved in the missile-target tactics.

In homing guidance systems, radar, infrared, or optical sensing (target-seeking) equipment in the missile itself measures the direction from missile to target. Measurements are more accurate from the missile's vantage point, and computations based on relative motion are simpler. However, because the missile must carry the additional load of a seeker and computer, the success of the system depends on the degree to which these subsystems can be miniaturized and simplified.

In an active homing system the missile's own radar illuminates the target, thus making the missile entirely autonomous, once the missile's radar has captured the target. In a passive homing system the missile carries only a receiver, which may be radar, infrared, or optical.

A beam-rider system is even simpler than the passive seeker; a radar receiver in the missile simply seeks to keep the missile in the beam of a launcher-based radar. The launcher station in turn keeps the radar beam centered on the target, so that the missile is guided along the beam to the target. *See* GUIDED MISSILE.

Guidance of long-range cruise (nonballistic) missiles is a problem of geographic navigation. In special cases a series of ground stations may form a command guidance system. Usually, however, the navigation system is carried in the missile, and guides it along a programmed course to a geographic destination. *See* AIRPLANE; BALLISTIC MISSILE; DRONE; MISSILE; NAVIGATION; SPACE NAVIGATION AND GUIDANCE. [R.H.C.]

Guided missile A pilotless, controlled-flight vehicle that is guided to a target by guidance and control equipment. This equipment may be carried in the missile vehicle itself, or guidance may be directed from the launch site. The term is generally reserved for aerodynamic, maneuverable missiles that may be guided to predetermined targets for military purposes.

Guided missiles are classified by launch/target mode such as air to air (AAM), air to surface (ASM), surface to air (SAM), surface to surface (SSM), and other possible modes. Missiles may be classified by range (short, medium, long) or by techniques related to tracking and guidance (radar, infrared heat seeker, optical or television, laser, radio, wire control command, acoustical). Some missiles make use of terrain following, which permits the missile to look at the terrain, compare it with a predetermined mapped route, and in effect fly a course as if by following a road map. *See* GUIDANCE SYSTEMS.

Guided missiles are generally self-propelled, and may use rocket motors (liquid or solid), air-breathing turbojet engines, ramjets, or various types of combined-cycle engines. For some missions, particularly air-to-surface missions, unpowered, gliding guided missiles may be used. *See* ELECTROMAGNETIC PROPULSION; ION PROPULSION; JET PROPULSION; ROCKET PROPULSION.

The kill mechanism for a missile consists of some form of explosive warhead and a system for detonation (fusing and arming). Warheads are typically either high-explosive or nuclear. Warheads may be exploded upon contact with the target, by command from an external source, by a proximity fuse that senses the target, by preset timers, and so on. *See* MISSILE. [M.L.Sp.]

Guild A group of species that utilize the same kinds of resources, such as food, nesting sites, or places to live, in a similar manner. Emphasis is on ecologically associated groups that are most likely to compete because of similarity in ecological niches, even though species can be taxonomically unrelated. The term was derived from the guild in human society composed of people engaged in an activity or trade held in common. *See* ECOLOGICAL COMMUNITY.

The guild concept focuses attention on the ways in which ecologically related species differ enough to permit coexistence, or avoid competitive displacement. For example, new places to live for some plants are provided by badger mounds in dense tall-grass prairie vegetation.

The guild is also commonly used as the smallest unit in an ecosystem in studies relating to environmental impact, wildlife management, and habitat classification. A representative species of a guild may be selected for study involving the uncertain assumption that environmental impact will influence this species in the same way as other guild members. *See* ECOSYSTEM. [P.W.P.]

Guinea worm infection An infection in humans with the nematode *Dracunculus medinensis*; also called dracunculosis or dracontiasis. The females develop in 8–12 months in the body cavities and migrate finally to the subcutaneous tissues, preferably of the limbs. A blister with a central papule forms at a spot where the worm's mouth approaches the epidermis. It bursts open, particularly upon contact with water, and numerous larvae are liberated. A small crustacean, the water flea (*Cyclops* and related genera), ingests the larvae and acts as an intermediate host. Humans become parasitized upon drinking well water containing these crustaceans.

The disease is and has been associated for centuries with religious rites. Various animals act as reservoirs, even in America. The disease is treated with some success with phenothiazine injections. Another method of treatment requires rolling the worm out inch by inch on a small stick. Prevention is by avoiding drinking water that contains infected crustaceans. [J.F.M.]

Gulf of California A deep, elongated trough in northwestern Mexico, 900 mi (1500 km) long and 60–120 mi (100–200 km) wide, separated from the Pacific Ocean by the mountainous peninsula of Baja California. A central constriction and a group of islands separate a shallow (<650 ft or 200 m) northern section from a deep region to the south which contains a series of basins 3000–12,000 ft (1000–3600 m) deep (see illustration).

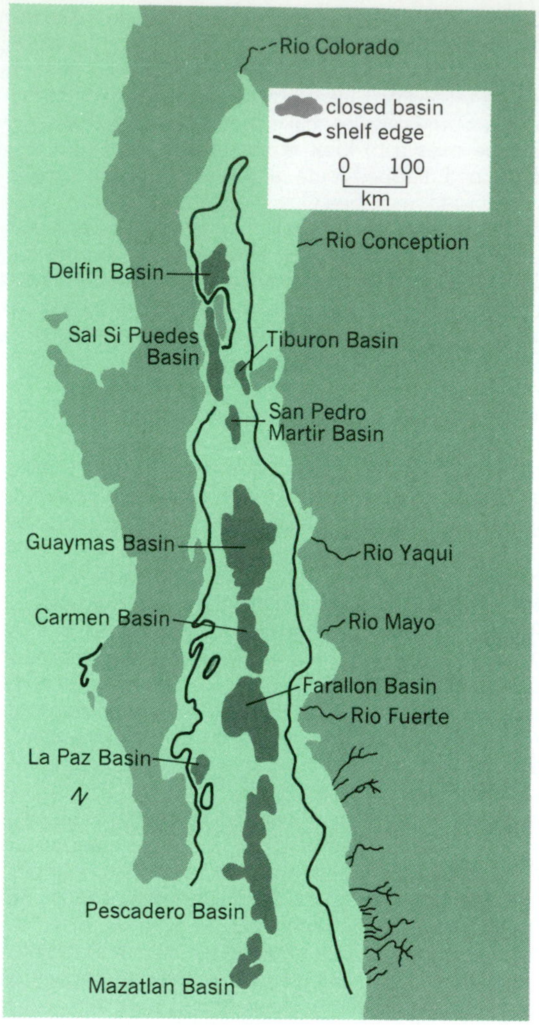

Principal relief features of Gulf of California. 1 km = 0.6 mi.

The Gulf is an arid region over most of its expanse. Baja California has an annual rainfall of 4.0–6.0 in. (10–15 cm); no permanent streams exist. On the eastern side rainfall increases from 4.0 in. (10 cm) in the northwest to 34 in. (85 cm) in the southeast; permanent streams occur only in the southern half.

The northern Gulf has sediments derived predominantly from the Colorado River. The southern Gulf receives sediments from the eastern side; a wide coastal plain and shelf indicate high rates of supply. Very little sediment is supplied from the peninsula, and the western shallow-water sediments consist of calcareous skeletal debris. [S.E.C.]

Gulf of Mexico A semienclosed marginal sea forming the southern marine boundary of the United States, the scene of extensive marine fisheries and important offshore gas, oil, and sulfur production. Covering an area of 650,000 mi^2 (1,700,000 km^2), with wide continental shelves on the northern and eastern boundaries, the basin has a large central area (Sigsbee Abyssal Plain) with depths in excess of 11,500 ft (3500 m).

The Florida shelf and the Campeche bank are carbonate areas with active reef formation currently taking place on some of the Florida Keys and the outer reefs of Campeche. The remainder of the Gulf has detrital sediments (mainly montmorillonite), deriving at present principally from the Mississippi and the Atchafalaya rivers. The Texas-Louisiana shelf and slope display a great number of salt domes similar to those occurring under the adjacent land areas. Associated petroleum pooling has made this a productive area for oil and gas recovery.

In the southeast the island of Cuba separates the two connections with the Atlantic—one via the Caribbean through the Yucatán Strait and the other through the Straits of Florida. The Yucatán current carries water northward in an intense flow with speeds typically 2–4 knots (103–206 cm/s) close to the western side of the Strait. This flow generally loops to the north and exists as the Florida Current, which is continuous with, and after augmentation becomes, the Gulf Stream. *See* CARIBBEAN SEA; GULF STREAM.

The major contributions of fresh water to the Gulf come from the Mississippi, Atchafalaya, and the lesser rivers of Texas. The low-salinity coastal waters hold to the shelf with a westward flow. Although the shallow Texas bays at times may be brackish, restricted circulation and local evaporation make them hypersaline over extended periods.

The most valuable of the food fisheries are for shrimp and oysters. Several high-quality fin fishes are also landed in quantity. There are large inshore and offshore sports fisheries.

[H.J.McL.]

Gulf Stream A great ocean current transporting about 70,000,000 tons (63,000,000 metric tons) of water per second (1000 times the discharge of the Mississippi River) northward from the latitude of Florida to the Grand Banks off Newfoundland. The Gulf Stream is thought of as a portion of a great horizontal circulation in the ocean, where each particle of water executes a closed circuit, sometimes moving slowly in midocean regions and other times rapidly in strong currents like the Gulf Stream. Thus the beginning and end of the Stream have arbitrary geographical limits (see illustration). *See* ATLANTIC OCEAN.

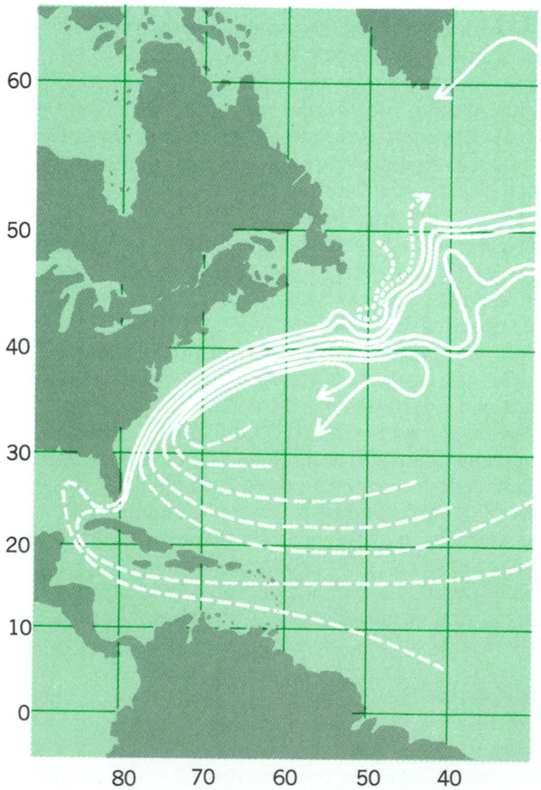

Sources (broken lines) and pattern (solid lines) of the Gulf Stream system. (*Adapted from C. O'D. Iselin, Papers Phys. Oceanogr. Meteorol., Mass. Inst. Technol. and Woods Hole Oceanogr. Inst., 4(4):1–99, 1936*)

The Gulf Stream is a narrow (62 mi or 100 km) and swift (up to 5 knots or 250 cm/s) eastward-flowing current jet which is embedded in a weaker and broader mean westward flow and which is surrounded by intense eddies. As it leaves the coast at Cape Hatteras, the Stream meanders from side to side like a river.

The near-surface Gulf Stream transports warm water from southern latitudes eastward to the Grand Banks, where the flow becomes broader and weaker, separating into several branches and eddies. About half the near-surface flow continues eastward across the Mid-Atlantic Ridge, and half recirculates southwestward, with part of the recirculation consisting of a countercurrent located south of the Stream.

The Gulf Stream is predominantly driven by the large-scale wind pattern, the westerlies in the north and the trades in the south. The winds exert a torque on the ocean that, due to the shape and rotation of the Earth, causes a large western-intensified gyre. Cold, deep water is formed in northern seas and flows southward as a western boundary current; warm water flows northward and replaces it. *See* Ocean circulation. [P.R.]

Gum A class of high-molecular-weight molecules, usually with colloidal properties, which in an appropriate solvent or swelling agent are able to produce gels at low dry-substance content. The molecules are either hydrophilic or hydrophobic. The term gum is applied to a wide variety of substances of gummy characteristics, and therefore cannot be precisely defined. *See* Gel.

Various rubbers are considered to be gums, as are many synthetic polymers, high-molecular-weight hydrocarbons, or other petroleum products. Chicle for chewing gum is an example of a hydrophobic polymer which is termed a gum but is not frequently classified among the gums. Quite often listed among the gums are the hydrophobic resinous saps that often exude from plants and are commercially tapped in balsam (gum balsam) and other evergreen trees (gum resin). Incense gums such as myrrh and frankincense are likewise fragrant plant exudates.

Usually, however, the term gum, as technically employed in industry, refers to plant polysaccharides or their derivatives. Modern usage of the term includes water-soluble derivatives of cellulose and derivatives and modifications of other polysaccharides which in the natural form are insoluble. Usage, therefore, also includes with gums the ill-defined group of plant slimes called mucilages. *See* Cellulose; Colloid; Polysaccharide.

Gums are used in foods as stabilizers and thickeners. They form viscous solutions which prevent aggregation of the small particles of the dispersed phase. In this way they aid in keeping solids dispersed in chocolate milk, air in whipping cream, and fats in salad dressings. Gum solutions also retard crystal growth in ice cream (ice crystals) and in confections (sugar crystals). Their thickening and stabilizing properties make them useful in water-base paints, printing inks, and drilling muds. Because of these properties they also are used in cosmetics and pharmaceuticals as emulsifiers or bases for ointments, greaseless creams, toothpastes, lotions, demulcents, and emollients. The adhesive properties of gums make them useful in the production of cardboard, postage stamps, gummed envelopes, and as pill binders. Other applications include the production of dental impression molds, fibers (alginate rayon), soluble surgery films and gauze, blood anticoagulants, plasma extenders, beverage-clarifying agents, bacteriological culture media, half-cell bridges, and tungsten-wire-drawing lubricants. [R.L.Wh.]

Gunsights Optical instruments which establish an optical line or axis for the purpose of aiming a weapon. The axis includes the observer's eye, a suitable mark in the instrument, and the target. Most gunsights employ as their basis either a telescope or a partially reflective mirror. *See* Mirror optics; Telescope.

A typical rifle sight consists of a terrestrial telescopic system having an objective, an eyepiece, an erector lens, and a reticle. Sometimes a field lens is employed to ensure uniform illumination. Aircraft gunsights are usually of the reflector type (also known as reflex sights) and employ in their simplest form a lamp, a reticle, a collimating lens, and a glass plate or partially reflecting mirror (see illustration). The collimator images the reticle pattern at infinity, and the mirror superimposes this image over the target area. Artillery sights can assume various forms, the simplest of which is the collimator sight, consisting of an objective having a reticle at its focus. When the eye is so placed as to receive light simultaneously from the target and the reticle, the latter appears superimposed on the former and a line of sight is established. *See* Lens (optics). [E.K.K.]

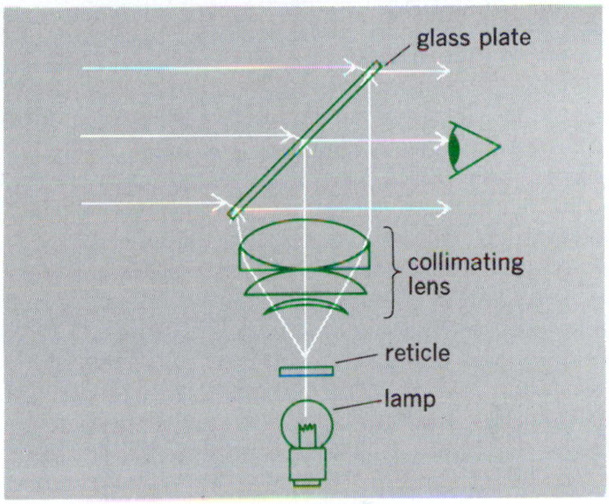

Reflex sight.

Gymnolaemata A class of bryozoans. Predominantly marine, gymnolaemates possess lophophores which are circular in basal outline and zooecia which are short, wide, vaselike or boxlike. Highly diverse in size and shape, most gymnolaemate colonies are small and delicate, but a few are large conspicuous growths. The individual zooids may be relatively isolated or loosely grouped side by side.

The gymnolaemates include several thousand species—mostly marine, some brackish, and a very few fresh-water—belonging to the two orders Ctenostomata and Cheilostomata. Appearing early in the Early Ordovician, when they included the possible ectoproct stem group, gymnolaemates remained quite inconspicuous until the Cretaceous, when they rose to the position of dominance among bryozoans which they still maintain. *See* Bryozoa; Cheilostomata; Ctenostomata. [R.J.Cu.]

Gymnostomatida An order of the Holotrichia which contains a large, widely distributed group of what are believed to be the most primitive ciliate protozoans. These organisms occur abundantly in sands of intertidal zones, as well as in the more usual fresh- and salt-water habitats. The body size is frequently large; ciliation is simple and plentiful; and the oral area lacks buccal ciliature, as the name of the order implies. *See* Holotrichia. [J.O.C.]

Gynogenesis The development of a fertilized egg through the action of the egg nucleus, without participation of the sperm nucleus. However, the spermatozoon penetrates the egg, as it does in normal fertilization, and usually contributes the division center (centrosome) for the establishment of a bipolar division apparatus.

Gynogenesis occurs as a regular phenomenon in nature in some nematodes, flatworms, and other invertebrates, and in some species of fishes. Gynogenesis also occurs spontaneously in species that usually have normal fertilization, and produces haploid embryos. *See* ANDROGENESIS; MEROGONY. [G.Fa.]

Gypsum A mineral with the chemical composition $CaSO_4 \cdot 2H_2O$. Gypsum is the commonest of the sulfate minerals. It occurs in five varieties: (1) rock gypsum; (2) gypsite, an impure earthy form; (3) alabaster, a massive, fine-grained, translucent variety; (4) satin spar, a fibrous silky form; and (5) selenite, transparent crystals. Crystals of gypsum are monoclinic, clear, white to gray, yellowish, and brownish in color, with well-developed cleavages. Luster is subvitreous to pearly. Hardness is 2 on Mohs scale and specific gravity is 2.3. *See* CALCIUM.

Gypsum is calcined in kilns at temperatures of 190–200°C (370–390°F) to remove part of the water of crystallization. The calcined product is known as plaster of paris. Gypsum and gypsum plaster products are chiefly used in the building industry and as a retarder in portland cement. Ground gypsum and anhydrite are used in the southern United States as soil conditioners to improve permeability. Gypsum and anhydrite, particularly the latter, may be made into ammonium sulfate fertilizer. Calcined gypsum is also used for industrial plasters, such as those used in pottery, molding, dentistry, and statuary. Pure white uncalcined gypsum, known as terra alba, is used as a filler in paper and paints and as a nutrient in growing yeast. Other uncalcined ground gypsum is used as a filler in insecticide. *See* ANHYDRITE; PLASTER OF PARIS.

Gypsum is of world-wide distribution. Thick commercial deposits of calcium sulfate occur in Nova Scotia and in Stassfurt, Germany. In the western United States extensive commercial gypsum deposits occur in California, Nevada, Utah, Texas, Iowa, Oklahoma, Kansas, Montana, Colorado, Arizona, South Dakota, Wyoming, and New Mexico. In the east, large mines are located in Michigan, New York, Ohio, and Indiana. [E.C.T.C.]

Gypsum plank A structural precast unit designed for use as the roof deck of industrial or other steel-frame buildings, and in some cases for the floor system as well. The four edges of the plank are bound with a tongue-and-groove steel form which, when fitted into an adjacent unit, produces a rigid steel I beam at the joint. The gypsum core is reinforced with welded galvanized-steel mesh. The planks are usually 2 in. (5 cm) thick, 15 in. (38 cm) wide, and 8–10 ft (2.4–3.0 m) in length. Specially designed clips fasten the planks to the supporting purlins and to an asphalt-impregnated felt built-up roofing used to resist the weather. The planks are fire-resistant and lightweight, are easily erected, and may be readily cut or drilled. *See* ROOF CONSTRUCTION. [C.M.A.]

Gyrator A linear, passive, two-port electric circuit element whose transmission properties are such that it is effectively a half wavelength longer for one direction of transmission than for the other direction of transmission. Thus a gyrator is a device that causes a reversal of signal polarity for one direction of propagation but not for the other. (A two-port element has a pair of input terminals and a pair of output terminals.) This device is novel, since it violates the theorem of reciprocity. *See* RECIPROCITY PRINCIPLE.

Until the early 1950s, all known linear passive electrical networks obeyed the theorem of reciprocity. However, several different types of nonreciprocal networks are now widely applied, principally at microwave frequencies. These devices are used to control the direction of signal flow and to protect or isolate components from undesired signals. One common application of a three-port nonreciprocal network, called a circulator, is to permit connection of a transmitter and a receiver to the same antenna. This is accomplished with minimum interference and virtually no power loss of either transmitted or received signal. *See* CONTINUOUS-WAVE RADAR.

Perhaps the first passive nonreciprocal system was an optical one proposed by Lord Rayleigh, making use of the rotation of the plane of polarization of light when it passed through a transparent material in the presence of a magnetic field. This phenomenon is called Faraday rotation.

The microwave analogy of Lord Rayleigh's device was proposed by C. L. Hogan. The nonreciprocal medium used is ferrite. In such a material, infinitesimal magnetic dipole moments which arise from the electronic structure of the material act gyroscopically when a steady magnetic field is applied. They precess about the applied field direction in a counterclockwise sense, thus permitting strong coupling to the component of a microwave-frequency magnetic field which is circularly polarized in the same sense. The component with the opposite sense of polarization is weakly coupled. Thus energy exchange between the magnetic dipoles and the microwave field is polarization-sensitive. *See* FERRIMAGNETISM; FERRITE. [F.J.R.]

Gyrocompass A north-seeking form of gyroscope used as a directional reference in navigation. Modern gyrocompasses are so reliable and so much more accurate than magnetic compasses that they are now used as the prime navigational instrument on nearly every ship and on major aircraft and missiles. *See* MAGNETIC COMPASS.

A gyrocompass combines the action of two devices, a pendulum and a gyroscope, to produce alignment with the Earth's spin axis. The principle is demonstrated with the model shown in the illustration, which consists of (1) a rapidly spinning, heavy gyro rotor, (2) a pendulous case which permits the rotor axle to nod up and down (angle θ), and (3) an outer gimbal which permits the axle to rotate in azimuth (angle ψ). For a gyroscope positioned at the Equator of the Earth, as the Earth rotates, the gimbal moves with it. So long as the rotor's spin axis is aligned with the Earth's axis, the gyro experiences no torque from Earth rotation. If there is misalignment, however, a sequence of restoring torques is initiated. *See* GYROSCOPE; PENDULUM.

In a shipboard installation the system must be mounted in a complete set of gimbals to isolate it from rolling, pitching, and yawing motions of the ship. Friction must be minimized.

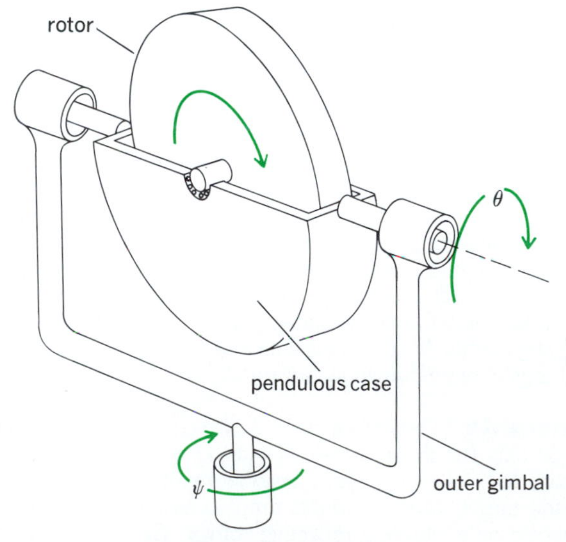

Gyrocompass model.

Moreover, Schuler tuning is employed to keep horizontal accelerations of the ship from producing false torques on the pendulum; the unique combination of gyro spin speed and pendulosity is chosen so that no acceleration of the instrument can disturb its vertical reference. *See* SCHULER PENDULUM. [R.H.C.]

For many years the use of gyrocompasses in aircraft was impractical because of their high speed and large, rapid changes in attitude. The north-south component of vehicle velocity produces an error which depends on the velocity magnitude. In order to correct for this error, the true ground speed of the aircraft must be determined and used to precess the gyro in the opposite direction. In addition, the direction of the local vertical must be determined accurately. These capabilities are contained in modern military aircraft which are equipped with gyroscopes mounted in conjunction with accelerometers to provide a stable inertial platform. They also are equipped with Doppler navigation radar which measures true velocity over the Earth's surface. *See* DOPPLER RADAR; INERTIAL GUIDANCE SYSTEM. [G.St.]

Gyrocotylidea An order of tapeworms of the subclass Cestodaria. All species are intestinal parasites of chimaeroid fishes and have an anterior end with an eversible proboscis and a posterior end with a ruffled adhesive organ. Typically, only one sexually mature worm is found in an individual host. The life history is incompletely known. *See* CESTODARIA. [C.P.R.]

Gyromagnetic effect An effect arising from the relation between the angular momentum and the magnetization of a magnetic substance. It is the effect which is exploited in the measurement of the gyromagnetic ratio of magnetic materials. The gyromagnetic effect is demonstrated by a simple experiment in which a freely suspended magnetic substance is subjected to a magnetic field. Upon a change in direction of the magnetic field, the magnetization of the substance must change. In order for this to happen, the atoms must change their angular momentum. Since there are no external torques acting on the system, the total angular momentum must remain constant. Thus the sample must acquire a mass rotation which may be measured. In this way, the gyromagnetic ratio may be determined. *See* GYROMAGNETIC RATIO. [E.A.; F.Ke.]

Gyromagnetic ratio The ratio of angular momentum to magnetic moment for atomic systems. This ratio is usually expressed in terms of the magnetomechanical factor g', as in Eq. (1). The ratio is written here in electromagnetic units; thus,

$$\frac{\text{Angular momentum}}{\text{Magnetic moment}} = \frac{2mc}{g'e} \qquad (1)$$

e/c and m are the charge and mass of the electron. The factor g' is sometimes loosely called the gyromagnetic ratio.

The magnetomechanical ratio is the inverse of the gyromagnetic ratio. It is usually denoted by γ and is equal to $g'e/2mc$. The magnetomechanical ratio of a substance identifies the origin of the magnetic moment. For example, for electron spin the angular momentum is $\frac{1}{2}\hbar$, where $\hbar$ is Planck's constant divided by 2π. The magnetic moment is the Bohr magneton $e\hbar/2mc$. Thus, the magnetomechanical ratio is given by Eq. (2). Since $\gamma = g'e/2mc$, for electron spin $g' = 2$. For orbital angular momentum, $\gamma = e/mc$ and $g' = 1$. The experimental values of g' for most ferromagnetic materials are in the neighborhood of 2, showing that the major contribution to the magnetization comes from the electron spin. In superconductors, on the other hand, the fact that $g' = 1$ shows that the diamagnetic currents which cause the Meissner effect are caused by electrons. *See* MEISSNER EFFECT; SUPERCONDUCTIVITY. [E.A.; F.Ke.]

$$\gamma = \frac{e\hbar/2mc}{\hbar 2} = \frac{e}{mc} \qquad (2)$$

Gyroscope An instrument, often called simply a gyro, that maintains an angular reference direction by virtue of a spinning mass. Gyroscopes are used in airplanes to "remember" the orientation of the horizon and the direction of north during maneuvers. Gyros measure the turning rate of gunsights to improve target tracking. The gyrocompass is a special north-seeking form of gyroscope developed to a high degree for shipboard use. *See* GYROCOMPASS.

Gyros are used in missiles, airplanes, ships, and torpedoes as the basic element in automatic steering systems. A set of gyroscopes may be used to stabilize instrument platforms in maneuvering vehicles. There are stringent accuracy requirements for gyroscopes used for the stable platforms in inertial navigation systems. All gyroscope applications depend upon a special form of Newton's law which states that a spinning body tends to react to a disturbing torque by precessing (rotating

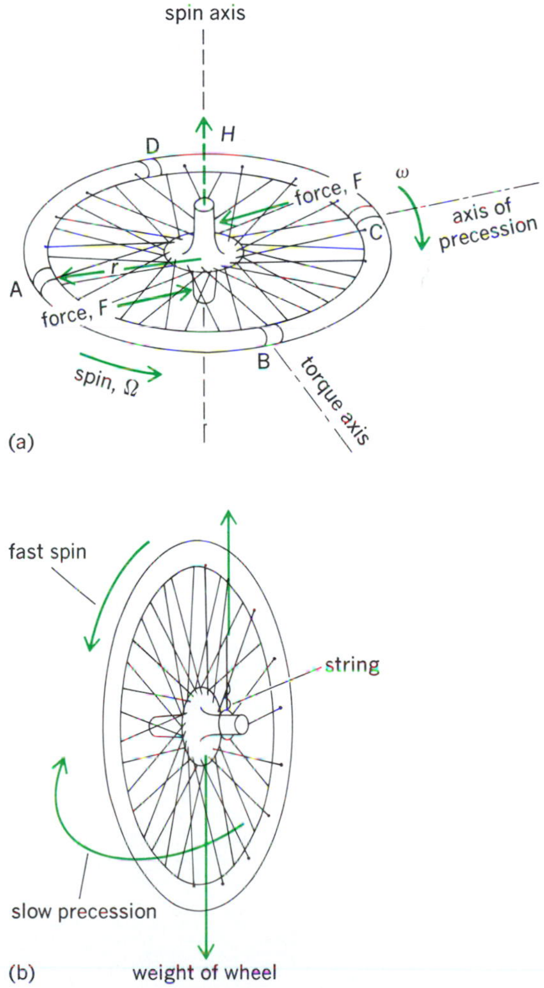

Gyroscope principle. (*a*) Relation of spin, torque, and precession axes. Force *F* applied to bicycle with spin Ω and angular momentum *H* alters motion of the rim, each portion of which passes successively through *A*, *B*, *C*, and *D*, resulting in precession ω. (*b*) Illustration of precession.

slowly) in a direction at right angles to the direction of the torque (see illustration).

The gyroscopic principle is used in instruments in several ways. In the free gyro, the spinning wheel is isolated from the airplane by gimbals. When the airplane banks, the gyro remains vertical and furnishes the pilot with an artificial horizon reference. *See* Automatic horizon.

The rate gyro is one of a class of instruments known as single-axis gyroscopes. It is mounted to the aircraft structure by only a single gimbal so that if the airplane rolls (about its fore-and-aft axis) the rate gyro is obliged to roll with it. This causes the gimbal to turn (precess) against the stiff spring by a small angle, and in this case the indication is the rate of roll. A set of three such gyros, mounted with input axes mutually orthogonal, will serve to determine completely the total angular velocity of the aircraft in space.

Families of gyroscopic instruments are based on either the free or the single-axis arrangement. [R.H.C.]

Gyrotron One of a family of microwave generators, also called cyclotron resonance masers, in which cyclotron resonance coupling between microwave fields and an electron beam in vacuum is the basis of operation. This type of coupling has the advantage that both the electron beam and the associated microwave structures can have dimensions which are large compared with a wavelength. Thus, cyclotron resonance masers are potentially greatly superior to conventional microwave tubes with respect to power capability at short wavelengths.

The development of these power sources is particularly significant for magnetically confined plasma fusion experiments. Microwave heating is considered an attractive method of supplying the energy needed to bring a reactor to ignition temperature, and gyrotrons provide a potential means of producing sufficient microwave power at the very short wavelength required. Gyrotrons also have potential application in millimeter-wave radar and communications systems. *See* Microwave tube; Nuclear fusion; Traveling-wave tube. [H.R.J.]

Hackberry A medium-sized to large tree, *Celtis occidentalis*. It occurs in the eastern half of the United States, except in the extreme south, and is characterized by corky or warty bark, by alternate, long-pointed serrate leaves unequal at the base, and by a small drupaceous fruit, with thin, sweet, edible flesh. It is used for furniture, boxes, and baskets, for shelterbelts, and as a shade tree. [A.H.G./K.P.D.]

Hadromerida An order of sponges of the class Demospongiae with monactinal megascleres that usually have a terminal knob at one end. Microscleres in the form of streptasters, asters, sigmas, or small spined diactinals may occur. Spongin is usually sparse in occurrence. In shape, hadromeridan sponges include radially symmetrical forms (see illustration) as well as encrusting, massive, or branching types. They occur in tidal and shallow waters of all seas and extend down to depths of at least 18,000 ft (5500 m). *See* Boring sponges; Demospongiae. [W.D.H.]

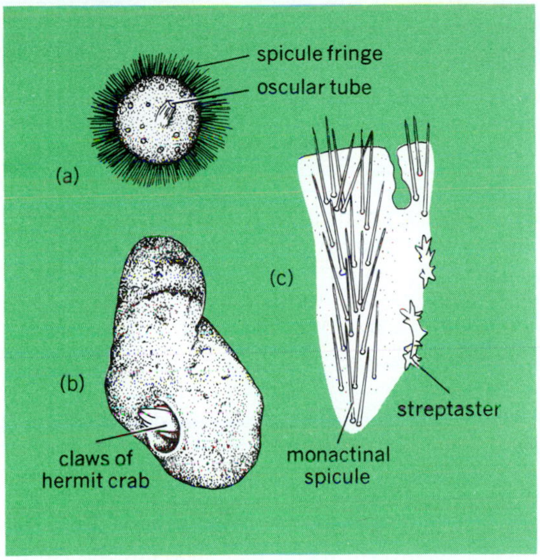

Hadromerine sponges. (*a*) *Radiella sol*, deep-sea species. (*b*) *Suberites ficus* living on shell occupied by hermit crab. (*c*) Spicule arrangement of *Spirastrella*.

Hadron The generic name of a class of particles which interact strongly with one another. Examples of hadrons are protons, neutrons, the π-, K-, and D-mesons, and their antiparticles. Protons and neutrons, which are the constituents of ordinary nuclei, are members of a hadronic subclass called baryons, as are strange and charmed baryons. Baryons have half-integral spin, obey Fermi-Dirac statistics, and are known as fermions. Mesons, the other subclass of hadrons, have zero or integral spin, obey Bose-Einstein statistics, and are known as bosons. The electric charges of baryons and mesons are either zero or ± 1 times the charge on the electron. Masses of the known mesons and baryons cover a wide range, extending from the pi-meson, with a mass approximately one-seventh that of the proton, to the upsilon-meson, with a mass about 10 times the proton mass. The spectrum of meson and baryon masses is not fully understood. *See* Baryon; Bose-Einstein statistics; Fermi-Dirac statistics; Meson; Neutron; Proton.

Much of the data relating to the properties of hadrons (both baryons and mesons) can be interpreted as if hadrons consist of more elementary constituents known as quarks and gluons. This conception of hadrons has widespread appeal in elementary particle physics. *See* Elementary particle; Gluons; Quarks. [A.K.M.]

Hadronic atom A hydrogenlike system that consists of a strongly interacting particle (hadron) bound in the Coulomb field and in orbit around any ordinary nucleus. The kinds of hadronic atoms that have been made and the years in which they were first identified include pionic (1952), kaonic (1966), Σ^--hyperonic (1968), and antiprotonic (1970). They were made by stopping beams of negatively charged hadrons in suitable targets of various elements, for example, potassium, zinc, or lead. The lifetime of these atoms is of the order of 10^{-12} s, but this is long enough to identify them and study their characteristics by means of their x-ray spectra. They are available for study only in the beams of particle accelerators. *See* Elementary particle; Hadron; Particle accelerator.

The hadronic atoms are smaller in size than their electronic counterparts by the ratio of electron to hadron mass. For example, in pionic calcium, atomic number $Z = 20$, the Bohr radius of the ground state is about 10 fermis (1 fermi = 1 femtometer = 10^{-15} m), and in ordinary calcium it is about 2500 fermis. Thus the atomic electrons are practically not involved in the hadronic atoms, and the equations of the hydrogen atom are applicable. The close approach of the hadrons to their host nuclei suggests that hadron-nucleon and hadron-nucleus forces will be in evidence, and this is one of the motivations for studying these new types of atoms. [C.E.W.]

Haemosporina A relatively small and generally rather compact group of protozoans in the subphylum Sporozoa. Authorities differ as to the group's taxonomic status; that assigned it by the Committee on Taxonomy and Taxonomic Problems of the Society of Protozoologists is followed here: a suborder of the order Eucoccida, subclass Coccidia, class Telosporea, subphylum Sporozoa. The Haemosporina are common protozoan parasites of vertebrates, and some of them are important as causes of illness and death. The best known of the group are the four species (genus *Plasmodium*) of malarial parasites of humans. *See* Malaria.

Transmission of these parasites is probably always effected in nature by the bite of some bloodsucking invertebrate. In the vertebrate host they reproduce asexually; sometimes this occurs in the tissues of certain internal organs, such as the lungs, liver, spleen, and brain; sometimes in the red blood cells (erythrocytes), or even in other types of blood and blood-forming cells; and often in both tissues and blood cells. The immature sex cells, gametocytes, always occur in erythrocytes or leukocytes (white blood cells). Gametocytes mature into gametes after ingestion by an intermediate host (arthropod). Fertilization ensues, with a subsequent period of development culminating in the production of numerous sporozoites. These tiny filamentous forms are infective for the vertebrate host. Since they can develop no further in the arthropod, infection in this host is self-limited, in the sense that no further buildup is possible; the insect is seldom harmed by the parasite. However, sporozoites may remain infective for a long time in the invertebrate host, perhaps as long as the insect lives. *See* SPOROZOA. [R.D.M.]

Hafnium A metallic element, symbol Hf, atomic number 72, and atomic weight 178.49. There are five naturally occurring isotopes. It is one of the less abundant elements in the Earth's crust.

Hafnium is a lustrous, silvery metal that melts at about 2222°C (4032°F). Reported values of the boiling point vary greatly, from about 2500 to about 5100°C (4530 to 9200°F). There are virtually no uses of the metal other than in control rods for nuclear reactors.

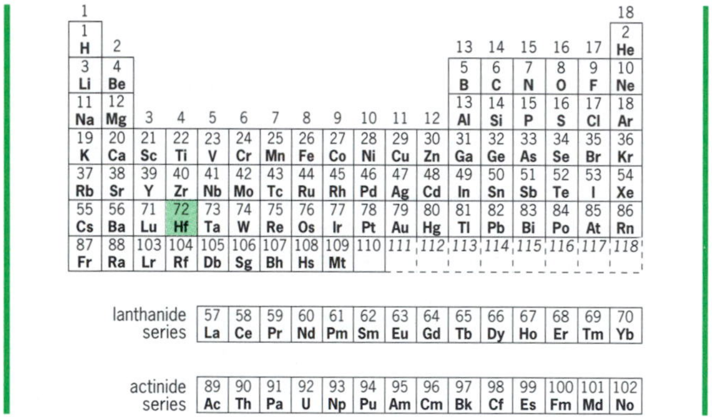

The chemistry of hafnium is almost identical with that of zirconium. The similarity of hafnium to zirconium is a consequence of the lanthanide contraction, which brings the ionic radii to very nearly identical values. Before (and since) the discovery of hafnium, this element was extracted with zirconium from its ores and passed with zirconium into all derivatives. Since the chemical properties are so similar, there has been no incentive to separate the hafnium except for making nuclear studies and components of nuclear reactors. *See* ZIRCONIUM. [W.B.B.]

Hail Precipitation composed of chunks or lumps of ice formed in strong updrafts in cumulonimbus clouds. Individual lumps are called hailstones. Most hailstones are spherical or oblong, some are conical, and some are bumpy and irregular. Diameters range from 0.2 to 6 in. (5 to 150 mm) or more. That is, the largest stones are grapefruit or softball size, and the smallest are pea size.

Very often hailstones are observed to be made of alternating rings of clear and white ice (see illustration). These rings indicate the growth processes of the hail. The milky or white portion of the growth occurs when small cloud droplets are collected by the hailstone and freeze almost instantaneously, trapping bubbles of air between the droplets and creating a milky appearance. The clear portion is formed when many droplets are collected so rapidly that a film of water spreads over the stone and freezes gradually, giving time for any trapped air bubbles to escape from the liquid.

The most favorable conditions for hail formation occur in the mountainous, high plains regions of the world. Hailstorms normally have relatively high, cool cloud bases and very strong updrafts within the clouds to carry the hailstones into the cooler regions of the cloud, where maximum growth occurs. Both small ice particles and supercooled liquid water (liquid water at temperatures below 32°F or 0°C) are needed for the ice particles to grow into hailstones. *See* CLOUD PHYSICS; PRECIPITATION (METEOROLOGY). [H.D.O.]

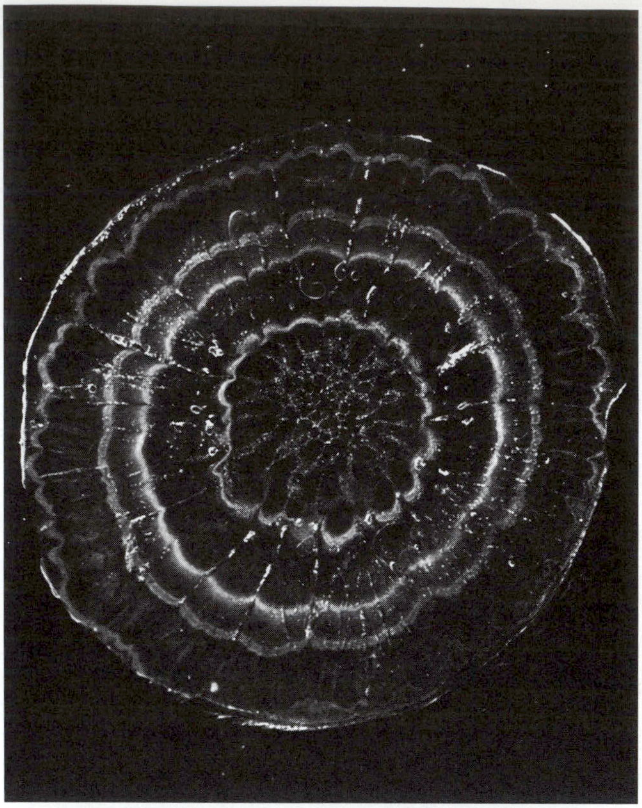

Cross section of a large hailstone showing the structure of alternating rings of clear and white ice. (*Alberta Research Council, Edmonton*)

Hair Nonliving, specialized epidermal derivatives characteristic only of modern mammals. However, it is now thought that hair was present in at least some therapsid reptiles. It consists of keratinized cells, tightly cemented together, which arise from the matrix at the base of a follicle. A follicle is a tubular epidermal downgrowth that penetrates into the dermis and widens into a bulb (the hair root) at its deep end. The follicle, together with a lateral outgrowth called the sebaceous gland, forms the pilosebaceous system. Rapid cell production in the matrix, and differentiation in the regions immediately above, produces a hair shaft which protrudes from the follicle mouth at the skin surface. *See* GLAND.

Hairs are not permanent structures but are continually replaced throughout the life of a mammal. In some species, for example, the rat, hamster, mouse, chinchilla, and rabbit, the replacement pattern is undulant, and waves of follicular activity can be traced across the body. In other species, for example, humans, cats, and guinea pigs, each follicle appears to cycle independently of others in the immediate area. *See* INTEGUMENT. [P.F.M.]

Hairworm An extremely elongate, nonsegmented worm (sometimes known as horsehair worm) belonging to the phylum Nematomorpha. The name is particularly applied to the approximately 200 species of the order Gordioidea, adults of which are found in smaller habitats of standing freshwater, such as wells, ponds, permanent swamp pools, and ditches.

Adult hairworms, usually with black or dark brown pigmentation, are mostly about 0.04 in. (1 mm) in diameter and range from 4 to 40 in. (10 cm to 1 m) in length.

In their unusual life cycle, hairworm larvae are parasitic in arthropods, while the adults are free-living with a nonfunctional gut and two long, cylindrical gonads. The sexes are separate, and adult males are much more active.

Hairworms are found all over the world, and the majority of species belong to the gordioid genera *Gordius*, *Paragordius*, *Chordodes*, *Parachordodes*, and *Gordionus*. There are five other species of nematomorphan worms in the marine planktonic genus *Nectonema*, and they are separated as the order Nectonematoidea. *See* NEMATOMORPHA. [W.D.R.-H.]

Half-life The time required for one-half of a given material to undergo chemical reactions; also, the average time interval required for one-half of any quantity of identical radioactive atoms to undergo radioactive decay.

The concept of the time required for all of the material to react is meaningless, because the reaction goes very slowly when only a small amount of the reacting material is left and theoretically an infinite time would be required. The time for half completion of the reaction is a definite and useful way of describing the rate of a reaction.

The specific rate constant k provides another way of describing the rate of a chemical reaction. This is shown in a first-order reaction, Eq. (1), where c_0 is the initial concentration and

$$k = \frac{2.303}{t} \log \frac{c_0}{c} \qquad (1)$$

c is the concentration at time t. The relation between specific rate constant and period of half-life, $t_{1/2}$, in a first-order reaction is given by Eq. (2). *See* CHEMICAL DYNAMICS. [F.D.]

$$t_{1/2} = \frac{2.303}{k} \log \frac{1}{1/2} = \frac{0.693}{k} \qquad (2)$$

The activity of a source of any single radioactive substance decreases to one-half in 1 half-period, because the activity is always proportional to the number of radioactive atoms present. For example, the half-period of ^{60}Co (cobalt-60) is $t_{1/2}$, = 5.3 years. Then a ^{60}Co source whose initial activity was 100 curies will decrease to 50 curies in 5.3 years. In 1 additional half-period this activity will be further reduced by the factor $\frac{1}{2}$. Thus, the fraction of the initial activity which remains is $\frac{1}{2}$ after one half-period, $\frac{1}{4}$ after two half-periods, $\frac{1}{8}$ after three half-periods, $\frac{1}{16}$ after four half-periods, and so on. The half-period is sometimes also called the half-value time or, with less justification, the half-life. *See* RADIOACTIVITY. [R.D.E.]

Halichondrida A small order of sponges of the class Demospongiae, subclass Ceractinomorpha, with a skeleton of diactinal or monactinal siliceous megascleres or both. Spongin is present in small amounts; microscleres are absent. A skinlike dermis is present and is often reinforced with spicules.

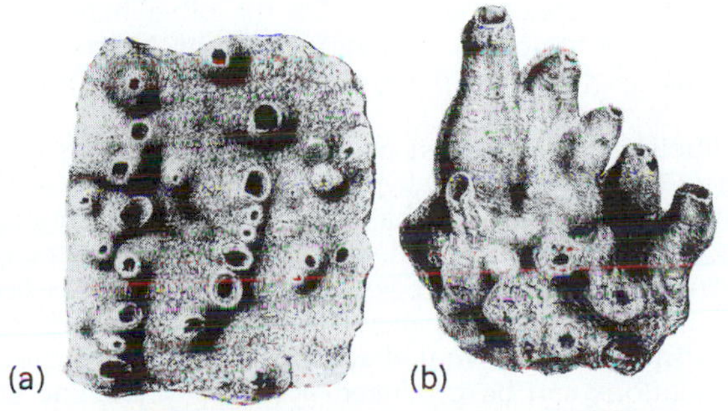

***Halichondria panicea*, shallow-water sponge. (*a*) Encrusting form. (*b*) Fistular form. (*After Bowerbank*)**

Halichondrid sponges are encrusting, massive, lobate, or branching in shape (see illustration). Halichondrids inhabit all seas, occurring chiefly in tidal areas and shallow waters of the continental shelf. Some species occur down to depths of at least 5000 ft (1500 m). Fossil species are unknown. *See* DEMOSPONGIAE. [W.D.H.]

Halide A compound of the type MX, where X may be fluorine, chlorine, bromine, iodine, or astatine, and M may be another element or organic radical. The compounds are considered derivatives of the hydrohalic acids, HX. All the halides form both covalent and ionic inorganic salts, coordination compounds, and many organic compounds.

Many properties show a gradual change as one compares the halides. The size of the ions, and the strength of the halide as a reducing agent, increases from F^- to I^-. It becomes more difficult to produce the element from the halide as one goes from I^- to F^-. The stability of complexes formed, and the solubilities of metal halides in water, increase going from I^- to F^-. *See* BROMIDE; CHLORIDE; FLUORIDE; HALOGENATED HYDROCARBON; HALOGENATION; IODIDE. [E.E.W.]

Halite The halite structure is one of the most fundamental in mineralogy. It is based on the cubic close-packing of the chloride anions with the sodium cations occupying all the available octahedral vacancies (see illustration). The structure forms the basis for many compounds such as manganosite, MnO; periclase, MgO; and galena, PbS. Fragments of the halite (rock-salt) structure are found in a large number of basic manganese arsenates and in the lead-rich sulfosalt minerals. [C.S.Hu./P.B.M.]

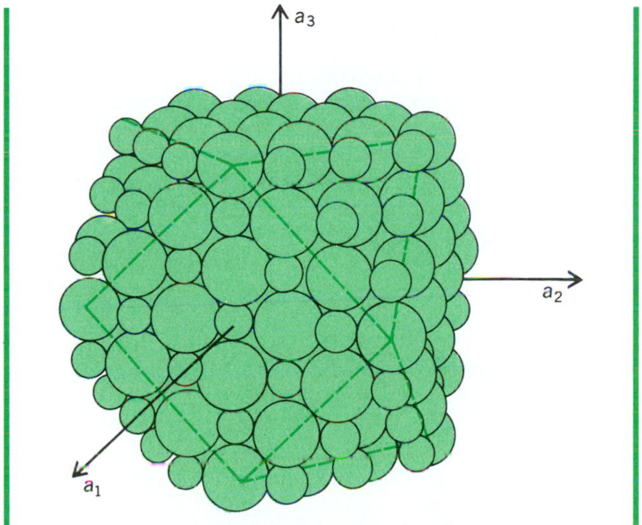

Packing model of halite, NaCl, with cuboctahedral outline (Na^+ small, Cl^- large). Arrows a_1, a_2, a_3 represent the directions of the edges of the cube of the unit cell. (*After C. S. Hurlbut, Jr., and C. Klein, Manual of Mineralogy, 19th ed., 1977*)

Hall effect The development of a transverse electric potential gradient in a current-carrying conductor placed in a magnetic field when the conductor is positioned so that the direction of the magnetic field is perpendicular to the direction of current flow. Analysis of the Hall effect yields important information on the band structure of metals and semiconductors and on the nature of the conductivity process. In semiconductor research, the magnitude of the Hall effect in simple cases provides a direct estimate of the concentration of charge carri-

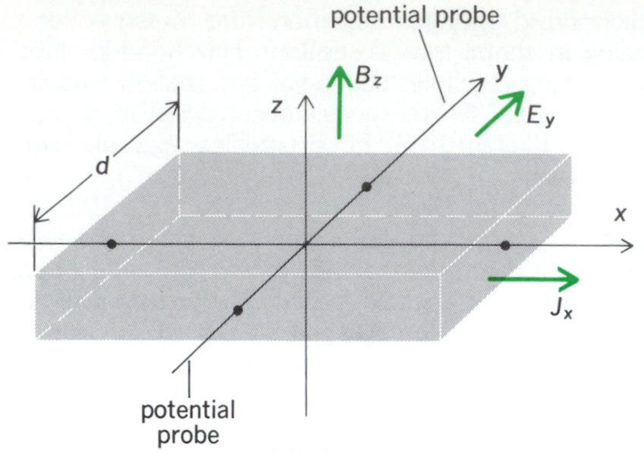

Hall effect.

ers. The Hall effect is one of the so-called galvanomagnetic effects. *See* GALVANOMAGNETIC EFFECTS.

The experimental arrangement for observing the Hall effect is shown schematically in the illustration. A voltage V_H appears between the sides of the specimen whenever the current density J_x and the magnetic field B_z are nonvanishing. The electric field $E_H = V_H/d = E_y$, where d is the width of the specimen using the coordinates of the illustration, is found to be proportional to the product of J_x and B_z. The quantities V_H and E_H are called the Hall voltage and Hall field, respectively. The Hall coefficient R is defined by the equation below.

$$R = E_y/J_xB_z$$

In semiconductors and metals, the sign and magnitude of R determine the sign and number of charge carriers, provided only one type of carrier (electrons or holes) is present. The Hall coefficient in semiconductors must be calculated with the aid of a two-band model when more than one type of carrier is responsible for charge transport. The Hall coefficient, especially in a multicarrier situation, is field-dependent. This field dependence provides valuable information on mobility ratios and relaxation times. *See* RELAXATION TIME OF ELECTRONS; SEMICONDUCTOR. [F.J.B.]

Halley's Comet The most famous of all comets; records of Halley's Comet appear at least as far back as 240 B.C. The comet's size, activity, and favorably placed orbit (with the perihelion roughly halfway between the Sun and the Earth's orbit) ensure its visibility to the naked eye at each apparition (see illustration). *See* COMET.

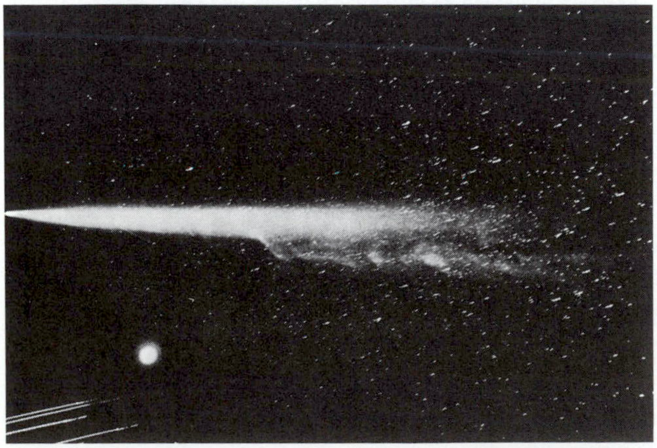

Halley's Comet on May 13, 1910. The planet Venus is in the lower part of the photograph. The tail stretched approximately 45° on this date. (*Lowell Observatory*)

This comet was the first to have its return predicted, a feat accomplished by Edmond Halley in 1705. He computed the orbits of several comets with Isaac Newton's then new gravitational theory. The orbits of comets observed in 1531, 1607, and 1682 were remarkably similar. Halley assumed that the sightings were of a single comet and predicted its return in 1758–1759. The prediction was verified, and the comet was named in his honor. The last perihelion passage of the comet was on February 9, 1986. In March 1986, six uncrewed spacecraft encountered Halley's Comet and produced data that have greatly enhanced the understanding of comets. The comet will return to its perihelion in 2061. [J.C.B.]

Halloysite A group of clay minerals. The major use of halloysite clay is in the manufacture of cracking petroleum catalysts. Pure halloysite clays are not abundant; hence their potential economic use has not been explored. *See* CLAY MINERALS; KAOLINITE.

There are two types of halloysite. Both are light-colored; one is porous and friable, while the other is dense, nonporous, and porcelainlike. The porous variety has the same chemical composition as kaolinite, whereas the nonporous hydrated form contains an added amount of water ($2H_2O$).

The halloysite minerals are made up of the same structure units as kaolinite; that is, layers composed of single silica tetrahedral and alumina octahedral units. They differ in the presence of a single layer of water molecules between each silicate layer in the hydrated form. Electron micrographs show that halloysite is composed of elongate particles which appear to be tubes. The tubes of the hydrated $4H_2O$ form appear to be made up of overlapping, curled-up sheets of the kaolinite type.

Halloysite is often found in hydrothermal deposits. It has also been reported in weathering products; however, it is a rare component of such materials. It is generally absent in sedimentary rocks, and it has been suggested that metamorphic processes would change halloysite to kaolinite. [F.M.W.; R.E.Gr.]

Hallucination An involuntary sensory experience of objectively nonexistent phenomena. The ability to distinguish hallucinations from perceptions requires contextual cues and the ability to use them.

Hallucinations can be elemental and meaningless (points of light, luminosities, noises, colors) or figural and identifiable (objects, sounds, voices). They may be static, repetitive, or erratic, or they may be developmental (sequential or unfolding). They may involve a clear sensorium (for example, the hallucinated event may be added to the perceived scene), or they may involve reduced consciousness (for example, the fusion of real and hallucinated events in the context of impaired attention and orientation, as in oneiric confusional states produced by neural degeneration or by acute alcohol withdrawal). They may be at least momentarily believed, as in the case of complex hallucinations superimposed on a context of otherwise normal awareness, consciousness, and feeling. They may be disbelieved, as in the case of psychomotor epileptic conditions that produce hallucinations in the context of strong feelings of recollection or dreamy unreality. On the other hand, reality testing may be completely suspended, as in oneiric confusional or normal (especially rapid-eye-movement or REM) dream states.

Hallucinations occur in abnormal conditions such as migraine headache, insulin hypoglycemia, high fever, schizophrenia (predominantly verbal hallucinations), psychomotor (temporal-lobe) epilepsy (visual, auditory, and gustatory-olfactory hallucinations), and narcolepsy (postural collapse, sleep-onset visual dream hallucinations, and postsleep paralysis, all suggesting an abnormal disinhibition of REM sleep). Hallucinations can be produced by acute or degenerative defects of subcortical and neocortical sensory projection mechanisms. In addition, hallucinations can be induced experimentally by sensory deprivation, by extended sleep deprivation, by direct electrical stimulation to sensory areas, or by psychoactive drugs (for example, LSD). *See* PERCEPTION. [D.B.C.]

Halo A system of luminous rings or arcs around the Sun or Moon, caused by refraction or reflection of light rays illuminating ice crystals floating in the air. These rays are seen as luminous circles around the Sun with a radius of either 22° (inner halo) or 46° (outer halo). Besides the inner and outer halos several similar arcs, luminous arcs, and bright spots are occasionally seen in the sky as a result of refraction or reflection of light rays illuminating ice crystals in the atmosphere. All these phenomena are usually considered parts of a more complex system, called the halo. The inner halo is, however, most frequently observed. *See* SUNDOG. [Z.S.]

Halocline A layer of seawater within which a marked salinity gradient is present. A permanent halocline of decreasing salinity usually coincides with the top of the permanent thermocline. The oceanic halocline furnishes significant clues to the origin and movement of water masses. Haloclines are extremely important in estuarine oceanography, however, as their depth, strength, and shape greatly influence circulation, mixing, and flushing of these embayments. Estuarine haloclines have strong positive gradients which are formed as fresh water flows seaward over seawater entering along the bottom. *See* ESTUARINE OCEANOGRAPHY; SEAWATER; THERMOCLINE. [J.J.S.]

Halocyprida An order of the subclass Myodocopa, class Ostracoda, characterized by biramous antennae with the endopod and exopod of similar size, reduction or absence of the seventh pair of appendages, an unpaired male copulatory organ, and the absence of a median eye. The taxon is subdivided into two suborders, Halocypridina and Cladocopina; each has both Recent and fossil representatives. Extant halocyprids are small ostracods, measuring less than 0.4 in (10 mm).

With the exception of one brackish-water representative, halocyprids are marine, cosmopolitan in all oceans, and found from surface waters to abyssal depths. Most are planktonic species, although a few are epibenthic or benthic, and they include filter feeders, carnivores, and possibly detritus feeders. *See* CRUSTACEA; MYODOCOPA. [P.A.McL.]

Haloform reaction The halogenation of acetaldehyde or a methyl ketone in aqueous basic solution. The reaction is characteristic of compounds that contain a CH_3CO— group linked to a hydrogen or to another carbon atom. The halogen is dissolved in aqueous base (sodium hydroxide or carbonate) to form an alkali hypohalite, which is the halogenating agent. This is added to the ketone. The reaction occurs in two phases: the methyl group is fully substituted by halogen to give an intermediate trihaloketone; the trihaloketone breaks down in the presence of alkali to produce the haloform and the sodium salt of a monocarboxylic acid. The haloforms which can be produced by this method are chloroform, bromoform, and iodoform.

The haloform reaction was formerly applied industrially, specifically in the manufacture of chloroform, but it has been virtually displaced by more economical methods. *See* CHLOROFORM; IODOFORM. [E.C.; M.G.G.]

Halogen elements The halogen family consists of the elements fluorine, F; chlorine, Cl; bromine, Br; iodine, I; and astatine, At. All the halogen elements except astatine exist in the Earth's crust and atmosphere. *See* HALOGEN MINERALS.

The halogens are the best-defined family of elements. They have an almost perfect gradation of physical properties. The increase in atomic weight from fluorine through iodine is paralleled by increases in density, melting and boiling points, critical temperature and pressure, heats of fusion and vaporization, and even in progressively deeper color (fluorine is pale yellow; chlorine, yellow-green; bromine, dark red; and iodine, deep violet).

Although all halogens generally undergo the same types of reactions, the extent and ease with which these reactions occur vary markedly. Fluorine in particular has the usual tendency of the lightest member of a family of elements to exhibit reactions not comparable to the other members. Each halogen must be considered individually, both in its preparation and in its reaction. *See* ASTATINE; BROMINE; CHLORINE; FLUORINE; HALIDE; HALOGENATION; IODINE. [R.J.C.; A.A.G.]

Halogen minerals Naturally occurring compounds containing a halogen as the sole or principal anionic constituent. There are over 70 such minerals, but only a few are common and can be grouped according to the following methods of formation.

1. Saline deposition by evaporation of seawater or salt lakes. Halite (rock salt), NaCl, is the most important of this type. Of the other minerals associated with halite, sylvite, KCl, and carnallite, $KMgCl_3 \cdot 6H_2O$, are the most important. *See* SALINE EVAPORITE.
2. Hydrothermal deposition. Fluorite, CaF_2, is the chief representative of this type. Cryolite, Na_3AlF_6, may be of primary deposition or may result from the action of fluorine-bearing solutions on preexisting silicates. *See* CRYOLITE; FLUORITE.
3. Secondary alteration. Chlorides, iodides, or bromides of silver, copper, lead, or mercury may form as surface alterations of ore bodies carrying these metals. The most common are cerargyrite, AgCl, and atacamite, $CU_2(OH)_3Cl$. *See* CERARGYRITE.
4. Deposition by sublimation. Halides formed as sublimation products about volcanic fumaroles include sal ammoniac, NH_4Cl; malysite, $FeCl_3$; and cotunnite, $PbCl_2$. At Mount Vesuvius, Italy, is the most noted occurrence of such minerals. *See* HALIDE.
5. Meteorites. Lawrencite, $FeCl_2$, has been found in iron meteorites. [C.S.H.]

Halogenated hydrocarbon One of a group of halogen derivatives of organic hydrogen- and carbon-containing compounds. The group includes monohalogen compounds (alkyl or aryl halides) and polyhalogen compounds that contain the same or different halogen atoms. The halogenated hydrocarbons may be classified under the general headings of alkyl halides, unsaturated (vinyl and allyl) halides, polyhalogenated compounds, and halogenated aromatic hydrocarbons.

The alkyl halides, especially bromides and iodides, are used extensively as research intermediates. Several are essential pharmaceutical intermediates. Alkyl chlorides are important industrially in the manufacture of other organic compounds, especially alcohols and esters.

The polyhalogenated hydrocarbons find many uses. They are generally chemically inert, and are widely used in many industrial processes as solvents, degreasers, and dry-cleaning agents. The polybromides, more reactive than the polychlorides, are frequently used in the synthesis of other compounds. The polyfluorides, often with attached chorine and bromine, are called freons and are widely used as refrigerants. The unsaturated polyhalogen compounds are also valuable in a variety of organic syntheses.

The halogenated aryl hydrocarbons are used in organic syntheses as solvents and as reactants. Generally, the chlorine compounds are more inert than the bromine and iodine compounds.

The aralkyl- or benzyl-type halides are more reactive than the aryl halides. They are used chiefly as chemical intermediates. *See* HALOGENATION. [E.C.; M.G.G.]

Halogenation A chemical reaction or process which results in the formation of a chemical bond between a halogen atom and another atom. Reactions resulting in the formation of halogen-carbon bonds are especially important. The halogenated compounds produced are employed in many ways, for example, as solvents, intermediates for numerous chemicals, plastic and polymer intermediates, insecticides, fumigants, sterilants, refrigerants, additives for gasoline, and materials used in fire extinguishers. *See* HALOGEN ELEMENTS.

Halogenation reactions can be subdivided in several ways, for example, according to the type of halogen (fluorine, chlorine, bromine, or iodine), type of material to be halogenated (paraffin, olefin, aromatic, hydrogen, and so on), and operating conditions and methods of catalyzing or initiating the reaction.

Halogenation reactions with elemental chlorine, bromine, and iodine are of considerable importance. Because of high exothermocities, fluorinations with elemental fluorine tend to have high levels of side reactions. Consequently, elemental fluorine is generally not suitable for direct fluorination. Two types of reactions are possible with these halogen elements, substitution and addition.

Substitution halogenation is characterized by the substitution of a halogen atom for another atom (often a hydrogen atom) or group of atoms (or functional group) on paraffinic, olefinic, aromatic, and other hydrocarbons. A chlorination reaction of importance that involves substitution is that between methane and chlorine. *See* SUBSTITUTION REACTION.

Addition halogenation involves a halogen reacting with an unsaturated hydrocarbon. Chlorine, bromine, and iodine react readily with most olefins; the reaction between ethylene and chlorine to form 1,2-dichloroethane is of considerable commercial importance, since it is used in the manufacture of vinyl chloride.

Addition reactions with bromine or iodine are frequently used to measure quantitatively the number of —CH═CH— (or ethylenic-type) bonds in organic compounds. Bromine numbers or iodine values are measures of the degree of unsaturation of the hydrocarbons.

Substitution halogenation on the aromatic ring can be made to occur via ionic reactions. The chlorination reactions with elemental chlorine are similar to those used for addition chlorination of olefins. *See* HALOGENATED HYDROCARBON. [L.F.A.]

Halophilism (microbiology) The phenomenon of demand for salt (sodium chloride) for growth and maintenance displayed by certain microbes. These microbes comprise a variety of different types: bacteria, molds, algae, amebas, and protozoans. They may be classified according to the salt concentration supporting best growth: extreme halophiles, best growth at 20–30% salt; moderate halophiles, best growth at 5–20% salt; and slight halophiles, best growth at 2–5% salt.

The halophilic bacteria are widely distributed in nature in waters of very high salinity, for example, the Dead Sea. In evaporation ponds for the production of solar salt, they are often present in the strongly saline brine in such large numbers that they impart a red color to the brine. [H.L.]

Haloragales An order of flowering plants, division Magnoliophyta (Angiospermae), in the subclass Rosidae of the class Magnoliopsida (dicotyledons). The order consists of 4 families, with less than 175 species in all. The Haloragales are herbs with perfect or often unisexual, more or less reduced flowers. Many of the species are aquatic. Pollen is commonly distributed by wind or water. The aquarium plant called parrot's feather (*Myriophylium*, family Haloragaceae) and the very large-leaved specimen plant *Gunnera* (family Gunneraceae) are well-known members of the Haloragales. *See* MAGNOLIOPSIDA; ROSIDAE. [A.Cr.]

Hamamelidae A small subclass of the class Magnoliopsida (dicotyledons) in the division Magnoliophyta (Angiospermae) consisting of 9 orders (Trochodendrales, Hamamelidales, Eucommiales, Urticales, Leitneriales, Juglandales, Myricales, Fagales, and Casuarinales), 23 families, and about 3400 species. They have strongly reduced, often unisexual flowers with poorly developed or no perianth. With the notable exception of some of the Urticales, they are all woody plants. Pollination is usually by wind. See articles on each order. *See also* FLOWER; MAGNOLIOPSIDA. [A.Cr.]

Hamamelidales A small order of flowering plants, division Magnoliophyta (Angiospermae), which gives its name to the subclass Hamamelidae in the class Magnoliopsida (dicotyledons). The family Hamamelidaceae contains about 100 species and the Platanaceae about 6 species; the other 4 families have only 2 species each. Within its subclass, the order is more advanced than the Trochodendrales in having vessels in the wood, but less advanced than the other orders in that the gynoecium consists either of separate carpels or of united carpels that open at maturity to release the seeds. Witch hazel (*Hamamelis*), sweet gum (*Liquidambar*), and the plane tree or sycamore (*Platanus*) are familiar members of the Hamamelidales. *See* HAMAMELIDAE; MAGNOLIOPSIDA; TROCHODENDRALES. [A.Cr.]

Hamilton's equations of motion The motion of a mechanical system may be described by a set of first-order ordinary differential equations known as Hamilton's equations. Because of their remarkably symmetrical form, they are often referred to as the canonical equations of motion of a system. They are equivalent to Lagrange's equations, but the fact that they are of first order and highly symmetrical makes them advantageous for general discussions of the motion of systems. *See* LAGRANGE'S EQUATIONS.

Hamilton's equations can be derived from Lagrange's equations. Let the coordinates of the system be q_j ($j = 1, 2, \ldots, f$), and let the dynamical description of the system be given by the lagrangian $L(q,\dot{q},t)$, where q denotes all the coordinates and a dot denotes total time derivative. Lagrange's equations are then given by Eq. (1). The momentum p_j canonically conjugate to q_j is defined by Eq. (2).

$$\frac{d}{dt}\frac{\partial L}{\partial \dot{q}_j} - \frac{\partial L}{\partial q_j} = 0 \qquad (1)$$

$$p_j = \frac{\partial L}{\partial \dot{q}_j} \qquad 2)$$

(The hamiltonian H is defined by Eq. (3). Then Hamilton's canonical equations are Eqs. (4).

$$H = \sum_{j=1}^{f} p_j\dot{q}_j - L\,(q,\dot{q},t) \qquad (3)$$

$$\dot{q}_j = \frac{\partial H(q,p,t)}{\partial p_j} \quad \dot{p}_j = -\frac{\partial H(q,p,t)}{\partial q_j} \qquad (4)$$

As they stand, Hamilton's equations are no easier to integrate directly than Lagrange's. Hamilton's equations are of great advantage in more general discussions, and they permit the making of canonical transformations which can lead to simplifications. *See* CANONICAL TRANSFORMATIONS.

The hamiltonian function H of classical mechanics is used to form the quantum-mechanical hamiltonian operator. [P.M.S.]

Hamilton's principle A variational statement known as Hamilton's principle forms a basis from which the equations of motion of a classical dynamical system may be deduced. Consider a mechanical system whose configuration is specified by f independent generalized coordinates $q_1, \ldots, q_f$. Let the configuration at time t_1 be given by the values $q_j(t_1)$ and that at time t_2 by the values $q_j(t_2)$. If the system is described dynamically by a lagrangian function $L(q,\dot{q},t)$ where q stands for all the f coordinates, then, according to Hamilton's principle, the trajectory or path of the system between the two configurations at the two given times is that which makes the value of the integral in the equation stationary relative to nearby paths between the same end points and taking the same time.

$$\Phi = \int_{t1}^{t2} L[q(t),\dot{q}(t),t]dt$$

Lagrange's equations may be derived from Hamilton's principle.

Hamilton's principle is invariant under any transformation of coordinates. It is also unaffected by the addition of a total time derivative to the lagrangian, because the integral of such a total derivative is independent of the path of integration. This allows the coordinates q_j and the momenta $p_j = \partial L/\partial \dot{q}_j$ to be subjected to arbitrary canonical transformations without changing the form of Lagrange's equations. *See* CANONICAL TRANSFORMATIONS; LAGRANGE'S EQUATIONS; LAGRANGIAN FUNCTION; MINIMAL PRINCIPLES. [P.M.S.]

Hamster The common name for any of 14 species of rodents in the family Cricetidae. The natural range of most of these species is Asia but a few, such as the common hamster (*Cricetus cricetus*), are found in Europe.

The common hamster is a solitary, aggressive animal with interesting burrowing and hoarding habits. The burrows are not deep, rarely more than 2 ft (0.6 m), and consist of a large central chamber with radiating side chambers for special purposes, such as for hoarding food, for living quarters, and for excretion. A so-called summer chamber is used for breeding. The hamster goes into its burrow in the autumn, closes off the entrance, and goes to sleep. It does not hibernate deeply and wakes up from time to time to eat. *See* HIBERNATION.

The golden hamster (*Mesocricetus auratus*) is extensively used as a laboratory animal for experimental purposes, especially in studies of the physiological aspects of hibernation. *See* RODENTIA. [C.B.C.]

Hand The terminal portion of the upper, or superior, extremity. Its bony framework consists of 8 carpal or wristbones arranged in 2 rows, 5 metacarpal or palm bones, and 14 phalanges or finger bones. These are held together, yet allowed considerable movement, by a complex system of ligaments.

Muscles of the forearm and hand act upon these bones, permitting a wide range of motion. The primate hand is characterized by the opposability of thumb and fingers, which permits grasping or prehension. In addition, fine voluntary control affords highly coordinated muscular action in humans. The hand contains many nervous receptors, mediating sensations of touch, pressure, and temperature, and thus serves as a primary sensory apparatus. [T.S.P.]

Haplopoda An order of carnivorous branchiopod crustaceans formerly included in the order Cladocera. Only one species, the fresh-water *Leptodora kindti*, is included.

The body is about 9 mm (0.4 in.) long. *Leptodora* is among the most transparent of multicellular freshwater animals. It swims slowly by means of enormous antennae and seizes its prey with its six pairs of segmented, grasping, thoracic limbs. The mandibles are styliform and of a type unique within the Branchiopoda. There is a single, median sessile eye. The carapace is reduced to a dorsal brood pouch and does not protect the body.

Parthenogenetic eggs and young are carried in the brood pouch in summer. In autumn, fertilized resting eggs are carried there which are shed freely, and the eggs overwinter. They hatch in spring as nauplii, a stage eliminated from the parthenogenetic phase of the life cycle.

Leptodora is widely distributed in the plankton of larger lakes of the Holarctic region. *See* BRANCHIOPODA. [G.Fr.]

Haplosclerida An order of sponges of the class Demospongiae, including species with a skeleton made up of a single category of siliceous megascleres embedded in spongin fibers or joined together in a network by a spongin cement. Microscleres (never asters) are present or absent. A modified dermal skeleton is absent. Many species form large upright branching colonies, the branches often being hollow.

Haplosclerid sponges inhabit all seas and are especially abundant in tidal and shallow waters of the continental shelf. Some species occur down to depths of 6600 ft (2000 m). The family Spongillidae is restricted to fresh water except for a few species which have secondarily invaded brackish water. *See* DEMOSPONGIAE. [W.D.H.]

Haplosporea A class of protozoa, often known as Acnidosporidia, in the subphylum Sporozoa. Haplosporea are distinguished from other similar groups by the production of spores lacking polar filaments. The spores are enclosed in a membrane, and each spore contains a single sporozoite.

Two orders make up the class according to the scheme proposed by the Committee on Taxonomy and Taxonomic Problems of the Society of Protozoologists. These are the Haplosporida and Sarcosporida. Haplosporida are parasites of invertebrates and primitive chordates (Ascidia). Sarcosporida are muscle parasites of vertebrates, particularly warm-blooded ones. *See* HAPLOSPORIDA; SPOROZOA. [R.D.M.]

Haplosporida A group of Protozoa usually regarded as an order within the class Haplosporea. The chief distinguishing characteristic of the Haplosporida, and the one from which their name is derived, is the production of uninucleate spores that lack polar capsules and polar filaments.

Haplosporida are mainly parasites of invertebrates, such as rotifers, annelids, crustacea, insects, and mollusks, but they also occur in the bodies of Ascidia (tunicates), which are primitive chordates. If the genus *Icthyosporidium* is properly included in the order, Haplosporida also parasitize fish. One species causes "neon" fish disease.

It is of considerable biological interest that some species of Haplosporida are hyperparasites, that is, parasitic on other parasites. They have been found in gregarines parasitizing annelids, and also in flukes. *See* HAPLOSPOREA; SPOROZOA. [R.D.M.]

Harbor Any body of water of sufficient depth for ships to enter and find shelter from storms or other natural phenomena. The modern harbor is a place where ships are built, launched, and repaired, as well as a terminal for incoming and outgoing ships. There are four principal classes of harbors: commercial, naval, fishery, and refuge for small craft. Most harbors are situated at the mouth of a river or at some point where it is easy to transfer cargoes inland by river barges, railroads, or trucks.

The term harbor is often confused with the term port. A port is a harbor with the necessary terminal facilities to expedite the moving of cargo and passengers at any stage of a journey. A good harbor must have a safe anchorage and a direct channel to open water, and must be deep enough for large ships. An efficient port must have enough room for docks, warehouses, and loading and unloading machinery.

Geographically, a port or harbor is usually limited to a comparatively small area of usable berthing space rather than an extended coastline.

There are three types of harbors: land-locked, natural harbors protected from the sea by a narrow inlet; unprotected harbors at which ships may dock even though unprotected from the hazards of changing tides, ocean waves, fogs, and ice; and artificial harbors carved out at sites where the natural features are unfavorable. The latter are fashioned by dredging and by constructing jetties, breakwaters, and sea basins to protect ships against unusually high or low tides. *See* COASTAL ENGINEERING. [B.J.W.]

Hardness scales Arbitrarily defined measures of the resistance of a material to indentation under static or dynamic load, to scratch, abrasion, or wear, or to cutting or drilling. Standardized tests compare similar materials according to the particular aspect of hardness measured by the test. Widely used tests for metals are Brinell, Rockwell, and Scleroscope tests, with modifications depending upon the size or condition of the material. Indentation tests compare species of wood or flooring materials, and abrasion tests serve as an index of performance of stones and paving materials.

Hardness tests are important in research and are widely used for grading, acceptance, and quality control of manufactured articles. The hardness designation or scale is associated with the test method or instrument used.

Resistance to scratching is defined by comparison with 10 selected minerals, which are numbered in the order of increasing hardness. This mineralogical scale, called Mohs scale, is 1, talc; 2, gypsum; 3, calcite; 4, fluorite; 5, apatite; 6, orthoclase; 7, quartz; 8, topaz; 9, corundum; and 10, diamond. Minerals lower in the scale are scratched by those with higher numbers.

Materials are differentiated qualitatively according to resistance to scratching or cutting by files especially selected for the purpose. Whether or not a visible scratch is produced on the material indicates its hardness in comparison with a sample of desired hardness.

Resistance to indentation by a hardened steel or tungsten carbide ball under specified load is the basis for Brinell hardness. Brinell hardness number (Bhn), expressed in kilograms per square millimeter, is obtained by dividing the load by the spherical surface area of the impression.

Indentation of a square-based diamond pyramid penetrator with an angle between opposite faces of 136° measures Vickers hardness. Vickers hardness number, also called diamond pyramid hardness, is equal to the load divided by the lateral area of the pyramidal impression.

Depth of indentation of either a steel ball or a 120° conical diamond with rounded point, called a brale, under prescribed load is the basis for Rockwell hardness. The depth of impression is indicated on a dial whose graduations represent the hardness number.

The pressure in kilograms per square millimeter required to embed a 0.75-mm (0.0295-in.) hemispherical diamond penetrator to a depth of 0.046 mm (0.0018 in.), producing an impression 0.36 mm (0.014 in.) in diameter, is the measure of Monotron hardness.

Height of rebound of a diamond-tipped weight or hammer falling within a glass tube from a height of 10 in. (25.4 cm) and striking the specimen surface measures Shore Scleroscope hardness.

Resistance to indentation over very small areas (as on small parts, the constituents of metal alloys, or for exploration of hardness variations) is called microhardness. [W.J.K./W.G.B.]

Hardy-Weinberg formula A basic mathematical relation used in population genetics. It gives the proportion of the various genotypes in a randomly mating population in terms of the frequencies of the genes. The formula was discovered independently in 1908 by G. H. Hardy, a British mathematician, and W. Weinberg, a German physician. *See* HUMAN GENETICS; POPULATION GENETICS.

In its simplest form the Hardy-Weinberg formula may be stated thus: If p is the proportion of gene A in the population and $q(= 1 - p)$ is the proportion of gene a, then after one generation of random mating the three genotypes AA, Aa, and aa will occur in the proportions p^2, $2pq$, and q^2. In other words the genotypes are given by the appropriate terms in the expansion of the binomial $(p + q)^2$. The extension to multiple alleles is direct.

The formula holds only for an infinite population and assumes random mating in the absence of significant mutation pressure or gene transfer between populations. However, it is an accurate approximation in many populations. *See* GENETICS. [J.F.Cr.]

Harmonic (periodic phenomena) A sinusoidal quantity having a frequency that is an integral multiple of the frequency of a periodic quantity to which it is related. *See* MODE OF VIBRATION; PARTIAL TONE.

A harmonic series of sounds is one in which the basic frequency of each sound is an integral multiple of some fundamental frequency. An ideal string (or air column) can vibrate as a whole or in a number of equal parts, and the respective periods of vibration are proportional to the lengths. These increasingly shorter lengths or periods form a harmonic series. *See* HARMONIC ANALYZER. [R.W.Y.]

Harmonic analyzer A device for separating and measuring the frequencies and amplitudes of the Fourier-series components of a complex periodic wave. A complete harmonic analysis would include finding the phase of each component. Most electrical analyzers find component amplitude only. *See* NONSINUSOIDAL WAVEFORM.

At low frequencies, harmonics may be measured with a wattmeter. The voltage to be analyzed is impressed upon the potential coil, and a current of adjustable frequency is passed through the current coil. By adjusting the frequency of the current to be close to that of the harmonic, a very slow beat can be produced and an accurate maximum reading can be obtained. *See* WATTMETER.

Two widely used types of instrument provide measurement of component amplitude at adjustable frequencies and nonmeasurement at other frequencies. The first of these, generally referred to as a distortion and noise meter, is designed to amplify uniformly over the frequency range of interest. The second type of instrument is a sound and vibration analyzer. Another widely used instrument employs a highly selective fixed-tuned filter in the heterodyne system. [M.G.F.]

Harmonic motion A periodic motion that is a sinusoidal function of time. It is often called simple harmonic motion (SHM). It is the simplest possible type of vibratory motion. The motion is symmetric about its midpoint, at which the velocity is greatest and the acceleration is zero. At the extreme displacements or turning points, the velocity is zero, and the acceleration is a maximum. The motion is characterized by a unique frequency (without overtones).

Harmonic motion may be present in very simple mechanisms. For example, if a wheel is rotating at a constant speed about a fixed axis, the projection on any fixed line of the motion of a point on the wheel is simple harmonic. Harmonic motion may also result from the response of a vibrating system to a periodic—in particular a sinusoidal—force. Harmonic motion is the typical motion of most simple systems that have been displaced from a position of stable equilibrium and then released, provided that the damping is negligible. The motion of a pendulum is approximately simple harmonic for small amplitudes. *See* PENDULUM.

The realization that atoms are continually vibrating in motions that are nearly harmonic is essential for understanding many properties of matter, including molecular spectra, heat capacity, and heat conduction. *See* DAMPING; FORCED OSCILLATION; HARMONIC OSCILLATOR; LATTICE VIBRATIONS; MOLECULAR STRUCTURE AND SPECTRA; PERIODIC MOTION; VIBRATION; WAVE MOTION. [J.M.Ke.]

Harmonic oscillator Any physical system that is bound to a position of stable equilibrium by a restoring force or torque proportional to the linear or angular displacement from this position. If such a body is disturbed from its equilibrium position and released, and if damping can be neglected, the resulting vibration will be simple harmonic motion, with no overtones. The frequency of vibration is the natural frequency of the oscillator, determined by its inertia (mass) and the stiffness of its restoring force.

The harmonic oscillator is not restricted to a mechanical system, but might, for example, be electric. Typical electronic oscillators, however, are only approximately harmonic.

If a harmonic oscillator, instead of vibrating freely, is driven by a periodic force, it will vibrate harmonically with the period of the force; initially the natural frequency will also be present, but any damping will eventually remove the natural motion. *See* DAMPING; FORCED OSCILLATION; HARMONIC MOTION.

In both quantum mechanics and classical mechanics, the harmonic oscillator is an important problem. It is one of the few rigorously soluble problems of quantum mechanics. The quantum-mechanical description of electromagnetic, electronic, mesonic, and other fields is usually carried out in terms of a (time) Fourier analysis. The individual Fourier components of noninteracting fields are independent harmonic oscillators. *See* ANHARMONIC OSCILLATOR. [J.M.Ke.]

Harmonic speed changer A mechanical-drive system used to transmit rotary, linear, or angular motion at high ratios and with positive motion. In the rotary version, shown schematically in the illustration, the drive consists of a rigid circular spline, an input wave generator, and a flexible spline. Any one of these can be fixed, used as the input, or used as the output. Any combination (fixed, driver, or driven) may be used. *See* SPLINES.

The advantages of this drive are (1) high-ratio gearing in small space, (2) speed reduction (or increase) in fixed ratio up to 1000:1 in one unit, (3) negligible wear of teeth, (4) balanced bearing loads, (5) negligible backlash, (6) efficiency of approximately 80% in a gear ratio of 400:1, and (7) adaptability to rotary-to-rotary, rotary-to-linear, and linear-to-linear drives. *See* GEAR DRIVE. [P.H.B.]

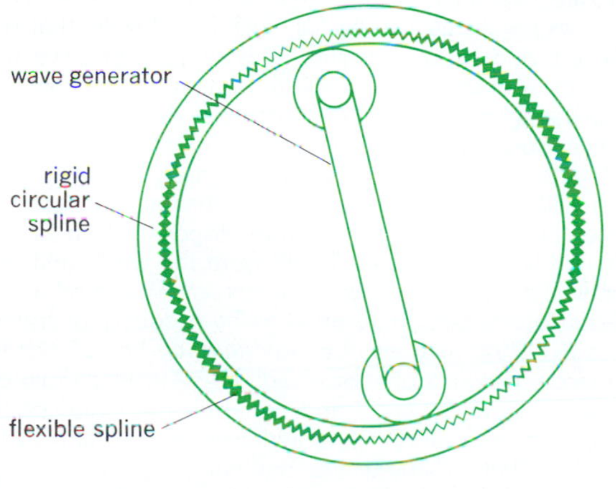

Rotary-to-rotary harmonic speed changer.

Harpacticoida An order of the Copepoda, with about 1700 species known. Their form is variable but is generally linear and more or less cylindrical (see illustration). These animals are minute and vary in length from 1/64 to 1/8 in. (0.4 to 3 mm). As a rule, the first thoracic segment is incorporated with the cephalothorax, and the last thoracic segment is included in the abdomen.

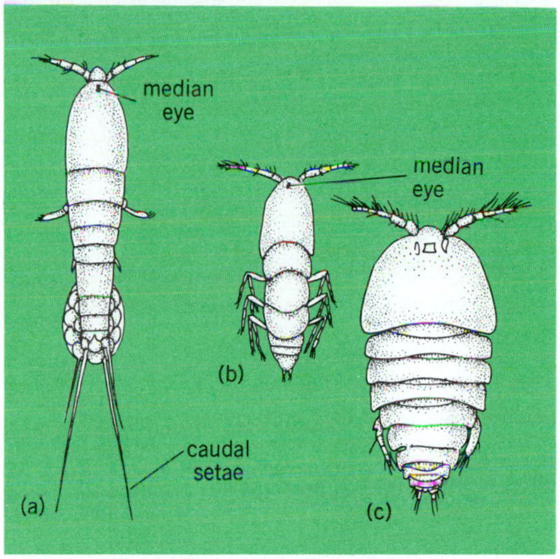

Some typical harpacticoids. (*a*) *Harpacticus chelifer.* (*b*) *Parategastes spaericus.* (*c*) *Alteutha interrupta.*

Harpacticoids have a worldwide distribution. They occur in all kinds of aquatic habitats, especially marine, but also among moss and leaves. In the sea they range from the shore to abyssal depths. In general, they are free-living and benthonic; some species are pelagic, parasitic, or commensal. *See* COPEPODA. [K.L.]

Harrowing Soil preparation for planting usually involves the pulling of an implement called a harrow over the plowed soil to break clods, level the surface, and destroy weeds. A wide variety of implements are classified as harrows; the most common kinds are the disk harrow, the spike-tooth harrow, the spring-tooth harrow, and the knife harrow. Previously the function of seedbed preparation was performed almost entirely by the implements classified as harrows. With the introduction of power, farming came to be performed in large part by field cultivators, rod weeders, rotary hoes, or treaders, subsurface packers, and various designs of rollers. Power-driven rotary tillers perform the function of both plowing and harrowing.

The spike-tooth is the oldest form of commercial harrow and consists of spikes or teeth (usually adjustable) extending downward from a frame. The teeth extend into the soil, and when the harrow is pulled forward, they cut through clods and break them. The teeth also stir and level the soil surface and kill weeds. This implement is most effective if used before clods dry; it is frequently attached behind the plow.

The spring-tooth harrow is similar to the spike-tooth type but has long curved teeth of spring steel. The spring action renders it suitable for rough or stony ground. It is particularly useful in bringing clods, roots of weeds, and obnoxious grasses to the surface for destruction, and to renovate and cultivate alfalfa fields. The knife harrow consists of a frame holding a number of knives which scrape and partly invert the surface to smooth it and

One type of disk harrow. (*Allis-Chalmers*)

destroy small weeds. The disk harrow is probably the most universally used type (see illustration). It cuts clods and trash effectively, destroys weeds, cuts in cover crops, and smooths and prepares the surface for other farming operations. The penetration of the disk harrow depends largely upon weight. [M.L.N.]

Hassium A chemical element, symbol Hs, atomic number 108. It was synthesized and identified in 1984 by using the Universal Linear Accelerator (UNILAC) at Darmstadt, West Germany, by the same team (led by P. Armbruster and G. Müzenberg) which first identified bohrium and meitnerium. The isotope ^{265}Hs was produced in a fusion reaction by bombarding a ^{208}Pb target with a beam of ^{58}Fe projectiles.

The discovery of bohrium and meitnerium was made by detection of isotopes with odd proton and neutron numbers. In this region, odd-odd nuclei show the highest stability against fission. Elements with an even atomic number are intrinsically less stable against spontaneous fission. The isotopes of hassium were expected to decay by spontaneous fission—which explains why meitnerium was synthesized before hassium.

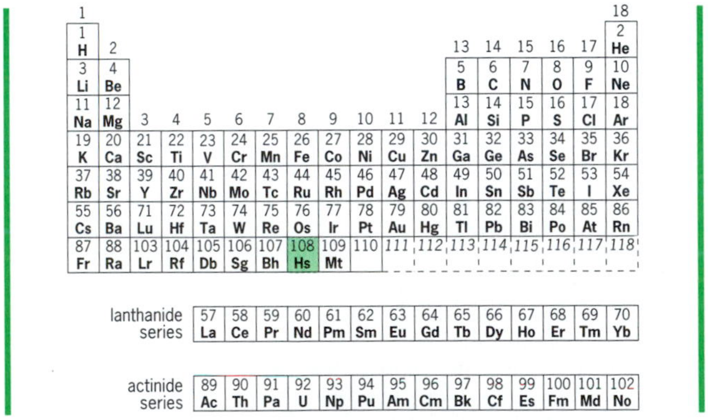
1 H
3 Li 4 Be
11 Na 12 Mg
19 K 20 Ca 21 Sc 22 Ti 23 V 24 Cr 25 Mn 26 Fe 27 Co 28 Ni 29 Cu 30 Zn 31 Ga 32 Ge 33 As 34 Se 35 Br 36 Kr
37 Rb 38 Sr 39 Y 40 Zr 41 Nb 42 Mo 43 Tc 44 Ru 45 Rh 46 Pd 47 Ag 48 Cd 49 In 50 Sn 51 Sb 52 Te 53 I 54 Xe
55 Cs 56 Ba 71 Lu 72 Hf 73 Ta 74 W 75 Re 76 Os 77 Ir 78 Pt 79 Au 80 Hg 81 Tl 82 Pb 83 Bi 84 Po 85 At 86 Rn
87 Fr 88 Ra 103 Lr 104 Rf 105 Db 106 Sg 107 Bh 108 Hs 109 Mt 110 111 112 113 114 115 116 117 118
5 B 6 C 7 N 8 O 9 F 10 Ne
13 Al 14 Si 15 P 16 S 17 Cl 18 Ar
2 He
lanthanide series 57 La 58 Ce 59 Pr 60 Nd 61 Pm 62 Sm 63 Eu 64 Gd 65 Tb 66 Dy 67 Ho 68 Er 69 Tm 70 Yb
actinide series 89 Ac 90 Th 91 Pa 92 U 93 Np 94 Pu 95 Am 96 Cm 97 Bk 98 Cf 99 Es 100 Fm 101 Md 102 No

As in the case of bohrium and meitnerium, the isotope was produced by fusion in a one-neutron deexcitation channel; in this case the compound system was ^{266}Hs. The reaction mechanism again was cold fusion. The isotope ^{265}Hs has a half-life of about 2 ms and decays by emission of an alpha particle of 10.36 MeV.

The elements bohrium, hassium, and meitnerium are stabilized by shell effects against spontaneous fission; this special stability may occur because these nuclei may prefer a sausage shape which is predicted to be energetically most favorable for them. *See* Bohrium; Meitnerium; Transuranium elements. [P.Ar.]

Hatchettite Mountain tallow, a yellow-white to yellow-green hydrocarbon occurring in Belgian coal seams; it is also called hatchettine. Hatchettite is translucent but darkens on exposure to air. It is soft, has no odor, is greasy to the touch, and consists of 85.5% carbon and 14.5% hydrogen. Its index of refraction is 1.47–1.50; it melts at 46–47°C (115–117°F), is sparingly soluble in alcohol or ether, decomposes in concentrated sulfuric acid, and has a specific gravity of 0.89–0.98. [I.A.B.]

Hausmannite A mineral with composition $Mn^{2+}Mn_2^{3+}O_4$. Hausmannite is most frequently massive-granular and possesses one perfect basal cleavage. The color is black, and streak dark brown. Hardness is 5.5 on Mohs scale, and the specific gravity is 4.81. It is an occasional ore of manganese, and it most frequently occurs in metamorphosed sedimentary manganese ore deposits, such as some small deposits in central Sweden and in the Central Provinces, India. [P.B.M.]

Hay fever An allergic disorder (also known as allergic rhinitis) of the nose and related structures due to sensitization by certain plant pollens, The term hay fever, although popularly used, is incorrect since hay is rarely implicated and fever is seldom present.

There is continuous or intermittent nasal obstruction accompanied by wheezing, sneezing, coughing, and the copious production of a watery discharge. The nasal mucosa becomes swollen and pale from vascular congestion and edema. Intense itching of the nose, mouth, and pharynx may occur, as well as excessive lacrimation, photophobia (light sensitivity), headache, irritability, insomnia, and gastric disturbances.

Skin testing and desensitization procedures are often helpful in identification of the inciting agent and in decreasing the severity of the reaction. Related forms of nasal allergy may be caused by exposure to other allergens such as foods, drugs, contactants, physical agents, and emotional crises in susceptible individuals. These may occur only in the nasal region but often are seen as part of a more generalized body reaction. *See* Hypersensitivity; Skin test. [N.K.M.]

Hazardous waste Any solid, liquid, or gaseous waste materials that, if improperly managed or disposed of, may pose substantial hazards to human health and the environment. Every industrial country in the world has had problems with managing hazardous wastes. Improper disposal of these waste streams in the past has created a need for very expensive cleanup operations. Efforts are under way internationally to remedy old problems caused by hazardous waste and to prevent the occurrence of other problems in the future.

A waste is considered hazardous if it exhibits one or more of the following characteristics: ignitability, corrosivity, reactivity, and toxicity. Ignitable wastes can create fires under certain conditions; examples include liquids, such as solvents, that readily catch fire, and friction-sensitive substances. Corrosive wastes include those that are acidic and those that are capable of corroding metal (such as tanks, containers, drums, and barrels). Reactive wastes are unstable under normal conditions. They can create explosions, toxic fumes, gases, or vapors when mixed with water. Toxic wastes are harmful or fatal when ingested or absorbed. When they are disposed of on land, contaminated liquid may drain (leach) from the waste and pollute groundwater. *See* Groundwater hydrology; Water pollution.

Hazardous wastes may arise as by-products of industrial processes. They may also be generated by households when commercial products are discarded. These include drain openers, oven cleaners, wood and metal cleaners and polishes, pharmaceuticals, oil and fuel additives, grease and rust solvents, herbicides and pesticides, and paint thinners.

The predominant waste streams generated by industries in the United States are corrosive wastes, spent acids, and alkaline materials used in the chemical, metal-finishing, and petroleum-refining industries. Many of these waste streams contain heavy metals, rendering them toxic. Solvent wastes are generated in large volumes both by manufacturing industries and by a wide range of equipment maintenance industries that generate spent cleaning and degreasing solutions. Reactive wastes come primarily from the chemical industries and the metal-finishing industries. The chemical and primary-metals industries are the major sources of hazardous wastes.

There is a growing acceptance throughout the world of the desirability of using waste management hierarchies for solutions to problems of hazardous waste. A typical sequence involves source reduction, recycling, treatment, and disposal. Source reduction comprises the reduction or elimination of hazardous waste at the source, usually within a process. Recycling is the use or reuse of hazardous waste as an effective substitute for a commercial product or as an ingredient or feedstock in an industrial process.

Treatment is any method, technique, or process that changes the physical, chemical, or biological character of any hazardous waste so as to neutralize such waste; to recover energy or material resources from the waste; or to render such waste nonhazardous, less hazardous, safer to manage, amenable for recovery, amenable for storage, or reduced in volume. Disposal is the discharge, deposit, injection, dumping, spilling, leaking, or placing of hazardous waste into or on any land or body of water so that the waste or any constituents may enter the air or be discharged into any waters, including groundwater.

There are various alternative waste treatment technologies, for example, physical treatment, chemical treatment, biological treatment, incineration, and solidification or stabilization treatment. These processes are used to recycle and reuse waste materials, reduce the volume and toxicity of a waste stream, or produce a final residual material that is suitable for disposal. The selection of the most effective technology depends upon the wastes being treated.

There are abandoned disposal sites in many countries where hazardous waste has been disposed of improperly in the past and where cleanup operations are needed to restore the sites to their original state. Cleaning up such sites involves isolating and containing contaminated material, removal and redeposit of contaminated sediments, and in-place and direct treatment of the hazardous wastes involved. As the state of the art for remedial technology improves, there is a clear preference for processes that result in the permanent destruction of contaminants rather than the removal and storage of the contaminating materials. [H.M.F.]

Head The region of the body consisting of the skull, its contents, and related structures. Its two principal parts are the cranium, or braincase, and the face.

The skin, hair, and subcutaneous tissues over the top of the skull are collectively known as the scalp. The regions of the cranium take their names from the underlying bones, for example, the temporal, parietal, frontal, and occipital regions.

The intracranial contents include the brain and uppermost portion of the spinal cord with their coverings (meninges), blood vessels, and the important cranial nerves, as well as the cerebrospinal fluid system. Many openings, or foramina, afford means of passage from within the skull for nerves and blood vessels. [T.S.P.]

Headache Pain within the head. It is probably the most common complaint for which people seek a physician's help. Headaches can be grouped into three primary categories: vascular, muscle-contraction, and organic.

Vascular headaches include classic and common migraine as well as cluster, toxic, and hypertensive headaches. All are caused by dilation of cerebral blood vessels. Constriction of the blood vessels may also occur in any part of the cerebral vasculature and cause the neurologic symptoms associated with some forms of vascular headache. Migraine affects one side of the head but may be bilateral. Nerologic symptoms, especially visual disturbances, are common. Cluster headache is the occurrence of migraines in groups or series. The cluster headache is characterized by its one-sided, excruciating attack that is usually localized around one eye. Other forms of vascular headache may be caused by systemic infection or fever, which causes dilation of the blood vessels. The ingestion of alcohol, poisons, or some medications used to treat hypertension or cardiac disease may produce adverse effects, including vascular headaches. *See* HYPERTENSION.

The most common form of headache is the muscle-contraction or tension headache. It is characterized by dull, constricting pain that can either occur intermittently or continue for days, months, or years. Muscle-contraction headaches usually affect both sides of the head and may be described as having a hat-band distribution of pain.

Very few headaches have an organic cause, such as brain tumor or aneurysm. Headache is not a prominent symptom of brain tumor: if present, headache will become progressively worse and constant, and it may not appear until late in the course of the tumor development. The headache associated with an aneurysm is usually mild until the aneurysm is at the point of rupture. If a patient complains of an exceptionally severe headache, organic disease, such as aneurysm, must be ruled out. *See* ANEURYSM.

Acute sinus headache is characterized by nasal congestion and fever. The headache is minimal in the morning and increases in severity through the day. Temporomandibular joint (TMJ) disease involves a faulty bite or misalignment of the teeth and can cause a headache. Eye conditions may also cause headache. The increased intraocular pressure of glaucoma, for example, may cause a headache, and so complaints of a recent onset of headache, particularly in the elderly, should prompt a screening for glaucoma. *See* GLAUCOMA; PAIN. [S.D.]

Health physics The science and profession that deals with problems of protection from the hazards of radiation or prevention of damage from exposure to this radiation while making it possible for humans to make full use of the various forms of energy. Initially health physics dealt only with ionizing radiations (alpha, beta, gamma, neutronic, mesonic, and so forth), but it has been extended to include nonionizing radiations (ultraviolet, visible, infrared, radio-frequency, microwave, long-wave, and sonic, ultrasonic, and infrasonic radiations). Health physics may be considered a border field of physics, biology, chemistry, mathematics, medicine, engineering, and industrial hygiene. *See* DOSIMETER; FILM BADGE; MONITORING OF IONIZING RADIATION; RADIATION BIOLOGY; RADIOMETRY. [K.Z.M.]

Hearing (human) The general perceptual behavior and the specific responses made in relation to sound stimuli. The perceptual behavior of hearing has many dimensions, including sound arousal, perceptual orientation, sound detection, sound discrimination, and perceptually motivated listening, such as music appreciation. Hearing behavior also involves the internal biochemical and physiological states associated with emotion, which add esthetic value and emotional tone to perception of speech, music, and noise. *See* PHONORECEPTION; PSYCHOACOUSTICS; SOUND.

Sound detection. There are two aspects of sound detection which should be considered: detection of a sound stimulus in an otherwise soundfree field, and detection of a sound stimulus in the presence of a second masking sound.

Limits of hearing. The frequency and intensity patterns which are perceived as sound have definite limits, which vary to some extent from one individual to another and under differ-

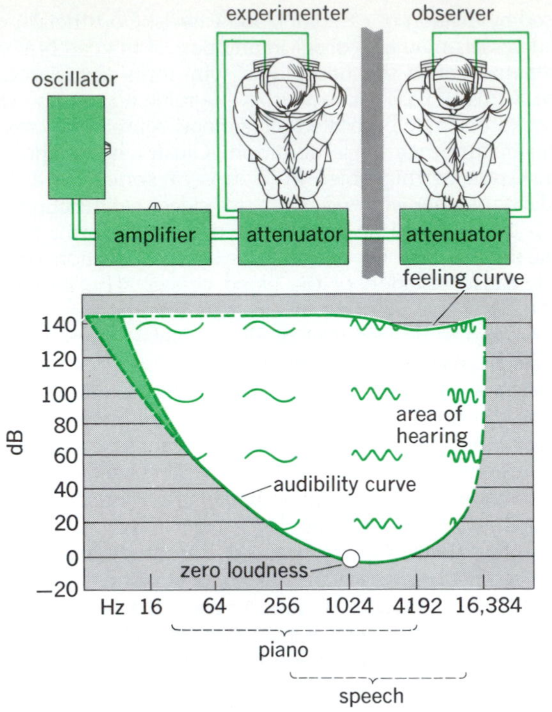

Fig. 1. Measuring the limits of hearing. Absolute thresholds are determined by method of adjustment.

ent conditions. These limits of hearing also vary markedly from one animal species to another. *See* AUDIOMETRY.

The limits of hearing are represented graphically as the area of hearing (Fig. 1). To obtain such a chart, two curves are plotted: the audibility curve, which is the lower curve in Fig. 1, and the upper curve of "pain" or "feeling." The audibility curve is determined by measuring the absolute threshold for a series of frequencies within the hearing range. The absolute threshold for a given frequency is the lowest intensity at which that frequency can be heard. The upper limit of the area of hearing is determined by increasing the intensities of tones until they produce a perception of pain, tickle, or feeling.

Figure 1 shows that human individuals normally perceive as sound those frequencies between about 16 Hz, or lower, and 20,000 Hz, with the greatest sensitivity around 1000–4000 Hz. That is, the audibility curve, showing absolute thresholds for different frequencies, dips lowest for frequencies of around 1000–4000 Hz. The range of sound intensities is expressed in terms of units called decibels. *See* DECIBEL.

The area of hearing varies with age as well as with injuries and other defects. Sensitivity to higher frequencies typically decreases with age. Some animals, such as the bat, mouse, rat, cat, and dog, have a frequency range beyond that of humans. *See* HEARING IMPAIRMENT.

Auditory masking. It is sometimes impossible to detect one tone in the presence of another tone or noise. The first tone is said to be masked by the second tone or noise. The masking effect is defined as the difference between the absolute threshold of the masked tone and the threshold of detection of this tone in the presence of the masking sound. In general, the masking effect is greater for tones near the masking sound in frequency, although tones higher in frequency than the masking sound are affected more than lower tones. The range of frequencies affected is greater, the greater the intensity of the masking sound, and the lower the frequency of the masking sound. *See* LOUDNESS; MASKING OF SOUND.

Sound discrimination. The differential threshold is a measure of ability to detect a difference between stimuli. The usual method for determining such thresholds for intensity and frequency of sounds is to introduce changes, either of intensity or frequency, into continuous sounds, and measure the smallest change that can be detected.

When white noise is used as a stimulus sound, the smallest change in intensity which can be detected remains fairly constant at slightly under 0.5 dB, above an intensity level of about 20 dB. Below 20 dB, the just-detectable change in intensity increases sharply. When pure tones are used as stimuli, a similar threshold of 0.5 dB or less is found above levels of about 40 dB for a 4000-Hz tone. The threshold is greater at lower intensities, and for tones of lower and higher frequencies. Since the decibel is a relative measure of intensity, one can say that the relative change in intensity which can be detected remains fairly constant above 20 dB for noise, and at somewhat higher intensity levels for pure tones.

The smallest change in frequency of pure tones which can be detected is fairly constant at about 2–3 Hz for frequencies up to 1000 Hz and at intensity levels over about 20 dB. Above 1000 Hz and 40 dB, the relative change in frequency which can be detected remains fairly constant, at about 0.2–0.4%.

In general, loudness increases with increased intensity, but is also a function of frequency. The audibility threshold in Fig. 1 is an equal loudness contour of tones that can just barely be heard. At different frequencies such equal loudness tones require different intensities.

Binaural hearing. Special characteristics of sound discrimination arise from binaural stimulation, or stimulation of the two ears.

The efficiency with which sound sources are localized in the environment is dependent on both learned cues and the ability of the auditory system to respond to slight differences in stimulation of the two ears. Success in judging distance depends on recognition of the sound, knowledge of how such a sound changes in loudness and timbre with distance, and corroboration from other sensory cues, especially visual ones. Even when other cues are eliminated, the direction of a sound which stimulates both ears can be judged fairly accurately except when it is exactly behind, above, or in front of the observer. Sounds to the right or left are localized in terms of differences at the two ears of intensity, time of arrival, or phase. Kinesthetic cues from head movements combine with binaural cues to enable observers to judge the elevation of sounds.

The ability to detect and discriminate sounds is increased with binaural stimulation. Tones heard with both ears sound louder than when heard with one. This summation effect accounts for the somewhat lower absolute thresholds found for binaural stimulation as opposed to monaural, and also for the lower intensity discrimination thresholds which are found for the two ears. Frequency discrimination is also better with binaural stimulation, especially for tones of very low and very high frequencies.

Response of auditory system. The ear (Fig. 2) is a system for transducing the mechanical energy of sound into electrical energy which triggers nerve impulses in the auditory nerve. As sound strikes the eardrum, it causes the three small bones of the middle ear to vibrate. These movements set up corresponding vibrations in the oval window, and thence in the fluids of the cochlea. *See* EAR.

The auditory organ of the inner ear, the cochlea, is an enclosed structure that contains three fluid-filled canals (helicotrema). The gelatinous-fibrous basilar membrane makes up the base of the cochlear canal (the central canal and smallest of the three). Upon this membrane is the organ of Corti and its hair cells, which are thought to be the true receptors of hearing. Proliferations from the fibers of the auditory nerve extend up the center of the cochlea and connect with these hair cells.

The ability to hear many different frequencies as distinct pitches is related to the ability of the cochlea to resolve these frequencies. Between about 200 and 2000 Hz, this resolution may be accomplished by differential response of the basilar

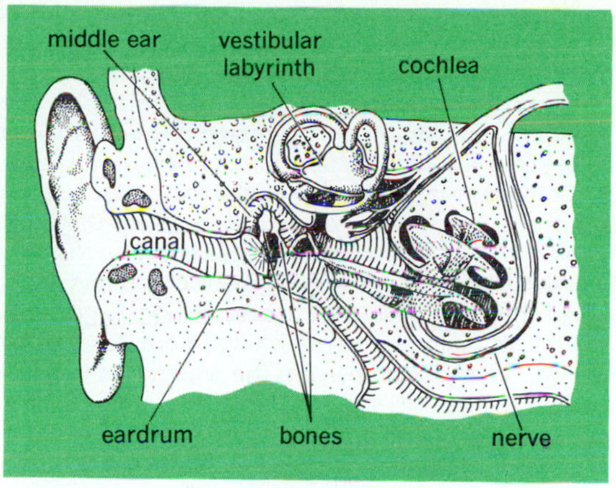

Fig. 2. Structure of the ear.

membrane. The cochlea has different resonance values at different points along its length, so that high tones cause the fluids and membrane to vibrate near the base, and low tones, near the apex. Vibrations of the oval window probably set up traveling waves in the fluids and membrane, which cause maximal vibrations at the place with a natural resonance value corresponding to the sound frequency. Below 200 Hz, the membrane vibrates as a whole, or nearly so.

Sensory nerve fibers, originating at the hair cells at all points along the organ of Corti, emerge from the cochlea to form the auditory nerve (VIIIth cranial nerve). This nerve enters the cochlear nucleus in the medulla oblongata, from which point connections are made with other centers in the midbrain, cerebellum, thalamus, and temporal lobe of the cortex. Discrimination of sound stimuli depends on differentiation of neural action at all levels. [K.U.S.]

Hearing aid An instrument used by a person who is hard of hearing to amplify sounds, particularly the sounds of speech. Modern electronic hearing aids are small enough to be worn comfortably and inconspicuously behind the ear, within the ear canal and concha, in the frames of spectacles, or in the clothing. A behind-the-ear aid is shown in the illustration. Essential components of a hearing aid are a microphone, an electronic amplifier, an earphone, and a plastic ear mold which serves to couple energy from the earphone to the eardrum either directly or through plastic tubes. Sometimes a vibrator is held by a spring headband behind the ear and delivers sound to the ear by bone conduction. *See* AMPLIFIER; EARPHONES; HEARING (HUMAN); HEARING IMPAIRMENT; MICROPHONE. [A.F.N.]

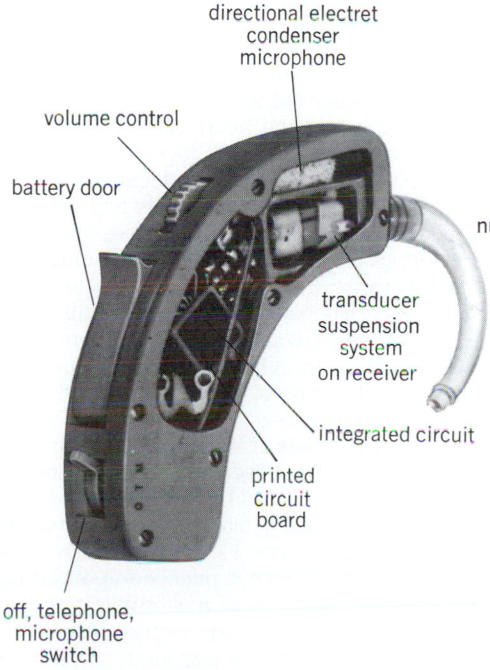

Behind-the-ear hearing aid with case open to show internal components. (*Beltone Electronics Corp.*)

Hearing impairment Impairment of hearing, referring to those with mild impairment, called hard of hearing, and those with severe impairment, called deaf. Hearing loss, both temporary and permanent, results from a broad range of etiologies.

Types of hearing impairment are based on time of onset and on the nature of the impairment. Congenital hearing loss is present from birth, and may vary in severity from mild to profound. Adventitious hearing loss occurs at any time during a lifetime and may have a course of either gradual, progressive development or of sudden onset. The three broad categories based on the nature of the impairment are: conductive hearing impairment, involving an impairment of the mechanism that conducts sound to the sense organ; sensorineural hearing impairment, caused by an abnormality of the sense organ in the inner ear or the auditory nerve; and central hearing impairment, which results from some injury or failure to function in the central nervous system. Combined conductive and sensorineural impairment is referred to as mixed hearing impairment. Modern diagnostic tests are making a number of new distinctions possible within these broad types of hearing impairment. *See* AUDIOMETRY; EAR.

Impedance of sound to the inner ear may be caused by congenital malformations, impaction by hardened earwax (cerumen), or infections in the external canal.

A very common cause of conductive hearing impairment is otitis media, infection of the middle-ear cavity, that may begin with a common cold, allergy, or upper respiratory infection. It is the most common cause of conductive hearing impairment in children. Another major cause of conductive impairment is the hereditary disease otosclerosis. Portions of the bony capsule surrounding the inner ear, normally the hardest bone in the body, are decalcified and replaced by a spongy, softer bone that grows and then hardens, often fixing the foot plate of the stapes at the point where the middle-ear conductive mechanism joins the inner ear. The sound reaching the inner ear is greatly attenuated. This progressive disease begins in youth, and hearing loss is first noted at adolescence or in the early twenties.

Other forms of hereditary hearing impairment are quite common in a population with severe congenital hearing impairment. Genetic causes are often suspected when no diagnosis can be made or the cause is otherwise unexplained, and especially when parents have a common ancestor.

Many drugs and chemicals are considered damaging to the inner ear, or ototoxic. Dihydrostreptomycin, kanamycin, and neomycin are definitely ototoxic, and both quinine and salicylates are highly suspect as causes of sensorineural impairment.

Acoustic trauma results from brief exposure to sudden explosive noises. A blast of noise, as from an explosion or gunshot, may permanently disorganize the organ of Corti, tearing pieces of it loose into the surrounding fluid of the inner ear. Exposure to any steady loud noise, such as that of factories, aircraft engines, or highly amplified music, usually results in tinnitus and short-term, high-frequency hearing loss referred to as temporary threshold shift. Eventually, with habitual exposure, permanent threshold shift will take place.

Hearing loss of old age, called presbycusis, is one of the most common causes of sensorineural impairment, particularly as life expectancy increases. Several aging processes contribute to presbycusis. Some changes take place in the middle ear with age, especially the loss of elasticity of connective tissue; some of the sensory cells of hearing are lost with age; or there may be a loss of neurons in the central nervous system, particularly with arteriosclerosis.

Prevention is the best treatment for hearing loss. Immunization for viral diseases, early medical intervention for upper-respiratory infections or earache, teaching children to blow their noses gently with both nostrils open, control of allergies, avoiding ototoxic drugs, and reduced exposure to loud noise are obvious prophylaxes. Routine hearing screening in pediatric offices, in well-baby clinics, and in schools, with appropriate medical follow-up, is also advisable.

Nonmedical aural rehabilitation for persons with hearing loss includes the judicious selection of personal electroacoustic hearing aids, auditory training and training to use amplification, instruction in speech reading, and conservation of speech by consciously relating speech movements to their tactile and kinesthetic sensations. *See* HEARING (HUMAN). [D.R.C.]

Heart (vertebrate) The muscular pumping organ of the cardiovascular system. The heart typically lies ventrally, near the anterior end of the trunk; it is ventral and medial to the gills in fish and at the base of the neck or in the chest region of tetrapods. In humans it is located behind the breastbone and ribs between the third and fifth costal cartilages. Its anterior portion or base is directed to the right and dorsally and is the area where the great vessels enter and leave the heart. The lower muscular portion ends in a blunt apex which lies behind the fifth costal cartilage on the left.

The muscular wall of the heart, the myocardium, is lined by an inner endocardium and is covered externally by membranous visceral pericardium. There are coronary arteries and veins to and from the heart, which has a specialized neuromuscular conducting system and autonomic nerve supply.

In fishes the heart is basically a simple tube which becomes subdivided into four successive chambers, the sinus venosus, atrium, ventricle, and conus arteriosus. Blood from the body enters the sinus and leaves the conus to go to the gills to be oxygenated. The ventricle supplies the main pumping force.

When lungs are introduced into the system in lungfish and tetrapods, the mixing of oxygenated and nonoxygenated blood becomes a problem. In brief, the sinus venosus and conus arteriosus disappear, becoming incorporated into the other chambers or the bases of the great vessels. At the same time the atrium and later the ventricle become divided into right and left chambers by a median septum.

In birds and mammals including humans (see illustration) the medial fibromuscular septum divides the heart into two lateral halves, each consisting of a thin-walled receiving chamber or atrium and a thicker, muscular pumping chamber or ventricle.

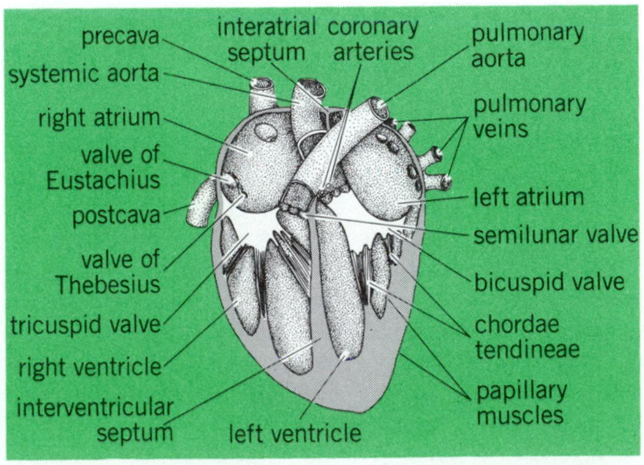

Internal structure of four-chambered mammalian heart, ventral view. (*After C. K. Weichert, Anatomy of the Chordates, 2d ed., McGraw-Hill, 1958*)

Blood enters the right atrium from the superior and inferior venae cavae which drain most of the body. It passes through the tricuspid valve to the right ventricle and is pumped to the lungs during systole, or contraction of the heart. Blood returns from the lungs by way of the pulmonary veins to the left atrium, passes into the left ventricle through the mitral valve, and during contraction is pumped out into the aorta. *See* CARDIOVASCULAR SYSTEM. [T.S.P.]

Heart disorders A group of structural and functional changes of the heart caused by diverse etiologic agents such as toxic chemicals, bacteria, viruses, fungi, parasites, and altered metabolism. These disorders are the leading cause of illness and death in the industrialized nations.

There are five general types of heart disorders based on the morphology of the heart changes: coronary heart disease, cardiomyopathies, rheumatic fever, infections, and congenital heart disease. Congestive heart failure is a common result of these several lesions. Congestive heart failure is a clinical syndrome resulting from diminished output of blood from the heart. The heart is unable to keep pace with the venous blood returning to it, resulting in the pooling of blood in the venous system with concomitant diminished filling of the arterial vessels. The basic derangement may be an impaired contractility of the heart muscle or an increased workload placed on the heart by extrinsic factors.

Coronary heart disease. Heart disorders associated with coronary vessel lesions are the product of insufficient blood supply to the heart muscle. There are three forms of coronary vessel disease: arteriosclerotic heart disease, myocardial infarction, and angina pectoris. With each there is insufficient blood flow through the narrowed lumen of a coronary artery, resulting in an insufficient oxygen supply to the heart muscle and subsequent muscular cell injury or death (necrosis). *See* ANGINA PECTORIS; ARTERIOSCLEROSIS; INFARCTION; THROMBOSIS.

Cardiomyopathies. Cardiomyopathies are a diverse group of heart disorders which have in common the myocardium as the major target of injury. They are divided into two subgroups: those which are inflammatory and those which are degenerative. *See* COR PULMONALE; DEGENERATIVE CARDIOMYOPATHY; ENDOCARDIAL FIBROELASTOSIS; HYPERTENSION; PERICARDITIS.

Rheumatic fever. Rheumatic fever is a systemic, nonsuppurative, often recurrent inflammatory process caused by group A beta hemolytic streptococci. Joints are the most frequent site of lesions; however, the most severe and important effect is injury to the heart. Most deaths from rheumatic fever occur long after the acute disease has subsided, and are the result of endocardial involvement at the heart valves. *See* RHEUMATIC FEVER.

Infections. Bacterial and mycotic endocarditis are inflammatory lesions of the lining of the chambers and valves of the heart caused by colonization of the heart valves by bacteria or fungi, and are serious life-threatening infections. Ordinarily the infection occurs in valves that have been previously damaged by rheumatic heart disease or congenital malformation. However, in a small proportion of cases there is no apparent preceding injury to the valve. Occasionally the underlying disease process is atherosclerosis of the valves or rarely syphilitic valvular disease. *See* ENDOCARDITIS.

Congenital heart disease. Congenital malformations of the heart occur in 0.4–1.0% of live births in the United States. A high percentage of miscarriages have evidence of heart malformations. The causes of congenital heart malformations are not established in most instances. However, there is some evidence that environmental chemicals may be important etiologic agents. The clinical effect of heart malformation is the alteration of the flow of blood. The large spectrum of anomalies of the heart are generally divided into two kinds: those in which the blood is shunted from the pulmonary blood flow with changes in the right atrium and ventricle, and those from the left atrium and ventricle affecting systemic blood flow. When blood is shunted from the pulmonary circuit without passing through the lungs, the blood is inadequately oxygenated and the blood appears blue or cyanotic (the cause of a "blue baby"). *See* CONGENITAL ANOMALIES; FETAL ALCOHOL SYNDROME. [N.K.M.]

Heartwater disease A septicemic, infectious disease of cattle, sheep, and goats in equatorial and southern Africa caused by *Cowdria ruminantium*, a rickettsialike pathogen. It is carried chiefly by the bont tick (*Amblyomma heberaeum*). Humans are not susceptible. *See* RICKETTSIALES. [C.B.P.]

Heartworms Long filarioid nematodes (*Dirofilaria immitis*) of the superfamily Filarioidea, which normally occur in the right heart chambers and adjacent vessels of the domestic dog. The progeny (microfilariae) begin to appear in the peripheral blood 7 to 9 months after infection. The microfilariae must complete their embryonation in mosquitoes. This has been demonstrated in over 70 mosquito species, but probably fewer than 10 are important vectors.

The first discernible pathology is the enlarged pulmonary artery with a thickening of the lining. Continuing infection results in the occlusion of most of the pulmonary capillary bed, cirrhosis, and kidney pathology. The pathology is commonly well advanced in chronic heartworm disease before labored breathing, coughing, and unexplained exhaustion, particularly on exercise, alerts the owner. This standard chemotherapy is a course of intravenous injections of thiacetarsamide. This adulticide is later followed by a microfilaricide, such as orally administered dithiazanine.

Prevention is obviously very important. Currently the only chemoprophylactic agent is diethylcarbamazine (DEC). It must be administered daily (orally) from the beginning of the mosquito season, throughout the mosquito season, and for 2 months thereafter. It is very risky to start this preventive medication unless the microfilariae have already been eliminated. *See* FILARIOIDEA. [G.F.O.]

Heat For the purposes of thermodynamics, it is convenient to define all energy while in transit, but unassociated with matter, as either heat or work. Heat is that form of energy in transit due to a temperature difference between the source from which the energy is coming and the sink toward which the energy is going. The energy is not called heat before it starts to flow or after it has ceased to flow. A hot object does contain energy, but calling this energy heat as it resides in the hot object can lead to widespread confusion. *See* ENERGY; INTERNAL ENERGY; TEMPERATURE; THERMODYNAMIC PRINCIPLES. [H.C.W.; W.A.S.]

Heat balance A particular form of an energy balance. The heat balance is generally useful in science but of special importance in process engineering. It is basic to the design and analysis of operating equipment since it provides a relationship between the energy terms in a process. As a simple example, the heat loss from a pipe carrying a hot fluid is difficult to measure directly but is easily calculated by a heat balance if the fluid properties at the ends of the pipe are known.

The conservation of energy requires that for any system the accumulation of energy within must equal the difference between energy entering and leaving. Energy terms are conveniently classified as: the heat Q transferred into the system across the boundary, the work W put into the system, and the energy of the material streams flowing to and from the system. The flowing streams carry internal energy U, potential energy, kinetic energy, and a flow work term pV which arises from forcing a volume of material V into or out of the system under the restraint of the pressure p.

Frequently, the kinetic energy, potential energy, and work terms are negligible or cancel out so that the energy balance simplifies to the equation below, where subscripts 1 and 2 re-

$$U_1 + p_1V_1 + Q = U_2 + p_2V_2 + \Delta E$$

fer to input and output streams, respectively, and ΔE represents accumulation of energy in the system. This simplified energy balance is generally referred to as a heat balance. [W.F.J.]

Heat balance, terrestrial atmospheric The (heat) energy exchange between the planetary Earth-plus-atmosphere system and space, consisting of the reception, distribution, and transformation of heat received from the Sun as well as the emission, distribution, and transformation of thermal (infrared) radiation arising from the Earth's surface and atmosphere. An area of 1 m^2 (10.76 ft^2) at the Earth's mean distance from the Sun and perpendicular to the Sun's rays would receive about 1390 watts (or joules per second) if there were no intervening atmosphere. This quantity is the solar "constant" that depends on the energy output of the Sun and the Earth's distance from it.

Since the area of a sphere is four times that of the circle it presents to parallel radiation, the average meter-squared area outside the Earth's atmosphere receives about 350 W of solar energy (see illustration). Thirty percent of this incident solar

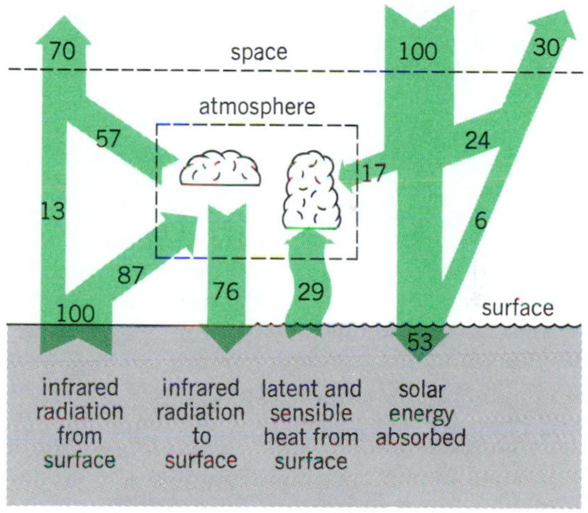

Average global distribution of incident solar energy and Earth infrared energy throughout the heat balance (350 W/m^2 is taken as 100 units).

energy is reflected and scattered back to space, mostly by clouds in the atmosphere; it is not available for use on Earth. This quantity, the percent reflected and scattered, is called the planetary albedo. The 70% of the solar energy that enters the planetary system is absorbed, mostly at the Earth's surface. The oceans absorb solar energy very well, and cover 70% of the planet's surface. This energy is partly transformed by plants, used to evaporate water, converted to thermal (infrared) radiation from the surface, and moved from the surface into the atmosphere by conduction and convection. The clouds and gases in the atmosphere also absorb some of the solar energy, thus evaporating clouds and heating the air. *See* GREENHOUSE EFFECT; SOLAR ENERGY; TERRESTRIAL RADIATION. [T.H.V.H.]

Heat capacity The quantity of heat required to raise a unit mass of homogeneous material one unit in temperature along a specified path, provided that during the process no phase or chemical changes occur, is known as the heat capacity of the material in question. Moreover, the path is so restricted that the only work effects are those necessarily done on the surroundings to cause the change to conform to the specified path. The path is usually at either constant pressure or constant volume.

In accordance with the first law of thermodynamics, heat capacity at constant pressure C_p is equal to the rate of change of enthalpy with temperature at constant pressure $(\partial H/\partial T)_p$. Heat capacity at constant volume C_v is the rate of change of internal energy with temperature at constant volume $(\partial U/\partial T)_v$. Moreover, for any material, the first law yields the relation shown in the equation below.

$$C_p - C_v = \left[P + \left(\frac{\partial U}{\partial V}\right)_T\right]\left(\frac{\partial U}{\partial T}\right)_P$$

See ENTHALPY; INTERNAL ENERGY; THERMODYNAMIC PRINCIPLES. [H.C.W.]

Heat exchanger A device used to transfer heat from a fluid flowing on one side of a barrier to another fluid (or fluids) flowing on the other side of the barrier.

When used to accomplish simultaneous heat transfer and mass transfer, heat exchangers become special equipment types, often known by other names. When fired directly by a combustion process, they become furnaces, boilers, heaters, tube-still heaters, and engines. If there is a change in phase in one of the flowing fluids—condensation of steam to water, for example—the equipment may be called a chiller, evaporator, sublimator, distillation-column reboiler, still, condenser, or cooler-condenser.

Heat exchangers may be so designed that chemical reactions or energy-generation processes can be carried out within them. The exchanger then becomes an integral part of the reaction system and may be known, for example, as a nuclear reactor, catalytic reactor, or polymerizer.

Heat exchangers are normally used only for the transfer and useful elimination or recovery of heat without an accompanying phase change. The fluids on either side of the barrier are usually liquids, but they may also be gases such as steam, air, or hydrocarbon vapors; or they may be liquid metals such as sodium or mercury. Fused salts are also used as heat-exchanger fluids in some applications.

Most often the barrier between the fluids is a metal wall such as that of a tube or pipe. However, it can be fabricated from flat metal plate or from graphite, plastic, or other corrosion-resistant materials of construction.

Heat exchangers find wide application in the chemical process industries, including petroleum refining and petrochemical processing; in the food industry, for example, for pasteurization of milk and canning of processed foods; in the generation of steam for production of power and electricity; in nuclear reaction systems; in aircraft and space vehicles; and in the field of cryogenics for the low-temperature separation of gases. Heat exchangers are the workhorses of the entire field of heating, ventilating, air-conditioning, and refrigeration. *See* CONDUCTION (HEAT); CONVECTION (HEAT); COOLING TOWER; DISTILLATION; EVAPORATOR; FURNACE; HEAT RADIATION; HEAT TRANSFER; TUBE-STILL HEATER; VAPOR CONDENSER. [R.F.Fr.]

Heat insulation Materials whose principal purpose is to retard the flow of heat. Thermal- or heat-insulation materials may be divided into two classes, bulk insulations and reflective insulations. The class and the material within a class to be used for a given application depend upon such factors as temperature of operation, ambient conditions, mechanical strength requirements, and economics.

Examples of bulk insulation include mineral wool, vegetable fibers and organic papers, foamed plastics, calcium silicates with asbestos, expanded vermiculite, expanded perlite, cellular glass, silica aerogel, and diatomite and insulating firebrick. They retard the flow of heat, breaking up the heat-flow path by the interposition of many air spaces and in most cases by their opacity to radiant heat.

Thermal conductivities of selected solids*

Material	Density, lb/ft³	Temp., °F(°C)	Conductivity (k),† Btu/(h)(ft²)(°F/in.)
Asbestos cement board	120	75(24)	4
Cotton fiber	0.8-2.0	75	0.26
Mineral wool, fibrous rock, slag, or glass	1.5-4.0	75	0.27
Insulating board, wood, or cane fiber	15	75	0.35
Foamed plastics	1.6	75	0.29
Glass			3.6-7.32
Hardwoods, typical	45	75	1.10
Softwoods, typical	32	75	0.80
Cellular glass	9	75	0.40
Fine sand (4/% moisture content)	100	40(4)	4.5
Silty clay loam (20/% moisture content)	100	40	9.5
Gypsum or plaster board	50	75	1.1

*From American Society of Heating, Refrigerating, and Air Conditioning Engineers, *Heating, Ventilating and Air Conditioning Guide*, 1959.

†Typical; suitable for engineering calculations. 1Btu/(h)(ft²)(°F/in.)=0.144 W/(m²)(°C/m).

Reflective insulations are usually aluminum foil or sheets, although occasionally a coated steel sheet, an aluminized paper, or even gold or silver surfaces are used. Refractory metals, such as tantalum, may be used at higher temperatures. Their effectiveness is due to their low emissivity (high reflectivity) of heat radiation.

The distinguishing property of bulk thermal insulation is low thermal conductivity; representative values are listed in the table. For a given thickness of material exposed to a given temperature difference, the rate of heat flow per unit area is directly proportional to the thermal conductivity of the material. *See* CONDUCTION (HEAT). [H.F.R.]

Heat pipe A device for transferring heat efficiently between two locations by using the evaporation and condensation of a fluid contained therein. Heat pipes have many applications in areas where reliable performance and low cost are of prime importance—for example, in electronics and heat exchangers. *See* HEAT EXCHANGER.

The heat pipe, the idea of which was first suggested in 1942, is similar in many respects to the thermosiphon. A large proportion of applications do not use heat pipes as strictly defined below, but employ thermosiphons (illus. *a*), sometimes known as gravity-assisted heat pipes. A small quantity of liquid is placed in a tube from which the air is then evacuated, and the tube is sealed. The lower end of the tube is heated, causing

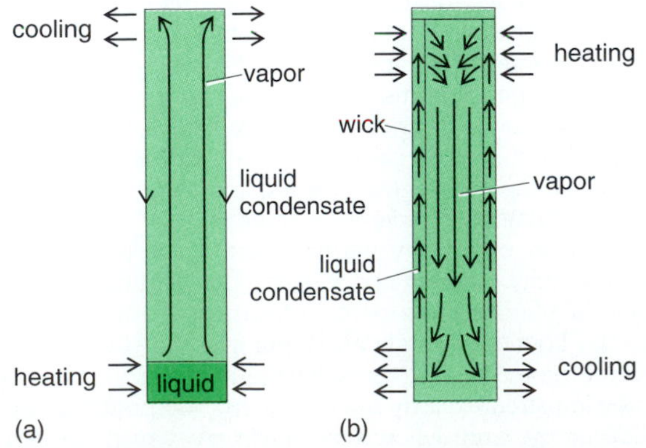

Heat transfer devices. (*a*) Thermosiphon. (*b*) Heat pipe; it can be in any position, not just vertical as shown.

liquid to vaporize and the vapor to move to the cooler end of the tube, where it condenses. The condensate is returned to the evaporator section by gravity. Since the latent heat of evaporation is generally high, considerable quantities of heat can be transported with a very small temperature difference between the two ends. Thus the structure has a high effective thermal conductance. The thermosiphon, also known as the Perkins tube, has been used for many years. A wide variety of working fluids have been employed, ranging from helium to liquid metals.

One limitation of the basic thermosiphon is that in order for the condensate to be returned by gravitational force to the evaporator region, the latter must be situated at the lowest point. The heat pipe is similar in construction to the thermosiphon, but in this case provision is made for returning the condensate against a gravity head. A wick, for example a few layers of fine gauze, is commonly used. This is fixed to the inside surface of the tube, and capillary forces return the condensate to the evaporator (illus. *b*). Since the evaporator position is not restricted, the heat pipe may be used in any orientation. If the heat pipe evaporator happens to be in the lowest position, gravitational forces will assist the capillary force. Alternative techniques, including centripetal forces and osmosis, may be used for returning the condensate to the evaporator.

Capillary forces are by far the most common form of condensate return employed, but a number of rotating heat pipes are used for cooling of electric motors and other rotating machinery. In some applications a mechanical pump is used to return condensate in two-phase run-around coil heat recovery systems. While this may be regarded as a retrograde step, it is a much more effective method for condensate return than reliance on capillary forces.

Applications are related to five principal functions of the heat pipe: separation of heat source and sink, temperature flattening, heat flux transformation, temperature control, and action as a thermal diode or switch. The two major applications, cooling of electronic components and heat exchange, can involve all of these features. In the case of electronics cooling and temperature control, all features can be important. In heat exchangers employing heat pipes, the separation of heat source and sink, and the action as a thermal diode or switch, are most significant. [D.A.Re.]

Heat pump The thermodynamic counterpart of the heat engine. A heat pump raises the temperature level of heat by means of work input. In its usual form a compressor takes refrigerant vapor from a low-pressure, low-temperature evaporator and delivers it at high pressure and temperature to a condenser (see illustration). The pump cycle is identical with the customary vapor-compression refrigeration system. *See* Refrigeration cycle.

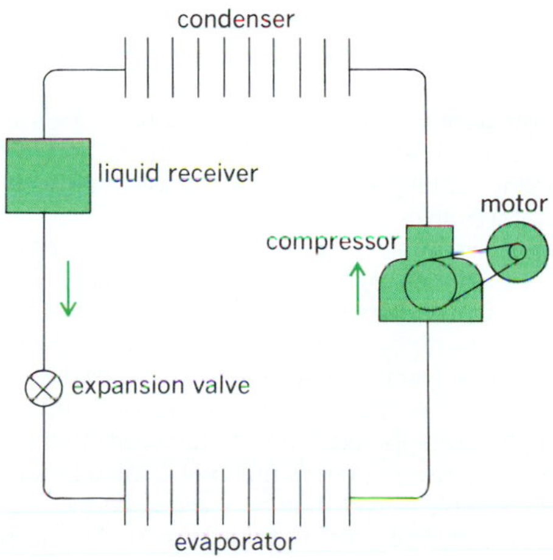

Basic flow diagram of heat pump with motor-driven compressor. For summer cooling, condenser is outdoors and evaporator indoors; for winter heating, condenser is indoors and evaporator outdoors.

This dual purpose is accomplished, in effect, by placing the low-temperature evaporator in the conditioned space during the summer and the high-temperature condenser in the same space during the winter. Thus, if 70°F (21°C) is to be maintained in the conditioned space regardless of the season, this would be the theoretical temperature of the evaporating coil in summer and of the condensing coil in winter. The actual temperatures on the refrigerant side of these coils would need to be below 70°F in summer and above 70°F in winter to permit the necessary transfer of heat through the coil surfaces. If the average outside temperatures are 100°F (38°C) in summer and 40°F (40°C) in winter, the heat pump serves to raise or lower the temperature 30° (17°C) and to deliver the heat or cold as required.

The heat pump is also used for a wide assortment of industrial and process applications such as low-temperature heating, evaporation, concentration, and distillation. [T.Ba.]

Heat radiation The energy radiated by solids, liquids, and gases as a result of their temperature. Such radiant energy is in the form of electromagnetic waves and covers the entire electromagnetic spectrum, extending from the radio-wave portion of the spectrum through the infrared, visible, ultraviolet, x-ray, and gamma-ray portions. From most hot bodies on Earth this radiant energy lies largely in the infrared region. *See* Electromagnetic radiation; Infrared radiation.

Radiation is one of the three basic methods of heat transfer, the other two methods being conduction and convection. *See* Conduction (heat); Convection (heat); Heat transfer.

A hot plate at 260°F (400 K) may show no visible glow; but a hand which is held over it senses the warming rays emitted by the plate. A temperature of more than 1300°F (1000 K) is required to produce a perceptible amount of visible light. At this temperature a hot plate glows red and the sensation of warmth increases considerably, demonstrating that the higher the temperature of the hot plate the greater the amount of radiated energy. Part of this energy is visible radiation, and the amount of this visible radiation increases with increasing temperature. A steel furnace at 2800°F (1800 K) shows a strong yellow glow. If a tungsten wire (used as the filament in incandescent lamps) is raised by resistance heating to a temperature of 4600°F (2800 K), it emits a bright white light. As the temperature of a substance increases, additional colors of the visible portion of the spectrum appear, the sequence being first red, then yellow, green, blue, and finally violet. The violet radiation is of shorter wavelength than the red radiation, and it is also of higher quantum energy. In order to produce strong violet radiation, a temperature of almost 5000°F (3000 K) is required. Ultraviolet radiation necessitates even higher temperatures. The Sun emits considerable ultraviolet radiation; its temperature is about 10,000°F (6000 K). Such temperatures have been produced on Earth in gases ionized by electrical discharges. The mercury-vapor lamp and the fluorescent lamp emit large amounts of ultraviolet radiation. Temperatures up to 36,000°F (20,000 K), however, are still much too low to produce x-rays or gamma radiation. A gas maintained at temperatures above 2×10^{6}°F (1×10^{6} K), encountered in nuclear fusion experiments, emits x-rays and gamma rays. *See* Nuclear fusion; Sun; Ultraviolet radiation.

A blackbody is defined as a body which emits the maximum amount of heat radiation. Although there exists no perfect blackbody radiator in nature, it is possible to construct one on the principle of cavity radiation. *See* Blackbody.

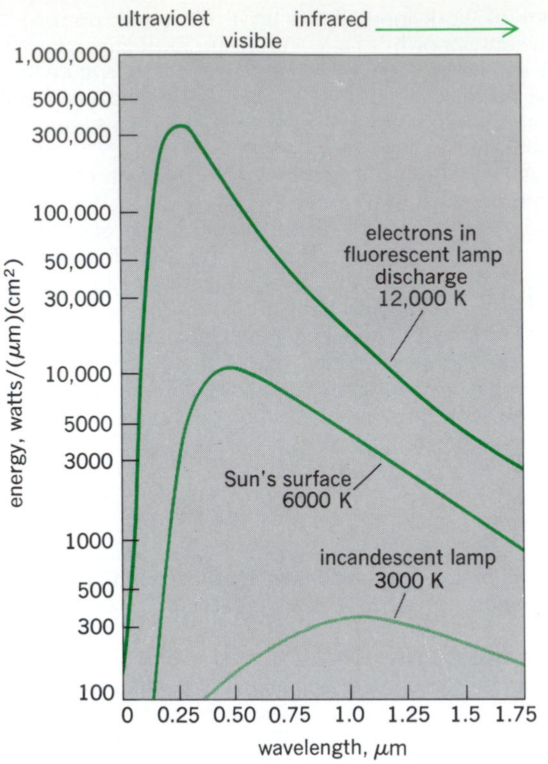

Graphs of Planck's law for various temperatures.

A cavity radiator is usually understood to be a heated enclosure with a small opening which allows some radiation to escape or enter. The escaping radiation from such a cavity has the same characteristics as blackbody radiation.

Kirchhoff's law correlates mathematically the heat radiation properties of materials at thermal equilibrium. It is often called the second law of thermodynamics for radiating systems. Kirchhoff s law can be expressed as follows: The ratio of the emissivity of a heat radiator to the absorptivity of the same radiator is a function of frequency and temperature alone. This function is the same for all bodies, and it is equal to the emissivity of a blackbody. A consequence of Kirchhoff s law is the postulate that a blackbody has an emissivity which is greater than that of any other body. *See* KIRCHHOFF'S LAWS OF ELECTRIC CIRCUITS.

Planck's radiation law represents mathematically the energy distribution of the heat radiation from 1 cm^2 of surface area of a blackbody at any temperature. Formulated by Max Planck early in the 20th century, it laid the foundation for the advance of modern physics and the advent of quantum theory. Equation (1) is the mathematical expression of Planck's radiation law,

$$R_\lambda = 37{,}418/\lambda^5[e^{14{,}388(\lambda T)} - 1] \qquad (1)$$

where R_λ is the total energy radiated from the body measured in watts per square centimeter per unit wavelength, at the wavelength λ. The wavelength in this formula is measured in micrometers. The quantity T is the temperature in kelvins, and e is the base of the natural logarithms. The illustration presents graphs of Planck's law for various temperatures and shows substances which attain these temperatures. It should be noted that these substances will not radiate as predicted by Planck's law since they are not blackbodies themselves.

The Stefan-Boltzmann law states that the total energy radiated from a hot body increases with the fourth power of the temperature of the body. This law can be derived from Planck's law by the process of integration and is expressed mathematically as Eq. (2), where R_T is the total amount of energy radiated from a blackbody in watts per square centimeter. When R_T is multiplied by the total emissivity, the total energy radiated from a real heat radiator is obtained. [H.G.S.; P.J.W.]

$$R_T = 5.670 \times 10^{-12}T^4 \qquad (2)$$

Heat transfer

Heat, a form of kinetic energy, is transferred in three ways: conduction, convection, and radiation. Heat transfer (also called thermal transfer) can occur only if a temperature difference exists, and then only in the direction of decreasing temperature. Beyond this, the mechanisms and laws governing each of these ways are quite different. *See* CONDUCTION (HEAT); CONVECTION (HEAT); HEAT RADIATION.

By utilizing a knowledge of the principles governing the three methods of heat transfer and by a proper selection and fabrication of materials, the designer attempts to obtain the required heat flow. This may involve the flow of large amounts of heat to some point in a process or the reduction in flow in others. All three methods operate in processes that are commonplace.

In industry, for example, it is generally desired to extract heat from one fluid stream and add it to another. Devices used for this purpose have passages for each of the two streams separated by a heat-exchange surface in the form of plates or tubes and are known as heat exchangers. The automobile radiator, the hot-water heater, the steam or hot-water radiator in a house, the steam boiler, the condenser and evaporator on either the household refrigerator or air conditioner, and even the ordinary cooking utensils in everyday use are all heat exchangers. *See* HEAT; HEAT EXCHANGER. [R.H.L.]

Heat treatment (metallurgy)

A procedure of heating and cooling a material without melting. Plastic deformation may be included in the sequence of heating and cooling steps, thus defining a thermomechanical treatment. Typical objectives of heat treatments are hardening, strengthening, softening, improved formability, improved machinability, stress relief, and improved dimensional stability. Heat treatments are often categorized with special names, such as annealing, normalizing, stress relief anneals, process anneals, hardening, tempering, austempering, martempering, intercritical annealing, carburizing, nitriding, solution anneal, aging, precipitation hardening, and thermomechanical treatment.

All metals and alloys in common use are heat-treated at some stage during processing. Iron alloys, however, respond to heat treatments in a unique way because of the multitude of phase changes which can be induced, and it is thus convenient to discuss heat treatments for ferrous and nonferrous metals separately. *See* IRON; STEEL.

Ferrous metals. Annealing heat treatments are used to soften the steel, to improve the machinability, to relieve internal stresses, to impart dimensional stability, and to refine the grain size.

Hardening treatments are used to harden steels by heating to a temperature at which austenite is formed and then cooling with sufficient rapidity to make the transformation to pearlite or ferrite unfavorable.

Some heat treatments are used to alter the chemistry at the surface of a steel, usually to achieve preferential hardening of a surface layer. Carburizing consists of subjecting the steel to an atmosphere of partially combusted natural gas which has been enriched with respect to carbon. In the nitriding treatment, nitrogen diffusing to the surface of the steel forms nitrides. Chromizing involves the addition of chromium to the surface by diffusion from a chromium-rich material packed around the steel or dissolved in molten lead. *See* SURFACE HARDENING OF STEEL.

Nonferrous metals. Many nonferrous metals do not exhibit phase transformations, and it is not possible to harden them by means of simple heating and quenching treatments as in steel. Unlike steels, it is impossible to achieve grain refinement by heat treatment alone, but it is possible to reduce the grain size by a combination of cold-working and annealing treatments.

Some nonferrous alloys can be hardened, but the mechanism is one by which a fine precipitate is formed, and the reaction is fundamentally different from the martensitic hardening reaction in steel. There are also certain ferrous alloys that can be precipitation hardened. However this hardening technique is used much more widely in nonferrous than in ferrous alloys. In titanium alloys, the β phase can transform in a martensitic reaction on rapid cooling, and the hardening of these alloys is achieved by methods which are similar to those used for steels. [B.L.A.]

Heating system An apparatus consisting of an energy source, a method of converting that energy to heat, and a transport system to convey the energy and heat to the point of use. Most heating systems include some manual or automatic method of controlling the heat output and delivery.

There are many sources of energy for use in heating. The earliest source, and still most common in developing countries, comprises wood and wood products such as paper, wood chips, and sawdust; peat is used in some cultures. Solar use for heating and electrical generation has become widespread, although economics discourages more general use. The generation of electrical energy requires the use of fossil fuels, water power, geothermal energy, or nuclear energy. Fossil fuels are used directly in furnaces and boilers. *See* ELECTRIC POWER SYSTEMS; ENERGY SOURCES; HEAT PUMP; SOLAR ENERGY.

The energy source is converted into heat by various means. Wood and fossil fuels are converted by burning, or the combustion process. Electrical energy can be converted directly into heat by resistance heaters. Solar energy requires collectors, with conversion to heat or electricity.

Many methods deliver heat to the point of use. Radiation systems take several forms. Cast-iron column radiators, using steam or hot water, have largely been superseded by convector radiators using steam, hot water, or electricity. Panel-type radiators are also used in ceilings or in floors. All require a piping or electrical distribution system. Forced-air warm-air heating, using electric motor–driven circulating fans, is common. *See* CENTRAL HEATING AND COOLING; COMFORT HEATING; DISTRICT HEATING; HOT-WATER HEATING SYSTEM; PANEL HEATING AND COOLING; RADIANT HEATING; SOLAR HEATING AND COOLING; STEAM HEATING; WARM-AIR HEATING SYSTEM. [R.W.Hai.]

Heavy minerals Minerals with a density above 2.9, which is the density of bromoform, the liquid used to separate heavy from light minerals. Heavy minerals are sometimes used in igneous petrology, but their main importance is in the study of sedimentary rocks. *See* SEDIMENTARY ROCKS.

If not modified by weathering or sedimentation processes, the heavy-mineral suites of sedimentary rocks, primarily sandstones, offer clues to the composition of the source rock. For example, amphibole and amphibole-epidote suites suggest an igneous or high-grade metamorphic source and augite suites a volcanic source.

Sediments can be characterized by their heavy-mineral association. The study of the distribution of heavy-mineral associations in a depositional basin permits the tracing of the distribution in space and time of sediment supply from various sources and assists in the unraveling of the erosional history of the source areas. Since frequently several sources supply sediments with different heavy-mineral suites to a single basin, the pattern of heavy-mineral associations can be complex, in particular when mixing occurs. Thus, heavy minerals are an aid in the study of the depositional history and paleogeography of a basin. In some cases they have been used successfully in correlating sedimentary strata. *See* STRATIGRAPHY. [T.H.V.A.]

Heavy water A form of water in which the hydrogen atoms of mass 1 (^1H) ordinarily present in water are replaced by deuterium (D or ^2H), the heavy stable isotope of hydrogen of mass 2. The molecular formula of heavy water is D_2O (or ^2H_2O). *See* DEUTERIUM.

Because the mass difference between ^1H and ^2H is the largest for any pair of stable (nonradioactive) isotopes in the periodic table, many of the physical and chemical properties of the pure isotopic species and their respective compounds differ to a significant extent. Selected physical properties of ^1H_2O and ^2H_2O are compared in the table.

Physical properties of ordinary and heavy water

Property	^1H_2O	^2H_2O (D_2O)
Molecular weight, ^{12}C scale	18.015	20.028
Melting point, °C	0.00	3.81
Normal boiling point, °C	100.00	101.42
Temperature of maximum density, °C	3.98	11.23
Density at 25°C, g/cm³	0.99701	1.1044
Critical constants		
Temperature, °C	374.1	371.1
Pressure, mPa	22.12	21.88
Volume, cm³/mol	55.3	55.0
Viscosity at 55°C, mPa·s	0.8903	1.107
Refractive index, n_D^{20}	1.3330	1.3283

Heavy water, judging from its higher melting and boiling points, its higher viscosity, and its surprisingly high temperature of maximum density, is a distinctly more structured liquid than is ordinary water. Heavy water is more extensively hydrogen-bonded, and the hydrogen bonds formed by ^2H are somewhat stronger than are those of ^1H.

The only large-scale use of heavy water in industry is as a moderator in nuclear reactors. Small amounts of heavy water are used to grow fully deuterated organisms, which serve as a source of fully deuterated compounds of biological importance. These are finding increasing use in research techniques such as small-angle neutron scattering, in high-resolution nuclear magnetic resonance spectroscopy of immobilized samples, and in the study of isotope effects. *See* ISOTOPE (STABLE) SEPARATION; NUCLEAR REACTOR; WATER. [J.J.K.]

Hederellida A marine bryozoan order in the class Stenolaemata. It is recorded from scattered localities in the Ordovician through Permian systems of the Paleozoic Era but is found prominently in Devonian strata. The colonies comprise tubular zooecia that commonly occur on brachiopod shells, echinoderm plates and columnals, and the outer surfaces of corals. Although usually encrusting, the colonies sometimes may be erect structures. The initial zooecium is a bulbous ancestrula, which gives rise to a tubular zooecium. Succeeding tubular zooecia bud one at a time from the lateral wall of the preceding zooecium. The tubular zooecia have perforated walls, and the distal opening of the zooecium is commonly sealed by a plate (possibly perforate). The zooecial wall apparently consisted of two layers, but it lacks the distinctive laminate structures of the order Trepostomata (Stenolaemata) and shows no additional thickening in the outer parts of the colony.

The simple, perforate tubular zooecia, in which the commonly oval zooecial opening is the same diameter as the zooecial tube, and the lack of accessory tubes or structures separate the hederellids from the Paleozoic cyclostomes. There are six recognized genera, including *Hederella*, *Reptaria*, *Hederopsis*, and *Clonopora*. [J.R.P.R.]

Hedgehog Members of the family Erinaceidae (order Insectivora), which includes 7 genera and 19 species of medium- to large-size animals generally characterized by spines on their back and sides. This family, however, is subdivided into two subfamilies, the spiny hedgehog group which includes the

Hedgehog (*Erinaceus europaeus*).

European hedgehog (*Erinaceus europaeus*); and the hairy hedgehog or gymnuras, long-tailed species which lack spines. The typical spiny hedgehog (see illustration) has well-developed eyes and a rudimentary to moderately long tail. In most species, there are five toes on each foot, although in some species these are reduced to four on the hindfeet. The spines can be erected by strong muscles, and serve as a protection for the naked or hairy belly, head, and limbs when the animal rolls itself into a ball. *See* INSECTIVORA. [C.B.C.]

Heidelberg man The European representative of the species *Homo erectus* which was ancestral to the later Neandertals and anatomically modern *H. sapiens*. Heidelberg man, or *Homo heidelbergensis* as it was originally called by O. Schoetensack, was formerly represented only by a largely intact lower jaw from a gravel pit near the village of Mauer, 16 km (10 mi) southeast of Heidelberg, Germany. The Mauer jaw has been dated on faunal and geological grounds to the early Middle Pleistocene, more than 400,000 years ago. Heidelberg man is now also represented by several other fossil remains, including teeth and the backs of craniums. *See* FOSSIL HUMAN. [E.T.]

Helicity (quantum mechanics) A fundamental quantized variable used in quantum mechanics to specify the relative orientations of spin and linear momentum of massless particles. It is a requirement of fundamental Dirac quantum mechanics that such particles have their spins aligned either parallel or antiparallel to their linear momentum. Particles having parallel alignment are arbitrarily assigned helicity +1; those having antiparallel alignment, −1. See MOMENTUM; SPIN (QUANTUM MECHANICS).

In a classic experiment on *K* electron capture by ^{152}Eu, M. Goldhaber, L. Grodzins, and A. Sunyar first showed that the neutrino emitted in the weak nuclear interaction had negative helicity—that its spin was aligned antiparallel to its momentum. An equivalent description of this situation is that these neutrinos are left-handed. Symmetry requires that antineutrinos be right-handed and have positive helicity. *See* ELECTRON CAPTURE; ELEMENTARY PARTICLE; NEUTRINO; QUANTUM MECHANICS; SYMMETRY LAWS (PHYSICS). [D.A.B.]

Helicobacter A genus of gram-negative bacilli whose members are spiral shaped, showing corkscrewlike motility generated by multiple, usually polar flagella. *Helicobacter* species require low concentrations of oxygen for maximum growth and produce the enzymes oxidase, catalase, and urease.

Different species of *Helicobacter* are found in the stomachs of different animals: *H. felis* in cats and dogs, *H. mustelae* in the domestic ferret, *H. nemestrinae* in the pigtailed macaque, and *H. acinonyx* in captive cheetahs with gastritis. *Helicobacter pylori,* found in the human stomach, is extremely common. In the United States and similarly developed countries, its prevalence increases at about 1% per year of age so that the majority of adults above age 50 are infected. In less developed countries, infection rates are dramatically higher, with up to 80% of children infected.

Helicobacter pylori is present in virtually all cases of chronic gastritis, which progresses slowly (years or decades) from asymptomatic to atrophic gastritis with impaired acid secretion. Virtually all individuals with duodenal ulcers are infected with *H. pylori*, which colonizes sites in the duodenum. The termination of treatment for duodenal ulcers leads to a high rate of recurrence of the ulcers, but ulcer treatment plus eradication of *H. pylori* from the stomach usually leads to a permanent cure. A significant proportion of individuals with *H. pylori*–associated atrophic gastritis develop intestinal-cell metaplasia in the stomach, a condition which is known to represent a precancerous state. *See* CANCER (MEDICINE); ULCER.

Helicobacter pylori virulence factors include its shape and its ability to rapidly move into and through the gastric mucous coating, which protects the organism from stomach acid; its surface-associated urease enzyme which neutralizes stomach acid near the organism; a cytotoxin; and a fibrillar adhesin which binds the organism to the surface of gastric epithelial cells. *Helicobacter pylori* can survive in large numbers in spite of antibodies which are secreted into the stomach and the host immune cell response (inflammatory response) characteristic of gastritis. [D.J.Ev.]

Helicoplacoidea A class of free-living, spindle- or pear-shaped plated echinozoans, known only from the Lower Cambrian (Olenellus Zone) of California. The skeletal plates formed a flexible test or shell, evidently carried on a muscular integument, which was contractile and was pleated in helical spirals. The oral and anal apertures are believed to have lain at either end of the body (see illustration). Helicoplacoids were apparently related to early kinds of sea cucumbers. *See* ECHINODERMATA; ECHINOZOA. [H.B.F.]

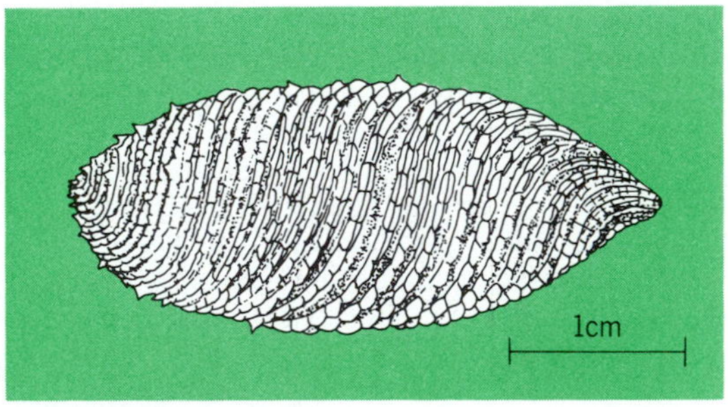

Helicoplacus curtisi (side view, reconstructed). (After J. W. Durham and K. E. Caster, *Science*, 140:820, 1963)

Helicopter An aircraft characterized by its large-diameter, powered, rotating blades. The helicopter can lift itself vertically by the reactive force generated as the rotating blades accelerate air downward. It can both lift and propel itself by accelerating air downward at an angle to the vertical. The helicopter is the most successful vertical takeoff and landing (VTOL) aircraft developed, by virtue of its relatively high efficiency in performing hovering and low-speed flight missions.

The key to understanding the operation and control of a helicopter lies in a knowledge of the forces and resultant motion of each rotor blade as momentum is imparted to the air. Unlike a fixed-wing aircraft, which derives its lift from the translational motion of the fuselage and airfoil-shaped wing relative to the air, the helicopter rotates its wings (or rotor blades) about a vertical shaft and thus is able to generate lift when the fuselage remains stationary.

Many different rotor arrangements have been used, and most of the early attempts at vertical flight were made with machines having multiple or coaxial counterrotating rotors. Most modern helicopters employ the single rotor or the tandem rotor configurations.

In addition to the selection of the number and location of the lifting rotors, designers have developed varied methods for attaching the blades to the rotor hub. Very early experiments conducted with the blades rigidly attached to the hub were unsatisfactory because of the excessive moments applied to the rotor mast. Based on the success achieved by the introduction of hinged attachments for the rotor blades, several configurations have been successfully manufactured. The teetering rotor used on two-bladed configurations has one central hinge which allows the blades to move in unison (one up, one down) like a seesaw. The gimbaled rotor is essentially equivalent to the teetering rotor and has been used on rotors with three or more blades. The articulated rotor has each blade attached to the hub by its own flapping hinge.

The growth of the helicopter industry in the United States is founded in the uses made by the armed forces. The technology which evolved to meet the needs of the military provided the base for an impressive growth in commercial applications. With such diverse operations as crop spraying, logging, construction, police and ambulance service, and passenger and corporate transportation, the industry has responded with a variety of commercial helicopters. *See* VERTICAL TAKEOFF AND LANDING (VTOL). [J.L.J.]

Helicosporida An order of protozoa in the class Myxosporidea (subphylum Cnidospora). It is characterized by production of spores with a relatively thick, single intrasporal filament and three uninucleate sporoplasms. *Helicosporidium parasiticum* is the only species in the order. The parasite infests the body cavity, fat bodies, and ganglia of mites (*Hericia hericia*) and the body cavity of fly larvae (*Dasyhelea obscura* and *Mycetobia pallipes*) found in the sap of horse chestnut and elm trees. *See* MYXOSPORIDEA. [R.F.N.]

Helimagnetism A property possessed by some metals, alloys, and salts of transition elements in which the atomic magnetic moments, at sufficiently low temperatures, are arranged in a spiral or helix. Simple antiferromagnets and ferromagnets can be considered as nonconical helimagnets with helical angles of 180 and 0°, respectively. In the same way, nonconical helimagnets may be considered as conical helimagnets with cone angle of 0°. Some typical helimagnets are listed in the table. The magnetic structures have been detected by neutron diffraction. [F.Ke.]

Some representative helimagnets

Substance	Magnetic structure	Temperature, K
MnO_2	Nonconical helix	$0 < T < 84$
$MnAu_2$	Nonconical helix	$0 < T < 363$
Dy	Nonconical helix	$85 < T < 179$
	Ferromagnet	$0 < T < 85$
$MnCr_2O_4$	Simple ferrimagnet	$18 < T < 43$
	Complex conical helix	$0 < T < 18$
Er	Conical helix	$0 < T < 20$
	Complex oscillation	$20 < T < 53$
	Sinusoidal antiferromagnet	$53 < T < 85$

Helion The doubly charged, positive ion of the lightest stable helium (He) isotope and the isobaric analog of mass 3 of the triton, the corresponding singly charged positive ion of the heaviest stable isotope of hydrogen. The helium-3 (^{3}He) atom occurs in nature with a fractional abundance of 0.014% but is more usually obtained as the product of the 12.4-year-half-life radioactive decay of tritium produced via the reaction ^{6}Li$(n,t)^4$He in neutron bombardment of ^{6}Li with neutrons. The helion has a mass excess of 14.931 MeV, a spin of ½, positive parity, and a magnetic dipole moment of −2.127624 nuclear magnetons. *See* HELIUM; TRITON.

The helion has been widely used as a nuclear projectile because of its large mass excess and because, through reactions such as (h,p), (h,t), and (h,α), it gives access to residual nuclei that can otherwise be studied only via reactions involving product neutrons and, consequently, lesser precision and greater difficulty. The h symbol for the helion has been recommended by the International Union of Pure and Applied Physics in parallel with the t symbol for the triton. *See* NUCLEAR REACTION; NUCLEAR STRUCTURE. [D.A.B.]

Helioseismology A technique for probing the interior of the Sun, using methods akin to terrestrial seismology. Because the Sun, although the nearest star by far, is a typical star, what can be learned of its interior through helioseismology is of broad importance to the stars in general.

Like terrestrial seismology, helioseismology entails the analysis of many "seismic" wave modes to determine the structure of the interior. However, although terrestrial seismic waves are initiated by a singular event such as an earthquake, waves within the Sun are continuously excited, probably by the turbulent convective motions in its outer layers. Thus the solar waves are always present at all points within the Sun and on its surface. The Sun is "ringing" like a bell, but not like one struck by a clapper; it vibrates more like a bell suspended in a sandstorm, continuously struck by tiny grains of sand. *See* SEISMOLOGY.

The solar waves are seen at the surface as up-and-down motions of the gases with a speed of about 0.3 mi/s (0.5 km/s) and a vertical displacement of about 30 mi (50 km). These waves are detected through the Doppler shift of the wavelength of absorption lines in the solar spectrum. The solar surface undulates up and down in a so-called five-minute oscillation. The oscillation is actually the superposition of as many as 10^7 individual modes of oscillation of the Sun as a whole, where each mode has its own characteristic frequency (near, but not exactly at, 0.003 cycle per second) and spatial pattern on the solar surface. Precise observations of the solar oscillations are difficult. A nearly continuous stream of data extending over days is needed to separate the many individual modes with nearly identical oscillation frequencies.

Helioseismology offers insight into the structure of the solar interior and also into its rotation. Waves propagating with or against the direction of rotation are carried by it, and their effective propagation speed and frequency are increased or decreased. The frequency shift for any mode depends on the average rotation rate within the resonant cavity for that mode, and comparison of the shift for many modes with different cavities makes it possible to determine how the rotation varies with depth. *See* SUN. [R.W.N.]

Heliozoia A subclass of the Actinopodea. Unlike Radiolaria, these protozoans have no central capsule. Most species

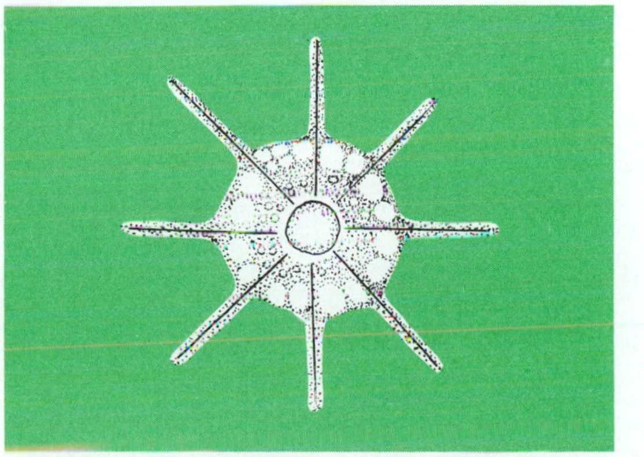

Actinophryida type of Heliozoia. (*After R. P. Hall, Protozoology, Prentice-Hall, 1953*)

live in fresh water. Pseudopodia may be either slender with an axial filament surrounded by cytoplasm (axopodia) or filamentous (filopodia). Certain floating species can roll along on the tips of their axopodia and also swim by moving their axopodia. Some species are naked; others have skeletal elements ranging from siliceous scales or spicules embedded in a gelatinous capsule to a reticulate chitinous skeleton often impregnated with silica. A centroplast may or may not be present. The subclass has three orders: Actinophryida (see illustration), Centrohelida, and Desmothoracida. *See* ACTINOPODEA. [R.P.H.]

Helium A gaseous chemical element, He, atomic number 2 and atomic weight 4.0026. Helium is one of the noble gases in group 0 of the periodic table. It is the second lightest element. The world's chief source of helium is a group of natural gas fields in the United States. *See* INERT GASES.

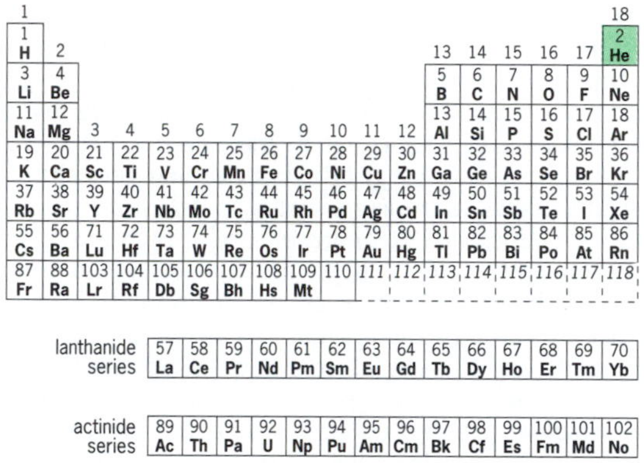
1 2 3 4 5 6 7 8 9 10 11 12 13 14 15 16 17 18
1 H 2 He
3 Li 4 Be 5 B 6 C 7 N 8 O 9 F 10 Ne
11 Na 12 Mg 13 Al 14 Si 15 P 16 S 17 Cl 18 Ar
19 K 20 Ca 21 Sc 22 Ti 23 V 24 Cr 25 Mn 26 Fe 27 Co 28 Ni 29 Cu 30 Zn 31 Ga 32 Ge 33 As 34 Se 35 Br 36 Kr
37 Rb 38 Sr 39 Y 40 Zr 41 Nb 42 Mo 43 Tc 44 Ru 45 Rh 46 Pd 47 Ag 48 Cd 49 In 50 Sn 51 Sb 52 Te 53 I 54 Xe
55 Cs 56 Ba 71 Lu 72 Hf 73 Ta 74 W 75 Re 76 Os 77 Ir 78 Pt 79 Au 80 Hg 81 Tl 82 Pb 83 Bi 84 Po 85 At 86 Rn
87 Fr 88 Ra 103 Lr 104 Rf 105 Db 106 Sg 107 Bh 108 Hs 109 Mt 110 111 112 113 114 115 116 117 118
lanthanide series 57 La 58 Ce 59 Pr 60 Nd 61 Pm 62 Sm 63 Eu 64 Gd 65 Tb 66 Dy 67 Ho 68 Er 69 Tm 70 Yb
actinide series 89 Ac 90 Th 91 Pa 92 U 93 Np 94 Pu 95 Am 96 Cm 97 Bk 98 Cf 99 Es 100 Fm 101 Md 102 No

Helium is a colorless, odorless, and tasteless gas. It has the lowest solubility in water of any known gas. It is the least reactive element and forms essentially no chemical compounds. The density and the viscosity of helium vapor is very low. Thermal conductivity and heat content are exceptionally high. Helium can be liquefied, but its condensation temperature is the lowest of any known substance. The properties of helium are given in the table. *See* LIQUID HELIUM.

Helium was first used as a lifting gas in balloons and dirigibles. This use continues for high-altitude research and for weather balloons. The principal use of helium is in inert gas—shielded arc welding. The greatest potential for helium use continues to emerge from extreme-low-temperature applications. Helium is the only refrigerant capable of reaching temperatures below 14 K (−434°F). The chief value of ultra-low temperature is the development of the state of superconductivity, in which there is virtually zero resistance to the flow of electricity. Other helium applications include use as a pressurizing gas in liquid-fueled rockets, in helium-oxygen breathing mixtures for divers, as a working fluid in gas-cooled nuclear reactors, and as a carrier gas for chemical analysis by gas chromatography.

Properties of helium

Property	Value
Atomic number	2
Atomic weight	4.0026
Melting point* at 25.2 atm pressure	−272.1°C (1.1 K)
Triple point (solid, helium I, helium II)	−271.37°C (1.78 K)
Triple point = λ-point (helium gas, helium I, helium II)	−270.96°C (2.19 K)
Boiling point at 1 atm pressure	−268.94°C (4.22 K)
Gas density at 0°C and 1 atm pressure, g/liter	0.17847
Liquid density at its boiling point, g/ml	0.1249
Solubility in water at 20°C, ml helium (STP)/1000 g water at 1 atm partial pressure of helium	8.61

*The melting point varies with the pressure.

Terrestrial helium is believed to be formed in natural radioactive decay of heavy elements. Most of this helium migrates to the surface and enters the atmosphere. The atmospheric concentration of helium (5.25 parts per million at sea level) could be expected to be higher. However, its low molecular weight permits helium to escape into space from the upper atmosphere at a rate roughly equal to its formation. Natural gases contain helium at concentrations higher than in the atmosphere. [A.W.F.]

Helix Any nonplanar curve all of whose tangents make the same angle with a fixed line. Other characteristic properties are that all principal normals are parallel to a plane and that the ratio of torsion to curvature is constant. If a helix has constant curvature (and hence constant torsion), it is a circular helix; it lies on a circular cylinder whose elements it cuts at a constant angle. *See* ANALYTIC GEOMETRY; DIFFERENTIAL GEOMETRY. [L.M.Bl.]

Helmholtz coils A pair of flat circular coils with equal numbers of turns and equal diameters, arranged with a common axis, and connected in series to have a common current (see illustration). The purpose of the arrangement is to obtain a

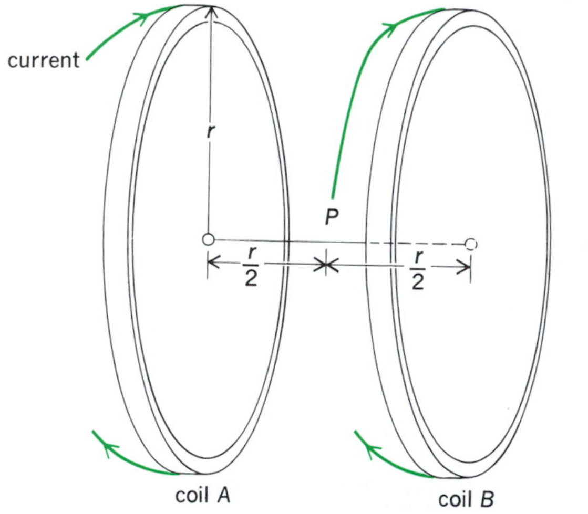

Helmholtz coils separated by distance *r*, resulting in a nearly uniform field of point *P*. (*After L. B. Loeb, Fundamentals of Electricity and Magnetism, 3d ed., 1947*)

magnetic field that is more nearly uniform than that of a single coil without the use of a long solenoid. The optimum arrangement is that in which the distance between the two coils is equal to the radius of one of the coils. *See* MAGNETIC FIELD; SOLENOID (ELECTRICITY). [K.V.M.]

Hematite The most important ore of iron, with composition α-Fe_2O_3. The crystals are thick tabular, usually flattened parallel to the base, and are frequently platy in habit. Hematite usually occurs as rouge-red earthy masses of finely divided particles. It is the major red-coloring agent in rocks and is a common interstitial cement in sediments. When mixed with quartzite or finely divided quartz, the mixture is called jasper, jaspilite, or taconite. Botryoidal masses are called "kidney ore," and splinters of these masses are "pencil ore."

The color is steel gray, blood red in thin fragments, and streak and powder are rouge red; hardness is 6 on Mohs scale and specific gravity is 5.25. The mineral is only weakly magnetic.

Hematite is the most widespread iron mineral. The most important ores are in low-to-medium-grade metamorphic rocks of sedimentary origin. Enormous beds occur in the Great Lakes region of the United States. Hematite also occurs in contact metamorphic and metasomatic deposits, often derived from the oxidation of magnetite and frequently associated with limestones.

Nearly every country in the world mines some hematite ore; the most important occurrences outside of the United States include India, Cuba, China, Chile, north African nations, and the Soviet Union. *See* IRON METALLURGY; REDBEDS. [P.B.M.]

Hematologic disorders Those disorders marked by aberrations in structure or function of the blood cells or the blood-clotting mechanism. Although many other diseases may be reflected by the blood and its constituents, the abnormalities of red cells, white cells, platelets, and clotting factors are considered to be primary hematologic disorders.

Red-cell abnormalities are principally represented by the anemias and polycythemias. In a wide variety of conditions the many forms of white cell present in the circulation, bone marrow, and lymphoid tissues of the body may be altered in form or number. The suffixes "-philia" and "-penia" denote increases and decreases, respectively, for the cells named. Leukopenia, neutrophilia, eosinophilia, and pancytopenia are examples of the wide range of possibilities. *See* ANEMIA; ERYTHROCYTOSIS; LEUKEMIA; LEUKOPENIA; POLYCYTHEMIA.

The hemorrhagic disorders result from a large number of known and unknown causes or contributing factors, often of a diverse nature. *See* HEMOPHILIA; HEMORRHAGE. [C.Ho.]

Hematopoiesis The process by which the cellular elements of the blood are formed.

The formation of these cells is one of the most active and important processes in the body. Most of the circulating cells live only for a short time and must be replaced in order to maintain life. For instance, in the human adult a red blood cell has a life of 120 days; 250 billion new red cells have to be produced daily to replace those that are destroyed.

Blood cells originate in the reticuloendothelial tissue, which is a loose, fibrous, highly vascularized mesh of fibers, endothelial cells, and macrophages. Within the spaces of the tissue are found the precursor (blast) cells of the definitive adult types. For the sake of convenience, the reticuloendothelial tissue is divided into two general but imprecise types: lymphoid and myeloid tissue.

Lymphoid tissue is primarily localized in the lymph nodes of the lymphatic system and is also in the spleen, thymus, and bone marrow. Several classes of white cells are produced, including the lymphocytes, macrophages, and monocytes. Myeloid tissue is normally limited in humans to the red bone marrow of the ribs, sternum, vertebrae, and proximal ends of the long bones of the body. It is concerned with the production of the erythrocytes (erythropoiesis), granular leukocytes, and megakaryocytes. Fragments of megakaryocytes form the blood platelets (thrombocytes). *See* BLOOD. [F.Wi.]

Hemiascomycetes A class of the subdivision Ascomycotina that includes the yeasts and yeastlike fungi. These are morphologically simple fungi; no ascocarp is formed, and the asci are produced free on the host or substrate. Asexual reproduction occurs by the formation of (budding) blastospores (or less frequently, by fission arthrospores). Two orders are recognized, the Endomycetales and the Taphrinales. *See* YEAST.

The vegetative body (thallus) of the Endomycetales may be either unicellular (true yeasts) or mycelial. In unicellular species, asci form when two vegetative cells fuse, and then the fused cell undergoes meiosis to form ascospores. In mycelial species, the hyphae are not very extensive. Sexual reproduction occurs when adjacent cells extend short lateral branches that fuse to form the asci.

The Endomycetales are common on substrates high in sugars, such as plant exudates, ripe fruits, and flower parts. Because they are microscopic, they are recognized mainly from cultures that have a homogeneous appearance and a characteristic odor. The most important genus is *Saccharomyces*; *S. cerevisiae* is the common bakery and brewery yeast, and *S. ellipsoideus* is used in winemaking. An important mycelial species is *Nematospora coryli*, which causes yeast spot disease of various crops.

The Taphrinales include the leaf curl disease fungi. The most widely recognized species are *Taphrina deformans*, cause of leaf curl of peach and almond trees, and *T. caerulescens*, cause of leaf blister of oaks. These fungi produce a well-developed mycelium in the host tissue but, when grown in culture, form only a yeastlike colony of single cells—a phenomenon known as dimorphism. Asci are produced when special binucleate hyphal cells beneath the host cuticle undergo nuclear fusion and the resulting diploid cell elongates to form an ascus on the leaf surface. The nucleus undergoes meiosis, and ascospores are formed. *See* ASCOMYCOTINA; EUMYCOTA; FUNGI. [R.T.Ha.]

Hemicellulose A term designating plant cell components which are made soluble by dilute alkali or which go into solution quite readily in hot dilute mineral acids with the formation of simple sugars. Hemicelluloses constitute about one-fourth of perennial plants and about one-third of annual plants. The term is usually applied to those polysaccharides in the cell wall of land plants which are extractable by dilute alkaline solutions. The term has also been used to include all the polysaccharide components of the cell wall other than cellulose. *See* CELL WALLS (PLANT); CELLULOSE.

Hemicelluloses extracted from different plant sources are rarely identical. In fact, many different hemicelluloses usually occur intermixed with each molecular type representing different degrees of polymerization. Because of this heterogeneity, few hemicelluloses have been isolated in a homogeneous state. Therefore, relatively little is known of the structure of these compounds that compose almost one-third of the carbohydrates in woody tissue.

D-Xylose is the dominant building unit of the hemicelluloses of most woods and annual plants. D-Mannose is also very abundant in hemicelluloses; the mannose content of softwoods is usually higher than that of hardwood. Often it occurs as a polymer, mannan, or in combination with D-glucose or D-galactose as a glucomannan, galactomannan, or galactoglucomannan. *See* MANNANS.

Hemicelluloses are important to the paper industry. In chemical wood pulps, hemicellulose is needed for satisfactory pulp quality. Its presence aids the swelling of the pulp, the bonding of the fibers, the bursting strength, tensile strength, tear resistance, folding endurance, opacity, and specific surface of the pulp sheet.

Hemicelluloses also serve as nutrients for yeasts, and they can be used for raw material in the production of furfural and ethyl alcohol. [R.L.Wh.]

Hemichordata A group of deuterostome animals that includes the classes Enteropneusta, Pterobranchia, and Planctosphaeroidea. The last, a monospecific class, is represented by *Planctosphaera pelagica*, a planktonic larva that occupies low depths and resembles the larval tornaria of Enteropneusta. *See* ENTEROPNEUSTA; PTEROBRANCHIA.

The hemichordates have a slender tubular diverticulum that projects forward into the protosome from the roof of the buccal cavity. That organ, also called a stomochord or buccal diverticulum, is a supporting axial rod of the protosome and resembles the notochord of Chordata, but it does not have the same position, structure, origin, or function, and so they are not homologous. *See* CHORDATA.

The hemichordates are not plentiful animals. All are marine species that live in a wide range of habitats and depths, from intertidal to abyssal, and show a worldwide distribution. They vary in size from a fraction of an inch to 7 ft or more. The members of the phylum differ widely. The enteropneusts are vermiform and solitary, whereas the pterobranchs are sacciform and colonial.

The hemichordates are a primitive group, having a tripartite body and coelom; their embryonic development resembles the echinoderms, with which they also share a primitive nervous system. The Hemichordata, therefore, may be a group at a low level of evolution, between echinoderms and chordates. [J.Ben.]

Hemicidaroida An extinct paraphyletic order of regular sea urchins (Echinoidea), identified by having a plain, unsculptured test and tubercles that are perforate and crenulate. They almost certainly include ancestors of later groups that developed imperforate and noncrenulate tuberculation and are thus paraphyletic. Two families are generally placed within the order, Hemicidaridae and Pseudodiadematidae. The oldest hemicidaroid is Late Triassic (Upper Norian) and the youngest comes from the Upper Cretaceous (Campanian). They appear to have been epifaunal grazers. *See* ECHINOIDEA. [A.Sm.]

Hemidiscosa An order of the subclass Amphidiscophora in the class Hexactinellida. These sponges are distinguished from the order Amphidiscosa in that the birotulates are hemidiscs with asymmetrical ends. *See* AMPHIDISCOPHORA; AMPHIDISCOSA. [W.D.H.]

Hemimorphite A mineral sorosilicate having the composition $Zn_4Si_2O_7(OH)_2 \cdot H_2O$; an ore of zinc. Crystals are usually colorless and the aggregates white, but in some cases there are faint shades of green, yellow, and blue. The mineral has a vitreous luster, a hardness of 4½ to 5 on Mohs scale, and a specific gravity of 3.45.

Hemimorphite has a wide distribution and has been mined in Belgium, Germany, Romania, England, Algeria, and Mexico. In the United States it is found at Sterling Hill, New Jersey; Friedensville, Pennsylvania; and Elkhorn Mountains, Montana. *See* SILICATE MINERALS. [C.S.Hu.]

Hemiptera An order of the class Insecta sometimes referred to as the Heteroptera. Both these names refer to the forewings, which are differentiated into a thickened basal area and a membranous apical region. These are the true bugs. Included in Hemiptera are such common insects as the bedbugs, stink bugs, plant bugs, lace bugs, and backswimmers.

The true bugs number about 25,000 species. They are known from all continents except Antarctica and occur on most islands. Hemiptera range in size from small aquatic and ground-inhabiting forms approximately 0.04–0.08 in. (1–2 mm) in length to giant water bugs 4 in. (100 mm) or more.

In habits, the true bugs range from strictly phytophagous types attached to a single host plant, to general predators on other insects, and even to specialized ectoparasites of bats. Many species are of economic importance as plant pests or vectors of disease. They occur in vegetation, on the ground, in and on the water, and in the nests of termites. Most water bugs depend on surface air held to the body by air spaces and hairs on the abdomen. As oxygen is depleted in the air bubble, it is replaced from the surrounding medium by diffusion.

Most Hemiptera are bisexual and oviparous, but parthenogenesis is known. Mating usually takes place on vegetation or on the ground, the pairing being end-to-end in stink bugs, squash bugs, chinch bugs, and similar species, and with the male above the female in most others.

Morphology. In Hemiptera the mouthparts are elongate and slender, forming a sucking mechanism. The beak arises from the anterior part of the head and the head is commonly directed forward or downward.

Hemiptera are further characterized (see illustration) by antennae, usually of four or five segments, a pair of compound eyes, and often two ocelli. The thorax consists of a prominent pronotum, a triangular mesothoracic scutellum, and a broad metathorax which is partly fused with the first abdominal segment. The mesothoracic wings, or hemelytra, overlap at their membranous apices when at rest. The hindwings are hitched to the forewings in flight by grooves and pegs. Wings are sometimes reduced to short pads and may be lacking in certain groups or even in members of a single species. The front legs are frequently enlarged and sometimes chelate in predacious forms, and the middle and hindlegs are adapted for swimming in some groups. The abdomen is of 10–11 segments, but only seven are commonly seen.

Subdivisions. Families of the Hemiptera and their distinguishing characteristics are listed in the table under the respective subdivisions.

Hydrocorisae. This subdivision contains nine families of water bugs with concealed antennae and without a bulbus ejaculatorius in the male. Many species are predaceous.

Corixoidea is a superfamily which contains the single family Corixidae, or water boatmen. Corixidae lack ocelli and have a unique type of mouthpart. Corixids swim with the dorsal side uppermost, using the oarlike middle and hind legs. Eggs are stalked and there are five nymphal instars. Adults sometimes fly to lights in great numbers.

In the superfamily Nepoidea, the Nepidae, or water scorpions, have a long breathing tube at the tip of the abdomen, through which they obtain air directly from the surface. The beak is short and stout to suck the juices of other insects on which they prey. Belostomatidae, or electric-light bugs, have short, straplike respiratory appendages at the tip of the abdomen. Giant water bugs of the genus *Lethocerus* are pests in fish ponds where they attack fry. They can inflict a painful bite when handled carelessly.

The superfamily Notonectoidea includes the Notonectidae or backswimmers: they swim ventral side uppermost, the hindlegs serving as oars. Breathing is facilitated by an air bubble, obtained by touching the tip of the abdomen to the surface.

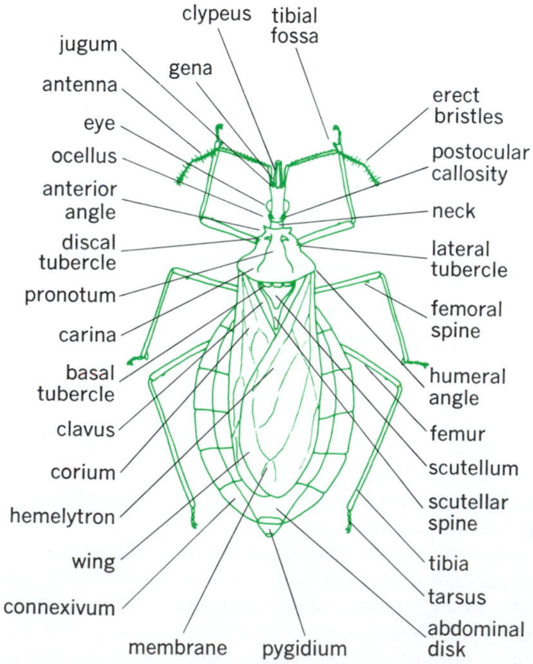

External anatomy of a bug.

Families of Hemiptera

Family	Common name	Distribution	Number of species
Subdivision Hydrocorisae			
Corixidae	Water boatmen	General	300
Nepidae	Water scorpions	General	170
Belostomatidae	Giant water bugs	General	140
Notonectidae	Backswimmers	General	170
Pleidae	None	General	20
Helotrephidae	None	Tropical	20
Naucoridae	Creeping water bugs	General	200
Gelastocoridae	Toad bugs	Tropical and subtropical	80
Ochteridae	Velvety shore bugs	Tropical and subtropical	20
Subdivision Amphibicorisae			
Gerridae	Water striders	General	300
Veliidae	Smaller water striders	General	200
Hydrometridae	Marsh treaders	General	50
Mesoveliidae	Water treaders	General	20
Hebridae	Velvety water bugs	General	40
Subdivision Geocorisae			
Superfamily Leptopodoidea			
Saldidae	Shore bugs	General	200
Leptopodidae	None	Tropical and subtropical	20
Leotichidae	None	Oriental	2
Superfamily Dipsocoroidea			
Dipsocoridae	None	General	
Schizopteridae	Jumping ground bugs	Tropical and subtropical	
Superfamily Cimicimorpha			
Cimicidae	Bat, bed, bird bugs	General	80
Anthocoridae	Flower bugs	General	300
Polyctenidae	Bat bugs	Tropical and subtropical	20
Miridae	Plant bugs	General	5000
Microphysidae	None	Palearctic	30
Plokiophilidae	None	Tropical	20
Nabidae	Damsel bugs	General	250
Tingidae	Lace bugs	General	700
Vianaidae	None	Neotropical	2
Thaumastocoridae	Palm bugs	Tropical	11
Superfamily Enicocephaloidea			
Enicocephalidae	Gnat bugs	General	300
Superfamily Reduvioidea			
Reduviidae	Assassin bugs	General	3500
Superfamily Aradoidea			
Aradidae	Flat bugs	General	800
Termitaphididae	Termite bugs	Tropical	10
Superfamily Pentatomorpha			
Idiostolidae	None	Chilean	2
Lygaeidae	Lygaeid bugs	General	2000
Thaumastellidae	None	Ethiopian	1
Colobathristidae	None	Tropical	70
Berytidae	Stilt bugs	General	100
Malcidae	None	Ethiopian, Oriental	30
Piesmatidae	Ash-gray leaf bugs	General but discontinuous	20
Pyrrhocoridae	Pyrrhocorid bugs	General	300
Largidae	None	General	100
Coreidae	Coreid bugs	General	2000
Rhopalidae	None	General	300
Stenocephalidae	None	Old World, Neotropical	20
Hyocephalidae	None	Australia	1
Pentatomidae	Stink bugs	General	2500
Phloeidae	Bark bugs	Neotropical	5
Plataspidae	None	Old World	400
Lestoniidae	None	Australia	1
Cydnidae	Ground or burrower bugs	General	600
Urostylidae	None	Oriental and Australian	50
Aphylidae (not placed)	None	Australian	2
Joppeicidae	None	Mediterranean	1

The minute bugs of the families Pleidae and Helotrephidae (superfamily Pleoidea) are suboval, with legs not fitted for rowing.

The creeping water bugs (superfamily Naucoroidea) are sometimes separated into the Naucoridae and Aphelocheiridae. They are suboval in body form. Respiration is either by an air bubble or by means of a plastron; in the Old World genus *Aphelocheirus*, the plastron consists of an ultramicroscopic hair pile which acts as a physical gill; the bug is thus able to remain submerged permanently.

In the superfamilies Gelastocoroidea and Ochteroidea, the Gelastocoridae, or toad bugs, and the Ochteridae (or Pelogonidae) are shore-line or mud inhabitants. Both have ocelli. The former are cryptically colored, resembling the sand or mud background. Ochterids are black with a silky sheen.

Amphibicorisae. This subdivision contains surface water bugs with antennae exposed and without a bulbus ejactulatorius in the male. Only the single, diverse superfamily Gerroidea has been proposed for the surface water bugs. All have conspicuous antennae, and the body is clothed with hydrofuge hairs. All Gerroidea are predacious.

Gerridae are the large water striders with long middle and hind legs and a median scent gland opening on the metasternum. The claws are inserted before the tips of the tarsi. The marine forms are always wingless but all others are polymorphic, with fully winged forms occurring together with short-winged or apterous types.

Veliidae are small water striders which have shorter legs and a longitudinal groove between the eyes. The claws are preapical. Like the Gerridae, these are pond inhabitants, stream-riffle bugs, and marine types, but the last are found only near shore in tropical reefs.

Hydrometridae are long, slender marsh treaders in which the head is longer than the thorax. The claws are apical.

Mesoveliidae and Hebridae are two small families which differ from others in having the well-developed ocelli and the single dorsal abdominal scent gland openings of the nymphs.

Geocorisae. This subdivision contains the land bugs with conspicuous antennae and an ejaculatory bulb in the male reproductive system. This subdivision can be divided into seven groups, each of which may be equivalent in rank to the Hydrocorisae and Amphibicorisae. Superfamilies that do not fit the above groupings are the Reduvioidea, the Saldoidea, the Aradoidea, and the Enicocephaloidea and Dipsocoroidea, of isolated position in the system.

The superfamily Saldoidea comprise the shore bugs which have three pairs of trichobothria on the vertex and a single nymphal scent gland opening. Dipsocoroidea is a group of minute ground inhabitants, of which the Schizopteridae live in leaf mold and the Dipsocoridae are predators on small insects under bark or in rotten wood or amid stones at the edge of streams.

Enicocephaloidea is a unique group in which the head is bilobed, the pronotum trilobed, and the wings completely membranous. They live under stones, beneath bark, and in leaf mold and are predators. Some species are wingless; others can cast off their wings.

Reduvioidea includes only the assassin bugs or conenose bugs of the family Reduviidae. Nearly all have a stridulatory furrow on the prosternum, which is scraped by the rostrum to produce a squeaking sound. Ocelli are generally present and the beak is three-segmented.

Among the Aradoidea are flat bugs of the family Aradidae and their specialized relatives, Termitaphididae. They lack ocelli and have four-segmented antennae. Most, if not all, are fungus feeders. The Termitaphididae have no wings. They are known from termite nests in the Old and New World tropics. Aradidae are nearly cosmopolitan and fully winged, brachypterous, stenopterous, or apterous.

Cimicimorpha is the largest group of the Geocorisae and is divided into three superfamilies, the Cimicoidea, Tingoidea, and Thaumastocoroidea.

1. Cimicoidea. Anthocoridae, or the flower bugs, are small predators on thrips, mites, and similar species in vegetation, in stored products, and in the nests of birds. The ocelli are distinct, and the front wings have a marginal fracture. Cimicidae contains the bedbugs, which lack ocelli, and have the short wings reduced to pads. The food consists of blood of birds and mammals. The Polyctenidae are bat ectoparasites which lack eyes and have ctenidia and strong claws. The Miridae family contains a majority of the species of Hemiptera. Included are plant bugs of both herbivorous and predacious type. Ocelli are lacking, the antennae are four-segmented, and there is only one dorsal abdominal scent gland opening in nymphs. Two small families, the Plokiophilidae and Microphysidae, are predacious. The Plokiophilids live in the webs of spiders and embiids.

Nabidae are the long "damsel" bugs which are slender predators on other insects. The ocelli are well developed, and the rostrum is four-segmented. The tropical Velocipedidae are related but differ in the broader form and darker coloration.

2. Tingoidea. This superfamily contains the lace bugs of the family Tingidae which have wings with many lacelike areolae. They lack ocelli, have four segments in the antennae and beak, and commonly have one or more bulbous or hoodlike elevations on the thorax. Pests are known on ornamental plants.

3. Thaumastocoroidea. The only family, the Thaumastocoridae, occurs in Australia and the New World tropics; it includes the royal palm bug of Florida.

Pentatomorpha includes the superfamilies Lygaeoidea, Pyrrhocoroidea, Coreoidea, and Pentatomoidea.

1. Lygaeoidea. This is the first of the superfamilies with ventral trichobothria (bristles) on the abdomen. The antennae are four-segmented. Ocelli are present. The Lygaeidae include the chinch bug, the false chinch bugs, and the milkweed bugs. Small families related to the lygaeids are the thread-legged Neididae (Berytidae), the small Piesmidae, and the tropical Colobathristidae. Some of the last have a unique method of stridulation with a filelike arch on the sides of the head.

2. Pyrrhocoroidea. This group includes the cotton stainers, which attack cotton bolls in the southwestern United States and over most of the tropics, and stout, dark bugs with a reddish border.

3. Coreoidea. The squash bugs and their relatives include the Coreidae, Rhopalidae, Alydidae, and Hyocephalidae. They have four-segmented antennae, a beak, distinct ocelli, and many veins in the membrane.

4. Pentatomoidea. This large group has marginal trichobothria. The antennae are usually five-segmented and the beak is four-segmented. Scutelleridae, or shield bugs, are not injurious in the United States. Cydnidae include the ground-burrowing bugs, which attack strawberries in sandy soil, and the negro bugs. Smaller families of little or no economic importance in the United States are the large tropical Tessaratomidae, the Acanthosomatidae, the Dinidoridae, the Aphylidae, the Urostylidae, the barklike Phloeidae, the shining, oval Plataspidae, and the Lestoniidae. The true stink bugs, Pentatomidae, include the black-and-red harlequin cabbage bug, and several green stink bugs on cotton and other crops. *See* INSECTA. [P.W.W.]

Hemispheric laterality The process that is responsible for the exchange of information between the two half brains. The human brain is a bilaterally symmetrical structure which is for the most part richly interconnected by two main bridges of neurons called the corpus callosum and anterior commissure. These structures can be surgically sectioned in humans in an effort to control the spread of epileptic seizures. Although there is no apparent change in everyday behavior of these patients, dramatic differences in cognitive function can be demonstrated under specialized testing conditions. Because of these studies it now can be said that in normal humans these cerebral commissures are largely responsible for behavioral unity; the neural mechanism keeps the left side of the body up to date with the activities of the right, and vice versa. *See* BRAIN. [M.S.G.]

Hemlock The genus *Tsuga* of the pine family, characterized by flattened needles with two white lines beneath the needlelike leaves, which have distinct short stalks. The cones are small and pendent.

Eastern hemlock (*T. canadensis*) occurs in eastern Canada, the Great Lakes states, and the Appalachians. Minutely toothed leaves are characteristic of this species. The wood is hard and strong, and is used for construction, boxes, crates, and paper pulp. The bark is one of the principal domestic sources of tannin. The eastern hemlock is a common ornamental tree.

Carolina hemlock (*T. caroliniana*), a species found in the southern Appalachians, has entire needles and is sometimes grown as an ornamental. The western hemlock (*T. hetero-*

phylla) grows in the extreme Northwest and in Alaska. Its needles resemble those of the eastern hemlock, but the white lines beneath are not so distinct. It is an important lumber tree, with uses similar to these of the eastern species. *See* PINALES.

[A.H.G./K.P.D.]

Hemoglobin The oxygen-carrying molecule of the red blood cells of vertebrates. This protein represents more than 95% of the solid constituents of the red cell. It is responsible for the transport of oxygen from the lungs to the other tissues of the body and participates in the transport of carbon dioxide in the reverse direction.

Each molecule of hemoglobin comprises four smaller subunits, called polypeptide chains. These are the protein or globin parts of hemoglobin. A heme group, which is an iron-protoporphyrin complex, is associated with each polypeptide subunit and is responsible for the reversible binding of one molecule of oxygen. The polypeptide chains and the heme are synthesized and combine together in nucleated red cells of the bone marrow. As these cells mature, the nuclei fragment and the cells, now called reticulocytes, begin to circulate in the blood. After sufficient hemoglobin has been formed in the reticulocyte, all nuclear material disappears and the cell is then called an erythrocyte, or red blood cell. Each hemoglobin molecule lasts as long as the red cell, which has an average life of 120 days. *See* PORPHYRIN.

Normal adult males and females have about 16 and 14 g, respectively, of hemoglobin per 100 ml of blood; each red cell contains about 29×10^{-12} g of hemoglobin. Red cells normally comprise 40–45% of the volume of whole blood.

The reversible combination of hemoglobin and oxygen can be represented by the reaction shown below. The equilibrium

$$\mathrm{Hb} + 4\mathrm{O_2} \overset{K_1}{\rightleftharpoons} \mathrm{HbO_2} + 3\mathrm{O_2} \overset{K_2}{\rightleftharpoons} \mathrm{Hb(O_2)_2} + 2\mathrm{O_2} \overset{K_3}{\rightleftharpoons} \mathrm{Hb(O_2)_3} + \mathrm{O_2} \overset{K_4}{\rightleftharpoons} \mathrm{Hb\,(O_2)_4}$$

constants for each step are not the same because an oxygen molecule on one heme group changes the affinity of the other hemes for additional oxygen molecules. This alteration in binding affinity during oxygenation is called heme-heme interaction and is due to small changes in the three-dimensional structure of the molecule.

Hemoglobin combines reversibly with carbon monoxide about 210 times more strongly than with oxygen. This strong affinity for carbon monoxide accounts for the poisoning effects of this gas.

Hemoglobin binds carbon dioxide by means of free amino groups of the protein but not by the heme group. The reversible combination with carbon dioxide provides part of the normal blood transport of this gas. Hemoglobin serves also as a buffer by reversible reactions with hydrogen ions. The acidic property of oxyhemoglobin is greater than deoxygenated hemoglobin. The extra binding of hydrogen ion by deoxyhemoglobin promotes the conversion of tissue carbon dioxide into bicarbonate ion and thus increases the amount of total carbon dioxide which can be transported by blood. *See* BLOOD; RESPIRATION. [R.T.J.]

Hemophilia A rare, hereditary blood disorder marked by a tendency toward excessive bleeding. It is almost entirely restricted to males, and is transmitted as a sex-linked mendelian recessive trait passing from an affected male through an unaffected or very mildly affected daughter to appear again in a grandson. Queen Victoria was a carrier, and several of her male descendants were affected. *See* HUMAN GENETICS; SEX-LINKED INHERITANCE.

Classical hemophilia (hemophilia A) is due to a deficiency of the antihemophilic factor or factor VIII, a clotting factor which is normally present in the blood in trace amounts and is essential for normal fibrin formation. Hemophilia B is a similar sex-linked bleeding disorder affecting males but characterized by a deficiency of another blood clotting factor, factor IX. It can only be distinguished from hemophilia A, which it closely resembles, by laboratory tests. Hereditary deficiencies of other clotting factors may give rise to bleeding disorders similar to hemophilia, but they are inherited as autosomal dominant or recessive characteristics.

Treatment consists in the intravenous administration of potent concentrates of factor VIII or IX prepared from human plasma and they are very effective in the management of hemophilia A and B, respectively. *See* BLOOD; HEMORRHAGE.

[C.Ho.]

Hemophilus A genus of gram-negative pleomorphic coccobacilli that are facultative anaerobes and almost invariably ferment carbohydrates and reduce nitrates. Most significantly, they require for propagation one or both of two growth factors: X (an iron prophyrin), and V (nicotinamide adenine dinucleotide [NAD] or one of its nucleoside precursors). The presence of both factors in blood accounts for the name *Hemophilus* ("blood-loving"); they are widely distributed in many plant, animal, and microbial cells. *See* BACILLACEAE.

All known members of the genus *Haemophilus* are obligate parasites of animals. Most of them are harmless commensal organisms living on the mucous membranes of the mouth, pharynx, and upper respiratory and lower genital tracts of humans and animals. The type species, *H. influenzae*, causes suppurative meningitis in infants, conjunctivitis (mainly in children), obstructive inflammation of the pharynx and larynx, inflammation of the middle ear, and various secondary infections, especially of the respiratory tract. *Hemophilus ducreyi* is the agent of a rather uncommon venereal disease, soft chancre. [E.L.B.]

Hemorrhage The escape of blood from within the vascular system. Hemorrhage may result from either trauma or disease of the vessel wall. In some cases there is no apparent break in the vessel wall, and the blood cells escape by a process called diapedesis.

The escape of blood following rupture of a vessel wall as a result of trauma is obvious and needs no explanation. The causes other than trauma can be divided into three main groups.

The first group consists of those conditions in which there is a chronic disease process affecting the vessel wall, such as arteriosclerosis or aneurysm formation. Either of these conditions, in association with an elevated blood pressure, can result in a break in the wall and subsequent hemorrhage. An infarct, or tissue death from any cause, may also result in hemorrhage.

The second group consists of those causes in which there is an acute process affecting the vessel wall, such as septicemia, bacterial toxins or focal colonies of bacteria in the vessel wall, poisoning by metals such as mercury or arsenic, or anoxia.

The third group consists of those hemorrhagic conditions which result from some defect in the blood itself. Under this heading are included leukemia, pernicious anemia, thrombocytopenia, and the clotting disorders. *See* ANEMIA; APOPLEXY; CIRCULATION DISORDERS; HEMATOLOGIC DISORDERS. [R.A.V.]

Hemorrhoids Soft, irregular venous abnormalities occurring in and above the anus. External hemorrhoids lie outside the anal ring and are commonly seen in two forms. The first is the acutely thrombosed hemorrhoid which gives great pain or tenderness. The second form occurs as part of a more extensive network of varicosities about the anal region. Symptoms vary greatly; often there is merely a slight bleeding tendency while at stool.

Internal hemorrhoids may be simple, hard-to-detect varicosities lying well within the anal canal, or they may be huge, protruding, blood-filled vessels that give intense pain. Severe hemorrhage may occur, as well as the passage of lesser amounts of bright red blood. [E.G.St./N.K.M.]

Hemp The fiber and the plant *Cannabis sativa*. It should not be confused with Manila hemp, which is not related to true hemp. Hemp contains the drug marijuana. *See* ABACA; MARIJUANA.

Hemp fiber, which for many years was the major raw material used in the manufacture of rope, now is used mostly in the production of small twines, linenlike fabrics and canvases, and, to some extent, in making special types of paper. *See* NATURAL FIBER.

Hemp is an annual crop, most of which is produced in Eastern Europe and mainland China, with some production in South Korea, Turkey, and Italy. [E.G.N.]

Henequen A fiber obtained from the leaves of *Agave fourcroydes*. It is produced only in Mexico, Cuba and El Salvador. Henequen is sometimes incorrectly called sisal, which is a closely related plant grown in Brazil and Africa. *See* SISAL.

The greatest quantity of henequen fiber goes into farm twine, followed by industrial tying twine and then light-duty rope. Padding for innerspring mattresses is made from the lowest grades of fiber and from flume tow, the short, tangled fiber that can be recovered from the cleaning operation. Henequen is exported as manufactured twine, rope, or padding, not as raw fiber. *See* NATURAL FIBER. [E.G.N.]

Heparin A highly sulfated mucopolysaccharide with blood anticoagulant activity, isolated from mammalian (chiefly beef) tissues. Heparin was first found in abundance in the liver, hence the name, but it is present in substantial amounts in the spleen, muscle, and lung as well. In the blood of most mammals, heparin is an antagonist to thrombin, prothrombin, and thromboplastin. It lessens the tendency of platelets to agglutinate. It is used in the treatment of venous thrombosis, embolism, myocardial infarction, and certain types of cerebral thrombosis. *See* POLYSACCHARIDE. [F.W.]

Hepatitis An inflammation of the liver caused by a number of etiologic agents, including viruses, bacteria, fungi, parasites, drugs, and chemicals. All types of hepatitis are characterized by distortion of the normal hepatic lobular architecture due to varying degrees of necrosis of liver cells, inflammation, and Kupffer cell enlargement and proliferation. There is usually some degree of disruption of normal bile flow, which causes jaundice. The severity of the disease is highly variable and often unpredictable. *See* LIVER.

The most common infectious hepatitis is of a viral etiology. The two well-recognized forms are infectious hepatitis (hepatitis A) and serum hepatitis (hepatitis B). Hepatitis A is orally acquired, has an incubation period of 15 to 45 days, occurs primarily in children and young adults, is usually a mild, often unrecognized, disease which has a duration of 5 to 8 weeks, usually leaves no residual effects, and has a very low mortality rate. Hepatitis B is parenterally transmitted, has an incubation period of 50 to 160 days, occurs at any age and is common in the elderly, is extremely common in hard-drug addicts, has an insidious onset and often produces a protracted disease, and has a mortality rate of up to 15%. It has been proven that many of these clinical concepts are inconstant.

Another frequently occurring form of hepatitis is caused by excessive ethyl alcohol intake and is referred to as alcoholic hepatitis. It usually occurs in chronic alcoholics and is characterized by fever, leukocytosis, and jaundice. *See* ALCOHOLISM.

Hepatitis can be caused by numerous commonly used drugs and can be pathogenetically related to a hypersensitivity allergic phenomenon or a dose-dependent phenomenon.

Chronic hepatitis is clinically defined in part as evidence of active liver disease for at least 3 consecutive months. The causes are multiple. It follows acute type B viral hepatitis in approximately 10% of the cases and is seen in association with the prolonged intake of certain drugs. Chronic hepatitis may also be found in young females, in whom it is possibly an autoallergic disease.

Clinical features of hepatitis include malaise, fever, jaundice, and serum chemical tests revealing evidence of abnormal liver function. In most mild cases of hepatitis, treatment consists of bedrest and analgesic drugs. In those individuals who develop a great deal of liver cell necrosis and subsequently progress into a condition known as hepatic encephalopathy, exchange blood transfusions are often used. This is done with the hope of removing or diluting the toxic chemicals thought to be the cause of this condition. [S.P.H.]

Herbarium A collection of plant specimens, pressed and mounted on paper or placed in liquid preservatives, and systematically arranged for subsequent consultation. Most of the great botanical gardens have such collections. *See* BOTANICAL GARDENS.

The chief value of a herbarium lies in the bringing together of large numbers of specimens of plants for comparative studies. It is important for botanists to have a knowledge of the appearance of living plants in their customary habitats, but without collections of preserved plants, such works as monographic treatments of various groups or manuals of the flora of various regions could not be prepared. [E.L.C.]

Some great herbaria of the world, with approximate number of specimens

Royal Botanical Gardens, Kew	7,000,000
V. L. Komarov Botanical Institute, Leningrad	5,000,000
British Museum (Natural History), London	4,000,000
Berlin-Dahlem, Botanischer Garten	4,000,000
Jardin des Plantes, Paris	5,000,000
Harvard University, Cambridge	3,500,000
Conservatoire et Jardin Botaniques, Geneva	4,000,000
Indian Botanic Garden, Calcutta	2,500,000
U.S. National Herbarium, Washington	3,000,000
N.Y. Botanical Garden, New York	3,000,000
Royal Botanic Garden, Edinburgh	2,000,000
National Herbarium, Melbourne	1,500,000
Missouri Botanical Garden, St. Louis	1,700,000

Herbicide Any chemical used to destroy or inhibit plant growth, especially of weeds or other undesirable vegetation. There are well over a hundred chemicals in common usage as herbicides. Many of these are available in several formulations or under several trade names. The variety of materials are conveniently classified according to the properties of the active ingredient as either selective or nonselective. Selective herbicides are those that kill some members of a plant population with little or no injury to others. Nonselective herbicides are those that kill all vegetation to which they are applied. Further subclassification is by method of application, such as preemergence (soil-applied before plant emergence) or postemergence (applied to plant foliage). Additional terminology sometimes applied to describe the mobility of post-emergence herbicides in the treated plant is contact (nonmobile) or translocated (mobile—that is, killing plants by systemic action).

A rapidly expanding use for nonselective herbicides is the destruction of vegetation before seeding in the practice of reduced tillage or no tillage. Some are also used to kill annual grasses in preparation for seeding perennial grasses in pastures. Additional uses are in fire prevention, elimination of highway hazards, destruction of plants that are hosts for insects and plant diseases, and killing of poisonous or allergen-bearing plants.

Preemergence or postemergence application methods derive naturally from the properties of the herbicidal chemical. The distinction between pre- and postemergence is not always clear-cut. For example, atrazine can exert its herbicidal action either following root absorption from a preemergence application or after leaf absorption from a postemergence treatment. [R.O.R.]

Herbig-Haro objects A small, bright, semistellar knot of nebular emission in one of the dark interstellar clouds of gas and dust from which stars form. Herbig-Haro objects are named for G. Herbig and G. Haro, who independently discovered them in the early 1950s. They range in size from 300 to 1000 astronomical units and can vary in intensity over periods of only a few years. Their spectra show the presence of the emission lines characteristically formed behind a radiative shock wave, and suggest masses only about a factor of 10 greater than the mass of the Earth. It is now believed that Herbig-Haro objects are manifestations of the mass-loss phenomenon associated with very young stars. *See* INTERSELLAR MATTER; MOLECULAR CLOUD.

Since the late 1970s, evidence of bipolar outflows has been detected in star-forming regions of the interstellar medium. In this dramatic phase of early stellar evolution, oppositely directed jets of high-speed gas are observed emanating from visible, pre-main-sequence T Tauri stars and, more frequently, from objects so young that their presence within their obscuring, parent clouds can be inferred only from infrared measurements. When these visible jets, which have speeds up to 400 km/s (250 mi/s), collide with the ambient interstellar gas, the violent heating and compression known as a shock wave results. Herbig-Haro objects are frequently the hot spots where the jets hit the surrounding material. Their luminous appearance is the result of excitation of the gas by the shock. *See* PROSTAR; T TAURI STAR.

However, not all Herbig-Haro objects are found at the terminal points of bipolar jets. Particularly when the source of the outflow is a much more luminous star than the Sun, Herbig-Haro objects can be scattered over a wide angular region rather than restricted to a well-defined outflow axis. Alternatively, several well-known Herbig-Haro objects with characteristic spectral emission lines are the brightest knots within individual jets, and probably reflect the presence of internal shocks. *See* STELLAR EVOLUTION. [A.I.S.]

Herbivory The consumption, without killing, of plants. Two broad types of herbivory can be recognized. The first is consumption by chewers, that is, animals with jaws that chew and make holes or gaps in leaves and stems, resulting in the removal of only parts of plants. The second is consumption by suckers, usually insects, which use a tubular tongue to extract liquid food from plant tissue. Herbivory usually involves a delicate balance in which the level of consumption is regulated so that only rarely is the plant severely damaged or killed. Occasionally, destructive outbreaks of caterpillars kill trees, but these tend to occur when a plant species is introduced to an area where it is not native and then cultivated as a monoculture. Also, many crop species have lost their natural antiherbivore defenses through breeding aimed at making them more palatable and acceptable as human food.

All parts of living plants are exploited by herbivores. Apart from the obviously mutualistic relationship between flowers and their pollinators, and between fruit and those fruit eaters that effect the dispersal of seeds, it is believed that herbivory is detrimental and that plants have evolved a variety of antiherbivore adaptations that restrict and regulate consumption. These adaptations include spines, prickles, hairs, tough and leathery leaf texture, and of particular importance, chemical compounds which play no part in normal growth and development of the plant but deter herbivores.

Grasses seem particularly well adapted to herbivory. They have a basal meristem which can survive all but the closest of grazing. Grasses and grazers first appeared in the Miocene, about 25 million years ago. Possibly, mutualism between these two groups is as intimate and inseparable as that between flowers and their pollinators. *See* ECOLOGICAL INTERACTIONS. [D.F.Ow.]

Hermaphroditism A condition in which components of both testes and ovaries are present in the same individual. Although true hermaphroditism is common among lower forms of animals such as annelids and mollusks, it is rare in humans. A more common condition in humans is pseudohermaphroditism, which simulates hermaphroditism. In female pseudohermaphroditism, or gynandry, the external sexual characteristics are in part or wholly of the male aspect, but internal female genitalia are present. In male pseudohermaphroditism, or androgyny, the individual has external sexual characteristics of female aspect, but has testes (usually undescended). *See* OVARY; REPRODUCTIVE SYSTEM DISORDERS; TESTIS. [S.P.P.]

Hernia An abnormal protrusion of a portion of the body through its containing wall. It is sufficiently common to have acquired a popular name, rupture.

Most hernias occur in areas of innate weakness in the body wall. Although such a weakness, and even an area of faulty closure of the wall, may have been present from birth, there is often no obvious bulge until late in life. After a major physical effort or a series of minor strains, various organs may suddenly protrude through the weakened area as a noticeable and often painful lump. Hernias in other instances occur through areas weakened by injury. Most frequently this is surgical incision, but violent stress or infection may also be responsible.

The most common type is the inguinal hernia. The weakness lies in the inguinal canal through which the vas deferens passes from the abdominal cavity to the scrotum. Umbilical hernia, through the defect left by the passage of umbilical cord structures during fetal life, is particularly common in infancy, and often closes off spontaneously. Femoral hernia, originating in the weakness produced by passage of the major arteries and veins from the abdomen into the legs, presents below the inguinal ligament and is common in women. In diaphragmatic hernia, where a large defect exists in the muscular diaphragm separating chest from abdomen, the abdominal organs rise into the chest cavity and may seriously impair breathing. Esophageal hiatus hernia is a small bulging of part of the stomach through the diaphragm around the esophagus. [R.N.B.]

Herpes A mild vesicular eruption of the skin or mucous membranes caused by a virus. The most familiar form is seen as cold sores or fever blisters.

Primary infections are chiefly inapparent; the commonest clinical form of the first attack is acute herpetic gingivostomatitis, a severe febrile disease with mouth eruptions. Recurrent attacks in persons with antibodies produce local lesions, often at mucocutaneous junctions. In some instances herpes simplex virus may infect the eye, producing acute keratitis which may permanently affect the cornea.

There are two types of herpes simplex virus, differing both antigenically and biologically: an oral type (type 1) and a genital type (type 2). Herpes simplex virus type 1 is probably more constantly present in humans than is any other virus. Primary infection occurs early; 70–90% of adults have antibodies. However, the virus may be carried throughout life. Transmission is by contamination from saliva or stools. Evidence indicates that type 2 is venereally transmitted. There is a high degree of association between prior infection with type 2 herpesvirus and the development of cancer of the uterine cervix. It is not yet known whether the relationship with cervical cancer is a causal one. *See* ANIMAL VIRUS; CYTOMEGALOVIRUS INFECTION. [J.L.Me.]

Herrings The common name for a family (Clupeidae) of about 70 genera of fishes in the order Clupeiformes. They are used extensively as food all over the world, and occur in all seas except the Arctic and Antarctic.

These fishes are the most primitive of the higher bony fishes. The fins have no supporting spines and are soft-rayed. There are usually four gill clefts, with the pectoral fins behind the gill openings. Scales are present on the body but absent on the head, and the swim bladder and lateral line may be missing.

The herring *Clupea harengus* has a circumpolar distribution. About eight other species of this genus are recognized, including the sprat or brisling (*C. sprattus*), which occurs in the Mediterranean and seas of western Europe, and the gizzard shad (*Dorosoma cepedianum*), which is a common species in the Potomac River. In Europe the herring is either salted, pickled, or smoked and cured as kippers. In Canada and the United States young herring are canned as "sardines."

The sardine, *Sardina pilchardus*, is a herring known commonly as the pilchard and is found along the European coasts in the Atlantic. The entire fish may be processed and preserved in oil, since the bones are soft and all parts are edible. Shads, herrings of the genus *Alosa*, occur in northern waters on both sides of the Atlantic. Anchovies comprise a family of herringlike fish, the Engraudidae, which together with the Clupeidae belong to the suborder Clupoidea. These fishes are found in the Mediterranean and range along the European coast as far north as Norway. *See* CLUPEIFORMES. [C.B.C.]

Hertzsprung-Russell diagram A graphical representation of the absolute magnitude and the surface temperature of stars. The ordinate is commonly the absolute visual magnitude. In the case of a cluster of stars, all of which may be assumed to be at the same distance, the apparent magnitude may be used in place of the absolute magnitude. For comparison with theory, the logarithm of the luminosity, in units of the Sun's luminosity, is used. The abscissa is the spectral type or the color index, both of which are related to the star's surface temperature. For comparison with theory, the logarithm of the effective temperature is used, the effective temperature being closely related to the surface temperature. When the absolute or apparent magnitude is plotted against a color index, the H–R diagram is also called a color-magnitude diagram. *See* MAGNITUDE (ASTRONOMY).

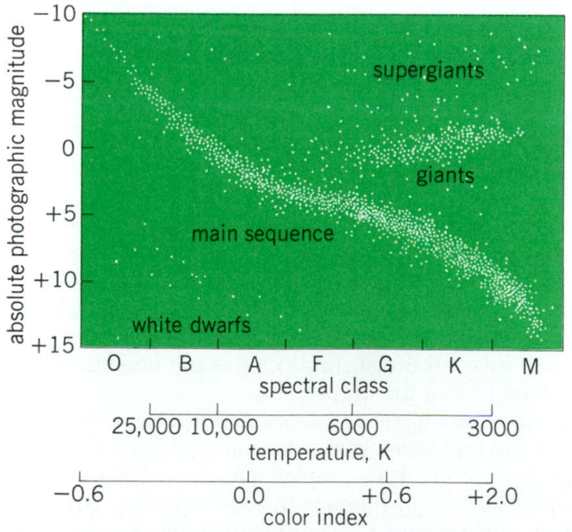

H–R diagram (schematic) representing the distribution of stars of known distance. (*After C. O. Abell, Exploration of the Universe, Holt, Rinehart and Winston, 1975*)

An H–R diagram of all stars with known absolute magnitudes and surface temperatures is shown in the illustration. By far the majority of the stars fall within a band running from the lower right to the upper left. These stars are called main-sequence stars. The region between the main sequence and the giants and supergiants, where few stars are found, is called the Hertzsprung gap.

The H–R diagram is useful for exhibiting the effects of stellar evolution. As a result of its evolution, a star's position in the H–R diagram changes with time. The time scales for noticeable changes in the position of a star in the H–R diagram generally are on the order of millions to billions of years. Because of stellar evolution, the H–R diagram is dynamic, rather than static, with stars moving into and out of each region. Those regions in which there are few stars, such as the Hertzsprung gap, are regions in which the stars are unstable and pass through relatively rapidly. Those regions where there are many stars, such as the main sequence, are those in which the stars stay for relatively long periods of time. *See* STAR; STELLAR EVOLUTION. [D.S.H.]

Hesperornithiformes An order of ancient extinct birds found from the Upper Cretaceous and frequently segregated in their own superorder, Odontognathae, on the basis of their possession of teeth. The hesperornithiforms were large, flightless, aquatic diving birds with the shoulder girdle and wings much reduced and the legs specialized for strong swimming. All aspects of the skeleton were typical of modern birds with the possible exception of teeth. The Hesperornithiformes almost certainly represent an evolutionary offshoot, not ancestral to any living birds; resemblances to loons and other modern diving birds are due to convergence. *See* AVES; ODONTOGNATHAE. [W.J.B.]

Heterochrony An evolutionary phenomenon that involves changes in the rate and timing of development. As animals and plants grow from their earliest embryonic stages to the adult, they undergo changes in shape and size. This life history of an individual organism is known as its ontogeny. The amount of growth that an organism experiences during its ontogeny can be more or less than its ancestor. This can apply to the organism as a whole or to specific parts.

Evolution can be viewed as a branching tree of modified ontogenies. Heterochrony that produces these changes in size and shape may be the link between genetics at one extreme and natural selection at the other.

If a character of one species in an evolutionary sequence undergoes less growth than its ancestor, the process is known as pedomorphosis. If it undergoes more growth, the process is known as peramorphosis. Each state can be achieved by varying the timing of onset, offset, or rate of development.

If development is stopped at an earlier growth stage in the descendant than in the ancestor (for example, by earlier onset of sexual maturity), ancestral juvenile features will be retained by the descendant adult (progenesis). If the onset of development of a particular structure is delayed in a descendant, the structure will develop less than in the ancestor (postdisplacement). The third process that produces pedomorphosis is neoteny, whereby the rate of growth is reduced.

For peramorphosis, development can start earlier in the descendant than in the ancestor (predisplacement); or the rate of development can be increased, thus increasing the allometric coefficient (acceleration); or development can be extended by a delay in the onset of sexual maturity (hypermorphosis).

As an organism grows, the number of cells that it produces increases. Ultimately, changes to rate and timing of growth are reflections of changes to the timing of onset and rate of cell development, and the balance between cell growth and cell death. Morphogens and growth hormones play a major role in controlling development in terms of initiation, rate of division, and migration. Therefore, changes to the timing of their expression affect the shape and size of the final adult structure. Inception of hormonal activity is under the control of genes that regulate the timing of its production. [K.J.McN.]

Heterocorallia A small, extinct, late Paleozoic order of fossil corals, known from Europe, North Africa, Asia, North America, and Australia, but limited to the Late Mississippian. They are found in calcareous shales and in limestones. Their very elongate skeletons, and the hooks on the outer septal edges of some, suggest a pseudoplanktonic existence attached to seaweeds. There are but rare indications of branching. *See* COELENTERATA. [D.H.]

Heterocyclic compounds Cyclic compounds in which the rings include at least one atom of an element different from the rest. Most types of heterocyclic compounds studied to date are organic compounds. An example of an organic heterocyclic compound is oxazoline (I); an example of an inorganic heterocyclic compound is the phosphonitrilic chloride (II). The smallest possible ring is three-membered, for example, ethylene oxide (III), but very large rings are possible, as in the crown ethers, for example, 18-crown-6 (IV). The cycle may contain only single bonds and is thus saturated; it may include one or more double bonds; or it may possess aromatic unsaturation characteristics of benzene, that is, it is heteroaromatic. Heterocyclic compounds can contain more than one ring, either heterocyclic or homocyclic.

Naturally occurring heterocyclic compounds are extremely common as, for example, most alkaloids, sugars, vitamins, DNA and RNA, enzymic cofactors, plant pigments, many of the components of coal tar, many natural pigments (such as indigo, chlorophyll, hemoglobin, and the anthocyanins), antibiotics (such as penicillin and streptomycin), and some of the essential amino acids (for example, tryptophan), and many of the peptides (such as oxytocin). Some of the most important naturally occurring high polymers are heterocyclic, including starch and cellulose. The major groups of natural products that are not mainly heterocyclic are the fats and most of the terpenes, steroids, and essential α-amino acids, though exceptions do exist.

Heterocyclic compounds may be named systematically. Many heterocycles, however, have nonsystematic names that are usually preferred by practicing chemists over the systematic ones. In the systematic approach to nomenclature the ring size is denoted by the appropriate stem. For example, three-membered saturated rings without nitrogen would have a name ending in -irane. The nature of the heteroatom is denoted by such prefixes as oxa-, thia-, or aza-, for oxygen, sulfur, or nitrogen, respectively. Thus, ethylene oxide (III) becomes oxirane. A five-membered unsaturated ring would have a name ending in -ole. A six-membered unsaturated ring containing nitrogen would have a name ending in -ine according to this scheme. Actually, the trivial names for many systems are commonly accepted, and the systematic names are not often used.

For details about specific heterocyclic systems *see* ACRIDINE; AZOLE; FURAN; HETEROCYCLIC POLYMER; IMIDAZOLE; INDOLE; ISOQUINOLINE; OXAZOLE; PYRAN; PYRAZOLE; PYRIDINE; PYRIMIDINE; PYRROLE; QUINOLINE; THIAZOLE; THIOPHENE. [R.A.A.]

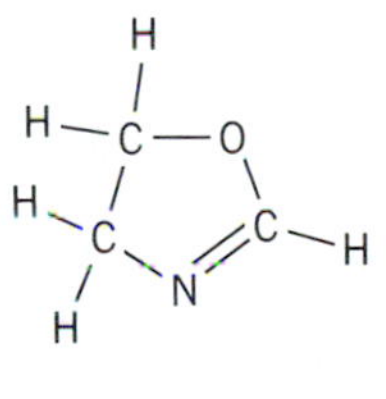

(I)

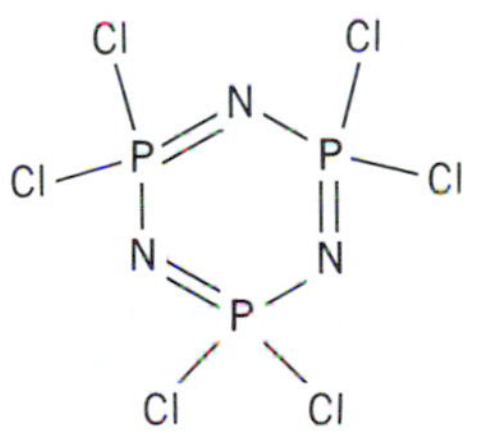

(II)

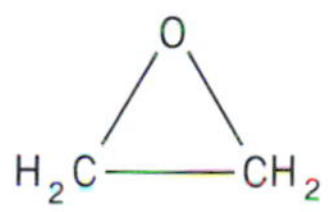

(III)

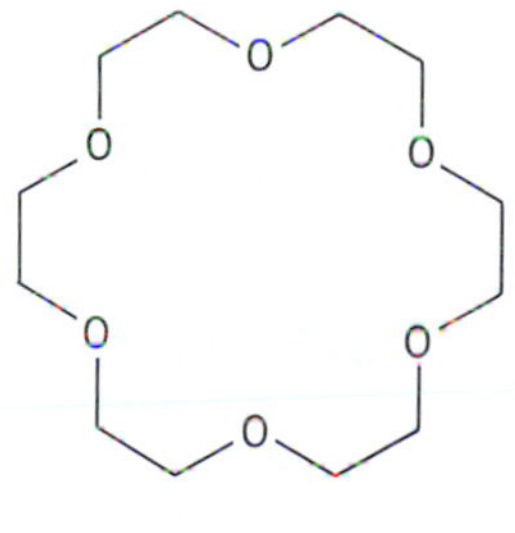

(IV)

Heterocyclic polymer Essentially, linear high polymers comprising heterocyclic rings, or groups of rings, linked together by one or more covalent bonds. As the search has continued for polymeric materials having useful properties at high temperatures (500°C or 930°F, or higher), much attention has been given to heterocyclic polymers. As a group such polymers are often both mechanically rigid and inherently resistant to thermal degradation. *See* POLYMER.

Some of these polymers form molecules in which the rings are fused together, as shown symbolically in the illustration (ladder polymers), and some form molecules in which fused rings are joined by single bonds (stepladder polymers). Similar considerations hold for simple aromatic systems, for example, linear polymers of benzene, but the heterocyclic systems have, in general, been more useful in application.

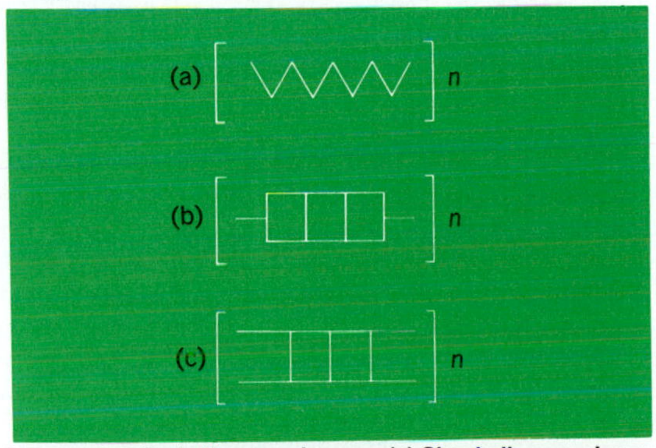

Structural units in linear polymers. (*a*) Simple linear polymer. (*b*) Stepladder polymer. (*c*) Ladder polymer.

Three heterocyclic polymers have been developed to the point of commercial availability; polyimides, polybenzimidazoles, and polybenzothiazoles. At least the first two appear to have established specialized markets.

Major applications for these rather expensive polymers are as metal-to-metal adhesives and as laminating resins for fibrous composites for structural applications in the aerospace industry. Other applications requiring both strength and resistance to oxidation at elevated temperatures have developed, including valve seats, bearings, and turbine blades. *See* POLYMERIC COMPOSITE. [J.A.M.]

Heterodyne principle The basic principle underlying the operation of a superheterodyne radio, television, or other receiver, wherein two alternating currents that differ in frequency are mixed in a nonlinear device to produce two new frequencies, corresponding to the sum and the difference of the input frequencies. Only the difference frequency is commonly used in a superheterodyne receiver, where it serves as the input to the intermediate-frequency amplifier. *See* RADIO RECEIVER; SUPERHETERODYNE RECEIVER. [J.Mar.]

Heterogeneous catalysis A chemical process in which the catalyst is present in a separate phase. In the usual case, the catalyst is a solid and the reactants and products are in gaseous or liquid phases. *See* CATALYSIS.

Heterogeneous catalysis proceeds by the formation and subsequent reaction of chemisorbed complexes which can be considered to be surface chemical compounds. A very simple case, where A → B is slow in the absence of catalyst, is shown in the following reaction:

$$\left.\begin{array}{l} {}^* + A \rightarrow {}^*A \\ A^* \rightarrow B^* \\ B^* \rightarrow B + {}^* \\ A \rightarrow B \end{array}\right\} \text{ chain propagating steps}$$

Reaction A → B is fast if the three preceding steps are fast. Here, * represents a catalytic site on the surface of the catalyst, $A^* \rightarrow B^*$ is called a surface reaction, $^* + A \rightarrow {}^*A$ represents the chemisorption of A, and $B^* \rightarrow B + {}^*$ represents desorption of B. *See* ADSORPTION.

With most sets of reactants, more than one reaction will be thermodynamically possible. The degree to which a given catalyst favors one reaction compared with other possible reactions is called the selectivity of the catalyst for the reaction. Two aspects of a catalyst are of particular importance: its selectivity, and its activity, which can be taken as the rate of conversion of reactants by a given amount of catalyst under specified conditions. Ideally, the rate will be proportional to the amount of catalyst.

The first important heterogeneous catalytic process to be used in the chemical industry was the manufacture of sulfuric acid from sulfur trioxide by the contact process in 1875. By the 1950s, heterogeneous catalytic processes had come to dominate the petroleum, petrochemical, and chemical industries. Today, about 70% of the crude oil refined in the United States is exposed to at least one heterogeneous catalytic process. Heterogeneous catalysis is a critical feature in energy conservation and interconversion, and is a key feature in the production of synthetic fuels from coal and oil shale. *See* COAL GASIFICATION; FISCHER-TROPSCH PROCESS.

Since catalytic activity will ordinarily be proportional to surface area, most catalysts are used in forms with large specific areas. Higher-area metal powders are often used for liquid-phase reactions in batch reactors. For example, finely divided nickel is used for the hydrogenation of unsaturated glycerides in the manufacture of margarine from vegetable oils. Supported catalysts are widely used. In these, the catalytic ingredient is dispersed in the internal porosity of such supports as silica gel, γ-alumina, and charcoals. These supports have large areas in the internal porosity, and their average pore diameters are 2–20 nm. Supported catalysts have the advantage that the area of the catalytic ingredient can be very large.

One important type of catalyst exposes strongly acidic sites in its internal porosity. Such catalysts are used to crack larger molecules of hydrocarbon into smaller ones in petroleum refining. Other catalysts, called dual-functional catalysts, have a hydrogenating catalytic ingredient on an acidic support. These are also of major importance in processing petroleum. *See* CATALYTIC REFORMING; CRACKING; HYDROCRACKING. [R.L.Bu.]

Heteronemertini An order of the class Anopla in the phylum Rhynchocoela, with an unarmed proboscis, a thick partly fibrous dermis, and a three-layered body musculature composed of outer longitudinal, median circular, and inner longitudinal strata. Cerebral organs, cephalic grooves and slits, and eyes are generally present. The alimentary system consists of a mouth, foregut, intestine with regular lateral diverticula but no cecum, and anus. Heteronemertini are mainly marine littoral in habit. *See* ANOPLA; RHYNCHOCOELA. [J.B.J.]

Heterophile antigen The serologic reactions of the tissue and blood-cell antigens of most animals are normally characteristic of the species. Significant serologic cross reactions usually occur only with antisera to the corresponding antigens of closely related species. The numerous groups of heterophile antigens—of which the Forssman antigens are the best studied—constitute significant exceptions. Heterophile antigens link the species hog-ox-human (blood group A), cat-horse, and dog-hog-cat-human, while several heterophile groups link otherwise diverse microorganisms. Similarities between antigens in mammalian hearts and the cell walls of the group A hemolytic streptococcus are also known. The cross reactions between the *Proteus* bacillus and the *Rickettsiae* are important in the diagnosis of typhus fever. *See* ANTIBODY; ANTIGEN; SEROLOGY. [M.J.Po.]

Heterophyiasis Presence, in the human small intestine, of the minute intestinal fluke *Heterophyes heterophyes*. The Nile Delta and Orient are endemic foci. Many reservoir hosts exist. Eggs in feces are ingested by snails, and cercariae finally emerge from these to encyst in the flesh of certain fish. Ingestion of the raw fish flesh by humans completes the cycle.

Large numbers of worms in humans damage the intestinal mucosa and cause diarrhea, colicky pains, and abdominal distress. Eggs may enter the lymph or blood to settle in the heart or brain, with serious complications. *See* MEDICAL PARASITOLOGY. [J.F.M.]

Heterosis Hybrid vigor or increase in size, yield, and performance found in hybrids, especially if the parents have previously been inbred. The application of heterosis has been one of the most important contributions of genetics to scientific agriculture in providing hybrid corn, and vigorous, high-yielding hybrids in other plants and in livestock. *See* BREEDING (ANIMAL); BREEDING (PLANT); GENETICS; MENDELISM.

There are two principal hypotheses to account for the association of size and vigor with heterozygosity, dominance and overdominance. The dominance hypothesis notes that any noninbred population carries a number of recessive genes that are harmful to a greater or lesser extent, but which are rendered ineffective by their dominant alleles. As they become homozygous through inbreeding, they exert their harmful effect. With hybridization, some of the detrimental recessives contributed to the hybrid by one parent are masked by dominant alleles from the other, and an increase in vigor is the result. The alternative hypothesis is that there are loci at which the heterozygote is superior in vigor to either homozygote. This, the overdominance hypothesis, also has the consequence that vigor is proportional to heterozygosity. The dominance hypothesis has been more widely accepted, but the two are very difficult to distinguish experimentally, and it is likely that overdominant loci are playing an appreciable role in heterosis, particularly in determining why one hybrid is better than another. *See* DOMINANCE. [J.F.Cr.]

Heterostraci An extinct group of ostracoderms or armored, jawless vertebrates (Agnatha). The armor has a distinctive microscopic structure, consisting of bone lacking any cavities for bone cells, surmounted by tubercles or ridges of dentine. Fragments of such armor from the Middle Ordovician of Australia and North America are the earliest remains definitely attributable to vertebrates. Heterostraci became more common toward the end of the Silurian and persisted through the Devonian. *See* AGNATHA.

The anterior part of the body was covered with plates. The posterior part of the body and the tail were covered with thick scales. There were no jaws, but the mouth was bounded behind by a number of small plates that may have been used for nibbling. [R.H.De.]

Heterotardigrada An order of the tardigrades; the majority of genera have widely varied structure. Cephalic appendages having a sensorial function are present, as well as cirrus lateralis and clava. This order of tardigrades is divided into two suborders, Arthrotardigrada and Echiniscoidea.

Members of the suborder Arthrotardigrada have toelike terminations of the legs. The tubular middle part of the leg telescopes into the broad proximal part. These animals are marine organisms found in sand or on algae.

In the suborder Echiniscoidea the legs terminate with claws. The middle part of the leg is partially retractable into the proximal part. At least the fourth pair of legs has a distinct fold. Frequently these animals are red because of the presence of carotenoid pigments. *See* CAROTENOID; TARDIGRADA. [Ev.M.]

Heterotrichida A large order of the Spirotrichia. The buccal ciliature is well developed, although in a number of families the somatic, or body, ciliature is really holotrichous in nature. Heterotrichs have become adapted to all sorts of habitats, including the digestive tracts of a variety of invertebrate and a few vertebrate hosts. Because of their size and amazing regenerative powers, a number of heterotrichs have been widely used in experimental research. Common genera include *Stentor*, *Blepharisma*, *Folliculina*, *Condylostoma*, *Climacostomum* (see illustration), and *Spirostomum*. *See* SPIROTRICHIA. [J.O.C.]

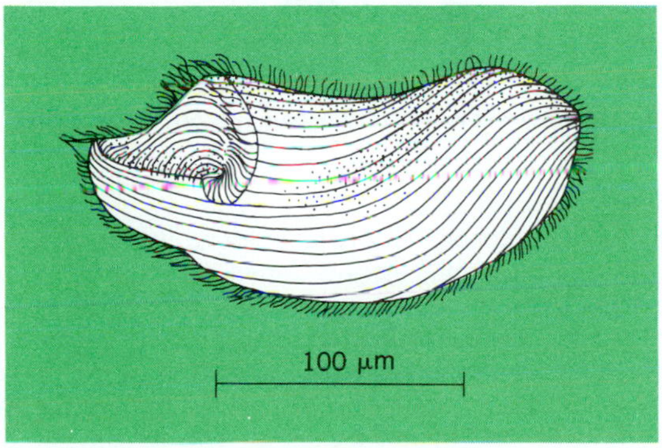

***Climacostomum*, an example of a heterotrich.**

Heulandite A mineral belonging to the zeolite family of silicates. It usually occurs in crystals with the prominent side being pinacoid, often having a diamond shape. There is perfect side pinacoid cleavage on which the luster is pearly; elsewhere the luster is vitreous. The hardness is 3½ to 4 on Mohs scale; specific gravity is 2.18–2.20. The mineral is usually white or colorless but may be yellow or red. *See* ZEOLITE.

Heulandite is essentially a hydrous calcium aluminum silicate, $Ca(Al_2Si_7O_{18})\cdot 6H_2O$. Heulandite is a secondary mineral found in cavities in basalts associated with other zeolites and calcite. Notable localities are in the Faeroe Islands, India, Nova Scotia, and West Paterson, New Jersey. [C.Fr.; C.S.Hu.]

Hexactinellida A class of the phylum Porifera which includes sponges with a skeleton made up basically of hexactinal siliceous spicules. Spongin is absent. Hexactinellida usually are radially symmetrical in form and have cylindrical, vase-, funnel-, or trumpet-shaped bodies. Some species are fanshaped and others exhibit a dendritic or labyrinthine mode of branching. Some species attach to a hard substratum directly by a basal attachment region, but many have basal tufts or mats of spicules which serve to anchor them in bottom sediments. The Hexactinellida are exclusively marine and occur in the deeper waters of all seas. Their bathymetric range extends from depths of about 80 ft (25 m) down to 28,000 ft (8500 m), but very few descend into the hadal zone, the faunal zone of the deep trenches.

The body wall has a cavernous structure penetrated by intercommunicating cavities across which stretch the soft parts in the form of filmlike trabeculae. Thimble-shaped flagellated chambers, arranged side by side in a single layer, are supported in the midst of the trabeculae. Five regions may be recognized in the body wall. The outer dermal membrane is often filmlike and is perforated by more or less round pores which serve for the entrance of water into the intertrabecular lacunae. The subdermal trabecular network, which is continuous with the dermal membrane, is dense peripherally, but the lacunae become more spacious toward the interior and form subdermal cavities. These are in contact with the outermost flagellated chambers. Fairly distinct inhalant canals may lead to deeper-lying flagellated chambers in species with thick walls and a folded chamber layer. The flagellated chambers are composed of choanocytes whose bases are continuous and form a reticular membrane perforated with holes which allow passage of water from the subdermal trabecular lacunae into the flagellated chambers. Each chamber opens widely through an apopyle into the subspongocoelic trabecular system, which in turn opens into the central spongocoel by way of a perforated membrane. Fairly distinct exhalant channels are often present in the trabecular system, spreading between the chamber layer and the spongocoel. The spongocoel opens through a large terminal osculum, covered over in some genera by a sieve plate, the meshwork of which is reinforced by spicules.

The skeleton is made up of megascleres and microscleres. Although many forms of spicules occur, most of them can apparently be derived from the triaxon or hexactin, in which three axes meet at right angles. The megascleres are also classified, according to their position in the body of the sponge, as dermal, parenchymal, and gastral.

Other animals often live in association with Hexactinellida. Sea anemones and hydroids live on some sponges of this class, and isopods and brittle stars are known to inhabit the cavernous interior regions.

Hexactinellida are well known as fossils because their rigid skeletons, often composed of a lattice of fused spicules, increase the chances of their being preserved intact. Sponges of this class are reported from the Cambrian Period upward. There is a notable peak of abundance of fossil remains in the Cretaceous Period. *See* PORIFERA. [W.D.H.]

Hexactinosa An order of the subclass Hexasterophora in the class Hexactinellida. The parenchymal megascleres in this order of sponges are united to form a rigid framework and consist wholly of simple hexactins which are arranged in parallel linear series. The members of each series are united one to another by a secondary envelope of silica. *See* HEXACTINELLIDA; HEXASTEROPHORA. [W.D.H.]

Hexagon A closed figure formed by the six line segments, or sides, that join in order six ordered points, or vertices, of a plane. In elementary geometry it is usually assumed that the sides of a hexagon do not cross and even that the figure bounds a convex region of the plane. The ancient Greek geometers investigated regular hexagons, that is, hexagons with all sides equal and with all angles included by adjacent sides equal to 120°. They found that a regular hexagon is obtained by applying the radius of a circle six times as a chord to the circumference, and that three regular hexagons completely fill up the plane about a point. *See* POLYGON. [L.M.Bl.]

Hexasterophora A subclass of sponges of the class Hexactinellida, in which the parenchymal microscleres are typically hexasters (small, six-rayed spicules, often with branched ends). This is a diverse assemblage of sponges commonly firmly fixed to the substratum by the base, less commonly anchored by means of basal spicule tufts or mats. The spicules of the body are sometimes free and unconnected, but the parenchymal megascleres are often fused to form a rigidly connected skeleton. The following orders are recognized: Hexactinosa, Lychniscosa, Lyssacinosa, and Reticulosa. *See* HEXACTINELLIDA; HEXACTINOSA; LYCHNISCOSA; LYSSACINOSA; PORIFERA; RETICULOSA. [W.D.H.]

Hibernation A general term applied to a condition of dormancy and torpor found in cold-blooded (poikilotherm) vertebrates and invertebrates. This rather universal phenomenon can be readily seen when body temperatures of poikilotherm animals drop in a parallel relation to ambient environmental temperatures. In the strict sense of the word, hibernation is a term which physiologists apply to relatively few species of mammals and birds.

Cold-blooded animals. Hibernation occurs with exposure to low temperatures and, under normal conditions, occurs principally during winter seasons when there are lengthy periods of low environmental temperatures. A related form of dormancy is estivation. Many animals estivate when exposed to prolonged periods of drought or during hot, dry summers. For all practical purposes, hibernation and estivation in animals are indistinguishable, except for the nature of the stimulus, which is either a cold or an arid environment.

There is no complete list of animals that hibernate; however, many examples can be found among the poikilotherms, both vertebrate and invertebrate. The poikilotherms are sometimes referred to as ectothermic, because their body temperatures are not internally regulated but follow the rise and fall of environmental temperatures. During hibernation and winter torpor, body temperatures reflect the environmental temperature, often to within a fraction of a degree. Among the classic examples of hibernators or estivators are reptiles, amphibians, and fishes among the vertebrates, and insects, mollusks, and many other invertebrates. [X.J.M.]

Warm-blooded vertebrates. Many mammals and some birds spend at least part of the winter in hiding, but remain no more drowsy than in normal sleep. On the other hand, some mammals undergo a profound decrease in metabolic rate and physiological function during the winter, with a body temperature near 0°C. This condition, sometimes known as deep hibernation, is the only state in which the warm-blooded vertebrate, with its complex mechanisms for temperature control, abandons its warm-blooded state and chills to the temperature of the environment. Between the drowsy condition and deep hibernation are gradations about which little is known. The bear, skunk, raccoon, and badger are animals which become drowsy in winter. The deep hibernators are confined to five orders of mammals: the marsupials, the Chiroptera or bats, the insectivores, the rodents, and, probably, the primates.

During a preparation period for hibernation, the animals either become fat, like the woodchuck, or store food in their winter quarters, like the chipmunk and hamster. Prior to hibernation, there is a general involution of the endocrine glands, but at least part of this occurs soon after the breeding season and is not directly concerned with hibernation. Animals such as ground squirrels become more torpid during the fall, even when kept in a warm environment, indicating a profound metabolic change which may be controlled by the endocrine glands.

The hibernator is capable of waking at any time, using self-generated heat, and this characteristic clearly separates the hibernating state from any condition of induced hypothermia. During the total period of hibernation, the hibernator spontaneously wakes from time to time, usually at least once a week. In the period of wakefulness the stored food is evidently eaten, but animals which do not store food rely on their fat for the extra energy during the whole winter. *See* HYPOTHERMIA. [C.P.L.]

Hickory Any species of the genus *Carya*, formerly known botanically as *Hicoria*. Hickories are mostly tall forest trees characterized by strong, terminal, scaly winter buds, pinnately compound leaves (see illustration), solid pith (not chambered), and fruit with an outer husk or exocarp which splits more or less readily into four parts, revealing a nut with a hard shell or endocarp.

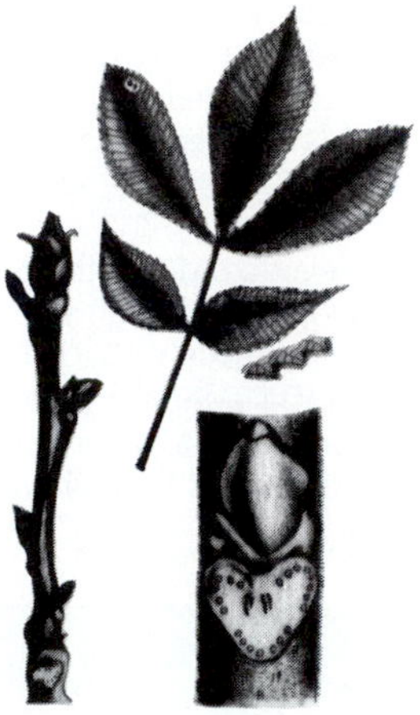

Twigs, buds, and leaves of shagbark hickory (*Carya ovata*).

The shagbark hickory (*C. ovata*) is found in the eastern half of the United States and adjacent Canada. It is the most important species because of the commercial value of its nuts, the hickory nuts of commerce, and of its wood. The pecan (*C. illinoensis*) is also a valuable species because of its commercially popular, thin-shelled, sweet nuts. Other species are the mockernut, shellbark, and pignut hickories. The remarkably tough and strong wood of all species makes it the world's best wood for tool handles. It is also used for parts of furniture, flooring, boxes, and crates, and for smoking meats. *See* FAGALES. [A.H.G./K.P.D.]

Hidden variables Additional variables or parameters that would supplement quantum mechanics. They would make it possible to unambiguously predict the result of a specific measurement on a single microscopic system. In contrast, pure quantum mechanics can only give probabilities for the various possible results of that measurement. Hidden variables would thus provide deeper insights into the quantum-mechanical probabilities. In this sense the relationship between quantum mechanics and hidden variables is analogous to the relationship between thermodynamics (for example, temperature) and statistical mechanics (the motions of the individual molecules). *See* STATISTICAL MECHANICS; THERMODYNAMIC PRINCIPLES.

In 1932 J. von Neumann provided an axiomatic basis for the mathematical methods of quantum mechanics. As a sidelight to this work, he rigorously proved from the axioms that any hidden variable theory was inconsistent with quantum mechanics. In 1966 J. S. Bell pinpointed the difficulty with von Neumann's proof—one of his axioms was fine for a pure quantum theory which makes statistical predictions, but the axiom was inherently incompatible with any hidden variable theory.

Hidden variable theories of the first kind are constructed so as to be self-consistent and to reproduce all the statistical predictions of quantum mechanics when the hidden variables are in an "equilibrium" distribution. Hidden variable theories of the second kind predict deviations from the statistical predictions of quantum mechanics, even for the "equilibrium" situations for which theories of the first kind agree with quantum mechanics. They are generally called local hidden variable theories because they are required to satisfy a locality condition. Intuitively, this seems to be a very natural condition. It says that an apparatus at one location should operate independently of any settings or actions of a second apparatus at a spatially separated location. In the strict Einstein sense of locality, the two apparatus must be independent during a time interval equal to the time required for a light signal to travel from one apparatus to the other.

Progress in local hidden variable theories has been made by examining correlations between measurements on two parts of a single system when the parts have become spatially separated. This is, in fact, the basis of a famous "thought" experiment (a hypothetical, idealized experiment in which the experimental results are deduced) introduced in 1935 by A. Einstein, B. Podolosky, and N. Rosen. They used this thought experiment to argue that quantum mechanics was not a complete theory. Although they did not refer to hidden variables as such, these would presumably provide the desired completeness.

New efforts were stimulated in 1952 when D. Bohm did the "impossible" by designing a hidden variable theory of the first kind. Bohm's theory was explicitly nonlocal, and this fact led Bell to reexamine the Einstein-Podolsky-Rosen experiment. He came to the remarkable conclusion that any hidden variable theory that satisfies the condition of locality cannot possibly reproduce all the statistical predictions of quantum mechanics. Bell's result can be put in the form of inequalities which must be satisfied by any local hidden variable theory but which may be violated by the statistical predictions of quantum mechanics under appropriate experimental conditions.

Due to the supplementary assumptions, small loopholes still remain, and experiments have been proposed to eliminate them. Nevertheless, the overwhelming evidence presently provided by experiments is against any theory that would supplement quantum mechanics with hidden variables and still retain the locality condition. Pure quantum theory has triumphed, and the remarkable Einstein-Podolsky-Rosen correlations have defied any reasonable classical kind of explanation. *See* QUANTUM MECHANICS. [E.S.F.]

Higgs boson An elementary scalar particle in the Glashow-Weinberg-Salam theory of electromagnetic and weak interactions. At present, there is no direct experimental evidence for its existence. It is closely associated with the origin of mass for all known elementary particles, as described in the Glashow-Weinberg-Salam theory. This gives the Higgs boson distinctive properties which shape the search for it. *See* ELECTROWEAK INTERACTION; WEINBERG-SALAM MODEL.

The mass of the Higgs boson is uncertain theoretically, being determined by the parameters of the scalar self-interactions. It is strongly suspected, however, that it is not possible for the Higgs mass to be far above 1 TeV/c^2 (where c is the speed of light).

The fact that the Higgs particle is closely related to the origin of mass endows it with special properties crucial in its production and detection. One consequence is that the heavy Higgs resonance is quite broad, with width approximately half the mass at 1 TeV/c^2, for example.

The first extensive searches for the Higgs boson probably will be carried out at the Large Electron Positron Storage Ring (LEP) and Stanford Linear Collider (SLC), e^+e^- machines with sufficient center-of-mass energy to operate at the Z resonance. If the expected number of Z's is produced, it should be possible to find the Higgs if its mass is below approximately 40 GeV/c^2. If LEP is upgraded to operate at a center-of-mass energy of approximately 200 GeV/c^2, it should be possible to produce the Higgs with mass up to approximately 80 GeV/c^2.

The next particle accelerators to extend the search will probably be hadron colliders (proton-proton or proton-antiproton). The proposed superconducting supercollider (SSC) should be able to search for the Higgs boson with mass greater than about twice the Z mass up to approximately 1 TeV/c^2. *See* ELEMENTARY PARTICLE; PARTICLE ACCELERATOR. [A.Ch.; M.So.]

High fidelity A term (often shortened to hi-fi) applied to sound reproduction, used to designate a state of performance in which realism is achieved. To achieve realism in sound reproduction, four fundamental conditions must be satisfied: (1) The frequency range must include without frequency discrimination all the audible components of the various sounds to be reproduced; (2) the volume range must permit noiseless and distortionless reproduction of the entire range of intensity associated with the sounds; (3) the reverberation characteristics of the original sound must be approximated in the reproduced sound; and (4) the spatial sound pattern of the original sound must be preserved in the reproduced sound. *See* DISK RECORDING; MAGNETIC RECORDING; QUADRAPHONIC SOUND SYSTEM; REVERBERATION; SOUND-REPRODUCING SYSTEMS; STEREOPHONIC SOUND. [H.F.O.]

High-power, short-pulse laser A pulsed laser whose output power has been greatly increased, with correspondingly shorter pulse durations, by the Q-switch technique. In this method, the optical path between one mirror and the amplification medium is blocked by a shutter. The medium is then excited beyond the degree ordinarily needed, but the shutter prevents laser action. At this time the shutter is abruptly opened and the stored energy is released in a giant pulse (1–100 MW peak power, lasting 1–30 nanoseconds for optically pumped solid-state lasers). Still higher peak powers can be obtained by passing this output through a traveling-wave laser amplifier (without mirrors). Peak powers in excess of 100 MW have been obtained in this way.

Still shorter, and higher-power, pulses can be generated by mode-locking techniques. A typical laser without mode locking usually oscillates simultaneously and independently at several closely spaced wavelengths. These modes of oscillation can be synchronized so that the peaks of their waves occur simultaneously at some instant. The result is a very short, intense pulse which quickly ends as the waves of different frequency get out of step. Mode-locked lasers have generated pulses shorter than 1 picosecond. *See* OPTICAL PULSES.

For the highest peak power, the output may be further intensified by additional stages of laser amplification. The beam diameter is increased by some optical arrangement, such as a telescope, so as to expand (dilute) the beam and thereby prevent damage to the laser material and optics.

Development of very large multistage lasers has been undertaken for research on thermonuclear fusion. *See* LASER; NUCLEAR FUSION; PULSED GAS LASER. [S.F.J.; A.L.S.]

High-pressure chemistry Chemistry at very high pressures, arbitrarily chosen to be above 10^4 bars (1 gigapascal), and mainly concerned with solid and liquid states. At 25°C (77°F) and 10^4 bars (1 GPa), nearly all ordinary gases are liquid or solid, and only a few liquids are not frozen; thus most high-pressure chemistry involves either higher temperatures, at which chemical reactions can occur at appreciable rates, or studies of internal arrangements in solids.

From 1 bar (10^2 kilopascals) to about 10^5 bars (10 GPa), normal low-pressure chemical behavior prevails, and only minor departures from the usual valence and coordination rules are found. However, many interesting changes in materials can be effected in this pressure range as atoms are forced into new bonding arrangements. From 10^5 to 10^9 bars (10 to 10^5 GPa), the energy added by compression becomes comparable with chemical bond energies, so that outer-shell electronic orbits are distorted and atoms and molecules change in character. A general tendency toward more metallic behavior is observed as the electrons become less strongly fixed to particular atoms, and chemical bonds may be broken. Upward of about 10^9 bars (10^5 GPa), the delocalization of electrons is extensive, and the material consists of a mixture of ions and electrons, so that chemical bonds are of little importance. The boundaries on these three pressure ranges are, of course, only approximate, and show some variation according to the temperature and the atoms involved.

The simplest effect of high pressure is the closer compression of atoms. The noble gases and alkali metals are quite compressible, whereas most oxides and the stronger metals are considerably stiffer. However, at a pressure exceeding about 10^5 bars (10 GPa), most of the easily compressed electronic clouds are tightened up, and the compressibilities of most substances approach each other.

Substances which consist of large molecules are easily stiffened or frozen by high pressures. The mobility of the molecules is sharply decreased by a sort of interlocking and tangling effect; thus for the substance to be sheared, chemical bonds must be broken, a process which requires considerable energy. This stiffening phenomenon limits the study of most reactions of organic molecules to low pressures because they are rather large and "freeze" easily, but yet are usually not stable enough to withstand the temperatures necessary for liquefaction or intermolecular reactions. [R.H.W.]

High-pressure physics High-pressure physics is concerned with the effects of high pressure on the properties of matter. Since most properties of matter are modified by pressure, the field of high-pressure physics encompasses virtually all branches of physics.

The "high" of high-pressure physics connotes experimental difficulty. At liquid-helium temperatures, pressures of several hundred bars are considered high. In general, however, the high-pressure range may be arbitrarily regarded as extending from about 10^3 bars upward to the present experimental limit. Prolonged static high pressures of as much as 5×10^5 bars (50 gigapascals) at room temperature have been achieved in certain types of apparatus. Transient pressures as high as about 10^7 bars (10^3 GPa) have been attained in shock waves produced by high explosives or by projectile impact.

The major effects of high pressure on matter include diminution of volume, phase transitions, changes in electrical, optical, magnetic, and chemical properties, increases in viscosity of liquids, and increases in the strength of most solids. In general solids are less compressible than liquids, and the compressibility of both solids and liquids decreases with increasing pressure. However, solids and liquids show wide individual variations in compressibility.

At high pressure many solids exhibit polymorphic phase changes, that is, a rearrangement of the atoms or molecules in the solid. There are no universally applicable rules governing the number of phase changes or the kind of phase change to be expected at high pressure, but there is a thermodynamic requirement that the phase that is stable at high pressure must have a smaller volume than the phase that is stable at low pressure. *See* THERMODYNAMIC PRINCIPLES.

Frequently, dramatic changes in physical properties result from phase changes. Ferromagnetic iron transforms to a paramagnetic form at pressures somewhat above 10^5 (10 GPa). In the same pressure range, the semiconducting element germanium transforms into a metallic phase that has an electrical conductivity greater than a million times that of the semiconductor.

Many phases that form at high pressure transform back to low-pressure phases as the pressure is released. However, some high-pressure phases may be retained in a metastable condition at low pressures, and some low-pressure phases can persist metastably at high pressure. Diamond, the high-pressure form of carbon, is thermodynamically unstable at room temperature and pressures below about 1.2×10^4 bars (1.2 GPa). Nonetheless diamond persists indefinitely as a metastable phase at low temperatures; it transforms to the stable form, graphite, only when heated to temperature in excess of 1000°C (1800°F) at low pressure. [R.K.L.; P.S.DeC.]

High-temperature chemistry The study of chemical phenomena occurring above 500 K (227°C or 440°F). High temperatures represent one of the important variables available to scientists for increasing the variety of possible chemical reactions over that expected for classical ground-state atoms and molecules. The relative population of excited rotational, vibrational, and electronic states can be enhanced by increasing the temperature and thus can effectively create new species and new mechanisms for reaction. The potentialities of this approach are well illustrated by the three laws of high-temperature chemistry: (1) At high temperatures everything reacts with everything. (2) The higher the temperature, the faster the reaction. (3) The products may be anything.

High temperatures also provide a common tie among the various options for energy production, conversion, or storage. For maximum thermodynamic efficiency, an energy production cycle should operate with a working fluid at as high a temperature as possible, and exhaust the spent fluid at as low a temperature as possible. Thus, in the combustion of coal to produce electric power or in the combustion of gasoline or diesel fuel to propel a car or an airplane, there is a need for materials of construction which allow operation of such devices at higher temperatures. *See* HIGH-TEMPERATURE MATERIALS.

It is convenient to discuss temperatures in terms of energy and to note that 11,500 K (20,200°F) corresponds to 1 electronvolt. In this sense, the particles emitted by radioactive nuclei or accelerated in cyclotrons and synchrotrons, which have energies in the keV, MeV, and BeV ranges, are effectively at temperatures of $\sim 10^7$ K, $\sim 10^{10}$ K, and $\sim 10^{13}$ K, respectively, and "high-energy physics" is synonymous with "ultrahigh-temperature chemistry."

Traditional high-temperature chemistry in the last several decades has been mainly concerned with phenomena in the range of 500–3000 K, although exotic flames can produce temperatures up to ~6000 K, shock waves can generate temperatures up to ~25,000 K, electric arcs can be operated in constricted modes to produce temperatures of ~50,000 K, and nuclear processes begin to occur at temperatures in the millions-of-degrees range. Laser excitation of selected energy states can produce species with effective temperatures in the range of 10^8 K. [J.L.M.]

High-temperature electronics The technology of electronic components capable of operating at high temperatures (above 572°F or 300°C). The need for electronic devices for geothermal well probes, planetary space probes, jet-engine controls, and nuclear power plant instruments is supplying the impetus for advancements in this area, since conventional silicon diodes, transistors, and integrated circuits will not function in this temperature range. The research has been carried out in three primary areas: new silicon devices, compound semiconductor devices, and integrated thermionic circuits. Some of these devices have the potential of extending the operational range of electronic circuits up to 1472°F (800°C).

Semiconductor devices can be divided into two categories: minority carrier devices, such as diodes and bipolar transistors, and majority carrier devices, such as the field-effect transistor (FET). Minority carrier devices made from silicon do not work well above 482°F (250°C). This is because the energy bandgap in silicon is only 1.1 electronvolt, and at temperatures above 250°C a large number of minority carriers are generated by thermal excitation.

Majority carrier devices in silicon (such as FETs) do not depend on minority carriers for their operation and therefore can be made to operate at higher temperatures. Enhancement-mode metal-oxide semiconductor FET (MOSFET) devices have been made to operate at 662°F (350°C). *See* JUNCTION DIODE; TRANSISTOR.

The basic need for a simple high-temperature rectifier diode and moderate power devices with gain have led to high-temperature research on compound semiconductors. Many of the group III–V compound semiconductors have energy bandgaps larger than silicon. For example, the semiconductor bandgap in gallium phosphide (GaP) is 2.2 eV. The thermally generated minority carrier density in GaP at 752°F (400°C) is small enough that it is possible to build minority carrier devices at high temperatures by using this and other wide-bandgap semiconductor materials.

A dramatic departure from semiconductor technology is the integrated thermionic circuit (ITC). This technology combines photolithographically defined subminiature thin-film metal patterns with planar vacuum-tube technology. The result is a technology that allows fabrication of active circuits with a density approaching that of present room-temperature silicon integrated circuits. [R.J.Ch.]

High-temperature materials A metal or alloy which serves above about 1000°F (540°C). More specifically, the materials which operate at such temperatures consist principally of some stainless steels, superalloys, refractory metals, and certain ceramic materials. The giant class of alloys called steels usually see service below 1000°F. The most demanding applications for high-temperature materials are found in aircraft jet engines, industrial gas turbines, and nuclear reactors. However, many furnaces, ductings, and electronic and lighting devices operate at such high temperatures.

In order to perform successfully and economically at high temperatures, a material must have at least two essential characteristics: it must be strong, since increasing temperature tends to reduce strength, and it must have resistance to its environment, since oxidation and corrosion attack also increase with temperature. *See* CORROSION; HIGH-TEMPERATURE CHEMISTRY.

High-temperature materials, always vital, have acquired an even greater importance because of developing crises in providing society with sufficient energy. The machinery which produces electricity or some other form of power from a heat source operates according to the basic Carnot cycle law, where the efficiency of the device depends on the difference between its highest operating temperature and its lowest temperature. Thus, the greater this difference, the more efficient is the device—a result giving great impetus to create materials that operate at very high temperatures. *See* CARNOT CYCLE; EFFICIENCY. [C.T.S.]

Highway engineering Highway planning, location, design, and maintenance. Before the design and construction of a new highway or highway improvement can be undertaken, there must be general planning and consideration of financing. As part of general planning it is decided what the traffic needs of the area will be for a considerable period, generally 20 years, and what construction will meet those needs. To assess traffic needs, the highway engineer collects and analyzes information about the physical features of existing facilities, the volume, distribution, and character of present traffic, and the changes to be expected in these factors. The highway engineer must determine the most suitable location, layout, and capacity of the new routes and structures. The detailed design is normally begun only when the preferred location has been chosen.

In selecting the best route careful consideration is given to the traffic requirements, terrain to be traversed, value of land needed for the right-of-way, and estimated cost of construction for the various plans. The photogrammetric method, which makes use of aerial photographs, is used extensively to indicate the character of the terrain on large projects, where it is most economical. On small projects ground-mapping methods are preferred. *See* PHOTOGRAMMETRY; SURVEYING; TOPOGRAPHIC SURVEYING AND MAPPING.

Detailed design of a highway project includes preparation of drawings or blueprints to be used for construction. These plans show, for example, the location, the dimensions of such elements as roadway width, the final profile for the road, the location and type of drainage facilities, and the quantities of work involved, including earthwork and surfacing. Soil studies are made to analyze gradation and physical properties of the soil. Selection of the type and thickness of roadway surfacing depends upon the loads to be accommodated. Designs for drainage structures, small bridges, and sometimes tunnels must be included in the overall planning. Other work involves planning rest stops on major expressway routes and designing barriers to control and reduce noise along busy routes. *See* PAVEMENT; TUNNEL.

Highway maintenance and operation are important aspects of highway engineering. Involved are the repair and upkeep of surfacing and shoulders, bridges and drainage facilities, signs, traffic control devices, guard rails, traffic striping on the pavement, retaining walls, and side slopes. Additional operations include ice control and snow removal. [A.N.C.]

Hilbert space Hilbert space is an abstract notion of great power and beauty which has been central to the development of mathematical analysis and forms the backdrop for many applications of analysis to science and engineering. Its essence lies in the fact that the objects of primary interest in analysis (namely, functions) enjoy geometrical properties which are in important ways analogous to the geometry of physical space. Thus the highly developed human visual and spatial intuition can lead to significant truths about functions. *See* EUCLIDEAN GEOMETRY.

Generalizations of euclidean space. Vectors (or directed line segments) in euclidean space have a rich structure. If certain of the desirable properties of euclidean space are isolated and adopted as postulates, a class of spaces is defined that may include spaces of functions that are of concern in analysis. During the early decades of the twentieth century it was gradually realized that it is possible to select a small number of properties in such a way that the resulting spaces possess virtually all the desirable features of euclidean space except those which are closely linked to finite dimensionality. These spaces are named after David Hilbert, who took a decisive step toward their introduction in 1906 when he proved the spectral theorem.

Role of Hilbert space. Hilbert's innovation occurred while he was investigating integral equations which arose from mathematical physics. These equations are continuous analogs of the systems of simultaneous linear equations which are encountered in elementary algebra, but the unknown entity, instead of being a finite set of numbers, is a function. Just as graphical methods give insight into the solution of a pair of simultaneous linear equations in two unkowns, so the development of Hilbert-space geometry had great consequences for the understanding of integral equations. Hilbert space is a truly fundamental mathematical structure which appears in widely disparate branches of pure and applied mathematics. A striking instance is quantum mechanics, where observable quantities are modeled by linear transformations of Hilbert space. *See* INTEGRAL EQUATION; LINEAR SYSTEMS OF EQUATIONS; OPERATOR THEORY; QUANTUM MECHANICS. [N.J.Y.]

Hill and mountain terrain Land surfaces characterized by roughness and strong relief. The distinction between hills and mountains is usually one of relative size or height, but the terms are loosely and inconsistently used.

Uplift of the Earth's crust is necessary to give mountain and hill lands their distinctive elevation and relief, but most of their characteristic features—peaks, ridges, valleys, and so on—have been carved out of the uplifted masses by streams and glaciers. Hill lands, with their lesser relief, indicate only lesser uplift, not a fundamentally different course of development. The features of hill and mountain lands are chiefly valleys and divides produced by sculpturing agents, especially running water and glacier ice. Local peculiarities in the form and pattern of these features reflect the arrangement and character of the rock materials within the upraised crustal mass that is being dissected.

Hill and mountain terrain occupies about 36% of the Earth's land area. The greater portion of that amount is concentrated in the great cordilleran belts that surround the Pacific Ocean, the Indian Ocean, and the Mediterranean Sea. Additional rough terrain, generally low mountains and hills, occurs outside the cordilleran systems in eastern North and South America, northwestern Europe, Africa, and western Australia. Eurasia is the roughest continent, more than half of its total area and most of its eastern portion being hilly or mountainous. Africa and Australia lack true cordilleran belts. The broad-scale pattern of crustal disturbance, and hence of rough lands, is now known to be related to the relative movements of a worldwide system of immense crustal plates. *See* MOUNTAIN; PLATE TECTONICS. [E.H.Ha.]

Hip The region of the junction of the thigh and trunk. The bony framework consists of the large, flaring hipbone, or pelvis, which contains the hip socket, or acetabulum. This receives the head of the thighbone, or femur. Large muscle groups surround the joint and pass from trunk to thigh to impart motion to each. In addition, the region is rich in blood vessels, nerves, and lymphatics, which are important in the supply of the lower limb. *See* JOINT (ANATOMY). [T.S.P.]

Hippopotamus The name for two species of even-toed (artiodactylid) ungulates which form the family Hippopotamidae. Both species occur in Africa; the great African hippopotamus (*Hippopotamus amphibius*) inhabits the rivers of tropical Africa, and the pygmy hippopotamus (*Choeropsis liberiensis*) lives near the rivers of western Africa but is more terrestrial.

The great African hippopotamus is the largest living artiodactylid. The ears are small and flexible, and the nostrils and eyes protrude so that they are out of the water as the animal floats. The skin is almost devoid of hair. The feet end in four toes enclosed in round hoofs. The pygmy hippopotamus is about the size of a large pig and is a more solitary species and does not live in large herds, as does the common species.

These animals migrate regularly, following the river upstream during the rainy season and downstream during the dry season to new pastures. They come onto land, especially at night, to feed on vegetation. The males of the common species occupy and maintain territories within which are small herds of females and juveniles. *See* ARTIODACTYLA. [C.B.C.]

Hirudinea A class of the annelid worms commonly known as leeches. These organisms are parasitic or predatory and have terminal suckers for attachment and locomotion. Most inhabit inland waters, but some are marine and a few live on land in damp places. The majority feed by sucking the blood of other animals, including humans.

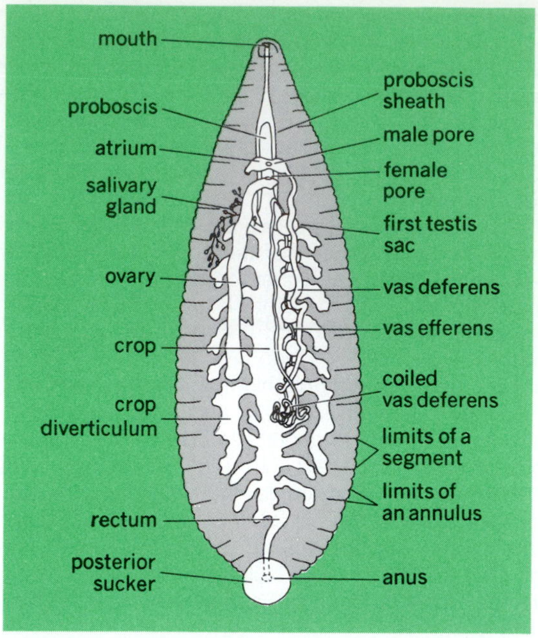

General structure of a leech. Male reproductive system is shown on the right, the female on the left. (*After K. H. Mann, A key to the British freshwater leeches, Freshwater Biol. Ass. Sci. Publ., 14:3–21, 1954*)

Leeches differ from other annelids in having the number of segments in the body fixed at 34, chaetae or bristles lacking, and the coelomic space between the gut and the body wall filled with packing tissue (see illustration). In a typical leech the first six segments of the body are modified to form a head, bearing eyes, and a sucker, and the last seven segments are incorporated into a posterior sucker.

The mouth of a leech opens within the anterior sucker, and there are two main methods of piercing the skin of the host to obtain blood: an eversible proboscis or three jaws, each shaped like half a circular saw, placed just inside the mouth. The process of digestion is very slow, and a meal may last a leech for 9 months. The carnivorous forms have lost most or all of their gut diverticula and resemble earthworms in having a straight, tubular gut. Leeches are hermaphroditic, having a single pair of ovaries and several pairs of testes.

The importance of leeches as a means of making incisions for the letting of blood or the relief of inflammation is declining, and in developed countries the bloodsucking parasites of mammals are declining, because of lack of opportunity for contact with the hosts. In other countries they are still serious pests. *See* ARHYNCHOBDELLAE; RHYNCHOBDELLAE. [K.H.M.]

Histamine A derivative of the amino acid histidine which is widely distributed in tissues. The effects produced by its release are manifold; almost every type of tissue or organ shows some response to it.

Histamine produces changes marked by loss of blood fluid and plasma protein from the vessels to adjacent tissue spaces. Histamine also causes either constriction or dilatation of the arteries and of portions of the circulatory system, depending on the animal species and other factors. This substance may also cause smooth muscle contraction, particularly of the uterus, the bronchioles, and the gallbladder.

Antihistamine compounds combat the general and specific effect of histamine release. These antihistamines are largely symptomatic in action and do not remove the inciting cause of the histamine release. *See* ANTIHISTAMINE. [N.K.M.]

Histidine An amino acid widely distributed in proteins. Histidine is present in relatively large amounts in hemoglobin. Histidine derivatives include the dipeptides carnosine and anserine, which are abundant in muscle, and ergothioneine which is abundant in erythrocytes and seminal fluid; however, the function of these compounds is unknown. Decarboxylation of histidine results in histamine. *See* AMINO ACIDS; HISTAMINE. [D.W.E.S.]

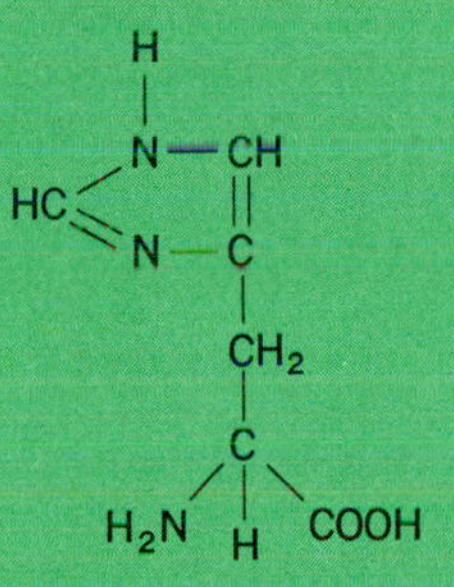

Histidine

Histochemistry A science that provides information about the distribution and activities of chemical components in tissues. Its techniques are largely extensions of those used in histology but make direct use of the chemical properties of tissue constituents to localize them in relation to the structure of the tissue.

Most histochemical procedures involve microscopic examination of tissue sections or smears of cells that have been treated with reagents that convert specific cell constituents into visible derivatives or, in the case of enzymes, treatment with substrate solutions which yield a visible reaction product at their sites of activity. In this context visible means colored, and therefore seen by ordinary light microscopy, or colorless but detectable by electron, fluorescence, interference, phase, polarizing, ultraviolet, or x-ray microscopy. [S.J.H.]

Histocompatibility The capacity to accept or reject a tissue graft. A set of genes, and the polymorphic cell surface antigens they control, provoke immunological rejection of grafts exchanged between genetically (and antigenically) dissimilar individuals. There are two categories of histocompatibility (H) genes (and antigens): major, which cause graft rejection in 2–3 weeks, and minor, where rejection usually takes more than a month. In mice the major histocompatibility gene complex (MHC) is designated the H-2 complex, and the antigens it controls, H-2 antigens. In humans the MHC is called the HLA complex.

Reactivity to foreign antigens is vested in a large population of lymphocytes, made up of many different clones, each with a different specificity of antigen-recognition capability. During the maturation of the immune system in fetal or neonatal animals, those clones with reactivity to self antigens are either deleted or suppressed, so that no destructive immune responses are permitted against self cells or tissues (self tolerance). *See* CELLULAR IMMUNOLOGY.

There are two broad classes of lymphocytes. T lymphocytes, which develop in the thymus, mediate graft rejection and "cell-mediated immunity" against viral and intracellular bacterial infections, but do not secrete free antibodies. B lymphocytes, which develop in bone marrow or the avian bursa of Fabricius, secrete free antibodies that neutralize bacterial toxins or viral infectivity, and promote phagocytosis and various other host defense reactions. The key point is that the need for self tolerance predicts that lymphocyte clones which recognize antigens very different from self antigens are likely to be more numerous than clones which recognize antigens closely related to self antigens. *See* TRANSPLANTATION BIOLOGY. [R.V.B.]

Histogenesis The developmental processes by which the definite cells and tissues which make up the body of an organism arise from embryonic cells. Among animals, the ectoderm, endoderm, and mesoderm, also known as the primary germ layers, provide the stem cells which gradually transform into distinctive kinds of cells and tissues. In the higher plants, meristematic cells, which occur wherever extensive growth takes place, provide the basis for tissue formation. *See* APICAL MERISTEM; EMBRYOLOGY; GERM LAYERS; HISTOLOGY; LATERAL MERISTEM. [C.B.C.]

Histology The study of the structure and chemical composition of tissues of animals and plants as related to their function. The primary aim is to understand how tissues are organized at all structural levels, including the molecular and macromolecular, the entire cell and intercellular substances, and the tissues and organs.

The four tissues of the animal body include cells and intercellular substances. They are (1) epithelium, in which the cells are generally closely applied to each other and separated by very little intercellular substance; (2) connective tissue, in which the cells are usually separated by greater amounts of intercellular substance, which may indeed form the great bulk of the tissue; (3) muscular tissue, whose cells are primarily concerned with contractility; and (4) nervous tissue, whose components are concerned primarily with rapid conduction of impulses. *See* CONNECTIVE TISSUE; EPITHELIUM; MUSCULAR SYSTEM; NERVOUS SYSTEM (VERTEBRATE).

The major fields of histological studies are morphological descriptions; developmental studies; histo- and cytophysiology; histo- and cytochemistry; and (5) fine (or submicroscopic) structure. [I.G.]

Histoplasmosis An infectious fungus disease primarily of humans, although it has been observed in animals. It is caused by a fungus, *Histoplasma capsulatum*, which has dimorphic properties. The infectious form is a mold whose spores are inhaled. As the organism invades tissues, the mold form is converted to a yeast form. This organism is endemic to the valleys of the Mississippi and Ohio rivers. *See* ASCOMYCOTINA; YEAST.

Histoplasmosis is primarily a disease of the lungs, usually in young children and in adults in their fifth or sixth decades. Approximately 50% of those individuals with the pulmonary form are asymptomatic. [L.D.H.]

Historadiography The technique for taking x-ray pictures of cells, tissues, or sometimes the whole animal or plant, if it is a small one. Soft x-rays, those with low penetrating power and relatively long wavelengths, are required for this type of picture. The best pictures are obtained when the tissues contain deposits of metallic elements which have a high absorption capacity for x-rays. *See* X-RAYS.

In applying the technique to tissues, a relatively thin section is placed against an x-ray film and irradiated with a beam of x-rays. When the film is developed, a picture of the object or section of tissue shows on the film. Another method attempts to focus the x-rays after they pass through the specimen. *See* X-RAY MICROSCOPE. [J.H.T.]

Hodgkin's disease A malignant lymphoid neoplasm, usually arising in lymph nodes characterized by morphological heterogeneity and bizarre giant tumor cells referred to as Reed-

Sternberg cells. The etiology of Hodgkin's disease is unknown, although current epidemiological data suggest an infectious (viral) etiology.

Persons with Hodgkin's disease usually seek medical advice because of enlarged painless lymph nodes. They may also have fever, weight loss, anorexia, pruritus, and anemia. The clinical extent of the disease is determined by a process of staging based on physical examination, various biopsies, and usually a laparotomy for examination of the spleen and liver.

Hodgkin's disease more commonly affects males than females, except for the nodular sclerosing variety, which occurs with equal frequency in both sexes. It generally is a disease of persons between the ages of 20 and 40, but it may affect the very young and the very old. Characteristically, persons with Hodgkin's disease exhibit a loss of cell-mediated immunity and become susceptible hosts for infection with a variety of microorganisms such as tubercle bacilli.

The treatment of Hodgkin's disease is dependent on its clinical stage and microscopic appearance. A combination of chemotherapy and radiation therapy is commonly used. *See* LYMPHATIC SYSTEM. [S.P.H.]

Hog cholera A highly contagious epizootic disease of pigs, also known as classical (or European) swine fever. The causative agent is a virus in the genus *Pestivirus*. This disease is the subject of statutory controls in a majority of countries, and has been eradicated from many areas, including the United States, Canada, Australasia, and parts of Europe. Clinically and pathologically, it closely resembles African swine fever, which is caused by an unrelated virus. *See* ANIMAL VIRUS.

Hog cholera can occur in European wild boar, but among domestic species only pigs are affected. Humans are not susceptible. The primary mode of transmission is by contact or proximity. Infected animals shed virus in all bodily secretions, including aerosols of respiratory mucus. The virus survival time in aerosol is short, and airborne transmission over long distances is not a factor. Virus may also be spread by contact with contaminated equipment and vehicles. It can survive for many months in frozen or refrigerated meat from infected pigs, and is not inactivated by mild forms of curing.

Pigs of any age may be affected. There are typically a high fever, loss of appetite, and dullness. Other symptoms include blotchy discoloration of the skin (particularly the extremities), incoordination and weakness of the hindquarters, constipation followed by diarrhea, gummed-up eyes, and coughing. Death occurs within 4–7 days, and the mortality is usually high.

The chronic form of disease is characterized by dullness, unthriftiness, capricious appetite, and variable degrees of coughing, diarrhea, and emaciation. There may be joint swellings and ulceration of the skin.

Strains vary in virulence, and hog cholera may still be suspected when milder signs occur in epizootic form. Low-virulence strains may produce few signs apart from reproductive failure in sows or congenital tremors in their offspring. [S.E.]

Hoisting machines Mechanisms for raising and lowering material with intermittent motion while holding the material freely suspended. Hoisting machines are capable of picking up loads at one location and depositing them at another anywhere within a limited area. In contrast, elevating machines move their loads only in a fixed vertical path, and monorails operate on a fixed horizontal path rather than over a limited area. *See* ELEVATING MACHINES; MONORAIL.

The principal components of hoisting machines are: sheaves and pulleys, for the hoisting mechanisms; winches and hoists, for the power units; and derricks and cranes, for the structural elements.

Sheaves and pulleys or blocks are a means of applying power through a rope, wire, cable, or chain. Sheaves are wheels with a grooved periphery that change the direction or the point of application of a force transmitted by means of a rope or cable. Pulleys are made up of one or more sheaves mounted in a frame, usually with an attaching swivel hook, eye, or similar device at one or both ends. Pulley systems are a combination of blocks. *See* BLOCK AND TACKLE; PULLEY.

Normally, winches are designed for stationary service, while hoists are mounted so that they can be moved about, for example, on wheel trolleys in connection with overhead crane operations. A winch is basically a drum or cylinder around which cordage is coiled for hoisting or hauling. The drum may be operated either manually or by power, using a worm gear and worm wheel, or a spur gear arrangement. A ratchet and pawl prevent the load from slipping; large winches are equipped with brakes, usually of the external band type.

A derrick is distinguished by a mast in the form of a slanting boom pivoted at its lower end and carrying load-supporting tackle at its outer end. In contrast, jib cranes always have horizontal booms. Derricks are standard equipment on construction jobs; they are also used on freighters for loading and unloading cargo, and on barges for dredging operations. Hoisting machines with a bridgelike structure spanning the area over which they operate are overhead-traveling or gantry cranes. *See* BULK-HANDLING MACHINES; DERRICK. [A.M.P.]

Holasteroida An order of irregular echinoids (sea urchins) of the superorder Atelostomata, with a strongly bilaterally symmetrical test. They have a small oval mouth lacking buccal notches that lies close to the anterior on the lower surface: a lantern is never present.

The 105 Recent species are divided into 13 genera. All are deep-sea forms. Their fossil record indicates that they were much more common and diverse in the past, with 61 genera divided into seven families, most of these being shallow-water forms. They are all deposit feeders, living either epifaunally or shallowly buried in unconsolidated substrata. *See* ATELOSTOMATA; ECHINODERMATA. [A.Sm.]

Hole states in solids Vacant electron energy states near the top of an energy band in a solid are called holes. A full band cannot carry electric current; a band nearly full with only a few unoccupied states near its maximum energy can carry current, but the current behaves as though the charge carriers are positively charged. *See* BAND THEORY OF SOLIDS.

The process of conduction in such a system may be visualized in the following way. An electron moves against an applied electric field by jumping into a vacant state. This transfers the position of the vacant state, or propagates the hole, in the direction of the field.

Hole conduction is important in many semiconductors, notably germanium and silicon. The occurrence of hole conduction in semiconductors can be favored by alloying with a material of lower valence than the "host." Semiconductors in which the conduction is primarily due to holes are called *p* type. *See* SEMICONDUCTOR. [J.C.]

Holectypoida An extinct order of primitive irregular sea urchins of the class (Echinoidea), which retain a functioning lantern throughout life. Although their periproct opens on the oral surface in the posterior interambulacrum, the test still shows considerable radial symmetry. The mouth is large, circular, and centrally positioned and is indented by sharp buccal notches. The holectypoids are distinguished from other primitive irregular echinoid groups by the presence of a fifth genital plate in the apical disc that may be perforated by a gonopore.

The approximately 130 nominal species are divided into nine genera and two families, Holectypidae and Discoididae. Holectypoids first appeared in the late Lower Jurassic and survived to the end of the Cretaceous (Maastrichtian), when they became extinct. They were mostly infaunal deposit feeders living in relatively coarse, permeable substrata. *See* ECHINODERMATA; ECHINOIDEA. [A.Sm.]

Holly The American species of holly (*Ilex opaca*) has evergreen leaves. It grows naturally in the eastern and southeastern United States close to the Atlantic and Gulf coasts, in the Mississippi Valley, and westward to Oklahoma and Missouri. It is best known for its bright red berries. The heartwood takes a high polish and is used for cabinet work and musical instruments; because it resembles ivory, it is sometimes used for keys for pianos and organs.

The English holly (*I. aquifolium*) is cultivated extensively in the extreme northwestern United States, but is not hardy in the northeastern states. Its spiny leaves are glossier than those of the American holly and have wavier margins. [A.H.G./K.P.D.]

Holmium A chemical element, Ho, atomic number 67, atomic weight 164.93, a metallic element belonging to the rare-earth group. The stable isotope ^{165}Ho makes up 100% of

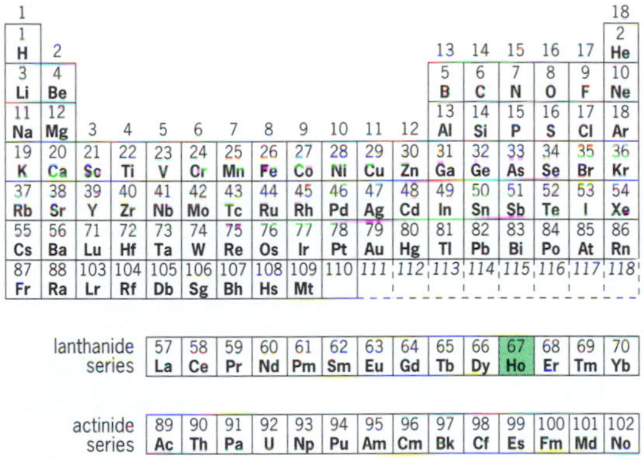

the naturally occurring element. The metal is paramagnetic, but as the temperature is lowered, it changes to antiferromagnetic and then to the ferromagnetic system. For properties of the metal *see* RARE-EARTH ELEMENTS. *See also* ANTIFERROMAGNETISM. [F.H.Sp.]

Holocene An imprecise term defining geologic deposits and their times and processes of formation, postdating the Ice

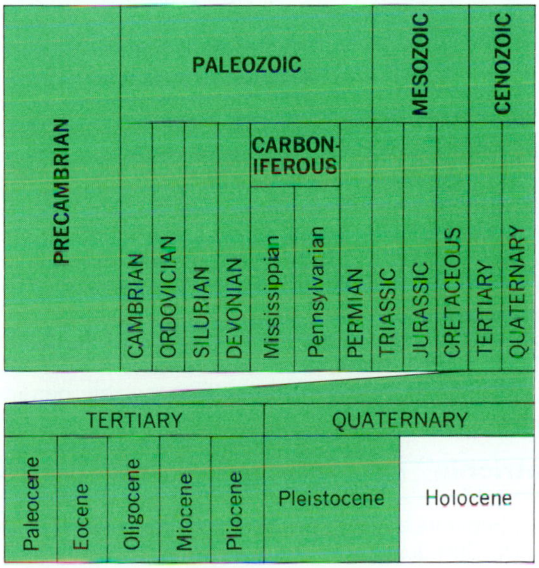

Age. Because the last deglaciation in middle and high latitudes was progressive, the boundary between the Holocene Epoch and the older Pleistocene Epoch has been set variously around a generally accepted value of 10,000 years. In its reference to this latest interval of geologic time, Holocene is essentially synonymous with Recent and Postglacial. *See* PLEISTOCENE.

The term is applied most commonly to sediments. Holocene strata represent virtually every environment of deposition, as they include all the sediments that are being deposited at present. Modern literature employs the term quite broadly as a time indicator for many other geological phenomena, such as uplift, ocean circulation, and volcanism. *See* QUATERNARY. [R.G.J.]

Holocephali One of two Recent subclasses of the cartilaginous fishes, or Chondrichthyes. The Holocephali, or chimaeras, differ from the other subclass, the Elasmobranchii, in having only four pairs of gill arches and gills that open to the exterior from a single pair of apertures; in the erectile dorsal fin and spine (see illustration); in the naked skin in adults; and in the absence of a cloaca and of ribs.

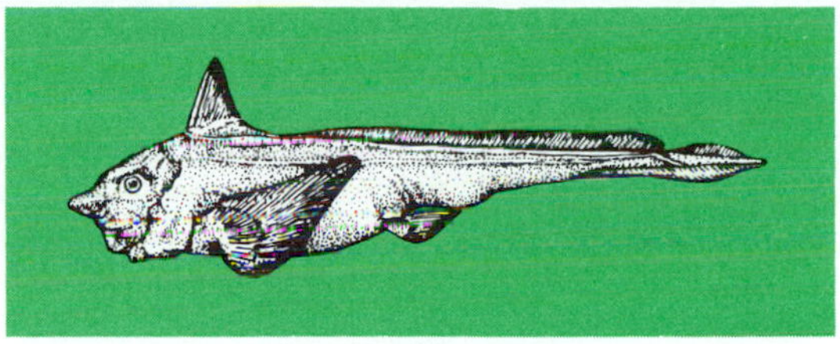

Deepwater chimaera (*Hydrolagus affinis*). (*After G. B. Goode and T. H. Bean, Oceanic Ichthyology, U.S. Nat. Mus. Spec. Bull. no. 2, 1895*)

Chimaeras date from the early Mesozoic. They are classified into a single order, the Chimaeriformes, one family, the Chimaeridae, four or five genera, and about 24 species. All chimaeras are marine, most living in deep water. They are of little economic importance. *See* CHONDRICHTHYES. [R.M.B.]

Holography A technique for recording, and later reconstructing, the amplitude and phase distributions of a coherent wave disturbance. Invented by Dennis Gabor in 1948, the process was originally envisioned as a possible method for improving the resolution of electron microscopes. While this original application has not proved feasible, the technique is widely used as a method of optical image formation, and in addition has been successfully used with acoustical and radio waves. *See* ACOUSTICAL HOLOGRAPHY.

The technique is accomplished by recording the pattern of interference between the unknown "object" wave of interest and a known "reference" wave (Fig. 1). In general, the object

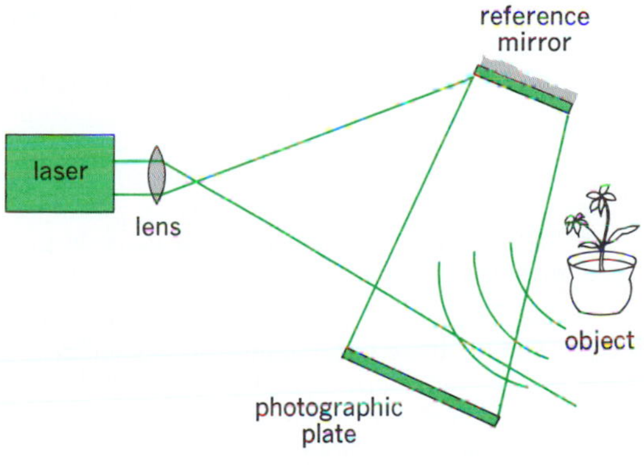

Fig. 1. Recording a hologram.

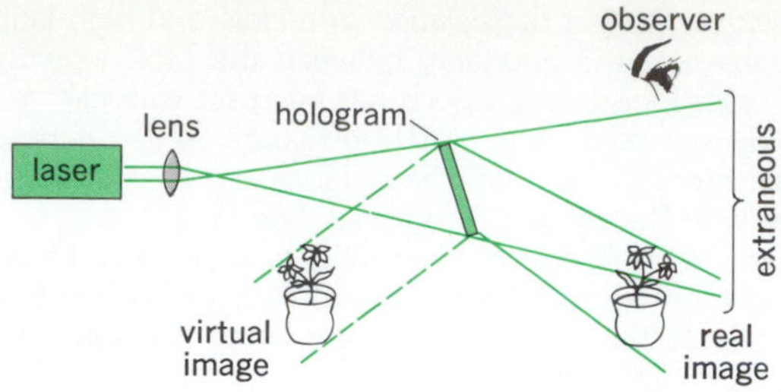

Fig. 2. Obtaining images from a hologram.

wave is generated by illuminating the (possibly three-dimensional) subject of concern with a highly coherent beam of light, such as supplied by a laser source. The waves reflected from the object strike a light-sensitive recording medium, such as photographic film or plate. Simultaneously a portion of the light is allowed to bypass the object, and is sent directly to the recording plane, typically by means of a mirror placed next to the object. Thus incident on the recording medium is the sum of the light from the object and a mutually coherent "reference" wave. *See* LASER.

The photographic recording obtained is known as a hologram (meaning a "total recording"); this record generally bears no resemblance to the original object, but rather is a collection of many fine fringes which appear in rather irregular patterns. Nonetheless, when this photographic transparency is illuminated by coherent light, one of the transmitted wave components is an exact duplication of the original object wave (Fig. 2). This wave component therefore appears to originate from the object (although the object has long since been removed) and accordingly generates a virtual image of it, which appears to an observer to exist in three-dimensional space behind the transparency. The image is truly three-dimensional in the sense that the observer's eyes must refocus to examine foreground and background, and indeed can "look behind" objects in the foreground simply by moving his or her head laterally.

The holographic technique has a number of unique properties which make it of great value as a scientific tool. Although the field is young, and new applications are continually emerging, certain important areas can be identified.

Holography has been demonstrated to offer the capability of several unique kinds of interferometry. This capability is a consequence of the fact that holographic images are coherent; that is, they have well-defined amplitude and phase distributions. Any use of holography to achieve the superposition of two coherent images will result in a potential method of interferometry. *See* INTERFEROMETRY.

Optical memories for storing large volumes of binary data in the form of holograms have been developed for commercial use. Such a memory consists of an array of small holograms, each capable of reconstructing a different "page" of binary data. When one of these holograms is illuminated by coherent light, it generates a real image consisting of an array of bright or dark spots, each spot representing a binary digit. *See* COMPUTER STORAGE TECHNOLOGY.

There has been interest in the use of holography for purposes of display of three-dimensional images. Applications have been found in the field of advertising, and there is increased use of holography as a medium for artistic expression. [J.W.Goo.]

Holostei One of three organizational levels (infraclasses) of the subclass Actinopterygii, or rayfin fishes. The holosteans are descended from the older Chondrostei and in turn are ancestral to the great mass of modern bony fishes, the Teleostei. *See* CHONDROSTEI; TELEOSTEI.

Holosteans made their first appearance in the Upper Permian as the order Semionotiformes; three additional orders arose in the Triassic Period, and the fifth and last order, the Aspidorhynchiformes, evolved in the Middle Jurassic. In the Jurassic and Lower Cretaceous, holosteans dominated actinopterygian fish life, but by the Late Cretaceous they had been largely replaced by teleosts.

Holosteans, although highly varied in body form, were structurally as well as temporally intermediate between chondrosteans, and teleosts, to which group they passed on substantial advances. In living holosteans the swim bladder is highly vascularized, and auxiliary aerial respiration is possible, a sometimes essential faculty in oxygen-poor waters of swamps. *See* ACTINOPTERYGII; AMIIFORMES; ASPIDORHYNCHIFORMES; PHOLIDOPHORIFORMES; PYCNODONTIFORMES; SEMIONOTIFORMES. [R.M.B.]

Holothuroidea A class of Echinozoa characterized by a cylindrical body and smooth leathery skin, and known as sea cucumbers. There are no arms, but a ring of five or more tentacles may surround the mouth, which is usually at one end of the body. There are no pedicellariae. Tube feet may be present or lacking. There are no ambulacral grooves. *See* ECHINOZOA.

Holothurians resemble worms because the pentamerous symmetry is largely concealed by a secondary bilateral symmetry, and the general absence of external spines distinguishes them from the other extant echinoderms (see illustration).

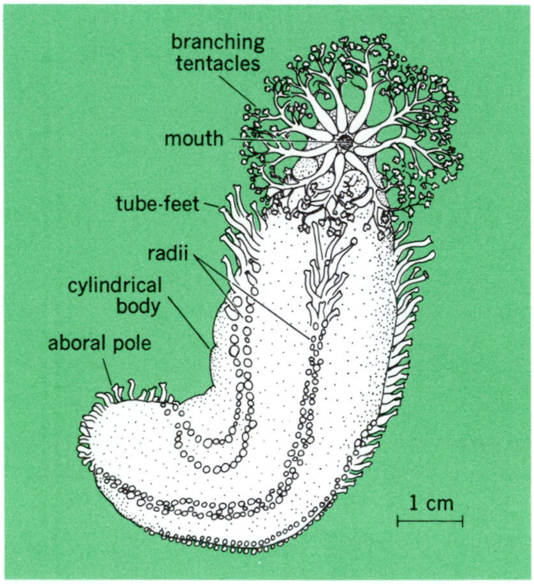

***Cucumaria*, a representative holothurian.**

The 1100 living species have been grouped in 170 genera arranged in six orders: the Dendrochirotida, Dactylochirotida, Aspidochirotida, Elasipodida, Molpadida, and Apodida. Colors vary widely; the most brilliant colors are found among the Synaptidae. Yellow, red, violet, and fawn tints occur, but many species are somber shades or black. *See* APODACEA.

Holothurians occur in all seas, from low-tide level down to the greatest depths explored. At depths below 5.5 mi (8.8 km) holothurians comprise 90% of the total mass of living matter, the rest being mainly starfishes. Two pelagic genera are known. [H.B.F.]

Holotrichia A major subclass of the class Ciliatea. These protozoans have a fairly uniform body ciliation, as the name implies. Separate articles appear on the groups listed in the following classification:

Subclass Holotrichia

Order: Gymnostomatida, Trichostomatida, Chonotrichida, Apostomatida, Astomatida, Hymenostomatida, Thigmotrichida

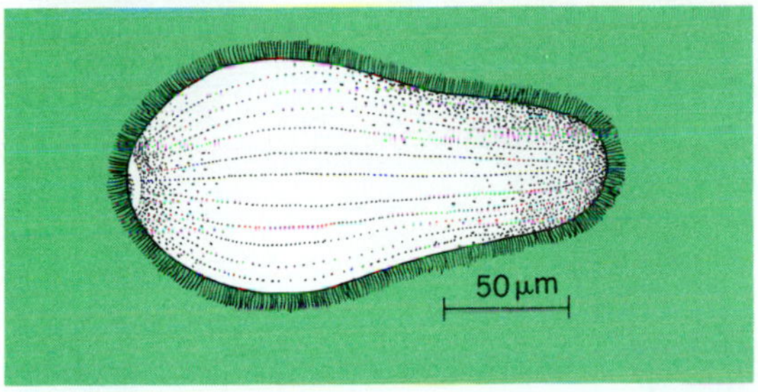

***Prorodon*, a primitive holotritch.**

The cilia are typically arranged in longitudinal rows over the body, although scattered exceptions exist. A mouth is often, although not always, present. The prototype of the Holotrichia is exemplified by the form portrayed in the illustration. *See* CILIATEA. [J.O.C.]

Homalozoa A subphylum of echinoderms, made up of members having a flattened theca or body lacking pentameral symmetry. Homalozoans (also called carpoids) include four extinct classes of relatively uncommon primitive echinoderms ranging in age from the Early or Middle Cambrian to the Late Carboniferous. Homalozoans have a flattened, asymmetrical to bilaterally symmetrical theca often composed of a marginal frame of large elongate plates surrounding top and bottom central areas that had numerous smaller plates and were probably flexible. All homalozoans were apparently mobile, benthic, detritus or suspension feeders that had adopted a flatfish way of life. *See* CARPOIDS; ECHINODERMATA. [J.Sp.]

Homeostasis In living substance, the maintenance of internal constancy and independence of the environment. The essence of living substance is that it differs chemically from the surrounding medium and yet maintains a dynamic equilibrium with its environment. Homeostasis occurs in all living cells and in the fluids and organ systems of multicellular organisms.

The cells of multicellular organisms are bathed in fluids which are kept relatively constant in ionic composition, osmotic composition, pH, level of sugar, and organic composition. Two patterns of homeostasis of internal state in the face of environmental change are recognized: (1) An animal may alter a given property with reference to the environment; the property conforms to the medium, and homeostasis consists in cellular adjustment such that metabolism continues in the altered state. For example, in cold-blooded animals the body temperature conforms to that of the environment. In many marine animals and endoparasites the body fluids conform in osmotic concentration with the medium. (2) An animal may regulate its internal state and maintain internal constancy despite an altered environment. Such regulation is achieved by a series of automatic feedback controls as environmental stress is applied; at some environmental limits, regulation fails and the animal cannot long survive. For example, warm-blooded animals maintain a constant body temperature over a range limited by heat and cold.

Two types of homeokinetic response are also distinguished: (1) Capacity adaptations are changes in cellular functions such that an intact animal can maintain normal activity over a range of internal states. (2) Resistance adaptations are changes which permit survival at extremes of some physical parameter. Resistance limits are narrow for intact integrated organisms, less restricted for isolated cells and tissues, and much wider for critical molecules such as proteins (denaturation) and lipids (melting). For example, many poikilothermic animals show little or no regulation of body temperature, but they alter energy-yielding enzyme systems to provide for nearly constant activity over a wide temperature range. Also, these animals can, by acclimation, become more or less resistant to heat or cold. Similar changes in thermal resistance occur in some homeotherms (temperature regulators). *See* HIBERNATION; HYPOTHERMIA; THERMOREGULATION. [C.L.P.]

Homing A process in navigation by which the destination is approached by keeping some navigational parameter constant. The parameter to be maintained constant is usually generated by the cooperative action of two sets of equipment, one on the vehicle and the other on the ground; that is, these equipments form a homing system.

Homing systems vary greatly in their complexity. One of the simpler systems is the radio compass, which operates with any suitably located radio beacon; the craft's heading is adjusted so that the bearing being read on the radio direction finder is kept constant. The loran system, on the other hand, is very complex. Specialized shipborne and ground-based installations are required to generate the parameter. In general, modern systems provide the homing feature while furnishing direct positional information. *See* LORAN. [P.C.S.]

Homoeosis The formation of an organ or part that differs from the original structure of the site. There is a basic difference between variants or mutants that arrest development and those which steer ontogeny onto an alternate course. This latter type of variation is most easily observed in those structures or organisms which are segmentally organized, and the variant is made evident by the transformation of one member of the segmental series into the likeness of some other member of that series. It is this kind of variant, or the transformation of one segmental identity into the likeness of some other segmental type that is defined as homoeosis.

The role of many homoeotic genes in development is to regulate the proper patterned expression of those events necessary to make a haltere as opposed to a wing or leg versus an antenna or proboscis. One can view this role in terms of some developmental switching mechanism. If the switch is working, a normal ontogenic pathway is followed. However, if the switch is broken (because of a mutation), an alternate course is taken. Importantly, a development is not arrested but is changed to a new and, in the proper place, normal pattern. It is not known how the products of the homoeotic genes regulate the normal developmental program. Based on the information available, the involvement could be at one of several levels. The gene products could serve to directly regulate the expression of large batteries of other genes, the products of which are necessary to construct the various anatomical portions of the organism. Alternatively, the products of homeostatic genes may be incorporated into the membranes of the cells in which they are present, and thereby secondarily alter the ability of those cells to react to chemical cues present in the developing organisms. *See* CELL DIFFERENTIATION; DEVELOPMENTAL BIOLOGY; DEVELOPMENTAL GENETICS; GENE ACTION; GENETIC ENGINEERING; MUTATION. [T.C.K.]

Homogeneous catalysis A process in which a catalyst is in the same phase as the reactant. A homogeneous catalyst is molecularly dispersed (dissolved) in the reactants, which are most commonly in the liquid state. Catalysis of the transformation of organic molecules by acids or bases represents one of the most widespread types of homogeneous catalysis. In addition, the catalysis of organic reactions by metal complexes in solution has grown rapidly in both scientific and industrial importance. *See* CATALYSIS. [D.F.]

Homoptera An order of the class Insecta related to the or-der Hemiptera. This is a major group of sucking insects, with more than 30,000 species, even though in Asia and Africa the number of undiscovered species probably still exceeds the discovered ones. Common examples are the cicadas, aphids, and leafhoppers. The group is difficult to characterize because of the large number and diverse forms of the species it contains.

The head of these insects is hypognathous or opisthognathous, the beak appearing to arise from the ventral posterior margin of the head or even from the prosternum. The gula is membranous or absent. As in the Heteroptera, the beak consists of two pairs of stylets, formed by the maxillae and the mandibles, ensheathed in the labium. The maxillary stylets fit together to form a double tube, one channel serving for the passage of food and the other for saliva.

Most winged species have four wings, but male scale insects have only two. In most forms both pairs of wings are membranous and transparent, but in some, the forewings are somewhat thickened and may then be either coriaceous and translucent, or opaque, and with or without an apical membranous area. When the insects are at rest, the forewings are usually held, rooflike, over the dorsum, with the apex of one of them slightly overlapping the apex of its complement (Fig. 1).

In most species metamorphosis is gradual, but in a few it is practically holometabolous. Adults and nymphs of most species are terrestrial, but a few species are subterranean in all stages and others are subterranean only in immature stages. A number of species are vectors of virus diseases of plants.

Series Coleorrhyncha. This group is characterized by the origin of the beak, formed at the anteroventral extremity of the face, and by the fact that the propleura form a sheath for the base of the beak. The hindwings are absent, and the forewings are held flat over the abdomen in repose. The flight function has been lost. They occur in Tasmania, New Zealand, and South America.

Series Auchenorrhyncha. This series and the Sternorrhyncha are the major groups of the Homoptera. In the Auchenorrhyncha the beak arises at the anteroventral extremity of the face and is not sheathed by the propleura. The Auchenorrhyncha includes a large number of species. A number of classifications have been proposed; the classification adopted here is a common one. It divides the series into the superfamily Fulgoroidea and the families Cicadidae, Cercopidae, Membracidae, and Cicadellidae. These families are not subordinate to the superfamily Fulgoroidea.

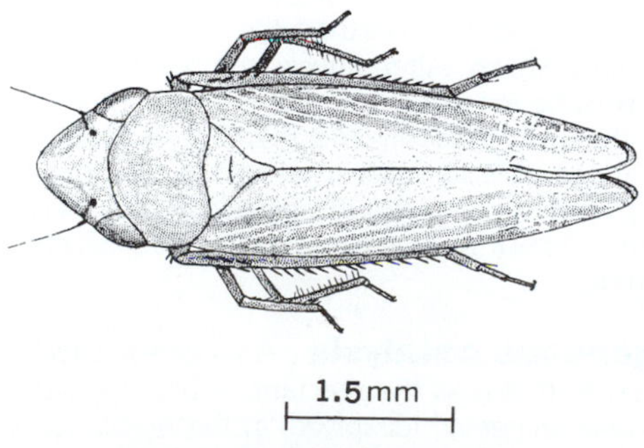

Fig. 1. A cicadellid, dorsal view.

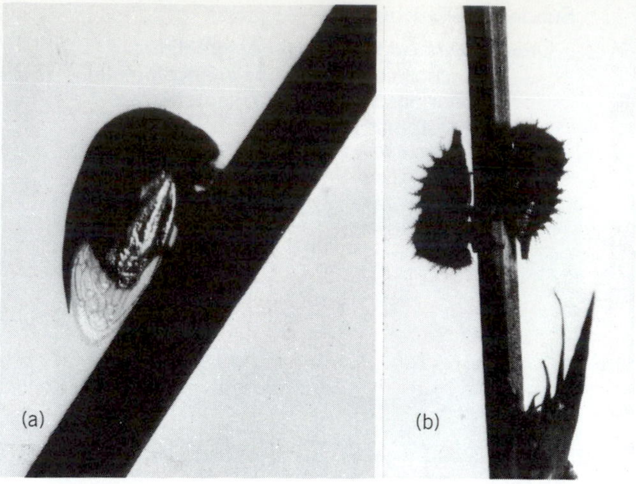

Fig. 2. Membracids on stems. (*a*) Adult. (*b*) Nymphs. (*Courtesy of C. H. Hanson*)

Superfamily Fulgoroidea includes insects commonly known as lantern flies. This group is subdivided into 20 families and includes many species which are important because of the economic damage they do while feeding or because they carry virus diseases of plants.

Included in the family Cicadidae are the cicadas, harvest flies, and jar flies. The insects in this family are probably better known to the layperson than any other homopterous family because of their large size and the strident songs of the males. Adults are usually found on trees or shrubs. At least in some species, the songs of the males assemble local populations.

Spittle bugs and froghoppers are common examples of the family Cercopidae. Insects in this group attract attention in the immature stages, during which they surround themselves with a mass of froth or spittle. One species, the meadow spittle bug, is very common in the temperate portion of the Northern Hemisphere, and its masses of spittle are familiar sights. Some species of Australia and the East Indies live within a calcium carbonate tube attached to stems or leaves.

The treehoppers (family Membracidae) are small to medium in size and seldom attract attention (Fig. 2*a*). Most of them feed on woody plants and are found on the stems in sunny locations. Frequently a number of specimens are arranged in a vertical row on the stem, all with their heads downward.

The greatest number of species occurs in the warmer regions of the world. In many species the enlarged pronotum has adornments, excrescences, and processes which are astonishing in appearance, some of them nearly as large as the remainder of the insect. The nymphs (Fig. 2*b*) leave the trees and feed on herbaceous plants, often occurring in great numbers in pastures. The adults return to woody plants before oviposition.

The leafhoppers are included in the large family Cicadellidae. These usually small insects are known to many people by sight but not by name, because of their common occurrence in great numbers at night near lights (Fig. 1). Probably the greatest number of species occurs in tropical areas, but the majority of these have not been described. Leafhoppers occasionally bite humans, but apparently they have never been seen taking blood. Several species have been found to be vectors of virus diseases of plants.

Series Sternorrhyncha. In this group of families, the beak appears to arise either between the fore coxae or behind them. The antennae are usually long, filamentous, and have no well-differentiated terminal setae. Wingless forms are common.

The Sternorrhyncha includes a large number of species, many of them of great economic importance. The winged forms are not strong flyers, but they are so light that they may be borne considerable distances by air currents.

The family Psyllidae are known as jumping plant lice. Its representatives resemble cicadas in appearance, but are much smaller. About 1000 species are known. Some psyllids produce severe damage to their food plants. This damage may result from the mere feeding by tremendous numbers of individuals, from the resulting yellowing or rolling of leaves, or from galls produced on the leaves. Indirect damage may result from the growth of fungi on leaves which have become coated with the sugary excrement, the honeydew, of the psyllids.

The whiteflies (family Aleyrodidae) are 0.28 in. (7 mm) or less in length and usually lightly covered with a white, powdery, waxy material which has led to their common name. Whiteflies directly damage plants by their feeding. Indirectly, damage results from spotting at the feeding site, growth of fungus on the excreted honeydew, or from increased susceptibility of leaves to winter damage.

Members of the large superfamily Aphidoidea have four wings, or none. The wings are usually membranous or whitish and opaque. The forewings are much larger than the hindwings. Honeydew is usually produced. This superfamily includes the families Aphididae and Chermidae. The true aphids (Aphididae) are very attractive to ants, and colonies of very small aphid species which otherwise might escape notice can often be located by observing the attending ants. In a few species, this relationship has progressed to the point where ants are necessary for the survival of the aphid species, as in the corn root aphid. Many aphid species are important because of damage done in feeding.

Chermidae is a small family of minute insects, the adelgids and phylloxeriids. Both winged and wingless forms occur. The grape phylloxera has been a severe pest of cultivated grapes and once threatened the entire wine industry of France.

The scale insects and mealy bugs (superfamily Coccoidea) are usually small. More than 4000 species have been described. In the males, the hindwings are reduced to clublike halteres. The wings are usually held flat over the back in repose, and the venation is greatly reduced. The females are wingless.

Scale insects injure the host plant by their feeding on leaves, stems, or roots, and a number of species are very important economically. Most species produce large quantities of honeydew. A few species have been shown to be vectors of virus diseases of cucumbers, tobacco, and cacao. *See* ENTOMOLOGY, ECONOMIC; INSECTA. [D.A.Y.]

Homosclerophorida An order of primitive sponges of the class Demospongiae, subclass Tetractinomorpha, with a skeleton consisting of equirayed, tetraxonid, siliceous spicules and their derivatives formed through reduction in number of rays.

Homosclerophorid sponges are mostly small in size and encrusting to massive in shape. They occur in tidal and shallow waters, down to depths of at least 1600 ft (500 m). *See* DEMOSPONGIAE. [W.D.H.]

Honey Dew melon A long-keeping cultivar of muskmelon, *Cucumis melo inodorus*, of the gourd family, Cucurbitaceae. Vines are vigorous and prolific and have large leaves and stems. The fruits are large, slightly oval, (diameter 6–7 in. or 15–18 cm), smooth, creamy yellow to ivory when ripe, and with little or no net. The flesh is thick, light green, tender, juicy, and very sweet with mild aroma and flavor; it contains 10% or more sugar when ripe and is rich in potassium and vitamin C, but not as rich in vitamin A as the orange-fleshed cantaloupe.

Honey Dew is of African origin, has been grown in France for many years, and was introduced into the United States in 1911. Most of the crop is harvested in the summer in California; almost all of the remaining crop is harvested in the spring in Texas, with less than 5% harvested in the fall in Arizona. *See* MUSKMELON; VIOLALES. [O.A.L.]

Honing The process of removing a relatively small amount of material from a surface by means of abrasive stones to obtain a desired finish or extremely close dimensional tolerance. Seldom is more than a few thousandths of an inch (0.001 in. = 0.025 mm) of stock removed. The abrading action of the fine grit stones occurs on a wide surface area rather than on a line of contact as in grinding. As it applies to machining, honing refers primarily to work done on cylindrical surfaces. In addition to metals and carbides, materials such as plastics, ceramics, and glass may be honed. Honing is done with manually operated equipment as well as with vertical and horizontal honing machines. *See* GRINDING. [A.H.T.]

Hookeriales An order of the mosses, many species of which occur in the tropics. The plants are small to robust, procumbent, and often flaccid, and frequently appear flattened. The stems are irregularly branched, and the cells of the peripheral layer are enlarged. The leaves vary in shape and size. They occur in many rows in some species, four rows in others, and appear to be in one plane (see illustration). *See* BRYOPSIDA. [W.H.W.]

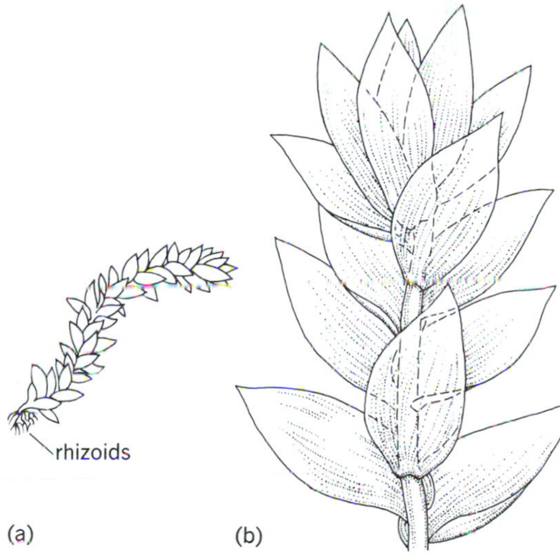

Hookeria acutifolia. **(*a*) A portion of an entire plant. (*b*) Apical portion enlarged. (*After W. H. Welch, Mosses of Indiana, Indiana Department of Conservation, 1957*)**

Hooke's law A generalization applicable to all solid materials, stating that stress is directly proportional to strain and expressed as

$$\frac{\text{Stress}}{\text{Strain}} = \frac{S}{\epsilon} = \text{constant} = E$$

where E is the modulus of elasticity, or Young's modulus, in pounds per square inch. The constant relationship between stress and strain applies only to stress below the proportional limit. *See* STRESS AND STRAIN; YOUNG'S MODULUS. [W.J.K./W.G.B.]

Hookworm A roundworm of the class Nematoda; an intestinal parasite of humans and other mammals. All species are included in the family Ancylostomidae and are phasmid nematodes; that is, there are phasmids, sensory (perhaps olfactory) structures, located posterior to the anal opening. Members of this family have a large mouth or buccal cavity armed either with chitinous cutting plates or teeth with which the worm attaches to the intestinal mucosa of the host.

Two species of hookworm are important to humans: the New World hookworm (*Necator americanus*) and the Old World hookworm (*Ancylostoma duodenale*). *See* HOOKWORM DISEASE; NEMATA. [C.B.C.]

Hookworm disease Microcytic hypochromic anemia in humans produced by the nematodes *Necator americanus* or *Ancylostoma duodenale* in the intestine. These nematodes suck blood from the small intestine. Although clinical manifestations in both cases are similar, *A. duodenale* infections are usually more severe and are characterized by severe eosinophilia and diarrhea. The host, if malnourished, cannot replace the blood lost, and anemia ensues wih all its symptoms. *See* ANEMIA; HOOKWORM.

The life history of *Ancylostoma* is similar to that of *Necator*. The eggs are eliminated in the host's feces. In areas of poor sanitation the soil becomes contaminated. With favorable temperature, humidity, shade, and aeration, eggs in soil embryonate in 24 h. A first-stage larva hatches and molts twice within 1 week to become infective. The larva actively penetrates the skin of humans. It then enters the bloodstream, and moves to the pulmonary alveoli. Pneumonia and other respiratory diseases may result. Penetration of the skin by infective hookworm larvae produces a dermatitis known as dew itch or ground itch. Migrating via trachea and esophagus to the duodenum, the parasite reaches adulthood in 2 months.

The infective larva of *A. braziliense*, found in dogs and cats, may infect the skin of humans and produce a creeping eruption that becomes contaminated with bacteria. The condition is known as larva migrans. The lesion may subsist for months. [J.F.M.]

Hop A plant (*Humulus lupulus*) belonging to the family Urticaceae. The plant is perennial, with the clockwise twining vine dying back to the ground each year. The male and female flowers are borne on separate plants. *See* URTICALES.

The mature hop cone—the hop of commerce—consists of papery bracts and bracteoles. As the cone matures, numerous golden-yellow granules (lupulin) develop on the bracteoles. These granules contain the resins and essential oils to which the hop owes its brewing value. *See* MALT BEVERAGE.

Hops are found primarily in the temperate areas of the world. In the United States the cultivated hop is grown in Washington, Oregon, Idaho, and California. [C.B.S.]

Hophornbeam The genus *Ostrya* of the birch family, represented in North America by two species. *Ostrya virginiana* is widely distributed in the eastern half of the United States and in the highlands of southern Mexico and Guatemala. It can be recognized by its fruit, which closely resembles that of the hop vine, and by its very scaly bark. The scales usually occur in narrow, more or less parallel, vertical strips. The leaves are sharply and doubly serrate. This is one of several trees known as ironwood because of its hard, strong wood; it is used for fence posts, tool handles, mallets, and other articles requiring hardness and strength. *Ostrya knowltonii* is a small rare tree of the southwestern United States. *See* FAGALES. [A.H.G./K.P.D.]

Hoplocarida A subclass of Crustacea, with a single extant order, Stomatopoda, commonly known as mantis shrimps. The Haplocarida was formerly included as a taxon within the Eumalacostraca, and disagreement regarding its independent origin still persists. The controversy is centered on the question of whether those elements of the eumalacostracan caridoid facies observed in haplocarids represent examples of homology or convergence. Investigations of fossil (Paleostomatopoda) and Recent Stomatopoda suggest that a distinct set of morphological features, sometimes referred to as a hoploid morphotype, clearly delineate the Hoplocarida from the Eumalacostraca. Morphological aspects of feeding observed in modern-day species, as well as features of the digestive system and abdominal musculature, tend to support the hypothesis of a distinct origin. *See* CRUSTACEA; STOMATOPODA. [P.A.McL.]

Hoplonemertini An order of the class Enopla of the phylum Rhynchocoela, characterized by possession of an elaborate armed proboscis consisting of an anterior thick-walled tube, a median portion armed with stylets, and a posterior blind tube. In some species the foregut (stomach) can be everted into the prey to achieve extracorporeal digestion prior to ingestion. There are two suborders, the Monostylifera and Polystylifera, separated on the number of stylers in the proboscis. Monostylifera include fresh-water, terrestrial, and symbiotic species, as well as the more common marine littoral forms. *See* ANOPLA; BDELLONEMERTINI; ENOPLA; RHYNCHOCOELA. [J.B.J.]

Horizon The visible horizon is the apparent boundary line between sky and earth or sea. The astronomical horizon is the great circle of the celestial sphere 90° from the zenith and the nadir. *See* ASTRONOMICAL COORDINATE SYSTEMS. [G.M.C.]

Hormone One of the chemical messengers produced by endocrine glands, whose secretions are liberated directly into the bloodstream and transported to a distant part or parts of the body, where they exert a specific effect for the benefit of the body as a whole. The endocrine glands involved in the maintenance of normal body conditions are pituitary, thyroid, parathyroid, adrenal, pancreas, ovary, and testis. However, these organs are not the only tissues concerned in the hormonal regulation of body processes. For example, the duodenal mucosa, which is not organized as an endocrine gland, elaborates a substance called secretin which stimulates the pancreas to produce its digestive juices. The placenta is also a very important hormone-producing tissue. See separate articles on the individual glands.

The hormones obtained from extracts of the endocrine glands may be classified into four groups according to their chemical constitution: (1) phenol derivatives, such as epinephrine, norepinephrine, thyroxine, and triiodothyronine; (2) proteins, such as the anterior pituitary hormones, with the exception of adrenocorticotropic hormone (ACTH), human chorionic gonadotropin, pregnant-mare-serum gonadotropin, and thyroglobulin; (3) peptides, such as insulin, glucagon, ACTH, vasopressin, oxytocin, and secretin; and (4) steroids, such as estrogens, androgens, progesterone, and corticoids. Hormones, with a few exceptions like pituitary growth hormone and insulin, may also be classified as either tropic hormones or target-organ hormones. The former work indirectly through the organs or glands which they stimulate, whereas the latter exert a direct effect on peripheral tissues. *See* ENDOCRINE SYSTEM (VERTEBRATE). [C.H.L.]

Horn (geology) A high mountain peak steepened by glaciation on all sides, and a remnant of a former alpine landscape deeply eroded by surrounding mountain glaciers. Horns may have three or four near-vertical faces, determined by the positions of cirque glaciers carving into the mountain below the faces. Horns occur in glaciated mountains all over the world, or as nunataks standing above ice sheet surfaces. Excavation of a cirque headwall, which may become the face of a horn above, prevents rock debris from accumulating and keeps the cliff clean. *See* CIRQUE; GLACIAL GEOLOGY.

The Matterhorn, located near Zermatt, Switzerland. (*Library of Congress Collections*)

A majestic four-sided horn, the Matterhorn rises to 14,684 ft (4477 m) on the Swiss-Italian frontier above Zermatt, Switzerland (*see* illustration). From the Swiss side, it appears as an isolated obelisk, but actually is at the end of a long arête. [S.E.Wh.]

Hornbeam The genus *Carpinus* of the birch family, represented in the United States by *C. caroliniana*, the American hornbeam or blue beech. Hornbeam is a small tree, and it has a smooth, steel-gray, fluted bark. It grows throughout the eastern half of the United States, especially in moist soil along banks of streams; it is sometimes called water beech. When mature, it is easily recognized by its peculiar bark, by the doubly serrate leaves resembling those of sweet birch, and by the small, pointed, angular winter buds with scales in four rows. The fruit is a small nutlet subtended by a three-lobed serrate bract. The wood is very hard, giving rise to the name ironwood.

The European hornbeam (*C. betulus*) is often cultivated in parks and estates. It can be distinguished by its larger size, larger winter buds, and larger three-lobed, almost entire fruiting bracts. *See* FAGALES. [A.H.G./K.P.D.]

Hornblende A general name given to the monoclinic calcium amphiboles that form extensive solid-solution series between the various metals in the generalized formula $(Ca,Na)_2(Mg,Fe,Al)_5(Al,Si)_8O_{22}(OH,F)_2$. Hornblende has a widespread occurrence in metamorphic, igneous (intrusive), and volcanic rocks, ranging from dominantly ferromagnesian rocks to granites. The term common hornblende refers to the middle-metamorphic-grade hornblendes found in schists. Common hornblende usually forms as long, prismatic needles, dark green to black in color. The hornblendes of the high grades of metamorphism are usually short, stubby prisms, black to brown in color, and high in iron and aluminum, with some sodium. *See* AMPHIBOLE; METAMORPHISM; SCHIST. [G.W.DeV.]

Hornfels A common name given to a class of metamorphic rocks produced by contact metamorphism, also known as hornstone. Hornfelses were originally sedimentary rocks. As magma intruded into the sediments, the heat given off induced recrystallization and effected a complete alteration of the primary sedimentary strata into hard, often flinty rocks. Their fine-grained, constituent minerals usually can be discerned only with the microscope. The chemical composition of a hornfels depends only upon the composition of the original sediment, and consequently the mineral composition of hornfelses is predetermined by the nature of the original sediment. *See* METAMORPHIC ROCKS; METAMORPHISM.

According to the chemical composition of the sediments, there will be 10 classes of silica-rich hornfelses:

1. Pure shale contains alumina and traces of ferromagnesia which form andalusite and cordierite (and an excess of silica which produces quartz).
2. A small addition of lime produces anorthite (in the actual hornfels represented by plagioclase, since some Na_2O is present) in addition to andalusite and cordierite.
3. More lime produces more anorthite, whereas andalusite disappears.
4. Next, hypersthene will form in addition to cordierite and anorthite.
5. With still more lime, the cordierite disappears, with the formation of more anorthite and hypersthene.
6. Lime is now so dominating that it combines with magnesia to form diopside in addition to anorthite and hypersthene.
7. Hypersthene disappears.
8. More lime reacts with anorthite to form grossularite in addition to diopside.
9. Anorthite is completely used up in the grossularite reaction.
10. Pure lime silicate, that is, wollastonite, occurs with diopside and grossularite. [T.F.W.B.]

Horology Measurement of the time dimension. In practice, horology is: the search for a steady or repetitive action; and the design of an instrument to perform that action and to indicate (read out) a measure of the action. Untit early in the 20th century, horology dealt with mechanical instruments, with effort distributed between improving accuracy and decreasing size of timepieces. Increasingly, however, electronic instruments have provided means for meeting these objectives. *See* ATOMIC CLOCK; CHRONOMETER; CLOCK; QUARTZ CLOCK; SUNDIAL; TIME; WATCH. [F.H.R.]

Horse A mammal which is an odd-toed ungulate (Perissodactyla) of the family Equidae. These animals have a single functional digit, the middle one, with the second and fourth digits present as vestigial structures. The wild horse (*Equus caballus*) is now represented by a single race, the Mongolian or Przewalski's horse (*E. c. przewalskii*), which survives in small numbers in central Asia. It stands only about 4.5 ft (1.4 m) high at the withers.

A mare may be bred at 1–2 years of age but usually not until 3 years old. One or two foals are born each year after a gestation period of about 11 months. The maximum life-span recorded for the horse is 60 years, although average maximum life-span is 35.

During their evolution the horses have increased in size from small terrierlike animals; the legs and feet increased in size and the original four digits gradually became reduced to one. There was a gradual change in the teeth as these animals developed into efficient browsers.

Throughout history many attempts were made to domesticate the various species of wild horses, but this accomplishment does not seem to have occurred before the Bronze Age.

Modern horses may be arbitrarily classified into three groups bred for specific activities; saddle and race horses, draft breeds, and harness horses. *See* PERISSODACTYLA. [C.B.C.]

Horseradish A hardy perennial crucifer, *Armoracia rusticana*, of eastern European origin belonging to the plant order Capparales. Horseradish is grown for its pungent roots, which are generally grated, mixed with vinegar and salt, and used as a condiment or relish. Production in the United States is limited to northern areas; Illinois, Wisconsin, and Missouri are important producing states. *See* CAPPARALES. [H.J.C.]

Horst A segment of the Earth's crust, generally long as compared to its width, that has been upthrown relative to the adjacent rocks (see illustration). Horsts range in size from those

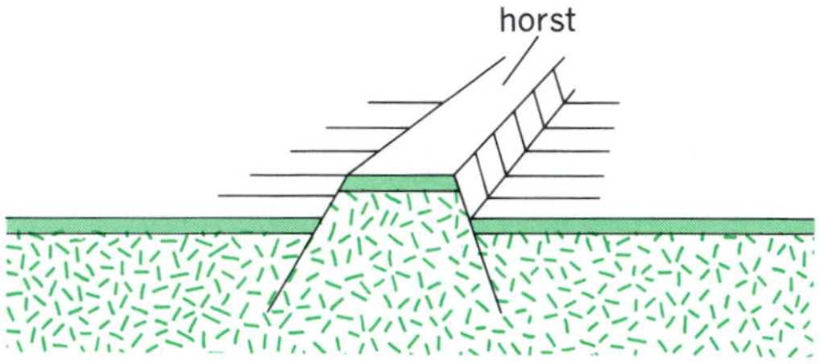

A simple horst with associated faults. (*After A.K. Lobeck, Geomorphology, McGraw-Hill, 1939*)

that have lengths and upward displacement of a few inches to those that are tens of miles long with upward displacements of thousands of feet. The faults bounding a horst on either side commonly have inclinations of 50–70° toward the downthrown blocks, and the direction of movement on these displacements indicates that they are gravity faults. These relationships suggest that horsts develop in regions where the crust has undergone extension. They may form in the crests of anticlines or domes, or may be related to broad regional warpings. *See* FAULT AND FAULT STRUCTURES. [P.H.O.]

Hospital infections Nosocomial infections; clinically apparent infections that develop while a person is hospitalized or that become apparent after a person is discharged from the hospital but are caused by microorganisms acquired during the stay there. Most nosocomial infections become clinically apparent while the individual is still hospitalized. However, as much as 25% of surgical wound infections, for example, appear after the person is discharged. Most nosocomial infections are probably not preventable, and many affect people whose resistance to infection is lowered by other means or by treatment for other diseases.

The four most frequently encountered types of nosocomial infection are, in descending order of frequency, urinary, wound and skin, respiratory, and bacteremic. *See* INFECTION. [B.D.]

Hot spots (geology) The surface manifestations of plumes, that is, columns of hot material, that rise from deep in the Earth's mantle. Hot spots are widely distributed around the Earth. One of their characteristics is an abundance of volcanic activity which persists for long time periods (greater than 1 million years). When the lithosphere (the rigid outer layer of the Earth) moves over a plume, a chain of volcanoes is left behind that progressively increases in age along its length. Hot spots are believed to be fixed with respect to each other and the deep mantle so that the age and orientation of these chains provide information on the absolute motions of the tectonic plates. *See* LITHOSPHERE; PLATE TECTONICS.

The Hawaiian-Emperor seamount chain in the central Pacific Ocean is a good example of a volcanic chain that was generated at a hot spot. The 3400-mi-long (5700-km) chain is made up mainly of tholeiitic lavas and ash tuff and pumice deposits. The lavas may have evolved from an initial submarine shield-building stage, through an explosive stage as they build up to sea level, and finally to a subaerial post-erosional stage. *See* LAVA; SEAMOUNT AND GUYOT.

Not all hot-spot volcanism is expressed in terms of highly lineated, multistage, volcanic chains. Aseismic ridges that extend up to or close to the axes of mid-oceanic ridges are another example of hot-spot volcanism. When a hot spot (for example, Iceland) is centered on the axis, pairs of ridges such as the Iceland-Faeroes Rise and the Greenland Rise are formed. Sometimes the plate (for example, Africa) has migrated off the hot spot (such as Tristan da Cunha), leaving behind ridge systems that no longer extend to the ridge axis (such as Rio Grande Rise and Western Walvis). *See* MID-OCEANIC RIDGE; VOLCANO; VOLCANOLOGY.

Another characteristic of hot spots is their association with broad swells in the Earth's topography. The Hawaiian hot-spot swell is believed to have been formed in response to either thermal or dynamic effects in an underlying mantle plume. The crustal and upper-mantle structure, which is constrained by seismic refraction data, shows that the oceanic crust is of uniform thickness beneath the swell. The long-wavelength correlation that is observed between the gravity anomaly and the topography (about 37 mGal mi^{-1} or 22 mGal km^{-1}) indicates that the mass excess of the swell is compensated by a low-density, high-temperature region below the crust. The uplift of hot-spot swells is believed to result from thermal perturbations in the underlying plume. The excess heights of swells suggest, on isostatic grounds, that temperature differences of about 450°F (250°C) occur between the plume and the surrounding mantle. Hot ascending plumes may raise the temperature of the overlying lithosphere, thereby thinning it.

Two classes of models have been proposed to explain hot-spot swells. In the reheating model, uplift is produced by thermal expansion that is confined to the conducting portion of the lithosphere (the thermal boundary layer). In the dynamic model, however, there is a contribution to the uplift that is produced by vertical normal stresses exerted to the seismically defined base of the lithosphere (the mechanical boundary layer) by convection.

The main distinguishing feature between the uplift models is that the reheating model predicts a higher heat flow than the dynamic model. Discrimination between these models therefore depends on how the subsidence history, heat flow, and long-term strength (which is controlled mainly by the temperature) differ from those for unperturbed lithosphere of the same age. [A.B.W.]

Hot-water heating system A heating system for a building in which the heat-conveying medium is hot water. A hot-water heating system consists essentially of water-heating or -cooling means and of heat-emitting means such as radiators, convectors, baseboard radiators, or panel coils. A piping system connects the heat source to the various heat-emitting units and includes a method of establishing circulation of the water or other medium and an expansion tank to hold the excess volume of water as it is heated and expands.

In a one-pipe system, radiation units are bypassed around a one-pipe loop. This type of system should only be used in small installations. In a two-pipe system (see illustration), radiation units are connected to separate flow and return mains, which may run in parallel or preferably on a reverse return loop, with no limit on the size of the system. In either type of system, circulation may be provided by gravity or pump.

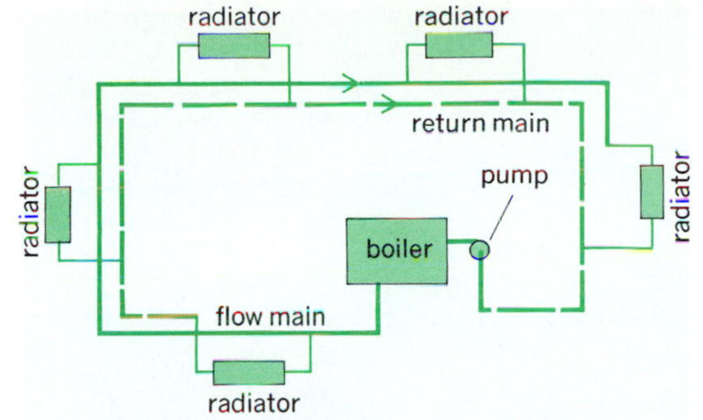

Two-pipe reverse return hot-water heating system.

One outstanding advantage of hot-water systems is the ability to vary the water temperature according to requirements imposed by outdoor weather conditions, with consequent savings in fuel. Radiation units may be above or below water heaters, and piping may run in any direction as long as air is eliminated. Hot water is admirably adapted to extensive central heating where high temperatures and high pressures are used and also to low-temperature panel-heating and -cooling systems. *See* COMFORT HEATING; DISTRICT HEATING; OIL BURNER. [E.L.W.]

Hot-wire anemometer A device for measuring air velocity, used principally in research on air turbulence and boundary layers. There are two types, known as constant-voltage and constant-temperature. The cooling of an electrically heated fine wire placed in a gas stream which alters wire resistance depends on the fluid velocity. With the constant-voltage type, as soon as the gas starts flowing, the hot wire cools off, the voltage across a Wheatstone bridge connected to the wire is kept constant, and the calibrated galvanometer shows a reading related to the velocity. This is useful only for very low velocities but is extremely sensitive—velocities down to 0.2 in./s (5 mm/s) are typical.

The constant-temperature, or constant-resistance, type is more useful. The resistance in the battery across the Wheatstone bridge is increased to such a value that the wire which was cooled by the stream is again brought to its original temperature. The current in the hot wire, read by a calibrated ammeter, gives a measure of the air velocity. Another important application is the measurement of turbulent fluctuations. These fluctuations show up as variations in wire resistance and are useful for boundary-layer studies. *See* BOUNDARY-LAYER FLOW; THERMOCOUPLE; WHEATSTONE BRIDGE. [H.Fo.]

Hubble constant The rate at which the velocity of recession of the galaxies increases with distance. Edwin Hubble established the existence of this motion observationally in 1929. *See* GALAXY, EXTERNAL.

The value of the expansion rate is still not agreed upon. The two values which seem to have the widest acceptance are 15 km per second per million light-years (1.6×10^{-18} s^{-1}) and 29 km per second per million light-years (3.1×10^{-18} s^{-1}).

If the expansion rate has been constant, then the age of the universe is the reciprocal of the Hubble constant. If, as seems more probable, the expansion has decelerated due to the gravitational attraction of the galaxies, the reciprocal of the Hubble constant gives an upper limit to the age, approximately 19 and 10×10^9 years respectively for the two values quoted above. W. A. Fowler gave the age of the universe as determined from the oldest globular clusters in the Galaxy as between 7 and 20×10^9 years, and from the abundances of the various radioactive element as between 10 and 15×10^9 years. It therefore appears that either value of the Hubble constant mentioned above is consistent with the other estimates of the age of the universe. *See* COSMOLOGY; STAR CLUSTERS. [L.C.G.]

Huebnerite A mineral with the chemical composition $MnWO_4$. Huebnerite is the manganese member of the wolframite solid-solution series. It commonly contains small amounts of iron. It occurs in monoclinic, short, prismatic crystals. Fracture is uneven. Luster varies from adamantine to resinous. Hardness is 4 on Mohs scale and specific gravity is 7.2. Huebnerite is transparent and yellowish to reddish-brown in color; streak is brown. *See* WOLFRAMITE. [E.C.T.C.]

Human-computer interaction The processes through which humans work with interactive computer systems. Some of these processes are readily observable in user interfaces to computer systems, which define the allowable communications between the computer and the user. Some of the interaction process occurs in the interpretation and generation of the communication, as the user and the computer form their responses to interface events. Human-computer interaction is also concerned with the design, evaluation, and implementation of interactive computer systems.

User interfaces include input devices, to transform users' physical actions to a form that the computer can process; output devices, to present messages from the computer system in a form which users can assimilate; a specification for the computer actions to respond to user inputs and to generate output presentations; and implementation of the computer actions in software and hardware. *See* COMPUTER PERIPHERAL DEVICES.

On the user side, interface processing begins with the user's goals in interacting with the computer. These can be refined into a sequence of actions to be taken by the computer, a sequence of inputs to the computer to cause these actions to take place, and physical actions to communicate the inputs. The outputs generated by the computer in response must then be perceived and interpreted. This interpretation may cause the user to continue the planned sequence of inputs or to bring it into better conformity with goals.

Developing human-computer interaction requires design of the processes on both sides of the interaction. The materials of the user interface, such as devices and processing specifications, are familiar to computer scientists as a design medium. However, the need to design user behavior as well implies additional knowledge and skills in applied cognitive science. Other contributing disciplines are ergonomics and graphics design. *See* HUMAN-FACTORS ENGINEERING.

As computing resources become less expensive and more widely available, human-computer interaction can offer increased communication bandwidth variety. User-interface devices may contain several display areas (windows) as views of different ongoing activities. Improvements in input techniques have made it easier to select items from displays, often using a hand-controlled pointing device (mouse). Use of multiple output media is also increasing, including sound, animations, and interactive video. *See* COMPUTER; COMPUTER GRAPHICS; DIGITAL COMPUTER; EMBEDDED SYSTEMS; VIRTUAL REALITY. [T.Ca.]

Human-factors engineering The area of knowledge dealing with the capabilities and limitations of human performance in relation to design of machines, jobs, and other modifications of the physical environment. Human-factors engineering seeks to ensure that humans' tools and environment are best matched to their physical size, strength, and speed and to the capabilities of the senses, memory, cognitive skill, and psychomotor preferences.

Human-factors engineering has also been termed human engineering, engineering psychology, applied experimental psychology, ergonomics, and biotechnology. It is related to the field of human-machine systems engineering but is more general, comprehensive, and empirical and not so wedded to formal mathematical models and physical analysis.

Among the problems of human-factors engineering are design of visual displays for ease and speed of interpretation; design of tonal signaling systems and voice communication systems for accuracy of communication; design of knobs and control handies and pedals; design of seats, work places, cockpits, and consoles in terms of humans' physical size, comfort, strength, and visibility. Human-factors engineering addresses problems of physiological stresses arising from such environmental factors as heat and cold, humidity, high and low atmospheric pressure, vibration and acceleration, radiation and toxicity, illumination or lack of it, and acoustic noise. Finally, the field includes psychological stresses of work speed and load and problems of memory, perception, decision-making, and fatigue. *See* HUMAN-MACHINE SYSTEMS. [T.B.S.]

Human genetics The study of human heredity and variation includes the physical basis (chromosomes and genes), the mechanisms whereby hereditary factors are transmitted from one generation to the next, the processes whereby genes determine characteristics, and the distribution of alternative genes and characteristics in different populations and races and the action of evolutionary forces upon them.

Chromosomes. The physical basis of human heredity is in all essential respects the same as in other organisms. Each cell contains in its nucleus threadlike chromosomes. Because of the nature of mitotic cell division, these chromosomes, barring some accident, are identical in number and kind with those present in the zygote formed by fusion of egg and sperm, and from which all other cells of the individual are derived. The chromosomes are nucleo-protein, and it is to be presumed that their deoxyribonucleic acid (DNA) is, as in other organisms, the actual chemical code of information transmitted via the germ cells from one generation to the next. The germ cells, formed by the meiotic divisions, carry one-half the chromosomes present in somatic cells and primordial germ cells; that is, the mature germ cells carry one chromosome of each kind instead of two. Fertilization restores the diploid chromosome number; each maternal chromosome is consequently matched by a similar paternal chromosome; and the corresponding genes at a particular site or locus in a given pair of chromosomes likewise form a pair. These paired genes, termed alleles, may be identical (homozygous) or different (heterozygous). *See* CHROMOSOME; MEIOSIS; MITOSIS.

The normal human chromosome number is 46, including 44 autosomes and 2 sex chromosomes. A diagram of the haploid chromosome set shows that the 12 largest chromosomes all have median, submedian, or subterminal centromeres (see illustration).

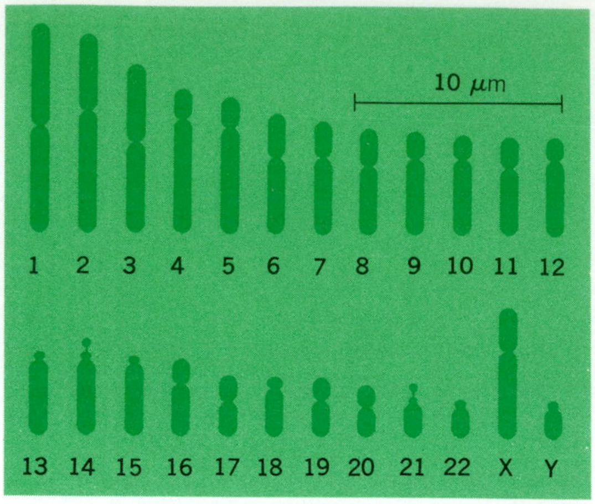

Drawing of a haploid set of human chromosomes. (*After E.H.Y. Chu and H. H. Giles, Human Chromosome complements in normal somatic cells in culture, Amer. J. Hum. Genet., 11(1):63–79, 1959*)

Among the smaller chromosomes there are 6, including the Y, with acrocentric, that is, nearly terminal, centromeres. The X is a large chromosome with a submedian centromere. Two smaller acrocentric chromosomes bear tiny satellites.

Sex is determined in the human species by a particular pair of chromosomes, represented in the female sex by two similar X chromosomes, and in the male sex by one X and one Y. Most of the Y chromosome is homologous with the proximal portion of the considerably larger X chromosome and crossing over might be expected to occur between the Y and this part of the X, because chiasmata are cytologically visible in this region. Eight or nine mutant loci have been assigned to this homologous portion of the X and Y chromosomes, on the basis of statistical evidence of their partial sex linkage. In order of closeness of linkage with sex, these are total color blindness, xeroderma pigmentosum, Oguchi's disease, spastic paraplegia, recessive epidermolysis bullosa, retinitis pigmentosa, hemorrhagic diathesis, and convulsive disorders. The validity of this partial sex linkage has also been questioned. *See* SEX-LINKED INHERITANCE.

The portion of the X chromosome present in two doses in females and in a single dose in males carries genes that show typical sex linkage, and of course must be linked with each other. The best-known case is the linkage between hemophilia and dichromatic (red-green) color blindness, which in some progenies of women heterozygous for both are closely linked. [H.B.G.]

Determination of inheritance pattern. The first step in all genetic analyses is to determine whether a character is inherited and if so, how it is inherited. Patterns of inheritance may be recognized only by the study of families. For genetic purposes, the term family means, as a minimum, parents and offspring; it may, however, include additional relatives.

By following the occurrence of certain characteristics over several generations it is possible to determine whether a characteristic is autosomal (dominant or recessive) or sex-linked (dominant or recessive). The inheritance is autosomal dominant when each affected individual has an affected parent. When affected males have affected sons, sex-linked dominance is excluded.

A simple type of autosomal inheritance concerns recessive characters such as albinism. An individual pedigree is not likely to contain sufficient information to permit the conclusion that inheritance of such a trait is recessive. Some features of pedigrees for rare recessive characters are that (1) the parents are usually normal (they are always normal if the character leads to early death, as in cystic fibrosis), (2) collateral and ancestral relatives are rarely affected, and (3) there may be an increased frequency of consanguineous matings. The last criterion is becoming of decreasing value in the United States.

Distinguishing features of pedigrees showing recessive sex linkage are that (1) affected males do not transmit the character (the gene) to their sons, (2) normal males have only normal descendants, (3) affected males other than sibs are related to each other via females, and (4) affected females do not occur. Exceptions may occur if the male in such a pedigree marries a carrier (heterozygous) female. In such a mating, affected males may have affected sons (the gene coming from the mother) and affected daughters, and normal males may have affected sons. The same features hold for sex-linked dominant characters. But, in addition, all daughters of affected males will be affected and normal females will have only normal descendants.

In the medicolegal uses of genetics, only genetic factors which are widespread in the population and which are inherited by the way of a simple, well-understood, and adequately tested genetic mechanism are useful. At present the only characters which fulfill these requirements are the ABO, MN, and Rh blood group systems. Accordingly, they are the only ones universally accepted for medicolegal use.

In cases of disputed paternity the question to be answered is whether this child of this mother could have this man as its father. If the child has an antigen not present in either its mother or alleged father, the man is excluded as the father; similarly, if the child does not have an antigen which the alleged father for genetic reasons had to transmit to his offspring, the man is excluded as the father. *See* BLOOD GROUPS. [A.G.S.]

Population analysis. A population, from the genetic viewpoint, constitutes a pool of genes with certain frequencies. The system by which the individuals mate determines how these genes are combined to form the genotypes. Two mating systems, panmixis or random mating, and assortative or nonrandom mating, are discussed in the following section.

Panmixis is a system of mating, also known as random mating, in which each member of one sex in a population is equally likely to mate with any member of the opposite sex. Such a population is said to be panmictic. A mating is then a random event, determined by chance. The frequency of certain types of matings is consequently determined by the frequency of the types of individuals in the population. The randomness of matings, however, refers only to certain characteristics of the mating individuals. The matings may be at random with respect to one characteristic (for example, blood groups) and yet not so with respect to some other characteristic (such as height). Random mating is not an absolute term and has meaning only with respect to specified characteristics of a given population. The result of panmixis in a large population with respect to one pair of autosomal genes (locus A, a, for example) is known as the Hardy-Weinberg. equilibrium. Another name is Castle-Hardy-Weinberg (CHW) law.

The Hardy-Weinberg law concerns the stability of the three genotypic (AA, Aa, aa) proportions in a large panmictic population in the absence of disturbing factors such as mutation. selection, migration, and sampling fluctuation. If a fraction p of all the genes at the Aa locus in the population is allele A and the rest ($q = 1 - p$) is allele a in both sexes, then a population consisting of $p^2(AA)$, $2pq(Aa)$, and q^2 (aa) will retain the same genotypic proportions in subsequent generations under the system of random mating. Such a population is said to be in an equilibrium state (of a neutral nature). The equilibrium law reflects the conservativeness of a mendelian population.

Assortative mating includes, in a broad sense, all systems of mating other than random mating. In the narrow but usual sense, it excludes the various systems of inbreeding (mating according to genetic relationship) and refers to nonrandom mating with respect to phenotypes. Positive assortative mating means that individuals of the same type tend to mate to a greater or lesser degree more often than they would by chance alone, and the matings between individuals of different types occur less often than dictated by chance alone. A preference based on phenotypic resemblance is involved. Familiar examples in humans are the positive correlations between spouses in such characteristics as color, stature, physique, and intelligence level, all of which are controlled to a certain extent by heredity. Positive assortative mating leads to greater parent-offspring correlation. If the degree of assortativeness in mating remains the same for a large number of generations, the population will approach an equilibrium condition. Moreover, the genotypic composition of an equilibrium population reached through assortative mating may be quite different from that reached through an inbreeding system. Negative assortative mating means that the individuals of unlike phenotypes tend to mate more often than expected by chance alone. Continued negative assortative mating of the same degree also leads to an equilibrium condition.

The difference between racial groups is usually found not in one but in many biological characteristics. A race, from the genetic viewpoint, is an aggregate of individuals who form more or less a mating group among themselves so that the genic content of this group differs from that of other mating groups. The fact that a marriage even between members of two different major races produces normal children who in tum also produce normal children shows that the chromosomes of all humans are homologous. Hence the racial differences must be attributed to the alleles of the various loci of the corresponding chromosomes. A particular allele may be present or absent in any race, but usually it occurs in all races. The racial differences arise from the differences in allelic frequencies. [C.C.L.]

Chromosome mapping. Human chromosome mapping, the localization of human genes to specific chromosomes or even regions of chromosomes, has undergone explosive growth since the development of somatic cell genetics. Combined with the more traditional methods of mapping from family studies and some other techniques, this has resulted in the mapping of more than 260 genes to the human autosomes. Although it has been estimated that less than 1% of all human genes have as yet been assigned to their respective chromosomes, it is now possible to speak of a human gene map. *See* GENETIC MAPPING; SOMATIC CELL GENETICS. [S.P.]

Human-machine systems The area of knowledge, also known as man-machine systems, dealing quantitatively with the communication and cooperation between people and machines in performing specific tasks. Human-machine systems engineering includes the experimentation techniques, the mathematical theory appropriate to handling data derived from experiments, and the engineering methods for applying this knowledge to the design of machines and systems. The pilot and the airplane, the driver and the automobile, the secretary and the typewriter, the talker and the telephone, the programmer and the computer are examples of human-machine systems. *See* SYSTEMS ENGINEERING.

Though it draws heavily on psychology and physiology, the field uses the language and concepts of engineering to analyze how people receive information through their vision, hearing, touch, and other senses, how they utilize this information to make decisions, and how they implement these decisions with their muscles. Often the response an individual makes in turn affects the new information which is displayed to the senses, thus completing a closed loop of information flow linking the person and machine.

As with pumps, gears, motors, and electronic circuits, engineers seek to describe the human operator in terms of definite input or stimulus quantities and output or response quantities and to connect these quantities by mathematical equations with a view to predicting the behavior of whole human-machine systems. The idea of predicting the behavior of a whole system from a knowledge of how the components are interconnected and what are the input-output equations of these components is at the root of what is commonly called systems theory.

Systems theory does not require understanding the inner workings of the components themselves; it is only necessary to know what that input-output relation is. This is why systems engineers often refer to their components as "black boxes." In the human-machine systems the engineer likewise need not understand the inner workings of the human nervous system—only what the stimulus-response relationship is in quantitative terms. The mathematical model of a person, moreover, need not correspond to the individual's anatomy or physiology in any way. It is only necessary that the black box representing the individual be connectable to those for the machine. Human-machine systems engineering is closely related to the field called human-factors engineering, which applies psychology and physiology to the design of machines but is not as concerned with compatible input-output equations for both humans and machine. *See* HUMAN-FACTORS ENGINEERING; SERVOMECHANISM. [T.B.S.]

Humidification The process of increasing the water-vapor content (humidity) of a gas. This process and its reverse operation, dehumidification, are important steps in air conditioning for human comfort and in many industrial operations. *See* AIR CONDITIONING; COMFORT HEATING; DEHUMIDIFIER; HUMIDITY.

Air (or other gas) can be humidified by direct injection of water vapor (steam) or, more commonly, by the evaporation of liquid water in contact with the airstream. When evaporation occurs, heat is required to provide the latent heat of vaporization. If no external source of heat is provided, either the water or the air, or both, will be cooled. The cooling of water by this process is the basis of operation for industrial cooling towers, whereas evaporative air coolers often used in hot, dry climates depend upon the air-cooling effect. In both these types of apparatus, humidification of the air occurs, although it is not the prime objective of the operation. In units designed primarily for humidification, the incoming air is usually heated to provide the latent heat of evaporation and to permit the air to leave the unit at controlled levels of both temperature and humidity. *See* COOLING TOWER. [A.L.K.]

Humidistat A controller that measures and controls relative humidity. A humidistat may be used to control either humidifying or dehumidifying equipment by the regulation of electric or pneumatic switches, valves, or dampers. Most methods for measuring humidity rely upon the swelling and shrinking of materials, such as human hair, silk, horn, goldbeater's skin, and wood, with increases and decreases in relative humidity.

An electronic humidistat includes a sensing element and a relay amplifier. The sensing element consists of alternate metal conductors on a small, flat plate with a plastic coating. An increase or decrease of the relative humidity causes a decrease or increase in the electrical resistance between the two sets of conductors; the change in resistance is measured by the relay amplifier. *See* HUMIDITY; PSYCHROMETRICS. [J.E.H./R.L.K.]

Humidity Atmospheric water-vapor content, expressed in any of several measures, especially relative humidity, absolute humidity, humidity mixing ratio, and specific humidity.

Relative humidity is the ratio, in percent, of the moisture actually in the air to the moisture it would hold if it were saturated at the same temperature and pressure. It is a useful index of dryness or dampness for determining evaporation, or absorption of moisture. *See* PSYCHROMETRICS.

Absolute humidity is the weight of water vapor in a unit volume of air expressed, for example, as grams per cubic meter or grains per cubic foot.

Humidity mixing ratio is the weight of water vapor mixed with unit mass of dry air, usually expressed as grams per kilogram. Specific humidity is the weight per unit mass of moist air and has nearly the same values as mixing ratio. [J.R.F.]

Humite A homologous series of magnesium nesosilicate minerals with general composition $Mg_{2n+1}(SiO_4)_n(F,OH)_2$. The known species include norbergite, chondrodite, humite, and clinohumite.

The minerals of the humite group have similar physical properties. The luster is resinous, and the color usually light yellow, brown, orange, or red. The pure synthetic Mg end members are colorless. Hardness is 6–6½ on Mohs scale, specific gravity is 3.1–3.2. These minerals are typically found in regionally crystallized marbles. Typical sources are the Grenville-age marbles in New York and Ontario, marble ejecta from Mount Vesuvius, and marbles from central Sweden. *See* SILICATE MINERALS. [P.B.M.]

Humus The amorphous, ordinarily dark-colored, colloidal matter in soil, representing a complex of the fractions of organic matter of plant, animal, and microbial origin that are most resistant to decomposition.

Humus consists of the combined residues of organic materials which have lost their original structure following the rapid decomposition of the simpler ingredients and includes synthesized cell substance as well as by-products of microorganisms. It is not a definite substance and is in a continual state of flux, disappearing by slow decomposition, and being constantly renewed by incorporation of residual matter. With a balance between these processes, humus, though not static, remains relatively uniform in nature and amount in a given soil. It constitutes a reservoir of stabilizing material which imparts beneficial physical, chemical, and biological properties to soil. Fertile soils are rich in humus.

Humus improves the texture of soils. It exerts a binding effect on sandy soils, and loosens the harder, clayey soils, thus increasing their porosity and permeability. It increases the moisture-holding capacity and improves the granular structure by cementing mineral particles into stable crumbs. This helps soils resist the pulverizing and eroding action of wind, water, and cultivation. As a storehouse of elements important to plants, humus functions as a regulator of soil processes by liberating gradually nutrients that would otherwise drain away. A soil rich in humus provides optimum conditions for the development of beneficial microorganisms and constitutes the best medium for growth of plants.

Peat is a type of humus that results from the decomposition of plant material under conditions of excessive moisture or in areas submerged in water. It is an organic deposit formed in marshes and swamps by the partial decomposition of countless generations of a variety of plants. *See* BOG; PEAT. [A.G.L.]

Hunger A term commonly used to refer to the subjective feelings that accompany the need for food; however, the study of this topic has come to include consideration of the overall control of food intake. More specifically, experimental work on the problem of hunger has been concerned with the sensory cues that give rise to feelings of hunger, the physiological mechanisms that determine when and how much food will be ingested, and the mechanism governing the selection of the food to be eaten.

Food consumption is basically controlled by the organism's nutritional status. Food deprivation leads to eating, and the ingestion of food materials terminates hunger sensations. Blood-sugar level can be used as a case in point. The concentration of blood sugar varies in a general way with the periodicity of the food cycle. Moreover, hyperglycemia (extremely high blood sugar) and hypoglycemia (extremely low blood sugar) have been observed to decrease and increase hunger respectively. Detailed analyses of normal life variations of blood sugar, however, reveal that the relation between the concentration of blood sugar and hunger is not sufficiently close for this single humoral factor to be able to control hunger in any simple and direct manner.

The evaluation of more local tissue utilization of food has proved a more promising approach to this problem. There is some evidence suggesting that the status of the liver is pivotal in the control of feeding. Depletion of liver glycogen stimulates feeding; its repletion terminates feeding in rats and rabbits.

Many stimuli that terminate feeding have been identified. Eating in food-deprived animals is inhibited by the reduction of either cellular water or of plasma fluid. It is also reduced by gastric distension and by infusing nutrients into the intestine and into the systemic, especially venous hepatic, circulation. Satiation produced by nutrient absorption from the intestine may be mediated, in part, by the gut hormone cholecystokinin. It is likely that the cholecystokinin is effective because it reduces the rate at which food passes through the stomach. [E.M.B.]

Huntington's disease A rare hereditary disease of the basal ganglia and cerebral cortex, resulting in choreiform (dancelike) movements, intellectual deterioration, and psychosis. The illness is strictly hereditary, as a dominant mendelian trait. All American patients are supposed to be descendants of two brothers with Huntington's disease, who migrated to the United States from England. No effective treatment is known. *See* HUMAN GENETICS; PSYCHOSIS. [F.C.R.]

Hurricane A tropical cyclone whose maximum sustained winds reach or exceed a threshold of 119 km/h (74 mi/h). In the western North Pacific ocean it is known as a typhoon. Many tropical cyclones do not reach this wind strength.

Maximum surface winds in hurricanes range up to about 200 mi (320 km) per hour. However, much greater losses of life and property are attributable to inundation from hurricane tidal surges and riverine or flash flooding than from the direct impact of winds on structures.

Tropical cyclones of hurricane strength occur in lower latitudes of all oceans except the South Atlantic and the eastern South Pacific, where combinations of cooler sea temperatures and prevailing winds whose velocities vary sharply with height prevent the establishment of a central warm core through a deep enough layer to sustain the hurricane wind system.

In the United States, property losses resulting from hurricanes have climbed steadily because of the increasing number of seashore structures. However, the loss of life, which has been huge in many storms, has decreased markedly. This is due mainly to the fact that warnings, aided by a more complete surveillance from aircraft and satellite, and extensive programs of public education, have become more accurate and more effective. Improvements in methodology for hurricane prediction have reduced the error in pinpointing hurricane landfall and have greatly reduced the probability of larger errors in prediction. *See* STORM; TROPICAL METEOROLOGY. [R.Sim.; J.Sim.]

Huygens' principle An assumption regarding the behavior of light waves, originally proposed by C. Huygens in the 17th century to explain the fact that light travels in straight lines and casts sharp shadows. Large-scale waves, such as sound waves or water waves, bend appreciably into the shadow. The special behavior of light may be explained by Huygens' principle, which states that "each point on a wavefront may be regarded as a source of secondary waves, and the position of the wavefront at a later time is determined by the envelope of these secondary waves at that time." Thus a wave WW originating at S is shown in the illustration at the instant it passes through an aperture. If a large number of circular secondary waves, originating at various points on WW, are drawn with the radius r representing the distance the wave would travel in time t, the envelope of these secondary waves is the heavily drawn circular arc $W'W'$. This represents the wave after t. If, as Huygens' principle requires, the disturbance is confined to the envelope, it will be 0 outside the limits indicated by points W'.

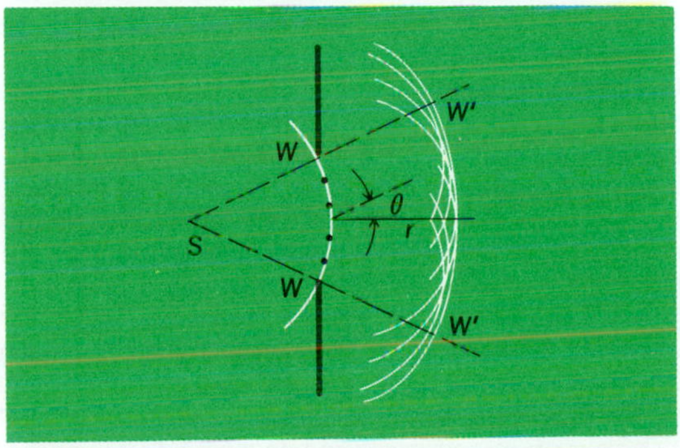

Huygens' principle: the construction for a spherical wave.

Careful observation shows that there is a small amount of light beyond these points, decreasing rapidly with distance into the geometrical shadow. This is called diffraction. *See* DIFFRACTION. [F.A.J./W.W.W.]

Hyades A group of stars, scattered through the "V" in the constellation of Taurus. An estimated 350 stars form the Hyades cluster. Because of the proximity of this cluster, which permits intensive observations, its main sequence stars are used as luminosity standards for the zero-age main sequence, from which the ages and distances of other clusters may be determined. However, the precise distance is still controversial. Reliable distance determinations fall between 140 and 150 light-years, but even a small error in the distance of the Hyades will magnify errors in distance determinations for more distant clusters. *See* STAR; STAR CLUSTERS. [H.S.H.]

Hyaluronic acid A polysaccharide which is an integral part of the gel-like substance of animal connective tissue; it supposedly serves as a lubricant and shock absorbent in the joints. Hyaluronic acid has also been isolated from umbilical cord, synovial fluid, skin, certain fowl tumors, and other sources. Treatment of this polysaccharide with the enzyme hyaluronidase, followed by acid hydrolysis, yields a disaccharide consisting of *N*-acetyl-D-glucosamine and D-glucuronic acid. This disaccharide appears to be the basic repeating structural unit that constitutes the hyaluronic acid molecule. *See* HYALURONIDASE; POLYSACCHARIDE. [W.Z.H.]

Hyaluronidase Any one of a family of enzymes, also known as hyaluronate lyases or spreading factors, produced by mammals, reptiles, insects, and bacteria, which catalyze the breakdown of hyaluronic acid. Some hyaluronidases also attack other similar polysaccharides. Since all liquefy the polysaccharide gel which fills the tissue spaces, they effectively accelerate diffusion so that injected, dissolved, or particulate matter (bacteria, viruses, toxins, or pigments) can diffuse through a larger volume of tissue. *See* HYALURONIC ACID.

The biological importance of the enzyme depends upon its source. That found in the culture filtrates of many strains of virulent bacteria permits the microorganisms to gain access to a larger volume of the host's tissue and, hence, to additional nutriment. That found in the venom of certain snakes and bees permits the toxin to produce more extensive damage to the victim. *See* ENZYME. [R.H.P.]

Hybrid dysgenesis A syndrome of abnormal traits that appears in the hybrids between certain strains of the fruit fly *Drosophila melanogaster*. The traits include partial sterility and greatly elevated rates of genetic mutations and chromosome rearrangements. Strains can be classified as P for paternally contributing or M for maternally contributing, so that only the hybrid sons and daughter of M females mated to P males show hybrid dysgenesis.

Hybrid dysgenesis is caused by the action of a family of transposable genetic elements, that is, segments of the genetic material (deoxyribonucleic acid, or DNA) with the special ability to move from one chromosomal site to another. Such elements range in size from a few hundred to a few thousand nucleotide pairs and typically occur in 10–100 genetic locations scattered throughout the chromosomes. Altogether, transposable elements are thought to compose 10–15% of the entire genetic complement of *Drosophila melanogaster* and are probably also common in most animal and plant species. The family of transposable elements that causes most cases of hybrid dysgenesis is called the P family, and the individual elements are called P factors because they occur only in the paternally contributing strains. *See* GENE; GENE ACTION; MUTATION; TRANSPOSONS. [W.R.E.]

Hydatellales An order of flowering plants, division Magnoliophyta (Angiospermae), in the subclass Commelinidae of the class Liliopsida (monocotyledons). The order consists of a single family with five species native to Australia, New Zealand, and Tasmania. The plants are small, submersed or partly submersed aquatic annuals with greatly simplified internal anatomy. The leaves are tufted at the base of the stem, and the inflorescence is a terminal head with two to several bracts, each subtending one to several reduced, unisexual flowers. These plants have sometimes been included within the Restionales, but the structural details of the ovary and seed set them apart. They are of no economic significance. *See* COMMELINIDAE; LILIOPSIDA; MAGNOLIOPHYTA; PLANT KINGDOM; RESTIONALES. [T.M.B.]

Hydrate A particular form of a solid compound which has water molecules associated with it An example is the hydrate of copper sulfate, $CuSO_4{\cdot}5H_2O$. Four water molecules are attached to the copper ion in the manner of coordination complexes, and the fifth water molecule is related to the sulfate and presumably held by hydrogen bonding.

Water can also be present in definite proportions in the crystal without being associated directly with the anion or cation. The water occupies a definite place in the crystal lattice.

Gas hydrates are compounds with a definite composition obtained when water and certain gases are chilled sufficiently. Chlorine, carbon dioxide, and acetylene all form such icy hydrates.

Some crystalline hydrates do not have definite proportions of water. Zeolites and similar silicate minerals, certain clays, and metallic oxides have variable proportions of water in their hydrated forms. *See* COORDINATION CHEMISTRY; HYDRATION. [F.W.]

Hydration The incorporation of molecular water into a complex with the molecules or units of another species. The complex may be held together by relatively weak forces or may exist as a definite compound. Many salts form solid hydrates when exposed to water vapor under certain conditions of temperature and pressure. Water is lost from these compounds when they are heated or when the water vapor pressure falls below a minimum value. Solids forming hydrates at low pressures are used as drying agents. *See* DELIQUESCENCE; DESICCANT; EFFLORESCENCE; HYDRATE; SOLUTION; SOLVATION. [F.J.J.]

Hydraulic accumulator A pressure vessel which oper-ates as a fluid source device or shock absorber. It is used to store fluid under pressure or to absorb excessive pressure increases. The hydraulic accumulator is an energy-efficient component, which allows the use of a smaller pump to achieve the same end results in terms of cylinder rod actuation speeds. In certain circuit designs, the accumulator will permit a pump motor to be completely shut down for an extended period of time while the accumulator supplies the necessary fluid to the circuit.

The operation of the hydraulic accumulator is induced by a pressurized gas (usually nitrogen), a spring, or a weighted plunger. The accumulator supplies fluid for actuator movement or to replace fluid lost by leakage. The gas-charged accumulator and the spring-type accumulator discharge their fluid into the system at pressures which are decreasing as the gas or spring expands. The weighted accumulator allows stored fluid to be discharged into the system at a constant pressure for the entirety of its downward stroke. *See* CONTROL SYSTEMS; HYDRAULICS; PUMP; SHOCK ABSORBER. [J.E.A.]

Hydraulic actuator A cylinder or fluid motor that converts hydraulic power into useful mechanical work. The mechanical motion produced may be linear, rotary, or oscillatory. Operation exhibits high force capability, high power per unit weight and volume, good mechanical stiffness, and high dynamic response. These features lead to wide use in precision control systems and in heavy-duty machine tool, mobile, marine, and aerospace applications. *See* CONTROL SYSTEMS; FLUIDICS.

Cylinder actuators provide a fixed length of straight-line motion. They usually consist of a tight-fitting piston moving in a closed cylinder. The piston is attached to a rod that extends from one end of the cylinder to provide the mechanical output. The double-acting cylinder (Fig. 1) has a port at each end of

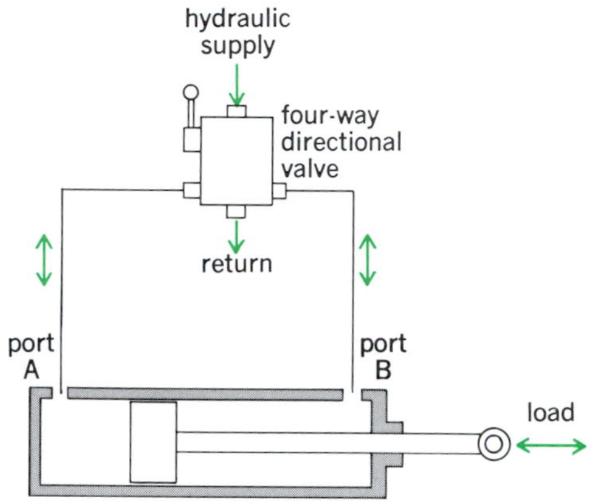

Fig. 1. Function of a hydraulic double-acting cylinder.

the cylinder to admit or return hydraulic fluid. A four-way directional valve functions to connect one cylinder port to the hydraulic supply and the other to the return, depending on the desired direction of the power stroke.

Limited-rotation actuators are used for lifting, lowering, opening, closing, indexing, and transferring movements by producing limited reciprocating rotary force and motion. Rotary actuators are compact and efficient, and produce high instantaneous torque in either direction. Figure 2 shows a piston-rack type of rotary actuator. Hydraulic fluid is applied to either the two end chambers or the central chamber to cause the two pistons to retract or extend simultaneously so that the racks rotate the pinion gear.

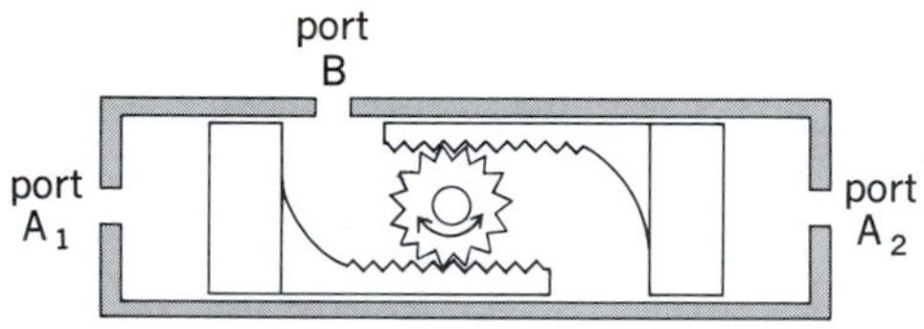

Fig. 2. Piston-rack type rotary actuator.

Rotary motor actuators are coupled directly to a rotating load and provide excellent control for acceleration, operating speed, deceleration, smooth reversals, and positioning. They allow flexibility in design and eliminate much of the bulk and weight of mechanical and electrical power transmissions.

Motor actuators are generally reversible and are of the gear or vane type. [C.M.]

Hydraulic gradient The slope along a closed water con-duit that measures the sum of elevation and pressure heads. Total head consists of these heads plus the velocity head. The elevation and pressure heads at a point can be measured directly by connecting a tube to a small hole in the conduit wall and observing the vertical height to which the water rises in the tube, called a piezometer. *See* BERNOULLI'S THEOREM. [D.R.F.H.]

Hydraulic jump An abrupt increase of depth in a free-surface liquid flow. A hydraulic jump is characterized by rapid flow and small depths on the upstream side, and by larger depths and smaller velocities on the downstream side. A jump can form only when the upstream flow is supercritical, that is, when the fluid velocity is greater than the propagation velocity c of a small, shallow-water gravity wave ($c = \sqrt{gh}$, where h is the depth). *See* OPEN CHANNEL. [D.R.F.H.]

Hydraulic press A combination of a large and a small cylinder connected by a pipe and filled with a fluid so that the pressure created in the fluid by a small force acting on the piston in the small cylinder will result in a large force on the large piston. The operation depends upon Pascal's principle, which states that when a liquid is at rest the addition of a pressure (force per unit area) at one point results in an identical increase in pressure at all points.

The principle of the hydraulic press is used in lift jacks, earth-moving machines, and metal-forming presses (see illustration). A comparatively small supply pump creates pressure

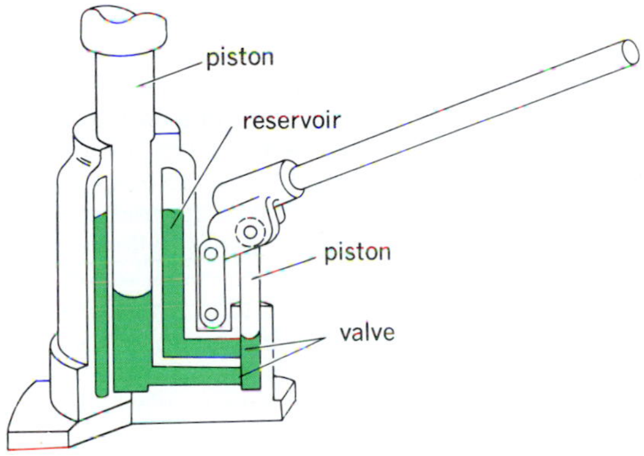

Hydraulic jack.

in the hydraulic fluid. The fluid then acts on a substantially larger piston to produce the action force. Heavy objects are accurately weighed on hydraulic scales in which precision-ground pistons introduce negligible friction. *See* MECHANICAL ADVANTAGE; SIMPLE MACHINE. [R.M.Ph.]

Hydraulic turbine A machine which converts the energy of an elevated water supply into mechanical energy of a rotating shaft. Most old-style waterwheels utilized the weight effect of the water directly, but all modern hydraulic turbines are a form of fluid dynamic machinery of the jet and vane type operating on the impulse or reaction principle and thus involving the conversion of pressure energy to kinetic energy. The shaft drives an electric generator, and speed must be of an acceptable synchronous value. *See* GENERATOR; IMPULSE TURBINE; PELTON WHEEL; TURBINE.

Efficiency of hydraulic turbine installations is always high, more than 85% after all allowances for hydraulic, shock, bearing, friction, generator, and mechanical losses. Material selection is not only a problem of machine design and stress loading from running speeds and hydraulic surges, but is also a matter of fabrication, maintenance, and resistance to erosion, corrosion, and cavitation pitting.

Pumped-storage hydro plants have employed various types of equipment to pump water to an elevated storage reservoir during off-peak periods and to generate power during on-peak periods where the water flows from the elevated reservoir through hydraulic turbines. *See* WATERPOWER. [T.Ba.]

Hydraulic valve lifter A device that eliminates the need for mechanical clearance in the valve train of internal combustion engines. Clearance is normally required to prevent the valve's being held open and destroyed as the valve train undergoes thermal expansion. However, clearance requires frequent adjustment and is responsible for much operating noise. The hydraulic lifter is a telescoping compression strut in the linkage between cam and valve, consisting of a piston and cylinder (see illustration). When no opening load

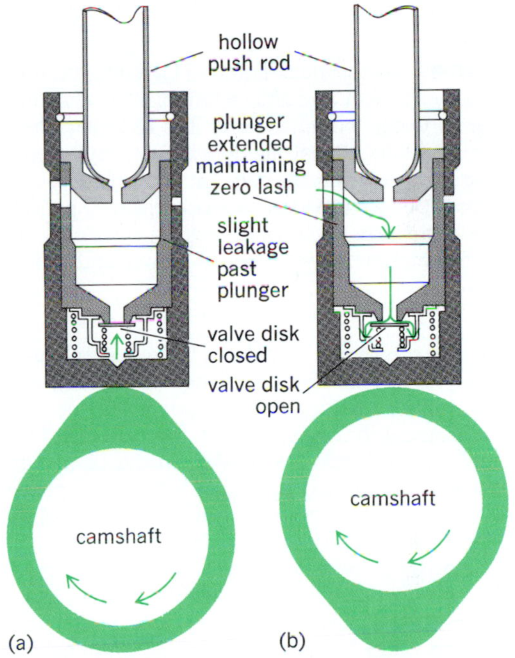

Positions of the hydraulic valve lifter, with engine valve (a) open and (b) closed. (*After W. H. Crouse, Automotive Mechanics, 5th ed., McGraw-Hill, 1965*)

exists, a weak spring moves the piston, extending the strut and eliminating any clearance. This action sucks oil into the cylinder past a check valve. The trapped oil transmits the valve-opening forces with little deflection. A slight leakage of oil during lift shortens the strut, assuring valve closure. The leakage oil is replaced as the spring again extends the strut at no load. *See* VALVE TRAIN.
[A.R.R.]

Hydraulics The behavior of water or other liquids, chiefly when in motion. As a part of fluid mechanics, hydraulics deals with the properties, behavior, and effects of all liquids at rest against, or in motion relative to, boundary surfaces or objects. Its laws also apply to bases when changes in density are small, that is, compressibility effects are negligible. *See* FLUID MECHANICS.

Common hydraulic applications encountered in civil engineering include the flow of water in pipes, canals, and rivers, in flood control, in land reclamation, and in hydroelectric power projects. Hydraulics also influences appurtenant structures and devices such as dams, dikes, locks, spillways, weirs, piers, ship hulls, gates, and valves. Mechanical and chemical engineering applications include the flow of gases, oils, or other liquids; lubrication; and fluid machinery such as pumps, turbines, propellers, fans, and fluid power transmission and control devices, including servomechanisms.

Characteristics of particular phenomena in hydraulics include fluid properties; the shapes, sizes, relative roughnesses and relative motions of the surfaces or objects involved; the relative times and distances (short or long); whether flow is closed as in a pipe or open as in a canal; and whether motion is in a relatively extensive fluid as a submarine deeply submerged or whether motion occurs where there is a free surface as a ship at sea.

Hydraulics deals with bulk fluids which are regarded as homogeneous; it is not concerned with molecular sizes. As distinct from solids, fluids, such as water, are unable to resist shearing forces while remaining in a state of equilibrium. Among fluids, a liquid mass will have a definite volume, varying only slightly with temperature and pressure, whereas a mass of gas will fill any space available to it, the pressure adjusting itself accordingly. *See* DENSITY; ELASTICITY; HYDROSTATICS; VISCOSITY. [W.A.]

Hydrazine A colorless liquid, H_2NNH_2 (boiling point 114°C or 237°F), with a musty, ammonialike odor. Physically it is similar to water, but chemically it is reducing, decomposable, basic, and bifunctional. Its derivatives range from simple salts to ring compounds, polymers, and coordination complexes. Major uses of hydrazine include such diverse applications as rocket fuels, corrosion inhibition in boilers, syntheses of biologically active materials and in rubber curing and foam-rubber production. [T.H.D.]

Hydride A compound containing hydrogen and another element. Many of the elements combine with hydrogen to form compounds, and even though hydrogen is less negative than the second element in the compound, they are called hydrides. For example, H_2S is called a hydride in the discussion here, although it is usually called hydrogen sulfide and properly so. Hydrides can be divided into three classes of compounds: covalent hydrides, saltlike hydrides, and metallic hydrides.

Covalent hydrides refer to compounds such as H_2O, H_2S, and NH_3, which are volatile. Covalent hydrides are formed from the nonmetals. Carbon hydrides would refer to natural gas, CH_4, and other hydrocarbons containing chains of carbons. Saltlike hydrides are ionic in nature, and nonvolatile. Lithium hydride, LiH, is a typical example. Metallic hydrides include compounds with simple formulas such as uranium hydride, UH_3, or palladium hydride, PdH_x. Some of these compounds are thought to contain the hydrogen in the holes in the metallic lattice. *See* HYDROCARBON; HYDROGEN; METAL HYDRIDES. [E.E.W.]

Hydrido complexes Complex hydrides containing a hydride ligand bonded to a central atom. The prefix hydro instead of hydrido is sometimes used. All are soluble in aromatic hydrocarbons. They are rather expensive, but their specific reducing powers make them attractive for synthesizing high-value products, such as pharmaceuticals, flavorings, fragrances, dyes, and insecticides. *See* COORDINATION COMPLEXES; METAL HY-DRIDES. [J.C.W.]

Hydroboration The process of producing organoboranes by the addition of diborane to unsaturated organic compounds. In ether solvents the addition of diborane to such molecules is exceedingly rapid and essentially quantitative. This reaction therefore makes the organoboranes readily available. Such organoboranes are finding increasing application as intermediates for organic synthesis.

Diborane is highly soluble in tetrahydrofuran, where it exists as the addition compound tetrahydrofuran-borane. Such solutions are often used for hydroboration, and merely involve bringing the two reactants together as indicated by reaction (1).

$$3RCH{=}CH_2 + C_4H_8O{:}BH_3 \rightarrow (RCH_2CH_2)_3B + C_4H_8O \quad (1)$$

Alternatively, sodium borohydride may be utilized to achieve hydroboration by the addition of boron trifluoride etherate. This is shown by reaction (2). Usually the organoborane is not

$$12RCH{=}CH_2 + 3NaBH_4 + 4(C_2H_5)_2O{:}BF_3 \rightarrow 4(RCH_2CH_2)_3B + 3NaBF_4 + 4(C_2H_5)_2O \quad (2)$$

isolated but is utilized in place, similar to applications of the Grignard reagent in synthesis.

Organoboranes are among the most versatile synthetic intermediates that are available to the organic chemist. The hydroboration reaction has made these intermediates readily available. *See* BORANE; ORGANIC CHEMICAL SYNTHESIS; ORGANOMETALLIC COMPOUND. [H.C.B.]

Hydrocarbon One of a group of chemical compounds composed only of hydrogen and carbon. The very large variety of hydrocarbons can best be divided into three classes (aliphatic, alicyclic, and aromatic), each of which may be further divided into a number of subclasses. *See* ALICYCLIC HYDROCARBON; ALIPHATIC HYDROCARBON; AROMATIC HYDROCARBON.

Of the natural sources of hydrocarbons, the largest are natural gas and petroleum. The composition of natural gas varies greatly depending on the area from which it is obtained.

Petroleum is a complex mixture of liquid and solid hydrocarbons, its composition also varying with the source. Its chief components are paraffinic hydrocarbons, saturated five- and six-carbon atom alicyclic hydrocarbons, and aromatic hydrocarbons. Olefins are usually absent. *See* NATURAL GAS; PETROLEUM.

Hydrocarbons are also obtained by the destructive distillation or carbonization of coal, producing coal gas and coal tar. Coal gas contains methane together with smaller amounts of ethane, ethylene, benzene, toluene, cyclopentadiene, and naphthalene. The normally liquid or solid hydrocarbons in the gas (namely the cyclopentadiene, benzene, toluene, and napthalene) can be recovered.

The volatile oils obtained from certain plants (caraway or dill), trees (pine), and citrus fruits (lemon or orange) contain members of a large group of hydrocarbons known as terpenes. *See* TERPENE. [L.S.]

Hydrocarbon resin Brittle or gummy materials prepared by the polymerization of several unsaturated constituents of coal tar, resin, or petroleum. The hydrocarbon resins are inexpensive and find many uses in rubber and asphalt formulations and in coating and calking compositions. Important classes are the coumarone-indene resins, the petroleum resins, and the polyterpene resins.

Coumarone and indene occur together in coal tar fractions. Polymerization may be readily effected in the presence of sulfuric acid to yield soft to hard, brittle polymers. The soft products are used as plasticizers and tackifiers in calking formulations and rubber compositions. The hard products are used to stiffen and strengthen synthetic rubber, and they also are used in floor coverings, roofing materials, varnishes, and metal paints, and in combination with other resins.

Petroleum resins are derived from mixtures of polymerizable dienes, for example, cyclopentadiene, and unsaturated compounds obtained from gasoline-refining operations. The properties and uses of the resins are generally similar to those of the coumarone-indene resins.

Polyterpenes are derived from the α- and β-pinene in turpentine. They polymerize in the presence of acid catalysts with the formation of soft to hard polymers of molecular weights up to about 2000. The properties and uses of the polyterpenes are generally similar to those of the other hydrocarbon resins described above. They are used in oil paints and adhesive formulations. *See* POLYMERIZATION. [J.A.M.]

Hydrocharitales An order of aquatic flowering plants, division Magnoliophyta (Angiospermae), in the subclass Alismatidae of the class Liliopsida (monocotyledons). The order consists of the single family Hydrocharitaceae, with about 100 species. Within the subclass the order is marked by its inferior, compound ovary with basically laminar placentation. The ovules are scattered over the walls of the individual carpels. The familiar aquarium plant, *Vallisneria spiralis*, called tape grass or eelgrass, is not a grass but belongs to the Hydrocharitaceae. *See* ALISMATIDAE; FLOWER; LILIOPSIDA. [A.Cr.]

Hydrocracking A catalytic, high-pressure process flexible enough to produce either of the two major light fuels—high octane gasoline or aviation jet fuel. It proceeds by two main reactions: adding hydrogen to molecules too massive and complex for gasoline and then cracking them to the required fuels. The process is carried out by passing oil feed together with hydrogen at high pressure (1000–2500 pounds per square inch gage or 7–17 megapascals) and moderate temperatures (500–750°F or 260–400°C) into contact with a bifunctional catalyst, comprising an acidic solid and a hydrogenating metal component. Gasoline of high octane number is produced, both directly and through a subsequent step such as catalytic reforming; jet fuels may also be manufactured simply by changing conditions with the same catalysts.

Generally, the process is used as an adjunct to catalytic cracking. Oils, which are difficult to convert in the catalytic process because they are highly aromatic and cause rapid catalyst decline, can be easily handled in hydrocracking, because of the low cracking temperature and the high hydrogen pressure, which decreases catalyst fouling. However, the most important components in any feed are the nitrogen-containing compounds, because these are severe poisons for hydrocracking catalysts and must be almost completely removed.

The products from hydrocracking are composed of either saturated or aromatic compounds; no olefins are found. In making gasoline, the lower paraffins formed have high octane numbers. The remaining gasoline has excellent properties as a feed to catalytic reforming, producing a highly aromatic gasoline which, with added lead, easily attains 100 octane number. Another attractive feature of hydrocracking is the low yield of gaseous components, such as methane, ethane, and propane, which are less desirable than gasoline. *See* GASOLINE.

The hydrocracking process is being applied in other areas, notably, to produce lubricating oils and to convert very asphaltic and high-boiling residues to lower-boiling fuels. *See* AROMATIZATION; CRACKING; HYDROGENATION. [C.P.B.]

Hydrodynamics That branch of continuum mechanics which treats of the laws of motion of an incompressible fluid and of the interactions of the fluid with its boundaries. For more detail on specific parts of the subject *see* FLUID FLOW; FLUID MECHANICS.

Fluids have many physical properties, including, besides thermodynamic and electromagnetic properties, the dynamical ones of density, viscosity, adhesion, and cohesion. Only the dynamical properties will be considered in this article, and furthermore, because capillary effects will not be treated, the last two properties will be of interest only insofar as they affect the inception of cavitation.

A flow field can be studied from either of two points of view, the lagrangian, in which one is concerned with the history of each fluid particle, and the eulerian, in which interest is focused, rather, on the velocity vectors associated with each point. *See* EULER'S EQUATIONS OF MOTION; LAGRANGE'S EQUATIONS.

A fruitful concept in hydrodynamics is that of the ideal or inviscid fluid. In such a fluid, across any element of area, the fluid on one side exerts a normal stress or pressure on the fluid on the other side, the magnitude of which is independent of the orientation of the element. Thus there is a scalar pressure field associated with the vector velocity field. A principal problem of hydrodynamics is to determine these fields corresponding to prescribed boundary conditions.

When viscosity is taken into account, the normal stresses across an area element are not, in general, independent of orientation, and in addition, tangential stresses are present. Instead of a scalar pressure field one must consider now a tensor stress field. These stresses, or their integrals, must be determined in order to solve the important problem of obtaining the force and moment acting on a body moving through the fluid. *See* NEWTONIAN FLUID; VISCOSITY.

Since the viscosity of the most common fluids, air and water, is very small, it is an excellent approximation to treat their flows as inviscid except at very low Reynolds numbers and in the neighborhood of the boundaries. Because the velocity of a fluid at a wall is zero relative to that of the boundary (nonslip condition), a so-called boundary layer in which viscous effects are important is present, within which appreciable shear stresses occur. Outside of this boundary layer the flow may be treated as inviscid, and to a good approximation, the pressure across the boundary layer may be assumed to be that in the inviscid flow at the outer edge of the boundary layer.

A simplified form of the Navier-Stokes equations, called the boundary-layer equations, yields laminar-flow solutions for the velocity profiles. It is known, however, that boundary-layer flows are often turbulent. A theory of the stability of the laminar boundary layer has been developed and current research is shedding considerable light on the processes whereby a disturbed laminar flow eventually breaks down into the typical random motions of turbulence, but the mechanisms are not yet clear. *See* BOUNDARY-LAYER FLOW; LAMINAR FLOW; NAVIER-STOKES EQUATIONS; TURBULENT FLOW. [L.L.]

Hydroelectric generator An electric rotating machine that transforms mechanical power from a hydraulic turbine or water wheel into electric power. Hydroelectric generators may have horizontal or vertical shafts, depending upon the turbine. The most common type in large ratings is the vertical-shaft, synchronous generator. The hydroelectric generator may be arranged to operate as a motor during periods of low power demand. In such a case the turbine serves as a pump, raising water to a high elevation for reuse to generate power at a period of peak load. Such a unit is termed a pump turbine. *See* ELECTRIC POWER GENERATION; GENERATOR; HYDRAULIC TURBINE; TURBINE; WATERPOWER. [L.T.R.]

Hydrofoil craft Marine craft of generally conventional hull form (as in planing or displacement craft), which are supported above the water by dynamic lift produced by winglike surfaces (hydrofoils) mounted below the hull. At low speed, these craft operate in the same manner as conventional hulls. As speed increases, they "take off" and "fly" supported by the hydrofoils. The lift principle employed by these foils in water is the same as that of an airfoil in air, although application of the principle entails consideration of the additional complexities caused by the air-water interface. *See* AIRFOIL.

The advantages which have encouraged the development of hydrofoil craft are increased efficiency, that is, a hydrofoil craft offers about half the resistance of a planing boat of similar weight and speed; and increased seakeeping ability.

There are two major types of hydrofoils: surface-piercing, which attain their stability by hydrodynamically balancing the amount of foil area above and below the water surface; and submerged-foil systems with autopilot which, by means of a height sensor and gyros, control the height and attitude of the craft above the water. The latter, more complicated system was devised to counter the effect of the orbital motion of the wave particles, which adversely affects surface-piercing systems, particularly in a following sea.   [T.M.B.]

Hydroformylation

An aldehyde synthesis process that falls under the general classification of a Fischer-Tropsch reaction but is distinguished by the addition of an olefin feed along with the characteristic carbon monoxide and hydrogen. In the oxo process for alcohol manufacture, hydroformylation of olefins to aldehydes is the first step. The second step is the hydrogenation of the aldehydes to alcohols. At times the term "oxo process" is used in reference to the hydroformylation step alone. In the hydroformylation step, olefin, carbon monoxide, and hydrogen are reacted over a cobalt catalyst to produce an aldehyde which has one more carbon atom than the feed olefin. As in the reaction below, the olefin conversion takes place by the addition of a formyl group (CHO) and a hydrogen atom across the double bond. *See* FISCHER-TROPSCH PROCESS.

$$R-CH=CH_2 + CO + H_2 \longrightarrow \begin{array}{l} R-CH_2-CH_2-CHO \\[1em] R-CH-CH_3 \\ \quad\quad | \\ \quad\quad CHO \end{array}$$

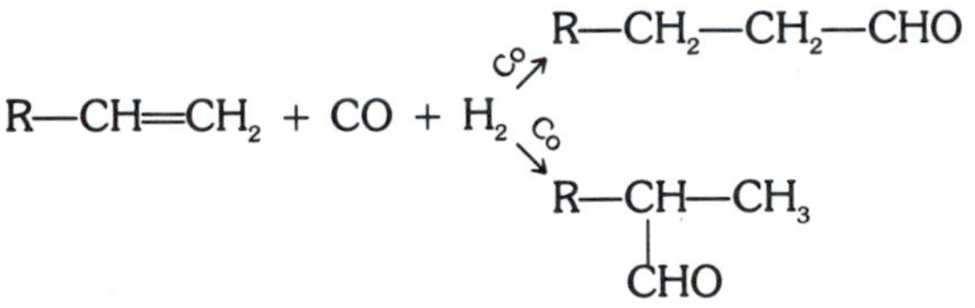

The aldehyde is then treated with hydrogen to form the alcohol. In commercial operations, the hydrogenation step is usually performed immediately after the hydroformylation step in an integrated system.

A wide range of carbon-number olefins, C_2–C_{16}, have been used as feeds. Propylene, heptene, and nonene are frequently used as feedstocks to produce normal and isobutyl alcohol, isooctyl alcohol, and primary decyl alcohol, respectively. Feed streams to oxo units may be single-carbon-number or mixed-carbon-number olefins.

The lower-carbon-number alcohols such as butanols are used primarily as solvents, while the higher-carbon-number alcohols go into the manufacture of plasticizers, detergents (surfactants), and lubricants.   [D.L.H.]

Hydrogen

The first chemical element in the periodic system. Under ordinary conditions it is a colorless, odorless, tasteless gas composed of diatomic molecules, H_2. The hydrogen atom, symbol H, consists of a nucleus of unit positive charge and a single electron. It has atomic number 1 and an atomic weight of 1.00797. The element is a major constituent of water and all organic matter, and is widely distributed not only

Properties of hydrogen	
Property	Value
Melting point	−259.2°C
Boiling point at 1 atm	−252.8°C
Density of solid at −259.2°C	0.0866 g/cm^3
Density of liquid at −252.8°C	0.0708 g/cm^3
Critical temperature	−240.0°C
Critical pressure	13.0 atm
Critical density	0.0301 g/cm^3

on the Earth but throughout the universe. There are three isotopes of hydrogen: protium, mass 1, makes up 99.98% of the natural element; deuterium, mass 2, makes up about 0.02%; and tritium, mass 3, occurs in extremely small amounts in nature but may be produced artificially by various nuclear reactions. *See* DEUTERIUM; ISOTOPE; TRITIUM.

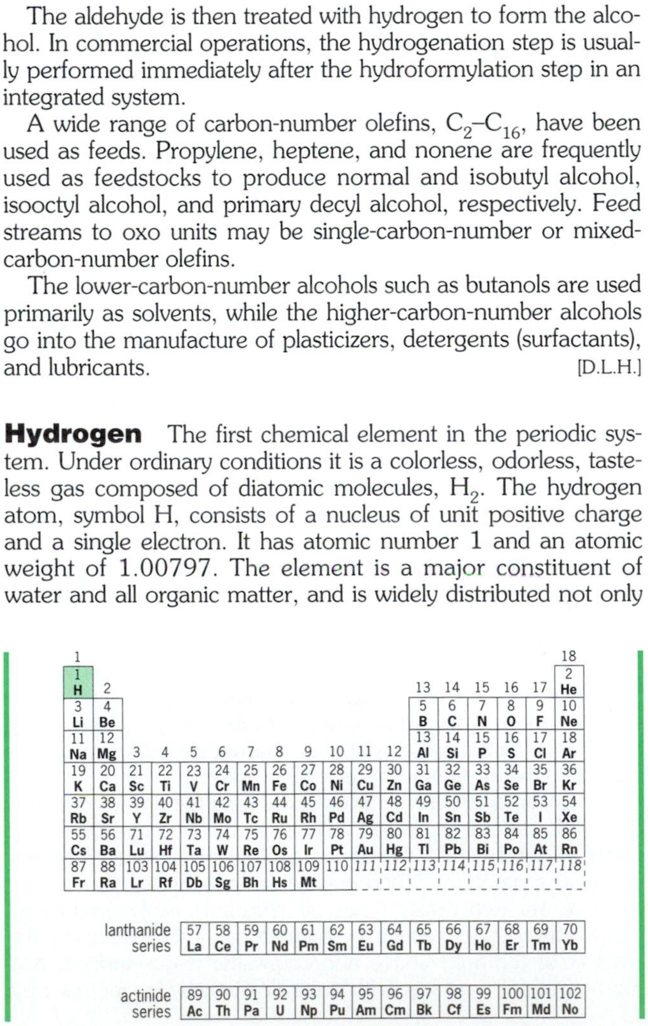

Uses. The largest single use of hydrogen is in the synthesis of ammonia. A rapidly expanding use for hydrogen is in petroleum-refining operations, such as hydrocracking and hydrogen treatment for removal of sulfur. Large quantities of hydrogen are consumed in the catalytic hydrogenation of unsaturated liquid vegetable oils to make solid fats. Hydrogenation is used in the manufacture of organic chemicals. Large quantities of hydrogen are used as a rocket fuel, in conjunction with oxygen or fluorine, and as a propellant for nuclear-powered rockets.

Properties. Ordinary hydrogen has a molecular weight of 2.01594. The gas has a density at 0°C and 1 atm of 0.08987 g/liter. Its specific gravity, compared to air, is 0.0695. Hydrogen is the lightest substance known. Some additional properties of hydrogen are listed in the table.

Hydrogen is somewhat more soluble in organic solvents than in water. Many metals adsorb hydrogen. The adsorption of hydrogen in steel may cause "hydrogen embrittlement," which sometimes leads to the failure of chemical processing equipment.

At ordinary temperatures hydrogen is a comparatively unreactive substance unless it has been activated in some manner, for example, by a suitable catalyst. At elevated temperatures it is highly reactive.

Although ordinarily diatomic, molecular hydrogen dissociates at high temperatures into free atoms. Atomic hydrogen is a powerful reducing agent, even at room temperature. It reacts with the oxides and chlorides of many metals, including silver, copper, lead, bismuth, and mercury, to produce the free metals. It reduces some salts, such as nitrates, nitrites, and cyanides of sodium and potassium, to the metallic state. It reacts with a number of elements, both metals and nonmetals, to yield hydrides such as NaH, KH, H_2S, and PH_3. With oxygen atomic hydrogen yields hydrogen peroxide, H_2O_2. With organic compounds atomic hydrogen reacts to produce a complex mixture of products. With ethylene, C_2H_4, for example, the products include ethane, C_2H_6, and butane, C_4H_{10}. The heat liberated when hydrogen atoms recombine to form hydrogen molecules is used to obtain very high temperatures in atomic hydrogen welding.

Hydrogen reacts with oxygen to form water. At room temperature this reaction is immeasurably slow, but is accelerated by catalysts, such as platinum, or by an electric spark, and then may take place with explosive violence.

With nitrogen, hydrogen undergoes an important reaction to give ammonia. Hydrogen reacts at elevated temperatures with a number of metals to give hydrides. The oxides of many metals are reduced by hydrogen at elevated temperatures either to the free metal or to lower oxides. Hydrogen reacts at room temperature with the salts of the less electropositive metals and reduces them to the metallic state. In the presence of a suitable catalyst hydrogen reacts with unsaturated organic compounds and adds to the double bond. *See* HYDROGENATION.

1																	18
1 H	2											13	14	15	16	17	2 He
3 Li	4 Be											5 B	6 C	7 N	8 O	9 F	10 Ne
11 Na	12 Mg	3	4	5	6	7	8	9	10	11	12	13 Al	14 Si	15 P	16 S	17 Cl	18 Ar
19 K	20 Ca	21 Sc	22 Ti	23 V	24 Cr	25 Mn	26 Fe	27 Co	28 Ni	29 Cu	30 Zn	31 Ga	32 Ge	33 As	34 Se	35 Br	36 Kr
37 Rb	38 Sr	39 Y	40 Zr	41 Nb	42 Mo	43 Tc	44 Ru	45 Rh	46 Pd	47 Ag	48 Cd	49 In	50 Sn	51 Sb	52 Te	53 I	54 Xe
55 Cs	56 Ba	71 Lu	72 Hf	73 Ta	74 W	75 Re	76 Os	77 Ir	78 Pt	79 Au	80 Hg	81 Tl	82 Pb	83 Bi	84 Po	85 At	86 Rn
87 Fr	88 Ra	103 Lr	104 Rf	105 Db	106 Sg	107 Bh	108 Hs	109 Mt	110	111	112	113	114	115	116	117	118

lanthanide series	57 La	58 Ce	59 Pr	60 Nd	61 Pm	62 Sm	63 Eu	64 Gd	65 Tb	66 Dy	67 Ho	68 Er	69 Tm	70 Yb
actinide series	89 Ac	90 Th	91 Pa	92 U	93 Np	94 Pu	95 Am	96 Cm	97 Bk	98 Cf	99 Es	100 Fm	101 Md	102 No

Principal compounds. Hydrogen is a constituent of a very large number of compounds containing one or more other elements. Such compounds include water, acids, bases, most organic compounds, and many minerals. Compounds in which hydrogen is combined with a single other element are commonly referred to as hydrides. For additional details on the compounds of hydrogen *see* ACID AND BASE; HYDRAZINE; HYDRIDE; HYDRIDO COMPLEXES; HYDROCARBON; HYDROGEN FLUORIDE; HYDROGEN PEROXIDE; WATER.

Preparation. A large number of methods may be used to prepare hydrogen gas. The choice of method is determined by such factors as the quantity of hydrogen desired, the purity required, and the availability and cost of raw materials. Among the processes frequently used are the reactions of metals with water or acids, the electrolysis of water, the reaction of steam with hydrocarbons or other organic materials, and the thermal decomposition of hydrocarbons. The principal raw materials for hydrogen production are hydrocarbons, such as natural gas, oil refinery gas, gasoline, fuel oil, and crude oil. [L.K.]

Hydrogen bomb A device in which an uncontrolled, self-sustaining thermonuclear fusion reaction is carried out in heavy hydrogen (deuterium) gas to produce an explosion. The hydrogen bomb is a fusion bomb. In a fusion reaction, the collision of two energy-rich nuclei results in a mutual rearrangement of their protons and neutrons to produce two or more reaction products, together with a release of energy. *See* NUCLEAR FUSION; THERMONUCLEAR REACTION.

For the hydrogen bomb reaction to become self-sustaining, a so-called critical temperature of about 3.5×10^7 K must be attained, possibly with the aid of the enormous temperatures developed by the explosion of a fission bomb. Once this temperature is achieved, the energy released in the initial reactions maintains the temperature, and the chain proceeds either until the supply of fusionable material is exhausted, or until sufficient expansion has taken place that the gas is cooled below the critical temperature. *See* ATOMIC BOMB; CHAIN REACTION (PHYSICS); NUCLEAR EXPLOSION. [K.W.P.]

Hydrogen bond The interaction which occurs when a hy-drogen atom, covalently bonded to an electronegative atom (as in A—H), interacts with another atom to form the aggregate A—H · · · Y. The shortest and strongest bond is indicated as A—H, while the secondary and weaker interaction is written as H · · · Y. Thus A—H is a proton donor, while (Y) is a proton acceptor which often contains lone pair electrons and can act as a base. The strongest hydrogen bonds are formed between the most electronegative (A) atoms such as fluorine, nitrogen, and oxygen which interact with (Y) atoms having electronegativity greater than that of hydrogen (C, N, 0, S, Se, F, Cl, Br, I). The weakest of hydrogen bonds are formed by acidic protons of C—H groups, as in chloroform and acetylene, and by olefinic and aromatic π-electrons acting as (Y).

The weaker the hydrogen bond, the shorter the lifetime of the complex it forms. An important aspect of weak hydrogen bond formation is that the different molecular aggregates which do form can be easily and reversibly transformed. Thus the small energy changes resulting in the rapid making and breaking of hydrogen bonds in biological systems are of great importance; for example, hydrogen bonding determines the configuration of the famous α-helix of DNA, and the structures of most proteins, thereby serving an important function in determining the nature of all living things. *See* CHEMICAL BONDING; DEOXYRIBONUCLEIC ACID (DNA); PROTEIN. [J.M.Wi.]

Hydrogen electrode A noble metal of large surface area covered with hydrogen gas in a solution of hydrogen ion saturated with hydrogen gas. The hydrogen gas may also be bubbled continuously over the noble metal; in this case, the affluent gas must be saturated with water vapor. Platinum is generally used as the noble metal. A large surface area is achieved by plating the surface with platinum sponge in a solution of chloroplatinic acid. In general, the noble metal is used in foil form and is welded to a wire sealed in the bottom of a hollow glass tube which is partially filled with mercury. Electrical contact to an external circuit is made through the mercury usually by a copper wire (see illustration).

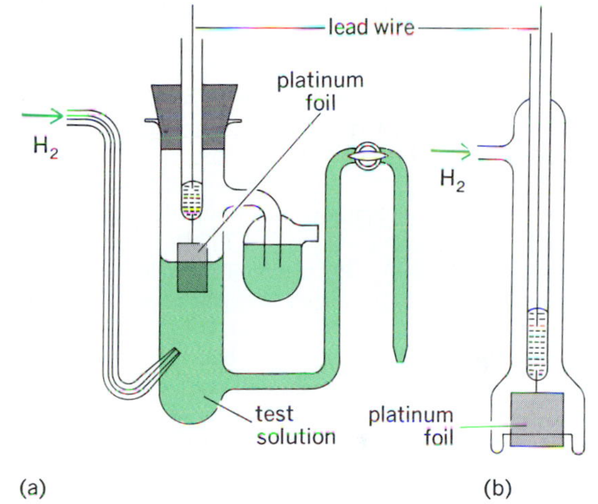

Hydrogen electrodes. (*a*) Stationary bubbling type and (*b*) dipping type.

The hydrogen electrode is reversible to hydrogen ion, and its electric potential is a logarithmic function of hydrogen ion activity. Because of this relation, this electrode is used to measure hydrogen ion activity. It cannot be used in reducing or oxidizing solutions or in solutions that poison the surface of the noble metal. Accordingly, it has limited application in the measurements of hydrogen-ion activity of solutions.

When the hydrogen-ion activity is unity and the hydrogen-gas pressure is at 1 atm (101 kilopascals), the potential of the hydrogen electrode is conventionally assigned a value of zero at all temperatures, and is used as a reference or standard electrode to which the potentials of all other electrodes are referred. All electrodes having a positive potential relative to the standard hydrogen electrode have greater reducing power than hydrogen; conversely, those of lower potential have a lower reducing power. Elements arranged in a series of their reducing power constitute what is called the electromotive-force series of the elements. *See* ELECTROCHEMICAL SERIES; ELECTRODE POTENTIAL; HYDROGEN ION. [W.J.H.]

Hydrogen fluoride The hydride of fluorine and the first member of the family of halogen acids. Anhydrous hydrogen fluoride is a mobile, colorless liquid that fumes strongly in air. It has the empirical formula HF, melts at −83°C, and boils at 19.8°C. The vapor is highly aggregated, and gaseous hydrogen fluoride deviates from perfect gas behavior to a greater extent than any other gaseous substance known. Aggregate formation in both the vapor and liquid phase arises from unusually strong hydrogen-bond interactions. *See* HYDROGEN BOND.

Anhydrous hydrogen fluoride is an extremely powerful acid, exceeded in this respect only by 100% sulfuric acid. Because anhydrous hydrogen fluoride is a superacid, many organic solutes dissolve in it to form stable carbonium ions. Alkali metal fluorides and silver fluoride dissolve readily in hydrogen fluoride to form conducting solutions. Anhydrous hydrogen fluoride dissolves a wide variety of organic compounds. Aqueous solutions of hydrogen fluoride (hydrofluoric acid) are relatively weakly acidic as compared to hydrochloric acid.

Hydrogen fluoride is a widely used industrial chemical. The largest use is in making fluorine-containing refrigerants (Freons, Genetrons). An increasingly important use of hydrogen fluoride is in the preparation of organic fluorocarbon compounds.

Both hydrogen fluoride and hydrofluoric acid cause unusually severe burns; appropriate precautions must be taken to prevent any contact of the skin or eyes with either the liquid or the vapor. *See* FLUORIDE; FLUORINE; HALOGENATED HYDROCARBON. [J.J.K.]

Hydrogen ion A proton combined with a number of water molecules. It is often written as H_3O^+ and called the hydronium ion. However, this species is best considered as an excess proton on a tetrahedral group of four water molecules and so would be designated as $H_9O_4^+$. For simplicity, it is most commonly written as H^+ (aq). *See* PROTON.

Since it is formed by the self-ionization of water, the hydrogen ion is present in all aqueous solutions. This formation also means that H^+(aq) is always found in the company of the hydroxide ion, OH^- (aq). The equilibrium relationship between the concentrations of these two species is a very important property of water. *See* IONIC EQUILIBRIUM.

The H^+(aq) and OH^- (aq) concentrations in pure water are equal to each other with a value of 10^{-7} mole/liter. Any aqueous solution with this concentration of H^+(aq) is called a neutral solution. If the H^+(aq) concentration is greater than 10^{-7} mole/liter, the solution is called acidic. Basic solutions are those in which the H^+(aq) concentration is less than 10^{-7} mole/liter. As the H^+(aq) concentration increases the OH^- (aq) concentration must decrease, and vice versa. In the most straightforward system, acids are substances that can donate an H^+(aq), and bases are substances that can accept one. *See* ACID AND BASE.

Hydrogen ion concentration determines the course of many chemical reactions that occur in living organisms and in the chemical industry. The control of hydrogen ion concentration is achieved in living organisms and in the laboratory by buffer systems. These are chemical mixtures designed to resist change in hydrogen ion concentration. *See* BUFFERS (CHEMISTRY).

Another property of the H^+(aq) is important in both theoretical and practical ways. H^+(aq) is the best conductor of electricity of any ion in aqueous solution. Its conductance at 77°F (25°C) is almost five times as large as the next-most-conducting ion. *See* ELECTROLYTIC CONDUCTANCE; HYDROGEN ELECTRODE.

The hydrogen ion concentration can vary over fourteen powers of 10. To avoid dealing with such exponentials, the concept of pH is used. Since all aqueous solutions contain both hydrogen ion and hydroxide ion, it is possible to define all degrees of acidity and basicity on the pH scale. *See* pH.

Two general methods are used for the determination of hydrogen ion concentrations. For relatively crude work, colorimetric methods are commonly used. These methods depend on the fact that certain natural and synthetic dyes have colors that depend on the hydrogen ion concentration. At times, paper is impregnated with such an indicator. In most precise work, a potentiometric method is used for the determination of hydrogen ion concentration. This method depends on an electrode whose potential is sensitive to hydrogen ion concentration. The only electrode commonly in use for practical pH measurements is the glass electrode. *See* ACID-BASE INDICATOR; COLORIMETRY; ION SELECTIVE MEMBRANES AND ELECTRODES; TITRATION; VOLUMETRIC ANALYSIS. [G.A.]

Hydrogen maser clock A type of atomic clock which employs maser oscillation of hydrogen atoms in a microwave cavity. In this active clock, hydrogen molecules are dissociated into a beam of hydrogen atoms. Those atoms in the upper hyperfine state are then focused into a Teflon-lined quartz bulb inside a cavity resonant at the hyperfine frequency of 1420 megahertz. The Teflon coating allows the atoms to bounce off the walls without much perturbation. As a result of this and the small size of the entrance hole, the atoms can be confined in the bulb for a long time, resulting in interaction times with the exciting radiation of the order of a second. This produces a very narrow line width. If the cavity losses are sufficiently low, maser oscillations take place. The hydrogen maser oscillator is the most stable oscillator known. Since the atoms do make collisions with the wall, there is a slight pulling from the free atom frequency. This results in an uncertainty of only about 2 parts in 10^{12}, so that the hydrogen maser makes an excellent clock. By reducing the hydrogen flux below the threshold for oscillation and providing 1420 MHz excitation, the apparatus can also be used as a passive clock. This greatly reduces frequency pulling effects due to cavity tuning changes and thus reduces the requirements for precise cavity temperature control at the expense of losing the spectral purity of the maser oscillation. *See* ATOMIC CLOCK; MASER. [L.S.C.]

Hydrogen peroxide A binary compound of hydrogen and oxygen, empirical formula H_2O_2, used mostly in dilute aqueous solutions as an oxidizing agent. Its most remarkable feature is its tendency to decompose readily into water and oxygen.

Anhydrous hydrogen peroxide is a clear, colorless liquid, of nearly the same viscosity and dielectric constant as water, but of greater density. Like water, it is strongly associated through hydrogen bonds. It boils at 150°C (300°F) with violent, sometimes explosive decomposition. Decomposition by light begins only in the near ultraviolet. As a solvent, hydrogen peroxide resembles water, except that acids and bases show much lower electrical conductivity. Although a fairly strong oxidant, it can act as a mild reducing agent, for example, with permanganates and perchromates.

Hydrogen peroxide is used mainly for bleaching cotton and other fibers, natural or synthetic. Increasing amounts are used in the pulp and paper industry. Its well-known cosmetic use as hair bleach consumes relatively little of the commercial 10% (30 volume) solution. In medicine it is useful for cleansing wounds and cuts, although its antiseptic action is rather slow. A limited but important use of the concentrated peroxide is for energy production in rockets, submarines (during submersion), airplanes (at takeoff), and the steering of space vessels. *See* BLEACHING; CHEMICAL FUEL.

Hydrogen peroxide, especially when concentrated, requires great care in handling and storing. When dropped on paper or wood, it can start a fire. Contact with the skin causes blotches that can be painful, but they disappear after a few hours without leaving traces. *See* HYDROGEN; OXYGEN; PEROXIDE. [P.A.G.]

Hydrogenation The chemical reaction of hydrogen with another substance, generally an unsaturated organic compound, and usually under the influence of temperature, pressure, and catalysts. There are several types of hydrogenation reactions. They include: (1) the addition of hydrogen to reactive molecules; (2) the incorporation of hydrogen accompanied by cleavage of the starting molecules (hydrogenolysis); and (3) reactions in which isomerization, cyclization, and so on, result.

Hydrogenation is synonymous with reduction in which oxygen or some other element (most commonly nitrogen, sulfur, carbon, or halogen) is withdrawn from, or hydrogen is added to, a molecule. When hydrogenation is capable of producing the desired reduction product, it is generally the simplest and most efficient procedure.

Hydrogenation is used extensively in industrial processes. Important examples are the synthesis of methanol, liquid fuels, hydrogenated vegetable oils, fatty alcohols from the corresponding carboxylic acids, alcohols from aldehydes prepared by the aldol reaction, cyclohexanol and cyclohexane from phenol and benzene, respectively, and hexamethylenediamine for the synthesis of nylon from adiponitrile. *See* Dehydrogenation; Fischer-Tropsch process; Hydroformylation; Hydrogen; Oxidation-reduction. [R.Le.]

Hydrography Now primarily the mapping of the ocean bottoms and the measuring of ocean currents; once broadly used to apply to studies of surface water of the Earth including measuring flow of streams and mapping of streams and lakes, as well as investigations of the ocean. The broader fields of hydrology and oceanography have absorbed much of the earlier scope of hydrography.

Today the term is used mainly with respect to charting oceans and estuaries and coastal waters for navigational purpose. Every maritime nation of any importance has established a hydrographic office as one of its regular bureaus. *See* Hydrology; Oceanography. [R.K.Li.]

Hydroida An order of the coelenterates which includes the fresh-water hydras, the attached and usually colonial hydroids, and many of the smaller jellyfish. It is the largest order of the class Hydrozoa.

The order Hydroida includes two principal suborders, Gymnoblastea and Calyptoblastea. The Gymnoblastea are those hydroids which lack protective cups around the hydranths and gonozoids. Jellyfish produced by these athecate hydroids are Anthomedusae. The Calyptoblastea include the hydroids with protective cups around the hydranths (hydrothecae) and around the gonozoids (gonothecae). Jellyfish of these thecate hydroids are called Leptomedusae. Two minor suborders are Limnomedusae and Chondrophora.

Anthomedusae are typically ovoid jellyfish, often with eyespots. Leptomedusae are usually flattened or saucer-shaped, have statocysts (sense organs of balance), and lack eyespots. The gonads of Anthomedusae are generally on the wall of the stomach just above the mouth, and those of Leptomedusae below the radial canals.

Young hydranths of gymnoblastic hydroids are small with five or more tentacles, but subsequently grow much larger and add more tentacles. Calyptoblastic hydranths, in contrast, emerge from a bud with a full complement of parts; they do not grow, live for only about a week, undergo regression and absorption by the colony, and are then replaced by new hydranths.

The fresh-water hydras are simple, motile polyps which do not produce colonies (see illustration). Buds separate from the parent and become individual polyps. Simple gonads develop on the body and there is no medusa. Hydras are sometimes included in the Gymnoblastea and sometimes placed in a suborder by themselves, the Hydrida.

Hydroids are species in which the polyp stage is usually dominant. Most hydroids are found near the shore attached to various supports such as rocks, wharves, boats, mussels, barnacles, worm tubes, crab and snail shells, and seaweeds. Sometimes the medusa stage is well developed and the hydroid stage may be lacking. Medusae are abundant both in coastal waters and in the open sea.

Hydras and hydroids have been used extensively in research on problems of growth, development, and regeneration. They have a high capacity for reorganization. Missing parts are quickly replaced, a bit of stem can produce a new hydranth, and completely disorganized masses of cells can reconstitute a new polyp. *See* Hydrozoa; Regeneration (biology). [S.Cr.]

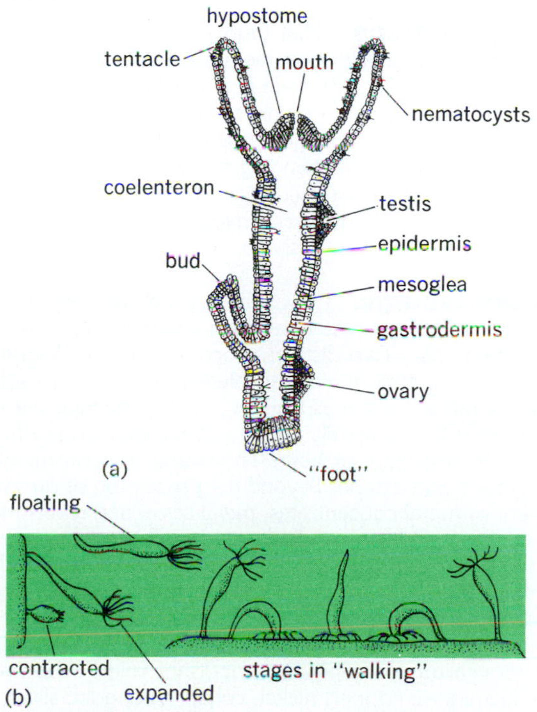

Hydra. **(*a*) Longitudinal section. (*b*) Movements. (*After T. I. Storer et al., General Zoology, 6th ed., McGraw-Hill, 1979*)**

Hydrokinematics Motion of a liquid apart from the cause of the motion. A liquid is treated as a continuum, the elements of which move along continuous paths. At any instant the flow pattern may be delineated by a family of streamlines which are everywhere tangent to the paths, but do not in general coincide with them unless the flow is steady. Complete specification of a flow field requires that the distributions of sources and vorticity be known. *See* Fluid flow; Hydrodynamics; Stokes stream function. [L.L.]

Hydrokinetics Forces produced by a liquid as a consequence of its motion. The mass of a liquid gives rise to inertia forces which are manifested as normal stresses or pressures when the liquid flows. The viscosity and cohesion of a liquid give rise to tangential shear forces and also affect the normal stresses. The interaction of these properties on the motion of the fluid is formally expressed in the equations of Navier and Stokes. Because of the difficulty of solving these equations in general, approximate solutions are often obtained by neglecting the less important properties in a particular flow situation. *See* Fluid flow; Hydrodynamics. [L.L.]

Hydrology The science that treats of the waters of the Earth: their occurrence, circulation, and distribution; their chemical and physical properties; and their reaction with the environment, including their relation to living things. The domain of hydrology embraces the full life history of water on the Earth.

Water in liquid and solid form covers most of the crust of the Earth. By a complex process powered by gravity and the action of solar energy, an endless exchange of water, in vapor, liquid, and solid forms, takes place between the atmosphere, the oceans, and the crust. Water circulates in the air and in the oceans as well as over and below the surface of land masses. This global scheme is called the hydrologic cycle, Although, general patterns of circulation are present in the atmosphere, the oceans, and the land masses, regional features are very irregular and random in their detail. Therefore, while causal relations underlie the overall process, important elements of chance affect local hydrological events. *See* Evapotranspiration; Terrestrial water.

Water is essential for all living things. Its adequate supply is a key factor for urban, agricultural, and industrial development. Water can also be the recipient of pollutants which degrade its quality for all uses, and may be a destructive agent when it inundates valleys and causes death and great damage. The rising of groundwater levels in agricultural lands may cause deterioration of the soils and loss of fertility by waterlogging and increased salinity. Erosion of soil by flowing waters, and ultimate deposition of the sediment in lakes, reservoirs, stream channels, and harbors are also serious problems. Thus, the means whereby natural waters may be captured and controlled are of utmost importance for the development of human economy. The study of hydrology provides the information necessary for determining those means. *See* EROSION; WATER CONSERVATION; WATER POLLUTION.

A number of field measurements are performed for hydrologic studies. Among them are the amount and intensity of precipitation; the quantities of water stored as snow and ice, and their changes in time; discharge of streams; rates and quantities of infiltration into the soil, and movement of soil moisture; rates of production from wells and changes in their water levels as indicators of groundwater storage; dissolved chemical elements and compounds in surface and ground waters; amounts of water transferred by evaporation and evapotranspiration to the atmosphere from snow, lakes, streams, soils, and vegetation; and sediment lost from the land and transported by streams. In addition, hydrology is concerned with research on the phenomena and mechanisms involved in all physical and biological components of the hydrogeologic cycle, with the purpose of understanding them sufficiently to permit quantitative predictions and forecasting. *See* GROUNDWATER HYDROLOGY; HYDROMETEOROLOGY; HYDROSPHERE; SURFACE WATER. [J.A.]

Hydrolysis Literally destruction, decomposition, or alteration of a chemical substance by water. In discussions of aqueous solutions of electrolytes the term hydrolysis is applied especially to reactions of cations (positive ions) with water to produce a weak base, or of anions (negative ions) to produce a weak acid. A salt of a weak acid or of a weak base, or of both a weak acid and weak base, is then said to be hydrolyzed. The degree of hydrolysis is the fraction of the ion which reacts with the water. The term solvolysis is employed for reactions of solutes with solvents in general.

A frequently quoted example is the hydrolysis of sodium acetate, which is highly dissociated in water, to yield sodium ion, Na^+, and acetate ion, $CH_3CO_2^-$. The acetate ion is represented below by the symbol A^-. Because acetic acid is a weak acid (it has a small dissociation constant), the acetate ion, A^-, combines with some of the hydrogen ion present in aqueous solutions, as shown in reaction (1).

$$A^- + H_3O^+ \rightleftharpoons HA + H_2O \quad (1)$$

The concentration of the hydrogen ion is involved in the equilibrium, reaction (2), between water and the hydrogen ion

$$2H_2O \rightleftharpoons H_3O^+ + OH^- \quad (2)$$

and the hydroxide ion. Because reaction (1) removes a product of reaction (2), each reaction proceeds to produce more of its products (the substances on the right side of each equation). The net result is reaction (3), which represents the reaction

$$A^- + H_2O \rightleftharpoons HA + OH^- \quad (3)$$

which is described as the hydrolysis of the acetate ion. When equilibrium is attained, the concentrations are such as to conform to the equilibrium constants of all three reactions (one of which is a mathematical consequence of the other two). *See* ACID AND BASE; HYDROGEN ION; HYDROLYTIC PROCESSES; IONIC EQUILIBRIUM. [T.F.Y.]

Hydrolytic processes Reactions of both organic and inorganic chemistry wherein water effects a double decomposition with another compound, hydrogen going to one component, hydroxyl to another, as in reactions (1)–(3). Although the

$$XY + H_2O \rightarrow HY + XOH \quad (1)$$

$$KCN + H_2O \rightarrow HCN + KOH \quad (2)$$

$$C_5H_{11}Cl + H_2O \rightarrow HCl + C_5H_{11}OH \quad (3)$$

word "hydrolysis" means decomposition by water, cases in which water brings about effective hydrolysis unaided are rare, and high temperatures and pressures are usually necessary. *See* HYDROLYSIS.

Hydrolytic reactions may be classified as follows: (1) hydrolysis with water alone; (2) hydrolysis with dilute or concentrated acid; (3) hydrolysis with dilute or concentrated alkali; (4) hydrolysis with fused alkali with little or no water at high temperature.

In the field of organic chemistry, the term "hydrolysis" has been extended to cover the numerous reactions in which alkali or acid is added to water. An example of an alkaline-condition hydrolytic process is the hydrolysis of esters, reaction (4), to

$$CH_3COOC_2H_5 + NaOH \rightarrow CH_3COONa + C_2H_5OH \quad (4)$$

produce alcohol. An example of an acidic-condition process is the hydrolysis of olefin to alcohol in the presence of phosphoric acid, reaction (5). The addition of acids or alkalies hastens

$$C_2H_4 + H_2O \xrightarrow{H_3PO_4} C_2H_5OH \quad (5)$$

such reactions even if it does not initiate the reaction. *See* ALCOHOLYSIS.

Perhaps some of the oldest and largest-volume hydrolysis technology is involved in soap manufacture. In the first step, glyceryl stearate acid, a fat, is hydrolyzed with water to yield stearic acid and glycerin. In the second step, the stearic acid is neutralized with caustic soda to give sodium stearate, the soap, and water. *See* FAT AND OIL.

Hydrolytic processes account for a huge product volume. Conversion of starch such as corn starch into maltose and glucose (sugar syrups) by treatment with hydrochloric acid is a major industry. Similarly, the production of furfural from pentosans of oat hulls or other cereal by-products such as corn cobs, rice hulls, or cottonseed bran is another commercial hydrolytic process. [D.L.H.]

Hydromechanics That branch of physics which deals with liquids, traditionally water, as a medium for the transmission of forces. A body immersed in a liquid experiences a vertical force proportional to the volume of liquid that it displaces. A fluid whose velocity changes produces forces as a result of changing pressure and of changing momentum. Such phenomena are applied in various machines. *See* ARCHIMEDES' PRINCIPLE; BERNOULLI'S THEOREM; BUOYANCY; HYDRAULIC PRESS; HYDRAULIC TURBINE; HYDRAULICS; HYDROSTATICS. [W.A.]

Hydrometallurgy The extraction and recovery of metals from their ores by processes in which aqueous solutions play a predominant role. Two distinct processes are involved in hydrometallurgy: putting the metal values in the ore into solution via the operation known as leaching; and recovering the metal values from solution, usually after a suitable solution purification or concentration step, or both. The scope of hydrometallurgy is quite broad and extends beyond the processing of ores to the treatment of metal concentrates, metal scrap and revert materials, and intermediate products in metallurgical processes. Hydrometallurgy enters into the production of practically all nonferrous metals and of metalloids, such as selenium and tellurium. Hydrometallurgical and pyrometallurgical processes complement each other. *See* LEACHING; PYROMETALLURGY.

Hydrometallurgy occupies an important role in the production of aluminum, copper, nickel, cobalt, zinc, gold, silver, platinum, selenium, tellurium, tungsten, molybdenum, uranium, zirconium, and other metals. *See* METALLURGY. [W.C.Co.]

Hydrometeorology The study of the occurrence, movement, and changes in the state of water in the atmosphere. The term is also used in a more restricted sense, especially by hydrologists, to mean the study of the exchange of water between the atmosphere and continental surfaces. This includes the processes of precipitation and direct condensation, and of evaporation and transpiration from natural surfaces. Considerable emphasis is placed on the statistics of precipitation as a function of area and time for given locations or geographic regions.

Water occurs in the atmosphere primarily in vapor or gaseous form. The average amount of vapor present tends to decrease with increasing elevation and latitude and also varies strongly with season and type of surface. Precipitable water, the mass of vapor per unit area contained in a column of air extending from the surface of the Earth to the outer extremity of the atmosphere, varies from almost zero in continental arctic air to about 6 g/cm^2 in very humid, tropical air.

Although a trivial proportion of the water of the globe is found in the atmosphere at any one instant, the rate of exchange of water between the atmosphere and the continents and oceans is high. Evaporation from the ocean surface and evaporation and transpiration from the land are the sources of water vapor for the atmosphere. Water vapor is removed from the atmosphere by condensation and subsequent precipitation in the form of rain, snow, sleet, and so on. The amount of water vapor removed by direct condensation at the Earth's surface (dew) is relatively small. *See* EVAPOTRANSPIRATION; HYDROLOGY; METEOROLOGY; PRECIPITATION (METEOROLOGY). [E.M.R.]

Hydrometer A direct-reading instrument for indicating the density, specific gravity, or some similar characteristic of liquids. Almost all hydrometers are made of a high-grade glass tubing. The main body is the float section in the bottom of which ballast, such as small shot, is secured. A small-diameter tube, the stem, extends from the upper end of the float section. Inside the stem is the scale, printed on heavy-grade paper, and well-secured within the stem so its position will not change. When the hydrometer is placed in a liquid, the stem extends vertically above the surface for a portion of its length.

Hydrometers may be classified according to the indication provided by graduations of the scale as follows: (1) density hydrometers, to indicate densities at a particular temperature, and usually for a particular liquid; (2) specific gravity hydrometers to indicate specific gravity of a liquid, with reference to water, at a particular temperature; (3) percentage hydrometers to indicate, at a particular temperature, the percentage of a substance such as salt, sugar, or alcohol dissolved in water (alcoholometers are an example); and (4) arbitrary scale hydrometers, indicating the density, specific gravity, or concentration of a liquid in terms of an arbitrarily defined scale, at a defined temperature. The last group includes the saccharimeter (indicates percentage of pure sucrose solutions); the Baumé hydrometer (measures specific gravity of liquids lighter than water); the lactometer (tests milk); and the barkometer (tests tanning extracts). *See* DENSITY; SPECIFIC GRAVITY. [H.S.B.]

Hydrophone A device which receives underwater sound waves and converts them to essentially equivalent electric waves. A hydrophone is the underwater analog of a microphone. Hydrophones are used in sonar apparatus, sonobuoys, and certain underwater weapons. *See* ACOUSTIC MINE; ACOUSTIC TORPEDO; MICROPHONE; SONAR; SONOBUOY; UNDERWATER SOUND. [H.F.O.]

Hydroponics One of several popular names given to the artificial culture of plants in inorganic salt solutions in open concrete tanks or other containers. The technique was devised in the mid-19th century as a means of studying mineral nutrients essential to the growth and development of plants, but for many years was only a laboratory procedure.

Mineral solution cultures are fully adequate for maximal growth if the essentials for plant development are provided, including light, air, favorable temperature, and a support for the plants. The size of such cultures has been increased and adapted to crop production. Means of root support have been provided by sand, gravel, and cinders; adequate aeration is maintained by pumps, by frequent drainage, or in other ways. Under these conditions, it is possible to obtain excellent crops in places where conventional methods of agriculture or gardening are unsuitable or impossible. Economic aspects of artificial culture, however, must be compared with ordinary soil cultivation in order to judge the feasibility of the soilless practice at a particular location. It is doubtful that nutriculture crop production will in a large measure supplant the usual agriculture method. [T.C.B.]

Hydroquinone A dihydric phenol ($C_6H_6O_2$; colorless crystals, melting point 169°C or 336°F) in which the two hydroxyl groups are in the para or 1,4 positions. The important relationship between hydroquinone and *p*-benzoquinone (commonly called simply quinone) is illustrated in the reaction below.

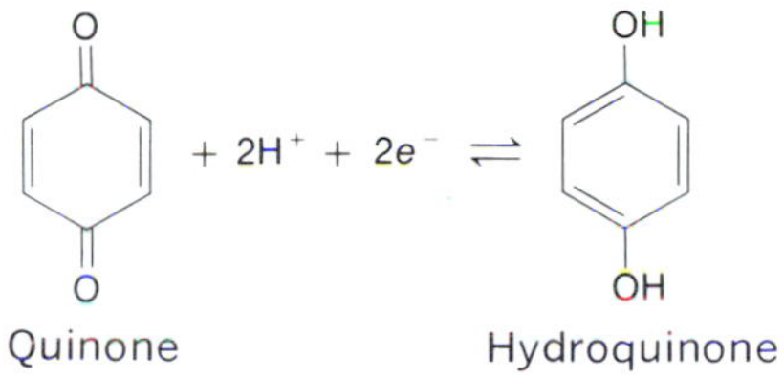

The facile oxidation of hydroquinone is the basis for the most important uses of the compound and its derivatives. As a photographic developer, hydroquinone reacts with light-activated silver bromide crystals to form black metallic silver (reduction product) and quinone (oxidation product). Various alkylated derivatives of hydroquinone and the monomethyl ether (hydroxyanisole) find use as preservatives and antioxidants. *See* PHENOL. [M.St.]

Hydrosphere Approximately 74% of the Earth's surface is covered by water, in either the liquid or solid state. The following waters, combined with minor contributions from groundwaters, constitute the hydrosphere: world oceans, fresh-water lakes, saline lakes and inland seas, rivers, soil moisture and vadose water, groundwater to depth of 4000 m, and icecaps and glaciers.

The oceans account for about 97% of the weight of the hydrosphere, while the amount of ice reflects the Earth's climate, being higher during periods of glaciation. The circulation of the waters of the hydrosphere results in the weathering of the landmasses. The rainwater falling on the continents, partly taken up by the ground and partly by the streams, acts as an erosive agent before returning to the seas.

The unique chemical properties of water make it an effective solvent for many gases, salts, and organic compounds. Circulation of water and the dissolved material it contains is a highly dynamic process driven by energy from the Sun and the interior of the Earth. Each component has its own geochemical cycle or pathway through the hydrosphere, reflecting the component's relative abundance, chemical properties, and utilization by organisms. The introduction of materials by humans has significantly altered the composition and environmental properties of many natural waters. *See* EROSION; GROUNDWATER HYDROLOGY; HYDROLOGY; LAKE; RIVER; SEAWATER; TERRESTRIAL WATER. [E.D.G.]

Hydrostatics The study of liquids at rest. In the absence of motion, there are no shear stresses; the internal state of stress at any point is determined by pressure alone. Hence, the pressure at a point is the same in all directions. Pressure acts normally to all boundary surfaces. For equilibrium under gravity, regardless of the shape of the containing vessel, the pressure is uniform over any horizontal cross section. Pressure varies with height or depth. Two different reference levels are used in measuring pressure. For many engineering purposes, gage pressure is used with pressure measured relative to atmospheric pressure as zero. For most scientific purposes, pressure is referred to true zero. Normal atmospheric pressure at sea level caused by the weight of the air above is approximately 101 kilopascals or 14.7 pounds per square inch absolute.

The buoyant force is the force exerted vertically upward by a fluid on a body wholly or partly immersed in it. Its magnitude is equal to the weight of the fluid displaced by the body. This value is also the vertical component of the fluid pressure force acting upward against the bottom of the body minus the fluid pressure force component (if any) acting vertically downward against the top of the body. If this buoyant force equals the weight of the body, the body will remain at the given level. If it exceeds the weight of the body, the latter will rise, and vice versa. The buoyant force as a single magnitude acts vertically upward through the center of buoyancy which is the center of gravity of the displaced fluid. *See* ARCHIMEDES' PRINCIPLE.

Pressure applied to a confined liquid is transmitted with equal intensity throughout the liquid and by it to all surfaces of the confining vessel or piping. Hence, a small force applied to a small area of a confined liquid can create a large force against a large area. If the small and large areas are pistons the device may be a hydraulic press or jack. Because the transmitting liquid is practically incompressible and its volume virtually constant, the linear movement of the large piston will be to that of the small piston in inverse proportion to their areas. The principle of multiplying a force by means of liquid pressure applies also to hydraulic brakes, power steering, control systems, and the like; the actuating force may be a pump instead of a small piston. *See* HYDRAULICS; HYDROMECHANICS. [W.A.]

Hydrothermal vent A hot spring on the ocean floor, where heated fluids exit. Most hydrothermal vents occur along the central axes of the Mid-Oceanic Ridge, an undersea mountain chain circling the Earth. Here new ocean floor is created as the sea floor on either side of the axis is transported outward. Vents of slightly different compositions also occur associated with undersea volcanoes or seamounts. Hydrothermal vent sites, or closely grouped clusters of vent deposits and exit ports, may cover areas from hundreds to thousands of square feet (tens to hundreds of square meters). Individual vent sites may be separated along mid-ocean ridges by more than a thousand miles (up to thousands of kilometers). *See* MID-OCEANIC RIDGE; MARINE GEOLOGY; SEAMOUNT AND GUYOT.

All of the hydrothermal vent sites occur in areas where quantities of magma exist below the sea floor. Cold seawater is drawn down into the oceanic crust toward the heat source. As the seawater is heated and reacts with surrounding rock, its composition changes. Sulfate and magnesium are major components of seawater lost during reaction; sulfide, metals, and gases such as helium and methane are major components gained. This modified seawater is known as hydrothermal fluid. Mixing of hydrothermal fluid with seawater leads to precipitation of minerals from solution, forming mineral deposits at the exit from the sea floor and so-called smoke, tiny mineral particles suspended in the rising plume of fluid. Black smoker vents are distinguished by the presence of such large quantities of minute mineral particles that the plumes become virtually opaque. *See* MAGMA.

Formation and outflow of hydrothermal fluid makes a major contribution to the concentration and balance of elements in the oceans by removing and adding elements to the original seawater composition. The quantities of elements added or removed from the oceans by hydrothermal venting around the world are comparable to quantities contributed by the worldwide flow of rivers into the oceans. Hydrothermal venting also represents a major flow of heat from the Earth's crust and a major mechanism for cooling of new oceanic lithosphere. *See* LITHOSPHERE; SEAWATER.

Perhaps the most striking feature of seafloor hydrothermal vents is their dense biologic communities. The specimens that live around the hydrothermal vents include new species, new genera, and even new families of animals. It is the first food chain discovered that does not rely on photosynthesis and the Sun as sources of food and energy, but instead on the hot fluid at the vents. The basic units of the food chain are bacteria that derive chemical energy from oxidizing the reduced chemical species in the hot waters. These bacteria, living in the vent fluids, in thick white and rust-colored mats around the vents, and in symbiotic relationships with other vent organisms, support a wide variety of organisms. [M.Go.]

Hydroxide One type of hydroxyl compound (OH-containing) whose solutions have a bitter taste, feel slippery to the touch, neutralize acids, and change the color of red litmus to blue. Hydroxides are also called bases and have a high concentration of OH^- ions in solution. When an oxide of an element, XO or X_2O, is dissolved in water, a hydroxyl compound of the type XOH is formed. Such species, XOH, can be ionized to give either OH^- ions or H^+ ions. The former are called bases and the latter acids. Ordinarily, when X is a metal, bases are obtained, for example, NaOH; and when X is a nonmetal, acids are obtained, for example, $S(OH)_6$, which loses two molecules of water and exists as H_2SO_4. Some insoluble oxides, Al_2O_3, for example, have both acid and basic properties. This class of compounds is said to be amphoteric.

Hydroxides are very important compounds industrially, sodium hydroxide being the most important. It is used in making other chemicals, rayon, petroleum products, and soap. *See* ACID AND BASE; HYDRATE. [E.E.W.]

Hydroxyl A chemical group in which oxygen and hydrogen are bonded and act as a single entity. In inorganic chemistry the hydroxyl group is known as the hydroxide ion (OH^-), and it is frequently bonded to metal cations, for example, sodium hydroxide (NaOH). In organic chemistry it frequently acts as a functional group, for example, in an alcohol (ROH, where R represents an alkyl group). *See* ACID AND BASE.

Many of the intermediate redox forms of dioxygen are toxic and damage important biomolecules. Much of this toxicity is thought to involve the generation and reactivity of hydroxyl (·OH), which is sometimes called the hydroxy radical. The most common means for producing hydroxyl is the reaction of a reducing agent with hydrogen peroxide (H_2O_2). Transition-metal ions, such as ferrous ion (Fe^{2+}), are the most common reducing agents for generating ·OH. *See* FREE RADICAL; OXYGEN TOXICITY.

Once generated, hydroxyl is a potent one-electron oxidant that forms the very stable OH^-ion, and it abstracts hydrogen atoms from organic molecules that contain C-H bonds to form the stronger O-H bond in water. The reaction of a radical, which contains an uneven number of electrons, with a molecule, which contains an even number of electrons paired in bonds, must generate a radical, because the number of electrons cannot change during the reaction. Thus, most reactions of radicals generate new radicals in processes called radical chain reactions. Reactions of radicals with molecules will continue to produce new radicals until other odd-electron species (such as transition-metal ions or other radicals) react with the radicals to produce even-electron molecules via termination reactions. *See* Chain reaction (chemistry); Transition elements. [H.H.T.]

Hydroxylation reaction One of several types of reactions used to introduce one or more hydroxyl groups into organic compounds. Unsaturated compounds are the most common starting materials. Monohydroxylation, the introduction of one hydroxyl group, is especially useful when isomeric alcohols cannot form, as in the conversion of ethylene to ethyl alcohol. Reactions with longer-chain olefins produce primarily secondary and branched-chain alcohols. Dihydroxylation, the introduction of two hydroxyl groups, is used to produce glycol. *See* Epoxidation; Ethylene glycol; Ethylene oxide; Glycol; Substitution reaction. [D.Sw.]

Hydroxyproline An amino acid occurring in certain proteins. Hydroxyproline, the 4-hydroxy analog of proline, shares many of the chemical and biological features of proline. Unlike proline, hydroxyproline is not found generally in proteins but is essentially limited to structural proteins of the collagen type of animals, and to a similar protein in the cell walls of higher plants. It constitutes about 10% of most collagen in vertebrates, but is present in higher content in certain invertebrate collagens. *See* Collagen; Protein.

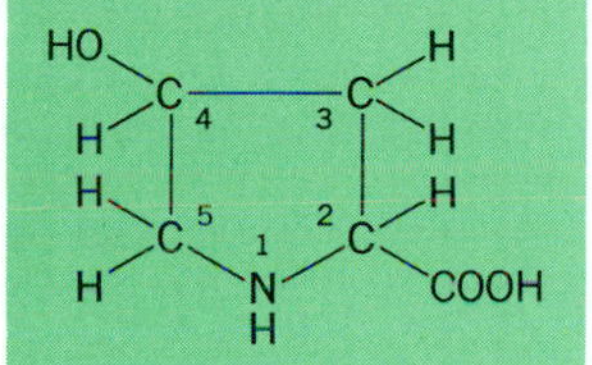

Hydroxyproline

In humans hydroxyproline is excreted in the urine, predominantly in small peptides. Hydroxyproline excretion in urine is increased during increased breakdown of collagen, and occurs in certain disorders of bone growth or bone destruction and in some metabolic disorders, for example, hyperthyroidism or hyperparathyroidism. *See* Amino acids; Proline. [E.Ad.]

Hydrozoa A class of the phylum Coelenterata which includes the fresh-water hydras, the marine hydroids, many of the smaller jellyfish, a few special corals, and the Portuguese man-of-war. The Hydrozoa may be divided into five orders: the Hydroida, Milleporina, Stylasterina, Trachylina, and Siphonophora. See separate article on each order.

The form of the body varies greatly among the hydrozoans. This diversity is due in part to the existence of two body types, the polyp and the medusa. A specimen may be a polyp, a medusa, a colony of polyps, or even a composite of the first two. Polyps are somewhat cylindrical, attached at one end, and have a mouth surrounded by tentacles at the free end. Medusae are free-swimming jellyfish with tentacles around the margin of the discoidal body.

In a representative life cycle, the fertilized egg develops into a swimming larva which soon attaches itself and transforms into a polyp. The polyp develops stolons (which fasten to substrates), stems, and other polyps to make up a colony of interconnected polyps. Medusae are produced by budding and liberated to feed, grow, and produce eggs and sperm.

Most hydrozoans are carnivorous and capture animals which come in contact with their tentacles. The prey is immobilized by poison injected by stinging capsules, the nematocysts. Most animals of appropriate size can be captured, but small crustacea are probably the most common food. *See* Coelenterata. [S.Cr.]

Hyena An African carnivore represented by three species of the family Hyaenidae. Hyenas resemble dogs superficially, but they are more closely related to the felids (cats). They are four-toed digitigrade (walking on the toes) animals with blunt, nonretractile claws that are used for digging and disinterring bodies. The gestation period is about 13 weeks, with two to four young composing the yearly litter. The maximum lifespan of this animal is about 25 years. *See* Carnivora. [C.B.C.]

Hyeniales An order of Devonian plants considered to be related to the Sphenopsida. The small dichotomously forked leaves tend to be borne in whorls. Some leaves bear terminal sporangia, but these appendages are neither aggregated into a tight cone nor separated by bracts. *See* Paleobotany. [H.P.B.]

Hygrometer An instrument for giving a direct indication of the amount of moisture in the air or other gas, the indication usually being in terms of relative humidity as a percentage that the moisture present bears to the maximum amount of moisture that could be present at the location temperature without condensation taking place. There are three major types of hygrometers: mechanical, electrical, and cold-spot or dew-point. *See* Dew point; Humidity.

In a simple mechanical type of hygrometer the sensing element is usually an organic material which expands and contracts with changes in the moisture in the surrounding air or gas. The material used most is human hair. As shown in the illustration, the bundle of hair is held under a slight tension by a spring, and a magnifying linkage actuates a pointer.

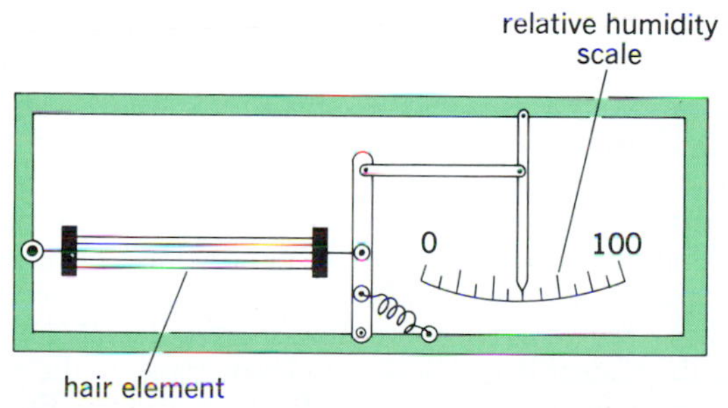

Hygrometer which uses hair as the sensing element. (*After D. M. Considine and S. D. Ross, Process Instruments and Controls Handbook, 2d ed., McGraw-Hill, 1974*)

In an electrical hygrometer the change in the electrical resistance of a hygroscopic substance is measured and converted to percent relative humidity.

In a third group of hygrometers, commonly called dew-point apparatus, the dew-point temperature is determined; this is the temperature at which the moisture in the gas is at the point of saturation, or 100% relative humidity. The usual procedure is to chill a polished surface until dew or a film of moisture just starts to appear and to measure the temperature of the surface. [H.S.B.]

Hymenomycetes A class of fungi in the subdivision Basidiomycotina. It is usually divided into two subclasses: Holobasidiomycetidae, delimited by nonseptate basidia and the absence of a yeast phase; and Phragmobasidiomycetidae, frequently with septate basidia and often forming a yeast phase. A typical hymenomycete produces a fruit body or basidiome with spore-bearing basidia organized in a membranelike layer called the hymenium. The fruit bodies are designed to allow for basidiospore discharge into air currents either directly off basidia or after falling from elaborate fertile surfaces. Falling spores, indicative of active discharge, accumulate in powdery masses called spore prints. The shape of the hymenium varies from lamellate (gilled, as in mushrooms), poroid (as in conk or bracket fungi), toothed (in hedge hog fungi), coralloid (coral fungi), labyrinthoid (daedaleoid fungi), wrinkled (merulioid fungi), or smooth to diffuse (corticioid fungi). Exceptional hymenomycetes may be aquatic, lack a mycelial phase, or lack a fruit body.

Antibiotics have been isolated from many species. Commercially grown edible species include the button mushroom (*Agaricus bisporus*), Shiitake (*Lentinula edodes*), Paddy Straw mushroom (*Volvariella volvacea*), and Wood Ear (*Auricularia polytricha*). Wild harvested species include the Matsutake (*Tricholoma matsutake* and *T. magnivelare*), chanterelles (*Cantharellus cibarius* and allies), and the King Bolete (*Boletus edulis*).

Most genera are either saprophytic (for example, *Agaricus* and *Polyporus*) or mycorrhizal with trees (*Albatrellus*, *Cortinarius*, *Ramaria*, and *Thelephora*). Others are parasites. *Heterobasidion* causes destructive tree diseases; *Rhizoctonia* and *Typhula* (snow molds) cause field crop losses; *Mycena citricolor* blights coffee leaves; and *Exobasidium*, an obligate plant pathogen, induces the formation of galls and leaf curls. Other notable pathogens include *Hohenbuehelia* and *Pleurotus*, which capture nematodes; *Serpula*, a major dry-rot agent; *Dictyonema*, a basidiolichen; and *Septobasidium*, which harnesses living scale insects. *See* BASIDIOMYCOTINA; EUMYCOTA; FUNGI; PLANT PATHOLOGY. [S.A.R.]

Hymenoptera The third largest order of insects, containing the sawflies, ants, wasps, bees, and related forms. Conservative estimates suggest that the world fauna may comprise well over 100,000 described species of this order, with many thousands still to be described. *See* INSECTA.

This order is of great importance to humans. Some members such as the sawflies, certain chalcidoids, and most cynipoids, feed during the larval stage on foliage or other plant tissues. Many species, such as the ichneumon flies, most chalcid flies, and wasps, are parasites or predators of other insects or spiders during their larval stage. Bees are indispensable in the pollination of many fruits, vegetables, and forage crops. *See* BEE.

Hymenoptera occur in all major faunal zones but are more abundant and have greater diversity of species in the tropical and temperate zones.

Adult Hymenoptera usually may be recognized by having two pairs of membranous wings with reduced venation, the hind pair smaller than the front pair, and by mouthparts formed for biting and often for lapping or sucking. In the higher forms, the abdomen is constricted basally, its first segment fused with the hind part of the thorax. Females always have an ovipositor modified for sawing, piercing, or stinging. Metamorphosis is complete.

The first four superfamilies of the Apocrita—the Ichneumonoidea, Chalcidoidea, Cynipoidea, and Proctotrupoidea—are commonly called the Parasitica, and the remaining superfamilies are known as the Aculeata. The Aculeata are stinging forms and the Parasitica are parasites of other insects. It is impossible to demarcate these two groups sharply because some Aculeata are parasites and some Parasitica are phytophagous. However, except for the phytophagous species of Parasitica, these insects lay their eggs in or on an insect or spider host while the Aculeata place theirs in nests with a provision of food. The table presents the major classification of the order as recognized in North America.

Morphology. The adult hymenopteran has a clearly differentiated head, thorax, and abdomen. Wings, when present, and legs are attached to the thorax.

Typically the head is so oriented that mouthparts are directed downward; however, all variations occur, and in some species the mouthparts are directed forward. The large compound eyes occupy much of the sides of the head, though they are reduced in size in many ants and some Parasitica. Three ocelli are typically present on the top of the head, but may be reduced or absent in wingless forms. The paired antennae arise from the face between the eyes, and they may be close to the mouthparts or removed from them.

The thorax consists of three segments, tightly fused together. Each segment bears a pair of legs, and each of the last two segments bears a pair of wings. In flight, the fore- and hindwings are joined by a row of tiny hooks along the fore margin of the posterior wing, which fit into the downfolded hind margin of the anterior wing. Flightless species with shortened, nonfunctional wings, or no wings at all occur in most major groups, except the sawflies and bees.

The abdomen primitively consists of 10 segments, though the number appears to be less because of modification or loss in the higher forms. The female ovipositor, or sting, is formed from processes of the eighth and ninth sterna.

In the Apocrita there is a pair of acid glands opening into a poison sac connected with the ovipositor. The secretion of these glands produces either a temporary paralysis when injected into their hosts by some Parasitica, or, usually, permanent paralysis when injected into their prey by aculeate wasps. Bees use their stings purely for defense. When a human is stung, the enzymes react with the tissues to release histamine. Death may occasionally result from anaphylactic shock, or from mechanical suffocation due to swelling of the lymphatic system. Medical assistance should be sought if severe swelling occurs following a sting, especially one on the face or throat.

Biology. Practically all hymenopterous adults are terrestrial forms, living in, on, or near the Earth's surface. A few species are secondarily aquatic, the adults swimming or walking under water to search out and parasitize aquatic or subaquatic hosts.

Most adults feed on plant nectar or honeydew secretions of various insects. A few sawflies prey on other insects. Some species of Parasitica and Aculeata imbibe body juices of the host or prey which they attack primarily for oviposition.

Mating takes place in a variety of situations, but it is always of rather short duration. Most species are represented by both males and females. Males are usually produced from unfertilized eggs and have half the normal number of chromosomes, while females are produced from fertilized eggs and have the normal number of chromosomes.

Hymenoptera exhibit complete metamorphosis during development and pass through an egg, larval, and pupal state. So far as is known, Hymenoptera always lay eggs. These are deposited in a protected situation on or near the supply of larval food. The hymenopterous egg is usually ovoid or sausage-shaped, and many species in some groups have stalked eggs.

Some species have only one generation a year in temperate zones, others have two, and many breed continually during the warmer months. Hymenoptera usually overwinter as prepupae but occasionally ants, social wasps, and some bees overwinter as adults or as larvae, as in some Parasitica.

Sexual dimorphism is often very marked. The two sexes of some species are so dissimilar that earlier students placed them in different genera or families. Even today there are many puzzles, and sexes have not been associated for many of the

Families of Hymenoptera

Classification	Common name	No. of species
Suborder Symphyta	Sawflies	1009
Superfamily Megalodontoidea		120
Xyelidae		33
Pamphiliidae	Web-spinning sawflies	87
Superfamily Tenthredinoidea		849
Pergidae		13
Argidae		32
Cimbicidae	Cimbicid sawflies	12
Diprionidae	Conifer sawflies	35
Tenthredinidae	Sawflies	757
Superfamily Siricoidea		28
Syntexidae		1
Siricidae	Horntails	15
Xiphydriidae		6
Orussidae		6
Superfamily Cephoidea		12
Cephidae	Stem sawflies	12
Suborder Apocrita		13,346
Superfamily Ichneumonoidea		3814
Stephanidae		7
Braconidae	Braconid wasps	1239
Ichneumonidae	Ichneumon flies	2568
Superfamily Chalcidoidea		2032
Mymaridae	Fairy flies	110
Trichogrammatidae	Minute egg parasites	39
Eulophidae		544
Elasmidae		17
Thysanidae		18
Eutrichosomatidae		2
Tanaostigmatidae		4
Encyrtidae		320
Eupelmidae		89
Eucharitidae		27
Perilampidae		31
Agaontidae	Fig insects	2
Torymidae	Torymids	181
Ormyridae		17
Pteromalidae		321
Eurytomidae	Seed and stem chalcids	203
Chalcididae	Chalcids	101
Leucospidae		6
Superfamily Cynipoidea	Gall wasps	877
Ibaliidae		6
Liopteridae		2
Figitidae		58
Cynipidae	Cynipids of gall wasps	811
Suborder Apocrita (cont.)		
Superfamily Proctotrupoidea		985
Evaniidae	Ensign flies	11
Gasteruptiidae		50
Pelecinidae	Pelecinid wasps	50
Vanhorniidae		1
Roproniidae		3
Heloridae		1
Proctotrupidae		54
Ceraphronidae		101
Diapriidae		304
Scelionidae	Scelionid wasps	272
Platygasteridae		182
Trigonalidae		5
Superfamily Bethyloidea		345
Chrysididae	Cuckoo wasps	124
Bethylidae		100
Sclerogibbidae		1
Dryinidae		120
Superfamily Scolioidea		643
Tiphiidae	Tiphiid wasps	185
Sierolomorphidae		2
Mutillidae	Velvet ants	409
Rhopalosomatidae		2
Scoliidae		26
Sapygidae		19
Superfamily Formicoidea		786
Formicidae	Ants	786
Superfamily Vespoidea		368
Vespidae	Hornets, yellow jackets, potter wasps	368
Superfamily Pompiloidea		279
Pompilidae	Spider wasps	279
Superfamily Sphecoidea	Fossorial wasps	1215
Ampulicidae		3
Sphecidae		1212
Superfamily Apoidea	Bees	3304
Colletidae	Colletid bees	149
Andrenidae	Andrenidae bees	852
Halictidae	Halictid and sweat bees	472
Melittidae		31
Megachilidae	Leafcutting bees	730
Apidae	Honeybees, bumblebees, and carpenter bees	1076

species having wingless females and winged males. Ordinarily the males are somewhat smaller than females, though the reverse is true in most species having wingless females. *See* SEXUAL DIMORPHISM; SOCIAL INSECTS. [K.V.K.]

Hymenostomatida An order of the Holotrichia which contains many species that often are of small size and fairly uniform ciliation. Primarily, these protozoans are of importance as the first possessors of a definite, though inconspicuous, buccal ciliature. This ciliature consists of an undulating membrane on the right side of the buccal cavity and an adoral zone of membranelles that is primitively composed of three membranelles on the left side. This tetrahymenal, or four-part, buccal ciliary apparatus is considered the fundamental condition from which the oral ciliature of many subsequent higher groups evolved. *See* CILIOPHORA.

The majority of hymenostomes are free-living fresh-water forms. *Paramecium* (see illustration) is the best-known genus of ciliates. It is a good-sized, widely distributed ciliate, and is a much-studied form. *Tetrahymena*, beginning to rival *Paramecium* as a favorite ciliate in much experimental work, owes its scientific popularity primarily to its ability to grow axenically, that is, free from all other organisms, in a chemically defined medium. *See* PROTOZOA. [J.O.C.]

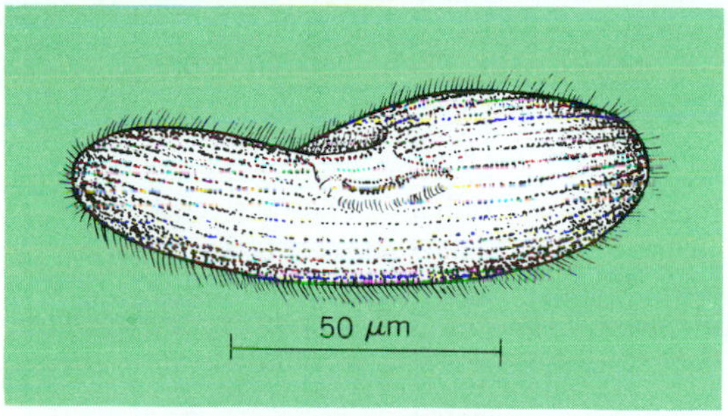

A representative *Paramecium*.

Hyperbola A curve cut from a cone or revolution by a plane that intersects both nappes of the cone and does not contain the apex. In analytic geometry it is shown (see illustration) that a hyperbola is the locus of points P in a plane, such

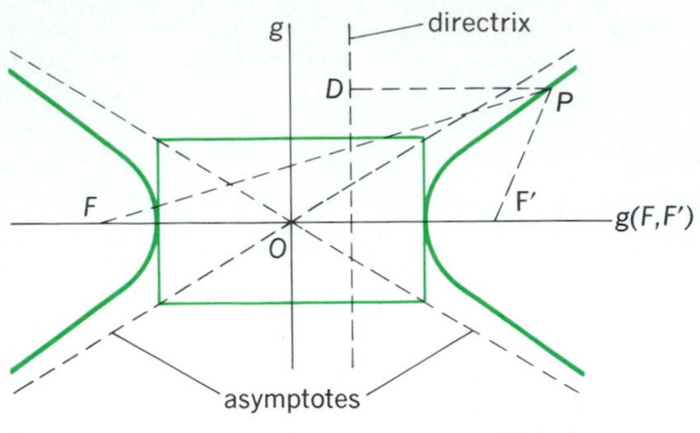

Hyperbola as a locus of points.

that $PF = \epsilon \cdot PD$, where PF and PD denote the distances of P from a fixed point F (focus) and a fixed line (directrix) of the plane, respectively, and ϵ is a constant, greater than 1. A hyperbola is also the locus of points P, the difference of whose distances from two fixed points F, F' (foci), $PF - PF'$, is a constant $2a$ that is less than the distance $2c$ between the foci. The curve is symmetric to the line $g(F,F')$ determined by F, F' and to O, their midpoint. It consists of two branches that are images of each other in the line g through O, perpendicular to $g(F,F')$. There are two lines through O, making equal angles with $g(F,F')$, and to each of which points on each branch get indefinitely close; that is, if point P traverses either branch of the hyperbola, its distance from these lines approaches zero. These lines are called asymptotes of the hyperbola. *See* ANALYTIC GEOMETRY; CONIC SECTION; HYPERBOLOID. [L.M.Bl.]

Hyperbolic function The hyperbolic sine and cosine of a real or complex variable z are defined by Eqs. (1). Both sinh z

$$\sinh z = \frac{e^z - e^{-z}}{2} \qquad \cosh z = \frac{e^z + e^{-z}}{2} \tag{1}$$

and cosh z have a period $2\pi i$ of e^z. From $De^z = de^z/dz = e^z$, Eqs. (2) are obtained.

$$D \sinh z = \cosh z \qquad D \cosh z = \sinh z \tag{2}$$

Since $e^z = \sum_{n=0}^{\infty} z^n/n!$ by definition, Eqs. (3) hold and the series converge for all z and yield relations (4). Thus rela-

$$\sinh z = \sum_{n=0}^{\infty} \frac{z^{2n+1}}{(2n+1)!} \qquad \cosh z = \sum_{n=0}^{\infty} \frac{z^{2n}}{(2n)!} \tag{3}$$

$$\sinh iz = i \sin z \qquad \cosh iz = \cos z \tag{4}$$

$$\sin iz = i \sinh z \qquad \cos iz = \cosh z$$

tions between the circular functions become hyperbolic (and vice versa) when z is replaced by iz.

The hyperbolic tangent and cotangent are defined by Eqs. (5). They have the period πi and have simple poles at the

$$\tanh z = \frac{\sinh z}{\cosh z} \qquad \coth z = \frac{\cosh z}{\sinh z} \tag{5}$$

zeros of cosh z and sinh z, respectively. Moreover Eqs. (6) hold true.

$$D \tanh z = \frac{1}{\cosh^2 z} \qquad D \coth z = -\frac{1}{\sinh^2 z} \tag{6}$$

By definition Eqs. (7) hold.

$$\operatorname{sech} z = \frac{1}{\cosh z} \qquad \operatorname{csch} z = \frac{1}{\sinh z} \tag{7}$$

See HYPERBOLA; TRIGONOMETRY. [L.Br.]

Hyperbolic navigation system A navigation system that produces hyperbolic lines of position through the measurement of the difference in times of transmission of radio signals from two or more synchronized transmitters at fixed points.

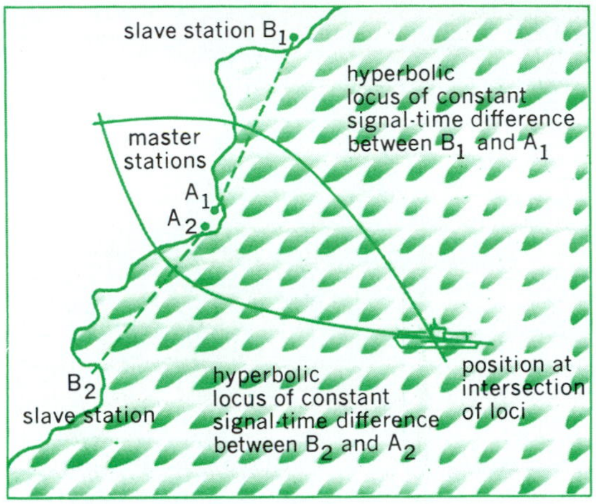

Two intersecting hyperbolic lines of position determining the location, or fix. (*ITT Laboratories*)

When synchronized signals are received from two transmitting stations, the difference in the times of arrival is constant on a hyperbola having the two transmitting stations as foci. The measured time difference locates the receiver on the hyperbolic line of position for that time difference. Another pair of transmitters provides another hyperbolic line of position. The intersection of the lines of position provides a navigational fix, as shown in the illustration. *See* DECCA; LORAN; OMEGA. [A.A.McK.]

Hyperboloid A quadric surface that has both elliptic and hyperbolic plane sections. It is symmetrical in each of a set of three mutually perpendicular planes that intersect pairwise in lines called axes of the hyperboloid. Two of these planes of symmetry intersect the hyperboloid in hyperbolas. If the third plane of symmetry does not intersect the hyperboloid, the latter consists of two separate but congruent parts and is called a hyperboloid of two sheets. If the third plane of symmetry inter-

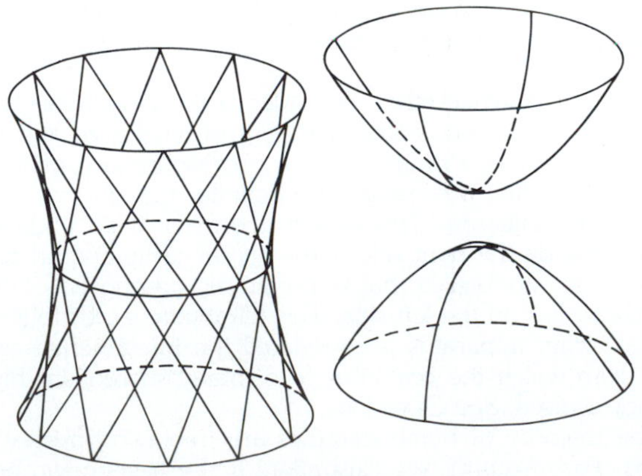
Hyperboloids of one and two sheets.

sects the hyperboloid, the curve of section is an ellipse (or circle) whose axes are also axes of the hyperboloid, and the latter is a connected surface called a hyperboloid of one sheet (see illustration). *See* ANALYTIC GEOMETRY; HYPERBOLA; QUADRIC SURFACE; SURFACE AND SOLID OF REVOLUTION. [J.S.F.]

Hypercharge A quantized attribute, analogous to electric charge, introduced in the classification of a subset of elementary particles—the so-called baryons—including the proton and neutron as its lightest members. As far as is known, electric charge is absolutely conserved in all physical processes. Hypercharge was introduced to formalize the observation that certain decay modes of baryons expected to proceed by means of the strong nuclear force simply were not observed. *See* ELECTRIC CHARGE.

Unlike electric charge, however, the postulated hypercharge was found not to be conserved absolutely; the weak nuclear interactions do not conserve hypercharge—and indeed can change hypercharge by ± 1 or 0 units.

When the known baryons are classified according to their electric charge and their hypercharge, they naturally group into octets, in the scheme first proposed by M. Gell-Mann and K. Nishijima. The quarks, hypothesized as the fundamental building blocks of matter, must have fractional hypercharge as well as electrical charge; the simplest quark model suggests values of $\frac{1}{3}$ and $\frac{2}{3}$, respectively. *See* BARYON; ELEMENTARY PARTICLE; QUARKS; SYMMETRY LAWS (PHYSICS); UNITARY SYMMETRY. [D.A.B.]

Hyperconjugation A phenomenon similar to conjugation in its formulation and manifestations, but the effects are weaker. It occurs when a CH_2 or CH_3 group (or in general an AR_2 or AR_3 group, where A may be any polyvalent atom and R any atom or radical) is adjacent to a multiple bond or to a group containing an atom with a lone π-electron, a π-electron pair or quartet, or a π-electron vacancy. Hyperconjugation can be sacrificial (this is, relatively weak) or isovalent (stronger).

Examples of molecules exhibiting first-order sacrificial hyperconjugation are:

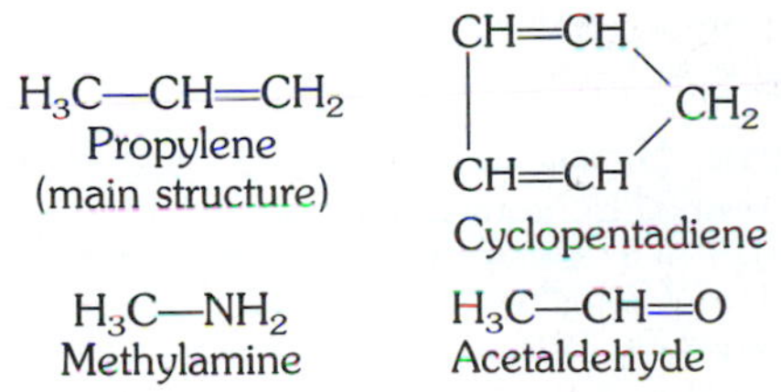

The analogy to conjugation is brought out by rewriting the main structures as

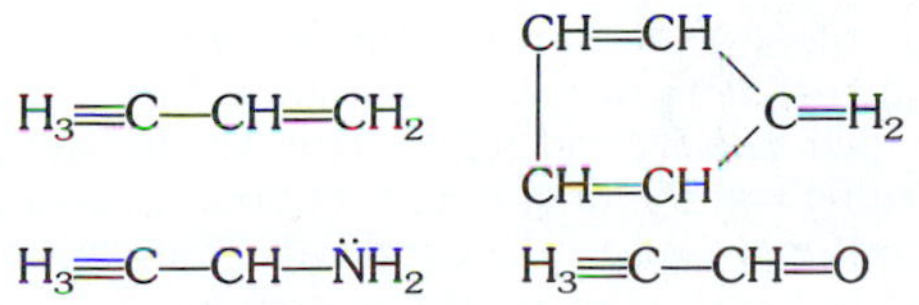

with minor resonance structures

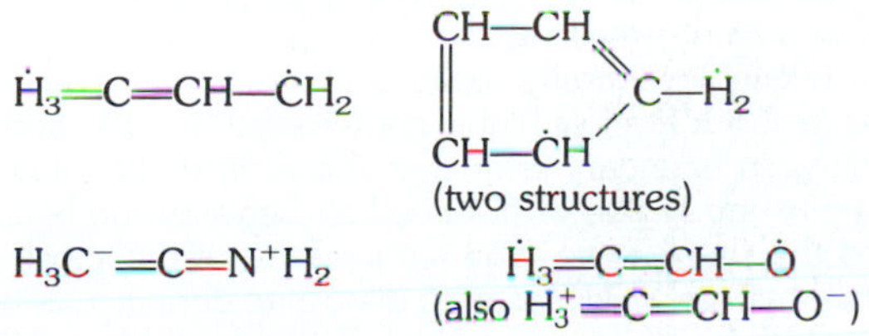

A still weaker effect is second-order sacrificial hyperconjugation, as in ethane (H_3≡C—C≡H_3; compare HC≡C—C≡CH) or in ethylene (H_2═C═C═H_2; compare H_2C═C═C═CH_2).

Examples of isovalent hyperconjugation are H_3≡C—C^+H_2 (with hyperisovalent resonance structure H_3^+═C═CH_2) and H_3≡C—$\dot{C}H_2$ (resonance structure $\dot{H}_3$═C═CH_2). The number of bonds is the same in the two resonance structures, just as in isovalent conjugation, but the second structure is energetically less favorable than the first structure; for this reason the resulting stabilizing resonance energy is less than in isovalent conjugation. [R.S.M.]

Hyperfine structure A closely spaced structure of the spectrum lines forming a multiplet component in the spectrum of an atom or molecule, or of a liquid or solid. In the emission spectrum for an atom, when a multiplet component is examined at the highest resolution, this component may be seen to be resolved, or split, into a group of spectrum lines which are extremely close together. This hyperfine structure may be due to a nuclear isotope effect, to effects related to nuclear spin, or to both. *See* ISOTOPE SHIFT; SPIN (QUANTUM MECHANICS).

The measurement of a hyperfine structure spectrum for a gaseous atomic or molecular system can lead to information about the nuclear magnetic and quadrupole moments, and about the atomic or molecular electron configuration. Important methods for the measurement of hyperfine structure for gaseous systems may employ an interferometer, or use atomic beams, electron spin resonance, or nuclear spin resonance. *See* ATOMIC STRUCTURE AND SPECTRA; ELECTRON PARAMAGNETIC RESONANCE (EPR) SPECTROSCOPY; INTERFEROMETRY; MAGNETIC RESONANCE; MOLECULAR BEAMS; NUCLEAR MAGNETIC RESONANCE (NMR); NUCLEAR MOMENTS. [L.D.R.]

Hypergeometric functions The analytic continuation of the function defined by the series in Eq. (1), where the shifted factorial $(a)_n$ is defined by Eq. (2). It satisfies differential equation (3), and for $|z| < 1$, Re $c >$ Re $b > 0$, is given by the integral representation of Eq. (4), where Γ represents the gamma function.

$$ {}_2F_1\,(a,b;c;z) = \sum_{n=0}^{\infty} \frac{(a)_n(b)_n}{(c)_n n!} z^n \qquad |z| < 1 \tag{1} $$

$$ (a)_n = a(a+1)\cdots(a+n-1) \qquad n = 1,2,\ldots \qquad (a)_0 = 1 \tag{2} $$

$$ z(1-z)y'' + [c - (a+b+1)z]y' - aby = 0 \tag{3} $$

$$ {}_2F_1\,(a,b;c;z) = \frac{\Gamma(c)}{\Gamma(b)\Gamma(c-b)} \int_0^1 (1-zt)^{-a} t^{b-1} (1-t)^{c-b-1} dt \tag{4} $$

See COMPLEX NUMBERS; GAMMA FUNCTION; SERIES.

The interest in hypergeometric functions comes from the many important functions which are special cases of the general hypergeometric function, the rich theory which has been developed for the general hypergeometric function, and the many times they occur in applications. Classically hypergeometric functions have arisen in science as solutions to differential equations. As discrete, rather than continuous, models of physical phenomena have become increasingly useful, hypergeometric functions have continued to arise as solutions to the equations governing these models. *See* DIFFERENTIAL EQUATION. [R.A.]

Hyperiidea A suborder of amphipod crustaceans. Most hyperiids can be recognized by the large eyes which cover nearly the entire surface of the head. The first maxillae and especially the maxillipeds are greatly reduced in comparison to the suborder Gammaridea. In the prehensile pereiopods, the claw is formed by the fifth and sixth segments rather than by the sixth and seventh segments, as in the Gammaridea. The second and third somites of the urosome are always fused, a condition rarely found in the Gammaridea. *See* AMPHIPODA; GAMMARIDEA.

The Hyperiidea are exclusively pelagic and marine. They are found in all the oceans, from the surface to great depths. Most are characteristic of oceanic rather than neritic waters, although there are some species that frequent inshore waters in the tropics. After the Copepoda and Euphausiacea, the Hyperiidea are the most abundant planktonic crustaceans. Some species are frequently found in association with other animals. Other species are truly free-living. Some hyperiids are luminescent; the whole body glows with a greenish-yellow light. *See* BIOLUMINESCENCE. [T.E.B.]

Hypermastigida An order of the Protozoa in the class Zoomastigophorea comprising the most complex flagellates, both structurally and in modes of division. All inhabit the alimentary canal of termites, cockroaches, and woodroaches. These organisms are multiflagellate. The nucleus is single and the organisms are plastic and slow-moving, generally ovoid to elongate. Flagella occur in spiral rows, in tufts, or over the enfire body. These flagellates vary from 15 to 350 micrometers in size. *See* PROTOZOA; ZOOMASTIGOPHOREA. [J.B.L.]

Hypernuclei Nuclei that consist of protons, neutrons, and one or more strange particles such as lambda particles. The lambda particle is the lightest strange baryon (hyperon); its lifetime is 2.6×10^{-10} s. Because strangeness is conserved in strong interactions, the lifetime of the lambda particle remains essentially unchanged in the nucleus also. Lambda hypernuclei live long enough to permit detailed study of their properties. *See* BARYON; ELEMENTARY PARTICLE; HYPERON; NUCLEAR STRUCTURE; STRONG NUCLEAR INTERACTIONS. [B.Po.]

Hyperon A collective name for any baryon with nonzero strangeness number s. The name hyperon has generally been limited to particles which are semistable, that is, which have long lifetimes relative to 10^{-22} s and which decay by photon emission or through weaker decay interactions. Hyperonic particles which are unstable (that is, with lifetimes shorter than 10^{-22} s are commonly referred to as excited hyperons. The known hyperons with spin $\frac{1}{2}\hbar$ (where $\hbar$ is Planck's constant divided by 2π) are Λ, Σ^-, Σ^0, and Σ^+ with $s = -1$, and Ξ^- and Ξ^0, with $s = -2$, together with the Ω^- particle, which has spin $\frac{3}{2}\hbar$ and $s = -3$. The corresponding antihyperons have baryon number $B = -1$, opposite strangeness s, and charge Q; they are all known empirically, except for $\overline{\Omega}^+$.

There is no deep distinction between hyperons and excited hyperons, beyond the phenomenological definition above. Indeed, the hyperon $\Omega(1672)^-$ and the excited hyperons $\Xi(1530)$ and $\Sigma(1385)$, together with the unstable nucleonic states $\Delta(1236)$, are known to form a unitary decuplet of states with spin $\frac{3}{2}\hbar$. *See* BARYON; ELEMENTARY PARTICLE; SYMMETRY LAWS (PHYSICS); UNITARY SYMMETRY. [R.H.D.]

Hyperplasia An increase in cell number. It should be distinguished from hypertrophy, which is an increase in the size of cells, although both bring about an increase in the size of an organ. *See* HYPERTROPHY.

Certain types of hyperplasia may be extremely difficult to differentiate from neoplasia, both grossly and microscopically. Not infrequently the mass of hyperplastic cells causes a localized swelling which grossly resembles a tumor. As in neoplasia, a rapid rate of proliferation is reflected in the large number of mitoses which may be evident. Hyperplasia differs from malignant neoplasia, however, in that the cells appear to be differentiating in an orderly fashion and the features of anaplasia are lacking. *See* ANAPLASIA; ATROPHY; NEOPLASIA. [W.F.P./N.K.M.]

Hypersensitivity An abnormal reactivity in humans and animals to allergens (substances foreign to the body) induced by exposure through injection, inhalation, ingestion, or contact. It is usually referred to as allergy (altered reactivity), and its commonly known clinical forms include asthma, hay fever, hives, infantile eczema, and the contact skin lesions caused by poison oak and poison ivy. Since the allergen may itself be nontoxic, the first experience with it occasions no difficulties. Second and subsequent exposures, however, may give rise to a variety of reactions of different degrees of severity, ranging from transient hives to death.

The allergen, or antigen, induces the allergic state and subsequently becomes responsible for the occurrence of reactions. Allergens vary in chemical nature, ranging from proteins and polysaccharides through simpler drugs and chemical compounds, even the salts of metals. *See* ANTIGEN.

Hypersensitivity may be classified into two categories, delayed and immediate. Delayed reactions begin after several hours, rather than in seconds or minutes as in immediate hypersensitivity. The two chief distinctiions between these two kinds of hypersensitivity are: (1) In delayed hypersensitivity lymphoid cells become activated to react with antigen by means of an effector substance not yet characterized. Such reactions may occur in any tissue where these cells and antigen meet (such as skin, lung, or kidney), and damage may extend to include all the structures that make up such organs. In an immediate hypersensitive reaction, circulating antibodies react with antigen. As a result of this combination, intermediary substances, such as histamine, are released and the antigen-antibody complexes may have directly injurious effects by their effect on certain specific tissues, notably blood-vessel walls and smooth muscle. (2) Since sensitive lymphoid cells react without relation to circulating antibodies, the delayed hypersensitive state cannot be transferred from the allergic subject to a normal one by injections of serum, as can immediate hypersensitivity. *See* ASTHMA; ATOPIC ALLERGY; HISTAMINE.

The steroids are useful on the basis of their anti-inflammatory effects in alleviating reactions in already-sensitive subjects. They are used clinically in severe atopic hypersensitivities while attempted desensitization is underway or when the state is intractable to immunologic therapy. Steroids are applicable also in delayed hypersensitivity when the lesions are so troublesome as to require alleviation before they subside spontaneously. *See* IMMUNOLOGY; STEROID. [S.R.]

Hypersonic flight Flight at speeds well above the local velocity of sound. By convention, hypersonic flight starts at about Mach 5 (five times the speed of sound) and extends upward in speed indefinitely. Supersonic vehicles also fly at speeds greater than the local speed of sound. However, when the Mach number is high, the flow field around an object exhibits a special behavior, which is worth studying separately from supersonic flight. This behavior is characteristic of hypersonic flight.

A body entering the Earth's atmosphere from space (for example, a meteorite, a ballistic reentry vehicle, or a spacecraft) has high velocity and hence large kinetic and potential energy. During reentry, drag forces act upon a reentry vehicle or spacecraft and cause it to decelerate, thus dissipating its kinetic and potential energy. The energy lost by the reentry vehicle is then transferred to the air within the flow field around the reentry vehicle. The flow field around the forward portion of a blunt-nosed vehicle (body of revolution or leading edge of a wing) generally exhibits (1) a distinct bow shock wave, (2) a shock layer of highly compressed hot gas, and (3) a highly sheared boundary layer over the surface. The flow field is defined as the region of disturbed air between the body surface and the shock wave. The temperature of the shock layer is so high that molecules begin to dissociate during collisions at about 4000°F (2500 K), or at Mach 7. At about Mach 12, gas in the stagnation region reaches a temperature of 6700°F (4000 K) and becomes ionized.

In general, at high hypersonic flight speed the characteristic temperature in the shock layer of a blunted body and in the boundary layer of a slender body is proportional to the square of the Mach number. Heat energy is then transferred from the hot gases to the vehicle surface by conduction and diffusion of chemical species in the boundary layer and by radiation from the shock layer near the nose. Heat energy is also radiated from the vehicle surface to space or to adjacent objects. One important problem confronting the designer of reentry vehicles or spacecraft is therefore to design a minimum-weight vehicle able to withstand large heat loads from adjacent hot-gas layers during reentry while retaining the ability to carry a given useful payload. *See* ATMOSPHERIC ENTRY. [S.-Y.C.]

Hypersonics A development of ultrasonics (also known as pretersonics) dealing with the production and utilization of sound waves of frequencies above 500 MHz. These frequencies are sufficiently high to approximate thermal waves, and they can interact directly with phonons. They also interact with electrons and holes and with spin waves in ferromagnetic materials. *See* HOLE STATES IN SOLIDS; PHONON; ULTRASONICS; WAVE MOTION.

A considerable development occurred in the field of hypersonics due to the production of thin-film transducers of cadmium sulfide or zinc oxide, the use of lasers in Brillouin scattering or Bragg reflection from sound waves, and the use of thermal pulses. They have provided considerable evidence on the interaction of acoustic waves with other thermal and electrical waves in the media. Frequencies as high as 2.5 terahertz have been obtained.

One application of hypersonics is the acoustic microscope. This was suggested as early as 1936 by S. Sokolov, but it was not until sound waves with very high frequencies with wavelengths on the order of those for optical wavelengths were obtained that acoustic microscopes were able to compete with optical means. *See* ACOUSTIC MICROSCOPE. [W.P.M.]

Hypertension High blood pressure. Blood pressure is expressed in two numbers: the higher number is the systolic blood pressure, which is the pressure exerted by the blood against the walls of the blood vessels while the heart is contracting. The lower number is the diastolic blood pressure, which is the residual pressure that exists between heart contractions, or while the heart is relaxing. Normal blood pressure provides sufficient blood flow to the vital organs, including the brain, heart, kidneys, intestine, and skeletal muscle.

It is not entirely accurate to think of high blood pressure as a distinct disease; high blood pressure appears to be both a disease and a risk factor for other diseases. At the highest end of the blood pressure distribution, there is an increased probability of premature death secondary to stroke, heart disease, or kidney failure. Lower on the distribution curve (for example, diastolic blood pressure of 90–104 mmHg, which is referred to as mild hypertension), the absolute risk of premature mortality is lower and continues to decline with further decreases in blood pressure. High blood pressure is thus a disease when its value is very high and a risk factor throughout its distribution. For diagnostic purposes, blood pressure is considered high when persistently above 140/90 mmHg.

Some cases of very high blood pressure are due to specific causes that may be surgically remediable. Most hypertension, however, results from the combination of a genetic predisposition and an environmental factor such as excessive sodium intake, sedentary habits, and stress. *See* ARTERIOSCLEROSIS; STRESS (PSYCHOLOGY).

High blood pressure can be controlled. Mild cases are treated by losing excess weight and reducing the intake of sodium and alcohol. More serious cases are treated with drugs such as diuretics, beta blockers, calcium antagonists, angiotensin-converting enzyme inhibitors, alpha blockers, and centrally acting compounds that affect regulatory centers in the brain. Treatment can usually assure a normal life. *See* CIRCULATORY SYSTEM; HEART DISORDERS. [M.Ho.]

Hypertrophy The increase in size of a cell, tissue, or organ. Hypertrophy of an organ or tissue is due either to an increase in the mass of the individual cells (true hypertrophy) or to an increase in the number of cells making up the tissue (hyperplasia) or to both. Certain organs such as the heart consist of cells which lack the ability to divide, and in such cases hypertrophy results in an increase in the size rather than in the number of the constituent cells. In other tissues such as the liver, lymph nodes, and adrenal cortex, hypertrophy of the organ is effected principally by an increase in the number of cells.

When the hypertrophy is the result of physiological rather than pathological stimuli, it is termed physiological hypertrophy. Examples are increased mass of skeletal muscles which may be produced by physical exercise, enlargement of the uterus in pregnancy, and increased number of circulating red blood cells found in people living at high attitudes. Compensatory hypertrophy is the increase in size of an organ or organ system following removal or destruction of a portion of it. *See* HYPERPLASIA. [W.F.P./N.K.M.]

Hyphochytriomycetes A class of the subdivision Mastigomycotina; a taxonomically small group of about 20 species of fungallike organisms characterized by the presence of a uniflagellate zoospore in the life cycle. In contrast to the Chytridiomycetes, the flagellum is anteriorly directed, with two rows of tripartite hairs. From similarities in ultrastructure and cellular chemistry and from molecular analysis, it has been determined that hyphochytrids are related to the biflagellated Oomycetes and heterokont algae, but they are nevertheless classified as a separate class within the Eumycota. *See* CHYTRIDIOMYCETES; OOMYCETES.

Two species, *Rhizidiomyces apophysatus* and *Hyphochytrium catenoides*, are widely distributed, possibly cosmopolitan, and occur as saprobes in soil and water. They have been grown in pure culture on defined media. *Rhizidiomyces apophysatus* has a unicellular thallus comprising a single zoosporangium with delicate rhizoids for anchorage and the absorption of nutrients. *Hyphochytrium catenoides* is filamentous with numerous intercalary hyphal swellings; under favorable conditions these swellings develop into zoosporangia. Sexual reproduction is unknown in these species and has not been well documented in the class. All species parallel the Chytridiomycetes in morphology and distribution, and they cannot be distinguished except by zoosporic structure. *See* EUMYCOTA; FUNGI; MASTIGOMYCOTINA. [D.J.S.B.]

Hyphomycetes A class of anamorphic (asexual or imperfect) fungi belonging to the Deuteromycotina. They lack locular fruit bodies (conidiomata), and so sporulation occurs on separate or aggregated hyphae, which may or may not be differentiated; the thallus consists of septate hyphae. About 1000 genera (with 550 synonyms) comprising more than 10,000 species are recognized. *See* DEUTEROMYCOTINA.

The Hyphomycetes, like other groups of deuteromycetes, is an artificial one composed almost entirely of anamorphic fungi of ascomycete affinity. Many are known anamorphs of Ascomycetes, although some have Basidiomycotina affinities because of their clamp connections or dolipore septa. Several of the latter are aquatic or aero-aquatic. Taxa are referred to as form genera and form species, because the absence of a sexual or perfect teleomorph state forces classification and identification by artificial rather than phylogenetic means. The unifying feature of the group is the production of conidia from superficial, exposed conidiogenous cells arising separately from vegetative hyphae or cells (mononematous), or incorporated on conidiophores that may be entirely separate or aggregated in cushionlike sporodochia or synnemata. Differences in insertion and arrangement of conidiogenous cells and conidiophores traditionally have been used to separate three orders. In the

Hyphomycetales they are solitary or at most fasciculate and tufted, in the Tuberculariales they are produced over the outer surface of a cushion-shaped–to–flattened conidioma (sporodochium), and in the Stilbellales they are united into a stipitate conidioma (synnema). *See* ASCOMYCOTINA; BASIDIOMYCOTINA.

Many Hyphomycetes can grow in synthetic media under artificial conditions. By using a variety of techniques that involve such factors as substrate composition, light requirements, and temperature, they can be induced to sporulate. The accurate identification of many groups, including yeasts, is based on nutritional tests, as in bacteriology.

The Hyphomycetes draw nourishment from living or dead organic matter and are adapted to grow, reproduce, and survive in a wide range of ecological situations. They can also be either stress-tolerant or combative. Their ability to colonize, decompose, and use substrates and to interact with or parasitize other organisms is a result of the enzymes, antibiotics, toxins, and other metabolites they produce, coupled with wide genetic diversity. They are extremely common in soils of all types and on leaf litter and other organic debris of both natural and manufactured origin.

The Hyphomycetes are found on or associated with fungi, including pathogens such as *Verticillium*, *Mycogone*, and *Cladobotryum*, and on lichens. Many are of medical importance, being associated with superficial (*Trichosporon*, *Exphiala*), cutaneous (*Microsporum*, *Trichophyton*, *Candida*), subcutaneous (*Fonsecaea*, *Petriellidium*), and systemic (*Histoplasma*, *Coccidioides*) infections. They are often opportunistic organisms that cause infections in immunocompromised patients. Hyphomycetes also cause economically important diseases in all types of vascular plants, especially agricultural and forestry crops, and function in a disseminative role. *See* MEDICAL MYCOLOGY; MYCOTOXIN; PLANT PATHOLOGY.

Hyphomycetes produce a wide variety of primary and secondary metabolites and are capable of effecting many different chemical and biochemical changes. By harnessing that capability in industrial processes, organic acids, enzymes, antibiotics, growth substances, alcohol, cheese, and Oriental foods, among others, can be produced and steroid transformation can take place. *See* FUNGI. [B.C.S.]

Hypnales An order of the true mosses (subclass Bryidae). Also known as the Hypnobryales, this order consists of 14 families and some 135 genera, primarily put together because they share a hypnoid peristome. The families show considerable sporophytic unity; it is gametophyte structure that determines family membership.

Hypnales are often known as feather mosses owing to their freely branched stems, with branches regularly or irregularly arranged in two rows. The plants form mats, especially in woodlands. The leaves are most commonly acute or acuminate, and the costa is generally short and double or nearly lacking. The sporophytes are lateral with elongate, smooth setae, and the capsules are generally inclined and asymmetric with well-developed double peristomes. *See* BRYIDAE; BRYOPHYTA; BRYOPSIDA. [H.Cr.]

Hypnosis A presumed altered state of consciousness in which the hypnotized individual is usually more susceptible to suggestion than in his or her normal state. In this context, a suggestion is understood to be an idea or a communication carrying an idea that elicits a covert or overt response not mediated by the higher critical faculties (that is, the volitional apparatus).

Hypnosis cannot be physiologically distinguished from the normal awake state of an individual, and for this reason its existence has been questioned by some investigators. There are few phenomena observed in association with hypnosis, if any, that are specific to the hypnotic state. Most are directly or indirectly produced by suggestions. Through suggestions given to hypnotized individuals, it is possible to induce alterations in memory, perception, sensation, emotions, feelings, attitudes, beliefs, and muscular state. Such changes can be, and usually are, incorporated into the complex behavior of the individual, resulting in amnesias and paramnesias, fuguelike conditions, paralysis, loss of sensory functions, changes in attention, personality alterations, hallucinatory and delusional behavior, and even physiological changes. Enhanced recall is sometimes possible. Although sometimes remarkable, the effects produced through hypnosis with the majority of individuals are much less spectacular than popularly believed. [A.M.W.]

Hypochlorite The negative ion, ClO^-, derived from hypochlorous acid, HClO. The hypochlorite ion is best known as an oxidizing agent and as a constituent of bleaching agents.

Hypochlorites are prepared by passing chlorine gas into alkaline solution and in this procedure the chlorine is simultaneously oxidized and reduced, as in the following reaction.

$$Cl_2 + 2NaOH \rightarrow NaCl + NaClO + H_2O$$

Sodium hypochlorite is extensively used as a bleach in the textile industry and in laundries and as a disinfectant in the home.

When chlorine is passed over slaked lime, a product called bleaching powder or chlorinated lime is obtained which approximates the formula CaCl(OCl). A related preparation, high-test hypochlorite (HTH), contains twice as much available chlorine as bleaching powder. *See* BLEACHING; CHLORINE; OXIDIZING AGENT. [E.E.W.]

Hypodermis The outermost cell layer of the cortex, also called the exodermis, of plants. It forms a prominent layer immediately under the epidermis in many but not all plants. Like the endodermis, it develops Casparian strips, suberin deposits, and cellulose deposits impregnated with phenolic or quinoidal substances. The hypodermis may produce substances that act as a barrier to the entry of pathogens, and in some plants it may function in the absorption of water and the selection of ions that enter the plant. *See* CORTEX (PLANT); ENDODERMIS. [D.S.V.F.]

Hypofluorous acid An unstable acid with the chemical formula HOF. Hypofluorous acid is a colorless gas at room temperature. At very low temperature it is a white solid, melting at −117°C (−179°F) to a pale yellow liquid that boils somewhat below room temperature. The compound reacts rapidly with water to form H_2O_2 in acid solution and O_2 in base. Most oxidizable substrates are attacked by HOF. Oxygen transfer from HOF provides a way of synthesizing a variety of compounds with isotopically labeled oxygen atoms. [E.H.Ap.]

Hypoplasia The failure of a tissue or organ to achieve complete development. Related but not synonymous terms are agenesis, the complete absence of a tissue, and aplasia, the virtual absence of a tissue or organ, only a remnant of the structure being present. Hypoplasia is usually apparent at birth. Examples of hypoplasia may be found in the kidney, the ovary, the testis, the heart, and in virtually every organ system. [N.K.M.]

Hypothermia The term which describes conditions of reduced body temperature in homeotherms, that is, those animals which routinely regulate and maintain body temperature (mammals and birds). Ordinarily the term hypothermia is not used in reference to conditions of reduced body temperature in poikilotherms (reptiles, amphibians, fishes, and invertebrates), cold-blooded animals which do not regulate and maintain body temperature. In those animals the effects of cold exposure are better termed cold torpor and loosely termed hibernation. *See* HIBERNATION.

An animal undergoing hypothermia reveals a number of dramatic biological events and conditions, for example, shivering, and reductions in body temperature and in respiration rate, and depression of cardiac activity. It is of note that many depressed physiological activities are comparable to processes of hibernation. If one assesses the true physiological meaning of these phenomena, it is evident that the animal has reduced its intake and transport of oxygen. Since the animal is still living in its immobile state resembling dormancy, a remarkable slowing of metabolic activity is demonstrated. This state of depressed metabolism means that an animal's tissues and organs have reduced requirements for oxygen. Oxygen needed for oxidative cellular metabolism is now required in small amounts since the body chemistry has slowed drastically. This reduction in the tissue-oxygen requirement is one of the most useful aspects of hypothermia.

In the practice of medicine the depressed metabolism of hypothermia has become an important tool in various forms of surgery and in the treatment of several illnesses. For example, cardiovascular surgery, where there is need for repair of heart valves and corrections of congenital defects, it has been found adequate to reduce body temperature from normal levels (37–38°C or 98.6–100.4°F) to about 30°C (86°F). This level of hypothermia can be achieved by surface cooling, in which the patient is placed in a modified ice bath or surrounded by ice blankets. The next level in surgical hypothermia is in bringing the body temperature down between 30 and 26°C (86 and 79°F). This is sometimes called moderate hypothermia, and at these levels it is necessary to cool the blood directly. The techniques usually involve cardiac bypass, where a portion of the circulation is shunted out of the body and blood-oxygen levels and flow rates are maintained by means of a highly sophisticated array of tubes, pumps, and oxygenators. Certain surgical procedures require deep levels of hypothermia, 25–15°C (77–59°F), with long periods of circulatory arrest. In these cases the heartbeat is markedly slowed, circulation is shunted and the body temperature is controlled outside of the body. *See* THERMOREGULATION. [X.J.M.]

Hypothesis A tentative supposition with regard to an unknown state of affairs, the truth of which is thereupon subject to investigation by any available method, either by logical deduction of consequences which may be checked against what is known, or by direct experimental investigation or discovery of facts not hitherto known and suggested by the hypothesis. [P.W.Br./H.Ma.]

Hypotrichida An order of Spirotrichia. These protozoans are commonly considered to represent the pinnacle of specialized development in the evolution of ciliates. Somatic ciliature of the ventral surface has been replaced by cirri; on the dorsal surface it is absent or represented by inconspicuous sensory bristles (see illustration). The adoral zone of membranelles is very prominently developed, and the buccal area may occupy a large part of the ventral surface of the body. The whole body is generally rigid. Hypotrichs occur ubiquitously in fresh- and salt-water habitats. *See* SPIROTRICHIA. [J.O.C.]

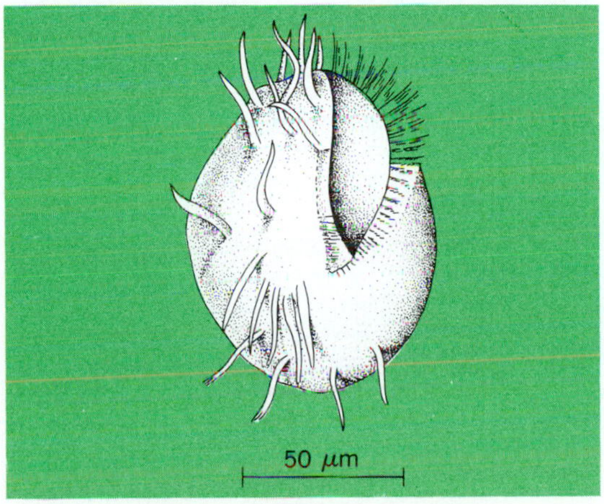

***Euplotes*, an example of a hypotrich.**

Hypoxia The failure of oxygen to gain access to, or to be utilized by, the body. Although the term anoxia is commonly used, a more precise term, hypoxia, is more often applicable because there is seldom a complete oxygen defect.

Oxygen deprivation may result from interference with some stage of the inspiration, lung diffusion, blood transport, cellular absorption, and final utilization by enzyme systems. A defect at any one or more of these major stages quite often induces a decreased ability of other related mechanisms to survive. This is seen most dramatically in any form of hypoxia in which the brain is deprived of the necessary oxygen for more than a few minutes. Nerve cell degeneration begins quickly, and although the original cause of hypoxia is removed, damage to the respiratory centers prevents resumption of breathing. *See* RESPIRATORY SYSTEM.

The term anoxia is used by many authorities to indicate an oxygen deficiency at the tissue level, and failure of cellular respiration may be designated histotoxic anoxia. There are other terms employed to differentiate the type of oxygen deficiency or the stage in the total respiratory process where defects occur. [E.G.St./N.K.M.]

Hyracoidea An order of ungulate mammals. Once abundant in the middle and late Tertiary faunas of Africa and reaching as far eastward as China during the Pliocene, they are represented today only by the small conies of Africa, Arabia, and Syria.

The short-tailed hyraxes keep their hindquarters in a crouch and are plantigrade, with the axis of symmetry of their feet going through the third digit. The strong hindlegs effectively propel the animals. The forefoot is four-fingered with only three functional fingers, and the hindfoot (pes) has three toes. The soles of the feet have soft, enlarged, elastic pads, which are kept moist by numerous glands allowing hyraxes to cling to rock surfaces and tree trunks. *See* DENTITION.

The living hyraxes (also called dassies or conies) are herbivorous. Ground-living forms live among rocks, and form colonies of up to 50 individuals. They are mainly active during the day. Some genera are solitary, nocturnal, and almost completely arboreal, rarely leaving the trees. *See* EUTHERIA; MAMMALIA. [F.S.S.]

Hysteresis A phenomenon wherein two (or more) physical quantities bear a relationship which depends on prior history. More specifically, the response Y takes on different values for an increasing input X than for a decreasing X.

If one cycles X over an appropriate range, the plot of Y versus X gives a closed curve which is referred to as the hysteresis loop. The response Y appears to be lagging the input X.

Hysteresis occurs in many fields of science. Perhaps the primary example is of magnetic materials where the input variable H (magnetic field) and response variable B (magnetic induction) are traditionally chosen. For such a choice of conjugate variables, the area of the hysteresis loop takes on a special significance, namely the conversion of energy per unit volume to heat per cycle. For mechanical hysteresis, it is customary to take the variables stress and strain, where the energy density loss per cycle is related to the internal friction. Thermal hysteresis is characteristic of many systems, particularly those involving phase changes, but here the hysteresis loops are not usually related to energy loss. *See* FERROELECTRICS; MAGNETIC HYSTERESIS; STRESS AND STRAIN; THERMAL HYSTERESIS. [H.B.H.; R.K.MacC.]

Hysteresis motor A synchronous motor without salient poles and without dc excitation which makes use of the hysteresis and eddy-current losses induced in its hardened-steel rotor to produce rotor torque. The stator and stator windings are similar to those of an induction motor and may be polyphase, shaded-pole, or capacitor type. The rotor is usually made up of a number of hardened steel rings on a nonmagnetic arbor. The hysteresis motor develops constant torque up to synchronous speed. The motor can, therefore, synchronize any load it can accelerate. These motors are built in small sizes, for instance, for electric clocks. *See* INDUCTION MOTOR; SYNCHRONOUS MOTOR. [L.V.B.]

Hysteriales (lichenized) An order of the Ascolichenes shared by the Ascomycetes, comprising those species with a so-called ascolocular structure. The hymenium consists of densely interwoven, branched pseudoparaphyses and irregularly scattered locules in which the asci are located.

The growth form of the Hysteriales is either crustose or fruticose. There are five or six families now included here. Arthoniaceae is the largest family, with two genera, *Arthonia* and *Arthothelium*, both widespread on bark, rarely on leaves, in temperate and tropical regions. The Roccellaceae is a family of conspicuous fruticose species that grow profusely on trees and rocks along the coastlines of Portugal, California, Baja California, and parts of western South America. These plants were collected by the ton in the Middle Ages for use as dyestuffs. These same lichens, on a smaller scale, now yield the dye used in litmus paper. *See* ASCOMYCOTINA. [M.E.H.]

I

Ice cream A commercial dairy food made by freezing while stirring a pasteurized mix of suitable ingredients. They may include but are not limited to milk fat, nonfat milk solids, or milk-derived ingredients; and sugars, corn syrup, water, flavoring, sometimes egg products, stabilizers, and emulsifiers. Air incorporated during the freezing process is also an important component.

Products classed as frozen dairy desserts include French ice cream, frozen custard, ice milk, sherbet, water ice, frozen dairy confection, dietary frozen dairy dessert, and mellorine or imitation ice cream.

In ice cream manufacture the basic ingredients are blended together; the liquid blend is called the mix. Then the mix is pasteurized, homogenized, cooled, and aged. The flavoring materials are usually added to the mix just before the freezing process, but fruits, nuts, and candies are not added until the ice cream is discharged from the freezer. *See* MILK. [W.S.A.]

Ice field A network of interconnected glaciers or ice streams, with common source area or areas, in contrast to ice sheets and ice caps. (An ice sheet is a broad, cakelike glacial mass with a relatively flat surface and gentle relief. Ice caps are properly defined as domelike glacial masses, usually at high elevation.) Being generally associated with terrane of substantial relief, ice-field glaciers are mostly of the broad-basin, cirque, and mountain-valley type. Thus, different sections of an ice field are often separated by linear ranges, bedrock ridges, and nunataks. [M.M.Mi.]

Ice island One of the massive bodies of floating ice in the Arctic Ocean, from 20 to 200 ft (6–60 m) thick, irregular in shape, and from a few square miles to 300 mi^2 (680 km^2) in area. Ice islands originate in the landfast ice along the high-latitude Northern shores of the Canadian archipelago and Greenland. This distinguishes them from the smaller and more rugged icebergs which originate in the glaciers of Greenland. The unbroken appearance of the ice islands contrasts greatly with the surrounding pack ice, which normally almost completely covers the Arctic Ocean with a maximum thickness of 20 ft (6 m).

Ice islands are tabular features without the pressure ridges found on pack ice. Long, shallow drainage channels are most evident in summer when they are filled with meltwater. The presence of rock piles on the surface and of dust and dirt layers within the ice, as well as plant and animal remains, testifies that the ice islands were close to land at one time. [K.H.]

Ice point The temperature at which liquid and solid water are in equilibrium under atmospheric pressure. The ice point is by far the most important "fixed point" for defining temperatures scales and for calibrating thermometers. A closely related point is the triple point, which has gained favor as the primary standard since it can be attained with great accuracy in a simple closed vessel, isolated from the atmosphere. *See* TEMPERATURE; TRIPLE POINT. [R.A.Bu.]

Iceberg A large mass of glacial ice broken off and drifted from parent glaciers or ice shelves along polar seas. Icebergs should be distinguished from polar pack ice which is sea ice, or frozen sea water, though rafted or hummocked fragments of the latter may resemble small bergs. *See* GLACIOLOGY; SEA ICE.

Arctic iceberg, eroded to form a valley or dry-dock type: grotesque shapes are common to the glacially produced icebergs of the North.

Icebergs are classified by shape and size. The terms used are arched, blocky, dome, pinnacled, tabular, valley, and weathered for berg description, and bergy-bit and growler for berg fragments ranging smaller than cottage size above water. The lifespan of an iceberg may be indefinite while the berg remains in cold polar waters, eroding only slightly during summer months. But under the influence of ocean currents, an iceberg that drifts into warmer water will disintegrate rapidly.

In the Arctic, icebergs (see illustration) originate chiefly from glaciers along Greenland coasts. It is estimated that a total of about 16,000 bergs are calved annually in the Northern Hemisphere, of which over 90% are of Greenland origin; but only about half of these have a size or source location to enable them to achieve any significant drift. No icebergs are discharged or drift into the North Pacific Ocean or its adjacent seas, except a few small bergs each year that calve from the piedmont glaciers along the Gulf of Alaska.

In the Southern Ocean, bergs originate from the giant ice shelves all along the Antarctic continent. These result in huge, tabular bergs or ice islands several hundred feet high and often over a hundred miles in length, which frequent the entire waters of the Antarctic seas. *See* ICE ISLAND. [R.P.D.]

Icebreaker A ship designed to break floating ice. Since the 1960s, potential resource development in the Arctic regions has led to the construction of icebreaking ships that can transit to all areas of the world, including the North Pole. Icebreakers provide the platforms from which polar science and research can be conducted on a year-round basis. *See* ANTARCTIC OCEAN; ARCTIC OCEAN.

Icebreakers may be classed as polar or subpolar, depending on their primary geographic area of operation. Polar icebreakers can operate independently in first-year, second-year, and multiyear ice in the Arctic or the Antarctic. Subpolar icebreakers operate in the ice-covered waters of coastal seas and lakes outside the polar regions. *See* MARITIME METEOROLOGY; SEA ICE.

The combined shapes of the bow, midbody, and stern constitute the hull form. The icebreaker's bow must be designed to break ice efficiently, and it consists of an inclined wedge that forms an angle of about 20° with the ice surface. As the icebreaker advances, the bow rides up on the edge of the ice until the weight of the ship becomes sufficiently large that the downward force causes the ice to fail. The broken ice pieces move under the hull, and the process is repeated. Traditionally, icebreakers have sloped sides of 5–20° to reduce the frictional resistance. The stern of the icebreaker is designed in a manner similar to the bow so that the ship can break ice while going astern. *See* SHIP DESIGN.

The hull of a polar icebreaker is built very strong so that it can withstand tremendous ice forces. Because frequent and high loads occur at very low air temperatures, specialty steels are used on polar icebreakers. It is common to have bow and stern plating of 1.5–2.0 in. (4–5 cm) in thickness.

Most icebreakers are powered with diesel engines. However, when the horsepower required for icebreaking is sufficiently great, gas turbines or nuclear steam turbines are needed. *See* MARINE ENGINE; SHIP POWERING AND STEERING.

Many auxiliary methods have been developed to improve the performance of icebreakers by reducing the friction between the hull and the ice. One widely used method is to roll or heel the ship from side to side. In addition, most icebreakers use a low-friction coating or a stainless-steel ice belt to reduce icebreaking resistance. [R.P.V.]

Ichneumon A member of the hymenopteran family Ichneumonidae. Included are more than 10,000 species of minute to large parasitic wasps which have a worldwide distribution. Ichneumons parasitize their victims by depositing eggs on or within the host. Endoparasitic forms undergo a complete metamorphosis in the host body with a number of instars or developmental stages; the adults emerge from the pupal or chrysalis stage of the host. *See* HYMENOPTERA; INSECTA. [C.B.C.]

Ichthyopterygia A subclass of extinct reptiles composed of predatory fish-finned and sea-swimming forms of the Mesozoic Era. All are much alike in body form: short-necked, streamlined, swordfishlike or porpoiselike in body, yet unrelated to either fishes or porpoises. A single order, Ichthyosauria, is included in this subclass. *See* REPTILIA. [C.L.C.]

Ichthyornithiformes An order of ancient fossil birds from the Upper Cretaceous marine chalk beds of the Niobrara formation. Two families are included: Ichthyornithidae, with seven species known from Kansas and Texas, and Apathornithidae, with one species each from Kansas and Wyoming. The Ichthyornithiformes possess all skeletal characteristics of modern birds (with some exceptions) and appear to be strong flying birds of inland seas, somewhat gull-like in their proportions and presumably in their habits. *See* AVES. [W.J.B.]

Ichthyostegalia An order comprising the oldest tetrapods yet found—labyrinthodont amphibians collected from Late Devonian fresh-water deposits of eastern Greenland. Two groups are distinguished, ichthyostegids and acanthostegids. The former includes the genus *Ichthyostega* (the "four-legged fish"), the bony parts of which have been reconstructed. The fishlike tail with dermal fin rays and the vestigial intracranial joint are holdovers of their piscine ancestors, the osteolepiform crossopterygians. By these features, in combination with pentadactyl limbs and otic notches for the eardrums, the ichthyostegalians bridge the gap between fish and tetrapods. The humerus is complex; the scales exhibit a trend toward elimination; and the chest skeleton is composed of large, broad ribs forming a trunk armor. *See* AMPHIBIA; LABYRINTHODONTIA. [E.J.]

Ideal aerodynamics The oldest branch of analytical aerodynamics, also referred to as ideal fluid dynamics. Although limited in its application, it is useful for explaining some airflow problems and for providing approximate answers. It is also still studied because many other analytical methods in aerodynamics presuppose a knowledge of this earlier work.

The use of a set of five basic simplifying assumptions characterizes ideal aerodynamics: (1) The flow is two-dimensional; that is, there exists a direction in which the flow characteristics do not change. (2) The flow is steady; that is, the flow characteristics do not change with time. (3) The flow is incompressible; that is, density does not change. (4) The fluid is nonviscous; that is, the effects of viscosity are ignored except, in some cases, indirectly. (5) The flow is irrotational; that is, the curl of the velocity vector is zero.

The assumption of constant density limits application to flows with a Mach number below about 0.6, depending on the accuracy desired. *See* AERODYNAMICS; LAPLACE'S IRROTATIONAL MOTION; FLUID FLOW; SUBSONIC FLIGHT. [F.McL.M.]

Idocrase A sorosilicate mineral of complex composition, with formula $Ca_{10}Al_4(Mg,Fe)_2Si_9O_{34}(OH)_4$, crystallizing in the tetragonal system; also known by the name vesuvianite. Crystals, frequently well formed, are usually prismatic with pyramidal terminations. The luster is vitreous to resinous; the color is usually green or brown but may be yellow, blue, or red. Hardness is $6\frac{1}{2}$ on Mohs scale; specific gravity is 3.35–3.45. *See* SILICATE MINERALS.

Idocrase is found characteristically in crystalline limestones resulting from contact metamorphism. Noted localities are Zermatt, Switzerland; Christiansand, Norway; River Vilui, Siberia; and Chiapas, Mexico. In the United States it is found at Sanford, Maine; Franklin, New Jersey; Amity, New York; and at many contact metamorphic deposits in western states. [C.S.Hu.]

Igneous rocks Those rocks which have congealed from a molten mass. They may be composed of crystals or glass or both depending on the conditions of formation. The molten matter from which they come is called magma; where erupted to the surface, it is commonly known as lava. Solidification of the hot rock melt occurs in response to loss of heat. Generated at depth the magma tends to rise. It commonly breaks through the Earth's crust and spills out on the Earth's surface or ocean floor to form volcanic or extrusive rocks. At the surface where cooling is rapid, fine-grained or glassy rocks are formed.

Where unable to reach the surface, magma cools more slowly, insulated by the overlying rocks; and a coarser texture develops. The resulting igneous rocks appear intrusive relative to adjacent rocks. In general, deeply formed (plutonic) rocks display the coarsest texture. Igneous rocks formed at shallow depths (hypabyssal) display features somewhat intermediate between those of volcanic and plutonic types. *See* MAGMA; PLUTON; VOLCANO; VOLCANOLOGY.

Texture refers to the mutual relation of the rock constituents within a uniform aggregate. It is dependent upon the relative amounts of crystalline and amorphous (glassy) matter as well as the size, shape, and arrangement of the constituents. Rock textures are highly significant; they shed light on the problem of rock genesis, and tell much about the conditions and environment under which the rock formed.

Schemes for classifying igneous rocks are numerous. Three principal methods of classification are used. (1) Megascopic schemes are based on the appearance of the rock-in-hand specimen or as seen with a magnifying glass (hand lens). Such schemes are useful in the field study of rocks. (2) Microscopic schemes (largely mineralogical) are employed in laboratory investigations where more detailed information is needed. (3)

Chemical schemes are very useful but have more limited application. The mineral content and texture of a rock generally tell much more about the rock's origin than does a bulk chemical analysis. Igneous rocks show great variations chemically, mineralogically, texturally, and structurally with few if any natural boundaries.

Plutonic rocks occur in large intrusive masses (batholiths, stocks, and other large plutons). They form at great depth and are often referred to as abyssal rocks. They are generated from large bodies of magma which cooled slowly.

Volcanic rocks are formed as lava flows or as pyroclastic rocks (heterogeneous accumulations of volcanic ash and coarser fragmental matter). They have solidified rapidly, and expanding gas bubbles formed by escaping volatiles frequently create highly porous rocks.

Hypabyssal rocks exhibit characteristics more or less intermediate between those of volcanic and plutonic types. They differ from volcanic rocks in that they are intrusive and generally free from glass and vesicular structures. They differ from plutonic rocks in that they occur in small bodies (dikes and sills) or in larger bodies formed at shallow depths (laccoliths) and they have textures characteristically resulting from more rapid cooling. *See* ANORTHOSITE; BASALT; DIABASE; FELDSPAR; FELDSPATHOID; GABBRO; GRANITE; GRANODIORITE; LABRADORITE; LEUCITE ROCK; MONZONITE; NEPHELINE SYENITE; PYROXENITE; QUARTZ; RHYOLITE; SYENITE; THERALITE; TRACHYTE.

Igneous rock-forming minerals may be classed as primary or secondary. The primary minerals are those formed by direct crystallization from the magma. Secondary minerals may form at any subsequent time. The principal primary minerals are relatively few and may be classed as light-colored (felsic) or dark-colored (mafic) varieties. Felsic is a mnemonic term for feldspathic minerals (feldspar and feldspathoids) and silica (quartz, tridymite, and cristobalite). Mafic is mnemonic for magnesium and iron-rich minerals (biotite, amphibole, pyroxene, and olivine). Felsic minerals are composed largely of silica, alumina, and alkalies. Mafics are rich in iron, magnesium, and calcium.

Secondary minerals include minerals formed by addition of material subsequent to solidification of the rock or by alteration of minerals already present in the rock. Alteration in which certain minerals become more or less reconstituted is common and widespread. The common alteration products derived from the essential primary minerals are as follows.

Primary mineral	*Secondary mineral*
Quartz	Not altered
Potash feldspar	Kaolinite, sericite
Plagioclase	Kaolinite, sericite (paragonite), epidote, zoisite, calcite
Nepheline	Cancrinite, analcite, natrolite
Leucite	Nepheline and potash feldspar
Sodalite	Analcite, cancrinite
Biotite	Chlorite, sphene, epidote, rutile, iron oxide
Hornblende	Actinolite, biotite, chlorite, epidote, calcite
Orthopyroxene	Antigorite, actinolite, talc
Clinopyroxene	Hornblende, actinolite, biotite, chlorite, epidote, antigorite
Olivine	Serpentine, magnetite, talc, magnesite

See PETROLOGY. [C.A.C.]

Ignition system The ignition system of an internal combustion engine initiates the chemical reaction between fuel and air in the cylinder charge. For maximum efficiency it is necessary to ignite the charge in each cylinder shortly before the piston reaches the top of its compression stroke. The best timing is determined experimentally for various engine speeds and loadings, and the ignition system is designed to provide this timing automatically. For smooth running, multicylinder engines are usually built so that the various pistons arrive at their firing top center positions in evenly spaced intervals. The sequence in which the cylinders reach their firing points depends upon the geometrical arrangement of the cylinders and crankshaft, and is called the firing order of the engine. The ignition system distributes the ignition impulse to the cylinders in this order. *See* INTERNAL COMBUSTION ENGINE.

A large class of engines, including most automobile engines, operate on a premixed charge of fuel vapor and air. In these engines the charge is ignited by passing a high-voltage electric current between two electrodes in the combustion chamber. The electrodes are incorporated in a removable unit. When a spark of sufficient energy jumps the gap between the electrodes, a self-propagating flame is produced in the fuel-air mixture which spreads rapidly throughout the charge. *See* COMBUSTION CHAMBER; DISTRIBUTOR; SPARK PLUG.

The electric energy for the spark is obtained from a storage battery in practically all spark-ignition engines. The battery, usually 6 or 12 volts, is part of a low-voltage primary circuit, which includes a switch, the primary winding of the ignition coil, a capacitor (or condenser), and breaker points. *See* STORAGE BATTERY.

The magneto system is similar to the battery system except that the voltage required to cause a flow of current in the primary winding is generated by the rotation of a set of permanent magnets, instead of being supplied by a battery. The magneto is often used with small compact installations such as lawnmowers, chain saws, and small outboard engines, where a battery would represent undesirable weight and bulk. [A.R.R.]

Iguanid A member of the reptilian family Iguanidae, in the order Squamata. The 400 species, found in the New World only, are adapted to various modes of life: terrestrial, arboreal, and fossorial (burrowing); some species are semiaquatic. They are mainly insectivorous, although there are herbivorous and omnivorous forms. The teeth are fixed to the inner edge of the jaws; the fleshy tongue is nonretractile; the body is compressed; the tail is long but rarely prehensile; the limbs have five clawed toes; and many species have dorsal or caudal spines, lappets, and crests. Many iguanids have the ability to change color and are commonly but erroneously referred to as chameleons. *See* REPTILIA; SQUAMATA. [C.B.C.]

Ileitis Regional ileitis is a sharply demarcated inflammation of the lower portion of the small intestine. Actually, other portions of the small and large intestine may be involved, so regional enteritis or regional enterocolitis may be preferred terms.

The disease is marked by intestinal obstruction, pain, cramps, diarrhea, and constipation. There is often associated weight loss or anemia. The cause is unknown, but faulty protein absorption, psychic factors, and abnormal lymph drainage have been implicated. Although regional ileitis is found most often in young adults, no age group is immune.

In contrast to regional ileitis, ordinary ileitis is any inflammation of the ileum from specific causes such as trauma or infections. *See* INTESTINE; INTESTINE DISORDERS. [E.G.St./N.K.M.]

Illiciales An order of flowering plants, division Magnoliophyta (Angiospermae), in the subclass Magnoliidae of the class Magnoliopsida (dicotyledons). The order consists of two families, Illiciaceae and Schisandraceae, with fewer than a hundred species in all. They are all woody, with scattered spherical cells containing volatile oils. The leaves are alternate and simple. The flowers are solitary or a few clustered together, regular, and hypogynous; the perianth has five to many segments that are not clearly differentiated into sepals and petals. The Illiciales are related to the Magnoliales and have sometimes been included in that order, but they are apparently more advanced in having fundamentally triaperturate pollen. *See* MAGNOLIIDAE; MAGNOLIOPSIDA. [T.M.Ba.]

Illite A general term for the mica-type clay minerals. Illite clays are used for the manufacture of structural clay products, such as brick and tile. Some degraded high plastic illites are used for bonding molding sands. *See* CLAY; CLAY MINERALS.

Illite is a common product of weathering if potash is present in the environment of alteration. It is a frequent constituent in many soil types, and may form in soils under certain conditions as a consequence of the addition of potash fertilizers. Illite is common in recent sediments and is particularly abundant in deep-sea clays. Because of its stability, illite is often found in ancient sediments. [F.M.W.; R.E.Gr.]

Illuminance The density of luminous flux incident on a surface. This word has been proposed by the Colorimetry Committee of the Optical Society of America to replace the term illumination. The definitions are the same. The symbol of illumination is E, and the equation is $E = dF/dA$, where A is the area of the illuminated surface and F is the luminous flux. *See* ILLUMINATION; PHOTOMETRY. [R.C.Pu.]

Illumination In a general sense, the science of the application of lighting. Radiation in the range of wavelengths of 0.38–0.76 micrometer produces the visual effect, commonly called light, by the response of the average human eye for normal (photopic) brilliance levels. Illumination engineering pertains to the sources of lighting and the design of lighting systems which distribute light to produce a comfortable and effective environment for seeing. In a specific quantitative sense, illumination is the combination of the spatial density of radiant power received at a surface and the effectiveness of that radiation in producing a visual effect. *See* ILLUMINANCE. [W.B.Bo.]

Ilmenite A rhombohedral mineral with composition $Fe^{2+}Ti^{4+}O_3$. The hardness is 5½ on Mohs scale, specific gravity 4.72. Color is black, and there is no cleavage. The mineral usually occurs massive or in thin plates. Two other minerals belonging to the ilmenite structure type are geikielite, $MgTiO_3$, and pyrophanite, $MnTiO_3$.

Ilmenite is the most abundant titanium mineral in igneous rocks and the most important ore of titanium. Important occurrences include the Ilmen Mountains, Russia (whence the name); Kragerø, Norway; and Allard Lake, Quebec, Canada. *See* TITANIUM. [P.B.M.]

Image processing The enhancement of images by systems in which the input into the system is an image and the output is also an image that has been improved in some way. The image may consist of a photographic negative, a medical thermogram, or a nuclear medicine image comprising a collection of dots, with the density of dots related to the distribution of a radioactive pharmaceutical in the body.

An image-processing system may have one of two goals: it either makes the image more pleasing or more useful to a human observer, or it actually performs some of the interpretation and recognition tasks usually performed by humans. The latter goal is often referred to as image analysis or pattern recognition.

Of the various types of hardware that can be used for image enhancement, the digital computer is the most generally useful and perhaps also the easiest to understand conceptually. Suppose that the initial image is a photographic transparency. The first step is to convert the density levels in the image into numbers that can be entered into the computer memory. This can be accomplished with a microdensitometer in which a small, rectangular aperture is used to isolate a single picture element, or pixel. The light transmitted through this element is measured with a photodetector, digitized, and stored. Then the aperture is moved over by its width, and the density of the adjacent picture element is recorded. In this way the entire picture is converted into a matrix of numbers suitable for computer manipulation. The most common example of computer manipulation is contrast enhancement, which is very useful if the input image is underexposed or depicts a low-contrast scene. A second example is a pseudocolor display, in which each gray level in the input is encoded as a color in the output. Once the computer manipulations are complete, the processed image may be displayed on a cathode-ray tube, a mechanical printer, or any other suitable device. *See* DIGITAL COMPUTER. [H.H.B.]

Imhoff tank A sewage treatment tank named after its developer, Karl Imhoff. Imhoff tanks differ from septic tanks in that digestion takes place in a separate compartment from that in which settlement occurs. The tank was introduced in the United States in 1907 and was widely used as a primary treatment process and also in preceding trickling filters. Developments in mechanized equipment have lessened its popularity, but it is still valued as a combination unit for settling sewage and digesting sludge. *See* SEPTIC TANK; SEWAGE. [W.T.I.]

Imidazole One of a group of organic heterocyclic compounds (also called iminazoles, glyoxalines, and 1,3-diazoles) containing a five-membered diunsaturated ring with two nonadjacent nitrogen atoms as part of the ring. Imidazole is a typical member of the group. *See* HETEROCYCLIC COMPOUNDS.

Imidazole itself is a water-soluble solid, mp 90°C (194°F), which is basic enough to form stable salts with both organic and inorganic acids. Imidazole is also weakly acidic, since the hydrogen at the 1 position may be replaced by metal. The low volatility of imidazole, bp 256°C (493°F), is indicative of considerable association by intermolecular hydrogen bonding. [W.J.Ge.]

Immittance The impedance or admittance of an alternating-current circuit. It is sometimes convenient to use the term immittance when referring to a complex number which may be either the impedance (ratio of voltage to current) or the admittance (ratio of current to voltage) of an electrical circuit. The units of impedance and admittance are, of course, different and so units cannot be assigned to an immittance. However, in certain theoretical work it may be necessary to deal with general functions which afterward will be specialized to become either an impedance or an admittance by the assignment of suitable units; in such cases it is convenient to refer to the functions as immittances. *See* ADMITTANCE; ALTERNATING CURRENT; ELECTRICAL IMPEDANCE. [J.O.S.]

Immune complex disease Local or systemic tissue injury caused by the vascular deposition of products of antigen-antibody interaction, termed immune complexes. Immune complex formation with specific antibodies causes the inactivation or elimination of potentially harmful consequences only when immune complexes deposit in tissues, inciting various mediators of inflammation, leading to injury and disease. When the reaction takes place in the extravascular fluids near the site of origin of the antigen (by injection, secretion, and such), focal injury can occur, as exemplified by the Arthus reaction or such conditions as experimental immune thyroiditis. Systemic disease may occur when soluble antigens combine with antibodies in the vascular compartment, forming circulating immune complexes that are trapped nonspecifically in the vascular beds of various organs, causing such clinical diseases as serum sickness or systemic lupus erythematosus with vasculitis and glomerulonephritis. The term immune complex disease usually signifies this systemic immune complex formation and vascular deposition. *See* AUTOIMMUNITY; IMMUNOLOGY. [E.H.C.; C.B.W.]

Immunity A state of resistance to an agent, the parasite, that normally produces an infection in one or more host species. Analogous immune responses also occur to tumors (cancers), in allergies and delayed hypersensitivity states, in autoimmune diseases, and in incompatible blood transfusions or tissue and organ transplantations. These responses are specifically lacking in immune deficiency diseases. Immunity is a complex resultant of many components, some native and hereditable, others acquired.

In acquired immunity both the invading parasite and the host may alter their offensive and resistive capacities with time. Many pathogenic organisms possess chemical components—enzymes, toxins, or surface antigens—which adversely affect the suceptible host. After contact with these substances (antigens) the host may, in turn, produce neutralizing or inhibiting substances (generally, specific antibodies), which check the infection.

Some organisms possess natural immunity. For example, animals, including humans, are generally insusceptible to even the most virulent plant pathogens, and plants are likewise insusceptible to animal pathogens.

Active immunity may be acquired upon recovery from an infection. Recovery from certain virus diseases, such as yellow fever, measles, and chickenpox, is usually attended by a lifelong immunity effective under the conditions of natural exposure. At the other extreme, immunity following recovery from the common cold is very short-lived or nonexistent. *See* CELLULAR IMMUNOLOGY; IMMUNOLOGY; VACCINATION. [H.P.T.]

Immunoassay An assay that quantifies antigen or antibody by immunochemical means. The antigen can be a relatively simple substance such as a drug, or a complex one such as a protein or a virus. *See* ANTIBODY; ANTIGEN.

The reactants are first mixed so that a varying quantity of one (A) is added to a constant amount of the other (B). The formation of an immune (antigen-antibody) complex is measured as a function of the varied reactant (A). The result is represented by a "standard curve" for reactant A. An unknown sample is tested by adding it to reactant B. The extent of the measured change is referred to the standard curve, and thereby is obtained the amount of reactant A which produces a comparable change. The amount is represented as the content of reactant A in the unknown sample. *See* IMMUNOFLUORESCENCE; IMMUNOLOGY; IMMUNONEPHELOMETRY; RADIOIMMUNOASSAY. [A.B.]

Immunochemistry A discipline concerned both with the structure of antibody (immunoglobulin) molecules and with their ability to bind an apparently limitless number of diverse chemical structures (antigens); with the structure, organization, and rearrangement of the genes coding for the immunoglobulin molecules; and with the structure and function of molecules on the surface of animal cells, such as the transplantation (histocompatibility) antigens, which recognize antibodies and the thymus-derived lymphocytes mediating the cellular immune response. *See* ANTIGEN; IMMUNOASSAY; IMMUNOGLOBULIN; RADIOIMMUNOASSAY; TRANSPLANTATION BIOLOGY. [W.H.K.; F.F.R.]

Immunoelectrophoresis A combination of the techniques of electrophoresis and immunodiffusion used to separate the components of a mixture of antigens and make them visible by reaction with specific antibodies. *See* ELECTROPHORESIS.

A medium such as agar is deposited on a convenient base, for example, a microscopic slide. A small well is cut in the medium. A test solution is deposited in the well, and the contained substances are separated by electrophoresis along one axis of the plate. A trough is then cut in the medium parallel to, but at some distance from, the line of the separated substances. The trough is filled with antiserum which contains antibodies to one or more of the separated substances. The antiserum and substances diffuse toward one another and, where they meet, form curvilinear patterns of precipitation. These can be seen directly in clear media or can be visualized after washing out unreacted materials and staining in opaque media. *See* IMMUNOASSAY. [A.B.]

Immunofluorescence A technique that uses a fluorochrome to indicate the occurrence of a specific antigen-antibody reaction. The fluorochrome labels either an antigen or an antibody. The labeled reactant is then used to detect the presence of the unlabeled reactant. The use of a labeled reactant (such as an antibody which both detects and indicates the antigen) to reveal the presence of an unlabeled one is termed direct immunofluorescence. The use of a labeled indicator antibody, which reacts with an unlabeled detector antibody that has previously reacted with an antigen, is termed indirect immunofluorescence. Substitution of a light meter for the human eye permits a quantitative measurement in immunofluorometry. *See* FLUORESCENCE; IMMUNOASSAY. [A.B.]

Immunogenetics A scientific discipline that uses immunological methods to study the inheritance of traits. Traditionally, immunogenetics has been concerned with moieties that elicit immune response, that is, with antigens (antigenic determinants). It has now broadened its scope to study also the genetic control of the individual's ability to respond to an antigen. *See* ANTIGEN.

The immunological methods used in immunogenetics are of two principal kinds, serological and histogenetical. In serological methods, antibodies are used to detect antigens, either in solution or on a cell surface. In histogenetical methods, immune cells (lymphocytes) are used to detect antigens on the surface of other cells. In modern immunogenetics research, the serological and histogenetical methods are combined with molecular methods in which the researcher isolates and works with the genes that code for the traits. This approach of going back and forth from classical to molecular methods has proved to be very successful and has led to the elucidation of several complex genetic systems. *See* ANTIBODY.

Animal immunogenetics relies heavily on the use of inbred, congenic, and recombinant inbred strains. Inbred lines result when individuals that are more closely related to each other than randomly chosen individuals mate together, for many generations. The advantage in working with inbred strains rather than outbred animals is that inbred strains restrict the variability of the conditions of an experiment. However, when two strains are compared and it is found that they respond differently to a treatment, it is not known to what gene this difference should be attributed. The strains may differ at as many genetic loci as two unrelated individuals in an outbred population do. To study the effect of single, defined genes, immunogeneticists have developed congenic lines. These lines always come in groups, the smallest group being a pair, which consists of a congenic line and its inbred partner strain. The two are homozygous at more than 97% of their loci (that is, they are inbred) and are identical except, ideally, at one locus—the locus that is to be studied. To find out whether two loci are on the same or on different chromosomes, two individuals that differ in the traits controlled by these loci are mated and then the F_1 hybrids are intercrossed. In the F_2, generation that results from this intercross, the genes assort either independently, if they are on different chromosomes, or nonrandomly, if they are the same chromosome—that is, when they are linked. Each time the strains are tested for linkage, this laborious procedure must be repeated. To avoid this repetition, immunogeneticists have prepared a "frozen" F_2 generation by establishing separate inbred lines from the different F_2 individuals. Such lines are called the recombinant inbred strains.

Contemporary immunogenetic research concentrates on two main categories of antigenic substances—those present in body fluids, primarily blood serum or plasma, and those expressed on surfaces of various cells. In the body-fluid antigens category, a prominent position is occupied by immunoglobulins. Although antibodies are usually used to detect antigens, they themselves may also serve as antigens, and antibodies can be produced against them. These antibodies against antibodies detect three principal kinds of antigenic determinants: isotypic, allotypic, and idiotypic. The main categories of cell-surface molecules studied by immunogenetical methods are blood-group antigens, histocompatibility antigens, tissue-restricted antigens, and receptors. Blood-group antigens are alloantigens found on erythrocytes. Histocompatibility antigens are antigens capable of inducing cellular immune responses and hence are detectable by histogenetical methods. Tissue-restricted antigens are expressed on some tissues but not on others and therefore serve as markers for cell sorting. Receptors are molecules that are capable of specifically interacting with certain other molecules. The interaction often leads to activation or inhibition of the receptor-bearing cell. *See* BLOOD GROUPS; GENETICS; HISTOCOMPATIBILITY; IMMUNOGLOBULIN; IMMUNOLOGY. [J.K.]

Immunoglobulin Any of a set of serum glycoproteins which have the ability to bind other molecules with a high degree of specificity. Molecules which are foreign or nonself are called antigens; and when an antigen is introduced into a vertebrate, the immunoglobulin induced is called an antibody. The critical property of this antibody is that it will combine specifically with the inducing antigen. *See* ANTIBODY; ANTIGEN.

Human myeloma proteins were the first structurally discrete immunoglobulins known, and they have played a central role in the elucidation of the antigenic, structural, and functional features of immunoglobulins. Myeloma proteins appear to result from the malignant expansion of a normal clone of antibody-producing cells; but since they are spontaneous or nonspecifically induced, they have no predictable antibody activity.

Immunoglobulins are heterogeneous with respect to charge, size, antigenicity, and function. Although charge heterogeneity was initially used to define the "γ-globulins," charge is the least useful of these criteria for precise characterization. The first separation of immunoglobulins by size was in 1937 when horse antipneumoccal antibody was separated into two different molecular weight groups: 900,000 and 150,000. The larger antibody has been named immunoglobulin macro (IgM) and the smaller immunoglobulin gamma (IgG), reflecting its electrophoretic mobility.

Antigenic analysis is the most practical approach in differentiating the classes and subclasses of immunoglobulins. The three most abundant classes in normal human serum are IgG, IgA, and IgM. Two other classes of antigenically distinct immunoglobulins, IgD and IgE, were first recognized as myeloma proteins because their concentration in normal human serum was too low to detect by the then available techniques.

IgG is the most abundant immunoglobulin in serum. The IgG molecule has a molecular weight of 150,000 and consists of two pairs of polypeptide chains which are covalently linked by disulfide bonds. One is called the Fab, for "fragment, antigen binding." The other fragment is crystallizable, contains no light chain, and does not bind antigen. This is called Fc, for "fragment, crystallizable."

IgA is the second most abundant immunoglobulin in human serum but is the primary immunoglobulin of secretions. The nature of IgA differs widely among various mammals and between serum and secretions.

Immunoglobulin macro or IgM is the largest immunoglobulin, with a molecular weight of about 900,000. Electron-microscopic studies show that the monomeric units are arranged like the spokes of a wheel, with the Fc portions at the hub and the antigen-binding Fab portions directed outward. IgM is the first immunoglobulin to appear during the primary immune response and is usually present only transiently.

IgD and IgE are present in minute amounts in normal human serum. No function has as yet been attributed to IgD. However, evidence suggests it may play a prominent role as a cell surface immunoglobulin and may serve as an antigen receptor. *See* IMMUNOLOGY. [J.D.C.]

Immunologic cytotoxicity The mechanism by which the immune system destroys or damages foreign or abnormal cells. Immunologic cytotoxicity may lead to complete loss of viability of the target cells (cytolysis) or an inhibition of the ability of the cells to continue growing (cytostasis). Immunologic cytotoxicity can be manifested against a wide variety of target cells, including malignant cells, normal cells from individuals unrelated to the responding host, and normal cells of the host that are infected with viruses or other microorganisms. In addition, the immune system can cause direct cytotoxic effects on some microorganisms, including bacteria, parasites, and fungi.

Immunologic cytotoxicity is a principal mechanism by which the immune response copes with, and often eliminates, foreign materials or abnormal cells. Cytotoxic reactions are frequently observed as a major component of an immune response that develops following exposure to foreign cells or microorganisms. In addition, there is increasing evidence that cytotoxic reactions represent a major mechanism for natural immunity and resistance to such materials. In most instances, cytotoxicity by immune components involves the recognition of particular structures on the target cells; also, the targets need to be susceptible to attack by the immune components. Some cells are quite resistant to immunologic cytotoxicity, and this appears to represent a major mechanism by which they can escape control by the immune system.

There are a variety of mechanisms for immunologic cytotoxicity. The two main categories are antibody- and cell-mediated cytotoxicity. Within cell-mediated cytotoxicity, there is a multiplicity of effector cell types and mechanisms that can be involved, including cytotoxic T lymphocytes, macrophage-mediated cytotoxicity, natural killer cells, granulocyte cytoxicity, and antibody-dependent cell-mediated cytotoxicity. *See* IMMUNOLOGY. [R.B.He.]

Immunological deficiency A state wherein the immune mechanisms are inadequate in their ability to perform their normal function, that is, the elimination of foreign materials (usually infectious agents such as bacteria, viruses, and fungi). Immune mechanisms are also responsible for the rejection of transplanted organs. These processes are accomplished by white blood cells known as lymphocytes, of which there are two major types, T lymphocytes (thymus-derived) and B lymphocytes (bone marrow-derived). *See* CELLULAR IMMUNOLOGY; TRANSPLANTATION BIOLOGY.

Immunological deficiency states result from a failure at any point in the complex set of interactions involving lymphocytes and immunoglobulins. In general, the diseases are due to absence of cell populations; failure of cells to mature; failure to secrete the products necessary for effective cell interactions; or failure of accessory cell populations or protein systems (for example, complement) which are necessary for the complete competence of B-cell immune function. Some of the diseases are carried on the X-chromosome and affect only males, being carried by females. *See* COMPLEMENT; SEX-LINKED INHERITANCE.

The prime symptom of immunodeficiency is an increased suceptibility to infections. Many of the organisms to which people are constantly exposed do not ordinarily have the capability to cause infections in immunocompetent individuals because these organisms are so weak that they cannot establish themselves in normals. In immunodeficients, however, they can cause fatal infections.

In general, immunodeficiency states are inherited. Immunodeficiency can also be acquired as a complication of other disease processes. One of the most common forms of deficiency is caused by aggressive treatment of leukemia. Another cause of induced immunodeficiency is seen with transplant rejection therapy. To prevent organ rejection, drugs which destroy lymphocytes must be administered. Certain viruses, such as the Epstein-Barr virus (EBV), which causes infectious mononucleosis, infect lymphocytes. Involvement of the lymphoid system is nearly always only temporary, but in a small number of individuals the virus cannot be eliminated, and a chronic infection of B cells leads to the loss of normal lymphocyte function. Another immunodeficiency disorder caused by virus infection is acquired immune deficiency syndrome (AIDS). Immunodeficiency can also be observed secondary to dietary deficiency. Two main varieties are seen. In protein-calorie malnutrition, serious deficiency primarily involving the T-cell system predisposes affected individuals to overwhelming infection by the agents of measles or tuberculosis. In the second variety, deficiency of single substances is the cause; the two most commonly observed deficiencies are those of zinc and biotin. *See* ACQUIRED IMMUNE DEFICIENCY SYNDROME (AIDS); TRANSPLANTATION BIOLOGY. [Ri.H.]

Immunological ontogeny The origin and development (ontogeny) of the lymphocyte system, from its earliest stages to the two major populations of mature lymphocytes: the thymus-dependent or T lymphocytes, and the thymus-independent or B lymphocytes. The T lymphocytes carry out those aspects of function which are called cell-mediated immunity, including graft rejection, elimination of tumor cells, and delayed hypersensitivity. B cells are responsible for humoral or antibody-mediated immunity. *See* CELLULAR IMMUNOLOGY; IMMUNITY.

For both systems, development or differentiation proceeds in discrete stages. In the first stage, pluripotent hematopoietic stem cells, which originate in the yolk sac in the embryo and then successively in fetal liver and fetal bone marrow, develop into precursor cells committed to becoming T or B cells. Hematopoiesis in human fetal liver begins at about 4 weeks of gestational age and in fetal bone marrow after 20 weeks. *See* HEMATOPOIESIS.

The thymus plays a strategic role in the development of T lymphocytes. Precursor cells are attracted into the thymus where, under the influence of this microenvironment, they undergo rapid proliferation and maturation. These maturing T cells also begin to express a variety of cell-surface markers, which parallels developing immunocompetence. From the thymus, the maturing T cells are exported to the peripheral lymphoid tissues. *See* THYMUS GLAND.

The earliest B cells identified in fetal liver are pre-B cells. As the cells mature, they express immunoglobulin M (IgM) and subsequently IgD on their surface. At this stage, the cell is a specific, competent B cell ready to interact with an antigen. In the course of B-lymphocyte differentiation, diversity of immunoglobulin classes is generated by an orderly switch from IgM to IgG to IgA with expression of the respective immunoglobulin on the cell surface. *See* ANTIGEN; IMMUNOGLOBULIN. [E.W.Ge.]

Immunological phylogeny The study of immunology and the immune system in evolution. All vertebrates can recognize and respond to nonself-molecular configurations on microorganisms, cells, or organic molecules by utilizing a complex recognition system termed the immune response. The presence of lymphocytes and circulating antibodies has been documented in all extant vertebrate species. However, the existence of induced, specific reactions directly homologous to the immune repertoire of vertebrates has not been clearly established in invertebrates.

The role of phagocytic cells in engulfing foreign pathogens has been documented in virtually all metazoan organisms. Phagocytic cells possess a limited capacity to discriminate self from nonself, and this is due in part to the presence of lectins (molecules capable of binding specifically to various sugars) on their surface. Although there is no evidence to suggest that invertebrate lectins and vertebrate immunoglobulins are homologous structures, sufficient diversity exists within lectins of certain species to indicate that these types of molecules and their cellular expression on phagocytes might serve as a primitive and universal recognition mechanism. *See* IMMUNOGLOBULIN; LECTINS.

All true vertebrates possess cells clearly recognizable as lymphocytes and can carry out T-cell functions, such as graft rejection, and show the capacity of B cells to synthesize and secrete immunoglobulins. True lymph nodes are not present in vertebrate species more primitive than mammals, but birds possess aggregates of lymphoid tissue probably serving a similar function. *See* CELLULAR IMMUNOLOGY; LYMPHATIC SYSTEM.

Humans possess five major classes or isotypes of immunoglobulin: IgG, IgM, IgA, IgE, and IgD. The IgM molecule is the first immunoglobulin to appear in ontogeny, and the first to appear in phylogeny. Immunoglobulins of cyclostomes, elasmobranchs (sharks and rays), and many teleost fishes consist only of IgM polymers. Immunoglobulins possessing heavy chains distinct from the μ-like heavy chains of those groups are present in some lungfish (Dipnoi) and in anuran amphibians (frogs and toads). Dipnoi have a low-molecular-weight non-IgM immunoglobulin (termed IgN). Birds possess IgM and IgA immunoglobutins, but also possess a non-IgM immunoglobulin similar to that of amphibians as their major immunoglobulin class. This immunoglobulin has been termed IgY. IgG immunoglobulins containing gamma chains clearly homologous to those of the humans and of true mammals are found only within the three subclasses of living mammals, namely, eutherians, metatherians (marsupials), and monotremes (for example, the echidna). *See* IMMUNOLOGICAL ONTOGENY.

Although the precise nature of the precursors of the specific elements of the immune system in evolution remains to be determined, the genetic and cellular events which lead to the capacity for specific immune recognition, diversification, and reactivity occurred early in vertebrate evolution. *See* IMMUNITY; IMMUNOLOGY. [J.J.Ma.]

Immunology The division of biological science concerned with the native or acquired resistance of higher living forms, or hosts, to infection with microorganisms. Immunology is an eclectic science drawing on many other branches of knowledge. The comprehensive study of infection and resistance may be divided, arbitrarily, into data relating to the pathogen, the host, and their interactions during infection. *See* ANTIBODY; ANTIGEN; DISEASE; IMMUNITY; INFECTION; PATHOGEN.

Immunology is also heavily concerned with assaying the immune status of the host through a variety of serological procedures, and in devising methods of increasing host resistance through prophylactic vaccination. There has also been much important investigation of induced resistance and tolerance to transplants of skin and organs, including tumors. *See* ACQUIRED IMMUNOLOGICAL TOLERANCE; BLOOD GROUPS; CELLULAR IMMUNOLOGY; HYPERSENSITIVITY; IMMUNITY; IMMUNOASSAY; INTERFERON; ISOANTIGEN; PHAGOCYTOSIS; PRECIPITIN; SEROLOGY; TRANSPLANTATION BIOLOGY; VACCINATION. [M.J.Po.]

Immunonephelometry An application of nephelometry to the quantification of antigen or antibody. The technique depends on the light-scattering properties of microparticles. The initial antigen-antibody complexes are macromolecular in size and do not scatter light. However, such complexes have the ability to aggregate. The size of the aggregates increases with time until they do scatter light. This process takes from several minutes to hours before it attains maximum, depending on various properties of the reactants. If other aspects are kept constant, including, for example, the concentration of one reactant, then both the rate at which the scatter of light increases and its maximal value increase with the concentration of the other reactant. Either the increase in the rate of scatter or its maximal value can be calibrated for various concentrations of a reactant to yield a concentration-response calibration curve. Unknown concentrations of that reactant can then be quantified by measuring the response and comparing it with the calibration curve. *See* ANTIBODY; ANTIGEN; IMMUNOASSAY. [A.B.]

Immunosuppression The natural or induced active suppression of the immune response, as contrasted with deficiency or absence of components of the immune system. Like many other complex biological processes, the immune response is controlled by a series of regulatory factors. A variety of suppressor cells play a role in essentially all of the known immunoregulatory mechanisms, such as: maintenance of immunological tolerance; limitation of antibody response to antigens of both thymic-dependent and thymic-independent types, as well as to antigens that stimulate reaginic antibody (antibodies involved in allergic reactions); generic control of the immune response; idiotype suppression; control of contact and delayed hypersensitivity; and antigenic competition. All of the major cell types [macrophages (monocytes), bone marrow-derived lymphocytes that normally mature into antibody-producing cells (B cells), and especially thymus-derived lymphocytes (T cells)] involved in the positive side of cellular interactions required for an immune response have also been found capable of functioning as suppressors in different regulatory systems. *See* IMMUNOLOGICAL DEFICIENCY.

Suppressor cells. Some suppressor functions are antigen- or carrier-specific. (A carrier is a molecule that can be chemically bound to another small molecule, called a hapten, in such a way that the combination induces an immune response that the hapten alone would not induce.) Others may not be carrier-specific, but may be specific for the type of response, such as immunoglobulin production but not delayed hypersensitivity. In the case of immunoglobulin production, the suppressor T cell may regulate the production of all immunoglobulin classes, a single class of immunoglobulins, or molecules that bind only a given antigen. Other suppressors may affect only cellular immunity and not humoral immunity. *See* CELLULAR IMMUNOLOGY; IMMUNOGLOBULIN.

Suppressor cells are critical in the regulation of the normal immune response. They play a role in regulating the magnitude and duration of the specific antibody response to an antigenic challenge. T cells are the major cells involved in immunosuppression, although activated phagocytic mononuclear cells are also significant as nonspecific suppressors in many systems. Helper T cells and suppressor T cells are different cell populations that are distinguished to a considerable extent by surface antigens that react with monoclonal antibodies or receptors for specific substances such as histamine. No single model explains the entire array of cellular suppressor phenomena. In different systems, T cells, macrophages, or even B cells may be the immediate targets of the suppressor cells and their secretions. Some suppression requires direct cell-cell interaction, whereas other suppression may be mediated by suppressor lymphokines. Both antigen-specific and antigen-nonspecific factors are known, and they may be secreted to act upon other cells, or especially in the case of antigen-specific factors, they may be integral parts of the cell membrane. The soluble immune-response suppressor factor, produced by activated T cells and then activated by monocytes, inhibits B-cell proliferation and immunoglobulin production in response to antigens. Macrophages also secrete suppressor factors, including prostaglandins that act on T cells and other soluble factors that are B-cell-specific. [K.S.W.; T.A.Wa.]

Immunosuppressors. Suppression of the immune response may be specific to a particular antigen or may be a response to a wide range of antigens encountered. The whole immune response may be depressed, or a particular population of immunologically active lymphocytes may be selectively affected. In some cases, the effect may be preferentially on T cells rather than B cells. If B cells are affected, it may be on a specific subclass of antibody-producing cells. Antigen-specific immunosuppression may be the result of deletion or suppression or a particular clone of antigen-specific cells, or the result of enhanced regulation of the immune response by antigen-specific suppressor cells. It can also be the result of increased production of anti-idiotypic antibody.

Nonspecific suppression of the immune response occurs in a number of rare immunological deficiency diseases of childhood. Acquired deficiency states affecting mainly T-cell function occur in states of malnutrition and in the presence of tumors. Acquired deficiencies may also occur secondary to a number of infectious diseases. The effect of total body irradiation on the lymphoreticular system also produces an acquired immunodeficiency. *See* ACQUIRED IMMUNE DEFICIENCY SYNDROME (AIDS).

There are a number of drugs that are capable of suppressing the immune response. The main stimulus for studies designed to identify these substances has been to devise means for controlling organ graft rejection. However, there has been considerable activity in looking for compounds that will suppress the immune response and reduce the inflammatory process in experimental models of rheumatoid arthritis. The ideal immunosuppressive drug should fulfill five main requirements: (1) There should be a wide margin of safety between a toxic and a therapeutic dose. (2) The drug should have a selective effect on lymphoid cells and not cause damage to the rest of the body. (3) If possible, this effect should be only on those cells which are involved in the specific immune process to be suppressed. (4) The drug should need to be administered for only a limited period until the immunological processes become familiar with the foreign antigen and begin to recognize it as part of "self." (5) The drug should be effective against immune processes once they have developed.

The result of any immune response is a balance between the action of effector cells mediating the phenomenon and suppressor cells regulating the response. Anything that reduces the regulatory function of suppressor cells will functionally increase the immune response. As suppressor cells are derived from rapidly turning-over precursor cells, and effector cells of T-cell-mediated immunity are derived from slowly dividing precursors, it is possible preferentially to depress the action of suppressor cells without affecting effector cells. This may be done by the use of alkylating agents given before immunization. It is likely that the chemotherapeutic effect of alkylating agents which are used extensively in the treatment of cancer in humans is partially due to these agents modifying the biological response to the tumor, producing an immunopotentiating action. *See* CHEMOTHERAPY; IMMUNITY; IMMUNOLOGY. [J.L.T.]

Immunotherapy The treatment of cancer by improving the ability of a tumor-bearing individual (the host) to reject the tumor immunologically. There are molecules on the surface of tumor cells, and perhaps in their interior, that are recognized as different from normal structures by the immune system and thus generate an immune response. The two components of the immune response are cell-mediated and antibody-mediated immunity, which must work in concert to overcome tumor cells. One type of thymus-derived lymphocyte (also called a cytotoxic T cell) can destroy tumor cells directly, while another recruits other white blood cells, the macrophages, that do the killing. Natural killer cells and perhaps other white blood cells may also participate. However, elements that normally regulate immunity, such as suppressor T cells, are stimulated excessively by the tumor, which leads to an immune response that is deficient and unable to reject the growing tumor. Thus the strategy of immunotherapy is to stimulate within or transfer to the tumor-bearing individual the appropriate antitumor elements while avoiding further stimulation of suppressor elements. *See* CELLULAR IMMUNOLOGY; IMMUNOLOGIC CYTOTOXICITY; IMMUNOSUPPRESSION.

There are four broad categories of immunotherapy: active, adoptive, restorative, and passive. Active immunotherapy attempts to stimulate the host's intrinsic immune response to the tumor, either nonspecifically or specifically. Nonspecific active immunotherapy utilizes materials that have no apparent antigenic relationship to the tumor, but have modulatory effects on the immune system, stimulating macrophages, lymphocytes, and natural killer cells. Specific active immunotherapy attempts to stimulate specific antitumor responses with tumor-associated antigens as the immunizing materials. Adoptive immunotherapy involves the transfer of immunologically competent white blood cells or their precursors into the host. Bone marrow transplantation, while performed principally for the replacement of hematopoietic stem cells, can also be viewed as adoptive immunotherapy. Restorative immunotherapy comprises the direct and indirect restoration of deficient immunological function through any means other than the direct transfer of cells. Passive immunotherapy means the transfer of antibodies to tumor-bearing recipients. This approach has been made feasible by the development of hybridoma technology, which now permits the production of large quantities of monoclonal antibodies specific for an antigenic determinant on tumor cells. *See* CANCER (MEDICINE); GENETIC ENGINEERING; IMMUNOLOGY; MONOCLONAL ANTIBODIES. [J.K.-M.; M.S.Mi.]

Impact A force, also known as impulsive force, which acts only during a short time interval but which is sufficiently large to cause an appreciable change in the momentum of the system on which it acts. The momentum change produced by the impulsive force is described by the momentum-impulse relation. For a discussion of this relation *see* IMPULSE (MECHANICS). *See also* COLLISION (PHYSICS). [P.W.S.]

Impedance matching The use of electric circuits and devices to establish the condition in which the impedance of a load is equal to the internal impedance of the source. This condition of impedance match provides for the maximum transfer of power from the source to the load. *See* ELECTRICAL IMPEDANCE.

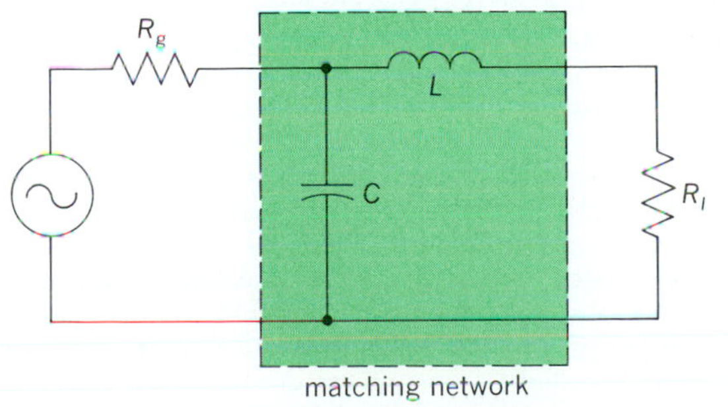

L-section impedance matching network with indirector *L* and capacitor *C* used in radio-frequency circuits. R_l = load resistance; R_g = generator resistance.

The maximum power transfer theorem of electric network theory states that at any given frequency the maximum power is transferred from the source to the load when the load impedance is equal to the conjugate of the generator impedance. When these conditions are satisfied, the power is delivered with 50% efficiency; that is, as much power is dissipated in the internal impedance of the generator as is delivered to the load. In general, the load impedance will not be the proper value for maximum power transfer. A network composed of inductors and capacitors may be inserted between the load and the generator to present to the generator an impedance that is the conjugate of the generator impedance (see illustration). [C.C.H.]

Impetigo A contagious skin disease characterized by superficial, thin-walled blisters which may develop into pustules. It is commonly seen in children but also found in adults. It is caused by either streptococcal or staphylococcal infections, or both. The disease is autoinoculable; that is, it can be quickly spread from an original site to any part of the body. No immunity is conferred by exposure. [E.G.St./N.K.M.]

Impsonite A black, naturally occurring carbonaceous material having specific gravity 1.10–1.25 and fixed carbon 50–85%. The origin of impsonite is not well understood, but it appears to be derived from a fluid bitumen that polymerized after it filled the vein in which it is found. *See* ASPHALT AND ASPHALTITE. [I.A.B.]

Impulse (mechanics) The integral of a force over an interval of time. For a force **F**, the impulse **J** over the interval from t_0 to t_1 can be written as Eq. (1). The impulse thus repre-

$$\mathbf{J} = \int_{t_0}^{t_1} \mathbf{F}\, dt \qquad (1)$$

sents the product of the time interval and the average force acting during the interval. Impulse is a vector quantity with the units of momentum.

The momentum-impulse relation states that the change in momentum of a mass m over a given time interval equals the impulse of the resultant force acting during that interval. The momentum change can be expressed in terms of the velocities $\mathbf{v}_1$ and $\mathbf{v}_0$ at times t_1 and t_0, respectively, giving Eq. (2).

$$\mathbf{J} = m(\mathbf{v}_1 - \mathbf{v}_0) \qquad (2)$$

See IMPACT; MOMENTUM. [P.W.S.]

Impulse generator An electrical apparatus which produces very short surges of high-voltage, or high-current, power. High impulse voltages are used to test the strength of insulators and of power equipment against lightning and switching surges. High-current impulses are produced by a discharge of capacitors connected in parallel. Such current surges may be used to magnetize permanent magnets or to produce the rising magnetic field in circular particle accelerators. [J.G.T.]

Impulse turbine A prime mover in which fluid (water, steam, or hot gas) under pressure enters a stationary nozzle where its pressure (potential) energy is converted to velocity (kinetic) energy. The accelerated fluid then impinges on the blades of a rotor, imparting its energy to the blades to produce rotation and overcome the connected rotor resistance. The impulse principle is basic to many turbines (see illustration).

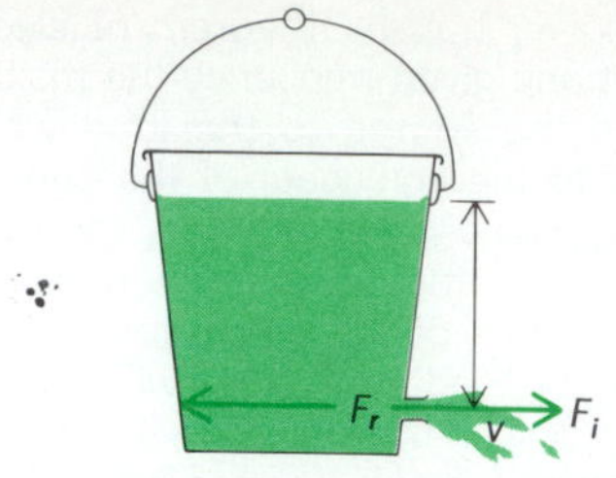

Water escaping through a nozzle near the base of a bucket, free to swing, illustrating the impulse (F_i) and reaction (F_r) forces of the jet issuing with a velocity V.

See Gas turbine; Hydraulic turbine; Pelton wheel; Reaction turbine; Steam turbine; Turbine propulsion. [T.Ba.]

Inadunata An extinct subclass of stalked Crinoidea comprising some 300 Paleozoic genera, ranging from the Upper Cambrian (Tremadocian) to the Permian. The arms (3–5 or more) were completely free and in no way incorporated into the calyx (see illustration). The calyx was either monocyclic or dicyclic, with conspicuous radials and basals, and all the thecal plates were sutured together into a rigid system. *See* Crinoidea; Echinodermata. [H.B.F.]

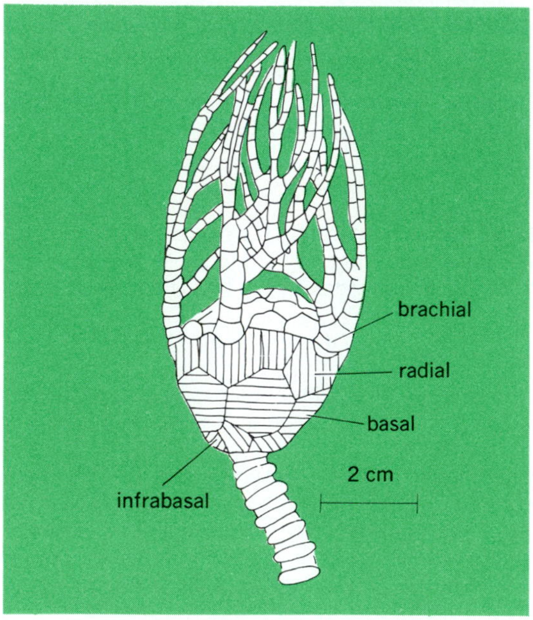

Cyathocrinus, a dicyclic inaduriate crinoid.

Inarticulata A class of phylum Brachiopoda. It is presently divided into four orders: Paterinida, Obolellida, Acrotretida, and Lingulida. *See* Acrotretida; Lingulida; Obolellida; Paterinida.

The two valves of the shell are typically not articulated and are held together only by soft tissue of the living animal. A few genera developed hinge mechanisms posteriorly. The posterior sector of one or both valves is commonly flattened. Internally, muscle scars and mantle canal impressions are usually the only features present; a median ridge or elevated septum may be developed, particularly in the brachial valve. In all inarticulates, the fibrous secondary layer characteristic of the Articulata is undeveloped. *See* Articulata (Brachiopoda); Brachiopoda. [A.J.R.]

Incandescence The emission of visible radiation by a hot body. A theoretically perfect radiator, called a blackbody, will emit radiant energy according to Planck's radiation law at any temperature. Prediction of the visual brightness requires additional consideration of the sensitivity of the eye, and the radiation will be visible only for temperatures of the blackbody which are above some minimum. The relation between brightness and temperature is plotted in the illustration. As shown, the minimum temperature for incandescence for the dark-adapted eye is about 390°C (730°F). Under these ideal observing conditions, the incandescence appears as a colorless glow. The dull red light commonly associated with incandescence of objects in a lighted room requires a temperature of about 500°C (930°F). *See* Blackbody; Heat radiation; Incandescent lamp; Vision. [H.W.Ru./G.R.H.]

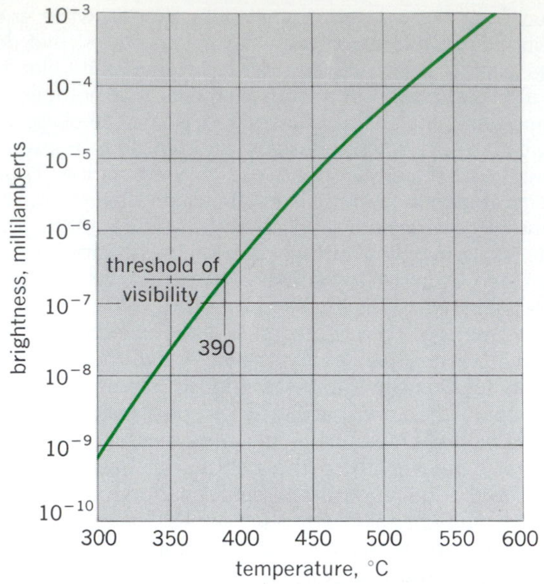

Relation between brightness of blackbody and termperture.

Incandescent lamp An electric lamp that produces light by heating a metallic filament to intense heat by passing an electric current through it. It is designed to produce light in the visible portion (at wavelengths of 380–760 micrometers) of the electromagnetic spectrum. The filament is prepared of special materials (commonly tungsten) and is enclosed in either an evacuated enclosure or one filled with an inert gas. In addition to radiation in the visible spectrum, infrared and ultraviolet energy are emitted, thus lowering the luminous efficiency of the lamp. When either of these radiations is accentuated, however, the lamp may be used as a source of that radiant energy. *See* Incandescence; Light.

The important parts of an incandescent lamp are the lamp enclosure or bulb, the filament, and the base. Standard lamps have various bulb shapes, bases, and filament constructions. The bulb may be clear, colored, inside-frosted, or coated with diffusing or reflecting material. Most lamps have soft-glass bulbs; hard glass is used when the lamp will be subjected to sudden and severe temperature changes.

Lamps are built for various voltage conditions, the most common being 115, 120, and 125 volts. High-voltage lamps are designed for ratings of 220–260 volts and low-voltage lamps for 6–64 volts. Lamps for use in 525–625 volt systems are designed to operate in groups of five in series across the line.

Incandescent lamps have been developed for many services. Most common are those used in general service and the miniature lamp. Special types have been developed for rough service applications, bake-oven use, severe vibration applications, showcase lamps, multiple lights (three-way lamp), sign lamps, spotlights, floodlights, and insect-control lamps. For other types of incandescent lamps *see* Arc lamp; Infrared lamp. [J.O.K.]

Incendiary One of a number of flammable materials and devices that are used to set fire to tactical and strategic targets, such as buildings, industrial installations, and fuel and ammunition dumps. Flame warfare extends also to antipersonnel use in the case of flamethrowers and fire bombs. [S.D.S.]

Inclined plane A plane surface inclined at an angle with the line of action of the force that is to be exerted or overcome. In the free-body diagram shown here, three forces act

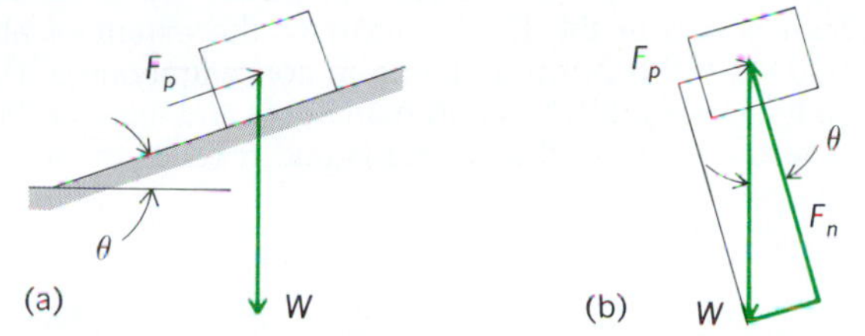

Weight resting on an inclined plane (*a*) with principal forces applied, and (*b*) their resolution into normal force.

on the object when no friction is present. The forces are its weight W, the force F_p parallel to the surface, and a force F_n normal to the surface. The summation of the forces acting in any direction on a body in static equilibrium equals zero; therefore, the summation of forces parallel to and forces normal to the surface are given by Eqs. (1) and (2). A force slightly great-

$$F_p - W \sin\theta = 0 \qquad (1)$$
$$F_n - W \cos\theta = 0 \qquad (2)$$

er than $W \sin\theta$ moves the object up the incline, but the inclined plane supports the greater part of the weight of the object. The principal use of the inclined plane is as ramps for moving goods from one level to another. *See* SIMPLE MACHINE. [R.M.Ph.]

Inclinometer An instrument for measuring magnetic dip, or inclination, which is defined as the angle between the magnetic field vector and the horizontal plane.

In the form of a simple dip needle, the instrument is a thin, pointed, mechanically balanced bar magnet mounted on the horizontal bearings. When oriented so that its plane of rotation is in the magnetic meridian, the needle aligns itself with the direction of the magnetic field and indicates on a circular scale the angle of dip. It is used principally in mine surveying where the minerals have a high content of magnetic materials. *See* MAGNETOMETER. [J.H.Ne.]

Inclusion blennorrhea Inclusion conjunctivitis of the newborn; an inflammation of the conjunctiva of the baby's eye caused by an agent of the PLT-Bedsonia group. The adult form of this disease is also called inclusion conjunctivitis.

So-called swimming-pool conjunctivitis is of the same etiology as conjunctivitis of the newborn. Infection of the newborn's eyes comes from the cervix of the mother, and since the reservoir is the genital tract, water of swimming pools is probably infected by discharges from the genital tracts of infected bathers. Chlorination of the water eliminates this source of adult conjunctivitis. [K.F.M.; J.S.]

Incompressible flow Fluid motion without change in density. For practical purposes liquids are assumed to flow incompressibly. At low velocity this is nearly the case; however, even for liquids, abrupt changes in velocity produce compression or rarefaction. In general, whenever fluid velocities are low (less than a fourth) relative to the rate of propagation of a pressure wave in the fluid, density variations during flow will be negligible and the flow will be effectively incompressible.

Important applications of incompressible flow theory include the aerodynamics of aircraft moving at low airspeeds, surface vehicles, and airflow in various industrial processes such as steelmaking. *See* AERODYNAMIC FORCE; BERNOULLI'S THEOREM; COMPRESSIBLE FLOW; FLUID FLOW; GAS DYNAMICS. [J.E.Ma.]

Index fossil The ancient remains and traces of a plant or animal that lived during a particular span of geologic time and that geologically dates the containing rocks. Index fossils are almost exclusively confined to sedimentary rocks which originated in such diverse environments as open oceans, tropical lagoons, coral reefs, beaches, lakes, and rivers. The fossil groups most useful as index fossils are generally marine and either floaters or open ocean swimmers, such as cephalopods, or bottom dwellers that had a floating or swimming stage in their life cycles, such as the medusa stage in the brachiopods. Such characteristics are necessary for rapid dispersal of newly evolved forms. On land, such mobile forms as the horses or wind-borne pollen and spores were relatively unrestricted by environmental barriers and became widely dispersed. All of these groups have provided biochronological zones of worldwide extent.

During the Cambrian Period (600,000,000 years before present) the oldest highly developed animals appeared; among them the trilobites provide the first important group of index fossils. Small plantlike floating colonial animals called graptolites have proved useful in correlating Ordovician (500,000,000 years B.P.) and Silurian (425,000,000 years B.P.) rocks. Ammonoids are a classic example of the internationally useful index fossil and are important beginning in the Devonian Period (405,000,000 years B.P.) and extending to the end of the Cretaceous Period (135,000,000 years B.P.). From the Pennsylvanian Period (310,000,000 years B.P.) fusulinids, a family of Foraminiferida, and pollen and spores from the coal forests are important indices. *See* BRACHIOPODA; CEPHALOPODA; CONODONT; FORAMINIFERIDA; FOSSIL; FUSULINACEA; GEOLOGICAL TIME SCALE; GRAPTOLITHINA; STRATIGRAPHIC NOMENCLATURE; STRATIGRAPHY. [C.C.]

Index mineral In metamorphic petrology, a characteristic mineral which by its presence in a rock indicates the mineral facies of the rock. An index mineral characterizes the mode of origin of a metamorphic rock in regard to temperature, pressure, and composition. A mineral such as quartz is not an index mineral because quartz is found in a great variety of rocks formed from low to high temperature and pressure. However, minerals with stability fields within restricted parts of the geologically important temperature-pressure range (by a given bulk composition of the containing rock) usually may be regarded as index minerals. In a crude way they are also geological thermometers and pressure gages. *See* GEOLOGIC THERMOMETRY; METAMORPHIC ROCKS. [T.F.W.B.]

Indian Ocean The smallest and geologically the most youthful of the three oceans. It differs from the Pacific and Atlantic oceans in two important aspects. First, it is landlocked in the north, does not extend into the cold climatic regions of the Northern Hemisphere, and consequently is asymmetrical with regard to its circulation. Second, the wind systems over its equatorial and northern portions change twice each year, causing an almost complete reversal of its circulation.

The eastern and western boundaries of the Indian Ocean are 147 and 20°E, respectively. In the southeastern Asian waters the boundary is usually placed across Torres Strait, and then from New Guinea along the Lesser Sunda Islands, across Sunda Strait and Singapore Strait.

The ocean floor is divided into a number of basins by a system of ridges. The largest is the Mid-Ocean Ridge, the greater part of which has a rather deep rift valley along its center. It lies like an inverted Y in the central portions of the ocean and ends in the Gulf of Aden. The Sunda Trench, stretching along

Java and Sumatra, is the only deep-sea trench in the Indian Ocean. East of the Mid-Ocean Ridge, deep-sea sediments are chiefly red clay; in the western half of the ocean, globigerina ooze prevails and, near the Antarctic continent, diatom ooze.

Atmospheric circulation over the northern and equatorial Indian Ocean is characterized by the changing monsoons. In the southern Indian Ocean atmospheric circulation undergoes only a slight meridional shift during the year. The surface circulation is caused largely by winds and changes in response to the wind systems. In addition, strong boundary currents are formed, especially along the western coastline, as an effect of the Earth's rotation and of the boundaries created by the landmasses.

North of 10°S the changing monsoons cause a complete reversal of surface circulation twice a year. In February, during the Northeast Monsoon, flow north of the Equator is mostly to the west and the North Equatorial Current is well developed. Its water turns south along the coast of Somaliland and returns to the east as the Equatorial Countercurrent between about 2 and 10°S. In August, during the Southwest Monsoon, the South Equatorial Current extends to the north of 10°S; most of its water turns north along the coast of Somaliland, forming the strong Somali Current. North of the Equator flow is from west to east and is called the Monsoon Current. Parts of this current turn south along the coast of Sumatra and return to the South Equatorial Current. During the two transition periods between the Northeast and the Southwest monsoons in April–May and in October, a strong jetlike surface current flows along the Equator from west to east in response to the westerly winds during these months. *See* OCEAN CIRCULATION.

Both semidiurnal and diurnal tides occur in the Indian Ocean. The semidiurnal tides rotate around three amphidromic points situated in the Arabian Sea, southeast of Madagascar, and west of Perth. The diurnal tide also has three amphidromic points: south of India, in the Mozambique Channel, and between Africa and Antarctica. It has more the character of a standing wave, oscillating between the central portions of the Indian Ocean, the Arabian Sea, and the waters between Australia and Antarctica. *See* TIDE. [K.W.]

Indicating gage A gage having contact points that move as they contact the part being inspected (see illustration). The movement is amplified on an indicator whose scale designates the limits of tolerance. Snap gages are commercially available with indicators instead of fixed contact points. Indicators are also used for inspecting out-of-roundness, centerline runout, or taper, by rotating the part on one axis or diameter and indicating the "run out" of another axis or diameter. [R.A.Bo.]

Electronic indicating gage being used to inspect a part. (*Federal Products Corp.*)

Indium A chemical element, In, atomic number 49, a member of group III and the fifth period of the periodic table. Indium has a relative atomic weight of 114.82.

Indium occurs in the Earth's crust to the extent of about 0.000001% and is normally found in concentrations of 0.1% or less. It is widely distributed in many ores and minerals but is largely recovered from the flue dusts and residues of zinc-processing operations.

1	2	3	4	5	6	7	8	9	10	11	12	13	14	15	16	17	18
1 **H**																	2 **He**
3 **Li**	4 **Be**											5 **B**	6 **C**	7 **N**	8 **O**	9 **F**	10 **Ne**
11 **Na**	12 **Mg**											13 **Al**	14 **Si**	15 **P**	16 **S**	17 **Cl**	18 **Ar**
19 **K**	20 **Ca**	21 **Sc**	22 **Ti**	23 **V**	24 **Cr**	25 **Mn**	26 **Fe**	27 **Co**	28 **Ni**	29 **Cu**	30 **Zn**	31 **Ga**	32 **Ge**	33 **As**	34 **Se**	35 **Br**	36 **Kr**
37 **Rb**	38 **Sr**	39 **Y**	40 **Zr**	41 **Nb**	42 **Mo**	43 **Tc**	44 **Ru**	45 **Rh**	46 **Pd**	47 **Ag**	48 **Cd**	49 **In**	50 **Sn**	51 **Sb**	52 **Te**	53 **I**	54 **Xe**
55 **Cs**	56 **Ba**	71 **Lu**	72 **Hf**	73 **Ta**	74 **W**	75 **Re**	76 **Os**	77 **Ir**	78 **Pt**	79 **Au**	80 **Hg**	81 **Tl**	82 **Pb**	83 **Bi**	84 **Po**	85 **At**	86 **Rn**
87 **Fr**	88 **Ra**	103 **Lr**	104 **Rf**	105 **Db**	106 **Sg**	107 **Bh**	108 **Hs**	109 **Mt**	110	*111*	*112*	*113*	*114*	*115*	*116*	*117*	*118*

lanthanide series	57 **La**	58 **Ce**	59 **Pr**	60 **Nd**	61 **Pm**	62 **Sm**	63 **Eu**	64 **Gd**	65 **Tb**	66 **Dy**	67 **Ho**	68 **Er**	69 **Tm**	70 **Yb**
actinide series	89 **Ac**	90 **Th**	91 **Pa**	92 **U**	93 **Np**	94 **Pu**	95 **Am**	96 **Cm**	97 **Bk**	98 **Cf**	99 **Es**	100 **Fm**	101 **Md**	102 **No**

Indium is used in soldering lead wires to germanium transistors and as a component of the intermetallic semiconductor used for germanium transistors. Indium arsenide, antimonide, and phosphide are semiconductors with unique properties. Other uses of indium are sleeve-type bearings to reduce corrosion and wear, glass-sealing alloys, and dental alloys. *See* GERMANIUM. [E.M.L.]

Indole The parent compound of a group of organic heterocyclic compounds containing the indole nucleus, which is a benzene ring fused to a pyrrole ring as in indole itself (I). The importance of the indole ring lies in its presence in a large number of naturally occurring compounds. *See* PYRROLE.

Indole can exist in two tautomeric forms, the more stable enamine form (I) and the 3-H-indole or imine form (II). Unsubstituted 3-H-indoles (sometimes called indolenines) and a structural isomer of indole, isoindole (III), are not stable, but have been shown to be reaction intermediates. They are isolable when properly substituted.

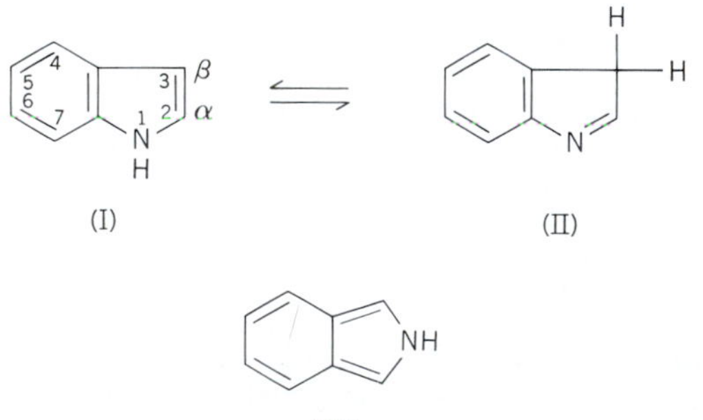

(I) (II) (III)

Indole (I) is a steam-volatile, colorless solid, melting point 52.5°C (126.5°F), boiling point 253°C (487°F). It is found in small amounts in coal tar, feces, and flower oils. Despite the presence of nitrogen, indole is not basic in the sense that it dissolves in aqueous acid or turns litmus blue. In fact, the hydrogen on the nitrogen is about as acidic as an aliphatic alcohol hydrogen. Indole is an aromatic compound and undergoes electrophilic substitutions much like benzene, although it is much more reactive than benzene. Its reactivity is comparable to that of phenol, and it undergoes a number of reactions similar to those of phenol. Indole itself reacts slowly with air and rapidly with most oxidizing agents to give intractable polymeric tars.

The indole alkaloids are a large group of substances containing the indole nucleus and can be isolated from plants. They contain a number of physiologically active materials, such as strychnine, reserpine, some forms of curare poison, and the rye-fungus drug, ergot. Lysergic acid diethylamide (LSD) is a synthetic, and not a naturally occurring, substance. There are many hundreds of other indole alkaloids. *See* ALKALOID; HETEROCYCLIC COMPOUNDS. [J.M.Bo.]

Inductance That property of an electric circuit or of two neighboring circuits whereby an electromotive force is induced (by the process of electromagnetic induction) in one of the circuits by a change of current in either of them. The term inductance coil is sometimes used as a synonym for inductor, a device possessing the property of inductance. *See* ELECTROMAGNETIC INDUCTION; ELECTROMOTIVE FORCE (EMF); INDUCTOR.

For a given coil, the ratio of the electromotive force of induction to the rate of change of current in the coil is called the self-inductance of the coil. An alternative definition of self-inductance is the number of flux linkages per unit current. Flux linkage is the product of the flux and the number of turns in the coil. Self-inductance does not affect a circuit in which the current is unchanging; however, it is of great importance when there is a changing current, since there is an induced emf during the time that the change takes place. For example, in an alternating-current circuit, the current is constantly changing and the inductance is an important factor.

The mutual inductance of two neighboring circuits is defined as the ratio of the emf induced in one circuit to the rate of change of current in the other circuit.

The International System (SI) unit of mutual inductance is the henry, the same as the unit of self-inductance. The same value is obtained for a pair of coils, regardless of which coil is the starting point.

The mutual inductance of two circuits may also be expressed as the ratio of the flux linkages produced in a circuit by the current in a second circuit to the current in the second circuit. *See* INDUCTANCE MEASUREMENT. [K.V.M.]

Inductance bridge A device for comparing inductances. The inductance bridge is a special case of an alternating-current impedance bridge. Just as the Wheatstone bridge is used to compare resistances, the impedance bridge is used to compare impedances which may contain inductance, capacitance, and resistance.

The inductance bridge shown in the illustration has resistors R_a and R_b as ratio arms and compares an unknown impedance Z_c to a standard consisting of R_d and L_d. If the standard L_d is variable, it and R_d are varied to reduce the detector (D) voltage to zero. This balances the bridge, and the equations of balance become

$$L_c/L_d = R_c/R_d = R_a/R_b$$

Theoretically the condition for balance is independent of frequency. In practice, the capacitance between turns and between layers of wire in the two coils will be different, and at high frequencies the bridge can be balanced at only one frequency at a time. *See* INDUCTANCE MEASUREMENT. [H.So./E.C.St.]

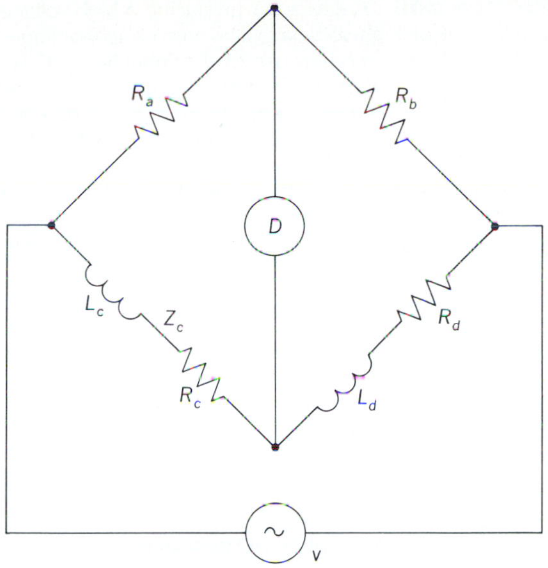

General inductance bridge.

Inductance measurement The determination of the electromagnetic parameter of inductance in an electric circuit. *See* INDUCTANCE.

Coils constructed so that their dimensions and consequently their inductance remain constant over long periods of time are used as inductance standards. If the dimensions are known, the inductance can be computed. Such an inductance is called a primary standard. Standards are usually maintained at constant temperature to keep the coil size and therefore the inductance from changing in value. When the inductance cannot be computed precisely, it may be measured by comparing it with a primary standard, and it would then be called a secondary standard. *See* INDUCTOMETER.

Impedance measurement is the determination of the total effect of a circuit element. The wire of which an inductor is wound has resistance, and there will be capacitance between turns. Thus every inductor exhibits resistance and capacitance as well as inductance. Also, every resistor exhibits some inductance and some capacitance, and every capacitor has some inductance and some resistance. Even if the capacitance of an inductor can be neglected, a measurement to determine the inductance is usually an impedance measurement giving the resistance as well as the inductance. *See* INDUCTANCE BRIDGE.

Measurement of distributed inductance of cables, lines, and waveguides is accomplished indirectly. For very low frequencies the cable currents are distributed uniformly throughout the conductors and the inductance per unit length is a maximum. At higher frequencies the skin effect causes the current to concentrate along the conductor surfaces. This concentration does not affect the magnetic field between the conductors, but it does tend to eliminate the field within each conductor itself. This reduction in magnetic field brings about a reduction in inductance. Consequently, as the frequency is increased the inductance decreases, approaching a fixed value.

The capacitance per unit length of a cable or open-wire transmission line is independent of frequency and can be measured at low frequencies. If direct measurements of attenuation and velocity of propagation are made, the inductance per unit length of transmission line may be computed at any frequency of operation. *See* ELECTRICAL MEASUREMENTS. [H.So./E.C.St.]

Induction coil A device for producing a high-voltage alternating current or high-voltage pulses from a low-voltage direct current. The largest modern use of the induction coil is in the ignition system of internal combustion engines, such as automobile engines. Devices of similar construction, known as vibrators, are used as rectifiers and synchronous inverters. *See* IGNITION SYSTEM; VIBRATOR.

The illustration shows a typical circuit diagram for an induction coil. The primary coil, wound on the iron core, consists of only a few turns. The secondary coil, wound over the primary, consists of a large number of turns.

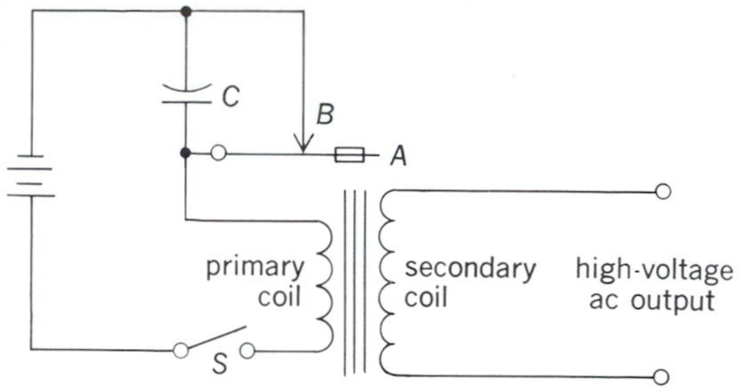

Typical circuit for an induction coil.

Induction coils of a different type are used in telephone circuits to step up the voltage from the transmitter and match the impedance of the line. The direct current in the circuit varies in magnitude at speech frequencies; therefore, no interrupter contacts are necessary. Still another type of induction coil, called a reactor, is really a one-winding transformer designed to produce a definite voltage drop for a given current. *See* REACTOR (ELECTRICITY). [N.R.B.]

Induction generator A nonsynchronous ac generator that is driven above synchronous speed by external sources of mechanical power. The construction of these machines is identical to that of induction motors. They can operate either as motors or generators, depending on whether the speed is below or above synchronous speed. In some frequency-changer sets, induction generators may operate as a motor part of the time. This is accomplished by coupling them to a two-speed synchronous machine capable of operating at system frequency as a motor at the higher speed and as a generator at the lower speed. Induction generators are not common in large sizes because of their poor power factor. They can deliver only leading currents. Moreover, they require a power supply from which to obtain their magnetizing current. This normally requires that they be operated in parallel with a synchronous source. Capacitors may be used to minimize the current taken from the source. Under such conditions the frequency of the induction generator is determined by the frequency of the synchronous source, and the output of the generator is determined by the mechanical input to its shaft. Induction generators do not require any dc excitation, and their control can be simple. Thus they are well suited to small, unattended hydroelectric units, where they are easily operated by remote control. *See* ALTERNATING-CURRENT GENERATOR; INDUCTION MOTOR. [L.T.R.]

Induction heating The heating of a nominally electrical conducting material by eddy currents induced by a varying electromagnetic field. The principle of the induction heating process is similar to that of a transformer. In the illustration, the inductor coil can be considered the primary winding of a transformer, with the workpiece as a single-turn secondary. When an alternating current flows in the primary coil, secondary currents will be induced in the workpiece. These induced currents are called eddy currents. The current flowing in the workpiece can be considered as the summation of all of the eddy currents.

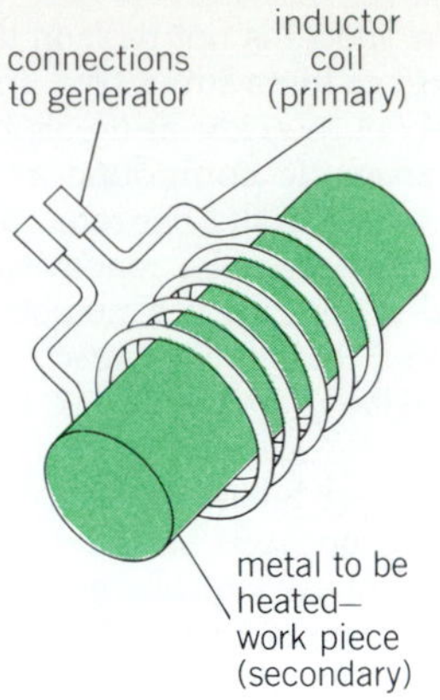

Basic elements of induction heating.

In the design of conventional electrical apparatus, the losses due to induced eddy currents are minimized because they reduce the overall efficiency. However, in induction heating, their maximum effect is desired. Therefore close spacing is used between the inductor coil and the workpiece, and high-coil currents are used to obtain the maximum induced eddy currents and therefore high heating rates. *See* CORE LOSS.

Induction heating is widely employed in the metalworking industry for a variety of industrial processes. While carbon steel is by far the most common material heated, induction heating is also used with many other conducting materials such as various grades of stainless steel, aluminum, brass, copper, nickel, and titanium products. *See* BRAZING; HEAT TREATMENT (METALLURGY); SOLDERING; WELDING AND CUTTING OF METALS.

The advantages of induction heating over the conventional processes (like fossil furnace or salt-bath heating) are the following: (1) Heating is induced directly into the material. It is therefore an extremely rapid method of heating. It is not limited by the relative slow rate of heat diffusion in conventional processes using surface-contact or radiant heating methods. (2) Because of skin effect, the heating is localized and the heated area is easily controlled by the shape and size of the inductor coil. (3) Induction heating is easily controllable, resulting in uniform high quality of the product. (4) It lends itself to automation, in-line processing, and automatic-process cycle control. (5) Startup time is short, and standby losses are low or nonexistent. (6) Working conditions are better because of the absence of noise, fumes, and radiated heat. *See* ELECTRIC HEATING. [G.F.B.]

Induction motor An alternating-current motor in which the currents in the secondary winding (usually the rotor) are created solely by induction. These currents result from voltages induced in the secondary by the magnetic field of the primary winding (usually the stator). An induction motor operates slightly below synchronous speed and is sometimes called an asynchronous (meaning not synchronous) motor. *See* ALTERNATING-CURRENT MOTOR.

Induction motors are the most common electric motors due to their simple construction, efficiency, good speed regulation, and low cost. Polyphase induction motors come in all sizes and find wide use where polyphase power is available. Single-phase induction motors are found mainly in fractional-horsepower (1 horsepower = 746 W) sizes, and those up to 25 hp are used where only single-phase power is available.

There are two principal types of polyphase induction motors: squirrel-cage and wound-rotor machines. The difference in these machines is in the construction of the rotor. The stator construction is the same and is also identical to the stator of a synchronous motor. Both squirrel-cage and wound-rotor machines can be designed for two- or three-phase current.

Single-phase induction motors display poorer operating characteristics than polyphase machines, but are used where polyphase voltages are not available. They are most common in small sizes ($\frac{1}{2}$ hp or less) in domestic and industrial applications. Their particular disadvantages are low power factor, low efficiency, and the need for special starting devices. [A.G.C.]

Inductive coordination The avoidance of inductive interference. Electric power systems, like almost everything run by electricity, depend on internal electric and magnetic fields; some of these fields find their way into the environment. The strongest of these fields can then induce voltages and currents in nearby devices and equipment and, in some cases, can interfere with the internal fields being used by electrical equipment in the vicinity. These induced voltages and currents, which are due to the coupling between the energized source and the electrical equipment, are called inductive interference. *See* ELECTRIC FIELD; ELECTROMAGNETIC INDUCTION; MAGNETIC FIELD.

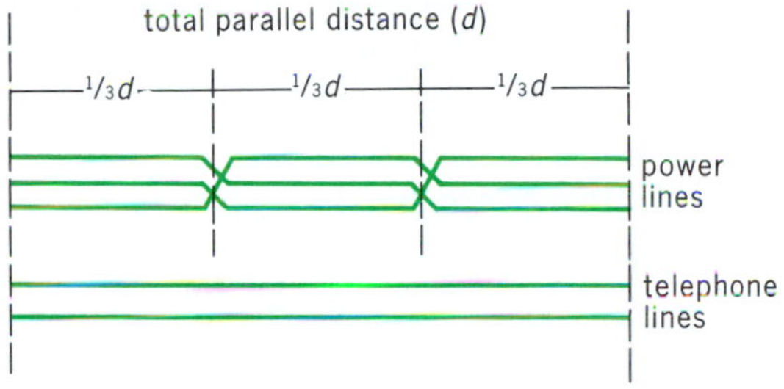

Transposition: the induced voltage and currents are canceled by transposing the power line as shown.

Overhead power lines cause practically all of the problems due to inductive coupling. For this reason and for safety considerations, power lines are restricted as far as possible to specific corridors or rights of way. Spacing between them and the requirements of their surroundings are considered and carefully calculated to minimize possible interference. These corridors are often shared by telephone lines, communication circuits, railroads, and sometimes trolley buses, each of which must be considered for possible inductive coupling.

Modern telephone and communication circuits are well shielded and rarely encounter interference from nearby power lines. However, where a long parallel exposure exists, inductive coupling can be reduced by balancing the operation of the power line and by transposition of power and communication lines (see illustration). Fences, long irrigation pipes, and large ungrounded objects within the right of way may experience considerable inductive coupling and must be grounded for safety. *See* ELECTRICAL INTERFERENCE; ELECTRICAL SHIELDING; GROUNDING; TRANSMISSION LINES. [A.A.M]

Inductive voltage divider An autotransformer that has its winding divided into 10 equal-turn sections so that when an alternating voltage V is applied to the whole winding, the voltage across each section is nominally $V/10$. The progressive voltages from one end to the section junctions are thus $V/10$, $2V/10$, $3V/10, \ldots$, and $9V/10$. *See* TRANSFORMER.

The division of voltage will be in error if there are differences of resistance and leakage inductance from section to section, and these errors will be significant if the differences are significant in relation to the input impedance of the winding. The most commonly used constructional technique for minimizing such errors is to make the winding by taking 10 equal lengths of insulated copper wire and twisting them into a "rope." The rope is wound onto a toroidal core made of thin, high-permeability, low-

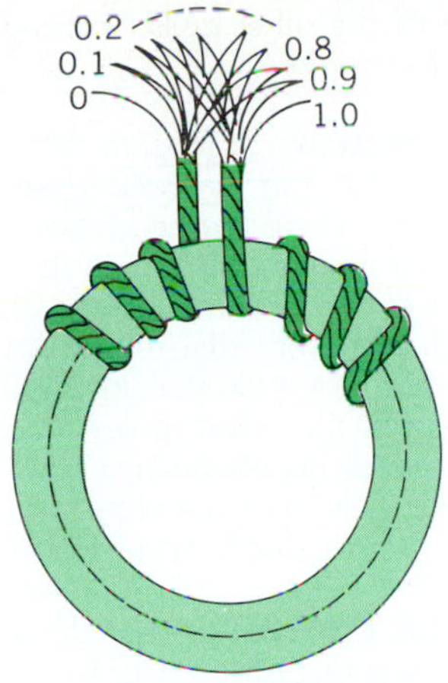

Rope winding on toroidal core making single-decade inductive voltage divider.

iron-loss magnetic material. The strands of the winding are then connected in series so that each strand forms one section of the 10-section divider (see illustration). The resistances of the sections are very closely equal since the strands were precut to be of the same length, and the leakage inductances are also closely equal and of very low value due to the close coupling given by this type of winding. The core form will have a low-reluctance magnetic path, giving a very high value of input impedance. By this means, division can be made accurately to a few hundredths of 1 part per million at low audio frequencies. Such units therefore find wide use as standards of ratio in the discipline of electrical measurements. *See* VOLTAGE MEASUREMENT. [T.A.D.]

Inductometer A coil of wire of known inductance. The inductance may be fixed, as in the case of primary standards; or the inductance may be adjustable by means of switches, or continuously variable by means of a movable-coil construction.

A standard self-inductance is constructed by winding a coil on a form having dimensions that are stable. Some materials dry out or absorb moisture over the years and consequently change continuously in size. The material on which a standard inductance is wound must not be subject to such change. Marble and some synthetics have been found suitable. A helical groove is cut in the material, and the coil is placed in this groove. *See* INDUCTANCE; INDUCTANCE BRIDGE; INDUCTANCE MEASUREMENT. [H.So./E.C.St.]

Inductor A device for introducing inductance into a circuit. The term covers devices with a wide range of uses, sizes, and types, including components for electric-wave filters, tuned circuits, electrical measuring circuits, and energy storage devices.

Inductors are classified as fixed, adjustable, and variable. All are made either with or without magnetic cores. Inductors without magnetic cores are called air-core coils, although the actual core material may be a ceramic, a plastic, or some other nonmagnetic material. Inductors with magnetic cores are called iron-core coils. A wide variety of magnetic materials are used, and some of these contain very little iron.

In fixed inductors coils are wound so that the turns remain fixed in position with respect to each other. Adjustable inductors have either taps for changing the number of turns desired, or consist of several fixed inductors which may be switched into various series or parallel combinations. Variable inductors are constructed so that the effective inductance can be changed. Means for doing this include (1) changing the permeability of a magnetic core; (2) moving the magnetic core, or part of it, with respect to the coil or the remainder of the core; and (3) moving one or more coils of the inductor with respect to one or more of the other coils, thereby changing mutual inductance. *See* INDUCTANCE. [B.L.R.; W.S.P.]

Inductor alternator A synchronous generator in which the field winding is fixed in magnetic position relative to the armature conductors. There are two types: the homopolar, in which the dc field coil is concentric with the shaft, and the heteropolar, in which the dc windings are distributed. In both types the ac windings are distributed and generate their induced voltage from the pulsation in the flux caused by the change in position of the salient poles on the rotor. Inductor alternators are used for high-frequency power and, in conjunction with static rectifiers, as a maintenance-free power source for ac excitation systems. *See* SYNCHRONOUS GENERATOR. [L.T.R.]

Industrial cost control A specific system or procedure for monitoring manufacturing costs to achieve profitability goals. The industrial cost control system should provide the framework for improving product profitability by identifying the variations between actual and target performance, discovering the causes of this variation, and eliminating the detrimental causes of cost variation.

In industrial cost control, performance of the control system is usually measured by the cost control performance of the factors of direct labor, material, and manufacturing burden or overhead. After creating the control accounts for these factors and after defining target costs for each account, those costs deviating from the desired standard can be quickly identified. Corrective action can then be taken to bring deviant costs back to the target level. Industrial cost control can be a simple and effective way to achieve goals of continued profitability. *See* INDUSTRIAL ENGINEERING; INSPECTION AND TESTING; INVENTORY CONTROL; MASS PRODUCTION; OPERATIONS RESEARCH; OPTIMIZATION; PERFORMANCE RATING; PERT; PROCESS ENGINEERING; PRODUCTION ENGINEERING; PRODUCTION METHODS; QUALITY CONTROL. [R.P.Lu.]

Industrial engineering As defined by the American Institute of Industrial Engineers, a branch of engineering "concerned with the design, improvement, and installation of integrated systems of people, material, equipment, and energy. It draws upon the specialized knowledge and skills in the mathematical, physical and social sciences together with the principles and methods of engineering analysis and design to specify, predict, and evaluate the results to be obtained from such systems."

The major activities of industrial engineers include work methods analysis and improvement, work measurement and the establishment of standards, job and workplace design, plant layout, materials handling, wage rates and incentives, cost reduction, suggestion evaluation, production planning and scheduling, inventory control, maintenance scheduling, equipment evaluation, assembly line balancing, systems and procedures, overall productivity improvement, and special studies—all done, almost exclusively in the early days, in manufacturing industries. As the technology evolved, additional activities were added to this list. These include machine tool analysis; numerically controlled machine installation and programming; computer analysis, installation, and programming; linear programming; queueing and other operations research techniques; simulations; management information systems; value analysis; human factors engineering; human/machine systems design; ergonomics; biomechanics; and the use of robots and automation. The industrial engineer of today is using computers more in production and test equipment, controls, and systems for analyses and special studies. Minicomputers and microprocessors are in wide and growing use. [A.G.Ho.]

Industrial meteorology The commercial application of weather information to the operational problems of business, industry, transportation, and agriculture in a manner intended to optimize the operation with respect to the weather factor. The weather information may consist of past weather records, contemporary weather data, predictions of anticipated weather conditions, or an understanding of physical processes which occur in the atmosphere. The operational problems are basically decisions in which weather exerts an influence.

Some of the applications of meteorological information are: professional meteorologists serving television or radio stations to explain to the public the changes in weather which either have occurred or are about to occur; forecasters working with airline dispatchers to utilize optimum flight paths and to avoid equipment tie-ups during lengthy adverse weather conditions; consultants who help utilities locate future power plant sites with minimal environmental impact; forecasters who advise shipping lines regarding ocean routing paths which avoid storms and decrease by a few hours the travel time for long-distance ocean freight movement; and consultant advisories which help industrial firms in the marketing of weather-sensitive products. In many cases the need for meteorological information is sufficiently important that a complete department within the operational segment of a company is established to serve the needs of that weather-affected firm. *See* WEATHER FORECASTING AND PREDICTION. [L.W.C.]

Industrial microbiology The study, utilization, and manipulation of those microorganisms (in the broad sense including bacteria, yeasts and other fungi, and some algae) capable of economically producing desirable substances, or changes in substances, and controlling unwanted microorganisms.

Since earliest times microorganisms have been used, intentionally or otherwise, as means of preserving foods or of making them more palatable, as in the making of such products as wine, cheese, butter, vinegar, and silage. They have also been used for such procedures as flax retting, removing mucilage from freshly harvested coffee beans, and copper leaching.

Microorganisms are used to produce chemical substances, which are then extracted from the culture mixture and purified. Included are (1) solvents such as ethanol, acetone, butanol, 2,3-butylene glycol, and other alcohols, such as glycerol, mannitol, erythritol, and polyhydric alcohols; (2) organic acids such as acetic, lactic, citric, gluconic, fumaric, propionic, itaconic, kojic, and gallic; (3) carbohydrates such as sorbose, fructose, dextrans, and other polysaccharides; (4) vitamins such as B_{12}, thiamine, and riboflavin; (5) amino acids such as lysine, arginine, and glutamic acid; (6) antibiotics such as penicillin, streptomycin, tetracycline, and chloramphenicol; (7) enzymes such as amylases, proteases, and glucose oxidases; and (8) nucleosides.

Microorganisms are used to produce desired changes in certain types of compounds or materials, such as in the microbiological transformations of steroids. More ancient are the uses of microorganisms in the textile industry with cotton, wool, and flax. More modern are the applications of bacteriological processes to the generation of electric power in fuel cells and to studies of human wastes as nutrients in closed space ecologies. In addition, microorganisms that are genetically altered are being used to produce insulin and interferon. [R.H.H.]

Industrial trucks Manually propelled (handtrucks) or powered carriers for transporting materials over level, slightly inclined, or slightly declined running surfaces. Some industrial trucks can lift and lower their loads, and others can also tier them. In any event, all such trucks maintain contact with the running surface over which they operate and, except when towed by a chain conveyor, follow variable paths of travel as distinct from conveying machines or monorails. *See* MATERIALS-HANDLING EQUIPMENT.

In industry the principle of handling materials in unit loads has developed in parallel with the increased use of powered industrial trucks, particularly the forklift type. These mobile mechanical handling aids have removed the limitations that existed when the weight and size of a load for movement and stacking depended mainly on the ability of a person to lift it manually. The unit-load principle of materials handling underlies the skid-platform and the pallet-forklift methods of operation. Both methods, especially the latter, have revolutionized handling techniques and equipment and even production equipment. [A.M.P.]

Industrial wastewater treatment A group of unit processes designed to separate, modify, remove, and destroy undesirable substances carried by wastewater from industrial sources. United States governmental regulations have been issued that involve volatile organic substances, designated priority pollutants, aquatic toxicity as defined by a bioassay, and in some cases nitrogen and phosphorus. As a result, sophisticated technology and process controls have been developed for industrial wastewater treatment. However, it is also necessary to implement better waste management and waste minimization practices that will result in reduction of waste loads and improved and more cost-effective operation of wastewater treatment facilities. *See* WATER POLLUTION.

Aerobic biological treatment is employed for the removal of biodegradable organics. Depending on the effluent requirements, the nature of the wastewater, and the availability of land, several process options are available. In the activated sludge process, the biomass generated is settled in a clarifier and recycled to the aeration basin in order to build a high concentration of biomass.

An aerated lagoon system is applicable (where large land areas are available) for treating nontoxic wastewaters, such as those generated by pulp and paper mills. Three basins in series are employed. The first basin is completely mixed in order to maximize biodegradation. In the second basin, at reduced power level, solids are permitted to settle and then undergo anaerobic degradation and stabilization. A third settling basin is usually employed to reduce effluent suspended solids.

Fixed-film processes include the trickling filter and the rotating biological contactor. In these processes, a biofilm is generated on a surface, usually plastic. As the wastewater passes over the film, organics diffuse into the film, where they are biodegraded. The trickling filter is primarily used as a pretreatment process for high concentrations of biodegradable organics in readily degradable wastewaters.

Anaerobic processes are sometimes employed before aerobic processes for the treatment of high-strength, readily degradable wastewaters. The primary advantages of the anaerobic process is low sludge production and the generation of energy in the form of methane (CH_4) gas. In the anaerobic process, organic materials are broken down to volatile fatty acids, primarily acetic acid, which in turn is broken down to methane and carbon dioxide (CO_2). High-rate processes include the anaerobic filter, in which the anaerobic organisms are grown on a support medium; the anaerobic contact process, which is similar to the activated sludge process and consists of a reactor, a clarifier, and a sludge recycle; and the upflow anaerobic sludge bed, which generates a granular sludge through which the wastewater is passed. Where land area is available, low-rate processes have the advantage of greater process stability in case of upset or shock loads (high concentrations of organics caused by spills or accidental dumps of wastewater). *See* BIODEGRADATION; SEWAGE DISPOSAL; SEWAGE TREATMENT.

Biological processes can remove only degradable organics. Nondegradable organics can be present in the influent wastewater or be generated as oxidation by-products in the biological process. Many of these organics are toxic to aquatic life and must be removed from the effluent before discharge. The most common technology to achieve this objective is adsorption on activated carbon. Granular carbon filters can be employed as a posttreatment in which the refractory organics are adsorbed on the carbon. *See* ACTIVATED CARBON; ADSORPTION.

Virtually all of the processes for industrial wastewater treatment generate a sludge that requires some means of disposal. In general, the processes for thickening and dewatering are the same as those in municipal wastewater treatment. Waste activated sludge is usually stabilized by aerobic digestion in which the degradable solids are oxidized by prolonged aeration. Ultimate disposal can be on land or by lagooning or incineration of dewatered material. *See* SEWAGE SOLIDS. [W.W Ec.]

Inert gases The inert gases, listed in the table, constitute group 18 of the periodic table of the elements. They are now better known as the noble gases, since stable compounds of xenon have been prepared. The noble gases are all monatomic.

All these gases occur to some extent in the Earth's atmosphere, but the concentrations of all but argon are exceedingly low. Argon is plentiful, constituting almost 1% of the air.

All the gases are colorless, odorless, and tasteless. They are all slightly soluble in water, the solubility increasing with increasing molecular weight. They can be liquefied at low temperatures, the boiling point being proportional to the atomic weight. All but helium can be solidified by reducing the temperature sufficiently, and helium can be solidified at temperatures of less than 2°F above absolute zero (0–1 K) by the application of an external pressure of 25 atm (2.5 megapascals) or more. *See* ARGON; HELIUM; KRYPTON; NEON; RADON; XENON. [A.W.F.]

The inert gases

Name	Symbol	Atomic number	Atomic weight
Helium	He	2	4.0026
Neon	Ne	10	20.183
Argon	Ar	18	39.948
Krypton	Kr	36	83.80
Xenon	Xe	54	131.30
Radon	Rn	86	(222)

Inertia That property of matter which manifests itself as a resistance to any change in the motion of a body. Thus when no external force is acting, a body at rest remains at rest and a body in motion continues moving in a straight line with a uniform speed (Newton's first law of motion). The mass of a body is a measure of its inertia. *See* MASS. [L.N.]

Inertia welding A welding process used to join similar and dissimilar materials at very rapid speed. It is, therefore, a very attractive welding process in mass production of good-quality welds. The ability to join dissimilar materials provides further flexibility in the design of mechanical components. The automotive and truck industry is the major user of this process.

Inertia welding is a type of friction welding which utilizes the frictional heat generated at the rubbing surfaces to raise the temperature to a degree that the two parts can be forged together to form a solid bond. The energy required for inertia welding comes from a rotating flywheel system built into the machine. Like an engine lathe, the inertia welding machine has a headstock and a tailstock. One workpiece held in the spindle chuck (usually with an attached flywheel) is accelerated rapidly, while the other is clamped in a stationary holding device of the tailstock. When a predetermined spindle speed is reached, the drive power is cut and the nonrotating part on the tailstock is pushed against the rotating part under high pressure. Friction between the rubbing surfaces quickly brings the spindle to a stop. At the same time the stored kinetic energy in the flywheel is converted into frictional heat which raises the temperature at the interface high enough to forge the two parts together without melting. *See* WELDING AND CUTTING OF METALS. [K.K.W.]

Inertial guidance system A self-contained system which can automatically determine the position, velocity, and attitude of a moving vehicle for the purpose of directing its future course. Based on prior knowledge of time, gravitational field, initial position, initial velocity, and initial orientation relative to a known reference frame, an inertial guidance system is capable of determining its present position, velocity, and orientation without the aid of external information. The generated navigational data are used to determine the future course for a vehicle to follow in order to bring it to its destination. Such systems have found application in the guidance and control of submarines, ships, aircraft, missiles, and spacecraft. *See* GUIDANCE SYSTEMS. [W.E.H.]

Inertial reference frame A frame of reference in which a body moves with constant velocity when free of impressed forces. It follows that one inertial frame cannot be accelerating with respect to another. If x_1, y_1, z_1 and x_2, y_2, z_2 are the coordinates of a particle P in parallel inertial frames 1 and 2, where 2 is moving with respect to 1 with velocity components v_x, v_y, v_z, then Eq. (1) holds. (Newtonian frames are understood here

$$\begin{aligned} x_2 &= x_1 - v_x t \\ y_2 &= y_1 - v_y t \\ z_2 &= z_1 - v_z t \end{aligned} \tag{1}$$

so that $t_2 = t_1$.) This is called a galilean transformation. Since the vs are independent of time, the component accelerations of P in 1 and 2 are the same, that is, Eq. (2) is valid.

$$A_{x2} = d^2x_2/dt^2 = d^2x_1/dt^2 = A_{x1} \text{ etc.} \tag{2}$$

See FRAME OF REFERENCE. [B.G.]

Inertial separator A device for utilizing the principle of momentum to separate larger particulate, primarily that above 10 micrometers in diameter, from a moving airstream. The larger particles tend to keep going in the same direction in which they entered the device, while the light-density air changes direction. Commonly offered in manifolded multiples of V-shaped pockets with louvered sides or in small-diameter cyclones, the inertial separators are used most frequently in hot, arid, desertlike regions where winds generate significant airborne dust and sand particles. A secondary bleed-air fan is used to discharge the separated particulate to the outside of the building. *See* AIR FILTER. [M.A.B.]

Infant diarrhea Diarrhea and its complications are the most important causes of infant death in most developing regions. The causes of the illness vary from dietary incompatabilities to intestinal infection. The most important infectious causes are, in approximate order of importance: rotavirus, the bacteria *Shigella* (causing dysentery) and *Salmonella*, the parasite *Giardia lamblia*, and enteropathogenic *Escherichia coli* bacteria (a common cause of hospital nursery outbreaks). Breast-feeding is associated with a decreased occurrence of diarrhea and represents a major means of preventing infantile diarrhea in the developing world. *See* DIARRHEA. [H.L.D.]

Infant respiratory distress syndrome A disorder usually affecting prematurely born infants and characterized by a rapid breathing rate, respiratory muscle retraction during inspiration and grunting sounds during expiration, and blood gas values reflecting oxygen deficiency, excessive carbon dioxide, and acidosis.

In order for newborn infants to adapt successfully to air breathing, the ventilatory surfaces of the lung must possess low surface tension, since high surface tension may result in alveolar collapse. The surface tension in the lung is kept low by a film of surfactant that is synthesized and secreted by the epithelial cells lining the alveoli. The lungs of premature infants, however, produce an insufficient amount of surfactant. Therefore the lungs tend to collapse after each breath, and the pulmonary blood vessels constrict causing blood from the right heart to bypass the lungs. Re-expanding collapsed lungs requires more work than normal inspiration; hence respiratory fatigue develops. The low oxygen levels caused by collapsed lungs and constriction of pulmonary blood vessels further suppress surfactant synthesis, and so a vicious cycle is created and progressive deterioration ensues. *See* LUNG.

In mild cases, there may be clinical improvement after a few days of illness, following which death rarely occurs. Untreated infants with more severe cases, however, experience progressive respiratory deterioration typically leading to death during the first 3 days of life.

Since infant respiratory distress syndrome is caused most commonly by premature birth, measures to prevent early birth are the most effective means of reducing risk. For the infant with respiratory distress syndrome, modern therapy has favorably altered the natural course. Treatment is directed toward the support of ventilation, circulation, acid-base balance, temperature control, and nutrition. [H.F.Kr.]

Infarction A process of vascular obstruction resulting in a region of dead tissue (necrosis) called an infarct. The usual cause is occlusion of an artery or vein by a thrombus or embolus and sometimes by severe arteriosclerosis; the development of the infarct depends to a great extent on the collateral circulation.

Infarcts commonly occur in the lungs, heart, brain, spleen, and kidneys. The common cause of infarcts of the heart is thrombosis of the coronary artery, usually secondary to arteriosclerosis. In addition, hemorrhage into an arteriosclerotic plaque can result in occlusion of the lumen. *See* EMBOLUS; HEART DISORDERS; THROMBOSIS. [R.A.V.]

Infection A term considered by some to mean the entrance, growth, and multiplication of a microorganism (pathogen) in the body of a host, resulting in the establishment of a disease process. Others define infection as the presence of microorganism in host tissues whether or not it evolves into detectable pathologic effects. The host may be a bacterium, plant, animal, or human being, and the infecting agent may be viral, rickettsial, bacterial, fungal, or protozoan. *See* EPIDEMIOLOGY; LYTIC INFECTION; MEDICAL BACTERIOLOGY; MEDICAL MYCOLOGY; MEDICAL PARASITOLOGY; PATHOGEN; VIRUS. [D.N.La.]

Infectious-disease control Since infection depends on factors in the host, the parasite, and the environment, there are various means by which infectious-disease control may be brought about. Immunization strengthens the defenses of the host; chemoprophylaxis is directed at the parasite; and quarantine and sanitary measures remove the parasite from the environment of susceptible people. *See* CHEMOTHERAPY; EPIDEMIOLOGY; SEWAGE DISPOSAL; VACCINATION. [C.W.]

Infectious mononucleosis A disease of children and young adults, characterized by fever and enlarged lymph nodes and spleen. EB (Epstein-Barr) herpesvirus is the causative agent.

Onset of the disease is slow and nonspecific with variable fever and malaise; later, cervical lymph nodes enlarge, and in about 50% of cases the spleen also becomes enlarged. The disease lasts 4–20 days or longer. Epidemics are common in institutions where young people live. EB virus infections occurring in early childhood are usually asymptomatic. In later childhood and adolescence, the disease more often accompanies infection—although even at these ages inapparent infections are common. *See* EPSTEIN-BARR VIRUS. [J.L.Me.]

Infectious myxomatosis A rapidly fatal viral disease of European rabbits, also known as rabbit pox, characterized by the formation of mucoid tumors in the skin and invasion of the blood and internal organs. It is spread by the bite of the mosquito. Rabbits in the area where the disease originated are immune. It has been successfully used as a method of extermination of unwanted rabbits in Australia where laboratory-infected rabbits have been released into the rabbit population. [A.E.Mo.]

Infectious papillomatosis A nonfatal disease of wild cottontail rabbits characterized by large, localized, fleshy tumors with horny surfaces, which occur on the neck and shoulders. It is caused by a virus, and natural infection occurs by inoculation of the skin surface as the rabbit enters its burrow. The disease is also known as Shope papilloma. [A.E.Mo.]

Infective dose The dose of microorganisms required to cause infection in 50% of the experimental animals; it is an important special case of the median effective dose and is also known as ID, or the median infective dose. *See* EFFECTIVE DOSE 50. [C.W.]

Infertility Inability to conceive or induce conception. An estimated 15% of couples in the United States are unable to conceive within one year of their first attempt at conception. Of all cases of infertility, 35% may be attributed to the male and 55% to the female; the remaining 10% is undetermined.

The principal cause of increasing rates of infertility is the postponement of pregnancy: adverse effects of increasing age on reproductive capacity include decreased conception rates and increased pregnancy losses. The increasing incidence of pelvic inflammatory disease is also thought to be a major cause. Pelvic infections lead to scar formation around the ovaries and the fallopian tubes, thereby impeding the transport of oocytes for fertilization. Ovulatory dysfunction accounts for approximately 20% of cases of infertility. Fallopian tube and uterine abnormalities account for 25% of the cases of infertile couples but such abnormalities can usually be corrected surgically.

Male infertility resulting from abnormal semen may be due to developmental defects, genitourinary infections, or varicocele. An evaluation of sperm count, motility, and morphology is helpful, but in most cases the cause of the abnormality remains undetermined. No treatment other than donor insemination is available for these couples. *See* PREGNANCY; REPRODUCTIVE SYSTEM DISORDERS. [R.E.Le.; R.D.K.]

Infinity The terms infinity and infinite have a variety of related meanings in mathematics. The adjective finite means "having an end," so infinity may be used to refer to something having no end. In order to give a precise definition, the mathematical domain of discourse must be specified.

Set theory provides a simple and basic example of an infinite collection—the class of natural numbers, or positive integers. A fundamental property of positive integers is that after each integer there follows a next one, so that there is no last integer. Now it is necessary in mathematics to treat the collection of all positive integers as an entity, and this entity is the simplest infinity, or infinite collection.

The term infinity appears in mathematics in a different sense in connection with limits of functions. For example, consider the function defined by $y = 1/x$. When x tends to 0, y approaches infinity, and the expression may be written as shown below.

$$\lim_{x \to 0} y = \infty$$

Precisely, this means that for an arbitrary number $a > 0$, there exists a number $b > 0$ such that when $0 < x < b$, then $y > a$, and when $-b < x < 0$, then $y < -a$. This example indicates that it is sometimes useful to distinguish $+\infty$ and $-\infty$. The points $+\infty$ and $-\infty$ are pictured at the two ends of the y axis, a line which has no ends in the proper sense of euclidean geometry.

In geometry of two or more dimensions, it is sometimes said that two parallel lines meet at infinity. This leads to the conception of just one point at infinity on each set of parallel lines and of a line at infinity on each set of parallel planes. With such agreements, parts of euclidean geometry can be discussed in the terms of projective geometry. For example, one may speak of the asymptotes of a hyperbola as being tangent to the hyperbola at infinity. [L.M.G.]

Inflammation The local response to injury, involving small blood vessels, the cells circulating within these vessels, and nearby connective tissue.

The early phases of the inflammatory response are stereotyped: A similar sequence of events occurs in a variety of tissue sites in response to a diversity of injuries. The response characteristically begins with hyperemia, edema, and margination of the circulating white blood cells. The white cells then migrate between the endothelial cells of the blood vessel into the tissue. The subsequent development of the inflammatory process is determined by factors such as type and location of injury, condition of the host, and the use of therapeutic agents. *See* EDEMA.

A local inflammatory response is usually accompanied by systemic changes: fever, malaise, and an increase in circulating leukocytes (leukocytosis). Such signals and symptoms are often helpful to the physician, first as clues to the presence of inflammation and later as an indication of its course.

Inflammation is basically a protective mechanism. The leakage of water and protein into the injured area brings humoral factors, including antibodies, into the locale and may serve to dilute soluble toxic substances and wash them away. The margination and migration of leukocytes brings these cells to the local site to deal with the injurious agent, dead or alive. [D.L.]

Inflammatory bowel disease Inflammatory bowel disease is a general term for two closely related conditions, ulcerative colitis and Crohn's disease. The diseases can affect the colon, distal, small intestine and sometimes other portions of the gastrointestinal tract as well as several sites outside the gastrointestinal tract. In 15–25% of cases limited to the colon, ulcerative colitis and Crohn's disease cannot be distinguished by clinical manifestations, x-ray examination, or even pathology. For this reason the broad term inflammatory bowel diseases is useful. The cause of these diseases is unknown.

Ulcerative colitis, an inflammatory condition limited to the colon, primarily affects the mucosa or lining of the colon. Marked inflammation gives rise to small ulcerations and microscopic abscesses that produce bleeding. The condition tends to be chronic, alternating between periods of complete remission and episodes of active and even life-threatening disease.

Crohn's disease, also known as regional enteritis, granulomatous colitis, and terminal ileitis, affects the colon and small intestine, and rarely the stomach or esophagus. Like ulcerative colitis, it is chronic and of unknown etiology. *See* DIGESTIVE SYSTEM. [L.A.Ka.]

Inflationary universe cosmology A theory of the evolution of the early universe, motivated by considerations from elementary particle physics as well as certain paradoxes of standard big bang cosmology, which asserts that at some early time the observable universe underwent a period of exponential expansion. During this inflationary epoch the scale of the universe increased by at least 28 orders of magnitude.

"Old" inflationary model. The suggestion of an inflationary period during the early universe, in connection with a specific model, was first made in 1980 by A. Guth. Guth reasoned that if grand unified symmetries are broken at some large energy scale, then a phase transition could occur in the early universe as the temperature cooled below the critical temperature where symmetry breaking occurs. According to the standard big bang model of expansion, the time at which this would occur would be about 10^{-30} after the initial expansion had begun. *See* GRAND UNIFICATION THEORIES; PHASE TRANSITIONS; SYMMETRY BREAKING; SYMMETRY LAWS (PHYSICS).

As Guth demonstrated, the effects of such a phase transition in the early universe could be profound. In order to calculate the dynamics of a phase transition it is necessary to follow the behavior of the relevant order parameter for the transition. This is done by determining the free energy of a system as a function of the order parameter, and following the changes in this energy as the temperature changes. The illustration shows a typical example of what might be expected for the case of symmetry breaking in grand unification theories. At some high temperature T the minimum of the relevant function is at zero value of the order parameter, the vacuum expectation value of a certain field. Thus the ground state of the system, which occurs when this energy is a minimum, will be the symmetric ground state. As the temperature is decreased, however, at a certain critical temperature T_c, a new minimum of the energy appears at a nonzero value of the order parameter. This is the symmetry breaking ground state of the system. *See* FREE ENERGY.

In the illustration it is seen that there is a barrier between the two minima. This means that classically the system cannot make a transition between the two states. However, it is a well-known property of quantum mechanics that the system can, with a certain very small probability, tunnel through the barrier and arrive in the new phase. Such a transition is called a first-order phase transition. Because the probability of such a tunneling process is small the system can remain for a long time in the symmetric phase before the transition occurs. This phenomenon is called supercooling. When the transition finally begins, "bubbles" of new phase locally appear throughout the original phase. As more and more bubbles form, and the original bubbles grow, eventually they combine and coalesce until all of the system is in the new phase. *See* POTENTIAL BARRIER; QUANTUM MECHANICS.

The metastable symmetric phase has a higher energy than the new lower energy symmetry breaking phase. Until the transition occurs, this means that the symmetric phase has associated with it a large constant energy density, independent of temperature. When this constant energy density is placed on the right hand side of Einstein's equations, where the energy density of matter appears, it is found that the resultant Hubble parameter describing expansion is a constant. Mathematically this implies that the scale size of the universe increases exponentially during this supercooling phase. This rapid expansion is what is referred to as inflation. *See* HUBBLE CONSTANT; RELATIVITY.

Successes of inflation. Guth pointed out that a period of exponential expansion at some very early time could solve a number of outstanding paradoxes associated with standard big bang cosmology.

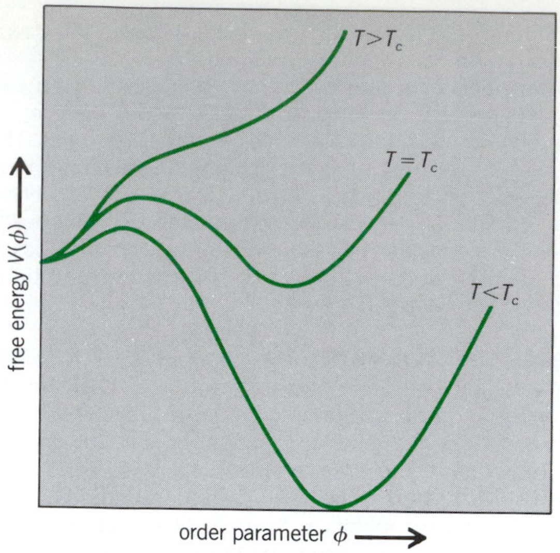

Free energy as a function of the order parameter for the case of a first-order transition.

Present observations seem to imply that either the present era is a unique time in the big-bang expansion, or else the initial conditions of expansion had to be fine-tuned to an incredible degree. Measurements of the observed expansion rate, combined with measurements of the observed mass density of the universe yield a value for the density parameter Ω which rapidly approaches zero for an open universe, infinity for a closed universe, and is exactly equal to one for a flat universe. All measurements of Ω yield values between about 0.1 and 2. However, theory suggests that once the value of Ω deviates even slightly from one, it very quickly approaches its asymptotic value far away from one for open or closed universes. Thus, it is difficult to understand why, after 10^{10} years of expansion, the value of Ω is now so close to one.

Inflation naturally explains why Ω should exactly equal one in the observable universe today. During inflation $\Omega(t)$ is driven arbitrarily close to one within the inflated region. If there are some 28 orders of magnitude of exponential expansion, then $\Omega(t)$ need not have been finely tuned to be close to one.

An equally puzzling problem which inflationary cosmology circumvents has to do with the observed large-scale uniformity of the universe. In particular, the 3-K microwave radiation background is known to be uniform in temperature to about one part in 10,000. However, in the standard big bang model the sources of this radiation observed coming from opposite directions in the sky were separated by more than 90 times the horizon distance (the distance a light ray could have travelled since the big bang explosion) at the time of emission. Since these regions could not possibly have been in physical contact, it is difficult to see why the temperature at the time of emission was so uniform in all directions. *See* COSMIC BACKGROUND RADIATION.

Inflation solves the horizon problem very simply. If the observed universe expanded by 28 orders of magnitude, then it originated from a region 10^{28} times smaller than the comparable region in the standard big-bang model. This makes it quite possible that at early times the entire observed universe was contained in a single horizon volume.

Unfortunately, the original inflationary scenario was fundamentally flawed. The phase transition began by the formation of bubbles of one phase nucleating amidst the initial metastable phase of matter. While these bubbles grow at the speed of light once they form, the space in between the bubbles is expanding exponentially. Thus, it is extremely difficult for bubbles to eventually occupy all of space, as is required for the completion of the transition.

"New" inflationary cosmology. In 1981, it was suggested that if the energy function (potential) of the illustration were slightly changed then it might be possible to maintain the successful phenomenology of the old inflationary model while avoiding its problems. In particular a special form of the potential was considered which is extremely flat at the origin. Such functions have essentially no barrier separating the metastable from the stable phase at low temperature. When the universe cooled down below the critical temperature, the order parameter could continuously increase from zero instead of tunnelling discretely to a large nonzero value. As long as the potential is sufficiently flat near the origin, however, it can take a long time before the order parameter approaches its value at the true minimum of the potential (a so-called slow-rollover transition). During this time the region of interest can again be expanding exponentially. Thus, in some sense a single bubble can undergo inflation in this scenario.

New inflation, too, is not without its problems. In order to have such a slow-rollover transition, the parameters of particle physics models must be finely tuned to some degree. No clear candidate model for new inflation has emerged from particle physics.

Inflationary models are attractive for many reasons. In particular they make definite predictions about the universe today. Specifically, if inflation is correct, the observable universe should have an exactly critical density now, and the anisotropy of the microwave background should have a specific magnitude, directly related to the observed clustering of galaxies. *See* BIG BANG THEORY; COSMOLOGY; UNIVERSE. [L.M.Kr.]

Inflorescence A flower cluster segregated from any other flowers on the same plant, together with the stems and bracts (reduced leaves) associated with it. Certain plants produce inflorescences, whereas others produce only solitary flowers. *See* FLOWER. [G.J.W.]

Influenza An acute respiratory viral infection, usually epidemic in occurrence. This disease has been one of the world's greatest killers; over 21 million deaths are attributed to the 1918–1919 pandemic. Three immunologic types of influenza virus are known. Group A appears to be subject to continual antigenic changes, group B changes to a lesser degree, and group C seems antigenically stable.

In human infection, the virus enters the respiratory tract in airborne droplets; within 1–2 days inflammation of the respiratory tract occurs, with fever, chills, and muscular aches. The illness is usually mild, and asymptomatic infections may occur. However, a more serious course or even death may result in older persons and those with chronic debilitating diseases. Bacterial complications may produce more severe illnesses, especially pneumonia. The lethal impact of an influenza epidemic is reflected in an increase in deaths due to cardiovascular and renal disease, as well as those due to pneumonia or to influenza itself.

Influenza occurs in successive epidemic waves and also in pandemics. The antigenic shifts of influenza viruses usually occur gradually, so that infection with strains widespread in one year confers some immunity against those prevalent in the next several years. However, sudden wide antigenic shifts do occur at 10–15-year intervals. *See* EPIDEMIC.

Vaccines made up of current strains, grown in embryonated eggs and inactivated by formalin or ultraviolet light, confer immunity to the homologous strains in the vaccine, but only relatively transient resistance to other strains. Since the vaccine is prepared from infected chick embryos, persons hypersensitive to eggs or egg products may experience severe allergic reaction. Attenuated live virus vaccines have been developed and used widely in the Soviet Union, and to a much lesser extent in Great Britain. One of the difficulties with the live vaccine is in obtaining a sufficiently high infection rate in those vaccinated. There is also the problem of antigenic shift that makes for obsolescence of the killed virus vaccines. *See* VACCINATION. [J.L.Me.]

Information systems engineering The process by which information systems are designed, developed, tested, and maintained. The technical origins of information systems engineering can be traced to conventional information systems design and development, and the field of systems engineering. Information systems engineering is by nature structured, iterative, multidisciplinary, and applied. It involves structured requirement analyses, functional modeling, prototyping, software engineering, and system testing, documentation, and maintenance.

Modern information systems solve a variety of data, information, and knowledge-based problems. In the past, most information systems were exclusively data-oriented; their primary purpose was to store, retrieve, manipulate, and display data. Application domains included inventory control, banking, personnel record keeping, and the like. The airline reservation system represents the quintessential information system of the 1970s. Since then, expectations as to the capabilities of information systems have risen considerably. Information systems routinely provide analytical support to users. Some of these systems help allocate resources, evaluate personnel, and plan and simulate large events and processes. The users expect information systems to perform all the tasks along the continuum shown in the illustration. *See* DATA-PROCESSING SYSTEMS.

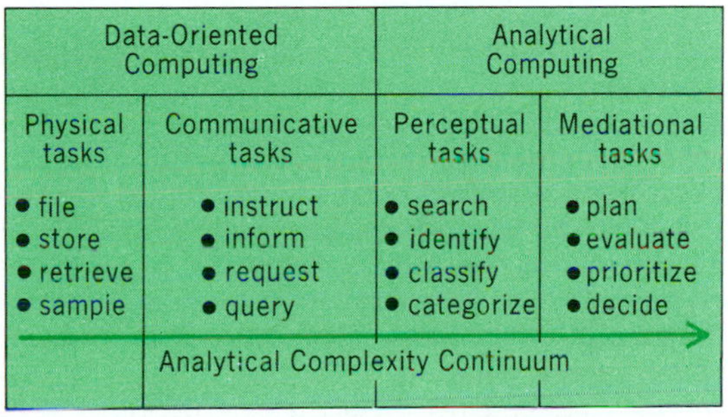

Data-Oriented Computing
Analytical Computing
Physical tasks
Communicative tasks
Perceptual tasks
Mediational tasks
• file • store • retrieve • sample
• instruct • inform • request • query
• search • identify • classify • categorize
• plan • evaluate • prioritize • decide
Analytical Complexity Continuum

Data-oriented and analytical computing, suggesting the range of information systems applications.

Systems engineering extends over the entire life cycle of systems, including requirement definitions, functional designs, development, testing, and evaluation. The systems engineer's perspective is different from that of the product engineer, software designer, or technology developer. The product engineer deals with detail, whereas the systems engineer takes an overall viewpoint. Systems engineering is based upon the traditional skills of the engineer combined with additional skills derived from applied mathematics, psychology, management, and other disciplines. The systems engineering process is a logical sequence of activities and decisions that transform operational needs into a description of system performance configuration. The process is by nature iterative and multidisciplinary. *See* SYSTEMS ENGINEERING. [S.J.A.]

Information theory A branch of communication theory devoted to problems in coding. A unique feature of information theory is its use of a numerical measure of the amount of information gained when the contents of a message are learned. Information theory relies heavily on the mathematical science of probability. For this reason the term information theory is often applied loosely to other probabilistic studies in communication theory, such as signal detection, random noise, and prediction. *See* ELECTRICAL COMMUNICATIONS; PROBABILITY.

In designing a one-way communication system from the standpoint of information theory, three parts are considered beyond the control of the system designer: (1) the source, which generates messages at the transmitting end of the system, (2) the destination, which ultimately receives the messages, and (3) the channel, consisting of a transmission medium or device for conveying signals from the source to the destination. The source does not usually produce messages in a form acceptable as input by the channel. The transmitting end of the system contains another device, called an encoder, which prepares the source's messages for input to the channel. Similarly the receiving end of the system will contain a decoder to convert the output of the channel into a form that is recognizable by the destination. The encoder and the decoder are the parts to be designed. In radio systems this design is essentially the choice of a modulator and a detector. *See* MODULATION.

A source is called discrete if its messages are sequences of elements (letters) taken from an enumerable set of possibilities (alphabet). Thus sources producing integer data or written English are discrete. Sources which are not discrete are called continuous, for example, speech and music sources. The treatment of continuous cases is sometimes simplified by noting that signal of finite bandwidth can be encoded into a discrete sequence of numbers.

The output of a channel need not agree with its input. For example, a channel might, for secrecy purposes, contain a cryptographic device to scramble the message. Still, if the output of the channel can be computed knowing just the input message, then the channel is called noiseless. If, however, random agents make the output unpredictable even when the input is known, then the channel is called noisy. *See* COMMUNICATIONS SCRAMBLING; CRYPTOGRAPHY.

Many encoders first break the message into a sequence of elementary blocks; next they substitute for each block a representative code, or signal, suitable for input to the channel. Such encoders are called block encoders. For example, telegraph and teletype systems both use block encoders in which the blocks are individual letters. Entire words form the blocks of some commercial cablegram systems. It is generally impossible for a decoder to reconstruct with certainty a message received via a noisy channel. Suitable encoding, however, may make the noise tolerable.

Even when the channel is noiseless, a variety of encoding schemes exists and there is a problem of picking a good one. Of all encodings of English letters into dots and dashes, the Continental Morse encoding is nearly the fastest possible one. It achieves its speed by associating short codes with the most common letters. A noiseless binary channel (capable of transmitting two kinds of pulse 0, 1, of the same duration) provides the following example. Suppose one had to encode English text for this channel. A simple encoding might just use 27 different five-digit codes to represent word space (denoted by #), A, B,..., Z; say # 00000, A 00001, B 00010, C 00011,..., Z 11011. The word #CAB would then be encoded into 00000000110000100010. A similar encoding is used in teletype transmission; however, it places a third kind of pulse at the beginning of each code to help the decoder stay in synchronism with the encoder. [E.N.G.]

Infrared astronomy The branch of astronomy that employs data in the heat radiation region of the electromagnetic spectrum to investigate the nature and physical characteristics of astronomical sources. The infrared region borders the visible region at a wavelength of 1 micrometer, and the radio region at 1000 μm. The spectral power of the heat radiation emitted by a blackbody reaches its maximum value at the point where the product of the wavelength and the absolute temperature is 2900 μm · K. *See* HEAT RADIATION; INFRARED RADIATION.

Observational technology. The Earth's atmosphere is opaque throughout most of the infrared, because of absorption by water (H_2O), carbon dioxide (CO_2), and occasionally a trace species like ozone (O_3). The illustration shows wavelength

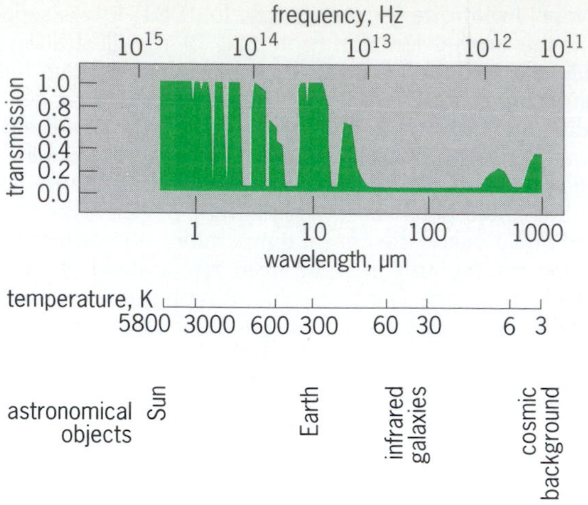

Location of infrared atmospheric windows in terms of frequency and wavelength. The temperatures of various blackbodies are shown where the peak in their spectral power distribution occurs. Types of astronomical objects with characteristic temperatures in each range are indicated.

regions in which the atmosphere is relatively transparent; observations can be made in these windows: near-infrared (1–3 μm), mid-infrared (3–30 μm), and (on those rare occasions that the atmosphere above is particularly dry) submillimeter (300–1000 μm). The atmosphere is opaque from the ground in the far-infrared (30–300 μm). However, at 45,000 ft (14 km) atmospheric transmission is about 0.80. Thus, far-infrared observations can be made from telescopes on aircraft, balloons, or spacecraft. *See* SUBMILLIMETER ASTRONOMY.

To the extent that the telescope and the atmosphere do not perfectly transmit astronomical infrared radiation, they themselves act like 300 K (80°F) sources of heat energy, generating a bright background. The contrast for observing Venus at noon with the unaided eye is about 600 times greater than for measuring Vega at 10 μm with even an optimized 98-in.-diameter (2.5-m) telescope. *See* TELESCOPE.

For both detectors and spectral filters, variety and capability are greater at shorter infrared wavelengths. Large-area detectors subdivided into several thousand picture elements (pixels) are rapidly becoming available. Mosaics of these large two-dimensional detectors have been assembled to make even larger focal planes. The two-dimensional detectors are generally mated to integrating, parallel-to-serial multiplexers. *See* OPTICAL DETECTORS.

Knowledge in infrared astronomy has been gained rapidly as a result of the success of several large-area sky surveys. The most sensitive was carried out by the *Infrared Astronomy Satellite* (*IRAS*), which carried a 24-in. (0.6-m) telescope cooled by liquid helium. More than 250,000 infrared sources were discovered. *See* CRYOGENICS; LIQUID HELIUM.

Celestial infrared sources. These include stars at wavelengths in the near-infrared region, and dust at longer wavelengths.

At near-infrared wavelengths, stars are the primary celestial sources of infrared radiation. Stars with surface temperatures less than 4000 K (6700°F), which is approximately the temperature of sunspots, emit most of their energy in the infrared. The majority of stars in the Milky Way Galaxy are this cool.

There are two large classes of stars cooler than the Sun. The first, and by far the most numerous, are simply low-mass stars that never reach the Sun's temperature. An object with a mass less than 0.08 times that of the Sun would never become a star. Intensive searches for these low-luminosity objects, called brown dwarfs, are motivated by the possibility that they could constitute a large fraction of the mass of the Milky Way Galaxy. *See* BROWN DWARF.

The second large class of stars cooler than the Sun comprises red giants, whose radii are about equal to the orbital radius of the Earth about the Sun. Studies of these stars, which can be detected at very large distances from the Sun, provide a powerful probe of the spatial distribution of solar-type stars throughout the Milky Way Galaxy. *See* STAR; STELLAR EVOLUTION.

Stars hotter than the Sun also appear to be relatively brighter at infrared wavelengths than at optical wavelengths. This reddening happens because interstellar space contains particles of dust. The *IRAS* image of the sky shows that the Milky Way Galaxy is more brilliant, flattened, and centrally concentrated than indicated from optical wavelength images. *See* INTERSTELLAR EXTINCTION; INTERSTELLAR MATTER; MILKY WAY GALAXY.

At wavelengths longer than about 3 μm, nonstellar sources of infrared radiation become relatively more prominent. The most important mechanism operating in the mid- and far-infrared wavelength range is emission by solid material.

Dust in the solar system, called zodiacal dust, reradiates the energy it absorbs from the Sun at mid-infrared wavelengths between 5 and 30 μm. If all stars form with planetary systems, the youngest ones should have more prominent zodiacal dust emission. An important result of the *IRAS* flight was the discovery of excess infrared emission from a nearby young star, Vega. If this excess emission is due to dust analogous to the zodiacal dust, then this discovery represents the first direct detection of a component of a planetary system associated with another star. *See* INTERPLANETARY MATTER.

Excess infrared radiation, attributable to thermal emission from dust, is also observed in the spectral energy distributions of red giants. However, the dust associated with these stars is condensed from gas that has drifted away from their outer, distended atmospheres. These stars play a key role in the chemical evolution of the Galaxy, since the matter they eject into the interstellar medium, enriched by nuclear burning in the stellar interior, is eventually incorporated into new stars.

At far-infrared wavelengths, the brightest celestial sources are the outer planets—Jupiter, Saturn, Uranus, and Neptune. Beyond the solar system, the brightest far-infrared sources are interstellar clouds. The largest of these are the most massive objects in the Milky Way Galaxy. In the interior of such a giant cloud, most external starlight is blocked out. This provides the conditions required (high density, low temperature) for density perturbations to undergo runaway collapse, resulting eventually in the formation of new stars. One of the most prodigious star-forming regions in the Galaxy is the nebula M17, whose large star population was discovered when a sensitive infrared camera was used to peer through the overlying layers of dust.

Half or more of the starlight of normal spiral galaxies is reprocessed by dust and emerges in the far-infrared. For some galaxies, almost all of the luminosity appears in the far-infrared. These galaxies must be in a transient stage since the star formation rates deduced from their infrared luminosities are so high that they would deplete their entire supply of interstellar gas and dust in a fraction of their lifetimes. Though infrared-bright galaxies are only a fraction of all galaxies, their luminosities are so great that they can be used to trace out the large-scale structure of the universe. *See* COSMOLOGY; GALAXY, EXTERNAL; UNIVERSE.

Several far-infrared sources observed by *IRAS* were shown to be distant galaxies with especially faint optical counterparts. Such large distances, coupled with their observed infrared brightness, means that these galaxies have luminosities as large as those of the most luminous quasars. Studies of these objects are motivated by the possibility that there may be an evolutionary link between these two classes of peculiar galaxies.

IRAS observations of quasars showed that many of them are brighter than expected in the infrared, especially at mid-infrared wavelengths. The discovery of infrared-bright quasars suggests that they may be both more numerous and more luminous than previously thought, deepening their mystery. *See* QUASAR. [S.Kl.; D.E.Kl.]

Infrared imaging devices Devices that convert an invisible infrared image into a visible image. The radiation available for imaging may be emitted from objects in the scene of interest (usually at the longer wavelengths called thermal radiation) or reflected. Reflected radiation may be dominated by sunlight or may be from controlled sources such as lasers used specifically as illuminators for the imaging device. The latter systems are called active, while those relying largely on emitted radiation are called passive. Active optical imaging systems were developed to achieve a nighttime aerial photographic capability, and work during World War 11 pushed such systems into the near-infrared spectral region. Development of passive infrared imaging systems came after the war, but only the advent of lasers allowed creation of active infrared imagers at wavelengths much longer than those of the photographic region. *See* INFRARED RADIATION; LASER.

Although developed largely for military purposes, infrared imaging devices have been valuable in industrial, commercial, and scientific applications. These range from nondestructive testing and quality control to earth resources surveys, pollution monitoring, and energy conservation. Infrared images from aerial platforms are used to accomplish "heat surveys," locating points of excessive heat loss. Calibration allows association of photographic tones with values of apparent (that is, equivalent blackbody) temperatures. Dark areas are "colder" than light ones. *See* NONDESTRUCTIVE TESTING; REMOTE SENSING. [G.J.Z.]

Infrared lamp A special type of incandescent lamp that is designed to produce energy in the infrared portion of the electromagnetic spectrum. The lamps produce radiant thermal energy which can be used to heat objects that intercept the radiation. An infrared lamp with a filament operating at 4000°F (2500 K) will release about 85% of its energy in the form of thermal radiant energy, about 15% as visible light, and a tiny fraction of a percent as ultraviolet energy. *See* HEAT RADIATION.

The lamps are supplied in three shapes. The most common shape, for general use, is that shown in illustration *a*, since the reflector unit is built in and needs only a suitable socket to form an infrared heating system. In industry, the lamp shown in illustration *b* is used with separate external reflectors designed to distribute the heat as desired.

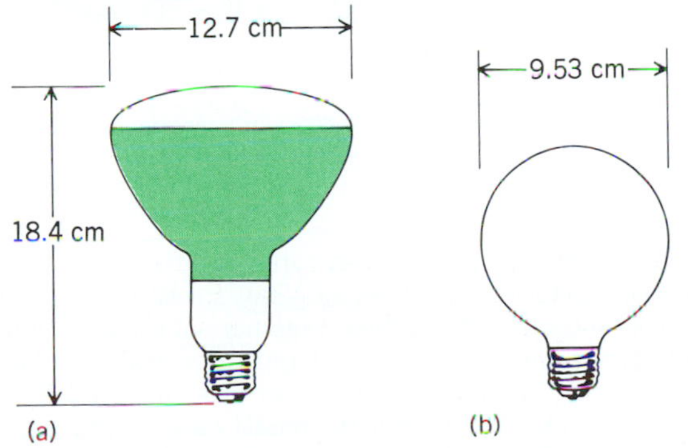

Shapes of infrared lamps. (*a*) Glass bulb with built-in reflector unit. (*b*) Glass bulb used with separate external reflector. 2.5 cm = 1 in.

The major advantage of infrared heating is that it is possible to heat a surface that intercepts the radiation without heating the air or other objects that surround the surface. Infrared lamps have many uses, including paint drying, evaporative drying, farm heating of animals, heating of food, control heating, and therapeutic heating of portions of the body to relieve muscle strains. *See* INCANDESCENT LAMP; INFRARED RADIATION; THERMOTHERAPY. [G.R.P.]

Infrared radiation Electromagnetic radiation in which wavelengths lie in the range from about 1 micrometer to 1 millimeter. This radiation therefore has wavelengths just a little longer than those of visible light and cannot be seen with the unaided eye. The radiation was discovered in 1800 by William Herschel.

An infrared source can be described by the spectral distribution of power emitted by an ideal body (a blackbody curve). This distribution is characteristic of the temperature of the body. A real body is related to it by a radiation efficiency factor or emissivity which is the ratio at every wavelength of the emission of a real body to that of a blackbody under identical conditions. The illustration shows curves for these ideal blackbodies radiating at a number of different temperatures. The higher the temperature, the greater the total amount of radiation. *See* EMISSIVITY.

Infrared detectors are based either on the generation of a change in voltage due to a change in the detector temperature resulting from the power focused on it, or on the generation of a change in voltage due to some photon-electron interaction in the detector material. This latter effect is sometimes called the internal photoelectric effect.

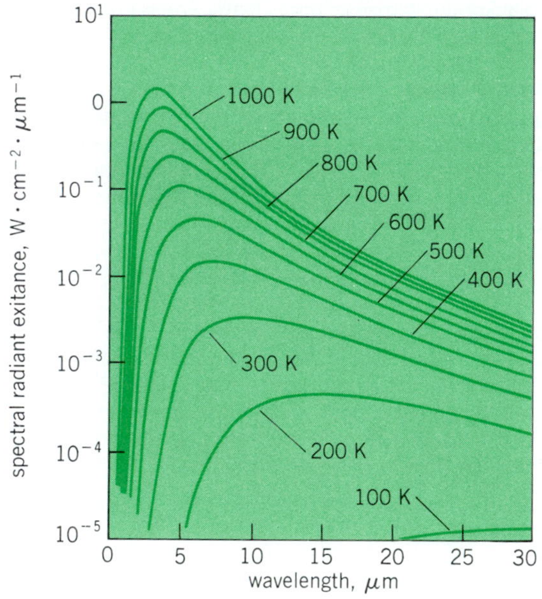

Radiation from blackbodies at different temperatures, shown on a logarithmic scale.

Infrared techniques have been applied in military, medical, industrial, meteorological, ecological, forestry, agricultural, chemical, and other disciplines. Weather satellites use infrared imaging devices to map cloud patterns and provide the imagery seen in many weather reports. Infrared imaging devices have also been used for breast cancer screening and other medical diagnostic applications. In most of these applications, the underlying principle is that pathology produces inflammation, and these locations of increased temperature can be found with an infrared imager. Airborne infrared imagers have been used to locate the edge of burning areas in forest fires. *See* INFRARED IMAGING DEVICES; REMOTE SENSING.

[W.L.Wo.]

Infrared spectroscopy The study of the interaction of material systems with electromagnetic radiation in the infrared region of the spectrum. The infrared region is valuable for the study of the structure of matter because the natural vibrational frequencies of atoms in molecules and crystals fall in the infrared range. Some gaseous molecules also have rotational frequencies in the far-infrared range, and certain frequencies corresponding to the energy levels of electrons in solids and in large molecules lie in the near infrared.

The infrared absorption spectrum of a molecule is highly characteristic, and often has been referred to as a molecular fingerprint. The spectrum can thus be used for molecular identification. Because the absorption of radiation at various infrared frequencies is quantitatively related to the number of absorbing molecules in a system, quantitative analysis is also possible.

The usual arrangement for measurement of an infrared spectrum consists of a source of continuous infrared radiation, an infrared spectrometer, and a detector. The infrared spectrometer disperses the radiation into a spectrum, and the detector converts the radiant energy into an electrical signal, which is amplified electronically and recorded by a chart recorder. To correct for variations caused by, among other factors, atmospheric absorption and the variation of source intensity with frequency, two spectra are recorded with a so-called double-beam spectrometer: one with a sample in the beam and one with the sample removed from the beam. The absorption by the sample as a function of frequency can then be computed from these two spectra. The comparison of the two spectra is obtained by rapid switching (10 times a second) of the beam from the infrared source back and forth between a path crossing the sample and a path passing by the sample.

The sharpness of the frequency and the high power per unit solid angle and unit spectral bandpass in a laser beam make it attractive for infrared spectroscopy at ultrahigh resolution. The main problem is the tuning of the laser frequency, that is, varying some parameter in the laser system so that its frequency may be varied continuously and accurately. Tunable infrared lasers have been used mainly to measure the energy levels due to the vibration and rotation of molecules in the gas phase at low pressures. By this means, very accurate values of such molecular parameters as internuclear distances, electric dipole moments, vibrational frequencies, internal force fields, and the like may be measured. *See* LASER SPECTROSCOPY.

Infrared spectra can be used to: identify pure chemical compounds by comparison of the spectrum of an unknown with previously recorded spectra of pure compounds; identify the constituents of mixtures; and show the presence of a group of atoms, a so-called functional group, in a molecule of unknown or doubtful structure. Also, since there is a linear quantitative relationship between the absorbance and the number of absorbing molecules, the quantitative analysis of mixtures by infrared means is feasible. Structures of molecules can be determined to varying degrees of refinement from infrared spectra. The infrared spectra of crystalline solids give information about modes of vibration of crystals, about hydrogen-bond vibrations when such bonds are present in them, and about electronic energy states in semiconductors and superconductors. *See* MOLECULAR STRUCTURE AND SPECTRA; SPECTROSCOPY.

[R.C.L.]

Infrasound Sound waves, particularly in the atmosphere, whose frequencies of pressure variation and of vibration are below the audible range, that is, lower than about 20 Hz. Earthquake and seismic waves are elastic waves which occur at infrasonic frequencies in the Earth's crust and in the oceans and seas. The physical laws of propagation in the atmosphere are essentially the same as for audible sound. The local speed of infrasound in air at ambient temperatures near 20°C (68°F) is about 340 m/s (1115 ft/s), the same as for audible sound.

At frequencies less than about 1.0 Hz, infrasound propagates through the atmosphere for distances of thousands of kilometers without substantial loss of energy. Sounds at these frequencies are almost always present at measurable intensities. Those of natural origin have many causes, including tornadoes, volcanic explosions, earthquakes, the aurora borealis, waves on the seas, large meteorites, and lightning discharges. When the wind blows, turbulent pressure fluctuations in the atmosphere occur at amplitudes up to tens of pascals, at infrasonic frequencies. People are unaware of these pressures via the sensation of hearing.

But sufficiently strong infrasound is "audible," contrary to simple acoustic tradition. However, there is no sensation of tone. Listeners variously describe audible infrasound as "pumping," "popping effect," or "chugging." The human body is particularly sensitive to vibrations and infrasound near 7 Hz, at which frequency there is an overall mechanical resonance of organs in the abdominal and chest cavities. Unpleasant physiological effects with pathological overtones occur when human subjects and animals are exposed to infrasound in the laboratory at sound pressure levels of 130 dB and greater. The equipment and machinery of modern technology can expose people to infrasound which is much stronger and occurs more often than in the natural environment. *See* SOUND. [R.K.C.]

Inhibitor (chemistry) A substance which is capable of stopping or retarding a chemical reaction. To be technically useful, such compounds must be effective in low concentrations, usually under 1%. The type of reaction which is most easily inhibited is the free-radical chain reaction. Vinyl polymerization and autoxidation are two important examples of the class. Another reaction type for which inhibitors have been found is corrosion, particularly in aqueous systems. *See* ANTIOXIDANT; CORROSION; FREE RADICAL; POLYMERIZATION. [L.R.M.]

Ink A dispersion of a pigment or a solution of a dye in a carrier vehicle, yielding a fluid, paste, or powder to be applied to and "dried" on a substrate. Writing, marking, drawing, and printing inks are applied by several methods to paper, metal, plastic, wood, glass, fabric, or other substrates. Often inks perform communicative, decorative, and even protective functions.

Printing ink can be classified according to its composition and texture, application, end use, and drying manner. Composition and texture may be any of the following types: oil or paste ink, solvent or liquid (aniline) ink, or specialty ink. The applications are letterpress, lithographic, flexographic, gravure, silk screen, stencil, duplicating, electrostatic, and jet. The end uses are news, publication, commercial, folding carton, book, corrugated box, paper bag, wrapper, label, metal container, plastic container, plastic film, foil, laminating, food insert, sanitary paper, and textile. The various drying manners are oxidizing, evaporating, penetrating, precipitating, polymerizing, reactive, gelling, and cold-setting or quick-setting.

Ultraviolet (radiation) cure inks are based on instant photopolymerization of monomers by a high-energy light source. They completely eliminate pollution from the web and spray powder on the sheet presses, and reduce energy consumption to 20% of that required by heat-set ovens previously employed in printing on paper or metal.

Inks developed for ball-point pens are Newtonian fluids of high tinctorial strength. These must be free of particles and premature drying so as to continue the feed to the paper without clogging. Rapid penetration into the paper accomplishes "drying." Stamp-pad or marking inks are impregnated into a cloth or foam rubber pad and transferred by pressure to a rubber type which is then stamped or impressed against the substrate. The inks must remain nondrying on the pad, yet rapidly penetrate into the stamped substrate. [B.V.Bu.]

Inland waterways The internal commercial water routes of the United States except for the Great Lakes. Transportation on the inland waterways of the United States accounts for 10% of the nation's total movement of commerce. Barges move 7.2%; the remaining 2.8% is moved in deep-draft vessels.

The 25,380-mi (40,845-km) network of commercially navigable inland channels is used primarily by barges which are either pushed by towboats or pulled on a hawser by tugboats. Where improvements of the channels have been made for barge transportation, a standard operating depth of 9 ft (2.74 m) is provided. Certain sections of these inland waterways, as has been indicated, are deep enough to support oceangoing vessels. The primary routes where both barges and oceangoing vessels operate are the Hudson River to Albany, the Mississippi River to Baton Rouge, the Houston Ship Channel, and the Columbia River to Portland.

The system of commercially navigable inland waterways of the United States is made up of a series of federal projects, with the exception of the New York State Barge Canal, which is a state project. Improvement, maintenance, and operation is the responsibility of the Army Corps of Engineers working with federal funds appropriated by the Congress. [B.B.C.]

Inoculation The process of introducing a microorganism or suspension of microorganism into a culture medium. The medium may be (1) a solution of nutrients required by the organism or a solution of nutrients plus agar; (2) a cell suspension (tissue culture); (3) embryonated egg culture; or (4) animals, for example, rat, mouse, guinea pig, hamster, monkey, birds, or human being. When animals are used, the purpose usually is the activation of the immunological defenses against the organism. This is a form of vaccination, and quite often the two terms are used interchangeably. Both constitute a means of producing an artificial but active immunity against specific organisms, although the length of time given by such protection may vary widely with different organisms. *See* CULTURE MEDIA; IMMUNITY; VACCINATION. [E.G.St./N.K.M.]

Inorganic chemistry The chemical reactions and properties of all the elements in the periodic table and their compounds, with the exception of the element carbon. The chemistry of carbon and its compounds falls in the domain of organic chemistry. The boundaries of inorganic chemistry with the other major areas of chemistry are not precisely defined, and it is often a matter of taste as to whether a particular topic is to be included in the field of inorganic chemistry or is to be considered physical or even organic chemistry. Investigations into theoretical inorganic chemistry or the study of problems in inorganic chemistry by quantitative and sophisticated physical methods may be considered either inorganic or physical chemistry quite arbitrarily. In similar fashion, organometallic compounds may be considered to be in the sphere of either inorganic or organic chemistry. To an increasing extent, the inorganic chemist is concerned with problems that once were considered the prerogative of physical chemists, organic chemists, or even biochemists. *See* BIOINORGANIC CHEMISTRY; CHEMICAL DYNAMICS; COORDINATION CHEMISTRY; GEOCHEMISTRY; NUCLEAR CHEMISTRY; ORGANOMETALLIC COMPOUND; PHYSICAL CHEMISTRY; SOLID-STATE CHEMISTRY. [J.J.K.]

Inorganic photochemistry The study of the light-induced behavior of various metal compounds. The physical and chemical properties of substances are generally altered by the absorption of light. Typical metal compounds have a characteristic number (coordination number) of molecules or ions (ligands) directly bonded to the metal center. For example a six-coordinate compound has the general formula ML_6^{n+}. Many of these compounds are colored, and much interest has been aroused by speculation that some metal compounds could mediate the transformation of solar radiation into useful chemical or electrical energy. *See* COORDINATION NUMBER.

The photochemistry of metal compounds has grown in concert with modern theories of the electronic structure of molecules and of chemical bonding in molecules. Photochemical studies are often designed to probe and test these theories. The range of pertinent studies spans most of the subdisciplines of chemistry and includes or bears on such topics as photophysics, the development of laser materials, catalysis, photosynthesis, oxidation-reduction chemistry, acid-base chemistry, organometallic chemistry, metalloenzyme chemistry, solid-state chemistry, and surface chemistry. *See* CHEMICAL BONDING; CHEMICAL DYNAMICS; COORDINATION CHEMISTRY; LASER PHOTOCHEMISTRY; PHOTOCHEMISTRY. [J.F.E.]

Inorganic polymer A giant molecule linked by covalent bonds but with an absence or near-absence of hydrocarbon units in the main molecular backbone; these may be included as pendant side chains. Carbon fibers, graphite, and so forth are considered inorganic polymers.

Some special characteristics of many inorganic polymers are a higher Young's modulus and a lower failure strain compared with organic polymers. Relatively few inorganic polymers dissolve in the true sense, or alternatively, if they swell, few can revert. Crystallinity and high glass transition temperatures are also much more common than in organic polymers. In highly cross-linked inorganic polymers, stress relaxation frequently involves bond interchange. *See* YOUNG'S MODULUS.

Inorganic polymers can be classified in a number of ways. Some are based on the composition of the backbone, such as the silicones (Si—O), the phosphazenes (P—N), and polymeric sulfur (S—S). Others are based on their connectivity, that is, the number of network bonds linking the repeating unit into the network. Thus the silicones based on R_2SiO, the phosphazenes based on NPX_2, and polymeric sulfur each have a connectivity of two, while boric oxide based on B_2O_3 has a connectivity of three, and amorphous silica based on SiO_2 has one of four (see illustration).

Among the well-known inorganic polymers are silicones, chalcogenide glasses, graphite, boron polymers, and silicate polymers. *See* BORON POLYMERS; CHALCOGENIDE GLASSES; GRAPHITE; SILICATE POLYMERS; SILICONE RESINS. [R.A.S.]

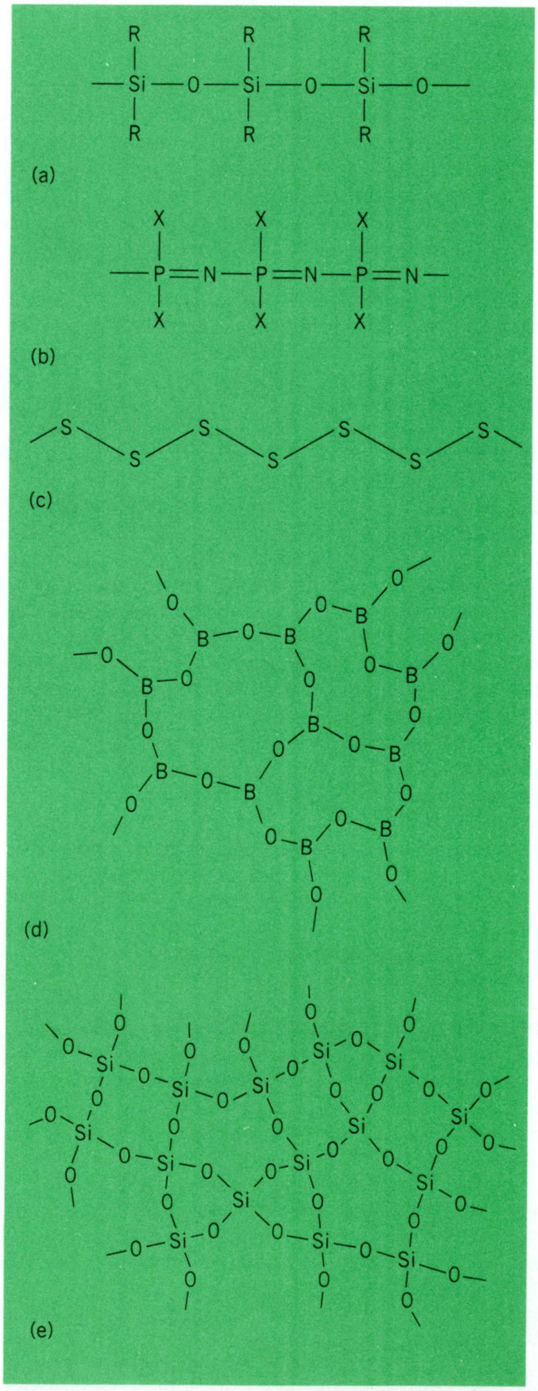

Polymers with varying connectivities: (*a*) siloxanes, two; (*b*) phosphazenes, two; (c) sulfur, two; (*d*) boric oxide, three; (*e*) amorphous silica, four. (*After N. H. Ray, Inorganic Polymers, Academic Press, 1978*)

Inositol A crystalline, water-soluble alcohol, often grouped with the vitamins. Inositol has the structural formula shown.

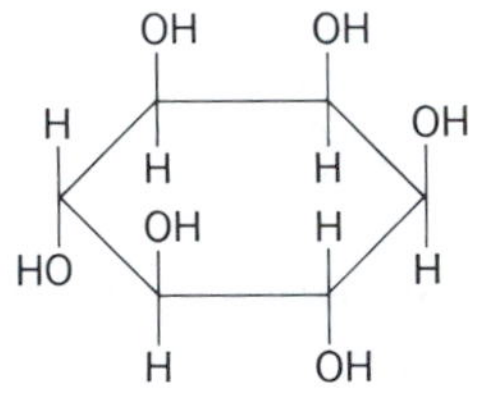

Almost nothing is known of its function in enzyme systems. It is metabolized as a carbohydrate, but its availability is limited in the presence of dietary calcium. There is no evidence of a human requirement for inositol in the diet, probably because a considerable amount is synthesized by intestinal bacteria. Yeast cells and mammalian tissues are capable of inositol synthesis. *See* CARBOHYDRATE. [F.W.]

Insect control, biological The term biological control was proposed in 1919 to apply to the use or role of natural enemies in insect population regulation. The enemies involved are termed parasites (parasitoids), predators, or pathogens. This remains preferred usage, although other biological methods of insect control have been proposed or developed, such as the release of mass-produced sterile males to mate with wild females in the field, thereby greatly reducing or suppressing the pests' production of progeny. Classical biological control is an ecological phenomenon which occurs everywhere in nature without aid from, or sometimes even understanding by, humans. However, humans have utilized the ecological principles involved to develop the field of applied biological control of insects, and the great majority of practical applications have been achieved with insect pests. Additionally, such diverse types of pest organisms as weeds, mites, and certain mammals have been successfully controlled by use of natural enemies.

Biological control offers several advantages over chemical control. Application costs are minimal and nonrecurring and there are no environmental pollution problems connected with biological control, whereas insecticides cause severe problems related to toxicity to humans, wildlife, birds, and fish, as well as causing adverse effects in soil and water. Biological control causes no upsets in the natural balance of organisms, but these upsets are common with chemical control; biological control is permanent, chemical control is temporary, usually one to many annual applications being necessary. *See* ENTOMOLOGY, ECONOMIC; INSECTICIDE. [P.DeB.]

Insect pathology A biological discipline embracing the general principles of pathology (disease in the broadest sense) as applied to insects. It refers to human observations and actions concerning the cause, symptomatology, gross pathology, histopathology, pathogenesis, and epizootiology of the diseases of insects; it is concerned with whatever can go wrong or become abnormal in an insect. A diseased insect may be suffering from an infectious disease caused by a microorganism or a noninfectious disease, such as a metabolic disturbance, a genetic abnormality, a nutritional deficiency, a physical or chemical injury, or injury caused by parasites or predators.

Insect pathology draws upon and contributes to the general field of microbiology and provides understanding of certain of the biological relationships existing between insects and microorganisms not pathogenic to them. Insect pathology may be viewed as a part of invertebrate pathology and a part of the branch of entomology. Insect pathology finds applications in agriculture, medicine, and biology generally. Microbial control, the use of microorganisms in biological control, is one area of applied insect pathology. Microorganisms are introduced to control insect pests for the protection of humans, animals, and agricultural crops. However, the suppression of disease in beneficial insects, such as the silkworm, honeybee, and ladybird beetle, is also of significant practical importance. *See* INSECT CONTROL, BIOLOGICAL; INVERTEBRATE PATHOLOGY. [J.D.Br.]

Insecta A class of the phylum Arthropoda sometimes called the Hexapoda. This is the largest class of animals, containing about 750,000 described species, but there are possibly as many as 5,000,000 actual species of insects. Like other arthropods, they have an external, chitinous covering. Fossil insects dating as early as the Paleozoic have been found throughout the world.

Classification. The class Insecta is divided into orders on the basis of the structure of the wings and the mouthparts, on the type of metamorphosis, and on various other characteristics. There are differences of opinion among entomologists as to the limits of some of the orders. The following are considered herein as orders of insects.

Subclass Apterygota
  Order: Thysanura—bristletails, silverfish
Subclass Pterygota
  Order: Ephemeroptera (Ephemerida)—mayflies
    Odonata—dragonflies and damselflies
    Blattaria (Dictyoptera)—cockroaches
    Mantodea—mantises
    Isoptera—termites
    Orthoptera—grasshoppers, katydids, crickets, walking sticks, cockroaches, and mantids
    Phasmatoptera
    Dermaptera—earwigs
    Embiidina (Embioptera)—webspinners
    Plecoptera—stoneflies
    Zoraptera—zorapterans
    Psocoptera—psocids
    Mallophaga—chewing lice
    Anoplura—sucking lice
    Thysanoptera—thrips
    Hemiptera—bugs
    Homoptera—cicadas, hoppers, aphids, whiteflies, scale insects
    Neuroptera—dobsonflies, fishflies, snakeflies, lacewings, antlions
    Coleoptera—beetles
    Strepsiptera
    Mecoptera—scorpionflies
    Siphonaptera—fleas
    Diptera—true flies
    Trichoptera—caddisflies
    Lepidoptera—butterflies and moths
    Hymenoptera—sawflies, ichneumons, chalcids, wasps, ants, bees

Morphology. Insects are usually elongate and cylindrical in form, and are bilaterally symmetrical. The body is segmented, and the ringlike segments are grouped into three distinct regions, the head, thorax, and abdomen. The head bears the eyes, antennae, and mouthparts; the thorax bears the legs and wings, when wings are present; the abdomen usually bears no locomotor appendages but often bears some appendages at its apex. Most of the appendages of an insect are segmented. The body wall serves not only as a covering, but also as a supporting structure to which many important muscles are attached.

A pair of compound eyes usually cover a large part of the head surface. In addition to having compound eyes, most insects also possess two or three simple eyes, the ocelli, usually located on the upper part of the head between the compound eyes. *See* EYE (INVERTEBRATE).

Insect mouthparts typically consist of labrum, or upper lip; a pair each of mandibles and maxillae; a labium, or lower lip; and a tonguelike structure, the hypopharynx. These structures are variously modified in different insect groups and are often used in classification and identification. The type of mouthparts an insect has determines how it feeds and what type of damage it is capable of causing. A cricket is said to have chewing mouthparts because it has heavily sclerotized mandibles that move sideways and can bite off and chew particles of food. A bedbug has long, slender, needlelike mandibles which can pierce the skin. Also, a pair of maxillae form a tube which is inserted through the puncture and used to suck blood.

Several forms of antennae are recognized, to which various names are applied; they are used extensively in classification. The antennae are usually located between or below the compound eyes and are often reduced to a very small size. They are sensory in function and act as tactile organs, organs of smell, and in some cases organs of hearing.

Insects are the only winged invertebrates, and their dominance as a group is probably due to their wings. Immature insects do not have wings, except in the mayflies. The wings are solid structures and may be likened to the two sides of a cellophane bag that have been pressed tightly together. The form and rigidity of the wing are due to the stiff chitinous veins which support and strengthen the membranous portion. At the base are small sclerites which serve as muscle attachments and produce consequent wing movement. The wings vary in number, size, shape, texture, and venation, and in the position at which they are held at rest. There is a common basic pattern of wing venation in insects which is variously modified and in general quite specific for different large groups of insects. Much of the classification of insects depends upon these variations. A knowledge of fossil insects depends largely upon the wings, because they are among the more readily fossilized parts of the insect body.

Internal anatomy. The intake of oxygen, its distribution to the tissues, and the removal of carbon dioxide are accomplished by means of an intricate system of tubes called the tracheal system. The principal tubes of this system, the trachea, open externally at the spiracles. Internally they branch extensively, extend to all parts of the body, and terminate in simple cells, the tracheoles. Many adaptations for carrying on respiration are known. Aquatic insects may have tracheal gills located either on the external portion of the body or in the rectal cavity. The larvae of certain insects, such as mosquitoes, have respiratory tubes at the posterior end of the body. Some aquatic insects with normal tracheal systems and active spiracles can go beneath the water and remain for extended periods of time by carrying with them a film of air on the surface of the body.

Insects possess an alimentary tract consisting of a tube, usually coiled, which extends from the mouth to the anus. It is differentiated into three main regions, the foregut, midgut, and hindgut.

Valves between the three main divisions of the alimentary canal regulate the passage of food from one region to another.

The excretory system consists of a group of tubes with closed distal ends, the Malpighian tubules. They vary in number from 1 to over 100, and extend into the body cavity. Various waste products are taken up from the blood by these tubules and passed out by way of the hindgut and anus.

The circulatory system of an insect is an open one, as compared with a closed system composed of arteries, veins, and capillaries, as in the vertebrates. The only blood vessel is a tube located dorsal to the alimentary tract and extending through the thorax and abdomen. The posterior portion of this tube is called the heart; the anterior part of the tube is called the dorsal aorta.

The nervous system consists of a brain, often called the supraesophageal ganglion, located in the head above the esophagus; a subesophageal ganglion, connected to the brain by two commissures that extend around each side of the esophagus; and a ventral nerve cord, typically double, extending posteriorly through the thorax and abdomen from the subesophageal ganglion. In the nerve cords there are enlargements, called ganglia. Typically, there is a pair to each body segment. From each ganglion of the chain, nerves extend to each adjacent segment of the body, and also extend from the brain to part of the alimentary canal.

Reproduction in insects is nearly always sexual, and the sexes are separate. Variations from the usual reproductive pattern occur occasionally. In many social insects, such as the ants and bees, certain females, the workers, may be unable to reproduce because their sex organs are undeveloped; in some insects, individuals occasionally occur that have characters of both sexes, called gynandromorphs. *See* SOCIAL INSECTS.

Metamorphosis. After insects hatch from an egg, they begin to grow and increase in size (growth) and will also usually change, to some degree at least, in form (metamorphosis) and often in appearance. The growth of an insect is accompanied by a series of molts, or ecdyses, in which the cuticle is shed and renewed.

The molt involves not only the external layers of the body wall, but also the cuticular linings of the tracheae, foregut, and hindgut; the cast skins often retain the shape of the insects from which they were shed. The shedding process begins with a splitting of the old cuticle, usually along the midline on the dorsal side of the thorax. This split grows and the insect eventually wriggles out of the old cuticle. The new skin, at first wrinkled and folded, remains soft and pliable long enough for the body to expand to its fullest capacity before hardening.

Insects have been grouped or classified upon the basis of the type of metamorphosis which they undergo. All entomologists, however, do not agree upon the same classification. These types of metamorphosis have been placed in two groups, under the heading of simple metamorphosis and complete metamorphosis. Some investigators include ametabolous, paurometabolous, and hemimetabolous types in simple metamorphosis, and consider complete metamorphosis to be synonymous with holometabolous. In order to cover the more detailed pattern, the following outline is used.

1. Ametabolous or primitive: No distinct external changes are evident with an increase in size.
2. Metabolous: Distinct changes, both in size and form, are evident.
   a. Paurometabolous: Direct metamorphosis that is simple and gradual; immature forms resemble the adult except in size and are referred to as nymphs.
   b. Hemimetabolous: Metamorphosis is incomplete; gills are present in aquatic larvae, or naiads.
   c. Holometabolous: Complete or indirect metamorphosis; stages in this development are the egg, larva, pupa, and adult, or imago. [D.M.DeL.]

Fossils. Fossil insects have been found in about 100 different rock formations throughout the world. One of the best known deposits is a shale bed in the town of Florissant, Colorado.

The study of fossil insects has made significant contributions to the understanding of the evolution of insects and of the relationships of existing orders and families. It has become apparent that the groups now living are but a small part of the total number that have existed on the Earth.

The fossil record, combined with morphological studies of living types, shows that insects have passed through three major evolutionary steps. The first of these was the development of wings as simple but functional outgrowths of the thoracic wall which could not be folded back over the abdomen when the insect was at rest. The second step was marked by the evolution of a complicated wing articulation, which enabled the wings to be folded back over the abdomen at rest, and by the presence of immature stages resembling the adults. The third step was the acquisition of a complicated postembryonic development with an immature stage (larva) very different from the adult. The fossils show that insects passed through these stages before the beginning of the Permian Period, some 225,000,000 years ago, when land-inhabiting vertebrates were only starting their evolutionary history. [F.M.C.]

Insecticide A material used to kill insects and related animals by disruption of vital processes through chemical action. Insecticides may be inorganic or organic molecules.

Insecticides are classified according to type of action as stomach poisons, contact poisons, residual poisons, systemic poisons, fumigants, repellents, attractants, insect growth regulators, or pheromones. Many act in more than one way. Stomach poisons are applied to plants so that they will be ingested as insects chew the leaves. Contact poisons are applied in a manner to contact insects directly, and are used principally to control species which obtain food by piercing leaf surfaces and withdrawing liquids. Residual insecticides are applied to surfaces so that insects touching them will pick up lethal dosages. Systemic insecticides are applied to plants or animals and are absorbed and translocated to all parts of the organisms, so that insects feeding upon them will obtain lethal doses. Fumigants are applied as gases, or in a form which will vaporize to a gas, so that they can enter the insects' respiratory systems. Repellents prevent insects from closely approaching their hosts. Attractants induce insects to come to specific locations in preference to normal food sources. Insect growth regulators are generally considered to act through disruption of biochemical systems or processes associated with growth or development, such as control of metamorphosis by the juvenile hormones, regulation of molting by the steroid molting hormones, or regulation of enzymes responsible for synthesis or deposition of chitin. Pheromones are chemicals which are emitted by one sex, usually the female, for perception by the other, and function to enhance mate location and identification; pheromones are generally highly species-specific. *See* FUMIGANT; PHEROMONE.

The use of inorganic insecticides has declined to almost nil. Probably the largest current uses are for inorganic arsenicals as constituents of formulations for wood impregnation and preservation. Organic insecticides began to supplant the arsenicals when DDT and lindane [1,1,1-trichloro-2,2-bis-(*p*-chlorophenyl)ethane] became available to the public during World War II. Historically, DDT and lindane were the two largest-volume organic insecticides. Since 1968, however, in the United States all nongovernmental and nonprescription uses of DDT have been canceled. Coincidentally, the use of lindane has declined both through development of resistant strains of insects and through official restriction.

Other chlorinated hydrocarbon insecticides which were available by or during the early 1970s included: TDE, methoxychlor, Dilan, chlordane, heptachlor, aldrin, dieldrin, endrin, toxaphene, endosulfan, Kepone, mirex, and Perthane. Registrations for all uses of TDE were canceled in 1971 because it is a metabolite of DDT. Chlordane, the first cyclopentadiene insecticide to achieve commercial status, was once the most effective chemical available for the control of cockroaches. As early as 1966, uses and registrations for aldrin and dieldrin were being curtailed. Most uses of aldrin and dieldrin were canceled by the U.S. Environmental Protection Agency (EPA) in 1975 because of their risks to human health and the environment, The principal remaining registered use for these two chemicals is for the control of subterranean termites. The production of aldrin has been stopped. Most uses of chlordane and heptachlor were similarly canceled in 1978. Registrations of other insecticides are being reviewed.

The development of organic phosphorus insecticides paralleled that of the chlorinated hydrocarbons. Since 1947, many thousands of organophosphorus compounds have been synthesized for evaluation as potential insecticides. A great diversity of activity against insects is found among organophosphorus insecticides. Unfortunately, many can be quite poisonous to humans and other warm-blooded animals, as well as to insects. Activity of organic phosphate insecticides results from the inhibition of the enzyme cholinesterase, which performs a vital function in the transmission of impulses in the nervous system.

The first commercial carbamate insecticide, carbaryl, was introduced during the mid-1950s and continues to enjoy broad use not only in the United States but worldwide. Carbamates, like organophosphates, are cholinergic. Several not only interfere with cholinesterase, but may also inhibit one or more enzymes known as aliesterases.

Diflubenzuron is another chemical with a new and different mode of action. It interferes with the action of chitin synthetase, an enzyme necessary for the synthesis of chitin, and as a consequence of that process, disrupts the orderly deposition of that vital element of the exoskeleton. Currently, it is registered for the control of the gypsy moth in hardwood forests of the eastern United States.

Pyrethrum is one of the oldest insecticides. Because it is highly active and quite safe for humans and animals, extensive effort was made during the period 1910–1945 to determine its chemical structure. These efforts culminated in the discovery of some of the most active insect control agents yet devised. The first product, allethrin, which was only partially synthetic, was introduced commercially during the early 1950s. Subsequent, entirely synthetic products have included: dimethrin, fenothrin, resmethrin, and tetramethrin.

Formulation of insecticides is extremely important in obtaining satisfactory control. Common formulations include dusts, water suspensions, emulsions, and solutions. Accessory agents, including dust carriers, solvents, emulsifiers, wetting and dispersing agents, stickers, deodorants or masking agents, synergists, and antioxidants, may be required to obtain a satisfactory product.

Proper timing of insecticide applications is important in obtaining satisfactory control. The application of insecticides should be correlated with the occurrence of the most susceptible or accessible stage in the life cycle of the pest involved. By and large, treatments should be made only when economic damage by a pest appears to be imminent. *See* INSECT CONTROL, BIOLOGICAL; MITICIDE; PESTICIDE. [G.F.L.]

Insectivora An order of mammals including such familiar forms as the hedgehogs, shrews, and moles. The Insectivora are of ancient origin and are difficult to define; they are united by primitive characters that, for the most part, are primitive for all placental mammals.

Although their name implies a diet of insects, most insectivores feed, or once fed, on a variety of invertebrate and even vertebrate prey; some were even frugivorous or omnivorous. Many insectivores have spines, notably the hedgehogs and tenrecs, and generally the senses of smell and hearing are far more acute than eyesight. Echolocation is known in shrews and tenrecs, and some shrews and *Solenodon* share the distinction of being the only mammals whose bite is poisonous, at least to small prey. Almost all insectivores are nocturnal or crepuscular, some are excellent tunnelers through the soil, and a few are aquatic.

Insectivores may be divided into two groups. The first group is composed of insectivores in which the intestinal cecum is lacking and the stapedial artery is a major source of blood supply to the brain. These are known as the Lipotyphia, or Insectivora proper. All living insectivores (except tree shrews, Tupaiidae) and most late Cenozoic, a few early Cenozoic, and possibly Cretaceous insectivores belong to this first group. The second group, sometimes called the Proteutheria, is generally more primitive and is primarily early Cenozoic in known occurrence. Living members have an intestinal cecum, and the blood supply to the brain is still primitively complex. The stapedial artery is not as important as in the lipotyphians. *See* DELTATHERIDIA; EUTHERIA; HEDGEHOG; MAMMALIA; MOLE (ZOOLOGY); SHREW; TENREC. [D.D.D./M.C.McK.]

Insectivorous plants Plants having variously modified, highly specialized leaves which capture and digest insects. The proteins of the digested insect bodies supply nitrogen, which otherwise may be unavailable to these plants in the places where they grow. Sometimes they are also called carnivorous (flesh-eating) plants. *See* NEPENTHALES; PITCHER PLANT; SECRETORY STRUCTURES (PLANT); SUNDEW; VENUS' FLYTRAP. [P.D.St./E.L.C.]

Insolation The amount of solar radiation which reaches a unit horizontal area of the Earth. It is the latitudinal variation of insolation which supplies energy for the general circulation of the atmosphere. Insolation outside the Earth's atmosphere depends on the angle of incidence of the solar beam and on the solar constant. The solar constant is the amount of energy which, in unit time, reaches a unit plane surface perpendicular to the Sun's rays outside the Earth's atmosphere when the Earth is at its mean distance from the Sun.

Latitudinal variation of insolation

Latitude	0°N	30°N	60°N	90°N
	(annual mean, ly/day*)			
Extraterrestrial	850	740	470	350
At surface				
With clear sky	570	520	320	220
With normal cloud cover	410	440	200	150
Total absorption (atmosphere and surface)	570	530	260	120

*Langleys per day (1 ly = 1 cal/cm^2).

Solar energy is modified by the air, the clouds, and the land and water surface of the Earth. The energy absorbed at the surface depends also on the surface albedo. The reflectivity of the solid Earth varies with ground cover. Forests absorb nearly all the energy incident on them. Fresh snow, on the other hand, reflects up to 90% of the incident energy. In middle latitudes, snow albedos may decrease to less than 40% with time, especially after warm spells have modified the surface. Grass, fields, and other surfaces will have intermediate albedos. When all the factors are considered, the absorption by the atmosphere and Earth's surface can be estimated, as shown in the last line of the table. It is interesting to note that the gradient of absorbed radiation, which is the driving force of the circulation, is similar to the gradient of the extraterrestrial insolation. *See* ALBEDO. [S.Fr.]

Inspection and testing Industrial activities which ensure that manufactured products, individual components, and multicomponent systems are adequate for their intended purpose. Complete inspection and testing programs cover evaluation of the product design prior to manufacturing, conformance to the design during manufacturing, and durability during subsequent storage, transport, and use. The terms inspection and testing are used interchangeably and jointly to describe closely related activities, but strictly speaking, inspection per se is the total activity dealing with: interpretation of product specifications; measurement and testing of the characteristics described in the specifications; methods of sampling and statistical analysis of results; and decisions on the disposition of the product evaluated. *See* GAGE; NONDESTRUCTIVE TESTING; QUALITY CONTROL.

The risks of occasionally making the wrong decisions in sampling inspection are an accepted part of industrial practice. The possibility of unsatisfactory product reaching the consuming public under such procedures is covered by guarantees which are now regulated by law. [J.A.Cl.]

Instinctive behavior A relatively complex response pattern which is usually present in one or both sexes of a given species. These responses have a genetic basis, are essentially unlearned, and are generally adaptive.

Instinctive behavior occurs when an animal has a particular internal state while it is in the presence of a specific external stimulation called a releaser or a sign stimulus. Neither the internal state nor the external stimulus alone is adequate for the elicitation of the response. Many animals show particular instinctive behaviors only during the mating season, when hormonal changes associated with sexual behavior sensitize specific portions of the central nervous system, which will then be active in the presence of the releaser. The external stimulus may be relatively simple or incredibly complex.

Within limits, the instinctive behaviors can be modified by learning. There is evidence, for example, that some predators learn to attack their prey at the back of the neck because when held in that position the prey cannot counterattack. *See* MIGRATORY BEHAVIOR; REPRODUCTIVE BEHAVIOR. [K.E.M.]

Instrument landing system (ILS) A continuous-wave amplitude-modulated, horizontally polarized fixed-beam system comprising the following aids to aircraft approaching a runway: (1) a very high-frequency localizer to give azimuth guidance with reference to the runway center line, with an operating frequency in the band 108–112 MHz; (2) an ultrahigh-frequency glide slope to give elevation guidance about an optimum approach angle, with an operating frequency in the band 328.6–335.4 MHz; and (3) two or three marker beacons operating at 75 MHz to give range with reference to the touchdown point. The system is shown in the illustration.

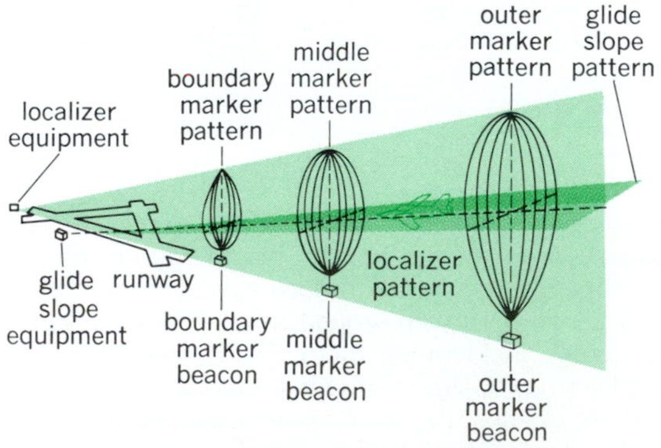

Guidance from fixed-beam low-approach system.

Since 1948, performance standards and recommended practices have been agreed on and formulated by the International Civil Aviation Organization (ICAO). In 1963, ICAO laid down three categories of ILS guidance criteria to standardize operations to lower approach visibility minima:

Category I. Operation down to a height of 60 m (200 ft) and 800 m (2600 ft) visibility.

Category II. Operation down to 30 m (100 ft) and 400 m (1300 ft) visibility.

Category III. Operation down to and along the surface of the runway unrestricted by cloud base and visibility conditions. *See* ELECTRONIC NAVIGATION SYSTEMS; PRECISION APPROACH RADAR (PAR). [F.H.T.]

Instrument science The systematically organized body of general concepts and principles underlying the design, analysis, and application of instruments and instrument systems.

Instruments are very diverse in function and form. They differ according to measurands, range of magnitudes and the dynamic variations of the measurand, required accuracy, and nature and environment of application. Equally diverse is the range of technologies that can be used to realize a particular instrument function. To enable this information to be handled in such a way that it can be usefully applied to the design, analysis, or application of instruments, it must be organized on the basis of a systematic framework of general concepts and principles. This framework is termed instrument science.

The basis of instrument science is: first, the consideration of instruments as members of the general class of information machines; and second, an analysis and synthesis of the instruments as systems, applying the principles of system engineering.

There exists a wide class of machines whose function is to acquire, process, and feed out information. This class includes measuring instruments, control apparatus, communication equipment, and computers. These machines function by the transformation of an input physical variable into an output variable in such a way that the output is functionally related to the input. Thus the output carries information about the input. This basic principle of functioning determines the general features of the analysis and synthesis of information machines. *See* COMPUTER; CONTROL SYSTEMS; ELECTRICAL COMMUNICATIONS.

Instruments and indeed other forms of information machines are conveniently analyzed and synthesized as systems. A system is a set of interconnected components functioning as a unit. There is a set of general principles and techniques for treating complex entities as systems. It is known as the systems approach, and is based on the decomposition of the complex whole into individual components. The individual components are considered in the first instance in terms of function rather than form.

Considering systems, such as instruments, as structures of functional building blocks makes clear that a very wide variety of instruments can be constructed from a much smaller variety of building blocks organized into a small variety of basic structures, such as a chain or loop connection. Further, it makes clear that apparently different systems—say, electrical and mechanical, or physical and biological—are essentially analogous. *See* INSTRUMENTATION; SYSTEMS ANALYSIS; SYSTEMS ENGINEERING. [L.Fi.]

Instrument transformer Electric transformers specifically designed for use in measurement and control circuits. The purpose of an instrument transformer is to convert primary voltages or currents to secondary values suitable for use with relays, meters, or other measuring equipment. A second purpose is to isolate the high-voltage primary circuit from the measurement circuit. Thus the construction of measuring devices and the protection of personnel using such devices are greatly simplified by the use of instrument transformers. For general discussion of transformers *see* TRANSFORMER.

There are two general types of instrument transformers: instrument potential transformers, which are used in voltage measurements, and instrument current transformers, which are used in current measurement. *See* ELECTRICAL MEASUREMENTS.

[I.F.K.]

Instrumental conditioning Learning based upon the consequences of behavior. For example, a rat may learn to press a lever when this action produces food. Instrumental or operant behavior is the behavior by which an organism changes its environment. The particular instances of behavior that produce consequences are called responses. Classes of responses having characteristic consequences are called operant classes; responses in an operant class operate on (act upon) the environment. For example, a rat's lever-press responses may include pressing with the left paw or with the right paw, sitting on the lever, or other activities, but all of these taken together constitute the operant class called lever pressing.

Consequences of responding that produce increases in behavior are called reinforcers. For example, if an animal has not eaten recently, food is likely to reinforce many of its responses. Whether a particular event will reinforce a response or not depends on the relation between the response and the behavior for which its consequences provide an opportunity. Some consequences of responding called punishers produce decreases in behavior. The properties of punishment are similar to those of reinforcement, except for the difference in direction.

The consequences of its behavior are perhaps the most important properties of the world about which an organism can learn, but few consequences are independent of other circumstances. Organisms learn that their responses have one consequence in one setting and different consequences in another. Stimuli that set the occasion on which responses have different consequences are called discriminative stimuli (these stimuli do not elicit responses; their functions are different from those that are simply followed by other stimuli). When organisms learn that responses have consequences in the presence of one but not another stimulus, their responses are said to be under stimulus control. For example, if a rat's lever presses produce food when a light is on but not when it is off and the rat comes to press only when the light is on, the rat's presses are said to be under the stimulus control of the light. *See* CONDITIONED REFLEX; SENSORY LEARNING. [A.C.Ca.]

Instrumentation Designing, manufacturing, and utilizing physical instruments or instrument systems for detection, observation, measurement, automatic control, automatic computation, communication, or data processing. The term instrumentation is also used for the ensemble of instruments and auxiliary equipment used in an experiment, test, or process, or in a plant, machine, or vehicle.

Since all branches of experimental science and technology depend on instrumentation, specialized instruments, with a corresponding body of knowledge and practice, have been developed separately in many fields. Thus chemical instrumentation, aeronautical instrumentation, medical instrumentation, optical instrumentation, and many other similar terms indicate areas of specialization in various industries or professions.

Instruments are sometimes classified according to the field or purpose of application, such as navigation instruments; according to their functions in instrument systems, such as detection, measurement, recording, computing, controlling, signal modification, or display; or according to the physical quantity or property that is to be measured or controlled by the instrument, such as flow, temperature, pressure, force, displacement, level, viscosity, acceleration, electrical quantities (voltage, current, resistance, and capacitance), or optical quantities (transmission, gloss, color, and brightness).

Other methods of describing or classifying instruments involve distinguishing operating principle or type (for example, an electron microscope of the transmission, emission, or scanning types), or specific range of applicability (for example, spectrometers for the infrared, visible, ultraviolet, or x-ray regions of electromagnetic radiation).

All instruments have inherent deficiencies. No instrument responds instantaneously to a change in the measurand; the lag is dependent on the natural frequency of the instrument system and its degree of damping (internal dissipation of energy). If the damping is too great or the natural frequency too low, the response lag may be excessive and the indication or output seriously in error during an interval of several times the natural period of the system. Conversely, if the damping is too small, the dynamic response may be excessive and the output may exhibit oscillations before attaining the desired value. In either event a detailed analysis of the instrument characteristics is essential to make appropriate corrections. *See* DAMPING.

Many instrument elements exhibit drift, a slow change in properties with time, resulting in a change in output. Many elements also exhibit the phenomenon of hysteresis; that is, they exhibit a response to increasing signals which is different from their response to decreasing signals. Environmental factors, such as temperature or vibration, may also affect performance of certain elements. The instrument maker must choose materials and operating principles which minimize these effects; the user must develop measurement techniques which either avoid these effects or correct for them. *See* HYSTERESIS.

An important aspect of instrumentation is the estimation of the uncertainties associated with measurements, and development of techniques to minimize them. Instruments in industrial use may be linked to the National Bureau of Standards through a calibration chain involving several other similar instruments, usually of successively better performance. *See* PHYSICAL MEASUREMENT. [W.A.Wi.]

Instrumentation amplifier A special-purpose linear amplifier, used for the accurate amplification of the difference between two (often small) voltages, often in the presence of much larger common-mode voltages, and having a pair of differential (usually high-impedance) input terminals, connected to sources V_{in1} and V_{in2}; a well-defined differential-mode gain A_{DM}; and a voltage output V_{out}, satisfying the relationship given in the equation below. It differs from an operational amplifier

$$V_{out} = A_{DM}(V_{in1} - V_{in2})$$

(op-amp), which ideally has infinite open-loop gain and must be used in conjunction with external elements to define the closed-loop transfer function. At one time built in discrete or hybrid form using operational amplifier and resistor networks, instrumentation amplifiers are readily available as inexpensive monolithic integrated circuits. Typical commercial amplifiers provide present gains of 1, 10, 100, and 1000. In some cases, the gain may be set to a special value by one or more external resistors. The frequency response invariably is flat, extending from 0 (dc) to an upper frequency of about 1 kHz to 1 MHz. *See* INTEGRATED CIRCUITS; OPERATIONAL AMPLIFIER; RESISTOR.

Instrumentation amplifiers are used to interface low-level devices, such as strain gages, pressure transducers, and Hall-effect magnetic sensors, into a subsequent high-level process, such as analog-to-digital conversion. *See* AMPLIFIER; DIFFERENTIAL AMPLIFIER; HALL EFFECT; PRESSURE TRANSDUCER; STRAIN GAGE. [B.Gi.]

Instrumented buoys Uncrewed floating structures (buoys) for the mounting, operation, data collection, and transmission of meteorological and oceanographic parameter-measuring systems. Instrumented buoys come in many different sizes and shapes, can be moored or drifting, and can transmit data back to ship or shore by radio directly or by satellite. Others record data in place for later recovery by ship and shore data reduction. Moored buoys can collect data in one location for long periods of time. Drifting buoys, driven by wind and current, collect data from large ocean areas over periods of up to 18 months.

Moored buoys are anchored in any water depth, and generally are used for long-term (months to years) routine measurement of wind speed and direction, barometric pressure, air temperature, sea-surface temperature, subsurface temperature, wave height and period, and wave spectral content. These buoys are moored by chain or by a combination of chain and synthetic line (nylon, for example) to bottom anchors. Instruments may be secured to or contained in the mooring lines at the depth of interest. Meteorological instruments generally are mounted at 33 ft (10 m) above the surface, a worldwide standard height; however, on smaller buoys they are mounted at 10–16 ft (3–5 m).

Drifting buoys are allowed to wander in response to wind, tide, and current. To be meaningful, the location of the data collection point needs to be known, and this information is provided by the satellite used to collect the data. Drifting buoys have developed as low-cost, easily deployed, lightweight (200–300-lb) platforms capable of making environmental measurements approaching the quality of moored buoys in data-sparse areas of the world's oceans. They can make routine measurements of wind speed and direction, barometric pressure, air and sea-surface temperature, and temperature versus depth to 2000 ft (600 m). With microprocessors as the hearts of these systems, like moored buoys, they can sample, process, and transmit data to a satellite. [W.O.R.]

Insulin The hormone, produced and secreted by the beta cells of the islets (insulae) of Langerhans of the pancreas, which regulates the use and storage of foodstuffs, especially the carbohydrates. When insufficient levels of insulin are produced, a state of diabetes occurs. Chemically insulin is a small, simple protein. The molecular weight is close to 6000, and the molecule consists of 51 amino acids, arranged in two linked polypeptide chains. Insulins from different species differ in their composition. These differences account for the fact that diabetics treated with animal insulins develop antibodies which may sometimes interfere with the action of the hormone. *See* CARBOHYDRATE METABOLISM; DIABETES; PANCREAS. [R.Lev.]

Intaglio (gemology) The name given to the type of carved gemstone in which the figure is engraved into the surface of the stone, rather than left in relief by cutting away the background, as in a cameo. Intaglios are almost as old as recorded history, for this type of carving was popular in ancient Egypt in the form of cylinders. Intaglios have been carved in a variety of gem materials, including emerald, crystalline quartz, hematite, and the various forms of chalcedony. *See* CAMEO. [R.T.L.]

Integer partition An integer partition of n is a nonincreasing sequence of positive integers whose sum is n; the number of these partitions is denoted by $p(n)$. Thus $p(4) = 5$, since 4, 3 + 1, 2 + 2, 2 + 1 + 1, and 1 + 1 + 1 + 1 all partition 4. No usable formula for $p(n)$ is known, but Euler discovered an elegant expression for the generating function: $p(x) = (1 - x)^{-1}(1 - x^2)^{-1} \cdots (1 - x^m)^{-1} \cdots$. Asymptotically, $p(n)$ obeys the formula below, where e^x is the exponential function which uses the base e. Thus $p(n)$ is eventually greater than any polynomial function of n but is eventually less than a^n for any $a > 1$.

$$p(n) \sim \left(\frac{1}{4n\sqrt{3}}\right) e^{x\sqrt{2n/3}}$$

Associated with an integer partition $n = a_1 + \cdots + a_k$ is an array of dots called its Ferrer's diagram, whose ith row contains a_i dots. Thus $4 = 3 + 1$ gives rise to the diagram. $\dot{\cdot}\cdot\cdot$.

A reflection of the diagram through its main diagonal interchanges rows and columns, converting $4 = 3 + 1$ into its dual partition $4 = 2 + 1 + 1$. This correspondence proves a number of identities, such as that the number of partitions of n into k parts equals the number of partitions of n which have greatest part k. *See* COMBINATORIAL THEORY; ENUMERATION OF ARRANGEMENTS. [T.Br.]

Integral equation An equation of the form typified by Eq. (1). The major problem is to decide when there is a function $\phi(x)$ which is a solution to the equation. Equations such as (1) arise from the analysis of ordinary differential equations. Integral equation (1) is an equation of the first kind if $a(x) \equiv 1$, and an equation of the second kind if $a(x) \equiv 0$.

$$\int_a^b K(x,y,\phi(y))dy + f(x) = a(x)\phi(x) \quad (1)$$

If $K(x,y,\phi) = 0$ for $x \leqslant y \leqslant b$, Eq. (1) is called a Voltera equation. Under mild assumptions the Volterra equation of the second kind can be solved by the method of successive approximations (also known as the method of Picard).

A special case of some interest is that of linear integral equations. The function $K(x,y,\phi)$ is a linear function of ϕ. The linear equations of the first and second kind are shown in Eqs. (2) and (3) respectively. The function $K(x,y)$ is called the kerne, and the complex number λ is called the parameter. The equation of the second kind, Eq. (3), is homogeneous if $f(x) \equiv 0$. In typical cases, the homogeneous equations will have only the trivial solution $\phi(x) \equiv 0$. For some values of the parameter λ, however, there will be nontrivial solutions. Such a value of the parameter λ is called a characteristic value for K, and the corresponding function ϕ is called a characteristic function for K. These concepts are related to corresponding concepts in linear algebra and operator theory. *See* EIGENVALUE (QUANTUM MECHANICS); MATRIX THEORY; OPERATOR THEORY.

$$f(x) = \int_a^b K(x,y\phi)(x)\, dy \quad (2)$$

$$\phi(x) = f(x) + \lambda \int_a^b K(x,y\phi)(y)\, dy \quad (3)$$

The integral equation, Eq. (2) or (3), is called a Fredholm equation, and $K(x,y)$ is called a Fredholm kernel if $\|K\| < \infty$, where $\|K\|$, the norm of K, is defined by Eq. (4). This is certainly true if K is continuous or even bounded. Even some infinite discontinuities are allowable. Fredholm equations of the second kind for which the parameter satisfies the inequality $|\lambda|\ \|K\| < 1$ can be shown to have unique solutions by the method of Picard. As a result there are no characteristic values for which $|\lambda| \leqslant 1/\|K\|$. In general, for any constant $M > 0$ there are only finitely many characteristic values which satisfy $|\lambda| \leqslant M$. Further, only finitely many linearly independent characteristic functions are associated with each characteristic value. *See* DIFFERENTIAL EQUATION; INTEGRAL TRANSFORM; INTEGRATION. [J.P.]

$$\|K\|^2 = \int_a^b \int_a^b |K(x,y)|^2 dxdy \quad (4)$$

Integral transform An integral relation between two classes of functions. For example, a relation such as Eq. (1) is said to define an integral transform. More generally, the integral may be a multiple integral, and the functions f, G, and ϕ

$$f(x) = \int_{-\infty}^{\infty} G(x,y)\phi(y)\, dy \quad (1)$$

may depend on a larger number of variables. Equation (1) is thought of as transforming a whole class, or space, of functions $\phi(y)$ into another class of functions $f(x)$. The function $G(x,y)$ is the kernel of the transform. One of the important uses of such a transform is based on the fact that a problem posed in one of the two spaces in question may be more easily solved in the other. For example, a differential equation to be solved for the function $\phi(y)$ may become an algebraic equation for the unknown function $f(x)$. *See* LAPLACE TRANSFORM.

The two basic problems for any integral transform are inversion and representation. In inversion the aim is to recover $\phi(y)$ from $f(x)$, the kernel $G(x,y)$ being known. That is, Eq. (1) is thought of as an integral equation (of the first kind) to be solved for the unknown function $\phi(y)$. A means of calculating $\phi(y)$ from $f(x)$ is called an inversion formula, and in its presence the transform achieves maximum utility, since explicit passage from each space to the other is thus assured. In representation the question is which functions $f(x)$ may be written or represented in the form (1). That is, one asks which functions $f(x)$ will make Eq. (1) solvable for $\phi(y)$. Usually this problem becomes more tractable when the solutions $\phi(y)$ are restricted to some subspace such as the class of positive or bounded functions. *See* INTEGRAL EQUATION.

An important special case of Eq. (1) is the convolution transform, when the kernel is a function of $(x - y)$, Eq. (2). An equivalent form of Eq. (2) is Eq. (3), since the change of vari-

$$f(x) = \int_{-\infty}^{\infty} G(x - y)\phi(y)\, dy \qquad (2)$$

$$F(x) \int_{0}^{\infty} K(xy)\Phi(y)\, dy \qquad (3)$$

able $x = e^t$, $y = e^{-u}$ carries Eq. (3) into Eq. (2) after a suitable change in notation. *See* INTEGRATION. [D.V.W.]

Integrated-circuit filter An electronic filter implemented as an integrated circuit, as contrasted with filters made by interconnecting discrete electrical components. The design of an integrated-circuit filter (also called simply an integrated filter) is constrained by the unavailability of certain types of components, such as piezoelectric resonators, that are often valuable in filtering. However, integrated filters can benefit from small size, close integration with other parts of a system, and the low cost of manufacturing very complex integrated circuits.

Filters have many applications. An important one is to smooth signal waveforms sufficiently to allow accurate sampling or to interpolate smoothly between given samples of a signal. Since the analog-to-digital converters that sample signals are usually made as integrated circuits (chips), it is often convenient to put the associated filters on the same chip. *See* ANALOG-TO-DIGITAL CONVERTER.

Passive filters, made by interconnecting inductors and capacitors, are not easily integrated because integrated-circuit inductors are usually of poor quality. This problem is less serious at frequencies above 1 GHz, so microwave filters can be passive. *See* MICROWAVE FILTER.

Because amplifiers are very cheap in integrated-circuit technology, active filters are widely implemented. The five main types of active filter—active-RC, MOSFET-C, transconductance-C, switched-capacitor (or switched-C), and active-RLC—are distinguished by their frequency-sensitive components. The switched-capacitor filters operate on samples of signals, while the other types operate without sampling (in continuous time). There is also a trend toward digital filters, which are easily integrated but require that analog signals be converted to digital form, which in turn requires filtering. *See* DIGITAL FILTER.

Discrete active-RC filters were widely used in the 1970s, and modern integrated filters are derived from them. The frequency-sensitive mechanism in active-RC filters is the charging of a capacitor C through a resistor R, giving a characteristic frequency $\omega_0 = 1/RC$ radians per second, at which the impedances of the resistor and capacitor are equal. Unfortunately, integrated-circuit manufacturing techniques do not control the product RC at all accurately, with variations of 20–50% being possible. This limits active-RC filtering to those applications where accuracy is unimportant, where external passive components are tolerable, or where tuning circuitry is available.

MOSFET-C filters replace the resistors of an active-RC filter with metal-oxide-semiconductor (MOS) transistors, in which a conducting channel along the surface can be enhanced or depleted by applying an electrical field from a gate electrode, thereby changing the resistance of the channel. The result is a tunable variant of an active-RC filter. *See* TRANSISTOR.

Transconductance-C filters combine the functions of the amplifier and the simulated resistor into a transconductance amplifier, whose output current (rather than output voltage) is proportional to its input voltage. Transconductance amplifiers can be very simple and hence are capable of high-frequency operation (up to approximately 1 GHz) but tend to have poor linearity when designed for high speeds.

A technique known as active-RLC filtering combines the ideas of active filtering with the use of physical inductors (made as spirals of metallization on the top layer of the chip). In this method, amplifiers, connected to simulate negative resistors, are used to enhance the performance of the inductors, whose losses can be modeled (to a first approximation) as being caused by a parallel positive resistance.

The primary advantage of switched-capacitor filters is that they can be very accurate, since critical frequencies are determined by the product of a clock frequency and a ratio of capacitors (rather than a single capacitor). Switched-capacitor filters are probably the most prevalent integrated filters. Most telephone systems, for example, use them to smooth signals before sampling them for digital transmission. *See* ELECTRIC FILTER; INTEGRATED CIRCUITS; SWITCHED CAPACITOR. [M.Sn.]

Integrated circuits Miniature electronic circuits produced within and upon a single semiconductor crystal, usually silicon. Integrated circuits range in complexity from simple logic circuits and amplifiers, about 1/20 in. (1.3 mm) square, to large-scale integrated circuits up to about 1/3 in. (8 mm) square. They contain tens of thousands of transistors and other components which provide computer memory circuits and complex logic subsystems such as microcomputer central processor units. *See* SEMICONDUCTOR.

Integrated circuits consist of the combination of active electronic devices such as transistors and diodes with passive components such as resistors and capacitors within and upon a single semiconductor crystal. The construction of these elements within the semiconductor is achieved through the introduction of electrically active impurities into well-defined regions of the semiconductor.

Generally, integrated circuits are not straightforward replacements of electronic circuits assembled from discrete components. They represent an extension of the technology by which silicon planar transistors are made. Because of this, transistors or modifications of transistor structures are the primary devices of integrated circuits. Methods of fabricating good-quality resistors and capacitors have been devised, but the third major type of passive component, inductors, must be simulated with complex circuitry or added to the integrated circuit as discrete components. *See* TRANSISTOR.

Integrated circuits can be classified into two groups on the basis of the type of transistors which they employ: bipolar integrated circuits, in which the principal element is the bipolar junction transistor; and metal oxide semiconductor (MOS) integrated circuits, in which the principal element is the MOS transistor. Both depend upon the construction of a desired pattern of electrically active impurities within the semiconductor body, and upon the formation of an interconnection pattern of metal films on the surface of the semiconductor. Bipolar circuits are generally used where highest logic speed is desired, and MOS for largest-scale integration or lowest power dissipation.

Bipolar integrated circuits. Complete digital integrated circuits generally contain tens to hundreds of inverters and gates interconnected as counters, arithmetic units, and other building blocks. Hundreds of such circuits may be fabricated on a single slice of silicon crystal. This feature of planar technology—simultaneous production of many circuits—is responsible for the economic advantages and wide use of integrated circuits.

A bipolar integrated circuit, representative of the simplest logic function, is illustrated schematically in Fig. 1. An inverter requires only a transistor and register, shown in cross section. Both the transistor and the resistor are formed within a separate pocket of *n*-type semiconductor surrounded on all sides by *p*-type regions. When the inverter circuit is operated, a reverse bias develops between the *n*-type pocket and its surroundings. The depletion region separating the *n*- and *p*-type regions has a very high resistance; consequently, the individual transistors and resistors are electrically isolated from each other, even though both of them have been formed within the same semiconductor crystal. *See* JUNCTION TRANSISTOR.

Integrated circuits based on amplifiers are called linear because amplifiers usually exhibit a linearly proportional response to input signal variations. However, the category includes memory sense amplifiers, combinations of analog and digital processing functions, and other circuits with nonlinear characteristics. Some digital and analog combinations include analog-to-digital converters, timing controls, and modems (data communications modulator-demodulator units). Linear circuits cannot yet match the power and frequency range of discrete transistors.

In the continuing effort to increase the complexity and speed of digital circuits, and the performance characteristics and versatility of linear circuits, a significant role has been played by the discovery and development of new types of active and passive semiconductor devices which are suitable for use in integrated circuits. Among these devices is the *pnp* transistor which, when used in conjunction with the standard *npn* transistors described earlier, lends added flexibility to the design of integrated circuits.

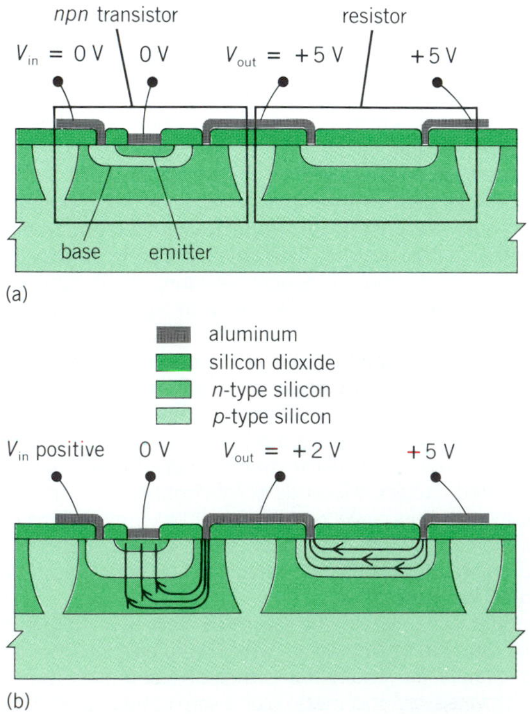

Fig. 1. Operation of bipolar inverter circuit. (*a*) Input voltage V_{in} is zero. (*b*) Positive input voltage applied. Arrows indicate direction of current flow.

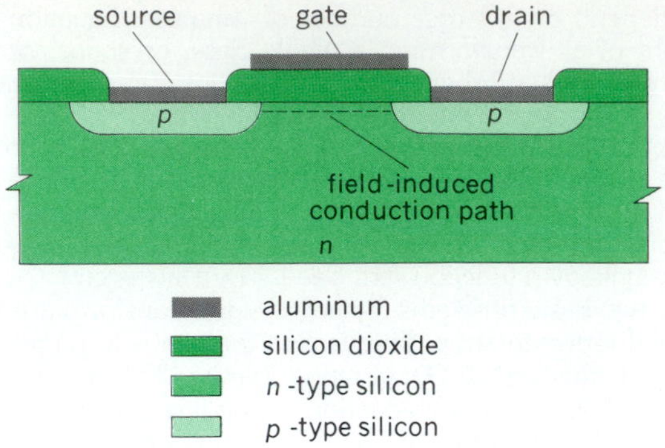

Fig. 2. Diagram of *p*-channel MOS field-effect transistor.

MOS integrated circuits. The other major class of integrated circuits is called MOS because its principal device is a metal oxide semiconductor field-effect transistor (MOSFET). It is more suitable for large-scale integration (LSI) than bipolar circuits because MOS transistors are self-isolating and can have an average size of less than a millionth of a square inch (6×10^{-4} mm^2). This has made it practical to use over 100,000 transistors per circuit. Because of this high density capability, MOS transistors are used extensively for high density RAMs, ROMs, and microprocessors.

Several major types of MOS device fabrication technologies have been developed. They are: metal-gate *p*-channel MOS, which uses aluminum for electrodes and interconnections (Fig. 2); silicon-gate *p*-channel MOS, employing polycrystalline silicon for gate electrodes and the first interconnection layer; *n*-channel MOS, which is usually silicon gate; and complementary MOS, which employs both *p*-channel and *n*-channel devices. Silicon gate *n*-channel and complementary MOS are the dominant technologies.

The basic principles of MOS technology can be illustrated with the metal-gate *p*-channel MOSFET in Fig. 2. It consists of two *p*-type diffused regions (such as can be obtained by the process used to form the base region of the *npn* transistors described earlier), separated by an *n*-type region which is under control of an electrode, called the gate. When there is no voltage applied to the gate, the two *p*-type regions, called source and drain, are electrically insulated from each other by the *n*-type region which surrounds them. When, however, a negative voltage is applied to the gate, the electric field induces a very thin *p*-type region at the surface of the silicon. This thin *p*-type region now connects the source and the drain, permitting the passage of current between them. The path is called a channel.

Both conceptually and structurally the MOS transistor is a much simpler device than the bipolar transistor. What kept this simple device from commercial utilization until 1964 is the fact that it depends on the properties of the semiconductor surface for its operation, while the bipolar transistor depends principally on the bulk properties of the semiconductor crystal. Hence MOS transistors became practical only when understanding and control of the properties of the oxidized silicon surface had been perfected to a very great degree. [R.Bu.; N.Be.; Y.E.-M.]

Integrated optics The study of optical devices, singly and in combinations, that are based on light transmission in waveguides, that is, structures that confine the propagating light to a region with one very small dimension (or sometimes two), of the order of the wavelength of the light. The principal motivation for these studies is to enable the combination of individual devices thus miniaturized, through waveguides or other means, into a functional optical system mounted on a small substrate. The resulting system is called an integrated optical circuit (IOC) by analogy with the semiconductor type of integrated circuit. An integrated optical circuit could, for example, include a laser, switches, polarizers, modulators, detectors, and so forth. An important use envisioned for integrated optical circuits is in connection with optical communications through the medium of glass fibers, which are themselves waveguides. Integrated optical circuits could be used in such a system as optical transmitters, repeaters, and receivers. *See* LASER; OPTICAL COMMUNICATIONS; OPTICAL FIBERS; OPTICAL MODULATORS; WAVEGUIDE. [E.M.Co.]

Integrated services digital network (ISDN) A generic term referring to the integration of communications services transported over digital facilities such as wire pairs, coaxial cables, optical fibers, microwave radio, and satellites. ISDN provides end-to-end digital connectivity between any two (or more) communications devices. Information enters, passes through, and exits the network in a completely digital fashion.

Since the introduction of pulse-code-modulation (PCM) transmission in 1962, the worldwide communications system has been evolving toward use of the most advanced digital technology for both voice and nonvoice applications. Pulse-code modulation is a sampling technique which transforms a voice signal with a bandwidth of 4 kHz into a digital bit stream, usually of 64 kilobits per second (kbps). *See* PULSE MODULATION.

Many aspects of telecommunications are improved with digital technology. For example, digital technology lends itself to very large-scale integration (VLSI) technology and its associated benefits of miniaturization and cost reduction. In addition, computers operate digitally. Digital transport provides for human-to-human, computer-to-computer, and human-to-computer interactions. The ISDN is capable of transporting voice, data, graphics, text, and even video information over the same equipment. *See* DATA COMMUNICATIONS; DIGITAL COMPUTER; INTEGRATED CIRCUITS.

The customer has access to a wide spectrum of communications services by way of a single access link. This is in contrast to existing methods of service access, which segregate services into specialized lines.

Associated with integrated access and ISDN is the concept of a standard interface. The objective of a standard interface is to allow any ISDN terminal to be plugged into any ISDN interface, resulting in terminal portability, flexibility, and ease in operation. *See* ELECTRICAL COMMUNICATIONS. [R.M.Wi.; A.E.J.]

Integration An operation of the infinitesimal calculus which has two aspects. The roots of one go back to antiquity, for Archimedes and other Greek mathematicians used the "method of exhaustion" to compute areas and volumes. A simple example of this is the approximation to the area of a circle obtained by inscribing a regular polygon of known area, and then repeatedly doubling the number of sides. The areas of the successive polygons are computable with the help of elementary geometry. The limit of the sequence of these areas gives the area of the circle. The area of each polygon can be regarded as being made up of the sum of the areas of triangles with vertices at the center of the circle, and so the process described is a constructive definition of an integral which is the limit of a sum. Modern definitions of integrals as limits of sums are discussed in this article.

The other aspect of integration is the process of finding antiderivatives, that is, for a given function $f(x)$ to find another function $g(x)$ whose derivative is $f(x)$. This aspect is related to the first by the fundamental theorem of integral calculus, so both processes are called integration.

In the early 19th century A. L. Cauchy gave a clear-cut definition of the definite integral for continuous functions and a proof of its existence. Later, G. F. B. Riemann discussed the integral for discontinuous functions and gave a necessary and sufficient condition for its existence. Thus the most generally used definition of the integral as the limit of a sum has come to be called the Riemann integral.

The precise definition of the Riemann integral for a real function f of one real variable x on a finite interval $a \leqq x \leqq b$ may be formulated as follows. Let P be a partition of the interval $[a,b]$ into n subintervals by points t_i, where $t_{i-1} < t_i$, $t_0 = a$, $t_n = b$, and consider a sum S of the form of Eq. (1), where

$$S = \sum_{i=1}^{n} f(x_i)(t_i - t_{i-1}) \qquad (1)$$

$t_{i-1} \leqq x_i \leqq t_i$. The sum S depends not only on the partition P but on the choice of the intermediate points x_i. It may happen that the sum S approaches a definite limit I when the maximum of the numbers $(t_i - t_{i-1})$ tends to zero, and in this case I is called the Riemann integral (or the definite integral) of f from a to b, and is denoted by the Leibnitzian symbol (2).

$$\int_a^b f(x)dx \qquad (2)$$

Also, f is said to be integrable on $[a,b]$. When f is a continuous function with positive values on the interval $[a,b]$, the integral has a simple geometrical interpretation as the area bounded by the x axis, the ordinates $x = a$ and $x = b$, and the graph of $y = f(x)$ (Fig. 1).

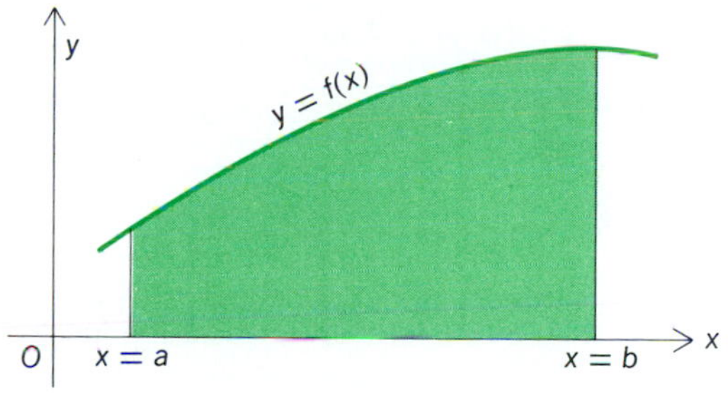

Fig. 1. Graph of $y = f(x)$.

It can be proved that a function f is integrable on $[a,b]$ if and only if the following two conditions are satisfied: f is bounded on $[a,b]$; and the set of points where f is discontinuous can be enclosed in a series (possibly infinite) of intervals, the sum of whose lengths is arbitrarily near to zero.

To develop the fundamental theorem of integral calculus let u be a variable in the interval $[a,b]$, on which f is integrable; then formula (3) defines a function of u which may be noted by

$$\int_a^u f(x)dx \qquad (3)$$

$I(u)$. If f is continuous on $[a,b]$, then f is integrable, and it is also true that $I(u)$ has a derivative $I'(u) = f(u)$. Now let h be any antiderivative of f, that is, $h'(u) = f(u)$ on $[a,b]$. Then $I'(u) - h'(u) = 0$, so $I(u) - h(u) = \text{constant} = -h(a)$, by the theorem of the mean for derivatives and the fact that $I(a) = 0$, so relation (4)

$$I = I(b) = h(b) - h(a) \qquad (4)$$

can be written. This is the fundamental theorem of integral calculus, and it shows that definite integrals may be calculated by the process of finding antiderivatives. For this reason antiderivatives are frequently called indefinite integrals and denoted by $\int f(x)\,dx$, and special methods of finding indefinite integrals for frequently occurring functions occupy a large part of elementary calculus. The standard notation for an indefinite integral of $f(x)$ is formula (5).

$$\int f(x)dx \tag{5}$$

The term improper integral refers to an extension of the notion of definite integral to cases in which the integrand is unbounded or the domain of integration is unbounded.

The concept called the Riemann integral can be extended to functions of several variables. The case of a function of two variables illustrates sufficiently the additional features which arise. To begin with, let $f(x,y)$ denote a real function defined on a rectangle of the form (6). Let P be a partition of R into n

$$R{:}a \leqq x \leqq b \qquad c \leqq y \leqq d \tag{6}$$

nonoverlapping rectangles R_i with areas A_i, and let (x_i,y_i) be a point of R_i. Define S by Eq. (7). In case the sum S tends to a

$$S = \sum_{i=1}^{n} f(x_i,y_i)A_i \tag{7}$$

definite limit I when the maximum diagonal of a rectangle R_i tends to zero, then f is said to be integrable over R, and the limit I is called the Riemann integral of f over R. It will be denoted here by the abbreviated symbol $\iint_R f$.

To define the integral of a function $f(x,y)$ over a more general domain D where it is defined, suppose that D is enclosed in a rectangle R, and define $F(x,y)$ by Eqs. (8). Then f is said to be

$$\begin{aligned} F(x,y) &= f(x,y) \text{ in } D \\ F(x,y) &= 0 \text{ outside } D \end{aligned} \tag{8}$$

integrable over D in case F is integrable over R, and Eq. (9) holds, by definition.

$$\iint_D f = \iint_R F \tag{9}$$

When the function $f(x,y)$ is continuous on D, and D is defined by inequalities of the form (10), where the functions $\alpha(x)$

$$a \leqq x \leqq b \qquad \alpha(x) \leqq y \leqq \beta(x) \tag{10}$$

and $\beta(x)$ are continuous, the double integral of f over D always exists, and may be represented in terms of two simple integrals by Eq. (11). In many cases, this formula makes possible the evaluation of the double integral.

$$\iint_D f = \int_a^b \left[\int_{\alpha(x)}^{\beta(x)} f(x,y)\, dy \right] dx \tag{11}$$

In the study of the "space" of real functions defined, for example, on the interval $a \leqq x \leqq b$, it is frequently useful to take formula (12) as the distance between the functions f

$$\int_a^b |f(x) - g(x)|\, dx \tag{12}$$

and g. This distance already has a meaning when f and g are Riemann-integrable, that is, bounded and not too discontinuous, in the sense specified for the Riemann integral. There is no generally useful extension of the concept of integral to apply to all real functions on $[a,b]$, but it is desirable to extend it to apply to the functions obtained from the continuous ones by certain limiting processes. In particular, it is desirable to have correspond to each sequence (f_n) of functions satisfying the Cauchy condition for convergence in terms of the distance (12), namely Eq. (13), a function g which is integrable (in the extended sense), and for which Eq. (14) holds. An extended

$$\lim_{\substack{m \to \infty \\ n \to \infty}} \int_a^b |f_m(x) - f_n(x)|\, dx = 0 \tag{13}$$

$$\lim_{n \to \infty} \int_a^b |f_n(x) - g(x)|\, dx = 0 \tag{14}$$

definition of integral having this property was given by H. L. Lebesgue.

F. Riesz devised an equivalent definition which can be stated quite simply, at least for the case of a bounded function $g(x)$. As a first definition, a point set S in the interval $[a,b]$ has measure zero in case it can be enclosed in a sequence (finite or infinite) of intervals, the sum of whose lengths is arbitrarily small. Also, a step function $f(x)$ is defined as one which is constant on each interval of a partition of $[a,b]$, as in Fig. 2. The Riemann

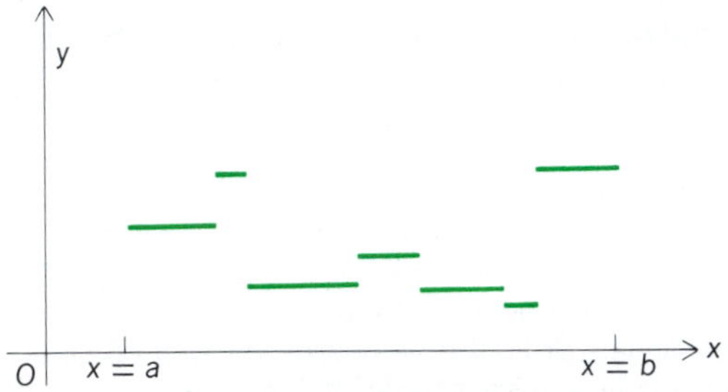

Fig. 2. Step function *f*(*x*).

integral of a step function is expressible as a finite sum. Then a bounded function $g(x)$ is integrable in Lebesgue's sense in case it is the limit of a uniformly bounded sequence of step functions $f_n(x)$, at each point of $[a,b]$ except those in a set S with measure zero, and by definition Eq. (15) can be written. In case

$$\int_a^b g(x)\, dx = \lim_{n \to \infty} \int_a^b f_n(x)\, dx \tag{15}$$

the step functions f_n in Eq. (15) are replaced by Lebesgue-integrable functions forming a uniformly bounded sequence, no new functions g are obtained.

The integral of Lebesgue is also defined for unbounded functions, but the points of infinite discontinuity do not need to be considered one by one. For each function $g(x)$ and each positive integer N let $g_N(x)$ denote the lesser of $g(x)$ and N. If each $g_N(x)$ is Lebesgue-integrable in the sense already defined, expression (16) forms a nondecreasing sequence, and so if ex-

$$\int_a^b g_N(x)\, dx \tag{16}$$

pression (16) is bounded, it tends to a finite limit, which is taken as the value of integral (17).

$$\int_a^b g(x)\, dx \tag{17}$$

If $\alpha(x)$ is a fixed function defined on the interval $[a,b]$, the sum S defined by Eq. (1) may be replaced by that in Eq. (18).

$$S = \sum_{i=1}^{n} f(x)_i\, [\alpha(t_i) - \alpha(t_{i-1})] \tag{18}$$

Then if the limit I of S exists in this case, it is called the Stieltjes integral of f with respect to α, and is denoted by integral (19).

$$\int_a^b f(x)\, d\alpha(x) \tag{19}$$

It has many of the properties of the Riemann integral, especially in case the function α is nondecreasing. *See* CALCULUS; FOURIER SERIES; SERIES. [L.M.G.]

Integument The skin, or outer covering, of the body. It is extremely diverse in its structure, varying both within species and from one species to another. It always consists of an outer epidermis and an innermost dermis (or corium), which rests on fatty connective tissue surrounding bone and muscle. The entire skin has a primary barrier function in protecting the soft internal tissues from physical trauma and bacteriological invasion, and in maintaining the internal salt and fluid levels. There are a variety of secondary functions which vary from species to species, namely, thermal insulation, sensory reception, scent secretion, respiration, attack and defense (claws, thick skins), and protective coloration. [P.F.M.]

Intelligence General mental ability due to the integrative and adaptive functions of the brain that permit complex, unstereotyped, purposive responses to novel or changing situations, involving discrimination, generalization, learning, concept formation, inference, mental manipulation of memories, images, words and abstract symbols, eduction of relations and correlates, reasoning, and problem solving.

Intelligence tests are diverse collections of tasks (or items), graded in difficulty. The person's performance on each item can be objectively scored (for example, pass or fail); the total number of items passed is called the raw score. Raw scores are converted to some form of scaled scores which can be given a statistical interpretation.

The first practical intelligence test for children, devised in 1905 by the French psychologist Alfred Binet, converted raw scores to a scale of "mental age," defined as the raw score obtained by the average of all children of a given age. Mental age (MA) divided by chronological age (CA) yields the well known intelligence quotient or IQ. When multipled by 100 (to get rid of the decimal), the average IQ at every age is therefore 100, with a standard deviation of approximately 15 or 16. Because raw scores on mental tests increase linearly with age only up to about 16 years, the conversion of raw scores to a mental-age scale beyond age 16 must resort to statistical artifices. Because of this problem and the difficulty of constructing mental-age scales which preserve exactly the same standard deviation of IQs at every age, all modern tests have abandoned the mental-age concept and the calculation of IQ from the ratio of MA to CA. Nowadays the IQ is simply a standardized score with a population mean of 100 and a standard deviation (σ) of 15 at every age from early childhood into adulthood. The middle 50%, considered "average," fall between IQs of 90 and 110. IQs below 70 generally indicate "mental retardation," and above 130, "giftedness." [A.R.J.]

Intelligent machine Any machine that can accomplish its specific task in the presence of uncertainty and variability in its environment. The machine's ability to monitor its environment allowing it to adjust its actions based on what it has sensed, is a prerequisite for intelligence. The term intelligent machine is an anthropomorphism in that intelligence is defined by the criterion that the actions would appear intelligent if a person were to do it. A precise, unambiguous, and commonly held definition of intelligence does not exist.

Examples of intelligent machines include industrial robots equipped with sensors, computers equipped with speech recognition and voice synthesis, self-guided vehicles relying on vision rather than on marked roadways, and so-called smart weapons, which are capable of target identification. These varied systems include three major subsystems: sensors, actuators, and control. The class of computer programs known as expert systems is included with intelligent machines, even though the sensory input and output functions are simply character-oriented communications. The complexity of control and the mimicking of human deductive and logic skills makes expert systems central in the realm of intelligent machines. *See* COMPUTER VISION; EXPERT SYSTEMS; GUIDANCE SYSTEMS; ROBOTICS; SPEECH RECOGNITION; VOICE RESPONSE.

Since the physical embodiment of the machine or the particular task performed by the machine does not mark it as intelligent, the appearance of intelligence must come from the nature of the control or decision-making process that the machine performs. Given the centrality of control to any form of intelligent machine, intelligent control is the essence of an intelligent machine. The control function accepts several kinds of data, including the specification for the task to be performed and the current state of the task from the sensors. The control function then computes the signals needed to accomplish the task. When the task is completed, this also must be recognized and the controller must signal the supervisor that it is ready for the next assignment (see illustration). *See* ADAPTIVE CONTROL; CONTROL SYSTEMS.

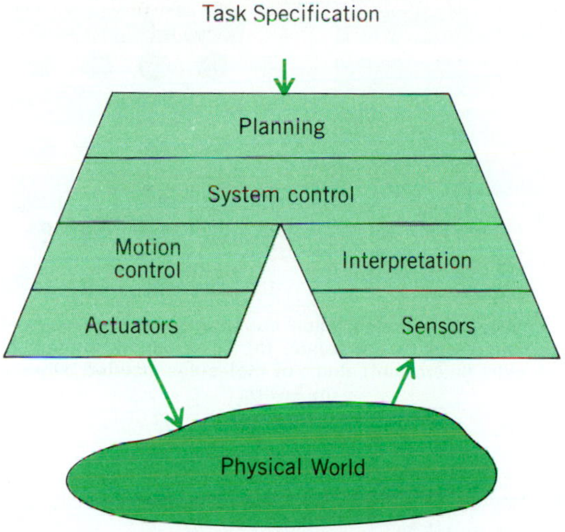

Flow of information and data in a typical intelligent machine.

Automatic, feedback, or regulatory systems such as thermostats, automobile cruise controls, and photoelectric door openers are not considered intelligent machines. Several important concepts separate these simple feedback and control systems from intelligent control. While examples could be derived from any of the classes of intelligent machines, robots will be used here to illustrate five concepts that are typical of intelligent control. (1) An intelligent control system typically deals with many sources of information about its state and the state of its environment. (2) An intelligent control system can accommodate incomplete or inconsistent information. (3) Intelligent control is characterized by the use of heuristic methods in addition to algorithmic control methods. (A heuristic is a rule of thumb, a particular solution or strategy to be used for solving a problem that can be used for only very limited ranges of the input parameters.) (4) An intelligent machine has a built-in knowledge base that it can use to deal with infrequent or unplanned events. (5) An algorithmic control approach assumes that all relevant data for making decisions is available. [J.F.Ja.]

Intercalation compounds Compounds that are formed by inserting extra atoms or molecules into a host structure, without disrupting the strong chemical bonds of the host material. For example, graphite is such a host material, in which the carbon atoms are covalently bonded to form layers one atom thick. The layers are held together only weakly by van der Waals forces. Lithium atoms can be inserted between the layers by immersing the graphite in liquid lithium at 400–750°F (200–400°C) to form a new structure with alternating, one-atom-thick layers of carbon and lithium (see illustration). Intercalation compounds that consist of a periodic structure of n layers of host material and one layer of intercalant can be formed from many layered hosts. These compounds are called nth-stage intercalation compounds. The example in the illustra-

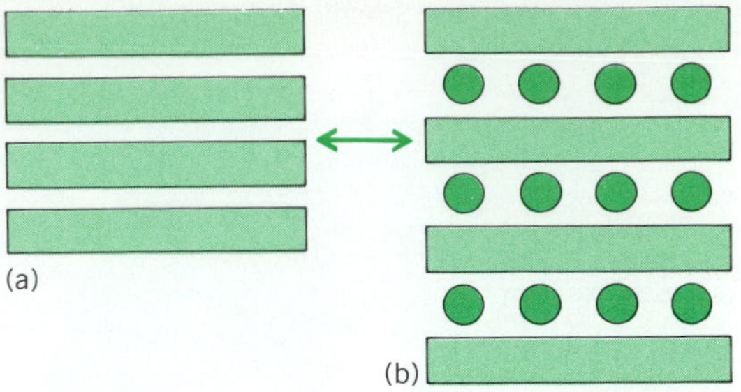

Formation of an intercalation compound. (*a*) Layered host compound, such as graphite. (*b*) Intercalation compound, with extra (intercalant) atoms or molecules inserted between the layers.

tion is a first-stage compound. *See* CHEMICAL BONDING; CRYSTAL; INTERMOLECULAR FORCES.

Since no strong chemical bonds are broken in the intercalation reaction, the kinetic barrier to reaction is usually small; that is, intercalation reactions proceed rapidly at low temperature, even at room temperature. Usually, the intercalation process is reversible, even at room temperature, if a chemical or electrochemical driving force is applied. This reversibility, coupled with the high rate of reaction that is possible at room temperature due to the large intercalant diffusivity between the layers, makes possible rechargeable high-energy-density batteries based upon the intercalation reaction. *See* BATTERY.

Hosts for insertion reactions include not only layered compounds (such as graphite, micas, clays, and transition-metal dichalcogenides) but also linear compounds, in which chains having strong bonding only along one direction are bonded weakly to each other, and also framework compounds, which have a structure with open channels into which intercalants smaller than the channel can be inserted. In the last case, the compounds are more frequently called insertion compounds. The physical properties of the host or the intercalant are often greatly changed by forming an intercalation compound. [F.J.D.]

Intercommunicating system A means of providing communication between two or more parties requiring either personal or business contact; also called an intercom. The calling parties are most often within the confines of the same building. The intercom may be a simple two-station system, such as the home "baby-sitter," or a multistation system, which is usually wired during construction of a home or professional building and provides, among other features, music, station bridging, exclusion, and push-to-talk. [W.Sch.]

Interface of phases The boundary between any two phases. Among the three phases, gas, liquid, and solid, five types of interfaces are possible: gas-liquid, gas-solid, liquid-liquid, liquid-solid, and solid-solid. The abrupt transition from one phase to another at these boundaries, even though subject to the kinetic effects of molecular motion, is statistically a surface only one or two molecules thick.

A unique property of the surfaces of the phases that adjoin at an interface is the surface energy which is the result of unbalanced molecular fields existing at the surfaces of the two phases. Within the bulk of a given phase, the intermolecular forces are uniform because each molecule enjoys a statistically homogeneous field produced by neighboring molecules of the same substance. Molecules in the surface of a phase, however, are bounded on one side by an entirely different environment, with the result that there are intermolecular forces that then tend to pull these surface molecules toward the bulk of the phase. A drop of water, as a result, tends to assume a spherical shape in order to reduce the surface area of the droplet to a minimum. *See* ADSORPTION; FLOTATION; FOAM; PHASE EQUILIBRIUM; SURFACE TENSION. [W.H.S.]

Interference filter An optical filter in which the wavelengths that are not transmitted are removed by interference phenomena rather than by absorption or scattering. In addition to being able to duplicate most of the spectral characteristics of absorption color filters, these devices can be made to transmit a very narrow band of wavelengths. They can thus be used as monochromators to examine a radiation source at the wavelength of a single spectrum line. For example, the solar disk can be observed in light of the hydrogen line Hα and thus the distribution of excited hydrogen over the disk can be determined. Most narrow-band interference filters are based on the Fabry-Perot interferometer. *See* INTERFERENCE OF WAVES; INTERFEROMETRY. [B.H.Bi.]

Interference microscope A microscope used for visualizing and measuring differences in phase or optical path in transparent or reflecting specimens. It is closely allied to the phase-contrast microscope. *See* PHASE-CONTRAST MICROSCOPE.

In the phase-contrast microscope, the final image is produced by separating the incident wave and the wave diffracted by the object, introducing a phase difference between them, and then recombining the altered waves. The final resultant wave has a lower amplitude than the incident wave, and a transparent object thus becomes visible. The same result is achieved in interference microscopy by a different method. The incident and diffracted waves are not separated, but interference is produced between the transmitted wave and another wave which originates from the same source. In Fig. 1, A represents the incident wave, B the wave transmitted by a perfectly transparent object. Since no energy is absorbed, B has the same amplitude as A, but is slightly delayed or altered in phase relative to A. The interfering wave is represented by the broken line. According to the Abbe theory the final image is formed by the interference or summation of all waves that pass through the optical system. If the interfering wave and the transmitted wave B are added together algebraically, the resultant wave in Fig. 1 will have a lower amplitude than the incident wave A, so that the transparent object will now appear to absorb light. The resultant wave can be changed by altering the phase and amplitude of the interfering wave, thus enabling the appearance of the image to be varied. Since two waves can only interfere if they have the same wavelength and are derived from a common source, it is usual to divide the light from the source into two parts by means of a beam-splitting device, such as a semireflecting mirror (Fig. 2). One beam is

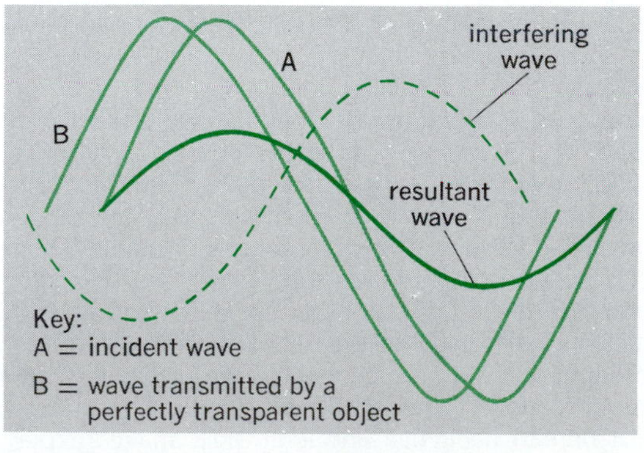

Fig. 1. Basic principle of interference microscopy. Interference is produced between B and another wave, shown by the broken line, derived from the same source. The resultant differs in amplitude from wave A so that the transparent object becomes visible.

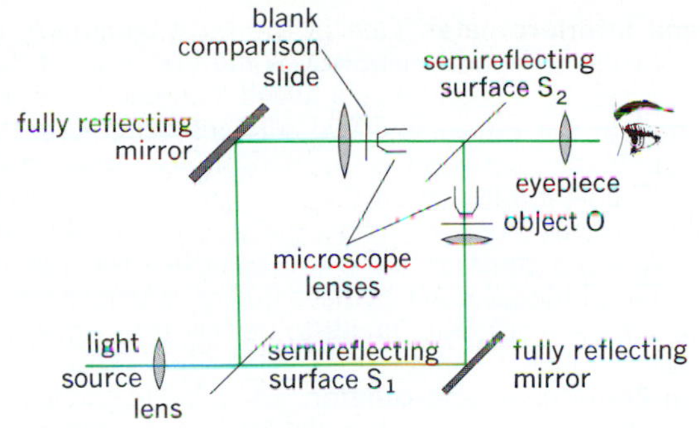

Fig. 2. A theoretical interference system for transparent specimens. The semireflecting mirror S_1 divides the light into two beams which follow similar paths. One beam passes through a microscope containing the object O, the other through a microscope containing a blank comparison slide. The beams are reunited at S_2 and are viewed through the eyepiece.

made to traverse the object; the other travels along a similar path, but does not pass through the object. When the two beams are recombined under suitable conditions, there is interference, and the differences in optical path introduced by various parts of the object can be seen as variations in intensity or color. *See* INTERFERENCE OF WAVES. [R.Ba.]

Interference of waves The process whereby two or more waves of the same frequency or wavelength combine to form a wave whose amplitude is the sum of the amplitudes of the interfering waves. The interfering waves can be electromagnetic, acoustic, or water waves, or in fact any periodic disturbance.

The most striking feature of interference is the effect of adding two waves in which the trough of one wave coincides with the peak of another. If the two waves are of equal amplitude, they can cancel each other out so that the resulting amplitude is zero. This is perhaps most dramatic in sound waves; it is possible to generate acoustic waves to arrive at a person's ear so as to cancel out disturbing noise. In optics, this cancellation can occur for particular wavelengths in a situation where white light is a source. The resulting light will appear colored. This gives rise to the iridescent colors of beetles' wings and mother-of-pearl, where the substances involved are actually colorless or transparent.

To observe interference with waves generated by atomic or molecular transitions, it is necessary to use a single source and to split the light from the source into parts which can then be recombined. In this case, the amplitude and phase changes occur simultaneously in each of the parts at the same time.

The simplest technique for producing a splitting from a single source was done by T. Young in 1801 and was one of the first demonstrations of the wave nature of light. In this experiment, a narrow slit is illuminated by a source, and the light from this slit is caused to illuminate two adjacent slits. The light from these two parallel slits can interfere, and the interference can be seen by letting the light from the two slits fall on a white screen. The screen will be covered with a series of parallel fringes. The location of these fringes can be derived approximately as follows: In the illustration, S_1 and S_2 are the two slits separated by a distance d. Their plane is a distance l from the screen. Since the slit S_0 is equidistant from S_1 and S_2, the intensity and phase of the light at each slit will be the same. The light falling on P from slit S_1 can be represented by Eq. (1) and from S_2 by Eq. (2), where f is the frequency, t the time, c the velocity of light; x_1 and x_2 are the distances of P from S_1 and S_2, and A_0 is the amplitude. This amplitude is assumed to be the same for each wave since the slits are close together, and x_1 and x_2 are thus nearly the same. The square of the amplitude or the intensity at P can then be written as Eq. (3).

$$A = A_0 \sin 2\pi f\left(t - \frac{x_1}{c}\right) \quad (1)$$

$$B = A_0 \sin 2\pi f\left(t - \frac{x_2}{c}\right) \quad (2)$$

$$I = 4A_0^2 \cos^2 \frac{2\pi f}{c}(x_1 - x_2) \quad (3)$$

In general, l is very much larger than y so that Eq. (3) can be simplified to Eq. (4).

$$I = 4A_0^2 \cos^2 \pi\left(\frac{yd}{l\lambda}\right) \quad (4)$$

Equation (4) is a maximum when Eq. (5) holds and a minimum when Eq. (6) holds, where n is an integer.

$$y = n\lambda \frac{l}{d} \quad (5)$$

$$y = (n + 1/2)\lambda \frac{l}{d} \quad (6)$$

Accordingly, the screen is covered with a series of light and dark bands called interference fringes. If the source behind slit S_0 is white light and thus has wavelengths varying perhaps from 400 to 700 nanometers, the fringes are visible only where $x_1 - x_2$ is a few wavelengths, that is, where n is small. At large values of n, the position of the nth fringe for red light will be very different from the position for blue light, and the fringes will blend together and be washed out.

The energy carried by a wave is measured by the intensity, which is equal to the square of the amplitude. The total energy falling on the screen is not changed by the presence of interference. The energy density at a particular point is, however, drastically changed. This fact is most important for those waves of the electromagnetic spectrum which can be generated by vacuum-tube oscillators. The sources of radiation or antennas can be made to emit coherent waves which will undergo interference. This makes possible a redistribution of the radiated energy. Quite narrow beams of radiation can be produced by the proper phasing of a linear antenna array. *See* ANTENNA (ELECTROMAGNETISM); INTERFEROMETRY. [B.H.Bi.]

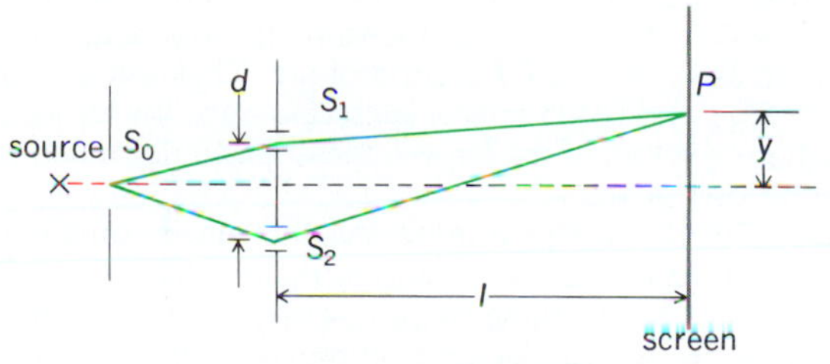

Young's two-slit interference.

Interferometry The design and use of optical interferometers. Optical interferometers based on both two-beam interference and multiple-beam interference of light are extremely powerful tools for metrology and spectroscopy. A wide variety of measurements can be performed, ranging from determining the shape of a surface to an accuracy of less than a millionth of an inch (25 nanometers) to determining the separation, by millions of kilometers, of binary stars. In spectroscopy, interferometry can be used to determine the hyperfine structure of spectrum lines. By using lasers in classical interferometers as well as holographic interferometers and speckle interferometers, it is possible to perform deformation, vibration, and contour measurements of diffuse objects that could not previously be performed. There are two basic classes of interferometers: division of wavefront and division of amplitude.

Michelson interferometer. The Michelson interferometer (Fig. 1) is based on division of amplitude. Light from an extended source S is incident on a partially reflecting plate (beam splitter) P_1. The light transmitted through P_1 reflects off mirror M_1 back to plate P_1. The light which is reflected proceeds to M_2 which reflects it back to P_1. At P_1, the two waves are again partially reflected and partially transmitted, and a portion of each wave proceeds to the receiver R, which may be a screen, a photocell, or a human eye. Depending on the difference between the distances from the beam splitter to the mirrors M_1 and M_2, the two beams will interfere constructively or destructively. Plate P_2 compensates for the thickness of P_1.

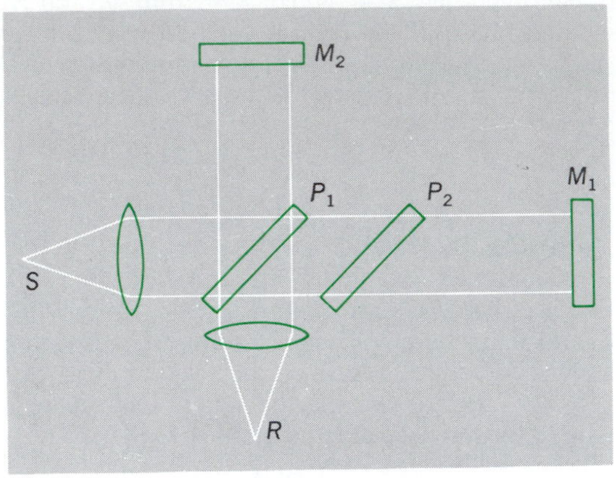

Fig. 1. Michelson interferometer.

The function of the beam splitter is to superimpose (image) one mirror onto the other. When the mirrors' images are completely parallel, the interference fringes appear circular. If the mirrors are slightly inclined about a vertical axis, vertical fringes are formed across the field of view. These fringes can be formed in white light if the path difference in part of the field of view is made zero. Just as in other interference experiments, only a few fringes will appear in white light.

Twyman-Green interferometer. If the Michelson interferometer is used with a point source instead of an extended source, it is called a Twyman-Green interferometer. The use of the laser as the light source for the Twyman-Green interferometer has made it an extremely useful instrument for testing optical components. The great advantage of a laser source is that it makes it possible to obtain bright, good-contrast, interference fringes even if the path lengths for the two arms of the interferometer are quite different. *See* LASER.

The Twyman-Green interferometer can be used to test a flat mirror. In this case, M_1 in Fig. 1 is a reference surface and M_2 is the flat surface being tested. If the test surface is perfectly flat, then straight, equally spaced fringes are obtained. Departure from the straight, equally spaced condition shows directly how the surface differs from being perfectly flat. A height change of half a wavelength will cause an optical path change of one wavelength and a deviation from fringe straightness of one fringe. Thus, the fringes give surface height information, just as a topographical map gives height or contour information.

The basic Twyman-Green interferometer can be modified to test concave-spherical mirrors. In the interferometer, the center of curvature of the surface under test is placed at the focus of a high-quality diverger lens so that the wavefront is reflected back onto itself. Likewise, a convex-spherical mirror can be tested. Also, if a high-quality spherical mirror is used, the high-quality diverger lens can be replaced with the lens to be tested.

Fizeau interferometer. One of the most commonly used interferometers in optical metrology is the Fizeau interferometer, which can be thought of as a folded Twyman-Green interferometer. In the Fizeau, the two surfaces being compared, which can be flat, spherical, or aspherical, are placed in close contact. The light reflected off these two surfaces produces interference fringes. For each fringe, the separation between the two surfaces is a constant. If the two surfaces match, straight, equally spaced fringes result. Surface height variations between the two surfaces cause the fringes to deviate from straightness or equal separation.

Mach-Zehnder interferometer. The Mach-Zehnder interferometer (Fig. 2) is a variation of the Michelson interferometer and, like the Michelson interferometer, depends on amplitude splitting of the wavefront. Light enters the instrument and is reflected and transmitted by the semitransparent mirror M_1. The reflected portion proceeds to M_3, where it is reflected through the cell C_2 to the semitransparent mirror M_4. Here it combines with the light transmitted by M_1 to produce interference. The light transmitted by M_1 passes through a cell C_1, similar to C_2, and is used to compensate for the windows of C_2. The major application of this instrument is in studying airflow around models of aircraft, missiles, or projectiles.

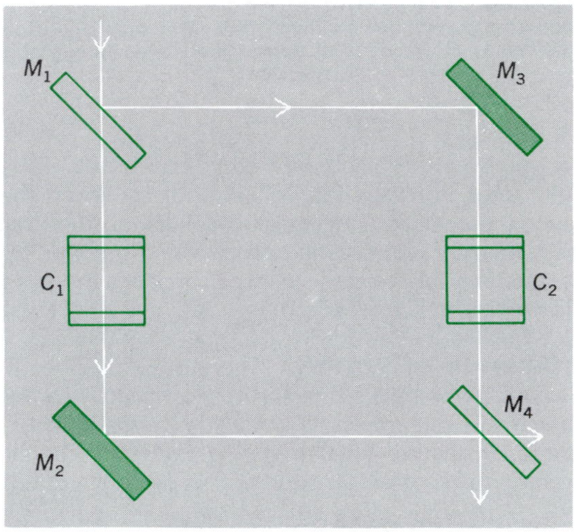

Fig. 2. Mach-Zehnder interferometer.

Shearing interferometers. In a lateral-shear interferometer a wavefront is interfered with a shifted version of itself. A bright fringe is obtained at the points where the slope of the wavefront times the shift between the two wavefronts is equal to an integer number of wavelengths. That is, for a given fringe the slope or derivative of the wavefront is a constant. For this reason a lateral-shear interferometer is often called a differential interferometer. Another type of shearing interferometer is a radial-shear interferometer. Here, a wavefront is interfered with an expanded version of itself. This interferometer is sensitive to radial slopes.

Michelson stellar interferometer. A Michelson stellar interferometer can be used to measure the diameter of stars which are as small as 0.01 second of arc. This task is impossible with a ground-based optical telescope since the atmosphere limits the resolution of the largest telescope to not much better than 1 second of arc.

Fabry-Perot interferometer. All the interferometers discussed above are two-beam interferometers. The Fabry-Perot interferometer is a multiple-beam interferometer since the two glass plates are partially silvered on the inner surfaces, and the incoming wave is multiply reflected between the two surfaces. The position of the fringe maxima is the same for multiple-beam interference as two-beam interference; however, as the reflectivity of the two surfaces increases and the number of interfering beams increases, the fringes become sharper.

Holographic interferometry. A wave recorded in a hologram is effectively stored for future reconstruction and use. Holographic interferometry is concerned with the formation and interpretation of the fringe pattern which appears when a wave, generated at some earlier time and stored in a hologram, is later reconstructed and caused to interfere with a comparison wave. It is the storage or time-delay aspect which gives the holographic method a unique advantage over conventional optical interferometry. *See* Holography.

Speckle interferometry. A random intensity distribution, called a speckle pattern, is generated when light from a highly coherent source, such as a laser, is scattered by a rough surface. The use of speckle patterns in the study of object displacements, vibration, and distortion is becoming of more importance in the nondestructive testing of mechanical components. *See* Nondestructive testing; Speckle. [J.C.Wy.]

Interferon A protein secreted by body cells when they are stimulated by viruses, bacteria, foreign cells, or antigenic macromolecules. It is the body's most rapidly produced defense against viruses. Interferon (IF) stimulates surrounding cells to produce other proteins which, in turn, may regulate viral multiplication, the immune response, and cell growth. Three different interferons have been identified: fibroepithelial, leukocyte, and immune. These are probably coded by three distinct cell genes. The three interferons have important differences in amino acid sequence, stimulus for induction, producer cell, and role in the body.

Interferon is under investigation as a treatment for viral infections, cancer, and autoimmune diseases. It functions in a number of animal phyla, ranging from fish to humans. Defense against viruses in plants is also related to interferon.

Interferon inhibits growth of many tumor cells more effectively than it inhibits growth of normal cells. Tumor inhibition by interferon may occur directly, through inhibition of tumor cell growth, or indirectly, through activation of cells that participate in the body's immune response (macrophages and lymphocytes).

Interferon is the most rapidly manufactured of the known body defenses against viruses. It appears within hours of viral infection and continues to be produced throughout the infection. A causal relationship between interferon and natural recovery from most viral infections is strongly supported by evidence developed since the discovery of interferon. [S.B.]

Interhalogen compounds The elements of the halogen family (fluorine, chlorine, bromine, and iodine) possess an ability to react with each other to form a series of binary interhalogen compounds (halogen halides) of general composition given by XY_n, where n can have the values 1, 3, 5, and 7, and where X is the heavier (less electronegative) of the two elements. All possible diatomic compounds of the first four halogens have been prepared. In other groups a varying number of possible combinations is absent. Although attempts have been made to prepare ternary interhalogens, they have been unsuccessful; there is considerable doubt that such compounds can exist. A list of known interhalogen compounds and some of their physical properties is given in the table. *See* Halogen elements.

The reactivity of the polyhalides reflects the reactivity of the halogens they contain. In general, they behave as strong oxidizing and halogenating agents. Most halogen halides (especially halogen fluorides) readily attack metals, yielding the corresponding halide of the more electronegative halogen. All halogen halides readily react with water. Such reactions can be quite violent and, with halogen fluorides, they may be explosive. They readily react with aliphatic and aromatic hydrocarbons and with oxygen- or nitrogen-containing compounds. [T.Su.]

Known interhalogen compounds

	XY	XY_3	XY_5	XY_7
	ClF	ClF_3	ClF_5	IF_7
mp	−154°C	−76°C	−103°C	
bp	−101°C	12°C	−14°C	4.77 (sublimes)
	BrF	BrF_3	BrF_5	
mp	≈−33°C	8.77°C	−62.5°C	
bp	≈20°C	125°C	40.3°C	
	IF	IF_3	IF_5	
mp	–	−28°C	10°C	
bp	–	–	101°C	
	BrCl	ICl_3*		
mp	≈−54°C	101°C		
bp	–	–		
	IC1†			
mp	27.2°C(α)			
bp	≈100°C			
	IBr			
mp	40°C			
bp	119°C			

*In the solid state the compound forms a dimer.
†Unstable β-modification exists, mp 14°C.

Interior ballistics The branch of applied physics which deals with the motions of projectiles while they are still inside the gun barrel or cannon, and are accelerated by the pressure of expanding gases resulting from the burning of the charge. The term is also used to refer to the similar events inside the casing of a solid-fuel rocket, in which case the exhaust blast is regarded as equivalent to the projectile.

If the propelling charge of the cartridge is changed into gas instantaneously, the gas pressure curve, from the instant of ignition to the instant the projectile leaves the muzzle, would be very nearly an adiabatic curve, quite similar to the curve showing the diminishing potential of a gravitational field. In reality, it takes some time for the powder charge to burn; furthermore, various resistance factors must be overcome. Therefore, the gas pressure in the gun barrel behind the projectile does not reach its maximum value until after the projectile has started moving inside the barrel. The pressure should reach a minimum value at the instant the projectile leaves the gun barrel (see illustration); hence, there is a definite relationship between powder charge, gas pressure, and barrel length. If the pressure is still high even when the projectile "unseals" the gun barrel, a portion of the propelling charge has been wasted. If the pressure drops too low before the projectile reaches the muzzle, the projectile is slowed down inside the barrel; in this case, shortening the barrel would actually improve the performance.

Even when the pressure drop has been properly calculated

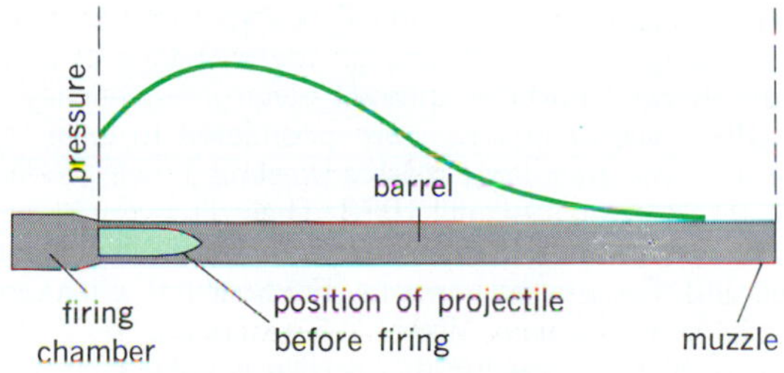

Gas pressure in gun barrel.

(and achieved), the exit velocity of the powder gases is still 2 to 2½ times the muzzle velocity of the projectile. This fact can be utilized to prevent or to minimize the recoil of the weapon by means of the muzzle brake first developed by Schneider-Creusot in France. The muzzle brake deflects the direction of motion of the gases by 120–180°, thereby creating what might be called a gas recoil opposite in direction to the projectile recoil. Recoilless guns eliminate recoil by discharging a quantity of the propelling gases through the rear. [W.L./D.Wo.]

Interleukin Any of a group of soluble molecules derived from lymphocytes (lymphokines) that has been purified, biochemically and functionally characterized, and genetically cloned. The first molecule to be given an interleukin designation is now referred to as interleukin 1 (IL-1), the second as IL-2, and so on. *See* LYMPHOKINES.

Interleukins act sequentially in an immune response. For example, IL-1 is produced by appropriately stimulated macrophages, such as those that have come into contact with bacteria. IL-1 can amplify the immune response because it acts on such a variety of target cells, rendering them hyperresponsive to secondary mediators; it acts on T cells, causing them to be responsive to IL-2, which induces their proliferation, and also acts on B cells, making them sensitive to signals from T cells, such as IL-4. The action of IL-1 on a variety of cell types induces the production of IL-6, which has its own spectrum of target cells and biological effects.

Many of the interleukins act synergistically with each other. For example, IL-3 elicits proliferation of myeloid progenitors, but the addition of IL-4 or IL-6 enhances the level of cell growth. IL-3 induces mast cell growth, which is further augmented by the addition of IL-4. The production of IgE antibodies, induced by IL-4, is potentiated by the addition of IL-5 or IL-6. *See* CELLULAR IMMUNOLOGY; IMMUNOLOGY. [J.F.Ma.; D.H.Ka.]

Intermediate-frequency amplifier A tuned amplifier employed in the amplification of the signals produced by the mixer in a radio receiver. Because the carrier frequency of the modulated signal from the mixer is essentially constant, the resonant frequency of the amplifier is fixed.

The proper design of the intermediate-frequency (i-f) amplifier is essential for good selectivity and reproduction of the original transmitted signal. If the amplifier is tuned too sharply, the high-frequency components of the modulating signal will be lost. To avoid this, stagger-tuning of the individual stages may be used. In a stagger-tuned amplifier the resonant frequency of each stage is slightly different from the carrier frequency, with the result that the gain is essentially constant over the bandwidth of the modulated signal. The gain decreases rapidly at frequencies outside this band.

The standard i-f frequency for broadcast radio receivers is 455 kHz; other frequencies used depend upon the application, such as television receivers or radar receivers. *See* AMPLIFIER. [H.F.K.]

Intermediate vector boson The fundamental particles that transmit the weak force. (An example of a weak interaction process is nuclear beta decay.) These elementary particles, which are also called *W* and *Z* particles, were discovered in 1983 in very high-energy proton-antiproton collisions. It is through the exchange of *W* and *Z* bosons that two particles interact weakly, just as it is through the exchange of photons that two charged particles interact electromagnetically. The intermediate vector bosons were postulated to exist in the 1960s; however, their large masses prevented their production and study at accelerators until 1983. Their discovery was a key step toward unification of the weak and electromagnetic interactions. *See* ELEMENTARY PARTICLE; FUNDAMENTAL INTERACTIONS; WEAK NUCLEAR INTERACTIONS; WEINBERG-SALAM MODEL.

The very high center-of-mass energies capable of producing the massive *W* and *Z* particles were achieved with collisions of protons and antiprotons at the laboratory of the European Organization for Nuclear Research (CERN) near Geneva, Switzerland. This was accomplished through conversion of the existing 4-mi-circumference (6-km) superproton synchrotron (SPS) into a proton-antiproton ($p\bar{p}$) collider. After injection into the SPS, the protons and antiprotons were accelerated together to an energy of 270 GeV each, and stored for several hours at this energy. This made a center-of-mass energy of 540 GeV per $p\bar{p}$ collision, which was about a factor of 10 higher than previously achieved with any other accelerator. *See* PARTICLE ACCELERATOR.

The charged boson (W^+ and W^-) mass is measured to be 80.9 ± 1.5 GeV/c^2, and the neutral boson (Z^0) mass is measured to be 95.6 ± 1.5 GeV/c^2. (For comparison, the proton has a mass of about 1 GeV/c^2.) There is an additional systematic error on these mass values of 3%, mainly due to calibration of the calorimeter. Prior to the discovery of the *W* and the *Z*, particle theorists had met with some success in the unification of the weak and electromagnetic interactions. Based on low-energy neutrino scattering data, which in the electroweak theory involves the exchange of virtual *W* and *Z* particles, theorists made predictions for the *W* and *Z* masses. The actual measured values are in agreement (within errors) with predictions. The discovery of the *W* and the *Z* particles at the predicted masses is an essential confirmation of the electroweak theory. [J.W.R.]

Intermetallic compounds Materials composed of two or more types of metal atoms, which exist as homogeneous, composite substances and differ discontinuously in structure from that of the constituent metals. They are also called, preferably, intermetallic phases. Their properties cannot be transformed continuously into those of their constituents by changes of composition alone, and they form distinct crystalline species separated by phase boundaries from their metallic components and mixed crystals of these components; it is generally not possible to establish formulas for intermetallic compounds on the sole basis of analytical data, so formulas are determined in conjunction with crystallographic structural information.

The term "alloy" is generally applied to any homogeneous molten mixture of two or more metals, as well as to the solid material that crystallizes from such a homogeneous liquid phase. Alloys may also be formed from solid-state reactions. In the liquid phase, alloys are essentially solutions of metals in one another, although liquid compounds may also be present. Alloys containing mercury are usually referred to as amalgams. Solid alloys may vary greatly in range of composition, structure, properties, and behavior. *See* ALLOY; NONSTOICHIOMETRIC COMPOUNDS; SEMICONDUCTOR; SOLID-STATE CHEMISTRY. [J.L.T.W.]

Intermolecular forces Attractive or repulsive interactions that occur between all atoms and molecules. These forces, which become significant at molecular separations of about 1 nm or less, are much weaker than forces associated with chemical bonds or electrostatic interactions of charged particles. They are important, however, since they are responsible for many of the physical properties of solids, liquids, and pressurized gases.

Intermolecular forces are responsible for many of the bulk properties of matter in all its phases. At increased pressures and sufficiently low temperatures the attractive forces between molecules in the gas will cause it to liquefy. The viscosity, surface tension, and diffusion of liquids are examples of physical properties which are a consequence of intermolecular forces. Repulsive forces prevent the molecules from approaching one another too closely and account for the high compressibility of liquids. Intermolecular forces between near and distant neighbors dictate the ordered molecular arrangements in crystalline solids. These forces also account for the elasticity of solids. Intermolecular forces also determine to an important extent the three-dimensional arrangement of biological molecules, polymers, and even smaller molecules. [G.E.E.]

Internal combustion engine A prime mover, the fuel for which is burned within the engine, as contrasted to a steam engine, for example, in which fuel is burned in a separate furnace. *See* ENGINE.

The most numerous of internal combustion engines are the gasoline piston engines used in passenger automobiles, outboard engines for motor boats, small units for lawn mowers, and other such equipment, as well as diesel engines used in trucks, tractors, earth-moving, and similar equipment. For other types of internal combustion engines *see* GAS TURBINE; ROCKET PROPULSION; ROTARY ENGINE; TURBINE PROPULSION.

The aircraft piston engine is fundamentally the same as that used in automobiles but is engineered for light weight and is usually air cooled. *See* RECIPROCATING AIRCRAFT ENGINE.

Characteristic features common to all commercially successful internal combustion engines include (1) the compression of air, (2) the raising of air temperature by the combustion of fuel in this air at its elevated pressure, (3) the extraction of work from the heated air by expansion to the initial pressure, and (4) exhaust. In 1862 Beau de Rochas proposed the four-stroke engine cycle as a means of accomplishing these conditions in a piston engine (see illustration). The engine requires two revolutions of the crankshaft to complete one combustion cycle. The first engine to use this cycle successfully was built in 1876 by N. A. Otto. *See* OTTO CYCLE.

Two years later Sir Dougald Clerk developed the two-stroke engine cycle by which a similar combustion cycle required only one revolution of the crankshaft. In 1891 Joseph Day simplified the two-stroke engine cycle by using the crankcase to pump the required air. Engines using this two-stroke cycle today have been further simplified by use of a third cylinder port which dispenses with the crankcase check valve used by Day. Such engines are in wide use for small units where fuel economy is not as important as mechanical simplicity and light weight. They do not need mechanically operated valves and develop one combustion cycle per crankshaft revolution. Nevertheless they do not develop twice the power of four-stroke cycle engines with the same size working cylinders at the same number of revolutions per minute (rpm). The principal reasons for this are (1) the reduction in effective cylinder volume due to the piston movement required to cover the exhaust ports, (2) the appreciable mixing of burned (exhaust) gases with the combustible mixture, and (3) the loss of some combustible mixture with the exhaust gases.

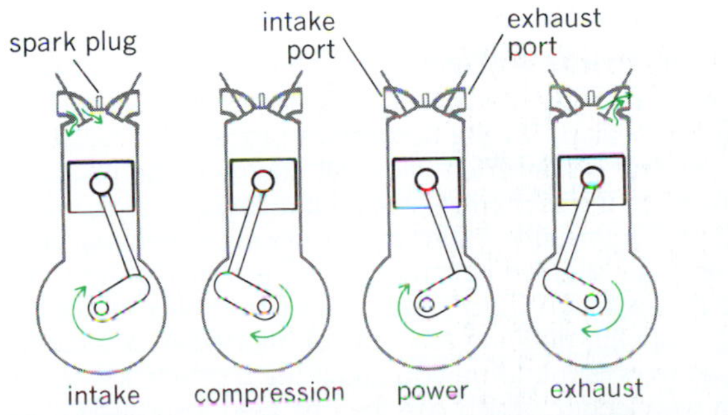

The four strokes of a conventional four-stroke engine cycle. For intake stroke the intake valve (left) has opened and the piston is moving downward, drawing air and gasoline vapor into the cylinder. In compression stroke the intake valve has closed and the piston is moving upward, compressing the mixture. On power stroke the ignition system produces a spark that ignites the mixture. As it burns, high pressure is created, which pushes the piston downward. For exhaust stroke the exhaust valve (right) has opened and the piston is moving upward, forcing the burned gases from the cylinder. (*After M. L. Smith and K. W. Stinson, Fuels and Combustion, McGraw-Hill, 1952*)

About 20 years after Otto first ran his engine, Rudolf Diesel successfully demonstrated an entirely different method of igniting fuel. Air is compressed to a pressure high enough for the adiabatic temperature to reach or exceed the ignition temperature of the fuel. Because this temperature is in the order of 1000°F (540°C), compression ratios of 12:1 to 20:1 are used commercially with compression pressures generally over 600 psi (4.1 megapascals). This engine cycle requires the fuel to be injected after compression at a time and rate suitable to control the rate of combustion. *See* DIESEL ENGINE; FUEL INJECTION.

There are many characteristics of the diesel engine which are in direct contrast to those of the Otto engine. The higher the compression ratio of a diesel engine, the less the difficulties with ignition time lag. Too great an ignition lag results in a sudden and undesired pressure rise which causes an audible knock. In contrast to an Otto engine, knock in a diesel engine can be reduced by use of a fuel of higher cetane number, which is equivalent to a lower octane number. *See* CETANE NUMBER; OCTANE NUMBER.

The larger the cylinder diameter of a diesel engine, the simpler the development of good combustion. In contrast, the smaller the cylinder diameter of the Otto engine, the less the limitation from detonation of the fuel.

High intake-air temperature and density materially aid combustion in a diesel engine, especially of fuels having low volatility and high viscosity. Some engines have not performed properly on heavy fuel until provided with a supercharger. The added compression of the supercharger raised the temperature and, what is more important, the density of the combustion air. For an Otto engine, an increase in either the air temperature or density increases the tendency of the engine to knock and therefore reduces the allowable compression ratio.

Diesel engines develop increasingly higher indicated thermal efficiency at reduced loads because of leaner fuel-air ratios and earlier cutoff. Such mixture ratios may be leaner than will ignite in an Otto engine. Furthermore, the reduction of load in an Otto engine requires throttling, which develops increasing pumping losses in the intake system. [N.MacC.]

Internal energy A characteristic property of the state of a thermodynamic system, introduced in the first law of thermodynamics. For a static, closed system (no bulk motion, no transfer of matter across its boundaries), the change in internal energy for a process is equal to the heat absorbed by the system from its surroundings minus the work done by the system on its surroundings. Only a change in internal energy can be measured, not its value for any single state. For a given process, the change in internal energy is fixed by the initial and final states and is independent of the path by which the change in state is accomplished. *See* THERMODYNAMIC PRINCIPLES. [P.J.B.]

Internal wave Wave motions of stably stratified fluids in which the maximum vertical motion takes place below the surface of the fluid. The restoring force is mainly due to gravity; when light fluid from upper layers is depressed into the heavy lower layers, buoyancy forces tend to return the layers to their equilibrium positions. Internal waves have been found in the atmosphere as lee waves (waves in the wind stream downwind from a mountain) and as waves propagated along an inversion layer (a layer of very stable air). They are also associated with wind shears at the lower boundary of the jet stream. In the oceans internal oscillations have been observed wherever suitable measurements have been made, but it is not completely certain that all of these oscillations are manifestations of internal waves rather than turbulent eddies. [C.S.C.]

International Date Line The 180° meridian, where each day officially begins and ends. A person traveling eastward, against the apparent movement of the Sun, gains 1 h for every 15° of longitude; a person traveling westward loses time at the

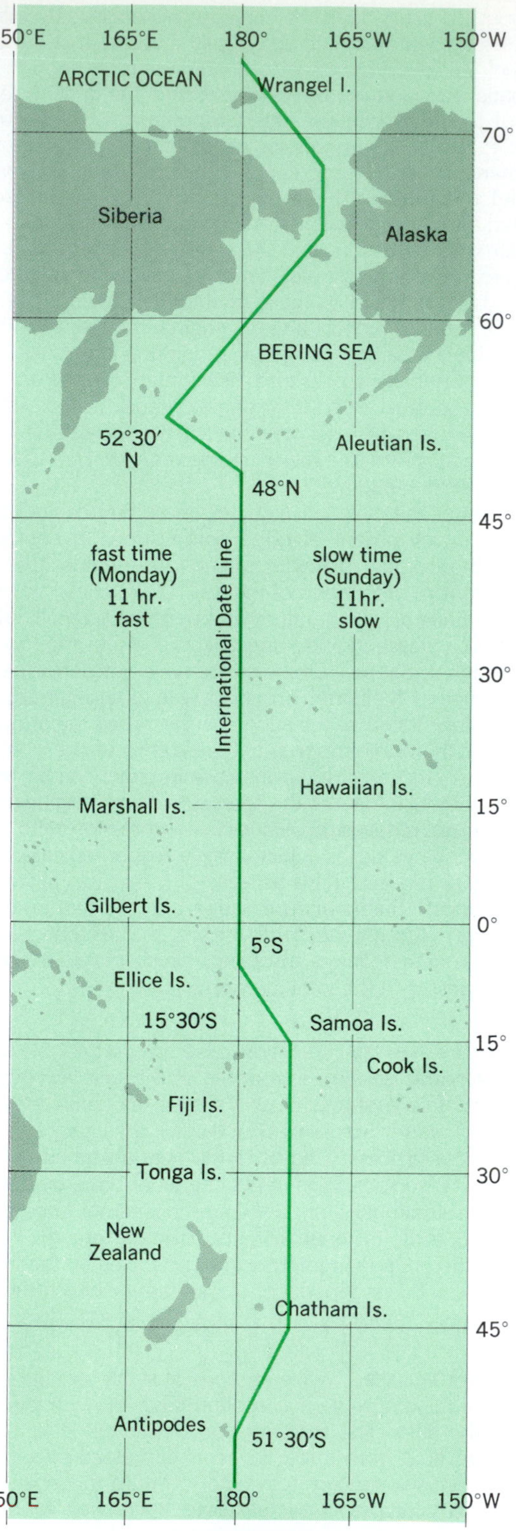

The International Date Line.

same rate. Two people starting from any meridian and traveling around the world in opposite directions at the same speed would have the same time when they meet, but would be 1 day apart in date. A traveler going west across the line loses a day; if it is Monday to the east, it will be Tuesday immediately across the International Date Line. In traveling eastward a day is gained; if it is Monday to the west of the line, it will be Sunday immediately across the line.

The 180° meridian is ideal for serving as the International Date Line. It is exactly halfway around the world from the zero, or Greenwich, meridian, from which all longitude is reckoned. It also falls almost in the center of the largest ocean; consequently there is the least amount of inconvenience as regards population centers. A few deviations in the alignment have been made (see illustration). [V.H.E.]

Interplanetary matter Gas and small solid particles in the space between the planets in the solar system. Most of the interplanetary gas originates in the Sun, flowing outward in the solar wind. The solid material originated in the asteroids and comets. *See* ASTEROID; COMET; INTERSTELLAR MATTER; SOLAR WIND. [P.G.M.]

Interplanetary propulsion Means of providing propulsive power for flight to the Moon or to a planet. A variety of different propulsion systems can be used. The space vehicles for these missions consist of a series of separate stages, each with its own set of propulsion systems. When the propellants of a given stage have been expended, the stage is jettisoned to prevent its mass from needlessly adding to the inertia of the vehicle. Although all propulsion systems actually used in interplanetary flight have been chemical rockets, several basically different propulsion systems have been studied. *See* ELECTROMAGNETIC PROPULSION; ION PROPULSION; ROCKET PROPULSION; ROCKET STAGING; SPACECRAFT PROPULSION. [G.P.S.]

Interpolation A process in mathematics used to estimate an intermediate value of one (dependent) variable which is a function of a second (independent) variable when values of the dependent variable corresponding to several discrete values of the independent variable are known.

Suppose, as is often the case, that it is desired to describe graphically the results of an experiment in which some quantity Q is measured, for example, the electrical resistance of a wire, for each of a set of N values of a second variable v representing, perhaps, the temperature of the wire. Let the numbers Q_i, $i = 1, 2, \ldots, N$, be the measurements made of Q and the numbers v_i be those of the variable v. These numbers representing the raw data from the experiment are usually given in the form of a table with each Q_i listed opposite the corresponding v_i. The problem of interpolation is to use the above discrete data to predict the value of Q corresponding to any value of v lying between the above v_i. If the value of v is permitted to lie outside these v_i, the somewhat more risky process of extrapolation is used. *See* EXTRAPOLATION. [K.S.K.]

Interstellar extinction Dimming of light from the stars due to absorption and scattering by grains of dust in the interstellar medium. In the absorption process the radiation disappears and is converted into heat energy in the interstellar dust grains. In the scattering process the direction of the radiation is altered. Interstellar extinction produces a dimming of the light from stars situated beyond interstellar clouds of dust. Measures of the radiation from pairs of stars of similar intrinsic properties but with differing amounts of interstellar extinction can be used to obtain information about the dependence of extinction on wavelength, which can then be used to provide clues about the nature of the interstellar dust grains. *See* SCATTERING OF ELECTROMAGNETIC RADIATION.

A detailed interpretation of this dependence and other data relating to interstellar dust suggests that the interstellar grains of dust range in size from about 0.01 to 1 μm and are composed of silicate grains and probably some type of carbon grain, and that the interstellar dust acquires coatings of water ice and ammonia ice in the densest regions of interstellar space. A comparison of interstellar extinction with the absorption by interstellar atomic hydrogen reveals that the dust contains about 1% of the mass of the interstellar medium. *See* INTERSTELLAR MATTER. [B.D.S.]

Interstellar matter The material (gaseous, molecular, and particulate) between the stars, constituting several percent of the total mass of the Galaxy. Being the reservoir from which new stars are born in the Galaxy, interstellar matter is of fundamental importance in understanding both the processes leading to the formation of stars, including the solar system, and ultimately the origin of life in the universe. Among the many ways in which interstellar matter is detected, perhaps the most familiar are attractive photographs of bright patches of emission-line or reflection nebulosity. However, these nebulae furnish an incomplete view of the large-scale distribution of material, because they depend on the proximity of one or more bright stars for their illumination. Radio observations of hydrogen, the dominant form of interstellar matter, reveal a widespread distribution throughout the thin disk of the Galaxy, concentrating in the spiral arms. Mixed in with the gas are small solid particles, called dust grains, of characteristic radius 0.1 micrometer. Although by mass the grains constitute less than 1% of the material, they have a pronounced effect through the extinction of starlight. Striking examples of this obscuration are the dark rifts seen in the Milky Way. On average, the density of matter is only one hydrogen atom per cubic centimeter (equivalently 10^{-24} g cm^{-3}), but because of the long path lengths over which the material is sampled, this tenuous medium is detectable. Radio and optical observations of other spiral galaxies show a similar distribution of interstellar matter in spiral arms in the galactic plane.

A hierarchy of interstellar clouds, concentrations of gas and dust, exists within the spiral arms. Many such clouds or cloud complexes are recorded photographically. However, the most dense, which contain interstellar molecules, are often totally obscured by the dust grains and so are detectable only through their infrared and radio emission. These molecular clouds contain the birthplaces of stars. *See* GALAXY, EXTERNAL; STELLAR EVOLUTION. [P.G.M.]

Intestine The portion of the digestive tract beyond the stomach. It is the region of the gut wherein digestion is completed through the combined action of bile, pancreatic juice, and secretions from the intestinal wall itself, and wherein the

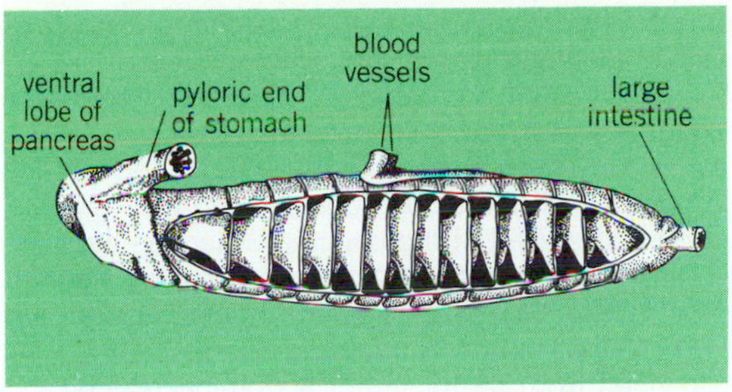

Fig. 1. Small intestine of shark (*Squalus acanthias*) cut open to show spiral valve.

products of digestion are finally absorbed. Two segments are generally distinguishable, a long slender small intestine and a wider terminal part, the large intestine. To keep absorptive surface adequate to meet the nutritive requirements of body mass, the ratio of the intestine length to body length is greater in large animals than in small animals. In addition, plant food appears to be difficult to digest, so that the intestine in herbivores is longer than in carnivores. The small intestine of a cow is about 150 ft (45 m) long and the larger intestine about 30 ft (9 m). The corresponding dimensions for humans are about 22 and 5 ft (6.6 and 1.5 m).

Small intestine. In elasmobranchs and some primitive fishes the mucous membrane of the rather short small intestine is thrown into a twisted fold, the spiral valve, which effectively increases surface area for absorption and lengthens the path which food must travel (Fig. 1). Among the higher vertebrates, especially mammals, the mucous membrane of the small intestine is thrown into numerous thin transverse folds, the plicae circulares, which greatly increase the internal surface of the intestine.

The villi, fingerlike outgrowths of the mucous membrane, are typical of the entire small intestine in birds and mammals. They are present in great numbers, and give the surface a soft velvety appearance. Humans have approximately 5,000,000 villi, which provide an absorptive surface area of about 100 ft^2 (10 m^2). Each villus is clothed with simple columnar epithelium and has a core of delicate fibrous tissue. Within each villus is a network of blood capillaries, a centrally located lymphatic capillary, or lacteal, and a few smooth muscle fibers. The muscle fibers provide for movements of the villi, which suggest a

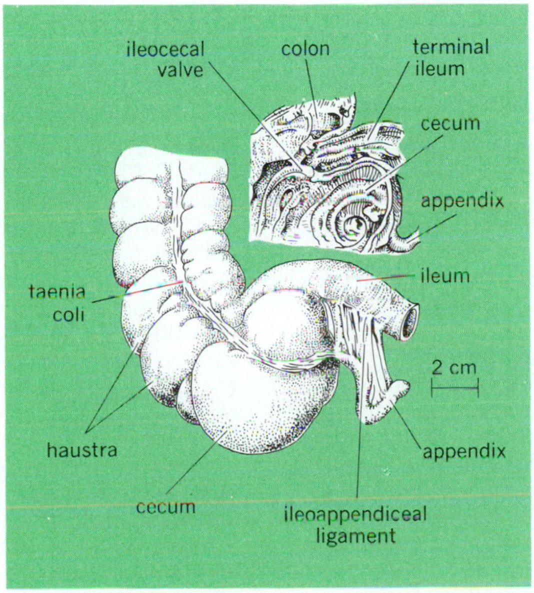

Fig. 2. Junction of ileum with large intestine in humans.

pumping action, and are believed to aid the forward movement of absorbed materials which have entered the lacteals.

The small intestine of mammals consists of three segments: duodenum, jejunum, and ileum. Transitions are gradual and boundaries rather arbitrary. The human duodenum is about 10 in. (25 cm) long, lies retroperitoneally, and makes a C-shaped bend around the head of the pancreas. Piercing the wall of the duodenum about 4 in. (10 cm) beyond the pylorus are the pancreatic and bile ducts. The jejunum in the average human is about 9 ft (2.7 m) long and the ileum about 13 ft (3.9 m). Both are suspended from the posterior abdominal wall by a mesentery, and together their loops fill much of the abdominal cavity. General structural features of the intestinal wall are similar throughout these two segments.

Large intestine. The large intestine in nonmammalian species is relatively short and often not clearly delimited from the small intestine. Its expanded lower end forms the cloaca, a passage which is shared with the reproductive and urinary systems. In reptiles and birds the cloaca absorbs substantial amounts of water, and the digestive and urinary waste products are discharged in a pastelike state.

The large intestine in mammals comprises four principal regions: cecum, appendix, colon, and rectum. At the opening of the small intestine into the side of the cecum, regurgitation is prevented by the ileocecal sphincter and a valvelike fold of the mucous membrane (Fig. 2). The human cecum lies in the light lower quadrant of the abdomen and is a relatively short, saccular structure terminating in the slender vermiform appendix. The colon in humans is about 5 ft (1.5 m) long, and consists of ascending, transverse, descending and sigmoid segments. The hepatic and splenic flexutres, which define the extent of the transverse colon, lie beneath the liver and spleen, respectively.

The two principal functions of the large intestine are concentration of the feces by absorption of water, and production of mucus to facilitate the forwarding of the feces.

The rectum, terminal segment of the large intestine, serves to hold the feces before discharge. In humans it is about 7 in. (17.5 cm) long and can be distended to a diameter of about 3 in. (7.5 cm). At its lower end the circular smooth muscle forms an internal (involuntary) sphincter. Below this is the external (voluntary) sphincter of striated muscle. *See* DIGESTIVE SYSTEM. [V.M.E.]

Intestine disorders

Intestine disorders Disorders of the intestine include congenital defects and vascular, mechanical, degenerative, inflammatory, and neoplastic disorders.

The most common congenital defect is Meckel's diverticulum, a pouch or tract which persists as the remnant of the fetal umbilical duct.

Vascular lesions include mesenteric thrombosis, a blocking of the arterial supply or venous drainage of the intestine by a thrombus or embolus. It is frequently associated with trauma, heart failure, or arteriosclerosis. *See* EMBOLUS; THROMBOSIS.

Common mechanical disorders are hernias, the protrusion of a loop of bowel through a defect or weakness of the abdominal wall, and peritoneal adhesions, resulting from recent or old peritonitis, which often lead to intestinal obstruction. *See* HERNIA.

The three degenerative lesions with the highest incidence are sprue, mucoviscidosis, and Whipple's disease, but all are relatively uncommon. The first is marked by fatty, foul-smelling, bulky stools (steatorrhea) and a multiplicity of other clinical symptoms. The second is an inborn deficit of pancreatic secretion that causes steatorrhea; other symptoms include salivary and sweat gland abnormality and asthma. Whipple's disease is a rare disease of middle age in which large fatty deposits accumulate in the intestinal walls and in the mesenteries.

Appendicitis is the most frequent provocation of surgery of the abdomen. It probably arises from a combination of causes, including infection and obstruction of the lumen, and also from parasitic infestations. Various chemicals and poisons may also produce inflammations of the small intestine (enteritis) or of the large bowel (colitis). Ulcerative colitis refers to a disease of unknown cause producing intestinal ulcerations, particularly in the colon.

Malignant tumors of the small intestine are not common, but the large intestine has a high incidence of carcinomas. Other malignancies are lymphomas, Hodgkin's disease, malignant melanomas, sarcomas, and leukemic infiltrations. *See* ONCOLOGY.

Benign tumors are represented by polyps, lipomas, myomas, and fibromas and usually occur in the colon, the small intestine rarely being involved. [E.G.St./N.K.M.]

Intra-Americas Sea

Intra-Americas Sea That area of the tropical and subtropical western North Atlantic Ocean encompassing the Gulf of Mexico, the Caribbean Sea, the Bahamas and Florida, the northeast coast of South America, and the juxtaposed coastal regions, including the Antillean Islands.

Meteorologically, the Intra-Americas Sea is a transition zone between truly tropical conditions in the south and a subtropical climate in the north. The Sea is also the region that either spawns or interacts with the intense tropical storms known locally as the West Indian Hurricane. Air flowing over the Sea acquires moisture that is the source of much of the precipitation over the central plains of North America. *See* HURRICANE; MONSOON METERorology; TROPICAL METERorology.

Ocean currents of the Intra-Americas Sea are dominated by the Gulf Stream system. Surface waters flow into the Sea through the passages of the Lesser Antilles, and to a lesser extent through the Windward Passage between Cuba and Haiti, and the Anegada Passage between Puerto Rico and Anguilla. These inflowing waters form the Caribbean Current, which flows westward and northward into the Gulf of Mexico through the Yucatán Channel. *See* OCEAN CIRCULATION.

River discharge from several South American rivers, notably the Orinoco and Amazon, drifts through the Intra-Americas Sea and carries materials thousands of kilometers from the deltas. The large deltas are heavily impacted by anthropogenic activities, but they remain the source of rich fisheries and plankton communities. Because of the small tidal range in the Sea, most deltas are wind-dominated geological features. *See* DELTA; EARTHQUAKE; MARINE FISHERIES; PLATE TECTONICS; TIDE.

[G.A.Ma.]

Intron

Intron In split genes, a portion that is included in ribonucleic acid (RNA) transcripts but is removed from within a transcript during RNA processing and is rapidly degraded. Split genes are those in which portions appearing in messenger RNAs (mRNAs) or in structural RNAs (exons) are not contiguous in a gene but are separated by lengths of deoxyribonucleic acid (DNA) encoding parts of transcripts that do not survive the maturation of RNA (introns). *See* EXON.

The number of introns in a gene varies greatly among genes, from 1 in the case of structural RNA genes to more than 30 in the yolk protein gene of the frog *Xenopus*. The lengths, locations, and compositions of introns also vary greatly among genes. However, in general, sizes and locations—but not DNA-sequence compositions—are comparable in homologous genes in different organisms. The implication is that introns became established in genes early in the evolution of eukaryotes, and while their nucleotide sequence is not very important, their existence, positions, and sizes are significant.

Introns allow genetic recombination to occur between the coding units rather than within them, thus providing a means of genetic evolution via wholesale reassortments of functional subunits or building blocks, rather than by fortuitous recombinations of actual protein-coding DNA sequences. *See* GENE; GENETIC CODE; RECOMBINATION (GENETICS). [P.M.M.R.]

Inventory control The process of managing the timing and the quantities of goods to be ordered and stocked, so that demands can be met satisfactorily and economically. Inventories are accumulated commodities waiting to be used to meet anticipated demands. Inventory control policies are decision rules that focus on the trade-off between the costs and benefits of alternative solutions to questions of when and how much to order for each different type of item.

The possible reasons for carrying inventories are: uncertainty about the size of future demands; uncertainty about the duration of lead time for deliveries; provision for greater assurance of continuing production, using work-in-process inventories as a hedge against the failure of some of the machines feeding other machines; and speculation on future prices of commodities. Some of the other important benefits of carrying inventories are: reduction of ordering costs and production setup costs (these costs are less frequently incurred as the size of the orders are made larger which in turn creates higher inventories); price discounts for ordering large quantities; shipping economies; and maintenance of stable production rates and work-force levels which otherwise could fluctuate excessively due to variations in seasonal demand.

The benefits of carrying inventories have to be compared with the costs of holding them. Holding costs include the following elements: cost of capital for money tied up in the inventories; cost of owning or renting the warehouse or other storage spaces; materials handling equipment and labor costs; costs of potential obsolescence, pilferage, and deterioration; property taxes levied on inventories; and cost of installing and operating an inventory control policy. Inventories, when listed with respect to their annual costs, tend to exhibit a similarity to Pareto's law and distribution. A small percentage of the product lines may account for a very large share of the total inventory budget (they are called class A items).

Continuous-review and fixed-interval are two different modes of operation of inventory control systems. The former means the records are updated every time items are withdrawn from stock. When the inventory level drops to a critical level called reorder point, a replenishment order is issued. Under fixed-interval policies, the status of the inventory at each point in time does not have to be known. The review is done periodically.

Uncertainties of future demand play a major role in the cost of inventories. That is why the ability to better-forecast future demand can substantially reduce the inventory expenditures of a firm. Conversely, using ineffective forecasting methods can lead to excessive shortages of needed items and to high levels of unnecessary ones.

Material requirements planning (MRP) systems (which are production-inventory scheduling softwares that make use of computerized files and data-processing equipment) are receiving widespread application. MRP systems have not yet made use of mathematical inventory theory. They recognize the implications of dependent demands in multiechelon manufacturing (which includes lumpy production requirements). Integrating the bills of materials, the given production requirements of end products, and the inventory records file, MRP systems generate a complete list of a production-inventory schedule for parts, subassemblies, and end products, taking into account the lead-time requirements. MRP has proved to be a useful tool for manufacturers, especially in assembly operations. [A.Do.]

Inverse scattering theory A theory whose objective is to determine the scattering object, or an interaction potential energy, from the knowledge of a scattered field. This is the opposite problem from direct scattering theory, where the scattering amplitude is determined from the equations of motion, including the potential. The equations of motion are usually linear (operator-valued) equations. *See* SCATTERING EXPERIMENTS (ATOMS AND MOLECULES); SCATTERING EXPERIMENTS (NUCLEI).

Inverse scattering theories can be divided into two types: (1) pure inverse problems, when the data consist of complete, noise-free information of the scattering amplitude; and (2) applied inverse problems, when incomplete data which are corrupted by noise are given. Many different applied inverse problems can be obtained from any pure inverse problem by using different band-limiting procedures and different noise spectra.

The difficulty of determining the exact object which produced a scattering amplitude is evident. It is often a priori information about the scatterer that makes the inversion possible.

Much of the basic knowledge of systems of atoms, molecules, and nuclear particles is obtained from inverse scattering studies using beams of different particles as probes. For the Schrödinger equation with spherical symmetry or in one dimension, there is an exact solution of the inverse problem.

A number of high-technology areas (nondestructive evaluation, medical diagnostics including acoustic and ultrasonic imaging, x-ray absorption and nuclear magnetic resonance tomography, radar scattering and geophysical exploration) use inverse scattering theory. Several classical waves including acoustic, electromagnetic, ultrasonic, x-rays, and others are used. *See* ACOUSTIC TOMOGRAPHY; BIOACOUSTICS; COMPUTERIZED TOMOGRAPHY; NONDESTRUCTIVE TESTING.

All of the inverse scattering technologies require the solution to ill-posed or improperly posed problems. A model equation is well posed if it has a unique solution which depends continuously on the initial data. It is ill posed otherwise. The ill-posed problems which are amendable to analysis, called regularizable ill-posed problems, are those which depend discontinuously upon the data. This destroys uniqueness, although solutions (in fact, many solutions) exist. [B.DeF.]

Inverse-square law Any law in which a physical quantity varies with distance from a source inversely as the square of that distance. When energy is being radiated by a point source (see illustration), such a law holds, provided the space between source and receiver is filled with a nondissipative, homogeneous, isotropic, unbounded medium. *See* POINT SOURCE.

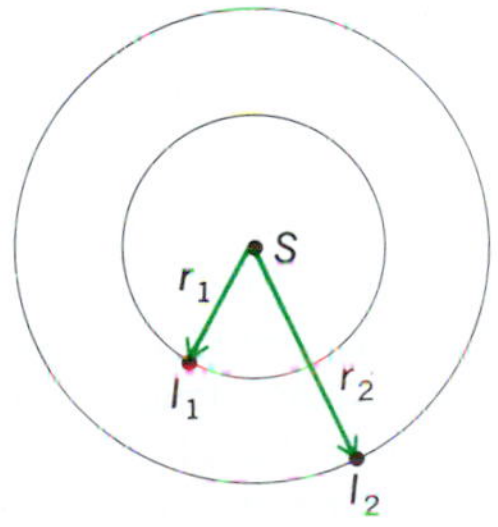

Point source S emitting energy of intensity I. The inverse-square law states that $I_1/I_2 = r_2^2/r_1^2$, where r = distance from S.

Similar reasoning shows that the same law applies to mechanical shear waves in elastic media and to compressional sound waves. The term is also used for static field laws such as the law of gravitation and Coulomb's law in electrostatics. [W.R.Sm.]

Invertebrate pathology All studies having to do with the principles of pathology as applied to invertebrates. It includes studies of the disease process in its broadest sense, that is, investigations of any departures from the normal state of health. Disease in an invertebrate is initiated by the same factors causing disease in vertebrates: infection with microbes; parasites; and noninfectious factors. Various approaches are used in invertebrate pathology. These include (1) studies of the cause of disease, the symptomatology, the resulting pathology on a gross or microscopic level, and pathophysiology; (2) defense mechanisms used by the host to prevent infection, to overcome microbes and parasites, or to repair damaged body parts by wound healing or regeneration; and (3) the effects and progress of infectious disease in a population (epizootiology). *See* INSECT PATHOLOGY. [P.T.J.]

Involute A term applied to a curve C' that cuts at right angles all tangents of a curve C (see illustration). Each curve C

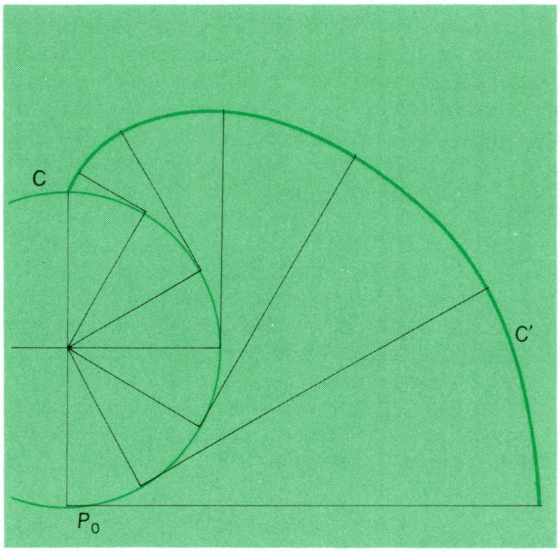

An involute C' of curve C.

has infinitely many involutes and the distance between corresponding points of any two involutes is constant. Let a length of string be coincident with a curve C, with one end fastened at a point P_0 of C. If the string is unwound, remaining taut, the other end of the string traces an involute C' of C. By varying the length of the string, all involutes of C are obtained. *See* ANALYTIC GEOMETRY. [L.M.Bl.]

Iodate A negative ion having the formula IO_3^-, and derived from iodic acid, HIO_3. Some salts such as $KH_2I_3O_9$ and $KH(IO_3)_2$ indicate that the acid may exist as polymers of the ion indicated by the empirical formula. Sodium and potassium iodates are the most important salts and are used in medicine. The iodates are more stable and are weaker oxidizing agents than bromates and chlorates. *See* BROMATE; CHLORATE; IODINE. [E.E.W.]

Iodide A compound which contains the iodine atom in 1− oxidation state and which is derived from hydriodic acid, HI. In comparing the iodide ion with the other halide ions, it should be pointed out that the iodides are more covalent, are the best reducing agents of the group, and form the least stable complexes. The aqueous solubilities of the metal iodides are much the same as the chlorides but in general a little lower.

Sodium or potassium iodide is added to table salt to prevent malfunction of the thyroid gland. Silver iodide is used in photographic films and papers. *See* HALIDE; HALOGENATED HYDROCARBON; IODINE. [E.E.W.]

Iodine A nonmetallic element, symbol I, atomic number 53, relative atomic mass 126.9045, the heaviest of the naturally occurring halogens. Under normal conditions iodine is a black,

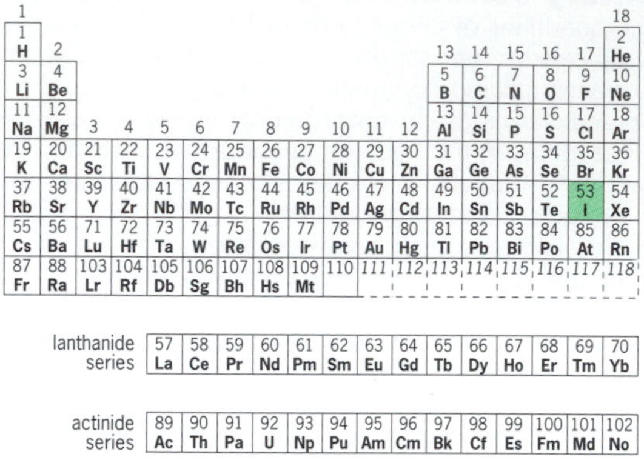
1 H
2 He
3 Li 4 Be
5 B 6 C 7 N 8 O 9 F 10 Ne
11 Na 12 Mg
13 Al 14 Si 15 P 16 S 17 Cl 18 Ar
19 K 20 Ca 21 Sc 22 Ti 23 V 24 Cr 25 Mn 26 Fe 27 Co 28 Ni 29 Cu 30 Zn 31 Ga 32 Ge 33 As 34 Se 35 Br 36 Kr
37 Rb 38 Sr 39 Y 40 Zr 41 Nb 42 Mo 43 Tc 44 Ru 45 Rh 46 Pd 47 Ag 48 Cd 49 In 50 Sn 51 Sb 52 Te 53 I 54 Xe
55 Cs 56 Ba 71 Lu 72 Hf 73 Ta 74 W 75 Re 76 Os 77 Ir 78 Pt 79 Au 80 Hg 81 Tl 82 Pb 83 Bi 84 Po 85 At 86 Rn
87 Fr 88 Ra 103 Lr 104 Rf 105 Db 106 Sg 107 Bh 108 Hs 109 Mt 110 111 112 113 114 115 116 117 118
lanthanide series 57 La 58 Ce 59 Pr 60 Nd 61 Pm 62 Sm 63 Eu 64 Gd 65 Tb 66 Dy 67 Ho 68 Er 69 Tm 70 Yb
actinide series 89 Ac 90 Th 91 Pa 92 U 93 Np 94 Pu 95 Am 96 Cm 97 Bk 98 Cf 99 Es 100 Fm 101 Md 102 No

lustrous, volatile solid; it is named after its violet vapor. *See* HALOGEN ELEMENTS.

The chemistry of iodine, like that of the other halogens, is dominated by the facility with which the atom acquires an electron to form either the iodide ion I^- or a single covalent bond —I, and by the formation, with more electronegative elements, of compounds in which the formal oxidation state of iodine is +1, +3, +5, or +7. Iodine is more electropositive than the other halogens, and its properties are modulated by: the relative weakness of covalent bonds between iodine and more electropositive elements; the large sizes of the iodine atom and iodide ion, which reduce lattice and solvation enthalpies for iodides while increasing the importance of van der Waals forces in iodine compounds; and the relative ease with which iodine is oxidized, Some properties of iodine are listed in the table. *See* ASTATINE; BROMINE; CHEMICAL BONDING; CHLORINE; FLUORINE.

Iodine occurs widely, although rarely in high concentration and never in elemental form. Despite the low concentration of iodine in sea water, certain species of seaweed can extract and accumulate the element. In the form of calcium iodate, iodine is found in the caliche beds in Chile. Iodine also occurs as iodide ion in some oil well brines in California, Michigan, and Japan.

The sole stable isotope of iodine is ^{127}I (53 protons, 74 neutrons). Of the 22 artificial isotopes (masses between 117 and 139), the most important is ^{131}I, with a half-life of 8 days. It is widely used in radioactive tracer work and certain radiotherapy procedures. *See* RADIOACTIVE TRACER.

Iodine exists as diatomic I_2 molecules in solid, liquid, and vapor phases, although at elevated temperatures (> 200°C or 390°F) dissociation into atoms is appreciable. Short intermolecular I…I distances in the crystalline solid indicate strong intermolecular van der Waals forces. Iodine is moderately soluble in nonpolar liquids, and the violet color of the solutions suggests that I_2 molecules are present, as in iodine vapor.

Although it is usually less vigorous in its reactions than the other halogens, iodine combines directly with most elements. Important exceptions are the noble gases, carbon, nitrogen, and some noble metals. The inorganic derivatives of iodine may be grouped into three classes of compounds: those with more electropositive elements, that is, iodides; those with other halogens; and those with oxygen. Organoiodine compounds fall into two categories: the iodides; and the derivatives in which iodine is in a formal positive oxidation state by virtue of bonding to another, more electronegative element. *See* GRIGNARD REACTION; HALOGENATED HYDROCARBON; HALOGENATION; IODIDE.

Some important properties of iodine

Property	Value
Electronic configuration	$[Kr]4d^{10}5s^25p^5$
Relative atomic mass	126.9045
Electronegativity (Pauling scale)	2.66
Electron affinity, eV	3.13
Ionization potential, eV	10.451
Covalent radius,—I,nm	0.133
Ionic radius, I^-, nm	0.212
Boiling point, °C	184.35
Melting point, °C	113.5
Specific gravity (20/4)	4.940

Iodine appears to be a trace element essential to animal and vegetable life. Iodide and iodate in sea water enter into the metabolic cycle of most marine flora and fauna, while in the higher mammals iodine is concentrated in the thyroid gland, being converted there to iodinated amino acids (chiefly thyroxine and iodotyrosines). They are stored in the thyroid as thyroglobulin, and thyroxine is apparently secreted by the gland. Iodine deficiency in mammals leads to goiter, a condition in which the thyroid gland becomes enlarged. *See* THYROID GLAND.

The bactericidal properties of iodine and its compounds bolster their major uses, whether for treatment of wounds or sterilization of drinking water. Also, iodine compounds are used to treat certain thyroid and heart conditions, as a dietary supplement (in the form of iodized salt), and for x-ray contrast media. *See* ANTIMICROBIAL AGENTS; ANTISEPTIC; SALT (FOOD).

Major industrial uses are in photography, where silver iodide is a constituent of fast photographic film emulsions, and in the dye industry, where iodine-containing dyes are produced for food processing and for color photography. *See* DYE; PHOTOGRAPHIC MATERIALS. [C.A.]

Iodoform A yellow, hexagonal solid with a penetrating odor also called triiodomethane, CHI_3. Its specific gravity is 4.08 and melting point 119°C (246°F). It is soluble in organic solvents and insoluble in water. It has weak bactericidal properties and exerts antiseptic action when applied to raw wounds, because of the liberation of free iodine. Iodoform also acts as an inhibitor for wound secretion, and this inhibits bacterial growth. Its chief use is in ointments for minor skin diseases. It is toxic when taken internally. *See* ANTIMICROBIAL AGENTS; HALOGENATED HYDROCARBON. [E.H.H.]

Ion An atom, or group of atoms, which by loss or gain of one or more electrons has acquired an electric charge. If the ion is formed from an atom of hydrogen or an atom of a metal, it is usually positively charged; if the ion is formed from an atom of a nonmetal or from a group of atoms, it is usually negatively charged. The number of electronic charges carried by an ion is called its electrovalence. The charges are denoted by superscripts which give their sign and number; for example, a sodium ion, which carries one positive charge, is denoted by Na^+; a sulfate ion, which carries two negative charges, by SO_4^{2-}. *See* ATOMIC STRUCTURE AND SPECTRA; CHEMICAL BONDING.

Salts are usually composed of orderly arrangements of ions which are not free to move easily in the solid. However, when the salt is fused or dissolved in water, the ions become free, and when an electric field is applied to the salt in solution, the positively charged cations move to the cathode and the negatively charged anions move to the anode. At the electrodes the ions lose their electric charge. This process is called electrolysis. *See* ELECTROLYSIS; IONIC CRYSTALS; SALT (CHEMISTRY); VALENCE. [T.C.W.]

Ion beam mixing A process in which bombardment of a solid with a beam of energetic ions causes the intermixing of the atoms of two separate phases originally present in the near-surface region. In the well-established process of ion implantation, the ions are incident instead on a homogeneous solid, into which they are incorporated over a range of depths determined by their initial energy. In the simplest example of ion beam mixing, the solid is a composite consisting of a substrate and a thin film of a different material (illustration *a*). Ions with sufficient energy pass through the film into the substrate, and this causes mixing of the film and substrate atoms (illustration *b*). If the ion dose is large enough, the original film will completely disappear (illustration *c*). This process may result in the impurity doping of the substrate, in the formation of an alloy or two-phase mixture, or in the production of a stable or metastable solid phase that is different from either the film or the substrate. *See* ION IMPLANTATION.

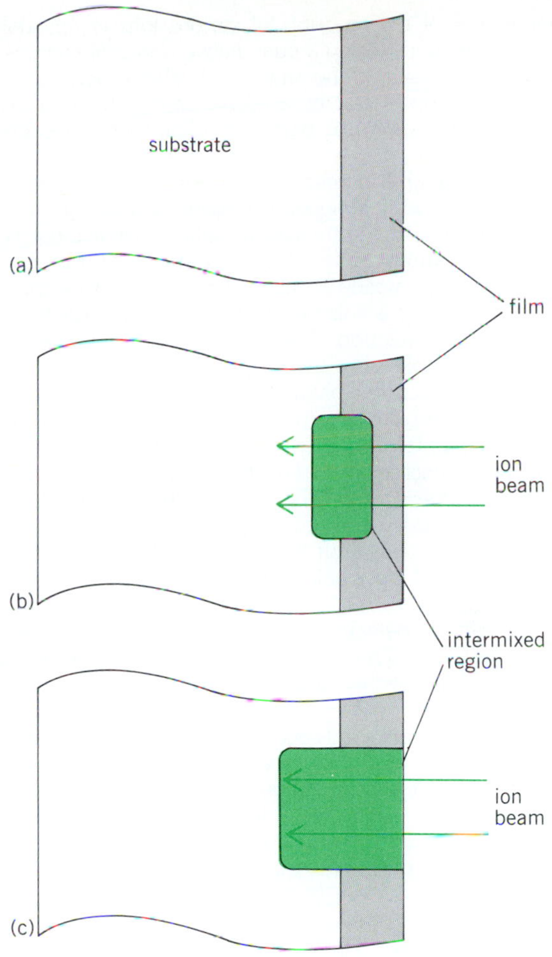

Ion beam mixing of film and substrate. (*a*) Before ion bombardment. (*b*) Partial intermixing. (c) Complete intermixing.

Like ion implantation, ion beam mixing is a solid-state process that permits controlled change in the composition and properties of the near-surface region of solids. Although not yet employed commercially, it is expected to be useful for such applications as the surface modification of metals and semiconductor device processing. In conjunction with thin-film deposition technology, ion beam mixing should make it possible to introduce many impurity elements at concentrations too high for ion implantation to be practical. [B.-Y.T.]

Ion chromatography A group of fast, efficient methods for separating chemical ions. Ion chromatography, a modern adaptation of ion-exchange chromatography, also permits measurement of the amounts of various ions in an analytical sample, thereby achieving a quantitative chemical analysis. The method is applicable to the analysis of either cations or anions. Very dilute samples can be analyzed, even down to the low parts-per-million (ppm) or parts-per-billion (ppb) concentration range.

In the typical arrangement used in ion chromatography, the eluant, which is a dilute aqueous solution of a salt or an acid, is pumped through the entire system until a chemical equilibrium has been attained in the ion-exchange column and the recorder shows a steady baseline signal. Next, a sample loop, which typically has a volume of 10 to 100 microliters, is filled with the sample solution. A valve switching arrangement then washes the sample from the loop and onto the ion-exchange column as a discrete "plug." An anion-exchange column is used to separate sample anions, and a cation-exchange column is used to separate cations.

On the ion-exchange column the sample ions are taken up by the exchange sites on the ion-exchange resin, and eluant ions that previously occupied these sites pass into solution. As each sample ion comes off the column and passes through the detector cell, the electronic signal from the detector changes and is recorded.

A recording of detector signal as a function of time is called a chromatogram. The amount of each ion in the analytical sample is determined by measurement of the height or area of each ion peak.

Ion chromatography has been used extensively for the determination of ions in drinking water, wastewater and process water, plating baths, detergents, foods, and many other applications. Analysis takes only a few minutes, and several ions can be determined in each chromatogram. *See* ANALYTICAL CHEMISTRY; CHROMATOGRAPHY; ION EXCHANGE; LIQUID CHROMATOGRAPHY. [J.S.Fr.]

Ion exchange The reversible exchange of ions of the same charge sign between a solution, usually aqueous, and an insoluble solid in contact with it. The phenomenon was first recognized in soils, where it accounts for the absorption and retention of water-soluble fertilizers. When potash (potassium chloride) is applied to soil, potassium ions are taken into the soil, while chemically equivalent amounts of sodium and calcium ions are released. If the potassium-loaded soil is now washed with a solution of common salt (sodium chloride), some of the potassium is released, while sodium takes its place. Water alone will not remove potassium ions from the soil, because there are no negative ions in the water to balance the charges.

A solid ion exchanger has an extended, open molecular framework that includes electrically charged, ionic groups. A cation exchanger exchanges positive ions and therefore has negative ions built into its framework. An anion exchanger has positive ions in its framework. The ions of the lattice are called the fixed ions; the smaller ions of opposite charge that can change places with ions in the solution are called counterions; small ions having the same charge as the fixed ions, which may enter the exchanger under certain conditions, are called co-ions.

Sometimes exchange occurs at the surface, as in glass and some clays, but more commonly the exchanger has a porous structure that may be crystalline, with channels, cavities, or layered spaces that let ions move in and out, or may be amorphous, like a cross-linked organic polymer. Solvent molecules enter these structures when the exchanger is in contact with solutions.

Ion exchange is common in nature. Many synthetic ion exchangers have been made which have important uses in chemical processing and chemical analysis. These include ion-exchange resins, macroporous resins, and synthetic inorganic exchangers (such as molecular sieves).

Ion exchange has numerous applications for industry and for laboratory research. Water conditioning provides the largest market for ion exchange. Hard water is softened by contact with a cation exchanger that carries sodium ions. Calcium and magnesium ions enter the exchanger and displace sodium ions, which do not cause hardness. In sugar refining, ion exchange removes salts but not nonelectrolytes. Salts present in raw sugar juices interfere with the crystallization of sugar.

Ion exchange is used very widely on the laboratory scale for analytical purposes. In modem high-performance liquid chromatography, ion exchange has a limited role; it has been used extensively to analyze mixtures of amino acids. Ion-exchange membranes are used as separators in electrolytic and voltaic cells. *See* DIALYSIS; ION-SELECTIVE MEMBRANES AND ELECTRODES; SALINE WATER RECLAMATION. [H.F.W.]

Ion-exchange chromatography A form of liquid chromatography based on selective ionic attractions between variously charged sample constituents and an ionized chromatographic matrix. There are two principal types of ion exchangers: cationic and anionic. The most commonly used ion exchangers consist of an organic polymeric backbone (usually a copolymer of styrene and divinylbenzene) with either acidic or basic exchange sites on its porous surface. The charged resins are capable of exchanging their cations or anions with those ions in the liquid phase which have a greater affinity for the matrix. Exchange interactions that take place during the passage of various ions through the column cause separation into discrete ionic zones. Ion-exchange chromatography is one of the widely employed forms of liquid chromatography. *See* CHROMATOGRAPHY; ION EXCHANGE; LIQUID CHROMATOGRAPHY. [M.V.N.]

Ion implantation A process of introducing impurities into the near-surface region of solids by directing a beam of energetic ions at the solid. When ions of sufficient energy are directed toward a crystal surface, they will penetrate the surface and slow to rest within the solid. The advantages of the implantation process are: precise control of the type of impurity to be introduced, the amount of impurity introduced, and the impurity distribution in depth. In addition, since any atoms can be added to any solid, mixtures can be formed which would not normally be found in nature, that is, systems which are not in thermal equilibrium. A major area of application of ion implantation has been to semiconductor device fabrication, where the process is a standard technique. Additional applications have been developed in metals, where unexpectedly beneficial effects on surface-sensitive properties have been found, and in insulators, and research in ion implantation has been undertaken for the purpose of developing scientific understanding and technological applications. *See* CRYSTAL; DIFFUSION IN SOLIDS; ION; SEMICONDUCTOR. [S.T.P.]

Ion propulsion Vehicular motion caused by reaction from the high-speed discharge of a beam of electrically equally charged minute particles. The charged particles, or ions, are accelerated in an electrostatic field produced within the vehicle that is propelled. Typical ion propellants include argon, sodium, xenon, cesium, and mercury.

Ion propulsion is characterized by high specific impulse but low thrust acceleration. Because high specific impulse means low propellant consumption, ion propulsion is attractive for a wide variety of applications. One functional category includes the use of ion thrusters on application satellites for orbit control, for station keeping (position maintaining of satellite in a given orbit), and attitude control. Another functional category involves the use of ion drives for cargo transports in Earth-Moon (geolunar) space. The third major functional application of ion propulsion is interplanetary transfer.

The power source for ion propulsion can be a nuclear reactor or a solar-radiation energy collector. In the first case, protection of the electrical equipment (and of personnel in the case of crewed vehicles) is required against neutron and γ-radiation; protection of the main radiator against radioactivity induced in the working fluid by the reactor is also necessary. If solar radiation is the power source, no radiation shield is required. Solar radiation can be used to provide electric power directly, through photovoltaic solar cells, or indirectly, through a solar heater in which the nuclear reactor is replaced by a solar-radiation collector and a heat exchanger.

Ion or electrostatic thrust devices consist of three components: an ionizer, which furnishes the ions or other charged particles; an accelerator, which provides the properly shaped electric or electromagnetic field for accelerating and properly shaping the charged-particle beam; and a neutralizer, usually an electron emitter, which neutralizes the exhaust beam after ejection. *See* ELECTROMAGNETIC PROPULSION; INTERPLANETARY PROPULSION; SPACECRAFT PROPULSION. [K.A.E.]

Ion-selective membranes and electrodes Membrane-based devices, involving permselective, ion-conducting materials, used for measurement of activities of species in liquids or partial pressures in the gas phase. Permselective means that ions of one sign may enter and pass through a membrane.

Ion-selective electrodes are classified mainly according to the physical state of the ion-responsive membrane material, and not with respect to the ions sensed.

Glass membrane electrodes are mainly used for hydrogen ion activity measurements. They predate the wider variety of membrane electrodes developed after 1960. Electrodes based on water-insoluble inorganic salts include sensors for F^-, Cl^-, Br^-, I^-, CN^-, SCN^-, S^{2-}, Ag^+, Cu^{2+}, Cd^{2+}, and Pb^{2+}. The compounds used are silver salts, mercury salts, sulfides of Cu, Pb, and Cd, and rare-earth salts. All of these are so-called white metals whose aqueous cations (except La^{3+}) are labile. Electrodes using liquid-ion exchangers are supported in the voids of inert polymers such as cellulose acetate, or in transparent films of polyvinyl chloride, and provide extensive examples of devices for sensing. Electrodes with chemical reactions interposed between the sample and the sensor surface permit a new degree of freedom in design of sensors for species which do not directly respond at an electrode surface. Two primary examples are the categories of gas sensors and of electrodes which use enzyme-catalyzed reactions. *See* CALOMEL ELECTRODE; ELECTRODE; ION EXCHANGE.

Electrodes for many species are, for the most part, commercially available. Applications may be batch or continuous. Important batch examples are potentiometric titrations with ion-selective electrode end-point detection, determination of stability constants of complexes and speciation identity, solubility and activity coefficient determinations, and monitoring of reaction kinetics, especially for oscillating reactions. Ion, selective electrodes serve as liquid chromatography detectors and as quality-control monitors in drug manufacture. Applications occur in air and water quality (soil, clay, ore, natural-water, water-treatment, sea-water, and pesticide analyses); medical and clinical laboratories (serum, urine, sweat, gastric-juices, extra-cellular-fluid, dental-enamel, and milk analyses); and industrial laboratories (heavy-chemical, metallurgical, glass, beverage, and household-product analyses). *See* ANALYTICAL CHEMISTRY; CHROMATOGRAPHY; TITRATION. [R.P.B.]

Ion sources Devices which produce positive or negative electrically charged atoms or molecules. *See* ION.

In general, ion sources fall into three major categories: those designed for positive-ion generation, those for negative-ion generation, and a highly specialized type of source designed to produce a polarized ion beam. The positive-ion source category may further be subdivided into sources specifically designed to generate singly charged ions and those designed to produce very highly charged ions.

Ion sources have acquired a wide variety of applications. They are used in a variety of different types of accelerators for nuclear research; have application in the field of fusion research; and are used for ion implantation, in isotope separators, in ion microprobes, as a means of rocket propulsion, in mass spectrometers, and for ion milling. *See* ION IMPLANTATION; ION PROPULSION; ISOTOPE (STABLE) SEPARATION; MASS SPECTROSCOPE; NUCLEAR FUSION; PARTICLE ACCELERATOR; SECONDARY ION MASS SPECTROMETRY (SIMS). [R.M.]

Ion-solid interactions Physical processes resulting from the collision of energetic ions, atoms, or molecules with condensed matter. These include elastic and inelastic backscattering of the projectile, penetration of the solid by the projectile, emission of electrons and photons from the surface, sputtering of neutral atoms and ions, production of defects in crystals, creation of nuclear tracks in insulating solids, and electrical, chemical, and physical changes to the irradiated matter resulting from the passage or implantation of the projectile.

When an energetic ion impinges upon the surface of condensed matter, it experiences a series of elastic and inelastic collisions with the atoms which lie in its path. These collisions occur because of the electrical forces between the nucleus and electrons of the projectile and those of the atoms which constitute the solid target. They result in the transformation of the kinetic energy of the projectile into internal excitation of the solid.

One of the most simple interactions occurs when the projectile collides with a surface atom and bounces back in generally the opposite direction from which it came. This process is known as backscattering. Its observation in 1911 led Ernest Rutherford to conclude that most of the matter in atoms is concentrated in a small nucleus. Now it is used as an analytical technique to measure the masses and locations of atoms on and near a surface. This technique for surface characterization is appropriately named Rutherford backscattering analysis, and is most commonly performed with alpha particles of about 2 MeV. Anther backscattering technique, known as ion-scattering spectrometry, uses projectiles with energies of perhaps 2 keV.

Although backscattering events are well enough understood to be used as analytical tools, they are relatively rare because they represent nearly head-on collisions between two nuclei. Far more commonly, a collision simply deflects the projectile a few degrees from its original direction and slows it somewhat, transferring some of its kinetic energy to the atom that is struck. Thus, the projectile does not rebound from the surface but penetrates deep within the solid, dissipating its kinetic energy in a series of grazing collisions.

The capacity of a solid to slow a projectile is called the stopping power, and is defined as the amount of energy lost by the projectile per unit length of trajectory in the solid. Stopping power is of central importance for many phenomena because it measures the capacity of a projectile to deposit energy within a thin layer of the solid and this energy drives secondary processes associated with penetration. In many insulating solids (including mica, glasses, and some plastics) the passage of an ion with a large electronic stopping power creates a unique form of radiation damage known as a nuclear track. When the substance is chemically etched, conical pits visible under an ordinary microscope are produced where ionizing particles have penetrated. The passage of single projectiles may thereby be observed. *See* PARTICLE TRACK ETCHING.

In the nuclear stopping region it is relatively likely that the projectile will transfer significant amounts of energy to individual target atoms. These atoms will subsequently strike others, and eventually a large number of atoms within the solid will be set in motion. This disturbance is known as a collision cascade. Collision cascades may cause permanent damage to materials, induce mixing of layers in the vicinity of interfaces, or cause sputtering if they occur near surfaces. *See* ION BEAM MIXING; SPUTTERING.

Ion implantation is used in the manufacture of integrated circuits and in the improvement of surface properties of metals. Ion-solid processes permit highly sensitive analyses for trace elements, the characterization of materials and surfaces, and the detection of ionizing radiation. Techniques employing them include secondary ion mass spectrometry (SIMS) for elemental analysis and imaging of surfaces, proton-induced x-ray emission (PIXE), ion-scattering spectrometry (ISS), and Rutherford backscattering analysis (RBS). They are also fundamental to the operation of silicon surface-barrier detectors which are used for the measurement of particle radiation, and of nuclear track detectors which are used in research as diverse as the dating of meteorites and the search for magnetic monopoles. *See* ACTIVATION ANALYSIS; BEAM-FOIL SPECTROSCOPY; ION IMPLANTATION; MAGNETIC MONOPOLES; PARTICLE DETECTOR; PROTON-INDUCED X-RAY EMISSION (PIXE); SECONDARY ION MASS SPECTROMETRY (SIMS); SURFACE PHYSICS. [R.A.We.]

Ion transport Movement of salts and other electrolytes in the form of ions from place to place within living systems.

In animals, ion transport may occur by any of several different mechanisms: electrochemical diffusion, active-transport requiring energy, or bulk flow as in the flow of blood in the circulatory system of animals or the transpiration stream in the xylem tissue of plants. The best-known system transporting ions actively is the sodium/potassium (Na/K) exchange pump which occurs in plasma membranes of virtually all cells. The physical basis of the Na/K pump has been identified as a form of the enzyme adenosinetriphosphatase (ATPase) which requires both Na^+ and K^+ for its activity of catalyzing the hydrolysis of adenosinetriphosphate (ATP). This reaction provides the energy required for the extrusion of Na^+ from cells in exchange for K^+ taken in, but the mechanism of coupling of the energy liberation with the transport process has not been established. [B.T.S.]

In plants, transport processes are involved in uptake and release of inorganic ions and in distribution of ions within plants, and thus determine ionic relations of plants.

In unicellular, filamentous, or simple thalloid algae, in mosses, in poorly differentiated aquatic higher plants, and experimentally in cell suspension cultures of higher plants, ion transport can be considered on the cellular level. In these systems all cells take up ions directly from the external medium. Intracellular distribution and compartmentation are determined by transport across other membranes within the cells.

Within tissues the continuous cell walls of adjacent cells form an apoplastic pathway for ion transport. A symplastic pathway is constituted by the cytoplasm extending from cell to cell via small channels crossing the cell walls. Transport over longer distances is important in organs (roots, shoots, leaves, fruits), which are composed of different kinds of tissues, and in the whole plant. Xylem and phloem serve as pathways for long-distance transport. Roots take up ions from the soil and must supply other plant organs. But there is also circulation within the plant; for example, potassium is readily transported both from root to shoot and in the opposite direction. *See* PHLOEM; XYLEM. [U.L.]

Ionic crystals A class of crystals in which the lattice-site occupants are charged ions held together primarily by their electrostatic interaction. Such binding is called ionic binding. Empirically, ionic crystals are distinguished by strong absorption of infrared radiation, good ionic conductivity at high temperatures, and the existence of planes along which the crystals cleave easily. *See* CRYSTAL.

Compounds of strongly electropositive and strongly electronegative elements form solids which are ionic crystals, for example, the alkali halides, other monovalent metal halides, and the alkaline-earth halides, oxides, and sulfides. Crystals in which some of the ions are complex, such as metal carbonates, metal nitrates, and ammonium salts, may also be classed as ionic crystals.

As a crystal type, ionic crystals are to be distinguished from other types such as molecular crystals, valence crystals, or metals. The ideal ionic crystal as defined is approached most closely by the alkali halides (see illustration). Other crystals often classed as ionic have binding which is not exclusively ionic but includes a certain admixture of covalent binding. Thus the term ionic crystal refers to an idealization to which real crystals correspond to a greater or lesser degree, and crystals exist having characteristics of more than one crystal type. *See* CHEMICAL BONDING.

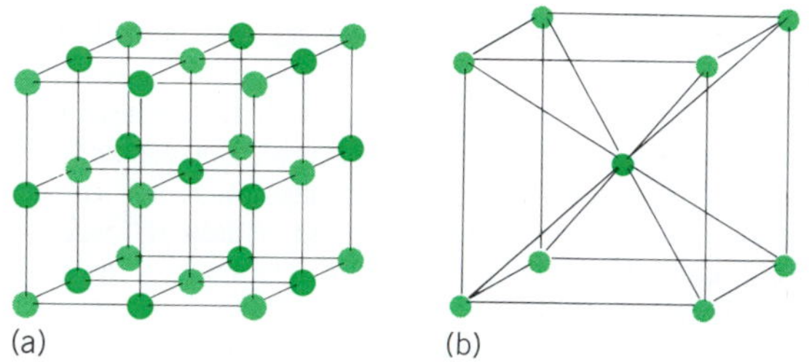

Lattices of (*a*) sodium chloride (darker circles are positive ions and lighter circles are negative ions) and (*b*) cesium chloride. (*After F. Seitz, The Modern Theory of Solids, McGraw-Hill, 1940*)

Ionic crystals, especially alkali halides, have played a very prominent role in the development of solid-state physics. They are relatively easy to produce as large, quite pure, single crystals suitable for accurate and reproducible experimental investigations. In addition, they are relatively easy to subject to theoretical treatment since they have simple structures and are bound by the well-understood Coulomb force between the ions. Being readily available and among the simplest known solids, they have thus been a frequent and profitable meeting place between theory and experiment. These same features of ionic crystals have made them attractive as host crystals for the study of crystal defects: deliberately introduced impurities, vacancies, interstitials, and color centers. *See* COLOR CENTERS. [B.G.D.]

Ionic equilibrium An equilibrium in a chemical reaction in which at least one ionic species is produced, consumed, or changed from one medium to another.

The wide variety of types of ionic equilibrium possible include: dissolution of an un-ionized substance, for example, the dissolution of hydrogen chloride (a gas) in water (an ionizing solvent), reactions (1)–(3); dissolution of a crystal in

$$HCl(g) \rightleftharpoons H^+ + Cl^- \qquad (1)$$

$$HCl(g) + H_2O \rightleftharpoons H_3O^+ + Cl^- \qquad (2)$$

$$HCl(g) + 4H_2O \rightleftharpoons H_9O_4^+ + Cl^- \qquad (3)$$

water, such as the dissociation of solid silver chloride, reaction (4); dissociation of a strong acid, for example, nitric acid, HNO_3, dissociates as it dissolves in water, as in reaction (5); dissociation of an ion in water, for example, the bisulfate ion, HSO_4^-, dissociates in water, as in reaction (6); and dissociation of water itself, is represented by reaction (7).

$$AgCl(crystal) \rightleftharpoons Ag^+ + Cl^- \quad (4)$$

$$HNO_3 + H_2O \rightleftharpoons H_3O^+ + NO_3^- \quad (5)$$

$$HSO_4^- + H_2O \rightleftharpoons H_3O^- + SO_4^{2+} \quad (6)$$

$$2H_2O \rightleftharpoons H_3O^+ + OH^- \quad (7)$$

See ACID AND BASE; CHEMICAL EQUILIBRIUM; HYDROLYSIS. [T.F.Y.]

Ionization The process by which an electron is removed from an atom, molecule, or ion. It is of basic importance to electrical conduction in gases and liquids. In the simplest case, ionization may be thought of as a transition between an initial state consisting of a neutral atom and a final state consisting of a positive ion and a free electron. In more complicated cases, a molecule may be converted to a heavy positive ion and a heavy negative ion which are separated. [G.H.M.]

Ionization chamber An instrument for detecting ionizing radiation by measuring the amount of charge liberated by the interaction of ionizing radiation with suitable gases, liquids, or solids.

While the gold leaf electroscope is the oldest form of ionization chamber, instruments of this type are still widely used as monitors of radiations by workers in the nuclear or radiomedical professions. However, for many purposes it is useful to measure the ionization pulse produced by a single ionizing particle. *See* ELECTROSCOPE; MONITORING OF IONIZING RADIATION.

The simplest form of a pulse ionization chamber consists of two conducting electrodes in a container filled with gas (see illustration). A battery, or other power supply, maintains an electric field between the positive anode and the negative cathode. When ionizing radiation penetrates the gas in the chamber—entering, for example, through a thin gastight window—this radiation liberates electrons from the gas atoms leaving positively charged ions. The electric field present in the gas sweeps these electrons and ions out of the gas, the electrons going to the anode and the positive ions to the cathode.

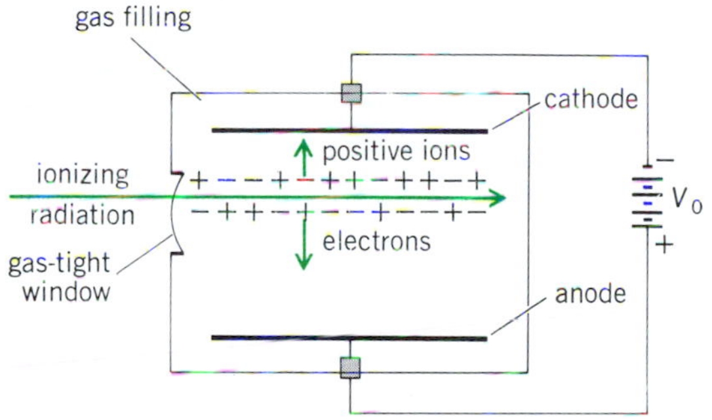

Parallel-plate ionization chamber.

In a chamber, such as that represented in the illustration, the current begins to flow as soon as the electrons and ions begin to separate under the influence of the applied electric field. The time it takes for the full current pulse to be observed depends on the drift velocity of the electrons and ions in the gas. Because the ions are thousands of times more massive than the electrons, the electrons always travel several orders of magnitude faster than the ions. As a result, virtually all pulse ionization chambers make use of only the relatively fast electron signal.

One of the most important uses of an ionization chamber is to measure the total energy of a particle or, if the particle does not stop in the ionization chamber, the energy lost by the particle in the chamber. In addition to energy information, ionization chambers are now routinely built to give information about the position within the gas volume where the initial ionization event occurred. This information can be important not only in experiments in nuclear and high-energy physics where these position-sensitive detectors were first developed, but also in medical and industrial applications.

Foremost among the other applications is the use of gas ionization chambers for radiation monitoring. Portable instruments of this type usually employ a detector containing approximately 60 in.^{3} (1 liter) of gas, and operate by integrating the current produced by the ambient radiation. Another application of ionization chambers is the use of air-filled chambers as domestic fire alarms. Yet another development in ion chamber usage is that of two-dimensional imaging in x-ray medical applications to replace the use of photographic plates.

Gaseous ionization chambers have also found application as total-energy monitors for high-energy accelerators. Such applications involve the use of a very large number of interleaved thin parallel metal plates immersed in a gas inside a large container.

Ionization chambers can be made where the initial ionization occurs, not in gases, but in suitable liquids or solids. In the solid-state ionization chamber (or solid-state detector) the gas filling is replaced by a large single crystal of suitably chosen solid material. In this case the incident radiation creates electron-hole pairs in the crystal, and this constitutes the signal charge. Silicon and germanium detectors have proved to be highly successful and have led to detectors that have revolutionized low-energy nuclear spectroscopy. The use of a liquid in an ionization chamber combines many of the advantages of both solid and gas-filled ionization chambers; most importantly, such devices have the flexibility in design of gas chambers with the high density of solid chambers. During the 1970s a number of groups built liquid argon ionization chambers and demonstrated their feasibility. [W.A.L.]

Ionization gage An instrument for measuring vacuum by ionizing the gas present and measuring the ion current. There are two types of ionization gages.

In the hot-filament ionization gage (see illustration), electrons emitted by a filament are attracted toward a positively charged grid electrode. Collisions of electrons with gas molecules produce ions, which are then attracted to a negatively charged electrode. The current measured at this electrode is directly proportional to the pressure or gas density.

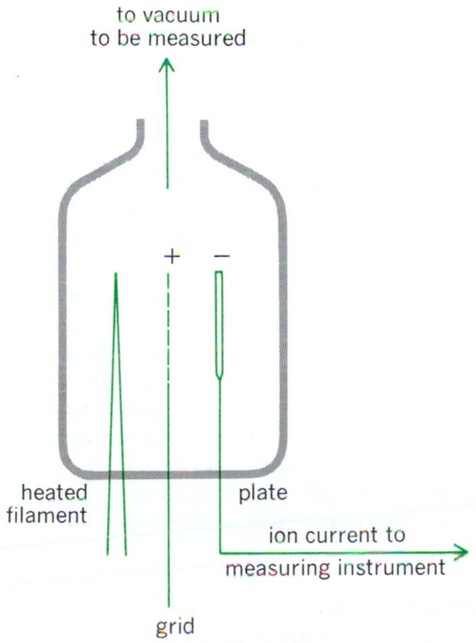

Hot-filament ionization gage.

In the cold-cathode (Philips or Penning) ionization gage, a high voltage is applied between two electrodes. Fewer electrons are emitted, but a strong magnetic field deflects the electron stream, increasing the length of the electron path which increases the chance for ionizing collisions of electrons with gas molecules. See VACUUM MEASUREMENT. [R.C.]

Ionization potential The potential difference through which a bound electron must be raised to free it from the atom or molecule to which it is attached. In particular, the ionization potential is the difference in potential between the initial state, in which the electron is bound, and the final state, in which it is at rest at infinity.

The ionization potential for the removal of an electron from a neutral atom other than hydrogen is more correctly designated as the first ionization potential. The potential associated with the removal of a second electron from a singly ionized atom or molecule is then the second ionization potential, and so on. [G.H.M.]

Ionophore A substance that can transfer ions from a hydrophilic medium, such as water, into a hydrophobic medium, such as hexane or a biological membrane, where the ions typically would not be soluble; also known as an ion carrier. The ions transferred are usually metal ions, for example, lithium (Li^+), sodium (Na^+), potassium (K^+), magnesium (Mg^{2+}), and calcium (Ca^{2+}); but there are ionophores that promote the transfer of other ions, such as ammonium ion (NH_4^+) or amines of biological interest. See ION.

There are two mechanisms by which ionophores promote the transfer of ions across hydrophobic barriers: ion-ionophore complex formation and ion channel formation. In complex formation, the ion forms a coordination complex with the ionophore in which there is a well-defined ratio (typically 1:1) of ion to ionophore. The ionophore wraps around the ion so that the ion exists in the polar interior of the complex, while the exterior is predominantly hydrophobic in character and as such is soluble in nonpolar media. The ion is coordinated by oxygen atoms present in the ionophore molecule through ion-dipole interactions. The ionophore molecule essentially acts as the solvent for the ion, replacing the aqueous solvation shell that normally surrounds the ion. See COORDINATION COMPLEXES.

Ionophores that act via ion channel formation are found in biological environments. The molecule forms a polar channel in an otherwise nonpolar cell membrane, allowing passage of small ions either into or out of the cell.

Naturally occurring ionophores fall into four classes, each of which has antibiotic activity: peptide, cyclic depsipeptide, macrotetrolide, and polyether ionophores. The biological activity of ionophore antibiotics is due to their ability to disrupt the flow of ions either into or out of cells. Under normal conditions, cells have a high internal concentration of potassium ions but a low concentration of sodium ions. The concentration of ions in the extracellular medium is just the reverse, high in sodium ions but low in potassium ions. This imbalance, which is necessary for normal cell function, is maintained by a specific transport protein (sodium-potassium adenosine triphosphatase) in the cell membrane that pumps sodium ions out of the cell in exchange for potassium ions. Ionophore antibiotics can disrupt this ionic imbalance by allowing ions to penetrate the cell membrane as ion-ionophore complexes or via the formation of ion channels. Gram-positive bacteria appear to be particularly sensitive to the effect of ionophores perturbing normal ion transport. See ANTIBIOTIC; ION TRANSPORT. [C.A.V.]

Ionosphere That part of the upper atmosphere which is sufficiently ionized by solar ultraviolet radiation so that the concentration of free electrons affects the propagation of radio waves. See IONIZATION; RADIO BROADCASTING; RADIO-WAVE PROPAGATION.

The ionosphere is structured in the vertical direction. It was first thought that discrete layers were involved, referred to as the D, E, F_1, and F_2 layers; however, rocket measurements have shown that the "layers" merge with one another to such an extent that they are now normally referred to as regions rather than layers. Since a vertically incident radio wave is reflected from the ionosphere at that level where the natural frequency of the plasma equals the radio frequency, it is possible to identify the electron concentration at the point of reflection in terms of the radio frequency. See ATMOSPHERE. [F.S.J.]

IR drop That component of the potential drop across a passive element (one which is not a seat of electromotive force) in an electric circuit caused by resistance of the element. This potential drop, by definition, is the product of the resistance *R* of the element and the current *I* flowing through it. The *IR* drop across a resistor is the difference of potential between the two ends of the resistor. See ELECTRICAL RESISTANCE; POTENTIALS. [J.W.St.]

Iridium A chemical element, Ir, atomic number 77, and atomic weight 192.2. Iridium in the free state is a hard, white metallic substance. See TRANSITION ELEMENTS.

Iridium has considerably less oxidation resistance than platinum or rhodium, but more so than ruthenium or osmium. A thin, adherent oxide film, IrO_2, forms above about 600°C

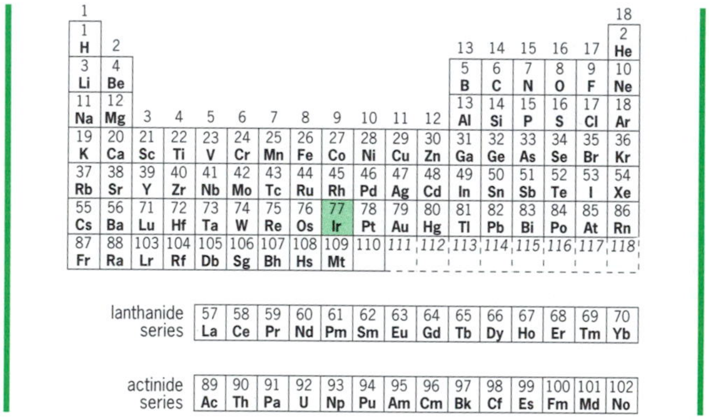

(1110°F). It is the only metal which can be used unprotected in air up to 2300°C (4170°F) with any degree of life expectancy. Iridium is not attacked by any acid, including aqua regia. Iridium has a strong tendency to form coordination compounds. The table gives values for some important properties of iridium.

Principal compounds include iridium trichloride, $IrCl_3$, a green, water-insoluble compound; sodium iridium(IV) chloride, $Na_2IrCl_6 \cdot 6H_2O$, a black, water-soluble crystalline solid; sodium

Some properties of iridium

Property	Value
Atomic weight, ^{12}C = 12.00000	192.2
Naturally occurring isotopes and percent abundance	192 (37.3%), 193 (62.7%)
Crystal structure	Face-centered cubic
Common chemical valence	3,4
Density at 25°C	22.55 g/cm³
Melting point	2447°C
Boiling point	4500°C
Specific heat at 0°C	0.0307 cal/g (0.0039 joule/g)

iridium(III) chloride, $Na_3IrCl_6 \cdot 12H_2O$, an olive-green, water-soluble crystalline solid; and ammonium iridium(IV) chloride, $(NH_4)_2IrCl_6$, a red-black, relatively insoluble crystalline solid.

Iridium's special properties have led to specialized applications, including crucibles for high-temperature growth of laser crystals, iridium-rhodium very-high-temperature thermocouples, and cladding applied over other materials. It is usually alloyed with platinum as the base, since a platinum–30% iridium alloy, for instance, is almost as corrosion-resistant as iridium and is great deal easier to fabricate. *See* Platinum; Rhodium. [H.J.A.]

Iron A chemical element, Fe, atomic number 26, and atomic weight 55.847. Iron is the fourth most abundant element in the crust of the Earth (5%). It is a malleable, tough, silver-gray, magnetic metal. It melts at 1540°C, boils at 2800°C, and has a density of 7.86 g/cm^3. The four stable, naturally occurring isotopes have masses of 54, 56, 57, and 58. The two main ores are hematite, Fe_2O_3, and limonite, $Fe_2O_3 \cdot 3H_2O$. Pyrites, FeS_2, and chromite, $Fe(CrO_2)_2$, are mined as ores for sulfur and chromium, respectively. Iron is found in many other minerals, and it occurs in groundwaters and in the red hemoglobin of blood.

The greatest use of iron is for structural steels; cast iron and wrought iron are made in quantity, also. Magnets, dyes (inks, blueprint paper, rouge pigments), and abrasives (rouge) are among the other uses of iron and iron compounds. *See* Cast iron; Iron alloys; Iron metallurgy; Stainless steel; Steel manufacture; Wrought iron.

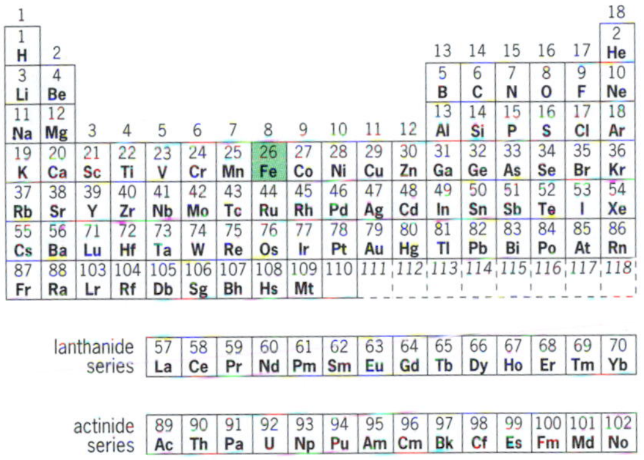
1
18
1 H
2
13 14 15 16 17
2 He
3 Li 4 Be
5 B 6 C 7 N 8 O 9 F 10 Ne
11 Na 12 Mg
3 4 5 6 7 8 9 10 11 12
13 Al 14 Si 15 P 16 S 17 Cl 18 Ar
19 K 20 Ca 21 Sc 22 Ti 23 V 24 Cr 25 Mn 26 Fe 27 Co 28 Ni 29 Cu 30 Zn 31 Ga 32 Ge 33 As 34 Se 35 Br 36 Kr
37 Rb 38 Sr 39 Y 40 Zr 41 Nb 42 Mo 43 Tc 44 Ru 45 Rh 46 Pd 47 Ag 48 Cd 49 In 50 Sn 51 Sb 52 Te 53 I 54 Xe
55 Cs 56 Ba 71 Lu 72 Hf 73 Ta 74 W 75 Re 76 Os 77 Ir 78 Pt 79 Au 80 Hg 81 Tl 82 Pb 83 Bi 84 Po 85 At 86 Rn
87 Fr 88 Ra 103 Lr 104 Rf 105 Db 106 Sg 107 Bh 108 Hs 109 Mt 110 111 112 113 114 115 116 117 118
lanthanide series
57 La 58 Ce 59 Pr 60 Nd 61 Pm 62 Sm 63 Eu 64 Gd 65 Tb 66 Dy 67 Ho 68 Er 69 Tm 70 Yb
actinide series
89 Ac 90 Th 91 Pa 92 U 93 Np 94 Pu 95 Am 96 Cm 97 Bk 98 Cf 99 Es 100 Fm 101 Md 102 No

There are several allotropic forms of iron. Ferrite or α-iron is stable up to 760°C (1400°F). The change of ß-iron involves primarily a loss of magnetic permeability because the lattice structure (body-centered cubic) is unchanged. The allotrope called γ-iron has the cubic close-packed arrangements of atoms and is stable from 910 to 1400°C (1670 to 2600°F). Little is known about δ-iron except that it is stable above 1400°C (2600°F) and has a lattice similar to that of α-iron.

The metal is a good reducing agent and, depending on conditions, can be oxidized to the 2+, 3+, or 6+ state. In most iron compounds, the ferrous ion, iron(II), or ferric ion, iron(III), is present as a distinct unit. Ferrous compounds are usually light yellow to dark green-brown in color; the hydrated ion, $Fe(H_2O)_6^{2+}$, which is found in many compounds and in solution, is light green. This ion has little tendency to form coordination complexes except with strong reagents such as cyanide ion, polyamines, and porphyrins. The ferric ion, because of its high charge (3+) and its small size, has a strong tendency to hold anions. The hydrated ion, $Fe(H_2O)_6^{3+}$, which is found in solution, combines with OH^-, F^-, Cl^-, CN^-, SCN^-, N_3^-, $C_2O_4^{2-}$, and other anions to form coordination complexes. *See* Coordination chemistry.

An interesting aspect of iron chemistry is the array of compounds with bonds to carbon. Cementite, Fe_3C, is a component of steel. The cyanide complexes of both ferrous and ferric iron are very stable and are not strongly magnetic in contradistinction to most iron coordination complexes. The cyanide complexes form colored salts. *See* Transition elements. [J.O.E.]

Iron alloys Solid solutions of metals, one metal being iron. A great number of commercial alloys have iron as an intentional constituent. Iron is the major constituent of wrought and cast iron and wrought and cast steel. Alloyed with large amounts of silicon, manganese, chromium, vanadium, molybdenum, niobium (columbium), selenium, titanium, phosphorus, or other elements, singly or sometimes in combination, iron forms the large group of materials known as ferroalloys that are important as addition agents in steelmaking. Iron is also a major constituent of many special-purpose alloys developed to have exceptional characteristics with respect to magnetic properties, electrical resistance, heat resistance, corrosion resistance, and thermal expansion. *See* Alloy; Ferroalloy; Steel. [H.E.McG.]

Iron metallurgy Extracting iron from ores and preparing it for use. Extraction involves the conversion of naturally occurring iron-bearing minerals into metallic iron. The term ironmaking is commonly used to include all of the industrial processes that convert raw materials into iron. The major process for the production of iron is the iron blast furnace. However, since 1970 the growth of alternative direct-reduction processes has been very significant. The principal difference between the blast furnace process and the direct-reduction processes is the temperature of operation. In the blast furnace, high operating temperatures enable the production of molten iron. At the lower operating temperatures of the direct-reduction processes, solid or sponge iron is produced. Most of the iron produced in the world is used in the production of steel. The remainder is converted to iron castings, ferroalloys, and iron powder. *See* Ferroalloy; Iron; Pressurized blast furnace; Pyrometallurgy; Steel manufacture. [G.R.St.P.]

Ironwood The name given to any of at least 10 kinds of tree in the United States, including the American hornbeam (*Carpinus caroliniana*), eastern hophornbeam (*Ostrya virginiana*), buckthorn bumelia (*Bumelia lycioides*), tough bumelia (*B. tenax*), buckwheat tree (*Cliftonia monophylla*), and swamp cyrilla or swamp ironwood (*Cyrilla racemiflora*). Leadwood (*Krugiodendron ferreum*), a native of southern Florida, has the highest specific gravity of all woods native to the United States and is also known as black ironwood. *See* Hophornbeam; Hornbeam. [A.H.G./K.P.D.]

Irregularia The name given by G. Cuvier in 1817 to an assemblage of echinoids in which the anus and periproct lie outside the apical system, the ambulacral plates remain simple, the primary radioles are hollow, and the rigid test shows more or less bilateral symmetry. The Irregularia are an artificial assemblage of similar but unrelated forms, and the name has no taxonomic validity. *See* Diadematacea; Echinoidea; Pygasteroida; Regularia. [H.B.F.]

Irrigation (agriculture) The artificial application of water to the soil to produce plant growth. Irrigation also cools the soil and atmosphere, making the environment favorable for plant growth. Water is applied to crops by surface, subsurface, sprinkler, and drip irrigation.

Surface irrigation includes furrow and flood methods. The furrow method is used for row crops. Corrugations or rills are small furrows used on close-growing crops. The flow, carried in furrows, percolates into the soil. Flow to the furrow is usually supplied by siphon tubes, spiles, gated pipe, or valves from

A side-roll sprinkler system which uses the main supply line (often more than 1000 ft, or 300 m, long) to carry the sprinkler heads and as the axle for wheels.

buried pipe. In the flood method, controlled flooding is done with border strips, contour or bench borders, and basins.

Subirrigation is a type of irrigation accomplished by raising the water table to the root zone of the crop or by carrying moisture to the root zone by perforated underground pipe. Either method requires special soil conditions for successful operation.

A sprinkler system consists of pipelines which carry water under pressure from a pump or elevated source to lateral lines along which sprinkler heads are spaced at appropriate intervals. Laterals are moved from one location to another by hand or tractor, or they are moved automatically. The side-roll wheel system, which utilizes the lateral as an axle (see illustration), is very popular as a labor-saving method.

Drip irrigation is a method of providing water to plants almost continuously through small-diameter tubes and emitters. It has the advantage of maintaining high moisture levels at relatively low capital costs. It can be used on very steep, sandy, and rocky areas and can utilize saline waters better than most other systems. The system has been most popular in orchards and vineyards, but is also used for vegetables, ornamentals, and for landscape plantings. *See* LAND DRAINAGE (AGRICULTURE); TERRACING (AGRICULTURE). [M.A.H.]

Irrotational flow Fluid flow in which each infinitesimal particle or element of the fluid preserves its original orientation. Because an element of fluid can be caused to rotate about its axis only by the application of viscous forces, irrotational flow is possible only for an ideal or nonviscous fluid. For fluids of low viscosity, such as air or water, irrotational flow may be approached in a free vortex. In a free vortex, a body of fluid rotates without the application of external torque because of angular momentum previously imparted to it. Examples are the rotation of fluid after leaving the impeller of a centrifugal pump, a tornado, or the rotation of water entering the drain of a tub. *See* FLUID FLOW; VORTEX. [R.L.D.]

Ischemic colitis A broad spectrum of responses occurring in individuals who have compromised circulation to the colon. Three interrelated factors must be altered before the condition is clinically apparent: the vessels supplying the colon must be compromised; the state of the microcirculation must be compromised; and the "virulence" of the bacteria in the lumen of the bowel must be changed. If minimal alterations are present, transient mucosal necrosis occurs, manifested by episodes of bloody diarrhea, with complete recovery of the patient. If more severe alterations occur, the extent of mucosal necrosis is greater. The latter presentation heals by fibrosis, which is clinically manifested by the presence of a colonic stricture. In its severest presentation, transmural involvement occurs with the development of gangrene of the colon. The ischemic process can be present at any level of the gastrointestinal tract, and affect any age group. Examples of the spectrum of ischemic bowel disease include ischemic colitis, acute necrotizing enterocolitis of premature infants, hemorrhagic necrosis of the gastrointestinal tract, and pseudomembranous, staphylococcal, uremic, and radiation enterocolitis and stress ulceration. [H.T.N.]

Isentropic flow The flow of a fluid is isentropic when its entropy is identical at all points in the flow. Isentropic flow can be approached for fluids flowing either in a duct or over the outside of a body. Because the entropy of the fluid is a thermodynamic property, similar to the enthalpy or energy of a fluid, the value of the entropy is fixed by the state of the fluid. For a pure substance, in the absence of external forces, entropy is a function of two independent properties. For example, in the absence of gravity, capillarity, electricity, and magnetism, the entropy of a single-phase fluid is a function of the pressure and temperature. *See* ENTHALPY; ENTROPY; ISENTROPIC PROCESS; THERMODYNAMIC PRINCIPLES. [P.E.Bl.]

Isentropic process In thermodynamics, a process involving change without any increase or decrease of entropy. Since the entropy always increases in a spontaneous process, one must consider reversible or quasistatic processes. During a reversible process the quantity of heat transferred is directly proportional to the system's entropy change. Systems which are thermally insulated from their surroundings undergo processes without any heat transfer; such processes are called adiabatic. Thus during an isentropic process there are no dissipative effects and the system neither absorbs nor gives off heat. For this reason the isentropic process is sometimes called the reversible adiabatic process. *See* ADIABATIC PROCESS; ENTROPY; THERMODYNAMIC PROCESSES. [P.E.Bl.]

Isentropic surfaces Surfaces along which the entropy and potential temperature of air are constant. Potential temperature, in meteorological usage, is defined by the relationship

$$\theta = T\left(\frac{1000}{P}\right)^{\frac{cp-cv}{cp}}$$

in which T is the air temperature, P is atmospheric pressure expressed in millibars, c_p is the specific heat of air at constant pressure, and c_v is the specific beat at constant volume. Since the potential temperature of an air parcel does not change if the processes acting on it are adiabatic (no exchange of heat between the parcel and its environment), a surface of constant potential temperature is also a surface of constant entropy. An advantage of representing meteorological conditions on isentropic surfaces is that there is usually little air motion through such surfaces, since thermodynamic processes in the atmosphere are approximately adiabatic. See ADIABATIC PROCESS. [F.S.]

Ising model A model which consists of a lattice of "spin" variables with two characteristic properties: (1) each of the spin variables independently takes on either the value $+1$ or the value -1; and (2) only pairs of nearest neighboring spins can interact. The study of this model in two dimensions forms the basis of the modern theory of phase transitions and, more generally, of cooperative phenomena.

A macroscopic piece of material consists of a large number of atoms, the number being of the order of Avogadro's number (approximately 6×10^{23}). Thermodynamic phenomena all depend on the participation of such a large number of atoms. Even though the fundamental interaction between atoms is short-ranged, the presence of this large number of atoms can, under suitable conditions, lead to an effective interaction between widely separated atoms. Phenomena due to such effective long-range interactions are referred to as cooperative phenomena. The simplest examples of cooperative phenomena are phase transitions. The most familiar phase transition is either the condensation of steam into water or the freezing of water into ice. Only slightly less familiar is the ferromagnetic phase transition that takes place at the Curie temperature, which, for example, is roughly 1043 K for iron. *See* CURIE TEMPERATURE; FERROMAGNETISM; PHASE TRANSITIONS. [B.M.McC.; T.T.W.]

Island biogeography Islands generally have fewer species of animals and plants than do comparable continental areas, at least if the islands lie some distance from continents. Often there are rather few genera, but some genera have many local species that sometimes exhibit strong adaptive radiation. The proportion of endemic species (found nowhere else) that are present generally increases with the degree of isolation and with the complexity of the habitat. Gigantism, flightlessness, and other unusual characteristics are relatively frequent.

Island floras and faunas are of far greater interest than either the number of species or their economic importance might seem to justify. How the plants and animals came to be on the islands, why so many of them are found nowhere else, why so many have special or strange forms, and where their ancestors came from, are fascinating questions which present problems of great scientific importance to the biologist. The theory of organic evolution emerged from studies of island faunas by Charles Darwin and Alfred Russel Wallace, and islands are still among the most advantageous places to study evolutionary processes. They also offer uniquely suitable sites for the investigation of the nature and functioning of ecosystems. *See* ECOLOGY; SPECIATION. [F.R.F.]

Isoantigen An immunologically active protein or polysaccharide present in some but not all individuals in a particular species. These substances initiate the formation of antibodies when introduced into other individuals of the species that genetically lack the isoantigen. Like all antigens, they are also active in stimulating antibody production in heterologous species. The ABO, MN, and Rh blood factors in humans constitute important examples. Consequently, elaborate precautions for typing are required in blood transfusion. *See* ANTIBODY; ANTIGEN; BLOOD GROUPS. [M.J.Po.]

Isobar (meteorology) A curve along which pressure is constant. Leading examples of its uses are in weather forecasting and meteorology. The most common weather maps are charts of weather conditions at the Earth's surface and mean sea level, and they contain isobars as principal information. Areas of bad or unsettled weather are readily defined by roughly circular isobars around low-pressure centers at mean sea level. Likewise, closed isobars around high-pressure centers define areas of generally fair weather. *See* AIR PRESSURE.

A principal use of isobars stems from the so-called geostrophic wind, which approximates the actual wind on a large scale. The direction of the geostrophic wind is parallel to the isobars, in the sense that if an observer stands facing away from the wind, higher pressures are to the person's right if in the Northern Hemisphere and to the left if in the Southern. Thus, in the Northern Hemisphere, flow is counterclockwise about low-pressure centers and clockwise about high-pressure centers, with the direction of the flow reversed in the Southern Hemisphere. *See* GEOSTROPHIC WIND; WEATHER MAP. [F.B.Sh.]

Isobar (nuclear physics) One of two or more atoms which have a common mass number A but which differ in atomic number Z. Thus, although isobars possess approximately equal masses, they differ in chemical properties; they are atoms of different elements. Isobars whose atomic numbers differ by unity cannot both be stable; one will inevitably decay into the other. *See* ELECTRON CAPTURE; RADIOACTIVITY. [H.E.D.]

Isobaric process A thermodynamic process in which the heat transfer to or from the gaseous system causes a volume change at constant pressure. This process can be illustrated by the expansion of a substance when it is heated. The system is then capable of doing an amount of work on its surroundings. The maximum work will be done when the external pressure of the surroundings on the system is equal to the pressure of the system. *See* ISOMETRIC PROCESS; POLYTROPIC PROCESS; THERMODYNAMIC PROCESSES. [P.E.Bl./W.A.S.]

Isobaric spin A quantum-mechanical variable or quantum number applied to hadrons, the strongly interacting fundamental particles, and compounds of hadrons (such as nuclear states) to facilitate consideration of the consequences of the charge independence of the strong (nuclear) forces. This variable is rarely labeled isobaric spin or isospin, and most commonly I-spin.

The many strongly interacting elementary particles (hadrons) and the compounds of these particles, such as nuclei, are observed to form sets or multiplers such that the members of the multiplet differ in their electromagnetic charge and electromagnetic properties but are otherwise almost identical. For example, the neutron and proton, with electric charges of zero and plus one fundamental unit (of the magnitude of the electronic charge), form a set of two such states. The pions, one with a unit of positive charge, one with zero charge, and one with a unit of negative charge, form a set of three. It appears that if the effects of electromagnetic forces and the closely related weak nuclear forces (responsible for beta decay) are neglected, leaving only the strong forces effective, the different members of such a multiplet are equivalent and cannot be distinguished in their strong interactions. The strong interactions are thus independent of the different electric charges held by different members of the set; they are charge-independent. *See* ELEMENTARY PARTICLE; FUNDAMENTAL INTERACTIONS; HADRON. [R.K.A.]

Isobryales An order of mosses in which the plants are slender to robust. In some families, they are strongly flattened, mostly with creeping primary stem, and irregularly to pinnately divided into erect, ascending, dendroid, or pendulous branches. The leaves are in two, three, or many rows, generally symmetrical. The sporophytes are pleurocarpous or cladocarpous. The capsule varies from immersed to exserted and is commonly smooth. *See* BRYOPSIDA. [W.H.W.]

Isocyanate One of a group of neutral derivatives of primary amines, formula R—N═C═O, that are obtained by reaction with excess phosgene and loss of hydrogen chloride as in the following:

$$RNH_2 + COCl_2 \rightarrow RNHCOCl \rightarrow R{-}N{=}C{=}O + HCl$$

Isocyanates hydrolyze in air to the unstable carbamic acids (RNHCOOH) which revert to the original amines by loss of carbon dioxide. Unhydrolyzed isocyanate then reacts with the amine to give a symmetrical urea, RNHCONHR. Normally this is an important reaction only because the nuisance must be avoided, but controlled hydrolysis of free isocyanate groups to liberate carbon dioxide is the basis of one type of foamed plastic. *See* AMINE; POLYURETHANE RESINS. [L.B.C.]

Isoelectric focusing A variation of electrophoresis in which the medium supports a pH gradient which includes the isoelectric pH of the species being studied. Many charged macromolecules have both positive and negative charges on their surfaces, and the electrophoretic mobility is related to the net excess of charge of one type or the other. As the pH becomes more acidic, the number of positive charges increases, and as the pH becomes more basic, the number of negative charges increases. For each molecule of this type, there is one pH, the isoelectric pH, at which the net charge on the surface is zero, so that the molecule does not move when an electric field is applied and thus has an electrophoretic mobility of zero. All molecules of a given isoelectric pH will migrate to the same region—hence the term isoelectric focusing. This method is particularly good for the analysis of microheterogeneity of protein species and other species which may differ slightly in their chemical content. *See* ELECTROPHORESIS; ISOELECTRIC POINT. [B.R.W.]

Isoelectric point The pH value of the dispersion medium of a colloidal suspension at which the colloidal particles do not move in an electric field; that is, the particles are electrophoretically inert. The isoelectric point is often employed to characterize colloidal material such as the proteins. However, a range of values is usually necessary, since the isoelectric point varies detectably with (1) the size of the particles, (2) the purity, and (3) the concentration of other than hydrogen ions. *See* COLLOID; ELECTROPHORESIS; PH; PROTEIN. [W.O.M.]

Isoelectronic sequence A term used in spectroscopy to designate the set of spectra produced by different chemical elements ionized in such a way that their atoms or ions contain the same number of electrons. The sequence in the table is an example. Since the neutral atoms of these elements each contain Z electrons, removal of one electron from scandium, two from titanium, and so forth, yields a series of ions all of which have 20 electrons. Isoelectronic sequences are useful in predicting unknown spectra of ions belonging to a sequence in which other spectra are known. *See* ATOMIC STRUCTURE AND SPECTRA. [F.A.J./W.W.W.]

Example of isoelectronic sequence

Designation of spectrum	Emitting atom or ion	Atomic number, Z
CaI	Ca	20
ScII	Sc^{+}	21
TiIII	Ti^{2+}	22
VIV	V^{3+}	23
CrV	Cr^{4+}	24
MnVI	Mn^{5+}	25

Isoetales A monotypic order of the plant division Lycopodiophyta in the class Isoetopsida, containing only one genus, *Isoetes*. These plants are called quillworts because in all 64 species the leaves are long and narrow with a spoonlike base, spirally arranged upon an underground cormlike structure. Most species are semiaquatic, although a few are terrestrial. They are confined to the cooler regions of the world. *See* LYCOPODIOPHYTA; STEM. [P.A.V.]

Isolaimoidea The only superfamily in the order Isolaimida, consisting of one family and genus. They are rather large for free-living soil nematodes (0.1–0.2 in. or 3–6 mm) and are found in seldom cultivated sandy soils. Some forms have anterior annulations, while others have posterior transverse rows of punctations. The diagnostic characteristics of this superfamily are presence of six hollow tubes around the oral opening and two whorls of six circumoral sensilla. Amphids are apparently absent, though some authors speculate that their function is taken over by the dorsolateral papillae of the second whorl. The triradiate stoma is elongate and has thickened walls anteriorly. The esophagus is clavate. The female ovaries are opposed, outstretched, and may have flexures. The gubernaculum, supporting the thickened male spicules, has two dorsal apophyses, and male preanal supplementary organs are papilloid. Paired caudal papillae are large in males, small in females. *See* NEMATA. [A.R.M.]

Isolated singularity If a function f is analytic at all points of a disc centered at a, except perhaps at the point a itself, then a is called an isolated singularity of the function f. There are three mutually exclusive possibilities for the nature of such a singularity:

1. It may happen that $|f(z)|$ remains bounded near the point a. In that case the function can be extended to an analytic function even at the point a by suitably choosing the value of $f(a)$; the point a is then called a removable singularity.

2. It may happen that $\lim_{z \to a} |f(z)| = \infty$. In that case a is a removable singularity for the function $g(z) = 1/f(z) = 1/g(z)$ where $g(z)$ is an analytic function vanishing at the point a. The function $f(z)$ is said to have a pole at the point a.

3. If neither of the two preceding cases arises, the function $f(z)$ is said to have an essential singularity at the point a. Singularities of this type can occur and do so frequently. *See* ANALYTIC FUNCTION. [R.C.G.]

Isoleucine An amino acid with the structure shown, considered essential for normal growth of animals. The biosynthesis of isoleucine occurs when pyruvate and α-ketobutyrate react to form α-aceto-α-hydroxybutyrate, which undergoes rearrangement and reduction to α,β-dihydroxy-β-methylvalerate. Dehydration to the α-keto acid and transamination complete the biosynthesis. *See* AMINO ACIDS. [E.A.Ad.]

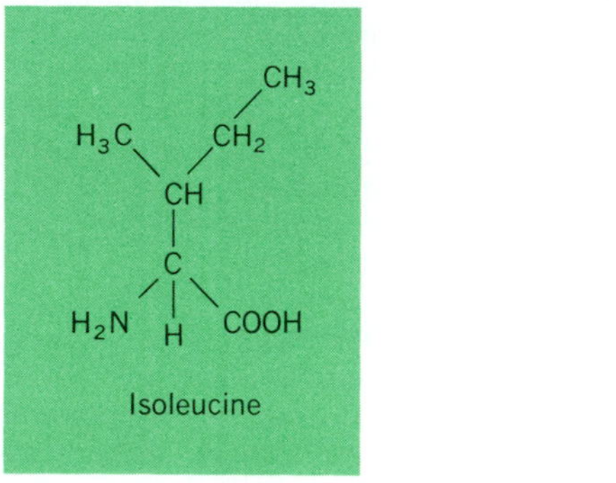

Isoleucine

Isomerization Rearrangement of the atoms within hydrocarbon molecules. Isomerization processes of practical significance in petroleum chemistry are (1) migration of alkyl groups, (2) shift of a single-carbon bond in naphthenes, and (3) double-bond shift in olefins. *See* ALKYLATION; AROMATIZATION; CRACKING; MOLECULAR ISOMERISM. [G.E.L.]

Isometric process A constant-volume thermodynamic process in which the system is confined by mechanically rigid boundaries. No direct mechanical work can be done on the surroundings by a system with rigid boundaries; therefore the heat transferred into or out of the system equals the change of internal energy stored in the system. This change in the internal energy, in turn, is a function of the specific heat and the temperature change in the system. See POLYTROPIC PROCESS. [P.E.Bl.]

Isomorphism (crystallography) A similarity of crystalline form between substances of similar composition. Two substances which are isomorphous have a similar chemical formula, an equal or nearly equal ratio of cation to anion radius, and comparable polarizabilities of their ions. Isomorphism is morphotropism in a narrower, more precise sense. Similarity in the macroscopic characteristics of isomorphous crystals becomes so close that extreme precision is needed to distinguish between them. See CRYSTAL. [W.C.D.]

Isopoda An order of malacostracan crustaceans characterized by a cephalon, or head, bearing one pair of maxillipeds in addition to the antennae, mandibles, and maxillae (see illustration). A carapace is lacking. The peraeon, or thorax, consists of seven distinct somites each bearing a pair of peraeopods, the legs. The pleon, or abdomen, has six somites. The first five pairs of pleonal appendages are foliaceous. The last pair is modified into hardened appendages called uropods and the last somite is fused with the telson into a pleotelson.

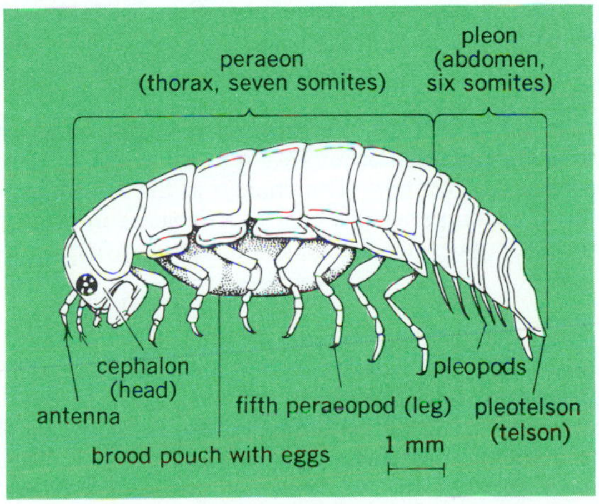

***Limnoria*, female. (After R. J. Menzies, *The comparative biology of the wood-boring isopod crustacean Limnoria*, Museum of Comparative Zoology, Harvard Coll. Bull., 112(5):363–388, 1954)**

The most familiar isopods are the terrestrial sow bugs or pill bugs. Many of the animals roll up into a compact ball when disturbed. Land isopods are usually found in moist environments, under decaying leaves and wood, and under rocks. The destructive marine wood-boring isopod *Limnoria*, the gribble, causes extensive damage to wharf piling in the United States, and one species is reported to attack treated timbers.

About 3000 species of isopods are known today, but it may be estimated that only one-half of the existing species have been described. See CRUSTACEA. [R.J.Me.]

Isoprene A five-carbon, conjugated diolefin, or diene, having the structure shown. It does not occur naturally but is obtained by the destructive distillation of gas oil, naphthas, and rubber.

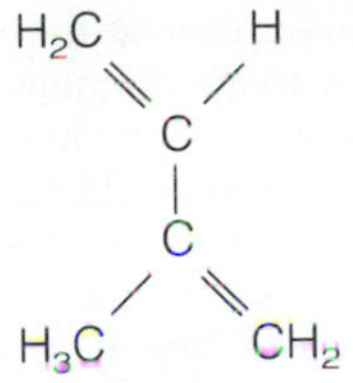

The terpenes may be regarded as multiples of the isoprene unit, C_5H_8. Indeed, isoprene may also be the foundation for other important plant products such as phytol, the sterols, and the carotenoids.

Isoprene is a mobile, colorless liquid boiling at 34.1°C (93.4°F), having a specific gravity of 0.862. It exhibits all the characteristic reactions of dienes of this type. Isoprene polymerizes readily to form dimers and high-molecular-weight resins. See TERPENE. [W.MOS.]

Isopropylphenol One of three isomeric compounds, having structures shown in the illustration. The nomenclature of the derivatives of these compounds is directly parallel to that of the cresol analogs.

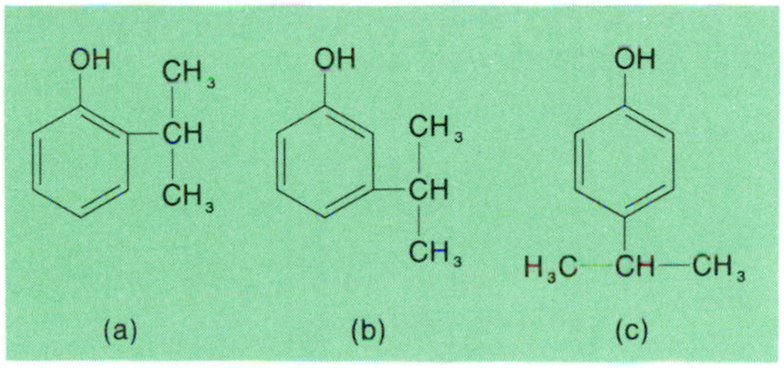

Structural formulas for (*a*) ortho, (*b*) meta, and (c) para isopropylphenol.

In many of the applications of which cresols have historically been used, one or another of the isopropylphenols can be used. As a by-product of an extremely large industrial manufacturing operation, that is, phenol from cumene, isopropylphenols are potentially plentiful and are free of many troublesome impurities commonly found in the cresols or cresylic acids of commerce, which are largely derived from coal tars or waste products of petroleum-refining operation. See CRESOL; PHENOL. [R.I.S.]

Isoptera An order of Insecta into which are grouped certain morphologically primitive forms commonly referred to as termites. They have a gradual metemorphosis, lacking true larval and pupal stages; the antennae are moniliform; the mouthparts are biting and prognathous; there are two pairs of subequal wings; and the abdomen is broadly joined to the thorax. All members of the Isoptera live in colonies. Termites perform a very important role in the breakdown of all types of cellulose debris.

Approximately 1900 species of Isoptera are recognized. Termites are often grouped into higher and lower forms. The higher termites include the family Termitidae, which contains about 80% of the species.

The mature termite (alate or imago) has membranous wings which extend beyond the end of the abdomen. In almost all termite species a second type of individual is produced in the colony. This is the soldier which lacks wings, is nonreproductive, and is variously modified for defense.

In the more advanced termites there is a third caste, the worker. Workers lack wings, are nonreproductive, and have mandibles which resemble those of the imagoes; they are usually blind. In addition to these definitive castes (alate, soldier, and worker), another type of individual may occur in the colony under certain circumstances. These individuals are the supplementary or replacement reproductives. Although the original pair (king and queen) may live for two or three decades, the life of the colony itself is not limited by their survival. If one or both are lost, other individuals in the colony become reproductive. *See* INSECTA; SOCIAL INSECTS; TERMITE. [F.M.We.]

Isopycnic The line of intersection of an atmospheric isopycnic surface with some other surface, for instance, a surface of constant elevation or of constant atmospheric pressure. An isopycnic surface is a surface in which the density of the air is constant. Such surfaces are also designated isosteric surfaces. On a surface of constant pressure, isopycnics coincide with isotherms, because on such a surface, density is a function solely of temperature. On a constant-pressure surface, isopycnics lie close together when the field is strongly baroclinic and are absent when the field is barotropic. *See* BAROCLINIC FIELD; BAROTROPIC FIELD; SOLENOID (METEOROLOGY). [F.S.]

Isoquinoline One of a group of organic compounds containing a benzene ring fused to the 3,4 positions of pyridine, having the structure shown. Isoquinoline is a representative

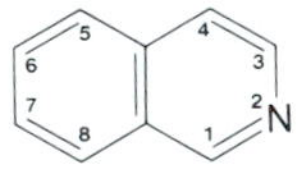

member of the group. Quinoline produced from coal tar contains approximately 1–4% of isoquinoline, and it is an important source of the latter material. Many plant alkaloids, especially those in the cactus, opium, and curare groups, are isoquinoline derivatives. *See* ALKALOID; HETEROCYCLIC COMPOUNDS; QUINOLINE.

Isoquinoline is a colorless, odorous, liquid with bp 243.3°C (469.9°F) and mp 26.5°C (79.7°F). Its stability to acid, base, or heat is high. With some exception, the reactions of isoquinoline have analogous counterparts in the reactions of quinoline. Substitution reactions such as nitration, sulfonation, bromination, and mercuration are observed. [W.J.Ge.]

Isostasy The principle that variations in the height of the Earth's surface are compensated by mass distributions beneath, leading to a state of purely hydrostatic pressure, independent of geographic position, at some moderate depth beneath sea level. For ideal local compensation, any column of given cross-sectional area through the Earth's crust, from the surface down to the level at which the hydrostatic state is reached, contains the same total mass. [G.D.G.]

Isotach A line along which the speed of the wind is constant. Isotachs are customarily represented on surfaces of constant elevation or atmospheric pressure, or in vertical cross sections. The closeness of spacing of the isotachs is indicative of the intensity of the wind shear on such surfaces. The term isovel is used synonymously with the term isotach. *See* WIND. [F.S.]

Isotachophoresis A variant of electrophoresis in which ionic species move with equal velocity in the presence of an electric field. The isotachophoretic condition is maintained when particles of different mobilities form boundaries within the solution so that the most mobile ions form the leading edge of the moving sample, followed by the less mobile ions in the order of their mobility. The reason that the ions with different mobility can move with the same velocity is that the electric field in each of the regions is inversely proportional to the mobility of the ionic species in order to maintain a constant current throughout the sample. Isotachophoresis has a number of important analytical and preparative applications based on its advantages of high resolving power, sensitivity, speed, and the ability to concentrate rather than to dilute the components which are being analyzed. *See* ELECTROPHORESIS. [B.R.W.]

Isothermal chart A map showing the distribution of air temperature (or sometimes sea surface or soil temperature) over a portion of the Earth's surface, or at some level in the atmosphere. On it, isotherms are lines connecting places of equal temperature. The temperatures thus displayed may all refer to the same instant, may be averages for a day, month, season, or year, or may be the hottest or coldest temperatures reported during some interval.

Isothermal charts are drawn daily in major weather forecasting centers; 5-day, 2-week, and monthly charts are used regularly in long-range forecasting; mean monthly and mean annual charts are compiled and published by most national weather services, and are presented in standard books on, for example, climate, geography, and agriculture. *See* AIR TEMPERATURE; WEATHER MAP. [A.Cou.]

Isothermal process A thermodynamic process which occurs with a heat addition or removal rate just adequate to maintain constant temperature. The change in the internal energy per mole U accompanying a change in volume in an isothermal process is given by the equation below, where T is the

$$U_2 - U_1 = \int_{V_1}^{V_2} \left[T\left(\frac{\partial P}{\partial T}\right)_V - P \right] dV$$

temperature, P is the pressure, and V is the volume per mole. The integral in the equation is zero for an ideal gas, and approximately zero for a condensed phase (solid or liquid) for which the volume changes vary little with pressure. Thus, in both these cases, $U_2 = U_1$. For real gases, the integral is nonzero, and the internal energy change is computed using the equation of state of the gas in the equation. *See* CHEMICAL THERMODYNAMICS; THERMODYNAMIC PROCESSES. [S.I.S.]

Isotone One of two or more atoms which display a constant difference $A - Z$ between their mass number A and their atomic number Z. Thus, despite differences in the total number of nuclear constituents, the numbers of neutrons in the nuclei of isotones are the same. The numbers of naturally occurring isotones provide useful evidence concerning the stability of particular neutron configurations. For example, the relatively large number (six and seven, respectively) of naturally occurring 50- and 82-neutron isotones suggests that these nuclear configurations are especially stable. On the other hand, from the fact that most atoms with odd numbers of neutrons are anisotonic, one may conclude that odd-neutron configurations are relatively unstable. *See* NUCLEAR STRUCTURE. [H.E.D.]

Isotope One of two or more nuclidic species of an element having identical number of protons (Z) in the nucleus but different number of neutrons (*N*). Isotopes differ in mass but chemically are the same element. All naturally occurring elements have radioactive isotopes, and the majority have at least one stable nuclide. Some elements which occur in nature, such as uranium, are radioactive but have isotopes with long half-lives. Of the 83 elements present on Earth in significant amounts, 20 possess only a single stable nuclide and are referred to as mononuclidic or anisotopic. The others have 2 to 10 stable isotopes.

Isotopic abundance usually refers to the isotopic composition of a naturally occurring terrestrial element. Some elements are observed to vary in isotopic composition. The variability ranges from a few per mil to a percent or two, although variations greater than this are observed in some samples. This variation occurs for several reasons. In the lighter elements—hydrogen, lithium, and boron, for instance—the isotopes differ enough in mass and to some extent in chemical reactivity that processes of distillation or chemical exchange between different chemical compounds of the element can produce significant differences in isotopic composition. Indeed, exchange reactions are used in the case of hydrogen, lithium, boron, carbon, nitrogen, and oxygen to separate isotopes of these elements on a relatively large scale.

Isotopes of certain elements possess unique or peculiar properties, and their separation or enrichment is desirable. Deuterium, ^{2}H, which occurs in an abundance of about 0.16% in terrestrial hydrogen, is useful for moderating neutrons in heavy-water reactors and is expected to form the fuel of fusion reactors in the future. Very large quantities of deuterated water have been produced by a variety of distillative, exchange, and electrolysis processes. The desirable fissionable isotope of uranium, ^{235}U, is enriched by gaseous diffusion in uranium hexafluoride gas in very large plants. *See* Deuterium; Heavy water; Isotope (stable) separation.

Of the biologically important elements, only carbon and hydrogen have radioisotopes of sufficiently long half-lives to be used in tracer studies in living organisms. Studies of metabolism, drug utilization, and other reactions in living organisms are best done with stable isotopes like ^{13}C, ^{15}N, ^{18}O, and ^{2}H. Compounds are tagged by introducing a high concentration of the isotope into the molecular structure, and the metabolized products are studied using the mass spectrometer to measure the altered isotopic ratios. *See* Radioactive tracer. [A.E.C.]

Isotope dilution techniques The introduction of radioisotopes into stable isotopes of an element in order to make volume, mass, and age measurements of the element. The technique may best be described by giving a simple example. If knowledge of the total amount of water present in a container is desired, a small amount of labeled water, D_2O or T_2O, is added to the H_2O in the container. For the purposes of the measurement, this amount can be considered to be 100%. After the labeled deuterium or tritium is completely mixed with the water in the container, a sample of the mixture of unit volume, perhaps 1 cm^3, is removed, and the amount of labeled water in it determined. If this amount is found to be a%, then the total amount of water in the container will be $100/a$.

In other applications the mass of an element may be determined. In this case the radioisotope is introduced and mixed, a sample is withdrawn, and both the amount of radioisotope and the amount of stable chemical present are measured. The ratio of these two amounts, called the specific activity, can be, expressed as % dose/mg. For example, sodium-24 is administered to a patient and the specific activity of sodium in the blood (the amount of labeled sodium divided by the amount of stable sodium) determined. This permits measurement of the total amount of exchangeable sodium within the body. [G.L.B.]

Isotope (stable) separation The physical separation of different stable isotopes of an element from one another. Many chemical elements always occur in nature as a mixture of several isotopes. The isotopes of any given element have identical chemical properties, but there are slight differences in their physical properties because of the differences in mass of the individual isotopes. Thus it is possible to separate physically the isotopes of an element to produce material of isotopic composition different from that which occurs in nature. Although these separation processes are all quite difficult and expensive to carry out, they are not inherently different from the usual operations employed in the chemical process industries. *See* Isotope.

The separation of isotopes is particularly important in the nuclear energy field because individual isotopes may have completely different nuclear properties. For example, uranium-235 is used as a fuel for nuclear chain reactors, heavy water (deuterium oxide) is used as a neutron moderator in nuclear chain reactors, and deuterium gas is a possible fuel for thermonuclear reactors. Separated isotopes are also used widely for research concerning the structure and properties of the nucleus. *See* Deuterium; Heavy water; Nuclear fuels; Radioisotope.

The process which is best suited for separating the isotopes of a given element depends upon the mass of the element and the desired quantity of separated material. Research quantities of separated isotopes are best prepared by electromagnetic separation in a mass spectrometer. When moderate quantities of a separated isotope are desired, thermal diffusion may be used.

In the large-scale separation of stable isotopes, the best processes are those which have the highest thermodynamic efficiencies. Reversible processes involving distillation and chemical exchange are best for separating the light isotopes such as deuterium. For heavy isotopes such as those of uranium, however, no appreciable separation is obtained by the reversible processes and some type of irreversible process such as gaseous diffusion must be used. [R.M.St.]

Isotope shift A small difference between the different isotopes of an element in the transition energies corresponding to a given spectral line transition. For a spectral line transition between two energy levels a and b in an atom or ion with atomic number Z, the small difference $\Delta E_{ab} = E_{ab}(A') - E_{ab}(A)$ in the transition energy between isotopes with mass numbers A', and A is the isotope shift. It consists largely of the sum of two contributions, the mass shift (MS) and the field shift (FS), also called the volume shift. The mass shift is customarily divided into a normal mass shift and a specific mass shift; each is proportional to the fractional mass difference $(A' - A)/A'A$. The normal mass shift is a reduced mass correction that is easily calculated for all transitions. The specific mass shift is produced by the correlated motion of different pairs of atomic electrons and is, therefore, absent in one-electron systems. The field shift is produced by the change in the finite size and shape of the nuclear charge distribution when neutrons are added to the nucleus. *See* Atomic structure and spectra; Nuclear structure. [P.M.K.]

Isotopic irradiation The subjection of a material to radiation from radioactive isotopes (radioisotopes). The radiation from radioisotopes produces essentially the same effect as the radiation from electron linear accelerators and other high-voltage particle accelerators, and the choice of which to use is based primarily on convenience and cost. Although the radioisotope radiation source does not require the extensive and complex circuitry necessary for a high-voltage radiation source, its radiation is always present and requires elaborate shielding for health protection and specialized mechanisms for bringing the irradiated objects into and out of the radiation beam. Further, the radiation output decreases with time according to the half-life of the radioisotope, which must therefore be replaced periodically. *See* PARTICLE ACCELERATOR.

The two main radioisotopes for industrial processing are cobalt-60 with a half-life of 5.261 years and an average gamma-ray energy of 1.25 MeV, and cesium-137 with a half-life of 30.3 years and a gamma-ray energy of 0.662 MeV. The application of industrial irradiation is increasing and includes crosslinking plastics, curing polymeric insulations on wires and cables, sterilizing medical devices, and food preservation.

Radioisotopes are used in the treatment of cancer by radiation. Encapsulated sources are used in two ways: the radioisotopes may be external to the body and the radiation allowed to impinge upon and pass through the patient (teletherapy), or the radiation sources may be placed within the body (brachytherapy). For teletherapy purposes cobalt-60 is the most commonly used isotope. Some cesium-137 irradiators have been built, but cesium-137 radiation is not as penetrating as that from cobalt-60. Radium-226 is now being replaced with cesium-137 sources as a brachytherapy encapsulated source. Iridium-192 is also being used for brachytherapy applications. Iodine-125 or gold-198 seeds are put directly in a tumor and permanently left in place.

For some medical applications the radioisotope is dispersed in the body; the most commonly used is iodine-131. When the radioisotope is administered either orally or intravenously in a highly purified form, it goes to the thyroid, where certain forms of thyroid disorders and cancers can be treated by the radiation. *See* RADIOACTIVITY AND RADIATION APPLICATIONS; RADIOISOTOPE; RADIOLOGY. [P.R.A.]

Isotropy (physics) A body is said to be isotropic if its physical properties are not dependent upon the direction in the body along which they are measured. A body displaying isotropy has only one refractive index, one dielectric constant, and so on. Most but not all liquids and aggregates made up of many small crystals randomly oriented in space are isotropic in all their properties. Depending on their symmetry, single crystals may or may not be isotropic with respect to a given property. *See* ANISOTROPY (PHYSICS); ELASTICITY. [D.T.]

Isozyme Many enzyme preparations, even after crystallization, still consist of a number of distinct proteins, each capable of catalyzing the same reaction. These different enzymes, originating from the same organism or even from the same cells, have been named isozymes. They can usually be separated from each other by electrophoretic or chromatographic techniques. For example, in many tissues the oxidation of lactic acid to pyruvic acid is found to be catalyzed by five different isozymes. These isozymes differ not only in electrophoretic mobility but also in some catalytic properties. *See* ENZYME. [D.W.]

Itch Itch sensations are believed to be mediated by a set of unmyelinated nerve fibers different from those serving pain. Itch can be evoked by mechanical, electrical, and chemical stimuli. Certain chemical stimuli are the most effective in producing itch, and these may be externally applied to the skin or released within the skin under certain conditions. A naturally occurring substance is histamine, which is released from mast cells in the skin following local injury or in response to foreign substances that trigger an allergic reaction. Some chemicals produce itch by causing histamine to be released in the skin, while others act independently to evoke itch and can do so in regions of skin previously depleted of histamine. *See* CUTANEOUS SENSATION; HISTAMINE; SENSATION; TOUCH. [R.L.]

Ixodides A suborder of the Acarina comprising the ticks. Ticks differ from mites, their nearest relatives, in their larger size and in having a pair of breathing pores, or spiracles, behind the third or fourth pair of legs. They have a gnathosoma, or head, which consists of a base, a pair of palps, and a rigid, elongated, ventrally toothed hypostome which anchors the parasite to its host. They also have a pair of protrusible cutting organs, or chelicerae, which permit the insertion of the hypostome. The 600 or so known species are all bloodsucking, external parasites of vertebrates including amphibians, reptiles, birds, and mammals. Ticks are divided into three families, Argasidae, Ixodidae (see illustration), and Nuttalliellidae.

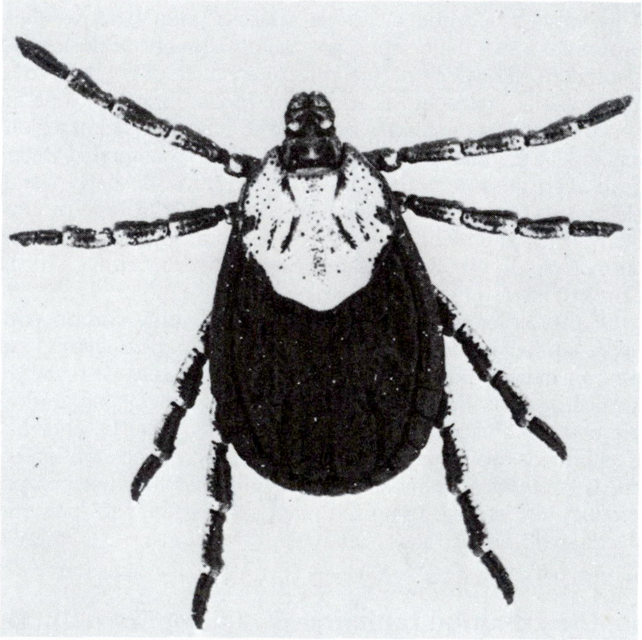

Ixodid tick, *Dermacentor andersoni*, female, enlarged to about 18 times natural size.

In contrast to argasids, Ixodidae have a scutum covering most of the dorsal surface of the male but only the anterior portion of females, nymphs, and larvae. They are known as the hard ticks. The sexes are thus markedly dissimilar.

The numerous tick-borne diseases of animals cause vast economic losses, especially in tropical and subtropical regions. Aside from carrying disease, several species are extremely important pests of humans and animals. Heavy infestations of ticks produce severe anemia in livestock, and even death, from loss of blood alone. *See* ACARI; FIÈVRE BOUTONNEUSE; Q FEVER; QUEENSLAND TICK TYPHUS; ROCKY MOUNTAIN SPOTTED FEVER; SIBERIAN TICK TYPHUS; SOUTH AFRICAN TICK-BITE FEVER; TICK FEVER; TICK PARALYSIS. [G.M.K.]

J

J particle An elementary particle with an unusually long lifetime and large mass which does not fit into any of the schemes for classifying the large number of previously known particles. The discovery in 1974 of J particles in proton-proton (p,p) and electron-positron (e^-,e^+) collisions was of great importance for elementary particle physics. *See* ELEMENTARY PARTICLE.

A family of particles has been identified which, by their emission and absorption of monochromatic gamma rays, transform into each other and into the J particle. The energy spectrum of the J-particle family is very similar to the positronium spectrum. The similarity between the J-particle spectrum and the positronium spectrum, together with the fact that the lifetime of the J particle is about 1000 to 10,000 times longer than that of other known elementary particles, indicates that the J particle is a bound state of a new type of quark and antiquark. This idea was first proposed by S. Glashow, who called it the charm quark. The J particle therefore is a charmonium state. *See* CHARM; POSITRONIUM; QUARKS. [S.C.C.T.]

Jade A name that may be applied correctly to two distinct minerals. The two true jades are jadeite and nephrite. In addition, a variety of other minerals are incorrectly called jade. Idocrase is called California jade, dyed calcite is called Mexican jade, and green grossularite garnet is called Transvaal or South African jade. The most commonly encountered jade substitute is the mineral serpentine. It is often called "new jade" or "Korean jade." The most widely distributed and earliest known true type is nephrite, the less valuable of the two. Jadeite, the most precious of gemstones to the Chinese, is much rarer and more expensive. *See* JADEITE; NEPHRITE. [R.T.L.]

Jadeite The monoclinic sodium aluminum pyroxene, $NaAlSi_2O_6$. Free crystals are rare. Jadeite usually occurs as dense, felted masses of elongated blades or as fine-grained granular aggregates. It has a Mohs hardness of 6.5 and a density of 3.25–3.35. It has a vitreous or waxy luster, and is commonly green but may also be white, violet, or brown.

Jadeite is always found in metamorphic rocks. It is associated with serpentine at Tawmaw, Burma; Kotaki, Japan; and San Benito County, California. It occurs in metasedimentary rocks of the Franciscan group in California and in Celebes. It is found in Tibet; Yunan Province, China; and Guatemala.

Jadeite is the more cherished of the two jade minerals, because of the more intense colors it displays. It is best known in the lovely intense green color resembling that of emerald. *See* JADE; PYROXENE. [L.Gr.]

Jahn-Teller effect A small distortion in the lattice structure of certain crystal defects and, in some cases, of entire crystals, and also in the structure of certain molecules, which reduces the symmetry and removes electronic degeneracy.

The Jahn-Teller effect is most commonly observed in the optical and microwave spectra of point defects and localized impurity centers in solids. Such a center has a certain point symmetry, that is, there is a group of coordinate transformations which leave the center and its surroundings unaltered. If this symmetry is high enough, some of the electronic states of the center will be orbitally degenerate. The Jahn-Teller theorem states that any such degenerate system is unstable against small distortions of the lattice framework which remove the degeneracy. *See* ATOMIC STRUCTURE AND SPECTRA; CRYSTAL; DEGENERACY (QUANTUM MECHANICS); MOLECULAR STRUCTURE AND SPECTRA. [M.D.S.]

Jamming Intentional generation of interfering signals by powerful transmitters as a countermeasure intended to block a communication or radar system or to impair its effectiveness appreciably. For example, radio broadcasts or radio messages can be jammed by beaming a more powerful signal on the same frequency at the area in which reception is to be impaired, using carefully selected noise modulation to give maximum impairment of intelligibility of reception. *See* ELECTRONIC WARFARE. [J.Mar.]

Japanning In the industrial sense, the finishing of metal objects with a black baking varnish that consists of a hard asphalt such as Gilsonite. Modern technology has made the practice obsolete. The term survives as an art form. Lacquers and high-gloss varnishes are hand-applied in layers of varying thickness for unusual effect to wood, ceramic, and metal creations in a variety of hues, often with filigree tracing of white or highlight color. *See* LACQUER; VARNISH. [C.Ma.; C.W.Si.]

Jasper An opaque, impure type of massive fine-grained quartz that typically has a tile-red, dark-brownish-red, brown, or brownish-yellow color. Jasper has been used since ancient times as an ornamental stone, chiefly of inlay work, and as a semiprecious gem material. *See* GEM; QUARTZ.

Jasper has a smooth conchoidal fracture with a dull luster. The specific gravity and hardness are variable; both values approach those of quartz. The color of jasper often is variegated in banded, spotted, or orbicular types. [C.Fr.]

Jaundice The yellow staining of the skin and mucous membranes associated with the accumulation of bile pigments, such as bilirubin, in the blood plasma. Bile pigments are the normal result of the metabolism of blood pigments, and are normally excreted from the blood into the bile by the liver. An increase in circulating bile pigments can, therefore, come about through increased breakdown of blood (hemolytic jaundice), through lack of patency of the bile ducts (obstructive jaundice), through inability or failure of the liver to clear the plasma (parenchymal jaundice), or through combinations of these. Jaundice occurs when the level of these circulating pigments becomes so high that they are visible in the skin and mucous membranes. *See* GALLBLADDER; LIVER.

Hemolytic jaundice results from certain morbid states that include various hemolytic anemias, hemolysis resulting from incompatible blood transfusion, severe thermal or electric injuries, or introduction of hemolytic agents into the bloodstream. Similar jaundice occurs in pulmonary infarction.

Chronic obstructive jaundice may be brought about through a variety of means. In the infant there may be a severe maldevelopment of the bile ducts, while in the adult obstructive jaundice is most commonly caused by impaction of a gallstone in the ducts, or benign and malignant tumors of the gallbladder, bile ducts, pancreas, or lymph nodes.

A wide variety of diseases exists in which part of the jaundice can be accounted for by actual damage to liver cells. This group comprises such conditions as inflammations of the liver, including viral hepatitis, Weil's disease, syphilis, parasitic infestations, and bacterial infections; toxic conditions, including poisoning and in a broader sense the toxemias associated with severe systemic diseases; tumorous conditions; and other miscellaneous conditions, the most common of which is congestive heart failure. [R.B.H.]

Java An island in Sundaland. Java is one of the richest agricultural islands of the tropics; it is 670 mi (1080 km) long and varies from 24 to 120 mi (39 to 190 km) in width, with an area of 48,800 mi^2 (132,800 km^2). The mountains, which are most prominent in the southern portion of the island, were formed during the late Tertiary folding. The junction of the fold mountains with the northern lowlands, which are an extension of the Sunda Shelf, represents a fundamental structural line that is parallel to the length of the island and to the main structural trends. Intense volcanic activity at the end of the Pliocene and in the Quaternary resulted in the formation of more than 20 volcanoes, of which 13 are active. Petroleum is found in a belt extending from Semarang through the island of Madura.

Lying close to the Equator, between 6 and 9°S latitude, the lowlands have high, even temperatures. Djakarta's mean monthly temperature ranges from 78 to 79.7°F (25.6 to 26.5°C), and its average relative humidity is 78% during the driest month. Occasional frosts occur at elevations above 4500 ft (1400 m). The heaviest rainfall is in the mountains and the western lowlands. Eastern Java has a pronounced dry season from July through September. *See* East Indies. [M.G.A.-C.]

Java man The Indonesian human fossils belonging to the species *Homo erectus*. The first remains of Java man were discovered in 1890–1891 by Eugene Dubois in central Java and consist of a skull cap, a fragment of a lower jaw, a few teeth, and a femur. Originally called "Pithecanthropus erectus" (meaning erect ape-man), these and related fossils were later recognized as sufficiently similar to modern humans to be called *Homo*. *See* Fossil human. [E.T.]

Jerboa The common name for 25 species of rodents which make up the family Dipodidae. All species occur in desert or semiarid regions of Asia; three species extend into North Africa. All of these animals are adapted for jumping, and the hindlegs and feet are extremely long (see illustration). The most common species is *Jaculus jaculus*, the lesser Egyptian jerboa. *See* Rodentia. [C.B.C.]

Typical jerboa, with long hindlegs and a tail that terminates in a tuft of hair.

Jet (gemology) A black, opaque material that takes a high polish. Jet has been used for many centuries for ornamental purposes. It is a compact variety of lignite coal. It has a refractive index of 1.66, a hardness of 3–4 on Mohs scale, and a specific gravity of 1.30–1.35. Jet is compact and durable, and can be carved or even turned on a lathe. The main source is Whitby, England, in hard shales. *See* Gem; Lignite. [R.T.L.]

Jet flow A local high-velocity stream in a relatively stationary surrounding fluid. Fluid may be caused to flow in a jet for various reasons. Propulsion of a body through a fluid may be by jet propulsion, a propeller, or a pump or compressor. A jet may be formed by flow out of a closed conduit, downstream from a propeller, or through an orifice or nozzle. The formation of a jet requires that a force be exerted on the fluid in the direction of the jet. An equal and opposite force is exerted on the machine causing the jet, and this is the propulsive force in the case of an airplane or a ship. *See* Jet propulsion. [V.L.S.]

Jet fuel Fuel blended from the light distillates fractionated from crude petroleum. All jet fuels must meet the stringent requirements of aircraft turbine engines and fuel systems, which demand extreme cleanliness and freedom from oxidation deposits in high-temperature zones. Combustors require fuels that atomize and ignite at low temperatures, burn with adequate heat release and controlled radiation, and neither produce smoke nor attack hot turbine parts. The operation of the aircraft in long-duration flights at high altitude imposes a special requirement of good low-temperature flow behavior. [W.G.D.]

Jet propulsion Propulsion of a body by means of force resulting from discharge of a fluid jet. This fluid jet issues from a nozzle and produces a reaction (Newton's third law) to the force exerted against the working fluid in giving it momentum in the jet stream. Turbojets, ramjets, and rockets are the most widely used jet-propulsion engines. *See* Ramjet; Turbojet.

In each of these propulsion engines a jet nozzle converts potential energy of the working fluid into kinetic energy. Hot high-pressure gas escapes through the nozzle, expanding in volume as it drops in pressure and temperature, thus gaining rearward velocity and momentum. This process is governed by the laws of conservation of mass, energy, and momentum and by the pressure-volume-temperature relationships of the gas-state equation. *See* Fluid dynamics; Jet flow; Nozzle. [J.W.Bl.]

Jet stream A relatively narrow, fast-moving wind current flanked by more slowly moving currents. Jet streams are observed principally in the zone of prevailing westerlies above the lower troposphere and in most cases reach maximum intensity, with regard both to speed and to concentration, near the tropopause. At a given time, the position and intensity of the jet stream may significantly influence aircraft operations because of the great speed of the wind at the jet core and the rapid spatial variation of wind speed in its vicinity. Lying in the zone of maximum temperature contrast between cold air masses to the north and warm air masses to the south, the position of the jet stream on a given day usually coincides in part with the regions of greatest storminess in the lower troposphere, though portions of the jet stream occur over regions which are entirely devoid of cloud. [F.S.]

Jet velocity The velocity of the engine exhaust gases relative to the exhaust nozzle. In ideal air-breathing cycles, it is assumed that the exhaust gas mass rate equals the inlet air mass rate, the mass of fuel burned being neglected, and that the exhaust gases are expanded to ambient pressure in the nozzle. Under these conditions the thrust of an engine is directly proportional to the airflow rate and to the difference between the jet and vehicle velocities; thrust is greatest at zero flight speed and becomes zero when the flight speed is the same as the jet velocity.

In the non-air-breathing rocket cycle, thrust of a rocket is independent of flight speed. Rather, at constant mass flow, thrust is directly proportional to nozzle exhaust velocity plus a pressure-times-area term. *See* PROPULSION; ROCKET PROPULSION. [R.R.H.]

Jetty A structure extending out from the shore at a river mouth or harbor entrance, located parallel to and at one or both sides of a navigation channel, at sites where littoral drift causes formation of bars across channel entrances. Jetties are used to keep a channel open by preventing littoral drift and by increasing velocity of flow in the channel. *See* COASTAL ENGINEERING; RIVER ENGINEERING. [E.J.Q.]

Jewel bearing A bearing used in quality timekeeping devices, gyros, and instruments, usually made of synthetic corundum (crystallized Al_2O_3) which is more commonly known as ruby or sapphire. The extensive use of such bearings in the design of precision devices is mainly due to the outstanding qualities of the material. Sapphire's extreme hardness imparts to the bearing excellent wear resistance, as well as the ability to withstand heavy loads without deformation of shape or structure. The crystalline nature of sapphire lends itself to very fine polishing and this, combined with the excellent oil- and lubricant-retention ability of the surface, adds to the natural low-friction characteristics of the material. Ruby has the same properties as sapphire. *See* ANTIFRICTION BEARING; GEM; GYROSCOPE; WATCH. [R.M.Sch.]

Johne's disease A specific enteritis, or inflammation of the gastrointestinal tract, of cattle, sheep, and deer, caused by the bacterium *Mycobacterium paratuberculosis*. The disease is spread by animals having eaten the droppings of sick animals. Animals that have Johne's disease must be destroyed because they rarely recover and present a focus of infection for other animals in the herd. *See* TUBERCULOSIS. [G.Mi.]

Joint (anatomy) The contact surface between two individual bones, also known as an articulation, whose detailed structure depends largely upon the movement between the bones. Articulations can be grouped according to the intervening tissues: diarthrosis, when cartilaginous articular pads are present and an articular cavity with a capsule separates the two bones; and synarthrosis, when the connecting material is continuous. The synarthrosis can be divided into three types: synchondrosis (cartilage between the bones), syndesmosis (collagen fibers between the bones), and synostosis (bone is continuous). Articulations may be freely movable, of limited movement (amphiarthrosis), or rigid (some sutures). Bones that are completely fused to one another are ankylosed, in which the articulation is obliterated.

The type of motion permitted by various joints may be restricted to one kind of movement or may be a combination of movements depending largely upon the configuration of the joint surfaces, intervening tissues, and ligaments associated with the articulation; in many joints, the ligaments are the major or sole limitations to the movements of bones. Some individuals possess a wider range of motion, particularly if the joints are malformed or diseased or if the ligaments vary in length or position. Moreover the ligaments provide most of the strength against tensile forces acting on the joint. *See* LIGAMENT; SKELETAL SYSTEM.

About one person in 16 who are over 15 years old has some form of joint disturbance. The most common conditions are forms of arthritis, which cause the inflammation or degeneration of joint structures. *See* ARTHRITIS; GOUT; RHEUMATISM. [W.J.B.]

Joint (geology) A fracture that traverses a rock and does not show any discernible displacement of one side of the fracture relative to the other. The term joint refers primarily to the actual fracture as represented by a fine line or trace marking the intersection of the fracture and rock surfaces. Commonly, however, the joint is represented superficially by a cleft or fissure resulting from weathering, by mechanical separation, or by one face of the fracture on an outcrop (see illustration).

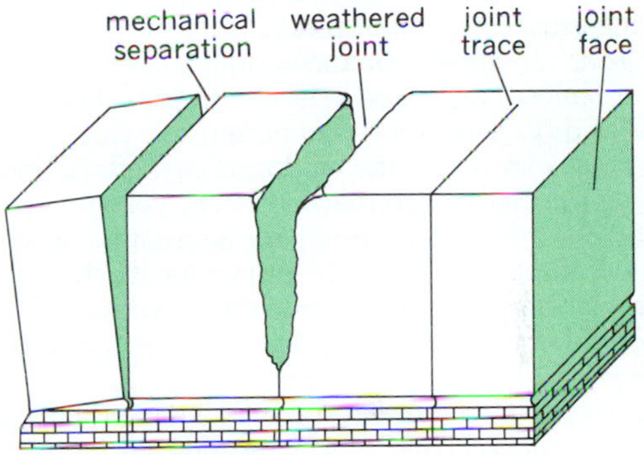

Aspects of a systematic joint, that is, a joint arranged in groups of regularly spaced planar fractures.

Joints are the most abundant structures in the Earth's crust and are present everywhere. They constitute structural inhomogeneities in rock bodies and as such are influential in determining details of form in topographic relief. Locally, systematic joints determine the azimuth of topographic features such as hills, valleys, stream courses, and cliff faces and underground features such as cave passages.

At present no one theory for the origin of systematic joints is accepted universally. Most geologists believe systematic joints are related genetically to folding, faulting, and uplift and attribute them to tangential compressional or tensional forces acting regionally or locally in the Earth's crust. [R.A.H.]

Joint (structures) The surface at which two or more mechanical or structural components are united. Whenever parts of a machine or structure are brought together and fastened into position, a joint is formed. *See* STRUCTURAL CONNECTIONS.

Mechanical joints can be fabricated by a great variety of methods, but all can be classified into two general types, temporary (screw, snap, or clamp, for example), and permanent (brazed, welded, or riveted, for example). [W.H.Cr.]

Jojoba *Simmondsia chinensis*, the only plant known to produce and store a liquid wax in its seed. The jojoba plant is native to the southwestern United States and Mexico. It is tolerant of some of the highest temperatures and most arid regions, and is being domesticated as a crop for hot low-rainfall regions around the world. A broadleaf evergreen shrub that is typically 3–10 ft (1–3 m) in height, it can grow as tall as 20 ft (6 m).

The seed-storage lipid of jojoba is a straight-chain ester. A majority of the wax molecules of jojoba are formed from acids and alcohols with 20 or 22 carbon atoms and one double bond. Many modifications can be made at the double bond, which results in the plant's versatility as an ingredient in a wide range of chemical products. Jojoba wax, used in cosmetics and lubricants, has the potential to serve as a basic feedstock if seed production costs are reduced. Jojoba is being developed simultaneously in many places around the world, and cultivation methods are variable and change rapidly. *See* FAT AND OIL (FOOD); WAX, ANIMAL AND VEGETABLE. [D.A.P.]

Josephson effect The passage of paired electrons (Cooper pairs) through a weak connection (Josephson junction) between superconductors, as in the tunnel passage of paired electrons through a thin dielectric layer separating two superconductors.

Quantum-mechanical tunneling of Cooper pairs through a thin insulating barrier (on the order of a few nanometers thick) between two superconductors was theoretically predicted by Brian D. Josephson in 1962. Josephson found that a current of paired electrons (supercurrent) would flow in addition to the usual current that results from the tunneling of single electrons. Josephson predicted that if the current did not exceed a limiting value (the critical current), there would be no voltage drop across the tunnel barrier. This zero-voltage current flow is known as the dc Josephson effect. Josephson also predicted that if a constant nonzero voltage were maintained across the tunnel barrier, an alternating supercurrent would flow through the barrier in addition to the dc current produced by the tunneling of unpaired electrons. This phenomenon is known as the ac Josephson effect. *See* TUNNELING IN SOLIDS.

Josephson pointed out that the magnitude of the maximum zero-voltage supercurrent would be reduced by a magnetic field. In fact, the magnetic field dependence of the magnitude of the critical current is one of the more striking features of the Josephson effect. Circulating supercurrents flow through the tunnel barrier to screen an applied magnetic field from the interior of the Josephson junction just as if the tunnel barrier itself were weakly superconducting. The screening effect produces a spatial variation of the transport current, and the critical current goes through a series of maxima and minima as the field is increased.

Josephson junctions, and instruments incorporating Josephson junctions, are used in applications for metrology at dc and microwave frequencies, frequency metrology, magnetometry, measurement of absolute temperatures below about 1 K, detection and amplification of electromagnetic signals, and other superconducting electronics such as high-speed analog-to-digital converters and computers. A Josephson junction, like a vacuum tube or a transistor, is capable of switching signals from one circuit to another; a Josephson tunnel junction is the fastest switch known. Josephson junction circuits are capable of storing information. Finally, because a Josephson junction is a superconducting device, its power dissipation is extremely small, so that Josephson junction circuits can be packed together as tightly as fabrication techniques permit. All the basic circuit elements required for a Josephson junction computer have been developed. *See* LOW-TEMPERATURE THERMOMETRY; SUPERCONDUCTING DEVICES; SUPERCONDUCTIVITY. [L.B.H.]

Joule's law A quantitative relationship between the quantity of heat produced in a conductor and an electric current flowing through it. As experimentally determined and announced by J. P. Joule, the law states that when a current of voltaic electricity is propagated along a metallic conductor, the heat evolved in a given time is proportional to the resistance of the conductor multiplied by the square of the electric intensity. Today the law would be stated as $H = RI^2$, where H is rate of evolution of heat in watts, the unit of heat being the joule; R is resistance in ohms; and I is current in amperes. This statement is more general than the one sometimes given that specifies that R be independent of I. Also, it is now known that the application of the law is not limited to metallic conductors. *See* ELECTRIC HEATING; OHM'S LAW. [L.G.H./J.W.St.]

Juglandales An order of flowering plants, division Magnoliophyta (Angiospermae), in the subclass Hamamelidae of the class Magnoliopsida (dicotyledons). The order consists of three families: the Juglandaceae with a little over 50 species, the Picrodendraceae with 3 species, and the Rhoipteleaceae with only 1 species. Within its subclass the order is sharply set off by its compound leaves. *Juglans* (walnut and butternut) and *Carya* (hickory, including the pecan, *C. illinoensis*) are familiar genera of the Juglandaceae. *See* HAMAMELIDAE; HICKORY; MAGNOLIOPSIDA. [A.Cr.]

Juncales An order of flowering plants, division Magnoliophyta (Angiospermae), in the subclass Commelinidae of the class Liliopsida (monocotyledons). The order consists of the family Juncaceae, with about 300 species, and the family Thurniaceae, with only 3. Within its subclass the order is marked by its reduced, mostly wind-pollinated flowers and capsular fruits with one to many anatropous ovules per carpel. The flowers have six sepals arranged in two more or less similar whorls, both sets chaffy and usually brown or green. *See* COMMELINIDAE; FLOWER; LILIOPSIDA. [A.Cr.]

Junction detector A device in which detection of radiation takes place in or near the depletion region of a reverse-biased semiconductor junction. The electrical output pulse is linearly proportional to the energy deposited in the junction depletion layer by the incident ionizing radiation. *See* CRYSTAL COUNTER; IONIZATION CHAMBER.

Introduced into nuclear studies in 1958, the junction detector, or more generally, the nuclear semiconductor detector, revolutionized the field. In the detection of both charged particles and gamma radiation, these devices typically improved experimentally attainable energy resolutions by about two orders of magnitude over that previously attainable. To this they added unprecedented flexibility of utilization, speed of response, miniaturization, freedom from deleterious effects of extraneous electromagnetic (and often nuclear) radiation fields, low-voltage requirements, and effectively perfect linearity of output response. They are now used for a wide variety of diverse applications. They are used for general analytical applications, giving both qualitative and quantitative analysis in the microprobe and the scanning transmission electron microscopes. They are used in medicine, biology, environmental studies, and the space program. In the last category they continue to play a very fundamental role, ranging from studies of the radiation fields in the solar system to the composition of extraterrestrial surfaces. *See* PARTICLE DETECTOR; SEMICONDUCTOR. [J.M.McK.]

Junction diode A semiconductor rectifying device in which the barrier between two regions of opposite conductivity type produces the rectification. Junction diodes are used in computers, radio and television, brushless generators, battery chargers, and electrochemical processes requiring high direct current and low voltage. Lower-power units are usually called semiconductor diodes, and the higher-power units are usually called semiconductor rectifiers. *See* SEMICONDUCTOR.

Junction diodes are classified by the method of preparation of the junction, the semiconductor material, and the general category of use of the finished device. By far the great majority of modern junction diodes use silicon as the basic semiconductor material. Germanium material was used in the first decade of semiconductor diode technology, but has given way to the all-pervasive silicon technology, which allows wider temperature limits of operation and produces stable characteristics more easily. Other materials are the group III–V compounds, the most common being gallium arsenide, which is used where its relatively large band-gap energy is needed.

In silicon units nearly all categories of diodes are made by

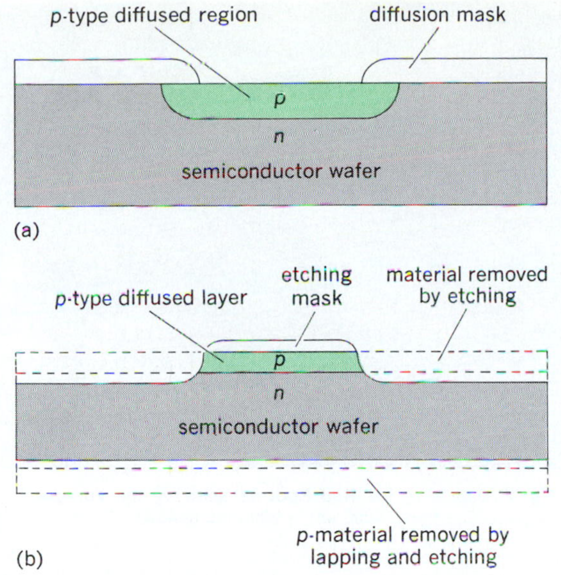

High-speed diffused silicon diodes. (*a*) Mesaless structure. (*b*) Mesa structure.

self-masked diffusion, as shown in illustration *a*. Exceptions are diodes where special control of the doping profile is necessary. In such cases, a variety of doping techniques may be used, including ion implantation, alloying with variable recrystallization rate, silicon transmutation by neutron absorption, and variable-impurity epitaxial growth. The mesa structure shown in illustration *b* is used for some varactor and switching diodes if close control of capacitance and voltage breakdown is required. *See* ELECTRONIC SWITCH; ION IMPLANTATION; RECTIFIER; SEMICONDUCTOR DIODE. [L.P.H.]

Junction transistor A transistor in which emitter and collector barriers are formed by *pn* junctions between semiconductor regions of opposite conductivity type. These junctions are separated by a distance considerably less than a minority-carrier diffusion length, so that minority carriers injected at the emitter junction will not recombine before reaching the collector barrier and therefore be effective in modulating the collector-barrier impedance. Junction transistors are widely used both as discrete devices and in integrated circuits. The discrete devices are found in the high-power and high-frequency applications. Silicon is the most widely used semiconductor material, although germanium is still used for some applications. *See* TRANSISTOR.

Most modern transistors are fabricated by the silicon self-masked planar double-diffusion technique. The structure of a planar diffused epitaxial transistor is shown in section in the illustration. In this structure both collector and emitter junctions are formed by diffusion of impurities from the top surface. In modern technology the base and emitter diffusions are carried out in two steps: a predeposition step, in which a very thin layer of heavily doped oxide is chemically deposited over the open surface of the silicon in the hole opened in the masking oxide; and a drive-in diffusion step, in which the deposited dopant is diffused into the silicon at a higher temperature than that used for the predeposition. The chemical predeposition step is being replaced by ion implantation directly through the oxide. *See* ION IMPLANTATION.

Silicon planar technology is used in fabricating integrated circuit chips. The general form of the transistor structure displayed in the illustration is used in integrated circuits. Such a structure is used for diodes as well as transistors since, for example, it is necessary only to connect the base and collector

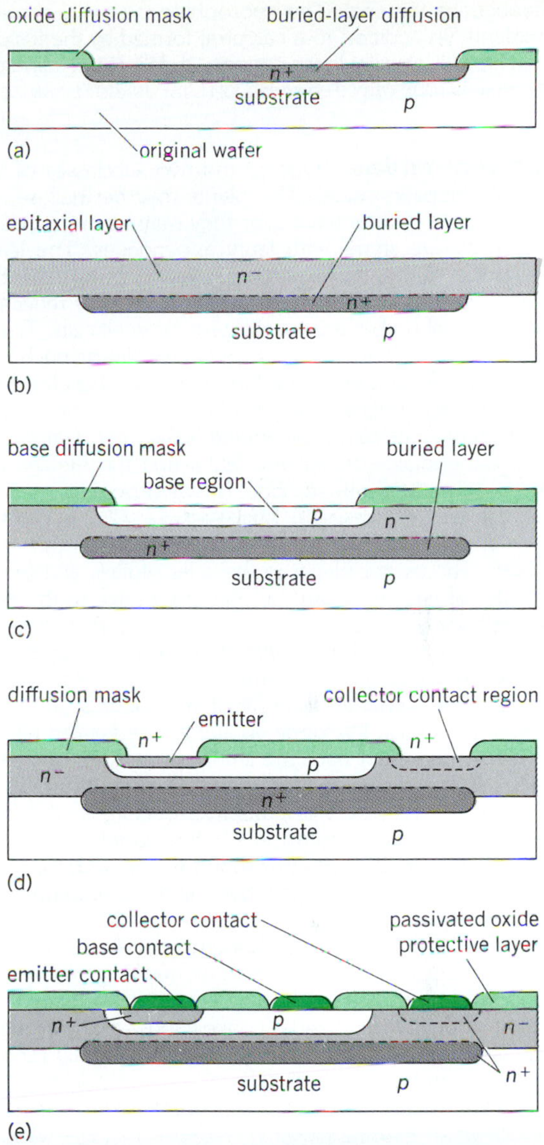

Double-diffused planar epitaxial transistor structure and method of fabrication. (*a*) Buried layer. (*b*) Epitaxial layer. (*c*) Collector junction formation. (*d*) Emitter junction. (*e*) Contact stripe placement.

contacts to use the collector junction as a diode. *See* INTEGRATED CIRCUITS. [L.P.H.]

Jungermanniales The largest order of liverworts, often called the leafy liverworts; it consists of 43 families. The leaves are in three rows, with the underleaves usually reduced or lacking. Other distinctive features include a perianth formed by a fusion of modified leaves, a short-lived seta, and a four-valved capsule. The leaves have an embryonic bilobed phase which may be lost on further development.

The plants of this order are dorsiventrally organized and leafy. They grow by means of an apical cell with three cutting faces, resulting in two rows of lateral leaves and a third row of underleaves which are generally reduced, and sometimes lacking. The stems lack a central strand. Rhizoids are usually present, all smooth. The leaves pass through a primordial two-lobed stage but may become two- to several-lobed, or unlobed (owing to obliteration of one primordial lobe). A midrib is lacking. Asexual reproduction by gemmae is common. Antheridia occur in leaf axils, sometimes also in axils of underleaves.

Archegonia are terminal. The sporophyte is usually protected by a perianth (in addition to a calyptra) formed by the fusion of leaves. The seta, usually long, consists of delicate, hyaline cells. The capsule is four-valved. *See* BRYOPHYTA; JUNGERMANNIIDAE.

[H.Cr.]

Jungermanniidae One of the two subclasses of liverworts (class Hepaticopsida). The plants may be thallose, with little or no tissue differentiation, or they may be organized into erect or prostrate stems with leafy appendages. The leaves, generally one cell in thickness, are mostly arranged in three rows, with the third row of underleaves commonly reduced or even lacking. Oil bodies are usually present in all cells. The rhizoids are smooth. The capsules, generally dehiscing by four valves, are usually elevated on a long, delicate, short-lived seta. The spore mother cells are deeply lobed.

The subclass consists of the orders Takakiales, Calobryales, and Jungermanniales, which are leafy, and the Metzgeriales, which are mostly thallose. *See* BRYOPHYTA; CALOBRYALES; JUNGERMANNIALES; METZGERIALES; TAKAKIALES.

[H.Cr.]

Jupiter The largest planet in the solar system, and the fifth in the order of distance from the Sun. It is visible to the naked eye, except for short periods when in near conjunction with the Sun. Usually it is the second brightest planet in the sky; only Mars at its maximum luminosity and Venus appear brighter. Jupiter is brighter than Sirius, the brightest star.

Through an optical telescope Jupiter appears as an elliptical disk, strongly darkened near the limb, and crossed by a series of bands parallel to the equator (Fig. 1). Even fairly small telescopes show a great deal of complex structure in the bands and disclose the rapid rotation of the planet. The details observed, however, do not correspond to the solid body of a planet but to clouds in its atmosphere, and the rotation period varies markedly with latitude.

Apart from the constantly changing details of the belts, some permanent or semipermanent markings have been observed to last for decades or even centuries, with some fluctuations in visibility. The most conspicuous, and permanent marking is the great Red Spot (Fig. 2), intermittently recorded since the middle of the 17th century and observed continually since 1878, when its striking reddish color attracted general attention. However, the vivid red coloration remains unexplained.

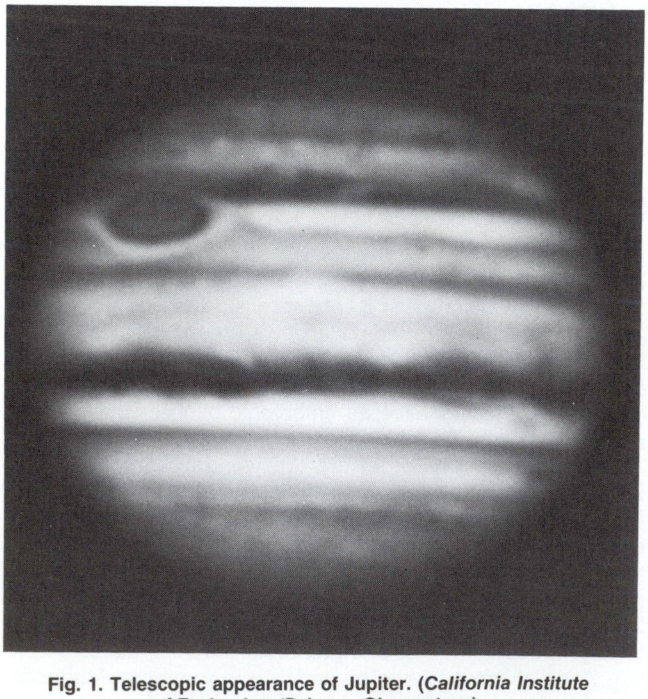

Fig. 1. Telescopic appearance of Jupiter. (*California Institute of Technology/Palomar Observatory*).

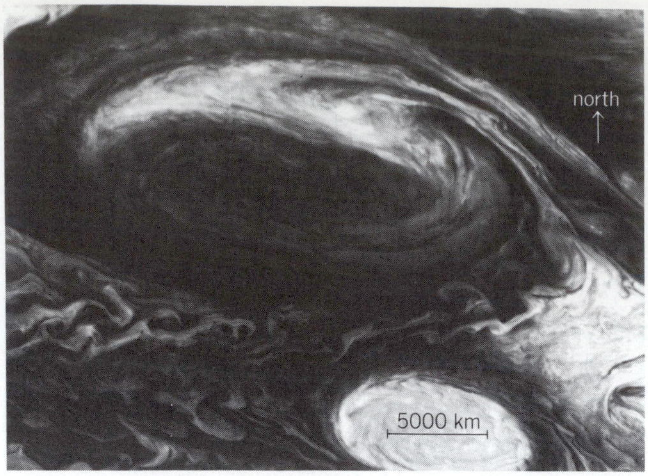

Fig. 2. Details of the Red Spot as seen by the *Voyager 1* flyby. 5000 km = 3000 mi. (*NASA*)

Atmosphere. The atmospheric pressure is probably a bit less than 3 atm (300 kilopascals). The atmosphere must be largely composed of hydrogen and helium. The radiometrically determined temperature of the visible disk of Jupiter, about -225 ± 55°F (130 ± 30 K), is in fairly good agreement with the value theoretically estimated from the assumption that the visible cloud layer is mainly composed of ice crystals of solidified ammonia (about -170°F or 160 K). *Pioneer 10* and *11* flybys showed that Jupiter's zones and belts consist of gases at different altitudes and temperatures. The coloring agents of the bands are unknown.

The Voyager probes showed that the Jovian clouds are affected by the same varieties of weather patterns that produce terrestrial weather, though on a much larger scale. Measurements of the velocities of surface features show that the Jovian atmosphere consists of regular patterns of alternately east and west wind velocities. While the major atmospheric features no doubt persist due to inertia, it is unknown what agent powers the Jovian weather system, though undoubtedly it comes from within the planet.

Interior composition and structure. Jupiter is primarily made up of liquid and metallic hydrogen. Its composition is similar to that of the Sun, namely 10 times as much hydrogen as helium (by numbers of atoms), and undoubtedly reflects the primordial composition of the solar system. Under the extreme pressures of the interior of the planet, the hydrogen would remain in a liquid metallic state, and this material is thought to make up the inner 60% of the radius of the planet. Above this layer it is speculated that there lies liquid hydrogen, with the pressure and temperature lowering sufficiently to become gaseous in nature at approximately 600 mi (1000 km) below the cloud tops. It seems likely that the core of Jupiter is made up of heavy iron-silicate molten material of several times the Earth's total mass and has a probable radius of a few thousand miles.

Jovian ring. *Voyager 2* discovered a faint ring about the planet on its closest passage. The brightest part of this ring is approximately 4000 mi (6000 km) in diameter, but the ring itself may possibly extend all the way to the planetary surface.

Satellites. Jupiter has 15 known satellites of which the four largest, I Io, II Europa, III Ganymede, and IV Callisto, discovered by Galileo in 1610, are by far the most important. The four Galilean satellites are of fifth and sixth stellar magnitudes and would be visible to the naked eye if they were not so close to the much brighter parent planet. All the others are faint telescopic objects. [E.M.H.]

Jurassic The system of rocks deposited during the middle part of the Mesozoic Era. The Jurassic System is normally underlain by the Triassic and overlain by the Cretaceous System. It was named after the Jura Mountains in Switzerland and consists mainly of sedimentary rocks. The duration of the Jurassic Period, which ended about 135,000,000 years ago, is estimated to be about 60,000,000 years.

Different opinions have been expressed on the configuration and position of continents and oceans during the Jurassic Period. The existence of large continents or land bridges in the region of present-day oceans is now denied by many authorities. A tentative reconstruction of Gondwanaland for the Callovian, a stage of the Middle Jurassic, is shown in the illustration.

The Precambrian shields, which form the oldest and most stable parts of the Earth's crust, were land areas subject to erosion during the Jurassic Period. Thus in North America the Canadian Shield was land bordered by seas both to the north and west, and the Scandinavian Shield was surrounded by seas in Late Jurassic times. Smaller areas underlain by older rocks were lands, the coastlines of which are indicated in many places by conglomerates and sandstones. See PRECAMBRIAN.

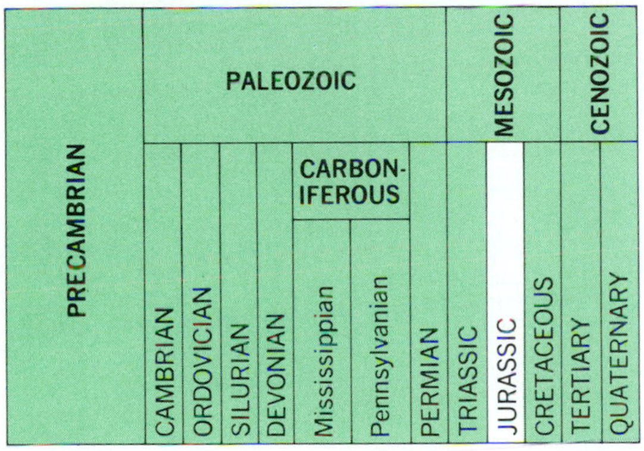

The Jurassic flora is characterized by ferns, which occur abundantly in certain regions, and by numerous gymnosperms. The presence of angiosperms is indicated by pollen grains. Calcareous algae built reefs in some marine areas.

Reptiles were represented by a great number of different forms. On the land the dinosaurs were the dominating forms. Skeletons of the first bird, *Archaeopterhyx*, were found in the Late Jurassic lithographic stones at Solenhofen, Bavaria. Practically all groups of modern fishes were present in the Jurassic. See DINOSAUR.

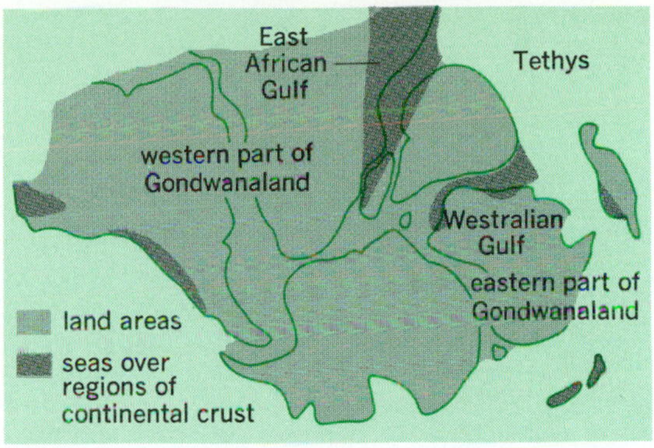

Tentative reconstruction of Gondwanaland for the Callovian. (After A. Hallam, *The bearing of certain palaeozoogeographic data on continental drift, Palaeogeography, Palaeoclimatology, Palaeoecology, vol. 3, 1967*)

The most important marine invertebrates of the Jurassic are the ammonites. Their short vertical and wide horizontal range and their appearance in almost every facies have made them the best guide fossils of the Jurassic. Land invertebrates are represented by such insects as flies, butterflies, and moths. Among the crustaceans, ostracods are known from both marine and fresh-water beds and are used for age determinations.

Iron ores of Jurassic origin are present in western Europe. Upper Jurassic coal is known from Scotland, the Lofoten Islands, and northeast Greenland. Highly bituminous rocks and oil shales occur in western Europe. Lime and cement are made from calcareous beds in England and Germany and bricks from the clays. Many good building stones from various beds are quarried in the Jurassic limestones of western Europe. Petroleum is obtained from various Jurassic horizons in the United States, Canada, Mexico, Brazil, France, Germany, Morocco, and Saudi Arabia. [H.Fr.]

Jute A natural fiber obtained from two Asiatic species, *Corchorus capsularis* and *C. olitorius*, of the plant family Malvaceae. These are tall, slender, half-shrubby annuals. The fibers are not very strong and deteriorate quickly in the presence of moisture, especially salt water. Jute is made into gunny, burlap bags, sacks for wool, potato sacks, covers for cotton bales, twine, carpets, rug cushions, curtains, and a linoleum base. Most of the commercial supply comes from plants grown in the Ganges and Brahmaputra valleys in Bangladesh and India. See MALVALES; NATURAL FIBER. [E.G.N.]

K

Kale Either of two cool-season biennial crucifers, *Brassica oleracea* var. *acephala* and *B. fimbriata*, of Mediterranean origin and belonging to the plant order Capparales. Kale is grown for its nutritious green curled leaves which are cooked as a vegetable. Distinct varieties (cultivars) are produced in Europe for stock feed. Kale is a minor vegetable in the United States. Virginia is an important producing state. *See* CAPPARALES. [H.J.C.]

Kaliophilite A rare mineral tectosilicate found in volcanic rocks high in potassium and low in silica. Kaliophilite is one of three polymorphic forms of $KAlSiO_4$. It crystallizes in the hexagonal system in prismatic crystals. The hardness is 6 on Mohs scale, and the specific gravity is 2.61. The principal occurrence of kaliophilite is at Monte Somma, Italy. *See* KALSILITE; SILICATE MINERALS. [C.S.Hu.]

Kalsilite A rare mineral found in volcanic rocks at Mafuru, in southwest Uganda. It is one of the three polymorphic forms of $KAlSiO_4$. The mineral is hexagonal. The specific gravity is 2.59. In index of refraction and general appearance in thin section it resembles nepheline and is difficult to distinguish from it. *See* KALIOPHILITE; SILICATE MINERALS. [C.S.Hu.]

Kame A round hill or small knoll of sand and gravel, a few meters to more than 100 m high (330 ft). It is one of a family of stratified glacial sediments formed by meltwater in contact with a disintegrating ice sheet. Melting stagnant ice produces large volumes of water carrying boulders, sand, silt, and clay into holes melted in the wasting glacier. When the ice melts completely, the sediment is left standing as small hills. Mixtures of loose rock debris (till) carried above in the melting ice may be lowered onto the kame. Kames are common wherever ice sheets melted, as in New England, New York, the midwestern United States, the British Isles, and Sweden. [S.E.Wh.]

Kanamycin An antibiotic produced by an actinomycete which has found utility in medicine. The producing organism is *Streptomyces kanamyceticus*. The antibiotic has bactericidal activity against a wide variety of gram-positive, gram-negative, and acid-fast bacteria. In general, its antimicrobial spectrum is similar to that of neomycin. Resistance to kanamycin develops slowly, and in stepwise fashion, except for the mycobacteria, which develop resistance rapidly. *See* DRUG RESISTANCE; NEOMYCIN. [J.Lei.]

Kangaroo The name for a number of Australian marsupials that are members of the family Macropodidae. This family also includes the wallabies. The kangaroos and their relatives occur principally in Australia, but are found in Tasmania and New Guinea as well.

Kangaroos have a long, thick tail that is used as a balancing organ, and enlarged hindlegs that are adapted for jumping in many species. The forelimbs are quite short, except in arboreal species such as the blacktree kangaroo (*Dendrolagus ursinus*) and its relatives, in which all four limbs are about the same length. The two largest species are the red kangaroo (*Macropus rufus*) and the great gray kangaroo (*M. giganteus*).

Kangaroos usually have one offspring each year. After the uterine gestation period of about 6 weeks, the very immature young is born and crawls into the marsupium. After an uninterrupted period of 2 months, it ventures out to find food and then returns to the safety of the marsupium. It may seek the protection of the pouch for up to 9 months. *See* MARSUPIALIA. [C.B.C.]

Kaolinite The principal mineral of the kaolinite group of clay minerals. Kaolinite is important in the ceramics industry because of its excellent firing properties and refractoriness. It is also used extensively as a filler in rubber products and for coating and filling paper products. *See* CERAMICS; CLAY MINERALS.

The kaolinite type of mineral forms under acid conditions at low temperatures and pressures. It can form under hydrothermal conditions or through the alteration of other clay minerals. Kaolinite is a principal component of lateritic-type soils. When calcium is present in the environment, the formation of kaolinite is retarded. The present-day marine environment does not favor the formation of kaolinite. The presence of a kaolinitic marine sediment is evidence of a kaolinitic source area, since this mineral does not form in the sea. It is also indicative of relatively rapid accumulation of material. [F.M.W.; R.E.Gr.]

Kapitza resistance A resistance to the flow of heat across the interface between liquid helium and a solid. The Kapitza resistance R_K is given by the equation below, where

$$R_K = A\Delta T/\dot{Q} \qquad (\text{cm}^2\ \text{K/W})$$

$\dot{Q}$ is the heat flow, A is the area of the interface, and ΔT is the temperature discontinuity across the interface. The inverse of this equation is known as the Kapitza conductance. The effect was discovered in 1941 by P. L. Kapitza.

Kapitza resistance is an important factor in most experiments at less than 2°F above absolute zero (1 K), since it hampers temperature equilibrium between refrigerator, sample, and thermometer. It is often the dominant thermal resistance at these temperatures. From a theoretical standpoint, Kapitza resistance provides information on the interaction of low-energy phonons and other excitations with each other and with surfaces. *See* CONDUCTION (HEAT); LIQUID HELIUM; LOW-TEMPERATURE PHYSICS; PHONON; QUANTUM SOLIDS. [R.R.]

Kapok tree Also called the silk-cotton tree (*Ceiba pentandra*), a member of the bombax family (Bombacaceae). The tree has a bizarre growth habit and produces pods containing seeds covered with silky hairs called silk cotton. It occurs in the American tropics, and has been introduced into Java, Philippine Islands, and Ceylon. The silk cotton is the commercial kapok used for stuffing cushions, mattresses, and pillows. *See* MALVALES. [P.D.St./E.L.C.]

Karman vortex street A double row of line vortices in a fluid. Under certain conditions a Karman vortex street is shed in the wake of bluff cylindrical bodies when the relative fluid velocity is perpendicular to the generators of the cylinder, as illustrated. This periodic shedding of eddies occurs first from

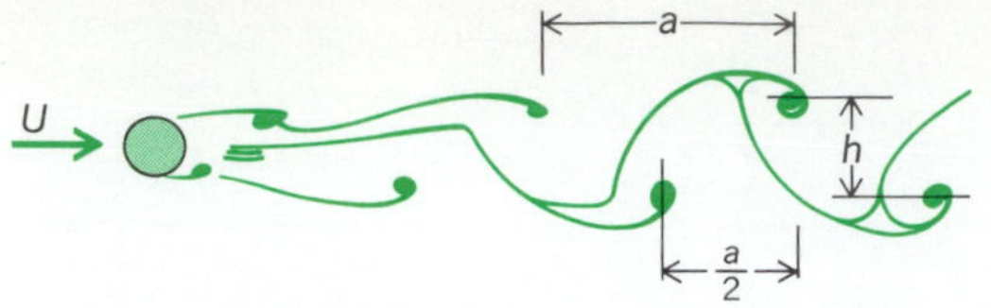

Karman vortex street. ***U*** **= stream speed;** ***a*** **= spacing between vortices;** ***h*** **= distance between two rows of vortices.**

one side of the body and then from the other, an unusual phenomenon because the oncoming flow may be perfectly steady. Vortex streets can often be seen, for example, in rivers downstream of the columns supporting a bridge. They can be created by steady winds blowing past smokestacks, transmission lines, bridges, missiles about to be launched vertically, and pipelines aboveground in the desert. *See* VORTEX. [A.E.Br.]

Karst topography Characteristic associations of minor, third-order, destructively developed land features resulting from subaerial and underground solution of limestone under conditions of humid climate. These pattern features are progressively carved into second-order structural forms, such as plains, plateaus, or even hilly and mountainous uplands, containing limestone layers at or near the surface. Virtually all limestone formations are products of biological or chemical precipitation of calcium carbonate, $CaCO_3$, from aqueous solution. They are therefore susceptible in varying degrees to resolution in surface waters and groundwaters, particularly when such waters are charged with carbon dioxide, CO_2, from vegetative sources or when the waters are under pressure. *See* LIMESTONE.

The sequence of changes in landforms of karst regions, from initiation of sinkholes through stages of their increase in number, enlargement, and coalescence, with collapse of caverns, and eventual wasting away of intervening hills, leaves broad-bottomed lowlands (uvalas) and scattered surviving elevations. Although this karst cycle is still subject to theoretical debate, the concept seems valid because different stages of landform development are identifiable in various karsts of differing ages. *See* CAVE; GEOMORPHOLOGY; WEATHERING PROCESSES. [J.H.B.]

Kata thermometer An instrument for measuring low air velocities in air-circulation problems. An alcohol thermometer with a large bulb is heated above 100°F (37.8°C), and its time to cool from 100 to 95°F (37.8 to 35.0°C), or some other interval above the ambient temperature, is measured. This time interval is a measure of the air current at the location. *See* AIR-VELOCITY MEASUREMENT. [H.Fo.]

Kelvin bridge A specialized version of the Wheatstone bridge network designed to eliminate, or greatly reduce, the effect of lead and contact resistance and thus permit accurate measurement of low resistance. The circuit shown in the illustration accomplishes this by effectively placing relatively high-resistance-ratio arms in series with the potential leads and contacts of the low-resistance standards and the unknown resistance. In this circuit R_A and R_B are the main ratio resistors, R_a, and R_b, the auxiliary ratio, R_x the unknown, R_s the standard, and R_y a heavy copper yoke of low resistance connected between the unknown and standard resistors.

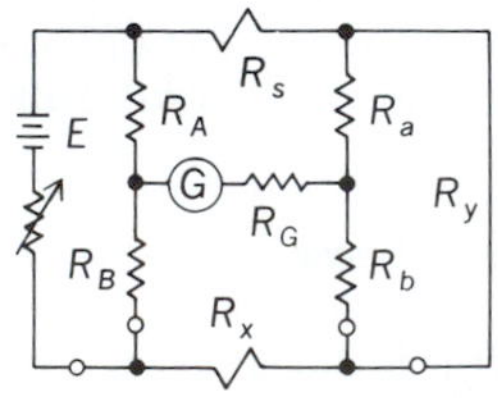

A Kelvin bridge used to measure an unknown low resistance.

As with the Wheatstone bridge, the Kelvin bridge for routine engineering measurements is constructed using both adjustable-ratio arms and adjustable standards. However, the ratio is usually continuously adjustable, over a short span, and the standard is adjustable in appropriate steps to cover the required range. *See* BRIDGE CIRCUIT; RESISTANCE MEASUREMENT; WHEATSTONE BRIDGE. [C.E.A.]

Kelvin's circulation theorem A theorem in fluid dynamics that pertains to an incompressible, inviscid fluid. A direct consequence of this theorem is a great simplification in understanding and analyzing a large class of fluid flows called irrotational flows.

Kelvin's theorem states that in an incompressible, inviscid fluid the circulation along a closed curve, always consisting of the same fluid particles, does not change with time. An important consequence of this theorem relates to fluid motions starting from rest or uniform motion; in such flows the circulation is initially zero for every possible closed curve and hence remains equal to zero thereafter according to Kelvin's theorem. *See* BERNOULLI'S THEOREM; FLUID FLOW; LAPLACE'S IRROTATIONAL MOTION. [A.E.Br.]

Kelvin's minimum-energy theorem A principle of fluid mechanics which states that the irrotational motion of an incompressible, inviscid fluid occupying a simply connected region has less kinetic energy than any other fluid motion consistent with the boundary condition of zero relative velocity normal to the boundaries of the region. *See* D'ALEMBERT'S PARADOX; FLUID FLOW; LAPLACE'S IRROTATIONAL MOTION. [A.E.Br.]

Kenaf An annual, short-day, herbaceous plant (*Hibiscus cannabinus*) of the Malvaceae family, cultivated for its stem fibers. The genus *Hibiscus* has approximately 200 species, of which one, roselle (*H. sabdariffa* var. *altissima*), also is occasionally referred to as kenaf.

Kenaf usually grows up to 15 ft (5 m) in height, and is cylindrical and either branched or unbranched. The stem is composed of two fibers, bast and core. The bast fibers, located in the bark, are long compared to the core fibers, produced in the stem interior. The leaves either are entirely heart shaped or display radiating lobes. Flowers are typically yellow with deep red centers. Wild forms of kenaf are found in east and central Africa, where for several centuries kenaf has been used for both fiber and food. Selection and breeding have developed varieties with higher fiber yields, improved disease resistance, and reduced branching.

Kenaf is grown commercially for fiber production in many areas of the world, with the largest producer being the People's Republic of China. In the United States, kenaf production is located in Texas, Louisiana, Mississippi, and California. Kenaf is used in the manufacturing of various paper and pulp products, and as poultry litter, potting soil amendments, chemical- and oil-spill absorbents, animal and horse bedding, and packing materials. Potential uses include the manufacturing of filters, particle boards, and insulation boards. Kenaf leaves, which contain 20–30% crude protein, also may have potential as a livestock feed source.

Kenaf can adapt to a wide range of climates and soils. However, because it cannot tolerate frost, planting should not occur at temperatures below 32°F (0°C). Optimum yields are generally obtained on well-drained soils with fertility levels to meet the nutritional requirements. Most varieties of kenaf are photoperiod sensitive, and vegetative growth increases until the daylight period becomes less than 12 h 30 min. Flowering is then initiated and the vegetative growth rate declines. Early planting maximizes yields by increasing the growing season. Kenaf is propagated by seed and must be replanted annually. Dense plant populations generally produce greater total and bast fiber yields, reduce weed populations, and improve harvesting efficiency. [C.G.C.]

Kennel cough A common, highly contagious respiratory disease of dogs, also known as canine infectious tracheobronchitis. Several different bacteria and viruses are usually associated with the disease. Symptoms are generally mild but may vary widely depending on the agent, the host, and environmental factors. The main feature of the disease is sudden onset of violent coughing in dogs that had a recent exposure to other, infected dogs. The disease is easily transmitted between dogs by droplets in the air or direct contact, and often occurs as outbreaks or as a seasonal infection. Most dogs completely recover within 2 weeks; however, chronic and severe forms of the disease sometimes occur.

Infectious agents commonly associated with the disease are the bacterium *Bordetella bronchiseptica* and canine parainfluenza virus. Each agent is capable of producing a mild form of the disease; however, most single-agent infections probably show no symptoms of disease. Several species of mycoplasmas have been isolated from the lower respiratory tract of dogs with kennel cough, but always in combination with another agent (for example, bordetella or canine parainfluenza virus). These mycoplasmas are normally found in the upper respiratory tract of healthy dogs. *See* BORDETELLA; MYCOPLASMAS.

Close contact with other dogs is usually required for transmission of kennel cough. Each of the viral agents of the disease, and possibly some of the mycoplasmas, has host ranges restricted to dogs. Because infections that show no symptoms of disease are also common, it is sometimes difficult to determine the source of the infection. Canine parainfluenza virus and *B. bronchiseptica* do not usually persist longer than a few weeks or a few months, respectively, in an individual dog.

Treatment of kennel cough is often unwarranted. However, antitussives, bronchodilators, and corticosteroids are used to relieve coughing, and antimicrobials are used to treat or prevent bronchopneumonia. The risk of acquiring kennel cough can be reduced by minimizing exposure to infectious agents. [D.A.Be.]

Kenyapithecus The genus name given by L. S. B. Leakey to several fossil Hominoidea from the Miocene deposits of Kenya. The older specimens, referred to as "Kenyapithecus africanus," are now included in the genus *Dryopithecus*, which includes most of the Miocene and Pliocene Pongidae, or apes. Moderate-sized with short faces, these primates are the East African representatives of the genus *Ramapithecus* and are considered by many to be the earliest known hominids. *See* FOSSIL HUMAN; RAMAPITHECUS. [E.T.]

Kepler's laws The three laws of planetary motion discovered by Johannes Kepler during the early years of the seventeenth century.

First law. The first law of Kepler states that a planet moves in an elliptical orbit around the Sun that is located at one of the two foci of the ellipse. An ellipse is one of the conic curves originally studied by Greek geometers. *See* CONIC SECTION; ELLIPSE.

In 1687, Isaac Newton demonstrated that any body, moving in an orbit around another body that attracts it with a force that varies inversely as the square of the distance between them, must move in a conic section. This path will be an ellipse when the velocity is below a certain limit in relation to the attracting force. Thus the first law is a general law that applies to all satellites held in orbit by an inverse-square force.

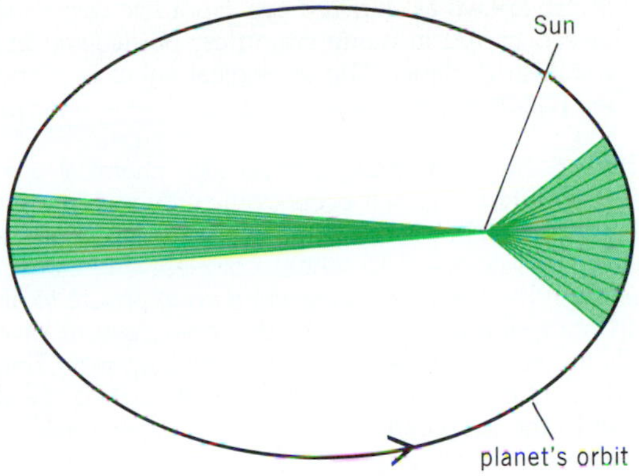

Demonstration of Kepler's first and second laws. The planet moves along an elliptical orbit at a nonuniform rate, so that the radius vector drawn to the Sun, which is located at one focus of the ellipse, sweeps out areas that are proportional to time. Thus, the planet would take equal times (corresponding to the equal areas) to traverse the unequal distances along the ellipse that correspond to the two shaded areas. The diagram greatly exaggerates the eccentricity of any orbital ellipse in the solar system.

Second law. Kepler needed some method by which to predict the locations of planets at given times, and this he provided with his second law. It states that the radius vector of the ellipse (the imaginary line between the planet and the Sun) sweeps out areas that are proportional to time (see illustration). A planet moves more swiftly when it is closer to the Sun and more slowly when it is farther removed.

Again Newton demonstrated the dynamic cause behind Kepler's second law. In this case, it is not restricted to forces that vary inversely as the square of the distance; rather it is valid for all forces of attraction between the two bodies. The second law expresses the principle of the conservation of angular momentum. *See* ANGULAR MOMENTUM.

Third law. Kepler's third law defines the relations that hold within the system of planets. It states that the ratio between the square of a planet's period (the time required to complete one orbit) to the cube of the mean radius (the average distance from the Sun during one orbit) is a constant. The four satellites that Galileo had discovered around Jupiter were found to obey the third law, as did the satellites later found around Saturn. Newton demonstrated once again that the third law is valid for every system of satellites around a central body that attracts them, as the Sun attracts the planets, with a force that varies inversely as the square of the distance. *See* GRAVITATION; ORBITAL MOTION. [R.S.We.]

Keratitis Any inflammation involving the cornea of the eye. It may involve the cornea alone (keratitis), or more commonly, both cornea and conjunctiva (keratoconjunctivitis). The various forms of keratitis and keratoconjunctivitis together constitute a major portion of all eye disease; keratitis is second only to cataract and equal to glaucoma as a cause of blindness. *See* KERATOCONJUNCTIVITIS.

Keratitis may be caused by mechanical injury, infectious agents, extremes of heat or cold, drying, radiation, corrosive or irritating chemicals, or allergy, or it may occur in association with diseases affecting the enfire body, such as a vitamin deficiency. Keratitis often accompanies other eye diseases, such as glaucoma. The degree of permanent impairment of vision is determined primarily by the extent of the injury rather than by the nature of the injurious agent. *See* GLAUCOMA. [W.R.A.]

Keratoconjunctivitis A viral disease characterized by an acute conjunctivitis, followed by keratitis, which in some cases leaves round, superficial opacities in the cornea for up to 2 years. It is also known as shipyard eye. Sporadic cases and epidemics have occurred in many countries, particularly in shipyards and industrial plants. The etiological agent is adenovirus type 8. *See* KERATITIS. [J.L.Me.]

Kernite A hydrated borate mineral with chemical composition $Na_2B_4O_6(OH)_2 \cdot 3H_2O$. It occurs only very rarely in crystals but is found most commonly in coarse, cleavable masses and aggregates. It is colorless to white; colorless and transparent specimens tend to become chalky white on exposure to air.

Boron compounds are used in the manufacture of glass, especially in glass wool used for insulation purposes. They are also used in soap, in porcelain enamels for coating metal surfaces, and in the preparation of fertilizers and herbicides. *See* BORATE MINERALS. [C.K.]

Kerogen A name given to the complex organic matter present in carbonaceous shales and oil shales. It is insoluble in all common solvents but on destructive distillation yields oil, gas, and acidic and basic compounds. Kerogen is formed by the conversion of plant and animal remains. It is the most common form of organic carbon on Earth, and it has been estimated that there is 1000 times as much kerogen as coal. *See* OIL SHALE. [I.A.B.]

Kerosine A refined petroleum fraction used as a fuel for heating and cooking, jet engines, lamps, and as a base for insecticides. Kerosine, known also as lamp oil, is recovered from crude oil by distillation. Specifications are established for specific grades of kerosine by government agencies and by refiners. For use in lamps, for example, a highly paraffinic oil is desired because aromatics and naphthenes give a smoky flame. In order to avoid atmospheric pollution, sulfur content must be low; and a minimum flash point of 100°F (38°C) is desirable to reduce explosion hazards. *See* DISTILLATE FUEL; JET FUEL. [H.C.R.]

Kerr effect Electrically induced birefringence that is proportional to the square of the electric field. When a substance (especially a liquid or a gas) is placed in an electric field, its molecules may become partly oriented. This renders the substance anisotropic and gives it birefringence, that is, the ability to refract light differently in two directions. This effect, which was discovered in 1875 by John Kerr, is called the electrooptical Kerr effect, or simply the Kerr effect. *See* ANISOTROPY (PHYSICS).

When a liquid is placed in an electric field, it behaves optically like a uniaxial crystal with the optical axis parallel to the electric lines of force. The Kerr effect is usually observed by passing light between two capacitor plates inserted in a glass cell containing the liquid. Such a device is known as a Kerr cell or optical Kerr shutter. Light passing through the medium normal to the electric lines of force (that is, parallel to the capacitor plates) is split into two linearly polarized waves.

In certain crystals there may be an electrically induced birefringence that is proportional to the first power of the electric field. This is called the Pockels effect. In these crystals the Pockels effect usually overshadows the Kerr effect, which is nonetheless present. In crystals of cubic symmetry and in isotropic solids (such as glass) only the Kerr effect is present. *See* ELECTROOPTICS. [M.A.D.]

Ketone One of a class of chemical compounds of the general formula

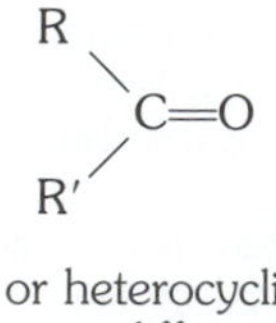

R and R′ are alkyl, aryl, or heterocyclic radicals. The groups R and R′ may be the same or different or incorporated into a ring as in cyclopentanone:

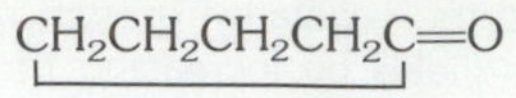

The ketones acetone and methyl ethyl ketone are used as solvents. Ketones are important intermediates in the syntheses of organic compounds.

By common nomenclature rules, the R and R′ groups are named, followed by the word ketone—for example, $CH_3CH_2COCH_2CH_3$ (diethyl ketone), $CH_3COCH(CH_3)_2$ (methyl isopropyl ketone), and $C_6H_5COC_6H_5$ (diphenyl ketone). The nomenclature of the International Union of Pure and Applied Chemistry uses the hydrocarbon name corresponding to the maximum number of carbon atoms in a continuous chain in the ketone molecule, followed by "-one," and preceded by a number designating the position of the carbonyl group in the carbon chain. The first two ketones above are named 3-pentanone and 3-methyl-3-butanone.

The lower-molecular-weight ketones are colorless liquids. Acetone and methyl ethyl ketone are miscible with water; the water solubility of the higher homologs decreases with increasing number of carbon atoms. Because of their characteristic odors, various ketones are of use in the flavoring and perfumery industry.

Addition to the carbonyl group is the most important type of ketone reaction. Ketones are generally less reactive than aldehydes in addition reactions. Methyl ketones are more reactive than the higher ketones because of steric group effects. [P.E.F.]

Ketosis An excessive accumulation of ketone bodies in the tissues, blood, and urine. Ketone bodies are formed by the oxidation of fatty acids by the liver. *See* KETONE.

When a disturbance in carbohydrate metabolism is present, ketone bodies are produced in excess. The ketone bodies, by virtue of their acid properties, hold an equivalent amount of base, thus producing a form of acidosis. This is seen clinically in certain cases of diabetes, in starvation, occasionally in hyperthyroidism, and in a variety of other disease states in which there is an absolute or relative decrease in carbohydrate availability. *See* DIABETES; THYROID GLAND DISORDERS.

In ketosis, the breath has a fruity odor, and there is usually evidence of dehydration; weakness, malaise, headache, nausea, and vomiting are common. [E.G.St./N.K.M.]

Key telephone system A system which can be installed on a customer's premises to provide access to more than one central-office line as well as other special line features such as wide area telecommunication service (WATS) and foreign exchange (FX). Many small businesses make use of key telephone systems (KTS), which may also provide communication between members of the same business community, either through the serving central office (CO) or private branch exchange (PBX) or an intercom system not associated with regular telephone lines.

The telephone instrument used with a key telephone system is called a key telephone set and is distinguished by the appearance of key switches, which provide the means for implementing most of the key system features. The sets are typically provided with 6, 10, or 20 keys (buttons) which provide access to the key station lines and functions. In addition, an instrument with provision for the pickup of 30 lines may be used as an attendant console. *See* SWITCHING SYSTEMS (COMMUNICATIONS); TELEPHONE SERVICE. [W.Sch.]

Kidney An organ involved with the elimination of water and waste products from the body. In vertebrates the kidneys are paired organs located close to the spine dorsal in the body cavity. They consist of a number of smaller functional units which are called urinary tubules or nephrons. The nephrons open to large ducts, the collecting ducts, which open into a ureter. The two ureters run backward to open into the cloaca or into a urinary bladder. In mammals, the kidneys are bean-

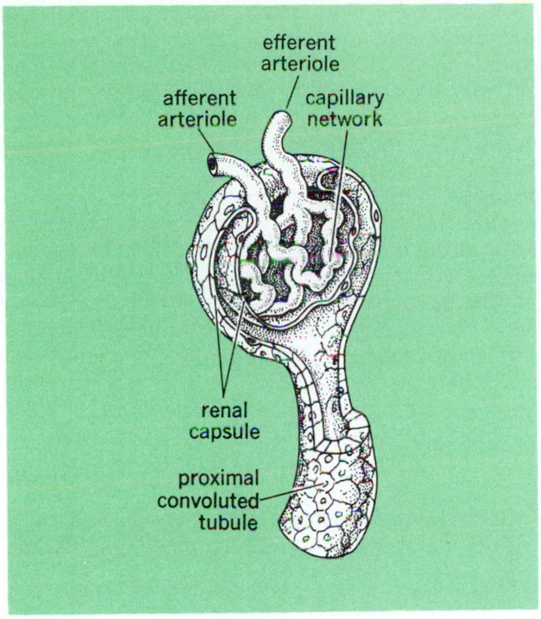

Nephron from frog kidney, dissected to show glomerulus within Bowman capsule.

shaped and found between the thorax and the pelvis. The number, structure, and function of the nephrons vary with evolution and, in certain significant ways, with the adaptation of the animals to their various habitats.

In its most primitive form, found only in invertebrates, the nephron has a funnel opening into the coelomic cavity followed by a urinary tubule leading to an excretory pore. In amphibians, some of the tubules have this funnel, but most of the tubules have a Bowman capsule (see illustration). In all higher vertebrates, the nephron has the Bowman capsule, which surrounds a tuft of capillary loops, called the glomerulus, constituting the closed end of the nephron. The inner epithelial wall of the Bowman capsule is in intimate contact with the endothelial wall of the capillaries. The wall of the capillaries, together with the inner wall of the Bowman capsule, forms a membrane ideally suited for filtration of the blood.

The blood pressure in the capillaries of the glomerulus causes filtering of blood by forcing fluid, small molecules, and ions through the membrane into the lumen of Bowman's capsule. This filtrate contains some of the proteins and all of the smaller molecules in the blood. As the filtrate passes down through the tubule, the walls of the tubule extract those substances not destined for excretion and return them to the blood in adjacent capillaries. Many substances which are toxic to the organism are moved in the opposite direction from the blood into the tubules. The urine thus produced by each nephron is conveyed by the collecting duct and ureter to the cloaca or bladder from which it can be eliminated.

In all classes of vertebrates the renal arteries deliver blood to the glomeruli and through a second capillary net to the tubules. The major blood supply to the kidney tubules comes, however, from the renal portal vein, which is found in all vertebrates except mammals and cyclostomes. Waste products from the venous blood can thus be secreted directly into the urinary tubules. *See* URINARY SYSTEM. [B.S.-N.]

Kidney disorders Disorders of the kidney are classified by both their etiology and their anatomic location along the nephron. Disorders other than those arising from tumors become evident when the kidney is unable to regulate the volume and composition of the extracellular fluids, resulting in edema and hypertension. Kidney disorders may be divided into the following categories: congenital, inflammatory, hereditary, infection-related, or associated with metabolic diseases (such as diabetes mellitus), toxins or drugs, circulatory collapse, and cancer.

The absence of kidneys at birth, which is quite rare, is incompatible with life. A slightly more common condition, however, is the presence of only one kidney or the presence of one kidney in an abnormal location. The most common congenital disorders are abnormalities in the peristaltic passage of urine, which normally flows from the kidneys into the ureters, where it is moved by a rhythmic peristaltic motion to the bladder. In many cases, diseases that inhibit the normal peristaltic flow of urine down the ureters because of a neuromuscular abnormality gradually run their course and do not require surgery. *See* CONGENITAL ANOMALIES.

The most common hereditary kidney disorder is adult polycystic kidney disease. It is an autosomal dominant disease that is found in all racial and ethnic groups and is characterized by the formation of cysts along the length of the nephron. As the cysts enlarge, the kidneys likewise enlarge—resulting in kidney failure in midadulthood. Alport's syndrome, a very rare genetic disease of the glomeruli, results in glomerular scarring and eventual renal failure within the second or third decade of life that cannot be reversed. *See* HUMAN GENETICS.

The kidneys are subject to infection from both local and remote sources and from a variety of causative organisms. The most common cause of pyelonephritis, a bacterial infection of the kidney, is obstruction of the bladder or ureters by a tumor or kidney stones that interrupt urine flow. The kidney may also be subject to infections originating at distant sites. Examples include bacterial products, such as endotoxins, that are released into the bloodstream and antibodies that are formed in response to bacterial invasion. *See* STREPTOCOCCUS.

The most common metabolic disorder to affect the kidney is diabetes mellitus. Of those persons with the insulin-dependent form of diabetes, 30–40% develop renal disease. Kidney dysfunction is one of the most life-threatening complications of diabetes. *See* DIABETES.

The two most common systemic diseases to affect the kidneys are systemic lupus erythematosus and vasculitis. Systemic lupus erythematosus is essentially a disease of young women and is characterized by the deposition of immunoglobulins and other immune substances in the glomeruli. Systemic vasculitis can affect either the medium-sized arteries or the arterioles. Involvement of the former leads to infarcts in various organs but seldom causes kidney failure. When the arterioles are affected, the lesions are diffuse, because the kidney is highly vascular and the arterioles, the smallest of the arteries, are abundant. Kidney failure often occurs.

The kidney is a major site for the excretion of waste products as well as heavy metals, the acids and alkalies of metabolism, and drugs. Because the urine becomes concentrated (100-fold) as it travels the length of the tubule, the concentration of otherwise-innocuous, or even therapeutic, substances may reach levels that are toxic to the kidneys. *See* TOXICOLOGY.

If systemic blood pressure drops to levels at which the kidneys are no longer perfused, the tubules become deprived of oxygen and suffer severe injury. The most common causes of underperfusion are blood loss and cardiac failure. If bloodflow to the kidneys is restored, epithelial cell damage is rapidly reversed and renal function returns to normal.

Cancer of the kidney occurs mainly in two age groups—the very young and those over 50 years of age. Tumors found in the young, the most common type being Wilms' tumor, are often bilateral, composed of fetal tissues, and may grow to a large size before detection. Some cancerous tumors of the kidney in children are responsive to radiation or chemotherapy. Tumors in the adult kidney, adenocarcinomas or hypernephromas, usually metastasize to the lungs and bones. *See* CANCER (MEDICINE); KIDNEY; URINARY SYSTEM. [G.S.]

Kiln A device or enclosure to provide thermal processing of an article or substance in a controlled temperature environment or atmosphere, often by direct firing, but occasionally by convection or radiation heat transfer. Kilns are used in many different industries, and the type of device called a kiln varies with the industry.

"Kiln" usually refers to an oven or furnace which operates at sufficiently high temperature to require that its walls be constructed of refractory materials. The distinction between a kiln and a furnace is often based more on the industry than on the design of the device. Generally the word "kiln" is used when referring to high-temperature treatment of nonmetallic materials such as in the ceramic, the cement, and the lime industries. When melting is involved as in steel manufacture, the term "furnace" is used, as in blast furnace and basic oxygen furnace. *See* FURNACE. [B.B.Cr.]

Kinematics That branch of mechanics which deals with the motion of a system of material particles without reference to the forces which act on the system. Kinematics differs from dynamics in that the latter takes these forces into account. *See* DYNAMICS.

For a single particle moving in a straight line (rectilinear motion), the motion is prescribed when the position of the particle is known as a function of the time. Plane kinematics of a particle is concerned with the specification of the position of a particle moving in a plane by means of two independent variables. The kinematics of a particle in space is concerned with the ways in which three independent coordinates may be chosen to specify the position of the particle at a given time, and with the relations between the first and second time derivatives of these coordinates and the components of velocity and acceleration of the particle.

Among the coordinate systems studied in kinematics are those used by observers who are in relative motion. In nonrelativistic kinematics the time coordinate for each such observer is assumed to be the same, but in relativistic kinematics proper account must be taken of the fact that lengths and time intervals appear different to observers moving relative to each other. *See* RELATIVITY. [H.C.Co./B.G.]

Kinetic energy The energy of a body due to its motion. A point particle with mass m and velocity v has kinetic energy given by Eq. (1). Any body in motion has Eq. (1) as its trans-

$$E = \tfrac{1}{2} mv^2 \qquad (1)]$$

lational kinetic energy. The velocity must be specified with respect to some inertial reference frame.

A rigid body can move as in Eq. (1), but in addition it can rotate about its center. The rotational kinetic energy E_R, for a body with moment of inertia I and angular frequency ω, in units of radians per second, is given by Eq. (2).

$$E_R = \tfrac{1}{2} I\omega^2 \qquad (2)]$$

The most general kinetic energy for a solid body in motion is given by Eq. (3).

$$E_k = \tfrac{1}{2} mv^2 + \tfrac{1}{2} I\omega^2 \qquad (3)]$$

See ROTATIONAL MOTION. [B.DeF.]

Kinetic theory of matter A theory which states that the particles of matter in all states of aggregation are in vigorous motion. In computations involving kinetic theory, the methods of statistical mechanics are applied to specific physical systems. The atomistic or molecular structure of the system involved is assumed, and the system is then described in terms of appropriate distribution functions. The main purpose of kinetic theory is to deduce, from the statistical description, results valid for the whole system. The distinction between kinetic theory and statistical mechanics is thus of necessity arbitrary and vague. Historically, kinetic theory is the oldest statistical discipline. Today a kinetic calculation refers to any calculation in which probability methods, models, or distribution functions are involved. *See* STATISTICAL MECHANICS.

Kinetic calculations are not restricted to gases, but occur in chemical problems, solid-state problems, and problems in radiation theory. Even though the general procedures in these different areas are similar, there are a sufficient number of important differences to make a general classification useful.

In classical ideal equilibrium problems there are no interactions between the constituents of the system. The system is in equilibrium, and the mechanical laws governing the system are classical. The basic information is contained in the Boltzmann distribution f (also called Maxwell or Maxwell-Boltzmann distribution) which gives the number of particles in a given momentum and positional range ($d^3x = dxdydz$, $d^3v = dv_x dv_y dv_z$, where x, y, and z are coordinates of position, and v_x, v_y, and v_z are coordinates of velocity). In Eq. (1) ϵ is the energy, $\beta =$

$$f(xyz, v_x v_y v_z) = Ae^{-\beta\epsilon} \qquad (1)$$

$1/kT$ (where k is the Boltzmann constant and T is the absolute temperature), and A is a constant determined from Eq. (2). The

$$\iint d^3x\, d^3v\, f = N \qquad \text{total number of particles} \qquad (2)$$

calculations of gas pressure, specific heat, and the classical equipartition theorem are all based on these relations. *See* BOLTZMANN STATISTICS.

Many important physical properties refer not to equilibrium but to nonequilibrium states. Phenomena such as thermal conductivity, viscosity, and electrical conductivity all require a discussion starting from the Boltzmann transport equation. If one deals with states that are near equilibrium, the exact Boltzmann equation need not be solved; then it is sufficient to describe the nonstationary situation as a small perturbation superimposed on an equilibrium state.

The basic classical procedure for arbitrary systems (systems with interactions taken into account) that allows the calculation of macroscopic entities is that using the partition function.

Classical nonideal nonequilibrium theory is the most general situation that classical statistics can describe. In general, very little is known about such systems.

There are quantum counterparts to the classifications just described. In a quantum treatment a distribution function is also used for an ideal system in equilibrium to describe its general properties. For systems of particles which must be described by symmetrical wave functions, such as helium atoms and photons, one has the Bose distribution, Eq. (3), where $\beta =$

$$f(v_x v_y v_z) = \frac{1}{(1/A)e^{\beta\epsilon} - 1} \qquad (3)$$

$1/kT$, and A is determined by Eq. (2). *See* BOSE-EINSTEIN STATISTICS.

For systems of particles which must be described by antisymmetrical wave functions, such as electrons, protons, and neutrons, one has the Fermi distribution, Eq. (4). The application to

$$f(v_x v_y v_z) = \frac{1}{(1/A)e^{\beta\epsilon} + 1} \qquad (4)$$

electrons as an (ideal) Fermi-Dirac gas in a metal is the basis of the Sommerfeld theory of metals. *See* FERMI-DIRAC STATISTICS; FREE-ELECTRON THEORY OF METALS; QUANTUM STATISTICS. [M.Dr.]

Kinetics (classical mechanics) That part of classical mechanics which deals with the relation between the motions of material bodies and the forces acting upon them. It is synonymous with dynamics of material bodies. *See* DYNAMICS.

Kinetics proceeds by adopting certain intuitively acceptable concepts which are associated with measurable quantities. These essential concepts and the measurable quantities used for their specification are as follows:

1. Space configuration refers to the positions and orientations of bodies in a reference frame adopted by the observer. It is expressed quantitatively by an arbitrarily chosen set of space coordinates, of which cartesian and polar coordinates are examples. All space coordinates rest on the notion of distance measurement.

2. Duration is expressed quantitatively by time measured by a clock or comparable mechanism.

3. Motion refers to change of configuration with time and is expressed by time rates of coordinate change called velocities and time rates of velocity change called accelerations. The classical assumption that coordinates behave as analytic functions of time permits representation of velocities and accelerations as first and second derivatives, respectively, of the space coordinates with respect to time.

4. Inertia is an attribute of bodies implying their capacity to resist changes of motion. A body's inertia with respect to linear motion is denoted by its mass.

5. Momentum is an attribute proportional to both the mass and velocity of a body. Momentum of linear motion is expressed as the product of mass and linear velocity.

6. Force serves to designate the influence exercised upon the motion of a particular body by other bodies, not necessarily specified. A quantitative connection between the motion of a body and the force applied to it is expressed by Newton's second law of motion, which is discussed later.

Distance, time, and mass are commonly regarded as fundamental, all other dynamical quantities being definable in terms of them.

A primary objective of classical kinetics is the prediction of the behavior of bodies which are subject to known forces when only initial values of the coordinates and momenta are available. This is accomplished by use of a principle first recognized by Isaac Newton. Newton's statement of the principle was restricted to the linear motion of an idealized body called a mass particle, having negligible extension in space.

The basic dynamical law set forth by Newton and known as his second law states that the time rate of change of a particle's linear momentum is proportional to and in the direction of the force applied to the particle. Stated analytically, Newton's second law becomes the differential equation, Eq. (1), in which m represents the particle's mass, v its velocity, F

$$\frac{d\,(mv)}{dt} = F \qquad (1)$$

the applied force, and t the time. Equation (1) provides a definition of force and of its units if units of mass, distance, and time have previously been adopted. The classical assumption of constancy of mass permits Eq. (1) to be expressed as Eq. (2),

$$ma = F \qquad (2)$$

where a represents the linear acceleration.

The behavior of systems composed of two or more interacting particles is treated by Newtonian dynamics augmented by Newton's third law of motion which states that when two bodies interact, the forces they exert on one another are equal and oppositely directed. The important laws of momentum and energy conservation are derivable for such systems (the latter only for forces of special type) and useful in solution of problems. *See* ACCELERATION; FORCE; GRAVITATION; HARMONIC MOTION; MASS; MOMENTUM; RIGID-BODY DYNAMICS; VELOCITY. [R.A.Fi.]

Kinetoplastida An order of the class Zoomastigophorea in the phylum Protozoa, also known as Protomastigida, containing a heterogeneous group of colorless flagellates possessing one or two flagella in some stage of their life cycle. These small organisms (5–89 μm in length) typically have pliable bodies. Some species are holozoic and ingest solid particles, while others are saprozoic and obtain their nutrition by absorption.

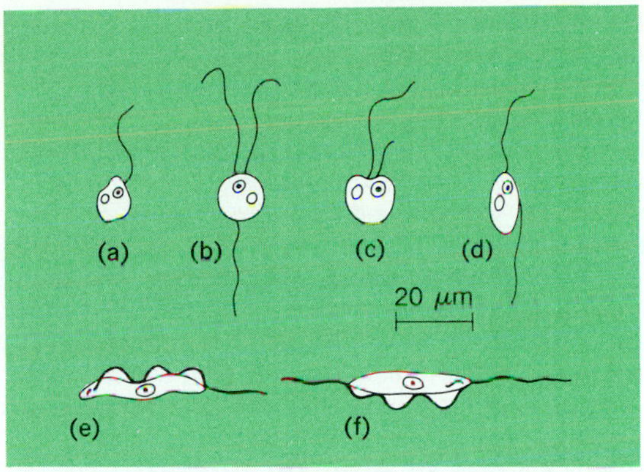

Representative genera of families of order Kinetoplastida. (*a*) *Oikomonas* (family Oikomonadidae), one anterior flagellum. (*b*) *Amphimonas* (family Amphimonadidae), two equally long anterior flagella. (c) *Monas* (family Monadidae), two unequally long anterior flagella. (*d*) *Bodo* (family Bodonidae), two unequally long flagella, one of them trailing. (e) *Trypanosoma* (family Trypanosomatidae), one flagellum with undulating membrane. (*f*) *Cryptobia* (family Cryptobiidae), two flagella, one free and one with undulating membrane.

There is disagreement on the division of the order into families. However, the five or more families can be divided into two general groups (see illustration). The first group contains simple organisms with no distinctive features save one or two flagella of equal or unequal length (includes the families Oikomonadidae, Amphimonadidae, Monadidae, and Bodonidae). The second group contains organisms which have an undulating membrane in addition to one or two flagella (includes Trypanosomatidae and Cryptobiidae). The most important family is the Trypanosomatidae, since it includes several species that infect humans and domestic animals with serious diseases, such as African sleeping sickness. *See* CILIA AND FLAGELLA; TRYPANOSOMATIDAE. [M.M.B.]

Kinorhyncha A phylum of free-living marine invertebrates less than 0.04 in. (1 mm) long. They are segmented and lack external ciliation (see illustration). Kinorhynchs are benthonic,

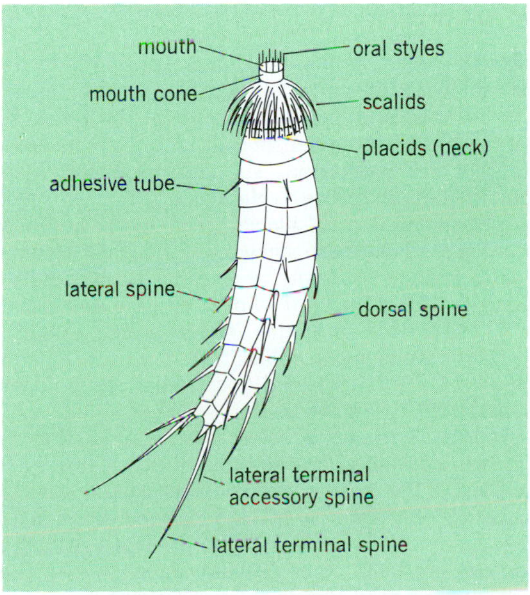

***Echinoderes* sp., a cyclorhagid kinorhynch (ventrolateral view).**

so-called because they generally dwell in mud or sand from intertidal to deep-sea habitats. Two orders are generally recognized, Cyclorhagida and Homalorhagida.

The body is covered by a transparent cuticle secreted by an underlying epidermis. The cuticle is molted only in the process of juvenile growth. Three body regions are recognized: a head segment, a neck segment and an 11-segment trunk. The head is completely retractable. When everted, it extends its five to seven circles of recurved spines called scalids. The neck consists of plates called placids which function to close the anterior opening of the trunk when the head is retracted. Most kinorhynchs have a pair of adhesive tubes on the ventral surface of the third or fourth segment. [R.P.Hi.]

Kirchhoff's laws of electric circuits Fundamental natural laws dealing with the relation of currents at a junction and the voltages around a loop. These laws are commonly used in the analysis and solution of networks. They may be used directly to solve circuit problems, and they form the basis for network theorems used with more complex networks.

One way of stating Kirchhoff's voltage law is: "At each instant of time, the algebraic sum of the voltage rise is equal to the algebraic sum of the voltage drops, both being taken in the same direction around the closed loop."

Kirchhoff's current law may be expressed as follows: "At any given instant, the sum of the instantaneous values of all the currents flowing toward a point is equal to the sum of the instantaneous values of all the currents flowing away from the point." *See* CIRCUIT (ELECTRICITY). [K.Y.T./R.T.W.]

Kite A tethered flying device that supports itself and the cable that connects it to the ground by means of the aerodynamic forces created by the relative motion of the wind. This relative wind may arise merely from the natural motions of the air or may be caused by towing the kite through the agency of its connecting cable.

The lifting force of all kites is produced by deflecting the air downward, the resulting change in momentum producing an upward force. To be successful, a kite must have an extremely low wing loading (weight/area) so that it can fly even on days when the wind velocity is not high. It must be completely stable, since the only controls available to the operator are the length of cable and the rate at which it is taken in or let out. Efficient design requires that its lift-to-drag ratio be as high as possible. *See* AERODYNAMIC FORCE; AERODYNAMICS. [D.C.H.]

Kiwifruit A vigorous deciduous fruiting vine (family Actinidiceae) that is native to central China, where it commonly grows in moist and sheltered areas on the forest edges. Kiwifruit requires both the female cultivar and a male pollenizer for successful fruit production. The kiwifruit industry depends on a single female cultivar, Hayward, the fruit having a creamy-white central core, black-brown seeds, and a bright translucent green outer flesh surrounded by a light-brown fuzzy skin. It is adapted to moderate climates in the temperate zone and requires 600–850 h of winter chilling (temperatures between 32 and 45°F or 0 and 7°C) to ensure uniform budbreak. Kiwifruit wood is susceptible to winter injury at temperatures below 14°F (−10°C), and flower buds can be damaged by frost below 29°F (−1.5°C).

The kiwifruit (*Actinidia deliciosa*) was introduced into cultivation in New Zealand with seed brought from China in 1904, making it one of the most recently domesticated fruiting plants. It appears that all vines now in New Zealand descended from one male and two female plants. The principal kiwifruit-growing countries are Italy, New Zealand, Japan, Chile, France, Greece, United States (California), and Australia.

Kiwifruit is a source of vitamin C, minerals such as potassium, calcium, and phosphorus, and dietary fiber. The primary use is for the fresh market, although culled fruit is processed into canned and frozen fruit slices, wine, jam, juice, and dried products. *See* FRUIT; THEALES. [J.K.Ha]

Klebsiella A genus of bacteria. Three species are recognized: *K. pneumoniae*, *K. ozaenae*, and *K. rhinoscleromatis*. *Klebsiella pneumoniae* may be found in soil, in water, and on plants. Frequently, it is found as a hospital-acquired bacterium which may colonize the throat or intestine or cause urinary tract infection, pneumonia, septicemia (mostly in debilitated patients), and meningitis (in newborns). *Klebsiella ozaenae* occurs in a certain type of chronic rhinitis called ozaena. *Klebsiella rhinoscleromatis* causes rhinoscleroma, a granulomatous disease of the upper respiratory tract occurring in Eastern Europe. *See* ENTEROBACTERIACEAE. [A.W.C.V.G.]

Klystron An evacuated electron-beam tube in which an initial velocity modulation imparted to electrons in the beam results subsequently in density modulation of the beam. A klystron is used either as an amplifier in the microwave region or as an oscillator.

For use as an amplifier, a klystron receives microwave energy at an input cavity through which the electron beam passes. The microwave energy modulates the velocities of electrons in the beam, which then enters a drift space. Here the faster electrons overtake the slower to form bunches. In this manner, the uniform current density of the initial beam is converted to an alternating current. The bunched beam with its significant component of alternating current then passes through an output cavity to which the beam transfers its ac energy. *See* MICROWAVE.

Klystrons may be operated as oscillators by feeding some of the output back into the input circuit. More widely used is the reflex oscillator in which the electron beam itself provides the feedback. The beam is focused through a cavity and is velocity-modulated there, as in the amplifier. The cavity usually has grids to concentrate the electric field in a short space so that the field can interact with a slow, low-voltage electron beam. Leaving the cavity, the beam enters a region of dc electric field opposing its motion, produced by a reflector electrode operating at a potential negative with respect to the cathode. The electrons do not have enough energy to reach the electrode, but are reflected in space and return to pass through the cavity again. The points of reflection are determined by electron velocities, the faster electrons going farther against the field and hence taking longer to get back than the slower ones. Reflex oscillators are used as signal sources from 3 to 200 GHz. They are also used as the transmitter tubes in line-of-sight radio relay systems and in low-power radars. [R.B.N.]

Knudsen number In fluid mechanics, the ratio l/L of the mean free path length l of the molecules of the fluid to a characteristic length L of the structure in the fluid stream. When the mean free path of the fluid particles is short relative to the size of the object being considered, the fluid can be treated as a continuum. If the path length between molecular encounters is comparable to or larger than a significant dimension of the flow region, the gas must be treated as consisting of discrete particles. *See* GAS DYNAMICS; STATISTICAL MECHANICS. [F.H.R.]

Koala A single species, *Phascolarctos cinereus*, which is a member of the family Phalangeridae in the mammalian order Marsupialia (pouch-bearing animals). It is a small animal that weighs from 11 to 17 lb (5 to 8 kg) when mature. They are restricted to eastern Australia. Not only do they have a specialized diet of eucalyptus leaves, but the leaves must be of a certain age from a specific species of tree, and the tree must grow upon a certain type of soil.

The koala breeds once each season, and the usual number of offspring is one. It remains in the pouch for 6 months; then it clings to the back of its mother and is carried around in this manner until 1 year old. See MAMMALIA; MARSUPIALIA. [C.B.C.]

Kohlrabi A cool-season biennial crucifer, *Brassica caulorapa* and *B. oleracea* var. *caulo-rapa*, of northern European origin belonging to the plant order Capparales. Kohlrabi is grown for its turniplike enlarged stem, which is usually eaten as a cooked vegetable (see illustration). A common cooked vegetable in Europe, especially Germany, kohlrabi is of minor importance in the United States. See CAPPARALES; TURNIP. [H.J.C.]

Kohlrabi (*Brassica caulorapa*), cultivar Early White Vienna. (*Joseph Harris Co., Rochester, New York.*)

Kondo effect The large anomalous increase in the resistance of certain dilute alloys of magnetic materials in nonmagnetic hosts as the temperature is lowered. This behavior is contrary to the decrease in electrical resistance with temperature observed in almost all other systems. It is believed that the interaction of the conduction electrons with the spin of the magnetic impurity creates rather long-range correlations in the electrons in the vicinity of the impurity and that this strong interaction tends to inhibit conductivity. This interaction is an exchange interaction similar to that in ferromagnets. See FERROMAGNETISM.

For temperatures above a few degrees absolute, however, thermal agitation tends to wash out these fairly delicate correlations, and on heating the sample further the contribution to the resistance diminishes. The temperature below which this effect predominates is referred to as the Kondo temperature, and it depends on the magnetic impurity and the host material.

It has been suggested that Kondo temperatures may range from millidegrees to hundreds of degrees, but for materials studied so far they are below about 30 K (−405°F). See CURIE TEMPERATURE.

The importance of studying the Kondo effect lies in the fact that it is a property of what is, presumably, the simplest possible magnetic system—a single magnetic atom in a nonmagnetic environment. It is hoped that an understanding of this system will provide a key to the far more complex and important problem of the electronic structure of magnetic materials themselves, one of the greatest challenges in physics. [M.Fo.]

Krebs cycle A sequence of enzymatic reactions that is involved in the oxidative metabolism of many organisms, especially those that carry out respiration. The cycle is also known as the citric acid cycle and the tricarboxylic acid cycle. Through this sequence, or pathway, the two-carbon acetyl unit, an important metabolic intermediate derived from a variety of substrates, can be oxidized to two molecules of CO_2. In aerobic metabolism, two molecules of oxygen are used for this oxidation. Organic acids containing four, five, and six carbons are formed in the course of the oxidation. These are consumed and regenerated continually through the operation of the entire sequence. Hence, this oxidative pathway is a catalytic mechanism which operates in a cyclic manner and is known as the citric acid cycle, because this tricarboxylic acid is one of the metabolic intermediates. The cycle is also named after H. A. Krebs, who recognized its functional role. The principal reactions of the Krebs cycle are outlined in the illustration. See METABOLISM.

The enzymatic reactions of the Krebs cycle are important, not only as a source of energy for living cells, but also as a

$$\underset{\text{Oxaloacetic acid}}{HOOC{-}CH_2{-}\overset{\overset{O}{\|}}{C}{-}COOH} \quad + \quad \underset{\text{Acetyl group of acetyl coenzyme A}}{CH_3\overset{\overset{O}{\|}}{C}{-}}$$

$$\downarrow (1)$$

$$\underset{\text{Citric acid}}{COOH{-}CH_2{-}COH(COOH){-}CH_2{-}COOH}$$

$$\downarrow (2)$$

$$\underset{\textit{cis}\text{-Aconitic acid}}{COOH{-}CH{=}C(COOH){-}CH_2{-}COOH}$$

$$\downarrow (3)$$

$$\underset{\text{Isocitric acid}}{COOH{-}CHOH{-}CH(COOH){-}CH_2{-}COOH}$$

$$\downarrow (4) \; -2[H]$$

$$CO_2 + \underset{\alpha\text{-Ketoglutaric acid}}{COOH{-}\overset{\overset{O}{\|}}{C}{-}CH_2{-}CH_2{-}COOH}$$

$$\downarrow (5) \; -2[H]$$

$$CO_2 + \underset{\text{Succinic acid}}{COOH{-}CH_2{-}CH_2{-}COOH}$$

$$\downarrow (6) \; -2[H]$$

$$\underset{\text{Fumaric acid}}{COOH{-}CH{=}CH{-}COOH}$$

$$\downarrow (7)$$

$$\underset{\text{Malic acid}}{COOH{-}CH_2{-}CHOH{-}COOH}$$

$$\downarrow (8) \; -2[H]$$

$$\underset{\text{Oxaloacetic acid}}{COOH{-}CH_2{-}\overset{\overset{O}{\|}}{C}{-}COOH}$$

Principal reactions of the Krebs cycle. Oxidative steps are indicated by the symbol −2[H].

source of essential metabolic intermediates which serve as starting materials for biosynthetic processes. [M.D.]

Kronig-Penney model An idealized, one-dimensional model of a crystal which exhibits many of the basic features of the electronic structure of real crystals. Consider the potential energy $V(x)$ of an electron shown in the illustration with an infi-

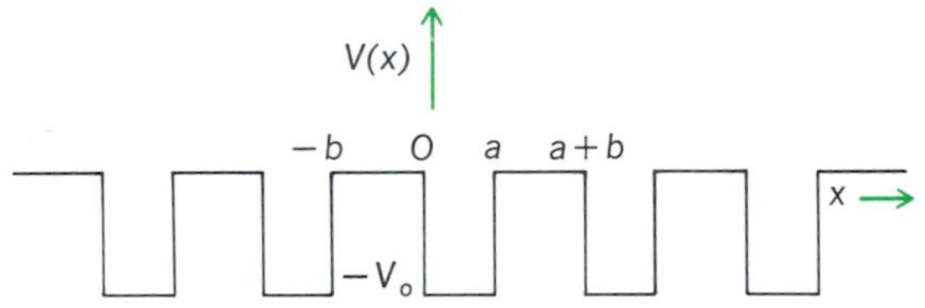

Potential energy which is assumed for the one-dimensional Kronig-Penney model.

nite sequence of potential wells of depth $-V_o$ and width a, arranged with a spacing b. The width and the curvatures of the allowed bands increase with energy. The Kronig-Penney model has been extended to include the effects of impurity atoms. *See* BAND THEORY OF SOLIDS. [J.C.]

Krypton A gaseous chemical element, Kr, atomic number 36, and atomic weight 83.80. Krypton is one of the noble

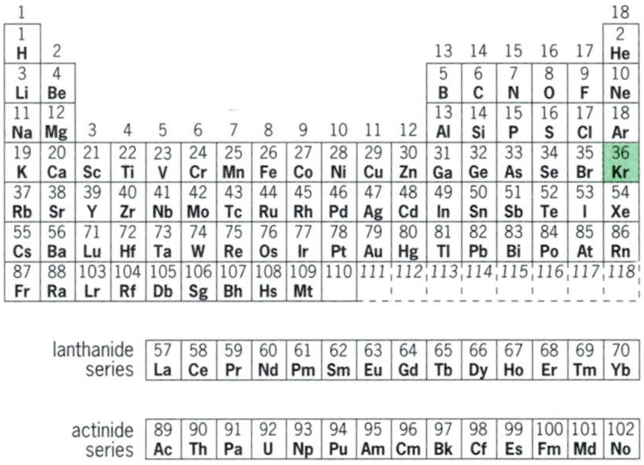
1 H
2 He
3 Li 4 Be
5 B 6 C 7 N 8 O 9 F 10 Ne
11 Na 12 Mg
13 Al 14 Si 15 P 16 S 17 Cl 18 Ar
19 K 20 Ca 21 Sc 22 Ti 23 V 24 Cr 25 Mn 26 Fe 27 Co 28 Ni 29 Cu 30 Zn 31 Ga 32 Ge 33 As 34 Se 35 Br 36 Kr
37 Rb 38 Sr 39 Y 40 Zr 41 Nb 42 Mo 43 Tc 44 Ru 45 Rh 46 Pd 47 Ag 48 Cd 49 In 50 Sn 51 Sb 52 Te 53 I 54 Xe
55 Cs 56 Ba 71 Lu 72 Hf 73 Ta 74 W 75 Re 76 Os 77 Ir 78 Pt 79 Au 80 Hg 81 Tl 82 Pb 83 Bi 84 Po 85 At 86 Rn
87 Fr 88 Ra 103 Lr 104 Rf 105 Db 106 Sg 107 Bh 108 Hs 109 Mt 110 111 112 113 114 115 116 117 118
lanthanide series
57 La 58 Ce 59 Pr 60 Nd 61 Pm 62 Sm 63 Eu 64 Gd 65 Tb 66 Dy 67 Ho 68 Er 69 Tm 70 Yb
actinide series
89 Ac 90 Th 91 Pa 92 U 93 Np 94 Pu 95 Am 96 Cm 97 Bk 98 Cf 99 Es 100 Fm 101 Md 102 No

gases in group 18 of the periodic table. Krypton is a colorless, odorless, and tasteless gas. The table gives some physical properties of krypton. The principal use for krypton is in filling electric lamps and electronic devices of various types. Krypton-argon mixtures are widely used to fill fluorescent lamps. *See* INERT GASES.

Physical properties of krypton

Property	Value
Atomic number	36
Atomic weight (atmospheric krypton only)	83.80
Melting point, triple point °C	−157.20
Boiling point at 1 atm pressure, °C	−153.35
Gas density at 0°C and 1 atm pressure, g/liter	3.749
Liquid density at its boiling point, g/ml	2.413
Solubility in water at 20°C, ml krypton (STP) per 1000 g water at 1 atm partial pressure krypton	59.4

The only commercial source of stable krypton is the air, although traces of krypton are found in minerals and meteorites.

A mixture of stable and radioactive isotopes of krypton is produced in nuclear reactors by the slow-neutron fission of uranium. It is estimated that about 2×10^{-8}% of the weight of the Earth is krypton. Krypton also occurs outside the Earth. [A.W.F.]

Kumquat Shrubs or small trees that are members of the genus *Fortunella*, which is one of the six genera in the group of true citrus fruits. Kumquats are believed to have originated in China and the Malay Peninsula, but are now widely grown in all citrus areas of the world. Of the several species the most common are *F. margarita*, which has oval-shaped fruit, and *F. japonica*, which has round fruit.

Kumquats, with their brilliant orange-colored fruits and dense green foliage, are highly ornamental and are most frequently grown for this reason. Kumquat fruits can be eaten whole without peeling; they are also used in marmalades and preserves and as candied fruits. *See* SAPINDALES. [F.E.G.]

Kutorginida An extinct order of brachiopods whose class assignment is uncertain. They occur in rocks of Early Cambrian and possibly also Middle Cambrian age. The shell has a primarily calcareous composition. The valves do not seem to be articulated by teeth and sockets. Both valves resemble articulate valves in having elongate muscle tracks and shelflike posterior regions (cardinal areas). *See* BRACHIOPODA. [M.W.F.]

Kwashiorkor disease A nutritional deficiency disease in infants and young children, caused primarily by diet in which protein is of poor quality or inadequate quantity, or both, and carbohydrate intake is normal or high. The disease has frequently been referred to as protein-calorie malnutrition or deficiency. Kwashiorkor occurs throughout the world, but especially among the underdeveloped areas. According to the World Health Organization, it is the most widespread and important dietary disease in the world today.

Clinical features consist of poor growth and development, edema, dyspigmentation of skin and hair, and psychological changes characterized by apathy and irritability. Pathological changes consist of enlarged fatty liver; atrophy of pancreas, salivary glands, and intestinal glands; and wasting of muscles. [H.Si.]

Kyanite A nesosilicate mineral, composition Al_2SiO_5, crystallizing in the triclinic system. Crystals are usually long and tabular and commonly occur in bladed aggregates. The specific gravity is 3.56–3.66. The hardness of kyanite, one of its interesting features, is 5 (Mohs scale) parallel to the length of the crystals but 7 across the length. The luster is vitreous to pearly, and the color is usually a shade of blue but may be white, gray, or green. *See* SILICATE MINERALS.

Kyanite crystals, some of gem quality, are found at St. Gothard, Switzerland. Kyanite has been mined in India and in the United States in Georgia and North Carolina for the manufacture of highly refractory porcelain. [C.S.Hu.]

Labradorite A plagioclase feldspar with composition range $Ab_{50}An_{50}$ to $Ab_{30}An_{70}$ ($Ab = NaAlSi_3O_8$; $An = CaAl_2Si_2O_8$). In some labradorite samples brilliant colors, much like those seen in oil films on water, result from the interference of light reflected at successive lamellar interfaces. *See* FELDSPAR; IGNEOUS ROCKS. [P.H.R.]

Labyrinthodontia An important subclass of fossil amphibians, first known in the Late Devonian, common in the Carboniferous and Permian, and persisting to the Late Triassic. They are named for infoldings of the enamel of their teeth, a primitive feature which is not universally present in labyrinthodonts and which is shared with rhipidistean crossopterygian fishes. It is generally accepted that the labyrinthodonts are descended from rhipidisteans. Labyrinthodonts are usually fairly large (up to 6.5 ft or 2 m long). Some were totally aquatic (in fresh water). Most were probably semiaquatic, while some (especially the Dissorophidae) appeared to be terrestrial, at least as adults. *See* AMPHIBIA. [R.E.DeM.]

Labyrinthulia Protozoa forming a subclass of Rhizopodea with obscure relationships to rest of the class. There is one order, Labyrinthulida. The mostly marine, ovoid to spindle-shaped, uninucleate organisms secrete a network of filaments (slime tubes) along which they glide, usually singly. This network inspired the name "net slime molds" sometimes applied to them. The mechanism of locomotion is unknown. *See* RHIZOPODEA. [R.P.H.]

Laccolith A geologic name for a body of igneous rock intruded into sedimentary rocks in such a way that the overlying strata have been notably lifted by the force of intrusion. In the most simple example the laccolith is circular or oval in plan, the subjacent sedimentary rocks and the floor of the laccolith are nearly horizontal, and the roof is dome-shaped with the sedimentary rocks arching over the dome. [J.A.N.]

Lacquer A fast-drying, hard, high-gloss surface coating. Lacquers are made by dissolving a cellulose derivative and other modifying materials in a solvent and adding pigment if desired. The cellulose derivative most commonly used is nitrocellulose, but a cellulose ester such as cellulose acetate or cellulose butyrate, or a cellulose ether such as ethyl cellulose is often used in formulation. Lacquers dry by evaporation of the solvent. They usually are applied by spray because of their rapid drying properties.

Lacquers have been used extensively as fast-drying, weather-resistant finishes for automobiles and as coatings for furniture and other factory-finished items. Their use diminished, however, as less expensive coating materials with improved properties appeared on the market. *See* SURFACE COATING. [C.R.Ma.; C.W.Si.]

Lacrimal gland A tubuloalveolar or acinous skin gland, also known as the tear gland. Two types occur among the vertebrates, the lacrimal proper and the Harderian. The eye glands drain into the nasal cavity by means of the lacrimal duct.

In land mammals the lacrimal gland proper is highly developed as a complex tubuloalveolar structure with several ducts which pour their copious fluid into the outer, upper part of the conjunctival sac or cavity. The tear substance washes across the eyeball and eventually passes through two small openings, one on the margin of each lid, into the lacrimal ducts. The latter converge to form the lacrimal sac, from which the nasolacrimal duct leads into the nasal passageway. Tears contain a considerable quantity of the common salt, sodium chloride. *See* EYE (VERTEBRATE); GLAND. [O.E.N.]

Lactam A cyclic amide that is the nitrogen analog of a lactone. For example, a γ-aminobutyric acid readily forms γ-butyrolactam (also known as 2-pyrrolidinone) upon heating, as in the reaction below. The tautomeric enol form of a lactam is

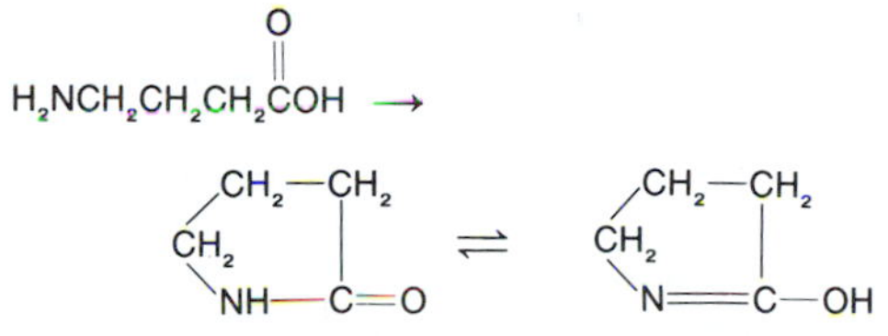

γ-Butyrolactam (keto form) — γ-Butyrolactim (enol form)

known as a lactim.

The δ-amino acids similarly form δ (six-membered-ring) lactams upon heating, but larger- and smaller-ring lactams must be made by indirect methods.

Several lactams are of considerable industrial importance. 2-Pyrrolidinone and 1-methyl-2-pyrrolidinone are made by heating γ-butyrolactone with ammonia and methylamine, respectively. They are useful specialty solvents. Vinylation of 2-pyrrolidinone with acetylene gives 1-vinyl-2-pyrrolidinone, which is polymerized to a substance commonly used in aerosol hair sprays.

The ß-lactam antibiotics comprise two groups of clinically important therapeutic agents, the penicillins and the cephalosporins. In both cases they contain a four-membered or ß-lactam ring which has its nitrogen atom and a carbon atom in common with another ring. Such substances are derived commercially from fermentation processes, followed usually by chemical manipulation of the functional groups. *See* AMINO ACIDS; ANTIBIOTIC; BIOCHEMICAL ENGINEERING; LACTONE. [P.E.F.]

Lactase An enzyme found in mammals, honeybee larvae, and some plants. It is a ß-galactosidase which hydrolyzes lactose to galactose and glucose. In mammals, lactase appears in the intestinal secretion from the intestinal villi, and exerts its effect on lactose in chyme. *See* ENZYME; GLUCOSE. [D.N.La.]

Lactate A salt or ester of lactic acid in which the acidic hydrogen of the carboxyl group has been replaced by a metal or

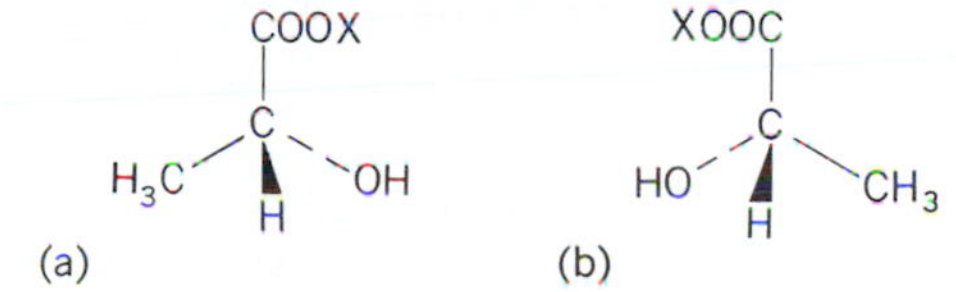

Optical isomers of lactic acid: (*a*) D(−) form; (*b*) L(+) form.

an organic radical (symbol X). The structural formulas of the two optical isomers are given in the illustration. Alkali metal salts have practical applications, for example, as blood coagulants in calcium therapy (calcium), as antiperspirants (aluminum), in treatment of anemia (iron), and as plasticizers (sodium). Esters are used as plasticizers and as solvents for lacquers, cellulose acetate, and cellulose nitrate. *See* ESTER; LACTIC ACID. [E.H.H.]

Lactation The function of the mammary gland providing milk nourishment to the newborn mammal. This process is under the control of the endocrine and nervous systems. It involves transformation of an inactive duct system to a lobuloalveolar glandular structure during pregnancy, cellular production of the components of milk (galactopoiesis), secretion into the ducts, and ejection under the stimulus of milking or suckling.

Lactation makes demands on the maternal regulation of calcium metabolism. Resorption of bone increases in lactating rats and women, and there is a marked increase in the absorption of calcium from the intestine. The elevated need for calcium results in an increased role for parathyroid hormone, calcitonin, and vitamin D in the regulation of the absorption and utilization of calcium. In humans a concomitant phenomenon frequently associated with lactation is amenorrhea. Consequently in some societies prolonged nursing is used as a birth control technique. *See* MAMMARY GLAND; MILK. [H.J.L.]

Lactic acid A slightly hygroscopic syrup with formula COOH—CHOH—CH_3 and a specific gravity of 1.294 at 25°C. The acid, also known as α-hydroxypropionic acid, contains one asymmetric carbon atom and exists as two optically active isomers that can be converted by an enzyme, racemase, into the racemic mixture. L(+)-Lactic acid, also called sarcolactic and paralactic acid, is metabolized by animals, whereas D(−)-lactic acid is not. The lactic acid of commerce is the racemic mixture, commonly known as *i*-lactic acid (optically inactive). Lactic acid is used in pharmaceutical preparations, in the textile and leather industry, in the production of inks, solvents, lacquers, and plastics, as the bread additive, calcium stearyl-2 lactylate, and as food preservative. [F.J.S.]

Lactobacillaceae A family of gram-positive, asporogenous, rod-shaped bacteria with the single genus *Lactobacillus*. They occur singly, in pairs, or in chains. They are primarily saprophytic, rarely pathogenic. The bacteria are generally known as sugar fermenters, producing lactic acid as a major product. They are found in fermenting food products, in the mouth, in the intestinal tract, and in body lesions. The strains usually are nonmotile and show neither reduction of nitrates, gelatin liquefaction, catalase production, nor surface growth in liquid media. The species are microaerophilic or anaerobic, inactive toward protein sources but fastidious in their requirements. [C.S.P.]

Lactone A cyclic, intramolecular ester derived from a hydroxy acid. Simple lactones are designated α, ß, γ, δ and so forth. Five- and six-membered lactones are very readily obtained by cyclization of a hydroxy acid or precursor as shown in reactions (1)–(3). Lactones with three- and four-membered rings are also known. *See* ESTER.

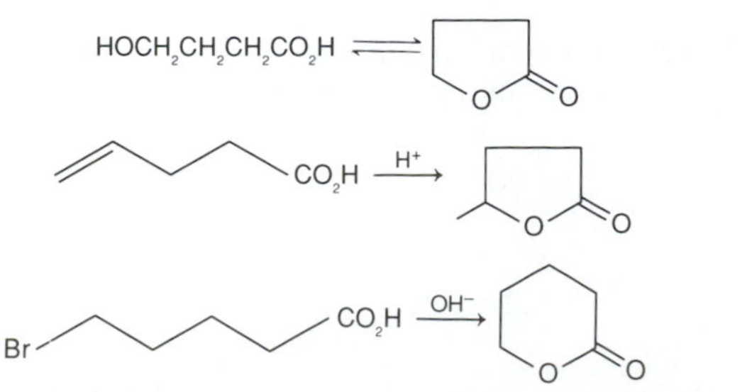

Lactones in various forms are found in numerous naturally occurring compounds. Unsaturated γ-lactone rings (I and II) are

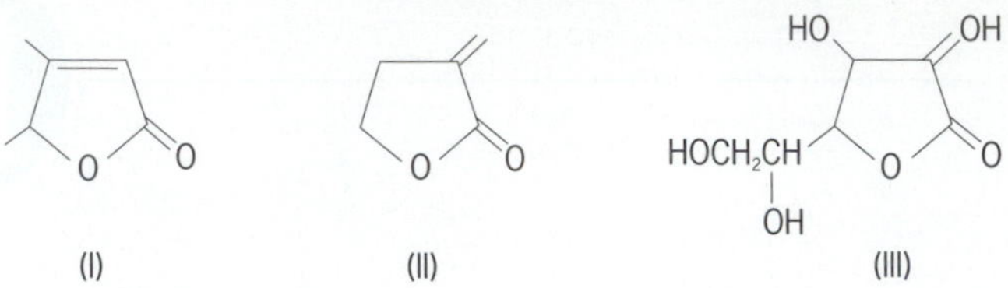

present in many components of essential oils. Ascorbic acid [vitamin C; (III)] is a carbohydrate lactone. *See* ASCORBIC ACID; ESSENTIAL OILS.

A major development in the chemistry of naturally occurring compounds has been the isolation from microorganisms of a number of macrocyclic lactones with rings containing from 12 to more than 30 atoms. These substances include immunosuppressive agents and antibiotics such as erythromycin. Although they represent some of the most complex organic compounds known, several of these macrocyclic lactones have been synthesized. *See* ANTIBIOTIC; ORGANIC CHEMICAL SYNTHESIS; ORGANIC CHEMISTRY. [P.E.F.; J.A.Mo.]

Lactose Milk sugar or 4-*O*-ß-D-galactopyranosyl-D-glucose. This reducing disaccharide is obtained as the α-D anomer (see formula, where the asterisk indicates a reducing group);

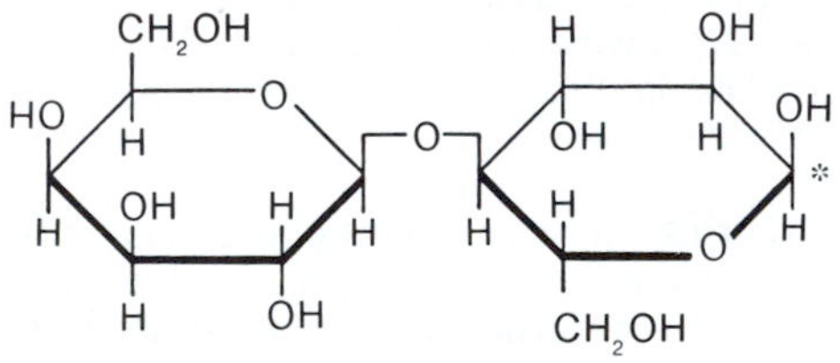

the melting point (mp) is 202°C (396°F). Lactose is found in the milk of mammals to the extent of approximately 2–8%. It is usually prepared from whey, which is obtained by a by-product in the manufacture of cheese. Upon concentration of the whey, crystalline lactose is deposited. [W.Z.H.]

Laevicaudata An order of fresh-water branchiopod crustaceans formerly included in the order Conchostraca. The body is up to about 7 mm (0.3 in.) in length; females have 12 trunk segments, males have 10. The trunk terminates in a feebly developed telson that lacks claws and is covered ventrally by an opercular lamella. The head is free, articulated with the trunk, and can be swung forward. It is covered by a headshield and has paired, sessile eyes. It bears large antennae that are used for swimming. The mandibles, while of the rolling, crushing type, have narrow masticatory surfaces and stout teeth.

Clumps of small drought-resistant eggs are carried by the female and shed at the next molt. They hatch as a late nauplius of distinctive type. The Laevicaudata are inhabitants of temporary water bodies and are almost worldwide in distribution. Although they can swim, they are usually associated with the bottom. *See* BRANCHIOPODA. [G.Fr.]

Lagomorpha The order of mammals including rabbits, hares, and pikas. Lagomorphs have two pairs of upper incisors (the second pair minute), and enamel surrounds the tooth, which does not form a sharp chisel. Motion of the jaw is vertical or transverse. Lagomorphs have three upper and two lower premolars, the earliest fossil rodents have one less of each. The tibia and fibula are fused, the fibula articulating with the calcaneum as in artiodactyls. There is a spiral valve in the cecum, and the scrotum is prepenial.

The order includes three families: Leporidae (rabbits and hares); Ochotonidae (pikas, whistling hares, or American coneys); and Eurymylidae, an extinct family from the Paleocene of Mongolia. *See* MAMMALIA; PIKA; RABBIT.

Leporidae are the most familiar members of the order. There are, in general, two kinds: rabbits (such as the American cottontail), which are relatively small, with shorter hindlegs, shorter ears, and short tails; and hares, larger forms with longer legs, ears, and tails. Rapid locomotion is by leaps, using the hindlegs, combined (especially in rabbits) with abrupt changes of direction. Both types occur in the same region, with rabbits inhabiting brush, scrub, or woods and hares living in open grassland. In North America, hares are usually called jackrabbits. [A.E.Wo.]

Lagrange's equations Equations of motion of a mechanical system for which a classical (non-quantum-mechanical) description is suitable, and which relate the kinetic energy of the system to the generalized coordinates, the generalized forces, and the time. If the configuration of the system is specified by giving the values of f independent quantities $q_1, \ldots, q_f$, there are f such equations of motion.

In their usual form, these equations are equivalent to Newton's second law of motion and are differential equations of the second order for the q's as functions of the time t. [P.M.S.]

Lagrangian function A function of the generalized coordinates and velocities of a dynamical system from which the equations of motion in Lagrange's form can be derived. The Lagrangian function is denoted by $L(q_1, \ldots, q_f; \dot{q}_1, \ldots \dot{q}_f; t)$. For systems in which the forces are derivable from a potential energy V, if the kinetic energy is T, the equation below holds.

$$L = T - V$$

See LAGRANGE'S EQUATIONS. [P.M.S.]

Lake An inland body of water, small to moderately large in size, with its surface exposed to the atmosphere. Most lakes fill depressions below the zone of saturation in the surrounding soil and rock materials. Generically speaking, all bodies of water of this type are lakes, although small lakes usually are called ponds, tarns (in mountains), and less frequently pools or meres. The great majority of lakes have a surface area of less than 100 mi^2 (259 km^2). More than 30 well-known lakes, however, exceed 1500 mi^2 (3885 km^2) in extent, and the largest freshwater body, Lake Superior, North America, covers 31,180 mi^2 (80,756 km^2). Most lakes are relatively shallow features of the Earth's surface. Because of their shallowness, lakes in general may be considered evanescent features of the Earth's surface, with a relatively short life in geological time.

Lakes differ as to the salt content of the water and as to whether they are intermittent or permanent. Most lakes are composed of fresh water, but some are more salty than the oceans. Generally speaking, a number of water bodies which are called seas are actually salt lakes; examples are the Dead, Caspian, and Aral seas. All salt lakes are found under desert or semiarid climates, where the rate of evaporation is high enough to prevent an outflow and therefore a discharge of salts into the sea.

Lakes with fresh waters also differ greatly in the composition of their waters. Because of the balance between inflow and outflow, fresh lake water composition tends to assume the composite dissolved solids characteristics of the waters of the inflowing streams—with the lake's age having very little influence. Under a few special situations, as crater lakes in volcanic areas, sulfur or other gases may be present in lake water, influencing color, taste, and chemical reaction of the water. *See* FRESH-WATER ECOSYSTEM; HYDROSPHERE; MEROMICTIC LAKE; SURFACE WATER.

Both natural and artificial lakes are economically significant for their storage of water, regulation of stream flow, adaptability to navigation, and recreational attractiveness. A few salt lakes are significant sources of minerals. *See* EUTROPHICATION. [E.A.A.]

Lamellibranchia The largest subclass of the class Bivalvia of the phylum Mollusca. The name refers to the two enormously enlarged gills, each composed of great numbers of parallel ctenidial filaments held together to form an extensive folded lamella. Each gill, although homologous both functionally and morphologically with gills of more primitive gastropods, is far more extensive than is required for the respiratory needs of the animal. Here the gill has become an organ of food collection: the most elaborate and efficient ciliary-feeding mechanism in the animal kingdom.

Subdivision of the Lamellibranchia is done most suitably into superfamilies, each with characteristic shell structure, dentition, ligament, and degree of union of the mantle lobes, involving, in many cases, formation of siphons to conduct the inhalant and exhalant water currents. The more important superfamilies include the Mytilacea (mussels), Ostreacea (true oysters), Cardiacea (true cockles), Tellinacea, Myacea (including "soft-shell" clams), Unionacea (fresh-water clams), Mactracea, and Adesmacea (or Pholadacea). *See* BIVALVIA; MOLLUSCA; SHIPWORM. [C.M.Y.]

Lamiales An order of flowering plants, division Magnoliophyta (Angiospermae), in the subclass Asteridae of the class Magnoliopsida (dicotyledons). The order consists of 5 families and more than 7800 species: Labiatae (about 3200 species; see illustration), Verbenaceae (about 2600 species),

***Monarda fistulosa*, an eastern North American species of wild bergamot, showing the characteristic opposite leaves and irregular corolla of the family Labiatae. (*A. W. Ambler, National Audubon Society*)**

and Boraginaceae (about 2000 species) make up the bulk of the order. Within its subclass the order is marked by its characteristic gynoecium, consisting of usually two biovulate carpels, each carpel divided between the ovules by a "false" partition, or with the two halves of the carpel seemingly wholly separate. Except in some of the more primitive species, the fruit usually consists of four separate or separating nutlets. In the Labiatae and most of the Boraginaceae, the four segments of the ovary are nearly or quite distinct from each other and are united chiefly by the gynobasic style. *See* ASTERIDAE; MAGNOLIOPSIDA. [A.Cr.]

Laminar flow Streamline flow of a viscous fluid which satisfies the Navier-Stokes equations of motion. In laminar flow, the fluid moves in layers without large irregular fluctuations. Laminar flow occurs at a low Reynolds number. This corresponds to the conditions of small velocities and dimensions of bodies, to very large viscosity, or to small density of the fluid.

Laminar flow plays an important role in several practical problems. The flow of oil in the bearings for lubrication is laminar. The theory of laminar flow shows that under great normal pressure, the oil in the bearing has only slight frictional resistance. Flow on the surface of modern aircraft and missiles flying at extremely high altitude may be laminar. Laminar boundary-layer flow determines the skin friction and the aerodynamic heating of these bodies. *See* NAVIER-STOKES EQUATIONS; REYNOLDS NUMBER; STREAMLINE FLOW. [S.-I.P.]

Laminariales An order of large brown algae (Phaeophyceae) commonly called kelps. Four families are recognized: Chordaceae, Laminariaceae, Alariaceae, and Lessoniaceae. Definitive features include a life history in which microscopic, filamentous, dioecious gametophytes alternate with a massive, parenchymatous sporophyte; growth of the sporophyte effected by meristems; and production of unilocular zoosporangia in extensive sori on blades of the sporophyte. *See* PHAEOPHYCEAE.

A mature sporophyte consists typically of a holdfast, stipe, and one or more blades. The holdfast usually comprises a cluster of rootlike structures (haptera); in a few species it is discoid or conical. The stipe, which varies in length from a few millimeters in *Hedophyllum* to more than 100 ft (30 m) in *Macrocystis* (giant kelp), may be branched or unbranched and may bear one or more blades.

Kelps are largely confined to cold waters of both hemispheres, displaying a peak of diversity, abundance, and luxuriance all around the North Pacific. Only *Lessonia* and *Ecklonia* (both Lessoniaceae) are characteristic of the Southern Hemisphere, but *Macrocystis* is bipolar, occurring abundantly in subantarctic waters. Kelps are lacking in Antarctica, where their ecological niche is filled by members of the brown algal family Desmarestiaceae. Some kelps are found in the intertidal zone, usually on rocks exposed to heavy waves, but most are subtidal.

The principal use of kelps is in the alginate industry. Various kelps are used extensively for human food in Japan. In China *Laminaria japonica* has been established as an important maricultural crop plant. Ecologically, such kelps as *Macrocystis* and *Nereocystis* are particularly important because they form protective canopies harboring special communities of smaller seaweeds, invertebrates, fishes, and the sea otter. *See* ALGAE; ALGINATE. [P.C.Si.; R.L.Moe]

Lamp An electric lamp is a device for converting electrical energy into illumination. The basic types include incandescent, vapor, fluorescent, and arc lamps. *See* ARC LAMP; FLUORESCENT LAMP; INCANDESCENT LAMP; MERCURY-VAPOR LAMP; VAPOR LAMP. [J.Mar.]

Lampridiformes An order of teleost fishes including the ribbonfishes, oarfishes, opah, and their allies, also known as the Allotriognathi. Although these fishes are diverse in form, they share characters that define this order. In most, the body is notably compressed, often ribbonlike (see illustration). The fins are composed of soft rays, or the dorsal has one or two anterior spines; the pelvic fin, if present, is thoracic, with 1–17 spineless rays; the pelvic girdle is inserted between the coracoids, or attached to them, or is lost; scales are small and cycloid or, commonly, absent; and the swim bladder is without a duct. Most species are large and colorful. They are all oceanic. *See* ACTINOPTERYGII. [R.M.B.]

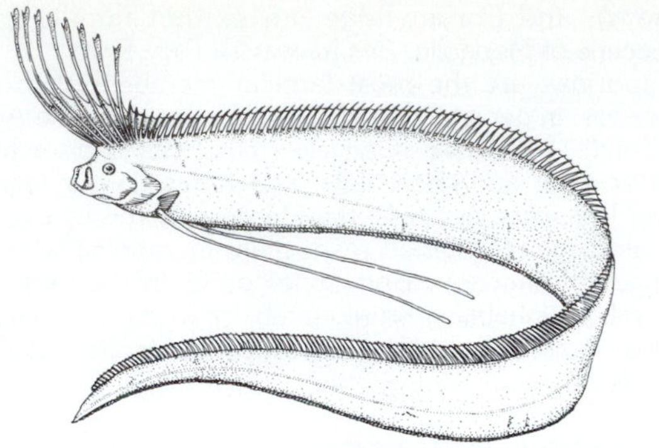

Oarfish (*Regalecus glesne*). (*After D. S. Jordan and B. W. Evermann, The Fishes of North and Middle America, U.S. Nat. Mus. Bull. no. 47, 1900*)

Lamprophyre Any of a heterogeneous group of gray to black, mafic igneous rocks characterized by a distinctive panidiomorphic and porphyritic texture in which abundant euhedral, dark-colored ferromagnesian (femic) minerals (dark mica, amphibole, pyroxene, olivine) occur in two generations—both early as phenocrysts and later in the matrix or groundmass—while felsic minerals (potassium feldspar, plagioclase, analcime, melilite) are restricted to the groundmass.

Many varieties of lamprophyre are known. Minettes, kersantites, vogesites, and spessartites are the most common and are sometimes collectively called calcalkaline lamprophyres. Camptonites and monchiquites are less common; alnoites are rare. These varieties, along with some others, are referred to as alkaline lamprophyres.

Lamprophyres are widespread but volumetrically minor rocks that apparently are restricted to the continents and are the last manifestation of igneous activity in a given area. They usually occur as subparallel or radial swarms of thin (~1.6–160 ft or ~0.5–50 m) dikes or, less commonly, sills, volcanic neck fillings, or diatremes, or, rarely, lava flows. *See* IGNEOUS ROCKS. [S.W.B.]

Land drainage (agriculture) The removal of water from the surface of the land and the control of the shallow groundwater table improves the soil as a medium for plant growth. The sources of excess water may be precipitation, snowmelt, irrigation water, overland flow or underground seepage from adjacent areas, artesian flow from deep aquifers, floodwater from channels, or water applied for such special purposes as leaching salts from the soil or for achieving temperature control.

The purpose of agricultural drainage can be summed up as the improvement of soil water conditions to enhance agricultural use of the land. Such enhancement may come about by direct effects on crop growth, by improving the efficiency of farming operations or, under irrigated conditions, by maintaining or establishing a favorable salt regime. Drainage systems are engineering structures that remove water according to the principles of soil physics and hydraulics. The consequences of drainage, however, may also include a change in the quality of the drainage water. Agricultural drainage is divided into two broad classes; surface and subsurface. Some installations serve both purposes. [J.N.Lu.]

Land reclamation The process by which seriously disturbed land surfaces are stabilized against the hazards of wind and water erosion. Surface mining for coal is responsible for almost one-half of the total land area disturbed in the United States. The drastic disturbance of the overburden severely changes the chemical and physical properties of the resulting spoils. These altered properties often create a hostile environment for seed germination and subsequent plant growth. Unless vegetative cover is established almost immediately, the denuded areas are subject to both wind and water erosion that pollute surrounding streams with sediment.

The removal of topsoil before mining and its replacement on the spoil surface after final grading have aided materially in the reclamation process. Surface grading techniques and seedbed preparation are very important in obtaining good vegetative cover for erosion and sedimentation control. Many plant species of economic importance can be used to produce hay, pasture, various horticultural crops, and major row crops. Commercial varieties of grasses that have shown promise on United States eastern strip mine spoils with moderate pH (5.0–6.0) include orchard grass, tall fescue, bromegrass, ryegrass, and timothy. Other varieties have shown promise on low-pH (4.5 or less) strip mine spoils. Most spoils in the western United States are returned to perennial grasses for eventual use by grazing livestock and wildlife. Almost any species of trees can be grown on strip mine areas as long as their nutrient and environmental needs are met. *See* SURFACE MINING. [O.L.B.]

Landing gear Those parts of an aircraft structure which serve to support the aircraft when it is not in the air. Tires, wheels, brakes, shock struts, drag struts, and miscellaneous equipment, such as retracting mechanisms, steering mechanisms, shimmy dampers, and doors, are the components of an aircraft landing gear. For special purposes the wheels and tires may be replaced with skids or skis. A landing gear is required on all piloted aircraft.

The most widely accepted arrangement is the tricycle landing gear. This arrangement places a nose gear well forward of the center of gravity on the fuselage and two main gears slightly aft of the center of gravity, with a sufficient distance between them to provide stability against rolling over during a yawed landing in a cross wind, or during ground maneuvers. The nose gear prevents nosing over when brakes are applied. The nose gear, which is always castered, is often equipped with power steering to allow ground maneuvering.

Another arrangement still used extensively is the tail-wheel landing gear. Two main-gear struts are located slightly ahead of the center of gravity, well-spaced laterally to provide lateral stability; a third wheel is mounted on the fuselage near the aft end of the airplane. It suffers in comparison with a tricycle gear because of increased problems associated with vision, ground maneuvering, and braking. [J.E.St.]

Landing ships and craft Combat vessels employed in amphibious operations to transport mobile equipment, amphibious vehicles, tanks, general cargo, and personnel, and to discharge them near or directly onto a beach. Landing ships are generally designated by the letters LS, landing craft by LC, and amphibious landing vehicles by LV. Landing craft, ranging from about 36 to 135 ft (11 to 41 m) in length, are not designed for long transoceanic voyages and are carried to the unloading area aboard landing ships. Landing ships, landing craft, and amphibious vehicles were first used during World War II, when a wide variety of types were developed. [C.J.P.]

Landscape architecture Landscape architecture was defined by Charles Eliot, one of America's first landscape architects, as the art of arranging and fitting land for human use and enjoyment. It is an applied art founded on the premise that use and beauty are compatible and that neither is complete without the other.

The services of the landscape architect, similar to those of a building architect, involve consultation; preparation of reports, plans, specifications, and estimates; assistance in letting contracts; and supervision of the work done by the contractors.

Projects undertaken by landscape architects include large parks and recreation areas, real estate subdivisions, highways and parkways, and landscaping for industries, institutions, cemeteries, and large private estates. [J.P.P.]

Landscape ecology The study of the ecological effects of spatial patterning of ecosystems. Specifically, landscape ecology examines the development and dynamics of spatial heterogeneity; interactions and exchanges across heterogeneous landscapes; the influences of spatial heterogeneity on biotic and abiotic processes; and the management of spatial heterogeneity. The consideration of spatial patterns distinguishes landscape ecology from traditional ecological studies, which frequently assume a spatially homogeneous system. *See* ECOSYSTEM.

The term landscape commonly refers to the landforms of a region considered together with its associated habitats at scales that range from acres to many square miles. For example, a rural landscape might include the forests, croplands, pastures, rivers, and towns within a region. Landscape ecology studies those large, heterogeneous areas as intact units with emphasis on ecological processes.

The discipline of landscape ecology arose from European traditions of regional geography and vegetation science. Many fields of study, including geography, ecology, landscape architecture, forestry, and regional planning, have contributed to the development of landscape ecology. In Europe, landscape ecology is well integrated into land-use planning and decision making.

Three useful landscape characteristics are structure, function, and change. Structure refers to the spatial patterns or relationships between distinctive ecosystems, that is, the distribution of energy, materials, and species in relation to the sizes, shapes, numbers, kinds, and configurations of components. Function encompasses the interactions between the spatial elements, that is, the flow of energy, materials, and organisms among the component ecosystems. Change refers to alteration in the structure and function of the ecological mosaic through time. *See* ECOLOGICAL COMMUNITY; ECOLOGICAL MODELING. [M.G.T.]

Landslide The perceptible downward sliding, falling, or flowing of masses of soil, rock, and debris (mixtures of soil and weathered rock fragments). Landslides range in size from a few cubic meters to over 10^9 m^3 (3.5×10^{10} ft^3), their velocities range from a few centimeters per day to over 100 m/s (330 ft/s), and their displacements may be several centimeters to several kilometers. *See* MASS WASTING.

The U.S. Highway Research Board classification divides landsliding of rock, soil, and debris, on the basis of the types of movement, into falls, slides, and flows. Other classifications consider flows, along with creep and other kinds of landslides, as general forms of mass wasting.

Falls occur when soil or rock masses free-fall through air. Falls are usually the result of collapse of cliff overhangs which result from undercutting by rivers or simply from differential erosion. Slides invariably involve shear displacement or failure along one or more narrow zones or planes. Internal deformation of the sliding mass after initial failure depends on the kinetic energy of the moving mass (size and velocity), the distance traveled, and the internal strength of the mass. Flows have internal displacement and a shape that resemble those of viscous fluids. Relatively weak and wet masses of shale, weathered rock, and soil may move in the form of debris flows and earthflows; water-soaked soils or weathered rock may displace as mudflows.

Mining and civil engineering works have induced myriads of

landslides, a few of them of a catastrophic nature. Open-pit mines and road cuts create very high and steep slopes, often quite close to their stability limit. Local factors (weak joints, fault planes) or temporary ones (surges of water pressure inside the slopes, earthquake shocks) induce the failure of some of these slopes. The filling of reservoirs submerges the lower portion of natural, marginally stable slopes or old landslides. Water lowers slope stability by softening clays and by buoying the lowermost portion, or toe, of the slope.

Advances in soil and rock engineering have improved the knowledge of slope stability and the mechanics of landsliding. Small and medium-sized slopes in soil and rock can be made more stable. Remedial measures include lowering the slope angle, draining the slope, using retaining structures, compressing the slope with rock bolts or steel tendons, and grouting. *See* ENGINEERING GEOLOGY; EROSION; SOIL MECHANICS. [A.S.N.]

Langevin function A mathematical function which is important in the theory of paramagnetism and in the theory of the dielectric properties of insulators. The analytical expression for the Langevin function is shown in the equation below. If

$$L(x) = \coth x - 1/x$$

$x \ll 1$, $L(x) \simeq x/3$. The paramagnetic susceptibility of a classical (non-quantum-mechanical) collection of magnetic dipoles is given by the Langevin function, as is the polarizability of molecules having a permanent electric dipole moment. *See* PARAMAGNETISM. [E.A.; F.Ke.]

Langmuir-Child law A law governing space-charge-limited flow of electron current between two plane parallel electrodes in vacuum when the emission velocities of the electrons can be neglected. It is often called the three-halves power law, and is expressed by the formula shown below.

$$j(\text{amperes/cm}^2) = \frac{\epsilon}{9\pi}\left(\frac{2e}{m}\right)^{1/2}\frac{V^{3/2}}{d^2}$$

$$= 2.33 \times 10^{-6}\,\frac{V\,(\text{volts})^{3/2}}{d\,(\text{cm})^2}$$

Here ϵ is the dielectric constant of vacuum, $-e$ the charge of the electron, m its mass, V is the potential difference between the two electrodes, d their separation, and j the current density at the collector electrode, or anode. The potential difference V is the applied voltage reduced by the difference in work function of the collector and emitter. The Langmuir-Child law applies, to a close approximation, to other electrode geometries as well. *See* SPACE CHARGE. [E.G.R.]

Lanolin The hydrous sheep's-wool wax (primarily cholesterol esters of higher fatty acids) derived as a by-product from the preparation of raw wool for the spinner. Lanolin is widely used as a base for emollients in cosmetics and shampoos. Perhaps because of its slight antiseptic effect and high resistance to rancidity it has continued wide use as a base for skin ointments. Actually, lanolin hinders absorption of the medication through the skin, as compared to other substances such as olive oil or lard. *See* WAX, ANIMAL AND VEGETABLE. [F.H.R.]

Lanthanide contraction The name given to an unusual phenomenon encountered in the rare-earth series of elements. The radii of the atoms of the members of this series decrease slightly as the atomic number increases. As the charge on the nucleus increases across the rare-earth series, all electrons are pulled in closer to the nucleus so that the radii of the rare-earth ions decrease slightly as the compounds go across the rare-earth series. Any given compound of the rare earths is very likely to crystallize with the same structure as any other rare earth. However, the lattice parameters become smaller and the crystal denser as the compounds proceed across the series. This contraction of the lattice parameters is known as the lanthanide contraction. *See* RARE-EARTH ELEMENTS. [F.H.Sp.]

Lanthanum A chemical element, La, atomic number 57, atomic weight 138.91. Lanthanum, the second most abundant element in the rare-earth group, is a metal. The naturally occurring element is made up of the isotopes ^{138}La and ^{139}La. The element occurs associated with other rare earths in mon-

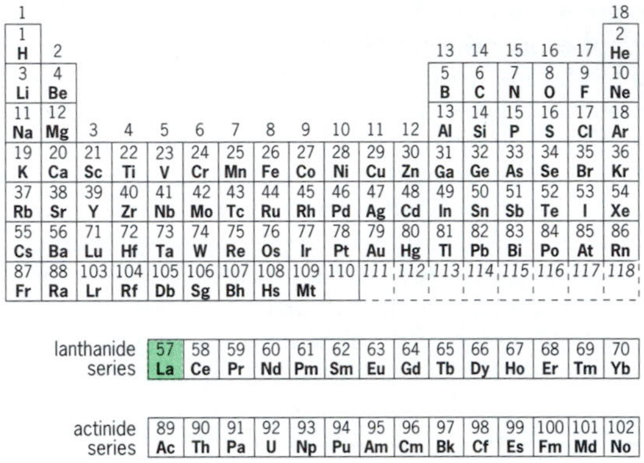

azite, bastnasite, and other minerals. It is one of the radioactive products of the fission of uranium, thorium, or plutonium. Lanthanum is the most basic of the rare earths. It is an important ingredient in glass manufacture. Lanthanum imparts a high refractive index to the glass and is used in the manufacture of expensive lenses. For other properties of the metal *see* RARE-EARTH ELEMENTS. [F.H.Sp.]

Laplace transform An integral transform extensively used by P. S. Laplace in the theory of probability. In simplest form it is expressed as the equation below. It is thought of as

$$f(s) = \int_0^\infty e^{-st}\phi(t)\,dt$$

transforming the determining function $\phi(t)$ into the generating function $f(s)$. The variable t is real, the variable s may be real or complex, $s = \sigma + i\tau$. As an example, if $\phi(t) = 1$ the integral converges for $\sigma > 0$, and $f(s) = 1/s$.

The Laplace transform is used for the solution of differential and difference equations, for the evaluation of definite integrals, and in many branches of abstract mathematics (functional analysis, operational calculus, and analytic number theory). *See* INTEGRAL TRANSFORM. [D.V.W.]

Laplace's differential equation Laplace's equation in two independent variables x and y is given as Eq. (1) and

$$\frac{\partial^2 u(x,y)}{\partial x^2} = \frac{\partial^2 u(x,y)}{\partial y^2} = 0 \qquad (1)$$

is of central importance in both pure mathematics and mathematical physics. A function $u(x,y)$ having continuous first and second partial derivatives and satisfying Laplace's equation in a neighborhood of a point is called harmonic at that point. If a plane piece of tinfoil has its edges kept at a temperature which varies from point to point but does not change with time, and if the flow of heat in the tinfoil is steady (that is, independent of the time), the temperature $u(x,y)$ at interior points of the foil is harmonic. Likewise Laplace's equation dominates the flow of electricity (the potential is similarly harmonic) and the flow of any incompressible fluid.

Many properties of Laplace's equation with two independent variables apply also in three or more dimensions. Thus, in three dimensions, a point distribution of matter of masses m_k at points (x_k, y_k, z_k) has a potential defined by Eq. (2), which is

$$u(x,y,z) \equiv \sum m_k[(x - x_k)^2 + (y - y_k)^2 + (z - z_k)^2]^{-1/2} \quad (2)$$

harmonic except in the points (x_k, y_k, z_k). Except at such points, the force (Newtonian law of gravitation) exerted by the distribution on a unit exploratory particle at (x,y,z) has the components $(\partial u/\partial x, \partial u/\partial y, \partial u/\partial z)$ and the component of the force in any direction is the directional derivative of $u(x,y)$ in that direction. *See* POTENTIALS; SPHERICAL HARMONICS. [J.L.W.]

Laplace's irrotational motion Laplace's equation for irrotational motion of an inviscid, incompressible fluid is partial differential equation (1), where x_1, x_2, x_3 are rectangular cartesian coordinates in an inertial reference frame, and Eq. (2)

$$\partial^2\phi \backslash \partial x_1{}^2 + \partial^2\phi \backslash \partial x_2{}^2 + 2^2\phi \backslash 2x_3{}^2 = 0 \quad (1)$$

$$\phi = \phi(x_1, x_2, x_3, t) \quad (2)$$

gives the velocity potential. The fluid velocity components, u_1, u_2, u_3 in the three respective rectangular coordinate directions are given by $u_i = \partial\phi/\partial x_i$, $i = 1,2,3$. More generally, in any inertial coordinate system, the equation is div (grad ϕ) = 0 and the velocity vector is $\mathbf{v}$ = grad ϕ. Irrotational motion implies that the fluid particles translate without rotation (like the cars on a ferris wheel). *See* FLUID FLOW. [A.E.Br.]

Laplacian The differential operator $\partial^2/\partial x^2 + \partial^2/\partial y^2 + \partial^2/\partial z^2$, in which the symbols x, y, z denote the variables of a rectangular cartesian coordinate system. The laplacian is frequently denoted by the symbol ∇^2 (read del square) in accordance with the fact that the laplacian of a scalar function $S(x,y,z)$ is the divergence of the gradient of S, that is, the equation below applies.

$$\partial^2 S/\partial x^2 + \partial^2 S/\partial y^2 + \partial^2 S/\partial z^2 = \nabla \cdot (\nabla S)$$

The laplacian operator is involved in some of the most fundamental equations of mathematical physics, namely, Laplace's equation ($\nabla^2 u = 0$), Poisson's equation, various wave equations, and the heat flow and diffusivity equations. *See* CALCULUS OF VECTORS; GAUSS' THEOREM; GRADIENT OF A SCALAR; GREEN'S THEOREM; WAVE EQUATION. [H.V.C.]

Lapping A precision abrading process used to bring a surface to a desired state of refinement or dimensional tolerance by removal of an extremely small amount of material. Lapping is accomplished by abrading a surface with a fine abrasive grit rubbed about it in a random manner. Usually less than 0.0005 in. (0.012 mm) of stock is removed. *See* GRINDING. [A.H.T.]

Larch A genus, *Larix*, of the pine family, with deciduous needles and short spurlike branches, which annually bear a crown of needles. The cones are small and persistent, varying by species in size, number, and form of the cone scales. The tamarack (*L. laricina*), also called hackmatack, is a native species. It has an erect, narrowly pyramidal habit, and grows in the northeastern United States, west to the Lake states, and across Canada to Alaska. The tough resinous wood is durable in contact with the soil and is used for railroad ties, posts, sills, and boats. Other uses include the manufacture of excelsior, cabinet work, interior finish, and utility poles. *See* PINALES.

The western larch (*L. occidentalis*), the most important and largest of all the species, grows in the northwestern United States and southeastern British Columbia. [A.H.G./K.P.D.]

Large systems control theory A branch of the theory of control systems concerned with the special problems that arise in the design of control algorithms (that is, control policies and strategies) for complex systems. Characteristically the systems have a great many inputs and outputs which interact in a complicated and usually imprecisely known manner. Some examples of such complex systems are the United States economy, the global telephone communication network, the electric power generation system for the western United States, large industrial plants, a modern military command, communication, control, and intelligence system, and a single advanced computer system composed of large arrays of interconnected microprocessors. *See* CONTROL SYSTEMS. [M.G.S.]

Larnite The alpha polymorph of calcium silicate (Ca_2SiO_4). Larnite is a mineral which crystallizes at high temperature. Its occurrences are practically confined to limestone or chalk zones in contact with semimolten basalts. At room temperature, larnite is metastable and inverts to its low-temperature polymorph calcio-olivine through shock. This leads to "fall," or disintegration of slags with time, and presents problems in the cement industry. The mineral is very rare, known from its type locality at Scawt Hill, County Antrim, Ireland, and from Crestmore, near Riverside, California. *See* SILICATE MINERALS. [P.B.M.]

Larynx The complex of cartilages and related structures at the opening of the trachea, or windpipe, into the pharynx, or throat. In humans and most other mammals, the signet-shaped cricoid cartilage forms the base of the larynx and rests upon the trachea. The thyroid cartilage, which forms the prominent Adam's apple ventrally, lies anterior to the cricoid. Dorsally there are paired pivoting cartilages, the arytenoids. Each is pyramid-shaped and acts as the movable posterior attachment for the vocal cords and the laryngeal muscles that regulate the cords. Two other small paired cartilages, the cuneiform and the corniculate, also lie dorsal to the thyroid cartilage. The epiglottis, a leaf-shaped elastic cartilage with its stem inserted into the thyroid notch, forms a lid to the larynx.

Laryngitis is an inflammation of the mucous membrane of the larynx always associated with hoarseness. Hoarseness is also a manifestation of paralysis of the recurrent laryngeal nerve of the vagus, which is easily damaged upon surgical removal of a goiter. [T.S.P.]

Laser A device that uses the principle of amplification of electromagnetic waves by stimulated emission of radiation and operates in the infrared, visible, or ultraviolet region. The term laser is an acronym for light amplification by simulated emission of radiation, or a light amplifier. However, just as an electronic amplifier can be made into an oscillator by feeding appropriately phased output back into the input, so the laser light amplifier can be made into a laser oscillator, which is really a light source. Laser oscillators are so much more common than laser amplifiers that the unmodified word "laser" has come to mean the oscillator, while the modifier "amplifier" is generally used when the oscillator is not intended. *See* MASER.

The process of stimulated emission can be described as follows. When atoms, ions, or molecules absorb energy, they can emit light spontaneously (as in an incandescent lamp) or they can be stimulated to emit by a light wave. This stimulated emission is the opposite of (stimulated) absorption, where unexcited matter is stimulated into an excited state by a light wave. If a collection of atoms is prepared (pumped) so that more are initially excited than unexcited, then an incident light wave will stimulate more emission than absorption, and there is net amplification of the incident light beam. This is the way the laser amplifier works.

A laser amplifier can be made into a laser oscillator by arranging suitable mirrors on either end of the amplifier. These

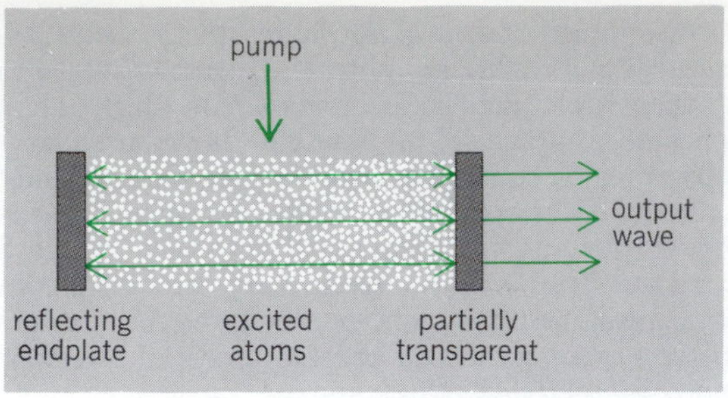

Structure of a parallel-plate laser.

are called the resonator. Thus the essential parts of a laser oscillator are an amplifying medium, a source of pump power, and a resonator. Radiation that is directed straight along the axis bounces back and forth between the mirrors and can remain in the resonator long enough to build up a strong oscillation. (Waves oriented in other directions soon pass off the edge of the mirrors and are lost before they are much amplified.) Radiation may be coupled out as shown in the illustration by making one mirror partially transparent so that part of the amplified light can emerge through it. The output wave, like most of the waves being amplified between the mirrors, travels along the axis and is thus very nearly a plane wave. *See* OPTICAL PUMPING.

In contrast to lasers, all conventional light sources are basically hot bodies which radiate by spontaneous emission. The overall wave produced by a conventional light source is a jumble of waves from the numerous individual atoms. The phase of the wave emitted by one atom has no relation to the phase emitted by any other atom, so that the overall phase of the light fluctuates randomly from moment to moment and place to place. The lack of correlation is called incoherence. *See* HEAT RADIATION.

In contrast to this, an ideal plane wave would have the same phase all across any wavefront, and the time fluctuations would be highly predictable (coherent). The output of a parallel-plate laser is very nearly such a plane wave and is therefore highly directional. This arises because in the laser oscillator atoms are stimulated to emit in phase with the stimulating wave, rather than independently, and the wave that builds up between the mirrors matches very closely the mirrors' surfaces. The output is powerful because atoms can be stimulated to emit much faster than they would spontaneously. It is highly monochromatic largely because stimulated emission is a resonance process that occurs most rapidly at the center of the range of wavelengths that would be emitted spontaneously. Since atoms are stimulated to emit in phase with the existing wave, the phase is preserved over many cycles, resulting in the high degree of time coherence of laser radiation. *See* COHERENCE; HEAT RADIATION.

The various types of lasers are classified according to their pumping (or excitation) scheme. The function of the pumping system is to maintain more atoms in the upper than in the lower state, thereby assuring that stimulated emission (gain transitions) will exceed (stimulated) absorption (loss transitions). This so-called population inversion ensures net gain (amplification greater than unity). The large classes of lasers include optically pumped lasers, gas-discharge lasers, pulsed gas lasers, chemical lasers, photodissociation lasers, nuclear lasers, gas-dynamic lasers, semiconductor lasers, free-electron lasers, and high-power, short-pulse lasers. See separate articles on each class.

The variety of technological uses for lasers has increased steadily since their appearance in 1960. Among the noticeable applications are those that utilize high-speed controllability of the tiny focal spot of a laser beam. For example, high-speed automatic scanners identify library cards, ski passes, and supermarket purchases and perform a variety of functions known as optical processing. Other uses for the laser beam's programmable control include information storage and retrieval (including three-dimensional holography and video disk reading), laser printing, micromachining, and automated cutting. Further applications involving high power include weaponry, laser welding, laser surgery (self-cauterizing), laser fusion, and materials processing. Optical communications utilize the laser's high frequency, which makes possible high information capacity. Bright laser beams are used by the construction industry to align straight excavations and for surveying. *See* CHARACTER RECOGNITION; CONTINUOUS-WAVE RADAR; GYROSCOPE; HOLOGRAPHY; INTEGRATED OPTICS; LASER PHOTOBIOLOGY; LASER PHOTOCHEMISTRY; LASER SPECTROSCOPY; OPTICAL COMMUNICATIONS; OPTICAL FIBERS; PHOTOACOUSTIC SPECTROSCOPY; VIDEO DISK RECORDING. [S.F.J.; A.L.S.]

Laser alloying A material processing method which utilizes the high power density available from focused laser sources to melt metal coatings and a portion of the underlying substrate. Since the melting occurs in a very short time and only at the surface, the bulk of the material remains cool, thus serving as an intimate heat sink. Large temperature gradients exist across the boundary between the melted surface region and the underlying solid substrate. The result is rapid self-quenching and resolidification.

For all laser sources, the exposure time (dwell time or pulse length) strongly influences the depth that will be melted. Longer exposure times result in deeper melting. Since deeper melting means a longer total time in the molten state, that means more time available for diffusion of the one or more alloying elements into the molten portion of the substrate. Deeper melting and longer melt times therefore result in more dilute surface alloys, while shallow melting and shorter melt times result in more concentrated surface alloys.

In making laser alloys, many other processing variables need to be considered. In addition to the exposure time, these include the laser power, the thickness of the film put down prior to laser melting, and in some instances the nature of the gaseous ambient during the laser processing. The processing variables are interrelated, and one variable cannot be freely changed without affecting another. Another consideration is that laser alloying is a liquid state–rapid quenching phenomenon. The near-surface region must be melted and yet vaporization avoided. Different minimum and maximum energy densities are thus defined for each laser exposure time. Laser alloying also involves very large temperature gradients and quenching from the liquid state. In this way it resembles other rapid-solidification technologies. The thermodynamic constraints which limit the conventional metallurgist do not necessarily apply. *See* ALLOY; LASER; METAL COATINGS. [C.W.D.]

Laser cooling The use of laser radiation to cool atoms and atomic ions to very low temperatures. The use of lasers to provide heat is familiar. Examples include surgery, welding, and inertial confinement fusion. However, laser light is also utilized to cool small specimens to very low temperatures, in some cases about 1 microkelvin. This cooling results from the mechanical momentum imparted to the atoms when they scatter light; by suitable arrangement of the frequency and position of the laser beam, the atoms can be made to scatter light only when scattering reduces their momentum.

In 1933 O. Frisch experimentally demonstrated the transfer of momentum from photons to atoms by deflecting a beam of sodium atoms with resonance radiation from a lamp. With the advent of tunable lasers, such effects can be much more pronounced. The narrow spectral width of lasers makes possible cooling by these mechanical forces. *See* ELECTROMAGNETIC RADIATION; LASER; LIGHT.

The simplest and commonest form of laser cooling is called Doppler cooling. It relies on the high spectral purity of lasers, the resonant nature of absorption of light by the atom, and the frequency shift of the light (as viewed by the atom) due to the Doppler effect. *See* DOPPLER EFFECT; QUANTUM MECHANICS.

The randomness in the times of absorption and in the direction of photon reemission act like random impulses on the atom that counteract the cooling effect. These random impulses, which cause heating, reach a balance with the cooling when the effective temperature T of the atoms reaches a minimum value. For many atoms, this temperature is near 1 mK or less.

In this rapidly developing field, new cooling schemes, also based on the conservation of momentum and energy in the photon-scattering process, are being developed. For example, one technique has produced cooling of neutral helium atoms to a temperature that approximately corresponds to the kinetic energy of an atom that is initially at rest and recoils from the emission of one photon ($T \approx 1\ \mu$K). In another technique, mercury ions bound in an electromagnetic trap have been cooled to the zero-point energy of motion for the confining potential.

Laser cooling is used in spectroscopy, collision studies, atom manipulation, and for studies of condensed matter. *See* PARTICLE TRAP. [D.J.Wi.]

Laser Doppler velocimeter A device that uses the Doppler effect to measure the velocity of fluid flow. When light is scattered from a moving object, a stationary observer will see a change in the frequency of the scattered light (Doppler shift) proportional to the velocity of the object. In the laser Doppler velocimeter this Doppler shift is used to measure the velocity of particles in a fluid. From the particle velocity the velocity of the fluid is inferred. A laser is used as light source because it is easily focused and is coherent and thereby enables the measurement of a frequency shift due to a Doppler effect. *See* DOPPLER EFFECT; FLOW MEASUREMENT; LASER. [M.Br.; L.P.E.]

Laser photobiology The interaction of laser light with biological molecules, and the applications to biology and medicine. *See* LASER.

Microirradiation is a useful technique for the study of cell function by alteration of a specific organelle or part of a cell. The laser beam is focused through the objective of a microscope onto the cell. Practically speaking, it is easy to obtain spots of about 1 micrometer in diameter. Ruby, neodymium, and argon lasers are used for this purpose.

Laser spectroscopy is used to probe biological processes in which very fast reactions are involved or to study structural changes of complex molecules. The two techniques used are flash photolysis (in the nanosecond and picosecond range) and Raman spectroscopy. *See* LASER SPECTROSCOPY; RAMAN EFFECT.

Continuous-wave lasers have been employed as a "light knife," that is as a surgical cutting and coagulation tool. Generally, CO_2 lasers, emitting in the infrared, are used for this purpose. When the laser energy is focused onto a tissue surface, a small volume of tissue is heated, and thus only this area is "cut off." An advantage of this procedure is that small capillaries are coagulated, preventing hemorrhage resulting from cut blood vessels. Argon lasers are the most commonly used for treating retinopathies, but also for glaucoma and cataract. Laser irradiation is used for removal of foreign pigments in the skin (tattoos), for treatment of vascular disorders ("wine marks"), and for removal of various pigmented skin lesions. [G.Mo.]

Laser photochemistry The branch of physical chemistry in which chemical reactions are induced, altered, or monitored by laser light.

The main advantages of a laser over a conventional light source (such as discharge or arc lamps) for study of photochemistry are threefold. First, laser light has a very small divergence angle. Second, laser light is exceptionally pure in color. Third, lasers are generally more powerful than conventional light sources. All in all, the laser provides the photochemist with a source of light between 10 and 20 orders of magnitude more spectrally bright than previously available.

One area of laser photochemistry is the study of single-collision chemical reaction dynamics. A natural and practical application of this research is the use of lasers to drive specific chemical reactions. One successful example of laser-induced chemistry is carbon dioxide laser irradiation of diborane (B_2H_6) to produce icosaborane ($B_{20}H_{16}$) in high yield; conventional pyrolysis generates a mixture of products in which $B_{20}H_{16}$ is not found.

Picosecond studies involve the use of state-of-the-art, ultrashort (10^{-12} s) laser pulses to probe the reaction dynamics of very fast processes. A sample is illuminated by the initial pump pulse followed by a staccato burst of probe pulses. By measurement of the absorption of the delayed probe pulses, the fast chemical processes initiated by the pump pulse can be monitored on an unprecedented time scale. *See* ULTRAFAST MOLECULAR PROCESSES.

A long-standing goal of the laser photochemist has been to manipulate the pathway of a photochemical process. One vigorously pursued approach to this problem involves using extremely intense lasers to induce dissociation of a specific bond in a molecule—in effect, molecular photosurgery.

Another large-scale application of laser photochemistry is the separation of isotopes. The laser preferentially dissociates molecules of a particular isotope; the isotopically enriched radical fragments then react with added scavengers, and can be removed. Application of this technique has achieved spectacular isotopic separations for a number of elements, most notably sulfur, boron, nitrogen, and hydrogen. *See* CHEMICAL DYNAMICS; ISOTOPE (STABLE) SEPARATION; LASER; PHOTOCHEMISTRY. [D.J.N.]

Laser-solid interactions Interactions of laser light with solids. The term usually refers to the thermal effects of absorption of high-intensity laser beams. For nonthermal laser interactions with matter *see* LASER PHOTOCHEMISTRY; LASER SPECTROSCOPY; NONLINEAR OPTICS.

The high power densities attainable with lasers allow melting and even vaporization of any solid material that is sufficiently opaque at a given wavelength or photon energy. This has led to a number of applications involving cutting and drilling of ceramics and other brittle materials, even diamonds. Welding of components from the smallest wires to huge steel plates is done commercially with high-power lasers. Metal alloying in surface regions is also a domain of lasers. *See* LASER ALLOYING; LASER WELDING.

Ion implantation has become a dominant method of introducing controlled quantities of impurities near the surface of silicon and other semiconductors. The implanted layers need a heat treatment to repair the displacement damage caused by bombardment with energetic ions and to move the implanted impurity ions into lattice locations where they replace host atoms and become electrically active. Laser heating is particularly suitable for annealing since only the implanted regions are heated. *See* ION IMPLANTATION.

Thin films of single-crystalline silicon over an insulating substrate are very attractive for high-speed integrated circuits. An important approach to the formation of such films is the controlled melting of thin polycrystalline layers deposited over fused silica substrates or over oxidized silicon wafers. Through a careful control of temperature gradients around the molten spot, by shaping the laser beam or patterning the film, single-crystalline regions can be obtained. The formation of silicon-on-insulator structures will lead in the future to three-dimensional circuits, with several levels of transistors on the same chip. *See* INTEGRATED CIRCUITS; LASER. [G.K.C.]

Laser spectroscopy Spectroscopy with laser light or, more generally, studies of the interaction between laser radiation and matter. Lasers have led to a rejuvenescence of classical spectroscopy, because laser light can far surpass the light from other sources in brightness, spectral purity, and directionality, and if required, laser light can be produced in extremely intense and short pulses. The use of lasers can greatly increase the resolution and sensitivity of conventional spectroscopic techniques, such as absorption spectroscopy, fluorescence spectroscopy, or Raman spectroscopy. Moreover, interesting new phenomena have become observable in the resonant interaction of intense coherent laser light with matter. Laser spectroscopy has become a wide and diverse field, with applications in numerous areas of physics, chemistry, and biology. *See* LASER; SPECTROSCOPY. [T.W.Ha.]

Laser welding Welding with a laser beam. The primary apparatus is the continuous-wave, convectively cooled CO_2 laser with either oscillator/amplifier (gaussian output beam) or unstable resonator (hollows output beam) optics. These lasers, available in output powers ranging from approximately 1000 to 15,000 W, have been used to demonstrate specific welding accomplishments in a variety of metals and alloys. Substantial advances in laser technology made possible the production of fully automated multikilowatt industrial laser systems which can be operated on a continuous production basis. These systems can be used for a variety of development programs and on-line production applications. *See* LASER; WELDING AND CUTTING OF METALS. [E.M.Br.]

Latent image An invisible image produced by a physical or chemical effect of light on the individual crystals (usually silver halide) of photographic emulsions. This image can be rendered visible by the process known as development. *See* PHOTOGRAPHY; PHOTOLYSIS. [R.H.N.]

Lateral line system Sensory organs, arranged in a linear series, which are found on the body surface of most fishes and some amphibians, and which respond to water movements caused by prey, predators, or members of the same species close to the body.

The basic component of each sensory organ is a sensory hair cell. At the external tip of the cell, two types of hair (cilia) project into a cylindrical cup-shaped structure (cupula). There is one long single hair with an internal structure containing an arrangement of filaments (a characteristic of motile cilia), and a bundle of shorter hairs with a simpler internal structure. At the base of the cell there are usually two connections to the nervous system: the afferent nerve carrying information to the brain, and the efferent nerve carrying information from the brain.

A neuromast organ consists of a group or field of several to many thousands of hair cells. The hairs from the component cells all project into one cupula, which follows the shape of the field, and so the cupulae may be flat plates or oval in cross section as well as cylindrical. The neuromast organs may be scattered in a random manner over the body surface (the free or superficial neuromast organs) or arranged in lateral lines over the head and along the body. The body lateral line is typically seen in the common food fishes running from the head along the flank to the tail.

The neuromast organs are displacement receptors, which are stimulated by movements of the surrounding water that might be caused by the swimming of predators, prey, or their own species, and the disturbances caused by their own swimming. The free neuromasts on the body surface are likely to be most sensitive.

Sound sources produce back-and-forth displacements of the water particles, which are particularly strong near the source. As long as the sound source is close enough and sufficiently intense, and the sound frequency is not too high, the hair cells should be able to respond to these displacements.

A striking characteristic of the lateral line is its length; no other sense organ is arranged in such a linear series. It ensures that the fish can make a simultaneous comparison of the stimuli reaching widely different parts of the body. In this way, the range of the source of stimulation can probably be ascertained, and much more accurate information can be gained about the direction of the source and probably its size. [J.H.S.B.]

Lateral meristem Strips or cylinders of dividing cells located parallel to the long axis of the organ in which they occur. Radial enlargement of the cells derived from these meristems increases the diameter of the organ. The lateral meristem is concerned with secondary growth in the sense that its meristematic activity adds cells to the primary body which was derived from the apical meristems. *See* APICAL MERISTEM. [V.I.C.]

Laterite Originally the name for the iron-rich weathering product of basalt in southern India. The term is now used in a compositional sense for weathering products composed principally of the oxides and hydrous oxides of iron, aluminum, titanium, and manganese. Clay minerals of the kaolin group are typically associated with, and are genetically related to, laterite. Laterites range from soft, earthy, porous material to hard, dense rock. Concretionary forms of varying size and shape commonly are developed. The color depends on the content of iron oxides and ranges from white to dark red or brown, commonly variegated. *See* BAUXITE; CLAY MINERALS; KAOLINITE; WEATHERING PROCESSES.

Mature lateritic soils lack fertility for most systems of agriculture. Savannas or parklike grasslands are typical on laterite. Clay, not laterite, is found beneath rainforests and jungle vegetation. [S.S.G.]

Lathe A machine for the removal of metal from a workpiece by gripping it securely in a holding device and rotating it under power against a suitable cutting tool. The tool may be moved radially or longitudinally in respect to the turning axis of the workpiece either manually or by attached power. Forms such as cones, spheres, and related concentric-shaped workpieces as well as true cylinders can be turned on a lathe. Machining operations such as facing, boring, and threading, which are variations of the turning process, can also be performed on a lathe.

Turning equipment may be classed as being either of the horizontal or vertical type referring to the turning axis of the workpiece in the machine. The basic engine lathe is primarily a manually operated machine. Filling the gap between the engine lathe and fully automatic turning equipment is the turret lathe. Fastest and best suited to high-quantity production is the automatic screw machine. [A.H.T.]

Latin square A square n by n matrix with entries from the set $N = \{0,1,\ldots,n-1\}$ so that each number occurs exactly once in each row and exactly once in each column. Two Latin squares are said to be orthogonal if when they are superposed the n^2 cells contain each of the n^2 pairs of numbers from N exactly once. For $n = 3$ one has the superposed orthogonal squares

$$\begin{matrix} 00 & 11 & 22 \\ 12 & 20 & 01 \\ 21 & 02 & 10 \end{matrix}$$

where the first digits form the first square, and the second digits the second square. This square may be used to design an agricultural experiment which tests the interaction of three varieties of grain with three types of fertilizers. A field is divided into nine plots in each of which the choice of grain is made according to the first digit and the choice of fertilizer according to the second digit. The Latin squares assure the even distribution of the varieties of grain and fertilizer in both directions, so that effects such as the variation of the soil are minimized, and the orthogonality allows the experimenter to try each fertilizer with each variety of grain.

For $n \leq 10$ each of these arrays can be viewed as a particular type of magic square whose entries are distinct two-digit numbers and such that each row and column sum is $11n(n - 1)/2$ (33 for $n = 3$ and 495 for $n = 10$).

If n is odd, it is easy to construct two orthogonal squares A and B. One takes $A = [a_{ij}]$ and $B = [b_{ij}]$ where $a_{ij} = i + j$, $b_{ij} = i + 2j$, reducing these sums by n or $2n$ if necessary to put them in the range $0,\ldots,n - 1$. If n is a multiple of 4, it is not much more difficult to make a construction. But for n of the form $4m + 2$ it is considerably harder. It is clearly impossible when $n = 2$, and, by an exhaustive trial, G. Tarry showed it to be also impossible for $n = 6$. Euler had conjectured in 1782 that no pair of orthogonal Latin squares existed for any n of the form $4m + 2$, but in 1959 orthogonal Latin squares were found for every $4m + 2$ greater than or equal to 10. *See* COMBINATORIAL THEORY. [T.B.]

Latite An aphanitic (not visibly crystalline) rock of volcanic origin, composed chiefly of sodic plagioclase and alkali feldspar with subordinate quantities of dark-colored minerals (biotite, amphibole, or pyroxene). Latite is intermediate between trachyte and andesite. *See* ANDESITE; TRACHYTE. [C.A.C.]

Latitude and longitude The latitude of a location specifies the angle between an imaginary line directed generally toward the center of the Earth and the Equator. The longitude measures the angle between the meridian (the plane defined by the Earth's axis and this local reference direction) and the plane of the Greenwich meridian.

Astronomical (or astronomic) latitude and longitude use the direction of gravity for the reference direction. This direction, known as the astronomical vertical, is perpendicular to the equipotential surface of the Earth's gravitational field at the location of the observer.

A particular geopotential surface approximating mean sea level in the open ocean is called the geoid. A mathematical surface in the form of an oblate ellipsoid may be constructed to approximate the geoid. The direction perpendicular to this reference ellipsoid at the observer's location is used as the reference direction in defining geodetic latitude and longitude. *See* GEODESY.

Geocentric latitude and longitude are defined by a reference direction which passes precisely through the center of mass of the Earth. These coordinates are determined mathematically from the geodetic latitude and longitude, assuming a fixed relationship between the center of the geodetic datum and the center of mass and knowing the mathematical shape of the ellipsoid. [D.D.McC.]

Lattice (mathematics) Lattice theory deals with properties of order and inclusion, much as group theory treats symmetry. As a generalization of Boolean algebra, lattice theory was first applied to algebraic number theory; however, it was later recognized as a major branch of mathematics, unifying various aspects of algebra, geometry, and functional analysis, as well as of set theory, logic, and probability. *See* BOOLEAN ALGEBRA; GROUP THEORY; SET THEORY.

The most basic concept of lattice theory is that of a partial ordering of a set S of elements $x, y, z,\ldots$. By this is meant a binary relation, usually denoted $\leqq$ (or $\geqq$), with the following properties:

(P1) $x \leqq x$ for all $x \in S$
(P2) If $x \leqq y$ and $y \leqq x$, then $x = y$
(P3) If $x \leqq y$ and $y \leqq z$, then $x \leqq z$

If $\leqq$ is any partial ordering of S, then its converse or dual $\geqq$ defined by statement (1), is also a partial ordering of S. This

$$x \geqq y \text{ if and only if } y \leqq x \qquad (1)$$

easily verified fact provides a fundamental Duality Principle, which is useful in many connections.

A lattice is defined as a partially ordered set in which any two elements x and y have a meet $x \cap y$ and a join $x \cup y$. These binary operations satisfy the four basic identities:

(L1) $x \cap x = x \cup x = x$
(L2) $x \cap y = y \cap x$ and $x \cup y = y \cup x$
(L3) $x \cap (y \cap z) = (x \cap y) \cap z$ and $x \cup (y \cup z) = (x \cup y) \cup z$
(L4) $x \cap (x \cup y) = x \cup (x \cap] y) = x$

The operations $\cap$ and $\cup$ are connected with the relation $\leqq$ by the condition that $x \leqq y$, $x \cap y = x$, and $x \cup y = y$ are three equivalent statements. Conversely, if L is an algebraic system with operations $\cap$ and $\cup$ satisfying (L1) to (L4) for all x, y, z, then the preceding condition defines $\leqq$ as a partial ordering of L, with respect to which $\cap$ and $\cup$ have the meanings defined above. [G.Bi.]

Lattice constant A parameter defining the unit cell of a crystal lattice, that is, the length of the edges of the cell and the angle between edges. If the unit cell is a cube, the side of it is the lattice constant, and it is usually in this sense that the term is used. *See* CRYSTALLOGRAPHY. [W.C.D.]

Lattice vibrations Periodic oscillation of the atoms in a crystal lattice about their equilibrium positions. As the crystal is heated, the amplitude of the vibrations increases. If the heating is continued, the temperature of the crystal eventually reaches a value at which the vibrations are so violent that the atoms break away from their lattice sites, and the solid melts. On the other hand, if the crystal is cooled to absolute zero, the amplitude of the vibrations does not subside entirely. A residual vibration of the atoms, which is quantum-mechanical in origin, remains. It is called the zero-point vibration.

Lattice vibrations are involved in many of the temperature-dependent phenomena of a solid. For example, the electrical resistance of a metal at room temperature arises primarily from the scattering of the conduction electrons by the vibrating atoms. The higher the temperature, and hence the more violent the vibrations of the atoms, the more the electrons are scattered, and the higher the electrical resistance becomes. The Bardeen-Cooper-Schrieffer theory of superconductivity postulates that subtle interactions between lattice vibrations and conduction electrons are mainly responsible for this phenomenon. *See* SUPERCONDUCTIVITY.

Lattice vibrations conduct heat through a crystal; thus a knowledge of the vibration mechanism is essential to an understanding of the heat conductivity of crystalline solids. [J.DeL.]

Launch complex The composite of facilities and support equipment needed to assemble, check out, and launch a rocket-propelled vehicle. The term usually is applied to the facilities and equipment required to launch larger vehicles for which a substantial amount of prelaunch preparation is needed. Small operational rockets may require similar but highly simplified resources on a much smaller scale. For these, the term launcher is usually used. *See* SPACE FLIGHT. [K.Kr.]

Laurales An order of flowering plants, division Magnoliophyta (Angiospermae), in the subclass Magnoliidae of the class Magnoliopsida (dicotyledons). The order includes eight families and about 250 species, but only the Lauraceae with 2000 species and the Monimiaceae with 450 species are large and diverse groups. Most members are tropical trees and shrubs, commonly with scattered spherical cells containing volatile oils. The leaves are usually simple and mostly entire. The flowers are often pollinated by beetles, and in most families show some degree of perigyny. The perianth parts are mostly distinct, but not clearly separable into petals and sepals. The stamens are well differentiated into filament and anther or are somewhat laminar. The pollen grains have two germinal apertures, or less frequently are inaperturate. The carpels, which vary in number, form separate but fully sealed pistils with but one functional ovule. Rarely (in family Gomortegaceae) the pistil consists of two or three united carpels.

This is one of the more archaic orders of flowering plants, clearly related to the Magnoliales but held to be more advanced in having mostly perigynous (or even epigynous) flowers, biaperturate or inaperturate pollen, and other morphological distinctions. Sometimes the order has been included within the Magnoliales. *See* AVOCADO; CAMPHOR TREE; MAGNOLIOPHYTA; MAGNOLIOPSIDA; PLANT KINGDOM; SASSAFRAS. [T.M.Ba.]

Lava Molten rock material that reaches the Earth's surface through volcanic vents and fissures; also, the igneous rock formed by consolidation of such molten material. Relatively rapid cooling at the Earth's surface may transform fluid lava into a dense-textured volcanic rock composed of tiny crystals or glass or both. Molten rock material below the Earth's surface, however, is usually known as magma and, upon cooling, gives rise to coarse-grained igneous rock, such as granite or gabbro.

During many volcanic eruptions the lava is so rapidly ejected that it is blown to bits by the explosive force of expanding gases. The small masses rapidly congeal and settle to the earth to form thick blankets of volcanic tuff and related pyroclastic rock. *See* IGNEOUS ROCKS; MAGMA; PYROCLASTIC ROCKS; TUFF; VOLCANIC GLASS; VOLCANO. [C.A.C.]

Law of the mean A theorem about derivatives, called the mean-value theorem of differential calculus. It is stated in this form: Suppose f is continuous for x from x_1 to x_2 inclusive, and suppose f has a derivative for each x between x_1 and x_2. Then there is some x of this kind for which the equation here

$$f'(x) = f(x_2) - \frac{f(x_1)}{x_2 - x_1}$$

holds true. This theorem enables one to prove that, if $f'(x)$ is always positive, then $f(x)$ increases as x increases. Also, if $f'(x)$ is always zero, then $f(x)$ remains constant as x changes. The first of these results is important for applications to the investigations of graphs. The second is virtually indispensable in the proof of the fundamental theorems of calculus and in establishing the uniqueness of the solutions of certain problems which involve antidifferentiation (the process of surmising what a function is from a knowledge of its derivative). Such problems abound in application of calculus to problems of motion. [A.E.Ta.]

Lawrencium A chemical element, symbol Lr, atomic number 103. Lawrencium, named after E. O. Lawrence, is the eleventh transuranium element; it completes the actinide series of elements. *See* ACTINIDE ELEMENTS; TRANSURANIUM ELEMENTS.

1	2	3	4	5	6	7	8	9	10	11	12	13	14	15	16	17	18
1 H																	2 He
3 Li	4 Be											5 B	6 C	7 N	8 O	9 F	10 Ne
11 Na	12 Mg											13 Al	14 Si	15 P	16 S	17 Cl	18 Ar
19 K	20 Ca	21 Sc	22 Ti	23 V	24 Cr	25 Mn	26 Fe	27 Co	28 Ni	29 Cu	30 Zn	31 Ga	32 Ge	33 As	34 Se	35 Br	36 Kr
37 Rb	38 Sr	39 Y	40 Zr	41 Nb	42 Mo	43 Tc	44 Ru	45 Rh	46 Pd	47 Ag	48 Cd	49 In	50 Sn	51 Sb	52 Te	53 I	54 Xe
55 Cs	56 Ba	71 Lu	72 Hf	73 Ta	74 W	75 Re	76 Os	77 Ir	78 Pt	79 Au	80 Hg	81 Tl	82 Pb	83 Bi	84 Po	85 At	86 Rn
87 Fr	88 Ra	103 Lr	104 Rf	105 Db	106 Sg	107 Bh	108 Hs	109 Mt	110	*111*	*112*	*113*	*114*	*115*	*116*	*117*	*118*

lanthanide series	57 La	58 Ce	59 Pr	60 Nd	61 Pm	62 Sm	63 Eu	64 Gd	65 Tb	66 Dy	67 Ho	68 Er	69 Tm	70 Yb
actinide series	89 Ac	90 Th	91 Pa	92 U	93 Np	94 Pu	95 Am	96 Cm	97 Bk	98 Cf	99 Es	100 Fm	101 Md	102 No

The nuclear properties of all the isotopes of lawrencium from mass 255 to mass 260 have been established. ^{260}Lr is an alpha emitter with a half-life of 3 min and consequently is the longest-lived isotope known. [A.Gh.]

Lawson criterion A necessary but not sufficient condition for the achievement of a net release of energy from nuclear fusion reactions in a fusion reactor. As originally formulated by J. D. Lawson, this condition simply stated that a minimum requirement for net energy release is that the fusion fuel charge must combust for at least enough time for the recovered fusion energy release to equal the sum of energy invested in heating that charge to fusion temperatures, plus other energy losses occurring during combustion. The Lawson criterion is to be thought of as only a rule of thumb for measuring fusion progress; detailed evaluation of all energy dissipative and energy recovery processes is required in order properly to evaluate any specific system. *See* NUCLEAR FUSION; PLASMA PHYSICS. [R.F.P.]

Lawsonite A metamorphic silicate mineral related chemically and structurally to the epidote group of minerals. Its composition is $CaAl_2(H_2O)(OH)_2[Si_2O_7]$. It possesses two perfect cleavages; crystals are orthorhombic prismatic to tabular, and colorless to pale blue; specific gravity is 3.1, and hardness is 6.5 on Mohs scale. *See* EPIDOTE; SILICATE MINERALS. [P.B.M.]

Layered complex In geology, an igneous rock body of large dimensions, 5–300 mi (8–480 km) across and as much as 23,000 ft (7000 m) thick, within which distinct subhorizontal stratification, or layering, is apparent and may be continuous over great distances, in some cases more than 60 mi (100 km). Although conspicuous layering may be found in other rocks of syenitic to granitic composition that are richer in silica, the great layered complexes of the world are, in an overall sense, of tholeiitic basaltic composition. (They may be viewed as intrusive analogs to continental flood basalts.) Indeed, their basaltic composition is of paramount significance to their origin. Only basaltic melts, originating in the mantle beneath the crust of the Earth, are both voluminous enough to occupy vast magma chambers and fluid enough for mineral layering to develop readily. The relatively low viscosity of basaltic melt is a consequence of its high temperature, 2100–2200°F (1150–1200°C), derived from the mantle source region, and its silica-poor, magnesium- and iron-rich (mafic) composition. *See* BASALT; EARTH; MAGMA.

Layered mafic complexes develop upon intrusion of large volumes of basaltic magma (120–24,000 mi^3 or 500–100,000 km^3) into more or less funnel-shaped (smaller complexes) or dish-shaped (larger complexes) chambers 3–5 mi (5–8 km) beneath the Earth's surface. It is widely held that such layering is dominantly produced by gravitational settling of early-formed (cumulus) crystals. These crystals begin to grow as the magma cools and, on reaching a critical size, begin to sink because of their greater density relative to that of the hot silicate melt. Although the sequential order of mineral crystallization can vary depending on subtle differences in magma chemistry, a classic sequence of crystallization from basaltic magma is olivine, $(Mg,Fe)SiO_4$; orthopyroxene, $(Mg,Fe)SiO_3$; clinopyroxene, $(Ca,Mg,Fe)SiO_3$; plagioclase $(Ca,Na)(Si,Al)_4O_8$.

While layers containing only one cumulus mineral may form under special circumstances, coprecipitation of two or three cumulus minerals—for example, olivine + orthopyroxene; orthopyroxene + plagioclase; or orthopyroxene + clinopyroxene + plagioclase—is more common. Under the influence of gravity and current movements in the cooling, tabular magma chamber, the cumulus minerals accumulate on the ever-rising floor of the chamber. The solid rock formed is known as a cumulate, a term that emphasizes its mode of origin and predominant content of cumulus minerals.

Lithologic layering within a complex is typically displayed on a variety of scales. On the broadest scale, a layered mafic complex may contain ultramafic cumulates rich in olivine and orthopyroxene at its base; mafic pyroxene- and plagioclase-rich cumulates at intermediate levels; and more evolved plagioclase-rich cumulates, or even granitic (granophyric) rocks, near its top.

Study of layered mafic complexes is of far more than academic interest because many of them host important deposits of chromium, copper, nickel, titanium, vanadium, and the platinum-group elements platinum, palladium, iridium, osmium, and rhodium. Each of these elements has a relatively high initial concentration in mafic magmas, but all must be dramatically concentrated within restricted layers to be recoverable economically. *See* IGNEOUS ROCKS. [G.K.Cz.]

Layout drawing A design drawing or graphical statement of the overall form of a component or device, which is usually prepared during the innovative stages of a design. Since it lacks detail and completeness, a layout drawing provides a faithful explanation of the device and its construction only to individuals such as designers and drafters who have been intimately involved in the conceptual stage. In a sense, the layout drawing is a running record of ideas and problems posed as the design evolves. In most cases the layout drawing ultimately becomes the primary source of information from which detail drawings and assembly drawings are prepared by other drafters under the guidance of the designer. *See* DRAFTING; ENGINEERING DRAWING. [R.W.M.]

Lazurite The chief mineral constituent in the ornamental stone lapis lazuli. Lazurite is a feldspathoid. It crystallizes in the isometric system, but well-formed crystals, usually dodecahedral, are rare. Most commonly, it is granular or in compact masses. The hardness is 5–5.5 on Mohs scale, and the specific gravity is 2.4–2.5. There is vitreous luster and the color is a deep azure, more rarely a greenish-blue. Lazurite is a tectosilicate, the composition of which is expressed by the formula $Na_4Al_3Si_3O_{12}S$.

Lapis lazuli is a mixture of lazurite with other silicates and calcite and usually contains disseminated pyrite. It has long been valued as an ornamental material. Localities of occurrence are in Afghanistan; Lake Baikal, Siberia; Chile; and San Bernardino County, California. *See* FELDSPATHOID; SILICATE MINERALS. [C.S.Hu.]

Leaching The removal of a soluble fraction, in the form of a solution, from an insoluble, permeable solid with which it is associated. The separation usually involves selective dissolving, with or without diffusion, but in the extreme case of simple washing it consists merely of the displacement (with some mixing) of one interstitial liquid by another with which it is miscible. The soluble constituent may be solid (as the metal leached from ore) or liquid (as the oil leached from soybeans).

Leaching is closely related to solvent extraction, in which a soluble substance is dissolved from one liquid by a second liquid immiscible with the first. Both leaching and solvent extraction are often called extraction. Because of its variety of applications and its importance to several ancient industries, leaching is known by a number of other names: solid-liquid extraction, lixiviation, percolation, infusion, washing, and decantation-settling. The liquid used to leach away the soluble material (the solute) is termed the solvent. The resulting solution is called the extract or sometimes the miscella.

Leaching processes fall into two principal classes: those in which the leaching is accomplished by percolation (seeping of solvent through a bed of solids), and those in which particulate solids are dispersed into the extracting liquid and subsequently separated from it. In either case, the operation may be a batch process or continuous. *See* EXTRACTION; FILTRATION; MASS-TRANSFER OPERATION; SOLVENT EXTRACTION. [S.A.M.]

Lead A chemical element, Pb, atomic number 82 and atomic weight 207.19. Lead is a heavy metal (specific gravity 11.34 at 16°C or 61°F), of bluish color, which tarnishes to dull gray. It is pliable, inelastic, easily fusible, melts at 327.4°C (621.3°F), and boils at 1740°C (3164°F). The normal chemical valences are 2 and 4. It is relatively resistant to attack by sulfuric and hydrochloric acids but dissolves slowly in nitric acid. Lead is amphoteric, forming lead salts of acids as well as metal salts of plumbic acid. Lead forms many salts, oxides, and organometallic compounds.

Industrially, the most important lead compounds are the lead oxides and tetraethyllead. Lead forms alloys with many metals and is generally employed in the form of alloys in most applications. Alloys formed with tin, copper, arsenic, antimony, bismuth, cadmium, and sodium are all of industrial importance. *See* LEAD ALLOYS; TETRAETHYLLEAD.

Lead compounds are toxic and have resulted in poisoning of workers from misuse and overexposure. However, lead poisoning is presently rare because of the industrial application of modern hygienic and engineering controls. The greatest hazard arises from the inhalation of vapor or dust. In the case of organolead compounds, absorption through the skin may become significant. Some of the symptoms of lead poisoning are headaches, dizziness, and insomnia. In acute cases there is usually stupor, which progresses to coma and terminates in death. The medical control of employees engaged in lead usage involves precise clinical tests of lead levels in blood and urine. With such control and the proper application of engineering control, industrial lead poisoning may be entirely prevented.

Lead rarely occurs in its elemental state. The most common ore is the sulfide, galena. The other minerals of commercial importance are the carbonate, cerussite, and the sulfate, anglesite, which are much more rare. Lead also occurs in various

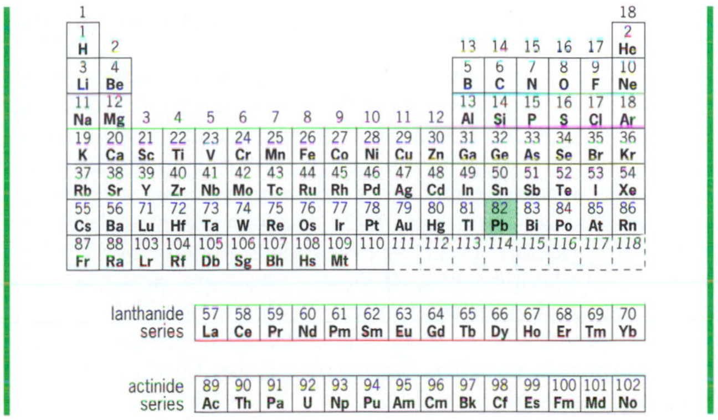

uranium and thorium minerals, arising directly from radioactive decay. Commercial lead ores may contain as little as 3% lead, but a lead content of about 10% is most common. The ores are concentrated to 40% or greater lead content before smelting. *See* LEAD METALLURGY.

The largest single use of lead is for the manufacture of storage batteries. Other important applications are for the manufacture of tetraethyllead, cable covering, construction, pigments, solder, and ammunition.

Organolead compounds are being developed for applications such as catalysts for polyurethane foams, marine antifouling paint toxicants, biocidal agents against gram-positive bacteria, protection of wood against marine borers and fungal attack, preservatives for cotton against rot and mildew, molluscicidal agents, anthelmintic agents, wear-reducing agents in lubricants, and corrosion inhibitors for steel.

Because of its excellent resistance to corrosion, lead finds extensive use in construction, particularly in the chemical industry. It is resistant to attack by many acids because it forms its own protective oxide coating. Because of this advantageous characteristic, lead is used widely in the manufacture and handling of sulfuric acid.

Lead has long been used as protective shielding for x-ray machines. Because of the expanded applications of atomic energy, radiation-shielding applications of lead have become increasingly important. *See* RADIATION SHIELDING.

Lead sheathing for telephone and television cables continues to be a sizable outlet for lead. The unique ductility of lead makes it particularly suitable for this application because it can be extruded in a continuous sheath around the internal conductors.

The use of lead in pigments has been a major outlet for lead but is decreasing in volume. White lead, $2PbCO_3 \cdot Pb(OH)_2$, is the most extensively used lead pigment. Other lead pigments of importance are basic lead sulfate and lead chromates. *See* PIGMENT.

A considerable variety of lead compounds, such as silicates, carbonates, and salts of organic acids, are used as heat and light stabilizers for polyvinyl chloride plastics. Lead silicates are used for the manufacture of glass and ceramic frits, which are useful in introducing lead into glass and ceramic finishes. Lead azide, $Pb(N_3)_2$ is the standard detonator for explosives. Lead arsenates are used in large quantities as insecticides for crop protection. Litharge (lead oxide) is widely employed to improve the magnetic properties of barium ferrite ceramic magnets. Also, a calcined mixture of lead zirconate and lead titanate, known as PZT, is finding increasing markets as a piezoelectric material. [H.S.; J.D.J.]

Lead alloys Substances formed by the addition of one or more elements, usually metals, to lead. Lead alloys may exhibit greatly improved mechanical or chemical properties as compared to pure lead. The major alloying additions to lead are antimony and tin. The solubilities of most other elements in lead are small, but even fractional weight percent additions of some of these elements, notably copper and arsenic, can alter properties appreciably.

Lead is used as a sheath over the electrical components to protect power and telephone cable from moisture. Lead alloy grids are used in the lead-acid storage battery (the type used in automobiles) to support the active material composing the plates. Chemical-resistant alloys are used extensively in many applications requiring resistance to water, atmosphere, or chemical corrosion. Lead bearing metals (babbitt metals) find frequent application in cast sleeve bearings, and are used extensively in freight-car journal bearings. Lead-base solder contains large amounts of tin with selected minor additions to provide specific benefits. *See* ALLOY; LEAD; LEAD METALLURGY; SOLDERING; TIN ALLOYS. [D.Wi.]

Lead metallurgy The extraction of lead from ore, its subsequent purification and processing, and its alloying with other metals to achieve desired properties.

There are many types and several specifications for lead. Primary lead is lead which is extracted and refined from ore. Secondary lead, recycled mostly from old storage batteries, can be refined to meet the same specifications as the primary metal, but the remelting, drossing, and recasting involved may have a negative impact on composition.

In the United States most lead ore is smelted and refined to a minimum purity of 99.85%. At and above this level of purity, four different grades of lead are recognized by the American Society for Testing and Materials: corroding, chemical, acid-copper, and common desilverized lead. The major differences among these grades are the allowable concentrations of copper, silver, and bismuth. Even trace amounts of these elements can have a significant effect on the properties or cost of the lead and justify having the four grades.

The first step in the beneficiation of ores to raise the lead content and to separate the lead from the zinc and iron minerals is concentrating. The standard processing begins with crushing the raw ore, followed by wet grinding, and then by mixing with flotation chemicals that collect the lead minerals in a froth, which is thickened and filtrated. The lead concentrate produced in this first processing step has a lead metal content of around 70%.

Before smelting, lead concentrates are frequently blended with high-grade raw ores or returned intermediates (flue dusts, limerock, and so forth) drawn from proportioning bins. These materials are pelletized so that a homogeneous and carefully sized smelter feed is provided. The feed is then sintered. The sinter product is charged into the top of a blast furnace. As the charge descends in the furnace, the molten metal flows to the bottom, from where it is withdrawn for refining. *See* LEAD. [A.L.P.]

Leaf A lateral appendage which is borne on a plant stem at a node (joint) and which usually has a bud in its axil. In most plants, leaves are flattened in form, although they may be nearly cylindrical with a sheathing base as in onion. Leaves usually contain chlorophyll and are the principal organs in which the important processes of photosynthesis and transpiration occur.

Morphology. A complete dicotyledon leaf consists of three parts: the expanded portion or blade; the petiole which supports the blades; and the leaf base. Stipules are small appendages that arise as outgrowths of the leaf base and are attached at the base of the petiole. The leaves of monocotyledons may have a petiole and a blade, or they may be linear in shape without differentiation into these parts; in either case the leaf base usually encircles the stem. The leaves of grasses consist of a linear blade attached to the stem by an encircling sheath.

Leaves are borne on a stem in a definite fixed order, or phyllotaxy, according to species (Fig. 1). For identification purposes, leaves are classified according to type (Fig. 2) and shape (Fig. 3), and types of margins (Fig. 4), tips, and bases (Fig. 5). The arrangement of the veins, or vascular bundles, of a leaf is called venation (Fig. 6). The main longitudinal veins are usually interconnected with small veins. Reticulate venation is most common in dicotyledons, parallel venation in monocotyledons.

Surfaces of leaves provide many characteristics that are used in identification. A surface is glabrous if it is smooth or free from hairs; glaucous if covered with a whitish, waxy material, or "bloom"; scabrous if rough or harsh to the touch; pubescent, a general term for surfaces that are hairy; puberulent if covered with very fine, downlike hairs; villous if covered with long, soft, shaggy hairs; hirsute if the hairs are short, erect, and stiff; and hispid if they are dense, bristly, and harshly stiff.

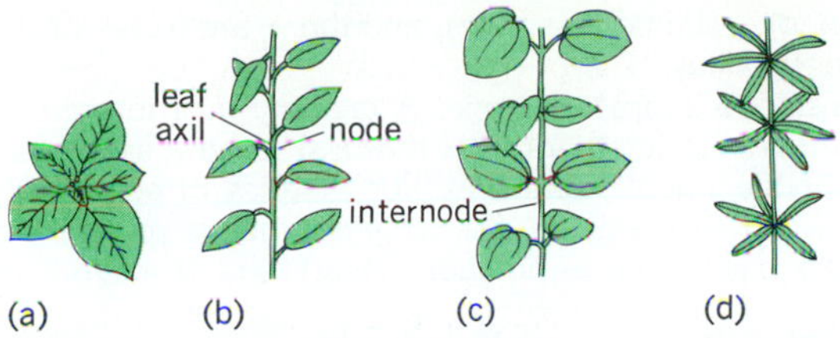

Fig. 1. Leaf arrangement. (*a*) Helical (top view). (*b*) Helical with elongated internodes (alternate). (*c*) Opposite (decussate). (*d*) Whorled (verticillate).

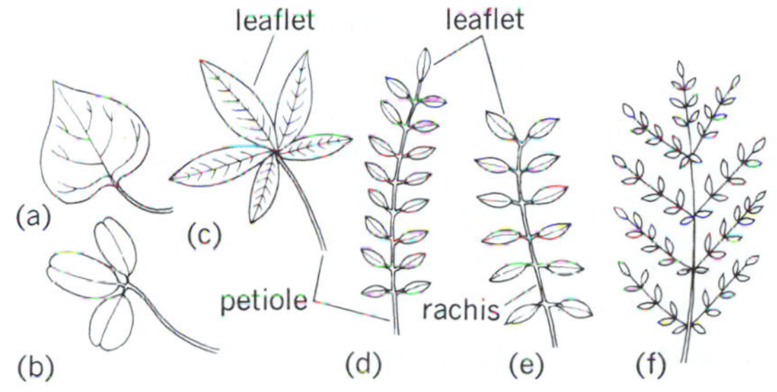

Fig. 2. Leaf types. (*a*) Simple. (*b*) Trifoliate. (*c*) Palmately compound. (*d*) Odd-pinnately compound. (*e*) Even-pinnately compound. (*f*) Decompound.

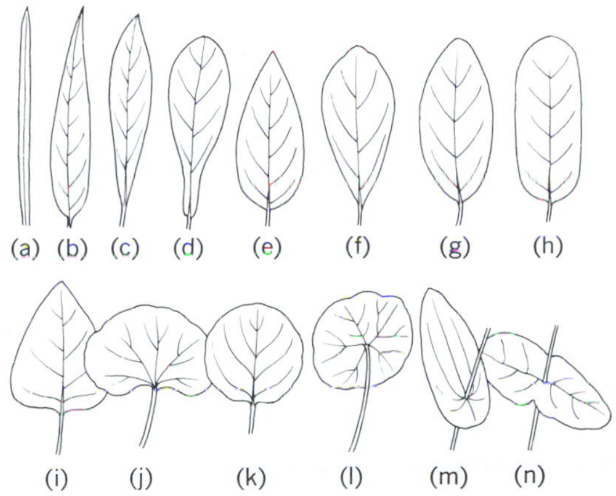

Fig. 3. Leaf shapes. (*a*) Linear. (*b*) Lanceolate. (*c*) Oblanceolate. (*d*) Spatulate. (*e*) Ovate. (*f*) Obovate. (*g*) Elliptic. (*h*) Oblong. (*i*) Deltoid. (*j*) Reniform. (*k*) Orbicular. (*l*) Peltate. (*m*) Perfoliate. (*n*) Connate.

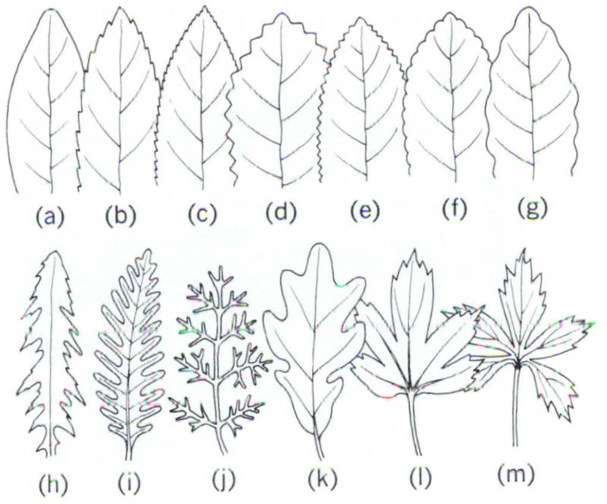

Fig. 4. Leaf margins of various types. (*a*) Entire. (*b*) Serrate. (*c*) Serrulate. (*d*) Dentate. (*e*) Denticulate. (*f*) Crenate. (*g*) Undulate. (*h*) Incised. (*i*) Pinnatifid. (*j*) Dissected. (*k*) Lobed. (*l*) Cleft. (*m*) Parted.

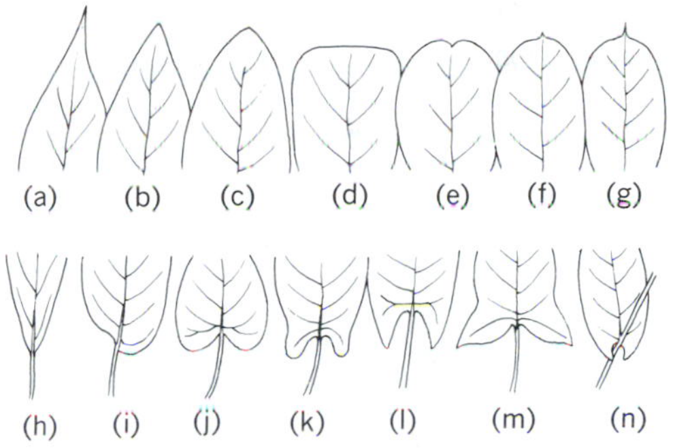

Fig. 5. Leaf tips and bases. (*a*) Acuminate. (*b*) Acute. (*c*) Obtuse. (*d*) Truncate. (*e*) Emarginate. (*f*) Mucronate. (*g*) Cuspidate. (*h*) Cuneate. (*i*) Oblique. (*j*) Cordate. (*k*) Auriculate. (*l*) Sagittate. (*m*) Hastate. (*n*) Clasping.

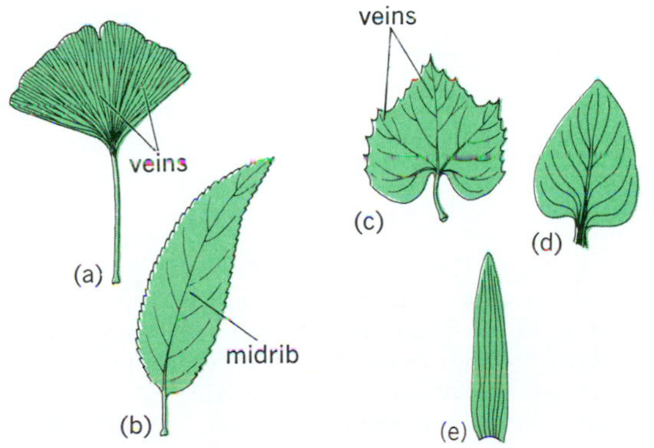

Fig. 6. Leaf venation. (*a*) Dichotomous. (*b*) Pinnate reticulate. (*c*) Palmate reticulate. (*d*) Parallel (expanded leaf). (*e*) Parallel (linear leaf).

The texture may be described as succulent when the leaf is fleshy and juicy; hyaline if it is thin and almost wholly transparent; chartaceous if papery and opaque but thin; scarious if thin and dry, appearing shriveled; and coriaceous if tough, thickish, and leathery.

Leaves may be fugacious, failing nearly as soon as formed; deciduous, failing at the end of the growing season; marcescent, withering at the end of the growing season but not falling until toward spring; or persistent, remaining on the stem for more than one season, the plant thus being evergreen. *See* DECIDUOUS PLANTS; EVERGREEN PLANTS.

Anatomy. The foliage leaf is the chief photosynthetic organ of most vascular plants. Although leaves vary greatly in size and form, they share the same basic organization of internal tissues and have similar developmental pathways. Like the stem and root, leaves consist of three basic tissue systems: the dermal tissue system, the vascular tissue system, and the ground tissue system. However, unlike stems and roots which usually have radial symmetry, the leaf blade usually shows dorsiventral symmetry, with vascular and other tissues being arranged in a flat plane.

Stems and roots have apical meristems and are thus characterized by indeterminate growth; leaves lack apical meristems, and therefore have determinate growth. Because leaves are more or less ephemeral organs and do not function in the structural support of the plant, they usually lack secondary growth and are composed largely of primary tissue only. *See* APICAL MERISTEM; ROOT (BOTANY); STEM.

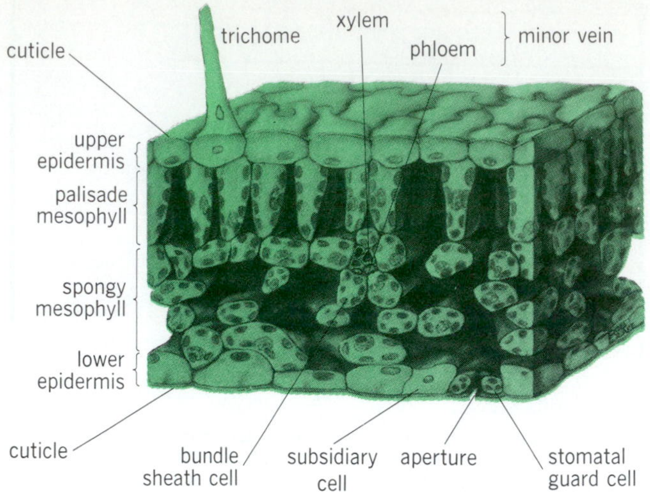

Fig. 7. Three-dimensional diagram of internal structure of a typical dicotyledon leaf.

The internal organization of the leaf is well adapted for its major functions of photosynthesis, gas exchange, and transpiration. The photosynthetic cells, or chlorenchyma tissue, are normally arranged in horizontal layers, which facilitates maximum interception of the Sun's radiation. The vascular tissues form an extensive network throughout the leaf so that no photosynthetic cell is far from a source of water, and carbohydrates produced by the chlorenchyma cells need travel only a short distance to reach the phloem in order to be transported out of the leaf (Fig. 7). The epidermal tissue forms a continuous covering over the leaf so that undue water loss is reduced, while at the same time the exchange of carbon dioxide and oxygen is controlled. *See* EPIDERMIS (PLANT); PARENCHYMA; PHLOEM; XYLEM. [N.G.D.]

Least-action principle Like Hamilton's principle, the principle of least action is a variational statement that forms a basis from which the equations of motion of a classical dynamical system may be deduced. Consider a mechanical system described by coordinates $q_1,\ldots, q_f$ and their canonically conjugate momenta $p_1,\ldots,p_f$. The action S associated with a segment of the trajectory of the system is defined by the equation below, where the integral is evaluated along the given segment

$$S = \int_c \sum_j p_j dq_j$$

c of the trajectory. The action is of interest only when the total energy E is conserved. The principle of least action states that the trajectory of the system is that path which makes the value of S stationary relative to nearby paths between the same configurations and for which the energy has the same constant value. The principle is misnamed, as only the stationary property is required. It is a minimum principle for sufficiently short but finite segments of the trajectory. *See* HAMILTON'S EQUATIONS OF MOTION; HAMILTON'S PRINCIPLE; MINIMAL PRINCIPLES. [P.M.S.]

Least-squares method A method of obtaining the best values (the ones with least error) of unknown quantities supposed to satisfy a system of linear equations of the form shown as notation (1), where $n > m$. Since there are more equations

$$\begin{array}{c} M_{11}a_1 + M_{12}a_2 + \cdots + M_{1m}a_m = b_1 \\ M_{21}a_1 + M_{22}a_2 + \cdots + M_{2m}a_m = b_2 \\ \cdots\cdots\cdots\cdots\cdots\cdots\cdots \\ M_{n1}a_1 + M_{n2}a_2 + \cdots + M_{nm}a_m = b_n \end{array} \qquad (1)$$

than unknowns, the system is said to be overdetermined. In the physical situation, the b_i are measured quantities, the M_{ij} are known (or assumed) quantifies, and the a_i are to be adjusted to their best values.

Consider a simple example. A quantity y of interest is supposed (perhaps for theoretical reasons) to be a linear function of an independent variable x. For a series of selected values $x_1, x_2, \ldots$ of x the values $y_1, y_2, \ldots$ of y are measured. The expected relation is shown as notation (2), and the problem is to find

$$\begin{array}{c} x_1\alpha + ß = y_1 \\ x_2\alpha + ß = y_2 \\ x_3\alpha + ß = y_3 \\ \cdots\cdots\cdots \end{array} \qquad (2)$$

the best values of α and ß, that is, respectively, the slope and intercept of the line which graphically represents the function. The best values of α and ß, in the least squares sense, are obtained by writing Eq. (3) and asserting that term

$$\eta_i = y_i - (x_i\alpha + ß) \qquad (3)$$

(4) shall be minimized with respect to α and ß; that is, that

$$\sum_{i}^{n} = 1\ \eta_i^2 \qquad (4)$$

Eqs. (5) hold. This leads to Eqs. (6) and (7), which may be

$$\frac{\partial}{\partial\alpha} \sum_{i=1}^{n} \eta_i^2 = 0 \qquad (5)$$

$$\frac{\partial}{\partial ß} \sum_{i=1}^{n} \eta_i^2 = 0$$

$$\alpha \sum_{i=1}^{n} x_i + nß - \sum_{i=1}^{n} y_i = 0 \qquad (6)$$

$$\alpha \sum_{i=1}^{n} x_i^2 + ß \sum_{i=1}^{n} x_i - \sum_{i=1}^{n} x_i y_i = 0 \qquad (7)$$

solved for α and ß. *See* CURVE FITTING. [McA.H.H.]

Lecanicephaloidea An order of tapeworms of the subclass Cestoda. All species are intestinal parasites of elasmobranch fishes. These tapeworms are distinguished by having a peculiar scolex divided into two portions. The lower portion is collarlike and bears four small suckers; the upper portion may be discoid or tentacle-bearing and is provided with glandular structures (see illustration). The scolex is usually buried in the intestinal wall of the host and may produce local pathology. The anatomy of the segments is very similar to that of the Proteocephaloidea. *See* CESTODA. [C.P.R.]

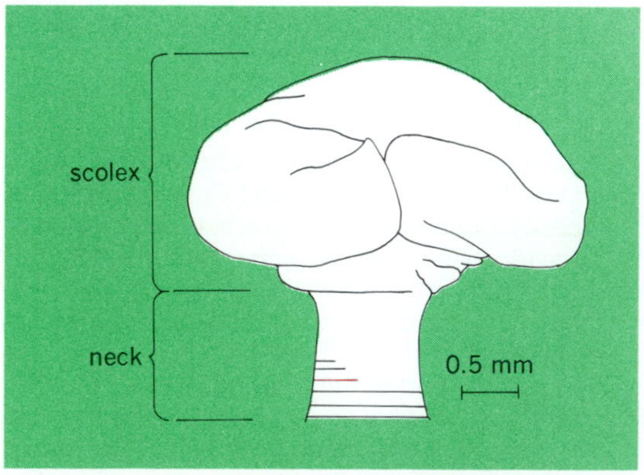

Anterior end of a lecanicephaloid tapeworm.

Lecanorales An order of the Ascolichenes, also known as the Discolichenes. Lecanorales is the largest and most typical order of lichens and parallels closely the fungal order Helotiales. The apothecia are open and discoid, with a typical hymenium and hypothecium. There are four growth forms—crustose, squamulose, foliose, and fruticose—all showing greater variability than any other order of lichens.

The Lecanorales is divided into 25 families, about 160 genera, and 8000–10,000 species. Family divisions are based on growth form of the thallus, structure of the apothecia, the species of symbiotic algae present, and spore characters. Species are separated by such characters as isidia, soredia, rhizines, and pores, and by chemistry. The larger families include: Cladoniaceae, Lecanoraceae, Lecideaceae, Parmeliaceae, Umbilicariaceae, and Usneaceae. [M.E.H.]

Le Chatelier's principle A description of the response of a system in equilibrium to a change in one of the variables determining the equilibrium.

For any chemical reaction equilibrium or phase equilibrium, an increase in temperature at constant pressure shifts the equilibrium in the direction in which heat is absorbed by the system. *See* CHEMICAL EQUILIBRIUM; PHASE EQUILIBRIUM.

For any reaction equilibrium or phase equilibrium, an increase in pressure at constant temperature shifts the equilibrium in the direction in which the volume of the system decreases.

For a reaction equilibrium in a dilute solution, addition of a small amount of a solute species that participates in the reaction will shift the equilibrium in the direction that uses up some of the added solute.

For an ideal-gas reaction equilibrium, addition at constant temperature and volume of a species that participates in the reaction will shift the equilibrium in the direction that consumes some of the added species. For an ideal-gas reaction equilibrium, addition at constant temperature and pressure of a species that participates in the reaction might shift the equilibrium to produce more of the added species or might shift the equilibrium to use up some of the added species; the direction of the shift depends on the reaction, on which species is added, and on the initial composition of the equilibrium mixture. [I.N.L.]

Lectins A group of proteins which possess the unique ability to agglutinate erythrocytes and other types of cells. Lectins are widely distributed in the plant kingdom, particularly among the legumes and to a lesser extent among the cereal grains. These substances have also been referred to as phytohemagglutinins, although the term phytolectins has been proposed in order to distinguish those lectins which are found in plants from those which are of animal or microbial origin.

All of the effects of lectins are believed to be a manifestation of the ability of lectins to bind to specific kinds of sugars present on the surface of cells. By virtue of this property, the lectins are an extremely useful tool, not only for the isolation and characterization of polysaccharides and glycoproteins, but also for probing the molecular structure of cell surfaces and the changes induced therein by chemical and biological agents.

There is some evidence to indicate that the lectins may serve to protect plants against certain pathogenic fungi and insect predators. Stimulation of pollen germination and seedling growth and regulation of cell wall extension are other physiological roles postulated for phytolectins. [I.E.Li.]

Lecythidales An order of flowering plants, division Magnoliophyta (Angiospermae), in the subclass Dilleniidae of the class Magnoliopsida (dicotyledons). The order consists of the single family Lecythidaceae, with about 450 species. They are tropical, woody plants with alternate, entire leaves, valvate sepals, separate petals, numerous centrifugal stamens, and a syncarpous, inferior ovary with axile placentation. Brazil nuts are the seeds of *Bertholletia excelsa*, a member of the Lecythidaceae. *See* BRAZIL NUT; DILLENIIDAE; MAGNOLIOPSIDA. [A.Cr.]

Leg (anatomy) The portion of the lower, or inferior, limb between the thigh and the foot. The leg bones are the tibia (a large, weight-supporting shinbone) and the slender, rodlike fibula. Compartmentalized muscle groups include functional sets which act on the foot, ankle, and thigh. Such muscles may either move the point to which they are attached or immobilize it to prevent movement, as when standing. [T.S.P.]

Legendre functions Solutions to the differential equation $(1 - x^2)y'' - 2xy' + \upsilon(\upsilon + 1)y = 0$.

The most elementary of the Legendre functions, the Legendre polynomial $P_n(x)$ Can be defined by the generating function in Eq. (1). More explicit representations are Eq. (2), and the hypergeometric function, Eq. (3). *See* HYPERGEOMETRIC FUNCTIONS.

$$(1 - 2xr + r^2)^{-1/2} = \sum_{n=0}^{\infty} P_n(x)r^n \qquad (1)$$

$$P_n(x) = \frac{(-1)^n}{2_n n!}\frac{d^n}{dx^n}(1 - x^2)^n \qquad (2)$$

$$P_n(x) = {}_2F_1[-n, n + 1; 1; (1 - x)/2] \qquad (3)$$

Generating function (1) implies Eq. (4). The function $(a^2 - 2ar\cos\theta + r^2)^{-1/2}$ represents the potential in an inverse

$$(a^2 - 2ar\cos\theta + r^2)^{-1/2} = \frac{1}{a}\sum_{n=0}^{\infty} P_n(\cos\theta)(r/a)^n \qquad (4)$$
$$0 < r < a$$

square field at a point P of a source at A, where r and a are the distances from P and A to a fixed point O, and θ is the angle between the segments PO and OA. *See* DIFFERENTIAL EQUATION; ORTHOGONAL POLYNOMIALS; POTENTIALS. [R.A.]

Legionnaire's disease A type of pneumonia caused by infection with the bacterium *Legionella pneumophila*. It is called Legionnaire's disease because it was first observed in an epidemic among those attending an American Legion convention in Philadelphia, Pennsylvania, in 1976. The initial symptoms are headache, muscle aches, and a generalized feeling of discomfort. The fever rises rapidly, reaching 102–105°F (38.9–40.6°C), and is usually accompanied by cough, shortness of breath, and chest pain. Abdominal pain and diarrhea are also often present. Ninety percent of Legionnaire's disease patients develop pneumonia, and approximately 15–20% of those die if they do not receive specific antibiotic treatment. Erythromycin is the antibiotic of choice.

The available epidemiologic evidence indicates that Legionnaire's disease is not contagious. Instead, it is thought that people become infected by inhalation of droplets of contaminated water, particularly the water that drifts down from cooling towers or evaporative condensers of air-conditioning systems. [J.E.McD.]

Legume A dry, dehiscent fruit derived from a single, simple pistil. When mature, it splits along both dorsal and ventral sutures into two valves. The term also designates any plant of the order Rosales that bears this type of fruit.

The family Leguminosae characteristically contains a single row of seeds attached along the lower or ventral suture of the fruit. The seeds are highly nutritious and several species of legumes furnish a large amount of food for both humans and animals. Some more common and important legumes are alfalfa, beans, clovers, kudzu, lespedeza, locust, peas, peanuts, soybeans, and vetch. *See* FRUIT; ROSALES. [P.D.St./E.L.C.]

Leguminous trees Trees of the legume family. Legumes commonly known to North Americans and Europeans are annuals such as soybean, peas, beans, and peanuts, and non-woody perennials such as clover or alfalfa. However, most legume genera occur in the subtropics and tropics in such diverse forms as 20-in.-tall (50-cm) annuals which may have large tubers, vines, shrubs, and 150-ft-tall (45-m) trees. Leguminous trees such as black locust (*Robinia pseudoacacia*) and honey locust (*Gleditsia triacanthyos*) occur in temperate regions; mesquite (*Prosopis* species) or *Acacia* species dominate much of the semiarid regions of North and South America, Africa, the Near East, and Australia; and large trees such as *Albizzia* species, *Dalberghia* species, and *Parkia speciosa* occur in the wet tropics. *See* LEGUME.

Managed orchards of selected strains of leguminous trees show potential for producing protein, carbohydrate, and fuel resources that require less water, nitrogen, and tillage than traditional crops. Leguminous trees can provide large quantities of livestock or human food in the form of pods, seeds, or leaf material. Tree- and shrub-based ecosystems such as forests typically lose less soil and soil nutrients than rowcrop agricultural systems.

Many of the tree legume woods are hard and durable with strikingly rich colors and prominent figure. Species with valuable wood include desert ironwood (*Olneya tesota*), mesquite (*Prosopis* species), screwbean (*Prosopis pubescens*), *Dalberghia*, *Pterocarpus*, and *Afrormosia*. Lumber of some of the wet humid tree legumes is more costly than African mahogany (*Khaya* species). [P.Fe.]

Leishmaniasis An infection caused by any one of three pathogenic hemoflagellates of the genus *Leishmania*, transmitted to humans by blood-sucking sand flies (*Phlebotomus*). Three diseases are distinguished: visceral leishmaniasis, known also as kala azar, caused by *L. donovani*; cutaneous leishmaniasis (Oriental sore, Baghdad boil), caused by *L. tropica*; and mucocutaneous leishmaniasis (espundia), caused by *L. braziliensis*. In cutaneous leishmaniasis the parasites invade only the skin and subcutaneous tissue, causing deep ulcers which upon healing confer a lasting immunity. It is widespread in the Near and Middle East, stretching to Asiatic Russia. Mucocutaneous leishmania affects the mucous membranes of the nose and mouth causing severe ulceration and disfigurement. It is prevalent in some areas of Central and South America. Visceral leishmaniasis is a severe, generalized, and often fatal infection affecting the organs rich in reticuloendothelial cells. There are endemic areas in India and China, in Central and North Africa, and along the Mediterranean littoral, where it affects mainly the young (infantile kala azar). [M.Yo.]

Leitneriales An order of flowering plants, division Magnoliophyta (Angiospermae) in the subclass Hamamelidae of the class Magnoliopsida (dicotyledons). The order consists of a single family, genus, and species (*Leitneria floridana*) of the southeastern United States. The plants are simple-leaved, dioecious shrubs with the flowers in catkins. *See* HAMAMELIDAE; MAGNOLIOPSIDA. [A.Cr.]

Lemming The name applied to 11 species of rodents in the subfamily Microtinae, family Muridae. These animals have a northern circumpolar distribution.

The Norway lemming (*Lemmus lemmus*) is found in the mountainous wastelands of northern Norway and Lapland. It is usually nocturnal and timid in its habits, except when a population explosion occurs with its resultant migration, of which the causative factors are not known. Cyclic variations in fertility may be a factor. Usually there are two litters of five offspring each year, but often four litters of two to eight offspring occur. *See* RODENTIA. [C.B.C.]

Lemniscate of Bernoulli A curve shaped like the figure eight (see illustration), referred to by Jacques Bernoulli in 1694. Let F_1, F_2 be points of a plane π with $F_1F_2 = 2a$, $a > 0$.

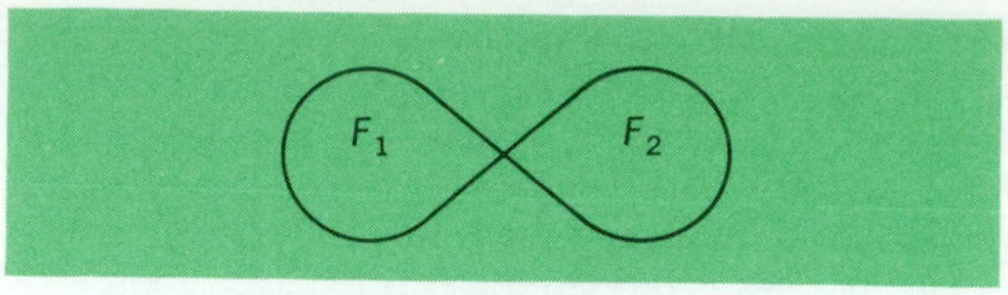

Curve known as lemniscate.

The locus of a point P of π which moves so that $PF_1 \cdot PF_2 = b^2$, where b is a positive constant, is called an oval of Cassini. The lemniscate is obtained when $b = a$. Its equation in rectangular coordinates is $(x^2 + y^2)^2 = a^2(x^2 - y^2)$ and in polar coordinates $\rho^2 = a^2 \cos 2\theta$. *See* ANALYTIC GEOMETRY. [L.M.Bl.]

Lemon The fruit *Citrus limon*. The yellow fruits are medium-sized and elongate with a prominent nipple. The lemon is more sensitive to cold than other major citrus fruits, and thus its commercial culture is restricted to areas with mild winter temperatures.

The lemon is grown primarily for its acid flavor. Lemon juice, very high in vitamin C, is used in beverages and has many culinary uses. It is also used widely in proprietary soft drinks. The principal by-products are citric acid from the juice and lemon oil from the peel. *See* ASCORBIC ACID.

Commercial lemon production developed first in Italy, mainly in Sicily. Italy is the largest producer, followed by California. Spain, Greece, and Argentina are also significant producers. [R.K.So.]

Lemur A primate of the family Lemuridae; all 16 species are indigenous to Madagascar. All lemurs have long tails, foxlike faces, and scent glands on the shoulder region and wrists. By constantly running its tail through its arms, the lemur creates a scent flag. Lemurs are largely dependent upon fruit as their diet; they also eat insects and eggs. A litter containing one or two young is born in the early spring after a gestation period of 2 months. The young cling to the underside of the mother for about the first 3 weeks and are then carried on her back for a period of time. *See* PRIMATES. [C.B.C.]

Length Extension in space. Length is one of the three fundamental physical quantities (the other two being mass and time), and therefore cannot be defined in terms of simpler quantities. It is measured by comparison with an arbitrary standard called the international meter, defined as 1,650,763.73 times the wavelength of the orange light emitted when a gas of pure krypton-86 is excited in an electrical discharge. [D.Wi.]

Lens (optics) A curved piece of ground and polished or molded material, usually glass, used for the refraction of light. Its two surfaces have the same axis. Usually this is an axis of rotation symmetry for both surfaces; however, one or both of the surfaces can be toric, cylindrical, or a general surface with double symmetry (see illustration). The intersection points of the symmetry axis with the two surfaces are called the front and back vertices and their separation is called the thickness of the lens. There are three lens types, namely, compound, single, and cemented. A group of lenses used together is a lens system. Such systems may be divided into four classes: telescopes, oculars (eyepieces), photographic objectives, and enlarging lenses.

Lens types. A compound lens is a combination of two or more lenses in which the second surface of one lens has the same radius as the first surface of the following lens and the two lenses are cemented together. Compound lenses are used instead of single lenses for color correction, or to introduce a surface which has no effect on the aperture rays but large effects on the principal rays, or vice versa. Sometimes the term compound lens is applied to any optical system consisting of more than one element, even when they are not in contact.

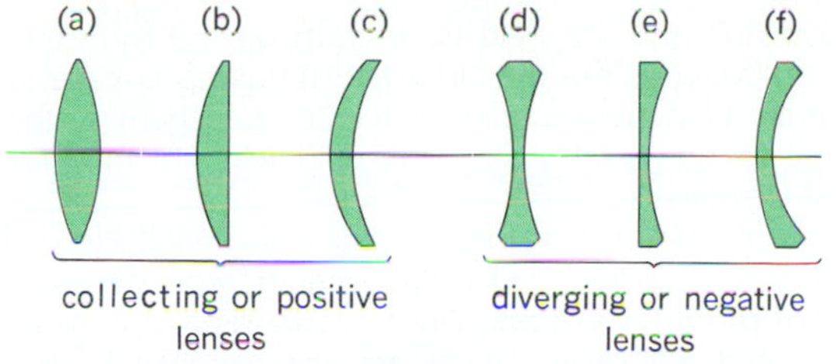

Common lenses. ***(a)*** **Biconvex.** ***(b)*** **Plano-convex.** ***(c)*** **Positive meniscus.** ***(d)*** **Biconcave.** ***(e)*** **Plano-concave.** ***(f)*** **Negative meniscus.** ***(After F. A. Jenkins and H. E. White, Fundamentals of Optics, 4th ed., McGraw-Hill, 1976)***

The diameter of a simple lens is called the linear aperture, and the ratio of this aperture to the focal length is called the relative aperture. This latter quantity is more often specified by its reciprocal, called the *f*-number. Thus, if the focal length is 50 mm and the linear aperture 25 mm, the relative aperture is 0.5 and the *f*-number is *f*/2. *See* FOCAL LENGTH.

A compound lens made of two or more simple thin lenses cemented together is called a cemented lens.

Lens systems. A lens system consisting of two systems combined so that the back focal point of the first (the objective) coincides with the front focal point of the second (the ocular) is called a telescope. Parallel entering rays leave the system as parallel rays. The magnification is equal to the ratio of the focal length of the first system to that of the second. *See* TELESCOPE.

A photographic objective images a distant object onto a photographic plate or film. The amount of light reaching the light-sensitive layer depends on the aperture of the optical system, which is equivalent to the ratio of the lens diameter to the focal length. The larger the aperture (the smaller the *f*-number), the less adequate may be the scene luminance required to expose the film. Therefore, if pictures of objects in dim light are desired, the *f*-number must be small. On the other hand, for a lens of given focal length, the depth of field is inversely proportional to the aperture.

In general, photographic objectives with large fields have small apertures; those with large apertures have small fields.

The basic type of enlarger lens is a holosymmetric system consisting of two systems of which one is symmetrical with the first system except that all the data are multiplied by the enlarging factor *m*. When the object is in the focus of the first system, the combination is free from all lateral errors even before correction. A magnifier in optics is a lens that enables an object to be viewed so that it appears larger than its natural size. The magnifying power is usually given as equal to one-quarter of the power of the lens expressed in diopters. *See* DIOPTER; MAGNIFICATION. [M.J.H.]

Lentil A semiviny annual legume with slender tufted and branched stems. The lentil plant (*Lens esculenta*) was one of the first plants brought under cultivation. The world's lentil production is centered in Asia, with nearly two-thirds of the production from India, Pakistan, Turkey, and Syria. Whitman and Spokane counties in Washington, and Latah, Benewah, and Nez Perce counties in Idaho grow about 95% of the lentils produced in the United States.

The seeds grow in short broad pods, each pod producing two or three thin lens-shaped seeds. Seed color varies from yellow to brown and may be mottled, although mottled seeds are not desirable for marketing.

Lentil seed is used primarily for soups but also in salads and casseroles. Lentils are more digestible than meat and are used as a meat substitute in many countries. *See* LEGUME. [K.J.M.]

Lenz's law A law of electromagnetism which states that, whenever there is an induced electromotive force (emf) in a conductor, it is always in such a direction that the current it would produce would oppose the change which causes the induced emf. If the change is the motion of a conductor through a magnetic field, the induced current must be in such a direction as to produce a force opposing the motion. If the change causing the emf is a change of flux threading a coil, the induced current must produce a flux in such a direction as to oppose the change.

Lenz's law is a form of the law of conservation of energy, since it states that a change cannot propagate itself. *See* CONSERVATION OF ENERGY; ELECTROMAGNETIC INDUCTION. [K.V.M.]

Leo The Lion, in astronomy, is a magnificent zodiacal constellation appearing during spring and early summer. It is the fifth sign of the zodiac. Leo is well defined and bears a close resemblance to the creature it represents (see illustration).

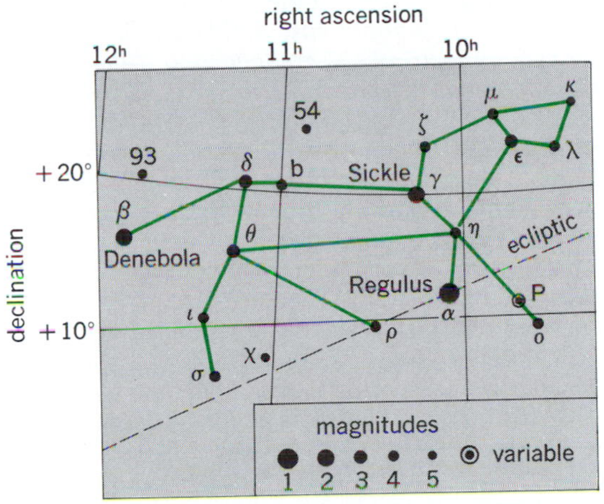

Line pattern of the constellation Leo. The grid lines represent the coordinates of the sky. Right ascension (E-W) in hours, and declination (N-S) in degrees, corresponding to the longitude and latitude of the Earth. The apparent brightness, or magnitudes, of the stars is shown by the sizes of the dots, which are graded by appropriate numbers as indicated.

Associated with this constellation are the famous Leonids shower of meteors, which can be seen radiating from Leo in November of each year and appearing especially brilliant at intervals of about 33 years. *See* CONSTELLATION. [C.-S.Y.]

Lepadomorpha A suborder of the Thoracica. These barnacles have a stalk or peduncle, which morphologically is the elongated, anterior, preoral region of the body, and a capitulum comprising the bivalved mantle enclosing the rest of the body (see illustration). The mantle folds are usually protected by

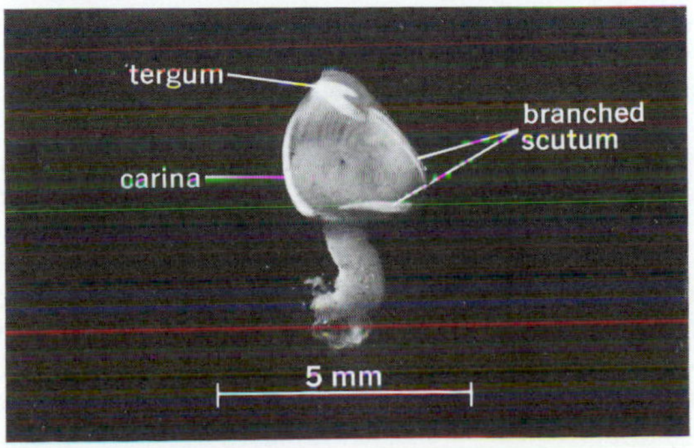

Morphology of *Octolasmis lowei*.

a varying number of calcareous plates. The peduncle also may be protected by small calcareous scales or granules. Caudal furca and filamentary appendages are often present. These barnacles are hermaphroditic, or with separate sexes.

Several families are distinguished, the most primitive being Scalpellidae. *See* THORACICA. [H.G.St.]

Lepidodendrales Giant club mosses, an order of extinct lycopods consisting primarily of arborescent forms, which constituted an important element of the Carboniferous flora. The group is characterized by branching, by spirally arranged microphyllous leaves, by the production of small amounts of secondary vascular tissues relative to the total bulk of the plant (several species are not known to have produced secondary tissues), and by heterospory (bearing spores of two types). In at least two genera heterospory attained a level of specialization approaching the seed habit. The order existed from Late Devonian through Permian times. Well-known genera are *Lepidodendron, Lepidophloios, Sigillaria, Bothrodendron, Stigmaria, Lepidostrobus, Lepidocarpon, Miadesmia,* and *Mazocarpon*. *See* LYCOPODIOPSIDA. [C.B.B.; H.W.P.]

Lepidolite A mineral of variable composition that is also called lithium mica and lithionite. Lepidolite is uncommon, occurring almost exclusively in structurally complex granitic pegmatites, commonly in replacement units. Lepidolite is a commercial source of lithium, commonly used directly in lithium glasses and other ceramic products.

It usually forms small scales or fine-grained aggregates. Its colors, pink, lilac, and gray, are a function of the Mn/Fe ratio. It is fusible at 2, yielding the crimson (lithium) flame. It has a perfect basal cleavage. Hardness is 2.5–4.0 on Mohs scale; specific gravity is 2.8–3.0. *See* MICA; SILICATE MINERALS. [E.W.H.]

Lepidoptera The order of scaly-winged insects, including the butterflies, skippers, and moths. This is one of the largest orders in the class Insecta. It has more than 100,000 species, about 10,000 in North America, and between 125 and 175 families. The adults have a covering of hairs and scales on the wings, legs, and body, and are often beautifully colored. With minor exceptions, the adults are also characterized by two pairs of membranous wings and sucking mouthparts, featuring a prominent, coiled proboscis. Butterflies and skippers usually fly in the daytime, and most moths are nocturnal. The adults usually take liquid food, such as nectar and juices of fruits. The caterpillars are almost always herbivorous.

Morphology. The most unusual feature of the head of the adult animal is the form of the mouthparts. The proboscis is extended by blood pressure created by the retraction of the stipites of the maxillae. Diagonal muscles within each proboscis unit cause the proboscis to coil. The liquid food is sucked up by means of a muscular pump, formed from the pharynx, buccal cavity, and cibarium, a food pocket of the mouth cavity. Ocelli are absent in many groups. Antennae are quite variable in form.

The prothorax is well developed in some lower groups but it is considerably smaller than the pterothoracic, or wing-bearing, segments, and is largely membranous in most Lepidoptera. The most prominent feature of the dorsum of the prothorax, in most groups, is a pair of protuberant sclerotized lobes.

The scales are very variable in form. Generally, they are flat, thin, sclerotized sacks, with striated outer surfaces. The vast spectrum of colors seen in the Lepidoptera can be grouped into two categories, pigmentary and structural colors. Pigmentary colors result from pigments which are present in the scales. Structural colors are the result of either fine surface ridges on the scales or layers within the cuticle, which interfere with or diffract the light. The structural colors are generally metallic or iridescent.

In most moths, the wings are coupled by a single spine formed by fused setae which project forward from the base of the hindwing and are held by a clasp on the forewing. The spine is known as the frenulum and the clasp is the retinaculum. In the Homoneura, there is a lobe, the jugum, at the base of the forewing, which engages the hindwing, or the frenulum, when it is present.

The form of the external genitalia, especially that of the male, has been widely used in the separation of species.

Developmental stages. Metamorphosis is complete. The larvae, commonly called caterpillars, are mandibulate and cylindrical, with short thoracic legs and a variable number of abdominal prolegs. They have one pair of thoracic and eight pairs of abdominal spiracles. Pupae are variable in form and have appendages that are usually firmly cemented down (obtect), though they are sometimes partly or completely free (exarate). Pupae often are enclosed in a silken cocoon.

Classification. Unfortunately, the classification of the Lepidoptera is the subject of considerable controversy; much will doubtless be resolved when more is known of the anatomy and life history of many groups.

A rather conservative classification has been used here. In many superfamilies only the more important families are mentioned. The table lists the important families.

Homoneura (Jugatae). Fore- and hindwings are similar in shape and venation. They are connected by a jugum and, sometimes, also by a frenulum. Mouthparts are mandibulate, with mandibles vestigial or absent, and the galeae forms a rudimentary proboscis. The female has a single genital opening. The pupae are free or partially free.

This small suborder, including less than 1% of the species in the order, contains a diverse group of primitive forms showing certain features in common with the Trichoptera, or caddis flies.

The superfamily Micropterygoidea includes one small family, the Micropterygidae, minute moths possessing toothed, functional mandibles and lacking even the most rudimentary proboscis. The larvae, which feed on mosses, are unusual in having eight pairs of abdominal prolegs.

Superfamily Eriocranioidea is a group of tiny moths; the mandibles are greatly reduced and untoothed. Three families, the Eriocraniidae, Neopseustidae, and Mnesarchaeidea, have been recognized within the superfamily. The leaf-mining larva lacks legs. The adults reportedly do not feed. The females have a piercing ovipositor. Superfamily Hepialoidea contains medium- to large-sized moths which possess rudimentary mouthparts. The larvae are borers. The rapid flying adults are mostly crepuscular, thus the common name swift, or ghost, moths. The only family of importance is the Hepialidae.

Heteroneura (Frenatae). Fore- and hindwings are markedly different in shape and venation. Usually they are connected by a frenulum and retinaculum. Mouthparts are formed for sucking or, rarely, are vestigial, Adults with functional mouthparts feed on nectar of flowers, juices of rotten fruits, and other liquids. The female usually has two genital openings. Pupae are usually obtect.

One family, the Incurvariidae, comprises the superfamily Incurvarioidea. The wings are covered with microscopic spines and the females have a single genital opening. The venation is almost complete.

Superfamily Nepticuloidea includes one family, the Nepticulidae. These tiny moths have wing spines and the females have a single genital opening, but they differ from the Incurvarioidea in having a reduced venation. The larvae are generally miners in leaves, bark, and rarely, in fruits.

Superfamily Cossoidea includes one family, the Cossidae, commonly called the carpenter or goat moths. These are heavy-bodied moths, with the abdomen extending well beyond the hindwings. Mouthparts are rudimentary except for labial palpi.

Superfamily Tineoidea contains 16–39 families; the number varies with the author. This is a very large group, of uncertain composition. These moths are of small size, usually with well-developed maxillary palpi. The labial palpi have a slender, pointed third segment. Venation may be reduced, and the wings may be divided into plumes.

The small, wide-winged moths of the superfamily Tortiicoidea belong to two families, the Olethreutidae and the Tortricidae. The maxillary palpi are vestigial or absent, and the third segment of the labial palpus is short and usually obtuse. The hair fringes of the wings are always shorter than the width of the wing.

The family contains a number of agriculturally undesirable species. Paramount among these is the codling moth (*Carpocapsa pomonella* L.), which is a very serious pest of apples and other fruits. The large genus *Laspeyresia* contains the interesting Mexican jumping bean moth (*L. saltitans* Westwood). The violent movements of the larvae of this moth are responsible for the action of the beans which they inhabit.

Tortricidae is a family which generally lacks the fringe of long hairs along the cubitus, characteristic of the Olethreutidae. The spruce budworm (*Choristoneura fumiferana* Clemens) is probably the most important injurious tortricid. In many places, especially in eastern Canada, it has defoliated vast areas of coniferous forest.

The superfamily Pyralidoidea is comprised of moths that are moderately small to medium-sized, long-legged, and slender-bodied. The maxillary palpi are usually well developed. Pyralididae is the second largest family of moths. They are small and medium-sized, and a wing expanse of 20–35 mm is not uncommon. The legs are usually long and slender. Pterophoridae is the family known as the plume moths. The wings are divided into featherlike plumes, of which there are usually two in the forewing and three in the hindwing. The moths lack maxillary palpi and have slender bodies and long legs. The larvae feed exposed or are borers.

Superfamily Zygaenoidea includes moderately small- to medium-sized moths that have complete venation, rudimentary palpi and, usually, a rudimentary proboscis. The wings are broad with short fringes. The larvae are short and more or less sluglike and are exposed feeders.

Superfamily Castnioidea includes one family, the Castniidae. They are large, diurnal, butterflylike moths with clubbed antennae, upright eggs, and boring larvae. A proboscis may be either present or absent. These moths are considered by some to be distantly related to the butterflies, but the resemblances may very well be due to convergence.

Size, distribution, and common names of some families of Lepidoptera

Classification	Common name	Distribution	No. of species*
Suborder Homoneura			
Micropterygidae	Micropterygids	Holarctic and Australia	3 (35)
Eriocraniidae	Ericocraniids	Holarctic	5 (20)
Mnesarchaeidae	Mnesarchaeids	New Zealand	
Hepialidae	Swift or ghost moths	Cosmopolitan	18 (200)
Suborder Heteroneura			
Incurvariidae	Yucca moths and relatives	Cosmopolitan	60
Nepticulidae	Serpentine leaf miners	Cosmopolitan	75
Cossidae	Goat or carpenter moths	Cosmopolitan	45
Aegeriidae	Clearwing moths	Cosmopolitan	120
Coleophoridae	Case bearers	Cosmopolitan	110 (900)
Gelechiidae	Gelechiids	Cosmopolitan	590 (3800)
Gracilariidae	Gracilariids	Cosmopolitan	235
Heliodinidae	Heliodinids	Cosmopolitan	21
Oecophoridae	Oecophorids	Cosmopolitan; largely Australian	225 (3000)
Orneodidae	Many-plume moths	Cosmopolitan	1
Psychidae	Bagworms	Cosmopolitan	25
Tineidae	Clothes moths and relatives	Cosmopolitan	135 (2500)
Yponomeutidae	Ermine moths	Cosmopolitan	65 (800)
Olethreutidae	Olethreutids	Cosmopolitan	715 (2500)
Tortricidae	Tortricids	Cosmopolitan	210 (1500)
Thyrididae	Window-winged moths	Tropical	10
Pyralididae	Pyralids; snout moths	Cosmopolitan	1135 (12,000)
Pterophoridae	Plume moths	Cosmopolitan	130
Eucleidae	Slug moths	Cosmopolitan	50 (900)
Megalopygidae	Flannel moths	Mostly American; a few African	11
Zygaenidae	Foresters and burnets	Palearctic, African, and Indo-Australian	—
Castniidae	Castniids	Neotropical and Indo-Australian	
Drepanidae	Hooktips	Holarctic	6
Geometridae	Measuring worms, loopers, cankerworms, carpets, waves, and pugs	Cosmopolitan	1200 (4000)
Uraniidae	Uraniids	Tropical	—
Sphingidae	Sphinx, hawk, or hummingbird moths	Cosmopolitan	106 (1000)
Lasiocampidae	Tent caterpillars, lappet moths	Cosmopolitan except New Zealand; mainly tropical	30 (1400)
Saturniidae	Giant silkworms	Cosmopolitan	60
Bombycidae	Silkworm and allies	Tropical	1 (introduced)
Arctiidae	Tiger moths	Cosmopolitan	200 (3600)
Lymantriidae	Tussock moths	Largely African and Indo-Malayan, but with important Holarctic species	27
Notodontidae	Prominents, puss moths	Cosmopolitan except New Zealand	120
Noctuidae	Noctuids, owlets, underwings, millers	Cosmopolitan	2700 (20,000)
Hesperiidae	Skippers, agave worms	Cosmopolitan	240 (3000)
Papilionidae	Swallowtails, bird-wings, parnassians	Cosmopolitan	27 (600)
Pieridae	Whites, sulfurs, orangetips	Cosmopolitan	61 (1000)
Nymphalidae	Four-footed butterflies	Cosmopolitan	211 (5000)
Libytheidae	Snout butterflies	Cosmopolitan	1 (17)
Lycaenidae	Blues, coppers, hairstreaks, metal marks	Cosmopolitan	138 (3500)

*The first figure is the number of described species in North America north of Mexico. This figure is reasonably accurate. The second figure, in parentheses, is a rough estimate of the number of described species in the world. It is difficult to postulate the actual total of Lepidoptera species from these figures since in some groups, such as the Papilionidae, probably more than 90% of the existing species have been described, while in others, such as many families of microlepidoptera, the figure may be well under 25%.

Members of the superfamily Geometroidea are small to large moths with reduced maxillary palpi and tympanal organs at the base of the abdomen. The frenulum may be present or absent.

Geometridae includes the measuring worms, loopers, and cankerworms, which make up a very large family of small- and medium-sized moths with slender bodies and relatively broad wings. The females are occasionally apterous. The larvae have the anterior prolegs reduced or absent; usually only those on segments 6 and 10 are well developed. They proceed with a characteristic looping motion, which is the basis for the scientific name. The larvae ordinarily are exposed feeders.

Sphinx, hawk, or hummingbird moths constitute the one family Sphingidae of the superfamily Sphingoidea. These medium-sized to very large, heavy-bodied moths have extremely rapid flight. The adults are mostly crepuscular or nocturnal, but a few genera are diurnal. The antennae are thickened and have a pointed apex. The proboscis is well developed and often extremely long. The wings are narrow, with the hindwing much shorter than the forewing. The larvae are external feeders and usually have a characteristic caudal horn. The pupa is in a cell in the ground or in a loose cocoon at the surface, and its long proboscis is often in a projecting case resembling a pitcher's handle.

Moths of the superfamily Saturnioidea are medium-sized to very large moths with the frenulum almost always reduced or absent. There is no tympanum and the mouthparts are usually reduced. The antennae are ordinarily pectinate, especially in the males.

Noctuoidea is a large, rather uniform superfamily of more than 20,000 species. Most of them are moderately large moths with reduced maxillary palpi. Tympanal organs are present in the metathorax.

The superfamily Hesperioidea includes one rather large family, the Hesperiidae. The skippers are small to moderately large, heavy-bodied, mostly diurnal insects with a clubbed antenna, which is bent, curved, or reflexed at the tip. The larvae have a prominent constriction, or neck, behind the head, and often live in leaves drawn together by silk. Those of the giant skippers are borers in yucca and agave. The pupa is usually enclosed in a slight cocoon.

Butterflies of the superfamily Papilionoidea are small to large diurnal insects with clubbed antennae, which are rounded at the tip and not bent or reflexed. The forewings always have two or more veins, which are stalked.

Biological aspects. The Lepidoptera are a group of insects on which much biological research remains to be done. A great deal is still unknown about the genetics, physiology, and ecology of this group. Moreover, butterflies and moths have proved useful as experimental animals in genetical research. The larvae of many species are injurious to certain crops, causing severe economic losses. Lepidoptera of all stages are subject to the attacks of a large number of predators, including birds, mammals, lizards, frogs, spiders, and certain other insects.

The Lepidoptera penetrate almost every section of the globe, with the major exception of Antarctica. Arctic and alpine tundra areas normally support a lepidopteran fauna which, although relatively poor in species, is rich in numbers. After rains, deserts are often alive with butterflies and moths. Tropical areas are by far the richest in species. One of the strangest habitats occupied by a lepidopteran is the hair of the neotropical three-toed sloth, where the sloth moth (*Bradypodicola hahneli*), a pyralid, passes its entire life cycle presumably feeding on algae which grow in the hair.

Migration is the most spectacular behavior occurring in the order. It is most frequent in the butterflies *Phoebis*, *Danaus*, *Libytheana*, and others, but is also known in moths such as *Chrysiridia*. Huge migratory swarms are frequently reported in many parts of the world, but this phenomenon, as well as the related communal roosting of adults, daily use of flyways, hilltopping, and so forth, is poorly understood.

Aggregations of butterflies at a mud puddle are a frequent sight, and moths have been observed "pumping," an act which consists of sipping water steadily from a puddle and ejecting it as a stream of drops or fine spray from the anus.

The phenomena of mimicry and protective resemblance are widespread in the Lepidoptera. Eyespot patterns in certain species elicit escape responses in passerine birds. Thus, they are an effective protective device. The appearance of dark forms of various moths in heavily industrialized, and thus heavily sooted, areas is a widespread phenomenon. This "industrial melanism" is doubtless due to shifting selection pressures and is one of the best-known examples of evolution in action. *See* BUTTERFLY; CATERPILLAR; INSECTA; MOTH; PROTECTIVE COLORATION. [P.R.E.]

Lepidosauria A subclass of reptiles, both living and extinct, in which the structure of the skull is characterized by two temporal openings (diapsid condition) on each side (see illustration). Lepidosauria differ from Archosauria, which are also diapsid, in that the bony arcades bordering their temporal openings may suffer reduction or loss, causing the apparent disappearance of one or both openings. They differ also in that no lepidosaur skull has any antorbital opening in front of the orbit, and in that their teeth are typically fused to the jaw (acrodont or pleurodont) rather than implanted in sockets (thecodont). *See* ARCHOSAURIA.

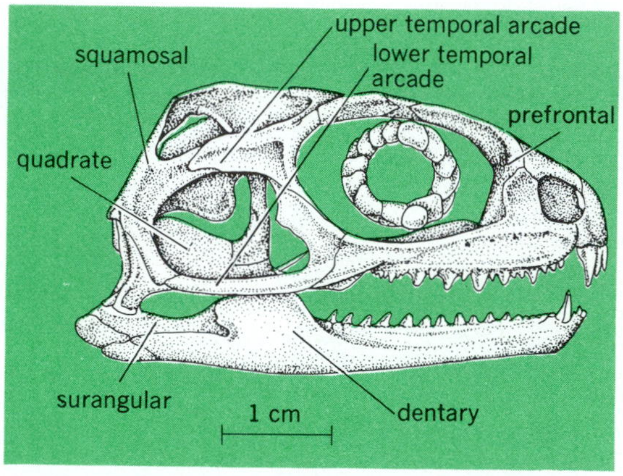

Skull of a living rhynchocephalian, *Sphenodon*, a typical diapsid. (After A. S. Romer, *Vertebrate Paleontology*, 3d ed., University of Chicago Press, 1966)

In the classification adopted here the Lepidosauria include three orders: Eosuchia, Rhynchocephalia, and Squamata. *See* EOSUCHIA; REPTILIA; RHYNCHOCEPHALIA; SQUAMATA. [A.J.C.]

Lepospondyli A term used to characterize Paleozoic amphibian groups in which the vertebral centra, basically spool-shaped, are formed by ossification directly around the notochord. In the Paleozoic, the larger amphibians were labyrinthodonts, but small lepospondyls appear to have been very abundant, particularly as water dwellers in the coal swamps. They are commonly considered to form three orders: Nectridea, Aistopoda, and Microsauria. The vertebral structure of the living amphibians is basically lepospondylous, and it is possible that part or all of the modern orders (unknown in the Paleozoic) have descended from lepospondyls, particularly the microsaurs. *See* AISTOPODA; AMPHIBIA; LABYRINTHODONTIA; MICROSAURIA; NECTRIDEA. [A.S.R.]

Leprosy A chronic infectious disease of humans, also known as Hansen's disease, of worldwide occurrence but most commonly found in tropical areas where the incidence may be as high as 5 per 1000 population. Infection appears most often to result from intimate contact with a leprous individual. The incubation period is long, ranging from a few months to several years. The presence of bacilli in peripheral nerves is a hallmark of leprosy. The most debilitating form of the disease, lepromatous leprosy, is characterized by multiple nodular lesions (lepromata) of the skin. Microscopically these lesions contain masses of rodlike microorganisms, leprosy bacilli. [L.B.]

Leptolepiformes An order of small ray-finned fishes of the Mesozoic that are important to the understanding of the early evolution of the Teleostei. The principal family included in this order, the Leptolepidae, has been reported from the Middle Triassic of Europe and ranges into the Cretaceous. Leptolepids represent the first teleosts as defined on the structure of the caudal skeleton. They were either derived from the earliest Pholidophoridae (known from the European Middle Triassic sediments) or shared a common ancestry with them. In size, body shape, fin position, structure of head and jaws, and general habitat, leptolepids are similar to the pholidophorids. In many ways these two families appear to parallel one another. *See* TELEOSTEI. [T.M.C.]

Lepton An elementary particle having no internal constituents which interacts through the electromagnetic, weak, and gravitational forces, but does not interact through the strong (nuclear) force. Leptons are very small, less than 10^{-18} m in size. This is less than 1/1000 the size of a nucleus and less than 10^{-8} the size of an atom. Leptons are much simpler than hadrons in both their structure and their behavior; hence leptons are considered to be more elementary than hadrons. At present the lepton family and the quark family are thought to lie at the same level of elementariness. *See* HADRON; QUARKS.

There are three known charged leptons: the electron (e), muon (μ), and tau (τ). Associated with each charged lepton is a neutral lepton called a neutrino. *See* ELECTRON; ELEMENTARY PARTICLE; NEUTRINO. [M.L.P.]

Leptospira A genus of spirochetes of the family Spirochaetacea. Recognition of two species, *Leptospira interrogans* and *L. biflexa*, has been proposed. Each comprises a large number of serological types (serovars). *Leptospira biflexa* serovars are apparently saprophytic organisms omnipresent in fresh surface waters and occasionally found in sea water. *Leptospira interrogans* comprises pathogenic leptospires which occur naturally in a large variety of wild and domesticated animals and cause acute, febrile, systemic disease in humans and other mammals. *See* LEPTOSPIROSIS; SPIROCHAETALES. [A.D.A.]

Leptospirosis An acute febrile disease of humans produced by spirochetes of many species of *Leptospira*. The incubation period is 6–15 days. Among the prominent features of the disease are fever, jaundice, muscle pains, headaches, hepatitis, albuminuria, and multiple small hemorrhages in the conjunctiva or skin. Meningeal involvement often occurs. The febrile illness subsides after 3–10 days. Fatal cases show hemorrhagic lesions in the kidney, liver, skin, muscles, and central nervous system. *See* SPIROCHAETALES.

Wild rodents are the principal reservoirs, although natural infection occurs in swine, cattle, horses, and dogs and may be transmitted to humans through these animals. Humans are infected either through contact with the urine or flesh of diseased animals, or indirectly by way of contaminated water or soil, the organisms entering the body through small breaks in the skin or mucous membrane. [T.B.T.]

Leptostraca The only extant order of the crustacean subclass Phyllocarida. The Leptostraca is represented by one fossil and a small number of living genera. These malacostracans are unique in having the carapace laterally compressed to such an extent that it forms a bivalvelike shell held together by a strong adductor muscle. The carapace covers only the thorax, leaving exposed the head, with its uniquely movable rostrum, stalked eyes, paired antennules, and antennae.

Leptostracans use the thoracopods to produce a feeding current and, in females, to form a brood pouch. That secondary brooding function suggests that egg-bearing females generally do not feed. Locomotion in leptostracans is accomplished by use of the first four pairs of pleopods. Most leptostracans are bottom dwellers, living on or slightly under the substrate, but one species is holopelagic, one inhabits hydrothermal vents, and still another is a marine cave dweller. *See* CRUSTACEA; PHYLLOCARIDA. [P.A.McL.]

Leptostromataceae A family of fungi of the order Sphaeropsidales. Although most species are saprophytes, some are fruit-tree pathogens. Pycnidia, the fruit bodies containing conidia, are black, shield-shaped, circular or oblong, and slightly asymmetrical; the opening of the pycnidium is long and slitlike. There are 80 genera and 375 species recognized. Many of these are conidial stages of Hemisphaeriales. *See* DEUTEROMYCOTINA. [N.F.Bu.]

Lespedeza A warm-season legume with trifoliate leaves, small purple pea-shaped blossoms, and one seed per pod. There are 15 American and more than 100 Asiatic species; two annual species and a perennial from Asia are grown as field crops in the United States. The American species are small shrubby perennials found in open woods and on idle land, rarely in dense stands; they are harmless weeds. *See* LEGUME. [P.T.]

Lethal dose 50 One special form of the effective dose 50, or median effective dose. The lethal dose 50, or median lethal dose (LD_{50}), is used when the response that is being investigated is the death of the experimental animal. The median lethal dose is therefore the dose which is fatal to 50% of the test animals. *See* EFFECTIVE DOSE 50. [C.W.]

Lethal gene A gene which brings about the death of the organism carrying it. Lethal genes constitute the most common class of mutations and are reflections of the fact that the fundamental function of genes is the control of processes essential to the growth and development of organisms. In higher diploid forms lethals are usually recessive and expressed only in homozygotes. Dominant lethals, expressed in heterozygotes, are rapidly eliminated and thus rarely detected. Recessive zygotic lethals are retained with considerable frequency in natural populations of cross-fertilizing organisms, while gametic lethals (those affecting normal functioning of eggs and sperm among animals, or the pollen and ovules of plants) are subject to stringent selection and are accordingly rare. *See* GENE; MOLECULAR BIOLOGY; MUTATION. [D.F.P.]

Lettuce A cool-season annual, *Lactuca sativa*, of Asian origin and belonging to the tribe Cichorium of the Compositae family. Lettuce is grown for its succulent leaves, which are eaten raw as a salad. Four varieties of this leading salad crop are head lettuce (*L. sativa* var. *capitata*), leaf or curled lettuce (*L. sativa* var. *crispa*) cos or romaine lettuce (*L. sativa* var. *longifolia*), and stem or asparagus lettuce (*L. sativa* var. *asparagina*). There are two types of head lettuce: butterhead, and crisphead or iceberg.

California raises more lettuce than any other state; Arizona, Florida, and Texas are next in importance. *See* ASTERALES. [R.G.Gr.]

Leucettida An order of the subclass Calcinea in the class Calcarea. These sponges have either a radiate arrangement of the flagellated chambers or a leuconoid structure. A distinct dermal membrane or cortex is present. The spongocoel is not lined with choanocytes; these cells are restricted to the flagellated chambers. Two families are recognized, the Leucascidae and Leucaltidae. *See* CALCAREA; CALCINEA. [W.D.H.]

Leucine An amino acid considered essential for normal growth of animals. In leucine biosynthesis, carboxyl and some

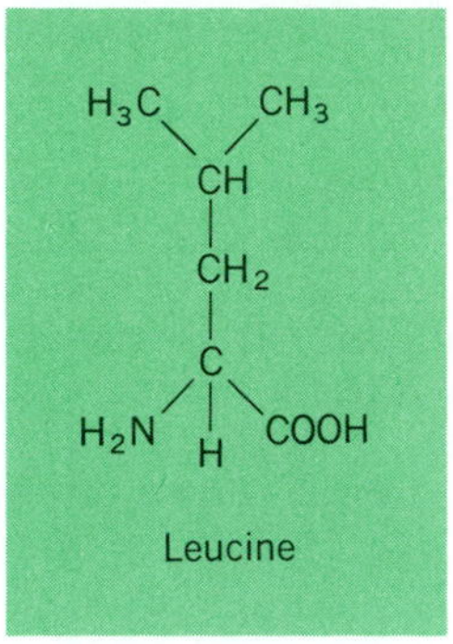

carbon atoms of leucine are derived from acetate; the rest of the molecule is furnished by α-ketoisovalerate, which arises in the biosynthesis of valine. *See* AMINO ACIDS. [E.A.Ad.]

Leucite A framework structure silicate of the feldspathoidal mineral group with the chemical composition $(K,Na)AlSi_2O_6$. Leucite found in the natural environment is usually nearly pure $KAlSi_2O_6$. The crystals are typically white, and the luster varies from dull to vitreous. Mohs hardness lies between 5.5 and 6. Density ranges from 2.45 to 2.5.

Crystals are common in the lavas of Nyiragongo volcano, Zaire. Other well-known areas include: West Kimberley, Australia; Mount Vesuvius, Italy; Highwood Mountains and Bearpaw Mountains, Montana; and the Leucite Hills, Wyoming. *See* SILICATE MINERALS. [W.Lu.]

Leucite rock Igneous rocks rich in leucite but lacking or poor in alkali feldspar. Those types with essential alkali feldspar are classed as phonolites, feldspathoidal syenite, and feldspathoidal monzonite. The group includes an extremely wide assortment both chemically and mineralogically.

The rocks are generally dark-colored and aphanitic types of volcanic origin. They consist principally of pyroxene and leucite and may or may not contain calcic plagioclase or olivine. Leucite rocks are rare. They occur principally as lava flows and small intrusives (dikes and volcanic plugs). *See* IGNEOUS ROCKS; LEUCITE. [C.A.C.]

Leucosoleniida An order of the subclass Calcaronea in the class Calcarea. These sponges have an asconoid structure. A true dermal membrane or cortex does not develop in this order, and the spongocoel is lined with choanocytes. Leucosoleniidae is the one family recognized. *See* CALCAREA; CALCARONEA. [W.D.H.]

Leukemia A disease of humans and animals characterized by generalized neoplastic proliferation of leukocytes, which are usually present in increased numbers in the bloodstream. As a result of uncontrolled proliferation, leukocytes or their precursors progressively replace the blood-forming tissues of the body, crowding out normal cellular elements so that anemia and thrombocytopenia ensue. Leukemic cells also tend to infiltrate certain other organs and tissues of the body, producing symptoms referable to the specific structures involved.

Leukemia includes a spectrum of clinical and cytological entities which may be subdivided according to three main criteria: the type of leukocyte affected; the degree of cytological differentiation; and the number of leukemic cells in the peripheral blood.

A great many factors have been suspected of causing leukemia, but few are supported by a substantial body of evidence. Probably the most firmly established causative factor for human leukemia is ionizing radiation. A number of chemical agents have been suspected of being leukemogenic, but proof is lacking, except for benzol, which seems capable of causing leukemia but only after years of heavy exposure. Studies on the etiology of leukemia have been particularly concentrated on possible viral agents. Also, two types of abnormality of the same pair of chromosomes (chromosome 21) have been found to be associated with leukemia.

The chief symptoms of leukemia are weakness and listlessness, severe anemia, hemorrhages into the skin or mucosal surfaces, and enlargement of spleen, liver, or peripheral lymph nodes. In a patient with symptoms suspiciously like those of leukemia, the usual first step in diagnosis is examination of the peripheral blood. Usually, however, additional diagnostic measures are utilized, the most important of which is examination of the bone marrow. In many cases biopsies of lymph nodes or liver are also obtained, and in some institutions puncture of the spleen in a manner similar to that used for the bone marrow has proven useful.

Leukemia is usually a fatal disease, but remarkable variations in duration of course and severity of symptoms are seen. These variations are to a large extent determined by the type of leukemia; however, within any given category of leukemia there are individual variations.

Several categories of drugs are in common use in the chemotherapy of leukemia, and new agents are being discovered at an impressive rate. Drugs in all the categories are capable of inducing remissions in individual cases. Each agent has a specific spectrum of effectiveness, and no one drug is useful in the treatment of all types of leukemia. These agents may be employed singly or in combination.

Bone marrow transplantation has become an accepted form of therapy for some cases of leukemia. Patients who receive bone marrow transplants are first treated with total-body irradiation and high-dose chemotherapy to destroy the leukemic cells (and also normal bone marrow cells).

Radiant energy in the form of conventional x-rays has a role in the therapy of leukemia; however, this therapy is limited by the fact that radiation of the entire body is highly toxic, and leukemia is a generalized disease. More useful, however, are certain radioactive isotopes which are selectively taken up by certain cell types, including leukemia cells. [J.B.B.]

Leukocidin A substance that destroys leukocytes, the white cells of blood. Leukocidins are extracellular substances which appear in the environment of certain species of growing bacteria. They are released into culture media, and their physical and chemical properties suggest that they are proteins. The most thoroughly studied leukocidins are those produced by staphylococci and streptococci. *See* STAPHYLOCOCCUS; STREPTOCOCCUS. [A.W.B.]

Leukopenia The condition of the blood characterized by a decrease in the number of circulating white cells, or leukocytes. It is not a disease in itself, but rather a symptom that may occur in the course of a great variety of diseases. Of the three kinds of white cells found in normal blood, it is the granulocytes that are largely affected in most forms of leukopenia, and these cells are the most important in resistance to infection. *See* BLOOD; INFECTION.

Leukopenia occurs as an incidental finding in a vast and heterogeneous group of pathologic conditions, especially infections. However, in most of the common bacterial infections the white count is elevated, reflecting the increased production of white cells. But even these infections, if overwhelming, may be associated with leukopenia if the demand for white cells is so great that the resources of the marrow are exhausted.

Leukopenia is regularly found in a group of diseases whose only common factor is enlargement of the spleen, for example, cirrhosis of the liver. Advanced and widespread cancer may produce leukopenia when it invades the bone marrow extensively and effectively crowds out the white-cell-producing tissues.

Leukopenia, sometimes severe and life-threatening, is regularly produced by certain physical and chemical agents when they are given in sufficient dosage. Among these are x-rays, benzene, arsenic, and a group of synthetic compounds used in the treatment of leukemia.

The only treatment that is undoubtedly useful is the administration of antibiotic drugs to control infection. These drugs have been life-saving in some cases, but their effectiveness is hampered by the almost complete lack of the patient's natural mechanisms of defense. [P.B.H.]

Level (communication and acoustics) Logarithm of the ratio of a given quantity to a reference quantity of the same kind. The base of the logarithm, the reference quantity, and the kind of level must be indicated. Frequently used levels, usually expressed in decibels, are power level, voltage level, and sound pressure level. Some other units of level are the neper (Np) and bel (B). *See* DECIBEL; NEPER. [R.W.Y.]

Level (surveying) An instrument for establishing a horizontal line of sight in order to find the difference of elevation between two points upon which a graduated rod is successively held to be sighted. The level usually consists of a telescope and attached spirit level, mounted on a tripod for rotation about a vertical axis (see illustration). The line of sight is fixed by cross hairs. Stadia hairs may be provided for convenience in keeping foresight and backsight distances balanced. The tubular spirit level vial is long, thus quite sensitive, with its upper interior surface being finely ground to a circular arc. A tangent to the center of the ground arc is parallel to the line of sight; thus the line of sight is horizontal (level) when the spirit bubble is centered. The instrument is leveled by manipulation of foot screws in the mounting. *See* SURVEYING. [B.A.B.]

High-precision tilting level. (*Wild Heerbrugg Instruments, Inc.*)

Level measurement The determination of the linear vertical distance between a reference point or datum plane and the surface of a liquid or the top of a pile of divided solids.

Liquid level measurement. Satisfactory measurements are possible only when the liquid is undisturbed by turbulence or wave action. When a liquid is too turbulent for the average level to be read, a baffle or stilling chamber is inserted in the tank or vessel to provide a satisfactory surface.

Stick, hook, and tape gages are used in open vessels where the surface of the liquid can readily be observed. The stick gage is a suitably divided vertical rod, or stick, anchored in the vessel so that the magnitude of the rise and fall of the liquid level may be observed directly. The hook gage provides a needle point, which is adjusted to produce a very tiny pimple in the liquid surface at the level reading, thereby minimizing the meniscus error. The tape gage reads the correct elevation when the point of a bob just touches the liquid surface.

Many forms of gage glass are available for the measurement of liquid level. Liquid in a tank or vessel is connected to the gage glass by a suitable fitting, and when the tank is under pressure the upper end of the glass must be connected to the tank vapor space. Thus the liquid rises to substantially the same height in the glass as in the tank, and this height is measured by suitable scale.

Various types of float mechanism are also used for liquid level measurement. The float, tape, and pulley gage provides an excellent method of measuring large changes in level with accuracy. It has the advantage that the scale can be placed for convenient reading at any point within a reasonable distance of the tank or vessel.

The change in buoyancy of a solid as its immersion in a liquid is varied is used to measure liquid level. This principle is used only when the densities of the liquid and vapor are substantially constant. Temperature changes will produce errors of greater magnitude than with the float mechanisms.

Hydrostatic head may also be used to measure liquid level. The pressure exerted by a column of liquid varies directly with its density as well as with its height, and thus this method of measurement requires that density be substantially constant. Densities of liquids vary with temperature; errors are therefore introduced with temperature changes, or the measuring element must be temperature compensated.

Electrode or probe systems are used in various forms for level indication and control. The number of electrodes and their design depend upon the characteristics of the liquid and the application. Fundamentally, a circuit through a relay coil is closed (or opened) when the liquid contacts a probe.

Capacitance-measuring devices can be used to measure levels of both dielectric (insulating) and conducting liquids. If the liquid being measured is a dielectric, one or two probes or rods, extending nearly to the bottom of the tank, are supported in an insulating mounting. The probes may be bare or covered with insulation. If the liquid is a conductor of electricity, only one probe is necessary but it must be covered with an insulating coating.

Nuclear level gages are used for difficult applications. Basically, all of the units involve a source of gamma (γ) radiation and a detector separated by the vessel or a portion of the vessel in which a liquid level varies. As the level rises, the detector receives less γ-radiation and thus the level is measured.

The sonic level detector is based on the time increment between the emission of a sound wave pulse and its reflection from liquid surface. The sound wave pulse is generated electronically, and its time in transit is measured very accurately by electronic means. If the speed of sound in the liquid or vapor is known accurately, the liquid level is known.

Solids level measurement. Solids level detectors are used to locate the top of a pile of divided solids in large vessels or processing equipment. The instruments are designed for the different solids handled, and the installation must be carefully made to ensure proper measurements. Because solids funnel, cone, and vary in average density with the particle size, shape, distribution, moisture content, and other factors, these detectors provide only an approximate indication of the volume present or the top of the pile. Solids level detectors are classified as continuous or fixed point. Continuous detectors provide a continuous measurement of the level over the range for which they were designed. Their output is an analog representation of the level of the solids. Fixed-point detectors indicate when a specific level has been reached and are used mainly for actuating alarm signals. By installing a number of these, however, at different points, the combined response can be made to approach that of a continuous detector. [H.S.B.]

Lever A pivoted rigid bar used to multiply force or motion, sometimes called the lever and fulcrum (see illustration). The

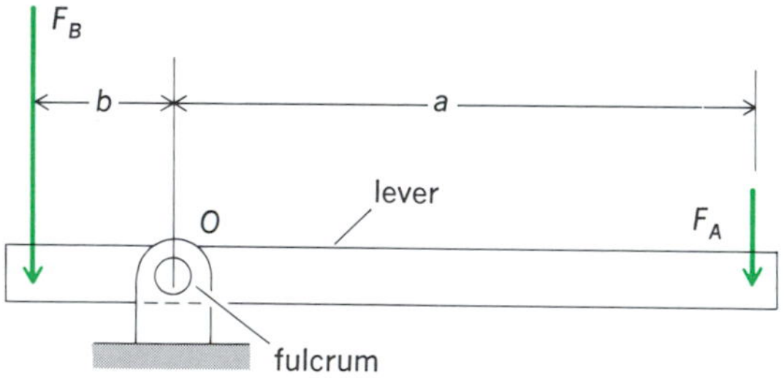

The lever pivots at the fulcrum.

lever uses one of the two conditions for static equilibrium, which is that the summation of moments about any point equals zero. The other condition is that the summation of forces acting in any direction through a point equals zero. *See* INCLINED PLANE.

If moments acting counterclockwise around the fulcrum of a lever are positive, then, for a frictionless lever, $F_B b - F_A a = 0$, which may be rearranged to give Eq. (1). If F_B represents the

$$F_B = \frac{a}{b} F_A \qquad (1)$$

output and F_A represents the input, the mechanical advantage, MA, is then given by Eq. (2).

$$\mathrm{MA} = \frac{F_B}{F_A} = \frac{a}{b} \qquad (2)$$

Applications of the lever range from the simple nutcracker and paper punch to complex multiple-lever systems found in scales and in testing machines used in the study of properties of materials. *See* SIMPLE MACHINE. [R.M.Ph.]

Libra The Balance, in astronomy, appearing during the spring. It is the seventh sign of the zodiac. The constellation consists of faint stars and is not conspicuous. It lies just west of the claws of Scorpius. The principal stars, α, ß, γ, and σ, out-

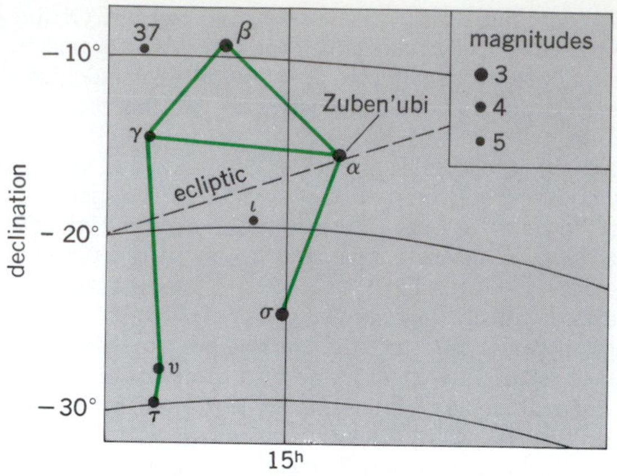

Line pattern of the constellation Libra. The grid lines represent the coordinates of the sky. The apparent brightness, or magnitude, of the stars is shown by the sizes of the dots, graded by appropriate numbers.

line a four-sided figure resembling a balance with beam and pans (see illustration). *See* CONSTELLATION. [C.-S.Y.]

Lichenes A group of organisms consisting of fungi and algae growing together symbiotically. The fungal component is usually an ascomycete or, less commonly, a basidiomycete.

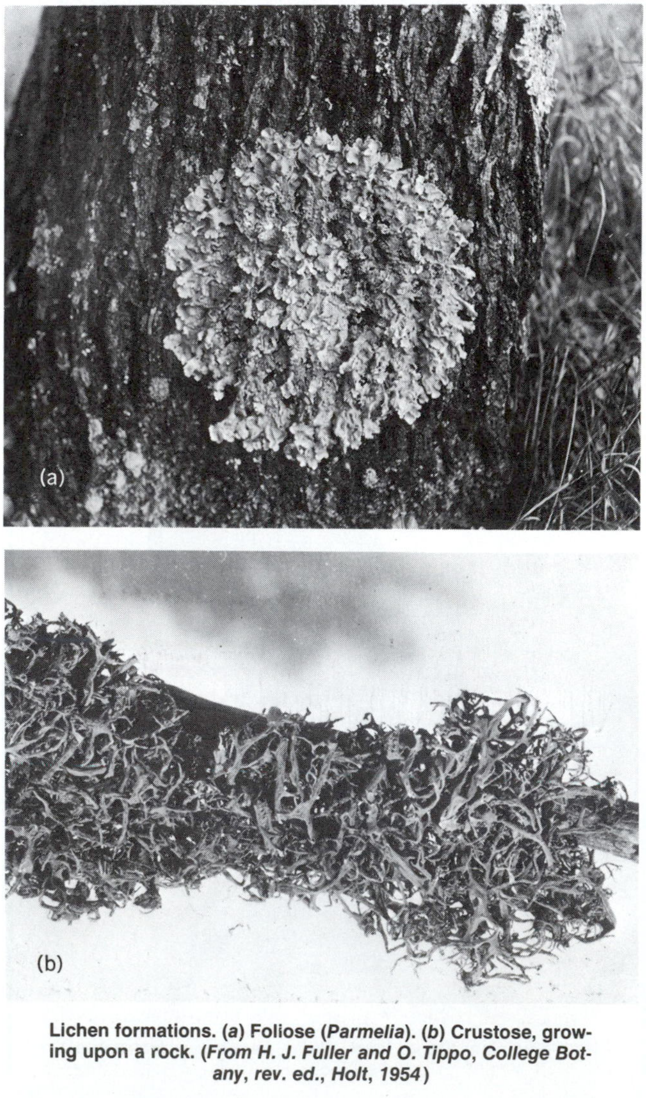

Lichen formations. (*a*) Foliose (*Parmelia*). (*b*) Crustose, growing upon a rock. (*From H. J. Fuller and O. Tippo, College Botany, rev. ed., Holt, 1954*)

The algal component is green or blue-green. When the two components grow together, they form a distinct, long-lived plant body (see illustration) that may be foliose (leaflike), crustose (crustlike), or fruticose (branched, shrubby, or bearded).

Because their dual nature was not discovered unti 1868, lichens were long classified as a separate division of the plant kingdom between the fungi and bryophytes. They are presently considered to be fungi, remarkable in their ability to parasitize yet live symbiotically with algae. Eventually the lichens will be assimilated among the orders and families of the true fungi, but since their relationships are so poorly known, they are still maintained out of necessity as a separate group. Three classes and five orders with nearly 18,000 species are recognized:

Class Ascolichenes
  Order: Pseudosphaeriales
    Hysteriales (Lichenized)
    Pyrenulales
    Lecanorales
    Caliciales
Class Basidiolichenes
Class Lichenes Imperfecti

See separate articles on the classes and orders. [M.E.H.]

Lidar The optical analog of radar. The term lidar is an acronym for light detection and ranging. Lidar systems employ intense pulses of light, typically generated by lasers, and large telescopes and sensitive optical detectors to receive the reflected pulses. They are most commonly used to measure the composition and structure of the atmosphere. The very narrow beamwidth, narrow linewidth, and ultrashort pulses of the laser make it possible to optically probe the atmosphere with exceptional sensitivity and resolution. When used to measure the range and velocity of hard targets, lidars are usually called laser ranging systems or laser radars. *See* LASER; OPTICAL DETECTORS; OPTICAL TELESCOPE; RADAR.

Ranging and altimeter systems. The most common lidar configuration is the monostatic system. The laser beam is either projected through the receiving telescope or propagates parallel to the optical axis of the telescope. If the system is designed for ranging or altimetry, the receiver measures the round-trip propagation time of the laser pulse between the lidar and the target.

The most sophisticated systems are used for ranging to retroreflector-equipped satellites and to the retroreflector arrays placed on the Moon by the Apollo astronauts. Accuracies of a few centimeters are achieved routinely. Data from these measurements are used to monitor geophysical phenomena such as continental drift, crustal dynamics, and the Earth's rotation rate. Airborne laser altimeters provide maps of surface topography, coastal water depth, forest canopies, sea ice distribution, volcanic landforms, impact craters, and ocean wave heights. *See* APPLICATIONS SATELLITES; REMOTE SENSING.

Atmospheric lidars. The targets of atmospheric lidars are either suspended dust and aerosols or gas molecules which are continuously distributed in the atmosphere along the propagation path of the laser beam. Atmospheric lidar systems are used to measure density profiles of the scatterers. Profile measurements are accomplished by pulsing the laser and then periodically sampling the detector output. The sampling process is called range gating.

Atmospheric lidars are classified according to the type of scattering mechanism exploited to make the measurement. Aerosol lidars measure scattering from atmospheric dust and aerosols. Rayleigh lidars are designed to measure the molecular scattered signal, which is proportional to atmospheric density. Resonance fluorescence lidars are used to measure the density profiles of specific molecular species such as sodium in the upper atmosphere. The differential absorption lidar (DIAL) measures species concentrations in the lower atmosphere using two lasers, one tuned to an absorption line of the species of interest and the other tuned just off the absorption line. Raman lidars measure the scattered signal at the Raman shifted wavelength, and have provided excellent measurements of atmospheric density, temperature, and water vapor concentration at altitudes below 10 km (6 mi). Doppler lidars are used to measure tropospheric winds. *See* DOPPLER EFFECT; DOPPLER RADAR; RAMAN EFFECT; SCATTERING OF ELECTROMAGNETIC RADIATION. [C.S.G.]

Lie detector A device intended to detect an involuntary physiological response that all persons exhibit when lying but never when telling the truth. Because there is no such specific lie response, actual "lie detector" tests used in the United States record breathing movements, blood pressure changes, and electrodermal responses on a polygraph while the respondent answers "yes" or "no" to a series of questions. From the recordings, one can determine whether "relevant" questions had greater impact on the respondent than the interpolated "control" questions. *See* ELECTRODERMAL RESPONSE. [D.T.L.]

Lie group A topological group with only countably many connected components whose identity component is open and is an analytic group. An analytic group or connected Lie group is a topological group with the additional structure of a smooth manifold such that multiplication and inversion are smooth. Many groups that arise naturally as groups of symmetries of physical or mathematical systems are Lie groups. The study of Lie groups has applications to analytic function theory, differential equations, differential geometry, Fourier analysis, algebraic number theory, algebraic geometry, quantum mechanics, relativity, and elementary particle theory. *See* GROUP THEORY; SMOOTH MANIFOLD; TOPOLOGY. [A.W.K.]

Ligament A strong, flexible connective tissue band usually found between two bony prominences. Most ligaments are composed of dense fibrous tissue formed by parallel bundles of collagen fibers. They have a shining white appearance and are pliable, strong, and noncompliant. A second kind of ligament, composed either partly or almost entirely of yellow elastic fibers, is extensible or compliant, thereby allowing the connected bones to move apart. *See* CONNECTIVE TISSUE; JOINT (ANATOMY). [W.J.B.]

Ligand A molecule with an affinity to bind to a second atom or molecule. This affinity can be described in terms of noncovalent interactions, such as the type of binding that occurs in enzymes that are specific for certain substrates; or of a mode of binding where an atom or groups of atoms are covalently bound to a central atom, as in the case of coordination complexes and organometallic compounds. Ligands of the latter type can be further distinguished by the nature of the orbitals used in bond formation. *See* ENZYME.

When a protein binds to another molecule, that molecule may be referred to as a ligand. The site where the ligand is bound is known as the binding or active site of the protein. In order for a molecule to be classified as a ligand for a protein, several weak interactions such as hydrophobic, van der Waals, and hydrogen bonding must take place simultaneously. Therefore, the binding of a ligand by a protein is generally quite specific. *See* CHEMICAL BONDING; COORDINATION CHEMISTRY. [T.J.Me.]

Light The term light, as commonly used, refers to the kind of radiant electromagnetic energy that is associated with vision. In a broader sense, light includes the entire range of radiation known as the electromagnetic spectrum. The branch of science dealing with light, its origin and propagation, its effects, and other phenomena associated with it is called optics. Spectroscopy is the branch of optics that pertains to the production and investigation of spectra. See OPTICS; SPECTROSCOPY.

Principal effects. The electromagnetic spectrum is a broad band of radiant energy which extends over a range of wavelengths running from trillionths of inches to hundreds of miles; wavelengths of visible light are measured in hundreds of thousandths of an inch. Arranged in order of increasing wavelength, the radiation making up the electromagnetic spectrum is termed gamma rays, x-rays, ultraviolet rays, visible light, infrared waves, microwaves, radio and TV waves, and very long electromagnetic waves. See ELECTROMAGNETIC RADIATION.

The fact that light travels at a finite speed or velocity is well established, both theoretically and experimentally. Numerous increasingly precise experiments have been conducted, leading to the currently accepted value of 299,792.458 km/s or 186,282.397 mi/s in a vacuum. In round numbers, the speed of light in vacuum or air may be said to be 186,000 mi/s or 300,000 km/s.

One of the most easily observed facts about light is its tendency to travel in straight lines. Careful observation shows, however, that a light ray spreads slightly when passing the edges of an obstacle. This phenomenon is called diffraction. The reflection of light is also well known. Reflection of light from smooth optical surfaces occurs so that the angle of reflection equals the angle of incidence, a fact that is most readily observed with a plane mirror. When light is reflected irregularly and diffusely, the phenomenon is termed scattering. The scattering of light by gas particles in the atmosphere causes the blue color of the sky. See DIFFRACTION; REFLECTION OF ELECTROMAGNETIC RADIATION.

The type of bending of light rays called refraction is caused by the fact that light travels at different speeds in different media—faster, for example, in air than in either glass or water. Refraction occurs when light passes from one medium to another in which it moves at a different speed. Familiar examples include the change in direction of light rays in going through a prism, and the bent appearance of a slick partially immersed in water. See REFRACTION OF WAVES.

In the phenomenon called interference, rays of light emerging from two parallel slits combine on a screen to produce alternating light and dark bands. This effect can be obtained quite easily in the laboratory, and is observed in the colors produced by a thin film of oil on the surface of a pool of water. Polarization of light is usually shown with Polaroid disks. Such disks are quite transparent individually. When two of them are placed together, however, the degree of transparency of the combination depends upon the relative orientation of the disks. It can be varied from ready transmission of light to almost total opacity, simply by rotating one disk with respect to the other. See INTERFERENCE OF WAVES; POLARIZED LIGHT.

When light is absorbed by certain substances, chemical changes take place. This fact forms the basis for the science of photochemistry.

Theory. Phenomena involving light may be classed into three groups: electromagnetic wave phenomena, corpuscular or quantum phenomena, and relativistic effects. The relativistic effects appear to influence similarly the observation of both corpuscular and wave phenomena. See RELATIVITY.

Wave phenomena. Interference and diffraction are the most striking manifestations of the wave character of light. Their fundamental similarity can be demonstrated in a number of experiments. The wave aspect of the entire spectrum of electromagnetic radiation is most convincingly shown by the similarity of diffraction pictures produced on a photographic plate, placed at some distance behind a diffraction grating, by radiations of different frequencies, such as x-rays and visible light. The interference phenomena of light are, moreover, very similar to interference of electronically produced microwaves and radio waves.

Polarization demonstrates the transverse character of light waves. Further proof of the electromagnetic character of light is found in the possibility of inducing, in a transparent body that is being traversed by a beam of plane-polarized light, the property of rotating the plane of polarization of the beam when the body is placed in a magnetic field. See FARADAY EFFECT.

The fact that the velocity of light had been calculated from electric and magnetic parameters (permittivity and permeability) was at the root of Maxwell's conclusion in 1865 that "light, including heat and other radiations if any, is a disturbance in the form of waves propagated...according to electromagnetic laws." Finally, the observation that electrons and neutrons can give rise to diffraction patterns quite similar to those produced by visible light has made it necessary to ascribe a wave character to particles. See ELECTRON DIFFRACTION; NEUTRON DIFFRACTION.

Corpuscular phenomena. In its interactions with matter, light exchanges energy only in discrete amounts, called quanta. This fact is difficult to reconcile with the idea that light energy is spread out in a wave, but is easily visualized in terms of corpuscles, or photons, of light.

The radiation from theoretically perfect heat radiators, called blackbodies, involves the exchange of energy between radiation and matter in an enclosed cavity. The observed frequency distribution of the radiation emitted by the enclosure at a given temperature of the cavity can be correctly described by theory only if one assumes that light of frequency ν is absorbed in integral multiples of a quantum of energy equal to $h\nu$, where h is a fundamental physical constant called Planck's constant.

When a monochromatic beam of electromagnetic radiation illuminates the surface of a solid (or less commonly, a liquid), electrons are ejected from the surface in the phenomenon known as photoemission, or the external photoelectric effect. It is found that the emission of these photoelectrons, as they are called, is immediate, and independent of the intensity of the light beam, even at very low light intensities. This fact excludes the possibility of accumulation of energy from the light beam until an amount corresponding to the kinetic energy of the ejected electron has been reached.

The scattering of x-rays of frequency ν_0 by the lighter elements is caused by the collision of x-ray photons with electrons. Under such circumstances, both a scattered x-ray photon and a scattered electron are observed, and the scattered x-ray has a lower frequency than the impinging x-ray. The kinetic energy of the impinging x-ray, the scattered x-ray, and the scattered electron, as well as their relative directions, are in agreement with calculations involving the conservation of energy and momentum. See COMPTON EFFECT; HEAT RADIATION; PHOTOELECTRICITY; PHOTON.

Quantum theories. The need for reconciling Maxwell's theory of the electromagnetic field, which describes the electromagnetic wave character of light, with the particle nature of photons, which demonstrates the equally important corpuscular character of light, has resulted in the formulation of several theories which go a long way toward giving a satisfactory unified treatment of the wave and the corpuscular picture. These theories incorporate, on one hand, the theory of quantum electrodynamics, first set forth by P. A. M. Dirac, P. Jordan, W. Heisenberg, and W. Pauli, and on the other, the earlier quantum mechanics of L. de Broglie, Heisenberg, and E. Schrödinger. Unresolved theoretical difficulties persist, however, in the higher-than-first approximations of the interactions between light and elementary particles.

Dirac's synthesis of the wave and corpuscular theories of light is based on rewriting Maxwell's equations in a Hamiltonian form resembling the Hamiltonian equations of classical mechanics. Using the same formalism involved in the transformation of classical into wave-mechanical equations by the introduction of the quantum of action $h\nu$, Dirac obtained a new equation of the electromagnetic field. The solutions of this equation require quantized waves, corresponding to photons. The superposition of these solutions represents the electromagnetic field. The quantized waves are subject to Heisenberg's uncertainty principle. The quantized description of radiation cannot be taken literally in terms of either photons or waves, but rather is a description of the probability of occurrence in a given region of a given interaction or observation. *See* HAMILTON'S EQUATIONS OF MOTION; QUANTUM ELECTRODYNAMICS; QUANTUM FIELD THEORY; QUANTUM MECHANICS; RELATIVISTIC QUANTUM THEORY; UNCERTAINTY PRINCIPLE. [G.W.S.]

Light amplifier In the broadest sense, a device which produces an enhanced light output when actuated by incident light. A simple photocell relay–light source combination would satisfy this definition. To make the term more meaningful, common usage has introduced two restrictions: (1) a light amplifier must be a device which, when actuated by a light image, reproduces a similar image of enhanced brightness; and (2) the device must be capable of operating at very low light levels without introducing spurious brightness variations (noise) into the reproduced image. The term is used synonymously with image intensifier. The light amplifier increases the brightness of an image which is below the visual threshold to a level where it can be readily seen with the unaided eye. It is, of course, impossible to see under conditions of complete darkness. Indeed, there is a fundamental lower limit of illumination under which an image of a given quality can be recognized. This limitation arises because of the corpuscular nature of light. *See* LIGHT; PHOTON.

In addition to their application for night vision, light amplifiers have been useful in many fields of science, such as astronomy, nuclear physics, and microbiology. [D.R.Co.]

Light curves Graphical descriptions of the light changes of variable stars. If the light changes occur in a regular manner, that is, if they are periodic or nearly so, a plot can be made of the brightness versus the phase. The phase is the time interval between an observation and the instant when the star was at some easily determined portion of the light curve. For eclipsing variables, this "epoch" will be at the time of greatest light loss when the star is faintest; for intrinsic variables the phase is usually reckoned from the time of maximum light when the star is brightest. *See* VARIABLE STAR. [F.B.W.]

Light-emitting diode A rectifying semiconductor device which converts electric energy into electromagnetic radiation. The wavelength of the emitted radiation ranges from the green to the near infrared, that is, from about 550 to over 1300 nanometers. Most commercial light-emitting diodes (LEDs) both visible and infrared, are fabricated from group III–V compounds. The most commonly used light-emitting diode is the red light-emitting diode, made of gallium arsenide-phosphide on gallium arsenide substrates.

Visible light-emitting diodes are used as solid-state indicator lights and as light sources for numeric and alphanumeric displays. Infrared light-emitting diodes are used in optoisolators and in optical fiber transmission in order to obtain the highest possible efficiency. *See* OPTICAL COMMUNICATIONS; OPTICAL FIBERS; OPTICAL ISOLATOR.

For discussion of the properties of *pn* junctions and light generation in solid-state devices *see* ELECTROLUMINESCENCE; JUNCTION DIODE; JUNCTION TRANSISTOR; LASER. [A.A.B.]

Light panel A surface-area light source that employs the principle of electroluminescence to produce light. Light panels are composed of two sheets of electrically conductive material, one a thin conducting backing and the other a transparent conductive film, placed on opposite sides of a plastic or ceramic sheet impregnated with a phosphor, such as zinc sulfide, and small amounts of compounds of copper or manganese. When an alternating voltage is applied to the conductive sheets, an electric field is applied to the phosphor. Each time the electric field changes, it dislodges electrons from the edges of the phosphor crystals. As these electrons fall back to their normal atomic state, they affect the atoms of the slight "impurities" of copper or manganese, and radiation of the wavelength of light is emitted. *See* ELECTROLUMINESCENCE.

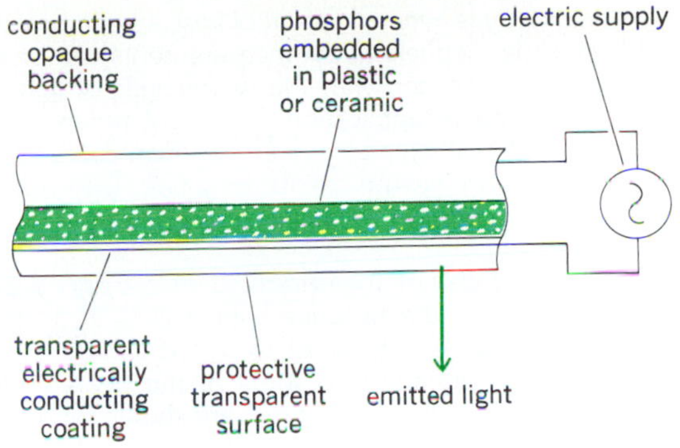

Simplified diagram of an electroluminescent cell.

In contrast to incandescent, vapor-discharge, and fluorescent lamps, which are essentially point or line sources of light, the electroluminescent light panel is essentially a surface source of light. Complete freedom of size and shape is a fascinating aspect of luminescent cells (see illustration). *See* ILLUMINATION. [W.B.Bo.]

Light-scattering photometry Optical methods used to measure the extent of scattering of light by particles suspended in fluids or by macromolecules in solution. Two different approaches are employed. The photometric measurement of the extent of attenuation of an incident light beam as it passes through the scattering medium is known as turbidimetry. The measurement of the intensity of light scattered to a detector which is not in the path of the incident light (often at right angles to it) is known as nephelometry—literally, the measurement of cloudiness.

Nephelometry and turbidimetry are also used for quantitative analytical chemical measurements. These methods were formerly considered relatively nonprecise and were used only to obtain approximate concentration information. With the advent of microprocessor-based instrumentation, however, it has been possible to overcome such prior limitations as nonlinearity of response with concentration, and these methods have become increasingly popular, particularly in the field of clinical chemistry. *See* SCATTERING OF ELECTROMAGNETIC RADIATION.

Turbidimetric analysis involves measurement of the intensity of light that is transmitted through a solution or suspension. For strongly turbid samples containing many particles or particles that are large compared with the wavelength of visible light, turbidimetry is the method of choice, and it is most often performed on a colorimeter or spectrophotometer. For a limited range of particle concentrations in suspensions, this method of measurement gives fairly good precision. *See* COLORIMETRY; SPECTROSCOPY.

The relative ease of discriminating the presence of scattered light from the dark background present in the absence of scattering makes nephelometric measurement an extremely sensitive tool for analytical purposes. For a relatively clear solution, where the light is only weakly scattered because of a low concentration of particles or the presence of particles which are extremely small compared with the wavelength of the incident light, nephelometry is generally the tool best suited to measurement. In principle, nephelometric measurements can be made at the detection angle and with the wavelength of light most suitable for a particular application. The most notable analytical application of nephelometry is in the quantitative analysis of specific human serum proteins. *See* IMMUNOASSAY; PHOTOMETRY. [J.C.St.; A.F.-T.C.]

Light-year A unit of measurement of astronomical distance. A light-year is the distance light travels in 1 sidereal year. One light-year is equivalent to 9.461×10^{12} km, or 5.879×10^{12} mi. Distances to some of the nearer celestial objects, measured in units of light time, are shown in the table.

Distances from the Earth to some celestial objects

Object	Distance from Earth (in light time)
Moon (mean)	1.3 s
Sun (mean)	8.3 min
Mars (closest)	3.1 min
Jupiter (closest)	33 min
Pluto (closest)	5.3 h
Nearest star (Proxima Centauri)	4.3 years
Andromeda Nebula (M31)	2,300,000 years

This unit, while useful for its graphic presentation of the enormous scale of stellar distances, is seldom used technically except in cosmology. *See* ASTRONOMICAL UNIT; PARALLAX (ASTRONOMY); PARSEC. [J.L.Gr.]

Lighthouse A distinctive structure, built on or near a shore, which exhibits a light of distinctive characteristics to serve as an aid to navigation. Lesser lights may be displayed from fixed structures called beacons or from floating buoys or lightships.

The characteristics of the lights displayed by lighthouses are given in light lists available to mariners and, in abbreviated form, on charts. Some lights have one or more sectors in which the light appears red, usually to warn of some danger in this sector. In other sectors most lights are white. *See* PILOTING. [A.B.M.]

Lightning The large spark produced by an abrupt discontinuous discharge of electricity through the air generally under turbulent conditions of the atmosphere.

Cumulonimbus clouds, which produce lightning, contain strongly upward-moving air and frequently extend to altitudes higher than 40,000 feet (12 km). Within these clouds, various factors act to cause the creation and separation of electric charge. The upper parts of thunderstorms are positively charged, whereas the central and lower parts are predominantly negative.

When cloud electrification has proceeded to the point where the potential gradient between adjacent charge centers, or between the cloud and ground, reaches the breakdown value, the large spark called lightning occurs. A breakdown potential of about 10,000 volts per centimeter has been suggested as an appropriate value for air containing water droplets. *See* CLOUD PHYSICS; GEOELECTRICITY; LIGHTNING AND SURGE PROTECTION; THUNDERSTORM. [L.J.B.]

Lightning and surge protection Means of protecting electrical systems, buildings, and other property from lightning and other high-voltage surges. From studies of lightning, two conclusions emerge: (1) Lightning will not strike an object if it is placed in a grounded metal cage. (2) Lightning tends to strike, in general, the highest objects on the horizon. *See* ATMOSPHERIC ELECTRICITY; LIGHTNING.

Lightning rod cone of protection. (*a*) Configuration of rods on a house. (*b*) Geometry of the principle.

One practical approximation of the grounded metal cage is the well-known lightning rod or mast. The effectiveness of this device is evaluated on the cone-of-protection principle. The protected area is the space enclosed by a cone having the mast top as the apex of the cone and tapering out to the base. If the radius of the base of the cone is equal to the height X of the mast, equipment inside this cone will rarely be struck. A radius equal to twice the height of the mast ($2X$) gives a cone of shielding within which an object will be struck occasionally. The cone-of-protection principle is shown in the illustration.

The probability that an object will be struck by lightning is considerably less if it is located in a valley. Therefore, electric transmission lines which must cross mountain ranges often will be routed through the gaps to avoid the direct exposure of the ridges.

There are a number of protective devices to limit or prevent lightning damage to electric power systems and equipment. The word protective is used to connote either one of two functions: the prevention of trouble, or its elimination after it occurs. Various protective means have been devised either to prevent lightning from entering the system or to dissipate it harmlessly if it does. Overhead ground wires and lightning rods are used to prevent lightning from striking the electrical system. Lightning arresters are protective devices for reducing the transient system overvoltages to levels compatible with the terminal-apparatus insulation. *See* SURGE SUPPRESSOR.

Immediate reclosure is a practice for restoring service after the trouble occurs by immediately reclosing automatically the line power circuit breakers after they have been tripped by a short circuit. The protective devices involved are the power circuit breaker and the fault-detecting and reclosing relays. *See* CIRCUIT BREAKER; ELECTRIC PROTECTIVE DEVICES. [G.D.B.]

Lignin A polymer found extensively in the cell walls of all woody plants, Lignin, one of the most abundant natural polymers, constitutes one-fourth to one-third of the total dry weight of trees. It combines with hemicellulose materials to help bind the cells together and direct water flow. *See* CELL WALLS (PLANT); HEMICELLULOSE; POLYMER.

Several methods have been devised for isolating lignin from wood. Some isolation methods are based on acid treatments in which the carbohydrate components (cellulose and hemicelluloses) are hydrolyzed to water-soluble materials. However, with such procedures, serious doubts exist as to whether the isolated lignin is representative of the "native" lignin. Enzymatic digestion of the carbohydrate in wood meal is a lengthy, tedious procedure but offers the greatest promise of leaving lignin unaltered during isolation.

Structural studies on lignin have been hampered by the random, cross-linked nature of the polymer. The relative proportions of the monomers which make up lignin (I and II) vary with

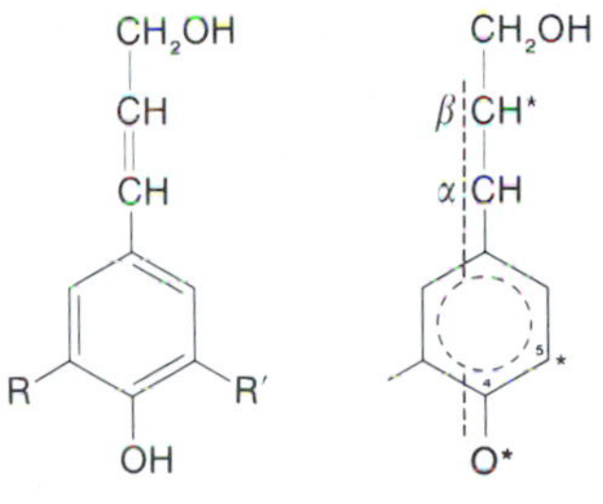

(I) (II)

the plant species. Lignin is formed in the plant by an enzymatic dehydrogenation of the monomers. The * designations shown on the structure (II) indicate the principal sites for coupling monomers.

In general, the markets for lignin products are not large or attractive enough to compensate for the cost of isolation and the energy derived from its burning. An exception is lignosulfonate, which is obtained during paper production either directly from sulfite pulping liquors or by sulfonation of acid-precipitated kraft lignin; its markets include a dispersant in carbon black slurries, clay products, dyes, cement, oil drilling muds, an asphalt emulsifier, a binder for animal feed pellets, a conditioner for boiler water or cooling water, and an additive to lead-acid storage battery plate expanders. *See* PAPER; WOOD CHEMICALS. [D.R.D.]

Lignite Soft and porous carbonaceous material intermediate between peat and subbituminous coal, with a heat value less than 8300 Btu/lb (19,300 kilojoules/kg) on a moist, mineral-matter-free basis. In North America, lignitic coals are classified as lignite A or lignite B, depending upon whether they have calorific value of more or less than 6300 Btu/lb (14,650 kilojoules/kg). Elsewhere, lignites are commonly called brown coals, with two major classes being recognized, namely, hard and soft brown coals.

The main uses of lignite have been for generation of electrical energy and for domestic and industrial heating. If shipped any great distance, lignite commonly requires drying. *See* COAL; PEAT. [J.A.Si.]

Lignumvitae A tree, *Guaiacum sanctum*, also known as holywood lignumvitae, which is cultivated to some extent in southern California and tropical Florida. Lignumvitae is native in the Florida Keys, Bahamas, West Indies, and Central and South America. It is an evergreen tree of medium size with abruptly pinnate leaves. The tree yields a resin or gum known as gum guaiac or resin of guaiac which is used in medicine. The very heavy black heartwood is used in bowling balls, blocks and pulleys, and parts of instruments. *See* SAPINDALES. [A.H.G./K.P.D.]

Liliales An order of flowering plants, division Magnoliophyta (Angiospermae), from whose name the names of the subclass Liliidae and class Liliopsida (monocotyledons) are derived. The order consists of some 13 families and nearly 7700 species, more than half of which belong to the family Liliaceae (4200 species). The Iridaceae, Dioscoreaceae, Agavaceae, and Smilacaceae are other good-sized families of the order. The ovary is either superior or inferior, the seeds are of ordinary number and structure, and most families and genera have septal nectaries in the flowers. *See* ASPARAGUS; GARLIC; LILIIDAE; LILIOPSIDA; ONION; SISAL. [A.Cr.]

Liliidae A subclass of the class Liliopsida (monocotyledons) of the division Magnoliophyta (Angiospermae), the flowering plants, consisting of 2 orders (Liliales and Orchidales), 17 families, and nearly 28,000 species. The Liliidae are syncarpous monocotyledons with both the sepals and the petals usually petaloid. The flowers generally have well-developed nectaries, and pollination is usually by insects or other animals. *See* FLOWER; LILIALES; LILIOPSIDA; ORCHIDALES. [A.Cr.]

Liliopsida One of the two classes which collectively make up the division Magnoliophyta (Angiospermae), the flowering plants. The Liliopsida, often known as Monocotyledoneae, or monocotyledons, embrace 4 subclasses (Alismatidae, Commelinidae, Arecidae, and Liliidae), 18 orders, 61 families, and about 55,000 species. *See* ALISMATIDAE; ARECIDAE; COMMELINIDAE; LILIIDAE.

All of the characters which collectively distinguish the Liliopsida from the Magnoliopsida (dicotyledons) are subject to exception, but most of the Liliopsida have parallel-veined leaves, and when the embryo is differentiated into recognizable parts, there is only a single cotyledon. The vascular bundles are generally scattered or borne in two or more rings, so that the stems and roots do not have a well-defined pith and cortex. Monocotyledons never have an intrafascicular cambium, and most of them have no secondary growth at all. The mature root system of monocots is wholly adventitious. The floral parts of monocots, when of definite number, are most often borne in sets of 3, seldom 4, never 5. The pollen is uniaperturate or of uniaperturate-derived type. *See* CORTEX (PLANT); FLOWER; LEAF; MAGNOLIOPSIDA; PITH; STEM. [A.Cr.]

Limburgite A dark, glass-rich igneous rock with abundant large crystals (phenocrysts) of olivine and pyroxene and with little or no feldspar. The glass is brown and usually alkali-rich. Phenocrysts are well formed. Limburgite is a rare rock and forms lava flows and small intrusive bodies (dikes, sills, and plugs). It is associated with basanite and alkali-basaltic rocks. *See* BASALT; IGNEOUS ROCKS. [C.A.C.]

Lime (botany) An acid citrus fruit, *Citrus aurantifolia*, usually grown in tropical or subtropical regions because of its low resistance to cold. The two principal groups of limes are the West Indian or Mexican and the Tahiti or Bearss. The fruit of West Indian lime is very small (walnut size) and strongly acid, and drops when fully colored. The Tahiti lime is seedless and its aroma is less pronounced.

Except in the United States, the commercial lime industry is restricted to the West Indian group. The major producing areas are India, Mexico, Egypt, and the West Indies. Commercial production of the Tahiti lime is largely confined to the United States. It is grown mainly in Florida, with some plantings in the warmer areas of southern California. *See* FRUIT; FRUIT, TREE. [R.K.So.]

Lime (industry) A general term for burned (or calcined) limestone, also known as quicklime, hydrated lime, and unslaked or slaked lime. Its predominant usage (90%) is as a basic industrial chemical. It still enjoys its traditional building uses. In order of decreasing size uses are: steel fluxing, water treatment, nonferrous metals (alumina, magnesium, copper, and others), pulp and paper, refractories, soil stabilization, sewage and trade waste treatment, chemicals, and glass manufacture. *See* LIMESTONE.

Lime is not a mineral; it is manufactured from a mineral—limestone, coral, oystershell, all being sources of calcium carbonate. Dolomite, a calcium-magnesium carbonate, is used to produce dolomitic (magnesium) lime. Only the purest types of stone or shell are used for lime. [R.S.B.]

Limestone A sedimentary rock composed dominantly of carbonate minerals, principally carbonates of calcium and magnesium. The chief minerals of limestones are calcite and aragonite, and in the dolomitic limestones, dolomite. Limestones are the most abundant of the nonclastic rocks. They are overwhelmingly the largest reservoir of the element carbon at or near the surface of the Earth. Much knowledge of invertebrate paleontology and consequently of the evolution of life and earth history comes from the fossils contained in them. Although the word limestone is used in the general sense above, specifically it refers to carbonate rocks dominated by the mineral calcite, $CaCO_3$. *See* ARAGONITE; CALCITE; LIME (INDUSTRY).

The textures of limestones reveal much about their origin. The chief textural elements in a great many limestones are fossils, whole or fragmental. The species represented and their state of preservation are a guide to the ecology and environment of the site of deposition. In addition, the nature of the fossil debris, whether complete articulated shells or fragments in all stages of disarticulation, fragmentation, and comminution, is a guide to the presence of bottom currents.

Some mechanically deposited limestones show well sorted particles of calcite cemented by clear intergranular calcite. If the particles are sand-size, the rock is termed calcarenite. Secondary textural features that are found in limestones include stylolites, cross-cutting veinlets filled with calcite, and replacement effects (typically that of rhombohedrons of dolomite replacing calcite).

Limestones may be referred to two types of origin, autochthonous and allochthonous. Autochthonous limestones have been formed in place by biogenic precipitation from the water of the environment, usually sea water; allochthonous limestones have been transported from the site of original precipitation by current action. *See* SEDIMENTARY ROCKS. [R.Si.]

Limiter circuit An electronic circuit used to prevent the amplitude of an electrical waveform from exceeding a specified level while preserving the shape of the waveform at amplitudes less than the specified level. The limiting action takes place by effectively shunting a normal load resistance with a much lower resistance at and above the specified limiting level.

Limiting action is usually accomplished by use of the highly nonlinear or "switching" voltage-current characteristic of an electronic device. A semiconductor diode used as a clamp is often the essential element in a limiter circuit. The action of such a diode may be understood with reference to the voltage-current characteristic of a typical semiconductor diode shown in the illustration. *See* CLAMPING CIRCUIT; SEMICONDUCTOR DIODE.

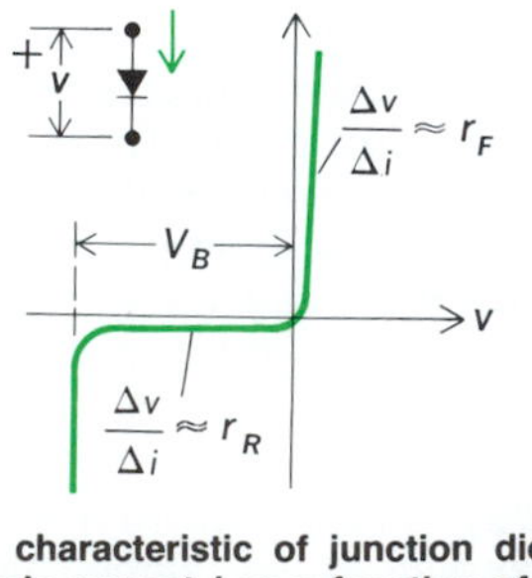

Voltage-current characteristic of junction diode shows instantaneous diode current *i* as a function of instantaneous voltage *v*. r_F = forward resistance; r_R = reverse-biased resistance; V_B = breakdown voltage; Δ denotes change.

Bipolar or field-effect transistors can also be used as limiters, in either the input or output circuits. Saturation characteristics of the output circuit of a field-effect transistor or bipolar transistor can also produce limiting action.

The term limiter is often defined in sufficiently broad terms to include the related operation of clipping, which also results in deletion of a portion of a waveform. Some authors even use the terms interchangeably. Limiters and clippers are often used together, rather than bidirectional limiters alone, where both negative and positive peaks of a waveform are to be removed. *See* CLIPPING CIRCUIT. [G.M.G.]

Limits and fits The extreme permissible values of a dimension are known as limits. The degree of tightness or looseness between two mating parts that are intended to act together is known as the fit of the parts. The character of the fit depends upon the use of the parts. Thus, the fit between members that move or rotate relative to each other, such as a shaft rotating in a bearing, is considerably different from the fit that is designed to prevent any relative motion between two parts, such as a wheel attached to an axle.

In selecting and specifying limits and fits for various applications, the interests of interchangeable manufacturing require that (1) standard definitions of terms relating to limits and fits be used; (2) preferred basic sizes be selected wherever possible to be reduce material and tool costs; (3) limits be based upon a series of preferred tolerances and allowances; and (4) a uniform system of applying tolerances (bilateral or unilateral) be used. *See* DESIGN STANDARDS; ENGINEERING DESIGN; MACHINE. [J.El.]

Limnology The study of lakes, ponds, rivers, streams, swamps, and reservoirs that make up inland water systems.

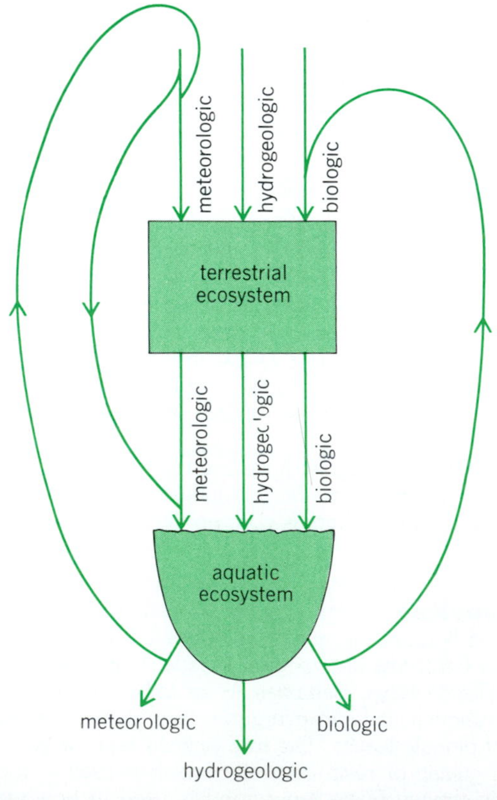

Diagrammatic model of the functional linkages between terrestrial and aquatic ecosystems. Vectors may be meteorologic, hydrogeologic, or biologic components moving nutrients or energy along the pathway shown.

Each of these inland aquatic environments is physically and chemically connected with its surroundings by meteorologic and hydrogeologic processes (see illustration).

Aquatic systems with excellent physical conditions for production of organisms and high nutrient levels may show signs of eutrophication. Eutrophic lakes are generally identified by large numbers of phytoplankton and aquatic macrophytes and by low oxygen concentrations in the profundal zone. *See* ECOLOGY; EUTROPHICATION; FRESH-WATER ECOSYSTEM; HYDROLOGY; LAKE; RIVER. [J.E.S.]

Limonite A field or generic term for natural hydrous iron oxides, the most common phase being the mineral goethite, α-FeO(OH). Limonite includes the so-called bog iron ores. It is the characteristic brown stain which coats rocks containing sulfide ores, such as pyrite and pyrrhotite, in the zone of weathering of these ores referred to as a gossan. It is formed by biogenic or inorganic precipitation in bog, spring, lacustrine, or marine deposits. [P.B.M.]

Limpet Name given to a variety of species of aquatic gastropod mollusks, all with a characteristic conical shell (see illustration) and a suckerlike foot. Limpets (like chitons) are well adapted for life on rocky surfaces exposed to wave action and, in the higher levels of the littoral zone, to alternating tidal submergence and aerial drying. All limpets move relatively slowly over rock or other hard surfaces, protecting themselves (against wave action or desiccation or predation) by clamping down the shell opening against the substrate by contraction of the enlarged shell muscles. There is never an operculum on the foot, and limpets are defenseless once detached.

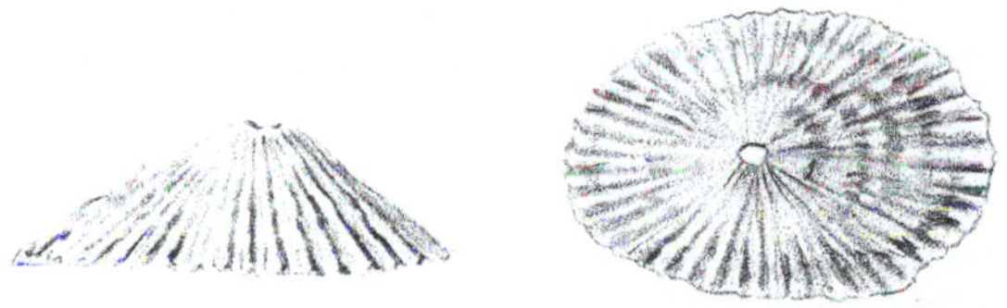

***Diodora aspera*, the rough-keyhole limpet of the Pacific coast of North America, viewed from the side and top.**

Limpets usually graze slowly and continuously by radular scraping of attached algae and diatoms from rock surfaces. The structural and functional adaptations of the limpet form have arisen in many distinct groups of gastropods. *See* GASTROPODA; MOLLUSCA. [W.D.R.-H.]

Linales An order of flowering plants, division Magnoliophyta (Angiospermae), in the subclass Rosidae of the class Magnoliopsida (dicotyledons). The order consists of three families: the Linaceae, with about 450 species; the Erythroxylaceae, with about 200 species; and the Humiriaceae, with about 50 species. They are simple-leaved herbs or woody plants, with hypogynous, regular, syncarpous flowers that have five to many stamens. *Linum usitatissimum* (the source of flax fibers and linseed oil) and *Erythroxylon coca* (the source of cocaine) are well-known members of the Linales. *See* COCA; FLAX; MAGNOLIOPSIDA; ROSIDAE. [A.Cr.]

Line In axiomatic geometry, a line frequently is taken as a primitive (undefined) concept: an element in line geometry; a class of points, in a geometry with point as an element, whose essential properties are stated as axioms. In analytic geometry, a line is identified with (or is the graph of) a linear equation $Ax + By + C = 0$, with $A, B \neq 0, 0$, in cartesian coordinates (x,y). As a derived concept, line is defined in terms of primitive notions such as distance or betweenness. *See* ANALYTIC GEOMETRY; EUCLIDEAN GEOMETRY; PLANE; POINT. [L.M.Bl.]

Line integral The line integral of a vector function **F** of position over a path C is represented by Eq. (1), where F_x, F_y,

$$\int \mathbf{F} \cdot d\mathbf{r} = \int_C F_x(x,y,z)\,dx + \int_C F_y(x,y,z)\,dy + \int_C F_z(x,y,z)\,dz \qquad (1)$$

F_z are the scalar components of **F** along the coordinate axes. The path C is supposed to be a curve, smooth at least in part, defined parametrically by equations of form (2) for each

$$x = x(p) \quad y = y(p) \quad z = z(p) \qquad (2)$$

smooth portion. The functions $F_x(x,y,z)$, etc., must be defined at all points of C.

When C is a closed curve, the line integral is called a circuit integral, and is written as notation (3).

$$\oint \mathbf{F} \cdot d\mathbf{r} \qquad (3)$$

See INTEGRATION. [McA.H.H.]

Line spectrum A discontinuous spectrum characteristic of excited atoms, ions, and certain molecules in the gaseous phase at low pressures. If an electric arc or spark between metallic electrodes, or an electric discharge through a low-pressure gas, is viewed through a spectroscope, images of the spectroscope slit are seen in the characteristic colors emitted by the atoms or ions present. *See* ATOMIC STRUCTURE AND SPECTRA; SPECTROSCOPY. [G.R.H.]

Linear algebra That branch of mathematics which deals with solutions of systems of linear equations and the related geometric notions of vector spaces and linear transformations. It is fundamental in the theory of the calculus of functions of several variables and hence is of great importance in the application of mathematics to physical and biological sciences, economics, and so on.

The word linear is derived from the fact that the equation of a line in two-dimensional analytic geometry has the form shown in Eq. (1) and a system of linear equations has the

$$ax + by = c \qquad (1)$$

corresponding form shown in Eq. (2), where $i = 1, 2, \ldots,$

$$a_{i1}x_1 + a_{i2}x_2 + \cdots + a_{in}x_n = b_i \qquad (2)$$

m. The a_{ij} and b_i are fixed quantities belonging to a specified field, for example, the field of real numbers, and solutions $(x_1,x_2,\ldots,x_n)$ are sought in the same field. [N.J.]

Linear programming A mathematical subject whose central theme is finding the point where a linear function defined on a convex polyhedron assumes its maximum or minimum value. The largest class of applications of linear programming occurs in business planning and industrial engineering. Its basic concepts are so fundamental, however, that linear programming has been used in almost all parts of science and social science where mathematics has made any penetration. Furthermore, even in cases where the domain is not a polyhedron, or the function to be maximized or minimized is not linear, the methods used are frequently adaptions of linear programming. *See* NONLINEAR PROGRAMMING; OPERATIONS RESEARCH.

A typical linear programming problem is to maximize expression (1), where $x_1,\ldots, x_n$ satisfy the conditions shown

$$c_1x_1 + \cdots + c_nx_n \qquad (1)$$

by notation (2). The a_{ij}, c_j, and b_i are constants; $x_1,\ldots, x_n$ are variables.

$$x_j \geqq 0$$
$$a_{11}x_1 + \cdots + a_{1n}x_n \leqq b_1$$
$$\cdots\cdots\cdots\cdots \quad (2)$$
$$a_{m1}x_1 + \cdots + a_{mn}x_n \leqq b_m$$

Sometimes the problem may be to minimize rather than to maximize; sometimes some of the variables may not be required to be nonnegative; sometimes some of the inequalities $a_{i1}x_1 + \cdots + a_{in}x_n \leqq b_i$ be reversed, or be equalities.

The linear inequalities, which the variables satisfy, correspond algebraically to the fact that the variable point $x = (x_1,\ldots,x_n)$ lies in a convex polyhedron. Hence, the principal mathematical bases for linear programming are the theory of linear equalities, a part of algebra, and the theory of convex polyhedra, a part of geometry.

If the function $c_1x_1 + \cdots + c_nx_n$ does not get arbitrarily large for points $x = (x_1,\ldots,x_n)$ on the convex polyhedron, then a maximum is attained at a vertex of the polyhedron.

The most popular method for solving linear programs, that is, the finding of the $x = (x_1,\ldots,x_n)$ which maximizes $c_1x_1 + \cdots + c_nx_n$, is the simplex method, developed by Dantzig in 1947. The method is geometrically a process of moving from vertex to neighboring vertex on the convex polyhedron, each move attaining a higher value of $c_1x_1 + \cdots + c_nx_n$, until the vertex yielding the greatest value is reached. Algebraically, the calculations are similar to elimination processes for solving systems of algebraic equations.

The popularity of the simplex method rests on the empirical fact that the number of moves from vertex to vertex is a small multiple (about 2–4) of the number of inequalities in most of the thousands of problems handled, although it has never been proved mathematically that this behavior is characteristic of "average" problems. [A.J.Ho.]

Linear system analysis The study of a system by means of a model consisting of a linear mapping between the system inputs (causes or excitations), applied at the input terminals, and the system outputs (effects or responses), measured or observed at the output terminals. A system is a set or an arrangement of things or devices so related or connected as to form a unity or organic whole. For example, the inputs or controlling factors of the national economy (the system) are government spending, interest rates, tax levels, and monetary policy. The outputs are the gross national product, unemployment rate, inflation rate, and the consumer price index. Similar correspondences can be established for most systems, ranging from the more common variety of physical systems to the less-well-defined socioeconomic variety whose structure and control represent some of the major problem areas of society. *See* SYSTEMS ANALYSIS; SYSTEMS ENGINEERING. [C.T.C.; P.E.B.]

Linear systems of equations Systems of mathematical equations of the form of system (1), where the a_{ij}, $i = 1, 2,\ldots, m$, $j = 1, 2,\ldots, n$, and the b_i, $i = 1, 2,\ldots, m$, are constants, or fixed numbers, and the x_i, $i = 1, 2,\ldots, n$ are called unknowns.

$$a_{11}x_1 + a_{12}x_2 + \cdots + a_{1n}x_n = b_1$$
$$a_{21}x_1 + a_{22}x_2 + \cdots + a_{2n}x_n = b_2$$
$$\cdots\cdots\cdots\cdots \quad (1)$$
$$a_{m1}x_1 + a_{m2}x_2 + \cdots + a_{mn}x_n = b_m$$

System (1) is referred to as m linear equations in n unknowns. A solution of system (1) is a set of numbers $c_1, c_2,\ldots, c_n$ such that when x_1 is replaced by c_1, x_2 by $c_2,\ldots, x_n$ by c_n, every equation of system (1) becomes a true equality. The problem posed by such a system of equations is to find criteria for the existence of a solution and, when solutions exist, to obtain systematic methods for finding the solutions.

Two linear systems of equations are equivalent if every solution of one of the systems is a solution of the other and vice versa.

If $m = n$ in system (1) and if the determinant $|A|$ of the matrix of the coefficients A, given by matrix (2), is not zero,

$$A = \begin{pmatrix} a_{11} & a_{12} & \cdots & a_{1n} \\ a_{12} & a_{22} & \cdots & a_{2n} \\ \cdots & \cdots & \cdots & \cdots \\ a_{m1} & a_{m2} & \cdots & a_{mn} \end{pmatrix} \quad (2)$$

then system (1) has a unique solution which can be obtained by Cramer's rule, which gives the values of the unknowns $x_1, x_2,\ldots, x_n$ as the ratio of two determinants. *See* DETERMINANT. Specifically, Cramer's rule gives

$$x_1 = \begin{vmatrix} b_1 & a_{12} & \cdots & a_{1n} \\ b_2 & a_{22} & \cdots & a_{2n} \\ \cdots & \cdots & \cdots & \cdots \\ b_n & a_{n2} & \cdots & a_{nn} \end{vmatrix} \Big/ |A|$$

$$x_2 = \begin{vmatrix} a_{11} & b_1 & a_{13} & \cdots & a_{1n} \\ a_{21} & b_2 & a_{23} & \cdots & a_{2n} \\ \cdots & \cdots & \cdots & \cdots & \cdots \\ a_{n1} & b_n & a_{n3} & \cdots & a_{nn} \end{vmatrix} \Big/ |A|$$

$$\cdots\cdots\cdots\cdots\cdots\cdots$$

$$x_n = \begin{vmatrix} a_{11} & \cdots & a_{1n-1} & b_1 \\ a_{21} & \cdots & a_{2n-1} & b_2 \\ \cdots & \cdots & \cdots & \cdots \\ a_{n1} & \cdots & a_{nn-1} & b_n \end{vmatrix} \Big/ |A|$$

In every case the determinant in the denominator is $|A|$, and for the value of the unknown x_i, the determinant in the numerator is the determinant of the matrix obtained from A by replacing the ith column of A by $b_1, b_2,\ldots, b_n$. If $|A| = 0$ in this case of m equations in n unknowns, the system may either be inconsistent or it may have infinitely many solutions. *See* EQUATIONS, THEORY OF; POLYNOMIAL SYSTEMS OF EQUATIONS. [R.A.Be.]

Linearity The relationship that exists between two quantities when a change in one of them produces a directly proportional change in the other. Thus, an amplifier has good linearity when doubling of its input signal strength always doubles the output signal strength. A transistor displays good linearity when doubling of the instantaneous base voltage serves to double the instantaneous value of collector current. [J.Mar.]

Lines of force Imaginary lines in fields of force whose tangents at any point give the direction of the field at that point and whose number through unit area perpendicular to the field represents the intensity of the field. The concept of lines of force is perhaps most common when dealing with electric or magnetic fields.

Electric lines of force are drawn to represent, or map, an electric field graphically in the space around a charged body. They are of great help in visualizing an electric field and in quantitative thinking about such a field. A magnetic field may also be represented by lines of force. Magnetic lines of force due to magnets originate on north poles and terminate on south poles, both inside and outside the magnet. *See* ELECTRIC FIELD; MAGNETIC FIELD. [R.P.Wi.]

Linewidth A measure of the width of the band of frequencies of radiation emitted or absorbed in an atomic or molecular transition. One of the dominant sources of electromagnetic radiation of all frequencies is transitions between two energy levels of an atomic or molecular system. The frequency of the radiation is related to the difference in the energy of the two levels by the Bohr relation, Eq. (1), where ν_0, is the frequency

$$\nu_0 = (E_1 - E_2)/h \quad (1)$$

of the radiation, h is Planck's constant, and E_1 and E_2 are the energies of the levels. This radiation is not monochromatic, but consists of a band of frequencies centered about ν_0 whose intensity $I(\nu)$ can be characterized by the linewidth. The linewidth is the full width at half height of the distribution function $I(\nu)$. The simplest case is for a transition from an excited state to the ground state for an atom or molecule at rest. For this case, the normalized distribution function is the lorentzian line profile given by Eq. (2). Here $\Delta\nu$ is the full width at half

$$I(\nu) = \frac{1}{\pi} \frac{\Delta\nu/2}{(\nu - \nu_0)^2 + (\Delta\nu/2)^2} \qquad (2)$$

maximum (FWHM). The FWHM is related to the lifetime τ of the excited level through Eq. (3). This is a manifestation of the

$$(\Delta\nu)(\tau) = \frac{1}{2\pi} \qquad (3)$$

quantum-mechanical uncertainty principle, and the linewidth $\Delta\nu$ is referred to as the natural linewidth. *See* ENERGY LEVEL (QUANTUM MECHANICS); QUANTUM MECHANICS; UNCERTAINTY PRINCIPLE.

Another major source of line broadening for atomic and molecular transitions is the Doppler shift due to thermal motion. For most situations the Doppler width is greater than the natural linewidth. *See* DOPPLER EFFECT.

A third major source of line broadening is collisions of the radiating molecule with other molecules. This broadens the line, shifts the center of the line, and shortens the lifetime of the radiating state.

For radiating atoms in a liquid or solid the width is usually dominated by the strong interaction of the radiator with the surrounding molecules. The net result is a broad line profile with a complex structure. *See* BAND THEORY OF SOLIDS. [F.M.P.]

Linguistics The science, that is, the general and universal properties, of language. The middle of the twentieth century saw a shift in the principal direction of linguistic inquiry from one of data collection and classification to the formulation of a theory of generative grammar, which focuses on the biological basis for the acquisition and use of human language and the universal principles that constrain the class of all languages. Generative grammar distinguishes between the knowledge of language (linguistic competence), which is represented by mental grammar, and the production and comprehension of speech (linguistic performance).

If grammar is defined as the mental representation of linguistic knowledge, then a general theory of language is a theory of grammar. A grammar includes everything one knows about a language; its phonetics and phonology (the sounds and the sound system), its morphology (the structure of words), its lexicon (the words or vocabulary), its syntax (the structure of sentences and the constraints on well-formed sentences), and its semantics (the meaning of words and sentences). *See* PSYCHOACOUSTICS; SPEECH; SPEECH PERCEPTION.

Linguistics is not limited to grammatical theory. Descriptive linguistics analyzes the grammars of individual languages; anthropological linguistics, or ethnolinguistics, and sociolinguistics focus on languages in relation to culture, social class, race, and gender; dialectologists investigate how these factors fragment one language into many. In addition, sociolinguists and applied linguists examine language planning, literacy, bilingualism, and second-language acquisition. Computational linguistics encompasses automatic parsing, machine processing, and computer simulation of grammatical models for the generation and parsing of sentences. If viewed as a branch of artificial intelligence, computational linguistics has as its goal the modeling of human language as a cognitive system. A branch of linguistics concerned with the biological basis of language development is neurolinguistics. The form of language representation in the mind, that is, linguistic competence and the structure and components of the mental grammar, is the concern of theoretical linguistics. The branch of linguistics concerned with linguistic performance, that is, the production and comprehension of speech (or of sign language by the deaf), is called psycholinguistics. Psycholinguists also investigate how children acquire the complex grammar that underlies language use. *See* ARTIFICIAL INTELLIGENCE; INFORMATION PROCESSING; PSYCHOLINGUISTICS. [V.A.F.]

Lingulida Inarticulate brachiopods forming an order that has persisted from the Early Cambrian to the present. One genus, *Lingula*, seems to be valid for both living animals and fossil species of the Ordovician. This implies an extremely low rate of evolution.

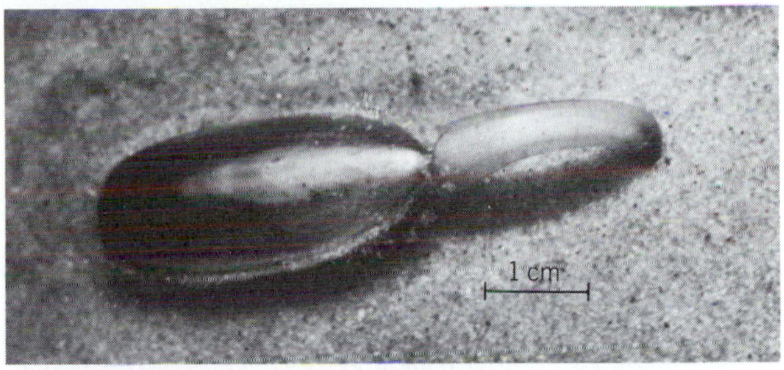

Living *Lingula reevi* from Kaneohe, Hawaii. (*Photograph by Susan C. Lum*)

The pedicle of lingulids is very large and extremely contractile. All other brachiopods have smaller pedicles, and some groups have none at all. The pedicle is used as a prop to aid the valves in shoveling through the sediment to make a burrow. When the animal is in its usual living position, it is anchored in place by a mass of sand cemented together by a secretion from the base of the pedicle (see illustration). *See* BRACHIOPODA; INARTICULATA. [C.S.H.]

Link An element of a mechanical linkage. A link may be a straight bar or a disk, or it may have any other shape, simple or complex. It is assumed, for simple analysis, to be made of unyielding material; that is, its shape does not change. The frame, or fixed member, of a linkage is one of the links.

Two links of a kinematic chain meet in a joint, or pair, by which these links are held together. Just as each joint is a pair having two elements, one from each link, so a link in a kinematic chain is the rigid connector of two or more elements belonging to different pairs. *See* LINKAGE (MECHANISM); MECHANISM. [D.P.Ad.]

Linkage (genetics) Failure of two or more genes to recombine at random as a result of their location on the same chromosome pair. Among the haploid products of a cell which has gone through meiosis, two genes located in the same chromosome pair remain in their two original combinations of alleles ("parental") unless an odd number of exchanges of homologous segments occurred within the interval bounded by their loci. The incidence of exchanges of homologous segments at meiosis is roughly proportional to the length of the chromosome segment between two loci. The percentage of recombinants thus provides an estimate of this length and a basis for constructing gene maps on which linked loci are arranged in linear order and spaced out in proportion to the recombination percentages between them. *See* MEIOSIS. [G.Po.]

Linkage (mechanism) A set of rigid bodies, called links, joined together at pivots by means of pins or equivalent devices. A body is considered to be rigid if, for practical purposes, the distances between points on the body do not change. Linkages are used to transmit power and information. They may be employed to make a point on the linkage follow a prescribed curve, regardless of the input motions to the link-

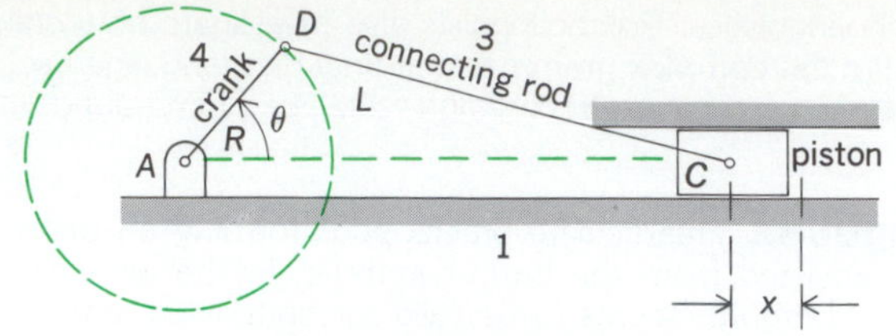

Slider crank mechanism.

age. They are also used to produce angular or linear displacement. *See* LINK; MECHANISM.

If the links are bars the linkage is termed a bar linkage. A common form of bar linkage is one for which the bars are restricted to a given plane, such as a four-bar linkage. A commonly occurring variation of the four-bar linkage is the linkage used in reciprocating engines (see illustration). Slider C is the piston in a cylinder, link 3 is the connecting rod, and link 4 is the crank. (Link 1 is the fixed base, A and D are pivots, R is the length of the crank, L is the length of the connecting rod, and θ denotes the angle of the crank.) This mechanism transforms a linear into a circular motion, or vice versa. The straight slider in line with the crank center is equivalent to a pivot at the end of an infinitely long link. *See* FOUR-BAR LINKAGE; PANTOGRAPH; SLIDER CRANK MECHANISM. [R.O.]

Linseed oil Oil expressed or extracted from flaxseed (*Linum usitatissimum*). Freshly expressed oil is a dark amber color and has a powerful odor. The crude product is filtered or centrifuged and decolorized, and as an article of commerce usually has a light yellow color similar to other refined vegetable oils. The chief acid of linseed oil is linolenic acid, which makes up well over half the fatty acids. Oleic acid and linoleic acid are present in lesser amounts.

Much linseed oil has been displaced from paint and varnish formulations by synthetic polymers, but development of new types of linseed oil emulsions for paints and for the treatment of concrete to protect it from winter damage has created greater markets than ever. In addition, much linseed oil is consumed in the formulation of putties and calking compounds, in printing inks, and in linoleum. Some linseed oil is incorporated into various resins and plastics, and small amounts are used in specialty lubricants, greases, and core oils for foundries. *See* DRYING OIL; SURFACE COATING. [F.W.]

Lip A fleshy fold above and below the entrance to the mouth of mammals. Even the lower vertebrates with movable jaws, such as fishes, amphibians, and most reptiles, have folds of skin external to the jaws and teeth. Nevertheless, only the typical mammals (marsupials and placental mammals) possess a highly developed type of true lips. These are separated by deep clefts from the jaws and are made mobile by containing muscles belonging to the facial group. The separating cleft or space constitutes the vestibule of the mouth. [L.B.A.]

Lipid One of a class of compounds which contains long-chain aliphatic hydrocarbons (cyclic or acyclic) and their derivatives, such as acids (fatty acids), alcohols, amines, amino alcohols, and aldehydes. The presence of the long aliphatic chain as the characteristic component of lipids confers distinct solubility properties on the simpler members of this class of naturally occurring compounds.

The lipids are generally classified into the following groups:

A. Simple lipids

1. Triglycerides or fats and oils are fatty acid esters of glycerol. Examples are lard, corn oil, cottonseed oil, and butter.
2. Waxes are fatty acid esters of long-chain alcohols. Examples are beeswax, spermaceti, and carnauba wax.
3. Steroids are lipids derived from partially or completely hydrogenated phenanthrene. Examples are cholesterol and ergosterol.

B. Complex lipids

1. Phosphatides or phospholipids are lipids which contain phosphorus and, in many instances, nitrogen. Examples are lecithin, cephalin, and phosphatidyl inositol.
2. Glycolipids are lipids which contain carbohydrate residues. Examples are sterol glycosides, cerebrosides, and plant phytoglycolipids.
3. Sphingolipids are lipids containing the long-chain amino alcohol sphingosine and its derivatives. Examples are sphingomyelins, ceramides, and cerebrosides.

Lipids are present in all living cells, but the proportion varies from tissue to tissue. The triglycerides accumulate in certain areas, such as adipose tissue in the human being and in the seeds of plants, where they represent a form of energy storage. The more complex lipids occur closely linked with protein in the membranes of cells and of subcellular particles. More active tissues generally have a higher complex lipid content; for example, the brain, liver, kidney, lung, and blood contain the highest concentration of phosphatides in the mammal. *See* FAT AND OIL; FAT AND OIL (FOOD); GLYCOLIPID; PHOSPHATIDE; SPHINGOLIPID; STEROID; TERPENE; TRIGLYCERIDE; VITAMIN; WAX, ANIMAL AND VEGETABLE. [R.H.G.]

Lipid metabolism The assimilation of dietary lipids and the synthesis and degradation of lipids; this article is restricted to mammals.

The principal dietary fat is triglyceride. This substance is not digested in the stomach and passes into the duodenum, where it causes the release of enterogastrone, a hormone which inhibits stomach motility. The amount of fat in the diet, therefore, regulates the rate at which enterogastrone is released into the intestinal tract. Fat, together with other partially digested foodstuffs, causes the release of hormones, secretin, pancreozymin, and cholecystokinin from the wall of the duodenum into the bloodstream.

Secretin causes the secretion of an alkaline pancreatic juice rich in bicarbonate ions, while pancreozymin causes secretion of pancreatic enzymes. One of these enzymes, important in the digestion of fat, is lipase. Cholecystokinin, which is a protein substance chemically inseparable from pancreozymin, stimulates the gallbladder to release bile into the duodenum. Bile is secreted by the liver and concentrated in the gallbladder and contains two bile salts, both derived from cholesterol: taurocholic and glycocholic acids. These act as detergents by emulsifying the triglycerides in the intestinal tract, thus making the fats more susceptible to attack by pancreatic lipase. In this reaction, which works best in the alkaline medium provided by the pancreatic juice, each triglyceride is split into three fatty acid chains, forming monoglycerides. The fatty acids pass across the membranes of the intestinal mucosal (lining) cells. Enzymes in the membranes split monoglyceride to glycerol and fatty acid, but triglycerides are reformed within the mucosal cells from glycerol and those fatty acids with a chain length greater than eight carbons: Short- and medium-chain fatty acids are absorbed directly into the bloodstream once they pass through the intestinal mucosa. *See* BILE ACID; CHOLESTEROL; DIGESTIVE SYSTEM; GALLBLADDER; LIVER; PANCREAS; TRIGLYCERIDE.

Obesity is a condition in which excessive fat accumulates in the adipose tissue. One factor responsible for this condition is excessive caloric intake. In starvation, uncontrolled diabetes, and many generalized illnesses the opposite occurs and the adipose tissue becomes markedly depleted of lipid. *See* ADIPOSE TISSUE; DIABETES; LIPID; METABOLIC DISORDERS; OBESITY. [M.A.R.]

Lipoprotein A macromolecule containing both lipid and protein. Lipoproteins are of fundamental importance in biology since lipids are generally poorly soluble or insoluble in the water-based environment of living systems. The formation of the lipid-protein complex solubilizes the lipid components in an aqueous medium, permitting transport and utilization. The ability to form lipoproteins is derived from the conformation, or three-dimensional structure, of the component proteins. The protein portion without its complement of lipid is called an apolipoprotein.

The protein part of any lipoprotein may be a single chain or multiple polypeptide chains. The lipid component is usually not covalently bound to the protein moiety, but in a few instances it may be. Consequently, for most forms of lipoproteins the lipid may be separated from the protein by extraction with organic solvents.

The relative amount of lipid and protein in different forms of lipoproteins is highly variable. A single molecule may contain one molecule of bound lipid or hundreds of molecules, depending upon the specific lipoprotein. In addition, the type of lipid present in the lipid-protein complex may vary a great deal. For example, low-density lipoprotein from human blood serum contains cholesterol, cholesterol esters, triglycerides, and phospholipids. *See* CHOLESTEROL; LIPID METABOLISM; PROTEIN.

Loosely defined, proteins embedded in membranes may also be lipoproteins. Such membrane proteins are generally insoluble in water or aqueous buffers but can be made soluble by the presence of detergents. The combination of protein and detergent is itself a form of lipoprotein since nearly all detergents have lipidlike properties. *See* DETERGENT. [L.J.Ba.]

Liposomes Aqueous compartments enclosed by lipid bilayer membranes; liposomes are also known as lipid vesicles. Lipid molecules consist of an elongated nonpolar (hydrophobic) structure with a polar (hydrophilic) structure at one end. When dispersed in water, they spontaneously form bilayer membranes, also called lamellae, which are composed of two monolayer sheets of lipid molecules with their nonpolar (hydrophobic) surfaces facing each other and their polar (hydrophilic) surfaces facing the aqueous medium. The membranes enclose a portion of the aqueous phase much like the cell membrane which encloses the cell; in fact, the bilayer membrane is essentially a cell membrane without its protein components.

Liposomes are often used to study the characteristics of the lipid bilayer. Properties of liposomes have been characterized by a variety of techniques: molecular organization by x-ray diffraction, nuclear magnetic resonance, electron paramagnetic resonance, and Raman spectroscopy; melting behavior (that is, crystal to liquid-crystal transition) by calorimetry; net electric surface charge by microelectrophoresis; size by light scattering and electron microscopy. *See* LIPID.

Liposomes have numerous uses as biochemical and biophysical tools: (1) as vehicles for the delivery of both water- and of oil-soluble materials to the cell; (2) as immunological adjuvants; (3) as substrates for the study of membrane properties such as rotational or translational diffusion in the plane of the membrane; and (4) as intermediates in the construction of bilayers large enough for the study of electrical properties of membranes. *See* CELL MEMBRANES. [R.C.MacD.; R.I.MacD.]

Liquefaction of gases Several processes have been developed for liquefying gases. Although significant differences in efficiency exist among them, difficulties encountered in reducing the more efficient to practice have made them all roughly competitive. Consequently, the choice of the refrigerating cycle to be used for liquefaction of a given gas is made largely on the basis of convenience. In large industrial liquefaction plants, however, a significant reduction in production costs may be realized by use of the more efficient cycles.

Although any thermodynamic process which results in a lowering of the temperature could in principle be utilized in the liquefaction of some gas, only four have found practical employment. These are, in order of increasing complexity, vapor compression; use of refrigerators to cool the gas at constant pressure until liquefaction occurs; utilization of the adiabatic performance of work by the gas in a cyclic process with regenerative cooling; and use of the Joule-Thomson process in a cycle with regenerative cooling. Combinations of these four processes have also often been found effective. [E.F.H.]

Liquefied natural gas (LNG) A product of natural gas which consists primarily of methane. Its properties are those of liquid methane, slightly modified by minor constituents. One property which differentiates liquefied natural gas (LNG) from liquefied petroleum gas (LPG) is the low critical temperature, about −100°F (−73°C). This means that natural gas cannot be liquefied at ordinary temperatures simply by increasing the pressure, as is the case with LPG; instead, natural gas must be cooled to cryogenic temperatures to be liquefied and must be well insulated to be held in the liquid state. *See* CRYOGENIC ENGINEERING; LIQUEFACTION OF GASES; LIQUEFIED PETROLEUM GAS (LPG). [A.W.F.]

Liquefied petroleum gas (LPG) A product of petroleum gases, principally propane and butane, which must be stored under pressure to keep it in a liquid state. At atmospheric pressure and above freezing temperature, these substances would be gases. Large quantifies of propane and butane are now available from the gas and petroleum industries. These are often employed as fuel for tractors, trucks, and buses and mainly as a domestic fuel in remote areas. Because of the low boiling point (−47.2 to 32°F or −44 to 0°C) and high vapor pressure of these gases, their handling as liquids in pressure cylinders is necessary. Owing to demand from industry for butane derivations, LPG sold as fuel is made up largely of propane.

LPG has a high octane rating, making it useful in engines having compression ratios above 10:1. Another factor of importance in internal combustion engines is that LPG leaves little or no engine deposit in the cylinders when it bums. *See* INTERNAL COMBUSTION ENGINE; PETROLEUM PRODUCTS. [M.Sou.]

Liquid A state of matter intermediate between that of crystalline solids and gases. Macroscopically, liquids are distinguished from crystalline solids in their capacity to flow under the action of extremely small shear stresses and to conform to the shape of a confining vessel. Liquids differ from gases in possessing a free surface and in lacking the capacity to expand without limit. On the scale of molecular dimensions liquids lack the long-range order that characterizes the crystalline state, but nevertheless they possess a degree of structural regularity that extends over distances of a few molecular diameters. In this respect, liquids are wholly unlike gases, whose molecular organization is completely random.

Liquids possess important transport properties, notably their capacity to transmit heat (thermal conductivity), to transfer momentum under shear stresses (viscosity), and to attain a state of homogeneous composition when mixed with other miscible liquids (diffusion). These nonequilibrium properties of liquids are well understood in macroscopic terms and are exploited in large-scale engineering and chemical-process operations. *See* GAS. [N.H.N.]

Liquid chromatography A form of chromatography involving passage of a liquid moving phase through a solid or a liquid stationary phase. The mobile phase is percolated through the column by means of either gravity, under pressure generated by a suitable pump, or centrifugal force. High-pressure pumps that generate up to several hundred atmospheres of inlet pressure are used in combination with chromatographic packings of only several micrometers of particle size. Since the viscosity of a liquid affects its flow at a given pressure, the pressure requirements vary for different liquids achieving identical flow. The two principal forms of liquid chromatography are ion-exchange and gel. *See* GEL CHROMATOGRAPHY; ION-EXCHANGE CHROMATOGRAPHY.

The mobile-phase composition is an important variable in chromatography. Sample solubility, competition between the sample and the mobile-phase molecules for available adsorbent surface, and solvation effects on the structure and function of some chromatographic materials can all influence the separation process. In liquid-liquid chromatography, the mobile phase and the stationary phase must be immiscible. Since immiscibility implies that the two phases must have significantly different chemical properties, only combinations of either a nonpolar mobile phase and a polar stationary phase or a polar mobile phase and a nonpolar stationary phase are feasible. *See* CHROMATOGRAPHY. [M.V.N.]

Liquid crystals A state of matter that mixes the properties of both the liquid and solid states. Liquid crystals may be described as condensed fluid states with spontaneous anisotropy. They are categorized in two ways: thermotropic liquid crystals, prepared by heating the substance, and lyotropic liquid crystals, prepared by mixing two or more components, one of which is rather polar in character (for example, water). *See* ANISOTROPY (PHYSICS).

Liquid crystals are a state of matter that combines a kind of long-range order (in the sense of a solid) with the ability to form droplets and to pour (in the sense of waterlike liquids). They also exhibit properties of their own such as the ability to form monocrystals with the application of a normal magnetic or electric field; an optical activity of a magnitude without parallel in either solids or liquids; and a temperature sensitivity which results in a color change in certain liquid crystals.

Liquid crystals have many applications. They are used as displays in digital wristwatches, calculators, panel meters, thermometers, and industrial products. They can be used to record, store, and display images which can be projected onto a large screen. They also have potential use as television displays. *See* ELECTRONIC DISPLAY. [G.H.B.]

Liquid helium Helium liquefies at a substantially lower temperature, 4.2 K (−452°F or −269°C), than any other substance; and below 2.2 K (−455.8°F) the liquid exhibits the extraordinary properties of superfluidity, most notably the ability to flow with complete absence of friction. Helium remains in a liquid state at absolute zero. All of these characteristics are due to the weakness of the attractive force between two helium atoms and to the small atomic mass, which according to the laws of quantum mechanics makes the atoms difficult to localize. *See* SUPERFLUIDITY.

Liquid helium was first produced in 1908 by H. Kamerlingh Onnes. Although extensive thermodynamic studies followed, the superfluid properties were not discovered until 1938. In addition to the common isotope of atomic weight 4, helium has a rare isotope of atomic weight 3 with quite different properties. In 1972 a superfluid transition in liquid ^3He was discovered at a temperature near 0.002 K (0.004°F above absolute zero, −459.67°F).

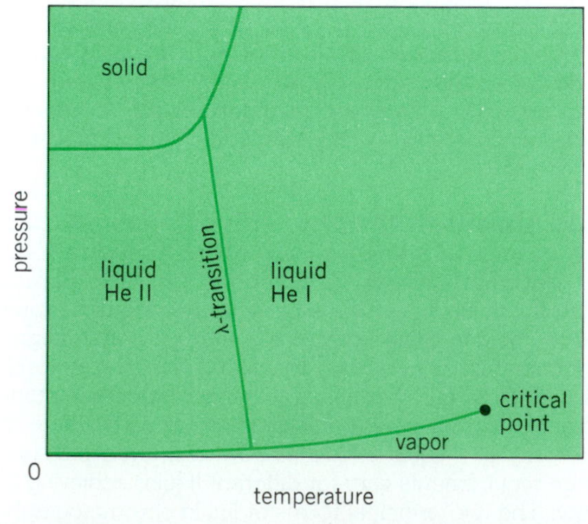

Phase diagram for ^4He. The critical point is at T = 5.20 K, P_c = 2.29 bars.

The phase diagram of ^4He (see illustration) shows several remarkable characteristics. Helium remains a liquid down to absolute zero unless a pressure that is greater than 25.3 bars (1 bar = 0.987 atm = 10^5 pascals) is applied. This is true despite the requirement of thermodynamics that the system have the same degree of order as a perfect crystal. A more subtle feature is a transition at the lambda line between two different liquid phases. This λ-transition is so named because the specific heat has a singularity resembling a backwards Greek lambda. There is no latent heat; such a transition is called second-order. The high-temperature liquid phase, called helium I, is a rather ordinary liquid. The λ-transition at 2.172 K (−455.76°F) (at vapor pressure) marks the onset of superfluidity, which is the characteristic property of the low temperature phase, helium II. [G.B.H.]

Liquid scintillation detector A particle detector in which the sensitive medium is a liquid scintillator. *See* SCINTILLATION COUNTER.

Liquid and plastic scintillators are the most extensively used forms of organic scintillator. They are noted for their fast response, for the ease with which they may be formed in arbitrary shapes, large or small, and for the economy they offer in achieving large, sensitive detecting volumes. Liquid scintillators also have the capacity to assimilate other liquids or substances and form homogeneous media, thereby providing an efficient and simple means for measuring the products of nuclear reaction or radioactive decay in those substances. *See* PARTICLE DETECTOR. [F.D.B.]

Lissajous figures Plane curves traced by a point which executes two independent harmonic motions in perpendicular directions, the frequencies of the motion being in the ratio of two integers. Such figures are widely used in frequency and phase measurements (see illustration). *See* HARMONIC MOTION.

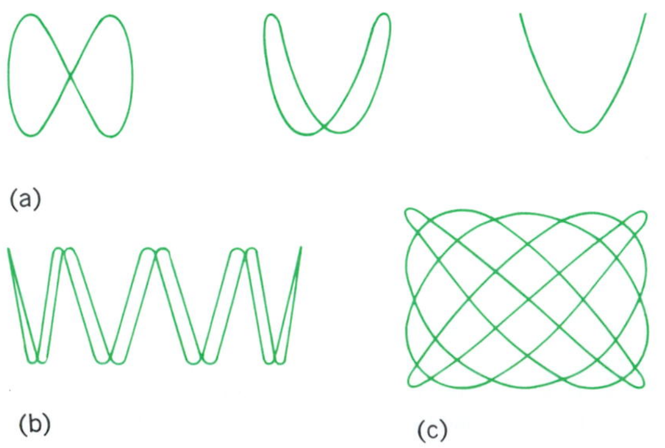

Typical Lissajous figures for ratios of vertical frequency to horizontal frequency. (*a*) 2:1, with various phase relations. (*b*) 8:1. (*c*) 5:4. (*After F. E. Terman, Radio Engineers' Handbook, McGraw-Hill, 1943*)

The cathode-ray oscilloscope furnishes the most important and practical means for the generation of the figures. The x-deflection plates of the tube are supplied with one alternating voltage, and the y-deflection plates with another. If the frequencies are incommensurable, the figure is not a closed curve and, except for very low frequencies, will appear as a patch of light because of the persistence of the screen. On the other hand, if the frequencies are commensurable, the figure is closed and strictly periodic; it is a true Lissajous figure, stationary on the screen and, if the persistence is sufficient, visible continuously as a complete pattern. *See* FREQUENCY MEASUREMENT; OSCILLOSCOPE; PHASE-ANGLE MEASUREMENT. [M.Gr.]

Lissamphibia The subclass of Amphibia including all living amphibians (frogs, toads, salamanders, and apodans). The other two subclasses are the Labyrinthodontia and the Lepospondyli. *See* LABYRINTHODONTIA; LEPOSPONDYLI.

Living amphibians are grouped together by possession of a unique series of characters, the most important of which are (1) pedicellate teeth, consisting of two segments, a crown and a pedicel; (2) an operculum-plectrum complex of the middle ear; (3) the papilla amphibiorum, a special sensory area of the inner ear; (4) green rods in the retina of the eye; (5) similar skin glands; and (6) a highly vascular skin used in respiration (cutaneous respiration). *See* AMPHIBIA; ANURA; URODELA. [R.E.]

Listeriosis A disease of humans and animals caused by the bacterium *Listeria monocytogenes*. The most common clinical manifestation of listeriosis in humans are meningitis and septicemia. These are observed most often in the newborn within the first month of life, or in debilitated adults who usually are 40 or more years old. [H.J.W.]

Litharenite A sandstone which contains greater than 25% detrital rock fragments and in which rock fragments are more abundant than feldspar grains. A sublitharenite contains more than 5% but less than 25% rock fragments. The terms litharenite (a contraction of lithic arenite) and sublitharenite refer to the two members of the lithic sandstone clan. Such sandstones are called graywacke, subgraywacke, lithic arenite, or sublithic arenite in other classifications.

Most rock fragments are susceptible to alteration during burial in a sedimentary basin: limestone grains dissolve, shale and schist grains compact, and volcanic grains (basalt, trachyte, and so on) react with groundwater to produce clay and zeolite minerals that precipitate in pores and cement the sand. These postdepositional changes make interpretation of the history of sandstones more difficult, and also influence the permeability of the rocks to migrating groundwater and petroleum. *See* GRAYWACKE; PETROGRAPHY; SANDSTONE; SUBGRAYWACKE. [E.F.McB.]

Lithium A chemical element, Li, atomic number 3, and atomic weight 6.939. Lithium heads the alkali metal family in the periodic table. In nature it is a mixture of the isotopes ^{6}Li and ^{7}Li. Lithium, the lightest solid element, is a soft, low-melting, reactive metal. In many physical and chemical properties it resembles the alkaline-earth metals as much as, or more than, it does the alkali metals. *See* ALKALI METALS; ALKALINE-EARTH METALS.

The major industrial use of lithium is in the form of lithium stearate as a thickener for lubricating greases. Other important uses of lithium compounds are in ceramics, specifically in porcelain enamel formulation; as an additive to give longer life and higher output in alkaline storage batteries; and in welding and brazing fluxes.

Lithium is a moderately abundant element and is present in the Earth's crust to the extent of 65 parts per million (ppm).

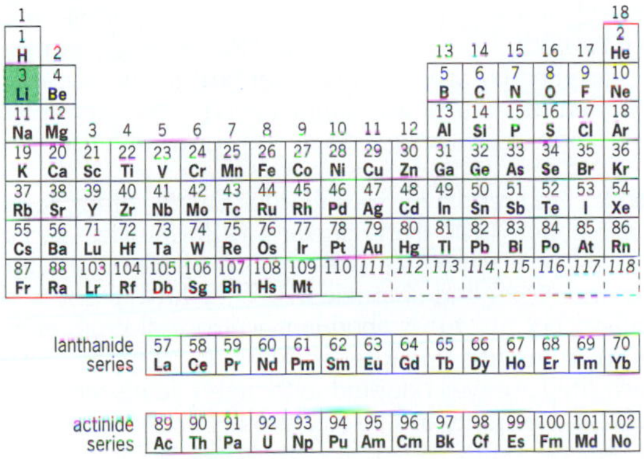
1 18
1 H 2 He
2 13 14 15 16 17
3 Li 4 Be 5 B 6 C 7 N 8 O 9 F 10 Ne
11 Na 12 Mg 3 4 5 6 7 8 9 10 11 12 13 Al 14 Si 15 P 16 S 17 Cl 18 Ar
19 K 20 Ca 21 Sc 22 Ti 23 V 24 Cr 25 Mn 26 Fe 27 Co 28 Ni 29 Cu 30 Zn 31 Ga 32 Ge 33 As 34 Se 35 Br 36 Kr
37 Rb 38 Sr 39 Y 40 Zr 41 Nb 42 Mo 43 Tc 44 Ru 45 Rh 46 Pd 47 Ag 48 Cd 49 In 50 Sn 51 Sb 52 Te 53 I 54 Xe
55 Cs 56 Ba 71 Lu 72 Hf 73 Ta 74 W 75 Re 76 Os 77 Ir 78 Pt 79 Au 80 Hg 81 Tl 82 Pb 83 Bi 84 Po 85 At 86 Rn
87 Fr 88 Ra 103 Lr 104 Rf 105 Db 106 Sg 107 Bh 108 Hs 109 Mt 110 111 112 113 114 115 116 117 118
lanthanide series 57 La 58 Ce 59 Pr 60 Nd 61 Pm 62 Sm 63 Eu 64 Gd 65 Tb 66 Dy 67 Ho 68 Er 69 Tm 70 Yb
actinide series 89 Ac 90 Th 91 Pa 92 U 93 Np 94 Pu 95 Am 96 Cm 97 Bk 98 Cf 99 Es 100 Fm 101 Md 102 No

This places lithium a little below nickel, copper, and tungsten, and a little above cerium and tin in abundance.

Noteworthy among lithium's physical properties are the high specific heat (heat capacity), large temperature range of the liquid phase, high thermal conductivity, low viscosity, and very low density. Lithium metal is soluble in liquid ammonia and is slightly soluble in the lower aliphatic amines, such as ethylamine. It is insoluble in hydrocarbons.

Lithium undergoes a large number of reactions with both organic, and inorganic, reagents. It reacts with oxygen to form the monoxide, Li_2O, and the peroxide, Li_2O_2. Lithium is the only alkali metal that reacts with nitrogen at room temperature to form a nitride, Li_3N, which is black. Lithium reacts readily with hydrogen at about 930°F (500°C) to form lithium hydride, LiH. The reaction of lithium metal with water is exceedingly vigorous. Lithium reacts directly with carbon to form the carbide, Li_2C_2. Lithium combines readily with the halogens, forming halides with the emission of light. While lithium does not react with paraffin hydrocarbons, it does undergo addition reactions with arylated alkenes and with dienes. Lithium also reacts with acetylenic compounds, forming lithium acetylides, which are important in the synthesis of vitamin A.

The most important lithium compound is lithium hydroxide. It is a white powder, and the material of commerce is actually lithium hydroxide monohydrate, $LiOH \cdot H_2O$. Lithium carbonate, $LiCO_3$, finds application in the ceramic industries and in medicine as an antidepressant. Both lithium halides, lithium chloride and lithium bromide, form concentrated brines with ability to absorb moisture over a wide temperature range; these brines are used in commercial air conditioning systems. [M.Si.]

Lithium primary cell A primary cell whose anode is composed of lithium. The lithium cell has a number of advantages over other primary cell systems, including high energy density, flat discharge characteristics, excellent service over a wide temperature range (as low as −40°C or −40°F), and good shelf life (up to 5 years without refrigeration). Nonaqueous solvents are used as the electrolyte because of the solubility of lithium in aqueous solutions. This cell is designed primarily for cardiac pacemakers and works with a typical current drain of 30 microamperes. *See* BATTERY; PRIMARY BATTERY; SOLID-STATE BATTERY. [J.D.; K.F.]

Lithosphere A term initially used in a general way to distinguish the rock phase of the Earth from the hydrosphere, biosphere, and atmosphere at the surface of the Earth. The term lithosphere has been used variously in recent geological literature to designate the crust (upper lithosphere), the crust plus the mantle (that is, the total silicate phase of the Earth), or the entire solid part of the Earth (crust + mantle + core).

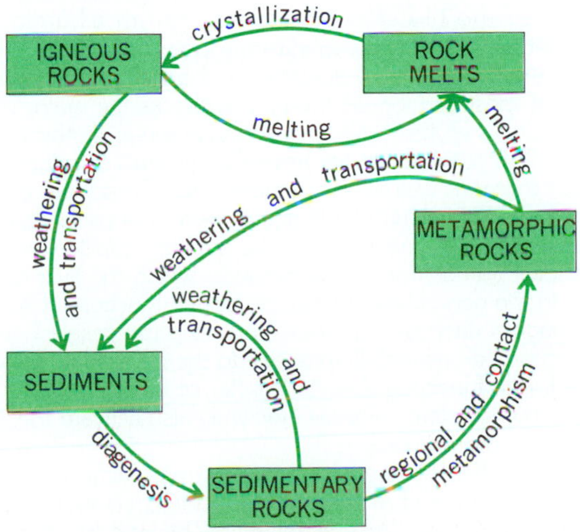

Important features of major geochemical cycle.

The lithosphere contains more than 99% of the mass of the Earth; it is therefore by far the most important reservoir of almost all of the chemical elements, and the processes in the lithosphere exert a dominant influence on many surface phenomena.

Erosion and chemical weathering are evidence of the pronounced disequilibrium between the atmosphere and the lithosphere. Chemical changes are taking place at various levels in the lithosphere and at the boundary between the lithosphere and the atmosphere, biosphere, and hydrosphere. The processes taking place at the Earth's surface can be observed directly and therefore are more completely understood than the processes taking place within and below the Earth's crust. Yet it is just the processes at depth that are of primary interest as a key to the history of the crustal evolution of the Earth.

Many of these processes participate in what has come to be known as the major geochemical cycle, whose most important aspects are shown in the illustration. The weathering of igneous, sedimentary, and metamorphic rocks at the lithosphere-atmosphere boundary produces sediments. These are buried and compacted at depth to produce sedimentary rocks. At more elevated temperatures or pressures or both, sedimentary rocks recrystallize to metamorphic rocks. At still higher temperatures partial or complete melting may produce rock melts, which cool to form igneous rocks. Exposure of these rocks to weathering processes then renews the cycle. *See* ATMOSPHERE; BIOSPHERE; DIAGENESIS; EARTH; HYDROSPHERE; MAGMA; METAMORPHISM; SEDIMENTOLOGY; WEATHERING PROCESSES. [H.D.Ho.]

Litopterna Hoofed herbivores confined to the Cenozoic of South America. The order was well represented from the Paleocene to the Pleistocene, and apparently arose on that continent from a condylarth ancestry. By later Paleocene time two main lines of descent were clearly demarcated. The Proterotheriidae displayed a remarkable evolutionary convergence with the horses in their dentition and in reduction of the lateral digits of their feet. In one group the foot was reduced to a single median toe by early Miocene time. The members of the Macraucheniidae were proportioned much as in the camels and by late Tertiary time had similarly lost the vertebral arterial canal of the cervical vertebrae. *See* CONDYLARTHRA; MAMMALIA. [R.H.T.]

Liver A large gland found in all vertebrates. It consists of a continuous parenchymal mass arranged to form a system of walls through which venous blood emanating from the gut must pass. This strategic localization between nutrient-laden capillary beds and the general circulation is associated with hepatic regulation of metabolite levels in the blood through storage and mobilization mechanisms controlled by liver enzymes.

Function. The liver is the sole source of such necessary constituents of the blood as fibrinogen, serum albumin, and cholinesterase (an enzyme which may prevent harmful buildup of the neuronal transmitter agent acetylcholine). In the embryonic stage of most vertebrates the liver serves as the major manufacturing site of erythrocytes, a process known as erythropoiesis. The liver also removes toxins from the systemic circulation and degrades them, as well as excess hormones. Particulate material may be removed through a phagocytic action of specialized cells (Kupffer cells) lining the lumen of the hepatic "capillary spaces," or sinusoids. In addition to the products which the liver delivers directly to the general circulation (endocrine function), it secretes bile through a duct system which, involving the gallbladder as a storage chamber, eventually passes into the duodenum (exocrine function). Bile functions as an emulsifier of fats to facilitate their digestion by fat-splitting lipases, and may also activate the lipase directly. *See* GALLBLADDER.

Anatomy. Knowledge of the precise connections between liver and intestine and of liver and systemic circulation provides the mechanical basis for understanding hepatic function. The human liver is a massive wedge-shaped organ divided into a large right lobe and a smaller left lobe. Its anterior surface underlies the diaphragm. The upper portion of the liver is partially covered ventrally by the lungs, whereas the lower portion overhangs the stomach and intestine. The entire liver is covered by Glisson's capsule, an adherent membranous sheet of collagenous and elastic fibers.

Venous blood from the intestine, and to a lesser extent from spleen and stomach, converges upon a short broad vessel, called the hepatic portal vein, which enters the liver through a depression in the dorsocaudal surface termed the porta hepatis. There the hepatic portal vein divides into a short right branch and a longer left branch. These vessels then ramify into the small branches which actually penetrate the functional parenchymal mass as the inner tubes of the portal canals.

The hepatic artery also enters at the porta hepatis and ramifies into smaller branches, which flank the portal venules within the portal canals. The branches of the portal vein and hepatic artery then empty into sinusoids, which are major regions of hepatovascular exchange. They communicate with small branches of the hepatic veins and, through the hepatic vein, the blood is returned to the heart by way of the vena cava.

The tiny bile canaliculi, which lie between grooves in adjacent parenchymal cells, communicate with tiny intralobular bile ducts. These intralobular bile ducts empty into increasingly larger interlobular bile ducts which lie within the portal canals and make up the third element of the so-called portal triad.

Although the biliary system is intimately associated with the portal system anatomically, even in terms of egress through the porta hepatis, it should be stressed that bile flow within this biliary system is in the opposite direction from that of blood flow. The hepatic duct arises in the region of the porta from the confluence of the largest bile ducts. [G.H.F.]

Liver disorders A heterogeneous group disease of particular importance because of the many essential physiological functions of the liver. A common condition characterized by accumulation of lipid (fat) within hepatocytes is fatty metamorphosis. The hepatocytes become swollen, and the liver can increase in size two to three times. The most common cause of a fatty liver in the United States is excessive alcohol intake; this is not due to nutritional deficiency but to changes produced in cellular enzymes by alcohol. *See* ALCOHOLISM.

Hepatitis is defined as inflammation of the liver. Viral hepatids is the most common type of hepatitis, and is usually caused by two distinct viruses called A and B. Alcoholic hepatitis is frequently seen in individuals consuming excessive quantities of ethyl alcohol. Cirrhosis is the term applied to widespread fibrosis in the liver, with nodular regeneration of hepatic parenchymal cells. *See* CIRRHOSIS; HEPATITIS; LIVER. [S.P.H.]

Living fossils Living species very closely resembling fossil relatives in most anatomical details. The term is a relative one and, applied loosely, could embrace nearly all extant animals and plants. In its more restricted usage, the term applies to living species with four additional characteristics: (1) truly close anatomical similarity to (2) an ancient fossil species—generally at least 100,000,000 years old; (3) living members of the group are represented by only a single or at best a few species, which are (4) often found in a very limited geographic area. Examples are horseshoe crabs, ginkgo trees, and coelacanth fish. [N.E.]

Lizard Any reptile in the suborder Sauria, order Squamata; there are about 3000 living species in 20 families. Included in the group are a variety of diverse forms such as the iguanas, geckos, monitors, and chameleons. They have a cosmopolitan distribution but are most abundant in tropical regions and are absent in Antarctica. Lizards are found in all types of habitats, to which they are well adapted; although a few species are herbivorous, the majority are insectivorous or carnivorous.

Although most species have paired limbs, some lizards have elongated bodies with reduced limbs and thus resemble snakes. Typical of the lizards are movable eyelids and a fused lower jaw. The tongue may be extensile, even prehensile, or thick and attached to the floor of the mouth with only the anterior edge extensile. Dentition is homodont, and the teeth are inserted on the jaw ridge (acrodont condition) or on the inner surface of the jaws (pleurodont). Lizards have epidermal scales which may be modified to form tubercles or reinforced with bony plate. *See* CHAMELEON; DENTITION; GECKO; GILA MONSTER; IGUANID; MONITOR; REPTILIA; SKINK; SQUAMATA. [C.B.C.]

Llama A member of the camel family (Camelidae) found only in South America. The llama is an artiodactyl with two toes on each foot. The upper lip is cleft and prehensile; the neck is long (see illustration). A single young is born after a gestation period of about 11 months. The maximum life-span is about 20 years. Like other members of the family, the llama is herbivorous.

The llama of South America.

Lama glama has been domesticated since the Inca civilization by the Peruvian Indians, who still keep large herds. Many interesting crosses have occurred among the different breeds in South America:

Llama (male) + alpaca (female) = huarizo
Alpaca + llama = misti; machurga
Vicuña + llama = llamavicuña
Guanaco + llama = llamahuanaco
Vicuña + alpaca = vacovicuna

While the fur provides material for rugs, rope, and cloth, the main use of this animal is as a beast of burden, especially in mining areas. *See* ALPACA; ARTIODACTYLA. [C.B.C.]

Loa loa The African eye worm, a filarioidean worm most commonly acquired by Caucasian immigrants, including missionaries, in Africa. Transmission is by daytime-feeding sylvan deer-flies, *Chrysops*. Fortunately, serious damage is rare even when the worm gets into the eye. The areas of pitting edema known as calabar swellings are painful and diagnostic. They commonly occur on the wrists, hands, arms, or orbital tissues. *See* FILARIOIDEA. [G.F.O.]

Load line A line drawn on the output characteristic curves of a vacuum tube or transistor, used to determine the operating range and the quiescent point (the operating point for zero signal input voltage or current). In the case of vacuum tubes, this graphical construction is necessary because the mathematical function relating plate current to plate voltage and grid voltage is not known analytically. Load lines may also be used with transistor circuits and other nonlinear devices to determine the quiescent point. *See* TRANSISTOR; VACUUM TUBE. [C.C.H.]

Loads, dynamic A force exerted by a moving body on a resisting member is called a dynamic or energy load. Loads suddenly applied with appreciable striking velocity, as when a falling weight strikes another body, are called impact loads. Such loads are produced by moving machine parts, cars on a bridge, airplane landing shock, falling weights, and other nonstationary loading conditions where forces are applied in relatively short time intervals. Dynamic load is expressed in terms of the amount of energy transferred or by an equivalent static load which has the same stress-producing effect on the member.

If a material or structure has sufficient elastic energy capacity as a resisting member to absorb completely the energy load through elastic action and recover completely from the deformation, it behaves elastically. *See* SHOCK ABSORBER.

An energy load exceeding the elastic energy capacity of the member produces inelastic deformation. Toughness is the energy absorbed per unit volume when the material is stressed to fracture. Capacity to dissipate energy overloads by plastic deformations is desirable in members subject to shock or impact. *See* STRESS AND STRAIN. [W.J.K./W.G.B.]

Loads, repeated Forces reapplied many times and causing varying stresses as the load changes. Repeated loads exist in crankshafts, axles, springs, piston rods, rails, bridge members, and many other machine and structural elements. Repeated loads considerably smaller than similarly applied static loads cause failure by progressive fracture. Designs for repeated loads depend on the maximum value of repeated stress which the material can resist, called the fatigue strength, and the magnitude of localized stress concentrations. *See* STRESS CONCENTRATION.

The maximum stress is called a stress concentration. High local stresses exist at bolt threads, abrupt change of shaft diameter, notches, holes, keyways, or fillet wells. Internal discontinuities, such as blowholes, inclusions, seams or cracks, ducts and knots in wood, voids in concrete, and variable stiffness of component constituents, cause stress concentrations. At points of contact of ball or roller bearings, gear teeth, or other local load applications, the stress may be greatly increased. *See* STRESS AND STRAIN. [J.B.S.]

Loads, transverse Forces applied perpendicularly to the longitudinal axis of a member. Transverse loading causes the member to bend and deflect from its original position, with internal tensile and compressive strains accompanying change in curvature.

Concentrated loads are applied over areas or lengths which are relatively small compared with the dimensions of the supporting member. Examples are a heavy machine occupying limited floor area, wheel loads on a rail, or a tie rod attached locally. Loads may be stationary or they may be moving, as with the carriage of a crane hoist or with truck wheels.

Distributed loads are forces applied continuously over large areas with uniform or nonuniform intensity. Closely stacked contents on warehouse floors, snow, or wind pressures are considered to be uniform loads. Variably distributed load intensities include foundation soil pressures and hydrostatic pressures.

Members subjected to bending by transverse loads are classed as beams. The span is the unsupported length. Beams

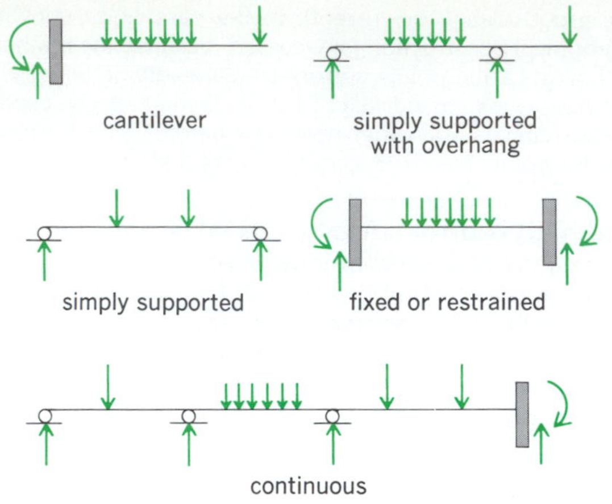

Types of beams.

may have single or multiple spans and are classified according to type of support, which may permit freedom of rotation or furnish restraint (see illustration). [J.B.S.]

Lobata An order of the Ctenophora in which the body is helmet-shaped. A pair of large lobe-shaped processes extend from the oral end in the pharyngeal plane. A pair of "auricles" are attached to the base of each process. The body is compressed, with the tentacular axis being shorter than the pharyngeal axis. The ribs are of unequal length, with the longest being the subpharyngeal pairs. The primary tentacles are degenerate and are replaced by secondary tentacles (see illustration).

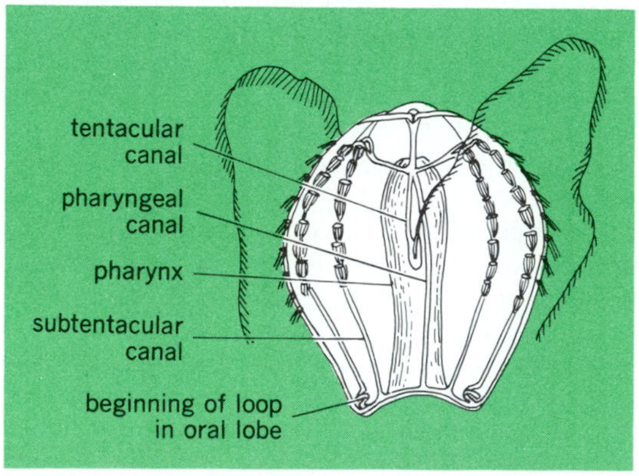

***Bolinopsis mikado*, of the Lobata.**

Examples of this order include *Bolinopsis*, *Ocyropsis*, *Mnemiopsis*, and *Leucothea*. *See* CTENOPHORA. [T.K.]

Lobosia A subclass of Rhizopodea characterized by lobopodia predominantly, although certain of these protozoan species also may form slender pseudopodia, or even develop several different kinds. Lobosia are divided into two orders, Amoebida and Arcellinida, which differ in presence or absence of a test. Pseudopodia emerge through the aperture of the test. *See* AMOEBIDA; ARCELLINIDA; PROTOZOA; SARCODINA; SARCOMAS-TIGOPHORA. [R.P.H.]

Lobster The name for a number of species of decapod crustaceans which are members of the family Homaridae and are commercially important because of their food value. Three species are commonly marketed: the European lobster (*Homarus vulgaris*), the Norwegian lobster (*Nephrops norvegicus*), and the American lobster (*Homarus americanus*). The American lobster is found in greatest abundance along the eastern coast of North America from Maine to New York, but ranges from Labrador to North Carolina. Migrations are chiefly to deeper waters in the fall, with a return to shallower waters in the spring. *See* DECAPODA (CRUSTACEA). [C.B.C.]

Local-area network A system consisting of a set of nodes that are interconnected by a set of links. The nodes and links cover a relatively small geographic area, ranging from a few feet to a mile. The nodes may be terminals, microcomputers, minicomputers, mainframes, printers, hard disks, or work stations. The links may be coaxial cable, twisted-pair wires, or fiber-optic cable. There are many reasons for developing and using local-area networks (LANs), including message communication, resource sharing, improved productivity, file transfer, application performance improvement, and combinations of these functions. *See* COAXIAL CABLE; COMMUNICATION CABLES; DATA-PROCESSING SYSTEMS; DIGITAL COMPUTER; MICROCOMPUTER; OPTICAL COMMUNICATIONS.

The three characteristics, not mutually exclusive, that differentiate local-area network architectures are control, topology, and communication technology. Control refers primarily to the techniques used for allowing access to the communications medium and resolving contention. There are three major control techniques: carrier sense, multiple access with collision detection (CSMA/CD); token passing; and centralized switching. In CSMA/CD systems, nodes are connected to a bus and have multiple, or concurrent, access to the communication medium (for example, coaxial cable). Nodes listen for existing transmissions to determine whether it is safe to transmit; hence, the terminology "carrier sense," or sensing the carrier. Token passing is a control technique that guarantees an upper limit on response time (that is, deterministic performance), but is more complex and expensive than CSMA/CD. This technique is used on rings and buses and uses a control message called the token for granting permission to a node to transmit. In a ring local-area network, for example, the token is rotated in the ring until a node that has data to transmit captures it. Centralized switching is used primarily in private automatic branch exchange (PABX) systems, which were first used for voice communication, and are challenging other types of local-area networks for market supremacy by emphasizing the integration of voice and data in one network.

Topology refers to the way in which nodes are connected. There are three major arrangements: bus, ring, and star. The control technique and the topology of a local-area network are mutually interdependent.

There are two major communication technologies: baseband and broadband. Baseband is unmodulated transmission; it transmits at the original frequency of the digital data. Broadband transmission is modulated and is based on community antenna television (CATV) technology. The PABX uses a different communication technology. Pulse-code modulation is used to quantize the analog voice signal into a binary bit stream for efficient digital transmission. This method is also compatible with integrating voice with data, since the data are already in digital form and can be merged with the digitized voice information. *See* DATA COMMUNICATIONS; PULSE MODULATION. [N.F.S.]

Local Group The cluster of galaxies to which the Milky Way Galaxy belongs. It is now generally accepted that most or all galaxies are clustered. The great clusters, which contain thousands of member galaxies, are rare, but can be recognized to very large distances. Clusters of smaller population are far more numerous. Most common are systems of only a few dozen or less galaxies; these are usually called groups of distinguish them from the richer clusters. The Local Group is such an entity. *See* COSMOLOGY.

The total membership of the Local Group is about two dozen. Those galaxies generally agreed to belong to the Local Group include three spirals: the Milky Way Galaxy; the Andromeda Galaxy, also known as Messier 31 (M31) and New General Catalog Number 224 (NGC 224); and the spiral in Triangulum—M33 or NGC 598. There are four irregular galaxies, including the nearest galaxian neighbors, the Large and Small Magellanic Clouds. Four Local Group galaxies are relatively small ellipticals. *See* ANDROMEDA GALAXY; GALAXY, EXTERNAL; MAGELLANIC CLOUDS. [G.O.A.]

Location fit Mechanical sizes of mating parts such that, when assembled, they are accurately positioned in relation to each other. Locational fits are intended only to determine the orientation of the parts. For normally stationary parts that require ease of assembly or disassembly, for parts that fit snugly, and for parts that move yet fit closely as in spigots, slight clearance is provided between parts. Where accuracy of location is important, transition fits are used. In these fits the holes and shafts are normally nearly the same diameter. For greater accuracy of location, the shafts are made slightly larger than the holes; such fits are termed location interference fits. *See* ALLOWANCE. [P.H.B.]

Lock-in amplifier An instrument for the measurement of periodic signals contaminated by noise and interference. The operation of a lock-in amplifier depends on the provision of a reference voltage that is synchronized to the signal to be measured. In a typical experimental application the reference is derived directly from the excitation source. *See* ELECTRICAL NOISE.

The signal channel of a lock-in amplifier contains a variable-gain alternating-current (ac) amplifier. The reference channel generates a precision square-wave voltage triggered from the applied reference, and incorporates a variable phase shifter. The variable phase shifter is used to adjust the relative phase of the signal and square-wave reference voltages appearing at the phase-sensitive detector. The phase-sensitive detector consists of a switching multiplier followed by a low-pass filter and will only detect (or rectify) signal voltages that are synchronous with the reference voltage. The direct-current (dc) component is transmitted to the final output by the low-pass filter and has a valve that depends on the signal amplitude and the phase difference between signal and reference. The user is able to maximize the response to a signal of interest by adjusting the reference phase from the front panel of the lock-in amplifier. *See* ELECTRIC FILTER; FUNCTION GENERATOR.

The response of the multiplier to asynchronous noise inputs is a fluctuating ac voltage with zero average value. The response to a synchronous signal is therefore quite unambiguous because there is no error due to rectified noise components.

Lock-in amplifiers operate over many decades of frequency without tuning or adjustment and are therefore suited to swept-frequency measurements. Because the detection process is phase-sensitive, the same instrument can be used in a range of measurements on signals and systems. Lock-in amplifiers are widely employed as precision vector voltmeters, selective microvoltmeters, frequency-response analyzers, and spectrum analyzers covering a frequency range from less than 1 Hz to several hundred kilohertz. *See* AMPLIFIER; SPECTRUM ANALYZER; VOLTMETER. [M.L.M.]

Locomotive A machine on flanged wheels, capable of propelling itself and other vehicles on railroad tracks by converting heat or electric energy into mechanical power. Most of the locomotive units in use on United States railroads are diesel-electric units. There are far fewer all-electric locomotives in service, and the steam locomotive has almost completely disappeared.

In addition to horsepower, locomotives are rated on the basis of tractive effort (measured at the rail) or drawbar pull (measured at the rear coupler of locomotive). A diesel or electric locomotive is designed to develop tractive effort equal to about 25% of the weight on its driving wheels when starting a train on clean dry rail. Greater tractive effort than this results in wheel slip. Three types of train resistance must be overcome by this tractive effort: resistance on grades, on curves, and during acceleration. *See* RAILROAD ENGINEERING. [W.D.E.]

Loculoascomycetes A class in the subdivision Ascomycotina; the Loculoascomycetes form a well-developed mycelium which bears the sexual (ascus) and asexual (conidium) states, and are distinguished from other ascomycetes by their method of ascocarp formation and their ascus structure. Ascocarp formation is ascolocular, a type of development in which certain cells of the vegetative hyphae undergo numerous divisions to form a small mass of homogeneous tissue (stroma). As the stroma enlarges, internal differentiation occurs to form the ascogenous cells, from which the asci will form. As the asci develop, they dissolve or crush the internal tissues of the stroma (now an ascostroma), creating a cavity (locule). As the ascostroma matures, a neck with a canal for ascospore discharge usually forms. An ascostroma may have one or several locules with asci. Species with a single locule are termed pseudothecia, and they often resemble the perithecia of the Pyrenomycetes. The asci have a two-layered wall (bitunicate). The outer layer (ectotunica) is thin and rigid, whereas the inner layer (endotunica) is thicker and elastic. At maturity, the ectotunica splits, allowing the endotunica to expand and forcibly discharge the eight ascospores. The ascospores may be hyaline or brown, but are usually multicelled.

Found primarily on living and dead plant tissues, members of this class include a number of important plant disease fungi, such as *Apiosporina morbosa* (black knot or cherry and other *Prunus* spp.); *Botryosphaeria rhodina* (stem diseases of various woody plants); *Cochliobolus heterostrophus* (leaf spot of corn and sorghum); *Guignardia bidwellii* (black rot of grapes); *Mycosphaerella berkeleyi* (leaf spot of peanut); *M. brassicicola* (ring spot of cabbage and other *Brassica* spp.); *M. fijiensis* (black leaf streak of banana); and *Venturia inaequalis* (apple scab). Sooty molds (*Capnodium*) often occur epiphytically on citrus fruits. *See* ASCOMYCOTINA; EUMYCOTA; FUNGI; PLANT PATHOLOGY. [R.T.Ha.]

Locust (forestry) A name commonly applied to two trees, the black locust (*Robinia pseudoacacia*) and the honey locust (*Gleditsia triacanthos*). Both of these commercially important trees have podlike fruits similar to those of the pea or bean.

The black locust is native in the Appalachian and the Ozark regions and is now widely naturalized in the eastern United States, southern Canada, and Europe. The wood is hard and durable in the soil and is used for fence posts, mine timbers, poles, and railroad ties.

The honey locust is native in the Appalachian and the Mississippi Valley regions, but is also widely naturalized in the eastern United States and southern Canada. The reddish wood is hard, strong, and coarse-grained and takes a high polish. Because it is durable in contact with soil, it is used for fence posts and railroad ties; it is also used for construction, furniture, and interior finish. [A.H.G./K.P.D.]

Loess An essentially unconsolidated, unstratified, calcareous silt. Most commonly it is homogeneous, permeable, buff to gray in color, and contains calcareous concretions and fossils. In natural and artificial excavations the loess maintains notably stable, vertical faces.

Although dominantly silt and often referred to as well sorted, loess is actually only moderately well sorted. The silt (and sand) grains are usually angular to subangular. Clay occurs as silt-size aggregates, as coatings on silt grains, and as interstitial filling. In texture and mineral composition, loess is comparable to the average mudstone or shale.

Loess is typically developed in areas peripheral to those covered by the last ice sheets in North America and Europe. Most of the loess seems to have been deposited during the latest (Wisconsin) glacial stage. *See* SEDIMENTOLOGY. [C.J.R.]

Logarithm An exponent of a suitably chosen positive number (base) larger than unity. Logarithms are of value in mathematical computation and in the equations and formulas used in expressing natural phenomena.

If $b^l = n$, where b is a positive number larger than unity, then l is called the logarithm of n to the base b and is written $l = \log_b n$. From this definition it follows at once that all positive numbers larger than unity have positive logarithms, all positive numbers smaller than unity have negative logarithms, and the logarithm of unity is equal to 0. Since $b^0 = 1$ irrespective of the value of b, it follows that the logarithm of unity is equal to 0. From the known properties of exponentials expressed by relations (1), it follows immediately that (1) the logarithm of a

$$b^{l1} \cdot b^{l2} = b^{l1 + l2} \qquad b^{l2} \div b^{l2} = b^{l1 - l2} \qquad (b^l)^m = b^{lm} \qquad \sqrt[m]{b^l} = b^{l/m} \tag{1}$$

product of two or more factors is equal to the sum of the logarithms of the factors; (2) the logarithm of the ratio of two numbers is equal to the difference between the logarithm of the numerator and the logarithm of the denominator; (3) the logarithm of the mth power of a number is the product of m and the logarithm of the number; and (4) the logarithm of the mth root of a number is equal to the logarithm of the number divided by m. *See* EXPONENT.

Although any positive number b larger than unity might have been chosen as the base of a system of logarithms, actually two numbers have been chosen in the construction of tables of logarithms, namely, the number $b = 10$ and the number $e = 2.718...$ defined as the sum of infinite series (2). The

$$1 + \frac{1}{1} + \frac{1}{1 \cdot 2} + \frac{1}{1 \cdot 2 \cdot 3} + \cdots + \frac{1}{1 \cdot 2 \cdot 3 \cdots n} + \cdots \tag{2}$$

system of logarithms to the base 10 is usually referred to as common logarithms; the system of logarithms to the base e is called natural logarithms. *See* E (MATHEMATICS).

The integral part of the common logarithm of a number larger than unity is one unit less than the number of digits before the decimal point. This integral part is called the characteristic; the decimal part is called the mantissa. Similarly, because the common logarithms of 0. 1, 0.01, and 0.001 are -1, -2, and -3, it follows that the logarithm of a number smaller than unity having p zeros after the decimal point may be expressed as the sum of a negative characteristic equal to $-(p + 1)$ and a positive mantissa. [A.N.L./S.Bo.]

Logging (forestry) The cutting and removal of the woody stem portions of forest trees. In effect, logging is the harvesting of timber from forests. Originally, logging produced only sawlog lengths for lumber sawmills, but with increased utilization of wood for other products such as pulp and paper, veneers, and poles and piles, the term logging includes practically all operations conducted in the forest in connection with felling trees, cutting the trunks to desired lengths, and transporting the merchantable material to processing mills or to shipping points. Most lumber log lengths are 12, 14 or 16 ft (3.6, 4.2, or 4.8 m), corresponding to the standard lengths of lumber used in general construction; or in multiples of these lengths, in which event final log lengths are cut at the mill. Pulpwood generally is cut in lengths of 4, 5, or 8 ft (1.2, 1.5, or 2.4 m), although sawlog lengths may be produced for efficient mechanized handling of larger diameters. For other products, various lengths are produced.

The steps usually followed in logging are: (1) selection of trees; (2) felling the trees; (3) bucking the trunks (cutting into appropriate lengths) and removing limbs and unmerchantable tops; (4) skidding or forwarding the bucked lengths to roadside or waterside; (5) loading on hauling vehicles, usually trucks; and (6) hauling to destination. *See* WOOD PRODUCTS. [A.E.W.]

Logic The subject that investigates, formulates, and establishes principles of valid reasoning.

The first attempt to investigate systematically acceptable modes of reasoning was made by Aristotle, in whose *Organon* reasoning was recognized as the subject of a special science. He formulated three basic "laws of thought": (1) the law of contradiction (no proposition is both true and false), (2) the law of excluded middle (each proposition is either true or false), and (3) the law of identity (each proposition implies itself). Advanced as his views of logic were, it seems doubtful that the idea of axiomatizing them ever occurred to him, despite the success of his contemporary Euclid in organizing geometry. That aspect of logic which enunciates or establishes valid reasoning (rather than merely investigating it) began with the publication in 1854 of George Boole's *An Investigation of the Laws of Thought*. The partly abstract treatment of logic presented in this work initiated the completely abstract developments that were to follow. In it the laws of thought are regarded as mere conventions which, like the postulates of euclidean geometry, might be modified or even rejected to create new logics. The only requirement that a "new" logic must satisfy is the one demanded of every deductive system—consistency. Just as different geometries are useful for different purposes, a logic that is appropriate in one environment might not be so in another. There are a growing number of competent individuals, for example, who consider that any logic containing Aristotle's law of excluded middle is not suitable for mathematics.

The unit of the propositional calculus is the proposition, which may be defined as the meaning of an indicative or declarative sentence, Each proposition has a truth-value; it is either true or false in the classical logic, but may assume other truth-values (for example, uncertain) in some of the new logics. It is convenient to use letters p, q, r, and so on to denote propositions. Propositions are combined by logical connectives to form other propositions. Chief among these are (1) negation (the negation of a proposition p, symbolized by p', may be formed verbally by writing "It is not the case that" before p); (2) conjunction (a binary operation symbolized by $\cdot$ or &, for example, $p \cdot q$, read "p and q"); (3) disjunction (a binary operation symbolized by + or $\vee$; for example, $p \vee q$, read "p or q"); (4) implication (a binary operation symbolized by $\supset$ or $\rightarrow$, for example, $p \rightarrow q$, read "p implies q"); and (5) equivalence (a binary operation denoted by $\equiv$; for example, $p \equiv q$, read "p is equivalent to q," means that each proposition implies the other). The proposition $p \cdot q$ asserts that both p and q are true. It is important to note that the disjunctive "or" is always used in the inclusive sense; that is, $p \vee q$ asserts that at least one, possibly both, of the propositions p and q is true. In addition, the meaning of "implication" in logic differs markedly from the ordinary sense of that term. The proposition $p \rightarrow q$ asserts merely that it is not the case that p is true and q is false; that is, in classical logic $p \rightarrow q$ means that either p is false or q is true. The logical connectives introduced above are examples of logical constants. They are not independent notions. Implication may be defined in terms of negation and conjunction, and this is true for all of the connectives. In fact, all the connectives can be defined in terms of one binary operation, the so-called Scheffer stroke function $p \mid q$, which may be read "p is false or q is false."

The logical connectives, the letters p, q, r, and so on, that stand for propositions (propositional variables), and the parentheses, brackets, and braces needed for punctuation are formal concepts. A propositional function is a combination of concepts, formal or nonformal, involving at least one variable, which is not a proposition but becomes one when all the variables are given specified values. Examples are (1) p and q, (2) $x^2 + y^2 = 1$, and (3) x is president of the United States. Unlike a proposition, a propositional function has no truth-value. On the other hand, the statement "For every integer x, $x^2 + 1 = 0$" is a proposition (not a propositional function) even though it contains a (bound) variable. A formal function is a propositional function that contains only formal concepts—for example $[(p \vee q) \cdot r] \supset s$. A truth function is a propositional function in which only propositional variables occur, and such that the truth value of each proposition obtained from the function by substituting specific propositions for the variables depends only on the truth values of those propositions. Examples are (1) $(p \cdot q) \vee r'$, (2) Napoleon was a great general and p (where p is a propositional variable). On the other hand, "Jones stated that p" is not a truth function.

Tautologies are formal truth functions such that every proposition obtained from them by substituting specific values (propositions) for the variables they contain has the truth value T (true). A truth function is contingent provided it assumes both truth values T and F (false)—here only the classical two-valued (aristotelian logic) truth functions are considered—and self-inconsistent provided it has only the value F. The character of a truth function is determined when it is shown to be self-inconsistent, contingent, or a tautology.

A truth function may be analyzed by means of a truth table, or matrix. Such tables are made first for the elementary truth functions p' $p \vee q$, $p \cdot q$, $p \supset q$ as follows:

(I)

p	p'
T	F
F	T

(II)

p	q	$p \vee q$
T	T	T
T	F	T
F	T	T
F	F	F

(III)

p	q	$p \cdot q$
T	T	T
T	F	F
F	T	F
F	F	F

(IV)

p	q	$p \supset q$
T	T	T
T	F	F
F	T	T
F	F	T

In every table each row of the last column gives that truth value of the function which is determined by those truth values (assigned to the variables) that are entered in the preceding columns of that row.

All tautologies can be deduced from a set of just four of them:

P1 $(p \vee p) \supset p$
P2 $q \supset (p \vee q)$
P3 $(p \vee q) \supset (q \vee p)$
P4 $(q \supset r) \supset [(p \vee q) \supset (p \vee r)]$

These simple tautologies were selected by Alfred North Whitehead and Bertrand Russell in their monumental work *Principia Mathematica* as the basic set from which all tautologies can be derived. This is analogous to what is done in an axiomatic geometry in which all the theorems of the geometry are deduced from a selected set of axioms or postulates that are assumed to be true.

It was proved by E. L. Post that postulates P1–P4 are consistent; that is, it is impossible to deduce from them truth functions X and X'. He also proved that the four postulates form an independent set; that is, no one of the postulates can be deduced from the others.

In nonaristotelian logics the principle of the excluded middle (that is, each proposition is either true or false) is not valid. There are, then, at least three truth values possible for a proposition, and for this reason such logics are known as many-valued. If "true" be interpreted in mathematics to mean "provable," and a proposition be called false provided its negation is provable, then the law of excluded middle is demonstrably false, since it has been shown that mathematics contains proposition p such that neither p nor its negation p' is provable. *See* BOOLEAN ALGEBRA; EUCLIDEAN GEOMETRY; MATHEMATICS; SCIENCE. [L.M.Bl.]

Logic circuits Electronic circuits which process information encoded as one of a limited set of voltage or current levels. Logic circuits are the basic building blocks used to realize consumer and industrial products that incorporate digital electronics. Such products include digital computers, video games, voice synthesizers, pocket calculators, and robot controls. *See* INTEGRATED CIRCUITS.

All logic circuits may be described in terms of three fundamental elements, shown graphically in the illustration. The NOT element has one input and one output; as the name suggests, the output generated is the opposite of the input in binary. In other words, a 0 input value causes a 1 to appear at the output; a 1 input results in a 0 output. (All signals are interpreted to be one of only two values, denoted as 0 and 1.)

The AND element has an arbitrary number of inputs and a single output. As the name suggests, the output becomes 1 if, and only if, all of the inputs are 1; otherwise the output is 0. The AND together with the NOT circuit therefore enables searching for a particular combination of binary signals.

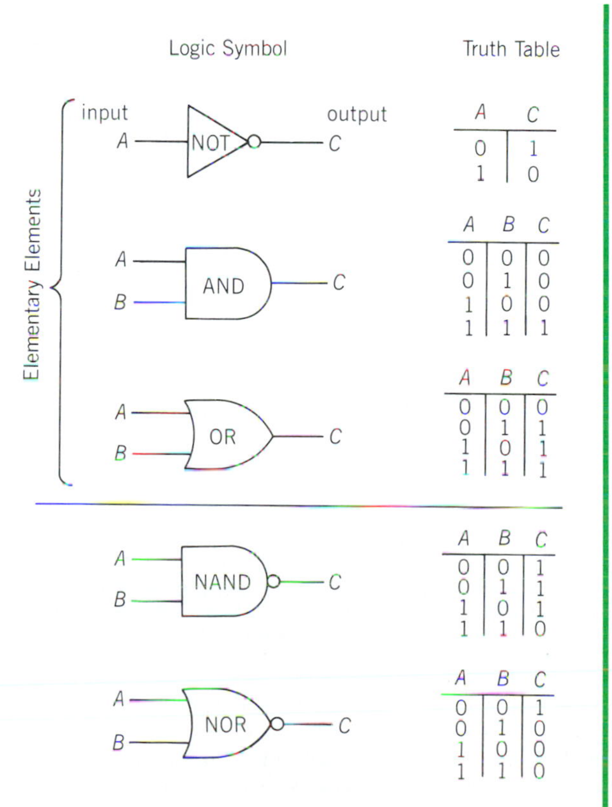

Logic elements.

The third element is the OR function. As with the AND, an arbitrary number of inputs may exist and one output is generated. The OR output is 1 if one or more inputs are 1.

The operations of AND and OR have some analogies to the arithmetic operations of multiplication and addition, respectively. The collection of mathematical rules and properties of these operations is called boolean algebra. *See* BOOLEAN ALGEBRA.

While the NOT, AND, and OR functions have been designed as individual circuits in many circuit families, by far the most common functions realized as individual circuits are the NAND and NOR circuits of the illustration. A NAND may be described as equivalent to an AND element driving a NOT element. Similarly, a NOR is equivalent to an OR element driving a NOT element.

As the names of the logic elements described suggest, logic circuits respond to combinations of input signals. Logic networks which are interconnected so that the current set of output signals is responsive only to the current set of input signals are appropriately termed combinational logic. An important further capability for processing information is memory, or the ability to store information. The logic circuits themselves must provide a memory function if information is to be manipulated at the speeds the logic is capable of. Logic circuit networks that include feedback paths to retain information are termed sequential logic networks, since outputs are in part dependent on the prior input signals applied and in particular on the sequence in which the signals were applied.

Several alternatives exist for the digital designer to create a digital system. Two common realizations are ready-made catalog-order devices, which can be combined as building blocks, and custom-designed devices. Gate-array devices comprise a two-dimensional array of logic cells, each equivalent to one or a few logic gates. Programmable logic arrays have the potential for realizing any of a large number of different sets of logic functions. In table look-up, the collection of input signals are grouped arbitrarily as address digits to a memory device. Finally, the last form of logic network embodiment is the microcomputer. *See* MICROCOMPUTER.

There are basically two logic circuit families in widespread use: bipolar and metal-oxide-semiconductor (MOS). *See* TRANSISTOR. [R.R.Sh; W.V.R.]

Lophiiformes The anglerfishes and their relatives. The first dorsal fin is reduced to a few flexible rays, of which the first is placed on top of the head and bears a terminal bulb or tassel and functions as a fishing lure (see illustration).

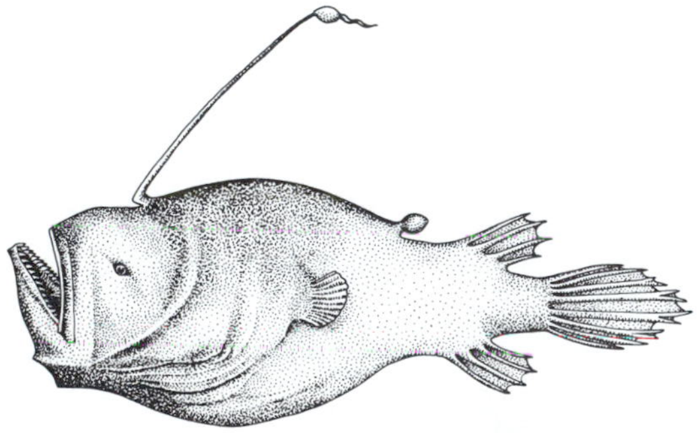

Anglerfish (*Cryptopsaras couesi*). (*After G. B. Goode and T. H. Bean, Oceanic Ichthyology, U.S. Nat. Mus. Spec. Bull. no. 2, 1895*)

Luminescent organs may be present, and in some deep-sea species the males are dwarfed and attached as ectoparasites on the females. There are 3 suborders, 15 families, nearly 60 genera, and about 195 known species. *See* ACTINOPTERYGII. [R.M.B.]

Lophophore The crown of tentacles which surrounds the mouth in the Bryozoa, Phoronida, and Brachiopoda. The numerous ciliated tentacles arise from a circular or horseshoe-shaped fold of the body wall. The tentacles are hollow outgrowths of the body wall, each containing fluid-filled extensions of the body cavity and extended hydraulically. The primary function of the lophophore is to gather food. On the tentacles are ciliary tracts which drive a current of food-particle-bearing water through the lophophore. While the lophophore is primarily a feeding organ, it may also play a role in reproduction, respiration, and larval locomotion. *See* BRACHIOPODA; BRYOZOA; PHORONIDA. [J.E.Wi.]

Loran A navigation system from which hyperbolic lines of position are determined by measuring the difference in times of arrival of pulses from widely spaced, synchronized transmitting stations. Since radio waves travel at the speed of light, a line of position represents a constant range difference from two transmitters.

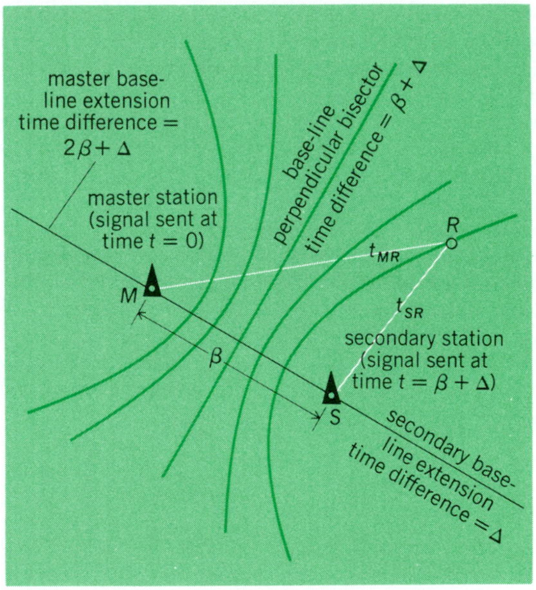

Principle of position fixing by hyperbolic navigation systems. (*After J. P. Van Etten, Navigation systems: Fundamentals of low and very-low-frequency hyperbolic techniques, Electr. Commun., 45(3):192–212, 1970*)

In the illustration, it is assumed that the master station M transmits a pulse signal at time $t = 0$, and that the secondary station S transmits a similar pulse signal Δ microseconds after receiving the master pulse signal. The secondary station receives the master signal at time $t = \beta$ microseconds, and therefore transmits its pulse at time $t = \beta + \Delta$ microseconds. A receiver at R measuring the difference in the times of arrival of the signals from the secondary and master stations measures the time difference $TD = (\beta + \Delta) + t_{SR} - t_{MR}$. The locus of all points with this common time difference is a hyperbola through the receiver position R. Similarly, every hyperbolic line of position is uniquely defined by a time difference. The intersection of two such lines of position gives a navigational fix, or location. Loran is used by commercial and military ships and aircraft. The name is derived from "long range navigation." *See* HYPERBOLIC NAVIGATION SYSTEM. [J.P.V.E.]

Lorentz transformations The family of mathematical transformations used in the special theory of relativity to relate the space and time variables of uniformly moving (inertial) reference systems. They were discovered in part by H. A. Lorentz during his studies of matter moving in an electromagnetic field and were formulated more completely by H. Poincaré and by Albert Einstein. If S and S' are two inertial reference frames with space-time coordinates (x,y,z,t) and (x',y',z',t'), respectively, the equations connecting these variables have the form ($x^0 = ct$, $x^1 = x$, $x^2 = y$, $x^3 = z$) of Eq. (1), where a^0, a^1, a^2, a^3

$$x'^{\alpha} = \sum_{\beta=0}^{3} L_{\beta}{}^{\alpha}(x^{\beta} - a^{\beta}) \quad (\alpha = 0, 1, 2, 3) \qquad (1)$$

are arbitrary real constants. Forming the two real matrices (2) and (3), the matrix L is subject to the condition shown

$$L = \begin{bmatrix} L_0{}^0 & L_1{}^0 & L_2{}^0 & L_3{}^0 \\ L_0{}^1 & L_1{}^1 & L_2{}^1 & L_3{}^1 \\ L_0{}^2 & L_1{}^2 & L_2{}^2 & L_3{}^2 \\ L_0{}^3 & L_1{}^3 & L_2{}^3 & L_3{}^3 \end{bmatrix}$$

$$\eta = \begin{bmatrix} 1 & 0 & 0 & 0 \\ 0 & -1 & 0 & 0 \\ 0 & 0 & -1 & 0 \\ 0 & 0 & 0 & -1 \end{bmatrix}$$

as Eq. (4), where L^{-1} and L_t are the inverse and transposed

$$L^{-1} = \eta L_t \eta \qquad (4)$$

matrices of L. *See* RELATIVITY. [E.L.Hi.]

Loricifera A phylum of multicellular invertebrates. These marine organisms are entirely meiobenthic; that is, they never exceed a maximum dimension of 400 micrometers and live in sediments ranging from deep-sea red clay to coarse sand or shell hash. They have some of the smallest known cells in the animal kingdom. Although they have been found throughout the world, only 10 species representing three genera and two families have been described: the first described species (*Nannaloricus mysticus*) from shallow coastal waters near Roscof, France; eight species (including the first loriciferan seen, *Pliciloricus enigmaticus*) off North Carolina; and a single species from the western Pacific. Well over 50 species thought to represent several additional genera are known but remain undescribed.

Adult loriciferans are bilaterally symmetrical. The body is regionated into a mouth cone which telescopically protrudes anteriorly from the center of a spherical, appendage-ringed (up to 400 rings may be present) head; a short, usually plated neck region; and a distinctively loricated (thick-cuticled) thorax-abdomen (see illus.). The head, along with its withdrawn mouth cone, may be inverted into the thorax-abdomen, which is then closed anteriorly by the cuticle of the neck region.

Loriciferans appear to be closely related to the Priapulida and perhaps less so to the Kinorhyncha. Certain structures of the mouth cone suggest affinities to the Tardigrada. Their presence may support the idea that proarthropods could have been developed from aschelminth ancestral stock. *See* ANIMAL KINGDOM; ARTHROPODA; KINORHYNCHA; PRIAPULIDA; TARDIGRADA. [R.P.H.]

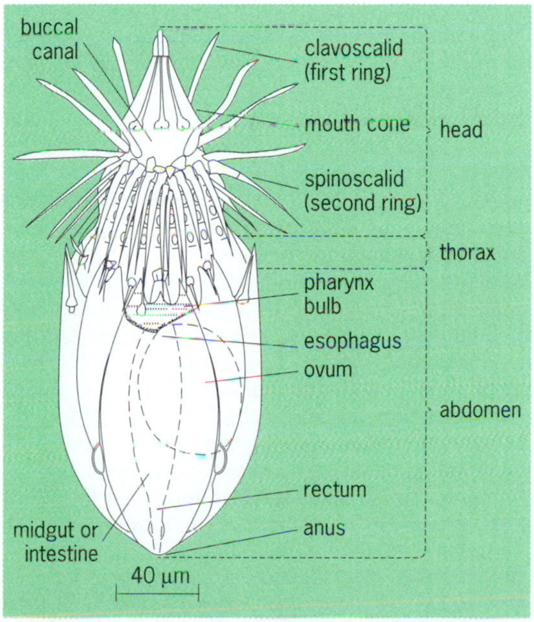

Ventral view of adult female loriciferan (*Nannaloricus mysticus*). (*After R. M. Kristensen, Loricifera, a new phylum with Aschelminthes characters from the meiobenthus, Z. Zool. Syst. Evolutionsforsch., 21(3):163–180, 1983*)

Loudness The perceptual intensity of sound. Loudness depends importantly on the physical intensity of sound, increasing when physical intensity increases and decreasing when physical intensity decreases. But loudness also depends on other physical properties of sound, such as frequency and duration. Sound waves with frequencies between 1000 and 5000 hertz are louder than sound waves that have the same intensity but lower or higher frequencies. The physical level of sound is expressed in decibels. *See* DECIBEL.

Because loudness depends on sound frequency as well as sound intensity or pressure, different stimuli with the same sound pressure level (SPL) may not be equally loud. One type of decibel scale, called the phon scale, overcomes this deficiency. The level of a sound in phons is the SPL in decibels of an equally loud 1000-Hz tone. Thus a 1000-Hz tone at 40-dB SPL has a level of 40 phons, as do all other sounds that equal its loudness, even though these other sounds may have SPLs much greater than 40 dB.

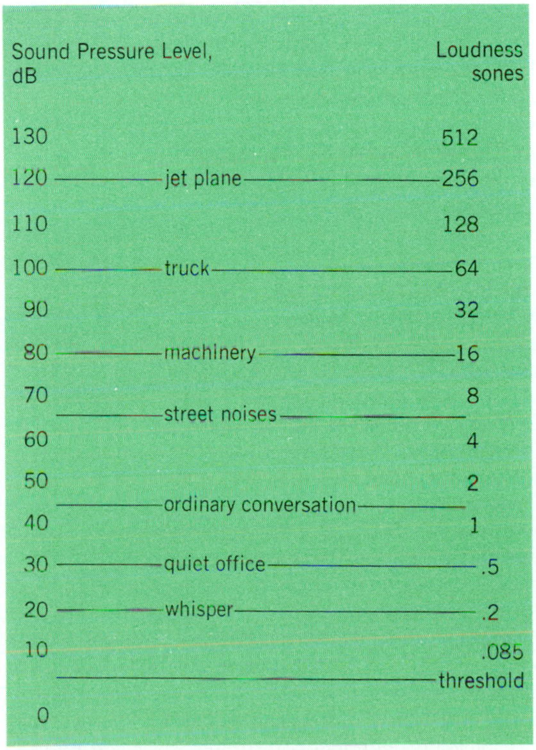

Decibel and loudness scales of common sounds.

The phon scale is often designated as a scale of loudness level. It is a scale of equal loudness, in that all equally loud sounds take the same level in phons, regardless of the SPL. Nevertheless, the phon scale is not a true scale of loudness, because it is a physical (decibel) scale. That is, a sound of 80 phons is not necessarily twice as loud as a sound of 40 phons. In order to determine a scale of perceived loudness, observers were asked to set the level of one sound to make it appear twice or half as loud as standard sounds. Observers were also asked to make numerical judgments of the degree of loudness. From such judgments a scale of loudness in sones was established. One sone is defined as the loudness of a 1000-Hz tone at 40-dB SPL heard with both ears. (When heard with one ear, loudness is half as great.) Above 40 dB, loudness in sones doubles with every 10-dB increase in the level of sound. Below 40 dB, loudness falls by half with decrements smaller than 10 dB (see illustration). *See* HEARING (HUMAN); SOUND; SOUND PRESSURE. [L.E.M]

Loudspeaker A device that converts electrical signal energy into acoustical energy, which it radiates into a bounded space, such as a room, or into outdoor space. The essential characteristic of the signal energy must be preserved in the energy-conversion process. The shorter term speaker is also used. The term unit, as in horn or driver unit, is applied to devices comprising a motor and diaphragm but lacking the acoustic element such as a horn to complete the speaker. The term speaker system is normally reserved for a plurality of speaker units and the associated accessories, horns, cabinets, and electrical networks required for an integrated system. *See* SOUND-REPRODUCING SYSTEMS.

Tests have shown that frequency ranges of 80–11,000 Hz for music and 150–8000 Hz for speech, if uniform throughout the listening area, give high-quality reproduction in an otherwise distortionless system. The pressure range requirements are set primarily by listener preference. *See* PSYCHOACOUSTICS.

Speakers are commonly classified by terms which describe their three important functional parts: the motor, diaphragm, and acoustic radiation-controlling element. The motor converts electrical energy into mechanical energy and couples the electrical signal source, commonly an amplifier, to the diaphragm. Speakers are also classified by their radiation-controlling elements into direct-radiator (hornless) and horn types. The direct-radiator types commonly employ baffles or enclosures. The diaphragm is the element which, vibrated by the motor, causes the air to vibrate; hence it couples the air load (radiation impedance) to the motor.

Direct-radiator speakers are the most widely used speakers. Of these, substantially all of the general-purpose and special-purpose low-frequency (woofer) speakers are of the moving-coil type. Electrostatic speakers are sometimes used as high-frequency speakers (tweeters) in broad-band systems and infrequently as general-purpose speakers.

Two or more identical speakers may be used to improve low-frequency efficiency and increase permissible input and output powers or, by proper relative orientation, improve the directional properties. Several substantially identical speakers may be used in combination to influence the directivity pattern. More commonly, the two or more speakers cover complementary frequency ranges. The more important advantages are (1) improved frequency response, (2) higher system efficiency, (3) improved directivity characteristic, (4) improved transient response, (5) reduced intermodulation, and (6) reduced frequency modulation, Coaxial units comprise two, or less commonly three, speaker units mounted on substantially the same axis in an integrated mechanical assembly. Appropriate wave filters or dividing networks control the amplitude and phase of the electrical signal reaching each unit.

In single-channel, or monophonic, systems speaker placement is largely controlled by the desirability of uniform response in the listening area. A corner location in a room is desirable, improving the low-frequency efficiency and reducing the angle the high-frequency radiation must cover. In stereophonic systems multiple speakers and signal channels are used to reproduce the spatial effects present in the original sound environment. The simpler conventional systems employ two channels and two speakers or speaker systems. These speakers are located symmetrically in the room, preferably along the shorter wall. In typical small listening rooms the speakers are 6–10 ft (2–3 m) apart. [H.S.K.]

Louse Any insect which is a member of the orders Anoplura (biting and sucking lice) or Mallophaga (biting or chewing lice). Lice are apterous (wingless) ectoparasites of birds and mammals. The leathery body is flattened dorsoventrally, with a head that is more or less distinct, and segmented legs (see illustration). The eggs, as they are laid, are attached to the hair or feathers of the host. *See* ANOPLURA; MALLOPHAGA. [C.B.C.]

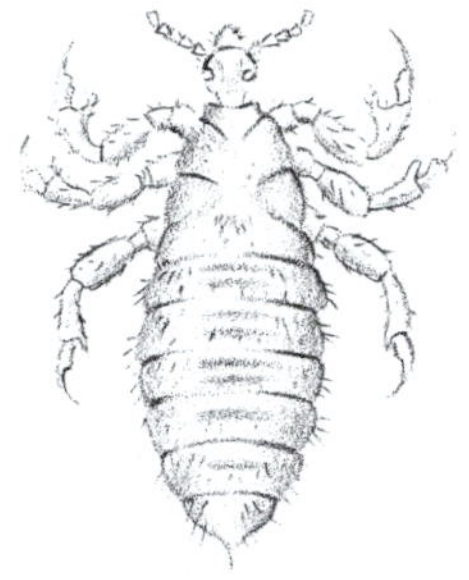

The human-head louse.

Low-level counting The measurement of very small amounts of radioactivity. The instruments used are special modifications of standard radioactivity measurement equipment, designed to yield greater detection efficiencies and lower background counting rates than those necessary for ordinary applications. These techniques are applied mainly to the measurement of long-lived natural radioactive isotopes and their daughter products, cosmic-ray-produced isotopes, and isotopes produced artificially during nuclear explosions. *See* RADIOACTIVITY.

Because of the statistical error associated with the measurement of radioactivity, accurate assays of samples with small amounts of radioactivity cannot be made in the presence of large amounts of radiation from other sources. The error can be decreased by reducing the background or by increasing the sample count rate. Thus most low-level counters are designed (1) to hold as large a sample as it is practical to obtain and to prepare; (2) to record with as high an efficiency as possible the radiations from the sample; and (3) to have as low a background counting rate as possible. The instruments which have been most widely used are internal gas counters, solid source beta-counters, and scintillation counters. *See* GEIGER-MÜLLER COUNTER; PARTICLE DETECTOR; SCINTILLATION COUNTER. [M.Stu.]

Low-pressure gas flow Flow of gases below atmospheric pressure, following the perfect gas laws, in pipes, fittings, and other common configurations. *See* GAS.

The laws of gas flow at subatmospheric pressures in the turbulent, transition, and laminar flow range are the same as the laws of gas flow at any pressures. The molecular flow range is characterized by a mean free path of molecules longer than the smallest linear dimension of the conduit. The laws of flow in this range are based on the assumption that the molecules have a Maxwell-Boltzmann distribution of velocities. *See* KINETIC THEORY OF MATTER; LAMINAR FLOW; MEAN FREE PATH; TURBULENT FLOW. [Z.C.D.]

Low-temperature acoustics The application of acoustics to research on the properties of condensed matter at low temperatures. Acoustic techniques are readily adaptable to the cryogenic environment and make possible many measurements of the structural and thermodynamic properties of materials at temperatures approaching absolute zero (0 K, which is −273°C). The study of sound propagation has also yielded major insights into the low-temperature phenomena of superconductivity in metals and superfluidity in liquid helium.

Solid materials. Acoustic measurements have been used to characterize the properties of a wide variety of solid-state materials, such as metals, dielectric crystals, amorphous solids, and magnetic materials. A measurement of the velocity of sound in a substance gives information on its elastic properties, while the attenuation of the sound characterizes the interaction of the lattice vibrations with the electronic and structural properties of the material. Ultrasonic frequencies, in the range from 20 kHz to 100 MHz and above, are commonly employed in these measurements because of the ease of generating and detecting the sound with piezoelectric quartz crystals.

Because the sound velocity effectively measures elastic constants, such measurements are used to characterize phase transitions in crystals where the structure of the lattice changes. The attenuation of sound in many crystals is due to defects and impurities in the crystal lattice and provides information on such structures. In a metal at very low temperatures, the dominant source of attenuation is the interaction of the sound with the conduction electrons. *See* CRYSTAL; LATTICE VIBRATIONS; PHASE TRANSITIONS; PHONON; SOUND ABSORPTION.

A large variety of magnetoacoustic effects are observed in metals and crystals. In these measurements, changes in the sound attenuation occur as the strength of a magnetic field applied to the sample is increased. One example is the phenomenon of nuclear acoustic resonance, resulting from the interaction of the nuclear spins in a crystal with vibrations of the lattice. There are also a number of other magnetoacoustic effects in metals which are useful in determining the orbits followed by the conduction electrons in the metal. *See* DE HAAS–VAN ALPHEN EFFECT.

Sound propagation is becoming increasingly useful for studying amorphous materials, in which there is no regular crystal structure. In materials such as silica glass (amorphous silicon dioxide, SiO_2), it has been found that at low temperatures there are only two quantum energy levels that are important. These levels correspond to two nearly equivalent arrangements of the atoms, with one arrangement having slightly higher energy. The stress from an imposed sound field can cause a transition from one arrangement to the other. If the relaxation rate back to the original configuration is comparable to the sound frequency, there will be a net absorption of energy from the sound wave. *See* AMORPHOUS SOLID.

When a metal is cooled below its superconducting transition temperature, there are striking changes in the attenuation of sound. At the transition some of the electrons near the Fermi surface begin to pair together, due to the attractive electron-phonon coupling. Once this occurs, the electrons can no longer exchange momentum with the lattice, and hence have zero resistance. This also means that the paired electrons no longer absorb energy from the sound wave, and the attenuation is from the remaining unpaired normal electrons. As the temperature is lowered well below the transition, the density of the unpaired electrons drops rapidly, and the attenuation becomes very small. *See* SUPERCONDUCTIVITY.

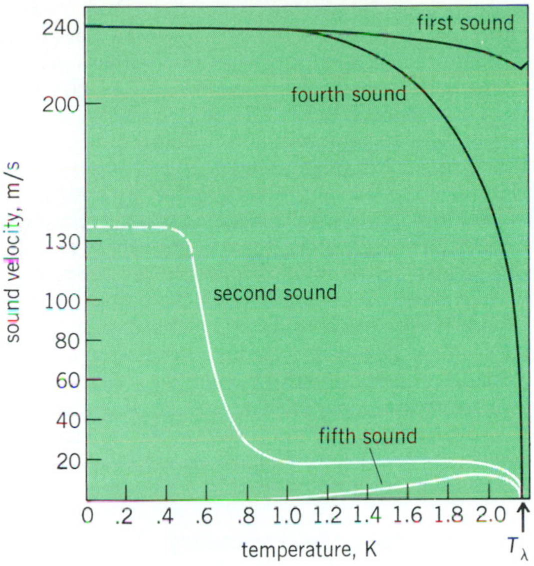

Velocity of the various types of sound in superfluid ^{4}He as a function of temperature.

Superfluid helium. Sound propagation has been extensively used to probe many of the unusual properties of superfluid helium. The novel features of the superfluid (zero viscosity and entropy) give rise to a rich variety of different types of sound which can propagate in the superfluid helium. Five distinct sound modes have been identified and observed experimentally. The sound velocities of a number of these modes are shown in the illustration as a function of temperature.

First sound is a pressure wave which propagates in the bulk liquid. It is quite similar to sound in ordinary fluids.

Second sound is an unusual type of wave: it is a temperature wave in the bulk superfluid. In this mode the normal fluid and superfluid move in opposite directions. This keeps the density constant, and hence there are no pressure oscillations in the wave (as in first sound); but because only the normal fluid carries entropy, there are oscillations in the entropy and thus in the temperature of the liquid. *See* SECOND SOUND.

Third sound is a wave which propagates in very thin films of helium. The third sound is a wave in which the thickness of the film varies, somewhat like waves in a tank of water. Because the films are so thin, only the superfluid can move, the normal fluid being immobilized by its viscosity.

Fourth sound is a pressure wave which propagates in superfluid helium when it is confined in a porous material such as a tightly packed powder. In such a situation the normal fluid is immobilized, and only the superfluid can flow freely because of its zero viscosity (the porous materials are often called superleaks for this reason). The fourth sound is analogous to first sound because it involves density and pressure oscillations and can be generated and detected by pressure transducers.

Fifth sound is a temperature wave which can propagate in helium confined in a superleak. It is analogous to second sound, except that again only the superfluid component can flow. *See* LIQUID HELIUM; SUPERFLUIDITY. [G.A.W.]

Low-temperature physics A study of the properties of gross matter at such low temperature that the quantum character of the substance becomes observable. In a general way, this implies temperatures of −452°F (4 K) and below. Quantum-mechanical coherence effects extend over readily measurable distances and give rise to superconductivity and superfluidity.

Zero-point energy "blows up" light atoms and greatly influences the bulk properties of the hydrogens and the heliums in the solid state. The existence of quantized vortices plays a large role in the basic behavior of superconductors and the liquid heliums. The existence of discrete atomic energy levels shows up via heat capacity anomalies (the Schottky effect) and various magnetic and nuclear phenomena. The "ortho-para conversion" is a striking thermal phenomenon in hydrogen compounds. Minima in electrical resistance (the Kondo effect) and thermal boundary resistances (Kapitza resistance) are other peculiarities under active study. The topic of cryogenics, most strictly interpreted, is the method of producing low temperatures. In addition to the considerable technology involved in refrigeration, there is a very important section of low-temperature physics concerned with temperature measurement and related instrumentation. *See* CRYOGENICS; KAPITZA RESISTANCE; KONDO EFFECT; LIQUID HELIUM; LOW-TEMPERATURE THERMOMETRY; QUANTIZED VORTICES; QUANTUM SOLIDS; SCHOTTKY ANOMALY; SUPERCONDUCTIVITY; SUPERFLUIDITY. [R.P.Hu.]

Low-temperature thermometry The measurement of temperature below 32°F (0°C). Very few thermometers are truly wide-range, and hence most of the conventional methods of thermometry tend to fail the further one moves below room temperature (see table). The defining instrument for the lower regions of the International Practical Temperature Scale, the platinum resistance thermometer, rapidly loses sensitivity below −405°F (30 K), and its official limit is set at −434.8l°F (13.81 K), the triple point of equilibrium hydrogen. This scale is based upon measurements of thermodynamic temperature made with the gas thermometer; the gas thermometer may be used down to about −456°F (2 K). *See* GAS THERMOMETRY.

The low-temperature region is unique in having available several different types of primary thermometers, all of which are quite practical. The least practical, perhaps, is the acoustic thermometer, which uses the property that, extrapolated to zero pressure, the speed of sound in a gas is proportional to $T^{1/2}$. This has been used in the range −456 to −424°F (2–20 K) as an alternative to, and check upon, the gas thermometer. The Johnson noise in a resistor can be used with particular advantage at low temperatures when allied with SQUID detector technology. *See* ELECTRICAL NOISE; SOUND.

Low-temperature thermometers

Thermometer	Temperature range, K
Thermocouples	
300 to 700 ppm Fe in Au/Ag + 0.37 at. % Au	1–25
Chromel/300 to 700 ppm Fe in Au	1–300
Chromel/Constantan	20–1100
Resistance thermometers	
Platinum (capsule)	4–500
Rhodium + 0.5 at. % Fe	0.5–300
Carbon	0.01–300
Germanium	0.01–30
Saturation vapor pressure thermometers	
Hydrogen	14–21
Helium-4	1.0–5.2
Helium-3	0.5–3.3
Magnetic thermometers	
Gadolinium metaphosphate, $Gd(PO_3)_3$	2–100
Cerous magnesium nitrate (CMN), $Ce_2Mg_3(NO_3)_{12} \cdot 24H_2O$; single crystal	0.003–4
CMN powder sphere or cylinder	0.002–4
Copper (and other nuclear paramagnets)	0.001–0.01
Gamma-ray anisotropy thermometers	
^{60}Co in hexagonal close-packed cobalt single crystal	0.002–0.04
^{54}Mn in iron	0.003–.03
^{54}Mn in nickel	0.004–0.045
Nuclear resonance thermometer	300 nK–2 K

In suitable systems it is possible to spatially orient atomic nuclei at very low temperatures, and if these nuclei are emitters of gamma rays, the emission pattern is anisotropic to a degree which is a measure of the thermodynamic temperature, Finally the magnetic susceptibility of suitable atomic nuclei may also be employed via the Curie law. Nuclear magnetic resonance or static SQUID-based techniques may be employed, but nuclear magnetic resonance is preferable in being unaffected by magnetic impurities, to which the second method falls hostage. *See* NUCLEAR MAGNETIC RESONANCE (NMR); SQUID; TEMPERATURE; TEMPERATURE MEASUREMENT; THERMOCOUPLE; THERMOMETER. [R.P.Hu.]

Lubricant A gas, liquid, or solid used to prevent contact of parts in relative motion, and thereby reduce friction and wear. In many machines, cooling by the lubricant is equally important. The lubricant may also be called upon to prevent rusting and the deposition of solids on close-fitting parts.

Crude petroleum is an excellent source of lubricants because a very wide range of suitable liquids, varying in molecular weight from 150 to over 1000 and in viscosity from light machine oils to heavy gear oils, can be produced by various refining processes (see table). In order to standardize on nomenclature for oils of differing viscosity, the Society of Automotive Engineers (SAE) has established viscosity ranges for the various SAE designations (see table).

Viscosity of oils for various applications

Application	Viscosity in centistokes at 25°C (77°F)	Primary function
Engine oils		
SAE 10W	60–90	Lubricate piston rings, cylinders, valve gear, bearings; cool piston; prevent deposition on metal surfaces
SAE 20	90–180	
SAE 30	180–280	
SAE 40	280–450	
SAE 50	450–800	
Gear oils		Prevent metal contact and wear of spur gears, hypoid gears, worm gears; cool gear cases
SAE 80	100–400	
SAE 90	400–1000	
SAE 140	1000–2200	
Aviation engine oils	220–700	Same as engine oils
Torque converter fluid	80–140	Lubricate, transmit power
Hydraulic brake fluid	35	Transmit power
Refrigerator oils	30–260	Lubricate compressor pump
Steam-turbine oil	55–300	Lubricate reduction gearing, cool
Steam cylinder oil	1500–3300	Lubricate in presence of steam at high temperatures

It is often desirable to add various chemicals to lubricating oils to improve their physical properties or to obtain some needed improvement in performance. These include viscosity-index improvers, pour-point depressants, antioxidants, antiwear and friction-reducing additives, and dispersants.

Synthetic lubricants may be superior to mineral lubricants in some applications. The main advantage of synthetics is that they have a greater operating range than a mineral oil. Included in this class are esters, containing oxidation inhibitors and sometimes mild extreme pressure additives, silicones, and the polyglycols, such as polypropylene and ethylene oxides.

The most useful solid lubricants are those with a layer structure in which the molecular platelets will readily slide over each other. Graphite, molybdenum disulfide, talc, and boron nitride possess this property. A unique type of solid lubricant is provided by the plastic polytetrafluoroethylene (PTFE). The principal difficulty encountered with the use of solid lubricants is that of maintaining an adequate lubricant layer between the sliding metal surfaces.

A lubricating grease is a solid or semifluid lubricant comprising a thickening (or gelling) agent in a liquid lubricant. Other ingredients imparting special properties may be included. An important property of a grease is its solid nature; it has a yield value. This enables grease to retain itself in a bearing assembly without the aid of expensive seals, to provide its own seal against the ingress of moisture and dirt, and to remain on vertical surfaces and protect against moisture corrosion, especially during shut-down periods. [R.G.L.]

Luminaire An electric lighting fixture, wall bracket, portable lamp, or other complete lighting unit designed to contain one or more electric lighting sources and associated reflectors, refractors, housing, and such support for these items as is necessary. *See* ILLUMINATION; INCANDESCENT LAMP; LAMP. [W.B.Bo.]

Luminance The luminous intensity of any surface in a given direction per unit of projected area of the surface viewed from that direction. The International Commission on Illumination defines it as the quotient of the luminous intensity in the given direction of an infinitesimal element of the surface containing the point under consideration, by the orthogonally projected area of the element on a plane perpendicular to the given direction. Simply, it is the luminous intensity per unit area. Luminance is also called photometric brightness.

Since the candela is the unit of luminous intensity, the luminance, or photometric brightness, of a surface may be expressed in candelas/cm^2, candelas/$in.^2$, and so forth.

The stilb is a unit of luminance (photometric brightness) equal to 1 candela/cm^2. It is often used in Europe, but the practice in America is to use the term candela/cm^2 in its place.

The apostilb is another unit of luminance sometimes used in Europe. It is equal to the luminance of a perfectly diffusing surface emitting or diffusing light at the rate of 1 lumen/m^2. *See* LUMINOUS INTENSITY; PHOTOMETRY. [R.C.Pu.]

Luminescence Light emission that cannot be attributed merely to the temperature of the emitting body. Various types of luminescence are often distinguished according to the source of the energy which excites the emission. When the light energy emitted results from a chemical reaction, such as in the slow oxidation of phosphorus at ordinary temperatures, the emission is called chemiluminescence. When the luminescent chemical reaction occurs in a living system, such as in the glow of the firefly, the emission is called bioluminescence. In the foregoing two examples part of the energy of a chemical reaction is converted into light. There are also types of luminescence that are initiated by the flow of some form of energy into the body from the outside. According to the source of the exciting energy, these luminescences are designated as cathodoluminescence if the energy comes from electron bombardment; radioluminescence or roentgenoluminescence if the energy comes from x-rays or from γ-rays; photoluminescence if the energy comes from ultraviolet, visible, or infrared radiation; and electroluminescence if the energy comes from the application of an electric field. By attaching a suitable prefix to the word luminescence, similar designations may be coined to characterize luminescence excited by other agents. Since a given substance can frequently be made to luminesce by a number of different external exciting agents, and since the atomic and electronic phenomena that cause luminescence are basically the same regardless of the mode of excitation, the classification of luminescence phenomena into the foregoing categories is only a matter of convenience, not of fundamental distinction.

When a luminescent system provided with a special configuration is excited, or "pumped," with sufficient intensity of excitation to cause an excess of excited atoms over unexcited atoms (a so-called population inversion), it can produce laser action. (Laser is an acronym for light amplification by stimulated emission of radiation.) This laser emission is a coherent stimulated luminescence, in contrast to the incoherent spontaneous emission from most luminescent systems as they are ordinarily excited and used. *See* LASER; OPTICAL PUMPING.

A second basis frequently used for characterizing luminescence is its persistence after the source of exciting energy is removed. Many substances continue to luminesce for extended periods after the exciting energy is shut off. The delayed light emission (afterglow) is generally called phosphorescence; the light emitted during the period of excitation is generally called fluorescence. In an exact sense, this classification, based on persistence of the afterglow, is not meaningful because it depends on the properties of the detector used to observe the luminescence. With appropriate instruments one can detect afterglows lasting on the order of a few thousandths of a microsecond, which would be imperceptible to the human eye. The characterization of such a luminescence, based on its persistence, as either fluorescence or phosphorescence would therefore depend upon whether the observation was made by eye or by instrumental means. These terms are nevertheless commonly used in the approximate sense defined here, and are convenient for many practical purposes. However, they can be given a more precise meaning. For example, fluorescence may be defined as a luminescence emission having an afterglow duration which is temperature-independent, while phosphorescence may be defined as a luminescence with an afterglow duration which becomes shorter with increasing temperature. For details on various types of luminescence *see* BIOLUMINESCENCE; CATHODOLUMINESCENCE; CHEMILUMINESCENCE; ELECTROLUMINESCENCE; FLUORESCENCE; PHOSPHORESCENCE; PHOTOLUMINESCENCE; THERMOLUMINESCENCE. [C.C.K.; J.H.S.]

Luminescence analysis Methods of chemical analysis in which analyte concentration is related to luminescence intensity or some other property of luminescence. Photoluminescence, particularly fluorescence, is the most widely used type of luminescence for chemical analysis. However, there are also important analytical applications of both chemiluminescence and bioluminescence.

Luminescence is light that accompanies the transition from an electronically excited atom or molecule to a lower energy state. The forms of luminescence are distinguished by the method used to produce the electronically excited species. When produced by absorption of incident radiation, the light emission is known as photoluminescence. Photoluminescence that is short-lived (10^{-8} s or less between excitation and emission) is known as fluorescence. Photoluminescence that is longer-lived (from 10^{-6} s all the way up to seconds) is known as phosphorescence. The reason for the difference in lifetime is that fluorescence involves an allowed, high-probability, transition while phosphorescence involves a forbidden, low-probability, transition. *See* FLUORESCENCE; LUMINESCENCE; PHOSPHORESCENCE; PHOTOLUMINESCENCE; SELECTION RULES (PHYSICS).

Chemiluminescence is observed when the electronically excited atom or molecule is formed as the product of a chemical reaction. If the light-producing reaction occurs in nature, such as the light emitted by fireflies, it is known as bioluminescence. *See* BIOLUMINESCENCE; CHEMILUMINESCENCE.

Photoluminescence excitation spectra are determined by measuring emission intensity at a fixed wavelength while vary-

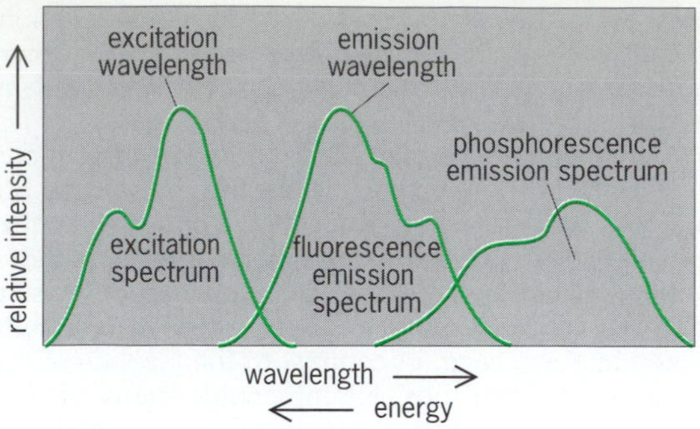

Typical photoluminescence excitation and emission spectra for molecules in solution.

ing the wavelength of the incident light used to produce the electronically excited species responsible for emission. The excitation spectrum is a measure of the efficiency of electronic excitation as a function of excitation wavelength. Photoluminescence emission spectra are determined by exciting at a fixed wavelength and varying the wavelength at which emission is observed. The illustration shows typical excitation and fluorescence and phosphorescence emission spectra for a molecule in solution. Between excitation and emission, electronically excited molecules normally lose some of their energy because of relaxation processes. As a consequence, the emission spectrum is at longer wavelengths, that is, at lower energy, than the excitation spectrum. Because the magnitude of the energy loss due to relaxation processes is greater for phosphorescence than for fluorescence, phosphorescence occurs at longer wavelengths than fluorescence. The spectra extend over a range of wavelengths, because in solution, molecules exist in a continuous distribution of vibrational and rotational energy levels. *See* ENERGY LEVEL (QUANTUM MECHANICS).

Observed luminescence intensities depend on three factors: (1) the number of electronically excited molecules or atoms produced by the excitation process; (2) the fraction of electronically excited molecules that emit light as they relax to a lower energy state; and (3) the fraction of the emitted luminescence that impinges on the detector and is measured. In the case of photoluminescence, the number of excited molecules is proportional to incident excitation intensity, the concentration of the luminescent species, and the efficiency with which the luminescence species absorbs the incident radiation. In the case of chemiluminescence and bioluminescence, the number of excited molecules depends on reactant concentrations and the efficiency with which the reaction pathway leads to production of the excited state. The dependence of intensity on concentration is the basis for chemical analysis based on luminescence.

By far the most important advantage of luminescence methods of chemical analysis is their ability to measure extremely low concentrations. This advantage arises because luminescence is measured relative to weak background signals. In contrast, methods based on the absorption of light require that a small difference between two large signals be measured. *See* SPECTROPHOTOMETRIC ANALYSIS. [W.R.Se.]

Luminous efficacy There are three ways this term can be used: (1) The luminous efficacy of a source of light is the quotient of the total luminous flux emitted divided by the total lamp power input. Light is visually evaluated radiant energy. Luminous flux is the time rate of flow of light. Luminous efficacy is expressed in lumens per watt. (2) The luminous efficacy of radiant power is the quotient of the total luminous flux emitted divided by the total radiant power emitted. This is always somewhat larger for a particular lamp than the previous measure, since not all the input power is transformed into radiant power. (3) The spectral luminous efficacy of radiant power is the quotient of the luminous flux at a given wavelength of light divided by the radiant power at that wavelength. A plot of this quotient versus wavelength displays the spectral response of the human visual system. It is, of course, zero for all wavelengths outside the range from 380 to 760 nanometers. It rises to a maximum near the center of this range. Both the value and the wavelength of this maximum depend on the degree of dark adaptation present. However, an accepted value of 683 lumens per watt maximum at 555 nanometers represents a standard observer in a light-adapted condition. *See* ILLUMINATION; LUMINOUS EFFICIENCY; LUMINOUS FLUX; PHOTOMETRY. [G.A.Ho.]

Luminous efficiency Visual efficacy of visible radiation, a function of the spectral distribution of the source radiation in accordance with the "spectral luminous efficiency curve," usually for the light-adapted eye or photopic vision, or in some instances for the dark-adapted eye or scotopic vision.

The spectral luminous efficiency of radiant flux is the ratio of luminous efficacy for a given wavelength to the value of maximum luminous efficacy. It is a dimensionless ratio. *See* ILLUMINATION; LUMINOUS EFFICACY; PHOTOMETRY. [G.A.Ho.]

Luminous energy The radiant energy in the visible region or quantity of light. It is in the form of electromagnetic waves, and since the visible region is commonly taken as extending 380–760 nanometers in wavelength, the luminous energy is contained within that region. It is equal to the time integral of the production of the luminous flux. *See* PHOTOMETRY. [R.C.Pu.]

Luminous flux The time rate of flow of light. It is radiant flux in the form of electromagnetic waves which affects the eye or, more strictly, the time rate of flow of radiant energy evaluated according to its capacity to produce visual sensation. The visible spectrum is ordinarily considered to extend from 380 to 760 nanometers in wavelength; therefore, luminous flux is radiant flux in that region of the electromagnetic spectrum. The unit of measure of luminous flux is the lumen. *See* PHOTOMETRY. [R.C.Pu.]

Luminous intensity The solid angular luminous flux density in a given direction from a light source. It may be considered as the luminous flux on a small surface normal to the given direction, divided by the solid angle (in steradians) which the surface subtends at the source of light. Since the apex of a solid angle is a point, this concept applies exactly only to a point source. The size of the source, however, is often extremely small when compared with the distance from which it is observed, so in practice the luminous flux coming from such a source may be taken as coming from a point. *See* CANDLEPOWER; PHOTOMETRY. [R.C.Pu.]

Luminous paint A type of paint that glows in the dark. Luminous paints may be either of the self-luminous type (energized by a radioactive salt) or of the type that requires preexcitation by an outside energy source such as light. Both types are made by incorporating luminescent material into the paint formulation. *See* LUMINESCENCE; PAINT. [C.C.K.; J.H.S.]

Lunar laser ranging An optical technique for monitoring the topocentric distance from Earth to specific points on the surface of the Moon with extremely high accuracy. While it is conceptually similar to terrestrial laser geodometry and artificial satellite applications of laser ranging (also called lidar, optical radar, or laser radar), there are significant differences in the techniques required in making the observations. The applica-

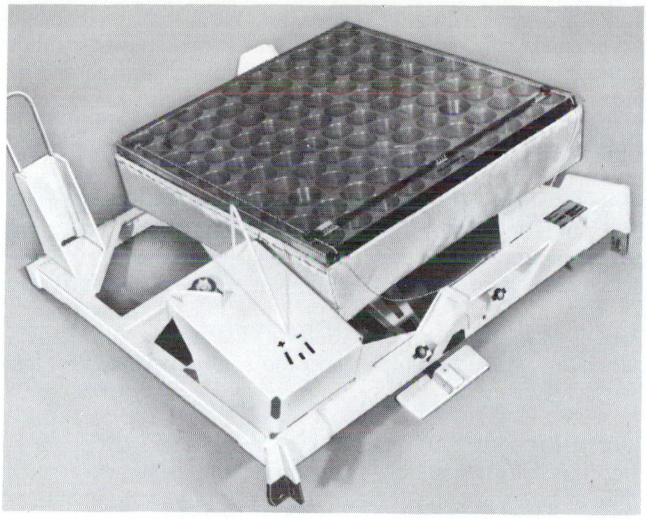

Laser retroreflector that was taken to the moon by *Apollo II*.

tions encompass the geometry and dynamics of both Earth and Moon. The features peculiar to lunar laser ranging, as contrasted with other laser applications, are related to the Moon's distance and motion.

The basic measurement consists of the precise time at which a light pulse is emitted from the laser, together with the elapsed time required for the pulse to travel from the transmitter to a known point on the Moon and return to an earth-bound detector. A precise definition of the lunar target is provided by using a retroreflector (see illustration), an array of mirrors whose design ensures that the incoming light will be reflected through almost exactly 180°, so that it will return to the emitting source. *See* LASER. [J.D.M.]

Lung Paired, air-filled respiratory sacs, usually in the anterior or anteroventral part of the trunk of most tetrapods. They lie within the coelom and are covered by peritoneum. In mammals they are within special chambers of the coelom known as pleural cavities and the peritoneum is termed pleura.

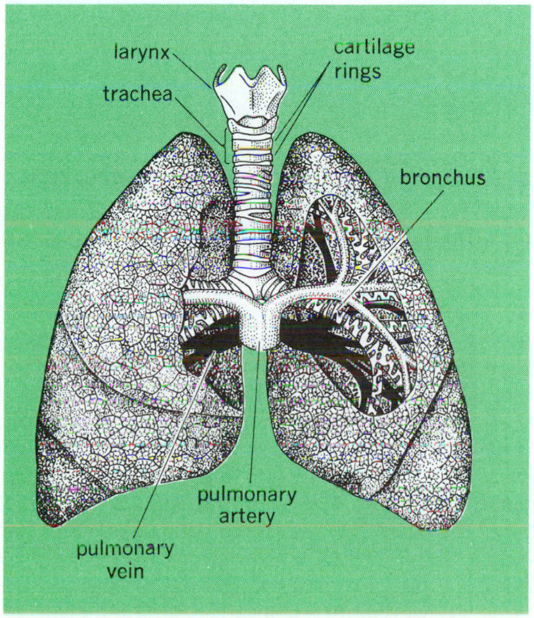

The human lung. (*After T. I. Storer and R. L. Usinger, General Zoology, 4th ed., McGraw-Hill, 1965*)

Amphibian lungs are often simple sacs, with only small ridges on the internal walls. In higher forms the lungs become more and more subdivided internally, thus increasing greatly the surface areas across which the respiratory exchange takes place. However, even in many reptiles the lungs may be quite simple. Birds have especially complex lungs with a highly differentiated system of tubes leading into and through them to the air sacs which are contained in many parts of the bird's body. Mammalian lungs are simpler, but in them the internal subdivision into tiny sacs or alveoli is extreme; there may be over 350,000,000 of them in one human lung.

In humans the two lungs lie within the chest, separated by the heart and mediastinum. The right lung has three lobes and the left lung two. A bronchus, an artery, and a vein enter each lung medially at the hilum; each branches again and again as it enters the lobules and smaller divisions of the lungs (see illustration). The terminal airways or bronchioles expand into small clusters of grapelike air cells, the alveoli. The alveolar walls consist of a single layer of epithelium and collectively present a huge surface. A small network of blood capillaries in the walls of the alveoli affords surfaces for the actual exchange of gases. *See* RESPIRATION; RESPIRATORY SYSTEM. [T.S.P.]

Lung disorders Inflammatory processes resulting from infections of bronchi and lungs are daily events in medical practice; scarring and tissue destruction as well as impaired pulmonary circulation are frequently encountered. Since the lungs literally filter the venous blood drainage from the entire body, they often become involved secondarily by blood-borne processes; many malignant tumors eventually metastasize to the lungs. More important still is the steadily rising incidence of primary malignancy of the lung and its controversial causative relationship to cigarette smoking and environmental pollution. *See* ONCOLOGY.

Although congestion and edema may occur separately, they are so frequently associated in the lungs that they may be discussed together. Factors predisposing to their development are increased pulmonary venous pressure, increased capillary permeability, decreased lymphatic drainage, and sodium retention. Heart failure is the most frequent cause of acute pulmonary congestion and edema. This is often a life-threatening situation that requires prompt medical attention. Shock, lung inflammations, and central nervous system disorders may also lead to acute pulmonary congestion and edema. *See* EDEMA.

Pulmonary embolism is the occlusion of one or more pulmonary arteries by a clot. The clot usually originates in the deep veins of the legs, but femoral, iliac, and pelvic veins may also be the site of origin. Predisposing factors in the development of vein occlusion with subsequent embolism are heart failure, postoperative convalescence, and malignant tumors. Massive pulmonary embolism is a life-threatening situation.

The adult secondary type of atelectasis is the collapse of pulmonary alveoli. This process can be caused by complete bronchial occlusion with subsequent alveolar air resorption or by increased intrapleural pressure produced by air or fluid in the pleural cavity or by deviation of the diaphragm.

Pneumonia is a general term denoting solidification (consolidation) of the lung tissue caused by inflammation. The disease may be caused by bacteria, viruses, chemical or physical agents, or any combination of these and varies greatly in pattern and severity. *See* PNEUMONIA.

Tuberculosis is caused by *Mycobacterium tuberculosis* of various types. Modern therapy is highly effective against tuberculosis. However, the disease must be considered serious, and it is not infrequent. Autopsy examinations have revealed active tuberculosis in patients that were not even suspected of having the disease. *See* TUBERCULOSIS.

Pneumoconioses are a group of pulmonary diseases caused by inhalation of specific dusts. Most induce alveolar fibrosis, sometimes with granuloma formation. Anthracosis is coal dust accumulation in lungs; silicosis is caused by silica particles; and asbestosis is a pneumoconiosis caused by asbestos inhalation. Other forms of pneumoconiosis include byssinosis (cotton dust), berylliosis (beryllium dust or vapor), and bagassosis (sugarcane fiber).

Emphysema is a common pulmonary disease, almost exclusively occurring in heavy cigarette smokers, characterized by destruction of lung tissue distal to the terminal bronchial. It is commonly associated with chronic bronchitis. *See* EMPHYSEMA.

Lung tumors constitute a heterogeneous group of neoplasms, most of which are carcinomas. Lung cancer is the most common tumor in men, and it is estimated that it may eventually surpass breast cancer as the most common female malignancy. *See* CANCER (MEDICINE); LUNG. [V.E.G.]

Lupus erythematosus One of the collagen diseases in which connective tissues show a similar form of biochemical and structural alteration. Each of the collagen diseases is a distinct clinical and pathologic entity and such a grouping exists more as a matter of convenience than because of a common etiologic factor. The causes are poorly understood, but hypersensitivity and nucleoprotein derangement appear to be involved. *See* AUTOIMMUNITY; HYPERSENSITIVITY.

Constitutional symptoms and signs (fever, weakness, fatigability, or weight loss) may be the first manifestation of illness and are often insidious. The joints are affected in 90% of the patients. Severe joint pain with little or no objective evidence of arthritis is most common. The heart is involved in over one-half of the cases and there may be heart failure. Lupus nephritis is very frequent and is the most serious feature of lupus. It usually progresses to subacute glomerulonephritis with uremia and usually hypertension, Anemia is also a very common feature of lupus. Lupus often mimics other diseases because of the wide possibility of tissue damage.

The disease is treated with high doses of corticosteroids which often exert a remarkable effect, especially in the early stages of the illness. [N.K.M.]

Lutetium A chemical element, Lu, atomic number 71, atomic weight 174.97, a very rare metal and the heaviest member of the rare-earth group. The naturally occurring element is made up of the stable isotope ^{175}Lu, 97.41%, and the long-life β-emitter ^{176}Lu with a half-life of 2.1×10^{10} years.

1	2	3	4	5	6	7	8	9	10	11	12	13	14	15	16	17	18
1 **H**																	2 **He**
3 **Li**	4 **Be**											5 **B**	6 **C**	7 **N**	8 **O**	9 **F**	10 **Ne**
11 **Na**	12 **Mg**											13 **Al**	14 **Si**	15 **P**	16 **S**	17 **Cl**	18 **Ar**
19 **K**	20 **Ca**	21 **Sc**	22 **Ti**	23 **V**	24 **Cr**	25 **Mn**	26 **Fe**	27 **Co**	28 **Ni**	29 **Cu**	30 **Zn**	31 **Ga**	32 **Ge**	33 **As**	34 **Se**	35 **Br**	36 **Kr**
37 **Rb**	38 **Sr**	39 **Y**	40 **Zr**	41 **Nb**	42 **Mo**	43 **Tc**	44 **Ru**	45 **Rh**	46 **Pd**	47 **Ag**	48 **Cd**	49 **In**	50 **Sn**	51 **Sb**	52 **Te**	53 **I**	54 **Xe**
55 **Cs**	56 **Ba**	71 **Lu**	72 **Hf**	73 **Ta**	74 **W**	75 **Re**	76 **Os**	77 **Ir**	78 **Pt**	79 **Au**	80 **Hg**	81 **Tl**	82 **Pb**	83 **Bi**	84 **Po**	85 **At**	86 **Rn**
87 **Fr**	88 **Ra**	103 **Lr**	104 **Rf**	105 **Db**	106 **Sg**	107 **Bh**	108 **Hs**	109 **Mt**	110	*111*	*112*	*113*	*114*	*115*	*116*	*117*	*118*

lanthanide series	57 **La**	58 **Ce**	59 **Pr**	60 **Nd**	61 **Pm**	62 **Sm**	63 **Eu**	64 **Gd**	65 **Tb**	66 **Dy**	67 **Ho**	68 **Er**	69 **Tm**	70 **Yb**
actinide series	89 **Ac**	90 **Th**	91 **Pa**	92 **U**	93 **Np**	94 **Pu**	95 **Am**	96 **Cm**	97 **Bk**	98 **Cf**	99 **Es**	100 **Fm**	101 **Md**	102 **No**

Lutetium, along with yttrium and lanthanum, is of interest to scientists studying magnetism. All of these elements form trivalent ions with only subshells which have been completed, so they have no unpaired electrons to contribute to the magnetism. Their radii with regard to the other rare-earth ions or metals are very similar so they form at almost all compositions either solid solutions or mixed crystals with the strongly magnetic rare-earth elements. Therefore, the scientist can dilute the magnetically active rare earths in a continuous manner without changing appreciably the crystal environment. *See* MAGNETOCHEMISTRY; RARE-EARTH ELEMENTS. [F.H.Sp.]

Lychee The plant *Litchi chinensis*, also called litchi, a member of the soapberry family (Sapindaceae). The fruit is a one-seeded berry. The thin, leathery, rough shell or pericarp of the ripe fruit is bright red in most varieties. Beneath the shell, completely surrounding the seed, is the edible aril or pulp.

It is a native of southern China, where it has been cultivated for more than 2000 years. It is grown in India, Union of South Africa, Hawaii, Burma, Madagascar, West Indies, Brazil, Honduras, Japan, Australia, and the southern United States. *See* SAPINDALES. [P.D.St/E.L.C.]

Lychniscosa An order of the subclass Hexasterophora in the class Hexactinellida. The parenchymal megascleres of these sponges are united to form a rigid framework. Examples of this order are *Aulocystis* and *Dactylocalyx*. *See* HEXACTINELLIDA; HEXASTEROPHORA. [W.D.H.]

Lycopodiales An order of the Lycopodiophyta with a fossil history extending back into the Paleozoic. These plants, commonly called club mosses, attained their greatest development during the Carboniferous Period and included several treelike forms comparable in size with the present-day pines. Today the order contains only two genera, *Lycopodium* and *Phylloglossum*, composed of small, herbaceous, evergreen plants with upright or trailing stems bearing numerous small, simple, spirally arranged leaves. The sporophylls (sporangium-bearing leaves) may be aggregated together into a cone or strobile (see illustration), or the sporangia may be borne singly on

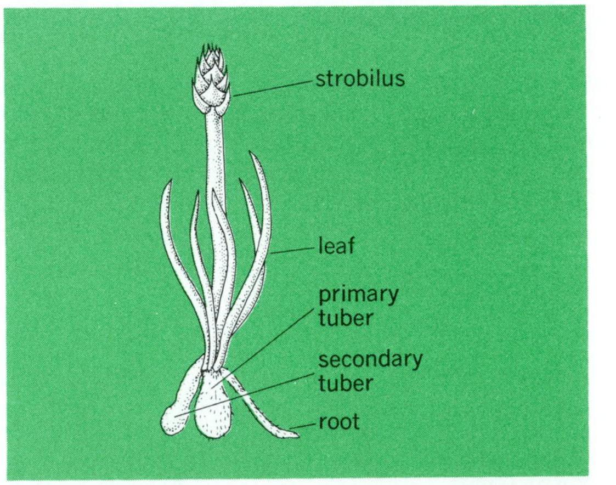

Fertile plant of *Phylloglossum drummondii*, twice natural size. (After A. W. Haupt, *Plant Morphology*, McGraw-Hill, 1953)

the upper surface of sporophylls scattered along the stem. The cones are often elevated on upright branches so that they occur above the leafy shoots. Because the cones resemble a club, these plants have been given the erroneous common name, club moss. *See* LYCOPODIOPHYTA. [P.A.V.]

Lycopodiophyta A division (formerly the subphylum Lycopsida) of the subkingdom Embryobionta (Embryophyta) having a long fossil history but now restricted to five living genera. The living members, commonly called club mosses, spike mosses, or quillworts, are significant in that the study of their structures and life cycles provides clues to the understanding of the many extinct species. Some of these plants, known only as fossils, were the dominant species of the extensive swampy forests of the Carboniferous. *See* LEPIDODENDRALES; LYCOPODIALES; LYCOPODIOPSIDA. [P.A.V.]

Lycopodiopsida One of the two classes of the division Lycopodiophyta, the other being the Isoetopsida; another name for this class is Lycopodineae. The plants in this class are commonly referred to as the lycopods. They have a very long history among vascular plants, dating back to the early Devonian and possibly Cambrian. The living members of this group are significant in that the study of their structures and life cycles provides clues to the understanding of the now extinct species of previous eras and the numerous changes which have occurred during the evolution of the present land flora. *See* PALEOBOTANY.

The two most common genera and the one having the widest distribution are *Lycopodium* and *Selaginella*. In *Lycopodium* the spores produced by the sporophyte are all alike (homosporous), whereas in *Selaginella* the spores are unlike (heterosporous). *See* ISOETALES; LEPIDODENDRALES; LYCOPODIALES; SELAGINELLALES; TRACHEOPHYTA. [P.A.V.]

Lyme disease A complex multisystem human illness caused by the tick-borne spirochete *Borrelia burgdorferi*. The disease usually begins with the characteristic skin lesion, erythema chronicum migrans, that may be accompanied by headache, stiff neck, fever, myalgias, arthralgias, malaise, fatigue, or swelling of the lymph nodes. These early clinical manifestations may last for several weeks and may be supervened by meningoradiculitis, meningoencephalitis, myocarditis, or migrating musculoskeletal pains. Still later, intermittent attacks of arthritis may occur. The arthritis may become chronic and result in the destruction of bone and cartilage in the large joints, especially the knees.

The fact that in the United States as well as in Europe many people contract Lyme disease without apparent exposure to ticks suggests that other hematophagous (blood-feeding) arthropods, such as biting flies and mosquitoes, serve as vectors of *B. burgdorferi*. Treatment with various antibiotics is effective. [W.Bu.]

Lymph node Aggregations of lymphoid tissue found along the course of lymphatic vessels. Lymph nodes vary in length from 1 mm to 2 cm (0.04 to 0.8 in.) or more, and in shape from spherical to flattened on one or more sides. A fibrous capsule surrounds each gland and sends supporting sheets, or trabeculae, through the lymphoid tissue (see illustration). The lymhoid tissue is composed of germinal centers of lymphocytes together with other cells, particularly those of the reticuloendothelial system. Lymph glands are found in superficial and deep sites, including the neck, axilla, groin, mediastinum, abdomen, pelvis, and intestinal wall (Peyer's patches). At least two functions are apparent: to supply lymphocytes to the circulation, and to remove bacteria and particulate matter carried to the nodes by the lymphatic vessels. *See* LYMPHATIC SYSTEM. [W.J.B.]

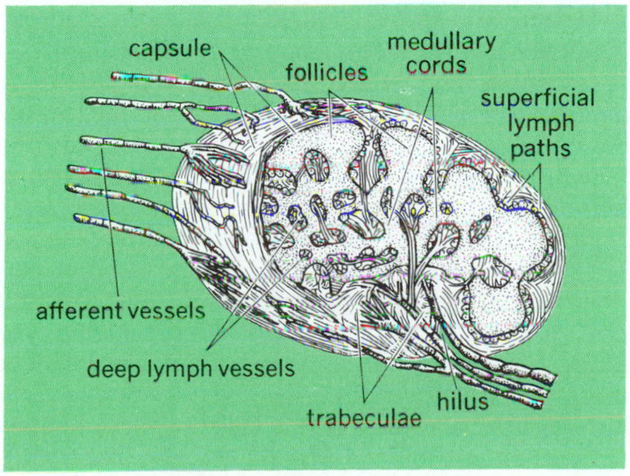

Diagrammatic sketch of a lymph node. (*After C. Toldt, An Atlas of Human Anatomy for Students and Physicians, vol. 1, 2d ed., Macmillan, 1941*)

Lymphadenitis An acute or chronic infection of the lymph nodes. Acute lymphadenitis may be caused by a variety of bacteria and toxic substances. Furthermore, lymph nodes draining an infected area show changes of acute inflammation. In comparatively mild reactions, white cells collect in the infected lymph nodes, and phagocytic activity is evident by ingestion by the inflammatory cells of bacteria, dead cells, tissue fragments, or any foreign material. With more severe involvement, pus collects in the nodes with abscess formation, and the infectious process often proceeds to neighboring lymph vessels, which are visible as red streaks leading to the nodes. The lymph nodes become enlarged and tender and the surrounding tissues are edematous. When pus formation and necrosis are marked, the lymph nodes open and drain on the body surface through sinus tracts. *See* EDEMA; PHAGOCYTOSIS. [F.A.C.]

Lymphangitis An infection of lymphatic channels which usually begins in an extremity and is frequently secondary to infection in the skin. The causative bacterium in many instances is a streptococcus. Several lymphatics draining the focus of infection in the skin become involved and these appear as red streaks that are firm, tender, and cordlike, especially if the lymph they contain coagulates. The infectious process usually extends only to the first chain of lymph nodes and then heals. In severe or untreated cases, the process reaches more distant lymph nodes, which may be destroyed by the inflammatory process, and finally enters the bloodstream, producing a bacteremia or infection of the blood. [F.A.C.]

Lymphatic system A system of vessels in the vertebrate body, beginning in a network of exceedingly thin-walled capillaries in almost all the organs and tissues except the brain and bones. This network is drained by larger channels, mostly coursing along the veins and eventually joining to form a large vessel, the thoracic duct, which runs beside the spinal column to enter the left subclavian vein at the base of the neck. The lymph fluid originates in the tissue spaces by filtration from the blood capillaries. While in the lymphatic capillaries it is clear and watery. However, at intervals along the larger lymphatic vessels, the lymph passes through spongelike lymph nodes, where it receives great numbers of cells, the lymphocytes, and becomes turbid.

The lymph nodes of mammals vary in number, size, form, and structure in different species. The amount of connective tissue of the lymph nodes, that is, the degree of development of the capsule and trabeculae, also varies in different mammals. Other lymphoid organs include the tonsils, thymus gland, and spleen, and in certain classes and groups of animals, structures which are confined to such groups, for instance, the bursa of Fabricius in the birds, a diverticulum from the lower end of the alimentary canal. *See* LYMPH NODE; SPLEEN; THYMUS GLAND; TONSIL.

The functions of the lymphatics are to remove particulate materials such as molecular proteins and bacteria from the tissues; to transport fat from the intestine to the blood; to supply the blood with lymphocytes; to remove excess fluid; also to return to the bloodstream the protein which has escaped from the blood capillaries. Basically, the composition of lymph closely resembles that of the plasma; lymph contains all of the types of protein found in plasma, but in lower concentration. The composition of lymph varies to some extent from one part of the body to another. Thus, the lymph from the liver contains more protein than that from the skin.

The lymph nodes serve as filtering-out places for foreign particles, including microorganisms, because the lymph comes into intimate contact with the many phagocytic cells of the sinusoids. These macrophages are of both the fixed and free wandering types. In addition to the phagocytic function, lymphoid tissue produces antibodies, although the actual process of antibody formation is not well understood. *See* CELLULAR IMMUNOLOGY; PHAGOCYTOSIS. [W.An.]

Lymphogranuloma venereum A venereal disease caused by a microorganism, a member of the PLT-Bedsonia (*Chlamydia*) group. It is also known as venereal bubo, The disease manifests itself as an enlargement of the inguinal lymph node and as such is seldom overlooked in the male; however, mild atypical or asymptomatic infections may be common. In the female the disease is more difficult to diagnose because the lymphadenopathy may be retroperitoneal. Fever commonly accompanies the development of the disease. Early diagnosis and treatment, supervision of infected persons, and public education are important in prevention. *See* VIRUS. [K.F.M.; J.S.]

Lymphokines The nonimmunoglobulin protein or glycoprotein molecules secreted by mitogen- or antigen-activated lymphocytes. Lymphokines enhance or suppress functions of other lymphoid cells of the immune system (for example, antibody formation by B cells, or graft rejection by T cells) and functions of inflammatory and foreign cells. They may also affect other biological functions; for example, resorption of bone is affected by the lymphokine known as osteoclast activating factor. Lymphokines may be considered to be the hormones of the immune defense system.

Lymphokines can be classified on the basis of presence or absence of antigenic specificity; nature of target cells (macrophages, polymorphonuclear leukocytes, lymphocytes, or other cells); or mode of action (inhibitory, stimulatory, or inflammatory). Antigen-specific lymphokines regulate immune responses exclusively to the antigen that induced their formation. These lymphokines bear determinants encoded by the immune response genes of the major histocompatibility complex (MHC). In contrast, antigen-nonspecific lymphokines act in a nonselective manner, and many of them, although not all, do not bear detectable MHC gene products. These lymphokines act on immunological responses of both B and T lymphocytes nonspecifically; they also modulate functions of the endothelium or inflammatory cells (macrophages, polymorphonuclear leukocytes, eosinophils, basophils, and others). Nonspecific lymphokines may also induce proliferation of hemopoietic cells, fibroblasts, or bone cells (repair mechanism), or they may destroy cells (for example, tumor cells can be destroyed by interferon). The lymphokines themselves are not preformed molecules present in resting lymphocytes but, rather, are synthesized and secreted as a result of activation of lymphoid cells by mitogens or antigens interacting with the cell membrane. Many, if not all, of the immune responses seem to be regulated by T lymphocytes mainly through cellular interactions via lymphokines. The T lymphocytes are thus able to initiate proper reaction against the antigen according to its nature, dose, and route of entry by turning on or off B cells, macrophages, or other T cells. *See* CELLULAR IMMUNOLOGY; HISTOCOMPATIBILITY; IMMUNOLOGY; IMMUNOSUPPRESSION. [D.W.Ka; R.Zu.]

Lymphoma A group of malignant tumors characterized by neoplastic proliferations of the various cell types endogenous to lymphoid tissue. Hodgkin's disease is considered to be a type of lymphoma. Lymphomas are classified on a morphological basis according to the cell of origin of the neoplasm, the degree of differentiation of the tumor cells, and the overall architecture of the involved lymphoid tissue. The etiology of lymphomas remains unknown.

The disease most commonly begins in the lymph nodes of the cervical (neck) and axillary (underarm) regions of the body; occasionally it originates in the lymphoid tissue of the intestine, and in Burkitt's lymphoma starts in the jaw. As the disease progresses, any organ in the body can become involved. Most commonly, the liver, spleen, and other lymph nodes become diseased as evidenced by their increase in size.

Most persons with lymphomas seek medical advice because of enlarged, nontender lymph nodes. These people usually have no other signs or symptoms, although occasionally they may be anemic and have a fever. Lymphomas are staged to determine the extent of involvement of the disease. Treatment is based on the stage of the disease and the tumor's microscopic appearance. The clinical course and prognosis of lymphomas are highly variable. When the disease is localized to a single group of lymph nodes in a well-defined region of the body, the therapy of choice is usually radiation. When the disease is disseminated, it is usually treated with a combination of drugs. Some long survivals and apparent cures have resulted from this chemotherapy. *See* HODGKIN'S DISEASE; LYMPHATIC SYSTEM. [S.P.H.]

Lyophilization Solvent removal from the frozen state by sublimation; commonly referred to as freeze-drying. Lyophilization is accomplished by freezing the material to be dried below its eutectic point and then providing the latent heat of sublimation. Precise control of heat input permits drying from the frozen state without product melt-back. In practical application, the process is accelerated and more precisely controlled under reduced pressure conditions. [S.C.T.]

Lyra The Lyre, in astronomy, a summer constellation, small but important. Lyra has a first-magnitude star, Vega, a navigational star and the most brilliant star in this part of the sky. Vega forms, with two faint stars to the east, an almost perfect equilateral triangle. The southern one in turn forms, with three brighter stars to the south, an approximate parallelogram. The resulting overall figure resembles a tortoise more than a stringed musical instrument. However, according to legend, Mercury made the first lyre from a turtleshell by placing strings across it. Hence the two different representations are not incompatible. *See* CONSTELLATION. [C.-S.Y.]

Lysenkoism A school of pseudoscience that flourished in the Soviet Union from the early 1930s to the mid-1960s, in violent opposition to traditional biology. The founder was T. D. Lysenko. He proclaimed a revolutionary fusion of agronomy and biological science, and therefore called his creation agrobiology, not Lysenkoism, the term that was used by his opponents.

A coherent outline of Lysenko's doctrines is hardly possible. The inheritance of acquired characters is often considered his central doctrine, though he came to it belatedly, as an offshoot of his original concept: "vernalization," or *iarovizatsiia*, a word that he coined. At first he used the word to describe the transformation of winter-habited wheat into spring habit as a result of moistening and chilling the seed before planting. He denounced the specialists who told him that the phenomenon had been observed and studied long before he put a new name on it, and he went on to extend the new term to almost any kind of seed treatment, and also to a stage in plant development that he claimed to have discovered. Lysenko came into conflict with scientific plant breeders and geneticists when he applied his concept of vernalization to hybridization and the selection of improved varieties. He came to endorse some of the crudest versions of the ancient belief in the inheritance of acquired characters. For example, he declared that domesticated plants are transformed into weeds by the hostile environment of poorly tended fields. *See* GENETICS; ORGANIC EVOLUTION; VERNALIZATION. [D.J.]

Lysin A term used to describe substances that will disrupt a cell, with the release of some of its constituents. Unless the damage is minor, this action leads to the death of the cell. Lysins vary in the range of host species whose cells they will attack and in their requirements for accessory factors for lysis; the immune lysins are strictest in their requirements. Erythrocytes are lysed by a wide variety of chemicals, including water and hypertonic salt solutions, which displace the osmotic pressure from that of isotonicity. They are also susceptible to surface-active substances, such as saponin. Many bacteria, such as the staphylococcus and the streptococcus, elaborate one or more hemolysins that will lyse erythrocytes from certain, although not all, species of animals. *See* LYTIC REACTION. [H.P.T.]

Lysine An amino acid constituent of proteins that contains a terminal amino group in addition to the usual α-amino group. Lysine must be supplied in the diet of animals and humans. It is widely distributed in proteins; however, it is absent in some such as zein (a grain protein). It is a precursor of 5-hydroxylysine, an amino acid occurring in only a few proteins such as collagen, and it is found in biocytin, a complexed form of the vitamin biotin, which is found in yeast. *See* AMINO ACIDS. [D.W.E.S.]

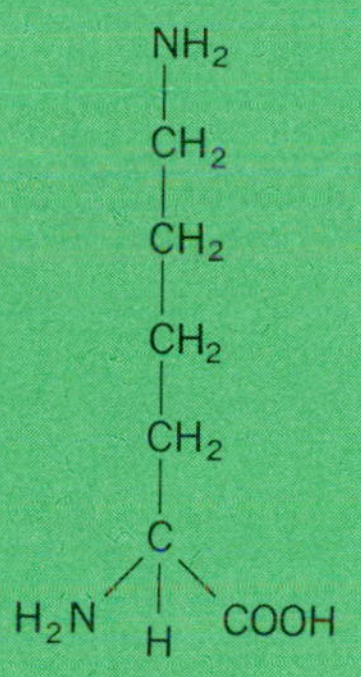

Lysine

Lysogeny Almost all strains of bacteria are lysogenic; that is, they have the capacity on rare occasions to lyse with the liberation of particles of bacteriophage (see illustration). Such particles can be detected by their ability to form plaques (colonies of bacteriophage) on lawns of sensitive (indicator) bacteria. The genetic determinant of the capacity of lysogenic bacteria to produce bacteriophage is a repressed phage genome (provirus) which exists in the bacterium in one of two states: (1) integrated into the bacterial chromosome (most cases), or (2) occupying some extra-chromosomal location (rare cases).

Bacteriophages which have the potential to exist as provirus are called temperate phages. When the provirus is integrated into the bacterial genome, it is called prophage. When the germinal substance (deoxyribonucleic acid or deoxyribonucleoprotein) of certain temperate phages enters a sensitive bacterium, the outcome may be death (lysis) for the bacterium as a result of phage multiplication, or it may result in the integration of the phage nucleic acid into the host genome (as a prophage), with the formation of a stable lysogenic bacterium. The lysogenic strain is designated by the name of the sensitive strain followed, in parentheses, by the strain of lysogenizing phage, for example, *Escherichia coli* (λ). Such a bacterium differs from its nonlysogenic ancestor in one very special way: It is immune to lysis by phage homologous to its carried prophage. *See* BACTERIOPHAGE. [L.B.]

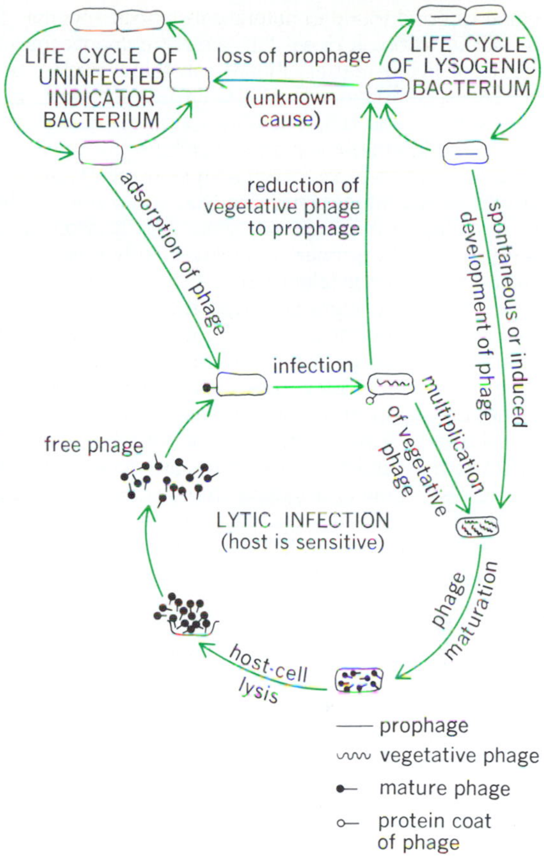

Life cycles of phage and bacterial host. (*After E. Jawetz, J. L. Melnick, and E. A. Adelberg, Review of Medical Microbiology, 2d ed., Lange, 1956*)

Lysorophia An order of elongated, swimming, and burrowing extinct lepospondylous amphibians. The single family Lysorophidae is best known from the upper Carboniferous and lower Permian genus *Lysorophus*. Fossils of this animal commonly are found in clusters of 100 or more individuals, which represent the burrowing, estivating, and possibly reproductive phases of the life cycle.

Body lengths of individuals found in these clusters range from about 3 to 30 in. (7.5 to 75 cm) and appear to represent different growth stages. The trunk consisted of about 100 vertebrae and was followed by a short flattened tail of about 15 more. Ribs were stout, long, and recurved, giving the trunk a robust snakelike appearance. Limbs were vestigial. The skull was small relative to the trunk. The jaws were set with a small number of sharp, conical teeth. Large branchial elements were present, suggesting the presence of external gills.

The closest living counterpart of *Lysorophus* is *Amphiuma*, a predaceous, aquatic urodele. Like *Amphiuma*, *Lysorophus* probably fed on small vertebrates and invertebrates, but the small gape of its mouth likely limited prey size severely. *Lysorophus* lived in areas subject to strong seasonality, to which its habit of estivation was an adaptation. *See* AMPHIBIA; LEPOSPONDYLI. [E.C.O.]

Lysosome A specialized cell part, of variable size and structure, surrounded by a single membrane and containing a mixture of hydrolytic (digestive) enzymes, most of which require an acid medium for optimal activity. Each enzyme splits a certain type of bond found in natural substances; together they have the ability to break down into small fragments the main constituents of living matter, that is, proteins, nucleic acids, polysaccharides, and lipids. Thus, the lysosomes make up the digestive system of the cell. They are found in most if not all animal cells, and in at least some plant cells.

There are two main types of lysosomes: primary lysosomes, packages of newly made enzymes that have not yet been involved in a digestive event; and secondary lysosomes, vacuoles derived from the primary lysosomes, within which digestion is taking place or has taken place.

Captured material is introduced into the cytoplasm of a cell as the content of a closed vesicle, called a phagosome or pinosome, surrounded by a membrane derived from the cell membrane. This vesicle later fuses with a lysosome in such a manner that their respective contents merge, while their membranes coalesce to form a single continuous boundary. The lysosomal enzymes and their substrates are brought together without any discontinuity affecting the membranes that surround them. Consequently, the enzymes cannot diffuse into the cytoplasm and attack the cell constituents indiscriminately. Even when lysosomes digest fragments of their own cells, they do so in a piecemeal and orderly fashion, without disruption of the cellular organization, at least under physiological conditions. Enzymes released from disrupted lysosomes can injure cells in pathological situations. [C.DeD.]

Lyssacinosa An order of the subclass Hexasterophora in the class Hexactinellida. In these sponges the parenchymal megascleres are typically free and unconnected but are sometimes secondarily united. *Asconema*, *Euplectella*, *Rossella*, and *Rhabdocalyptus* are examples of this order. *See* HEXACTINELLIDA; HEXASTEROPHORA. [W.D.H.]

Lytic infection Infection of a bacterium by a bacteriophage with subsequent production of more phage particles and lysis, or dissolution, of the cell. The viruses responsible are commonly called virulent phages. Lytic infection is one of the two major bacteriophage-bacterium relationships, the other being lysogenic infection. *See* BACTERIOPHAGE; LYSOGENY. [P.B.C.]

Lytic reaction A term used in serology to describe a reaction that leads to the disruption or lysis of a cell. The best example is the lysis of sheep red blood cells by specific antibody and complement in the presence of Ca^{2+} by (calcium ion) and Mg^{2+} (magnesium ion), a reaction that forms the indicator system of the standard Wassermann test for syphilis, as well as other complement-fixation reactions. In this example lysis results in the release of cellular hemoglobin into the medium; the reaction may be followed by visual or instrumental estimation of the decreased cell turbidity or the increased color of the medium due to the free hemoglobin. The initiation of lysis by complement can apparently proceed after the attachment of only one molecule of IgM or two molecules of IgG antibody to the red blood cell. IgM and IgG are both immunoglobulins. *See* ANTIBODY; COMPLEMENT; COMPLEMENT-FIXATION TEST; SEROLOGY. [H.P.T.]

M

Macadamia nut The fruit of a tropical evergreen tree, *Macadamia ternifolia*, native to Queensland and New South Wales and now grown commercially in Australia and Hawaii. The trees bear many small white or pinkish flowers in drooping racemes, each of which may mature from 1 to 20 fruits. These consist of a leathery outer husk (pericarp) which splits along one side at maturity, freeing the very hard-shelled, nearly round seed or nut. Two types of nuts are recognized, the most

***Macadamia integrifolia*. (*a*) Mature nuts. (*b*) Nuts without husks. (*c*) Nuts in husk showing method of dehiscence. (*After R. A. Jaynes, ed., Handbook of North American Nut Trees, Humphrey Press, 1969*)**

important commercially having a smooth shell and the other having a rough shell and sometimes referred to another species, *M. integrifolia* (see illustration). [L.H.MacD.]

Mach number In fluid mechanics, the ratio of the free stream velocity to the velocity of sound in the fluid at the same condition, such as temperature and pressure. Mach number is also the ratio of the inertia force of the fluid to the force of compressibility or the elastic force. In most fluid systems compressibility effects become important for values of the Mach number greater than about 0.3. A body moving through a fluid at a velocity less than sonic is preceded by a region of gradually varying density and pressure that controls the flow around the body. At Mach numbers equal to or greater than unity the gradual transition of pressure cannot exist, and shock waves, or regions of abruptly altered pressure and density, form at critical sections on or near the surface of the body and extend outward. *See* SHOCK WAVE; STREAMLINING. [G.Mu.]

Machine A combination of rigid or resistant bodies having definite motions and capable of performing useful work. The term mechanism is closely related but applies only to the physical arrangement that provides for the definite motions of the parts of a machine. For example, a wristwatch is a mechanism, but it does no useful work and thus is not a machine. Machines vary widely in appearance, function, and complexity from the simple hand-operated paper punch to the ocean liner, which is itself composed of many simple and complex machines. *See* MACHINERY; SIMPLE MACHINE. [R.M.Ph.]

Machine key Generally, a device used to prevent relative rotation of a shaft and the member to which it is connected, such as the hub of a gear, pulley, or crank. Many types of keys (see illustration) are available, and the choice in any installation depends on such factors as power requirements, tightness of fit, stability of connection, and cost.

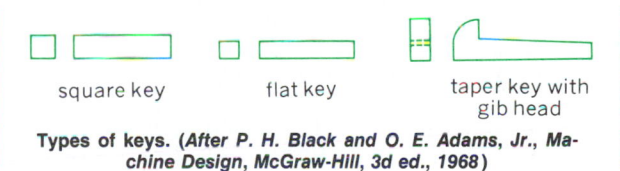

Types of keys. (*After P. H. Black and O. E. Adams, Jr., Machine Design, McGraw-Hill, 3d ed., 1968*)

Square keys are common in general industrial machinery. Flat keys are used where added stability of the connection is desired, as in machine tools. Square or flat keys may be of uniform cross section or they may be tapered. In tapered keys the width is uniform and the height of the key tapers. Tapered keys may have gib heads to facilitate removal. Other types of keys have been developed for special applications. [P.H.B.]

Machine tools Tools used to modify the shapes of materials in specific, controlled ways, such as by drilling holes, turning diameters, grinding radii, and performing many more operations on almost any type of rigid or semirigid material. The materials they can handle include the ferrous and nonferrous metals, wood, and plastic.

Machine tools allow a cutting tool to be brought into contact repeatedly with the workpiece material. In some cases the tool is fed into the stationary material, and in other cases the material is fed into the stationary tool. At each pass of the cutting tool, a small chip of the material is removed. Sawing, milling, turning, and drilling are all examples of chip-producing machining. Grinding also produces chips, but these chips are very fine and appear to be powder. *See* DRILLING MACHINE; GRINDING MILL; MILLING MACHINE; SAWING; TURNING (WOODWORKING).

The material and its desired shape determine the most effective machining technique. Drilling produces holes of moderate diameter with reasonably good dimensional control. If a more precise hole diameter is desired, a subsequent reaming operation is performed. Large-diameter holes are usually bored on lathes or machining centers. *See* BORING; REAMER.

Broaching is used for rapid production of slots of small or medium size. The slots may be almost any shape, depending on the shape of the broaching cutter. *See* BROACHING.

Cylindrical parts are usually produced by turning on lathes. Lathes are effective for large stock-removal operations to moderate tolerances. *See* LATHE.

All machining operations have optimum speeds and feeds. These are determined by carefully weighing the economics of tool life, required production rates, and operator skill and attentiveness. Speed is the rate at which the cutting edges passes against the workpiece. Feed is the amount of stock that will be removed by each cutting edge of the cutting tool. *See* MANUFACTURING PROCESSES; NUMERICAL CONTROL; TOOLING. [J.R.C.B.]

Machinery A group of parts arranged to perform a useful function. Normally some of the parts are capable of motion; others are stationary and provide a frame for the moving parts. The terms machine and machinery are so closely related as to be almost synonymous; however, machinery has a plural implication, suggesting more than one machine. Common examples of machinery include automobiles, clothes washers, and airplanes; machinery differs greatly in number of parts and complexity.

Some machinery simply provides a mechanical advantage for human effort. Other machinery performs functions that no human being can do for long-sustained periods. *See* MACHINE; MECHANICAL ENGINEERING; SIMPLE MACHINE. [R.S.S.]

Machining An operation that changes the shape, surface finish, or mechanical properties of a material by the application of special tools and equipment. Machining almost always is a process where a cutting tool removes material to effect the desired change in the workpiece. Typically, powered machinery is required to operate the cutting tools. *See* PRODUCTION METHODS.

Although various machining operations may appear to be very different, most are very similar: they make chips. These chips vary in size from the long continuous ribbons produced on a lathe to the microfine sludge produced by lapping or grinding. These chips are formed by shearing away the workpiece material by the action of a cutting tool. Cylindrical holes can be produced in a workpiece by drilling, milling, reaming, turning, and electric discharge machining. Rectangular (or nonround) holes and slots may be produced by broaching, electric discharge machining, milling, grinding, and nibbling. Cylinders may be produced on lathes and grinders. Special geometries, such as threads and gears, are produced with special tooling and equipment utilizing the turning and grinding processes mentioned above. Polishing, lapping, and buffing are variants of grinding where a very small amount of stock is removed from the workpiece to produce a high-quality surface.

In almost every case, machining accuracy, economics, and production rates are controlled by the careful evaluation and selection of tooling and equipment. Speed of cut, depth of cut, cutting-tool material selection, and machine-tool selection have a tremendous impact on machining. In general, the more rigid and vibration-free a machining tool is, the better it will perform. Jigs and fixtures are often used to support the workpiece. Since it relies on the plastic deformation and shearing of the workpiece by the cutting tool, machining generates heat that must be dissipated before it damages the workpiece or tooling. Coolants, which also acts as lubricants, are often used.

To increase the life and speed of cutting tools, they are often coated with a thin layer of extremely hard material such as titanium nitride or zirconium nitride. These materials, which are applied over the cutting edges, provide excellent wear resistance. They are also brittle, so they rely on the toughness of the underlying cutting tool to support them. Coated tools are more expensive than conventional tools, but they can often cut at much higher rates and last significantly longer. When used properly on sufficiently rigid machine tools, they are far more economical than conventional tooling. *See* METAL COATINGS. [J.R.C.B.]

Mackerel A fish which is a member of the order Perciformes, family Scombridae. There are about 50 carnivorous species found in the middle layer or near the surface of tropical and temperate seas. Mackerel are characterized by a long slender body, pointed head, and large mouth.

Scomber scombrus, the common mackerel, is an important fish commercially. It is a migratory species found on both sides of the North Atlantic. The Pacific mackerel (*Pneumatophorus diego*) is also an important commercial fish but differs from the common mackerel in having a swim bladder. The American Spanish mackerel (*Scomberomorus maculatus*) is a choice food fish. *See* PERCIFORMES. [C.B.C.]

McLeod gage A type of instrument used to measure vacuum by application of the principle of Boyle's law.

A known volume of a gas whose pressure is to be measured is trapped by raising the level of a fluid (mercury or oil) by means of a plunger, by lifting a reservoir, by using pressure, or by tipping the apparatus. As the fluid level is further raised, the gas is compressed into the capillary tube (see illustration).

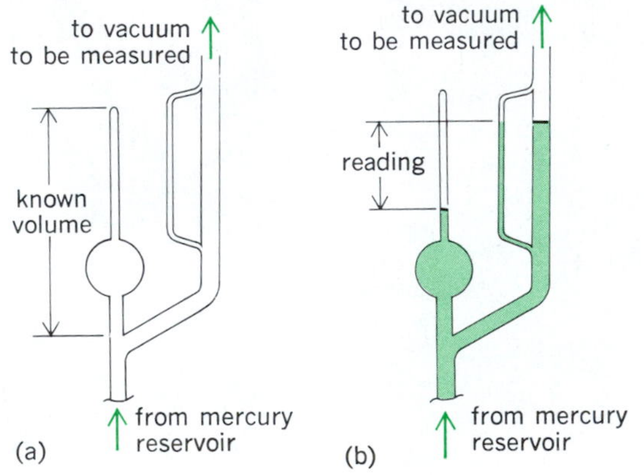

McLeod gage. (*a*) Filling (charging) position. (*b*) Measuring position.

Obeying Boyle's law, the compressed gas now exerts enough pressure to support a column of fluid high enough to read. Readings are somewhat independent of the composition of the gas under pressure. *See* VACUUM MEASUREMENT. [R.C.]

Macrodasyida An order of the phylum Gastrotricha. They inhabit marine or brackish waters, seldom fresh waters. Some do not exceed 0.5 mm (0.02 in.) in length, and most are not more than 1–1.5 mm (0.04–0.06 in.); these live in clean to detritus-rich marine sands of littoral or sublittoral areas. All have front and rear groups of adhesive tubes; most also have tubes along their sides. Some have cuticular thickenings or scales or hooks. *Turbanella* and *Tetranchyroderma* are the most common and most abundant of macrodasyids, with numbers of 50–100 per cubic centimeter of sand (800–1600 per cubic inch) not being unusual. *See* GASTROTRICHA. [W.D.Hu.]

Macroevolution The larger course of evolution by which the categories of animal and plant classification above the species have been evolved from each other and have differentiated into the forms within each. Only that part of evolution for which concrete evidence exists will be considered.

The organisms observed, alive or as fossils, can be classified into large groups or phyla. Within each of these the body is organized in variations of a single fundamental type. Such groups are the annelids, mollusks, echinoderms, and other phyla. Macroevolution is the process by which evolution has taken place within these phyla. In some cases evidence can be given that two or more such phyla have earlier evolved from a single ancestral group.

The evidence for the reality of macroevolution is derived from many sources. Paleontology provides evidence from fossils of the course of evolution. Fossils preserved at intervals of geological time may be compared and deductions made of their relationships. Thus, phylogenetic trees may be constructed. Studies of comparative morphology and embryology give evidence of many kinds from the similarities and dissimilarities of living animals. Geographical distribution of forms, together with the known history of the lands, gives data which should be in agreement with beliefs concerning the evolution of the organisms. *See* PALEONTOLOGY.

Speciation or microevolution has continued throughout evolution and provides its background. But the question remains whether the course of change in the long progress of macroevolution has been controlled in other ways besides the mechanisms considered in speciation. Another question for discussion is whether macroevolution is merely long-continued speciation. *See* SPECIATION.

Throughout the course of evolution, change has occurred during the life of organisms. These organisms must always be competent to survive in competition with other organisms. The whole biology evolves; evolution is a phenomenon in natural history. Also, change of habit and habitat normally accompanies the evolution of structure. It is probably without meaning to ask which of these precedes the other.

Types of change. In all evolution one may distinguish two types of change, cladogenesis and phyletic evolution. Cladogenesis is the type of evolution associated with attered habit and habitat, usually in populations of the species separated from the rest. Adaptation to the new conditions results in division of the originally homogeneous species into distinguishable parts. Phyletic evolution is the gradual evolution of a population in its environment without division into isolated parts. This consists partly in improved adaptation to the conditions, partly in readaptation when the conditions change, as they frequently do in natural environments. Usually change in environmental conditions is gradual, but sudden and irreversible change sometimes occurs and organisms often spread into new habitats where conditions differ from those to which they are adapted. The organisms must then adapt quickly to the new conditions if they are to survive. In such circumstances evolution will be rapid under the changed selection due to the new conditions. This process is known as quantum evolution and is considered as a special although extreme case of phyletic evolution. It may lead to new groups on any taxonomic level. The distinctions between these types of evolution are illustrated diagrammatically in the illustration. The evolutionary changes in the organisms are similar in all the types.

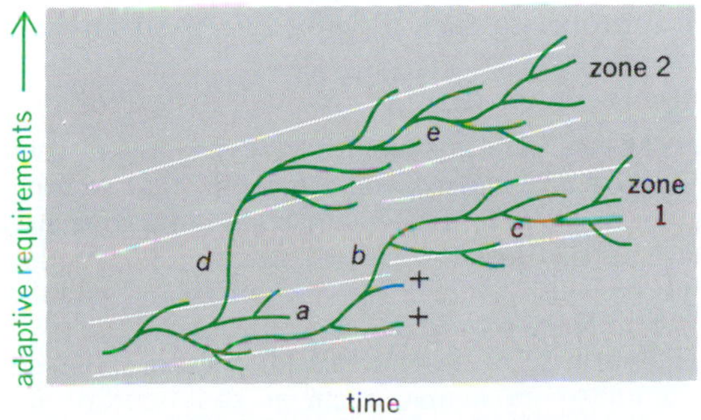

Diagram of the main types of evolution: *a* represents phyletic evolution in an adaptive zone or niche (zone 1); *b* indicates sudden alteration of conditions in this zone, with the death (+) of most of the lines inhabiting it and the readaptation of one line to the new conditions, that is, quantum evolution; *c* shows phyletic evolution in the new conditions; *d* shows migration to a new niche (zone 2, cladogenesis); *e* indicates phyletic evolution in this zone. Rapid (tachytelic) evolution occurs at *b* and *d* and horotelic evolution at *a*, *c*, and *e*. (*Modified from G. G. Simpson, Tempo and Mode in Evolution, Columbia University Press, 1944*)

Adaptive radiation. The course of evolution differs between successful groups, such as the vertebrates or the insects, which have enlarged their areas of dominance with time and evolved into many smaller groups, and other groups, such as many forms among the lower invertebrates, which show no progressive evolution and have often survived through long ages with little change.

The knowledge of vertebrate evolution is the most detailed. When the evolution of this successful group is examined it is found that a limited number of dominant types (several groups of fishes, amphibians, reptiles, birds, and mammals) have followed each other, each in a new mode of vertebrate life, and that each was evolved from the previous dominant group. Further, each dominant group divided soon after its appearance into a large number of subsidiary types adapted to more restricted modes of life within the range of the larger group. This diversification into different adaptive zones is known as adaptive radiation. In adaptive radiation the greater part of the changes of form are due to alterations of the relative sizes of the parts of the body by modifications of their growth rates during development and not to the evolution of new organs. In the mammals the radiating lines appear today as the orders of bats, primates, rodents, and so on. All of these can be traced back to near the beginning of the dominance of the mammals. Each line continues to radiate throughout its successful history. Formation of new organs, which are not modifications of organs previously present, occurs in the radiation but accounts for only a small part of the changes. The development of horns in the mammals is an example.

The evolution of a major dominant group from a preceding group, for example, the mammals or birds from the reptiles, or the amphibians from the fishes, demands much more fundamental changes than those of adaptive radiation. The whole form and function of the body is reorganized. It takes a considerable time, for example, about 40,000,000 years in the mammals and birds, and proceeds by successive alterations of one feature after another toward the new type. It occurs in one of the radiating lines of the preceding group, while its other lines, though they continue to radiate, clearly remain members of the preceding group. *See* ORGANIC EVOLUTION. [G.S.C.]

Macroscelidea A mammalian order, the elephant shrews and their allies, which consists of a single African family, Macroscelididae. They are mostly hopping and scampering mammals, divided among several subfamilies, some of which resemble rodents and ungulates in certain features. Characteristic of all these animals is an intestinal cecum and the arrangement of bones in the ear region. The long proboscis of the first living macroscelidids to be discovered led to the name elephant shrews. *See* MAMMALIA. [M.C.McK.]

Madelung constant A numerical constant α_M in terms of which the electrostatic energy U of a three-dimensional periodic crystal lattice of positive and negative point charges q_+, $-q_-$, N in number, is given by the equation below, where d is

$$U = -\frac{1}{2}\frac{Nq_+q_-}{d}\alpha_M$$

the nearest-neighbor distance between positive and negative charges and N is large. Knowledge of such electrostatic energies as given by the Madelung constant is of importance in the calculation of the cohesive energies of ionic crystals and in many other problems in the physics of solids. *See* IONIC CRYSTALS.

Madelung constants for some common ionic crystals

Crystal structure	Madelung constant, α_M
Sodium chloride, NaCl	1.7476
Cesium chloride, CsCl	1.7627
Zinc blende, α-ZnS	1.6381
Wurtzite, β-ZnS	1.641
Fluorite, CaF_2	5.0388
Cuprite, CuO_2	4.1155
Rutile, TiO_2	4.816
Anatase, TiO_2	4.800
Corundum, Al_2O_3	25.0312

The Madelung constants for a number of common ionic crystal structures are given in the table. For these cases d is chosen as the nearest-neighbor distance. *See* CRYSTAL. [B.G.D.]

Magellanic Clouds Of all easily observable external galaxies, the one nearest to the Milky Way galactic system. The distance of the Large Magellanic Cloud (LMC) is 55,000 parsecs (180,000 light-years or 1.70×10^{18} km or 1.05×10^{18} mi); that of the Small Magellanic Cloud (SMC) is 63,000 parsecs (205,000 light-years or 1.94×10^{18} km or 1.21×10^{18} mi); these distances are uncertain by about 10%. Because of their proximity, many types of galactic objects, such as bright main sequence stars, supergiants, classical cepheid variables, novae, and planetary nebulae can be studied. *See* CEPHEIDS; GALAXY, EXTERNAL; VARIABLE STAR.

Both Clouds are visible to the naked eye. E. P. Hubble classified them as irregular galaxies, although there is now evidence that supports the hypothesis that the Large Magellanic Cloud (at least) may be a barred spiral. The Large Magellanic Cloud is much more massive than the Small. [L.H.Al.]

Magic numbers Numbers of neutrons or protons in nuclei which correspond to particularly stable structures and closed shells. In nature the magic numbers for both neutrons and protons are 2, 8, 14, 20, 50, and 82. The next neutron magic numbers are 126 and 184; the next proton magic numbers are expected to be 114 and perhaps 164. Nuclei having magic numbers of both neutrons and protons have been found to have spherical equilibrium shapes and special stability; they have anomalously low capture probabilities for additional neutrons and protons respectively. *See* NUCLEAR STRUCTURE. [D.A.B.]

Magic square An n by n matrix of (not necessarily distinct) nonnegative integers all of whose row and column sums are equal to a prescribed number x. The number of such magic squares is 1 when $n = 1$, $x + 1$ when $n = 2$, and in general is a polynomial of the form $a_m x^m + a_{m-1}x^{m-1} + \cdots + a_1 x + 1$, where $m = (n - 1)^2$. Although the coefficients a_i are not given explicitly in this formula, for each n they may be computed by the method of undetermined coefficients, so that a computer search of magic squares for small values of x will also determine the number of such squares for all values of x. *See* COMBINATORIAL THEORY; ENUMERATION OF ARRANGEMENTS; MATRIX THEORY. [T.Br.]

Magma The hot material, partly or wholly liquid, from which igneous rocks form. Besides liquids, solids and gas may be present in magma. Most observed magmas are silicate melts with associated crystals and gas, but some inferred magmas are carbonate, phosphate, oxide, sulfide, and sulfur melts.

Strictly, any natural material which contains a finite proportion of melt (hot liquid) is a magma. However, magmas which contain more than about 60% by volume of solids generally have finite strength and fracture like solids.

Hypothetical, wholly liquid magmas which develop by partial melting of previously solid rock and segregation of the liquid into a volume free of suspended solids and gas are called primary magmas. Hypothetical, wholly liquid magmas which develop by crystallization of a primary magma and isolation of rest liquid free of suspended solids are called parental (or secondary) magmas. Although no unquestioned natural examples of either primary or parental magmas are known, the concepts implied by the definitions are useful in discussing the origins of magmas.

Bodies of flowing lava and natural volcanic glass prove the existence of magmas. Such proven magmas include the silicate magmas corresponding to such rocks as basalt, andesite, dacite, and rhyolite as well as rare carbonate-rich magmas and sulfur melts. Oxide-rich and sulfide-rich magmas are inferred from textural and structural evidence of fluidity as well as mineralogical evidence of high temperature, together with the results of experiments on the equilibrium relations of melts and crystals. *See* IGNEOUS ROCKS; LAVA.

Magma is presumed to underlie regions of active volcanism and to occupy volumes comparable in size and shape to plutons of eroded igneous rocks. However, it is not certain that individual plutons existed wholly as magma at one time; therefore, actual bodies of magma may be smaller than plutons. *See* PLUTON.

Diverse origins are probable for various magmas. Basaltic magmas because of their high temperatures probably originate within the mantle several tens of kilometers beneath the surface of the Earth. Rhyolitic magmas may originate through crystallization of basaltic magmas or by melting of crustal rock. Intermediate magmas may originate within the mantle or by crystallization of basaltic magmas, by melting of appropriate crustal rock, and also by mixing of magmas or by assimilation of an appropriate rock by an appropriate magma. *See* PETROGRAPHIC PROVINCE; VOLCANO. [A.T.A.]

Magnesite The mineral form of magnesium carbonate, $MgCO_3$. Iron, manganese, and cobalt may replace some magnesium in magnesite. Magnesite has the same structure as calcite. It is usually massive and white, but iron impurities may give it a brownish tint. The specific gravity is 3 and the hardness is 4 on Mohs scale.

Magnesite may be associated with peridotites, soapstones, and dolomites, or it may occur as sedimentary deposits. Deposits are in Austria, Manchuria, California, Nevada, and Washington. Magnesite is a source of magnesia. *See* CARBONATE MINERALS; MAGNESIUM. [R.I.Ha.]

Magnesium A metallic chemical element, Mg, in group 2 of the periodic system, atomic number 12, atomic weight 24.312. Magnesium is silvery white and extremely light in weight. The specific gravity is 1.74, and the density is 1740 kg/m^3 (0.063 lb/in.^{3} or 108.6 lb/ft^3). Because of this lightness

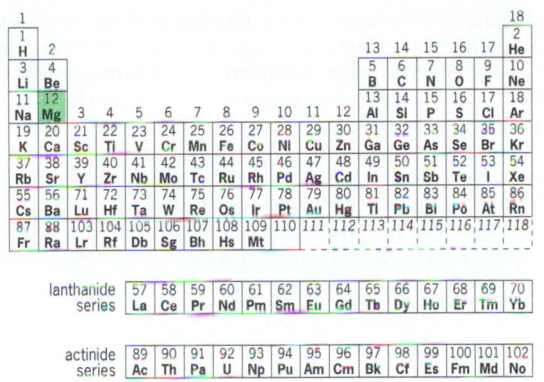

1	2	3	4	5	6	7	8	9	10	11	12	13	14	15	16	17	18
1 **H**																	2 **He**
3 **Li**	4 **Be**											5 **B**	6 **C**	7 **N**	8 **O**	9 **F**	10 **Ne**
11 **Na**	12 **Mg**											13 **Al**	14 **Si**	15 **P**	16 **S**	17 **Cl**	18 **Ar**
19 **K**	20 **Ca**	21 **Sc**	22 **Ti**	23 **V**	24 **Cr**	25 **Mn**	26 **Fe**	27 **Co**	28 **Ni**	29 **Cu**	30 **Zn**	31 **Ga**	32 **Ge**	33 **As**	34 **Se**	35 **Br**	36 **Kr**
37 **Rb**	38 **Sr**	39 **Y**	40 **Zr**	41 **Nb**	42 **Mo**	43 **Tc**	44 **Ru**	45 **Rh**	46 **Pd**	47 **Ag**	48 **Cd**	49 **In**	50 **Sn**	51 **Sb**	52 **Te**	53 **I**	54 **Xe**
55 **Cs**	56 **Ba**	71 **Lu**	72 **Hf**	73 **Ta**	74 **W**	75 **Re**	76 **Os**	77 **Ir**	78 **Pt**	79 **Au**	80 **Hg**	81 **Tl**	82 **Pb**	83 **Bi**	84 **Po**	85 **At**	86 **Rn**
87 **Fr**	88 **Ra**	103 **Lr**	104 **Rf**	105 **Db**	106 **Sg**	107 **Bh**	108 **Hs**	109 **Mt**	110	*111*	*112*	*113*	*114*	*115*	*116*	*117*	*118*

lanthanide series	57 **La**	58 **Ce**	59 **Pr**	60 **Nd**	61 **Pm**	62 **Sm**	63 **Eu**	64 **Gd**	65 **Tb**	66 **Dy**	67 **Ho**	68 **Er**	69 **Tm**	70 **Yb**
actinide series	89 **Ac**	90 **Th**	91 **Pa**	92 **U**	93 **Np**	94 **Pu**	95 **Am**	96 **Cm**	97 **Bk**	98 **Cf**	99 **Es**	100 **Fm**	101 **Md**	102 **No**

combined with alloy strength suitable for many structural uses, magnesium has long been known as industry's lightest structural metal.

With a density only two-thirds that of aluminum, magnesium is used in countless applications where weight saving is an important consideration. The metal also has, however, many desirable chemical and metallurgical properties which account for its extensive use in a variety of nonstructural applications.

Magnesium is very abundant in nature, occurring in substantial amounts in many rock-forming minerals such as dolomite, magnesite, olivine, and serpentine. In addition, magnesium is also found in sea water, subterranean brines, and salt beds. It is the third most abundant structural metal in the Earth's crust, exceeded only by aluminum and iron.

Some of the properties of magnesium in metallic form are listed in Table 1. Magnesium is very active chemically. It will actually displace hydrogen from boiling water, and a large number of metals can be prepared by thermal reduction of their salts and oxides with magnesium. The metal will combine with most nonmetals and with practically all acids. Magnesium reacts only slightly or not at all with most alkalies and many organic chemicals, including hydrocarbons, aldehydes, alcohols, phenols, amines, esters, and most oils. As a catalyst, magnesium is useful for promoting organic condensation, reduction, addition, and dehalogenation reactions. It has long been used for the synthesis of complex and special organic compounds by the well-known Grignard reaction. Principal alloying ingredients include aluminum, manganese, zirconium, zinc, rare-earth metals, and thorium. *See* MAGNESIUM ALLOYS.

Table 1. Physical properties of primary magnesium (99.9% pure)

Property	Value
Atomic number	12
Atomic weight	24.312
Atomic volume, cm^3/g-atom	14.0
Crystal structure	Close-packed hexagonal
Electron arrangement in free atoms	(2) (8) 2
Mass numbers of the isotopes	24, 25, 26
Percent relative abundances of ^{24}Mg, ^{25}Mg, ^{26}Mg	77, 11.5, 11.5
Density, g/cm^3 at 20°C	1.738
Specific heat, cal/g/°C at 20°C (1 cal = 4.2 joules)	0.245
Melting point, °C	650
Boiling point, °C	1110±10

Magnesium compounds are used extensively in industry and agriculture. Table 2 lists the major magnesium compounds and indicates some of their more significant applications. [W.H.Gr.; S.C.E.]

Table 2. Principal magnesium compounds and uses

Compound	Uses
Magnesium carbonate	Refractories, production of other magnesium compounds, water treatment, fertilizers
Magnesium chloride	Cell feed for production of metallic magnesium, oxychloride cements, refrigerating brines, catalyst in organic chemistry, production of other magnesium compounds, flocculating agent, treatment of foliage to prevent fire and resist fire, magnesium melting and welding fluxes
Magnesium hydroxide	Chemical intermediate, alkali, medicinal
Magnesium oxide	Insulation, refractories, oxychloride and oxysulfate cements, fertilizers, rayon-textile processing, water treatment, papermaking, household cleaners, alkali, pharmaceuticals, rubber filler catalyst
Magnesium sulfate	Leather tanning, paper sizing, oxychloride and oxysulfate cements, rayon delustrant, textile dyeing and printing, medicinal, fertilizer ingredient, livestock-food additive, ceramics, explosives, match manufacture

Magnesium alloys The most important alloying ingredients used in magnesium alloys are aluminum, zinc, manganese, silicon, zirconium, rare-earth metals, and thorium. The specific gravity of magnesium alloys ranges from 1.74 to 1.83. It has led to a great many structural applications in the aircraft, transportation, materials-handling, and portable-tool and -equipment industries. Magnesium alloys are commonly used in the form of die castings, which account for 75% of their usage in structural applications. *See* MAGNESIUM. [T.E.L.]

Magnet A piece of ferromagnetic material whose domains have been aligned sufficiently so that it produces a net magnetic field outside itself and can experience a net torque when placed in a magnetic field produced by some other source. For a discussion of ferromagnetic domains *see* FERROMAGNETISM.

If a bar of magnetically hard ferromagnetic material is magnetized to saturation in the magnetic field of a current flowing in a coil of wire, and the current is then turned off, the magnetic flux density in the bar drops down the hysteresis curve to a point below the retentivity point. The flux density goes below this point because the field becomes negative due to the demagnetizing field of the magnet's poles. The important point here is that the bar magnet does retain an appreciable magnetization and, if the material is well chosen, this magnetization will remain more or less permanently, provided the magnet is not heated, dropped, exposed to too strong a magnetic field, or in general, treated too roughly. Such a magnet is called a permanent magnet. *See* MAGNETIC HYSTERESIS.

A keeper is a piece of soft (easily magnetized) ferromagnetic material (such as soft iron) which (when the magnet is not in use) extends from one pole to the other, and through which the magnetic flux lines between the poles are concentrated. A keeper is used especially with U-shaped or horseshoe magnets. Its presence decreases the demagnetizing effect of the poles and thus helps to keep the magnet strong. *See* ELECTROMAGNET; MAGNETIC FIELD; MAGNETIC FLUX; MAGNETIC MATERIALS; MAGNETISM; MAGNETIZATION; MAGNETOSTATICS. [R.P.Wi.]

Magnet wire Insulated copper or aluminum wire used in the coils of all types of electromagnetic machines and devices. It is single-strand wire insulated with enamel, varnish, cotton, glass, asbestos, or combinations of these. To meet the immense variety of uses and to gain competitive advantage, a great number of kinds of enamel and fiber insulations are widely available. *See* ELECTRICAL INSULATION; MAGNET. [P.L.A.; C.J.Her.]

Magnetic bubble memory A memory storage technology which uses localized magnetized regions to store information. Magnetic bubble memory is the integrated-circuit analogy to rotating magnetic memories such as disks and tapes.

Magnetic bubble memories employ materials that are easily magnetized in one direction but are hard to magnetize in the orthogonal direction. The most commonly used materials are magnetic garnets which are deposited on a nonmagnetic substrate to form the basis of a bubble memory chip. A thin film of one of these materials, with the easy direction perpendicular to the surface, will allow only two natural directions of magnetization—up or down in the easy direction. When no external magnetic field is applied, the magnetic garnet film forms serpentine patterns of upward and downward magnetization. When a magnetic field is applied by sandwiching the film between two permanent magnets, cylindrical-shaped magnetic domains are formed in place of the serpentine structure. These cylinders are called magnetic bubbles and have a magnetization pointing in the opposite direction of the surrounding area. In effect, magnetic bubbles are magnetic islands in a magnetic sea of opposite magnetic polarity.

A bit of storage is represented as the presence or absence of a bubble at a given storage location. There are several methods of defining the storage locations, but the most common way is to use the chevron patterns. These chevrons are made of a soft ferromagnetic material such as permalloy and are deposited on the surface of the magnetic garnet. There can be one bubble for each chevron.

The chevrons also serve as the path for bubble movement when the information is accessed. To organize the bubble memory bits, the chevron patterns are deposited to form loops along which the bubbles can travel. The simplest organization is one single, large loop. The drawback is that the bubbles must travel the entire length of the loop before the information can be retrieved, and this is time-consuming. Most bubble memory chips now use a more sophisticated architecture called the major/minor loop (see illustration). In these multiple storage loops, the length and the time taken to traverse a minor loop is much smaller than for one major, large loop.

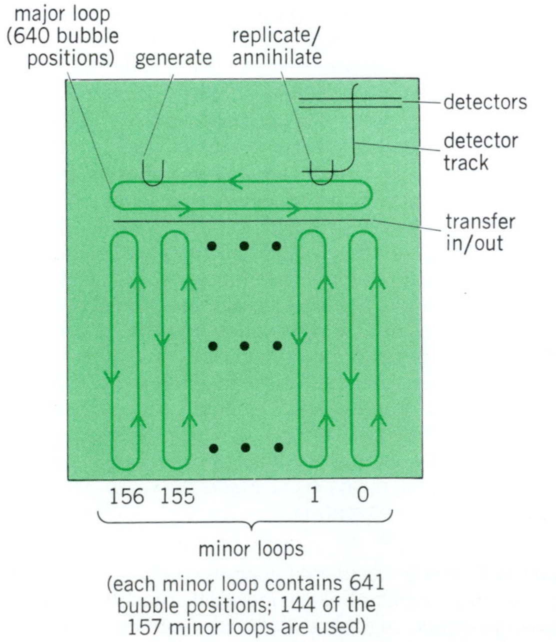

Diagram of the major/minor loop architecture of a 100,000-bit magnetic bubble memory chip. With multiple storage loops, the time to traverse a minor loop is much reduced as compared with one major, large loop.

The information is retrieved by waiting until the desired bubbles are at the top of the minor loops. At that point one bubble from each minor loop is transferred to the major loop. This "bubble train" traverses the major loop, during which time the information can be read or new information can be substituted.

To move the bubbles along the loops, a rotating magnetic field must be generated. For each rotation of the magnetic field each bubble moves synchronously from one chevron to the next. In other words, the magnetic bubbles move within the material, but no physical motion takes place. *See* COMPUTER STORAGE TECHNOLOGY; INTEGRATED CIRCUITS. [J.E.J.]

Magnetic circuits Closed paths of magnetic flux; also a design method using such paths to compute the magnetic field of a core geometry that is often encountered, for instance, high-permeability flux-path segments (and their associated air gaps), each segment having reasonably definite length and area. Examples of magnetic circuits are transformer cores (see illustration), relay frames, and iron parts of electrical machinery. The magnetic-field equations for these devices look so similar to dc circuit equations that they are called magnetic circuits. *See* MAGNETIC FLUX. [M.G.F.]

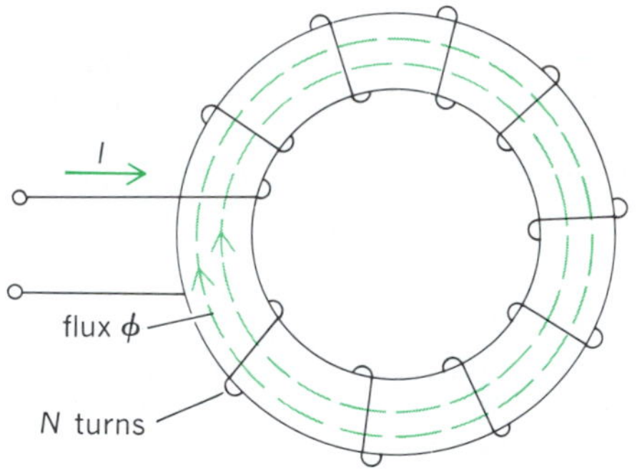

Diagram of a toroidal magnetic circuit; *l* is length. (*After A. E. Fitzgerald, D. E. Higginbotham, and A. Grabel, Basic Electrical Engineering, McGraw-Hill, 1967*)

Magnetic compass A compass depending for its directive force upon the attraction of the Earth's magnetism for a magnet free to turn in any horizontal direction. A compass is an instrument used for determining horizontal direction.

The magnetic compass operates on the principle that like magnetic poles repel each other whereas unlike poles attract each other. The Earth has internal magnetism similar to that which would result from a short, powerful bar magnet at the center of the Earth. The lines of force connecting the two poles are vertical at two points on the surface of the Earth, called magnetic poles, and horizontal along the magnetic equator, a line approximating a great circle nearly midway between the magnetic poles. The horizontal component of the Earth's magnetic field varies from a maximum on or near the magnetic equator to zero at the magnetic poles.

A simple magnetic compass consists of a magnetized needle mounted so as to be free to align itself with the horizontal component of the Earth's field. A magnetic compass is unreliable or inoperative in the vicinities of the magnetic poles. [A.B.M.]

Magnetic ferroelectrics Materials that display both magnetic order and spontaneous electric polarization. Research on these materials has enabled considerable advances to be made in understanding the interplay between magnetism and ferroelectricity. The existence of both linear and higher-order coupling terms has been confirmed, and their consequences studied. They have given rise, in particular, to a number of magnetically induced polar anomalies and have even provided an example of a ferromagnet whose magnetic moment per unit volume is totally induced by its coupling via linear terms to a spontaneous electric dipole moment.

Most known ferromagnetic materials are metals or alloys. Ferroelectric materials, on the other hand, are nonmetals by definition. It therefore comes as no surprise to find that there are no known room-temperature ferromagnetic ferroelectrics. In fact, there are no well-characterized materials which are known to be both strongly ferromagnetic and ferroelectric at any temperature.

Somewhat unaccountably, antiferromagnetic ferroelectrics are also comparative rarities in nature. Nevertheless, a few are known, and among them the barium-transition-metal fluorides are virtually unique in providing a complete series of isostructural examples. They have the chemical composition $BaXF_4$ in which X is a divalent ion of one of the 3*d* transition metals, manganese, iron, cobalt, or nickel. These materials are orthorhombic and all spontaneously polar (that is, pyroelectric) at room temperature. For all except the iron and manganese materials, which have a higher electrical conductivity than the others, the polarization has been reversed by the application of an electric field, so that they are correctly classified as ferroelectric. Long-range antiferromagnetic ordering sets in at temperatures somewhat below 100 K (−280°F). Structurally the materials consist of XF_6 octahedra which share corners to form puckered *xy* sheets which are linked in the third dimension *z* by the barium atoms. *See* CRYSTAL.

The importance of these magnetic ferroelectrics is the opportunity they provide to study and to separate the effects of a variety of magnetic and nonmagnetic excitations upon the ferroelectric properties and particularly upon the spontaneous polarization. Measurements are often made via the pyroelectric effect, which is the variation of polarization with temperature. This effect is an extremely sensitive indicator of electronic and ionic charge perturbations in polar materials. Through these perturbations the effects of propagating lattice vibrations (phonons), magnetic excitations (magnons), electronic excitations (excitons), and even subtle structural transitions can all be probed with precision. *See* PYROELECTRICITY.

Of all the X ions present in the series $BaXF_4$, the largest is Mn^{2+}. As the temperature is reduced from room temperature, the fluorine cages contract and eventually the divalent manganous ion becomes too big for its cage, precipitating a complicated structural transition at 250 K (−10°F). One of the more interesting effects of this phase transition is that it produces a lower-temperature phase with a crystal symmetry low enough to support the existence of the linear magnetoelectric effect, a linear coupling between magnetization and polarization. Below the antiferromagnetic transition at 26 K in $BaMnF_4$ this linear coupling produces a canting of the antiferromagnetic sublattices through a very small angle (of order 0.2 degree of arc). The result is a spontaneous, polarization-induced magnetic moment. At low temperatures $BaMnF_4$ is therefore technically a weak ferromagnet, although the resultant magnetic moment is extremely small, and it is more usually referred to as a canted antiferromagnet. This is the only well-categorized example of pyroelectrically driven ferromagnetism. *See* ANTIFERROMAGNETISM; FERROELECTRICS; FERROMAGNETISM; MAGNETISM. [M.E.L.]

Magnetic field A condition existing in the vicinity of a magnetic body (or a current-carrying medium) whereby the magnetic forces due to the body (or current) are detectable. In a magnetic field, there is a force on a moving charge in addition to the electrostatic (Coulomb) forces between charges, or there is a force on a magnetic pole. The description of the field may be in terms of magnetic induction (flux density) or in terms of magnetic field strength. *See* MAGNETIC INDUCTION.

A body can be magnetized by bringing it into a magnetic field due to currents or magnets. Except in the case of a ring magnetized along its circumference, the field associated with a magnetized body extends to the region surrounding the body. The external effect usually appears in limited regions of the body called poles. A magnetized bar of iron has two poles, one at either end; and from the fact that the bar will set itself in an approximate north-south direction in the Earth's field, it appears that there are two kinds of poles. The pole that is at the north end of the bar is called a north-seeking pole; that at the south end is called a south-seeking pole. The two poles at the ends are merely indications of the continuous magnetization within the body. An indication of the validity of this statement is the fact that when a bar magnet is broken into two parts, two new poles appear at the break, and the orientation of the poles in each fragment is the same as it was in the original magnet. *See* MAGNETIZATION.

It is observed that magnetic poles exert forces on each other and upon moving charges in the region near the poles. There is a field near the pole, and the pole may be considered as the cause of that field.

If the poles of a magnetized body are small enough that they may be considered point poles, the force that one pole exerts upon another is found to be proportional to the product of the pole strengths and inversely proportional to the square of the distance between them. This statement is called Coulomb's law of magnetostatics. *See* MAGNET; MAGNETOSTATICS. [K.V.M.]

Magnetic flux Lines used to represent the magnetic induction **B** in a magnetic field. *See* MAGNETIC FIELD; MAGNETIC INDUCTION.

Lines of flux used to represent the field in magnitude and direction are selected so that they are parallel to **B** at each point, and the number of lines of flux per unit area of a surface perpendicular to the field is equal to the magnitude of the magnetic induction, The total number of lines of induction through a surface is the magnetic flux Φ. Magnetic induction is the flux per unit area or flux density. [K.V.M.]

Magnetic flux linkage The sum of the number of webers of magnetic flux linking each of the individual turns of a coil under the influence of a magnetic field. The concept of magnetic flux linkage simplifies the analysis of electromagnetic systems. The flux linkage λ of a coil of *N* turns may be expressed mathematically by the equation below,

$$\lambda = \sum_{i=1}^{N} \Phi_i$$

where Φ_i is the webers of flux linking the *i*th turn of the coil. *See* MAGNETIC FIELD; MAGNETIC FLUX. [G.McP.]

Magnetic hysteresis The lagging of magnetization behind magnetizing force as the magnetic condition of a ferromagnetic material is changed.

When a ferromagnetic sample that is initially demagnetized is subjected to a continuously increasing magnetizing force *H*, the relation between *H* and flux density *B* is shown by the normal magnetization curve *Oab* of the illustration. The point *a* indicates the magnetic condition as the increasing magnetic intensity has reached H_1. If *H* is increased to a maximum value H_2 and then decreased again to H_1, the decreasing flux density does not follow the path of increase, but decreases at a rate less than that at which it rose. This lag in the change of *B* behind the change of *H* is called hysteresis. If the value of *H* is further reduced from H_1 to zero, *B* is not reduced to zero but to a value B_r. The specimen has retained a permanent magnet-

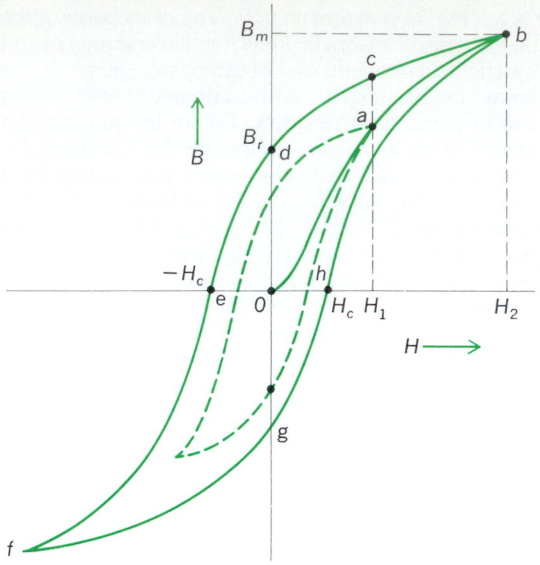

Hysteresis loop; symbols explained in text.

ism. This ordinate B_r is called the retentivity or remanence. The value of B may be reduced to zero at e by reversing the direction of H and increasing its value to H_c. This value H_c is called the coercive force or coercivity.

As H is increased in the negative direction, the magnetization proceeds along the curve of the illustration until at f the values of B and H are the same as those at b, but opposite in direction. When the reverse changes in H are made, the magnetization changes along the curve $fghb$. This entire loop $bdefghb$ is called a hysteresis loop. If the hysteresis loop starts from another point on the normal magnetization curve, such as a, there will be a smaller hysteresis loop entirely within the larger, such as the broken-line loop of the illustration. *See* MAGNETIZATION. [K.V.M.]

Magnetic induction

A vector quantity that is used as a quantitative measure of a magnetic field. It is defined in terms of the force on a charge moving in the field by Eq. (1), where

$$B = \frac{F}{qv \sin \theta} \quad (1)$$

B is the magnitude of the magnetic induction and F is the force on the charge q, which is moving with speed v in a direction making an angle θ with the direction of the field. The direction of the vector quantity B is the direction in which the force on the moving charge is zero.

The magnetic induction may also be expressed in terms of the force F on a current element of length l and current I, as in Eq. (2). The meter-kilogram-second (mks) unit of magnetic

$$B = \frac{F}{Il \sin \theta} \quad (2)$$

induction is derived from this equation by expressing the force in newtons, the current in amperes, and the length in meters. The unit of B is thus the newton/ampere-meter.

Magnetic flux density is the magnetic flux per unit area through a surface perpendicular to the magnetic induction. Magnetic flux density and magnetic induction are equivalent terms. *See* MAGNETIC FIELD; MAGNETIC FLUX. [K.V.M.]

Magnetic instruments

Instruments designed for the measurement of magnetic field strength or magnetic flux density, depending on their principle of operation.

Hall-effect instruments. Often called gaussmeters, these instruments measure magnetic field strength. They have a useful working range from 10 A/m to 2.4 MA/m (0.125 oersted to 30 kilooersteds). When a magnetic field, H_z, is applied in a direction at right angles to the current flowing in a conductor (or semiconductor), a voltage proportional to H_z is produced across the conductor in a direction mutually perpendicular to the current and the applied magnetic field. This phenomenon is called the Hall effect. The output voltage of the Hall probe is proportional to the Hall coefficient, which is a characteristic of the Hall-element material, and is inversely proportional to the thickness of this material. For a sensitive Hall probe, the material is thin with a large Hall coefficient. The semiconducting materials indium arsenide and indium antimonide are particularly suitable. *See* HALL EFFECT.

Fluxgate magnetometer. This instrument is used to measure low magnetic field strengths. It is usually calibrated as a gaussmeter with a useful range of 0.2 millitesla to 0.1 nanotesla (2 gauss to 1 microgauss).

Fluxmeter. This instrument is designed to measure magnetic flux. A fluxmeter is a form of galvanometer in which the torsional control is very small and heavy damping is produced by currents induced in the coil by its motion. This enables a fluxmeter to accurately integrate an emf produced in a search coil when the latter is withdrawn from a magnetic field, almost independently of the time taken for the search coil to be moved. *See* GALVANOMETER; MAGNETIC FLUX.

Electronic charge intergrators. Often termed an integrator or gaussmeter, an electronic charge integrator, in conjunction with a search coil of known effective area, is used for the measurement of magnetic flux density. Integrators have almost exclusively replaced fluxmeters because of their independence of level and vibration. The instrument (see illus.) consists of a

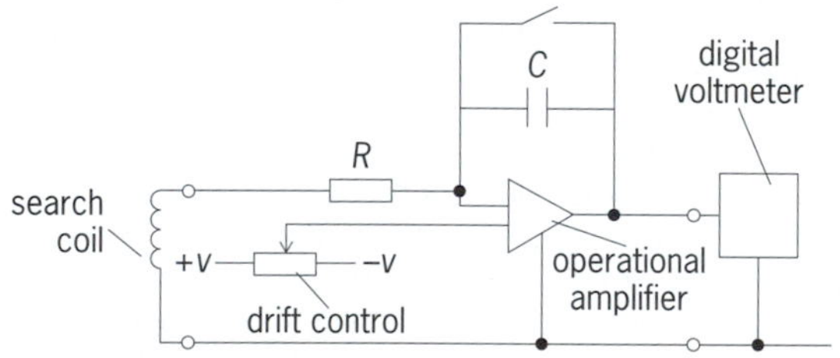

Arrangement of an electronic charge integrator.

high-open-loop-gain (10^7 or more) operational amplifier with a capacitive feedback and resistive input. *See* OPERATIONAL AMPLIFIER.

Rotating-coil gaussmeter. This instrument measures low magnetic field strengths and flux densities. It comprises a coil mounted on a nonmagnetic shaft remote from a motor mounted at the other end. The motor causes the coil to rotate at a constant speed, and in the presence of a magnetic field or magnetic flux density a voltage is induced in the search coil. The magnitude of the voltage is proportional to the effective area of the search coil and the speed of rotation. *See* MAGNETIC FIELD; MAGNETIC INDUCTION; MAGNETOMETER. [A.E.D.]

Magnetic lens

A magnetic field with axial symmetry capable of converging beams of charged particles of uniform velocity and of forming images of objects placed in the path of such beams. Magnetic lenses are employed as condensers, objectives, and projection lenses in magnetic electron microscopes, as final focusing lenses in the electron guns of cathode-ray tubes, and for the selection of groups of charged particles of specific velocity in velocity spectrographs.

Magnetic lenses may be formed by solenoids or helical coils of wire traversed by electric current, by axially symmetric pole pieces excited by a coil encased in a high-permeability material such as soft iron, or by similar pole pieces excited by permanent magnets. In the last two instances the armatures and pole pieces serve to concentrate the magnetic field in a narrow region about the axis.

Magnetic lenses are always converging lenses. Their action differs from that of electrostatic lenses and glass lenses in that they produce a rotation of the image in addition to the focusing action. For the simple uniform magnetic field within a long solenoid the image rotation is exactly 180°. Thus a uniform magnetic field forms an erect real image of an object on its axis. [E.G.R.]

Magnetic levitation Contactless, frictionless support and transportation of objects or vehicles by the use of magnets. Magnetic levitation provides vehicles that do not rely on friction for propulsion, guidance, or braking and are therefore very quiet.

The technique relies upon the controlled use of magnetic forces to balance gravitational forces and hold the vehicle clear of the track it follows. Magnets also control the vehicle laterally to prevent contact with the track. In designing a magnetic levitation system, there are two choices: magnets can attract the track, pulling up toward a rail, or repel the track, pushing up from a surface.

Electromagnetic systems. Most applications in England, Japan, and Germany use an attraction system with direct-current electromagnets working in conjunction with a steel rail. The magnetic flux in the rail is either axial or transverse to the direction of travel. The transverse system requires simple magnets but a heavy and expensive rail—a combination with poor lateral characteristics. Axial flux designs give good lateral characteristics, but a more difficult and less flexible magnet design is required. *See* ELECTROMAGNET; MAGNET; MAGNETIC FLUX.

Electrodynamic systems. The second option in the choice of system is to use repulsion. Electromagnets are always used with a virtual magnet induced in aluminum coils or an aluminum and steel track. This is an inherently stable technique, with the vehicle resting on a magnetic cushion. Using induction techniques, the Japanese have developed magnets and track to operate vehicles at very high speeds, about 310 mi/h (500 km/h), with a clearance of about 12 in. (300 mm). In order to achieve the very strong magnetic fields necessary to support a vehicle at this distance, superconducting materials are used in the windings. *See* SUPERCONDUCTING DEVICES. [M.G.Po.]

Magnetic materials A term generally applied to materials exhibiting ferromagnetism. The magnetic properties of all materials make them respond in some way to a magnetic field. Most materials are diamagnetic or paramagnetic and show almost no response. The materials which are most important to magnetic technology are ferromagnetic and ferrimagnetic materials. *See* FERRIMAGNETISM; FERROMAGNETISM; MAGNETISM.

Magnetic materials are further classified as soft or hard according to the ease of magnetization. Soft materials are used in devices in which change in the magnetization during operation is desirable, sometimes rapidly, as in ac generators and transformers. Hard materials are used to supply a fixed field either to act alone, as in a magnetic separator, or to interact with others, as in loudspeakers and instruments. *See* FERRITE; IRON ALLOYS; MAGNETIZATION. [A.C.B.]

Magnetic moment The relationship between a magnetic field and the torque exerted on a magnet, a current loop, or a charge that is moving in the field.

When a magnet is placed in a magnetic field of strength H, there is a torque L exerted on the magnet by the field. The torque is a maximum when the axis of the magnet is perpendicular to the field. The ratio of the torque for this position to the strength of the field is called the magnetic moment M of the magnet. *See* MAGNET.

If a flat coil of wire of N turns and area A, in which there is a current I, is placed in a magnetic field of flux density B, the coil experiences a torque L given by Eq. (1), where θ is the

$$L = NIAB \sin \theta \qquad (1)$$

angle between the field and the normal to the plane of the coil. The torque is maximum when $\theta = 90°$, that is, when the plane of the coil is parallel to the field. The ratio of the maximum torque to the flux density B is the magnetic moment of the coil, as shown in Eq. (2). If a charge is spinning, there is a

$$M = \frac{L}{B} = NIA \qquad (2)$$

charge in motion and thus an electric current. The spin is equivalent to a tiny current loop which has a magnetic moment. Atomic nuclei also possess magnetic moments. *See* NUCLEAR MOMENTS. [K.V.M.]

Magnetic monopoles Magnetic charges; hypothetical entities designed to put magnetism on an equal footing with electricity. In this article, gaussian units are used throughout.

The name monopole or one-pole for a magnetic charge is meant as a contrast to dipole or two-pole. A magnetic dipole is a system producing a magnetic field which looks like that of two equal but opposite magnetic charges close together, when the field is measured at a large distance from the dipole. Such a field can be produced by a loop of wire carrying an electric current. In this case, it is obviously impossible to tear the dipole apart and get two magnetic monopoles.

Electric fields are produced by electric charges according to the formula $\mathbf{E}(\mathbf{r}) = q\mathbf{r}/r^3$, where q is a point charge at the origin of coordinates, $\mathbf{r}$ is the position of observation at distance r from q, and $\mathbf{E}$ is the resulting field at $\mathbf{r}$. They exert forces on point charges $\mathbf{F} = q\mathbf{E}$, where $\mathbf{F}$ is the force, q is the charge, and $\mathbf{E}$ is the electric field from all other charges. A magnetic field is produced by moving electric charges. The magnetic field about a wire carrying electric current is a well-known example. In turn, magnetic fields exert forces on charges according to $\mathbf{F} = q(\mathbf{v}/c) \times \mathbf{B}$, where $\mathbf{v}$ is the velocity, $\mathbf{B}$ is the magnetic field, and c is the speed of light. This force only acts on a moving charge and is perpendicular to the direction of motion and to the magnetic field. It is natural to speculate that there may be magnetic charges which act as sources of magnetic fields in the same way that electric charges produce electric fields. Two such magnetic charges would obey the famous rules that like charges repel and unlike charges attract each other. *See* ELECTRON. [A.S.G.]

Magnetic permeability A factor, characteristic of a material, that is proportional to the magnetic flux density (magnetic induction) B produced in the material by a magnetic field divided by the intensity of the field H. Permeability is usually represented by the Greek letter μ.

Consider a solenoid that has been bent into a circular form so that the ends are joined together. This winding is called a toroid (see illustration). For a closely wound toroid, almost all the flux is in the interior. The intensity of the magnetic field inside the toroid is given by Eq. (1), where l is the mean cir-

$$H = \frac{NI}{l} \qquad (1)$$

cumference of the toroid and I is the current in the toroid coil. *See* MAGNETIC FIELD; SOLENOID (ELECTRICITY).

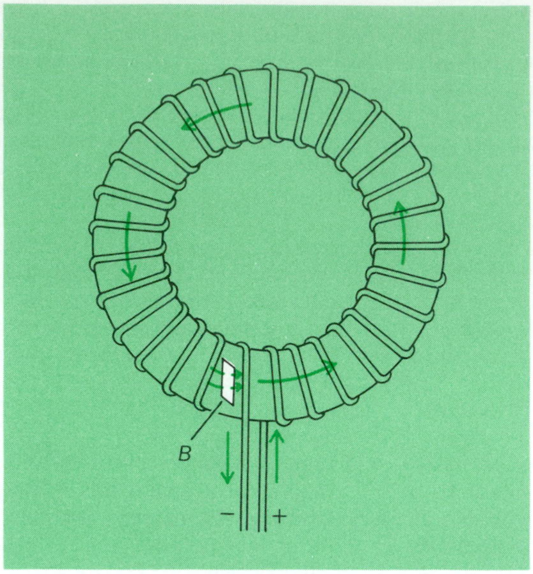

A solenoid wound in the form of a toroid.

The flux density B within the toroid is given, from Ampère's law, by Eq. (2) if the toroid is in empty space. Then Eq. (3)

$$B_0 = \mu_0 \frac{NI}{l} \tag{2}$$

$$\mu_0 = \frac{B_0}{H} \tag{3}$$

holds. If a medium takes the place of the empty space within the toroid, the value of B changes for the same value of H. The ratio of B to H, given in Eq. (4), is called the absolute

$$\mu = \frac{B}{H} \tag{4}$$

permeability of the medium. *See* AMPÈRE'S LAW.

It is convenient to define another quantity, called the relative permeability, μ_r, as the ratio of the permeability μ of the material to the permeability μ_0 of empty space, as in Eq. (5).

$$\mu_r = \frac{\mu}{\mu_0} \tag{5}$$

Relative permeability is a pure number, and independent of the system of units used. The permeability of free space has the numerical value of $4\pi \times 10^{-7}$ henry per meter. *See* ELECTRICAL UNITS AND STANDARDS.

Materials may be classified in terms of their relative permeabilities. Diamagnetic materials have values of μ_r, a little less than unity; paramagnetic materials have μ_r, a little greater than unity. Ferromagnetic materials are those that have relative permeabilities considerably greater than unity, which are variable, depending upon the value of H and the previous magnetic history. For some ferromagnetic materials, the maximum value of μ_r may be in the thousands. *See* DIAMAGNETISM; FERROMAGNETISM; MAGNETIC MATERIALS; PARAMAGNETISM. [K.V.M.]

Magnetic reception (biology) Sensitivity to magnetic stimuli. A great variety of biological effects resulting from exposure to fields many thousands of times more intense than Earth's have been reported. Among these are changes in plant growth rates, retardation of embryo development, changes in enzyme activity, alterations of tumor growth, and other indications of stress.

Evidence of magnetic detection has been given for honeybees and other kinds of insects, including termites, beetles, and fruit flies (*Drosophila*); for certain bacteria; and for birds. Most of the evidence for magnetic detection by birds has come from studies of their migratory and homing behavior. Results strongly suggest that birds possess a magnetic compass, that is, they can determine compass bearings from the geomagnetic field. Birds' sensitivity to magnetic stimuli is roughly similar to less than 10^{-4} gauss. The physical mechanism for magnetic detection by living organisms is unknown, though a variety of possibilities have been put forward. [W.T.K.]

Magnetic recording The technique of storing electric signals as a magnetic pattern on a moving magnetic surface. In magnetic recording a time-varying electric current produces a corresponding time-varying magnetic field in a recording head. This field then lays down a spatially varying pattern of magnetization in a sheet or strip of magnetic material moving past the head. When this recorded pattern moves past a playback head, a time-varying electric output is generated in the head corresponding to the original electric signal. A common form of magnetic recorder is diagrammed in Fig. 1. Magnetic tape from the supply reel moves past an erase head which removes any previously recorded pattern and leaves the tape in a demagnetized condition. An electric signal fed to the recording head lays down the desired new pattern on the tape. This recorded pattern induces an electric output as it moves over the playback head. Since the magnetization pattern remains in the tape until erased or otherwise altered, the tape may be rewound onto the supply reel and replayed as often as desired, with the erase and recording heads inactivated. The motor-driven capstan and pinch roller pull the tape over the heads at a uniform, controlled speed during recording and playback.

Magnetic recorders have become indispensable in innumerable areas of modern society. Audio recorders are sold by the millions as consumer products in reel-to-reel, cartridge, and cassette formats. Tape recorders are central in professional sound-recording studios and in radio and television broadcasting. They serve as mass-storage memories in computer systems and are used as instrumentation recorders for collecting data in research, industry, and military applications. They are employed in satellites and spacecraft for delayed transmission of data to Earth stations.

Principles of operation. Regardless of the nature of the application and of the mechanical configuration of the recorder, all magnetic recorders rely on the same basic magnetic behavior for their operation.

Thin plastic tapes are the medium now in universal use in tape recorders. The magnetic coating consists of very tiny particles dispersed in a thermosetting resin that is spread evenly on the surface of the plastic as a viscous liquid. The most frequently used materials for the magnetically active particles are iron oxide and chromium dioxide. Each of the particles in a tape coating is a tiny permanent magnet embedded in the resin so it cannot move.

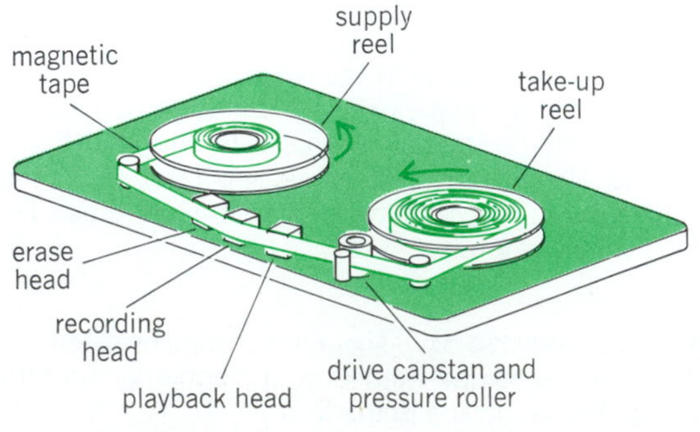

Fig. 1. Common form of magnetic tape recorder.

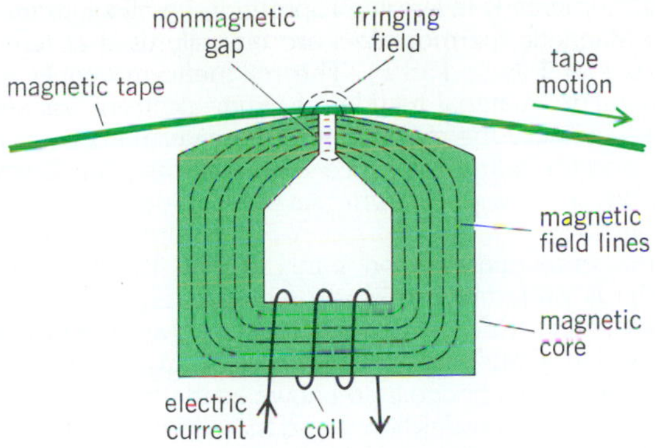

Fig. 2. Schematic representation of a magnetic recording head.

Magnetic heads. While magnetic heads may differ widely in geometrical details, they are basically of the form sketched in Fig. 2, whether they are recording or playback heads. Frequently, the same head is used for both functions. As indicated in the figure, a coil of wire is wound on a core of material that very readily carries magnetic flux (that is, a high-permeability material). A nonmagnetic gap is formed in the core at some point. When an electric current passes through the coil, magnetic field lines associated with the current follow paths around the high-permeability core. At the gap, some of the lines spread outside the core to form a fringing field. The recording medium passes through the fringing field above the gap and is magnetized there during recordings. *See* MAGNETIZATION.

Digital recording. Magnetic recording is naturally adapted to the use of digital signals consisting of binary states represented by ones and zeros. The two digital states are made to correspond to the two polarities of magnetic saturation of the recording medium. Since only saturation is involved, there is no concern with linearity of recording and high-frequency bias is not needed, nor is erasure of the recording medium. While the most widespread use of digital recording is in computers, it is finding increasing use in instrumentation recorders, satellite and telecommunication recorders, machine-tool controllers, electronic games, and audio and video recorders.

Audio recorders. The earliest significant use of magnetic recording was in the audio-frequency domain, and today audio tape recorders greatly outnumber other types. Machines for professional use in recording studios operate at standard tape speeds of 15 in./s (38.1 cm/s) and 30 in./s (76.2 cm/s), since short-wavelength losses are reduced at these speeds. By contrast, the cassette recorders sold in the mass consumer market have a tape speed of only $1\frac{7}{8}$ in./s (4.8 cm/s). Certain other recorders requiring only low-grade voice quality, such as dictation machines and recorders for long-term monitoring purposes, operate at even slower tape speeds.

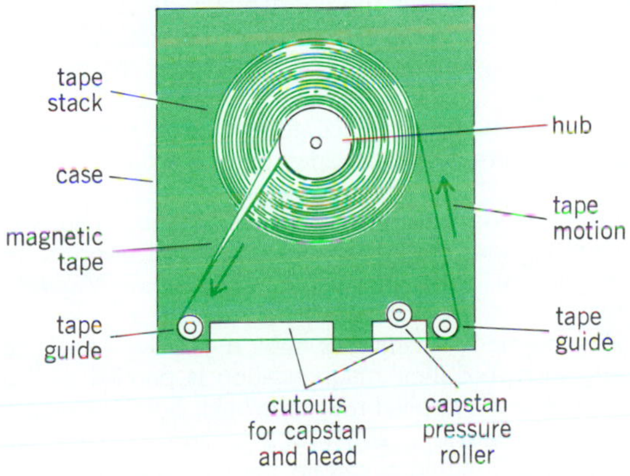

Fig. 3. Diagram of a Stereo-8 audiotape cartridge.

Of the numerous attempts to eliminate the inconvenience of threading tape through its path from the supply reel to the take-up reel in a recorder, only two have received widespread acceptance. They are the endless-loop cartridge and the coplanar-hub cassette. In the cartridge, the tape is wound on a single hub. Tape is drawn out of the wound stack at the hub, pulled past a playback head by a capstan, and returned to the outside of the stack, with the two ends of the length of tape being spliced together to form an endless loop (Fig. 3). In an audio cassette the tape is wound in a stack on a hub, with one end of the tape being permanently attached to the hub. The other end of the tape is permanently attached to an identical hub in the same plane as the first and symmetrically located in the flat plastic case. The tape path between the hubs is determined by guide posts in the case, and the tape travels from one hub to the other.

Video recorders. The purpose of video tape recorders is to record and play back television signals. The television signal frequency-modulates a carrier which is recorded on the magnetic tape, with frequencies in excess of 15 MHz being involved. The high head-to-tape speeds required for recording these high frequencies are obtained by mounting the heads on a rapidly rotating wheel that sweeps the heads more or less transversely across the tape from edge to edge.

Recorders for computers. From the inception of electronic computers, magnetic tape recorders have been an indispensable element in the hierarchy of computer memories. During the 1960s and 1970s, magnetic disk files assumed an equally important function in the hierarchy. All recording associated with computers employs digital signals. Since the loss of even a single bit in a digital number can have near-catastrophic results, much effort goes into assuring precision and reliability in the manufacture and fabrication of tapes, disks, heads, and mechanical structures involved in the recorders. *See* COMPUTER STORAGE TECHNOLOGY. [J.G.W.]

Magnetic relaxation The relaxation or approach of a magnetic system to an equilibrium or steady-state condition as the magnetic field is changed. This relaxation is not instantaneous but requires time. The characteristic times involved in magnetic relaxation are known as relaxation times. Relaxation has been studied for nuclear magnetism, electron paramagnetism, and ferromagnetism.

Magnetism is associated with angular momentum called spin, because it usually arises from spin of nuclei or electrons. The spins may interact with applied magnetic fields, the so-called Zeeman energy; with electric fields, usually atomic in origin; and with one another through magnetic dipole or exchange coupling, the so-called spin-spin energy. Relaxation which changes the total energy of these interactions is called spin-lattice relaxation; that which does not is called spin-spin relaxation. (As used here, the term lattice does not refer to an ordered crystal but rather signifies degrees of freedom other than spin orientation, for example, translational motion of molecules in a liquid.) Spin-lattice relaxation is associated with the approach of the spin system to thermal equilibrium with the host material; spin-spin relaxation is associated with an internal equilibrium of the spins among themselves. *See* MAGNETISM; SPIN (QUANTUM MECHANICS). [C.P.S.]

Magnetic resonance A phenomenon exhibited by the magnetic spin systems of certain atoms whereby the spin systems absorb energy at specific (resonant) frequencies when subjected to alternating magnetic fields. The magnetic fields must alternate in synchronism with natural frequencies of the magnetic system. In most cases the natural frequency is that of precession of the bulk magnetic moment of constituent atoms or nuclei about some magnetic field. Because the natural frequencies are highly specific as to their origin (nuclear magnetism, electron spin magnetism, and so on), the resonant method makes possible the selective study of particular features of interest. For example, it is possible to study weak nuclear magnetism unmasked by the much larger electronic paramagnetism or diamagnetism which usually accompanies it.

Nuclear magnetic resonance (that is, resonance exhibited by nuclei) reveals not only the presence of a nucleus such as hydrogen, which possesses a magnetic moment, but also its interaction with nearby nuclei. It has therefore become a most powerful method of determining molecular structure. The detection of resonance displayed by unpaired electrons, called electron paramagnetic resonance, is also an important application. *See* ELECTRON-NUCLEAR DOUBLE RESONANCE; MAGNETISM; NUCLEAR MAGNETIC RESONANCE (NMR); PARAMAGNETIC RESONANCE. [C.P.S.]

Magnetic separation methods All materials possess magnetic properties. Substances that have a greater permeability than air are classified as paramagnetic; those with a lower permeability are called diamagnetic. Paramagnetic materials are attracted to a magnet; diamagnetic substances are repelled. Very strongly paramagnetic materials can be separated from weakly or nonmagnetic materials by the use of low-intensity magnetic separators. Minerals such as hematite, limonite, and garnet are weakly magnetic and can be separated from nonmagnetics by the use of high-intensity separators.

Magnetic separators are widely used to remove tramp iron from ores being crushed, to remove contaminating magnetics from food and industrial products, to recover magnetite and ferrosilicon in the float-sink methods of ore concentration, and to upgrade or concentrate ores. Magnetic separators are extensively used to concentrate ores, particularly iron ores, when one of the principal constituents is magnetic. *See* MAGNETIC MATERIALS; MAGNETISM; MECHANICAL SEPARATION TECHNIQUES; ORE DRESSING. [F.D.DeV.]

Magnetic susceptibility The magnetization of a material per unit applied field. It describes the magnetic response of a substance to an applied magnetic field. *See* MAGNETISM; MAGNETIZATION.

All ferromagnetic materials exhibit paramagnetic behavior above their ferromagnetic Curie points. The general behavior of the susceptibility of ferromagnetic materials at temperatures well above the ferromagnetic Curie temperature follows the Curie-Weiss law. The paramagnetic Curie temperature is usually slightly greater than the temperature of transition. *See* CURIE TEMPERATURE; CURIE-WEISS LAW; FERROMAGNETISM.

Most paramagnetic substances at room temperature have a static susceptibility which follows a Langevin-Debye law. Saturation of the paramagnetic susceptibility occurs when a further increase of the applied magnetic field fails to increase the magnetization, because practically all the magnetic dipoles are already oriented parallel to the field. *See* PARAMAGNETISM.

The susceptibility of diamagnetic materials is negative, since a diamagnetic substance is magnetized in a direction opposite to that of the applied magnetic field. The diamagnetic susceptibility is independent of temperature. Diamagnetic susceptibility depends upon the distribution of electronic charge in an atom and upon the energy levels. *See* DIAMAGNETISM.

The susceptibility of antiferromagnetic materials above the Néel point, which marks the transition from antiferromagnetic to paramagnetic behavior, follows a Curie-Weiss law with a negative paramagnetic Curie temperature. [E.A.; F.Ke.]

Magnetic thermometer A thermometer whose operation is based on Curie's law, which states that the magnetic susceptibility of noninteracting (that is, paramagnetic) dipole moments is inversely proportional to absolute temperature. Magnetic thermometers are typically used at temperatures below 1 K (−458°F). The magnetic moments in the thermometric material may be of either electronic or nuclear origin. Generally the magnetic thermometer must be calibrated at one (or more) reference temperatures. *See* ELECTRON; CURIE-WEISS LAW; MAGNETIC MOMENT; NUCLEAR MOMENTS; PARAMAGNETISM.

In the temperature region 1 mK to 1 K, the thermometric material is preferably an electronic paramagnet, typically a nonconducting hydrous rare-earth salt such as cerium magnesium nitrate (CMN) $[2Ce(NO_3)_3 \cdot 3Mg(NO_3)_2 \cdot 24H_2O)]$. The mutual-inductance bridge (also known as the Hartshorn bridge) has been the most widely employed measuring circuit for precision electronic paramagnetic thermometry.

The much weaker nuclear paramagnets are used for thermometry only in the ultralow-temperature region. The nuclear thermometer loses adequate sensitivity for calibration purposes above 50–100 mK unless it is operated in a high polarizing field (*H* greater than 0.1 tesla). It can be utilized as a self-calibrating primary thermometer if a measurement of the spin-lattice relaxation time is conducted in parallel to the temperature determination. Nuclear paramagnetism is most conveniently monitored with nuclear magnetic resonance techniques. *See* LOW-TEMPERATURE THERMOMETRY; MAGNETIC RELAXATION; NUCLEAR MAGNETIC RESONANCE (NMR). [M.K.]

Magnetic thin films Sheets of magnetic material with thicknesses of a few micrometers or less, used in the electronics industry. Magnetic films can be single-crystalline, polycrystalline, or amorphous in the arrangement of their atoms. Applications include magnetic bubble technology, magnetoresist sensors, thin film heads, and recording media. *See* MAGNETIC BUBBLE MEMORY.

Both ferro- and ferrimagnetic films are used. The ferromagnetic films are usually transition-metal-based alloys. For example, permalloy is a nickel-iron alloy. The ferrimagnetic films, such as garnets or the amorphous films, contain transition metals such as iron or cobalt and rare earths. The ferrimagnetic properties are advantageous in bubble applications where a low overall magnetic moment can be achieved without a significant change in the Curie temperature. They are also useful for magnetooptic applications. *See* FERRIMAGNETISM; FERROMAGNETISM; MAGNETIC MATERIALS; MAGNETISM; MAGNETIZATION. [P.C.]

Magnetism Magnetism comprises those physical phenomena involving magnetic fields and their effects upon materials. Magnetic fields may be set up on a macroscopic scale by electric currents or by magnets. On an atomic scale, individual atoms cause magnetic fields when their electrons have a net magnetic moment as a result of their angular momentum. A magnetic moment arises whenever a charged particle has an angular momentum. It is the cooperative effect of the atomic magnetic moments which causes the macroscopic magnetic field of a permanent magnet. *See* MAGNET; MAGNETIC FIELD; MAGNETIC MOMENT.

Materials can be grouped according to their magnetic behavior, although there is some overlap among groups.

1. Diamagnetic substances have a negative magnetic susceptibility; they are magnetized in a direction opposite to that of an applied magnetic field. *See* DIAMAGNETISM; MAGNETIC SUSCEPTIBILITY.

2. Paramagnetic substances have a positive magnetic susceptibility such that their magnetization is parallel and usually proportional to the applied magnetic field. *See* PARAMAGNETISM.

3. A ferromagnetic substance is one with net atomic moments which, within a certain temperature range, tend to line up in such a way that there can exist a net magnetization in

the absence of an applied field. This category includes all ferromagnets and ferrimagnets and some helimagnets. At sufficiently high temperatures all these substances become paramagnetic. *See* CURIE TEMPERATURE; FERRIMAGNETISM; FERROMAGNETISM; HELIMAGNETISM.

4. An antiferromagnetic substance is one with net atomic moments which, within a certain temperature range, tend to line up in such a way that there can exist no net magnetization in the absence of an applied field. This category includes all antiferromagnets and some helimagnets. *See* ANTIFERROMAGNETISM; MAGNETIC MATERIALS; MAGNETIC PERMEABILITY; MAGNETIC RESONANCE; MAGNON. [E.A.; F.Ke.]

Magnetite A cubic mineral and member of the spinel structure type with composition $[Fe^{3+}]^{IV}[Fe^{2+}Fe^{3+}]^{VI}O_4$. The color is opaque iron-black and streak black, the hardness is 6 (Mohs scale), and the specific gravity is 5.20. The habit is octahedral, but the mineral usually occurs in granular to massive form, sometimes of enormous dimensions. Magnetite is a natural ferrimagnet, but heated above 1072°F (578°C; the Curie temperature) it becomes paramagnetic.

The major magnetic ore of iron, magnetite may be economically important if it occurs in sufficient quantities. The most spectacular ore body occurs at Kiruna in northern Sweden. Other important occurrences are in Norway, Russia, and Canada. *See* IRON; SPINEL. [P.B.M.]

Magnetization The process of becoming magnetized; also the property and in particular the extent of being magnetized. Magnetization has an effect on many of the physical properties of a substance. Among these are electrical resistance, specific heat, and elastic strain. *See* MAGNETOCALORIC EFFECT; MAGNETORESISTANCE; MAGNETOSTRICTION.

The magnetization **M** of a body is caused by circulating electric currents or by elementary atomic magnetic moments, and is defined as the magnetic moment per unit volume of such currents or moments. In the mks system of units, **M** is measured in webers per square meter.

The magnetic induction or magnetic flux density **B** is given by the equation, where **B** and **M** are in webers/m², **H**, the

$$\mathbf{B} = \mu_0(\mathbf{H} + \mathbf{M}) \text{ (mks)}$$

applied magnetic field, is in ampere-turns/m, and μ_0, the permeability of free space, is defined as $4\pi \times 10^{-7}$ henry/m, that is, weber/(ampere-turn)(m). *See* ELECTRICAL UNITS AND STANDARDS; MAGNETIC INDUCTION.

The topic of magnetization is generally restricted to materials exhibiting spontaneous magnetization, that is, magnetization in the absence of **H**. All such materials will be referred to as ferromagnets, including the special category of ferrimagnets. A ferromagnet is composed of an assemblage of spontaneously magnetized regions called domains. Within each domain, the elementary atomic magnetic moments are essentially aligned, that is, each domain may be envisioned as a small magnet. An unmagnetized ferromagnet is composed of numerous domains, oriented in some fashion.

The process of magnetization in an applied field **H** consists of growth of those domains oriented most nearly in the direction of **H** at the expense of others, followed by rotation of the direction of magnetization against anisotropy forces. *See* FERROMAGNETISM.

On removal of the field **H**, some magnetization will remain, called the remanence $\mathbf{M}_r$.

Curves, sometimes called *B-H* curves, are used to describe magnetic materials. They are plotted with **H** as abscissa and with either **M** or **B** as ordinate. In the illustration, $\mathbf{B}_r$ is the remanent induction ($\mathbf{B}_r = \mathbf{M}_r$); **H**c is the coercive force, or reverse field required to bring the induction **B** back to zero;

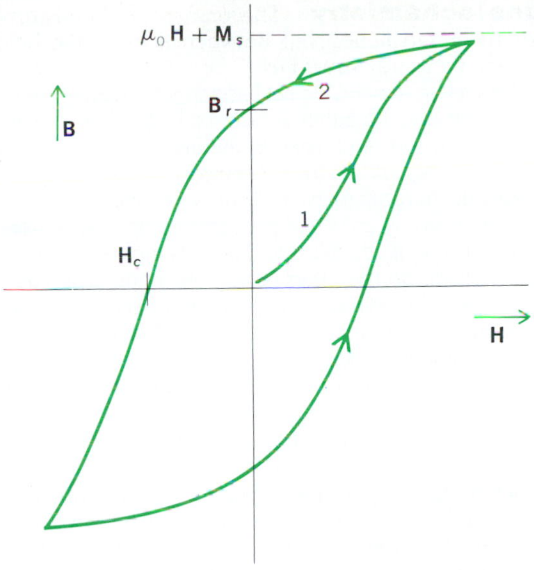

Magnetization curves.

and $\mathbf{M}_s$ is the saturation magnetization, or magnetization when all domains are aligned. The saturation magnetization is equal to the spontaneous magnetization of a single domain, except that it is possible to increase this magnetization slightly by application of an extremely large field. Saturation magnetization is temperature dependent, and disappears completely above the Curie temperature T_c where a ferromagnet changes into a paramagnet. *See* CURIE TEMPERATURE.

The irreversible nature of magnetization is shown most strikingly by the fact that the path of demagnetization does not retrace the path of magnetization—path 2 of the illustration does not retrace path 1. There is a tendency for the magnetization to show hysteresis, that is, to lag behind the applied field, and the loop in the illustration is called a hysteresis loop. *See* MAGNETIC HYSTERESIS. [E.A.; F.Ke.; K.V.M.]

Magneto A type of permanent-magnet alternating-current generator frequently used as a source of ignition energy on tractor, marine, industrial, and aviation engines. *See* ALTERNATING-CURRENT GENERATOR.

Modern induction-type magnetos consist of a permanent-magnet rotor and stationary low- and high-tension windings, also called the primary and secondary windings. The energy output of a magneto is obtained as a result of a rapid rate of change of flux through the stationary windings. The primary winding has comparatively few turns and the secondary winding has many thousand turns of fine wire. One end of the secondary winding is connected to an end of the primary winding and grounded to the frame of the magneto. The primary winding is closed on itself through a breaker mechanism actuated by a cam on the magneto shaft. The breaker is mechanically set to interrupt the primary circuit each time the flux through the winding is changing at its greatest rate. The sudden collapse of the primary current induces a very high voltage in the secondary winding. *See* IGNITION SYSTEM; INTERNAL COMBUSTION ENGINE. [R.T.W.]

Magnetocaloric effect The reversible change of temperature accompanying the change of magnetization of a ferromagnetic or paramagnetic material. This change in temperature may be of the order of 1°C (2°F), and is not to be confused with the much smaller hysteresis heating effect, which is irreversible. *See* THERMAL HYSTERESIS. [E.A.; F.Ke.]

Magnetochemistry The branch of chemistry which studies the interrelationship between a magnetic held and atomic and molecular structures.

A substance in a magnetic field acquires an intensity of magnetization which may be either smaller or larger than that induced in a vacuum by the same field. In the first case, the substance is said to be diamagnetic. In the second case, the substance may be paramagnetic, ferromagnetic, or antiferromagnetic.

Diamagnetism, a universal property of matter, is usually of the order of magnitude 10^{-6} to 10^{-5}. Temperature-dependent paramagnetism, on the other hand, arises only when an atom, ion, or molecule possesses a permanent magnetic moment either in the ground state or in an excited state. A permanent magnetic moment is the result of the presence of one or more unpaired electrons. Paramagnetic susceptibilities are of the order of magnitude 10^{-4} to 10^{-3}.

A substance composed of atoms with permanent magnetic moments which are very near to one another (for example, iron metal) may display ferromagnetism. This phenomenon occurs when large numbers of the atoms with permanent magnetic moments interact so that their individual moments align in a parallel fashion, giving rise to a large resultant moment.

On the other hand, a similar substance (for example, manganese metal) may display antiferromagnetism. Here, the magnetic moments align in an antiparallel fashion, thus largely canceling the individual magnetic moments of the atoms. Parallel versus antiparallel alignment depends, among other factors, upon interatomic distances. *See* ANTIFERROMAGNETISM; ATOMIC STRUCTURE AND SPECTRA; ELECTRON PARAMAGNETIC RESONANCE (EPR) SPECTROSCOPY; MAGNETIC RESONANCE; MAGNETISM; MOLECULAR STRUCTURE AND SPECTRA. [D.M.Gr.]

Magnetohydrodynamic power generator A system for the generation of electrical power through the interaction of a flowing, electrically conducting fluid with a magnetic field. As in a conventional electrical generator, the Faraday principle of motional induction is employed, but solid conductors are replaced by an electrically conducting fluid. The interactions between this conducting fluid and the electromagnetic field system through which power is delivered to a circuit are determined by the electrical magnetohydrodynamic (MHD) equations, while the properties of electrically conducting gases or plasmas are established from the appropriate relationships of plasma physics. Major emphasis has been placed on MHD systems utilizing an ionized gas, but an electrically conducting liquid or a two-phase flow can also be employed. *See* ELECTROMAGNETIC INDUCTION; PLASMA PHYSICS.

When combined with a steam turbine system to serve as the high-temperature or topping stage of a binary cycle, an MHD generator has the potential for increasing the overall plant thermal efficiency to around 50%, and values higher than 60% have been predicted for advanced systems. This follows from the increased efficiency made available by the higher source temperature. Thus, the MHD generator is a heat engine or electromagnetic turbine which converts thermal energy to a direct electrical output via the intermediate step of the kinetic energy of the flowing working fluid. *See* ENGINE; MAGNETOHYDRODYNAMICS; THERMODYNAMIC PROCESSES. [W.D.J.]

Magnetohydrodynamics The science that deals with the dynamics or motion of a fluid interacting with a magnetic field. The fluid must be a good conductor of electricity and hence can be a liquid metal or, more usually, an ionized gas or plasma. (Magnetohydrodynamics is alternately called hydromagnetics or magnetogas dynamics.) Magnetohydrodynamics is important in the development of controlled thermonuclear reactors. Other applications include simulation of hypersonic flight conditions, ionic thrust for outer-space propulsion, space-vehicle braking upon reentry to the atmosphere, high-energy particle accelerators, microwave generators, thermionic energy-conversion devices, application of thin metallic coatings, and the study of cosmic and upper atmospheric phenomena.

The conducting fluid and magnetic field interact through electric currents that flow in the fluid. The currents are induced as the conducting fluid and the magnetic field lines move across each other. In turn, the currents influence both the magnetic field and the motion of the fluid. Qualitatively, the magnetohydrodynamic interactions tend to link the fluid and the field lines so as to make them move together.

Although some early work in magnetohydrodynamics was concerned with liquid metals, a much wider interest has developed in phenomena which involve ionized gases or plasmas. A large electrical conductivity in a plasma requires a high density of high-energy electrons. This can occur for a plasma either in thermal equilibrium at relatively high temperatures, from a few electronvolts upward (1 electronvolt is equivalent to a temperature of 11,600 K or 20,900°F), or in a nonequilibrium situation where the ions and molecules remain at a low temperature and the electrons are supplied with energy by an external source such as a microwave generator or ultraviolet radiation.

Plasmas are encountered in interstellar space, in the Sun and stars, and in the upper atmosphere, as well as in human-made devices into which energy is fed from electrical, chemical, or nuclear sources. Strong shock waves forming ahead of a blunt object traveling with hypersonic velocities through low-density air may heat the air sufficiently to ionize it. *See* MAGNETIC FIELD. [R.L.]

Magnetometer Primarily, an instrument for measuring the intensity of a magnetic field. As related to the Earth's field, magnetometers are classed as absolute or relative, depending on whether they are capable of calibration without direct comparison with a standard magnetic instrument. Most types of magnetometers must be standardized by measuring the intensity of a field whose strength is accurately determined by other means, thus providing a relative measurement. [J.H.Ne.]

Magnetomotive force The magnetomotive force (mmf) around a magnetic circuit is the work per unit magnetic pole required to carry the pole once around the circuit. It is the analog of electromotive force. *See* ELECTROMOTIVE FORCE (EMF); MAGNETIC CIRCUITS.

It is expressed mathematically in the equation below,

$$\text{mmf} = \oint H \cos \theta \, ds$$

where $H \cos \theta$ is the component of magnetic field strength in the direction of a length of path ds. The line integral is taken around any closed path in the field. *See* MAGNETIC FIELD. [K.V.M.]

Magneton A unit of magnetic moment used for atomic, molecular, or nuclear magnets. *See* MAGNETIC MOMENT.

The Bohr magneton μ_B has the value of the classical magnetic moment of the electron, which can theoretically be calculated as shown in Eq. (1), where e and m are the electronic

$$\mu_B = \mu_0 = \frac{e\hbar}{2mc} = 9.2741 \times 10^{-2} \text{ erg/gauss}$$
$$= 9.2741 \times 10^{-24} \text{ joule/tesla} \quad (1)$$

charge and mass, $\hbar$ is Planck's constant divided by 2π, and c is the velocity of light. *See* ELECTRON.

The magnetic moment of an atom or molecule results from contributions from both the orbital angular momentum of the atomic electrons and the electronic moments themselves (attributed to the electron spin). When certain groups of atoms are compared, their moments show simple ratios. Observation of this fact led to the definition of the Weiss magneton (before the Bohr magneton) on a purely experimental basis as the unit for these moments. Its value is given by Eq. (2).

$$\mu_W = 1.853 \times 10^{-21} \text{ erg/gauss}$$
$$= 1.853 \times 10^{-24} \text{ joule/tesla} \qquad (2)$$

The nuclear magneton is obtained from the Bohr magneton by replacing m by the proton mass; it is thus 1836.15 times smaller than the Bohr magneton. The neutron, which has no net charge, has a magnetic moment of the same order (but opposite sign) as the proton. *See* NEUTRON; NUCLEAR MOMENTS. [McA.H.H.]

Magnetooptic Kerr effect A magnetooptic effect that deals with the changes that are produced in the optical properties of a reflecting surface of a ferromagnetic substance when the substance is magnetized. In a typical case this will result in elliptically polarized light appearing in reflection, when the ordinary rules of metallic reflection would give only plane-polarized light. *See* MAGNETOOPTICS. [G.H.Di.; W.W.W.]

Magnetooptics That branch of physics which deals with the influence of a magnetic field on optical phenomena. Considering the fact that light is electromagnetic radiation, an interaction between light and a magnetic field would seem quite plausible. It is, however, not the direct interaction of the magnetic field and light that produces the known magnetooptic effects, but the influence of the magnetic field upon matter which is in the process of emitting or absorbing light. *See* COTTON-MOUTON EFFECT; FARADAY EFFECT; MAJORANA EFFECT; MAGNETOOPTIC KERR EFFECT; VOIGT EFFECT; ZEEMAN EFFECT. [G.H.Di.; W.W.W.]

Magnetoresistance The change of electrical resistance produced in a current-carrying conductor or semiconductor on application of a magnetic field H. Magnetoresistance is one of the galvanomagnetic effects. It is observed both with H parallel to and transverse to the current flow. The change of resistance usually is proportional to H^2 except in very large fields, where it becomes proportional to H. *See* GALVANOMAGNETIC EFFECTS. [E.A.; F.Ke.]

Magnetosphere A comet-shaped cavity or bubble around the Earth, carved in the solar wind. This cavity is formed because the Earth's magnetic field represents an obstacle to the solar wind, which is a supersonic flow of plasma blowing away from the Sun. As a result, the solar wind flows around the Earth, confining the Earth and its magnetic field into a long cylindrical cavity with a blunt nose. Since the solar wind is a supersonic flow, it also forms a bow shock a few earth radii away from the front of the cavity. The boundary of the cavity is called the magnetopause. The region between the bow shock and the magnetopause is called the magnetosheath. The Earth is located about 10 earth radii from the blunt-nosed front of the magnetopause. The long cylindrical section of the cavity is called the magnetotail, which is on the order of a few thousand earth radii in length, extending approximately radially away from the Sun. *See* MAGNETIC FIELD; SOLAR WIND; SUN.

The magnetosphere has been extensively explored by a number of satellites carrying sophisticated instruments. The satellite observations have indicated that the cavity is not an empty one, but is filled with plasmas of different characteristics. The Earth's dipolar magnetic field is considerably deformed by these plasmas and the electric currents generated by them. *See* VAN ALLEN RADIATION.

All other magnetic planets, such as Mercury, Jupiter, and Saturn, have magnetospheres which are similar in many respects to the magnetosphere of the Earth. [S.-I.A.]

Magnetostatics The study of magnets and the fields produced by magnets. It should not be confused with the science of electromagnetism, which is concerned with the study of magnetic fields produced by currents. When considering magnetostatics, one looks upon magnet poles as seats of magnetism, and Coulomb's law of force between point poles in empty space is of fundamental importance. This law, expressed in rationalized meter-kilogram-second (mks) units, and using the Sommerfeld proposal, is written as the equation, below, where $\mu_0 = 4\pi \times 10^{-7}$ weber/ampere-meter is the

$$F = \frac{\mu_0}{4\pi} \frac{m_1 m_2}{r^2}$$

permeability of empty space; m_1 is the pole strength of one point pole and m_2 of the other, with pole strengths expressed in ampere-meters; F is the force in newtons; and r is the distance in meters between the point poles. *See* MAGNET; MAGNETIC FIELD. [R.P.Wi.]

Magnetostriction The change of length of a ferromagnetic substance when it is magnetized. More generally, magnetostriction is the phenomenon that the state of strain of a ferromagnetic sample depends on the direction and extent of magnetization. The phenomenon has an important application in devices known as magnetostriction transducers. *See* FERROMAGNETISM.

The magnetostrictive effect is exploited in transducers used for the reception and transmission of high-frequency sound vibrations. Nickel is often used for this application. *See* SONAR; ULTRASONICS. [E.A.; F.Ke.]

Magnetron The oldest of a family of crossed-field microwave electron tubes wherein electrons, generated from a heated cathode, move under the combined force of a radial electric field and an axial magnetic field. By its structure a magnetron causes moving electrons to interact synchronously with traveling-wave components of a microwave standing-wave pattern in such a manner that electron potential energy is converted to microwave energy with high efficiency. Magnetrons have been used since the 1940s as pulsed microwave radiation sources for radar tracking. Because of their compactness and the high efficiency with which they can emit short bursts of megawatt peak output power, they have proved excellent for installation in aircraft as well as in ground radar stations. In continuous operation, a magnetron can produce a kilowatt of microwave power which is appropriate for rapid microwave cooking. *See* ELECTRON TUBE.

The magnetron is a device of essentially cylindrical symmetry (see illustration). On the central axis is a hollow cylindrical cath-

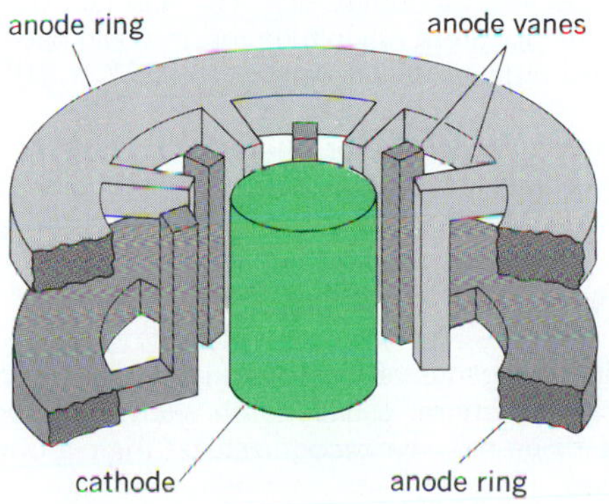

An interdigital-vane anode circuit and cathode, indicating the basic cylindrical geometry of the magnetron. (*After G. D. Sims and I. M. Stephenson, Microwave Tubes and Semiconductor Devices, Blackie and Son, London, 1963*)

ode. The outer surface of the cathode carries electronemitting materials, primarily barium and strontium oxides in a nickel matrix. Such a matrix is capable of emitting electrons when current flows through the heater inside the cathode cylinder. *See* VACUUM TUBE.

At a radius somewhat larger than the outer radius of the cathode is a concentric cylindrical anode. The anode serves two functions: (1) to collect electrons emitted by the cathode and (2) to store and guide microwave energy. The anode consists of a series of quarter-wavelength cavity resonators symmetrically arranged around the cathode. *See* CAVITY RESONATOR.

A radial dc electric field (perpendicular to the cathode) is applied between cathode and anode. This electric field and the axial magnetic field (parallel and coaxial with the cathode) introduced by pole pieces at either end of the cathode provide the required crossed-field configuration. [R.J.Co.]

Magnification A measure of the effectiveness of an optical system in enlarging or reducing an image. For an optical system that forms a real image, such a measure is the lateral magnification *m*, which is the ratio of the size of the image to the size of the object. If the magnification is greater than unity, it is an enlargement; if less than unity, it is a reduction.

The angular magnification is the ratio of the angles formed by the image and the object at the eye. In telescopes the angular magnification (or, better, the ratio of the tangents of the angles under which the object is seen with and without the lens, respectively), can be taken as a measure of the effectiveness of the instrument.

Magnifying power is the measure of the effectiveness of an optical system used in connection with the eye. The magnifying power of a spectacle lens is the ratio of the tangents of the angles under which the object is seen with and without the lens, respectively. The magnifying power of a magnifier or an ocular is the ratio of the size under which an object would appear seen through the instrument at a distance of 10 in. or 250 mm (the distance of distinct vision) divided by the object size. *See* LENS (OPTICS); OPTICAL IMAGE. [M.J.H.]

Magnitude (astronomy) The astronomical scale of brightness of stars and planets. A difference of five magnitudes between two objects corresponds to a factor of 100 in their ratio brightness. The basis of this system is that what the eye notes as equal intervals of brightness are actually equal ratios. One magnitude corresponds to a light ratio of $(100)^{1/5} = 2.512$. Magnitudes should be defined monochromatically, or at least over a narrow and well-defined range of wavelengths; the subscript λ indicates the band used. The zero point of the magnitude system is defined so that the brightest stars are near zero.

The energy received from a star depends on its distance *r* as well as on its intrinsic luminosity. The absolute magnitude is defined as the apparent magnitude a star would have if located at a standard distance of 10 parsecs (3.0857×10^{17} m). *See* PARSEC.

A wide variety of multicolor photoelectric systems are in use to measure color; correlations between such color and brightness measures are defined by the results for a group of standard stars. The color of a star is given, in general, by a color index, on a particular base line. [J.L.Gr.]

Magnolia A genus of trees with large, chiefly white flowers, and simple, entire, usually large alternate leaves. In the winter the twigs may be recognized by their aromatic odor when bruised.

The most important species commercially is *Magnolia acuminata*, commonly called cucumber tree, which grows in the Appalachian and Ozark mountains. The fruit is red when ripe and resembles a small cucumber in shape.

The wood of the magnolia is similar to that of the tulip tree and is rather soft, but it is of such wide natural dimensions that it is valued for furniture, cabinetwork, flooring, and interior finish.

Magnolia species occur naturally in a broad belt in the eastern United States and Central America, with a similar region in eastern Asia and the Himalayas. *See* MAGNOLIALES. [A.H.G./K.P.D.]

Magnoliales The nomenclaturally typical order of the subclass Magnoliidae in the class Magnoliopsida (dicotyledons) of the division Magnoliophyta (Angiospermae), the flowering plants. The Magnoliales are noteworthy as a loose aggregation of the most primitive existing families of angiosperms. The order consists of 19 families and about 5600 species. The Lauraceae (about 2200 species) and Annonaceae (about 2100 species) are the largest families of the order, while the family Magnoliaceae is much smaller, with only a little over 200 species (see illustration). Within its subclass the order is characterized by the presence of spherical "ethereal oil" cells and by a well-developed perianth of separate tepals. All members of the order are woody. *See* AVOCADO; CAMPHOR TREE; MAGNOLIA; MAGNOLIOPSIDA; SASSAFRAS; TULIP TREE. [A.Cr.]

Flower of the tulip tree (*Liriodendron tulipifera*). (Photography by F. E. Westlake, from National Audubon Society)

Magnoliidae A subclass of the class Magnoliopsida (dicotyledons) in the division Magnoliophyta (Angiospermae), the flowering plants. The subclass consists of 6 orders, 36 families, and more than 11,000 species. The Magnoliidae are the most primitive subclass of flowering plants. In general, they have a well-developed perianth, which may or may not be differentiated into sepals and petals. *See* ARISTOLOCHIALES; FLOWER; MAGNOLIALES; MAGNOLIOPSIDA; NYMPHAEALES; PAPAVERALES; PIPERALES; RANUNCULALES. [A.Cr.]

Magnoliophyta A division of seed plants consisting of about 220,000 widely distributed species, or about two-thirds of all known kinds of plants. Because the young seeds, called ovules, are enclosed in an ovary, the Magnoliophyta are often called angiosperms (Greek angeion, vessel, plus sperma, seed), in contrast to the gymnosperms (division Pinophyta), which lack an ovary and have "naked" ovules. They further differ in that most angiosperms have vessels (specialized water-conducting tubes) in the xylem, whereas most gymnosperms do not. Angiosperms range from tiny herbs to gigantic trees, and they dominate the land vegetation of most of the Earth. They also occur in lakes and ponds and streams, either rooted to the bottom or free-floating, but only a very few are truly marine, and these are rooted to the bottom near the shore of the ocean.

The angiosperms may be characterized as vascular plants with roots, stems, and leaves, usually with well-developed vessels in the xylem and with companion cells in the phloem. The central cylinder has leaf gaps or scattered vascular bundles; the ovules are enclosed in an ovary; and the female gametophyte is reduced to a few-nucleate embryo sac without an archegonium. The male gametophyte is reduced to a tiny pollen grain that gives rise to a pollen tube containing a tube nucleus and two sperms; one sperm fuses with the egg in the embryo sac to form a zygote, and the other fuses with two nuclei of the embryo sac to form a triple fusion nucleus that is typically the forerunner of the endosperm of the seed. *See* LEAF; PHLOEM; POLLEN; ROOT (BOTANY); SEED; STEM; XYLEM.

The angiosperms are often called flowering plants because the flower is a characteristic feature not found in other groups of plants. *See* FLOWER; REPRODUCTION (PLANT).

The angiosperms are usually considered to be the most highly evolved division of the subkingdom Embryobionta. Their highly specialized and relatively efficient conducting tissues, combined with the protection of their ovules in an ovary, give them a competitive advantage over most other groups of land plants in most regions. This general advantage is reflected in their abundance and in their great diversification, with different species adapted to a wide range of ecological niches. *See* EMBRYOBIONTA.

The Magnoliophyta consist of two classes, the Magnoliopsida and Liliopsida. The Magnoliopsida usually have two cotyledons in the embryo of the seed and are commonly called dicotyledons, or dicots; the Liliopsida usually have embryos with a single cotyledon and are commonly called monocotyledons, or monocots. *See* LILIOPSIDA; MAGNOLIOPSIDA.

All the important staples of the human diet are either angiosperms or are derived from animals (except for fish) which feed largely on angiosperms. In addition to food, they provide clothing and shelter and furnish raw materials for countless industrial processes. *See* PLANT KINGDOM; TRACHEOPHYTA. [A.Cr.]

Magnoliopsida One of the two classes of flowering plants which collectively make up the division Magnoliophyta (Angiospermae). The Magnoliopsida, often known as Dicotyledoneae or dicotyledons, embrace 6 subclasses, 56 orders, 293 families, and about 165,000 species. *See* ASTERIDAE; CARYOPHYLLIDAE; DILLENIIDAE; HAMAMELIDAE; MAGNOLIIDAE; ROSIDAE.

All of the characters which collectively distinguish the Magnoliopsida from the Liliopsida (monocotyledons) are subject to exception, but in general the Magnoliopsida have two cotyledons and net-veined leaves. The vascular bundles are typically borne in a ring (or cylinder) enclosing a pith. Increase in thickness of stems and roots, after the primary tissues have matured, results from meristematic activity of a cambial layer which passes through the vascular bundles. In about half of the species of the group, the cambium of the stem forms a continuous cylinder which produces a new layer of wood (secondary xylem) and bark (secondary phloem) each growing season for year after year. Such plants become trees or shrubs.

It is widely agreed that the most primitive existing angiosperms belong to the dicotyledons (especially the order Magnoliales) and that most of the characters which distinguish the monocotyledons as a group are derived rather than primitive. *See* LILIOPSIDA; MAGNOLIALES; MAGNOLIOPHYTA; PLANT KINGDOM. [A.Cr.]

Magnon A quasi-particle which is introduced to describe small departures from complete magnetic ordering in ferro-, ferri-, antiferro-, and helimagnetic arrays. *See* ANTIFERROMAGNETISM; FERRIMAGNETISM; FERROMAGNETISM; HELIMAGNETISM.

A sinusoidal variation of that angular momentum which is associated with magnetism (mostly spin momentum of the electrons) is called a spin wave. A magnon is a quantized spin wave, much as a phonon is a quantized lattice-vibration wave or a photon is a quantized electromagnetic wave.

The low-temperature behavior of ordered magnetic materials is very accurately described in terms of the statistical excitation of magnons. In particular, the departure of ferromagnetic magnetization from complete saturation is found to fall off proportionally to $T^{3/2}$ as the temperature T increases from absolute to zero to about half the Curie temperature.

The energy of a spin wave, or magnon, increases with increasing wave vector $\mathbf{k}$, that is, with decreasing wavelength of the ripple. The reason is that the so-called exchange forces which produce magnetic ordering are very short-range, causing only neighboring spins strongly to resist alignment.

The precessions of interacting spins are very complicated and, in general, the equations of motion contain nonlinear terms. These become important for the large-amplitude precessions induced by intense microwave fields. Large-amplitude precessions can also break up into acoustic vibrations, provided the acoustic eigenfrequencies equal the difference between the microwave (pump) frequency and that of some magnetostatic mode. This is called magnetoacoustic resonance. [F.Ke.]

Mahogany A hard, red or yellow-brown wood which takes a high polish and is extensively used for furniture and cabinetwork. The West Indies mahogany tree (*Swietenia mahagoni*), a native of tropical regions in North and South America, is a large evergreen tree with smooth pinnate leaves. Together with other species it yields the world's most valuable cabinet wood. In the United States it occurs naturally only in the extreme southern tip of Florida, but it is planted elsewhere in the state as an ornamental and shade tree. *See* GERANIALES. [A.H.G./K.P.D.]

Maillard reaction A nonenzymatic chemical reaction involving condensation of an amino group and a reducing group, resulting in the formation of intermediates which ultimately polymerize to form brown pigments (melanoidins). The reaction was named for the French biochemist Louis-Camille Maillard. It is of extreme importance to food chemistry, especially because of its ramifications in terms of food quality. *See* AMINE; REACTIVE INTERMEDIATES.

There are three major stages of the reaction. The first comprises glycosylamine formation and rearrangement *N*-substituted-1-amino-1-deoxy-2-ketose (Amadori compound). The second phase involves loss of the amine to form carbonyl intermediates, which upon dehydration or fission form highly reactive carbonyl compounds through several pathways. The third phase occurring upon subsequent heating involves the interaction of the carbonyl flavor compounds with other constituents to form brown nitrogen-containing pigments (melanoidins). These are highly desirous compounds in certain foods browned by heating in the presence of oxygen.

The Maillard reaction is considered undesirable in some biological and food systems. The interaction of carbonyl and amine compounds might damage the nutritional quality of proteins by reducing the availability of lysine and other essential amino acids and by forming inhibitory or antinutritional compounds. The reaction is also associated with undesirable flavors and colors in some foods, particularly dehydrated foods. *See* AMINO ACIDS; CARBONYL; LYSINE. [M.E.B.]

Majorana effect A magnetooptic effect that deals with optical anisotropy of colloidal solutions. The effect probably is caused by the orientation of the particles in the magnetic field. *See* MAGNETOOPTICS. [G.H.Di., W.W.W.]

Malachite A basic carbonate of copper with the chemical formula $Cu_2(OH)_2(CO_3)$. Large quantities of malachite have been found in Siberia and used for ornamental stone. It has been mined as an ore of copper near Kolwezi, Republic of the Congo.

Malachite is monoclinic but usually occurs in massive forms or in bundles of radiating fibers. It is invariably green. It has a specific gravity of 4.05 and a hardness of 3.5–4 on Mohs scale. *See* CARBONATE MINERALS; COPPER. [R.I.Ha.]

Malacostraca The largest and most diversified class of the Crustacea; includes the shrimps, lobsters, crabs, sow bugs, beach hoppers, and their allies. The shell or carapace may be large, small, vestigial, or absent; the tail or abdomen is long or short; the eyes are generally set on movable stalks but may be sessile or even coalesced. Despite this diversity, the unity of the group is demonstrated by the following characteristics which all share. The maximum number of appendages is 19 pairs. The trunk limbs are sharply differentiated into a thoracic series of eight pairs and an abdominal series of six pairs. The female genital duct always opens at the level of the sixth thoracic segment, whereas those of the male open at the level of the eighth.

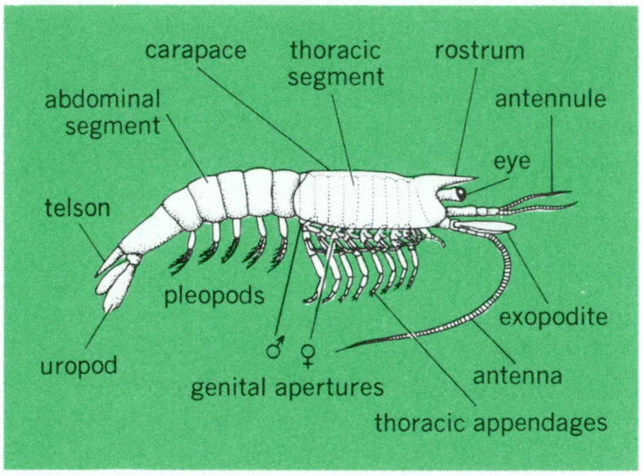

Caridoid facies. (*After E. R. Lankester, ed., A Treatise on Zoology, pt. 7. fasc. 3, A. and C. Black, 1909*)

Malacostraca are divided into three subclasses, the Phyllocarida, Hoplocarida, and Eumalacostraca. Central to the classification of the Eumalacostraca has been the concept of the "caridoid facies" (see illustration). This term refers to a series of morphological attributes generally common to the four orders, Syncarida, Pancarida, Peracarida, and Eucarida. *See* EUCARIDA; EUMALACOSTRACA; HOPLOCARIDA; PERACARIDA; PHYLLOCARIDA; SYNCARIDA. [P.A.McL.]

Malaria A group of diseases caused by hemosporidian blood parasites of the genus *Plasmodium*, transmitted to humans by the bite of *Anopheles* mosquitoes.

Malarias have been known since ancient times. Their association with swamps and marshes has long been known, and the names given to these fevers—malaria (bad air), paludism (fever of the swamps), and Sumpfenfieber—bear witness to it. The worldwide prevalence of malaria, its great impact upon the health of the population of warm climates, and its toll in mortality and morbidity have made it the outstanding single global health problem.

With the bite of an infected mosquito, a few hundred parasites are introduced into the body. They find their way to the liver, where they develop in hepatic cells and grow to schizonts, each containing several thousand minute parasites. From the liver, the released parasites find their way to the bloodstream to attack red blood cells. Their synchronic development in erythrocytes and their division and simultaneous bursting produce the typical malaria attack. Together with this parasitic growth, other, more resistant, forms appear in the blood, destined to undergo sexual development (gametocytes). When taken up by a female *Anopheles* during her blood meal, the fertilized gametes undergo a process of sporogony in the midgut of their insect vector and subsequently penetrate the salivary glands as invasive sporozoites; from there they reach their human host in the act of the mosquito bite.

Until 1930 quinine was the only specific remedy available. Since 1930 a number of synthetic chemotherapeutic compounds have been used in the therapy and prophylaxis of malaria: atabrine, chloroquine, primaquine, paludrine, pyrimethamine, and others.

A global program to eradicate malaria under the auspices of the World Health Organization has been underway since the end of World War II. By applying DDT and other contact insecticides, it has been possible to eliminate malaria from Greece, Italy, and a number of endemic countries. *See* HAEMOSPORINA. [M.Yo.]

Mallophaga A comparatively small order of insects numbering perhaps 3000 known species which are commonly called the bird lice or, more correctly, the biting lice. Most of them occur among the feathers of birds. Only a comparatively small number, perhaps 300 species, occur on mammals. The group can be distinguished from the sucking lice by the fact that they always possess mandibles. Unlike sucking lice, they never have claws enlarged or modified to close about a hair or a feather. The significant difference between biting and sucking lice is the fact that the Mallophaga rarely transmit any disease of their host, while the sucking lice are well-known vectors of organisms that cause certain diseases.

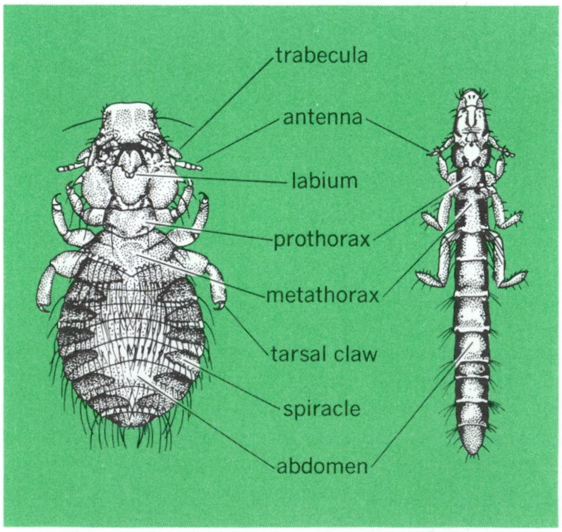

Morphology of two typical mallophagans.

The group may be technically defined as follows (see illustration): relatively small insects, much flattened and permanently without wings. The antennae are five-segmented; mandibles are distinctly developed and their apices cross. The prothorax is developed as a distinct segment, while the mesothorax and metathorax at times are closely fused. Legs possess either one or two terminal claws. The ovipositor is greatly reduced in size.

Like the sucking lice, the Mallophaga are closely adapted to life upon a single host species or, at most, upon a group of closely related hosts. The physical adaptation is not as close as in the sucking lice. There is enough adaptation, however, to preclude the ready passage of a mallophagan from one host species to another, particularly at times other than in the nest or when the hosts are in close bodily contact. The result is that they are usually passed from an animal to its offspring as a sort of racial inheritance, and the insects live as upon an "island" which is the host. Because of this, the classification of the Mallophaga reflects in a way the classification of their hosts. *See* ANOPLURA; INSECTA. [D.M.DeL.]

Malnutrition A compromised state of nutritional health resulting from inadequate intake of food or inability to assimilate and metabolize nutrients properly. Malnutrition may arise as a primary malady, due to insufficient intake of food or of any of many nutrients essential for the body to grow and maintain good health; this is by far the most common type of malnutrition. Malnutrition may also develop as a secondary consequence of many underlying disease states and psychic disorders. *See* NUTRITION.

Primary malnutrition may be further complicated when the available food is of compromised quality. Individuals and large populations suffer from inadequate food supplies throughout the nonindustrialized areas of the world: as many as one-half to two-thirds of the Earth's population have too little food.

An individual's stage of growth is a major determinant of vulnerability to malnutrition. The groups of individuals that are experiencing most rapid growth (infants, preschool children, pregnant women, and adolescents) are at highest risk. Lack of a specific essential nutrient is manifested most prominently in the body's most rapidly growing and dividing cells. These include cells of the skin and hair, the mucous membranes lining the gastrointestinal tract, and the blood cells. Thus lesions of the skin, gastrointestinal symptoms, and anemia are commonly encountered in the malnourished person. *See* KWASHIORKOR DISEASE.

Disease states are commonly associated with malnutrition in individuals who are ill for significant periods. A variety of diseases may be responsible for markedly restricted intakes of food, called secondary malnutrition. The ill person may lose interest in ingesting an adequate diet, even though nourishing and attractive food is available. Repeated episodes of vomiting may make food intake impossible. Mental illness may be expressed as an abnormal aversion to food as in anorexia nervosa. The alcoholic patient may misuse his or her economic resources and not have food available or ignore the necessity of eating. *See* ALCOHOLISM; METABOLIC DISORDERS.

Disease of the gastrointestinal tract may cause symptoms of secondary malnutrition, even though there is a normal dietary intake. Inability to digest food, or to absorb the digested nutrients, will eventually lead to malnutrition. Infectious diseases, as well as cancers and other malignant neoplasms, increase the metabolic activity of the body, raising the demand for certain essential nutrients and food energy. [N.J.S.]

Malt beverage A fermented beverage produced from grain. Beer is a generic term used to describe alcoholic beverages made from cereal grains, especially barley, in the form of malt. Ale, lager, porter, and stout are different kinds of beer made by recognizably similar processes. The United States is the largest producer of beer in the world.

The manufacture of beer is a complex natural process of three general parts: the preparation of barley by germination, or the malting process; the actual digestion of barley (now malt) starch to produce a solution of sugars (called wort) and the adjustment of flavor with hops, which are the brewhouse processes; and the fermentation of these sugars by yeast to yield alcohol, carbon dioxide gas, and flavor compounds to produce beer. [M.J.L.]

Maltase An enzyme which breaks down (hydrolyzes) the disaccharide maltose into glucose. Maltase has been found in the pancreas, intestine, liver, kidney, and blood serum of animals. It has also been found in the bacteria, fungi, and monocotyledonous and dicotyledonous plants. *See* CARBOHYDRATE METABOLISM; ENZYME. [D.N.La.]

Maltose An oligosaccharide, known as malt sugar, a reducing disaccharide (see illustration). It is fermentable by yeast in the presence of D-glucose.

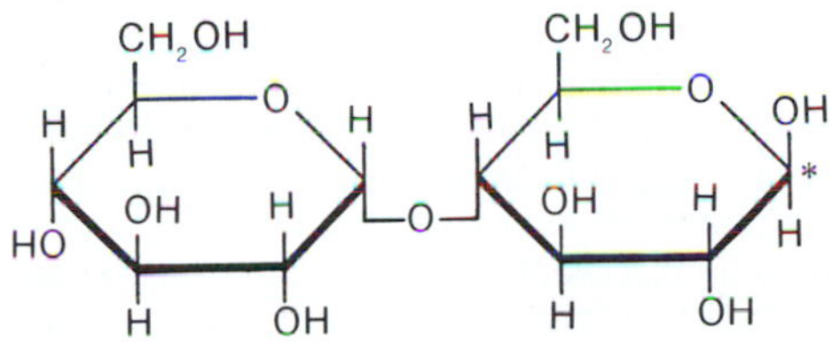

Formula for maltose (α form; * indicates reducing group).

The action of animal (salivary and pancreatic) as well as plant (germinating cereals, sweet potato) amylases on starch, dextrin, and glycogen produces maltose as the main end product. Maltose is hydrolyzed by acids and the enzyme maltase to two molecules of D-glucose. *See* GLUCOSE; MALTASE; OLIGOSACCHARIDE. [W.Z.H.]

Malvales An order of flowering plants, division Magnoliophyta (Angiospermae), in the subclass Dilleniidae of the class Magnoliopsida (dicotyledons). The order consists of 6 families and more than 3500 species. The Malvaceae, with more than 1500 species, and the Sterculiaceae, with about 1000 species, are the largest families of the order. The Malvales have hypogynous flowers with valvate calyx, mostly separate petals that are often convolute in bud, usually numerous and centrifugal stamens that are often connate by their filaments, and a basically syncarpous pistil with axile (rarely parietal) placentation and often numerous carpels.

Cotton (*Gossypium*), the linden tree (*Tilia*), and such garden ornamentals as hollyhock (*Althaea rosea*) and rose of Sharon (*Hibiscus syriacus*) are familiar members of the Malvales. *See* BALSA; BASSWOOD; COLA; COTTON; DILLENIIDAE; FLOWER; JUTE; MAGNOLIOPSIDA; OKRA. [A.Cr.]

Mammalia The vertebrate class Mammalia contains thousands of living and fossil animals that can be characterized by a complex of skeletal, soft anatomical, and physiological features.

In living mammals the allantois and amnion are present during embryonic development. As a rule hair covers the body, but it can be secondarily reduced in many orders such as cetaceans, proboscideans, perissodactyls, and sirenians. The fur cover, in addition to serving as protection, also insulates to retain body heat. The four-chambered heart (two ventricles and two auricles) keeps the circulation of the lungs separate from the body circulation, thus resulting in more efficient transport of oxygen to body tissues. The left aortic arch is dominant. The live-born young (monotremes are born from reptilelike eggs) feed on milk produced by the specialized skin glands of the female. The relatively large brain of mammals is the result of selection for an improved memory and for increased intelligence. *See* FETAL MEMBRANE; HAIR; LACTATION; MAMMARY GLAND; THERMOREGULATION.

The teeth of mammals are highly diagnostic even on the species level, and an understanding of dental evolution is the chief key to comprehending mammalian phylogeny. The longest and most crucial span of mammalian evolution, that during the Jurassic and Cretaceous (about 180,000,000–63,000,000 years ago) is deciphered mainly on the basis of cheek teeth (molars). Tracing the various homologous cusps and crests of different mammalian groups enables scientists to establish relatively firm evolutionary ties between groups. *See* DENTITION; TOOTH.

As a rule mammals have seven cervical vertebrae. During growth of the skeleton most mammalian long bones and some of the vertebrae, unlike those in reptiles, have ossified terminal ends (epiphyses) separated by the growing cartilage layer from the middle part of the bone. The cartilage layer in which growth occurs ossifies when growth is completed. The epiphyses enable mammals to have strongly articulating joints while the individual is still growing. Mammals tend to have ribs only in the thoracic region. These ribs are firmly attached to the sternum, forming the lung enclosing rib cage, covered posteriorly by the muscular diaphragm. *See* SKELETAL SYSTEM.

Oral digestion, control of internal body temperatures, a more efficient circulatory system, superior kidneys to excrete wastes of an increased metabolism, a skeletal and muscular system capable of more rapid locomotion, and in general, more efficient functioning by all the bodily organs rendered the mammals competitively superior to the reptiles. Instead of laying eggs like their reptile ancestors, mammals evolved development of the young inside the mother, thus protecting the offspring during their most vulnerable stage of development. Postnatal care of the young probably evolved concurrently with internal development. As a result of sustaining high metabolic activity, a greater awareness of the environment became necessary. Mammals developed keener, more sensitive noses and ears, and to coordinate this increased sensory information the brain became larger and more complex. *See* DIGESTIVE SYSTEM; KIDNEY; MUSCULAR SYSTEM; REPRODUCTIVE SYSTEM.

The mammals are placed in three subclasses as given in the following classification scheme.

Class Mammalia
  Subclass Prototheria
    Order Monotremata
  Subclass Allotheria
    Order Multituberculata
  Subclass Theria
    Infraclass Triconodonta
      Order: Triconodonta
        Docodonta
    Infraclass Pantotheria
      Order: Pantotheria
        Symmetrodonta
    Infraclass Metatheria
      Order Marsupialia
    Infraclass Eutheria
      Order: Insectivora
        Deltatheridia
        Dermoptera
        Tillodontia
        Taeniodonta
        Chiroptera
        Macroscelidea
        Primates
        Carnivora
        Condylarthra
        Pantodonta
        Dinocerata
        Pyrotheria
        Proboscidea
        Sirenia
        Desmostylia
        Hyracoidea
        Embrithopoda
        Notoungulata
        Astrapotheria
        Xenungulata
        Litopterna
        Perissodactyla
        Artiodactyla
        Edentata
        Pholidota
        Tubulidentata
        Cetacea
        Rodentia
        Lagomorpha

See separate articles on each group. [F.S.S.]

Mammary gland One of the distinguishing and unique anatomical structures of Mammalia, designed to secrete a nutrient for the newborn suitable to bridge the transition from intra- to extrauterine life. The most primitive mammal, the duckbill or platypus, resembles reptiles in laying eggs. Its mammary glands are unique in lacking nipples, which are present in all other mammals. The milk oozes out of two mammary gland areas and is lapped up by the young. *See* MILK.

The mammary glands are considered skin glands because their embryonic origin stems from the proliferation of the ectoderm which gives rise to the skin. At birth, the mammary gland consists of the nipple and one or more ducts arising from primary and secondary sprouts. Until the approach of puberty, little further extension and sprouting of the duct system occurs. Then, through the stimulus of estrogen and recurring estrous cycles, considerable duct growth occurs to form fatty pads or breast tissue.

When animals become pregnant, the growth of the duct system continues with the development of many side branches. At the sides and ends of these branched ducts, a new type of spherical cellular growth occurs, termed alveoli, transforming the tubular gland into a typical compound tubulo-alveolar gland. The alveoli may be compared to individual grapes and the bunch of grapes to a lobule, each of which is supported and surrounded by connective tissue. The alveoli are lined internally by epithelial cells capable of secreting milk into the central expandable cavity called the lumen. During mammary gland lobule-alveolar growth, the lumen of the alveoli do not show because milk is not secreted.

Lobule-alveolar growth is stimulated during the first two-thirds of pregnancy by the simultaneous secretion of estrogen and progesterone, first by the ovary and later by the placental membranes as well. It has been shown that additional lobule-alveolar growth occurs during early lactation, stimulated by anterior pituitary hormones or their target glands. *See* OVARY; PITUITARY GLAND.

During the final third of pregnancy, the cells of the mammary gland begin to secrete a fluid called colostrum. The colostrum accumulates in the glands and causes a gradual enlargement of the breast or udder. At the time of parturition the mammary glands are fully prepared to secrete. The functional activity of the glands is controlled by the lactogenic hormone secreted by the anterior pituitary. *See* LACTATION. [C.W.T.]

Mammography A radiological technique used to image breast tissues. Conventional mammographs (plain film) and xeromammographs are used most commonly, with ultrasound, computerized tomography, and enhanced mammography offering supplementary information. Both plain-film mammography and xeromammography use x-rays of low energy which are effective in imaging the breast skin, parenchyma, fat, and calcifications. An x-ray beam exiting the breast contains structural information that can be recorded by either photochemical or conductive processes. The photochemical system of conventional mammography uses film containing silver halide crystals, whereas the conductive system of xeromammography relies on a charged selenium plate to record the radiographic image. Both methods can also be used to image the lactiferic ducts after they have been injected with a radiopaque dye (contrast material). This examination is called galactography or ductography. *See* RADIOGRAPHY. [N.-F.B.]

Mandarin A name used to designate a large group of citrus fruits in the species *Citrus reticulata* and some of its hybrids. This group is variable in the character of trees and fruits since the term is used in a general sense to include many different forms, such as tangerines, King oranges, Temple oranges, tangelos (hybrids between grapefruit and tangerine), Satsuma oranges, and Calamondin, presumably a hybrid between a mandarin and a kumquat. *See* KUMQUAT; ORANGE; TANGERINE.

Although tangerines are the most extensively planted of the mandarin group, others, particularly the Temple orange, the Murcott orange, and the tangelos, of which there are several varieties, are important commercial fruits in the United States. *See* FRUIT; FRUIT, TREE; SAPINDALES. [F.E.G.]

Mandibulata A subphylum of the phylum Arthropoda. The Mandibulata, also known as the Antennata, includes the classes Crustacea, Insecta, Chilopoda, Diplopoda, Symphyla, and Pauropoda. The mandibulate arthropods possess a pair of mandibles, or jaws, which characterize the group. *See* CHILOPODA; CRUSTACEA; DIPLOPODA; INSECTA; PAUROPODA; SYMPHYLA. [C.B.C.]

Manganese A chemical element, Mn, of atomic number 25 and atomic weight 54.9380. Manganese is one of the transition elements of the first long period of the periodic table, falling between chromium and iron. It has certain properties in common with both of these metals. Although relatively little known or used in its pure form, it has great practical importance in the manufacture of steel.

1	2	3	4	5	6	7	8	9	10	11	12	13	14	15	16	17	18
1 H																	2 He
3 Li	4 Be											5 B	6 C	7 N	8 O	9 F	10 Ne
11 Na	12 Mg											13 Al	14 Si	15 P	16 S	17 Cl	18 Ar
19 K	20 Ca	21 Sc	22 Ti	23 V	24 Cr	25 Mn	26 Fe	27 Co	28 Ni	29 Cu	30 Zn	31 Ga	32 Ge	33 As	34 Se	35 Br	36 Kr
37 Rb	38 Sr	39 Y	40 Zr	41 Nb	42 Mo	43 Tc	44 Ru	45 Rh	46 Pd	47 Ag	48 Cd	49 In	50 Sn	51 Sb	52 Te	53 I	54 Xe
55 Cs	56 Ba	71 Lu	72 Hf	73 Ta	74 W	75 Re	76 Os	77 Ir	78 Pt	79 Au	80 Hg	81 Tl	82 Pb	83 Bi	84 Po	85 At	86 Rn
87 Fr	88 Ra	103 Lr	104 Rf	105 Db	106 Sg	107 Bh	108 Hs	109 Mt	110	*111*	*112*	*113*	*114*	*115*	*116*	*117*	*118*

lanthanide series	57 La	58 Ce	59 Pr	60 Nd	61 Pm	62 Sm	63 Eu	64 Gd	65 Tb	66 Dy	67 Ho	68 Er	69 Tm	70 Yb
actinide series	89 Ac	90 Th	91 Pa	92 U	93 Np	94 Pu	95 Am	96 Cm	97 Bk	98 Cf	99 Es	100 Fm	101 Md	102 No

Manganese melts at 1244 ± 3°C (2271 ± 5.4°F) and boils at 2095°C (3803°F). In the solid state it exists in four allotropic modifications designated α, ß, γ, and δ. Manganese in either the α or ß modifications at room temperature is both hard and brittle and cannot be plastically deformed to any significant extent. Ductility is, however, associated with the simpler structure of γ-manganese.

Manganese readily oxidizes in air to form a brown oxide coating. It also readily oxidizes at elevated temperatures. In this respect its behavior is more akin to its neighbor in the periodic table of higher atomic number, iron, than its neighbor with lower atomic number, chromium. *See* TRANSITION ELEMENTS. [A.H.Su.]

Manganese is a fairly reactive metal. Although the massive metal is somewhat slow to react, the powdered metal reacts with ease and, in some cases, quite vigorously. When it is heated in air or oxygen, powdered manganese forms a red oxide, Mn_3O_4. With water, at room temperature, hydrogen and manganese(II) hydroxide, $Mn(OH)_2$, are formed. In the case of acids, because manganese is such a reactive metal, hydrogen is liberated along with the formation of a manganese(II) salt. Manganese reacts at elevated temperatures with the halogens, sulfur, nitrogen, carbon, silicon, phosphorus, and boron.

In its many different compounds, manganese has oxidation states of 1+ through 7+. The most common oxidation states are 2+, 4+, and 7+. All the compounds, except those containing Mn^{II}, are deeply colored. For example, potassium permanganate, $KMnO_4$, forms aqueous solutions that are reddish-purple; potassium manganate, K_2MnO_4, forms deep-green solutions.

The compounds of manganese have many uses in industry. Manganese dioxide is used as a drying agent or catalyst in paints and varnishes, as a decolorizer in glass manufacturing, and in dry cells. Potassium permanganate is used for bleaching purposes, for decolorization of oils, and as an oxidizing agent in preparative and analytical chemistry. *See* DRY CELL; OXIDIZING AGENT. [W.W.We.]

Manganese nodules Concentrations of manganese and iron oxides found on the floors of many oceans. The origin of these potato-shaped metal-rich deposits has been elucidated; their complex growth histories are revealed by the textures of nodule interiors shown in the illustration.

Marine manganese nodules from certain regions are significantly enriched in nickel, copper, cobalt, zinc, molybdenum, and other elements so as to make them important reserves for these strategic metals.

Although manganiferous nodules and crusts have been sampled or observed on most sea floors, attention has focused on the nickel-plus-copper-rich nodules (2–3 wt% metals) from the north equatorial Pacific in a belt stretching from southeast Hawaii to Baja California, as well as the high-cobalt nodules from seamounts in the Pacific Ocean. Manganese nodules from the Atlantic Ocean and from higher latitudes in the

Reflected-light photograph of the polished surface of a sectioned manganese nodule showing the complex growth history of the concretionary deposit (diameter 1.6 in. or 4 cm).

Pacific Ocean have significantly lower concentrations of the minor strategic metals. However, surveys of the Indian Ocean have revealed metal-enrichment trends comparable to those found in the Pacific Ocean nodules; high Ni + Cu-bearing nodules are found adjacent to the Equator.

Microchemical analyses have revealed that chemical differences exist between the outermost top (exposed to sea water) and bottom (immersed in sediment) layers of manganese nodules. Surfaces buried in underlying sediments are generally higher in Mn, Ni, and Cu contents, compared to the more Fe + Co-rich surfaces exposed to sea water. Episodic rolling-over of a nodule accounts for fluctuating concentrations of Mn, Fe, Ni, Cu, Co, and other metals across sectioned manganese nodules. [R.G.Bu.]

Manganese oxide minerals Minerals that contain manganese (Mn) and oxygen (O) or the hydroxyl ion (OH) as principal components. Over 20 manganese oxide minerals have been identified. They can be broadly categorized by the primary oxidation state of manganese in the mineral as tetravalent (Mn^{4+}), trivalent (Mn^{3+}), or divalent (Mn^{2+}) manganese oxides. Tetravalent and trivalent manganese oxides occur in widespread continental and marine environments. They are the primary constituents of manganese nodules and crusts that occur in vast quantities on the ocean floors. Manganese oxide minerals are of economic importance as a source of manganese for the manufacture of steel, and some are used as the cathodic material in dry-cell batteries. *See* HYDROXYL; MANGANESE.

The basic structural unit for many of the manganese oxide minerals is an octahedron formed from a manganese cation surrounded by six oxygens (MnO_6). Octahedra in these structures are linked to one another by sharing either corners (one oxygen atom) or edges (two oxygen atoms). The octahedra are then linked in some minerals to form chains of different widths, and in other minerals to form layers or sheets of octahedra. *See* CRYSTAL.

The manganese oxides are black and opaque with the exception of manganosite (emerald green) and pyrochroite (colorless to pale green or blue). Both manganosite and pyrochroite become black on exposure to air. Most of the manganese oxides have a specific gravity of 4–5 and a hardness of 5–7 or less. Some of the minerals have a wide range of hardness. Many of the manganese oxides commonly are poorly crystalline and occur in irregular masses or grains. Many manganese oxides are intimately intergrown with other manganese oxides or other minerals on a fine scale. *See* HAUSMANNITE; MANGANITE; PYROLUSITE. [S.T.]

Manganite A mineral having composition MnO(OH) and crystallizing in the orthorhombic system in prismatic crystals with deep vertical striations. The hardness is 4 on Mohs scale, and the specific gravity is 4.3. The luster is metallic and the color iron black. Fine crystals have been found in the Harz Mountains; in Cornwall, England; and in the United States at Negaunee, Michigan. Manganite is a minor ore of manganese. *See* MANGANESE. [C.S.Hu.]

Mango A tree (*Mangifera indica*) of the family Anacardiaceae that originated in the Indo-Burma region and is now grown throughout the world. The mango is a medium to large evergreen tree; it produces a dense, round canopy, with leaves which are reddish brown when young and dark green when mature.

In the United States, mangoes are grown only in Florida and on a small scale in Hawaii, usually as backyard trees. Their greatest importance is in India, where almost 2,000,000 acres (800,000 hectares) are grown constituting 75% of the world area devoted to mango production. The ripe fruit is eaten raw as a dessert or used in the manufacture of juice, jams, jellies, and preserves. Unripe fruit can be made into pickles or chutneys. Mangoes are a good source of vitamins A and C. *See* FRUIT, TREE; SAPINDALES. [R.M.Wa.]

Mangrove A swamp forest of low to tall trees and some shrubs, commonly associated with some salt marsh herbs. Swamp forests occur along the borders of many tropical shores where wave action is not intense and mud and peat are deposited. Most plants of this community are halophytes that are well adapted to salt water and fluctuations of tide level. Some have stilt or prop roots to help hold them on the shifting sediments and others have erect root structures (pneumatophores) that crop out above the surface (see illustration).

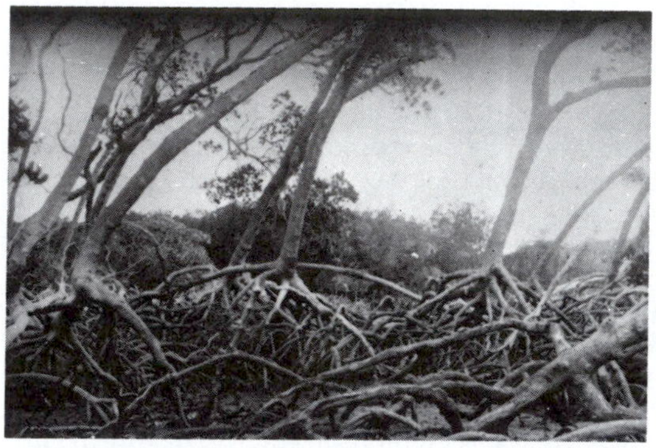

Interior of mangrove swamp, Newton Island. ***(From J. A. Steers, Salt marshes, Endeavour, 18(70):75–82, 1959)***

The mangrove community often develops as a distinct halosere. In such areas it is zoned from open water landward in a series of different species. The landward zone species develop on sediments and peats which were initially deposited in the seaward zone. These changes due to deposition in the swamp often extend the coast outward and form incipient islands in shallow, quiet waters. The swamps when dense also afford some protection against erosion resulting from violent storms. Thus, mangrove swamps have a significant geologic role. *See* ECOLOGICAL SUCCESSION; ECOLOGY.

The composition of mangrove swamps is strikingly different in two regions. Only four (Western) species predominate in the Americas and along the west coast of Africa, but over 10 (Eastern) species are frequent in swamps of eastern Africa, Asia, and the western Pacific region. The genera belong to a number of families. They are examples of convergent adaptations to their saline habitat. [J.H.D.]

Manic-depressive psychosis A severe disturbance of affect (feeling or emotion); hence it is referred to as an affective psychosis. It is characterized by extreme and pathological elation alternating with severe dejection, both of which may last for months or years. There may be normal intervals between these states, and cases are known in which only the manic or the depressive attack occurs once or repeatedly. *See* PSYCHOSIS.

The etiology of the disorder is obscure. Endocrine and constitutional factors probably play a role, although specific data to prove this are lacking. Hereditary factors, as shown in research on twins, are of importance. [F.C.R.]

Manifold (mathematics) An *n*-dimensional manifold is a connected, locally compact space with a countable basis, each point of which has a neighborhood homeomorphic to euclidean *n*-space. Examples are simple closed curves (one-dimensional manifold) and surface of a sphere or a torus (two-dimensional manifolds). A fundamental problem of topology (solved only for $n = 1,2$) is to classify *n*-dimensional manifolds

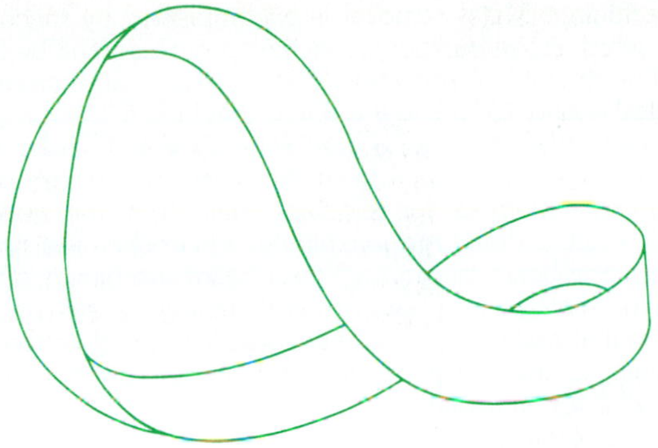

Möbius band, of interest in study of manifolds.

into types such that two manifolds are homeomorphic if and only if they belong to the same type. If a long rectangular ribbon is given half a turn and the two ends glued together, the resulting surface is called a Möbius band (see illustration). A manifold is orientable provided no portion of it is homeomorphic to a Möbius band; otherwise, it is nonorientable. *See* TOPOLOGY. [L.M.Bl.]

Mannans A group of polysaccharides composed chiefly or entirely of D-mannose units. Mannans occur in the thickened cell walls of palm seeds as reserve polysaccharides, which disappear during germination. The structural formula below

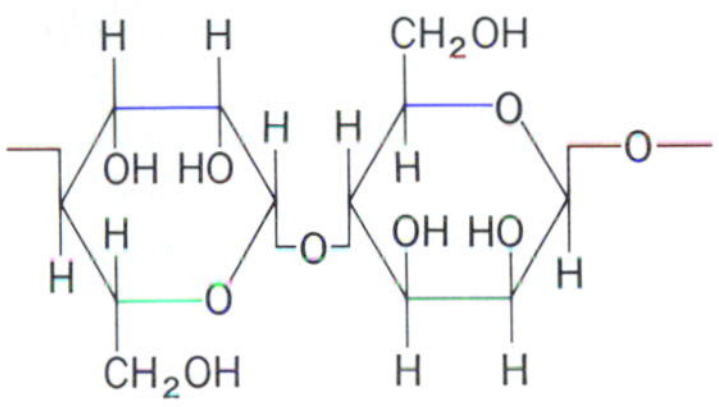

shows that the molecule is linear and is somewhat similar to cellulose. Glucomannans and galactoglucomannans are generally prominent components of coniferous woods, but they occur in only very small quantities in deciduous woods. *See* CELL WALLS (PLANTS); POLYSACCHARIDE. [R.L.Wh.]

Manometer A double-leg liquid-column gage used to measure the difference between two fluid pressures. Micromanometers are precision instruments which typically measure from very low pressures to 50 mm of mercury (6.7 kilopascals). The barometer is a special case of manometer with one pressure at zero absolute. *See* BAROMETER.

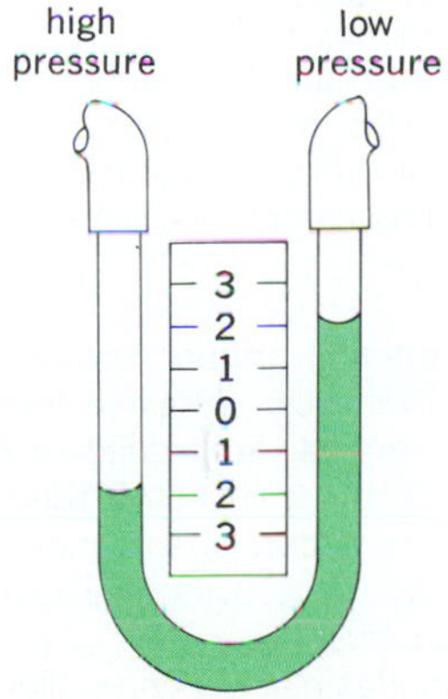

U-tube manometer.

The various types of manometers have much in common with the U-tube manometer, which consists of a hollow tube, usually glass, a liquid partially filling the tube, and a scale to measure the height of one liquid surface with respect to the other (see illustration). If the legs of this manometer are connected to separate sources of pressure, the liquid will rise in the leg with the lower pressure and drop in the other leg. The difference between the levels is a function of the applied pressure and the specific gravity of the pressurizing and fill fluids.

A well-type manometer has one leg with a relatively small diameter, and the second leg is a reservoir. The cross-sectional area of the reservoir may be as much as 1500 times that of the vertical leg, so that the level of the reservoir does not change appreciably with a change of pressure. Mercurial barometers are commonly made as well-type manometers.

The inclined-tube manometer is used for gage pressures below 10 in. (250 mm) of water differential. The leg of the well-type manometer is inclined from the vertical to elongate the scale. Inclined double-leg U-tube manometers are also used to measure very low differential pressures. *See* PRESSURE MEASUREMENT. [J.H.Z.]

Manufactured fiber One of a number of textile fibers and filaments manufactured from chemical substances of natural, inorganic, or synthetic origin. Although the nomenclature has varied in the past, the correct technical usage makes a distinction based on the origin of the raw material. Accordingly, rayon is a manufactured fiber but not a synthetic fiber. Nylon is a synthetic fiber that also falls into the broader classification of manufactured fibers. In addition to names based on their chemical composition, manufactured fibers are customarily known by the manufacturer's name or trademark, or by terms referring to characteristic properties, for example, thermoplastic fibers, or manufacturing process, for example, wet spun fibers and viscose rayon.

Manufactured fibers, with the exception of the inorganic ones, are composed of long, chainlike molecules called linear polymers. The arrangement (orientation) of these polymer chains with respect to the fiber axis, as well as the manner in which they fit or pack together (crystallinity), is controlled both by their chemical nature and by the manufacturing steps. In turn, these three variables—chemical nature, orientation, and crystallinity—establish the fiber properties, such as strength, flexibility, resilience, and durability. Fiber properties are important factors in the appearance, comfort, and wear of garments, home furnishings, and other textile applications.

To spin filaments, the starting material is made fluid by melting or dissolving a solvent, extruded through small holes, and then solidified by cooling (melt spinning), by chemical treatment (wet spinning), or by evaporation of the solvent (dry spinning). The filaments are produced in various sizes, and they range in diameter from finer than a human hair to large bristles used for toothbrushes.

Filaments are combined to form yarns of two different types: continuous filament yarns and spun yarns. Spun yarns are made from short fibers called staple. These are cut in lengths from 1 to 4½ in. (2.5 to 11.4 cm), and are held together in the yarn by crimp and twist in a manner similar to that used for natural fibers. The choice of yarn depends on the intended purpose; fabrics of spun-staple yarns are usually more bulky and less smooth and lustrous than fabrics of continuous filament yarns.

The fiber types of major importance in the United States are classified by composition as follows: (1) cellulosic, composed of regenerated cellulose, cellulose diacetate, and cellulose triacetate; (2) synthetic, composed of polyamide, polyester, polyacrylic, polyvinyl, and polyolefin resins; and (3) inorganic, composed of glass and metal. Numerous natural sources other than cellulose have been tried. Modern sources, including petroleum, natural gas, coke, acetylene, and related chemicals, yield the monomers, or starting units, for addition polymerization (polyvinyls, acrylics, dinitriles, and olefins such as polyethylene and polytetrafluoroethylene), and for condensation polymerization (polyamides, polyesters, and polyurethanes). *See* POLYMER; TEXTILE. [R.E.S.]

Manufacturing engineering A function of industry that deals with all aspects of the production process. The function of manufacturing engineering is to develop and optimize the production process. This can best be understood by considering the total process through which a designer's concept becomes a marketable product: (1) The product designer (or product design department) conceptualizes a product. Drawings and one or more prototypes of this product are produced, usually by manual methods, but increasingly by computer-aided design and manufacturing. (2) The finalized prototype and its part drawings are released to the manufacturing engineering department, which starts designing and building an economically justifiable process by which the product will be produced. (3) When the manufacturing process developed by manufacturing engineering has been thoroughly tried and proved workable, it is turned over to the production department, which assumes responsibility for product manufacture.

In brief, the manufacturing engineering department bridges the gap between product design and full production. It discharges responsibilities in a number of subordinate fields of engineering: process engineering, tool engineering, material handling, plant engineering, and standards and methods. It must be noted that this list of component disciplines, while typical, is not universal. In virtually all manufacturing facilities the composition of subordinate engineering functions is adjusted to meet the needs of the product to be made. In all cases, however, the objective of manufacturing engineering is the same: to design and develop the manufacturing process in the most economical manner possible. *See* COMPUTER-AIDED DESIGN AND MANUFACTURING; ENGINEERING DESIGN; INDUSTRIAL ENGINEERING; METHODS ENGINEERING; PROCESS ENGINEERING; PRODUCTION ENGINEERING; TOOLING. [D.B.D.]

Manufacturing processes The various operations that create useful products from raw materials. Manufacturing processes include the creation of raw stock from which parts can be made; the operations for changing the shape and configuration of the raw stock and fabricating piece parts; operations for improving the performance of the parts in service; operations for joining and assembling parts into a final product; plating and other surface finishing operations; and the inspection and testing to assure the quality and performance of the product in service. By far the greatest effort is in processing metal, although nonmetals including wood, plastics, and composites constitute an increasing part of manufactured products.

The main source of metals is from ores. Once extracted, the metal may be cast into ingots for subsequent use, or it may be refined further to produce finished structural parts. The purified metal may also be ground into a finer powder from which parts can be made by sintering in a mold under controlled temperature and pressure. *See* METAL; METALLURGY.

Metal castings and ingots are made by pouring molten metal into preformed molds. This process is the ultimate way for forming metal into intricate shapes. Most metal castings are made in sand molds. The use of computers and automatic molding machines has led to better and more consistent control of all the factors in sand molds. Other techniques include vacuum molding and the full-mold process. In the former, an airtight seal is made around the sand in the mold and then a vacuum is created so that air pressure will hold the sand in place. The full-mold process uses expanded polystyrene foams for the pattern. The foam pattern is embedded in the sand, and when the metal is poured into the mold, it vaporizes the foam and takes its place. *See* METAL CASTING.

Metal removal processes performed in order to size a part or to improve surface properties come under the general heading of machining. Metal removal is accomplished by mechanical work, which is the traditional machining process, and by the so-called nontraditional processes that use electrical, thermal, and chemical energy to loosen the bonds that hold the metal together. In mechanical removal a specially shaped and hardened tool is used to cut away metal from the workpiece in the form of chips. Other methods for removing metal from the workpiece include electric discharge machining, electrochemical machining, electron-beam machining, laser-beam machining, plasma-beam machining, and photochemical etching. Each method has features that make it especially adapted for special applications. Machining is one of the manufacturing operations that are most successful for automation applications. *See* MACHINE TOOLS; NUMERICAL CONTROL.

A large share of manufactured parts are made by metal-forming operations. The source material is generally sheet metal from which the blanks (or stock) for making a part are cut. The blank can then be radically deformed in a variety of ways to give the desired shape. Common examples of products manufactured by using forming operations are food and beverage cans, metal cookware, coins, piping, and auto-body sheet metal. *See* SHEET-METAL FORMING.

Forging is a process of hammering a billet of metal into the proper shape and form. The raw stock is hammered by tough, hardened dies to produce a part with good dimensional control requiring only finishing by machining and polishing. *See* FORGING; METAL FORMING.

Powder metallurgy is an important manufacturing process, because of the possibilities for mixing a broad range of ingredients and then combining them by sintering. This method imparts special properties to the end product, and excellent dimensional control and desirable surface finishes can be obtained. *See* METAL MATRIX COMPOSITES; POWDER METALLURGY; SINTERING.

A composite is a combination of two or more distinct materials that are brought together to yield a material having properties superior to the separate constituents. A fiber may be incorporated into the matrix by winding, by weaving, and by layers—for example, an automotive tire—or it may be chopped into short lengths and thoroughly mixed with the matrix material. *See* COMPOSITE MATERIAL.

Joining is a term that covers a wide variety of methods for attaching different parts to form a unit. Among the more prevalent methods are welding, brazing, soldering, adhesive bonding, and mechanical fasteners. *See* ADHESIVE BONDING; BOLT; JOINT (STRUCTURE); SCREW; SCREW FASTENER; WELDING AND CUTTING OF METALS.

Assembly may require up to 50% of the labor, time, and cost for manufacturing a product; and it will be a principal factor in the quality and performance of the product. There are a variety of assembly methods that have developed over the years, for example the assembly line characterized by automobile production. Automated assembly has great potential for reducing the cost of production. However, assembly is the most difficult manufacturing operation to automate. Instances where parts can be properly oriented, delivered, and assembled by mechanical means yield good results in automation. Jobs requiring perception and intelligence do not. Robots with computer vision and with the ability to modify their work pattern to accommodate different situations hold promise for expanding automated assembly. *See* COMPUTER VISION; ROBOTICS. [R.E.Be.]

Manure Any plant and animal residue that may contain excreta of animals. Manures, classified into animal manure, plant manure, and compost, are added to the soil in various stages of decomposition. In the soil they undergo further degradation through the action of soil microorganisms, raising the soil fertility level and improving soil texture by increasing humus content. *See* HUMUS.

Animal manure includes plant residues like straw, used as litter, as well as solid and liquid excreta. Compost consists of plant and animal residues allowed to rot before being applied to soil. Green manure is plant material in the form of a growing crop plowed into the soil. Because the object of manuring is to increase the supply of nitrogen, leguminous crops are grown and then plowed into the soil. *See* SOIL MICROBIOLOGY. [A.G.L.]

Map projections Methods of transforming a spherical representation of the Earth's surface to a nonspherical surface, usually a plane. These methods are useful because globes and curved-surface reproductions are cumbersome, expensive, and difficult to measure upon. An infinite number of methods is possible. Each method deforms the geometric relationships among points on the sphere in some way, since a sphere and a plane are not applicable surfaces. Consequently, directions, distances, areas, and angular relationships on the Earth are never completely recreated on a map.

Cylindrical projections. These result from symmetrical transfer of the spherical surface to a tangent or intersecting cylinder. True or correct scale will obtain along the great circle of tangency or the two homothetic small circles of intersection. If the axis of the cylinder is made parallel to the axis of the Earth, the parallels and meridians will appear as right, perpendicular lines. Points on the Earth equally distant from the tangent great circle (Equator) or small circles of intersection (parallels equally spaced on either side of the Equator) will have equal scale departure. The pattern of deformation will therefore parallel the parallels, as change in scale occurs in a direction perpendicular to the parallels. If the cylinder is turned 90° with respect to the Earth's axis, the projection is said to be transverse, and the pattern of deformation will be symmetric with respect to a great circle through the poles. If the turn of the cylinder is less than 90°, an oblique projection results (Fig. 1).

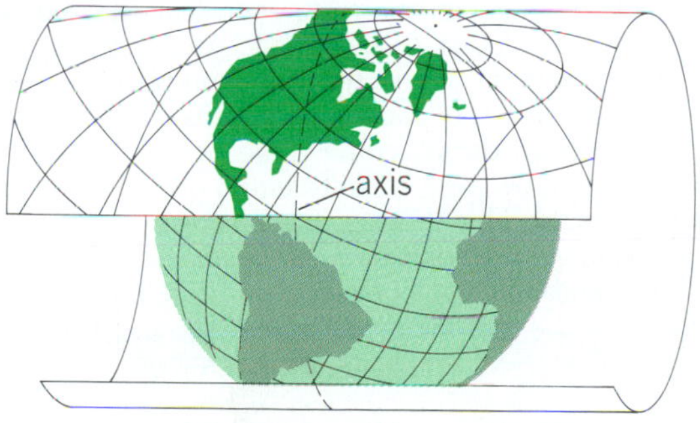

Fig. 1. Oblique Mercator projection on a cylinder. (*After American Oxford Atlas, Oxford, 1951*)

Conic projections. Transfer to a tangent or intersecting cone is the basis of these projections. True scale will obtain along one or two small circles in the same hemisphere. Conic projections are usually arranged with the axis of the cone parallel to the Earth's axis. Consequently, meridians appear as radiating straight lines and parallels as concentric arcs which intersect with the meridians at right angles. Conical patterns of deformation parallel to parallels, that is, scale departure will be uniform along any parallel (Fig. 2).

Azimuthal projections. These result from the transfer to a tangent or intersecting plane established perpendicular to a right line passing through the center of the Earth. All geometrically developed azimuthal projections are transferred from some point on this line. Points on the Earth equidistant from the point of tangency or the center of the circle of intersection have equal scale departure. Hence the pattern of deformation is circular and concentric to the aforementioned center. All azimuthal projections, whether geometrically or mathematically derived, have two aspects in common: (1) All great circles that pass through the center of the projection appear as straight lines. (2) All azimuths from the center are truly displayed (Fig. 3). [A.H.Ro.]

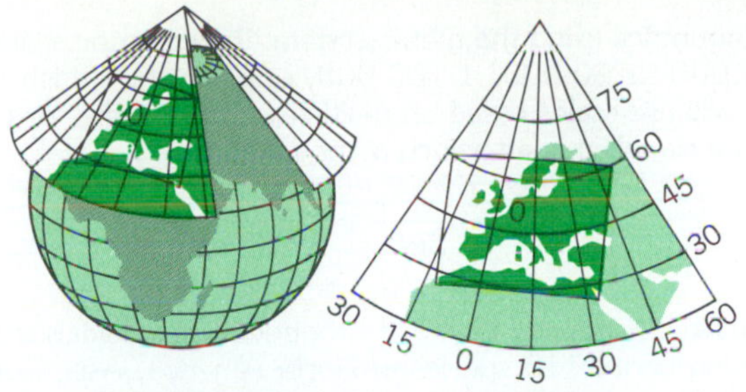

Fig. 2. Conical projection, based on origin 45°N, 10°E. (*After American Oxford Atlas, Oxford, 1951*)

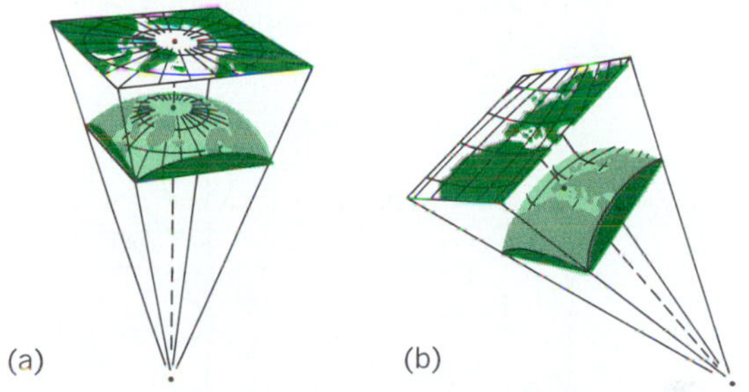

Fig. 3. Azimuthal projections with tangent planes have been raised above sphere surface. (*a*) Polar azimuthal. (*b*) Oblique azimuthal. (*After American Oxford Atlas, Oxford, 1951*)

Map scale The relation between the length of a line on a map to the corresponding line in nature. Map scale is expressed in three ways:

1. Representative fraction, as

$$\frac{1}{R} = \frac{\text{length on map}}{\text{length in nature}}$$

Thus, if on a map 1 in. represents 10 mi, it is expressed by the representative fraction 1/633,600 or 1:633,600, as there are 63,360 in. in 1 mi.

2. Miles-per-inch scale indicates the number of miles represented and is shown on maps as "Scale: 1 in. to 10 mi."

3. Graphic scale is a graduated line that portrays the actual length of miles or kilometers as they appear on the map. Graphic scales have the advantage of remaining true after the map has been enlarged or reduced in reproduction (see illustration).

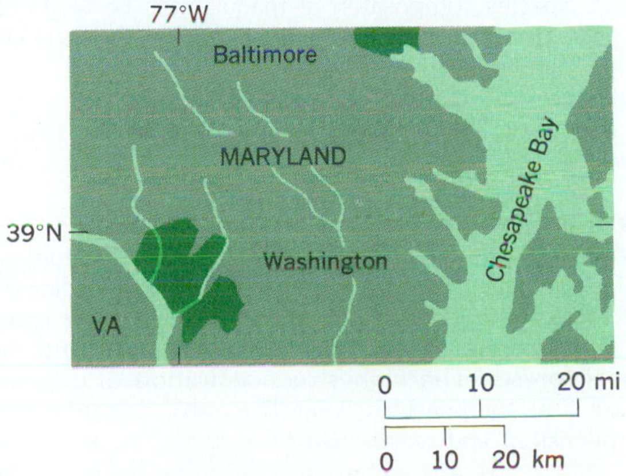

A sketch map and its graphic scales.

In countries using the metric system, the common scales are 1:25,000, 1:50,000, 1:100,000, and so on. British maps generally use scales based on multiples of 1:63,360; for example, the so-called quarter-inch maps which mean

$$\frac{1}{4 \times 63{,}360} = \frac{1}{253{,}440}$$

[E.Ra./W.C.]

Maple A genus, *Acer*, of broad-leaved, deciduous trees including about 115 species in North America, Asia, Europe, and North Africa. This genus is characterized by simple, opposite, usually palmately lobed (rarely pinnate) leaves, generally inconspicuous flowers, and a fruit consisting of two long-winged samaras or keys (see illustration).

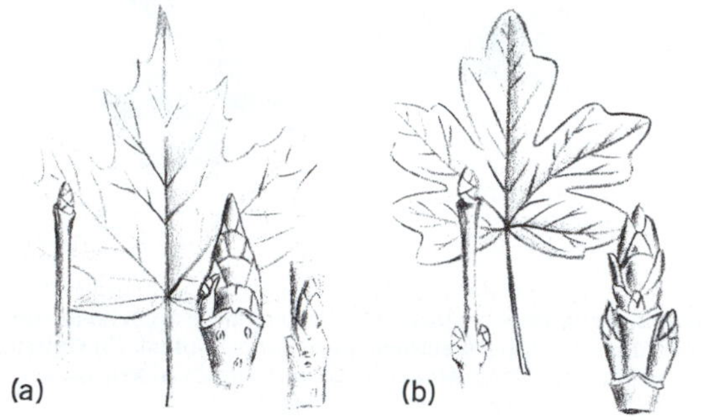

Characteristic maple leaves, twigs, and buds. (*a*) Sugar maple (*Acer saccharum*). (*b*) Hedge maple (*A. campestre*).

The most important commercial species is the sugar or rock maple (*A. saccharum*), called hard maple in the lumber market. This tree grows in the eastern half of the United States and adjacent Canada. It can be recognized by its gray furrowed bark, sharp-pointed scaly winter buds, and symmetrical oval outline of the crown.

Maples rank third in the production of hardwood lumber. Hard maple is used for flooring, furniture, boxes, crates, woodenware, spools, bobbins, motor vehicle parts, veneer, railroad ties, and pulpwood. It is the source of maple sugar and syrup and is planted as a shade tree. *See* SAPINDALES. [A.H.G./K.P.D.]

Maquis A type of vegetation composed of shrubs, or "scrub," usually not exceeding 10 ft (3 m) in height, the majority having relatively small, hard, leathery, often spiny or needlelike drought-resistant leaves and occurring in areas with a Mediterranean climate. In this climate precipitation occurs principally in winter and relative humidity is low and evaporation high in summer. The life form of the plants together with the climatic regime, not the species composition, characterizes maquis. Species composition of maquis may be very uniform over wide areas, or may vary considerably locally, and between widely separated areas.

Californian chaparral is closely related ecologically to maquis and is included in the concept by some authorities. *See* CHAPARRAL. [R.K.G.]

Marattiales An ancient order of ferns which in some respects is intermediate between the orders Ophioglossales and Polypodiales. Whereas their fossil remains indicate they were formerly worldwide and abundant, they now comprise only 7 genera and about 150 species limited to the humid tropical forests. The highest concentration of genera and species is in southeastern Asia. They are mostly large ferns with persistent leaf bases. *See* OPHIOGLOSSIDAE; POLYPODIALES. [P.A.V.]

Marble A term applied commercially to any limestone or dolomite taking polish. Marble is extensively used for building and ornamental purposes. *See* DOLOMITE; LIMESTONE.

In petrography the term marble is applied to metamorphic rocks composed of recrystallized calcite or dolomite. Schistosity, often controlled by the original bedding, is usually weak except in impure micaceous or tremolite-bearing types. Calcite (marble) deforms readily by plastic flow even at low temperatures. Therefore, granulation is rare, and instead of schistosity there develops a flow structure characterized by elongation and bending of the grains concomitant with a strong development of twin lamellae. *See* METAMORPHIC ROCKS; MINERALOGY; SCHIST.

Pure marbles attaining 99% calcium carbonate, $CaCO_3$, are often formed by simple recrystallization of sedimentary limestone. Dolomite marbles are usually formed by metasomatism. *See* CALCITE; DOLOMITE; METASOMATISM. [T.F.W.B.]

Marcasite A mineral having composition FeS_2 and crystallizing in the orthorhombic system. Marcasite frequently has a radiating structure and may be globular or stalactitic. There is poor prismatic cleavage. The hardness is 6–6.5 on Mohs scale and the specific gravity is 4.89. The luster is metallic and the color pale bronze-yellow to nearly white on a fresh fracture. Marcasite and pyrite are dimorphous; both have the composition FeS_2. Because marcasite is whiter, it is called white iron pyrite.

Marcasite is found in metalliferous deposits associated with lead and zinc ores, as replacement deposits in limestone, and in concretions in clays and shales. The nodular and lenticular masses in coal known as brasses are in part marcasite and in part pyrite. *See* PYRITE. [C.S.Hu.]

Marchantiales An order of the liverwort subclass Marchantiidae. Characteristic features include the differentiation of upper and lower tissues of the gametophyte, ventral scales, and rhizoids of two kinds. The sporophyte is considerably reduced, and the capsule dehisces irregularly. The archegonia, though dorsal, come to be pendant from an elevated receptacle because of differential growth resulting in decurved margins. The stalks of the receptacles are modified branches often with rhizoids in one, two, or rarely four furrows along their length. The order consists of 12 families grouped in two suborders, the more complex Marchantiineae and the simplified Ricciineae. *See* BRYOPHYTA; MARCHANTIIDAE. [H.Cr.]

Marchantiidae One of the two subclasses of liverworts (class Hepaticopsida). The gametophytes are ribbonlike or rosette-shaped thalli, usually showing considerable internal tissue differentiation. The rhizoids may be both smooth and internally pegged on the same thalli. Oil bodies, if present, are restricted to scattered cells that lack chloroplasts. The antheridia are usually ovoid, and the archegonia usually consist of six rows of cells. The sporophytes are generally reduced.

The subclass differs from the thallose Metzgeriales of the subclass Jungermanniidae in the intemal differentiation of the thallose gametophytes, the smooth and pegged rhizoids, and the variation in the way the capsules dehisce. The subclass is divided into three orders, the Marchantiales, the Monocleales, and the Sphaerocarpales. *See* BRYOPHYTA; JUNGERMANNIIDAE; MARCHANTIALES; METZGERIALES; MONOCLEALES; SPHAEROCARPALES. [H.Cr.]

Margarine An emulsified fatty food product used as a spread and a baking and cooking fat, consisting of an aqueous phase dispersed in the fat as a continuous phase. Developed originally as a butter substitute, margarine is now considered a food in its own right and is manufactured in forms unknown to butter, such as plastic, soft, or fluid. However, margarine is colored, flavored, fortified with vitamins, and otherwise formulated to have the same or similar taste, appearance, and nutritional value as butter. *See* BUTTER.

Currently, margarine produced in the United States must contain not less than 80% fat. The fats and oils must be edible but may be from any vegetable or animal carcass source, natural or hydrogenated. The required aqueous phase may be water, milk, or solutions of dairy or vegetable protein, and must be pasteurized. Vitamin A must be added to yield a finished margarine with not less than 15,000 international units per pound (0.45 kg). Optional ingredients include salt or potassium chloride for low-sodium diets, nutritive sweeteners, fatty emulsifiers, antioxidants, preservatives, edible colors, flavors, vitamin D, acids, and alkalies. *See* FAT AND OIL (FOOD). [T.J.W.]

Marijuana The Spanish name for the dried leaves and flowering tops of the hemp plant, *Cannabis sativa* (Cannabinaceae). The narcotic ingredients allegedly have stimulating effects, and after smoking two or three cigarettes, the smoker often has a feeling of well-being and increased power and ability. After excessive amounts of the drug, illusions are often common, as well as pleasing, fanciful hallucinations. Sometimes the excessive user experiences disorientation and even delirium. *See* HEMP; URTICALES. [P.D.St./E.L.C.]

Marine boiler A steam boiler designed to suit the marine environment and generally arranged to supply steam to the main propulsion machinery, ship's service electric generators, feed-pump drivers, and other auxiliary services.

Marine boilers are usually of the two-drum water-tube type with water-cooled furnaces, superheaters, desuperheaters, and heat recovery equipment of the economizer or air-heater type. The majority of ships are fitted with two boilers, although some large passenger ships may have three or more. Some cargo ships are fitted with only one boiler, and in some of these cases a smaller auxiliary boiler may be fitted for emergency or in-port steaming use.

Marine boilers are generally arranged for oil firing. Oil is used extensively because of its simplicity of handling and storing and the fact that it can be stored in spaces that often cannot be used for carrying cargo. However, the shortage of fuel oil, coupled with its ever-increasing cost, is bringing back coal as a fuel for marine boilers. *See* BOILER; MARINE MACHINERY; STEAM-GENERATING UNIT. [R.P.G.]

Marine containers Standardized rectangular boxes for the transport of marine cargo. Since 1960, the ocean transportation of general cargo or freight has undergone a revolutionary technological change. The innovation was containerization—the development of standard marine cargo containers for consolidating packages into units of interchange between ships, docks, trucks, and railcars, and the development of special ships and handling systems to transport these containers at sea. This innovation brought economies of scale to marine cargo-handling operations, introduced capital-intensive processes to the labor-oriented stevedoring tasks, reduced cargo theft and damage, reduced the time a ship spent in port, and provided the means for efficient intermodal transport of cargoes.

Basic to the change to containerization was a new approach to loading and unloading cargoes. Instead of using nets or slings to lift individual bales, boxes, sacks, or pallets of cargo in and out of the ship's holds, the new systems employ standard cargo containers and special handling equipment to place the containers aboard (see illustration). The containers are loaded or stuffed at an inland factory or terminal and usually are never opened until they reach their ultimate destination. The ship carrying these containers becomes an extension of a truckline or a railroad.

Container crane and straddle carrier.

Intermodability, moving containerized cargo by more than one mode of transport without the need for intermediate reloading, is the essence of this technology. The containers can be moved to the ship via regular highway trailers or trailer chassis, or by flatbed railcar or rail piggyback. Once at the marine terminal, they can be lifted or rolled on the ships. Upon arrival at the discharge port, all land and water modes are again available to move them, with cargo undisturbed, to the consignee. [J.C.C.]

Marine ecology Marine ecology comprises the ecology of the world's oceans with their shores and estuaries. Individual organisms, groups of organisms, and the ocean environments in which they exist constitute ecological systems, termed ecosystems. Examples vary in size from a particular seaweed ecosystem, to a tidepool ecosystem, to an estuarine bay ecosystem. The size limits, or boundaries, of a particular ecosystem that distinguish it from others may depend upon physical barriers to community dispersal, such as a submarine mountain; environmental factors that restrict the area of the biotic community, such as the salinity factor; the productivity cycles within the community; or the reproductive and dispersal potentials of the community. *See* ECOSYSTEM.

The nonliving, or abiotic, materials in a marine ecosystem cycle comprise not only a variety of water-soluble inorganic nutrient salts, such as the phosphates, nitrates, and sulfates of calcium, potassium, and sodium and the dissolved gases oxygen and carbon dioxide, but also organic compounds, such as the various amino acids, vitamins, and growth substances. Most of the solid material dissolved in the sea originated from the weathering of the crust of the Earth.

The marine ecosystem can be divided into two large areas, the pelagic and benthic divisions. Each of these consists of various zones. The pelagic division embodies all the waters of the oceans and their adjacent salt-water bodies. It is divisible into the neritic zone, extending offshore to the edge of the continental shelf to a depth of over 650 ft (200 m), and the oceanic zone, embracing the remaining offshore waters.

The benthic division of the marine environment embraces the entire ocean floor, both coastal bottom and deep-sea bottom, properly termed the littoral and deep-sea systems, respectively. The littoral system consists of the eulittoral zone and the sublittoral zone. The eulittoral zone extends from high-tide level to about 200 ft (60 m), the approximate depth below which attached plants do not grow abundantly. This is probably the richest zone of the marine ecosystem in respect to numbers and kinds of organisms, as well as in variety of ecological types and habitat modifications. The sublittoral zone terminates at depths varying in different latitudes between about 650 to 1300 ft (200 to 400 m), depending upon light and temperature factors that modify the distribution of benthic animals and plants.

The deep-sea system of the benthic division is subdivided into an upper part, the archibenthic zone or, more meaningfully, the continental deep-sea zone, extending from the edge of the continental shelf (650–1300 ft or 200–400 m) to depths of about 2600–3600 ft (800–1100 m), and the abyssal-benthic zone that embraces the remainder of the benthic deep-sea system. The zones have little or no light, relatively constant conditions of salinity and temperature, and steadily decreasing numbers and kinds of organisms. [C.L.N.]

Marine engine An engine that propels a waterborne vessel. Even in small craft the marine engine must have the following characteristics: reliability, light weight, compactness, fuel economy, low maintenance, long life, relative simplicity for operating personnel, ability to reverse, and ability to operate steadily at low or cruising speed. *See* Boat propulsion; Marine machinery; Ship powering and steering.

Steam is a common type of propulsion for large ships. The diesel engine has gained wide acceptance in foreign merchant ships, but in the United States the majority of seagoing vessels use steam propulsion.

The gas turbine generally consists of an axial compressor discharging compressed air to a combustion chamber where fuel is burned, adding heat. The products of combustion at high temperature and pressure then pass through a gas turbine that drives the compressor and load. Generally, the term "gas turbine" is applied to the entire plant. *See* Gas turbine.

Both diesel and gasoline internal combustion engines are used in marine applications. Many moderate- and low-power marine installations use automotive or locomotive engines designed for variable load and intermittent service. High-power marine propulsion units normally are called on to operate continuously under load.

The turboelectric type of drive, comprising one or more steam turbine generators and ac propulsion motors, is also used for ship propulsion. The diesel-electric type of drive, composed of one or more dc diesel generator sets and often a double-armature propulsion motor, is used in tugs, dredges, Coast Guard cutters, and icebreakers, where maneuvering and a wide range in propeller speed are necessary. [J.T.H.]

Marine engineering The engineering discipline concerned with the ship design process. As such, it is closely related to naval architecture. In the development of a new ship design the marine engineer is responsible for the selection of the main propulsion plant and all of its auxiliaries. These include steam turbines, boilers, diesel engines, gas turbines, or other alternative combinations for driving the propellers. The electrical plant for ship hotel services, pumps, and so forth, is normally part of the responsibility of the marine engineer, as well as fuel systems with their associated piping and pumps, and refrigeration and air conditioning.

The marine engineer plans the machinery layout, does the heat balance calculation, and in essence designs the whole propulsion system. The design of the propeller is sometimes handled by the naval architect and sometimes by the marine engineer. However, most large ship propellers are designed by specialists. *See* Boat propulsion; Marine boiler; Marine engine; Marine machinery; Marine refrigeration; Naval architecture; Propeller (marine craft); Ship design; Ship powering and steering; Ship propulsion reactor. [R.B.C.]

Marine fisheries The harvest of animals and plants from the ocean to provide food and recreation for people, food for animals, and a variety of organic materials for industry. It is now generally agreed that the world catch is approaching a maximum, which may be less than 100,000,000 metric tons per year. If methods can be devised to harvest smaller organisms not heretofore used because they have been too costly to catch and process, it has been estimated that the yield could perhaps be increased severalfold. Russia is said to have succeeded in developing an acceptable human food product from Antarctic krill. *See* Fishery. [J.L.McH.]

Marine geology The study of the portion of the Earth beneath the oceans. More than 70% of the Earth's surface is covered by marine waters. Of the oceanic area (3.61×10^8 km^2 or 1.39×10^8 mi^2) approximately 3×10^8 km^2 (1.2×10^8 mi^2) is contributed by the deep-sea floor; the remaining 6×10^7 km^2 (2.3×10^7 mi^2) represents the submerged margins of the continents. The distribution of elevations on the Earth is shown in Fig. 1.

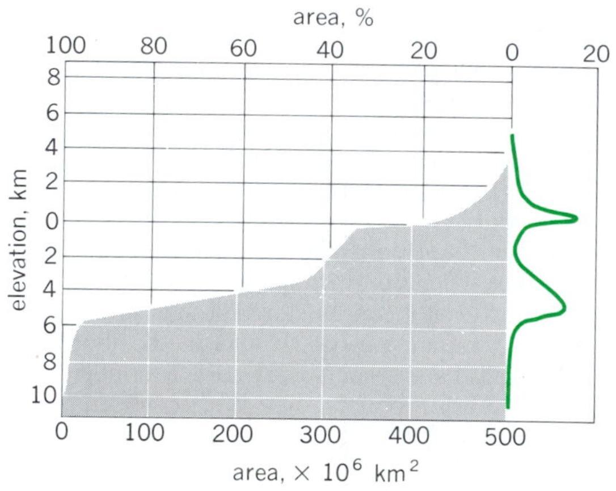

Fig. 1. Hypsographic curve showing area of Earth's solid surface above any given level of elevation or depth. Curve at the right shows frequency distribution of elevations and depths for 2-km intervals. 1 km = 0.6 mi.

Physiographic provinces. The Earth relief lies at two dominant levels (Fig. 1); one, within a few hundred meters of sea level, represents the normal surface of the continental blocks; the other, between 4000 and 5000 m (2.5 and 3.0 mi) below sea level, and comprising more than 50% of the Earth's surface, represents the ocean-basin floor. The topographic provinces beneath the sea can be included under three major morphologic divisions: continental margin, ocean-basin floor, and mid-oceanic ridge.

Continental margins. The continental margin includes those provinces associated with the transition from continent to ocean floor. The continental margin in the Atlantic and Indian oceans is generally composed of continental shelf, continental slope, and continental rise. The continental slopes are cut by many submarine canyons. Submarine alluvial fans extend out from the seaward ends of the larger canyons. *See* Continent; Continental margin; Submarine canyon.

The continental margin can be divided into three categories of provinces. Category I includes the continental shelf, marginal plateaus, and shallow epicontinental seas, all slightly submerged portions of the continental block. Category II includes the continental slope, marginal escarpments, and the landward slopes of marginal trenches, all expressions of the outer edge of the continental block. Category III includes the continental rise, the ridge-basin complex, and the ridge-trench complex.

Ocean-basin floor. Excluding the marginal trenches and mid-oceanic ridges, the ocean-basin floor includes the deepest portions of the ocean. Approximately one-third of the Atlantic and three-fourths of the Pacific fall under this heading. The ocean-basin floor can be divided into three categories of provinces; the abyssal floor, oceanic rises, and seamounts and seamount groups.

The abyssal floor includes the broad, deep areas of the central portion of the ocean. An abyssal plain is a smooth portion of the deep-sea floor where the gradient of the bottom does not exceed 1:1000. Abyssal plains adjoin all continental rises and can be distinguished from the continental rise by a distinct change in bottom gradient. At their seaward edge, most of the abyssal plains gradually give way to abyssal hills, which are thought to represent tectonic or volcanic relief of a type identical with that buried beneath the abyssal plains. Abyssal plains are also found in the marginal trenches, marginal basins, and in epicontinental marginal seas. Features of exactly the same morphology and origin are found in some lakes. Of similar origin are archipelagic aprons, which spread out from the base of oceanic islands. *See* MARINE SEDIMENTS; OCEANIC ISLANDS.

Oceanic rises are areas slightly elevated above the abyssal floor which do not belong to the continental margin or the mid-oceanic ridges. In contrast to the mid-oceanic ridges, oceanic rises are nonseismic; their relief is more subdued, and they are asymmetrical in cross section.

A seamount is any submerged peak more than 900 m (500 fathoms) high. Seamounts are distributed through all the physiographic provinces of the oceans. Seamounts sometimes occur randomly scattered but more often lie in linear rows. Virtually all conical seamounts are believed to be extinct or active volcanoes. *See* SEAMOUNT AND GUYOT.

Mid-Oceanic Ridge. The middle third of the Atlantic, Indian, and South Pacific oceans is occupied by a broad, fractured swell known as the Mid-Oceanic Ridge. In the Atlantic, it is known as the Mid-Atlantic Ridge; in the southern Indian Ocean, the Mid-Indian Ridge; in the Arabian Sea, the Carlsberg Ridge and Murray Ridge; and in the South Pacific, the Easter Island Ridge. [B.C.He.]

Underlying structure. Because approximately 70% of its surface is covered by the oceans, the typical structure of the Earth is found in the oceanic and not in the land areas. Statistical examination shows that most of the Earth's solid surface is either at the elevation of the ocean floors or at the elevation of the continents. The anomalous areas, those of extreme or of intermediate elevation, are long, narrow features—the mountain ranges, island arcs, deep-sea trenches, and continental margins. With the exception of a few intermediate areas, such as the Red Sea, the crustal structure of the ocean basins is distinctly different from the crustal structure of the continents, and it appears that this has been the case for most of the Earth's history.

Ocean basins and continental margins. Figure 2 is a structure section across an Atlantic type (rifted) margin based on an interpretation of geologic and geophysical data within the framework of plate tectonics and sea-floor spreading. On the continent and continental shelf, the rocks beneath the sedimentary layers are mainly of the acidic type, such as granite, gneiss, or schist. Beneath the rocks is an intermediate layer (layer 2) believed to be gabbroic or basaltic. The mantle is probably composed of ultramafic rocks, such as peridotite,

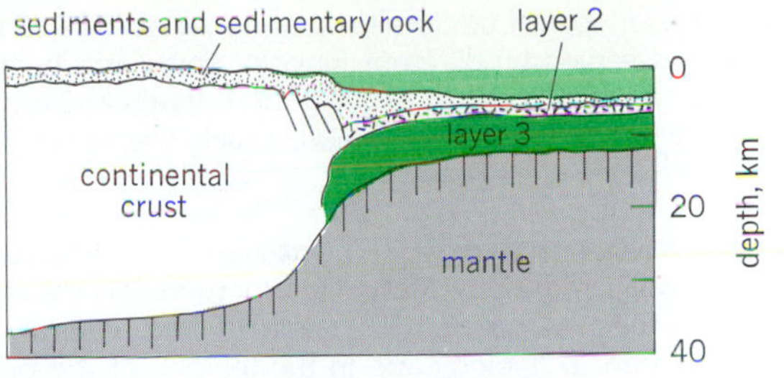

Fig. 2. Structure section across Atlantic-type (rifted) margin. 1 km= 0.6 mi.

enstatite, or eclogite. This is the most prominent layer in the Earth, extending from near the surface approximately halfway to the center. There are some variations in the upper parts of the mantle between continental and oceanic areas, but the most apparent difference is in the thickness and composition of the crust. The continental crust is six or seven times thicker than the oceanic crust (layer 3) and contains almost all of the acidic rocks, such as granites, whereas the oceanic crust is almost entirely composed of basic rock. *See* EARTH.

The thickness of the crust of the continental shelves is intermediate between that of the continents and oceans, and its composition is continental. In some places, such as parts of the east coast of North America and South America, the continental shelf and continental slope are broad areas where erosion of the continental masses has resulted in the deposition of many thousands of meters of sediment during the past several million years. In other areas which do not receive much sediment, notably the west coast of the Americas, there is only a narrow continental shelf.

Submarine ridges. There are two types of oceanic ridges, seismic and aseismic. The mid-ocean ridge system is of the former type and is the largest single feature on the Earth. Over its entire length the ridge is characterized by a narrow zone of shallow-focus earthquake epicenters. Usually associated with this zone are a rift valley and a large positive magnetic anomaly. In addition, parallel linear bands of alternating positive and negative magnetic anomalies are found on the flanks in many areas. The sediment cover is typically thin and in some places in the crest region is entirely absent in a zone 100–500 km (60–300 mi) wide. On the flanks an appreciably thicker cover is found; in some places it covers the basement rock more or less uniformly, while in others it is collected in pockets separated by exposed basement peaks or ridges.

Deep-sea trenches. Deep-sea trenches are important structural features associated with some continental margins, island arcs, earthquake belts, and areas of volcanic activity. Most of them are confined to the margins of the Pacific Ocean, although there are trenches in the Atlantic and Indian oceans. In the bottom are layers of sediments which are generally thickest landward of the trench axes. The great depth of the underlying dense layers causes a pronounced deficiency of gravity, a characteristic feature of all deep-sea trenches. Earthquake foci tend to lie on inclined planes dipping landward from the trenches. *See* MARINE SEDIMENTS. [M.Ew.; J.Ew.]

Marine machinery The engines, blowers, and other moving devices used on ships, boats, and related vessels. Practically all machinery used on ships has shore-based counterparts. Machinery is often developed for marine use prior to shore development and use. The marine environment places many demands on marine machinery. These include an extra measure of reliability, performance on a rolling and pitching platform, and limited space and weight requirements, as well as anticorrosion properties. These factors make necessary the frequent use of special designs for marine service. In addition, the purposes served by different types of ships and boats call for machinery of a number of types and characteristics. *See* BOAT PROPULSION; MARINE BOILER; MARINE ENGINE; PROPELLER (MARINE CRAFT); SHIP POWERING AND STEERING. [J.T.H.]

Marine microbiology The study of the microscopic organisms living in the sea. Marine microorganisms were probably among the first (cellular) living entities and have played a tremendous part in geology and in the biology of the oceans. They are still of paramount importance as transformers of organic and inorganic substances and as food for other, often larger, organisms.

Marine microbes may be considered as (1) autotrophs and (2) heterotrophs or phagotrophs. Autotrophs live on inorganic materials and use carbon dioxide as the sole carbon source, obtaining their energy from sunlight (phototrophs) or from chemical reactions (chemoautotrophs). The heterotrophs and phagotrophs (bacteria, heterotrophic algae and fungi, and protozoa) require a source of carbon more complex than carbon dioxide; this they obtain, along with energy, by breaking down or ingesting other microorganisms, plants, and animals.

Marine microorganisms are the food of a large number of marine animals including many crustaceans of the zooplankton, molluscans, and plant-eating fish. [E.J.F.W.]

Marine mining The process of recovering mineral wealth from sea water and from deposits on and under the sea floor. While mineral resources to the value of trillions of dollars do exist in and under the oceans, their exploitation is not simple. Many environmental problems must be overcome and many technical advances must be made before the majority of these deposits can be mined in competition with existing land resources.

The mineral resources of the marine environment are of three basic types: the dissolved minerals of the ocean waters; the unconsolidated mineral deposits of marine beaches, continental shelf, and deep-sea floor; and the consolidated deposits contained within the bedrock underlying the seas. As with land deposits, the initial stages preceding the production of a marketable commodity include discovery, characterization of the deposit to assess its value and exploitability, and mining, including beneficiation of the material to a salable product. [M.J.Cru.]

Marine navigation The process of directing a watercraft to a destination in a safe and expeditious manner. From a known present position, a course is determined which avoids dangers, and on this course estimates are made of time schedules. The task is to periodically effect en route checks and to make required adjustments.

The method used will depend on the type of vessel and on its role or mission. The devices available range from a simple compass to a host of sophisticated electronic systems. In all cases, the navigator must plan and prepare by setting instruments in order and by checking for predictable current and tidal effects and hazards to navigation en route. This preparation includes having the latest correct charts and reviewing pertinent sections of sailing directions, light lists, and tide and current tables.

The methods used to fulfill the requirements of navigation come under one of the following broad categories: dead reckoning, piloting, celestial navigation, and electronic navigation. The first three categories have become somewhat standardized; the fourth category, on the other hand, has been under constant and innovative development. *See* CELESTIAL NAVIGATION; DEAD RECKONING; ELECTRONIC NAVIGATION SYSTEMS; PILOTING.

With the unprecedented expansion of the world's merchant fleets and the ever-increasing number of very large deep-draft tankers with poor maneuvering characteristics and high-speed containerships, come additional hazards to shipping through collisions at sea. Hence, the international maritime community is constantly seeking ways to put to use available technology to provide the navigator with the tools necessary to cope with the problem of collision avoidance.

Radar can only present certain information and this information must be used properly and prudently by the navigator in traffic situations. To avoid this condition of task overload on the navigator and to reduce or eliminate the human error in calculations or interpretation, collision avoidance aids or systems (CAS), also known as automatic radar plotting aids (ARPA), of varying degrees of sophistication have been developed.

Integrated navigation-conning systems have the basic function of helping reduce the number of accidents between ships by having the computer correlate most of the data available during crowded conditions and thus help unburden the navigator by presenting this data in more usable form for better and quicker decision making.

In a program established for the purpose of reducing the number of collisions and strandings and at the same time safeguarding the environment, the U.S. Coast Guard, under the authority of the Ports and Waterways Safety Act of 1972, has instituted vessel traffic services (VTS) in a number of heavily trafficked United States ports. Generally the service provides marine traffic management of an advisory nature, but in an especially hazardous situation it may be necessary for the VTS to exercise emergency control of vessel movements. *See* NAVIGATION. [A.E.F.]

Marine refrigeration The application of refrigeration on shipboard for the preservation of perishables in transit and for the preservation of foodstuffs to be used by the passengers and crew. Marine refrigeration is also used for maintaining certain cargo products in liquid form that would otherwise evaporate when stored at or near ambient conditions. Refrigeration is also required during warm weather for air-conditioning the quarters for passengers and crew. For air conditioning, similar but larger compressors than those used for ship's stores are generally employed, and for larger systems chilled water is circulated through the coils situated in various fan rooms throughout the quarters area. *See* AIR CONDITIONING; REFRIGERATION. [R.P.G.]

Marine sediments The accumulation of minerals and organic remains on the sea floor. Marine sediments vary widely in composition and physical characteristics as a function of water depth, distance from land, variations in sediment source, and the physical, chemical, and biological characteristics of their environments. The study of marine sediments is an important phase of oceanographic research and, together with the study of sediments and sedimentation processes on land, constitutes the subdivision of geology known as sedimentology. *See* OCEANOGRAPHY.

Traditionally, marine sediments are subdivided on the basis of their depth of deposition into littoral 0–66 ft (0–20 m), neritic 66–660 ft (20–200 m), and bathyal 660–6600 ft (200–2000 m) deposits. This division overemphasizes depth. More meaningful, although less rigorous, is a distinction between sediments mainly composed of materials derived from land, and sediments composed of biological and mineral material originating in the sea. Moreover, there are significant and general differences between deposits formed along the margins of the continents and large islands, which are influenced strongly by the nearness of land and occur mostly in fairly shallow water, and the pelagic sediments of the deep ocean far from land.

Sediments of continental margins. These include the deposits of the coastal zone, the sediments of the continental shelf, conventionally limited by a maximum depth of 330–660 ft (100–200 m), and those of the continental slope. Because of large differences in sedimentation processes, a useful distinction can be made between the coastal deposits on one hand (littoral), and the open shelf and slope sediments on the other (neritic and bathyal). Furthermore, significant differences in sediment characteristics and sedimentation patterns exist between areas receiving substantial detrital material from land, and areas where most of the sediment is organic or chemical in origin.

Coastal sediments include the deposits of deltas, lagoons, and bays, barrier islands and beaches, and the surf zone. The zone of coastal sediments is limited on the seaward side by the depth to which normal wave action can stir and transport sand, which depends on the exposure of the coast to waves and does not usually exceed 66–100 ft (20–30 m); the width of this zone is normally a few miles. The sediments in the coastal zone are usually land-derived. The material supplied by streams is sorted in the surf zone; the sand fraction is transported along the shore in the surf zone, often over long distances, while the silt and clay fractions are carried offshore into deeper water by currents. Consequently, the beaches and barrier islands are constructed by wave action mainly from material from fairly far away, although local erosion may make a contribution, while the lagoons and bays behind them receive their sediment from local rivers.

The types and patterns of distribution of the sediments are controlled by three factors and their interaction: (1) the rate of continental runoff and sediment supply; (2) the intensity and direction of marine transporting agents, such as waves, tidal currents, and wind; and (3) the rate and direction of sea level changes. The balance between these three determines the types of sediment to be found. *See* DELTA; ESTUARINE OCEANOGRAPHY.

On most continental shelves, equilibrium has not yet been fully established and the sediments reflect to a large extent the recent rise of sea level. Only on narrow shelves with active sedimentation are present environmental conditions alone responsible for the sediment distribution. Sediments of the continental shelf and slope belong to one or more of the following types: (1) biogenic (derived from organisms and consisting mostly of calcareous material); (2) authigenic (precipitated from sea water or formed by chemical replacement of other particles, for example, glauconite, salt, and phosphorite); (3) residual (locally weathered from underlying rocks); (4) relict (remnants of earlier environments of deposition, for example, deposits formed during the transgression leading to the present high sea level stand); (5) detrital (products of weathering and erosion of land, supplied by streams and coastal erosion, such as gravels, sand, silt, and clay).

Much of the fine-grained sediment transported into the sea by rivers is not permanently deposited on the self but kept in suspension by waves. This material is slowly carried across the shelf by currents and by gravity flow down its gentle slope, and is finally deposited either on the continental slope or in the deep sea. If submarine canyons occur in the area, they may intercept these clouds, or suspended material, channel them, and transport them far into the deep ocean as turbidity currents. If the canyons intersect the nearshore zone where sand is transported, they can carry this material also out into deep water over great distances. *See* CONTINENTAL MARGIN; REEF.

[T.H.V.A.]

Deep-sea sediments. In general, classifications are difficult to apply because so many deep-sea sediments are widely ranging mixtures of two or more end-member sediment types. However, they can be divided into biogenic and nonbiogenic sediments.

Biogenic sediments, those formed from the skeletal remains of various kinds of marine organisms, may be distinguished according to the composition of the skeletal material, principally either calcium carbonate or opaline silica. The most abundant contributors of calcium carbonate to the deep-sea sediments are the planktonic foraminiferids, coccolithophorids and pteropods. Organisms which extract silica from the sea water and whose hard parts eventually are added to the sediment are radiolaria, diatoms, and to a lesser degree, dilicoflagellates and sponges. The degree to which deep-sea sediments in any area are composed of one or more of these biogenic types depends on the organic productivity of the various organisms in the surface water, the degree to which the skeletal remains are redissolved by sea water while setting to the bottom, and the rate of sedimentation of other types of sediment material. Where sediments are composed largely of a single type of biogenic material, it is often referred to as an ooze, after its consistency in place on the ocean floor.

The nonbiogenic sediment constituents are principally silicate materials and, locally, certain oxides. These may be broadly divided into materials which originate on the continents and are transported to the deep sea (detrital constituents) and those which originate in place in the deep sea, either precipitating from solution (authigenic minerals) or forming from the alteration of volcanic or other materials. The coarser constituents of detrital sediments include quartz, feldspars, amphiboles, and a wide spectrum of other common rock-forming minerals. The finer-grained components also include some quartz and feldspars, but belong principally to a group of sheet-silicate minerals known as the clay minerals, the most common of which are illite, montmorillonite, kaolinite, and chlorite. The distributions of several of these clay minerals have yielded information about their origins on the continents and, in several cases, clues to their modes of transport to the oceans.

[P.E.Bi.]

Maritime meteorology The moderating, energizing, and redistributing effects that bodies of water have on air temperature, weather, and climate by releasing energy and heat to the atmosphere in a number of characteristics ways.

To a greater extent than is commonly realized, the climates of the oceans determine the climates of the Earth. The oceans constitute the prime source of energy for all water-vapor-induced atmospheric processes. The latent heat liberated by the condensation of water vapor evaporated from the oceans supplies the energy for tropical cyclone formation; and such water vapor from the oceans is otherwise a primary stage in the energy cycle of the general atmospheric circulation. *See* ATMOSPHERE; STORM.

The oceans affect weather and climate, then, because they are a tremendous reservoir for the storage of solar energy. This energy is absorbed during periods of the year (or in regions) of surplus solar energy. It is then stored or transported by ocean currents, to be released to the atmosphere at times (or in regions) of deficiency in solar energy. In this sense the oceans become equalizers or moderators of climates. *See* CONTINENTALITY (METEOROLOGY).

[W.C.J.]

Marjoram The aromatic herb *Majorana hortensis*, a common plant in Mediterranean areas. The spicy camphoraceous odor of marjoram has long been cherished as an addition to a wide variety of foods; in the Middle Ages marjoram was used as an air freshener. Marjoram is in the mint family (Lamiaceae) and is a close relative of European or Greek oregano, with which it is often confused. There is still controversy concerning the proper taxonomic classification of this plant. Some authors place it in the genus *Origanum*, while others continue to separate it into its own genus *Majorana*. *See* OREGANO.

Marjoram is a small perennial (1–2 ft or 30–60 cm tall), and has ovate leaves to 1 in. (2.5 cm) long. The leaves are slightly hairy, as are the erect somewhat woody stems. Majoram flowers are white to very light lavender or pink in color but are very small and usually go unnoticed; the entire flower spike or inflorescence is, however, easily noticeable.

The dried and fresh leaves are used as flavoring for meats (sausage), vegetables, cheeses, poultry stuffing blends, and sauces, especially tomato-based sauces. *See* LAMIALES; SPICE AND FLAVORING. [S.Kir.]

Marl An argillaceous, nonindurated calcium carbonate deposit that is commonly gray or blue-gray. It is somewhat friable, and in some respects resembles chalk, with which it is interbedded in some localities. It is formed in some fresh-water lakes, partially by the action of some aquatic plants. The clay content of marls varies, and all gradations between small amounts of clay (marly limestones) and large amounts (marly clay) are found. Marlstone, or marlite, is an indurated rock of the same composition as marl. The marlstones are not fissile but blocky and massive with subconchoidal fracture. *See* LIMESTONE; SEDIMENTARY ROCKS. [R.Si.]

Marmoset Any of 10 species of primate which are members of the family Callithricidae and occur in South America. They are the most primitive of all primates, having claws rather than nails, a nonprehensile tail, and litters of more than two.

The pygmy marmoset (*Cebuella pygmaea*), found in the forests of the Amazon, is the smallest of all living monkeys. In some species, such as the common marmoset (*Callithrix jacchus*), the male carries the young about and returns them to the female only at feeding time. The diet is varied and consists of insects, fruits, seeds, and small vertebrates. *See* PRIMATES. [C.B.C.]

Marmot Several species of rodents of the genus *Marmota* which are members of the squirrel family, Sciuridae. These mammals are found in Europe, Asia, and North America.

One of the best-known marmots is the alpine marmot (*Marmota marmota*). It has a broad round head, a short furry tail, prominent eyes, and medium-sized ears; it is covered with thick, coarse, dull-colored fur. Other marmots are the red marmot (*M. caudata*) of Central Asia, the red-bellied marmot (*M. flaviventris*), ranging from New Mexico to British Columbia, and the woodchuck, or groundhog (*M. monax*), of North America. All species seem to prefer high altitudes except the woodchuck, which lives in woodlands and on farms. In many ways the marmots resemble the prairie dog. *See* PRAIRIE DOG; RODENTIA. [C.B.C.]

Mars The planet fourth in distance from the Sun. It is visible to the naked eye as a bright red star, except for short periods near its conjunctions with the Sun. Having a mean (synodic) period between oppositions of 780 days, Mars is in opposition with the Sun every other year. *See* PLANET.

In a telescope, Mars usually appears as a bright reddish disk marked by complex, semipermanent dark regions and variable white polar caps. Scientists have long known of atmospheric phenomena, variable polar caps of ice and clouds, and large bright and dark regions on the surface with dimensions of several hundred kilometers or more. But a multitude of other small surface formations with characteristic scales less than 125 mi (200 km) manifest their existence by the optical illusion of "canals" and "oases."

Surface. The Martian surface, which initially resembled the heavily cratered lunar uplands, has been modified extensively by the processes of volcanism, faulting, and the effects of wind and water. The northern hemisphere tends to be characterized by large volcanic lava plains, whereas the southern hemisphere appears to be older and more heavily cratered.

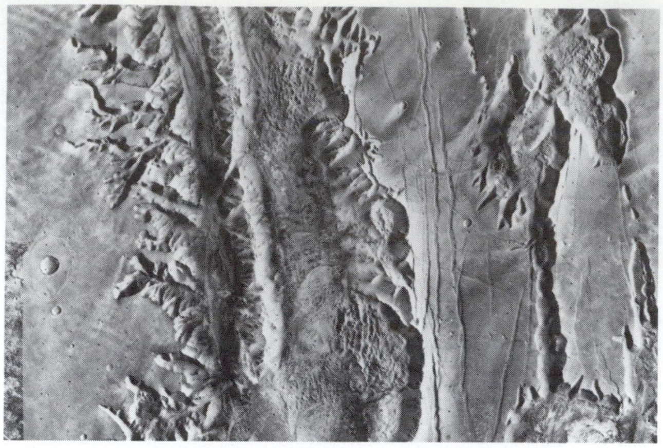

Fig. 1. Part of western end of Valles Marineris, showing canyon walls; photographed by *Viking Orbiter 1.* (*NASA*)

Among the more impressive features on the Martian surface are several enormous shield volcanoes, larger than any mountains on Earth. The largest and perhaps the youngest is Olympus Mons, which is nearly 1375 mi (600 km) across at its base and stands approximately 16 mi (26 km) above the surrounding terrain. Three other large shield volcanoes, Ascraeus Mons, Pavonis Mons, and Arsia Mons, lie along the nearby Tharsis Ridge. In shape and structure the Martian shield volcanoes bear a strong resemblance to their Hawaiian counterparts, At the summit of each shield is a complex of calderas, collapsed craterlike features that were once vents for lava.

Perhaps the most spectacular features on the Martian surface are the huge canyons located primarily in the equatorial regions. Valles Marineris, actually a system of canyons, extends for over 3000 mi (5000 km) along the equatorial belt. (Fig. 1). In places, the canyon complex is as much as 300 mi (500 km)

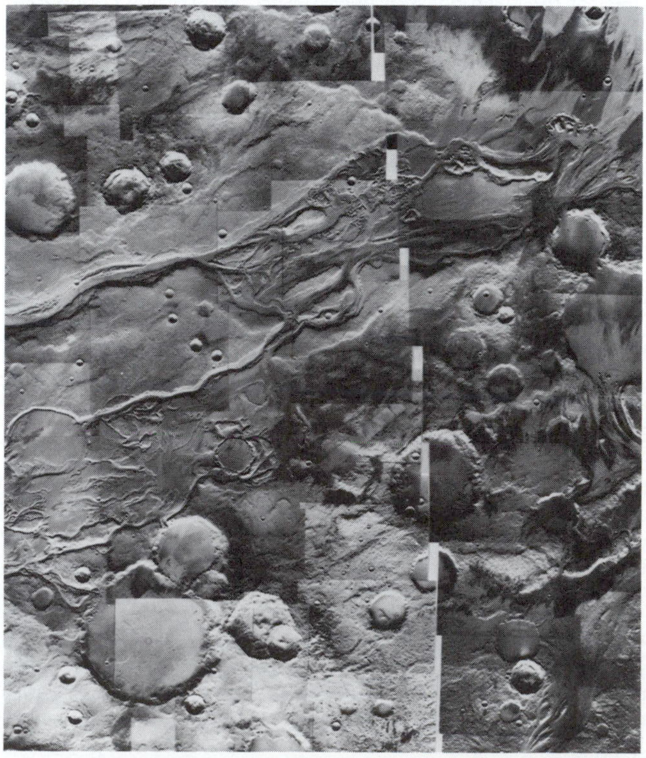

Fig. 2. Channel system in the Mangala Vallis region of Mars; photographed by *Viking Orbiter 1.* (*NASA*)

wide and drops to more than 4 mi (6 km) below the surrounding surface. Dwarfing the Grand Canyon, Valles Marineris is comparable in size to the great East African Rift Valley.

An astonishing class of features on the Martian surface is a widespread network of channels that bear a very strong resemblance to dry river beds (Fig. 2). Ranging from broad, sinuous features nearly 40 mi (60 km) wide, to small, narrow networks less than 300 ft (100 m) wide, the channels present the strong impression that they were created by water erosion.

The seasonal cycle of growth and decay of the bright polar caps has long been taken as evidence of the presence of water on Mars. However, evidence from the Mariner spacecraft identified carbon dioxide as the principal constituent of the polar snow, with lesser amounts of water ice also present. Formed during autumn and winter by condensation and deposition of the icy mist covering the polar regions, the polar caps at the end of winter cover a vast area extending down to latitude 60° in the southern hemisphere and 70° in the northern hemisphere. The residual caps, which never completely disappear, are believed to be composed almost entirely of water ice.

Atmosphere. The Martian atmosphere is very thin. It is composed principally of carbon dioxide, but contains nitrogen, argon, oxygen, and a trace of water vapor totaling about 5%. Both carbon dioxide and ice form clouds in the Martian atmosphere.

Localized dust storms appear quite frequently on Mars. Because of the extreme thinness of the Martian atmosphere, wind velocities greater than 90–110 mi/h (40–50 m/s) are needed to set surface dust grains in motion, although they can then be carried to high altitudes by winds of lesser velocity. Some dust storms develop with such intensity that their total extent may be hemispheric or even global.

Viking spacecraft. In 1976 two American spacecraft—the Viking landers—successfully dropped onto the surface of Mars. The spacecraft collected more than 1400 photographs and large quantities of other data concerning the character, composition, and organic content of the surface, and the composition and climatic patterns of the lower atmosphere. Three biological investigations and a gas chromatograph/mass spectrometer attempted to settle the question of whether life existed on the planet. The latter instrument detected no organic compounds at all (within the limits of its sensitivity). One of the trio of biological experiments gave a positive result—the release of oxygen from a soil sample when humidified—but there is no consensus among the researchers as to the source of this reaction.

Satellites. Mars has two small satellites, Phobos and Deimos. Neither satellite is massive enough to be gravitationally contracted to a spherical shape. Both satellites are saturated with impact craters. Both satellites have the same low albedo, approximately 0.05. This low reflectivity suggests that the Martian satellites may have had their origin in the asteroid belt. *See* ASTEROID. [J.K.B.]

Marsileales A small order of heterosporous, leptosporangiate ferns (division Polypodiophyta) which grow in water or wet places and are rooted to the substrate. The leaves arise on long stalks from the rhizome and typically have floating blades with four leaflets, suggesting a four-leaved clover. The order contains a single family and only 3 genera, about 50 species in all, most of them belonging to the widespread genus *Marsilea* (water clover). *See* POLYPODIOPHYTA. [A.Cr.]

Marsupialia An order of animals, long considered the only order of the mammalian infraclass Metatheria.

The marsupials are characterized by the presence of a pouch (marsupium) in the female, a skin pocket whose teat-bearing abdominal wall is supported by epipubic bones. The young are born in the embryonic state and crawl unaided to the marsupium, where they attach themselves to the teats and continue their development. In a few species the pouch is vestigial or has disappeared completely.

A wide variety of terrestrial adaptations are found among the 82 genera of living marsupials. Some spectacular examples of evolution convergent upon placental modes of life are exemplified by the Australian marsupial "moles" (*Notoryctes*), "wolves" (*Thylacinus*), and "flying squirrels" (*Petaurus*). For the most part, the adaptive radiation of the marsupials has paralleled that of the placentals, yielding many ecological analogs, such as kangaroos and wallabies, which are the counterparts of the placental deer and antelopes, although these groups differ widely in structure. *See* ANTEATER; EUTHERIA; KANGAROO; KOALA; MAMMALIA; METATHERIA; OPOSSUM. [R.H.T.]

Marten The name applied to seven species of carnivores which are members of the family Mustelidae. This family also includes the skunk, weasel, otter, badger, and wolverine.

The American marten (*Martes americana*) inhabits the cooler forests of North America. Its brown pelt, known as the American sable, is highly prized. It is small and has scent glands, as do most members of the family. Other species are the pine marten (*M. martes*), yellow-throated marten (*M. flavigula*), stone marten (*M. foina*), Japanese marten (*M. melampus*), and South Indian yellow-throated marten (*M. gwatkinsi*). These are all found in Asia; the pine and stone martens occur in Europe as well.

The adult weighs a maximum of 3 lb (1.4 kg), and for the size of the animal, the gestation period of about 38 weeks is unusually long. This is apparently because these species, as many other members of the family do, have delayed implantation of the fertilized egg; that is, after fertilization there is a lapse of several months before embryonic activation is initiated. Mating occurs in the summer and the litter of three or four is born the following spring. *See* BADGER; CARNIVORA; FISHER; MAMMALIA; OTTER; SABLE; SKUNK; WEASEL; WOLVERINE. [C.B.C.]

Maser A device for coherent amplification or generation of electromagnetic waves by use of excitation energy in resonant atomic or molecular systems. The word is an acronym for *m*icrowave *a*mplification by *s*timulated *e*mission of *r*adiation. The device uses an unstable ensemble of atomic or molecular particles which may be stimulated by an electromagnetic wave to radiate excess energy at the same frequency and phase as a stimulating wave, thus providing coherent amplification. Masers, however, are not limited to the microwave region; this type of amplification has been extended to include a frequency range from audio to infrared or optical frequencies. Maser-type amplifiers and oscillators are also sometimes referred to as molecular, or quantum-mechanical, since they involve processes on a molecular scale, and since some types cannot be adequately described by classical mechanics, but show characteristic quantum-mechanical phenomena.

Maser amplifiers can have exceptionally low noise, and come close to effectively amplifying a single quantum of radiation in the microwave region. Their inherently low noise makes maser oscillators that use very narrow atomic or molecular resonances extremely monochromatic, providing a basis for frequency standards. *See* ATOMIC CLOCK.

Because of their low noise and high sensitivity, maser amplifiers are particularly useful for reception and detection of very weak signals in radio astronomy, microwave radiometry, long-distance radar, and long-distance microwave communications. They also provide research tools for very sensitive amplification or detection of electromagnetic radiation.

Thermodynamic equilibrium of an ensemble of particles—such as atoms, molecules, electrons, or nuclei—which have discrete energy levels and which may radiate electromagnetic energy, requires that the number n_1 of particles in a lower

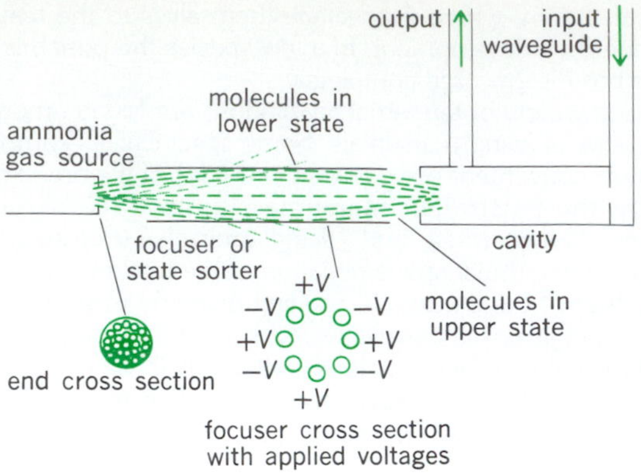

Schematic of beam-type maser.

level 1 be related to the number n_2 in an upper level 2 by the Boltzmann distribution condition $n_1/n_2 = e^{(E1 - E2)kT}$, where E_1 and E_2 are the respective energies of the two levels; k is Boltzmann's constant; and T is the absolute temperature, a positive number. Thermodynamic equilibrium also requires that phases of oscillation of the particles, or relative phases of quantum-mechanical wave functions for the various states, be random. A violation of either condition can result in instabilities which may release electromagnetic radiation. The frequency ν of radiation released is characteristically given by $h\nu = E_2 - E_1$, where h is Planck's constant. *See* BOLTZMANN STATISTICS.

The particles may be stimulated by an electromagnetic wave to make transitions from the lower to the upper level, absorbing energy from the wave, or from the upper to the lower level, imparting energy to the wave, and thereby increasing the wave amplitude coherently. Stimulated transitions from the upper to the lower state and those from the lower to the upper state are equally probable. For equilibrium at any positive temperature, the Boltzmann distribution requires that n_1 be greater than n_2. Therefore, there is a net absorption of energy from the wave, because particles which absorb are more numerous than those which emit. If the condition $n_2 > n_1$ occurs, the system may be said to have a negative absolute temperature, because the Boltzmann condition is fitted only by a negative value of T. If there are not too many counterbalancing losses from other sources, this condition allows a net amplification, because particles which emit energy are more numerous than those which absorb.

Gas masers. The beam-type maser (see illustration) was the first type of maser to be suggested. Ammonia gas issues from a small orifice into a vacuum system to form a molecular beam. Molecules in the lower of the two states are deflected away from the axis of the state sorter or focuser by inhomogeneous electric fields which act on their dipole moments. Those molecules in the upper state are deflected toward the axis and sent into the microwave-resonant cavity. If losses in the cavity walls and coupling holes are sufficiently small, or if the number of molecules is sufficiently large, amplification or oscillation will occur. This maser is particularly useful as a frequency or time standard because of the relative sharpness and invariance of resonances of the ammonia beam.

Solid-state masers. Solid-state masers usually involve the electrons of paramagnetic atoms or molecules in a static or slowly varying magnetic field. In the simplest case, the two-level solid-state maser, only one electron on each molecule is affected. The energy of the electron is quantized into two levels, according to whether the magnetic moment, associated with the electron spin, is parallel or antiparallel to the magnetic field. At thermal equilibrium there are more magnetic moments parallel than antiparallel to the field. This situation may be reversed by interchanging the two populations of particles.

Optical and infrared masers. Optical and infrared masers utilize a variety of principles for excitation of the nonequilibrium distribution. All involve multimode cavities, usually consisting of two optically flat plates between which the radiation is reflected many times through the excited medium. Such masers are particularly valuable as oscillators, producing coherent light that is extremely monochromatic and can be focused either to a very intense small spot or radiated in a remarkably parallel beam. *See* LASER. [C.H.T.; J.P.Go.]

Masking of sound Interference with the audibility of a sound caused by the presence of another sound. More specifically, the number of decibels (dB) by which the intensity level of a sound (signal) must be raised above its threshold of audibility, to be heard in the presence of a second sound (masker), is called the masking produced by the masker on the signal. The masker and the signal may be identical or may differ in frequency, complexity, or time.

Four major efforts are to be noted when both the masker and signal are pure tones and are presented simultaneously. First, the higher the level of the masker, the greater the masking. Second, masking is greatest when the frequency of the masker is in the vicinity of that of the signal. Third, the masking caused by a tone is much greater on frequencies higher than that of the masker than on frequencies below it. Fourth, in addition to the two tones themselves, other tones are heard which do not exist except in the listener's hearing. These tones are caused by nonlinear effects in the hearing mechanism.

The most widely studied complex masking sound is random noise which has energy at all frequencies and is said to be flat if the level for each 1-Hz bandwidth of the noise is the same. When random-flat noise is used to mask a pure tone, only a narrow frequency band (critical band) of the noise centered at the tonal frequency causes masking. When noise masks speech, either the detectability of speech or speech intelligibility can be measured. The level for speech intelligibility is about 10–14 dB higher than for speech detectability. *See* ACOUSTIC NOISE; HEARING (HUMAN); SOUND. [W.A.Y.]

Masonry A built-up construction or combination of building units of such materials as clay, shale, concrete, glass, gypsum, or stone, set in mortar or plain concrete. Common usage of the term masonry implies the use of cut stone set in mortar and patterned conventionally. In modern engineering practice, masonry includes unit construction not only of natural stone but also of brick and tile—burned-clay manufactured products. Plant-made bricks and blocks of gypsum, glass, concrete and sand-lime mixes are also regularly used in unit masonry construction.

Unit masonry may be plain (unreinforced) or reinforced by the use of steel mesh or small bars placed in the mortar beds as soon as the mortar is in place, or through vertical cavities in bricks or tile, in accordance with recognized design procedures. Such reinforcement gives to unit masonry increased resistance to tensile and shear forces, thus enhancing its value in resisting lateral forces such as those from seismic action. Multistory buildings of bearing-wall masonry construction are found throughout the world. [R.F.L.]

Mass The quantitative or numerical measure of a body's inertia, that is, of its resistance to being accelerated.

Because it is often necessary to compare masses of such dissimilar bodies as a sample of sugar, a sample of air, an electron, and the Moon, it is necessary to define mass in terms of a property that not only is inherent and permanent but is also universal in that it is possessed by all known forms of matter. All matter possesses two properties, gravitation and inertia. The property of gravitation is that every material body attracts every other material body. The property of inertia is that every material body resists any attempt to change its motion. A body's motion is said to change if the body is accelerated, that is, if it increases or decreases its speed or changes the direction of its motion. Because of its inertia a body cannot be accelerated unless a force is exerted on it. The greater the inertia of a body, the less will be the acceleration produced by a given force. See GRAVITATION; INERTIA.

The present definition of mass is in terms of inertia. The masses of two bodies are compared by applying equal forces to the bodies and measuring their accelerations. For example, the two bodies may be allowed to collide. According to Newton's third law, each body will then experience an equally strong force. If there are no external forces, and if a_1 and a_2 are the measured accelerations of the two bodies, the ratio of the masses of the two bodies is by definition given by the following equation:

$$\frac{m_1}{m_2} = \frac{a_2}{a_1}$$

This equation gives only ratios of masses; it is therefore necessary to designate the mass of some one body as the standard mass to which the masses of all other bodies can be compared. The body that has been chosen for this purpose is a cylinder of platinum-iridium alloy. It is known as the international standard of mass; its mass is called 1 kilogram (kg), and it is kept at the International Bureau of Weights and Measures near Paris, France. Replicas of the standard mass, kept at various national laboratories, are periodically compared with this standard.

Einstein's special theory of relativity predicts that the inertia of a body should increase if the energy of the body increases. This prediction has been conclusively verified experimentally. It follows that the mass of a body will increase if, for example, the body gains speed (addition of kinetic energy), or its temperature rises (addition of heat energy), or the body is compressed (addition of elastic energy). See CONSERVATION OF MASS. [L.N.]

Mass defect The difference between the mass of an atom and the sum of the masses of its individual components in the free (unbound) state. The mass of an atom is always less than the total mass of its constituent particles; this means, according to Albert Einstein's well-known formula, that an energy of $E = mc^2$ has been released in the process of combination, where m is the difference between the total mass of the constituent particles and the mass of the atom, and c is the velocity of light.

The mass defect, when expressed in energy units, is called the binding energy, a term which is perhaps more commonly used. See NUCLEAR BINDING ENERGY. [W.W.W.]

Mass flow rate meter An instrument that measures the mass of fluid flowing through a pipe, duct, or conduit, per unit time. There have been many attempts to develop practical industrial mass flowmeters. The most consistently satisfactory method of determining mass flow has been through calculation using volume flow and density determined either by measurement or inference from pressure and temperatures. Only a few of the mechanical types will be mentioned here.

In one type, the fluid in the pipe is made to rotate at a constant speed by a motor-driven impeller. The torque required by a second, stationary impeller to straighten the flow again is a direct measurement of the mass flow.

A second type, described as gyroscopic/Coriolis mass flowmeter, employs a C-shaped pipe and a T-shaped leaf spring as opposite legs of a tuning fork. An electromagnetic forcer excites the tuning fork, thereby subjecting each moving particle within the pipe to a Coriolis-type acceleration. The resulting forces angularly deflect the C-shaped pipe an amount that is inversely proportional to the stiffness of the pipe and proportional to the mass flow rate within the pipe. The angular deflection (twisting) of the C-shaped pipe is optically measured twice during each cycle of the tuning-fork oscillation (oscillating at natural frequency). A digital logic circuit converts the timing of the signals into a mass-flow-rate signal. See CORIOLIS ACCELERATION.

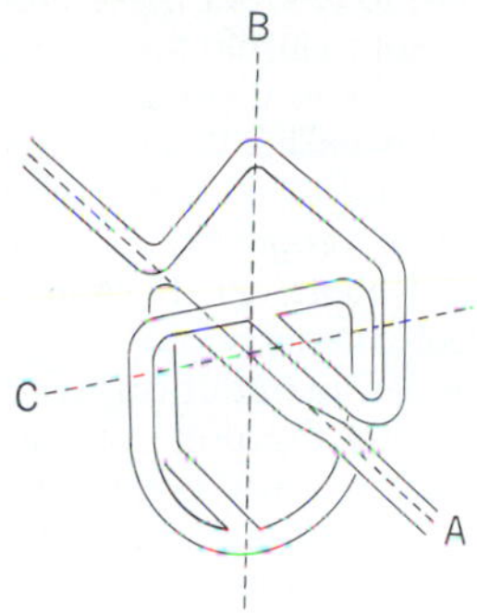

Gyroscopic flowmeter.

Another type is based on the principle of the gyroscope (see illustration). The flowing stream is sent through a pipe of suitable shape so that the mass of material flowing corresponds to a gyro wheel rotating about axis C. The entire gyrolike assembly is rotating about axis A, and a torque is produced about axis B proportional to the angular momentum of the simulated gyro wheel. This torque is therefore directly proportional to the mass rate of flow. The pretzellike configuration of pipe is introduced to eliminate centrifugal and other extraneous effects. See FLOW MEASUREMENT; GYROSCOPE. [M.Br.; L.P.E.]

Mass-luminosity relation The relation, observed or predicted by theory, between the quantity of matter a star contains (its mass) and the amount of energy generated in its interior (its luminosity). Because of the great sensitivity of the rate of energy production in a stellar interior to the mass of the star, the mass-luminosity relation provides an important test of theories of stellar interiors.

For a family of stars with different masses but with the same mixture of chemical elements uniformly distributed throughout the stellar volumes, there will be a unique mass-luminosity relation. Since most of the stars in the solar neighborhood have about the same chemical composition, the observed relation, obtained from binary stars for which masses and luminosities can be observationally evaluated, conforms reasonably well with theory. See BINARY STAR; STAR. [D.M.P.]

Mass number The mass number A of an atom is the total number of its nuclear constituents, or nucleons, as the protons and neutrons are collectively called. The mass number is placed before and above the elemental symbol, thus ^{238}U. The mass number gives a useful rough figure for the atomic mass; for example, ^{1}H = 1.00814 atomic mass units (amu), ^{238}U = 238.124 amu, and so on. See ATOMIC NUMBER. [H.E.D.]

Mass production The manufacture of products in large quantities. The products may be machine-made individual parts, assembled items, or process items such as chemicals, coal, oil, cement, or cereals.

Eli Whitney, one of the foremost inventors of the early 1800s, is credited with being the father of mass production through his development of the concept of interchangeable manufacture. Whitney invented machines, tools, gages, jigs, and fixtures so that all parts could be made to close dimensional tolerances. Today it is difficult to realize that until about 1920 if anyone broke an automobile axle there was usually a wait of at least a week for the new one to be sent out from the factory.

In order for an item to warrant mass production, there must be a large, stable market to justify the type of specialized equipment required for economical production. Even this concept has evolved to what is called flexible manufacturing, which involves control of many operations, whether machining or assembly, by computers, programmable controllers, or numerical control systems. Robots can be quite easily "trained" to change from the requirements of one design to another.

The effects of mass production are not all technical. Automation, by lowering the unit cost of a product enables more people to buy the product, creates a need for new maintenance services, calls for more raw materials, and increases shipping and handling jobs both to and from the factory, with consequent overall increases in employment. *See* OPERATIONS RESEARCH; PRODUCTION ENGINEERING; ROBOTICS. [R.A.Li.]

Mass spectrometry An analytical technique for identification of chemical structures, determination of mixtures, and quantitative elemental analysis, based on application of the mass spectrometer. Organic and inorganic molecular structure determination is based on the fragmentation pattern of the ion formed when the molecule is ionized; further, because such patterns are distinctive, reproducible, and additive, mixtures of known compounds may be quantitatively analyzed. Quantitative elemental analysis of organic compounds requires exact mass values from a high-resolution mass spectrometer; trace analysis of inorganic solids requires a measure of ion intensity as well. *See* MASS SPECTROSCOPE.

Organic compounds. For analysis of organic compounds the principal methods of ion production are electron impact, chemical ionization, field ionization, and field desorption.

In electron impact, a gaseous sample of a molecular compound is ionized with a beam of energetic electrons and part of the energy is transferred to the ion formed by the collision:

$$A—B—C + e \rightarrow A—B—C^{+} + e + e$$

For most molecules the production of cations is favored over the production of anions by a factor of about 10^4. The cation corresponding to the simple removal of the electron is commonly called the molecular ion and normally will be the ion of greatest m/e ratio in the spectrum. In the ratio m is the mass of the ion in atomic mass units and e is the charge of the ion measured in terms of the number of electrons removed (or added) during ionization. Occasionally the ion is of vanishing intensity, and sometimes it collides with another molecule to abstract a hydrogen or another group. In these cases an incorrect assignment of the molecular ion may be made unless further tests are applied. Proper identification gives the molecular weight of the sample.

The remaining techniques were devised generally to circumvent the problem of the weak or vanishingly small intensity of a molecular ion. In chemical ionization, the ions to be analyzed are produced by transfer of a heavy particle (H^+, H^-, or heavier) to the sample from ions produced from a reactant gas. Field ionization and field desorption techniques are used for less volatile material. The sample is ionized when it is in a very high field gradient (several volts per angstrom) near an electrode surface. In field ionization, the sample is admitted as a gas and ionized within a few angstroms of the surface. In field desorption, the sample is coated on the metal surface. *See* IONIZATION POTENTIAL; LASER SPECTROSCOPY; PLASMA CHROMATOGRAPHY.

Mixtures. For a given instrument, the mass spectrum of a compound serves as a fingerprint of a compound under standard conditions; it varies with temperature, ionizing voltage, and parameters of the construction of the instrument. A mixture of gases may be analyzed because spectra are additive.

A further powerful technique for analysis of mixtures containing unknowns is mass spectral analysis of gas chromatographic effluents either by direct hookup of the gases exiting from the chromatograph, including carrier gas, to the introduction system of the spectrometer or by analysis of trapped effluent fractions.

Similarly, mass spectrometers may be coupled to liquid chromatographs; the solvent may be removed by evaporation of the solution on a moving belt which next passes into the ion source and is then heated to evaporate the sample, or by freezing solvent on liquid-nitrogen-cooled fingers surrounding the source; or the solvent may be used itself as a reagent gas for chemical ionization. Ion cyclotron resonance spectroscopy has become of special importance because of the development of Fourier-transform ion cyclotron resonance techniques. *See* GAS CHROMATOGRAPHY; LIQUID CHROMATOGRAPHY.

Inorganic solids. The analysis of solid inorganic samples can be made either by vaporization at very high temperatures or by volatilization of the sample surface so that particles are atomized and ionized with a high-energy spark. The wide range of energies given to the particles requires a double-focusing mass spectrometer for analysis.

Secondary ion mass spectrometry is most commonly used for surface analysis. A primary beam of ions accelerated through a few kilovolts is focused on a surface; ions are among the products sputtered from this surface, and they may be directly analyzed in a quadrupole filter. *See* SECONDARY ION-MASS SPECTROMETRY (SIMS); SPECTROSCOPY; TRACE ANALYSIS. [M.M.Bu.]

Mass spectroscope An instrument used for determining the masses of atoms or molecules found in a sample of gas, liquid, or solid. It is analogous to the optical spectroscope, in which a beam of light containing various colors (white light) is sent through a prism to separate it into the spectrum of colors present. In a mass spectroscope, a beam of ions (electrically charged atoms or molecules) is sent through a combination of electric and magnetic fields so arranged that a mass spectrum is produced (see illustration). If the ions fall on a photographic plate which after development shows the mass spectrum, the instrument is called a mass spectrograph; if the spectrum is allowed to sweep across a slit in front of an electrical detector which records the current, it is called a mass spectrometer.

Mass spectroscopes are used in both pure and applied science. Atomic masses can be measured very precisely. Because of the equivalence of mass and energy, knowledge of nuclear

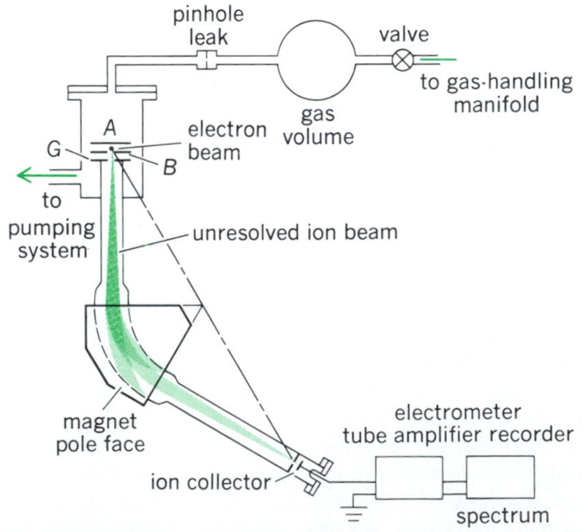

Schematic drawing of mass spectrometer tube. Ion currents are in the range 10^{-10} to 10^{-15} ampere and require special electrometer tube amplifiers for their detection. In actual instruments the radius of curvature of ions in a magnetic field is 4–6 in. (10–15 cm).

structure and binding energy of nuclei is thus gained. The relative abundances of the isotopes in naturally occurring or artificially produced elements can be determined.

Empirical and theoretical studies have led to an understanding of the relation between molecular structure and the relative abundances of the fragments observed when a complex molecule, such as a heavy organic compound, is ionized. When a high-resolution instrument is employed, the masses of the molecular or fragment ions can be determined so accurately that identification of the ion can frequently be made from the mass alone. *See* MASS SPECTROMETRY.

Because chemical compounds may have mass spectra as unique as fingerprints, mass spectroscopes are widely used in industries such as oil refineries, where analyses of complex hydrocarbon mixtures are required. *See* ATOMIC WEIGHT; BETA PARTICLES. [A.O.N.]

Mass-transfer operation An operation in chemical engineering that involves transfer of material from one phase to another, or from one place to another within a single phase. Mass transfer may occur for various purposes, for example, to effect a chemical reaction, to obtain a separation of components, or to obtain uniform distribution of material within a phase. The most common types are (1) fluid-fluid transfer, occurring typically in absorption, distillation, and liquid-liquid extraction, and (2) fluid-solid transfer, occurring typically in crystallization and gas adsorption. *See* ADSORPTION; COUNTERCURRENT TRANSFER OPERATIONS; CRYSTALLIZATION; DIALYSIS; DIFFUSION IN GASES AND LIQUIDS; DISTILLATION; GAS ABSORPTION OPERATIONS; ION EXCHANGE; LEACHING; SOLVENT EXTRACTION. [C.R.W.]

Mass wasting A generic term for downslope movement of soil and rock, primarily in response to gravitational body forces. Mass wasting is distinct from other erosive processes in which particles or fragments are carried down by the internal energy of wind, running water, or moving ice and snow.

The stability of slope-making materials is lost when their shear strength (or sometimes their tensile strength) is overcome by shear (or tensile) stresses, or when individual particles, fragments, and blocks are induced to topple or tumble. The shear and tensile strength of earth materials depends on their mineralogy and structure. Processes that generally decrease the strength of earth materials include one or more of the following: structural changes, weathering, groundwater, and meteorological changes. Stresses in slopes are increased by steepening, heightening, and external loading due to static and dynamic forces. Processes that increase stresses can be natural or result from human activities. Although other classifications exist, these movements can be conveniently classified according to their velocity into two types: creep and landsliding. *See* SOIL MECHANICS.

Geologically, creep is the imperceptible downslope movement at rates as slow as a fraction of millimeter per year; its cumulative effects are ubiquitously expressed in slopes as the downhill bending of bedded and foliated rock, bent tree trunks, broken retaining walls, and tilted structures. There are two varieties of geologic creep. Seasonal creep is the slow, episodic movement of the uppermost several centimeters of soil, or fractured and weathered rock. It is especially important in regions of permanently frozen ground. Rheologic creep, sometimes called continuous creep, is a time-dependent deformation at relatively constant shear stresses of masses of rock, soil, ice, and snow. This type of creep affects rock slopes down to depths of a few hundred meters, as well as the surficial layer disturbed by seasonal creep. Continuous creep is most conspicuous in weak rocks and in regions where high horizontal stresses (several tens of bars or several megapascals) are known to exist in rock masses at depths of 330 to 660 ft (100 to 200 m).

Landsliding includes all perceptible mass movements. Three types are generally recognized on the basis of the type of movement: falls, slides, and flows. Falls involve free-falling material; in slides the moving mass displaces along one or more narrow shear zones; and in flows the distribution of velocities within the moving mass resembles that of a viscous flow. *See* LANDSLIDE.

Mass wasting is an important consideration in the interaction between humans and the environment. Deforestation accelerates soil creep. Engineering activities such as damming and open-pit mining are known to increase landsliding. On the other hand, enormous natural rock avalanches have buried entire villages and claimed tens of thousands of lives. [A.S.N.]

Massif A block of the Earth's crust commonly consisting of crystalline gneisses and schists, the textural appearance of which is generally markedly different from that of the surrounding rocks. Common usage indicates that a massif has limited areal extent and considerable topographic relief. Structurally, a massif may form the core of an anticline or may be a block bounded by faults or even unconformities. In any case, during the final stages of its development a massif acts as a relatively homogeneous tectonic unit which to some extent controls the structures that surround it. Numerous complex internal structures may be present; many of these are not related to its development as a massif but are the mark of previous deformations. [P.H.O.]

Mastigomycotina Zoosporic fungi; an artificial grouping of fungi with all members, except a few advanced ones, producing free-living zoospores in some part of their life cycle. The thallus may be unicellular, multicellular, or mycelial with many reproductive bodies. The thallus has a chitinaceous or cellulosic wall. In a few endoparasitic species the thallus is naked (lacks a wall). Forms of sexual reproduction are many and varied.

The three classes are distinguished by the type of flagellation: single whiplash (Chytridiomycetes), single tinsel (Hyphochytridiomycetes), or biflagellate (Oomycetes).

Distribution is cosmopolitan; many are aquatic, occurring in fresh water or brackish water, and a few are marine. They are also abundant in soils, and some have part of their life cycles on vascular plants. Many are saprobes, but some are parasitic on insects or fish, and some are destructive plant pathogens. [D.J.S.B.]

Mastigophora A superclass of the Protozoa also known as the Flagellata. The common morphological flagellate type is spherical to cylindrical on an anteroposterior axis. Despite their common morphological plan, based on the flagellum as a means of locomotion, Flagellata are very diverse in shape, colony formation, internal structure, external shell or test, color, physiology, reproduction, and choice of environment.

The Mastigophora exhibit marked plantlike characteristics, so that texts and references in botany always treat at least some of them. Some workers include all colorless flagellates in the algae; however, flagellates display many distinctly animal features and sometimes are treated as Protozoa. Thus, flagellates are regarded as a link between the plant and animal kingdoms. *See* PHYTAMASTIGOPHOREA; ZOOMASTIGOPHOREA.

Mastigophora possess one common feature, locomotion by means of one or more whiplike protoplasmic extrusions termed flagella. The flagellate body is typically monaxial and elongate to some degree. Cells are practically spherical in *Monas*, but ovoid, cordiform, pyriform, fusiform, acicular, tubular, or flattened cells are more common. Flattened species typically glide along a surface rather than swim. A particular shape is normally maintained by *Ochromonas*, whereas *Mastigamoeba* and some others form pseudopodia, and certain species (*Euglena agilis*) undergo frequent distortions termed euglenoid or metabolic changes of shape.

Structurally the flagellate cell is not simple. The cytoplasm is sometimes quite vacuolated (*Collodictyon* or *Trepomonas*) and occasionally colored, although color is generally confined to chromatophores when these are present. One or more contractile vacuoles may be located near the flagellar base.

Most flagellates are single-celled but colony formation is frequent. Cells may be naked (*Oicomonas*), enclosed in a thick cellulose cell wall or pellicle (*Euglena spirogyra*), or in a chitinous, calcareous, or silicious test, the lorica or shell (*Trachelomonas*, *Distephanus speculum*).

Flagellates adjust to wide ranges of pH, osmotic pressure, light, and temperature but optima are found upon investigation. Some flagellate species occur in both fresh and sea water, but cannot be transferred directly from fresh to sea water. Almost any aqueous ecological niche contains Flagellata.

Flagellata are near the base of the food pyramid, probably next to bacteria. Their varied synthetic abilities, fast reproductive rate, and huge numbers in both fresh and salt water compensate for their usual small size. The oceanic blooms which kill fish and other animals are principally flagellates, but green flagellates reaerate polluted water. They often cause tastes and odors in potable water, but they readily attack organic matter in natural water as do bacteria, They include many parasites dangerous to humans and animals and are themselves frequently parasitized. *See* PROTOZOA. [J.B.L.]

Mastitis Inflammation of the mammary gland. This condition is most frequently caused by infection of the gland with bacteria that are pathogenic for this organ. It has been described in humans, cows, sheep, goats, pigs, horses, and rabbits. Mastitis causes lactating women to experience pain when nursing the child, it damages mammary tissue, and the formation of scar tissue in the breast may cause disfigurement.

The mammary gland is composed of a teat and a glandular portion. The gland has defensive mechanisms to prevent and overcome infection with bacteria. Nonspecific defense mechanisms include teat duct keratin, lactoferrin, lactoperoxidase, and complement. Specific defense mechanisms are mediated by antibodies and include opsonization, direct lysis of pathogens, and toxin neutralization. Milk contains epithelial cells, macrophages, neutrophils, and lymphocytes. To induce mastitis, a pathogen must first pass through the teat duct to enter the gland, survive the bacteriostatic and bactericidal mechanisms, and multiply. Bacteria possess virulence factors such as capsules and toxins which enable them to withstand these protective mechanisms. *See* LACTATION.

When bacteria multiply within the gland, there is a release of inflammatory mediators and an influx of neutrophils. The severity of mammary infection is classified according to the clinical signs. In humans, infection occurs during lactation, with clinical episodes most frequent during the first 2 months of lactation. In acute puerperal mastitis the tissue becomes hot, swollen, red, and painful, and a fever may be present. In the absence of treatment, this may progress to a pus-forming mastitis, with the development of breast abscesses.

In acute puerperal mastitis of humans, suitable antibiotics are administered by the intravenous or intramuscular route, while in the abscess form surgical drainage is provided in addition to antibacterial therapy. Penicillins, cephalosporins, and erythromycin are administered locally into the infected gland after milking for 1–2 days. Additional antibiotics are given systemically, that is, intravenously or intramuscularly, in severe cases of mastitis, and also to improve bacteriologic cure rates. *See* ANTIBIOTIC. [N.L.N.]

Match A short piece of wood or other material at the end of which has been placed a substance that can be ignited by friction. Matches are of two types, strike-anywhere and safety. The strike-anywhere match, also called the kitchen match, contains potassium chlorate ($KClO_3$) and phosphorus sesquisulfide (P_4S_3) as the main ingredients, with added quantities of iron oxide, zinc oxide, and glue. In addition, it includes ground glass to create friction. Safety matches involve two separate compositions, the first coated on the matchstick and the second on the side of the container. The first contains potassium chlorate, potassium dichromate, manganese dioxide, sulfur, iron oxide, ground glass, and glue. The second contains red phosphorus, antimony sulfide, and an abrasive. The heat produced by the ignition of a minute amount of the red phosphorus when the match is struck is sufficient to ignite the match head. The wood or cardboard of the matchstick is coated with paraffin to give a longer and more ready flame and is impregnated with borax to prevent afterglow. *See* COMBUSTION; FLAME; PHOSPHORUS; PYROTECHNICS. [W.E.Go.]

Material requirements planning A formal computerized approach to inventory planning, manufacturing scheduling, supplier scheduling, and overall corporate planning. The material requirements planning (MRP) system provides the user with information about timing (when to order) and quantity (how much to order), generates new orders, and reschedules existing orders as necessary to meet the changing requirements of customers and manufacturing. The system is driven by change and constantly recalculates material requirements based on actual forecast orders. It makes adjustments for possible problems prior to their occurrence, as opposed to traditional control systems which looked at more historical demand and reacted to existing problems. *See* MANUFACTURING ENGINEERING.

The logic of the material requirements planning system is based on the principle of dependent demand, a term describing the direct relationship between demand for one item and demand for a higher-level assembly part or component. For example, the demand for the number of wheel assemblies on a bicycle is directly related to the number of bicycles planned for production; further, the demand for tires is directly dependent on the demand for wheel assemblies. In most manufacturing businesses, the bulk of the raw material and in-process inventories are subject to dependent demand. Dependent demand quantities are calculated, while independent demand items are forecast. Independent demand is unrelated to a higher-level item which the company manufactures or stocks. Generally, independent demand items are carried in finished goods inventory and subject to uncertain end customer demand. Spare parts or replacement requirements for a drill press are an example of an independent demand item.

By use of the computer, material requirements planning is able to manipulate massive amounts of data to keep schedules up to date and priorities in order. The technological advances in computing and processing power, the benefits of on-line capabilities, and reduction in computing cost make computerized manufacturing planning and control systems such as material requirements planning powerful tools in operating modern manufacturing systems productively. *See* INDUSTRIAL ENGINEERING; INVENTORY CONTROL; SYSTEMS ENGINEERING. [L.C.Gi.]

Materials handling The loading, moving, and unloading of materials. The hundreds of different ways of handling materials are generally classified according to the type of equipment used. For example, the International Materials Management Society has classified equipment as (1) conveyor, (2) cranes, elevators, and hoists, (3) positioning, weighing, and control equipment, (4) industrial vehicles, (5) motor vehicles, (6) railroad cars, (7) marine carriers, (8) aircraft, and (9) containers and supports. *See* MATERIALS-HANDLING EQUIPMENT. [R.Mu.]

Materials-handling equipment Devices used for handling materials in an industrial distribution activity. The equipment moves products as discrete articles, in suitable containers, or as solid bulk materials which are relatively free-flowing.

Many different types of machines result from combinations and permutations of the following factors: (1) The route over which the product is moved may be fixed or variable; (2) the path of travel may be horizontal, inclined, declined, or vertical; (3) motion may be imparted to the product manually, by the force of gravity, by air pressure, by vacuum, by vibration, or by power-actuated components of the machine; (4) the motion may be continuous or intermittent (reciprocating); and (5) the product may be supported or carried suspended during the handling operation. Based upon their most common characteristics, materials-handling machines can be grouped into six broad categories, listed in the following cross references. *See* BULK-HANDLING MACHINES; CONVEYING MACHINES; ELEVATING MACHINES; HOISTING MACHINES; INDUSTRIAL TRUCKS; MONORAIL. [A.M.P.]

Materials science and engineering A multidisciplinary field concerned with the generation and application of knowledge relating to the composition, structure, and processing of materials to their properties and uses. The field encompasses the complete knowledge spectrum for materials ranging from the basic end (materials science) to the applied end (materials engineering). It forms a bridge of knowledge from the basic sciences (and mathematics) to various engineering disciplines.

The study of metallic materials constitutes a major division of the materials science and engineering field. Most metals have a crystalline structure of closely packed atoms arranged in an orderly manner. In general they are good electrical and thermal conductors. Many are relatively strong at room temperature and retain good strength at elevated temperatures. Metals and alloys are often cast into the nearly final shape in which they will be used (castings). Ferrous metals and alloys contain iron as their major metallic element; nonferrous metals and alloys contain elements other than iron as their major metallic element. *See* ALLOY; ALUMINUM ALLOYS; CAST IRON; COPPER ALLOYS; IRON ALLOYS; METAL; METAL CASTING; NICKEL ALLOYS; STAINLESS STEEL; STEEL; ZINC ALLOYS.

The study of ceramic materials forms a second major division of the field of materials science and engineering. Ceramics are inorganic materials consisting of metallic and nonmetallic elements chemically bonded together. Most ceramic materials have high hardness, high-temperature strength, and good chemical resistance; however, they tend to be brittle. Ceramics in general have low electrical and thermal conductivities, which makes them useful for electrical and thermal insulative applications. Most ceramic materials can be classified into three groups: traditional ceramics, technical ceramics, and glasses. *See* CERAMICS; GLASS; PORCELAIN.

The study of polymeric materials forms a third major division of materials science and engineering. Most of these materials consist of carbon-containing long molecular chains or networks. Structurally, most of them are noncrystalline, but some are partly crystalline. The strength and ductility of polymeric materials vary greatly. Most polymers have low densities and relatively low softening or decomposition temperatures. Many are good thermal and electrical insulators. Polymeric materials have replaced metals and glasses for many applications. *See* POLYMER.

A fourth major division of materials science and engineering comprises the study of composite materials. A composite material is a mixture of two or more materials that differ in form and chemical composition and are essentially insoluble in each other, and most are produced synthetically by combining various types of fibers with different matrices to increase strength, toughness, and other properties. Three important types of composite materials have polymeric, metallic, or ceramic matrices. *See* COMPOSITE MATERIAL; METAL MATRIX COMPOSITES.

In addition to metallic, ceramic, polymeric, and composite materials, materials science and engineering is also concerned with the research and development of other special classes of materials that are based on applications. Some major types of these materials are electronic materials, optical materials, magnetic materials, superconducting materials, dielectric materials, nuclear materials, biomedical materials, and building materials. *See* INTEGRATED CIRCUITS; MAGNETIC MATERIALS; OPTICAL FIBERS; SUPERCONDUCTING DEVICES; SUPERCONDUCTIVITY. [W.F.S.]

Maternal behavior The pattern of care given an offspring by its mother. Many species reproduce generation after generation without receiving or providing any parental care. Insects and fish commonly produce vast numbers of offspring that they neither feed nor defend, resulting in the loss of many offspring to predators and to other hazards. Their great numbers, however, ensure that some will survive and reproduce. Also, if the young of a species are self-sufficient at birth or if they mature very rapidly after birth, they can often survive and reproduce with little parental care.

Parental behavior is most highly developed in species that produce only a few offspring at once, mature slowly, and have complex behavior patterns that can be learned only through extended practice. Parental behavior can involve as little as hiding the young and never seeing them again or it can involve years of feeding, defending, and teaching. In most species that care for their young after birth, the female does most or all of the work. The biological explanation for this arises from the fact that a male can father vast numbers of offspring, whereas a female can bear only a few. Each offspring that fails to survive or reproduce can represent a significant proportion of the female's lifetime reproductive output. Thus, the females that pass their genes on to subsequent generations in greatest numbers are not those that bear the most young but those that invest more in caring for the few young they produce. Active care of the young by the male is most common when males of a species limit themselves to one or a few female partners. *See* REPRODUCTIVE BEHAVIOR.

Factors that influence the onset of maternal behavior include physiological and psychological factors related to pregnancy, hormonal influences associated with childbirth, the behavior of the newborn, and cultural factors such as learning and social traditions. Although hormones play necessary roles in recovery from pregnancy and in milk production, there is no evidence that they play a critical role in any other aspect of human maternal behavior. Cultural factors and the behavior of the newborn are the primary cues that prompt human mothers to care for their young. [E.Wa.; G.Pos.]

Mathematical biology In its broadest sense, mathematical biology encompasses all applications of mathematics, computer technology, and quantitative theorizing to biological systems and the underlying processes within them.

Mathematical biology arose in the application of mathematical equations to biological situations involving chemical or physical processes already understood quantitatively by the methods of classical physics and chemistry. An example is the process of diffusion, by which oxygen enters the bloodstream from the lungs and subsequently passes into the tissues to be utilized in energy production, a fundamental phenomenon found in all living systems. An understanding of all biological processes in which diffusion plays a role is substantially improved by the fact that mathematical equations describing diffusion have been available to physicists and chemists for some time and can be adapted to the particular circumstances involved in biology.

As a result of the extensive amount of research, mathematical models of biological systems have become prominent and sometimes even decisive tools in the development of understanding of living systems. These models range in scope from the submicroscopic domain of intracellular biochemical reactions and metabolism to the treatment of the problems of population growth; they range in interest from highly abstract problems arising from control and communication processes in the brain to more immediate practical problems such as computer diagnosis of human diseases. In a disciplinary sense they range through anatomy, physiology, pharmacology, psychology, clinical medicine, genetics, taxonomy, botany, and a variety of other areas. A large number of mathematical models of the heart, brain, kidney, lungs, bloodstream, bone, and endocrine and muscular systems have been developed. *See* BIOELECTRIC MODEL.

Thus, with the computer the ability of the mathematician to treat component atomic processes quantitatively can be extended to the construction of very complex systems of these component processes. With computers, in other words, the process of reductionism, whereby biological systems are reduced to simpler physical and chemical processes, has been reversed, and biological phenomena can be treated at an appropriate level of complexity. [G.P.M.]

Mathematical ecology The application of mathematical theory and technique to ecology. The earliest studies in ecology were by naturalists interested in organisms and their relationships to the environment. Such investigations continue to this day as an active and central part of the subject, and have focused attention on understanding the ecological and evolutionary relationships among species. A large mathematical literature has arisen as an aid to the formalization of concepts and ideas; the underlying goal has been to understand patterns of nature. For the most part, such approaches are retrospective, designed to help in understanding how current ecological relationships developed, and to place that development within appropriate evolutionary context. *See* ECOLOGY.

The first important mathematical efforts in ecology were made by P. F. Verhulst and others to describe the dependence of population growth rate on population size and to infer the consequences of such relations. Models of the type developed by Verhulst are called nonlinear, because the graph of growth rate versus population size is not a straight line, but typically bends over due to increased competitive effects.

The most basic models of growth are iterations, in which population size is recomputed at consecutive times in accordance with a basic underlying growth model. Such a model can be expressed in mathematical form by the difference equation [Eq. (1)], where $N(t)$ is the population size at time t,

$$N(t + h) - N(t) = F(N(t))h \qquad (1)$$

$N(t + h)$ is the size h units later, and the function F defines the growth rule. The time period for which the population growth rate remains fixed determines the characteristics of growth. If the time period h is short, a population governed by such a law typically will come into balance with its environment. However, if the time period is too long, the adjustment of the growth rate is delayed and the population may undergo oscillations, continuing to increase long after the resources have been depleted, and then declining until resources recover. When h is small, a powerful approximation to Eq. (1) can be obtained by using differential calculus; in particular, the limiting form is the differential equation [Eq. (2)]. If F is continuous,

$$\frac{dN}{d}t = F(N) \qquad (2)$$

solutions to Eq. (2) never exhibit oscillations, since they approximate the situation for small h.

Models of the sort just described are the simplest possible—they ignore all complications due to the fact that a population is composed of a heterogeneous assemblage of individuals, and further that it interacts with other populations and with its abiotic environment. Thus, although these simple models have found application in applied areas such as fishery science, in which one seeks to develop models relating one year's fish population size to the previous year's spawning stock, it is well recognized that more precise answers require consideration of various complications. [S.A.L.]

Mathematical geography The branch of physical geography that deals with attributes of the Earth, such as its size, shape, and movements, and with those relations to the heavenly bodies, especially the Sun, that affect the features of the face of the Earth.

The Earth is a nearly spherical body, 24,901.57 mi (40,075.19 km) in circumference, rotating on an axis, and revolving in an orbit around the Sun. The equatorial diameter is 7,926.41 mi (12,756.33 km) and the polar diameter is 7,899.84 mi (12,713.56 km). Its shape may be better classed as an oblate spheroid. *See* EARTH; GEODESY.

Rotation and revolution are the two major motions of the Earth. The Earth rotates on an axis, the ends of which are the poles, making one complete rotation in 24 h, with a speed at the Equator of about 1000 mi/h (464 m/s). The rotation is from west to east (in Earth-bound terms), making night and day, and the impression that heavenly bodies rise in the east and set in the west.

The rotating Earth revolves in an orbit around the Sun, making one complete revolution in 365 days, 5 h, 48 min, 45.15 s, or about $365\frac{1}{4}$ days. *See* INTERNATIONAL DATE LINE; TIME.

The positions and changes of the Earth's axis are most important factors in Earth-Sun relationships. As the Earth moves through its orbit, the axis of rotation is constantly inclined at about $66\frac{1}{2}°$ (66.55°) from the plane of the ecliptic, or $23\frac{1}{2}°$ (23.45°) from a line perpendicular to the ecliptic. In addition to maintaining its inclination with the ecliptic, the polar axis of rotation always points to the same spot in the heavens throughout its yearly revolution around the Sun. *See* PRECESSION OF EQUINOXES.

From June 21, when the Sun's rays are vertical at the Tropic of Cancer (summer solstice), summer continues in the Northern Hemisphere until September 23, when the Sun's rays strike directly over the Equator (autumnal equinox). Fall begins with the Sun overhead at the Equator and continues until December 22 (winter solstice), when the Sun has reached its southern zenithal position at the Tropic of Capricorn, $23\frac{1}{2}°$ south. Winter finds the overhead Sun's rays moving back toward the Equator, which is attained on March 21 (vernal equinox). Spring is that period when the Sun is again overhead north of the Equator, from March 21 to June 21. *See* EQUINOX; NORTH POLE; SOLSTICE; SOUTH POLE.

Although less important to humans than Earth-Sun relationships, the Moon plays a significant role as the major force in the production and timing of tides. The tide is due essentially to the gravitational pull of the Moon; the greatest force will be on the Earth right under the Moon and at a directly opposite point on the other side of the Earth. The Sun also exerts gravitational pull on the Earth's waters and produces tidal effects, but these are much less marked than the lunar tides. *See* TIDE. [V.H.E.]

Mathematical physics The area of physics aimed at deducing the consequences of the more established physical theories by relying mainly on the method of mathematical solution, presuming that the basic laws of physics are known. A fruitful approach is possible largely because there is close analogy between the mathematical problems arising in different fields of theoretical physics. The same partial differential equations are encountered in many different contexts.

Through World War II, the main technique of mathematical physics was the analytical mathematical solution of problems. Since World War II, high-speed computing machines have become increasingly important and have made numerical solutions possible for many problems where the analytical technique did not work.

The term mathematical physics is sometimes used synonymously with theoretical physics. *See* THEORETICAL PHYSICS. [H.A.Be.]

Mathematical software Software that is designed to solve equations, analyze experimental data, evaluate integrals, and solve similar computational problems. In the early days of computing, users entirely wrote their own software, including the software for these problems. As scientific computation has become more demanding, however, users have depended increasingly on software written by specialists. There are now software libraries covering a full range of mathematical problems, domain-specific libraries, and entire systems of interactive mathematical computation.

A common form of mathematical software is the subroutine library. From such a library the user can select subroutines and easily incorporate them into a program, thus saving time and usually yielding a more accurate and efficient program. Library subroutines are grouped into areas such as roots of polynomials, evaluation of integrals, solution of ordinary differential equations, solution of partial differential equations, solution of linear equations, statistical computations, and special functions. There may be a dozen or more subroutines within an area, each treating a different subproblem. A domain-specific library covers one of these areas. Most libraries are written in Fortran, a language commonly used in scientific computation. *See* PROGRAMMING LANGUAGES.

An interactive system allows commands to be typed in the form of a mathematical expression in a notation similar to Fortran. While the operands can be scalars, they also can be matrices and vectors, thus making it easy to do matrix arithmetic and to solve matrix equations. Some of these systems allow algebraic manipulations and give the analytic solution of problems; they are sometimes called symbolic algebra systems. *See* SYMBOLIC COMPUTATION.

The precision of arithmetic and the range of representable numbers are important computer parameters affecting mathematical software. The design of mathematical software must take these arithmetic parameters into account so as to achieve the most accurate result allowed on the machine being used, and so as to avoid failure of the computation because of exceeding the range of representable numbers. Mathematical software libraries provide functions for obtaining arithmetic parameters for a given machine. *See* DIGITAL COMPUTER; PROGRAMMING LANGUAGES; SOFTWARE. [L.D.F.]

Mathematics Mathematics is frequently encountered in association and interaction with astronomy, physics, and other branches of natural science, and it also has deep-rooted affinities to the humanities. It is a realm of knowledge entirely unto itself, and one of considerable scope.

Relation to science. Mathematics is not a branch of natural science itself. It does not deal with phenomena and objects of the external world and their relations to each other but, strictly speaking, only with objects and relations of its own imagery. One can practice meaningful mathematics without being concerned with science at all, and philosophical attempts to reduce all origin of mathematics to utilitarian motives are wholly unconvincing. However, mathematics is the language of science in a deep sense. Mathematics is an indispensable medium by which and within which science expresses, formulates, continues, and communicates itself. And just as the language of true literacy not only specifies and expresses thoughts and processes of thinking but also creates them in turn, so does mathematics not only specify, clarify, and make rigorously workable concepts and laws of science, but also at certain crucial instances becomes an indispensable constituent of their creation and emergence as well.

Creative formulas. A formula is a string of mathematical symbols subject only to certain general rules of composition. To a working mathematician a string of symbols is a formula if it is something worth remembering. Much mathematics is concentrated in and propelled by certain formulas of unusual import.

Foundations—mathematical logic. A prime demand on mathematics is that it be deductively rigorous, and a traditional model for intended rigor is Euclid's presentation of mathematical assertions in theorems. A theorem is a proposition which has been proved, excepting certain first theorems called axioms, which are admitted without proof; and to prove a theorem means to obtain it from other theorems by certain procedures of deduction or inference. It had long been commonplace that each branch of mathematics was based on its own axioms, but during the 19th century, mathematicians arrived at the insight that even the same branch might have alternate axioms. Specifically, there were envisaged alternate versions of two- and three-dimensional geometry, the axiom varied being the axiom on parallels. It was also recognized that a set of axioms becomes mathematically possible if it is logically consistent, that is, if one cannot deduce from the axioms to theorems one of which, as a proposition, is the negation of the other. *See* EUCLIDEAN GEOMETRY.

At the same time certain developments led to the realization that not only the axioms but the rules of inference themselves might be, and even ought to be, subjected to variations. Now if axioms and rules of inferences are both viewed as subject to change, it is customary to speak of a mathematical system or also a formal system, and, of course, an irreducible first requirement is that the system be consistent after the manner just stated. Consistency alone is a somewhat negative property. There is a further property, called completeness, which is more positive, and which, if present, is very welcome. A system is complete if for any proposition which can be formulated it either can be proved that it holds or that its negation holds.

Some of the developments that led to doubt as to whether the traditional rules of inference are inviolate were the following.

1. G. Boole had found in 1854 that the classical Aristotelian connectives "and," "or," "negation of" for propositions follow rules similar to those which the operations addition, multiplication, "the negative of" obey in ordinary algebra (Boole's algebra of propositions); his conclusions took from rules of inference the status of untouchability.

2. G. Cantor, the founder of the theory of sets and operations between them, defined a set (intuitively or naively) as the collection of all objects having a certain property which is verbally expressible. Especially, "the set of all sets" is again a set and it has the peculiarity of being a set which contains itself as one of its elements. But this leads to the following contradictory situation (Russell's paradox): Divide the totality of all possible sets into two categories. A set shall belong to category I if it does not contain itself as an element, and to category II if it does contain itself as an element. Now form the set M whose elements are the sets of category I. It can now be reasoned by deductive steps admissible in Cantor's own theory that the set M cannot belong to either of the two categories, although the original division into categories did assign each set to one of them.

3. In 1904 E. Zermelo formulated the following axiom of choice: Given any family of nonvacuous sets $\{S\}$, no matter

how (infinitely) large the family may be, it is possible to choose simultaneously an element $x = x_s$ from each given set S and thus to consider the set M consisting of precisely these elements. By the use of this axiom some striking theorems in classical mathematics could be proved which, without the use of the axiom, seemed to be logically out of reach entirely. Mathematicians began to wonder whether a theorem based on the axiom of choice is indeed valid or, at any rate, whether it has the same level of validity as one without it, and as a consequence, theorems employing the axiom of choice were frequently labeled as such. *See* SET THEORY.

Some of the doubts were resolved eventually; the most striking results are the following ones of K. Gödel. (1) Any consistent mathematical system which is sufficient for classical arithmetic must be incomplete. (2) Any such system remains consistent if one adds to it the axiom of choice, so that working mathematics cannot disprove the axiom of choice. In 1963 P. Cohen showed that the axiom cannot be proved, either. *See* LOGIC.

Constructiveness. Some mathematicians object to mere existence proofs, and they demand that any proof also be constructive. The interpretations of this demand differ widely. Some proofs closely approach what a practical mathematician welcomes; if, for instance, a theorem asserts the existence of a number or a function, then the proof must also embody a procedure for actual computation of the solution, approximately, at least. Other versions are little more than the negative requirement that certain combinations of inference be avoided. There are also views which combine both; the best known among the last is the intuitionist view. It firmly demands a certain kind of constructiveness, which, however, does not necessarily guarantee the calculation by present-day computing machines. However, the actual stricture by which intuitionism became widely known is that proof by contradiction is not admissible. Proof by contradiction is also called proof by double negation, and it is equivalent to the Aristotelian law of the excluded middle. It assumes tentatively that the proposition to be proved is false and from this assumption deduces a contradiction to a previously established theorem.

Space in mathematics. If geometry is the mathematics of space, then, in a superficial sense, all mathematics began with geometry, because apparently it began with measurements of figures: length, area, volume, and size of angles. It did not concern itself with questions of shape but with clarifying and deciding when figures are equal or substantially equal with regard to form. The first true theory of geometry was that of the Greeks, whose primary concern was study of the basic concept of equality of figures—their congruence and similarity—and the Greeks were so determined to dissociate their theory from the preceding phase of merely making measurements that Euclid's extensive work, for instance, avoids to a fault any kind of actual measurements. But, for all its lofty purposes, Greek geometry was too rigid and circumscribed to be able really to cope with the mathematical problem of space. Geometry did not progress further until, with the advent of coordinate systems, introduced by Descartes and his predecessors, a better mathematics of space could be initiated.

If a cartesian coordinate system is etched into two- or three-dimensional euclidean space, then the space becomes a point set, each point being a pair (x^1, x^2) or a triple (x^1, x^2, x^3) of real numbers, and any figure a suitable subset of it. This is a deliberate process of arithmetization of space which unifies space and number at the base. It does not hamper geometry in its task of pursuing problems of shape but instead aids it. In the cartesian plane, two figures are similar if the points of one can be obtained from the points of the other by means of a transformation, Eqs. (1), where Eq. (2) applies and where, for some $\rho < 0$, Eq. (3) can be written. The similarity is a congru-

$$y^1 = a^1 + \alpha_1{}^1 x^1 + \alpha_2{}^1 x^2$$
$$y^2 = a^2 + \alpha_1{}^2 x^1 + \alpha_2{}^2 x^2 \qquad (1)$$

$$\alpha_1{}^1 \cdot \alpha_1{}^2 + \alpha_2{}^1 \cdot \alpha_2{}^2 = 0 \qquad (2)$$

$$(\alpha_1{}^1)^2 + (\alpha_2{}^1)^2 = (\alpha_1{}^2)^2 + (\alpha_2{}^2)^2 = \rho^2 \qquad (3)$$

ence if, and only if, $\rho = 1$ (orthogonal transformation). Now this analytic representation of congruence and similarity suggests a geometric examination of the most general linear transformations, Eqs. (1), which are nonsingular, that is, for which the determinant $|\alpha_q{}^p| \neq 0$. They were virtually unknown to the Greeks, although they highlight the axiom of parallels of Euclid's geometry. A one-to-one transformation of the cartesian plane is such a linear transformation if, and only if, it carries a straight line onto a straight line and parallel straight lines into parallel straight lines.

The family of all linear transformations constitutes a transitive group, and the subfamily of orthogonal transformations is already a transitive group. F. Klein made the pronouncement, which is generally accepted, that there arises a geometry on a space if on the space there is given a transitive group of transformations; two figures are considered equal whenever one figure can be carried into the other figure by one of the transformations.

The arithmetization of space led to a purely mathematical creation of n-dimensional space, euclidean and other, for any integer dimension n, by defining its points generally as n-tuples of real numbers $(x^1,\ldots, x^n)$ with suitable definitions for various geometrical relations between such points. The best-known application of this was the four-dimensional space of the theory of relativity, but actually multidimensional geometry had been playing a part in physics before that. If a mechanical system involves M mass points, it was customary in effect to introduce the space of dimension $n = 3M$, whose points are the states of the system, that is, the n-tuples of coordinates $\{x_m{}^1, x_m{}^2, x_m{}^3\}$, $m = 1,\ldots, M$, at any one time point. Also, if there are restraints operative in the system, then the Lagrange-Hamilton theory suitably reduced the dimension of the space by the use of the free parameters of the system instead of the original n coordinates themselves. The use of free parameters spread from mechanical systems to other systems in physics and chemistry, and all so-called equations of state are geared to this. Finally, in quantum theory a state of a system has infinitely many coordinates and the infinitely dimensional space representing it is a Hilbert space. Also, partly under the influence of Hilbert space, mathematicians have become fascinated with infinitely dimensional spaces in general. They are being studied intensively, and large parts of mathematics are being pressed into these new frames of reference.

The arithmetization of space is also reflected in the ever-widening use of graphs and charts. *See* ALGEBRA; ANALYTIC GEOMETRY; CALCULUS OF VECTORS; COORDINATE SYSTEMS; GEOMETRY; PROBABILITY STATISTICS; TOPOLOGY. [S.Bo.]

Matrix isolation A technique for providing a means of maintaining molecules at low temperature for spectroscopic study. This method is particularly well suited for preserving reactive species in a solid, inert environment. Elusive molecular fragments, such as free radicals that may be postulated as important controlling intermediates for chemical transformations used in industrial reactions, high-temperature molecules that are in equilibrium with solids at very high temperatures, and molecular ions that are produced in plasma discharges or by high-energy radiation all can be examined by using absorption (infrared, visible, and ultraviolet), electron-spin resonance, and laser-excitation spectroscopes.

The experimental apparatus for matrix isolation experiments is designed with the method of generating the molecular tran-

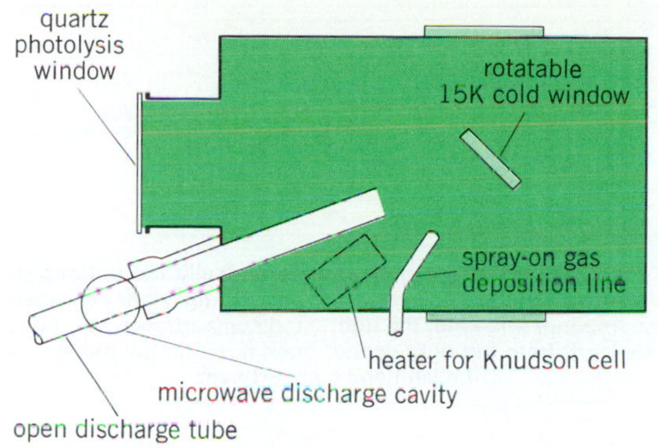

Vacuum-vessel base cross section for matrix photoionization experiments.

sient and performing the spectroscopy in mind. The illustration shows the cross section of a vacuum vessel used for absorption spectroscopic measurements. The matrix sample is introduced through the spray-on line; argon is the most widely used matrix gas, although neon, krypton, xenon, and nitrogen are also used. The reactive species can be generated in a number of ways: mercury-arc photolysis of a trapped precursor molecule through the quartz window, evaporation from a Knudsen cell in the heater, chemical reaction of atoms evaporated from the Knudsen cell with molecules deposited through the spray-on line, and vacuum-ultraviolet photolysis of molecules deposited from the spray-on line by radiation from discharge-excited atoms flowing through the tube. For laser excitation studies, the sample is deposited on a tilted copper wedge which is grazed by the laser beam, and light emitted or scattered at approximately 90° is examined by a spectrograph. In electron-spin resonance studies, the sample is condensed on a sapphire rod that can be lowered into the necessary waveguide and magnet.

The matrix isolation technique enables spectroscopic data to be obtained for reactive molecular fragments, many of which cannot be studied in the gas phase. [L.A.]

Matrix mechanics A formulation of quantum theory in which the operators are represented by time-dependent matrices. *See* MATRIX THEORY.

Matrix mechanics is not useful for obtaining quantitative solutions to actual problems; on the other hand, because it is concisely expressed in a form independent of special coordinate systems, matrix mechanics is useful for proving general theorems. [E.G.]

Matrix theory A matrix is a rectangular array of numbers, or other elements, of the form below. The array A is an m by

$$A = \begin{pmatrix} a_{11} & a_{12} & \cdots & a_{1n} \\ a_{21} & a_{22} & \cdots & a_{2n} \\ \cdots & \cdots & \cdots & \cdots \\ a_{m1} & a_{m2} & \cdots & a_{mn} \end{pmatrix}$$

n matrix with m rows and n columns, and the size of A is said to be m by n. The rows of a matrix are always numbered from the top down and the columns from left to right. The position of each element in the array is given by its subscripts; that is, a_{ij} is the element in the ith row and jth column. Since every element of A is represented by a_{ij}, as i takes on the values 1, 2,..., m and j the values 1, 2,..., n, a_{ij} is called the typical element of A, and the compact notation $A = (a_{ij})$ is used when the size of A is given.

Matrices have application as computational devices in such widely diversified fields as economics, psychology, statistics, engineering, physics, and mathematics. In mathematics, matrices are useful tools in the study of linear systems of algebraic equations, linear differential equations, linear mappings and transformations, and bilinear and quadratic forms.

If $m = n$ in (1), A is called a square matrix of order n. If $m = 1$, A is a row matrix, and if $n = 1$, A is a column matrix. The elements a_{ij} of A, for which $i = j$, are the principal diagonal elements. A diagonal matrix is a matrix such that $a_{ij} = 0$ if $i \neq j$, and a scalar matrix is a square diagonal matrix with equal diagonal elements. An identity matrix is a scalar matrix in which the common diagonal element is the number 1. An n by n identity matrix is denoted by I_n.

Matrices can be regarded as generalized numbers, and their utility in applications depends on the possibility of combining them in certain definite ways. The matrix operations of addition, subtraction, and multiplication are defined in terms of these same operations for the elements, and they satisfy some, but not all, of the rules of ordinary algebra. In discussing these matrix operations it is assumed that the elements of the matrices are numbers.

Two matrices $A = (a_{ij})$ and $B = (b_{ij})$ are equal if they have the same size m by n and a_{ij} for all i, j. Matrices $A = (a_{ij})$ and $B = (b_{ij})$ of the same size m by n are added by adding correspondingly placed elements; that is, $A + B = C = (c_{ij})$ is an m by n matrix where $c_{ij} = a_{ij} + b_{ij}$ for all i, j. The m by n matrix with 0 in every position is denoted by 0, and is called a null matrix. The matrix $-A = (-a_{ij})$ is the negative of the matrix $A = (a_{ij})$, and $A + (-A) = 0$. Subtraction of m by n matrices is defined by $B - A = B + (-A) = (b_{ij} - a_{ij})$.

A matrix $B = (b_{ij})$ is conformable with respect to a matrix $A = (a_{ij})$ if B has size n by q and A has size m by n; that is, B has the same number of rows as A has columns. The matrix product AB is defined only when B is conformable with respect to A. The product $C = AB$ is an m by q matrix and the element in the i, j position of C is obtained by multiplying the n elements in the ith row of A into the n elements in the jth column of B, term by term, and adding these products. Thus if

$$A = \begin{pmatrix} 2 & 0 & -1 \\ 4 & 1 & \tfrac{1}{2} \end{pmatrix} \quad \text{and} \quad B = \begin{pmatrix} 5 \\ 1 \\ -2 \end{pmatrix}$$

then

$$AB = \begin{pmatrix} 12 \\ 20 \end{pmatrix}$$

If A and B are square matrices of the same size, then both AB and BA are defined. Unlike the case for numbers, it may happen for matrices that $AB \neq BA$ and $AB = 0$ with $A \neq 0$ and $B \neq 0$.

The product of a matrix A and a number a is called a scalar product and is obtained by multiplying every element a_{ij} of A by a.

The transpose of an m by n matrix A is the n by m matrix B which has as its ith row the ith column of A and as its jth column the jth row of A for all i, j. If the transpose of a matrix A is denoted by A', and B is conformable with respect to A, then $(AB)' = B'A'$. A matrix A is symmetric if $A = A'$. A symmetric matrix is necessarily square.

A square n by n matrix is nonsingular if the determinant of A is not zero. Otherwise A is singular. A nonsingular matrix A has a unique inverse, that is, a matrix A^{-1} such that $AA^{-1} = A^{-1}A = I_n$. The inverse of a nonsingular matrix is easily described but is difficult to compute for matrices of large size. Many important applications require the calculation of the inverse, and numerical methods are used to approximate the elements of the inverse. *See* DETERMINANT. [R.A.Be.]

Matter (physics) The substance composing bodies perceptible to the senses. The distinguishing properties of matter are gravitation and inertia. Any entity exhibiting these properties when at rest is matter. Although electromagnetic radiation also possesses these properties to some extent, this radiation always moves with the speed of light. All material bodies have mass, which is a measure of inertia; every material body near the Earth's surface has weight, which is a measure of the Earth's gravitational attraction for the body. *See* GRAVITATION; INERTIA; MASS; WEIGHT. [D.Wi.]

Matthiessen's rule An empirical rule which states that the total resistivity of a crystalline metallic specimen is the sum of the resistivity due to thermal agitation of the metal ions of the lattice and the resistivity due to the presence of imperfections in the crystal. This rule is a basis for understanding the resistivity behavior of metals and alloys at low temperatures.

The resistivity of a metal results from the scattering of conduction electrons. Lattice vibrations scatter electrons because the vibrations distort the crystal. Imperfections such as impurity atoms, interstitials, dislocations, and grain boundaries scatter conduction electrons because in their immediate vicinity the electrostatic potential differs from that of the perfect crystal. [F.J.B.]

Maxillopoda A class of Crustacea whose application is gaining moderately widespread use, despite the fact that its validity is not universally accepted by researchers. The constituent subclasses of the Maxillopoda also remain unsettled.

The class Maxillopoda was proposed for those taxa with six thoracic somites (with some exceptions), a well-developed mandibular palp in adults, well-developed maxillules and maxillae adapted for filtering in filter feeders, and the lack of gnathobases on the appendages of the thorax. The Recent taxa included were the Copepoda, Branchiura, Mystacocarida, and provisionally, the Cirripedia. Subsequently the Cirripedia (subdivided into Cirripedia sensu stricto and Ascothoracida) were unequivocally incorporated. The Ostracoda are now included by some but excluded by others. More recent suggestions have included the Tantulocarida, and even the probably noncrustacean Pentastomida, within the Maxillopoda. *See* BRANCHIURA; CIRRIPEDIA; COPEPODA; CRUSTACEA; OSTRACODA; PENTASTOMIDA; TANTULOCARIDA. [P.A.McL.]

Maxwell's demon An imaginary being whose action appears to contradict the second law of thermodynamics, which identifies the natural direction of change with the direction of increasing entropy. There has always been a certain degree of discomfort associated with the acceptance of the law, particularly in relation to the time reversibility of physical laws and the role of molecular fluctuations. In 1867, J. C. Maxwell considered, in this connection, the action of "a finite being who knows the paths and velocities of all the molecules by inspection." This being was later referred to as a demon by Lord Kelvin, and the usage has been generally adopted. *See* ENTROPY; THERMODYNAMIC PRINCIPLES; TIME, ARROW OF.

The activity of Maxwell's demon can be modeled by a trapdoor in a partition between two regions full of gas at the same pressure and temperature. The trapdoor needs to be restrained by a light spring to ensure that it is closed unless it is struck by molecules traveling from the left (see illus.). Its hinging is such that molecules traveling from the right cannot open it. The essential point of Maxwell's vision was that molecules striking the trapdoor from the left would be able to penetrate into the right-hand region but those present on the right would not be able to escape back into the left-hand region. Therefore, the initial equilibrium state of the two regions, that of equal pressures, would be slowly replaced by a state in which the two regions acquired different pressures as molecules accumulated in the right-hand region at the expense of the left-hand region. Only a slightly more elaborate mechanical arrangement is needed to change the apparatus to one in which the temperatures of the two regions move apart. In each case, the demonic trapdoor appears to be contriving a change that is contrary to the second law, for an implication of that law is that systems in either mechanical equilibrium (at the same pressure) or thermal equilibrium (at the same temperature) cannot spontaneously diverge from equilibrium.

As frequently occurs in science, the resolution of a paradox or the elimination of an apparent conflict with a firmly based law depends on a detailed analysis of the proposed arrangement. Numerous analyses of this kind have shown that the activities of Maxwell's demon do not in fact result in the overthrow of the second law. [P.W.A.]

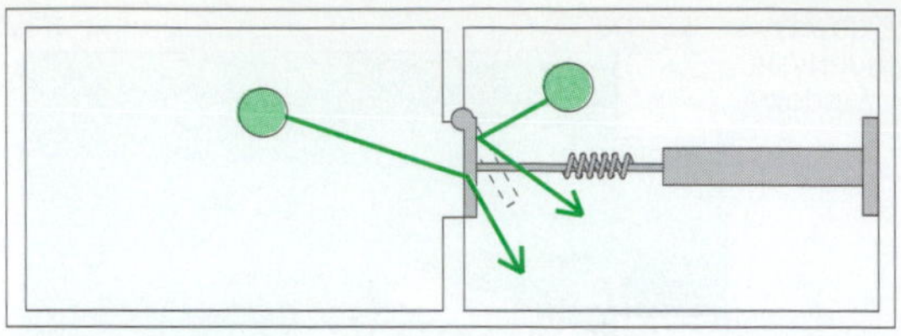

Type of device that emulates mechanically the actions of Maxwell's demon. Molecules traveling to the right can open the trapdoor and enter the right-hand compartment, but those striking it from the right cannot open it, so do not move into the left-hand compartment.

Maxwell's equations Four differential equations proposed by James Clerk Maxwell in 1864 as the basis of the theory of electromagnetic waves. They may be written, in vector notation, as Eqs. (1)–(4), where **D** is the electric displacement, **B** the magnetic flux density, **E** the electric field strength or intensity, **H** the magnetic field strength or intensity, ρ the charge density, and **i** the current density.

$$\nabla \cdot \mathbf{D} = \rho \qquad (1)$$

$$\nabla \cdot \mathbf{B} = 0 \qquad (2)$$

$$\nabla \times \mathbf{E} = -\frac{\delta \mathbf{B}}{\delta t} \qquad (3)$$

$$\nabla \times \mathbf{H} = \mathbf{i} + \frac{\delta \mathbf{D}}{\delta t} \qquad (4)$$

The first equation states that electric flux lines, if they end at all, will do so on electric charges. The second states that magnetic flux lines never terminate. The third is a form of Faraday's law of induction, which states that the rate of change of the magnetic flux threading a circuit equals the electromotive force or line integral of **E** around the circuit. The fourth integral is based partially on A. M. Ampère's experiments on steady currents which show that the line integral of the magnetic intensity **H** (or $\mathbf{B}/\mu$, where μ is the permeability) around a closed curve equals the current encircled. *See* DISPLACEMENT CURRENT; EQUATION OF CONTINUITY; STOKES' THEOREM. [W.R.Sm.]

Mean effective pressure A term commonly used in the evaluation for positive displacement machinery performance which expresses the average net pressure difference in pounds per square inch (psi) on the two sides of the piston in engines, pumps, and compressors. It is also known as mean pressure and is abbreviated as mep or mp.

In an engine (prime mover) it is the average pressure which urges the piston forward on its stroke. In a pump or compressor it is the average pressure which must be overcome, through the driver, to move the piston against the fluid resistance.

The criterion of mep is a vitally convenient device for the evaluation of a reciprocating engine, pump, or compressor design as judged by initial cost, space occupied, and deadweight. See COMPRESSOR; DIESEL CYCLE; THERMODYNAMIC CYCLE; VAPOR CYCLE. [T.Ba.]

Mean free path The average distance traveled between two similar events. The concept of mean free path is met in all fields of science and is classified by the events which take place. The concept is most useful in systems which can be treated statistically, and is most frequently used in the theoretical interpretation of transport phenomena in gases and solids, such as diffusion, viscosity, heat conduction, and electrical conduction. The types of mean free paths which are used most frequently are for elastic collisions of molecules in a gas, of electrons in a crystal, of phonons in a crystal, and of neutrons in a moderator. See KINETIC THEORY OF MATTER. [W.D.W.]

Measles An acute, highly infectious viral disease, with cough, fever, and maculopapular rash. It is of worldwide endemicity.

The virus enters the body via the respiratory system, multiplies there, and circulates in the blood. Prodromal cough, sneezing, conjunctivitis, photophobia, and fever occur, with Koplik's spots in the mouth. A rash appears after 14 days' incubation and persists 5–10 days. Serious complications may occur in 1 of every 15 persons; these are mostly respiratory (bronchitis or pneumonia), but neurological complications are also found. Encephalomyelitis occurs, but it is rare. Permanent disabilities may ensue for a significant number of persons.

Live attenuated virus vaccine can effectively prevent measles; vaccine-induced antibodies persist for years. In unvaccinated populations, immunizing infections occur in early childhood during epidemics which recur after 2–3 years' accumulation of susceptible children. Transmission is by coughing or sneezing. Measles is spread chiefly by children during the catarrhal prodromal period; it is infectious from the onset of symptoms until a few days after the rash has appeared. By the age of 20 over 80% of persons have had measles. Second attacks occur but are very rare. Treatment is symptomatic. [J.L.Me.]

Measure A reference sample used in comparing lengths, areas, volumes, masses, and the like. The measures employed in scientific work are based on the international units of length, mass, and time—the meter, the kilogram, and the second—but decimal multiples and submultiples are commonly employed. Prior to the development of the international metric system, many special-purpose systems of measures had evolved and many still survive, especially in Great Britain and the United States. See METRIC SYSTEM; PHYSICAL MEASUREMENT; TIME; UNITS OF MEASUREMENT; WEIGHT. [D.Wi.]

Measure theory A branch of mathematical analysis connected with the theory of integration. In order to discuss this subject, a formal definition of the term measure must be given.

Let X be an arbitrary set. Let m be a fixed collection of subsets of X satisfying the following conditions:

1. $\phi \in m$. (ϕ is the empty set. The symbol $\in$ indicates that ϕ is an element of m).

2. If $A \in m$, then $A^c \in m$. (A^c is the complement of A. It consists of those elements of X which do not belong to A.)

3. If $A_1, A_2, \ldots \in m$, then

$$\bigcup_{k=1}^{\infty} A_k \in m \tag{1}$$

(This set is the union of $A_1, A_2, \ldots$. It consists of those elements of X which belong to at least one of the sets $A_1, A_2, \ldots$.)

In this situation the collection m is called a σ-algebra. Here the term algebra refers to the various set operations (complementation, union, intersection), and the prefix σ to the fact that countably many such operations can be performed with sets in m and still result in sets in m. For example, the three properties mentioned above imply another property:

4. If $A_1, A_2, \ldots \in m$, then

$$\bigcap_{k=1}^{\infty} A_k \in m \tag{2}$$

(This set is the intersection of $A_1, A_2, \ldots$. It consists of those elements of X which belong to all of the sets $A_1, A_2, \ldots$.) See SET THEORY.

Now suppose that X is a set and m is a particular σ-algebra of subsets of X. A measure μ is a function which assigns to each set in m a certain nonnegative real number (or $+\infty$) and which satisfies the following conditions:

1. $\mu(\phi) = 0$.

2. If A and $B \in m$ and are disjoint ($A \cap B = \phi$), then Eq. (3) holds.

$$\mu(A \cup B) = \mu(A) + \mu(B) \tag{3}$$

3. If $A_1, A_2, \ldots \in m$, then Eq. (4) holds.

$$\mu\left(\bigcup_{k=1}^{\infty} A_k\right) \leq \sum_{k=1}^{\infty} \mu(A_k) \tag{4}$$

Various other properties can be derived from these conditions, such as:

4. If A and $B \in m$ and $A \subset B$ (that is, A is a subset of B), then $[\mu](A) \leq \mu(B)$.

5. If $A_1, A_2, \ldots \in m$ and are mutually disjoint ($A_i \cap A_j = \phi$ if $i \neq j$), then Eq. (5) holds.

$$\mu\left(\bigcup_{k=1}^{\infty} A_k\right) = \sum_{k=1}^{\infty} \mu(A_k) \tag{5}$$

The last property is of crucial importance and is called the countable additivity property of the measure μ.

Measure theory has a great number of important applications. Undoubtedly, the most important is the application to integration. See INTEGRATION. [B.F.J.]

Measured daywork A tool used primarily in manufacturing facilities as a control device to measure productive output in relation to labor input within a specific time period. The measurement of the work content is accomplished through the use of time standards which are usually the result of a stopwatch time study, predetermined time standards (methods-time measurement, the work-factor system), or some other form of work-measurement technique designed to measure tasks of labor under normal and average conditions.

A measured daywork plan is similar to incentive pay plans inasmuch as in both plans, time standards are used as a device to measure operator performance and also for various forms of management planning. However, in a measured daywork plan, worker income is based on a fixed hourly rate established by management, and is usually affected only by job classification, shift premiums, and overtime adjustments. Because of the fixed hourly rate in a measured daywork plan, there is little incentive for a worker to exceed a normal or standard level of performance or productivity. On the other hand, time standards are more readily acceptable, and become less an item of contention to the employee and bargaining unit (union).

The term daywork as used in industry denotes a fixed hourly rate that is not raised or lowered by varying worker performance levels. The hourly rate for a particular job should be a fair one relative to other jobs in the shop, and should also be comparable to rates of pay for similar jobs in the industrial community.

Once measured labor time standards have been established for shop operations, in addition to evaluating operator performance and identifying labor costs, new-product costs can also be determined prior to release to production, worker-power planning and scheduling can be done, equipment capacity requirements can be identified, and planning and make/buy decisions can be facilitated. *See* PERFORMANCE RATING; PRODUCTIVITY; WAGE INCENTIVES; WORK MEASUREMENT. [D.S.]

Mechanical advantage Ratio of the force exerted by a machine (the output) to the force exerted on the machine, usually by an operator (the input). The term is useful in discussing a simple machine, where it becomes a figure of merit. It is not particularly useful, however, when applied to more complicated machines, where other considerations become more important than a simple ratio of forces. *See* EFFICIENCY; SIMPLE MACHINE. [R.M.Ph.]

Mechanical alloying A materials-processing method that involves the repeated welding, fracturing, and rewelding of a mixture of powder particles, generally in a high-energy ball mill, to produce a controlled, extremely fine microstructure. The mechanical alloying technique allows alloying of elements that are difficult or impossible to combine by conventional melting methods. In general, the process can be viewed as a means of assembling metal constituents with a controlled microstructure. If two metals will form a solid solution, mechanical alloying can be used to achieve this state without the need for a high-temperature excursion. Conversely, if the two metals are insoluble in the liquid or solid state, an extremely fine dispersion of one of the metals in the other can be accomplished. The process of mechanical alloying was originally developed as a means of overcoming the disadvantages associated with using powder metallurgy to alloy elements that are difficult to combine. *See* CRUSHING AND PULVERIZING; POWDER METALLURGY; SOLID SOLUTION.

Some oxides are insoluble in molten metals. Mechanical alloying provides a means of dispersing these oxides in the metals. Examples are nickel-based superalloys strengthened with dispersed thorium oxide or yttrium oxide (Y_2O_3). These superalloys have excellent strength and corrosion resistance at elevated temperatures, making them attractive candidate materials for use in applications such as jet-engine turbine blades, vanes, and combustors. A number of other potential applications for mechanical alloying material are being explored, including powders for coating applications, alloys of immiscible systems, amorphous alloys, intermetallics, cermets, and organic-ceramic-metallic material systems in general. *See* AMORPHOUS SOLID; CERMET; HIGH-TEMPERATURE MATERIALS; INTERMETALLIC COMPOUNDS; METAL COATINGS.

Liquids or solid immiscible systems are difficult to process by conventional pyrometallurgy; mechanical alloying provides a route to obtain a homogeneous distribution in the solid phase. *See* PYROMETALLURGY. [F.H.Fr.]

Mechanical classification A sorting operation in which mixtures of particles of mixed sizes, and often of different specific gravities, are separated into fractions by the action of a stream of fluid. Water is ordinarily used as the sorting fluid, but other liquids or air or other gases may be used (see illustration).

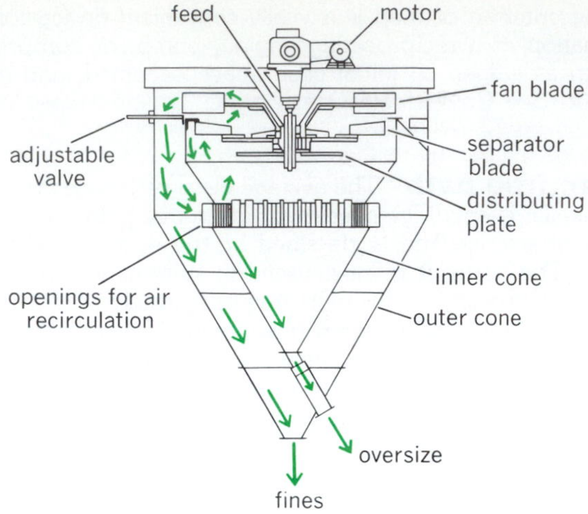

The double-cone air separator. *(After W. L. McCabe and J. C. Smith, Unit Operations of Chemical Engineering, McGraw-Hill, 3rd ed., 1975)*

The main objective of classification is to separate the particles according to size. This function is identical to that of screening, but classification is applicable to smaller particles, especially those that are undersize. For small particles it is more economical than screening. In classification the oversize and undersize are called sands and slimes, respectively.

Material also may be mechanically classified by specific gravity, a method that separates substances differing in chemical composition. This is called hydraulic separation. Such classification is based on the fact that, in a fluid, particles of the same specific gravity but of different size or shape settle at different constant speeds. Large, heavy, round particles settle faster than small, light, needlelike ones. If the particles also differ in specific gravity, the speed of settling is further affected. This is the basis for the separation of particles by kind rather than by size alone. *See* FLOTATION; MECHANICAL SEPARATION TECHNIQUES; UNIT OPERATIONS. [W.L.McC.]

Mechanical engineering One of several recognized fields of engineering. To grasp the meaning of mechanical engineering, it is desirable to take a close look at what engineering really is. The Engineers' Council for Professional Development has defined engineering as the profession in which a knowledge of the mathematical and physical sciences gained by study, experience, and practice is applied with judgment to develop ways to utilize economically the materials and forces of nature for the progressive well-being of mankind. It is a profession in which study in mathematics and science is blended with experience and judgment for the production of useful things.

Formal training of a mechanical engineer includes mastery of mathematics through the level of differential equations. Training in physical science embraces chemistry, physics, mechanics of materials, fluid mechanics, thermodynamics, statics, and dynamics. *See* ENGINEERING; MACHINERY; MASS PRODUCTION; TECHNOLOGY. [R.S.S.]

Mechanical impedance For a system executing simple harmonic motion, the mechanical impedance is the ratio of force to particle velocity. If the force is that which drives the system and the velocity is that of the point of application of the force, the ratio is the input or driving-point impedance. If the velocity is that at some other point, the ratio is the transfer impedance corresponding to the two points.

Mechanical impedance is a complex quantity. The real part, the mechanical resistance, is independent of frequency if the dissipative forces are proportional to velocity; the imaginary part, the mechanical reactance, varies with frequency, becoming zero at the resonant and infinite at the antiresonant frequencies of the system. *See* FORCED OSCILLATION; HARMONIC MOTION. [M.Gr.]

Mechanical rectifier A device which uses a synchronously operated mechanical switch to convert a single-phase or polyphase alternating voltage to a direct voltage. Single-phase mechanical rectifiers are made for small current output and are normally called vibrators. For large values of power where low voltages (less than 600 volts) are desired, the polyphase mechanical rectifier is used. This low-voltage device has higher efficiency than electronic rectifiers, which have appreciable voltage drop across the arc. These devices are not commonly used in power system applications. *See* RECTIFIER; VIBRATOR. [A.G.C.]

Mechanical separation techniques A group of laboratory and production operations whereby the components of a polyphase mixture are separated by mechanical methods into two or more fractions of different mechanical characteristics. The separated fractions may be homogeneous or heterogeneous, particulate or nonparticulate.

Types of mechanical separator

Materials separated	Separators
Liquid from liquid	Settling tanks, liquid cyclones, centrifugal decanters, coalescers
Gas from liquid	Still tanks, deaerators, foam breakers
Liquid from gas	Settling chambers, cyclones, electrostatic precipitators, impingement separators
Solid from liquid	Filters, centrifugal filters, clarifiers, thickeners, sedimentation centrifuges, liquid cyclones, wet screens, magnetic separators
Liquid from solid	Presses, centrifugal extractors
Solid from gas	Settling chambers, air filters, bag filters, cyclones, impingement separators, electrostatic and high-tension precipitators
Solid from solid	
By size	Screens, air and wet classifiers, centrifugal classifiers
By other characteristics	Air and wet classifiers, centrifugal classifiers, jigs, tables, spiral concentrators, flotation cells, dense-medium separators, magnetic separators, electrostatic separators

The techniques of mechanical separation are based on differences in phase density, in phase fluidity, and in such mechanical properties of particles-as size, shape, and density; and on such particle characteristics as wettability, surface charge, and magnetic susceptibility. Obviously, such techniques are applicable only to the separation of phases in a heterogeneous mixture. They may be applied, however, to all kinds of mixtures containing two or more phases, whether they are liquid-liquid, liquid-gas, liquid-solid, gas-solid, solid-solid, or gas-liquid-solid.

Methods of mechanical separations fall into four general classes: (1) those employing a selective barrier such as a screen or filter cloth; (2) those depending on difference in phase density alone (hydrostatic separators); (3) those depending on fluid and particle mechanics; and (4) those depending on surface or electrical characteristics of particles. A wide variety of separation devices have been devised and are in use. The more important kinds of equipment are listed in the table, grouped according to the phases involved. *See* CENTRIFUGATION; CLARIFICATION; DUST AND MIST COLLECTION; ELECTROSTATIC PRECIPITATOR; FILTRATION; FLOTATION; MAGNETIC SEPARATION METHODS; MECHANICAL CLASSIFICATION; SCREENING; SEDIMENTATION (INDUSTRY); THICKENING. [S.A.M.]

Mechanical vibration The continuing motion, repetitive and often periodic, of a solid or liquid body within certain spatial limits. Vibration occurs frequently in a variety of natural phenomena such as the tidal motion of the oceans, in rotating and stationary machinery, in structures as varied in nature as buildings and ships, in vehicles, and in combinations of these various elements in larger systems. The sources of vibration and the types of vibratory motion and their propagation are subjects that are complicated and depend a great deal on the particular characteristics of the systems being examined. Further, there is strong coupling between the notions of mechanical vibration and the propagation of vibration and acoustic signals through both the ground and the air so as to create possible sources of discomfort, annoyance, and even physical damage to people and structures adjacent to a source of vibration.

Mass-spring-damper system. Although vibrational phenomena are complex, some basic principles can be recognized in a very simple linear model of a mass-spring-damper system (see illustration). Such a system contains a mass M, a spring

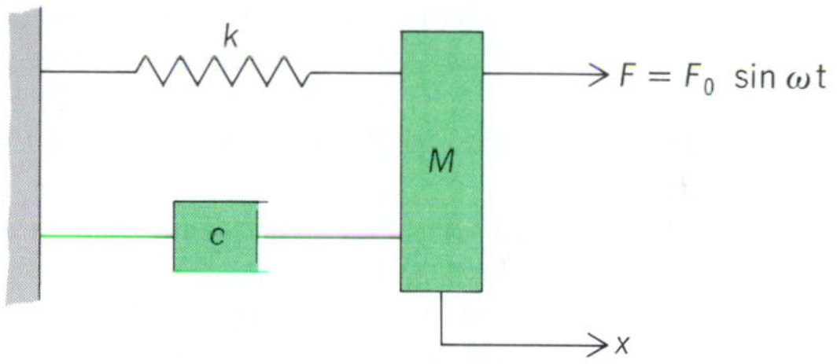

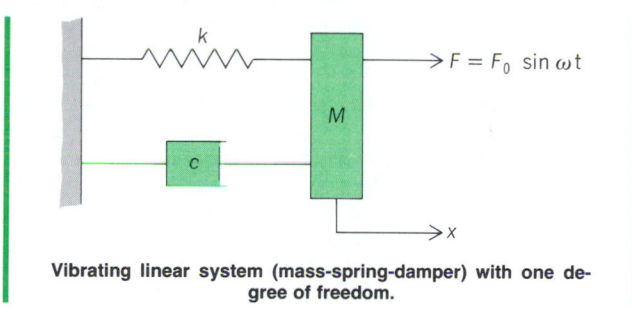

Vibrating linear system (mass-spring-damper) with one degree of freedom.

with spring constant k that serves to restore the mass to a neutral position, and a damping element which opposes the motion of the vibratory response with a force proportional to the velocity of the system, the constant of proportionality being the damping constant c. This damping force is dissipative in nature, and without its presence a response of this mass-spring system would be completely periodic. *See* DAMPING.

Complex systems. The foregoing model of the linear spring-mass-damper system contains within it a number of simplifications that do not reflect conditions of the real world in any obvious way. These simplifications include the periodicity of both the input and, to some extent, the response; the discrete nature of the input, that is, the assumption that it is temporal in nature with no reference to spatial distribution; and the assumption that only a single resonant frequency and a single set of parameters are required to describe the mass, the stiffness, and the damping. The real world is far more complex. Many sources of vibration are not periodic. These include impulsive forces and shock loading, wherein a force is suddenly applied for a very short time to a system; random excitations, wherein the signal fluctuates in time in such a way that its amplitude at any given instant can be expressed only in terms of a probabilistic expectation; and aperiodic motions, wherein the fluctuation in time may be some prescribed nonperiodic function or some other function that is not readily seen to be periodic.

Sources of vibration. There are many sources of mechanical and structural vibration that the engineer must contend with in both the analysis and the design of engineering systems. The most common form of mechanical vibration problem is motion induced by machinery of varying types, often but not always of the rotating variety. Other sources of vibration include: ground-borne propagation due to construction; vibration from heavy vehicles on conventional pavement as well as vibratory signals from the rail systems common in many metropolitan areas; and vibrations induced by natural phenomena, such as earthquakes and wind forces. Wave motion is a source of vibration in mechanical and structural systems associated with offshore structures.

Effect of vibrations. The most serious effect of vibration, especially in the case of machinery, is that sufficiently high alternating stresses can produce fatigue failure in machine and structural parts. Less serious effects include increased wear of parts, general malfunctioning of apparatus, and the propagation of vibration through foundations and buildings to locations where the vibration of its acoustic realization is intolerable either for human comfort or for the successful operation of sensitive measuring equipment. *See* ACOUSTIC NOISE; SOUND; VIBRATION; WEAR. [C.L.D.; J.P.D.H.]

Mechanics In its original sense, mechanics refers to the study of the behavior of systems under the action of forces. Mechanics is subdivided according to the types of systems and phenomena involved.

An important distinction is based on the size of the system. Those systems that are large enough can be adequately described by the newtonian laws of classical mechanics; in this category, for example, are celestial mechanics and fluid mechanics. On the other hand, the behavior of microscopic systems such as molecules, atoms, and nuclei can be interpreted only by the concepts and mathematical methods of quantum mechanics.

Mechanics may also be classified as nonrelativistic or relativistic mechanics, the latter applying to systems with material velocities comparable to the velocity of light. This distinction pertains to both classical and quantum mechanics.

Finally, statistical mechanics uses the methods of statistics for both classical and quantum systems containing very large numbers of similar subsystems to obtain their large-scale properties. *See* CLASSICAL FIELD THEORY; CLASSICAL MECHANICS; DYNAMICS; FLUID MECHANICS; QUANTUM MECHANICS; STATICS; STATISTICAL MECHANICS. [B.G.]

Mechanism Classically, a mechanical means for the conversion of motion, the transmission of power, or the control of these. Mechanisms are at the core of the workings of many machines and mechanical devices. In modern usage, mechanisms are not always limited to mechanical means. In addition to mechanical elements, they may include pneumatic, hydraulic, electrical, and electronic elements. In this article, the discussion of mechanism is limited to its classical meaning. *See* MACHINE.

Most mechanisms consist of combinations of a relatively small number of basic components. Of these, the most important are cams, gears, links, belts, chains, and logical mechanical elements. The last include such devices as ratchets, trips, detents, and interlocks. In order to understand how any mechanism works, their degree of freedom, structure, and kinematics must be considered. *See* BELT DRIVE; CAM MECHANISM; CHAIN DRIVE; ESCAPEMENT; GEAR; LINKAGE (MECHANISM); RATCHET.

Degree of freedom is conveniently illustrated for mechanisms with rigid links. The discussion is limited to mechanisms which obey the general degree-of-freedom equation,

$$F = \lambda(l - j - 1) + \sum f_i$$

where F = degree of freedom of mechanism, l = number of links of mechanism, j = number of joints of mechanism, f_i = degree of freedom of relative motion at ith joint, σ = summation symbol (summation over all joints), and λ = mobility number (the most common cases are $\lambda = 3$ for plane mechanisms and $\lambda = 6$ for spatial mechanisms). *See* DEGREE OF FREEDOM (MECHANICS).

The kinematic structure of a mechanism refers to the identification of the joint connection between its links. Just as chemical compounds can be represented by an abstract formula and electric circuits by schematic diagrams, the kinematic structure of mechanisms can be usefully represented by abstract diagrams. The structure of mechanisms for which each joint connects two links can be represented by a structural diagram, or graph, in which links are denoted by vertices, joints by edges, and in which the edge connection of vertices corresponds to the joint connection of links; edges are labeled according to joint type, and the fixed link is identified as well. Thus the graph of the slider-crank mechanism of illustration *a* is as shown in illustration *b*. In this figure the circle around vertex 1 signifies that link 1 is fixed.

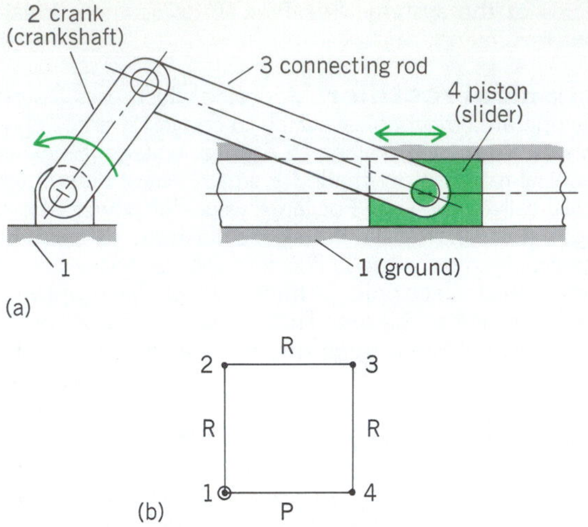

Slider-crank mechanism. (*a*) Mechanism. (*b*) Graph of mechanism. *R* = pin joint; *P* = sliding joint.

Kinematics is divided into kinematic analysis (analysis of a mechanism of given dimensions) and synthesis (determination of the proportions of a mechanism for given motion requirements). It includes the investigation of finite as well as infinitesimal displacements, velocities, accelerations and higher accelerations, and curvatures and higher curvatures in plane and three-dimensional motions. *See* KINEMATICS.

The design of mechanisms involves many factors. These include their structure, kinematics, dynamics, stress analysis, materials, lubrication, wear, tolerances, production considerations, control and actuation, vibrations, critical speeds, reliability, costs, and environmental considerations. Modern trends in the design of mechanisms emphasize economical design analysis by means of computer-aided design techniques. *See* COMPUTER-AIDED DESIGN AND MANUFACTURING. [F.F.]

Mechanoreceptors Receptors that provide the organism with information about such mechanical changes in the environment as movement, tension, and pressure. In higher animals receptors are actually the only means by which information of the surroundings is gained and by which reactions to environmental changes are stated. *See* SENSATION.

In primitive forms of life, such as unicellular organisms and sponges, the receptor and effecter functions are built into the same cell. The cell is directly excited by and reacts to the environmental stimulus. At higher levels of organization, the receptor separates from the effector and appears specialized in function and structure in the receptor cell of the primitive nerve system of actinians (sea anemones). Stimuli of the environment excite the receptor cell, and this excitation is conveyed to another specialized cell, the effector. In another step of phylogenetic development, a third element appears, the motor nerve cell, which serves as a link between the receptor and effector. This arrangement appears clearly in coelenterates and is carried through in its essential features up to the vertebrates.

Mechanoreceptors are excited by mechanical disturbances of their surroundings through deformation of their structure, through pressure or tension, or through a combination of these. In general, little energy is required for mechanical stimuli to cause a detectable excitation in mechanoreceptors.

From a physical point of view, mechanoreceptors are energy transducers; they convert mechanical into electrical energy, which in turn triggers the nerve impulse. Deformation leads to a sequence of events which may be summarized by the following scheme:

$$\text{Mechanical stimulus} \rightarrow \text{Generator current} \rightarrow \text{Nerve impulse (action potential)}$$

The generator current is the earliest detectable sign of excitation. The most salient characteristic of the generator current is its graded nature; its amplitude increases continuously, without visible steps, if the stimulus strength is progressively increased. When the generator current reaches a certain critical amplitude, an all-or-nothing potential is discharged in the sense organ which may then propagate as an all-or-nothing nerve impulse along the afferent axon of the receptor. *See* NERVOUS SYSTEM (INVERTEBRATE); NERVOUS SYSTEM (VERTEBRATE). [W.R.L.]

Mecoptera A small order of insects called the scorpion flies. Characteristic of the adult insect is the peculiar prolongation of the head into a beak, which bears chewing mouthparts. They are small to medium in size. The insects either have two pairs of large, net-veined wings of equal size, often with dark areas, or have short and aborted wings. The legs are long and slender. In some species, the male abdomen has a terminal enlargement which is held recurved over the back so that he resembles a scorpion, thus the common name. The Mecoptera are found in moist habitats within densely wooded areas. The adults are omnivorous but feed chiefly on small insects. *See* INSECTA. [B.E.R.]

Medical bacteriology The study of bacteria which affect human health, especially those which produce diseases known as infections, such as typhoid fever, plague, and tuberculosis. Sometimes called pathogens, certain bacteria are able to establish themselves in or on the body, where they grow and multiply, producing local or general functional upsets of the person thus infected. Medical bacteriologists study the whole process of infection but are primarily concerned with the role played by bacteria. They isolate suspected bacteria, relate them to the disease at hand, analyze their pertinent characteristics, study genetic changes which involve pathogenicity, and weigh other phases of prevention or therapy.

The identification of pathogenic bacteria is frequently used to establish the nature of a disease in a patient or for testing a response to antibiotics. It consists of searching for characteristics which will quickly and efficiently distinguish pathogenic bacteria from those with which they might be confused. Characteristics fall into four natural groups: morphology, metabolism, pathogenicity, and antigenicity. One or more of these groups may be necessary to establish identification. *See* PATHOGEN.

Bacteria may be members of a pathogenic species and still vary sharply in virulence. A strain or culture of bacteria which infects readily and produces severe effects is called virulent. One which, though similar in all other known ways, does not infect is called avirulent. A virulent organism which has been reduced in some degree from its expected virulence is termed attenuated. *See* VIRULENCE.

Whereas virulence focuses on the bacteria, immunity refers to variations in receptiveness (susceptibility) on the part of the infected person. Virulent bacteria may affect two persons quite differently. Immunity is a complex constitutional state which varies in degree both as to individual mechanisms and as to each specific infection. *See* IMMUNITY. [M.S.M.]

Medical chemical engineering The application of chemical engineering to medicine, frequently involving mass transport and separation processes, especially at the molecular level. Materials engineering, an area of specialization associated with chemical engineering, is frequently involved. *See* CHEMICAL ENGINEERING.

Major commercially available systems, devices, and products which have had significant input from chemical engineers include: hemodialyzers (the artificial kidney), peritoneal dialyzers, heart-lung machines (blood oxygenators), plasmapheresis equipment, controlled-release polymers (releasing drugs into living tissue), and synthetic suture material, which will dissolve at a programmed rate. The most important requirements for many of these systems and products are that there will be no unintended components dissolving out of the materials into blood or tissue and that the surface will not cause gross destruction of, or loss of, the cellular or molecular elements of tissue. [E.W.Me.]

Medical control systems Systems that adapt and utilize for medical application the methodologies, techniques, and devices developed for the control of physical systems. The concept of control usually suggests an autonomous or semiautonomous process that measures the difference between actual and desired performance of a system and, according to an analytically derived algorithm, manipulates energy to affect the system behavior to minimize the difference. Two types of application to medicine are apparent. The first incorporates the clinician (human operator) in the control loop. The second involves the development of an autonomous process for the control of one or more physiological variables. In the sequence leading to control action, namely, measurement—analysis—control, it is possible to make measurements of many medical indicators, expressed in physical terms, with almost arbitrary accuracy, speed, and precision.

Considerable advances have been made in the development of mechanized and semiautomated methods for measurement and analysis, the necessary precursors to effective control. Physiological monitoring during surgery, intensive therapy, or coronary care are representative examples. Another important area of measurement relates to the mechanization of routine medical diagnostic procedures.

Computer-based mathematical models are becoming available as decision aids to the clinician in diagnosing and treating disease. From the many physiological and clinical measures available, the clinician selects a set of observations for the particular patient. These are fed into the diagnostic subsystem. The resultant diagnostic statements and raw clinical observations are then fed to a decision maker which selects the next action by the clinician, either to carry out further tests or to start treatment. Again medical data are used as in diagnosis, but here local treatment objectives are employed. The treatment selected is fed to the relevant medical effector which applies the relevant responses to patient and environment.

The field of artificial organs, prostheses, and ortheses (orthopedic braces or devices), is an area where examples of direct-feedback control are being implemented. Developments in cardiac pacemakers imply that the autonomous regulation of heart rate may soon replace the current fixed-rate pacer and the variable-rate device that must be adjusted by the wearer. Other examples are the myoelectric-feedback control of artificial limbs and active devices that route neural signals around damaged pathways to operate distal members. *See* CONTROL SYSTEMS; PROSTHESIS. [E.R.Ca.]

Medical imaging A medical specialty that uses x-rays, gamma rays, high-frequency sound waves, and magnetic fields to produce images of organs and other internal structures of the body. In diagnostic radiology the purpose is to detect and diagnose disease, while in interventional radiology, imaging procedures are combined with other techniques to treat certain diseases and abnormalities. *See* RADIOLOGY.

Film x-ray studies, the most common radiologic procedures, are made up of still pictures of the various organs and tissues in the body. In these procedures, x-rays are passed through the body to expose the photographic film that is placed on the opposite side of the body. The changes in film density that result from exposure allow the radiologist to distinguish between normal and abnormal tissue and to diagnose many different disease types. *See* X-RAYS.

Fluoroscopy is a dynamic x-ray imaging technique that produces a moving image over time. It is essential for evaluating organ movement such as the beating of the heart or movement of the diaphragm. The gastrointestinal series and the barium enema are the most common fluoroscopic studies. These procedures begin with the administration of a barium mixture either by ingestion or by an enema that fills the stomach or large intestine. The barium mixture, known as a contrast medium, like dense tissues, blocks the x-ray beam. Fluoroscopy then reveals the location of the barium-coated lining of the stomach and intestine and enables the radiologist to observe as they contract and distend.

Angiography is the radiologic study of blood vessels. Because arteries and veins are not normally visible in conventional x-ray studies, an iodinated compound, which is opaque to the x-ray, must be injected into the bloodstream. An arteriogram is an x-ray study of the arteries; a venogram is an x-ray study of the veins. Arteriography is most often used to show the presence and extent that arteries have become clogged and narrowed by arteriosclerosis, which can lead to strokes and heart attacks.

Computed tomography (CT), also called computed axial tomography (CAT), is a scanning technique that combines computer and x-ray technologies. The computer constructs a two-dimensional anatomic image that represents a cross-sectional slice through the body. Three-dimensional images can be generated by using special computer software. These are especially useful in planning reconstructive orthopedic or plastic surgery. *See* COMPUTERIZED TOMOGRAPHY.

Ultrasound imaging, or sonography, is a diagnostic imaging procedure that uses high frequency sound waves instead of ionizing radiation. During an ultrasound examination, a lightweight transducer is placed on the patient's skin over the part to be imaged. The transducer produces sound waves that penetrate the skin to reach tissues and organs. When the sound waves strike specific tissue surface, echoes are produced. The echoes are detected by the transducer and are then electronically converted into an anatomic image that is displayed on a video screen. The image can also be recorded on film or videotape. Ultrasound imaging is commonly used in obstetrics to monitor the position and development of the fetus and also to detect any fetal abnormalities or problems in the pregnancy. Ultrasound is also used to show problems in other internal structures, including the gallbladder, kidney, and heart. Doppler ultrasound can monitor blood flow through veins and arteries. It is commonly used to study kidney transplants and blood flow to the brain and also to diagnose blocked arteries. *See* MEDICAL ULTRASONIC TOMOGRAPHY.

Magnetic resonance imaging (MRI) is a diagnostic procedure that uses a large, high-strength magnet, radio-frequency signals, and a computer to produce images. The technique of MRI is extremely useful in evaluating diseases of the brain and spine. It is also used to evaluate joints, bone and soft tissue abnormalities, as well as abnormalities of the chest, abdomen, and pelvis. *See* NUCLEAR MAGNETIC RESONANCE (NMR).

Nuclear medicine imaging studies use radioactive compounds called radionuclides of radiopharmaceuticals that emit gamma rays, or some emit beta particles. The chemicals are formulated so that they collect temporarily in the parts of the body to be studied. For most nuclear imaging studies, the radionuclide is injected into the patient and the images are taken with a gamma camera suspended above the patient who lies on a table. The camera detects the gamma rays emitted from the radionuclide in the patient's body and uses this information to produce an image that shows the distribution of the radionuclide within the body. The image is recorded on film and is called a scintigram or scan. Scintigrams of the heart and bone are the two most common nuclear medicine examinations.

The single-photon emission computed tomography (SPECT) examination uses a computer to obtain two-dimensional images that are thin slices of internal organs such as the heart, brain, and liver. The SPECT images can display organs with much greater detail than conventional scintigrams.

Positron emission tomography (PET) is a more refined radiologic technique that is used to study the metabolic activity inside an organ. The technique has been shown to be useful in the study of brain-related disorders, such as epilepsy and Alzheimer's disease, and of the vitality of heart tissue.

Interventional radiology combines imaging procedures with various injection and catheter techniques to treat tumors, blockages, bleeding vessels, and other abnormalities without extensive surgery. Among the more common interventional procedures is angioplasty, which is used to treat blocked or narrowed arteries. *See* RADIOGRAPHY. [M.Lo.; R.DelF.]

Medical mycology The study of fungi (molds, yeasts) having pathogenic properties for humans. Some of these fungi are called opportunistic pathogens because they can cause disease only in those individuals whose immunological systems have been compromised either through administration of certain drugs or because of some other disease state such as a malignancy. There are fungi that can cause disease in a noncompromised host and are considered to be primary pathogens, even though they may also cause an opportunistic mycotic disease. *See* OPPORTUNISTIC INFECTIONS.

Most fungi pathogenic for humans reside in soil as saprophytes. Humans are not a natural environment for them and when infected by these organisms are considered to be accidental hosts. *See* ACTINOMYCOSIS; ASPERGILLOSIS; BLASTOMYCOSIS; CANDIDIASIS; CHROMOMYCOSIS; COCCIDIOIDOMYCOSIS; DERMATOPHYTOSIS; HISTOPLASMOSIS; NOCARDIOSIS. [L.D.H.]

Medical parasitology The branch of medical microbiology which deals with the relationships between humans and those animals which live in or on them.

The most important human parasites fall into four phyla: the Protozoa, Nemata, Platyhelminthes, and Arthropoda. *See* ACARI; ANOPLURA; CESTODA; DIPTERA; MALLOPHAGA; NEMATA; SIPHONAPTERA.

Different organisms show extreme diversity in their localization within the host body. Some are exclusively intracellular, most are extracellular only, and some are both. Nearly every organ, tissue, and fluid of the body can be invaded: Trypanosomes live in the blood, lymph nodes, and cerebrospinal fluid; filaria live in the lymph channels and eye fluids; other worms inhabit the intestine, bronchi, and bile ducts; and a variety of organisms occur in the heart, brain, liver, spleen, and glands of internal secretion. In these sites the effect on the tissue varies from complete destruction to almost complete harmlessness, and the same parasite at different times may have either of these effects.

Also, the relationships between microorganisms and humans may be permanent or temporary, the microorganism being known, accordingly, as either an obligatory or a facultative parasite. For diseases due to protozoa *see* AMEBIASIS; BALANTIDIASIS; CHAGAS' DISEASE; LEISHMANIASIS; MALARIA; PNEUMOCYTOSIS; TRICHOMONIASIS. For diseases due to heiminths *see* ASCARIASIS; CLONORCHIASIS; FASCIOLIASIS; FASCIOLOPSIASIS; GUINEA WORM INFECTION; HETEROPHYIASIS; HOOKWORM DISEASE; PARAGONIMIASIS; PINWORM INFECTION; SCHISTOSOMIASIS; TAPEWORM DISEASE; TRICHINOSIS. [M.Yo.]

Medical ultrasonic tomography A mapping or imaging technique, used to obtain clinically useful information about the structure and functioning of tissues and organs, in which acoustic pulses are emitted from an acoustoelectric transducer, and echoes are received from acoustic impedance discontinuities along the assumed line-of-sight axial propagation path. A number of different modes of operation have emerged, each having areas of usefulness.

The A (amplitude) mode uses acoustic pulse emissions and echo reception along a single line-of-sight axial propagation path, and thus provides a one-dimensional mapping. This mode of operation cannot provide identification of structural features. It is, however, a most accurate method of measuring time delays and, therefore, distances between echo-producing structures or distances of structures from transducers, provided the speed of sound propagation in the medium is known.

The M mode of operation is used to display the movement of time-varying, echo-producing structures by intensity-modulating the trace as it is swept slowly across the oscilloscope screen in a direction at right angles to the fast time-base sweep. This mode of operation is used extensively in diagnosing disorders of the heart. *See* ECHOCARDIOGRAPHY; HEART DISORDERS.

For a two-dimensional picture to be obtained, the line-of-sight propagation path must be scanned and the position and direction of the path monitored and used to form a two-dimensional picture. Typically, the B-mode display is formed by moving the transducer so that the line-of-sight path remains in a single plane. The time-base trace of the cathode-ray oscilloscope screen is moved to correspond, in position and direction, to the ultrasonic line-of-sight propagation path, and echoes are displayed as intensity modulations of the trace.

The C (constant-range) mode provides a two-dimensional image display at constant time delay, and presumed constant distance, from the ultrasonic transducer. The scanning is arranged so that the point at constant depth along the propagation path (beam axis) traverses a plane. *See* MEDICAL IMAGING; ULTRASONICS. [F.Du.]

Medical waste Any solid waste that is generated in the diagnosis, treatment, or immunization of human beings or animals, in research pertaining thereto, or in the production or testing of biologicals. Since the development of disposable medical products in the early 1960s, the issue of medical waste has confronted hospitals and regulators. Previously, reusable products included items such as linens, syringes, and bandages; they were sterilized or disinfected prior to reuse, and the principal waste product was limited to human pathological tissue.

Most hazardous substances are described by their relevant properties, such as corrosive, poison, or flammable. Medical waste was originally defined in terms of its infectious properties, and thus it was called infectious waste. However, given the difficulty of identifying pathogenic organisms in waste that might cause disease, it has become standard practice to define medical waste by types or categories. While definitions differ somewhat under different regulations, in the United States the Centers for Disease Control cite four categories of infective wastes that should require special handling and treatment: laboratory cultures and stocks, pathology wastes, blood, and items with sharp points such as needles and syringes (sharps). These categories, of necessity, require that the generator of these wastes exercise judgment in identifying the material to be included.

Usually, the regulations mandate that, once identified, medical waste be separated from other waste as close to the source of generation as possible. After separation, the medical waste must be rendered noninfectious. Various treatment options are available, performed either at the source of generation or by a commercial entity off-site. The treatment technologies currently used for medical waste include incineration, sterilization, chemical disinfection, and microwave. *See* HAZARDOUS WASTE. [R.A.Sp.]

Medicine The field of science devoted to healing. Many subdivisions exist and more ramifications appear almost daily. Included in the area of medicine are the clinical specialties of surgery, pediatrics, psychiatry, obstetrics, and others. Internal medicine is the specialization which deals with internal diseases of a nonsurgical nature.

Related to the clinical specialties, particularly in regard to medical education and research, are the basic medical sciences. These include, among others, anatomy, physiology, psychology, pharmacology, biochemistry, and microbiology. Midway between the basic and the clinical sciences lies pathology, the study of the structural and functional alterations caused by diseases or abnormal states.

An important area in all specialties is preventive medicine and public health. This form of medicine supplies a necessary link with the community, state, or large geographic region in matters of prevention, mass treatment, and statistical appraisals of health matters. It is also concerned with socioeconomic factors related to physical and mental well-being. *See* PUBLIC HEALTH.

Socialized medicine is that form which exists under the direct control and financing of the state. The National Health Service of Great Britain is the best-known example, and other systems exist.

Other subdivisions of medicine, with names that are largely self-explanatory, include veterinary, legal, tropical, and military medicine. *See* FORENSIC MEDICINE.

Although medicine is based primarily upon scientific information and method, an important feature is the relationship between the physician and the patient. It is at this point that the necessary scientific background of medicine gives way to the art of healing. *See* SURGERY. [E.G.St./N.K.M.]

Mediterranean Sea The Mediterranean Sea lies between Europe, Asia Minor, and Africa. It is completely landlocked except for the Strait of Gibraltar, the Bosporus, and the Suez Canal. The Mediterranean is conveniently divided into an eastern basin and a western basin, which are joined by the Strait of Sicily and the Strait of Messina.

The total water area of the Mediterranean is 965,900 mi^2 (2,501,000 km^2), and its average depth is 5040 ft (1536 m). The greatest depth in the western basin is 12,200 ft (3719 m), in the Tyrrhenian Sea. The eastern basin is deeper, with a greatest depth of 18,140 ft (5530) m in the Ionian Sea about 55 km off the Greek mainland. The Atlantic tide disappears in the Strait of Gibraltar. The tides of the Mediterranean are predominantly semidiurnal. [J.Ly.]

Meiosis A special type of cell division that occurs in diploid or polyploid tissues and brings about a reduction in chromosome number. It is usual for the chromosome number to be halved. It is therefore the opposite of fertilization, for meiosis produces the nuclei which themselves, or their distant mitotic descendants, fuse to form a zygote. *See* MITOSIS.

Meiosis compensates for the doubling of the chromosome number brought about at fertilization by reducing it again before the next fertilization. Meiosis occurs immediately prior to gamete formation in most animals. There are usually a further two or three mitotic divisions before gametes are formed in higher plants. *See* GAMETOGENESIS; REPRODUCTION (ANIMAL); REPRODUCTION (PLANT).

First division. Meiosis is a continuous sequence of events, but for ease of description and reference certain readily recognizable stages have been named (see illustration). The first division involves the separation of homologs that have paired with each other.

Prophase 1. There are five stages delineated for the sequence of events occurring during prophase 1.

1. Leptotene. This is the earliest stage; the chromosomes appear as long thin threads and are differentiated along their length into beadlike structures named chromomeres. In homologous chromosomes these chromomeres correspond in size, number, and position, and together with the position of centromeres, nucleolar organizers, and heterochromatic regions, by pachytene they enable the cytologist to recognize particular chromosomes.

2. Zygotene. Here homologous chromosomes begin to pair in a highly specific way. The pairing, or synapsis, may begin at any place along the length of the chromosome, but more usually it starts near the ends. The precision of pairing is such that homologous chromomere pairs with homologous chromomere. While the pairing is taking place, continued shortening, thickening, and coiling result in the homologs being twisted about each other so that there is close alignment along their entire length. When pairing is complete, the diploid complement is reduced to the haploid number of bivalent chromosomes or bivalents.

3. Pachytene. The term refers to the more obvious thickening of the chromosomes which reflects their paired nature and the fact that chromosome duplication has occurred.

4. Diplotene. The attraction between the homologous chromosomes observed during zygotene gives way to repulsion. The bivalents are however held together by the points of exchange which, when opened out by the repulsion, appear as crosses or chiasmata. It is also clear that each bivalent consists of four strands at this stage, each chromosome having duplicated to form two chromatids.

5. Diakinesis. Intense internal coiling of the sister chromatids now occurs and this results in greater shortening and thickening than has been seen hitherto. One consequence is that the chromatid pairs on either side of a chiasma twist through approximately 90° at the point of exchange, revealing the characteristic loops and half loops. The nucleoli and the nuclear membrane break down.

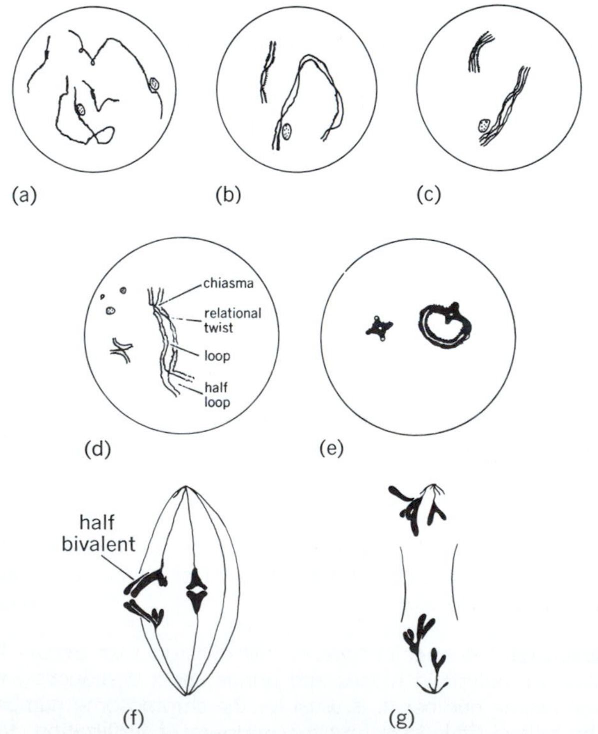

Stages of the first nuclear division of meiosis. (*a–e*) Prophase, comprising (*a*) leptotene, (*b*) zygotene, (*c*) pachytene, (*d*) diplotene, and (*e*) diakinesis. (*f*) Metaphase 1. (*g*) Anaphase 1. Two pairs of homologous chromosomes are shown; one of these pairs carries nucleolar organizers; the other has heterochromatic segments.

Prometaphase 1. Following the breakdown of the nuclear membrane, a spindle is formed. Sister kinetochores of each chromosome face the same pole (syntelic orientation). Unpaired chromosomes (univalents) which sometimes occur when pairing or chiasma formation fails may show either amphitelic or syntelic behavior. Also, the orientation of one bivalent on the spindle is independent of the others. Following this autoorientation, congression occurs and this leads to full metaphase.

Metaphase 1. In the first metaphase of meiosis the centromeric regions of the homologs come to lie equidistant on either side of the equator; the actual distance is dependent upon the position of the proximal chiasma. There is considerable tension on the bivalents at this stage, so that the position of the chiasmata may have slipped toward the chromosome ends.

Anaphase 1. In meiosis this duplication does not occur until the anaphase of the second nuclear division. Instead, the homologs (each consisting of a chromatid pair) are separated by the contraction of the spindle fibers pulling them to opposite poles of the spindle. In meiosis there is often an elongation of the spindle at late anaphase 1, which ensures that the products of the first division are well separated before the onset of the second.

Telophase 1 and interkinesis. These stages, which may not occur in some organisms, involve the regrouping of the haploid sets of chromatid pairs into organized nuclei. There may be a relaxation of the intense coiling begun at diakinesis.

Second division. The second division is mechanically the same as a normal mitotic division, but at metaphase sister chromatids are held together only at the centromere, so that chromosomes appear X- or V-shaped. Chromosomes are present in a haploid instead of a diploid number. At anaphase 2, centromeres of sister chromatids move to opposite poles where the end result is four haploid nuclei, and the reduction in chromosome number has been accomplished.

Significance. Apart from the reduction of chromosome number, there are two other major consequences of meiosis. The independent orientation of bivalents at metaphase 1 ensures random assortment of whole chromosomes, while chiasma formation results in nonsister chromatids exchanging lengths of chromosome material, thereby recombining genes which they carry, leading to the breakup of linked allelic combinations. As a result, it is highly unlikely that any of the meiotic products of an individual will be exactly the same as any other, unless it contains devices for restricting crossing-over autoorientation. *See* CELL DIVISION; GENE. [G.G.N.]

Meissner effect The expulsion of magnetic flux from the interior of a superconducting metal when it is cooled in a magnetic field to below the critical temperature, near absolute zero, at which the transition to superconductivity takes place. It was discovered by Walther Meissner in 1933, when he measured the magnetic field surrounding two adjacent long cylindrical single crystals of tin and observed that at −452.97°F (3.72 K) the Earth's magnetic field was expelled from their interior. This indicated that at the onset of superconductivity they became perfect diamagnets. This discovery showed that the transition to superconductivity is reversible, and that the laws of thermodynamics apply to it. The Meissner effect forms one of the cornerstones in the understanding of superconductivity, and its discovery led F. London and H. London to develop their phenomenological electrodynamics of superconductivity. *See* DIAMAGNETISM; THERMODYNAMIC PRINCIPLES.

The magnetic field is actually not completely expelled, but penetrates a very thin surface layer where currents flow, screening the interior from the magnetic field.

The Meissner effect is subject to limitations. Full diamagnetism is not observed in polycrystalline samples, and the effect is not observed in impure samples or samples with certain geometrics, such as a round flat disk, with the magnetic field parallel to the axis of rotation. *See* SUPERCONDUCTIVITY. [H.W.M.]

Meitnerium A chemical element, symbol Mt, atomic number 109. The decay of meitnerium is observable in the fusion reaction between ^{58}Fe and ^{209}Bi. This experiment was performed in 1982 at Darmstadt, West Germany, by a team led by P. Armbruster and G. Müzenberg. This same German team had discovered bohrium (atomic number 107), using similar techniques.

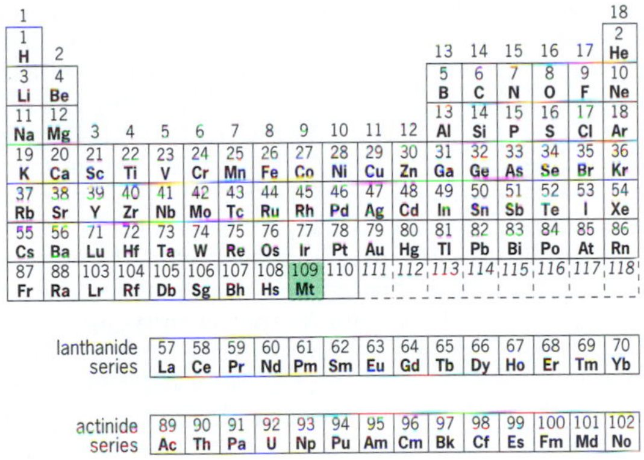

1 2 3 4 5 6 7 8 9 10 11 12 13 14 15 16 17 18
1 H
2 He
3 Li 4 Be
5 B 6 C 7 N 8 O 9 F 10 Ne
11 Na 12 Mg
13 Al 14 Si 15 P 16 S 17 Cl 18 Ar
19 K 20 Ca 21 Sc 22 Ti 23 V 24 Cr 25 Mn 26 Fe 27 Co 28 Ni 29 Cu 30 Zn 31 Ga 32 Ge 33 As 34 Se 35 Br 36 Kr
37 Rb 38 Sr 39 Y 40 Zr 41 Nb 42 Mo 43 Tc 44 Ru 45 Rh 46 Pd 47 Ag 48 Cd 49 In 50 Sn 51 Sb 52 Te 53 I 54 Xe
55 Cs 56 Ba 71 Lu 72 Hf 73 Ta 74 W 75 Re 76 Os 77 Ir 78 Pt 79 Au 80 Hg 81 Tl 82 Pb 83 Bi 84 Po 85 At 86 Rn
87 Fr 88 Ra 103 Lr 104 Rf 105 Db 106 Sg 107 Bh 108 Hs 109 Mt 110 111 112 113 114 115 116 117 118
lanthanide series 57 La 58 Ce 59 Pr 60 Nd 61 Pm 62 Sm 63 Eu 64 Gd 65 Tb 66 Dy 67 Ho 68 Er 69 Tm 70 Yb
actinide series 89 Ac 90 Th 91 Pa 92 U 93 Np 94 Pu 95 Am 96 Cm 97 Bk 98 Cf 99 Es 100 Fm 101 Md 102 No

In a sequence of bombardments of ^{209}Bi targets with beams of ^{50}Ti, ^{54}Cr, and ^{58}Fe, the compound systems ^{259}Db, ^{263}Bh, and ^{267}Mt have been produced. Out of 10^{11} nuclear encounters, only one apparently leads to production of an atom of meitnerium. Nevertheless, the probability of producing the observed event randomly is 10^{-18}. Even with only one atom found, the existence of meitnerium can be considered as highly probable.

Y. T. Oganessian and his team at Dubna Laboratories in Russia repeated the Darmstadt experiment in 1984 with a tenfold higher irridiation dose. The nuclide ^{246}Cf, an alpha emitter with a half-life of 1.5 days, the seventh member of the decay chain, was separated chemically. Seven alpha decays were registered compatible with the decay energy and half-life of ^{246}Cf. Moreover, one fission event from ^{258}Rf was seen. The cross section given is 3×10^{-40} m^2. The formation of the isotope ^{266}Mt thus has been confirmed indirectly. *See* BOHRIUM; NUCLEAR FISSION; NUCLEAR REACTION; RADIOACTIVITY; TRANSURANIUM ELEMENTS. [P.Ar.]

Melanterite A mineral having composition $FeSO_4 \cdot 7H_2O$. Melanterite occurs mainly in green, fibrous or concretionary masses, or in short, monoclinic, prismatic crystals. Luster is vitreous, hardness is 2 on Mohs scale, and specific gravity is 1.90.

Melanterite is a common secondary mineral derived from oxidation and hydration of iron sulfide minerals such as pyrite and marcasite. Its occurrence is widespread. It is not an ore mineral. *See* MARCASITE; PYRITE. [E.C.T.C.]

Melilite A complete solid solution series ranging from gehlenite, $Ca_2Al_2SiO_7$, to akermanite, $Ca_2MgSi_2O_7$, often containing appreciable Na and Fe. The Mohs hardness is 5–6, and the density increases progressively from 2.94 for akermanite to 3.05 for gehlenite. The luster is vitreous to resinous, and the color is white, yellow, greenish, reddish, or brown. Akermanite-rich varieties occur in thermally metamorphosed siliceous limestones and dolomites, but more gehlenite-rich ones result if A1 is present. Melilites are found instead of plagioclase in silica-deficient, feldspathoid-bearing basalts. *See* SILICATE MINERALS. [L.Gr.]

Melioidosis An endemic disease of animals and humans. The causative organism, *Pseudomonas pseudomallei*, exists primarily as a saprophyte in the soil and water of rice-growing areas of southeastern Asia.

The multiple forms of melioidosis in humans present confusing symptoms. A 2–5 day fulminating, pulmonary, septicemic disease is seldom diagnosed before death. An indolent pulmonary form with thin-walled abscesses resembles tuberculosis except for chills and fever. A chronic systemic form produces widespread semigranulomatous abscesses in many organs. [W.R.Mi.]

Melting point The temperature at which a solid changes to a liquid. For pure substances, the melting or fusion process occurs at a single temperature, the temperature rise with addition of heat being arrested until melting is complete.

Melting points reported in the literature, unless specifically stated otherwise, have been measured under an applied pressure of 1 atm (10^5 pascals), usually 1 atm of air. (The solubility of air in the liquid is a complicating factor in precision measurements.) Upon melting, all substances absorb heat, and most substances expand; consequently an increase in pressure normally raises the melting point. A few substances, of which water is the most notable example, contract upon melting; thus, the application of pressure to ice at 32°F (0°C) causes it to melt. Large changes in pressure are required to produce significant shifts in the melting point.

For solutions of two or more components, the melting process normally occurs over a range of temperatures, and a distinction is made between the melting point, the temperature at which the first trace of liquid appears, and the freezing point, the higher temperature at which the last trace of solid disappears, or equivalently, if one is cooling rather than heating, the temperature at which the first trace of solid appears. *See* PHASE EQUILIBRIUM; SOLUTION; SUBLIMATION; TRIPLE POINT. [R.L.S.]

Membrane distillation A separation method in which a nonwetting, microporous membrane is used with a liquid feed phase on one side of the membrane and a condensing, permeate phase on the other side. Separation by membrane distillation is based on the relative volatility of various components in the feed solution. The driving force for transport is the partial pressure difference across the membrane. Separation occurs when vapor from components of higher volatility passes through the membrane pores by a convective or diffusive mechanism. *See* CONVECTION (HEAT).

Membrane distillation shares some characteristics with another membrane-based separation known as pervaporation, but there also are some vital differences. Both methods involve direct contact of the membrane with a liquid feed and evaporation of the permeating components. However, while membrane distillation uses porous membranes, pervaporation uses nonporous membranes.

Membrane distillation systems can be classified broadly into two categories: direct-contact distillation and gas-gap distillation. These terms refer to the permeate or condensing side of the membrane; in both cases the feed is in direct contact with the membrane. In direct-contact membrane distillation, both sides of the membrane contact a liquid phase; the liquid on the permeate side is used as the condensing medium for the vapors leaving the hot feed solution. In gas-gap membrane distillation, the condensed permeate is not in direct contact with the membrane.

Potential advantages of membrane distillation over traditional evaporation processes include operation at ambient pressures and lower temperatures as well as ease of process scale-up. *See* CHEMICAL SEPARATION TECHNIQUES; MASS-TRANSFER OPERATION; MEMBRANE SEPARATIONS; SALINE WATER RECLAMATION. [S.S.K.; N.N.L.]

Membrane mimetic chemistry The study of processes and reactions whose developments have been inspired by the biological membrane. Faithful modeling of the biomembrane is not an objective of membrane mimetic chemistry. Rather, only the essential components of natural systems are recreated from relatively simple, synthesized molecules. (The term membrane mimetic is more restrictive than the term biomimetic. Biomimetic chemistry is directed at the mechanistic elucidation of biochemical reactions and at the development of new compounds modeled on specific biological systems.) *See* CELL MEMBRANES.

Various surfactant aggregate systems have been used in membrane mimetics.

Surfactants (detergents) contain distinct hydrophobic (apolar) and hydrophilic (polar) regions. Depending on the chemical structure of their hydrophilic polar head groups, surfactants can be neutral, positively charged, or negatively charged. *See* DETERGENT.

Aqueous micelles are spherical aggregates, 4–8 nanometers in diameter, formed dynamically from surfactants in water above a characteristic concentration, the critical micelle concentration. *See* MICELLE.

Monomolecular layers are formed by spreading naturally occurring lipids or synthetic surfactants, dissolved in volatile solvents, over water in a trough. The polar head groups of the surfactants are in contact with water, the subphase, while their hydrocarbon tails protrude above it. *See* MONOMOLECULAR FILM.

Other systems used in membrane mimetics are multilayer assemblies (Langmuir-Blodgett films), bilayer lipid membranes, and vesicles prepared by sonication from naturally occurring lipids. *See* SONOCHEMISTRY.

Membrane mimetic chemistry has become a versatile chemical tool. Applications of compartmentalization of reactants in membrane mimetic systems involve altered reaction rates, products, stereochemistries, and isotope distributions. Monolayers and organized multilayers can be employed profitably as molecular electronic devices. Opportunities also exist for using different surfactant aggregates with polymeric membranes for the control and regulation of reverse osmosis and ultrafiltration. *See* SURFACTANT; ULTRAFILTRATION. [J.H.Fe.]

Membrane separations Processes for separating mixtures by using thin barriers (membranes) between two miscible fluids. A suitable driving force across the membrane, for example concentration or pressure differential, leads to the preferential transport of one or more feed components.

Membrane separation processes are classified under different categories depending on the materials to be separated and the driving force applied: (1) In ultrafiltration, liquids and low-molecular-weight dissolved species pass through porous membranes while colloidal particles and macromolecules are rejected. The driving force is a pressure difference. (2) In dialysis, low-molecular-weight solutes and ions pass through while colloidal particles and solutes with molecular weights greater than 1000 are rejected under the conditions of a concentration difference across the membrane. (3) In electrodialysis, ions pass through the membrane in preference to all other species, due to a voltage difference. (4) In reverse osmosis, virtually all dissolved and suspended materials are rejected and the permeate is a liquid, typically water. (5) For gas and liquid separations, unequal rates of transport can be obtained through nonporous membranes by means of a solution and diffusion mechanism. Pervaporation is a special case of this separation where the feed is in the liquid phase while the permeate, typically drawn under subatmospheric conditions, is in the vapor phase. (6) In facilitated transport, separation is achieved by reversible chemical reaction in the membrane. High selectivity and permeation rate may be obtained because of the reaction scheme. Liquid membranes are used for this type of separation. *See* DIALYSIS; DIFFUSION IN GASES AND LIQUIDS; ION-SELECTIVE MEMBRANES AND ELECTRODES; MASS-TRANSFER OPERATION; OSMOSIS; TRANSPORT PROCESSES; ULTRAFILTRATION. [N.N.L; S.S.K.]

Memory The recollection of past events or the performance of previously learned skills without practice after the passage of time are evidence of memory function. In order to say that an organism remembers, the present behavior of that organism must be influenced by its past experience. However, this is not a sufficient condition since it should include changes brought about by physical growth or fatigue. For one to say that memory has occurred, the organism must compare a given stimulus with a stimulus which has occurred previously, and its behavior must reflect this comparison. The concept of memory thus indicates the formation of some kind of associative link between two stimuli separated in time. For this to occur the first stimulus must produce a persistent change in the organism. This change is often referred to as the memory trace or engram. To demonstrate that this change has occurred, the performance of the organism must be altered in some measurable way. In experimental studies of memory, the organism is presented with controlled stimuli, and in some quantitative way the behavioral output is measured after the passage of time. Such studies are designed both to define the psychological laws of memory and to investigate the neurological substrate in which memory occurs. [A.G.; M.E.J.]

Mendelevium A chemical element, Md, atomic number 101, the twelfth member of the actinide series of elements. Mendelevium does not occur in nature; it was discovered and is prepared by artificial nuclear transmutation of a lighter element. Known isotopes of mendelevium have mass numbers from 248 to 258 and half-lives from a few seconds to about 55 days. They are all produced by charged-particle bombardments of more abundant isotopes. The amounts of mendelevium which are produced and used for studies of chemical and nuclear properties are usually less than about a million atoms; this is of the order of a million times less than a weighable

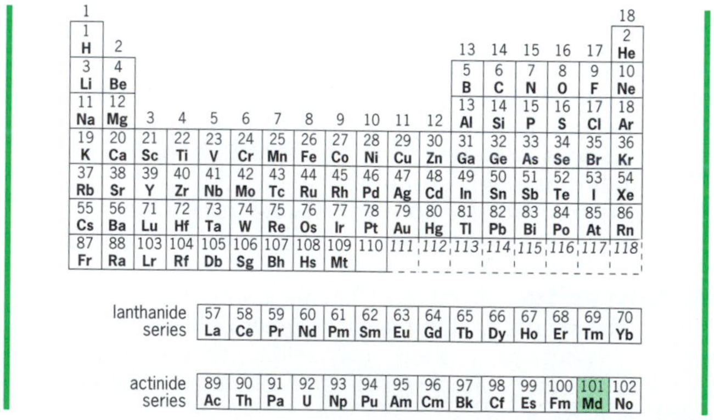

amount. Studies of the chemical properties of mendelevium have been limited to a tracer scale. The behavior of mendelevium in ion-exchange chromatography shows that it exists in aqueous solution primarily in the 3 + oxidation state characteristic of the actinide elements. However, it also has a dipositive (2+) and a monopositive (1+) oxidation state. *See* ACTINIDE ELEMENTS; TRANSURANIUM ELEMENTS. [G.T.S.]

Mendelism The basic laws of inheritance discovered in 1865 by Gregor Mendel, a monk in the Augustinian monastery at Brünn, Moravia. Mendel's laws were formulated to explain the basis for results of his experiments on the inheritance of seven pairs of alternative characteristics in the garden pea. He studied each character separately by making hybrids between plants which were constant for one of a pair of alternative traits, such as round versus wrinkled seeds, green versus yellow pods, and tall versus short stems. In the first hybrid generation Mendel noted that each character of the hybrid resembles that of one of the parental forms. These characters he called dominant, and the contrasting hidden ones, recessive.

Law of segregation. In a second generation, bred by self-pollinating the hybrid, Mendel observed that both the dominant and the recessive character appeared unaltered and in the proportion of three dominant to one recessive. He extended his experiments to subsequent generations and found that plants showing a recessive characteristic bred true, whereas among the dominants one bred true to two that bred like the original hybrid. Mendel formulated the following hypothesis to account for these results. The characters are controlled in their inheritance by units—now called genes—which occur paired in each parent and are separated so that they occur singly in the mature reproductive cells, the gametes. The units become paired once again following the union of gametes in the formation of the next generation. When the paired units are dissimilar a hybrid results. In the hybrid the paired units remain unaltered and subsequently are segregated into approximately equal numbers of gametes. These recombine randomly with reference to the particular units they contain in forming the next generation. *See* GENE; RECOMBINATION (GENETICS).

Diagrammatically this may be represented as follows:

Parent generation:

AA (dominant) aa (recessive)

Gametes of parents:

A a

First hybrid generation:

Aa (resembles dominant parent)

Gametes of hybrid:

$$(\tfrac{1}{2}\,\mathrm{A} - \tfrac{1}{2}\,\mathrm{a}) \times (\tfrac{1}{2}\,\mathrm{A} + \tfrac{1}{2}\,\mathrm{a})$$

Eggs Pollen

Second hybrid generation:

$$\tfrac{1}{4}\,\mathrm{AA} + \tfrac{1}{4}\mathrm{Aa} + \tfrac{1}{4}\,\mathrm{Aa} + \tfrac{1}{4}\,\mathrm{aa}$$

$$\tfrac{3}{4}\text{ dominant: }\tfrac{1}{4}\text{ recessive}$$

Law of independent assortment. Mendel also investigated the offspring of hybrids involving several alternative characters. To cite one experiment, he pollinated a seed parent of round yellow peas (*AA* and *BB*, both dominant characters) with pollen from a parent plant of wrinkled green peas (*aa* and *bb*). The first hybrid generation formed round yellow seeds and the second hybrid generation yielded seeds of four sorts in the following numbers: 315 round and yellow, 101 wrinkled and yellow, 108 round and green, and 32 wrinkled and green. Mendel adduced that this ratio resulted from independent assortment of the hereditary units controlling these two characters. That is, in the formation of the gametes of the hybrid one-half contain *A* and one-half *a*; similarly one-half contain *B* and one-half *b*. Each pair is sorted out independently of the other so that the hybrid produces four kinds of gametes in approximately equal numbers:

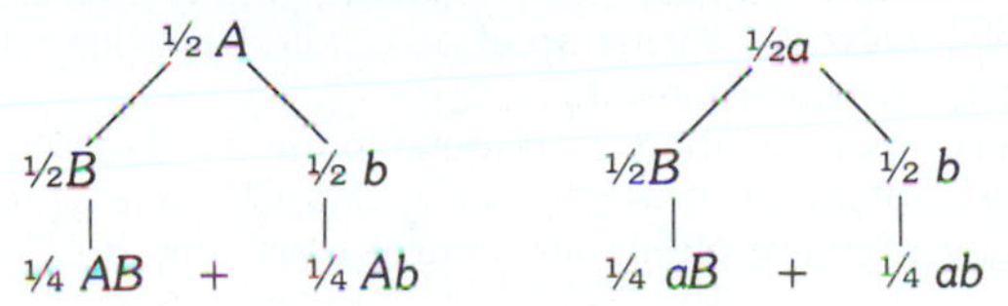

The four kinds of gametes unite at random to form nine different genetic types in the second generation. With dominance, only four differently appearing types are observed.

Thus Mendel's hypothesis predicts four differently appearing types in the proportions 9 round, yellow seeds: 3 wrinkled, yellow seeds: 3 round, green seeds: 1 wrinkled, green seed. *See* GENETICS. [H.H.SM.]

Meninges In mammals, the three membranes that cover the brain and spinal cord: the dura mater, the arachnoid membrane, and the pia mater. The outermost, the dura mater, is a tough, fibrous, double-layered structure that is adherent to the skull. The inner layer of the dura mater sends separating sheets between the cerebral hemispheres and between the cerebrum and cerebellum. It also contains large venous sinuses and forms sheaths for nerves leaving the skull. The middle layer, the arachnoid, is a delicate serous layer loosely investing the brain. Below this is the spongy subarachnoid cavity which contains the circulating cerebrospinal fluid. The innermost layer, the pia mater, is a vascular layer which closely follows each convolution of the brain. Together the meninges furnish protection, blood supply, drainage, and cerebrospinal channels for the brain. *See* NERVOUS SYSTEM (VERTEBRATE). [T.S.P.]

Meningitis An inflammation of the meninges and brain, caused by a variety of agents, including viral, bacterial, and protozoan. It is a common disease which may result in serious complications. *See* MENINGOCOCCUS.

Meningitis usually results from a previous infection of the bloodstream, but sometimes it results from direct extension of an infection of a contiguous site, such as the mastoid or the sinuses. A fracture of the skull or a congenital defect of the meninges may contribute; a respiratory infection frequently precedes meningitis.

Diagnosis is based upon clinical signs, such as headache, vomiting, convulsions, fever, and in the case of the meningococcus, a rash. Diagnosis is confirmed by examination of a sample of spinal fluid withdrawn by lumbar puncture. [W.B.]

Meningococcus A major human pathogen belonging to the bacterial genus *Neisseria*, and the cause of meningococcal meningitis and meningococcemia. The official designation is *N. meningitidis*. The meningococcus is a gram-negative, aerobic, nonmotile diplococcus. It is fastidious in its growth requirements and is very susceptible to adverse physical and chemical conditions.

Humans are the only known natural host of the meningococcus. Transmission occurs by droplets directly from person to person. Fomites and aerosols are probably unimportant in the spread of the organism. The most frequent form of host-parasite relationship is asymptomatic carriage in the nasopharynx.

The most common clinical syndrome caused by the meningococcus is meningitis, which is characterized by fever, headache, nausea, vomiting and neck stiffness and has a fatality rate of 15% (higher in infants and adults over 60). Disturbance of the state of consciousness quickly occurs, leading to stupor and coma. Many cases also have a typical skin rash consisting of petechiae or purpura. *See* MENINGITIS. [R.Go.]

Menopause The irreversible cessation of regular monthly uterine bleeding in the adult human female, marking the end of her ability to become pregnant. The mean age of menopause (occurrence of the last menses) in the United States is 51 plus or minus 2.5 years. Menopause probably occurs because the ovary runs out of eggs and the cyclic rise and fall of brain and ovarian hormones designed to prepare the uterus to receive and nourish pregnancy no longer occurs.

Menopause is one event in the climacteric, the period of time during which the reproductive capability slows down and finally stops. The biochemical hallmark of this period is a reduction in estrogen production by the ovary. Some estrogen continues to be produced by the adrenal gland and the fatty tissues throughout the body, but this amount is very small compared with premenopausal levels. A whole host of psychological problems have been attributed to estrogen deprivation, but well-documented proof of these relationships is lacking. *See* HORMONE; MENSTRUATION. [G.Gu.]

Menstruation Periodic sloughing of the uterine lining in women of reproductive age. Menstrual bleeding indicates the onset of the menstrual cycle, which lasts an average of 27–30 days, although ranges of 21–60 days have been recorded.

The principal control mechanism is a complex hormonal regulation. The follicle-stimulating hormone (FSH) of the anterior pituitary gland stimulates the developing ovarian follicle, which secretes increasing amounts of estrogens during the first half of the menstrual cycle. The estrogenic hormones cause a proliferation and thickening of the uterine mucosa together with blood vessel and gland enlargement. During the middle of the cycle the ovum is extruded (ovulation) from the ovarian follicle. The remaining empty follicle becomes the corpus luteum, formed by the ingrowth of connective tissue, which begins to secrete progesterone to prepare the uterus for pregnancy. If the ovum is not fertilized, the corpus luteum degenerates, progesterone secretion is thus greatly reduced, and the vascularized mucosa of the uterus is sloughed off. *See* ESTROGEN; PROGESTERONE; UTERUS.

The onset of menstruation (menarche) occurs between the ages of 9 and 16, with the majority of females beginning at 12–14 years. Duration of the flow and cycle length may be quite variable during these early years. Gradually, a degree of regularity is established but the flow may last 3–7 days; this, as well as the copiousness of the flow, may vary from one period to another. Cessation of menses (menopause) occurs predominantly in women 49–53 years old. *See* MENOPAUSE.

There is no menstrual bleeding during pregnancy. This amenorrhea sometimes continues during nursing until weaning; however, ovulation and conception are still possible at this time. *See* ESTRUS; LACTATION; PREGNANCY. [S.P.P.]

Mental deficiency A condition, also known as mental retardation, which is characterized by intellectual retardation, social inadequacy, and persistent dependency. This condition is evidence of an arrest in development, and is apparent from birth or an early age. The causes range from genetic factors to metabolic dysfunction, specific injury to the central nervous system, and environmental factors.

Mental deficiency is divided into a number of subgroups on the basis of cause, degree of manifest intellectual retardation, and ability to be trained. From the aspect of cause, the mentally deficient are classified into nine groups: infections and intoxications; trauma or physical agent; metabolism or nutrition; gross brain disease (postnatal); unknown prenatal influences; chromosomal abnormality; gestational disorders; following psychiatric disorder; and environmental influences.

Mentally retarded individuals are often classified in terms of their intelligence quotient (IQ). The highest group, designated as "mild," in general has IQ limits of 55–70. The next lowest group in degree of retardation, with IQ limits of 40–54, is known as "moderate." The lowest groups, with demonstrable IQs generally below 39, are known as "severe" and "profound," and include approximately 3% of the known cases. *See* INTELLIGENCE.

Classification terms more commonly used in public schools, and which roughly correspond to those referred to above, are educable, trainable, and multihandicapped. The educable are those who, by special training, can be expected to attain relative independence. The trainable are those who, with special training, are capable of a minor degree of self-help and social participation, but in a sheltered environment. [H.Le.]

Menthol A monocyclic, saturated, secondary terpene alcohol with the formula below. Menthol contains three asym-

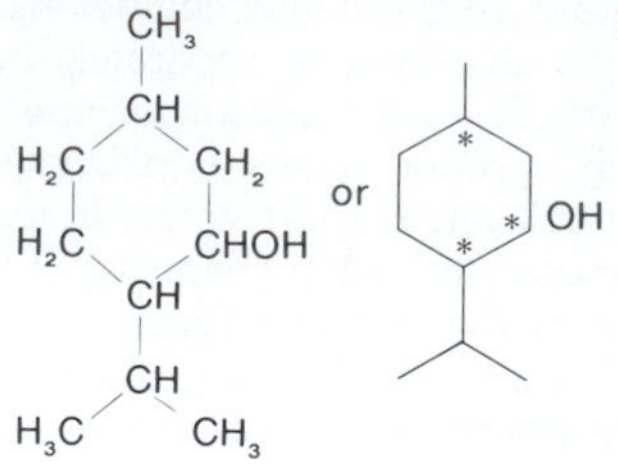

metric carbon atoms (starred in the formula). It can exist in four externally compensated and eight optically active forms. The only forms encountered in nature are the *l*-menthol and *d*-neomenthol. Commercial *l*-menthol is isolated principally from the oil of *Mentha arvensis*. It possesses a distinct peppermint flavor and gives the impression of cooling the mouth and skin. *See* TERPENE. [W.Mos.]

Mercaptan One of a group of organosulfur compounds which are also called thiols or thio alcohols and which have the general structure RSH. Aromatic thiols are called thiophenols, and biochemists often refer to thiols as sulfhydryl compounds. The unpleasant odor of volatile thiols causes them to be classed as stenches, but the odors of many solid thiols are not unpleasant.

Mercaptans (1) form salts with bases, (2) are easily oxidized to disulfides and higher oxidation products such as sulfonic acids, (3) react with chlorine (or bromine) to form sulfenyl chlorides (or bromides), and (4) undergo additions to unsaturated compounds, such as olefins, acetylenes, aldehydes, and ketones. The insoluble mercury salts (mercaptides) are used to isolate and identify mercaptans. *See* ORGANOSULFUR COMPOUND. [N.K.]

Merchant ship A power-driven ship employed in commercial transport on the oceans and large inland bodies of water such as the Great Lakes. The relatively small craft used for inland waterway transportation are not commonly referred to as ships.

Commodities transported by water are classified as breakbulk, unitized, or bulk (dry or liquid) cargoes. Generally, water cargo transportation is cheaper per ton-mile than land or air transportation; approximately 90% of the United States overseas trade revenue is waterborne, while the rest is airborne. *See* MARINE CONTAINERS.

A passenger ship, as defined by International Safety of Life at Sea (SOLAS) rules, carries more than 12 passengers on international voyages. Passenger vessels that also transport cargo are called passenger-cargo ships. [A.M.D'A.]

Mercury (element) A chemical element, Hg, atomic number 80 and atomic weight 200.59. Mercury is a silverwhite liquid at room temperature (melting point −38.89°C or −37.46°F); it boils at 357.25°C (675.05°F) under atmospheric pressure. It is a noble metal that is soluble only in oxidizing solutions. Solid mercury is as soft as lead. The metal and its compounds are very toxic. With some metals (gold, silver, platinum, uranium, copper, lead, sodium, and potassium, for example) mercury forms solutions called amalgams. *See* AMALGAM; TRANSITION ELEMENTS.

In its compounds, mercury is found in the 2+, 1+, and lower oxidation states, for example, $HgCl_2$, Hg_2Cl_2, or $Hg_3(AsF_6)_2$. Often the mercury atoms are doubly covalently bonded, for example, Cl—Hg—Cl or Cl—Hg—Hg—Cl. Some mercury(II) salts, for example, $Hg(NO_3)_2$ or $Hg(ClO_4)_2$, are quite soluble in water and dissociate normally. The aqueous solutions of these salts react as strong acids because of hydrolysis. Other mercury(II) salts, for example, $HgCl_2$ or $Hg(CN)_2$, also dissolve in water, but exist in solution as only slightly dissociated molecules. There are compounds in which mercury atoms are bound directly to carbon or nitrogen atoms, for example, $H_3C—Hg—CH_3$ or $H_3C—CO—NH—Hg—NH—CO—CH_3$. In complex compounds, for example, $K_2(HgI_4)$, mercury often has three or four bonds.

Metallic mercury is used as a liquid contact material for electrical switches, in vacuum technology as the working fluid of diffusion pumps, for the manufacture of mercury-vapor rectifiers, thermometers, barometers, tachometers, and thermostats, and for the manufacture of mercury-vapor lamps. It finds application for the manufacture of silver amalgams for tooth fillings in dentistry. Of importance in electrochemistry are the standard calomel electrode, used as the reference electrode for the measurement of potentials and for potentiometric titrations, and the Weston standard cell. *See* Calomel electrode.

Mercury is commonly found as the sulfide, HgS, frequently as the red cinnabar and less often as the black metacinnabar. A less common ore is the mercury(I) chloride. Occasionally the mercury ore contains small drops of metallic mercury. *See* Cinnabar.

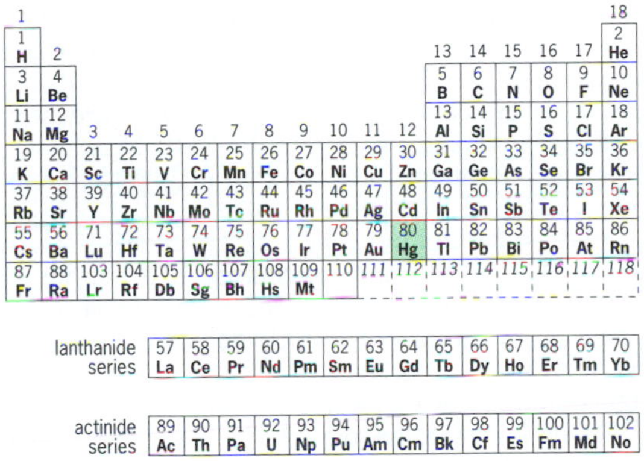

The surface tension of liquid mercury is 484 dynes/cm, six times greater than that of water in contact with air. Hence, mercury does not wet surfaces with which it is in contact. In dry air metallic mercury is not oxidized. After long standing in moist air, however, the metal becomes coated with a thin layer of oxide. In air-free hydrochloric acid or in dilute sulfuric acid, the metal does not dissolve. Conversely, it is dissolved by oxidizing acids (nitric acid, concentrated sulfuric acid, and aqua regia). [K.B.]

Mercury (planet) The planet nearest to the Sun. It is visible to the naked eye shortly after sunset or shortly before sunrise when it is near its greatest angular distance from the Sun. The diameter is 3031±1 mi (4878±2 km). The mass is 0.055 times the mass of the Earth. Mercury has the highest density of any planet except Earth, indicating that iron and nickel are its main chemical constituents. The surface of the planet is composed of silicate rock, heavily cratered from exposure to impacting meteoritic debris. There is no atmosphere, so that the surface retains the record of nearly the entire past geological history. Understanding of Mercury was greatly advanced in 1974 and 1975 when the United States spacecraft *Mariner 10* made three passes close to the planet.

On Mercury, the maximum surface temperature varies with distance from the Sun, at perihelion reaching 700 K (800°F) at the equator. At night, however, the unshielded surface cools rapidly to below 110 K (−340°F), producing the greatest diurnal variations in temperature in the solar system. A meter below the surface, the temperature hardly varies from an average value of about 350 K (170°F). *Mariner 10* obtained photographic coverage of nearly half of Mercury, with 20% of the surface imaged at a resolution of 1 km or better. The face of Mercury shows evidence of internal (endogenic) geologic processes as well as the scars of impacts (see illustration).

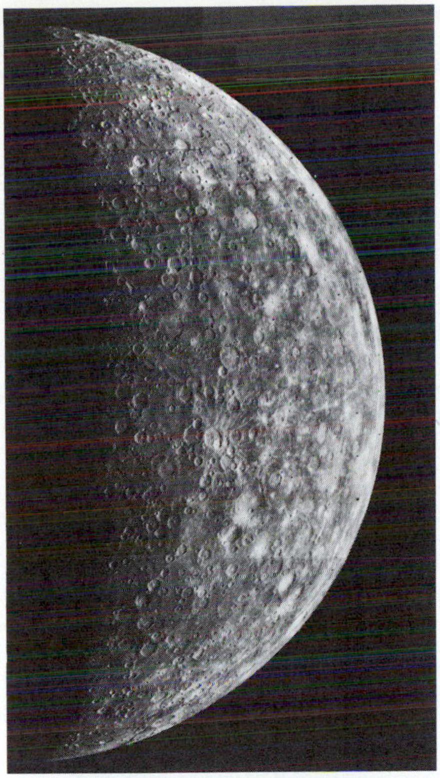

Mercury at quarter phase as seen by the approaching *Mariner 10* spacecraft in March 1974. (*Jet Propulsion Laboratory*)

Much of the surface appears flooded with lava, in the same way as the lunar maria, probably as the result of extensive past volcanism. Also many great compressional scarps that cut across other topographic features may have been caused by shrinkage of the planet as its interior cooled. [D.Mo.]

Mercury battery A primary dry-cell battery consisting of a zinc anode, a cathode of mercuric oxide (HgO) mixed with graphite, and an electrolyte of potassium hydroxide (KOH) saturated with zinc oxide (ZnO). With carefully purified materials and balanced amounts of ZnO and HgO, the cell has very low self-discharge and makes efficient use of the active materials. In some cells which require long-term continuous drains, for example, in hearing aid use, MnO_2 is added to the HgO.

Contained in a steel can, the active materials are separated by a porous material which prevents migration of conducting particles from the mercuric oxide pellet. The electrolyte is completely absorbed in the active materials, separator, and absorbent materials. The steel can serves as the contact to the HgO. A metal top with a concentric neoprene grommet closes off the top of the can and serves as the contact to the zinc.

The ampere-hour capacity of mercury cells is relatively unchanged with variation of discharge schedule and to some extent with variation of discharge current. Outstanding features of mercury cells include flat discharge curve, small variation in capacity with intermittent or continuous discharge, shelf life of several years, and good high-temperature characteristics. *See* DRY CELL; PRIMARY BATTERY; RESERVE BATTERY. [J.D.]

Mercury-vapor lamp A vapor or gaseous discharge lamp in which the arc discharge takes place in mercury vapor. This lamp is widely used for roadway and all other forms of illumination, and as a source of ultraviolet radiation for industrial applications.

The arc discharge takes place in a transparent tube of fused silica, or quartz, and this quartz tube is usually mounted inside a larger bulb of glass (see illustration). The outer bulb reduces the ultraviolet radiation of the inner arc tube, encloses and protects the mount structure, and can be coated with phosphors that greatly improve the color of the light emitted.

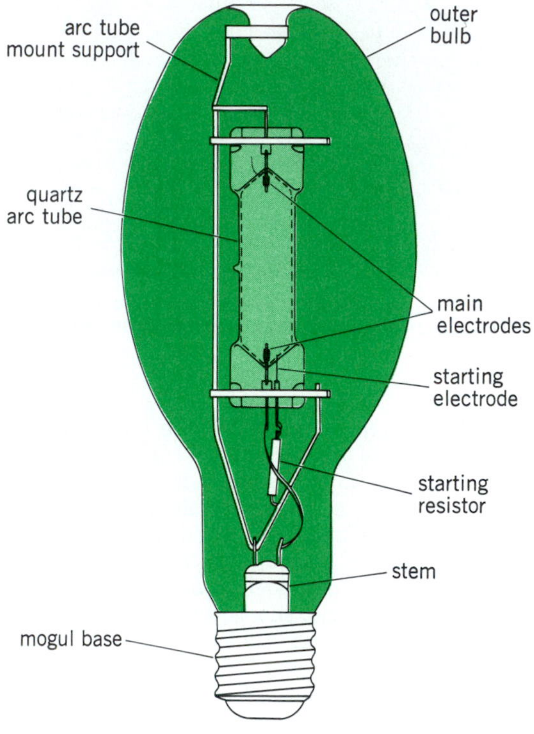

High-pressure mercury-vapor lamp.

The arc tube has electrode assemblies mechanically sealed into each end, At one end of the arc tube, a smaller starting electrode is located close to the main electrode. The starting electrode is connected through a high resistance to the opposite electric polarity of the adjacent main electrode. The arc tube itself is filled with argon gas and a small amount of pure mercury before being sealed.

Radiation from the mercury arc is confined to four specific wavelengths in the visible portion of the spectrum and several strong lines in the ultraviolet. The visible radiation from clear mercury lamps provides light with a distinct blue-green appearance and poor color rendition. Red objects, for example, look brown or black, and human skin looks unattractive. For this reason, most mercury lamps have a phosphor color-correcting coating, and clear bulbs are used only where appearance of colors is secondary to some gain in efficiency. *See* ILLUMINATION; LAMP; ULTRAVIOLET LAMP; VAPOR LAMP. [T.F.N.]

Meridian That half of a great circle on Earth that passes through points having the same longitude and terminates at the North and South poles. The meridian from which longitudes are measured is the one passing through Greenwich, near London, England; it is called the prime meridian.

The celestial meridian is a great circle on the celestial sphere passing through the two celestial poles and the observer's zenith. Two branches of it are distinguished, each extending from pole to pole, the upper branch containing the zenith and the lower the nadir. *See* ASTRONOMICAL COORDINATE SYSTEMS. [G.M.C.]

Merogony The normal or abnormal development of a part of an egg following cutting, shaking, or centrifugation of the egg before or after fertilization. Depending on the character of the nucleus present in the egg fragment, the following types of merogony are recognized: (1) diploid merogony, the nucleus being the normal fusion product of the egg and sperm nuclei and thus diploid; (2) andromerogony, development with the haploid sperm nucleus; (3) gynomerogony, development of a fragment of a fertilized egg with the haploid egg nucleus; (4) parthenomerogony, development of a nucleated fragment of an unfertilized egg following parthenogenetic stimulation. *See* ANDROGENESIS; GYNOGENESIS; PARTHENOGENESIS. [G.Fa.]

Meromictic lake A lake whose water is permanently stratified and therefore does not circulate completely throughout the basin at any time during the year. Normally lakes in the temperate zone mix completely during the spring and autumn when water temperatures are approximately the same from top to bottom. In meromictic lakes there are no periods of overturn or complete mixing because seasonal changes in the thermal gradient are either small or overridden by the stability of a chemical gradient, or the deeper waters are physically inaccessible to the mixing energy of the wind. Most commonly, the vertical stratification is stabilized by a chemical gradient in meromictic lakes.

The upper stratum of water in a meromictic lake is mixed by the wind and is called the mixolimnion. The bottom, denser stratum, which does not mix with the water above, is referred to as the monimolimnion. The transition layer between these strata is called the chemocline.

Of the hundreds of thousands of lakes on the Earth, only about 120 are known to be meromictic. In general, meromictic lakes in North America are restricted to: sheltered basins that are proportionally very small in relation to depth and that often contain colored water, basins in arid regions, and isolated basins in fiords. *See* FRESH-WATER ECOSYSTEM; LAKE; LIMNOLOGY. [G.E.Li.]

Merostomata A class of the phylum Arthropoda, subphylum Chelicerata. Merostomes are aquatic chelicerates, characterized by abdominal appendages bearing respiratory organs. Most merostomes are extinct; only the horseshoe crabs, comprising four species and three genera (*Carcinoscorpio* and *Tachypleus* of eastern Asia and *Limulus polyphemus* of eastern North America) survive.

The body of a merostome consists of a prosoma, or head, which lacks antennae, has a pair of compound eyes and a pair of simple median eyes, and bears the chelicerae (pincers) and five pairs of uniramous walking legs with gnathobases for mastication. The opisthosoma, or trunk, consists of 12 or fewer segments which may be freely articulating or partly or entirely fused into a solid shield; the opisthosoma bears the respiratory appendages. The telson (tail) is a solid, usually spikelike, structure. *See* CHELICERATA; LIVING FOSSILS. [N.E.]

Mesogastropoda The largest and most diverse order of gastropod mollusks in the subclass Prosobranchia. Mesogastropods are all single-gilled and thus monotocardiac, that is, with only one auricle and asymmetry of other cardiac, renal, and genital structures. Mesogastropods with a wide variety of marine lifestyles are found burrowing in soft muds and living exposed to air near high-tide level. There are also estuarine, fresh-water, and terrestrial species in the order. The anatomical asymmetry makes possible a functional separation of genital from renal ducts, which in turn allows the development of internal fertilization, of large eggs, of ovoviviparity, and of true viviparity (all features advantageous for the evolution of nonmarine stocks).

Mesogastropods are structurally diverse, and most classifications divide them into 13 superfamilies, 4 of which number several thousand species and encompass some of the world's most abundant, ubiquitous, and cosmopolitan marine snails. *See* GASTROPODA; PROSOBRANCHIA. [W.D.R.-H.]

Meso-ionic compound A member of a class of five-membered ring heterocycles (and their benzo derivatives) which possess a sextet of π electrons in association with the atoms composing the ring but which cannot be represented satisfactorily by any one covalent or polar structure.

Two main types, depending formally upon the origin of the electrons in the π system, have been identified; they are exemplified by compounds (I) and (II). In structure (I) the nitrogen and

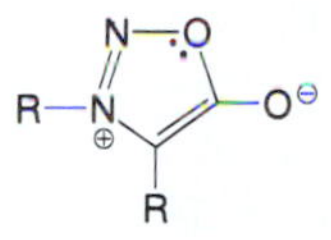

(I)

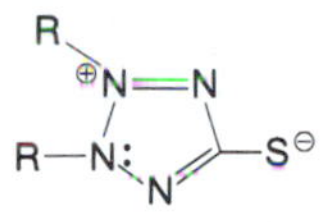

(II)

oxygen atoms, 1,3 to each other, are shown as donating two electrons each to the total of eight electrons in the whole π system, whereas in structure (II) the two middle nitrogen atoms, 1,2 to each other, are the two-electron donors.

The term "satisfactorily" in the definition refers to the fact that the charge in the ring cannot be associated exclusively with one ring atom. Thus, these compounds are in sharp contrast with other dipolar structures, such as ylides, and such compounds are not considered meso-ionic. *See* YLIDE.

There has been considerable interest in the pharmacological activity of meso-ionic compounds, and derivatives have shown a variety of antibiotic, anthelminthic, antidepressant, and anti-inflammatory properties. *See* HETEROCYCLIC COMPOUNDS. [J.P.Fr.]

Mesometeorology That portion of meteorology comprising the knowledge of intermediate-scale atmospheric phenomena, that is, in the size range of approximately 1–1200 mi (2–2000 km) and with time periods typically, but not always, less than 1 day. Unlike the larger weather systems on synoptic scales (the scales resolved by current weather reporting station networks) which typically produce significant changes over periods of days, most mesoscale phenomena have interdiurnal periods (less than 1 day), and consequently their changes are often more startling. In addition to time and space criteria for defining mesoscale, dynamical considerations can be used.

For observing mesoscale phenomena over midlatitude land masses, the average spacing between atmospheric sounding stations is about 180–360 mi (300–600 km) and soundings are taken twice each day. Consequently, only the largest mesoscale phenomena, with wavelengths greater than about 600 mi (1000 km), are routinely observed (resolved) by this network. Information with higher time-and-space resolution is available from aircraft observations and networks of radar stations, profilers, and satellite imagery. The profiler is a ground-based hybrid observing system of vertically pointing radar and microwave radiometry. The remote-sensing platforms often show that mesoscale weather systems are distinct components of larger synoptic-scale cyclones (low-pressure systems) and anticyclones (high-pressure systems).

Although the satellite imagery and radar data provide extensive areal coverage and clearly reveal the presence of the mesoscale systems, they do not provide measurements of certain atmospheric parameters (such as temperature, moisture, and pressure) in a form in which mesoscale structures and circulations can be readily quantified and understood. Thus, while mesoscale phenomena can be "observed," they cannot be studied and predicted (in the conventional manner) as easily as synoptic-scale systems. *See* METEOROLOGICAL SATELLITES; RADAR METEOROLOGY; WEATHER FORECASTING AND PREDICTION.

Mesoanalysis is the analysis of meteorological data in a manner that reveals the presence and characteristics of mesoscale phenomena. Because sounding stations are so widely spaced, only the largest of mesoscale systems can be resolved by the free-air (sounding) data. On the other hand, the density of surface observing stations is often satisfactory for identifying mesoscale features or circulations. Probably the most common application of mesoanalysis is for forecasting convective (thunderstorm) weather systems. *See* THUNDERSTORM.

Modern weather-forecasting techniques rely heavily upon predictions made from computers. These predictions, commonly called numerical model forecasts, require as input the three-dimensional initial state of the atmosphere. Normally, this initial condition is produced from the previous forecast and the most recent observations from the network of atmospheric soundings. For many mesoscale phenomena, it has been possible to develop relationships between the large-scale (synoptic) environment and the occurrence of particular types of mesoscale events. By using these relationships, the prediction of the synoptic-scale environment by the numerical models is then used to infer the likelihood of specific mesoscale events. Using satellite, radar, and conventional surface observations, the onset of an event is readily detected and appropriate adjustments to local forecasts are implemented. These adjustments usually come in the form of very short-term forecasts, commonly called nowcasts, and typically are valid for only about 3 h. However, depending upon the particular mesoscale phenomena, longer-term (3–12 h) forecasts sometimes are possible. *See* METEOROLOGY; NOWCASTING; STORM DETECTION. [J.Mi.F.]

Meson The generic name for any hadronic particle with baryon number zero. Such particles were first envisaged in 1935 by H. Yukawa. In 1937 C. Anderson established the existence of positively and negatively charged μ mesons (now known as muons) of mass 105.6 MeV in cosmic radiations, but these were soon shown to have very weak coupling with nucleons, that is, to be nonhadronic and therefore not "mesons," as this name is used today. The nuclear force mesons turned out to be the π mesons (pions), of mean mass about 138 MeV, first identified in cosmic radiation by C. F. Powell in 1947. *See* BARYON; COSMIC RAYS; HADRON; LEPTON.

In 1948, studies of the cosmic radiation indicated the existence of heavier mesons (masses about 495 MeV), now known as K mesons (kaons). There are four K mesons, K^+, K^- and two neutral particles, K_s^0 and K_L^0, and strong K^+, K^-, and K_L^0 beams are obtainable from higher-energy (≥30 GeV) proton accelerators. Since the $\pi^\pm$, $K^\pm$, and $K_{L,S}^0$ mesons have relatively long lifetimes (about 10^{-8} to 10^{-10} s), they are referred to as semistable.

In 1961 the ω meson of mass 783 MeV was discovered through its rapid decay to three pions in a bubble chamber investigation. It is highly unstable, with lifetime 6.6×10^{-23} s, as corresponds to the natural width 10.1 MeV observed in its mass value. Such a short lifetime implies that the ω meson decays through hadronic interactions. Many such highly unstable heavy mesons are not established, with lifetimes shorter than 10^{-22} s, decaying hadronically to lighter mesons, and more continue to be discovered.

In 1974 the J/ψ meson of mass 3097 MeV was discovered. Although it decays to mesons, its natural width is only 0.063 MeV (lifetime 10^{-20} s). Also, it decays electromagnetically, yielding an e^+e^- or $\mu^+\mu^-$ pairs, at a rate comparable with those for mesonic final states. This new phenomenon gave the first indication of the existence of heavy new quarks. In consequence, its study has often been termed the "new physics." *See* J PARTICLE; QUARKS.

Hadrons are now considered to be composite, consisting of spin-$\frac{1}{2}$ quarks (q), corresponding antiquarks ($\bar{q}$), and some number of gluons (g), the last being the quanta of the intermediate field which binds the quarks and antiquarks to form hadrons. Baryon number $B = +\frac{1}{3}$ holds for a quark q, $B = -\frac{1}{3}$ for antiquark $\bar{q}$, while $B = 0$ holds for a gluon. In this view, the simplest possibility is that each meson is a quark-antiquark (q-$\bar{q}$) pair bound together by the gluon field, and this model accounts quite well for most of the known mesons and their properties. *See* ELEMENTARY PARTICLE; GLUONS. [R.H.D.]

Mesosauria An order of extinct aquatic reptiles, also known as Proganosauria, of latest Carboniferous or earliest Permian time, about 280,000,000 years ago. The best-known genus is *Mesosaurus*. Like many aquatic reptiles, mesosaurs have a very long snout. The teeth are numerous and long, and appear very delicate. They may have served for filter feeding on soft-bodied invertebrates.

Mesosaurs, an early offshoot of the Carboniferous "stem reptiles" or captorhinomorphs, were the first reptiles to invade marine waters, but apparently soon became extinct, leaving no descendants. *See* REPTILIA. [R.L.C.]

Mesoscopic physics A subdiscipline of condensed-matter physics that focuses on the properties of solids in a size range intermediate between bulk matter and individual atoms or molecules. The size scale of interest is determined by the appearance of novel physical phenomena absent in bulk solids and has no rigid definition; however, the systems studied are normally in the range of 100 nanometers (the size of a typical virus) to 1000 nm (the size of a typical bacterium). Other branches of science, such as chemistry and molecular biology, also deal with objects in this size range, but mesoscopic physics has dealt primarily with artificial structures of metal or semiconducting material which have been fabricated by the techniques employed for producing microelectronic circuits. Thus mesoscopic physics has a close connection to the fields of nanofabrication and nanotechnology. Three categories of new phenomena in such systems are interference effects, quantum size effects, and charging effects. *See* ARTIFICIALLY LAYERED STRUCTURES; NANOSTRUCTURE; QUANTIZED ELECTRONIC STRUCTURE (QUEST); SEMICONDUCTOR HETEROSTRUCTURES.

Interference effects. In the mesoscopic regime, scattering from defects induces interference effects which modulate the flow of electrons. The experimental signature of mesoscopic interference effects is the appearance of reproducible fluctuations in physical quantities. For example, the conductance of a given specimen oscillates in an apparently random manner as a function of experimental parameters (see illus.). However, the same pattern may be retraced if the experimental parameters are cycled back to their original values; in fact, the patterns observed are reproducible over a period of days.

Quantum size effects. Another prediction of quantum mechanics is that electrons confined to a particular region of space may exist only in a certain set of allowed energy levels. The spacing between these levels increases as the confining region becomes smaller. One striking phenomenon which arises from these quantum size effects is the steplike increase of the conductance of electrons flowing through a constriction of several hundred nanometers' width. *See* ENERGY LEVEL (QUANTUM MECHANICS).

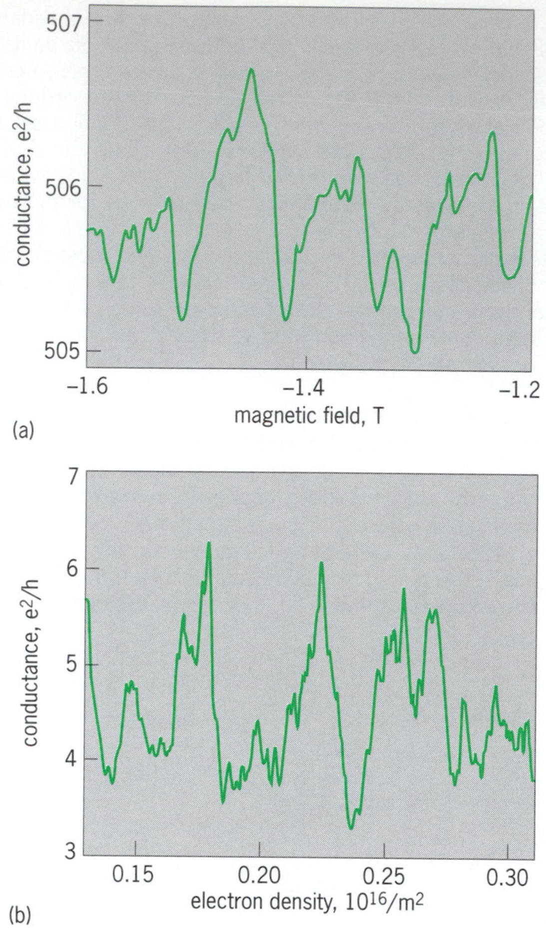

Conductance of 2000-nm gold wire as a function of magnetic field measured at a temperature of 0.04 K. The pattern observed is reproducible over a period of days. (*After R. A. Webb and S. Washburn, Quantum interference fluctuations in disordered materials, Phys. Today, 41(12):46–55, December 1988*).

Another mesoscopic system that shows quantum size effects consists of isolated islands of electrons that may be formed at the appropriately patterned interface between two different semiconducting materials. The electrons typically are confined to disk-shaped regions termed quantum dots. The confinement of the electrons in these systems changes their interaction with electromagnetic radiation significantly. *See* QUANTIZED ELECTRONIC STRUCTURE (QUEST).

Charging effects. Isolated mesoscopic solids such as quantum dots or metallic grains on an insulating substrate also show novel effects associated with the discreteness of the charge on the electron. Devices known as single-electron transistors (SETs) are by far the most sensitive electrometers (instruments for measuring electrical charge) presently known. *See* ELECTROMETER; TRANSISTOR. [A.D.S.]

Mesosphere An atmospheric layer stretching from the stratopause, at an altitude of about 30 mi (50 km), to the mesopause at about 53 mi (85 km). Like the troposphere, from which it is separated by the stratosphere, it is a layer in which the temperature generally decreases with height. Many of the atmospheric variations encountered in the mesosphere are linked to complicated processes in the underlying layers.

Special features of the mesosphere include a region of extremely strong winds centered near 40 mi (65 km), tidal oscillations and gravity waves of appreciable amplitude, meteors, and noctilucent clouds. The clouds are observed in middle and high latitudes near an altitude of 50 mi (80 km), in summer, when the Sun is at a small angle below the horizon. A light-scattering cloud prevails near the mesopause, over the summer pole. Meteors, especially numerous near 59 mi (95 km), commonly burn up before reaching the stratopause. Also in the domain of the mesosphere is the greater part of the D region, the lowest ionized layer of the atmosphere. *See* Atmosphere; Ionosphere; Meteor; Noctilucent clouds; Stratosphere. [R.S.Q.]

Mesostigmata A suborder of the Acarina (also called Parasitiformes), commonly referred to as the mites. Mesostigmata are characterized by a single pair of breathing pores, or stigmata, located laterally in the middle of the idiosoma between the second and third, or third and fourth, legs. A two- or three-tined palpal claw and a tritosternum, which is located ventrally on the idiosoma just behind the gnathosoma, are also characteristic for most mesostigmatid mites.

Many of the Mesostigmata are predacious on other small arthropods or are scavengers and fungus-feeders that make up an important segment of the soil fauna. Some members of the family Phytoseiidae are beneficial to humans as predators of the destructive spider mites, Tetranychidae. Other forms are parasitic on insects and other invertebrates.

Terrestrial arthropods are the only invertebrate hosts affected by mesostigmatid mites. As a rule, phoresy is the predominant reason for the association. Vertebrate hosts are restricted to mammals, birds, and reptiles. The Mesostigmata that parasitize them are of two main sorts: They are either internal parasites of the respiratory passages, or they are ectoparasitic nidicoles; that is, they reproduce in the nests of their hosts. *See* Acari; Phoresy. [J.H.C.; R.W.St.]

Mesozoa A division of the animal kingdom sometimes ranked as intermediate between the Protozoa and the Metazoa. The Mesozoa comprise two orders of small, wormlike organisms, the Dicyemida and the Orthonectida. Both are parasitic in marine invertebrates. The body consists of a single layer of ciliated cells enclosing one or more reproductive cells. These body cells are rather constant in number and arrangement for any given species. The internal cells do not correspond to the endoderm of other animals, as they have no digestive function. The life cycles are complex, involving both sexual and asexual generations (metagenesis). *See* Animal kingdom. [B.H.McC.]

Mesozoic In geology, the system of rocks younger than the Paleozoic and older than the Cenozoic; also, the geologic era during which those rocks were formed. The era is now believed to have extended from about 225,000,000 to 64,000,000 years ago. The Mesozoic is commonly referred to as the Age of Reptiles inasmuch as these animals dominated the land, sea, and air. The era is divided into three systems or periods which are, from oldest to youngest, Triassic, Jurassic, and Cretaceous. *See* Cretaceous; Jurassic; Triassic. [W.A.Co.]

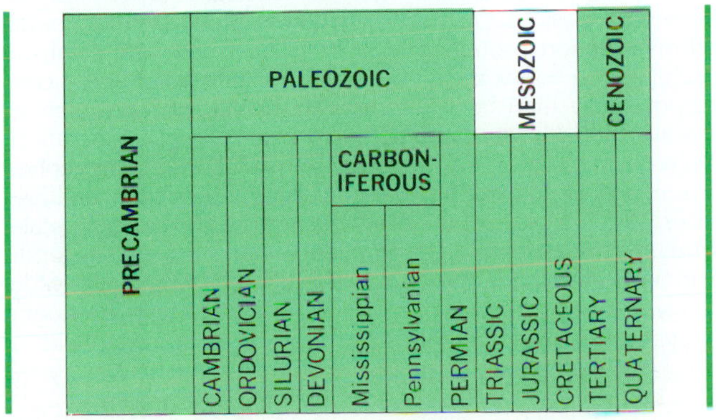

Messier catalog An early listing of nebulae and star clusters. Charles Messier (1730–1817) was primarily interested in discovering comets, so he compiled a list of objects that might be confused with comets in small telescopes. His first catalog was published in the Royal Academy of Sciences of France in 1771. The compilation contains both open and globular star clusters, diffuse galactic nebulae, so-called planetary nebulae, and external galaxies.

The New General Catalog (NGC) and the supplementary Index Catalogs (IC) published by J. L. E. Dreyer at the end of the 19th century were far more extensive than that of Messier. Consequently the NGC numbers are generally used for all except a few objects whose Messier numbers had become firmly established. *See* Galaxy, external; Nebula; Star clusters. [L.H.Al.]

Metabolic disorders Disorders which involve an alteration in the normal metabolism of carbohydrates, lipids, proteins, water, and nucleic acids. Deviations in normal metabolic processes are evidenced by various syndromes and diseases. *See* Malnutrition; Metabolism.

Abnormal protein metabolism. The diseases associated with abnormal protein metabolism have been classified as follows: (1) diseases associated with increased production of proteins, (2) those associated with decreased production of proteins, (3) diseases associated with the production of abnormal proteins, and (4) diseases associated with the excretion of unusual amounts of amino acids.

Hyperproteinemia is an increase of protein, usually one type of plasma protein. When it occurs, more β- or γ-globulins are produced while fewer total proteins, including α-globulins and albumins, are synthesized. The diseases usually associated with hyperglobulinemia are multiple myeloma, kala-azar, Hodgkin's disease, lymphogranuloma inguinale, sarcoidosis, liver cirrhosis, and amyloid. The mechanisms which stimulate the production of proteins, particularly of one type, are not known, but the plasma cell has been implicated. *See* Cirrhosis; Hodgkin's disease.

A decreased amount of protein, hypoproteinemia, may be the result of a lack of essential amino acids for protein synthesis, a metabolic block, or other interference with the normal synthesis mechanism. Increased excretion of protein, particularly in chronic renal disease with a loss of albumin in the urine (albuminuria), is another common cause of hypoproteinemia. Kwashiorkor is the best example of hypoproteinemia resulting from dietary deficiency. *See* Hepatitis; Kwashiorkor disease.

Another group of abnormal proteins have been associated with the hemoglobins. Hemoglobin functions in the transport of oxygen and carbon dioxide in the blood. Several different hemoglobins in red blood cells in mammals have been described, one of which produces the disease called sickle-cell anemia. *See* Hemoglobin.

Amino aciduria is a group of disorders in which there appears to be an increase in the amount of amino acids excreted in the urine. They are due to abnormal protein metabolism and result either from an overflow mechanism where the concentration of amino acids in the serum surpasses the renal threshold of the glomerular membrane, or from defective absorption of amino acids in the renal tubules. The overflow amino acidurias include phenylketonuria, alkaptonuria, and liver disease. *See* Phenylketonuria; Protein metabolism.

Abnormal nucleic acid metabolism. Since nucleic acids are closely associated with chromosomes and the process of cell division as well as with protein synthesis, they are one of the most important of cell constituents. They have been shown to be very susceptible to injury by various agents, and interruption of division is one of the earliest signs of cell damage. A disturbance in nucleic acid metabolism is of extreme importance to many other aspects of cell metabolism. The whole field of virus infection of mammalian cells is undoubtedly largely concerned with a disturbance in nucleic acid or nucleoprotein metabolism. *See* GOUT; LUPUS ERYTHEMATOSUS.

Abnormal lipid metabolism. Obesity in general may be attributed to an excess intake of calories and only rarely may glandular deficiencies such as hypopituitarism and hypothyroidism be implicated as its cause. Increased caloric intake results in an increase in the size and number of fat cells distributed throughout normal lipid deposits and many organs of the body. *See* OBESITY; PITUITARY GLAND DISORDERS; THYROID GLAND DISORDERS.

Hyperlipemia, an excess of lipid in the blood, is usually secondary to uncontrollable diabetes, hypothyroidism, biliary cirrhosis, or lipid nephrosis. A primary type of lipemia is caused by a genic defect. There is marked delay in the clearing of ingested fat so that a milky serum with elevation of all the lipid fractions, especially of the neutral fat, occurs. Severe sclerosis, and occasionally xanthomatosis, is common in this disease in early adulthood. Another primary hyperlipemia, called idiopathic hypercholesteremia, has also been described. Sclerosis of all the vessels is marked, with a high incidence of myocardial infarct.

There are small tumors composed of fat cells called lipomas which are commonly found in the subcutaneous tissue and which occasionally become malignant liposarcomas.

There are four common major diseases which are associated with accumulations of lipid in and around damaged cells. These are, in order of importance, arteriosclerosis, myocardial infarct, cirrhosis of the liver, and nephrosis. *See* ARTERIOSCLEROSIS; CIRRHOSIS; INFARCTION.

There is another group of diseases in which lipid accumulates because of a disturbance in lipid metabolism, not dependent as far as is known on external stimuli. These are the so-called lipid storage diseases which are in some ways very similar to the glycogen storage disease. In all of these diseases a large accumulation of lipids appears in many cells, but particularly in the reticuloendothelial cells of the lymph nodes, liver, spleen, and bone marrow. The lipid seems to be distributed throughout the cell and is not well localized. *See* LIPID METABOLISM.

Abnormal carbohydrate metabolism. There are five pathological states which are reasonably well elucidated in terms of deficient mechanisms or in terms of the specific carbohydrate involved: diabetes mellitus, glycogen storage disease (von Gierke's disease), Hurler-Pfaundler's disease, galactosemia, and malignant neoplasm. The end effects of these diseases have been known for many years; the causes still remain obscure. *See* DIABETES.

In glycogen storage disease extraordinarily large amounts of glycogen are deposited in essential organs such as the heart, liver, kidney, and skeletal muscle. The symptoms of the disease arise when the accumulation of glycogen interferes with the normal cells' metabolism. When this occurs in the heart, the muscle fibers appear split by large amounts of glycogen, muscle contractility is impaired, and heart failure and death may eventually result.

In Hurler-Pfaundler's disease there is excess storage of a mucopolysaccharide in the spleen, liver, and other tissues. The symptoms, as in glycogen storage disease, result from interference with the cells' function. The enzymatic deficiency in this disease is still unknown.

Galactosemia is a disease of the newborn resulting from an inherited deficiency of the enzyme concerned with the conversion of galactose to glucose, phosphogalactose transferase. The principal food of newborns is milk which contains large quantities of lactose. Lactose is usually broken to galactose and glucose, and the glucose is quickly utilized. The galactose, however, must first be converted to glucose. If the galactose is not converted, it accumulates in cells and interferes with their normal functions. *See* CARBOHYDRATE METABOLISM. [D.W.Ki.]

Metabolism All the physical and chemical processes by which living, organized substance is produced and maintained and the transformations by which energy is made available for use by an organism.

In defining metabolism, it is customary to distinguish between energy metabolism and intermediary metabolism, although the two are, in fact, inseparable. Energy metabolism is primarily concerned with overall heat production in an organism, while intermediary metabolism deals with chemical reactions within cells and tissues. In general, the term metabolism is interpreted to mean intermediary metabolism. *See* ENERGY METABOLISM.

Metabolism thus includes all biochemical processes within cells and tissues which are concerned with their building up, breaking down, and functioning. The synthesis and maintenance of tissue structure generally involves the union of smaller into larger molecules. This part of metabolism, the building of tissues, is termed anabolism. The process of breaking down tissue, of splitting larger protoplasmic molecules into smaller ones, is termed catabolism. Growth or weight gain occurs when anabolism exceeds catabolism. On the other hand, weight loss results if catabolism proceeds more rapidly than anabolism, as in periods of starvation, serious injury, or disease. When the two processes are balanced, tissue mass remains the same.

The metabolism of the three major foodstuffs, carbohydrates, fats, and proteins, is intimately interrelated, so any clearcut division of the three is arbitrary and inaccurate. Thus the metabolism of protoplasm is concerned with all three of these foodstuffs. The metabolic pathways of carbohydrates, fats, and proteins cross at many points; thus certain pathways of metabolism are shared in common by fragments of these different classes of foodstuffs.

Some of the metabolic processes of the protoplasm of both plant and animal cells occur along common pathways; carbohydrate metabolism in plants is similar in many details to carbohydrate metabolism in animals. Therefore the study of metabolism in any organism is, in a sense, the study of metabolism in all protoplasm. *See* CARBOHYDRATE METABOLISM; KREBS CYCLE; LIPID METABOLISM; PROTEIN METABOLISM. [M.B.McC.]

Metachlamydeae An artificial group of flowering plants, division Magnoliophyta (Angiospermae), recognized in the Englerian system of classification, consisting of those families of dicotyledons (Magnoliopsida) in which the petals are characteristically fused, forming a sympetalous corolla. This is in contrast to the Archichlamydeae, which have separate petals or lack petals. The group has also been called Sympetalae. *See* ARCHICHLAMYDEAE; MAGNOLIOPHYTA. [A.Cr.]

Metal An electropositive chemical element. Physically, a metal atom in the ground state contains a partially filled band with an empty state close to an occupied state. Chemically, upon going into solution a metal atom releases an electron to become a positive ion. Consequently in biotic systems metal atoms function prominently in ionic transport and electron exchange. In bulk a metal has a high melting point and a correspondingly high boiling temperature; except for mercury, metals are solid at standard conditions. Direct observation shows a metal to be relatively dense, malleable, ductile, cohesive, highly conductive both electrically and thermally, and lustrous. When crystals of the elements are classified along a scale from plastic to brittle, metals fall toward the plastic end. Furthermore, molten metals mixed with each other over wide ranges of proportions form, upon slowly cooling, homogeneous close-packed crystals. In contrast, a metal mixed with a nonmetal completely combines into a homogeneous crystal only in one or a few discrete stoichiometric proportions.

For detailed discussions of metals, in particular, exceptions to generic behavior, see separate articles on each metal. *See also* ELEMENTS; PERIODIC TABLE. [F.H.R.]

Metal-base fuel A fuel containing a metal of high heat of combustion as a principal constituent. High propellant performance in either a rocket or an air-breathing engine is obtained when the heat of combustion of the fuel is high. Chemically, high heats of combustion are attained by the oxidation of the low-atomic-weight metals in the upper left-hand corner of the periodic table. The generally preferred candidates are lithium, beryllium, boron, carbon, magnesium, and aluminum.

The metallized additive can be used in either a liquid or solid propellant. When the pure metal is added to liquid fuels, an emulsifying or gelling agent is employed which maintains the particles in uniform suspension. When used in composite solid rocket propellants, the metal powder is usually mixed with the oxidizer and unpolymerized fuel, and the propellant is then processed in the usual way.

Early compounding of metallized propellants employed the free metal itself. As a result of extensive research in metalloorganic compounds, however, several classes of metallic compounds have been employed. Two major reasons for the use of such compounds are that the solubility of the metal in the fuel can be realized, resulting in a homogeneous propellant, and performance higher than that for the pure metal can be obtained in some cases.

The major classes of metallic compounds of interest as high-performance propellants include the hydrides, amides, and hydrocarbons. Additional classes include mixtures either of two metals or of two chemical groups, such as an amine hydrocarbon.

Most metallic fuels are costly and many, because of particle-size requirements or synthesis in a specific compound, are in limited supply. The combustion gases all produce smoky exhausts which may be objectionable in use. Engine development problems are also increased because of the appearance of smoke and deposits in the engine. [D.Al.]

Metal carbonyl A compound of a metal combined with carbon monoxide (CO). The structure involves a coordinate link between the metal atom and the CO group, with each CO donating a pair of electrons to the metal. Metal carbonyls are usually low-melting crystalline solids, and several are liquids at room temperature. They are generally soluble in organic solvents but insoluble in water. The primary decomposition products are the metal and carbon monoxide. Metal carbonyls are quite reactive. [H.S.]

Metal casting The introduction of molten metal into a cavity or mold where, upon solidification, it becomes an object whose shape is determined by mold configuration. Casting offers several advantages over other methods of metal forming: it is adaptable to intricate shapes, to extremely large pieces, and to mass production; it can provide parts with uniform physical and mechanical properties throughout; and, depending on the particular material being cast, the design of the part, and the quantity being produced, its economic advantages can surpass other processes.

Two broad categories of metal-casting processes exist: ingot casting (which includes continuous casting) and casting to shape. Ingot castings are produced by pouring molten metal into a permanent or reusable mold. Following solidification these ingots (or bars, slabs, or billets, as the case may be) are then further processed mechanically into many new shapes. Casting to shape involves pouring molten metal into molds in which the cavity provides the final useful shape, followed only by machining or welding for the specific application. There are four basic types of these casting processes: sand, permanent-mold, die, and centrifugal. [K.R.]

Metal cluster compounds A compound in which two or more metals aggregate so as to be within bonding distance of one another. Heterogeneous metal clusters of the transition metals deposited on supports such as alumina, silica, and zeolites have been widely used in catalysis. However, research involving soluble metal cluster compounds in homogeneous solution indicates that they can mimic the heterogeneous counterpart. Since solution chemistry is more amenable to characterization, the soluble clusters should lead to an intimate understanding of catalytic mechanisms and to the design of very selective catalysts.

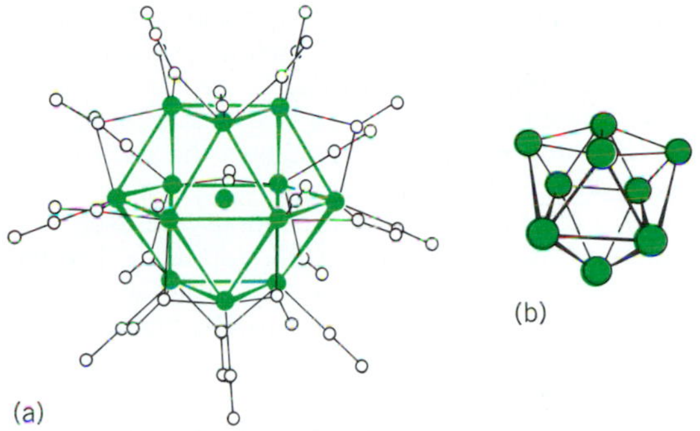

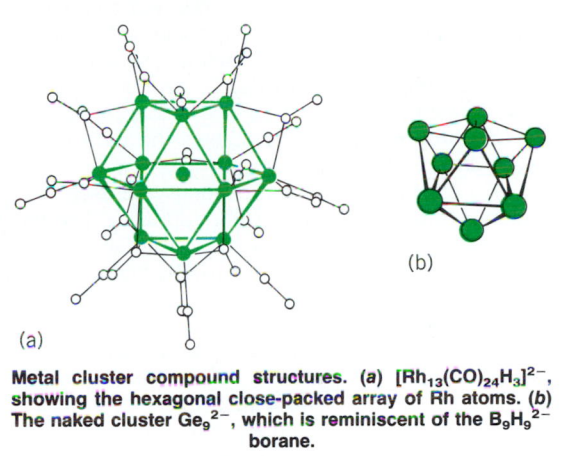

Metal cluster compound structures. ***(a)*** **$[Rh_{13}(CO)_{24}H_3]^{2-}$, showing the hexagonal close-packed array of Rh atoms.** ***(b)*** **The naked cluster Ge_9^{2-}, which is reminiscent of the $B_9H_9^{2-}$ borane.**

Metal clusters in solution fall into two general categories: those encumbered with a coordination sphere of chemically bound ligands (ligated clusters; see illustration), and those surrounded only by a sphere of solvent molecules (the so-called naked clusters). Clusters of the first class are found primarily in the case of transition metals and are critical to catalysis since the substrate undergoing catalytic transformation must be chemically bound to the cluster some time during the catalytic cycle. *See* HETEROGENEOUS CATALYSIS; HOMOGENEOUS CATALYSIS. [R.W.Ru.]

Metal coatings Thin films of material bonded to metals in order to add specific surface properties, such as corrosion or oxidation resistance, color, attractive appearance, wear resistance, optical properties, electrical resistance, or thermal protection. This article discusses various methods of applying either metallic coatings or nonmetallic coatings, such as vitreous enamel and ceramics, and the conversion of surfaces to suitable reaction-product coatings. For other methods for the protection of metal surfaces *see* CLADDING; ELECTROLESS PLATING; ELECTROPLATING OF METALS; LACQUER; PAINT.

Hot-dipped coatings of low-melting metals provide inexpensive protection to the surfaces of a variety of steel articles. Thoroughly cleaned work is immersed in a molten bath of the coating metal. The coating consists of a thin alloy layer together with relatively pure coating metal that adheres to the work as it is withdrawn from the bath.

Sprayed coating permits the coating of assembled steel structures to obtain corrosion resistance, the building up of worn machine parts for rejuvenation, and the application of highly refractory coatings with melting points in excess of 3000°F (1650°C).

Cementation coatings are surface alloys formed by diffusion of the coating metal into the base metal, producing little dimensional change. Parts are heated in contact with powdered coating material that diffuses into the surface to form an alloy coating, whose thickness depends on the time and the temperature of treatment.

In vapor deposition a thin specular coating is formed on metals, plastics, paper, glass, and even fabrics. Coatings form by condensation of metal vapor originating from molten metal, from high-voltage discharge between electrodes (cathode sputtering), or from chemical means such as hydrogen reduction or thermal decomposition (gas plating) of metal halides.

Immersion coatings are produced either by direct chemical displacement or for thicker coatings by chemical reduction (electroless coating). Metal ions plate out of solution onto the workpiece.

Vitreous enamel coatings are glassy but noncrystalline coatings for attractive durable service in chemical, atmospheric, or moderately high-temperature environments. In wet enameling a slip is prepared of a water suspension of crushed glass, flux, suspending agent, refractory compound, and coloring agents or opacifiers. The slip is applied by dipping or flow coating; it is then fired at a temperature at which it fuses into a continuous vitreous coating. Dry enameling is used for castings, such as bathtubs. The casting is heated to a high temperature, and then dry enamel powder is sprinkled over the surface, where it fuses. *See* FRIT.

Essentially crystalline, ceramic coatings are used for high-temperature protection above 1100°C (2000°F). The coatings may be formed by spraying refractory materials such as aluminum oxide or zirconium oxide, or by the cementation processes for coatings of intermetallic compounds such as molybdenum disilicide. *See* CERMET.

Surface-conversion coatings provide an insulating barrier of low solubility formed on steel, zinc, aluminum, or magnesium without electric current. The article to be coated is either immersed in or sprayed with an aqueous solution, which converts the surface into a phosphate, an oxide, or a chromate.

Anodic coatings of protective oxide may be formed on aluminum or magnesium by making them the anode in an electrolytic cell. If permanent color is required, the coating is impregnated with a dye before sealing. *See* CORROSION. [W.W.Br.]

Metal forming Manufacturing processes by which parts or components are fabricated from metal stock. In the specific technical sense, metal forming involves changing the shape of a piece of metal. In general terms, however, it may be classified roughly into five categories: mechanical working, such as forging, extrusion, rolling, drawing, and various sheet-forming processes; casting; powder and fiber metal forming; electroforming; and joining processes. *See* DRAWING OF METAL; EXTRUSION; FORGING; METAL CASTING; METAL ROLLING; POWDER METALLURGY; SHEET-METAL FORMING. [S.Ka.]

Metal halide lamp A high-pressure discharge lamp that is enclosed in a quartz envelope containing metal halides, usually iodides, and produces high-efficacy white light. These lamps are widely used for sports stadiums, roadways, commercial interiors, and industrial applications. The singular lamp feature is the compact geometry and high efficacy of nearly white light. *See* LUMINOUS EFFICACY.

The metal halide lamp requires a high voltage in order to start. Once the arc discharge is ignited, the internal vapor pressure begins to increase until it reaches a preset value, usually around 5 atm (500 kilopascals). These lamps are always connected to an auxiliary power supply called the ballast which supplies the proper voltage and current for starting and operating the lamps. The light output gradually increases over approximately 2 min as the various ingredients begin to vaporize and emit light. The light from the arc discharge comes from the metal components of the iodide compounds, which are typically a mixture of sodium, thallium, indium, scandium, dysprosium, and occasionally tin iodide. A combination of these metals produces a pleasing white light of very high efficiency, between 80 and 120 lumens per watt or even higher. *See* ARC DISCHARGE.

Special ingredients can be used inside these lamps to make them suitable for plant growth, photocopying, and scientific usages. Certain types of compact metal halide lamps are ideal for use with movie and slide projection. In these cases, the ingredients chosen (usually indium) are used because they produce a small, intense high-brightness arc. The spectra of these lamps are nearly continuous and approach the spectra found in an incandescent lamp. *See* CINEMATOGRAPHY; INCANDESCENT LAMP; OPTICAL PROJECTION SYSTEMS.

In cases where a more diffuse light source is desired, the outer jacket of the lamp can be coated with a white phosphor in order both to diffuse the light and to change the color of the output and in some cases actually improve the efficiency. [G.H.R.]

Metal hydrides A compound in which hydrogen is bonded chemically to a metal or metalloid element. The compounds are classified generally as ionic, transition metal, and covalent hydrides. Covalent hydrides are of two subtypes, binary and complex. Certain hydrides have achieved a position of modest industrial importance, but most are of theoretical interest only.

Under extreme conditions such as in electric discharges, many metals form volatile, short-lived transient hydrides of the general formula type MH. Although some of these can be prepared experimentally, most are observed only by their spectra. They are important in studying molecular bonding. The action of atomic hydrogen at low temperatures forms surface films of unstable hydrides with many metals. *See* HYDRIDO COMPLEXES; HYDROGEN. [J.C.W.]

Metal matrix composite A material consisting of two or more substances, one of which is a metal or metallic alloy, and which exhibits its own distinctive behavior. The latter criterion sets a composite apart from conventional metallic materials such as steels and nonferrous alloys, both of which contain two or more constituents. For example, the stiffness or strength of a composite may be higher than that of the individual components, or radically different from either.

Typically, metal matrix composites consist of wires, filaments, fibers (fine wires or filaments), whiskers, or thin sheets of a strong, relatively brittle material embedded in the weaker but more ductile metallic constituent. This combination and particular distribution of the constituents result in enhanced stiffness and strength coupled with ductility and toughness (the ability to withstand impact or shock loading). The need for a fine-scale distribution of the reinforcement, and the role of the matrix in stress transfer, crack propagation, and protection of the reinforcement are considerations which apply to metal matrix composites. The reinforcement may be either continuous, that is, extending the full length of the composite, or discontinuous, consisting of chopped lengths having a particular ratio of length to diameter (aspect ratio).

The overall property combinations of metal matrix composites are particularly attractive in aerospace and marine engineering, propulsion, and transportation. Unlike polymer matrix composites, metal matrix composites can be utilized at both ambient temperature and high temperatures. *See* ALLOY; COMPOSITE MATERIAL; POLYMERIC COMPOSITE. [A.La.]

Metal rolling Reducing or changing the cross-sectional area of a workpiece by the compressive forces exerted by rotating rolls. The original material fed into the rolls is usually an ingot from a foundry. The largest product in hot rolling is called a bloom; by successive hot- and then cold-rolling operations the bloom is reduced to a billet, slab, plate, sheet, strip, and foil, in decreasing order of thickness and size. The initial breakdown of the ingot by rolling changes the coarse-grained, brittle, and porous structure into a wrought structure with greater ductility and finer grain size.

A schematic presentation of the rolling process, in which the thickness of the metal is reduced as it passes through the rolls, is shown in illustration *a*. The speed at which the metal moves during rolling changes, as shown in illustration *b*, to keep the

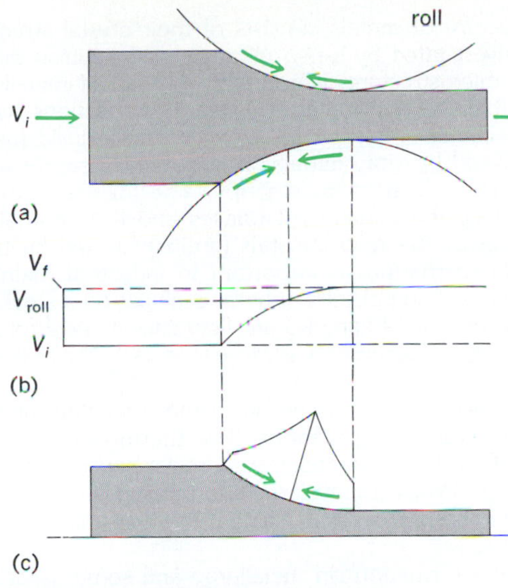

The rolling process. (*a*) Direction of friction forces in the roll gap. (*b*) Velocity distribution. (*c*) Normal pressure acting on the strip in the roll gap. V_i = initial velocity, V_f = final velocity, V_{roll} = velocity during rolling operation.

volume rate of flow constant throughout the roll gap. Hence, as the thickness decreases, the velocity increases; however, the surface speed of a point on the roll is constant, and there is therefore relative sliding between the roll and the strip. The normal pressure distribution on the roll and hence on the strip is of the form shown in illustration *c*. Because of its particular shape this pressure distribution is known as the friction hill.

A great variety of roll arrangements and equipment are used in rolling. The proper reduction per pass in rolling depends on the type of material and other factors; for soft, nonferrous metals, reductions are usually high, while for high-strength alloys they are small. Requirements for roll materials are mainly strength and resistance to wear. Common roll materials are cast iron, cast steel, and forged steel. [S.Ka.]

Metallic-disk rectifier A rectifier that consists of one or more disks of metal in contact with coatings or layers of a semiconductor material. Alternating current is changed into pulsating direct current by the rectifying action that occurs at the junction interface between a metal disk and its mating-semiconductor layer. The most common examples are the selenium and copper oxide rectifiers. *See* SEMICONDUCTOR RECTIFIER. [J.Mar.]

Metallic electrode arc lamp An arc lamp whose positive electrode is solid copper, and whose negative electrode is formed of magnetic iron oxide with titanium as the light-producing element and other chemicals to control steadiness and vaporization. In this type of lamp, light is produced by luminescent vapor introduced into the arc by conduction from the cathode. These lamps are limited to direct-current use. A regulating mechanism adjusts the arc. *See* ARC LAMP. [J.O.K.]

Metallic glasses Alloys having amorphous or glassy structures. A glass is a solid material obtained from a liquid which does not crystallize during cooling. It is therefore an amorphous solid, which means that the atoms are packed in a more or less random fashion similar to that in the liquid state. The word glass is generally associated with the familiar transparent silicate glasses containing mostly silica and other oxides of aluminum, magnesium, sodium, and so on. These glasses are not metallic; they are electrical insulators and do not exhibit ferromagnetism. In 1960 the first true metallic glass, $Au_{80}Si_{20}$, an alloy containing 80 at. % Au and 20 at. % Si, was obtained. *See* ALLOY; AMORPHOUS SOLID. [P.E.D.]

Metalloacid elements The chemical elements with the following atomic numbers and names: 23, vanadium, V; 41, niobium, Nb; 73, tantalum, Ta; 24, chromium, Cr; 42, molybdenum, Mo; 74, tungsten, W; 25, manganese, Mn; 43, technetium, Tc; and 75, rhenium, Re. These elements are the subgroup members of groups V, VI, and VII, respectively, of the periodic table. In the elemental state all are metals of relatively high density, high melting point, and low volatility. The classification metalloacid elements refers to the fact that their oxides react with water to give somewhat acidic solutions, in contrast to the more typical behavior of the oxides of other metals which yield basic solutions. *See* PERIODIC TABLE; TRANSITION ELEMENTS. [B.B.Cu.]

Metallocenes A group of organometallic compounds, biscyclopentadienyl metals, which generally possess a sandwich structure (see illustration).

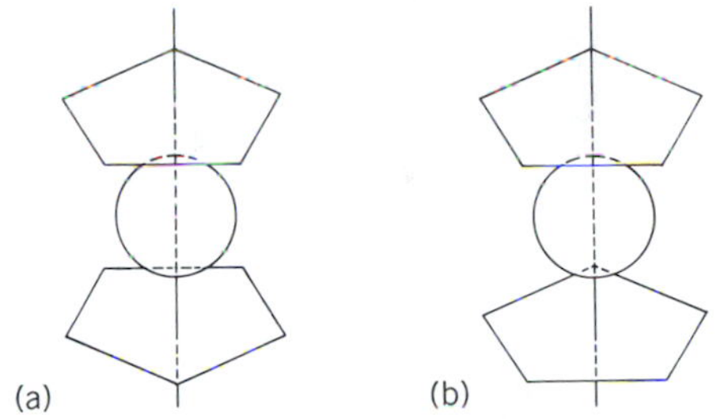

Metallocene structures. (*a*) Staggered. (*b*) Eclipsed.

Almost all transition metals have been found to form metallocenes. According to the center metal atom, metallocenes are called ferrocenes (Fe), ruthenocenes (Ru), and so on. In a brief definition, both cyclopentadienyl groups bind with the metal atom in π fashion (termed pentahapto, abbreviated h^5).

Metallocenes have been used for coating or plating metal by their vapor-phase decomposition. Applications of metallocenes are versatile and have potential for further developments. *See* FERROCENE; ORGANOMETALLIC COMPOUND. [M.Ts.]

Metallography The study of the structure of metals and alloys by various methods, especially optical and electron microscopy. The term metallography once designated the sci-

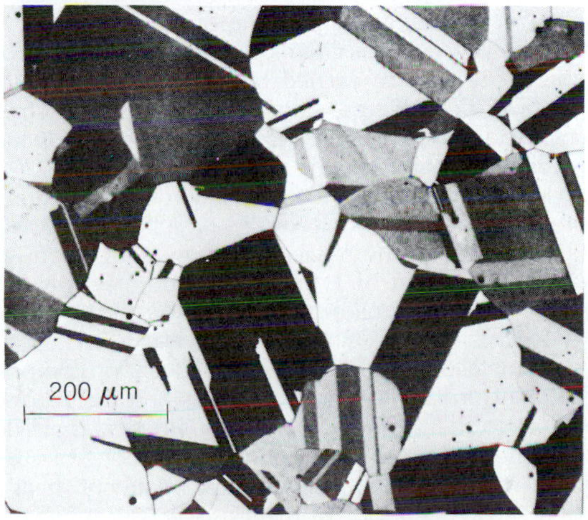

Photomicrograph of the structure of annealed brass (70% Cu, 30% Zn). (*Courtesy of W. R. Johnson*)

ence relating the properties of metals to their structure, but this science is now called physical metallurgy. *See* METALLURGY.

The optical microscopy of metals is conducted with reflected light on surfaces suitably prepared to reveal structural features (see illustration); a resolution of about 200 nanometers and a linear magnification of at most 2000 × can be obtained. Electron microscopy is generally carried out either with replicas of specimen surfaces or with thin foils prepared from bulk specimens. The replica technique provides a resolution of detail as small as 5 nm and magnifications of up to approximately 50,000 × . The thin-foil technique provides a resolution of better than 1 nm and a useful magnification of 500,000 × . *See* ELECTRON MICROSCOPE; OPTICAL MICROSCOPE.

Metallography serves both research and industrial practice. Optical microscopy has long been a standard method for the investigation of the phase constitution of metallic systems and the microstructure of metals. Phase transformations and the mechanisms by which metals deform plastically are becoming better understood through the application of electron microscopy. Metallographic methods are used in industry for the control of production processes, especially heat treatments. These methods are frequently indispensable in the analysis of the causes of service failures of metallic objects. Metallographic methods have been adapted to the examination of nonmetallic materials, especially ceramics, semiconductors, and minerals. [M.B.B.; J.B.V.S.]

Metalloid An element which exhibits the external characteristics of a metal but behaves chemically both as a metal and as a nonmetal. Arsenic and antimony, for example, are hard crystalline solids that are definitely metallic in appearance. They may, however, undergo reactions that are characteristic of both metals and nonmetals. However, only when this dualistic chemical behavior is very marked and the external appearance metallic is the element commonly called a metalloid. *See* METAL; NONMETAL. [F.J.J.]

Metallurgy The technology and science of metallic materials. Metallurgy as a branch of engineering is concerned with the production of metals and alloys, their adaptation to use, and their performance in service. As a science, metallurgy is concerned with the chemical reactions involved in the processes by which metals are produced and the chemical, physical, and mechanical behavior of metallic materials.

The field of metallurgy may be divided into process metallurgy (production metallurgy, extractive metallurgy) and physical metallurgy. In this system metal processing is considered to be a part of process metallurgy and the mechanical behavior of metals a part of physical metallurgy.

Process metallurgy, the science and technology used in the production of metals, employs some of the same unit operations and unit processes as chemical engineering. These operations and processes are carried out with ores, concentrates, scrap metals, fuels, fluxes, slags, solvents, and electrolytes. Different metals require different combinations of operations and processes, but typically the production of a metal involves two major steps. The first is the production of an impure metal from ore minerals, commonly oxides or sulfides, and the second is the refining of the reduced impure metal, for example, by selective oxidation of impurities or by electrolysis. *See* ELECTROMETALLURGY; HYDROMETALLURGY; IRON METALLURGY; ORE DRESSING; PYROMETALLURGY; STEEL MANUFACTURE.

Physical metallurgy investigates the effects of composition and treatment on the structure of metals and the relations of the structure to the properties of metals. Physical metallurgy is also concerned with the engineering applications of scientific principles to the fabrication, mechanical treatment, heat treatment, and service behavior of metals. *See* ALLOY; HEAT TREATMENT (METALLURGY).

The structure of metals consists of their crystal structure, which is investigated by x-ray, electron, and neutron diffraction, their microstructure, which is the subject of metallography, and their macrostructure. Crystal imperfections, which provide mechanisms for processes occurring in solid metals, are investigated by x-ray diffraction and metallographic methods, especially electron microscopy. The microstructure is determined by the constituent phases and the geometrical arrangement of the microcrystals (grains) formed by those phases. Macrostructure is important in industrial metals. It involves chemical and physical inhomogeneities on a scale larger than microscopic. Examples are flow lines in steel forgings and blowholes in castings. *See* METALLOGRAPHY; X-RAY DIFFRACTION.

Phase transformations occurring in the solid state underlie many heat-treatment operations. The thermodynamics and kinetics of these transformations are a major concern of physical metallurgy. Physical metallurgy also investigates changes in the structure and properties resulting from mechanical working of metals.

For more information on metallurgy and some associated techniques see articles on individual metals and their metallurgy. *See also* ELECTROPLATING OF METALS; METAL COATINGS; METAL FORMING. [M.B.B.]

Metamerism The serial repetition of parts along the length of the body axis in bilaterally symmetrical animals. The successive subdivisions are called metameres, somites, or segments, hence the synonym, segmentation. Common examples are the muscles and spinal nerves in the human body and in the body and tail of many mammals, snakes and lizards, salamanders, and fishes. It also occurs in other chordates, and in arthropods and annelid worms. *See* ANIMAL SYMMETRY. [H.L.H.]

Metamict state The amorphous state of substances that have lost their original crystalline structure because of the radioactivity of uranium or thorium. The process of rendering crystalline substances partly or wholly amorphous is known as metamictization. Minerals whose crystal structure has been disrupted by this process are known as metamict minerals. Some minerals which show crystal form are nevertheless structurally amorphous but have attained this state in a different manner than glasses or rigid gels. This third class of amorphous substances is known as metamict (mixed otherwise).

The more significant features of metamict minerals are:

1. Anomalous optical behavior. Many metamict substances are heterogeneous, being partly isotropic and partly anisotropic (doubly refractive).

2. Pyrognomic behavior. The minerals readily become incandescent on heating, although this varies greatly.

3. Glasslike properties, The minerals lack cleavage but exhibit conchoidal fracture, some being particularly brittle.

4. Density. The density is increased by heating.

5. Reconstitution of crystalline structure by heating.

6. Resistance to attack by acids. It is increased by heating.

7. Presence of uranium (U) or thorium (Th). In addition the presence of rare earths and related elements has been emphasized by some observers. *See* RARE-EARTH ELEMENTS.

8. Existence of some minerals in both the crystalline and metamict states. In these cases little, if any, chemical difference can be found.

9. Amorphous behavior when exposed to x-rays. This is regarded as the crucial test. Some traces of x-ray diffraction may be observed, though the minerals have become optically quite isotropic. *See* X-RAY DIFFRACTION. [P.M.H.]

Metamorphic rocks Preexisting rock masses in which new minerals, textures, or structures are formed at higher temperatures and greater pressures than those normally present at the Earth's surface. *See* IGNEOUS ROCKS; SEDIMENTARY ROCKS.

Two groups of metamorphic rocks may be distinguished; cataclastic rocks, formed by the operation of purely mechanical forces; and recrystallized rocks, or the metamorphic rocks properly so called, formed under the influence of metamorphic pressures and temperatures. Cataclastic rocks are mechanically sheared and crushed. They represent products of dynamometamorphism, or kinetic metamorphism. Chemical and mineralogical changes generally are negligible. The rocks are characterized by their minute mineral grain size. Each mineral grain is broken up along the edges and is surrounded by a corona of debris or strewn fragments. *See* METAMORPHISM.

Metamorphic rocks, properly so called, are recrystallized rocks. The laws of recrystallization are not the same as those of simple crystallization from a liquid, because the crystals can develop freely in a liquid, but during recrystallization the new crystals are encumbered in their growth by the old minerals. Consequently, the structures which develop in metamorphic rocks are distinctive and of great importance, because in many ways they reflect the physiochemical environment of recrystallization and thereby the genesis and history of the metamorphic rock.

The metamorphic minerals may be arranged in an idioblastic series (crystalloblastic series) in their order of decreasing force of crystallization as follows: (1) sphene, rutile, garnet, tourmaline, staurolite, kyanite; (2) epidote, zoisite; (3) pyroxene, hornblende; (4) ferrogmagnesite, dolomite, albite; (5) muscovite, biotite, chlorite; (6) calcite; (7) quartz, plagioclase; and (8) orthoclase, microcline. Crystals of any of the listed minerals tend to assume idioblastic outlines at surfaces of contact with simultaneously developed crystals of all minerals of lower position in the series.

Igneous magma at high temperature may penetrate into sedimentary rocks, it may reach the surface, or it may solidify in the form of intrusive bodies (plutons). Heat from such bodies spreads into the surrounding sediments, and because the mineral assemblages of the sediments are adjusted to low temperatures, the heating-up will result in a mineralogical and textural reconstruction known as contact metamorphism. *See* CONTACT AUREOLE; PLUTON.

The effects produced do not depend only upon the size of the intrusive. Other factors are amount of cover and the closure of the system, composition and texture of the country rock, and abundance of gaseous and hydrothermal magmatic emanations. The heat conductivity of rocks is so low that gases and vaporous emanations become chiefly responsible for transportation and transfer of heat into the country rock.

Crystalline schists, gneisses, and migmatites are typical products of regional metamorphism and mountain building. If sediments accumulate in a slowly subsiding geosynclinal basin, they are subject to down-warping and deep burial, and thus to gradually increasing temperature and pressure. They become sheared and deformed, and a general recrystallization results. However, subsidence into deeper parts of the crust is not the only reason for increasing temperature. It is not known what happens at the deeper levels of a live geosyncline, but obviously heat from the interior of the Earth is introduced regionally and locally, partly associated with magmas, partly in the form of "emanations" following certain main avenues, determined by a variety of factors. From this milieu rose the lofty mountain ranges of the world, with their altered beds of thick sediments intercalated with tuffs, lava, and intrusives, all thrown into enormous series of folds and elevated to thousands of feet. Thus were born the crystalline schists with their variants of gneisses and migmatites. *See* GEOSYNCLINE; OROGENY.

Well-defined series of mineral facies have been singled out. Sedimentary rocks of the lowest metamorphic grade have recrystallized to give rocks of the zeolite facies. At slightly higher temperatures the greenschist facies develops—chlorite, albite, and epidote being characteristic minerals. A higher degree of metamorphism produces the epidote-amphibolite facies, and a still higher degree the true amphibolite facies in which hornblende and plagioclase mainly take the place of chlorite and epidote. Representative of the highest regional metamorphic grade is the granulite facies, in which most of the stable minerals are water-free, for example, pyroxenes and garnets. Any sedimentary unit will recrystallize according to the rules of the several mineral facies, the complete sequence of events being a progressive change of the sediment by deformation, recrystallization, and alteration in the successive stages: greenschist facies→epidote-amphibolite facies→amphibolite facies→granulite facies. *See* FACIES (GEOLOGY); GRANULITE.

The normal continental crust is entirely made up of metamorphic rocks; where thermal, mechanical, and geochemical equilibrium prevails, there are only metamorphic rocks. Border cases of this normal situation occur in the depths where ultrametamorphism brings about differential melting and local formation of magmas. When equilibrium is restored, these magmas congeal and recrystallize to (metamorphic) rocks. At the surface, weathering processes oxidize and disintegrate the rocks superficially and produce sediments as transient products. Thus the cycle is closed; petrology is without a break. All rocks that are found in the continental crust were once metamorphites. *See* AMPHIBOLITE; EPIDIORITE; EPIDOSITE; GREISEN; MARBLE; MICA SCHIST; MIGMATITE; PHYLLITE; PHYLLONITE; QUARTZITE; SCAPOLITE; SERPENTINITE; SOAPSTONE. [T.F.W.B.; R.C.N.]

Metamorphism The alterations and transformations in preexisting rock masses effected by temperature and pressure, but excluding changes produced by weathering and sedimentation. The changes may include the production of new minerals, structures, or textures, or all three. They give a distinctive new character to the rock as a whole, but they do not involve the loss of individuality of a rock mass, such as changes brought about by fusion. Quantitatively, the metamorphic rocks, including gneisses and migmatites, are the most important group of rocks in the crust of the continents. *See* METAMORPHIC ROCKS; GNEISS; MIGMATITE.

Different kinds of metamorphism may be defined according to genetic criteria, such as the geologic processes that were assumed to have caused the metamorphism, or the physical and chemical conditions that appear to have been predominant in determining the course of metamorphism. Using these criteria three general kinds of metamorphism are noted below.

1. Dislocation, mechanical, or dynamic metamorphism is the result of pressure (or stress) along dislocations in the Earth's crust. The deformed rocks commonly show marked zones of extremely fine-grained rocks, such as mylonites, whose structures are determined by crushing and movement of the grains without important recrystallization of old, or growth of new, minerals. This type of metamorphism is local and restricted in occurrence.

2. Contact or thermal metamorphism occurs in response to increased temperature induced by adjacent intrusions of magma. Chemical reconstitution of the rocks is due to magmatic exhalation; other conditions, such as confining pressure, exert subordinate influence.

3. Regional metamorphism, the most widespread type, is brought about by an increase in both temperature and pressure in orogenic regions, which are vast segments of the crust represented by the folded mountain ranges. Heat and pressure are mainly consequences of downwarping and deep burial. Pressure is also generated by shearing stresses accompanying the orogenic movements. *See* OROGENY. [T.F.W.B./R.C.N.]

Metaplasia A type of epithelial or connective-tissue cellular reaction in which a pathway of cell specialization (differentiation) is selected other than the usual, resulting in the replacement of the original specialized type of tissue by another type. For example, the trachea is normally lined by mucus-secreting columnar epithelium. Following prolonged irritation, the primordial cells of the respiratory epithelium produce daughter cells that differentiate into stratified squamous epithelium.

Physiologic metaplasia occurs in normal developmental processes, such as in the ciliated mucous columnar epithelium in the esophagus during embryonic development. It is replaced by a stratified squamous epithelium by the time of birth. Pathologic metaplasia represents those types of changes of epithelial or connective tissue cells associated with alteration of the cell environment by toxic agents, bacteria, chronic inflammation, parasites, and avitaminosis A. Frequent sites of pathologic epithelial metaplasia are the gallbladder, urinary bladder, oral mucosa, uterine cervix, and stomach. Pathologic connective-tissue metaplasia, such as bone formation in fibrous scars, blood-vessel walls, or cartilage, is also frequently seen. *See* ANAPLASIA; HYPERPLASIA; HYPOPLASIA; NEOPLASIA. [N.K.M.]

Metasomatism A process by which the chemical composition of rocks is altered. Every transformation of minerals by chemical replacement is designated as metasomatism, and in this sense most of the newly formed constituents of the metamorphic rocks would be metasomatic minerals. *See* METAMORPHIC ROCKS; METAMORPHISM; MINERAL.

Metasomatism is fundamentally a change, atom for atom, in a rock mineral. It postulates a reaction between a solid mineral and a disperse or fluid phase. Two types of metasomatism are recognized: mineral metasomatism and rock metasomatism. The mechanism of the mineral metasomatism in a metamorphic rock is the same as that of the rock metasomatism of large bodies, but in mineral metasomatism the fluid phase, which is generated of and by the metamorphic rock, would cause only a relocation of the chemical elements within the rock without changing the chemical composition of the whole rock. In the metasomatic rock, however, the fluid is introduced into the rock and acts metasomatically, thereby changing the initial chemical composition of the rock. Thus the mineral metasomatism taking place in a metamorphic rock may be considered as a metabolism of rock, that is, a relocation of the chemical elements which have to migrate over only short distances; whereas true metasomatism of rocks should be restricted to only those replacements by which the chemical composition of the whole rock, as well as that of the individual minerals, is essentially changed through an immigration of foreign chemical elements and an expulsion of other elements over long distances, up to several kilometers. Metasomatism in this sense would include, for example, the alteration of limestone into dolomite. Distinction may also be made between mutative, additive, and expulsive metasomatism. *See* DOLOMITE ROCK.

Metasomatism is widespread in silicate rocks. In contact zones of igneous intrusions the replacements may affect not only the intruded rocks but also the crystallized magma from which the fluids are derived. Such magmatic fluids are often gaseous, giving rise to the so-called pneumatolytic metasomatism with formation of skarn rocks. *See* PNEUMATOLYSIS; SKARN. [T.F.W.B.]

Metastable state In quantum mechanics, an excited stationary energy state whose lifetime is unusually long. *See* EXCITED STATE; STATIONARY STATE.

The stationary states of atoms and nuclei, or indeed of any material system, ordinarily are computed while ignoring the interactions between matter and the electromagnetic radiation field. When these interactions are included, a system in an excited state always has some probability of emitting radiation in the form of one or more photons, and thereby decaying to a lower or less excited state. For a metastable state, the probability of making such a transition is unusually small. *See* NUCLEAR ISOMERISM. [E.G.]

Metatheria An infraclass of therian mammals including a single order, the Marsupialia. The Metatheria are distinguished from the Eutheria (the placental mammals) by numerous characters. The braincase is small, the angular process of the mandible is inflected, and a pair of marsupial bones articulates with the pelvis. Almost all living marsupials have a pouch on the belly of the female in which the young are carried after birth. The early marsupials were unable to compete with the more progressive later placental forms and died out except in South America and Australia, where they were isolated by water barriers. *See* EUTHERIA; MAMMALIA; THERIA. [D.D.D./F.S.S.]

Metazoa The multicellular or tissue animals that make up all of the animal kingdom except the one-celled or acellular Protozoa. The cells of the metazoan body are organized in layers or groups as tissues or organ systems variously specialized for performing different bodily functions. The cellular organization into tissues varies widely in different groups of animals and in the parts of any one animal. In some instances the cells lose their boundaries after growth. [T.I.S.]

Meteor Observable phenomenon resulting from entry of a particle of matter, called a meteoroid, into the Earth's atmosphere at a speed between 7 and 45 mi/s (11 and 72 km/s). Meteoric material originates in the solar system and is attributed to asteroidal and cometary debris.

The mass of individual particles extends from micrometer-size dust specks to complete asteroids with weights in the thousands of tons. These extreme cases represent special fields of study. *See* METEORITE.

Meteoroids become visible at altitudes of from 50 to 75 mi (80 to 120 km), depending on their velocities. At these altitudes the energy transferred to meteoroids by collisions with atmospheric atoms and molecules becomes sufficient to vaporize the solids. Further collisional interactions of the vaporization products with the air produce luminosity and ionization. The detectable portions of trajectories, during which the meteoroids are luminous, are called trails. These phenomena may be observed by visual, photographic, telescopic, or radar techniques.

The meteor orbit for the few tens of kilometers of trail may almost always be described as a straight line. This straight-line trajectory is a great circle as seen projected on the celestial sphere. If the same meteor is observed from several different locations, the extension, backward along the trail, of the great circle observed at each station intersects the same point on the celestial sphere. This is the meteor's radiant point and is determined solely by the direction of motion of the meteoroid. In the same way, if a number of meteors traveling parallel to one another are observed from a single location, the backward extensions of their trails intersect at the common radiant point. Meteors with parallel motion and identical velocities must have identical orbits, and, very likely, identical origins. Groups of meteors with these characteristics are called meteor showers.

Apart from the elements they contain, visual meteors may bear no similarity to meteorites. There is good evidence that most meteoroids fragment as they pass through the atmosphere, and some evidence that nearly complete disruption of certain meteoroids occurs whenever the dynamic pressures on the meteoroids exceed 0.5 pound per square inch (3.4 kilopascals). The destruction of an entire meteoroid by so small a pressure, and not by thermal effects, is indicative of extreme fragility. Other observations indicate that the density of meteoroids may possibly be less than that of water. It is presumed that the low density and extreme fragility are due to porosity of the meteoric material. [R.E.McC.]

Meteorite A natural, nonterrestrial object which survives passage through the Earth's atmosphere and arrives as a solid at the Earth's surface. As the meteorite passes through the atmosphere, it may be viewed as a falling star, with all of the optical brilliance and auditory effects characteristic of meteors. However, only an extremely small fraction of the meteors survive their atmospheric entry. Hence, meteorites represent rare events, but meteors can be seen on almost any clear night if moonlight or the lights of a city do not overilluminate the nighttime sky. *See* METEOR.

Meteorites consist mostly of metal and silicate material. They can be classified as irons, stony irons, or stones, depending on the relative proportions of the iron and the silicates. There are only about 2500 meteorites known. These do not include the tiny micrometeorites that have been found in oceanic sediments and in high-altitude dust. Micrometeorites include both interplanetary dust and some of the numerous bits of debris that are generated when larger objects pass through the atmosphere. *See* MICROMETEORITE.

Meteorites represent samples of the primitive interplanetary rubble that orbits the Sun but did not aggregate into a planetary-sized object. This rubble consists of bodies that range from hundreds of kilometers in diameter to tiny particles of interplanetary dust. Collisions between these bodies are frequent, particularly in the thickly populated region between the orbits of Mars and Jupiter. The larger bodies that are located between the orbits of Mars and Jupiter are called asteroids. More than 1600 asteroids are known, and they appear to have the same range of densities and chemical compositions that are observed in meteorites. *See* ASTEROID.

Comets are another possible source of meteorites. Cometary debris is probably the source of many meteors, including the more spectacular meteor showers, but there is no compelling evidence that comets are the source of any known meteorites. *See* COMET; SOLAR SYSTEM.

The impact of a large meteorite, one weighing more than a few hundred tons, releases so much energy that the meteorite explodes. This produces a depression called a meteorite crater. There was little interest in terrestrial meteorite craters prior to the observation that impact craters are responsible for many of the surface features of the Moon and Mars. [O.K.M.]

Meteorological balloon A balloon used to carry radiosondes and other lightweight instruments aloft, also known as a sounding balloon. It is an extensible balloon made from natural or synthetic rubber and filled with either hydrogen or helium. Hundreds of these balloons are launched daily by meteorological stations throughout the world in support of the upper-air program of the World Meteorological Organization, as well as local weather forecasting. These flights are vertical sounding probes of the atmosphere for measuring pressure, temperature, humidity, and wind velocity. They can carry up to 40 lb (18 kg) of useful payload. High-altitude types carrying much lighter payloads may exceed 25 mi (40 km) before expanding to the bursting point. [W.R.N.]

Meteorological optics A branch of atmospheric physics or physical meteorology in which optical phenomena occurring in the atmosphere are described and explained.

Meteorological optics can be divided into two main parts. In the first part, the atmosphere is considered as a continuous medium in which the speed of propagation of electromagnetic waves (light) varies with the density. In the second part, attention is given to the presence of particulate matter in the atmosphere, such as air molecules, dust and haze particles, cloud and raindrops, and ice crystals, and to the phenomena resulting from the interaction of light with these particles. In a third, minor part, applications to measurement of atmospheric properties are considered.

Several technical problems can be solved by direct application of the results of meteorological optics, such as the amount of illumination of objects on the ground or high in the atmosphere and the conditions of their visibility under different atmospheric conditions. [Z.S.]

Meteorological radar Radar equipment for observing precipitation, clouds, and other atmospheric phenomena. The most extensive application for such equipment is in short-term prediction of rain or severe weather at ground locations and in studies of severe storms, precipitation development, and the spatial structure of precipitation patterns. *See* CLOUD; PRECIPITATION (METEOROLOGY); RADAR METEOROLOGY.

The deterioration of resolution with increasing range is often a serious limitation in quantitative studies. Moreover, ranges beyond 180 mi (300 km) are generally not very useful even in qualitative observations, At that range a horizontally directed radar beam is nearly 4 mi (6 km) above the ground and therefore above many rain clouds. Although large thunderstorms may extend to 6 mi (10 km) or higher and thus be detected at more distant ranges, quantitative measurements are usually limited to ranges within 120 mi (200 km). Radars operating at wavelengths shorter than about 3 cm (frequencies higher than 10 GHz) are limited to still closer ranges, because appreciable amounts of energy are lost in transit to and from a distant object owing to attenuation by intervening precipitation and absorption by atmospheric water vapor and oxygen. At wavelengths of 10 cm and longer, these losses are negligible for most purposes.

The principal quantity determined from the observations is the spatial distribution of the radar reflectivity of meteorological scatterers. Ordinarily this is accomplished by rotating the antenna in azimuth and elevation in a synchronized manner that allows full, three-dimensional scanning of the region around the radar. The reflectivity at a given point is determined from the power of the signal reflected from the scatterers in the measurement volume centered at the point. Because these scatterers are not fixed but move relative to one another, the signal fluctuates in time and only the average power is related to the reflectivity.

Distinct from the radars employed for cloud and precipitation observations, high-atmosphere meteorological radar is used for investigating winds and the thermal structure of the high atmosphere. Operating at wavelengths ranging from about a half meter to several meters (in the VHF and UHF bands of the electromagnetic spectrum), these radars much of the time are able to detect echoes from clear air. Scattering is accounted for, not by hydrometeors or other particulate matter, but by refractive index inhomogeneities in the optically clear air. *See* RADAR. [R.R.R.]

Meteorological rockets Synoptic exploration of the stratospheric circulation through use of small rocket systems. The data obtained from rocket investigations has produced dramatic changes in the scientific view of this frontier region of the atmosphere, with a resulting alteration of the structural concepts into an atmospheric model which is primarily characterized by intense dynamics. *See* STRATOSPHERE.

The Meteorological Rocket Network (MRN) has roughly doubled the observed atmospheric volume and its conception and operation has epitomized the international cooperation. The MRN began at North American missile ranges in 1959, was expanded into a global network in 1968, and now includes 56 sites. [W.L.W.]

Meteorological satellites Earth-orbiting spacecraft carrying a variety of instruments for measuring visible and invisible radiation from the Earth and its atmosphere. The instruments of a single polar-orbiting satellite can view the

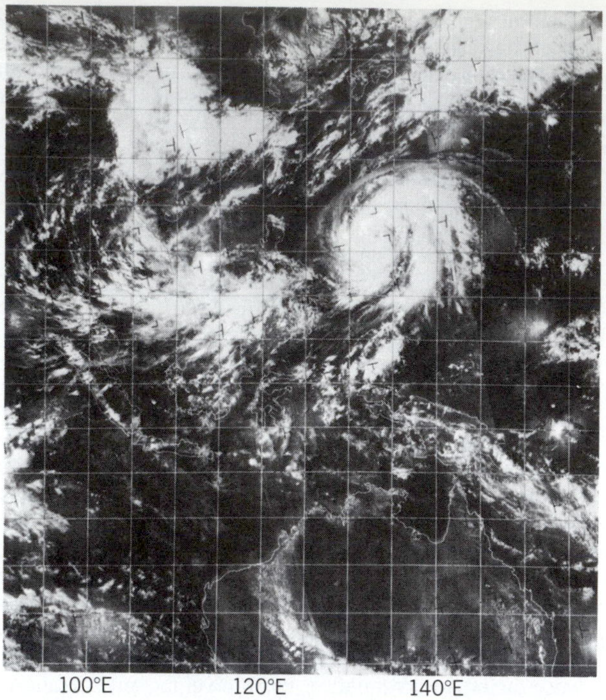

***ESSA 7* montage of about 28 pictures taken Oct. 17, 1968, transposed to Mercator projection. Typhoon Gloria appears above picture center. (*NOAA*)**

entire Earth twice each 24 h—a striking contrast to the small fraction of the global atmosphere that is sampled by the combination of all other methods by all nations. *See* WEATHER FORECASTING AND PREDICTION.

The space age of meteorology began when *Tiros 1* (Television and Infrared Observation Satellite) was launched on April 1, 1960. From that point on the United States launched a large number of satellites which produced a prodigious flow of weather data. Most satellites provide information for routine analysis; others are intended for research and development.

Pictures from orbiting spacecraft are assembled with a sophisticated communication-computer system. Pictures are tape-recorded while the vehicle transits remote regions. Once within (FM) radio range of a receiving station the tape is read out and the signals sent by land lines to the National Environmental Satellite Service near Washington, D.C. There they are fed into a high-speed computer for assembling and mapping. Pictures displayed for photographing on a cathode-ray tube are overlaid with latitude-longitude lines and coastlines. These are fitted to the picture by a computer that calculates their shape and placement from the satellite message, and from the known height and position of the camera at a picture time (see illustration). *See* WEATHER MAP.

While visible pictures are irreplaceable for some purposes, the 1970s witnessed an accelerated use of infrared images and data. Infrared radiation, heat energy, is highly dependent on the temperature of the radiating material, a characteristic that makes it possible to deduce a myriad of information. Visible and infrared images exhibit an important difference. Thick clouds, regardless of their altitude, reflect most of the sunlight that falls on their upper surface so that they are very bright when photographed from space. On the other hand, low clouds, lying in the warmer atmosphere, and high clouds, in the cold upper atmosphere, radiate at quite different intensities and therefore exhibit different brightnesses on film.

Ocean-surface temperatures, critical to both meteorology and oceanography, can be deduced from infrared measurements in cloudless regions. Sophisticated computer programs analyze infrared data from all ocean areas each day, deducing which measurements were made in cloud-free areas, using only these for computing sea-surface temperatures. Other sensors respond to water vapor in the atmosphere, while those measuring a broad spectrum of wavelengths observe the total heat lost to space. This debit side of the Earth's heat budget is a critical variable for long-range forecasting. [L.F.Hu.]

Meteorology The science concerned with the atmosphere and its phenomena. Meteorology is primarily observational; its data are generally "given." The meteorologist observes the atmosphere—its temperature, density, winds, clouds, precipitation, and other characteristics—and aims to account for its observed structure and evolution (weather, in part) in terms of external influence and the basic laws of physics. *See* ATMOSPHERE.

Synoptic meteorology is the branch of meteorology comprising the knowledge of atmospheric phenomena connected with the weather, applied mainly in weather forecasting, and based on data acquired by the synoptic method. This method involves the study of weather processes through representations of atmospheric states determined by synchronous observations at a network of stations, most of which are at least 6 mi (10 km) apart. By international agreement, the data taken from the Earth's surface and aloft at certain international hours of observation are inserted on weather maps, upper-air maps, vertical cross sections, time sections, and sounding diagrams with international symbols according to fixed rules. These crude representations are then analyzed and critically evaluated in accordance with the knowledge of existing structure models in the (lower) atmosphere, in order to ascertain the best approximation to a three-dimensional image of the true atmospheric state at the hour of observation. From one such representation, or a series of them, future atmospheric states are then derived with the aid of empirical knowledge of their behavior and by application of the theoretical results of dynamic meteorology. *See* WEATHER MAP.

The branch of meteorology known as dynamical meteorology is the science of naturally produced motions in the atmosphere and of the related distributions in space of pressure, density, temperature, and humidity. Based on thermodynamic and hydrodynamic theories, it forms the main scientific basis of weather forecasting and climatology. *See* CLIMATOLOGY; WEATHER FORECASTING AND PREDICTION. [T.Be.]

Metering orifice A wall opening with sharp inlet edges, through which fluid flows, the wall thickness being less than an eighth of the area of the opening. An orifice serves basically to control or to meter the rate of fluid flow. It may be used as an instrument for the measurement of fluid flow, including oils, air, gases, steam, and other vapors, or as an element in a machine to limit the flow of fluid. Common shapes are sharp-edge, with bevel facing out, and square. Often, fuel enters a chamber through an orifice, as in a carburetor. *See* FLOW MEASUREMENT. [R.E.Sp.]

Methane A member of the alkane or paraffin series of hydrocarbons with the formula shown below. Methane is called

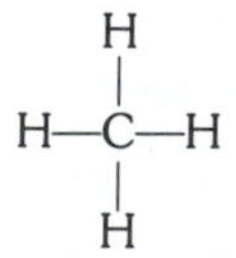

marsh gas because it forms by anaerobic bacterial decomposition of vegetable matter in swampy land. Coal miners know it as firedamp because mixtures with air are combustible. It is a major constituent of natural gas (50–90%) and of coal gas. It forms in large amounts in sewage disposal processes, especially in anaerobic digestion. As a liquid it freezes at −182.6°C (−296.7°F) and boils at −161.6°C (−258.9°F).

In addition to its use as a fuel, methane is important as a source of organic chemicals and of hydrogen. Its reaction with steam at high temperatures in the presence of catalysts yields carbon monoxide and hydrogen (synthesis gas), which can be catalytically converted to liquid alkanes (Fischer-Tropsch process) or to methanol and other alcohols. *See* ALKANE; FISCHER-TROPSCH PROCESS; HYDROFORMYLATION; NATURAL GAS. [L.S.]

Methane-producing bacteria Bacteria characterized by the unique ability to produce the gaseous hydrocarbon methane (CH_4) as the major product of metabolism. These bacteria are found in nature wherever plant material is undergoing anaerobic decomposition. Although methane-producing bacteria are among the most oxygen-sensitive bacteria known, they are widespread in habitats such as the digestive tracts of animals (intestine, rumen, cecum), sediments, swamps, marshes, sanitary landfills, sewage sludge digesters, and decaying heartwood of certain trees. In these habitats other bacteria consume the available oxygen, thus making it possible for anaerobic zones of low reducing potential to be established. Methane-producing bacteria are terminal organisms in the microbial metabolic food chain in which organic matter is decomposed. However, the amount of methane formed in covered sanitary landfills is large enough that fills should be safely vented. *See* METHANE. [R.S.W.]

Methanol The simplest alcohol (alkanol). Methanol has the formula shown below. Formerly obtained from wood as a co-

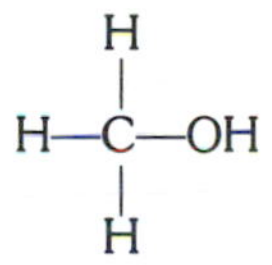

product of charcoal production, it was called wood spirit or wood alcohol.

Methanol is a colorless, poisonous liquid with essentially no odor and very little taste. Its lack of odor is an important hazard in handling, and often strong-smelling wood alcohol or kerosine is added so that workers will become aware of leaks or spills. Methanol boils at 64.7°C (149°F). It is miscible with water and most organic liquids, including gasoline. It is extremely flammable, burning with a nearly invisible blue flame.

Methanol has become a major feedstock for the synthesis of acetic acid, and large amounts of methanol are consumed in the production of methyl esters. Methanol itself is employed as an extractant and solvent for many substances, and is sometimes blended with gasoline in cold weather to reduce condensation problems. *See* ALCOHOL. [F.W.]

Methionine An amino acid considered essential for normal growth of animals. Methionine is an important methyl group donor in transmethylation reactions. For this purpose it must first be activated by adenosine triphosphate (ATP), forming *S*-adenosylmethionine. This compound readily transfers its methyl group to suitable acceptors, leaving *S*-adenosylhomocysteine. Active methyl groups originate during methionine biosynthesis. A hydroxymethyl group is reduced to a methyl group through the mediation either of coenzyme B_{12} or of tetrahydrofolic acid. *See* AMINO ACIDS. [E.A.Ad.]

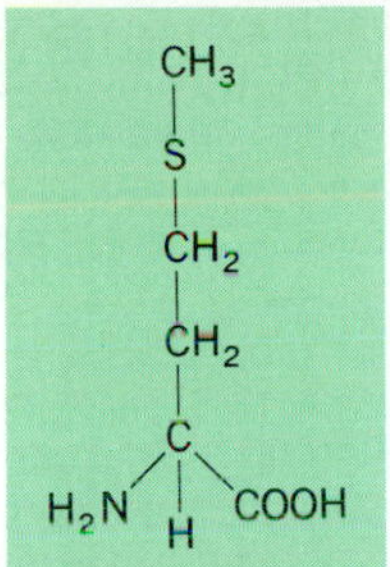

Methionine

Methods engineering A technique used by progressive management to improve productivity and reduce costs in both direct and indirect operations of manufacturing and nonmanufacturing business organizations. Methods engineering is applicable in any enterprise wherever human effort is required. It can be defined as the systematic procedure for subjecting all direct and indirect operations to close scrutiny in order to introduce improvements that will make work easier to perform and will allow work to be done smoother in less time, and with less energy, effort, and fatigue, with less investment per unit. The ultimate objective of methods engineering is profit improvement. *See* OPERATIONS RESEARCH; PRODUCTIVITY.

Methods engineering includes five activities: planning, methods study, standardization, work measurement, and controls. Methods engineering, through planning, first identifies the amount of time that should be spent on a project so as to get as much of the potential savings as is practical. Invariably the most profitable jobs to study are those with the most repetition, the highest labor content (human work as distinguished from mechanical or process work), the highest labor cost, or the longest life-span. Next, through methods study, methods are improved by observing what is currently being done and then by developing better ways of doing it. The standardization phase includes the training of the operator to follow the standard method. Then the number of standard hours in which operators working with standard performances can do their job is determined by measurement. Finally, the established method is periodically audited, and various management controls are adjusted with the new time data. The system may include a plan for compensating labor that encourages attaining or surpassing a standard performance. [B.W.N.]

Metric system A system of units used in scientific work throughout the world and employed in general commercial transactions and engineering applications in most of the developed nations of the world except for Great Britain and the United States. The basic units of the metric system define length (meter), mass (kilogram), and time (second).

The chief advantage of the metric system is that it is based on standards that have been accepted by international agreement, and it therefore provides a common basis for all scientific measurements. A second advantage of the metric system lies in the fact that only decimal multiples and submultiples of the fundamental length and mass units and of other derived units are employed. *See* PHYSICAL MEASUREMENT; TIME; UNITS OF MEASUREMENT. [D.Wi.]

Metzgeriales An order of liverworts in the subclass Jungermanniidae. Twelve families make up the Metzgeriales, which are also known as the Anacrogynae because of archegonia produced behind the growing apex. The gametophyte plant body is flat, elongated, and usually thallose, with no tissue differentiation or surface pores; less commonly there is a stem with two rows of leaves. Capsules dehisce by valves. *See* BRYOPHYTA; JUNGERMANNIIDAE. [H.Cr.]

Mica Any one of a group of hydrous aluminum silicate minerals with platy morphology and perfect basal (micaceous) cleavage. The most common micas are muscovite, $KAl_2(AlSi_3O_{10})(OH)_2$; paragonite, $NaAl_2(AlSi_3O_{10})(OH)_2$; phlogopite, $K(Mg,Fe)_3(AlSi_3O_{10})(OH)_2$; biotite, $K(Fe,Mg)_3(AlSi_3O_{10})(OH)_2$; and lepidolite, $K(Li,Al)_{2.5-3.0}(Al_{1.0-0.5}Si_{3.0-3.5}O_{10})(OH)_2$. Ca, Ba, Rb, and Cs can sometimes substitute for Na and K; Mn, Cr, and Ti for Mg, Fe, and Li; and F for OH.

Mica is commonly found as small flakes or lamellar plates without a crystal outline. Muscovite and biotite sometimes occur in thick books, tabular prisms with a hexagonal outline that can be up to several feet across. The prominent basal cleavage is a consequence of the layered crystal structure. Thin cleavage sheets of micas, particularly muscovite and phlogopite, are flexible, elastic, tough, and translucent to transparent (isinglass). They have low electrical and thermal conductivity and high dielectric strength.

Micas have Mohs hardnesses of 2–3 and specific gravities of 2.8–3.2. They have a vitreous to pearly luster. Muscovite is colorless to pale shades of brown, green, or gray. Paragonite is colorless to pale yellow. Phlogopite is pale yellow to brown. Biotite is dark green, brown, or black. Lepidolite is most often pale lilac, but it can also be colorless, pale yellow, or pale gray.

Commercial mica is of two main types: sheet, and scrap or flake. Sheet muscovite, mostly from pegmatites, is used as a dielectric in capacitors and vacuum tubes in electronic equipment. Lower-quality muscovite is used as an insulator in home electrical products such as hot plates, toasters, and irons. Scrap and flake mica is ground for use in coatings on roofing materials and waterproof fabrics, and in paint, wallpaper, joint cement, plastics, cosmetics, well drilling products, and a variety of agricultural products. *See* CAPACITOR; ELECTRIC INSULATOR; SILICATE MINERALS. [L.Gr.]

Mica schist A widely distributed group of rocks of medium to high metamorphic grade, composed essentially of mica and quartz and exhibiting a foliated or schistose structure which is easily revealed by the parallel orientation of the mica flakes. Varieties are biotite schists, biotite-sericite schists, biotite-chlorite schists, and albite-biotite schists. Of lower metamorphic grade and of fine grain are phyllites or chlorite-sericite schists, without biotite but otherwise similar to mica schists, into which they pass by gradual transitions. *See* MICA; PHYLLITE.

Mica schists are most widely distributed in Precambrian areas and in the younger eroded mountain ranges where they usually represent metamorphic equivalents of aluminum-rich sediments (clays and argillites). However, tuffs and certain acid igneous rocks (rhyolites, granites) may form part of the primary constituents. *See* METAMORPHIC ROCKS. [T.F.W.B.]

Micelle A colloidal aggregate of a unique number (50→100) of amphipathic molecules, which occurs at a well-defined concentration called the critical micelle concentration. In polar media such as water, the hydrophobic part of the amphiphiles forming the micelle tends to locate away from the polar phase while the polar parts of the molecule (head groups) tend to locate at the polar micelle solvent interface. A micelle may take several forms, depending on the conditions and composition of the system, such as distorted spheres, disks, or rods (see illustration). Micelles are formed in nonpolar media such as benzene, where the amphiphiles cluster around small water droplets in the system, forming an assembly known as a reversed micelle.

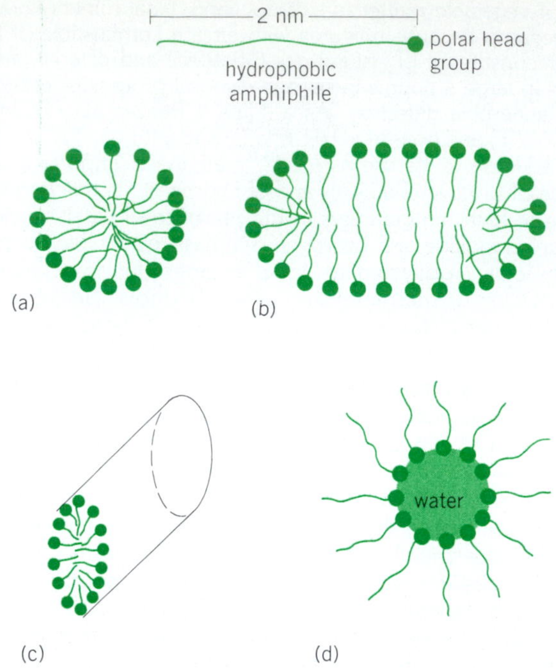

Form of an amphiphile and several forms of micelle: (*a*) spherical, (*b*) disk, (*c*) rod, and (*d*) reversed.

Micellar systems have the unique property of being able to solubilize both hydrophobic and hydrophilic compounds. They are used extensively in industry for detergency and as solubilizing agents. *See* DETERGENT; SOAP. [J.K.T.]

Microbial biofilm An adhesive substance, the glycocalyx, and the bacterial community which it envelops at the interface of a liquid and a surface. When a liquid is in contact with an inert surface, any bacteria within the liquid are attracted to the surface and adhere to it. In this process the bacteria produce the glycocalyx. The bacterial inhabitants within this microenvironment benefit as the biofilm concentrates nutrients from the liquid phase. However, these activities may damage the surface, impair its efficiency, or develop within the biofilm a pathogenic community that may damage the associated environment. Microbial fouling or biofouling are the terms applied to these actual or potential undesirable consequences. Microbial fouling affects a large variety of surfaces under various conditions. Microbial biofilms may form wherever bacteria can survive; familiar examples are dental plaque and tooth decay. *See* PERIODONTAL DISEASE; TOOTH DISORDERS.

The process of microbial fouling involves the secretion by the biofilm inhabitants of metabolites that damage the surface or the production of organic or inorganic deposits upon the surface. The consequences of microbial fouling are (1) physical damage to the surface as a result of microbial growth and metabolite activity; (2) the reduction in the proper function of the surface because of the presence of the biofilm; and (3) the creation of a reservoir of potential pathogens within the biofilm. These consequences are not mutually exclusive. Microbial fouling in such places as drinking water, food industries, marine environments, and industrial water systems is of great concern, and it is therefore carefully monitored. *See*

CORROSION; FOOD MICROBIOLOGY; WATER PURIFICATION; WATER SUPPLY ENGINEERING. [H.L.-Sc.; J.W.C.]

Microbial degradation The removal of organic compounds from natural ecosystems by bacteria; also referred to as biodegradation. Microorganisms occupy a niche or habitat because various chemical compounds are present there that serve as potential food or energy sources for the bacteria. Hence, the degradation or utilization of these compounds by certain bacterial strains allows for the concomitant removal or destruction of chemical compounds that are hazardous to other organisms and persistent in nature. Microbial degradation can be either primary or ultimate. Primary biodegradation (biotransformation) represents an alteration in structure so that the physical and chemical properties of the compound are changed. Ultimate biodegradation (mineralization), which is measured through the use of radiolabeled substrates, results in breakdown of the chemical into carbon dioxide.

Some strains of bacteria are unable to completely metabolize certain compounds, and instead they utilize a readily assimilable compound for energy and only partially utilize the more refractive compound. This is known as the principle of cometabolism, which is useful in biodegradation. Normally, microbial populations in soil or sewage readily degrade naturally occurring aromatic compounds, and often attack those aromatic compounds that have been substituted with chlorine atoms. Thus, in environments contaminated with polychlorinated biphenyl compounds (PCBs), degradation is often slow, but cometabolism processes become valuable for more rapid removal of these compounds. *See* POLYCHLORINATED BIPHENYLS.

Bioremediation is the active use of microorganisms to decompose toxic substances. Older methods of incineration, deep-well injection, and ocean dumping for cleaning up the environment have limitations and often result in additional pollution. Bioremediation through wastewater treatment or land farming, on the other hand, offers a natural method of destroying toxic chemicals without environmental damage. Approaches to developing new strains of bacteria involve chemostat isolation, chemostat selective matings, and recombinant DNA techniques. Biodegradation can be effective in cleaning up hazardous waste sites, water, and soil. Leakage of hydrocarbons into the environment from above- and belowground storage tanks is a major problem, and the applications of different commercial strains of bacteria have been successful in breaking down some of these compounds. *See* BIOLEACHING; HAZARDOUS WASTE; INDUSTRIAL MICROBIOLOGY. [P.A.Va.]

Microbial ecology The study of interrelationships between microorganisms and their living and nonliving environments. Microbial populations are able to tolerate and to grow under varying environmental conditions, including habitats with extreme environmental conditions such as hot springs and salt lakes. Understanding the environmental factors controlling microbial growth and survival offers insight into the distribution of microorganisms in nature, and many studies in microbial ecology are concerned with examining the adaptive features that permit particular microbial species to function in particular habitats.

Within habitats some microorganisms are autochthonous (indigenous), filling the functional niches of the ecosystem, and others are allochthonous (foreign), surviving in the habitat for a period of time but not filling the ecological niches. Because of their diversity and wide distribution, microorganisms are extremely important in ecological processes. The dynamic interactions between microbial populations and their surroundings and the metabolic activities of microorganisms are essential for supporting productivity and maintaining environmental quality of ecosystems. Microorganisms are crucial for the environmental degradation of liquid and solid wastes and various pollutants and for maintaining the ecological balance of ecosystems—essential for preventing environmental problems such as acid mine drainage and eutrophication. *See* ECOSYSTEM; EUTROPHICATION.

The various interactions among microbial populations and between microbes, plants, and animals provide stability within the biological community of a given habitat and ensure conservation of the available resources and ecological balance. Interactions between microbial populations can have positive or negative effects, either enhancing the ability of populations to survive or limiting population densities. Sometimes they result in the elimination of a population from a habitat. *See* RHIZOSPHERE.

The transfer of carbon and energy stored in organic compounds between the organisms in the community forms an integrated feeding structure called a food web. Microbial decomposition of dead plants and animals and partially digested organic matter in the decay portion of a food web is largely responsible for the conversion of organic matter to carbon dioxide. *See* BIOMASS; FOOD WEB.

Only a few bacterial species are capable of biological nitrogen fixation. In terrestrial habitats, the microbial fixation of atmospheric nitrogen is carried out by free-living bacteria, such as *Azotobacter*, and by bacteria living in symbiotic association with plants, such as *Rhizobium* or *Bradyrhizobium* living in mutualistic association within nodules on the roots of leguminous plants. In aquatic habitats, cyanobacteria, such as *Anabaena* and *Nostoc*, fix atmospheric nitrogen. The incorporation of the bacterial genes controlling nitrogen fixation into agricultural crops through genetic engineering may help improve yields. Microorganisms also carry out other processes essential for the biogeochemical cycling of nitrogen. *See* BIOGEOCHEMISTRY; NITROGEN CYCLE; NITROGEN FIXATION.

The biodegradation (microbial decomposition) of waste is a practical application of microbial metabolism for solving ecological problems. Solid wastes are decomposed by microorganisms in landfills and by composting. Liquid waste (sewage) treatment uses microbes to degrade organic matter, thereby reducing the biochemical oxygen demand (BOD). *See* ESCHERICHIA; SEWAGE TREATMENT; WATER PURIFICATION. [R.M.A.]

Microbiological methods Methods customarily employed in the laboratory for the study of microorganisms include techniques to describe, enumerate, identify, and study the physiological processes of microscopic forms of life.

Microscopes that provide approximately a thousandfold magnification are used to observe bacteria. Lesser magnification is required for yeasts, molds, algae, and protozoa, but much higher magnification (up to several-hundred-thousandfold) as provided by electron microscopes is utilized for visualization of viruses. Electron microscopy is also used to view thinly sliced sections of bacteria and other organisms in order to study details of their internal structure. *See* ELECTRON MICROSCOPE.

Special staining methods are used to provide information concerning the chemical nature of the surface and some of the subcellular structures of microorganisms. Bacteria are separated into two large groups (gram-negative or gram-positive) by virtue of their staining by sequential application of basic dyes and a decolorizing agent that make up Gram's stain. *See* GRAM'S STAIN; STAIN (MICROBIOLOGY).

Counting of microorganisms generally involves inoculation of a series of tenfold dilutions into sterile culture agar media that are presumed to be adequate for the initiation of growth and subsequent multiplication of each living organism in the sample being counted. Each living microbial cell produces a mass of cells (called a colony) that is visible to the unaided eye. A count of these colonies multiplied by the reciprocal of the dilution inoculated into the dish provides a quantitative measure of the microbial population of the original specimen. *See* CULTURE MEDIA.

Identification and other studies of a microorganism normally must be preceded by its isolation and cultivation of a pure culture. This is usually accomplished by transferring some of the cells from a colony growing on an agar medium to a sterile liquid medium. (A fully grown culture, after one day of incubation, may exceed a billion cells per milliliter.) Characterization of a pure culture of a microorganism consists primarily of determinations of various physiological capabilities of this large population of cells. *See* PURE CULTURE.

Research studies on microorganisms may involve all of the methods of modern cellular and molecular biology. Basic studies of the genetic processes of microorganisms have yielded much knowledge of fundamental genetics. Biochemical studies of cellular processes frequently involve the use of both physical and chemical methods to separate and purify specific intracellular components from homogeneous populations of microbial cells. The physical structure of a preparation may be examined by electron microscopy and by a variety of other sophisticated physicochemical methods. Biochemical activity may be studied by the methods of enzyme chemistry. [J.B.E.]

Microbiology The science and study of microorganisms. These include the protozoa, algae, fungi, bacteria, viruses, and rickettsiae. *See* ALGAE; BACTERIA; FUNGI; PROTOZOA; VIRUS.

Microbiology at first developed largely as an applied science with emphasis on the role of microorganisms as pathogens and their association with the spoilage of food. Little attention was paid to the essential biological properties of the microorganisms themselves. However, the studies on their physiology and biochemistry, carried out in the 1920s, pointed to the fundamental unity underlying the behavior of all living organisms. The enormous diversity encountered among microorganisms and the specificity of their physiological patterns helped make them favorite objects for biochemical investigations. This, in turn, stimulated an interest in all sorts of microorganisms, with the result that microbiology became acknowledged as an important branch of biological science.

For medical and sanitary aspects of microbiology *see* EPIDEMIOLOGY; IMMUNOLOGY; MEDICAL BACTERIOLOGY; MEDICAL MYCOLOGY; MEDICAL PARASITOLOGY; RICKETTSIOSES; VIRUS. For food products spoilage *see* FOOD MICROBIOLOGY; MILK. For the transformation of microorganisms in water and soil *see* SOIL MICROBIOLOGY; WATER MICROBIOLOGY. For industrial aspects *see* INDUSTRIAL MICROBIOLOGY. For methodology *see* MICROBIOLOGICAL METHODS. [C.B.V.N.]

Microbiota (human) Microbial flora harbored by normal, healthy individuals. A number of microorganisms have become adapted to a particular site or ecologic niche in or on their host. Some are normal residents that are regularly found, and if disturbed will rapidly reestablish themselves; others are transient microorganisms that may colonize the host for short periods but are unable to permanently colonize. The normal fetus is sterile, but during and after birth the infant is exposed to an increasing number of microorganisms. Subsequently, those organisms best adapted to survive and colonize particular sites establish themselves and become predominant. Physiologic factors such as the availability of nutrients, temperature, moisture, pH, oxidation-reduction potential, and resistance to local antibacterial substances play an important role in determining the ability of a microorganism to become established at a particular site. The normal indigenous microbial flora is exceedingly complex, consisting of many different species of bacteria, fungi, viruses, and protozoa. The great majority of these commensal and symbiotic organisms are bacteria and fungi.

The indigenous microorganisms play an important role by protecting the normal host from invasion by microorganisms with a greater potential for causing disease. They compete with the pathogens for essential nutrients and for receptors on host cells by producing bacteriocins and other inhibitory substances, making the environment inimical to colinization by pathogens.

In the healthy individual the morphologic integrity of the body surface provides a very effective first line of defense. The intact skin is an efficient physical barrier that can be penetrated by very few microorganisms. The secretion of specific antimicrobial substances and bactericidal fatty acids by the sebaceous glands also retards microbial invasion.

Mucosal surfaces also provide a mechanical barrier in the respiratory, gastrointestinal, and genitourinary tracts. These surfaces are bathed in secretions with antimicrobial activity. In the respiratory tract, mechanical cleansing is accomplished by the cough and mucociliary action. Recurrent infections of the sinuses, middle ear, bronchial tract, and lungs occur in individuals who have an impairment of ciliary activity. These infections are usually caused by *Staphylococcus pneumoniae* and *Haemophilus influenzae*, the more virulent pus-forming organisms found in the nasopharynx. Defects in ciliary activity also cause bacterial respiratory infections in cigarette smokers and heavy alcohol drinkers. *See* STAPHYLOCOCCUS.

Once the natural barriers of the skin and mucous membranes are breached, the next major line of defense is the polymorphonuclear leukocytes. Individuals with disorders of these phagocytic cells have an increased incidence of serious infections with their indigenous microflora.

The complement system is another nonspecific mechanism of the body for the elimination of invading microorganisms. Complement proteins in conjunction with organs of the reticuloendothelial system (spleen, liver, and bone marrow) play a key role in the removal of encapsulated bacteria from the bloodstream. Splenectomized individuals and those with a nonfunctioning spleen because of sickle cell disease have an increased incidence of fulminating infections caused by *S. pneumoniae*, *H. influenzae*, *Neisseria meningitidis*, and recently recognized unusual organisms. *See* COMPLEMENT. [H.P.W.]

Microcline The name for a triclinic potassium-rich feldspar. The microcline structure is built up of rather pure $KAlSi_3O_8$. Microcline is usually twinned with typical crosshatching, indicating that it usually grew (stably or metastably) as sanidine (monoclinic) and subsequently changed during geological time into the triclinic modification by passing through a state called normal orthoclase. Microcline is typically found in granites and pegmatites. Green microcline is known as amazon stone or amazonite. *See* FELDSPAR. [F.H.L.]

Micrococcaceae A family of gram-positive cocci, all producing the enzyme catalase (splitting hydrogen peroxide to water and oxygen) and with a respiratory metabolism. Endospores are not formed. Bright yellow, red, green, and purple pigments are formed by many of the organisms in this group. The family contains three genetically nonrelated genera: *Micrococcus*, *Staphylococcus*, and *Planococcus*. *See* STAPHYLOCOCCUS. [M.M.]

Microcomputer A digital computer whose central processor unit (CPU) is a microprocessor (see illustration). A microcomputer can be a 4-bit-, 8-bit-, or 16-bit-word machine depending upon the word size of its microprocessor. A microcomputer used to be a 4-bit- or 8-bit-word machine, and a minicomputer a 16-bit-word machine. Because of the advance of integrated circuit technology, even though minicomputers are still basically 16-bit-word computers, their physical size, weight, and cost have been reduced considerably. On the other hand, the microcomputer has expanded from 8-bit- to 16-bit-word machines with sophisticated supporting hardware and

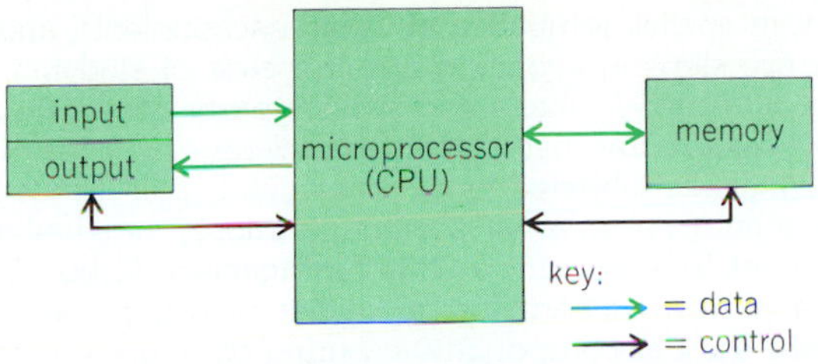

Basic elements of a microcomputer.

software. They can interface with mass storage media such as cassette tapes and floppy disks. Furthermore, in addition to machine and assembly programming languages, microcomputers can be programmed using high-level programming languages, such as BASIC, FORTRAN, or Pascal, which once were available only on minicomputers. Hence, the difference between microcomputers and minicomputers has become very small. *See* COMPUTER STORAGE TECHNOLOGY; DIGITAL COMPUTER.

Based on their sizes and applications, microcomputers are classified into three categories: one-card (or minimum) microcomputers; personal (home or business) microcomputers; and microcomputer development systems. Because of their low cost, the use of personal microcomputers in education, home entertainment, personal record keeping, and business has become very popular. And, they can have peripherals such as keyboard and cathode-ray-tube display, audio cassette tape recorders, disk drives, and printers. A complete microcomputer system that is used to test both the software and hardware of other microcomputer-based systems is called a microcomputer development system. Such a system provides hardware and software support from initial program development to debugging the prototype system.

A microcomputer is generally used for a dedicated task as part of a system. Because of its low cost, the microcomputer has begun to replace the minicomputer in many dedicated systems, such as data acquisition systems, processor control systems, data-processing systems, and remote-terminal control systems. *See* MICROPROCESSOR. [S.C.L.]

Microfilming The technique of making, on a reduced scale, a photographic film record of documents, printed matter, and the like, which can be enlarged for reading. Reductions may range from 8 to 40 times, although in extreme cases a reduction of 60 times has been used. *See* PHOTOCOPYING PROCESSES; PHOTOGRAPHY. [W.Cl.]

Micromanipulation The techniques and science of microdissection, microvivisection, microisolation, and microinjection. This field of knowledge is also known as micrurgy. Applications of micrurgy are many, especially in biology, cancer research, microchemistry, biophysics, colloid technology, and metallurgy.

The basic instruments required for micrurgy are (1) micromanipulators or micropositioners which convert the relatively crude hand movements to the exquisitely fine movements of the microtools; (2) microtools such as needles, pipes, hooks, loops, electrodes, scalpels, or forceps; and (3) microscopes and accessories. [M.J.K.]

Micrometeorite A meteorite less than about 0.1 mm in diameter. These bodies are usually called micrometeoroids while still in space. Micrometeorites can be expected to survive the atmospheric encounter since their size permits them to radiate the tremendous heat energy generated by friction with the atmosphere before it can cause vaporization. The Earth collects micrometeorites at a rate of about 10 tons (9 metric tons) per day. Those larger than micrometeorite size are destroyed during atmospheric entry and produce the visible phenomena known as meteors. Studies have shown that micrometeoroids yield important clues to the nature, origin, and evolution of the solar system and contribute significantly to the physical and chemical nature of the Earth's upper atmosphere. *See* METEOR; METEORITE. [R.K.S.]

Micrometeorology A branch of atmospheric dynamics and thermodynamics which deals primarily with the interaction between atmosphere and ground; the interchange of masses, momentum, and energy at the earth-air interface; in short, with the lower boundary conditions of atmospheric processes. Micrometeorology, along with other branches of meteorology, is also concerned with investigating and predicting the transport and dispersion of pollution from such sources as smokestacks and automotive exhausts in the lower atmosphere. Micrometeorology typically deals with the transport (or flux) of air properties in the vertical direction (conduction and convection) and the vertical variation of these fluxes. *See* AIR POLLUTION; SEAWATER; SOIL.

Micrometeorological research is important in that it supplies detailed information about the physical processes in the region of the atmosphere where life is most abundant. It closes the gaps of information from synoptic networks. It produces results useful for applications in various fields such as climatology, oceanography, soil physics, agriculture, biology, chemical warfare, and air pollution. Among the branches of atmospheric physics, micrometeorology is the one whose subjects are most amenable to fairly complete experimental description, and to testing of the theoretical models. *See* AGRICULTURAL METEOROLOGY; CLIMATOLOGY; CROP MICROMETEOROLOGY; INDUSTRIAL METEOROLOGY. [H.H.L.]

Micrometer A precision instrument used to measure small distances and angles. A common use is on a machinist's caliper, as in the illustration. *See* CALIPER.

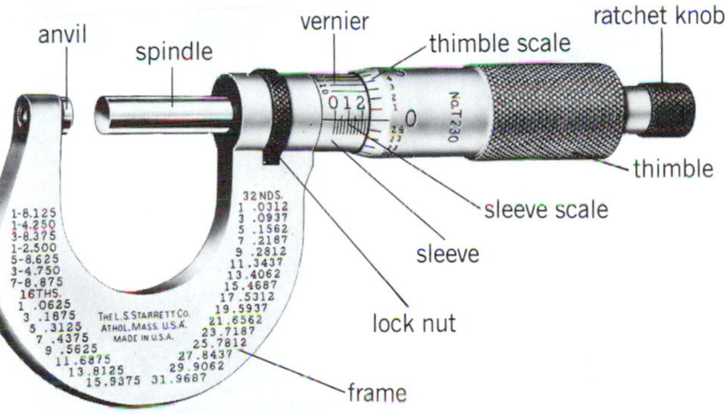

Machinist's outside caliper with micrometer reading 0.250 in. (*L. S. Starrett Co.*)

The spindle of the caliper is an accurately machined screw, which is rotated by the thimble or the ratchet knob until the object to be measured is in contact with both spindle and anvil. The ratchet slips after correct pressure is applied, ensuring consistent, accurate gaging. The number 1 on the sleeve represents 0.1 in.; the smallest divisions are 0.025 in. A vernier scale allows accurate reading to 0.0001 in. *See* VERNIER. [F.H.R.]

Microorganisms A heterogeneous assemblage of simple organisms, consisting of the protozoa, algae, fungi, rickettsiae, viruses, and bacteria. All these forms have in common a relatively simple organization which sets them apart from true plants and true animals: They are either unicellular or, if multicellular, their tissues are relatively undifferentiated. They are presumed to be evolutionary offshoots of the ancestral groups from which the plant and animal kingdoms have evolved.

The belated discovery of microorganisms and the failure at first to recognize their unique properties led to their arbitrary classification within the plant and animal kingdoms. This practice is still in use, although many biologists are now aware of the misconceptions which it engenders. *See* BACTERIA; FUNGI; MICROBIOLOGY; PROTOZOA; VIRUS. [E.A.Ad.]

Micropaleontology A branch of paleontology dealing with the fossilized microscopic organic remains (microfossils) of the geologic past, their structure, biology, phylogenetic relations, and distribution in space and time. The study of these microfossils has become an independent scientific field largely because: (1) The size of these fossils requires special methods for collection and examination. (2) Their abundance in geologic formations makes it possible to analyze their spatial distribution and the rates of morphological changes during the course of evolution by means of statistical methods which can be used only under exceptional circumstances in the study of larger fossils. (3) Microfossils have become indispensable tools in certain branches of applied geology, especially in the exploration for oil-bearing strata, because countless numbers of these minute fossils may be obtained from small pieces of subsurface rock recovered from drill holes. (4) The diversity of microfossils, their wide spatial distribution in varied environments, and their distinctive steps in evolution and the ease of studying them have contributed to make micropaleontology one of the most actively studied branches of the earth sciences.

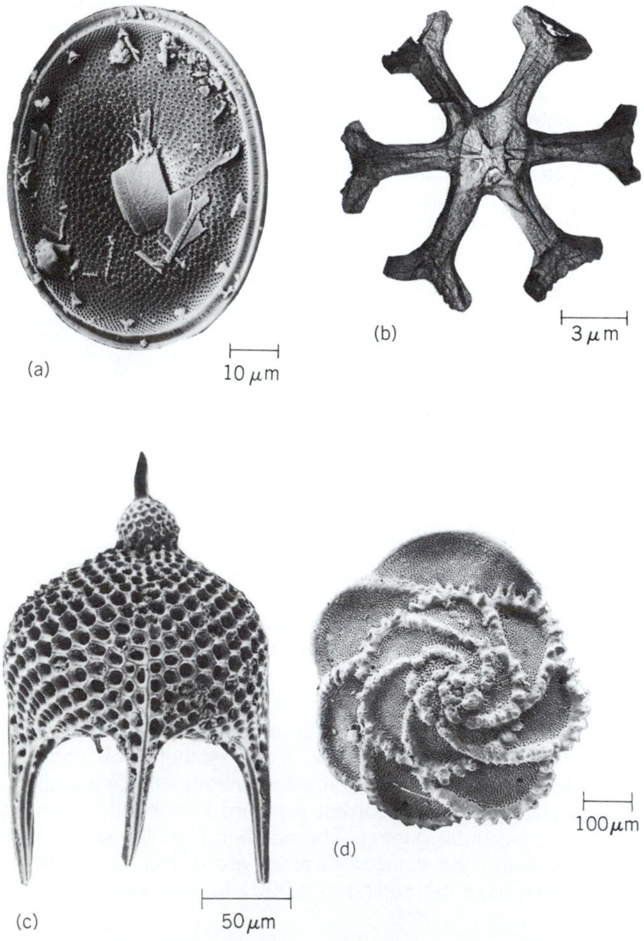

Representative microfossils. (*a*) Diatom; (*b*) asterolith, a calcareous nannoplankton; (c) radiolarian; and (*d*) planktonic foraminiferan.

The material subjected to micropaleontological studies forms a spectrum from primitive plants to advanced vertebrates (see illustration). The only prerequisite for organisms to become the subject of micropaleontological studies is their possession of resistant skeletal components ensuring their preservation in sedimentary strata as fossilized remains even after biological, chemical, or mechanical processes have destroyed the organisms' soft parts.

Most major groups of organisms incorporate, besides organic compounds, hard resistant materials that serve for structural support or protection. The more common substances found among the microfossils are calcium carbonate, silicon dioxide (or silica), calcium phosphate in the form of the mineral apatite (typical of bones and teeth), sporonine (principal constituent of pollen and spore walls), and various complex organic compounds. [T.S.]

Microphone An electroacoustic device containing a transducer which is actuated by sound waves and delivers essentially equivalent electric waves. Modern conventional microphones may be classified as pressure, gradient, combination pressure-gradient, and wave types. *See* TRANSDUCER.

In the pressure microphone the electrical response is caused by variations in sound pressure. Pressure microphones are inherently nondirectional (omnidirectional), because pressure is a scalar and not a vector quantity.

A pressure-gradient microphone is one in which the electrical response corresponds to some function of the difference in pressure between two points in space. In general, when the distance between the two points is small compared to the wavelength, the pressure gradient corresponds to the particle velocity in the sound wave. A velocity microphone is one in which this condition holds.

The wave microphone consists of a system in which the directivity depends upon some type of wave interference. The most common wave microphones are line, reflector, and lens types. Line and reflector microphones are highly directional in the speech-frequency range and may therefore be used for long-distance pickup of speech under conditions of high ambient noise and excessive reverberation. High-sensitivity transducers are employed in line and reflector microphones to provide an adequate signal-to-noise ratio.

The hot-wire microphone consists of a fine wire heated by the passage of an electric current. The cooling due to the motion of air past the wire causes a change in electrical resistance of the wire. In a sound wave the particle velocity cools the wire. There are also minor cooling effects produced by the sound pressure. The change in resistance due to the passage of a sound wave may be used to detect its presence. However, the frequency of the electrical output is twice the frequency of the sound wave because the wire is cooled equally by both positive and negative particle velocities. Therefore, this microphone cannot be used for the reproduction of sound. [H.F.O.]

Microphonics The transformation of the energy of mechanical vibration into an electrical signal. The term microphonics is most commonly applied to the production of an undesirable electrical output by an electronic device undergoing mechanical vibration.

Under certain conditions, electrical wires and cables are microphonic in that they produce undesirable electrical outputs in response to mechanical excitation. This is most commonly encountered with coaxial cables. In addition, microphonic effects have been noted in ceramic capacitors, variable resistors, point-contact transistors and diodes, and switches.

In order to avoid microphonics, special consideration is given to the electromechanical design of many electronic components. In addition, electronic equipment designed for use in high-vibration environments, such as aircraft radios, are often mounted with vibration isolators to reduce the mechanical excitation of the equipment. *See* SHOCK ISOLATION. [D.N.K.]

Microprocessor A central processor unit (CPU) on a single integrated circuit (IC) chip. The physical size of a typical microprocessor wafer is around 0.2 × 0.2 in. (5 × 5 mm), but may vary from one microprocessor to another. The illustration is a magnified view of a typical microprocessor. Based on its

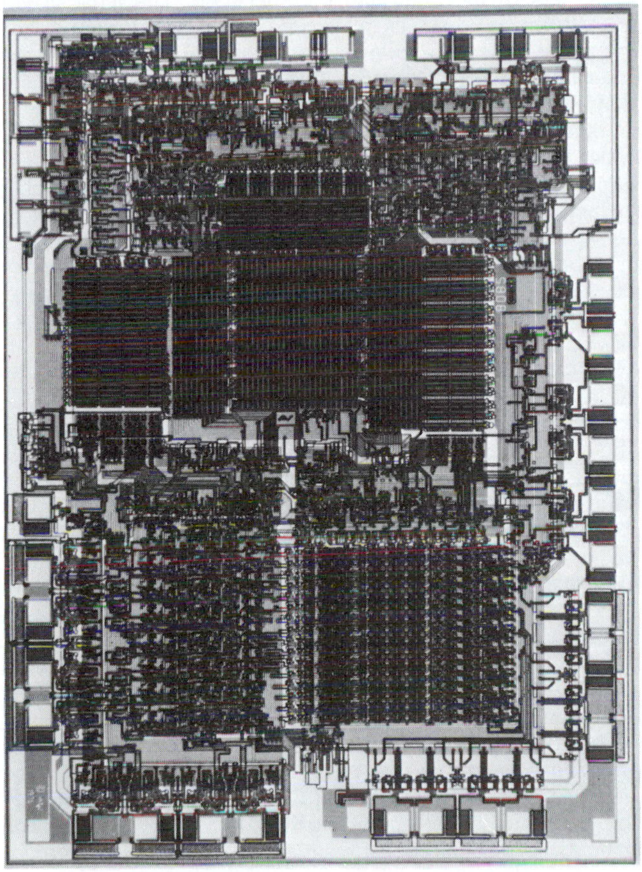

A typical microprocessor (Intel 8085). (*Intel Corp.*)

word size (bits), the commercially available microprocessors can be classified as 4-bit, 8-bit, 16-bit, or bit-sliced. The physical size and the number of pins of a microprocessor chip are proportional to its word size.

Each microprocessor has its own architecture, which encompasses the general layout of its major components, the principal features of these components, and the manner in which they are interconnected. However, the major components of all microprocessors are the same: (1) the clock; (2) control unit, comprising the program counter (PC), instruction register (IR), processor status word (PSW), and stack pointer (SP); (3) control memory; (4) bus control; (5) working register; (6) arithmetic/logic unit (ALU); (7) internal memory or stack. They are also called the hardware of a microprocessor. Besides the hardware, a microprocessor has its instruction set or software. According to the type of operations that the instructions perform, an instruction set can be subdivided into group operations for arithmetic, data transfer, branching, logic, and input/ output (I/O). A set of logically related instructions stored in memory is referred to as a program. The microprocessor "reads" each instruction from memory in a logically determinate sequence, and uses it to initiate processing actions.

The main difference between the microprocessor digital system design and the hard-wired logic digital system design is that the former uses the microprocessor to replace hard-wired logic by storing program sequences in the read-only memory (ROM) rather than implementing these sequences with gates, flip-flops, counters, and so on. After the design is completed, modifications may be made by simply changing the program in the ROM. The microprocessor digital system design is widely used because of the following advantages:

1. Manufacturing costs of industrial products can be significantly reduced.
2. Products can get to the market faster, providing a company with the opportunity to increase product sales and market share.
3. Product capability is enhanced, allowing manufacturers to provide customers with better products, which can frequently command a higher price in the marketplace.
4. Development costs and time are reduced.
5. Product reliability is increased, while both service and warranty costs are reduced. *See* SEMICONDUCTOR MEMORIES.

Application areas include: industrial sequence controllers; machine tool controllers; point-of-sale terminals; intelligent terminals; instrument processors; traffic light controllers; weather data collection systems; and process controllers. *See* CONTROL SYSTEMS; DIGITAL COMPUTER; INTEGRATED CIRCUITS; MICROCOMPUTER. [S.C.L.]

Micropygoida An order of regular echinoids belonging to the Diadematacea, established for the two recognized species of *Micropyga*. They have an aulodont lantern with grooved teeth. Test plating is imbricate, and ambulacra are composed of trigeminate compound plates in which upper and lower elements are reduced to demiplates. Pore pairs are biserially arranged in ambulacral columns, and there are unique umbrellalike aboral tube feet. *Micropyga* is a deep-water echinoid, found between 480 and 4290 ft (150 and 1340 m) depth in the Indo-West Pacific. There are no fossils that can be placed in this taxon with certainty, but it is possible that the Upper Jurassic *Pedinothuria* may belong here. *See* DIADEMATACEA; ECHINODERMATA. [A.B.S.]

Microradiography The process of producing enlarged radiographic images on photographic film. Microradiography is applicable to small metallurgical, biological, and other specimens with gross structures not easily visible on conventional 1:1 radiographs. Historadiography is a term used for the microradiography of biological specimens, and hence is a branch of histology. Microradiography is a type of x-ray microscopy. *See* HISTORADIOGRAPHY; OPTICAL MICROSCOPE; RADIOGRAPHY; X-RAY MICROSCOPE.

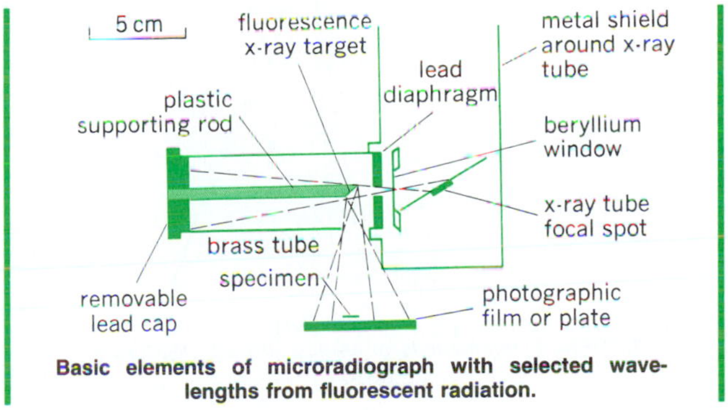

Basic elements of microradiograph with selected wavelengths from fluorescent radiation.

An enlarged radiographic image can be obtained in two ways: (1) by projection microradiography, also termed shadow microscopy, in which the specimen is placed at various distances from the film to secure inherent enlargement of divergent beams, and (2) by contact microradiography, in which the specimen is in contact with a fine-grained photographic emulsion so that the absorption image is photographically enlarged or viewed with a microscope.

In microradiography it is convenient to be able to generate a wide range of essentially monochromatic beams, instead of relying upon a series of x-ray tubes with different targets. It is possible to generate such beams at will by exciting secondary fluorescent radiation from a variety of elements with a single high-intensity source of primary rays. The experimental arrangement is shown in the illustration. Obviously the fluorescent beams are much lower in intensity than primary rays directly used, but exposure times are by no means prohibitive, and there is a large gain in homogeneity of the beam and hence in sharpness and resolution of the microradiographic image. *See* X-RAY TUBE.

Since microradiography is a combination of radiography and micrography, analyses derived from it are at three levels: (1) for structure, which is most common and simplest; (2) for mass distribution; and (3) for mass-chemical analysis. The last two are accomplished best with thin biological specimens and with monochromatic beams of long wavelength. [G.L.Cl.]

Microsauria A diverse order of small, extinct amphibians, known only from the Pennsylvanian and lower Permian of North America and Europe. They range from obligatorily aquatic, perennibranchiate genera with lateral-line canal grooves, to fully terrestrial lizardlike forms. Several families include long-bodied, possibly burrowing, species. Limbs are always retained, and the tail is never specialized as a swimming organ. Microsaurs are recognized by the possession of a broad, strap-shaped occipital condyle, and no more than a single bone in the temporal series. The trunk vertebrae are spool-shaped.

The specific origin of the group remains unknown. Although intermediate forms are not known, microsaurs appear to be the most probable group of Paleozoic amphibians from which two modern amphibian orders, the apodans and the salamanders, have evolved. *See* AMPHIBIA; LEPOSPONDYLI. [R.L.C.]

Microscope An instrument used to obtain an enlarged image of a small object. The image may be seen, photographed, or sensed by photocells or other receivers, depending upon the nature of the image and the use to be made of the information of the image.

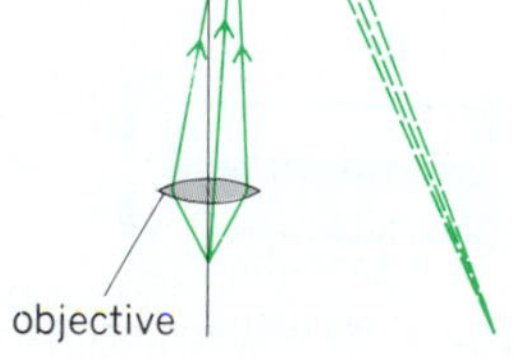

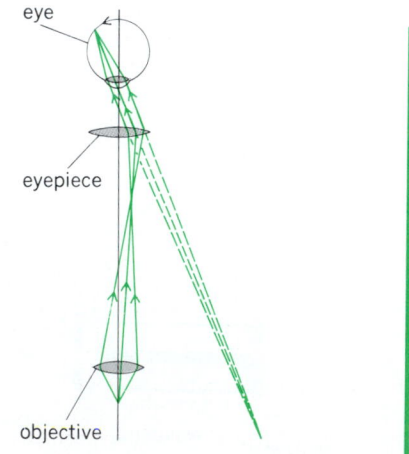

Compound microscope diagram. ***(After F. A. Jenkins and H. E. White, Fundamentals of Optics, 4th ed., McGraw-Hill, 1976)***

A simple microscope, hand lens, or magnifier usually is a round piece of transparent material, ground thinner at the edge than at the center, which can form an enlarged image of a small object. Commonly, simple microscopes are double convex or planoconvex lenses, or systems of lenses acting together to form the image.

The compound microscope utilizes two lenses or lens systems. One lens system forms an enlarged image of the object and the second magnifies the image formed by the first. The total magnification is then the product of the magnifications of both lens systems (see illustration).

The typical compound microscope consists of a stand, a stage to hold the specimen, a movable body-tube containing the two lens systems, and mechanical controls for easy movement of the body and the specimen. The lens system nearest the specimen is called the objective; the one nearest the eye is called the eyepiece or ocular. A mirror is placed under the stage to reflect light into the instrument when the illumination is not built into the stand. For objectives of higher numerical aperture than 0.4, a condenser is provided under the stage to increase the illumination of the specimen. Various optical and mechanical attachments may be added to facilitate the analysis of the information in the doubly enlarged image. *See* ELECTRON MICROSCOPE; FLUORESCENCE MICROSCOPE; INTERFERENCE MICROSCOPE; LENS (OPTICS); OPTICAL MICROSCOPE; PHASE-CONTRAST MICROSCOPE; REFLECTING MICROSCOPE; X-RAY MICROSCOPE. [O.W.R.]

Microsporidea A class of Cnidospora characterized by the production of minute spores with a single intrasporal or one or two intracapsular filaments and a single sporoplasm. The spore membrane of these protozoans is usually a single piece. Microsporidians are mainly intracellular parasites of arthropods and fishes. Microsporida is the only order of this class. *See* CNIDOSPORA. [R.F.N.]

Microtechnique The art of preparing objects for examination under the microscope and of preserving objects so prepared. Few objects yield useful information if examined without such preparation, which may involve, in addition to preliminary preservation, hardening, rendering transparent, selective coloration of parts, and cutting into thin slices.

The four types of microscope slide commonly made are wholemounts, smears, squashes, and sections. The last three methods are merely devices to make thinner, or smaller, objects unsuitable for the first method. In all four methods, the objects are permanently preserved in a mounting medium between a glass slide, about 0.04 in. (1 mm) thick, and a glass cover slip 0.04 to 0.008 in. (1 to 0.2 mm) thick. Preliminary fixation and preservation, staining, and the final mounting media are common to all four types. [P.Gr.]

Microwave An electromagnetic wave which has a wavelength in the centimeter range. Microwaves occupy a region in the electromagnetic spectrum which is bounded by radio waves on the side of longer wavelengths and by infrared waves on the side of shorter wavelengths (see illustration). There are no sharp boundaries between these regions except by arbitrary definition. In the microwave region, a further delineation is sometimes made with such names as decimeter, centimeter, or millimeter waves.

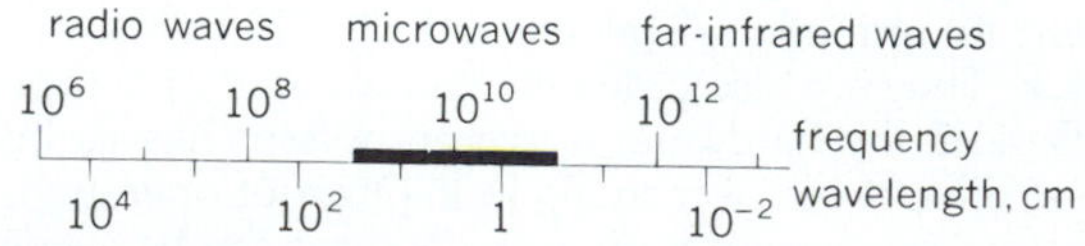

A portion of the electromagnetic spectrum.

All modern microwave generators are electronic devices which produce continuous-wave oscillations of a single tunable frequency. Some important microwave generators are known as klystrons, magnetrons, and traveling-wave oscillators. Their power outputs range from microwatts to thousands of kilowatts, depending upon the type and design of the generator and the operating frequency. *See* KLYSTRON; MAGNETRON; MICROWAVE SOLID-STATE DEVICES; MICROWAVE TUBE; TRAVELING-WAVE TUBE.

Circuit elements. Any particular grouping of physical elements which are arranged or connected together to produce certain desired effects on the behavior of microwaves is known as a microwave circuit.

A waveguide is a circuit element which constrains or guides the propagation of microwaves along a path defined by the physical construction of the guiding element. It can be, for instance, a coaxial cable having an outer conductor of annular cross section coaxially placed with respect to an inner conductor. By prevailing usage, however, a microwave waveguide usually means a hollow metallic tube which can confine and guide the propagation of microwaves.

A microwave attenuator is a device that causes the field intensity of a wave to decrease by absorbing a part of the incident power. A phase changer is a device that causes the field intensity to shift its phase without attenuating its amplitude.

A microwave detector is a device that can demonstrate the presence of a microwave by a specific effect that the wave produces. One of the most effective means of detection makes use of a nonlinear device which converts the microwave field intensity into either a direct current or a low-frequency alternating current. A silicon crystal making a pinpoint contact with a tungsten wire is perhaps the most commonly used detector for microwaves.

Another type of microwave detector in common use is called a bolometer. It is a device whose resistance changes sensitively with temperature, the latter being a function of the absorbed microwave power. *See* BOLOMETER.

A microwave wavemeter is usually made of a tunable resonant cavity which measures the free-space wavelength (or frequency) of a wave by the position of tuning at resonance. *See* CAVITY RESONATOR.

Transmission. A microwave transmitter is similar in all its principal aspects to an ordinary radio transmitter. It consists of a microwave generator, a power amplifier (if necessary), a circuit containing all necessary elements, means of modulation to impart some form of information or program to the waves, and an antenna network to send the waves out into space. Since World War II, klystron amplifiers have come into practical use at high powers, and traveling-wave amplifiers at low powers. *See* ANTENNA (ELECTROMAGNETISM).

Methods involving amplitude modulation (AM), frequency modulation (FM), or phase modulation (PM) can be used to transmit information. Transmission by pulsed waves is an efficient method of sending an extraordinarily large peak power at the expense of a small average power. Even apart from the power considerations, pulse transmission is the most effective way of securing echo signals by reflection, as in radar. *See* AMPLITUDE MODULATION; FREQUENCY MODULATION; PHASE MODULATION; PULSE MODULATION.

Propagation in space. In space, a wave tends to travel along a straight line in the absence of obstructions. However, if the transmitted beam either wholly or partially hits the Earth's surface or any physical obstacle, it will be scattered, with the particular phenomenon variously described as reflection, refraction, or diffraction. A microwave has too high a frequency to be reflected by the ionosphere at angles near vertical incidence. However, for microwaves slanted toward the horizon, reflection by the ionosphere becomes possible. *See* MICROWAVE OPTICS.

Advantages. From the point of view of radio communication, a great advantage of microwaves lies in the immense spaciousness of useful frequencies. For instance, the frequency difference between the S-band (wavelength around 10 cm) and K-band (wavelength around 1 cm) is roughly 20,000 MHz, about 100 times the combined frequency range of present-day radio broadcasting, radio communication, and television. Another advantage of microwaves is related to the high directivity and resolving power of microwave wavelengths. Narrow microwave beams can be readily formed by antennas of physically convenient sizes.

Applications. The most important practical application of microwaves is radar. In fact, microwave techniques have been developed mainly under the incentive of radar work. Next to radar the use of microwaves as carrier waves in relay links for multichannel transmission of telephone, telegraph, and television has already become practically important.

Microwave spectroscopy has become an established science chiefly because of the availability of microwave instruments and techniques. Another important advance came with the development of the so-called atomic clocks, which use microwave resonance interactions with either cesium atoms or ammonia molecules. The introduction of the ammonia maser led to the development of the solid-state maser, which is virtually a noiseless amplifier. One of the most important applications of the maser is in the field of radio astronomy, which is in its own right a fertile ground for microwave application. In the field of nuclear physics instrumentation, microwave electronics has been applied in a most significant manner to high-energy linear accelerators and similar devices. *See* ATOMIC CLOCK; MASER; MICROWAVE SPECTROSCOPY; PARTICLE ACCELERATOR; RADIO ASTRONOMY. [C.K.J.]

Microwave filter A two-port component used to provide frequency selectivity in satellite and mobile communications, radar, electronic warfare, metrology, and remote-sensing systems operating at microwave frequencies (1 GHz and above). Microwave filters perform the same function as electric filters at lower frequencies, but differ in their implementation because circuit dimensions are on the order of the electrical wavelength at microwave frequencies. Thus, in the microwave regime, distributed circuit elements such as transmission lines must be used in place of the lumped-element inductors and capacitors used at lower frequencies. This can make microwave filter design more difficult, but it also introduces a variety of useful coupling and transmission effects that are not possible at lower frequencies.

The majority of modern microwave filters are designed by using the insertion-loss method, whereby the amplitude response of the filter is approximated by using network synthesis techniques that have been extended to accommodate microwave distributed circuit elements. A general four-step procedure is followed: determination of filter specifications, design of a low-pass prototype filter, scaling and transforming the filter, and implementation (conversion of lumped elements to distributed elements).

Microwave filters are implemented in many ways. Waveguide cavity band-pass filters have very low insertion loss, making them preferred for frequency multiplexing in satellite communication systems. Coaxial low-pass filters, made with sections of coaxial line with varying diameters, are compact and inexpensive. Planar filters in microstrip or stripline form (see illus.) are important for integration with hybrid or monolithic microwave integrated circuits. While planar filters are usually more cost effective than waveguide versions, their insertion loss is usually greater. Computer-aided design procedures are used in the synthesis of more sophisticated amplitude and phase responses, and active microwave devices (field-effect transistors) are used to provide filters with gain or tunable

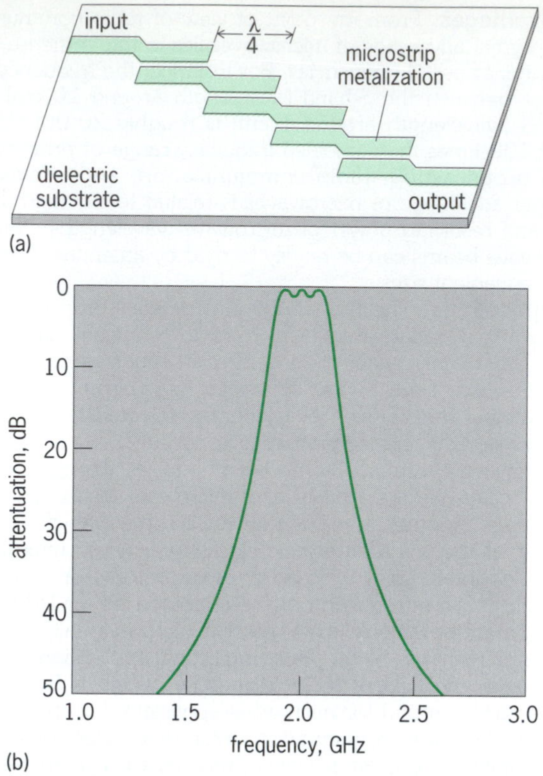

Third-order (four-section) microwave band-pass filter. (*a*) Layout of the parallel coupled stripline filter. (*b*) Calculated frequency response of the filter, with center frequency of 2 GHz and 0.5-dB equal-ripple passband.

response characteristics. *See* COAXIAL CABLE; COMPUTER-AIDED DESIGN AND MANUFACTURING; ELECTRIC FILTER; MICROWAVE; MICROWAVE SOLID-STATE DEVICES; TRANSMISSION LINES; WAVEGUIDE. [D.M.Po.]

Microwave free-field standards The means for setting up electromagnetic fields of precisely determined intensity at microwave frequencies in unbounded regions of space. Such standards are used to evaluate field probes and antennas for measuring field strength and power density. The standardization of these devices is necessary before they are used for determining the performance of radar and communications systems or for assessing such systems for health and safety risks or electromagnetic compatibility. *See* ELECTROMAGNETIC RADIATION; MICROWAVE.

Antennas used in free-field standards are usually either half-wave dipoles or waveguide horns. The half-wave dipole is a collinear device with a length of approximately one-half of the free-space wavelength of the radiated wave. The gain and pattern characteristics of a pyramidal horn can be calculated to accuracies of about ± 0.2 dB; however, the reflections at the throat and aperture discontinuities significantly influence the gain-frequency characteristic and calibration is usually necessary. *See* WAVEGUIDE.

The ideal environment for making measurements on antennas is a large unobstructed volume which is free of reflecting objects and electromagnetically interfering signals—that is, a free-space condition. A practical solution is to use an anechoic chamber to set up simulated free-space conditions in a bounded environment. Low reflection of electromagnetic signals from the walls of such a chamber is achieved by the use of an electromagnetic wave absorbent layer covering all of the reflecting surfaces within the room or chamber, the outer shell of which is a metallic structure to give shielding against interfering signals encroaching on the test region.

Clear open sites can be used at low frequencies, or for high-gain antennas where far-field measurements require such large distances that enclosed or anechoic environments are impractical. However, since electromagnetic waves are strongly reflected by a ground plane, this effect must be accommodated, and one of two courses can be adopted. One is to do measurements as far above the ground as possible. The alternative is to make use of the ground plane by working close to it—if necessary, enhancing its reflection with a metal ground plane or grid—and by making allowance for the reflection in the analysis of performance. *See* ANECHOIC CHAMBER.

Probably the single most important parameter of a standard antenna, when considered for metrological applications, is the boresight or maximum gain, and a number of methods of determining this parameter have been devised. The three-antenna method involves measurement of the transmission between two polarization-matched antennas. Only the product of the antenna gains can be obtained from this measurement; however, with three antennas the measured combinations will yield the gain of each antenna uniquely. The extrapolation method involves determining the transmission characteristic between two antennas as the transmission path is increased through about 4 to 10 Rayleigh distances. With good metrology it is possible to get a sufficiently accurate characterization to allow extrapolation to the true far-field range. Antenna metrology has increasingly concentrated on near-field scanning techniques with the objective of improving antenna characterization. In this method a probe antenna is used to sample the magnitude and phase, for orthogonal polarizations, of the radiated fields over a well-defined surface, which can be a plane, a cylinder, or a sphere, a few wavelengths from the antenna under test.

The essential requirement for the calibration of devices for measuring power flux density or field strength is the creation of a substantially plane wave of known power density which encompasses the effective aperture of the device to be tested. This is effected by launching a known power through an antenna or transverse-electromagnetic cell of known characteristics and calculating the field strength or power density from the appropriate equation. [R.W.Ye.]

Microwave landing system (MLS) A system of microwave signals used to guide an aircraft during final approach and landing. *See* MICROWAVE.

Microwave landing systems (MLS) use ground-originated radio transmissions that provide precise positional information referenced to a single runway. The ground facilities that provide the required signals are (1) a localizer station, sited at the far end of the runway that radiates signals used to measure azimuth angles; (2) a glide path station, sited to the side of the runway near the touchdown zone which radiates signals used to measure elevation angles; and (3) a distance-measuring transponder usually sited at the localizer station that responds to radioed interrogations from the aircraft so that a distance based on elapsed propagation time can be determined in the aircraft. *See* DISTANCE-MEASURING EQUIPMENT.

The limitation of the instrument landing system (ILS) that inspired its replacement is fundamentally due to the frequency bands in which its constituent subsystems operate (108–112 MHz, 328.6–335.4 MHz, and 75 MHz). At these frequencies the antenna arrays required to form relatively confined energy lobes are large and often difficult to site. Thus, replacements for the ILS use the microwave frequencies where complex beams can be achieved with antennas of reasonably small apertures. *See* ANTENNA (ELECTROMAGNETISM).

The International Civil Aviation Organization (ICAO) standardization process was directed to a system concept called the time reference scanning beam (TRSB). The system has two hundred 300-kHz-wide channels in the frequency band from

5.030 to 5.091 GHz. The angular guidance is conveyed by a scanning beam with an angular velocity of 20,000° per second. It sweeps through the coverage sector in a "to" and then "fro" direction. The measured time interval between the "to" and "fro" beam passages is the measure of the angle at the reception point. To achieve the high angular velocity of the beam, the antennas employed for these functions are electronically scanned arrays.

A significant feature of the TRSB-type MLS is its message capability. This capability provides a means of transmitting to a receiving aircraft a complete range of information relating to the special characteristics of the particular ground system. The geometric relationships of the various elements of the system are also transmitted. These messages will allow sophisticated receiving and processing equipment in the aircraft to adapt the approach and landing maneuvers to a wide variety of conditions. *See* AIR NAVIGATION; ELECTRONIC NAVIGATION SYSTEMS. [F.B.P.]

Microwave measurements A collection of techniques particularly suited for development of devices and monitoring of systems where physical size of components varies from a significant fraction of an electromagnetic wavelength to many wavelengths. *See* MICROWAVE.

Virtually all microwave devices are coupled together with a transmission line having a uniform cross section. The concept of traveling electromagnetic waves on that transmission line is fundamental to the understanding of microwave measurements. *See* MICROWAVE TRANSMISSION LINES.

At any reference plane in a transmission line there are considered to exist two independent traveling electromagnetic waves moving in opposite directions. One is called the forward or incident wave, and the other the reverse or reflected wave. The electromagnetic wave is guided by the transmission line and is composed of electric and magnetic fields with associated electric currents and voltages. Any one of these parameters can be used in considering the traveling waves, but the measurements in the early development of microwave technology made principally on the voltage waves led to the custom of referring only to voltage. One parameter in very common use is the voltage reflection coefficient Γ, which is related to the incident, V_i, and reflected, V_r voltage waves by Eq. (1).

$$\Gamma = \frac{V_r}{V_i} \tag{1}$$

Impedance. The voltage reflection coefficient Γ is related to the impedance terminating the transmission line and to the impedance of the line itself. If a wave is launched to travel in only one direction on a uniform reflectionless transmission line of infinite length, there will be no reflected wave. The input impedance of this infinitely long transmission line is defined as its characteristic impedance Z_0. An arbitrary length of transmission line terminated in an impedance Z_0 will also have an input impedance Z_0. *See* ELECTRICAL IMPEDANCE.

If the transmission line is terminated in the arbitrary complex impedance load Z_L, the complex voltage reflection coefficient Γ_L at the termination is given by Eq. (2).

$$\Gamma = \frac{Z_L - Z_0}{Z_L + Z_0} \tag{2}$$

Even when there is no unique expression for Z_L and Z_0 such as in the case of hollow uniconductor waveguides, the voltage reflection coefficient Γ has a value because it is simply a voltage ratio. In general, the measurement of microwave impedance is the measurement of Γ. Both amplitude and phase of Γ can be measured by direct probing of the voltage standing wave set up along a transmission line by the two opposed traveling waves, but this is a slow technique.

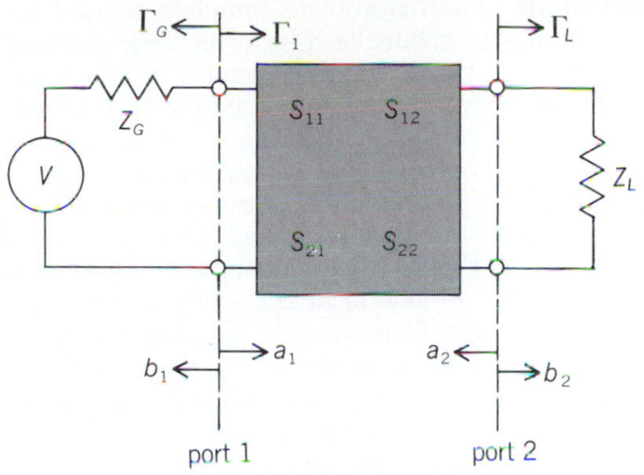

A two-port inserted between a load and a generator. S_{nm} are the scattering coefficients of the two-port.

Directional couplers have been used for many years to perform much faster swept frequency measurement of the magnitude of Γ, and more recently the use of automatic network analyzers under computer control has made possible rapid, accurate measurements of amplitude and phase of Γ over very broad frequency ranges. *See* DIRECTIONAL COUPLER.

Power. A required increase in microwave power is expensive whether it be the output from a laboratory signal generator, the power output from a power amplifier on a satellite, or the cooking energy from a microwave oven. To minimize this expense, absolute power must be measured. Most techniques involve conversion of the microwave energy to heat energy which, in turn, causes a temperature rise in a physical body. This temperature rise is measured and is approximately proportional to the power dissipated. The whole device can be calibrated by reference to low-frequency electrical standards and application of appropriate corrections. *See* RADIOMETRY.

The power sensors are simple and can be made to have a very broad frequency response. A power meter can be connected directly to the output of a generator to measure available power P_A, or a directional coupler may be used to permit measurement of a small fraction of the power actually delivered to the load.

Scattering coefficients. While the measurement of absolute power is important, there are many more occasions which require the measurement of relative power which is equivalent to the magnitude of voltage ratio and is related to attenuation. Also there arises frequently the need to measure the relative phase of two voltages. Measurement systems having this capability are referred to as vector network analyzers, and they are used to measure scattering coefficients of multiport devices. The concept of scattering coefficients is an extension of the voltage reflection coefficient applied to devices having more than one port. The most simple is a two-port. Its characteristics can be specified completely in terms of a 2 × 2 scattering matrix, the coefficients of which are indicated in the illustration. The incident voltage at the reference plane of each port is defined as a, and the reflected voltage is b. Voltages a and b are related by matrix equation (3), where (S_{nm}) is the scattering matrix of the junction. Writing Eq. (3) out for a two-port device gives Eqs. (4) and (5). Examination of Eq. (4) shows,

$$(b_n) = (S_{nm})(a_m) \tag{3}$$

$$b_1 = S_{11}a_1 + S_{12}a_2 \tag{4}$$

$$b_2 = S_{21}a_1 + S_{22}a_2 \tag{5}$$

for example, that S_{11} is the voltage reflection coefficient looking into port 1 if port 2 is terminated with a Z_0 load ($a_2 = 0$). *See* MATRIX THEORY.

Heterodyne. The heterodyne principle is used for scalar attenuation measurements because of its large dynamic range and for vector network analysis because of its phase coherence. The microwave signal at frequency f_s is mixed with a microwave local oscillator at frequency f_{LO}, in a nonlinear mixer. The mixer output signal at frequency $f_s - f_{LO}$ is a faithful amplitude and phase reproduction of the original microwave signal but is at a low, fixed frequency so that it can be measured simply with low-frequency techniques. One disadvantage of the heterodyne technique at the highest microwave frequencies is its cost. Consequently, significant effort has been expended in development of multiport network analyzers which use several simple power detectors and a computer analysis approach which allows measurement of both relative voltage amplitude and phase with reduced hardware cost. *See* HETERODYNE PRINCIPLE.

Noise. Microwave noise measurement is important for the communications field and radio astronomy. The measurement of thermal noise at microwave frequencies is essentially the same as low-frequency noise measurement, except that there will be impedance mismatch factors which must be carefully evaluated. The availability of broadband semiconductor noise sources having a stable, high, noise power output has greatly reduced the problems of source impedance mismatch because an impedance-matching attenuator can be inserted between the noise source and the amplifier under test. *See* ELECTRICAL NOISE; ELECTRICAL NOISE GENERATOR; MICROWAVE NOISE STANDARDS.

Use of computers. The need to apply calculated corrections to obtain the best accuracy in microwave measurement has stimulated the adoption of computers and computer-controlled instruments. An additional benefit of this development is that measurement techniques that are superior in accuracy but too tedious to perform manually can now be considered. For a discussion of attenuation in microwave circuits *see* ATTENUATION (ELECTRICITY). [R.F.Cl.]

Microwave noise standards Electrical noise generators which produce calculable noise intensities at microwave frequencies, and which are used to calibrate other noise sources by using comparison methods. Noise standards are based upon the blackbody or thermal radiator and generate noise power according to Planck's radiation law. The practical realization of a blackbody in the microwave region consists of a microwave absorber with unity absorptivity. This can be achieved by using a transmission line terminated in its characteristic impedance, or in microwave terminology a matched termination. *See* HEAT RADIATION; MICROWAVE TRANSMISSION LINES; TRANSMISSION LINES.

The range of sources which require calibration and the desire to obtain low uncertainties dictate that microwave thermal noise standards are required with temperatures both above and below the ambient temperature. Sources have been developed with temperatures in the range from 4 to 1300 K (−452° to 1900°F). The low temperatures are normally achieved by immersion of the matched termination in a cryogenic liquid of which liquid nitrogen (77 K or − 321°F) is the most common. Standards for measurement of high-temperature sources have the termination in a heated oven. A transition section supports the temperature gradient from the thermal termination to the ambient temperature output which connects to the measurement system. *See* MICROWAVE; MICROWAVE MEASUREMENTS; RADIOMETRY. [M.W.S.]

Microwave optics The study of those properties of microwaves which are analogous to the properties of light waves in optics. The fact that microwaves and light waves are both electromagnetic waves, the major difference being that of frequency, already suggests that their properties should be alike in many respects. But the reason microwaves behave more like light waves than, for instance, very low-frequency waves for electrical power (50 or 60 Hz) is primarily that the microwave wavelengths are usually comparable to or smaller than the ordinary physical dimensions of objects interacting with the waves.

As is the case with light, a beam of microwaves propagates along a straight line in a perfectly homogeneous infinite medium. This phenomenon follows directly from a general solution of the wave equation in which the direction of a wave normal does not change in a homogeneous medium. *See* WAVE EQUATION.

With some modification the laws of reflection and refraction can be applied to the propagation of microwaves inside a dielectric-filled metallic waveguide. Another interesting application is associated with the microwave analog of total internal reflection in optics. A properly designed dielectric rod (without metal walls) can serve as a waveguide by totally reflecting the elementary plane waves. Still another case of interest is that of a microwave tens. *See* ANTENNA (ELECTROMAGNETISM); REFLECTION OF ELECTROMAGNETIC RADIATION; REFRACTION OF WAVES; WAVEGUIDE.

In an analogous manner to light, a microwave undergoes diffraction when it encounters an obstacle or an opening which is comparable to or somewhat smaller than its wavelength. *See* MICROWAVE. [C.K.J.]

Microwave reflectometer A form of directional coupler that is used for measuring the power flowing in both directions in a waveguide. A pair of single-detector couplers appropriately positioned on opposite sides of the waveguide can be used for this purpose, with one detector positioned to monitor transmitted power and the other detector positioned to measure power reflected back from a discontinuity in the line. Each coupler receives a constant small fraction of the energy flowing in one direction in the waveguide. *See* RADIO-FREQUENCY IMPEDANCE MEASUREMENTS. [J.Mar.]

Microwave solid-state devices Semiconductor devices for the generation, amplification, detection, and control of electromagnetic energy in the frequency range of about 1–100 GHz. Devices that generate or amplify microwave energy are referred to as active microwave devices. The most important active solid-state microwave devices are microwave transistors, transferred electron (Gunn) devices, avalanche diodes, and tunnel diodes. Devices that detect or control microwave energy are referred to as passive microwave devices. The most important passive solid-state microwave devices include PIN diodes, varactor diodes, and point-contact and Schottky-barrier detector diodes. *See* MICROWAVE; MICROWAVE TUBE; SEMICONDUCTOR DIODE; TRANSISTOR.

Active devices. At the lower microwave frequencies, silicon bipolar transistors and gallium arsenide (GaAs) field-effect transistors (FETs) are the most widely used active solid-state devices. At the higher microwave frequencies, the usefulness of transistors is severely limited by high-frequency effects, and two-terminal negative resistance devices such as transferred-electron devices, avalanche diodes, and tunnel diodes predominate.

Most modern silicon bipolar microwave transistors are epitaxial diffused *npn* structures (Fig. 1). The basic principles of operation of bipolar microwave transistors are the same as those of low-frequency bipolar transistors. However, because of the higher frequencies involved, the requirements on dimensions, process control, packaging, heat sinking, and radio-frequency (rf) circuitry are much more severe for microwave transistors than for low-frequency transistors.

Microwave GaAs FETs are a type of microwave transistor that provides improved performance over silicon microwave transistors. Low-noise, low-power, and medium-power GaAs FETs are commercially available.

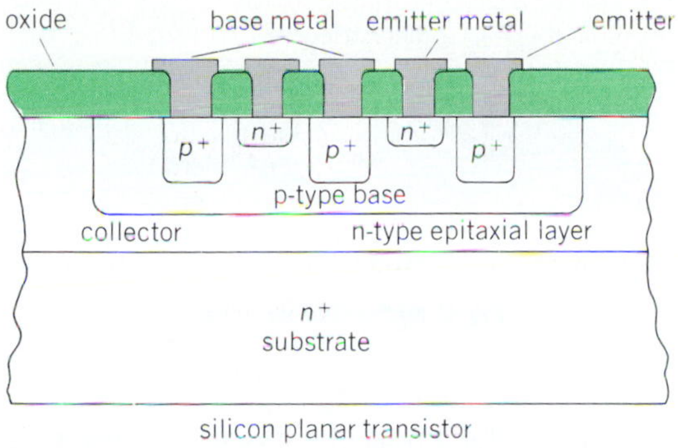

Fig. 1. Cross section of a planar silicon bipolar-power microwave transistor. (*After H. Sobol and F. Sterzer, Microwave power sources, IEEE Spectrum, 9(4):20–33, 1972*)

In transferred-electron devices (TEDs), negative resistance is achieved by taking advantage of the negative differential mobility of electrons in certain *n*-type III–V compounds, particularly GaAs. Structurally, TEDs are particularly simple semiconductor devices. They consist merely of a bar of transferred electron material with ohmic cathode and anode contacts.

Figure 2 shows graphs of electron drift velocity versus electric field for silicon and GaAs. The drift velocity for silicon exhibits a "normal" behavior, increasing monotonically with increasing electric field. In GaAs, however, the drift velocity decreases with increasing field above about 3 kV/cm. The negative differential mobility of GaAs is caused by the transfer of electrons from high- to low-mobility energy bands, hence the name "transferred electron devices." *See* SEMICONDUCTOR.

Avalanche diodes are junction devices that produce a negative resistance by appropriately combining impact avalanche breakdown and charge-carrier transit time effects. Avalanche breakdown in semiconductors occurs if the electric field is high enough for the charge carriers to acquire sufficient energy from the field to create electron-hole pairs by impact ionization. Transit time effects occur if the time the charge carriers spend traversing the diode becomes an appreciable fraction of an rf period. The two most important classes of avalanche diodes are IMPATT diodes (impact avalanche and transit time diodes) and TRAPATT diodes (trapped-plasma avalanche-triggered transit diodes).

Tunnel diodes were first described by L. Esaki in 1958. They are heavily doped *pn* junctions that exhibit a negative differential resistance over part of their current-voltage characteristics. The negative resistance is caused by processes (quantum-mechanical tunneling) that are so fast that there are no transit time effects even at the highest microwave frequencies. *See* TUNNEL DIODE.

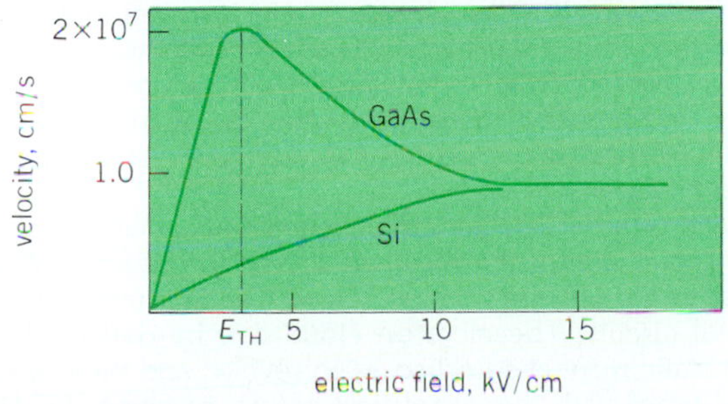

Fig. 2. Curves of electron drift velocity versus electric field for silicon and gallium arsenide. (*After H. Sobol and F. Sterzer, Microwave power sources, IEEE Spectrum, 9(4):20–33, 1972*)

Passive devices. The oldest passive microwave device is the point-contact diode, a diode that was widely used in World War II radars. More modern passive microwave solid-state devices include Schottky-barrier diodes, PIN diodes, and varactor diodes.

Point-contact and Schottky-barrier diodes are majority-carrier rectifiers that are used as the nonlinear element in passive microwave-frequency down-converters and in rectifiers. The rectification characteristics of these diodes are similar to those of *pn* junction diodes, but unlike *pn* junction diodes they do not exhibit any minority-carrier charge storage capacitance. This lack of charge-storage capacitance makes point-contact and Schottky-barrier diodes particularly attractive for down-converter applications, since variations in capacitance are in general undesirable in down-conversion. In point-contact diodes, a pointed tungsten wire is forced down on a semiconductor wafer (usually silicon). In Schottky-barrier diodes, the rectifying junction is formed at the interface of a deposited metal layer and a semiconductor crystal, usually silicon or gallium arsenide. *See* POINT-CONTACT DIODE.

PIN diodes are junction diodes in which the *p* and *n* regions are separated by a relatively thick high-resistivity intrinsic layer. These diodes are widely used in microwave switches, electronically variable microwave attenuators, and microwave limiters. *See* JUNCTION DIODE.

A varactor (variable-capacitance) diode is a *pn* junction diode that is specifically designed for applications that make use of the voltage dependence of the diode depletion-layer capacitance. *See* VARACTOR. [F.St.]

Microwave spectroscopy

Microwave spectroscopy The study of the interaction of matter and electromagnetic radiation in the microwave region of the spectrum. *See* MICROWAVE; SPECTROSCOPY.

The interaction of microwaves with matter can be detected by observing the attenuation or phase shift of a microwave field as it passes through matter. These are determined by the imaginary or real parts of the microwave susceptibility (the index of refraction). The absorption of microwaves may also trigger a much more easily observed event like the emission of an optical photon in an optical double-resonance experiment or the deflection of a radioactive atom in an atomic beam. *See* MOLECULAR BEAMS.

At room temperature, the relative population difference between the states involved in a microwave transition is a few percent or less. The population difference can be close to 100% at liquid helium temperatures, and microwave spectroscopic experiments are often performed at low temperatures to enhance population differences and to eliminate certain line-broadening mechanisms. The population differences between the states involved in a microwave transition can also be enhanced by artificial means. When the molecules or atoms with inverted populations are placed in an appropriate microwave cavity, the cavity will oscillate spontaneously as a maser (microwave amplification by stimulated emission of radiation). *See* MASER.

The magnetic dipole and electric quadrupole interactions between the nuclei and electrons in atoms and molecules can lead to energy splittings in the microwave region of the spectrum. Thus, microwave spectroscopy has been used extensively for precision determinations of spins and moments of nuclei. *See* HYPERFINE STRUCTURE; MOLECULAR BEAMS; NUCLEAR MOMENTS.

The rotational frequencies of molecules often fall within the microwave range, and microwave spectroscopy has contributed a great deal of information about the moments of inertia, the spin-rotation coupling mechanisms, and other physical properties of rotating molecules. *See* MOLECULAR STRUCTURE AND SPECTRA.

The magnetic resonance frequencies of electrons in fields of a few thousand gauss (a few tenths of a tesla) lie in the microwave region. Thus, microwave spectroscopy is used in the study of electron-spin resonance or paramagnetic resonance. *See* ELECTRON PARAMAGNETIC RESONANCE (EPR) SPECTROSCOPY; MAGNETIC RESONANCE.

The cyclotron resonance frequencies of electrons in solids at magnetic fields of a few thousand gauss (a few tenths of a tesla) lie within the microwave region of the spectrum. Microwave spectroscopy has been used to map out the dependence of the effective mass on the electron momentum.

For other applications *see* ATOMIC CLOCK; COSMIC BACKGROUND RADIATION; RADIO ASTRONOMY. [W.Hap.]

Microwave transmission lines Structures used for transmission of electromagnetic energy at microwave frequencies from one point to another. A transmission line may be defined more precisely as a system of material boundaries forming a continuous path from one place to another and capable of directing the transmission of electromagnetic energy along this path. *See* TRANSMISSION LINES.

The structures in most widespread use as microwave transmission lines are coaxial lines and hollow-pipe waveguides. Other structures, including striplines, open-wire lines, and dielectric rods, are used in special applications.

Coaxial lines are widely used at lower frequencies, and are satisfactory for many applications at microwave frequencies. In a coaxial line (Fig. 1) the electromagnetic waves are transmitted through the dielectric medium bounded by two conducting, coaxial cylinders. Because of skin effect, the currents in the conductors are concentrated in the surfaces of the conductors bounding the dielectric medium.

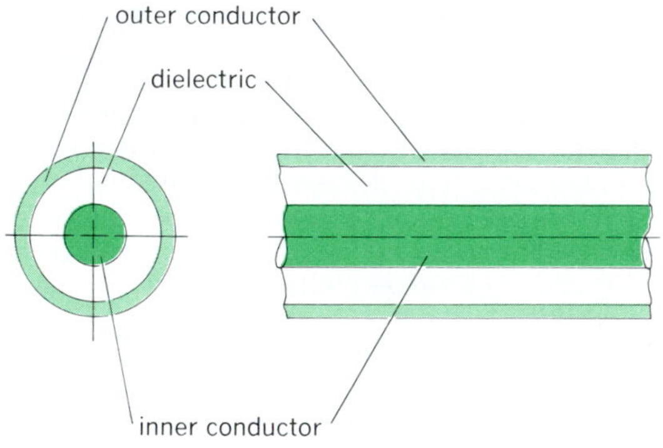

Fig. 1. Coaxial transmission line.

The hollow-pipe waveguide and other types of transmission lines are increasingly preferred at the higher frequencies. Electromagnetic waves are transmitted through the interior of hollow metal pipes, and electric currents flow on the inner surfaces. In contrast to coaxial lines, hollow-pipe waveguides are characterized by cutoff frequencies; that is, for a waveguide of given dimensions, the operating frequency must be higher than a critical cutoff frequency for energy to propagate. Waveguides therefore become impractically large at lower frequencies, and the dimensions are reasonably small only in the microwave range.

Dielectric rods and tubes can also be used as microwave transmission lines, as can single-conductor lines, which are sometimes coated with dielectric. These structures can have very low attenuation. Striplines (Fig. 2), which consist of a metal strip supported above a metal plane or between two metal planes, have somewhat higher attenuation than hollow-pipe waveguides. They are, however, frequently used where structures of unusual compactness and physical simplicity are desired.

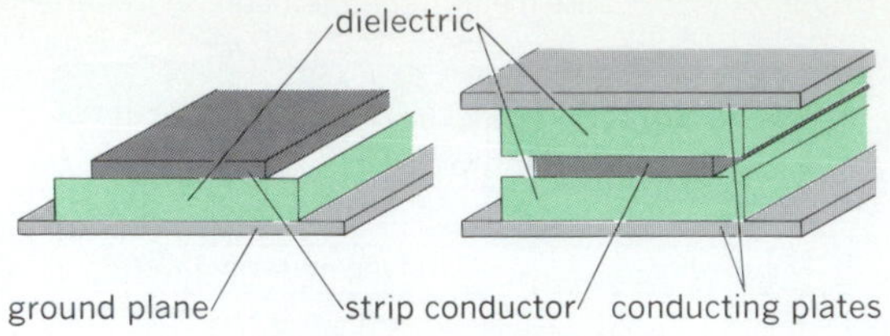

Fig. 2. Microwave striplines.

An important characteristic of a transmission line is its characteristic impedance, defined as the input impedance to a uniform line of infinite length. If the terminal or load impedance of a transmission line is equal to the characteristic impedance of the line, all of the energy in the wave traveling toward the line is absorbed in the load impedance, and the line is said to be matched. If the load impedance is not equal to the characteristic impedance, part of the traveling wave is reflected.

For most applications a matched transmission line is an optimum condition, as it minimizes the undesired losses of energy in the line itself and maximizes the energy delivered to the load impedance. Frequently, however, the load impedance does not match the characteristic impedance of the transmission line, and various matching networks and devices must then be employed. *See* IMPEDANCE MATCHING; MICROWAVE. [T.Mor.]

Microwave tube A high-vacuum tube designed for operation in the frequency region from approximately 3000 to 300,000 MHz. Two considerations distinguish a microwave tube from vacuum tubes used at lower frequencies: the dimensions of the tube structure in relation to the wavelength of the signal that it generates or amplifies, and the time during which the electrons interact with the microwave field. *See* VACUUM TUBE.

In the microwave region wavelengths are in the order of centimeters; resonant circuits are in the forms of transmission lines that extend a quarter of a wavelength from the active region of the microwave tube. With such short circuit dimensions the internal tube structure constitutes an appreciable portion of the circuit. For these reasons a microwave tube is made to form part of the resonant circuit. Leads from electrodes to external connections are short, and electrodes are parts of surfaces extending through the envelope directly to the external circuit that is often a coaxial transmission line or cavity. *See* CAVITY RESONATOR; TRANSMISSION LINES.

At microwaves the period of signal is in the range of 0.001–1 nanosecond. Only if transit time is less than a quarter of the signal period do significant numbers of electrons exchange appreciable energy with the signal field. Transit time is reduced in several ways. Electrodes are closely spaced and made planar in configuration, and high interelectrode voltages are used.

Tubes designed by the foregoing principles are effective for wavelengths from a few meters to a few centimeters. At shorter wavelengths different principles are necessary. To obtain greater exchange of energy between the electron beam and the electromagnetic field several alternative designs have proved practical.

Instead of collecting the electron beam at a plate formed by the opposite side of the resonant circuit, the beam is allowed to pass into a field-free region before reacting further with an external circuit. The electron cloud can be deflected by a strong static magnetic field so as to revolve and thereby react several times with the signal field before reaching the plate. *See* KLYSTRON; MAGNETRON.

Instead of producing the field in one or several resonant circuits, the field can be supported by a distributed structure along which it moves at a velocity comparable to the velocity of electrons in the beam. The electron beam is then directed close to this structure so that beam and field interact over an extended interval of time. *See* BACKWARD-WAVE TUBE; MICROWAVE; TRAVELING-WAVE TUBE. [F.H.R.]

Mictacea A proposed order of the Peracarida established for two small crustacean species, *Hirsutia bathyalis* and *Mictocaris halope*. The two species share many features common to other peracaridans but differ sufficiently to justify their assignment to a distinct order with two monotypic families. Common peracaridan features include a brood pouch formed by basal lamellae of the perepods (oöstegites) in the female; a small movable process (lacinia mobilis) on the mandible; free thoracic somites not fused to a carapace shield; a single maxilliped of typical peracarid form; and partially immobile pereopodal basal segments.

Mictocaris is a cave-dwelling species, whereas *Hirsutia* has been found only in soft muddy sediment of the deep sea. Nothing is known about the feeding habits of these crustaceans, but with its spined first pereopod and fossorial second, *Hirsutia* is thought to be carnivorous. In contrast, the feeding appendages in *Mictocaris* resemble those of thermosbaenaceans, which scrape food particles from the substrate.

Mictacea appear to be most closely related to the Thermosbaenacea, Spelaeogriphacea, and Mysidacea. *See* CRUSTACEA; MYSIDACEA; PERACARIDA; SPELAEOGRIPHACEA; THERMOSBAENACEA. [P.A.McL.]

Middle-atmosphere dynamics The motion of that portion of the atmosphere that extends in altitude roughly from 10 to 100 km (6 to 60 mi). The Earth's climate is determined by a balance between incoming solar and outgoing Earth thermal radiative energy, both of which must necessarily pass through the middle atmosphere. The lower portion, the stratosphere, contains many greenhouse gases (ozone, water vapor, carbon dioxide, methane, nitrous oxide, chlorofluorocarbons, and others); and it is predicted to cool at the same time as the lower atmosphere is warmed by the greenhouse effect. The middle atmosphere is also a focus for effects of emissions from proposed commercial fleets of stratospheric aircraft. In addition to the chemistry involved, dynamical transport modeling and measurements are needed to predict the widespread transport of these important trace gases and emissions over the globe. *See* ATMOSPHERE; STRATOSPHERE; TERRESTRIAL RADIATION.

The stratosphere constitutes the lower part of the middle atmosphere, from about 10 to 50 km (6 to 30 mi) altitude; from about 50 to 80 km (30 to 48 mi) or so lies the mesosphere. The location of the base of the stratosphere (called the tropopause) depends on meteorological conditions, varying on average from about 10 km (6 mi) in altitude at the poles to about 16 km (10 mi) at the Equator.

Atmospheric gravity waves result from combined gravitational and pressure gradient forces. Typical characteristics are transverse polarization, vertical wavelengths of 0.1–0 km (0.06–6 mi), horizontal wavelengths of 1–100 km (0.6–60 mi) or more, and periods in the range of 5 min to several hours. These waves may be excited by airflow over orography (mountains) as standing lee waves, by growing clouds, and by large-scale storm complexes in the lower atmosphere; and then they propagate up into the middle atmosphere.

Planetary-scale Rossby waves are large and slowly moving waves affected by the Coriolis effect due to the Earth's rotation. Rossby waves are common at middle latitudes in winter, where they can propagate up into the middle atmosphere from excitation regions below. Near the Equator, hybrid Rossby-gravity waves and also Kelvin waves (a special class of eastward-propagating internal gravity waves having no north-south velocity component) have been observed in the middle atmosphere. *See* CORIOLIS ACCELERATION.

A variety of global-scale normal-mode oscillations are also found in the middle atmosphere, prominent examples being wave 1, westward-moving waves with periods of about 5 and 16 days, and wave 3, a westward-moving feature with a period of about 2 days. Another observed oscillation in the middle atmosphere has been found with periods in the range 1–2 months (propagating up from the lower atmosphere).

Waves resulting from fluid dynamical instabilities are also observed: medium-scale (waves 4–7) eastward-moving waves, which are actually the tops of tropospheric storm systems, can dominate the circulation of the summer Southern Hemisphere lower stratosphere. The medium-scale waves have periods of 10–20 days. *See* DYNAMIC INSTABILITY; DYNAMIC METEOROLOGY. [J.L.Sta.]

Midnight sun A phenomenon observed in the polar zones of the Earth near the time of the summer solstice, when the Sun remains visible above the horizon at midnight and reaches its minimum altitude without setting. The midnight sun is a consequence of the inclination of the rotation axis of the Earth, by which the Earth presents in turn each pole to the Sun for 6 months. [G.DeV.]

Mid-Oceanic Ridge An interconnected system of broad submarine rises totaling about 60,000 km (36,000 mi) in length, the longest mountain range system on the planet. The origin of the Mid-Oceanic Ridge is intimately connected with plate tectonics. Wherever plates move apart sufficiently far and fast for oceanic crust to form in the void between them, a branch of the Mid-Oceanic Ridge will be created. In plan view the plate boundary of the Mid-Oceanic Ridge comprises an alternation of spreading centers (or axes or accreting plate boundaries) interrupted or offset by a range of different discontinuities, the most prominent of which are transform faults. As the plates move apart, new oceanic crust is formed along the spreading axes, and the ideal transform fault zones are lines along which plates slip past each other and where oceanic crust is neither created nor destroyed. *See* PLATE TECTONICS; TRANSFORM FAULT.

Separation of plates causes the hot upper mantle to rise along the spreading axes of the Mid-Oceanic Ridge; partial melting of this rising mantle generates magmas of basaltic composition that segregate from the mantle and rise in a narrow zone at the axis of the Mid-Oceanic Ridge to form the oceanic crust. The partially molten mantle "freezes" to the sides and bottoms of the diverging plates to form the mantle lithosphere that, together with the overlying "rind" of oceanic crust, comprises the lithospheric plate. At the axis of the Mid-Oceanic Ridge the underlying column of crust and mantle is hot and thermally expanded; this thermal expansion explains why the Mid-Oceanic Ridge is a ridge. With time, a column of crust plus mantle lithosphere cools and shrinks as it moves away from the ridge axis as part of the plate. The gentle regional slopes of the Mid-Oceanic Ridge therefore represent the combined effects of sea-floor spreading (divergent plate motion) and thermal contraction. *See* LITHOSPHERE; MAGMA.

The height and thermal contraction rate of the ridge crest are relatively independent of the rate of sea-floor spreading; thus the width and regional slopes of the Mid-Oceanic Ridge depend primarily on the rate of plate separation (spreading rate). Where the plates are separating at 2 cm (0.8 in.) per year, the Mid-Oceanic Ridge has five times the regional slope but only one-fifth the width of a part of the ridge forming where the plates are separating at 10 cm (4 in.) per year. One consequence of the relation between the width and plate separation rate of the Mid-Oceanic Ridge is that more ocean water is displaced, thereby raising sea level, during times of globally faster plate motion. *See* SEA-LEVEL FLUCTUATIONS.

Although the Mid-Oceanic Ridge exhibits little systematic depth variation along much of its length, there are several bulges (swells) of shallower sea floor. For reasons not well understood, the sea-floor bulges are more prominent along parts of the Mid-Oceanic Ridge where the rate of plate separation (spreading rate) is slower, for example, along the northern Mid-Atlantic Ridge and the Southwest Indian Ridge.

The axis of the Mid-Oceanic Ridge—that is, the active plate boundary between two separating plates—is a narrow zone only a few kilometers wide, characterized by frequent earthquakes, intermittent volcanism, and scattered clusters of hydrothermal vents where seawater, percolating downward and heated by proximity to hot rock, is expelled back into the ocean at temperatures as high as 350°C (660°F). Surrounding such vents are deposits of hydrothermal minerals rich in metals, as well as exotic animal communities including, in some vent fields, giant tubeworms and clams. *See* HYDROTHERMAL VENT; MARINE GEOLOGY; VOLCANO. [P.R.V.]

Migmatite Rocks originally defined as of hybrid character due to intimate mixing of older rocks (schist and gneiss) with granitic magma. Now most plutonic rocks of mixed appearance, regardless of how the granitic phase formed, are called migmatites. Commonly they appear as veined gneisses.

Several modes of origin have been proposed. (1) Granitic magma may be intercalated between thin layers of schist (lit-par-lit injection) to form a banded rock called injection gneiss. (2) The granitic magma may form in place by selective melting of the rock components. (3) The granitic layers may develop by metamorphic differentiation (redistribution of minerals in solid rock by recrystallization). (4) The granitic layers may represent selectively replaced or metasomalized portions of the rock. *See* GRANITIZATION; METAMORPHISM; METASOMATISM. [C.A.C.]

Migratory behavior Regularly occurring, oriented seasonal movements of individuals of many animal species. The term migration is used to refer to a diversity of animal movements, ranging from short-distance dispersal and one-way migration to round-trip migrations occurring on time scales from hours (the vertical movements of aquatic plankton) to years (the return of salmon to their natal streams following several years and thousands of kilometers of travel in the open sea).

Many temperate zone species, including many migrants, are known to respond physiologically to changes in the day length with season (photoperiodism). For example, many north temperate organisms are triggered to come into breeding condition by the interaction between the lengthening days in spring and their biological clocks (circadian rhythms). Similar processes, acting through the endocrine system, bring animals into migratory condition.

To perform regular oriented migrations, animals need some mechanism for determining and maintaining compass bearings. Animals use many environmental cues as sources of directional information. Work with birds has shown that species use several compasses.

Many species of vertebrates and invertebrates possess a time-compensated Sun compass. With such a system, the animal can determine absolute compass directions at any time of day; that is, its internal biological clock automatically compensates for the changing position of the Sun as the Earth rotates during the day. Many arthropods, fish, salamanders, and pigeons can perceive the plane of polarization of sunlight, and may use that information to help localize the Sun even on partly cloudy days.

Only birds that migrate at night have been shown to have a star compass. Unlike the Sun compass, it appears not to be linked to the internal clock. Rather, directions are determined by reference to star patterns which seem to be learned early in life.

Evidence indicates that several insects, fish, a salamander, certain bacteria, and birds may derive directional information from the weak magnetic field of the Earth. *See* MAGNETIC RECEPTION (BIOLOGY).

Wind provides directional information to birds, but because it changes with time, it cannot yield compass information directly. Nocturnal songbird migrants nearly always fly with a following wind; in part this is because they select a night with favorable winds to initiate migration.

Three beach-dwelling amphipod crustaceans seem to possess a time-compensated lunar compass in addition to a Sun compass.

Many kinds of animals show the ability to return to specific sites following a displacement. The phenomenon can usually be explained by familiarity with landmarks near "home" or sensory contact with the goal. For example, salmon are well known for their ability to return to their natal streams after spending several years at sea. Little is known about their orientation at sea, but they recognize the home stream by chemical cues (olfactory) in the water. The young salmon apparently imprint on the odor of the stream in which they were hatched. Current evidence indicates that birds imprint on or learn some feature of their birthplace, a prerequisite for them to be able to return to that area following migration. On its first migration, a young bird appears to fly in a given direction for a programmed distance. Upon settling in a wintering area, it will also imprint on that locale and will thereafter show a strong tendency to return to specific sites at both ends of the migratory route.

Only in birds can an unequivocal case be made for the existence of true navigation, that is, the ability to return to a goal from an unfamiliar locality in the absence of direct sensory contact with the goal. This process requires both a compass and the analog of a map. Present evidence suggests that the map is not based on information from the Sun, stars, landmarks, or magnetic field. Other possibilities such as olfactory, acoustic, or gravitational cues are being investigated, but the nature of the navigational component of bird homing remains the most intriguing mystery in this field. [K.P.A.]

Military aircraft Aircraft that are designed or modified for highly specialized military applications. Balloons, gliders, airplanes, and helicopters have been used for military purposes.

Basic types of military aircraft include bombers, fighters, transports, patrol, trainers, and reconnaissance and observation aircraft (Figs. 1 and 2). Bombers are usually characterized by relatively long range, low maneuverability, and large weapon-carrying capability. Bombers may be equipped to

Fig. 1. Grumman F-14 fleet interceptor, variable sweep fighter. (*Grumman Aircraft Co.*)

Fig. 2. Lockheed C-5A global strategic transport/cargo aircraft. (*Lockheed-Georgia Aircraft Co.*)

deliver conventional or nuclear weapons in day, night, or adverse weather.

Fighters are relatively short-range, highly maneuverable, fast airplanes designed to destroy enemy aircraft and to attack ground targets. They can carry machine guns, cannons, rockets, guided missiles, and bombs, depending upon the missions. Some fighters called fighter bombers can carry conventional or nuclear weapons several hundreds of kilometers behind enemy lines to strike priority ground targets.

Transport aircraft carry troops and war supplies. Many are adaptations of airplanes used by commercial airlines. Cargo-carrying aircraft have been developed especially for the military services. These include the world's largest aircraft, capable of moving large quantities of people and material rapidly to distant points. A special-purpose type of transport airplane is the aerial tanker. By means of special refueling fittings together with other equipment, fighters, bombers, and helicopters can take on fuel to fill their tanks while in flight.

Reconnaissance aircraft are used to observe the enemy visually or by radar or photography. These aircraft may be specially designed or may be modified from a basic fighter or bomber type.

Fig. 3. General Dynamics AGM-109 cruise missile. (*General Dynamics/Convair*)

Helicopters deserve special mention in military use. They are unexcelled for rescue work and for delivery of people and material to otherwise inaccessible areas. *See* HELICOPTER; VERTICAL TAKEOFF AND LANDING (VTOL).

Cruise missiles are small, long-range, uncrewed aircraft (Fig. 3). Although termed missiles, they have more of the features of aircraft than of missiles. Cruise missiles have wings and a turbine engine, giving them ranges of a few hundred miles to over 1200 mi (2000 km). They are autonomous, using inertial navigation systems for en route navigation with periodic updates from space or terrain match correlation schemes. Cruise missiles permit the crewed aircraft to stand off from the target and strike it with the small, low-flying cruise missiles. *See* MISSILE.

[L.M.N.]

Military satellites Artificial Earth satellites used for a variety of military purposes. Many, if not most, crewed and uncrewed satellites successfully orbited by the various countries of the world have been either specifically military in nature or usable for military purposes.

Russia has launched almost twice as many satellites as the United States. Virtually all of the uncrewed satellites that Russia has sent into Earth orbit have been launched by the military. A simple comparison of numbers is not sufficient to draw conclusions about the two nations' space programs because of differences in the effectiveness of particular kinds of flight hardware.

The major functions of military satellite missions are: communications, navigation, meteorology, surveillance, geodesy, nuclear test and radiation detection, and research and technology. Future space shuttle programs will include all or some of the above. *See* APPLICATIONS SATELLITES; COMMUNICATIONS SATELLITE; METEOROLOGICAL SATELLITES; SATELLITE NAVIGATION SYSTEMS; SPACE SHUTTLE.

[H.W.Br.]

Milk The lacteal secretion, practically free from colostrum, obtained by the complete milking of one or more healthy cows and containing not less than 8.25% milk solids (not fat) and not less than 3.25% milk fat. Among mammals, humans utilize milk as a source of food. The dairy cow supplies the vast majority of milk for human consumption, particularly in the United States; however, milk from goats, water buffalo, and reindeer is also consumed in other countries. Without qualification, the general term milk refers to cow's milk.

Average composition of milk is 87.2% water, 3.7% fat, 3.5% protein, 4.9% lactose, and 0.7% ash. Whole milk and skim milk are classified as excellent sources of calcium, phosphorus, and riboflavin because 10% of the daily nutritional requirement is supplied by not over 100 kcal (420 kilojoules). These two beverages are also classified as good sources of protein and thiamine; and whole milk is a good source of vitamin A. To be classified as good, the source must contribute 10% of a nutrient in not over 200 kcal (840 kilojoules). Milk is a good source of protein rich in all the essential amino acids.

Processing. Most raw milk collected at farms is pumped from calibrated and refrigerated stainless steel tanks into tank trucks for delivery to processing plants. The actual processing of raw milk begins with either separation or clarification. These machines are essentially similar except that in the clarifier the cream and skim milk fractions are not separated. Many processers have units called standardizer-clarifiers which separate only a small fraction of the fat from the raw whole milk; the amount of fat removed can be regulated. This facilitates the production of milk of standard fat content even though that in the raw product may vary.

Milk is rendered free of pathogenic bacteria by pasteurization. This is accomplished in a manner so that every particle of milk is heated to a specified temperature and held at that temperature for a specified time. *See* PASTEURIZATION.

Fat globules in fluid milk products are broken by homogenization into sizes that are 2 micrometers or less and thus are relatively unaffected by gravitational forces. The U.S. Public Health Service specifies that the fat content of the upper 6 in. (100 ml) of a quart of homogenized milk that has been undisturbed for 48 h cannot differ by more than 10% from that of the remainder.

Most milk is fortified with 400 international units (IU) of vitamin D per quart, and some skim milk is fortified also with 2000 IU of vitamin A per quart. These vitamins as concentrates are added either by automatic dispensing into a continuous flow of milk prior to pasteurization or as a single quantity in a batch operation.

Products. Many fermented or cultured products are produced from milk. These fermentations require the use of bacteria that ferment lactose or milk sugar.

Cultured buttermilk consists of skim milk or low-fat milk which is pasteurized at 180°F (82°C) for 30 min, cooled to 72°F (22°C), and inoculated with an active starter culture containing *Streptococcus lactis* and *Leuconostoc citrovorum*. The mixture is incubated at 70°F (21°C) and cooled when acidity is developed to approximately 0.8%. This viscous product is then agitated, packaged, and cooled. The desired flavor is created by volatile acids and diacetyl; the latter is produced by *L. citrovorum*.

One of the oldest fermented milks known is yogurt. Yogurt is prepared using whole or low-fat milk with added nonfat milk solids. The milk is heated to approximately 180°F (82°C) for 30 min, homogenized, cooled to 115°F (46°C), inoculated with an active culture, and packaged. Yogurt cultures are mixtures of *Streptococcus thermophilus* and *Lactobacillus bulgaricus* in a 1:1 ratio. Balance of these organisms in the culture is important for production of a quality product.

Concentrated and dried milk products. To reduce costs of transportation and handling, either part or all of the water is removed from milk. Moreover, the partly dehydrated milk can either be sterilized or dried to permit unrefrigerated storage for prolonged periods. Many different milk products (such as dry whole milk, evaporated milk, and condensed milk) are produced for these specific reasons. The composition of some of these is controlled by standards of identity and some by request of the commercial buyer. [R.L.Br.]

Milky Way Galaxy The large disk-shaped aggregation of stars, gas, and dust in which the solar system is located. The term Milky Way is used to refer to the diffuse band of light visible in the night sky emanating from the Milky Way Galaxy. Although the two terms are frequently used interchangeably, Milky Way Galaxy, or simply the Galaxy, refers to the physical object rather than its appearance in the night sky.

The Milky Way Galaxy contains about 2×10^{11} solar masses of visible matter. (The Sun's mass is 4.4×10^{30} lb or 2×10^{30} kg.) Roughly 97% is in the form of stars, and about 3% is in the form of interstellar gas. The gas both inside the stars and in the interstellar medium is primarily hydrogen and helium, with a small admixture of all of the heavier atoms. The mass of dust is about 1% of the interstellar gas mass and is therefore an insignificant fraction of the total mass of the Galaxy. The presence of this dust, however, limits the view from the Earth in the plane of the Milky Way Galaxy to a small fraction of the Galaxy's diameter in most directions. *See* INTERSTELLAR MATTER.

The Milky Way Galaxy contains four major structural subdivisions: the nucleus, the bulge, the disk, and the halo. The Sun is located in the disk about half way between the center and the distance at which the disk of stars becomes undetectable. The currently accepted value of the distance of the Sun from the galactic center is 8.5 kiloparsecs, although some measurements suggest that the distance may be as small as 7 kpc. *See* PARSEC.

The nucleus of the Milky Way Galaxy is an ill-defined region within a few tens of parsecs of the geometric center. The nucleus is the source of very energetic activity detected by means of radio waves, infrared radiation, and gamma radiation, some of which is generally believed to be powered by matter interacting with a black hole at the very center of the Milky Way. No unambiguous evidence has yet been found, however, to confirm the presence of a black hole. *See* BLACK HOLE; GAMMA-RAY ASTRONOMY; INFRARED ASTRONOMY; RADIO ASTRONOMY.

The bulge is a spheroidal distribution of stars centered on the nucleus that extends to a distance of about 3 kpc from the center. It contains a relatively old population of stars, very nearly as old as the Milky Way itself. *See* STAR.

The disk consists of a disk of stars and a disk of gas. The disk of stars can be identified to about 17 kpc from the center of the Galaxy, and the disk of gas can be identified to distances of at least 22 kpc from the center. The disk is very thin and has the relative proportions of a long-playing phonograph record. It is the location of the spiral arms that are characteristic of most disk-shaped galaxies, and are the sites of present-day star formation. *See* GALAXY, EXTERNAL; STELLAR EVOLUTION.

The halo is a rarefied spheroidal distribution of stars that is very nearly devoid of interstellar gas and dust and that surrounds the disk. The stars found in the halo are the oldest stars in the Galaxy. The stars are found individually as so-called field stars as well as in globular clusters, that is, spherical clusters of up to about a million stars with very low abundances of elements heavier than helium. The extent of the halo is not well determined, but globular clusters with distances of about 40 kpc from the center have been definitely identified. Dynamical evidence suggests that the halo contains nonluminous matter in some unknown form, commonly referred to as dark matter. The dark matter, if it exists, contains most of the mass of the Galaxy, dominating that even in the form of stars. *See* COSMOLOGY; STAR CLUSTERS.

Evidence suggests that the ages of the oldest stars in the Milky Way are within about 10% of the age of the universe as a whole. The Milky Way Galaxy presumably formed from a very large (about 10^{15} times the mass of the Sun) cloud of gas of hydrogen and helium that was essentially devoid of all heavier elements. The material in this gas cloud formed the Virgo supercluster, and one portion of it formed the Milky Way and the Local Group. The first generation of stars are those that make up the population in the halo. Presumably, as the gas cloud collapsed under the influence of its self-gravity, viscous dissipation in the gas allowed the protogalaxy to collapse along the rotation axis, forming the disk at a later evolutionary stage than the halo. That part of the disk that has not yet been used up in the process of star formation continues to form stars. [L.Bl.]

Milleporina An order of the class Hydrozoa of the phylum Coelenterata. These are the "string corals" of shallow tropical seas. Their structure is similar to that of hydroids except for the addition of a calcareous exoskeleton (illustration *a*). Because of this skeleton, they resemble true corals (Anthozoa). The skeleton is covered by a thin layer of tissue, is penetrated by interconnecting tubes, and is perforated by tiny holes through which the bodies of the "coral animals," or polyps, are extended (illustration *b*).

The polyps are of two types (illustration *c*): nutritive gastrozooids with tentacles and mouth; and protective polyps, which are long and armed with stinging cells but have no mouth.

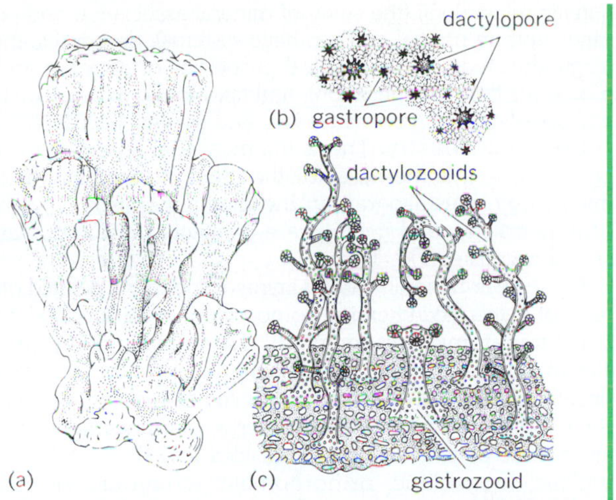

Milleporina. ***(a)*** **Piece of dry** ***Millepora*****, showing typical flabellate shape.** ***(b)*** **Same, magnified, showing pores. (c) Polyps of** ***Millepora*****. (*****After L. H. Hyman, The Invertebrates, vol. 1, McGraw-Hill, 1940*****)**

Millepores produce medusae, or jellyfish, in which sex cells develop. *See* HYDROZOA. [S.Cr.]

Millerite A mineral having composition NiS and crystallizing in the hexagonal system. Millerite usually occurs in hairlike tufts and radiating groups of slender to capillary crystals. The hardness is 3–3.5 (Mohs scale) and the specific gravity is 5.5. The luster is metallic and the color pale brass yellow. Millerite is found in many localities in Europe, notably in Germany and Czechoslovakia. In the United States it is found with pyrrhotite at the Gap Mine, Lancaster County, Pennsylvania; with hematite at Antwerp, New York; and in geodes in limestone at Keokuk, Iowa. In Canada large cleavable masses are mined as a nickel ore in Lamotte Township, Quebec. [C.S.Hu.]

Millet A common name applied to at least five related members of the grass family grown for their edible seeds: foxtail millet (*Setaria italica*), proso millet (*Panicum miliaceum*), pearl or cat-tail millet (*Pennisetum typhoideum*), Japanese barnyard millet (*Echinochloa frumentacea*), raggee or finger millet (*Eleusine coracana*), and koda millet (*Paspalum scrobiculatum*).

As a crop for human food, pearl millet is grown widely in the tropics and subtropics in regions of limited rainfall where there is a growing season of 90 to 120 days. The naked seeds are yellowish to whitish in color and about the size of wheat grain. The dried grain is usually pulverized to make a meal or flour and then cooked in soups, in porridge, or as cakes. [H.B.S.]

Milling machine A machine for the removal of metal by feeding a workpiece through the periphery of a rotating circular cutter. The multitoothed cutter of a milling machine produces a milled surface as each revolving tooth removes a portion of metal from the passing workpiece.

Milling machines may generally be classed as horizontal or vertical, depending on the power spindle's axis of rotation. A movable, horizontal table holds the workpiece and carries it through the path of the cutter teeth (see illustration). On the universal mill, the table may also be rotated horizontally; on machines with a universal head, the cutter may be tilted to the desired angle. [A.H.T]

Part of a horizontal milling machine. (***Brown and Sharpe Manufacturing Co.*****)**

Mineral A naturally occurring substance with a characteristic chemical composition expressed by a chemical formula. Although organic substances such as coal and oil are usually listed under mineral resources, they are not minerals, being complex mixtures without definite chemical formulas. Minerals constitute an extremely important natural resource. Most metals and inorganic chemicals, and many other products essential to civilization, are derived from minerals. Both forests and farms are dependent upon soils, which are composed chiefly of minerals.

Occurrence. Minerals may occur as individual crystals or they may be disseminated in some other mineral or rock. Most minerals at the face of the earth occur in rocks. Minerals may be attached to the surface of some opening. Geodes are cavities in a rock which are lined with a mineral, or in some cases, completely filled. The term vug applies to irregular openings containing ore minerals. Veins are fissures of varying size which have been filled with one or more minerals. The term lode is used for a vein or group of veins in a definite area. Gangue minerals are the worthless minerals associated with a valuable mineral or ore. The term ore is usually restricted to minerals from which a metal is obtained. A placer is a concentration of relatively heavy and durable minerals which have been transported and redeposited in a stream bed where the water velocity is lowered. *See* GEODE.

Names. Minerals usually have both a chemical name and a mineral name. Thus lead sulfide occurring in nature is called galena, and sodium chloride is called halite. Most modern mineral names end in "ite." Some have a chemical connotation, as molybdenite, MoS_2, and zincite, ZnO. Physical properties are the basis for the names magnetite, graphite (to write), rhodonite (rose color), and cryolite (ice stone). Some minerals have variety names, as amethyst, agate, and jasper for quartz. There are also colloquial names, such as fool's gold for pyrite.

Classification. Minerals are classified first with respect to chemical composition, and then so far as possible by isomorphism, or similarity of crystalline form. In general the cation is of less significance than the anion. The major groups in mineral classification are:

Native elements	Sulfate type
Sulfide and sulfo minerals	Chromates
Oxides and hydrated oxides	Molybdates
Halogen minerals	Tungstates
Nitrates	Phosphate type
Carbonates	Arsenates
Borates	Vanadates
	Silicates

See BORATE MINERALS; CARBONATE MINERALS; HALOGEN MINERALS; NATIVE ELEMENTS; NITRATE MINERALS; SILICATE MINERALS.

Groups. Any complete classification of minerals must be based on chemical composition. However, limited groups of minerals may be of interest for various special purposes. Such groups may be based on origin, type of occurrence, certain physical properties, or use.

Primary minerals are those formed from the magma, including the pegmatite and hydrothermal phases. Others are secondary. Rock-forming minerals are those which make up the great bulk of the igneous, sedimentary, and metamorphic rocks. Thus quartz, feldspar, and mica are found in granite and are rock-forming minerals. They are also called essential minerals, in contrast to the occasional tiny crystals of pyrite, zircon, or apatite which might occur in granite, and which are called accessory minerals.

Classes of minerals are those which make up a certain chemical group, as carbonates, sulfates, or oxides.

Isomorphous groups are those whose members are strictly isomorphous, as the garnet group. *See* ISOMORPHISM (CRYSTALLOGRAPHY).

Families of minerals include certain closely related minerals which have close chemical and physical similarities, but are not necessarily isomorphous, as the feldspars or pyroxenes.

Economic minerals are those which are of economic importance, and include both metallic (ore minerals) and nonmetallic minerals, as cryolite and sulfur, and gem minerals.

Clay minerals have certain common physical and chemical properties. They are fine-grained, plastic when wet, and become hard when dried or fired. They are chiefly hydrous silicates of aluminum. *See* CLAY MINERALS.

Stable minerals are those resistant to both chemical and mechanical weathering, being both insoluble and hard. Heavy minerals are those which collect in placers and beach sands because of their higher specific gravities, or those which can be separated in the laboratory by gravity methods. Detrital minerals are rock fragments which are essentially unaltered. Authigenic minerals are those generated in the place of formation, and allogenic minerals are those which have been transported. *See* AUTHIGENIC MINERALS.

Mineral associations refer to minerals formed by the same process and hence closely associated. Mineral sequence refers to a series of associated minerals formed at successive stages.

A mineral suite is a general term which may apply to a group of associated minerals in one deposit; a representative group from a certain locality; or a group of specimens showing variations, as in color or form, in a single mineral species. *See* MINERALOGY. [L.S.R.]

Mineralogy The science which concerns the study of natural inorganic substances, whether of terrestrial or extraterrestrial origin, called minerals. Mineralogy is a science that cannot be easily defined. It is most properly a branch of inorganic chemistry, but the discipline concentrates on the origin, description, and classification of minerals. *See* MINERAL.

Thus, four main categories may be considered: crystal chemistry (composition and atomic arrangement of minerals); paragenetic mineralogy (the study of mineral association and occurrence both in natural and synthetic systems); descriptive mineralogy (the study of the physical properties of minerals and the means for their identification); and taxonomic mineralogy (mineral classification, systematization, and nomenclature).

Crystal chemistry. This is the most vital aspect of mineralogy because it is the basis for the other studies. The fields of mineralogy, crystallography, inorganic chemistry, geochemistry, petrology, and geology are connected in the domain of crystal chemistry (Fig. 1).

A particular mineral, called a mineral species, is defined on the basis of a specified chemical composition and a specified crystal structure (atomic arrangement). These two criteria provide almost sufficient knowledge for characterization of a mineral since in principle all other properties can be derived from them.

Crystalline substances, that is, periodic arrangements of matter in three dimensions, can be divided into six crystal systems: triclinic, monoclinic, orthorhombic, tetragonal, hexagonal (including the trigonal and rhombohedral subdivisions), and cubic. These crystal systems can be considered as structure cells, each of which contains a certain integral number of atoms of a substance. Figure 2 depicts these systems and offers the criteria for distinguishing them. *See* CRYSTALLOGRAPHY.

Any crystalline substance has a certain integral number of its essential chemical formula units loosely called molecules) in its structure cell. Thus, each unique atomic position in the structural unit can be occupied by a particular kind (or kinds) of atom(s). Consider the ideal cell formula $(A_a)(B_b)\ldots(P_p)$. Each of the parentheses specifies a unique atomic position. The capital letters specify the element present and the small subscripts the number of times it occurs in the cell. If the small subscripts have a factor in common, it is factored out and what remains is the formula unit. The ideal formula is further defined in terms of the atomic element which occurs in excess of 50 mole % within each of the parentheses. The structure type, along with the ideal formula unit, defines a mineral species.

This strict definition of a species is required since a particular mineral may have a range of compositions. The range of compositions is called a series. The ideal limiting compositions are

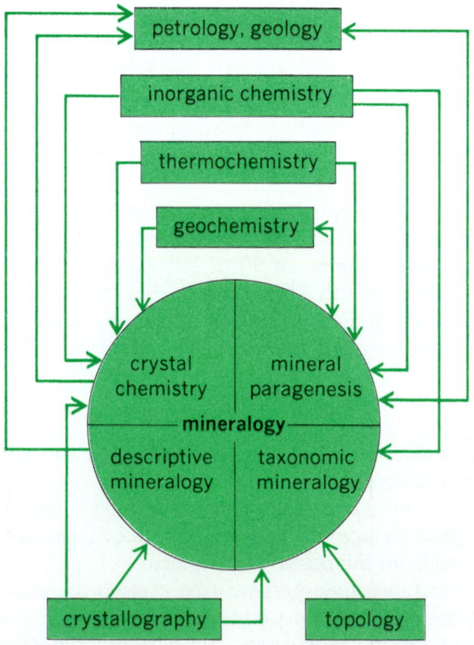

Fig. 1. Diagram showing the transmission of information between mineralogy and some other sciences. Arrows imply the direction of information.

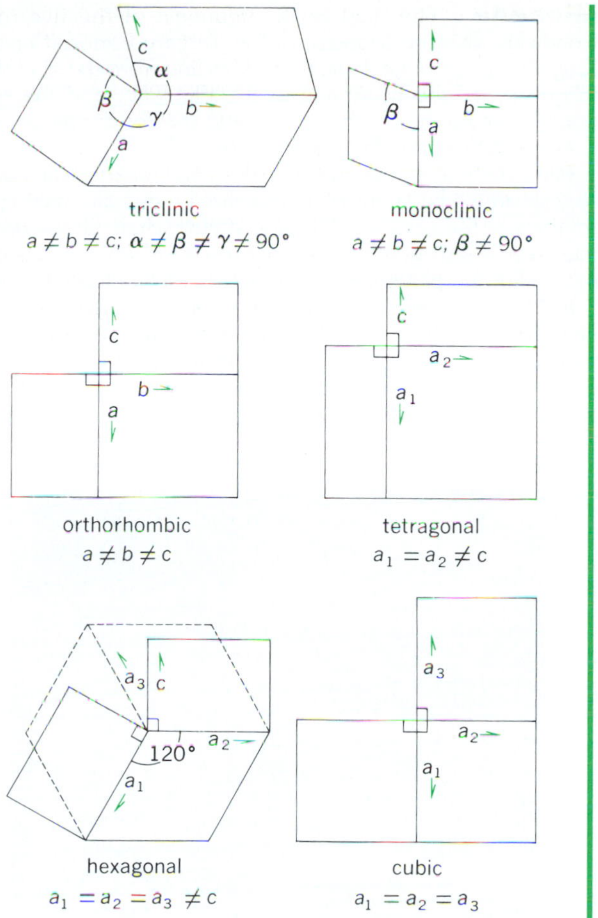

Fig. 2. The cell shapes of the six crystal systems shown as their principal projections.

called end members and each of the end members has a specific name.

Paragenetic mineralogy. Paragenetic mineralogy is the study of mineral paragenesis, or the association and order of crystallization of minerals. The problem may concern mineral association within a single hand specimen or may embrace a much larger region, such as an entire ore body, in which case many representative specimens are judiciously collected. This study usually accompanies the analysis of the general geological structures within and around the ore body, such as the bedding, folding, and faulting. Included among the important aspects of paragenetic mineralogy are ore mineralogy, the mineralogy of a sequence of phases crystallized from a parent magma, the sequence of minerals crystallized in a vein, and so forth. *See* GEOLOGY; PETROLOGY.

Mineral paragenesis is usually considered in relative time and the absolute difference in time between the oldest and youngest minerals is often not known. Absolute age differences can be obtained in some instances, for example, by lead isotope age dating of a sequence of crystallized lead-bearing minerals. *See* ROCK AGE DETERMINATION.

Descriptive mineralogy. Mineral recognition directly by the senses is very subjective and requires considerable experience. Gross features of a mineral such as color, form, hardness, and specific gravity are important criteria for identification in the field where a well-equipped laboratory is usually not available. More objective criteria such as the optical properties and x-ray powder diffraction spectra of a mineral require specialized equipment, but the results are usually certain since these data are known for most mineral species and are extensively tabulated. Older methods such as fusibility, flame tests, and blowpipe analysis have been largely abandoned.

Taxonomic mineralogy. There are approximately 3000 distinct mineral species known to science, About 60 new mineral species are discovered each year. For a new species to be properly defined, the chemical analysis, structure cell and space group, crystal morphology, powder pattern, optical data, all physical properties, and paragenesis must be given as completely as possible. [P.B.M.]

Minimal principles

In the treatment of physical phenomena, it can sometimes be shown that, of all the processes or conditions which might occur, the ones actually occurring are those for which some characteristic physical quantity assumes a minimum value. These processes or conditions are known as minimal principles. The application of minimal principles provides a powerful method of attacking certain problems that would otherwise prove formidable if approached directly from first principles.

One simple minimal principle asserts that the state of stable equilibrium of any mechanical system is the state for which the potential energy is a minimum. Other general theorems of classical dynamics that are related to minimal principles are Hamilton's principle and the principle of least action. *See* HAMILTON'S PRINCIPLE; LEAST-ACTION PRINCIPLE. [D.Wi.]

Minimal surfaces

A branch of mathematics belonging to the calculus of variations, differential geometry, and geometric measure theory. A surface, interface, or membrane is called minimal when it has assumed a geometric configuration of least area among those configurations into which it can readily deform. Soap films spanning wire frames or compound soap bubbles enclosing volumes of trapped air are common examples. *See* DIFFERENTIAL GEOMETRY; MEASURE THEORY.

Geometrically, the mean curvature of a surface S at a point is the difference between the maximum upward curvature there and the maximum downward curvature; in particular, a surface of zero mean curvature has such principal curvatures equal and opposite and hence typically appears "saddle-shaped." It turns out that S is a minimal surface; that is, it cannot be perturbed to less area leaving its boundary fixed, provided the mean curvature is zero at each of its points; such a surface could occur, for example, as a soap film spanning a wire frame. The corresponding minimal surface equation is partial differential equation. If, alternatively, S were part of a soap bubble enclosing trapped air, the mean curvature of S would be proportional to the difference in air pressure between the two sides of S. In the calculus of variations, typically area-minimizing properties of minimal surfaces are emphasized. In differential geometry, minimal surfaces are defined as surfaces of zero mean curvature; surfaces of constant mean curvature are also extensively studied.

The two-dimensional surface is dominant in determining shape whenever the energy of a system is changed significantly by a displacement or a change in area of the surface. Such surfaces include the interfaces between crystals in a typical rock or metal, the film of soapy water between the air cells in a soap froth, the membrane separating the cells in living tissue, and the cracks separating basalt columns. Minimization of surface area plays a role in determining the shape of many living organisms. *See* FOAM; GRAIN BOUNDARIES. [F.J.A.]

Mining

The taking of minerals from the earth, including production from surface waters and from wells. Usually the oil and gas industries are regarded as separate from the mining industry. The term mining industry commonly includes such functions as exploration, mineral separation, hydrometallurgy, electrolytic reduction, and smelting and refining, even though these are not actually mining operations. *See* IRON METALLURGY; METALLURGY; ORE DRESSING.

Mining is broadly divided into three basic methods: opencast, underground, and fluid mining. Opencast, or surface mining, is done either from pits or gouged-out slopes or by strip mining, which involves extraction from a series of successive parallel trenches. Dredging is a type of strip mining, with digging done from barges. Hydraulic mining uses jets of water to excavate material.

Underground mining involves extraction from beneath the surface, from depths as great as 10,000 ft (3 km), by any of several methods.

Fluid mining is extraction from natural brines, lakes, oceans, or underground waters; from solutions made by dissolving underground materials and pumping to the surface; from underground oil or gas pools; by melting underground material with hot water and pumping to the surface; or by driving material from well to well by gas drive, water drive, or combustion. Most fluid mining is done by wells. In another type of well mining, insoluble material is washed loose by underground jets and the slurry is pumped to the surface. *See* OPEN-PIT MINING; PETROLEUM ENGINEERING; PLACER MINING; SURFACE MINING; UNDERGROUND MINING. [E.Ju.]

Mink Any of three species of aquatic carnivorous mammals that are members of the family Mustelidae in the genus *Mustela*. They are found in the forested areas of North America, Europe, and Siberia. The mink is an excellent swimmer, aided by its slightly webbed hindfeet, and feeds on crayfish, frogs, snakes, and fish. The body is elongate and the legs are short (see illustration).

The American mink (*Mustela vison*) is the largest species of mink.

There is one litter each year, with three or four young being born in April or May after a gestation period of about 8 weeks. The young are mature the following season. The pelts of these animals are valuable; many mink farms have been established throughout the world. This animal is a natural predator on the muskrat, over whose numbers it exerts a control. *See* BADGER; CARNIVORA; MARTEN; OTTER; SKUNK; WEASEL. [C.B.C.]

Minnow A fresh-water fish in the family Cyprinidae, order Cypriniformes. There are over 200 species in North America, but as a group they are more abundant in tropical Africa and southeastern Asia.

This family is highly variable in morphological characteristics. Pharyngeal teeth are usually present, the body may be naked or covered with cycloid scales, there is a large swim bladder, and most species have spineless fins. Sexual dimorphism is common, and breeding males may be brightly colored or have nuptial tubercles on the head. The majority of species are small in size; many other small or immature species of fishes are erroneously called minnows. *See* CYPRINIFORMES. [C.B.C.]

Miocene The next to the youngest of the five major worldwide divisions (epochs) of the Tertiary Period (Cenozoic Era); the epoch of geologic time extending from the end of the Oligocene to the beginning of the Pliocene; and the rocks (series) formed during this epoch and the fossils therein. *See* CENOZOIC; OLIGOCENE; PLIOCENE; TERTIARY.

Miocene time embraces two widespread reexpansions of the sea, separated by an interval of localized mountain making and shiftings of the seaways. Miocene strata include all the normal marine and terrestrial deposits. Extrusive volcanic rock is common; Miocene plutonic rocks are locally (Philippines) exposed.

In Miocene seas certain gastropod families suddenly diversified and multiplied, and other well-established gastropods persisted; large scallops of Oligocene inspiration, and clams expanded; oysters attained great size; cake urchins and sand dollars diversified and multiplied; diatoms, corals, cephalopods,

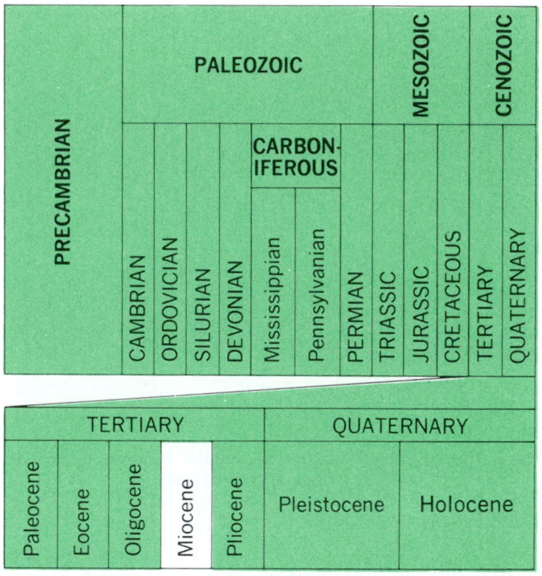

crustacea, insects, and large-toothed sharks were locally important; and offshore small foraminifers proliferated. Mammalian marine adaptations (seals, sea lions, walruses, and seacowlike extinct desmostylids) appeared, or became modernized (sea cows, dugongs, and whales).

On land, except for rodents, all but two or three recognized mammal families already existed. Eurasian faunas included deer, hyenas, the earliest giraffes, and the bovines. A huge bear-dog, *Amphicyon*, was conspicuous, and mastodonts spread into North America for the first time. As grasslands spread, grazing life became dominant over browsing life; horses and camels, especially, evolved and spread rapidly. Australian, South American, and Madagascar mammal faunas remained isolated.

Ancestral redwoods flourished in China and California. Responding to a progressive cooling of climates through the Cenozoic, floras shifted equatorward along the north-south–trending mountain ranges bordering the Pacific, though species piled up where strandlines interfered (Malaysia and the Caribbean). Where mountains run east-west through the Old World, however, a Miocene Mediterranean flora was trapped and is ancestral to the Recent flora upon which human agricultural, high civilizations first developed. [R.M.Kl.]

Mira The first star recognized to be a periodic variable. Called "The Wonderful" by the ancients, Mira was discovered by Fabricius in 1596 and has a period of 332 ±9 days. It is the prototype of long-period variables, and like most of its class, changes considerably from epoch to epoch, At times the maximum is as bright as second magnitude, and sometimes it is barely fifth magnitude. Mira (*o* Ceti) has a faint companion discovered after its presence was predicted from a study of the spectrum. *See* VARIABLE STAR. [M.W.M.]

Mirage A name for a variety of unusual images of distant objects seen as a result of the bending of light rays in the atmosphere during abnormal vertical distribution of air density. If the air closer to the ground is much warmer than the air above, the rays are bent in such a way that they enter the observer's eyes along a line lower than the direct line of sight. The object is then seen below the horizon, the inferior mirage. If the air closer to the ground is much colder than the air above, the rays are bent in the opposite direction, arriving at the observer's eyes above the line of sight; the object then seems to be elevated or floating in the air, the superior mirage. Mirages can be seen most frequently along an overheated highway surface; the inferior mirage of the sky gives the impression of water reflection over a wet pavement, which disappears upon a closer viewing. [Z.S.]

Mirror optics The use of plane or curved reflecting surfaces for the purpose of reverting, directing, or forming images. An optical surface which specularly reflects the largest fraction of the incident light is called a reflecting surface. Such surfaces are commonly fabricated by polishing of glass, metal, or plastic substrates, and then coating the surface of the substrate with a thin layer of metal, which may be covered in addition by a single or multiple layers of thin dielectric films. The law of reflection states that the incident and reflected rays will lie in the plane containing the local normal to the reflecting surface and that the angle of the reflected ray from the normal will be equal to the angle of the incident ray from the normal. *See* GEOMETRICAL OPTICS.

The formation of images in the plane mirrors is easily understood by applying the law of reflection. The illustration shows the formation of the image of a point formed by a plane mirror. Each of the reflected rays appears to come from a point image located a distance behind the mirror equal to the distance of the object point in front of the mirror. In the illustration, the face of the observer can be considered as a set of points, each of which is imaged by the plane mirror. Since the observer is viewing the facial image from the object side of the mirror, the face will appear to be reversed left for right in the virtual image formed by the mirror. The illustration also indicates the redirection of light by a plane mirror, in that a viewer who cannot observe the object point directly can observe the virtual image of the point formed by the mirror. A simple optical device which is based on this principle is the simple mirror periscope, which uses two mirrors to permit viewing of scenes around an obstacle. *See* PERISCOPE.

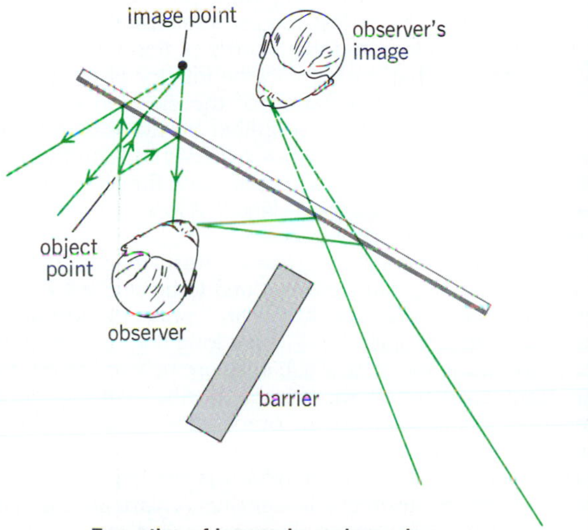

Formation of images by a plane mirror.

A curved mirror, either spherical or conic in form, will produce a real or virtual image in much the same manner as a lens, but generally with reduced aberrations. There will be no chromatic aberrations since the law of reflection is independent of the color or wavelength of the incident light. *See* ABERRATION (OPTICS); OPTICAL IMAGE.

Both concave and convex spherical mirrors are commonly encountered. Convex mirrors are commonly used as wide-angle rearview mirrors in automobiles or on trucks. A common application of concave mirrors is the magnifying shaving mirror frequently found in bathrooms.

A spherical mirror will form an image which is not perfect, except for particular conjugate distances. The use of a mirror which has the shape of a rotated conic section, such as a parabola, ellipsoid, or hyperboloid, will form a perfect image for a particular set of object-image conjugate distances and will have reduced aberrations for some range of conjugate relations. The most familiar applications for conic mirrors are in reflecting telescopes. *See* OPTICAL PRISM; OPTICAL SURFACES; OPTICAL TELESCOPE; REFLECTION OF ELECTROMAGNETIC RADIATION; TELESCOPE. [R.R.S.]

Misophrioida A small but evolutionarily significant order of the subclass Copepoda, containing only a few species. Misophrioids have two subdivisions of the body, the anterior prosome and posterior urosome, articulated immediately behind the fifth thoracic somite. This feature immediately distinguishes misophrioids from calanoids. Like calanoids, some misophrioids have a dorsal heart, a character that distinguishes the taxon from both cyclopoids and harpacticoids. The misophrioids have four unique characteristics: a carapacelike posterior extension of the head region (cephalosome) that encloses the first leg-bearing segment; the absence of the nauplius eye in all life stages; the retention of the antennary gland as the functional excretory system of adults; and a single naupliar developmental stage.

The unique characteristics of the misophrioids are interpreted as having resulted from the adaptation to a bathypelagic mode of life and gorging as a feeding strategy. However, species of three newly described genera also exhibit a number of characteristics that approach those attributed to a hypothetical copepod ancestor. *See* CALANOIDA; COPEPODA; CRUSTACEA; CYCLOPOIDA; HARPACTICOIDA. [P.A.McL.]

Missile An object capable of being projected or hurled, usually with the intent of striking some distant object. More particularly, a missile is usually a weapon that is self-propelled after leaving the launching device. The term thus excludes projectiles fired from guns as well as free-failing bombs. The propulsion criterion is not an absolute test, however, since gliding bombs, especially those guided after launch, are also frequently classed as missiles.

There are two primary categories of modern missiles; unguided and guided. All missiles, to be effective, must be directed in some sense, but those subject to no further control after leaving the launching device are usually classed as unguided.

Typical of unguided missiles are the ground-launched and air-launched free-flying rockets used in enormous number in World War II.

The growth of the sciences of instrumentation, electronics, and automatic control has led to the development of devices for the guidance of missiles in flight. Certain targets contrast with their surroundings by emitting or reflecting radiation in a distinctive manner. (Airplanes and ships are typical of the targets falling into this class.) The direction, and sometimes the distance, of targets of this kind can frequently be sensed by radiation receiving instruments. These instruments, located either in the missile or on the ground, can be used to guide the missile continuously toward the target.

Other targets cannot be readily distinguished from their surroundings by nonhuman means. Guided missiles used against these are frequently directed to the predetermined geographical location of such targets. Guidance errors are likely to be substantial, particularly as the distance from the guidance station to the target becomes great. In these cases, use of nuclear warheads becomes mandatory on economic grounds.

Missiles are classified in many ways. A common classification is according to the medium from which the missile is launched and to which it is directed. Thus, there are surface-to-air missiles, surface-to-surface missiles, air-to-surface missiles, and so on. They may also be classified according to range.

Missiles are also classified according to flight profile. The two categories are aerodynamic missiles (sometimes called cruise missiles) and ballistic missiles. Cruise missiles usually have wings or enlarged fins to give lift and maneuverability. A ballistic missile has no wings. It must be aimed sufficiently high to permit it to fall freely under the influence of gravity until it reaches the target. *See* BALLISTIC MISSILE.

High-speed missiles present problems of protecting sensitive components from the high temperatures produced by aerodynamic drag and sometimes from the high inertial forces produced by high accelerations. Only during the reentry phase is heat protection necessary.

The shape of the reentering body is sometimes quite blunt. This shape causes a greater fraction of the kinetic energy of the missile to be transferred to the surrounding atmosphere. However, blunt bodies slow down greatly before striking the ground, and they also have large radar reflectivity. These qualities are undesirable if antiballistic missiles are defending the target. As a consequence, the trend is toward sharper cones, accepting the greater heat transfer by providing greater protection. The ablative heat shield has almost entirely superseded the heat-sink type. *See* NOSE CONE. [R.C.Tr.]

Mississippian A large division of late Paleozoic geologic time, varyingly considered to rank as an independent period or as a main subdivision (termed epoch) of the Carboniferous Period. It is named Mississippian for a succession of highly fossiliferous marine strata consisting largely of limestones found along the Mississippi River between southeastern Iowa and southern Illinois. The rocks formed during the term defined as Mississippian are classed as a geologic system by most American geologists, and this practice is increasingly accepted outside of North America even though foreign usage favors Lower Carboniferous. Widespread marine transgression of the continents, accompanied especially by accumulation of limestones, characterized Mississippian time. Locally, this was preceded or succeeded by mountain building. *See* CARBONIFEROUS. [R.C.Mo.]

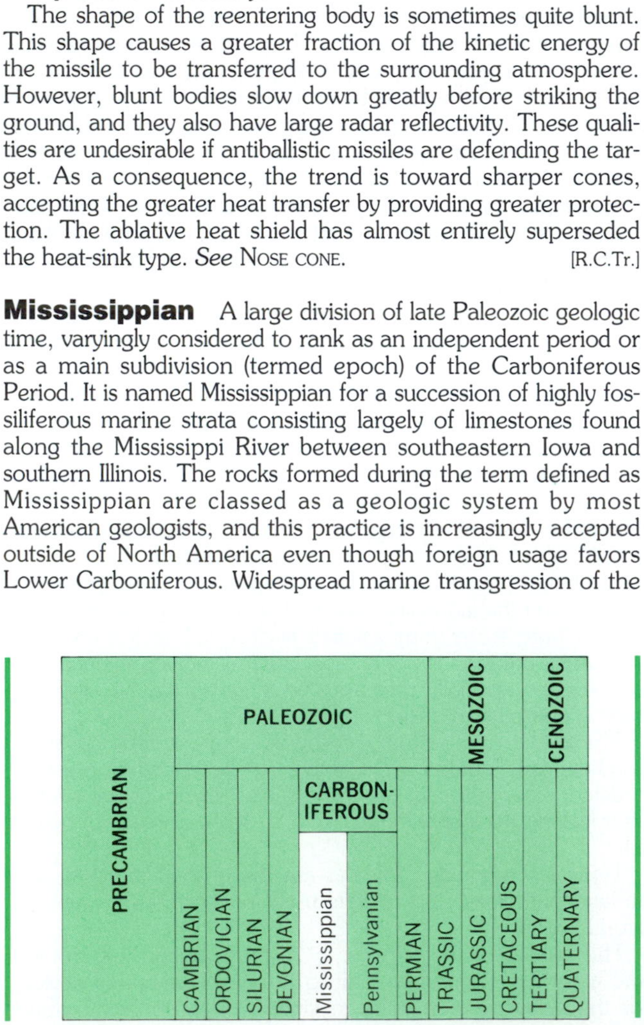

Mistletoe The name given to several species of the mistletoe family (Loranthaceae). The true mistletoe of Europe is *Viscum album*, and among the early nations this was an important ceremonial plant, which probably accounts for the origin of the custom of kissing under the mistletoe. In the United States, the common representative of the group is *Phoradendron flavescens*. All of the mistletoes are green hemiparasites; that is, they obtain water and minerals from the host plant but manufacture their own food. *See* SANTALALES. [P.D.St./E.L.C.]

Miticide A chemical useful for killing phytophagous and parasitic members of the order Acari of the class Arachnida. Although these animals are not insects, their control has become a part of economic entomology. Several species of mites are parasitic, but a considerable number are economically important as plant pests. Miticides are chemicals used specifically for the control of plant-feeding mites. *See* ACARI; INSECTICIDE; PESTICIDE. [G.F.L.]

Mitochondria Cell structures, sometimes called chondriosomes, which are minute cytoplasmic organelles found in all cells. They are clearly visible in the living cell when examined with the dark-field or phase-contrast microscope, and can be identified with the bright-field microscope when specifically stained by Janus green, applied supravitally. They may appear as small spherical granules, short rods, or long filaments. Spherical forms range from 0.2 to more than 2.0 micrometers in diameter and filamentous forms may vary from 1 to 7 micrometers in length.

Mitochondria may change form in different physiological conditions and are sensitive indicators of cell injury. In living cells they are in constant motion, executing slow flexuous movements as they move from one part of the cell to another. As a rule they maintain their characteristic shape during these movements, but occasionally filamentous forms break up into short rods or spheres. In rare instances they elongate and thrust out lateral branches which coalesce temporarily with other parts of the same mitochondrion to form annular or plexiform (network) structures. New mitochondria arise only from preexisting mitochondria by a process of budding or fission.

In electron micrographs of thin sections of cells the mitochondria are seen to be limited by two dense lines $\sim$5 nanometers thick and $\sim$5 nanometers apart (Fig. 1). The outermost of these is smooth and continuous, whereas the inner turns inward at more or less regular intervals along the length of each mitochondrion to form a series of thin folds, the cristae mitochondriales, that project into the interior of the organelle. According to the interpretation of the electron microscopic images, the mitochondrion is limited by two separate membranes, the innermost of which is unfolded to form the cristae. Although internal membranes usually take the form of folds or cristae, they may instead form slender villous or tubular projections called tubuli mitochondriales.

In negatively stained, fragmented fresh mitochondria, the inner surface of the organelle is found to be studded with small particles 9 nm in diameter which are connected to the membrane by a slender stalk 3.5–4.0 nm long (Fig. 2). These particles, called inner membrane subunits, are believed to be sites of important enzymes of the mitochondrion, but are not preserved by routine methods of preparation of tissues for electron microscopy.

The interior of the mitochondrion is occupied by an amorphous matrix of appreciable density. A variable number of dense particles $\sim$50 nm in diameter, called matrix granules,

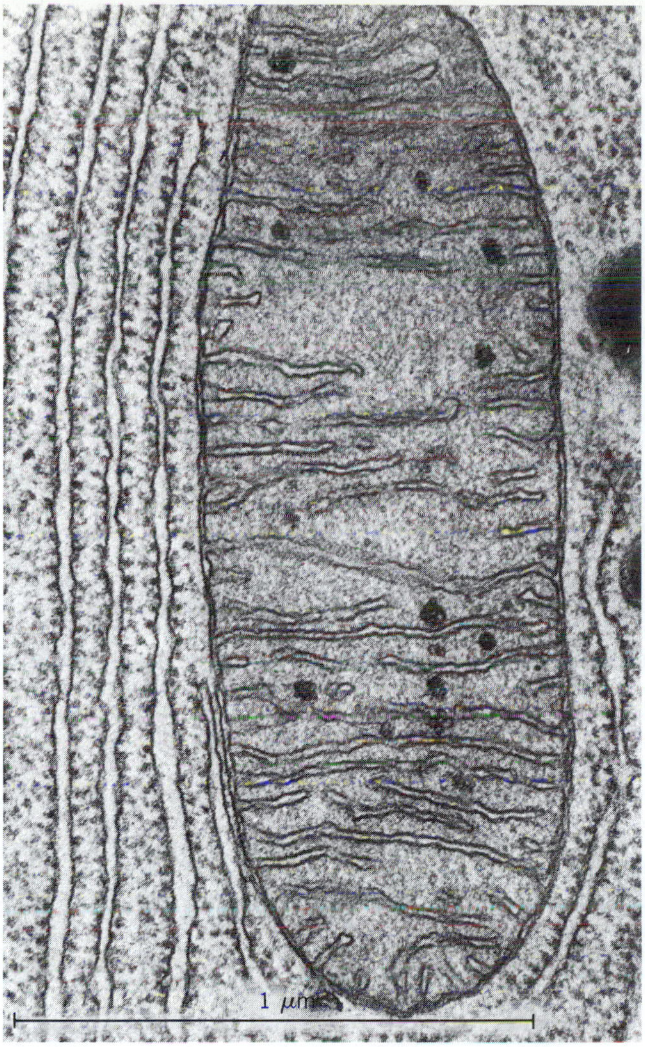

Fig. 1. Electron micrograph of mitochondrion from pancreatic acinar cell, showing typical cristae projecting into a moderately dense matrix containing matrix granules. (*Courtesy of K. R. Porter*)

are randomly distributed in the matrix between the cristae, and are perhaps concerned with regulating the interval ionic environment of the mitochondrion. Also present are smaller granules ~12 nm in diameter that contain ribonucleic acid (RNA). Deoxyribonucleic acid (DNA) has also been found, and this appears to reside in slender filaments that can be visualized in the matrix in electron micrographs of the mitochondria in some tissues.

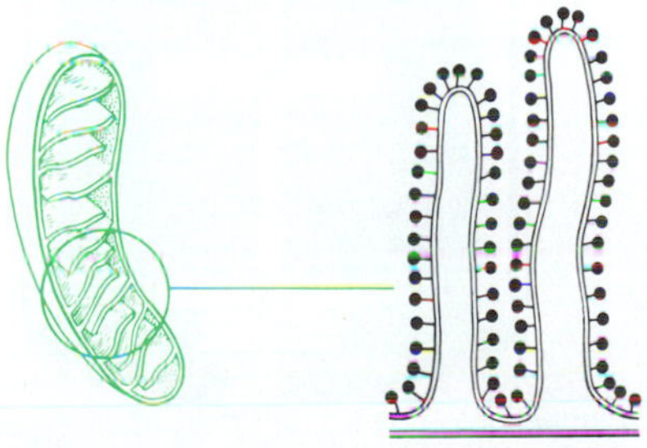

Fig. 2. Schematic representation of two cristae, showing the inner membrane subunits which are not preserved in the usual preparation of tissue for electron microscopy. These minute particles are believed to contain the enzymes of oxidative phosphorylation. Relative size is exaggerated.

Mitochondria contain highly integrated enzyme systems providing energy for cell metabolism. Among the most important of these is the tricarboxylic acid cycle and its associated respiratory enzymes. These produce the energy-rich compound adenosinetriphosphate which is essential for many vital cellular activities, including cell motility and muscular contraction. *See* ADENOSINETRIPHOSPHATE (ATP); CELL (BIOLOGY); CELL MEMBRANES; KREBS CYCLE; PROTEIN. [D.W.F.]

Mitosis A division of the cell nucleus in which a spindle and chromosomes are involved. The process results in two daughter nuclei which are genetically identical to each other and to the cell from which they were derived. It involves the exact duplication of the chromosome threads, followed by a precise separation of the products of the duplication to each of the daughter cells. This is brought about by a series of mechanical movements during which the entire nucleus and cytoplasm are temporarily reorganized. Mitosis is the characteristic mode

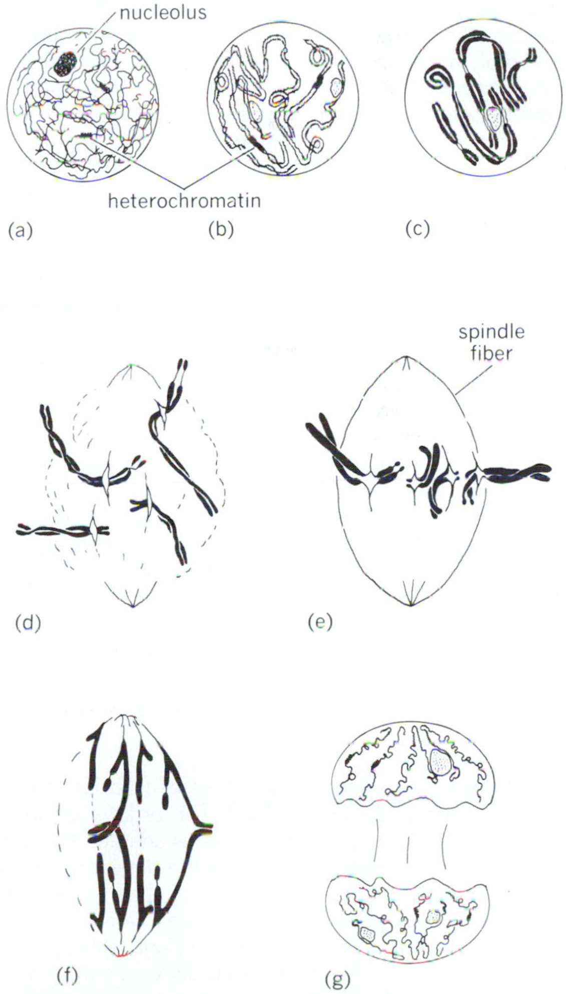

Mitosis. (*a*) Interphase: deoxyribonucleic acid (DNA) replicates; the chromosomes duplicate. (*b*) Early and (*c*) late prophase: chromosomes appear, condense, and spiralize; the phase is terminated by disappearance of nucleoli and disruption of nuclear membrane. (*d*) Prometaphase: spindle is organized; centromeres attach; congression of chromatid pairs to spindle equator occurs. (*e*) Metaphase: chromatid pairs are stationary at equator of spindle. (*f*) Anaphase: regions of special cycle of division divide; spindle fibers contract, pulling sister chromatids to opposite poles. (*g*) Telophase: reorganization of two interphase nuclei occurs. (*Drawn after the style of K. R. Lewis and B. John, Chromosome Marker, Churchill, 1963*)

of division of eukaryotic cells, although other less precise forms of division are known, for example, amitosis. *See* CELL DIVISION.

Mitotic cycle. The first indication that a cell is about to divide is the appearance within the nucleus of long chromosome threads (see illustration). At the outset these are divided longitudinally into pairs of sister chromatids, the consequence of reduplication of the chromosomal DNA during the preceding interphase. During prophase the chromosomes condense and spiralize, becoming shorter and thicker; sister chromatids often show relational coiling. The spindle apparatus forms, and at metaphase centromeres congress at the equator of the spindle. In anaphase, sister chromatids, led by their centromeres, pass to opposite poles. Following reorganization of nuclei in telophase, the cell may undergo cytokinesis, which in animal cells normally involves constriction and ultimate division of the cytoplasm, while in plant cells a new wall is laid down by the phragmoplast in the cytoplasm between the daughter nuclei. In the absence of cytokinesis, multinucleate cells or coenocytes may be formed.

The spindle apparatus is shaped somewhat like a football and in many species, especially in animals, is associated with centriolar asters at each of its poles. When fully formed at metaphase, the spindle is a discrete, distinct structure, and when suitably illuminated, can be seen without difficulty in the normal living cell.

Significance. The chromosomes are permanent structures capable of accurate self-duplication which, in turn, depends upon the exact replication of the individual DNA molecules composing the chromosomes. Before replication, the chromosomes consist of two subunits of DNA, corresponding to the two chains of the DNA double helix. During interphase, the replication of DNA involves the separation of the complementary strands over a short distance, followed by the synthesis of complementary strands on each of the parental strands. During mitosis the products of DNA and chromosome duplication are distributed equally between daughter nuclei. After the nucleus has divided, the cytoplasm divides, producing two cells from the original one. Each daughter nucleus should then carry exactly the same heritable information as the other and the parent nucleus. *See* CYTOKINESIS.

Mitotic crossing over may well be an important method of redistributing genetic material in organisms which do not have a sexual stage, for example, imperfect fungi. Linkage groups could be broken up by a cycle of diploidization, recombination, and nonmeiotic haploidization, a sequence which has been named the parasexual cycle by Pontecorvo. In this way recombination of whole chromosomes and parts of chromosomes can occur. *See* CHROMOSOME; DEOXYRIBONUCLEIC ACID (DNA). [G.G.N.]

Mitteniales An order of true mosses (subclass Bryidae) found in Australia and New Zealand. The order consists of a single species. *Mittenia plumula*, which is adapted for growth in caves or cavelike places. Branches of the persistent protonema consist of spherical cells which reflect light from a backing of chloroplasts, thus providing a glow. The stems are simple and erect, and the leaves are 2–4-ranked and oblonglingulate and blunt or apiculate. The midrib ends above the midleaf, and the cells are short and smooth.

The order is remarkably similar to the Schistostegales of North Temperate distribution in protonema and habitat, but the other features of gametophyte and sporophyte are quite different. A distinctive feature is the double peristome with 32 segments derived in part from the outermost cell walls of the endothecium (that is, the inner portion of the embryonic capsule). *See* BRYIDAE; BRYOPHYTA; BRYOPSIDA; SCHISTOSTEGALES. [H.Cr.]

Mixer A device having two or more signal inputs, usually adjustable, and one common output. It is used in audio amplifiers to combine the outputs of individual microphones and other audio-signal sources linearly and in desired proportions to produce one audio output signal. In television, a mixer serves similarly to combine the outputs of two or more television cameras or other video signal sources. The mixer stage in a superheterodyne receiver combines the incoming modulated radio-frequency (rf) signal with the signal of a local rf oscillator to produce a modulated intermediate-frequency (i-f) signal. Crystal diodes are widely used as mixers in radar and other microwave equipment. [J.Mar.]

Mixing A common operation to effect distribution, intermingling, and homogeneity of matter. Actually the operation is called agitation, with the term mixing being applicable when the goal is blending, that is, homogeneity. Other processes, such as reaction, mass transfer (includes solubility and crystallization), heat transfer, and dispersion, are also promoted by agitation. The type, extent, and intensity of agitation determine both the rates and adequacy of a particular process result. The agitation is accomplished by a variety of equipment.

Most liquid mixing is done by rotating impellers in vertical cylindrical vessels. A typical impeller-type liquid mixer with a variety of features is shown in the illustration. The internal features, including the vessel itself, are considered as a whole, that is, as the agitated system. The forces applied by the impeller develop overall circulation or bulk flow. Superimposed on this flow pattern, there is molecular diffusion, and if turbulence is present, also turbulent eddies. These provide micromixing. Solids, granular to powder, are mixed in a variety of contrivances.

Solids of different density and size are mixed in tumblers (a double cone turning end on end) or with agitators (a helical ribbon rotating in a horizontal trough). The duration of mixing is an important additional variable because classification and sep-

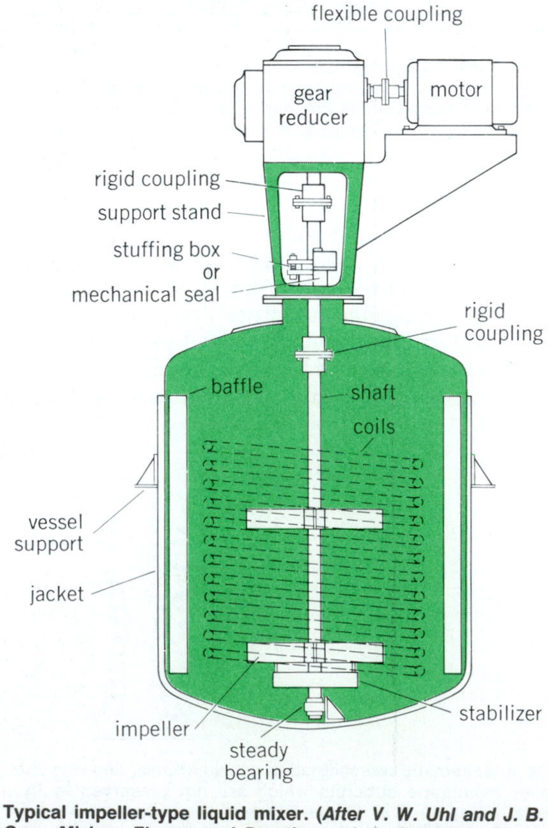

Typical impeller-type liquid mixer. (*After V. W. Uhl and J. B. Gray, Mixing: Theory and Practice, vol. 2, Academic Press, 1967*)

aration often occur after attainment of the desired distribution if the operation is carried on too long. [V.W.U.]

Mixture An aggregate composed of two or more distinct chemical components which retain their identities regardless of the degree to which they have become mingled. The substituents may be present in any proportion and may form a homogeneous or heterogeneous system. There is no chemical interaction between the components of a mixture, and it is possible, at least theoretically, to separate them by physical means. Attractive or repulsive forces between the different substituents, however, may affect the gross characteristics of the aggregate. For example, the total volume of a mixture of several liquids may not be the sum of the separate volumes. [F.J.J.]

Mizar A multiple star in the constellation Ursa Majoris. Mizar (ζ Ursa Majoris) is located at the bend in the handle of the Big Dipper. A very wide visual double, Mizar (the horse) and Alcor (the rider), is visible to the naked eye. Mizar is also a telescopic double star; both components are also spectroscopic binary stars. Mizar was the first spectroscopic binary to be discovered (1889). *See* BINARY STAR; URSA MAJOR. [J.L.Gr.]

Möbius inversion A useful counting technique which exploits the relationship between a pair of functions f and g defined on a lattice L. Suppose that g satisfies Eq. (1), and that

$$g(s) = \sum_{s \le t} f(t) \quad (1)$$

the values of g are known; then the values of f can be found if the calculation in Eq. (1) can be "inverted" to define f in terms of g. The principle of Möbius inversion states that there is a function $\mu(s,t)$, called the Möbius function of the lattice, defined for pairs of elements (s,t) in L and which depends only on the lattice, such that Eq. (2) holds.

$$f(s) = \sum_{s \le t} \mu(s,t)g(t) \quad (2)$$

See COMBINATORIAL THEORY; ENUMERATION OF ARRANGEMENTS. [T.B.]

Mode of vibration A characteristic manner in which vibration occurs. In a freely vibrating system, oscillation is restricted to certain characteristic frequencies; these motions are called normal modes of vibration.

An ideal string, for example, can vibrate as a whole with a characteristic frequency $f = (1/2L)\sqrt{T/m}$, where L is the length of string between rigid supports, T the tension, and m the mass per unit length of the string. The displacements of different parts of the string are governed by a characteristic shape function. The frequency of the second mode of vibration is twice that of the first mode. Similarly, modes of higher order have frequencies that are integral multiples of the fundamental frequency.

Because the frequencies are in the ratios 1:2:3..., the modes of vibration of an ideal string are properly called harmonics. Not all vibrating bodies have harmonic modes of vibration, however. *See* HARMONIC (PERIODIC PHENOMENA); VIBRATION. [R.W.Y.]

Model theory The theory underlying the design and use of models to predict the characteristics of any system. It is particularly valuable when the desired system, or prototype, is large and complex. A model can be built, tested, and modified at comparatively low cost. If the model is properly designed, the results obtained from it may be used with a high degree of confidence in predicting the performance of the prototype. Models are thus widely used in the design of bridges, dams, unusual buildings, aircraft, river and harbor works, agricultural machinery, and a variety of network systems.

If the results obtained from a model are to be applicable to the prototype, a rigorous set of conditions on the model must be satisfied or else the deviations from these conditions must be taken into account in the predictions of the prototype behavior. The similitude relationships that must exist between model and prototype may be developed conveniently by using dimensional analysis. *See* DIMENSIONAL ANALYSIS.

The first step is to identify all of the independent variables in the system which will influence the magnitude of the quantity that is to be predicted. A set of dimensionless variables is then formed from these variables, using dimensional analysis. These dimensionless variables, or pi terms, are grouped in an equation, called the characteristic equation, in the form of Eq. (1),

$$\pi_1 = f(\pi_2, \pi_3, \pi_4, \ldots, \pi_s) \quad (1)$$

in which π_1 is the dimensionless term containing the variable to be predicted, and the other π terms are dimensionless quantities based on the independent variables. A similar equation may be written for the model, which involves the same variables as the prototype; this is shown in Eq. (2). If the two

$$\pi_{1m} = f(\pi_{2m}, \pi_{3m}, \pi_{4m}, \ldots, \pi_{sm}) \quad (2)$$

systems involve the same phenomenon, the forms of the two functions will be identical. If the model is so designed that Eqs. (3) hold, it follows that Eq. (4) is true. Therefore, if the design

$$\pi_{2m} = \pi_2, \pi_{3m} = \pi_3, \pi_{4m} = \pi_4, \ldots, \pi_{sm} = \pi_s \quad (3)$$

$$\pi_1 = \pi_{1m} \quad (4)$$

equations, Eqs. (3), are satisfied in the construction and operation of the model, the prediction equation, Eq. (4), may be used to predict the magnitude of π_1 and hence the desired variable in the prototype, from the measured magnitude of π_{1m} in the model.

In some situations two design conditions may lead to conflicting requirements that cannot be satisfied simultaneously. In situations of this kind the design of the model is based on the predominant number, and the influence of the other number or numbers must be distorted. [G.Mu.]

Modem A device that converts the digital signals produced by terminals and computers into the analog signals that telephone circuits are designed to carry. Despite the availability of several all-digital transmission networks, the analog telephone network remains the most readily available facility for voice and data transmission. Since terminals and computers transmit data using digital signaling, whereas telephone circuits are designed to transmit analog signals used to convey human speech, a device is required to convert from one to the other in order to transmit data over telephone circuits. The term modem is a contraction of the two main functions of such a unit, modulation and demodulation. The device is also called a data set. *See* INTEGRATED SERVICES DIGITAL NETWORK (ISDN); MODULATION.

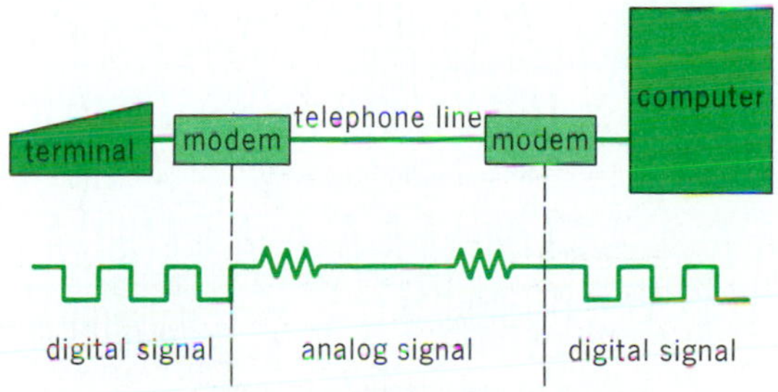

Signal conversion performed by modems. A modem converts a digital signal to an analog tone (modulation) and reconverts the analog tone into its original digital signal (demodulation).

In its most basic form a modem consists of a power supply, transmitter, and receiver. The power supply provides the voltage necessary to operate the modem's circuitry. The transmitter section contains a modulator as well as filtering, wave-shaping, and signal control circuitry that converts digital pulses (often input as a direct-current signal with one level representing a digital one and another level a digital zero) into analog, wave-shaped signals that can be transmitted over a telephone circuit. The receiver section contains a demodulator and associated circuitry that is used to reverse the modulation process by converting the received analog signals back into a series of digital pulses (see illustration). *See* DATA COMMUNICATIONS; DEMODULATOR; ELECTRIC FILTER; ELECTRICAL COMMUNICATIONS; ELECTRONIC POWER SUPPLY; MODULATOR; WAVE-SHAPING CIRCUITS. [G.He.]

Modulation The process or result of the process whereby a message is changed into information-bearing signals that not only unambiguously represent the message but also are suitable for propagation over the transmitting medium to the receiver. Demodulation or detection is the process whereby, in response to the received wave, either the original message or information pertaining to the original message is made available in the form desired.

Modulation implies bandwidth occupancy. For any signal to change in a way that cannot be predicted implies necessarily that the signal occupy a nonzero band of frequencies. For example, the spoken word occupies a band from a few hundred to several thousand hertz.

Defined broadly, modulation is the process or result of the process whereby some parameter of one wave is varied in accordance with another. As is customary in the treatment of modulation, the word "wave" is used as a generic term intended to include such concepts as signal, voltage, current, pressure, displacement, and the like, whether these are constant or changing.

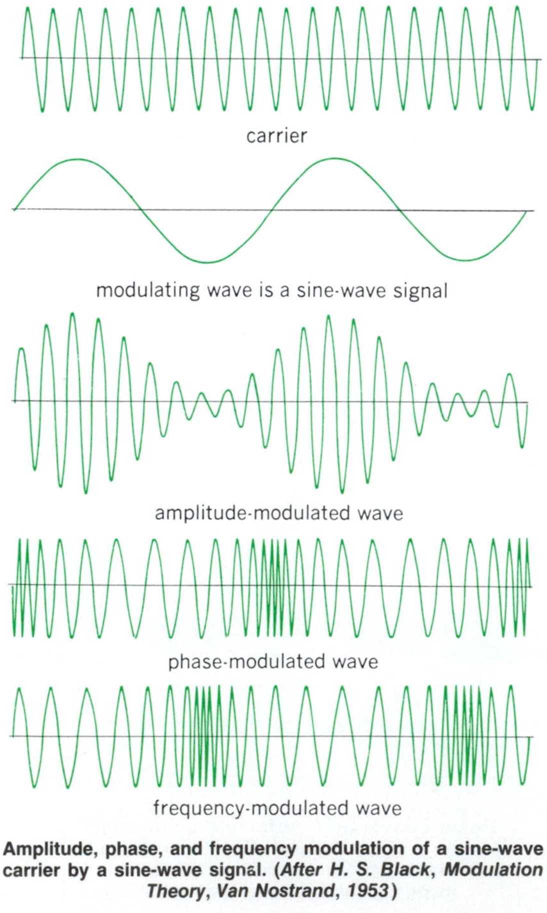

Amplitude, phase, and frequency modulation of a sine-wave carrier by a sine-wave signal. (*After H. S. Black, Modulation Theory, Van Nostrand, 1953*)

This broad definition of modulation implies three fundamental concepts: modulating wave, carrier, and modulated wave. A modulating wave changes some parameter of the wave to be modulated; a carrier is a wave suitable for modulation by the modulating wave; a modulated wave has some parameter changed in accordance with the modulating wave.

Common forms of modulation are amplitude modulation, angle modulation (which includes frequency modulation and phase modulation), and pulse modulation (see illustration). *See* AMPLITUDE MODULATION; ANGLE MODULATION; FREQUENCY MODULATION; PHASE MODULATION; PULSE MODULATION. [H.S.Bl.]

Modulator Any device or circuit by means of which a desired signal is impressed upon a higher-frequency periodic wave known as a carrier. The process is called modulation. The modulator may vary the amplitude, frequency, or phase of the carrier. *See* MODULATION.

There are many ways to accomplish amplitude modulation, but in all cases a nonlinear element or device must be employed. The modulating signal controls the characteristics of the nonlinear device and thereby controls the amplitude of the carrier. *See* AMPLITUDE MODULATION; AMPLITUDE-MODULATION DETECTOR; AMPLITUDE MODULATOR.

The frequency modulator usually changes the effective capacitance or inductance in the frequency-determining LC circuit of the oscillator. However, other techniques can be used. For example, a multivibrator can be used to generate carrier frequencies up to a few megahertz, and the multivibrator frequency can be modulated by controlling the base, gate, or grid bias supply voltage. *See* FREQUENCY MODULATION; FREQUENCY-MODULATION RADIO; FREQUENCY MODULATOR. [C.L.A.]

Mohair The long, lustrous hair of the Angora goat, which originated in the area around Ankara (Angora), Turkey. Mohair is a smooth, strong, durable, and resilient fiber. It enhances softness and luster in fabrics. Mohair absorbs dye evenly and brilliantly, retains color well, and permits unusual decorative effects. It is mainly used as an apparel fiber but may be used in upholstery, draperies, wigs, hairpieces, and rugs. Leather produced from the skin is useful for gloves, purses, and novelties. *See* WOOL. [C.E.T.]

Moho (Mohorovičić discontinuity) A seismic discontinuity at the base of the Earth's crust inferred from travel time curves indicating that seismic waves undergo a sudden increase in velocity. Immediately below the discontinuity the velocity of seismic compressional waves increases to a little over 5 mi/s (8 km/s). The discontinuity is named after its discoverer, the Yugoslavian geophysicist A. Mohorovičić. Such a discontinuity seems to exist every place that geophysicists seismologically investigate the Earth's structure. Under the ocean basins its depth is generally 6–7 mi (10–12 km); under the continents it is usually at a depth of 20–22 mi (33–35 km). It usually lies deeper under mountain masses and shallower under lowlands and plains, sloping upward near coastal margins. Islands and island arcs are also marked by a downward bend of the Moho under the ocean basins. *See* CONTINENT; SEISMOLOGY. [G.G.L.]

Moiré pattern When one family of curves is superposed on another family of curves, a new family called the moiré pattern appears.

To produce moiré patterns, the lines of the overlapping figures must cross at an angle of less than about 45°. The moiré lines are then the locus of points of intersection. The illustra-

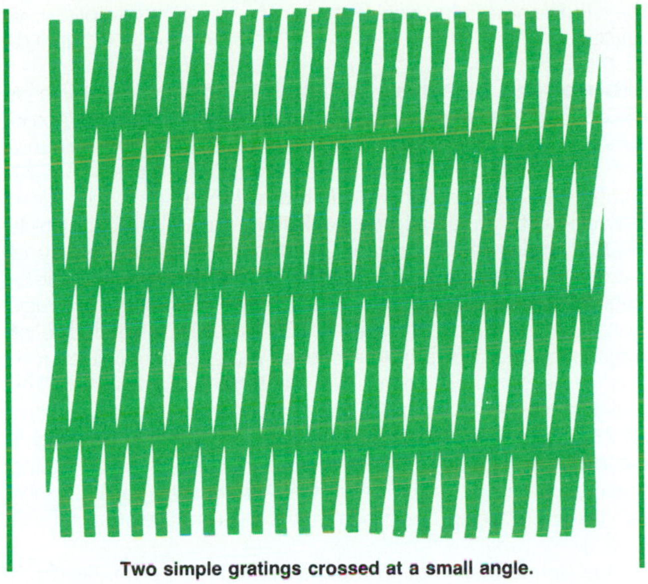

Two simple gratings crossed at a small angle.

tion shows the case of two identical figures of simple gratings of alternate black and white bars of equal spacing. When the figures are crossed at 90°, a checkerboard pattern with no moiré effect is seen. At crossing angles of less than 45°, however, one sees a moiré pattern of equispaced lines, the moiré fringes. The spacing of the fringes increases with decreasing crossing angle. This provides one with a simple method for measuring extremely small angles (down to 1 second of arc). As the angle of crossing approaches zero, the moiré fringes approach 90° with respect to the original figures.

Even when the spacings of the original figures are far below the resolution of the eye, the moiré fringes will still be readily seen. This phenomenon provides a means of checking the fidelity of a replica of a diffraction grating. *See* DIFFRACTION GRATING.

Moiré techniques are widely used in the stress analysis of metals, in the examination of large optical surfaces, in investigating aberrations of lenses, and in determining a refractive index gradient (for example, that of sugar molecules diffusing into water). [G.O.]

Moisture-content measurement Measurement of the ratio or percentage of water present in a gas, a liquid, or a solid (granular or powdered) material. Nearly all materials contain free water, the relative amount being dependent upon the physical and chemical properties of the material. The primary purpose of determining and maintaining moisture contents within specified limits can usually be traced to economic factors, trade practices, or legal requirements.

Moisture content has a number of synonymous terms, many of which are specific to certain industries, types of product, or material. The water content in solid, granular, or liquid materials is usually referred to as moisture content on either the wet or dry basis; the wet basis is common to most industries. Specifically, moisture content on the wet basis refers to the quantity of water per unit weight or volume of the wet material. A weight basis is preferred. The textile industry uses the dry basis for moisture content of textile fibers. Often referred to as regain moisture content, the dry basis or regain refers to the quantity of water in a material expressed as a percentage of the weight of the bone-dry (thoroughly dried) material.

The moisture content in air is referred to as humidity, either absolute or relative. Absolute humidity is the number of pounds of water vapor associated with 1 lb (0.5 kg) of dry air, also called just humidity, Relative humidity is the ratio, usually expressed as a percentage, of the partial pressure of water vapor in the actual atmosphere to the vapor pressure of water at the prevailing temperature. Relative humidity is customarily reported by the U.S. Weather Bureau because it essentially describes the degree of saturation of the air. However, air which is saturated (100% RH) at 50°F (10°C) is quite dry (19% RH) when heated to 100°F (38°C). A changing basis of this type is not convenient for many purposes such as computations used in air conditioning, combustion, or chemical processing; therefore absolute units, such as dew point or grains of water per pound of dry air, are more acceptable. Dew point is the temperature at which a given mixture of air and water vapor is saturated with water vapor. *See* DEW POINT; HUMIDITY.

Gases. The measurement of water content in gases and mixtures of air and gases is important in industry. A number of commercially manufactured instruments are available for these measurements; their principles of operation include condensation, used in dew- or fog-point indicators; dimensional change, used by hygrometers; thermodynamic equilibrium, used by wet-bulb psychrometers; and absorption methods, which serve as the basic principle for gravimetric and electric conductivity or dielectric types. *See* HYGROMETER; PSYCHROMETER; PSYCHROMETRICS.

The importance of humidity in relation to personal comfort is well known. The air conditioning industry produces equipment to maintain comfortable conditions of temperature and humidity. Considerable industrial air conditioning is also done for process reasons. Control of humidity is also important in the preservation of materials, especially those which are hygroscopic, and in the storage of food products.

Liquids and solids. The development of instrumentation for the measurement and control of moisture in liquids and solids has been due in a large part to the great need that exists in many processes where the control of a precise moisture is critical. The desirability of a specific moisture content in a product during its preliminary manufacturing process is often required. In general, however, the rigid control of moisture content occurs most frequently in the final product to assure its quality and the fulfillment of legal or trade practices for the individual product.

Instruments suitable for the measurement of moisture content may be classified as periodic and continuous. In general, only those instruments offering continuous measurement are practical for the automatic control of moisture content in a product. The periodic instrument types are generally automated versions of conventional laboratory moisture-analysis procedures.

Moisture measuring instruments may also be classified by operating principle. Those instruments employing electrical conductivity (either dc or ac), absorption of electromagnetic energy (radio-frequency regions), electrical capacitance (dielectric constant change), and infrared energy radiations are more readily adapted to continuous measurements inasmuch as the response of these instruments to moisture changes is very fast. Those instruments employing automatic oven drying, chemical titrations, equilibrium hygrometric methods, distillation methods, and so forth are usually of the intermittent, or periodic, type. [L.E.C.]

Mole (chemistry) One mole is the mass numerically equal (in grams) to the relative molecular mass of a substance. It is the amount of a substance that contains the same number of molecules as there are atoms in 0.012 kilogram of carbon-12. The mole is an individual unit of mass, that is, it relates only to a given substance. If the relative molecular mass is μ, 1 mole = μ grams, and the M of one mole is μ g/mol.

Moles of different elements or compounds contain the same number of molecules. This number, called the Avogadro number N_0, is 6.023×10^{23}.

Since they contain the same number of molecules, moles of

ideal gases occupy the same volume at the same temperature and pressure. At 0°C and 1 atmosphere (101,325 pascals) pressure, this volume, called the molar volume, is 22.4 liters. *See* AVOGADRO NUMBER; GAS; RELATIVE MOLECULAR MASS. [T.C.W.]

Mole (medicine) A local growth of the skin which is usually rounded, raised, and pigmented. Related to birthmarks, it is correctly called a nevus. Nevi may be present at birth or may develop later. They are almost universally present but vary greatly in size, color, and distribution, as well as in components. Most nevi are soft, but some are corrugated and warty. Those present at birth represent a form of birthmark cellular in nature, in contrast to vascular birthmarks such as port-wine stains.

All nevi may become malignant, but controversy exists over which nevi are precursors of malignant melanomas, malignant tumors, of melanin-producing cells. Mechanical irritation, trauma, and inadequate removal may predispose nevi to malignant changes. Any nevus which shows a change in color, size, or hardness should be referred to a physician. [N.K.M.]

Mole (zoology) A mammal belonging to the family Talpidae. There are 19 species distributed on all continents except Australia. Moles are insectivores, feeding mainly on earthworms and insect larvae, and are highly specialized for their burrowing habits.

The body is stout and cylindrical and with a short neck (see illustration). The eyes and external ears are small or vestigial, and the conical head terminates in a long naked muzzle. The forelimbs are especially adapted for digging, having very powerful muscles and a spadelike bony structure. Mating occurs in the spring; after a gestation period of 30–40 days a litter of 3–7 young are born.

The European mole, *Talpa europaea*.

Moles are solitary animals and rarely come aboveground. Molehills are seen where temporary burrows are made, but permanent runs are made by compressing the earth so there are no molehills to betray their presence. Each mole lives in its own fortress constructed as a central chamber around which are two circular passages, one at a higher and the other at a lower level, connected by short passages. The upper passage connects with the central chamber and the lower one leads to the main exit. There is also an emergency exit leading to the main exit from below the central chamber. *See* INSECTIVORA; MAMMALIA. [C.B.C.]

Molecular adhesion A particular manifestation of intermolecular forces which causes solids or liquids to adhere to each other. The term is usually used with reference to adhesion between two different materials, in contrast to the term cohesion, which refers to the action of intermolecular forces whereby solid and liquid substances are held to substantially constant volume. Thus adhesive forces are not different in kind from cohesive forces. The behavior of different substances in contact is greatly affected by the relative strength of adhesive forces between different materials or cohesive forces between the same materials. *See* COHESION (PHYSICS); INTERMOLECULAR FORCES. [E.U.C.]

Molecular beams Well-directed streams of atoms or molecules in vacuum. Utilization of molecular beams is a cornerstone technique in the investigation of molecular structure and interactions. Molecular beams are usually formed at sufficiently low particle density for the interaction of one beam molecule with another to be negligible. This ensemble of truly isolated molecules is available for the spectroscopic study of molecular energy levels using photon probes from the radio-frequency to optical portions of the electromagnetic spectrum. Some of the best-determined fundamental knowledge of physics comes from spectroscopic molecular-beam experiments. Beyond this, beams can be applied as probes of the multifaceted nature of gases, plasmas, surfaces, and even the structure of solids. An application intermediate in complexity is the study of molecular interactions by means of two colliding beams, where one might be a beam of charged particles such as ions or electrons. *See* SCATTERING EXPERIMENTS (ATOMS AND MOLECULES).

One simple means of forming a beam is to permit gas from an enclosed chamber to escape through a small orifice into a second chamber maintained at high vacuum by means of large pumps (illustration *a*). A useful number of molecules passes forward along the horizontal axis of the apparatus. A well-collimated beam is then formed by requiring that those molecules entering the test chamber where an experiment is to be performed pass not only through the orifice but also through a second small hole separating the collimating and test chambers.

If higher velocities are desired, then a charge-exchange beam system can be used. In this scheme (illustration *b*), ions are produced by some ionizing process such as electron impact on atoms within a gas discharge. Since the ions are electrically charged, they can be accelerated to the desired velocity and focused into a beam using electric or magnetic fields. The last step in neutrally charged beam formation is to pass the ions through a neutralizing gas where electrons from the gas molecules are transferred to the beam ions in charge-exchange molecular collisions. *See* ION SOURCES.

Much of molecular spectroscopy involves the absorption or emission of light by molecules in a gas sample. The frequency of the light photon is proportional to the separation of molecu-

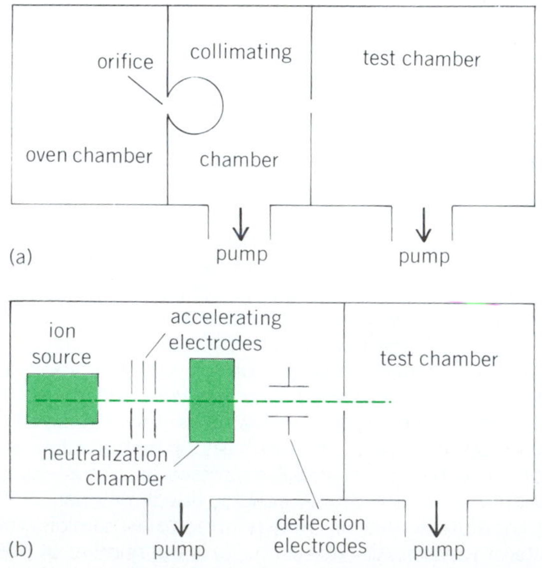

Schematic diagrams of systems for producing molecular beams. (*a*) Conventional oven-beam system. (*b*) Charge-exchange beam system.

lar energy levels involved in the spectroscopic transition. However, the molecule density in typical gas samples is so high that the energy levels are slightly altered by collisions between molecules, with the transition frequency no longer characteristic of the free molecule. The use of low-density molecular beams with their sensitive detection techniques can reduce this collision alteration problem, with the result that atomic properties can be measured to accuracies of parts per million or even better. If the very simplest atoms or molecules are employed, then the basic electromagnetic interactions holding the component electrons and nuclei together can be precisely studied. This is of great importance to fundamental physics, since theoretical understanding of electromagnetic interactions through quantum electrodynamics represents the most successful application of quantum field theory to elementary particle physics problems. *See* QUANTUM ELECTRODYNAMICS; QUANTUM FIELD THEORY.

The development of tunable, strong laser sources of single-frequency light beams has added another dimension to molecular-beam experiments. With laser radiation resonantly tuned to excite a molecule from its normal ground state to one of its infinite number of vibrationally, rotationally, and electronically excited states, the number of possible studies and applications of excited molecular beams becomes enormous. *See* LASER SPECTROSCOPY; MOLECULAR STRUCTURE AND SPECTRA; NUCLEAR STRUCTURE. [J.E.B.]

Molecular biology In its broadest context, molecular biology is that part of biology which attempts to interpret biological events in terms of the molecules in the cell. The subject matter merges indistinguishably with older and more traditional areas of biology, such as biochemistry and genetics. In a general way molecular biology may be described as that subject which grew out of the merging of biochemistry and genetics, since in this discipline special emphasis is placed on the molecular basis of genetics and of macromolecular replication.

Molecular biophysics is a related discipline that analyzes various biological phenomena on their most fundamental level, that is, on the level of the behavior of molecules. Because the primary emphasis is on a molecular explanation or description, one of the major tools in this discipline is chemistry, which is the science of molecules. On the other hand, many of the techniques used in molecular biophysics are taken directly from physics, such as the use of x-ray diffraction or nuclear magnetic resonance to study molecular structure. Thus molecular biophysics can be characterized as a discipline which utilizes the selective aspect of both chemistry and physics to study phenomena of biology on the molecular level. *See* BIOCHEMISTRY; BIOPHYSICS; GENETICS.

It is possible to describe a biological system in terms of various types of flux or flow in the living cell. The most obvious is a flow of matter, in which substances are taken up by the cell and a series of transformations are carried out within it; for example, glucose is converted into carbon dioxide and water. The determination of the detailed pathways of this conversion has been the traditional subject matter of biochemistry. In intermediary metabolism, the interconversions of a large variety of organic and inorganic molecules are described in great detail. A general conclusion reached in this type of investigation is the fact that a special role is played by the macromolecular proteins, which are polymers of amino acids. Proteins form the enzymes which have catalytic activity, and in a general way their presence or absence governs the metabolic pathways in living cells. *See* PROTEIN.

Another type of flux or flow in living tissue is that of energy. Associated with the flow of matter into cells and its conversion into other types of molecules is a conversion of chemical energy. For example, energy is contained in the chemical bonds of the glucose molecule, and the controlled breakdown of glucose into carbon dioxide and water leads to a controlled release of this chemical energy and its conversion into chemical energy associated with other types of molecules. Chemical energy is necessary to carry out many types of work in living cells, such as the work of chemical synthesis in which large molecules are built up through polymerization. Other types of work include mechanical work or osmotic work in which molecules are made to move against concentration gradients within the cell. The flow of chemical energy through the cell has been an integral part of intermediary metabolism, and a large part of biochemistry consists of describing metabolic pathways in terms of both molecular and energy conversions which are carried out within the cell. *See* ENERGY METABOLISM.

There is a third type of flux which is of essential importance in understanding the total molecular activity of the cell. This is the flow of information. By information is meant the directions or blueprints by means of which the cell knows which types of molecules to make. Since the control of intermediary metabolism is dependent upon the presence or absence of the protein enzymes, the information flow may be regarded as the system in the cell whereby the directions for assembling particular protein molecules are stored and transmitted through the cell. There is thus a system for controlling the metabolic activities of cells. This feature is characteristic of living matter, and it sharply divides such systems from nonliving material. In some ways, it may be said that the core of molecular biology is the study of the molecular representation of information in living cells and its manner of distribution and reproduction. *See* DEOXYRIBONUCLEIC ACID (DNA); RIBONUCLEIC ACID (RNA). [Al.R.]

Molecular cloud A large and relatively dense cloud of cold gas and dust in interstellar space from which new stars are born. Molecular clouds consist primarily of molecular hydrogen (H_2) gas, with temperatures in the range 10–100 K. Molecular hydrogen is not directly observable under most conditions in molecular clouds. Therefore, almost all current knowledge about the properties of molecular clouds has been deduced from observations of trace constituents, mostly simple molecules such as carbon monoxide (CO), which have strong emission lines in the centimeter- and millimeter-wavelength portions of the electromagnetic spectrum. Molecular clouds are widespread in the plane of the Milky Way Galaxy, with the greatest concentration of clouds lying in a broad ring encircling the galactic center. *See* MILKY WAY GALAXY; RADIO ASTRONOMY.

Most molecular gas in the Milky Way Galaxy is found in giant molecular clouds which are about 100,000 times the mass of the Sun. The closest and best-studied giant molecular clouds lie at a distance of about 450 parsecs (1 parsec $= 3 \times 10^{13}$ km or 2×10^{13}mi) in the constellation Orion where, over the last 10^7 years, they have given birth to several groups of hot and massive stars, as well as several thousand lower-mass stars. *See* ORION NEBULA.

Smaller molecular clouds, ranging from one to several thousand times the mass of the Sun, tend to form only low-mass Sun-type stars. A category of even smaller molecular clouds lies far from the plane of the Milky Way Galaxy and is too small to form stars. *See* PROTOSTAR.

Molecular clouds typically have average gas densities ranging from about 10 molecules per cubic centimeter on large scales to more than 10^6 molecules per cubic centimeter in cloud cores. The clouds have a very complex internal structure consisting of clumps and filaments of dense gas surrounded by interclump gas of much lower density.

About 100 different chemical species have been so far identified within molecular clouds, indicating that there is a rich chemistry taking place. About 1% of the mass of molecular clouds is in the form of interstellar dust grains that absorb starlight, making molecular clouds opaque at visible and ultraviolet wavelengths. Therefore, most nearby clouds can be seen in silhouette against the background of stars. *See* INTERSTELLAR EXTINCTION; INTERSTELLAR MATTER. [J.Ba.]

Molecular distillation A process by which substances are distilled in high vacuum at the lowest possible temperature and with the least damage to their composition. In a conventional distillation carried out at atmospheric pressure, molecules that leave a liquid surface as a vapor in their movement toward a condensing surface travel in straight paths until they collide with a molecule of residual gas (usually air), with another vapor molecule, or with the surface of the distilling vessel. These collisions either return the vapor to the liquid or greatly slow its rate of travel from the liquid to the condenser.

If the residual gas is virtually removed from a distillation vessel (for example, to pressures approximating 0.001 mmHg), and if the condensing surface is located close to the evaporating liquid, then conditions are favorable for molecular distillation (see illustration).

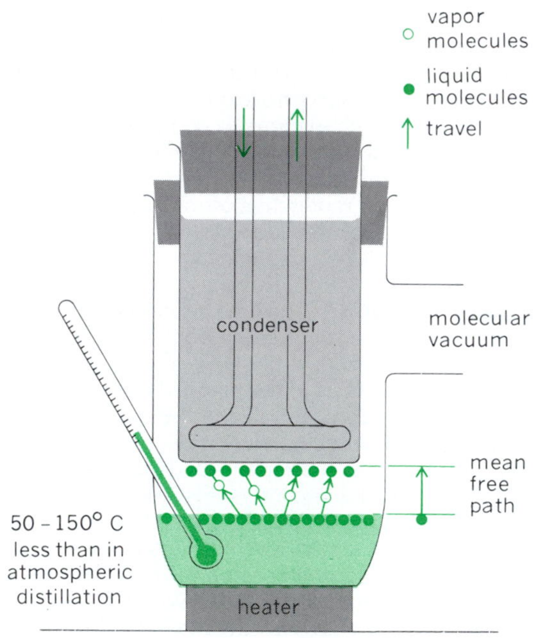

Schematic diagram illustrating distillation at molecular distillation conditions.

Materials composed of compounds with molecular weights in the approximate range of 400 to 1200 (there are exceptions) are effectively distilled by the molecular distillation process. Since this method can distill these compounds at temperatures 90–270°F (50–150°C) lower than by any other means, their thermal decomposition is at a minimum, which is very important when dealing with heat-sensitive materials. Also, any tendency toward oxidation during distillation is eliminated, since there is no atmospheric oxygen present.

At molecular distillation pressure, evaporation takes place at all temperatures, for there is no boiling point as there is in atmospheric pressure or in conventional vacuum distillation. Since evaporation is slow at lower temperatures, the liquid temperature is usually raised until the rate of evaporation meets practical requirements.

Molecular distillation is widely used in the laboratory for analytical work, as well as in the chemical industry for the improvement of plasticizers, silicone fluids, flavors and fragrances, pharmaceuticals, and edible products. In the latter category, high-purity monoglycerides used in baked goods and confectionery, as well as cosmetics, are purified by molecular distillation. [E.S.Ba.]

Molecular genetics In classical genetics the gene was regarded as a unit of information which directed the expression of certain features, such as the color of flowers or eyes or the shape of a leaf. This view of the gene has changed considerably in molecular genetics, since the gene is now recognized as the information necessary to assemble a protein molecule, which, in turn, regulates the production of a pigment in the eye or the growth of a leaf cell. The protein itself is a complicated molecule; accordingly, the gene today is seen to have a great deal of structure as compared to that visualized in classical genetics. The basic unit of information is that piece of deoxyribonucleic acid which makes the messenger ribonucleic acid which codes for a particular polypeptide chain. *See* GENE.

One of the characteristic features of the molecules that contain genetic information is the fact that they are capable of self-replication but that this replication process is not entirely perfect. Occasionally during DNA replication an incorrect base is inserted into the newly synthesized DNA chain. When this DNA replicates further, it produces altered DNA and organisms which differ slightly from those produced by the original strain. This is the phenomenon of mutation, and it is the mechanism by which genetic information changes. Mutants make messenger RNA molecules with a modified sequence, and they are expressed in the formation of a protein molecule which differs slightly from that of the original molecule. This modification may be trivial: The substitution of one amino acid for another may leave the protein largely unchanged. In other circumstances the modified protein molecule may have some chemical advantage in the cell over the original molecule; the mutation may thus produce a cell with a selective advantage over the original cell, better able to survive in its environment. *See* DEOXYRIBONUCLEIC ACID (DNA); MUTATION; PROTEIN.

Genetic information is encoded linearly in the DNA molecule. An important activity in molecular genetics is that of determining the sequence of genes found in the DNA. This activity has been very successful in small organisms such as bacteria or viruses in which genetic maps have been made relating positions in the bacterial genome to the production of specific proteins. Such genetic maps are useful in the study of the overall phenomena of cell division and gene expression. *See* GENETIC MAPPING.

Genes are often expressed in functionally related groups. For example, if a bacterial cell is exposed to a new kind of sugar molecule which it must utilize in order to grow, the cell often has the capacity to adapt to the new type of nutrient by forming proteins capable of metabolizing this particular type of sugar. A great deal of study has been carried out on the mechanisms which control gene expression. This work has shown the existence of regulator genes that act in such a way as to modify the expression of other genes. Thus not all genes in the bacterial cell are structural in the sense that they define proteins used in enzymatic activities. Other types of genes exist which control the expression of structural genes. *See* GENETICS. [A.R.]

Molecular isomerism The property of compounds (isomers) which have the same molecular formula but different physical and chemical properties. The difference in properties is caused by a difference in molecular structure (that is, molecular architecture). A typical example is dimethyl ether, CH_3OCH_3, a chemically quite inert gas which condenses at −24°C, and ethyl alcohol, CH_3CH_2OH, a liquid of substantial chemical reactivity which boils at 78°C; both compounds have the molecular formula C_2H_6O.

Isomers may be classified as constitutional isomers or stereoisomers. Constitutional isomers differ in constitution or connectedness, relating to the question as to which atoms are linked to which others and how. Dimethyl ether and ethanol (Fig. 1) are constitutional isomers. In dimethyl ether each car-

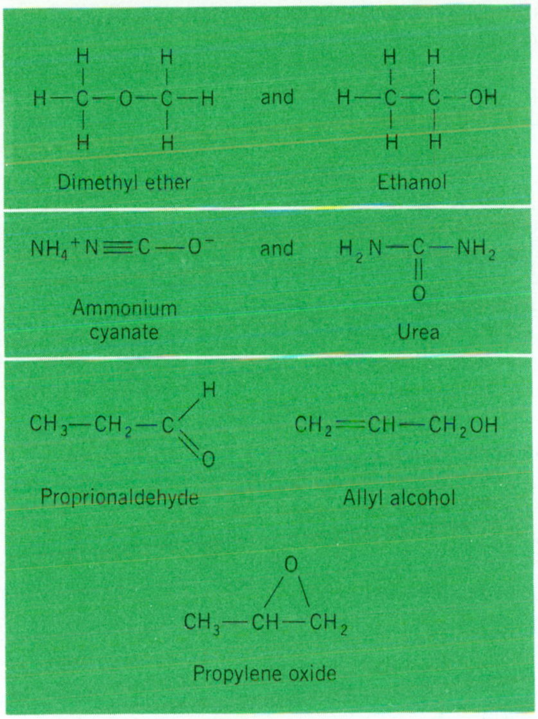

Fig. 1. Functional isomers.

bon is connected to three hydrogen atoms and the one oxygen atom; the two carbon atoms are thus equivalent. In ethyl alcohol (ethanol) one carbon is linked to three hydrogen atoms and the other carbon; the second carbon is linked to the first carbon, two hydrogens, and the oxygen atom which, in turn, is linked to the sixth hydrogen atom; the two carbon atoms are not equivalent. Stereoisomers, in contrast, have the same constitution but differ in the three-dimensional array of the atoms in space, called configuration.

Constitutional isomers. Constitutional isomers have been subdivided into functional isomers, positional isomers, and chain isomers.

Functional isomers (Fig. 1) differ in functional group, that is, the group (or groups) most material in determining chemical behavior. In the third example shown in Fig. 1 (propionaldehyde) the three compounds all correspond to the molecular formula C_3H_6O, but the first one has an aldehyde function, the second combines a double bond with an alcohol function, and the third one has an epoxide function.

Positional isomers (Fig. 2) have the same functional group but differ in its position along a chain or in a ring. Closely related are chain isomers which also have the same functional group or groups but differ in the shape of the carbon chain (Fig. 3*a*); quite similar are ring isomers (Fig. 3*b*) which differ in the size of one or more rings. Ring and chain isomers together are sometimes called skeletal isomers.

Stereoisomers. Compounds which have not only the same molecular formula but also the same constitution (connectivity of atoms) but which differ in the disposition of the atoms in space are called stereoisomers. Stereoisomers, in turn, are subdivided into two types: those that are mirror images of each other, called enantiomers, and those which are not mirror images, called diastereomers or diastereoisomers.

Enantiomers are unique in that they always come in pairs. Either a molecule is superposable with its mirror image, in which case it does not have an enantiomer, or it is not superposable with its mirror image, in which case it has one and only one enantiomer (since an object can have only one mir-

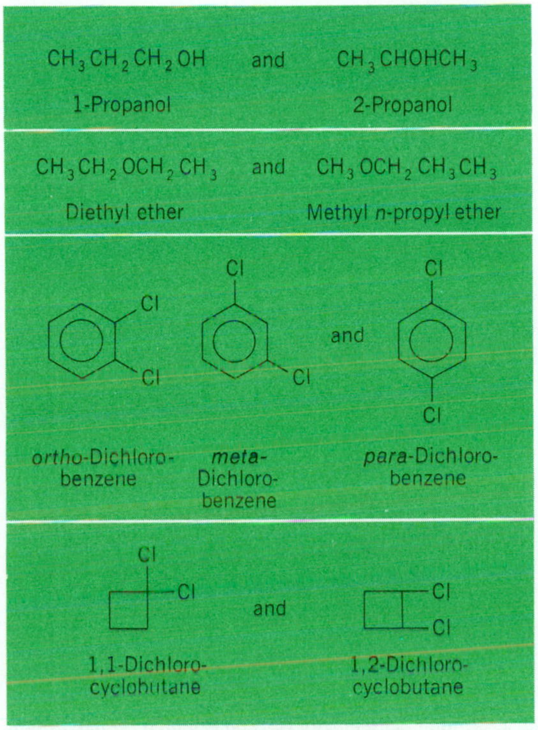

Fig. 2. Positional isomers.

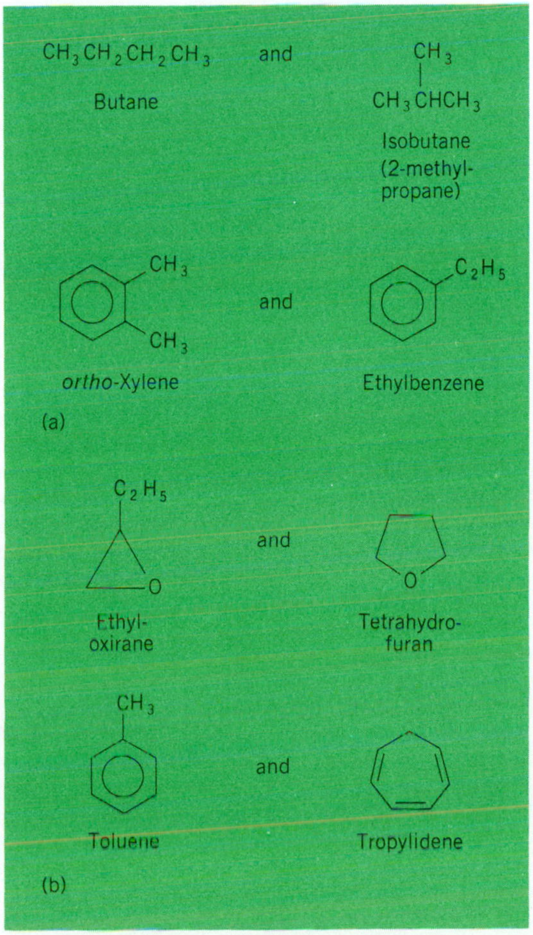

Fig. 3. Skeletal isomers. (*a*) Chain isomers. (*b*) Ring isomers.

ror image). Molecules which are not superposable with their mirror images are called chiral; those which are so superposable are called achiral. Enantiomers are much more alike than are other sets of isomers (constitutional isomers or diastereomers); thus they have the same melting point, boiling point, free energy, spectral properties, x-ray diffraction pattern, and so on.

Diastereomers have the same constitution but different spatial arrangement and are not mirror images. They resemble constitutional isomers in that there may be more than two isomers in a set and that their physical, energetic, and spectral properties are generally quite distinct. *See* CONFORMATIONAL ANALYSIS; OPTICAL ACTIVITY; STEREOCHEMISTRY; TAUTOMERISM. [E.L.E.]

Molecular mechanics A non-quantum-mechanical way of computing structures, energies, and some properties of molecules. While quantum mechanics treats electrons explicitly, molecular mechanics treats electrons implicitly. For this reason, molecular mechanics calculations are much faster than quantum calculations, and this method is heavily used by scientists who need to quickly determine shapes of molecules. The goal of molecular mechanics is to build a computer model of reality. This is done with potential energy functions. Parameters are then selected for those potential functions so that molecules whose structures, energies, and properties are known can be reproduced with a specified degree of precision. Once that has been accomplished, the molecular mechanics model may be used to compute structures, energies, and properties of unknown molecules. *See* ENERGY.

Molecular mechanics considers molecules as a collection of atomic masses held together by "sticky" forces. These forces represent the electrons holding the atoms together to form molecules. A simple conception is to consider the atoms as balls connected by springs. In this model, as in all real molecules, the atoms migrate toward their lowest-energy, most stable positions. Anything that moves the atoms from their equilibrium positions increases the internal energy of the molecule. This energy, like that of a mechanical spring model, is the potential energy. If one of the bonds were elongated (or compressed) by pulling (or pushing) it from its resting length, the potential energy would increase, and a restoring force would return it to equilibrium position. The magnitude of this restoring force depends on the strength of the spring connecting the masses. This, in turn, is proportional to the strength of the bond connecting the atoms.

Molecular mechanics has several limitations. Most important, it is necessary to have access to a dataset of known, related molecules to use for parametrization. Thus the discovery of completely new molecular species with this computational method is not possible. Also, molecular mechanics cannot account for bond making or bond breaking (the essence of chemical reactions) because of the way it implicitly treats electrons. Finally, the motionless structure computed is not very realistic, so many scientists implement molecular dynamics to achieve better model molecules. *See* CHEMICAL BONDING; COMPUTATIONAL CHEMISTRY. [K.B.L.]

Molecular orbital theory A quantum-mechanical (mathematical) model for the electronic structure and chemical bonding of a molecule. Each electron is associated with its particular wave function ψ, which spans the entire molecule. Each molecular orbital, as these wave functions are called, is definable by certain quantum numbers which govern its shape and energy. In addition to the spatial quantum numbers, each electron possesses a spin quantum number which has a value of either $\pm\frac{1}{2}$. *See* QUANTUM NUMBERS.

It is reasonable that the form of the molecular orbital in the vicinity of a particular atom should be similar to that of an atomic orbital located on that atom, since the forces on the electron will be chiefly those due to the nucleus of the atom and other electrons near that nucleus. Hence, a frequently used approximation is to express the molecular orbital as a linear combination of atomic orbitals which include all the atomic centers. Thus, Eq. (1) can be written, where the ϕ are

$$\psi = \sum_i C_i \phi_i \tag{1}$$

normalized atomic orbitals located on the various atoms and the C_i are coefficients which determine the contribution of the atomic orbitals to the molecular orbital. The set of atomic orbitals ϕ_i used to construct the molecular orbital is called the basis set for the molecular orbital.

The energy of the system is given by Eq. (2), where $\mathcal{H}$ is the

$$E = \int \psi^* \mathcal{H} \psi \, d\tau \Big/ \int \psi^* \psi \, d\tau \tag{2}$$

hamiltonian operator which takes into account all the kinetic and potential energy contributions, and $d\tau$ is the volume element for the electron. Since the energy will depend upon the coefficients C_i, application of the variation principle will specify those coefficients for which the energy is a minimum. This approach requires the solution of the secular determinant, Eq. (3), where S_{ij} is the overlap integral, $\int \phi_i^* \phi_j \, d\tau$, and H_{ij} =

$$|H_{ij} - ES_{ij}| = 0 \tag{3}$$

$\int \phi_i^* \mathcal{H} \phi_j \, d\tau$. Evaluation of the secular determinant results in n values of the energy E, where n is the number of ϕ functions used to construct the molecular orbital. Each energy value will generate its own set of coefficients C_i, so there will be a set of n molecular orbitals, ψ_n, one associated with each energy E_n. Therefore, the number of molecular orbitals will be identical with the number of atomic orbitals used. *See* MATRIX MECHANICS; QUANTUM CHEMISTRY.

Applied to the simple system H_2^+, in which only 1s orbitals on each of the hydrogen atoms A and B are used to construct the molecular orbitals, the theory results in two molecular orbitals, Eqs. (4) and (5).

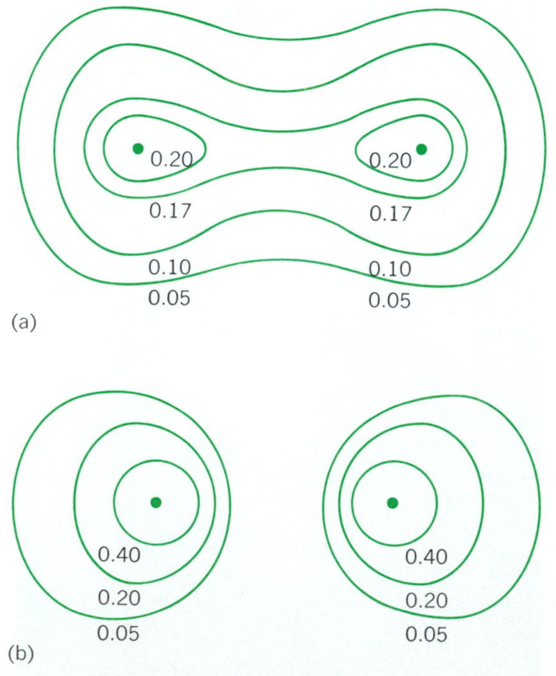

Fig. 1. Contour diagrams for H_2^+. (*a*) Bonding orbital. (*b*) Antibonding orbital.

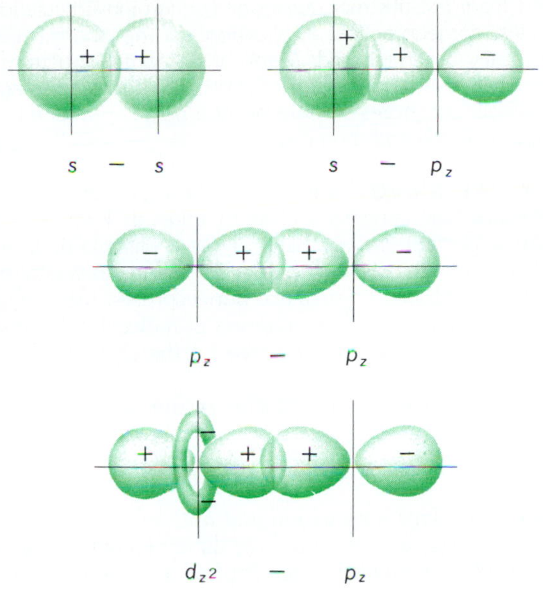

Fig. 2. Atomic orbitals capable of σ-bonding.

$$\psi(b) = \frac{1}{\sqrt{2}}\,\phi_A(1s) + \frac{1}{\sqrt{2}}\,\phi_B(1s) \qquad (4)$$

$$\psi(a) = \frac{1}{\sqrt{2}}\,\phi_A(1s) - \frac{1}{\sqrt{2}}\,\phi_B(1s) \qquad (5)$$

Contour diagrams of the electron densities $\psi^*\psi$ for the two functions are given in Fig. 1. The numerical values on the diagrams are for $\psi^*\psi$ = constant evaluated in atomic units. In the case of the bonding orbital ψ_b, the electron density is concentrated between the two nuclei while the antibonding orbital ψ_a possesses a nodal plane in this region and the charge is pushed away from the nuclei. Note that both orbitals have axial symmetry along the line connecting the two nuclei. Such orbitals are called sigma (σ) orbitals and can result from combinations of atomic orbitals whose charge densities prior to bonding lie along the line. Some typical examples of such orbitals are shown in Fig. 2.

Molecular orbitals, bonding and antibonding, can also be constructed by linear combinations of atomic orbitals such that charge densities occur above and below the line connecting the two nuclei, but a nodal plane exists along the line. For example, combinations of p_x orbitals on adjacent atoms can result in bonding and antibonding orbitals as illustrated in Fig. 3. The resultant orbitals are called pi (π) orbitals. The p_y orbitals on each of the atoms in a diatomic molecule are capable of analogous combinations. In addition, π-bonding can occur between *p* and *d* or between *d* and *d* orbitals. Generally, π-interactions are not as strong as σ-interactions, so that the bonding orbital is not lowered nor the antibonding orbital raised in energy as much in π-bonding compared to σ-bonding. *See* CHEMICAL BONDING; MOLECULAR STRUCTURE AND SPECTRA. [R.F.F.]

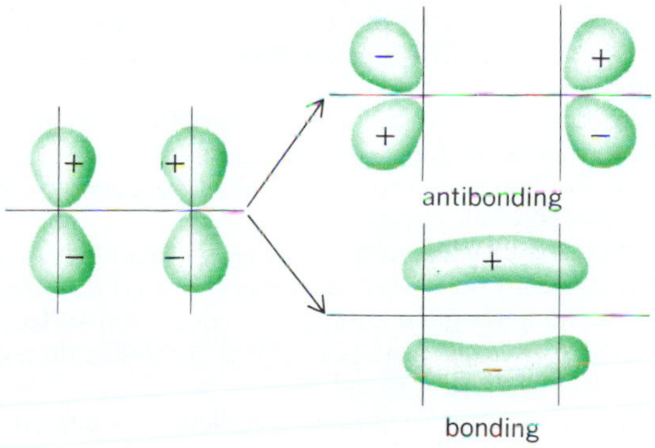

Fig. 3. π-Bonding and antibonding *p* orbitals.

Molecular pathology Pathology is a branch of biology that deals with the nature of disease. Molecular pathology is a segment of the more general discipline and deals specifically with the study of the bases and mechanisms of disease on a molecular or chemical level. The field includes the study of protein alterations, metabolic disorders, and chromosome aberrations. *See* CHROMOSOME; METABOLIC DISORDERS; PROTEIN.

The major problems that face molecular pathology involve the elucidation of the mechanism of cellular injury, including cell death, cellular regeneration, malignant transformation of cells to cancerous tissue, multicellular interactions (including the immune response), autoimmune disease and inflammation, renal disease, and birth defects. Molecular pathology also must unwind the mysteries of aging, degenerative disease, and central nervous system dysfunction, including mental illness. *See* AGING; AUTOIMMUNITY; CONGENITAL ANOMALIES; DEATH; ONCOLOGY; PATHOLOGY. [E.A.S.]

Molecular physics The study of the physical properties of molecules. Molecules possess a far richer variety of physical and chemical properties than do isolated atoms. This is attributable primarily to the greater complexity of molecular structure, as compared to that of the constituent atoms. Molecules also possess additional energy modes because they can vibrate; that is, the constituent nuclei oscillate about their equilibrium positions and rotate when unhindered. These modes give rise to additional spectroscopic properties, as compared to those of an atom; molecular spectroscopy in the optical, infrared, and microwave regions is one of the physical chemist's most powerful means of identifying and understanding molecular structure. Molecular spectroscopy has also given rise to the rapidly growing field of molecular astronomy.

Molecular physics is primarily concerned with the study of properties of isolated molecules, as contrasted to the more general study of molecular reactions, which is the domain of physical chemistry. Such properties, in addition to the broad field of spectroscopy, include electron affinities (for the formation of molecular negative ions); polarizabilities (the "distortability" of the molecule along its various symmetry axes by external electric fields); magnetic and electric multipole moments, attributable to the distributions of electric charge; currents and spins of the molecule; and the (nonreactive) interactions of molecules with other molecules, atoms, and ions. *See* COSMOCHEMISTRY; INFRARED SPECTROSCOPY; INTERMOLECULAR FORCES; MICROWAVE SPECTROSCOPY; MOLECULAR BEAMS; MOLECULAR STRUCTURE AND SPECTRA; SPECTROSCOPY. [B.B.]

Molecular recognition The ability of biological and chemical systems to distinguish between molecules and regulate behavior accordingly. How molecules fit together is fundamental in disciplines such as biochemistry, medicinal chemistry, materials science, and separation science. A good deal of effort has been expended in trying to evaluate the underlying intermolecular forces. The weak forces that act over short distances (hydrogen bonds, van der Waals interactions, and aryl stacking) provide most of the selectivity observed in biological chemistry and permit molecular recognition. The recognition event initiates behavior such as replication in nucleic acids, immune response in antibodies, signal transduction in receptors, and regulation in enzymes. Most studies of recognition in organic chemistry have been inspired by these biological phenomena. It has been the task of bioorganic chemistry to develop systems

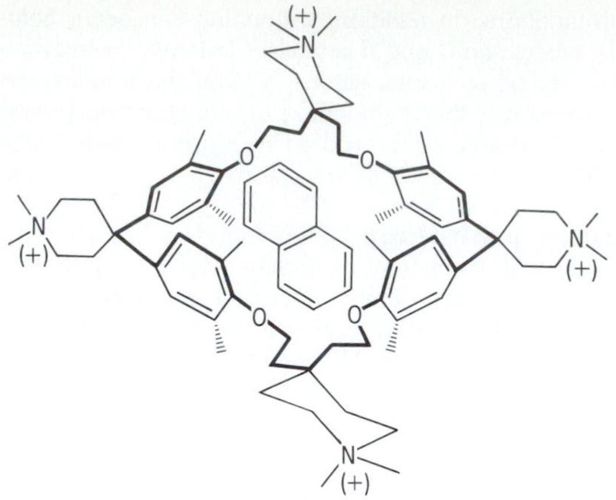

(I)

capable of such complex behavior with molecules that are comprehensible and manageable in size, that is, with model systems. *See* ANTIBODY; CHEMORECEPTION; ENZYME; HYDROGEN BOND; INTERMOLECULAR FORCES; NUCLEIC ACID; SYNAPTIC TRANSMISSION.

The advantage of cyclic structures lies in their ability to restrict conformation or flexibility. A rigid matrix of binding sites, that is, preorganized sites, is usually associated with high selectivity in binding. A flexible matrix tends to accept several binding partners. Although sacrificing selectivity, this has the advantage of transmitting conformational information and is relevant to biological signaling events. *See* CONFORMATIONAL ANALYSIS.

Macrocyclic (crown) ethers can bind and transport ions and imitate biological processes involving macrolides. Large ring structures that are lined with oxygen present an inner surface which is complementary to the spherical outer surface of positively charged ions. *See* CROWN ETHERS.

Cyclophane-type structures offer considerable rigidity because of the aromatic nuclei. Binding forces between host and guest are largely hydrophobic. A typical system is a cyclophane-naphthalene complex (I), in which a naphthalene guest is bound by a water-soluble cyclophane derivative. Other macrocyclic structures include the cyclodextrins and hybrid structures assembled from macrocyclic subunits. *See* AROMATIC HYDROCARBON; COORDINATION COMPLEXES; FERROCENE; NAPHTHALENE.

Because the encircling of larger, more complex molecules with macrocycles poses structural problems, other molecular shapes have been explored. Cleft molecules offer advantages in this regard. The principle underlying these systems involves the shape of the small organic target molecules: convex in surface and bearing functional groups that diverge from their centers. Accordingly, designing a trap for such targets requires molecules of a concave surface in which functional groups converge. This complementarity is also a feature of the immune system: the "hot spots" of an antigen tend to be convex, whereas the binding sites of the antibody are concave.

Systems featuring a cleft have been developed to bind adenine derivatives and other heterocyclic systems through chelation, as shown in (II). *See* CHELATION.

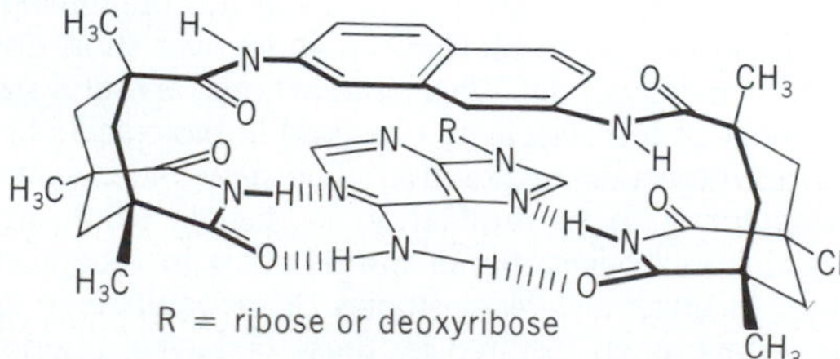

(II)

Apart from the abstract questions concerning articulation of molecules, some practical applications in the pharmaceutical industry may be envisioned. Many of the target structures are biologically active, and the use of synthetic sequestering agents for metabolic substrates can represent a novel approach to biochemical methods and drug delivery. [J.Reb.]

Molecular sieve Any one of the crystalline metal aluminosilicates belonging to a class of minerals known as zeolites. An important characteristic of the zeolites is their ability to undergo dehydration with little or no change in crystal structure. The dehydrated crystals are honeycombed with regularly spaced cavities interlaced by channels of molecular dimensions which offer a very high surface area for the absorption of foreign molecules.

The basic formula for all crystalline zeolites can be represented as

$$M_{2/n}O{:}Al_2O_3{:}xSiO_2{:}yH_2O$$

where M represents a metal ion and *n* its valence. The crystal structure consists basically of a three-dimensional framework of SiO_4 and AlO_4 tetrahedrons (see illustration). The tetrahedrons

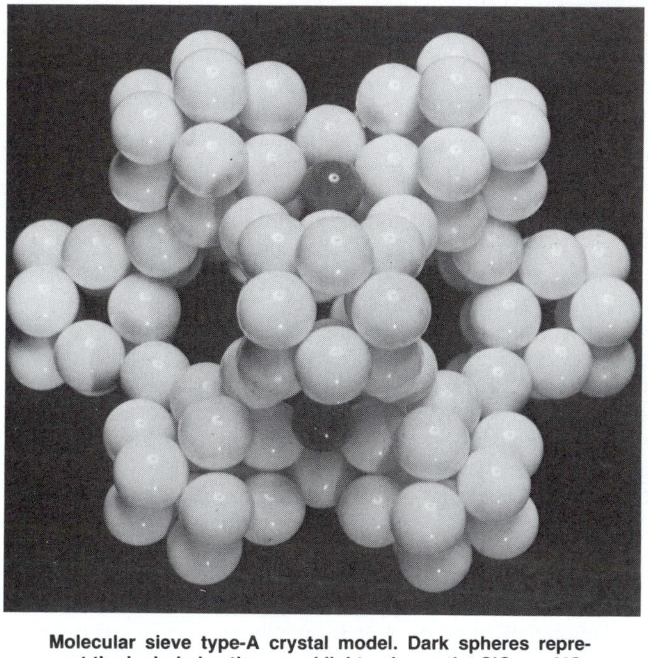

Molecular sieve type-A crystal model. Dark spheres represent the included cations, and light spheres the SiO_4 or AlO_4 tetrahedrons.

are cross-linked by the sharing of oxygen atoms, so that the ratio of oxygen atoms to the total of silicon and aluminum atoms is equal to 2. The electrovalence of the tetrahedrons containing aluminum is balanced by the inclusion of cations in the crystal. One cation may be exchanged for another by the usual ion-exchange techniques. The size of the cation and its position in the lattice determine the effective diameter of the pore in a given crystal species.

The properties of molecular sieves as adsorbents which distinguish them from nonzeolitic adsorbents are (1) the relatively strong coulomb fields generated by the adsorption surface and (2) the uniform pore size; the pore size is controlled, in a given crystal species, by the associated cation.

The basic characteristics of molecular sieves are utilized commercially in several production and research applications. Their absorption properties make them useful for drying, purification, and separations of gases and liquids. Conversely, molecular sieves can be preloaded with chemical agents, which are thereby isolated from the reactive system in which they are dispersed until released from the adsorbent either thermally or by displacement by a more strongly adsorbed compound. They are also used as cation exchange media and as novel catalysts and catalyst supports. *See* ADSORPTION; GAS CHROMATOGRAPHY; ION EXCHANGE; ZEOLITE. [R.L.Ma.]

Molecular structure and spectra Until the advent of quantum theory, ideas about the structure of molecules evolved gradually from analysis and interpretation of the facts of chemistry. Chemists developed the concept of molecules as built from atoms in definite proportions, and identified and constructed (synthesized) a great variety of molecules. Later, when the structure of atoms as built from nuclei and electrons began to be understood with the help of quantum theory, a beginning was made in seeing why atoms can combine in definite ways to form molecules; also, infrared spectra began to be used to obtain information about the dimensions and the nuclear motions (vibrations) in molecules. However, a fundamental understanding of chemical bonding and molecular structure became possible only by application of the present form of quantum theory, called quantum mechanics. This theory makes it possible to obtain from the spectra of molecules a great deal of information about the nature of molecules in their normal as well as excited states, and about dissociation energies and other characteristics of molecules. *See* CHEMICAL BONDING.

Molecular sizes. The size of a molecule varies approximately in proportion to the numbers and sizes of the atoms in the molecule. Simplest are diatomic molecules. These may be thought of as built of two spherical atoms of radii r and r', flattened where they are joined. The equilibrium value R_e of the distance R between their nuclei is then smaller than the sum of the atomic radii. However, the nuclei of atoms in two different molecules cannot normally approach more closely than a distance $r + r'$; r and r' are called the van der Waals radii of the atoms.

To describe a polyatomic molecule, one must specify not merely its size but also its shape or configuration. For example, carbon dioxide (CO_2) is a linear symmetrical molecule, the O—C—O angle being 180°. The H—O—H angle in the nonlinear water (H_2O) molecule is 105°. Many molecules which are essential for life contain thousands or even millions of atoms. Proteins are often coiled or twisted and cross-linked in ways which are important for their biological functioning.

Dipole moments. Most molecules have an electric dipole moment. In atoms, the electron cloud surrounds the nucleus so symmetrically that its electrical center coincides with the nucleus, giving zero dipole moment; in a molecule, however, these coincidences are disturbed, and a dipole moment usually results.

Thus, when the atoms of HCl come together, there is some shifting of the H-atom electron toward the Cl. A complete shift would give H^+Cl^-, which would constitute an electric dipole of magnitude eR_e, where e is the electronic charge. But in fact the dipole moment is only 0. 17 eR_e. This is because the actual electronic shift is only fractional. For further discussion *see* ELECTRONEGATIVITY.

Molecular polarizability. In the preceding consideration of dipole moments, the discussion has been in terms of atoms and molecules free from external forces. An electric field pulls the electrons of an atom or molecule toward it and pushes the nuclei away, or vice versa. This action creates a small induced dipole moment, whose magnitude per unit strength of the field is called the polarizability.

Molecular energy levels. The states of motion of nuclei and electrons in a molecule, or of electrons in an atom, are restricted by quantum mechanics to special forms with definite energies. The state of lowest energy is called the ground state; all others are excited states. In analogy to water levels, one speaks of energy levels. Excited states exist only momentarily, following an electrical or other stimulus. *See* QUANTUM CHEMISTRY; QUANTUM MECHANICS.

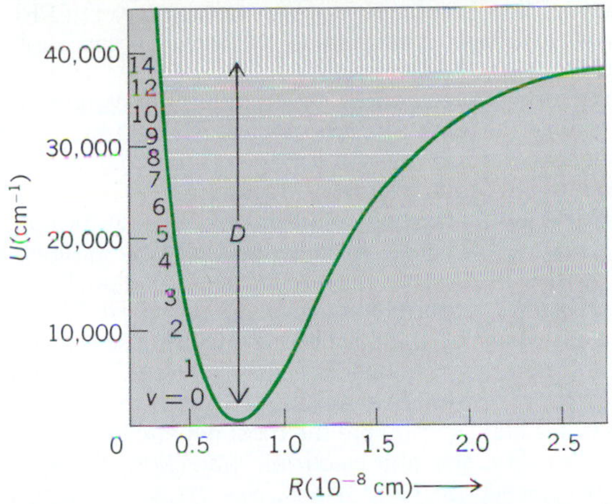

$U(R)$ curve of ground electronic state of H_2 with vibrational levels and dissociation continuum. D indicates the dissociation energy. Maximum v here is 14. (After G. Herzberg, *Molecular Spectra and Molecular Structure*, vol. 1, 2d ed., Van Nostrand, 1950)

Excitation of an atom consists of a change in the state of motion of its electrons. Electronic excitation of molecules can also occur, but alternatively or additionally, molecules can be excited to discrete states of vibration and rotation.

In a diatomic vibration, R varies periodically above and below R_e. The possible vibration energies are related to the potential curve, which shows how the energy of attraction $U(R)$ of the atoms varies with R. The $U(R)$ curve and vibrational levels for the ground electronic state of H_2 are shown in the illustration. For stable (attractive) $U(R)$ curves, the vibrational levels decrease in spacing as υ increases, until finally, as the spacing approaches zero, a maximum υ is reached; in the illustration, this is 14. After a small gap, a dissociation continuum of energy levels then sets in. Here the atoms have enough mutual kinetic energy to fly apart.

The total energy of any molecule can be written as Eq. (1).

$$E = E_{el} + E_{\upsilon} + (E_r + E_{fs} + E_{hfs} + E_{ext}) \qquad (1)$$

Both the electronic energy E_{el} and vibration energy E_{υ} can be discrete or continuous. The quantities E_r, E_{fs}, and E_{hfs} denote rotational, fine-structure, and hyperfine-structure energies. The last two appear as small or minute splittings of the rotation levels. The spacings ΔE of adjacent discrete levels of each type are usually in the order given in notation (2). The E_{ext} term in

$$\Delta E_{el} \gg \Delta E_{\upsilon} \gg \Delta E_r \gg \Delta E_{fs} \gg \Delta E_{hfs} \qquad (2)$$

Eq. (1) refers to additional fine structure which appears on subjecting molecules to external magnetic fields (Zeeman effect) or electric fields (Stark effect). *See* FINE STRUCTURE (SPECTRAL LINES); HYPERFINE STRUCTURE; STARK EFFECT; ZEEMAN EFFECT.

Polyatomic molecules have much more complicated patterns of vibrational and (usually) rotational energy levels than diatomic molecules.

Molecular spectra. The frequencies $c\nu$ (c = speed of light) of electromagnetic spectra obey the Einstein-Bohr equation, Eq. (3),

$$hc\nu = E' - E'' \qquad (3)$$

where h is Planck's constant. Molecular emission spectra accompany jumps in energy from higher to lower levels; absorption spectra accompany jumps from lower to higher levels.

Molecular spectra can be classified as fine-structure or low-frequency spectra, rotation spectra, vibration-rotation spectra, and electronic spectra. Low-frequency spectra are discussed elsewhere. *See* Electron paramagnetic resonance (EPR) spectroscopy; Magnetic resonance; Microwave spectroscopy; Molecular beams; Spectroscopy.

Transitions between energy levels differing only in rotational state give rise to pure rotation spectra. These typically consist of a sequence of lines spaced almost equidistantly, and lying in the far infrared or the microwave region.

Spectra involving only vibrational and rotational state changes consist of bands which lie mainly in the infrared. Each band consists of two sets of closely spaced rotational lines, one on each side of a central frequency. Vibration-rotation absorption bands of liquids and solutions are widely used in chemical analysis. Here the rotational structure is blurred out, and only an "envelope" is seen. *See* Infrared spectroscopy.

Electronic band spectra are the most general type of molecular spectra. For any one electronic transition, the spectrum consists typically of many bands. *See* Atomic structure and spectra; Intermolecular forces; Molecular weight; Raman effect; Resonance (molecular structure); Scattering experiments (atoms and molecules); Valence. [R.S.M.]

Molecular weight The sum of the atomic weights of all the atoms in a molecule. Atomic weights (and therefore molecular weights) are relative weights arbitrarily referred to an assigned atomic weight of 12.0000 exactly for the most abundant isotope of carbon, ^{12}C. *See* Atomic weight.

The amount of a substance which has a weight in grams equal to the molecular weight of the substance is called a gram-mole or a gram-molecular weight. The number of molecules in a gram-mole is exceedingly large (6.02×10^{23}) and is called the Avogadro number. *See* Avogadro number.

The ultimate analysis of a substance provides precise knowledge of formula weight, namely, the sum of the atomic weights of the atoms in the simplest formula for the substance. The molecular weight is an exact multiple of the formula weight. For example, analysis shows that the simplest formula of benzene is CH and that the formula weight, the sum of the atomic weights of carbon and hydrogen, is 13.02. Each benzene molecule is a hexagonal structure of true formula C_6H_6 and molecular weight 78.11. Some substances, such as sodium chloride, contain no discrete molecules and are best assigned formula weights but no molecular weights. Known molecules range in molecular weight from two (hydrogen) to several billion.

Molecular weights are determined by a variety of methods. Historically, molecular-weight determinations were essential in establishing the stoichiometry which forms the basis for the fundamental laws of chemistry; today, they are made to interpret chemical reactions, to aid in determining molecular structure and molecular shapes, to give a better understanding of solvent-solute interaction, and to provide data for the design and control of a great variety of useful industrial processes. [J.R.A.]

Molecule A molecule may be thought of either as a structure built of atoms bound together by chemical forces or as a structure in which two or more nuclei are maintained in some definite geometrical configuration by attractive forces from a surrounding swarm of negative electrons. Besides chemically stable molecules, short-lived molecular fragments called free radicals can be observed under special circumstances. *See* Chemical bonding; Free radical; Molecular structure and spectra. [R.S.M.]

Mollicutes The class of bacteria comprising the mycoplasmas, with the single order Mycoplasmatales, and two families, Mycoplasmataceae and Acholeplasmataceae.

The organisms are gram-negative; though motility has been described for a few species, they are usually nonmotile; they are nonsporing; serum is required for growth.

Several *Mycoplasma* species are important pathogens of humans and animals, while others are common saprophytes of the mucous membranes of the lower genital tract and the oropharynx. *Mycoplasma pneumoniae*, previously regarded as a virus, and known under the name of the Eaton agent, causes cold agglutin-associated primary atypical pneumonia. *Mycoplasma hominis*, which is a common inhabitant of the lower genital tract, appears to be a potential pathogen. *Mycoplasma salivarium* and *M. orale* are common members of the normal oral microflora. The role of *M. fermentans*, found rarely in the human genital tract, is obscure. The same holds true for the T-mycoplasmas (T for tiny) that are very common inhabitants of the genital tracts of males and females, and of which at least seven different serotypes are known. Two varieties of the type species *M. mycoides* are the etiological agents of contagious pleuro-pneumonia of cattle and goats.

Considerable interest has been shown in mycoplasmas as possible pathogens of plants. In plant diseases such as yellows-type disease, corn stunt, and alfalfa witches, mycoplasma-like structures have been demonstrated in great numbers by electron microscopy of thin sections of both diseased plant tissues and insect vectors. [E.A.F.]

Mollusca A major phylum of the animal kingdom comprising an extreme diversity of external body forms (oysters, clams, chitons, snails, slugs, squid, and octopuses among others), all based on a remarkably uniform basic plan of structure and function. The phylum name is derived from *mollis*, meaning soft, referring to the soft body within a hard calcareous shell, which is usually diagnostic. Soft-bodied mollusks make extensive use of ciliary and mucous mechanisms in feeding, locomotion, and reproduction.

The Mollusca constitute a successful phylum; there are probably over 110,000 living species of mollusks, a number second only to that of the phylum Arthropoda, and more than double the number of vertebrate species. More than 99% of living molluscan species belong to two classes: Gastropoda (snails) and Bivalvia. Ecologically, these two classes can make up a dominant fraction of the animal biomass in many natural communities, both marine and fresh-water. Certain bivalve species are the most abundant marine benthic animals, and computations have suggested that one of these may encompass a larger "standing crop" biomass of animal tissue than any other single animal species on the planet.

The phylum Mollusca is divided into seven distinct extant classes, three of which (Gastropoda, Bivalvia, and Cephalopoda) are of major significance in terms both of species numbers and of ecological bioenergetics, and one extinct class. An outline of their classification follows.

Class Monoplacophora (mainly fossil;
  but one living genus, *Neopilina*)
Class Aplacophora
Class Polyplacophora
Class Scaphopoda
Class Rostroconchia (fossil only)
Class Gastropoda
  Subclass: Prosobranchia
    Opisthobranchia
    Pulmonata
Class Bivalvia (or Pelecypoda)
  Subclass: Protobranchia
    Lamellibranchia
    Septibranchia
Class Cephalopoda (or Siphonopoda)
  Subclass: Nautiloidea (Tetrabranchia)
    Ammonoidea (fossil only)
    Coleoidea (Dibranchia)

Functional morphology. The unique basic plan of the Mollusca involves the different modes of growth and of functioning of the three distinct regions of the molluscan body (see illustration). These are: the head-foot with some nerve concentrations, most of the sense organs, and all the locomotory organs; the visceral mass (or hump) containing organs of digestion, reproduction, and excretion; and the mantle (or pallium) hanging from the visceral mass and enfolding it and secreting the shell. In its development and growth, the head-foot shows a bilateral symmetry with an anterioposterior axis of growth. Over and around the visceral mass, however, the mantle-shell shows a biradial symmetry, and always grows by marginal increment around a dorsoventral axis. It is of considerable functional importance that a space is left between the mantle-shell and the visceral mass forming a semi-internal cavity; this is the mantle cavity or pallial chamber within which the typical gills of the mollusk, the ctenidia, develop. This mantle cavity is almost diagnostic of the phylum; it is primarily a respiratory chamber housing the ctenidia, but with alimentary, excretory, and genital systems all discharging into it.

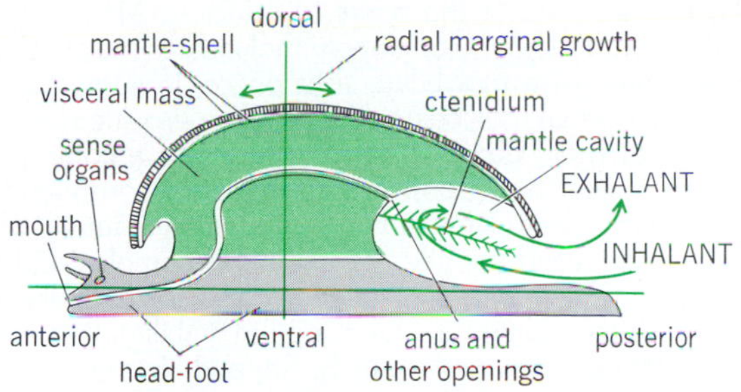

Generalized model of a stem mollusk (or archetype) in side view. Water circulation through the mantle cavity, gills (ctenidia), and pallial complex is from ventral inhalant to dorsal exhalant. (*After W. D. Russell-Hunter, A Life of Invertebrates. Macmillan, 1979*)

The shell is always underlain by the mantle, a fleshy fold of tissues which has secreted it. The detailed structure of the shell and of the mantle edge (with three functionally distinct lobes) is also consistent throughout the Mollusca. The shell is made up of calcium carbonate crystals enclosed in a meshwork of tanned proteins. It is always in three layers.

Each of the eight classes of the Mollusca has a characteristic body form and shell shape. Two classes are enormous (Gastropoda and Bivalvia), one of moderate extent (Cephalopoda), the others being minor by comparison. The Gastropoda constitute a diverse group with the shell usually in one piece. This shell may be coiled as in typical snails—that is, helicoid or turbinate—or it may form a flattened spiral, or a short cone as in the limpets, or it may be secondarily absent as in the slugs. Most gastropods are marine, but many are found in fresh waters and on land; in fact, they are the only successful nonmarine mollusks.

The cardiac structures of mollusks are also closely linked to the pallial complex. If there is a symmetrical pair of ctenidia, there will be a symmetrical pair of auricles on either side of the muscular ventricle of the heart; if one ctenidium, one auricle; if four ctenidia, four auricles. Body fluids in mollusks are almost all blood, and body cavities are almost all hemocoel. The respiratory pigment is usually hemocyanin in solution, so that neither circulatory efficiency nor blood oxygen-carrying capacity is high.

Uniquely molluscan is the use of cilia in "sorting surfaces," which can segregate particles into different size categories and send them to be disposed of in different ways in several parts of the organism. In a simpler type of sorting surface, the epithelium is thrown into a series of ridges and grooves, the cilia in the grooves beating along them and the cilia on the crests of the ridges beating across them. Thus, fine particles impinging on the surface can be carried in the direction of the grooves, while larger particles are carried at right angles. Such sorting surfaces occur both externally on the feeding organs and internally in the gut of many mollusks. For example, on the labial palps of bivalves, they are used to separate the larger sand grains (which are rejected) from the smaller microorganisms which then pass to the mouth.

The range in levels of complexity of molluscan nervous systems is comparable to that found in the phylum Chordata. The four-strand nervous system with one pair of tiny ganglia found in chitons is not dissimilar to the neural plan in turbellarian flatworms. In contrast, the nervous system and sense organs of a cephalopod like an octopus are equaled and exceeded only by those of some birds and mammals.

In all primitive mollusks, the sexes are separate, and external fertilization follows the spawning of eggs and sperm into the sea. In more advanced mollusks, eggs are larger (and fewer), fertilization may become internal (with complex courtship and copulatory procedures), and larval stages may be sequentially suppressed. A remarkably large number of mollusks (including many higher snails) are hermaphroditic. Although some are truly simultaneous hermaphrodites, many more show various kinds of consecutive sexuality. Most often the male phase occurs first, and these species are said to show protandric hermaphroditism.

Distributional ecology. Mollusks are largely marine. The extensive use of ciliary and mucous mechanisms in feeding, locomotion, reproduction, and other functions demands a marine environment for the majority of molluscan stocks. Apart from a small number of bivalve genera living in brackish and fresh waters, all nonmarine mollusks are gastropods.

Despite the soft, hydraulically moved bodies and relatively permeable skins typical of all mollusks, some snails are relatively successful as land animals, although they are largely limited to more humid habitats. The primary physiological requirements for life on land concern water control, conversion to air breathing, and temperature regulation.

In the sea, all classes of mollusks are found, and all habitats have mollusks. Protobranchiate bivalves are found at depths of over 30,000 ft (9000 m). Although ecologically cephalopod mollusks are limited to the sea, there are sound reasons for claiming modem cephalopods as the most highly organized invertebrate animals. The functional efficiencies of jet propulsion and of massive brains in squid, cuttlefish, and octopuses have not been paralleled in their other physiological systems.

In addition to the extreme diversity of external body form exhibited by different mollusks, they show a remarkable diversity in their ecological distribution and life styles. However, the basic molluscan plan of structure and function always remains recognizable. *See* APLACOPHORA; BIVALVIA; CEPHALOPODA; GASTROPODA; LAMELLIBRANCHIA; MONOPLACOPHORA; POLYPLACOPHORA; SCAPHOPODA; SNAIL. [W.D.R.-H.]

Molluscum contagiosum A skin disease caused by a virus and characterized by the appearance of small discrete lesions in groups on the face, arms, or genitalia. The lesions are firm and pearly white with a sharply indented central core, and yield an infectious filtrate which produces the disease when inoculated into human volunteers. The disease, which may be epidemic in children, occurs in all ages and is worldwide in distribution. [A.E.Mo.]

Molybdenite A mineral having composition MoS_2. Molybdenite is the chief ore of molybdenum. It crystallizes in the hexagonal system, but crystals are rare and when found are hexagonal plates. It is commonly in scales or foliated masses. The mineral has a greasy feel. The hardness is 1.5 (Mohs scale) and the specific gravity is 4.7. The luster is metallic and the color lead gray. Molybdenite and graphite have long been confused because of their nearly identical physical properties. They can be distinguished by the streak left on glazed paper, black for graphite and green for molybdenite. Molybdenite has been used as a lubricant.

Molybdenite occurs in various places in Norway, Sweden, Australia, England, China, and Mexico. In the United States, molybdenite is found in small amounts at many localities but the most important occurrence is at Climax, Colorado. *See* MOLYBDENUM. [C.S.Hu.]

Molybdenum A chemical element, Mo, atomic number 42, and atomic weight 95.95; one of the transition elements. A silver-gray metal with a density of 10.22 g/cm^3 (5.907 oz/in.^{3}), molybdenum melts at 2610°C (4730°F). The table gives selected physical properties for metallic molybdenum. *See* TRANSITION ELEMENTS.

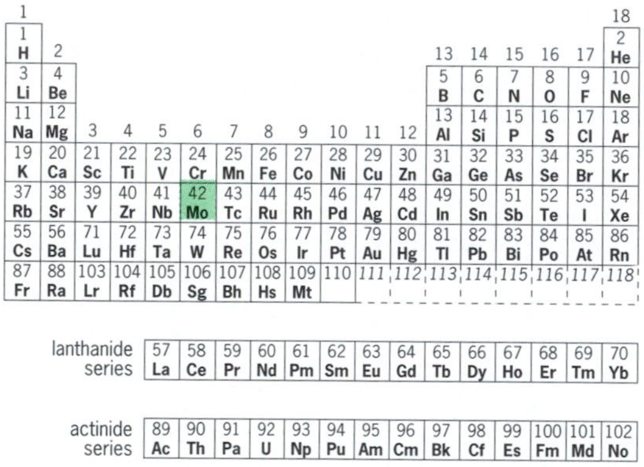

Molybdenum is found in many parts of the world, but relatively few deposits are rich enough to warrant recovery costs. Most molybdenum comes from mines where its recovery is the primary objective of the operation. The remainder is recovered as a by-product of certain copper-mining operations. *See* MOLYBDENITE.

Molybdenum forms compounds in which it displays valence states of 0, 2+, 3+, 4+, 5+, and 6+. Molybdenum as an ionizable cation has not been observed, but cationic species, such as molybdenyl, MoO_2^{2+} are known to exist. The chemistry of molybdenum is extremely complex and, with the exception of the halides and chalcogenides, very few simple compounds are known.

Selected physical properties of metallic molybdenum

Property	Value*
Melting point	2610°C
Heat of fusion	6.7 kcal/mol
Boiling point	5560°C
Heat of vaporization	117.4 kcal/mol
Heat capacity (298.16–1800 K)	5.48 + 1.30 × 10^{-3}T (in cal/deg mol)
Specific heat (20°C)	0.064 cal/g°C
Thermal conductivity	
200°C	0.298 cal/s/cm^2/cm/°C
1100°C	0.239 cal/s/cm^2/cm/°C
2200°C	0.206 cal/s/cm^2/cm/°C
Mean linear expansion coefficient	
20–150°C	5.43 × 10^{-6}/°C
20–1600°C	6.65 × 10^{-6}/°C
Electrical conductivity (0°C)	34% IACS* (58.62 microhm-cm)

*1 cal = 4.184 joules.
†International Annealed Copper Standard.

Molybdenum dioxide and trioxide are the most common and most stable oxides; other reported oxides are metastable and are essentially laboratory species.

Molybdic acid, H_2MoO_4 (or $MoO_3 \cdot H_2O$) forms a series of stable, normal salts of the types $M_2^{2+}MoO_4$, $M^{2+}MoO_4$, and $M_2^{3+}(MoO_4)_3$. Polymeric or isopoly molybdates may be formed by acidification of a molybdate solution or, in some cases, by heating the normal molybdates. Hydrogen peroxide reacts with a number of molybdates to form a series of anionic peroxy compounds. Another group of molybdenum compounds are the heteropoly electrolytes, a large, fundamental family of salts and free acids, each member containing a complex and high-molecular-weight anion. Molybdenum also forms halides and oxyhalides, representing a wide range in stability, and a series of homologous compounds with S, Se, and Te that are somewhat similar to the oxides. [H.Mo.]

Molybdenum alloys Solid solutions of molybdenum and other metals. Two alloys are considered commercially important, TZM Moly and Moly-30W.

TZM Moly alloy [0.5% titanium (Ti), 0.1% zirconium (Zr), and 99.4% molybdenum (Mo)] has been the choice for applications that demand superior hot strength and creep resistance, or resistance to recrystallization and softening in comparison to pure molybdenum. This alloy is amenable to a pair of additional metallurgical strengthening or hardening mechanisms compared to warm- or cold-work strain hardening for pure molybdenum: (1) solid-solution hardening by dissolution of titanium, zirconium, and carbon in the molybdenum matrix for microstructural stability, and (2) dispersion hardening by fine precipitation of complex spheroidal carbides to improve high-temperature strength and creep resistance.

An alloy of 70% molybdenum and 30% tungsten (W), Moly-30W has a high melting point of 5125°F (2830°C), 400°F (222°C) higher than pure molybdenum. This allows higher service temperatures without incipient melting or serious metal erosion. However, more significant uses developed because Moly-30W exhibits exceptional corrosion resistance to chemical attack by high-purity molten zinc, or even zinc vapors. *See* ALLOY; MOLYBDENUM. [R.W.Bu.]

Moment of inertia A relation between the area of a surface or the mass of a body to the position of a line. The

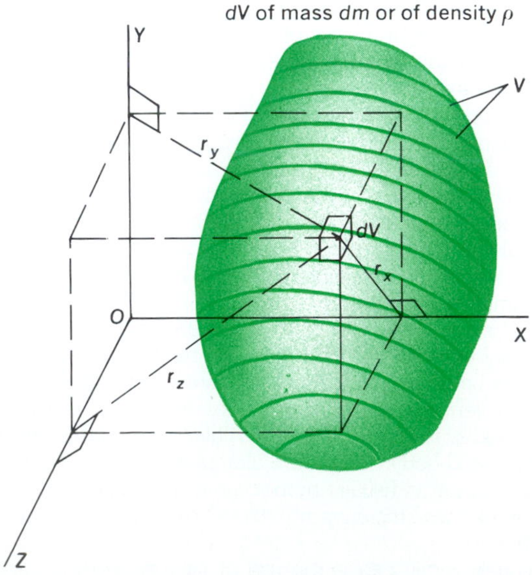

Moment of inertia of a volume.

analogous positive number quantities, moment of inertia of area and moment of inertia of mass, are involved in the analysis of problems of statics and dynamics respectively.

The moment of inertia of a figure (area or mass) about a line is the sum of the products formed by multiplying the magnitude of each element (of area or of mass) by the square of its distance from the line. The moment of inertia of a figure is the sum of moments of inertia of its parts.

For a body of mass distributed continuously within volume V, the movement of inertia of the mass about the X axis is given by either $I_X = \int r_x^2\, dm$ or $I_x = \int r_x^2 \rho\, dV$, where dm is the mass included in volume element dV at whose position the mass per unit volume is ρ (see illustration). Similarly $I_Y = \int r_y^2 \rho\, dV$ and $I_Z = \int r_z^2 \rho\, dV$.

The moments of inertia of a figure about lines which intersect at a common point are generally unequal. The moment is greatest about one line and least about another line perpendicular to the first one. A set of three orthogonal lines consisting of these two and a line perpendicular to both are the principal axes of inertia of the figure relative to that point. If the point is the figure's centroid, the axes are the central principal axes of inertia. The moments of inertia about principal axes are principal moments of inertia. *See* CENTROIDS (MATHEMATICS); PRODUCT OF INERTIA; RADIUS OF GYRATION. [N.S.F.]

Momentum Linear momentum is the product of the mass and the linear velocity of a body. It is defined by Eq. (1),

$$\mathbf{P} = m\mathbf{v} \qquad (1)$$

where m is the mass and $\mathbf{v}$ is the linear velocity. Since linear momentum is the product of a scalar and a vector quantity, it is a vector and hence has both magnitude and direction.

According to the general statement of Newton's second law, for a force $\mathbf{F}$, a momentum $\mathbf{P}$, and a time t, Eq. (2) holds.

$$\mathbf{F} = d\mathbf{P}/dt \qquad (2)$$

Thus Newton's second law involves the time rate of change of momentum. Changes of momentum are important in collision processes. *See* COLLISION (PHYSICS).

When a group of bodies is subject only to forces that members of the group exert on one another, the total momentum of the group remains constant. *See* ANGULAR MOMENTUM; CONSERVATION OF MOMENTUM; IMPULSE (MECHANICS). [P.W.S.]

Monazite A phosphate (mineral) of the cerium metals, $(Ce,La,Y,Th)(PO_4)$. Ordinarily lanthanum, La, is present in about 1:1 ratio with cerium. Small amounts of the yttrium, Y, earths substitute for Ce and La. Thorium substitutes for Ce and La and generally ranges up to 10% ThO_2. A series of monazite minerals ranging up to 30% ThO_2 probably exists. Thorium-free monazite is rare. Uranium, U, in small amounts has been reported.

Monazite crystallizes in the monoclinic system. Crystals are prismatic and generally minute but occasionally large. Colors range from white through shades of yellow, green, and brown.

Monazite occurs in many regions of the world, but the major production comes from placer deposits in Idaho, South Carolina, and Florida in the United States, and in India, Brazil, and the Union of South Africa. Recovered monazite concentrates are sources of thorium, thorium compounds, and cerium metals. *See* CERIUM; RADIOACTIVE MINERALS; RARE-EARTH ELEMENTS. [W.R.Lo.]

Mongoose The name for about 39 species of carnivorous mammals which are members of the family Viverridae. This family also includes the civets and genets. Mongooses are restricted in their distribution to the warmer regions of the Old World, ranging from the Mediterranean into Africa and Southeast Asia. These are plantigrade animals about the size of a cat and have a long slender body, short legs, nonretractile claws, and scent glands. Some species are fair climbers even though the claws are nonretractile. Mongooses are predators, feeding on snakes, frogs, fishes, and crabs, and they are especially fond of bird and crocodile eggs. The mongoose cannot legally be brought into the United States because of its destructive habits. *See* CARNIVORA; CIVET. [C.B.C.]

Monhysterida An order of nematodes in which, generally, the stoma is funnel shaped and lightly cuticularized; however, in some families the stoma is spacious and heavily cuticularized, and is armed with protrusible teeth. The amphids vary from simple spirals to circular forms. Usually the second and third circlets of cephalic sensilla are combined, but in some taxa the third circlet of four distinct setae is separate. The normal pattern of distribution is often disrupted by numerous cervical setae. The near-cyclindrical esophagus is sometimes swollen posteriorly. The cuticle may be smooth or may have annuli or ornamentation. When the annuli are distinct, the somatic setae may be long and in four to eight longitudinal rows. The female gonads are outstretched and either single or paired.

There are three monhysterid superfamilies: The Linhomoeoidea and Siphonolaimoidea are primarily marine forms; feeding habits among the groups are unknown. The Monhysteroidea are free-living nematodes found in all environments, from marine waters to fresh waters and soil; feeding habits are unknown. *See* NEMATA. [A.R.M.]

Monitor Any of 27 species of the reptilian family Varanidae in the order Squamata, distributed in Africa, India, Malaysia, and Australia. They are carnivorous and voracious and are the largest and heaviest known lizards. They have long, slender forked tongues which can be protruded well beyond the mouth or withdrawn into a sheath at its base. The head is mobile and can be turned in all directions, The tail, which is round in terrestrial species and flattened in aquatic species, serves as a weapon. Depending on their habitat, these reptiles eat fish, snakes, amphibians, lizards, mammals, and birds.

The largest of all known lizards, the Komodo dragon (see illustration) was not discovered until 1912, when it was found inhabiting the small island of Komodo in the Dutch East Indies. It feeds on small deer and wild pigs, as well as members of its own species and dead animals. It has been protected, since it is

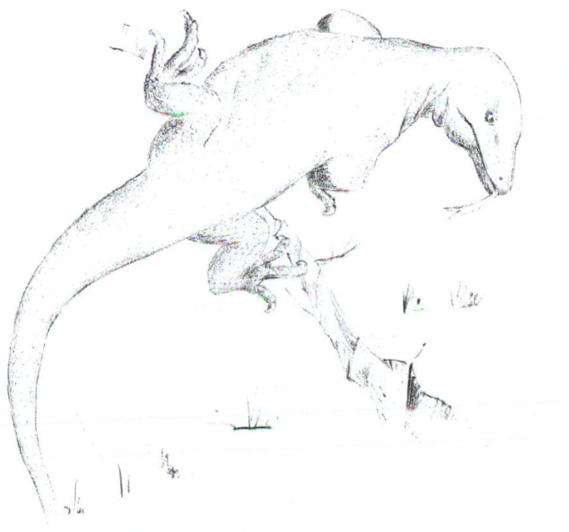

The Komodo dragon (*Varanus griseus*).

considered to be a relict of extinct giant lizards. It is an extremely active and savage species, much feared by the natives. *See* Lizard; Reptilia; Squamata. [C.B.C.]

Monitoring of ionizing radiation The use of meters and special techniques to determine the absorbed dose of ionizing radiation received by individuals; also, the use of meters and other devices to determine the type of radiation, its energy spectrum and direction, and the absorbed dose in the various areas and inside the human body. Monitoring frequently is divided into personnel monitoring, building surveys, and area monitoring.

Personnel monitoring involves those operations directly associated with the measurement and recording of the absorbed dose received by the individual. Personnel monitoring includes the issuing of dosimeters; the reading, maintenance, and calibration of these devices; the keeping of exposure records; and personal contacts with individuals to determine the causes of exposure and to recommend procedures to limit the recurrence of exposures. *See* Dosimeter; Film badge; Health physics.

Building surveys are made with many types of survey or monitoring instruments which, for the most part, can be grouped into three classes: Geiger-Müller counters, scintillation counters, and ionization chambers. This equipment is used to measure the dose rate and accumulated dose in various work areas and to estimate the surface contamination on floors, walls, furniture, and equipment. One of the hazards of greatest concern in many types of work with radioactive materials is the inhalation of airborne dusts, fumes, and gases. As a consequence, various types of equipment, such as air filters, precipitators, impingers, and charcoal collectors, have been developed to make collections of airborne contamination for radioisotope analysis. *See* Geiger-Müller counter; Ionization chamber; Scintillation counter.

Area monitoring is concerned with the measurement of the buildup and spread of radioactive contamination in the air, water, and soil outside work areas. Many instruments used for building surveys are used in area monitoring. [K.Z.M.]

Monkey An adaptive or evolutionary grade among the primates, represented by members of two of the three modern anthropoid superfamilies. The New World, platyrrhine monkeys (Ceboidea) and Old World, catarrhine forms (Cercopithecoidea) probably reached a monkey level of adaptation independently some time after their separation from a common ancestor, perhaps 50 million years ago. The term "monkey" is not indicative of taxonomic or phylogenetic relationship: the closest relatives of the cercopithecoids are not the ceboid monkeys but the Old World apes and the humans.

The Ceboidea comprise two families, while the Cercopithecoidea are today considered to comprise only one family, with two subfamilies. A modern classification of the suborder Anthropoidea follows.

Suborder Anthropoidea
- Infraorder Platyrrhini
  - Superfamily Ceboidea
    - Family Cebidae
      - Subfamily Cebinae (capuchin and squirrel monkeys)
      - Subfamily Callitrichinae (marmosets and tamarins)
    - Family Atelidae
      - Subfamily Atelinae (howler and spider monkeys)
      - Subfamily Pitheciinae (saki, owl, and titi monkeys)
- Infraorder Catarrhini
  - Superfamily Parapithecoidea (extinct Egyptian monkeys)
  - Superfamily Cercopithecoidea
    - Family Cercopithecidae
      - Subfamily Cercopithecinae (cheek-pouched monkeys, macaques, baboons, guenons, and mangabeys)
      - Subfamily Colobinae (leaf eaters, langurs, and colobus monkeys)
  - Superfamily Hominoidea (gibbons, great apes, and humans)

Monkeys are hard to characterize as a group because of their great diversity, and because much of the discussion reflects a comparison with the apes. Both monkeys and apes contrast with the prosimian grade in that they are typically large, diurnal animals that live in social groups. Monkeys differ from apes in their possession of a tail, a smaller brain, quadrupedal pronograde posture, and a usually longer face. They are generally smaller than apes, but large monkeys outweigh gibbons. Like almost all primates, monkeys are pentadactyl, with nails rather than claws on the digits in most cases. They have pectoral mammary glands and well-developed vision. Monkeys are primarily vegetarian and inhabit forested tropical or subtropical regions of Africa, Asia, and South America. The differences between the New and Old World monkeys are summarized in the table.

Major contrasts between New and Old World monkeys and special features of each

Ceboidea (New World species)	Cercopithecoidea (Old World species)
Nose platyrrhine (nasal septum wide, nostrils open to sides)	Nose catarrhine (septum narrow, nostrils open downward)
Tail long, prehensile in atelines and *Cebus* only	Tail short to long, nonprehensile
3 premolar teeth in each quadrant	2 premolars in each quadrant
24 deciduous, 36 permanent teeth (4 fewer in Callitrichinae) I 2/2 C 1/1 P 3/3 M 3/3 (M 2/2 in Callitrichinae)	20 deciduous, 32 permanent teeth I2/2 C1/1 P2/2 M3/3
	Ischial callosities present
	Cheek pouches in Cercopithecinae
Jaws and teeth lightly built in Cebidae; more robust in Atelidae, with deep lower jaw	Sacculated stomach in Colobinae
Fingers and toes with curved nails (clawlike in Callitrichinae)	All nails tend to be flattened
Big toe opposable, thumb not fully so and sometimes reduced in Cebidae	Thumb and big toe opposable, thumb reduced in Colobinae

Old World species are found throughout all the warmer regions of the eastern Hemisphere, except Australia and Madagascar. Many of the familiar monkeys are included in this family, such as the rhesus macaque, Barbary "ape," mangabey, baboon, and mandrill.

New World species or ceboids occupy forested areas from southern Mexico to Argentina. They are divided into two main groups, or families (their major characteristics are given in the table). All are arboreal, including a few with prehensile tails; there is no living form, nor any evidence of a fossil form, that has come to the ground habitually. Familiar monkeys included in this group are marmosets, capuchins, sakis, howler monkeys, and spider and wooly monkeys. *See* Primates. [E.D.]

Monoamine oxidase A mitochondrial enzyme which oxidatively deaminates intraneuronal biogenic amines, some of which are important neurotransmitters in the peripheral and central nervous system. Monoamine oxidase inactivates a biogenic amine by converting it to an aldehyde, which may be further oxidized to an acid or reduced to an alcohol. This enzyme has become increasingly important for the investigation of psychiatric disorders, and their pharmacological treatment. *See* Enzyme.

The function of monoamine oxidase is to inactivate catecholamines (dopamine and norepinephrine), and the indoleamine (serotonin), which are, along with acetylcholine, the most important biogenic amine neurotransmitters in the mammalian brain. Removal of these amines from the synaptic cleft through active and passive uptake into the nerve cell terminal is thought to be the major route of their inactivation under ordinary conditions. Some of this amine is then reused for neural transmission, but most of the rest is metabolized by monoamine oxidase. Monoamine oxidase is important to the metabolism of newly synthesized biogenic amines which are not stored or immediately released. Therefore, when monoamine oxidase activity is markedly reduced through the administration of one of its inhibitors, the concentrations of its substrates will increase until they are sufficiently high to inhibit further synthesis (end-product inhibition). It has been well established that clinical effects of monoamine oxidase inhibition in humans (for example, antidepressant action) require at least 85% inhibition of the enzyme because of the excess amount of monoamine oxidase present in the brain. *See* ACETYLCHOLINE; NORADRENERGIC SYSTEM; SEROTONIN; SYNAPTIC TRANSMISSION.

There has been considerable interest in the role of brain monoamine oxidase activity in a variety of psychiatric disorders, especially schizophrenia, depression, and alcoholism. Because of theories that schizophrenia might be due to increased dopaminergic activity while depression might be due to decreased serotonergic or noradrenergic activity or both, it has been of interest to determine if monoamine oxidase activity might be altered in brain tissue or platelets in either of these conditions. *See* AFFECTIVE DISORDERS; ALCOHOLISM; BRAIN; SCHIZOPHRENIA. [H.Y.M.]

Monocleales An order of liverworts of the subclass Marchantiidae, consisting of a single genus (*Monoclea*). The gametophyte is among the largest of all liverworts; its thallus consists of homogeneous cells except for scattered oil cells. The rhizoids are smooth and both thick-walled and thin-walled. The sex organs are grouped but not elevated: the antheridia are sunken in cavities and grouped into receptacles, while the archegonia are enclosed in groups by involucres. The very long, massive seta considerably elevates the capsules which dehisce spoonlike by one slit. The lobing of spore mother cells is unique in the subclass Marchantiidae. *See* BRYOPHYTA; MARCHANTIIDAE. [H.Cr.]

Monoclonal antibodies Antibody proteins that bind to a specific target molecule (antigen) at one specific site (antigenic site). Monoclonal antibodies are produced by a laboratory culture of cells (a cell line), grown from a single antibody-secreting cell. Development of the technology for making monoclonal antibodies has had an impact in both science and medicine; applications include experiments in basic immunology, medical diagnostics, and therapeutics.

A cell line that produces monoclonal antibodies is created by a combination of immunological and tissue culture techniques. First, a foreign substance (antigen) is injected into a mouse. The mouse immune system reacts to the antigen by producing antibodies that bind to it. Immunization with a complex antigen results in the production of a number of different antibodies binding to its different conformational features (antigenic sites). In the spleen each antibody-producing cell makes only one type of antibody, which reacts with a single antigenic site. An immunized mouse will have many antibodies in the serum component of its blood (antiserum) which react with many different features of a complex antigen. In 1975 G. Köhler and C. Milstein developed a technique for immortalizing spleen cells in laboratory culture and simultaneously retaining their antibody secretion ability. This allowed the isolation and culture of individual antibody-secreting cells and the subsequent production of monoclonal antibodies.

The ability to produce monoclonal antibodies has revolutionized immunology and the use of immunological techniques in other disciplines. Previously, antibodies could be isolated only from serum of an immunized animal, and analysis of an antibody response reflected the combination of activities and properties of all the different antibodies in that antiserum. But by making monoclonal antibodies from an immunized animal, it has been possible to analyze individual molecular properties of antibodies produced in a single immune response and to understand more about the mechanism by which antibodies are generated. *See* ANTIBODY; ANTIGEN; IMMUNOLOGY; TISSUE CULTURE. [F.M.B.]

Monogenea A subclass of the Trematoda which are ectoparasites of the gills, skin, and orifices of fishes and, less frequently, of the esophageal tracts and bladders of amphibians and turtles. They have conspicuous anterior and posterior holdfasts, the latter usually armed. The terminal genitalia are frequently sclerotized. The group is characterized by sexual reproduction, direct development, and a single host in the life cycle.

The most widely used classification employs two orders, the Monophisthocotylea, in which the posthaptor is without discrete multiple suckers or clamps, and the Polyopisthocotylea, with suckers or clamps on the posthaptor.

Body shapes of the various genera are distinctive, sometimes bizarre, as in *Vallisia*, which is sickle-shaped. Paired external suckers or buccal cavity suckers and adhesive glands occur anteriorly. The posterior holdfast is either solid and armed with central anchors and marginal hooks (Gyrodactyloidea), sucker-shaped with anchors and hooks (Capsaloidea; see illustration), or solid and bearing suckers or clamps (Polyopisthocotylea).

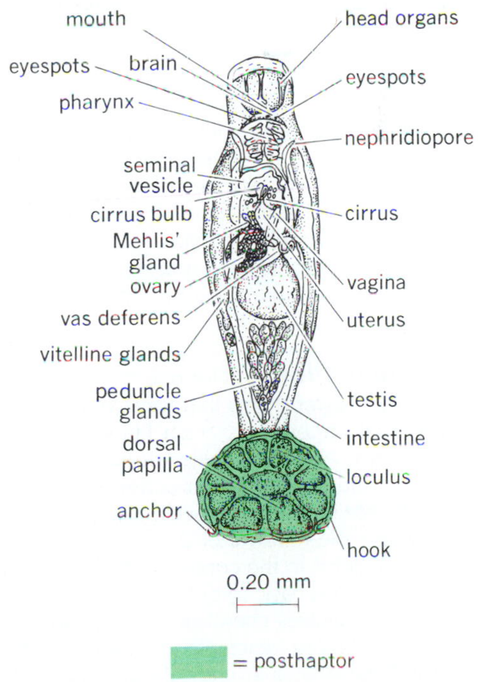

A monogeneid of the superfamily Capsaloidea, *Heterocotyle aetobatis* from the spotted eagle ray, ventral view.

Monogenea usually have direct development involving simple metamorphosis from the ciliated larval stage to the nonciliated juvenile. Juvenile anchors and hooks may be retained or replaced by adult suckers or clamps. Cross-fertilization or, perhaps less frequently, self-fertilization of hermaphroditic individuals resulting in egg capsules which hatch on the host or in its environment is most common. *See* TREMATODA. [W.J.Ha.]

Monogononta A class of the phylum Rotifera which contains the majority of species in this invertebrate class. The organisms of this order are characterized by the presence of a single gonad in both males and females. There is a striking degree of sexual dimorphism, with the males being small and degenerate. The order is made up of three suborders: Ploima, Flosculariacea, and Collothecacea.

In suborder Ploima there is an exceptional diversity of form, varying from soft-bodied wormlike rotifers to species with variously ornamented, loricate shells (see illustration). Most of the free-swimming benthonic and pelagic rotifers belong to this suborder. Locomotion is by the ciliated corona.

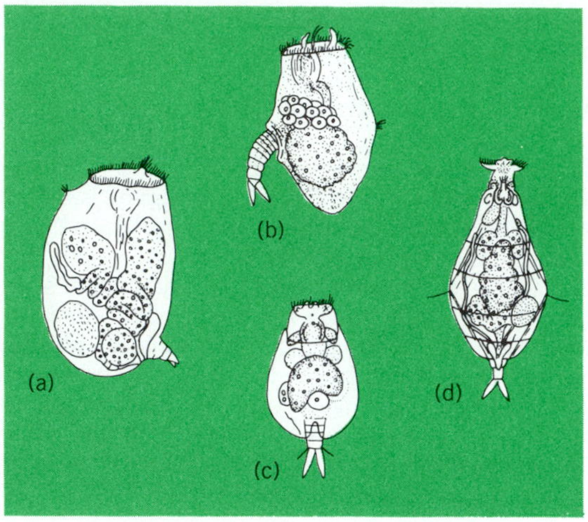

Ploimates. (*a*) *Asplanchnopus* sp. (*b*) *Ploesoma* sp. (*c*) *Euchlanis dilata.* (*d*) *Notomata copeus.*

The suborder Flosculariacea contains the spectacular sessile rotifers formerly known as melicertaceans of the family Flosculariidae, as well as a number of equally notable free-swimming forms included in the family Testudinellidae.

The suborder Collothecacea contains but a single family, the Collothecidae, made up of five genera. Most species of Collothecidae are sessile, and many are encased in gelatinous tubes. *See* ROTIFERA. [E.H.A.]

Monomolecular film A film one molecule thick; often referred to as a monolayer. Films that form at surfaces or interfaces are of special importance. Such films may reduce friction, wear, and rust, or may stabilize emulsions, foams, and solid dispersions. Thin films on water surfaces reduce evaporation losses, which are important in arid regions throughout the world. Nevertheless, the removal of thin films of contaminants is one of many problems in the control of pollution. The broad field of catalysis, which is basic to petroleum refining and many chemical industries, involves chemical reactions that are accelerated in the thin films of reactants at interfaces. Moreover, thin films containing proteins, cholesterol, and related compounds constitute biological membranes, the internal interfaces that control the complex processes of life. *See* CATALYSIS; CELL MEMBRANES.

Monolayers on solids, or at liquid interfaces, may be formed by adsorption from the adjacent bulk phases; the process may show high specificity for particular chemical species. Measurements of the extent of adsorption have historically provided information on the composition and structure of monolayers formed in this way. A variety of surface-sensitive instrumental techniques, such as diffraction and scattering of low-energy electrons, neutrons, and ions, and spectroscopy of adsorbed species, have been brought to bear to obtain information about the structure of the surface layer and chemical perturbations in it. *See* ADSORPTION; SPECTROSCOPY.

In order to form spread monolayers which are sufficiently stable to study, a substance must combine low solubility and volatility with some moiety which attracts it to the liquid surface; for films on water, this generally means one or more polar functional groups. Totally nonpolar substances, such as the higher-molecular-weight paraffin hydrocarbons, will not spread on water (although they can spread on liquids of very high surface tension, such as mercury). Typical among the large group of substances which do form insoluble monolayers on water are the long-chain fatty acids and their derivatives such as glycerides, sterols, and many lipid substances of biological origin, including the fat-soluble vitamins and natural pigments such as chlorophyll. Many polar synthetic polymers, including polyvinyl acetate and polymethyl methacrylate, can be made to spread as monolayers on water; so can many proteins, because their tertiary structure unfolds at the air-water interface. *See* COLLOID; EMULSION; FOAM; INTERFACE OF PHASES; LUBRICANT; SURFACE TENSION; WATER CONSERVATION. [G.L.G.]

Mononchida An order of nematodes having a full complement of cephalic sensilla on the lips in two circlets of 6 and 10. The amphids are small and cuplike, and are located just posterior to the lateral lips; the amphidial aperture is either slitlike or ellipsoidal. The stoma is globular and heavily cuticularized, and is derived primarily from the cheilostome. The stoma bears one or more massive teeth that may be opposed by denticles in either transverse or longitudinal rows. The esophagus is cylindrical conoid, with a heavily cuticularized luminal lining. The excretory system is atrophied. Males have ventromedial supplements and paired spicules. The gubernaculum may possess lateral accessory pieces. Females have one or two ovaries. Caudal glands and a spinneret are common; however, they may be degenerate or absent.

There are three mononchid superfamilies: The Mononchoidea contain some of the most common and easily recognized free-living nonparasitic nematodes that occur in soils and fresh waters throughout the world. The closely related Bathyodontoidea are inhabitants of soil or fresh water and prey on small microorganisms. The nonparasitic Monochuloidea comprise both soil and fresh-water species, all of which are predators of microfauna. *See* NEMATA. [A.R.M.]

Monoplacophora A class of the phylum Mollusca. Although fossil monoplacophorans had been known since the end of the nineteenth century, it was the discovery of a living species in deep water off Costa Rica in the 1950s that led to universal acceptance of the class. Monoplacophorans are bilaterally symmetrical, univalved mollusks that vanish from the fossil record at the end of the Paleozoic, about 240 million years ago. The living species, such as *Neopilina galatheae*, are rare and inhabit deep water, which may explain their absence from Mesozoic and Cenozoic rocks. Living monoplacophorans have a limpet-shaped shell, a circular foot attached by pairs of retractor muscles, and several gills on each side of the body. *See* GASTROPODA; MOLLUSCA; POLYPLACOPHORA. [B.Ru.]

Monopulse radar A radar that obtains a complete measurement of the target's angular position from a single echo pulse. Together with the range measurement performed with the same pulse, the target position in three dimensions is determined completely. Usually a train of echo pulses is then employed to make a large number of repeated measurements and produce a refined estimate, but this is not intrinsically necessary. An important advantage of a monopulse tracking radar over one employing conical scan is that the instantaneous angular measurements are not subject to errors caused by target scintillation. An additional advantage of monopulse tracking as compared to conical scan is that no mechanical action is required in monopulse. *See* RADAR. [R.I.B.]

Monorail A distinctive type of materials-handling machine that provides an overhead, normally horizontal, fixed path of travel in the form of a trackage system and individually propelled hand or powered trolleys which carry their loads suspended freely with an intermittent motion. Because monorails operate over fixed paths rather than over limited areas, they differ from overhead-traveling cranes, and they should not be confused with such overhead conveyors as cableways. *See* BULK-HANDLING MACHINES; MATERIALS-HANDLING EQUIPMENT. [A.M.P.]

Monosaccharide A class of simple sugars containing a chain of 3–10 carbon atoms in the molecule, known as polyhydroxy aldehydes (aldoses) or ketones (ketoses). They are very soluble in water, sparingly soluble in ethanol, and insoluble in ether. The number of monosaccharides known is approximately 70, of which about 20 occur in nature. The remainder are synthetic. The existence of such a large number of compounds is due to the presence of asymmetric carbon atoms in the molecules. Aldohexoses, for example, which include the important sugar glucose, contain no less than four asymmetric atoms, each of which may be present in either D or L configuration. The number of stereoisomers rapidly increases with each additional asymmetric carbon atom.

A list of the best-known monosaccharides is given below:

Trioses:	$CH_2OH \cdot CHOH \cdot CHO$, glycerose (glyceric aldehyde)
	$CH_2OH \cdot CO \cdot CH_2OH$, dihydroxy acetone
Tetroses:	$CH_2OH \cdot (CHOH)_2 \cdot CHO$, erythrose
	$CH_2OH \cdot CHOH \cdot CO \cdot CHO$, erythrulose
Pentoses:	$CH_2OH \cdot (CHOH)_3 \cdot CHO$, xylose, arabinose, ribose
	$CH_2OH \cdot (CHOH)_2 \cdot CO \cdot CH_2OH$, xylulose, ribulose
Methyl pentoses (6-deoxyhexoses):	
	$CH_3(CHOH)_4 \cdot CHO$, rhamnose, fucose
Hexoses:	$CH_2OH \cdot (CHOH)_4 \cdot CHO$, glucose, mannose, galactose
	$CH_2OH \cdot (CHOH)_3 \cdot CO \cdot CHOH$, fructose, sorbose
Heptoses:	$CH_2OH \cdot (CHOH)_5 \cdot CHO$, glucoheptose, galamannoheptose
	$CH_2OH \cdot (CHOH)_4 \cdot CO \cdot CH_2OH$, sedoheptulose, mannoheptulose

Aldose monosaccharides having 8, 9, and 10 carbon atoms in their chains have been synthesized. *See* CARBOHYDRATE; KETONE; OPTICAL ACTIVITY; STEREOCHEMISTRY. [W.Z.H.]

Monosodium glutamate The single sodium salt of glutamic acid used in foods to accentuate flavors. It is also known as MSG. Molecular structure is represented below.

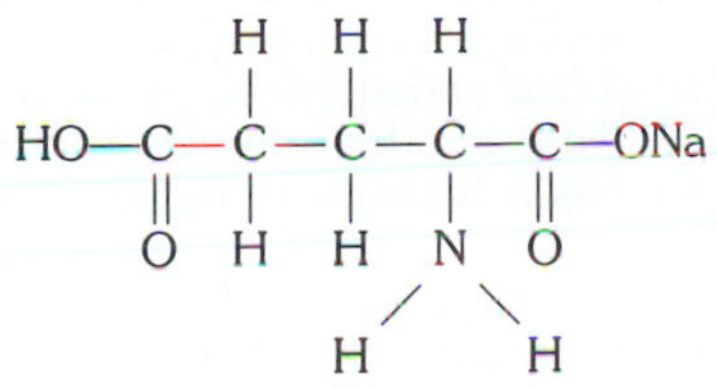

The crystal form available in commerce is the monohydrate, with structure as represented plus one molecule of water of hydration. *See* GLUTAMIC ACID.

Originally produced from seaweed in the Orient, it is now made principally from cereal glutens, such as those of wheat, corn, and soybeans, from solutions evolved in the manufacture of beet sugar, and by microbiological fermentation of carbohydrates. The two raw materials used for the greater proportion of commercial production are wheat gluten and desugared beet-sugar molasses.

Monosodium glutamate is recognized as a standard of identity ingredient in several commercial food preparations. Its principal use is in the preparation of canned and dried soups, but it also enters into the production of some meat, vegetable, fowl, and fish products. It is the so-called secret ingredient used by many of the famous restaurant and hotel chefs. [P.D.V.M.]

Monotremata The single order of the mammalian subclass Prototheria. Two living families, the Tachyglossidae and the Ornithorhynchidae, make up this unusual order of quasi-mammals, or mammallike reptiles.

The Tachyglossidae comprise the echidnas (spiny anteaters), which have relatively large brains with convoluted cerebral hemispheres. The known genera, *Tachyglossus* and *Zaglossus*, are terrestrial, feeding on termites, ants, and other insects. They are capable diggers, both to obtain food and to escape enemies. Like hedgehogs, they can erect their spines and withdraw their limbs when predators threaten. Commonly one egg, but occasionally two or even three, is laid directly into the marsupium (pouch) of the mother where it is incubated for up to 10 days. Species of *Tachyglossus* live in rocky areas, semideserts, open forests, and scrublands. They are found in Australia, Tasmania, New Guinea, and Salawati Island. Species of *Zaglossus* are found in mountainous, forested areas.

The duck-billed platypus, comprising the Ornithorhynchidae, has a relatively small brain with smooth cerebral hemispheres. The young have calcified teeth, but in the adult these are replaced by horny plates which form around the teeth in the gums. The snout, as the name indicates, is duck-billed. The semiaquatic platypus is a capable swimmer, diver, and digger. Two eggs are usually laid by the female into a nest of damp vegetation. After incubating the eggs for about 10 days the female leaves, returning only when the eggs are hatched. The platypus is found in Australia and Tasmania in almost all aquatic habitats. *See* MAMMALIA; PROTOTHERIA. [F.S.S.]

Monsoon meteorology The study of the structure and behavior of the atmosphere in those areas of the world that have monsoon climates. In lay terminology, monsoon connotes the rains of the wet summer season that follows the dry winter. However, for mariners, the term monsoon has come to mean the seasonal wind reversals.

In true monsoon climates, both the wet summer season that follows the dry winter and the seasonal wind reversals should occur. Winds from cooler oceans blow toward heated continents in summer, bringing warm, unsettled, moisture-laden air and the season of rains, the summer monsoon. In winter, winds from the cold heartlands of the continents blow toward the oceans, bringing dry, cool, and sunny weather, the winter monsoon.

Based on these criteria, monsoon climates of the world include almost all of the Eastern Hemisphere tropics and subtropics, which is about 25% of the surface area of the Earth. The areas of maximum seasonal precipitation straddle or are adjacent to the Equator. Two of the world's areas of maximum precipitation (heavy rainfall) are within the domain of the monsoons: the central and south African region, and the larger south Asia-Australia region. The monsoon surface winds emanate from the cold continents of the winter hemisphere, cross the Equator, and flow toward and over the hot summer-hemisphere land masses.

India presents the classic example of a monsoon climate region, with an annual cycle that brings southwesterly winds and heavy rains in summer (the Indian southwest monsoon) and northeasterly winds and dry weather in winter (the north-east winter monsoon).

Like all weather systems on Earth, monsoons derive their primary source of energy from the Sun. About 30% of the Sun's energy that enters the top of the atmosphere is transmitted back to space by cloud and surface reflections. Little of the remainder is absorbed directly by the clear atmosphere; it is absorbed at the Earth's surface according to a seasonal cycle. The opposition of seasons in the Northern and Southern hemispheres leads to a slow movement of surface air across the Equator from winter hemisphere to summer hemisphere, forced by horizontal pressure gradients and vertical buoyancy forces resulting from differential seasonal heating. Such a seasonally reversing rhythm is most pronounced in the monsoon regions. *See* ALBEDO; ATMOSPHERE; HEAT BALANCE, TERRESTRIAL ATMOSPHERIC; INSOLATION; METEOROLOGY; TROPICAL METEOROLOGY. [J.S.Fe.]

Monstrilloida A small, aberrant order of the crustacean subclass Copepoda. It comprises two families whose members, as larvae, are parasitic on invertebrates, particularly polychaete worms and prosobranch mollusks. Monstrilloids are characterized by a total absence of mouthparts and gut in the free-swimming, nonfeeding adult phase. Antennae are lacking, but antennules are usually well developed. Thoracopods number four pairs, and females carry two egg sacs on a pair of ovigerous spines. *See* COPEPODA; CRUSTACEA. [P.A.McL.]

Monte Carlo method A technique for estimating the solution x of a numerical mathematical problem by means of an artificial sampling experiment. The estimate is usually given as the average value, in a sample, of some statistic whose mathematical expectation is equal to x. In most of the useful applications, the mathematical problem itself arises in a problem of probability, either in physics or in operational research.

The importance of the method arises primarily from two sources: the practical need to solve equations that are too complicated to solve by analytic methods alone, and the increased importance of all numerical methods because of the advent of the electronic computer.

The main advantage of Monte Carlo is that many numerical problems are in practice too complicated to solve in any other method. A familiar example is the estimation of the probability of winning a game of pure chance: Sometimes the only reasonably simple method of estimation is to play the game several times. There are also numerical problems that can be solved by deterministic methods but are logically simpler to solve approximately by the Monte Carlo method. Sometimes poor approximations are satisfactory because the aim is merely to determine the strategic variables of a problem. This is likely to be a fruitful technique in mathematical economics. Another situation where a poor approximation is satisfactory occurs when there is available an iterative method of calculation, that is, a method of successive approximation, which converges to the right answer in a reasonable time provided that the first trial solution is not too far from the truth. The Monte Carlo method may then be used for obtaining a first trial solution.

The main disadvantage of the Monte Carlo method is that for each extra decimal place required it is necessary to multiply the sample size by 100. To calculate π to five decimal places by throwing a needle would require about 10^{10} throws. *See* OPERATIONS RESEARCH; PROBABILITY; QUEUEING THEORY; STATISTICS; STOCHASTIC PROCESS. [I.J.G.]

Month Any of the several units of time based on the revolution of the Moon around Earth.

The calendar month is one of the 12 arbitrary periods into which the calendar year is divided.

The synodic month is the average period of revolution of the Moon with respect to the Sun, the same as the average interval between successive full moons. Its duration is 29.531 days.

The tropical month is the period required for the mean longitude of the Moon to increase 360°, or 27.322 days.

The sidereal month, 7 seconds longer than the tropical month, is the average period of revolution of the Moon with respect to a fixed direction in space.

The anomalistic month, 27.555 days in duration, is the average interval between closest approaches of the Moon to Earth.

The nodical month, 27.212 days in duration, is the average interval between successive northward passages of the Moon across the ecliptic. *See* CALENDAR; TIME. [G.M.C.]

Montmorillonite A group name for all clay minerals with an expanding structure, except vermiculite, and also a specific mineral name for the high alumina end member of the group. *See* CLAY MINERALS; VERMICULITE.

Montmorillonite clays have wide commercial use. The high colloidal, plastic, and binding properties make them especially in demand for bonding molding sands and for oil-well drilling muds. They are also widely used to decolorize oils and as a source of petroleum cracking catalysts. *See* CLAY.

Members of the montmorillonite group of clay minerals vary greatly in their modes of formation. Alkaline conditions and the presence of magnesium particularly favor the formation of these minerals. Several important modes of occurrence are in soils, in bentonites, in mineral veins, in marine shales, and as alteration products of other minerals. Recent sediments have a fairly high montmorillonite content. *See* BENTONITE; MARINE SEDIMENTS. [F.M.W.; R.E.Gr.]

Monzonite A phaneritic (visibly crystalline) plutonic rock composed chiefly of sodic plagioclase (oligoclase or andesine) and alkali feldspar (microcline, orthoclase, usually perthitic), with subordinate amounts of dark-colored (mafic) minerals (biotite, amphibole, or pyroxene). Monzonite is more or less intermediate between syenite and diorite. *See* SYENITE. [C.A.C.]

Moon The Earth's natural satellite. United States and Soviet spacecraft have obtained lunar data and samples, and American astronauts have orbited, landed upon, and roved upon the Moon.

The Earth and Moon now make one revolution about their barycenter, or common center of mass (a point about 4670 km from the Earth's center), in $27^d\ 7^h\ 43^m\ 11.6^s$. This sidereal period is slowly lengthening, and the distance (now about 60.27 earth radii) between centers of mass is increasing, because of tidal friction in the oceans of the Earth.

The Moon's present orbit is inclined about 5° to the plane of the ecliptic. As a result of differential attraction by the Sun on the Earth-Moon system, the Moon's orbital plane rotates slowly relative to the ecliptic (the line of nodes regresses in an average period of 18.60 years) and the Moon's apogee and perigee rotate slowly in the plane of the orbit (the line of apsides advances in a period of 8.850 years). Looking down on the system from the north, the Moon moves counterclockwise. It travels along its orbit at an average speed of nearly 0.6 mi/s (1 km/s) or about 1 lunar diameter per hour.

As a result of the Earth's annual motion around the Sun, the direction of solar illumination changes about 1° per day, so that the lunar phases do not repeat in the sidereal period given above but in the synodic period, which averages $29^d\ 12^h\ 44^m$.

When the lunar line of nodes coincides with the direction to the Sun and the Moon happens to be near a node, eclipses can occur. *See* ECLIPSE.

The relation between the Moon's shape and its mass distribution is very important to theories of lunar origin and the history of the Earth-Moon system. By radio altimetry, Apollo confirmed that the Moon's surface on the far side is higher on the average than the near side; that is, the center of mass is offset from the center of figure. The offset is about 2 km (1.2 mi) toward the Earth. These observations suggest that the Moon's crust is thicker on the far side than on the near side.

The Moon's small size and low mean density result in surface gravity too low to hold a permanent atmosphere, and therefore it was to be expected that lunar surface characteristics would be very different from those of Earth. However, the bulk properties of the Moon are also quite different—the density alone is evidence of that. The Moon is too small to have compressed its silicates into a metallic phase by gravity; therefore, if it has a dense core at all, the core should be of nickel-iron. Available data suggest that the Moon's iron core may have a diameter of at most a few hundred kilometers.

As can be seen from the Earth with the unaided eye, the Moon has two major types of surface: the dark, smooth maria and the lighter, rougher highlands. Photography by spacecraft shows that, for some unknown reason, the Moon's far side consists mainly of highlands (Fig. 1). Both maria and highlands are covered with craters of all sizes. Numerous different types of craters can be recognized. Most prominent at full moon are the bright ray craters whose grayish ejecta appear to have traveled for hundreds of miles across the lunar surface. Observers have long recognized that some erosive process has been and may still be active on the Moon. Bombardment of the airless Moon by meteoritic matter and solar particles, and extreme temperature cycling, are now considered the most likely erosive agents, but local internal activity is also a possibility.

The lunar mountains, though very high (26,000 ft or 8000 m), are not extremely steep, and lunar explorers see rolling rather than jagged scenery. Though a widespread network of fault traces is visible, there is no evidence on the Moon of the great mountain-building processes seen on the Earth.

Basins on the Moon's near side, namely, Imbrium, Serenitatis, and Crisium, appear fully flooded. These were maria created by giant impacts, followed by subsidence of the ejecta and (probably much later) upwelling of lava from inside the Moon. Examination of small variations in Lunar Orbiter motions has revealed that each of the great circular maria is the site of a positive gravity anomaly (excess mass). The old argument about impact versus vulcanism as the primary agent in forming the lunar relief appears to be entering a new, more complicated phase with the confirmation of extensive flooding of impact craters by lava on the Moon's near side, while on the far side, where the crust is thicker, the great basins remain mostly empty.

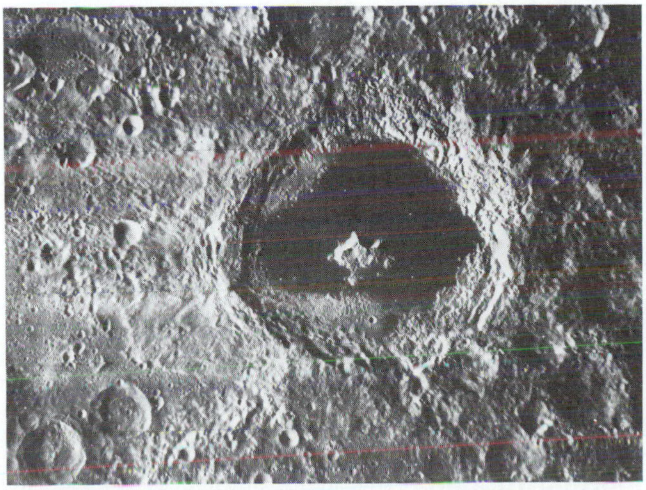

Fig. 1. Mare Tsiolkovski on far side of Moon. Crater, partly flooded by dark mare material, is about 120 mi (200 km) across. (*Langley Research Center, NASA*)

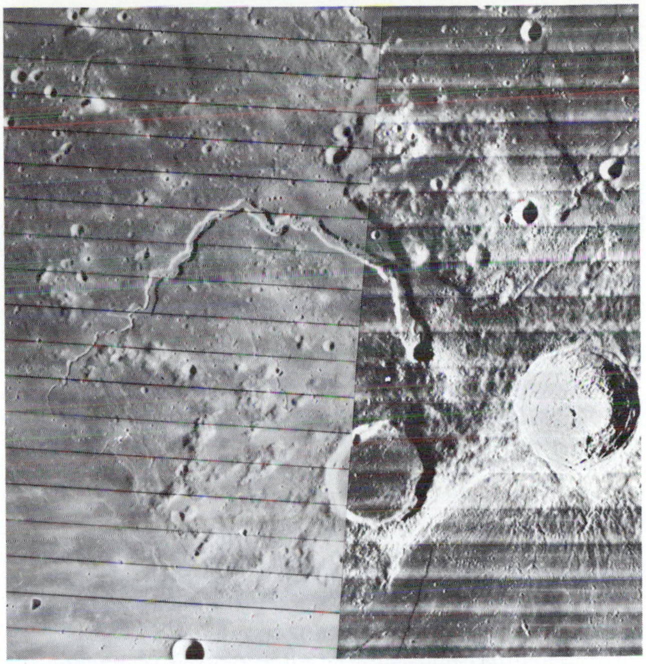

Fig. 2. Schroeter's Valley, a large sinuous rille, and craters Herodotus and Aristarchus. Aristarchus, brightest of the ray craters, is about 25 mi (40 km) across and 2000 ft (3600 m) deep. (*Langley Research Center, NASA*)

In some of the Moon's mountainous regions bordering on the maria are found sinuous rilles (Fig. 2). These winding valleys were shown in Lunar Orbiter pictures to have an exquisite fineness of detail. No explanation for them yet offered has proved entirely convincing.

The Moon seems to be totally covered, to a depth of at least tens of meters, by a layer of rubble and soil with very peculiar optical and thermal properties. This layer is called the regolith. The observed optical and radio properties all point to a highly porous or underdense structure for at least the top few millimeters of the lunar surface material. A dark-gray, fine soil appears

Fig. 3. Lunar soil and rocks, and the trenches made by *Surveyor 7*. (*Jet Propulsion Laboratory, NASA*)

to mantle the entire Moon, softening most surface contours and covering everything except occasional fields of rocks (Fig. 3). This soil, with a slightly cohesive character like that of damp sand and a chemical composition similar to that of some basic silicates on the Earth, is a product of the radiation, meteoroid, and thermal environment at the lunar surface. [J.D.Bu.]

Moose An even-toed ungulate (Artiodactyla) which is a member of the deer family, Cervidae. *Alces alces* is the largest member of the family and ranges in the boreal forested areas throughout North America and in northern Eurasia. The moose is known as the elk in Europe and is believed by some authorities to be a race of the American moose (*A. americana*). The legs are long, making the animal well-adapted for its feeding habits of wading for aquatic plants and browsing on trees and bushes. During the rutting season in the early fall, the male gathers a number of cows together, and mating takes place. After a gestation period of about 37 weeks, one or two calves are born. *See* ARTIODACTYLA. [C.B.C.]

Moraine An accumulation of glacial debris, usually till, with distinct surface expression related to some former ice front position. End moraine, the most common form, is an uneven ridge of till built in front of or around the terminus of a glacier margin, and reflects some degree of equilibrium between rate of ice motion, supply of rock debris at the ice front, temperature of the glacier base, and shape and resistance of underlying bedrock (see illustration).

End moraine in Pennsylvania (*U.S. Geological Survey*).

If an end moraine represents the farthest forward position a glacier ever moved, it is a terminal moraine. It demonstrates a steady-state condition for a period of time within the ice body where constant forward motion is balanced by frontal melting; and a continual supply of debris, as on an endless conveyor belt, is brought forward to the glacier terminus. If the ice front then melts farther back than it moves forward, till is spread unevenly over the land as ground moraine. If a retreatal position of steady-state equilibrium is maintained again, a recessional moraine may be constructed.

Drumlins, produced by glacier streamlining of ground moraine, are probably the best-known moraine forms. *See* DRUMLIN. [S.E.Wh.]

Mordant A substance or combination of substances that facilitates the fixing of a dye to a fiber. A mordant enables the production of a more permanent and often deeper color. Metallic salts or hydroxides are most frequently used as mordants.

Certain mordants act directly on the fiber, making it more susceptible to the dye. Fabrics are then pretreated with the mordant before exposure to the dye. Other mordants function through the formation of a complex with the dye. The complex acts as the dyeing agent. Mordant and dye in this case are exposed simultaneously to the fabric. *See* DYE; DYEING. [F.J.J.]

Mormonilloida The smallest of seven orders of Copepoda, containing two species in a single genus; their position in the copepod hierarchy has not been determined. Mormonilloids are free-living copepods that, in body form, closely resemble certain cyclopoids. However, they differ significantly in possessing mouthparts typical of the Calanoida. The presence of both gymnoplean and podoplean characteristics suggests a phylogenetic relationship with the Misophrioida. The lack of a heart, the small number of antennal segments, and the absence of fifth thoracic legs immediately distinguish mormonilloids from misophrioids. *See* CALANOIDA; MISOPHRIOIDA.

Both mormonilloid species are pelagic, primarily occurring in a depth range of 1350–2300 ft (410–700 m) in the eastern North Atlantic. Males have never been observed. *See* COPEPODA; CRUSTACEA. [P.A.McL.]

Morphactin One of a group of fluorene-carboxylic acid derivatives that are morphogenetically active substances in higher plants. The most active compounds, used as herbicides, are 9-hydroxyfluorene-(9)-carboxylic acid (fluorenol) and especially the 2-chloro derivative.

Morphactins affect the adult organs of plants very little, although leaf abscission or shedding of the stem apex may be caused. Much more frequently it is the new growth that is affected: Stem internodes are shortened, leaf laminae are reduced, and apical dominance (suppression of lateral shoots by the apical bud) is deranged, so that branching is increased. Plants that have developed under the influence of morphactin show, therefore, a dwarf bushy habit with numerous morphological anomalies; but, protected from competition with other plants, they normally survive and may bloom (though sometimes with deformed flowers) and even form fruit.

Other noticeable effects are retardation of germination, strong inhibition of lateral root formation, inhibition of the nodule formation on the roots of legumes, stopping of the spreading of rosette plants, change in the number and placing of flower buds, alternation in flower sex expression, induction of parthenocarpy, and abnormal development of embryos within the seeds. [H.Z.]

Morphine The most important opium alkaloid. High-quality opium contains 10–15% morphine (sometimes as high as 25%) along with smaller amounts of codeine, thebaine, papaverine, narcotine, and other bases. Its formula is $C_{17}N_{19}NO_3$, and it has the structure depicted below.

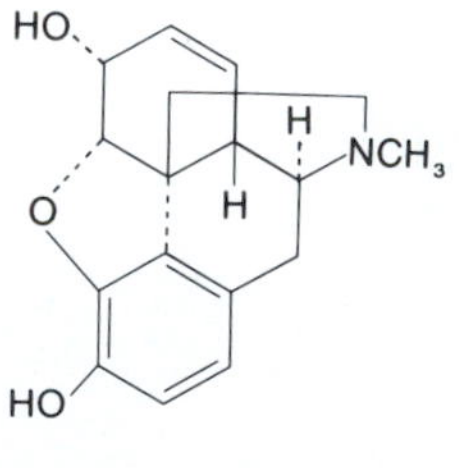

Clinically, the great usefulness of the drug lies in its depressant action, which causes a raising of the threshold of pain, sleepiness, and euphoria. While repeated use sometimes leads to addiction, morphine is an excellent analgesic and has certain well-defined and indispensable uses in medicine and surgery. *See* ALKALOID; ANALGESIC. [F.W.]

Morphotropism Similarity of structure, axial ratios, and angles between faces of one or more zones in crystalline substances whose formulas can be derived one from another by substitution. Examples of morphotropic substances are compounds of the type M_2SO_4, where M stands for Li, Na, K, Rb, Cs, Ag, Tl, or NH_4. Structurally, this is explained as follows. The crystal structure is essentially determined by the large SO_4 groups. The oxygen ions form a tetrahedron, with the sulfur in its center. The other ions are in cavities in between four or six of the oxygen ions according to their dimensions. Organic compounds also show morphotropism. No definite general rules, however, can be given for its occurrence. *See* COORDINATION NUMBER; CRYSTAL; CRYSTALLOGRAPHY; ISOMORPHISM (CRYSTALLOGRAPHY). [W.C.D.]

Mortar A plastic mixture of solids and water used to bond masonry units together. Mortar exhibits hardening characteristics (or sets) as reactions of the cementitious materials occur with water.

The principal solids in mortar are sand and cementitious materials, such as hydrated lime (or slaked quicklime) and portland cement. The latter is absent in a pure lime mortar. The portland cement reacts rapidly with water, giving the mortar its initial set. Lime reacts more slowly, both with carbon dioxide (CO_2) from the atmosphere to form $CaCO_3$, and with sand (SiO_2) to form calcium silicates. Mortar thus hardens with age.

Lime is used in mortars mainly because of the good workability it imparts, leading to well-filled joints. *See* LIME (INDUSTRY). [J.A.P.]

Mosaicism The coexistence of two or more genetically distinct cell populations derived originally from a single zygote, Mosaics may arise at any stage of development, from the two-cell stage onward, or in any tissue which actively proliferates thereafter. The phenomenon is commonly observed in many species of animals and plants and may be caused by somatic mutation or chromosomal nondisjunction. An individual animal or plant may exhibit mosaicism, or it may occur in a culture of a single cell- or tissue-type obtained from an individual.

Chromosome nondisjunction is probably the principal cause of chromosomal aberration, which in turn may lead to the development of mosaicism. During cell division, the two sister chromatids usually separate completely, each chromatid going to opposite poles of the cell guided by the spindle apparatus. In some cases, one chromatid will fail to completely separate, or it may lag behind. This nondisjunction will lead to the presence of both sister chromatids in the same daughter cell instead of one in each of the daughter cells (Fig. 1). Somatic crossing-over leads to the production of a recombinant mosaic where chromosome segments, with their corresponding blocks of genes, are exchanged between homologous chromosomes during mitosis. The occurrence of this process leads to mosaicism, mostly manifested as spots (clones of variant cells) on the cuticle of insects or on leaves, petals, or stamen hairs. *See* CHROMOSOME.

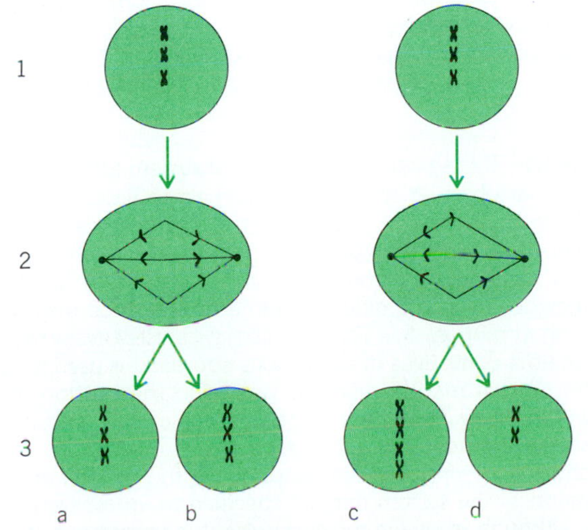

Fig. 1. Mosaicism caused by chromosome nondisjunction. (1) Two cells with an identical chromosome complement are in the process of cellular division. (2) In the cell on the left, the sister chromatids have separated normally, whereas in the other cell, one chromatid has remained in the equatorial zone. (3) Two normal daughter cells (a and b) are produced as a result of a normal mitotic division. Two aneuploid cells (c and d) result from the abnormal division. If c and d are viable, a tissue mosaicism may result with a mixture of normal cells and some aneuploid cells (c, d, or c and d).

Fig. 2. Bilateral gynandromorphism in the moth *Abraxas grossulariata*. The left side shows the typical male wing patterning, whereas the right side shows a female form.

Sex chromosome mosaicism, the presence of a mixture of cell populations with different X and Y chromosome constitutions, is not uncommon and is often seen in individuals with ovarian dysgenesis. The presence of a significant proportion of chromosomally abnormal cells in any such mosaic will tend to lead to a clinically expressed syndrome. The proportion of each constituent clone may vary from tissue to tissue, but is relatively stable in each individual site throughout adult life. Sex mosaics (gynandromorphs) are particularly striking where a difference in the secondary sexual characteristics exists between the normal sexes. For example, in a butterfly with bilateral gynandromorphism, the left side may show the characteristic wing color and pattern of the male, and the light wing, typical female patterning (Fig. 2). *See* CHIMERA; GENETICS; SEX-LINKED INHERITANCE. [A.W.W.]

Mosquito Any member of the dipterous subfamily Culicinae in the family Culicidae. These insects are slender and fragile, with long legs, a long slender abdomen, and narrow wings. As with other members of the Diptera, distinct halteres (balancing organs) are present. Mouthparts, adapted for piercing and sucking, are well developed in the females, since they alone are blood feeders, and weakly developed in the males, which feed on nectar, ripe or decaying fruit, and other sweet substances. The head is small with large compound eyes, but no ocelli (simple eyes) are present. Antennae are long and beadlike in the female, and plumose or feathery in the male.

The larvae of all mosquitoes are aquatic and, although they vary in structural details, all have a breathing tube or similar structure which allows them to obtain air while floating just below the water surface. Some larvae are predacious, though generally food consists of organic matter present in the water. The pupae, which are also aquatic, differ radically from the larvae in morphology, and eventually metamorphose into the adult, or imago, stage. All three stages are important in identification of the species.

In addition to being biting pests, these insects form an important group in medical entomology, since they are vectors of many diseases of humans. Of the 90 genera and subgenera of mosquitoes, three are commonly known because of their medical importance: *Culex*, the common house mosquito; *Aedes*, the yellow-fever mosquito; and *Anopheles*, the malaria mosquito. These common names are misleading, since various species of each genus may transmit a variety of diseases. *See* DIPTERA; MALARIA; YELLOW FEVER. [C.B.C.]

Mössbauer effect Recoil-free gamma-ray resonance absorption. The Mössbauer effect, also called nuclear gamma resonance fluorescence, has become the basis for a type of spectroscopy which has found wide application in nuclear physics, structural and inorganic chemistry, biological sciences, the study of the solid state, and many related areas of science.

The fundamental physics of this effect involves the transition (decay) of a nucleus from an excited state of energy E_e to a ground state of energy E_g with the emission of a gamma ray of energy E_γ. If the emitting nucleus is free to recoil, so as to conserve momentum, the emitted gamma ray energy is $E_\gamma = (E_e - E_g) - E_r$, where E_r is the recoil energy of the nucleus. The magnitude of E_r is given classically by the relationship $E_r = E_\gamma^2/2mc^2$ where m is the mass of the recoiling atom and c is the speed of light. Since E_r is a positive number, the E_γ will always be less than the difference $E_e - E_g$, and if the gamma ray is now absorbed by another nucleus, its energy is insufficient to promote the transition from E_g to E_e.

In 1957 R. L. Mössbauer discovered that if the emitting nucleus is held by strong bonding forces in the lattice of a solid, the whole lattice takes up the recoil energy, and the mass in the recoil energy equation given above becomes the mass of the whole lattice. Since this mass typically corresponds to that of 10^{10} to 10^{20} atoms, the recoil energy is reduced by a factor of 10^{-10} to 10^{-20}, with the important result that $E_r \approx 0$ so that $E_\gamma = E_e - E_g$; that is, the emitted gamma-ray energy is exactly equal to the difference between the nuclear groundstate energy and the excited-state energy. Consequently, absorption of this gamma ray by a nucleus which is also firmly bound to a solid lattice can result in the "pumping" of the absorber nucleus from the ground state to the excited state. *See* ENERGY LEVEL (QUANTUM MECHANICS); EXCITED STATE; GAMMA RAYS; GROUND STATE.

In a typical Mössbauer experiment the radioactive source is mounted on a velocity transducer which imparts a smoothly varying motion (relative to the absorber, which is held stationary), up to a maximum of several centimeters per second, to the source of the gamma rays. These gamma rays are incident on the material to be examined (the absorber). Some of the gamma rays are absorbed and reemitted in all directions, while the remainder of the gamma rays traverse the absorber and are registered in an appropriate detector.

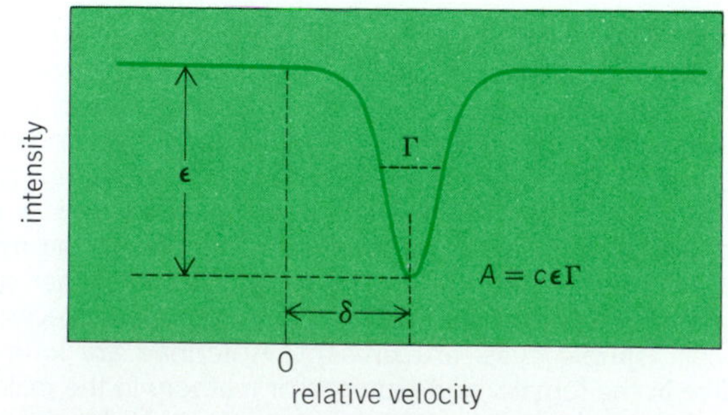

Mössbauer spectrum of an absorber which gives an unsplit resonance line. The spectrum is characterized by a position δ, a line width Γ, and an area *A* related to the effect magnitude ε.

A typical display of a Mössbauer spectrum, which is the result of many repetitive scans through the velocity range of the transducer, is shown in the illustration. In certain nuclides the Mössbauer resonance line displays splitting that arises from the coupling of the nuclear electric quadrupole moment with the electric field gradient or of the nuclear magnetic dipole moment with the magnetic field at the nucleus, providing information on the magnitude of these interactions.

Mössbauer effect experiments have been used to elucidate problems in a very wide range of scientific disciplines. Applications include the measurement of nuclear magnetic and quadrupole moments and of excited-state lifetimes involved in the nuclear decay process; study of the chemical consequences of nuclear decay; study of the nature of magnetic interactions in iron-containing alloys and of the dependence of the magnetic field in these alloys on various parameters; study of the effects of high pressure on chemical properties of materials; investigation of the relationship between chemical composition and structure on the one hand and the superconductive transition on the other; investigation of the structure of compounds; and study of the structure and bonding properties of metal atoms in complex biological molecules. [R.H.He.]

Moth An insect of the order Lepidoptera, which also includes the butterflies. Moths are principally nocturnal, heavy-bodied insects. The wings are folded in a rooflike position, wrapped around the body, or held in a horizontal position. The antennae have varied forms and shapes, and are rarely enlarged at the tip. The majority of adults have mouthparts adapted for sucking nectar from flowers. In a few species the mouthparts are vestigial; these species do not feed as adults.

Moths have a complete metamorphosis during development. The eggs are attached to plants where the wormlike larvae, known as caterpillars, develop. They are of considerable economic importance, since they are phytophagous, and many are serious pests of cultivated plants; also, some feed on fabrics, others on stored foods such as grains. Most caterpillars spin a cocoon in which they pupate and from which, after a period of rest or diapause, the imago or adult emerges. *See* LEPIDOPTERA. [C.B.C.]

Motion If the position of a material system as measured by a particular observer changes with respect to time, that system is said to be in motion with respect to the observer. Absolute motion, then, has no significance, and only relative motion may be defined; what one observer measures to be at rest, another observer in a different frame of reference may regard as being in motion. *See* FRAME OF REFERENCE; RELATIVE MOTION.

The time derivatives of the various coordinates used to specify the system may be used to prescribe the motion at any instant of time. How the motion develops in subsequent instants is then determined by the laws of motion. In classical dynamics it is supposed that in principle the motion and configuration of the system may be specified to an arbitrary precision, although in quantum mechanics it is recognized that the measurement of the one disturbs the other.

The most general theory of motion that has yet been developed is quantum field theory, which combines both quantum mechanics and relativity theory, as well as the experimentally observed fact that elementary particles can be created and annihilated. *See* DEGREE OF FREEDOM (MECHANICS); DYNAMICS; EULER'S EQUATION OF MOTION; HAMILTON'S EQUATIONS OF MOTION; HARMONIC MOTION; KINEMATICS; KINETICS (CLASSICAL MECHANICS);

Lagrange's equations; Newton's laws of motion; Oscillation; Periodic motion; Quantum field theory; Quantum mechanics; Rectilinear motion; Relativity; Rotational motion. [H.C.Co./B.G.]

Motor An electric rotating machine that converts electric energy into mechanical energy. Because of its many advantages, the electric motor has largely replaced other motive power in industry, transportation, mines, business, farms, and homes. Electric motors can meet a wide range of service requirements—starting, accelerating, running, braking, holding, and stopping a load. They are available in sizes from a small fraction of a horsepower (1 hp = 0.75 kilowatt) to many thousands of horsepower, and in a wide range of speeds. The speed may be fixed (or synchronous), constant for given load conditions, adjustable, or variable. Many motors are self-starting and reversible. *See* Electric rotating machinery.

Electric motors may be alternating-current (ac) or direct-current (dc). There are many types of each. Although ac motors are more common, dc motors are unexcelled for applications requiring simple, inexpensive speed control or sustained high torque under low-voltage conditions. *See* Alternating-current motor; Direct-current motor; Induction motor; Repulsion motor; Synchronous motor; Universal motor.

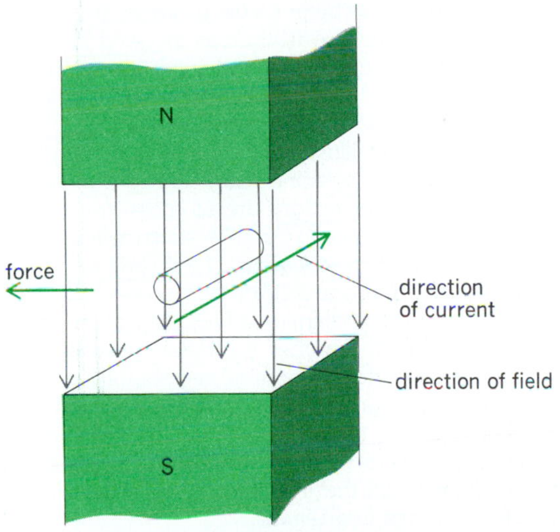

Relative directions of field flux, current, and force.

When a conductor located in a magnetic field carries current, a mechanical force is exerted upon it (see illustration). This force has the value shown in Eq. (1), where i is the current in amperes, B is the magnetic density in webers per square meter, and l is the conductor length in meters. The illustration shows the relative directions of current, field, and force. The force reverses with either current or field reversal, but not when both are reversed. The torque T is the product of this force and the rotor radius.

$$F = Bil \qquad \text{newtons} \qquad (1)$$

If the conductor moves in the direction of F, an electromotive force (emf) e is generated which opposes the current (motor action). If the conductor is moved against F, this emf will assist the current (generator action). Its value is shown in Eq. (2), where v is the velocity of the conductor across the flux in meters per second.

$$e = vBl \qquad \text{volts} \qquad (2)$$

The product ei represents the power converted, in watts, shown in Eqs. (3) and (4), which are the bases for the emf and output formulas of dynamo machinery. Many machines will operate as either a motor or a generator, but they should be designed for the particular service. *See* Generator. [A.F.P.]

$$\text{Motor output} = ei - \text{rotative loss} \qquad (3)$$

$$\text{Generator shaft input} = ei + \text{rotative loss} \qquad (4)$$

Motor-generator set A motor and one or more generators, with their shafts mechanically coupled, used to convert an available power source to another desired frequency or voltage. The motor of the set is selected to operate from the available power supply; the generators are designed to provide the desired output.

The principal advantage of a motor-generator set over other conversion systems is the flexibility offered by the use of separate machines for each function. Since a double energy conversion is involved, electrical to mechanical and back to electrical, the efficiency is lower than in most other conversion methods. *See* Generator; Motor. [A.R.E.]

Motor systems Those portions of nervous systems that regulate and control the contractile activity of muscle and the secretory activity of glands. Muscles and glands are the two types of organ by which an organism reacts to its environment; together they constitute the machinery of behavior. Cardiac muscle and some smooth muscle and glandular structures can function independently of the nervous system but in a poorly coordinated fashion. Skeletal muscle activity, however, is entirely dependent on neural control. Destruction of the nerves supplying skeletal muscles results in paralysis, an inability to move. The somatic motor system includes those regions of the central nervous system involved in controlling the contraction of skeletal muscles in a manner appropriate to environmental conditions and internal states. *See* Gland; Muscle.

Skeletal muscle. The nerve supply to skeletal muscles of the limbs and trunk is derived from large nerve cells called motoneurons, whose cell bodies are located in the ventral horn of the spinal cord. Muscles of the face and head are innervated by motoneurons in the brainstem. The axons of the motoneurons traverse the ventral spinal roots (or the appropriate cranial nerve roots) and reach the muscles via peripheral nerve trunks. In the muscle, the axon of every motoneuron divides repeatedly into many terminal branches, each of which innervates a single muscle fiber. The region of innervation, called the neuro-

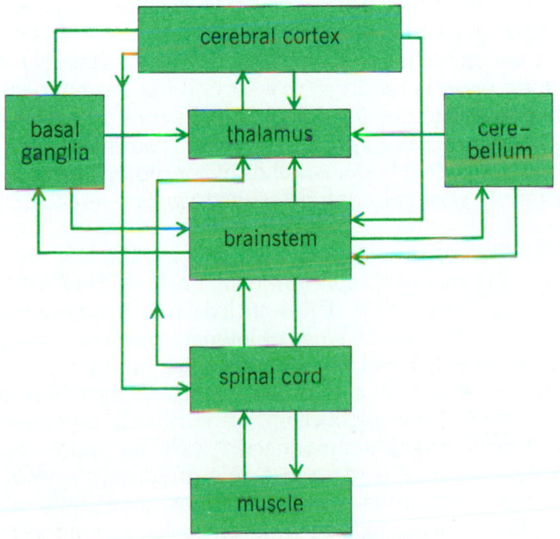

Schematic diagram of the major components of the vertebrate motor systems. Arrows indicate the main neural connections between regions.

muscular junction, or motor end plate, is a secure synaptic contact between the motoneuron terminal and the muscle fiber membrane. *See* SYNAPTIC TRANSMISSION.

Since synaptic transmission at the neuromuscular junction is very secure, an action potential in the motoneuron will produce contraction of every muscle fiber that it contacts. For this reason, the motoneuron and all the fibers it innervates form a functional unit, called the motor unit. The number of muscle fibers in a single motor unit may be as small as six (for intrinsic eye muscles) or over 700 (for motor units of large limb muscles). For a given muscle the average number of muscle fibers per motor unit can be determined by dividing the total number of fibers in the muscle by the number of motoneurons in its pool. The typical innervation ratio for limb muscles is on the order of 120–150. In general, muscles involved in delicate rapid movements have fewer muscle fibers per motor unit than large muscles concerned with gross movements.

Components of skeletal motor system. Motoneurons are activated by nerve impulses arriving through many different neural pathways. Some of their neural input originates in peripheral receptor organs located in the muscles themselves, or in receptors in skin or joints. Many muscle receptors discharge in proportion to muscle length or tension; such receptors have relatively potent connections to motoneurons, either direct monosynaptic connections or relays via one or more interneurons. Similarly, stimulation of skin and joints, particularly painful stimulation, can strongly affect motoneurons. Such simple segmental pathways constitute the basis for spinal reflexes. The other major source of input to motoneurons arises from supraspinal centers. The illustration shows the main nervous system centers involved in controlling the input to motoneurons.

Segmental circuits. At the spinal level, the motoneurons and muscles have a close reciprocal connection. Afferent connections from receptors in the muscles return sensory feedback to the same motoneurons which contract the muscle. Connections to motoneurons of synergist and antagonist muscles are sufficiently potent and appropriately arranged to subserve a variety of reflexes. In animals with all higher centers removed, these segmental circuits may function by themselves to produce simple reflex responses. Under normal conditions, however, the activity of segmental circuits is largely controlled by supraspinal centers. Descending tracts arise from two major supraspinal centers: the cerebral cortex and the brainstem.

Brainstem. The brainstem, which includes medulla and pons, is a major and complex integrating center which combines signals descending from other higher centers, as well as afferent input arising from peripheral receptors. The descending output from brainstem neurons affects motor and sensory cells in the spinal cord. Brainstem centers considerably extend the motor capacity of an animal beyond the stereotyped reflex reactions mediated by the spinal cord. In contrast to segmental reflexes, these motor responses involve coordination of muscles over the whole body. Another major motor function of the brainstem is postural control, exerted via the vestibular nuclei of the ear.

Besides neurons controlling limb muscles, the brainstem also contains a number of important neural centers involved in regulating eye movements. These include motor neurons of the eye muscles and various types of interneurons that mediate the effects of vestibular and visual input on eye movement.

Cerebellum. Another important coordinating center in the motor system is the cerebellum, an intricately organized network of cells closely interconnected with the brainstem. The cerebellum receives a massive inflow of sensory signals from peripheral receptors in muscles, tendons, joints, and skin, as well as from visual, auditory, and vestibular receptors. Higher centers, particularly the cerebral cortex, also provide extensive input to the cerebellum via pontine brainstem relays. The integration of this massive amount of neural input in the cerebellum somehow serves to smooth out the intended movements and coordinate the activity of muscles. Without the cerebellum, voluntary movements become erratic, and the animal has difficulty accurately terminating and initiating responses. The output of the cerebellum affects primarily brainstem nuclei, but it also provides important signals to the cerebral cortex.

Basal ganglia. At another level of motor system are the basal ganglia. These massive subcortical nuclei receive descending input connections from all parts of the cerebral cortex. Their output projections send recurrent information to the cerebral cortex via the thalamus, and their other major output is to brainstem cells.

Cerebral cortex. At the highest level of the nervous system is the cerebral cortex, which exerts control over the entire motor system. The cerebral cortex performs two kinds of motor function: certain motor areas exert relatively direct control over segmental motoneurons, via a direct corticospinal pathway, the pyramidal tract, and also through extrapyramidal connections via supraspinal motor centers. The second function, performed in various cortical association areas, involves the programming of movements appropriate in the context of sensory information, and the initiation of voluntary movements on the basis of central states. Cortical language areas, for example, contain the circuitry essential to generate the intricate motor patterns of speech. Limb movements to targets in extrapersonal space appear to be programmed in parietal association cortex. Such cortical areas involved in motor programming exert their effects via corticocortical connections to the motor cortex, and by descending connections to subcortical centers, principally basal ganglia and brainstem.

As indicated in the illustration, the motor centers are all heavily interconnected, so none really functions in isolation. In fact, some of these connections are so massive that they may form functional loops, acting as subsystems within the motor system. For example, most regions of the cerebral cortex have close reciprocal interconnections with underlying thalamic nuclei, and the corticothalamic system may be considered to form a functional unit. Another example is extensive connection from cerebral cortex to pontine regions of the brainstem, controlling cells that project to the cerebellum, which in turn projects back via the thalamus to the cerebral cortex. Such functional loops are at least as important in understanding motor coordination as the individual centers themselves. *See* BRAIN; NERVOUS SYSTEM (VERTEBRATE). [E.E.F.]

Motor vehicle A trackless, self-propelled vehicle for land transportation of people or commodities or for moving materials. Passenger and commercial vehicles have comparable components but separate identities, primarily because of their different functions. Passenger vehicles are built for the comfort and protection of occupants, whereas commercial vehicles are designed more ruggedly to withstand daily abuse and are available in an infinite variety of forms. *See* AUTOMOBILE; BUS; EARTHMOVER; TRACTOR.

Passenger vehicles include automobiles, recreation vehicles and all-terrain vehicles such as campers and motor homes with sleeping accommodations, motor bikes, motorcycles, scooters, trail bikes, dune buggies, and snowmobiles. Included in the broad range of body types and configurations of commercial vehicles are ambulances, airport limousines, taxicabs, buses, vans, tankers, light- and heavy-duty highway and industrial trucks, tractors, and self-propelled farm equipment, off-highway construction and excavation units, and various military vehicles.

A typical motor vehicle begins to operate when the ignition switch is turned on. Battery current is released so that the starter motor can crank the engine, as the carburetor delivers an explosive fuel mixture to the engine cylinders.

Simultaneously, the ignition system sends high-tension current to the spark plugs to ignite the fuel. Connecting rods transmit energy from the pistons to the crankshaft, while the camshaft (geared to the crankshaft) times intake or exhaust valve operation and firing intervals for spark plugs. Power impulses from each piston rotate the crankshaft, which is a source of motion for the vehicle. Power plant energy is transmitted to the power train, consisting of the clutch, transmission, drive shaft, and rear axle. In front- or rear-drive vehicles the power train is known as a transaxle when its elements are located in the front or rear. Engine output to the wheels, controlled by the driver, is governed by the power train and the accelerator for required speed and direction. *See* AUTOMOTIVE ENGINE; AUTOMOTIVE TRANSMISSION; AXLE; DIFFERENTIAL.

In addition to the major components (power plant and power train), automotive vehicles have several operating systems vital to proper functioning. The fuel system includes the fuel tank, fuel lines leading to the fuel pump, filter, and carburetor, with an air cleaner mounted on the carburetor to supply adequate clean air.

Conventional carburetors and intake manifolds have been replaced in some automobiles by fuel-injection systems that have a metering computer which controls fuel flow to each cylinder. This system produces better combustion and less fuel waste than conventional systems, which distribute fuel to all cylinders. The distribution is based on air and engine temperatures and manifold pressure and speed required, with automatic controls for all operating conditions. *See* CARBURETOR; FUEL INJECTION.

There are three basic sources of pollution from exhaust systems: fuel tank vapor, crankcase fumes, and exhaust fumes. Although conventional intake and exhaust systems remain in use, emission-control equipment and no-lead gasoline are now required in new vehicles. Positive crankcase ventilation (PCV) uses engine suction to draw crankcase and fuel tank fumes, as well as additional quantities of air, from the air cleaner into the system. Normal fuel and air mixtures for these added combustibles are controlled by the PCV valve; a vapor separator returns liquid to the fuel tank, or vapor to the PCV unit. The exhaust manifold has an air-injection system driven by an air pump belted to the crankshaft, to create afterburn of unexploded hot engine exhaust gas for more complete combustion. A catalytic converter inserted in the exhaust pipe line ahead of the muffler breaks down noxious nitrogen oxides, hydrocarbons, and carbon monoxide into harmless or inert matter.

At the heart of the whole electrical system are the ignition switch, battery, alternator, voltage regulator, and battery discharge sensor; without them none of the ancillary circuits would function. The ignition switch controls a half-dozen circuits which use these components; the starting, charging, and ignition circuits, lighting system, accessories, and warning sensors that keep the driver informed of changes in electrical, mechanical, fuel, oil, and other operating conditions that require correction. *See* AUTOMOTIVE ALTERNATOR; AUTOMOTIVE VOLTAGE REGULATOR; BATTERY; IGNITION SYSTEM.

Power plants are cooled by water or a solution of water and permanent antifreeze for year-round protection against boiling over in summer and freezing in winter. The system includes a water pump and fan driven by the engine, water jackets around cylinders in the engine head and block, a radiator to dissipate heat from the coolant, and a thermostat to control operating temperatures. Also included in cooling systems are vehicle heaters for passenger comfort, as well as a transmission oil cooling element. *See* ENGINE COOLING.

Several types of steering systems are used for manual or hydraulic power-assist control when the vehicle is in motion, with provisions for manual operation in the event of failure of the power unit. The introduction of power steering reduced the manual effort needed, and subsequently, variable-ratio steering has produced faster vehicle response when full right or left turns are negotiated. *See* AUTOMOTIVE STEERING; POWER STEERING.

Suspension systems are the means by which passenger and vehicle are protected from road shock due to irregular street and highway surfaces. Shock absorbers are designed for air or hydraulic operation, or for either type with auxiliary spring-loaded units for heavy-duty usage in conjunction with suspension systems. Mounting methods may be axle to frame or suspension unit to frame, or within the suspension unit itself. *See* AUTOMOTIVE SUSPENSION; SHOCK ABSORBER.

Conventional four-wheel drum brake systems have progressed to disk brakes; some imports are equipped with four-wheel disk brakes. *See* AUTOMOTIVE BRAKE; BRAKE.

Two lubricating systems are used in passenger and commercial vehicles to reduce friction and wear of moving engine and chassis parts. The engine lubrication system includes an oil pan and sump for oil supply and return, oil pump, and filter; oil level is measured by a dipstick and checked periodically for replenishment. Lubricant is also applied to the fittings of strategic working parts of the chassis at specified points and at prescribed service intervals for satisfactory vehicle operation.

[P.A.O.]

Motorboating A form of oscillation that occurs at a very low audio frequency in a system, circuit, or component. It is caused by excessive amount of audio feedback at low frequencies. This oscillation is a succession of pulses; when these occur in a circuit that is feeding a loudspeaker, the pulses result in putt-putt sounds resembling those made by a motorboat.

[J.Mar.]

Mountain A feature of the Earth's surface that rises high above its base and has generally steep slopes and a relatively small summit area. Commonly the features designated as mountains have local heights measurable in thousands of feet, lesser features of the same type being called hills, but there are many exceptions. *See* HILL AND MOUNTAIN TERRAIN.

Mountains rarely occur as isolated individuals. Instead they are usually found in roughly circular groups or massifs, such as the Olympic Mountains of northwestern Washington, or in elongated ranges, like the Sierra Nevada of California. An array of linked ranges and groups, such as the Rocky Mountains, the Alps, or the Himalayas, is a mountain system. North America, South America, and Eurasia possess extensive cordilleran belts, within which the bulk of their higher mountains occur. *See* CORDILLERAN BELT; MASSIF; MOUNTAIN SYSTEMS.

As a rule, mountains represent portions of the Earth's crust that have been raised above their surroundings by upwarping, folding, or buckling, and have been deeply carved by streams or glaciers into their present surface form. Some individual peaks and massifs have been constructed upon the surface by outpourings of lava or eruptions of volcanic ash. *See* OROGENY.

[E.H.Ha.]

Mountain meteorology The effects of mountains on the atmosphere, ranging over all scales of motion, including very small (such as turbulence), local (for instance, cloud formations over individual peaks or ridges), and global (such as the monsoons of Asia and North America).

The most readily perceived effects of a mountain, or even of a hill, are related to the blocking of air flow. When there is sufficient wind, the air either goes around the obstacle or over it, causing waves in the flow similar to those in a river washing over a boulder. Since ascending air cools by adiabatic expansion, the saturation point of water vapor may be reached in such waves as they form over an obstacle, and a cloud then forms in the ascending branch of the wave motion. Such a cloud dissipates in the descending branch where adiabatic warming takes place. The shapes and amplitudes of these lee waves (they form over and to the lee of mountains) depend not only on the thermal stability and on the vertical wind shear in the overlying atmosphere but also on the shape of the underlying terrain. *See* CLOUD; CLOUD PHYSICS; WAVE (PHYSICS).

On a grander scale, mountain ranges, such as the Sierras of North and South America, place an obstacle in the path of the westerly winds (that is, winds from the west), which generally prevail in middle latitudes. Such a blockage tends to generate a high-pressure region upwind from the mountains (this may be viewed as air piling up as it prepares to jump the hurdle), and a low-pressure area downwind. Thus, there is a stronger push against the mountains on the high-pressure western side than on the low-pressure eastern side. The net effect is the slowing down of the atmospheric flow (mountain torque). *See* TORQUE.

Less subtle than mountain torque effects are the large-scale meanders that develop in the global flow patterns once they have been perturbed, mainly by the North and South American Andes and by the Plateau of Tibet and its Himalayan mountain ranges. These meanders in the large-scale flow are known as planetary waves. They appear prominently in the pressure patterns of hemispheric or global weather maps. *See* WEATHER MAP.

The major monsoon circulations interact with the global circulation, shaped in part by sea-surface temperature anomalies in the equatorial Pacific. The various aspects of mountain meteorology, therefore, have to be viewed within the larger picture. There is a continuous interaction between the weather effects on all space and time scales generated by the mountains and the weather patterns that prevail elsewhere on the Earth. *See* METEOROLOGY; MONSOON METEOROLOGY; WIND. [E.R.R.]

Mountain systems Long, broad, linear to arcuate belts in the Earth's crust where extreme mechanical deformation and thermal activity have been (or are being) concentrated.

Mountain systems in the general sense occur both on continents and in ocean basins, but the geological properties of the systems in continental as opposed to oceanic settings are distinctly different. The mechanical strain in classical, continental mountain systems is expressed in the presence of major folds, faults, and intensive fracturing and cleavage. Thermal effects are in the form of vast volcanic outpourings, intruded bodies of igneous magma, and metamorphism. Uplift and deformation in young mountain systems are conspicuously displayed in the physiographic forms of topographic relief. Where mountain building is presently taking place, the dynamics are partly expressed in warping of the land surface and significant shallow or deep earthquake activity. Locations of ancient mountain systems in continental regions now beveled flat by erosion are clearly disclosed by the presence of highly deformed, intruded, and metamorphosed rocks.

Two basic classes of oceanic mountain systems exist. A world-encircling oceanic rift mountain system has been built along the extensional tectonic boundary between plates diverging at rates of 0.8–2.4 in. or 2–6 cm per year from the mid-oceanic ridges. This rift mountain system is exposed to partial view in Iceland. The second type, island arc mountain systems, occur in oceanic basins where the crust dives downward at trench sites, thus underthrusting adjacent oceanic crust. *See* MARINE GEOLOGY; OCEANIC ISLANDS.

The classical, conspicuous mountain systems of the Earth occur at the continent/ocean interface, for this is the site where plate convergence has led to major sedimentation, subduction of oceanic crust under continents, collision of island arc mountain systems with continents, and head-on collision of continents. *See* OROGENY; PLATE TECTONICS. [G.H.D.]

Mouse The name associated with any species of animals which are members of the families Muridae, Heteromyidae, Cricetidae, and Zapodidae in the order Rodentia. Some of the more common species are listed in the table. Many species are used for research in both biology and medicine. In addition to their use in studying the mechanisms of genetics, they are important in the study of carcinogenesis, effects of drugs, and virology. They are also important experimental animals in studying cell physiology, such as for cell and tissue culture research. The familiar white mouse is an albino form of the house mouse and is used extensively in research.

The common house mouse is one of the oldest known species of domestic rodent pests. It usually has a maximum lifespan of 4 years with four to six litters of four to eight young each per year. The gestation period is about 3 weeks. These rodents begin to breed at 3 months of age. Adults have a pointed snout, compact body, and an equally long tail. The ears are fairly large, as are the legs. While omnivorous, they have a preference for grains and other vegetable foods. Of the 44 species known, only one species occurs in the United States, and it has become wild in some parts of the country. *See* RODENTIA. [C.B.C.]

Classification and representative species of mice

Families and subfamilies	Examples
Family: Heteromyidae	
Subfamily: Perognathinae	Pocket and kangaroo mice
Subfamily: Heteromyinae	Spiny pocket mice
Family: Cricetidae	
Subfamily: Cricetinae	Climbing mice, harvest mice, water mice, white-footed mice, pygmy mice
Family: Muridae	
Subfamily: Murinae	Striped mice, house mice, spiny mice, harvest mice, field mice, forest mice
Subfamily: Dendromurinae	African tree mice
Family: Zapodidae	Jumping mice

Moving-coil frequency meter An instrument that employs one or more pivoted moving coils and resonant circuits to measure frequency.

The illustration shows one design of a moving-coil resonant-type instrument. Two coils, A and B, are placed at right angles to form a moving element and are supplied through a resonant circuit. The coils are tuned to different frequencies, coil A to a frequency slightly below the lowest scale point and coil B an equal amount above the upper scale point. A fixed coil C divided into two parts carries the sum of the currents of the two circuits of coils A and B. When the frequency of the applied voltage equals the midscale frequency the currents in the two resonant circuits are practically equal and the phase lead of one coil equals the phase lag of the other with respect to the fixed-coil current. At frequencies near the low end of the scale,

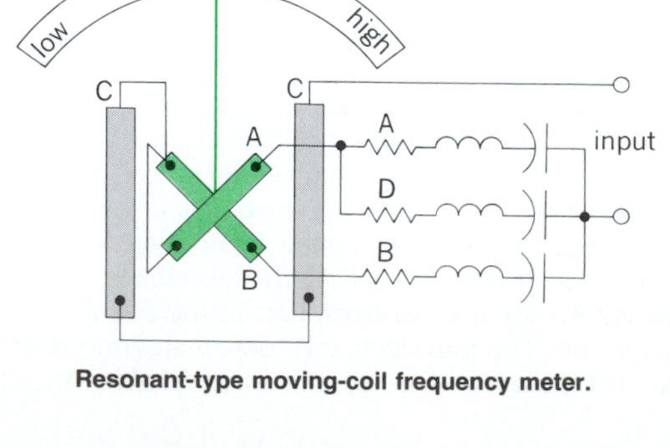

Resonant-type moving-coil frequency meter.

the moving element comes to rest with coil A approaching a position parallel to the fixed coil. As the circuit of coil A approaches resonance, the current in coil A is relatively high and in phase with the fixed-coil current and the current in coil B leads with respect to the fixed-coil current. At frequencies in the higher range, coil B approaches resonance and takes an equilibrium position parallel to the plane of the fixed coil. If the supply frequency is very low, the impedance of the circuit A and B being equal, the pointer will indicate a fictitious midscale reading. To rectify this misleading measurement, a third coil D is introduced in parallel with coil A and resonates at frequencies much less than the lowest scale point. *See* FREQUENCY MEASUREMENT. [E.B.C.]

Moving-iron frequency meter An instrument in which the movement of a pivoted iron vane or needle under the influence of current-carrying coils is used to measure frequency.

A resonant-type frequency meter operating on the moving-iron principle is shown in Fig. 1. Two field coils are mounted opposite to each other and are connected so that their fluxes oppose. If i_1 and i_2 are moving-coil currents, the resultant field current will be $i_1 - i_2$. The moving system which lies between the field coils is composed of an iron vane of magnetic material centered within the armature coils and rigidly attached to the shaft. Both the field coils are connected to series-resonance circuits in which resonance is produced below the operating range for one circuit and above the operating range for the other.

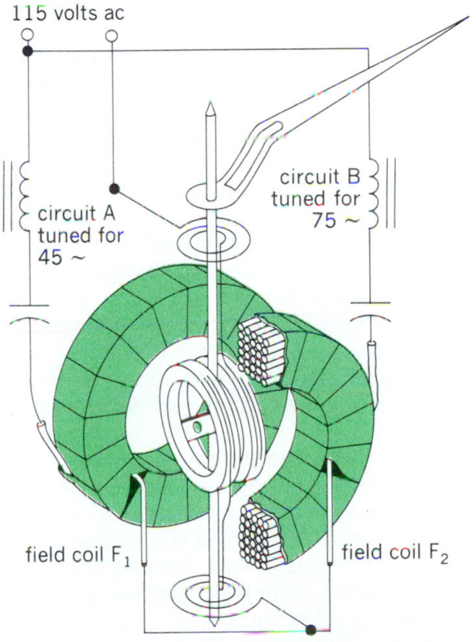

Fig. 1. Resonant-type frequency meter employing the moving-iron principle. (*General Electric Co.*)

Deflecting torque is produced by the reaction of the armature, or moving coil, field with the in-phase component of the field coils. In addition to this deflecting torque there is a countertorque produced by the action of the field flux in the magnetic vane of the moving element. Equilibrium occurs when the magnitude of the countertorque is equal to the deflecting torque, which causes the pointer to take a position indicating the frequency.

Another design of a moving-iron frequency meter, but of the ratiometer type, is shown in Fig. 2. Two coils A and B are arranged at right angles to each other. Coil A is in series with a

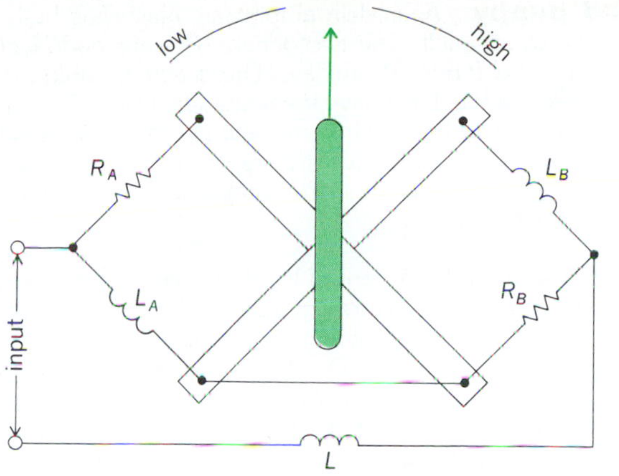

Fig. 2. Ratiometer-type Weston frequency meter which uses the moving-iron principle.

resistance R_A, and this combination is in parallel with an inductance L_A. Similarly, coil B is in series with L_B, and the combination is in parallel with R_B. The complete circuit thus acts as a bridge network. When the bridge is balanced by suitable values of parameters, the currents in the two coils are equal and the moving element, a pivoted soft-iron needle, will indicate the mean-frequency position as shown in the diagram. *See* FREQUENCY MEASUREMENT. [E.B.C.]

Moving-target indication A method of presenting pulse-radar echoes in a manner that discriminates in favor of moving targets and suppresses stationary objects. Moving-target indication (MTI) is almost a necessity when moving targets are being sought over a region from which the ground clutter echoes are very strong. The most common presentation of the output of a radar with MTI is a plan-position indicator (PPI) display. The moving targets appear as bright echoes, while ground clutter is suppressed. *See* RADAR. [J.M.C.]

Mucilage A naturally occurring, high-molecular-weight (200,000 and up), organic plant product of unknown detailed structure. The term is loosely used, often interchangeably with the term gum. Chemically, mucilage is closely allied to gums and pectins but differs in certain physical properties. Although gums swell in water to form sticky, colloidal dispersions and pectins gelatinize in water, mucilages form slippery, aqueous colloidal dispersions. Mucilages are formed in normal plant growth within the plant by mucilage-secreting hairs, sacs, and canals, but they are not found on the surface as exudates as a result of bacterial or fungal action after mechanical injury, as are gums. Mucilages occur in nearly all classes of plants in various parts of the plant, usually in relatively small percentages, and are not infrequently associated with other substances, such as tannins. The chief industrial sources of mucilages are Icelandic and Irish moss, linseed, locust bean, slippery elm bark, and quince seed. *See* ADHESIVE; GUM; PECTIN. [E.H.H.]

Mucormycosis A severe infection caused by ubiquitous molds of the order Mucorales which occurs only in people who have deficient immune responses. The mold is found in soil and on fruits and may grow on bread. While most people have contact with these fungi without becoming infected, certain individuals with immune defects develop invasive infections which result in disease. The types of disease that result from Mucorales infection are limited. The mold first invades the nasal passages and then spreads directly through sinuses and the orbit to the brain or via a major blood vessel. One common finding is arterial clotting with tissue necrosis or venous clotting with obstruction. [D.Ar.]

Mud puppy A urodele amphibian which has both lungs and gills as an adult. The two genera, *Proteus* and *Necturus*, make up the family Proteidae. These aquatic animals have small lidless eyes. They have vomerine teeth as well as palatine and pterygoid teeth, which in this animal are not separate. The vertebrae are amphicoelous, that is, biconcave. Fertilization is internal, and there is a sperm receptacle in the female. *See* AMPHIBIA; URODELA. [C.B.C.]

Mud volcano A conical landform composed of clay and silt with variable admixtures of sand and rock fragments, the whole resulting from eruption of wet mud and impelled upward by fluid or gas pressure. The form of such features depends upon the ratio of mineral matter to the fluids and gases in the mobile material. Associated features include mud pits, developed where the mobile material contains excessive amounts of fluid and gas, and pitch lakes, formed where the bitumens of petroleum residues are at least as voluminous as the clay particles in the viscous mud. Minor amounts of petroleum residues are often found in mud volcanoes and associated features, but pitch lakes are comparatively rare phenomena. [H.H.Su.]

Muffler A device used to attenuate sound being propagated in conjunction with a moving stream of fluid (usually gas). Mufflers may be placed in two general classifications, depending upon the mode by which sound energy is removed from the gas stream. Those devices which attenuate by reflecting sound back to the source are called nondissipative, or reactive. Devices which absorb sound energy as the gas passes through them are called dissipative devices.

Reactive types are predominantly used in the intake and exhaust of automotive and other reciprocating engines and compressors. Reflection is provided through the use of acoustic filters, resonators, and change in direction by bends in the pipe confining the gas stream. These mufflers are particularly useful for low-frequency applications and for those installations where high temperatures or flammable gases restrict the use of dissipative materials. *See* ACOUSTIC FILTER; ACOUSTIC RESONATOR.

Dissipative types are used in a wide range of applications where low pressure drop and high attenuation at predominantly middle and high frequencies are required. Many different configurations and types of materials can be employed, depending upon the individual application. In a simple ventilating system for a home or office the structure might consist of lining the interior walls of the duct with a blanket of glass fiber material of the order of 1 in. (25 mm) in thickness. On the other hand, a large wind tunnel or jet-engine test cell exhaust might require a complicated design of ducts, absorbent baffles in a parallel stack, cylinders of absorbing material mounted as a spatial array, or one of a number of proprietary designs combining these elements. *See* SOUND ABSORPTION. [W.J.G.]

Mulberry A genus (*Morus*) of trees characterized by milky sap and simple, often lobed, alternate leaves. White mulberry (*M. alba*) was introduced into the United States from China during the 19th century as a source of food for silkworms. The silkworm project was unsuccessful, but the trees remained and are common in cities and on the borders of forests. Red mulberry (*M. rubra*) grows in the eastern half of the United States and in southern Ontario. The wood is used for fence posts, furniture, interior finish, agricultural implements, and barrels. *See* URTICALES. [A.H.G./K.P.D.]

Mule The mule is a hybrid sired by a male ass (*Equus asinus*) out a female horse (*E. caballus*). The opposite cross, very seldom made, produces the hinny (a hybrid between a stallion and a female ass). The mule and the hinny are usually sterile, but two authenticated cases of mules producing living progeny are known. Male mules, often called horse mules, are almost always castrated to make them more tractable as work animals. Mules are noted for their endurance, surefootedness, and ability to stand hard work in hot weather. They can safely be self-fed in lots or corrals, whereas horses cannot. Usually steady and free from nervous excitability, mules can be handled by inexperienced or careless farm labor. [J.M.K.]

Multiaccess computer A computer system in which computational and data resources are made available simultaneously to a number of users. Users access the system through terminal devices, normally on an interactive or conversational basis. A multiaccess computer system may consist of only a single central processor connected directly to a number of terminals (that is, a star configuration), or it may consist of a number of processing systems which are distributed and interconnected with each other as well as with the user terminals.

The primary purpose of multiaccess computer systems is to share resources. The resources being shared may be simply the data-processing capabilities of the central processor, or they may be the programs and the data bases they utilize. The earliest examples of the first mode of sharing are the general-purpose, time-sharing, computational services. Examples of the latter mode are airlines reservation systems in which it is essential that all ticket agents have immediate access to current information. *See* DATABASE MANAGEMENT SYSTEMS.

System components. The major hardware components of a multiaccess computer system are terminals or data entry/display devices, communication lines to interconnect the terminals to the central processors, a central processor, and on-line mass storage. Terminals may be quite simple, providing only the capabilities for entering or displaying data, or they may have an appreciable amount of "local intelligence" to support simple operations like editing of the displayed text without requiring the involvement of the central processor. The interconnecting communication lines can be provided by utilizing the common-user telephone system or by obtaining leased, private lines from the telephone company or a specialized carrier.

System operating requirements. A multiaccess system must include the following functional capabilities: (1) multiline communications capabilities that will support simultaneous conversations with a reasonably large number of remote terminals; (2) concurrent execution of a number of programs with the ability to quickly switch from executing the program of one user to executing that of another; (3) ability to quickly locate and make available data stored on the mass storage devices while at the same time protecting such data from unauthorized access.

The ability of a system to support a number of simultaneous sessions with remote users is an extension of the capability commonly known as multiprogramming. In order to provide such service, certain hardware and software features should be available in the central processor. Primary among these is the ability to quickly switch from executing one program to another while protecting all programs from interference with one another.

Memory sharing is essential to the efficient operation of a multiaccess system. A popular memory management technique is the utilization of paging. The program is broken into a number of fixed-size increments called pages. Similarly, central memory is divided into segments of the same size called page frames. (Typical sizes for pages and page frames are 512 to 4096 bytes.) Under the concept known as demand paging, only those pages that are currently required by the program are loaded into central memory.

Software capabilities. The control software component of most interest to an interactive user is the command interpreter. This routine interacts directly with users, accepting requests for service and translating them into the internal form required by the remainder of the operating system, as well as controlling all interaction with the system.

The capability to page the memory as outlined above can be utilized to provide users with the impression that each has available a memory space much larger than is actually assigned. Such a system is said to provide a virtual memory environment. Similarly, the ability of the operating system to quickly change context from one executing program to another will result in users' receiving the impression that each has an individual processor. *See* DATA-PROCESSING SYSTEMS; DIGITAL COMPUTER. [P.H.E.]

Multilevel control theory An approach to the control of large-scale systems based on (1) decomposition of the complex overall control problem into simpler and more easily managed subproblems, and (2) coordination of the subproblems so that overall system objectives and constraints are satisfied.

The controllers are organized in a multilevel hierarchical structure according to three basic criteria: functional decomposition, plant decomposition, and temporal decomposition.

In the functional decomposition approach the overall control problem is partitioned into a nested set of generic control functions.

In the plant decomposition approach the controlled system (plant) is partitioned into subsystems along lines of weak interaction. Each subsystem has its own (first-level) controller which acts to satisfy local objectives and constraints. A second-level controller (coordinator) influences the actions of the local controllers to compensate for subsystem interactions so that overall objectives and constraints are satisfied.

The multilevel structure induces an ordering with respect to time scale; specifically, the mean period of control action tends to increase as one proceeds from a lower to a higher level of the hierarchy. This motivates the concept of a temporal control hierarchy wherein a control or decision-making problem is partitioned into subproblems based on the different time scales relevant to the associated action functions. These time scales reflect such factors as the response time of the plant, the band-width characteristics of the disturbance inputs, and trade-off considerations relating the benefit of control action to its cost. The temporal hierarchy may embrace a broad spectrum of control/decision-making activities, including process control, production control, scheduling, and planning functions. These may range in time scale from seconds to weeks to years. *See* CONTROL SYSTEMS. [I.L]

Multimedia technology Synergistic union of digital video, audio, computer, information, and telecommunication technologies. Information services have traditionally been available in several ways. On the educational front, information acquisition, authoring, composition, publishing, and delivery have been predominantly in printed form (as journals, newspapers, and books). By contrast, entertainment has been produced and delivered via television and radio broadcasts, predominantly in analog form. Advances in hardware and software have computerized the processes of information acquisition, authoring, composition, and publishing. Large-capacity storage disks and high-speed fiber-optic networks enable the integration of diverse media, such as text, audio, and video, during storage and delivery. These advances have stimulated the development of multimedia technologies in which information in different forms (for example, journals, newspapers, books, movies, encyclopedias) are digitally composed and published on multimedia platforms, stored on high-capacity servers, and made available via broadband networks. *See* COMPACT DISK; LOCAL-AREA NETWORKS; OPTICAL RECORDING.

Multimedia technologies have had a wide-ranging impact on both education and entertainment, and offer innovative applications. For instance, digital multimedia libraries, rather than being a central storehouse for information, can function as rapid-access mechanisms to geographically distributed electronic databases. Such "libraries without walls" provide ubiquitous access to hundreds of millions of items, including books, lectures, historical documents, photos, maps, laws, cartoons, and software, and enable self-paced education that accommodates different learning styles. Phone lines provide two-way video communication. Users' sites equipped with fiber-optic networks can support hundreds of entertainment channels, including teleshopping channels and pay-per-view video channels.

The key issue for multimedia information management is the enormous size of digital video. Compression techniques help reduce the sizes of multimedia information by factors of tens to hundreds. The MPEG standard for video compression strives to reduce the digital video rate to about 1.2 megabits/s so as to enable storage and retrieval on CD-ROMs. *See* COMPUTER. [P.V.R.]

Multimeter A common term for a volt-ohm-milliammeter, also called an analyzer or circuit analyzer. A much less common usage applies to a self-contained test instrument containing two or more single- or multi-scale indicating instruments for measuring simultaneously two or more electrical quantities. *See* VOLT-OHM-MILLIAMMETER. [I.F.K./E.C.St.]

Multiple cropping A farming technique for growing several food crops on the same field in one year. Multiple cropping is most widely used in the tropical regions of South and Central America, Africa, and Asia. The advantages of multiple cropping in the densely populated sections of these regions enables farmers to produce on their own small farms the food needed to feed their families. Often four or five different food crops can be grown in succession on the same land in one calendar year. Such practices make it possible to support very dense populations. [R.Br.]

Multiple proportions, law of This law states that, when two elements combine together to form more than one compound, the weights of one element that unite with a given weight of the other are in the ratio of small whole numbers. The law can be illustrated by the composition of the five oxides of nitrogen. One gram of nitrogen is combined with 2.85 g of oxygen in nitrogen pentoxide, N_2O_5; with 2.28 g in nitrogen dioxide, NO_2; with 1.71 g in nitrogen trioxide, N_2O_3; with 1.14 g in nitric oxide, NO; and with 0.57 g in nitrous oxide, N_2O. These numbers are in the simple ratio of 5:4:3:2:1. *See* DEFINITE COMPOSITION, LAW OF. [T.C.W.]

Multiple sclerosis A disease which affects otherwise healthy young adults with multiple lesions (demyelinated plaques) of the central nervous system; blindness, urinary incontinence, incoordination, dysarthria (slurred speech), numbness, and paralysis are common. These signs may develop rapidly or slowly and may disappear just as mysteriously. Complete recovery from the acute episode occurs in about two-thirds of all cases, so that evaluation of any treatment is difficult. Although most cases are characterized by repeated remissions and exacerbations, about 25% are steadily and slowly progressive without the typical ups and downs. The average duration of illness is about 27 years, so that it frequently puts a severe psychological and economic stress upon the patient and the family. [E.C.A.; C.-M.S.]

Multiplexing The means or result of the means whereby a number of independent messages, each unambiguously represented by information-bearing signals, may be sent and received over a common transmitting medium. Viewed broadly, a multiplex system is any part of a communication system in which two or more distinct signal channels are combined.

Multiplexing is generic to most present-day systems of telecommunications. Many types of multiplexing are possible. Typical examples are time division, frequency division, and phase discrimination. *See* ELECTRICAL COMMUNICATIONS. [H.S.Bl.]

Multiplication One of the fundamental operations of arithmetic and algebra. The symbol × is commonly employed in arithmetic to denote multiplication. Because of its resemblance to the letter *x*, it is rarely used in algebra, where multiplication is frequently denoted by a dot (as in $a \cdot b$) or, most often, merely by juxtaposition of letters (for example, *ab*), Multiplication of numbers (real or complex) is associative, $a(bc) = (ab)c$; commutative, $ab = ba$; and distributive with respect to addition, $a(b + c) = ab + ac$; but the term has been extended to denote binary operations on many other kinds of objects, and these operations need not possess all the properties of ordinary multiplication listed above (for example, multiplication of matrices is not commutative). *See* ADDITION; ALGEBRA; DIVISION; NUMBER SYSTEMS; SUBTRACTION. [L.M.Bl.]

Multipole radiation Gamma rays, internal conversion electrons, or positron-electron pairs of defined characteristics emitted from an atom when the nucleus makes a transition between two energy states. The multipole order is the number of units of angular momentum removed by the radiation. This number is not necessarily equal to the difference between the spins of the nucleus in its initial and final states because the nuclear spin direction may change. Thus a quadrupole radiation will result when a state of spin 2 makes a transition to a state of spin 0, but a transition from a state of spin 2 to one of spin 1 may also result in quadrupole radiation if there is an appropriate change in nuclear spin direction.

From multipole radiation measurements, the static and dynamic properties of nuclear energy states may be determined, and this information may be used in theories of nuclear structure.

In addition to the energy and angular momentum values, a third characteristic of a nuclear state is the parity. There are two classes of multipole radiation, the electric and the magnetic, and the designation of a given radiation, depends upon both the angular momentum change and whether the parities of the initial and final states are the same or different. *See* NUCLEAR RADIATION; PARITY (QUANTUM MECHANICS); SPIN (QUANTUM MECHANICS). [D.E.A.]

Multiprocessing An organizational technique in which a number of processor units are employed in a single computer system to increase the performance of the system in its application environment above the performance of a single processor of the same kind. In order to cooperate on a single application or class of applications, the processors share a common resource. Usually this resource is primary memory, and the multiprocessor is called a primary memory multiprocessor. A system in which each processor has a private (local) main memory and shares secondary (global) memory with the others is a secondary memory multiprocessor, sometimes called a multicomputer system because of the looser coupling between processors. The more common multiprocessor systems incorporate only processors of the same type and performance and thus are called homogeneous multiprocessors; however, heterogeneous multiprocessors are also employed. A special case is the attached processor, in which a second processor module is attached to a first processor in a closely coupled fashion so that the first can perform input/output and operating system functions, enabling the attached processor to concentrate on the application workload. *See* COMPUTER STORAGE TECHNOLOGY; OPERATING SYSTEM.

Multiprocessor systems may be classified into four types: single instruction stream, single data stream (SISD); single instruction stream, multiple data stream (SIMD); multiple instruction stream, single data stream (MISD): and multiple instruction stream, multiple data stream (MIMD). Systems in the MISD category are rarely built. The other three architectures may be distinguished simply by the differences in their respective instruction cycles:

In an SISD architecture there is a single instruction cycle; operands are fetched in serial fashion into a single processing unit before execution. Sequential processors fall into this category.

An SIMD architecture also has a single instruction cycle, but multiple sets of operands may be fetched to multiple processing units and may be operated upon simultaneously within a single instruction cycle. Multiple-functional-unit, array, vector, and pipeline processors are in this category. *See* SUPERCOMPUTER.

In an MIMD architecture, several instruction cycles may be active at any given time, each independently fetching instructions and operands into multiple processing units and operating on them in a concurrent fashion. This category includes multiple processor systems in which each processor has its own program control, rather than sharing a single control unit.

MIMD systems can be further classified into throughput-oriented systems, high-availability systems, and response-oriented systems. The goal of throughput-oriented multiprocessing is to obtain high throughput at minimal computing cost in a general-purpose computing environment by maximizing the number of independent computing jobs done in parallel. High-availability multiprocessing systems are generally interactive, often with never-fail real-time online performance requirements.

The goal of response-oriented multiprocessing (or parallel processing) is to minimize system response time for computational demands. *See* COMPUTER SYSTEMS ARCHITECTURE; CONCURRENT PROCESSING; FAULT-TOLERANT SYSTEMS; REAL-TIME SYSTEMS. [P.C.Pa.]

Multituberculata The single order of the nominally mammalian subclass Allotheria. This extremely successful order of mammals is known to have flourished on northern continents from Late Jurassic time until the end of the Eocene, finally dying out about 40,000,000 years ago. The only mammalian herbivores of the Mesozoic, these animals differentiated by Late Cretaceous time into a number of distinct phyla which persisted generally through the Paleocene but which, during the Eocene, succumbed to competition from condylarths, primates, and finally rodents. Three suborders of the order Multituberculata are known, Plagiaulacida, Ptilodontoidea, and Taeniolabidoidea. *See* ALLOTHERIA. [M.C.McK.]

Multivariable control The control of systems characterized by multiple inputs, which are usually referred to as the controls; or by multiple outputs, which are often the measured variables to be controlled (see illustration); or by both multiple inputs and outputs. Automobiles, chemical processing plants, aerospace vehicles, biological systems, and the national economy are all examples of multivariable systems which require and

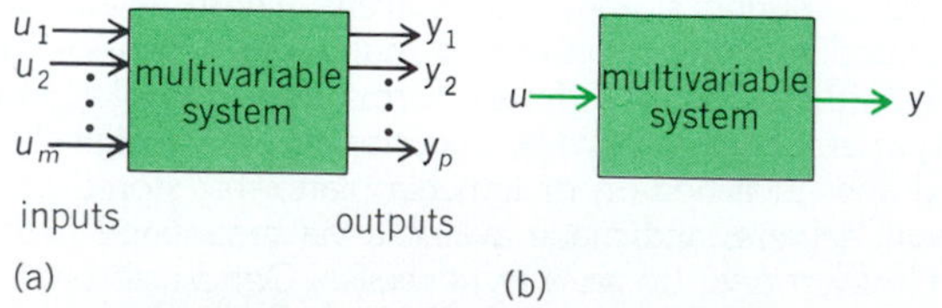

Two multivariable system representations. (*a*) Input and output variables shown separately as $u_1, u_2, \ldots, u_m$ and $y_1, y_2, \ldots, y_p$. (*b*) Input and output variables shown in vector form as *u* and *y*.

receive some form of regulation or control, be it mathematically contrived or not. In many cases, control of such systems is implemented without a previously given design or explicit mathematical model, such as in the manual control of relatively simple motorized vehicles, or in a person's control of arm and leg movements. In more complex tasks, however, such as the manual manipulation of an inherently unstable helicopter, automatic control devices (autopilots) are employed to assist the operator.

The control objectives to be achieved in the multivariable case include those control objectives which are generally sought in the scalar, that is, single input/output case, and also depend on the particular application involved. In particular, stability is usually a primary concern in the design of any control system, along with various measures of the degree of stability. While controller failure in the scalar case can lead to catastrophic failure of the overall system, such need not be true in the multivariable case due to the interactions between various input/output pairs. More specifically, in certain applications it may be possible to design a multivariable controller "driven" by all p outputs which performs satisfactorily even when feedback information from one or more outputs is lost. The integrity of such a multivariable controller would be much better than that of its scalar counterpart.

In many control applications, a primary objective is that of tracking, or ensuring that the system output or outputs track a desired input or inputs with little or no steady-state errors. In addition, simultaneous regulation of external disturbances is also often desired; that is, ensuring that any external disturbances affect the plant outputs as little as possible. It is also obvious that simplicity of controller design is a highly desirable design objective both in the scalar and multivariable cases. One final design objective unique to the multivariable case is that of minimizing or eliminating the interaction between loops, which is often referred to as decoupling; that is, perfect decoupling would imply a controlled system where each input affects only one output or, otherwise stated, a system whose compensated transfer matrix is diagonal and nonsingular. *See* CONTROL SYSTEMS; PROCESS CONTROL. [W.Wo.]

Multivibrator A form of relaxation oscillator consisting normally of two or more active devices, such as transistors, interconnected by electric networks. In a multivibrator a portion of the output voltage or current of each active device is applied to the input of the other with such magnitude and polarity as to maintain the devices alternately conducting over controllable periods of time. The transition time of each device from one state to the other is extremely short. As a consequence the voltage waveform from the output of each of the devices is essentially rectangular in form.

Multivibrators are classified by the manner in which the reversal-of-state action of each device is initiated and by the method of control of the time interval in each state, whether from external sources or from the decay of voltage across a capacitor in circuits containing *RC* time constants within the multivibrator itself.

Symmetrical bistable multivibrator. In bistable multivibrators either of the two devices may remain conducting, with the other nonconducting, until the application of an external pulse. Such a multivibrator is said to have two stable states.

The original form of bistable multivibrator made use of vacuum tubes and was known as the Eccles-Jordan circuit. It was sometimes called a flip-flop or binary circuit because of the two alternating output voltage levels. The junction field-effect transistor (JFET) circuit (Fig. 1) is a modern version of the Eccles-Jordan circuit. Its resistance networks between positive and negative supply voltages are such that, with no current flowing to the drain of the first JFET, the voltage at the gate of the second is slightly negative, zero, or limited to, at most, a slightly

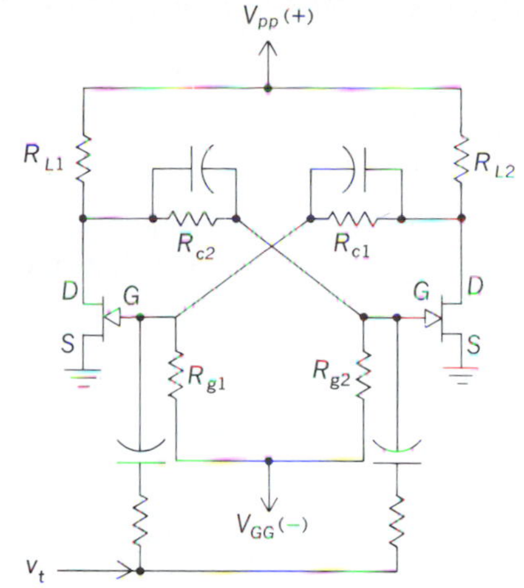

Fig. 1 Bistable multivibrator. *D* = drain; *S* = source; *G* = gate; V_t = input (trigger) voltage. The *V*'s indicate voltages, and the *R*'s, resistances.

positive value. The resultant current in the drain circuit of the second JFET causes a voltage drop across the drain load resistor; this drop in turn lowers the voltage at the gate of the first JFET to a sufficiently negative value to continue to reduce the drain current to zero. This condition of the first device OFF and second ON will be maintained as long as the circuit remains undisturbed.

If a fast negative pulse is then applied to the gate of the ON transistor, its drain current decreases and its drain voltage rises. A fraction of this rise is applied to the gate of the OFF transistor, causing some drain current to flow. The resultant drop in drain voltage, transferred to the gate of the ON transistor, causes a further rise at its drain. The action is thus one of positive feedback, with nearly instantaneous transfer of conduction from one device to the other. There is one such reversal each time a pulse is applied to the gate of the ON transistor. Normally pulses are applied to both transistors simultaneously so that whichever device is ON will be turned off by the action.

A bipolar transistor counterpart of the JFET bistable multivibrator with *npn* bipolar transistors may be used. The base of the transistor corresponds to the gate, the emitter to the source, and the collector to the drain. Insulated-gate field-effect transistors (IGFETs) may also be used in multivibrators.

Unsymmetrical bistable circuits. Bistable action can be obtained in the emitter- or source-coupled circuit with one of the set of cross-coupling elements removed (Fig. 2). In this case, regenerative feedback necessary for bistable action is obtained by the one remaining common coupling element, leaving one emitter or gate free for triggering action. Biases can be adjusted such that device 1 is ON, forcing device 2 to be OFF. In this case, a pulse can be applied to the free input in such a direction as to reverse the states. Alternatively, device 1 may initially be OFF with device 2 ON. Then an opposite polarity pulse is required to reverse states. Such an unsymmetrical bistable circuit, historically referred to as the Schmitt trigger circuit, finds widespread use in many applications.

Monostable multivibrator. A monostable multivibrator has only one stable state. If one of the normal active devices is in the conducting state, it will remain so until an external pulse is applied to make it nonconducting. The second device is thus made conducting and remains so for a duration dependent upon *RC* time constants within the circuit itself.

Astable multivibrator. The astable multivibrator has capacitance coupling between both of the active devices and therefore has no permanently stable state. It will generate a periodic

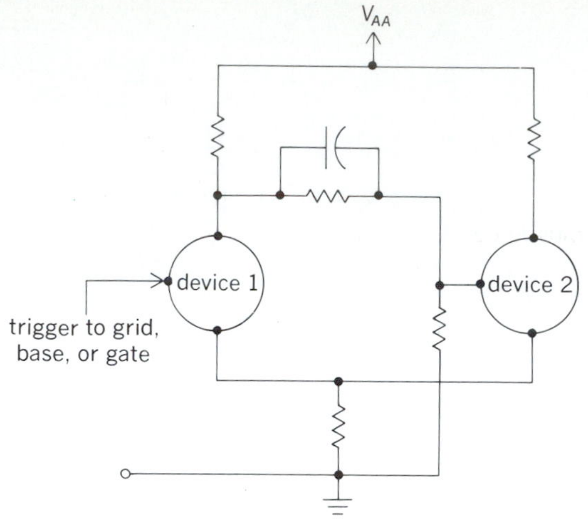

Fig. 2. Unsymmetrical bistable multivibrator.

rectangular waveform at the output with a period equal to the sum of the OFF periods of the two devices.

Astable multivibrators, although normally free-running, can be synchronized with input pulses recurrent at a rate slightly faster than the natural recurrence rate of the device itself. If the synchronizing pulses are of sufficient amplitude, they will bring the internal waveform to the conduction level at an earlier than normal time and will thereby determine the recurrence rate.

Logic gate multivibrators. Multivibrators may be formed by using two cross-coupled logic gates, with the unused input terminals used for triggering purposes. Such circuits are called flip-flops. *See* LOGIC CIRCUITS; TRANSISTOR. [G.M.G.]

Mumps An acute contagious viral disease, characterized chiefly by enlargement of the parotid glands (parotitis).

Besides fever, the chief signs and symptoms are the direct mechanical effect of swelling on glands or organs where the virus localizes. One or both parotids may swell rapidly, producing severe pain when the mouth is opened. In orchitis, the testicle is inflamed but is enclosed by an inelastic membrane and cannot swell; pressure necrosis produces atrophy, and if both testicles are affected, sterility may result. The ovary may enlarge, without sequelae.

An attenuated live virus vaccine can induce immunity without parotitis. It is recommended particularly for adults exposed to infected children, for students in boarding schools and colleges, and for military troops. [J.L.Me.]

Muonium An exotic atom, Mu or (μ^+e^-), formed when a positively charged muon (μ^+) and an electron are bound by their mutual electrical attraction. It is a light, unstable isotope of hydrogen, with a muon replacing the proton. Muonium has a mass 0.11 times that of a hydrogen atom due to the lighter mass of the muon, and a mean lifetime of 2.2 microseconds, determined by the spontaneous decay of the muon ($\mu^+ \rightarrow e^+\nu_e\bar{\nu}_\mu$). Muonium is formed when beams of μ^+ produced in particle accelerators are stopped in certain nonmetallic targets.

Since muonium is a system consisting only of leptons, it serves as a testing ground for the theory of quantum electrodynamics (QED), which describes the electromagnetic interaction between particles. Muonium chemistry and muonium spin rotation (MSR) are two developing subfields which seek to understand the chemical and physical behavior of a light hydrogen isotope in matter and to probe the structure of materials. *See* POSITRONIUM; QUANTUM ELECTRODYNAMICS. [P.O.E.]

Murine typhus fever An infectious disease closely related to epidemic typhus fever. This generally milder, febrile disease with a more gradual onset, shorter duration, and lower mortality than the epidemic form is caused by *Rickettsia typhi* (synonyms *manchuriae*, *mooseri*). *See* EPIDEMIC TYPHUS FEVER; RICKETTSIALES; RICKETTSIOSES.

The primary disease cycle occurs between rats and their fleas with sporadic involvement of humans. The vector is chiefly the tropical rat flea *Xenopsylla cheopis*, though other fleas can be infected. As in epidemic typhus, contamination by the feces of the vector, rather than its bites, is responsible for transmission. *See* SIPHONAPTERA.

Because of the primary cycle in commensal rats and their fleas, prevention and control are directed to them. The sporadic human incidence does not indicate need for a vaccination program, and protection of personnel in food-handling establishments in urban and suburban areas, where most cases occur, is best accomplished by rat control. [C.B.P.]

Muscle The tissue in the body in which cellular contractility has become most apparent. Almost all forms of protoplasm exhibit some degree of contractility, but in muscle fibers specialization has led to the preeminence of this property. In vertebrates three major types of muscle are recognized: smooth, cardiac, and skeletal.

Smooth muscle. Smooth muscle, also designated visceral and sometimes involuntary, is the simplest type. These muscles consist of elongated fusiform cells which contain a central oval nucleus. The size of such fibers varies greatly, from a few microns up to 0.02 in. (0.5 mm) in length. These fibers contract relatively slowly and have the ability to maintain contraction for a long time. Smooth muscle forms the major contractile elements of the viscera, especially those of the respiratory and digestive tracts, and the blood vessels. Smooth muscle fibers in the skin regulate heat loss from the body. Those in the walls of various ducts and tubes in the body act to move the contents to their destinations, as in the biliary system, ureters, and reproductive tubes.

Smooth muscle is usually arranged in sheets or layers, commonly oriented in different directions. The major physiological properties of these muscles are their intrinsic ability to contract spontaneously and their dual regulation by the autonomic nerves of the sympathetic and parasympathetic systems. *See* AUTONOMIC NERVOUS SYSTEM.

Cardiac muscle. Cardiac muscle has many properties in common with smooth muscle; for example, it is innervated by the autonomic system and retains the ability to contract spontaneously. Presumably, cardiac muscle evolved as a specialized type from the general smooth muscle of the circulatory vessels. Its rhythmic contraction begins early in embryonic development and continues until death. Variations in the rate of contraction are induced by autonomic regulation and by many other local and systemic factors.

The cardiac fiber, like smooth muscle, has a central nucleus, but the cell is elongated and not symmetrical. It is a syncytium, a multinuclear cell or a multicellular structure without cell walls. Histologically, cardiac muscle has cross-striations very similar to those of skeletal muscle, and dense transverse bands, the intercalated disks, which occur at short intervals. *See* HEART (VERTEBRATE).

Skeletal muscle. Skeletal muscle is also called striated, somatic, and voluntary muscle, depending on whether the description is based on the appearance, the location, or the innervation. The individual cells or fibers are distinct from one another and vary greatly in size from over 6 in. (15 cm) in length to less than 0.04 in. (1 mm). These fibers do not ordinarily branch, and they are surrounded by a complex membrane, the sarcolemma. Within each fiber are many nuclei;

thus it is actually a syncytium formed by the fusion of many precursor cells.

The transverse striations of skeletal muscle form a characteristic pattern of light and dark bands within which are narrower bands. These bands are dependent upon the arrangement of the two sets of sliding filaments and the connections between them. *See* MUSCLE PROTEINS; MUSCULAR SYSTEM. [W.J.B.]

Muscle proteins Specialized proteins in muscle cells are the building blocks of the structures constituting the moving machinery of muscle. They are disposed in myofilaments which are discernible by electron microscopy. These myofilaments are of two kinds, and their regular arrangement within the cell gives the striated pattern to skeletal muscle fibers (see illustration). It is recognized that the sliding of the two sets of

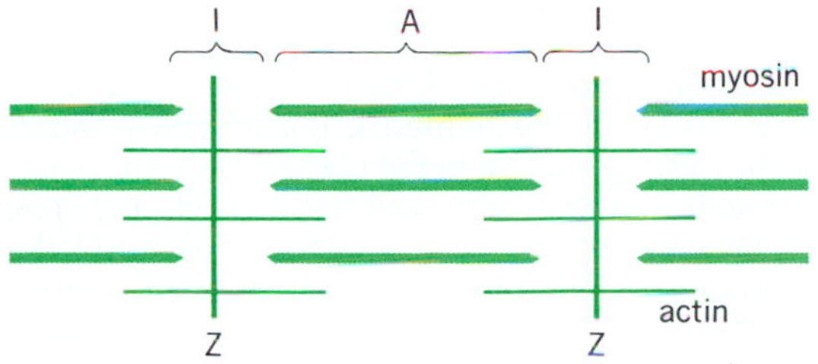

Banding pattern of a sarcomere, showing arrangement of myosin and actin molecules. A, I, and Z indicate bands.

filaments relative to each other is the molecular basis of muscle contraction. To understand the ultimate mechanism that causes the movement of these filaments (relative to each other), it is necessary to consider the features of the individual molecules making up these filaments. Evidence has been accumulating that practically all nonmuscle cells, although lacking the filaments of muscle, contain proteins similar to those found in muscle; these proteins are likely to be involved in cell motility and in determining properties of cell membranes.

Molecules of myosin, amounting to about 60% of the total muscle protein, are arranged in filaments occupying the central zone of each segment (sarcomere) or the fibril, the A band. Myosin is an elongated molecule made up of two intertwined heavy peptide chains (molecular weight of about 200,000) whose ends form two separate globular structures. The intertwined portion forms a rigid rod; each of the two globular portions (heads) contains a center capable of combining with, and splitting off, the terminal phosphate of adenosine triphosphate (ATP); ATP is the ultimate source of energy for muscle contraction. Each globular head can combine with actin.

The chief constituent of the filaments originating in the Z band of each sarcomere is actin. Actin filaments are made up of globular units in a double helix; there are 13 to 15 of these units in each strand for every complete turn of this helix.

These two proteins, or rather protein complexes, are associated with the actin filaments. Tropomyosin is an α-helical protein containing two intertwined polypeptide chains that extend over about seven globular actin units. Each tropomyosin molecule is associated with one troponin molecule, which in turn consists of three different types of subunits. In contrast to the tropomyosin molecule, the troponin complex and its subunits are thought to be essentially spherical in shape. Both tropomyosin and troponin are required to make the interaction of actin and myosin sensitive to calcium ions.

A great deal of information has come to light concerning the process by which the interaction between actin and myosin is regulated in the living cell. In higher organisms troponin and tropomyosin participate in this regulation. In the presence of these proteins the combination of actin and myosin cannot take place. However, if a small amount of ionized calcium is present, the inhibitory effect of the tropomyosin-troponin system is reversed, and the interaction with actin and myosin becomes fully effective.

This regulation involves a movement of tropomyosin molecules within the thin filaments from positions in which they block the combination of actin with the myosin moiety. Within the muscle cell the amount of calcium available is regulated by certain membranous structures, collectively the sarcoplasmic reticulum, that appear to have the ability to store calcium in the resting state. Upon excitation, the outer membrane of the muscle cell undergoes a change in permeability. This change spreads inside the muscle through the so-called transverse tubular system and causes a release of calcium within the cell. The calcium ions thus released permit the interaction of actin and myosin, leading to the liberation of chemical energy from ATP and its transformation into mechanical work. *See* MOTOR SYSTEMS; MUSCLE. [J.Ge.]

Muscovite One of the mica group of minerals with the composition $K_2Al_4(Si_6Al_2)O_{20}(OH)_4$. It is also called white, or potash, mica. Muscovite occurs in some granites and is especially abundant in pegmatites, the chief commercial deposits. It is a widespread constituent of slates, phyllites, schists, and gneisses. Physical properties include easy perfect basal cleavage to sheets that are flexible and elastic; specific gravity of 2.7–3.1; and hardness of 2–2.5 on Mohs scale. Colors in sheets are in shades of brown, green, and yellow; viewed microscopically, muscovite is colorless. *See* MICA; SILICATE MINERALS. [E.W.H.]

Muscular dystrophy A group of muscle diseases which are hereditary and characterized by progressive muscle weakness and wasting. The etiology of the various muscular dystrophies is unknown. Muscular dystrophy is a primary disease of the muscle cells themselves, and not the result of abnormality in their innervation. In some types of muscular dystrophy the disease appears to be restricted to skeletal muscles alone, but in some others cardiac muscle may also be affected, and in still others the skeletal muscle involvement appears to be part of a more generalized disorder, with abnormalities in other organ systems as well.

The most common forms of muscular dystrophy are the sex-linked recessive Duchenne's pseudohypertrophic muscular dystrophy and the dominantly inherited myotonic dystrophy. Other, less common forms include a milder form of sex-linked pseudohypertrophic muscular dystrophy (Becker's type), dominantly inherited facioscapulohumeral muscular dystrophy, and recessively inherited limb-girdle muscular dystrophy. The ocular myopathies appear to include several different disease entities, all of which have as a common feature the progressive paralysis of extraocular muscles.

As the basic defect in any of the muscular dystrophies is unknown, treatment is purely symptomatic. Unfortunately, no treatment has been shown to alter the course of the muscular dystrophies. [S.M.Su.]

Muscular system The muscular system consists of muscular cells, the contractile elements with the specialized property of exerting tension during contraction, and associated connective tissues. The three morphologic types of muscles are voluntary muscle, involuntary muscle, and cardiac muscle. The voluntary, striated, or skeletal muscles are involved with general posture and movements of the head, body, and limbs. The involuntary, nonstriated, or smooth muscles are the muscles of the walls of hollow organs of the digestive, circulatory, respiratory, and reproductive systems, and other visceral structures.

Cardiac muscle is the intrinsic muscle tissue of the heart. *See* MUSCLE.

Anatomy. Muscle groups are particularly distinct in elasmobranchs and other primitive fishes, and they are generally defined on the basis of their embryonic origin in these animals. Two major groups of skeletal muscles are recognized, somatic (parietal) muscles, which develop from the myotomes, and branchiomeric muscles, which develop in the pharyngeal wall from lateral plate mesoderm. The somatic musculature is subdivided into axial muscles, which develop directly from the myotomes and lie along the longitudinal axis of the body, and appendicular muscles, which develop within the limb bud from mesoderm derived phylogenetically as buds from the myotomes.

The vertebrate muscular system is the largest of the organ systems, making up 35–40% of the body weight in humans. The movement of vertebrates is accomplished exclusively by muscular action, and muscles play the major role in transporting materials within the body. Muscles also help to tie the bones of the skeleton together and supplement the skeleton in supporting the body against gravity. *See* SKELETAL SYSTEM.

Most of the axial musculature is located along the back and flanks of the body, and this part is referred to as trunk musculature. But anteriorly the axial musculature is modified and assigned to other subgroups. Certain of the occipital and neck myotomes form the hypobranchial muscles, and the most anterior myotomes form the extrinsic ocular muscles.

The hypaxial musculature of tetrapods can be subdivided into three groups: (1) a subvertebral (hyposkeletal) group located ventral to the transverse processes and lateral to the centra of the vertebrae, (2) the flank muscles forming the lateral part of the body wall, and (3) the ventral abdominal muscles located on each side of the midventral line. The subvertebral musculature assists the epaxial muscles in the support and movement of the vertebral column. Most of the flank musculature takes the form of broad, thin sheets of muscle that form much of the body wall and support the viscera. The midventral hypaxial musculature in all tetrapods consists of the rectus abdominis, a longitudinal muscle on each side of the midline that extends from the pelvic region to the anterior part of the trunk.

The hypobranchial musculature extends from the pectoral girdle forward along the ventral surface of the neck and pharynx to the hyoid arch, chin, and into the tongue. It is regarded as a continuation of part of the hypaxial trunk musculature.

Limb muscles are often classified as intrinsic if they lie entirely within the confines of the appendage and girdle, and extrinsic if they extend from the girdle or appendage to other parts of the body. In fishes, movements of the paired fins are not complex or powerful and the appendicular muscles in the strictest sense are morphologically simple. In terrestrial vertebrates, the limbs become the main organs for support and locomotion, and the appendicular muscles become correspondingly powerful and complex. The muscles are too numerous to describe individually, but they can be sorted into dorsal and ventral groups, because tetrapod muscles originate embryonically in piscine fashion from a dorsal and a ventral premuscular mass within the limb bud. In general, the ventral muscles, which also spread onto the anterior surface of the girdle and appendage, act to protract and adduct the limb and to flex its distal segments; the dorsal muscles, which also extend onto the posterior surface of the girdle and appendage, have the opposite effects (retraction, abduction, and extension). The limb muscles also serve as flexible ties or braces that can fix the bones at a joint and support the body.

Flight in birds has entailed a considerable modification of the musculature of the pectoral region. As one example, the ventral adductor muscles are exceedingly large and powerful, and the area from which they arise is increased by the enlargement of the sternum and the evolution of a large sternal keel. Not only does a ventral muscle, the pectoralis, play a major role in the downstroke of the humerus, but a ventral muscle, the supracoracoideus, is active in the upstroke as well.

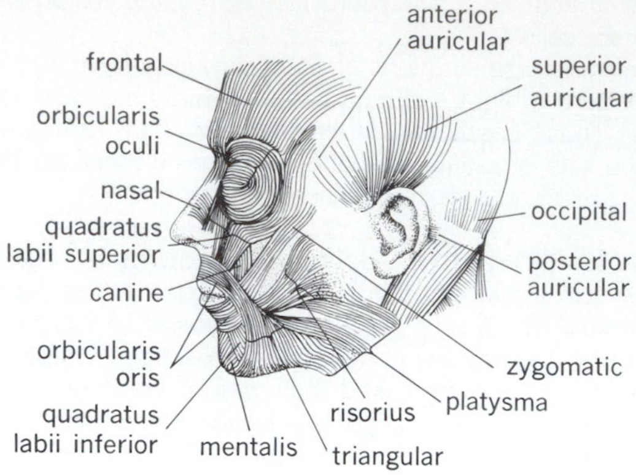

Fig. 1. Human facial muscles. (*After H. W. Rand, The Chordates, Blakiston, 1950*)

In a number of terrestrial vertebrates, particularly amniotes, certain of the more superficial skeletal muscles of the body have spread out beneath the skin and inserted into it. These may be described as integumentary muscles. Integumentary muscles are particularly well developed in mammals and include the facial muscles and platysma, derived from the hyoid musculature, and often a large cutaneous trunci (Fig. 1). The last is derived from the pectoralis and latissimus dorsi and fans out beneath the skin of the trunk. The twitching of the skin of an ungulate is caused by this muscle.

Muscle mechanics. Many of the bones serve as lever arms, and the contractions of muscles are forces acting on these arms (Fig. 2). The joint, of course, is the fulcrum and it is at one end of the lever. The length of the force arm is the perpendicular distance from the fulcrum to the line of action of the muscle; the length of the work arm is the perpendicular distance from the fulcrum to the point of application of the power generated in the lever. Compactness of the body and physiological properties of the muscle necessitates that a muscle attach close to the fulcrum; therefore, the force arm is considerably shorter than the work arm. Most muscles are at a mechanical disadvantage, for they must generate forces greater than the work to be done, but an advantage of this is that a small muscular excursion can induce a much greater movement at the end of the lever. *See* SKELETAL SYSTEM.

Slight shifts in the attachments of a muscle that bring it toward or away from the fulcrum, and changes in the length of the work arm, can alter the relationship between force and amount or speed of movement.

In general, the force of a muscle is inversely related to the amount and speed of movement that it can cause. Certain pat-

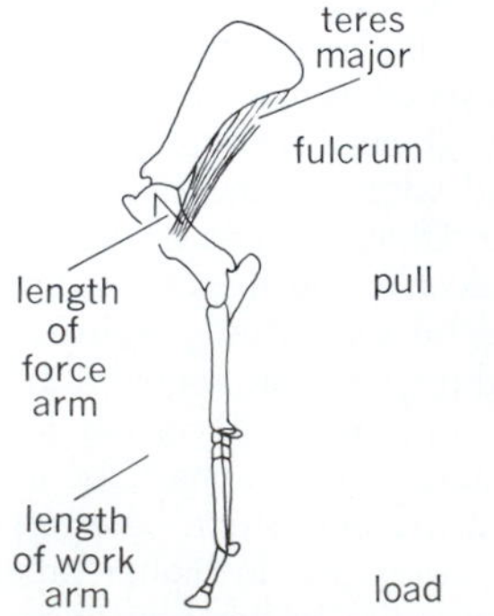

Fig. 2. A typical vertebrate lever system.

terns of the skeleton and muscles are adapted for extensive, fast movement at the expense of force, whereas others are adapted for force at the expense of speed. In the limb of a horse, which is adapted for long strides and speed, the muscles that move the limb insert close to the fulcrum and the appendage is long. This provides a short force arm but a very long work arm to the lever system (Fig. 2). In the front leg of a mole, which is adapted for powerful digging, the distance from the fulcrum to the insertion of the muscles is relatively greater and the length of the appendage is less, with the result that the length of the force arm is increased relative to the length of the work arm. [W.F.W.]

Muscular system disorders Disorders affecting skeletal (voluntary) muscle. The normal functioning of the skeletal muscle is dependent not only on the integrity of the muscle fibers themselves, but also on that of the motor cortex, the pyramidal tract, and the extrapyramidal system (including the cerebellum). It also depends on innervation by the motoneurons of the brainstem and the spinal cord. In addition, the proper functioning of the other organ systems, such as the endocrine system, and variations in the concentration of various electrolytes may also affect muscle function.

Damage to the motor cortex or the pyramidal tract produces the type of weakness seen in humans after a stroke or spinal cord injury. Although the paralyzed limb may initially be flaccid (hypotonic), spasticity (hypertonia) eventually develops. Despite the weakness, muscle atrophy is usually not striking. When the extrapyramidal system or the cerebellum is the site of damage, instead of weakness there are uncontrolled movements, difficulty with coordination, or both. In either of these situations there is no characteristic change in the muscle, either grossly or microscopically. At most, atrophy of type 2 fibers is seen. *See* ATROPHY.

Motoneuron damage. With damage to the spinal motoneuron or its axon, there is flaccid weakness of the muscle with proportionate wasting. Direct involvement of the spinal motoneuron was typically seen in poliomyelitis, but is now seen more commonly in the progressive spinomuscular atrophies of infancy and childhood, and in amyotrophic lateral sclerosis (ALS) in adults. Spontaneous twitching of groups of muscle fibers (fasciculation) innervated by the same motoneuron (motor unit) is frequently seen in these disorders. *See* POLIOMYELITIS.

Progressive diseases of unknown cause may occur in infancy Werdnig-Hoffmann disease) or later in childhood (Kugelberg-Wielander disease). Both diseases result from degeneration of the motoneurons and appear to be inherited in an autosomal recessive pattern. In the infantile form, the baby is often floppy from birth with generalized weakness, a poor cry, and difficulty in sucking and breathing. Many children succumb in early childhood. In the later childhood forms, the rate of progression is slower and the outlook better. In these children weakness is more marked in the proximal muscles of the limbs.

Amyotrophic lateral sclerosis is a progressive disease of unknown cause, in which both the brainstem and spinal motoneurons, as well as the corticospinal tracts, undergo degeneration. This is a relatively rapid progressive disease, with death usually occurring within 3 years of diagnosis due to swallowing difficulty and respiratory failure. Although some cases of ALS appear to be inherited, most cases occur sporadically.

Neuromuscular junction. The most common example of disease at the neuromuscular junction is myasthenia gravis. Other, less common diseases are the myasthenic Eaton-Lambert syndrome and botulism. *See* MYASTHENIA GRAVIS.

Myopathy. Abnormalities of the muscle itself (myopathy) obviously result in muscle weakness, and muscle diseases fall into two large groups: those with a genetic basis and those which are nongenetic.

Congenital myopathies and the muscular dystrophies constitute the genetic muscle diseases. Congenital myopathies are characterized by a generalized weakness which is present at birth. The weakness is usually not progressive and often improves with time. Muscle fiber necrosis, a characteristic of the muscular dystrophies, is not seen in the congenital myopathies. *See* MUSCULAR DYSTROPHY.

Nongenetic or acquired diseases are all characterized by rapidly progressive weakness of the proximal muscles of the limbs, and so resemble the limb-girdle form of muscular dystrophy in the distribution of weakness. They are often included as inflammatory myopathies. Individuals with these disorders have difficulty arising from a recumbent or sitting position, in climbing stairs, and in lifting heavy objects onto a shelf. They also often have tender and painful muscles, may be febrile, and may have other manifestations of a systemic illness.

Metabolic diseases. A number of metabolic diseases have been associated with muscle symptoms. These include thyroid diseases and certain endocrine diseases, particularly Cushing's syndrome, which are either spontaneous or secondary to therapeutically administered adrenocorticosteroid hormones. No specific histologic changes have been described in the muscle in these disorders. Muscular symptoms are also seen with some of the glycogen storage diseases. Muscle weakness and paralysis may also be associated with alterations in the level of serum potassium.

Myotonia. A delayed relaxation of the muscle after forceful contraction (myotonic) is another symptom of muscle disease. This phenomenon is a characteristic feature of myotonic dystrophy and of congenital myotonia. Myotonia is usually present from early life and is often associated with muscle hypertrophy, but muscle weakness is not a feature of this disease. The muscle shows no definite histologic changes. *See* MUSCULAR SYSTEM. [S.M.Su.]

Mushroom A fungus belonging to the basidiomycetous order Agaricales, or the fruiting body (basidiocarp) of such a fungus. The generalized life history of a mushroom begins with the germination of a basidiospore which produces a primary

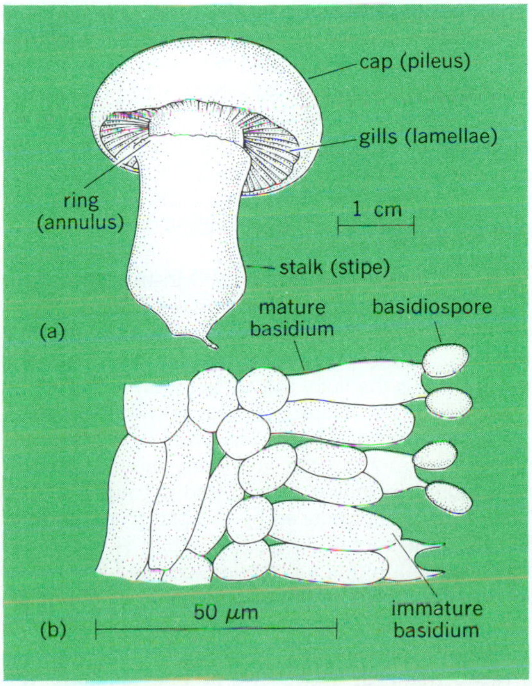

***Agaricus bisporus*, the common cultivated mushroom. (*a*) Basidiocarp. (*b*) Cross section of gill, showing basidia in various stages of development.**

mycelium composed of filaments (hyphae) with one nucleus per cell. If two compatible primary mycelia fuse, a secondary mycelium with two nuclei in each cell develops. This underground secondary mycelium, which may be extensive and perennial, forms at its periphery small masses of compacted hyphae which enlarge and differentiate into immature basidiocarps, or mushroom buttons (see illustration). These buttons enlarge rapidly, burst through the soil and become mature basidiocarps (see illustration). Basidia form on plates of tissue (gills) underneath the cap of the mushroom. Each basidium produces basidiospores (usually four) which are shot off and dispersed by wind. Since mushrooms develop near the margin of the circular mycelium, they sometimes form "fairy rings." *See* BASIDIOMYCOTINA; FUNGI.

Basidiocarps of some mushrooms are prized as food; others are poisonous. The term toadstool is commonly applied to basidiocarps of poisonous species. The only way to distinguish edible from poisonous species is by identification. [R.M.P.]

Musical instruments Instruments for producing musical sounds have long been classified as woodwinds, brass, percussion, or strings; to these must be added electrical and electronic instruments. In a sense, all these instruments implement and extend the capability of the original musical instrument, the singing voice. The classes mentioned are useful in grouping instruments in a general way for the kinds of sounds they produce, even though woodwind instruments are not necessarily made of wood, nor are brass instruments always made of metal.

Woodwind instruments. Woodwind instruments are distinguished primarily by the fact that the effective length of the vibrating air column is shortened by opening lateral side holes in succession. Two distinctly different means of generating the sound are employed. For the flute, and its half-size version the piccolo, the player blows across the embouchure hole near one end in such away as to cause periodic puffs of air to enter the tube; after a turbulent turning these puffs excite the air column longitudinally. This method of excitation leaves the tube acoustically open in the sense that the contained air vibrates much as it does in a simple tube with both ends open to the atmosphere.

For the double-reed oboe or bassoon, the player holds between the lips a pair of thin reeds (pieces of cane appropriately thinned, shaped, and bound together) that beat against each other to change the player's breath to puffs of air. For clarinets and saxophones, a single reed attached to a mouthpiece by a ligature functions in a similar way. The portion of the mouthpiece (the lay) against which the reed beats must be appropriately curved; the character of the sound is modified somewhat by the volume of the mouthpiece as well as by the shape and material of the reed.

For both the single- and double-reed instruments, the reeds vibrate under the influence of sound waves reflected back from the distant end of the air column and allow the puffs of air to enter when the sound pressure within the instrument is large. Thus (in contrast to the flute) the air vibration at the reed end is that associated with an acoustically closed end.

Brass instruments. The typical brass instrument consists of a cup-shaped mouthpiece, a slightly tapered mouthpipe, cylindrical tubing including valves, and a roughly hyperbolic bell. Puffs of air are introduced by the player via vibrating lips stretched over the mouthpiece. The action is comparable to that of the clarinet, in that the mouthpiece end is nearly closed acoustically. The length of the air column is increased by tubing switched in by use of valves, either piston or rotary: a common arrangement is such that the first valve lowers the intonation by two semitones, the second by one semitone, and the third by three semitones. For a given length of tubing, different tones are produced by tensioning the lips to excite different modes of vibration whose frequencies are approximately in the ratios 2:3:4:5:6:8.

Percussion instruments. Instruments such as the timpani (kettledrums) and xylophone are called percussion instruments because the sound is initiated by a blow. Two kinds of sound producers are involved: a membrane under tension, associated with a cavity that can influence the frequency of vibration, as in the case of the timpani; and a rigid bar or plate vibrating transversely, whose frequency is little affected by any resonator that may be attached. Some percussion instruments give a well-defined sound that excites a sensation of definite pitch, such as does a church bell; others, such as drums, cymbals, and triangles, are useful primarily for rhythm effects.

Stringed instruments. For the guitar and harp, strings are set into vibration by plucking; for the other stringed instruments the vibration is usually initiated and maintained by bowing. The frequency of vibration is primarily established by the length, tension, and mass per unit length of the string. A string vibrates not only at the lowest (fundamental) frequency, but also at the same time at higher frequencies which tend toward integer multiples of the fundamental frequency. The sound radiated from the instrument is thus complex.

The radiation of sound from a stringed instrument is enhanced by a resonator consisting of an almost closed air cavity. Some of the energy of the vibrating string is transmitted via the bridge to the walls of the cavity. In a carefully constructed violin the resonances of the air cavity and its vibrating walls are distributed in frequency in order to afford a relatively uniform response throughout the playing range of the instrument.

Keyboard instruments. Instruments such as the celesta, pipe organ, accordion, and piano are usually put in a group called keyboard instruments, because the respective vibrating bars, pipes, reeds, and strings in these instruments are selected by use of keys in a keyboard. The celesta and piano could also be described as percussion instruments, because hammers strike the bars and strings; the pipe organ and the accordion, with its wind-driven free reeds, are wind instruments. By its multiple keyboards (and pedal board) the pipe organ puts under the control of a single player thousands of sources whose distinctive sounds can be reproduced on command.

Electrical and electronic types. Musical instruments of the kinds already described become quasi-electrical instruments by the addition of a microphone to pick up the airborne sound, an amplifier, and a loudspeaker. Alternately, a vibration pickup can be used to generate an electrical signal directly from the vibration of a string; this is the case in the electric guitar and electric piano. The signal can be modified electrically before being radiated as sound from a loudspeaker. Since the resonance box or sounding board is not needed for these electrical stringed instruments, the damping is very slight and the vibration can last for a long time so that relatively sustained sounds can be obtained if desired. Rigid vibrators, such as reeds, rods, bars, or tubes, with associated pickups are also exploited as electrical musical instruments. *See* ACOUSTIC RESONATOR; RESONANCE (ACOUSTICS AND MECHANICS); SOUND-REPRODUCING SYSTEMS; SYMPATHETIC VIBRATION; VIBRATION. [R.W.Y.]

Musk-ox An even-toed ungulate, *Ovibos moschatus*, which is a member of the family Bovidae in the mammalian order Artiodactyla. This single species is the northernmost representative of the family, ranging through the tundra areas and snowfields of Canada and Alaska, as well as Greenland.

The musk-ox derives its name from the musky odor it emits. It is a stoutly built animal (see illustration) and has a coat of long dense hair that is resistant to the extreme cold of the windswept treeless tundra. They do not hibernate and are usually found in herds of 20–100 animals huddled together for warmth. As protection against its natural enemy, the wolf, the musk-ox will form a circle with the young inside. Since a cow

The musk-ox (*Ovibos moschatus*).

produces a single calf every 2 years, the numbers have been so reduced that they are now protected by the Canadian government. *See* ARTIODACTYLA. [C.B.C.]

Muskeg A term derived from Chippewan Indian for "grassy bog." In North America ecologists apply diverse usage, but most include peat bogs or tussock meadows, with variable woody vegetation such as spruce or tamarack. Plant remains accumulate when trapped in the water or in media such as sphagnum moss which inhibit decay. The peat might accrue indefinitely or might approach a steady state of raised bogs, blanket bogs, or forest if input became balanced by erosion or by loss as methane and CO_2.

Organic terrain occurs on every continent. The largest expanses of it are in Russia (especially western Siberia) and Canada. Typical bog or spruce-larch muskeg develops in cool temperature and subarctic to arctic lands. In the subtropics and the tropics there is considerable organic terrain, as in Paraguay, Uruguay, and Guyana in South America. Thus, despite differences in climate or floristics, the phenomenon of peat formation persists if local aeration or biochemical conditions hinder decay. Despite climatic and biotic differences, gross peat structure categories seem comparable the world over. *See* BOG; PEAT. [N.W.R.]

Muskmelon The edible fruit of *Cucumis melo*, belonging to the gourd family, Cucurbitaceae, as do other vine crops such as cucumber, watermelons, pumpkin, and squash. The muskmelon appears to be indigenous to Africa. *See* CANTALOUPE; HONEY DEW MELON; PERSIAN MELON; VIOLALES.

The plants are annual, trailing vines, with three to five runners. The runners produce short fruiting branches, which bear the perfect flowers and later the fruits. Muskmelons maturing on the vine without becoming overripe are superior in quality to those harvested immature. The sugar content, flavor, and texture of the fresh flesh improves very rapidly as the fruit approaches maturity. When mature, the melon is sweet, averages 6 to 8% sugar, and has a slight to distinctly musky odor and flavor, depending upon cultivar and environment. The flesh is rich in potassium, in vitamin C and, when deep orange, also in vitamin A. [F.W.Z.]

Muskrat The largest member of the rodent subfamily Microtinae in the family Muridae, it is also known as the musquash. Also included in this subfamily are the lemmings and voles. *Ondatra zibethica*, the single species, is about the size of a rabbit. It is economically important for its fur, which is shiny, soft, and thick. It received its name because of its inguinal glands which produce a characteristic musky odor. The muskrat is omnivorous, eating various types of vegetation, such as reeds and tuberous roots of aquatic plants, and animals, such as mussels, insects, and fish. These animals become active in March, when mating occurs. The female has two or three litters of 5–9 young each year, and the last litter of the season remains in the parental abode through winter. *See* LEMMING; RODENTIA. [C.B.C.]

Mustard Any one of a number of annual crucifer species of Asiatic origin belonging to the plant order Capparales. Mustards eaten as greens are *Brassica juncea*, *B. juncea* var. *crispifolia*, and *B. hirta*. Table mustard and oils are obtained from *B. nigra*. Important production centers for mustard greens are in the South, where the crop is popular. Montana and the West Coast states are important sources of mustard seed. *See* CAPPARALES. [H.J.C.]

Mutagens and carcinogens Mutagens are agents capable of altering the genetic material of an organism, leading to inherited differences; carcinogens are agents that cause cancer in animals, including humans.

Occupational skin cancer among workers exposed to mineral oils and tars was well known over a century ago; the first pure chemical compounds (carcinogens) that could be shown to induce tumors in experimental animals were isolated from coal tar and synthesized in the laboratory in 1930. Since then an enormous effort has been expended on the study of chemical carcinogenesis (the induction of cancer with chemicals). Well over 3000 compounds are carcinogenic in animals, and of these a few are known to be active also in humans. For example, the high incidence of bladder cancer among former workers in dyestuff factories is directly attributable to their daily contact with aromatic amines such as 2-napthylamine and benzidine, both carcinogens. The great majority of chemical carcinogens are organic compounds belonging to a number of diverse structural types (polycyclic hydrocarbons, aromatic amines, nitrosamines, and so on). Some natural products, such as the aflatoxins which can occur in peanuts and certain grains as a result of contamination with the common yellow mold *Aspergillus flavus*, and a few inorganic materials, notably asbestos, are also carcinogenic. *See* AFLATOXIN; ASBESTOS.

Under the correct conditions almost all carcinogens are mutagens. This positive correlation is probably the best indication that the hypothesis that cancers are caused by somatic mutations is essentially correct.

It has slowly become apparent that a high proportion of cancers are caused by environmental factors, and therefore are preventable, at least in theory. This conclusion has been reached largely as a result of cancer epidemiology. For example, the incidence of stomach cancer is much higher in Japan than in the United States, but that among Japanese who have emigrated to California tends toward that of the indigenous population. This becomes noticeable in the second and third generations, suggesting that although the disease has an environmental cause, this is connected with cultural differences rather than some general factor to which the whole Japanese population in California is exposed.

There is no mystery about one major disease, lung cancer; all the epidemiological evidence agrees that this is largely caused by the cultural habit of cigarette smoking. Cigarette tar is readily shown to be mutagenic and carcinogenic in appropriate tests. *See* CANCER (MEDICINE); MUTATION; ONCOLOGY. [M.M.C.]

Mutation Any alteration capable of being replicated in the genetic material of an organism. When the alteration is in the nucleotide sequence of a single gene, it is referred to as gene mutation; when it involves the structures or number of the chromosomes it is referred to as chromosome mutation, or rearrangement. Mutations may be recognizable by their effects on the phenotype of the organism (mutant).

Gene mutations. Two classes of gene mutations are recognized: point mutations and intragenic deletions. Two different types of point mutation have been described. In the first of these, one nucleic acid base is substituted for another. The second type of change results from the insertion of a base into, or its deletion from, the polynucleotide sequence. These mutations are all called sign mutations or frame-shift mutations because of their effect on the translation of the information of the gene. *See* NUCLEIC ACID.

More extensive deletions can occur within the gene which are sometimes difficult to distinguish from mutants which involve only one or two bases. In the most extreme case, all the informational material of the gene is lost.

A single-base alteration, whether a transition or a transversion, affects only the codon or triplet in which it occurs. Because of code redundancy, the altered triplet may still insert the same amino acid as before into the polypeptide chain, which in many cases is the product specified by the gene. Such DNA changes pass undetected. However, many base substitutions do lead to the insertion of a different amino acid, and the effect of this on the function of the gene product depends upon the amino acid and its importance in controlling the folding and shape of the enzyme molecule. Some substitutions have little or no effect, while others destroy the function of the molecule completely.

Single-base substitutions may sometimes lead not to a triplet which codes for a different amino acid but to the creation of a chain termination signal. Premature termination of translation at this point will lead to an incomplete and generally inactive polypeptide.

Sign mutations (adding or subtracting one or two bases to the nucleic acid base sequence of the gene) have a uniformly drastic effect on gene function. Because the bases of each triplet encode the information for each amino acid in the polypeptide product, and because they are read in sequence from one end of the gene to the other without any punctuation between triplets, insertion of an extra base or two bases will lead to translation out of register of the whole sequence distal to the insertion or deletion point. The polypeptide formed is at best drastically modified and usually fails to function at all. This sometimes is hard to distinguish from the effects of intragenic deletions. However, whereas extensive intragenic deletions cannot revert, the deletion of a single base can be compensated for by the insertion of another base at, or near, the site of the original change. *See* GENE; GENETIC CODE.

Chromosomal changes. Some chromosomal changes involve alterations in the quantity of genetic material in the cell nuclei, while others simply lead to the rearrangement of chromosomal material without altering its total amount. *See* CHROMOSOME.

Origins of mutations. Mutations can be induced by various physical and chemical agents or can occur spontaneously without any artificial treatment with known mutagenic agents.

Until the discovery of x-rays as mutagens, all the mutants studied were spontaneous in origin; that is, they were obtained without the deliberate application of any mutagen. Spontaneous mutations occur unpredictably, and among the possible factors responsible for them are tautomeric changes occurring in the DNA bases which alter their pairing characteristics, ionizing radiation from various natural sources, naturally occurring chemical mutagens, and errors in the action of the DNA-polymerizing and correcting enzymes.

Spontaneous chromosomal aberrations are also found infrequently. One way in which deficiencies and duplications may be generated is by way of the breakage-fusion-bridge cycle. During a cell division one divided chromosome suffers a break near its tip, and the sticky ends of the daughter chromatids fuse. When the centromere divides and the halves begin to move to opposite poles, a chromosome bridge is formed, and breakage may occur again along this strand. Since new broken ends are produced, this sequence of events can be repeated. Unequal crossing over is sometimes cited as a source of duplications and deficiencies, but it is probably less important than often suggested.

In the absence of mutagenic treatment, mutations are very rare. In 1927 H. J. Muller discovered that x-rays significantly increased the frequency of mutation in *Drosophila*. Subsequently, other forms of ionizing radiation, for example, gamma rays, beta particles, fast and thermal neutrons, and alpha particles, were also found to be effective. Ultraviolet light is also an effective mutagen. The wavelength most employed experimentally is 253.7 nm, which corresponds to the peak of absorption of nucleic acids.

Some of the chemicals which have been found to be effective as mutagens are the alkylating agents which attack guanine principally although not exclusively. The N7 portion appears to be a major target in the guanine molecule, although the O^6 alkylation product is probably more important mutagenically. Base analogs are incorporated into DNA in place of normal bases and produce mutations probably because there is a higher chance that they will mispair at replication. Nitrous acid, on the other hand, alters DNA bases in place. Adenine becomes hypoxanthine and cytosine becomes uracil. In both cases the deaminated base pairs differently from the parent base. A third deamination product, xanthine, produced by the deamination of guanine, appears to be lethal in its effect and not mutagenic. Chemicals which react with DNA to generate mutations produce a range of chemical reaction products not all of which have significance for mutagenesis.

Significance of mutations. Mutations are the source of genetic variability, upon which natural selection has worked to produce organisms adapted to their present environments. It is likely, therefore, that most new mutations will now be disadvantageous, reducing the degree of adaptation. Harmful mutations will be eliminated after being made homozygous or because the heterozygous effects reduce the fitness of carriers. This may take some generations, depending on the severity of their effects. Chromosome alterations may also have great significance in evolutionary advance. Duplications are, for example, believed to permit the accumulation of new mutational changes, some of which may prove useful at a later stage in an altered environment.

Rarely, mutations may occur which are beneficial: Drug yields may be enhanced in microorganisms; the characteristics of cereals can be improved. However, for the few mutations which are beneficial, many deleterious mutations must be discarded. Evidence suggests that the metabolic conditions in the treated cell and the specific activities of repair enzymes may sometimes promote the expression of some types of mutation rather than others. *See* DEOXYRIBONUCLEIC ACID (DNA). [B.J.K.]

Myasthenia gravis

Myasthenia gravis A disease resulting from an abnormality in neuromuscular transmission, characterized by a fluctuating degree of muscle weakness. The weakness is usually aggravated by activity, and there is partial or complete restoration of strength after a period of rest or the administration of anticholinesterase medications. *See* ACETYLCHOLINE.

It has been shown that the basic defect in myasthenia gravis is a reduction in the number of acetylcholine receptor sites in the postsynaptic membrane of the neuromuscular junction. It has also been shown that many myasthenic patients have immunoglobulins in the serum that partially block acetylcholine receptors. *See* AUTOIMMUNITY; IMMUNOGLOBULIN.

Abnormalities have been demonstrated in the thymus gland and skeletal muscle in myasthenia gravis. There is an increased incidence of thymoma in myasthenia gravis, and in those without a thymoma, hyperplasia of the germinal centers is a common finding in the thymus gland.

Although the disease affects young women more commonly, usually in the third decade, it can occur in either sex at any age. In the majority of persons, weakness affects muscles of head, neck, and limbs (generalized myasthenia), but in some the weakness is restricted to the muscles of the eyes (ocular myasthenia), in which case the disease is usually benign.

The standard treatment for myasthenia gravis has been the use of longer-acting anticholinesterase agents; thymectomy and immunosuppressive drugs are reserved for those patients with generalized myasthenia that does not respond sufficiently to these agents. *See* SYNAPTIC TRANSMISSION. [S.M.Su.]

Mycobacterial diseases Humans are subject to infection by species of *Mycobacterium*, a genus of acid-fast, nonmotile, rod-shaped bacteria, some of which may produce disease. Studies completed in many countries demonstrate that mycobacterial infections with no recognizable disease are common; in many areas reactions to tuberculin very often are due to infections with mycobacteria other than tubercle bacilli; and a number of mycobacterial diseases, some of which mimic tuberculosis, are prevalent in some countries and very rare in others. Their epidemiological relationships are strikingly different from those of tuberculosis. Some mycobacterial diseases are characteristically localized in the superficial, cooler parts of the body. Each species of *Mycobacterium* is distinct in structure and in biochemical and other reactions (see table).

Principal mycobacterial diseases of humans

Disease	Species	Disease character
Tuberculosis	*Mycobacterium tuberculosis*	Lung or other tissue destruction, cough, wasting
Kansasii disease	*M. kansasii*	Tuberculosislike
Battey disease	*M. intracellulare, M. avium*	Tuberculosislike
Scrofula	*M. scrofulaceum*	Gland swelling in neck, children, or other
Habana disease	*M. simiae*	Pulmonary or other
Szulgai disease	*M. szulgai*	Tuberculosislike or other
Xenopi disease	*M. xenopi*	Tuberculosislike or other
Leprosy	*M. leprae*	Skin, nerve, bone of face, and elsewhere
Swimming pool granuloma	*M. marinum*	Ulcer at site of broken skin
Tropical ulcer	*M. ulcerans*	Dermal and subcutaneous ulcers
Fortuitum disease	*M. fortuitum, M. chelonei*	Prosthesis or wound contaminant, rare lung or other disease

Although these species are demonstrably distinct by immunological tests, prominent cross reactions occur; diagnostic serologic tests with patients' sera have not been available for identification of the diseases. *See* IMMUNOLOGY; LEPROSY; SEROLOGY; SKIN TEST; TUBERCULOSIS. [E.H.R.]

Mycology A field of microbiology concerned with the study of fungi, those microorganisms included in the division Thallobionta (Thallophyta) of the plant kingdom. Fungi lack photosynthetic pigments and include such familiar plants as the molds, yeasts, and bacteria. *See* FUNGI; THALLOBIONTA.

Medical mycology is a specialized field of study concerned with fungi that cause disease. Practically all of the pathogenic fungi affecting humans belong to either the Schizomycetes or the Fungi Imperfecti; a few species are Phycomycetes. From a clinical viewpoint two major categories of fungus disease are recognized: those which cause superficial skin infections, the dermatophytes, and those which may produce deep invasive infections. *See* DEUTEROMYCOTINA; MEDICAL MYCOLOGY. [E.G.St./R.E.K.]

Mycorrhizae Joint or dual organs of absorption formed when healthy absorbing organs (roots, rhizomes, or thalli) of many, perhaps most, higher plants are inhabited by symbiotic fungi.

Mycorrhizae are of several types of structure and physiology as expected from the fact that the host species may belong to the Bryophyta, Polypodiophyta, Pinophyta, or Magnoliophyta, and the fungi to Basidiomycetes, Ascomycetes, Phycomycetes, and to the imperfect fungi. Two main structural types are usually recognized: (1) Ectomycorrhizae (ectotrophic mycorrhizae), in which the fungus forms a mantle or sheath about the host organ and its hyphae only penetrate between the cells, and (2) Endomycorrhizae (endotrophic mycorrhizae), in which the fungal hyphae penetrate into the cells of the host, inhabit them for a time, and are usually eventually eliminated by digestion. Some mycorrhizae seem to be intermediate between these two structural groups.

The work on the physiology of mycorrhizae has sufficed to show that although most kinds of these organs are more efficient in nutrient absorption than uninfected roots, they may be divided into two groups of carbon nutrition—two groups which do not conform to the structural groups. In all ectomycorrhizae, and in the endomycorrhizae of green Ericaceae and those endomycorrhizae with Endogonaceae (phycomycetes) as fungi (vesicular-arbuscular mycorrhizae), the symbiotic system depends for carbon on photosynthesis by the host. In endomycorrhizae caused by fungi with septate hyphae (Basidiomycetes and imperfect fungi), such as occur in those members of the Orchidaceae, Ericales, and other taxa which have a saprophytic phase or are saprophytes in the adult stage, the symbiotic system is supplied with carbon compounds by the fungus. [J.L.H.]

Mycotoxin Any of the mold-produced substances that may be injurious to vertebrates upon ingestion, inhalation, or skin contact. The diseases they cause, known as mycotoxicoses, need not involve the toxin-producing fungus. Diagnostic features characterizing mycotoxicoses are: the disease is not transmissible; drug and antibiotic treatments have little or no effect; in field outbreaks the disease is often seasonal; the outbreak is usually associated with a specific foodstuff; and examination of the suspected food or foodstuff reveals signs of fungal activity. *See* AFLATOXIN; ERGOTISM; MEDICAL MYCOLOGY. [A.Ci.]

Myelin A highly refractile fatty sheath surrounding many nerve fibers. Myelin is composed of alternating dense and light strata repeating radially at a period of 16–18 nanometers. Every other dense stratum is less dense and often appears as a double line. The very dense strata are called major dense lines, and the less dense ones intraperiod lines.

The main function of the myelin sheath is to facilitate conduction of the nerve action potential. The action potential is a propagated depolarization of the axon membrane that extends throughout the length of the axon. This spreads as a flow current carried by ions from point to point along the axon surface. Ions are exchanged with the gap between the axon and the Schwann cell membrane. In nonmyelinated nerve fibers these ions rapidly diffuse into the extracellular space, and propagation of the action potential is relatively slow, less than 3.3 ft/s (1 m/s). In myelinated nerve fibers the ions have no free pathway to the extracellular space since the membranes of the mesaxons are closely apposed and the gap between the axon and the Schwann cell membrane at the node is also sealed off. The only place where there is free access to the extracellular space is at the nodes of Ranvier. The consequence of this is that current flows primarily through the axon membrane at the node and jumps via the extracellular space to the next node to complete the circuit. [J.D.Ro.]

Myiasis The infestation of vertebrates by the larvae, or maggots, of numerous species of flies. These larvae may invade different parts of the bodies of these animals or may appear externally. The Diptera of medical and veterinary importance are largely confined to the families Oestridae, Calliphoridae, and Sarcophagidae.

In cutaneous myiasis, the larvae are found in or under the skin. There may be a migration of some species of these larvae through host tissues, resulting in a swelling with intense itching. Such a condition is known as larva migrans, or creeping eruption, and may require surgical treatment.

Intestinal myiasis in humans is usually the result of accidentally swallowing the eggs or larvae of these flies. It occurs commonly in many herbivores who ingest the eggs when feeding on contaminated herbage. The larvae settle in the stomach or intestinal tract of the animal host.

Cavity, or wound, myiasis occurs when the larvae invade natural orifices, such as the nasopharynx, vulva, and sinuses, or artificial openings such as wounds. External myiasis includes infestation by those maggots which are blood feeders. *See* DIPTERA; MEDICAL PARASITOLOGY. [C.B.C.]

Mylonite A rock that has suffered extreme mechanical deformation and granulation but has remained chemically unaltered. In the early stages of this process, cataclasites or flaser rocks are produced. These rocks are characterized by the fact that the nature of the parent rock is easily recognized. The mylonites, however, are altered beyond recognition. In spite of their pulverized condition they are hard, coherent, and often black and glassy-looking rocks, because the crushing took place under confining pressures so high that the comminuted particles have been welded tightly together. *See* METAMORPHIC ROCKS. [T.F.W.B.]

Myocarditis Cardiomyopathies characterized by a prominent infiltration of inflammatory cells interstitially between the myocardial fibers. Most of these inflammatory cardiomyopathies are secondary to systemic disease and thus are termed secondary inflammatory cardiomyopathies. Primary myocarditis is rare and the causes are not well understood. The secondary myocarditis is associated with systemic bacterial, viral, fungal, and parasitic diseases, such as typhoid, diphtheria, scarlet fever, coxsackie virus, influenza, polio, measles toxoplasmosis, and trichinosis. Irrespective of the cause, the morphological appearance of inflammatory myocarditis is similar. Grossly the heart is normal; however, microscopically there is an interstitial inflammatory infiltrate with edema. The cellular infiltrate consists of patches of eosinophils, giant cells, and some lymphocytes. *See* HEART DISORDERS. [N.K.M.]

Myodocopa One of two extant subclasses of Ostracoda. The subclass comprises two superorders, the Myodocopida and Halocyprida. Myodocopods possess a heart. The valves of the carapace are unequal and often only weakly calcified. However, they may be smooth, strongly ribbed, or provided with nobs or spines. Five to seven pairs of appendages are present. *See* HALOCYPRIDA; MYODOCOPIDA.

Myodocopods are all marine,, although a few do inhabit brackish waters, and they are represented in all oceans from the surface to abyssal depths. Because of the development of the antennae, most species of myodocopids can swim; however, most are epibenthic or benthic inhabitants.

From a phylogenetic standpoint, the Myodocopa are represented in the fossil record by Entomoconchacea and Entomozoacea (Ordovician?-Carboniferous); however, none of the surviving major taxa can be considered primitively close to the ancestral ostracod. *See* CRUSTACEA; OSTRACODA. [P.A.McL.]

Myodocopida An order of the subclass Myodocopa (Ostracoda). In contrast to other living ostracods, nearly all myodocopids have a carapace with a permanent anterior aperture when the valves are otherwise drawn together. Lateral eyes are well developed, each with several lenses, and a frontal organ (organ of Bellonci) projects from the center of the forehead. Through the permanent aperture or gape, the antennae project for locomotion. Some myodocopid forms are pelagic but most are benthic; in the former, the carapace may be very thin, containing little or no calcite. *See* MYODOCOPA.

The order includes the single extant superfamily Cypridinoidea, which has a carapace that is usually calcified and round-backed, with the rostrum curved downward. The largest living ostracod, *Gigantocypris agassizi*, is a swimming species of the Cypridinacea with an extremely thin subglobular carapace reaching 0.92 in. (23 mm) in length. From the standpoint of appendage anatomy, it has been postulated that the Myodocopida are the more primitive of the two orders, separating from the Halocyprida in the pre-Silurian era. *See* CRUSTACEA; OSTRACODA. [P.A.McL.]

Myriangiales An order of parasitic fungi of the class Ascomycetes which produce their asci at various levels in uniascal locules within stromata, which are masses of tightly interwoven somatic hyphae. The stroma weathers away and exposes the asci which contain the spores. Examples of Myriangiales are *Elsinoë veneta* and *E. fawcetii*, which cause raspberry and citrus anthracnose, respectively, and *Myriangium duriaei*, which parasitizes scale insects. *Piedraia hortai*, which attacks human hair causing black piedra, has also been placed here. *See* ASCOMYCOTINA; FUNGI; MEDICAL MYCOLOGY; PLANT PATHOLOGY. [C.J.A.]

Myricales An order of flowering plants, division Magnoliophyta (Angiospermae), in the subclass Hamamelidae of the class Magnoliopsida (dicotyledons). The order consists of the single family Myricaceae, with about 50 species. Within its subclass the order is marked by its simple, resinous-dotted, aromatic leaves and unilocular ovary with two styles and a single ovule. The plants are trees or shrubs, and the flowers are much reduced and borne in catkins. The fruit is a small, waxy-coated drupe or nut. Several species of *Myrica* are occasionally cultivated as ornamentals. *See* FLOWER; HAMAMELIDAE; MAGNOLIOPSIDA. [A.Cr.]

Myrtales An order of flowering plants, division Magnotiophyta (Angiospermae), in the subclass Rosidae of the class Magnoliopsida (dicotyledons). The order contains 13 families and about 9000 species. The two largest families are the Melastomataceae, with about 4000 species, and the Myrtaceae, with about 3000. The Onagraceae, with about 650 species, are common in temperate regions, especially of western North America, in contrast to the larger families of the order, which are chiefly tropical.

The Myrtales usually have opposite, simple, entire leaves and perigynous to epigynous flowers with a compound pistil and most commonly axile placentation. Allspice and cloves are obtained from *Pimenta dioica* and *Syzygium aromaticum*, respectively, which are trees belonging to the Myrtaceae. *Eucalyptus* is another genus of the same family. *See* ALLSPICE; CLOVE; EUCALYPTUS; MAGNOLIOPSIDA; ROSIDAE. [A.Cr.]

Mysidacea An order of free-swimming Crustacea. Mysidacea contains two suborders, the Lophogastrida and Mysida. Most species are marine, many are euryhaline, and some are fresh-water. They are known from inshore and surface waters down to depths of over 23,000 ft (7000 m) [deepest record 23,430–23,960 ft or 7100–7260 m in a closing net], attaining a much greater size in the greater depths.

The adult body consists of 19 somites; 5 head, 8 thoracic, and 6 abdominal with a tail fan composed of uropods and a telson (see illustration). Each somite bears one pair of functionally modified, biramous appendages. The carapace envelops most of the thorax and is fused dorsally with not more than 4 of the anterior thoracic somites. The eyes are pedunculate and movable.

Locomotion is effected by a paddling action of the pleopods in lower forms. In higher groups thoracic exopods set up backwardly directed currents which propel the animal forward. The tail fan and abdomen assist in swimming and can produce springing movements, the "escape mechanism." Walking may be accomplished by the thoracic endopods.

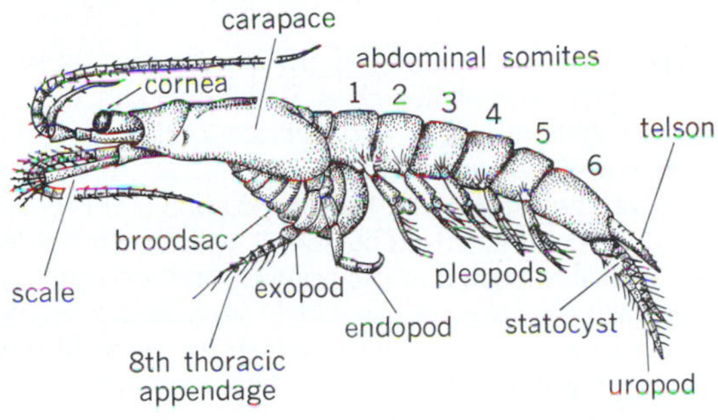

Diagram of a mysid, showing caridoid facies.

Mysids are not combatant, relying on agility, camouflage, sheltering under rocks or weeds, or burrowing in the sea floor to escape enemies. Six species, all blind and representing the three most primitive genera of Mysida, are cavernicolous in fresh or brackish subterranean waters. Nine species of *Heteromysis*, almost the most highly developed genus of mysids, are known to live as commensales. No parasitic forms are known.

Pelagic species in upper waters are usually transparent and, except for the cornea, are colorless. Most littoral species have a chromatophore system whereby color changes are made to harmonize with the background.

During daylight most mysids remain feeding at, or near, the bottom, migrating at dusk toward upper levels where the young are liberated. Toward dawn, adults descend again, while the juveniles usually remain nearer the surface until mature. *See* MALACOSTRACA; PERACARIDA. [O.S.T.]

Mystacocarida An order of primitive Crustacea. The three species are in the genus *Derocheilocaris*. The body is wormlike and about 0.2 in. (5 mm) long. The cephalothorax bears first antennae, second antennae, mandibles, first maxillae, and second maxillae. Maxillipeds are on a separate segment, and four additional free thoracic segments bear platelike appendages. The six abdominal segments are without appendages. The labrum is enormous, mouthparts and nervous system are primitive, and the genital pore is on the thorax. *See* CRUSTACEA. [R.W.P.]

Myxiniformes The hagfishes, sometimes called Myxinoidea, an order of eellike, jawless vertebrates (class Agnatha), often grouped with the Petromyzontida as Cyclostomata. They resemble the Petromyzontida in a number of respects. However, they differ from Petromyzontida in having the nasal opening at the tip of the snout and leading to the pharynx so that it can be used in respiration; in having barbels around the mouth; in having 6–15 pairs of gill pouches which open to the exterior either separately or by a common aperture; in having only one pair of semicircular canals; in having degenerate eyes lacking lens and iris and not visible externally; in having a continuous dorsal-caudal-ventral fin with fin rays only in the caudal portion; in having both male and female organs in one individual, though only one matures; and in being exclusively marine.

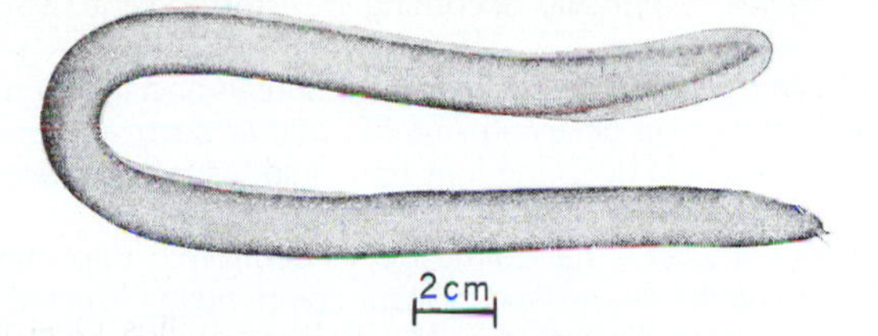

Atlantic hagfish (*Myxine glutinosa*). (*After G. B. Goode, U.S. Nat. Mus. Bull. no. 27, 1884*)

Hagfishes live on or near muddy bottoms in fairly deep waters, where they feed largely on dead or disabled fishes and polychaete worms, into which they bore with their rasplike tongues.

Hagfishes are divided into three genera: *Myxine* (see illustration), known in both the Atlantic and Pacific oceans; *Paramyxine*, known from Japan; and *Bdellostoma*, known in the Pacific. *See* AGNATHA; CYCLOSTOMATA (CHORDATA); PETROMYZONTIDA. [R.H.De.]

Myxobacterales An order of gliding bacteria. The Myxobacterales or myxobacters are also known as the slime bacteria. They are found mainly in the surface layers of soil, rotting wood, compost, and manure.

The vegetative cells are flexible, slender rods 1–2 micrometers wide and 3–5 micrometers long. The cells are not enclosed by a rigid cell wall; thus, they are not as refractile as the cells of true bacteria and are somewhat difficult to see microscopically. On a surface the cells display a characteristic gliding motility, although neither flagella nor other motility organelles are present. Because of this, motility myxobacterial colonies are not limited like bacterial colonies but are, rather, rapidly spreading, thin "swarms," often with irregular, veinlike thickenings. The cells tend to arrange themselves in parallel rows, forming small tongues and islets at the edge of the swarm.

The fruiting bodies formed by the myxobacters are unique among the bacteria. After a period of vegetative growth on a suitable medium, the cells of the swarm begin to aggregate toward several centers. The aggregating myxobacters form tongues of parallel rows of cells which converge on the center. Arriving at the center, the cells heap up on themselves and become transformed into resting cells called microcysts. The fruiting body may be a more or less formless mass of microcysts embedded in a mass of slime, or it may become variously branched and twisted, bearing clusters of cysts at the tips of the branches. Mature fruiting bodies are usually pigmented. [W.R.S.]

Myxomycetes A group of microorganisms belonging to the Myxomycota classified with the fungi and given the class name Myxomycetes because of the funguslike reproductive phase, or sometimes classified with the protozoans (one-celled animals) and given the class name Mycetozoa because of the animallike assimilative phase. Evolutionary origins are obscure and controversial, but many believe the Myxomycetes are descendants of protozoan ancestors.

The reproductive phase is characterized by sedentary funguslike fruiting structures containing one to many spores. The assimilative phase is initiated by spore germination that gives rise to one or more naked myxamoebae or swarm cells which may serve as gametes. Fusion of gametes forms the zygote, which develops into a free-living, multinucleate, acellular mass of protoplasm, the plasmodium, exhibiting protoplasmic streaming and amoeboid movement. The mode of nutrition is animallike, with ingestion of particulate matter (usually bacteria) by engulfment. The plasmodium undergoes a series of developmental stages, eventually becoming transformed into various types of fructifications.

Most slime molds are universally distributed and live in moist and dark places on decaying organic matter such as standing dead trees, and old decaying logs and dead twigs and leaves on the forest floor.

The Myxomycetes have little direct economic importance, but their beautiful design and brilliant colors have attracted and fascinated many collectors. *See* MYXOMYCOTA. [H.W.K.]

Myxomycota Organisms treated as classes—Acrasiomycetes (cellular slime molds), Myxomycetes (plasmodial, acellular, or true slime molds), Protosteliomycetes, and Plasmodiophoromycetes (endoparasitic slime molds)—and classified with the fungi, following the rules of botanical nomenclature. These same groups may be classified in the kingdom Protista at various taxonomic ranks, following zoological nomenclature. These slime mold groups are not necessarily closely related, but they all do have primitive amebalike colonial stages somewhere in their life cycles. *See* FUNGI. [H.W.K.]

Myxosporida An order of the protozoan class Myxosporidea (subphylum Cnidospora). It is characterized by the production of spores with one or more valves and polar capsules, and by possession of a single sporoplasm with or without an iodinophilous vacuole. Myxosporidians are mainly parasites of fishes. They infect all parts of the body, including the heart and brain, and often induce considerable pathological changes in the host tissue.

Infection begins with the ingestion of the spore by a host fish. The digestive fluids cause the polar filaments to be extruded, and at the same time the sporoplasm is released from the spore (see illustration). The sporoplasm, or amebula, reaches the specific site of infection directly through the gut wall or by way of the bloodstream. The amebula becomes a trophozoite when it starts feeding on the host tissues. The trophozoite then goes through a series of nuclear divisions and, by a process of budding, gives rise to a number of cells, each of which eventually develops into a sporont. A sporont is a monosporoblast if one spore is produced and a pansporoblast if two or more spores are formed. The sporont undergoes a series of nuclear divisions, in which the number of nuclei produced will determine the number of spores and polar capsules to be formed. That is, in every spore one nucleus is involved in the formation of each valve and each polar capsule. Two nuclei become the gametic nuclei, which then fuse to form the zygotic nucleus of the sporoplasm. *See* CNIDOSPORA; MYXOSPORIDEA. [R.F.N.]

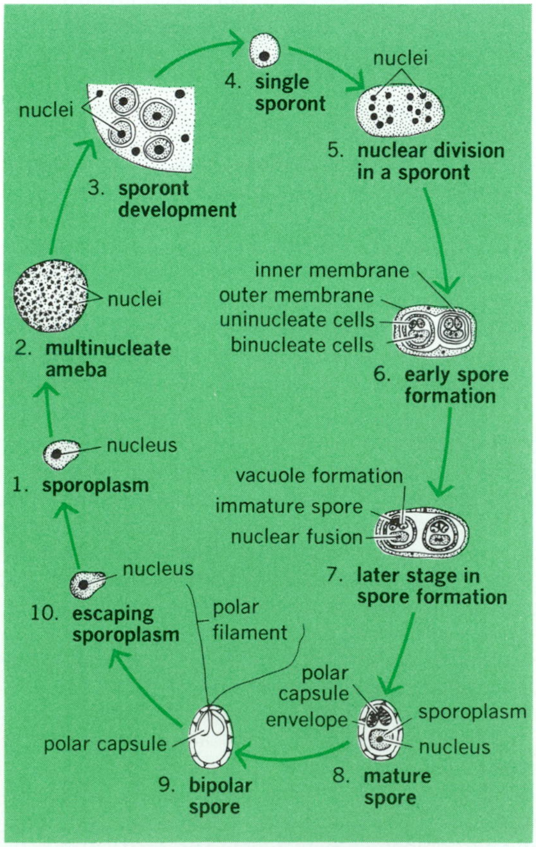

Life cycle of *Myxobolus*, a fish parasite.

Myxosporidea A class of the protozoan subphylum Cnidospora. Members of this class, which includes the orders Myxosporida, Actinomyxida, and Helicosporida, are parasites in fish, a few amphibians, and invertebrates. The Myxosporida are divided into two suborders, the Unipolarina and Bipolarina.

Unipolarina is characterized by spores with one to six (never five) polar capsules located at the anterior end, except in some genera in which the capsules are widely separated or located in the central part of the spore but in which the polar filament is attached near the anterior end. The Unipolarina contains nine or more families.

Bipolarina, containing a single family with three genera, is characterized by the presence of one capsule at or near each end of a fusiform or ellipsoid spore. *See* ACTINOMYXIDA; CNIDOSPORA; HELICOSPORIDA; MYXOSPORIDA. [R.F.N.]

Myxovirus A group of animal viruses that originally included the human influenza viruses, mumps virus, and Newcastle disease virus. Its name was based on a special affinity of these members for certain mucins, in a reaction concerned with the adsorption and penetration of virus into the cell. A number of other viruses have been found also to share characteristics of the group, particularly in their morphology, although not all have a demonstrated affinity for mucins. The group now includes influenza and fowl plague viruses, and the paramyxovirus subgroup, in which mumps, measles, parainfluenza, respiratory synctial (RS) disease, and Newcastle disease viruses have been placed. Rinderpest and canine distemper viruses are related to measles virus.

The myxoviruses range from 80 to 300 nanometers in size; they possess a coiled inner helix of ribonucleoprotein surrounded by an ether-sensitive lipoprotein envelope studded with projections.

All myxoviruses except canine distemper, rinderpest, and RS viruses agglutinate red blood cells. Such hemagglutination by the virus, and the development in the patient of specific antibodies which inhibit hemagglutination, are properties useful in diagnosis. *See* ANIMAL VIRUS; INFLUENZA; MEASLES; MUMPS; NEWCASTLE DISEASE; PARAINFLUENZA VIRUS; PARAMYXOVIRUS. [J.L.Me.]

Myzostomaria An aberrant group of Polychaeta recognized for four families, each with one genus. Myzostomidae with *Myzostomum* comprise about 130 species from worldwide areas. Mesomyzostomidae with *Mesomyzostomum* has two species, from Japan and the Aru Islands. Protomyzostomidae with *Protomyzostomum* is known for three species, from Japan and the Murman Sea, and Stelechopidae with *Stelechopus* has a single species, from Crozet Island, Antarctic Ocean. All are either external or internal parasites of echinoderms, chiefly crinoids, and hence from deep water. Most are greatly depressed, broad and very small, measuring at most a few millimeters. The separation of this group of polychaetes into families and species is based on external and internal characters. *See* ANNELIDA; POLYCHAETA. [O.H.]

N

Nacreous clouds Clouds whose coloration and iridescence suggest mother-of-pearl (nacre); they are an occasional phenomenon of the lower stratosphere (12–15 mi or 20–25 km) during the winter in high latitudes and have been reported from Scotland, Norway, Alaska, and Antarctica. Nacreous clouds are generally stationary, and therefore they are attributed to wave motions; standing waves, often in the lee of mountain ranges, provide the mechanism for abrupt lifting of air with accompanying condensation at such low temperatures as −112 to −130°F (−80 to −90°C). The clouds may display rather pure colors across the rainbow range and may appear to be very bright, even dazzling, shortly after sunset. Unlike the much higher noctilucent clouds of summer, nacreous clouds are also plainly seen in daylight. The optical phenomena of nacreous clouds indicate that they are probably composed of spherical particles (of ice) about 1 to 2 micrometers in diameter. *See* NOCTILUCENT CLOUDS; STANDING WAVE; STRATOSPHERE. [E.Ke.]

Nailing The driving of nails in a manner that will position and hold two or more members, usually of wood, in a desired relationship to each other. The contact pressures between the surfaces of the nails and the surrounding wood fibers hold the nails in position. Some types of nails are shown in the illustration.

Factors that determine the strength and efficiency of a nailed joint are (1) the type of wood, (2) the nail used, (3) the conditions under which the nailed joint is used, and (4) the number of nails. In general, hard, dense woods hold nails better than soft woods. The better the resistance of a nail to direct withdrawal from a piece of wood, the tighter the joint will remain. To increase resistance to withdrawal or loosening, nails may be coated, etched, spirally grooved, annularly grooved, or barbed, as illustrated.

Blunt-pointed nails are often used to prevent the wood from splitting. Using nails of a smaller diameter also tends to prevent splitting but requires a greater number of nails per joint. Beeswax is sometimes applied to nail points to make them drive more easily, but it also reduces the holding power of the nail. [A.H.T.]

Najadales An order of aquatic and semiaquatic flowering plants, division Magnoliophyta (Angiospermae), in the subclass Alismatidae of the class Liliopsida (monocotyledons). The order consists of 8 families and a little more than 200 species. The Potamogetonaceae, with about 90 species, are the largest family of the order, and the name Potamogetonales is sometimes used instead of Najadales for the group. The Najadales are Alismatidae in which the perianth, when present, is not differentiated into evident sepals and petals. Usually the flowers are not individually subtended by bracts. The Zosteraceae of this order are unique among flowering plants, in that they grow submersed in the ocean, albeit in shallow water near the shore. *Zostera marina*, or eelgrass, is a common member of the family. *See* ALISMATIDAE; LILIOPSIDA. [A.Cr.]

Nanochemistry The study of the synthesis and characterization of materials in the nanoscale size range (1 to 10 nanometers). These materials include large organic molecules, inorganic cluster compounds, and metallic or semiconductor particles. The synthesis of nanoscale inorganic materials is important because the small size endows these particles with unusual structural and optical properties that may find application in catalysis and electrooptical devices. Approaches to the synthesis of these materials have focused on constraining the reaction environment through the use of surface-bound organic additives, porous glasses, zeolites, clays, or polymers. The use of synthetic approaches that are inspired by the biological processes result in the deposition of inorganic materials such as bones, shells, and teeth (biomineralization). This biomimetic

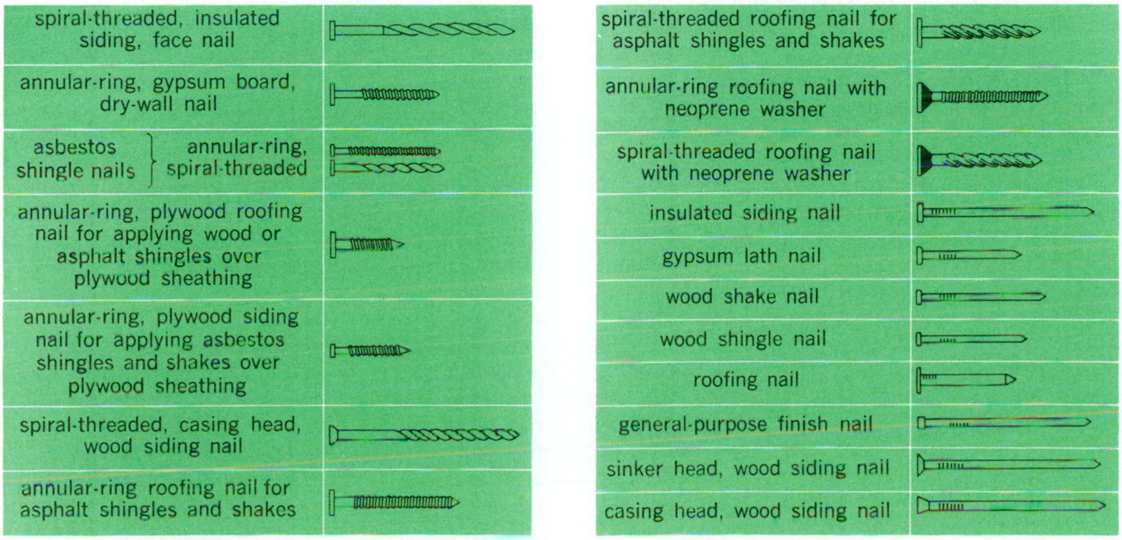

Special- and general-purpose nails.

approach involves the use of assemblies of biological molecules that provide nanoscale reaction environments in which inorganic materials can be prepared in an organized and controlled manner. Examples of biological assemblies include phospholipid vesicles and the polypeptide micelle of the iron storage protein, ferritin. *See* MICELLE.

Vesicles are bounded by an organic membrane that provides a spatial limit on the size of the reaction volume. If a chemical reaction is undertaken in this confined space that leads to the formation of an inorganic material, the size of the product will also be constrained to the dimensions of the organic host structure. Provided that the chemical and physical conditions are not too severe to disrupt the organic membrane, these supramolecular assemblies may have advantages over inorganic hosts such as clays and zeolites because the chemical nature of the organic surface can be systematically modified so that controlled reactions can be accomplished. *See* SUPRAMOLECULAR CHEMISTRY.

One problem encountered with the use of phospholipid vesicles is their sensitivity to changes in temperature and ionic strength. Procedures have been developed in which the biomolecular cage of the iron storage protein, ferritin, has been used as a nanoscale reaction environment for the synthesis of inorganic materials. In the simplest approach the native iron oxide core is transformed into another material by chemical reaction within the protein shell. [S.Ma.]

Nanostructure A material structure assembled from a layer or cluster of atoms with size of the order of nanometers. Interest in the physics of condensed matter at size scales larger than that of atoms and smaller than that of bulk solids (mesoscopic physics) has grown rapidly since the 1970s, owing to the increasing realization that the properties of these mesoscopic atomic ensembles are different from those of conventional solids. As a consequence, interest in artificially assembling materials from nanometer-sized building blocks arose from discoveries that by controlling the sizes in the range of 1–100 nm and the assembly of such constituents it was possible to begin to alter and prescribe the properties of the assembled nanostructures. *See* MESOSCOPIC PHYSICS.

Nanostructured materials are modulated over nanometer length scales in zero to three dimensions. They can be assembled with modulation dimensionalities of zero (atom clusters or filaments), one (multilayers), two (ultrafine-grained overlayers or coatings or buried layers), and three (nanophase materials), or with intermediate dimensionalities (see illus.).

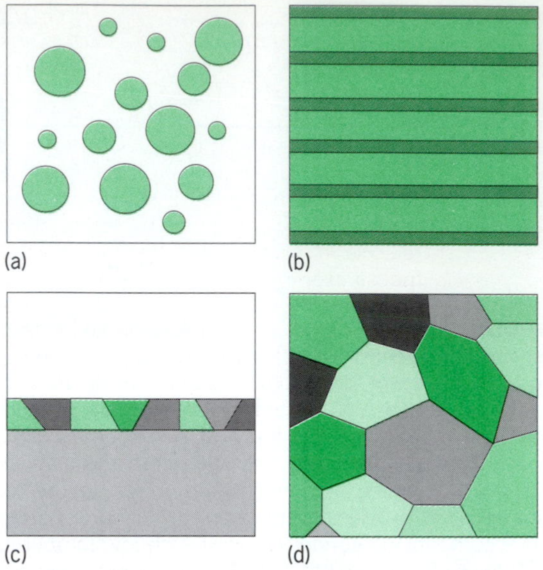

Schematic of four basic types of nanostructured materials, classified according to integral modulation dimensionality. (*a*) Dimensionality 0: clusters of any aspect ratio from 1 to infinity. (*b*) Dimensionality 1: multilayers. (*c*) Dimensionality 2: ultrafine-grained overlayers (coatings) or buried layers. (*d*) Dimensionality 3: nanophase materials. (*After R. W. Siegel, Nanostructured materials: Mind over matter, Nanostruct. Mat., 3:1–18, 1993*)

Multilayers and clusters. Multilayered materials have had the longest history among the various artificially synthesized nanostructures, with applications to semiconductor devices, strained-layer superlattices, and magnetic multilayers. Recognizing the technological potential of multilayered quantum heterostructure semiconductor devices helped to drive the rapid advances in the electronics and computer industries. A variety of electronic and photonic devices could be engineered by utilizing the low-dimensional quantum states in these multilayers for applications in high-speed field-effect transistors and high-efficiency lasers, for example. Subsequently, a variety of nonlinear optoelectronic devices, such as lasers and light-emitting diodes, have been created by nanostructuring multilayers. *See* ARTIFICIALLY LAYERED STRUCTURES; LIGHT-EMITTING DIODE; SEMICONDUCTOR HETEROSTRUCTURES; TRANSISTOR.

The advent of beams of atom clusters with selected sizes allowed the physics and chemistry of these confined ensembles to be critically explored, leading to increased understanding of their potential, particularly as the constituents of new materials, including metals, ceramics, and composites of these materials. A variety of carbon-based clusters (fullerenes) have also been assembled into materials of much interest. In addition to effects of confinement, interfaces play an important and sometimes dominant role in cluster-assembled nanophase materials, as well as in nanostructured multilayers. *See* ATOM CLUSTER; CERAMICS; FULLERENE.

Synthesis and properties. A number of methods exist for the synthesis of nanostructured materials. They include synthesis from atomic or molecular precursors (chemical or physical vapor deposition, gas condensation, chemical precipitation, aerosol reactions, biological templating), from processing of bulk precursors (mechanical attrition, crystallization from the amorphous state, phase separation), and from nature (biological systems). Generally, it is preferable to synthesize nanostructured materials from atomic or molecular precursors, in order to gain the most control over a variety of microscopic aspects of the condensed ensemble; however, other methodologies can often yield very useful results. *See* VAPOR DEPOSITION. [R.W.Si.]

Nanotechnology Systems for transforming matter, energy, and information, based on nanometer-scale components with precisely defined molecular features. The term nanotechnology has also been used more broadly to refer to techniques that produce or measure features less than 100 nanometers in size; this meaning embraces advanced microfabrication and metrology. Although complex systems with precise molecular features cannot be made with existing techniques, they can be designed and analyzed. Studies of nanotechnology in this sense remain theoretical, but are intended to guide the development of practical technological systems.

Nanotechnology based on molecular manufacturing requires a combination of familiar chemical and mechanical principles in unfamiliar applications. Molecular manufacturing can exploit mechanosynthesis, that is, using mechanical devices to guide the motions of reactive molecules. By applying the conventional mechanical principle of grasping and positioning to conventional chemical reactions, mechanosynthesis can provide an unconventional ability to cause molecular changes to occur at precise locations in a precise sequence. Reliable positioning is required in order for mechanosynthetic processes to construct objects with millions to billions of precisely arranged atoms.

Mechanosynthetic systems are intended to perform several basic functions. Their first task is to acquire raw materials from an externally provided source, typically a liquid solution containing a variety of useful molecular species. The second task is to process these raw materials through steps that separate molecules of different kinds, bind them reliably to specific sites, and then (often) transform them into highly active chemical species, such as radicals, carbenes, and strained alkenes and alkynes. Finally, mechanical devices can apply these bound, active species to a workpiece in a controlled position and orientation and can deposite or remove a precise number of atoms of specific kinds at specific locations.

Several technologies converge with nanotechnologies, the most important being miniaturization of semiconductor structures, driven by progress in microelectronics. More directly relevant are efforts to extend chemical synthesis to the construction of larger and more complex molecular objects. Protein engineering and supramolecular chemistry are active fields that exploit weak intermolecular forces to organize small parts into larger structures. Scanning probe microscopes are used to move individual atoms and molecules. *See* MOLECULAR RECOGNITION; MONOMOLECULAR FILM; NANOSTRUCTURE; SCANNING TUNNELING MICROSCOPE; SUPRAMOLECULAR CHEMISTRY. [E.Dr.]

Naphtha Any one of a wide variety of volatile hydrocarbon mixtures. They are sometimes obtained from coal tar but are more often derived from petroleum. Physical properties vary widely. The initial boiling point may be as low as 27°C (80°F), and end points may reach 260°C (500°F). Boiling ranges are sometimes as narrow as 11°C (20°F) or as wide as 110°C (200°F). Products sold as naphthas find their greatest use as solvents, thinners, or carriers.

There is a fairly sharp differentiation between aliphatic and aromatic naphthas. Aliphatic naphthas are relatively low in odor and toxicity and tend, also, to be low in solvent power. The aromatic naphthas are highly solvent. Their main components are toluene and xylenes; benzene is less desirable because of the extreme toxicity of its vapors. *See* PETROLEUM PRODUCTS. [J.K.R.]

Naphthalene A colorless crystalline aromatic hydrocarbon, $C_{10}H_8$, with the familiar odor of mothballs, melting point 80.1°C (176°F), boiling point 218°C (424°F). It is almost insoluble in water but soluble in nearly all organic solvents. Structurally it is best represented as two benzenoid rings fused together.

One gallon of coal tar yields approximately one pound of naphthalene (approximately 120 grams per liter). It is present also in certain fractions obtained by the reforming or steam cracking of certain petroleum distillates and is recovered commercially on a large scale.

Of the naphthalene consumed, the majority is converted to phthalic anhydride. Aside from a few percent used in the manufacture of mothballs, the remainder is converted to naphthalene compounds, which are used as dye intermediates, tanning agents, and surface-active agents. *See* AROMATIC HYDROCARBON; POLYNUCLEAR HYDROCARBON. [C.K.B.]

Narcotic A drug which diminishes the awareness of sensory impulses, especially pain, by the brain. This action makes narcotics useful therapeutically as analgesics. While they are the most powerful pain-relieving agents available, their use is complicated by a number of undesirable side actions. *See* ANALGESIC.

All of the generally used narcotics are in some way related to opium, and the term opiate is sometimes used interchangeably with the term narcotic. Opium is a gummy exudate obtained from the unripe seed capsules of the opium poppy. Crude opium contains over a dozen alkaloids, all of which have been isolated and identified as to their structural chemistry. From this knowledge chemists have developed a number of synthetic chemical compounds, some of which have important advantages over the naturally occurring alkaloids. Therapeutically important natural alkaloids are morphine, codeine, and papaverine. Among the important synthetic narcotics are meperidine (Demerol), dihydromorphine (Dilaudid), oxymorphone (Numorphan), alphaprodine (Nisentil), anileridine (Leritine), piminodine (Alvodine), levorphanol (Levo-Dromoran), methadone (Dolophine), and phenazocine (Prinadol). *See* ALKALOID; OPIATES; POPPY.

Nalorphine (Nalline) is a narcotic antagonist and is used in the treatment of acute overdosage from narcotics; it is dangerous to drug addicts. Heroin is a highly addicting narcotic, and is so dangerous in this regard that the drug has been completely banned by both federal and state laws under all circumstances.

Pharmacology. The pharmacology of narcotics is generally similar to that of morphine, the principal narcotic used for its analgesic effects. Differences among them lie in the potency of their action and in the degree and variety of the side actions which they produce. Effects are those of analgesia, accompanied by a state of euphoria characterized by drowsiness and a change of mood from anxiety and tension to calmness and equanimity. It should be remembered that whatever narcotic is used, the effects are dose-related, and in higher doses all narcotics produce deep sleep and eventually general depression of all brain functions. Death from overdosage is due to depression of the respiratory centers with resultant failure of respiration.

The predominant pharmacological effect of morphine (and the other narcotics) is on the central nervous system. From the standpoint of its medicinal use, its most important action is relief of pain. Along with its valuable medicinal use morphine produces a great many undesirable side actions; the most frequent are depressed respiratory activity, the production of nausea and vomiting, and the inhibition of defecation and urination.

Drug dependence. All narcotics have the potential for producing dependence and addiction when used repeatedly over a period of time. Drug dependence results from compulsive, continued use of the drug, and is characterized by one or more of the following conditions: habituation, tolerance, or addiction.

Like any other habit pattern, habitual use of a drug can develop. Common examples are the use of nicotine in the form of cigarettes, or caffeine in the form of coffee or tea. Such habituation is generally regarded as innocuous.

Repeated ingestion of a drug in which the effect produced by the original dose no longer occurs results in tolerance. To produce the original effect, it is necessary to increase the dose.

When the body develops a dependence for the drug, addiction occurs. If the drug is suddenly stopped after a period of frequent use, a withdrawal syndrome develops, which is characterized by physical pain and widespread body reactions. The addict comes to dread the development of such painful and distressing reactions, and is trapped into continuing the drug.

All narcotics can produce habituation, tolerance, and addiction to a greater or less degree. Addiction to codeine is relatively rare but possible. Addiction to heroin develops rapidly, and this narcotic is therefore exceedingly dangerous. *See* DRUG ADDICTION. [J.M.Di.]

Native elements Those elements which occur in nature uncombined with other elements. Aside from the free gases of the atmosphere there are about 20 elements that are found as minerals in the native state. These are divided into metals, semimetals, and nonmetals. Gold, silver, copper, and platinum are the most important metals and each of these has been found abundantly enough at certain localities to be mined as an ore. Rarer native metals are others of the platinum group, lead, mercury, tantalum, tin, and zinc. Native iron is found sparingly both as terrestrial iron and meteoric iron.

The native semimetals can be divided into (1) the arsenic group, including arsenic, antimony, and bismuth; and (2) the tellurium group, including tellurium and selenium.

The native nonmetals are sulfur, and carbon in the forms of graphite and diamond. Native sulfur is the chief industrial source of that element. [C.S.Hu.]

Natrolite A fibrous or needlelike mineral belonging to the zeolite family of silicates. Most commonly it is found in radiating fibrous aggregates. The hardness is 5–5½ on Mohs scale, and the specific gravity is 2.25. The mineral is white or colorless with a vitreous luster that inclines to pearly in fibrous varieties. The chemical composition is $Na_2(Al_2Si_3O_{10}) \cdot 2H_2O$, but some potassium is usually present substituting for sodium.

Natrolite is a secondary mineral found lining cavities in basaltic rocks. Its outstanding locality in the United States is at Bergen Hill, New Jersey. *See* ZEOLITE. [C.Fr.; C.S.Hu.]

Natural fiber A fiber obtained from a plant or animal. For many centuries, wool and flax were the chief fibers for apparel in temperate climates, and cotton in the tropics.

Vegetable fibers. The basic biological unit of vegetable fibers is the plant cell, with a length frequently more than a thousand times its diameter. The cotton fiber is a single cell. The bast and hard fiber strands are composed of many overlapping, parallel fiber cells held together in a fiber bundle that may be as long as the leaf or stem from which it comes.

Cotton. The cotton plant is a member of the mallow family (Malvaceae). The grading of cotton fibers is based on various considerations, but the length of fiber is the most important. The most important universal standard grades are determined mainly by staple length. For example, while the Sea Island cotton extends up to 2 in. (5 cm) in staple length, Indian varieties range from ⅞ to ⅝ in. (2.2 to 1.7 cm). Yarn quality depends on inherent fiber properties of strength, length, length uniformity, and fineness. In order to ensure a desired uniformity and product quality, batches of raw fiber, different in staple length and other characteristics, are processed by blending together.

After weaving, the resulting cotton fabric may receive further processing, which can also be complex, especially because modern textile technology can change cotton fabric more than any other material. Bleaching, dyeing, and printing can be considered among these, and with the introduction of chemical finishes, cotton can be given many properties, such as durable press, crease resistance, water repellence, and flame retardance. *See* BLEACHING; COTTON; DYEING.

Bast fibers. These fibers which come from the phloem or bast of dicotyledonous plants are also called soft fibers. The greatest quantity of these fibers goes into fabrics, but they have other uses such as thread, twine, or paper. The more widely used bast fibers are jute, flax, hemp, and ramie.

Flax is grown mostly in northern and eastern Europe. The fiber is separated from the stem of the plant by a system of retting and scutching. It is used primarily for linen fabrics, but many other products are made from it. The line (long, parallel) fiber is used principally for fine damask table linen, apparel fabrics, sheeting, lace, thread, and special twine. The tow (short, tangled fiber produced in the scutching and in the hackling operations) and some of the coarser grades of line fiber are used for toweling, canvas, mail bags, meat wraps, twine, and for high-grade papers such as cigarette, bond, and currency. The fiber from flax raised for seed is also used for high-grade papers (but not for textiles). *See* FLAX.

Hemp comes from the stems of the plant *Cannabis sativa* and should not be confused with Manila hemp, Mauretius hemp, sunn hemp, and others that are not related to true hemp. The fiber is produced mainly in eastern Europe and central Asia. It is removed from the stem by retting and scutching. For centuries, hemp was the principal fiber used for marine cordage until replaced by abaca and sisal. It is still used extensively for twine and for many of the same products as flax. *See* HEMP.

Ramie fiber is removed from the stem of the plant (*Bohmeria nivea*) by a beating and scraping operation that leaves it in ribbons. These ribbons are dried and later chemically degummed. Thus, ramie is not retted as other bast fibers are. The world's largest producer is believed to be mainland China. The greatest use for ramie fiber is in fabric and twine. Also made from ramie are thread, gas mantles, belting, canvas, packing for propeller-shaft bearings, and knitting yarn for blends with wool and synthetic fibers. *See* RAMIE.

Jute fiber is obtained from the stems of white jute (*Corchorus capsularis*) and from tossa and daisee (*C. olitorius*). The fiber is removed from the stem by retting and then stripping the retted fiber from the stems by hand. Jute is grown mainly in Bangladesh and India, but Burma, Nepal, and Brazil produce sizable amounts. Its suitability for uses requiring strong, bulky fabric or twine and its low cost compared to competitive fibers have made jute second only to cotton among the natural fibers in world consumption. *See* JUTE.

Hard fibers. Hard fibers, also known as leaf or cordage fibers, come from the leaves of monocotyledonous plants. The fiber of the leaves of many species is used, but the three that are consumed in greatest quantity are abaca, sisal, and henequen.

Frequently known as Manila hemp, abaca comes from the pseudostem (modified leaf petioles that look like a tree trunk) of *Musa textilis*. Most abaca is grown in the Philippines, but other countries have produced it from time to time. Abaca is traditionally used in marine cordage and in other applications that require a strong, water-resistant rope. However, abaca is now being used more for pulp which goes into tea bags, mimeograph mats, air filters, sausage casings, and other products. *See* ABACA.

The sisal plant (*Agave sisalana*) is grown mainly in Africa and Brazil, while henequen (*A. fourcroydes*) is grown mainly in Mexico and Cuba. These are cordage fibers removed from the leaves by defibering. The greatest use for *Agave* fibers is for twine and rope. Upholstery padding, especially that used in innerspring mattresses, is made from the lower grades of fiber. It is also used for pulping, plastic reinforcement, and patio rugs. *See* HENEQUEN; SISAL. [E.G.N.]

Animal fibers. Many animal species produce fibrous structures in a variety of forms and with different functionalities. These include the open webs of spiders, the more closed cocoons of moths, the fibrous part of bird feathers, and body hairs found on species from insects to mammals, including humans. The common characteristic of these fibrous structures is that they are all denatured proteins. Silk and mammalian hairs have achieved commercial and technical importance, although the utility of feathers is due largely to their fibrous qualities.

Mammalian fibers. Mammalian fibers are used by humans in the form of fur, all types of textiles, fiber fills, ropes, brushes, wigs, and so on. All mammalian fibers are closely related. Many species produce a double coat, containing both long, coarse, and somewhat sparse guard hairs, and a shorter, fine, and dense underfur. *See* HAIR.

Mammalian hairs can reversibly sorb more water than any other commercially used fibers, with the exception of some regenerated cellulose. Part of the comfort factor of wool-based textiles is due to this fact. Mammalian fibers are more elastically extensible than any other natural fibers. In water they can be repeatedly stretched to 30%, with full recovery of their length and strength.

The most important commercial animal fiber is wool. Some

of the more important fiber qualities are fineness, length, crimp, color, handle, soundness, and cleanness. The finest and most valuable wool is produced by the Australian Merino sheep, developed from a Spanish breed. Medium- and coarse-variety fibers are produced by other breeds. *See* SHEEP; WOOL.

Mohair is produced by the Angora goat mostly in Turkey, South Africa, and the United States. Luster is its most important quality. The Cashmere goat in Asia grows fur fibers which are the source of cashmere; softness is the most important characteristic. The fine underfur is the valuable component of all goat hairs. *See* CASHMERE; MOHAIR.

The fiber known as camel originates from the Bactrian two-humped species from Asia. For textile purposes only the fine underfur is collected by plucking or shearing. The alpaca and vicuna are South American members of the camel family. The wild vicuna grows a shorter fleece but the finest fibers; the softest wool type materials are made from it. *See* ALPACA; VICUNA.

The Angora rabbit hair is very fine, and relatively long. It is used mostly for blending with wool for soft fabric effects both in woven and knitted textiles. The fur hair of common rabbits is still very fine but shorter. Their most important use is in felted hats. Other animal fibers from rodents, pigs, and horses have only limited uses in felts, brushes, and fiber fills.

Silk. Although it has lost ground to new synthetics, silk is still an important luxury fiber. It is the secretory product of the larval stage of many moth varieties. The cultivated breed is the *Bombyx mori*. The worm, before changing into the chrysalis stage, spins a cocoon around itself from the silk fiber. A whole cocoon is made up from a single fiber. Silk has a few unique properties: it is the finest animal fiber; it has no cellular structure since it was never a part of body tissue; and it is a continuous filament. The most important qualities of silk fabrics are softness, smoothness, luster, and drape. These are based mostly on the small fiber diameter and lack of surface texture. *See* SILK; TEXTILE; TEXTILE CHEMISTRY; TEXTILE PRINTING. [W.S.T.]

Natural gas A combustible gas that occurs in porous rock of the Earth's crust and is found with or near accumulations of crude oil. Being in gaseous form, it may occur alone in separate reservoirs. More commonly it forms a gas cap, or mass of gas, entrapped between liquid petroleum and impervious capping rock layer in a petroleum reservoir. Under greater pressure it is intimately mixed with, or dissolved in, crude oil.

Typical natural gas consists of hydrocarbons having a very low boiling point. Methane makes up approximately 85% of the typical gas. Ethane may be present in amounts up to 10%; and propane up to 3%. Butane, pentane, hexane, heptane, and octane may also be present. While normal hydrocarbons having 5–10 carbon atoms are liquids at ordinary temperatures, they may be present in the vapor form in natural gas. Carbon dioxide, nitrogen, helium, and hydrogen sulfide may also be present.

Natural gas has no distinct odor. Its main use is for fuel, but it is also used to make carbon black, natural gasoline, certain chemicals, and liquefied petroleum gas. Propane and butane are obtained in processing natural gas. *See* LIQUEFIED PETROLEUM GAS (LPG); PETROLEUM PRODUCTS. [M.T.H.]

Natural language processing Computer analysis and generation of natural language text. The goal is to enable natural languages, such as English, French, or Japanese, to serve either as the medium through which users interact with computer systems such as database management systems and expert systems (natural language interaction), or as the object that a system processes into some more useful form such as in automatic text translation or text summarization (natural language text processing). *See* DATABASE MANAGEMENT SYSTEMS.

In the computer analysis of natural language, the initial task is to translate from a natural language utterance, usually in context, into a formal specification that the system can process further. Further processing depends on the particular application. In natural language interaction, it may involve reasoning, factual data retrieval, and generation of an appropriate tabular, graphic, or natural language response. In text processing, analysis may be followed by generation of an appropriate translation or a summary of the original text, or the formal specification may be stored as the basis for more accurate document retrieval later. Given its wide scope, natural language processing requires techniques for dealing with many aspects of language, in particular, syntax, semantics, discourse context, and pragmatics.

The first aspect of natural language processing, and the one that has perhaps received the most attention, is syntactic processing, or parsing. Syntactic processing is important because certain aspects of meaning can be determined only from the underlying structure and not simply from the linear string of words. A second phase of natural language processing, semantic analysis, involves extracting context-independent aspects of a sentence's meaning. Given that most natural languages allow people to take advantage of discourse context, their mutual beliefs about the world, and their shared spatio-temporal context to leave things unsaid or say them with minimal effort, the purpose of a third phase of natural language processing, contextual analysis, is to elaborate the semantic representation of what has been made explicit in the utterance with what is implicit from context. A fourth phase of natural language processing, pragmatics, takes into account the speaker's goal in uttering a particular thought in a particular way—what the utterance is being used to do. *See* ARTIFICIAL INTELLIGENCE. [B.W.]

Nautical Almanac A book published annually by the governments of the principal maritime nations which contains the astronomical data required for navigation by observations of celestial objects. An almanac (from the Arabic word for "calendar") generally contains calendrical tables of astronomical phenomena, or ephemerides, and may also include diverse information of nonastronomical character. The United States *Nautical Almanac* has been published since 1855.

With the *Nautical Almanac*, and a chronometer or wireless time signals to determine Greenwich Mean Time (GMT), the navigator needs only the means of measuring positions of celestial objects with respect to the local vertical as defined by the physical horizon or by an artificial horizon. Such an instrument is the mariner's sextant. The angular distance between the local vertical and the celestial object is equivalent to the angular distance on the surface of the Earth between the observer and that point on the Earth where the celestial object is directly overhead, that is, the ground point of the celestial object. *See* SEXTANT.

The purpose of the tables in the *Nautical Almanac* is to allow the navigator to derive the ground point of the celestial object for any moment of time. By measuring the distance from two or more ground points and solving the navigational triangle for each observation, the navigator establishes the ship's position uniquely. *See* CELESTIAL NAVIGATION.

A similar publication but with slightly reduced tabular accuracy is especially designed to facilitate celestial air navigation. *See* AIR ALMANAC; EPHEMERIS. [R.L.Du.]

Nautiloidea A group of externally shelled cephalopods, represented by the single extant genus *Nautilus*. The formal designation of this group as a subclass is now generally used only for those externally shelled cephalopods that resemble *Nautilus* in having completely coiled shells (and then the subclass includes *Nautilus* itself). In living forms, the basic structural plan includes a shell consisting of a septate phragmocone, a living chamber, and a siphuncle. In fossil nautiloids, this simple pattern is modified in great variety with respect to shell form and size, structure and size of the siphuncle, and the large number of devices to counteract the buoyancy of the phragmocone. The shape of fossil nautiloids may deviate in many ways from the simple *Nautilus* model.

Fossil nautiloids are found on all continents, including Antarctica. Especially noteworthy are the rich Ordovician and Silurian faunas of North America, northern Europe, Czechoslovakia, and central and southern China. Coiled nautiloids almost certainly moved around by jet propulsion like *Nautilus* and lived close to the sea floor, at moderate to intermediate depths in many different environments. Other types may have included agile swimmers as well as slow-moving benthic adaptations. *See* AMMONOIDEA; CEPHALOPODA. [C.T.]

Naval architecture An engineering discipline which in-volves the design of ships, boats, drilling rigs, submarines, and other floating or submerged craft. The naval architect is responsible for the overall concept of a new ship and specifically responsible for the design of the hydrodynamic shape of the hull, the structure, and the general arrangement, including the cargo-handling system. On the other hand, the propulsion plant and all its auxiliaries are the responsibility of the marine engineer.

The education of a naval architect is often combined with that of a marine engineer, so that both students receive some of the same training. In small ship-design offices, the naval architect–marine engineer may handle both hull and machinery design. The education of a naval architect also covers much of the same fundamental engineering and mathematics as that of the mechanical engineer, and the same types of degree programs are offered. *See* MARINE ENGINEERING; SHIP DESIGN. [R.B.C.]

Naval armament A general term which covers the ordnance and control systems used in naval ships and by naval aircraft. It includes a wide spectrum of weapons designed for use against targets in the air, on land or sea, or under the surface of the ocean. The weaponry used by naval forces thus runs from nuclear warheads to small arms and includes weapons that are intended for use against a particular type of target as well as the more general-purpose weapons.

Fig. 1. ASROC missile being launched from destroyer. (*Official U.S. Navy photograph*)

Fig. 2. Polaris missile being fired from a submerged submarine. (*Official U.S. Navy photograph*)

Naval armament may be air-launched, surface-launched, or submarine-launched. It includes guns, guided missiles, rockets, bombs, depth charges, torpedoes, and mines.

Guided missiles. Naval missiles may be launched from ships, submarines, or aircraft. Research has been aimed at producing missiles which can be used, say, by planes as well as by surface ships, or at modifying missiles designed for one type of work for another use. Advanced radars and computerized control systems can increase missile capabilities.

ASROC (Fig. 1) and SUBROC (submarine rocket) are antisubmarine weapons, launched respectively by surface ships and submarines. The weapon is aimed by a shipboard computer using target information obtained by sonar. The unguided rocket is fired from a launcher (ASROC) or a torpedo tube (SUBROC) and follows a ballistic trajectory through the air to the target's predicted position.

Polaris (Fig. 2), Poseidon, and Trident are solid-fueled ballistic missiles launched from submerged submarines. They are ejected from their vertical launch tubes by compressed air or by a gas-steam system. The missile is guided by a self-contained inertial-navigation system to the correct point in space for an unguided (ballistic) trajectory to its target. Poseidon, with greater range and a multiple warhead, is replacing Polaris in most of the Navy's nuclear-powered ballistic-missile submarines

(SSBN). Trident also has a multiple warhead, with a considerably greater range than Poseidon. *See* Guided missile; Missile.

Guns. Though missiles are widely used by ships and aircraft, guns continue to be an integral part of the naval arsenal. While missiles are superior for long-range attack and for defense against supersonic planes and missiles at high altitudes and long ranges, the opposite is often true for such missions as shore bombardment, gunfire support of land forces, and defense against small craft. Renewed attention has been given to lighter guns, with high rates of fire, and to quick-reaction control systems for close-in defense against aircraft and cruise missiles.

Bombs. Bombs of many types and sizes, from small antipersonnel weapons to 2000-pound bombs designed for blast effect, as well as nuclear weapons, can be carried by naval aircraft. "Smart bombs" can be guided to their target from a standoff distance. Homing bomb system (HOBOS) types are television-guided adaptations of standard "dumb" bombs with added guidance systems and controllable tail surfaces. Laser-guided bombs (LGB) use laser homing. The antipersonnel-antimateriel (APAM) bomb contains a cluster of small bomblets. If these hit a hard surface, they have a shaped-charge effect. If they strike a soft surface, such as earth or sand, they rebound for a fragmentation airburst.

Torpedoes. Torpedoes travel underwater under their own power to intercept and destroy surface ships and submarines. Modern naval torpedoes are fast, far-ranging, and armed with a heavy explosive warhead. Torpedoes may be homing (guided acoustically to their target) or nonhoming (following a preset course). Torpedoes may be launched by aircraft, submarines, or surface warships. *See* Acoustic torpedo.

Mines. A mine is a thin-cased, non–self-propelled weapon filled with high explosive and planted underwater, where it is designed to explode when struck or closely approached by a ship. Mines can be contact type (fired by actually striking the hull of a passing ship) or influence type (detonated by the close approach of a ship). Influence mines thus are much harder to sweep than contact mines. Mines can be planted by surface ships, submarines, or aircraft. Specially designed minehunters and minesweepers use a variety of detection and clearance devices to locate and dispose of mines. *See* Acoustic mine; Naval ship. [J.C.R.]

Naval ship

A ship designed primarily for use in warfare, either directly in combat operations or to provide services and support such operations. United States naval ships can be categorized as either combatants or noncombatants. They also can be classified by their performance capabilities and by their intended mission. Specific military features in warfare areas such as antiair warfare (AAW), antisubmarine warfare (ASW), and surface warfare (SUW) are performance attributes; whereas mission groupings include battleships, cruisers, destroyers, frigates, aircraft carriers, mine craft, patrol craft, amphibious ships, auxiliary ships, service craft, and high-speed or high-performance ships.

Combatant ships. This category includes battleships, cruisers, destroyers, frigates (formerly destroyer escorts), patrol craft, and aircraft carriers. Cruisers and aircraft carriers now are the capital ships of the United States fleet.

The primary purpose of a cruiser is to provide antiaircraft and antisubmarine protection to an aircraft carrier task force or group. These ships are multimission warships suited for independent operations. Nuclear cruisers (Fig. 1) generally are deployed with nuclear-powered aircraft carriers. This mix provides the high speed and unlimited endurance necessary to exploit the full capabilities of the task group.

The DD 963 Spruance Class destroyer is primarily an antisubmarine warfare ship, designed to detect and destroy submarines. Internal changes to the hull and equipment have made this destroyer the quietest ship in the world. The USS Charles F. Adams Class (DDG 2) of guided-missile destroyers went through a modernization which brought it up to the fighting capability required to counter the latest known offensive threats.

Fig. 1. Nuclear-powered guided-missile cruiser USS *Virginia*, CGN 38, serving the fleet since 1976. (*Official U.S. Navy photograph*)

Another group of combatants is called frigates. The frigate is generally a single-purpose ship which provides the fleet with adequate escort vessels.

The largest warships are aircraft carriers (Fig. 2) which, through the aircraft they support, can project power many miles from their position. An aircraft carrier is a floating mobile air station. Its flight deck provides the runways, its island is the control tower, the large hangar area below the flight deck contains maintenance and repair shops, and deep in the hull are tanks for aircraft fuel storage and magazines for stowing aircraft weapons and ammunition.

Patrol ships constitute a naval combatant ship subcategory which augments surface forces, particularly in coastal areas and narrow seas where surveillance is impractical for conventional ships. Technical interest is highest in high-performance patrol craft which strive for maximum capability in the smallest package.

The ships used for the transport, landing, support, and control of assault troops collectively constitute the amphibious warfare force. With the exception of the ships, boats, and craft that actually ground on the beach, the amphibious ships used in World War II and Korea were conversions of commercial ships. With the advent of the helicopter and the successful adaptation of it for landing assault troops, the nature of

Fig. 2. Conventionally propelled aircraft carrier, USS *John F. Kennedy* (CV67). (*Official U.S. Navy photograph*)

amphibious warfare changed and with the change came new ships. *See* LANDING SHIPS AND CRAFT.

Mine warfare ships constitute a subcategory of combatant naval ships for laying or for locating, removing, and neutralizing naval mines. Emphasis in postwar years has not been on mine laying but on the latter functions, which are loosely called minesweeping. Mine countermeasures support ships carry, in addition to the mine warfare command functions, a number of minesweeping launches and minesweeping helicopters.

Auxiliary ships. Auxiliary ships are those ships which provide services and support naval operations. Other floating configurations which provide general services are not auxiliary ships but are service craft. Included in the auxiliaries are the tenders for destroyers and submarines and the ships that replenish the fleet with supplies of ammunition, oil, naval stores, and combat support items. Also included are ships for cargo, troops, and various fields of research involving submarines, hydrofoils, oceanography, and surveying. [W.H.Hu.; D.J.W.]

Navier-Stokes equations Three scalar partial differential equations that describe conservation of momentum for the motion of a viscous, incompressible fluid. They may be expressed vectorially as one equation,

$$\rho \frac{\partial \mathbf{v}}{\partial t} + \rho(\mathbf{v} \cdot \nabla)\mathbf{v} = - \nabla p + \rho \mathbf{f} + \mu \nabla^2 \mathbf{v}$$

where ρ is fluid density, $\mathbf{v}$ is fluid velocity vector, p is fluid pressure, $\mathbf{f}$ is body force (such as gravity) per unit mass, μ is fluid viscosity coefficient, and t is time. These equations, together with the continuity relation, $\nabla \cdot \mathbf{v} = 0$, and suitable boundary conditions determine the flow field; for example, $\mathbf{v}$ and p are determined as functions of position in space and of time. One of these boundary conditions is that of no slip at the surface of a body; that is, the fluid immediately at the body surface "sticks" to it and thus has the same velocity as the surface itself. *See* NEWTONIAN FLUID.

Few mathematical solutions to this complicated set of nonlinear partial differential equations are known, except for simple geometrics. The importance of viscosity in determining the flow depends on the relative size of the body. Approximations to the Navier-Stokes equations for small Reynolds number give good results. For $Re \gg 1$, the effects of viscosity are confined to a thin layer near the surface of bodies in the fluid. *See* BOUNDARY-LAYER FLOW; D'ALEMBERT'S PARADOX; FLUID FLOW; REYNOLDS NUMBER. [A.E.Br.]

Navigation The process of directing the movement of a craft from one place to another. Usually modern navigation is considered to encompass four procedures—piloting, position fixing, homing, and dead reckoning. Some consider celestial navigation to be a separate activity from other position-fixing methods. If this classification is accepted, the number of procedures increases to five. *See* AIR NAVIGATION; CELESTIAL NAVIGATION; DEAD RECKONING; MARINE NAVIGATION; NAVIGATION INSTRUMENTS; PILOTING; TERRESTRIAL COORDINATE SYSTEMS. [A.B.M.]

Navigation instruments Measuring devices used in ships and aircraft to determine geographic position. The accuracy of such devices has improved steadily since the beginnings of travel, with occasional spectacular new inventions suddenly enlarging the navigator's domain. Notable milestones are the magnetic compass, the sextant, the gyrocompass, radio systems, inertial guidance, and satellite navigation. *See* ELECTRONIC NAVIGATION SYSTEMS; GYROCOMPASS; INERTIAL GUIDANCE SYSTEM; MAGNETIC COMPASS; SATELLITE NAVIGATION SYSTEMS; SEXTANT. [R.H.C.]

Neandertals A group of late archaic humans from Europe, the Near East, and central Asia that immediately preceded the first modern humans in those regions. The Neandertals are included by some within the species *Homo sapiens*, recognizing their close affinities to modern humans; others place them in their own species, *Homo neanderthalensis*, emphasizing the differences between them and modern humans.

The first recognized Neandertal remains were found in the Neander Valley near Dusseldorf, Germany, in 1856. Since then the remains of several hundred Neandertals have been discovered. Since the Neandertals were the first humans to bury their dead, a number of largely complete skeletons are preserved, providing detailed knowledge of their biology.

In the early twentieth century, when Neandertals were the only archaic humans known, they were reconstructed as semihuman, dull-witted, and brutish. Hence their popular image was that of the archetypical cavemen. They are now recognized as relatively recent members of the human lineage; they lived between about 125,000 and 32,000 years ago, as compared with earlier members of the genus *Homo* who extend back more than 2 million years. The Neandertals share many features with modern humans both anatomically and behaviorally. Yet, a number of important contrasts between them and more recent humans are recognized.

Physically, the Neandertals were about the same height as most modern humans, on the average 5 ft 5 in. (166 cm), but they were much more heavily built. They had heavy necks, broad and muscular shoulders, and extremely muscular arms, hands, and legs. Estimates of their strength show them to have been about twice as strong as athletic modern humans. Their leg bones show a marked thickening of their shafts, which is indicative of both marked strength and endurance—a necessary part of their survival.

The position of the Neandertals in modern human ancestry remains controversial. Whatever the extent to which Neandertals can be claimed to be ancestors of modern humans, they represent the most recent phase of premodern humans, one in which people were less efficient than modern humans at hunting and gathering, and compensated for their cultural limitations with biological attributes such as tremendous strength, large front teeth, and thermal adaptations. Yet they exhibited the beginnings of many of the attributes of modern humans. They were very successful for about 100,000 years, but they were eventually replaced by humans who were better able to exploit their environments. *See* FOSSIL HUMAN. [E.T.]

Nearshore sedimentary processes The processes that shape the shore features of coastlines and begin the mixing, sorting, and transportation of sediments and runoff from land, particularly those interactions among waves, winds, tides, currents, and land that relate to the waters, sediments, and organisms of the continental shelf and nearshore areas. *See* MARINE SEDIMENTS.

The energy for nearshore processes comes from the sea and is produced by the force of winds blowing over the ocean, by the gravitational attraction of Moon and Sun acting on the mass of the ocean, and by various impulsive disturbances at the atmospheric and terrestrial boundaries of the ocean. These forces produce waves and currents that transport energy toward the coast. The configuration of the land mass and adjacent shelves modifies and focuses the flow of energy and determines the intensity of wave and current action in coastal waters. Rivers and winds transport erosion products from the land to the coast, where they are sorted and dispersed by waves and currents. *See* OCEAN CIRCULATION; OCEAN WAVES.

The dispersive mechanisms operative in the nearshore waters of oceans, bays, and lakes are all quite similar, differing only in intensity and scale, variables that are determined primarily by the nature of the wave action and the dimensions of the surf zone. The most important mechanisms are the orbital motion of the waves, which is the basic mechanism by which wave energy is expended on the shallow sea bottom, and the currents of the nearshore circulation system that produce a continuous interchange of water between the surf zone and offshore. The dispersion of water and sediments near the coast and the formation and erosion of sandy beaches are some of the more common manifestations of nearshore processes.

Erosional and depositional nearshore processes play an important role in determining the configuration of coastlines. Erosion is usually dominant off headlands and along coastal sections backed by alluvium and other unconsolidated material, whereas deposition is most common along indentations between headlands. The overall effect of such processes is usually a straightening and smoothing of the coastline. However, this is not always the case; differential wave erosion may cause a rapid erosion of material between headlands and thus cause irregularities in the coastline. *See* COASTAL LANDFORMS.

Whether deposition or erosion will be predominant in any particular place depends upon a number of interrelated factors: the amount of available beach sand and the location of its source; the configuration of the coastline and of the adjoining ocean floor; and the effects of wave, current, wind, and tidal action. The establishment and persistence of natural sand beaches are often the result of a delicate balance among a number of these factors, and any changes, natural or human-made, tend to upset this equilibrium. [D.L.I.]

Nebula Originally, any fixed, extended, and usually fuzzy luminous object seen in a telescope. Nebulae are now distinguished from star clouds that can be resolved into individual stars, but earlier workers were unable to differentiate between white nebulae, which are stellar systems so remote as to show no individual stars, and gaseous or diffuse nebulae in the Milky Way Galaxy. *See* STAR CLOUDS.

Extragalactic nebulae are stellar systems comparable with the Milky Way Galaxy or the Magellanic Clouds in size and number of stars, and are more properly termed external galaxies. *See* GALAXY, EXTERNAL.

This article deals with gaseous nebulae. This class of objects includes diffuse nebulae which contain dust and gas of the interstellar medium, excited and caused to fluoresce by embedded stars. Gaseous nebulae are members of the Milky Way galactic system, and small compared with its overall dimensions.

Diffuse nebulae range from huge masses of relatively high surface brightness, such as the Orion Nebula, down to faint, milky structures a hundred times less dense that are detectable only with long exposures and special filters. Diffuse nebulae may contain both dust and gas or may be purely gaseous, such as the California Nebula. The nebulosity found in the Pleiades and elsewhere consists of dust with no luminous gas, although it is probable that there is also a great quantity of neutral hydrogen. *See* ORION NEBULA.

Variable-brightness nebulae are associated with abnormal variable stars and are frequently fan-shaped in appearance. Nebulae of this type are associated with T Tauri variables, which are believed to be stars in the process of formation.

Planetary nebulae are so denoted because they often show small greenish disks in the telescope, not unlike the images of the planets Uranus and Neptune. The energy emitted by planetary nebulae is derived from the rich ultraviolet radiation of stars embedded within them.

The detonation of a star in a supernova event causes the ejection of the outer layers into the surrounding interstellar medium. In early stages as in the Crab Nebula, the radiating material consists of ejecta from the star. In the later stages this rapidly moving material is slowed down as it mixes with the surrounding dust and gas of the interstellar medium. Supernovae remnants characteristically emit nonthermal radio-frequency emission, whereby they are often detected in nearby galaxies as well as in the Milky Way system. *See* CRAB NEBULA; SUPERNOVA. [L.H.Al.]

Nectarine A smooth-skinned, fuzzless form of peach, *Prunus persica*. The nectarine's lack of pubescence is a simple recessive genetic characteristic. Classically, the fruits were thought of as being somewhat smaller, softer, and richer in flavor than those of the peach. More recently developed cultivars, however, approximate fresh-market peaches in size and firmness but are not usually superior in flavor.

California is practically the sole commercial producer of nectarines. There is a considerable number of plantings in irrigated areas in south-central Washington. *See* FRUIT; FRUIT, TREE; PEACH; ROSALES. [L.F.H.; C.H.B.]

Nectridea An order of extinct (Carboniferous and early Permian) lepospondylous amphibians characterized by vertebrae in which large fan-shaped hemal arches grow directly downward from the middle of each caudal centrum; the neural arches of the tail, and sometimes of the trunk as well, have a similar shape. Most nectridians can be clearly separated into two types. In one, represented by *Urocordylus*, the skull is long and slender, the body and tail elongated, and the limbs reduced. In the other type, the head and trunk are broad and

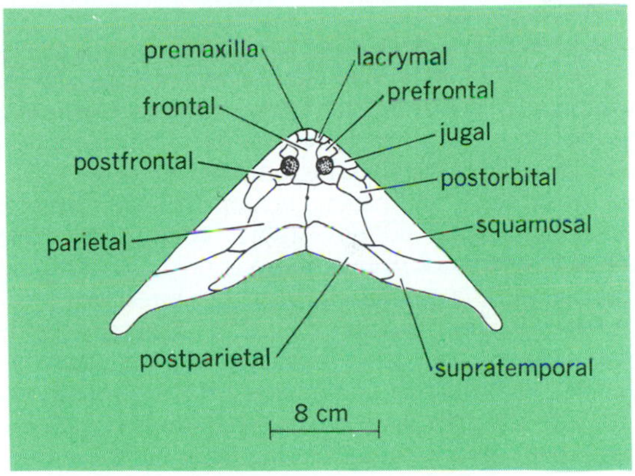

Skull of a lower Permian nectridian, *Diplocaulus*. (*After E. H. Colbert, Evolution of the Vertebrates, 2d ed., 1969*)

flattened and the black corners of the skull table taper into long slender "horns" (see illustration). *See* AMPHIBIA; LEPOSPONDYLI. [A.S.R]

Negative ion An atomic or molecular system with an excess of negative charge. Negative ions, also called anions, are formed in attachment processes in which an additional electron is captured by an atom or molecule. Negative ions were first reported in the early days of mass spectrometry. It was soon learned that even a small concentration of such weakly bound, negatively charged systems had an appreciable effect on the electrical conductivity of gaseous discharges. Negative ions now play a major role in a number of areas of physics and chemistry involving weakly ionized gases and plasmas. Applications include accelerator technology, injection heating of thermonuclear plasmas, material processing, and the development of tailor-made gaseous dielectrics. In nature, negative ions are known to be present in tenuous plasmas such as those found in astrophysical and aeronomical environments. The absorption of radiation by negative hydrogen ions in the solar photosphere, for example, determines the Sun's spectral distribution. *See* ION; ION SOURCES; PLASMA PHYSICS. [D.J.Pe.]

Negative-resistance circuits Circuits or devices whose static (dc) current-voltage characteristic at one or more ports has a range in which the slope is negative. Figure 1 shows two typical negative-resistance characteristics in which a positive increment of voltage is accompanied by a negative increment of current. Strictly, the term also includes circuits whose static current-voltage characteristics do not have negative-slope ranges, but in which the input impedance or admittance at one or more ports has a negative real component. The negative resistance may arise from inherent properties of a device or from the proper choice of electrical circuit configuration and operating conditions.

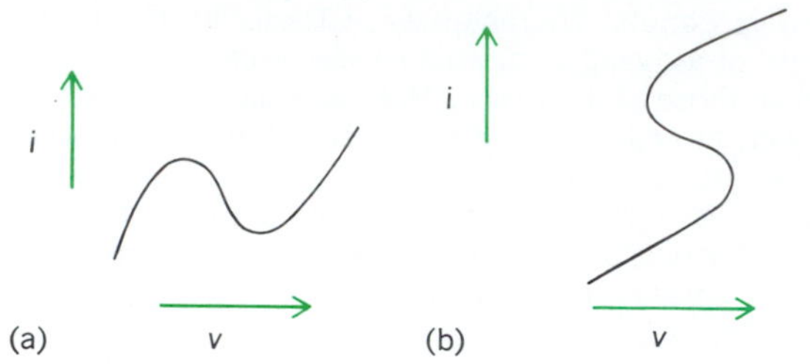

Fig. 1. Examples of (*a*) voltage-stable and (*b*) current-stable current-voltage characteristic curves.

Negative-resistance circuits and the circuits based upon them find many applications in counting or information storage, in the generation of pulses or alternating voltages, in instrumentation, in the control of current and power, and in physical realization of idealized (lossless) or hypothetical circuits.

Basic circuits. A port having a characteristic of the form of Fig. 1*a* is said to be voltage-stable (voltage-controllable) because at any value of voltage there is only one value of current, but over a range of current there are three values of voltage, between which transition may take place spontaneously or under excitation. Similarly, a port having a characteristic of the form of Fig. 1*b* is said to be current-stable (current-controllable) because for each value of current there is only one value of voltage, but over a range of voltage there are three values of current, between which transition may take place. In general, negative-resistance circuits may have both voltage-stable and current-stable ports.

Figure 2 shows one type of negative-resistance circuit. The indicated direction of input current Δi resulting from a positive increment of input voltage Δv is that which would obtain if the input resistance of the port were positive (dissipative). If the voltage amplification A of the amplifier is positive and greater than unity, the output voltage $A\Delta V$ is of the same sign as the input voltage and greater in magnitude. The current Δi_r through the resistance R is then in the direction shown. If this current exceeds the current Δi_i into the amplifier input, Δi must be opposite in direction to that shown. In other words, the positive increment of impressed voltage produces a negative increment of current, and the slope of the characteristic relating current i with voltage v is negative. Because the range of input voltage over which the amplifier has an amplification greater than unity is limited, the negative-slope portion of the characteristic is also limited with respect to voltage, and the characteristic is the voltage-stable form shown in Fig. 1*a*. A similar analysis shows that a voltage-stable port can be formed across the output terminals of the amplifier and that current-stable ports can be formed in place of resistor R or by opening the connection between the lower input and output terminals. Voltage-stable and current-stable ports can in general also be formed within the amplifier itself. *See* AMPLIFIER.

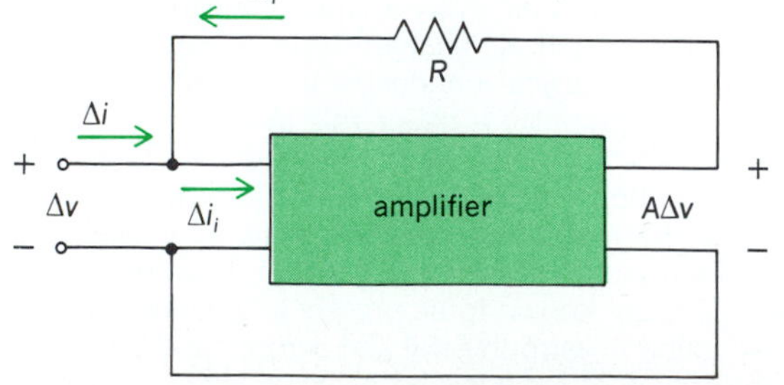

Fig. 2. Example of a circuit that has a voltage-stable current-voltage characteristic.

The two essentials of a negative-resistance circuit are amplification and positive feedback; any circuit modification that tends to increase the open-loop amplification of the amplifier and feedback network is favorable to the production of negative resistance. In order that negative resistance can be obtained under static conditions, it is necessary that the amplifier be capable of amplifying increments of direct voltage. *See* FEEDBACK CIRCUIT.

Bistable, astable and monostable circuits, and sine-wave oscillators can be formed from negative resistance circuits. *See* MULTIVIBRATOR; OSCILLATOR.

Negative-resistance devices. Two examples of current-stable negative-resistance devices are the *pnpn* transistor and the unijunction transistor; two examples of voltage-stable devices are the tunnel (Esaki) diode and the Gunn diode. An advantage of negative-resistance devices over circuits that incorporate two transistors to achieve negative resistance is the simplicity of bistable, monostable, and oscillator circuits in which they are used. *See* MICROWAVE SOLID-STATE DEVICES; TRANSISTOR; TUNNEL DIODE. [H.J.R.]

Negative temperature The property of a thermodynamical system which satisfies certain conditions and whose thermodynamically defined absolute temperature is negative. The essential requirements for a thermodynamical system to be capable of negative temperature are: (1) the elements of the thermodynamical system must be in thermodynamical equilibrium among themselves in order for the system to be described by a temperature at all; (2) there must be an upper limit to the possible energy of the allowed states of the system; and (3) the system must be thermally isolated from all systems which do not satisfy both of the first two requirements, that is, the internal thermal equilibrium time among the elements of the system must be short compared to the time during which appreciable energy is lost to or gained from other systems.

The second condition must be satisfied if negative temperatures are to be achieved with a finite energy. Most systems do not satisfy this condition; for example, there is no upper limit to the possible kinetic energy of a gas molecule. Systems of interacting nuclear spins, however, have the characteristic that under suitable circumstances they can satisfy all three of the above conditions, in which case the nuclear spin system can be at negative absolute temperature. *See* KINETIC THEORY OF MATTER; STATISTICAL MECHANICS.

The transition between positive and negative temperatures is through infinite temperature, not absolute zero; negative absolute temperatures should therefore not be thought of as colder than absolute zero, but as hotter than infinite temperature. *See* ABSOLUTE ZERO; TEMPERATURE. [N.F.R.]

Neisseriaceae A family of aerobic, gram-negative cocci and coccobacilli. The most predominant cellular arrangement is in pairs; when the cells are spheroid, the adjacent sides are flattened. Many of the known species are parasitic, and several cause disease in humans. Neisseriaceae contains four genera: *Neisseria*, *Branhamella*, *Moraxella*, and *Acinetobacter*. [M.M.]

Nemata A phylum of unsegmented worms. A classification of nematodes follows:

Phylum Nemata
  Class Adenophorea (Aphasmidia, Aphasmidea)
    Subclass Enoplia
      Order: Enoplida
        Isolaimida
        Mononchida
        Dorylaimida
        Trichocephalida
        Mermithida
        Muspiceida
    Subclass Chromadoria
      Order: Araeolaimida
        Chromadorida
        Desmoscolecida
        Monhysterida
  Class Secernentea (Phasmidia, Phasmidea)
    Subclass Rhabditia
      Order: Rhabditida
        Strongylida
        Ascaridida
    Subclass Spiruria
      Order: Spirurida
        Camallanida
    Subclass Diplogasteria
      Order: Diplogasterida
        Aphelenchida
        Tylenchida

Diagnosis. The Nemata are unsegmented or pseudosegmented (any superficial annulation limited to the cuticle) bilaterally symmetrical worms with a basically circular cross section. The body is covered by a noncellular cuticle. The cylindrical body is usually bluntly rounded anteriorly and tapering posteriorly. The body cannot be easily divided into head, neck, and trunk or tail, although a region posterior to the anus is generally referred to as the tail. The oral opening is terminal (rarely subterminal) and followed by the stoma, esophagus, intestine, and rectum which opens through a subterminal anus. Females have separate genital and digestive tract openings. In males the tubular reproductive system joins posteriorly with the digestive tract to form a cloaca. The sexes are separate and the gonads may be paired or unpaired. Females may be oviparous or ovoviviporous.

Adult nematodes are extremely variable in size, ranging from less than 0.012 in. (0.3 mm) to over 26 ft (8 m). Nematodes are generally colorless except for food in the intestinal tract or for those few species which have eyespots.

Life cycle. Reproduction among nematodes is either amphimictic or parthenogenetic (rarely hermaphroditic). After the completion of oogenesis the chitinous egg shell is formed and a waxy vitelline membrane forms within the egg shell; in some nematodes the uterine cells deposit an additional outermost albuminoid coating. Upon deposition or within the female body, the egg proceeds through embryonation to the eellike first- or second-stage larva, but following eclosion the larva proceeds through four molts to adulthood. This represents a direct life cycle, but among parasites more diversity occurs.

Distribution. Nemata comprise the third largest phylum of invertebrates, being exceeded only by Mollusca and Arthropoda. In sheer numbers of individuals they exceed all other metazoa. As parasites of animals they exceed all other helminths combined. Nematodes have been recovered from the deepest ocean floors to the highest mountains, from the Arctic to the Antarctic, and in soils as deep as roots can penetrate. [A.R.M.]

Nematicide A type of chemical used to kill plant-parasitic nematodes. Nematicides may be classed as soil fumigants or soil amendments, space fumigants, surface sprays, or dips. Soil treatments are commonly used because most plant-pathogenic species spend part or all of their life cycle in the soil, in or about the roots of plants. Nematicides may be liquids, gases, or solids, but on a field scale, liquids are most practical. *See* NEMATA; PESTICIDE. [D.J.R.]

Nematomorpha A phylum of worms that was formerly considered to be a class of the phylum Aschelminthes; commonly called the hairworms, and closely allied to the nematodes. The adults are free-living in aquatic habitats, while the juveniles are parasitic in arthropods. The nematomorphs are found all over the world. They are divided into two classes, the Nectonematoidea and Gordioidea, with a total of 225 species. *See* NEMATA.

The body is long and slender with a maximum length of 5 ft (1.5 m) and a diameter of 0.02–0.12 in. (0.5–3 mm). The females are longer than the males. The posterior end may be rounded with a terminal cloaca, or it may form two or three lobes in a forklike structure. The body color is yellowish, brown, or almost black. The body wall consists of three layers: an outer, rather thick fibrous cuticle; an epidermis consisting of a single layer of cells; and innermost, a muscle layer with longitudinal fibers only.

The body cavity extends the length of the body. It may be filled with tissue so that only minor spaces are left around the digestive system and the gonads.

The sexes are always separate, and the gonads are paired and stringlike extending the length of the body. During copulation the male coils itself around the female and places a drop of sperm near the cloacal opening of the female. The sperm cells actively enter the seminal receptacle. The eggs are laid in water in strings, and the adults die after egg laying. When hatched, the larvae swim to an aquatic arthropod. They penetrate the body wall of the host by means of their characteristic proboscis, which is armed with hooks and three long stylets. The gradual development in the host lasts some months without any metamorphosis. When they are mature, the worms leave the host. [B.J.Mu.]

Nematophytales An enigmatic group of fossil plants, in mid-Silurian to lower Upper Devonian rocks, composed of intertwined, branching tubes of two sizes: 10–50 micrometers in diameter, and 1–10 micrometers. Although they are referred to the algae by some authors, the occurrence of this group in inland swamps, coastal plain deposits, and marine deposits close to shore indicates that they were terrestrial organisms, unrelated to any known groups, perhaps at an intermediate level between algae and bryophytes. *See* PALEOBOTANY. [H.P.B.]

Neodymium A metallic chemical element, Nd, atomic number 60, atomic weight 144.24. Neodymium belongs to the rare-earth group of elements. The naturally occurring element includes the six isotopes. The oxide, Nd_2O_3, is a light-blue powder. It dissolves in mineral acids to give reddish-violet solutions. For properties of the metal *see* RARE-EARTH ELEMENTS.

The salts have found application in the ceramic industry for coloring glass and for glazes. The glass is particularly useful in goggles used by glass blowers, since it absorbs the intense yel-

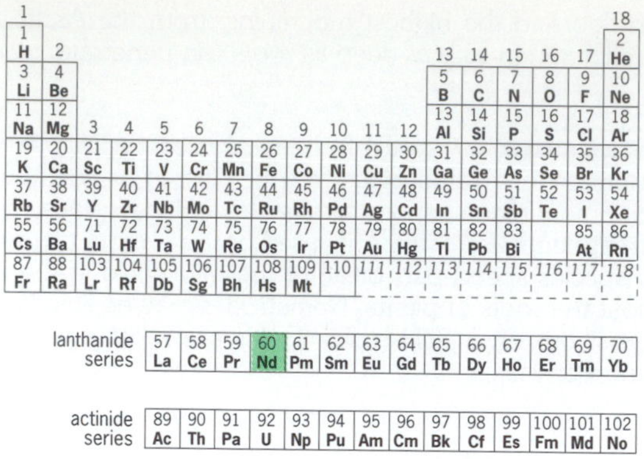

low D line of sodium present in the flame. The element has found commercial application in the manufacture of lasers. [F.H.Sp.]

Neogastropoda An order of gastropods, also known as the Stenoglossa, which contains the most highly developed snails (see illustration). Respiration is by means of ctenidia. The nervous system is concentrated, an operculum is usually present, and the sexes are separate. All families in this order are marine. The family Muricidae contains the rock snails, which are all predatory; Buccinidae, most abundant in northern seas, contains the whelks in the genus *Buccinum*, which are frequently used for fish bait; other families, such as the Volutidae, Olividae, and Harpidae, are prized by shell collectors for their beautiful coloration. The family Conidae contains the poisonous cone shells, which have caused several deaths by their poisonous bite. *See* GASTROPODA. [W.J.C.]

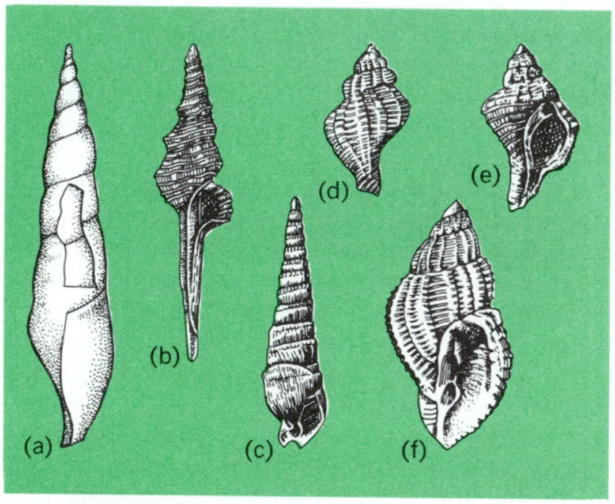

Neogastropoda. (*a*) Ordovician species of *Subulites*, widespread Ordovician and Silurian genus. (*b*) *Falsifusus*, early Tertiary genus. (c) Miocene species of familiar existing *Terebra* (Tertiary-Recent). (*d, e*) *Urosalpinx* (Tertiary-Recent). (*f*) *Cancellaria* (Tertiary-Recent). (*After R. R. Shrock and W. H. Twenhofel, Principles of Invertebrate Paleontology, 2d ed., McGraw-Hill, 1953*)

Neognathae One of the two recognized superorders making up the subclass Neornithes of the class Aves. They are characterized as flying birds with fully developed wings and sternum with a keel, caudal vertebrae fused into a pygostyle, and absence of teeth in both jaws, or modifications of these conditions in secondary flightless birds.

This superorder includes all living birds and all known fossil birds since the Late Cretaceous; only the ancestral Jurassic *Archaeopteryx* and the specialized Cretaceous *Hesperonis* and its allies do not belong to the Neognathae. *See* ARCHAEORNITHES; AVES; ODONTOGNATHAE; RATITES. [W.J.B.]

Neognathostomata A superorder of Echinoidea, subclass Euechinoidea. These invertebrates are characterized by having a rigid, exocyclic test and a lantern or jaw apparatus developed sometime during the life history and usually persisting into the adult stage. The included orders are the Holectypoida, Clypeasteroida, and Cassiduloida. *See* CASSIDULOIDA; CLYPEASTEROIDA; ECHINODERMATA; ECHINOIDEA; EUECHINOIDEA; HOLECTYPOIDA. [H.B.F.]

Neogregarinida An order of the protozoan subclass Gregarinia, class Telosporea, subphylum Sporozoa. All gregarines are parasites of the digestive tract and body cavity of invertebrates or lower chordates; their large, mature trophozoites (vegetative stages) live outside the host's cells. The Neogregarinida are thought to be relatively advanced gregarines which live in insects. There are only about 29 species of about 12 genera, and 4 families. *See* GREGARINIA. [N.D.L.]

Neolampadoida A group of small, deep-water cassiduloid echinoids with neotenous characteristics, treated as an order by some workers; possibly polyphyletic. The presence of bourrelets and phyllodes, the elongate first ambulacral plates, the undifferentiated tuberculation, and undifferentiated posterior interambulacral plating all indicate their relationships lie with cassiduloids. The only character shared by members of this group is the lack of petals (they have simple ambulacral pores only). Other characteristics, such as apical disc plating, are varied, indicating at least two independent origins from shallow-water cassiduloids.

There are seven genera, each monospecific. Five are living today and are usually found at depths of 430–1280 ft (135–400 m). A Miocene species and an Upper Eocene species are also known. *See* ECHINODERMATA. [A.B.S.]

Neolithic The period of prehistoric culture in which people made some of their stone tools by grinding the edges to create smooth surfaces. The term emerged in the mid-nineteenth century. Stone axes were early examples of Neolithic tools. Chipped stone tools, typical of the earlier Paleolithic period, were also made by Neolithic peoples. Later in the nineteenth century other terms were added: Mesolithic (between Paleolithic and Neolithic), and a Copper Age between Neolithic and the Bronze Age. Thus three successive Stone Ages (Paleolithic, Mesolithic, and Neolithic) were assumed, followed by three Metal Ages (Copper, Bronze, and Iron). Civilization, marked by the emergence of political states, civic buildings, and usually writing, followed the Neolithic. Later archeological study, still continuing, has shown that cultural evolution has not always followed exactly this proposed pattern. *See* PALEOLITHIC.

The Neolithic began only some time after the end of the last Ice Age, some 11,000 years before present, but did not begin everywhere at the same time; indeed, in some areas the spread of civilization engulfed peoples who had never experienced the Neolithic. By tradition, prehistorians have usually limited their use of the term Neolithic to Europe and the Near East, and some have abandoned the use entirely. However, the term has frequently been expanded to indicate a cultural pattern that had some agriculture and associated villages as the most complex social and physical entities known to the inhabitants. The Neolithic originated independently with the beginnings of agriculture in at least five areas: the Near East, the Hwang-Ho valley of northern China, southeastern Asia, Meso-America, and Peru. All later agriculture was seemingly derived by cultural diffusion from one of these areas, as that of Europe and Africa developed by diffusion from the Near East.

Prehistorians and ethnologists have not usually used the term Neolithic for cultures in the Western Hemisphere.

However, at the time of European discovery the Americas were occupied mainly by the civilized cultures of Meso-America and western South America or by village farmers. Only the natives of Canada, of much of the western United States, and of the southern quarter of South America were hunters and gatherers; all other native Americans lived mainly by agriculture, and the noncivilized agriculturalists should certainly be termed Neolithic. [C.A.Re.]

Neomeniomorpha A subclass of creeping, vermiform mollusks in the class Aplacophora. They are covered by a spicular integument and recognized by the presence of a ventral groove within which lies a narrow foot and by the absence of on oral shield. Neomenioids range in size from less than 0.08 to 12 in. (2 mm to 300 mm) and are found from subtidal areas to the abyss, at depths over 16,000 ft (5000 m). There are 23 families with 70 genera and 193 species worldwide.

Neomenioids creep by means of their ciliated foot along a track of sticky mucus produced from a ciliated, reversible pedal pit at the anterior end of the pedal groove. Anterior to the pedal pit, the head end is held above the substratum and freely moved.

All neomenioids are hermaphroditic. A barrel-shaped, nonfeeding larva called a pericalymma either is brooded or swims by means of a ciliated cellular test within which the animal develops; metamorphosis through loss or resorption of the test occurs within 10 days. *See* Aplacophora; Mollusca. [A.H.S.]

Neomycin A colorless antibiotic produced by certain strains of *Streptomyces fradiae* in submerged culture. Chemically, the structure of neomycin, $C_{23}N_{46}N_6O_{13}$, is analogous to that of streptomycin. Neomycin is active against a great variety of gram-positive cocci and rods, gram-negative rods, and acid-fast bacteria, notably the tuberculosis organism. It is not active against anaerobic bacteria, fungi, most protozoa, rickettsiae, and viruses.

Clinically, neomycin is used topically in the treatment of bacterial infections of the skin, and orally in the treatment of bacterial infections of the digestive tract. It can also be used systemically in the treatment of deep-seated bacterial infections that are resistant to other, less toxic antibiotics. *See* Antibiotic; Streptomycin. [S.A.W.]

Neon A gaseous chemical element, Ne, with atomic number 10 and atomic weight 20.183. Neon is a member of the family of noble gases. The only commercial source of neon is the Earth's atmosphere, although traces of neon are found in natural gas, minerals, and meteorites. *See* Inert gases.

1	2	3	4	5	6	7	8	9	10	11	12	13	14	15	16	17	18
1 H																	2 He
3 Li	4 Be											5 B	6 C	7 N	8 O	9 F	10 Ne
11 Na	12 Mg											13 Al	14 Si	15 P	16 S	17 Cl	18 Ar
19 K	20 Ca	21 Sc	22 Ti	23 V	24 Cr	25 Mn	26 Fe	27 Co	28 Ni	29 Cu	30 Zn	31 Ga	32 Ge	33 As	34 Se	35 Br	36 Kr
37 Rb	38 Sr	39 Y	40 Zr	41 Nb	42 Mo	43 Tc	44 Ru	45 Rh	46 Pd	47 Ag	48 Cd	49 In	50 Sn	51 Sb	52 Te	53 I	54 Xe
55 Cs	56 Ba	71 Lu	72 Hf	73 Ta	74 W	75 Re	76 Os	77 Ir	78 Pt	79 Au	80 Hg	81 Tl	82 Pb	83 Bi	84 Po	85 At	86 Rn
87 Fr	88 Ra	103 Lr	104 Rf	105 Db	106 Sg	107 Bh	108 Hs	109 Mt	110	111	112	113	114	115	116	117	118

lanthanide series	57 La	58 Ce	59 Pr	60 Nd	61 Pm	62 Sm	63 Eu	64 Gd	65 Tb	66 Dy	67 Ho	68 Er	69 Tm	70 Yb
actinide series	89 Ac	90 Th	91 Pa	92 U	93 Np	94 Pu	95 Am	96 Cm	97 Bk	98 Cf	99 Es	100 Fm	101 Md	102 No

Considerable quantities of neon are used in high-energy physics research. Neon fills spark chambers used to detect the passage of nuclear particles. Liquid neon can be utilized as a refrigerant in the temperature range about 25 to 40 K (−416 to −387°F). Neon is also used in some kinds of electron tubes, in Geiger-Müller counters, in spark-plug test lamps, and in warning indicators on high-voltage electric lines. A very small wattage produces visible light in neon-filled glow lamps; such lamps are used as economical night and safety lights. *See* Neon glow lamp.

Neon is colorless, odorless, and tasteless; it is a gas under ordinary conditions. Some of the other properties of neon are given in the table. Neon does not form any chemical compounds in the ordinary sense of the word; there is only one atom in each molecule of gaseous neon. [A.W.F.]

Physical properties of neon

Property	Value
Atomic number	10
Atomic weight (atmospheric neon only)	20.183
Melting point, °C	−248.6
Boiling point at 1 atm pressure, °C	−246.1
Gas density at 0°C and 1 atm pressure, g/liter	0.8999
Liquid density at its boiling point, g/ml	1.207
Solubility in water at 20°C, ml neon (STP)/1000 g water at 1 atm partial pressure neon	10.5

Neon glow lamp A low-wattage lamp often used as an indicator light or as an electronic circuit component. The neon lamp usually consists of a pair of electrodes sealed within a bulb containing neon gas at a low pressure. Some of the smaller bulbs are equipped with wire leads that are connected directly into the electrical supply circuit; others are equipped with conventional bases that vary with the size of the lamps (see illustration).

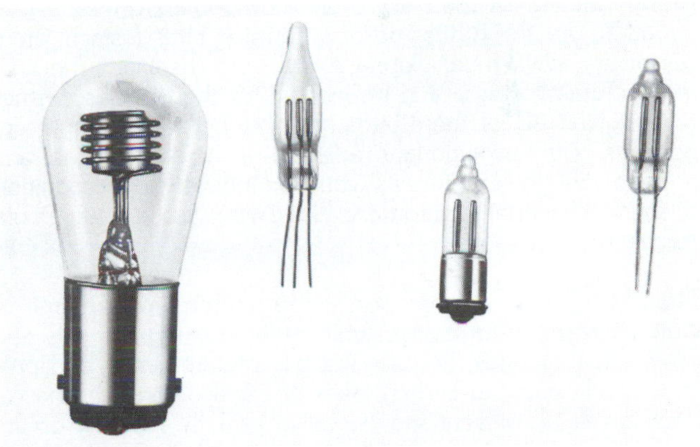

Some examples of glow lamps.

Electrodes sealed in a neon atmosphere will emit electrons if a sufficient voltage difference is impressed across them. In glow lamps the electrodes are usually treated to emit electrons freely. With a sufficiently high voltage between electrodes, the velocity of electron flow is high enough to ionize the neon nearest the negative electrode (cathode). The neon then emits a reddish-orange glow similar to the color of neon sign tubing. With direct current the glow is restricted to the immediate vicinity of the negative electrode. With alternating current, both electrodes act alternately as cathodes, and the glow appears alternately at both surfaces. At usual frequencies, the alternations occur so rapidly that both electrodes appear to glow constantly. In dc circuits, the voltage across the electrodes may be reduced significantly, once the lamp has started, without causing the lamp to go out. [A.M.]

Neoplasia An abnormal proliferation of somatic cells resulting in a mass of tissue called a neoplasm or tumor. Neoplasia literally means new growth. The involved cells are characterized by an abnormality which allows them to escape the normal regulatory events. This anomaly is passed on to daughter cells.

The mass of tissue which forms the neoplasm can be divided into two components. One is the parenchyma, which consists of the abnormal, proliferating cells. The other is the stroma, which is connective tissue providing support. It includes the vascular tissues. The relative amount of parenchyma and stroma varies in different types of tumors.

Neoplasia can result in a tumor whose parenchyma have a high degree of differentiation. This means that in appearance and function the cells are very like normal tissue. These neoplasms typically remain localized and generally can be successfully removed. They are referred to as benign. Conversely, there are neoplasms whose parenchyma have varying degrees of abnormalities in structure and behavior. They tend to invade adjacent tissue. Small pieces may, by various means, locate in other places in the body where they form secondary tumors called metastases. These kinds of neoplasms are referred to as malignant or cancerous.

Thus, there is a whole range of neoplastic responses. The type of tumor produced is partially dependent upon the cell type involved in the change and the position in the body. Most cells are capable of undergoing neoplastic changes. *See* ONCOLOGY; TUMOR. [N.K.M.; C.Qu.]

Neornithes The subclass of the class Aves containing all known birds except the Jurassic fossil *Archaeopteryx*, which alone constitutes the subclass Archaeornithes. Some authorities divide the Neornithes into two superorders: Odontognathae, containing only the fossil order Hesperornithiformes; and Neognathae, containing the remaining orders. Additional superorders that have been segregated by other authors are: Ichthyornithes, for the fossil order Ichthyornithiformes; Palaeognathae, for the Ratites or large running birds (ostrich, emu, cassowary, kiwi, moa, elephant bird, and rhea) and the tinamous (Tinamiformes); and Impennes, for the Sphenisciformes or penguins. All of these birds are now believed to represent portions of the evolutionary radiation of the Neognathae, and not separate phyletic lines as would be indicated by recognition of these additional superorders. *See* AVES; NEOGNATHAE; ODONTOGNATHAE. [K.C.P.]

Neoteny A phenomenon among some salamanders, in which larvae of large size, while still retaining the gills and other larval features, become sexually mature, mate, and produce fertile eggs. In certain lakes of Mexico, only the neotenous larvae are present and are called axolotls. Neoteny occurs in certain species of the family Ambystomidae. *See* PEDOGENESIS. [T.I.S.]

Nepenthales An order of flowering plants, division Magnoliophyta (Angiospermae), in the subclass Dilleniidae of the class Magnoliopsida (dicotyledons). The order (also known as Sarraceniales) consists of 3 well-marked small families: the Droseraceae, with about 90 species; the Nepenthaceae, with about 80 species; and the Sarraceniaceae, with only about 16. The plants characteristically grow in waterlogged soils which are deficient in available nitrogen. They are herbs or shrubs with alternate, simple leaves that are modified for catching insects, from which they absorb nitrogenous nutrients. The pitcher plants (*Sarracenia*), sundew (*Drosera*), and Venus' flytrap (*Dionaea muscipula*) are well-known members of the order. *See* DILLENIIDAE; MAGNOLIOPSIDA. [A.Cr.]

Neper A unit of attenuation used in transmission-line theory. On a uniform transmission line having waves traveling in only one direction, the magnitudes of voltage E and of current I decrease with distance x traveled, as given by Eq. (1), where

$$\frac{E}{E_0} = \frac{I}{I_0} = \epsilon^{-\alpha x} \qquad (1)$$

E_0, I_0, and α are constants. The attenuation in nepers between the points where E_0 and I_0 are measured and where E and I are measured is given by Eq. (2), in which ln denotes the natural

$$\alpha x = \ln \frac{E_0}{E} = \ln \frac{I_0}{I} \qquad (2)$$

(or napierian) logarithm. One neper equals 8.686 dB, the decibel being the practical unit of attenuation. *See* DECIBEL; TRANSMISSION LINES. [E.W.K.]

Nepheline A mineral of variable composition: in its purest state, $NaAlSiO_4$; often nearly $Na_3K(AlSiO_4)_4$; but generally $(Na,K,\square,Ca,Mg,Fe^{2+},Mn,Ti)_8(Al,Si,Fe^{3+})_{16}O_{32}$, where □ represents vacant crystallographic sites, and Ca, Mg, Fe^{2+}, Mn, Ti, and Fe^{3+} are usually present in only minor or trace amounts. The most important variations in nepheline composition are due to crystalline solution of $KAlSiO_4$ (the mineral kalsilite), and substitution of □ for K. *See* SILICATE MINERALS.

The salient physical properties of nepheline are: a Mohs scale hardness of 5.5–6.0; a specific gravity between 2.56 and 2.67; a typically dark gray, light gray, or white color, but it can also be colorless (nepheline is colorless in petrographic thin section); and a vitreous or greasy luster. Nepheline occurs as simple hexagonal prisms or, more commonly, as isolated shapeless grains or irregular polycrystalline masses.

Nepheline is the most abundant feldspathoid mineral; it occurs in a wide variety of SiO_2-deficient (quartz-free) and alkali-rich volcanic, plutonic, and metamorphic rocks. In volcanic rocks, nepheline occurs chiefly as a primary mineral in phonolites, kenytes, and melilite basalts, and it is the characteristic mineral of nephelinites.

Both "pure" (processed) nepheline and nepheline syenite are used as raw materials for the manufacture of glass, various ceramic materials, alumina, pottery, and tile. *See* FELDSPATHOID; IGNEOUS ROCKS; NEPHELINE SYENITE; NEPHELINITE. [J.G.B.]

Nepheline syenite A phaneritic (visibly crystalline) plutonic rock with granular texture, composed largely of alkali feldspar (orthoclase, or microcline, usually perthitic), nepheline, and dark-colored (mafic) minerals (biotite, soda-amphibole, and soda-pyroxene). Nepheline syenites have many features in common with syenites into which they grade, but chemically, mineralogically, and texturally they are much more variable. *See* SYENITE.

The alkali feldspar is soda-rich and usually exhibits perthitic texture (intergrown potash and soda feldspars). Nepheline, a gray mineral with a greasy appearance, is also a major constituent. It may be highly altered and is commonly converted to bright-yellow cancrinite. The most common accessory minerals are sphene, zircon, apatite, ilmenite, and magnetite.

Nepheline syenite and related rocks are rare and generally occur in small bodies (dikes, sills, laccoliths, stocks, and small irregular plutons). Only a few large bodies are known. Three of the largest bodies are in southern Greenland, Pilaansberg in South Africa, and Kola Peninsula, Russia. *See* IGNEOUS ROCKS. [C.A.C.]

Nephelinite A dark-colored, aphanitic (very finely crystalline) rock of volcanic origin, composed essentially of nepheline (a feldspathoid) and pyroxene.

The texture is usually porphyritic with large crystals (phenocrysts) of augite and nepheline in a very-fine-grained matrix. Augite phenocrysts may be diopsidic or titanium-rich and may be rimmed with soda-rich pyroxene (aegirine-augite). Microscopically the matrix is seen to be composed of tiny crystals or grains of nepheline, augite, aegirite, and sodalite with occasional soda-rich amphibole, biotite, and brown glass. Accessories usually include magnetite, ilmenite, apatite, sphene, and perovskite.

Nephelinite and related rocks are very rare. They occur as lava flows and small, shallow intrusives. A great variety of these feldspathoidal rocks is displayed in Kenya. *See* Feldspathoid; Igneous rocks. [C.A.C.]

Nephrite One of the amphibole group of rock-forming minerals. The minutely fibrous structure of nephrite makes it exceedingly durable. It occurs in a variety of colors, mostly of low intensity, including medium and dark green, yellow, white, black, and blue-gray. Nephrite has a hardness of 6 to 6½ on Mohs scale, a specific gravity near 2.95, and refractive indices of 1.61 to 1.64. Nephrite occurs throughout the world; important sources include Russia, New Zealand, Alaska, several provinces of China, and a number of states in the western United States. *See* Amphibole; Gem; Jade. [R.T.L.]

Nephritis A general term referring to an inflammatory process affecting the kidney. Included in this broad category are diseases caused by specific infectious agents such as viruses, spirochetes, bacteria, fungi, or parasites. A large group of common diseases affecting the kidney have no known cause, but may be associated with diseases elsewhere in the body such as streptococcal pharyngitis, lupus erythematosus, or arteriosclerosis. Finally, there are those conditions which are seen following exposure to toxins such as heavy metals, carbon tetrachloride, and certain drugs. *See* Glomerulonephritis; Pyelonephritis. [G.E.Str.]

Neptune The outermost of the four giant planets, a near twin of Uranus in size, mass, and composition. Its discovery in 1846 within a degree from the theoretically predicted position was one of the great achievements of celestial mechanics.

Through a small telescope, Neptune appears as a tiny greenish disk. Its linear diameter of 30,900 mi (49,400 km) is very similar to that of Uranus. The mass of Neptune is 17.21 times the mass of Earth, corresponding to a mean density of 1.66, somewhat above that of its sister planet. This suggests that the enrichment of heavy elements (when compared with Jupiter and Saturn, which exhibit a solar mixture) is somewhat greater in Neptune than in Uranus, but both must still be composed primarily of hydrogen and helium.

The temperature that this distant object would assume as a result of heating by solar radiation can be calculated as −380°F (45 K). But temperatures measured near a wavelength of 13 micrometers are very much higher, over −280°F (100 K). At radio frequencies, which permit observations of radiation from the lower atmosphere, one finds temperatures increasing with depth.

Neptune has two known satellites. The larger and brighter one, Triton, was found visually by W. Lassell in 1846; Nereid was discovered in a photographic search by G. P. Kuiper in 1949. [T.C.O.]

Neptunium A chemical element, symbol Np, atomic number 93. Neptunium is a member of the actinide or 5*f* series of elements. It was synthesized as the first transuranium element in 1940 by bombardment of uranium with neutrons to produce neptunium-239. The lighter isotope ^{237}Np, a long-lived alpha emitter with half-life 2.14×10^6 years, is particularly important chemically.

Neptunium metal is ductile, low-melting (637°C or 1179°F), and in its alpha form is of high density, 20.45 g/cm³ (11.82 oz/in.³). The chemistry of neptunium may be said to be intermediate between that of uranium and plutonium. Neptunium metal is reactive and forms many binary compounds, for example, with hydrogen, carbon, nitrogen, phosphorus, oxygen, sulfur, and the halogens. *See* Actinide elements; Nuclear chemistry; Transuranium elements. [R.A.Pe.]

1	2	3	4	5	6	7	8	9	10	11	12	13	14	15	16	17	18
1 H																	2 He
3 Li	4 Be											5 B	6 C	7 N	8 O	9 F	10 Ne
11 Na	12 Mg											13 Al	14 Si	15 P	16 S	17 Cl	18 Ar
19 K	20 Ca	21 Sc	22 Ti	23 V	24 Cr	25 Mn	26 Fe	27 Co	28 Ni	29 Cu	30 Zn	31 Ga	32 Ge	33 As	34 Se	35 Br	36 Kr
37 Rb	38 Sr	39 Y	40 Zr	41 Nb	42 Mo	43 Tc	44 Ru	45 Rh	46 Pd	47 Ag	48 Cd	49 In	50 Sn	51 Sb	52 Te	53 I	54 Xe
55 Cs	56 Ba	71 Lu	72 Hf	73 Ta	74 W	75 Re	76 Os	77 Ir	78 Pt	79 Au	80 Hg	81 Tl	82 Pb	83 Bi	84 Po	85 At	86 Rn
87 Fr	88 Ra	103 Lr	104 Rf	105 Db	106 Sg	107 Bh	108 Hs	109 Mt	110	*111*	*112*	*113*	*114*	*115*	*116*	*117*	*118*

lanthanide series	57 La	58 Ce	59 Pr	60 Nd	61 Pm	62 Sm	63 Eu	64 Gd	65 Tb	66 Dy	67 Ho	68 Er	69 Tm	70 Yb
actinide series	89 Ac	90 Th	91 Pa	92 U	93 Np	94 Pu	95 Am	96 Cm	97 Bk	98 Cf	99 Es	100 Fm	101 Md	102 No

Nerve A group of nerve fibers coursing together as a bundle in the peripheral nervous system. The individual fibers are covered by Schwann cells, many of which contain large amounts of myelin, which makes the nerve appear shiny white. The nerve fibers with their Schwann cell sheaths are held together by connective tissue. In most nerves, some of the fibers are sensory (carrying information to the central nervous system) and some are motor (carrying information from the central nervous system to peripheral glands and muscles). When both sensory and motor fibers are in a nerve, it is called a mixed nerve. *See* Myelin.

In the central nervous system (brain and spinal cord) a group of nerve fibers running together is called a tract. Glial cells, not Schwann cells, form the sheaths of tract fibers, and there is no connective tissue holding the bundle together. Whereas most nerves are mixed, there is functional segregation in the central nervous system so that most tracts have only one functional type of fiber. *See* Motor systems; Nervous system (vertebrate). [D.B.W.]

Nervous system (invertebrate) All multicellular organisms have a nervous system, which may be defined as assemblages of cells specialized by their shape and function to act as the major coordinating organ of the body. Nervous tissue underlies the ability to sense the environment, to move and react to stimuli, and to generate and control all behavior of the organism. Compared to vertebrate nervous systems, invertebrate systems are somewhat simpler and can be more easily analyzed. Invertebrate nerve cells tend to be much larger and fewer in number than those of vertebrates. They are also easily accessible and less complexly organized; and they are hardy and amenable to revealing experimental manipulations. However, the rules governing the structure, chemistry, organization, and function of nervous tissue have been strongly conserved phylogenetically. Therefore, although humans and the higher vertebrates have unique behavioral and intellectual capabilities, the underlying physical-chemical principles of nerve cell activity and the strategies for organizing higher nervous systems are already present in the lower forms. Thus neuroscientists have taken advantage of the simpler nervous systems of invertebrates to acquire further understanding of those processes by which all brains function. *See* Nervous system (vertebrate).

Invertebrate and vertebrate nerve cells differ more in quantity, or degree, than in qualitative features. Aside from differences in size and numbers, the most striking difference is that invertebrate neurons have a unipolar shape, whereas most vertebrate neurons are multipolar. An additional general contrast between invertebrate and vertebrate nervous systems is that invertebrates tend to have more neurons displaced to the periphery (outside the central nervous system) and to perform more integrative and processing functions in the periphery. Vertebrates perform almost all their integration within the central nervous system, using interneurons. Invertebrate nervous systems also seem to have a greater potential for regrowth, regeneration, or repair after damage than do vertebrate nerve cells. Many invertebrates continue to add new nerve cells to their ganglia with age; vertebrates, in general, do not. Only vertebrate neurons have myelin sheaths, a specialized wrapping of glial membrane around axons, increasing their conduction speed. Invertebrates tend to enhance conduction velocity by using giant axons, particularly for certain escape responses. *See* MYELIN. [J.E.Bla.]

Nervous system (vertebrate) A coordinating and integrating system which functions in the adaptation of an organism to its environment. An environmental stimulus causes a response in an organism when specialized structures, receptors, are excited. Excitations are conducted by nerves to effectors which act to adapt the organism to the changed conditions of the environment. In animals, humoral correlation is controlled by the activities of the endocrine system.

The central nervous system consists of the brain and spinal cord; most peripheral nerves are connected to the spinal cord. Peripheral nerves that are connected directly to the brain are called cranial nerves. *See* BRAIN; CENTRAL NERVOUS SYSTEM; CRANIAL NERVE; SPINAL CORD.

The nervous system is divided into the cerebrospinal (also called voluntary or somatic) system and the autonomic (also called visceral or involuntary) system. The cerebrospinal system controls all voluntary muscle motions, while the operation of organs (heart muscle and so on) is controlled by the autonomic system. The autonomic system is divided into the parasympathetic system, which plays an inhibitory role, and the sympathetic system, which is responsible for visceral reflexes, especially in emergency situations. *See* AUTONOMIC NERVOUS SYSTEM; PARASYMPATHETIC NERVOUS SYSTEM; SYMPATHETIC NERVOUS SYSTEM. [A.He.]

Nervous system disorders A satisfactory classification of diseases of the nervous system should include not only the type of reaction (congenital malformation, infection, trauma, neoplasm, vascular diseases, and degenerative, metabolic, toxic, or deficiency states) but also the site of involvement (meninges, peripheral nerves or gray or white matter of the spinal cord, brainstem, cerebellum, and cerebrum). This article will consider mostly the various types of reactions and only incidentally the sites affected. Clinical signs and symptoms of dysfunction of the nervous system depend not only on the type and location of lesions but also on the severity and rapidity of development of the lesion.

Malformations. The central nervous system develops as a hollow neural tube by the fusion of the crests of the neural groove, beginning in the cervical area and progressing rostrally and caudally, the last points to close being termed the anterior and posterior neuropores. If the anterior neuropore fails to close (about 24 days of fetal age), anencephaly develops. The poorly organized brain is exposed to amniotic fluid and becomes necrotic and hemorrhagic, leaving a froglike head on an infant who usually dies within hours after birth. The eyes, brainstem, cerebellum, and spinal cord are usually normal. *See* CONGENITAL ANOMALIES.

If the posterior neuropore fails to close (about 26 days of fetal age), the lumbosacral neural groove is exposed to amniotic fluid. Without the protective coverings of meninges, spines, muscles, and skin, the nervous tissue becomes partially necrotic and incorporated in a scar. Such a meningomyelocele is readily infected unless buried surgically within a few hours after birth. In addition, in about 95% of such infants hydrocephalus (a dilatation of the ventricles) occurs. The hydrocephalus can usually be adequately treated by shunting the ventricular fluid into the venous system or peritoneal cavity.

Infections. Infections of the nervous system may occur through a defect in the normal protective coverings caused by certain congenital malformations, as mentioned above, but also through other defects as the result of trauma, especially penetrating wounds or fractures opening into the paranasal sinuses or mastoid air cells. Subsequent infection of the nervous system may be the major complication of such "open head" injuries.

Viral infections vary widely geographically, generally related to the necessity for intermediate hosts and vectors (animal reservoirs) by which the virus is spread from animals to humans, or from one human to another. Poliomyelitis is primarily an intestinal infection which occasionally spreads to the nervous system. Herpes zoster has a similarly restricted preference for infecting sensory nerve cells and producing an acute skin eruption in the distribution of the affected sensory cells. Herpes simplex is closely related to herpes zoster, resides in the trigeminal or sacral sensory nerve cells, and intermittently produces eruptions in the distribution of these cells: "fever blisters" in and around the mouth in type I herpes, or similar blisters in the genital area in type II herpes. *See* HERPES; POLIOMYELITIS.

Complications of infections of the meninges include hydrocephalus, cranial nerve paralyses, and focal destructions of neural tissue.

Inflammations. Certain viruses, such as lymphocytic choriomeningitis and mumps, frequently produce a meningitis in humans from whose cerebrospinal fluid the virus is relatively easily grown. Certain other viruses, such as measles and varicella, occasionally produce meningitis or encephalomyelitis, but the cerebrospinal fluid does not contain the virus, and the hypothesis is that an allergic reaction is responsible. *See* MENINGITIS.

Allergy to one's own tissue elements is an interesting possibility that has evoked many experimental approaches. One of the best-studied examples is experimental allergic encephalomyelitis. Two diseases, multiple sclerosis affecting the central nervous system and the Landry-Guillain-Barré syndrome affecting the peripheral nervous system, are considered likely candidates eventually to be related to experimental allergic encephalomyelitis and experimental allergic neuritis, respectively. *See* MULTIPLE SCLEROSIS.

Vascular diseases. Vascular diseases of the nervous system are commonly called strokes, a term which emphasizes the suddenness of onset of neurological disability, as though the individual were struck down by a blow to the head. Such a cataclysmic onset is characteristic of vascular diseases, since the nerve cell can function without nutrients for only a matter of seconds and will die if not renourished within several minutes. *See* HEMORRHAGE.

Small and large strokes may be suffered clinically as the result of small or large hemorrhages or ischemic episodes occurring in or around the brain. One of the common ways the brain reacts to these injuries is by swelling. If the person survives the first critical week after a massive stroke with swelling, the neurologic deficit itself is rarely fatal, but the accompanying prolonged convalescence may contribute to the development of pneumonia or pyelonephritis, which may be fatal.

Degenerative diseases. Degenerative, metabolic, toxic, and deficiency states include the largest numbers of both common and rare diseases of the nervous system. Since neurons in the brain may be destroyed after birth and cannot be replaced, mental deterioration, deafness and blindness, incoordination and adventitious movements, and other neurologic signs so typical of these disorders are generally not reversible even if the basic metabolic defect is corrected. *See* HUNTINGTON'S DISEASE; METABOLIC DISORDERS; PARKINSON'S DISEASE.

Neoplasms. Neoplasms of the nervous system can be conveniently divided into primary and metastatic, the primary into gliomas and others, and the metastatic into bronchogenic and others. These four groups each account for about 25% of all intracranial neoplasms. [E.C.A.; C.-M.S.]

Network (mathematics) A collection of points (nodes), together with arcs joining certain pairs of nodes which may represent communications links between various locations, roads joining certain cities, and so on. Numbers representing, for example, the capacity or length of the arcs are frequently assigned. Problems considered may be of a combinatorial nature (finding the number of possible connections between two nodes) or of an optimization nature (finding the shortest path, or maximizing the amounts shipped between two nodes). [G.Ow.]

Neural crest A strip of ectodermal material in the early vertebrate embryo inserted between the prospective neural plate and epidermis. After closure of the neural tube the crest cells migrate into the body and give rise to parts of the neural system: the main part of the visceral cranium, the mesenchyme, the chromaffin cells, and pigment cells. The fact that mesenchyme arises from this ectodermal organ was directly contrary to the doctrine of the specificity of the germ layers. Neural crest no doubt exists in all vertebrate groups, including the cyclostomes. It has been most thoroughly studied in amphibians and the chick. *See* GERM LAYERS. [S.H.]

Neural network An information-processing device that consists of a large number of simple nonlinear processing modules, connected by elements that have information storage and programming functions. The field of neural networks is an emerging technology in the area of machine information processing and decision making. This technology makes extensive use of the terminology and methods of artificial intelligence. The main thrusts are toward highly innovative machine and algorithmic architectures, radically different from those that have been employed in conventional digital computers, and toward overcoming the major technological shortcomings and inherent limitations imposed by the traditional information-processing machines. The information-processing elements and components of neural networks, inspired by neuroscientific studies of the structure and function of the human brain, are conceptually simple. Neural networks have created an unusual amount of interest in the engineering and industrial communities by opening up new directions for research on commercial and military applications. *See* ARTIFICIAL INTELLIGENCE; NEUROBIOLOGY.

Automated information processing is achieved by means of modules that in general involve four functions: input/output (getting in and out of the machine), processing (executing prescribed specific information-handling tasks), memory (storing information), and connections between different modules providing for information flow and control. Neural networks contain a very large number of simple processing modules. This contrasts with traditional digital computers, which contain a small number of complex processing modules that are rather sophisticated in the sense that they are capable of executing very large sets of prescribed arithmetic and logical tasks (instructions). In conventional digital computers, the four functions listed above are carried out by separate dedicated machine units. In neural networks information storage is achieved by components which at the same time effect connections between distinct machine units (see illustration). *See* COMPUTER SYSTEMS ARCHITECTURE; DIGITAL COMPUTER.

The information-processing properties of neural networks depend mainly on two factors: the network topology (the scheme used to connect elements or nodes together), and the algorithm (the rules) employed to specify the values of the weights connecting the nodes. While the ultimate configuration and parameter values are problem-specific, it is possible to classify neural networks, on the basis of how information is stored or retrieved, in three broad categories: neural networks behaving as learning machines with a teacher; neural networks behaving as learning machines without a teacher; and neural networks behaving as associative memories. Within these three categories, several generic models have found important applications, and still others are under intensive investigation. [N.DeC.]

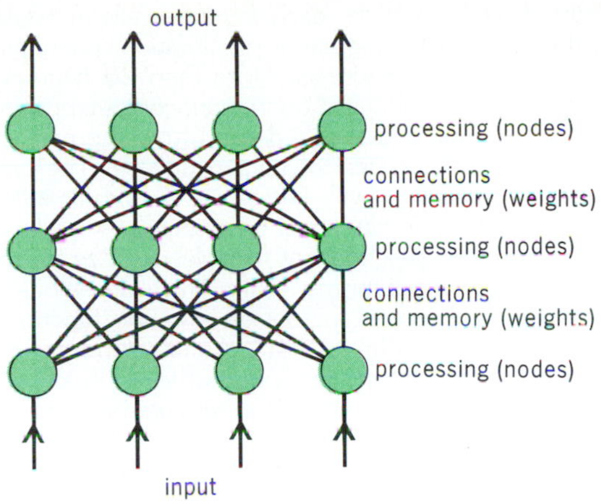

Architecture of multilayer neural network.

Neurobiology The development and function of the nervous system, with emphasis on how nerve cells generate and control behavior. The major goal of neurobiology is to explain at the molecular level how nerve cells differentiate and develop their specific connections and how nerve networks store and recall information. Ancillary studies on disease processes and drug effects in the nervous system also provide useful approaches for understanding the normal state by comparison with perturbed or abnormal systems. The functions of the nervous system may be studied at several levels: molecular, subcellular (organelle), cellular, simple multicellular interacting systems, complex systems, and higher functions (whole animal behavior). *See* ENDORPHINS; NERVOUS SYSTEM (INVERTEBRATE); NERVOUS SYSTEM (VERTEBRATE); NEURON. [J.R.B.]

Neurohypophysis hormone Extracts of the neurohypophysis, or posterior lobe of the pituitary, exhibit the following biological properties when administered to mammals: (1) The extracts affect blood pressure by constricting the blood vessels leading from the arteries (pressor action); (2) they stimulate the contraction of smooth muscle and specifically of the musculature of the uterus (oxytocic action); (3) they cause ejection of milk from the mammary gland; and (4) they cause inhibition of water excretion through the kidneys (antidiuretic action). These effects are due to the presence of two peptide hormones, oxytocin and vasopressin.

From a quantitative standpoint, vasopressin is the more versatile of the two, since it exercises appreciable uterine-stimulating, avian vasodepressor, and milk-ejecting actions in addition to its pressor and antidiuretic effects. Oxytocin has a slight but definite pressor and antidiuretic action in addition to its own characteristic effects. These overlapping biological activities of the two hormones reflect their close structural relationship. *See* HORMONE; PITUITARY GLAND. [C.H.L.]

Neuroimmunology The study of basic interactions among the nervous, endocrine, and immune systems during development, homeostasis, and host defense responses to injury. In its clinical aspects, neuroimmunology focuses on diseases of the nervous system, such as myasthenia gravis and multiple sclerosis, which are caused by pathogenic autoimmune processes, and on nervous system manifestations of immunological diseases, such as primary and acquired immunodeficiencies. *See* AUTOIMMUNITY; IMMUNOLOGICAL DEFICIENCY.

Neuroimmune interactions are dependent on the expression of at least two structural components: immunocytes must display receptors for nervous system-derived mediators, and the mediators must be able to reach immune cells in concentrations sufficient to alter migration, proliferation, phenotype, or secretory or effector functions. More than 20 neuropeptide receptors have been identified on immunocompetent cells.

It has been found that stimuli derived from the nervous system could affect the course of human disease. The onset or progression of tumor growth, infections, or chronic inflammatory diseases, for example, could be associated with traumatic life events or other psychosocial variables such as personality types and coping mechanisms. More direct indications of the influence of psychosocial factors on immune function have been provided by findings that cellular immunity can be impaired in individuals who are exposed to unusually stressful situations, such as the loss of a close relative. *See* CELLULAR IMMUNOLOGY.

During responses to infection, trauma, or malignancies, cells of the immune system produce some cytokines in sufficiently high quantities to reach organs that are distant from the site of production. These cytokines include predominantly the products of mononuclear phagocytes, including interleukin 1, tumor necrosis factor, and interleukin 6. They are known to act on the bone marrow, liver, and connective tissue cells as well as the nervous system. The last activity presumably contributes to the achievement of optimal adjustment of the organism during defense responses. Fever is the classic example of changes in nervous system function that are induced by products of the immune system: interleukin 1, which is produced by monocytes after stimulation by certain bacterial products, binds to receptors in the hypothalamus and evokes changes via the induction of prostaglandins. Interleukin 1 also induces slow-wave sleep. Both fever and sleep may be regarded as protective behavioral changes. *See* INTERLEUKIN.

Cytokines and neuropeptides, which were previously thought to be exclusively produced in the immune or nervous tissues, can actually be products of both systems. Some of the pluripotent cytokines that were historically the first to be detected in cells of the immune system and that regulate growth and differentiation of diverse cell types are now also found to be produced within the nervous system. Interleukin 1 and interleukin 6, which are predominantly products of mononuclear phagocytes, are also produced by glioblastoma and possibly microglial cells. It is likely that these mediators are produced during host responses to brain injury and serve functions similar to those in other organs to regulate clearance of damaged tissue and its repair. Conceivably, these interleukins may also play a role during development. The demonstration of interleukin-1 immunoreactivity in fibers of the central nervous system raises the possibility that this cytokine might also act as a neurotransmitter. *See* ENDOCRINE SYSTEM (VERTEBRATE); IMMUNOLOGY; NERVOUS SYSTEM (VERTEBRATE); NEUROSECRETION.

[M.L.; W.Ku.; P.V.]

Neuron A nerve cell: the functional unit of the nervous system. Structurally, the neuron is made up of a cell body or soma and one or more long processes: a single axon and dendrites. The cell body contains the nucleus and usual cytoplasmic organelles with an exceptionally large amount of rough endoplasmic reticulum, called Nissl substance in the neuron. The longest cell process is the axon, which is capable of transmitting propagated nerve impulses. There may be none, one, or many dendrites composing part of a neuron. If there is no dendrite, it is a unipolar neuron; with one dendrite, it is a bipolar neuron; if there is more than one dendrite, it is a multipolar neuron. In most neurons only the axon propagates nerve impulses; the dendrites and somas are also irritable but do not propagate nerve impulses. *See* NERVOUS SYSTEM (VERTEBRATE). [D.B.W.]

Neuroptera An order of delicate insects having endopterygote development, chewing mouthparts, and soft bodies. Included are the insects commonly termed lacewings, ant lions, dobsonflies, and snake flies. The order consists of about 25 families and is widely distributed. *See* ANT LION.

The adults have long, slender antennae and usually four similar wings, although the front pair is generally slightly larger than the hind pair. The adults of most species are strongly attracted to lights. The larvae are aggressive predators. The larvae of lacewings are especially destructive to aphids, scale insects, and mites. *See* ENDOPTERYGOTA; INSECTA. [F.M.C.]

Neurosecretion The synthesis and release of hormones by neurons (nerve cells). Such neurons are specifically called neurosecretory cells, and their products are called neurohormones. Like conventional (nonglandular) neurons, neurosecretory cells are able to receive signals from other neurons and to generate and conduct impulses. There is some evidence that these impulses trigger the release of the neurohormones. Because of the glandular function of neurosecretory cells, they are able to send messages to structures that are distant from their axonal terminals, whereas conventional neurons can only send messages locally. Neurosecretory cells, like conventional neurons, have an axon, Nissl granules, and neurofibrillae. Furthermore, like conventional neurons, neurosecretory cells of vertebrates have dendrites, but the neurosecretory cells of invertebrates may lack dendrites. *See* NEURON.

In organisms with a well-developed circulatory system, the neurohormones are typically sent by the vascular route, whereas in lower invertebrates that lack an organized circulatory system, the neurohormones simply diffuse to the target. At one time there was thought to be a sharp line of distinction between the nervous and endocrine systems. However, experimentation revealed not only that the nervous and endocrine systems interact in many ways, but also that some neurons are by themselves actually capable of producing hormones. The fact that some neurons are hormone producing shows how closely related the nervous and endocrine systems really are. In fact, there is a definite trend toward treating both of these systems merely as the two components of a single coordinating system called the neuroendocrine system. *See* ENDOCRINE SYSTEM (INVERTEBRATE); ENDOCRINE SYSTEM (VERTEBRATE); NERVOUS SYSTEM (VERTEBRATE). [M.F.]

Neurotic disorders Mental disorders characterized by symptoms such as phobias, obsessive thoughts, and compulsive actions, or by losses of specific bodily functions. A neurotic disorder, termed neurosis in the psychoanalytic literature, is distinguished from more severe mental disturbances by a continued ability to recognize reality, and from more diffuse character disorders by the relatively specific nature of the symptoms. Neurosis is quite common and treatable through a range of psychological and biological methods.

The three classic neuroses were identified by S. Freud. The first is phobia (anxiety hysteria in Freud's terminology), which is characterized by unreasonable fear of common objects or situations. The second classic neurosis is conversion (conversion hysteria), in which there is loss of function in part of the body that generally does not correspond to an anatomic or physiologic disorder. The third of the original neuroses, obsessive-compulsive neurosis, is characterized by intrusive, repetitive thoughts of a disturbing nature, and a compelling need to perform ritualized, repetitive, and apparently senseless acts. *See* OBSESSIVE-COMPULSIVE DISORDER; PHOBIC REACTION; PSYCHOSOMATIC DISORDERS.

Historically, neurosis was described and defined by Freud, and the dominant theories of its origin and treatment were psychoanalytic. There have been several generations of psychoanalytic theories, but one generally adopted by classical psychoanalysts holds that each neurotic symptom represents a conflict between an unacceptable impulse and the prohibition against that impulse. In the analytic theory, neurotic symptoms are formed when a frustration in current life prevents direct satisfaction of a wish. The recommended therapy is classical psychoanalysis, or psychoanalytic psychotherapy. This is aimed at making conscious to the patient the nature of the desire that is frustrated, as well as the prohibitions that prevent its direct satisfaction. *See* ANXIETY STATES; NEUROBIOLOGY; PSYCHOANALYSIS; PSYCHOTHERAPY. [M.M.]

Neutral currents Exchange currents which carry no electric charge and mediate certain types of electroweak interactions. The discovery of the neutral-current weak interactions and the agreement of their experimentally measured properties with the theoretical predictions were of great significance in establishing the validity of the Weinberg-Salam model of the electroweak forces.

The electroweak forces come in three subclasses: the electromagnetic interactions, the charged-current weak interactions, and the neutral-current weak interactions. The electromagnetic interaction is mediated by an exchanged photon γ. Since the photon carries no electric charge, there is no change in charge between the incoming and the outgoing particles. The charged-current weak interaction is mediated by the exchange of a charged intermediate boson, the W^+, and thus, for example, an incoming neutral lepton such as the ν_μ is changed into a charged lepton, the μ^-. In the neutral-current weak interactions, the exchanged intermediate boson, the Z^0, carries no electric charge (hence the name neutral-current interaction), and thus for example, an incident neutral lepton, such as the ν_μ, remains an outgoing neutral ν_μ. *See* ELECTRON; INTERMEDIATE VECTOR BOSON; LEPTON; NEUTRINO; PHOTON.

The neutral-current interactions were experimentally discovered in 1973, and have since been extensively studied, in neutrino scattering processes. Very important information about the properties of the neutral currents have been obtained by studying the interference effects between the electromagnetic and the neutral-current weak interactions in the scattering of polarized electrons on deuterium. Parity violating effects in atomic physics processes due to the neutral weak currents have been observed, and predicted parity-violating nuclear effects have been searched for. *See* ELEMENTARY PARTICLE; FUNDAMENTAL INTERACTIONS; PARITY (QUANTUM MECHANICS); SYMMETRY LAWS (PHYSICS); WEAK NUCLEAR INTERACTIONS; WEINBERG-SALAM MODEL. [C.B.]

Neutralization reaction (immunology) A procedure in which the chemical or biological activity of a reagent or a living organism is inhibited, usually by a specific neutralizing antibody. As an example, the lethal or the dermonecrotic actions of diphtheria toxin on animals may be completely neutralized by an equivalent amount of diphtheria antitoxin.

Antibodies to bacterial, snake-venom, and other enzyme preparations regularly precipitate them from solution so that the supernates are devoid of enzyme activity; however, the neutralization of activity in the precipitate may range from complete to negligible. *See* IMMUNOLOGY; NEUTRALIZING ANTIBODY; SEROLOGY. [H.P.T.]

Neutralizing antibody An antibody that reduces or abolishes some biological activity of a soluble antigen or of a living microorganism. Thus, diphtheria antitoxin is a neutralizing antibody that, in adequate amounts, abolishes the pathological effects of diphtheria toxin in animals. This is only one characteristic; the other general properties of the antibody are those of the immunoglobulin family (IgG, IgA, or IgM) to which it belongs. *See* ANTIBODY; IMMUNOGLOBULIN. [H.P.T.]

Neutrino An elementary particle designated by the Greek symbol ν, with zero rest mass and zero electric charge. It is classified as a lepton. The neutrino is the only known elementary particle that has only weak interactions. Thus, the neutrino is a unique tool in the study of weak forces; for this reason, ν interactions have been the subject of active study at large particle accelerators. *See* ELEMENTARY PARTICLE; LEPTON; WEAK NUCLEAR INTERACTIONS.

The existence of the neutrino was postulated in 1930 to explain the apparent nonconservation of energy in beta-decay process $n \rightarrow p + e^- + \bar{\nu}$. It was not until 1953 that the existence of the neutrino was experimentally verified, by observation of the interactions caused by free neutrinos. Quantitative studies of ν interactions resulted in the discovery that there are two distinct neutrinos—the electron neutrino ν_e, associated with beta decay, and the muon neutrino ν_μ, associated with pion decay, $\pi^+ \rightarrow \mu^+ + \nu_\mu$. From the absence of neutrinoless double beta decay, it can be inferred that ν_e is not identical to its antiparticle $\bar{\nu}_e$. It is thus believed that there are four distinct neutrinos, namely, ν_e, ν_μ, $\bar{\nu}_e$, and $\bar{\nu}_\mu$. *See* RADIOACTIVITY.

In the late 1970s a new heavy charged lepton, the tau ($\tau^\pm$), was discovered to exist in addition to the previously known charged leptons, the electron ($e^\pm$) and the muon ($\mu^\pm$). There is some theoretical speculation that two new additional types of neutrinos, the tau neutrino (ν_τ) and the anti-tau neutrino ($\bar{\nu}_\tau$), are associated with this new heavy lepton.

Considerations of angular momentum conservation in pion decay establish the spin of the neutrino to be ½ in units of $h/2\pi$ where h is Planck's constant. There has been some experimental indication and theoretical speculation that the mass of at least some of the neutrinos is not identically zero but has some finite, although very small, value. This possibility has fundamental cosmological implications since astrophysicists believe that there are a very large number of neutrinos in the universe, and if the neutrinos had a nonzero mass they would constitute a significant fraction of the total mass of the universe. *See* SPIN (QUANTUM MECHANICS).

According to the rules of quantum mechanics, a massless spin-½ particle such as the neutrino can have its spin lined up along its direction of motion or opposite to its direction of motion; these are called the helicity states, right-handed and left-handed respectively. Thus, in general, such a particle has four components—the particle and its antiparticle, right-handed and left-handed. It was, however, predicted by the two-component theory of the neutrino, and shown by experiment, that only two of the four components exist for the neutrinos—left-handed neutrinos and right-handed antineutrinos. *See* HELICITY (QUANTUM MECHANICS); RELATIVISTIC QUANTUM THEORY. [C.B.]

Neutron An elementary particle having approximately the same mass as the proton, but lacking a net electric charge. It is indispensable in the structure of the elements, and in the free state it is an important reactant in nuclear research and the propagating agent of fission chain reactions.

Neutrons and protons are the constituents of atomic nuclei. The number of protons in the nucleus determines the chemical nature of an atom, but without neutrons it would be impossible for two or more protons to exist stably together within nuclear dimensions. The protons, being positively charged, repel one another by virtue of their electrostatic interactions. The presence of neutrons weakens the electrostatic repulsion, without weakening the nuclear forces of cohesion. In light nuclei the resulting balanced, stable configurations contain protons and neutrons in almost equal numbers, but in heavier elements the neutrons outnumber the protons; in ^{238}U, for example, 146 neutrons are joined with 92 protons. Only one nucleus, ^{1}H, contains no neutrons. For a given number of protons, neutrons in several different numbers within a restricted range often yield nuclear stability—and hence the isotopes of an element. *See* ISOTOPE; NUCLEAR STRUCTURE; PROTON.

Free neutrons have to be generated from nuclei, and since they are bound therein by cohesive forces, an amount of energy equal to the binding energy must be expended to get them out. Nuclear machines, such as cyclotrons and electrostatic generators, induce many nuclear reactions when their ion beams strike target material. Some of these reactions release neutrons, and these machines are sources of high neutron flux. Neutrons are released in the act of fission, and nuclear reactors are unexcelled as intense neutron sources. *See* NUCLEAR BINDING ENERGY; NUCLEAR REACTION; NUCLEAR REACTOR.

Having no electric charge, neutrons interact so slightly with atomic electrons in matter that energy loss by ionization and atomic excitation is essentially absent. Consequently they are vastly more penetrating than charged particles of the same energy. The main energy-loss mechanism occurs when they strike nuclei. As with billiard balls, the most efficient slowing-down occurs when the bodies that are struck in an elastic collision have the same mass as the moving bodies; hence the most efficient neutron moderator is hydrogen, followed by other light elements: deuterium, beryllium, and carbon. The great penetrating power of neutrons imposes severe shielding problems for reactors and other nuclear machines, and it is necessary to provide walls, usually of concrete, several feet in thickness to protect personnel. *See* RADIATION DAMAGE TO MATERIALS; RADIATION SHIELDING; RADIOLOGY.

Free neutrons are themselves radioactive, each transforming spontaneously into a proton, an electron (β^- particle), and an antineutrino. This instability is a reflection of the fact that neutrons are slightly heavier than hydrogen atoms. Neutrons are, individually, small magnets. This property permits the production of beams of polarized neutrons, that is, beams of neutrons whose magnetic dipoles are aligned predominantly parallel to one direction in space. Neutrons spin with an angular momentum of ½ in units of $h/2\pi$, where h is Planck's constant. The negative sign which is attached to the magnetic moment indicates that the magnetic moment vector and the angular momentum vector are oppositely directed. *See* MAGNETON; NUCLEAR MOMENTS; NUCLEAR ORIENTATION; SPIN (QUANTUM MECHANICS).

When neutrons are completely slowed down in matter, they have a maxwellian distribution in energy that corresponds to the temperature of the moderator with which they are in equilibrium. The maxwellian distribution has a tail extending to very low energies, and a few neutrons (about 10^{-11} of the main neutron flux) at this extreme have energies less than 5×10^{-8} attojoule (3×10^{-7} eV), and hence velocities of less than about 7 m/s. The de Broglie wavelength of these ultracold neutrons is greater than 50 nanometers, which is so much larger than interatomic distances in solids that they interact with regions of a surface rather than with individual atoms, and as a result they are reflected from polished surfaces at all angles of incidence. Ultracold neutrons can be expected to be important in basic physics and to have applications in studies of surfaces and of the structure of inhomogeneities and magnetic domains in solids. *See* ANTINEUTRON; ELEMENTARY PARTICLE; THERMAL NEUTRONS.

[A.H.Sn.]

Neutron-antineutron oscillations Periodic transitions between the state of a neutron and the state of an antineutron. In 1970 the possible existence of neutron-antineutron oscillations and of other processes violating the conservation of baryon number B, which until then appeared to be a well-established law of nature, was first explicitly suggested by V. A. Kuz'min. In 1979 S. L. Glashow again raised this problem in the framework of theories unifying strong, weak, and electromagnetic interactions. These theories allow for processes violating baryon number conservation. Glashow pointed out that neutron-antineutron ($n\bar{n}$) transitions (with change in baryon number $\Delta B = 2$) would be mediated by a particle having a mass of the order of 10^6 GeV, intermediate between the mass of the $W^{\pm}$, Z^0 bosons (10^2 GeV) and the mass of the particle unifying the three interactions (10^{15} GeV). The search for $n\bar{n}$ oscillations would then also provide information on a range of energies not accessible to particle accelerators for some years to come. *See* BARYON; FUNDAMENTAL INTERACTIONS; GRAND UNIFICATION THEORIES; INTERMEDIATE VECTOR BOSON; SYMMETRY LAWS (PHYSICS).

Neutrons oscillating inside a nucleus would occasionally annihilate with emissions of π mesons, and this would be another process that would make matter unstable in addition to proton decay ($\Delta B = 1$). Direct measurements of neutron-antineutron oscillations consist in allowing an intense beam of slow neutrons to travel in a vacuum for as long a time as possible, in a region where the magnetic field of the Earth has been reduced, and looking for antineutron annihilations on a target. *See* ANTINEUTRON; NEUTRON.

[G.Fa.]

Neutron diffraction The phenomenon associated with the interference processes which occur when neutrons are scattered by the atoms within solids, liquids, and gases. The use of neutron diffraction as an experimental technique is relatively new compared to electron and x-ray diffraction, since successful application requires high thermal-neutron fluxes, which can be obtained only from nuclear reactors. These diffraction investigations are possible because thermal neutrons have energies with equivalent wavelengths near 0.1 nanometer and are therefore ideally suited for interatomic interference studies.

In the scattering of neutrons by atoms, there are two important interactions. One is the short-range, nuclear interaction of the neutron with the atomic nucleus. This interaction produces isotropic scattering because the atomic nucleus is essentially a point scatterer relative to the wavelengths of thermal neutrons. Strong resonances associated with the scattering process prevent any regular variation of the nuclear scattering amplitudes with atomic number. The other important process for the scattering of neutrons by atoms is the interaction of the magnetic moment of the neutron with the spin and orbital magnetic moments of the atom. *See* SCATTERING EXPERIMENTS (ATOMS AND MOLECULES); SCATTERING EXPERIMENTS (NUCLEI).

Since the nuclear scattering amplitudes for neutrons do not vary uniformly with atomic number, there are certain types of chemical structures which can be investigated more readily by neutron diffraction than by x-ray diffraction. Moreover, since neutron scattering is a nuclear process, when the scattering amplitude of an element is not favorable for a particular investigation, it is frequently possible to substitute an enriched isotope which has scattering characteristics that are markedly different. The most significant application of neutron diffraction in chemical crystallography is the structure determination of composite crystals which contain both heavy and light atoms, and the most important compounds in this general classification are the hydrogen-containing substances.

The interaction of the magnetic moment of the neutron with the orbital and spin moments in magnetic atoms makes neutron scattering a unique tool for the study of a wide variety of magnetic phenomena, because information is obtained on the magnetic properties of the individual atoms in a material. This interaction depends on the size of the atomic magnetic moment and also on the relative orientation of the neutron spin and of the atomic magnetic moment with respect to the scattering vector and with respect to each other. Consequently, detailed information can be obtained on both the magnitude and orientation of magnetic moments in any substance which displays magnetic properties.

The investigation of antiferromagnetic and ferrimagnetic substances is one of the most important applications of the neutron diffraction technique, because detailed information on the magnetic configuration in these systems cannot be obtained by other methods.

One of the most important uses of inelastic neutron scattering is the study of thermal vibrations of atoms about their equilibrium positions, because lattice vibration quanta, or phonons, can be excited or annihilated in their interactions with low-energy neutrons. The measurements provide a direct determination of the dispersion relations for the normal vibrational modes of the crystal and do not require the large corrections necessary in similar x-ray investigations. These measured dispersion relations furnish the best experimental information available on interatomic forces that exist in crystals. *See* ELECTRON DIFFRACTION; MAGNON; NEUTRON SPECTROMETRY. [M.K.W.]

Neutron optics A title by analogy of certain phases of neutron physics in which the wave character of neutrons dominates and leads to behavior similar to that of light. Neutrons can be reflected at small glancing angles from plane surfaces; they show various scattering phenomena with similarity to light, and they can be diffracted by crystals. Although they can also be polarized, the analogy with light is in this case invalid because the polarization of neutrons depends upon their possession of a constant magnetic moment, which light waves lack. *See* NEUTRON; NEUTRON DIFFRACTION. [A.H.Sn.]

Neutron spectrometry A generic term applied to experiments in which neutrons are used as the probe for measuring excited states of nuclides and for determining the properties of these states. The term neutron spectroscopy is also used. The strength of the interaction between a neutron and a target nuclide can vary rapidly as a function of the energy of the incident neutron, and it is different for every nuclide. At particular neutron energies the interaction strength for a specific nuclide can be very strong; these narrow energy regions of strong interactions are called resonances (see illustration). The strength of the interaction, expressing the probability that an interaction of a given kind will take place, can be considered as the effective cross-sectional area σ presented by a nucleus to an incident neutron.

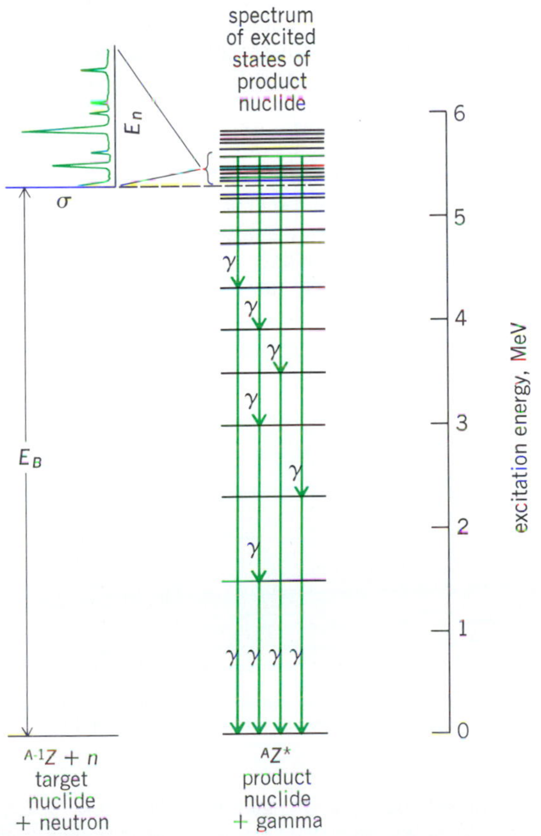

Energy-level diagram for the product nucleus $^{A}Z^{*}$ with mass number A and charge number Z. The asterisk emphasizes that the product nucleus is in an excited state, from which it returns to ground state by emitting gamma rays (γ). Excitation energy is the sum of the energy of the neutron E_n and the binding energy E_B of the neutron which has been added to the target nuclide.

Neutron spectroscopy can be carried out by two different techniques (or a combination): (1) by the use of a time-pulsed neutron source which emits neutrons of many energies simultaneously, combined with the time-of-flight technique to measure the velocities of the neutrons; this time-of-flight technique can be used for neutron measurements from 10^{-3} eV to about 200 MeV; (2) by the use of a beam of nearly monoenergetic neutrons whose energy can be varied in small steps approximately equal to the energy spread of the neutron beam; however, useful "monoenergetic" neutron sources are not available from about 10 eV to about 10 keV.

Neutron spectroscopy has yielded a mass of valuable information on nuclear systematics for almost all nuclides. The distribution of the spacings between nuclear levels and the average of these spacings have provided valuable tests for various nuclear theories. The properties of these levels, that is, the probabilities that they decay by neutron or gamma-ray emission, or by fission, and the averages and distribution of these probabilities have stimulated much theoretical effort.

In addition, knowledge of neutron cross sections is fundamental for the optimum design of thermal fission power reactors and fast neutron breeder reactors, as well as fusion power reactors now in the conceptual stage. Cross sections are needed for nuclear fuel materials such as ^{235}U or ^{239}Pu, for fertile materials such as ^{238}U, for structural materials such as iron and chromium, for coolants such as sodium, for moderators such as beryllium, and for shielding materials such as concrete. *See* NUCLEAR STRUCTURE; REACTOR PHYSICS. [J.A.H.]

Neutron star A star containing about 1½ solar masses of material compressed into a volume approximately 10 km (6 mi) in radius (1 solar mass = 2×10^{30} kg, or 4.4×10^{30} lb). Neutron stars are one of the end points of stellar evolution and are probably the final states of medium-mass and possibly of high-mass stars. The density of neutron star material is 10^{14} to 10^{15} times the density of water and exceeds the density of matter in the nuclei of atoms. Neutron stars are pulsars (pulsating radio sources) if they rotate sufficiently rapidly and have strong enough magnetic fields. *See* PULSAR; STELLAR EVOLUTION.

Neutron stars play a role in astrophysics which extends beyond their status as strange, unusual types of stellar bodies. The interior of a neutron star is a cosmic laboratory in which matter is compressed to densities which are found nowhere else in the universe. Neutron stars in double-star systems can cause the emission of x-rays when matter flows toward the neutron star, swirls around it, and heats up. Neutron stars are probably formed in supernova explosions. A few pulsars are found in double-star systems, and careful timing of the pulses they emit can test Einstein's general theory of relativity. *See* GRAVITATION; RELATIVITY; SUPERNOVA.

If Einstein's theory of gravitation is the correct one, a neutron star with a mass larger than some limiting value will collapse catastrophically, because its internal pressure will be insufficient, and become a black hole. The exact value of this limiting mass is not known precisely, but lies between 3 and 5 solar masses. *See* BLACK HOLE.

Most of the interior of a neutron star consists of matter which is almost entirely composed of neutrons. In the bulk of the star, this matter is in a superfluid state, where circulation currents can flow without resistance. This material is under pressure, since it must be able to support the tremendous weight of the overlying layers at each point in the neutron star. This pressure, called degeneracy pressure, is caused by the close packing of the neutrons rather than by the motion of the particles. As a result, neutron stars can be stable no matter what the internal temperature is, because the pressure that supports the star is independent of temperature. *See* NEUTRON; SUPERFLUIDITY. [H.L.Sh.]

New Britain The largest island of the Bismarck Archipelago. Also known as Neu Pommern, New Britain is 370 mi (595 km) long and 90 mi (145 km) at its maximum width, and it has an area of about 13,000 mi² (34,000 km²). Some of the mountains of the rugged range are over 7000 ft (2100 m) and are active volcanoes. Rabaul, the capital of the Bismarcks, is located in the caldera of an active volcano, one of whose walls has collapsed and is open to the sea. *See* EAST INDIES. [M.G.A.-C.]

New Guinea The second largest island in the world. West Irian, together with eastern New Guinea and Papua, is nearly 1500 mi (2400 km) long and 390 mi (630 km) wide at its maximum, with an area of approximately 340,000 mi² (880,000 km²).

The southern lowland area is part of the Sahul Shelf. The high mountains of the central part of the island are a Tertiary fold belt with a long geosynclinal history. Peaks of the central mountains rise to more than 16,000 ft (4900 m) in the Nassau and Orange ranges. Since the snow line is at about 14,500 ft (4400 m), many of these are snow-capped, although within 5° of the Equator. North of the central mountains there is a broad valley with a coastal range to its north. Important deposits of minerals have been found: gold, petroleum, coal, iron ore, tin, copper, platinum, manganese, osmiridium, and others.

Heavy rains are received from the two annual monsoons. These rains, with the high temperatures (the mean monthly high in December is 78°F or 25.6°C and the low in August is 74°F or 23.3°C), result in a tropical rainforest over most of the island, except where savannas have been culturally induced. *See* EAST INDIES. [M.G.A.-C.]

New Ireland New Ireland, or Neu Mecklinberg, has an area of 3340 mi² (8650 km²); it is 200 mi (320 km) long, with an average width of 20 mi (32 km). It is very mountainous but lacks active volcanoes. *See* EAST INDIES. [M.G.A.-C.]

New Zealand A landmass in the Southern Hemisphere, bounded by the South Pacific Ocean to the north, east, and south and the Tasman Sea to the west, with a total land area of 103,883 mi² (269,057 km²). The exposed landmass represents about one-quarter of a subcontinent, with three-quarters submerged. This long, narrow, mountainous country, oriented northeast to southwest, consists of two main islands, North Island and South Island, surrounded by a much greater area of crust submerged to depths reaching 1.2 mi (2 km).

South Island lowlands are either alluvial plains as in Otago, Southland, and Nelson, or glacial outwash fans as in Westland and Canterbury. North Island lowlands such as Hawke's Bay, Wairarapa, and Manawatu are alluvial; the Waikato, Hauraki, and Bay of Plenty lowlands occupy structural basins that contain large volumes of reworked volcanic debris from the central volcanic region. The alluvial lowlands of both main islands form the most agriculturally productive areas of the country. *See* PLAINS.

The climate of New Zealand is influenced by three main factors: a location in latitudes where the prevailing airflow is westerly; an oceanic environment; and the mountain chains, which modify the weather systems as they pass eastward, causing high rainfalls on windward slopes and sheltering effects to leeward.

Weather is determined mostly by series of anticyclones and troughs of low pressure that produce alternating periods of settled and variable conditions. Westerly air masses are occasionally replaced by southerly airstreams, which bring cold conditions with snow in winter and spring to areas south of 39°S, and northerly tropical maritime air, which brings warm humid weather to the north and east coasts. *See* METEOROLOGY.

Rainfall on land is 16–470 in. (400–12,000 mm) per year, with the highest rainfall being on the western windward slopes of the mountains, and the lowest on the eastern basins in the lee of the Southern Alps in Central Otago and south Canterbury. Annual rain days are at least 130 for most of North Island, but on South Island the totals are far more variable, with over 200 occurring in Fiordland, 180 on the west coast, and fewer than 80 in Central Otago. Summer droughts are relatively common in Northland, and in eastern regions of both islands. *See* DROUGHT; PRECIPITATION (METEOROLOGY).

Droughts, springtime air frosts, and hailstorms are the major common climatic hazards for the farming industry, but floods associated with prolonged intense rainstorms are the major general hazard.

The economy is heavily dependent on the natural resources soil, water, and plants. New Zealand has few exploitable minerals, but possesses a climate generally favorable for agriculture, pastoral farming, renewable forestry, and tourism. With a small population (3.4 million), much of its manufacturing is concerned with processing produce from the land and surrounding seas, and supplying the needs of those industries.

Because of its high relief and its location on an active crustal plate boundary in the zone of convergence between Antarctic air masses and tropical air masses, New Zealand is prone to high-intensity and high-frequency natural hazards—earthquakes, volcanic eruptions, large and small landslides, and floods. [M.J.Se.]

Newcastle disease An epizootic viral disease of fowls, with respiratory, gastrointestinal, and central nervous system involvement. It may be transmitted to humans who work with fowls, usually appearing as a conjunctivitis; recovery is complete in 10–14 days. The virus is related to other paramyxoviruses in hemagglutination characteristics, as well as in size and internal structure. *See* MYXOVIRUS; PARAMYXOVIRUS. [J.L.Me.]

Newtonian fluid A fluid in which the state of stress at any point is a linear function of the time rate of strain at that point. The fluid thus bears a direct analogy to a hookean solid, for which the state of stress is a linear function of the strain. Many gases and liquids are closely newtonian over a wide range of pressures and temperatures.

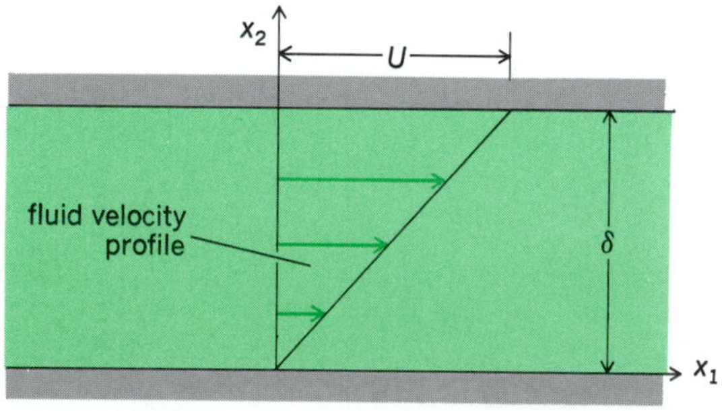

Newtonian fluid. Top plate moves relative to bottom one to produce Couette flow of intervening viscous fluid.

The simplest example of newtonian fluid flow is Couette flow, the low-speed steady motion of a viscous fluid between two infinite plates moving parallel to each other with relative velocity U, as illustrated. The shear stress in the fluid is constant and equals $\mu(\partial u_1/\partial x_2) = \mu(U/\delta)$, where δ is the distance between the plates, μ is the viscosity, u_i are velocity components of fluid, and x_i are rectangular cartesian coordinates with $i = 1, 2, 3$. *See* FLUID FLOW; FLUIDS. [A.E.Br.]

Newton's laws of motion Three fundamental principles which form the basis of classical, or newtonian, mechanics. They are stated as follows:

First law: A particle not subjected to external forces remains at rest or moves with constant speed in a straight line.

Second law: The acceleration of a particle is directly proportional to the resultant external force acting on the particle and is inversely proportional to the mass of the particle.

Third law: If two particles interact, the force exerted by the first particle on the second particle (called the action force) is equal in magnitude and opposite in direction to the force exerted by the second particle on the first particle (called the reaction force).

The newtonian laws have proved valid for all mechanical problems not involving speeds comparable with the speed of light and not involving atomic or subatomic particles. *See* DYNAMICS; FORCE; KINETICS (CLASSICAL MECHANICS). [D.Wi.]

Niacin A vitamin also known as nicotinic acid, a component of the vitamin B complex. It is a white water-soluble powder stable to heat, acid, and alkali, with the structure shown below.

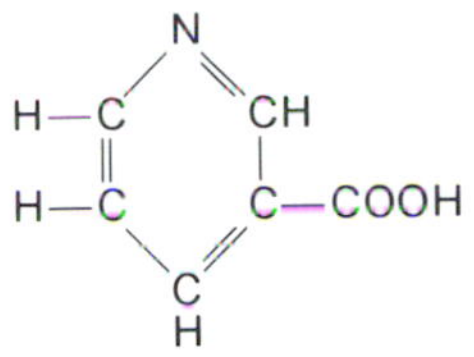

Many animals, including humans, are capable of synthesizing niacin in varying degrees from the amino acid tryptophan. Niacin is widely distributed in foods. Yeasts, wheat germ, and meats, particularly organ meats, are rich sources of the vitamin. Some foods such as milk are relatively poor sources of niacin but contain generous quantities of tryptophan. *See* TRYPTOPHAN.

Niacin-deficiency disease is known as pellagra and is particularly prevalent among the poor people whose diet is largely corn. The recommended dietary allowance for niacin is 6.6 niacin equivalents per 1000 kcal. *See* PELLAGRA; VITAMIN. [S.N.G.]

Nibbling The cutting of material by the action of a reciprocating punch. The nibbler takes repeated small bites as the work is passed beneath it. The workpiece must be backed up by a support or die.

Nibbling machines are constructed with considerable distance, or throat, between the punch and its supporting upright. This distance, plus the use of a round punch which allows the workpiece to be moved about, permits the cutting of irregular shapes. Duplicate pieces may be made by using templates as guides for the punch. [A.H.T.]

Niccolite A minor ore of nickel. Niccolite is a mineral having composition NiAs. Crystals are rare, and niccolite usually occurs in massive aggregates with metallic luster and pale copper-red color. Because of the color, not the composition, it is called copper nickel. The hardness is 5.5 on Mohs scale and the specific gravity is 7.78. It is found in vein deposits with cobalt and silver minerals as in the silver mines of Saxony, Germany, and Cobalt, Ontario, Canada. *See* NICKEL; PYRRHOTITE. [C.S.Hu]

Nickel A chemical element, Ni, atomic number 28, a silver-white, ductile, malleable, tough metal. The atomic mass of naturally occurring nickel is 58.71.

1	2	3	4	5	6	7	8	9	10	11	12	13	14	15	16	17	18
1 H																	2 He
3 Li	4 Be											5 B	6 C	7 N	8 O	9 F	10 Ne
11 Na	12 Mg											13 Al	14 Si	15 P	16 S	17 Cl	18 Ar
19 K	20 Ca	21 Sc	22 Ti	23 V	24 Cr	25 Mn	26 Fe	27 Co	28 Ni	29 Cu	30 Zn	31 Ga	32 Ge	33 As	34 Se	35 Br	36 Kr
37 Rb	38 Sr	39 Y	40 Zr	41 Nb	42 Mo	43 Tc	44 Ru	45 Rh	46 Pd	47 Ag	48 Cd	49 In	50 Sn	51 Sb	52 Te	53 I	54 Xe
55 Cs	56 Ba	71 Lu	72 Hf	73 Ta	74 W	75 Re	76 Os	77 Ir	78 Pt	79 Au	80 Hg	81 Tl	82 Pb	83 Bi	84 Po	85 At	86 Rn
87 Fr	88 Ra	103 Lr	104 Rf	105 Db	106 Sg	107 Bh	108 Hs	109 Mt	110	*111*	*112*	*113*	*114*	*115*	*116*	*117*	*118*

lanthanide series	57 La	58 Ce	59 Pr	60 Nd	61 Pm	62 Sm	63 Eu	64 Gd	65 Tb	66 Dy	67 Ho	68 Er	69 Tm	70 Yb
actinide series	89 Ac	90 Th	91 Pa	92 U	93 Np	94 Pu	95 Am	96 Cm	97 Bk	98 Cf	99 Es	100 Fm	101 Md	102 No

Nickel consists of five natural isotopes having atomic masses of 58, 60, 61, 62, 64. Seven radioactive isotopes have also been identified, having mass numbers of 56, 57, 59, 63, 65, 66, and 67.

Most commercial nickel goes into stainless steel and other corrosion-resistant alloys. Nickel is also important in coins as a replacement for silver. Finely divided nickel is used as a hydrogenation catalyst. *See* NICKEL ALLOYS.

Nickel is a fairly plentiful element, making up about 0.008% of the Earth's crust and 0.01% of the igneous rocks. Appreciable quantities of nickel are present in some kinds of meteorite, and large quantities are thought to exist in the Earth's core. Two important ores are the iron-nickel sulfides, pentlandite and pyrrhotite $(Ni,Fe)_xS_y$; the ore garnierite, $(Ni,Mg)SiO_3 \cdot nH_2O$, is also commercially important. Nickel occurs in small quantities in plants and animals. It is present in trace amounts in sea water, petroleum, and most coal.

Nickel metal is of moderate strength and hardness (3.8 on Mohs scale). When viewed as very small particles, nickel appears black. The density of nickel is 8.90 times that of water at 20°C (68°F). Nickel melts at 1455°C (2651°F) and boils at 2840°C (5144°F). Nickel is only moderately reactive. It resists alkaline corrosion and does not burn in the massive state, although fine nickel wires can be ignited. Nickel is above hydrogen in the electrochemical series, and it dissolves slowly in dilute acids, releasing hydrogen. In metallic form nickel is a moderately strong reducing agent.

Nickel is usually dipositive in its compounds, but it can also exist in the oxidation states 0, 1+, 3+, and 4+. Besides the simple nickel compounds, or salts, nickel forms a variety of coordination compounds or complexes. Most compounds of nickel are green or blue because of hydration or other ligand bonding to the metal. The nickel ion present in water solutions of simple nickel compounds is itself a complex, $[Ni(H_2O)_6]^{2+}$. [W.E.C.]

Nickel alloys Combinations of nickel with other metals. Nickel-base alloys may be melted in open-hearth, electric-arc, or induction furnaces in air, under inert gas, or in vacuum. Casting may also be done under these same ambient conditions.

Nickel 211 and Duranickel alloy 301 are essentially binary alloys with 4.75% manganese and 4.5% aluminum, respectively. A characteristic use of nickel 211 is as wire for sparkplug electrodes. Duranickel alloy 301 is well suited to the manufacture of springs and diaphragms.

Monel alloy 400 contains about two-thirds nickel and one-third copper and is the oldest of the commercial nickel-base alloys, dating from about 1905. Nickel-chromium binary alloys are used primarily in specialty high-temperature service. Other nickel alloys contain combinations of other metals in varying quantities. *See* NICKEL. [E.N.S.; G.Sm.]

Nickel metallurgy The extraction and refining of nickel from its ores. Nickel's properties of strength, toughness, and resistance to corrosion have been used to advantage in alloys since ancient times. Although nickel ranks twenty-fourth in order of abundance of the elements, there are relatively few nickel deposits of commercial importance. Nickel ores are of two generic types, sulfides and laterites. Explorations of the ocean bottoms have revealed vast deposits of manganese oxide nodules which contain significant values of nickel, copper, and cobalt.

Selection of processes for nickel extraction is largely determined by the type of ore to be treated. Sulfide ores are amenable to concentration by such methods as flotation or magnetic separation. The state of combination of nickel in the lateritic ores usually precludes such enrichment, thus requiring treatment of the total ore.

Sulfide ores are first crushed and ground to liberate the mineral values and then subjected to froth flotation or magnetic separation to concentrate the valuable constituents and reject the gangue or rock fraction. The nickel concentrate is treated by pyrometallurgical processes. The major portion undergoes partial roasting in multihearth or fluidized-bed furnaces to eliminate about half of the sulfur and to oxidize the associated iron. The hot calcine, plus flux, is smelted in natural gas–coal-fired reverberatory furnaces operating at about 1200°C to produce a furnace matte, enriched in nickel, and a slag for discard. The furnace matte is transferred to converters and blown with air in the presence of more flux to oxidize the remaining iron and associated sulfur, yielding Bessemer matte containing nickel, copper, cobalt, small amounts of precious metals, and about 22% sulfur. The molten Bessemer matte is cast into 25-ton (22.5-metric ton) molds in which it undergoes controlled slow cooling. After crushing and grinding, the metallics are removed magnetically and treated in a refining complex for recovery of metal values.

The bulk of the nickel originating from lateritic ores is marketed as ferronickel. The process employed is basically simple and involves drying and preheating the ore usually under reducing conditions. The hot charge is then further reduced and melted in an electric-arc furnace, and the crude metal is refined and cast into ferronickel pigs. A substantial amount of nickel is produced from lateritic ores by the nickel sulfide matte technique. In this process the ore is mixed with gypsum or other sulfur-containing material such as high-sulfur fuel oil, followed by a reduction and smelting operation to form matte. The molten furnace matte is upgraded in either conventional or top-blown rotary converters to a high-grade matte, which can be further refined by roasting and reduction to a metallized product. *See* NICKEL; PYROMETALLURGY. [A.I.]

Nicotinamide adenine dinucleotide (NAD) An organic coenzyme and one of the most important components of the enzymatic systems concerned with biological oxidation-reduction reactions. It is also known as NAD, diphosphopyridine nucleotide (DPN) coenzyme I, and codehydrogenase I. NAD is found in the tissues of all living organisms. *See* COENZYME.

The nicotinamide, or pyridine, portion of NAD can be reduced chemically or enzymatically with the formation of reduced or hydrogenated NAD (NADH). NAD functions as the immediate oxidizing agent for the oxidation, or dehydrogenation, of various organic compounds in the presence of appropriate dehydrogenases, which are specific apoenzymes, or protein portions of the enzyme. In the dehydrogenase reactions one hydrogen atom is transferred from the substrate to NAD, while another is liberated as hydrogen ion.

NAD and its reduced form, NADH, serve to couple oxidative and reductive processes and are constantly regenerated during metabolism. Hence, they serve as catalysts and NAD is referred to as a coenzyme. In some enzymatic reactions a different coenzyme, triphosphopyridine nucleotide is required. Dehydrogenases are generally quite specific with respect to the coenzyme which they can utilize. *See* ENZYME; NICOTINAMIDE ADENINE DINUCLEOTIDE PHOSPHATE (NADP). [M.D.]

Nicotinamide adenine dinucleotide phosphate (NADP) A coenzyme and an important component of the enzymatic systems concerned with biological oxidation-reduction systems. It is also known as NAD, triphosphopyridine nucleotide (TPN), coenzyme II, and codehydrogenase II. The compound is similar in structure and function to nicotinamide adenine dinucleotide (NAD). It differs structurally from NAD in having an additional phosphoric acid group esterified at the 2′ position of the ribose moiety of the adenylic acid portion. In biological oxidation-reduction reactions the NADP molecule becomes alternately reduced to its hydrogenated form (NADPH) and reoxidized to its initial state (see illustration). *See* CARBOHYDRATE METABOLISM; COENZYME; ENZYME. [M.D.]

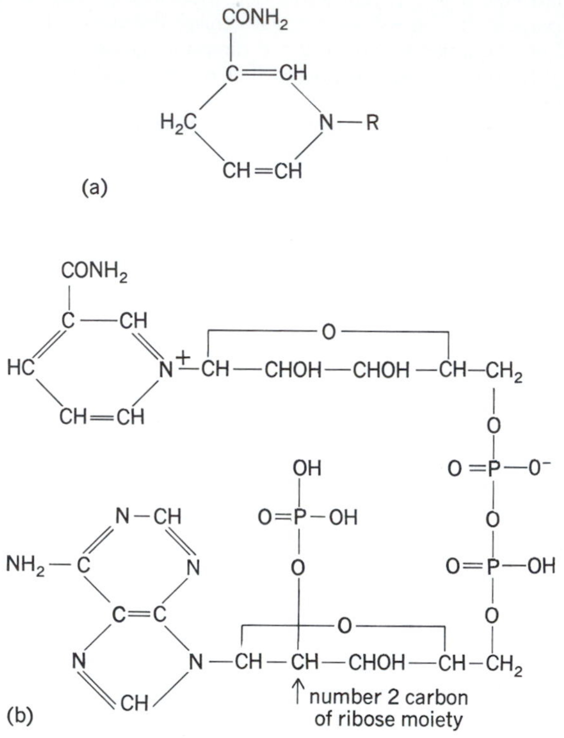

Triphosphopyridine nucleotide. (*a*) The reduced form of the nicotinamide portion of TPNH. (*b*) The oxidized form of the molecule (TPN).

Nicotine A liquid alkaloid from the dried leaves of tobacco. Present in *Nicotiana tabacum* and *N. rustica*, nicotine is recovered by extraction of cigarette wastes. Chemically it is $C_{10}H_{14}N_2$, a colorless to amber oil which takes up moisture from the air very avidly. It has a strong tobacco odor and intensely bitter, burning taste. Quite soluble in water, nicotine forms salts with mineral acids, such as hydrochloride, hydroiodide, and sulfate. The sulfate is used as an agricultural insecticide.

Nicotine is highly toxic. Symptoms include extreme nausea, vomiting, evacuation of intestine and urinary bladder, mental confusion, and convulsions. *See* ALKALOID. [F.W.]

Niobium A chemical element, Nb, atomic number 41 and atomic weight 92.906. In the United States this element was originally called columbium. The metallurgists and metals industry still use this older name.

Most niobium is used in special stainless steels, high-temperature alloys, and superconducting alloys such as Nb_3Sn. Niobium is also used in nuclear piles.

Niobium metal has a density of 8.6 g/cm³ (5.0 oz/in.³) at

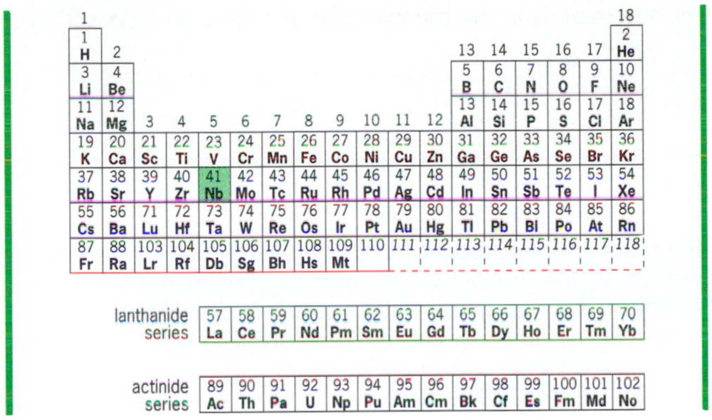

20°C (68°F), a melting point of 2468°C (4474°F), and a boiling point of 4927°C (8900°F). Metallic niobium is quite inert to all acids except hydrofluoric, presumably owing to an oxide film on the surface. Niobium metal is slowly oxidized in alkaline solution. It reacts with oxygen and the halogens upon heating to form the oxidation state V oxide and halides, with nitrogen to form NbN, and with carbon to form NbC, as well as other elements such as arsenic, antimony, tellurium, and selenium.

The oxide Nb_2O_5, melting point 1520°C (2768°F), dissolves in fused alkali to yield a soluble complex niobate, $Nb_6O_{19}^{8-}$. Normal niobates such as NbO_4^{3-} are insoluble. The oxide dissolves in hydrofluoric acid to give ionic species such as $NbOF_5^{2-}$ and $NbOF_6^{3-}$, depending on the fluoride and hydrogen-ion concentration. The highest fluoro complex which can exist in solution is NbF_6^-. [E.M.L.]

Nippotaeniidea An order of tapeworms of the subclass Cestoda. The few known species are intestinal parasites of Eurasian fresh-water fishes. The head bears a single terminal sucker. The segmental anatomy shows relationships to the Pseudophyllidea and Cyclophyllidea. The life history is unknown. It is probable that this order is related to the proteocephalids. *See* CESTODA; CYCLOPHYLLIDEA; PSEUDOPHYLLIDEA. [C.P.R.]

Niter A potassium nitrate mineral with chemical composition KNO_3. Niter crystallizes in the orthorhombic system, generally in thin crusts and delicate acicular crystals; it occurs in massive, granular, or earthy forms. It is brittle; hardness is 2 on Mohs scale; specific gravity is 2.109. The luster is vitreous, and the color and streak are colorless to white. *See* NITRATE MINERALS.

Niter is commonly found, usually in small amounts, as a surface efflorescence in arid regions and in caves and other sheltered places. Niter occurs associated with soda niter in the desert regions of northern Chile, and in similar occurrences in Italy, Egypt, Russia, the western United States, and elsewhere. [G.Sw.]

Nitrate The negative ion NO_3^-, derived from nitric acid, HNO_3. Because almost all metallic nitrates are water-soluble, they are not found in nature but are produced from nitric acid. One notable exception to this is the impure sodium nitrate, called Chile saltpeter, which occurs in large amounts because of the aridity and very small rainfall of the Chilean coastal plain.

Because nitrates contain nitrogen in its highest oxidation state (5+), the ion is a useful oxidizing agent. Because of this property, nitrates are often constituents of matches and explosives. Nitrates are an important source of nitrogen in fertilizers. *See* FERTILIZER; NITRIC ACID; NITROGEN. [E.E.W.]

Nitrate minerals These minerals are few in number and with the exception of soda niter are of rare occurrence. Normal anhydrous and hydrated nitrates occurring as minerals are soda niter, $NaNO_3$; niter, KNO_3: ammonia niter, NH_4NO_3; nitrobarite, $Ba(NO_3)_2$; nitrocalcite, $Ca(NO_3)_2 \cdot 4H_2O$; and nitromagnesite, $Mg(NO_3)_2 \cdot 6H_2O$. In addition there are three known naturally occurring nitrates containing hydroxyl or halogen, or compound nitrates. They are gerhardtite, $Cu_2(NO_3)(OH)_3$; buttgenbachite, $Cu_{19}(NO_3)_2Cl_4(OH)_{32} \cdot 3H_2O$; and darapskite, $Na_3(NO_3)(SO_4) \cdot H_2O$. *See* NITER; SODA NITER.

The natural nitrates are for the most part readily soluble in water. For this reason they occur most abundantly in arid regions, particularly in South America along the Chilean coast. *See* FERTILIZER; NITROGEN. [G.Sw.]

Nitration An important chemical conversion (unit process) in which a nitro group (—NO_2) is introduced into a hydrocarbon compound. Three types of nitration are common, and these are classified according to their chemical structure:

1. C-nitration in which the nitro group attaches itself to a carbon atom, as indicated in reaction (1).

$$>C{-}H + HNO_3 \rightarrow\ >C{-}NO_2 + H_2O \qquad (1)$$

2. O-nitration (really esterification) in which an O-N bond is formed to produce a nitrate, as seen in reaction (2).

$$>C{-}OH + HNO_3 \rightarrow\ >C{-}O{-}NO_2 + H_2O \qquad (2)$$

3. N-nitration, in which there is formation of a N-N bond, such as shown by reaction (3).

Nitration of aromatics is employed commercially for the production of numerous nitroaromatics. Explosives such as TNT, tetryl, ammonium picrate, and cyclonite (RDX) are produced by nitration. Olefins can be nitrated with nitric acid to produce nitroolefins, nitro alcohols, and the subsequent nitronitrate esters. *See* NITRIC ACID; NITRO AND NITROSO COMPOUNDS; NITROPARAFFIN; OXIDIZING AGENT. [L.F.A.; R.N.S.]

Nitric acid A strong mineral acid having the formula HNO_3. Pure nitric acid is a colorless liquid with a specific gravity of 1.52 at 25°C (77°F); it freezes at −47°C (−53°F). Nitric acid is used in the manufacture of ammonium nitrate and phosphate fertilizers, nitro explosives, plastics, dyes, and lacquers. The principal commercial process for the manufacture of nitric acid is the Ostwald process, in which ammonia, NH_3, is catalytically oxidized with air to form nitrogen dioxide, NO_2. When the dioxide is dissolved in water, 60% nitric acid is formed. Production of 90–100% nitric acid is based on processes such as the reaction of sulfuric acid with sodium nitrate (an older method of nitric acid manufacture), dehydration of 60% acid, and oxidation of nitrogen dioxide in a solution of dilute nitric acid. *See* AMMONIA; NITROGEN. [F.J.J.]

Nitride A binary compound of nitrogen with elements less electronegative than nitrogen. Common practice, however, is to exclude azides (such as NaN_3), in which specialized bonding exists among the nitrogen atoms, and the binary compounds with hydrogen, the halogens, and the oxygen group elements. With this limitation, essentially solid nitrides will be encountered.

Direct reaction of alkali metals with nitrogen yields azides, which can be decomposed by heating to form the nitrides Li_3N, Na_3N, K_3N, Rb_3N, and Cs_3N. These nitrides decompose to nitrogen gas and the elements in the vicinity of 400°C

(752°F); they react with water vapor to liberate ammonia (NH_3) and form the metal hydroxides.

The nitrides of the alkaline earth elements (Be_3N_2, Mg_3N_2, Ca_3N_2, Sr_3N_2, and Ba_3N_2) may be prepared by direct reaction of the element with nitrogen or with ammonia at elevated temperatures. The other metals also form nitrides except for the elements Ag, Au, Hg, Tl, Sn, Pb, Sb, and Bi. *See* AZIDE; NITROGEN; OXIDE. [R.K.E.]

Nitrile One of a group of organic chemical compounds of general formula RC≡N. A nitrile is named from the acid to which it can be hydrolyzed by adding the suffix -onitrile to the acid stem, for example, acetonitrile from acetic acid. An alternative system names the group attached to CN, thus CH_3CN is also named methyl cyanide. In more complex structures the CN group is named as a substituent, cyano.

Industrially, nitriles are formed by heating carboxylic acids with ammonia and a dehydration catalyst under pressure. For the preparation of acrylonitrile, which is used on a large scale in the plastics industry, a vapor-phase catalytic ammoxidation of propylene has been developed. *See* ACRYLONITRILE; AMINE; CARBOXYLIC ACID. [P.E.F.]

Nitrite The negative ion NO_2^- derived from the unstable nitrous acid HNO_2. Because of the intermediate oxidation state of nitrogen in nitrites (3+), the ion can act as either an oxidizing or reducing agent. Most nitrites are water-soluble. Sodium and potassium nitrites are quite stable and find extensive use in dyestuff manufacture and organic synthesis. *See* NITRATE; NITROGEN. [E.E.W.]

Nitro and nitroso compounds Nitro compounds are derivatives of organic hydrocarbons having one or more —NO_2 groups with nitrogen-to-carbon bonding. They differ from the oxygen-linked nitrites, which are esters. The group lacks enough electrons to form double bonds with both oxygens. However, both oxygens react alike; hence the bond is regarded as a resonance hybrid of single and double bonds.

Aromatic nitro compounds have been used chiefly as dye intermediates, explosives, and pharmaceuticals. They are formed readily by the reaction of aromatic compounds with nitric acid; H is replaced by the —NO_2 group, for example,

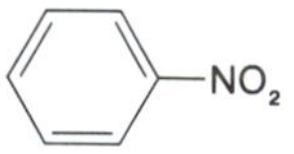

Aliphatic nitro compounds are prepared with difficulty and have grown in importance only since the development of vapor-phase nitration of hydrocarbons with nitric acid vapors at 420°C (788°F).

Nitroso compounds contain the —NO group attached to carbon or nitrogen. Many are unstable intermediates, for example, nitrosobenzene formed during the reduction of nitrobenzene. *See* NITRATION. [A.L.H.]

Nitroaromatic compound A member of the class of organic compounds in which the nitro group (—NO_2) is attached directly to the cyclic, aromatic nucleus. The prototypal compound is nitrobenzene. It is prepared by the reaction of benzene with nitric acid in the presence of sulfuric acid, as shown in the reactions below. The most significant use of

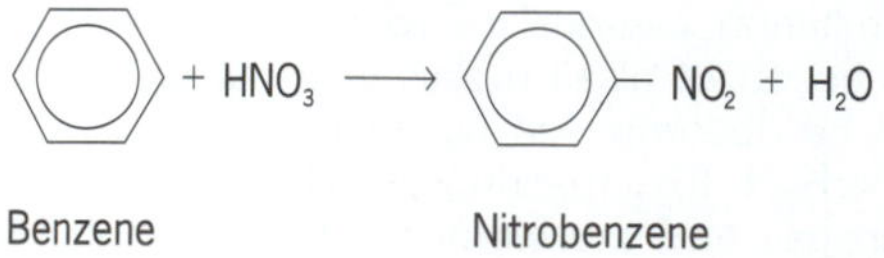

Benzene Nitrobenzene

nitrobenzene is in the manufacture of aniline (I). About 97% of

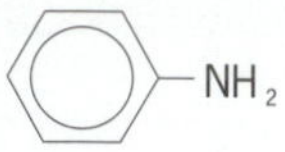

(I)

the nitrobenzene produced in the United States is converted to aniline, which is used in the manufacture of plastics, rubber additives, dyes, drugs, and other products. *See* BENZENE.

Second to nitrobenzene in commercial importance are the mononitrotoluenes, particularly the ortho and para isomers, (II) and (III), respectively. Reaction of toluene with a mixture of

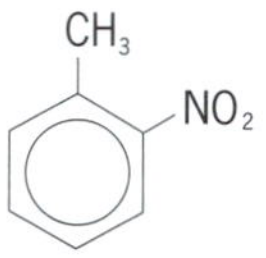

(II)

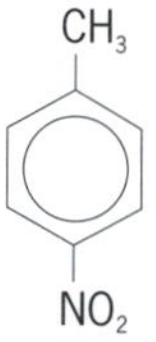

(III)

nitric and sulfuric acid at about 40°C (104°F) gives a high yield of a mixture of the isomers, which are separated by a combination of fractional distillation and crystallization. The nitrotoluenes are important intermediates in the preparation of dyes, rubber chemicals, and agricultural chemicals.

2,4,6-Trinitrotoluene (TNT; IV) is a military explosive that is

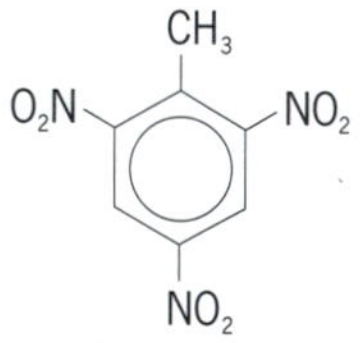

(IV)

stable, nonhygroscopic, and relatively insensitive to impact, friction, shock, and electric spark. It is produced by nitration of toluene in successive stages at progressively higher temperatures and concentrations of acid.

Although literally thousands of other aromatic ring compounds, including the heterocyclics, have been converted to their nitro derivatives, few such compounds have achieved any significant industrial importance. *See* AROMATIC HYDROCARBON; NITRATION. [P.E.F.]

Nitrobacteriaceae A family of gram-negative chemolithotrophic bacteria which live autotrophically, derive energy from the oxidation of either ammonia to nitrite, or nitrite to nitrate, and obtain carbon for growth from carbon dioxide. They are the agents of nitrate formation in nature. Seven genera are recognized. Four genera oxidize ammonia to nitrite: *Nitrosomonas*, *Nitrosococcus*, *Nitrosospira*, and *Nitrosolobus*. Three genera oxidize nitrite to nitrate: *Nitrobacter*, *Nitrospina*, and *Nitrococcus*. *See* CHEMOLITHOTROPHIC BACTERIA; NITROGEN CYCLE; SOIL MICROBIOLOGY. [J.Me.]

Nitrofuran One of a family of yellow crystalline compounds possessing antimicrobial activity, characterized by a nitro group in the 5-position of a 2-substituted furan ring. Most nitrofurans have a broad antibacterial spectrum, while relatively few exhibit clinically useful antifungal and antiprotozoan activity.

Nitrofurans of importance in medicine include nitrofurazone, mainly used as a topical antibacterial agent; furazolidone, used primarily for its local antitrichomonal action in the vagina; nifuroxime, a topical antifungal agent; and nitrofurantoin, used particularly for genitourinary tract infections. *See* CHEMOTHERAPY. [J.J.J.]

Nitrogen A chemical element, N, atomic number 7, atomic weight 14.0067. Nitrogen is a gas under normal conditions.

Molecular nitrogen is the principal constituent of the atmosphere (78% by volume of dry air), in which its concentration is a result of the balance between the fixation of atmospheric nitrogen by bacterial, electrical (lightning), and chemical (industrial) action, and its liberation through the decomposition of organic materials by bacteria or combustion. In the combined

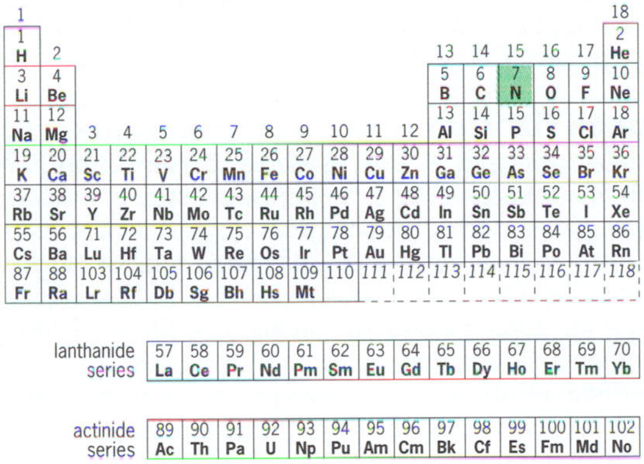

state, nitrogen occurs in a variety of forms. It is a constituent of all proteins (both plant and animal) as well as of many other organic materials. Its chief mineral source is sodium nitrate.

Because of the importance of nitrogen compounds in agriculture and the chemical industry, much of the industrial interest in elementary nitrogen has been in processes for converting elemental nitrogen into nitrogen compounds. Nitrogen is also used for filling bulbs of incandescent lamps and, in general, wherever a relatively inert atmosphere is required.

Nitrogen, as it occurs in nature, consists of two isotopes, ^{14}N and ^{15}N, in the abundance ratio of 99.635 to 0.365. In addition the radioactive isotopes ^{12}N, ^{13}N, ^{16}N, and ^{17}N have been made by a variety of nuclear reactions. At standard temperature and pressure, elemental nitrogen exists as a gas with a density of 1.25046 g per liter. The molecular formula is N_2. Some physical properties of elemental nitrogen are listed in Table 1.

Elemental nitrogen has a low reactivity toward most common substances at ordinary temperatures. At high temperatures, molecular nitrogen reacts with chromium, silicon, titanium, aluminum, boron, beryllium, magnesium, barium, strontium, calcium, and lithium to form nitrides; with O_2 to form NO; and at moderately high temperatures and pressures in the presence of a catalyst, with hydrogen to form ammonia. Above 1800°C (3270°F), nitrogen, carbon, and hydrogen combine to form hydrogen cyanide.

Table 1. Properties of nitrogen

Property	Value
Heat of transformation (α–β)	54.71 cal/mole
Heat of fusion	172.3 cal/mole
Heat of vaporization	1332.9 cal/mole
Critical temperature	126.26 ± 0.04 K
Critical pressure	33.54 ± 0.02 atm
Density: α-form	1.0265 g/ml at −252.6°C
β-form	0.08792 g/ml at −210.0°C
Liquid	1.1607−0.0045*T* (*T* = abs temp)

When molecular nitrogen is subjected to the actions of a condensed electrode discharge or to a high-frequency electrodeless discharge, it is partially changed to an activated, unstable condition, from which on standing it returns to its normal state with the emission of a golden-yellow afterglow.

The elements of the nitrogen family exhibit in their compounds three principal oxidation states, −3, +3, and +5, though other oxidation states are also exhibited. All the elements of the nitrogen family form hydrides of the type shown in the formula below and +3 oxides, +5 oxides, +3 halides

$$\begin{array}{c} \mathrm{H} \\ \mathrm{H}:\ddot{\mathrm{M}}:\mathrm{H} \\ \cdot\cdot \end{array}$$

(MX_3), and, except for nitrogen and bismuth, + 5 halides (MX_5). Nitrogen is the most electronegative element of the nitrogen family. Table 2 lists the principal classes of inorganic

Table 2. Compounds of nitrogen

Oxidation state	Examples
+5	N_2O_5, HNO_3, nitrates, NO_2X
+4	$N_2O_4 \rightleftharpoons 2NO_2$
+3	N_2O_3, HNO_2, nitrites, NOX, NX_3
+2	NO, Na_2NO_2, nitrohydroxylamates
+1	N_2O, $H_2N_2O_2$, hyponitrites
0	N_2
−1/3	HN_3, azides
−1	NH_2OH, hydroxylammonium salts
−2	NH_2NH_2, hydrazinium salts, hydrazides
−3	NH_3, ammonium salts, amides, imides, nitrides

nitrogen compounds. Thus, in addition to the typical oxidation states of the family (−3, +3, and +5), nitrogen forms compounds with a variety of additional oxidation states. *See* AMINE; AMMINE; AMMONIA; HYDRAZINE; NITRIC ACID; NITRIDE. [H.H.S.]

Compounds containing the dinitrogen molecule, N2, bound to a metal are called nitrogen complexes or dinitrogen complexes. Outstanding in their ability to form coordination compounds with nitrogen are a number of metals which belong to the group VIII transition metal family. For each metal of this group, several nitrogen complexes have been identified. Nitrogen complexes of these metals occur in low oxidation states, such as Co(I) or Ni(O). The other ligands present in these complexes besides N_2 are usually of a type known to stabilize low oxidation states; phosphines appear to be particularly prominent bonding partners in this respect. The illustration shows the structure of a typical N_2 complex, elucidated by a crystal structure determination. The N-N bond axis in this complex is

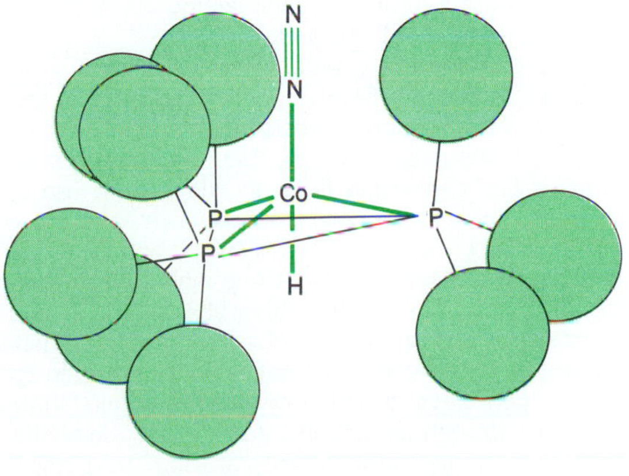

Structure of a coordination compound with N_2 (colored circles represent phenyl groups).

aimed, within the limits of experimental error, directly toward the position of the metal atom. *See* COORDINATION CHEMISTRY; COORDINATION COMPLEXES. [H.B.]

Nitrogen cycle The collective term given to the natural biological and chemical processes through which inorganic and organic nitrogen are interconverted. It includes the process of ammonification, ammonia assimilation, nitrification, nitrate assimilation, nitrogen fixation, and denitrification.

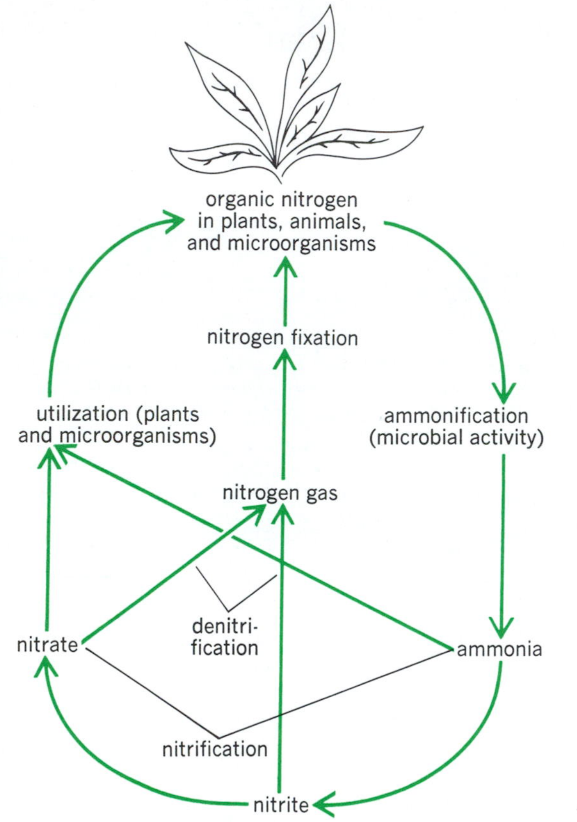

Diagram of the nitrogen cycle.

Nitrogen exists in nature in several inorganic compounds, namely N_2, N_2O, NH_3, NO_2^-, and NO_3^-, and in several organic compounds such as amino acids, nucleotides, amino sugars, and vitamins. In the biosphere, biological and chemical reactions continually occur in which these nitrogenous compounds are converted from one form to another. These interconversions are of great importance in maintaining soil fertility and in preventing pollution of soil and water.

An outline showing the general interconversions of nitrogenous compounds in the soil-water pool is presented in the illustration. There are three primary reasons why organisms metabolize nitrogen compounds: (1) to use them as a nitrogen source, which means first converting them to NH_3, (2) to use certain nitrogen compounds as an energy source such as in the oxidation of NH_3 to NO_2^- and of NO_2^- to NO_3^-, and (3) to use certain nitrogen compounds (NO_3^-) as terminal electron acceptors under conditions where oxygen is either absent or in limited supply. The reactions and products involved in these three metabolically different pathways collectively make up the nitrogen cycle.

There are two ways in which organisms obtain ammonia. One is to use nitrogen already in a form easily metabolized to ammonia. Thus, nonviable plant, animal, and microbial residues in soil are enzymatically decomposed by a series of hydrolytic and other reactions to yield biosynthetic monomers such as amino acids and other small-molecular-weight nitrogenous compounds. These amino acids, purines, and pyrimidines are decomposed further to produce NH_3 which is then used by plants and bacteria for biosynthesis, or these biosynthetic monomers can be used directly by some microorganisms. The decomposition process is called ammonification.

The second way in which inorganic nitrogen is made available to biological agents is by nitrogen fixation (this term is maintained even though N_2 is now called dinitrogen), a process in which N_2 is reduced to NH_3. Since the vast majority of nitrogen is in the form of N_2, nitrogen fixation obviously is essential to life. The N_2-fixing process is confined to prokaryotes (certain photosynthetic and nonphotosynthetic bacteria). The major nitrogen fixers (called diazotrophs) are members of the genus *Rhizobium*, bacteria that are found in root nodules of leguminous plants, and of the cyanobacteria (originally called blue-green algae). *See* NITROGEN FIXATION. [L.E.Mo.]

Nitrogen fixation The conversion of dinitrogen (N_2) from the atmosphere into a form available to plants and hence to animals and humans. This is an important step in the terrestrial nitrogen cycle. Both chemical and biological processes contribute, to turn over some 2×10^8 metric tons of nitrogen per year over the planet as a whole: industrial fertilizer production accounts for about one-third; roughly two-thirds of the global input of fixed nitrogen arises from the biological process.

Chemical fixation. Three basic processes are used that give different products containing fixed nitrogen. All require a considerable input of thermal or electrical energy.

In arc processes air is passed through an electric arc and about 1% nitric oxide is formed, which can be converted to nitrates by established chemical procedures. In the widely used Haber-Bosch process, hydrogen is generated from natural gas or petroleum, mixed with air, and burned to yield a nitrogen-hydrogen mixture. This reacts (under 100–800 atm pressure [1 atm = 10^5 pascals] at 200–700°C or 390–1290°F with a metal oxide catalyst) to give ammonia. In the cyanamide process, calcium cyanamide is generated by heating calcium carbide in nitrogen; when moistened, it hydrolyzes to give urea and ammonia. These are the major industrial processes for obtaining nitrogenous fertilizer. *See* AMMONIA; CYANAMIDE; ELECTROCHEMICAL PROCESSES; FERTILIZER.

Biological fixation. Only prokaryotes (bacteria and blue-green algae) fix nitrogen. Nitrogen-fixing microbes fall into two main groups, free-living and symbiotic.

The free-living microbes can be subclassified. The aerobes require air for growth and fix nitrogen where air is present. Nitrogen-fixing, free-living facultative bacteria are widespread: the bacterial genera *Klebsiella* and *Bacillus* include many strains that can fix nitrogen, but they do so only in the absence of air. They grow perfectly well in air if prefixed nitrogen such as ammonium salt is present, but they do not then fix nitrogen. Some *Clostridium* strains, some methanogenic bacteria, and some sulfate-reducing bacteria are free-living nitrogen-fixing anaerobes; none grows in the presence of air, whether they fix nitrogen or not. Finally there are the free-living phototrophs: many blue-green algae fix nitrogen in air; many photosynthetic bacteria can fix nitrogen, but only when growing anaerobically. The blue-green algae are ecologically the most important of the free-living, nitrogen-fixing organisms, particularly in regard to the productivity of arid, tundra, and waterlogged regions (such as rice paddies).

The best-known symbiotic bacteria belong to the genus *Rhizobium*. Species of *Rhizobium* colonize the roots of leguminous plants and stimulate the formation of nodules, within which they fix nitrogen aerobically. Rhizobial symbioses are of great agricultural importance in soybean, pulse, and similar legume crops. Of comparable ecological importance are the symbiotic endophytes of nonlegumes, such as *Alnus* (alder), *Casuarina* (Australian pine), or *Ceanothus*, which are found in root nodules, where they fix nitrogen. *See* SOIL MICROBIOLOGY. [J.Po.]

Nitroparaffin One of a series of aliphatic compounds with relatively high boiling points and high molecular polarity. Such compounds are used as solvents for cellulose derivatives, waxes, and many substances of high molecular weight. Their chief use is in the synthesis of other organic compounds. The commercially available nitroparaffins include nitromethane, nitroethane, 1-nitropropane, and 2-nitropropane.

Polynitroparaffins such as 2,2-dinitropropane increase the octane number in diesel fuels and also reduce the smoke nuisance in diesel motors. *See* NITRATION. [L.B.C.]

Nobelium A chemical element, No, atomic number 102. Nobelium is a synthetic element produced in the laboratory. It

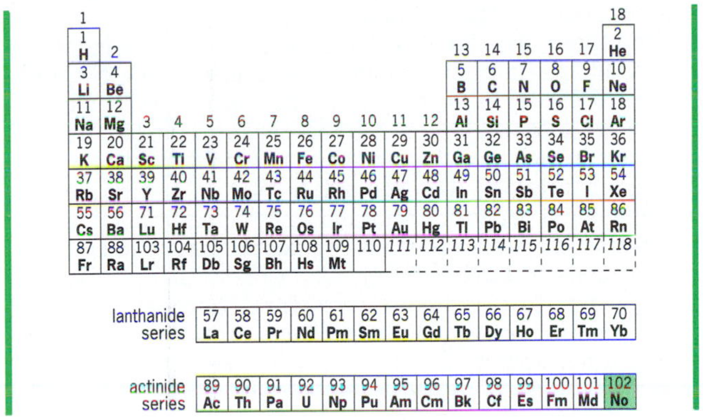

1
18
1 H
2
2 He
13 14 15 16 17
3 Li 4 Be
5 B 6 C 7 N 8 O 9 F 10 Ne
11 Na 12 Mg
3 4 5 6 7 8 9 10 11 12
13 Al 14 Si 15 P 16 S 17 Cl 18 Ar
19 K 20 Ca 21 Sc 22 Ti 23 V 24 Cr 25 Mn 26 Fe 27 Co 28 Ni 29 Cu 30 Zn 31 Ga 32 Ge 33 As 34 Se 35 Br 36 Kr
37 Rb 38 Sr 39 Y 40 Zr 41 Nb 42 Mo 43 Tc 44 Ru 45 Rh 46 Pd 47 Ag 48 Cd 49 In 50 Sn 51 Sb 52 Te 53 I 54 Xe
55 Cs 56 Ba 71 Lu 72 Hf 73 Ta 74 W 75 Re 76 Os 77 Ir 78 Pt 79 Au 80 Hg 81 Tl 82 Pb 83 Bi 84 Po 85 At 86 Rn
87 Fr 88 Ra 103 Lr 104 Rf 105 Db 106 Sg 107 Bh 108 Hs 109 Mt 110 111 112 113 114 115 116 117 118
lanthanide series
57 La 58 Ce 59 Pr 60 Nd 61 Pm 62 Sm 63 Eu 64 Gd 65 Tb 66 Dy 67 Ho 68 Er 69 Tm 70 Yb
actinide series
89 Ac 90 Th 91 Pa 92 U 93 Np 94 Pu 95 Am 96 Cm 97 Bk 98 Cf 99 Es 100 Fm 101 Md 102 No

decays by emitting an alpha particle, that is, a doubly charged helium ion. Only atomic quantities of the element have been produced to date. Nobelium is the tenth element heavier than uranium to be produced synthetically. It is the thirteenth member of the actinide series, a rare-earth-like series of elements. *See* ACTINIDE ELEMENTS: RADIOACTIVITY; RARE-EARTH ELEMENTS; TRANSURANIUM ELEMENTS. [P.R.F.]

Nocardia A genus of gram-positive, nonmotile, aerobic, pathogenic or saprophytic, mycelium-forming bacteria belonging to the so-called nocardioform actinomycetes. The principal natural habitat of the nocardiae is the soil, where they are abundant and widely distributed and probably take part in the remineralization of organic matter. *Nocardia* species also occur in aquatic habitats, and have been implicated in the biodeterioration of rubber joints in water and sewage pipes. Populations of *Nocardia* species have been isolated from foam on the surface of aeration tanks in activated-sludge sewage-treatment plants.

Nocardial infections, usually referred to as nocardiosis and actinomycetoma, are usually acquired from the environment by inhalation or aspiration of dust particles contaminated with the nocardiae or by traumatic inoculation of the organisms into the skin or bloodstream. Antibacterial drugs form the therapeutic basis for treating nocardial infections. *See* ANTIBIOTIC; NOCARDIOSIS. [K.P.S.]

Nocardiosis An actinomycetous disease of humans caused by several species of *Nocardia*. The lesions are purulent and granulomatous, usually involving the subcutaneous tissues, although a pulmonary form is also recognized. The latter form of nocardiosis is the most serious, often metastasizing to the viscera and central nervous system. *Nocardia asterioides*, the primary etiological agent, is easily found in the soil. [L.D.H.]

Nociceptors Cutaneous receptors that have bare nerve endings in or beneath the skin and are selectively responsive to noxious stimuli described as painful when applied to human skin. There are different sets of thinly myelinated nerve fibers and unmyelinated fibers whose endings are differentially responsive to only noxious thermal stimuli (hot or cold, or both) or only to noxious mechanical stimuli (such as pricking or pinching the skin). Other types of nociceptors are less specific, and the most common of these is one which responds to any noxious stimulus, whether it be chemical, thermal, or mechanical, and thus is called the polymodal nociceptor. Following injury to the skin, such as that produced by a superficial burn, the threshold responses to mechanical or thermal stimuli of many nociceptors may be greatly lowered. In such a state of heightened sensitivity to pain, termed hyperalgesia, normally innocuous stimulation such as gentle rubbing or warming can evoke a sensation of pain. *See* CUTANEOUS SENSATION. [R.La.M.]

Noctilucent clouds High-altitude clouds which are seen at the mesopause and are still illuminated by the Sun when it is already well below the observer's horizon during astronomical twilight. Noctilucent clouds occur only during the summer months at latitudes from about 45 to 70°. Generally, they resemble high cirrus clouds with pronounced band or wave structures and are usually so tenuous that stars can be seen through them; they are always pure white to blue-white.

It is now agreed that the cold temperatures are a necessary, but not a sufficient, condition for noctilucent-cloud formation. It is also generally accepted that the clouds are composed of ice crystals or water droplets which condense from water vapor brought up to the mesopause by the turbulence present during the arctic summer. *See* CLOUD PHYSICS. [R.K.S.]

Noeggerathiales An incompletely known and poorly de-fined group of vascular plants whose geologic range extends from Upper Carboniferous to Triassic. Their taxonomic status and position in the plant kingdom are uncertain since morphological evidence (because of the paucity of the fossil record) does not make it possible to place the group confidently in any recognized major subdivision of the vascular plants. The Noeggerathiales have been proposed in the evolutionary scheme in two remotely related groups of vascular plants, the Pteropsida and the Sphenopsida. *See* PALEOBOTANY; PLANT KINGDOM. [E.S.B.]

Nomograph A graphical relationship between a set of variables that are related by a mathematical equation or law. The fundamental principle involved in the construction of a nomographic or alignment chart consists of representing an equation containing three variables, $f(u,v,w) = 0$, by means of three scales in such a manner that a straight line cuts the three scales in values of u, v, and w, satisfying the equation. The cutting line is called the isopleth or index line. Numbers may be quickly and easily read from the scales of such a chart even by one unfamiliar with the construction of the chart and the equation involved. The illustration shows such an example. Assume that it is desired to find the value of E when $D = 2$ and $Q = 50$. Lay a straightedge through 50 on the Q scale and through 2 on the D scale and read 11.8 at its intersection with the E scale. As another example, it might be desired to know what value or values of D should be used if E and Q are required to be 10 and 60, respectively. A straightedge through $E = 10$ and $Q = 60$ cuts the D scale in two points, $D = 2.8$ and $D =$

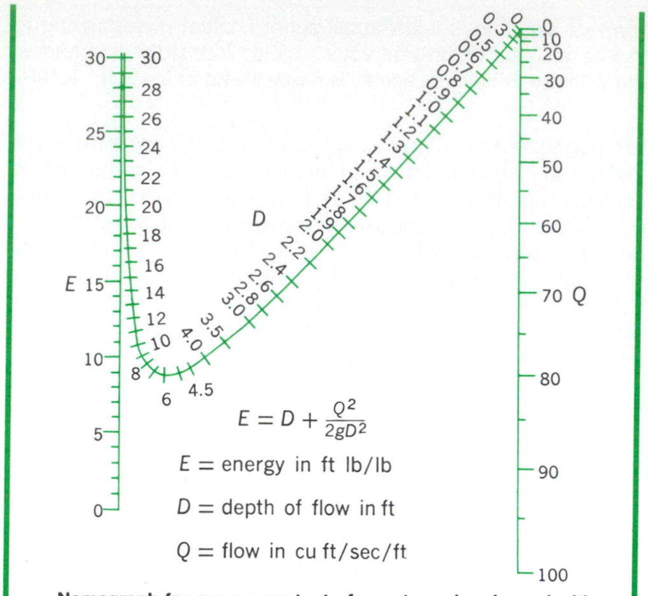

Nomograph for energy content of a rectangular channel with uniform flow.

9.4. This is equivalent to finding two positive roots of the cubic equation $D^3 - 10D^2 + 56.25 = 0$. It is assumed that $g = 32$ ft/s^2 in this equation. [R.D.D.]

Nondestructive testing The use of tests to examine an object or material to detect imperfections, determine properties, or assess quality without changing its usefulness. Nondestructive testing is a term generally applied to certain tests that are utilized in industrial operations, although some of the same technologies are used in the medical field, such as radiography (x-ray) and ultrasonics.

Some nondestructive tests involve complex high-technology systems, while others, which employ the same basic test principle, are easily applied by unskilled operators. Nearly every form of energy is used in nondestructive tests, including all wavelengths of the electromagnetic spectrum, as well as a broad range of frequencies of vibrational mechanical energy. Although there are many methods used for nondestructive testing, six are applied widely in industry: visual-optical, liquid-penetrant, magnetic-particle, eddy-current, ultrasonic, and radiographic.

Visual-optical method. This method can be applied to the examination of surfaces and transparent materials. A visual test might be as simple as an inspector looking at an object to check color or surface quality. At the other extreme, laser beams and complex image-recognition equipment can be used in automated scanning systems to perform an inspection. Endoscopes are rigid or flexible (fiber-optic) devices used to inspect internal surfaces of objects. They can be very small in diameter (⅛ in. or 3 mm) and usually include their own illumination source.

Liquid-penetrant method. Testing with liquid penetrants is useful for locating imperfections that are open to the surface of a wide variety of nonporous materials. To perform the test, a thin oillike liquid containing a dye is applied to the material and it moves into small openings or cracks. After the excess liquid on the surface is carefully removed and the object is dried, a developer is applied to the surface. This material is typically a fine powder, such as talc, usually in suspension in a liquid. It acts like a blotter and pulls the liquid penetrant up out of surface imperfections. The liquid tends to spread in the developer, thereby enlarging the indication.

Magnetic-particle method. Magnetic-particle testing is used for detecting discontinuities at or near the surface in magnetic materials. Steel and cast iron parts such as those used in automobiles, aircraft, and railroad equipment are subjected to magnetic-particle testing to detect processing or fatigue cracks. To perform the test, the test object is properly magnetized, and then finely divided magnetic particles are applied to its surface. When the object is properly oriented to the induced magnetic field, a discontinuity creates a leakage field which attracts and holds the particles, forming a visible indication.

Eddy-current method. Eddy current tests are based upon correlation between electromagnetic properties and physical or structural properties of a test object. Eddy currents are induced in metals whenever they are brought into an ac magnetic field. These eddy currents create a secondary magnetic field, which opposes the inducing magnetic field. The presence of discontinuities or material variations alters eddy currents, thus changing the apparent impedance of the inducing coil or of a detection coil. Coil impedance indicates the magnitude and phase relationship of the eddy currents to their inducing magnetic field current. Conditions such as heat treatment, composition, hardness, phase transformation, case depth, cold working, strength, size, thickness, cracks, seams, and inhomogeneities are indicated by eddy-current tests.

Ultrasonic method. Ultrasonic waves can detect and locate structural discontinuities or differences and measure the thickness of a variety of materials. To perform a contact test, a transducer is placed on an object to which a coupling material has been applied, and the results are interpreted from cathode-ray-tube indications. To accomplish this, an electric pulse is generated in a test instrument and transmitted through a cable to a transducer, which converts the electric pulse into mechanical vibrations. These low-energy-level vibrations are transmitted through a cable to a transducer, which converts the electric pulse into mechanical vibrations. These low-energy-level vibrations are transmitted through a coupling material into the test object, where the ultrasonic energy is attenuated, scattered, or reflected to indicate conditions within the object. Reflected or transmitted sound energy is usually reconverted to electrical energy by the same transducer and returned to the test instrument. The received energy is then usually displayed on a cathode-ray tube. The presence, position, and amplitude of echoes indicate conditions of the test-object material. *See* ULTRASONICS.

Radiographic method. Radiography can be used to reveal internal structure in industrial materials and assemblies, just as it does for a doctor or dentist who inspects parts of the body. The most widely used inspection approach involves the transmission of x-rays through an object to produce an image on x-ray film. This inspection method can be applied to many materials and assemblies. Proper placement of components in assemblies, detection of porosity, inclusions, or cracks, and differentiation of different materials, thicknesses, or densities are among the useful results obtained by radiography. *See* RADIOGRAPHY. [D.D.Do.]

Nonelectrolyte A chemical compound which does not conduct electricity either when fused or when dissolved in a solvent such as water. The bonding in nonelectrolytes is covalent. Most organic compounds, with the exception of the acids and amines, are nonelectrolytes, as are many inorganic compounds such as the halides of the nonmetals. *See* CHEMICAL BONDING; ELECTROLYTE; SOLUTION. [T.C.W.]

Nongonococcal urethritis (NGU) Human urethral inflammation not associated with common bacterial pathogens. A significant cause of nongonococcal urethritis may be bacteria of the Bedsonia (*Chlamydia*) group, particularly trachoma–inclusion conjunctivitis (TRIC) or lymphogranuloma venereum agents. The infective agent is in most cases the agent of inclusion conjunctivitis, which depends upon venereal transmission and a genital-tract reservoir for its perpetuation. The disease may be a severe purulent urethritis, but it is more often mild with a remitting course. In the female the same agent causes cervicitis. *See* LYMPHOGRANULOMA VENEREUM. [K.F.M.]

Nonlinear acoustics That portion of physics devoted to study of the behavior of nonlinear sonic and ultrasonic disturbances and the various phenomena associated with their propagation. The word nonlinear implies that nonlinear differential equations are necessary for an adequate mathematical description of the phenomena. Many nonlinear phenomena are observed only at high acoustical amplitudes, and thus are referred to as finite-amplitude effects. The word macrosonics also is used in this connection. Subjects of investigation in nonlinear acoustics include waveform distortion of finite-amplitude sonic and ultrasonic waves, weak shock-wave propagation, collinear and noncollinear wave interactions of sound with sound, parametric phenomena in sound waves, acoustic radiation pressure, acoustic streaming, and cavitation.

The definitive modern work relating theory and experiment for finite-amplitude sound waves in fluids (gases and liquids) was carried out in 1960 by W. Keck and R. T. Beyer. The same approach was used in 1965 by M. A. Breazeale and J. Ford to describe the propagation of a finite-amplitude longitudinal wave in a solid. For an isotropic solid or along the principal crystallographic directions in a cubic crystal, they showed that the propagation can be described by a differential equation whose solution takes the same form as that describing finite-amplitude wave propagation in liquids or gases. By identifying the nonlinearity parameter, it thus is possible to describe gases, liquids, and solids by differential equations having the same mathematical form. Much of the recent progress in nonlinear acoustics has been in measuring the nonlinearity parameters of gases, liquids, and solids.

The basic equation of nonlinear acoustics can be derived either from hydrodynamics or from nonlinear elasticity, depending upon the medium under consideration. The end result is a nonlinear differential equation appropriate to the description of finite-amplitude distortion of waves in gases, liquids, and solids.

An initially sinusoidal wave will distort as it propagates. Its progression is indicated in the illustration, where the waveform of the particle velocity u is shown at various values of the distance a from a sinusoidal driver. At the discontinuity distance $a = L$, the wave front is predicted to have a vertical tangent. Beyond $a = L$, the theoretical solution begins to predict a multivalued function, a physically unrealistic situation. The actual physical situation is that if the dissipation is small enough, the waveform becomes a repeated shock wave. *See* SHOCK WAVE.

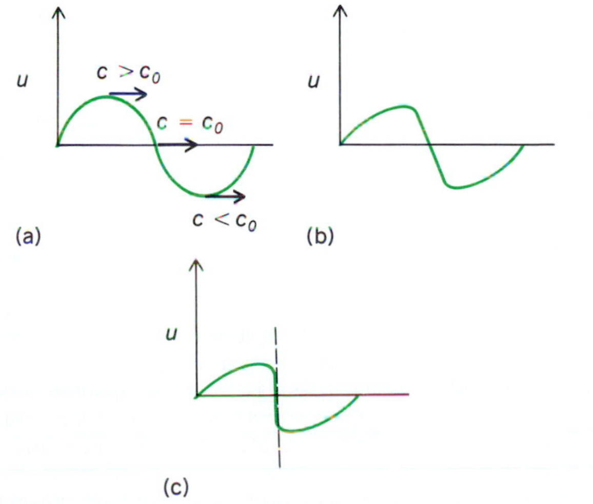

Progressive distortion of an initially sinusoidal finite-amplitude wave. (*a*) *a* = 0. (*b*) *a* > 0. (*c*) *a* = *L*.

Other than the nonlinear distortion of an initially sinusoidal ultrasonic wave, there are other effects which arise from the existence of nonlinear terms in the differential wave equation. The sum-and-difference-frequencies $\omega_1 \pm \omega_2$ arising from simultaneous interaction of two frequencies ω_1 and ω_2, as well as the radiation pressure observed in both solids and liquids, are examples of such nonlinear effects. In addition, a number of cavitation phenomena can be described only in terms of nonlinear differential equations. The wide range of nonlinear effects has given rise to a correspondingly wide range of applications of nonlinear acoustics.

Nonlinear acoustics has made possible the measurement of physical quantities that heretofore had not been determined. Values of the nonlinearity parameters of a number of fluids and solids are available, and have proved to be virtually independent of temperature, especially for the [100] direction. This fact is very significant in improving the understanding of the relationship between the nonlinearity of a solid measured by acoustical techniques and anharmonicity of the interatomic potential function determined from thermal expansion, neutron scattering, or other techniques.

The term streaming is given to the bulk flow of fluid that results whenever a sound wave is present in the medium. For high-intensity waves in fluids the effects of streaming can be very noticeable. Aluminum particles suspended in xylol can be used to demonstrate the streaming phenomenon, in which the nonlinear acoustical radiation force gives rise to a circulation of the particles, whose path can be followed with a time exposure.

A transducer driven by two frequencies ω_1 and ω_2 can exhibit the directionality inherent in the high-frequency propagation, while allowing the low-frequency $\omega_2 - \omega_1$ to be propagated. *See* PARAMETRIC ARRAYS.

In the course of developing the acoustic microscope, it was observed that the resolution of the instrument was limited by the wavelength of the acoustical signal propagating through the medium under investigation. In some instances, advantage can be taken of the fact that the wavelength of the second harmonic is half that of the fundamental and hence is capable of correspondingly higher resolution. *See* ACOUSTIC MICROSCOPE.

Several acoustic positioners are being developed for high-temperature containerless processing of materials in the microgravity environment of space. Acoustic forces associated with high-intensity sound fields can be used to move and manipulate objects. In all the methods for containerless processing the sample material is freely suspended in a furnace by acoustic force fields arising from nonlinearity of the medium. The suspended sample then is heated above its melting point and resolidified. During the thermal cycle the sample may be processed in a number of ways and its properties can be measured. Acoustic levitation also can be observed in fluids. In this case, the positioning of bubbles that have been induced by cavitation of the fluids has been observed. *See* ACOUSTIC LEVITATION; CAVITATION; SOUND. [M.A.Br.]

Nonlinear control theory The theory of control systems that do not have the property of superposition, that is, of systems in which some or all of the outputs are not linear functions of the inputs. All physical systems, although in many cases modeled on linear systems, exhibit some degree of nonlinearity. An electronic amplifier will suffer output saturation if the input exceeds its allowable limits, producing the nonlinear characteristic shown in Fig. 1. As a result of backlash, the input-output characteristic of a gear train exhibits hysteresis, which can significantly degrade the performance of a closed-loop control system. A relay, with ideal characteristic as shown in Fig. 2, is widely used in practice because of its high reliability and low cost. *See* HYSTERESIS.

The behavior of nonlinear control systems is quite different from that of linear control systems. A stable nonlinear system may sustain a fixed-amplitude oscillation without the applica-

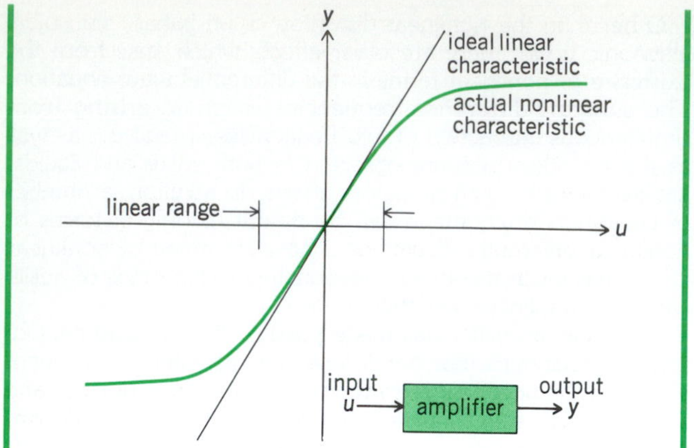

Fig. 1. Input/output characteristic of amplifier showing saturation.

tion of an input. The unforced response of a nonlinear system may be entirely different for different initial conditions. The state trajectory (the path followed by the state vector in the course of time) of a nonlinear system may exhibit jump discontinuities and possess two or more equivalent states (states that are physically identical even though the values of the state variables describing them are different). Frequencies can exist in the output that are not present in the input. These phenomena do not occur in linear control systems.

Most nonlinear control systems can be described by a set of first-order differential equations of the form in Eq. (1), where x

$$x = f(x,u) \tag{1}$$

is the n-dimensional state vector and u the m-dimensional input vector. Except in special cases, the quantitative analysis of Eq. (1) is carried out on a digital computer. *See* DIFFERENTIAL EQUATION; LINEAR ALGEBRA.

The state x_e is called an equilibrium state of the system described by Eq. (1) if Eq. (2) is satisfied. At an equilibrium

$$f(x_e,0) = 0 \tag{2}$$

state, $\dot{x}_e = 0$, and if no input is applied the state trajectory will remain indefinitely at the equilibrium state. One aspect of the stability of nonlinear control systems is defined in terms of its equilibrium states. An equilibrium state is said to be stable if, for any initial state in the neighborhood of x_e, the resulting trajectory remains in the vicinity of x_e. If, for all such initial states, the resulting trajectories all approach x_e as time tends to infinity, the equilibrium state is said to be asymptotically stable. These definitions characterize the local behavior of the state

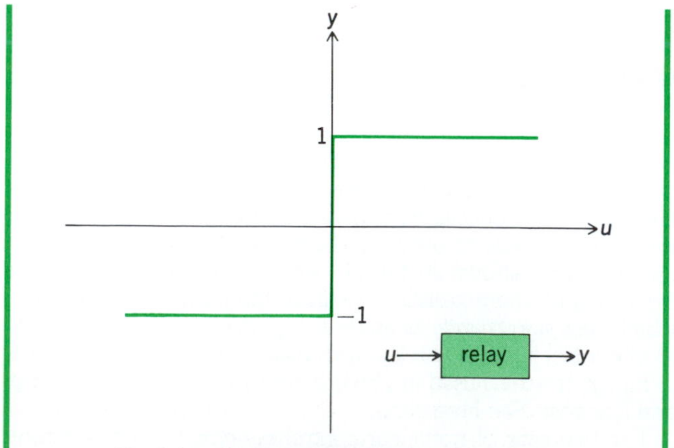

Fig. 2. Input/output characteristic of ideal relay.

trajectory in the vicinity of equilibrium points. If the properties hold for any initial state, the equilibrium state is said to be stable or asymptotically stable in the large. *See* CONTROL SYSTEMS; RELAY CONTROL SYSTEMS. [P.E.B.; C.T.C.]

Nonlinear optical devices Devices based on a class of optical effects that result from the interaction of electromagnetic radiation from lasers with nonlinear materials. Nonlinear means that the effect depends on the intensity of the light. Nonlinear effects are due to the nonlinear contribution to the polarization of the medium, which can be expressed as a power series expansion in the incident electric field E by the equation below, where $\chi^{(1)}$ is the linear, and $\chi^{(2)}$ and $\chi^{(3)}$ are

$$P = \epsilon_0(\chi^{(1)} E + \chi^{(2)} E^2 + \chi^{(3)} E^3 + \cdots)$$

second- and third-order susceptibilities, respectively. *See* ELECTRIC SUSCEPTIBILITY.

Nonlinear optical devices can be classified roughly into two categories: $\chi^{(2)}$ devices that generate light at new frequencies and $\chi^{(3)}$ devices that process optical signals. The first category covers second-harmonic generators and optical parametric oscillators, while the second category contains four-wave-mixing beam deflectors, phase-conjugate mirrors, etalon switches and logic devices, and waveguide couplers. *See* OPTICAL PHASE CONJUGATION.

Generation of laser frequencies in the ultraviolet region of the spectrum is of considerable importance in many applications such as laser fusion and laser isotope separation. Second-harmonic generators and optical parametric oscillators have produced tunable coherent radiation for these and other applications. *See* ISOTOPE (STABLE) SEPARATION; LASER; NUCLEAR FUSION.

Nonlinear optical devices may be used in digital signal processing and optical computing. The parallel nature of light, which may make the simultaneous operation of many devices possible, together with the high speed of optical components, promises very large bandwidths for number crunching computation and information processing. Many counterparts of electronic circuit elements such as transistors and gates have already been demonstrated by using optical technology. Methods for optical interconnections within electronic computers have been considered.

When a nonlinear material is sandwiched between two partially reflecting mirrors, a nonlinear Fabry-Perot etalon is formed. Normally, a thin semiconductor slab is used to provide the nonlinearity. Such a semiconductor etalon is capable of performing logic operations, switching, and controlling one light beam with another. Etalon devices are most suitable for parallel processing because many beams could be focused on a single etalon, thereby defining a two-dimensional array of pixels. Employing parallel processors in optical computing could considerably enhance the speed of computations. *See* COMPUTER SYSTEMS ARCHITECTURE; CONCURRENT PROCESSING; OPTICAL INFORMATION SYSTEMS; SUPERCOMPUTER.

When light travels in a medium with an index of refraction larger than the surrounding material, light traveling at a shallow enough angle is trapped in the higher-index medium by total internal reflection. The guided-wave devices confine the light over a long interaction length (a fraction of an inch or few millimeters), thereby permitting high intensities in the guided region. Also, the long interaction length permits the use of materials with weaker nonlinearities to obtain the phase shift (of the order of 180°) needed for logic devices. *See* OPTICAL FIBERS; REFLECTION OF ELECTROMAGNETIC RADIATION.

One method that has been assessed for optical interconnection is the use of holographic elements to diffract the light to the desired locations. However, conventional holograms are not programmable and would have to be replaced every time a different set of pixels was to be accessed. Four-wave mixing is one of the avenues that has been investigated to perform real-time programmable holography. *See* HOLOGRAPHY; INTEGRATED OPTICS; NONLINEAR OPTICS. [N.Pe.; H.M.G.]

Nonlinear optics A field of study concerned with the interaction of electromagnetic radiation and matter in which the matter responds in a nonlinear manner to the incident radiation fields. The nonlinear response can result in intensity-dependent variation of the propagation characteristics of the radiation fields or in the creation of radiation fields at new frequencies. Nonlinear effects can take place in solids, liquids, gases, and plasmas, and may involve one or more electromagnetic fields as well as internal excitations of the medium. The wavelength range of interest generally coincides with the spectrum covered by lasers, extending from the far infrared to the vacuum ultraviolet, but some nonlinear interactions have been observed at wavelengths extending into the x-ray range. Historically, nonlinear optics precedes the laser, but most of the work done in the field has made use of the high powers available from lasers. *See* LASER.

Nonlinear materials. Nonlinear effects of various types are observed in all materials. It is convenient to characterize the response of the medium mathematically by expanding it in a power series in the electric and magnetic fields of the incident optical waves. The linear terms in such an expansion give rise to the linear index of refraction, linear absorption, and the magnetic permeability of the medium, while the higher-order terms give rise to nonlinear effects. The coefficients in the expansion that describe the strength of the nonlinear interactions are now linear susceptibilities.

In general, nonlinear effects associated with the electric field of the incident radiation dominate over magnetic interactions. Effects involving the electric dipole response of the medium, termed the polarization, are generally the most important of the electric field interactions. Generally the magnitudes of the nonlinear susceptibilities decrease rapidly as the order of the interaction increases. *See* ELECTRIC SUSCEPTIBILITY; POLARIZATION OF DIELECTRICS.

Intensity-dependent effects. Nonlinear polarization components at the same frequencies as those in the incident waves can result in effects that change the index of refraction or the absorption coefficient, quantities that are constants in linear optical theory. Materials that are transparent at low optical intensities can have their absorption increase at high intensities. This effect involves simultaneous absorption of multiple numbers of photons from one or more incident waves.

In materials which have a strong linear absorption at the incident frequency, the absorption can decrease at high intensities. This effect, termed saturable absorption, was observed long before lasers were available. Saturable absorbers are useful in operating Q-switched and mode-locked lasers.

Intensity-dependent changes in the refractive index can affect the propagation characteristics of a laser beam. For many materials the index of refraction increases with increasing optical intensity. If the laser beam has a profile that is more intense in the center than at the edges, the profile of the refractive index corresponds to that of a positive lens, causing the beam to focus. This effect, termed self-focusing, can cause an initially uniform laser beam to break up into many smaller spots with diameters of the order of a few micrometers and intensity levels that are high enough to damage many solids. In other materials the refractive index decreases as the optical intensity increases. The resulting nonlinear lens causes the beam to defocus, an effect termed self-defocusing.

Phase conjugation. The same interactions that give rise to the intensity-dependent effects can also cause nonlinear interactions between waves that are at the same frequency but are otherwise distinguishable, for example, by their direction of polarization or propagation. Termed degenerate four-wave mixing, this interaction gives rise to a number of effects such as amplification of a weak probe wave and generation of a fourth wave that propagates in a direction opposite to that of a probe wave. In one of the most interesting aspects of this interaction, the backward wave has the same distribution of phase variations as the probe wave but with their sense reversed. This effect, termed phase conjugation or time reversal, can allow the correction of phase distortions of light beams caused by propagation through inhomogeneous media, by aberrated optics, or by mode dispersion in optical waveguides. It has important applications in the area of image transmission through optical fibers and in various forms of optical processing and information transfer.

Nonlinear spectroscopy. The variation of the nonlinear susceptibility near the resonances that correspond to sum- and difference-frequency combinations of the input frequencies forms the basis for various types of nonlinear spectroscopy which allow study of energy levels that are not normally accessible with linear optical spectroscopy. Nonlinear spectroscopy can be performed with many of the interactions discussed earlier. Many types of four-wave mixing interactions can also be used in nonlinear spectroscopy. *See* LASER SPECTROSCOPY; RAMAN EFFECT.

Inelastic scattering. Light can scatter inelastically from fundamental excitations in the medium, resulting in the production of radiation at new frequencies that differ from the incident frequencies by an amount that is equal to the frequency of the fundamental excitation. The difference in photon energy between the incident and scattered fields is taken up or given up by the medium.

Coherent effects. Another class of effects involves excitations of the medium in which the phase of the atomic wave function is preserved. These effects are generally observed only for short light pulses, of the order of several nanoseconds or less. In one interaction, termed self-induced transparency, a pulse of light of the proper shape, magnitude, and duration can propagate unattenuated in a medium which is otherwise absorbing. [J.F.R.]

Nonlinear physics The study of situations where, in a general sense, cause and effect are not proportional to each other; in other words, if the measure of what is considered to be the cause is doubled, the measure of its effect is not simply twice as large. Many examples have been known in physics for a long time, and they seemed well understood. Over the last few decades, however, physicists have noticed that this lack of proportionality in some of the basic laws of physics often leads to unexpected complications, if not to outright contradictions. Thus, the term nonlinear physics refers more narrowly to these developments in the understanding of physical reality.

Linearity in nonlinear systems. When a large number of particles starts out in a condition of stable equilibrium, the result of small external forces is well-coordinated vibrations of the whole collection, for example, the vibrations of a violin string, or of the electric current in an antenna. Each collective motion acts like an independent oscillator, each with its own frequency. In more complicated systems, many vibrational modes can be active simultaneously without mutual interference. A large dynamical system is, therefore, described in terms of its significant degrees of freedom, thought to be only loosely coupled. The motion of any part of the whole becomes multiperiodic; for example, a water molecule has bending and stretching vibrations with different frequencies, and both are coupled with the rotational motion of the whole molecule. *See* ANTENNA (ELECTROMAGNETISM); DEGREE OF FREEDOM (MECHANICS); MOLECULAR STRUCTURE AND SPECTRA; VIBRATION.

Failure of perturbation theory. H. Poincaré discovered at the end of the nineteenth century that for many problems this perturbation theory is not entirely satisfactory. He showed, in the case of the Moon's motion around the Earth, that the disturbance by the Sun is strong enough that this standard mathematical procedure fails. The main culprits are resonances, which occur when the frequencies of different degrees of freedom are combined through their nonlinear coupling. A key nonperturbative phenomenon is known to engineers as phase lock: When different frequencies arise in simple multiples of one another, the whole dynamical system falls into a dynamical trap; and for a continuous range of initial conditions, the interaction changes the frequencies of the individual degrees of freedom sufficiently so as to "lock" the motion into the resonance. *See* ASTEROID; PHASE-LOCKED LOOPS; RESONANCE (ACOUSTICS AND MECHANICS).

KAM theorem. In the 1950s, A. N. Kolmogoroff provided a first account of how the addition of a weak coupling generates chaotic regions and islands in phase space. This problem was later worked out in detail by V. Arnold and J. Moser to yield the KAM theorem. This theorem gives detailed information about the loss of the regular structure as the strength of the coupling increases. It does not say anything, however, about the trajectories in the newly created areas of chaotic behavior. These further investigations are the main goal of such fields as chaos or complexity. The impact of Poincaré's general arguments and the KAM theorem reaches into every area of nonlinear physics. The oldest among the areas is hydrodynamics, where the phenomenon of turbulent flow has so far resisted any effective control. This is what makes weather prediction so difficult. Signal propagation along the nerves and transmission of pulses through synaptic connections are other well-known nonlinear processes. *See* CHAOS; NEUROBIOLOGY; NONLINEAR ACOUSTICS; NONLINEAR OPTICS. [M.C.G.]

Nonlinear programming

An area of applied mathematics concerned with finding the values of the variables which give the smallest or largest value of a specified function in the class of all variables satisfying prescribed conditions. The function which is to be optimized is called the objective function, and the functions defining the prescribed conditions are referred to as the constraint functions or constraints. This general problem is called the nonlinear programming problem. The study of the theoretical and computational aspects of the nonlinear programming problem is called nonlinear programming, mathematical programming, or optimization theory. When there are no constraints, the nonlinear programming problem is said to be unconstrained; otherwise the problem is said to be constrained. If the objective function and the constraint functions are linear, the nonlinear programming problem is said to be a linear programming problem. When the objective function is quadratic and the constraints are linear, the nonlinear programming program is said to be a quadratic programming problem. *See* LINEAR PROGRAMMING.

The general nonlinear programming problem can be stated as below, where $x = (x_1, \ldots, x_n)$ are the variables of the prob-

$$\begin{aligned} &\text{minimize } f(x) \\ &\text{subject to } g_i(x) \geq 0, \qquad i = 1, \ldots, m \\ &\qquad\qquad h_j(x) = 0, \qquad j = 1, \ldots, p \end{aligned}$$

lem, f is the objective function, g_i are the inequality constraints, and h_j are the equality constraints. By changing f to $-f$, a maximization problem is transformed to a minimization problem. Moreover, an inequality constraint in the form $g_i(x) \geq 0$ is equivalent to $-g_i(x) \leq 0$. Consequently the format of the problem above can handle both types of inequality constraints and both minimization and maximization problems.

Many of the quantitative problems in business, economics, and engineering design can be expressed as nonlinear programming problems. General computational methods have been designed and implemented on large digital computers. While considerable progress has been made in both the design and algorithms and computer technology, there are still many practical large nonlinear programs which cannot be solved today. Research is aimed at taking advantage of the structure of the particular problem in question. One important example of structure is sparseness. Sparseness would occur if the nonlinear programming problem had many variables, but the objective function and each constraint function involved only relatively few variables. *See* OPERATIONS RESEARCH; OPTIMIZATION; PROCESS CONTROL. [R.A.T.]

Nonmetal

The elements are conveniently, but arbitrarily, divided into metals and nonmetals. The nonmetals do not conduct electricity readily, are not ductile, do not have a complex refractive index, and in general have high ionization potentials.

If the periodic table is divided diagonally from upper left to lower right, all the nonmetals are on the right-hand side of the diagonal. Examples of elements which do not fit neatly into this useful but arbitrary classification are tin, which exists in two allotropic modifications, one definitely metallic and the other with many properties of a nonmetal, and tellurium and antimony. Such elements are called metalloids. *See* METAL; METALLOID; PERIODIC TABLE. [T.C.W.]

Non-newtonian fluid

A fluid whose flow behavior departs from that of an ideal newtonian fluid. In a newtonian fluid the rate of shear in the fluid under isothermal conditions is proportional to the corresponding stress at the point under consideration. A non-newtonian fluid is one for which this rather strict requirement is not met. There are several ways in which real fluids can depart from the relatively simple newtonian condition, but it is better to think in terms of non-newtonian regions rather than non-newtonian fluids because many fluids exhibit both newtonian and non-newtonian behavior, depending on the conditions of flow, and cannot be strictly classified as being exclusively one type or the other. *See* NEWTONIAN FLUID.

Non-newtonian fluids can be broadly classified into three types as follows:

1. Time-independent fluids for which the rate of shear at any point in the fluid is some function of the shear stress at that point and depends on nothing else. (A newtonian fluid is a special case of a fluid in this category.)
2. More complex time-dependent fluids for which the relationship between shear stress and shear rate depends on the time the fluid has been sheared, that is, on its previous history.
3. Fluids which have the characteristics of both viscous liquids and elastic solids and exhibit partial elastic recovery after deformation, the so-called viscoelastic fluids.

Time-independent fluids. These fluids may be conveniently divided into three distinct types. These types, the flow curves for which are shown in Fig. 1, are as follows.

Bingham plastics exhibit a yield stress which is the stress that must be exceeded before flow starts. Thereafter the rate-of-shear curve is linear. This behavior is found in such materials as toothpaste, oil paints, and drilling muds. There are other materials which also exhibit a yield stress, but the flow curve is thereafter not linear. These are usually called generalized Bingham plastics.

Pseudoplastic fluids show no yield value, but the ratio of shear stress to the rate of shear, which may be termed the "apparent viscosity," falls progressively with shear rate. This phenomenon of shear thinning is characteristic of suspensions of asymmetric particles or solutions of polymers such as cellulose derivatives.

Dilatant fluids are similar to pseudoplastics in that they show no yield stress, but the apparent viscosity for dilatant materials increases with increasing rates of shear. Such "shear thicken-

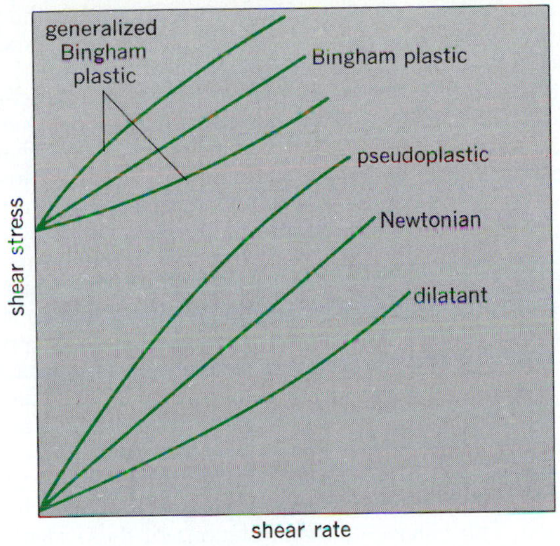

Fig. 1. Non-newtonian flow curves.

ing" is observed with suspensions of solids at high solids content (approaching point of tightest packing). Corn-flour pastes can be dilatant.

Time-dependent fluids. These fluids may be subdivided into two classes—thixotropic fluids and rheopectic fluids—according to whether the shear stress decreases or increases with time when the fluid is sheared at a constant rate.

The structure breakdown of thixotropic materials is a function of the duration of shear as well as of the rate of shear. Consider a thixotropic material which is sheared at a constant rate after a period of rest. The structure will be progressively broken down, and the apparent viscosity will decrease with time. The rate of breakdown of structure during shearing at a given rate will depend on the number of linkages available for breaking and therefore must decrease with time. Also, the simultaneous rate of reformation of structure will increase with time as the number of possible new structural linkages increases. Eventually a state of dynamic equilibrium is reached when the rate of buildup of structure equals the rate of breakdown. This equilibrium position depends on the rate of shear and moves toward greater breakdown at increasing rates of shear.

Rheopectic fluids are fluids for which the structure builds up on shearing, and this phenomenon can be regarded as the reverse of thixotropy—in fact, rheopectic behavior is often referred to as antithixotropy. Such materials are comparatively rare, and few if any are of industrial significance.

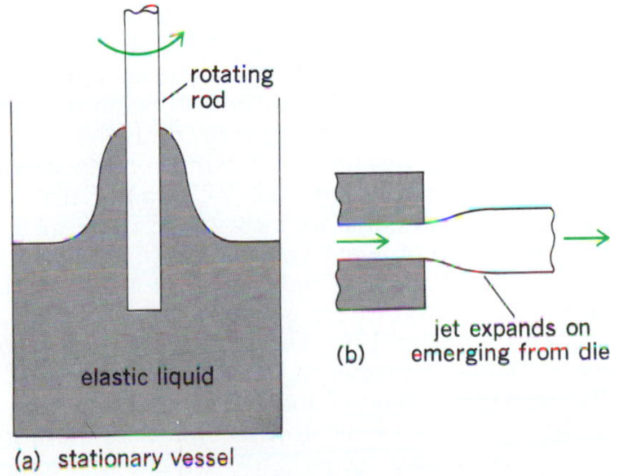

Fig. 2. Viscoelastic effects at a free surface. (*a*) Weissenberg effect. (*b*) Die-swell effect.

Viscoelastic fluids. A viscoelastic material is one which possesses both viscous and elastic properties; that is, although the material might be viscous, it exhibits a certain elasticity of shape and is capable of storing energy of deformation.

This type of behavior, which is typical of solutions of macromolecules and molten polymers, is analogous to mechanical springs and dashpots in series. Real materials behave as though they comprised a large number of such spring-dashpot combinations, each with its own relaxation time defined as the ratio of viscosity to the corresponding rigidity modulus in the combination.

Flow. The flow behavior of non-newtonian fluids cannot be predicted quantitatively except in relatively simple situations. In the case of time-independent non-newtonian fluids the behavior under laminar flow in tubes can be treated rigorously, but turbulent flow is not well understood. Although time-dependent fluids, particularly of the thixotropic type, are of considerable commercial importance, there is little progress of engineering significance on problems where the changes of fluid properties with time are important. The steady flow of viscoelastic fluids can be reasonably well predicted, but when unsteady conditions exist, such as the flow through valves and fittings and in the whole field of turbulent flow, the elastic properties of the fluid are of major importance and quantitative treatments are not yet well developed. At a free surface, viscoelastic effects can manifest themselves in the form of the Weissenberg effect, in which an elastic liquid tends to climb up a rotating rod, as shown in Fig. 2*a*, or as the swelling of a jet on extrusion from a die, as in Fig. 2*b*. These phenomena are the result of the normal stresses set up in viscoelastic fluids under shear. *See* FLUID FLOW; FLUIDS. [J.Har.; W.L.Wi.]

Nonsinusoidal waveform

Nonsinusoidal waveform The representation of a wave that does not vary in a sinusoidal manner. Electric circuits containing nonlinear elements, such as electron tubes, iron-core magnetic devices, and transistors, commonly produce nonsinusoidal currents and voltages. When these are repetitive functions of time, they are called nonsinusoidal electric waves. Oscillograms, tabulated data, and sometimes mathematical functions for segments of such waves are often used to describe the excursions throughout one cycle. A cycle corresponds to 2π electrical radians and covers the time interval T seconds in which the wave repeats itself.

These electric waves can be represented by a constant term, the average or dc component, plus a series of harmonic terms in which the frequencies of the harmonics are integral multiples of the fundamental frequency. The fundamental frequency f_1, if it does exist, has the time span $T = 1/f_1$ seconds for its cycle. The second-harmonic frequency f_2 then will have two of its cycles within T seconds, and so on.

The series of terms stated above is known as a Fourier series and can be expressed in the form of the equation below, where

$$y(t) = B_0 + C_1 \sin(\omega t + \phi_1) + \cdots + C_n \sin(n\omega t + \phi_n) + \cdots$$
$$= \sum_{n=0}^{\infty} C_n \sin(n\omega t + \phi_n)$$

$y(t)$, plotted over a cycle of the fundamental, gives the shape of the nonsinusoidal wave. The radian frequency of the fundamental is $\omega = 2\pi f_1$, and n is an integer. C_1 is the amplitude of the fundamental ($n = 1$), and succeeding C_n's are the amplitudes of the respective harmonics having frequencies corresponding to $n = 2, 3, 4$, and so on, with respect to the fundamental. The phase angle of the fundamental with respect to a chosen time reference axis is ϕ_1, and the succeeding ϕ_n's are the phase angles of the respective harmonics. See Fourier series.

The equation for $y(t)$ shows all its separate components, which, in general, include an infinite number of terms. In practical problems the first several terms usually yield an approximate result sufficiently accurate for portrayal of the actual wave. The degree of accuracy desired in representing faithfully the actual wave determines the number of terms that must be used in any computation. [B.L.R.; W.S.P.]

Nonstoichiometric compounds Chemical compounds in which the relative number of atoms is not expressible as the ratio of small whole numbers, hence compounds for which the subscripts in the chemical formula are not rational (for example, $Cu_{1.987}S$). Sometimes they are called berthollide compounds to distinguish them from daltonides, in which the ratio of atoms is generally simple. Nonstoichiometry is a property of the solid state and arises because a fraction of the atoms of a given kind may be (1) missing from the regular structure (for example, $Fe_{1-\delta}O$), (2) present in excess over the requirements of the structure (for example, $Zn_{1+\delta}O$), or (3) substituted by atoms of another kind (for example, $Bi_2Te_{3\pm\delta}$). The resulting materials are generally of variable composition, intensely colored, metallic or semiconducting, and different in chemical reactivity from the parent stoichiometric compounds from which they are derived.

Nonstoichiometry is best known in the binary compounds of the transition elements, particularly the hydrides, oxides, chalcogenides, pnictides, carbides, and borides. It is also well represented in the so-called insertion or intercalation compounds, in which a metallic element or neutral molecule has been inserted in a stoichiometric host. Nonstoichiometric compounds are important in some solid-state devices (such as rectifiers, thermoelectric generators, and photodetectors) and are probably formed as chemical intermediates in many reactions involving solids (for example, heterogeneous catalysis and metal corrosion).

The simplest way to classify nonstoichiometric compounds is to consider which element is in excess and how this excess is brought about. A classification scheme largely based on this distinction but which also includes some examples of ternary systems is as follows.

Binary compounds:
- I. Metal nonmetal ratio greater than stoichiometric
  - (*a*) Metal in excess, for example, $Zn_{1+\delta}O$
  - (*b*) Missing nonmetal, for example $UH_{3-\delta}$, $WO_{3-\delta}$
- II. Metal: nonmetal ratio less than stoichiometric
  - (*a*) Metal-deficient, for example, $Co_{1-\delta}O$
  - (*b*) Nonmetal in excess, for example, $UO_{2+\delta}$
- III. Deviations on both sides of stoichiometry, for example, $TiO_{1\pm\delta}$

Ternary compounds (insertion compounds):
- IV. Oxide "bronzes," for example, $M_\delta WO_3$, $M_\delta V_2O_5$
- V. Intercalation compounds, for example, $K_{1.5+\delta}MoO_3$, $Li_\delta TiS_2$

Excluded from consideration are the recognized impurity materials, such as $Na_{1-2x}Ca_xCl$, which are best considered as conventional solid solutions wherein ions of one kind and perhaps vacancies have replaced an equivalent number of ions of another kind. [M.J.Si.]

Noradrenergic system A neuronal system that is responsible for the synthesis, storage, and release of the neurotransmitter norepinephrine. Norepinephrine consists of a single amine group and a catechol nucleus and is therefore referred to as a monoamine or catecholamine. Norepinephrine exists in both the central and peripheral nervous systems, but it is the central nervous system norepinephrine that has been associated with several brain functions, including sleep, memory, learning, and emotions. The major clustering of norepinephrine-producing neuron cell bodies in the central nervous system is the locus coeruleus. This center, located in the pons but with extensive projections throughout the brain, produces more than 70% of all brain norepinephrine.

The effects of norepinephrine after its release are achieved by its binding with specific adrenergic receptors located on the neuronal plasma membrane. This binding initiates a chain of events (the effector system) in the target cell which are mediated through cyclic adenosinemonophosphate. Once norepinephrine is released, it is inactivated by reuptake or by enzymatic degradation. Reuptake, the major mechanism of inactivation, involves the active transport of norepinephrine back across the presynaptic membranes, and it is then stored in vesicles for future use. Enzymatic degradation occurs primarily by two enzymes, monoamine oxidase, which is located in the synapse and on intraneuronal mitochondria, and catechol-*O*-methyl-transferase, which is located in the synapse and on postsynaptic neuronal membranes. *See* MONOAMINE OXIDASE.

Certain medications achieve their affect by altering various stages of synthesis, storage, release, and inactivation of norepinephrine. The behavioral manifestations of these alterations have led to a better understanding of norepinephrine's role in various psychiatric disorders. It has been suggested that altered noradrenergic function may be of etiologic significance in depressive illness, anxiety disorder, and schizophrenia. *See* AFFECTIVE DISORDERS; SCHIZOPHRENIA. [D.S.C.; A.Bre.]

Normal (mathematics) A term generically synonymous with perpendicular, which often refers specifically to a line that goes through a point P of a curve C and is perpendicular to the tangent to C at P. If curve C is not a plane curve, all normal lines of C at point P on C lie in a plane, the normal plane of C at P. *See* ANALYTIC GEOMETRY. [L.M.Bl.]

North America The northern of the two continents of the New World or Western Hemisphere, extending from narrow parts in the tropics to progressively broader portions in middle latitudes and Arctic polar margins. This protuberance of the Earth's crustal shell stands with some 9.363×10^6 mi^2 (2.426×10^7 km^2) of land above the sea and ranks third in size among the continents. The continental mass upstands with deep submarine continental scarps toward the ocean basins at a marginal break or change in slope which appears at varying shallow depths of continental submarine shelf. These shelf areas of shallow sea bottom extend offshore for distances up to several hundred miles. Most of this outline of North American physical geography emphasizes physiography, especially land surface character, but it also includes some aspects of continental shelves considered as coastal-zone character. This focus is intended to present a basis for understanding other environmental and human features with which the land character is commonly interrelated. *See* CONTINENT.

Location. North America, the continent with adjacent islands, traditionally includes Greenland and the West Indies. The North American continent extends from southeastern Panama (6°6′N) at its isthmian frontier with South America to the northern tip of the Boothia Peninsula (72°10′N), northeast of the magnetic pole. This makes an extreme north-south distance of approximately 4554 mi (7327 km). If the Arctic islands are included, this distance increases to 5272 mi (8483 km); the northern tip of Grant Land on Ellesmere Island is at 82°30′N. The maximum longitudinal spread is from easternmost Greenland (18°W) to the westernmost of the Aleutian Islands (172°E), a total of 170°. Canada and the United States, exclusive of Alaska, lie between longitude 52° and 142°W. Since the narrower portion of the continent lies equatorward of 30°N latitude, relatively little of its area is in the tropics, whereas the greatest area occurs in mid-latitudes with considerable portions extending into the subpolar and polar regions. *See* ARCTIC AND SUBARCTIC ISLANDS; WEST INDIES.

Physiographic provinces. The land surface is the base upon which all the other geographical elements are placed. North America can be reasonably divided into some six broad physiographic divisions, based on their general geologic structure and subsequent landform development. These are, in turn, further subdivided on the basis of more localized variations in topography.

Canadian Shield. The Canadian Shield, also known on the mainland as the Laurentian Upland province, is the core around which the rest of North America has been built. In general, this province occupies the northeast quarter of North America, being defined by the St. Lawrence Gulf and River, the upper Great Lakes, and the series of large lakes in central Canada extending north from the United States border, including Winnipeg, Athabaska, Great Slave, and Great Bear lakes. The two extensions of this region into the United States are the Adirondack Mountains in northern New York State and the Superior Highland circling the western end of Lake Superior.

The Canadian Shield is composed of ancient crystalline rocks which have been subjected to geological processes such as volcanism, folding, faulting, warpage, and erosion. Extensive erosion over long periods has resulted in a surface that has been somewhat smoothed and lowered in relief so it resembles an undulating plain, except along the southern and eastern borders. Along the Atlantic Coast there is an escarpment that has been serrated by the tongues of mountain glaciers into fiords, making this area the most rugged section of the province. *See* FIORD; GEOLOGICAL TIME SCALE.

The Canadian Shield is one of the great mineralized areas of the world, containing a variety of both ferrous and nonferrous ores. Detailed prospecting and utilization of the minerals are generally limited to the outer border of this province, especially along the south fringe where economical transportation is provided.

Appalachian Highlands. A great variety of highlands and associated lowlands in Eastern United States and southeastern Canada are considered as composing the Appalachian Highlands, making it a very complex region. It is generally divided into the New England–Maritime Canada province and the Appalachian province, which in turn is further subdivided into older and newer sections.

Rocks ranging from ancient crystalline to Recent sediments are found in the same geographic area in various sections of this highland region. The crystalline rocks predominate in the northern and eastern portions, whereas the northwest and west are characterized by sediments of Paleozoic time.

Geologic structures of almost every type, which have been subjected to a variety of erosional and depositional processes, are found somewhere in this region. Doming, folding, faulting, and tilting, as well as volcanic activities, contribute to the variety of surface features. Major uplifts followed by long periods of erosion have left what might otherwise have been a very high and rugged mountain area as a generally subdued highland with maximum elevation only slightly exceeding 6000 ft (1800 m) in a few summits. Even the peaks are smoothed and rounded by erosional forces. The northeast-southwest trend of the province from Newfoundland to northern Alabama appears in most of the individual structures.

This highland area is rich in mineral resources which tend to be localized. The newer Appalachians, as a whole, have the greatest bituminous coal reserve in the world and are the site of the first developed petroleum field in the United States. Elsewhere, minor coal deposits are found in Nova Scotia; iron ores in Newfoundland, New England, New Jersey, and Alabama; lead, zinc, and wolframite in New Jersey; asbestos in Quebec; limestone, granite, marble, and slate in various localities; and gold in North Carolina.

Southeastern coastal plain. A distinctive coastal plain margins North America along the coastal zone of the Atlantic Ocean and the Gulf of Mexico. This plain varies in width but is nearly unbroken from Cape Cod and Long Island to Yucatan in southeastern Mexico. In most parts, very flat seaward margins of low-lying and most recent sedimentary deposits are followed inland by a tendency for a coastwise banding of features. A landward increase in elevation and erosional dissection also tends to contribute toward rolling and irregular plains with local relief up to a very few hundreds of feet. A gradual gentle uplift or upwarp of the geologically recent sedimentary rock layers means that they are little disturbed from their original horizontal attitude and dip slightly seaward and that such regions have most uplift inland toward bordering backland areas of differing character. Erosion, therefore, bares lower and somewhat older sedimentary layers progressively in a landward direction. Landward-facing cuestas or bands of low hills may develop on more-resistant rock exposures, whereas lower relief and occasional flat intervales develop by differential erosion in exposures of less-resistant rock layers. The farthest inland of these, commonly developed in Cretaceous clays, are termed inner lowland. The landward coastal plain boundary is commonly the inner margin of the Cretaceous days, and the border between the coastal plain layers and the oldland or backland rocks is known as the fall line or fall zone. *See* ESCARPMENT; FALL LINE.

A predominance of land subsidence marks the region from North Carolina to Cape Cod with many embayments and produces shorelines irregular in contrast to most other parts of recent upraisal or stillstand of the land relative to the sea. A strongly marked variance from north to south in climatic characteristics with an especially steep temperature gradient is reflected in such aspects as natural vegetation and length of growing season and agricultural production, as well as the pattern contrasts in winter resorts of Florida and the Gulf Coast versus the definite winter influences of the New York and New Jersey coastal zones.

Within the essential unity of this great coastal plain form and structure, six or seven subdivisions are useful in outlining the regional variations characterizing its long extent: Northern embayed, Sea Island, Peninsular Florida, East Gulf section, Mississippi alluvial plain, West Gulf, and Yucatan section.

Arctic margin plains. Two plains units that are commonly considered parts of the interior North American lowlands are the Arctic slope of Alaska and the Mackenzie plains of Canada. They open out to Arctic Ocean shores from which the polar ice pack recedes for an unpredictable duration during the late part of the warm season. These plains are subject to arctic cold and long hours of darkness during the low-sun period, and conversely to long hours of daylight with short but surprisingly warm periods during the high sun. Both are strongly marked in their lower parts by small lakes and ponds, polygonal ground, and other features of permanently frozen ground. They are also subject to the widespread flooding of lower and flatter parts which characterizes subarctic regions, especially when the headwaters thaw earlier than the ice breakup in the main streams to the northward. *See* PERMAFROST.

Interior provinces. Interior North America is so predominantly plains land, although it contains a few highland parts, that other physical attributes commonly contribute as much to regional differences as do structure and surface form of the plains. The aspects of great area and huge distances often cause the extent and character of climate and its closely associated problems of natural vegetation and soil to make the greatest regional physical contrasts in many parts of the plains. Recent (Pleistocene) glaciation north of the Ohio and Missouri river valleys leaves an impress on patterns of drainage, surface forms, and even soil materials quite different from areas not recently affected by glaciers.

The interior provinces include the interior low plateaus, the Ozark province, the Ouachita province, the Central Lowland province, and the Great Plains province.

Cordilleran North America. A large proportion of western North America is a great north-south highland of mountain masses, ranges, and systems and of elevated basins and plateaus. With the Central American ranges and mountains of somewhat different structure and axial orientation, the North American Cordillera continues from those of Andean South America, the great mountain rim of the Pacific. These Pacific-encircling mountains turn westward in Alaska and swing toward the Asian shores of the Pacific in the great arcuate festoons of the Aleutian, Kuril, and Japanese islands.

The consequences in the physical and human geography of the continent are pronounced but somewhat in contrast with those of South America. The great orographic barrier made difficult an articulation of settlement and development between the eastern regions and the western highlands and coastal zones.

Although bold in plan and extent, the Cordilleran lands are diverse in component highland parts and commonly intricate in local details of highland relief. One classification uses three or four levels of category to outline the highland characteristics. The largest parts may be designated as three systems: Rocky Mountain system, ranges extending in various parts on the eastern margin of the Cordilleran region from New Mexico to Arctic shores of Alaska; intermontane plateau system, westward of the Rockies and extending even further (from southern Mexico to Alaska); and Pacific mountain system, from the Aleutians to peninsular Lower California.

Central America. This narrowest part of the North American mainland has predominantly mountainous landforms, but it is geologically and structurally different from the great ranges, basins, and plateaus of western Cordilleran North America. It is perhaps best considered as a composite of at least five rather separate and different parts, more closely related to the Antillean mountain system. [C.V.C.]

North Pole That end of the Earth's axis which points toward the North Star, Polaris (Alpha Ursae Minoris). It is the geographical pole where all meridians converge, and should not be confused with the north magnetic pole, which is in the Canadian Archipelago. The North Pole's location falls near the center of the Arctic Sea.

The North Pole has phenomena unlike any other place except the South Pole. For 6 months the Sun does not appear above the horizon, and for 6 months it does not go below the horizon. As there is a long period (about 7 weeks) of continuous twilight before March 21 and after September 23, the period of light is considerably longer than the period of darkness. [V.H.E.]

North Sea A flooded portion of the northwest continental margin of Europe occupying an area of over 190,000 mi^2 (500,000 km^2). The North Sea is the scene of extensive marine fisheries and important offshore oil and gas. In the south, water is shallower than 160 ft (50 m), but north of 58°N it deepens gradually to 650 ft (200 m) at the top of the continental slope. A band of deep water of up to 1300 ft (400 m) extends around the south and west coast of Norway and is known as the Norwegian Trench.

The nontidal residual current circulation of the southern North Sea is mainly determined by wind velocity, but in the north, well-defined non-wind-driven currents have been identified, especially in summer. Two of these currents bring in water from outside the North Sea; one flows through the channel between Orkney and Shetland (the Fair Isle current), and the other follows the continental slope north of Shetland and merges with the Fair Isle current southwest of Norway before entering the Skagerrak.

In some areas of the North Sea, especially in the shallow parts of the south, and also within parts of the permanent inflowing currents, phytoplankton production persists throughout the summer, due to the continuous mixing of nutrients into the surface waters. There is a rich diversity of zooplankton within the North Sea. Copepods are of particular importance in the food web. There are many commercially important fish stocks, notably herring, mackerel, haddock, cod, whiting, saithe, and plaice. [H.D.D.]

Northwest Passage The northern sea route between the Atlantic and Pacific oceans through the Canadian Archipelago. The entire route is frozen over in winter. As the ice cap retreats northward in summer, two routes are opened. The eastern approach through Davis and Hudson straits is navigable in June, at which time the partially melted ice floes drift into the Atlantic Ocean on the Labrador Current. This approach remains open until November.

In the Canadian Archipelago the ice is landlocked. These passages remain closed until the ice melts. The tortuous southern route, close along the Canadian mainland coast, is usually passable from mid-August to mid-October. The more direct northern route, through Lancaster Sound and the Barrow and McClure straits, has never been found entirely ice-free. This passage has only been made by Wind-class icebreakers. *See* ARCTIC AND SUBARCTIC ISLANDS; BERING SEA; SEA ICE. [J.P.T.]

Nose The nasal cavities and the structures surrounding and associated with them. The nose functions primarily as the organ of smell and in most tetrapods also assumes a respiratory function, forming the anterior end of the air passage through which air is drawn in and in which it is warmed and moistened.

In humans the nasal cavities are triangular openings that pass from the external nares back to the dorsal part of the pharynx (see illustration). The lateral walls are composed principally of portions of the ethmoid and sphenoid bones and projections of three turbinate bones, or conchae, on each side. The floor of the nose is formed by the palate, which is also the roof of the mouth. The nasal cavities are lined with respiratory epithelium, which also lines the paranasal sinuses. The latter are cavities in the frontal, ethmoid, sphenoid, and maxillary bones which communicate with the nasal passages. The external nose consists of the two nasal bones that form the bony bridge and two pairs of lower nasal cartilages. These together with the tightly adherent skin determine the individual shape and size of the human nose. *See* OLFACTION. [T.S.P.]

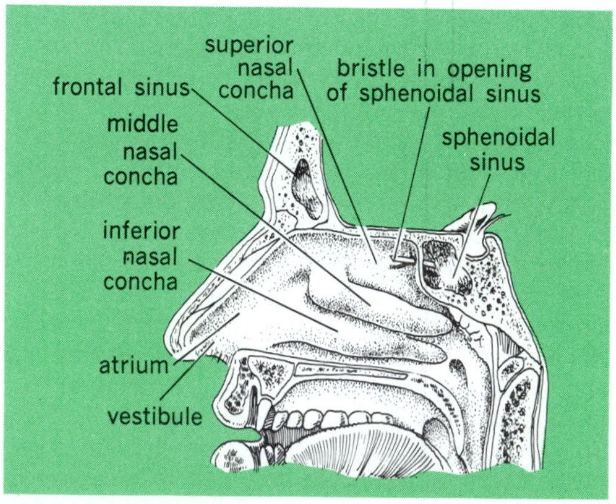

Human nose. ***(After W. J. Hamilton et al., Textbook of Human Anatomy, Macmillan, 1956).***

Nose cone The forward portion of a spacecraft that is designed for atmospheric entry. Nose cones are utilized for intercontinental ballistic missiles and spacecraft such as Apollo and space shuttles. The nose cone is required to withstand heating encountered during atmospheric entry, maintain the structural integrity of the spacecraft, prevent overheating of the payload, and usually maintain the aerodynamic characteristics of the spacecraft.

Even for a properly designed shape, it is inevitable that some fraction of the spacecraft's initial kinetic energy will finally reach the nose cone in the form of heat. The design of the heat shield for the nose cone is a complex procedure, which is highly dependent on the heating level. There are a variety of surface-protection or cooling systems which have been used. Generally these systems consist of heat sinks of various types: the absorption of heat by virtue of a material's sensible heat capacity, latent heat capacity, or chemical heat capacity. That is, heat absorption is accomplished by a temperature rise, a phase change, or a chemical reaction. Aerodynamic lift is employed by such vehicles as reusable space shuttles to lower heating rates so that the nose cone material can radiate away much of the incident heating.

Ablation is used to provide surface protection. The designer can divert heat from the spacecraft by allowing the nose cone's outer layer of material to melt, vaporize, or sublime. While large ablation rates provide excellent thermal protection, the resulting change in profile due to surface recession can adversely change the aerodynamic characteristics of the spacecraft. The designer must account for this change. *See* SPACE SHUTTLE; SPACECRAFT STRUCTURE. [P.R.N.]

Notacanthiformes An order of actinopterygian fishes known also as the Heteromi (spiny eels or notacanths) and the Lyopomi (halosaurs). The body is elongated, tapers posteriorly, and has no caudal fin (see illustration).

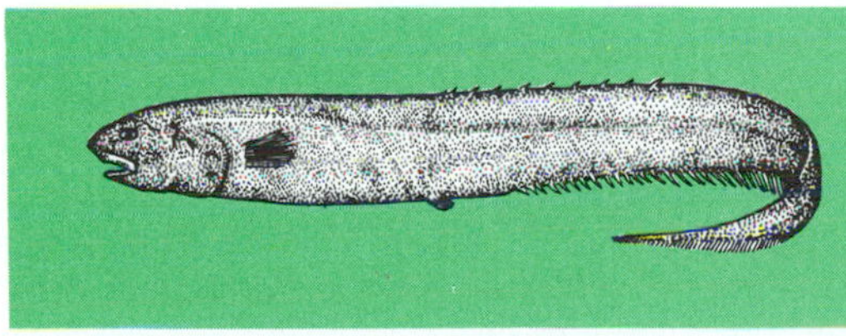

Spiny eel (*Notacanthus nasus*). (After D. S. Jordan and B. W. Evermann, *The Fishes of North and Middle America*, U.S. Nat. Mus. Bull. no. 47, 1900)

This small order, which has a history extending back to the Upper Cretaceous, includes 3 families, about 8 Recent genera, and about 25 species. Spiny eels and halosaurs inhabit deep seas of all oceans; some have photophores. They are like true eels (Anguilliformes) in that they lack a firm suspension of the pectoral girdle from the skull, but some have fin spines like the perciform fishes. *See* ACTINOPTERYGII; ANGUILLIFORMES; PHOTOPHORE GLAND. [R.M.B.]

Nothosauria An order of extinct aquatic diapsid reptiles in the infraclass Sauropterygia, known from the Triassic System of Europe, North America, and Asia. Nothosaurs represent a primitive grade of evolution within the infraclass, which also includes the Jurassic and Cretaceous plesiosaurs.

Nothosaurs range approximately 3–13 ft (1–4 m) in length. The neck is long, with at least 15 cervical vertebrae. Like aquatic lizards and crocodiles, early nothosaurs probably swam primarily by lateral undulation of the trunk and tail, with the limbs held to the side to reduce drag. Advanced nothosaurs show modifications of the front limbs to act as paddles; the rear limbs are somewhat reduced.

Nothosaurs are distinguished from other aquatic reptiles by the closure of the openings in the palate that occur in more primitive reptiles, and by the covering of the base of the braincase. The ventral surface of the shoulder girdle is characterized by a wide gap between the base of the scapulae and the coracoids, behind a stout transverse bar formed by the interclavicle and the blades of the clavicles. As in plesiosaurs, part of the scapula is superficial to the clavicle, a reversal of the relationship of these bones in other reptiles. Primitive nothosaurs resemble primitive lepidosauromorphs in the pattern of the skull roof but have lost the lower temporal bar. *See* DIAPSIDA; PLESIOSAURIA; REPTILIA. [R.L.C.]

Notomyotina A suborder of Phanerozonida (subclass Asteroidea) in which the upper marginals alternate in position with the lower marginals to impart a degree of flexibility to the arm, and each of the tube feet has a terminal sucking disk. Paxillae are present on the upper surface. Each arm usually contains a pair of dorsal muscles, whose contraction enables the arm to be turned upward over the disk. These are mainly deep-water forms. *See* ASTEROIDEA; ECHINODERMATA. [H.B.F.]

Notostraca An order of crustaceans of moderate size (20–90 mm) generally referred to the Branchiopoda. They are called the tadpole shrimps. The cylindrical trunk consists of 25–44 body segments, and the number varies slightly within a species. The first 11 somites each bear one pair of legs. Behind these are a varying number of segments with an increasing number of legs; a varying number of legless segments; and finally a telson with two narrow, cylindrical, caudal filaments (cercopods). The dorsal shield is rounded and flattened dorsoventrally, and usually covers only part of the animal. The animals feed mainly on detritus, and the food is moved to the mouth in a ventral groove. The young hatch in the nauplius larval stage.

The group contains only two genera, *Triops*, or *Apus*, and *Lepidurus* (see illustration), and about a dozen species. They live in nonpermanent waters, sometimes occurring in great numbers, and are distributed all over the world. *See* BRANCHIOPODA. [F.Li.]

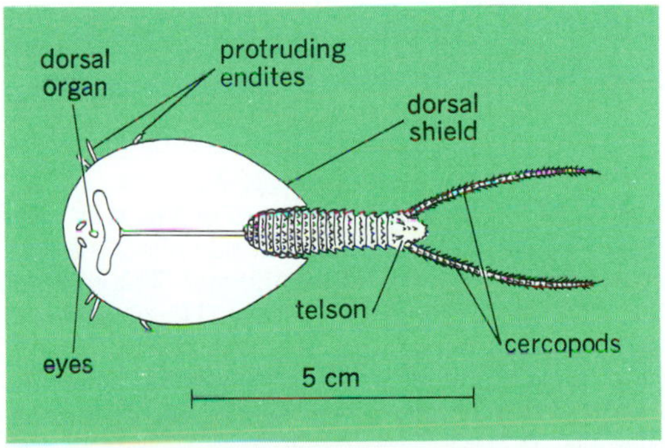

***Lepidurus arcticus*, female, dorsal aspect.**

Notoungulata An order of dominant, hoofed herbivores of the Cenozoic of South America that are abundantly represented in Paleocene through Pleistocene nonmarine sedimentary rocks of that continent. Diverging from a primitive condylarth ancestry at an early date, they radiated into a wide diversity of forms, some of which were convergent with Northern Hemisphere ungulates.

Notoungulates were characterized by a skull with an expanded temporal region due to the presence of a large sinus in the squamosal and no postorbital bar. The feet were primitive, with five toes (three or two in some advanced forms), and the weight was borne mainly by the third digit. *See* EUTHERIA. [R.H.T.]

Nova The brightening of a seemingly undistinguished star within a few days to thousands or even hundreds of thousands of times its previous luminosity. These events, which occur about 10 times a year in a large spiral galaxy like the Milky Way, are the classical novae (short for novae stellae, or new stars, so called because the pre-flare-up stars are too faint to have been seen in the days before telescopes). A much rarer and still more violent class of stellar flare-up is given the name supernova. *See* SUPERNOVA.

The classical novae are defined and distinguished from other kinds of explosive stellar events by the following observed properties: (1) brightening by more than a factor of 4000 in a few days; (2) noticeable fading in less than a few years (more often weeks to months); (3) a complex, changing spectrum, showing emission lines of common gases at high temperature; (4) ejection of gas at speeds of 100 to 5000 km/s (60 to 3000 mi/s); and (5) no historical record of a similar previous event involving that particular star.

The single most important step in understanding nova explosions occurred in 1964, when Robert Kraft showed that most previously observed novae had occurred in close (that is, stars not much farther apart than their own sizes) binary systems, containing one hot, compact star (like a white dwarf) and one cooler, more extended star (like a main sequence or subgiant star). The systems typically have large amounts of stray gas and masses near 1 solar mass (2×10^{30} kg). *See* BINARY STAR; STAR; WHITE DWARF STAR. [V.T.]

Novobiocin A moderately broad-spectrum acid antibiotic produced by strains of *Streptomyces niveus* and *S. spheroides*. It is a dibasic acid with the structure below.

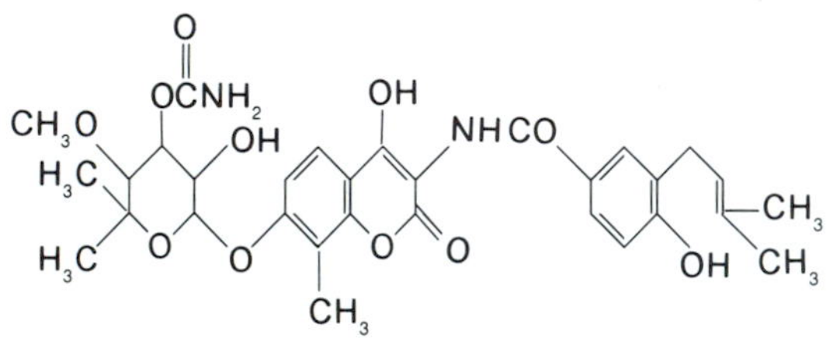

The antibiotic is relatively nontoxic unless it is given for more than 1 week. The most frequent side effect is a skin rash which may subside with continued therapy. The rash may or may not recur upon subsequent dosage. Novobiocin is highly active against staphylococci and certain strains of *Proteus*. It is also active against pneumococci, streptococci, *Brucella*, and certain strains of *Escherichia*, *Aerobacter*, and *Pseudomonas*. *See* ANTIBIOTIC. [G.M.S.]

Nowcasting A form of very short-range weather forecasting. The term nowcasting is sometimes used loosely to refer to any area-specific forecast for the period up to 12 h ahead that is based on very detailed observational data. However, nowcasting should probably be defined more restrictively as the detailed description of the current weather along with forecasts obtained by extrapolation up to about 2 h ahead. Useful extrapolation forecasts can be obtained for longer periods in many situations, but in some weather situations the accuracy of extrapolation forecasts diminishes quickly with time as a result of the development or decay of the weather systems. *See* METEOROLOGY; WEATHER FORECASTING AND PREDICTION.

Nowcasting involves the use of very detailed and frequent meteorological observations, especially remote-sensing observations, to provide a precise description of the "now" situation from which very short-range forecasts can be obtained by extrapolation. Of particular value are the patterns of cloud, temperature, and humidity which can be obtained from geostationary satellites, and the fields of rainfall and wind measured by networks of ground-based radars. These kinds of observations enable the weather forecaster to keep track of smaller-scale events such as squall lines, fronts and thunderstorm clusters, and various terrain-induced phenomena such as land-sea breezes and mountain-valley winds. Meteorologists refer to these systems as mesoscale weather systems because their scale is intermediate between the large, synoptic-scale cyclones, and the very small or microscale, features such as boundary-layer turbulence. *See* MESOMETEOROLOGY; METEOROLOGICAL SATELLITES; RADAR METEOROLOGY. [K.A.B.]

Nozzle A projecting opening that directs the flow of fluid into an open space. Some nozzles maintain the fluid in a jet; an example is the needle nozzle that directs water against the buckets of an impulse turbine. Other nozzles disperse the fluid in an atomized mist; an example is the cone nozzle that sprays liquid fuel into a combustion chamber. The nozzle may be an integral part of a machine, as the nozzle in a steam turbine, or it may be a separate interchangeable piece as on a fire truck.

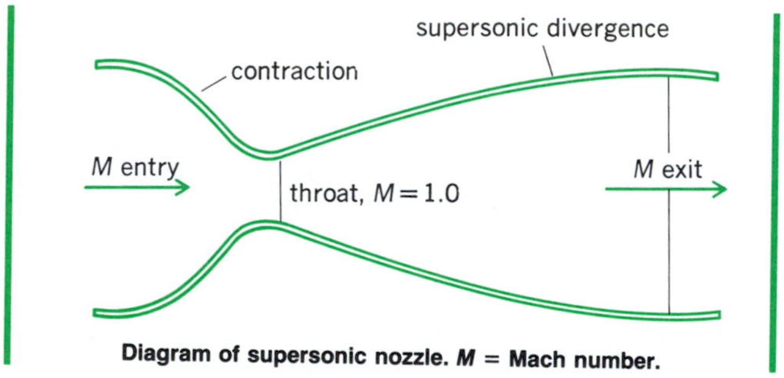

Diagram of supersonic nozzle. *M* = Mach number.

As used in a wind tunnel, a nozzle increases fluid velocity but with the added requirement that the higher-velocity stream be uniform and parallel. Physically the nozzle consists of a contracting section. If the final fluid velocity is to be supersonic, a divergent portion downstream of the contraction is also required (see illustration). The region at the minimum section is called the throat. [F.D.K.]

Nuclear battery A battery that converts the energy of particles emitted from atomic nuclei into electric energy. Two basic types have been developed: (1) A high-voltage type, in which a beta-emitting isotope is separated from a collecting electrode by a vacuum or a solid dielectric, provides thousands of volts but the current is measured in picoamperes (pA); (2) a low-voltage type gives about 1 volt with current in microamperes (μA).

In the high-voltage type, a radioactive source is attached to one electrode, emitting charged particles. The source might be strontium-90, krypton-85, or hydrogen-3 (tritium), all of which are pure beta emitters. An adjacent electrode collects the emitted particles. A vacuum or solid dielectric separates the source and the collector electrodes. The principal use of the high-voltage battery is to maintain the voltage of a charged capacitor. The current output of the radioactive source is sufficient for this purpose.

Three different concepts have been employed in the low-voltage type of nuclear batteries: (1) a thermopile, (2) the use of an ionized gas between two dissimilar metals, and (3) the two-step conversion of beta energy into light by a phosphor and the conversion of light into electric energy by a photocell. *See* BATTERY. [J.D.; L.R.; K.F.]

Nuclear binding energy The amount by which the mass of an atom is less than the sum of the masses of its constituent protons, neutrons, and electrons expressed in units of energy. This energy difference accounts for the stability of the atom. In principle, the binding energy is the amount of energy which was released when the several atomic constituents came together to form the atom. Most of the binding energy is associated with the nuclear constituents (protons and neutrons), or nucleons, and it is customary to regard this quantity as a measure of the stability of the nucleus alone. *See* NUCLEAR STRUCTURE.

A widely used term, the binding energy (BE) per nucleon, is defined by the equation below, where ${}_ZM^A$ represents the

$$\text{BE/nucleon} = \frac{[ZH + (A - Z)n - {}_ZM^A]c^2}{A}$$

mass of an atom of mass number A and atomic number Z, H and n are the masses of the hydrogen atom and neutron, respectively, and c is the velocity of light. The binding energies of the orbital electrons, here practically neglected, are not only small, but increase with Z in a gradual manner; thus the BE per nucleon gives an accurate picture of the variations and trends in nuclear stability.

The binding energy, when expressed in mass units, is known as the mass defect, a term sometimes incorrectly applied to quantity $M - A$, where M is the mass of the atom. [H.E.D.]

The term binding energy is sometimes also used to describe the energy which must be supplied to a nucleus in order to remove a specified particle to infinity, for example, a neutron, proton, or α-particle. A more appropriate term for this energy is the separation energy. This quantity varies greatly from nucleus to nucleus and from particle to particle. [D.H.W.]

Nuclear chemical engineering The branch of chemical engineering that deals with the production and use of radioisotopes, nuclear power generation, and the nuclear fuel cycle. A nuclear chemical engineer requires training in both nuclear and chemical engineering. As a nuclear engineer, he or she should be familiar with the nuclear reactions that take place in nuclear fission reactors and radioisotope production, with the properties of nuclear species important in nuclear fuels, with the properties of neutrons, gamma rays, and beta rays produced in nuclear reactors, and with the reaction, absorption, and attenuation of these radiations in the materials of reactors. *See* BETA PARTICLES; GAMMA RAYS; NEUTRON; NUCLEAR FUELS.

As a chemical engineer, he or she should know the properties of materials important in nuclear reactors and the processes used to extract and purify these materials and convert them into the chemical compounds and physical forms used in nuclear systems. *See* CHEMICAL ENGINEERING; NUCLEAR REACTOR.

Aspects of nuclear reactors of concern to nuclear chemical engineers include production and purification of the uranium dioxide fuel, production of the hafnium-free zirconium tubing used for fuel cladding, and control of corrosion and radioactive corrosion products by chemical treatment of coolant. A chemical engineering aspect of heavy-water reactor operation is control of the radioactive tritium produced by neutron activation of deuterium. Aspects of liquid-metal fast-breeder reactors of concern to nuclear chemical engineers include fabrication of the mixed uranium dioxide-plutonium dioxide fuel, purity control of sodium coolant to prevent fouling and corrosion, and reprocessing of irradiated fuel to recover plutonium and uranium for recycle. *See* NUCLEAR FUEL CYCLE; PLUTONIUM; URANIUM. [M.Be.]

Nuclear chemistry An interdisciplinary field that, in general, encompasses the application of chemical techniques to the solution of problems in nuclear physics. The discovery of the naturally occurring radioactive elements and of nuclear fission are classical examples of the work of nuclear chemists.

Although chemical techniques that are employed in nuclear chemistry are essentially the same as those in radiochemistry, these fields may be distinguished on the basis of the aims of the investigation. Thus, a nuclear chemist utilizes chemical techniques as a tool for the study of nuclear reactions and properties, whereas a radiochemist utilizes the radioactive properties of certain substances as a tool for the study of chemical reactions and properties. For the application of radioactive tracers to chemical problems *see* RADIOCHEMISTRY. For the chemical effects of radiation on various systems *see* RADIATION CHEMISTRY.

The chemical identification of radioactive nuclides and the determination of their nuclear properties has been one of the major activities of nuclear chemists. Such studies have produced an extensive array of radioisotopes, and present studies are concerned mainly with the more difficult identification of nuclides of very short half-life. Nuclear chemical investigations led to the discovery of the synthetic radioactive elements which do not have any stable isotopes and are not formed in the natural radioactive series (technetium, promethium, astatine, and the transuranium elements). Other major areas of nuclear chemistry include studies of nuclear structure and spectroscopy and of the probability and mechanisms of various nuclear reactions. *See* NUCLEAR FISSION; NUCLEAR REACTION; NUCLEAR STRUCTURE. [E.P.S.]

Nuclear engineering That branch of engineering that deals with the production and use of nuclear energy. It is concerned with the development, design, construction, and operation of power plants which convert energy produced by fission or fusion to other useful forms such as heat or electrical energy. Development of these unique sources of energy requires novel solutions to difficult mechanical, electrical, and materials problems. Because many of the components and systems operate in the presence of intense high-energy radiation, special problems that are generated by the interaction of radiation with various materials are encountered. Such problems are unique to nuclear engineering. Training of nuclear engineers places special emphasis on this area. *See* DECONTAMINATION OF RADIOACTIVE MATERIALS; HEALTH PHYSICS; MONITORING OF IONIZING RADIATION; NONDESTRUCTIVE TESTING; NUCLEAR FISSION; NUCLEAR FUEL CYCLE; NUCLEAR FUELS; NUCLEAR FUELS REPROCESSING; NUCLEAR FUSION; NUCLEAR POWER; NUCLEAR REACTOR; RADIATION DAMAGE TO MATERIALS; RADIATION SHIELDING; RADIOACTIVE WASTE MANAGEMENT; REACTOR PHYSICS. [W.Ke.]

Nuclear explosion An explosion for which the energy is produced by a nuclear transformation (either fission or fusion).

Although the terms atomic explosion and atomic weapon are occasionally used, the energy is derived in processes that involve the nuclear forces of matter rather than the atomic forces; hence, nuclear explosion and nuclear weapon are more appropriate designations. The energy released in a nuclear explosion is usually stated in terms of the equivalent energy release of a given mass of TNT using convenient units such as kiloton (KT) or megaton (MT).

The early nuclear explosions—such as the Alamogordo event (Trinity) or the Hiroshima, Japan, bomb—were about 20 KT. Tactical fission weapons in the sub-KT range have been announced by the United States. Fusion weapons are sometimes called hydrogen bombs or thermonuclear weapons, terms that describe their constituents and explosive mechanisms.

The prompt energy release of a nuclear explosion vaporizes the constituents and products, as well as the bomb container. These extremely hot gases are also at extremely high pressure and, following the explosion, exert enormous forces on their surrounding medium, such as air, earth, or water, thus initiating a complex series of effects which depend upon the surrounding medium. Four types of bursts are distinguished: (1) air, (2) underground, (3) underwater, and (4) very high altitude (exoatmospheric). *See* ATOMIC BOMB; HYDROGEN BOMB; NUCLEAR FISSION; NUCLEAR FUSION; RADIOACTIVE FALLOUT. [H.Br.]

Nuclear fission An extremely complex nuclear reaction representing a cataclysmic division of an atomic nucleus into two nuclei of comparable mass. This rearrangement or division of a heavy nucleus may take place naturally (spontaneous fission) or under bombardment with neutrons, charged particles, gamma rays, or other carriers of energy (induced fission). Although nuclei with mass number A of approximately 100 or greater are energetically unstable against division into two lighter nuclei, the fission process has a small probability of occurring, except with the very heavy elements. Even for these elements, in which the energy release is of the order of 200 megaelectronvolts, the lifetimes against spontaneous fission are reasonably long. *See* NUCLEAR REACTION.

Liquid-drop model. The stability of a nucleus against fission is most readily interpreted when the nucleus is viewed as being analogous to an incompressible and charged liquid drop with a surface tension. Long-range Coulomb forces between protons act to disrupt the nucleus, whereas short-range nuclear forces, idealized as a surface tension, act to stabilize it. The degree of stability is then the result of a delicate balance between the relatively weak electromagnetic forces and the strong nuclear forces. Although each of these forces results in potentials of several hundred megaelectronvolts, the height of a typical barrier against fission for a heavy nucleus, because they are of opposite sign but do not quite cancel, is only 5 or 6 MeV. Investigators have used this charged liquid-drop model with great success in describing the general features of nuclear fission and also in reproducing the total nuclear binding energies. *See* NUCLEAR BINDING ENERGY; NUCLEAR STRUCTURE; SURFACE TENSION.

Shell corrections. The general dependence of the potential energy on the fission coordinate representing nuclear elongation or deformation for a heavy nucleus such as ^{240}Pu is shown in Fig. 1. The expanded scale used in this figure shows

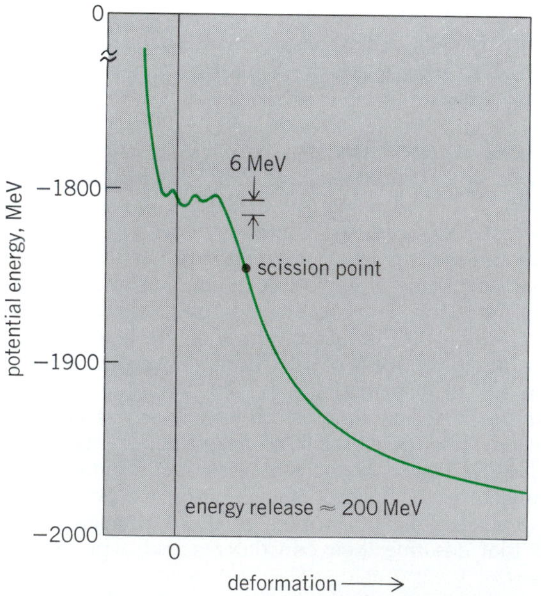

Fig. 1. Plot of the potential energy in MeV as a function of deformation for the nucleus ^{240}Pu. (*After M. Bolsteli et al., New calculations of fission barriers for heavy and superheavy nuclei, Phys. Rev., 5C:1050–1077, 1972*)

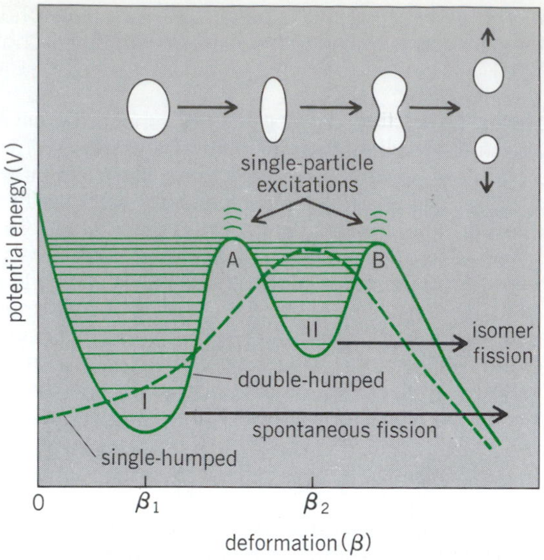

Fig. 2. Schematic plots of single-humped fission barrier of liquid-drop model and double-humped barrier introduced by shell corrections. Humps at *A* and *B* result in minima in potential energy at deformation of β_1 and β_2. States in these wells are designated class I and class II states respectively. (*After J. R. Huizenga, Nuclear fission revisited, Science, 168:1405–1413, 1979*)

the large decrease in energy of about 200 MeV as the fragments separate to infinity. It is known that ^{240}Pu is deformed in its ground state, which is represented by the lowest minimum of −1813 MeV near zero deformation. This energy represents the total nuclear binding energy when zero of potential energy is the energy of the individual nucleons at a separation of infinity. The second minimum to the right of zero deformation illustrates structure introduced in the fission barrier by shell corrections, that is, corrections dependent upon microscopic behavior of the individual nucleons, to the liquid-drop mass. Although shell corrections introduce small wiggles in the potential-energy surface as a function of deformation, the gross features of the surface are reproduced by the liquid-drop model. Since the typical fission barrier is only a few megaelectronvolts, the magnitude of the shell correction need only be small for irregularities to be introduced into the barrier. This structure is schematically illustrated for a heavy nucleus by the double-humped fission barrier in Fig. 2, which represents the region to the right of zero deformation in Fig. 1 on an expanded scale. The fission barrier has two maxima and a rather deep minimum in between. For comparison, the single-humped liquid-drop barrier is also schematically illustrated. The transition in the shape of the nucleus as a function of deformation is schematically represented in the upper part of the figure.

Experimental consequences. The observable consequences of the double-humped barrier have been reported in numerous experimental studies. In the actinide region more than 30 spontaneously fissionable isomers have been discovered between uranium and berkelium, with half-lives ranging from 10^{-11} to 10^{-2} s. These decay rates are faster by 20 to 30 orders of magnitude than the fission half-lives of the ground states, because of the increased barrier tunneling probability (see Fig. 2). Several cases in which excited states in the second minimum decay by fission are also known. Normally these states decay within the well by gamma decay; however, if there is a hindrance in gamma decay due to spin, the state (known as a spin isomer) may undergo fission instead.

Fission probability. The cross section for particle-induced fission $\sigma(y,f)$ represents the cross section for a projectile y to react with a nucleus and produce fission, as shown by the equation below. The quantities $\sigma_R(y)$, Γ_f, and Γ_t are the total

$$\sigma(y,f) = \sigma_R(y)\,(\Gamma_f/\Gamma_t)$$

reaction across sections for the incident particle y, the fission width, and the total level width, respectively where $\Gamma_t = \Gamma_f + \Gamma_n + \Gamma_y + \cdots$ is the sum of all partial-level widths. All the quantities in the above equation are energy-dependent.

When the incoming neutron has low energy, the likelihood of reaction is substantial only when the energy of the neutron is such as to form a compound nucleus in one or another of its resonance levels. The requisite sharpness of the "tuning" of the energy is specified by the total level width Γ. The nuclei ^{233}U, ^{235}U, and ^{239}Pu have a very large cross section to take up a slow neutron and undergo fission because both their absorption cross section and their probability for decay by fission are large. The probability for fission decay is high because the binding energy of the incident neutron is sufficient to raise the energy of the compound nucleus above the fission barrier. The very large, slow neutron fission cross sections of these isotopes make them important fissile materials in a chain reactor. *See* CHAIN REACTION (PHYSICS); REACTOR PHYSICS.

Postscission phenomena. After the nuclear fragments are separated, they are further accelerated as the result of the large Coulomb repulsion. The initially deformed fragments collapse to their equilibrium shapes, and the excited primary fragments lose energy by evaporating neutrons. After neutron emission, the fragments lose the remainder of their energy by gamma radiation, with a lifetime of about 10^{-11} s. The variation of neutron yield with fragment mass is directly related to the fragment excitation energy. Minimum neutron yields are observed for nuclei near closed shells because of the resistance to deformation of nuclei with closed shells. Maximum neutron yields occur for fragments that are "soft" toward nuclear deformation.

After the emission of the prompt neutrons and gamma rays, the resulting fission products are unstable against ß-decay. For example, in the case of thermal neutron fission of ^{235}U, each fragment undergoes on the average about three ß-decays before it settles down to a stable nucleus. For selected fission products (for example, ^{87}Br and ^{137}I) ß-decay leaves the daughter nucleus with excitation energy exceeding its neutron binding energy. The resulting delayed neutrons amount, for thermal neutron fission of ^{235}U, to about 0.7% of all the neutrons given off in fission. Though small in number, they are quite important in stabilizing nuclear chain reactions against sudden minor fluctuations in reactivity. *See* DELAYED NEUTRON; NEUTRON; THERMAL NEUTRONS. [J.R.Hu.]

Nuclear fuel cycle The steps by which fissionable (for example ^{233}U, ^{235}U, ^{239}Pu) and fertile (for example, ^{138}U, ^{232}Th) materials are prepared for use in, and recycled or discarded after discharge from, the nuclear reactor. These steps include mining of uranium- or thorium-bearing ore and milling of the ore to form concentrates. The uranium concentrate is converted to the volatile uranium hexafluoride (UF_6) that is used in the separation of isotopes to produce uranium enriched in the fissile ^{235}U. Another part of the fuel cycle is the fabrication of the enriched uranium into fuel assemblies. After the fuel has liberated the desired amount of heat in the reactor, the spent assemblies are reprocessed to separate the remaining fissionable and fertile material (uranium and plutonium) from the nuclear wastes. Other steps in the fuel cycle include the various transportation operations that move materials from one step to another, often connecting plants many hundreds of miles apart. Finally, waste management includes the treatment, storage, and disposal of radioactive wastes from the many other parts of the fuel cycle. The fuel cycle for a thorium-based reactor requires, in addition to the uranium fuel cycle steps outlined above, the mining, milling, and purification of thorium and the reprocessing of thorium-containing fuel into its components, which include unused thorium, uranium in the form of fissionable ^{233}U, and nuclear waste. *See* NUCLEAR FUELS REPROCESSING: NUCLEAR REACTOR; RADIOACTIVE WASTE MANAGEMENT; THORIUM; URANIUM. [M.J.St.]

Nuclear fuels The fissionable and fertile elements and isotopes used as the sources of energy in nuclear reactors. Although many heavy elements can be made to fission by bombardment with high-energy alpha particles, protons, deuterons, or neutrons, only neutrons can provide a self-sustaining reaction.

The number of neutrons released in the fission process varies from one per many fissions for elements just beyond the fission point (silver) to two or more per fission for the heavier elements, such as thorium and uranium. Even in such elements, neutron capture by the nucleus accompanied by the release of excess energy in the form of a gamma ray occurs in many cases, rather than nuclear fission. This reduces the number of neutrons available for further fission. The ratio of neutron capture to neutron fission varies from nucleus to nucleus and changes with the energy of the bombarding neutrons. Only a few isotopes of the heavy elements have a higher probability of fission than capture. These fissionable isotopes, ^{233}U, ^{235}U, and ^{239}Pu, are the only materials that can sustain the fission reaction and are therefore called nuclear fuels. *See* NUCLEAR REACTOR; THORIUM; URANIUM. [R.L.Be.]

Nuclear fuels reprocessing The treatment of spent reactor fuel elements to recover fissionable and fertile material. Spent fuel is usually discharged from reactors because of chemical, physical, and nuclear changes that make the fuel no longer efficient for the production of heat, rather than because of the complete depletion of fissionable material. Therefore, discharged fuel usually contains fissionable material in sufficient amounts to make its recovery attractive. In the case of breeder reactors, in order to take advantage of the characteristic of the breeder reactor to produce more fissionable material than is used, reprocessing of fuel must be done to recover the fissile material bred into part of the fuel. If fertile material is also contained in the fuel, it is ordinarily recovered and purified during fuel reprocessing. Purification of the valuable constituents consists of the removal of fission products and extraneous structural material present in the fuel. *See* NUCLEAR FUELS; NUCLEAR REACTOR.

There are several basic steps involved in fuel reprocessing. After fuel has been discharged from a nuclear reactor, it is common practice to store the fuel submerged in 15–20 ft (4.6–6.1 m) of water (for cooling and radiation-shielding purposes) for a period of 50–200 days; this allows the short-lived fission products to decay radioactively. Following the cooling period, the fuel is mechanically cut or disassembled into convenient sizes.

The specific steps next undertaken depend upon the particular reprocessing method employed to separate the desired products from each other, from fission products, and from extraneous structural materials. Although many separation methods exist, the one based upon solvent extraction principles is most frequently used for fuel reprocessing.

The operating experience to date with fuel-reprocessing plants has shown them to be relatively safe in spite of hazards from radiation and nuclear criticality, as well as other hazards of a more conventional nature. *See* HEALTH PHYSICS; RADIOACTIVE WASTE MANAGEMENT. [M.J.St.]

Nuclear fusion One of the primary nuclear reactions, the name usually designating an energy-releasing rearrangement collision which can occur between various isotopes of low atomic number. *See* NUCLEAR REACTION.

Interest in the nuclear fusion reaction arises from the expectation that it may someday be used to produce useful power, from its role in energy generation in stars, and from its use in the fusion bomb. Since a primary fusion fuel, deuterium, occurs naturally and is therefore obtainable in virtually inexhaustible supply, solution of the fusion power problem would permanently solve the problem of the present rapid depletion of chemically valuable fossil fuels. As a power source, the lack of radioactive waste products from the fusion reaction is another argument in its favor as opposed to the fission of uranium. *See* HYDROGEN BOMB; NUCLEAR FISSION.

In a nuclear fusion reaction the close collision of two energy-rich nuclei results in a mutual rearrangement of their nucleons (protons and neutrons) to produce two or more reaction products, together with a release of energy. The energy usually appears in the form of kinetic energy of the reaction products, although when energetically allowed, part may be taken up as energy of an excited state of a product nucleus. In contrast to neutron-produced nuclear reactions, colliding nuclei, because they are positively charged, require a substantial initial relative kinetic energy to overcome their mutual electrostatic repulsion so that reaction can occur. This required relative energy increases with the nuclear charge Z, so that reactions between low-Z nuclei are the easiest to produce. The best known of these are the reactions between the heavy isotopes of hydrogen, deuterium, and tritium. *See* DEUTERIUM; TRITIUM.

Nuclear fusion reactions can be self-sustaining if they are carried out at a very high temperature. That is to say, if the fusion fuel exists in the form of a very hot ionized gas of stripped nuclei and free electrons termed a plasma, the agitation energy of the nuclei can overcome their mutual repulsion, causing reactions to occur. This is the mechanism of energy generation in the stars and in the fusion bomb. It is also the method envisaged for the controlled generation of fusion energy. *See* PLASMA PHYSICS.

The cross sections (effective collisional areas) for many of the simple nuclear fusion reactions have been measured with high precision. It is found that the cross sections generally show broad maxima as a function of energy and have peak values in the general range of 0.01 barn (1 barn $= 10^{-24}$ cm^2) to a maximum value of 5 barns, for the deuterium-tritium (D-T) reaction. The energy releases of these reactions can be readily calculated from the mass difference between the initial and final nuclei or determined by direct measurement.

Some of the important simple fusion reactions, their reaction products, and their energy releases are:

$$\mathrm{D} + \mathrm{D} \rightarrow \mathrm{He}^3 + n + 3.25\ \mathrm{MeV}$$

$$\mathrm{D} + \mathrm{D} \rightarrow \mathrm{T} + p + 4.0\ \mathrm{MeV}$$

$$\mathrm{T} + \mathrm{D} \rightarrow \mathrm{He}^4 + n + 17.6\ \mathrm{MeV}$$

$$\mathrm{He}^3 + \mathrm{D} \rightarrow \mathrm{He}^4 + p + 18.3\ \mathrm{MeV}$$

$$\mathrm{Li}^6 + \mathrm{D} \rightarrow 2\mathrm{He}^4 + 22.4\ \mathrm{MeV}$$

$$\mathrm{Li}^7 + \mathrm{p} \rightarrow 2\mathrm{He}^4 + 17.3\ \mathrm{MeV}$$

If it is remembered that the energy release in the chemical reaction in which hydrogen and oxygen combine to produce a water molecule is about 1 eV per reaction, it will be seen that, gram for gram, fusion fuel releases more than 1,000,000 times as much energy as typical chemical fuels. [R.F.P.]

Nuclear isomerism The existence of excited states of atomic nuclei with unusually long lifetimes. If the lifetime of a specific excited state is unusually long, compared with the lifetimes of other excited states in the same nucleus, the state is said to be isomeric. The definition of the boundary between isomeric and normal decays is arbitrary, and the term is therefore used loosely. *See* EXCITED STATE; PARITY (QUANTUM MECHANICS); SPIN (QUANTUM MECHANICS).

The predominant decay mode of excited nuclear states is by γ-ray emission. The rate at which this process occurs is determined largely by the spins, parities, and excitation energies of the decaying state and of those to which it is decaying. In particular, the rate is extremely sensitive to the difference in the spins of initial and final states and to the difference in excitation energies. Both extremely large spin differences and extremely small energy differences can result in a slowing of the γ-ray emission by many orders of magnitude, resulting in some excited states having unusually long lifetimes and therefore being termed isomeric.

In addition to spin isomers, two other types of isomers have been identified. The first of these arises from the fact that some excited nuclear states represent a drastic change in shape of the nucleus from the shape of the ground state. In many cases this extremely deformed shape displays unusual stability, and states with this shape are therefore isomeric. A particularly important class of these shape isomers is observed in the decay of heavy nuclei by fission, and the study of such fission isomers has been the subject of intensive effort. *See* NUCLEAR FISSION.

A more esoteric form of isomer has also been observed, the so-called pairing isomer which results from differences in the microscopic motions of the constituent nucleons in the nucleus. A state of this type has a quite different character from the ground state of the nucleus, and is therefore also termed isomeric. *See* NUCLEAR STRUCTURE. [R.Bet.]

Nuclear laser A device in which laser action in a gas is excited by the fast-moving ions produced in nuclear fusion. These fusion products excite and ionize the gas atoms, and make it possible to convert directly from nuclear to optical energy. *See* LASER. [S.F.J.; A.L.S.]

Nuclear magnetic resonance (NMR) A phenomenon exhibited by a large number of atomic nuclei which is based upon the existence of nuclear magnetic moments which are associated with quantized nuclear spins. These nuclear moments, when placed in a magnetic field, give rise to distinct nuclear Zeeman energy levels between which spectroscopic transitions can be induced by radio-frequency radiation. Plots of these transition frequencies, called spectra, furnish important information about molecular structure and sample composition.

Spinning nuclei with nonzero angular momentum behave like tiny magnets. When placed in an external magnetic field, these tiny nuclear magnets assume certain allowed orientations (Zeeman energy levels) with respect to it. The nuclei can be reoriented by adding energy of the exact transition frequency.

As the separation between the nuclear Zeeman levels is directly proportional to the strength of the perturbing magnetic field, the transition frequency can be varied for a given nucleus by merely changing the applied magnetic field. In this regard, nuclear magnetic resonance (NMR) spectroscopy is unlike other spectroscopic methods, in which the investigator is unable to control the frequency of the spectral transition. Thus an NMR spectrum may be secured by varying the magnetic field to bring the separation of the Zeeman levels into correspondence with a constant irradiating frequency; or the alternative experimental method may be used, in which a constant magnetic field is employed and the irradiating frequency is varied over the range of spectroscopic frequencies. *See* ZEEMAN EFFECT.

It is this unique field-frequency relationship that makes possible the application of NMR spectroscopy in molecular studies. Although identical nuclei have the same frequency dependence upon the magnetic field, a difference in the chemical environment can modify the applied magnetic field, so that all such nuclei in the same sample do not experience the same net magnetic field, and thus they do not resonate at exactly the same frequency. The corresponding spectral shift in the transition frequencies between two such chemically nonequivalent nuclei is referred to as the chemical shift. Being directly proportional to the total applied field, this parameter is recorded in the relative units of parts per million (ppm). *See* SPECTROSCOPY. [G.Sl.]

Nuclear magnetic resonance flowmeter An instrument that uses the phenomenon of nuclear magnetic resonance to measure the velocity of fluid flow. Nuclei of the flowing fluid are resonated by a radio-frequency field superimposed on an intense permanent magnetic field. A detector downstream measures the amount of decay of the resonance and thereby senses the velocity of the fluid. The most effective fluids are hydrocarbons, fluorocarbons, and water-bearing liquids because this resonance is most pronounced in fluorine and hydrogen nuclei. *See* FLOW MEASUREMENT; NUCLEAR MAGNETIC RESONANCE (NMR). [M.Br.; L.P.E.]

Nuclear medicine A subspecialty of medicine based primarily on the use of radioactive substances in medical diagnosis, treatment, and research. In a typical examination a radioactive tracer is given, usually by injection into an arm vein, and the distribution of the radioactive substance within the body or a part of the body is portrayed in a series of nuclear images. The imaging is based on the emission of gamma rays by the tracer substance that pass out of the body and are recorded by a scintillation camera. The scintillation camera is a radiation detection device consisting of a large sodium iodide crystal (nearly 1 m in diameter in some cases) which detects the gamma rays that interact with the crystal and locates where on the detector face the interaction has occurred. These interactions are used to produce a picture or image of where the gamma rays originated within the body. The methods are related to those in routine x-ray examinations, except that gamma rays are emitted from the body to provide the diagnostic information, rather than being transmitted through the body. Nuclear images depend on specific radioactive elements or compounds labeled with radioactive elements being selectively concentrated in an organ, making it possible to obtain pictures that provide information about regional function within the organ. *See* GAMMA-RAY DETECTORS; SCINTILLATION COUNTER.

Nuclear imaging techniques make it possible to examine functions of the human body noninvasively, in ways that at times surpass the perception of pathologists at the autopsy table and surgeons at the operating table. The body can be examined as a series of physiologic processes, not simply as static structures. *See* RADIOACTIVE TRACER; RADIOISOTOPE; RADIOLOGY. [H.N.W.]

Nuclear molecule A quasistable entity of nuclear dimensions formed in nuclear collisions and comprising two or more discrete nuclei that retain their identities and are bound together by strong nuclear forces. Whereas the stable molecules of chemistry and biology consist of atoms bound through various electronic mechanisms, nuclear molecules do not form in nature except possibly in the hearts of giant stars; this simply reflects the fact that all nuclei carry positive electrical charges, and that under all natural conditions the long-range electrostatic repulsion prevents nuclear components from coming within the grasp of the short-range attractive nuclear force which could provide molecular binding. But in energetic collisions this electrostatic repulsion can be overcome. *See* NUCLEAR STRUCTURE. [D.A.B.]

Nuclear moments Intrinsic properties of atomic nuclei: electric moments result from deviations of the nuclear charge distribution from spherical symmetry; magnetic moments are a consequence of the intrinsic spin and the rotational motion of nucleons within the nucleus. The classical definitions of the magnetic and electric multipole moments are written in general in terms of multipole expansions. *See* NUCLEAR STRUCTURE; SPIN (QUANTUM MECHANICS).

In special cases nuclear moments can be measured by direct methods involving the interaction of the nucleus with an external magnetic field or with an electric field gradient produced by the scattering of high-energy charged particles. In general, however, nuclear moments manifest themselves through the hyperfine interaction between the nuclear moments and the fields or field gradients produced by either the atomic electrons' currents and spins, or the molecular or crystalline electronic and lattice structures. *See* HYPERFINE STRUCTURE. [N.Ko.]

Nuclear orientation The directional ordering of an assembly of nuclear spins I with respect to some axis in space. Under normal conditions nuclei are not oriented; that is, all directions in space are equally probable. For a system of nuclear spins with rotational symmetry about an axis, the degree of orientation is completely characterized by the relative populations a_m of the $2I + 1$ magnetic sublevels $m(= I, I - 1, \ldots, -I)$.

Nuclear orientation can be achieved in various ways. The most obvious way is to modify the energies of the $2I + 1$ magnetic sublevels so as to remove their degeneracy and thereby change the populations of these sublevels. The spin degeneracy can be removed by a magnetic field interacting with the nuclear magnetic dipole moment, or by an inhomogeneous electric field interacting with the nuclear electric quadrupole moment. Significant differences in the populations of the sublevels can be established by cooling the nuclear sample to low temperatures. This means of producing nuclear orientation is called the static method. In contrast, there is the dynamic method, which is related to optical pumping in gases. There are other ways to produce oriented nuclei; for example, in a nuclear reaction such as the capture of polarized neutrons (produced by magnetic scattering) by unoriented nuclei. *See* DYNAMIC NUCLEAR POLARIZATION; OPTICAL PUMPING.

Oriented nuclei have been used to measure nuclear properties, for example, magnetic dipole and electric quadrupole moments, spins, parities, and mixing ratios of nuclear states. Oriented nuclei have been used to examine some of the fundamental properties of nuclear forces, for example, nonconservation of parity in the weak interaction. Measurement of hyperfine fields, electric-field gradients, and other properties relating to the environment of the nucleus have been made by using oriented nuclei. Nuclear orientation thermometry is one of the few sources of a primary temperature scale at low temperatures. Oriented nuclear targets used in conjunction with beams of polarized and unpolarized particles have proved very useful in examining certain aspects of the nuclear force. *See* LOW-TEMPERATURE THERMOMETRY; NUCLEAR MOMENTS; NUCLEAR STRUCTURE; PARITY (QUANTUM MECHANICS). [H.Mar.]

Nuclear physics The discipline involving the structure of atomic nuclei and their interactions with each other, with their constituent particles, and with the whole spectrum of elementary particles that is provided by very large accelerators. The nuclear domain occupies a central position between the atomic range of forces and sizes and those of elementary-particle physics, characteristically within the nucleons themselves. As the only system in which all the known natural forces can be studied simultaneously, it provides a natural laboratory for the testing and extending of many fundamental symmetries and laws of nature. Containing a reasonably large, yet manageable number of strongly interacting components, the nucleus also occupies a central position in the universal many-body problem of physics. *See* ATOMIC NUCLEUS; ATOMIC STRUCTURE AND SPECTRA; ELEMENTARY PARTICLE; SYMMETRY LAWS (PHYSICS).

Nuclear physics is unique in the extent to which it merges the most fundamental and the most applied topics. Its instrumentation has found broad applicability throughout science, technology, and medicine; nuclear engineering and nuclear medicine are two very important areas of applied specialization. *See* NUCLEAR ENGINEERING; NUCLEAR RADIATION (BIOLOGY); RADIOLOGY.

Nuclear chemistry, certain aspects of condensed matter and materials science, and nuclear physics together constitute the broad field of nuclear science; outside the United States and Canada elementary particle physics is frequently included in this more general classification. *See* ANALOG STATES; COSMIC RAYS; FUNDAMENTAL INTERACTIONS; ISOTOPE; NUCLEAR CHEMISTRY; NUCLEAR FISSION; NUCLEAR FUSION; NUCLEAR ISOMERISM; NUCLEAR MOMENTS; NUCLEAR REACTION; NUCLEAR REACTOR; NUCLEAR SPECTRA; NUCLEAR STRUCTURE; PARTICLE ACCELERATOR; PARTICLE DETECTOR; RADIOACTIVITY; SCATTERING EXPERIMENTS (NUCLEI); WEAK NUCLEAR INTERACTIONS. [D.A.B.]

Nuclear power Power derived from fission or fusion nuclear reactions. More conventionally, nuclear power is interpreted as the utilization of the fission reactions in a nuclear power reactor to produce steam for electrical power production, for ship propulsion, or for process heat. Fission reactions involve the breakup of the nucleus of heavy-weight atoms and yield-energy release which is more than a millionfold greater than that obtained from chemical reactions involving the burning of a fuel. Successful control of the nuclear fission reactions provides for the utilization of this intensive source of energy, and with the availability of ample resources of uranium deposits, significantly cheaper fuel costs for electrical power generation are attainable. Safe, clean, economic nuclear power has been the objective both of the federal government and of industry's programs for research, development, and demonstration. Critics of nuclear power seek a complete ban or at least a moratorium on new commercial plants. *See* NUCLEAR FISSION.

The inherent dangers associated with nuclear power which involves unprecedented quantities of radioactive materials, including possible wide-scale use of plutonium, have been recognized, and extensive programs for safety, ecological, and biomedical studies, research, and testing have been integral with the advancement of the engineering of nuclear power. Criteria for siting a nuclear power station involve thorough investigation of the region's geology, seismology, hydrology, meteorology, demography, and nearby industrial, transportation, and military facilities. Also included are emergency plans to cope with fires or explosions, and radiation accidents arising from operational malfunctions, natural disasters, and civil disturbances. *See* NUCLEAR REACTOR; RADIOACTIVE WASTE MANAGEMENT. [H.S.I.]

Nuclear quadrupole resonance A selective absorption phenomenon observable in a wide variety of polycrystalline compounds containing nonspherical atomic nuclei when placed in a magnetic radio-frequency field. Nuclear quadrupole resonance (NQR) is very similar to nuclear magnetic resonance (NMR), and was originated as an inexpensive (no stable homogeneous large magnetic field is required) alternative way to study nuclear moments. It later gained a modest popularity. *See* MAGNETIC RESONANCE; NUCLEAR MAGNETIC RESONANCE (NMR).

In the simplest case, for example, ^{35}Cl in solid Cl_2, NQR is associated with the precession of the angular momentum I (and the nuclear magnetic dipole moment μ) of the nucleus, depicted in the illustration as a flat ellipsoid of rotation, around the symmetry axis (taken as the z axis) of the Cl_2 molecule fixed in the crystalline solid. The precession, with constant angle θ between the nuclear axis and symmetry axis of the molecule, is due to the torque which the inhomogeneous molecular electric field exerts on the nucleus of electric quadrupole moment eQ. The absorption occurs classically when the frequency of the rf field and that of the precessing motion of the angular momentum coincide.

NQR spectra have been observed in the approximate range 1–1000 MHz. Most of the NQR work has been on molecular

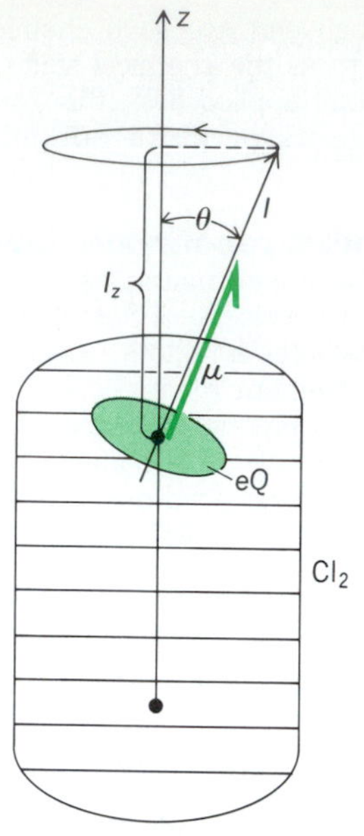

Interaction of ^{35}Cl nucleus with the electric field of a Cl_2 molecule.

crystals. For such crystals the coupling constants found do not differ very much from those measured for the isolated molecules in microwave spectroscopy. The most precise nuclear information which may be extracted from NQR data are quadrupole moment ratios of isotopes of the same element. If values for the axial gradient of the molecular electric field can be estimated from atomic fine structure data, then fair values of the quadrupole moment may be obtained. However, it has also proved very productive to use the quadrupole nucleus as a probe of bond character and orientation and crystalline electric fields and lattice sites, and much data has been accumulated in this area. *See* MICROWAVE SPECTROSCOPY. [H.De.]

Nuclear radiation All particles and radiations emanating from an atomic nucleus due to radioactive decay and nuclear reactions. Thus the criterion for nuclear radiations is that a nuclear process is involved in their production. The term was originally used to denote the ionizing radiations observed from naturally occurring radioactive materials. These radiations were alpha rays (energetic helium nuclei), beta rays (negative electrons), and gamma rays (electromagnetic radiation with wavelength much shorter than visible light). *See* ALPHA PARTICLES; BETA PARTICLES; GAMMA RAYS.

Nuclear radiations have traditionally been considered to be of three types based on the manner in which they interact with matter as they pass through it. These are the charged heavy particles with masses comparable to that of the nuclear mass (for example, protons, alpha particles, and heavier nuclei), electrons (both negatively and positively charged), and electromagnetic radiation. For all of these, the interactions with matter are considered to be primarily electromagnetic. The behavior of mesons and other particles is intermediate between that of the electron and heavy charged particles.

A striking difference in the absorption of the three types of radiations is that only heavy charged particles have a range. That is, a monoenergetic beam of heavy charged particles, in passing through a certain amount of matter, will lose energy without changing the number of particles in the beam. Ultimately, they will be stopped after crossing practically the same thickness of absorber. For electromagnetic radiation (gamma rays) and neutrons, on the other hand, the absorption is exponential. The difference in behavior reflects the fact that charged particles are not removed from the beam by individual interactions, whereas gamma radiation photons (and neutrons) are removed. Electrons exhibit a more complex behavior. *See* ELECTRON; NUCLEAR REACTION. [D.G.K.]

Nuclear radiation (biology) Nuclear radiations are used in biology because of their common property of ionizing matter. This makes their detection relatively simple, or makes possible the production of biological effects in any living cell. *See* NUCLEAR RADIATION.

All ionizing radiations produce biological changes, directly by ionization or excitation of the atoms in the molecules of biological entities, such as in chromosomes, or indirectly by the formation of active radicals or deleterious agents, through ionization and excitation, in the medium surrounding the biological entities. Ionizing radiation, having high penetrating power, can reach the most vulnerable part of a cell, an organ, or a whole organism, and is thus very effective. The radiations emitted by radioactive isotopes of the various elements are used in biological research. *See* AUTORADIOGRAPHY; HISTORADIOGRAPHY.

The medical uses of nuclear radiations may be divided into three distinct classes:

1. The radiations, which are principally x-rays, are used to study the anatomical configuration of body organs, usually for the purpose of detecting abnormalities as an aid in diagnosis.
2. The radiations are used for therapeutic purposes to produce biological changes in such tissues as tumors.
3. The radiations are used as a simple means of tracing a suitable radioactive substance through different steps in its course through the body, in the study of some particular physiological process. *See* RADIOLOGY. [G.F./E.H.Q.]

Nuclear reaction A process that occurs as a result of interactions between atomic nuclei when the interacting particles approach each other to within distances of the order of nuclear dimensions ($\simeq 10^{-12}$ cm). While nuclear reactions occur in nature, understanding of them and use of them as tools have taken place primarily in the controlled laboratory environment. In the usual experimental situation, nuclear reactions are initiated by bombarding one of the interacting particles, the stationary target nucleus, with nuclear projectiles of some type, and the reaction products and their behaviors are studied.

Types of nuclear interaction. As a generalized nuclear process, consider a collision in which an incident particle strikes a previously stationary particle, to produce an unspecified number of final products. If the final products are the same as the two initial particles, the process is called scattering. The scattering is said to be elastic or inelastic, depending on whether some of the kinetic energy of the incident particle is used to raise either of the particles to an excited state. If the product particles are different from the initial pair, the process is referred to as a reaction.

The most common type of nuclear reaction, and the one which has been most extensively studied, involves the production of two final products. Such reactions can be observed, for example, when deuterons with a kinetic energy of a few megaelectronvolts are allowed to strike a carbon nucleus of mass 12. Protons, neutrons, deuterons, and alpha particles are observed to be emitted, and reactions (1)–(4) are responsible. In these

$$ {}^{2}_{1}\mathrm{H} + {}^{12}_{6}\mathrm{C} \rightarrow {}^{2}_{1}\mathrm{H} + {}^{12}_{6}\mathrm{C} \quad (1)$$

$$ {}^{2}_{1}\mathrm{H} + {}^{12}_{6}\mathrm{C} \rightarrow {}^{1}_{1}\mathrm{H} + {}^{13}_{6}\mathrm{C} \quad (2)$$

$$ {}^{2}_{1}\mathrm{H} + {}^{12}_{6}\mathrm{C} \rightarrow {}^{0}_{1}n + {}^{13}_{7}\mathrm{N} \quad (3)$$

$$ {}^{2}_{1}\mathrm{H} + {}^{12}_{6}\mathrm{C} \rightarrow {}^{4}_{2}\mathrm{He} + {}^{10}_{5}\mathrm{B} \quad (4)$$

equations the nuclei are indicated by the usual chemical symbols; the subscripts indicate the atomic number (nuclear charge) of the nucleus, and the superscripts the mass number of the particular isotope. These reactions are conventionally written in the compact notation $^{12}\mathrm{C}(d,d)^{12}\mathrm{C}$, $^{12}\mathrm{C}(d,p)^{13}\mathrm{C}$, $^{12}\mathrm{C}(d,n)^{13}\mathrm{N}$, and $^{12}\mathrm{C}(d,\alpha)^{10}\mathrm{B}$, where d represents deuteron, p proton, n neutron, and α alpha particle. In each of these cases the reaction results in the production of an emitted light particle and a heavy residual nucleus. If the residual nucleus is formed in an excited state, it will subsequently emit this excitation energy in the form of gamma rays or, in special cases, electrons. The residual nucleus may also be a radioactive species, in which case it will undergo further transformation in accordance with its characteristic radioactive decay scheme. *See* RADIOACTIVITY.

Nuclear cross section. In general one is interested in the probability of occurrence of the various reactions as a function of the bombarding energy of the incident particle. The measure of probability for a nuclear reaction is its cross section. Consider a reaction initiated by a beam of particles incident on a region which contains N atoms per unit area (uniformly distributed), and where I particles per second striking the area result in R reactions of a particular type per second. The fraction of the area bombarded which is effective in producing the reaction products is R/I. If this is divided by the number of nuclei per unit area, the effective area or cross section $\sigma = R/IN$. This is referred to as the total cross section for the specific reaction, since it involves all the occurrences of the reaction. The dimensions are those of an area, and total cross sections are expressed in either square centimeters or barns (1 barn $= 10^{-24}$ cm^2). The differential cross section refers to the probability that a particular reaction product will be observed at a given angle with respect to the beam direction. Its dimensions are those of an area per unit solid angle (for example, barns per steradian).

Reaction mechanism. Various reaction models have been extremely successful in describing certain classes or types of nuclear reaction processes. In general, all reactions can be classified according to the time scale on which they occur, and the degree to which the kinetic energy of the incident particle is converted into internal excitation of the final products. A large fraction of the reactions observed has properties consistent with those predicted by two reaction mechanisms which represent the extremes in this general classification. These are the mechanisms of compound nucleus formation and direct interaction.

Compound nucleus formation is envisioned to take place in two distinct steps. In the first step the incident particle is captured by (or fuses with) the target nucleus, forming an intermediate or compound nucleus which lives a long time ($\simeq 10^{-16}$ s) compared to the approximately 10^{-22} s it takes the incident particle to travel past the target. During this time the kinetic energy of the incident particle is shared among all the nucleons, and all memory of the incident particle and target is lost. The compound nucleus is always formed in a highly excited unstable state, is assumed to approach themodynamic equilibrium involving all or most of the available degrees of freedom, and will decay, as the second step, into different reaction products, or through so-called exit channels. The essential feature of the compound nucleus formation or fusion reaction is that the probability for a specific reaction depends on two independent probabilities: the probability for forming the compound nucleus, and the probability for decaying into that specific exit channel.

Some reactions have properties which are in striking conflict with the predictions of the compound nucleus hypothesis. Many of these are consistent with the picture of a mechanism where no long-lived intermediate system is formed, but rather a fast mechanism where the incident particle, or some portion of it, interacts with the surface, or some nucleons on the surface, of the target nucleus. These direct reactions are assumed to involve only a very small number of the available degrees of freedom. Most direct reactions are of the transfer type where one or more nucleons are transferred to or from the incident particle as it passes the target, leaving the two final partners either in their ground states or in one of their many excited states. Such transfer reactions are generally referred to as stripping or pick-up reactions, depending on whether the incident particle has lost or acquired nucleons in the reaction.

Inelastic scattering is also a direct reaction. Whereas the states preferentially populated in transfer reactions are those of specific single-particle or shell-model structure, the states preferentially excited in inelastic scattering are collective in nature. *See* NUCLEAR STRUCTURE; SCATTERING EXPERIMENTS (NUCLEI). [D.G.K.]

Nuclear reactor A system utilizing nuclear fission in a controlled and self-sustaining manner. Neutrons are used to fission the nuclear fuel, and the fission reaction produces not only energy and radiation, but also additional neutrons. Thus a neutron chain reaction ensues. A nuclear reactor provides the assembly of materials to sustain and control the neutron chain reaction, to appropriately transport the heat produced from the fission reactions, and to provide the necessary safety features to cope with the radiation and radioactive materials produced by the operation of the nuclear reactor. *See* CHAIN REACTION (PHYSICS); NUCLEAR FISSION.

Nuclear reactors are used in a variety of ways as sources for energy, for nuclear radiations, and for special tests and feasibility demonstrations. The generation of electric energy by a nuclear power plant requires the use of heat to produce steam or to heat gases to drive turbogenerators. Direct conversion of the fission energy into useful work is possible, but an efficient process has not yet been realized to accomplish this. Thus, in its operation the nuclear power plant is similar to the conventional coal-fired plant, except that the nuclear reactor is substituted for the conventional boiler. *See* NUCLEAR POWER.

Fuel and moderator. The fission neutrons are released at very high energies and are called fast neutrons. The average kinetic energy is 2 MeV, with a corresponding neutron speed of 1/15 the speed of light. Neutrons slow down through collisions with nuclei of the surrounding material. This slowing-down process is made more effective by the introduction of light-weight materials, called moderators, such as heavy water (deuterium oxide), ordinary (light) water, graphite, beryllium, beryllium oxide, hydrides, and organic materials (hydrocarbons). Neutrons that have slowed down to an energy state in equilibrium with the surrounding material are called thermal neutrons. The probability that a neutron will cause the fuel material to fission is greatly enhanced at thermal energies, and thus most reactors utilize a moderator for the conversion of fast neutrons to thermal neutrons.

With suitable concentrations of the fuel material, neutron chain reactions also can be sustained at higher neutron energy levels. The energy range between fast and thermal is designated as intermediate. Fast reactors do not have moderators and are relatively small. Reactors have been built in all three categories.

Only three isotopes—uranium-235, uranium-233, and plutonium-239—are feasible as fission fuels, but a wide selection of materials incorporating these isotopes is available.

Heat removal. The major portion of the energy released by the fissioning of the fuel is in the form of kinetic energy of the fission fragments, which in turn is converted into heat through the slowing down and stopping of the fragments. For the heterogeneous reactors this heating occurs within the fuel elements. Heating also arises through the release and absorption of the radiations from the fission process and from the radioactive materials formed. The heat generated in a reactor is removed by a primary coolant flowing through the reactor.

Reactor coolants. Coolants are selected for specific applications on the basis of their heat-transfer capability, physical properties, and nuclear properties.

Water has many desirable characteristics. It was employed as the coolant in the first production reactors and most power reactors still utilize water as the coolant. In a boiling-water reactor (see illustration) the water is allowed to boil and form steam that is piped to the turbine. In a pressurized-water reactor the coolant water is kept under increased pressure to prevent boiling, and transfers heat to a separate stream of water in a steam generator, changing that water to steam. *See* COOLING TOWER; RADIOACTIVE WASTE MANAGEMENT; VAPOR CONDENSER.

For both boiling-water and pressurized-water reactors, the water serves as the moderator as well as the coolant. Both light and heavy water are excellent neutron moderators, although heavy water (deuterium oxide) has a neutron-absorption cross section approximately 1/500 that for light water. The most serious limitation of water as a coolant for power reactors is its high vapor pressure. The high pressure necessary for water-cooled power reactors imposes severe design problems. *See* NUCLEAR REACTION.

Gases are inherently poor heat-transfer fluids as compared with liquids because of their low density. This situation can be improved by increasing the gas pressure; however, this introduces other problems. Helium is the most attractive gas (it is chemically inert and has good thermodynamic and nuclear properties) and has been selected as the coolant for the development of high-temperature gas-cooled reactor (HTGR) systems, in which the gas transfers heat from the reactor core to a steam generator. Gases are capable of operation at extremely high temperature, and they are being considered for special process applications and for direct-cycle gas-turbine applications.

Diphenyl and terphenyl possess good neutron-moderating properties and have lower vapor pressures than water. Organic coolants are noncorrosive and relatively inexpensive. Their major disadvantage is dissociation or decomposition under irradiation.

The alkali metals have excellent heat-transfer properties and extremely low vapor pressures at temperatures of interest for power generation. Sodium is the most attractive because of its relatively low melting point and high heat-transfer coefficient. It is also abundant, commercially available in acceptable purity, and relatively inexpensive. It is not particularly corrosive, provided low oxygen concentration is maintained. Its nuclear properties are excellent for fast reactors. In the liquid metal fast breeder reactor sodium in the primary loop collects the heat generated in the core and transfers it to a secondary sodium loop in the heat exchanger, from which it is carried to the steam generator.

Molten salts have been used as reactor coolants because they have favorable high-temperature properties and because mixtures of salts containing fuel permit fluid fuel-coolant systems.

Fluid flow and hydrodynamics. Because heat removal must be accomplished as efficiently as possible, considerable attention must be given to the fluid-flow and hydrodynamic characteristics of the system.

The heat capacity and thermal conductivity of the fluid at the temperature of operation have a fundamental effect upon the design of the reactor system. The heat capacity determines the mass flow of the coolant required. The fluid properties (thermal conductivity, viscosity, density, and specific heat) are important in determining the surface area required for the fuel to permit transfer of the heat generated at reasonable temperature differences. This, in turn, affects the design of the fuel—in particular, the amount and arrangement of the fuel elements. These factors combine to establish the pumping characteristics of the system because pressure drop and coolant-temperature rise are directly related.

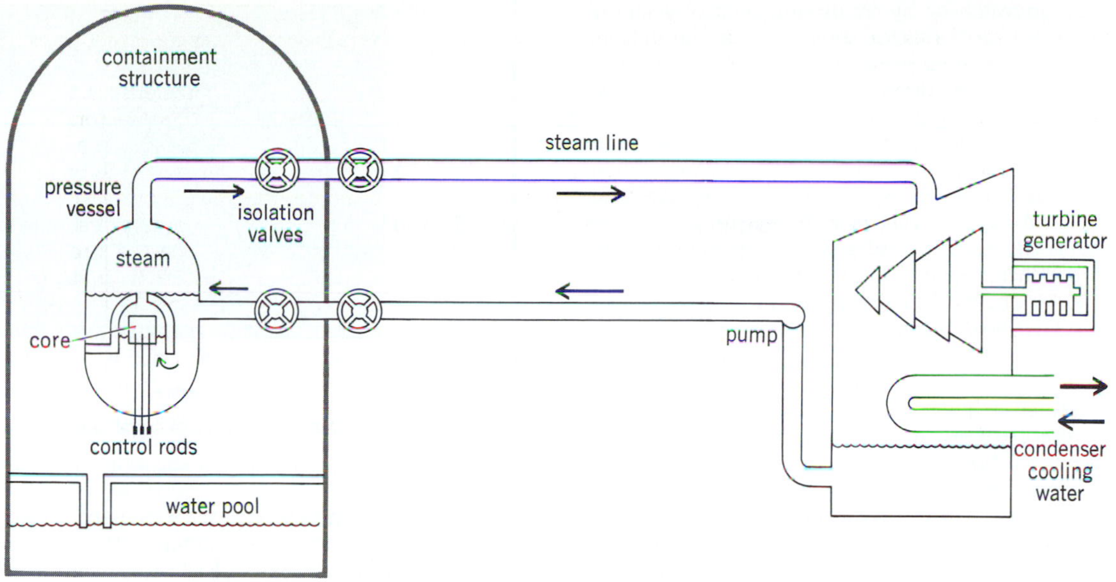

Boiling-water reactor (BWR). ***(Atomic Industrial Forum, Inc.)***

Thermal stress. The temperature of the reactor coolant increases as it circulates through the reactor. Fluctuations in power level or in coolant flow rate result in variations in the temperature rise. A reactor is capable of very rapid changes in power level, particularly, reduction in power level. Therefore, reactor-coolant systems must be designed to accommodate the temperature transients that may occur because of rapid power changes. In addition, they must be designed to accommodate temperature transients that might occur as a result of a coolant-system malfunction, such as pump stoppage. The consequent temperature stresses induced in the various parts of the system are superimposed upon the thermal stresses that exist under normal steady-state operations.

Coolant-system components. The development of reactor systems has necessitated concurrent development of special components for reactor coolant systems. These have been required even for systems employing conventional coolants, such as water or air. Because of the hazard of radioactivity, leak-tight systems and components are a prerequisite to safe, reliable operation and maintenance. Special problems are introduced by many of the fluids employed as reactor coolants.

More extensive component developments have been required for the alkali metals which are chemically very active and are extremely poor lubricants. Centrifugal pumps employing unique bearings and seals must be specially designed. Liquid metals are excellent electrical conductors, and, in some special cases, electromagnetic-type pumps have been developed. These pumps are completely sealed, contain no moving parts, and derive their pumping action from electromagnetic forces imposed directly on the fluid.

Core design. A typical reactor core for a power reactor consists of the fuel element rods supported by a grid-type structure inside a vessel.

Structural materials employed in reactor systems must possess suitable nuclear and physical properties and must be compatible with the reactor coolant under the conditions of operation. The most common structural materials employed in reactor systems are aluminum, stainless steel, and zirconium alloys. Aluminum and zirconium alloys have favorable nuclear properties, whereas stainless steel has favorable physical properties. Aluminum is widely used in low-temperature reactors; zirconium and stainless steel are used in high-temperature reactors. Zirconium is relatively expensive, and its use is therefore confined to applications where neutron absorption is critical.

Heterogeneous reactors maintain a separation of fuel and coolant by cladding the fuel. The cladding material must be compatible with both the fuel and the coolant.

The cladding materials must also have favorable nuclear properties. The neutron-capture cross section is most significant because the parasitic absorption of neutrons by these materials reduces the efficiency of the nuclear fission process. Aluminum is a very desirable material in this respect; however, its physical strength and corrosion resistance in water decrease very rapidly above about 300°F (149°C). Zirconium has very favorable neutron properties, and in addition can be made reasonably corrosion-resistant in high-temperature water. It is extensively used for water-cooled power reactors. Stainless steel is used for the fuel cladding in fast reactors.

Control. The control of reactors requires the measurement and adjustment of the critical condition. A reactor is critical when the rate of production of neutrons equals the rate of consumption in the system. The neutrons are produced by the fission process and are consumed in a variety of ways, including absorption to cause fission, nonfission capture in fissionable materials, capture in fertile materials, capture in structure or coolant, and leakage from the reactor. A reactor is subcritical (power level decreasing) if the number of neutrons produced is less than the number consumed. The reactor is supercritical (power level increasing) if the number of neutrons produced exceeds the number consumed.

Reactors are controlled by adjusting the balance between neutron production and neutron consumption. Normally, neutron consumption is controlled by varying the absorption or leakage of neutrons; however, the neutron-generation rate can be controlled by varying the amount of fissionable material in the system.

The reactor control system requires the movement of neutron-absorbing rods (control rods) in the reactor under very exacting conditions. They must be arranged to increase reactivity (increase neutron population) slowly and under absolute control. They must be capable of reducing reactivity, both rapidly and slowly.

Control drives are normally electromechanical devices that impart linear or swinging motion to the control rods. Normal operation of the control drives can be accomplished manually by the reactor operator or by automatic control systems. Reactor scram (very rapid reactor shutdown) can be initiated automatically by one or more system scram-safety signals, or it can be started manually by depressing a scram button convenient to the operator in the control room.

Applications. Reactor applications include production of fissionable fuels (plutonium and uranium-233); mobile, stationary, and packaged power plants; research, testing, teaching-demonstration, and experimental facilities; space and process heat; dual-purpose designs; and special applications. The potential use of reactor radiation for sterilization of food and other products, for chemical processes, and for high-temperature applications has been recognized. [H.S.I.]

Nuclear spectra The distribution in energy or momentum of radiations emitted by radioactive nuclei or in a nuclear reaction; also, the graphical display of data from instruments used to measure such radiations (for example, magnetic spectrometers and scintillation detectors). *See* NUCLEAR RADIATION.

Nuclear spectra are widely used in both basic and applied research. For the former, it is usually the information which can be inferred from studies of such spectra rather than the radiations themselves which are of primary interest. From experimental determinations of the type of radiation, its energy, and the changes that it causes in angular momentum and parity, one is able to deduce information pertaining to the static and dynamic properties of the initial and final nuclear configurations. This, in turn, enables one to gain insight into the structure of nuclei as well as the forces between nucleons. In applied research, it is the radiations themselves or the effects which they produce that are of most importance. *See* NUCLEAR STRUCTURE.

The distribution in energy (momentum) of radiations emitted during transitions between nuclear configurations can be discrete or continuous. Transitions which give rise to discrete spectra are those in which a single type of radiation is emitted. The emitted particle then has a specific energy determined by the two nuclear configurations and type of radiation involved. However, if two or more radiations are emitted during a transition, the energy will be shared among them. In this case, the energy of each particle emitted can take on a continuum of values between well-defined limits. This then gives rise to a continuous spectrum. *See* NUCLEAR REACTION; RADIOACTIVITY.

[D.J.Ho.]

Nuclear structure The atomic nucleus is at the center of the atom and contains 99.975% of the total mass of the atom. Its average density is about 3×10^{11} kg/cm^3; its diameter is about 10^{-12} cm, and thus much smaller than the diameter of the atom, which is about 10^{-8} cm.

The nucleus is composed of protons and neutrons. The number of protons is usually denoted by Z, while that of neutrons is denoted by N. The number of protons is equal to the number of electrons in the atom. Since the proton is positively charged, the electron is negatively charged, and the neutron is neutral, the atom as a whole is neutral under normal conditions. The total number of protons and neutrons in a nucleus is called the mass number and denoted by $A = N + Z$. Nuclei having the same proton number but different neutron number are called isotopes. Nuclei having the same neutron number but different proton number are called isotones. Finally, nuclei with the same mass number are called isobars. *See* ATOMIC STRUCTURE AND SPECTRA; ELECTRON; NEUTRON; PROTON.

Properties. The bulk properties of nuclei include their sizes, density distributions, and masses. The charge distribution can be quite accurately described as a bell-shaped distribution (called a Woods-Saxon distribution), shown in the illustration for nuclei of cobalt and bismuth.

The nucleus is a quantum-mechanical system. Its detailed

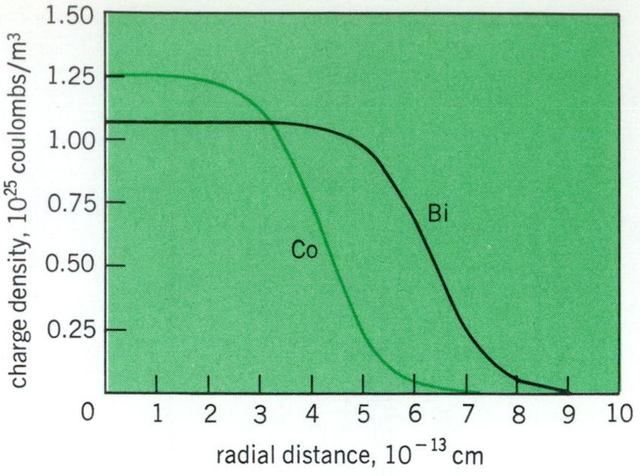

Charge density in nuclei of cobalt and bismuth.

properties are best described by a diagram, called the energy-level diagram, in which its quantum states are listed, together with the expectation values of all measurable quantities. Three of these are particularly important: the energy, the angular momentum, and the parity. In addition, other properties, such as magnetic and electric moments, are often measured. *See* ANGULAR MOMENTUM; ENERGY LEVEL (QUANTUM MECHANICS); PARITY (QUANTUM MECHANICS).

Nuclear models. A number of models have been developed to account for the properties of nuclei.

The basic model for the description of nuclear properties is the shell model. The strong nuclear force binds together protons and neutrons, called by the single name nucleons. Each individual nucleon moves in the average potential generated by all the others. This potential has the same shape as the nuclear matter distribution of the illustration, but upside down. The states of a single nucleon in the average potential cluster together into layers or shells, much like the single-particle states in atoms. The closed shells appear at nucleon numbers 2, 8, 20, 28, 50, 82, 126. The numbers at which closed shells occur are called magic numbers. *See* MAGIC NUMBERS; SPIN (QUANTUM MECHANICS).

Closed-shell nuclei behave, in many respects, as inert. Most properties of nuclei can be described by considering as active only nucleons outside the closed shells. These are called valence nucleons. In addition to the average potential in which they move, it is assumed that there is a residual interaction between valence nucleons. When the number of valence nucleons is small (up to about 4), the effects of the residual interaction can be calculated easily.

When the number of valence nucleons is large (greater than 5–6), some striking regularities develop in the level scheme. These regularities indicate the presence of collective features in nuclei. The collective features arise from two special properties of the residual interaction between valence nucleons: first, its pairing property which tends to pair off nucleons, one with angular momentum up and one with angular momentum down, to give zero resultant; and second, a quadrupole property which favors a distortion of the nuclear surface in the shape of an ellipsoid.

The number of energy levels of a nucleus below an excitation energy of 2 MeV is usually small. These levels can be described by using the shell model and allowing only for excitations of the valence nucleons from some single-particle levels to others nearby in energy. At higher excitation energies, the number of observed states increases considerably. These states arise from the excitation of the valence nucleons to higher single-particle levels and from excitations of nucleons from the closed shells to the valence shell. Because of their large number, it is no longer possible to describe properties of individual states. Thus, statistical methods are employed to describe average properties of states of this sort.

Simple collective modes of excitation occur even at higher excitation energy. Unlike the low-lying collective modes, these are no longer necessarily associated with pairing and quadrupole correlations. They arise from some coherent excitation of nucleons to shells other than the valence shell. These modes are called giant resonances, and are observed in inelastic scattering of strongly interacting particles off nuclei (α-scatttering, for example), in inelastic scattering of electrons off nuclei, and in photoabsorption. *See* GIANT NUCLEAR RESONANCES.

Nuclear forces. The nucleons which form the nucleus are held together by the strong nuclear forces. The forces which act between two nucleons when they are free (not inside the nucleus) have been studied by performing scattering experiments of one nucleon on the other, and by analyzing the properties of the deuteron, the only bound state of two nucleons. Because the proton and the neutron share many properties, it is convenient to consider them as two different states of the same particle, the nucleon. In order to distinguish protons from neutrons, the nucleon is given a sort of intrinsic angular momentum, called isospin. *See* ISOBARIC SPIN.

Nuclear structure calculations require the knowledge of the residual interaction between nucleons. This interaction is different from the free nucleon-nucleon interaction because of the modifications introduced by the presence of the other nucleons in the nucleus. The interaction inside nuclei is called the effective interaction, and its relationship to the free interaction is not fully understood. One property of the effective interaction, its charge independence, arises from the charge independence of the free nucleon-nucleon interaction. *See* ANALOG STATES; NUCLEAR REACTION; NUCLEAR SPECTRA; STRONG NUCLEAR INTERACTIONS. [F.I.]

Nucleation The formation within an unstable, supersaturated solution of the first particles of precipitate capable of spontaneous growth into large crystals of a more stable solid phase. These first viable particles, called nuclei, may either be formed from solid particles already present in the system (heterogeneous nucleation) or be generated spontaneously by the supersaturated solution itself (homogeneous nucleation). *See* SUPERSATURATION.

Nucleation is significant in analytical chemistry because of its influence on the physical characteristics of precipitates. Processes occurring during the nucleation period establish the rate of precipitation, and the number and size of the final crystalline particles. *See* COLLOID; FLOCCULATION; PRECIPITATION (CHEMISTRY). [D.H.K.; L.Go.]

Nucleic acid An acidic, chainlike biological macromolecule consisting of multiply repeated units of phosphoric acid, sugar, and purine and pyrimidine bases. Nucleic acids as a class are involved in the preservation, replication, and expression of hereditary information in every living cell. There are two types of nucleic acid: deoxyribonucleic acid (DNA) and ribonucleic acid (RNA). *See* DEOXYRIBONUCLEIC ACID (DNA); RIBONUCLEIC ACID (RNA). [E.Jo.]

Nucleon A collective name for a proton or a neutron. Protons and neutrons are the main constituents of the nuclei of atoms and have considerable similarity between themselves. They have the same spin, the same statistics, approximately the same mass, and, through the process of beta decay, can transform into each other. Thus it is convenient to have a common term to designate them both.

Occasionally neutrons and protons are considered as two states of a single particle called the nucleon, the two states being distinguished by the special value of an internal variable which can then assume only two values and is called the third component of the isotopic spin. *See* ISOBARIC SPIN; NEUTRON; PROTON. [E.Se.]

Nucleoprotein A generic term for any member of a large class of proteins associated with nucleic acid molecules. Nucleoprotein complexes occur in all living cells and in viruses, where they play vital roles in reproduction and protein synthesis.

Classification of the nucleoproteins depends primarily upon the type of nucleic acid involved—deoxyribonucleic acid (DNA) or ribonucleic acid (RNA)—and on the biological function of the complex. Deoxyribonucleoproteins (complexes of DNA and proteins) constitute the genetic material of all organisms and of many viruses. They function as the chemical basis of heredity and are the primary means of its expression and control. Most of the mass of chromosomes is made up of DNA and proteins whose structural and enzymatic activities are required for the proper assembly and expression of the genetic information encoded in the molecular structure of the nucleic acid. *See* DEOXYRIBONUCLEIC ACID (DNA).

Ribonucleoproteins (complexes of RNA and proteins) occur in all cells as part of the machinery for protein synthesis. This complex operation requires the participation of messenger RNAs (mRNAs), amino acid transfer RNAs (tRNAs), and ribosomal RNAs (rRNAs), each of which interacts with specific proteins to form functional complexes called polysomes, on which the synthesis of new proteins occurs. *See* RIBONUCLEIC ACID (RNA).

In simpler life forms, such as viruses which infect animal and plant cells and bacteriophages which infect bacteria, most of the mass of the viral particle is due to its nucleoprotein content. The material responsible for the hereditary continuity of the virus may be DNA or RNA, depending on the type of virus, and it is usually enveloped by one or more proteins which protect the nucleic acid and facilitate infection. *See* BACTERIOPHAGE; VIRUS. [V.G.A.]

Nucleosome The fundamental nucleohistone structural subunit of eukaryotic chromosomes. Nucleosomes (or ν-bodies) are particles about 10 nanometers in diameter, composed of a complex of DNA and histones, periodically positioned on a long DNA molecule like beads on a string. They represent the first level of DNA packaging in a hierarchy of organization that progresses up to the chromosomes visible in a light microscope. Nucleosomes appear to be universal among eukaryotic organisms. More than 90% of the DNA in any particular cell nucleus is contained in this form of packaging.

Neither the histones nor the nucleosomes have any apparent enzymatic or catalytic activity. At the minimum, a major function of the nucleosome is to package the DNA into a more compact form (leading to a sevenfold contraction of DNA length), at the same time permitting most of the DNA considerable exposure to the nuclear milieu. *See* CHROMOSOME; DEOXYRIBONUCLEIC ACID (DNA). [D.E.O.; A.L.O.]

Nucleosynthesis Theories of the origin of the elements involve synthesis with nucleons (neutrons and protons), the elementary building blocks of the nucleus. The nuclear theory, comprising nine distinct processes, is commonly called nucleosynthesis. Any acceptable theory of nucleosynthesis must lead to an understanding of the cosmic abundance of the elements observed in the solar system, stars, and the interstellar medium. The curve of these abundances is shown in the illustration.

Observations of the expanding universe and of the 3 K (−454°F) background radiation lend credence to the belief that the universe originated in a primordial event known as the "big bang" some 1–2 × 10^{10} years ago. Absence of stable mass 5 and mass 8 nuclei preclude the possibility of synthesizing the major portion of the nuclei of masses greater than 4 in those first few minutes of the universe, when density and temperature

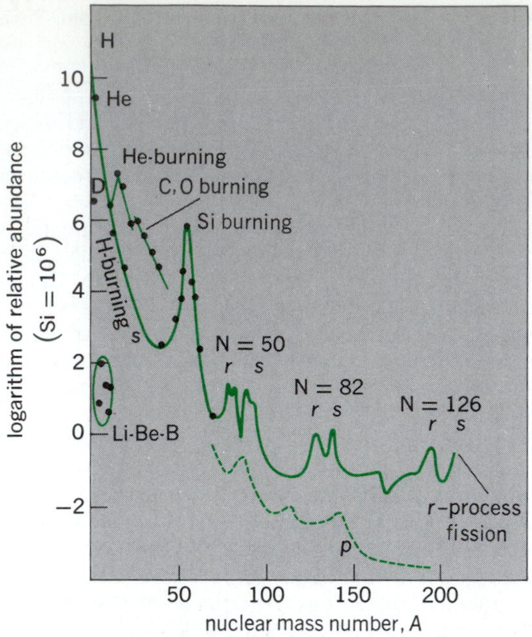

Schematic diagram of cosmic abundance as a function of nuclear mass number *A*, based on data of H. E. Suess and H. C. Urey. Predominant nucleosynthetic processes are indicated. (*Adapted from E. M. Burbidge et al., Synthesis of elements in stars, Rev. Mod. Phys., 29:547–650, 1957*)

were sufficiently high to support the necessary nuclear reactions to synthesize elements. *See* BIG BANG THEORY; COSMOLOGY.

The principal source of energy in stars is certainly nuclear reactions which release energy by fusion of lighter nuclei to form more massive nuclei. There is considerable evidence that nucleosynthesis is going on in stars, and has been doing so for billions of years.

Hydrogen burning, the first of the nine processes of nucleosynthesis, converts hydrogen nuclei into helium. In stars of 1.2 or less solar masses, hydrogen burning proceeds by the proton-proton chain. Such reactions occurring during the big bang are believed to be responsible for the helium and deuterium, and possibly some of the ^{7}Li observed today. In more massive stars where temperatures are equal to or greater than 2×10^7 K, hydrogen burning is accomplished through proton captures by carbon, nitrogen, and oxygen nuclei, in the carbon-nitrogen-oxygen (CNO) cycles, to form ^{4}He. *See* CARBON-NITROGEN-OXYGEN CYCLES; NUCLEAR FUSION; PROTON-PROTON CHAIN.

When the hydrogen fuel is exhausted in the central region of the star, the core contracts and its temperature and density increase. Helium then becomes the fuel for further energy generation and nucleosynthesis. The basic reaction in this thermonuclear phase is the three-alpha process in which three ^{4}He nuclei (three alpha particles) fuse to form a carbon nucleus of mass 12 (^{12}C).

Upon exhaustion of the helium supply, if the star has an initial mass of at least 7.5 solar masses, gravitational contraction of the core can lead to a temperature of about 9×10^8 °F (5×10^8 K), where it becomes possible for two ^{12}C nuclei to overcome their high mutual Coulomb-potential barrier and fuse to form ^{20}Ne, ^{23}Na, and ^{24}Mg.

Carbon burning is immediately followed by a short-lived stage, sometimes referred to as neon burning, in which ^{20}Ne is photodisintegrated. The eventual result is that most of the carbon from helium burning becomes oxygen, which supplements the original oxygen formed in helium burning. This is all followed by the fusion of oxygen nuclei at significantly higher temperatures.

Silicon burning commences when the temperature reaches 5×10^9 °F (3×10^9 K). In this phase, photodisintegration of ^{28}Si and other intermediate-mass nuclei in the neighborhood of $A = 28$ produces copious supplies of protons, alpha particles, and neutrons. Capture of these by other intermediate-mass nuclei synthesizes nuclei up to a mass A of about 60, and results in the buildup of the iron-group peak near $A = 56$.

Because neutrons are neutral particles, their capture is not affected by the Coulomb barrier that inhibits charged-particle capture by the massive nuclei. If the number of neutrons per seed nucleus is small, so that time intervals between neutron captures are long compared to the beta-decay lifetimes of unstable nuclei that are formed, the *s*-process (slow process) takes place. The seed nuclei are probably primarily those in the iron peak, but neutron processes probably alter some of the lighter-mass nuclei as well. In this slow process, after neutron capture a nucleus that is unstable because of excess neutrons decays by electron and antineutrino emission to a stable isobar which may capture another neutron, leading to a chain of captures and decays to synthesize nuclei of masses up to about $A = 209$, when alpha decay becomes a deterrent to further synthesis by the *s*-process. The *s*-process is believed to occur in red giant stars.

The *r*-process occurs when a large neutron flux allows rapid captures of neutrons. Only in the massive explosions of supernovae are such large neutron fluxes conceivable. With such rapid successive captures, it is possible to proceed through unstable nuclei before they can either beta- or alpha-decay. In this manner it is possible to synthesize nuclei all the way into the transuranic elements.

The *p*-process produces proton-rich heavier elements probably by modifying a small fraction of the *s*-process and *r*-process nuclei through (p,γ) (p,n), and (γ,n) reactions.

Spallation of more abundant nuclei such as carbon, nitrogen, and oxygen by protons can account for the low-abundance nuclides ^{6}Li, ^{9}Be, ^{10}B, and ^{11}B and for some ^{7}Li. This breaking up of the more massive nuclei to form the light nuclei is called the *l*-process. The most probable site for the *l*-process is in the interstellar medium, where high-energy galactic cosmic rays bombard the interstellar gas. *See* NUCLEAR REACTION; STELLAR EVOLUTION. [G.R.C.]

Nucleotide A cellular constituent that is one of the building blocks of ribonucleic acids (RNA) and deoxyribonucleic acid (DNA). In biological systems, nucleotides are linked by enzymes in order to make long, chain-like polynucleotides of defined sequence. The order or sequence of the nucleotide units along a polynucleotide chain plays an important role in the storage and transfer of genetic information. Many nucleotides also perform other important functions in biological systems. *See* ADENOSINETRIPHOSPHATE (ATP); COENZYME; CYCLIC NUCLEOTIDES; DEOXYRIBONUCLEIC ACID (DNA); NICOTINAMIDE ADENINE DINUCLEOTIDE (NAD); NUCLEIC ACID; RIBONUCLEIC ACID (RNA).

Nucleotides are generally classified as either ribonucleotides or deoxyribonucleotides. Both classes consist of a phosphorylated pentose sugar that is linked via an *N*-glycosidic bond to a purine or pyrimidine base. The combination of the pentose sugar and the purine or pyrimidine base without the phosphate moiety is called a nucleoside. *See* PURINE; PYRIMIDINE. [E.P.G.]

Nuclide A species of atom that is characterized by the constitution of its nucleus, in particular by its atomic number Z and its neutron number $A - Z$, where A is the mass number. The total number of stable nuclides is approximately 275. About a dozen radioactive nuclides are found in nature, and in addition, hundreds of others have been created artificially. [H.E.D.]

Nudibranchia An order of the gastropod subclass Opisthobranchia containing about 2500 living species of carnivorous sea slugs. They occur in all the oceans, at all depths, but reach their greatest size and diversity in warm shallow seas.

In this, the largest order of the Opisthobranchia, there is much evidence indicating polyphyletic descent from a number of long-extinct opisthobranch stocks. In all those lines which persist to the present day, the shell and operculum have been discarded in the adult form. In many of them the body has quite independently become dorsally papillate. In at least two suborders these dorsal papillae have acquired the power to nurture nematocysts derived from their coelenterate prey so as to use them for the nudibranch's own defense. In other cases such papillae (usually called cerata) have independently become penetrated by lobules of the adult digestive gland, or they may contain virulent defensive glands or prickly bundles of dagger-like calcareous spicules. *See* OPISTHOBRANCHIA. [T.E.T.]

Number systems Integral numbers may be represented as linear combinations of powers of any convenient and arbitrarily chosen base. The choice of the base is not always made on a rational basis; number systems have been based on 5, 6, 10, and 60. More recently, systems based on 2 and 8 have proved quite useful in computer applications. The duodecimal number system, in which numbers are represented as linear combinations of powers of 12, has certain advantages because 12 has the factors 1, 2, 3, 4, 6, and 12.

Decimal system. Every positive integer is uniquely a polynomial in 10 with coefficients, called digits, taken from 0, 1,..., 9. The fact that

$$205714 = 4 + 1 \cdot 10 + 7 \cdot 10^2 + 5 \cdot 10^3 + 0 \cdot 10^4 + 2 \cdot 10^5$$

is nearly always lost in present-day teaching, and in the hurried application of ordinary arithmetic, In fact, numbers are likely to be thought of as merely an orderly arrangement of decimal digits.

The operations of addition and multiplication consist of the corresponding operations with polynomials, together with rules that serve to keep the results inside the system so that they can be used in future operations. Subtraction introduces negative numbers that may be handled by introducing a special digit called a sign digit with its own rules of combination, or by introducing complementation in which the digits of a number are subtracted from 9, except for the last nonzero digit which is subtracted from 10.

Division is a process that can be carried out only rarely with absolute exactness in the decimal system, the process usually being nonterminating. This introduces the notion of infinite decimal expansions and the more or less theoretical operations with such numbers. In practice, truncation and rounding are used as in $2/3 = .66667$, with consequent errors and departure from the axioms of arithmetic.

Binary system. In the binary system every positive integer is the sum of distinct powers of 2 in just one way. Thus $434 = 2^8 + 2^7 + 2^5 + 2^4 + 2^1$, and this is expressed by writing 110110010. The digits corresponding to 2^0, 2^2, 2^3, and 2^6 are zero, since these powers do not occur in 434. The first dozen integers are written as follows:

1	1	4	100	7	111	10	1010
2	10	5	101	8	1000	11	1011
3	11	6	110	9	1001	12	1100

The great advantage of the binary system lies in the fact that there are only two kinds of binary digits, or "bits," namely 0 and 1. This not only gives a simplified arithmetic but provides a language in which to treat two-valued functions or bistable systems. Among its disadvantages is the fact that the binary system requires nearly three times as many digits to represent a given number as does the familiar decimal system.

Digital computers invariably use the binary system. The so-called decimal computers code the decimal digits into binary form, while the purely binary machines use full binary arithmetic.

Arithmetic in the binary system is remarkably simple. For addition, only $1 + 1 = 10$ is needed, while the multiplication table reduces to $1 \cdot 1 = 1$. Such simple operations are readily performed electronically with extreme rapidity and reliability.

The binary system is useful not only to represent numbers but also to record and process information. In fact the unit of information is a binary digit. For example, given a set S of objects and a property P, it is possible to record which objects have the property P, and which do not, by assigning a binary position to each object of S and recording there a 1 or 0 according as the property P is, or is not, possessed by the corresponding object. A binary computer, with its ability to extract and examine a given binary digit, can use this compact method of storing information.

Octal system. To write a number in the octal system, once it has been expressed in the binary system, one merely groups the binary digits by threes, beginning at the binary point and working to the left and right. The octal system with its eight digits 0, 1,..., 7 affords a convenient way of condensing the lengthier display of the binary system. Arithmetic in the octal system resembles the familiar decimal arithmetic. The octal system requires only 10% more digits than the decimal system to represent the same amount of information. Some computing systems use base 16, in which case binary information is handled in sets of four bits. [D.H.L.]

Number theory The primary objects of number theory are the properties and mutual relations of the natural numbers 1, 2, 3, ... and more generally of the integers, which include also the zero and the negative integers. The integers form a ring, that is, a domain within which addition, subtraction, multiplication (but not necessarily division) can always be carried out. The numbers can be classified in many ways: for example, odd and even numbers, square numbers, prime numbers, perfect numbers.

Elementary number theory. The main concern of this branch of number theory is divisibility. A number d is called a divisor of n (in symbols: d/n) if there exists an integer t such that $n = dt$. A prime number is a number that has only 1 and itself as divisors. Euclid proved that there is no last prime number in the sequence of integers. Indeed, if 2, 3, 5,..., p are known prime numbers, then the number shown in Eq. (1) is

$$N = 2 \cdot 3 \cdot 5 \cdot \cdots \cdot p + 1 \qquad (1)$$

divisible by none of them and is therefore either a new prime number itself or is divisible by prime numbers not used in the construction of N. The fundamental theorem of number theory states that a number n can be factored into prime numbers in only one way when the order of the prime factors is disregarded.

A perfect number is defined as a number equal to the sum of its proper divisors, or divisors smaller than the number. Euclid showed that $(2^n - 1)2^{n-1}$ is a perfect number if $2^n - 1$ is a prime number, a so-called Mersenne prime. If $2^n - 1$ is a prime, then n must itself be a prime. At present 36 Mersenne primes and thus 36 perfect numbers are known, the first of which are 6, 28, and 496.

Eratosthenes (in the 3d century B.C.) showed how all prime numbers up to some large integer x can be found if only the primes p up to $\sqrt{x}$ are already known. It is indeed sufficient to strike out from the list of all integers, starting with 2, the multiples $2p$, $3p$,..., etc., of all the primes up to $\sqrt{x}$. The remaining integers are precisely all the primes up to x.

If $a - b$ is divisible by m, then a is called congruent b modulo m, as shown in symbols in formula (2). This relation is an

$$a \equiv b \pmod{m} \qquad (2)$$

equivalence; that is, it is reflexive, symmetric, and transitive, and defines, therefore, equivalence classes of numbers congruent to each other, called residue classes. There are evidently m residue classes modulo m. The number of those residue classes which contain only numbers coprime to m is called Euler's function $\varphi(m)$.

Diophantine equations. There are infinitely many integral solutions for the pythagorean equation, Eq. (3). The solutions

$$x^2 + y^2 = z^2 \qquad (3)$$

are all contained in Eqs. (4), where u and v are positive integers.

$$x = u^2 - v^2 \quad y = 2uv \quad z = u^2 + v^2 \qquad (4)$$

Pierre de Fermat made the famous statement that Eq. (5)

$$x^n + y^n = Z^n \qquad (5)$$

is not solvable in integers for any integral exponent $n > 2$. Besides $n = 4$, only prime exponents n need to be considered.

Single equations, or systems of equations in more unknowns than the number of equations, but with added conditions, such as allowing only solutions in integers, are called diophantine. Equations (3) and (5) are examples of diophantine equations.

Algebraic number theory. The results concerning Fermat's last theorem are obtained through the methods of algebraic number theory. Karl F. Gauss carried over the concepts of number theory to the ring $R[i]$ of all complex integers $a + bi$, where a and b are ordinary integers. The law of unique prime factorization is preserved in this ring. Ordinary prime numbers $p \equiv 3 \pmod 4$ are also prime numbers in $R[i]$, whereas $2 = -i(1 + i)^2$ and the primes $p \equiv 1 \pmod 4$ are split $p = (a + bi)(a - bi)$. An algebraic number field $R(\theta)$ of degree n is generated by the root θ of an algebraic equation $F(x) = 0$ of degree n, having rational coefficients. A number α in this field is called an (algebraic) integer if it satisfies an algebraic equation with rational integer coefficients, the highest of which is 1. The algebraic integers in an algebraic number field form again an integral domain. However, the prime factorization is not necessarily unique, as the example $21 = 3 \cdot 7 = (1 + 2\sqrt{-5})(1 - 2\sqrt{-5}) = (4 + \sqrt{-5})(4 - \sqrt{-5})$ shows, where each product is irreducible in the ring $R[\sqrt{-5}]$. Uniqueness is restored through the introduction of ideals.

Analytic number theory. For certain problems of number theory, the methods of analysis, that is, of calculus and function theory, have to be used. The most famous problem of this sort is the number $\pi(x)$ of prime numbers $p \leqq x$. About 1793, Gauss conjectured the relation in Eq. (6). This equation was

$$\lim_{x \to \infty} \frac{\pi(x) \log x}{x} = 1 \qquad (6)$$

proved a century later. The methods were created by G. F. B. Riemann who investigates the function $\zeta(s)$ defined by the series shown in Eq. (7). Here $s = \sigma + it$ is a complex variable;

$$\zeta(s) = \sum_{n=1}^{\infty} \frac{1}{n^s} \qquad (7)$$

the series in Eq. (7) is convergent for $\sigma > 1$. By a method of analytic continuation, the function can be defined in the whole complex plane of s. It is a meromorphic function with only a simple pole of residue 1 at $s = 1$. It can be shown that $\zeta(s)$ has no zeros for $\sigma = 1$. This fact and the pole at $s = 1$ suffice to establish Eq. (6). Riemann also proposed the still unproved "Riemann hypothesis": All zeros of $\zeta(s)$ have a real part not exceeding ½. See COMPLEX NUMBERS; NUMBER SYSTEMS; ZERO.

[H.B./E.Gr.]

Numerical analysis The development and analysis of computational methods (and ultimately of program packages) for the minimization and the approximation of functions, and for the approximate solution of equations, such as linear or nonlinear (systems of) equations and differential or integral equations. Originally part of every mathematician's work, the subject is now often taught in computer science departments because of the tremendous impact which computers have had on its development. Research focuses mainly on the numerical solution of (nonlinear) partial differential equations and the minimization of functions.

Numerical analysis is needed because answers provided by mathematical analysis are usually symbolic and not numeric; they are often given implicitly only, as the solution of some equation, or they are given by some limit process. A further complication is provided by the rounding error which usually contaminates every step in a calculation (because of the fixed finite number of digits carried).

Even in the absence of rounding error, few numerical answers can be obtained exactly. Among these are (1) the value of a piece-wise rational function at a point and (2) the solution of a (solvable) linear system of equations, both of which can be produced in a finite number of arithmetic steps. Approximate answers to all other problems are obtained by solving the first few in a sequence of such finitely solvable problems. A typical example is provided by Newton's method: A solution c to a nonlinear equation $f(c) = 0$ is found as the

$$\text{limit } c = \lim_{n \to \infty} x_n,$$

with x_{n+1} being a solution to the linear equation

$$f(x_n) f'(x_m)(x_{n+1} - x_n) = 0,$$

that is, $x_n + 1 = x_n - f(x_n)/f'(x_n)$, $n = 0, 1, 2, \ldots$. Of course, only the first few terms in this sequence $x_0, x_1, x_2, \ldots$ can ever be calculated, and thus one must consider when to break off such a solution process and how to gauge the accuracy of the current approximation.

An otherwise satisfactory computational process may become useless, because of the amplification of rounding errors. A computational process is called stable to the extent that its results are not spoiled by rounding errors. The extended calculations involving millions of arithmetic steps now possible on computers have made the stability of a computational process a prime consideration.

Interpolation and approximation. Polynomial interpolation provides a polynomial p of degree n or less which uniquely matches given function values $f(x_0), \ldots, f(x_n)$ at corresponding distinct points $x_0, \ldots, x_n$. The interpolating polynomial p is used in place of f, for example in evaluation, integration, differentiation, and zero finding. Accuracy of the interpolating polynomial depends strongly on the placement of the interpolation points, and usually degrades drastically as one moves away from the interval containing these points (that is, in case of extrapolation). See EXTRAPOLATION.

When many interpolation points (more than 5 or 10) are to be used, it is often much more efficient to use instead a piecewise polynomial interpolant or spline. Suppose the interpolation points above are ordered, $x_0 < x_1 < \cdots < x_n$. Then the cubic spline interpolant to the above data, for example, consists of cubic polynomial pieces, with the ith piece defining the interpolant on the interval $[x_{i-1}, x_i]$ and so matched with its neighboring piece or pieces that the resulting function not only matches the given function values (hence is continuous) but also has a continuous first and second derivative.

Interpolation is but one way to determine an approximant. In full generality, approximation involves several choices: (1) a set P of possible approximants, (2) a criterion for selecting from P a particular approximant, and (3) a way to measure the approximation error, that is, the difference between the function f to be approximated and the approximant p, in order to judge the quality of approximation.

Solution of linear systems. Solving a linear system of equations is probably the most frequently confronted computational task. It is handled either by a direct method, that is, a method which obtains the exact answer in a finite number of steps, or by an iterative method, or by a judicious combination of both. Analysis of the effectiveness of possible methods has led to a workable basis for selecting the one which best fits a particular situation.

Direct methods require a number of operations which increases with the cube of the number of unknowns. Some types of problems arise wherein the matrix of coefficients is sparse, but the unknowns may number several thousand; for these, direct methods are prohibitive in computer time required. One frequent source of such problems is the finite difference treatment of partial differential equations. A significant literature of iterative methods exploiting the special properties of such equations is available. For certain restricted classes of difference equations, the error in an initial iterate can be guaranteed to be reduced by a fixed factor, using a number of computations that is proportional to $n \log n$, where n is the number of unknowns. Since direct methods require work proportional to n^3, it is not surprising that as n becomes large, iterative methods are studied rather closely as practical alternatives.

Differential equations. Classical methods yield practical results only for a moderately restricted class of ordinary differential equations, a somewhat more restricted class of systems of ordinary differential equations, and a very small number of partial differential equations. The power of numerical methods is enormous here, for in quite broad classes of practical problems relatively straightforward procedures are guaranteed to yield numerical results, whose quality is predictable. *See* DIFFERENTIAL EQUATION. [C.DeB.]

Numerical control The method of controlling machines by the application of digital electronic computers and circuitry. Machine movements that are controlled by cams, gears, levers, or screws in conventional machines are directed by computers and digital circuitry in computer numerical control (CNC) machines.

Computer numerical control provides very flexible and versatile control over machine tools. Most machining operations require that a cutting tool be fed at some speed against a workpiece. In a conventional machine such as a turret lathe, the turning tool is mounted on a slide with hand-operated infeed and crossfeed slides. The operator manually turns a crank that feeds the cutting tool into the workpiece (infeed) to the desired diameter. Another crank then moves the turning tool along the longitudinal axis of the machine and produces a cylindrical cut along the workpiece. The feed rate of the turning tool is sometimes controlled by selecting feed gears. These gears move the axis slide at the desired feed. A CNC machine replaces the hand cranks and feed gears with servomotor systems. *See* LATHE; MACHINE TOOLS; SERVOMECHANISM.

Computer numerical controls allow the desired cut depths and feed rates to be "dialed in" rather than controlled by cranks, cams, and gears. This provides precise, repeatable machine movements that can be programmed for optimal speeds, feeds, and machine cycles. All cutting-tool applications, whether on a lathe, drill press, or machining center, have optimum speeds and feeds, which are determined by carefully weighting the economics of tool life, required production rates, and operator attentiveness. With computer numerical control these parameters are set once, and then they are repeated precisely for each subsequent machine cycle.

Machines with numerical control (NC) became possible in the 1960s, when the development of the transistor made electronic servosystems practical. Earlier NC cutting tools were programmed with punched tape. This technique, while an impressive and productive technology, lacked the sophistication and flexibility that have become available with the development of more advanced systems. *See* COMPUTER; TRANSISTOR.

Computer numerical control advanced further with the advent of inexpensive computers in the early 1980s. Instead of being programmed to move to discrete, precalculated locations, the machines can be programmed to perform complex mathematical calculations that determine the locations of the machine's axes. Programming has become simpler, more powerful, and more user-friendly. It has even become possible to program machine tools graphically at the machine while machining is taking place. *See* PROGRAMMABLE CONTROLLERS.

Machine systems designed for computer numerical control use servosystems to control the position of cutting tools. A typical servosystem is composed of six parts: a lead screw (ball screw) and lead nut, an electric servomotor, a servoamplifier, a position-feedback device, a velocity-feedback device, and a computer.

Lead screws are precision-ground screws. To prevent any backlash in the screw system, a special lead nut utilizing hundreds of ball bearings tracks the screw.

Servomotors are specially constructed electric motors designed to operate at very low angular velocities. Although they sometimes drive the lead screw through a gear reduction system, they are usually directly coupled to the lead screw. This direct coupling minimizes any backlash or positioning errors introduced by gears. The motor is required to provide full torque at low, or zero, revolutions per minute (rpm). Traditionally, servomotors have been direct-current types. The electrical requirements for alternating-current motors are more complex than for direct-current motors, but the motors themselves are more reliable, especially in hostile environments. *See* ALTERNATING-CURRENT MOTOR; DIRECT-CURRENT MOTOR; GEAR.

The servoamplifier powers the servomotor; it receives a signal voltage from the computer that indicates the desired machine axis direction and speed. A tachometer is mounted on the servomotor and measures the motor's rpm. The output of the tachometer is fed into the servoamplifier, which examines the rpm of the servomotor and varies the voltage and current supplied to the motor to keep it turning at the desired rate. *See* TACHOMETER.

Position feedbacks are provided by resolvers or encoders. These allow the computer to know where the machine's axes are at any time, and to calculate how far the axes must move to reach their programmed positions.

The computer provides overall control of the entire system. It examines the part program, determines the desired positions of the machine axes, then provides special signals to the servoamplifiers to move the servomotors and thereby move the machine's axes. The computer monitors the position feedback provided by the resolvers or encoders. In addition, the computer handles various so-called housekeeping tasks such as displaying the axes' locations on the cathode-ray-tube. The computer also monitors the program to ensure that dangerous functions, such as opening the chuck in the middle of a cut, have not been programmed.

Computer numerical control machines are used mainly when flexibility is required or variable and complex part geometries must be created. They are used to produce parts in lot sizes of a few pieces to several thousand. Extremely large manufacturing lot sizes frequently call for more product-specific machines, which can be optimized for large production runs.

It is often assumed that a machine controlled by a computer is inherently more accurate than an equivalent manual machine, this is not necessarily true. In fact, CNC machines are sometimes less accurate than manual machines. The CNC machines rely on precision-ground slides and screws to position accurately, and normal machine wear can degrade this accuracy. The computer is usually unable to detect or compensate for this wear, while a skilled mechanist can. *See* COMPUTER-AIDED DESIGN AND MANUFACTURING; CONTROL SYSTEMS; TOOLING. [J.R.C.B.]

Numerical indicator tube Any electron tube capable of visually displaying numerical figures. Some varieties also display alphabetical characters and commonly used symbols. In many electronic circuits and equipments it is desirable to indicate numbers or characters. Such a tube can be one of a set which displays, for instance, the magnitude and polarity of voltage in digital form. This is similar to electrical clocks which display the time in numerical form.

A simple and extensively used form is a cold-cathode gas tube in which there is a series of cathodes which light up because of the cathode glow that surrounds any cathode in a gas discharge. A common anode is used. The desired cathode is selected by any suitable switching scheme. The cathodes are made by photoetching the desired shape from a suitable metal. They are insulated and stacked one above the other, so that any particular number can be read when the corresponding cathode is energized. *See* ELECTRICAL CONDUCTION IN GASES; GAS TUBE. [R.W.W.]

Numerical taxonomy The grouping by numerical methods of taxonomic units based on their character states. The application of numerical methods to taxonomy, dating back to the rise of biometrics in the late 19th century, has received a great deal of attention with the development of the computer and computer technology. Numerical taxonomy provides methods that are objective, explicit, and repeatable, and is based on the ideas first put forward by M. Adanson in 1763.

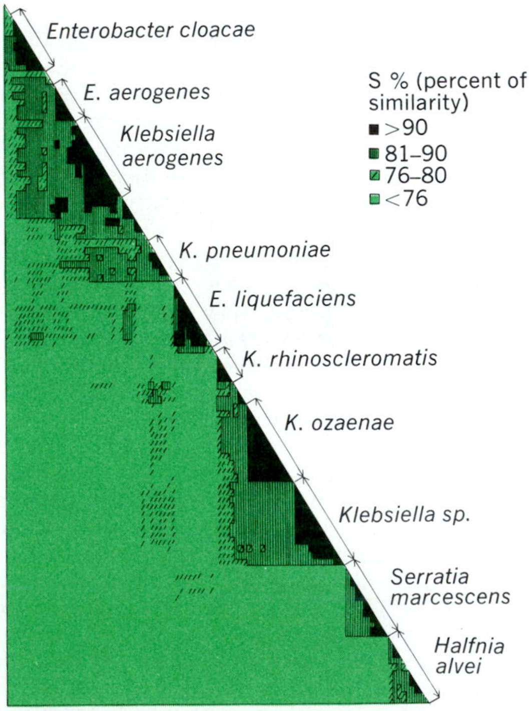

Similarity matrix showing relationships between clusters of strains of enteric bacteria. (*After R. Johnson et al., Numerical taxonomy study of the Enterobacteriaceae, Int. J. System. Bacteriol., 25(1):12–37, 1975*)

These ideas, or principles, are that the ideal taxonomy is composed of information-rich taxa based on as many features as possible, that *a priori* every character is of equal weight, that overall similarity between any two entities is a function of the similarity of the many characters on which the comparison is based, and that taxa are constructed on the basis of diverse character correlations in the groups studied. *See* TAXON.

In the early stages of development of numerical taxonomy, phylogenetic relationships were not considered. However, numerical methods have made possible exact measurement of evolutionary rates and phylogenetic analysis. Furthermore, rapid developments in the techniques of direct measurement of the homologies of deoxyribonucleic acid (DNA), and ribonucleic acid (RNA), between different organisms now provide an estimation of "hybridization" between the DNAs of different taxa and, therefore, possible evolutionary relationships. Thus, research in numerical taxonomy often includes analyses of the chemical and physical properties of the nucleic acids of the organisms, the data from which are correlated with phenetic groupings established by numerical techniques. *See* PHYLOGENY; TAXONOMIC CATEGORIES.

Several formulas have been suggested for calculating similarity between organisms. Resemblance is expressed by coefficients of similarity usually ranging between unity and zero, the former for perfect agreement and the latter for no agreement. Coefficients of dissimilarity (distance) can also be employed, in which case the range is between zero and an undefined positive value, the former for identity and the latter for maximal distance or disparity. Resemblance coefficients are usually tabulated in the form of a matrix, with one coefficient for every pair of taxonomic entities. The similarity index is the simplest of the coefficients employed to calculate similarity: $S = n_s/(n_s + n_d)$, where n_s represents the number of positive features shared by two strains being compared and n_d represents the number of features positive for one strain and negative for the other (and vice versa). An example of an S-value matrix is shown in the illustration.

A variety of clustering methods has been employed in numerical taxonomy, including sequential, agglomerative, hierarchic, and nonoverlapping clustering. Dendrograms are most commonly employed to represent taxonomic structure resulting from cluster analysis. These have the advantage of being readily interpretable as conventional taxonomic hierarchies. Dendrograms are, in general, distinguished into phenograms representing phenetic relationships and cladograms representing evolutionary branching sequences. [R.R.C.]

Nummulites An extinct genus (properly known as *Camerina*) of the relatively large (up to 1.4 in. or 35 mm in diameter) unicellular protozoans of the order Foraminiferida. Individuals of this genus developed a lenticular, discoidal, planispirally coiled, involute, multichambered test (shell) composed of calcium carbonate. They inhabited shallow, warm marine waters of the tropical zone during the early Tertiary Period. *See* FORAMINIFERIDA.

The pyramids of Egypt are made of blocks of an Eocene nummulitic limestone. The numerous species are used in geologic correlation. [W.S.C.]

Nursing The application of the principles of physical, biological, and social sciences in the physical and mental care of people, sick and well. Nursing includes therapy as directed by the doctor; physical and emotional care; and patient and family education in rehabilitation, health maintenance, and disease prevention. When several persons are simultaneously involved in the nursing care of a person or persons, nursing may also include the application of the principles and skills involved in management. *See* MEDICINE. [R.Sl.]

Nut (engineering) In mechanical structures, an internally threaded fastener. Plain square and hexagon nuts for bolts and screws are available in three degrees of finish: unfinished, semifinished, and finished. There are two standard weights: regular and heavy. For specific applications, there are other standard forms such as jam nut, castellated nut, slotted nut, cap nut, wing nut, and knurled nut.

Wing and knurled nuts are designed for applications where a nut is to be tightened or loosened by using finger pressure only. *See* SCREW FASTENER. [W.J.L.]

Nutation (astronomy and mechanics) In mechanics, the term nutation refers to a bobbing or nodding up-and-down motion of a spinning rigid body, such as a top or a gyroscope, as it precesses about its vertical axis. Astronomical nutation refers to irregularities in the precessional motion of the equinoxes caused by the varying torque applied to the Earth by the Sun and Moon. Astronomical nutation, which is sometimes called nutational wandering of the terrestrial poles, should not be confused with nutation as defined in mechanics; the latter is present even if the source of the torques is unvarying. *See* PRECESSION. [R.E.Bo.]

Nutmeg A delicately flavored spice obtained from the nutmeg tree (*Myristica fragrans*), a native of the Moluccas, or Spice Islands. The tree is a dark-leafed evergreen, and is a member of the nutmeg family (Myristicaceae). The golden-yellow, mature fruits resemble apricots (see illustration). They gradually lose moisture and when completely ripe, the husk (pericarp) splits open, exposing the shiny brown seed which is the nutmeg of commerce. Nutmeg oil is used in medicine, perfumery, and dentifrices, and in the tobacco industry. *See* MAGNOLIALES. [P.D.St./E.L.C.]

Mature nutmeg (*Myristica fragrans*) fruits. (*USDA*)

Nutrition The science of nourishment, including the study of the nutrients that each organism must obtain from its environment in order to maintain life and reproduce. Mammals need for their nutrition (aside from water and oxygen) a highly complex mixture of chemical substances, including amino acids; carbohydrates; certain lipids; a great variety of minerals; and vitamins. *See* AMINO ACIDS; CARBOHYDRATE METABOLISM; LIPID METABOLISM; PROTEIN METABOLISM; VITAMIN.

One of the bases for current and continued interest in nutrition is the fact that individuals who have differing genetic backgrounds have differing nutritional needs, considered quantitatively; for this reason various human ills may arise because the individuals concerned do not get all of the nutrients in amounts compatible with their own distinctive requirements. *See* DISEASE; MALNUTRITION; METABOLIC DISORDERS. [R.J.Wi.]

Nymphaeales An order of flowering plants, division Magnoliophyta (Angiospermae), in the subclass Magnoliidae of the class Magnoliopsida (dicotyledons). The order consists of three families, the Nymphaeaceae, Nelumbonaceae, and Ceratophyllaceae, with scarcely a hundred species in all. They are all aquatic herbs without cambium or vessels. Water lilies (*Nymphaea*; see illustration) and the lotus lily (*Nelumbo*) are familiar members of the order. *See* MAGNOLIIDAE; MAGNOLIOPSIDA. [A.Cr.]

A common eastern American species of water lily (*Nymphaea odorata*). The large and indefinite number of tepals, stamens, and carpels is characteristic of the family and a large part of the order. (*Photograph by Hugh Spencer, National Audubon Society*)

Nystatin An antifungal antibiotic useful in the therapy of a wide variety of nonsystemic fungal infections and also as an ingredient in animal feeds for enhanced growth rates with poultry and swine. It is active against a wide range of yeasts and other fungi, including *Candida*, *Aspergillus*, *Penicillium*, *Botrytis*, and others, but it is without activity against bacteria. Chemically it is a polyene. It is produced biosynthetically by fermentation with a strain of *Streptomyces noursei*. [R.E.B.]

Oak A genus (*Quercus*) of trees, some of which are shrubby, with about 200 species, mainly in the Northern Hemisphere. About 50 species are native in the United States. All oaks have scaly winter buds, usually clustered at the ends of the twigs, and single at the nodes. The fruit is a nut (acorn). The leaves are simple and usually lobed.

Oaks furnish the most important hardwood lumber in the United States. Principal uses are for charcoal, barrels, building construction, flooring, railroad ties, mine timbers, boxes, crates, vehicle parts, ships, agricultural implements, caskets, woodenware, fence posts, piling, and veneer. Oak is also used for pulp and paper products. *See* Fagales. [A.H.G./K.P.D.]

Oasis An isolated fertile area, usually limited in extent and surrounded by desert. The term was initially applied to small areas in Africa and Asia typically supporting trees and cultivated crops with a water supply from springs and from seepage of water originating at some distance. However, the term has been expanded to include areas receiving moisture from intermittent streams or artificial irrigation systems. Thus the floodplains of the Nile and Colorado rivers can be considered vast oases, as can arid irrigated areas.

Among the native plants restricted to waste areas and abandoned plots are a wide variety of phreatophytes, including tamarisk (*Tamarix* sp.), cattail (*Typha latifolia*), sedges, and rushes; farther away many halophytes are found where drainage is poor; and beyond this fringe, desert vegetation takes over. [W.G.McG.]

Oats An agricultural crop (genus *Avena*) grown for its grain and straw in most countries of the temperate zones of the world. In general, oats are a cool-season crop which requires a moist climate. They grow well on both light and heavy soils if sufficient moisture and fertility nutrients are available.

Oat grain usually contains 10–16% protein, which makes it especially good for rations for young livestock and for human food. Dehulled oat seeds are used extensively for making breakfast cereals, such as porridge made from rolled oats. In general, oat grain contains adequate quantities of minerals and B vitamins for normal diet. The fat content is low. Oat hulls are used to make furfural, an important chemical in nylon manufacturing. [K.J.F./J.A.Sh.]

Obesity The presence of an excessive amount or abnormal distribution of fat in an individual. This tendency toward obesity is enhanced by rich diets, lack of regular exercise, and the rejection of certain fundamental health rules. In addition, emphasis has been placed upon the relationship between the overindulgence in food and tension-producing situations. People are said to eat too much because they are anxious, shy, insecure, or hostile, or have some other need for which eating can at least partly compensate. *See* Neurotic disorders.

Studies in nutrition and metabolism indicate that, although such factors certainly may play a role in many cases, more subtle differences occur among individuals. As a person ages, the metabolic rate tends to slow down somewhat; more important, however, there are changes in hormonal balance which affect fat utilization by the body. *See* Hormone; Nutrition.

Obesity occasionally appears in relation to specific disease processes. These include certain disorders of the hypothalamus or injury to it resulting from changes in adjacent structures, such as in the formation of some pituitary tumors. Froehlich's syndrome, characterized by a feminine pattern of obesity and sexual dysfunction, is thought to result from associated damage to the neighboring hypothalamus, although it was formerly believed to be the result of hypopituitarism. The influence of hereditary factors cannot be overlooked because it has been established that obesity and other body characteristics tend to be transmitted from one generation to the next. *See* Human genetics. [E.G.St./N.K.M.]

Object-oriented programming A computer programming methodology that focuses on data items rather than processes. Traditional software development models assume a top-down approach. A functional description of a system is produced and then refined until a running implementation is achieved. Data structures (and file structures) are proposed and evaluated based on how well they support the functional models.

The object-oriented approach focuses first on the data items (entities, objects) that are being manipulated. The emphasis is on characterizing the data items as active entities which can perform operations on and for themselves. Then, how system behavior is implemented through the interaction of the data items is described.

The essence of the object-oriented approach is the use of abstract data types, polymorphism, and reuse through inheritance.

Abstract data types define the active data items described above. A traditional data type in a programming language describes only the structure of a data item. An abstract data type also describes operations that may be requested of the data item. It is the ability to associate operations with data items that makes the items active. The abstract data type makes operations available without revealing the details of how the operations are implemented, preventing programmers from becoming dependent on implementation details. The definition of an operation is considered a contract between the implementor of the abstract data type and the user of the abstract data type. *See* Abstract data types.

Polymorphism in the object-oriented approach refers to the ability of a programmer to treat many different types of objects in a uniform manner by invoking the same operation on each object. Because the objects are instances of abstract data types, they may implement the operation differently as long as they fulfill the agreement in their common contract.

A new abstract data type (class) can be created in object-oriented programming simply by stating how the new type differs from some existing type. A feature that is not described as different will be shared by the two types, constituting reuse through inheritance. Inheritance is useful because it replaces the practice of copying an entire abstract data type in order to change a single feature. *See* Computer; Programming languages. [J.J.Shi.]

Obolellida A small order of inarticulate brachiopods which are predominantly Early Cambrian but which are known in Middle Cambrian rocks; they are among the oldest known representatives of the phylum. *See* INARTICULATA.

Superficially, they show the closest similarity to obolids, which belong to the order Lingulida. Like the obolids, they are biconvex and subcircular to elongate oval in outline and have marginal beaks. They differ, however, in a number of features, most conspicuously in shell composition; the shell of the Obolellida is of calcium carbonate in contrast to the calcium phosphate shell of the obolids. *See* BRACHIOPODA; LINGULIDA. [A.J.R.]

Obsessive-compulsive disorder A type of neurosis. The characteristic symptom pattern of the obsessive compulsive is the irrational persistence of ideation and enactment which is recognized as irrational but which the person cannot alter in spite of his efforts. Seemingly irrational and fragmentary phrases, admonitions, and melodies may haunt the person in an unshakable way in the case of obsessive symptoms and, in compulsive behavior, the person is driven to perform acts whose relevance for his problems remains utterly obscure to him, for example, repeated handwashing, sometimes to a self-injurious point. *See* NEUROTIC DISORDERS. [J.S.Br.; W.Mis.]

Obsidian A volcanic glass, usually of rhyolitic composition, formed by rapid cooling of viscous lava. The color is jet-black because of abundant microscopic, embryonic crystal growths (crystallites) which make the glass opaque except on thin edges. Iron oxide dust may produce red or brown obsidian.

Obsidian usually forms the upper parts of lava flows. Well-known occurrences are Obsidian Cliffs in Yellowstone Park, Wyoming; Mount Hekla, Iceland; and the Lipari Islands off the coast of Italy. *See* IGNEOUS ROCKS; VOLCANIC GLASS. [C.A.C.]

Occultation The apparent disappearance of a star or planet behind the surface of the Moon, or of a satellite behind the disk of the parent planet. Occultations by the Moon are observed to determine the position of the Moon at the time of the phenomenon. They are also used for precise determinations of longitude. Because of the eastward motion of the Moon against the background of stars, immersion (or disappearance) of the star or planet occurs at the eastern limb, and emersion (or reappearance) at the western limb. A single occultation is visible only from a certain region of Earth. Outside that region, the effect of parallax causes the star or planet to remain clear of the lunar disk. [S.D.G.]

Ocean circulation The general circulation of the ocean. The term is usually understood to include large-scale, nearly steady features, such as the Gulf Stream, as well as current systems which change seasonally but are persistent from one year to the next, such as the Davidson Current, off the northwestern United States coast, and the equatorial currents in the Indian Ocean. A great number of energetic motions have periods of a month or two and horizontal scales of a few hundred kilometers—a very-low-frequency turbulence, collectively called eddies. These are transient features, but locally have all the characteristics of currents. They are ubiquitous features in the ocean, although their energy varies widely, being highest near Western Boundary Currents. Energetic motions are also concentrated near the local inertial period (24 h, at 30° latitude) and at the periods associated with tides (primarily diurnal and semidiurnal). *See* TIDE.

The greatest single driving force for currents, as for waves, is the wind. Furthermore, the ocean absorbs heat at low latitudes and loses it at high latitudes. The resultant effect on the density distribution is coupled into the large-scale wind-driven circulation. Some subsurface flows are caused by the sinking of surface waters made dense by cooling or high evaporation. *See* OCEAN WAVES.

Surface currents. Except in western boundary currents, and in the Antarctic Circumpolar Current, the system of strong surface currents is restricted mainly to the upper 330–660 ft (100–200 m) of the sea. The mid-latitude anticyclonic gyres, however, are coherent in the mean well below 3300 ft (1000 m). The average speeds of the open-ocean surface currents remain mostly below 20 cm/s (0.4 knot). Exceptions to this are found in the western boundary currents, such as the Gulf Stream, and in the Equatorial Currents of the three oceans, all of which have velocities of 1–2 m/s (2–4 knots).

Deep circulation. The deep circulation results in part from the wind stress and in part from the internal pressure forces which are maintained by the budgets of heat, salt, and water. Both groups of forces are dependent upon atmospheric influences. Apart from Coriolis and frictional forces, the topography of the sea bottom exercises a decisive influence on the course of deep circulation.

The deep circulation in the oceans is more difficult to perceive than the circulation in the marginal seas. In addition to the internal pressure forces, determined by the distribution of density and the piling up of water by the wind, there are also the influences of Coriolis forces and large-scale turbulence. There are areas in tropical latitudes in which the surface water, as a result of strong evaporation, has a relatively high density. In thermohaline convection, the water sinks while flowing horizontally until it reaches a density corresponding to its own, and then spreads out horizontally. In this way the colder and deeper levels of the oceans take on a layered structure consisting of the so-called bottom water, deep water, and intermediate water. [W.Stu.]

Ocean waves The irregular moving bumps and hollows on the ocean surface. Winds blowing over the ocean, in addition to producing currents, create surface water undulations called waves or a sea. The characteristics of these waves (or the state of the sea) depend on the speed of the wind, the length of time that it has blown, the distance over which it has blown, and the depth of the water. If the wind dies down, the waves that remain are called a dead sea. Waves can travel hundreds and thousands of miles from where they were generated into areas where the wind is light. These waves are called swell. *See* SEA STATE. For a discussion of wave characteristics *see* CAPILLARY WAVE; INTERNAL WAVE; NEARSHORE SEDIMENTARY PROCESSES; SEICHE; STORM SURGE; TSUNAMI.

Wave spectral components first grow linearly in resonance with turbulent pressure components advected by the winds of the atmosphere. The components then grow exponentially by extracting energy from the wind profile. As saturation at a particular frequency is reached, the formation of whitecaps limits further growth. Although not accepted by all scientists, there is considerable evidence that, if the wind blows long enough at constant velocity over a large enough area, the spectrum of the waves will depend only on wind speed.

The rate at which waves are generated depends very much on whether or not there is a background spectrum present when the wind speed increases. With no background, a fairly long time is required to generate a fully developed sea. With a background, the time required is some variable fraction, often one-half or less, of the time required from a truly zero initial condition. Variable background conditions imply that the spectrum at all points over the ocean must be tracked accurately for accurate determination of changes. [W.J.P.; G.N.]

Instruments designed to measure ocean waves can be grouped into two classifications: those that sense the elevation of the surface water, and those that sense the subsurface pressure fluctuations generated by the waves. Surface elevation sensing gages include (1) surface float devices mechanically connected to a recording mechanism, (2) devices that record the buoyant force on a vertical cylinder, (3) recorders actuated by vertical accelerometers mounted in surface buoys, (4) inverted echo sounders (fathometers) fixed below the surface to echo off the water surface, (5) accurate absolute altimeter recorders operated in aircraft flying at a fixed elevation, (6) stereophotographs, and (7) electrical elements whose resistance or capacitance is a function of the elevation of the water surface. [F.E.S.]

Oceanic birds Any species of birds which are capable of living for long periods of time away from land and obtaining food from the open sea, that is, pelagic birds. They come to the land only during the breeding season, The avian order Procellariformes contains the largest number and variety of pelagic forms, all of which live beyond the continental edge, the home range of the true deep-sea species. These birds depend on a diet of various aquatic forms, such as mollusks and planktonic crustaceans, rather than upon fish. They are adapted to their salt-water environment because of the large nasal glands known as salt glands, which control salt secretion and maintain a physiological balance. *See* PROCELLARIIFORMES.

Most species of oceanic birds are long-lived and have a long period of maturation and a low reproductive rate; many species produce a single egg each year. The young are helpless when hatched and are fed by regurgitation. Included in this group are the albatrosses, fulmars, petrels, and shearwaters. [C.B.C.]

Oceanic islands Those islands which rise from the deep-sea floor rather than from shallow continental shelves. Most islands in gulfs and seas that fringe the great ocean basins are geologically similar to the nearby continents. On the other hand, almost all islands that rise from the ocean basins are volcanoes with or without coral reef and, geologically, bear little relation to the continents. Volcanic islands are only the tops of much larger undersea volcanoes, most of which are associated with great submarine structures, such as submarine ridges and fractures in the Earth's crust (see illustration). *See* MARINE GEOLOGY.

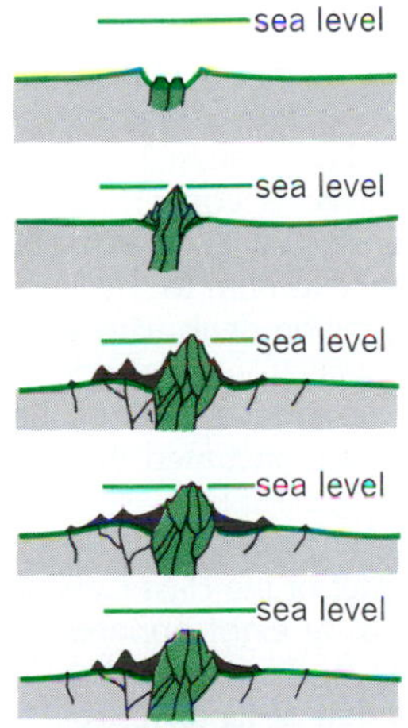

Sequence of development of submarine volcano or ridge. Center areas are initial volcanic extrusion; black areas are later deposits of volcanic material and erosional debris.

On the deep-sea floor, volcanoes begin as lava flows from fissures in the Earth's crust under 2 to 3 mi (3 to 5 km) of water. Gradually they build upward through the water, but about nine-tenths of the submarine volcanoes become inactive and stop their growth before they reach the sea surface. The others burst from the deep water into a new realm, where wave and subaerial erosion combat their upward growth. At first the volcanoes tend to produce ash and cinders, which are easily eroded. Gradually as the pile becomes broader, volcanoes rise above the waves and more resistant fluid lava flows build a solid island. Where several nearby volcanoes merge together, as in Hawaii, a great island may form. *See* VOLCANO; VOLCANOLOGY.

Volcanoes are active for no more than a few million years, however, and inactive volcanoes are inevitably worn down to shallow submarine banks by erosion which never stops. In addition to these worn-down volcanoes, drowned former islands called guyots or tablemounts have been discovered in all the ocean basins, mostly at depths of 1000–7000 ft (300–2000 m). *See* SEAMOUNT AND GUYOT. [E.L.H.; H.W.Me.]

Oceanographic platforms Supporting structures (stations) for instruments used in sensing and recording the various parameters of the oceanic environment. The platforms come in assorted shapes and sizes, ranging from moored buoys to fixed offshore towers and from surface survey ships to deep-diving submersibles. The former group collects a history of data over long periods of time in a specific location, whereas the latter group collects data over a large area, usually for a specific instant of time. *See* OCEANOGRAPHIC VESSELS; UNDERSEA VEHICLES. [J.J.Hr.]

Oceanographic vessels The largest and most basic tools used in the scientific study and exploration of the oceans. The term oceanographic vessels includes both conventional and special-purpose research ships and specialized research vehicles, such as deep submersibles and moored buoys. Uncrewed platforms and equipment, such as buoys, are commonly considered instruments rather than vehicles. *See* OCEANOGRAPHIC PLATFORMS; UNDERSEA VEHICLES.

The primary purpose of a research ship is to carry scientists and technicians and their equipment to locations at sea where their investigations are to take place. Conventional oceanographic ships are used for multidiscipline research. Special-purpose oceanographic vessels, such as research submarines, crewed buoys, fisheries research ships, special ships to tend and carry deep-submergence vehicles (see illustration), research drilling ships, and routine survey ships, are often limited by the special requirements of their primary tasks from being economically suitable for multidiscipline operation in oceanographic investigations. [J.Le.]

Deep submersible tender *Lulu* operated by Woods Hole Oceanographic Institution.

Oceanography The scientific study and exploration of the oceans and seas in all their aspects, including the sediments and rocks beneath the seas; the interaction of sea and atmosphere; the body of seawater in motion and subject to internal and external forces; the living content of the seas and sea floors and the behavior of these organisms; the chemical composition of the water; the physics of the sea and sea floor; the origin of ocean basins and ancient seas; and the formation and interaction of beaches, shores, and estuaries. Hence oceanography consists of the marine aspects of several disciplines and branches of science: geology, meteorology, biology, chemistry, physics, geophysics, geochemistry, fluid mechanics, and in its more theoretical aspects, applied mathematics. It is also an environmental science which describes and attempts to explain all processes in the ocean, and the interrelation of the ocean with the solid and gaseous phases of the Earth and with the universe. *See* MARINE ECOLOGY; MARINE GEOLOGY; OCEAN CIRCULATION; OCEAN WAVES; SEAWATER; TIDE.

Indirect benefits of oceanography arise from the application of marine meteorology to weather prediction, not only over the sea areas but also over the land. The study of marine geology and marine ecology aid in the understanding of the character of oil-bearing sedimentary rocks found on land. [W.A.N.]

Octagon A figure formed by the eight line segments (sides) that join in order eight ordered points (vertices) of a plane. In elementary geometry it is assumed that the sides do not cross and, usually, even that the finite region of the plane bounded by the sides is convex (convex octagon). A regular octagon has each two of its sides congruent, and each angle made by adjacent sides equal to 135° (see illustration). The area of a regular octagon of side a is $2(\sqrt{2} + 1)a^2$. *See* POLYGON; REGULAR POLYTOPES. [L.M.Bl.]

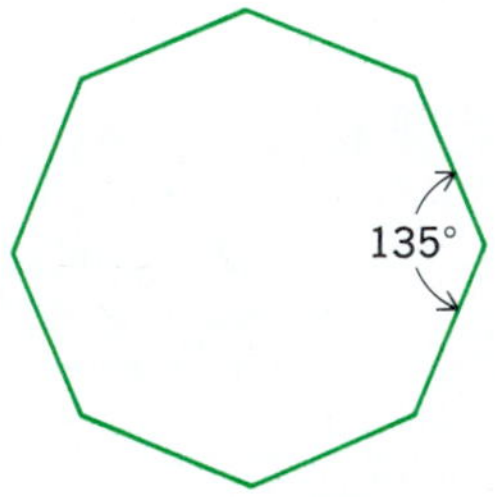

Regular octagon.

Octahedron In geometry, a solid with 8 faces. The faces of a regular octahedron are all congruent equilateral triangles. These 8 triangles (see illustration), have a total of 6 vertices and 12 edges—the vertices and edges of the octahedron. The solid is dual to the cube, which has 8 vertices, 6 faces, and 12 edges. The centers of the faces of a regular octahedron are the vertices of a cube; conversely, the centers of the faces of a cube are the vertices of a regular octahedron. The volume of an octahedron with edge a is $a^3\sqrt{2}/3$. *See* POLYHEDRON; REGULAR POLYTOPES. [L.M.Bi.]

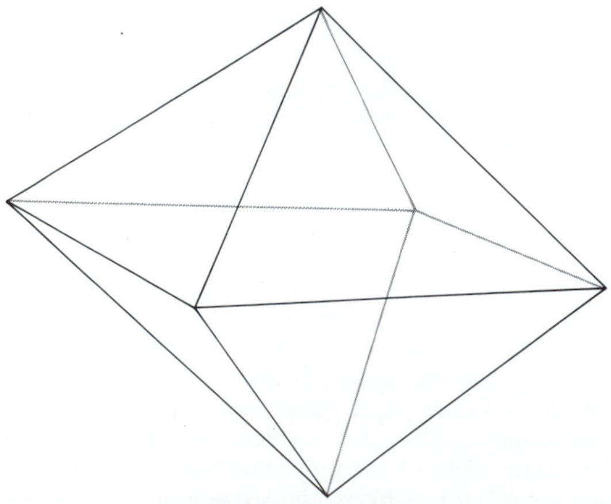

Regular octahedron.

Octane An alkane having the formula C_8H_{18}. The 18 possible isomers range in boiling point from 99.3°C or 2170°F (2,2,4-trimethylpentane) to 125.6°C or 258.1°F (*n*-octane).

The branched-chain octanes are obtained commercially by the catalytic alkylation of isobutane with the *n*-butylenes or isobutylene. They have high antiknock rating and are therefore valuable constituents of gasoline. *See* ALKANE; GASOLINE; OCTANE NUMBER. [L.S.]

Octane number A standard laboratory measure of a fuel's ability to resist knock during combustion in a spark-ignition engine, A single-cylinder four-stroke engine of standardized design is used to determine the knock resistance of a given fuel by comparing it with that of primary reference fuels composed of varying proportions of two pure hydrocarbons, one very high in knock resistance and the other very low. A highly knock-resistant isooctane (2,2,4-trimethylpentane, C_8H_{18}) is assigned a rating of 100 on the octane scale, and normal heptane (C_7H_{16}), with very poor knock resistance, represents zero on the scale. Octane number is defined as the percentage of isooctane required in a blend with normal heptane to match the knocking behavior of the gasoline being tested. *See* SPARK KNOCK; OCTANE.

For fuels with a rating higher than 100 octane, the rating is usually obtained by determining the amount of tetraethyllead compound that needs to be added to pure isooctane to match the knock resistance of the test fuel. *See* ANTIKNOCK AGENT. [J.C.L.]

Octave The interval between two sounds having a basic frequency ratio of 2:1; also, the interval in pitch between two tones such that one tone may be regarded as duplicating at the next higher pitch the basic musical import of the other tone. The first definition given is concerned with physical oscillations, and in this sense, the interval in octaves between any two frequencies is the logarithm to the base 2 of the frequency ratio. The second definition deals with sensations. Historically, a ratio of 2:1 in frequency was considered to correspond to the pitch interval of an octave, but psychological research of the 20th century has demonstrated that this is not necessarily the case. *See* PITCH; SCALE (MUSIC). [R.W.Y.]

Octopoda An order of the class Cephalopoda (subclass Coleoidea), characterized by eight appendages that encircle the mouth, a saclike body, and an internal shell that is much modified or reduced from that of its ancestors. One or two rows of suckers without chitinous rings occur along the eight highly flexible contractile arms. The approximately 200 species of octopods include shallow-water forms like the common octopus, *Octopus vulgaris*; open-ocean species like the paper argonaut, *Argonauta argo*; and deep-sea forms with fins like the flapjack devilfish, *Opisthoteuthis californica*. Two suborders divide the Octopoda into those with paddle-shaped fins on the body and tendrillike cirri on the arms (Cirrata) and those without fins or cirri (Incirrata). *See* CEPHALOPODA; COLEOIDEA. [C.F.E.R.]

Oculosida An order of Radiolaria. Pores of these protozoans are restricted to certain areas in the central capsule, which may be a single or double membrane in different groups. The order is subdivided on the basis of arrangement of the pores. In the Monopylina (Nassellarina) pores lie at one pole of the single-layered capsule. In the Tripylina (Phaeodorina) the major opening (astropyle) usually contains a perforated plate. A common feature is the presence, near the astropyle, of olive-colored material (possibly it is partially digested food) sometimes known as phaeodium. *See* RADIOLARIA. [R.P.H.]

Odonata An order of the class Insecta known as dragonflies. There are probably less than 3000 species known throughout the world. The order is divided into the Anisoptera, or true dragonflies, and the Zygoptera, or damselflies.

The young inhabit ponds, streams, and marshes; the adults fly over these localities or adjacent land. The adult structure is unique, characterized by a head with large compound eyes and wings with clear or transparent membranes traversed by networks of veins (see illustration); the male has accessory genital organs possessed by no other insects.

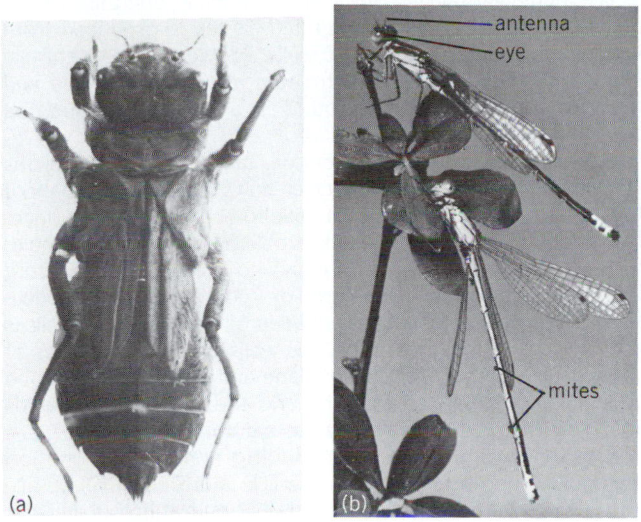

Typical Odonata. (*a*) Dragonfly nymph. (*b*) Adult damselflies.

There are three general stages in the life history: the egg, the nymph (see illustration), sometimes called naiad, and the adult. Development is usually slow, often requiring 3–5 years. Rarely is there more than one generation a year in the northern range. Adults may live for an extended period in summer. Eggs are laid by insertion into plant stems, either beneath the water or just above the surface. Others are dropped directly into the water and sink to the bottom, where they hatch and the nymphs develop.

Dragonflies constitute one of the oldest insect orders; they can be traced back through fossil records to the Carboniferous and Permian. Surprisingly few changes have occurred since then, although the order has diversified and specialized to counter competition and to avoid enemies; for example, wing venation is more complex and body structure more varied. [P.G.]

Odontognathae One of the two superorders making up the subclass Neornithes, or true birds. It was originally erected to include fossil birds that could fly or that were secondarily flightless but still possessed teeth; it was regarded therefore as an evolutionary step between *Archaeopteryx* and fully evolved modern birds. It is now considered to contain only a simple order, Hesperornithiformes, of which the best-known family is the Hesperornithidae containing several species from the Upper Cretaceous of North America. *See* ARCHAEORNITHES; AVES; HESPERORNITHIFORMES; NEORNITHES. [W.J.B.]

Odontostomatida An order of the Spirotrichia which represents a minor group of small, bizarre-looking protozoan species. The odontostomes are compressed laterally and possess very little ciliature. Even the adoral zone of membranelles is reduced in prominence. These ciliates are found in sewage disposal environs and other fresh- or salt-water habitats which have a very low oxygen content. *See* CILIOPHORA; SPIROTRICHIA. [J.O.C.]

Oedogoniales An order of filamentous fresh-water green algae (Chlorophyceae) with unique morphological features including (1) an elaborate method of cell division that results in the accumulation of apical caps, (2) zoospores and antherozoids with a subapical crown of flagella, and (3) a highly specialized type of oogamy. There is a single family, Oedogoniaceae, comprising three genera. *See* CHLOROPHYCEAE.

Oedogonium has the largest number of species (several hundred) and is the most common of the three genera. Its unbranched filaments are initially attached by a holdfast cell to submerged vegetation, stones, or wood, usually in permanent ponds or pools, but at maturity they may form free-floating masses. The cells are cylindrical, each containing a reticulate chloroplast with numerous pyrenoids.

Vegetative multiplication by fragmentation is common. In asexual reproduction, zoospores with a subapical crown of up to 120 flagella are formed singly within a cell. In sexual reproduction, there is a highly specialized interplay between female and male elements. The egg is a metamorphosed protoplast of an enlarged spherical cell (oogonium). Antherozoids, with a subapical crown of about 30 flagella, are produced in pairs or tetrads in very small discoid antheridia. The two types of sex organs may occur on the same filament (homothallic species), or on different filaments (heterothallic species), but in either case certain species have an indirect development of antherozoids. These species, termed nannandrous in distinction to those with direct development (macrandrous), form short cylindrical cells which initially appear like antheridia, but in which the protoplast metamorphoses into a single swarmer bearing a subapical crown of flagella. [P.C.Si.; R.L.Moe]

Oegophiurida An order of Ophiuroidea comprising eight families. Seven of these are represented by Paleozoic fossils, but one has a living representative, *Ophiocanops*. Oegophiurids appeared in the Ordovician and were apparently the first ophiuroids in which the ambulacral ossicles fused to form the typical ophiuroid arm vertebrae. Dorsal and ventral arm plates are lacking, and the ambulacral groove remains as in their somasteroid ancestors. The gastric ceca extend into the arms, and the gonads are located there too, arranged in series. *See* OPHIUROIDEA. [A.C.C.]

Ohmmeter A small, portable instrument using a microammeter and associated circuitry to measure resistance by the voltmeter-ammeter method. Additional circuits are usually included to measure alternating and direct-current volts and amperes, and the instrument is called a volt-ohm-milliammeter, or multimeter. *See* AMMETER; RESISTANCE MEASUREMENT; VOLT-METER. [C.E.A.]

Ohm's law The direct current flowing in an electrical circuit is directly proportional to the voltage applied to the circuit. The constant of proportionality R, called the electrical resistance, is given by the equation below, in which V is the applied voltage

$$V = RI$$

and I is the current. Numerous deviations from this simple, linear relationship have been discovered. *See* CONDUCTIVITY; ELECTRICAL RESISTANCE. [C.E.A.]

Oil and gas, offshore Oil and natural gas prospecting and exploitation on the continental shelves and slopes. Exploration on the more than 8×10^6 mi^2 (20.7×10^6 km^2) of the world's continental shelves and slopes, lying between the shore and 1000-ft (300-m) water depth includes exploration or drilling or both off the coasts of more than 75 nations.

For many years petroleum companies stopped at the water's edge or sought and developed oil and gas accumulations only in the shallow seas bordering onshore producing areas. Exploration deeper under the sea did not begin in earnest until the world's burgeoning appetite for energy sources, coupled with a lessening return from land drilling, provided the incentives for the huge investments required for drilling in the open sea. There has been a steady annual increase in the number of wells drilled in deep water beyond the continental shelves, although the costs of developing the actual producing systems are enormous.

The underwater search has been made possible only by vast improvements in offshore technology. A wide variety of rig platforms has evolved, some designed to cope with specific hazards of the sea and others for more general work. The world's mobile platform fleet can be divided into four main groupings: self-elevating platforms, submersibles, semisubmersibles, and floating drill ships.

The most widely used mobile platform is the self-elevating, or jack-up, unit (see illustration). It is towed to location, where

Offshore self-elevating drilling platform. (*Marathon Oil Co., Findlay, Ohio*)

the legs are lowered to the sea floor, and the platform is jacked up above wave height. These self-contained platforms are especially suited to wildcat and delineation drilling.

The submersible platforms are towed to location and then submerged to the sea bottom. They are very stable and can operate in areas with soft sea floors. Difficulty in towing is a disadvantage, but this is partially offset by the rapidity with which they can be raised or lowered, once on location. Semisubmersibles are a version of submersibles. They can work as bottom-supported units or in deep water as floaters.

Floating drill ships are capable of drilling in 60-ft (18-m) to abyssal depths. They are built as self-propelled ships or with a ship configuration that requires towing.

The move of exploration into the open hostile sea has required not only the development of drilling vessels but a host of auxiliary equipment and techniques. A whole new industrial complex has developed to serve the offshore industry. Of particular interest is the development of diving techniques and submersible equipment to aid in exploration and the completion of wells. Economics will eventually require sea-bottom completions in which robots will make the necessary pipe and well connections. [G.R.S.]

Oil and gas field exploitation In the petroleum industry, a field is an area underlain without substantial interruption by one or more reservoirs of commercially valuable oil or gas, or both. A single reservoir (or group of reservoirs which cannot be separately produced) is a pool. Several pools separated from one another by barren, impermeable rock may be superimposed one above another within the same field. Pools have variable areal extent. Any sufficiently deep well located within the field should produce from one or more pools. However, each well cannot produce from every pool, because different pools have different areal limits.

Development of a field includes the location, drilling, completion, and equipment of wells necessary to produce the commercially recoverable oil and gas in the field.

General considerations. Oil and gas production necessarily are intimately related, since approximately one-third of the gross gas production in the United States is produced from wells that are classified as oil wells. However, the naturally occurring hydrocarbons of petroleum are not only liquid and gaseous but may even be found in a solid state, such as asphaltite and some asphalts.

Where gas is produced without oil, the production problems are simplified because the product flows naturally throughout the life of the well and does not have to be lifted to the surface. However, there are sometimes problems of water accumulations in gas wells, and it is necessary to pump the water from the wells to maintain maximum, or economical, gas production. The line of demarcation between oil wells and gas wells is not definitely established. Most gas wells produce quantities of condensable vapors, such as propane and butane, that may be liquefied and marketed for fuel, and the more stable liquids produced with gas can be utilized as natural gasoline.

Production methods in producing wells. The common methods of producing oil wells are (1) natural flow; (2) pumping with sucker rods; (3) gas lift; (4) hydraulic subsurface pumps; (5) electrically driven centrifugal well pumps; and (6) swabbing.

However, most wells are not self-flowing and various lifting methods must be employed. Approximately 90% of the wells made to produce by some artificial lift method in the United States are equipped with sucker-rod–type pumps. In these the pump is installed at the lower end of the tubing string and is actuated by a string of sucker rods extending from the surface to the subsurface pump. The two common variations are mechanical and hydraulic long-stroke pumping. Other lifting mechanisms are the gas lift, hydraulic subsurface pumps, swabs, bailers, jet pumps, and sonic pumps. *See* OIL AND GAS WELL COMPLETION.

Production instruments. The commoner and more important instruments required in petroleum production operations are the following:

1. Gas meters, which are generally of the orifice type, are designed to record the differential pressure across the orifice, and the static pressure.
2. Recording subsurface pressure gages small enough to run down 2-in. (3-cm) ID (inside diameter) tubing are used extensively for measuring pressure gradients down the tubing of flowing wells, recording pressure buildup when the well is closed in, and measuring equilibrium bottom-hole pressures.
3. Subsurface samplers designed to sample well fluids at various levels in the tubing are used to determine physical properties.
4. Oil meters of various types are utilized to meter crude oil flowing to or from storage.
5. Dynamometers are used to measure polished-rod loads.

6. Liquid-level gages and controllers are used. They are similar to those used in other industries, but with special designs for closed lease tanks. [R.L.Ch.]

Oil and gas well completion The operations that prepare for production of a well drilled to an oil or natural gas reservoir. Various problems of well casing during and at the end of drilling are related to modes of completing connection between the proper reservoirs and the surface. Tubing inside the casing and valves and a pumping unit at the surface must deliver reservoir products to the surface at a controlled rate. Variations in reservoir and overlying formations may require special techniques to keep out water or sand or to increase the production rate. *See* OIL AND GAS FIELD EXPLOITATION; OIL AND GAS WELL DRILLING.

Casing. Oil and gas wells are walled with steel tubing (casing) which is cemented in place. Tubing lengths, called joints, are threaded and coupled so that they may be joined together in a continuous string in the well bore. Properly placed and cemented in the hole, casing protects fresh-water reservoirs from contamination, supports unconsolidated rock formations, maintains natural separation of formations, aids in the prevention of blowouts and waste of reservoir energy, and acts as a conduit for receiving pipe of smaller diameters through which the well effluent may be brought to the surface under controlled conditions. It may be necessary to set many strings of casing in one hole before reaching the objective (Fig. 1).

Well hole–reservoir connection. Reservoir conditions determine the type of completion technique to be followed: (1) barefoot completion, (2) preperforated liner, or (3) casing set through and perforated.

Barefoot completion is frequently used when the character of the producing rock is such that it does not require supplemental support or screening. With this method, the production casing is seated above the producing section and the formation contents enter the bore hole from the bare or unlined producing stratum or strata, hence the term barefoot.

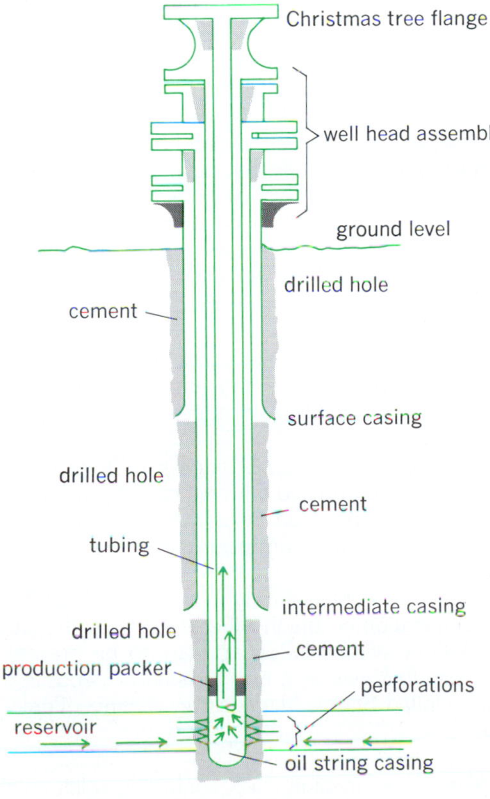

Fig. 1. Casing detail; casing strings in an oil well.

Liner-type completion is similar to a barefoot completion except that the open portion of the hole is cased with a preperforated section of casing called a liner.

Gun perforating is a method of forming holes through the casing and into a formation from within a well bore. The two more popular methods are bullet perforating, as with a rifle, and jet perforating, as with a torch.

Production-flow control. A steel tube, smaller in diameter than the casing, serves as a production flow line within the well. The tubing is run inside the casing and is either suspended in the hole or set on a production packer at or near the producing interval. The top of the tubing string terminates at the surface in a sealing element in the wellhead assembly to which the so-called Christmas tree is attached. A manifold constructed of steel valves and fittings, placed on top of the casings protruding above the surface, is called a Christmas tree (Fig. 2). Its purpose is to maintain the well under proper control, to receive the formation products under pressure, and to control the rate of daily production from the reservoir and direct it into a pipeline, generally at reduced pressures, to the oil-gathering station.

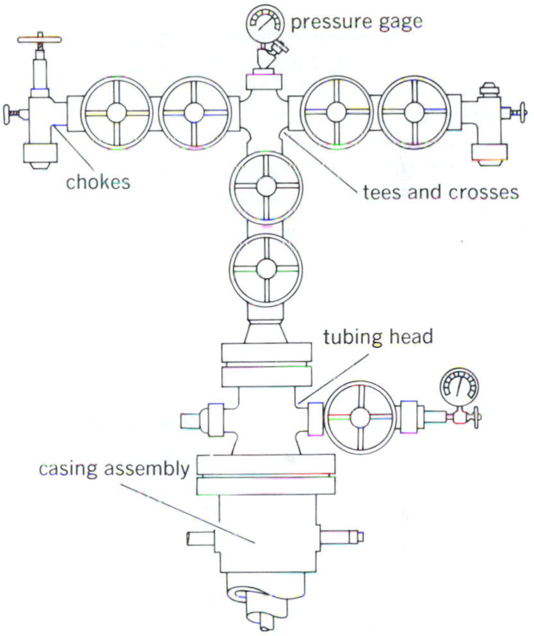

Fig. 2. Typical layout of a Christmas tree.

Most oil wells require secondary means of removing the reservoir product. The most common of several methods is the pumping unit. *See* PETROLEUM-ENHANCED RECOVERY.

Production-stimulation techniques. Any method designed to increase the production rate from a reservoir is defined as production stimulation. Three of the methods used are acidizing, fracturing, and employing explosives.

In acidizing, varied volumes of hydrochloric acid are used in limestone and dolomite or other acid-soluble formations to dissolve the existing flow-channel walls and enlarge them. Formation fracturing is a hydraulic process aimed at the parting of a desired section of formation. Selected grades of sand or particles of other materials are added to the fracturing fluid in varied quantities. These particles pack and fill the fracture, acting as a propping agent to hold it open when the applied pressure is released. Explosives are used to remove reservoir-blocking material from the reservoir face and to create fractures in the rock to increase production. *See* PETROLEUM RESERVOIR ENGINEERING. [H.S.Br.]

Oil and gas well drilling The drilling of holes for exploration and extraction of crude oil and natural gas. Deep holes and high pressures are characteristics of petroleum drilling not commonly associated with other types of drilling. In general, it becomes more difficult to control the direction of the drilled hole as the depth increases, and additionally, the cost per foot of hole drilled increases rapidly with the depth of the hole. Drilling-fluid pressure must be sufficiently high to prevent blowouts but not high enough to cause fracturing of the bore hole. Formation-fluid pressures are commonly controlled by the use of a high-density clay-water slurry, called drilling mud. The primary disadvantage in the use of drilling muds is the relatively low drilling rate which normally accompanies high bottom-hole pressure. Drilling rates can often be increased by using water to circulate the cuttings from the hole; when feasible, the use of gas as a drilling fluid can lead to drilling rates as much as 10 times those attained with mud. *See* OIL AND GAS WELL COMPLETION; PETROLEUM GEOLOGY; ROTARY TOOL DRILL; TURBODRILL; WELL LOGGING. [J.B.C.]

Oil burner A device for converting fuel oil from a liquid state into a combustible mixture. A number of different types of oil burners are in use for domestic heating. These include sleeve burners, natural-draft pot burners, forced-draft pot burners, rotary wall flame burners, and air-atomizing and pressure-atomizing gun burners. The most common and modern type that handles 80% of the burners used to heat United States homes is the pressure-atomizing-gun-type burner shown in the illustration.

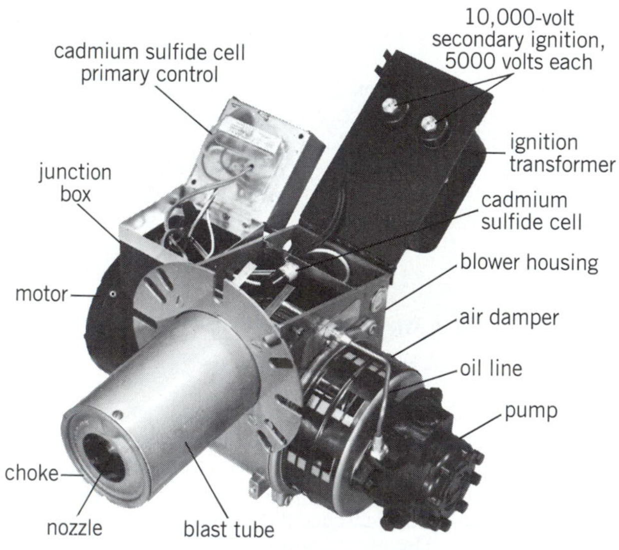

An oil burner of the pressure-atomizing type. (*Automatic Burner Corp.*)

The sleeve burner, commonly known as a range burner because of its use in kitchen ranges, is the simplest form of vaporizing burner. The natural-draft pot burner relies on the draft developed by the chimney to support combustion. A modification of this burner is the forced-draft pot burner which supplies its own air for combustion and does not rely totally on the chimney. The rotary wall flame burners have mechanically assisted vaporization. The gun-type burner uses a nozzle to atomize the fuel so that it becomes a vapor, and burns easily when mixed with air.

The oil burner is used for a wide assortment of heating, air conditioning, and processing applications. Oil burners heat commercial buildings such as hospitals, schools, and factories. Air conditioners using the absorption refrigeration system have been developed and fired with oil burners. Oil burners are used to produce CO_2 in greenhouses to accelerate plant growth. They also produce hot water for many commercial and industrial applications. *See* AIR COOLING; COMFORT HEATING; HOT-WATER HEATING SYSTEM; OIL FURNACE. [R.A.K.]

Oil field model A small-scale and commonly simplified replica of subsurface conditions of interest and value in petroleum prospecting and oil-field development. The term model has been applied by geologists to a simplified diagram and by mathematical physicists to a formal analysis with special boundary conditions and related attributes. These and somewhat more complex physical models have value in transmitting concepts and relationships to the nonspecialist, but they are generally designed by geologists and petroleum engineers to aid investigation of a particular problem. To do so, the model is scaled to size, shape, and like attributes, or according to forces upon it, relative to its archetype.

Physical models present the earth features in a readily visible fashion easier to grasp than mathematical analysis. For a discussion of mathematical models *see* PETROLEUM RESERVOIR MODELS. [P.W.]

Oil field waters Waters of varying mineral content which are found associated with petroleum and natural gas or have been encountered in the search for oil and gas. They are also called oil field brines, or brines. They include a variety of underground waters, usually deeply buried, and have a relatively high content of dissolved mineral matter. These waters may be (1) present in the pore space of the reservoir rock with the oil or gas, (2) separated by gravity from the oil or gas and thus lying below it, (3) at the edge of the oil or gas accumulation, or (4) in rock formations which are barren of oil and gas. Brines are commonly defined as water containing high concentrations of dissolved salts. Potable or fresh waters usually are not considered oil field waters but may be encountered, generally at shallow depths, in areas where oil and gas are produced.

Probably the most important geological use of oil field water analyses is their application to the quantitative interpretation of electrical and neutron well logs, particularly micrologs. *See* PETROLEUM GEOLOGY; WELL LOGGING. [P.McG.]

Oil furnace A combustion chamber in which oil is the heat-producing fuel. Fuel oils, having from 18,000 to 20,000 Btu/lb (42–47 megajoules/kg), which is equivalent to 140,000 to 155,000 Btu/gal (39–43 megajoules/liter), are supplied commercially. The lower flash-point grades are used primarily in domestic and other furnaces without preheating. Grades having higher flash points are fired in burners equipped with preheaters. *See* FUEL OIL.

Domestic oil furnaces with automatic thermostat control usually operate intermittently, being either off or operating at maximum capacity. *See* OIL BURNER. [F.H.R.]

Oil sand A loose to consolidated sandstone or a porous carbonate rock, impregnated with a heavy asphaltic crude oil, too viscous to be produced by conventional methods; also known as tar sand or bituminous sand. *See* ASPHALT AND ASPHALTITE.

Oil sands are distributed throughout the world but the largest proven accumulation occurs in Alberta, Canada (the Athabasca deposit). A large accumulation appears to be present in the Orinoco Basin in Venezuela, and far smaller deposits occur in Russia, the United States, Madagascar, Albania, Trinidad, and Romania. [G.R.G.]

Oil shale A sedimentary rock containing solid, combustible organic matter in a mineral matrix. The organic matter, often called kerogen, is largely insoluble in petroleum solvents, but decomposes to yield oil when heated. Although "oil shale" is used as a lithologic term, it is actually an economic term referring to the rock's ability to yield oil. No real minimum oil yield or content of organic matter can be established to distinguish oil shale from other sedimentary rocks. Additional names given to oil shales include black shale, bituminous shale, carbonaceous shale, coaly shale, cannel shale, cannel coal, lignitic shale, torbanite, tasmanite, gas shale, organic shale, kerosine shale, coorongite, maharahu, kukersite, kerogen shale, and algal shale. [J.W.S.; H.B.J.]

Okra A warm-season annual, *Hibiscus esculentus*, of Ethiopian origin. Okra, also called gumbo, is grown for its immature pods (see illustration), which are generally used for preparing soups but are also eaten as a freshly cooked vegetable. It is a member of the order Malvales and is related to cotton. Georgia, Florida, and Louisiana are important producing states. *See* MALVALES. [H.J.C.]

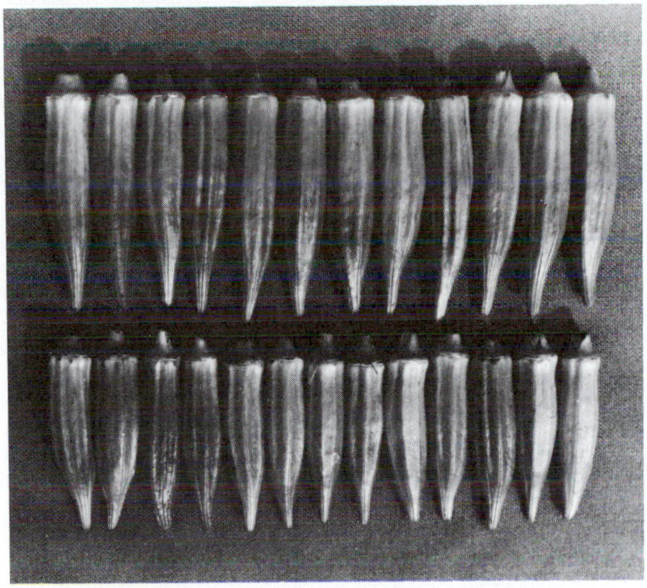

Okra pods. (*Asgrow Seed Co., subsidiary of The Upjohn Co.*)

Olbers' paradox The riddle of cosmic darkness. The obvious explanation for the darkness of the night sky, that the Sun is on the other side of the Earth, does not account for the fact that, in space far from any star, the universe is full of darkness and not of light.

In a boundless universe of stars, with no interstellar absorption, every line of sight from the eye must eventually intercept the surface of a star. If most stars are similar to the Sun, the sky at every point should shine as bright as the Sun's disk. The sky (or celestial sphere) is 180,000 times larger than the Sun's disk, and the starlight incident on the Earth should therefore be 180,000 times more intense than sunlight, which obviously is not the case. Hermann Bondi resurrected the riddle of cosmic darkness in 1952 and attributed it to the nineteenth-century astronomer Wilhelm Olbers, although, as is now known, it had previously been discussed by Edmund Halley and other astronomers.

Edgar Allan Poe suggested in 1848 that the universe is not old enough for the light from very distant stars to have reached the Earth; this was investigated by Lord Kelvin in 1901. Modern calculations confirm Kelvin's results: light travels at approximately 186,000 mi/s (300,000 km/s) and, in a static universe 10–20 $\times$ 10^9 years old, stars cannot shine long enough for their light to reach the Earth from regions sufficiently distant for the visible stars to cover the entire sky. This means that stars cannot shine long enough to fill the universe with radiation in equilibrium with their surfaces. Clearly, if the sky at night is dark in a static universe of finite age, then in an expanding universe of similar age the night sky is even darker because of the redshift. *See* BIG BANG THEORY; COSMOLOGY; UNIVERSE. [E.H.]

Oleate A salt or ester of oleic acid, whose formula is given below, in which the acid hydrogen of the carboxyl group is

$$CH_3(CH_2)_7CH{=}CH(CH_2)_7C({=}O){-}OH$$

replaced by a metal or an organic radical. Oleates occur in nature chiefly as the glyceryl ester, found in substantial amounts in animal and vegetable fats. A few of the simpler esters have commercial applications in textile, leather, cosmetics, and pharmaceuticals. Alkali metal oleates are water-soluble and are the chief components of toilet and laundry soaps. *See* CARBOXYLIC ACID; DETERGENT; ESTER; SOAP. [E.H.H.]

Olefin sulfide One of a number of organosulfur compounds, also called episulfides or thiiranes, that are the sulfur analogs of olefin oxides. The homologous series is represented by the structural formula shown below. The episulfides are far

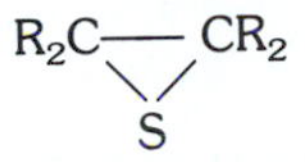

less known than the epoxides. They undergo some reactions which are analogous to those of epoxides, for example, ring opening with HCl, H_2S, RSH, and alcohols, ROH. The tendency of olefin sulfides to polymerize (to polysulfides) is quite marked, however, and frequently complicates their use in other reactions. *See* ORGANOSULFUR COMPOUND. [N.K.]

Olfaction One of the chemical senses. In mammals the olfactory receptors are located in the upper parts of the nasal cavity. In humans they are located in a small patch in each nasal cavity, consisting of about 0.4 in.^{2} (2.5 cm^2) of mucous membrane, colored with a yellow pigment. The olfactory cells are long and ovoid in shape, terminating at the distal end in several delicate hairs projecting into and possibly through the mucous covering the nasal epithelium. The proximal end consists of fine unmyelinated nerve fibers which pierce the bony cribriform plate to enter directly the olfactory bulb of the brain. *See* BRAIN.

Nerve endings of the trigeminal nerve fibers, utilized in common chemical sensitivity, are widely distributed throughout the nasal and olfactory area. Many odorants, for example, ammonia, stimulate both the olfactory and free nerve endings.

Most odorous compounds are organic. Both the arrangement and structure of the molecule as well as the presence of certain groups within the molecule appear to influence odor. In spite of the relative inaccessibility of the olfactory end organs, odor materials can be detected at extremely low concentrations. It has been estimated that olfaction is 10,000 times more sensitive than taste. Threshold concentrations for well-known odorants, such as ethylmercaptan, have been cited in the range of 4×10^{-8} mg/liter of air.

There is extensive literature on theories of olfaction, ranging from the purely chemical to the purely physical. The factors considered are the chemical or physical nature of an adequate stimulus and the nature of the receptor; for example, the receptor may be composed of only a few basic kinds of endings or many specialized different endings. Though many theories are provocative, there is, as yet, insufficient ground for accepting one theory over the other. *See* CHEMICAL SENSES. [C.P.]

Oligocene The third of the five major worldwide divisions (epochs) of the Tertiary Period (Cenozoic Era); the epoch of geologic time extending from the end of the Eocene to the beginning of the Miocene. Oligocene also denotes the series of rocks formed during the epoch and the fossils therein. *See* CENOZOIC; EOCENE; MIOCENE; TERTIARY.

Marine faunal provinces were fragmented in the Oligocene. Conversely, some isolated Northern Hemisphere landmasses became reconnected, permitting interchange of some mammals (opossums, anthracotheres, chalicotheres, tapirs, rhinocerids). Cool-water marine shellfish moved equatorward, certain pelecypod and gastropod stocks arriving for the first time in habitats they still occupy today. The marine planktonic foraminiferal faunas of the preceding Eocene have been shown to have suffered an entire eclipse, having become restricted to a single lineage.

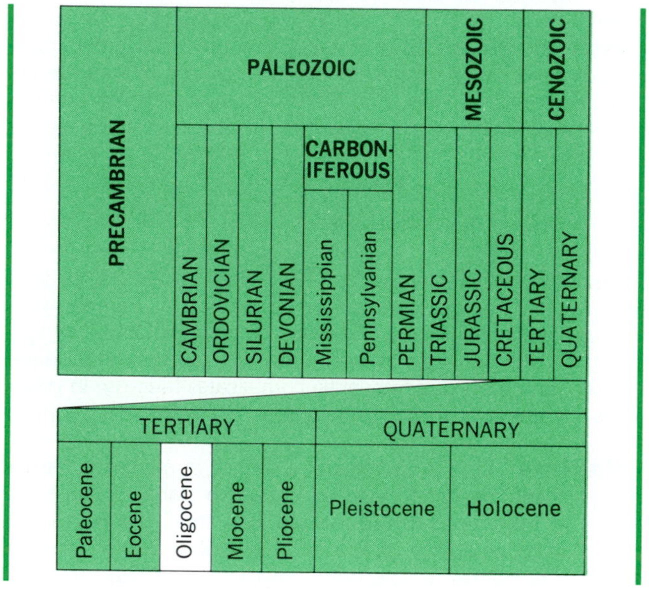

With the rise of grasses the earlier huge browsers, such as the brontotheres and other archaic mammals, made their last stand. Hyracoids appeared and have left a lone relic descendant coney or "rock rabbit." Primitive primates (lemurs) and insectivores carried on in isolated Madagascar. Ten living genera of birds already existed. The first New and Old World monkeys and Great Apes appeared. Carnivora modernized (dogs, cats, saber-toothed cats) alongside their disappearing creodont ancestors. Beavers, pangolins, three-toed horses, and pig, peccary, and tapir families appeared and other hoofed mammals diversified. *See* PALEOBOTANY; PALEONTOLOGY. [R.M.Kl.]

Oligochaeta A class of the phylum Annelida including worms such as the earthworms. There are 21 families with over 3000 species. These animals exhibit both external and in-ternal segmentation. They usually possess setae which are not borne on parapodia. Oligochaetes are hermaphroditic. The gonads are few in number and situated in the anterior part of the body, the male gonads being anterior to the female go-nads. The gametes are discharged through special ducts, the oviducts and sperm ducts. A clitellum is present at maturity. There is no larval stage during development.

The oligochaetes are primarily fresh-water and burrowing terrestrial animals. A few are marine and several species occur in the intertidal zone.

Oligochaetes are cylindrical, elongated animals with the anterior mouth usually overhung by a fleshy lobe, the prostomium, and the anus terminal. The body plan is that of a tube within a tube. Externally, the segments are marked by furrows. The setae or bristles are borne on most segments. Other external features are the pores of the reproductive systems opening on certain segments, the openings of the nephridia, and in many earthworms dorsal pores which open externally from the coelom. Some aquatic species have extensions of the posterior part of the body which function as gills.

The oligochaetes have been used in studies of physiology, regeneration, and metabolic gradients. Some aquatic forms are important in studies of stream pollution as indicators of organic contamination. Earthworms are important in turning over the soil and reducing vegetable material into humus. It is likely that fertile soil furnishes a suitable habitat for earthworms, rather than being a result of their activity. *See* ANNELIDA. [P.C.H.]

Oligoclase A plagioclase feldspar with a composition ranging from $Ab_{90}An_{10}$ to $Ab_{70}An_{30}$, where Ab = $NaAlSi_3O_8$ and An = $CaAl_2Si_2O_8$. If Fe_2O_3 is present as thin flakes oriented parallel to certain structurally defined planes, such oligoclase is called aventurine or sunstone. *See* FELDSPAR; IGNEOUS ROCKS. [F.H.L.]

Oligonucleotide A deoxyribonucleic acid (DNA) or ribonucleic acid (RNA) sequence composed of two or more covalently linked nucleotides. Oligonucleotides are classified as deoxyribooligonucleotides or ribooligonucleotides. Fragments containing up to 50 nucleotides are generally termed oligonucleotides, and longer fragments are called polynucleotides. *See* DEOXYRIBONUCLEIC ACID (DNA); RIBONUCLEIC ACID (RNA).

A deoxyribooligonucleotide consists of a 5-carbon sugar called deoxyribose joined covalently to phosphate at the 5′ and 3′ carbons of this sugar to form an alternating, unbranched polymer. A ribooligonucleotide consists of a similar repeating structure where the 5-carbon sugar is ribose. Chemically synthesized oligonucleotides of predetermined sequence have proven to be very useful for studying a large number of biochemical processes. In the 1960s, these compounds were used to decipher the genetic code. Later, chemically prepared deoxyoligonucleotides were joined to form genes for transfer RNAs. Gene synthesis from synthetic deoxyoligonucleotides is now routinely used to prepare genes and modified genes for proteins having potential clinical applications. Oligonucleotides have also been used to diagnose genetic disorders and bacterial or viral infections. *See* GENE; GENETIC CODE; GENETIC ENGINEERING; NUCLEIC ACID. [M.H.C.]

Oligopygoida An order of irregular echinoids in the superorder Neognathostomata resembling clypeasteroids but lacking the accessory ambulacral pores characteristic of that group. Oligopygoids have well-developed petals, and there are characteristic small demiplates present below the petals. The apical disk is monobasal and the mouth oval and usually deeply sunken. Oligopygoids have a lantern, which closely resembles that of clypeasteroids, and their lantern muscle-attachment structure are a mixture of ambulacral and interambulacral processes.

There are two gena, *Oligopygus* and *Haimea*, containing about 25 species, all from the middle and upper Eocene of the Caribbean and Gulf of Mexico regions. They were probably infaunal deposit feeders like present-day laganiids. *See* ECHINODERMATA. [A.B.S.]

Oligosaccharide A sugar composed of 2 or more monosaccharide units. Those sugars containing up to 6 units, many of which occur in nature, have been isolated as crystalline compounds. Fragments, obtained by controlled hydrolysis of various polysaccharides with acid and consisting of monosaccharides up to 10 units, are also termed oligosaccharides. *See* MONOSACCHARIDE.

The oligosaccharides may be considered as glycosides in which a hydroxyl (OH) group of one monosaccharide is condensed with the reducing group of another, with the loss of $n - 1$ molecules of water (n = number of monosaccharide residues). This condensation process is shown below.

$$C_6H_{12}O_6 + C_6H_{12}O_6 - H_2O = C_{12}H_{22}O_{11} \quad \text{Disaccharide}$$

$$3C_6H_{12}O_6 - 2H_2O = C_{18}H_{32}O_{16} \quad \text{Trisaccharide}$$

If two sugar units are joined in this manner, a disaccharide results; a linear array of three monosaccharides thus joined by glycosidic bonds is a trisaccharide, and so forth. On the basis of the number of constituent monosaccharide units, the oligosaccharides are classified as disaccharides, trisaccharides, tetrasaccharides, and so on. No sharp distinction can be drawn between the oligosaccharides and polysaccharides; it is chiefly a matter of the latter's possessing higher molecular weights. *See* POLYSACCHARIDE.

The monosaccharide units of an oligosaccharide may be alike, as in maltose, which on hydrolysis gives two molecules of D-glucose, or different, as in sucrose or raffinose. Sucrose consists of D-glucose and D-fructose, and raffinose consists of D-glucose, D-fructose, and D-galactose residues. *See* MALTOSE; SUCROSE. [W.Z.H.]

Oligotrichida A minor order of the Spirotrichia. If somatic ciliature is present, it is sparse. The bodies are round in cross section, and the adoral zone of membranelles is often highly developed at the anterior, or oral, end of the organism. Species are found in fresh- and salt-water habitats, and none occurs as a parasite. *Halteria* (see illustration) has long bristles which are used in a kind of jumping movement. *See* CILIOPHORA; SPIROTRICHIA. [J.O.C.]

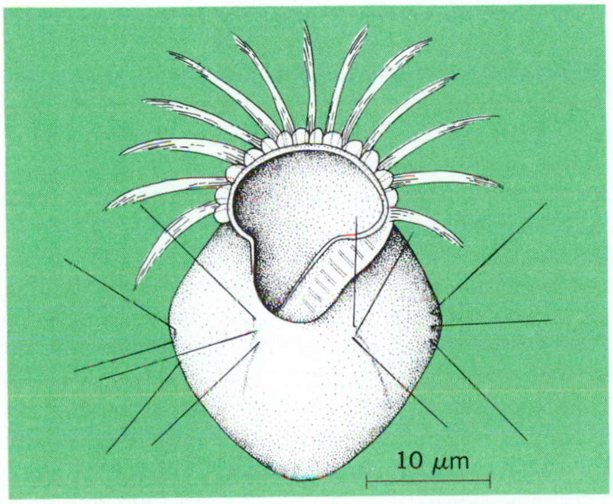

***Halteria*, an example of an oligotrichid.**

Olive Olive fruits, which are produced by a small to medium-sized evergreen tree (*Olea europaea*), can be eaten, after processing, as table olives, or can be extracted for oil that is used on salads, for cooking, for body lotions, or for medicinal purposes. The olive tree is a historically ancient cultivated plant, having been domesticated by early civilizations in the eastern Mediterranean regions. Olive culture later spread to all the Mediterranean countries and subsequently to South America, California, South Africa, and Australia.

Olive fruits are harvested for table olives in the autumn, when the fruits change from green to straw or to a slightly red color. Raw fruits contain a bitter glucoside which makes them inedible, but treatment with an alkali such as sodium hydroxide (lye) neutralizes the bitterness. The lye must subsequently be leached out of the fruits with water. In another method most of the bitterness can be removed by leaching with salt water. A lactic acid fermentation process is widely used in the Mediterranean countries to preserve the olives.

For oil production, the fruits are harvested in midwinter when they have become black and have reached their maximum oil content—15 to 25% of the fresh weight, depending on the variety. In producing olive oil, the freshly harvested fruits including pits are ground, after which this material is placed in burlap or cloth bags and placed under high pressure. Oil and water are extracted and transferred to tanks, where the oil rises to the top and is removed. *See* FAT AND OIL (FOOD). [H.T.Ha.]

Olivine A name given to a group of magnesium-iron silicate minerals crystallizing in the orthorhombic system. The luster is vitreous and the color olive-green, giving rise to the name olivine. Hardness is 6½–7 on Mohs scale; specific gravity is 3.27–3.37, increasing with increase in iron content. *See* SILICATE MINERALS.

Olivine is a nesosilicate with composition $(Mg,Fe)_2SiO_4$. It comprises a complete solid solution series from the pure iron member fayalite, Fe_2SiO_4, to the pure magnesium member forsterite, Mg_2SiO_4. Minerals of intermediate composition have been given their own names but are usually designated simply as olivine.

Olivine occurs chiefly as a rock-forming mineral in igneous rocks. Although it may be present in granites and other light-colored rocks, it is found chiefly in the dark rocks such as gabbro, basalt, and peridotite. The rock lunite is composed almost completely of olivine. Olivine is one of the first minerals to form upon crystallization of a magma. It is believed that this early-formed olivine accumulated through the process of magmatic differentiation to form the large dunite masses.

At a few localities, notably on St. John's Island in the Red Sea and in Burma, olivine is found in transparent crystals. These are cut into gemstones which go under the name of peridot. Olivine is a major constituent of many stony meteorites. [C.S.Hu.]

Omega A global radio navigation system developed by the U.S. Navy. The development of the system was the result of a Department of Defense requirement for a worldwide, all-weather, continuous radio aid to navigation. It provides position information by measuring the phase difference between signals radiated by a network of transmitting stations. Omega utilizes eight strategically located terrestrial very-low-frequency transmitting stations. The locus of constant phase difference for any pair of stations is a hyperbolic line of position. Geographical position is obtained by plotting or computing the intersection of two or more such hyperbolic lines of position. Omega provides moderate position-fixing accuracy (2–4 nautical miles or 3.7–7.4 km) anywhere on the globe.

Omega is a hyperbolic navigation system using phase comparison of very-low-frequency (10 to 14 kHz) continuous-wave radio signals. The system can be used by aircraft, ships, and land vehicles, and also by submarines operating at moderate antenna depths of about 40 to 50 ft (12 to 15 m). All Omega transmitters are synchronized in phase and transmit on each of four common navigation frequencies: 10.2, 11.05, 11⅓, and 13.6 kHz. Each station also has its own unique frequency on which it transmits, with all transmissions on a time-shared basis according to the Omega signal format.

Omega receivers vary in complexity according to the degree of performance to be achieved and the number of automatic features included in each design. The simplest semiautomatic equipment may include only provisions for measuring phase difference at 10.2 kHz. However, most receivers are able to make phase difference measurements on multiple frequencies (for example, 10.2 and 13.6 kHz, or 10.2, 13.6, and 11⅓ kHz). The most advanced receivers utilize all four navigation frequencies and, in addition, receive and utilize the unique frequencies for acquisition and tracking purposes.

Differential Omega—in which a receiver located in a known and fixed geographical position monitors the Omega signals and broadcasts corrections on a real-time basis to mobile users in the vicinity—can improve position-fixing accuracy to better than ½ nautical mile (0.9 km) within a range of 50 mi (80 km) of the monitor receiver. Such an implementation has been established along the coast of France by the French government. *See* ELECTRONIC NAVIGATION SYSTEMS; HYPERBOLIC NAVIGATION SYSTEM. [J.P.V.E.]

Omphacite The pale to bright green monoclinic pyroxene found in eclogites and related rocks. Omphacites are essentially members of the solid solution series between jadeite ($NaAlSi_2O_6$) and diopside ($CaMgSi_2O_6$). The density ranges from 3.16 to 3.43 g/cm^3; hardness is 5–6.

Omphacite is stable only at the relatively high pressures of the blueschist and eclogite facies of metamorphism, where it is associated with minerals such as glaucophane and lawsonite or pyropic garnet, respectively. In such environments it also occurs in veins, either on its own or with quartz. *See* GLAUCOPHANE; PYROXENE. [T.J.B.H.]

Onchocerciasis A disease caused by the microfilariae worm *Onchocerca volvulus* in the subcutaneous lymphatics. It is characterized by subcutaneous nodules which are most conspicuous where the skin lies closely over bony structures, via cranium, pelvic girdle, joints, and shoulder blades. When they are on the head, the microfilariae reach the eyes. Ocular disturbances vary from mild transient bleary vision to total and permanent blindness. This is a major problem in the provinces of Oaxaca and Chiapas in Mexico and the nearby Pacific slopes of Guatemala, where infection is an occupational hazard and associated with coffee growing. The blackfly is the vector.

Termed "river blindness," onchocerciasis also affects field workers in the Ivory Coast, Ghana, Mali, Togo, and Upper Volta. *See* FILARIOIDEA. [G.F.O.]

Oncholaimoidea A superfamily of nematodes in the order Enoplida. These are principally marine and brackish-water forms with alleged predaceous and carnivorous feeding habits. Generally, the stoma is armed with one dorsal tooth and two subventral teeth, and its wall may be further fortified with transverse rows of small denticles. The stoma is divisible into two parts; in some species the stoma of the adult male is collapsed or indistinct. The cephalic sensilla are in two whorls, one circumoral and composed of six papilliform sensilla. The second whorl combines the ancestral two whorls of six and four into a single whorl of ten sentiform sensory organs. In some forms these sensilla are papilliform. The cylindrical-toconoid esophagus may exhibit a series of muscular bulbs posteriorly. The cuticle is generally smooth, and over the length of the body there are scattered sensory setae or papillae. *See* NEMATA. [A.R.M.]

Oncofetal antigens Antigens that are commonly present both in fetal tissue during early development of life and in adult tissue when cancer occurs. Immunization of animals such as rabbits or goats with these substances induces their immune system to produce antibodies; therefore, these substances are antigenic.

These antigens, primarily protein in nature and produced by cancer cells, are the products of one or more genes that normally are expressed only during fetal development and then are repressed in adult life. Production of these proteins as a result of activation of the controlled genes by a yet unknown mechanism in association with cancer has been called retrogenetic expression. Very minute but significant changes of these fetal antigens in body fluids can serve in detecting the early oncogenic process and in monitoring the efficacy of, and in developing new modalities of, cancer treatment.

Although several oncofetal antigens have been investigated, two of these, alpha fetoprotein (AFP) and carcinoembryonic antigen (CEA), have received the most attention.

Alpha fetoprotein is a normal embryonic product during fetal development. Alpha fetoprotein in the fetus is primarily synthesized in the liver, yolk sac, and gastrointestinal tract. In the adult, alpha fetoprotein in the serum is elevated in individuals with primary hepatocellular carcinoma and with teratocarcinomas of the ovary or testes. As a tumor marker, alpha fetoprotein has been used for large-scale screening in Africa, China, and Japan.

Since its discovery in 1965, the carcinoembryonic antigen of human digestive cancer has been the most studied oncofetal antigen. This glycoprotein was initially detected by immunoprecipitation techniques in the embryonic and fetal gut, pancreas, and liver in the first two trimesters of gestation, and also in entodermally derived digestive-system cancers. As an oncofetal antigen, carcinoembryonic antigen is less valuable as a cancer test than initially reported. Although it is generally accepted that the carcinoembryonic antigen assay should not be used as a screening test for cancer in the general population, it does have an adjunctive role in diagnostic procedures.

A few other less defined oncofetal antigens have been investigated. The fetal antigens associated with pancreatic cancer, named pancreatic oncofetal antigen and pancreas cancer–associated antigen, have been described. These pancreatic oncofetal antigens are found in extracts of the fetal pancreas, adult human pancreatic tumors and sera from individuals with carcinoma of the pancreas, and to a lesser degree in persons with other malignancies. *See* ANTIGEN; ONCOLOGY. [T.M.Ch.]

Oncogenes Genes whose activities contribute to neoplasia. They comprise genes introduced into the cell by infectious agents, and cell genes whose activities are abnormal as a result of mutation or unusual expression. *See* GENE; NEOPLASIA.

Most, if not all, neoplasms are due to permanent alterations in one or a few cells that grow to form a tumor. These changes, because of their usual irreversibility, are thought to result from mutations, which either destroy or change the functions of particular genes or alter the regulation of gene expression (transcription and translation). Since higher organisms contain thousands of genes in every cell, the task of finding the few that are presumed to be altered in neoplasia has been difficult. Nonetheless, molecular biologists identified a group of genes, collectively known as oncogenes, that seem to be affected by the mutagenic events that lead to cancer and whose activity appears important to the cancer cell. *See* DEOXYRIBONUCLEIC ACID (DNA); GENE ACTION; MUTAGENS AND CARCINOGENS.

Oncogenes were discovered by virologists. Virus infection can lead to neoplasia by a variety of mechanisms in many species. Many retroviruses containing ribonucleic acid (RNA) carry transforming genes (known as viral oncogenes: *v-onc*) that are closely related to genes in normal cells called protooncogenes (*c-onc*). Protooncogenes are found in both simple and complex eukaryotic organisms, and it is suspected that they are involved in the basic biochemical activities of the cell. *See* ANIMAL VIRUS; TUMOR VIRUSES.

The discovery of *v-onc* genes suggested that their *c-onc* ancestors might also be implicated in neoplasia. Research has shown the following ways in which *c-onc* activity can be subverted in neoplasia, thus converting a protooncogene into a cell oncogene: (1) Acquisition by a tumor virus (transduction). (2) Nearby insertion of tumor virus DNA (insertional mutagenesis). (3) Rearrangements of chromosomal segments containing *c-onc* DNA (translocation, deletion). (4) Increase in number of chromosomal segments containing *c-onc* DNA (amplification). (5) Chemical change (mutation) in specific nucleotides composing *c-onc* DNA. *See* CANCER (MEDICINE); ONCOLOGY. [J.Wy.]

Oncology The study of the causes, development, characteristics, and treatment of tumors. The word tumor as commonly used is synonymous with neoplasm. Tumors are new growths of masses of tissue cells in excess of the needs of the tissue from which the cell masses are derived. *See* NEOPLASIA; TUMOR.

Benign tumors are usually slow-growing, localized, and circumscribed. Microscopically they have a relatively orderly architecture, and the cells generally resemble adult body cells. In contrast, malignant tumors, regardless of their appearance or tissue of origin, tend to grow rapidly, to invade surrounding tissues, and to spread to distant sites in the body.

Though cancers may start at any age, they occur with progressively increasing frequency in age groups beyond 40. In men the five most common sites for cancer are skin, lungs, prostate, stomach and colon. In women they are breast, uterus, skin, colon, and stomach. Three types of cancer (leukemia, neuroblastoma, and embryonal carcinoma of the kidneys) are significant contributors to infant mortality. *See* CANCER (MEDICINE).

Etiology. Etiology is the study of the causes of the transformation of normal cells to neoplastic ones. The transformation involves a change in cellular heredity. The daughter cells of a neoplastic cell inherit their altered form and behavior from their parent cell and pass the change on to subsequent generations of cells. Normally, cells apportion their energy between proliferation and specialized function to meet the needs of the body. Neoplastic cells subordinate specialized functions to proliferation. The change from normal to neoplastic usually involves a stepwise series of chemical, morphological, and functional changes. Carcinogenesis is the term applied to the process.

The causes of the transformation of normal cells to neoplastic can be divided into two general groups, genetic and environmental. For the vast majority of cancers the relative importance of inherited and environmental factors in the genesis of cancers remains unknown. The general genetic makeup of individuals may enhance or diminish their susceptibility to exogenous agents. A whole population may be exposed to a similar dose of an agent, yet only some individuals develop cancer. Different races and strains of individuals have a varying susceptibility to the development of cancer.

General estimates suggest that approximately 80–90% of the neoplasms in human beings have an exogenous environmental agent as a prime causative factor. A major opportunity to control cancer in the human rests with improving the environment. More than 400 agents are known, and many others are suspected. The agents include viruses, chemicals, radiation, parasites, fungi, and bacteria.

Viruses that cause neoplasms (oncogenic viruses) occur in many species of animals. About two-thirds of oncogenic viruses are of the ribonucleic acid (RNA) type, and one-third are DNA viruses. The usual viral infections encountered, such as influenza, kill the cells attacked, and consequently destroy the virus too, and therefore no lasting effects are produced. *See* TUMOR VIRUSES.

Several radiations have been proved to be potent carcinogens, although the mechanism of action on the cell is not established. Whether the radiation directly alters the DNA of the genome or whether it alters some other aspect of cell metabolism remains moot.

The existence of cancer-producing chemicals in the environment was first conclusively shown by Sir Percival Potts in 1775, when he astutely observed that cancers of the scrotum occurred frequently in chimney sweeps. The chemical substance involved was isolated 140 years later when coal tars, painted on the ears of rabbits, produced carcinomas. Subsequently, numerous carcinogenic hydrocarbons have been extracted from coal tar. Among the known carcinogenic compounds are the polycyclic aromatic hydrocarbons, such as 7,12-dimethyl benz(a)anthracene or benzo(a)pyrene; aromatic amines such as dimethyl amino azobenzene ("butter yellow"); urethane; and alkylating agents such as epoxides and beta-propriolactone. Some food contaminants and additives, estrogens, several metals, and asbestos fiber have carcinogenic effects. Air pollution and tobacco are known causes of lung cancer. *See* MUTAGENS AND CARCINOGENS.

Tumor biology. Benign and malignant tumors have some attributes in common. Both may begin at almost any site within the body. One cell or a small group of cells may begin to proliferate in an autonomous manner. Consequently, tumors are primarily composed of living cells, and they are deranged descendants of body cells. Though more or less resembling the cell of origin, tumor cells usually vary from the normal in size, shape, and staining properties and have defective physical and spatial relationships with the surrounding tissue. Most significant is their accentuated proliferative vigor. This is a basic feature of all tumors, for without it the prime characteristic, growth, would be lacking.

Benign tumors are composed of cells that usually closely resemble the tissue of origin and differ from normal tissue primarily in general architecture and excess of cells. Malignant tumors (cancers), on the other hand, have in addition to some similarities with benign tumors many unique features. The three cardinal properties of malignancy are anaplasia, invasion, and metastasis. These properties refer to the appearance and behavior of cancer cells, in which they usually differ greatly from benign tumor cells. They reflect the independence of cancer cells, that is, the ability to grow at sites other than the locus of origin.

Anaplasia is the structural aberration of cancer cells which results in a more primitive (embryonic) appearance and diminished or absent adult function. Their size, shape, staining properties, and spatial relationships to one another are usually markedly deranged. Two consequences of the development of autonomy by neoplastic cells are invasion and metastasis. These newly acquired properties of malignant cells are responsible for destructive growth and spread to the detriment or death of the host. Some cancers appear to be independent of regulation from their inception and are able to invade and metastasize immediately. Others undergo a stepwise series of precancerous changes before becoming autonomous (malignant). Sometimes this may take many years. Invasion precedes metastasis in tumor progression to autonomous growth. Individual or small groups of cancer cells, no longer dependent on their site of origin for survival, are capable of infiltrating the surrounding host tissues and remaining viable.

Classification. Because of their widely differing growth patterns, each type of benign and malignant tumor has a different implication of illness and life expectancy.

Carcinoma. The table is a list of common tumors based on the tissue of origin, cell type, and the presence or absence of malignant features. Those tumors derived from the internal and external body surface coverings and their derivatives (such as the skin; sweat glands; breast; and lining of the respiratory, gastrointestinal, urinary, and genital systems) are epithelial tumors. Malignant tumors derived therefrom are called carcinomas. When a carcinoma takes origin from the skin or compa-

Tumor classification

Tissue of origin	Benign tumor	Malignant tumor
Epithelial tissues		
Skin	Papilloma	Epidermoid carcinoma
Glands	Adenoma	Adenocarcinoma
Nervous tissue		
Nerve cells	Neuroma	Neuroblastoma
Supportive cells	Glioma	Glioblastoma
Pigment cells	Melanoma	Melanocarcinoma
Placenta	Chorionic adenoma	Chorionic carcinoma
Connective tissues		
Adult	Fibroma	Fibrosarcoma
Embryonic	Myxoma	Myxosarcoma
Cartilage	Chondroma	Chondrosarcoma
Bone	Osteoma	Osteogenic sarcoma
Fat	Lipoma	Liposarcoma
Smooth muscle	Leiomyoma	Leiomyosarcoma
Lymphoid tissue	Giant follicle lymphoma	Lymphosarcoma
Mixed tissues		
Ovary or testis	Teratoma	Teratocarcinoma
Salivary gland	Mixed tumor	Mixed tumor

rable epithelium, it is called epidermoid carcinoma. Benign skin tumors are called papillomas. When malignancies are derived from glandular epithelium, they are called adenocarcinomas. Adenomas are benign glandular tumors. For example, a benign tumor of bronchial glands is called a bronchial adenoma. A malignancy derived from these glands is called bronchial adenocarcinoma. Though remotely descended from embryonal ectoderm, tumors of the nervous system have many features peculiar to themselves and represent a separate subgroup. For the same reason tumors derived from the placental epithelium also form a separate subgroup.

Sarcoma. The term sarcoma is applied to malignancies arising from the derivatives of embryonal mesoderm, such as connective tissues, bone, muscle, cartilage, blood vessels, and blood cells. A benign fibrous connective tissue tumor is a fibroma, whereas a malignant one is a fibrosarcoma. A major subgroup is made up of the tumors composed of blood cells, the lymphomas and leukemias. Mixed tumors and teratomas comprise a third major category of tumors. They are composed of mixtures of cells of the above categories. In all three categories the Greek word root "oma," meaning tumor, is applied to the end of the name of the involved tissue cell type. The table lists only a few representative tumors in each category.

Basal cell carcinoma. One of the most common tumors, the so-called basal cell carcinoma, is not included in the table because of the difficulty of classification. It begins as a small nodule in the skin which ulcerates and it has a ragged, dirty, irregular appearance and gradually increases in size. The tumor remains local, rarely if ever metastasizing, but will invade, compress, and destroy surrounding tissues. It erodes through cartilage and bone; hence the term rodent ulcer is sometimes applied. Therefore the invasive properties of the tumor are consistent with malignancy. However, its failure to metastasize suggests a benign nature.

In contrast, epidermoid carcinoma is a truly malignant skin tumor. It may arise wherever the stratified squamous type of epithelial cells may be found, such as skin, mouth, esophagus, bronchus, and uterine cervix. Beginning as an elevated, papillary, hard nodule in the epithelium, it more or less rapidly infiltrates the deeper connective and adipose tissues and spreads to the regional lymph nodes. If the primary tumor is not treated, metastases may occur to internal organs, principally lungs or liver.

Adenocarcinoma. Adenocarcinomas may arise wherever glandular tissue is found. They frequently arise in the breast, rectum, colon, stomach, pancreas, prostate, and uterus. Breast adenocarcinoma, derived from the lining of breast ducts, begins as a small, single, firm nodule which has an irregular border. It can be moved about only with difficulty and to a limited degree.

Unlike breast carcinomas, which are often noticeable when they are small, adenocarcinomas of the colon or rectum may achieve considerable size before being discovered. They tend to infiltrate and metastasize with vigor.

At the other extreme, adenocarcinomas of the prostate are exceedingly common in elderly men, but they uncommonly cause symptoms or spread outside the prostate gland.

Neuroblastoma, a malignant tumor derived from nervous system cells, is the most common cancer of infants. Occasionally the tumor is present at birth.

Lymphoma-leukemia tumors. The lymphoma-leukemia group of tumors has many unique features. Delineation into benign and malignant forms is difficult. *See* LEUKEMIA; LYMPHOMA.

Diagnosis and treatment. The most important tumor diagnostic tool is the oldest and simplest, a careful physical examination. Tumors being space-occupying masses, they are frequently detected by direct vision or palpation. Cancers of the skin and female breast are among the most frequently occurring types.

Diagnostic x-rays are another means of visualizing tumors. Blood chemistry studies and blood counts may at times reveal clues of the existence of a cancerous process. Because noncancerous lesions such as inflammations and degenerations may visibly resemble cancer, the diagnosis of cancer is not proved until a biopsy is done.

Exfoliative cytology is used extensively as an aid to the early diagnosis of cancer. Cells that flake off from various body surfaces are stained and observed under the microscope. When the Papanicolaou smear from the uterus is observed microscopically, the exfoliated normal cells surround several epidermoid carcinoma cells. This technique has been most effective in the early diagnosis of carcinoma of the cervix of the uterus.

The three major types of weapons for the treatment of cancer available to physicians are surgery, radiation, and chemotherapy. Complete surgical excision of a benign tumor is usually sufficient to effect a cure. Malignant tumors can also be cured if they have not infiltrated adjacent tissues or metastasized by the time of excision. Radiation therapy, including x-ray and radioactive metals (cobalt, radium, phosphorus, iodine, and others), has application in the treatment of many types of cancers. Strong beams of radiation are focused on the tumor, causing the death of susceptible cancer cells and scar formation. Thus radiation which may produce cancer at one dosage level can destroy it at another. Whether radiation or surgery or both are used in the treatment of a particular neoplasm depends on many variables, including the exact type of neoplasm and its size, location, and rate of spread and the general conditions of the patient. [N.K.M.]

Onion A cool-season biennial, *Allium cepa*, of Asiatic origin and belonging to the plant order Liliales. The onion is grown for its edible bulbs.

Related species are leek (*A. porrum*), garlic (*A. sativum*), Welsh onion (*A. fistulosum*), shallot (*A. ascalonicum*), and chive (*A. schoenoprasum*).

Onion varieties (cultivars) are classified mainly according to pungency (mild or pungent) and use (dry bulbs or green bunching). Bulbs may be white, red, or yellow. Varieties differ markedly in their keeping quality and in their response to length of day. Hybrid varieties, with increased disease resistance, longer storage life, and improved quality, are rapidly displacing older varieties. Texas, New York, and California are important producing states. *See* LILIALES. [H.J.C.]

Oniscoidea A suborder of the Isopoda which contains the terrestrial members of these crustaceans. They are popularly known as sow bugs, slaters, wood lice, or, in the case of those that roll themselves into a ball, pill bugs. Land isopods commonly occur under rocks, loose bark, leaf mold, and similar moist places. They abound in humid tropical and warm-temperate regions, particularly in the Old World. A few are practically cosmopolitan and have probably been accidentally transported by humans with plants, soil, or building materials. When abundant in gardens or greenhouses, they sometimes do considerable damage by gnawing on plants.

The body is either flattened dorsoventrally or in pill bugs is highly vaulted like an armadillo. Its three subdivisions, the head, thorax, and abdomen, are broadly joined (see illustration). The lateral margins generally form a continuous oval outline, except in types with the abdomen abruptly narrower than the thorax. The surface may be smooth or variously sculptured with tubercles, ridges, or spines. Sexual dimorphism is rare.

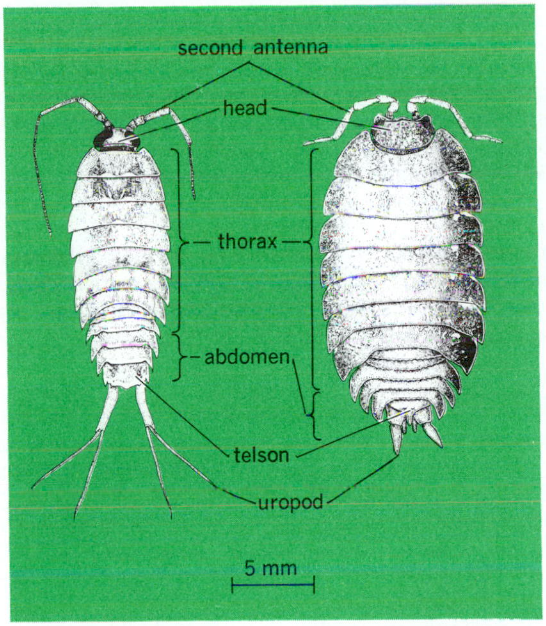

Genera of oniscoideans. (*a*) *Ligia*. (*b*) *Porcellio*.

Six fused segments, including the first thoracic somite, make up the head. It bears two pairs of antennae (the first being vestigial), two sessile compound eyes, and four pairs of mouthparts. The thorax has seven free segments, each bearing a pair of similar seven-jointed walking legs. The abdomen has five segments plus a terminal telson. Its appendages are biramous and include five pairs of platelike pleopods and one pair of uropods. *See* ISOPODA. [M.A.M.]

Onychophora A phylum of terrestrial animals, frequently referred to as Peripatus. The 80 or more species, divided into 13 genera, live in the tropics of both hemispheres and in the subtropical regions of the Southern Hemisphere. They undoubtedly represent the only survivors of a much more diversified phylum.

The adults of Onychophora species vary considerable in size of the individuals of a species. The body is annulated and covered with minute papillae, elongate, and wormlike, with 25–45 pairs of short, ventral legs that are no longer than the thickness of the body. Most species are gray, tan, or brown, but some are green, others purple, and a few are red or reddish-orange. They may be striped or speckled.

The body is distinctly segmented internally, which corresponds with the pairs of legs externally. The main body cavity is a hemocoel, quite different from the simple coelom of the Annelida. The heart is a tube, open at both ends, with paired openings or ostia along the sides of the tube. The excretory system consists of a series of metanephridia in each segment except the antennal and mandibular segments. A small bladder is present in each leg. This empties through a nephridial duct to a small nephropore between the ventral pads on the foot.

A pair of oral papillae, located one on each side of the mouth opening, squirt out a fluid toward prospective prey. This hardens into a sticky net which tangles the legs of the prey, mostly small, soft-bodied insects, but sometimes even large grasshoppers. The onychophoran then places its mouth on the body of the prey and exerts suction. The mandibles are extended from inside the mouth cavity. These paired structures are directed backward and are used to penetrate the integument of the prey and shred the tissue beneath. Digestive fluids are secreted into the hole. Somewhat liquefied tissue is then sucked up by the host and swallowed. [R.H.A.]

Onychopoda A specialized order of brachiopod crustaceans formerly included in the order Cladocera. The body is up to about 12 mm (0.5 in.) in length, but much of the length of the longest species is made up by a caudal process.

The head and thorax are short, as is the abdomen in some species, but in others it is drawn out into a long caudal process. A carapace is present but is reduced to a dorsal brood pouch, leaving the body naked. A large median compound eye occupies much of the head.

Onychopods swim actively by means of their antennae—the antennules are small and sensory—and seize their food with their four pairs of grasping trunk limbs. Most are predators, but detritus is also eaten by some species. The mandibles are stoutly denticulate.

Reproduction is mostly by parthenogenesis (males are unknown in some species from the Caspian Sea); eggs and young are carried in the brood pouch. Sexual reproduction gives rise to freely shed resistant eggs that overwinter in temperate zone species. Onychopods occur in the sea and fresh water and are worldwide in distribution, but fresh-water species occur only in the Holarctic temperate zone. There is a remarkable group of endemic species in the Ponto-Caspian region. *See* BRACHIOPODA. [G.Fr.]

Onyx Banded chalcedonic quartz, in which the bands are straight and parallel, rather than curved, as in agate. Unfortunately, in the colored-stone trade, gray chalcedony dyed in various solid colors such as black, blue, and green is called onyx, with the color used as a prefix. Because the color is permanent, the fact that it is the result of dyeing is seldom mentioned.

The natural colors of true onyx are usually red or brown with white, although black is occasionally encountered as one of the colors. When the colors are red-brown with white or black, the material is known as sardonyx; this is the only kind commonly used as a gemstone. Its most familiar gem use is in cameos and intaglios. *See* CAMEO; CHALCEDONY; INTAGLIO (GEMOLOGY). [R.T.L]

Oogenesis The sequence of events that leads to the formation of the female gamete. Oogenesis commences with proliferation of primordial germ cells which give rise to oogonia, and it terminates with the mature oocyte (egg). The oocytes develop from oogonia and enter a growth phase that is initiated by the follicle-stimulating hormone (FSH), produced by the anterior portion of the pituitary gland, or adenohypophysis. Since the hormone stimulates the production of gametes, it is referred to as a gonadotropin. In insects the production of oocytes is stimulated by the juvenile hormone fabricated by the gland known as the corpus allatum. In other forms the agent responsible for the growth of oogonia is unknown. Growth of the oocyte is manifested by an increase in its volume and often by a change in its shape. The growth phase also encompasses a period during which the oocyte must differentiate into a mature egg, that is, an egg that possesses the capability, upon fertilization, of producing an embryo. *See* VITELLOGENESIS. [E.An.]

Oolite Any of the small, more or less spherical particles commonly found in limestones and dolomites. Most oolites are 0.5–1.0 mm (0.02–0.04 in.) in diameter, but their size range is much greater. Oolites show varying degrees of departure from sphericity; some may be ellipsoidal, others may be appreciably flattened or distorted. The term oolite has been used to denote both the small, spherical bodies and the rock composed of an aggregate of these bodies. Some geologists prefer to call the particles oolite and the rock by its common lithologic name, prefixed by the word oolitic, for example, oolitic limestone. *See* DOLOMITE; LIMESTONE; SEDIMENTARY ROCKS. [R.Si.]

Oomycetes A class of fungi in the subdivision Mastigomycotina. They comprise a group of heterotropic, funguslike organisms that are classified with the zoosporic fungi (Mastigomycotina) but in reality are related to the heterokont algae. They are distinguished from other zoosporic fungi by the presence of biflagellate zoospores. Some taxa are nonzoosporic. Asexual reproduction involves the release of zoospores from sporangia; in some taxa the sporangium germinates with outgrowth of a germ tube. Sexual reproduction occurs when an oogonial cell is fertilized by contact with an antheridium, resulting in one or more oospores.

Oomycetes are cosmopolitan, occurring in fresh and salt water, in soil, and as terrestrial parasites of plants. Many species can be grown in pure culture on defined media. There are four orders: The Saprolegniales and Leptomitales are popularly known as water molds. Some species are destructive fish parasites. Many Lagenidiales are parasites of invertebrates and algae. The Peronosporales are primarily plant parasites attacking the root, stem, or leaf, and include some of the more destructive plant pathogens. *See* EUMYCOTA; FUNGI; MASTIGOMYCOTINA. [D.J.S.B.]

Opal A natural hydrated form of silica. There are many different varieties of opal, but the best known are those which are highly prized as gemstones. Precious opal displays the property of opalescence, a fine play of spectral colors resulting from the interference of light rays within the stone. Fire opal shows intense orange-to-red reflections against a yellow-to-orange body color. Black opal has a black background against which the colors are displayed. Common opal is milk white, yellow, green, or red, but without opalescence. The variety hyalite is clear and colorless with a globular surface. *See* OPALESCENCE.

Opal has a hardness of 5–6 on Mohs scale. The specific gravity varies from 1.9 to 2.2, depending upon the water content. Opal is found in cavities in igneous and sedimentary rocks and in fossil wood in which it is the petrifying substance. Its largest deposits are in sedimentary beds as diatomite which result from the accumulation of tiny opalean tests of diatoms. Such a deposit is at Lompoc, California. *See* SILICATE MINERALS. [C.S.Hu.]

Opalescence The milky iridescent appearance of a dense transparent medium when the system (or medium) is illuminated by polychromatic radiation in the visible range, such as sunlight. Slight changes in the rainbowlike color of the system can occur, depending on the scattering angle, that is, the angle between the directions of incident radiation and of observation.

Opalescence is a general term which applies to the optical phenomenon of intense scattering in the visible range of the electromagnetic radiation by a system with strong local optical inhomogeneities. The iridescence, or rainbowlike display of interference of colors, arises because the intensity of scattered light is approximately proportional to the reciprocal fourth power of the wavelength of incident light (Rayleigh's law). *See* SCATTERING OF ELECTROMAGNETIC RADIATION. [B.C.]

Opalinata A superclass of Sarcomastigophora, including the single order Opalinida, with the single family Opalinidae. This group was formerly classified as ciliates because of the uniform covering of "cilia"; these structures are actually short flagella. There is no cytostome, no contractile vacuole, and no cytoproct. The body size is large, some forms having a length greater than 2000 micrometers. All species are found in the large intestine or rectum of cold-blooded vertebrate hosts. *See* SARCOMASTIGOPHORA. [J.O.C.]

Opaque medium A medium which is impervious to rays of light, that is, not transparent to the human eye. By extension, a medium may be described as opaque if it does not transmit infrared waves or other regions of the electromagnetic spectrum, such as the x-ray, ultraviolet, and microwave regions. The property of zero transmittance does not necessarily imply total reflectance; that is, opacity can result both from reflection and from absorption of incident rays. *See* ABSORPTION. [M.G.M.]

Open channel A covered or uncovered conduit in which liquid (usually water) flows with its top surface bounded by the atmosphere. Typical open channels are rivers, streams, canals, flumes, reclamation or drainage ditches, sewers, and water-supply or hydropower aqueducts.

Open-channel flow is classified according to steadiness, a condition in relation to time, and to uniformity, a condition in relation to distance. Flow is steady when the velocity at any point of observation does not change with time; if it changes from instant to instant, flow is unsteady. At every instant, if the velocity is the same at all points along the channel, flow is uniform; if it is not the same, flow is nonuniform. Nonuniform flow which is steady is called varied; nonuniform flow which is unsteady is called variable.

Flow occurs from a higher to a lower elevation by action of gravity. If the phenomenon is short, wall friction is small or negligible, and gravity shapes the flow behavior. Gravity phenomena are local; they include the hydraulic jump, flow over weirs, spillways, or sills, flow under sluices, and flow into culvert entrances.

If the phenomenon is long, friction shapes the flow behavior. Friction phenomena include flows in rivers, streams, canals, flumes, and sewers. [W.A.]

Open circuit A condition in an electric circuit in which there is no path for current between two points; examples are a broken wire and a switch in the OPEN, or OFF, position. *See* CIRCUIT (ELECTRICITY).

Open-circuit voltage is the potential difference between two points in a circuit when a branch (current path) between the points is open-circuited. Open-circuit voltage is measured by a voltmeter which has a very high resistance (theoretically infinite), such as a vacuum-tube voltmeter. [C.F.G.]

Open-pit mining The extraction of ores of metals and minerals by surface excavations. This method of mining is applicable for near-surface deposits which do not have a ratio of overburden (waste material that must be removed) to ore that would make the operation uneconomic. Where these criteria are met, large-scale earth-moving equipment can be used to give low unit mining costs.

Most open-pit mines are developed in the form of an inverted cone, with the base of the cone on the surface. Exceptions

Roads and rounds that have developed open cuts on Cerro Bolivar in Venezuela. (*U.S. Steel Corp.*)

are open-pit mines developed in hills or mountains (see illustration). The walls of most open pits are terraced with benches to permit shovels, front-end loaders, or bucket-wheel excavators to excavate the rock and to provide access for trucks, trains, or belts to transport the rock out of the pit. Numerous benches or terraces in the pit permit a number of areas to be worked at one time and are necessary for giving slope, typically less than a 2-to-1 slope, to the sides of the pit and for ore blending. Using multiple benches also permits a balanced operation in which much of the overburden is removed from upper benches at the same time that ore is being mined from lower benches.

The principal operations in mining are drilling, blasting, loading, and hauling. These operations are usually required for both ore and overburden. Occasionally the ore or the overburden is soft enough that drilling and blasting are not required.

[C.D.B.]

Operating system A set of programs to control and coordinate the operation of hardware and software in a computer system. Operating systems typically consist of two sets of programs. Control programs direct the execution of user application programs and support programs. They also supervise the location, storage, and retrieval of data (including input and output to various peripherals) and allocate the resources of the computer system to the different tasks entered into the system. Processing programs include applications programs, utility programs, system diagnostics, and language translators, such as compilers and interpreters. The growth of microcomputers has resulted in the development of operating systems specifically designed for them. *See* MICROCOMPUTER; MICROPROCESSOR.

There are five general types of operating systems:

1. Real-time. Processing functions are executed within narrow time limits, as in updating a data bank or reading and evaluating input from a sensor or transducer.
2. Serial or batch. Tasks or jobs are run one at a time according to a schedule determined by the operating system.
3. Multiprogramming. Two or more jobs may be run concurrently in the same system.
4. Time sharing. Multiple users have simultaneous access to the system. A user is not aware that others have access to the system; the impression is that of exclusive use of the system. *See* MULTIACCESS COMPUTER.
5. Multiprocessing. Two or more processing units are coupled together to execute a single job. *See* DIGITAL COMPUTER; MULTIPROCESSING.

[H.L.HE.]

Operational amplifier A voltage amplifier that amplifies the differential voltage between a pair of input nodes. For an ideal operational amplifier (also called an op amp), the amplification or gain is infinite.

Most existing operational amplifiers are produced on a single semiconductor substrate as an integrated circuit. These integrated circuits are used as building blocks in a wide variety of applications. *See* INTEGRATED CIRCUITS.

Although an operational amplifier is actually a differential-input voltage amplifier with a very high gain, it is almost never used directly as an open-loop voltage amplifier in linear applications for several reasons. First, the gain variation from one operational amplifier to another is quite high and may vary by ± 50% or more from the value specified by the manufacturer. Second, other nonidealities such as the offset voltage make it impractical to stabilize the dc operating point. Finally, performance characteristics such as linearity and bandwidth of the open-loop operational amplifier are poor. In linear applications, the operational amplifier is almost always used in a feedback mode.

A block diagram of a classical feedback circuit is shown in illus. *a*. The transfer characteristic, often termed the feedback gain A_f of this circuit, is given by Eq. (1). In the limiting case, as

$$\frac{X_o}{X_i} = A_f = \frac{A}{1 + A\beta} \tag{1}$$

A becomes very large, the feedback gain is approximated by Eq. (2).

$$A_f \simeq \frac{1}{\beta} \tag{2}$$

See FEEDBACK CIRCUIT.

An operational amplifier is often used for the amplifier designated A in this block diagram. Since A_f in the limiting case is independent of A, the exact gain characteristics of the operational amplifier become unimportant provided the gain is large. Although linear applications of the operational amplifier extend well beyond the simple feedback block diagram of illus. *a*, the applications invariably involve circuit structures with feedback that make the characteristics of the circuit nearly independent of the exact characteristics of the operational amplifier. Such circuits are often termed active circuits.

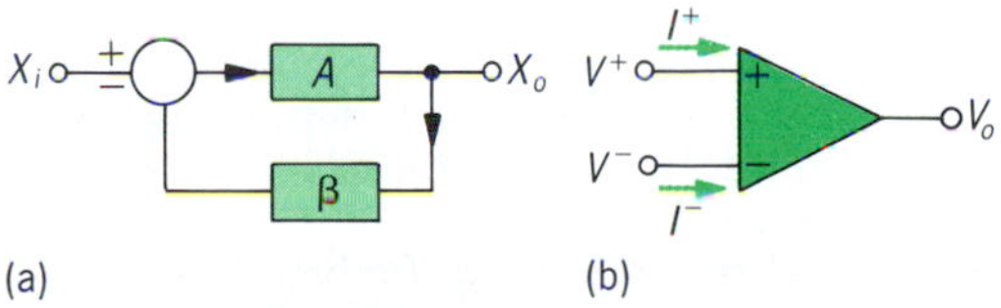

Basic circuits. (*a*) Classical feedback circuit. (*b*) Operational amplifier symbol typically used in circuit diagrams.

The commonly used operational amplifier symbol is shown in illus. *b*. In this circuit, the output voltage is related to the gain A of the operational amplifier by Eq. (3), where A is very

$$V_0 = A(V^+ - V^-) \tag{3}$$

large and the input currents I^+ and I^- are nearly zero. *See* AMPLIFIER; CIRCUIT (ELECTRONICS).

[P.M.VanP.]

Operations research The application of scientific methods and techniques to decision-making problems. A decision-making problem occurs where there are two or more alternative courses of action, each of which leads to a different and sometimes unknown end result. Operations research is also used to maximize the utility of limited resources. The objective is to select the best alternative, that is, the one leading to the best result.

To put these definitions into perspective, the following analogy might be used. In mathematics, when solving a set of simultaneous linear equations, one states that if there are seven unknowns, there must be seven equations. If they are independent and consistent and if it exists, a unique solution to the problem is found. In operations research there are figuratively "seven unknowns and four equations." There may exist a solution space with many feasible solutions which satisfy the equations. Operations research is concerned with establishing the best solution. To do so, some measure of merit, some objective function, must be prescribed.

In the current lexicon there are several terms associated with the subject matter of this program: operations research, management science, systems analysis, operations analysis, and so forth. While there are subtle differences and distinctions, the terms can be considered nearly synonymous. *See* SYSTEMS ENGINEERING.

Methodology. The success of operations research, where there has been success, has been the result of the following six simply stated rules: (1) formulate the problem; (2) construct a model of the system; (3) select a solution technique; (4) obtain a solution to the problem; (5) establish controls over the system; and (6) implement the solution.

The first statement of the problem is usually vague and inaccurate. It may be a cataloging of observable effects. It is necessary to identify the decision maker, the alternatives, goals, and constraints, and the parameters of the system. A statement of the problem properly contains four basic elements that, if correctly identified and articulated, greatly eases the model formulation. These elements can be combined in the following general form: "Given (the system description), the problem is to optimize (the objective function), by choice of the (decision variable), subject to a set of (constraints and restrictions)."

In modeling the system, one usually relies on mathematics, although graphical and analog models are also useful. It is important, however, that the model suggest the solution technique, and not the other way around.

With the first solution obtained, it is often evident that the model and the problem statement must be modified, and the sequence of problem-model-technique-solution-problem may have to be repeated several times. The controls are established by performing sensitivity analysis on the parameters. This also indicates the areas in which the data-collecting effort should be made.

Implementation is perhaps of least interest to the theorists, but in reality it is the most important step. If direct action is not taken to implement the solution, the whole effort may end as a dust-collecting report on a shelf.

Mathematical programming. Probably the one technique most associated with operations research is linear programming. The basic problem that can be modeled by linear programming is the use of limited resources to meet demands for the output of these resources. This type of problem is found mainly in production systems, but is not limited to this area. *See* LINEAR PROGRAMMING.

Stochastic processes. A large class of operations research methods and applications deals with stochastic processes. These can be defined as processes in which one or more of the variables take on values according to some, perhaps unknown, probability distribution. These are referred to as random variables, and it takes only one to make the process stochastic.

In contrast to the mathematical programming methods and applications, there are not many optimization techniques. The techniques used tend to be more diagnostic than prognostic; that is, they can be used to describe the "health" of a system, but not necessarily how to "cure" it. *See* PROBABILITY; QUEUEING THEORY; STOCHASTIC PROCESS.

Scope of application. There are numerous areas where operations research has been applied. The following list is not intended to be all-inclusive, but is mainly to illustrate the scope of applications: optimal depreciation strategies; communication network design; computer network design; simulation of computer time-sharing systems; water resource project selection; demand forecasting; bidding models for offshore oil leases; production planning; classroom size mix to meet student demand; optimizing waste treatment plants; risk analysis in capital budgeting; electric utility fuel management; optimal staffing of medical facilities; feedlot optimization; minimizing waste in the steel industry; optimal design of natural-gas pipelines; economic inventory levels; optimal marketing-price strategies; project management with CPM/PERT/GERT; air-traffic-control simulations; optimal strategies in sports; optimal testing plans for reliability; optimal space trajectories. *See* DECISION THEORY; GERT; INVENTORY CONTROL; PERT. [W.G.L.]

Operator theory At one level of abstraction an operator is simply a function whose arguments and values are real- (or complex-) valued functions of one or more real variables; in more naive terms an operator is a rule for converting such real- (or complex-) valued functions into others. The following are simple examples: (i) the operator which takes each differentiable real-valued function of one variable into its derivative; (ii) the operator which takes each twice-differentiable function f of one variable into expression (1); (iii) the operator which takes each twice-differentiable function f of three variables into expression (2); and (iv) the operator which takes the continuous function f of one real variable into the function g where relation (3) holds.

$$\left(\frac{df}{dx}\right)^2 + x^2\frac{d^2f}{dx^2} \qquad (1)$$

$$\frac{\partial^2 f}{\partial x^2} + \frac{\partial^2 f}{\partial y^2} + \frac{\partial^2 f}{\partial z^2} \qquad (2)$$

$$g(x) \equiv \int_0^1 \sqrt{x+y}\, f(y)\, dy \qquad (3)$$

Since an operator is a function, the usual functional notation is applicable. $L(f)$ may be used to denote the result of operating on f with the operator L. The set of all functions f for which $L(f)$ is defined is called the domain of L, and the set of all functions g such that $L(f) = g$ for some f in the domain of L is called the range of L. It is obvious that solving a differential or integral equation is equivalent (in many ways) to solving an operator equation $L(f) = g$, where g and L are given and it is required to find f. Moreover, the operator concept can be very useful both in theory and practice, producing a great variety of illuminating insights. *See* DIFFERENTIAL EQUATION; INTEGRAL EQUATION.

In large part the fruitfulness of the operator concept can be traced to two sources. One of these is the possibility of adding and multiplying operators in such a way that many, though not all, of the laws of ordinary algebra hold. The other is the fact that the ranges and domains of operators behave in many respects like ordinary space and, indeed, may be regarded as contained in infinite dimensional generalizations of the familiar three-dimensional space of solid geometry. This makes it possible to think of an operator as a geometrical transformation and to exploit one's spatial intuition. [G.W.M.]

Operator training The specialized education of an organization's employees in the general knowledge, specific skills, and overall mental attitude required to do their jobs effectively. It is regarded as an essential function of an enterprise. As science advances and technology becomes more complex, competent and continuous training increases in importance.

Operator training can be of new employees and can include introduction to the company, orientation to the work situation, indoctrination in the rules and regulations of the firm, instruction in specific job skills, knowledge of particular subjects, and attitude improvement. Training can also be directed to existing employees to prepare them for job transfers or promotions, to maintain or upgrade their existing skills levels, to keep them current on new tools, techniques, or technologies, to aid in their personal or professional growth, or to improve their morale and attitudes. Training can also be of employees who have become injured, aged, or infirm in some way so that they can no longer do the job for which they were originally trained.

Productive industrial work requires the presence of at least six elements: (1) proper tools, machines, and equipment; (2) proper raw materials and supplies; (3) an operator skilled in the work to be done; (4) an environment suitable for the above three elements to operate; (5) correct instructions clearly communicated to the operator; and (6) motivation on the operator's part to perform.

A distinction may be made between training and development. Training may be said to be the imparting to the operator the correct skills, knowledge, and attitude to do the present or immediately prospective job. Development may be said to be the more general education and upgrading of the employee in subjects not as much required in the present job but anticipated to be needed in future work, usually an advancement or a promotion. Human resources development (HRD) is a term used by some firms to encompass these functions. [V.M.A.]

Operon A contiguous linear array of genes whose expression is regulated as a unit. Genes in an operon are preceded by two regulatory sites in the chromosome: the promoter and the operator.

An operon is characterized by several features as follows.

1. The genes of an operon encode proteins that have related functions, such as the successive enzymatic steps in the degradation of the sugar lactose or in the biosynthesis of the amino acid histidine. Presumably, such genes are organized into an operon so that their expression can be coordinated.

2. The genes of an operon are contiguous to one another on the chromosome, adjacent to the promoter and operator sites.

3. The protein products of the genes of an operon are always produced in a fixed ratio. For example, if a given operon encodes three proteins, A, B, and C, the ratios of the rates of synthesis of each of these proteins to the other is constant. A doubling in the rate of production of A is accompanied by a doubling in the rate of production of B and C. Thus, all of the genes of an operon are turned on and off together.

4. Genes are expressed by transcription of the information stored in the DNA into a short-lived nucleic acid molecule called messenger RNA. The mRNA is translated in a complex series of reactions to yield a protein product that carries out a specific cellular function. All of the genes of an operon are transcribed into one large mRNA molecule, which is translated into the individual protein products.

5. A certain class of mutations in the genes of an operon affects the expression not only of the gene in which the mutation resides, but also of genes located to one side of the mutated gene. These mutations, termed polar mutations, reduce the amount of mRNA corresponding to genes distal to the mutated site.

6. The genes of an operon are all regulated by the same mechanism. Therefore, mutations that alter this common mechanism affect all genes of the operon equally. *See* CHROMOSOME; DEOXYRIBONUCLEIC ACID (DNA); GENE; GENE ACTION; RIBONUCLEIC ACID (RNA). [F.G.C.; H.M.J.]

Ophioglossidae A subclass of the class Polypodiopsida known as the adder's-tongue ferns. It is a small group with only 3 genera and about 80 species. Two genera, *Ophioglossum* and *Botrychium*, are widely distributed in tropical and temperate regions and have about the same number of species; the third genus, *Helminthostachys*, is represented by a single species confined to southeastern Asia and Polynesia. These are considered the most primitive of the present-day ferns. No fossils have been reported for this group.

The chromosome number is high in the species of *Ophioglossum*. *Ophioglossum petiolatum*, a tropical species, has a chromosome count of over 1000, the largest number observed in a naturally occurring species of vascular plants. *See* POLYPODIOPSIDA. [P.A.V.]

Ophiolite A distinctive assemblage of mafic plus ultramafic rocks generally considered to be fragments of oceanic lithosphere that have been tectonically transported into continental margins and island arcs. A complete idealized ophiolite sequence from bottom to top includes the following units: (1) an ultramafic tectonite complex composed mostly of multiply-deformed harzburgite, dunite, and minor chromitite; (2) a plutonic complex of layered mafic-ultramafic cumulates at the base grading upward to massive gabbro, diorite, and plagiogranite; (3) a mafic sheeted-dike complex; and (4) an extrusive section of massive and pillowed lavas and intercalated sediments. Virtually all described ophiolites are younger than 5×10^8 years; most lack complete sections and are dismembered and fragmented. The estimated original thickness is variable, ranging from about 1 mi (2 km) to more than 5 mi (8 km).

Throughout the world, ophiolites typically occur as long, narrow belts in two distinct geographic settings: (1) Those in the Alpine-Mediterranean region (Tethyan ophiolites) were formed in small ocean basins that were surrounded by older attenuated continental crust. (2) Those in western North America and the Circum-Pacific (Cordilleran) region seem to have formed in interarc basins. These Cordilleran ophiolites are generally incomplete, metamorphosed, or dismembered, but they commonly form the basement rocks for many North American terranes. *See* CONTINENTAL MARGIN; LITHOSPHERE; PLATE TECTONICS; STRUCTURAL GEOLOGY; TECTONOPHYSICS. [J.G.L.]

Ophiurida An order of Ophiuroidea in which the vertebrae articulate by means of ball-and-socket joints, and the arms, which do not branch, move mainly from side to side and do not coil in the vertical plane. The disk and arms are usually sheathed in regularly arranged plates. These are disposed in four series on the arms, namely, one dorsal, one ventral, and two lateral. There is a single madreporite. The order embraces most of the known genera of brittle stars and includes 13 families. *See* OPHIUROIDEA. [H.B.F.]

Ophiuroidea A subclass of the Asterozoa, known as the brittle stars, in which the arms are usually clearly demarcated from a central disk and perform whiplike locomotor movements, and the tube feet are nonsuctorial sensory tentacles. In all existing ophiuroids the ambulacral plates fuse together in pairs to form articulating joints termed vertebrae, and the ambulacral groove is converted into an internal epineural canal.

There are about 1900 extant species referred to 230 genera, arranged to form 3 orders: Oegophiurida, Phrynophiurida, and Ophiurida. There is also one Paleozoic order, the Stenurida. *See* OEGOPHIURIDA; OPHIURIDA; PHRYNOPHIURIDA; STENURIDA.

Ophiuroids are usually five-armed, with a few species being regularly six- or seven-armed. In some Euryalae the arms may branch repeatedly; these are the so-called basket fishes. Tropical species are often patterned in contrasting colors, but most ophiuroids tend to match their environment. Some species are luminescent, although not constantly so.

Ophiuroids occur in all the oceans from low-tide level downward, often in dense populations which number millions to the hectare. Six families range below a depth of 2 mi (3.2 km); the genera *Ophiura*, *Amphiophiura*, and *Ophiacantha* range below 4 mi (6.4 km). The shallow-water forms hide among algae, under stones, or within sponges or bury the disk in sand or mud, leaving only the arms protruding. Deep-water forms lie in or on the bottom material or adhere to corals or cidarids. [H.B.F.]

Opiates Drugs derived from opium, the dried juice of the oriental poppy seed. The pharmacologically active substances, which constitute approximately 25% of the extract, are the alkaloids morphine, codeine, and papaverine. The newer synthetic compounds which resemble morphine in their action are called opioids.

The principal effect of opium and opioids is to relieve pain. Even today morphine remains the best analgesic. It also assuages anxiety and causes slight drowsiness, relaxation, and a euphoric state of mind. These psychic effects are so agreeable that many troubled individuals seek solace by ingesting, smoking, or injecting opiates. *See* MORPHINE.

Codeine has an action similar to morphine, but its analgesic effects are less. Papaverine has almost no analgesic action, and is used as an antispasmodic to relieve vascular spasm and undesirable contraction of smooth muscle. *See* ANALGESIC; DRUG ADDICTION; NARCOTIC. [R.D.A.]

Opilioacariformes The smallest and most primitive of the three orders of Acari, comprising only one suborder, the Notostigmata, and one family, the Opilioacaridae. These mites are characterized by the possession of a pretarsus, or apotele, on the pedipalp, with prominent claws, which are usually relatively unmodified and paired like those on the legs. There are two pairs of eyes on the propodosoma, above the second pair of legs. *See* ACARI. [J.H.C.]

Opiliones An order of the class Arachnida. Individual members of the Opiliones or Phalangida, popularly known as harvestmen or daddy-longlegs, typically possess an unsegmented cephalothorax broadly joined to a segmented abdomen, paired chelate chelicerae, paired palpi, four pairs of segmented legs, and a pair of simple eyes. In length the bodies range from less than 0.04 to over 0.6 in. (1 to over 15 mm); respiration is by tracheae; some forms possess scent glands at the anteriolateral portions of the cephalothorax.

Worldwide in distribution, harvestmen vary greatly in appearance. The majority of temperate forms have relatively small bodies and enormously elongated legs. Tropical forms are often somewhat flattened and have relatively shorter legs.

Three suborders are recognized: the Cyphophthalmi, small mitelike forms; the Laniatores, often with flattened bodies, mainly tropical in distribution with some adapted to cave life; and the Palpatores, typically the long-legged forms of temperate areas. *See* ARACHNIDA. [C.J.Go.; M.L.G.]

Opisthobranchia A subclass in the class Gastropoda containing about 4000 living species, arranged in nine orders, including the herbivorous Aplysiomorpha (sea hares) and Sacoglossa and the carnivorous Thecosomata (sea butterflies) and Nudibranchia (sea slugs). Primitive members of many of the orders show adaptations for burrowing beneath sand or mud; more advanced members are always active surface-living or pelagic forms. This trend is accompanied by a decrease in the importance of the shell and operculum for passive defense. These are replaced by more dynamic chemical (some species secrete decinormal sulfuric acid through the skin if annoyed), physical (daggerlike calcareous epidermal spicules), or biological (redirected nematocysts derived from coelenterate prey) defensive mechanisms.

The adult shells of the primitive opisthobranchs living today are often strongly developed and sometimes colorful; a more typical opisthobranch shell is fragile, inflated, and egg-shaped. In rather more advanced forms, the external shell has a very wide gape, and in animals like *Berthella* the widely gaping shell is wholly internal, covered by the mantle. The most varied shells are found in the Sacoglossa. In the highest sacoglossans, some bullomorphs and aplysiomorphs, and in all the nudibranchs, the true shell is completely lost after larval metamorphosis. *See* GASTROPODA; MOLLUSCA; NUDIBRANCHIA; SACOGLOSSA. [T.E.T.]

Opisthocomiformes A small order of birds that contains the single family Opisthocomidae. Its one species, the hoatzin, is restricted to South America and has frequently been included in the Galliformes as a suborder. One group of scientists claims that it is simply a cuckoo, Cuculidae, and not a very aberrant one. In the absence of definite evidence of its relationship, it will be treated here as a distinct order. *See* CUCULIFORMES; GALLIFORMES.

The hoatzin is a unique bird, often considered rather reptilian because of the habits of the nearly unfeathered young, which leave the nest soon after birth to crawl about trees by using the clawed wings as well as the feet. The hoatzins are also adept swimmers, capable of surviving a fall into water. They are medium-sized birds of dark brown plumage streaked with white; a distinct crest of reddish-brown feathers tops the head, which is small and has a short, stout bill. The wings are long and rounded, the legs short with strong toes; the long hallux is on the same level as the anterior toes. Hoatzins are weak fliers, but they are arboreal, clambering about in trees to feed on leaves and fruit. Hoatzins are found in the forests of northern South America, especially along rivers and streams.

The hoatzins are known in the fossil record only by the Miocene *Hoazinoides* from Colombia, a typical form that reveals nothing about the evolutionary history of the order. *See* AVES. [W.J.B.]

Opossum Any member of the family Didelphidae in the order Marsupialia which includes about 65 species found in the New World. These mammals are arboreal and are mainly omnivorous.

The common opossum (*Didelphis marsupialis*) is an extremely adaptable mammal that ranges from Argentina through Central America into the United States. The uterine gestation period is quite short, about 12 days, and the young are born no larger than a bee. They then make their way through the mother's fur to the pouch (marsupium), where they remain and continue to develop for as long as 100 days. There may be two litters of 9–12 each year. The opossum is a sedentary animal and the adults, living in solitary pairs, rarely leave their territory. *See* MARSUPIALIA. [C.B.C.]

Opportunistic infections Infections that owe their emergence to generalized or local defects in the host's defense mechanisms. In general, chances for acquiring an infection in humans are proportionate to the number of microbes at the site of infection, proportionate to their virulence, and inversely proportionate to the strength of the host's defenses.

One possible classification of defects in the host's defense mechanisms would list: defects in the production of antibodies; defects in cell-mediated immunity; defects in phagocytosis; and defects in certain local defense factors. Few human diseases are associated with only one such defect. In most diseases underlying opportunistic infection, multiple mechanisms are at work to affect host defenses. One form of immunosuppression, for instance, is treatment with steroid hormones; it affects cell-mediated immunity, phagocytosis, and (to a lesser degree) antibody production. Another form of immunosuppression, cytotoxic chemotherapy, causes the number of phagocytes to drop, diminishes antibody production, and causes damage to mucous membranes.

Among other conditions favoring opportunistic infections are burns, alcoholism, chronic lung diseases, chronic heart failure, cirrhosis of the liver, renal failure, cystic fibrosis, diabetes mellitus, hypoparathyroidism, sickle-cell disease, leukemias, lymphomas, and various defects in phagocytic mechanisms, as well as the status after splenectomy. Certain manipulative procedures such as catheterization, intravenous cannulation, and surgery predispose as well, often to local infections. *See* CHEMOTHERAPY; INFECTION. [A.W.C.V.G.]

Opsonin A term used in serology and immunology to refer to a substance that enhances the phagocytosis of bacteria by leukocytes. Opsonin is generally synonymous with the bacteriotropin of F. Neufeld and coworkers (1904–1905), a relatively thermostable antibody, increased in amount during specific immunization, that renders the corresponding bacterium more susceptible to phagocytosis. There is evidence that this action can be promoted to some extent by antibody alone, but that it is substantially increased by the further addition of the thermolabile complement system. *See* AGGLUTINATION REACTION; ANTIBODY; LYTIC REACTION; NEUTRALIZATION REACTION (IMMUNOLOGY); PHAGOCYTOSIS; PRECIPITIN; SERUM. [H.P.T.]

Optical activity The effect of asymmetric compounds on polarized light. To exhibit this effect, a molecule must be nonsuperimposable on its mirror image, that is, must be related to its mirror image as the right hand is to the left hand. An optically active compound and its mirror image are called enantiomers or optical isomers (see illustration). Enantiomers differ

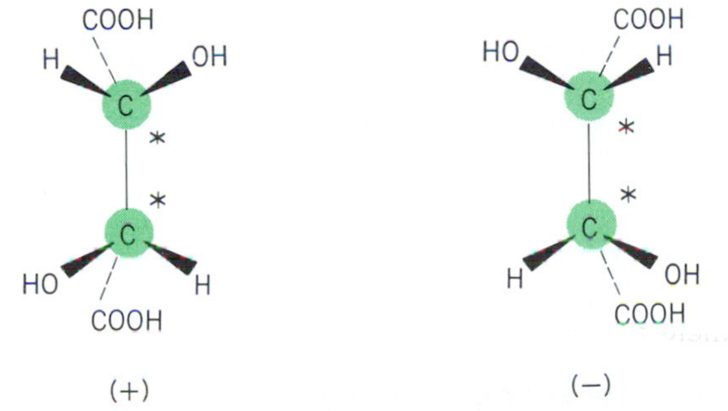

Enantiomers of tartaric acid.

only in their geometric arrangements; they have identical chemical and physical properties. The right-handed and left-handed forms of a molecule can be distinguished only by their optical activity or by their interactions with other asymmetric molecules. Optical activity can be used to probe other aspects of molecular geometry, as well as to identify which enantiomer is present and its purity.

The physical basis of optical activity is the differential interaction of asymmetric substances with left versus right circularly polarized light. If solids and substances in strong magnetic fields are excluded, optical activity is an intrinsic property of the molecular structure and is one of the best methods of obtaining structural information from a sample in which the molecules are randomly oriented. The relationship between optical activity and molecular structure results from the interaction of polarized light with electrons in the molecule. Thus the molecular groups that contribute most directly to optical activity are those that have mobile electrons which can interact with light. Such groups are called chromophores, since their absorption of light is responsible for the color of objects. For example, the chlorophyll chromophore makes plants green. *See* FARADAY EFFECT; MAGNETO-OPTICS; POLARIZED LIGHT; STEREOCHEMISTRY.

Optical activity is measured by two methods, optical rotation and circular dichroism. The optical rotation method depends on the different velocities of left and right circularly polarized light beams in the sample. The velocities are not measured directly, but both beams are passed through the sample simultaneously. This is equivalent to using plane-polarized light. The differing velocities of the left and right circularly polarized components yield a rotation of the plane of polarization. Circular dichroism is the difference in absorption of left and right circularly polarized light. Since this difference is about a millionth of the absorption of either polarization, special techniques are needed to determine it accurately. Circular dichroism is reported as a difference in absorption, or as an ellipticity (a measure of the elliptical polarization of the emergent beam). [V.M.]

Optical bistability A phenomenon in which an optical device can exist in either of two stable states. Optical bistability is an expanding field of research because of its potential application to all-optical logic and because of the interesting physical phenomena it encompasses. A bistable optical device can function as a variety of logic devices. It can have two stable

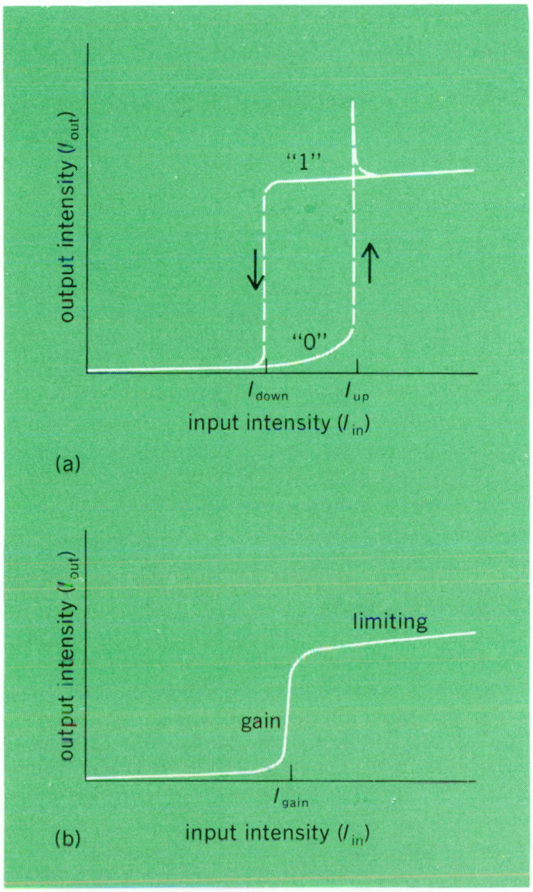

Transmission of a typical bistable optical device, under conditions of (*a*) bistability (memory) and (*b*) high ac gain (optical transistor, discriminator, or limiter).

output states labeled 0 and 1 for the same input, as shown in illustration *a*. Thus it can serve as an optical memory element. Under slightly modified operating conditions, the same device can exhibit the optical transistor characteristic of illustration *b*. For input intensities close to I_{gain}, small variations in the input light are amplified, in much the same way that a vacuum tube triode or transistor amplifies electrical signals. The characteristic of illustration *b* can also be used as a discriminator; inputs above I_{gain} are transmitted with far less attenuation than those below. Finally, there is limiting action above I_{gain}: large changes in the input hardly change the output. There is hope that bistable devices will revolutionize optical processing, switching, and computing. Since the first observation of optical bistability in a passive, unexcited medium in 1974, bistability has been observed in many different materials. Current research is focused on optimizing these devices and developing smaller devices of better materials which operate faster, at higher temperatures, and with less power.

Optical bistability has also attracted the attention of physicists interested in fundamental phenomena. In fact, a bistable device often consists of a nonlinear medium with an optical resonator; it is thus similar to a laser except that the medium is unexcited (except by the coherent light incident on the resonator). Such a device constitutes a simple example of a strongly coupled system of matter and radiation. Counterparts of many of the phenomena studied in lasers such as fluctuations, regenerative pulsations, and optical turbulence can be observed under better-controlled conditions in passive bistable systems. *See* LASER.

The transmission of information as signals impressed on light beams traveling through optical fibers is replacing electrical transmission over wires. With optical pulses and optical transmission, the missing component of an all-optical signal processing system is an optical logic element in which one light beam or pulse controls another. Because of the high frequencies of optical electromagnetic radiation, such all-optical systems have the potential for subpicosecond switching and room-temperature operation. *See* INTEGRATED OPTICS; OPTICAL COMMUNICATIONS. [H.M.G.]

Optical communications The transmission of speech, data, pictures, or other information by light. An information-carrying light wave signal originates in a transmitter, passes through an optical channel, and enters a receiver which reconstructs the original information. The ensemble of these three items constitutes an optical communication system.

Optical communication is one of the most advanced forms of communication by electromagnetic waves. Light waves are generally understood to occupy the part of the electromagnetic spectrum with wavelengths in the range 0.2–100 micrometers. The advantages sought in using light waves instead of electromagnetic waves with longer wavelengths are threefold:

1. Because of its short wavelength, light can be focused into narrower beams or confined in smaller waveguides than radiation of longer wavelengths.
2. The information-carrying capacity is potentially greater than that of longer-wavelength radiation.
3. The highest transparency for electromagnetic radiation achieved in any solid material is that of silica glass in the wavelength range 1–2 μm. This transparency is orders of magnitude higher than that of any other solid material in any other part of the spectrum. *See* ELECTROMAGNETIC RADIATION; LIGHT.

Until the 1960s, the only available sources generated light from a multitude of independent atomic radiators. Such light cannot be focused effectively to form a narrow beam or to propagate along the axis of a waveguide. The demonstration of the first laser removed this shortcoming and made feasible in the optical wavelength region all the communication techniques formerly limited to microwaves and other electromagnetic waves of longer wavelengths. *See* LASER.

Systems. Although optical fiber systems presently appear as the most significant outgrowth of intense exploration in the field of optical communications, they were not the only or first optical systems considered or utilized. A free-space optical communications channel exists, for example, between orbiting satellites. For satellite-to-satellite communications, the region between the information source and information use is ideal, essentially a vacuum, which does not distort or attenuate the light beam. For this application, the laser is the best optical source because of its ability to radiate spatially coherent light. For satellite-to-Earth communications and other communication through air, the Earth's atmosphere strongly influences the light transmission between information source and destination. In neither case would the transmission reliability be considered satisfactory for most communication purposes. On the other hand, the probability of outage of an atmospheric propagation system can be arbitrarily low if the transmission path is short. For example, optical communication links confined to rooms or buildings for the remote control of television sets or video recorders have become quite popular. *See* OPTICAL FIBERS.

Optical transmitters. The preferred sources for use in optical communication systems are semiconductor diodes which generate light in forward-biased operation. The simplest device is the light-emitting diode which radiates in all directions from a fluorescent area located in the diode junction. Since fibers accept only light entering the core at a small angle with respect to the axis, only a small part of the light radiated from a light-emitting diode can be captured and transmitted by the fiber. *See* LIGHT-EMITTING DIODE.

In the case of the semiconductor laser, two ends of the junction plane are furnished with mirror surfaces which form an optical resonator and enhance the light bouncing back and forth between the mirrors by stimulated emission. As a result, the light emitted through the partially transparent end mirrors is well collimated within a narrow cone of angles. A large fraction of the light can be captured and transmitted by optical fibers. In practical semiconductor lasers, the semitransparent mirrors are simply formed by careful cleaving of two sides of the junction.

Optical receivers. The simplest photodetectors are designed to capture light in a reverse-biased junction where it generates electron-hole pairs. The electrical carriers are swept out by the internal electrical field and induce a photocurrent in the external circuit. Greater receiver sensitivity may be achieved by multiplying the photo-generated carriers in the diode junction. This can be done by virtue of an internal electrical field that is sufficiently high to cause avalanche multiplication of free carriers. *See* OPTICAL DETECTORS. [D.C.G.]

Optical detectors Devices that generate a signal when light is incident upon them. The signal may be one observed visually in reflected or transmitted light, or it may be electrical.

One of the most widely used optical detectors is photographic film. This method of detection is based on a photochemical process in particles embedded in the film. This fairly sensitive technique is more effective than others in detecting or recording, or both, the large amount of information usually contained in an image or a complicated optical spectrum. *See* PHOTOGRAPHY.

In many cases, fast detector response is needed to detect rapid changes in incident light; furthermore, quite often an electrical output signal is desired. In such cases, the optimum detector is one based on either external or internal photoemission.

In an externally photoemitting device, the light incident upon a surface causes an emission of electrons from that surface into a vacuum. These electrons can be amplified in number by using electric and magnetic fields to cause them to impinge on other surfaces, thereby producing secondary electrons. These detectors are limited to wavelengths in the visible region or shorter, but are capable of a sensitivity and speed of response higher than any other type. See PHOTOMULTIPLI-ER; PHOTOTUBE; TELEVISION CAMERA TUBE.

In an internally photoemitting device, incident light produces free charge carriers within a body and is detected through the effect of these charge carriers on the electrical impedance of the body. In its simplest and most widely used form, the body is a photoconductor whose impedance is high in the absence of incident light. Such detectors are less sensitive and slower than external photoemitters, but they require no vacuum envelope and operate at low voltages. They also have the advantage that some of them are sensitive to all wavelengths shorter than about 40 micrometers, far into the infrared. See OPTICAL COMMUNICATIONS; OPTICAL MODULATORS; PHOTOCONDUCTIVE CELL; PHOTOCONDUCTIVITY; PHOTODIODE; SEMICONDUCTOR DIODE. [J.K.G.]

Optical fibers Flexible transparent fiber devices, used for either image or data transmission, in which light propagates by total internal reflection. The optical fiber, or light guide, has a core of material with a refractive index higher than that of the surrounding cladding material. Fiber properties and requirements for image transfer, in which information is continuously transmitted over relatively short distances, are quite different than for data transfer where on-off pulses of light are used to transmit information over much longer distances.

There are three basic types of light guides. Multimode, stepped refractive index profile fibers (illustration *a*) are typically used for conventional image transfer as well as for short-distance data transmission. The number of rays or "modes" of light which are guided is determined by the core size and core/clad refractive index difference. In such a fiber, an initially sharp pulse made up of many modes broadens as it travels long distances through the fiber, since high-angle modes have a longer distance to travel than low-order modes. This limits the data transmission rate and distance because it determines how closely input pulses can be spaced without overlap at the output end.

A graded index multimode fiber (illustration *b*), where the core refractive index decreases with increased radial distance, can be used to minimize pulse broadening due to mode dispersion. Light rays travel more slowly near the core center than near the edge, resulting in the speed of the high-order modes approximating the speed of the low-order modes. This type of fiber is suitable for intermediate-distance, intermediate-data-rate transmission systems.

In the single-mode fiber type (illustration *c*), with low refractive index difference and small core size, the effect of mode dispersion on pulse dispersion is eliminated since only one mode is guided. This fiber is useful for long-distance, high-data-rate applications. See OPTICAL COMMUNICATIONS; REFLECTION OF ELECTROMAGNETIC RADIATION; REFRACTION OF WAVES; WAVEGUIDE. [S.R.N.]

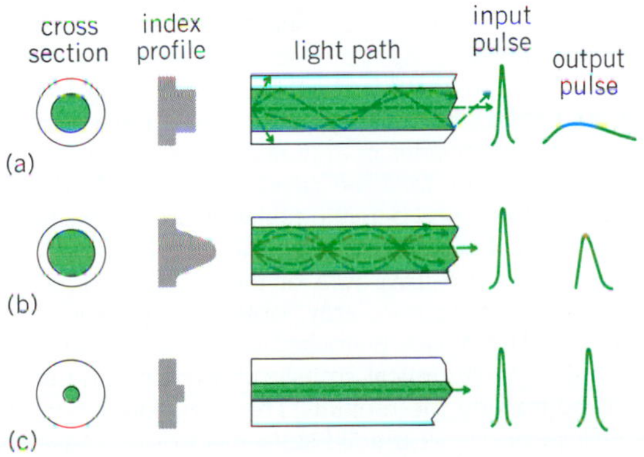

Types of optical fiber. (*a*) Multimode, stepped refractive index profile. (*b*) Multimode, graded index. (*c*) Single-mode, stepped index.

Optical flat A disk of high-grade quartz glass approximately ¾ in. (2 cm) thick, having at least one side ground and polished with a deviation in flatness usually not exceeding 50 nanometers all over, and a surface quality of 5 microfinish or less. When two surfaces of this quality are placed lightly together so that the air is not wrung out from between them, they are separated by a film of air and actually touch at only one point. This point is the vertex of a wedge of air separating the two pieces.

If parallel beams of light pass through the flat, part will be reflected against the surface being inspected, while part will be reflected directly back through the flat. Because the distance between the surfaces is constantly increasing along the angle, the beams reflected from the flat and the beams reflected from the workpiece will alternately reinforce and interfere with each other, producing a pattern of alternate light and dark bands (see illustration). Each succeeding full band from a point of contact means the distance between surfaces is one wavelength thicker. If the light is relatively monochromatic, the wavelength is known. Red with a wavelength of 295 nm is commonly used. Thus a definite relationship is established between lineal measurement and light waves. Optical flats are used for two general purposes, determination of surface contour and comparison of lineal measurement. [R.A.B.]

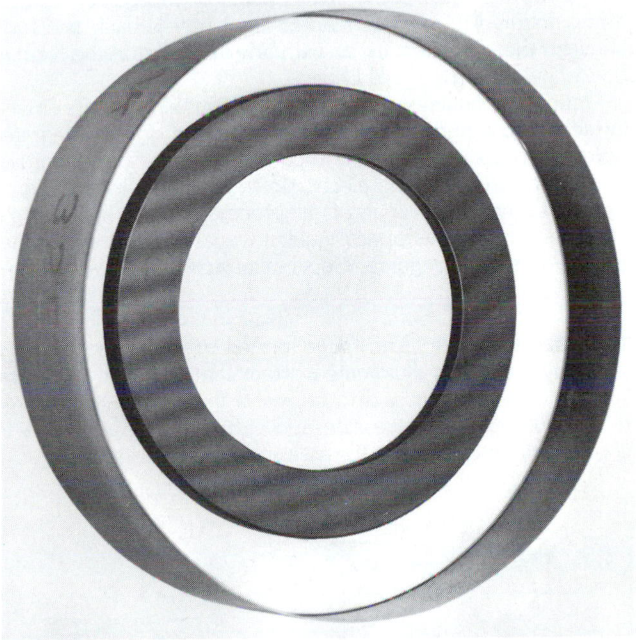

Optical flat being used to determine flatness of seal ring. Interference bands on seal ring face show lines of constant depth. (*Van Keuren Co.*)

Optical guided waves Optical-frequency electromagnetic waves confined within an optical waveguide, a structure designed to carry such waves from one place to another somewhat as a pipe carries water. Optical waveguides confine light by the method of total internal reflection. (The terms optical and light are used here in the broadest sense to include visible and near-infrared electromagnetic radiation.) The demonstrations of the first semiconductor laser and the first low-loss glass optical fiber initiated a technological revolution. *See* Reflection of electromagnetic radiation.

Because of the high data rates that can be achieved, the transmission of information in the form of optical guided waves confined within an optical-fiber waveguide has become the preferred method for the telecommunications industry. The optical fiber consists of two concentric glass cylinders. The inner core region is made of a glass that has a slightly higher index of refraction than the outer cladding region, as required for total internal reflection to occur. Core diameters in the range of 4–50 micrometers are used routinely to carry near-infrared optical radiation. Pulses of light with wavelengths between 0.8 and 1.55 μm are injected into the fiber on the near end. The presence or absence of a pulse is interpreted as a one or a zero by a receiver on the far end, typically many miles (kilometers) away. *See* Optical communications; Optical fibers.

Devices such as semiconductor lasers make use of optical guided waves over distances much shorter than those spanned by optical fibers. In a semiconductor laser, a planar waveguide also serves as an electrically driven amplifying medium. Gallium arsenide (GaAs), aluminum gallium arsenide (AlGaAs), or indium gallium arsenide phosphide (InGaAsP) are common examples of semiconductor laser materials. When mirrors are formed on the ends of the semiconductor waveguide, typically a few hundred micrometers apart, light travels back and forth between those mirrors just as with any laser, but in the form of optical guided waves. *See* Laser.

Optical waveguides are the basic constituents of an emerging technology known variously as integrated optics, integrated optoelectronics, or photonic integrated circuits. This technology integrates optical and electronic components into or on an optical waveguide. The aim is to process and manipulate light while it is trapped as optical guided waves within the confines of the optical waveguide. *See* Integrated optics; Waveguide.

[D.G.H.]

Optical image The image formed by the light rays from a self-luminous or an illuminated object that traverse an optical system. The image is said to be real if the light rays converge to a focus on the image side and virtual if the rays seem to come from a point within the instrument (see illustration).

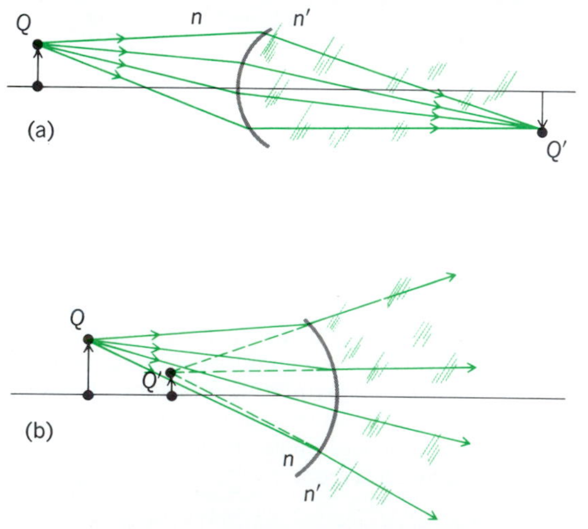

Optical images. (*a*) Real image. Rays leaving object point *Q* and passing through the refracting surface separating media *n* and *n′* are brought to a focus at the image point *Q′*. (*b*) Virtual image. Rays leaving *A* and refracted by the concave surface separating *n* and *n′* appear to be coming from the virtual image point *Q′*. As the rays are diverging, they cannot be focused at any point. (*Modified from F. A. Jenkins and H. E. White, Fundamentals of Optics, 3d ed., McGraw-Hill, 1957*)

The optical image of an object is given by the light distribution coming from each point of the object at the image plane of an optical system. The ideal image of a point according to geometrical optics is obtained when all rays from an object point unite in a single image point. However, diffraction theory teaches that even in this case the image is not a point but a minute disk. *See* Diffraction.

From the standpoint of geometrical optics, if this most desirable type of image formation cannot be achieved, the next best objective is to have the image free from all but aperture errors (spherical aberration). In this case the light distribution in the image plane is still circular, resembling the point image; there is a true coordination of object point and image, although the image may be slightly unsharp. If the aperture errors are small, or if the image is viewed from a distance, such an image formation may be very satisfactory. *See* Aberration (optics).

Asymmetry and deformation errors may be very disturbing if not held in check, because the light distribution of the image of a point in this case has a decidedly undesirable shape. [M.J.H.]

Optical information systems Devices that use light to process information. Optical information systems or processors consist of one or several light sources; one- or two-dimensional planes of data such as film transparencies, various lenses, and other optical components; and detectors. These elements can be arranged in various configurations to achieve different data processing functions. As light passes through various data planes, the light distribution is spatially modulated proportional to the information present in each plane. This modulation occurs in parallel in one or two dimensions, and the processing is performed at the speed of light. The high-speed and parallel-processing features of optical information processors, together with their small size and low power dissipation, are their attractive features. The processing speed is limited in practice by the rate at which new planes of data can be introduced into the system and the rate at which processed data can be removed from the output detectors.

Practical systems employ real-time and reusable spatial light modulators rather than film. These devices include various liquid, ferroelectric, magnetooptic, electrooptic, and acoustooptic crystals to convert optical data (for example, an ambient scene) or electrical data into a form suitable for spatially modulating the light passing through the system. (Usually, coherent laser light is employed.) Lenses, mirrors, computer-generated holograms, holographic optical elements, and fiber optics are used to manipulate and control the light as it passes through the system and thus to provide a quite versatile collection of architectures for diverse applications. Optical information processors can be grouped into three major classes: optical image processors, optical signal processors, and optical computers. The architecture of the optical system and the type of processing functions performed in each case differ considerably. *See* Acoustooptics; Electrooptics; Holography; Laser; Magnetooptics; Optical fibers.

[D.Ca.]

Optical isolator A very small four-terminal electronic circuit element that includes in an integral package a light emitter, a light detector, and, in some devices, solid-state electronic circuits. The emitting and detecting devices are so positioned that the majority of the emission from the emitter is optically coupled to the light-sensitive area of the detector. The device is also known as an optoisolator, optical-coupled isolator, and optocoupler. The device is housed in an integral opaque package so that the only optical emission impinging on the detector is that produced by the emitter. This configuration of components can perform as a solid-state electronic transformer or relay, since an electronic input signal causes an electronic output signal without any electrical connection between the input and the output terminals.

Optical isolators are particularly useful to electronic circuit and systems designers because two circuits often have a large voltage difference between them and yet it is necessary to transfer a small signal between them without changing the basic voltage level of either.

The optical emitter most commonly selected for use in optical isolators is the gallium arsenide light-emitting diode (LED), which emits in the infrared or near-infrared regions of the spectrum. The other input devices commonly used are gas-discharge and incandescent lamps. These emitters tend to be somewhat larger and slower than LEDs, but are used when their particular electronic characteristics are desired, and in conjunction with detectors that have a spectral peak in the visible portion of the spectrum. *See* INCANDESCENT LAMP; LIGHT-EMITTING DIODE.

The light detectors that are used in the construction of optical isolators include light-dependent resistors (such as photocells), light-sensitive devices that generate a voltage without any electrical input (such as photovoltaic devices), light-sensitive devices that switch from one state to another (such as phothyristors), and light-sensitive devices that modify a voltage or current (such as phototransistors, photodiodes, and photodetector-amplifier combinations). *See* CIRCUIT (ELECTRONICS); OPTICAL DETECTORS. [R.D.Co.]

Optical materials Generally, all substances used to reflect, refract, filter, polarize, modulate, detect, or disperse infrared, visible, or ultraviolet radiation or light; more specifically, the transparent materials usually used for windows and lenses. These materials are often some formulation of glass, a plastic in either bulk or thin-film form, or crystals either single or in aggregates. The most important optical properties are the degree and spectral region of transparency, the value of the refractive index n over the same spectral region, and the uniformity of the sample. Desirable materials are usually hard, strong, and relatively insensitive to temperature variations. *See* REFRACTION OF WAVES; SCATTERING OF ELECTROMAGNETIC RADIATION.

The optical material most widely used is optical glass, which is available in a wide range of refractive indices and dispersions and differs from ordinary glass in its freedom from imperfections. It must be free from unmelted particles or "stones," from bubbles, and from chemical inhomogeneity, which gives rise to regions of variable refractivity known as cords or striae. *See* GLASS.

Because of certain limitations of glasses, single-crystal or polycrystalline aggregates are used for special applications. The most useful of these are of cubic symmetry, so that the refractive index and other physical properties are isotropic.

The most significant properties of plastics (compared to glass) are their softness and molecular complexity. In the visible spectrum they can be used for relatively cheap lenses but must be protected from abrasion. They can be used in the infrared at relatively short and very long wavelengths.

For the ultraviolet region of the spectrum, the main refractors are LiF, CaF_2, and SiO_2. The region short of this is known as the vacuum ultraviolet, where the atmosphere and all known materials are opaque. In the very far infrared most of the alkali halides "open up" and transmit reasonably well. Many plastics are useful if not thicker than several thousandths of an inch. These are used primarily as windows in the form of membranes to maintain a prescribed interior atmosphere. [W.L.Wo.]

Optical methods of chemical analysis Methods that deal with the measurement of the extent of the interaction of light (electromagnetic radiation) with matter. Both the manner of interaction and the type of electromagnetic radiation utilized vary widely. The common wavelength units are the angstrom (Å), the micrometer (μm), and the nanometer (nm), which have values as shown below.

$$1\ \text{Å} = 10^{-10}\ \text{m}$$
$$1\ \mu\text{m} = 10^{-6}\ \text{m}$$
$$1\ \text{nm} = 10^{-9}\ \text{m}$$

Frequency units are usually given in cm^{-1}, called wave numbers of kaysers, so that the frequency which corresponds to 1 μm is equal to a value of 10,000 cm^{-1}.

A broader interpretation of optical methods also includes the corresponding techniques using higher-energy x-rays and lower-energy microwaves. For discussions of the principal methods *see* FLAME PHOTOMETRY; INFRARED SPECTROSCOPY; LUMINESCENCE ANALYSIS; MICROWAVE SPECTROSCOPY; POLARIMETRIC ANALYSIS; REFRACTOMETRIC ANALYSIS; SPECTROCHEMICAL ANALYSIS; SPECTROPHOTOMETRIC ANALYSIS; X-RAY FLUORESCENCE ANALYSIS. [J.N.L.]

Optical microscope An instrument used to obtain an enlarged image of a small object. In general, a compound microscope consists of a light source, a condenser, an objective, and an ocular or eyepiece, which can be replaced by a recording device such as a photoelectric tube or a photographic plate. The optical microscope is limited by the wavelengths of the light used and by the materials available for manufacturing the lenses. Some optical errors, for instance, curvature of field and lateral color, are not generally corrected in the objective but neutralized in the ocular. *See* ABERRATION (OPTICS); LENS (OPTICS).

Magnifying power. The magnifying power of a compound microscope is the product of the magnification of the objective and the magnifying power of the eyepiece. The latter is computed like that of any magnifier. The magnification of the objective is equal to the distance from the second focal point to the image formed by the objective, divided by the focal length. An objective of 18-mm focal length thus has a power of 10 ×. It is customary to specify objectives in terms of magnifying power instead of focal length. The distance mentioned is called the optical tube length (generally 180 mm), and is to be distinguished from the mechanical tube length, which is the length of the mechanical tube itself. *See* MAGNIFICATION.

Catadioptric systems. Catadioptric systems have been developed for microscopes. Their great advantage is their comparatively small chromatic aberration. A type consisting of a single element is shown in Fig. 1*a*. Pure mirror systems have no color aberrations. All microscopic work in the ultraviolet region is done with catadioptric systems. A type designed for this purpose is sketched in Fig. 1*b*.

Condensers. An external auxiliary lens is used to condense the light from a light source so that the object is brightly and

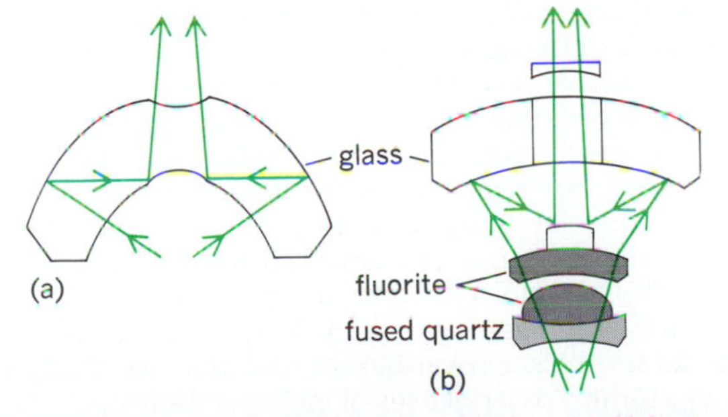

Fig. 1. Two types of catadioptric objective. (*a*) Maksutov type. (*b*) 53×, NA 0.72, ultraviolet objective, designed by Gray. Glass elements in the latter serve purely as reflectors. (*Photographic Service Department, Kodak Research Laboratory*)

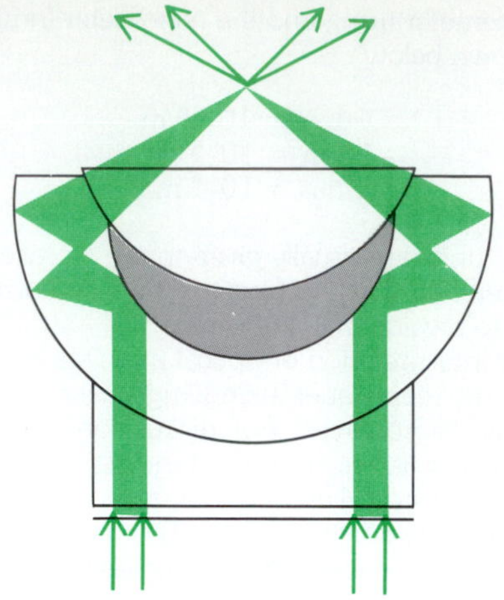

Fig. 2. Cardioid condenser. Shaded meniscus area is air space; other areas are portions of condenser through which light passes. (*Photographic Service Department, Kodak Research Laboratory*)

uniformly illuminated. The usual purpose of a condenser system is to make sure that as much light as possible coming from the object goes through an optical system. Condensers are used in macroscopic projection, in which an illuminated film or slide is imaged with the help of a projection objective or magnifier. In microscope systems, they are used to direct the light from a light source so that the rays from any object point fill most of the entrance pupil.

In microscope practice, it is sometimes advantageous to increase the contrast of small objects by making them appear as bright objects on a dark background. This is achieved by arranging the condenser so that direct light is cut off and only light diffracted or dispersed from the object enters the microscope. The cardioid condenser sketched in Fig. 2 is an example of such a dark-field condenser. [M.J.H.]

Light microscope. The mirror, condenser, oculars, and body tube of the light microscope are frequently known as the optical train. The stand, stage, and adjustments comprise the mechanical part of the microscope.

A mirror is usually attached to the substage of the microscope to reflect light along the optic axis of the microscope. When no condenser is used, the concave mirror is used because it concentrates more light on the specimen; a plane mirror is used with a condenser.

The objective is the basic part of the microscope; it forms the image that is again enlarged by the eyepiece. Photomicrographic objectives are designed to produce a flat image with little distortion. For convenience, two to five objectives can be mounted on a revolving nosepiece to be parfocal and parcentric, so that the specimen remains almost in focus at the center of the field as the objectives are changed.

The commonly used Huygenian ocular has a fairly flat field with marked pincushion distortion. Compensating oculars complete the color correction for apochromatic objectives and have less distortion, but they do have curvature of field. *See* EYEPIECE.

The monocular body tube may be of adjustable length. American microscopes are designed for a mechanical tube length of 160 mm (6.3 in.) and a cover glass thickness of 0.18 mm (0.007 in.). The draw tube is lengthened for thinner and shortened for thicker cover glasses to correct for the spherical aberrations from cover glasses of incorrect thickness.

Binocular bodies are designed for the use of both eyes. Most binocular bodies use prisms to reflect one-half of the light to each eye. Because each eye sees the same field, these binocular bodies do not give stereoscopic vision. The binocular is often longer than the monocular body and the proper tube length is maintained with a compensating lens.

Inverted microscope. The inverted microscope has the body of the microscope, including the objective and the ocular, below the stage and the illumination above the stage for transmitted light. The inverted microscope is especially useful for the examination of surfaces. Large and awkward specimens can be moved over the stage more readily than with the usual microscope. The inverted microscope is also useful for microdissection and the observation of hanging-drop preparations.

Comparison microscope. The comparison microscope is an arrangement of two microscopes connected by a special viewing ocular so that the field of one microscope is seen at one side of a vertical dividing line and the field of the other microscope on the opposite side of the dividing line; or it may be a projection type of microscope in which the image is compared with a template or known pattern.

Dissecting microscope. Dissecting microscopes are of two types. The simplest is a magnifying glass mounted on a support above a glass plate, used for the dissection of materials.

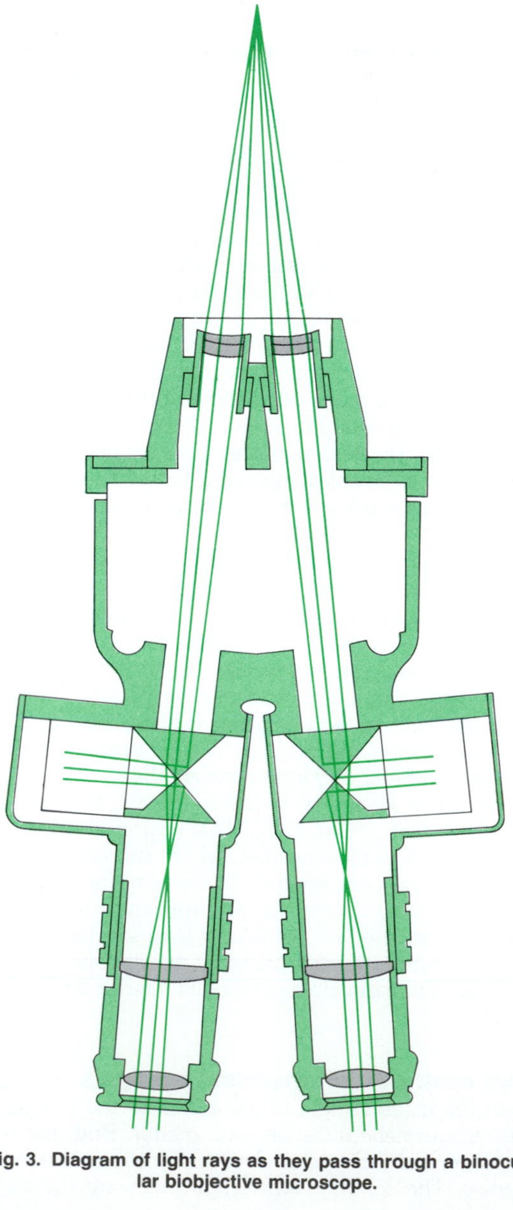

Fig. 3. Diagram of light rays as they pass through a binocular biobjective microscope.

The more usual dissecting microscope, often called a Greenough microscope, is a stereoscopic microscope composed of two separate microscopes fastened together and used as a single unit on one stand (Fig. 3). This is a truly stereoscopic instrument because the right eye sees the specimen from the ght side and the left eye from the left side. Prisms are usually included in the body tube to erect the image; thus movements of the specimen are direct and are not reversed as with the monobjective microscope.

Metallurgical microscope. The metallurgical microscope is a laboratory microscope with a focusing stage and a vertical illuminator, used primarily for the examination of metal surfaces. *See* MICROSCOPE. [O.W.R.]

Optical modulators Devices that serve to vary some property of a light beam. The direction of the beam may be scanned as in an optical deflector, or the phase or frequency of an optical wave may be modulated. Most often, however, the intensity of the light is modulated.

Rotating or oscillating mirrors and mechanical shutters can be used at relatively low frequencies (less than 10^5 Hz). However, these devices have too much inertia to operate at much higher frequencies. At higher frequencies it is necessary to take advantage of the motions of the low-mass electrons and atoms in liquids or solids. These motions are controlled by modulating the applied electric fields, magnetic fields, or acoustic waves in phenomena known as the electrooptic, magnetooptic, or acoustooptic effect, respectively. *See* ACOUSTOOPTICS; ELECTROOPTICS; KERR EFFECT; MAGNETOOPTICS. [I.P.K.]

Optical phase conjugation A process that involves the use of nonlinear optical effects to precisely reverse the direction of propagation of each plane wave in an arbitrary beam of light, thereby causing the return beam to exactly retrace the path of the incident beam. The process is also known as wavefront reversal or time-reversal reflection. The unique features of this phenomenon suggest widespread application to the problems of optical beam transport through distorting or inhomogeneous media.

Optical phase conjugation is a process by which a light beam interacting in a nonlinear material is reflected in such a manner as to retrace its optical path. As the illustration shows, the image-transformation properties of this reflection are radically different from those of a conventional mirror. The incoming rays and those reflected by a conventional mirror (illustration *a*) are related by inversion of the component of the *k*-vector or wave vector *k* normal to the mirror surface. Thus a light beam can be arbitrarily redirected by adjusting the orientation of a conventional mirror. In contrast, a phase-conjugate reflector (illustration *b*) inverts the vector quantity $\vec{k}$ so that, regardless of the orientation of the device, the reflected conjugate light beam exactly retraces the path of the incident beam. This retracing occurs even though an aberrator (such as a piece of broken glass) may be in the path of the incident beam. Looking into a conventional mirror, one would see one's own face, whereas looking into a phase-conjugate mirror, one would see only the pupil of the eye.

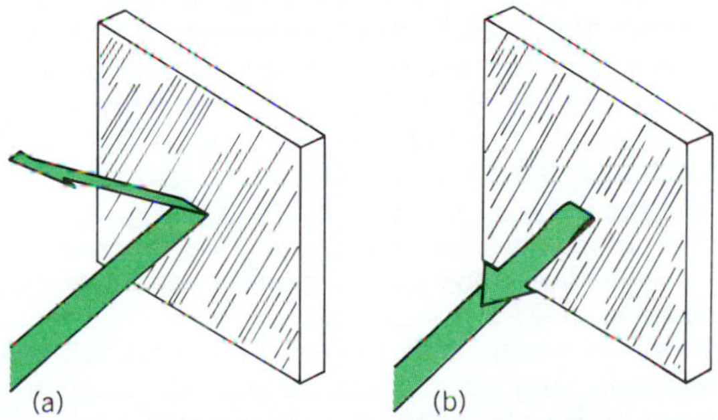

Comparison of reflections (*a*) from a conventional mirror and (*b*) from an optical phase conjugator. (*After V. J. Corcoran, ed., Proceedings for the International Conference on Laser '78 for Optical and Quantum Electronics, STS Press, McLean, Virginia, 1979*)

These remarkable image-transformation properties (even in the presence of a distorting optical element) open the door to many potential applications in areas such as laser fusion, atmospheric propagation, fiber-optic propagation, image restoration, real-time holography, optical data processing, nonlinear microscopy, laser resonator design, and high-resolution nonlinear spectroscopy. *See* HOLOGRAPHY; LASER; NONLINEAR OPTICS; OPTICAL COMMUNICATIONS. [R.A.F.; B.J.F.]

Optical prism An optical system consisting of two or more usually plane surfaces of a transparent solid or embedded liquid at an angle with each other. Prisms are used for deviating light. Since the amount of deviation depends on the refractive index of the prism, which varies with wavelength, prisms can also be used for dispersing light, and they can therefore be used to separate white light into its monochromatic parts. Prisms can also be used instead of mirrors for deviating light, with the added advantage that the reflecting surfaces are protected against corrosion. *See* DISPERSION (RADIATION); MIRROR OPTICS; REFRACTION OF WAVES.

Dispersing prisms deviate light of different wavelengths by different amounts. To increase the dispersion, several prisms with their refracting edges parallel can be used. The Rayleigh prism, shown in illustration *a*, is an example of such a system. By using a prism made of a material, such as flint glass, that has a high dispersion, and adding one or more prisms made of a material having a low dispersion, such as crown glass (illustration *b*), the deviation can be neutralized without neutralizing the dispersion to give a direct-vision prism system. The arrangement shown in illustration *b* is known as the Amici prism system. By using a similar arrangement but adjusting the angle so that the dispersion, but not the deviation, is neutralized, it is possible to make a prism system that is achromatic over a small part of the spectrum, like an achromatic lens. *See* BINOCULARS; DIOPTER; GEOMETRICAL OPTICS; LENS (OPTICS); OPTICAL MATERIALS; PERISCOPE; RESOLVING POWER (OPTICS). [M.J.H.]

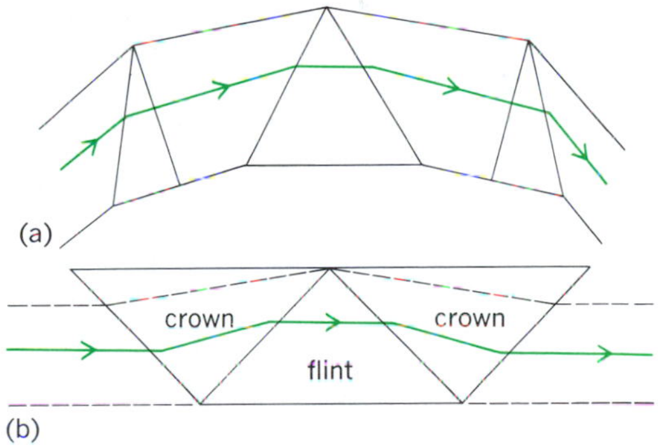

Some types of dispersing prisms. (*a*) Rayleigh prism system. (*b*) Amici direct-vision system consisting of a flint-glass prism and two crown-glass prisms.

Optical projection systems Optical projection is the process whereby a real image of a suitably illuminated object is formed by an optical system in such a manner that it can be viewed, photographed, or otherwise observed. Essential equipment in an optical projection system consists of a light source, a condenser, an object holder, a projection lens, and (usually) a

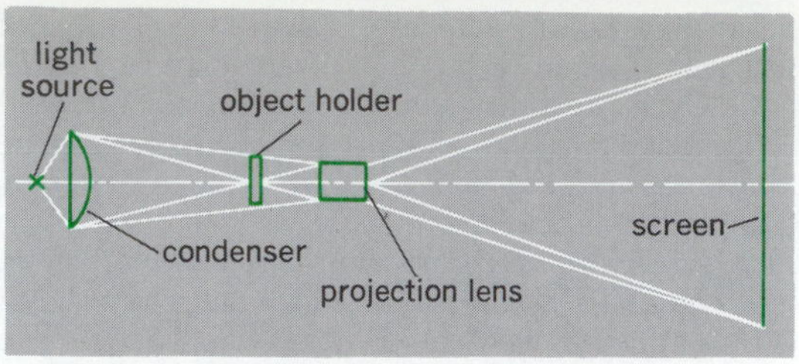

Fig. 1. A simple optical projection system.

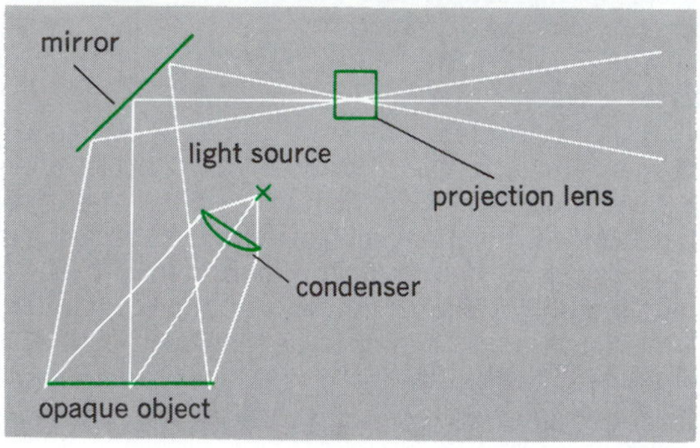

Fig. 2. An epidiascope, or system for projecting an image of an opaque object.

screen on which the image is formed (Fig. 1). For some important applications of optical projection *see* CINEMATOGRAPHY.

The luminance of the image in the direction of observation will depend upon (1) the average luminance of the image of the light source as seen through the projection lens from the image point under consideration, (2) the solid angle subtended by the exit pupil of the projection lens at this image point, and (3) the reflective or transmissive characteristics of the screen. Usually it is desirable to have this luminance as high as possible. Therefore, with a given screen, tens, and projection distance, the best arrangement is to have the light source imaged in the projection lens, with its image filling the exit pupil as completely and as uniformly as possible.

The object is placed between the condenser and the projection lens. If transparent, it can be inserted directly in the light beam; however, it should be positioned, and the optical system should be so designed that it does not vignette (cut off) any of the image of the light source in the projection lens. If the object is opaque, an arrangement known as an epidiascope (Fig. 2) is used. [A.J.H.]

Optical pulses Short flashes of light, used to isolate moments of time. Spark or flash photography can freeze the most rapid movement of macroscopic objects with flashes of 10^{-7} s in duration. High-speed flash lamps and electronics play an important role in the study of fast photophysical and photochemical processes with a resolution of 10^{-10} s. In 1966 a laser technique was used to generate the first optical pulses of less than 10 picoseconds in duration. Pulses as short as a tenth of a picosecond have been produced with the continuous dye laser system. *See* LASER; STROBOSCOPIC PHOTOGRAPHY.

The more frequencies or modes oscillating in a laser, the shorter the optical pulse that can be generated. Organic dyes are ideal for generating pulses, because dyes are capable of oscillating over a broad range of frequencies or colors. To make a pulse, each of these oscillating frequencies must maintain a fixed phase relationship with one another to be coherently added to produce a short optical pulse. This process is called mode locking. [C.V.S.]

Optical pumping The process of causing strong deviations from thermal equilibrium populations of selected quantized states of different energy in atomic or molecular systems by the use of optical radiation (that is, light of wavelengths in or near the visible spectrum), called the pumping radiation.

Optical pumping is vital for light amplification by stimulated emission in an important class of lasers. For example, the action of the ruby laser involves the fluorescent emission of red light by a transition from an excited level E_2 to the ground level E_1. In this case E_2 is relatively high above E_1 and the equilibrium population of E_2 is practically zero. Amplification of the red light by laser action requires that number of atoms N_2 exceed N_1 (population inversion). The inversion is accomplished by intense green and violet light from an external source which excites the chromium ion in the ruby to a band of levels, E_3 above E_2. From E_3 the ion rapidly drops without radiation to E_2, in which its lifetime is relatively long for an excited state. Sufficiently intense pumping forces more luminescent ions into E_2 by way of the E_3 levels than remain in the ground state E_1, and amplification of the red emission of the ruby by stimulated emission can then occur. *See* LASER. [W.W.]

Optical recording The process of recording sound signals on a photosensitive medium so that they may be reproduced at a subsequent time. Photographic film has been widely used as the photosensitive medium, but in the late 1970s development of another kind of medium, the so-called optical disk, was undertaken. The introduction of the laser as a light source greatly improves the quality of reproduced sound signals. The pulse code modulation (PCM) techniques make it possible to obtain extremely high-fidelity reproduction of sound signals in optical disk recording systems.

Optical film recording. Optical film recording is also termed motion picture recording or photographic recording. A sound motion picture recording system consists basically of a modulator for producing a modulated light beam and a mechanism for moving a light-sensitive photographic film relative to the light beam and thereby recording signals on the film corresponding to the electrical signals (Fig. 1). A sound motion picture reproducing system is basically a combination of a light source, optical system, photoelectric cell, and mechanism for moving a film carrying an optical record by means of which the recorded photographic variations are converted into electrical signals of approximately like form. *See* CINEMATOGRAPHY; SOUND RECORDING.

Laser-beam film recording. An optical film system which utilizes a laser as a light source and a combination of an acoustooptical modulator (AOM) and an acoustooptical deflector (AOD) instead of a galvanometer was developed in the late 1970s. A 100-kHz pulse width modulation (PWM) circuit converts the audio input signal into a PWM signal. The laser beam is made to continuously scan the sound track area at right angles to the direction of the film transport. This is done by means of the AOD, which in turn is driven by a 100-kHz sawtooth signal. Simultaneously, the laser beam is pulse-width-modulated by means of the AOM, which is driven by a 100-kHz PWM signal. The AOD scanning signal and the AOM pulse-width-modulated signal combine and generate the variable-area sound track exposure on the film (Fig. 2). Because the actual distance between successive scans is very small, the traces of successive scans are fused into a pattern of variable-area recording. *See* PULSE MODULATION.

PCM optical sound disk player. An optical disk allows a color television program of about 30 minutes' duration to be reproduced from a recording on a phonographlike disk 12 in. (30 cm) in diameter. Owing to its wide frequency range characteristics, up to several MHz, it can also serve to record or reproduce digitized (pulse code modulation or PCM) audio signals. The PCM optical sound disk player has the following

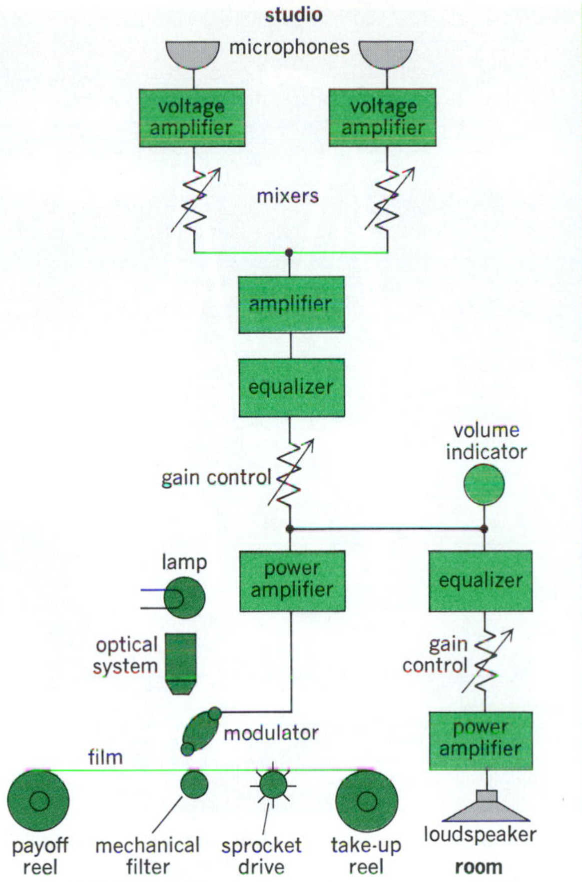

Fig. 1. Schematic arrangement of apparatus in a complete optical sound motion-picture recording system.

superior characteristics which are not available from the conventional disk: (1) wide-frequency-range (up to 20 kHz) audio signals; (2) extremely large dynamic range; and (3) the complete absence of various distortions inherent in conventional records such as tracing distortion, pinch effect, and flutter and wow. These are direct results of PCM techniques applied to audio signal recording.

In the recording system, an originating disk is obtained by coating a thin layer of evaporated metal on one side of a glass disk. The audio signal is passed through a PCM coding circuit for conversion into a digital signal which is then frequency-modulated. This frequency-modulated signal is then applied to

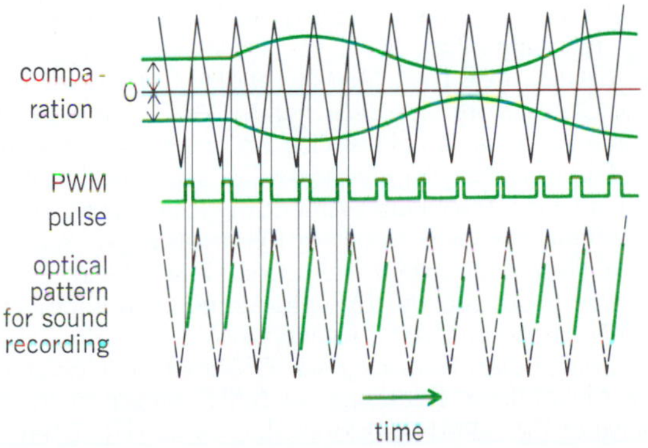

Fig. 2. Schematics of the sound track waveform generation of the laser sound recorder. (*After T. Taneda et al., A high quality optical sound recording system using a scanned laser beam, SMPTE J., 89:95–97, 1980*)

the optical modulator which chops the continuous beam of light from the argon laser. The laser-beam intensity corresponding to the frequency modulation is then focused into a spot of about 0.8 micrometer diameter by the optical condenser. The information is thus recorded as a series of pits in thin metal film on the disk.

The PCM laser sound disk player rotates the disk at the correct speed. Its optical system directs and focuses a low power (1 mW) helium-neon laser beam to a small read spot on the surface of the disk and then collects the reflected optical energy and directs it to a single photodetector. The received energy at the photodetector changes according to the presence of recorded pits. Its electronic circuits process the received signal into digitized form, and the subsequent digital-to-analog converter converts the digital signals into the audio output signals with high fidelity. *See* DIGITAL-TO-ANALOG CONVERTER. [H.D.]

Optical rotatory dispersion A term used to describe the change in rotation as a function of wavelength experienced by linearly polarized light as it passes through an optically active substance. *See* OPTICAL ACTIVITY.

In all materials the rotation varies with wavelength. The variation is caused by two quite different phenomena. The first accounts in most cases for the majority of the variation in rotation and should not strictly be termed rotatory dispersion. It depends on the fact that optical activity is actually circular birefringence. In other words, a substance which is optically active transmits right circularly polarized light with a different velocity from left circularly polarized light.

In addition to this pseudodispersion which depends on the material thickness, there is a true rotatory dispersion which depends on the variation with wavelength of the indices of refraction for right and left circularly polarized light. *See* POLARIZED LIGHT. [B.H.Bi.]

For wavelengths that are absorbed by the optically active sample, the two circularly polarized components will be absorbed to differing extents. This unequal absorption is known as circular dichroism. Circular dichroism causes incident linearly polarized light to become elliptically polarized. *See* ABSORPTION.

Optical rotatory dispersion and circular dichroism are closely related, just as are ordinary absorption and dispersion. If the entire optical rotatory dispersion spectrum is known, the circular dichroism spectrum can be calculated, and vice versa.

In order for a molecule (or crystal) to exhibit circular birefringence and circular dichroism, it must be distinguishable from its mirror image. An object that cannot be superimposed on its mirror image is said to be chiral, and optical rotatory dispersion and circular dichroism are known as chiroptical properties.

Most biological molecules have one or more chiral centers and undergo enzyme-catalyzed transformations that either maintain or reverse the chirality at one or more of these centers. Still other enzymes produce new chiral centers, always with a high specificity. These properties account for the fact that optical rotatory dispersion and circular dichroism are widely used in organic and inorganic chemistry and in biochemistry. *See* ENZYME; STEREOCHEMISTRY.

In the absence of magnetic fields, only chiral substances exhibit optical rotatory dispersion and circular dichroism. In a magnetic field, even substances that lack chirality rotate the plane of polarized light, as shown by M. Faraday. Magnetic optical rotation is known as the Faraday effect, and its wavelength dependence is known as magnetic optical rotatory dispersion. In regions of absorption, magnetic circular dichroism is observable. *See* FARADAY EFFECT; MAGNETICOOPTICS. [R.W.Wo.]

Optical surfaces Many common optical instruments contain optical materials bounded by spherical surfaces with a wide range of curvatures. A surface that is not a sphere is called an aspherical surface. Examples of such aspherical surfaces with symmetry of rotation that are occasionally used in optical instruments are conic sections (ellipsoids, hyperboloids, and paraboloids). Cylindrical and toroidal lenses are used in anamorphotic systems.

Since it is much easier to grind and polish spherical surfaces than aspherical ones, most optical systems consist of lenses that are bounded by two spherical surfaces. These surfaces are arranged so that their centers lie on a line known as the axis of the system. The point of intersection of the surface with the axis is called the vertex. A plane surface is generally considered as a special case of a spherical surface which has its center at infinity and which can therefore be said to have an infinite radius.

Aspherical surfaces are difficult to manufacture. When many identical aspherical elements are to be made in series, special methods of grinding and polishing must be used. For instance, templates are used for spectacle lenses and for paraboloidal and ellipsoidal mirrors. Where extreme correction is desired, for instance, in astronomical telescopes, retouching by hand is the proper procedure. For condensers and for other lenses where extreme accuracy is not needed, a molding process is used. Large paraboloidal mirrors for searchlights and ellipsoidal mirrors for arc lamps used to project motion pictures are made by a "dropping" process. A sheet of plate glass is laid on a suitable concave mold and heated until it softens and can be sucked into the mold. *See* Geometrical optics; Lens (optics); Optical telescope. [M.J.H.]

Optical telescope An instrument that collects light energy from a distant source and focuses it into an image that can then be studied by a number of different techniques. This definition of an optical telescope must be narrowed somewhat. Satellite-borne telescopes operated to study celestial x-rays or ultraviolet radiation and ground-based telescopes used to study radio radiation can be called optical telescopes in the sense that they operate according to the principles of geometrical optics. Ground-based telescopes are used to study radiation from celestial objects in the wavelength range from the Earth's atmospheric cutoff in the near ultraviolet to the infrared, that is, from 300 to 1000 nanometers. Such instruments may be classified as (1) large astronomical telescopes, used to study the nature of astronomical objects themselves, and (2) astronomical transit instruments, used to study positions and motions of astronomical objects and in the accurate determination of time. *See* Astronomical transit instrument; Geometrical optics; Radio telescope; Telescope.

There are basically three types of optical systems in use in astronomical telescopes; refracting systems whose main optical elements are lenses which focus light by refraction; reflecting systems, whose main imaging elements are mirrors which focus light by reflection; and catadioptric systems, whose main elements are a combination of a lens and a mirror. The most notable example of the last type is the Schmidt camera. *See* Schmidit camera.

Astronomers seldom use large telescopes for visual observations. Instead, they record their data for future study. Modern developments in photoelectric imaging devices are supplanting photographic techniques for many applications. The great advantages of detectors such as charge-coupled devices is the fact that they are very sensitive, and the images can be read out onto a computer-compatible magnetic tape or disk for immediate analysis. See Astronomicawl photography; Charge-coupled devices.

Light received from most astronomical objects is made up of radiation of all wavelengths. The spectral characteristics of the radiation emitted from or reflected by a body may be extracted by special instruments called spectrographs. *See* Astronomical spectroscopy.

For many applications the Earth's atmosphere limits the effectiveness of larger telescopes. The most obvious deleterious effect of the Earth's atmosphere is image scintillation and motion, collectively known as "poor seeing." On the very best nights at ideal observing sites, the image of a star will be spread out over a 0.25-arc-second seeing disk; on an average night, the seeing disk may be between 0.5 and 2.0 seconds of arc. Telescopes larger than about 20 in. (50 cm) are built, not for resolution, but for light-gathering power, which depends on the area of the primary mirror or objective lens. Once again, however, the atmosphere limits what can be observed.

The 158-in. (4.01-m) Mayall reflector of the Kitt Peak National Observatory. The dark tube at the upper end of the telescope is the prime focus cage. The large horseshoe bearing floats on oil pads, seen near each end of the mounting walkway. (*Kitt Peak National Observatory photograph*)

The 200-in. (5.08-m) Hale telescope at Palomar Mountain, California, was completed in 1950. The primary mirror of this reflector is 200 in. in diameter with a 40-in. (1.02-m) hole in the center. The Hale telescope was the first that was large enough to have a prime-focus cage where the observer could sit inside the main tube.

The 158-in. (4.01-m) Mayall reflector at the Kitt Peak National Observatory was dedicated in 1973 (see illustration). The prime focus has a field of view six times greater than that of the Hale reflector. An identical telescope was subsequently installed at Cerro Tololo Inter-American Observatory, in Chile.

In 1989 the European Southern Observatory put into operation their New Technology Telescope. The 141-in. (3.58-m) mirror was produced by the spincasting technique, where molten glass is poured into a rotating mold.

The Smithsonian Astrophysical Observatory/University of Arizona Multi-Mirror Telescope, or MMT, is one of the most innovative of the newer telescopes. It uses six 6-ft (1.8-m) mirrors working together to provide light-gathering power equivalent to a 15-ft (4.5-m) mirror.

The Keck Telescope on Mauna Kea, Hawaii, completed in 1993, is the largest of the segmented mirror telescopes to be put into operation. The telescope itself is a fairly traditional design. However, its primary mirror is made up of 36 individual hexagonal segments mosaiced together to form a single 386-in. (10-m) mirror. Electronic sensors built into the edges of the segments monitor the relative positions of the segments, and feed the results to a computer-controlled actuator system.

The Hubble Space Telescope (HST), which was launched into Earth orbit by the space shuttle on April 24, 1990, is a 94.5-in. (2.4-m) Cassegrain reflector, which feeds light into a number of scientific instruments. These instruments can see objects many times fainter than those achieved by prior technology. After several years of operation, the instruments are changed in orbit and replaced by innovative instruments. *See* SATELLITE ASTRONOMY. [R.D.Ch.]

Optical tracking instruments A family of optical instruments used for precise time-correlated observation of distant airplanes, missiles, and artificial satellites, all of which travel at apparent velocities much greater than those of most astronomical objects. The instruments supply permanent engineering records for the determination of spatial position, missile attitude, structural behavior, and performance of specific mechanisms during test flights. These observations enable engineers to correct design, improve performance, and collect scientific data from missiles at extreme distances and altitudes.

The instruments used fall by function into two classes, those which determine spatial position and those which record engineering events. Tracking telescopes are the basic engineering-event recording systems, while cinetheodolites and ballistic cameras are used for the precise determination of spatial position.

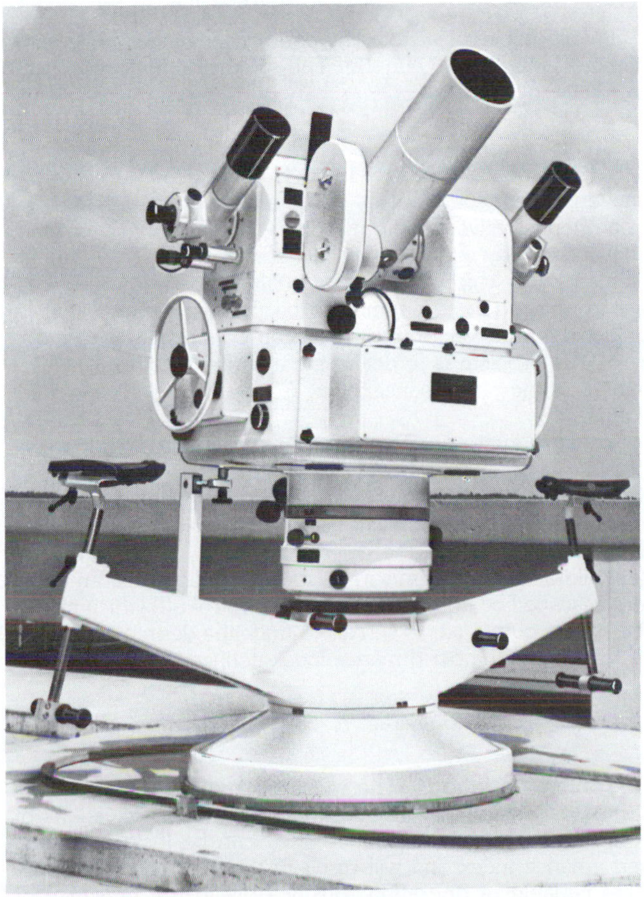

Contraves Model E cinetheodolite with 60- and 120-in. (1.5- and 3.0-m) objective. Tracking is by servos with either slaved, single, or dual-operator control. Laser automatic tracking is being added to the instrument for single-station laser-radar operation. (*Fecker Systems Division, Owens-Illinois and Contraves AG*)

Spatial position determination can be considered the determination of the position of a moving target using dynamic adaptations of the methods of civil surveying. The classical techniques utilize a minimum of four instruments on a precisely measured baseline to locate a moving target. Each instrument records the data for computing the direction of the line of sight to the target for each instant of time. This information, in the form of analog or digital elevation and azimuth angles, the times of observation, and the known location of each instrument, is used to triangulate for the location of the missile as a function of time.

Cinetheodolite. Most of the spatial position work is performed by cinetheodolites (see illustration), which are surveying theodolites having 35-mm motion-picture cameras with 60- to 240-in. (1.5- to 6. 1-cm) focal-length lenses substituted for the surveyor's eye and telescope. The system of cameras is synchronized up to a maximum of 30 frames/s from a master control station for simultaneous exposure as the cinetheodolites follow the moving missile. Each photograph records the elevation angle, the azimuth angle, the missile image, and the reticle lines which define the instrumental axis.

The rapid evolution of laser systems led to the development of a laser automatic-tracking range-finding digital cinetheodolite, which is an optical radar capable of single station solution of the target location. The requirements for automatic tracking and range-finding have led to the development of a cinetheodolite with 1-second accuracy.

Ballistic cameras. These are fixed-axis, wide-angle, photographic-plate cameras, capable of more precise spatial position determination by recording on one plate multiple exposures of the missile against a stellar background. Use of a static system and precisely cataloged star positions decreases necessity for long-term mechanical stability and accuracy, allowing ballistic cameras to achieve 2–5 seconds angular accuracy. Pyrotechnic flares or electronic stroboscopic lamps at the missile are used to indicate the missile positions against the night sky.

Tracking telescopes. These are long-focal-length telescopes mounted to track missiles in flight precisely while collecting missile performance data. In the normal aided-tracking mode of operation, the operator observes the missile through a tracking sight and controls the motion of the telescope with a control knob. The system must track at rates up to 10°/s with a precision of 1–2 minutes of arc if the system resolution of $\frac{1}{2}$ second is not to be lost because of relative motion of the image during the exposure.

The precise tracking of modern tracking telescopes has permitted the use of slit spectrographs for infrared and visible-light spectroscopy. *See* ASTRONOMICAL PHOTOGRAPHY; CAMERA; LENS (OPTICS). [G.A.E.]

Optically pumped laser A device in which the population inversion necessary for laser action is achieved by concentrating light (for example, from a flash lamp, the Sun or another laser) onto the amplifying medium.

Many such lasers are three-level lasers; that is, ground-state atoms are excited by absorption of light to a broad upper-energy stage, from which they quickly relax to the emitting state. Laser action occurs as they are stimulated to emit radiation and so return to the original ground state. Solid three-level lasers usually make use of ions of a rare-earth element, such as neodymium, or of a transition metal, such as chromium, dispersed in a transparent crystal or glass. The ions have broad absorption bands in which pumping radiations can be absorbed. Thus broad-band white light can be used to excite the atoms.

Many rare-earth ion lasers use a fourth level above the

ground state. This level serves as a terminal level for the laser transition, and is kept empty by rapid nonradiative relaxation to the ground state. This means that, relative to three-level systems, a population inversion is easier to maintain, and therefore such materials require relatively low pumping light intensity for laser action.

Liquid lasers have structures generally like those of optically pumped solid-state lasers, except that the liquid is generally contained in a transparent cell. Some liquid lasers make use of rare-earth ions in suitable dissolved molecules, while others make use of organic dye solutions. The dyes can lase over a wide range of wavelengths, depending upon the composition and concentration of the dye or solvent. Thus tunability is obtained throughout the visible, and out to a wavelength of about 0.9 micrometer. Fine adjustment of the output wavelength can be provided by using a diffraction grating or other dispersive element in place of one of the laser mirrors.

In several infrared regions, tunable laser action can be obtained by using an infrared gas laser to pump a semiconductor crystal in a magnetic field, giving amplification by stimulating spin-flip Raman scattering from the electrons in the semiconductor. Tuning is achieved by varying the magnetic field. *See* LASER. [S.F.J.; A.L.S.]

Optics Narrowly, the science of light and vision; broadly, the study of the phenomena associated with the generation, transmission, and detection of electromagnetic radiation in the spectral range extending from the long-wave edge of the x-ray region to the short-wave edge of the radio region. This range, often called the optical region or the optical spectrum, extends in wavelength from about 1 nanometer to about 1 millimeter. *See* GEOMETRICAL OPTICS; METEOROLOGICAL OPTICS; PHYSICAL OPTICS; VISION.

The discoveries of the experimentalists of the early 17th century formed the basis of the science of optics. The statement of the law of refraction, the development of the astronomical telescope, observations of diffraction, and the principles of the propagation of light all came in this relatively short period. The publication of Isaac Newton's *Opticks* in 1704, with its comprehensive and original studies of refraction, dispersion, interference, diffraction, and polarization, established the science.

In the early 19th century many productive investigators established the transverse-wave nature of light. The relationship between optical and magnetic phenomena led to the crowning achievement of classical optics—the electromagnetic theory of J. C. Maxwell. Maxwell's theory, which holds that light consists of electric and magnetic fields propagated together through space as transverse waves, provided a general basis for the treatment of optical phenomena. In particular, it served as the basis for understanding the interaction of light with matter and, hence, as the basis for treatment of the phenomena of physical optics. *See* ELECTROMAGNETIC RADIATION; LIGHT; MAXWELL'S EQUATIONS.

In the 20th century optics has been in the forefront of the revolution in physical thinking caused by the theory of relativity and especially by the quantum theory.

The science of optics finds itself in a position that is satisfactory for practical purposes but less so from a theoretical standpoint. The theory of Maxwell is sufficiently valid for treating the interaction of high-intensity radiation with systems considerably larger than those of atomic dimensions. The modern quantum theory is adequate for an understanding of the spectra of atoms and molecules and for the interpretation of phenomena involving low-intensity radiation, provided one does not insist on a very detailed description of the process of emission or absorption of radiation. However, a general theory of relativistic quantum electrodynamics valid for all conditions and systems has not been worked out.

The development of the laser has been an outstanding event in the history of optics. The theory of electromagnetic radiation from its beginnings was able to comprehend and treat the properties of coherent radiation, but the controlled generation of coherent monochromatic radiation of high power was not achieved in the optical region until the work of C. H. Townes and A. L. Schawlow in 1958 pointed the way. Many achievements in optics, such as holography and interferometry over long paths, have resulted from the laser. *See* HOLOGRAPHY; INTERFEROMETRY; LASER. [R.C.L.]

Optimal control (linear systems) A branch of modern control theory which deals specifically with the problem of determining optimal control for linear systems when the performance index to be minimized depends quadratically on the system's variables.

Although the class of problems encompassed by linear optimal control theory may at first seem restrictive, its results have found wide application in many different control problems for both linear and nonlinear systems and, because of the linear nature of the resulting optimal control laws, are particularly simple to implement.

Regulator problem. The simplest type of linear optimal control problem with a quadratic cost is the so-called regulator problem. Its solution contains all of the essential characteristics of linear optimal control problems in general and establishes a framework from which more advanced topics in the field can be examined. Consider the system described by state equation (1), where $x(t)$ is the n-dimensional state vector and $u(t)$ is the

$$\dot{x} = A(t)x + B(t)u \qquad x(t_0) \text{ specified} \tag{1}$$

p-dimensional input vector. The performance index to be minimized, J, is defined by Eq. (2). The symmetric matrices S_1 and

$$J = \tfrac{1}{2}\, x^T(t_1)S_1x(t_1) + \tfrac{1}{2}\int_{t_0}^{t_1} x^TQ(t)x + u^TR(t)u\,dt \tag{2}$$

$Q(t)$, which penalize the system for large excursions in both the final and intermediate values of the system state, are assumed to be non-negative definite; that is, x^TS_1x and x^TQx are nonnegative for any vector x with real components. The symmetric matrix $R(t)$ which reflects the penalty associated with excessive control effort is assumed to be positive definite; that is, $u^TRu>0$ for any vector u with real components not all equal to zero. The problem is to determine a control vector $u(t)$ which minimizes the performance index J subject to constraint equation (1).

Tracking systems. The optimal regulator problem is a special case of the more general problem of determining a control which causes the system output or state to track a desired trajectory as closely as possible without undue control effort. In the regulator problem the desired trajectory is $x(t) \equiv 0$. The general tracking problem can be stated as follows: Given a system described by Eq. (1), determine a control $u(t)$ which causes the system output $y(t) = C(t)x(t)$ to track the desired output $z(t)$ by minimizing the performance index defined by Eq. (3) where $e(t) \equiv z(t) - y(t)$.

$$J = \tfrac{1}{2}e^T(t_1)S_1e(t_1) + \tfrac{1}{2}\int_{t_0}^{t_1} e^TQ(t)e + u^TR(t)u\,dt \tag{3}$$

Problem variations. The regulator and tracking problems can be extended to include a number of special variations of practical importance. Penalties on u can be included in the performance index and will result in dynamic feedback structures capable of operating satisfactorily in the presence of slowly varying plant disturbances. Controller constraints such as piecewise constant gains or a requirement that the system employ output feedback lead to optimization problems which are solvable by numerical search techniques. Requiring the control input to assume only those values which are constrained in magnitude (for example, $|u(t)| \leq 1$) leads to a control law which is nonlinear over most of the state space (u assumes values on the extremity of the constraint set) and linear in the vicinity of the state space origin. This type of system is said to use dual-mode control.

The extension of the linear optimal control problem to include the effects of random noise in the plant and measurement equations yields a feedback structure which contains an optimal state estimator followed by the linear feedback structure associated with the deterministic regulator problem. *See* CONTROL SYSTEMS; LINEAR SYSTEM ANALYSIS; OPTIMAL CONTROL THEORY; STOCHASTIC CONTROL THEORY. [P.E.B.]

Optimal control theory An extension of the calculus of variations for dynamic systems with one independent variable, usually time, in which control (input) variables are determined to maximize (or minimize) some measure of the performance (output) of a system while satisfying specified constraints. Theory is conveniently divided into two parts: optimal programming, where the control variables are determined as functions of time for a specified initial state of the system, and optimal feedback control, where the control variables are determined as functions of the current state of the system.

Examples of optimal control problems are: (1) determining paths of vehicles between two points to minimize fuel or time, and (2) determining feedback control logic for vehicles or industrial processes to keep them near a desired operating point in the presence of disturbances with acceptable control magnitudes.

Dynamic systems are conveniently divided into two categories: continuous dynamic systems, where the control and state variables are functions of a continuous independent variable, such as time or distance, and discrete dynamic systems, where the independent variable changes in discrete increments. Many discrete systems are discretized versions of continuous systems; the discretization is often made so that (1) the system can be analyzed or controlled by digital computers (or both), or (2) measurements of continuous outputs are made at discrete intervals of time (sampled-data systems) in order to share data transmission channels.

A large class of interesting optimal control problems can be described as follows: the dynamic system is described by a set of coupled first-order ordinary differential equations of the form of Eq. (1), where x (the state vector) represents the state vari-

$$\dot{x} = f(x,u,t) \tag{1}$$

ables of the system (such as position, velocity, temperature, voltage, and so on); u (the control vector) represents the control variables of the system (such as motor torque, control-surface deflection angle, valve opening, and so on); t is time; and f represents a set of functions of x, u, and t. The performance index J, which one desires to minimize, is a scalar function of the final time t_f and the final state $x(t_f)$ plus an integral from the initial time t_0 to the final time t_f of a scalar function of the state and control vectors and time, as given in Eq. (2). Possible con-

$$J = \varphi[x(t_f),t_f] + \int_{t_0}^{t_f} L[x(t),u(t),t]dt \tag{2}$$

straints include: (*a*) specified vector functions constraining the initial time t_0 and the initial state $x(t_0)$, as in Eq. (3); (*b*) speci-

$$\alpha[x(t_0),t_0] = 0 \tag{3}$$

fied vector functions constraining the final time and final state, as in Eq. (4); (*c*) specified vector functions constraining the

$$\psi[x(t_f),t_f] = 0 \tag{4}$$

control variables [inequality (5)]; (*d*) specified vector functions

$$C[u(t),t] \leq 0 \tag{5}$$

constraining both control and state variables [inequality (6)];

$$CS[x(t),u(t),t] \leq 0 \tag{6}$$

and (*e*) specified vector functions constraining only the state variables [inequality (7)]. Constraints *a* and *b* are equality con-

$$S[x(t),t] \leq 0 \tag{7}$$

straints, whereas constraints *c*, *d*, and *e* are inequality constraints. *See* LINEAR SYSTEM ANALYSIS.

Classes of problems. There are several important classes of optimal control problems.

Bang-bang control is a type of control in which the control variables are either at their maximum or at their minimum values, never in between. This type of control is optimal, for example, in certain minimum-fuel problems where the control variables enter linearly into system equations (1), the performance index in Eq. (2), and the constraints in Eq. (5). *See* NONLINEAR CONTROL THEORY.

Bang-zero-bang control is a type of control in which the control values are at their maximum values, zero, or their minimum values. This type of control is optimal, for example, in certain minimum-fuel problems where the control variables enter linearly into system equations (1) and the constraints in Eq. (5), and the performance index depends on the magnitude (but not the sign) of the control variables.

Linear-quadratic problems are problems in which system equations (1) are linear and the performance index in Eq. (2) is quadratic in the states x and the control u. *See* OPTIMAL CONTROL (LINEAR SYSTEMS).

Neighboring optimum feedback control (NOFC) about a nominal optimal path is a viable alternative to a "complete" optimal feedback control solution. NOFC constitutes an approximation to the optimal feedback control solution in a region neighboring the nominal optimal path.

Stochastic optimal control theory. This takes into consideration, in an ensemble average sense, random disturbances on the dynamic system, random initial conditions, and random errors in measurements. A control design is sought that maximizes (or minimizes) the performance criterion on the average and satisfies the constraints within certain tolerances, also on the average. *See* ESTIMATION THEORY; STOCHASTIC CONTROL THEORY; STOCHASTIC PROCESS.

Differential games. These are two-sided optimal control problems. Typically, control system A tries to minimize the performance criterion, while control system B tries to maximize it. This may lead to a minimax feedback strategy in which both systems do their best, taking into account the other controller's intelligence in seeking the opposite goal. *See* CONTROL SYSTEMS; DECISION THEORY; GAME THEORY. [A.E.Br.]

Optimization In its most general meaning, the efforts and processes of making a decision, a design, or a system as perfect, effective, or functional as possible. Formal optimization theory encompasses the specific methodology, techniques, and procedures used to decide on the one specific solution in a defined set of possible alternatives that will best satisfy a selected criterion. Because of this decision-making function the term optimization is often used in conjunction with procedures which more appropriately belong in the more general domain of decision theory. Strictly speaking, formal optimization techniques can be applied only to a certain class of decision problems known as decision making under certainty. *See* DECISION THEORY.

Concepts and terminology. Conceptually, the formulation and solution of an optimization problem involves the establishment of an evaluation criterion based on the objectives of the optimization problem, followed by determination of the optimum values of the controllable or independent parameters that will best satisfy the evaluation criterion. This determination is accomplished either objectively or by analytical manipulation of the so-called criterion function, which relates the effects of the independent parameters on the dependent evaluation criterion parameter. In most optimization problems there are a number of conflicting criteria and a compromise must be reached by a trade-off process which makes relative value judgments among the conflicting criteria. Additional practical considerations encountered in most optimization problems include so-called functional and regional constraints on the parameters. The former represents physical or functional interrelationships which exist among the independent parameters (that is, if one is changed it causes some changes in the others); the latter limits the range over which the independent parameters can be varied.

Optimization techniques work only for a specific system or configuration which has been described to a point where all criteria and parameters are defined within the system and are isolated and independent from other parameters outside the boundaries of the defined system. Formal optimization is not a substitute for creativity in that it depends on a clear definition of the system to be optimized, and even though an optimum solution for a given system configuration is obtained, this does not guarantee that a better solution is not available.

Problem formulation. Definition is a critical part of problem formulation and consists of:

(1) A description of the system configuration to be optimized, including definition of system boundaries to an extent that system parameters become isolated and independent of external parameters.

(2) Definition of a single, preferably quantitative, parameter which will serve as the overall evaluation criterion for the specific optimization problem. This is the dependent parameter (it depends upon the choice of the optimum solution) which measures how well the solution satisfies the desired objectives of the problem and this is the parameter which will be maximized or minimized to satisfy the objectives.

(3) Definition of controlled or independent parameters that will have an effect on the criterion. These are the parameters whose values determine the value of the criterion parameter, and these should include all controllable parameters which influence the criterion and are within the boundaries of the defined problem.

The most critical aspect of formulating a formal optimization problem is the establishment of a satisfactory criterion function that describes the behavior of the evaluation criterion as a function of the independent parameters. The criterion function, often also referred to as the pay-off or objective function, usually takes the form of a penalty or cost function (which attempts are made to minimize) or a merit, benefit or profit function (which attempts are made to maximize by choosing the optimum values of the independent variables). In order to apply formal solution methods, the criterion function should be expressed graphically or analytically.

Physical principles of operation which govern the relationship among the various independent parameters of the problem represent functional constraints. Most optimization problems involve practical limits on the range over which each parameter or function of the parameters can be varied. They represent the regional constraints.

Solution of optimization problems. The reliability and the sophistication of the solutions of optimization problems increase as the problem becomes better defined. In the early stages of optimization of a complex problem, the considerations may be primarily objective judgments based on the optimizer's judgments of the relevant parameters. As alternative subsystem and system configurations evolve through a process of conceptualization, analysis and evaluation, and elimination of undesirable system configurations, a clearer configuration of parts of the problem can be defined, together with the relevant criteria, parameters, and constraints. Procedures for defining competing systems or alternate strategies and formulating the problem in order to apply formal optimization techniques are also common to operations research, systems analysis, and systems design. See GAME THEORY: METHODS ENGINEERING; MONTE CARLO METHOD; PRODUCTION ENGINEERING; PRODUCTION PLANNING. [I.P.]

Oral glands Glands located in the mouth that secrete fluids to moisten and lubricate the mouth and food and may initiate digestive activity, and that may perform other specialized functions. Fishes and aquatic amphibians have only solitary mucus-secreting cells in the epithelium of the mouth cavity. Multicellular glands first appeared in land animals to keep the mouth moist and make food easier to swallow. Some glands of terrestrial amphibians have a lubricative secretion; others serve to make the tongue sticky for use in catching insects. Some frogs secrete a serous fluid that contains ptyalin, a digestive enzyme. The oral glands of reptiles are much the same, but are more distinctly grouped. In poisonous snakes and the single poisonous lizard, the Gila monster, certain oral glands of the serous type are modified to form venom. Also many of the lizards have glands that are mixed in character, containing both mucous and serous cells. Oral glands are poorly developed in crocodilians and sea turtles. Birds bolt their food, yet grain-eaters have numerous glands, some of which secrete ptyalin.

All mammals except aquatic forms are well supplied with oral glands. There are numerous small glands, such as the labial glands of the lips, buccal glands of the cheeks, lingual glands of the tongue, and palatine glands of the palate. Besides these, there are larger paired sets in mammals that are quite constant from species to species and are commonly designated as salivary glands (see illustration). The parotid gland, near each ear,

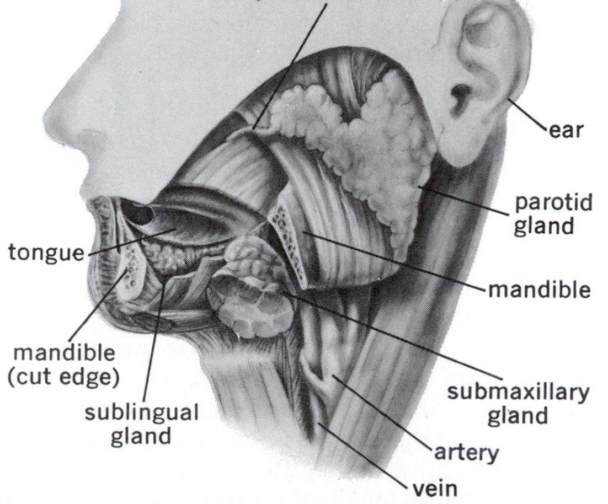

The salivary glands, shown by a partial dissection of the head. ***(After J. C. Brash, ed., Cunningham's Textbook of Anatomy, 9th ed., Oxford, 1951)***

discharges into the vestibule. The submaxillary or submandibular gland lies along the posterior part of the lower jaw; its duct opens well forward under the tongue. The sublingual gland lies in the floor of the mouth. It is really a group of glands, each with its duct. See GLAND. [L.B.A.]

Orange The sweet orange (*Citrus sinensis*) is the most widely used species of citrus fruit and commercially is the most important. The sour or bitter oranges, of lesser importance, are distinct from sweet oranges and are classified as a separate species, *C. aurantium*. The United States is the largest producer of oranges, followed by Spain, Italy, and Brazil. The orange is also a major crop in several other countries.

Sweet orange fruit is consumed fresh or as frozen or canned juice. A large portion of the crop, particularly in the United States, is used as frozen concentrate. After the juice is extracted, the peel and pulp are used for cattle feed. Peel oil is used in perfumes and flavoring, and citrus molasses is used as a livestock feed.

The sweet orange tree is a moderately vigorous evergreen with a rounded, densely foliated top. The fruits are round or somewhat elongate and usually orange-colored when ripe. They can be placed in four groups: the common oranges, acidless oranges, pigmented oranges, and navel oranges. They may also be distinguished on the basis of early midseason, and late maturity. *See* FRUIT; FRUIT, TREE. [R.K.So.]

Orangutan The largest species of great ape, *Pongo pygmaeus*, occurring outside Africa, in the low-lying forest areas of Borneo and Sumatra. This primate may grow to more than 4 ft (1.2 m) in height and weigh about 150 lb (68 kg), the male being larger than the female. The arms are longer than the legs. The canines are more fanglike than those of humans, and the orangutan is distinguished from humans by a large laryngeal cavity which appears as a large pouch below the chin.

These animals are slow and not very agile on the ground or in trees. Their diet consists mainly of fruit. Adult males live apart from the females except during the mating season. At this time a temporary tree nest is built to be used by a solitary adult or a female and young. Both sexes are sexually mature at 8–10 years of age. A single young is born each year after a gestation period of not quite 8 months. Males assume no responsibility in caring for the young.

Of all the anthropoids, this animal has a head which has developed in such a manner that there is a high, arched forehead like that of humans. There is some fear that this animal faces extinction; estimates place the present population at less than 5000. *See* PRIMATES. [C.B.C.]

Orbital motion In astronomy the motion of a material body through space under the influence of its own inertia, a central force, and other forces. Johann Kepler found empirically that the orbital motions of the planets about the Sun are ellipses. Sir Isaac Newton, starting from his laws of motion, proved that an inverse-square gravitational field of force requires a body to move in an orbit that is a circle, ellipse, parabola, or hyperbola.

Two bodies revolving under their mutual gravitational attraction, but otherwise undisturbed, describe orbits of the same shape about a common center of mass. The less massive body has the larger orbit. In the solar system, the Sun and Jupiter have a center of mass just outside the visible disk of the Sun. For each of the other planets, the center of mass of Sun and planet lies within the Sun.

For this reason, it is convenient to consider only the relative motion of a planet of mass m about the Sun of mass M as though the planet had no mass and moved about a center of mass $M + m$. The orbit so determined is exactly the same shape as the true orbits of planet and Sun about their common center of mass, but it is enlarged in the ratio $(M + m)/M$. *See* CENTER OF MASS; PLANET.

Orbital velocity v of a planet moving in a relative orbit about the Sun may be expressed by Eq. (1) where a is the semimajor axis, and r is the distance from the planet to the Sun.

$$v^2 = G(M + m)\left(\frac{2}{r} - \frac{1}{a}\right) \qquad (1)$$

In the special case of a circular orbit, $r = a$, and the expression becomes Eq. (2).

$$v^2 = \frac{G(M + m)}{a} \qquad (2)$$

When the eccentricity of an orbit is exactly unity, the length of the major axis becomes infinite and the ellipse degenerates into a parabola. The expression for the velocity then becomes Eq. (3). This parabolic velocity is referred to as the velocity of escape, since it is the minimum velocity required for a particle to escape from the gravitational attraction of its parent body. *See* ESCAPE VELOCITY.

$$v^2 = G(M + m)\left(\frac{2}{r}\right) \qquad (3)$$

Eccentricities greater than unity occur with hyperbolic orbits. Because in a hyperbola the semimajor axis a is negative, hyperbolic velocities are greater than the escape velocity.

Parabolic and hyperbolic velocities seem to be observed in the motions of some comets and meteors. Aside from the periodic ones, most comets appear to be visitors from cosmic distances, as do about two-thirds of the fainter meteors. It is possible that many of the "parabolic" comets are actually moving in elliptical orbits of extremely long period. The close approach of one of these visitors to a massive planet, such as Jupiter, could change the velocity from parabolic to elliptical if retarded, or from parabolic to hyperbolic if accelerated. It is possible that many of the periodic comets, especially those with periods under 9 years, have been captured in this way. *See* COMET; GRAVITATION; PERTURBATION (ASTRONOMY). [R.L.Du.]

Orchid Any member of the orchid family (Orchidaceae), one of the largest families of plants, with 450 genera and perhaps 20,000 species. Orchids have a wide distribution, being most abundant in tropical forests where the majority are epiphytes. In the temperate and arctic regions, the genera are terrestrial. Unusual features of this group are the complex and highly specialized flowers, the minute microscopic seeds having no endosperm, and the great number of seeds in an orchid capsule. The extraordinary beauty of the flowers makes orchids the basis of a multimillion-dollar floral industry. Vanilla is obtained from the pods of *Vanilla planifolia*, and the tubers of some Asiatic species are collected, dried, and marketed as salep, which is used both as a medicine and a food. *See* ORCHIDALES; VANILLA. [P.D.St./E.L.C.]

Orchidales An order of flowering plants, division Magnoliophyta (Angiospermae), in the subclass Liliidae of the class Liliopsida (monocotyledons). The order consists of 4 families: the Orchidaceae, with about 20,000 species; the Burmanniaceae, with about 130 species; the Cordiaceae, with 10 species; and the Geosiridaceae, with 1 species. The Orchidales are mycotrophic, sometimes nongreen Liliidae with very numerous tiny seeds that have an undifferentiated embryo and little or no endosperm. The ovary is always inferior and apparently lacks the septal nectaries found in many Liliatae, although other kinds of nectaries are present. *See* FLOWER; LILIIDAE; LILIOPSIDA; ORCHID. [A.C.]

Ordovician The second-oldest period in the Paleozoic Era. The rocks deposited during this time interval (these are termed the Ordovician System) overlie those of the Cambrian and underlie those of the Silurian. The Ordovician Period was about 60,000,000 to 65,000,000 years long, and lasted from about 495,000,000–500,000,000 to about 435,000,000 years ago.

The type area for the Ordovician System is those parts of Wales and England that include rocks bearing the fossils that composed the second of the three major Lower Paleozoic faunas cited by C. Lapworth. The Ordovician System in Britain was divided to five major units called series, each distinguished by a unique fossil fauna. The Ordovician System may be recog-

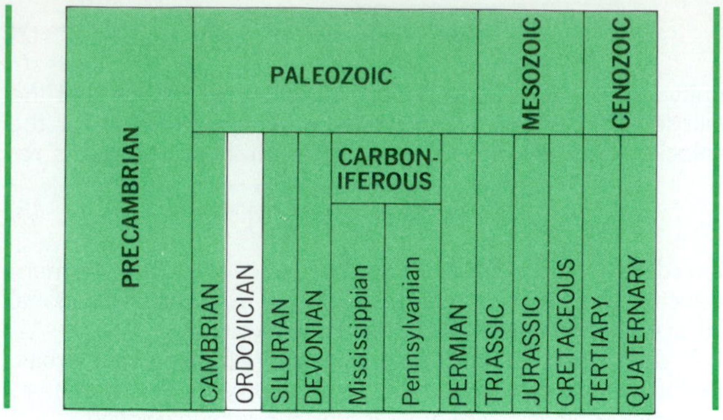

nized in nearly all parts of the world, because the fossils that characterize it are so broadly defined and so widely distributed that some may be found wherever fossiliferous Ordovician rocks occur.

The present-day continental areas of Asia, Australia, North America, and part of Argentina in South America appear to have been mostly under shallow continental shelf seas that were in tropical climates during the Ordovician. Ordovician rocks over much of these areas are typified by considerable thicknesses of carbonate (lime) rocks that accumulated in shallow subtidal and intertidal environments similar to those seen today in the Bahamas, Bermuda, and the Florida Keys.

Life abounded in Ordovician shelf seas and in the surface waters over shelf margins and ocean basins. Those environments in which carbonates accumulated included many different types of organisms that lived on the sea floor. Most of the animals were invertebrates that took foods from suspension in the surrounding sea water. Bottom-dwelling marine animals were closely spaced in many shallow tropical environments, suggesting that nutrients were plentiful in them.

Late Ordovician glacial deposits are found across nearly all of northern Africa. Glacial beds of apparent Late Ordovician age are present in several localities in South America as well. Remanent magnetism and sedimentary rock evidence suggests that a pole, probably the South Pole, was located in northern Africa and that continental-type glaciation occurred about it in the Late Ordovician.

Significant tectonic events took place during the mid-Ordovician. Uplifts took place across most of the areas that had been under shallow shelf seas. Mid-Ordovician uplift in northern Africa appears to have been a precursor to glaciation. Much of mid-Ordovician North America became land, and a karst topography with a relief of as much as 430 ft (130 m) developed on exposed carbonate rocks. A tectonic highland (Taconica) formed along the margin of what is today eastern North America, probably as a result of plate collision. *See* PALEOGEOGRAPHY; PALEOZOIC; PLATE TECTONICS. [W.B.N.B.]

Ore and mineral deposits Ore deposits are naturally occurring geologic bodies that may be worked for one or more metals. The metals may be present as native elements, or, more commonly, as oxides, sulfides, sulfates, silicates, or other compounds. The term ore is often used loosely to include such nonmetallic minerals as fluorite and gypsum. The broader term, mineral deposits, includes, in addition to metalliferous minerals, any other useful minerals or rocks. Minerals of little or no value which occur with ore minerals are called gangue. Some gangue minerals may not be worthless in that they are used as by-products; for instance, limestone for fertilizer or flux, pyrite for making sulfuric acid, and rock for road material. *See* MINERAL.

Mineral deposits that are essentially as originally formed are called primary or hypogene. The term hypogene also indicates formation by upward movement of material. Deposits that have been altered by weathering or other superficial processes are secondary or supergene deposits. Mineral deposits that formed at the same time as the enclosing rock are called syngenetic, and those that were introduced into preexisting rocks are called epigenetic.

The distinction between metallic and nonmetallic deposits is at times an arbitrary one since some substances classified as nonmetals, such as lepidolite, spodumene, beryl, and rhodochrosite, are the source of metals. The principal reasons for distinguishing nonmetallic from metallic deposits are practical ones, and include such economic factors as recovery methods and uses.

The Earth's crust consists of igneous, sedimentary, and metamorphic rocks. The table gives the essential composition of the crust and shows that 10 elements make up more than 99% of the total. Of these, aluminum, iron, and magnesium are industrial metals. The other metals are present in small quantities, mostly in igneous rocks.

Elemental composition of Earth's crust based on igneous and sedimentary rocks*

Element	Weight, %	Atom, %	Volume, %
Oxygen	46.71	60.5	94.24
Silicon	27.69	20.5	0.51
Titanium	0.62	0.3	0.03
Aluminum	8.07	6.2	0.44
Iron	5.05	1.9	0.37
Magnesium	2.08	1.8	0.28
Calcium	3.65	1.9	1.04
Sodium	2.75	2.5	1.21
Potassium	2.58	1.4	1.88
Hydrogen	0.14	3.0	

*From T. F. W. Barth, *Theoretical Petrology*, copyright © 1952 by John Wiley and Sons, Inc. (recalculated from F. W. Clarke and H. S. Washington, 1924).

Most mineral deposits are natural enrichments and concentrations of original material produced by different geologic processes. Economic considerations, such as the amount and concentration of metal, the cost of mining and refining, and the market value of the metal, determine whether an ore is of commercial grade. [A.F.H.]

Ore dressing Treatment of ores to concentrate their valuable constituents (minerals) into products (concentrate) of smaller bulk, and simultaneously to collect the worthless material (gangue) into discardable waste (tailing). The fundamental operations of ore-dressing processes are the breaking apart of the associated constituents of the ore by mechanical means (severance) and the separation of the severed components (beneficiation) into concentrate and tailing, using mechanical or physical methods which do not effect substantial chemical changes.

Comminution is a single- or multistage process whereby ore is reduced from run-of-mine size to that size needed by the beneficiation process. The process is intended to produce individual particles which are either wholly mineral or wholly gangue, that is, to produce liberation. Since the mechanical forces producing fracture are not susceptible to detailed control, a class of particles containing both mineral and gangue (middling particles) are also produced. Comminution is divided into crushing (down to 6- to 14-mesh) and grinding (down to micrometer sizes).

Screening is a method of sizing whereby graded products are produced, the individual particles in each grade being of nearly the same size. In beneficiation, screening is practiced for two reasons: as an integral part of the separation process, for example, in jigging; and to produce a feed of such size and size range as is compatible with the applicability of the separation process. *See* SCREENING.

Beneficiation consists of two fundamental operations: the determination that an individual particle is either a mineral or a gangue particle (selection); and the movement of selected particles via different paths (separation) into the concentrate and tailing products. When middling particles occur, they will either be selected according to their mineral content and then caused to report as concentrate or tailing, or be separated as a third product (middling), which is reground to achieve further liberation. *See* FLOTATION; LEACHING; MECHANICAL SEPARATION TECHNIQUES.

Separation is achieved by subjecting each particle of the mixture to a set of forces which is usually the same irrespective of the nature of the particles excepting for the force based upon the discriminating property. This force may be present for both mineral and gangue particles but differing in magnitude, or it may be present for one type of particle and absent for the other. As a result of this difference, separation is possible, and the particles are collected in the form of concentrate or tailing.

Magnetic separation utilizes the force exerted by a magnetic field upon magnetic materials to counteract partially or wholly the effect of gravity. Thus under the action of these two forces, different paths are produced for the magnetic and nonmagnetic particles. *See* MAGNETIC SEPARATION METHODS. [M.D.Ha.]

Oregano A herb, also known as wild marjoram. The dried leaves of several species of aromatic plants are known as oregano; thus oregano is a common name for a general flavor and aroma rather than the name of a specific plant.

European (*Origanum vulgare*) and Greek (*O. hervacleoticum*) oregano are both in the mint family (Lamiaceae). Mexican oregano is obtained primarily from plants of *Lippia graveolens*. These small aromatic shrubs in the verbena family grow wild in Mexico. Origanum oil used in perfumery is steam-distilled primarily from Spanish oregano, *Thymus capitatus*. *See* LAMIALES.

European oregano can be distinguished by its strong piquant character and tall growth with dark, broad leaves; it is a perennial erect herb 2–3 ft tall (0.6–1 m) with pubescent stems, ovate dark green leaves, and white or purple flowers. Native to southern Europe, southwest Asia, and the Mediterranean countries, European oregano is usually found growing in the dry, rocky, calcareous soils of the mountain regions. Greece, Italy, Spain, Turkey, and the United States are the primary sources of European oregano.

Dried oregano leaves are used as a culinary herb in meat and sausage products, salads, soups, Mexican foods, and barbeque sauces. The essential oil of oregano is used in food products, cosmetics, and liqueurs. *See* MARJORAM; SPICE AND FLAVORING. [S.Kir.]

Organic chemical synthesis The preparation of a desired organic compound by altering the molecular structure of another compound in a rational and predetermined manner. This process may require one or more individual chemical reactions wherein each reaction results in a discrete modification of molecular structure. The reactions in organic synthesis may require the use of catalysts or inorganic reagents and frequently involve the combination of two or more molecules of different substances.

Adaptation of chemical reactions to the stepwise construction of complex compounds is one of the most important facets of organic chemistry from both a theoretical and a practical point of view. An almost unlimited number of different organic compounds are theoretically capable of existence. Although many organic compounds occur in nature, only a small fraction of the theoretically possible compounds have been isolated from natural sources. Consequently, desired organic compounds that do not occur naturally or cannot be obtained economically from natural sources must be synthesized. The synthesis of organic compounds forms the basis of the plastics and synthetic fiber industries. Many valuable dyes, flavors, fragrances, vitamins, and drugs are also prepared by synthesis. Organic synthesis is also used to confirm the structures of organic compounds that have been postulated on the basis of evidence derived from chemical behavior and physical measurements. *See* CONDENSATION REACTION; ORGANIC CHEMISTRY; ORGANIC REACTION MECHANISM; POLYMERIZATION; STEREOCHEMISTRY; STERIC EFFECT (CHEMISTRY); SUBSTITUTION REACTION. [R.A.Kr.]

Organic chemistry The chemistry of the compounds of carbon. Organic chemistry owes its peculiar and important position to the fact that carbon, almost alone among the elements, is capable of uniting with itself indefinitely to form compounds. Other elements, notably boron, display similar tendencies, but carbon forms a far greater number of compounds. Secondly, carbon, almost without exception, displays a constant valence of 4. On these two principles the science of organic chemistry is built. The number of carbon compounds that are theoretically capable of existence is staggering, and this very fact poses problems of major magnitude in connection with nomenclature, molecular structure, and arrangement in space of the atoms comprising organic molecules. *See* CARBON.

Organic compounds in general differ from inorganic compounds by the nature of the bonds by which the component atoms of a molecule are united. The valence bonds of most inorganic compounds are of the ionic or electrovalent type. In contrast, covalent or electron pair bonds occur in some inorganic compounds, such as ammonia, and in almost all carbon compounds. Covalent bonds can be altered to give them some ionic character. In general, bonds range from those that are essentially covalent to those that are entirely ionic. There is no sharp boundary between ionic and covalent bonds. However, covalent bonds do share certain characteristics which cause striking differences between the physical and chemical properties of inorganic compounds, such as salts, and those of typical organic compounds. In general, organic compounds possess relatively low melting points, can frequently be distilled, are sparingly soluble or insoluble in water but soluble in nonaqueous solvents, and with a few exceptions do not conduct electricity.

Over the years certain conventions for expressing molecular structures of organic compounds have been accepted. Standard practice represents a shared electron bond involving a single pair of electrons (a single bond) by a single line. Double bonds, involving two pairs of shared electrons, are represented by a double line, and triple bonds, involving three pairs of shared electrons, are represented by a triple line. *See* CHEMICAL BONDING; CONJUGATION (CHEMISTRY); RESONANCE (MOLECULAR STRUCTURE).

Further simplification is often achieved by omission of bond lines when the meaning is obvious, as illustrated below with the hydrocarbon propene.

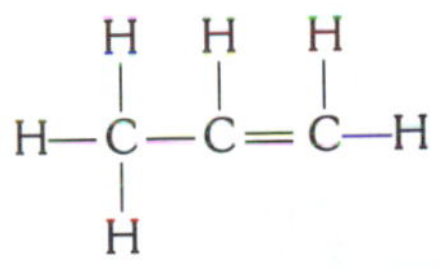

$H_3C—CH{=}CH_2$

$H_3CCH{=}CH_2$

An organic compound containing only single bonds between carbon atoms is said to be saturated; a compound containing one or more multiple bonds between carbon atoms is said to be unsaturated.

The ability of carbon to combine with itself leads to the existence of series of compounds, the formulas of which differ by a constant increment. Such a series, known as a homologous series, is illustrated as follows by the alkanes, each member of

CH_4 — Methane

CH_3CH_3 or C_2H_6 — Ethane

$CH_3CH_2CH_3$ or C_3H_8 — Propane

which differs from the next lower member by the increment CH_2. *See* CHEMICAL STRUCTURES.

Classification of organic compounds. Because of the great number and variety of organic compounds (well over 1,000,000 are known), some systematic classification scheme and systematic method of nomenclature for dealing with them becomes mandatory. The problem is complex, and no completely satisfactory system has yet been devised.

Carbon atoms may combine in the form of long chains, either straight or containing branches, or they may combine to form rings. Further, since carbon is also capable of covalent bonding with other atoms, incorporation of such so-called hetero atoms into carbon rings is possible. This situation furnishes the basis for a broad classification of organic compounds into three main groups depending upon the arrangement of the carbon atoms and the presence or absence of atoms other than carbon in the cyclic compounds. Under these terms of reference, organic compounds may be classified as acyclic compounds, which contain no ring structural arrangements of the constituent atoms; carbocyclic compounds, which contain one or more rings consisting solely of carbon atoms; and heterocyclic compounds, the rings of which contain one or more atoms other than carbon. These principles are illustrated by the following:

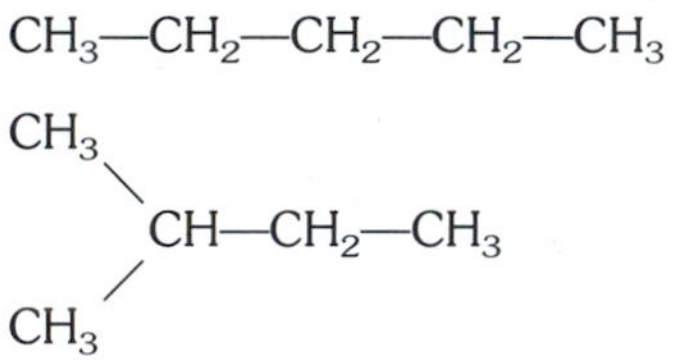

Acyclic

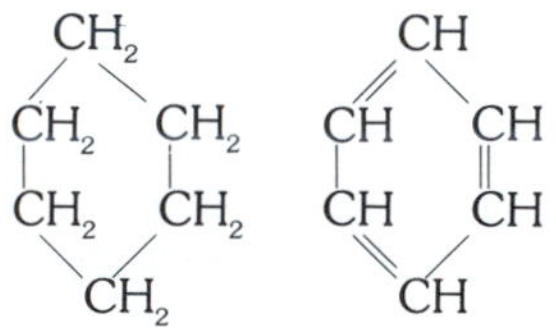

Carbocyclic

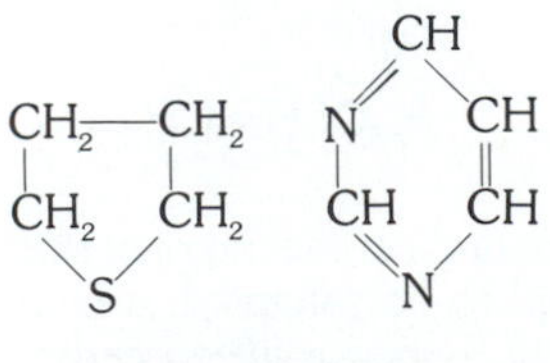

Heterocyclic

Compounds which contain only carbon and hydrogen are known as hydrocarbons. In the hydrocarbons, one or more of the hydrogen atoms, at least in principle, may be replaced by any other atom or group of atoms capable of entering into a covalent bond. If a group of atoms replaces a hydrogen atom of a given hydrocarbon it is considered a functional group since its presence changes the chemical properties of the substituted compound from those of the parent substance (usually the saturated hydrocarbon). The common functional groups are:

Name	*Structure*
Halo (chloro, bromo, etc.)	—Cl,—Br
Hydroxyl	—OH
Aldehyde	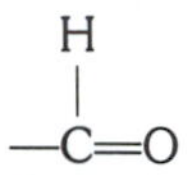
Carboxyl	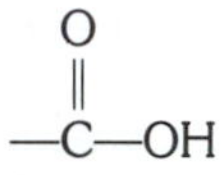
Ketone	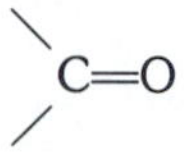
Ether	—O—
Amino	$—NH_2$
Cyano	—C≡N
Thiol or mercapto	—SH
Sulfonic acid	$—SO_3H$

Nomenclature of organic compounds. When the molecular structure of a compound becomes known, assignment of a logical systematic name becomes possible, at least in principle. After many attempts to develop a rational system of nomenclature, the matter was finally placed in the hands of a committee of the International Union of Pure and Applied Chemistry. From the efforts of this committee, an international system, the IUPAC system, is emerging, by which organic compounds can be named in a logical manner. However, a certain amount of nationalism persists and practice does not uniformly conform to the IUPAC rules.

The system of nomenclature for acyclic hydrocarbon derivatives, carbocyclic compounds, and heterocyclic compounds is outlined below.

Acyclic hydrocarbons. The acyclic hydrocarbons are subdivided on the basis of the presence or absence of double or triple bonds. Acyclic compounds containing no multiple bonds are known as alkanes or paraffins; those containing one or more double bonds are known as alkenes or olefins; and those containing one or more triple bonds are known as alkynes or acetylenes. *See* ALIPHATIC HYDROCARBON; ALKANE; ALKENE; ALKYNE.

Acyclic hydrocarbon derivatives. In general, the same principles which govern the naming of branched-chain alkanes, alkenes, and alkynes hold. The parent compound is that which contains the longest continuous straight carbon chain.

For halogen compounds, the prefixes fluoro, chloro, bromo, and iodo are used with the name of the parent alkane:

$CH_3CHClCH_2CH_3$ is 2-chlorobutane

$CH_3CHBrCH_2CH_2CH_2Br$ is 1,4-dibromopentane

Hydroxyl derivatives are commonly known as alcohols, and many of them are named as such. CH_3OH is methyl alcohol, and CH_3CH_2OH is ethyl alcohol.

Aldehydes, particularly the lower members of the series, are frequently named by replacing the terminal -ic of the related acid by -aldehyde. Ketones are named by three methods. The lower symmetrical ketones are named by replacing the terminal -ic of the acid which would yield them on pyrolysis by -one. For naming mixed ketones, the word ketone is prefixed by the names of the radicals joined to the carbonyl carbon atom. In the Geneva system the terminal -e of the parent alkane is replaced by -one. The carbonyl group position must be shown by a number. *See* ALDEHYDE; KETONE.

The carboxyl function is characteristic of organic acids. Since many acids were originally isolated from natural products, trivial names associated with their sources are common. For example, HCOOH, formic acid, is named from the Latin *formica* (ant). Branched-chain acids are named frequently as derivatives of acetic acid, CH_3COOH, or as derivatives of the parent straight-chain acid.

Ethers are generally named by reference to the alkyl radicals present and addition of the word ether; $CH_3CH_2—O—CH_3$ is ethyl methyl ether. By the Geneva system they are named as alkoxy derivatives of the parent hydrocarbon. *See* ETHER.

Amines are alkyl derivatives of ammonia. They are classified into primary, secondary, or tertiary amines according to whether 1, 2, or 3 hydrogen atoms of ammonia have been replaced by organic radicals. They are conveniently named by using the names of the radicals attached to nitrogen as prefixes to the word amine. *See* AMINE.

Carbocyclic compounds. The carbocyclic compounds are subdivided into the aromatic compounds and the alicyclic compounds according to certain structural features and chemical properties. *See* ALICYCLIC HYDROCARBON; AROMATIC HYDROCARBON.

The aromatic carbocyclic compounds are characterized structurally by an alternating system of single and double bonds. The parent aromatic hydrocarbons are almost always known by trivial names (such as benzene or phenol) many of which indicate the source from which they were originally obtained. For convenience, it is not customary to write the individual carbon and hydrogen atoms in the structural formulas of aromatic hydrocarbons. Rather, each angular position is assumed to represent a carbon atom, and the presence of sufficient hydrogen atoms to bring the carbon atoms to the quadrivalent state is also assumed. Typical examples are as follows:

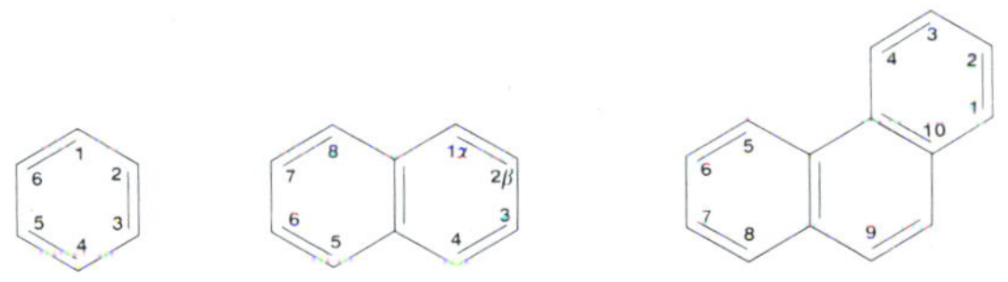

Benzene, C_6H_6 — Naphthalene, $C_{10}H_8$ — Phenanthrene, $C_{14}H_{10}$

The problem of nomenclature of the alicyclic compounds is formidable and is complicated by the occurrence of structural, positional, and geometrical isomerism, as well as stereoisomerism. Definitive rules for naming and numbering many of the polycyclic compounds have been formulated by the International Union of Pure and Applied Chemistry. *See* ALICYCLIC HYDROCARBON; MOLECULAR ISOMERISM.

Heterocyclic compounds. One or more carbon atoms (with or without accompanying hydrogens) in a carbocyclic compound may be replaced by a hetero atom (sometimes in combination with hydrogen). A systematic method which can be applied to the nomenclature of any heterocyclic compound has been developed. In this the related aromatic hydrocarbons are taken as reference standards, the nature of the hetero atom(s) is indicated by a prefix (oxa- for O, thia- for S, and aza- for N), and the position of the hetero atom or atoms is indicated by a number corresponding to the accepted numbering of the aromatic hydrocarbons. Thus

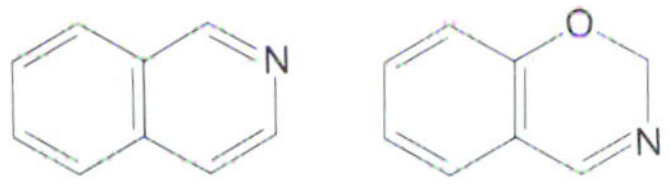

Trivial name:	Isoquinoline	None
Systematic name:	2-Azanaphthalene	1-Oxa-3-aza-2*H*-naphthalene

In compounds carrying more than one hetero atom, numbers are selected so that oxygen bears the lowest number, followed by sulfur and, finally, by nitrogen. The most highly unsaturated compound is taken as the parent. Derivatives in which one or more double bonds are saturated are referred to as dihydro or tetrahydro derivatives of the parent. *See* HETEROCYCLIC COMPOUNDS. [R.C.E.]

Organic conductor An organic substance with low electrical resistance. Two major classes of organic conductors are charge-transfer compounds and conducting polymers.

Charge-transfer compounds. The search for organic conductors in the early 1970s led to the observation of metallic-like electrical conduction in well-ordered molecular crystals and the discovery of many new phenomena such as the stabilization of charge-density waves and spin-density waves, new mechanisms for electronic transport, organic superconductivity, and new states of matter produced under strong magnetic fields. Most of these phenomena are due to the low dimensionality (one or two dimensions) of the electron gas in the charge-transfer compounds where they were observed. Among these properties, superconductivity has created much interest since the zero-resistance state is now observed at temperatures as high as 10 K (−442°F) in certain organic superconductors. *See* SUPERCONDUCTIVITY.

Charge-transfer compounds are two-component materials containing anionic and cationic species originating by charge transfer between donor and acceptor entities; these may be two organic molecules or an organic molecule with an inorganic ion. Tetrathiafulvalene-tetracyanoquinodimethane (TTF-TCNQ) is the prototype of charge-transfer organic crystals. Its crystal structure exhibits piled-up segregated columns of donor TTF and acceptor TCNQ molecules (illus. *a*). In the solid state, the amount of charge transferred from donor to acceptor is determined by the overall crystal stability.

Another class of organic conductors is exemplified by radical cation salts such as $(TMTSF)_2X$ where the organic molecule is tetramethyltetraselenafulvalene and X is an inorganic anion. In this class of materials (Bechgaard salts) the molecules display a zigzag packing along the stacking axis (illus. *b*), where one positive charge (hole) is shared between two organic molecules.

The strong overlap between electron clouds of neighboring

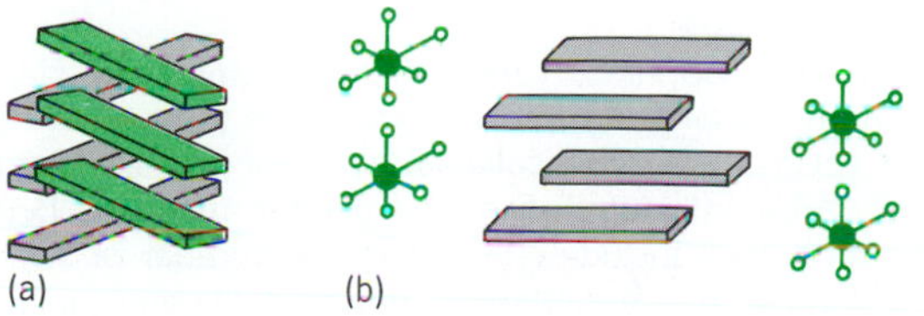

Stacking of charge-transfer compounds. (*a*) Segregated stacking of TTF-TCNQ-like materials. (*b*) Typical zigzag stacking of the $(TMTSF)_2X$ series.

molecules along the stacks spreads the partially filled molecular electronic states into an energy band 0.5–1 eV wide. This bandwidth is large enough to allow electron delocalization among all molecules on a given stack and to promote electrical conduction similar to that in metal crystals. *See* Band theory of solids; Conduction (electricity); Delocalization; Molecular orbital theory.

Conducting polymers. Polymeric materials are typically considered as insulators. However, research since the late 1970s has led to the discovery of polymeric materials with extremely high conductivity, approaching that of copper. The prospect of materials combining the properties of plastics and metals or semiconductors has led to a search for applications, made attractive because improved polymers no longer suffer from such drawbacks as low stability, processing difficulties, and brittleness. Most conducting polymers can be switched reversibly between conductive and nonconductive states, with the result that their conductivities can span an enormous range. This switching is accomplished through oxidation-reduction (redox) chemistry, the conductivity being sensitive to the degree of oxidation of the polymer backbone. This property distinguishes conducting polymers from metals and semiconductors and is the basis of many existing and potential applications. In addition, certain polymers become conducting upon oxidation or reduction and thus can exhibit *p*- or *n*-type conduction. *See* Oxidation-reduction; Polymer; Semiconductor.

In a conducting polymer an oxidant removes electrons from the π-electron system of the polymer, creating radical cations that, at high concentrations, dimerize to form cation pairs known as bipolarons. Charge-balancing counterions are concomitantly incorporated between polymer chains. The overall process is referred to as doping, and the counteranion (or countercation in the case of reduction) is the dopant.

Considerable progress has been made on the theory of important parameters such as oxidation potentials, band gaps, and band widths, often with good agreement with experiment. Such work is important for the design of new conductive polymers with specific properties. *See* Solid-state chemistry; Solid-state physics. [G.E.W.]

Organic evolution The modification of living organisms during their descent, generation by generation, from common ancestors. Organic, or biological, evolution is to be distinguished from other phenomena to which the term evolution is often applied, such as chemical evolution, cultural evolution, or the origin of life from nonliving matter. Organic evolution includes two major processes: anagenesis, the alteration of the genetic properties of a single lineage over time; and cladogenesis, or branching, whereby a single lineage splits into two or more distinct lineages that continue to change anagenetically.

Anagenesis consists of change in the genetic basis of the features of the organisms that constitute a single species. Populations in different geographic localities are considered members of the same species if they can exchange members at some rate and hence interbreed with each other, but unless the level of interchange (gene flow) is very high, some degree of genetic difference among different populations is likely to develop. The changes that transpire in a single population may be spread to other populations of the species by gene flow. *See* Species concept.

Almost every population harbors several different alleles at each of a great many of the gene loci; hence many characteristics of a species are genetically variable. All genetic variations ultimately arise by mutation of the genetic material. Broadly defined, mutations include changes in the number or structure of the chromosomes and changes in individual genes, including substitutions of individual nucleotide pairs, insertion and deletion of nucleotides, and duplication of genes. Many such mutations alter the properties of the gene products (ribonucleic acid and proteins) or the timing or tissue localization of gene action, and consequently affect various aspects of the phenotype (that is, the morphological and physiological characteristics of an organism). Whether and how a mutation is phenotypically expressed often depends on developmental (epigenetic) events. *See* Gene; Genetic code; Mutation; Ribonucleic acid (RNA).

Natural selection is a consistent difference in the average rate at which genetically different entities leave descendants to subsequent generations; such a difference arises from differences in fitness (that is, in the rate of survival, reproduction, or both). In fact, a good approximate measure of the strength of natural selection is the difference between two such entities in their rate of increase. The entities referred to are usually different alleles at a locus, or phenotypically different classes of individuals in the population that differ in genotype. Thus selection may occur at the level of the gene, as in the phenomenon of meiotic drive, whereby one allele predominates among the gametes produced by a heterozygote. Selection at the level of the individual organism, the more usual case, entails a difference in the survival and reproductive success of phenotypes that may differ at one locus or at more than one locus. As a consequence of the difference in fitness, the proportion of one or the other allele increases in subsequent generations. The relative fitness of different genotypes usually depends on environmental conditions.

Different alleles of a gene that provides an important function do not necessarily differ in their effect on survival and reproduction; such alleles are said to be neutral. The proportion of two neutral alleles in a population fluctuates randomly from generation to generation by chance, because not all individuals in the population have the same number of surviving offspring. Random fluctuations of this kind are termed random genetic drift. If different alleles do indeed differ in their effects on fitness, both genetic drift and natural selection operate simultaneously. The deterministic force of natural selection drives allele frequencies toward an equilibrium, while the stochastic (random) force of genetic drift brings them away from that equilibrium. The outcome for any given population depends on the relative strength of natural selection (the magnitude of differences in fitness) and of genetic drift (which depends on population size).

The great diversity of organisms has come about because individual lineages (species) branch into separate species, which continue to diverge. This splitting process, speciation, occurs when genetic differences develop between two populations that prevent them from interbreeding and forming a common gene pool. The genetically based characteristics that cause such reproductive isolation are usually termed isolating mechanisms. Reproductive isolation seems to develop usually as a fortuitous by-product of genetic divergence that occurs for other reasons (either by natural selection or by genetic drift). *See* Speciation.

A frequent consequence of natural selection is that a species comes to be dominated by individuals whose features equip them better for the environment or way of life of the species. Such features are termed adaptations. Although many features of organisms are adaptive, not all are, and it is a serious error to suppose that species are capable of attaining ideal states of adaptation. Some characteristics are likely to have developed by genetic drift rather than natural selection, and so are not adaptations; others are side effects of adaptive features, which exist because of pleiotropy or developmental correlations.

Higher taxa are those above the species level, such as genera and families. A taxon such as a genus is typically a group of species, derived from a common ancestor, that share one or more features so distinctive that they merit recognition as a separate taxon. The degree of difference necessary for such recognition, however, is entirely arbitrary: there are often no sharp limits between related genera, families, or other higher taxa, and very often the diagnostic character exists in graded steps among a group of species that may be arbitrarily divided into different higher taxa. Moreover, a character that in some groups is used to distinguish higher taxa sometimes varies among closely related species or even within species. In addition, the fossil record of many groups shows that a trait that takes on very different forms in two living taxa has developed by intermediate steps along divergent lines from their common ancestor; thus the inner ear bones of mammals may be traced to jaw elements in reptiles that in turn are homologous to gill arch elements in Paleozoic fishes.

The characteristics of a species evolve individually or in concert with certain other traits that are developmentally or functionally correlated. Because of this mosaic pattern of evolution, it is meaningful to speak of the rate of evolution of characters, but not of species or lineages as total entities. Thus in some lineages, such as the so-called living fossils, many aspects of morphology have evolved slowly since the groups first came into existence, but evolution of their deoxyribonucleic acid and amino acid sequences has proceeded at much the same rate as in other lineages. Every species, including the living fossils, is a mixture of traits that have changed little since the species' remote ancestors, and traits that have undergone some evolutionary change in the recent past. The history of life is not one of progress in any one direction, but of adaptive radiation on a grand scale: the descendants of any one lineage diverge as they adapt to different resources, habitats, or ways of life, acquiring their own specialized features as they do so. There is no evidence that evolution has any goal, nor does the mechanistic theory of evolutionary processes admit of any way in which genetic change can have a goal or be directed toward the future. However, for life taken as a whole, the only clearly discernible trend is toward ever-increasing diversity. [D.J.Fu.]

Organic geochemistry The study of the composition and processes of origin, movement, and alteration of biologically derived carbonaceous substances in rocks, waters, and the atmosphere. These may be highly dispersed organic materials in fine-grained rocks (including petroleum source rocks and oil shales); continuous phases (petroleum) occupying pore networks in reservoir rocks; or solid layers of mainly carbonaceous material, for example, coals. See COAL; OIL SHALE; PETROLEUM.

Another branch of organic geochemistry deals with the prebiologic or abiogenic formation of organic substances. For example, inorganic gases can be converted to essentially organic molecules when subjected to various forms of energy, particularly radiation. See GEOCHEMISTRY. [R.D.McI.]

Organic nomenclature A system by which a unique and unambiguous name or other designation is assigned to a given organic molecular structure. A set of rules adopted by the International Union of Pure and Applied Chemistry (IUPAC) is the basis for a standardized name for any organic compound. Common or nonsystematic names are used for many compounds that have been known for a long time. The latter names have the advantage of being short and easily recognized, as in the examples below. However, in contrast to systematic names, common or trivial names do not convey information from which the structure can be written by reference to prescribed rules.

In the IUPAC system, a name is formed by combination of a parent alkyl chain or ring with prefixes and suffixes to denote substituents. For aliphatic compounds, the parent name is a stem that denotes the longest straight chain in the structure with the ending -ane. The first four members of the alkane series are methane, ethane, propane, and butane. From five carbons on, the names follow the Greek numerical roots, for example, pentane and hexane. Changing the ending -ane to -yl gives the name of the corresponding radical or group; for example, CH_3CH_2—is the ethyl radical or ethyl group. Branches on an alkyl chain are indicated by the name of a radical or group as a prefix. The location of a branch or other substituent is indicated by number; the chain is numbered from whichever end results in the lowest numbering. See ALKANE.

Double or triple bonds are indicated by the endings -ene or -yne, respectively. The configuration of the chain at a double bond is denoted by *E*- when two similar groups are on opposite sides of the plane bisecting the bond, and *Z*- when they are on the same side. See ALKENE; ALKYNE.

A compound containing a functional group is named by adding to the parent name a suffix characteristic of the group. If there are two or more groups, a principal group is designated by suffix and the other(s) by prefix.

The same general principles apply to cyclic compounds. For alicyclic rings, the prefix cyclo- is followed by a stem indicating the number of carbon atoms in the ring, as is illustrated in the structure below. In bicyclic compounds, the total number of

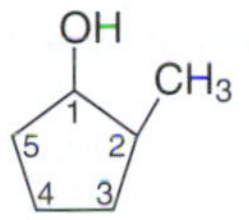

2-Methylcyclopentanol

carbons in the ring system is prefixed by bicyclo- and numbers in brackets which indicate the number of atoms in each connecting chain. See ALICYCLIC HYDROCARBON.

The names of aromatic hydrocarbons have the ending -ene, which denotes a ring system with the maximum number of noncumulative double bonds. Each of the simpler polycyclic hydrocarbons has a different parent name. To number the positions in a polycyclic aromatic ring system, the structure must first be oriented with the maximum number of rings arranged horizontally and to the right. The system is then numbered in clockwise sequence starting with the atom in the most counterclockwise position of the upper right-hand ring, omitting atoms that are part of a ring fusion. See AROMATIC HYDROCARBON.

Basis of nomenclature of heterocyclic compounds

Heteroatom	Prefix	Ring size	Suffix
O	ox(a)-	4	-ete
S	thi(a)-	5	-ole
N	az(a)-	6	-ine
		7	-epine

Systematic names for rings containing a heteroatom (O, S, N) are based on a combination of a prefix denoting the heteroatom(s) and a suffix denoting the ring size, as indicated in the table. See CHEMICAL STRUCTURES; CHEMICAL SYMBOLS AND FORMULAS; HETEROCYCLIC COMPOUNDS; ORGANIC CHEMISTRY. [J.A.Mo.]

Organic photochemistry A branch of chemistry that deals with light-induced changes of organic material. Because it studies the interaction of electromagnetic radiation and matter, photochemistry is concerned with both chemistry and physics. In considering photochemical processes, therefore, it is also necessary to consider physical phenomena that do not involve strictly chemical changes, for example, absorption and emission of light, and electronic energy transfer.

Because several natural photochemical processes were known to play important roles (namely, photosynthesis in plants, process of vision, and phototropism), the study of organic photochemistry started very early in the twentieth century. However, the breakthrough occurred only after 1950 with the availability of commercial ultraviolet radiation sources and modern analytical instruments for nuclear magnetic resonance (NMR) spectroscopy and (gas) chromatography.

In general, most organic compounds consist of rapidly interconvertible, nonseparable conformers, because of the free rotation about single bonds. Each conformer has a certain energy associated with it, and its own electronic absorption spectrum. This is one cause of the broad absorption bands produced by organic compounds in solution. The equilibrium of the conformers may be influenced by the solvent and the temperature. According to the Franck-Condon principle, which states that promotion of an electron by the absorption of a photon is much faster than a single vibration, each conformer will have its own excited-state configuration. *See* CONFORMATIONAL ANALYSIS; FRANCK-CONDON PRINCIPLE; MOLECULAR STRUCTURE AND SPECTRA.

Because of the change in the pi-bond order upon excitation of the substrate and the short lifetime of the first excited state, the excited conformers are not in equilibrium and each yields its own specific photoproduct. Though different conformers may lead to the same photoproduct and one excited conformer may lead to several photoproducts, a change in solvent, temperature, or wavelength of excitation influences the photoproduct composition. This is especially true with small molecules; larger molecules with aromatic groups are less sensitive for small wavelength differences.

The influence of wavelength is also important when the primary photoproduct also absorbs light and then gives rise to another photoreaction. Excitation with selected wavelengths or monochromatic light by use of light filters or a monochromator, respectively, may then be profitable for the selective production of the primary product. Similarly, irradiation at a low temperature is helpful in detecting a primary photoproduct that is unstable when heated (thermolabile). SEE COLOR FILTER.

[W.H.L.]

Organic reaction mechanism A detailed description of an overall organic process, R → P, in which the participants and their interconversion rates are characterized. All of the steps from reactants R via possible intermediates I to products P are included, as in reactions (1). To characterize the reaction

$$R \underset{-1}{\overset{1}{\rightleftharpoons}} I_1 \qquad I_1 \underset{-2}{\overset{2}{\rightleftharpoons}} I_2 \qquad I_2 \underset{-3}{\overset{3}{\rightleftharpoons}} P \qquad (1)$$

path, it is possible to use a variety of experimental techniques and invoke theoretical and quasitheoretical exclusions.

From an energy viewpoint, a mechanism may be represented schematically in a free energy (G) and reaction coordinate ($Q^{\ddagger}$) diagram (Fig. 1). Energy minima are associated with observable species (R, P, I), while energy maxima are associated with fleeting transients or activated complexes (AC). Here all of the participants as well as barrier heights, for example, $\Delta G_1^{\ddagger} = G_1^{\ddagger} - G(R)$ for step 1 or $\Delta G_{-2}^{\ddagger} = G_2^{\ddagger} - G(I_2)$ of Eqs. (1), are indicated and $Q^{\ddagger}$ relates in a general way to the progress of a multistep reaction. Unfortunately, this view of a mechanism is somewhat akin to reconstructing a race from snapshots of the horses at the start and finish lines.

In a more comprehensive picture, chemical changes among species may be regarded as taking place on a hypersurface representing the changes in potential energy V as a function of atom configurations or a related set of orthogonal coordinates

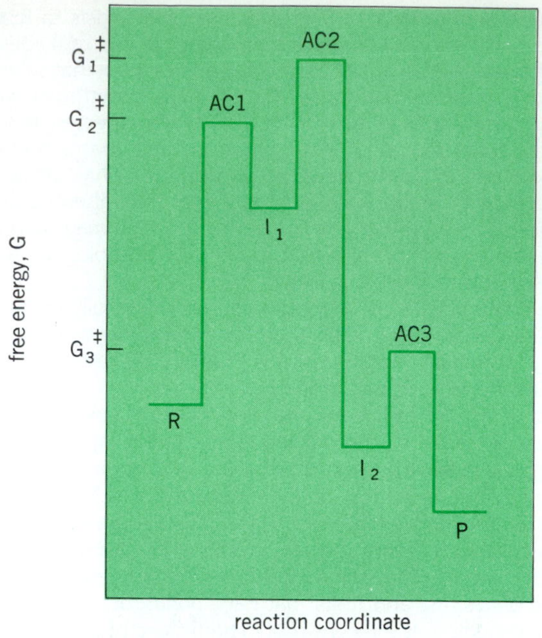

Fig. 1. Progress of a general organic reaction, Eq. (1), in terms of free energy and reaction coordinate.

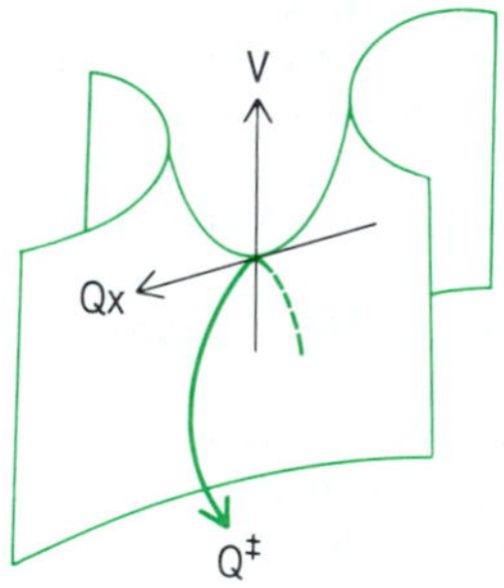

Fig. 2. Potential energy *V* in the AC region.

(Q_i). Figure 2 is a portion of this surface which focuses on the change $I_1 \rightleftharpoons I_2$. Q_x stands for all of the coordinates except $Q^{\ddagger}$; the activated complex is identified with a saddle point or one dimensional maximum along $Q^{\ddagger}$. While there is an energy barrier to overcome, the energy is minimized along $Q^{\ddagger}$.

Individual steps by which changes occur are termed elementary. In such a reaction, one or two—rarely more—molecules become the energized aggregate labeled activated complex. To illustrate, reactions (2) and (3) show the activated complexes with partial bonds.

Unimolecular

$$HCCl_3 \rightarrow (HCl_2C\ .\ .\ .\ Cl) \rightarrow HCl_2C\cdot + Cl\cdot \qquad (2)$$

Bimolecular

$$HO^- + H_3Cl \rightarrow (HO\ .\ .\ .\ CH_3\ .\ .\ .\ I)^- \rightarrow HOCH_3 + I^- \qquad (3)$$

Special techniques are employed to determine the elementary step or steps involved in a reaction. At the outset it is necessary to decide what the process is, and determine that it does in fact take place. *See* CHEMICAL DYNAMICS. [S.I.M.]

Organoactinides Organometallic compounds of the actinides—elements 90 and beyond in the periodic table. Both the large sizes of actinide ions and the presence of 5*f* valence orbitals are unique features which differ distinctly from most, if not all, other metal ions.

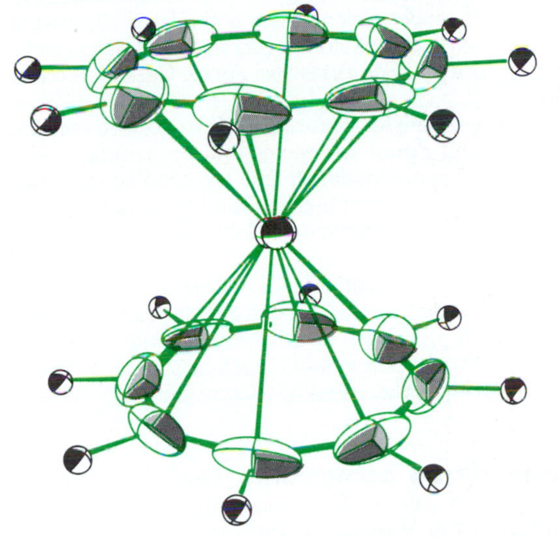

Molecular structure of $U(C_8H_8)_2$ determined by single-crystal x-ray diffraction. (*After K. O. Hodgson and K. N. Raymond, Inorg. Chem., 12:458, 1973*)

Organometallic compounds have been prepared for all actinides through curium (element 96), although most investigations have been conducted with readily available and more easily handled natural isotopes of thorium (Th) and uranium (U). Organic groups (ligands) which bind to actinide ions include both π- and σ-bonding functionalities. The importance of this type of compound reflects the ubiquitous character of metal-carbon two-electron sigma bonds in both synthesis and catalysis. *See* CATALYSIS; CHEMICAL STRUCTURES.

The molecular structures of a number of organoactinides have been determined by single-crystal x-ray and neutron-diffraction techniques. In almost all cases the large size of the metal ion gives rise to unusually high (as compared to a transition-metal compound) coordination numbers. That is, a greater number of ligands or ligands with greater spatial requirements can be accommodated within the actinide coordination sphere. The sandwich complex bis(cyclooctatetraenyl)-uranium (uranocene), an example of this latter type, is shown in the illustration. *See* ACTINIDE ELEMENTS; COORDINATION CHEMISTRY; COORDINATION NUMBER; METALLOCENES; NEUTRON DIFFRACTION; ORGANOMETALLIC COMPOUND; X-RAY DIFFRACTION. [T.J.M.]

Organoarsenic compound A derivative of arsenic containing at least one organic radical attached through carbon to the arsenic atom by means of a covalent bond. All organoarsenic derivatives can be derived schematically by formal substitution or oxidation from the primary, secondary, and tertiary arsines. This is shown in the accompanying display. Depending on the state of oxidation, organoarsenic compounds are derivatives of trivalent or pentavalent arsenic. A special type of organoarsenic compounds which do not have an equivalent in the organophosphorus series are the arseno compounds, RAs═AsR.

All organoarsenic compounds exhibit physiological activity. The high mammalian toxicity of the trivalent derivatives makes a number of them potential chemical warfare agents. Several of the less toxic derivatives were used as chemotherapeutics for

$RAsH_2$ Alkylarsine	R_2AsH Dialkylarsine	R_3As Trialkylarsine
↓	↓	↓
$RAsO$ Alkyl-arsenoxide	$(R_2As)_2O$ Dialkyl-arsinoxide	R_3AsO Trialkyl-arsinoxide
↓	↓	
$RAs(OH)_2$ Alkylarsonous acid	R_2AsOH Dialkylarsinous acid	
↓	↓	
$RAs(O)(OH)_2$ Alkylarsonic acid	$R_2As(O)OH$ Dialkylarsinic acid	

combating certain infectious diseases before the discovery of the antibiotics. *See* INSECTICIDE; ORGANOPHOSPHORUS COMPOUND. [F.W.H./T.C.S.]

Organometallic compound One of a group of substances containing carbon-metal bonds. This definition can be broadened to include all compounds of elements that possess metallike properties and that are combined chemically with one or more carbon atoms. In this respect, metal alkoxides, chelates, salts of organic acids, and related compounds are not discussed in this article. *See* CHELATION.

Organometallic compounds have found many commercial applications. The use of tetraethyllead as a gasoline antiknock additive is a good example. Organometallic compounds have played an important role in the development of silicone polymers, polyvinylchloride, and polyethylene. *See* POLYOLEFIN RESINS; TETRAETHYLLEAD.

A systematic classification of all known organometallic compounds is very difficult. For the sake of convenience, however, they can be divided into three main classes. The more electropositive metals of groups 1 and 2 of the periodic table form organometallic compounds that are nonvolatile, usually poorly soluble in organic solvents, and essentially ionic in nature. Lithium, beryllium, and magnesium compounds are much less ionic than those of the remaining metals, however. The metals (excepting transition metals) and metalloids of groups 13, 14, 15, and 16 form organometallic compounds which are mainly volatile, soluble in organic solvents, and principally covalently bonded. The transition metals constitute a third main group, possessing bonding to aromatic or aromatic-type groups which is mainly of a special *d*-orbital, or "sandwich," type. *See* CHEMICAL BONDING; FERROCENE; METALLOCENES; ORGANOARSENIC COMPOUND; ORGANOPHOSPHORUS COMPOUND.

The reaction of metals or metalloids with organic halides is widely used in the synthesis of organometallic compounds. Another means of preparing organometallic compounds involves the reaction of highly reactive metals with hydrocarbons that contain active hydrogen atoms. The displacement of a metal in an organometallic compound by a more reactive metal to produce a new organometallic compound has been widely used, although reactions of this type are usually reversible.

The reaction of organometallic compounds with metal salts has found extensive application in the synthesis of many types of organometallic compounds. Examples include the reactions of alkylsodium compounds, alkyllithium compounds, and Grignard reagents with halides of elements of groups 13, 14, 15, and 16. Other reactions include methods by which an organometallic compound of a given metal is used to prepare other organometallic derivatives of the same metal. *See* GRIGNARD REACTION. [M.D.R.]

Organophosphorus compound One of a series of derivatives of phosphorus that have at least one organic (alkyl or aryl) group attached to the phosphorus atom linked either directly to a carbon atom or indirectly by means of another element (for example, oxygen). The mono-, di-, and trialkylphosphines (and their aryl counterparts) can be regarded formally as the parent compounds of all organophosphorus compounds.

Considering the large number of organic groups that may be joined to phosphorus as well as the incorporation of other elements in these materials, the number of combinations is practically unlimited. A vast family in itself is composed of the heterocyclic phosphorus molecules, in which phosphorus is one of a group of atoms in a ring system.

Some organophosphorus compounds have been used as polymerization catalysts, lubricant additives, flameproofing agents, plant growth regulators, and insecticides. Organophosphorus compounds were made during World War II for use as chemical warfare agents in the form of nerve gases (Sarin, Trilon 46, Soman, and Tabun). *See* PHOSPHORUS. [S.E.Cr.]

Organoselenium compound One of a group of compounds that contain both selenium (Se) and carbon (C) and frequently other elements as well, for example, halogen, oxygen (O), sulfur (S), or nitrogen (N). Organoselenium compounds have become common in organic chemistry laboratories, where they have numerous applications, particularly in the area of organic synthesis. Organoselenium compounds formally resemble their sulfur analogs and may be classified similarly. For instance, selenols, selenides, and selenoxides are clearly related to thiols, sulfides, and sulfoxides. Despite structural similarities, however, sulfur and selenium compounds are often strikingly different with respect to their stability, properties, and ease of formation. *See* ORGANIC CHEMICAL SYNTHESIS; ORGANOSULFUR COMPOUND; SELENIUM.

Selenols have the general formula RSeH, where R represents either an aryl or alkyl group. They are prepared by the alkylation of hydrogen selenide (H_2Se) or selenide salts, as well as by the reaction of Grignard reagents or organolithium compounds with selenium, as in reactions (1).

$$H_2Se\ (\text{or } HSe^-) \xrightarrow{RX}$$

$$\underset{\text{Selenol}}{RSeH} \xleftarrow[2.\ H_3O^+]{1.\ Se} \underset{\text{Grignard reagent}}{RMgX} \quad \text{or} \quad \underset{\text{Organolithium compound}}{RLi} \qquad (1)$$

Selenols are stronger acids than thiols. They and their conjugate bases are powerful nucleophiles that react with alkyl halides (R′X; R′ = alkyl group) or similar electrophiles to produce selenides (RSeR′); further alkylation yields selenonium salts, as shown in reactions (2), where R″ represents an alkyl

$$\underset{\text{Selenol}}{RSeH} \xrightarrow{R'X} \underset{\text{Selenide}}{RSeR'} \xrightarrow{R''X} \underset{\text{Selenonium salt}}{R{-}\overset{\overset{\displaystyle R''}{|}}{Se^+}{-}R'} + X^- \qquad (2)$$

group that may be different from the alkyl group R′. Both acyclic and cyclic selenides are known. *See* ALKYLATION; ELECTROPHILIC AND NUCLEOPHILIC REAGENTS; GRIGNARD REACTION; REACTIVE INTERMEDIATES.

Diselenides (RSeSeR) are usually produced by the aerial oxidation of selenols; they react in turn with chlorine or bromine, yielding selenenyl halides or selenium trihalides. Diselenides are easily oxidized to seleninic acids or anhydrides by reagents such as hydrogen peroxide or nitric acid.

Selenoxides are readily obtained from the oxidation of selenides with hydrogen peroxide (H_2O_2) or similar oxidants. Selenoxides undergo facile elimination to produce olefins and selenenic acids, a process known as syn-elimination.

Selenocarbonyl compounds tend to be considerably less stable than their thiocarbonyl or carbonyl counterparts because of the weaker double bond between the carbon and selenium atoms. Selenoamides, selenoesters and related compounds can be isolated, but selenoketones (selones) and selenoaldehydes are highly unstable. *See* ALDEHYDE; AMIDE; CHEMICAL BONDING; ESTER; KETONE.

The charge-transfer complexes formed between certain organoselenium donor molecules and appropriate acceptors such as tetracyanoquinodimethane are capable of conducting electric current. Selenium-containing polymers are also of interest as organic conductors. *See* COORDINATION COMPLEXES; ORGANIC CONDUCTOR.

Selenium is an essential trace element, and its complete absence in the diet is severely detrimental to human and animal health. The element is incorporated into selenoproteins such as glutathione peroxidase, which acts as a natural antioxidant. *See* BIOINORGANIC CHEMISTRY; COORDINATION CHEMISTRY; PROTEIN. [T.G.B.]

Organosilicon compound One of a group of compounds in which silicon (Si) is bonded to an organic functional group (R) either directly or indirectly via another atom. Formally, all organosilanes can be viewed as derivatives of silane (SiH_4) by the substitution of hydrogen (H) atoms. The most common substituents are methyl (CH_3; Me) and phenyl (C_6H_5; Ph) groups. However, tremendous diversity results with cyclic structures and the introduction of heteroatoms. *See* SILICON.

Organosilicon compounds are not found in nature and must be prepared in the laboratory. The ultimate starting material is sand (silicon dioxide, SiO_2) or other inorganic silicates, which make up over 75% of the Earth's crust. The useful properties of silicone polymers were identified in the 1940s; widespread interest in organosilicon chemistry followed.

The chemistry of organosilanes can be explained in terms of the fundamental electronic structure of silicon and the polar nature of its bonds to other elements. Silicon appears in the third row of the periodic table, immediately below carbon (C) in group IV, and has many similarities with carbon. However, it is the fundamental differences between carbon and silicon that make silicon so useful in organic synthesis and of such great theoretical interest. *See* CARBON.

In the majority of organosilicon compounds, silicon follows the octet rule and is 4-coordinate. This trend can be explained by the electronic configuration of atomic silicon ($3s^23p^23d^0$) and the formation of sp^3-hybrid orbitals for bonding. Unlike carbon ($2s^22p^2$), however, silicon is capable of expanding its octet and can form 5- and 6-coordinate species with electronegative substituents, such as the 6-coordinate octahedral dianion, $[Me_2SiF_4]^{2-}$. *See* COORDINATION CHEMISTRY; ELECTRON CONFIGURATION; VALENCE.

Silicon is a relatively electropositive element that forms polar covalent bonds ($Si^{\delta+}{-}X^{\delta-}$) with carbon and other elements, including the halogens, nitrogen, and oxygen. The strength and reactivity of silicon bonds depend on the relative electronegativities of the two elements. For example, the strongly electronegative elements fluorine (F) and oxygen (O) form bonds that have tremendous thermodynamic stabilities. *See* CHEMICAL THERMODYNAMICS; ELECTRONEGATIVITY.

Before 1981, multiply bonded silicon compounds were identified only as transient species both in solution and in the gas phase. However, when tetramesityldisilene ($Ar_2Si{=}SiAr_2$, where Ar = 2,4,6-trimethylphenyl) was isolated, it was found to be a crystalline, high-melting-point solid having excellent stability in the absence of oxygen and moisture.

Numerous reactive species have been generated and characterized in organosilicon chemistry. Silylenes are divalent silicon species (SiR_2, where R = alkyl, aryl, or hydrogen), analogous to carbenes in carbon chemistry. The dimerization of two silylene units gives a disilene. Silicon-based anions, cations, and radicals are important intermediates in the reactions of silicon. *See* CHEMICAL BONDING; FREE RADICAL; REACTIVE INTERMEDIATES.

The direct process for the large-scale preparation of organosilanes has provided a convenient source of raw materials for the development of the silicone industry. The process produces a mixture of chloromethylsilanes from elemental silicon and methyl chloride in the presence of a copper (Cu) catalyst, as in the reaction below.

$$\text{Me—Cl} + \text{Si} \xrightarrow{\text{Cu catalyst}} \text{Cl}_2\text{SiMe}_2 + \text{other Cl}_n\text{SiMe}_{4-n}$$

In spite of concentrated research efforts in this area, the synthetic methods available for the controlled formation of silicon-carbon bonds remain limited to only a few general reaction types. These include the reaction of organometallic reagents with silanes, catalytic hydrosilylation of multiple bonds, and reductive silylation.

The role of silicon in organic synthesis is quite extensive, and chemists exploit the unique reactivity of organosilanes to accomplish a wide variety of transformations. Silicon is usually introduced into a molecule to perform a specific function and is then removed under controlled conditions.

Polysilanes are organosilicon compounds that contain either cyclic arrays or linear chains of silicon atoms. The isolation and characterization of both cyclosilanes (with up to 35 silane units) and high-molecular-weight linear polysilanes (with up to 3000 silane units) have demonstrated that silicon is capable of extended chain formation (catenation). Polysilanes contain only silicon-silicon bonds in their backbone, which differentiates them from polysiloxanes (silicones), which contain alternating silicon and oxygen repeat units. *See* SILICONE RESINS.

Polysilanes have found applications as photoresists in microlithography, charge carriers in electrophotography, and photoinitiators for vinyl polymerization. Polysilanes also function as preceramic polymers for the manufacture of silicon carbide fibers. *See* ORGANIC CHEMICAL SYNTHESIS. [H.Y.]

Organosulfur compound One of a group of substances which contain both carbon and sulfur. Oxygen, nitrogen, the halogens, and phosphorus are also often present. Thousands of such compounds are well known. *See* SULFUR.

Many medicinal and natural products are organosulfur compounds, for example, the penicillins, the sulfa drugs, insulin, and amino acids such as cysteine and methionine. The utilizations of organosulfur compounds involve enormous commercial developments in medicinals, detergents, sulfide rubbers and polymers, sulfur dyes, solvents, and agricultural chemicals (herbicides, fungicides, and insecticides). Besides their practical uses, studies of organosulfur compounds helped to establish fundamental theory on the nature of valence bonding, of molecular geometry, and of reaction mechanisms of organic chemistry.

The most common classes of organosulfur compounds are: mercaptans, RSH; thiophenols, ArSH; disulfides, RSSR′, where R may be an aliphatic, aromatic, or heterocyclic radical; sulfides, including cyclic sulfides, RSR; sulfenyl chlorides, RSCl; sulfoxides; sulfones; sulfonyl chlorides, RSO_2Cl; sulfonamides, $R—SO_2NH_2$; thioaldehydes and thioketones, RC(H)═S and $R_2C═S$; sulfinic acids, RSO_2H; sulfonic acids, RSO_3H; esters of sulfonic acids, RSO_2OR'; thiol esters and thio acids. *See* DIMETHYL SULFOXIDE; HETEROCYCLIC COMPOUNDS; MERCAPTAN; PETROLEUM PROCESSING; SULFONAMIDE; SULFONE; SULFONIC ACID; THIO COMPOUNDS; THIOETHER. [N.K.]

Oriental vegetables Oriental vegetables are very important in Asian countries, but are considered as minor crops in the United States and Europe. However, in recent years there has been an increased interest in these crops because of their unusual flavors and textures and in some cases their high nutritional values. Some of the more common ones are described below.

Chinese cabbage, celery cabbage, napa, or pe-tsai (*Brassica campestris*, pekinensis group; *B. rapa*, pekinensis group; *B. pekinensis*) belongs to the mustard (Cruciferae) family, and is a biennial leafy plant but is grown as an annual. The harvested part is a head which is composed of broad crinkled leaves with a very wide, indistinct, white midrib. The outer leaves are pale green, and the inner leaves of the head are blanched. It is a good salad vegetable because of the mild flavor and crisp texture.

There are many varieties of pak choy, bok choi, Chinese mustard, or celery mustard (*Brassica campestris*, chinensis group; *B. rapa*, chinensis group; *B. chinensis*). The crop is grown as an annual. Pak choy does not form a head like most varieties of Chinese cabbage, but forms a celerylike stalk of tall, dark green leaves, with prominent white veins and long white petioles (see illustration). The base of the petiole may be expanded and spoon-shaped; the blade of the leaf is smooth, and not crinkled like that of Chinese cabbage.

The large, long, white radish (*Raphanus sativus*, longipinnatus group) is often called oriental winter radish, daikon, Chinese winter radish, lobok, and lob paak. Radish is a dicotyledonous herbaceous plant grown for its long, enlarged roots.

Edible podded peas, China peas, sugar peas, or snow peas (*Pisum sativum*, macrocarpon group) belong to the Leguminosae family. The primitive forms of pea are slightly bitter and have a tough seed coat, which allows for long dormant periods. In the garden pea the pod swells first as the seeds enlarge, but in the edible podded pea the immature seeds bulge the pod, which demarks the developing seeds. Unlike the regular garden

Pak choy or Chinese mustard (*Brassica campestris*). (*University of California Agricultural Experiment Station*)

peas which have tough fibery seed pods, the edible podded peas were selected for the tender pods and not for the seeds.

Yard-long bean or asparagus bean (*Vigna sinensis*, sesquipedelis group) belongs to the Leguminosae family. It is an annual climbing plant. It is a relative of the cow pea (*V. unguiculata*).

Mung bean or green gram (*Phaseolus aureus*) sprouts have been used by the Chinese for their remarkable healing qualities for thousands of years. Only recently has the Western world recognized the value of sprouted leguminous seeds (mung, soy, and alfalfa sprouts). The unsprouted mung bean seeds contain negligible amounts of vitamin C (ascorbic acid), whereas the sprouted seeds contain 20 mg per 100 g of sprouts, which is as high as tomato juice.

Jicama (*Pachyrrizus erosus*), also called the yam bean, is indigenous to Mexico and Central America. It belongs to the pea (Leguminosae) family. The crop is grown for its enlarged turnip-shaped root, which is eaten raw or cooked, has a crisp texture, and is sweetish in taste.

Chinese winter melon, wax gourd, winter gourd, white gourd, ash gourd, Chinese preserving melon, ash pumpkin, or tung kwa (*Benincasa hispida*; *B. cerifera*) is a viny annual cucurbit. The flesh of mature fruits is used in making Chinese soups. It can be eaten raw or made into sweet preserves similar to citron or watermelon rind preserves.

Balsam pear, alligator pear, bitter gourd, bitter melon, bitter cucumber, or fu kwa (*Momordica charantia*) is an annual herbaceous vine. The heart-shaped to cylindrical fruits are extremely bitter; some of the bitterness is removed before cooking by peeling and steeping in salt water. Immature fruits are less bitter than mature ones.

Chinese okra or angled loofa (*Luffa acutangula*) is called zit kwa by the Chinese; it is a close relative of *L. cylindrica*, known as loofa, sponge gourd, or dishcloth gourd. Chinese okra is an annual climbing vine grown for the immature fruits. The thoroughly mature fruits can be made into vegetable sponge.

Water spinach, water convolvulus, swamp cabbage, kang kong, or shui ung tsoi (*Ipomoea aquatica*) is a perennial semiaquatic plant grown for its long, tender shoots. It is a relative of the sweet potato (*I. batatas*), but does not produce an enlarged root. [M.Y.]

Orion The Warrior, in astronomy, and undoubtedly the finest of all constellations in the sky. Orion is a winter group near the celestial equator. Four of the most prominent stars, α, γ, β, and κ, form a huge crude rectangle (see illustration). The group is pictured as the figure of a warrior, holding a shield with his left hand and swinging a club with his raised right arm ready to strike the charging Bull. Betelgeuse (meaning armpit), one of the largest stars known, is the red star at the right shoulder, Bellatrix is at the left shoulder, and Rigel, the blue-white star, is at the left leg. Three bright stars in a straight line in the middle of the rectangle represent the warrior's belt. The center star is Alnilam. *See* CONSTELLATION; ORION NEBULA; TAURUS. [C.-S.Y.]

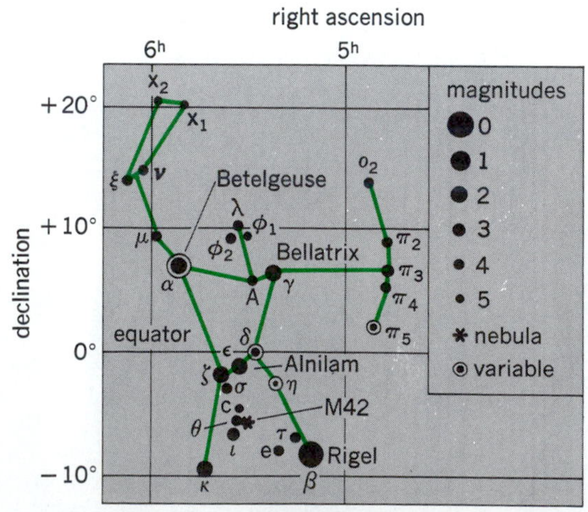

Line pattern of the constellation Orion. M42 designates the Orion Nebula. The grid lines represent the coordinates of the sky. The apparent brightness, or magnitude, of the stars is shown by the size of the dots, graded by appropriate numbers.

Orion Nebula A portion of a vast complex of ionized hydrogen, cool molecular clouds, and solid grains which is found in the club (or sword) of the constellation of Orion, and which shines because of radiation received from hot stars embedded within it (see illustration). Designated Messier 42 or NCG 1976, it is the brightest and perhaps one of the densest ionized hydrogen (HII) regions known. *See* ORION.

Orion Nebula, NGC 1976.

Radio-frequency and infrared observations have shown the Orion Nebula to be extremely intricate, with dense cool clouds in which numerous, often complex molecules abound. The hot Trapezium stars, which excite the nebula, are very young and only recently formed from the interstellar medium. This and various other lines of evidence show the luminous Orion Nebula to be very young, less than a million years old, and possibly only a few tens of thousands of years of age. Radio observations indicate that the Orion Nebula and the entire complex of hot stars, gas, and grains are surrounded by an expanding shell of neutral hydrogen. *See* INTERSTELLAR MATTER; NEBULA. [L.H.Al.]

Ornithischia One of two orders of extinct reptiles popularly known as dinosaurs, the other being the Saurischia. The two orders are distinguished by various anatomical differences, but the shape of the pelvis is the primary distinction, tetraradiate or superficially "birdlike" in the Ornithischia and triradiate in the Saurischia. The ordinal names refer to these distinctive conditions: "bird-hipped" and "reptile-hipped." Like the saurischians, ornithischians probably originated from the Thecodontia in Early or Middle Triassic times. Ornithischian remains are known from rock strata ranging from Middle Triassic to latest Cretaceous in age and have been found on all continents except Antarctica. On the basis of tooth shape, all ornithischians are judged to have been herbivores. Most kinds had peculiar beaks or bills, featuring a unique predentary bone at the front of the lower jaw, for plucking vegetation. Many possessed highly specialized batteries of teeth for crushing, grinding, or slicing plant tissues. *See* SAURISCHIA; THECODONTIA.

Traditional classifications recognize four suborders: Ornithopoda, Stegosauria, Ankylosauria, and Ceratopsia. A fifth suborder, Pachycephalosauria, was suggested as a result of new discoveries in Mongolia. *See* DINOSAUR. [J.H.O.]

Ornithopter An aircraft whose wings beat like those of a bird, bat, or insect. Models are recorded as early as 400 B.C. (Archytas). Since the mid-19th century some have been flown successfully; a number of machines of person-carrying size have been flown for short distances. *See* FLIGHT.

The beating surfaces may be nonlifting (that is, propelling only), in which case they are properly called fins, or they may be wings which lift as well as propel. [J.L.G.F.P.]

Orogeny The process by which mountainous tracts such as the Alpine-Himalayan, Appalachian, and Cordilleran orogenic belts are formed, also known as orogenesis. Characteristically, orogenic belts are long and linear or arcuate, with distinctive zones of sedimentary, deformational, and thermal patterns that are, in general, parallel but asymmetric to the belts. Orogenic belts have complex internal geometry, involving extensive mass transport of very dissimilar rock sequences of dominantly marine sediments and continental crust (sial) and oceanic crust and mantle (ophiolite suite). Intensive deformation and metamorphism of a particular orogenic belt is relatively short-lived when compared to the time during which much of the sedimentary rock of the belt was deposited. Typically, orogenic belts are sites of abnormally thick accumulations of sedimentary and volcanic rock and severe deformation and thermal alterations. *See* MOUNTAIN SYSTEMS.

Seismological and oceanographic studies have revealed a globe-encircling, seismically active belt of mountains, called oceanic ridges, and trenches with associated volcanic island arcs. Models propose that orogeny is a consequence of the evolution of oceanic ridges and trenches and continental drift. Although oceanic ridges and trench–island arc systems are mountainous, they are not orogenic belts in the classical sense which restricts orogenic belts to continental masses. Orogeny results from interactions of this global, continuously evolving system of oceanic ridges and trenches, according to concepts of sea-floor spreading or, currently, lithosphere plate tectonics.

Fundamental concepts of orogeny based on plate tectonics are that continental margins of the Atlantic type, which develop during ocean opening, are converted to island arc–cordilleran-type (Andean) orogenic belts as lithosphere is consumed in subduction zones, below trenches, along the continental margin during closing of an ocean. A collision-type (Himalayan) orogenic belt is superimposed on a cordilleran belt upon final ocean closing, as opposing continental margins suture. *See* CORDILLERAN BELT; OCEANIC ISLANDS; PLATE TECTONICS.

The plate tectonics model indicates that mountain building results from two fundamental mechanisms. Island arc–cordilleran orogeny, for the most part thermally driven, occurs along leading lithosphere plate edges, above subduction zones. Continent–island arc or continent-continent collision orogeny is for the most part mechanically driven and occurs subsequently to island arc development or during final closing of an ocean or both. Actual orogenic belts are usually the result of complex combinations of these basic mechanisms. [J.M.B.; J.F.D.]

Orpiment A mineral having composition As_2S_3. Crystals are small, tabular, and rarely distinct; the mineral occurs more commonly in foliated or columnar masses. The hardness is 1.5–2 (Mohs scale) and the specific gravity is 3.49. The luster is resinous and pearly on the cleavage surface; the color is lemon yellow. Orpiment is found in Romania, Peru, Japan, and Russia. In the United States it occurs at Mercer, Utah; Manhattan, Nevada; and in deposits from geyser waters in Yellowstone National Park. *See* ARSENIC. [C.S.Hu.]

Orthida An order of articulate brachiopods which includes the oldest known representatives of the class. They first appeared in Early Cambrian times and became especially prolific and diverse in the Ordovician. Subsequently, the group diminished in relative abundance during Silurian and Late Paleozoic time, finally becoming extinct during the Permian. The majority of the order belong to the suborder Orthidina. Characteristically, these are biconvex, finely ribbed shells with a straight hinge line and well developed interareas on both valves. *See* ARTICULATA (BRACHIOPODA); BRACHIOPODA. [A.J.R.]

Orthoclase A name for potassium feldspar, $KAlSi_3O_8$, that usually contains some sodium feldspar (up to about 50 mole % $NaAlSi_3O_8$) either in solid solution or exsolved as relatively pure $NaAlSi_3O_8$. The latter can be present as albite or analbite or in structural states intermediate between albite and analbite. *See* ALBITE; FELDSPAR. [F.H.L.]

Orthoester A trialkyl derivative of a nonexistent orthoacid, $RC(OH)_3$, with general formula $RC(OR)_3$. The structure of the orthocarbonates is $C(OR)_4$. Orthoesters are liquids with ethereal odors and are stable to aqueous alkalies but sensitive to acid hydrolysis. By far the most common and most important is ethyl orthoformate, $HC(OC_2H_5)_3$, although ethyl orthoacetate and higher members are known. Orthoformate esters are frequently used in synthetic organic chemistry, for example, in preparing aldehydes from Grignard reagents. *See* ESTER; GRIGNARD REACTION. [E.B.R.]

Orthogonal polynomials A special case of orthogonal functions that arise in many physical problems (often as the solutions of differential equations), in the study of distribution functions, and in certain other situations where one approximates fairly general functions by polynomials. *See* PROBABILITY.

Each set of orthogonal polynomials is defined with respect to a particular averaging procedure. The average value of a suitable function f is denoted by $E\{f\}$. An example is shown in Eq. (1). In general an averaging procedure has the form shown in Eq. (2), a Stieltjes integral, where σ is a distribution function,

$$E\{f\} = \frac{1}{2}\int_{-1}^{1} f(x)\,dx \qquad (1)$$

$$E\{f\} = \int_{-\infty}^{\infty} f(x)\,d\sigma(x) \qquad (2)$$

that is, an increasing function with $\sigma(-\infty) = 0$ and $\sigma(+\infty) = 1$.

Two functions f and g are said to be orthogonal with respect to a given averaging procedure if $E\{f\bar{g}\} = 0$ where the bar denotes complex conjugation. By the system of orthogonal polynomials associated with the averaging procedure is meant a sequence $P_0, P_1, P_2 \ldots$ of polynomials P_n having exact degree n, which are mutually orthogonal, that is, $E\{P_m\bar{P}_n\} = 0$ for $m \neq n$. This last condition is equivalent to the statement that each P_n is orthogonal to all polynomials of degree less than n. Thus P_n has the form $P_n(x) = a_0 + a_1x + a_2x^2 + \cdots + a_nx^n$ where $a_n \neq 0$ and is subject to the n conditions $E\{x^k P_n\} = 0$ for $k =$

0, 1,..., $n-1$. This gives n linear equations in the $n+1$ coefficients of P_n, leaving one more condition, called a normalization, to be imposed. The method of normalization differs in different references. *See* POLYNOMIAL SYSTEMS OF EQUATIONS. [C.S.He.]

Orthonectida An order of Mesozoa. The orthonectids parasitize various marine invertebrates as multinucleate plasmodia. The plasmodia multiply by fragmentation. Eventually they give rise asexually, by polyembryony, to sexual males and females. Commonly only one sex arises from a given plasmodium.

These sexually mature forms escape as minute ciliated organisms. Structurally they are composed of a single layer of ciliated epithelial cells surrounding an inner mass of sex cells. The ciliated cells are disposed in rings around the body.

After insemination the eggs develop in the female and form ciliated larvaea. When liberated, these larvae invade new individuals of their host and then disaggregate, liberating germinal cells which give rise to new plasmodia. *See* MESOZOA. [B.H.McC.]

Orthoptera A heterogeneous order of generalized insects with gradual metamorphosis, chewing mouthparts, and four wings. The front pair of wings is thickened and leathery, while the second pair is thin, membranous, and folded fanwise beneath the first. Many forms have reduced wings or none at all. Most Orthoptera are plant eaters. Some live in marshy areas, but they are not aquatic in the true sense.

Orthoptera are among the noisiest of insects, possessing both well-developed sound-producing and sound-perceiving organs. The Locustidae produce sound by rubbing a series of tiny pegs on the inner faces of the femora against the outer wings. The auditory organs are on either side of the base of the abdomen. In most of the remaining Orthoptera sound is produced by rapidly rubbing the wings together, and the auditory organs are located at the bases of the front tibiae.

Habitats of the Orthoptera are variable. Representatives of most families are found on the surface of the ground. However, the Grylloblattidae occur under stones or in loose soil, the Tridactilidae in sand or mud, the Mantidae in grasses or herbaceous cover. Members of these families are not of great economic importance.

The number of families included in this order varies because the species differ so in shape and form. Some of the families are described below.

The walkingsticks and leaf insects are examples of the family Phasmidaea. Many are remarkable mimics of dead twigs, being long, slender, wingless, and brownish in color (illustration *a*). Occasionally they become so abundant that they defoliate trees.

The family Blattidae includes the cockroaches and are flattened insects with a large, scalelike pronotum which overhangs the head. Some cockroaches are wingless, but most are winged. A few species live in human dwellings and are among the most abundant and persistent of household pests. *See* COCKROACH.

The family Mantidae includes the praying mantids, characterized by a long slender prothorax bearing a pair of large grasping legs, and a freely moving head with large eyes (illustration *b*). Mantids are predacious, clutching the prey with the large forelegs and bringing it to the mouth.

The true grasshoppers and the migratory locusts are members of the family Locustidae. The antennae are usually less than half the length of the body. The hindlegs are adapted for jumping (illustration *c*). To this group belong the majority of crop-destroying Orthoptera.

Members of the family Tetrigidae are tiny grasshoppers, called grouse locusts or pygmy grasshoppers, which superficially resemble the Locustidae. The front wings, however, are reduced to small scalelike structures, and the pronotum extends backward over the whole body.

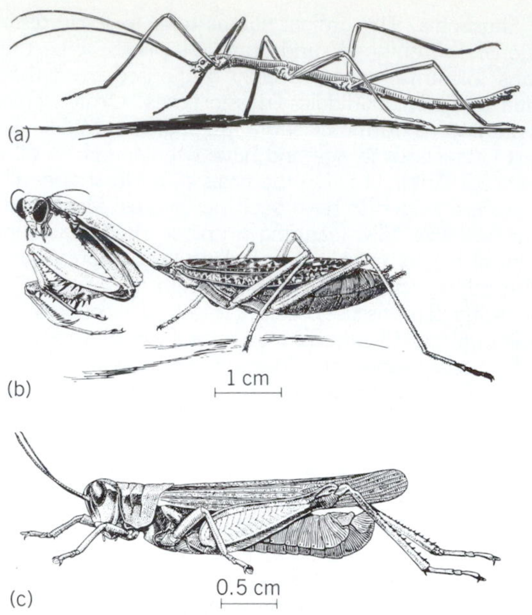

Orthoptera. (*a*) Walkingstick (*Diapheroma femorata*). (*b*) Praying mantid (*Stagomantis carolina*). (c) Grasshopper (*Melanoplus mexicanus*). (*Illinois Natural History Survey*)

The family Tettigoniidae consists of the longhorn or green grasshoppers. The antennae are usually as long as, or longer than, the body, and most of the common species are green. The hindlegs are fitted for jumping. Most of the Tettigoniidae are active at night. To this group belong the katydids, meadow grasshoppers, cave or camel crickets, and the Mormon cricket.

The true crickets belong to the family Gryllidae. These are usually rather chunky, dark-colored insects with long antennae.

The mole crickets, insects highly specialized for fossorial existence, are in the family Gryllotalpidae. They burrow just beneath the soil, pushing up long ridges as do moles. *See* INSECTA. [H.B.M.]

Orthoquartzite A rock (also known as quartzose sandstone) composed almost entirely (over 95%) of detrital quartz grains.

The minor accessory minerals present in orthoquartzite are only the most stable ones, zircon and tourmaline. The texture of orthoquartzites tends to be distinctive. Typically, the grains are extremely well rounded and tend to have high sphericity as well. The particles of orthoquartzites are bound together by precipitated mineral cements. The most abundant mineral cement is quartz. In some orthoquartzites the silica cement is chert or opal. Not all orthoquartzites are well cemented; many are friable and almost unindurated. Carbonate minerals are also important as cementing agents in orthoquartzites.

Typically, orthoquartzites are associated with fossiliferous limestone and calcareous shale beds. Orthoquartzites are most abundant in thin sedimentary sections, primarily in the interior of continents, but may also be found as the early deposits in some geosynclines. *See* QUARTZ; SEDIMENTARY ROCKS. [R.Si.]

Orthorhombic pyroxene A group of minerals having the general chemical formula $XYSi_2O_6$, in which the Y site contains Fe or Mg and the X site contains Fe, Mg, Mn, or a small amount of Ca (up to about 3%). The end members of this solid solution series are enstatite ($Mg_2Si_2O_6$) and ferrosilite ($Fe_2Si_2O_6$). Names used for intermediate members of the series are enstatite, bronzite, hypersthene, and orthoferrosilite.

Many of the physical and optical properties of orthopyroxene are strongly dependent upon composition, and especially upon the Fe-Mg ratio. In hand specimens, orthopyroxene can be distinguished from amphibole by its characteristic 88° cleavage angles, and from augite by color—augite is typically green to black, while orthopyroxene is more commonly brown, especially on slightly weathered surfaces.

Orthopyroxene is a widespread mineral in metamorphic rocks. It is characteristic of granulite facies metamorphism, in both mafic and feldspathic gneisses. Orthopyroxene occurs in many basalts and gabbros, particularly those of tholeiitic composition, and many meteorites, but is notably absent from most alkaline igneous rocks. The greatest abundance of orthopyroxene is in ultramafic rocks, especially those in large layered intrusions. *See* PYROXENE. [R.J.Tr.]

Orthotrichales An order of the true mosses (subclass Bryidae) consisting of five families and 23 genera. The plants grow in mats or tufts in relatively exposed places, on trunks of trees and on rock. They are rather freely branched and prostrate, or may be sparsely forked and erect-ascending; the habit is more or less pleurocarpous, but sporophytes may be produced at the ends of leading shoots or branches. The leaves are generally oblong and broadly pointed, with a strong midrib. The capsules are often ribbed, and the peristome, if present, has a poor development of endostome. The calyptrae are often hairy. *See* BRYIDAE; BRYOPHYTA; BRYOPSIDA. [H.Cr.]

Oscillation Any effect that varies in a back-and-forth or reciprocating manner. Examples of oscillation include the variations of pressure in a sound wave and the fluctuations in a mathematical function whose value repeatedly alternates above and below some mean value.

The term oscillation is for most purposes synonymous with vibration, although the latter sometimes implies primarily a mechanical motion. The alternating current and the associated electric and magnetic fields are referred to as electric (or electromagnetic) oscillations.

If a system is set into oscillation by some initial disturbance and then left alone, the effect is called a free oscillation. A forced oscillation is one in which the oscillation is in response to a steadily applied periodic disturbance.

Any oscillation that continually decreases in amplitude, usually because the oscillating system is sending out energy, is spoken of as a damped oscillation. An oscillation that maintains a steady amplitude, usually because of an outside source of energy, is undamped. *See* ANHARMONIC OSCILLATOR; DAMPING; FORCED OSCILLATION; HARMONIC OSCILLATOR; MECHANICAL VIBRATION; OSCILLATOR; VIBRATION. [J.M.Ke.]

Oscillator An electronic circuit that converts energy from a direct-current source into a periodically varying electrical output. If the output voltage is a sine-wave function of time, the generator is called a sinusoidal, or harmonic, oscillator. Only sinusoidal oscillators are discussed in this article. If the output waveform contains abrupt changes in voltage, such as occur in a pulse or square wave, the device is called a relaxation oscillator. *See* RELAXATION OSCILLATOR; WAVE- SHAPING CIRCUITS.

The fundamental laws governing sinusoidal oscillators are the same for all oscillator circuits. These basic concepts are illustrated in Fig. 1. The amplifier provides an output voltage v_o as a consequence of an external input signal voltage v_s. The voltage v_o is applied to a circuit called a feedback network whose output is v_f. If the feedback voltage v_f were made identically equal to the input voltage v_s, and if the external input were disconnected and the feedback voltage connected to the amplifier input terminals 1 and 2, the amplifier would continue to provide the same output voltage v_o as before. This requires that the instantaneous values of v_f and v_s be exactly equal at all times. Since no restriction was made on the waveform, it need not be sinusoidal.

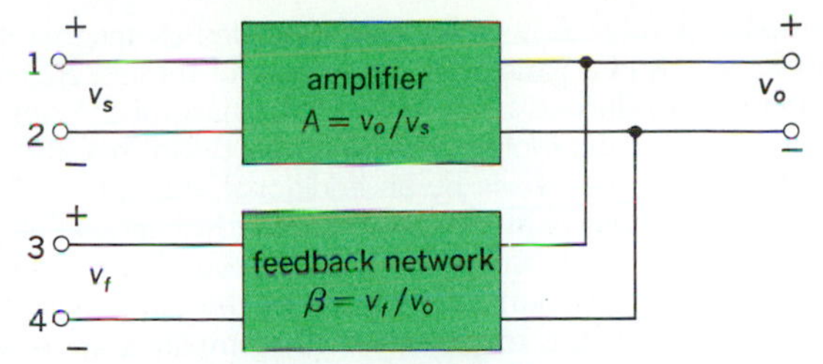

Fig. 1. An amplifier and feedback network which is not yet connected to form a closed loop.

If the entire circuit operates linearly and the amplifier or feedback network or both contain reactive elements, the only periodic wave that will preserve its form is the sinusoidal waveform, and such a circuit will be a sinusoidal oscillator. For sinusoidal oscillators the condition where v_s equals v_f requires that amplitude, phase, and frequency of v_s and v_f be identical. The phase shift introduced in a signal while being transmitted through a reactive network is invariably a function of the frequency, and there is usually only one frequency at which v_f and v_s are in phase. Therefore, a sinusoidal oscillator operates at the frequency for which the total phase shift of the amplifier and feedback network is precisely zero (or an integral multiple of 2π). The frequency of a sinusoidal oscillator, provided the circuit oscillates at all, is therefore determined by the condition that the loop phase shift is zero.

Another condition, which must clearly be met if the oscillator is to function, is that the magnitude of v_s and v_f must be identical. If the amplifier has a voltage amplification, or gain, A, then v_o equals Av_s. The fraction of the voltage v_o applied to the input through the feedback network is called the feedback factor β. Therefore, the equations below are obtained. If v_f is to

$$v_f = \beta v_o \qquad v_f = \beta A v_s$$

equal v_s, then βA must equal 1. βA is called the loop gain.

An oscillator will not function if, at the oscillator frequency, the magnitude of the product of the gain of the amplifier and the feedback factor of the feedback network is less than unity. The condition of unity loop gain is called the Barkhausen criterion.

If βA is less than unity, the removal of the external generator will immediately result in a cessation of oscillations. An oscillator in which the loop gain is exactly unity is an abstraction that is completely unrealizable in practice. A practical oscillator always has a βA somewhat larger than unity (say 5%) to ensure that, with incidental variations in transistor and circuit parameters, βA does not fall below unity.

The phase-shift oscillator (Fig. 2) exemplifies the principles set forth above. A field-effect transistor (FET) amplifier of con-

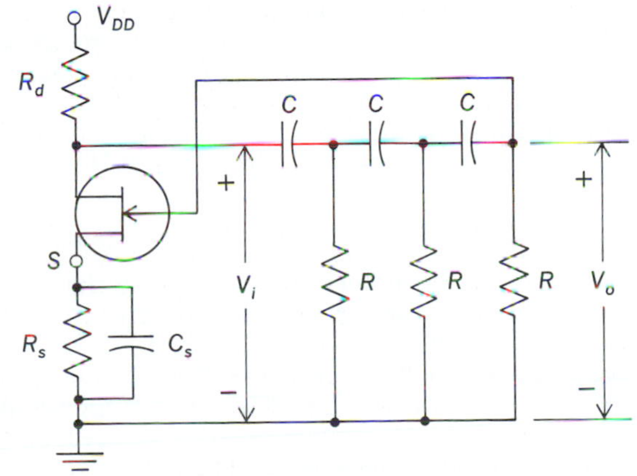

Fig. 2. An FET phase-shift oscillators.

ventional design is followed by three cascaded arrangements of a capacitor *C* and a resistor *R*, the output of the last *RC* combination being returned to the gate. The phase of the signal is shifted 180° by the amplifier, and the network of resistors and capacitors shifts the phase by an additional amount. At some frequency the phase shift introduced by the *RC* network is precisely 180°, and the total phase shift around the circuit is exactly zero. At this particular frequency the circuit will oscillate, provided that the magnitude of the amplification is sufficiently large. The phase-shift oscillator is particularly suited to frequencies from several hertz to several hundred kilohertz and so includes the range of audio frequencies. [E.S.Y.]

Oscillatory reaction A chemical reaction in which some composition variable of a chemical system exhibits regular periodic variations in time or space. It is a basic tenet of chemistry that a closed system moves inexorably toward an unchanging state called chemical equilibrium. That motion can be described by the monotonic increase of entropy if the system is isolated, and by the monotonic decrease of Gibbs free energy if the system is constrained to constant temperature and pressure. *See* CHEMICAL EQUILIBRIUM; ENTROPY; GIBBS FUNCTION.

The species taking part in a chemical reaction can be classified as reactants, products, or intermediates. The concentrations of reactants decrease. Intermediates are formed by some steps and destroyed by others. If there is only one intermediate, and if its concentration is always much less than the initial concentrations of reactants, this intermediate attains a stable steady state in which the rates of formation and destruction are virtually equal. Some oscillations require at least two intermediates which interact in such a way that the steady state of the total system is unstable to the minor fluctuations present in any collection of molecules. The concentrations of the intermediates may then oscillate regularly, although the oscillations must disappear before the inevitable monotonic approach to equilibrium.

The systems whose chemistries are best understood all involve an element that can exist in several different oxidation states. An example is the so-called Belousov-Zhabotinsky reaction. A strong oxidizing agent (bromate) attacks an organic substrate (such as malonic acid), and the reaction is catalyzed by a metal ion (such as cerium) that can exist in two different oxidation states.

As long as bromide ion (Br) is present, it is oxidized by bromate (BrO_3^-), as in reaction (1).

$$BrO_3^- + 2Br^- + 3H^+ \rightarrow 3HOBr \qquad (1)$$

When bromide ion is almost entirely consumed, the cerous ion (Ce^{3+}) is oxidized, as in reaction (2). Reaction (2) is inhibited

$$BrO_3^- + 4Ce^{3+} + 5H^+ \rightarrow HOBr + 4Ce^{4+} + 2H_2O \qquad (2)$$

by Br^-, but when the concentration of bromide has been reduced to a critical level, reaction (2) accelerates autocatalytically until bromate is being reduced by Ce^{3+} many times as rapidly as it is by Br^- when reaction (3) is dominant.

The hypobromous acid (HOBr) brominates the organic substrate to form bromomalonic acid (BrMA), as in reaction (3).

$$2Ce^{4+} + BrMA \rightarrow 2Ce^{3+} + Br^- + \text{oxidized organic matter} \qquad (3)$$

Reaction (3) creates the bromide ion necessary to shut off fast reaction (2) and throw the system back to dominance by slow reaction (1).

As other redox oscillators become understood, they fit the same pattern of a slow reaction destroying a species that inhibits a fast reaction that can be switched on autocatalytically; the fast reaction then generates conditions to produce the inhibitor again. *See* OXIDATION-REDUCTION. [R.M.No.]

Oscillograph A measurement device for determining waveform by recording the instantaneous values of a quantity such as voltage as a function of time, It consists of three major components: (1) a primary detector for sensing the instantaneous values of the quantity; (2) the timing system for introducing a time scale on the record; and (3) some means for recording the waveform. *See* WAVEFORM.

The electromagnetic oscillograph and the cathode-ray oscillograph are the two basic forms of oscillographs in common use.

Electromagnetic oscillographs are either direct-writing oscillographs or light-beam oscillographs. The best known of the direct-writing forms is the electrocardiograph. In this device the primary detector is a galvanometer with a multiturn moving coil. Attached to the moving coil is a pen arm, which traces an ink record on a continuously moving paper chart. The heart beats at about one cycle per second (1 Hz) and at this low frequency the pen follows the vibrations nicely and draws a graph in ink that can be read immediately. *See* GALVANOMETER.

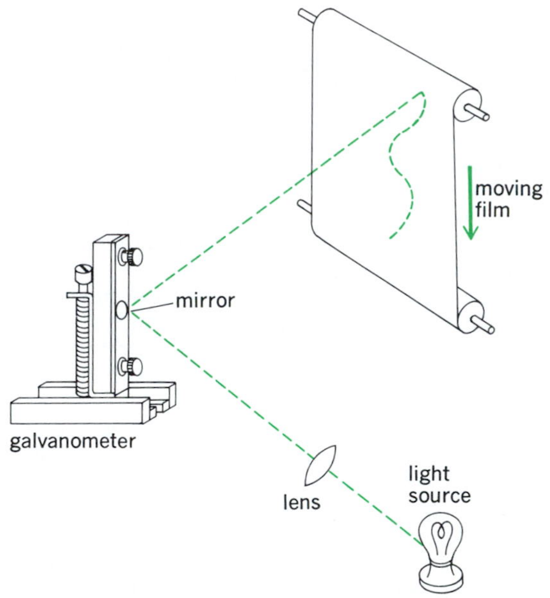

Bifilar electromagnetic oscillograph with light-beam, photographic-film recorder. (*General Electric Co.*)

The light-beam oscillograph, usable up to 500 Hz, is a more accurate and more commonly used type of oscillograph. The essential components of a light-beam-type magnetic oscillograph are shown in the illustration. The galvanometer has a moving coil, usually with a single U-shaped turn (bifilar type), but multiturn coils are sometimes used. A tiny mirror attached to the moving coil is the only mass added to the moving system for recording purposes. A beam of light, focused to a point of light by the lens, is reflected from the mirror onto a photographic film that moves at constant speed. The lateral position of the point where the light impinges on the film is a function of the position of the mirror and, therefore, the instantaneous value of the current flowing in the moving coil.

The low-frequency limitations (500 Hz) of the relatively large masses of the electromagnetic oscillograph element are eliminated in the cathode-ray oscillograph. This device is a cathode-ray oscilloscope combined with a camera that makes a photographic record of the screen image. The terms cathode-ray oscillograph and cathode-ray oscilloscope are commonly interchanged. Strictly speaking, the oscillograph contains means for producing records, whereas the oscilloscope does not. *See* OSCILLOSCOPE. [I.F.K./E.C.St.]

Oscilloscope An electronic instrument which produces a luminous plot on a fluorescent screen showing the relationship of two or more variables. In most cases it is an orthogonal (x,y) plot with the horizontal axis being a linear function of time. The vertical axis is normally a linear function of voltage at the signal input terminal of the instrument.

The primary advantage of a cathode-ray oscilloscope over other forms of plotting devices is its speed of response. Commercially available instruments in the general-purpose category can display frequencies as high as 100 MHz while special high-speed oscilloscopes can respond as high as 2000 MHz. The horizontal linear time axis of one general-purpose oscilloscope may be varied in 25 calibrated ranges from a slow speed of 10 cm in 50 seconds to a high of 10 cm in 0.2 microseconds. The same oscilloscope can record on photographic film a single trace at a rate of 250 cm/microsecond.

In its normal form the cathode-ray oscilloscope is made up of five basic elements:

1. The cathode-ray tube and associated controls in focus, intensity or brightness, and astigmatism. *See* CATHODE-RAY TUBE.
2. The vertical or signal amplifier and its associated devices such as input terminal, attenuators, position control, and ac or dc amplifier operation selector. *See* AMPLIFIER; VOLTAGE AMPLIFIER.
3. Horizontal-axis time-base circuits, frequently called the sweep generator. Included in this group are the sweep-base control, trigger or synchronizing circuits, and usually a circuit for turning on the cathode-ray tube beam only when the sweep is going in the left-to-right direction on the screen. *See* SWEEP GENERATOR.
4. Auxiliary facilities, such as amplitude or time calibrators and repetition rate generators.
5. Power supplies furnishing the correct operating voltages for the above circuits. *See* OSCILLOGRAPH. [H.V.]

Osmium A chemical element, Os, atomic number 76, and atomic weight 190.2. Osmium is a hard white metal of rare natural occurrence.

Osmium, like the other platinum metals, is catalytically active. Osmium tetroxide is used as an organic reagent and as

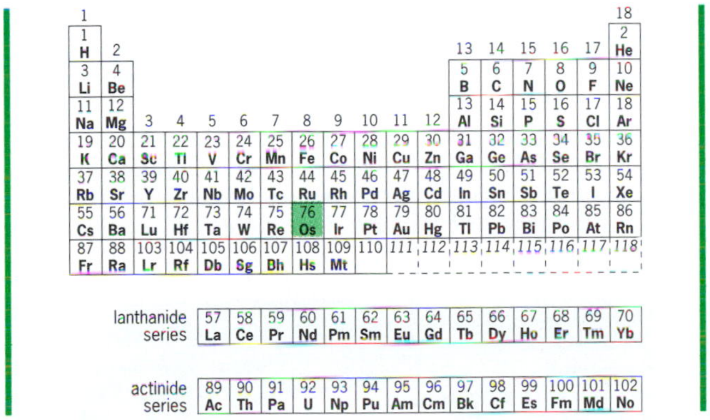

a stain for tissue in microscopy. Osmium alloyed with rhodium, ruthenium, iridium, or platinum is used in fountain-pen nibs, phonograph needles, electrical contacts, and instrument pivots. *See* PLATINUM.

The chemistry of osmium is very complicated because of the many valences exhibited by the element and the tendency of each of these to form numerous complex ions. Osmium is a very hard metal; it and its alloys are wear-resistant. Pure osmium and alloys in which it predominates are unworkable, so that they must either be used in the cast form or fabricated by

Physical properties of osmium

Property	Value
Atomic weight (^{12}C = 12.00000)	190.2
Naturally occurring isotopes, with percent abundance in parentheses	184(0.018)
	186(1.59)
	187(1.64)
	188(13.3)
	189(16.1)
	190(26.4)
	192(41.0)
Crystal structure	Close-packed hexagonal
Common chemical valence	4, 6, 8
Density at 25°C, g/cm^3	22.59
Melting point, °C	3045
Boiling point, °C	5020
Specific heat at 0°C, cal/g	0.0309

powder metallurgy. The table gives values of some physical properties of osmium.

Osmium tetrachloride, $OsCl_4$, is a black solid which is insoluble in nonoxidizing acids. Osmium tetroxide, OsO_4, is a very pale-yellow crystalline solid with a melting point of 40°C and a boiling point of 130°C; it is the most important osmium compound. This very poisonous compound is soluble in water and carbon tetrachloride. It is a powerful oxidizing agent. [H.J.A.]

Osmoregulatory mechanisms Physiological mechanisms for the maintenance of an optimal and constant level of osmotic activity of the fluid within and around the cells, considered to be most favorable for the initiation and maintenance of vital reactions in the cell and for maximal survival and efficient functioning of the entire organism.

The actions of osmoregulatory mechanisms are, first, to impose constraints upon the passage of water and solute between the organism and its surroundings and, second, to accelerate passage of water and solute between organism and surroundings. The first effect requires a change of architecture of membranes in that they become selectively permeable and achieve their purpose without expenditure of energy. The accelerating effect, apart from requiring a change of architecture of cell membranes, requires expenditure of energy and performance of useful osmotic work. Thus substances may be moved from a region of low to a region of higher chemical activity. Such movement can occur in opposition to the forces of diffusion of an electric field and of a pressure gradient, all of which may act across the cell membrane. It follows that there must be an energy source which is derived from the chemical reactions of cellular metabolism and that part of the free energy so generated must be stored in molecules which are driven across the membrane barrier. Active transport is the modern term for such processes. *See* CELL MEMBRANES; OSMOSIS. [W.A.B.; T.P.S.]

Osmosis The transport of solvent through a semipermeable membrane separating two solutions of different solute concentration. The solvent diffuses from the solution that is dilute in solute to the solution that is concentrated.

The flow of liquid through such a barrier may be stopped by applying pressure to the liquid on the side of higher solute concentration. The applied pressure required to prevent the flow of solvent across a perfectly semipermeable membrane is called the osmotic pressure and is a characteristic of the solution. The walls of cells in living organisms permit the passage of water and certain solutes, while preventing the passage of other solutes, usually of relatively high molecular weight. These walls act as selectively permeable membranes, and allow osmosis to occur between the interior of the cell and the surrounding media. *See* EDEMA; OSMOREGULATORY MECHANISMS; SOLUTION. [F.J.J.]

Ostariophysi A superorder of actinopterygian fishes comprising two orders, Siluriformes and Cypriniformes. The characters of the Ostariophysi agree in the main with the Salmoniformes, the generalized basal order from which most teleosts have evolved, but with one notable difference: The anterior four or five vertebrae are much modified as a protective encasement for a series of bony ossicles that form a chain connecting the swim bladder with the inner ear, the so-called Weberian apparatus. Other characters include an abdominal pelvic fin, if present, usually with many rays and a more or less horizontal pectoral fin placed low on the side. Usually there are no fin spines, but one or two spines derived from consolidated soft rays may be found in dorsal, anal, or pectoral fins, and a spinelike structure is present in the adipose fin of certain catfishes. The superorder Ostariophysi consists of 2 well-marked orders, about 34 families, and approximately 5000 species. *See* CYPRINIFORMES; SILURIFORMES. [R.M.B.]

Osteichthyes The bony fishes, one of the three classes of Recent fishlike vertebrates. It includes most of the familiar fishes. The Osteichthyes are similar to the Chondrichthyes, or cartilaginous fishes, and contrast with the living Agnatha in having jaws, paired nostrils, true teeth, paired pelvic and pectoral fins and girdles (unless lost secondarily), three semicircular canals, and bony scales (unless lost or modified). Separation of the Osteichthyes from the Paleozoic Placodermi and Acanthodii is more difficult because these groups agree in most basic vertebrate features. The Osteichthyes contrast with the Chondrichthyes in having a bony skeleton (some Recent bony fishes possess a largely cartilaginous skeleton), a swim bladder (at least primitively), a true gill cover, and mesodermal ganoid, cycloid, or ctenoid scales (sharks possess dermal denticles, or placoid scales) which are sometimes modified or lost. Fertilization is usually external, but if it is internal, the intromittent organ is not derived from pelvic-fin claspers, Most often a modified anal fin or a fleshy tube or sheath functions in sperm transfer. *See* AGNATHA; CHONDRICHTHYES; COPULATORY ORGAN; SCALE (ZOOLOGY); SWIM BLADDER.

The Recent fauna of the class Osteichthyes includes 3 subclasses, 32 orders, about 357 families, roughly 3570 genera, and probably about 17,600 species. A classification of the Osteichthyes follows. Equivalent names are given in parentheses; orders known only as fossils are preceded by asterisks. For more detailed information see separate articles on each group (the order Dipteriformes is discussed in the article Dipnoi).

Class Osteichthyes
- Subclass Actinopterygii
  - Infraclass Chondrostei
    - Order: *Palaeonisciformes
      - Polypteriformes (Cladistia)
      - Acipenseriformes
  - Infraclass Holostei
    - Order: Semionotiformes (Protospondyli, Ginglymodi, and Lepisostei)
      - *Pyconodontiformes
      - Amiiformes (Halecomorphi)
      - *Aspidorhynchiformes
      - *Pholidophoriformes (Halecostomi)
  - Infraclass Teleostei
    - Order: *Leptolepiformes
      - Elopiformes (Isospondyli in part)
      - Anguilliformes (Apodes and Saccopharyngiformes or Lyomeri)
      - Notacanthiformes (Lyopomi and Heteromi)
      - Clupeiformes
      - Osteoglossiformes (Isospondyli in part and Mormyriformes)
      - Salmoniformes (Myctophiformes, or Iniomi, and Haplomi)
      - Cetomimiformes (Cetunculi)
      - Ctenothrissiformes
      - Gonorynchiformes
      - Cypriniformes (Heterognathi, Eventognathi, and Gymnonoti)
      - Siluriformes (Nematognathi)
      - Percopsiformes (Salmopercae and Amblyopsiformes)
      - Batrachoidiformes (Haplodoci)
      - Gobiesociformes (Xenopterygii)
      - Lophiiformes (Pediculati)
      - Gadiformes (Anacanthini)
      - Atheriniformes (Beloniformes or Synentognathi, Cyprinodontiformes or Microcyprini, and Phallostethiformes)
      - Beryciformes (Berycomorphi)
      - Zeiformes (Zeomorphi)
      - Lampridiformes (Allotriognathi)
      - Gasterosteiformes (Thoracostei and Solenichthyes)
      - Pegasiformes (Hypostomides)
      - Synbranchiformes (Synbranchii)
      - Perciformes (Acanthopterygii or Percomorphi)
      - Pleuronectiformes (Heterosomata)
      - Tetradontiformes (Plectognathi)
- Subclass Crossopterygii
  - Order: *Osteolepiformes (Rhipidistia)
    - Coelacanthiformes (Coelacanthini)
- Subclass Dipnoi (Dipneusti)
  - Order: Dipteriformes

[R.M.B.]

Osteoglossiformes An order of soft-rayed, actinopterygian fishes that includes the mooneyes, featherbacks, mormyrids or elephantfishes, and bonytongues.

Osteoglossiforms have the primary bite between the well-toothed tongue and the roof of the mouth, usually the strongly toothed parasphenoid but occasionally the endopterygoids. The mouth is bordered by the premaxilla, which in some forms is fused to its mate from the opposite side, and the maxilla. A unique feature is the presence of paired, usually bony rods at the base of the second gill arch.

Osteoglossiforms represent one of the basal stocks among the teleosts. The Ichthyodectidae and two related families are well represented in Cretaceous marine deposits, and *Allothrissops* and *Pachythrissops* of the Jurassic and Cretaceous probably belong here. Recent members may be classified in 2 suborders, 6 families, 22 genera, and about 135 species. All

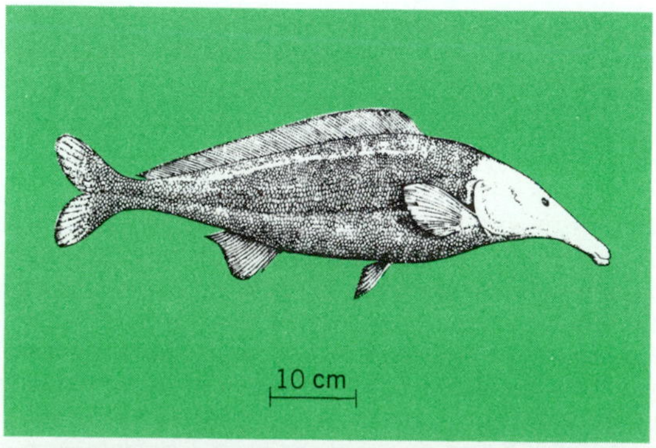

African elephantnose (*Mormyrus proboscirostris*). (*After G. A. Boulenger, Catalogue of Fresh Water Fishes of Africa in the British Museum, Natural History, vol. 1, 1909*)

inhabit fresh water, and all are tropical except for the two species of Hiodontidae or mooneyes of North America, where they have a history to the Eocene.

Numerically the Osteoglossiformes are dominated by the Mormyridae, African river and lake fishes of varied size and form. In some the snout is very blunt and rounded, in others it is elongated (see illustration), the source of the vernacular name elephantfish. The related family Gymnarchidae consists of a single species. Scientifically, mormyrids and gymnarchids are of especial interest in that all are electrogenic. Modified muscles in the caudal peduncle generate and emit a continuous electric pulse. The discharge frequency varies, being low at rest and high when the fish is under stress. The mechanism operates like a radar device, since the fish is alerted whenever an electrical conductor enters the electromagnetic field surrounding it. *See* ACTINOPTERYGII; ELECTRIC ORGAN (BIOLOGY). [R.M.B.]

Osteomyelitis Inflammation of a bone and its adjacent bone marrow, usually resulting from infection. Pyogenic (pus-forming) organisms, such as the hemolytic staphylococci, are the most frequent agents in acute cases. Chronic forms occur and may be induced by tuberculosis, syphilis, and other microorganisms.

The long tubular bones of the arms and legs are most often involved. Bone destruction can become extensive if prompt diagnosis and treatment is not initiated. The acute phase is marked by a pus-forming inflammatory reaction. A persistent osteomyelitis tends to become chronic and shows combined reparative attempts mingled with continued tissue breakdown. The acute clinical course is marked by fever, chills, leukocytosis, and local symptoms of inflammation such as pain, swelling, and redness of the affected part. [E.G.St./N.K.M.]

Osteoporosis A metabolic bone disease involving a reduction in bone mass per unit volume to a level leading to fracture, especially of the spine, forearm, and hip. Perhaps 10 million individuals in the United States suffer from osteoporosis, both primary (postmenopausal in the elderly female, senile in the elderly male) and secondary (associated with other diseases such as Cushing's disease). The individual with osteoporosis develops fractures of the three areas noted above and has associated pain; spinal curvature ("dowager's hump") and height loss is common when such fractures occur in the spine.

In secondary osteoporosis, the causes of bone mass loss and fractures may be readily identifiable, such as an excess of cortisone in Cushing's disease; the precise cause of primary osteoporosis is more difficult to define. There are a number of abnormalities of bone cells and of skeletal hormones that occur in an aging individual, but none is unique and specific for an aging osteoporotic individual. Nevertheless, a number of risk factors are noted to contribute to the occurrence of osteoporosis, particularly postmenopausal osteoporosis. Among these factors are inadequate calcium intake during life and particularly after menopause, inadequate exercise, and loss of the female hormone estrogen after a natural or surgical (hysterectomy) menopause. Estrogens protect against loss of bone due to their apparent inhibition of parathyroid hormone, a hormone which simulates bone resorption. Other risk factors are leanness and small stature. *See* CALCIUM METABOLISM; ESTROGEN; PARATHYROID HORMONE.

Osteoporosis therapy must be divided into symptomatic therapy and therapy for the underlying disease. Symptomatic therapy is usually related to previous skeletal fractures and includes pain relief, short-term limitation of activity, and back bracing when appropriate. Dietary calcium or oral calcium supplements, estrogen, calcitonin, the experimental agent diphosphonate, and a form of vitamin D which is produced in the kidney and increases calcium absorption function to slow bone resorption and bone loss. Assessing the response (beneficial or detrimental) to therapy may be accomplished by monitoring of bone mass. [C.H.Ch.]

Osteostraci An order of extinct jawless vertebrate fishes, also called Cephalaspida, of the class Agnatha, known from the Middle Silurian to Upper Devonian of Europe, Asia, and North America. They were mostly small, about 2 in. to 2 ft (5 to 60 cm) in length. The head and part of the body were encased in a solid armor of bone, and the posterior part of the body and the tail were covered with thick scales. Some early forms lacked paired fins, though most possessed flaplike pectoral fins. One or two dorsal fins were present. Their depressed shape and the position of the eyes on the top of the head suggest that Osteostraci were bottom dwellers (see illustration). The underside of the throat region was covered by small plates and there was a small mouth in front. *See* AGNATHA. [R.H.DE.]

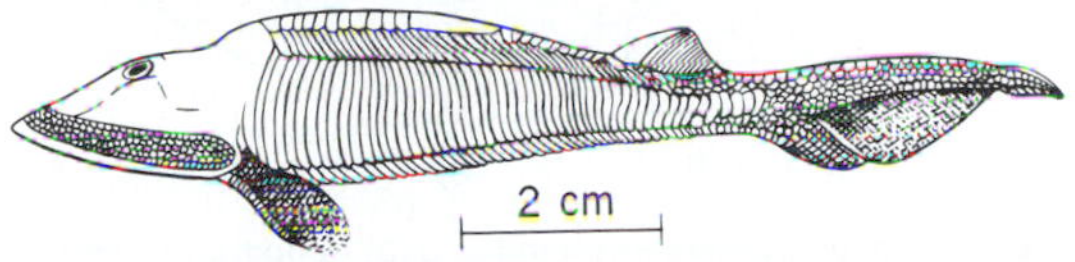

The ostracoderm *Hemicyclaspis*, a cephalaspid, a Lower Devonian jawless vertebrate. (*After E. H. Colbert, Evolution of the Vertebrates, Wiley, 1955*)

Ostracoda A major taxon of the Crustacea containing small bivalved animals, with most between 0.04 and 0.08 in. (1 and 2 mm). They inhabit aquatic environments in nearly all parts of the world. Semiterrestrial species have been described from moss and leaf-litter habitats in Africa, Madagascar, Australia, and New Zealand, and from vegetable debris of marine origin in the Kuril Archipelago. Of the more than 2000 species extant, none is truly parasitic and most are free-living. However, a few fresh-water and marine forms live commensally on other animals. Most ostracods are scavengers, some are herbivorous, and a few are predacious carnivores.

Knowledge of the morphology of the Ostracoda is based primarily on relatively few selected species. The two valves, sufficient to enclose the rest of the animal, are joined dorsally along a hinge, which may vary from a simple juncture to a complex series of teeth and sockets. From the dorsal part of the carapace, the elongate body is suspended as a pliable sac, with lateral flaps of hypodermis extending between the lamellae of the valves. It is unsegmented, but reinforced by chitinous processes for rigidity in the vicinity of the appendages. There is no true abdominal region.

Ostracods have separate sexes, although many species are parthenogenetic and lack males. Most ostracods lay their eggs on the substrate or on vegetation, but some transfer them to the posterior space within the carapace, where they hatch and the young brood is retained for a time. *See* CRUSTACEA; PALEOCOPA. [P.A.McL.]

Otter Members of the family Mustelidae, a large group of carnivores which also includes the martens, weasels, badgers, and skunks. Otters are more completely adapted to aquatic life than other members of the family, having a long thin body, short legs, and somewhat flattened head. The small ears possess a membrane which closes the ear canal when the animal dives. The feet have five webbed toes with nonretractile claws. The tail is broad and flattened, being used along with body movements for swimming. The short thick fur is impervious to water. Like other members of the family all have well-developed perianal scent glands. *See* CARNIVORA; MAMMALIA. [C.B.C.]

Otto cycle The basic thermodynamic cycle for the prevalent automotive type of internal combustion engine. The engine uses a volatile liquid fuel (gasoline) or a gaseous fuel to carry out the theoretic cycle shown in the illustration. The cycle consists of two isentropic (reversible adiabatic) phases interspersed between two constant-volume phases. The theoretic cycle should not be confused with the actual engine built for such service as automobiles, motor boats, aircraft, lawn mowers, and other small self-contained power plants.

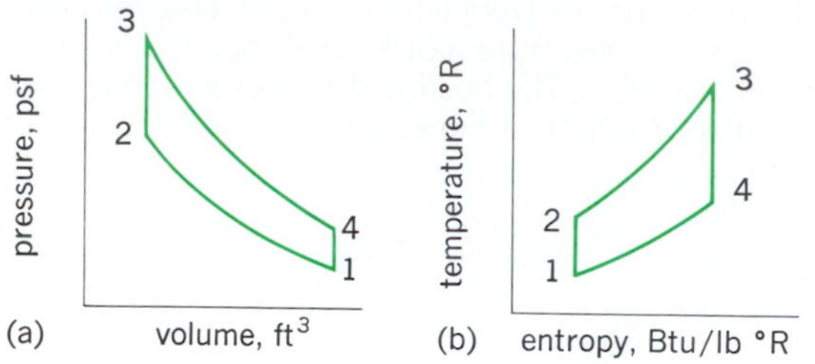

Diagrams of (*a*) pressure-volume and (*b*) temperature-entropy for Otto cycle.

The thermodynamic working fluid in the cycle is subjected to isentropic compression, phase 1–2; constant-volume heat addition, phase 2–3; isentropic expansion, phase 3–4; and constant-volume heat rejection (cooling), phase 4–1.

The Otto cycle is represented in many millions of engines utilizing either the four-stroke principle or the two-stroke principle. Evidence indicates that actual Otto engines offer peak efficiencies (25±%) at compression ratios of 15±. Above this ratio, efficiency falls. The most probable explanation is that the extreme pressures associated with high compression cause increasing amounts of dissociation of the combustion products. This dissociation, near the beginning of the expansion stroke, exerts a more deleterious effect on efficiency than the corresponding gain from increasing compression ratio. *See* Brayton cycle; Carnot cycle; Diesel cycle; Internal combustion engine; Thermodynamic cycle. [T.Ba.]

Ovarian disorders Ovarian disorders range from the fairly common physiological cystic enlargements of the ovaries to neoplasms of malignant character which may be life-threatening and require more urgent attention.

Developmental abnormalities are expressed either as sexual infantilism or as the increasingly complex variety of morphological, hormonal, and chromosomal parameters associated with sexual identity. Gonadal agenesis refers to bilateral rudimentary "streak" gonads and infantile external genitalia as seen in, for example, Turner's syndrome. The ovaries contain stroma only, without germ cells (ova). Ovarian "dysgenesis" includes chromosomal mosaics, true hermaphrodites who possess both ovarian and testicular tissue (ovotestis), and pseudohermaphrodites.

The ovary normally contains numerous microscopic cysts (Graafian follicles). After ovulation the cystic space may become filled with blood (a hemorrhagic corpus luteum). Not infrequently these physiological cysts enlarge which may result in a surgical procedure and excision of the cyst to exclude a neoplastic process.

The polycystic ovary syndrome includes bilateral polycystic ovaries, primary amenorrhea or oligomenorrhea, infertility, obesity, and hirsutism. There is a wide spectrum of clinical and pathological abnormality in this type of disorder. Amenorrhea or anovulatory bleeding and infertility are characteristic symptoms. The clinical and pathologic changes are due to a prolonged anovulatory state and unopposed estrogen stimulation. The ovaries are typically enlarged, with numerous subcapsular follicular cysts.

Endometriosis (ectopic endometrial glandular and stromal tissue) can also produce a cystic mass in the ovary (endometrioma) filled with old, brown, clotted blood ("chocolate cyst"). Spontaneous rupture of endometrial cysts and other hemorrhagic physiological cysts of the ovary may produce peritoneal irritation and acute abdominal symptoms.

The large number of often bewildering patterns of ovarian neoplasms have resulted in many different classifications. Those based on hormone secretion have become particularly imprecise. Unfortunately, most ovarian malignancies are discovered at an advanced stage, since ovarian neoplasms in general tend to be asymptomatic until they produce a palpable mass. Malignant neoplasms are "staged" I–IV to determine the extent of growth beyond the ovary. The stage establishes a prognosis and most effective mode of treatment. This can be done accurately only by surgical examination of the abdominal cavity (laparotomy). *See* Cancer (medicine); Ovary. [M.Ek.]

Ovary A part of the reproductive system of all female vertebrates. Although not vital to individual survival, the ovary is vital to perpetuation of the species. The function of the ovary is to produce the female germ cells or ova, and in some species to elaborate hormones that assist in regulating the reproductive cycle.

The ovaries develop as bilateral structures in all vertebrates, but adult asymmetry is found in certain species of all vertebrates from the elasmobranchs to the mammals.

The ovary of all vertebrates functions in essentially the same manner. However, ovarian histology of the various groups differs considerably. Even such a fundamental element as the ovum exhibits differences in various groups. *See* Ovum.

The mammalian ovary is attached to the dorsal body wall. The free surface of the ovary is covered by a modified peritoneum called the germinal epithelium. Just beneath the germinal epithelium is a layer of fibrous connective tissue. Most of the rest of the ovary is made up of a more cellular and more loosely arranged connective tissue (stroma) in which are embedded the germinal, endocrine, vascular, and nervous elements.

The most obvious ovarian structures are the follicles and the corpora lutea. The smallest, or primary, follicle consists of an oocyte surrounded by a layer of follicle (nurse) cells. Follicular growth results from an increase in oocyte size, multiplication of the follicle cells, and differentiation of the perifollicular stroma to form a fibrocellular envelope called the theca interna. Finally, a fluid-filled antrum develops in the granulosa layer, resulting in a vesicular follicle.

The cells of the theca interna hypertrophy during follicular growth and many capillaries invade the layer, thus forming the endocrine element that is thought to secrete estrogen. The other known endocrine structure is the corpus luteum, which is primarily the product of hypertrophy of the granulosa cells remaining after the follicular wall ruptures to release the ovum. Ingrowths of connective tissue from the theca interna deliver capillaries to vascularize the hypertrophied follicle cells of this new corpus luteum; progesterone is secreted here. *See* Estrogen; Estrus; Menstruation; Progesterone. [K.L.D.]

Overdrive An automotive device supplied as special equipment and containing a step-up planetary gear arrangement located between the transmission and propeller shaft. Overdrive permits the propeller shaft to be driven at transmission output shaft speed, or faster than transmission output shaft speed when the overdrive comes into operation. For a given forward speed of the car, overdrive gives lower engine speed, quieter operation, and reduced gasoline consumption.

The overdrive contains two major components, a planetary

gear system and a free-wheeling device (or overrunning clutch). The overrunning clutch has two functions. It locks the transmission mainshaft and the overdrive output shaft together when the planetary gear system is inactive, providing direct drive through the overdrive. The second function is to permit the output shaft to overrun the transmission mainshaft when the planetary gear train is in action, providing overdrive. The overrunning clutch consists of an outer shell attached to the output shaft. A circular inner member attached to the transmission shaft carries a series of cams or flats; a hardened steel roller rides on each cam. When the clutch is in action, the rollers wedge between the cams and outer shell so that the cams drive the outer shell. During overrunning, the outer shell rotates faster than the cams and the rollers move from loaded contact with the cams to disengage the clutch. *See* AUTOMOTIVE TRANSMISSION; CLUTCH; PLANETARY GEAR TRAIN. [F.R.McF.]

Overtone (music and acoustics) An upper partial tone. As a consequence of the distinction "upper," a sixth overtone, for example, is the same as the seventh partial tone, so that numbering tends to become confused. The word overtone has been used as a synonym for partial, harmonic, or mode of vibration, and the word tone has still other meanings in music. It seems desirable, therefore, not to use overtone at all, but to select one of the other words that has a more specific meaning. *See* HARMONIC (PERIODIC PHENOMENA); PARTIAL TONE; TONE (MUSIC AND ACOUSTICS). [R.W.Y.]

Overvoltage The difference between electrode potential under electrolysis conditions and the thermodynamic value of the electrode potential in the absence of electrolysis for the same experimental conditions; sometimes called overpotential. It is expressed in volts and is often quoted in absolute value. Understanding of overvoltage phenomena is essential in industrial electrochemistry, since they are often a major factor in determining the efficiency of electrode processes. *See* DECOMPOSITION POTENTIAL; ELECTROCHEMICAL PROCESS; ELECTRODE POTENTIAL; ELECTROLYSIS.

The overvoltage for a given electrode reaction depends on electrolysis conditions such as current density (current per unit area), concentration of electrolyzed substance, temperature, presence of foreign electrolytes, nature of the solvent, and absorption of foreign substances on the electrode. The overvoltage for given electrolysis conditions varies greatly from one electrolyzed substance to another—from less than 1 mV to a few volts.

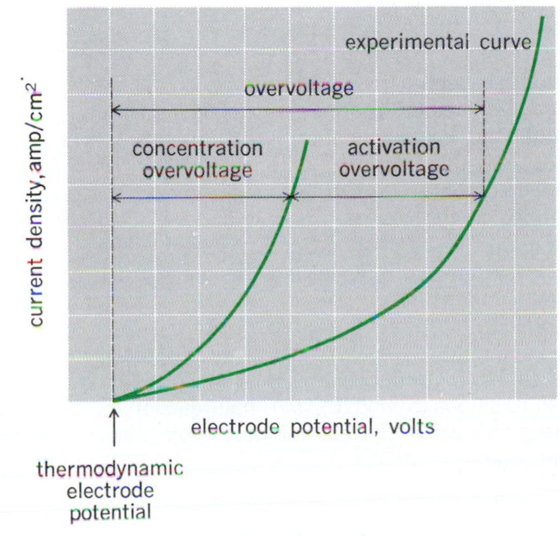

Current-potential curve.

Overvoltage values are determined from the variations of electrode potential with current density by the following method. The electrolytic cell is connected to a direct-current power supply, and the potential of the electrode being studied is measured against a reference electrode for different current densities. Results are plotted as a current-potential curve (see illustration), and the overvoltage is read on this curve.

Concentration overvoltage and activation overvoltage are the two components into which the overvoltage can generally be decomposed. Concentration overvoltage arises from local variations of the concentration of reactants or products of electrolysis or both near the electrode. The thermodynamic potential calculated for the concentrations at the electrode surface under electrolysis conditions is different from the potential in the absence of electrolysis. The difference between these two potentials is the concentration overvoltage. Activation overvoltage results from the relative slowness of electrode reactions. An energy barrier must be overcome by the substances involved in an electrode reaction just as for purely chemical reactions. Activation overvoltage corresponds to this additional consumption of electrical energy. [P.De.]

Ovum The egg or female sex cell. Strictly speaking, the term refers to this cell when it is ready for fertilization, but it is often applied to earlier or later stages. Confusion is avoided by using qualifying adjectives such as immature, ripe, mature, fertilized, or developing ova. The mature ova are generally spheroidal and large. The number of ova produced at one time varies in different animals, from millions in many marine animals that spawn into the surrounding sea water to about a dozen or less in mammals in which adaptations for internal nourishment of the developing embryo and care of the young are highly developed.

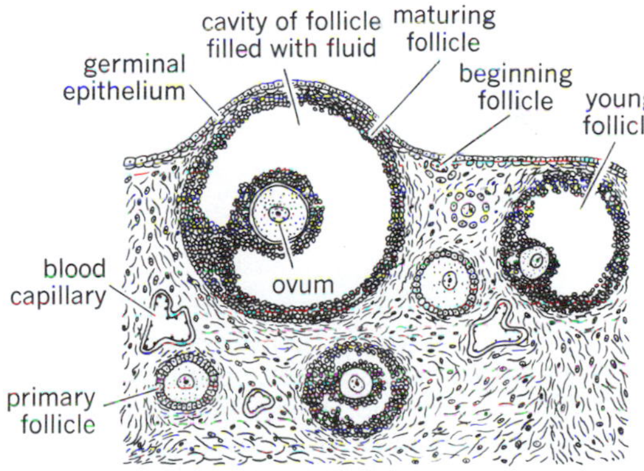

Section of a mammalian ovary.

In the ovary the immature ovum is associated with follicle cells through which it receives material for growth. In mammals, as the egg matures, these cells arrange themselves into a structure known as the Graafian, or vesicular, follicle, consisting of a large fluid-filled cavity into which the ovum, surrounded by several layers of cells, projects from the layer of follicle cells that constitutes the inner wall (see illustration). The fluid contains estrogenic female sex hormone secreted by cells in an intermediate layer of the follicular wall.

Yolk, or deutoplasm, is essentially a food reserve in the form of small spherules, present to a greater or lesser extent in all eggs. It accounts largely for the differences in size of eggs. Eggs are classified according to the distribution of yolk. In the isolecithal type there is a nearly uniform distribution through the cytoplasm, as in most small eggs. The yolk in telolecithal eggs is increasingly concentrated toward one pole, as in the large eggs of fish, amphibians, reptiles, and birds. Centrolecithal, or centrally located, yolk occurs in eggs of insects and cephalopod mollusks. *See* GAMETOGENESIS; OOGENESIS. [A.T./H.L.H.]

Oxalate A salt or ester of oxalic acid with the formula

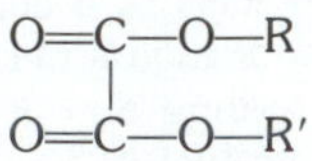

where R and R′ are alkyl groups or metallic ions attached to the acid to produce a series of salts and esters. In general, alkali-metal salts are water-soluble; others are insoluble. Many salts have practical applications, for example, in pyrotechnics (sodium), blueprinting (potassium-iron), and analytical procedures (ammonium). Diethyl oxalate is used as a solvent, as a dyestuff intermediate, and in plastics. *See* ESTER; OXALIC ACID. [E.H.H.]

Oxalic acid A white solid acid melting with decomposition at 189.5°C (373.1°F). Oxalic acid (ethanedioic acid), with the structure

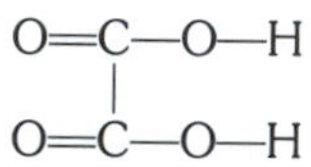

is the first of a series of dicarboxylic acids. Its salts (oxalates) are prevalent in nature, for example, KHC_2O_4 in plants of the oxalis family (wood sorrel) and CaC_2O_4 in eucalyptus bark. *See* OXALATE.

The acid is used as a bleaching agent for rust and ink stains, in textile and leather production, and as monoglyceryl oxalate in the manufacture of allyl alcohol and formic acid. [E.B.R.]

Oxazole One of a group of organic heterocyclic compounds containing oxygen and nitrogen in the 1 and 3 positions of a five-membered diunsaturated ring. The formula

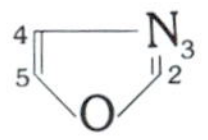

shows the structure and numbering system for a typical member of the group, 1,3-oxazole, considered the parent compound. This compound is a colorless, volatile, weakly basic liquid (bp 69–70°C or 156–158°F), with an odor resembling that of pyridine. Oxazole is miscible with water and organic solvents. Mineral acids form salts that tend to dissociate in water. Oxazoles show appreciable resistance to disruption by heat, by acid, and by alkali. The nucleus is susceptible to oxidation, the 4,5 position being the usual point of attack. Hydrogenation over a platinum catalyst or with sodium and alcohol gives tetrahydro derivatives (oxazolidines) or ring-cleavage products. *See* AZOLE; HETEROCYCLIC COMPOUNDS. [W.J.Ge.]

Oxidation process A process in which oxygen is caused to combine with other molecules. The oxygen may be used as elemental oxygen, as in air, or in the form of an oxygen-containing molecule which is capable of giving up all or part of its oxygen. Oxidation in its broadest sense, that is, an increase in positive valence or removal of electrons, is not considered here if oxygen itself is not involved. *See* OXIDATION-REDUCTION.

Most oxidations occur with the liberation of large amounts of energy in the form of either heat, light, or electricity. The stable ultimate products of oxidation are oxides of the elements involved. These oxidations occur in nature as corrosion, decay, and respiration and in the deliberate burning of matter such as wood, petroleum, sulfur, or phosphorus to oxides of the constituent elements.

The principal variables to be considered and controlled in any partial oxidation are temperature, pressure, reaction time (or contact time), nature of catalyst, if any, mole ratio of oxidizing agent, and whether the substance to be oxidized is to be kept in the liquid or vapor phase. Only a narrow range of conditions unique to each substance being oxidized and each product desired will give satisfactory yields. It is also essential to maintain conditions outside the range of spontaneous ignition, to avoid explosive mixtures or the accidental accumulation of unstable peroxides, and to choose materials which not only can resist the environmental conditions but also which do not have adverse catalytic effects or otherwise interfere with the desired reaction. *See* COMBUSTION. [I.E.L.]

Oxidation-reduction An important concept of chemical reactions which is useful in systematizing the chemistry of many substances. Oxidation can be represented as involving a loss of electrons by one molecule and reduction as involving an absorption of electrons by another. Both oxidation and reduction occur simultaneously and in equivalent amounts during any reaction involving either process.

Oxidation number. The oxidation state is a concept which describes some important aspects of the state of combination of the elements. An element in a given substance is characterized by a number, the oxidation number, which specifies whether the element in question is combined with elements which are more electropositive or more electronegative than it is. It further specifies the combining capacity which the element exhibits in a particular combination. A scale of oxidation numbers is defined by assigning to an oxygen atom in an ion such as SO_4^{2-} the value of 2−. That for sulfur as 6+ then follows from the requirement that the sum of the oxidation numbers of all the atoms add up to the net charge on the species. The value of 2− for oxygen is not chosen arbitrarily. It recognizes that oxygen is more electronegative than sulfur, and that when it reacts with other elements it seeks to acquire two more electrons, by sharing or outright transfer from the electropositive partner, so as to complete a stable valence shell of eight electrons.

Although oxidation number is in some respects similar to valence, the two concepts have distinct meanings. In the substance H_2, the valence of hydrogen is 1 because each H makes a single bond to another H, but the oxidation number is 0, because the hydrogen is not combined with a different element. *See* VALENCE.

When the oxidation number of an atom in a species is increased, the process is described as oxidation, no matter what reagent produces it; when a decrease in oxidation number takes place, the process is described as reduction, again without regard to the identity of the reducing agent. The term oxidation has been generalized to imply combination of an element with an element more electronegative than itself.

Reactions. In an oxidation-reduction reaction, some element decreases in oxidation state and some element increases in oxidation state. The substances containing these elements are defined as the oxidizing agents and reducing agents, and they are said to be reduced and oxidized, respectively. The processes in question can always be represented formally as involving electron absorption by the oxidizing agent and electron donation by the reducing agent. For example, reaction (1) can be regarded as the sum of the two partial processes, or half-reactions, (2) and (3). Similarly, reaction (4) consists of the two half-reactions (5) and (6), with half-reaction (5) being taken five times to balance the electron flow from reducing agent to oxidizing agent.

$$2Fe^{3+} + 2I^- \rightarrow 2Fe^{2+} + I_2 \quad (1)$$

$$2I^- \rightarrow I_2 + 2e^- \quad (2)$$

$$2Fe^{3+} + 2e^- \rightarrow 2Fe^{2+} \quad (3)$$

$$16H^+ + 2MnO_4^- + 10I^- \rightarrow 8H_2O + 2Mn^{2+} + 5I_2 \quad (4)$$

$$2I^- \rightarrow I_2 + 2e^- \quad (5)$$

$$16H^+ + 2MnO_4^- + 10e^- \rightarrow 2Mn^{2+} + 8H_2O \quad (6)$$

Each half-reaction consists of an oxidation-reduction couple; thus, in half-reaction (6) the reducing agent and oxidizing agent making up the couple are manganous ion, Mn^{2+}, and permanganate ion, MnO_4^-, respectively; in half-reaction (5) the reducing agent is I^- and the oxidizing agent is I_2. The fact that MnO_4^- reacts with I^- to produce I_2 means that MnO_4^- in acid solution is a stronger oxidizing agent than is I_2. Because of the reciprocal relation between the oxidizing agent and reducing agent comprising a couple, this statement is equivalent to saying that I^- is a stronger reducing agent than Mn^{2+} in acid solution. Reducing agents may be ranked in order of tendency to react, and this ranking immediately implies an opposite order of tendency to react for the oxidizing agents which complete the couples. In the list below some common oxidation-reduction couples are ranked in this fashion:

Strong reducing agent ↑ Increasing reducing power; Weak reducing agent		Weak oxidizing agent ↑ Increasing oxidizing power; Strong oxidizing agent
Mg	=	$Mg^{2+} + 2e^-$
Zn	=	$Zn^2 + 2e^-$
H_2	=	$2H^+ + 2e^-$
Cu	=	$Cu^2 + 2e^-$
I^-	=	$\frac{1}{2}I_2 + e^-$
Fe^{2+}	=	$Fe^{3+} + e^-$
Br^-	=	$\frac{1}{2}Br_2 + e^-$
Cl^-	=	$\frac{1}{2}Cl_2 + e^-$
$4H_2O + Mn^{2+}$	=	$MnO_4^- + 8H^+ + 5e^-$

See ELECTROCHEMICAL SERIES; ELECTRONEGATIVITY; OXIDIZING AGENT. [H.Ta.]

Oxide A binary compound of oxygen with another element. Oxides have been prepared for essentially all the elements except the noble gases. Often, several different oxides of a given element can be prepared; a number exist naturally in the Earth's crust and atmosphere: silicon dioxide (SiO_2) in quartz; aluminum oxide (Al_2O_3) in corundum; iron oxide (Fe_2O_3) in hematite; carbon dioxide (CO_2) gas; and water (H_2O).

Most elements will react with oxygen at appropriate temperature and oxygen pressure conditions, and many oxides may thus be directly prepared. Most metals in massive form react with oxygen only slowly at room temperatures because the first thin oxide coat formed protects the metal. The oxides of the alkali and alkaline-earth metals, except for beryllium and magnesium, are porous when formed on the metal surface, and they provide only limited protection to the continuation of oxidation, even at room temperatures. Gold is exceptional in its resistance to oxygen, and its oxide (Au_2O_3) must be prepared by indirect means. The other noble metals, although ordinarily resistant to oxygen, will react at high temperatures to form gaseous oxides.

Oxides may be classified as acidic or basic according to the character of the solution resulting from their reactions with water. The nonmetal oxides generally form acid solutions and the metal oxides generally form alkaline solutions. *See* ACID AND BASE; EQUIVALENT WEIGHT; OXYGEN. [R.K.E.]

Oxidizing agent A participant in a chemical reaction that absorbs electrons from another reactant. In the process a component atom of this substance undergoes a decrease in oxidation number. In this action as an oxidizing agent, the substance undergoes reduction.

A measure of the effectiveness of a reagent as an oxidizing agent is its reduction potential. This is, in electrochemical terms, the equivalent of the free-energy change for the reduction process. The element with the highest reduction potential (and, therefore, the strongest oxidizing agent) is fluorine, F_2. The practical effectiveness of a given oxidizing (or reducing) agent will depend upon both the thermodynamics and the available kinetic pathway for the reaction process. *See* CHEMICAL THERMODYNAMICS.

Substances that are widely used as oxidizing agents in chemistry include ozone (O_3), permanganate ion (MnO_4^-), nitric acid (HNO_3), as well as oxygen itself. Organic chemists have empirically developed combinations of reagents to carry out specific oxidation steps in synthetic processes. The action of molecular oxygen as an oxidizing agent may be made more specific by photochemical excitation to an excited singlet electronic state. [F.J.J.]

Oxime One of a group of chemicals derived from aldehydes (RCH=NOH, aldoximes) or ketones (RR′C=NOH, ketoximes) used for isolation and identification of carbonyl compounds. In general, they are easily purified and have characteristic melting points. The properties of certain oximes have made them industrially important. Oximes have received considerable attention because of their stereochemistry and their participation in the Beckmann rearrangement. *See* ALDEHYDE; KETONE; STEREOCHEMISTRY. [L.B.C.]

Oximetry A technique that employs an oximeter, a photoelectric photometer, to measure the oxygenated fraction of the hemoglobin. This fraction is usually expressed in percent, which is referred to as the "oxygen saturation of blood."

One type of oximeter is designed to measure the oxygen saturation of blood circulating in a particular tissue of an intact animal or human. The tissue most commonly studied is the cartilaginous pinna of the ear, and the instrument used for this purpose is called an ear oximeter.

Another type of oximeter is designed to measure the oxygen saturation of blood during, or shortly after, its withdrawal from various sites in the vascular system. Such a device usually is designated a cuvette oximeter. [E.H.W.]

Oxygen A gaseous chemical element, O, atomic number 8, and atomic weight 15.9994. Oxygen is of great interest because it is the essential element both in the respiration process in most living cells and in combustion processes. It is the most abundant element in the Earth's crust. About one-fifth (by volume) of the air is oxygen. Uncombined gaseous

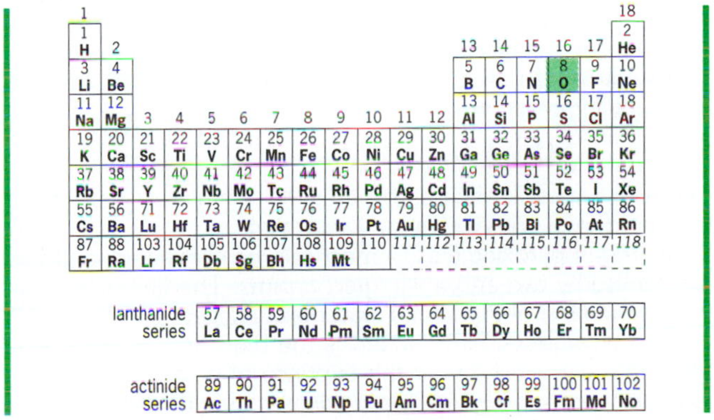

oxygen usually exists in the form of diatomic molecules, O_2, but oxygen also exists in a unique triatomic form, O_3, called ozone. *See* OZONE.

Oxygen is separated from air by liquefaction and fractional distillation. The chief uses of oxygen in order of their importance are (1) smelting, refining, and fabrication of steel and other metals; (2) manufacture of chemical products by controlled oxidation; (3) rocket propulsion; (4) biological life support and medicine; and (5) mining, production, and fabrication of stone and glass products.

Under ordinary conditions oxygen is a colorless, odorless, and tasteless gas; it condenses to a pale blue liquid. Oxygen is one of a small group of slightly paramagnetic gases, and it is the most paramagnetic of its group. Liquid oxygen is also slightly paramagnetic. Some data on oxygen and some properties of its ordinary form, O_2, are listed in the table.

Properties of oxygen

Property	Value
Atomic number	8
Atomic weight	15.9994
Triple point (solid, liquid, and gas in equilibrium)	−218.80°C (54.35 K)
Boiling point at 1 atm pressure	−182.97°C (90.18 K)
Gas density at 0°C and 1 atm pressure, g/liter	1.4290
Liquid density at the normal boiling point, g/ml	1.142
Solubility in water at 20°C, ml oxygen (STP) per 1000 g water at 1 atm partial pressure of oxygen	30

Practically all chemical elements except the inert gases form compounds with oxygen. Among the most abundant binary compounds of oxygen are water, H_2O, and silica, SiO_2, the latter being the chief ingredient of sand. Among compounds containing more than two elements, the most abundant are the silicates, which constitute most of the rocks and soil. Other widely occurring compounds are calcium carbonate (limestone and marble), calcium sulfate (gypsum), aluminum oxide (bauxite), and the various oxides of iron which are mined as a source of iron. *See* OXIDE. [A.W.F.]

Oxygen toxicity A toxic effect in a living organism caused by a species of oxygen. Oxygen has two aspects, one benign and the other malignant. Those organisms that avail themselves of the enormous metabolic advantages provided by dioxygen (O_2) must defend themselves against its toxicity. The complete reduction of one molecule of O_2 to two of water (H_2O) requires four electrons; therefore, intermediates must be encountered during the reduction of O_2 by the univalent pathway. The intermediates of O_2 reduction, in the order of their production, are the superoxide radical (O_2^-), hydrogen peroxide (H_2O_2), and the hydroxyl radical ($HO\cdot$). *See* OXYGEN; SUPEROXIDE CHEMISTRY.

The intermediates of oxygen reduction, rather than O_2 itself, are probably the primary cause of oxygen toxicity. It follows that defensive measures must deal with these intermediates. The superoxide radical is eliminated by enzymes that catalyze the following reaction.

$$O_2^- + O_2^- + 2H^+ \rightleftharpoons H_2O_2 + O_2$$

These enzymes, known as superoxide dismutases, have been isolated from a wide variety of living things.

Hydrogen peroxide (H_2O_2) must also be eliminated, and this is achieved by two enzymatic mechanisms. The first of these is the dismutation of H_2O_2 into water and oxygen, a process catalyzed by catalases. The second is the reduction of H_2O_2 into two molecules of water at the expense of a variety of reductants, a process catalyzed by peroxidases. *See* ENZYME.

The multiplicity of superoxide dismutases, catalases, and peroxidases, and the great catalytic efficiency of these enzymes, provides a formidable defense against O_2^- and H_2O_2. If these first two intermediates of O_2 reduction are eliminated, the third ($HO\cdot$) will not be produced. No defense is perfect, however, and some $HO\cdot$ is produced; therefore its deleterious effects must be minimized. This is achieved to a large extent by antioxidants, which prevent free-radical chain reactions from propagating. *See* ANTIOXIDANT; CHAIN REACTION (CHEMISTRY); FREE RADICAL; PEPTIDE.

The apparent comfort in which aerobic organisms live in the presence of an atmosphere that is 20% O_2 is due to a complex and effective system of defenses against this peculiar gas. Indeed, these defenses are easily overwhelmed, and overt symptoms of oxygen toxicity become apparent when organisms are exposed to 100% O_2. For example, a rat maintained in 100% O_2 will die in 2 to 3 days. [I.Fr.]

Oxymonadida An order of class Zoomastigophorea in the phylum Protozoa. These are colorless flagellate symbionts in the digestive tract of the roach *Cryptocercus* and of certain termites. They are xylophagous; that is, they ingest wood particles taken in by the host. Seven or more genera of medium or large size have been identified, the organisms varying from pyriform to ovoid in shape. At the anterior end a pliable necklike rostrum attaches the organism to the host intestinal wall, but they are sometimes free. They can be either uni- or multinucleate. *See* ZOOMASTIGOPHOREA. [J.B.L.]

Oxytetracycline A crystalline, amphoteric, broad-spectrum antibiotic elaborated by the actinomycete *Streptomyces rimosus*. The antibiotic is known commercially as Terramycin and is produced by fermentation processes.

Oxytetracycline is used against infectious diseases in both human and veterinary medicine. It has broad-spectrum activity against bacteria, spirochetes, rickettsiae, large viruses, and certain protozoa and metazoa. Oxytetracycline has agricultural application as an animal growth stimulant, in poultry feed to increase egg production, as a food preservative, and as a crop-spray ingredient. *See* ANTIBIOTIC. [C.R.S.]

Oxyurina A major group of the Nematoda, also known as the Oxyurata. It is one of the two suborders of Ascaridida. In one modern system of classification, it has essentially the composition and characteristics outlined under Oxyuroidea. In another, some of the component groups such as the Heterakidae, Cosmocercidae, and Kathlaniidae are assigned instead to the suborder Ascaridina. *See* ASCARIDIDA; NEMATA. [J.T.L.]

Ozone A powerfully oxidizing allotropic form of the element oxygen. The ozone molecule contains three atoms (O_3). Ozone gas is decidedly blue, and both liquid and solid ozone are an opaque blue-black color, similar to that of ink. *See* OXYGEN.

Some properties of ozone are given in the table. Ozone has a characteristic, pungent odor familiar to most persons

Some properties of ozone

Property	Value
Density of the gas at 0°C, 1 atm pressure	2.154 g/liter
Density of the liquid	
−111.9°C	1.354 g/ml
−183°C	1.573 g/ml
Boiling point at 1 atm pressure	−111.9°C
Melting point of the solid	−192.5°C

because ozone is formed when electrical apparatus produces sparks in air. Ozone is irritating to mucous membranes and toxic to human beings and lower animals.

Ozone is a more powerful oxidizing agent than oxygen, and oxidation with ozone takes place with evolution of more heat and usually starts at a lower temperature than when oxygen is used. In the presence of water, ozone is a powerful bleaching agent, acting more rapidly than hydrogen peroxide, chlorine, or sulfur dioxide. *See* OXIDIZING AGENT.

Ozone is utilized in the treatment of drinking-water supplies. Odor- and taste-producing hydrocarbons are effectively eliminated by ozone oxidation. Iron and manganese compounds which discolor water are diminished by ozone treatment. Compared to chlorine, bacterial and viral disinfection with ozone is up to 5000 times more rapid.

Ozone occurs to a variable extent in the Earth's atmosphere. Near the Earth's surface the concentration is usually 0.02–0.03 ppm in country air, and less in cities except when there is smog. At vertical elevations above 13 mi (20 km), ozone is formed by photochemical action on atmospheric oxygen. Maximum concentration of 5×10^{12} molecules/cm^3 (more than 1000 times the normal peak concentration at Earth's surface) occurs at an elevation of 19 mi (30 km). *See* ATMOSPHERIC OZONE. [A.W.F.]

Ozonolysis A process which uses ozone to cleave unsaturated organic bonds. Generally, ozonolysis is conducted by bubbling ozone-rich oxygen or air into a solution of the reactant. The reaction is fast at moderate temperatures. Intermediates are usually not isolated but are subjected to further oxidizing conditions to produce acids or to reducing conditions to form alcohols or aldehydes. An unsymmetrical olefin is capable of yielding two different products whose structures are related to the groups substituted on the olefin and the position of the double bond.

Before World War I, ozonolysis was applied commercially to the preparation of vanillin from isoeugenol. The only modern application of the technique in the United States is in the manufacture of azelaic and pelargonic acids from oleic acid. *See* ALKENE; OZONE. [R.K.Ba.]

P

Pacific islands All of the islands of the open Pacific Ocean plus the Philippines on the west. As thus defined, the islands number nearly 10,000, but they are scattered over an area of 64,000,000 mi^2 (166,000,000 km^2), roughly one-third of the Earth's surface. The total island land area is 170,000 mi^2 (440,000 km^2), or about 0.25% of the Pacific Ocean.

Most Pacific islands rise from the deep sea. All the large islands and many small ones bear volcanic high points rising hundreds, or even thousands, of feet above sea level. The great majority of the islands, however, are low, many being tiny islets on or close to surface reefs. Islands of all sizes are concentrated in the southwest Pacific, particularly in the tropical latitudes.

The typical Pacific islands lie far from continental shores and are of volcanic origin. Some are not as isolated as they appear on maps, because bathymetric charts show that they are high points on submarine ridges of one sort or another. Many other islands are in linear groups that are independent of midoceanic ridges. Still other islands are arranged in arcuate groups, either continental arcs such as the Philippines or oceanic arcs such as the Marianas and Palau. *See* OCEANIC ISLANDS. [H.S.L.]

Pacific Ocean The Pacific Ocean has an area of 6.37×10^7 mi^2 (1.65×10^8 km^2) and a mean depth of 14,000 ft (4280 m). It covers 32% of the Earth's surface and 46% of the surface of all oceans and seas, and its area is greater than that of all land areas combined. Its mean depth is the greatest of the three oceans and its volume is 53% of the total of all oceans. Its greatest depths in the Marianas and Japan trenches are the world's deepest, more than 6 mi (10 km).

The two major wind systems driving the waters of the ocean are the westerlies which lie about 40–50° lat in both hemispheres (the "roaring forties") and the trade winds from the east which dominate in the region between 20°N and 20°S. These give momentum directly to the west wind drift (flow to the east) in high latitudes and to the equatorial currents which flow to the west. At the continents there is flow of water from one system to the other and huge circulatory systems result. *See* OCEAN CIRCULATION; SOUTHEAST ASIAN WATERS.

The swiftest flow (greater than 2 knots) is found in the Kuroshio Current near Japan. It forms the northwestern part of a huge clockwise gyre whose north edge lies in the west wind drift centered at about 40°N, whose eastern part is the south-flowing California Current, and whose southern part is the North Equatorial Current.

Equatorward of 30° lat heat received from the Sun exceeds that lost by reflection and back radiation, and surface waters flowing into these latitudes from higher latitudes (California and Peru currents) increase in temperature as they flow equatorward and turn west with the Equatorial Current System. They carry heat poleward and transfer part of it to the high-latitude cyclones along the west wind drift. The temperature of the equatorward currents along the eastern boundaries of the subtropical anticyclones is thus much lower than that of the currents of their western boundaries at the same latitudes. The highest temperatures (more than 82°F or 28°C) are found at the western end of the equatorial region. Along the Equator itself somewhat lower temperatures are found. The cold Peru Current contributes to its eastern end, and there is apparent upwelling of deeper, colder water at the Equator.

Upwelling also occurs at the edge of the eastern boundary currents of the subtropical anticyclones. When the winds blow strongly equatorward (in summer) the surface waters are driven offshore, and the deeper colder waters rise to the surface and further reduce the low temperatures of these equatorward-flowing currents. *See* UPWELLING.

The limiting temperature in high latitudes is that of freezing. Ice is formed at the surface at temperatures slightly less than 30°F (−1°C) depending upon the salinity; further loss of heat is retarded by its insulating effect. The ice field covers the northern and eastern parts of the Bering Sea in winter, and most of the Sea of Okhotsk, including that part adjacent to Hokkaido (the north island of Japan). Summer temperatures, however, reach as high as 43°F (6°C) in the northern Bering Sea and as high as 50°F (10°C) in the northern part of the Sea of Okhotsk. *See* BERING SEA.

Pack ice reaches to about 62°S from Antarctica in October and to about 70°S in March, with icebergs reaching as far as 50°S. *See* ICEBERG; SEA ICE.

Surface waters in high latitudes are colder and heavier than those in low latitudes. As a result, some of the high-latitude waters sink below the surface and spread equatorward, mixing mostly with water of their own density as they move, and eventually become the dominant water type in terms of salinity and temperature of that density over vast regions.

The most conspicuous water masses formed in the Pacific are the Intermediate Waters of the North and of the South Pacific, which on the vertical sections include the two huge tongues of low salinity extending equatorward beneath the surface from about 55°S and from about 45°N. The southern tongue is higher in salinity and density and lies at a greater depth. [J.L.Re.]

Packet switching A software-controlled means of directing digitally encoded information in a communication network from a source to a destination. Switching and transmission are the two basic functions that effect communication on demand from one point to another in a communication network, an interconnection of nodes by transmission facilities. Each node functions as a switch in addition to having potentially other nodal functions such as storage or processing.

Switched (or demand) communication can be classified under two main categories: circuit-switched communication and store-and-forward communication. Store-and-forward communication, in turn, has two principal categories: message-switched communication (message switching) and packet-switched communication (packet switching).

In circuit switching, an end-to-end path of a fixed bandwidth (or speed) is set up for the entire duration of a communication or call. In store-and-forward switching, the message, either as a whole or in parts, transits through the nodes of the network one node at a time. As the name implies, the entire message, or a part of it, is stored at each node and then forwarded to the next.

In message switching, the switched message retains its integrity as a whole message at each node during its passage through the network from the source node to the destination node. For very long messages, this requires large buffers (or storage capacity) at each node. Also, in situations where one or more intermediate nodes are involved, the constraint of receiving the very last bit to the next node may result in unacceptable delays. Packet switching can be viewed as a technique which breaks a large message into fixed-size, small packets and then switches these packets through the network as if they were individual messages. This approach reduces the need for large nodal buffers and "pipelines" the resources of the network so that a number of nodes can be active at the same time in switching a long message, reducing significantly the transit delay of the original message through the network. One important characteristic of packet switching is that network resources are consumed only when data are actually sent.

The practicality and efficiency of a geographically large network using packet switching was first demonstrated in the early 1970s by ARPANET, developed by the Advanced Research Project Agency of the U.S. Department of Defense. A number of subsequent public packet networks have been developed.

Data communication (or computer communication) is the primary application for packet networks. Computer communication traffic characteristics are fundamentally different from those of voice traffic. Data traffic is usually bursty, lasting from several milliseconds to several minutes or hours. The holding time for data traffic is also widely different from one application to another. These characteristics of data communication make packet switching an ideal choice for most applications. *See* DATA COMMUNICATIONS; SWITCHING SYSTEMS (COMMUNICATIONS).

[P.K.V.]

Packing A seal usually used for high pressure as in steam and hydraulic applications. The motion between parts may be infrequent as in valve stems, or continual as in pump or engine piston rods. There is no sharp dividing line between seals and packing; both are dynamic pressure resistors under motion. Diverse materials are used for packing such as impregnated fiber, rubber, cork, or asbestos compounds. In packings, it is necessary that the surface finish of the contacting metal part be smooth for long life of the material. *See* PRESSURE SEAL.

[P.H.B.]

Pain Pain, especially in its acute form, is usually a reflection of a tissue-damaging or potentially tissue-damaging stimulus. There is a transmission system that conveys this information to the central nervous system. This phenomenon is called nociception. Pain is more complex than other sensory systems such as vision or hearing because it not only involves the transfer of sensory information to the nervous system, but produces suffering which then leads to aversive corrective behavior. In certain disease states, defects in the transmission system can of themselves generate false information to the nervous system, as though tissue damage were occurring in the periphery. An example of this is phantom limb pain, in which the individual often has a crushing type of pain in a foot that has been amputated.

Acute pain such as occurs with broken bones and other significant injuries is almost inevitably accounted for by the phenomenon of nociception and is probably a purely neurophysiological event. However, the more pain becomes a chronic phenomenon, the more such influences as psychological factors and behavior become part of the expression of pain.

Acute pain is a useful warning system. There are specific nerve paths for conducting this sensation (see illustration). Pain receptors in the skin and other tissues are nerve terminals which lack any special characteristics, and they are probably triggered by a chemical stimulus when potential tissue damage occurs. There appear to be two types of terminals: one responds to many types of painful stimuli, whereas the other specifically responds to either mechanical or thermal energy. When the terminals are stimulated, the pain (that is, nociception message) is carried along specific small sensory fibers called A-delta and C fibers. The A-delta fibers are larger and transmit the "first pain" or "fast pain." The smaller C fibers transmit a secondary dull continuous pain. These nerve fibers were traditionally believed to enter the spinal cord through the dorsal root, but it now seems that many also enter through the ventral root into the spinal cord.

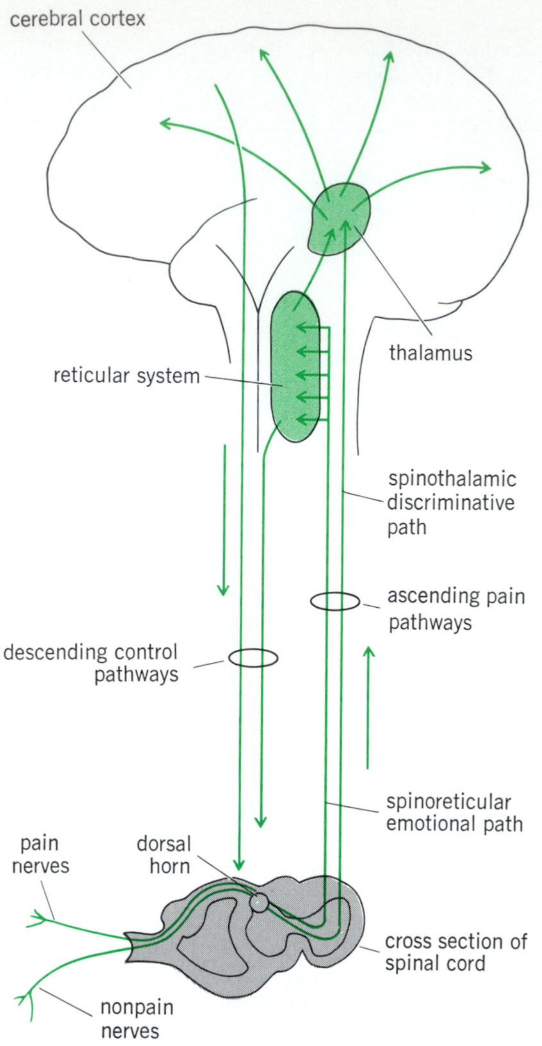

Neurophysiology of incoming pain. Sensation from peripheral receptors travels along specific pain nerves, and is modulated throughout the spinal cord and brain.

Having entered the spinal cord, these fibers relay in the dorsal horn of the spinal gray matter, an area of considerable regulation and modulation of the incoming pain stimulus which is influenced by other incoming sensory stimuli; that is, touch or pressure sensations can suppress the transmission of signals in the small pain fibers. This helps to explain why when a person is hurting, the pain can be reduced by rubbing the affected part, and this phenomenon forms the basis of some of the treatment strategies of stimulation-produced analgesia. In addition, the incoming pain signal in the spinal cord is also modulated by descending signals from the brain. At times of anxiety, these pain signals may be augmented. From these relay stations in the dorsal horn, the pain signal is carried by two nerve paths up to the brain. The classical pathway is the spinothalamic tract, on the side of the spinal cord opposite to the incoming stimulus, and this leads to the posterior part of the thalamus in the brainstem, and from there nerve paths radiate the pain sensation to many parts of the cerebral cortex, where the pain is appreciated. In addition to this direct path, there is also a diffuse ascending path known as the spinoreticular tract which relays to many of the basal ganglia in the brain, and from there to areas of the brain connected with motivational and affective behavior such as the hippocampus and the cingulate gyrus. It is possible that narcotic analgesics exert some of their action on this ascending spinoreticular tract because these drugs tend to reduce the suffering aspects of pain, but still preserve many of the discriminative qualities so that individuals can still feel the pain, but it does not bother them so much. *See* ANALGESIC; NARCOTIC.

Certain parts of the brainstem around the central canal appear to exert a strong inhibitory effect on incoming pain signals. Stimulation of these areas probably releases endorphins, which are morphinelike substances produced by the body and liberated at various sites on the incoming pain path to suppress these signals. *See* ENDORPHINS. [T.M.M.]

Paint A fluid, with viscosity, drying time, and flowing properties dictated by formulation, normally consisting of a vehicle or binder, a pigment, a solvent or thinner, and a drier, which may be applied in relatively thin layers and which changes to a solid in time. The change to a solid may or may not be reversible, and may occur by evaporation of the solvent, by chemical reaction, or by a combination of the two.

In modern technology, paint is classified in three major categories because of differing performance requirements: architectural paints, commercial finishes, and industrial coatings. A fourth category is artistic media.

Architectural paints are air-drying materials applied by brush or spray to architectural and structural surfaces and forms for decorative and protective purposes. Materials are classified by formulation type as solvent-thinned and water-thinned. The drying mechanism of solvent-thinned paints predominantly may be by solvent evaporation, oxidation, or a combination of the two. Solvent-thinned paints which dry essentially by solvent evaporation rely on a fairly hard resin as the vehicle. Resins include shellac, cellulose derivatives, acrylic resins, vinyl resins, and bitumens. In paints that dry by oxidation, the vehicle is usually an oil or an oil-based varnish. Water-thinned paints may be subdivided into those in which the vehicle is dissolved in water and those in which it is dispersed in emulsion form. Paints with water-soluble vehicles include the calcimines, in which the vehicle is glue, and casein paints, in which the vehicle is casein or soybean protein. Materials formed by emulsion polymerization are described as a latex, and products are called latex paints. *See* DRIER (PAINT); DRYING OIL; ENAMEL; POLYMER.

Commercial finishes include air-drying or baking-cured materials applied by brush, spray, or magnetic agglomeration to kitchen and laundry appliances, automobiles, machinery, and furniture and used as highway marking materials. Industrial coatings are subdivided by their intended service: corrosion-resistant coatings, high-temperature coatings, and coatings for immersion service. *See* PIGMENT; SURFACE COATING.

[C.R.Ma.; C.W.Si.]

Palaeacanthocephala An order of Acanthocephala, the adults of which are parasitic worms found in fishes, aquatic birds, and mammals. They have the following characteristics. The nuclei of the hypodermis are fragmented and the chief lacunar vessels are lateral. The males have usually 2–7 cement glands. The ligament sac in the female breaks down so that the eggs develop in the body cavity. Proboscis hooks occur in long rows and spines are present on the body of some species. Species which commonly occur in vertebrates are *Leptorhynchoides thecatus* and *Corynosoma* (see illustration). *See* ACANTHOCEPHALA. [D.V.Mo.]

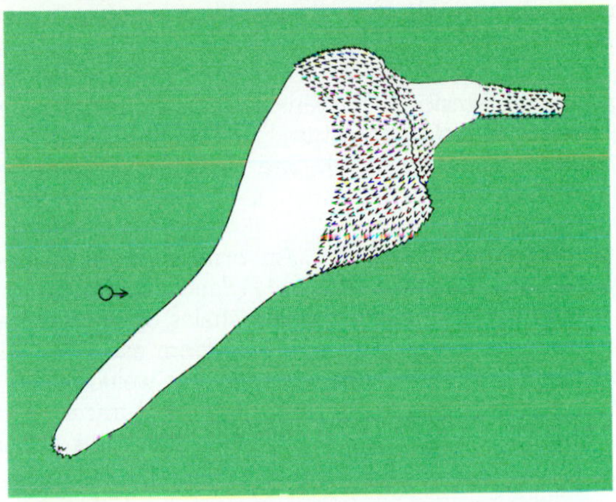

***Corynosoma reductum*. (After H. J. Van Cleave, *Acanthocephala of North American Mammals*, University of Illinois Press, 1953)**

Palaeonemertini A rarer order of the class Anopla in the phylum Rhynchocoela, characterized by an unarmed proboscis, a thin gelatinous dermis, and either a two-layered or three-layered body musculature. Many members show primitive features, such as a peripherally located nervous system and the absence of ocelli, ciliated grooves, and intestinal diverticula. Cerebral organs, if present, are generally simple. *See* ANOPLA; RHYNCHOCOELA. [J.B.J.]

Palaeonisciformes A large order of chondrostean fish which includes the earliest known and most primitive ray-finned fishes and has a long history from the Lower Devonian, or earlier, to the Cretaceous. Many genera and species, both marine and fresh-water, have been described, but the majority are poorly understood. There are two principal Paleozoic suborders: the fusiform Palaeoniscoidei and their deep-bodied derivatives, the Platysomoidei.

Palaeoniscoids are small to medium-sized fishes with a heavily ossified exoskeleton. The body is usually covered with thick rhombic scales surfaced with enamel and articulated in oblique rows by a peg and socket mechanism. Posteriorly the body is tapered to a graceful point. This slender tapered portion is turned upward in heterocercal fashion to form the dorsal or axial lobe of the caudal fin.

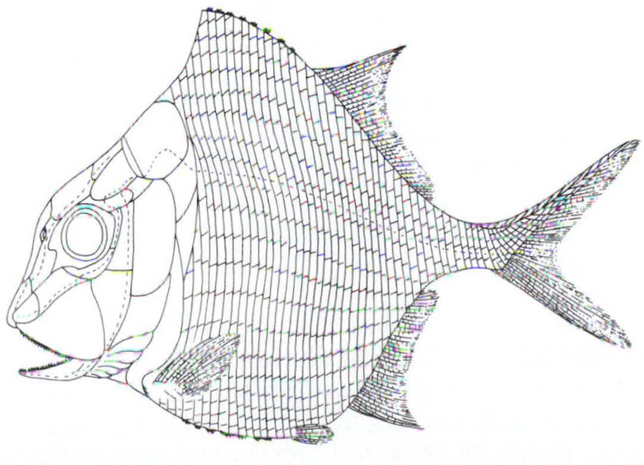

***Platysomus parvulus*, an Upper Carboniferous platysomid from Britain which attained a length of 5 in. (13 cm). (Modified from R. H. Traquair, *On the structure and affinities of the Platysomidae*, Trans. Roy. Soc. Edinburgh, vol. 29, 1879)**

Typical platysomids have a laterally compressed, rhombic-shaped body with an angle at the front of the long dorsal and anal fins, small pelvic fins or none, a short lower jaw, and an upright jaw suspension. The flank scales are much deeper than long; the tail is heterocercal and forked; the facial profile is very steep (see illustration). *See* ACANTHODII; CHONDROSTEI; CROSSOPTERYGII; OSTEICHTHYES. [T.M.C.]

Palaeospondyloidea An ordinal name assigned to the single, tiny, problematic fish *Palaeospondylus*, which is known only from Middle Devonian shales of restricted extent in Caithness, Scotland. This animal seldom exceeded 2 in. (5 cm) in length. The skeleton consists of a well-calcified skull, vertebral column, caudal fin, and occasional traces of paired pelvic fins. The relationships of *Palaeospondylus* are uncertain. [R.H.De.]

Palate The roof of the mouth in those vertebrates whose mouth cavity and nasal passages are wholly or partially separate.

The palate of mammals consists of two portions. The hard palate, more anterior in position, underlies the nasal cavity, whereas the soft palate hangs like a curtain between the mouth and nasal pharynx. The hard palate has an intermediate layer of bone. The oral surface of the hard palate is a mucous membrane, covered with a stratified squamous epithelium.

The soft palate is a backward continuation from the hard palate. Its free margin connects on each side with two folds of mucous membrane, the palatine arches, enclosing a palatine tonsil. In the midline the margin extends into a fingerlike projection named the uvula. The oral side of the soft palate continues as the covering of the hard palate, and the submucosa contains pure mucous glands.

Besides separating the nasal passages from the mouth, the hard palate is a firm plate, against which the tongue crushes and manipulates food. The soft palate, at rest, is pendant. In sucking, swallowing, or vomiting it is raised to separate the oral from the nasal portion of the pharynx. The closing action also occurs in speech, except for certain consonants requiring nasal resonance. *See* SPEECH. [L.B.A.]

Paleobiochemistry The study of chemical processes used by organisms that lived in the geological past. Most information on the nature of life in the geological past comes from the study of fossils; a record of biochemical processes that occurred can be found in the organic molecules of sedimentary rocks and fossils. The organic matter in fossil fuel deposits and finely dispersed in shales and limestones represents the debris of cells which have been chemically altered to a more stable form. A comparison of the molecular structure of these preserved organic compounds with that of components of living cells enables the researcher to identify similarities and dissimilarities between past and present biochemistry.

Paleobiochemical studies have shown that a number of the common chemical processes used by living organisms today have been in use for a very great length of time. Paleobiochemical techniques are also used on materials returned from extraterrestrial sources to determine whether life exists outside of Earth. *See* PALEOECOLOGY; PALEONTOLOGY. [T.C.H.]

Paleobotany The study of fossil plants and vegetation of the geologic past. As a branch of paleontology, it combines a knowledge of both geology and botany. Its materials are the fossilized remains of prehistoric plants preserved in the rocks, including fossil leaves, seeds, pollen, spores, and wood, and occasionally fruits and flowers. From such materials the paleobotanist attempts to reconstruct whole plants as they actually grew, as well as the ancient forests of which they were a living part. The fossil plant record also shows the succession of plants which have inhabited the Earth. Paleobotany involves not only collecting, describing, and naming fossil plants, but also aims at their interpretation in terms of the evolutionary history of plant groups, the relationships between extinct and living forms, the reconstruction of the environmental conditions under which the ancient forests lived, and the relative geologic ages of the rocks in which the fossils were found. *See* PALEONTOLOGY; PALYNOLOGY.

Most fossil plants are found in shales and fine-grained siltstones, which are the lithified muds of ancient deposits. The best collections are obtained from rocks originally laid down as deposits in floodplains, lakes, swamps or bogs, and coastal lagoons or estuaries. Volcanic ash, the lithified form of which is known as tuff, is also an excellent medium for the rapid burial and subsequent preservation of plant remains. Much knowledge of ancient plants is derived from a study of coal itself and of the fossil plants that occur both in coal and in the roof shales and underclays associated with coal beds. *See* COAL; COAL BALLS; COAL PALEOBOTANY; PETRIFACTION. [E.Do.]

Paleoceanography The study of the history of the circulation, chemistry, biogeography, fertility, and sedimentation of the oceans. It is a quantitative field that uses the geologic record to construct and test rigorous models of the behavior of the past ocean. It is this emphasis that distinguishes the field from the study of marine facies which, as old as geology itself, has made use of the sparse oceanic record preserved on the continents.

Paleoceanography relies on three separate data sets: reconstruction of the configuration of continents and oceans and of the topography of the ocean floor, a temporal framework usually biostratigraphic in nature, and sediment data bearing on ocean chemistry, fertility, circulation, sediment input, and so on. The geographic reconstruction is furnished by plate tectonics from paleomagnetic and sea-floor spreading data. *See* PLATE TECTONICS.

Environmental parameters are derived from the mineralogical, textural, chemical, and paleontological properties of ocean sediments. Calcium carbonate content is a key to the nature of the bottom waters; organic matter content informs about the level of oxygenation; silica and phosphate hold clues to the fertility of the surface water. Floral and faunal assemblages can, with careful statistical work, yield valuable numerical data on salinity, temperature, and a few other variables for times not too far in the past. Oxygen and carbon isotopes similarly provide salinity and temperature data. *See* INDEX MINERAL; MARINE SEDIMENTS; PALEONTOLOGY. [T.H.V.A.]

Paleocene The oldest of five major worldwide divisions (epochs) of the Tertiary Period (Cenozoic Era); the epoch of geologic time extending from the end of the Cretaceous Period (Mesozoic Era) to the Eocene Epoch. *See* TERTIARY.

Paleocene time encompassed two intervals of regional uplift and marine regression, separated by an interval of marked changes in the distribution of temporarily expanding seas. The preceding Cretaceous was brought to a close by a shrinking of the world's previous seaways, giving rise to local fresh-water deposition, to extensive terrestrial deposits, and to widespread erosion through regional uplifts. True mountain-making movements (orogenies) were few though locally intense. Areas of known continuous marine deposition from Cretaceous to Paleocene are scattered and restricted.

Paleocene marine life was characterized by the spread of pelecypods, gastropods, echinoids, and foraminifers already present but rare in earlier seas. Great limestone-forming foraminifers dominated topical shallow clear marine waters. At a rather sharply defined biostratigraphic horizon (end of the Cretaceous) termed the time of "the Great Dying," previously

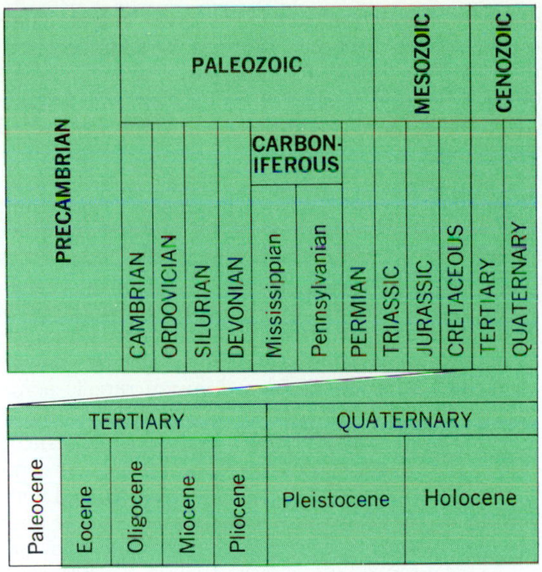

dominant cephalopod mollusks (true belemnites and ammonites) and marine reptiles (ichthyosaurs, plesiosaurs, and mososaurs) became suddenly extinct, as did the huge dinosaur reptiles on land. The outstanding feature of Paleocene land life was the appearance, proliferation, and explosive evolution of small archaic placental mammals (insectivores, creodont carnivores, condylarths, ancestral rodents, and primates), while an older marsupial mammal fauna became isolated in Australia. In the closing phases of the Paleocene, mammalian faunas indicate reestablishment of some intercontinental land connections. *See* PALEONTOLOGY. [R.M.Kl.]

Paleoclimatology That aspect of the study of climates which deals with ancient, long-term patterns of weather, usually treated on a global or regional scale. Paleoclimatology uses "climatic indicators" that are identified and analyzed by geologists; this is one aspect of paleogeography, the study of past environments. The science of paleoclimatology thus requires knowledge of both geology and atmospheric physics. *See* CLIMATOLOGY; PALEOGEOGRAPHY.

Examples of paleoclimatic indicators are: fossil palm leaves discovered in Spitsbergen and Antarctica suggesting tropical conditions there at one time; the distinctive grooving of rock surfaces by glacier ice in India and the Sahara Desert; bones of mammoths found in New York State. Continental drift and plate tectonics explain some of the anomalies, but not all.

Both modern and ancient climates are dependent upon two variables, planetary conditions and extraplanetary controls. The extraterrestrial irregularities are due, on the one hand, to the slight changes in the relationship of the Earth to the Sun, and, on the other, to extremely subtle variations of the Sun's emissions. Variables within the planet itself depend on such things as the amount of dust thrown into the atmosphere by volcanic eruptions, changes in the height of mountains, and changes in the oceanographic circulation.

The Earth's environments have passed through both repetitive events and evolutionary, noncyclic stages of development. The principal cyclic control is probably the rotation of the Galaxy, which takes place about once every 200–250 × 10^6 years. Its effects are not yet well understood, but it almost certainly affects the Earth's gravitational field and many dependent factors, not the least of which is climate.

It is known that although there have been very large climatic changes in the past, they could not have surpassed certain limits because the fossil record of extinct life shows that favorable metabolic environments (about 25±10°C or 77±18°F persisted in spite of all revolutionary events. The geologic record shows that major extinctions and new speciation tend to coincide in time with major events in paleogeography and paleomagnetic and climatic history. The specific mechanisms of all these changes constitute one of the major challenges of science. *See* EXTINCTION (BIOLOGY); SPECIATION. [R.W.F.]

Paleocopa An order of extinct ostracods, also called Palaeocopida. The outstanding characteristic is a long, straight hinge (see illustration). These ostracods lack a frontal opening and a duplicature. Lobes and sulci, vertically elongate, are conspicuous in many genera and are conventionally designated from the anterior end of the valve as L1, S1, L2, S2. Of these,

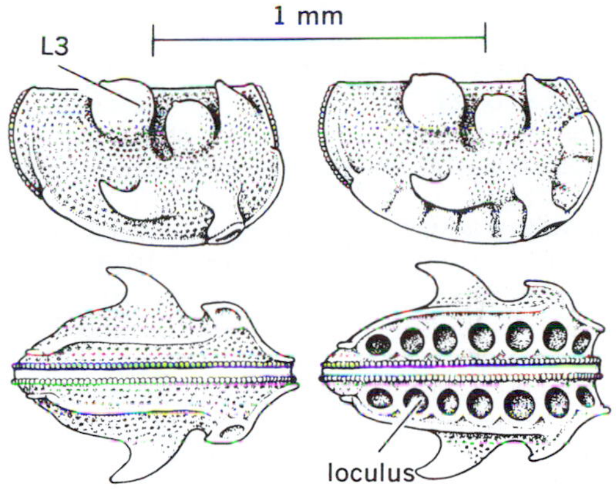

Abditoloculina pulchra **Kesling, male (left) and female carapaces in lateral view (top row) and ventral view. Devonian of New York; the bulbous L3 is typical of the family Hollinidae.**

S2 is the most developed sulcus and is present in all lobate forms; it marks the position of the internal adductor muscle scars. Soft parts have never been found preserved, making comparison with living groups uncertain. The order is divided into suborders Beyrichicopina and Kloedenellocopina. *See* OSTRACODA. [R.V.K.]

Paleoecology The ecology of the geologic past, a study of the relations of fossil organisms to each other and to the environments in which they lived. Paleoecology is a study based on inferences and interpretations and on the basic assumption that the animals and plants of the past lived under essentially the same environmental conditions as do living relatives. In studying the record as preserved in the rocks, the paleoecologist attempts, as does the ecologist, to make a complete list of the animals and plants that inhabited a particular area and to obtain an estimate of their relative abundance. *See* ECOLOGY.

Some paleoecological interpretations are based upon data obtained from the fossils themselves, others on features of the rocks that contain the fossils; in many cases data from both sources are used. The shape of a cephalopod shell or other swimming or crawling form may suggest a mode of locomotion; the thin and fragile shells of certain foraminifers or gastropods may point clearly to a pelagic existence. Corals and sedentary or burrowing pelecypods and brachiopods may be found in position of growth, indicating that they probably were buried in the place where they lived. When the members of a fossil assemblage are compared with living relatives whose life habits are known, it may become apparent that the fossil assemblage is a mixed one with representatives from more than one environment. Such an interpretation may be confirmed if some members of the group exhibit signs of breakage or wear or if only the lighter unattached valves of sedentary forms are present. Studies of the isotopic composition of the shells may indicate the temperature or some other aspect of the past environment.

The rock containing the fossils may indeed be the major source of interpretative data. The texture of the enclosing sediments may reveal much about the site of deposition. In water-laid sediments the occurrence of graded beds, that is, beds in which the texture grades upward from coarse to fine, may point to landslides or some form of turbidity current that threw the sediment into suspension before permitting final settlement. The presence of certain types of ripple marks, mud cracks, rain prints, or other sedimentary structures may aid in determining the nature of the original environment. Other interpretations may be based on a high content of organic matter or lime carbonate, or on the presence of oolites, glauconite, or nodules of phosphorite or manganese oxides. *See* SEDIMENTARY ROCKS.

Combinations of the above-mentioned types of data may give a suggestion or a clear indication, in the case of water-laid sediments, about such features as type of bottom, nearness to land, depth, agitation, and turbidity. *See* PALEONTOLOGY. [H.S.L.]

Paleogeography The geography of the ancient past. Because all the events of geologic history have taken place in the context of a geography different from that of the modern world, paleogeography has always been of central concern in studying geology. It encompasses all aspects of geographic information. Since the advent of plate tectonics, it has become apparent that certain geologic features are related to the shift of continental blocks and the opening and closing of ocean basins. Geography changes constantly through time. The correct interpretation of geologic history depends on the compilation of paleogeographic reconstructions of the geologic past. *See* CONTINENTAL DRIFT; PLATE TECTONICS.

Paleogeography provides the framework for all the geologic events that have affected the surface of the Earth. It is also vital in specialized studies, such as the distribution of organisms and community patterns in paleontology. Paleogeography also has practical value in the search for resources by providing reconstructions that can be used to locate promising regions in which to search for undiscovered resources.

Paleogeography incorporates information from many geologic disciplines. Geophysics, petrology, sedimentology, stratigraphy, and paleontology each provide different types of data. Past continental positions are determined through the use of paleomagnetism. The azimuth direction of the remanent magnetism of a rock indicates the direction to the magnetic north or south pole at the time the rock formed and therefore the original orientation of the area, which may have moved or rotated since the rocks formed. The inclination of the remanent magnetic field can be used to determine the ancient latitude. Latitude can also be inferred from the presence of climatically sensitive sediments. Longitude in the geologic past is difficult to determine because there are no preserved measures of absolute longitude as there are for latitude. However, reasonably accurate relative longitudes have been determined for most continents through time by a combination of simple spatial constraints, biogeographic relationships, and the apparent pattern of plate motion. *See* PALEOCLIMATOLOGY; PALEOMAGNETISM. [R.A.B.]

Paleoindian The term used to describe the culture of the earliest inhabitants of the Americas. Most archeologists agree that humans entered the New World from Asia via the Bering Land Bridge, but the time of the earliest immigration is still unknown. Two schools of thought have emerged concerning this question. One group of scholars believes that humans entered North America at the end of the Wisconsin glaciation (about 13,000 years ago) and that Clovis culture is the earliest recognized cultural tradition in the New World. The other group of scholars contends that the first immigration took place during the middle to early Wisconsin glacial periods (about 20,000–60,000 years ago) or earlier, and suggests that pre-Clovis cultural evidence has been or will be discovered. *See* GLACIAL EPOCH.

Two major technologies of the Asian Paleolithic are relevant to the question of the origins of the earliest Americans. The first is a crude pebble core-and-flake technology which has been in existence in southeast Asia since the Lower Paleolithic (about 1 million years ago). Some scholars suggest that peoples using this technology may have migrated across Siberia into North America prior to 20,000 years ago. Clovis technology is thought to have been developed later by one of these groups of people, with the idea diffusing rapidly across the landscape. The second technology is that of the Upper Paleolithic found in the Dyuktai culture of Siberia (about 30,000–12,000 years ago). Dyuktai technology consisted primarily of microblades, microblade cores, burins, and unifacially flaked artifacts. Dyuktai people hunted large ice age mammals such as the mammoth throughout northeast Asia, and are thought to have moved into Alaska around 12,000 years ago. *See* PALEOLITHIC.

After the Dyuktai peoples were ensconced in the extreme eastern edge of Beringia, the glaciers began to melt and the ice age ameliorated. This caused the sea levels to rise, which separated Alaska from Siberia and isolated these peoples from their Asian homeland. At the same time, the glaciers straddling the Rocky Mountains receded, opening an ice-free corridor which allowed access southward into the rest of the Western Hemisphere. It is thought that the Dyuktai people who remained in Alaska developed into the American Paleoarctic Tradition, which has a nearly identical artifact assemblage. The people who migrated southward during this time may have developed Clovis technology. [De.S.]

Paleolithic The prehistoric period when people made stone tools exclusively by chipping or flaking. John Lubbock proposed and defined the term Paleolithic, or Old Stone Age, in 1865, and also defined a subsequent Neolithic, or New Stone Age, during which some stone tools were formed by polishing or grinding. Later archeologists altered these definitions; to many today the Paleolithic is the period during which human beings lived entirely by hunting and gathering, while the Neolithic is the following interval during which plant and animal domestication was introduced. To other archeologists, the Paleolithic is simply a time interval, roughly equivalent to the Pleistocene Epoch, while the Neolithic comprises the early part of the succeeding Holocene (or Recent) Epoch. *See* GEOLOGICAL TIME SCALE; NEOLITHIC.

It is impossible to devise a rigorous, global definition of the Paleolithic or any other cultural stage, because artifact technology and economic practices have changed independently at different times in different parts of the world. Hence, Paleolithic will be used here informally to refer to the time interval between the earliest appearance of stone tools, more than 2 million years B.P. (before present), and the end of the last glacial period, 12,000–10,000 years B.P.

The oldest artifacts found so far come from sites in Ethiopia, Kenya, and Tanzania where they are dated to between 2.5 and 1.6 million years B.P. They comprise crude flakes and the modified pebbles and stone chunks from which the flakes were struck. Approximately 1.6–1.5 million years B.P., at least some people in East Africa began to manufacture the bifacially flaked tools known to archeologists as hand axes. The flaking of the first hand axes was crude, but as time passed their flaking tended to become more refined. By 1 million years B.P. hand-ax makers had spread through most of Africa and the Near East, and by 750,000 years B.P. they had reached Europe.

The Early Stone Age/Lower Paleolithic apparently persisted until sometime between 200,000 and 130,000 B.P., the exact time perhaps depending on the place. In sub-Saharan Africa the Early Stone Age was succeeded by the Middle Stone Age, while in North Africa and Eurasia the Lower Paleolithic was followed by the Middle Paleolithic. Most Middle Stone Age/Middle Paleolithic assemblages lack hand axes. The principal Middle Stone Age/Middle Paleolithic tools are well-made stone flakes, often modified by edge flaking ("retouch") into types called sidescrapers, knives, denticulates (serrate-edged pieces), and so forth. In Europe, Middle Paleolithic people were the Neandertals, *Homo sapiens neanderthalensis*. Neandertals also lived in the Near East during the early Middle Paleolithic, but were apparently supplanted by anatomically modern people (*Homo sapiens sapiens*) during the latter part (?after 70,000–60,000 B.P.). Outside of Europe and the Near East, bones representing Middle Paleolithic/Middle Stone Age people are rare, very fragmentary, or both. *See* NEANDERTALS.

In Europe, the Near East, and North Africa, the Middle Paleolithic was followed by the Upper Paleolithic, perhaps 50,000 years B.P. in the Near East, adjacent North Africa, and eastern Europe, but only 35,000 years B.P. in western Europe. The Upper Paleolithic and Later Stone Age are difficult to characterize artifactually, because wherever they are well known, they exhibit great variability in time and space. The amount of artifactual change through time and space during the Upper Paleolithic/Later Stone Age far exceeds that in earlier periods and suggests an ability to innovate that earlier people did not possess, perhaps because of biological (neurological) limitations. Conventionally, the Upper Paleolithic is said to end with the end of the Last Ice Age, 12,000–10,000 years B.P., although Upper Paleolithic artifact types and economic practices continued on for many millennia throughout much of Eurasia. In sub-Saharan Africa, the Later Stone Age persisted widely into the first millennium before Christ, and in places, particularly in southern Africa, Later Stone Age people were still present at time of first European contact, just a few centuries ago. *See* ANTHROPOLOGY; ARCHEOLOGY; FOSSIL HUMAN.

[R.G.K.]

Paleomagnetism The study of the direction and intensity of the Earth's magnetic field through geologic time. Paleomagnetism has been, and continues to be, an important tool in unraveling the past movements of the Earth's tectonic plates. By studying the records of the ancient magnetic field left in rocks, earth scientists are able to learn how the continental and oceanic plates have moved relative to the Earth's spin axis and relative to one another. In addition, the global reference frame of the Earth's magnetic field provides a very useful basis for temporal correlation of rocks on a local or global geographic scale (magnetostratigraphy). *See* GEOMAGNETISM; PLATE TECTONICS.

Many rocks acquire remanent magnetizations at or about the time they are formed. These magnetizations are nearly always parallel to the direction of the Earth's magnetic field at the locality where the rock formed. *See* ROCK MAGNETISM.

In paleomagnetic studies a suite of carefully oriented samples spanning a time interval long enough to average magnetic secular variations is collected. For magnetostratigraphy, an ordered suite of samples spanning the stratigraphic section of interest are collected. The samples are taken to the laboratory, where they are cut into small upright cylinders and their magnetization is measured by using a sensitive magnetometer.

The end product of the laboratory experiments is a suite of magnetization vector directions from the collected samples. These directions are specified by the inclination I, the angle that the magnetization vector makes with the horizontal, and the declination D, the angle that the projection of the magnetization vector upon a horizontal plane makes with true north, reckoned positive clockwise from north. Provided that the sample collection represents a sufficiently long time span to average out secular variation, representative mean D and I values and an associated uncertainty in direction may be calculated by using statistical techniques. The mean declination and inclination, together with the inclination-latitude relationship mentioned earlier and some elementary spherical trigonometry, allow the calculation of a representative paleomagnetic pole from the rock unit. By connecting paleomagnetic poles of different ages in an ordered time sequence, an apparent polar wander path (APWP) may be constructed for a particular tectonic plate (see illustration). The APWP specifies the displacement history of a plate or continent with respect to the spin axis, and can be directly compared with APWPs from other plates or continents to determine whether relative movements have occurred.

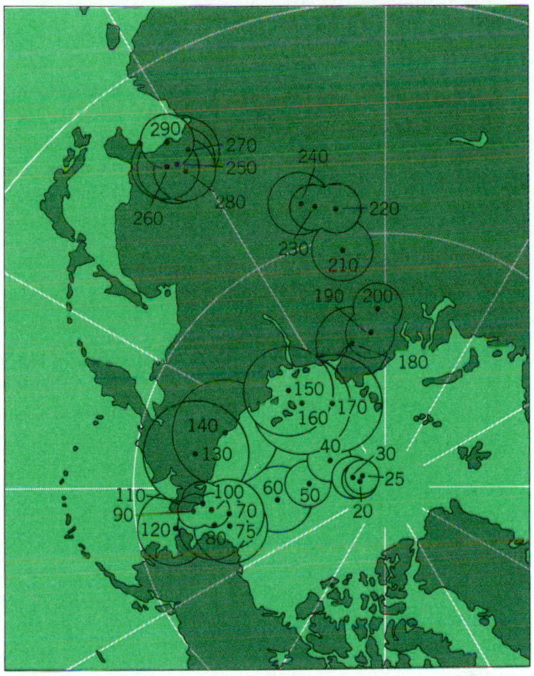

Apparent polar wander path for North America for Late Carboniferous time to the present. The numbers represent time in millions of years before present. Circles encompass standard error. *(After E. Irving, Paleopoles and paleolatitudes of North America and speculations about displaced terranes, Can. J. Earth Sci., 16:669–694, 1979)*

The end product of a magnetostratigraphic study is a set of normal (N) and reversed (R) magnetizations from the stratigraphic section under investigation. The positioning and frequency of occurrence of these N-to-R and R-to-N transitions is highly diagnostic in many cases, and by using these data together with other local geologic information, such as the position of major unconformities, one stratigraphic section can be correlated with another over considerable distances. The method can also be used over intracontinental and intercontinental distances. However, because the field has only two possible states (N or R), correlation over longer distances where tectonics and sedimentation rates may vary is correspondingly less accurate.

[M.O.McW.]

Paleontology The study of animal history as recorded by fossil remains. Paleontology lies on the boundary between two disciplines, biology and geology.

The fossil record includes a very diverse class of objects ranging from molds of microscopic bacteria in rocks more than 3×10^9 years old to unaltered bones of human fossils in ice-age gravel beds formed only a few thousand years ago. Quality of preservation ranges from the occasional occurrence of soft parts (skin and feathers, for example) to barely decipherable impressions made by shells in soft mud that later hardened to rock. *See* FOSSIL; MICROPALEONTOLOGY.

The most common fossils are hard parts of various animal groups. Thus the fossil record is not an accurate account of the complete spectrum of ancient life but is biased in overrepresenting those forms with shells or skeletons. Fossilized worms are extremely rare, but it is not valid to make the supposition that worms were any less common in the geologic past than they are now. *See* EDIACARAN FAUNA.

The data of paleontology consist not only of the parts of organisms but also of records of their activities: tracks, trails, and burrows. Dinosaur footprints, for example, are very common in the Connecticut Valley. Even chemical compounds formed only by organisms can, if extracted from ancient rocks, be considered as part of the fossil record. Artifacts made by humans, however, are not termed fossils, for these constitute the data of the related science of archeology, the study of human civilizations. *See* ARCHEOLOGY; PALEOBIOCHEMISTRY.

Although life was not present on the Earth when it first formed, the raw materials necessary for its natural development were available. Electrical discharges and ultraviolet radiation induced the components of the original atmosphere to form complex organic compounds which later combined to form the prototype of a living cell. The oldest recognizable organisms so far discovered are bacterialike cells from South African cherts dated at 3.1×10^9 years old. By the time the Gunflint cherts of Canada were deposited 2×10^9 years ago, complex algae had evolved. With the exception of a few forms found in rocks just slightly older than 6×10^8 years, invertebrate animals are not found until the base of the Cambrian Period, 6×10^8 years ago. It is not known why so many groups of invertebrates made their first appearance as fossils in rocks of the same age; perhaps the base of the Cambrian marks a time during which animals first developed hard parts. The first vertebrate fossils, primitive fishes, occur in Lower Ordovician rocks. Vertebrates invaded the land with the evolution of amphibians in the Late Devonian, about 3.5×10^8 years ago. Reptiles evolved soon afterward in Mississippian times; the first true mammals are Jurassic in age. Humans are newcomers, a product of the last few million years. Humans belong to a single species, a natural product of an unplanned evolution; yet from a geological perspective, humans have altered the Earth far more than it had changed in any comparable time since life evolved. *See* PALEOBOTANY. [S.J.G.]

Paleoseismology The study of geological evidence for past earthquakes. This scientific discipline has contributed greatly to modern understanding of the nature of earthquakes. The patterns of earthquakes, in both space and time, evolve over centuries and millennia and cannot be discovered by modern instruments. Knowledge of these patterns is important for understanding the physics of earthquakes and for forecasting future destructive earthquakes.

In certain natural environments, the features related to ancient earthquakes are preserved in the landforms and superficial layers of the Earth's surface. Geologists use this paleoseismological evidence to extend the short historical and instrumental record of earthquakes into ancient centuries and millennia. Such paleoseismological studies have clarified the earthquake record of many parts of the world, including the midcontinent and east coast of the United States, northern Africa, southern Europe, China, Japan, Indonesia, and New Zealand.

The geological preservation of ancient earthquakes has enabled scientists to compare modern earthquakes with those of the past. In 1983, for example, a sparsely populated region of Idaho was struck by a magnitude-7.3 earthquake. Subsequent investigations revealed a fresh, 30-km-long (18-mi) fault scarp running along the western base of the lofty Lost River Range. Inspection of the fresh escarpment revealed that it is surmounted by a more subdued, vegetated escarpment of nearly identical length and height. Excavations across this ancient fault scarp showed that it had formed about 5000 years earlier during an event very similar to the 1983 earthquake. This is one of several examples of what paleoseismologists call a characteristic earthquake: apparently, some earthquakes are nearly identical repetitions of their predecessors. *See* EARTHQUAKE; SEISMOLOGY. [K.Si.]

Paleosol A soil that formed or began forming on a landscape in the geologic past, that is, an ancient soil. There are three kinds of paleosols: relict, buried, and exhumed.

Relict soils formed on preexisting landscapes but subsequently were not buried under younger sediments. Buried soils are soils formed on preexisting landscapes and subsequently buried by younger sediment or rock. These soils crop out in excavations, such as stream banks or road cuts. Exhumed paleosols are soils that were buried but have been reexposed on the present land surface by erosion of the covering mantle. These soils occupy appreciable parts of the present land surface and are juxtaposed geographically to other soils more in harmony with the present environment.

Most paleosols are similar in their morphologies and properties to soils on the present land surface but are not necessarily analogs of the soils of the same area or region. Analyses of paleosols and comparison with present surface soils and their environments aid in reconstructing possible environments of landscapes of the geologic past. *See* SOIL. [R.V.R.]

Paleozoic A major segment of geologic time, representing the time of deposition of the oldest sedimentary rocks on Earth which contain identifiable fossils and whose strata can be correlated worldwide. Paleozoic strata commonly occur across all continents as thousands of feet of limestone or dolomite with less shale and sandstone, and in long, narrow geosynclinal troughs around the continents filled with even thicker silty sandstones, dark shale, and chert.

The era is divided into systems: Cambrian (beginning almost 6×10^8 years ago), Ordovician, Silurian, Devonian, Carboniferous (Mississippian and Pennsylvanian in North America), and Permian, ending about 2.25×10^8 years ago. These European-defined rock systems were based on changes in fossil life and were conceived to mark events in geologic history such as major rises and falls of sea level over the continents, mountain-building episodes, and times of intrusion of molten magma (granite) in folded belts peripheral to the continental plates. These major events of Earth history are now believed to have been caused by extensive lateral movements of continental plates opening and closing great oceans on the globe's surface. *See* PLATE TECTONICS.

Although a proto-Atlantic Ocean was present early in the Paleozoic, it had essentially disappeared by the end of the Paleozoic as the Baltic-Russian and African continents moved westward relative to North America and impinged on this continent. Late in the Paleozoic the North American plate was completely surrounded by uplifted mountain chains formed as the earlier subsiding troughs were squeezed against it.

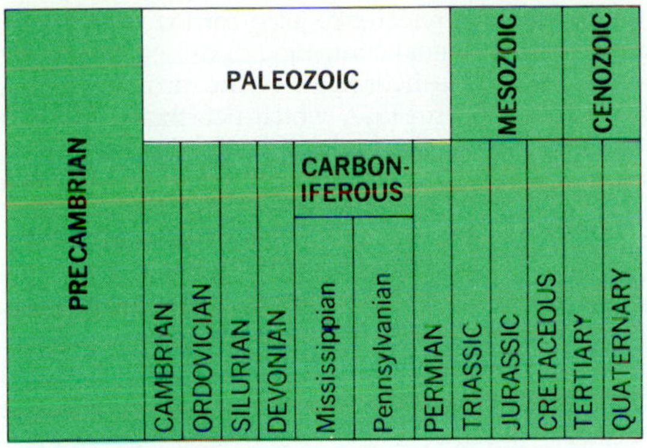

Paleozoic life was quite different from that represented by the fossils of later strata. Major groups of dominant Paleozoic invertebrate fossils such as brachiopods, trilobites, nautiloid cephalopods, and bryozoans which dwelt in the tropical seas of that time are now either extinct or very rare. The middle of the Paleozoic saw the rise of fishes, followed in Carboniferous time by amphibians. The humid equatorial climate of the Northern Hemisphere Carboniferous resulted in vast coal-producing swamps, and much oil and gas were produced by dark marine basinal shales and collected in limestone bordering shallow continental seas. Reptiles up to several meters in length dominated vertebrate life in the Permian. During this immense span of time the most ancient forms of animal and plant life had evolved or become extinct, setting the stage for modem plants, reptiles, mammals, and shellfish.

In Permian time, climate all over the world became more extreme, fluctuating between humid and dry, and in more arid times permitting deposition of redbeds and gypsum and rock salt which precipitated as vast inland saline seas evaporated. In marked contrast, a great southern continent composed of a united Africa, South America, Antarctica, Australia, and India was mostly ice-covered toward the end of the Paleozoic. The Paleozoic thus closed after 3.5×10^8 years with major changes in world climate, intense and widespread movement of continental plates, and attendant episodes of mountain building. *See* CAMBRIAN; CARBONIFEROUS; DEVONIAN; ORDOVICIAN; SILURIAN. [J.L.Wi.]

Palladium A chemical element, Pd, atomic number 46, and atomic weight 106.4. Palladium is a white and very ductile metal which resembles platinum and follows it in importance and abundance. *See* PLATINUM.

Palladium supported on carbon or alumina is used as a catalyst in a number of chemical processes including hydrogenation in liquid and gas-phase reactions. Perhaps the largest single use of pure palladium is for low-current electrical contacts. Palladium is unique in the number of metals with which it can be alloyed, and it generally produces ductile solid solutions.

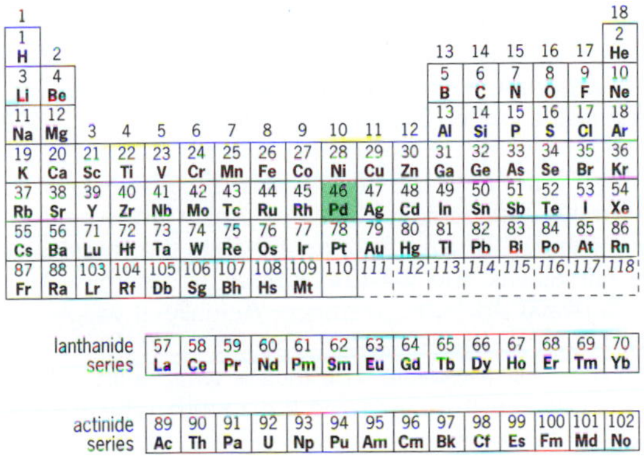

Palladium is soft and ductile and may be fabricated into fine wire and thin sheet. On heating at temperatures up to about 800°C (1472°F), palladium forms a visible oxide tarnish, PdO. This oxide is thin and adherent, not tending to scale or flake off. Above 800°C (1472°F), the oxide dissociates and the metal is bright if cooled quickly to room temperature. The values of some important properties are given in the table.

Some properties of palladium

Property	Value
Atomic weight ^{12}C = 12.00000	106.4
Naturally occurring isotopes and % abundance	102 (.96)
	104 (10.97)
	105 (22.23)
	106 (27.33)
	108 (26.71)
	110 (11.81)
Crystal structure	Face-centered cubic
Common chemical valence	2,4
Density at 25°C, g/cm^3	12.01
Melting point, °C	1554
Boiling point, °C	2900
Specific heat at 0°C, cal/g	0.0584

Hydrogen is readily absorbed by palladium and will diffuse through heated palladium at a relatively rapid rate. This property is used in hydrogen purifiers, which pass hydrogen but no other gas. Palladium in ordinary atmospheres is resistant to tarnish but will tarnish slightly in outdoor exposure to sulfur-contaminated atmospheres. At room temperature palladium is resistant to hydrofluoric, phosphoric, perchloric, acetic, hydrochloric, and sulfuric acids as gases, but may be attacked by some of them at 100°C (212°F).

Palladium chlorides and related compounds are the most important. Palladium chloride, $PdCl_2$, is used in electrodeposition; it and related chlorides are used in refining cycles and as sources of pure palladium sponge in thermal decomposition processes. Palladium monoxide, PdO, and dihydroxide, $Pd(OH)_2$, are used as sources of palladium catalysts. Sodium tetranitropalladate, $Na_2Pd(NO_2)_4$, and other complex salts are used as the basis of plating baths. [H.J.A.]

Palm Any member of the Arecaceae (Palmae), a large family of monocotyledons (Liliopsida), with about 3500 species. Most palms are trees with a slender, unbranched trunk and a terminal crown of large leaves that are folded between the veins, like a fan. As with other monocotyledonous trees, the woodiness of the stem derives not from the presence of great amounts of xylem but from the lignification of other tissues. A few palms have climbing stems with scattered leaves, and some others have very short stems so that the leaves arise at or near the surface of the ground.

It is said that there are more than 1000 uses for palms and their products. The coconut palm (*Cocos nucifera*) and the date palm (*Phoenix dactylifera*) are the two most important palms. *See* ARECALES; COCONUT; LILIOPSIDA. [A.CR.]

Palmitate A salt (soap) or ester of palmitic acid,

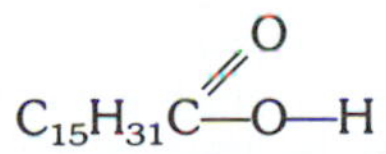

in which the acidic hydrogen has been replaced by a metal or an organic radical. Palmitates occur in nature, chiefly as the glyceryl esters, found in substantial amounts in animal and vegetable fats. The esters of long-chain alcohols are known as waxes.

Simple esters have limited uses, chiefly in plastics. Alkali metal palmitates are soluble in water and, with the similar stearates and oleates, are the major components of toilet and laundry soaps. Other metal soaps are used in lubrication greases, pharmaceuticals, and cosmetics, and as waterproofing agents and fungicides. *See* DETERGENT; ESTER; FAT AND OIL; SOAP. [E.H.H.]

Palpigradi An order of rare arachnids comprising 21 known species from tropical and warm temperate regions. American species occur in Texas and California. All are minute, whitish, eyeless animals, varying from 0.027 to 0.112 in. (0.68 to 2.8 mm) in length, that live under stones, in caves, and in other moist, dark places. The elongate body terminates in a slender, multisegmented flagellum set with setae. In a curious reversal of function, the pedipalps, the second pair of head appendages, serve as walking legs. The first pair of true legs, longer than the others and set with sensory setae, has been converted to tactile appendages which are vibrated constantly to test the substratum. *See* ARACHNIDA. [W.J.Ger.]

Palynology The scope of palynology has enlarged from the study of spores and pollen to include other microscopic organisms and microscopic fragments of megascopic plants and animals that are observed in sediments. The modern palynologist is concerned with advanced techniques of the compound microscope, with transmission and scanning electron microscopy, and with infrared, fluorescent, and ultraviolet microscopy. *See* MICROSCOPE.

Palynology can be divided into two fields, neopalynology and paleopalynology. The former is concerned with extant microorganisms and disassociated microscopic parts of megaorganisms; the latter is concerned with their fossil counterparts. *See* MICROPALEONTOLOGY; POLLEN.

Fossil spores and pollen occur in many types of rock. They are most abundant in organic sediments such as peat and coal. They are absent or rare only in the coarse elastic sediments, in some of the chemically precipitated rocks, and in rocks or soils that have been oxidized or metamorphosed after lithification. [L.R.Wi.]

Pancreas A composite gland in most vertebrates, containing both exocrine cells—which produce and secrete enzymes involved in digestion—and endocrine cells, arranged in separate islets which elaborate at least two distinct hormones, insulin and glucagon, both of which play a role in the regulation of metabolism, and particularly of carbohydrate metabolism. *See* CARBOHYDRATE METABOLISM.

Anatomy. The pancreas of mammals shows large variations. The extremes are the unique, massive pancreas of humans, and the richly branched organ of the rabbit. Usually, the main duct, the duct of Wirsung, opens into the duodenum very close to the hepatic duct. In humans, the pancreas weighs about 2.5 oz (70 g). It can be divided into head, body, and tail. Accessory pancreases are frequently found anywhere along the small intestine, in the wall of the stomach, and in Meckel's diverticulum.

The exocrine portion of the pancreas shows tubuloalveolar glands. Each terminal alveolus is called an acinus. The various acini have central cavities, which open into intralobular ducts through narrow intercalated tubes. The interlobular ducts anastomose and ultimately form the main duct of Wirsung. The activity of the acini is stimulated by secretin as well as by pilocarpine.

The endocrine portion shows cellular masses called islands or islets of Langerhans, in which the cellular cords or masses are more or less isolated by irregular spaces filled with connective tissue and blood capillaries. The two main types of cells are the alpha and the beta cells.

Between the grapelike exocrine portion with its ducts and the islands of Langerhans, it is possible to observe connective tissue septa, numerous blood vessels, and nerves.

Physiology. The pancreatic juice carried to the duodenum is a slightly alkaline liquid containing trypsinogen, which, when activated, causes the hydrolysis of the proteins into amino acids, amylase, and maltase, which act on the glucides, and lipase, which causes the hydrolysis of fatty substances. The intense stimulation of the pancreatic secretion after ingestion of food is considered to be the result of a nervous reflex originating in the mouth, and also of direct introduction of acids and fats into the duodenum, causing the liberation of a hormone called secretin into the bloodstream to stimulate the exocrine secretion.

In 1889 J. Von Mering and O. Minkowsky were the first to remove completely the pancreas of a dog, thereby causing true diabetes. F. Banting and C. Best (1922) prepared pancreatic extracts which were able to prevent the lethal effects of pancreatectomy. The same effect was obtained with extracts from pancreas in which, after ligature of the duct of Wirsung, the exocrine portion of the gland had disappeared. *See* DIABETES.

The alpha cells and beta cells in the islets are the sources of two hormones, insulin from the beta, and glucagon, also known as the hyperglycemic factor, from the alpha. The former is a hormone which influences carbohydrate metabolism, enabling the organism to utilize sugar. The latter accelerates the conversion of liver glycogen into glucose. Glucagon elevates the blood sugar level, and its effects are the opposite of those of insulin, so that the two hormones together maintain the sugar metabolism of the body in balance. When the level of sugar in the blood becomes too low, the secretion of glucagon is stimulated. *See* GLUCAGON; INSULIN. [E.L.V.C.]

Pancreas disorders The pancreas is affected by a variety of congenital and acquired diseases. Because of its dual functional role, diseases may develop in the exocrine portion of the pancreas or in the endocrine portion.

The most frequent congenital lesion of the pancreas is more appropriately designated as a developmental abnormality—ectopic or aberrant pancreas. Ectopic pancreas can be found anywhere within the gastrointestinal tract, but is more frequent in the stomach and duodenum.

Cystic fibrosis (mucoviscidosis) is a systemic disease in which mucus secretion is altered so that a viscid mucus is produced. The disease is inherited as a mendelian recessive. Cystic fibrosis affects all exocrine glands, including the acinar portion of the pancreas. Production of altered mucus leads to dilation of the exocrine ducts (cystic), destruction of acinar tissue, and replacement of the destroyed tissue by fibrous connective tissue (fibrosis). The islets are not affected by this disease. Elevation in secretion of sodium and chloride in sweat is also common.

Acute hemorrhagic pancreatitis is a serious disease of unknown etiology which causes sudden liberation of activated pancreatic enzymes that digest the pancreatic parenchyma. The digestive process leads to dissolution of fat and production of calcium soaps. In addition, rupture of pancreatic vessels occurs with resultant hemorrhage and shock. This disease is associated with bilary tract disease, especially gallstones (cholelithiasis), and with alcoholism.

Diabetes mellitus is the principal disease associated with the endocrine portion of the pancreas and is the fifth leading cause of death in the United States. *See* DIABETES. [H.T.N.]

Panda The name for two species of carnivores in the raccoon family, Procyonidae, both of which are found in Asia. The red panda (*Ailurus fulgens*) inhabits the forested regions of the Himalayas and western China. It has thick fur which gives it a heavy stocky appearance. Actually, it weighs an average of 10–12 lb (4.5–5.5 kg). The fur is black on the lower parts and red on the back. The face is white with a dark stripe from each eye to the corner of the mouth. The red panda is primarily a vegetarian, feeding on lichens, roots, and bamboo shoots. It is arboreal, nesting in a hollow tree or in the fork of a tree. The giant panda (*Ailuropoda melanoleuca*) superficially resembles a bear and may be a true member of the Ursidae. Little is known about the natural history of this animal; its diet of bamboo restricts it to the bamboo forests located in cold, damp mountainous areas of central China. The adults weigh 200–300 lb (90–135 kg) and are somewhat smaller than the brown bear. *See* CARNIVORA. [C.B.C.]

Pandanales An order of flowering plants, division Magnoliophyta (Angiospermae), in the subclass Arecidae of the class Liliopsida (monocotyledons). The order contains only the single pantropical family Pandanaceae, with nearly 900 species. They are more or less arborescent, sparingly branched monocotyledons with numerous, long, firm, narrow, parallel-veined leaves that usually have spiny margins. The flowers are small, very numerous, unisexual, and usually crowded into a dense spadix. *See* ARECIDAE; LILIOPSIDA. [A.Cr.]

Panel heating and cooling A system in which the heat-emitting and heat-absorbing means is the surface of the ceiling, floor, or wall panels of the space which is to be environmentally conditioned. The heating or cooling medium may be air, water, or other fluid circulated in air spaces, conduits, or pipes within or attached to the panel structure. For heating only, electric current may flow through resistors in or on the panels. *See* ELECTRIC HEATING.

Heat energy is transmitted from a warmer to a cooler mass by conduction, convection, and radiation. The output from heating surfaces comprises both radiation and convection components in varying proportions. In panel heating systems, especially the ceiling type, the radiation component predominates. *See* RADIANT HEATING.

When a panel system is used for cooling, the dew-point temperature of the ambient air must remain below the surface temperature of the heat-absorbing panels to avoid condensation of moisture on the panels. Panel cooling effectively prevents the disagreeable feeling of cold air blown against the body and minimizes the occurrence of summer colds. *See* COMFORT HEATING; HOT-WATER HEATING SYSTEM. [E.L.W./R.Ko.]

Pantodonta An extinct order of mammals which included the first large land animals of the Tertiary of the Northern Hemisphere. They filled some of the ecologic niches left open by the extinction of the dinosaurs and were well adapted to a life in swamps. Their feet had five toes, usually with broad hooves, but *Titanoides* had powerful digging claws. *Pantolambda*, from the middle Paleocene of North America, is the most primitive pantodont. *See* MAMMALIA. [M.C.McK.]

Pantograph A four-bar parallel linkage, with no links fixed, used as a copying device for generating geometrically similar figures, larger or smaller in size, within the limits of the mechanism. In the illustration the curve traced by point T will be similar to that generated by point S. This similarity results because points T and S will always lie on the straight line $\overline{OTS}$; triangles $\overline{OBS}$ and $\overline{TCS}$ are always similar because lengths $\overline{OB}$, $\overline{BS}$, $\overline{CT}$, and $\overline{CS}$ are constant and $\overline{OB}$ is always parallel to $\overline{CT}$. Distance $\overline{OT}$ always maintains a constant proportion to distance $\overline{OS}$ because of the similarity of the above triangles. Numerous modifications of the pantograph as a copying device have been made. *See* FOUR-BAR LINKAGE.

A second use of the pantograph geometry is seen in the collapsible parallel linkage used on electric locomotives and rail cars to keep a current-collector bar or wheel in contact with an overhead wire. Two such congruent linkages in planes parallel to the train's motion are affixed securely on the top of the locomotive with joining horizontal members perpendicular to each other. The uppermost member collects the current, and powerful springs thrust the configuration upward with sufficient pressure normally to make low-resistance contact from wire to collector. [D.P.Ad.]

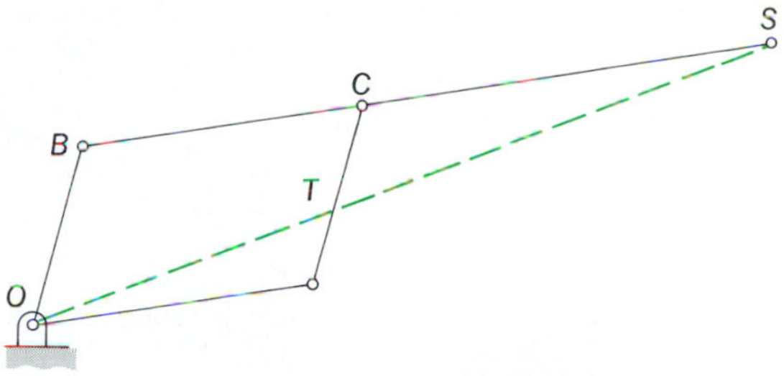

Similar triangles of a pantograph.

Pantothenic acid A member of the B vitamin group with the structural formula

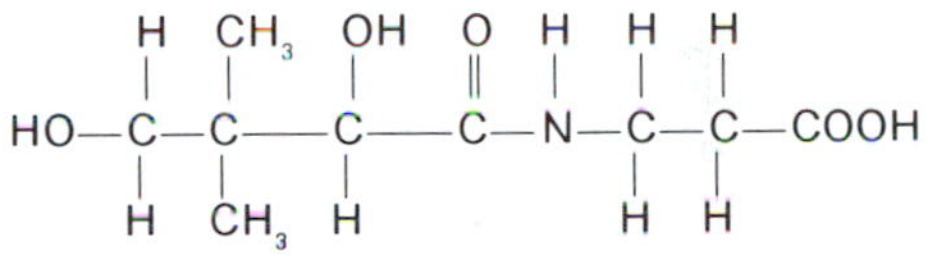

It is a light-yellow, viscous oil which is readily soluble in H_2O. Pantothenic acid is widely distributed, and liver, kidneys, fresh green vegetables, and egg yolks are among its best sources. Losses of the vitamin during cooking are minimal, as it is present in stable conjugated form in food.

No definite pathologic lesions due to a specific pantothenic acid deficiency have been reported in humans. The factors affecting the requirement for pantothenic acid are probably similar to those altering the needs for the other B vitamins. Pantothenic acid has been reported to improve the reactions of young men to stress. The widespread occurrence of pantothenic acid assures protection against the deficiency state under most conditions. Most Americans eat 10 mg of pantothenic acid per 2500 cal of good diet. The daily requirement is probably about 3–5 mg. *See* VITAMIN. [S.N.G.]

Pantotheria The most primitive of the three intraclasses of the mammalian subclass Theria. The early members of this infraclass retained many reptilian features of the jaws, but by the end of the Jurassic they were fully mammalian and gave rise to the marsupials and placentals at about the beginning of the Cretaceous. *See* EUTHERIA; METATHERIA; THERIA. [M.C.McK.]

Papaverales An order of flowering plants, division Magnoliophyta (Angiospermae), subclass Magnoliidae of the class Magnoliopsida (dicotyledons). The order consists of only 2 families, Papaveraceae and Fumariaceae, with about 300 species each. Within its subclass, the order is marked by its syncarpous gynoceium, parietal placentation, and only two (seldom three) sepals. Most of the species are herbaceous, and many of them contain isoquinoline alkaloids similar to those in the order Ranunculales. The Papaveraceae, with regular flowers, numerous stamens, and a well-developed latex system, include the poppies (*Papaver* and related genera; see illustration), bloodroot, (*Sanguinaria*), and celandine (*Chelidonium*). The Fumariaceae, with four or six stamens, irregular flowers that usually have some of the petals spurred or saccate, and no latex system, include the bleeding heart (*Dicentra spectabilis*)

Oriental poppy (*Papaver orientale*). (*Photograph by John H. Gerard, from National Audubon Society*)

and some other common ornamentals. *See* MAGNOLIIDAE; MAGNOLIOPSIDA; POPPY; RANUNCULALES. [A.Cr.]

Paper The term "paper" is traditionally applied to felted or matted sheets of cellulose fibers, formed on a fine wire screen from a dilute water suspension, and bonded together as the water is removed and the sheet is dried. In present-day usage, paper may also include sheet materials produced from other types of fibers (particularly mineral or synthetic) formed and bonded by other means.

Raw materials. Of the many raw materials used by the paper industry, cellulose fibers occupy the dominant position although other fibrous materials, particularly some of the synthetic high polymers, are assuming increasing importance.

Wood is the primary source of cellulose fibers for papermaking. Before use, the wood must be reduced to the fibrous state; this operation is called pulping. Commercial pulping operations are of three principal types: mechanical, full chemical, and semichemical.

Mechanical pulping involves the reduction of wood to the fibrous state by purely mechanical means. Logs, of either 2- or 4-ft (0.6- or 1.2-m) length, are ground into pulp by means of large revolving grindstones. Except for a few water-soluble components, all the constituents of the wood are present in the ground wood pulp.

Full chemical pulping employs chemical reagents to effect a separation of the cellulose fibers from the other wood components. Wood chips are cooked with suitable chemicals, usually at elevated temperatures and pressures. The object is to dissolve the lignin and other extraneous compounds, leaving the cellulose intact and in fibrous form. The kraft or sulfate process is the most extensively employed. Here the active pulping ingredients are sodium hydroxide and sodium sulfide, in a strongly alkaline solution. Almost any wood species can be pulped by this process.

Semichemical pulping includes several processes in which mild chemical action, followed by mechanical attrition, is used. These processes are employed largely, but not exclusively, on deciduous wood species.

Other sources of cellulose fibers are significant, although of secondary importance. Cotton rags were for many years the principal raw material of the industry. Grasses, reeds, and agricultural residues find considerable use principally in Europe and the Far East. The reuse of waste papers is common, particularly for the lower grades of paper products.

Manufacture. Paper manufacture is largely a mechanical operation, although the chemical and physicochemical aspects are all-important in determining the final sheet properties. The tendency of cellulose fibers to bond together, when dried from a water suspension, is basic to papermaking technology.

Unmodified cellulose fibers, as obtained from pulping and bleaching operations, are generally unsuited for papermaking. They must first be refined; the refining operation is conducted mechanically in beaters or refiners.

During refining, the pulp fibers are separated, crushed, frayed, fibrillated, and cut. They imbibe water and swell, becoming more flexible and more pliable. Their capacity to bond with one another on drying is greatly enhanced, partly through modification of the fiber surfaces and partly because of the creation of new surface area.

A variety of additive materials are introduced to the papermaking pulps during stock preparation. Fillers, such as clays, titanium dioxide, or calcium carbonate, are used for the control of sheet opacity and for other secondary reasons. Many paper grades are sized to impart a degree of water resistance; rosin is the most commonly used sizing agent. Dyestuffs are used extensively for color control, even with white sheets. Other additives, such as wet-strength agents, deflocculating agents, and defoamers, are used as they are needed.

The continuous paper machine converts a very dilute aqueous suspension of fibers and other ingredients into a dry sheet of paper at speeds which may vary from a few feet to over $\frac{1}{2}$ mi (about 1 m to 0.8 km) per minute. The machine may be one city block in length; even though the operations are complex, the principles involved are relatively few and straightforward.

The Fourdrinier machine is one of the two major sheet-forming devices in widespread use. It consists in essence of a continuously moving wire belt or screen, to which the dilute papermaking slurry is fed and from which the wet, formed sheet is removed continuously. The slurry issues through the slice onto the wire, as a jet of uniform and proper thickness, speed, and consistency.

The wet sheet at about 20% consistency is removed from the wire and transferred to a supporting woolen felt, which carries it through a series of press rolls. Here more water is expressed mechanically, and the sheet progresses to the drier section at about 33% consistency. The remaining water is removed by evaporation.

The second major type of paper machine, the cylinder machine, differs from the Fourdrinier only in the forming part. Here, in place of the moving wire, one or a series of rotary cylindrical filters is used. Each screen-covered cylinder is mounted in a vat and operates partially submerged in the dilute papermaking slurry being supplied to it. As the cylinder revolves, water drains through the screen to the interior of the cylinder, and the wet sheet is formed. This sheet is removed at the top of the cylinder.

Many grades of paper receive some type of treatment after formation and drying in order to enhance certain desirable characteristics. Machine calendering is one of the most common operations. Here the sheet is passed through the nips formed by a series of steel rolls, one held on top of the other. Machine coating is practiced extensively on book and magazine papers. The coating is often clay, contained in a high-solids aqueous suspension which includes suitable adhesives and other ingredients.

Paper products. The products of the paper industry are extremely varied. Throughout the centuries paper has retained its traditional uses, but the phenomenal growth of the industry during the past few decades has been due largely to new applications and uses of paper and paper-based materials. These new applications for the most part involve converted products. Two common machine converting operations were mentioned above, Supercalendering, embossing, coating, waxing, laminating, impregnating, saturating, corrugating, and printing are but a few of the common operations. The corrugated paperboard shipping container is used widely, and represents one of the largest single uses of paper. Multiwall sacks are used extensively for handling granular materials. Food packaging, with its trend toward the small unit packages, has led to extensive paper utilization, with the paper often coated, waxed, resin-impregnated, or combined with other foils and films. Building papers and boards constitute a sizable fraction of the industry output. *See* CELLULOSE; WOOD CHEMICALS. [R.P.Wh.]

Paprika A type of pepper, *Capsicum annuum* (order Polemoniales), with nonpungent flesh, grown for its long red fruit. It is of American origin, but is most popular in Hungary and adjacent countries. Seeds are removed from the mature fruit, and the flesh is dried and ground to prepare the dry condiment commonly referred to as paprika. California is the only important producing state in the United States. *See* PEPPER; SOLANALES. [H.J.C.]

Parablastoidea A small class of primitive blastozoan echinoderms containing three genera found in the early Middle Ordovician in eastern Canada; northeastern, eastern, south-central, and western United States; and near Leningrad, Russia. Parablastoids have a bud-shaped theca or body with well-developed pentameral symmetry. A stem with one-piece columnals attached the theca to the sea floor, suggesting that parablastoids were attached, medium- to high-level suspension feeders. Although they converged on blastoids in thecal design and way of life, parablastoids had differences in their plating, ambulacra, and respiratory structures to indicate a separate origin and history. That is the justification for assigning parablastoids and blastoids to different classes. *See* CRINOZOA; ECHINODERMATA. [J.Sp.]

Parabola A member of the class of curves that are intersections of a plane with a cone of revolution. It is obtained (see illustration) when the cutting plane is parallel to an element of the cone. *See* CONIC SECTION.

In analytic geometry the parabola is defined at the locus of points (in a plane) equally distant from the fixed point F (focus) and a fixed line (directrix) not through the point. It is symmetric about the line through F perpendicular to the directrix.

The curve has numerous other properties of interest in both pure and applied mathematics. For example, the trajectory of an artillery shell, assumed to be acted upon only by the force of gravity, is a parabola.

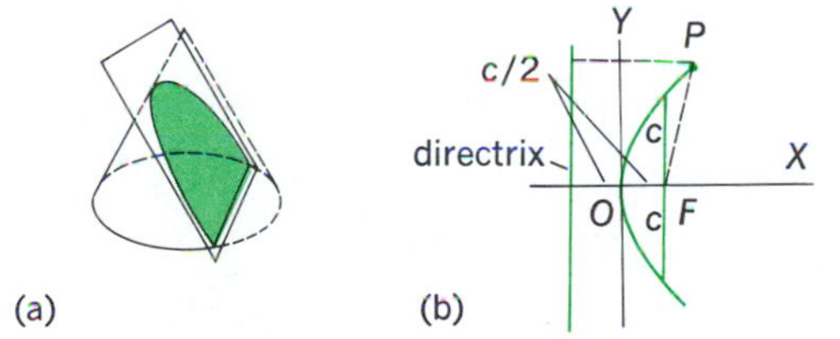

Parabola as (*a*) conic section and (*b*) locus of points.

Archimedes found the area bounded by an arc of a parabola and its chord; for example, the area bounded by the parabola $y^2 = 2cx$ and its latus rectum (the chord through F perpendicular to the axis) is $\frac{2}{3}c^2$. *See* PARABOLOID. [L.M.Bl.]

Paraboloid A quadric surface (a surface having an equation of second degree in three variables) which has an axis such that every section of the surface by a plane parallel to the axis is a parabola or straight line, whereas every section by a plane perpendicular to the axis is a central conic with center on the axis. If these central conics are all ellipses (or circles or a single point), the paraboloid is called an elliptic paraboloid (see illustration); and in particular if the ellipses are circles, the paraboloid is a paraboloid of revolution. Otherwise, these sections normal to the axis are hyperbolas (or a pair of lines through the axis), and the paraboloid is a hyperbolic paraboloid.

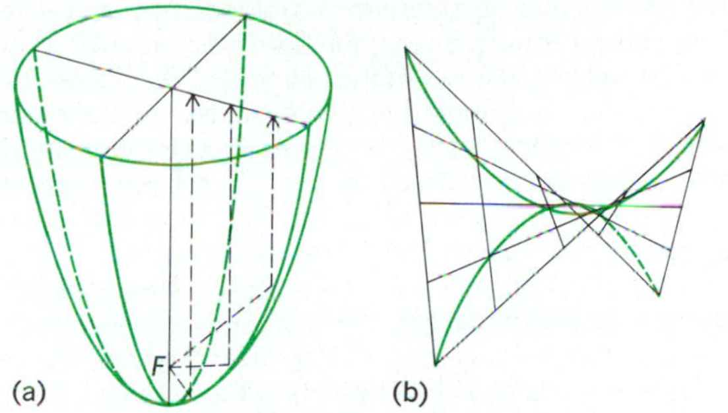

Two types of paraboloid. (*a*) Paraboloid of revolution, and (*b*) hyperbolic paraboloid.

Paraboloids of revolution are used as reflecting surfaces in automobile headlights and reflecting telescopes. Rays parallel to the axis are all reflected through a single point, called the focus, and rays from the focus are reflected parallel to the axis. *See* ANALYTIC GEOMETRY; MIRROR OPTICS; PARABOLA; QUADRIC SURFACE; SURFACE AND SOLID OF REVOLUTION. [J.S.F.]

Parachute A flexible, lightweight structure, generally intended to retard the passage of an object within or through atmosphere by materially increasing the resistive surface. A parachute is a decelerator or air-braking device in the general form of an oblate hemisphere. The parachute is the only suitably demonstrated device for emergency descent from aircraft. It comprises a canopy and cords, which form the suspension and attachment between canopy and object. A parachute canopy is a membrane which relies upon pressure differential across it to maintain its inflated shape. The differential is created by entrapment of an air mass on the inside and movement of the air on the outside. [S.E.W.]

Paraffin A term used variously to describe either a waxlike substance or a group of compounds. The former use pertains to the high-boiling residue obtained from certain petroleum crudes. It is recovered by freezing out on a cold drum and is purified by crystallization from methyl ethyl ketone. Paraffin wax is a mixture of 26- to 30-carbon alkane hydrocarbons; it melts at 52–57°C (126–135°F). Microcrystalline wax contains compounds of higher molecular weight and has a melting point as high as 90°C (190°F). The name paraffin was formerly used to designate a group of hydrocarbons—now known as alkanes. *See* ALKANE. [A.L.H.]

Paragonimiasis Presence of the fluke *Paragonimus westermani* encapsulated in the lungs or other tissues of humans. The Orient is the classic endemic focus, but cases of human infection have occurred in the Americas. The disease is connected with eating raw or pickled fresh-water crustaceans. Eggs of the fluke are passed in sputum or feces, embryonate in water, and hatch into the ciliated miracidium, which penetrates into snails (*Melania* sp.). Cercariae finally emerge and encyst in the crustacean. When the crustacean is ingested by humans, the metacercariae migrate by the way of the abdominal cavity and the diaphragm to the lungs. Symptoms resemble those of various lung diseases, such as tuberculosis, bronchiectasis, and cancer. *See* DIGENEA. [J.F.M.]

Parainfluenza virus A member of the paramyxovirus subgroup of the myxoviruses which are associated with a variety of respiratory illnesses. The virus particles range in size from 90 to 200 nanometers, agglutinate red blood cells, and (like the influenza viruses) contain a receptor-destroying enzyme. They differ from the influenza viruses in their large size, their possession of the larger ribonucleoprotein helix characteristic of the paramyxoviruses, their tendency to lyse as well as agglutinate erythrocytes, and their generally poor growth in eggs. In culture, these viruses grow well in primary monkey or human epithelial cells. *See* Animal virus; Influenza; Myxovirus; Paramyxovirus. [J.L.Me.]

Parallax (astronomy) The difference in direction to a celestial object from two separated points of observation. The distance between observational points is the base line. The base line may be provided by the diurnal motion of Earth around its axis, in which case the parallax is diurnal or geocentric; it may be provided by the motion of the Earth in its orbit about the Sun, in which case the parallax is annual or heliocentric; or it may be provided by the motion of the solar system within the local galaxy, in which case the parallax is secular, as illustrated.

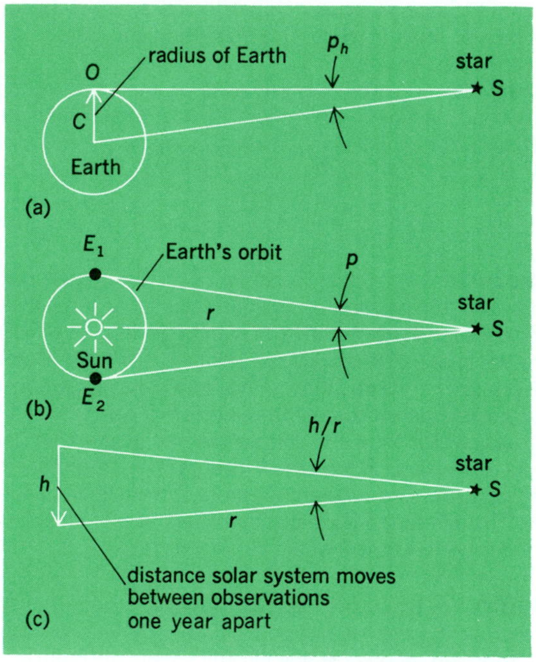

Three types of parallax. (*a*) Geocentric or horizontal. (*b*) Annual. (*c*) Secular.

The size of Earth produces a geocentric parallax noticeable in observations of objects within the solar system. Horizontal parallax p_h is defined as the angle between the directions to the object S at the horizon as seen from the center C of Earth and from the observer's location O on the surface of Earth, The mean equatorial horizontal parallax refers to Earth's equatorial radius (3963 mi or 6378 km) as seen from the mean distance of the celestial object.

The size of the Earth's orbit produces a heliocentric parallax noticeable in observations of the nearer stars. This parallax plays a fundamental role in the determination of the distances to stars and is of basic importance in the study of the physical properties of stars. Stellar parallax or annual parallax is the maximal angle subtended by one astronomical unit at the star's location.

The motion of the solar system seems attractive for the purpose of measuring stellar distances, because time alone provides an indefinite extension of the base line. However, there are severe limitations because of the motions of the stars themselves; hence only values of the average parallax for groups of stars, called mean secular parallaxes, may be measured. The method has proved important to extending geometric knowledge of the Milky Way system, in a statistical fashion, by measuring average distances up to several thousand light-years for groups of stars. [P.V.deK.]

Parallel circuit An electric circuit in which the elements, branches (elements in series), or components are connected between two points with one of the two ends of each component connected to each point. The illustration shows a simple parallel circuit. In more complicated electric networks

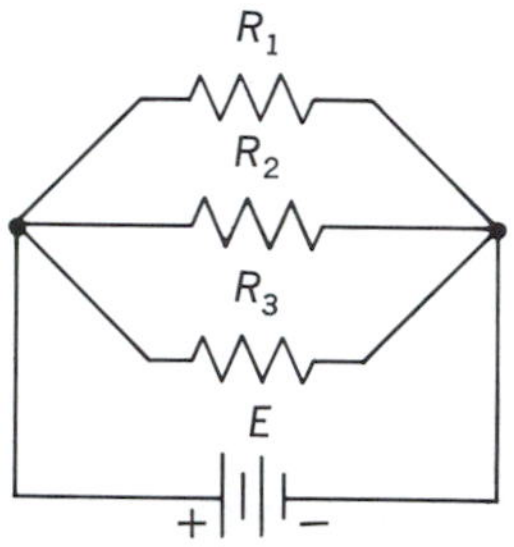

Schematic of a parallel circuit. *E* is a battery; R_1, R_2, and R_3 are resistors.

one or more branches of the network may be made up of various combinations of series or series-parallel elements. *See* Circuit (electricity).

In a parallel circuit the potential difference (voltage) across each component is the same. However, the current through each branch of the parallel circuit may be different. For example, the lights and outlets in a house are connected in parallel so that each load will have the same voltage (120 volts) but each load may draw a different current (0.5 ampere in a 60-watt lamp and 10 amperes in a toaster). [C.F.G.]

Parallelepiped A polyhedron having six faces that are parallel in pairs. Each face is a parallelogram (see illustration). If adjacent edges are perpendicular, the parallelepiped is called a rectangular parallelepiped—or in common speech, a rectangular box. The formula $V = Bh$ means that the volume of any

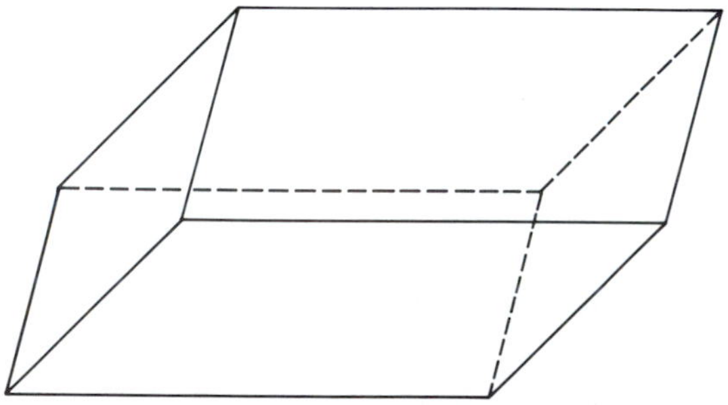

Diagram of a parallelepiped.

parallelepiped is equal to the product of its base (that is, the area of a face) times the altitude (that is, the distance from the base to the parallel face). For a rectangular parallelepiped, the volume is equal to the product of the lengths of three edges that meet at a vertex. *See* Polyhedron; Prismatoid and prismoid; Solid (geometry). [J.S.F.]

Parallelogram A quadrilateral, or four-sided polygon, having (1) opposite sides parallel or (2) opposite sides equal

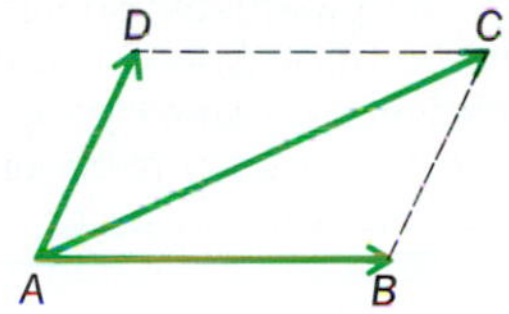

Diagram showing addition of vectors.

and coplanar, or (3) a pair of opposite sides equal and parallel. Its opposite angles are equal, its adjacent angles are supplementary, and its diagonals bisect each other. Its area is the product of its base and altitude. The parallelogram law for the addition of vectors states that, if two vectors are represented by the directed segments $\overrightarrow{AB}$ and $\overrightarrow{AD}$, their vector sum is represented by the directed diagonal $\overrightarrow{AC}$ of the parallelogram *ABCD* (see illustration). *See* QUADRILATERAL; RECTANGLE; SQUARE. [J.S.F.]

Paralysis agitans A syndrome of the central nervous system caused by diffuse pathology of the brain, resulting in tremor, rigidity, and other symptoms.

From an etiological viewpoint, the syndrome may be divided into paralysis agitans (Parkinson's disease), postencephalitic parkinsonism, and symptomatic paralysis agitans in combination with liver disease (Wilson's disease) and arteriosclerosis. The pathological lesions of all these disorders are similar and involve degenerative lesions in the basal ganglia of the cerebral hemispheres and in the cerebral cortex. *See* PARKINSON'S DISEASE.

The course is progressive, but decades may pass before patients are severely disabled. Therapy is symptomatic and includes drugs such as belladonna root and its various components, hyoscine, atropine, stramonium, and scopalamine. [F.C.R.]

Paramagnetic resonance Magnetic resonance arising from electrons in paramagnetic substances or from electrons in paramagnetic centers in diamagnetic substances. For applied fields of several thousand gauss (several tenths of a tesla) the electron paramagnetic resonance (EPR) experiments are done at microwave frequencies, commonly at 3-cm or at 1-cm wavelengths. In some instances, nuclear resonance apparatus has been used with correspondingly lower applied fields. The most sensitive apparatus detects approximately 10^{12} electron spins for a line 1 gauss (10^{-4} tesla) broad, a sensitivity far greater than obtained by nonresonant methods, such as those utilizing paramagnetic susceptibility. *See* ELECTRON PARAMAGNETIC RESONANCE (EPR) SPECTROSCOPY; MAGNETIC RESONANCE. [C.P.S.]

Paramagnetism A property exhibited by substances which, when placed in a magnetic field, are magnetized parallel to the field to an extent proportional to the field (except at very low temperatures or in extremely large magnetic fields). Paramagnetic materials always have permeabilities greater than 1, but the values are in general not nearly so great as those of ferromagnetic materials. Paramagnetism is of two types, electronic and nuclear.

The following types of substances are paramagnetic:

1. All atoms and molecules which have an odd number of electrons. According to quantum mechanics, such a system cannot have a total spin equal to zero; therefore, each atom or molecule has a net magnetic moment which arises from the electron spin angular momentum. Examples are organic free radicals and gaseous nitric oxide.

2. All free atoms and ions with unfilled inner electron shells and many of these ions when in solids or in solution. Examples are transition, rare-earth, and actinide elements and many of their salts. This includes ferromagnetic and antiferromagnetic materials above their transition temperatures. For a discussion of these materials *see* ANTIFERROMAGNETISM; FERRIMAGNETISM; FERROMAGNETISM.

3. Several miscellaneous compounds including molecular oxygen and organic biradicals.

4. Metals. In this case, the paramagnetism arises from the magnetic moments associated with the spins of the conduction electrons and is called Pauli paramagnetism.

Relatively few substances are paramagnetic. Aside from the Pauli paramagnetism found in metals, the most important paramagnetic effects are found in the compounds of the transition and rare-earth elements which have partially filled 3*d* and 4*f* electron shells respectively.

Electronic paramagnetism arises in a substance if its atoms or molecules possess a net electronic magnetic moment. The magnetization arises because of the tendency of a magnetic field to orient the electronic magnetic moments parallel to itself.

Nuclear paramagnetism arises when there is a net magnetic moment due to the magnetic moments of the nuclei in a substance. Nuclear magnetic moments are about 10^3 times smaller than electron magnetic moments. As a result, nuclear paramagnetism produces effects 10^6 times smaller than electron paramagnetic or diamagnetic effects. *See* DIAMAGNETISM; MAGNETIC RESONANCE; NUCLEAR MOMENTS. [E.Ad.; F.Ke.]

Paramecium A single-celled protozoan belonging to family Parameciidae of the order Holotrichia which have cilia that usually cover the entire body surface. There are about nine species. Some species are large and have a cosmopolitan distribution. The majority of species occur in fresh water, especially in stagnant pools, and are divided into two morphologically distinct groups: aurelia and bursaria. Included in the first group are elongate, cigar-shaped organisms which appear to be nearly circular in cross section, and which have a rounded posterior end, and a temporary structure known as the cytopyge or anal opening located on the side. The bursaria group contains organisms which are short and broad, flat in cross section, with a blunt posterior end and a nearly terminal cytopyge. The beating cilia are the locomotor organelles which propel paramecia in a spirally rotating manner.

Asexual reproduction by transverse binary fission is characteristic. In conjugation, there is a temporary fusion of cells with an exchange of nuclear material. Conjugation alters the genetics of the exconjugants since there is an exchange of nuclear material involving two different physiological variants or mating types. Conjugation only occurs between different strains. *See* HOLOTRICHIA; REPRODUCTION (ANIMAL). [C.B.C.]

Parameter An auxiliary variable, functions of which give the coordinates of a curve or surface. The coordinates of a curve are functions of one parameter. A curve in 3-space has parametric equations (1).

$$x = f(t) \quad y = g(t) \quad z = h(t) \qquad (1)$$

The coordinates of a surface are functions of two parameters, shown in Eqs. (2).

$$x = f(u,v) \quad y = g(u,v) \quad z = h(u,v) \qquad (2)$$

An arbitrary constant in an equation is also called a parameter. Variations in the values of the parameter generate a system of equations which may represent a family of curves or surfaces. Such families are called one-parameter, two-parameter, and so on, according to the number of independent parameters. *See* PARAMETRIC EQUATION. [L.Br.]

Parametric amplifier A highly sensitive low-noise amplifier for ultrahigh-frequency and microwave radio signals, utilizing as the active element an inductor or capacitor whose reactance is varied periodically at another microwave or ultrahigh frequency. A varactor diode is most commonly used as the variable reactor. Amplification of weak signal waves occurs through a nonlinear modulation or signal-mixing process which produces additional signal waves at other frequencies. This process may provide negative-resistance amplification for the applied signal wave and increased power in one or more of the new frequencies which are generated. *See* VARACTOR.

There are several possible circuit arrangements for obtaining useful parametric amplification. The two most common are the up-converter and the negative-resistance amplifier. In both types, the pump frequency is normally much higher than the input-signal frequency. In the up-converter, a new signal wave is generated at a higher power than the input wave. In the negative-resistance device, negative resistance is obtained for the input-signal frequency, causing an enhancement of signal power at the same frequency. *See* NEGATIVE-RESISTANCE CIRCUITS.

The most important advantage of the parametric amplifier is its low level of noise generation. The parametric amplifier finds its greatest use as the first stage at the input of microwave receivers where the utmost sensitivity is required. Its noise performance has been exceeded only by the maser. Maser amplifiers are normally operated under extreme refrigeration using liquid helium at about 4 K above absolute zero (− 452°F). The parametric amplifier does not require such refrigeration but in some cases cooling to very low temperatures has been used to give improved noise performance that is only slightly poorer than the maser. *See* AMPLIFIER; MASER. [M.E.Hi.]

Parametric arrays Arrays of sources (or receivers) of sound formed by variation of appropriate parameters of the propagation medium. Normally, these parameters are the local sound speed and the particle velocity which vary because of the presence of large-amplitude pump, or primary, sound waves.

The usual parametric source configuration simply consists of a directional transducer (often a plane piston or planar array) driven at two frequencies near the transducer resonance, forming a dual-frequency sound beam called the primary beam. Because sound-wave propagation is not a completely linear process, signals at new frequencies are formed effectively through the interaction of sound with sound as the beam progresses and are generated along the length of the primary beam. The lowest of these new frequencies is the difference of the two primary frequencies, and so the primary beam acts as an end-fire array of sources at the difference frequency. The effective length of the array will be determined by the attenuation of the primary beam, which occurs either as a result of small-signal absorption or, for sufficiently high primary amplitudes, as a result of nonlinear losses due to the generation of harmonics of the primary frequencies and other intermodulation components, such as the sum-frequency component.

Most applications of parametric sources have been to underwater acoustics, but their use in air, as well as in other media, may be expected. Because the effective length of a parametric source can be made quite long in practice, it is possible to generate highly directional difference-frequency beams, and because the primary amplitude is shaded very gradually along the length of the array, these beams can be made practically side-lobe-free, in contrast to the beams from conventional acoustic sources. As a result, echoes from a parametric source exhibit practically no reverberation, whereas conventional echoes may be obscured by reverberation from reflection of the side lobes. Thus, the parametric source may be expected to be useful in reverberation-limited situations where one desires a narrow beam from a small projector. Such applications include precision fathometry, subbottom profiling, echo ranging, communications, and Doppler navigation logs. In order to obtain the advantages of a parametric source, however, one must be willing to tolerate low efficiency and low search rate. *See* DIRECTIVITY; ECHO SOUNDER; SONAR; SOUND; UNDERWATER SOUND; UNDERWATER TRANSDUCER. [M.B.M.]

Parametric equation A type of mathematical equation used, typically, to represent curves in a plane or in space of three dimensions, In principle, however, there is no limitation to any particular number of dimensions. A parameter is actually an independent variable. In elementary analytic geometry a curve in the xy plane is often studied, in the first instance, as the locus of an equation $y = F(x)$ or $G(x,y) = 0$. The form $y = F(x)$ is not adequate for the complete representation of certain curves, whereas the form $G(x,y) = 0$ may be adequate. The circle $x^2 + y^2 - 16 = 0$ affords an example. But the form $G(x,y) = 0$ is not always convenient. The parametric form $x = f(t)$, $y = g(t)$ is often the most convenient; moreover, it is often the naturally occurring form of representation of the curve. For the circle $x^2 + y^2 - 16 = 0$, one possible parametric representation is $x = 4 \cos t$ and $y = 4 \sin t$.

A pair of equations $x = f(t)$, $y = g(t)$, where f and g are continuous functions defined for some interval of values of t, for example, $a \leq t \leq b$, is said to define a parametric curve. If one thinks of t as time, the equations define the motion of the point (x,y) as t increases from a to b. Clearly the path can cross itself, double back on itself, or the point may even remain motionless.

A parametric surface in space of three dimensions is defined by $x = f(u,v)$, $y = g(u,v)$, $z = h\ (u,v)$, where f, g, h are continuous functions of the two parameters, u,v. *See* ANALYTIC GEOMETRY; CALCULUS; PARTIAL DIFFERENTIATION. [A.E.Ta.]

Paramo A biological community, essentially a grassland, covering extensive high areas in equatorial mountains of the Western Hemisphere. Geographically, paramos are limited to the Northern Andes and adjacent mountains. Paramos occur in alpine regions above timberline and are controlled by a complex of climatic and soil factors peculiar to mountains near the Equator. The richly diverse flora and the fauna of the paramos are adapted to severely cold, mostly wet conditions. Humans have found some paramos suitable for living and use. Since paramos were defined originally in Spanish-speaking countries, no English equivalent has been given to the term. [H.G.B.]

Paramyxovirus A subgroup of myxoviruses that includes the viruses of mumps, measles, parainfluenza, respiratory syncytial (RS) disease, and Newcastle disease. Like influenza viruses, the paramyxoviruses are ribonucleic acid (RNA)–containing viruses and possess an ether-sensitive lipoprotein envelope. *See* ANIMAL VIRUS; MEASLES; MUMPS; MYXOVIRUS; NEWCASTLE DISEASE; PARAINFLUENZA VIRUS. [J.L.Me.]

Paranoia A mental disease characterized by logically systematized delusions of persecution. The chief symptoms are systematic, consistent, and persistent delusions of persecution, far exceeding character traits of hypersensitivity, suspicions and querulousness, irascibility, and pettiness which are found in paranoid personalities. The hostility, seclusiveness, and cynicism of paranoiacs make contact with them difficult. The cause of paranoia is not clearly established. *See* PARANOID STATE; PSYCHOSIS. [F.C.R.]

Paranoid state A severe and chronic psychosis with predominant delusions of persecution. The delusions, in contrast to those of paranoia, are inconsistent, illogical, and often accompanied by hallucinations. Otherwise, the personality does not show the disintegration which occurs in schizophrenia. However, some schools of psychiatry do not separate this mental disorder from the schizophrenias. *See* PARANOIA; SCHIZOPHRENIA. [F.C.R.]

Parapertussis An infection of the human respiratory tract caused by the bacillus *Bordetella parapertussis.* It resembles mild pertussis, from which it can be distinguished only by bacteriological methods. The disease is widespread and is often confused with mild whooping cough. The course of the disease is about 3 weeks and complications are rare. Only symptomatic care is usually necessary. [W.L.Br.]

Parasitic castration Destruction of the reproductive organs by parasites. Direct castration occurs when parasites penetrate the sex organs and feed on them, as trematodes do in the gonads of mollusks. Indirect castration results when parasites cause the gonads to atrophy. Some parasitic barnacles cause the regression of gonads of crabs but do not actually invade the gonads. Changes in the secondary sex characters commonly result.

In plants, the term parasitic castration has been employed to describe the action of smut fungi in the grain of wheat and the conversion of stamens and carpels into petals in plants infested by nematodes. *See* RHIZOCEPHALA; SMUT (MICROBIOLOGY). [P.G.Re.]

Parasitic oscillation An undesired oscillation which may occur in any type of circuit, such as an audio-, video-, or radio-frequency amplifier, oscillator, modulator, or pulse waveform generating circuit. For example, it often happens that with no apparent input signal to an amplifier an output voltage of considerable magnitude is obtained. The amplifier may be oscillating because some part of the output is inadvertently being fed back into the input. This feedback may result from the output impedance of the power supply. If feedback does occur through the power supply impedance, the oscillations can usually be stopped by the use of appropriately placed decoupling networks. *See* AMPLIFIER; FEEDBACK CIRCUIT; OSCILLATOR.

Feedback may also occur through the interelectrode capacitance from grid to plate of a tube, through lead inductances, stray wiring, and other paths, which are often difficult to determine exactly. Parasitic oscillations represent a waste of power, a distortion of the desired waveform, or a complete malfunctioning of the circuit. Hence these oscillations must be eliminated. This can usually be accomplished by a change in circuit parameters, a rearrangement of wiring, some additional bypassing or shielding, a change of tube or transistor, the use of an rf (radio-frequency) inductor in the plate circuit, the use of rf chokes in series with the filament lead, and so on. *See* ELECTRICAL NOISE; ELECTRICAL SHIELDING; GROUNDING. [J.Mi.]

Parasitology A branch of biology which deals with those organisms, plant or animal, which have become dependent on other living creatures. The essential criterion of parasitism is dependency, the loss of freedom to live an independent existence; all degrees of dependency obtain, from transitory symbiosis to complete helplessness. There is a direct correlation between the degree of dependency and the extent of parasitism. Plants are primarily free-living and consequently parasitism is far more prevalent in the animal kingdom. Here it is both extensive and intensive, since few, if any, species are free from attack and since the parasites are often present in enormous numbers. Parasitism has appeared in every phylum of animals, where the individuals themselves are parasites or serve as hosts to parasitic forms. Some parasites are temporary or facultative, but most are obligatory, that is, unable to survive apart from their hosts. Parasitism involves a gradual and progressive adaptation on the part of the parasite, and recovery of an independent status becomes increasingly difficult.

Host animals are more or less seriously affected by parasitic infestation, but if well nourished, they may not manifest patent illness. If undernourished, the host becomes increasingly debilitated and the development or activity of the sexual organs is inhibited; parasitic castration is the ultimate result. Parasites may occlude ducts, produce lesions that permit bacterial invasion, or produce substances toxic to the host. *See* MEDICAL PARASITOLOGY; PARASITIC CASTRATION. [H.W.S.]

Parasympathetic nervous system A portion of the autonomic system. It consists of two neuron chains, but differs from the sympathetic nervous system in that the first neuron has a long axon and synapses with the second neuron near or in the organ innervated. In general, its action is in opposition to that of the sympathetic nervous system, which is the other part of the autonomic system. It cannot be said that one system, the sympathetic, always has a excitatory role and the other, the parasympathetic, an inhibitory role; the situation depends on the organ in question. However, it may be said that the sympathetic system, by altering the level at which various organs function, enables the body to rise to emergency demands encountered in flight, combat, pursuit, and pain. The parasympathetic system appears to be in control during such pleasant periods as digestion and rest. The alkaloid pilocarpine excites parasympathetic activity while atropine inhibits it. *See* AUTONOMIC NERVOUS SYSTEM; SYMPATHETIC NERVOUS SYSTEM. [D.B.W.]

Parathyroid gland An endocrine organ usually associated with the thyroid gland in the neck region and possessed by all vertebrates except the fishes. In response to lowered serum calcium concentration, a hormone is produced which promotes bone destruction and inhibits the phosphorus-conserving activity of the kidneys.

In humans, there are typically four glands situated as shown in the illustration; however, the number varies between three and six, with four appearing about 80% of the time. Variations in the positioning of the glands along the craniocaudal axis

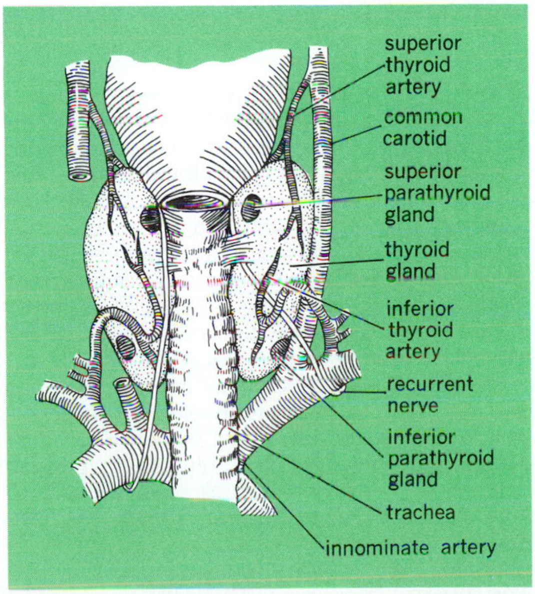

Common positions of human parathyroid glands on the posterior aspect of the thyroid. (*After W. H. Hollinshead, Anatomy of the endocrine glands, Surg. Clin. N. Amer., 21(4):1115–1140, 1952*)

occur but, excepting parathyroid III which may occasionally be found upon the anterior surface of the trachea, the relation to the posterior surface of the thyroid is rarely lost. [W.E.D.]

The site of action of the parathyroid hormone has not been finally settled. Three viewpoints have been proposed: the renal (phosphaturic) theory, the bone theory, and the citrate theory.

The renal theory is based on the finding that hypoparathyroid subjects respond to parathyroid hormone by a prompt urinary excretion of phosphate. By directly acting on the kidney, the hormone promotes the elimination of phosphate, resulting in a lowering of serum phosphate, withdrawal of phosphorus from the skeleton, and mobilization of calcium, thus elevating the levels of serum calcium.

The bone theory proposes that the primary effect of parathyroid hormone is stimulation of osteoclastic activity of bone. The preponderance of evidence supports the view that the parathyroid glands function in calcium homeostasis through a primary effect of the hormone on bone.

According to the citrate theory, the hormone causes an increase of the concentration of citrate ion in the serum which, in turn, exerts a powerful solubilizing action on hydroxylapatite and on bone minerals. None of these theories has been established definitively. It appears likely that several or all of these mechanisms may be operative simultaneously. *See* ENDOCRINE SYSTEM (VERTEBRATE); PARATHYROID HORMONE. [C.D.T.]

Parathyroid hormone The only known secretory product of the parathyroid glands. It is a polypeptide of 76 amino acids. The hormone is of prime importance in regulating calcium and phosphate metabolism, and thereby bone turnover. It is released into the bloodstream whenever the calcium concentration in blood plasma falls below normal levels. The released hormone acting upon three organs, kidney, bone, and intestinal tract, brings about the restoration of a normal blood calcium concentration. The hormone increases the absorption of calcium and phosphate in the intestine, retention of calcium and excretion of phosphate by the kidney, and resorption of both calcium and phosphate from bone; and inhibits the formation of new bone. The net result of all these changes is to elevate plasma calcium and depress plasma phosphate. Thus a negative feedback relationship exists between the action of parathyroid hormone, the blood calcium level, and the rate of parathyroid hormone secretion. *See* BONE; CALCIUM METABOLISM; PARATHYROID GLAND; PHOSPHATE METABOLISM. [H.Ras.]

Parazoa A name proposed for a subkingdom of animals which includes the sponges. Erection of a separate subkingdom for the sponges implies that they originated from protozoan ancestors independently of all other Metazoa. This theory is supported by the uniqueness of the sponge body plan and by peculiarities of fertilization and development. Much importance is given to the fact that during the development of sponges with parenchymella larvae, the flagellated external cells of the larva take up an internal position as choanocytes after metamorphosis, whereas the epidermal and mesenchymal cells arise from what was an internal mass of cells in the larva. These facts suggest that either the germ layers of sponges are reversed in comparison with those of other Metazoa or the choanocytes cannot be homologized with the endoderm of other animals. Either interpretation supports the wide separation of sponges from all other Metazoa to form the subkingdom Parazoa or Enantiozoa. *See* METAZOA; PORIFERA. [W.D.H.]

Parenchyma A ground tissue chiefly concerned with the manufacture and storage of food. The primary functions of plants, such as photosynthesis, assimilation, respiration, storage, secretion, and excretion—those associated with living protoplasm—proceed mainly in parenchymal cells. Parenchyma is frequently found as a homogeneous tissue in stems, roots, leaves, and flower parts. Other tissues, such as sclerenchyma, xylem, and phloem, seem to be embedded in a matrix of parenchyma; hence the use of the term ground tissue with regard to parenchyma is derived. The parenchymal cell is one of the most frequently occurring cell types in the plant kingdom. *See* PLANT ANATOMY; PLANT PHYSIOLOGY.

Typical parenchyma occurs in pith and cortex of roots and stems as a relatively undifferentiated tissue composed of polyhedral cells that may be more or less compactly arranged and show little variation in size or shape. The mesophyll, that is, the tissue located between the upper and lower epidermis of leaves, is a specially differentiated parenchyma called chlorenchyma because its cells contain chlorophyll in distinct chloroplastids.

This chlorenchymatous tissue is the major locus of photosynthetic activity and consequently is one of the more important variants of parenchyma. Specialized secretory parenchymal cells are found lining resin ducts and other secretory structures. *See* PHOTOSYNTHESIS; SECRETORY STRUCTURES (PLANT). [R.L.Hu.]

Paresis, general An inflammatory and degenerative disease of the brain caused by an infection with *Treponema pallidum*. It is also called syphilitic meningoencephalitis. The disease is characterized by psychotic behavior with a profound and progressive intellectual and emotional deterioration and neurological changes. The disease, unless treated, takes a rapid downhill course and results in death. *See* PSYCHOSIS; SYPHILIS. [F.C.R.]

Paresthesia One of several designations given to abnormally intense and disagreeable pain arising from apparently trivial stimulation. Other terms used to identify the same symptom are spontaneous pain, overreaction or overresponse, paradoxic pain, hyperpathia, protopathic pain, dysesthesia, and central pain. Indeed, there are more different names available than there are good descriptions of the pain. *See* PAIN. [F.A.G.]

Pareto's law A law (sometimes called the 20–80 rule) describing the frequency distribution of an empirical relationship fitting the skewed concentration of the variate-values pattern. The phenomenon wherein a small percentage of a population accounts for a large percentage of a particular characteristic of that population is an example of Pareto's law. When the data are plotted graphically, the result is called a maldistribution curve. To take a specific case, an analysis of a manufacturer's inventory might reveal that less than 15% of the component part items account for over 90% of the total annual usage value.

The mathematics required to calculate and graph the curve of Pareto's law is simple arithmetic. It should be noted, however, that the calculations need not be done in all cases. It may suffice to merely make a rough approximation of a situation in order to determine whether or not Pareto's law is present and whether benefits may subsequently accrue. [V.M.A.]

Parity (quantum mechanics) A physical property of a wave function which specifies the wave function's behavior under simultaneous reflection of all spatial coordinates of the wave function through the origin, that is, when x is replaced by $-x$, y by $-y$, and z by $-z$. If the wave function ψ satisfies Eq. (1), it is said to have even parity. If, on the other hand, Eq. (2) holds, the wave function is said to have odd parity. These two expressions can be combined in Eq. (3), where

$$\psi(x,y,z) = \psi(-x, -y, -z) \qquad (1)$$

$$\psi(x,y,z) = -\psi(-x, -y, -z) \qquad (2)$$

$$\psi(x,y,z) = P\psi(-x, -y, -z) \qquad (3)$$

$P = \pm 1$ is a quantum number having only the two values $+1$ (designated as even parity) and -1 (odd parity). The physical property defined by P is quantized and is called parity. More precisely, parity is defined as the eigenvalue of the operation of space inversion. Parity is a concept that has meaning only for waves and therefore has no meaning in classical particle physics.

Parity does have a meaning for the Schröinger wave function of a particle in quantum mechanics. It likewise has meaning for the wave function of any system. *See* QUANTUM MECHANICS.

The conservation of parity is a consequence of the inversion symmetry of space. Thus, parity would be conserved if the statement of physical laws were independent of the handedness of the coordinate system used. Of course, the fact that most people are right-handed is not a physical law but an accident of evolution; there is nothing in the laws of physics which favors a right-handed over a left-handed human. The same holds for optically active organic compounds, such as the amino acids. However, the statement that the neutrino is left-handed *is* a physical law.

All the strong interactions between elementary particles (for example, nuclear forces) and the electromagnetic interactions are symmetrical to inversion, so that parity is conserved by these interactions. As far as is known, only the ß-interactions (which involve neutrinos) and the other weak interactions are not symmetrical to inversion and do not conserve parity. The weak interactions contribute little to all processes except the decays of elementary particles (including ß-decay of nuclei), so that in all other processes parity is very nearly conserved. *See* ELEMENTARY PARTICLE; FUNDAMENTAL INTERACTIONS.

Atomic and nuclear energy states are characterized by a definite parity (which may be different for different energy states of the same nucleus), and the conservation of parity has an important bearing on atomic and nuclear reactions.

One of the selection rules which derives from parity conservation is the following: The same spin-zero boson cannot decay into both two π-mesons and three π-mesons, because these final states have opposite parities, even and odd respectively. But the positive K-meson is observed to do just this: It has both the K_2 and K_3 decay modes, and its spin is zero as deduced from the distribution of momenta in the K_3 mode. Thus the conclusion is that parity is not conserved in this decay. In 1956, T. D. Lee and C. N. Yang made the bold hypothesis that parity also is not conserved in ß-decay. They reasoned that the magnitude of the ß-decay coupling is about the same as the coupling which leads to decay of the K-meson, so these decay processes may be manifestations of a single kind of coupling. Also, there is a very natural way to introduce parity nonconservation in ß-decay, namely by assuming a restriction on the possible states of the neutrino (two-component theory). *See* NEUTRINO. [C.J.G.]

Parkinson's disease A chronic, progressive disorder of the central nervous system, marked by slow movement and muscular rigidity, weakness, and tremor at rest.

The typical patient is wide-eyed, staring, has a masklike face, walks with short shuffling steps, the body bent forward and the arms held stiffly at the sides. Tremor is marked when a limb is at rest and may become so pronounced that involuntary "pill-rolling" movements of the fingers and thumb appear. The tremor usually disappears upon movement, to be replaced by a cogwheel type of motion. Unless other brain areas have been affected by the precipitating agent, no mental changes are seen.

The disorder is slowly progressive and leads to increasing incapacity. Treatment has been largely symptomatic and directed toward the relief of pain and cramps and the development of the best mental and occupational situation possible under the circumstances. Developments in neurosurgery and chemotherapy have produced dramatic results in certain cases. Excellent reversal of the Parkinson state has been achieved by treatment with L-dopa. [N.K.M.]

Parrot Any member of the single family Psittacidae, which comprises the order Psittaciformes. There are about 316 species in 82 genera and all are essentially similar, with the exception of a few extreme forms. In addition to the forms commonly known as parrots, the family includes the parakeets, cockatoos, macaws, and lories.

One of the outstanding features of this group is the bill, which is short, stout, and strongly hooked and is used, along with the feet, in climbing. The degree of mobility of the parrots' upper jaw exceeds that seen in most other birds. These birds are fruit and seed eaters and have a well-developed crop. The wings are strong and rounded and while most species can fly rapidly for short distances, a few species have almost lost the ability to fly. Parrots are typically birds of forested areas and are distributed in large numbers throughout the tropics in Africa, Asia, South America, and Australia. Australian parakeets live on the ground principally, but perch and nest in trees. Plumage is brightly colored and most commonly green, but many species are multicolored.

Many species of this family are an important source of an infection of humans known as psittacosis, which has a high incidence among bird breeders and handlers. *See* PSITTACIFORMES; PSITTACOSIS. [C.B.C.]

Parsec A unit of measure of astronomical distances. One parsec is equivalent to 3.084×10^{13} kilometers, or 1.916×10^{13} miles. There are 3.26 light-years in 1 parsec. The parsec is defined as the distance at which the semimajor axis of Earth's orbit around the Sun (1 astronomical unit) subtends 1 second of arc. Thus, because the angle is small, the equation below holds. A parsec is then 206,265 astronomical units. At a

$$\frac{1 \text{ astronomical unit}}{1 \text{ parsec}} = 1 \text{ second} = \frac{1}{206{,}265}$$

distance of 1 parsec, the parallax is 1 second of arc. The nearest star is about 1.3 parsecs distant; the farthest known galaxy is several billion parsecs. *See* PARALLAX (ASTRONOMY). [J.L.Gr.]

Parsley A biennial, *Petroselinum crispum*, of European origin belonging to the plant order Umbellales. Parsley is grown for its foliage and is used to garnish and flavor foods. It contains large quantities of vitamins A and C. Two types, plain-leafed and curled, are grown for their foliage; Hamburg parsley (*P. crispum* var. *tuberosum*), also called turnip-rooted parsley, is grown for its edible parsniplike root. *See* APIALES. [H.J.C.]

Parsnip A hardy biennial, *Pastinaca sativa*, of Mediterranean origin belonging to the plant order Umbellales. The parsnip is grown for its thickened taproot and is used primarily as a cooked vegetable. Exposure of mature roots to low temperatures, not necessarily freezing, improves the quality of the root by favoring the conversion of starch to sugar. *See* APIALES. [H.J.C.]

Parthenocarpy The development of fruit without fertilization of the ovule. There are two types of natural parthenocarpy. Vegetative parthenocarpy is fruit set and development without the stimulation of either pollination or fertilization, so that it is endogenously controlled. Stimulative parthenocarpy requires pollination, but not fertilization, of the ovule for fruiting to occur. Examples of fruit in the first category are oriental persimmons, cultivated banana, Washington Navel orange, and pineapple; and in the second category, Black Corinth grape.

Three generic classes of chemical regulators are involved in parthenocarpy: auxins, cytokinins, and gibberellins. Depending on the stage of fruiting, each chemical regulator may have an independent action or may be dependent on another regulator for a fruit to respond and proceed through a normal developmental sequence. *See* AUXIN; CYTOKININS; FRUIT; GIBBERELLIN. [R.H.B.]

Parthenogenesis A special type of sexual reproduction in which an egg develops without entrance of a sperm. It is common among rotifers, plant lice or aphids, thrips, many ants, bees, and wasps, and some crustaceans. Males are unknown in certain thrips and rotifers. Queen honeybees produce drones, or males, by parthenogenesis but also lay fertilized eggs that yield females (the workers) and queens. Aphids have successive generations of parthenogenetic females in spring and summer, then produce both sexes by parthenogenesis. These later mate; the females lay fertilized eggs that hatch in spring as females, and parthenogenesis begins. [T.I.S.]

Partial differentiation A mathematical operation performed on functions of more than one variable. In this article only two or three variables are considered; however, the principles apply to functions of n variables, for any positive integer $n > 1$. If $z = f(x,y)$, the partial derivative $\partial z/\partial x$ is defined as the derivative of $f(x,y)$ with respect to x, y being regarded as fixed; that is,

$$\frac{\partial z}{\partial x} = \lim_{h \to 0} \frac{f(x+h,y) - f(x,y)}{h}$$

Another notation for $\partial z/\partial x$ is $f_1(x,y)$. The other first partial derivative is $\partial z/\partial y$, also written $f_2(x,y)$. For values at particular points the notation is

$$\left(\frac{\partial z}{\partial x}\right)_{(a,b)} = f_1(a,b)$$

In the case of a function of three variables, $f(x,y,z)$, the expression is

$$\frac{\partial f}{\partial z} = f_3(x,y,z)$$

The second derivatives of $f(x,y)$ are given by

$$f_{11}(x,y) = \frac{\partial}{\partial x}\left(\frac{\partial f}{\partial x}\right) \qquad f_{12}(x,y) = \frac{\partial}{\partial y}\left(\frac{\partial f}{\partial x}\right)$$

$$f_{21}(x,y) = \frac{\partial}{\partial y}\left(\frac{\partial f}{\partial y}\right) \qquad f_{22}(x,y) = \frac{\partial}{\partial y}\left(\frac{\partial f}{\partial y}\right)$$

It can happen that $f_{12}(x,y) \neq f_{21}(x,y)$, but this will not happen in common practice, especially with elementary functions. If f_1, f_2, f_{12}, f_{21} are defined in neighborhood of (a,b), and if f_{12}, f_{21} are continuous at (a,b), then $f_{12}(a,b) = f_{21}(a,b)$. In addition, there are more delicate theorems relating to this matter.

The notion of the differentiability of a function is fundamental in the theory of partial differentiation. The requirement that $f(x,y)$ be differentiable is not the same as the requirement that $f_1(x,y)$ and $f_2(x,y)$ both exist; it is a more inclusive requirement. The geometric meaning of f being differentiable at (a,b) is that the surface defined by $z = f(x,y)$ has a tangent plane not parallel to the z axis when $x = a$, $y = b$. In analytic terms the condition is that if

$$\epsilon = f(a+h, b+k) - f(a,b) - f_1(a,b)h - f_2(a,b)k$$

then

$$\lim_{(h,k)\to(0,0)} \frac{\epsilon}{|h| + |k|} = 0$$

A sufficient condition that f be differentiable at (a,b) is that the partial derivatives f_1, f_2 be defined at all points near (a,b), and continuous at (a,b).

The prime importance of the differentiability concept is that the differentiability property is needed in proving the chain rule for functions of several variables. This rule asserts that a differentiable function of a differentiable function is differentiable, and the rule tells how to compute partial derivatives of the composite function. For example, if $x = f(s,t)$, $y = g(s,t)$, where f and g are differentiable, and if $z = F(x,y)$, where F is differentiable, then the composite function is $G(s,t) = F[f(s,t),g(s,t)]$. Then $z = G(s,t)$ is differentiable as a function of s and t, and

$$\frac{\partial G}{\partial s} = \frac{\partial F}{\partial x}\frac{\partial f}{\partial s} + \frac{\partial F}{\partial y}\frac{\partial g}{\partial s}$$

$$\frac{\partial G}{\partial t} = \frac{\partial F}{\partial x}\frac{\partial f}{\partial t} + \frac{\partial F}{\partial y}\frac{\partial g}{\partial t}$$

These equations, expressing the formal part of the chain rule, are often written in the form

$$\frac{\partial z}{\partial s} = \frac{\partial z}{\partial x}\frac{\partial x}{\partial s} + \frac{\partial z}{\partial y}\frac{\partial y}{\partial s}$$

$$\frac{\partial z}{\partial t} = \frac{\partial z}{\partial x}\frac{\partial x}{\partial t} + \frac{\partial z}{\partial y}\frac{\partial y}{\partial t}$$

See CALCULUS; DIFFERENTIATION; PARAMETRIC EQUATION. [A.E.Ta.]

Partial tone A simple sinusoidal component of a complex tone. It may be part of a complex physical oscillation; alternatively, a partial tone (or partial) is a component of a sound sensation, distinguished as a simple tone that cannot be further analyzed by the ear and that contributes to the timbre of the complex sound. *See* HEARING (HUMAN); TONE (MUSIC AND ACOUSTICS).

The physical sound from a violin string, for example, is usually composed of a number of partials. Each of these partials of the airborne sound results from vibration in a number of equal parts that takes place when the string vibrates as a whole. Such characteristic vibrations by aliquot parts are also called partials; they are components of the complex motion of the string. Each such characteristic vibration pattern can also occur individually, however. Thus it seems preferable to call these characteristic motions modes of vibration rather than partials, to make it clear that they can exist independently without being parts of a complex motion. *See* MODE OF VIBRATION; VIBRATION.

If a string is bowed steadily, the frequencies of the partials of the resulting complex tone will be integral multiples of the lowest (fundamental) frequency, and the partials may properly be called harmonics. If, however, the same string is struck or plucked and then allowed to vibrate freely, the frequencies of the corresponding modes of vibration are, in general, no longer exactly in the ratios of integers, and the partials and modes of vibration are inharmonic. *See* HARMONIC (PERIODIC PHENOMENA). [R.W.Y.]

Particle accelerator An electrical device which accelerates charged atomic or subatomic particles to high energies. The particles may be charged either positively or negatively. If subatomic, the particles are usually electrons or protons and, if atomic, are charged ions of various elements and their isotopes throughout the entire periodic table of the elements.

Accelerators that produce various subatomic particles at high intensity have many practical applications in industry and medicine as well as in basic research. Electrostatic generators, pulse transformer sets, cyclotrons, and electron linear accelerators are used to produce high levels of various kinds of radiation that in turn can be used to polymerize plastics, provide bacterial sterilization without heating, and manufacture radioisotopes which are utilized in industry and medicine for direct treatment of some illnesses as well as research. They can also be used to provide high-intensity beams of x-rays with extreme penetrating power that can be used for cancer therapy, as well as for x-ray radiographic determination of flaws and structural problems in heavy industrial steel castings and other types of structures. *See* ISOTOPIC IRRADIATION; NONDESTRUCTIVE TESTING; RADIATION BIOLOGY; RADIATION CHEMISTRY; RADIOACTIVITY AND RADIATION APPLICATIONS; RADIOGRAPHY; RADIOLOGY.

Particle accelerators fall into two general classes—electrostatic accelerators that provide a steady dc potential, and varieties of accelerators that employ various combinations of time-varying electric and magnetic fields.

Electrostatic accelerators. Electrostatic accelerators in the simplest form either accelerate the charged particle from the source of high voltage to ground potential or from ground potential to the source of high voltage. The maximum energy available from this kind of accelerator is limited by the ability of the voltage generator to provide some maximum high voltage. *See* ELECTROSTATIC ACCELERATOR.

Time-varying field accelerators. In contrast to the high-voltage-type accelerator which accelerates particles in a continuous stream through a continuously maintained potential, the time-varying accelerators must necessarily accelerate particles in small discrete groups or bunches.

An accelerator that varies only in electric field and does not use any magnetic guide or turning field is customarily referred to as a linear accelerator or linac. In the simplest version of this kind of accelerator, the electrodes that are used to attract and accelerate the particles are connected to a radio-frequency (rf) power supply or oscillator in such a way that alternate electrodes are of opposite polarity. In this way each successive gap between adjacent electrodes is alternately accelerating and decelerating. If these acceleration gaps are appropriately spaced to accommodate the increasing velocity of the accelerated particle, the frequency can be adjusted so that the particle bunches are always experiencing an accelerating electric field as they cross each successive gap. In this way modest voltages can be used to accelerate bunches of particles indefinitely, limited only by the physical length of the accelerator construction. Some of the larger linacs are over a mile in length.

As accelerators are carried to a higher and higher energy, a linac eventually reaches some practical construction limit because of length. This problem of extreme length can be circumvented conveniently by accelerating the particles in a circular path maintained by either static or time-varying magnetic fields. Accelerators utilizing steady magnetic fields as guide paths are usually referred to as cyclotrons or synchrocyclotrons, and are arranged to provide a steady magnetic field over relatively large areas that allow the particles to travel in a circular orbit of gradually increasing diameter as they increase in energy. *See* MAGNETIC FIELD.

Practical limitations of magnet construction have kept the size of circular proton accelerators with static magnetic fields to the vicinity of 100 to 1000 MeV. For even higher energies, up to 400 GeV per nucleon in the maximum-size proton accelerator in operation, it is necessary to vary the magnetic field as well as the electric field in time. In this way the magnetic field can be of a minimal practical size, which is still quite extensive for a 400-GeV accelerator. This circular magnetic containment region or "race track" is injected with relatively low-energy particles that can coast around the magnetic ring when it is at minimum field strength. The magnetic field is then gradually increased to stay in step with the higher magnetic rigidity of the particles as they are gradually accelerated with a time-varying electric field. *See* SYNCHROTRON.

Superconducting magnets. The study of the fundamental structure of nature and all associated basic research require an ever increasing energy in order to allow finer and finer measurements on the basic structure of matter. Since the voltage-varying and magnetic-field-varying accelerators also have limits to their maximum size in terms of cost and practical construction problems, the only way to increase particle energies even further is to provide higher-varying magnetic fields through superconducting magnet technology, which can extend electromagnetic capability by a factor of 4 to 5. *See* SUPERCONDUCTING DEVICES.

Storage rings. Beyond this limit the only other possibility is to accelerate particles in opposite directions and arrange for them to collide at certain selected intersection regions around the accelerator. The main technical problem is to provide adequate numbers of particles in the two colliding beams so that the probability of a collision is moderately high. Such storage ring facilities are in operation for both electrons and protons, and design and construction of much larger ones have been undertaken. *See* STORAGE RINGS.

Collective accelerators. Development of a completely different kind of accelerator, the collective-effect accelerator, has been undertaken in several countries. The idea is to accelerate a cloud or bunch of electrons containing one or a few protons or heavier atoms within the electrostatic well established by the cloud structure so that the electrical forces providing the containment are sufficiently strong to drag the heavy particles along. *See* ELECTRON RING ACCELERATOR. [H.E.W.]

Particle detector A device used to detect and measure radiations characteristically emitted in nuclear processes, including γ- or x-rays, lightweight charged particles (electrons or positrons), nuclear constituents (neutrons, protons, and heavier ions), and subnuclear constituents such as mesons. The device is also known as a radiation detector. Detectors are essential tools for the discovery of radioactive minerals, for all studies of the structure of matter at the atomic, nuclear, and subnuclear levels, and for protection from the effects of radiation. They have also become important practical tools in the analysis of materials using the techniques of neutron activation and x-ray fluorescence analysis.

A convenient way to classify radiation detectors is according to their mode of use: (1) For detailed observation of individual photons or particles, a pulse detector is used to convert each such event (that is, photon or particle) into an electrical signal. (2) To measure the average rate of events, a mean current detector, such as an ion chamber, is often used. (3) Position-sensitive detectors are used to provide information on the location of particles or photons in the plane of the detector. (4) Track-imaging detectors image the whole three-dimensional structure of a particle's track. (5) The time when a particle passes through a detector or a photon interacts in it is measured by a timing detector. Such information is used to determine the velocity of particles and when observing the time relationship between events in more than one detector.

The ionization produced by a charged particle is the effect most commonly employed in a particle detector. In the basic type of gas ionization detector, an electric field applied between two electrodes separates and collects the electrons and positive ions produced in the gas by the radiation to be measured. Track-imaging detectors rely on a secondary effect of the ionization along a particle's track to reveal its structure.

In semiconductor detectors, a solid replaces the gas. The "insulating" region (depletion layer) of a reverse-biased *pn* junction in a semiconductor is employed. *See* BUBBLE CHAMBER; CLOUD CHAMBER; IONIZATION CHAMBER; JUNCTION DETECTOR.

In addition to producing free electrons and ions, the passage of a charged particle through matter temporarily raises electrons in the material into excited states. When these electrons fall back into their normal state, light may be emitted and detected as in the scintillation detector. *See* LIQUID SCINTILLATION DETECTOR; SCINTILLATION COUNTER.

Neutral particles, such as neutrons, cannot be detected directly by ionization. Consequently, they must be converted into charged particles by a suitable process and then observed by detecting the ionization caused by these particles.

Although ionization detectors dominate the field, a number of detector types based on other radiation-induced effects are used. Notable examples are: (1) transition radiation detectors, which depend on the x-rays and light emitted when a particle passes through the interface between two media of different refractive indices; (2) track detectors, in which the damage caused by charged particles in plastic films and in minerals is revealed by etching procedures; (3) thermo- and radiophotoluminescent detectors, which rely on the latent effects of radiation in creating traps in a material or in creating trapped charge; and (4) Cerenkov detectors, which depend on measurement of the light produced by passage of a particle whose velocity is greater than the velocity of light in the detector medium. *See* CERENKOV RADIATION. [F.S.G.]

Particle flow Particle flow is important to many industrial processes, including the pneumatic conveying of solids, transport of solids in liquids (slurries), removal of particulates from gas streams for pollution control, combustion of pulverized coal, and drying of particulates in the food and pharmaceutical industries. The vast majority of flow problems in industrial design involve the flow of gas or liquids with suspended solids. *See* DRYING; FLUIDIZED-BED COMBUSTION; PIPELINE.

A key parameter in fluid-particle flows is the Stokes number, which is the ratio of the response time of a particle to a time characteristic of a flow system. Particle response time is the time that a particle takes to respond to a change in carrier flow velocity. If the Stokes number is small (say, less than 0.1), the particles have sufficient time to respond to the change in fluid velocity, so the particle velocity approaches the fluid velocity. However, if the Stokes number is large (say, greater than 10), the particles have little time to respond to the varying fluid velocity and the particle velocity shows little change.

The relative concentration of the particles in the fluid is referred to as loading. The loading may be defined in several ways, such as the ratio of particle mass flow to fluid mass flow. Many industrial applications involve highly loaded particle flows.

If the particle loading is small, the fluid will affect the particle properties (velocity, temperature, and so forth), but the particles will not influence the fluid properties. This is referred to as one-way coupling. If the conditions are such that there is a mutual interaction between the particles and fluid, the flow is two-way-coupled. Two-way coupling effects are reduced with increasing Stokes number because the particles undergo less acceleration.

If the particle motion is controlled by the action of the fluid on the particle, the flow is termed dilute. But if the particle concentration is sufficiently high, the particles will collide with each other and their motion will be dependent on particle-particle collisions; the flow is then regarded as dense. *See* FLUID FLOW; PARTICULATES. [C.T.Cr.]

Particle track etching A technique of selective chemical etching to reveal tracks of heavy nuclear particles in a wide variety of solid substances. Developed in order to see fossil particle tracks in extraterrestrial materials, the technique finds application in many fields of science and technology.

An etchable track is produced if the charged particle has a sufficiently high radiation-damage rate and if the damaged region in the solid is permanently localized. Thus only highly ionizing particles are detectable; only nonconductors record tracks; and radiation-sensitive plastics can detect lighter particles than can radiation-insensitive minerals and glasses. The conical shape of the etched track depends on the ratio of the rate of etching along the track to the bulk etching rate of the solid.

The lunar surface, meteorites, and other objects exposed in space have been irradiated by charged particles from a variety of sources in the Sun and the Galaxy. Comparison of fossil particle tracks in lunar rocks and meteorites with spacecraft measurements of present-day radiations has established that solar flares and galactic cosmic rays have not changed over the last 2×10^7 years—the typical time a lunar rock exists before being shattered by impacting interplanetary debris.

Studies of tracks in a piece of glass from the *Surveyor 3* spacecraft after a 2.6-year exposure on the lunar surface, and of tracks in plastic detectors exposed briefly above the Earth's atmosphere in rockets, have led to the surprising discovery that the Sun preferentially ejects heavy elements in its flares rather than an unbiased sample of its atmosphere. The existence of galactic cosmic rays with atomic number greater than 30 was discovered in 1966 when fossil particle tracks were first studied in meteorites. Several particles heavier than uranium have been detected, indicating that cosmic rays originate in sources where synthesis has proceeded explosively beyond uranium. *See* COSMIC RAYS.

Unique advantages of etched-track detectors in nuclear and elementary particle physics are their ability to distinguish heavy-particle events in a large background of lightly ionizing radiation and their ability to detect individual rare events by a specialized technique such as electric-spark scanning or ammonia penetration through etched holes. These advantages have permitted such advances as the measurement of very long fission half-lives and the discovery of ternary fission. *See* NUCLEAR FISSION; SUPERTRANSURANICS; TRANSURANIUM ELEMENTS.

The spontaneous fission of ^{238}U, present as a trace-element purity, gives tracks that can be used to date terrestrial samples ranging from rocks to human artifacts. Because fission tracks are erased in a particular mineral at a well-defined temperature, one can use the apparent fission-track ages as a function of distance from the heat source to measure the thermal (tectonic) history of regions. *See* FISSION TRACK DATING.

Filters are produced by irradiating thin plastic sheets with fission fragments and then etching holes to the desired size. Uses include biological research, wine filtration, and virus sizing. A uranium exploration method relies on a survey of radon emanation, as measured by alpha-particle tracks in plastic detectors, to locate promising locations in which to drill. Plastic detectors are also used in conjunction with a beam of high-energy heavy ions to take radiographs of cancer patients that reveal details not detectable in x-rays. [P.B.P.]

Particle trap A device used to confine charged or neutral particles where their interaction with the wall of a container must be avoided. Electrons or protons accelerated to energies as high as 1 teraelectronvolt (10^{12} electronvolts) are trapped in magnetic storage rings in high-energy collision studies. Other forms of magnetic bottles are designed to hold dense hot plasmas of hydrogen isotopes for nuclear fusion. At the other end of the energy spectrum, ion and atom traps can store isolated atomic systems at temperatures below 1 millikelvin. Other applications of particle traps include the storage of antimatter such as antiprotons and positrons (antielectrons) for high-energy collision studies or low-energy experiments. *See* ANTIMATTER; ANTIPROTON; NUCLEAR FUSION; PARTICLE ACCELERATOR; PLASMA PHYSICS; POSITRON.

Charged-particle traps. Charged particles can be trapped in a variety of ways. An electrostatic (Kingdon) trap is formed from a thin charged wire. The ion is attached to the wire, but its angular momentum causes it to spiral around the wire in a path with a low probability of hitting the wire.

A magnetostatic trap (magnetic bottle) is based on the fact that a charged particle with velocity perpendicular to the magnetic field lines travels in a circle, whereas a particle moving parallel to the field is unaffected by it. In general, the particle has velocity components both parallel and perpendicular to the field lines and moves in a helical spiral. In high-energy physics, accelerators and storage rings also use magnetic forces to guide and confine charged particles. A tokamak has magnetic field lines configured in the shape of a torus, confining particles in spiral orbits. Another type of bottle uses a magnetic mirror.

The radio-frequency Paul trap uses inhomogeneous radio-frequency electric fields to confine particles, forcing them to oscillate rapidly in the alternating field (see illustration). If the amplitude of oscillation (micromotion) is small compared to the trap dimensions, the trap may be thought of as increasing the (kinetic) energy of the particle in a manner that is a function of the particle position. The particle moves to the position of minimal energy and is therefore attracted to the center of the trap where the oscillating electric fields are weakest.

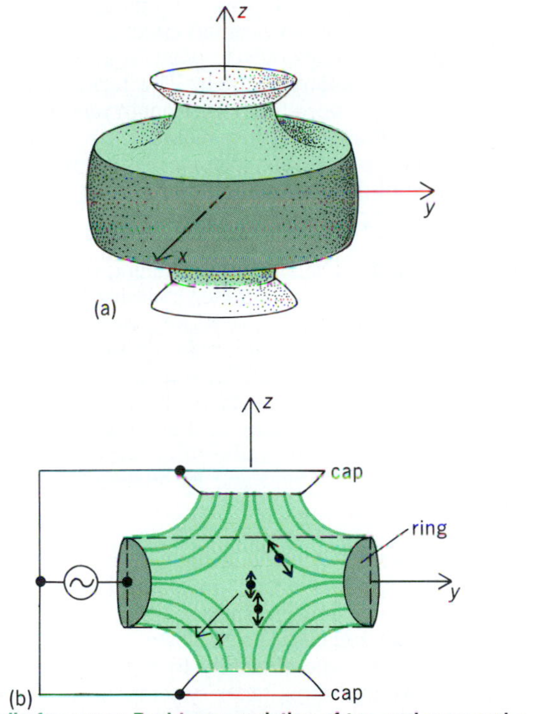

Radio-frequency Paul trap consisting of two end caps and a ring electrode. (*a*) Cutaway view (*after G. Kamas, ed., Time and Frequency Users's Manual, National Bureau of Standards Technical Note 695, 1977*). (*b*) Cross section, showing the amplitude of the instantaneous oscillations for several locations in the trap.

The Penning trap, with the same electrode configuration as in the illustration, uses a combination of static electric and magnetic fields instead of oscillating electric fields.

Neutral-particle traps. Uncharged particles such as neutrons or atoms are manipulated by higher-order moments of the charge distribution such as the magnetic or electric dipole moments.

Magnetic traps of neutral particles use the fact that atoms usually have a magnetic dipole moment $\boldsymbol{\mu}$ on which the gradient of a magnetic field **B** exerts a force. The atom can be in a state whose magnetic energy $W = -\boldsymbol{\mu} \cdot \mathbf{B}$ increases or decreases with the field strength, depending on whether the moment is antiparallel or parallel to the field. A magnetic field cannot be constructed with a local maximum in a current-free region, but a local minimum is possible, allowing particles seeking a weak field to be trapped.

Laser traps use the strong electric fields **E** of the laser beam to induce an electric dipole moment **p** on the atom. A laser field tuned below the atomic resonance polarizes the atom in phase with the driving field; the instantaneous dipole moment points in the same direction as the field. Thus the energy of the atom $W = -\mathbf{p} \cdot \mathbf{E}$ is lowered if it is in a region of high laser intensity. The high-intensity trapping region is formed simply by focusing the beam of a laser. *See* LASER.

Magnetooptic hybrid traps use, instead of the dipole forces induced by the laser field, the scattering force that arises when an atom absorbs photons. *See* LASER COOLING. [S.Ch.]

Particulates Solids or liquids in a subdivided state. Because of this subdivision, particulates exhibit special characteristics which are negligible in the bulk material. Normally, particulates will exist only in the presence of another continuous phase, which may influence the properties of the particulates. A particulate may comprise several phases. The table categorizes particulate systems and relates them to commonly recognized designations. *See* ALLOY; EMULSION; FOAM; GEL.

Fine-particle technology deals with particulate systems in which the particulate phase is subject to change or motion, and is concerned with those particles which are tangible to human senses, yet small compared to the human environment—particles that are larger than molecules but smaller than gravel. Fine particles are in abundance in nature (as in rain, soil, sand, minerals, dust, pollen, bacteria, and viruses) and in industry (as in paint pigments, insecticides, powdered milk, soap, powder, cosmetics, and inks). Particulates are involved in such undesirable forms as fumes, fly ash, dust, and smog and in military strategy in the form of signal flares, biological and chemical warfare, explosives, and rocket fuels.

Many of the characteristics of particulates are influenced to a major extent by the particle size. For this reason, particle size has been accepted as a primary basis for characterizing particulates. However, with anything but homogeneous spherical particles, the measured "particle size" is not necessarily a unique property of the particulate but may be influenced by the technique used. Consequently, it is important that the techniques used for size analysis be closely allied to the utilization phenomenon for which the analysis is desired.

Types of particulate systems

System			Hydrosol	Aerosol	Powder
Continuous Phase		Solid	Liquid	Gas	None (or gas)
Dispersed or particulate phase	Gas	Sponge	Foam	—	—
	Liquid	Gel	Emulsion	Mist Spray Fog Rain	—
	Solid	Alloy	Slurry Suspension	Fume Dust Snow Hail	Single phase; Multiphase (ores, flour)

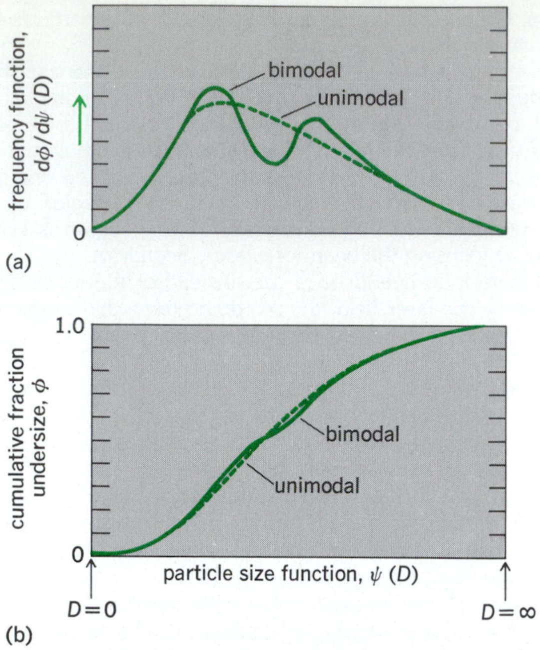

Methods for representing size distribution. (a) Frequency distribution. (b) Cumulative distribution.

Size is generally expressed in terms of some representative, average, or effective dimension of the particle. The most widely used unit of particle size is the micrometer (μm). Another common method is to designate the screen mesh that has an aperture corresponding to the particle size. The screen mesh normally refers to the number of screen openings per unit length or area; several screen standards are in general use.

Particulate systems are often complex. Primary particulates may exist as loosely adhering (as by van der Waals forces) particles called flocs or as strongly adhering (as by chemical bonds) particulates called agglomerates. Primary particles are those whose size can only be reduced by the forceful shearing of crystalline or molecular bonds. *See* CHEMICAL BONDING; INTERMOLECULAR FORCES.

Mechanical dispersoids are formed by comminution, decrepitation, or disintegration of larger masses of material, as by grinding of solids or spraying of liquids, and usually involve a wide distribution of particle sizes. Condensed dispersoids are formed by condensation of the vapor phase (or crystallization of a solution) or as the product of a liquid- or vapor-phase reaction; these are usually very fine and often relatively uniform in size. Condensed dispersoids and very fine mechanical dispersoids generally tend to flocculate or agglomerate to form loose clusters of larger particle size.

Most real systems are composed of a range of particle sizes. The two common general methods for representing size distribution graphically are shown in the illustration. The frequency distribution (illustration *a*) gives the fraction of particles $d\phi$ (on whatever basis desired) that lie in a given narrow size range dD as a function of the average size of the range (or of some function of the average size). A cumulative distribution (illustration *b*) is the integral of the frequency curve. It gives the fraction ϕ of the particles that are smaller or larger than a given size D. *See* INTEGRATION; STATISTICS.

If a particle suspended in a fluid is acted upon by a force, it will accelerate to a terminal velocity at which the resisting force due to fluid friction just balances the applied force. If a particle falls under the action of gravity, this velocity is known as the terminal gravitational settling velocity.

Particles suspended in a fluid partake of the molecular motion of the suspending fluid and hence acquire diffusional characteristics analogous to those of the fluid molecules. This random zigzag motion of the particles, commonly known as brownian motion, is obvious under the microscope for particles smaller than 1 μm. *See* BROWNIAN MOVEMENT; DIFFUSION IN GASES AND LIQUIDS. [C.E.La.]

Pascal's law A law of physics which states that a confined fluid transmits externally applied pressure uniformly in all directions. More exactly, in a static fluid, force is transmitted at the velocity of sound throughout the fluid. The force acts normal to any surface. This natural phenomenon is the basis of the pneumatic fire, balloon, hydraulic jack, and related devices. *See* HYDROSTATICS. [K.Arn.; R.S.R.]

Paschen-Back effect An effect on spectral lines obtained when the light source is placed in a very strong magnetic field, first explained by F. Paschen and E. Back in 1921. In such a field the anomalous Zeeman effect, which is obtained with weaker fields, changes over to what is, in a first approximation, the normal Zeeman effect. The term "very strong field" is a relative one, since the field strength required depends on the particular lines being investigated. It must be strong enough to produce a magnetic splitting that is large compared to the separation of the components of the spin-orbit multiplet. *See* ATOMIC STRUCTURE AND SPECTRA; ZEEMAN EFFECT. [F.A.J./W.W.W.]

Passeriformes The order of perching birds (an inadequate name used for lack of a better inclusive English name), containing about 5300 species, or more than half of the living species of birds. It comprises two major divisions: a somewhat more primitive group collectively known as suboscines, usually divided into three suborders (Eurylaimi, Tyranni, and Menurae); and the large suborder Oscines or Passeres, the songbirds. Separation of the Oscines from the suboscines is based chiefly on the structure of the syrinx and its attached muscles. *See* AVES; PERCHING BIRDS. [K.C.P.]

Passive radar A technique for surveillance, mapping, navigation, and guidance that employs the reception of microwave-frequency energy radiated by warm bodies, or reflected from other sources. *See* MICROWAVE.

Passive radar has some similarity to both passive infrared systems and active (radiating) radar systems. Like infrared systems, passive radar emits no energy of its own, and therefore does not give away its position or existence. Target resolution is inferior to that of active radar and infrared systems. However, the ability of a passive radar system to discriminate between different targets and backgrounds can be better than that of either active radar or infrared.

The ability of a passive radar to discriminate between different objects depends primarily on (1) the apparent temperature differential between the objects (this takes into account emissivity and reflectivity), (2) the grazing angle between the antenna beam and the object, (3) antenna polarization, (4) antenna beam width, and (5) the minimum detectable signal level of the receiver. The grazing angle, or angle of incidence, is a factor when viewing smooth objects, such as bodies of water. Polarization is most pronounced at small grazing angles. The effect of polarization varies with the object, providing an additional method of discriminating between objects at the same temperature. *See* HEAT RADIATION; RADAR; REFLECTION OF ELECTROMAGNETIC RADIATION. [P.J.K.]

Pasteurella A genus of gram-negative, nonmotile, nonsporulating, facultatively anaerobic coccobacillary to rod-shaped bacteria which are parasitic and often pathogens in many species of mammals, birds, and reptiles. It was named to honor Louis Pasteur in 1887. Genetic studies have shown that *Pasteurella*, together with *Haemophilus* and *Actinobacillus*, constitute a family, Pasteurellaceae.

The genus contains at least 10 species. *Pasteurella multocida* causes hemorrhagic septicemia in various mammals and fowl cholera, and is occasionally transmitted to humans, mainly in rural areas. Human pasteurellosis may include inflammation in bite and scratch lesions, infections of the lower respiratory tract and of the small intestine, and generalized infections with septicemia and meningitis. *Pasteurella canis* and *P. stomatis* may cause similar, though generally less severe, infections in humans after contact with domestic or wild animals. Although drug-resistant *Pasteurella* strains have been encountered, human *Pasteurella* infections are as a rule readily sensitive to the penicillins and a variety of other chemotherapeutic agents. *See* ANTIBIOTIC; DRUG RESISTANCE. [W.Ma.]

Pasteurization The application of mild heat for a specified time to a liquid food or beverage to enhance its keeping properties and to destroy any harmful microorganisms present. For milk, the times and temperatures employed are based upon the thermal tolerance of *Mycobacterium tuberculosis*, one of the most heat-resistant of non-spore-forming pathogens. A temperature of either 143°F (61.7°C) for 30 min or of 161°F (71.7°C) for 15 s is employed. Vegetative cells of most bacteria are killed by the heat treatment, but endospores are unaffected. [C.F.N.]

Patent Common designation for letters patent, which is a certificate of grant by a government of an exclusive right with respect to an invention for a limited period of time. A United States patent confers the right to exclude others from making, using, or selling the patented subject matter in the United States and its territories. Portions of those rights deriving naturally from it may be licensed separately, as the rights to use, to make, to have made, and to lease. Any violation of this right is an infringement.

An essential substantive condition which must be satisfied before a patent will be granted is the presence of patentable invention or discovery. To be patentable, an invention or discovery must relate to a prescribed category of contribution, such as process, machine, manufacture, composition of matter, plant, or design. In the United States there are different classes of patents for different members of these categories. [D.W.B.]

Paterinida A small order of inarticulated brachiopods that is relatively common in Cambrian rocks and ranges into beds of Middle Ordovician age. Paterinids characteristically have a thin shell of calcium phosphate. Both valves are convex, the pedicle valve usually being the higher, and both typically have well-developed propareas. Internally, the lateral margins of the delthyrium are commonly thickened, but other features are poorly developed and the muscle scars are extremely faint. *See* BRACHIOPODA; INARTICULATA. [A.J.R.]

Pathogen Any agent capable of causing disease. The term pathogen is usually restricted to living agents, which include viruses, rickettsia, bacteria, fungi, yeasts, protozoa, helminths, and certain insect larval stages. *See* DISEASE.

Pathogenicity is the ability of an organism to enter a host and cause disease. The degree of pathogenicity, that is, the comparative ability to cause disease, is known as virulence. The terms pathogenic and nonpathogenic refer to the relative virulence of the organism or its ability to cause disease under certain conditions. This ability depends not only upon the properties of the organism but also upon the ability of the host to defend itself (its immunity) and prevent injury. The concept of pathogenicity and virulence has no meaning without reference to a specific host. For example, gonococcus is capable of causing gonorrhea in humans but not in lower animals. *See* MEDICAL MYCOLOGY; MEDICAL PARASITOLOGY; PLANT PATHOLOGY; PLANT VIRUSES AND VIROIDS; VIRULENCE. [D.N.La.]

Pathology The study of the etiologies, mechanisms, and manifestations of disease. Techniques and knowledge gained from other disciplines, including anatomy, physiology, microbiology, biochemistry, and histology, are utilized. The information obtained from the study of pathology is necessary prior to developing methods with which to control and prevent disease.

With the light microscope it became possible to correlate the observed signs and symptoms in an individual with cellular changes. In its early stages pathology was very descriptive. Diseases were understood and categorized in part, by how gross and microscopic anatomy was altered. In the last half of the 19th century, by using this approach to pathology, coupled with microbiological techniques, it was learned that the major causes of human death were biotic agents: protozoans, bacteria, viruses, and fungi. Infectious diseases took a heavy toll in human lives. Better sanitation and public health measures were instrumental in controlling these diseases, and the production of antibiotics and immunization procedures further reduced their importance. It is now apparent that all diseases reflect changes at the molecular level. Scientists are beginning to understand what these biochemical alterations are in some diseases.

There are many branches of pathology. Divisions are made depending upon focus of interest. Clinical pathology is concerned with diagnosis of disease. As medicine has expanded, subspecialties such as surgical pathology and neuropathology have developed. Experimental pathology attempts to study disease mechanisms under controlled conditions. General pathology covers all areas, but in less detail, and serves in medical education.

A relatively new area of pathology is environmental pathology, which deals with disease processes resulting from physical and chemical agents. At present, the leading causes of death have environmental agents as the known or suspected major etiologic factors; these diseases include heart disease, atherosclerosis, and cancer. It is believed that with understanding, many such diseases, like those produced in response to biotic agents, can be brought under control. *See* DISEASE. [N.K.M.; C.Qu.]

Pathotoxin A chemical of biological origin, other than an enzyme, that plays an important causal role in a plant disease. Most pathotoxins are produced by plant pathogenic fungi or bacteria, but some are produced by higher plants, and one has been reported to be the product of an interaction between a plant and a bacterial pathogen. Some pathogen-produced pathotoxins are highly selective in that they cause severe damage and typical disease symptoms only on plants susceptible to the pathogens that produce them. Others are nonselective and are equally toxic to plants susceptible or resistant to the pathogen involved. A few pathotoxins are species-selective, and are damaging to many but not all plant species. In these instances, some plants resistant to the pathogen are sensitive to its toxic product. *See* PLANT PATHOLOGY. [H.Wh.]

Pattern formation (biology) The mechanisms that ensure that particular cell types differentiate in the correct location within the embryo and that the layers of cells bend and grow in the correct relative positions. Pattern formation is one of four processes that underlie development, the others being growth, cell diversification, and morphogenesis. *See* ANIMAL GROWTH; ANIMAL MORPHOGENESIS; CELL DIFFERENTIATION; PLANT GROWTH; PLANT MORPHOGENEIS.

Pattern formation is the creation of a predictable arrangement of cell types in space during embryonic development. The types of patterns of cell types found in animals and plants can be conveniently described as simple or complex. Simple patterns involve the spatial arrangement of identical or equivalent structures such as bristles on the leg of a fly, hairs on a person's head, or leaves on a plant. Such equivalent patterns are thought to be produced by mechanisms that are the same or very similar in the fly and the plant. Complex patterns are those that are made up of parts that are not equivalent to one another. In the vertebrate limb, for example, the structure of the arm is different at each level, with one bone (humerus) in the upper arm, two bones (radius and ulna) in the lower arm, and a complex set of bones making up the wrist and the hand. How are such nonequivalent parts patterned during development? The theoretical framework that allows a basis for understanding how such patterns arise is called positional information. Two stages exist in the positional information framework. First, a cell must become aware of its position within a developing group, or field, of cells. This specification of cellular position requires a mechanism by which each cell within a field can obtain a unique value or address. The second component is the interpretation of the positional address by a cell to manifest a particular cell type by the expression of a particular set of genes. *See* DEVELOPMENTAL BIOLOGY; EMBRYONIC DIFFERENTIATION. [N.Ho.]

Pattern recognition The construction of algorithms capable of processing the range of input patterns of interest to yield a set of output patterns such that the occurrence of each input causes a uniquely identifiable output pattern to occur.

It is essentially a decoding process in which discriminations are made. In simpler cases tree methods of serial segmentation find value. Ideally, such measures enable one to divide remaining patterns at each decision node into two disjoint patterns, similar in principle to the basis for binary decoding.

Another approach makes use of as many measures as possible that appear to be related to the desired discriminations. A search is then made for an appropriate differential weighting of these measures so as to obtain maximal spacing between patterns. This approach is sometimes referred to as factor analysis. *See* AUTOMATA THEORY. [N.B.R.]

Pauropoda A class, and perhaps the most obscure group, of the Myriapoda. They are pale creatures, no more than 0.04–0.08 in. (1–2 mm) in length, inhabiting damp situations in leaf litter, under bark, stones and debris, and in humus and similar detritus. Apparently very widely distributed as a class, they have been undiscovered only in deserts and in the arctic and antarctic regions.

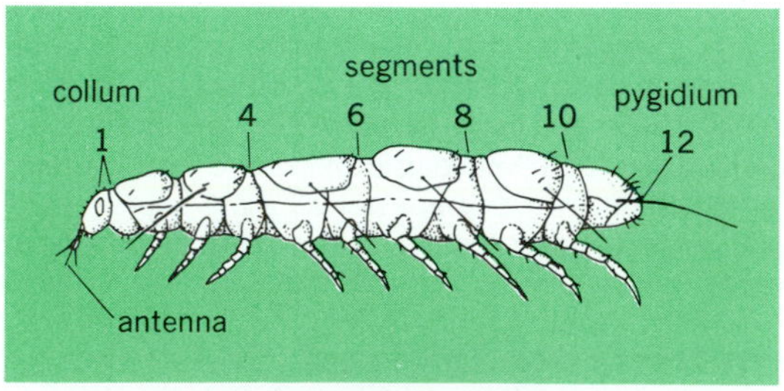

***Pauropus silvaticus.* (After R. E. Snodgrass, A Textbook of Arthropod Anatomy, Cornell University Press, 1952)**

Like millipedes, they are progoneate and have one pair of maxillae, and their trunk segments display a certain degree of amalgamation. Their peculiar bifurcate antennae and adult complement of 12 trunk segments with 9 pairs of functional legs are distinctive within the myriapod complex (see illustration). All pauropods lack eyes, spiracles, tracheae, and a circulatory system.

The class currently consists of 2 families with less than 10 genera; there are probably fewer than 60 species known. [R.E.Cr.]

Pavement An artificial surface laid over the ground to facilitate travel. A pavement's ability to support loads depends primarily upon the magnitude of the load, how often it is applied, the supporting power of the soil underneath, and the type and thickness of the pavement structure. Before the necessary thickness of a pavement can be calculated, the volume, type, and weight of the traffic (the traffic load) and the physical characteristics of the underlying soil must be determined.

Once the grading operation has been completed and the subgrade compacted, construction of the pavement can begin. Pavements are either flexible or rigid. Flexible pavements, which are composed of aggregate (sand, gravel, or crushed stone) and bituminous material (see illustration), have less resistance to bending than do rigid pavements, which are made of concrete. Both types can be designed to withstand heavy traffic. Selection of the type of pavement depends, among other things, upon (1) estimated construction costs; (2) experience of the highway agency doing the work with each of the two types; (3) availability of contractors experienced in building each type; (4) anticipated yearly maintenance costs; and (5) experience of the owner in maintenance of each type. *See* CONCRETE; HIGHWAY ENGINEERING. [A.N.C.]

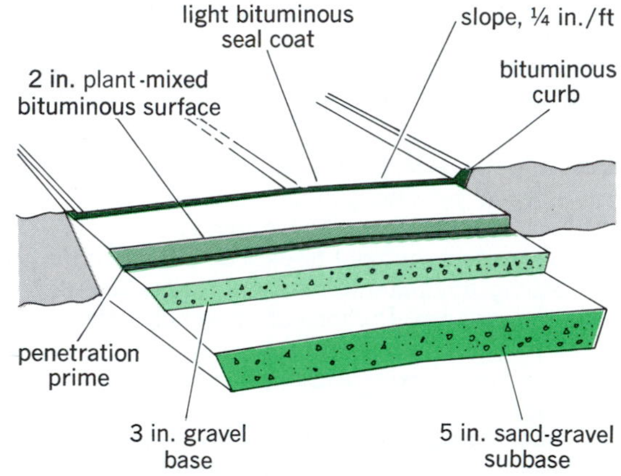

Flexible pavement design for a city collector street with maximum traffic load of 5 tons (4.5 metric tons) per axle. Right-of-way is 60 ft (18 m) wide and the pavement width is 38 ft (11.6 m). Berms or boulevards at the sides are sloped in order to drain toward the street. 1 in. = 2.5 cm.

Pawl The driving link or holding link of a ratchet mechanism, also called a click or detent. In the illustration the driving pawl at *A*, forced upward by lever *B*, engages the teeth of the ratchet wheel and rotates it counterclockwise. Holding pawl *C* prevents clockwise rotation of the wheel when the pawl at *A* is making its return stroke, Pawl and ratchet are an open, upper pair.

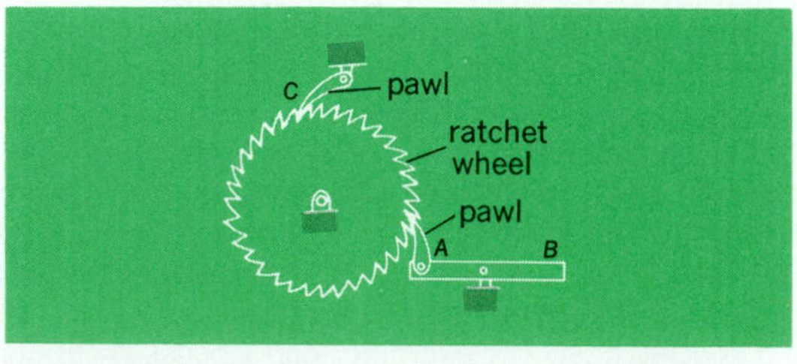

Holding and driving pawls with a ratchet wheel.

Driving and holding pawls likewise engage rack teeth on the plunger of a ratchet lifting jack, such as those supplied with automobiles. A ratchet wheel with a holding pawl only, acting as a safety brake, is fastened to the drum of a capstan, winch, or other powered hoisting device.

A double pawl can drive in either direction or be easily reversed in holding. A cam pawl prevents the wheel from turning clockwise by a wedging action while permitting free counterclockwise rotation. This technique is used in the automobile hill-holder to prevent the vehicle's rolling backward. *See* ESCAPEMENT. [D.P.Ad.]

Paxillosida An order of sea stars and members of the class Asteroidea. The name is derived from the club-shaped plates, or paxillae, that form the sea star's upper skeletal surface and that have tiny spinelets or granules covering their tips. Paxillosida encompasses six families, the largest being the Astropectinidae, Luidiidae, and Porcellanasteridae. The Ctenodiscidae, Goniopectinidae, and Radiasteridae are represented by comparatively few members. Astropectinids and luidiids are primarily predators of mollusks and other echinoderms. The former are found over a wide range of depths, whereas the latter live in relatively shallow water. Porcellanasterids are deep-water asteroids that swallow sediment in bulk as they bury themselves to a level just below the surface.

In addition to the presence of paxillae, paxillosidans are characterized by a number of unusual features that have been thought to indicate a primitive phylogenetic position among living asteroids. Although the tube feet of most asteroids have suckered disks, those of most paxillosidans are pointed. The digestive system of most asteroids is relatively complex and complete, terminating in an anus, whereas that of the paxillosidans is simple, saclike, and lacking an anus in some members. Most asteroids can extrude their stomach during feeding, but that ability is limited in paxillosidans. A brachiolarian larval stage has been recognized in the development of most asteroids, yet that stage is thought to be absent from paxillosidans. *See* ASTEROIDEA; ECHINODERMATA. [D.B.B.]

Pea The pea is one of the oldest cultivated crops. It is a native to western Asia from the Mediterranean Sea to the Himalaya Mountains. It appears to have been carried to Europe as early as the time of the lake dwellers of prehistoric times. Peas were introduced into China from Persia about A.D. 400; they were introduced into the United States in very early Colonial days.

Garden peas (*Pisum sativum*) have wrinkled seed coats at maturity when dry; field peas (*P. arvense*) have a smooth seed coat. Both types are annual leafy plants. Each leaf bears three pairs of leaflets and ends in a slender tendril. Five to nine round seeds are enclosed in a pod about 3 in. (7.5 cm) long. Seed color varies from white to cream, green, yellow, or brown. Smooth-seeded varieties may be harvested fresh for freezing or canning, or harvested dry as edible peas. Dry peas may be split or ground and prepared in various ways, such as for split-pea soup.

Wisconsin, Washington, Minnesota, Oregon, Illinois, New York, Pennsylvania, Utah, and Idaho lead in the production of peas harvested green. [K.J.M.]

Peach A native of China, the peach (*Prunus persica*) is, adapted to the temperate zone where winter temperature does not go below − 15°F (−26°C). Therefore, commercial culture is confined to the less rigorous parts of the temperate zone and to areas where nearby bodies of water may exert a modifying influence on climate. In Europe the peach is adapted to Italy, Spain, and southern France. Principal peach-growing areas in North America are (1) central California on the Pacific Coast, with some production in Washington, Oregon, Utah, and Colorado; (2) Maryland, Delaware, and southern New Jersey; (3) Georgia and the Carolinas; (4) Tennessee, Kentucky, and southern Illinois and Indiana; and (5) southern New England, the Niagara region of Canada, western New York, and western Michigan. *See* FRUIT; FRUIT, TREE. [H.B.T.]

Peanut A self-pollinated, one- to six-seeded legume which is cultivated throughout the tropical and temperate climates of the world. The oil, expressed from the seed, is of high quality, and a large percentage of the annual world production is used for this purpose. In the United States some 65% goes into the cleaned and shelled trade, the end products of which are roasted or salted peanuts, peanut butter, and confections. *See* ROSALES.

Botanically, peanuts may be divided into three main types, Virginia, Spanish, and Valencia, based on branching order and pattern and the number of seeds per pod. The USDA Marketing Standards includes an additional type, Runner, which refers to the small-seeded Virginia type produced in Georgia and Alabama. *See* SEED.

The peanut's most distinguishing characteristic is the yellow flower, which resembles a butterfly (papilionaceous) and is borne above ground. The pod, a one-loculed legume, splits under pressure along a longitudinal ventral suture. Pod size varies, and seed weight varies from 0.08 to 0.2 oz (0.2 to 5 g). The number of seeds per pod usually is two in the Virginia type, two or three in the Spanish, and three to six in the Valencia. *See* LEGUME. [A.Pe.]

Pear A fruit native to western Asia or nearby Europe and of very old culture, having been known nearly 1000 years B.C. It spread across Europe and was extremely popular in Belgium and France during the 18th and 19th centuries. Early settlers throughout North America made extensive efforts to grow pears, but they do best in an equable, dry climate with sufficient summer heat to develop good quality. Over 95% of United States commercial production is in the interior valleys of California, Oregon, and Washington.

Nearly all United States pear production is of the European pear (*Pyrus communis*) which includes the varieties Bartlett (known as Williams in Europe), d'Anjou, Bosc, and Comice. Other major types are *P. serotina*, the Oriental sand pear, with roundish shape and gritty flesh; hybrids between *P. communis* and *P. serotina*, represented by Kieffer and Leconte; and *P. nivalis*, the snow pear, grown in Europe for cider or perry (a fermented liquor). *See* FRUIT; FRUIT, TREE. [R.P.L.]

Pearl Any mollusk-formed calcareous concretion that displays an orient and is lustrous. There are two major groups of bivalved mollusks in which gem pearls may form: the saltwater pearl oyster (*Pinctada*), and a number of genera of fresh-water clams. Usually, jewelers refer to salt-water pearls as Oriental pearls, regardless of their place of discovery, and to those from fresh-water bivalves as fresh-water pearls.

Between the body mass and the valves of the mollusk extends a curtainlike tissue called the mantle. In order for a pearl to form, a tiny object such as a parasite or a grain of sand must work through the mantle. When this happens, secretion of nacre around the invading object builds a pearl within the body of the mollusk. Whole pearls form within the body mass of the mollusk, in contrast to blister pearls, which form as protrusions on the inner surface of the shell. Edible oysters produce lusterless concretions, but never pearls.

The substitute for natural pearls, to which the name cultured pearl has been given, is usually made by inserting a large bead into a mollusk to be coated with nacre. [R.T.L.]

Peat A dark-brown or black residuum produced by the partial decomposition and disintegration of mosses, sedges, trees, and other plants that grow in marshes and other wet places. Forest-type peat, when buried and subjected to geological influences of pressure and heat, is the natural forerunner of most coal. Moor peat is formed in relatively elevated, poorly drained moss-covered areas, as in parts of Northern Europe. *See* COAL; HUMUS. [G.H.C.]

Pebble mill A tumbling mill that grinds or pulverizes materials without contaminating them with iron. Because the pebbles have lower specific gravity than steel balls, the capacity of a given size shell with pebbles is considerably lower than with steel balls. The lower capacity results in lower power consumption. The shell has a nonmetallic lining to further prevent iron contamination, as in pulverizing ceramics or pigments (see illustration). Selected hard pieces of the material being ground can be used as pebbles to further prevent contamination. *See* TUMBLING MILL. [R.M.H.]

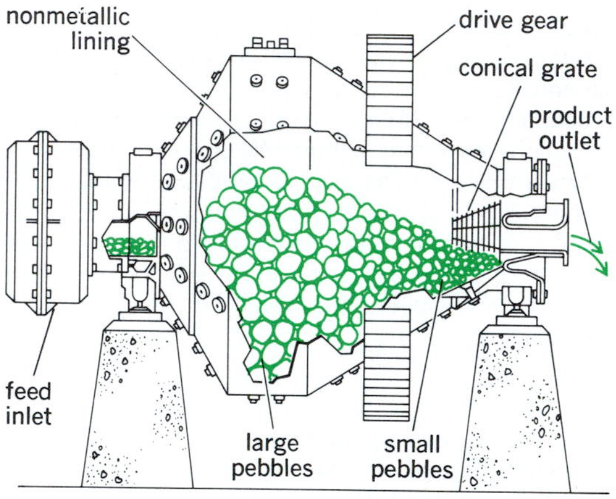

Diagrammatic sketch of a conical pebble mill.

Pecan A large tree (*Carya illinoensis*) of the family Juglandaceae, and the nut from this tree. Native to valleys of the Mississippi River and tributaries as far north as Iowa, to other streams of Texas, Oklahoma, and northern and central Mexico, this nut tree has become commercially important throughout the southern and southwestern United States and northern Mexico. [R.H.S.]

Pectin A group of polysaccharides occurring in the cell walls and intercellular layers of all land plants. They are extractable with hot water, dilute acid, or ammonium oxalate solutions. Pectins are precipitated from aqueous solution by alcohol and are commercially used for their excellent gel-forming ability.

Commercially, the primary source of pectin is the peel of citrus fruits such as lemon and lime, although orange and grapefruit may be used. A secondary source is apple pomace and sunflower heads.

Pectin is widely used in the food industry, principally in the preparation of gels. It is used as a base for jelly and as a stabilizer in some dairy products and frozen desserts, such as sherbet, and also as edible protective coatings for sausages, almonds, candied dried fruit, and soft dates. *See* GEL; POLYSACCHARIDE. [R.L.Wh.; J.R.D.]

Pectolite A mineral inosilicate with composition $Ca_2NaSi_3O_8(OH)$. The hardness is 5 on Mohs scale, and the specific gravity is 2.75. The mineral is colorless, white, or gray with a vitreous to silky luster. Pectolite is found in the United States at Paterson, Bergen Hill, and Great Notch, New Jersey. *See* SILICATE MINERALS. [C.S.Hu.]

Pediculosis Human infestation with lice. There are two biological varieties of the human louse, *Pediculus humanus*, var. *capitis* and var. *corporis*, each showing a strong preference for a specific location on the human body. *Pediculus humanus capitis* colonizes the head and *P. h. corporis* lives in the body-trunk region.

These lice are wingless insects which are ectoparasites. Their mouthparts are modified for piercing skin and sucking blood. The terminal segments of their legs are modified into clawlike structures which are utilized to grasp hairs and clothing fibers.

Lice are important vectors of human diseases. Their habit of sucking blood and their ability to crawl rapidly from one human to another transmit such diseases as typhus (rickettsial) and epidemic relapsing fever (spirochetal). The body fluids and feces of infected lice transmit these diseases. *See* EPIDEMIC TYPHUS FEVER; LOUSE. [R.Su.]

Pedinoida An order of Diadematacea, making up those genera which possess solid spines and a rigid test. The ambulacra show typical diadematoid structure, and the tubercles are noncrenulate. The single known family, Pedinidae, includes 15 genera, ranging from the Late Triassic onward, though only one genus, *Caenopedina*, survives today. *See* DIADEMATACEA. [H.B.F.]

Pedogenesis Reproduction by larval or immature animals, especially the parthenogenetic reproduction by larvae of certain gall midges. *Miastor* is a typical example in which this phenomenon occurs. *See* NEOTENY. [T.I.S.]

Pedology Defined narrowly, a science that is concerned with the nature and arrangement of horizons in soil profiles; the physical constitution and chemical composition of soils; the occurrence of soils in relation to one another and to other elements of the environment such as climate, natural vegetation, topography, and rocks; and the modes of origin of soils. Pedology so defined does not include soil technology, which is concerned with uses of soils.

Broadly, pedology is the science of the nature, properties, formation, distribution, and function of soils, and of their response to use, management, and manipulation. The first definition is widely used in the United States and less so in other countries. The second definition is worldwide. *See* SOIL; SOIL MECHANICS. [R.W.S.]

Peening A metal-finishing operation, also called shot peening, in which small steel shot is thrown against a piece such as a cutting tool. The impact of the shot on the work plastically deforms the surface to a depth of a few thousandths of an inch, producing residual compressive stress. The material is thus made more resistant to fatigue failure. Surface hardness of the material is also increased slightly by the cold working produced by the shot. The shot is hurled at high velocity upon the work by centrifugal force or by an air blast. [A.H.T.]

Pegasiformes The sea moths or sea dragons, a small order of peculiar actinopterygian fishes also known as the Hypostomides. The body is encased in a broad, bony framework anteriorly and has bony rings posteriorly, simulating the seahorses and pipefishes (see illustration). Unlike those fishes, which have the mouth at the tip of the produced snout, sea moths have enlarged nasal bones which form a rostrum that projects well forward of the small, toothless mouth. The greatly expanded horizontal pectoral fin belies

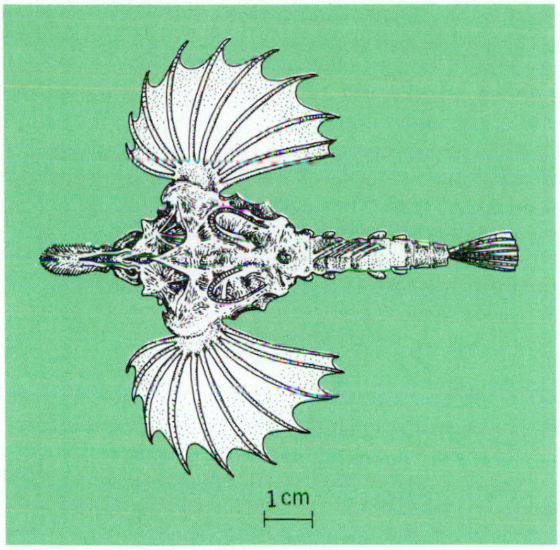

Sea moth (*Pegasus draconis*). (*After D. S. Jordan and J. O. Snyder, vol. 24, Leland Stanford University Contributions to Biology, 1901*)

its appearance and does not function in aerial gliding. The pelvic fin is abdominal and consists of a slender spine and one or two long rays. The short, opposed dorsal and anal fins have no spines. There is no swim bladder. *See* GASTEROSTEIFORMES.

There is a single family, Pegasidae, with one genus, *Pegasus*, and four or five species. They live amidst vegetation on Indo-Pacific shores from East Africa to Japan, Australia, and Hawaii. There is no fossil record. *See* ACTINOPTERYGII. [R.M.B.]

Pegasus The Winged Horse, in astronomy, an autumnal constellation. Pegasus is usually identified by the four bright stars α, ß, γ, and α (Alpha Andromedae) situated on the corners of a large square known as the Great Square in Pegasus (see illustration). The star Alpheratz at the northeastern corner

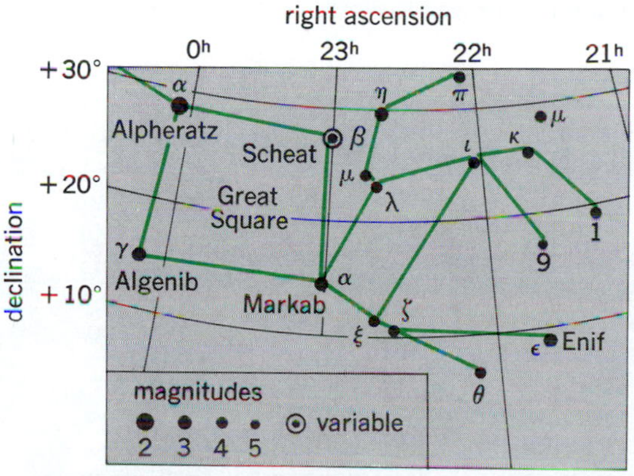

Line pattern of the constellation Pegasus. The grid lines represent the coordinates of the sky. The apparent brightness, or magnitude, of the stars is shown by the sizes of the dots, graded by appropriate numbers.

of the square is really in the constellation Andromeda. *See* CONSTELLATION. [C.-S.Y.]

Pegmatite Exceptionally coarse-grained and relatively light-colored crystalline rock composed chiefly of minerals found in ordinary igneous rocks. Extreme variations in grain size also are characteristic, and close associations with dominantly fine-grained aplites are common. Pegmatites are widespread and very abundant where they occur, especially in host rocks of Precambrian age, but their aggregate volume in the Earth's crust is small. Many pegmatites have been economically valuable as sources of clays, feldspars, gem materials, industrial crystals, micas, silica, and special fluxes, as well as beryllium, bismuth, lithium, molybdenum, rare-earth, tantalumniobium, thorium, tin, tungsten, and uranium minerals. *See* APLITE; IGNEOUS ROCKS.

Essential minerals (1) in granitic pegmatities are quartz, potash feldspar, and sodic plagioclase; (2) in syenitic pegmatites, alkali feldspars with or without feldspathoids; and (3) in diorite and gabbro pegmatities, soda-lime or lime-soda plagioclase. Varietal minerals such as micas, amphiboles, pyroxenes, black tourmaline, fluorite, and calcite further characterize the pegmatites of specific districts. Accessory minerals include allanite, apatite, beryl, garnet, magnetite, monazite, tantalite-columbite, lithium tourmaline, zircon, and a host of rarer species. [R.H.J.]

Peking man The East Asian representatives of *Homo erectus*. The name is in reference to the site of Choukoutien near Peking (Beijing), China, where most of the fossils have been found. The fossils depict hominids similar to the other *H. erectus* samples. They had brains, on the average, about two-

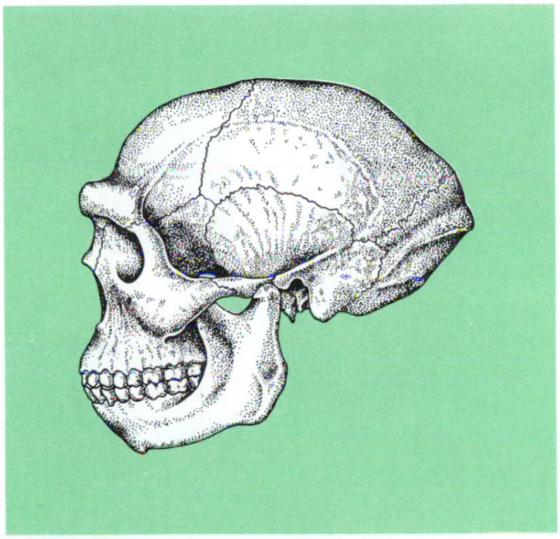

Reconstruction of female *Sinanthropus* skull, showing the heavy browridges. (*After M. F. Ashley Montagu, An Introduction to Physical Anthropology, 2d ed., Charles C. Thomas, 1951*)

thirds of modern human size, large faces with massive brows and jaws (see illustration), and limbs functionally modern but exceptionally strongly built. *See* FOSSIL HUMAN. [E.T.]

Pelecaniformes An order of aquatic birds characterized by having all four toes joined by webs. Several fossil and six living families are usually recognized. The latter include the pelicans (Pelecanidae), cormorants (Phalacrocoracidae), anhingas or snakebirds (Anhingidae), frigate birds or man-o'-war birds (Fregatidae), gannets and boobies (Sulidae), and tropic birds (Phaethontidae). All are primarily fish eaters, but methods of catching prey vary widely. *See* AVES. [K.C.P.]

Pellagra A disease resulting from severe deficiency of the vitamin B complex. Niacin deficiency is the principal dietary defect in pellagra; the low tryptophan content of some foods also plays a role in this disease.

Skin, gastrointestinal, and neurologic symptoms may occur. Redness, scaling, and brownish discoloration of the skin, particularly in areas exposed to sunlight, are common lesions. Marked enlargement of the tongue and drooling are prominent symptoms. The mucous membranes of the mouth, eyes, urethra, and vagina may show swelling and have a bright red, smooth appearance. Gastrointestinal symptoms may be vague or may take the form of severe nausea, vomiting, and bloody diarrhea. Early nervous system involvement is displayed by the neurasthenic syndrome characterized by restlessness, anxiety, and insomnia. Organic psychoses may follow continued deficiency and are marked by memory loss, confusion, and a variable pattern of affective behavior, such as depression or paranoia. Actual delirium, limb rigidity, and certain uncontrolled reflexes mark the severe case. *See* NIACIN; VITAMIN. [E.G.St./N.K.M.]

Pelletron accelerator An electrostatic particle accelerator that utilizes charging chains consisting of steel cylinders joined by links of solid insulating material such as nylon (see illustration). The metal cylinders are charged as they leave a pulley at ground potential, and the charge is removed as they pass over a pulley in a high-potential terminal.

Charging chain of 14-MV tandem Pelletron at Australian National University, Canberra.

Pelletrons range in maximum operating voltages from 200 kV to 25 MV and have beam current capabilities ranging from a few microamperes up to 0.8 milliampere. Machines to give 1 MV or more are enclosed in pressure tanks and are insulated by sulfur hexafluoride (SF_6) gas at pressures up to about 8 atm (800 kilopascals).

A large number of electrostatic accelerators of the belt-charged Van de Graaff type are in operation throughout the world with voltage capabilities of 10 MV and below. Many of the 10-MV Van de Graaff accelerators have been upgraded to 13 MV. Because of belt trouble above 10 MV, most of the upgraded machines have been converted to Pelletron charging. *See* VAN DE GRAAFF GENERATOR.

Commonly, two-stage (tandem) accelerators have utilized a double-ended column extending through the enclosing tank from one end to the other, with the high-potential terminal at the center. Negative ions enter at one end, are stripped of two or more electrons at the terminal, and receive a second acceleration as they proceed in a straight line through the high-energy accelerating tube to the other end of the accelerator. In a folded tandem accelerator the column extends in only one direction from the terminal, and a 180° magnet directs ions through a second accelerating tube. *See* PARTICLE ACCELERATOR. [R.G.H.]

Pelmatozoa A division of the Echinodermata made up of those forms which are anchored to the substrate during at least a part of the life history. Formerly treated as a formal unit of classification with the rank of subphylum, pelmatozoans are now realized to be a heterogeneous assemblage of forms with similar habits but dissimilar ancestry, their common features having arisen by convergent evolution. Most pelmatozoan echinoderms are members of the subphylum Crinozoa, but some echinozoans also exhibit a sedentary, anchored life, with modifications for such existence. *See* CRINOZOA; ECHINODERMATA; ECHINOZOA; ELEUTHEROZOA. [H.B.F.]

Peltier effect A phenomenon discovered in 1834 by J. C. A. Peltier, who found that at the junction of two dissimilar metals carrying a small current the temperature rises or falls, depending upon the direction of the current. In view of experiments, which establish that the rate of intake or output of heat is proportional to the magnitude of the current, it can be shown that an electromotive force resides at a junction. Electromotive forces of this type are called Peltier emf's. *See* SEEBECK EFFECT; THERMOELECTRICITY; THOMSON EFFECT. [J.W.St.]

Pelton wheel An impulse type of hydraulic turbine. In the impulse turbine, pressure of the water supply is converted into velocity by a nozzle. The water jet then impinges on the buckets of the turbine wheel or runner. In a Pelton wheel the buckets have a splitter in the middle to divide the water jet and cause it to flow across the cupped faces of the buckets and emerge at the sides (see illustration). The water jet thus imparts

A 12-in. (30-cm) Pelton wheel. (*Photograph by R. L. Daugherty and P. Kyropoulas*)

its kinetic energy to the buckets. Pelton wheels are usually operated from high-head sources. *See* HYDRAULIC TURBINE. [F.H.R.]

Pelycosauria An extinct order of primitive, mammallike reptiles of the subclass Synapsida. They are characterized by a temporal fossa that lies low on the side of the skull. The group is known from rocks of the Upper Carboniferous and lower and middle Permian. Three suborders are included: Ophiacodonta, Edaphosauria, and Sphenacodontia. Late in the early Permian the sphenacodonts gave rise to more advanced mammallike reptiles, the therapsids. *See* SYNAPSIDA; THERAPSIDA. [E.C.O.]

Pendulum A rigid body mounted on a fixed horizontal axis, about which it is free to rotate under the influence of gravity. The period of the motion of a pendulum is virtually independent of its amplitude and depends primarily on the geometry of the pendulum and on the local value of g, the acceleration of gravity. Pendulums have therefore been used as the control elements in clocks, or inversely as instruments to measure g.

Motion. In the schematic representation of a pendulum shown in the illustration, O represents the axis and C the center of mass. The line OC makes an instantaneous angle θ with the vertical. In rotary motion of any rigid body about a fixed axis, the angular acceleration is equal to the torque about the axis divided by the moment of inertia I about the axis. If m represents the mass of the pendulum, the force of gravity can be considered as the weight mg acting at the center of mass C.

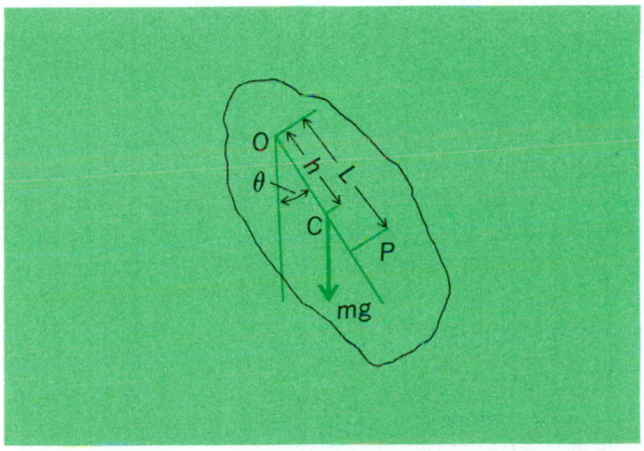

Schematic diagram of a pendulum. *O* represents the axis, *C* is the center of mass, and *P* the center of oscillation.

If the amplitude of motion is small, the motion is simple harmonic. The period T, time for a complete vibration (for example, from the extreme displacement right to the next extreme displacement right), is given by Eq. (1). *See* HARMONIC MOTION.

$$T = 2\pi\sqrt{I/mgh} \tag{1}$$

The actual form of a pendulum often consists of a long, light bar or a cord that serves as a support for a small, massive bob. The idealization of this form into a point mass on the end of a weightless rod of length L is known as a simple pendulum. An actual pendulum is sometimes called a physical or compound pendulum. In a simple pendulum the lengths h and L become identical, and the moment of inertia I equals mL^2. Equation (1) for the period becomes Eq. (2).

$$T = 2\pi\sqrt{L/g} \tag{2}$$

Center of oscillation. Equation (2) can be used to define the equivalent length of a physical pendulum. Comparison with Eq. (1) shows that Eq. (3) holds. The point P on line OC

$$L = I/mh \tag{3}$$

of the illustration, whose distance from the axis O equals L, is called the center of oscillation. Points O and P are reciprocally related to each other in the sense that if the pendulum were suspended at P, O would be the center of oscillation.

Types. Kater's reversible pendulum is designed to measure g, the acceleration of gravity. It consists of a body with two knife-edge supports on opposite sides of the center of mass as at O and P (and with at least one adjustable knife-edge). If the pendulum has the same period when suspended from either knife-edge, then each is located at the center of oscillation of the other, and the distance between them must be L, the length of the equivalent simple pendulum. The value for g follows from Eq. (2).

The ballistic pendulum is a device to measure the momentum of a bullet. The pendulum bob is a block of wood into which the bullet is fired. The bullet is stopped within the block and its momentum transferred to the pendulum. This momentum is determined from the amplitude of the pendulum swing. *See* BALLISTICS.

The spherical pendulum is a simple pendulum mounted on a pivot so that its motion is not confined to a plane. The bob then moves over a spherical surface. A Foucault pendulum is a spherical pendulum suspended so that its plane of oscillation is free to rotate. Its purpose is to demonstrate the rotation of the Earth. *See* FOUCAULT PENDULUM; SCHULER PENDULUM. [J.M.Ke.]

Penguin Any member of the avian order Sphenisciformes which contains the single family Spheniscidae. There are about 18 species, some of these having definite geographical races, included in 6 genera (fossil forms known since the Tertiary from Antarctic remains are excluded).

These flightless marine birds range throughout the Southern Hemisphere and are more highly adapted for their marine existence than any other group of birds. The feathers, unlike those of other birds, cover the body densely. The structurally modified wings do not fold, and function as flippers. The tail is short, the bill is variable in shape and form, the feet are short and webbed, and the legs are set far back on the body. The different species vary from 1 to 4 ft (0.3 to 1.2 m) in height. Penguins move through the water by using the flippers for propulsion and the tail and feet for direction; they swim as fast as seals underwater. On shore these bipedal animals move in a rapid but ungainly manner. Their diet consists principally of crustaceans, fish, and mollusks, such as squid. Penguins are gregarious and nest in large colonies. They are monogamous, pairing for life. The number of eggs varies from one to three, and incubation behavior is variable with the species. *See* SPHENISCIFORMES. [C.B.C.]

Penicillin One of the beta-lactam antibiotics, all of which possess a four-ring beta-lactam structure fused with a five-membered thiazolidine ring. These antibiotics are nontoxic and kill sensitive bacteria during their growth stage by the inhibition of biosynthesis of their cell wall mucopeptide. *See* PLANT CELL.

The antibiotic properties of penicillin were first recognized by A. Fleming in 1928 from the serendipitous observation of a mold, *Penicillium notatum*, growing on a petri dish agar plate of a staphylococcal culture. The mold produced a diffuse zone which lysed the bacterial cells. Commercial production of penicillin came from the pioneer work of E. Chain and H. W. Florey in 1938. Penicillin (as penicillin G) was made available to the allied troops in Europe in the latter part of World War II.

Penicillin is produced from the fungal culture *P. chrysogenum* that was isolated from a moldy cantaloupe. The biosynthesis of penicillin is known in detail, and all the enzymes involved in the formation of this secondary metabolite have been isolated and purified.

The fermented penicillin G and penicillin V are susceptible to destruction by an enzyme (beta-lactamase) produced by certain bacteria which makes them resistant. The penicillins methicillin, oxacillin, nafcillin, cloxacillin and dicloxacillin are resistant to hydrolysis by beta-lactamases and are used to treat staphylococcal infections. Cloxacillin and dicloxacillin are used orally. Ampicillin and amoxicillin are penicillins with extended spectra, and they are effective against many gram-negative bacteria. They are used mainly orally against streptococci and other respiratory-tract pathogens, including *Haemophilus influenzae*, in the treatment of sinusitis, bronchitis, and pneumonia. They are used extensively in pediatrics and against *Listeria monocytogenes* and *Salmonella* spp. *See* ANTIBIOTIC; DRUG RESISTANCE; SALMONELLOSES; STREPTOCOCCUS.

Penicillin G, the most commonly fermented penicillin, is produced by the addition of a precursor, phenylacetic acid, to the growing culture. Use of phenoxyacetic acid as a precursor produces penicillin V. Both penicillins are recovered by extraction into organic solvents at acid pH, and precipitation as their potassium or sodium salt. Penicillin G is generally given by injection against penicillin-sensitive streptococci such as pneumococci (meningitis), and in treatment of endocarditis and gonorrhea. Penicillin V is acid stable and is usually given orally. It is effective in the treatment of upper respiratory infections and periodontal work. *See* GONORRHEA; MENINGITIS. [D.A.Lo.]

Penis The male organ of copulation, or phallus. In mammals the penis consists basically of three elongated masses of erectile tissue. The central corpus spongiosum (corpus urethrae) lies ventral to the paired corpora cavernosa. The urethra runs along the underside of the spongiosum and then normally rises to open at the expanded, cone-shaped tip, the glans penis, which fits like a cap over the end of the penis. Loose skin encloses the penis and also forms the retractable foreskin, or prepuce.

Erection of the penis is caused by nervous stimulation resulting in engorgement of the spiral helicine arteries and the plentiful venous sinuses of the organ. In most mammals other than Primates the penis is retracted into a sheath when not in use.

In submammalian forms the penis is not as well developed. Crocodilians, turtles, and some birds have a penis basically like that of mammals, although not as well developed, lying in the floor of the cloaca. When erected, it protrudes from the rudimentary. Other vertebrates lack a penis, although various functionally comparable organs may be developed such as the claspers on the pelvic fins of sharks and the gonopodia on the anal fins of certain teleost fishes. *See* COPULATORY ORGAN. [T.S.P.]

Pennatulacea An order of the subclass Alcyonaria, commonly called the sea pens. These animals lack stolons and live with their bases embedded in the soft substratum of the sea. The colony consists of a distal rachis bearing many polyps and a polypless proximal peduncle, whose terminal end sometimes expands to form a bladder. The colony of *Pennatula* looks like a feather being formed of numerous secondary polyps which arise from leaf-shaped lateral expansions of the very elongated primary axial or terminal polyp. In the other form of colony the polyps arise directly from the primary one, as in *Veretillum* (see illustration) and *Renilla* and *Cavenularia*. The colony has a horny unbranched axial skeleton composed of pennatulin and some calcium carbonate and phosphate, but

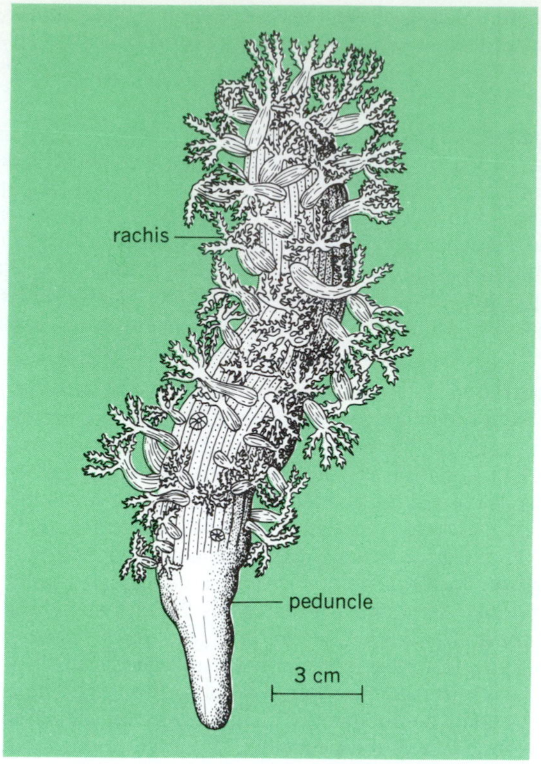

***Veretillum cynomorium*.**

Cavenularia, including the luminous species, has a rudimentary one and *Renilla* lacks a skeleton. *See* ALCYONARIA. [K.At.]

Pennsylvanian A major division of late Paleozoic time, considered either as an independent period or as the younger subperiod of the Carboniferous. In North America the Pennsylvanian has been widely recognized as a geologic period and derives its name from a thick succession of mostly nonmarine, coal-bearing strata in Pennsylvania. Radiometric ages place the beginning of the period at approximately 320,000,000 years ago and its end at about 280,000,000 years ago. In northwestern Europe, strata of nearly equivalent age are commonly designated as Upper Carboniferous and in eastern Europe as Middle and Upper Carboniferous. *See* CARBONIFEROUS.

During Pennsylvanian time the paleoequator extended across North America from southern California to Newfoundland, through northwestern Europe into the Ukrainian region of eastern Europe, and across parts of northern China. This was a time of extensive coal deposition in a tropical belt that appears to have included areas from 15 to 20° north and south of the

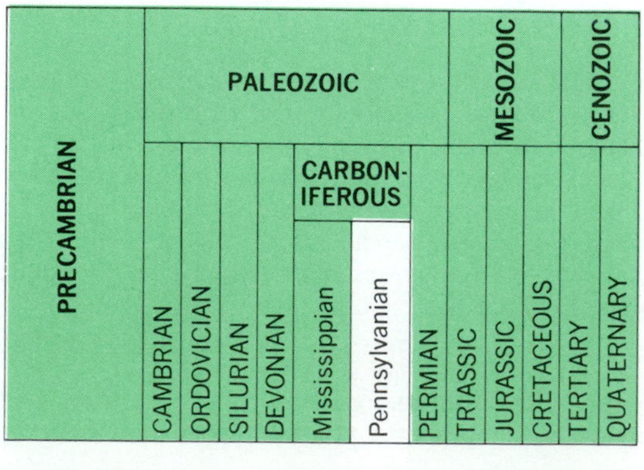

paleoequator. Coal of this age is abundant and relatively widespread and has great economic importance.

Petroleum is commonly trapped in nearshore marine deposits of Pennsylvanian age, particularly in carbonate banks near the edge of shelves, in longshore bars and beaches, in reefs and mounds, and at unconformities associated with transgressive-regressive shore lines. Many of these traps contribute significantly to petroleum production.

Pennsylvanian paleogeography changed significantly during the period as the supercontinent Pangaea gradually was formed by the joining together of Gondwana and Laurasia. The result was an extensive orogeny, or mountain-building episode, which supplied the vast amounts of sediments that make up most of the Pennsylvanian strata in the eastern and midwestern part of the United States.

Evidence in the form of well-developed tree rings, less diverse fossil floras and faunas, and glacial deposits indicates that temperate and glacial conditions were common in nonequatorial climatic belts during Pennsylvanian time. Climatic fluctuations during the period caused significant increases and decreases in the amount of water that was temporarily stored in the glaciers in Gondwana and contributed to eustatic changes of sea level. [C.A.R.]

Pentagon A geometric figure formed by the five line segments, or sides, that join in order five ordered points, or vertices, of a plane. In elementary geometry it is supposed that the sides do not cross, and even that the figure bounds a convex region of the plane (convex pentagon). *See* POLYGON; REGULAR POLYTOPES. [L.M.Bl.]

Pentamerida An order of articulate brachiopods whose genera are first known in rocks of Middle Cambrian age and which became extinct in the Devonian. *See* ARTICULATA (BRACHIOPODA).

Prior to the Late Ordovician the order was represented only by the suborder Syntrophiidina, a group that was never abundant but which continued into Early Devonian times. The early syntrophiidines differed little from the orthaceans, their presumed ancestors, although a strong dorsal median fold was characteristic. In later forms the posterior margin was curved and interareas became reduced; internally one of the most typical features of the order was developed, a spondylium. This spoon-shaped structure is formed by coalescence of the dental plates and carried some or all of the ventral musculature. Changes also occurred in the brachial valve, notably in improved definition of the sockets and elaboration of the brachiophore bases, which commonly extended to the floor of the valve. *See* BRACHIOPODA; ORTHIDA. [A.J.R.]

Pentastomida A class of bloodsucking arthropods, parasitic in the respiratory organs of vertebrates, that frequently are referred to as the Linguatulida or tongue worms. The adults are vermiform, with a short cephalothorax and an elongate, annulate abdomen that may be cylindrical or flattened.

The class is divided into two orders: the Cephalobaenida, a more primitive group, and the Porocephalida, a more specialized one. The first has six-legged larvae and the other, four-legged larvae. The mitelike form of the larvae, with short stumpy legs, demonstrates relationship to the arthropods. Characteristic arthropod features include the presence of (1) jointed appendages in the larvae; (2) stigmata or breathing pores in the body wall; (3) specialized reproductive organs, especially those of the male; and (4) ecdysis or molting of larvae and nymphs. More than 50 species have been described.

Human infection occurs frequently in Africa and the Orient, where humans are an accidental intermediate host of the nymphal form. The liver is a common site of infection, and large numbers of larvae may produce serious and even fatal effects. *See* ARTHROPODA; CEPHALOBAENIDA; POROCEPHALIDA. [H.W.S.]

Pentlandite A mineral having composition $(Fe,Ni)_9S_8$. Pentlandite is the major ore of nickel. It is usually massive, showing a well-defined octahedral parting. The hardness is 3.5–4 (Mohs scale) and the specific gravity varies from 4.6 to 5.0, depending on the ratio of iron to nickel; greater amounts of iron cause an increase in the specific gravity. The luster is metallic and the color yellowish bronze. Pentlandite is found at many localities in small amounts, but its chief occurrence is at Sudbury, Ontario, where it is mined on a large scale. *See* NICKEL. [C.S.Hu.]

Pepper The garden pepper, *Capsicum annuum* (family Solanaceae), is a warm-season crop originally domesticated in Mexico. It is usually grown as an annual, although in warm climates it may be perennial. This species includes all peppers grown in the United States except for the "Tabasco" pepper (*C. frutescens*), grown in Louisiana. Other cultivated species, *C. chinense*, *C. baccatum*, and *C. pubescens*, are grown primarily in South America. Some 10–12 strictly wild species also occur in South America. Peppers are grown worldwide, especially in the more tropical areas, where the pepper is an important condiment.

Sweet (nonpungent) peppers, harvested fully developed but still green, are widely used in salads or cooked with other foods. Perfection pimento, harvested red ripe, is used for canning. Paprika is made from ripe red pods of several distinct varieties; the pods are dried and ground. *See* PAPRIKA; PIMENTO.

The ripe color of most varieties is red, a few varieties are orange-yellow, and in Latin America brown-fruited varieties are common. Nutritionally, the mature pepper fruit has three to four times the vitamin C content of an orange, and is an excellent source of vitamin A. *See* ASCORBIC ACID; VITAMIN A. [P.G.S.]

Peppermint The mint species *Mentha piperita* (family Lamiaceae), a sterile interspecific hybrid believed to have occurred in nature from the hybridization of fertile *M. spicata*. Peppermint oil is obtained by steam distillation from the partially dried hay. The main uses of peppermint oil are to flavor chewing gum, confectionery products, toothpaste, mouthwashes, medicines, and as a carminative in certain medical preparations for the alleviation of digestive disturbances. [M.J.M.]

Pepsin A proteolytic enzyme found in the gastric juice of mammals, birds, reptiles, and fish. It is formed from a precursor, pepsinogen, which is found in the stomach mucosa. Pepsinogen is converted to pepsin either by hydrochloric acid, naturally present in the stomach, or by pepsin itself. *See* ENZYME.

Pepsin is prepared commercially from the glandular layer of fresh hog stomachs. It is a part of the crude preparation known as rennet, which is used to curdle milk in preparation for cheese manufacture. Pepsin is also used for a variety of other applications in food manufacturing; to modify soy protein and gelatin, thereby providing whipping qualities; to modify vegetable proteins for use in nondairy snack items; to make precooked cereals into instant hot cereals; and to prepare animal and vegetable protein hydrolysates for use in flavoring foods and beverages. [M.So.]

Peptic ulcer A sharply defined ulceration of the upper gastrointestinal tract, characterized by loss of the mucous lining and variable penetration into or through the organ wall. The loss of the mucosa, ordinarily covered by a mucous secretion, lays bare the musculomembranous wall. The crater formed is covered with an exudate lying over the raw tissue, except in chronic ulcers where the pit consists largely of scar tissue. The exact causes are obscure, but emotional tension and certain psychological patterns are frequently present in affected individuals.

Most ulcer patients show an increased acidity of the stomach or an abnormal production of acid during times when the stomach is usually quiescent, as at night. Both the action of the acid on raw surfaces and the resulting muscular spasms are thought to be associated with the typical ulcer pain. Scarring, constriction, perforation, and hemorrhage are common complications. [E.G.St./N.K.M.]

Peptide A compound that is made up of two or more amino acids joined by covalent bonds which are formed by the elimination of a molecule of H_2O from the amino group of one amino acid and the carboxyl group of the next amino acid. Peptides larger than about 50 amino acid residues are usually classified as proteins. Glutathione is the most abundant peptide in mammalian tissue. Hormones such as oxytocin (8), vasopressin (8), glucagon (29), and adrenocorticotropic hormone (39) are peptides whose structures have been deduced; in parentheses are the numbers of amino acid residues for each peptide.

For each step in the biological synthesis of a peptide or protein there is a specific enzyme or enzyme complex that catalyzes each reaction in an ordered fashion along the biosynthetic route. However, it is noteworthy that, although the biological synthesis of proteins is directed by messenger RNA on cellular structures called ribosomes, the biological synthesis of peptides does not require either messenger RNA or ribosomes. *See* AMINO ACIDS; PROTEIN; RIBONUCLEIC ACID (RNA); RIBOSOMES. [J.M.M.]

Peracarida A superorder of the class Crustacea, subclass Malacostraca. Common examples of the orders are the opossum shrimps, aquatic sow bugs, and side-swimmers. The Peracarida includes the orders Amphipoda, Cumacea, Isopoda, Mysidacea, Spelaeogriphacea, and Tanaidacea. These orders share a number of characters, the most notable being that the young develop within a thoracic marsupium, which they leave at a late stage of development.

Other features which distinguish the Peracarida from other groups of Malacostraca are the following: The protopodite of the second antenna usually consists of three segments. The thoracic legs are flexed between the fifth and sixth segments. The heart is generally elongate, extending through the greater part of the thoracic region; in the isopods, it may extend into, or lie entirely in, the abdomen, where respiratory exchange occurs. Spermatozoa are usually filiform, in contrast to their spherical or vesicular form in the Eucarida and Hoplocarida. *See* AMPHIPODA; CRUSTACEA; CUMACEA; ISOPODA; MYSIDACEA; SPELAEOGRIPHACEA; TANAIDACEA. [T.E.B.]

Perbromate The negative ion BrO_4^-. The perbromate ion is tetrahedral, like the perchlorate and meta periodate ions. Perbromic acid is a very strong acid, and its potassium, rubidium, and cesium salts have relatively low solubilities in water. Aqueous solutions of the alkali perbromates are completely stable, while perbromic acid itself is stable up to concentrations of 6 *M*. Under ordinary conditions perbromates are very sluggish oxidizing agents, being among the least reactive of all the oxyhalogen compounds. *See* PERCHLORATE. [E.H.Ap.]

Percent A ratio comparison of two quantities expressed by using 100 equal parts, or hundredths; symbolized %. There are three major uses of percent: part of a whole, rate, and comparison of any two quantities.

Part of a whole. The basic idea of percent is as a ratio that shows a part of a whole. The technical name for the whole is base.

If 89 out of 100 problems are correct, the part-whole comparison shows 89/100 or 89% correct. If the whole is not already divided into equal parts, an equivalent ratio to 100 is found.

Certain ratios are easy to express as hundredths. When 3 baskets are made in basketball out of 10 attempts, an equivalent ratio using 100 is found from which the percent is obvious: 3/10 = 30/100 = 30%.

For 35 hits out of 126 times at bat, the ratio 35/126 is not easily expressed as a ratio using 100. Hundredths will be more obvious by dividing 35 by 126, and then reading the number of hundredths to find the percent: $35 \div 126 \approx 0.278$ or 27.8%. The percent is obtained by moving the decimal point two places to the right.

Rate. While percent always means a comparison to 100, percent can show a rate of so many per 100, not so many out of 100. A sales tax of 6% means a rate of 6 cents for each dollar, and this 6 cents is in addition to the dollar. The tax amount can often be calculated mentally by multiplying the 6% rate by the number of dollars.

Interest paid or interest received is done by using rates. Simple interest at a rate of 6% means \$6 per \$100 for a full year. If interest is calculated monthly on the unpaid balance and the yearly rate is 18%, the monthly rate is approximately 18% ÷ 12 or about 1.5%.

With compound interest, the amount of interest is added each compounding period, and this total amount is subject to compounding for the next period. For example, \$1.00 invested at 8% will be worth 108% of \$1.00 at the end of one year, or \$1.08. The worth at the end of the second year is 108% of \$1.08, or $(1.08)^2$. At this same rate, after 10 years the compounded value is $(1.08)^{10} = 2.1589247$, or \$2.15, using 1.08 as a factor for 10 times.

Comparing any two quantities. Percent is used to compare any two quantities, but special care must be given to the base for the comparison. Comparison of city A with a population of 42,000 people and city B with 67,000 people will depend on the base. A compared to B is 42,000/67,000, 62.7%; A is 62.7% of B. B compared to A is 67,000/42,000, 1.595, or 159.5%; B is 159.5% of A. *See* ARITHMETIC. [J.N.P.]

Perception Those subjective experiences of objects or events that ordinarily result from stimulation of the receptor organs of the body. This stimulation is transformed or encoded into neural activity (by specialized receptor mechanisms) and is relayed to more central regions of the nervous system where further neural processing occurs. Most likely, it is the final neural processing in the brain that underlies or causes perceptual experience, and so perceptionlike experiences can sometimes occur without external stimulation of the receptor organs, as in dreams.

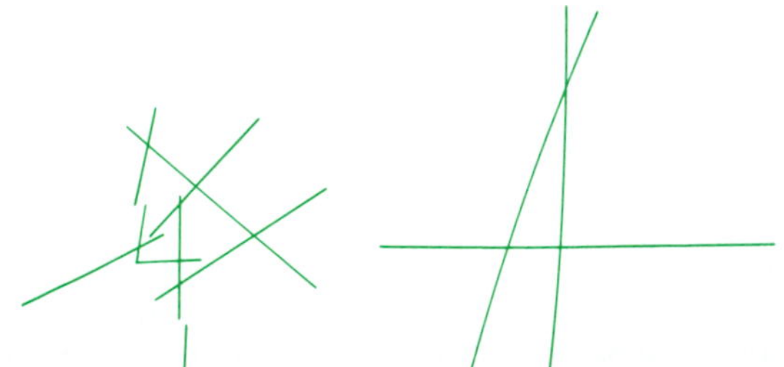

Fig. 1. Perceptual organization. (*a*) The figure of a four is immediately and spontaneously perceived despite the presence of other overlapping and adjacent lines. (*b*) The four, although physically present, is not spontaneously perceived and is even difficult to see when one knows it is there.

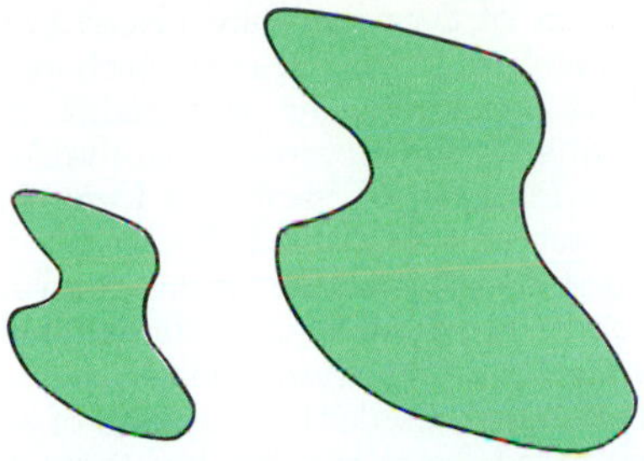

Fig. 2. Transposition of form; the two shapes clearly look the same despite the difference in size.

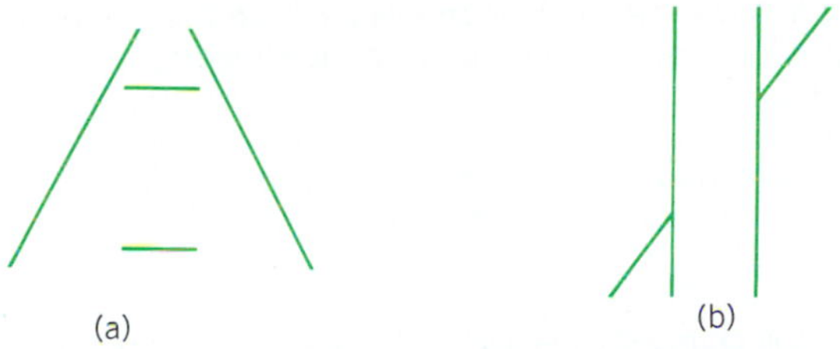

Fig. 4. Geometrical illusions. (*a*) The Ponzo illusion in which the two horizontal lines of equal length appear unequal. (*b*) The Poggendorff illusion in which the two oblique line segments are aligned with one another (that is, are collinear) but appear to be misaligned.

In contemporary psychology, interest generally focuses on perception or the apprehension of objects or events, rather than simply on sensation or sensory process (such as vision or hearing). While no sharp line of demarcation between these topics exists, it is fair to say that sensory qualities are generally explicable on the basis of mechanisms within the receptor organ, whereas object and event perception entails higher-level activity of the brain. *See* HEARING (HUMAN); SENSATION; VISION.

Since objects or events are not experienced only through vision, the term perception obviously applies to other sense modalities as well. Certainly things and their movement may be experienced through the sense of touch. Such experiences derive from receptors in the skin (tactile perception), but more importantly, from the positioning of the fingers with respect to one another when an object is grasped, the latter information arising from receptors in the muscles and joints (haptic or tactual perception). The position of the parts of the body are also perceived with respect to one another whether they are stationary (proprioception) or in motion (kinesthesis), and the position of the body is experienced with respect to the environment through receptors sensitive to gravity such as those in the vestibular apparatus in the inner ear. Auditory perception yields recognition of the location of sound sources and of structures such as melodies and speech. Other sense modalities such as taste (gustation), smell (olfaction), pain, and temperature provide sensory qualities but not perceptual structures as do vision, audition, and touch, and thus are usually dealt with as sensory processes. *See* OLFACTION; PAIN; PROPRIOCEPTION; TEMPERATURE SENSES; TOUCH.

Constancy. By and large, these perceptual properties of objects remain remarkably constant despite variations in distance, slant, and retinal locus caused by movements of the observer. This fact, referred to as perceptual constancy, is perhaps the hallmark of perception and more than any other, serves to characterize the field of perception.

Examples of perceptual constancy are: size (except at very great distances, an object appears the same size whether seen nearby or far away, although the size of its image on the retina can be very different); shape (a circle seen from the side is perceived as a circle, although it appears as an ellipse on the retina); orientation (objects appear to keep the same orientation in space, independently of the orientation of the observer's head); and position (a fixed object remains perceived as stationary even when its image on the retina moves because of eye or head movements).

Motion perception. Perception is not only of objects but of events as well. The kind of event that has been investigated in great detail is the perception of motion. As already noted, perceived movement cannot simply be explained by the motion of an object's retinal image since image motion caused by observer or eye movement does not lead to perceived object movement. Moreover, an object tracked by smooth-pursuit eye movements will appear to move, although in that case there is essentially no motion of the object's image over the retina. Similarly, an afterimage will appear to move during eye movement even in a completely darkened room. This illusory effect is analogous to Emmert's law. Where ordinarily the movement of the retinal image caused by the moving eye is computed to signify "no object motion," thus yielding position constancy (since the image motion and eye motion are equal in magnitude), the same computational rule must signify "object motion" in the case of the afterimage.

Form perception. One question requiring study is why a triangular object appears triangular when a triangular image of the object is present on the retina. Before attempting to answer this question, it is important first to make certain distinctions. Form perception means the experience of a shaped region in the field. Recognition means the experience that the shape is familiar. Identification means that the function or meaning or category of the shape is known. For those who have never seen the shape before, it will be perceived but not recognized or identified. For those who have, it will be perceived as a certain familiar shape and also identified. Recognition and identification obviously must be based on past experience, which means that through certain unknown processes, memory contributes to the immediate experience that one has of the triangle, giving the qualities of familiarity and meaning.

The figure of a 4 in Fig. 1*a* is seen as one unit, separate from other units in the field, even if these units overlap. This means that the parts of the figure are grouped together by the perceptual system into a whole, and these parts are not grouped with the parts of other objects. This effect is called perceptional organization. There are other problems about form perception that remain to be unraveled. For example, the

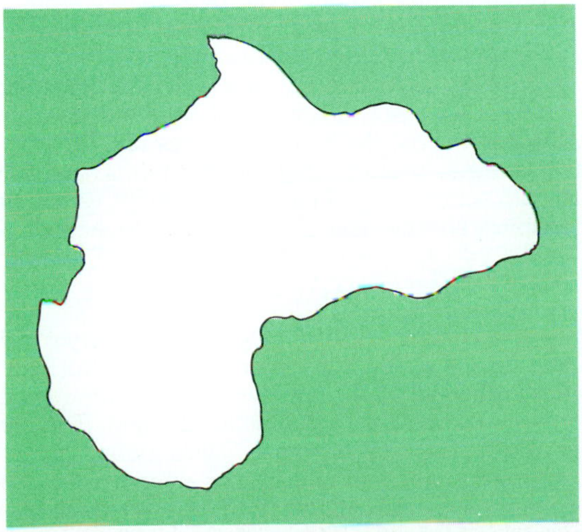

Fig. 3. This figure does not look like anything recognizable. If, however, it is tilted 90° clockwise, it will be recognized as an outline map of Africa.

size of a figure can vary, as can its locus on the retina or even its color or type of contour, without affecting its perceived shape (Fig. 2).

A further fact about form perception is that it is dependent upon orientation (Fig. 3). It is a commonplace observation that printed or written words are difficult to read when inverted, and faces look very odd or become unrecognizable when upside down. Simple figures also look different when their orientation is changed: a square looks like a diamond when tilted by 45°.

Geometrical illusions. Related to the topic of form perception is the misperception of the size or direction of parts of figures that constitutes many of the geometric illusions. In an illusion figure, one particular part is perceived to be either longer or shorter than another part, although they are objectively equal (Fig. 4*a*); or the direction of a contour is perceived to be different from that of another contour although they are the same (Fig. 4*b*). For reasons still not understood, the background or context of the rest of the figure affects these parts.

Innate or learned? A central problem is whether the perception of properties such as form and depth or the achievement of veridical perception as in the constancies is innately determined or is based on past experience. By "innate" it is meant that the perception is the result of evolutionary adaptation and thus is present at birth or when the necessary neural maturation has occurred. By "past experience" it is meant that the perception in question is the end result of prior exposure to certain relevant patterns or conditions, a kind of learning process. Despite centuries of discussion of this problem, and considerable experimental work, there is still no final answer to the question. It now seems clear that certain kinds of perception are innate, but equally clear that past experience also is a determining factor. *See* EXTRASENSORY PERCEPTION (ESP); INTELLIGENCE. [I.R.]

Perching birds Members of the order Passeriformes, which includes about 5300 species, approximately half of the living species of birds. There are two major divisions of the order, Oscines and Suboscines.

The foot is not webbed and has four toes, with the first toe or hallux pointing backward, enabling the bird to firmly grasp a branch or twig and perch. The resonating chamber of the syrinx has a number of membranes whose vibrations are controlled by muscles. The form of the bill varies with the type of food (seed, insects, nectar, or prey) used by the bird.

Species that live on insects frequently breed in high latitudes and migrate longer distances than other members of the order. There are some species that do not migrate at all. Some species are found in regions which are practically treeless, such as desert or mountain regions, while other species have adapted to cultivated areas, such as orchards and farms.

The members of the order have a cosmopolitan distribution and, except in polar regions, make up the predominant avifauna of a region. They exhibit a correspondingly wide variety of breeding habits, nesting sites, number of eggs, and types of nests.

Suboscines. This primitive group is divided into the suborders Eurylaimi, Tyranni, and Menurae. The broadbills comprise the single family Eurylaimidae of the suborder Eurylaimi. Examples are the long-tailed broadbill (*Psarisomus dalhousiae*), indigenous in Asia; the green broadbill (*Calystomena viridis*) of Borneo, and the wattled broadbill (*Eurylaimus sleerii*) of the Philippines. The woodcreepers, wedgebills, and woodhewers make up the family Dendrocolaptidae of the suborder Tyranni. They range throughout Mexico into South America. In the same range are the ovenbirds of the family Furnariidae in the suborder Tyranni. The small antbirds of the family Formicariidae, suborder Tyranni, are confined to Central and South America.

The 400 species of flycatchers are placed in two families: Muscicapidae (members of the Oscines) which includes the Old World flycatchers or fantails; and Tyrannidae, the tyrant flycatchers. This latter group is confined to the Americas. The manakins, family Pipridae, of South and Central America are known for their elaborate courtship behavior.

Menurae is a small suborder that includes only four species of Australian birds, two species of lyrebirds (Menuridae), and two species of scrubbirds (Atrichornithidae).

Oscines. This large suborder is composed of about 40 families commonly called songbirds. Among these are the common species found in the United States: larks, swallows, wrens, martins, waxwings, crows, warblers, thrushes, finches, vireos, blackbirds, orioles, and sparrows. *See* PASSERIFORMES. [C.B.C.]

Perchlorate A compound which contains chlorine in the 7 + oxidation state and which is derived from perchloric acid, $HClO_4$. Perchlorates are more stable than chlorates, chlorites, or hypochlorites but are nevertheless excellent oxidizing agents. On heating, perchlorates decompose into potassium chloride, KCl, and oxygen gas. Because of their oxidizing properties, perchlorates find use in explosives and as oxidizing agents in the laboratory. *See* CHLORINE; HYPOCHLORITE; OXIDIZING AGENT; PERBROMATE. [E.E.W.]

Perciformes The typical spiny-rayed fishes, also known by the ordinal names Acanthopteri and Percomorphi. This is the largest order of vertebrates; the approximately 7500 species include 41% of all fishes. Perciformes include a diversity of structural types and sizes. The characters of the Perciformes include fin spines, usually present; a pelvic fin which, if present, is usually thoracic or jugular in position; the pelvic girdle usually attached to the cleithra, sometimes connected by ligaments; the pelvic fin usually with a spine and 5 soft rays, the latter occasionally reduced; the pectoral fin base more or less vertical, usually placed well up on the side; a swim bladder without a duct; scales usually ctenoid, sometimes secondarily cycloid, absent, or variously modified; and the caudal fin with 17 principal rays (15 branched) or fewer.

Perciform fishes dominate the modern vertebrate life of the oceans and have done so throughout the Cenozoic. The group first appeared in the Upper Cretaceous, after which it underwent a rapid adaptive radiation; many of the basic structural types, as well as most major perciform derivatives such as the Pleuronectiformes and Tetraodontiformes, were present in the Eocene, A few families of perciforms have been notably successful in fresh water. Other families have effectively adapted to life in the deep seas, and still others have become specialized for pelagic existence. It is in the shore areas, the offshore banks, the coral reefs, the coastal beaches and lagoons, and the intertidal zone, however, that the perciforms have attained their ultimate achievement. Here the enormous variety attests to the adaptive effectiveness of the group. *See* ACTINOPTERYGII; OSTEICHTHYES. [R.M.B.]

Percopsiformes A small order of actinopterygian fishes that is also known as the Salmopercae and Amblyopsiformes. The order is thought to be remotely related to the codfishes (Gadiformes) and toadfishes (Batrachoidiformes). The characters (see illustration) include single, ray-supported dorsal and anal fins, each usually with one to four anterior spines; pelvic fin, if present, subabdominal in position, with three to eight soft rays; pelvic girdle, if present, attached to the postcleithra; swim bladder without a duct; and body covered with cycloid or ctenoid scales. *See* BATRACHOIDIFORMES; GADIFORMES.

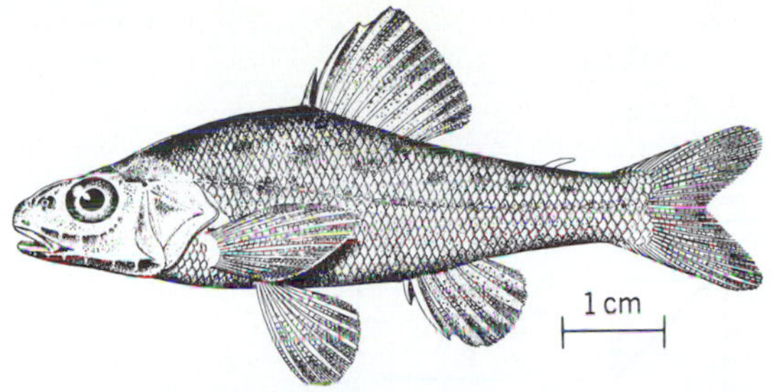

Sand roller (*Percopsis transmontana*). (*After D. S. Jordan and B. W. Evermann, The Fishes of North and Middle America, U.S. Nat. Mus. Bull. 47, 1900*)

The order comprises three families, five genera, and eight Recent species of North American fresh-water fishes. *See* ACTINOPTERYGII; TELEOSTEI. [R.M.B.]

Performance rating A procedure for determining the value for a factor which will adjust the measured time for an observed task performance to a task time that one would expect of a trained operator performing the task, utilizing the approved method and performing at normal pace under specified workplace conditions. Normal time (ultimately subjectively based) is the time that a trained worker requires to perform the specified task under defined workplace conditions, employing the assumed philosophy of "a fair day's work for a fair day's pay."

The performance rating process is concerned with determining normal pace during the work portion of an average day and must, therefore, consider the fatigue recovery aspects of allowance (nonwork) times occurring during the day. The following two equations relate factors in determining how much time a worker will be allowed per unit of output:

$$\text{Standard time} = \text{normal time} \times \text{allowances}$$
$$\text{Normal time} = \text{observed time} \times \text{rating factor}$$

If the observed time for a task is adjusted by the performance rating factor to determine normal time, and allowance time is added for nonwork time, the standard time will represent the allowed time per unit of production.

The most commonly employed rating technique throughout the history of stopwatch time study, including the present, is referred to as pace rating. A properly trained employee of average skill is time-studied while performing the approved task method under specified work conditions. Rating consists only of determining the relative pace (speed) of the operator in relation to the observer's concept of what normal pace should be for the observed task, including consideration of expected allowances to be applied to the standard. *See* HUMAN-FACTORS ENGINEERING; METHODS ENGINEERING; WORK MEASUREMENT. [P.E.H.]

Pericarditis An inflammation of the pericardium. This occurs secondary to many general clinical diseases, but is usually minor in degree and is not the more significant part of the disease process. Primary pericarditis occurs when the inflammation of the pericardium dominates the disease process. Most primary pericarditis is of unknown cause. Known causes of paricarditis include bacterial, tubercular, viral, and mycotic infections, as well as deranged metabolic processes such as uremia, hypercholesterolemia, or neoplastic diseases. Pericarditis can accompany acute myocardial infarction, or it may occur associated with hypersensitivity diseases, (rheumatic fever, rheumatoid arthritis, polyarteritis, and so on). *See* HEART (VERTEBRATE); HEART DISORDERS. [N.K.M.]

Pericycle As commonly defined, the outer boundary of the stele of plants. Originally it was interpreted as a band of cells between the phloem and the innermost layer (endodermis) of the cortex. Such pericycle is commonly found in roots and, in lower vascular plants, also in stems. In higher vascular plants, however, a distinct layer of cells may not be present between the phloem and the cortex. The pericycle, if present, may be composed of parenchyma or sclerenchyma cells with relatively thin or heavily thickened walls. It may be one to several layers in radial dimensions.

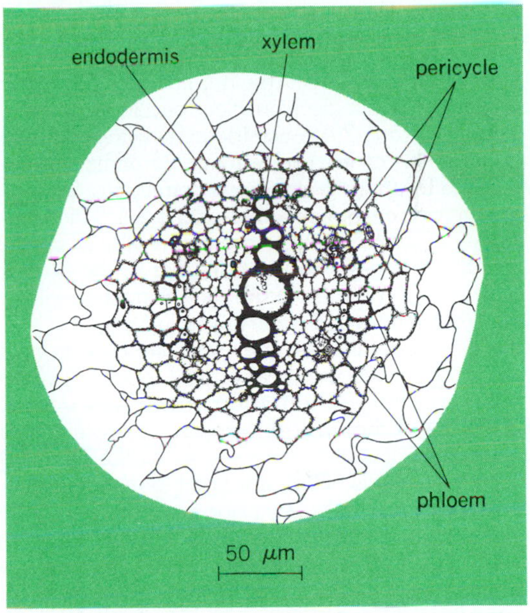

Transection of central part of sugarbeet. (*From K. Esau, Hilgardia, 9(8), 1935*)

Primordia of branch roots commonly arise in the pericycle in seed plants, most frequently outside the xylem ridges (see illustration). The first cork cambium may also arise in the pericycle of those roots that have secondary vascular tissues. In roots, a part of the vascular cambium itself (that outside the primary xylem ridges) originates from pericycle cells. *See* CORTEX (PLANT); ENDODERMIS; LATERAL MERISTEM; PARENCHYMA; PHLOEM; ROOT (BOTANY); SCLERENCHYMA; STEM; XYLEM. [V.I.C.]

Pericyclic reaction Concerted (single-step) processes in which bond making and bond breaking occur simultaneously (but not necessarily synchronously) via a cyclic (closed-curve) transition state. Although a given reaction may appear formally to be pericyclic, it cannot be assumed to be a concerted process. In each case, the detailed mechanism of the reaction must be established experimentally. Pericyclic reactions can be promoted either by heat or by light; the stereochemistry of the reaction is determined by the mode of activation employed and the number of electrons that are delocalized in the transition state. *See* CHEMICAL BONDING; PHYSICAL ORGANIC CHEMISTRY; STEREOCHEMISTRY.

Four types of pericyclic reactions that are frequently encountered in organic chemistry are electrocyclic processes, cycloadditions, sigmatropic shifts, and cheletropic reactions.

Electrocyclic processes are reactions that involve their cyclization across the termini of a conjugated π-system with concomitant formation of a new σ-bond or the microscopic reverse. The sequence of steps involved in the forward reaction must be the same, in the reverse order, as that in the reverse direction when the forward and reverse reactions are carried out under identical conditions. This statement is known as the principle of microscopic reversibility.

The effect of the mode of activation upon the stereochemistry of an electrocyclic process is shown in the reaction (1),

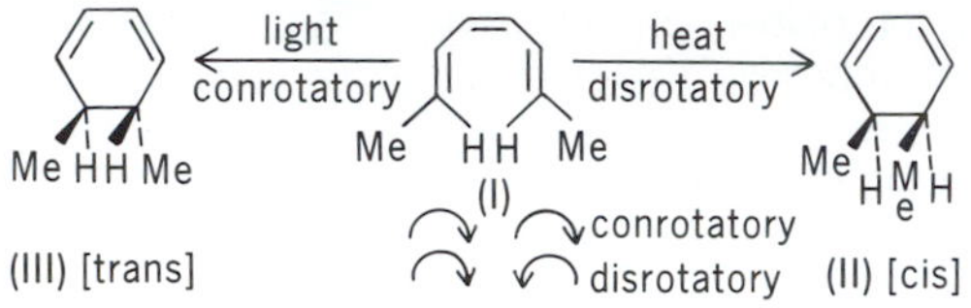

(1)

where Me = methyl, for the hexatrienecyclohexadiene interconversion (a six-electron electrocyclic process). Thus, when *trans,cis,trans*-2,4,6-octatriene [structure (I)] is heated, disrotatory motion of the two terminal 2*p* orbitals occurs; that is, they rotate in opposite directions thereby resulting in exclusive formation of *cis*-5,6-dimethylcyclohexa-1,3-diene (II). The corresponding photochemical process results in conrotatory motion of the termini in structure (I); that is, the two terminal 2*p* orbitals rotate in the same direction thereby yielding *trans*-5,6-dimethylcyclohexa-1,3-diene (III) exclusively.

Cycloadditions occur when two (or more) π-electron systems react under the influence of heat or light to form a cyclic compound with concomitant formation of two new σ-bonds that join the termini of the original π-systems. The stereochemistry of this reaction is classified with respect to the two molecular planes of the reactants. Thus, if σ-bond formation occurs from the same face of the molecular plane across the termini of one of the component π-systems, the reaction is said to be suprafacial on that component. If instead σ-bond formation occurs from opposite faces of the molecular plane, the reaction is said to be antarafacial on that component. This distinction is illustrated in reaction (2) for two thermal processes, where the symbol ≠ indicates the structure of the transition state. Reaction (2*a*) shows the Diels-Alder [4 + 2] cycloaddition of butadiene (IV) to ethylene (V), a six-electron pericyclic reaction in which additions across the termini of the diene (four-electron component) and dienophile (two-electron component) both occur suprafacially. Reaction (2*b*) shows a [14 + 2] cycloaddition in which σ-bond formation occurs suprafacially on the two-electron component [tetracyanoethylene (VI)] and antarafacially on the fourteen-electron component [heptafulvalene (VII)].

Sigmatropic shifts involve migration of a σ-bond that is flanked at either (or both) ends by conjugated π-systems. Either one or both ends of the σ-bond may migrate to a new location within the one or more flanking π-systems.

Cheletropic reactions involve extrusion of a fragment via concerted cleavage of two σ-bonds that terminate at a single atom or the reverse process. Cheletropic fragmentations may be either linear or nonlinear [reaction 3].

R. B. Woodward and R. Hoffmann introduced an application of molecular orbital theory that permits prediction of rates and products of pericyclic reactions. They utilized symmetry properties of molecular orbitals to estimate relative energies of diastereoisomeric transition states for structurally similar pericyclic reactions.

In an alternative theoretical approach to understanding pericyclic reactions, the transition state is examined directly, and attempts to estimate the degree of electronic stabilization

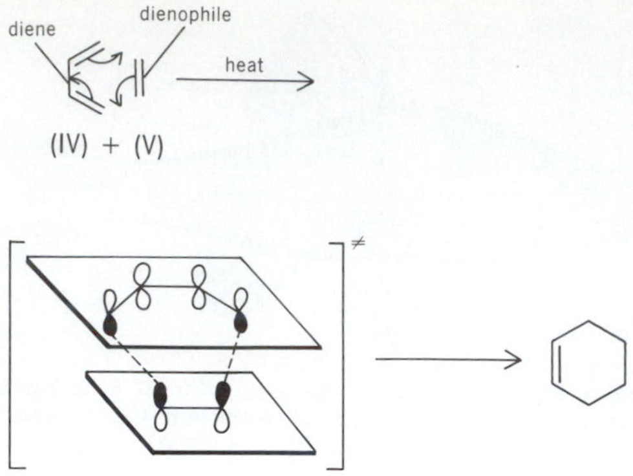

(2*a*)

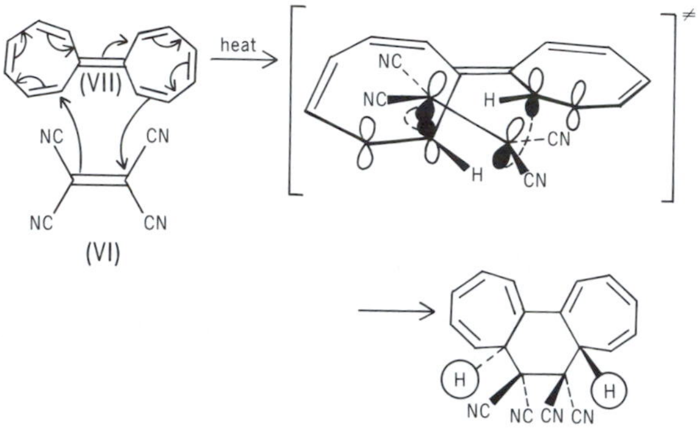

(2*b*)

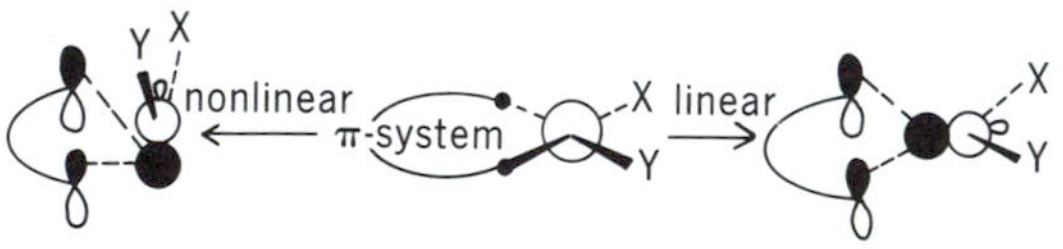

(3)

(allowedness) or destabilization (forbiddenness) inherent in that transition state are made. One such approach emphasizes the importance of frontier orbitals (highest-occupied–lowest-unoccupied molecular orbitals) in determining the course of a pericyclic reaction. *See* Diastereoisomer; Delocalization; Diels-Alder reaction; Electron configuration; Molecular orbital theory; Organic reaction mechanism; Woodward-Hoffmann rule.

[A.P.M.]

Periderm A group of tissues which replaces the epidermis in the plant body. Its main function is to protect the underlying tissues from desiccation, freezing, heat injury, mechanical destruction, and disease. Although periderm may develop in leaves and fruits, its main function is to protect stems and roots. The fundamental tissues which compose the periderm are the phellogen, phelloderm, and phellem.

The phellogen is the meristematic portion of the periderm and consists of one layer of initials. These exhibit little variation in form, appearing rectangular and somewhat flat in cross and radial sections, and polygonal in tangential sections.

The phelloderm cells are phellogen derivatives formed inward. The number of phelloderm layers varies with species, season, and age of the periderm. In some species, the periderm lacks the phelloderm altogether. The phelloderm consists of living cells with photosynthesizing chloroplasts and cellulosic walls.

The phellem, or cork, cells are phellogen derivatives formed outward. These cells are arranged in tiers with almost no intercellular spaces except in the lenticel regions. After completion of their differentiation, the phellem cells die and their protoplasts disintegrate. The cell lumens remain empty, excluding a few species in which various crystals can be found. The remarkable impermeability of the suberized cell walls is largely due to their impregnation with waxes, tannins, cerin, friedelin, and phellonic and phellogenic acids.

Lenticels are loose-structured openings that develop usually beneath the stomata and that facilitate gas transport through the otherwise impermeable layers of phellem. See BARK; SCLERENCHYMA. [Y.W.; H.Wi.]

Peridotite An igneous rock consisting of more than 90% of millimeter- to centimeter-sized crystals of olivine, pyroxene, and hornblende, with olivine predominant. Other minerals are mainly plagioclase, chromite, and garnet. Most of the volume of the Earth's mantle probably is peridotite. See OLIVINE; PYROXENE.

Peridotites have three principal modes of occurrence corresponding approximately to their textures: (1) Peridotites with well-formed olivine crystals occur mainly as layers in gabbroic complexes. (2) Peridotite nodules in alkaline basalts and diamond pipes generally have equigranular textures, but some have irregular grains. (3) Peridotite also occurs on the walls of rifts in the deep sea floor and as hills on the sea floor, some of which reach the surface.

Peridotite is an important rock economically. Pure olivine rock (dunite) is quarried for use as refractory foundry sand and refractory bricks used in steelmaking. The sulfides associated with peridotites are common ores of nickel and platinoid metals. The chromite bands commonly associated with peridotites are the world's major ores of chromium. See IGNEOUS ROCKS. [A.T.A.]

Perigee The point nearest the Earth in the orbit of the Moon or of an artificial satellite. At perigee the Moon is 5% closer to Earth than at its mean distance. Because, on the average, the Moon and Sun subtend nearly equal angles, a solar eclipse near perigee lasts about 5 min; an eclipse near apogee is annular. See ECLIPSE; MOON; PERIHELION. [G.P.K.]

Perihelion In astronomy, that point at one extremity of the major axis of the elliptical, parabolic, or hyperbolic orbit of a planet or comet about the Sun where the planet or comet is closest to the Sun. The instant when a planet or comet is at perihelion is referred to as the time of perihelion passage. For Earth this occurs about January 3. See ORBITAL MOTION. [R.L.Du.]

Period (periodic phenomena) The time interval of a single repetition of a varying quantity of a motion or phenomenon which repeats itself regularly. The period is the reciprocal of the frequency. Waves which have regularly repeated time-varying quantities are termed periodic. See FREQUENCY (WAVE MOTION); PERIODIC MOTION; WAVE MOTION. [W.J.G.]

Period doubling A scenario for the transition of a natural process from regular motion to chaos. Various natural processes develop in time in a way that depends upon prevailing environmental details. A quantity that specifies the particular state of the environment of a process is called a parameter, and is taken as a fixed value over the course of development of the process.

It is a frequent natural occurrence for a process to have a regular and easily describable motion for some range of parameters, but to have complex, irregular, and difficult-to-describe motions for other ranges of parameters. In the context of fluid flow, the latter circumstance is termed turbulence. In a more general context it is called chaos (which includes fluid turbulence but presages an underlying generality). See FLUID FLOW; TURBULENT FLOW.

Sometimes, as the environmental parameters are varied, a process may systematically exhibit more irregular motions, turning over into chaotic motion beyond some parameter value. In analogy to the phenomenology of phase transitions, this circumstance is termed a transition to chaos. There are a variety of qualitatively different transitions to chaos, each termed a scenario. Period doubling is one frequently encountered scenario leading to chaos for which a full theoretical account exists. Since it occurs in a wide variety of processes of significantly divergent physical characters (for example, fluid-flow, chemical reactions, and electronic devices), it is sensible to consider it as a phenomenon in its own right. See PHASE TRANSITIONS.

In order to observe this scenario, it is sufficient that all but one parameter is held fixed. Over some range of this varied parameter (it shall be defined to increase over the range of investigation) the motion is observed to be periodic. Above a certain value of the parameter the motion grows more complicated (a bifurcation has occurred): after the amount of time T for which the motion exactly repeated itself just prior to the bifurcation, the motion now slightly fails to do so, exactly repeating, however, after another T seconds. That is, the period has doubled from T to $2T$. As the parameter is further increased, the error to repeat after the first half of the new period systematically increases. A still further increase of parameter produces another bifurcation resulting in a new doubling of the period: the motion slightly fails to repeat after two roughly periodic cycles, exactly doing so after four. As the parameter is further increased, there are successive period-doubling bifurcations, more and more closely spaced in parameter value until at a critical value the doubling has occurred an infinite number of times, so that the motion is now no longer periodic and hence of a more complex character than had yet been encountered. Unpredictably complex motions occur for values of the parameter above its critical value, although ranges of parameter still exist for which the system exhibits new periodic motions. Indeed any period-doubling system exhibits the same sequence of truly chaotic motion and interspersed periodicities as its parameter increases. Thus there is a strong degree of qualitatively universal behavior for all systems experiencing this scenario.

However, there is also a precise quantitative universality. That is, without knowing the system (or its equations) essentially all measurable quantities can be predicted: By looking at the data alone, it would not be possible to guess the physical system responsible for that data. Thus, reminiscent of thermodynamics, questions can be posed and answered in a general manner that bypasses the specific mechanisms governing any particular system. [M.J.F.]

Periodate A salt which contains iodine in the 7 + oxidation state and which is derived from periodic acid. Periodic acid is known in three forms: metaperiodic acid, HIO_4; dimesoperiodic acid, $H_4I_2O_9$; and paraperiodic acid, H_5IO_6. The corresponding salts are also known. Only the periodates of sodium and potassium are important. See IODINE. [E.E.W.]

Periodic motion Any motion that repeats itself identically at regular intervals. If $x(t)$ represents the displacement of any coordinate of the system at time t, a periodic motion has the property defined by the following equation for every value of

$$x(t + T) = x(t)$$

the variable time t. The fixed time interval T between repetitions, or the duration of a cycle, is known as the period of the motion. *See* PERIOD (PERIODIC PHENOMENA).

The motion of the escapement mechanism of a watch, the motion of the Earth about the Sun, and the more complicated motion of the crankshaft, piston rods, and pistons in an engine running at uniform speed are all examples of periodic motion.

The vibration of a piano string after it is struck is a damped periodic motion, not strictly periodic according to the definition. Although the motion very nearly repeats itself, and with a fixed repetition time, each successive cycle has a slightly smaller amplitude. *See* DAMPING; HARMONIC MOTION; VIBRATION; WAVE MOTION. [J.M.Ke.]

Periodic table A list of elements (atoms) ordered along horizontal rows according to atomic number (the number of electrons in an atom and also the charge on its nucleus). In the periodic table (see illus.), the rows are arranged so that elements with nearly the same chemical properties occur in the same column (group), and each row ends with a noble gas (closed-shell element that is generally inert). For chemists, the position of atoms in the periodic table provides the most powerful guide for classifying the expected properties of molecules and solids made from these particular atoms. *See* INERT GASES.

The origin of the periodic table was explained in the 1920s in terms of the basic physical laws (quantum mechanics) obeyed by the electrons of an atom. Thus, the rows in the periodic table correspond to the shell number, n, and groups correspond to a particular electronic configuration designated by the number and type of electrons in the outermost shell. These electrons govern chemical properties and are known as valence electrons. *See* ELECTRON CONFIGURATION; QUANTUM MECHANICS; VALENCE.

Additional information from the physical laws of atoms can be incorporated into the periodic table and can greatly enhance its organizing capability. For example, configuration energy adds a third dimension to the periodic table. The configuration energy is defined in terms of the ionization energy (I), the energy required to remove an electron from an atom.

Besides enhancing the organizing capability of the periodic table, the concept of configuration energy explains many long-standing puzzles about the table itself. It explains the existence of the metalloid band of elements (configuration energy is nearly constant in this band) and why these elements divide the metals from the nonmetals. Elements possessing configuration energies with magnitudes greater than those of the metalloids are nonmetals; those with lower configuration energies are metals. *See* METAL; NONMETAL.

The lack of numerical or analytic connection between the traditional two-dimensional periodic table and methods used to predict the structure and reactivity of molecules and solids has long reduced the table's usefulness. However, configuration energy, introduced as a new dimension of the periodic table, is just the average atomic energy level, and simultaneously the average density of states, for the atoms out of which the mole-

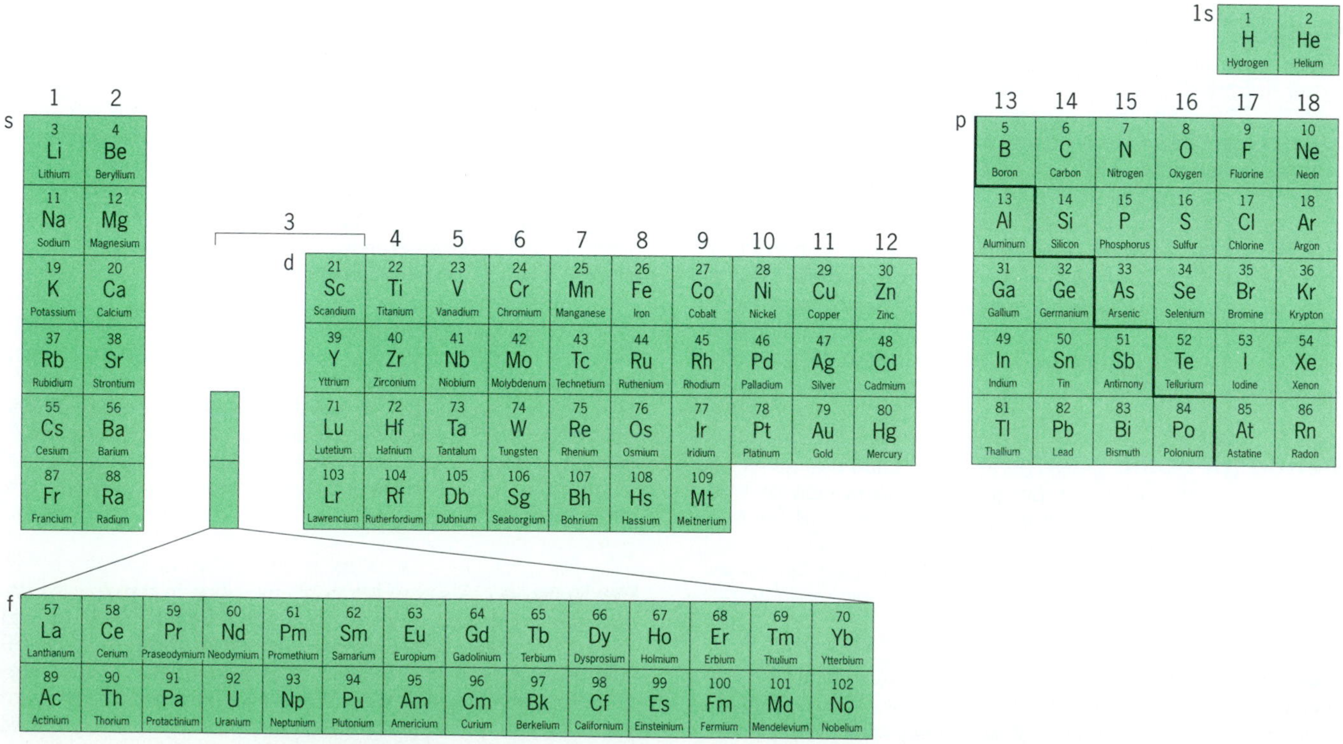

Two-dimensional periodic table. The atomic numbers are listed above the symbols identifying the elements. The heavy line separates metals from nonmetals. The letters s, p, d, and f indicate types of electrons in motion around the nucleus. (*After J. Emsley, The Elements, Oxford University Press, 1989*)

cular-orbit–energy-level diagrams and energy bands in solids are constructed, thereby tying the periodic table directly to present-day research techniques. *See* Band theory of solids; Energy level (quantum mechanics); Molecular orbital theory; Molecular structure and spectra. [L.C.A.]

Periodontal disease An inflammatory lesion caused by bacteria affecting the tissues housing the roots of the teeth. The disease, sometimes called pyorrhea, increases in prevalence and severity with increasing age, and it is the principal cause of tooth loss in adult humans throughout the world. When only the gum tissue or gingiva is affected, the disease is called gingivitis, but when the process extends into the deeper structures it is known as periodontitis. The diseased tissues appear abnormally red and slightly swollen, and they tend to bleed, sometimes profusely, when the teeth are brushed. In some cases the gums may become thickened and scarred, and they may recede, exposing the root surface. As the disease advances, the attachment of the gum to the tooth is lost, creating a periodontal pocket, a large portion of the gum is destroyed, and the bone surrounding the roots is resorbed. The teeth become loose, abscesses form, and extraction is required.

Both gingivitis and periodontitis are caused by bacteria that form plaques on the surfaces of the teeth at the gingival sulcus or pocket. These plaques may contain 250 or more separate microbial species. Plaques of any microbial composition can cause gingivitis, but specific bacteria appear to be necessary for induction of periodontitis. Among the bacteria involved in periodontitis are various species of *Porphyromonas*, *Bacteroides*, *Actinobacillus*, *Eikenella*, *Fusobacterium*, *Wolinella*, and other less well-characterized species. Spirochetes are present in active lesions, but their role remains unclear. The bacteria extend apically along the interface between the tooth root and the gingival tissue and causes periodontal pockets to form.

The principal features of the pathogenesis of periodontitis have been described. The lesions begin as an acute inflammatory response followed by a dense accumulation of lymphoid cells. There is a net loss of collagen in the area nearest the junctional epithelium and periodontal pocket, with scarring and fibrosis of the connective tissues at more distant sites. The junctional epithelium is converted into an ulcerated pocket epithelium, the alveolar bone housing the tooth roots is resorbed, and the periodontal ligament is destroyed. Products released by infiltrating leukocytes, including prostaglandins, interleukins and collagenase, and other hydrolytic enzymes, are involved in tissue destruction.

Bacterial colonization and extension activate several host defense mechanisms. The most effective of these is the accumulation of functional neutrophilic granulocytes between the surface of the plaque and the gingival tissue. These cells tend to counter and limit microbial extension. The bacteria appear to invade the periodontal connective tissues, where they induce immunopathologic and other destructive inflammatory reactions in the host, and these lead, in major part, to the observed tissue destruction. Periodontal destruction is episodic, with periods of exacerbation characterized by highly acute inflammation, followed by periods of quiescence.

Although bacteria are essential for induction of the disease, predisposing factors are also important, though their elucidation is not complete. Individuals who manifest functionally abnormal neutrophilic granulocytes or monocytes are unusually susceptible to the severe early onset forms of periodontitis. The leukocyte abnormality appears to be genetically transmitted. The early-onset forms have been designated as prepubertal, juvenile, and rapidly progressive periodontitis; adult periodontitis has a later onset and does not seem to be related to leukocyte abnormalities. Some persons with acquired immune deficiency syndrome (AIDS) manifest a highly destructive, unique form of periodontitis. Other predisposing conditions include unusually stressful situations and periods of hormone imbalance occurring at puberty, during pregnancy, and in some women taking birth control drugs.

Good daily oral hygiene practices, including vigorous brushing of the exposed surfaces of the teeth and use of dental floss, interproximal brushes, and other devices to clean between the teeth, constitute the most effective measures to prevent periodontal disease. Basic ingredients of treatment of existing disease include bringing the infection under control and establishing conditions which preclude reinfection. All of the microbial deposits must be removed from the crown and root surfaces. In individuals with severe forms of periodontitis, these procedures may be supplemented by use of antibiotics either systematically or directly into the pocket. These procedures usually lead to reduction of the inflammation and to some shrinkage in the gums, but the periodontal pockets remain. Based on the traditional view that treated pockets may become reinfected and the disease may continue to spread, surgical treatment may be performed with the aim of reducing pocket depth and restoring normal tissue contours. Alternatively, regenerative procedures use various grafting materials, including freeze-dried decalcified bone or bone substitutes and guided tissue regeneration. To perform guided tissue regeneration, flaps are opened in the gingival tissue and the root surfaces are thoroughly cleaned; a porous membrane is placed around the tooth, covering the bone defect with or without placing grafts, and flaps covering the membranes are sutured into place. The membrane permits the wound site to become populated with cells having the capacity to generate new bone, cementum, and periodontal attachment. *See* Tooth disorders. [R.C.P.]

Perischoechinoidea A subclass of Echinoidea lacking stability in the number of columns of plates that make up the ambulacra and interambulacra. The ambulacral columns vary from 2 to 20, the interambulacral from 1 to 14. Of the four included orders, three are exclusively Paleozoic; the other, Cidaroida, includes both Paleozoic and extant members and is probably ancestral to all other surviving echinoids. *See* Bothriocidaroida; Cidaroida; Echinocystitoida; Echinoidea. [H.B.F.]

Periscope An optical instrument that permits viewing along a displaced or deflected axis, providing an observer with the view from a position which may be inaccessible or dangerous. Periscopes range in complexity from the simple unit-power tank periscope to the complex multielement submarine periscope.

The tank periscope, intended to protect the user from bullets, employs a pair of plane, parallel, reflecting surfaces (either mirrors or prisms), so arranged in a mount that the path of light through the instrument forms a crude letter *Z*. If powers greater than unity are desired or if the periscope is to be used for sighting, a terrestrial telescope can be added to the periscope. *See* Telescope.

In the submarine periscope, it is necessary to employ a telescope system having a wide field of view and uniform illumination across a field which can be fitted into a long, narrow tube whose length-to-diameter ratio may be 50 or greater. This is achieved by utilizing a plurality of lenses so spaced along the length of the tube as to cause the incoming principal rays from the edge of the field to be deviated from side to side within the tube (see illustration). In general, the greater the number of lenses, the wider the field of view.

Various modifications of the basic optical systems described here are employed as viewing periscopes in military aircraft

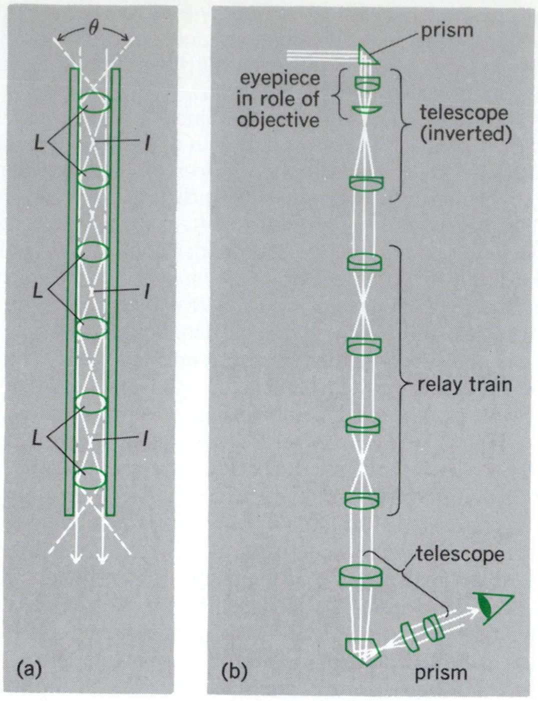

Periscopic relay train. (*a*) Showing lenses *L*, inversions *i*, and angle of view θ. (*b*) Between a pair of facing telescopes in a submarine periscope.

and as viewing devices in particle accelerators and nuclear reactors. The cystoscope and endoscope are slender, sometimes mechanically flexible periscopes used for visual examination and photography of body cavities inaccessible to direct observation; an entirely different basis for the design of such instruments is in the use of bundles of optical fibers. *See* OPTICAL FIBERS. [E.K.K.]

Perissodactyla An order of exclusively herbivorous mammals, including the living horses, zebras, asses, tapirs, and rhinoceroses, and their diverse and abundant relatives. The feet are mesaxonic, that is, with the axis going through the third toe; the femur has a third trochanter for muscle attachment; and the astragalus has a characteristic flat head and short neck. Perissodactyls probably derive from a phenacodontid condylarth. The earliest known forms, the primitive horse *Hyracotherium* (formerly called Eohippus, the "dawn horse") and the tapir *Homogalax*, occur in early Eocene sediments of North America and Europe.

The explanation for the very successful radiation of early perissodactyls probably lies in their adaptations for fast running. Many later perissodactyls, however, particularly some of the rhinoceroses and brontotheres, developed elephantine, columnar legs, in accord with their large to gigantic size.

The order is divided into 3 suborders with 13 families. A classification scheme for the order follows:

Order Perissodactyla
  Suborder Hippomorpha
    Superfamily Equoidea
      Family Equidae
      Family Palaeotheriidae
    Superfamily Brontotherioidea
      Family Brontotheriidae
  Suborder Ancylopoda
    Superfamily Chalicotherioidea
      Family Eomoropidae
      Family Chalicotheriidae
  Suborder Ceratomorpha
    Superfamily Tapiroidea
      Family Isectolophidae
      Family Lophialetidae
      Family Depertellidae
      Family Lophiodontidae
      Family Helaletidae
      Family Tapiridae
    Superfamily Rhinoceratoidea
      Family Hyracodontidae
      Family Amynodontidae
      Family Rhinoceratidae

See MAMMALIA. [F.S.S.]

Peritoneum The membranous lining of the coelomic, especially the abdominal, cavity, which surrounds most of the organs. It is composed mainly of flattened epithelial cells that produce a small amount of watery, or serous, fluid. In the embryo the coelomic wall is lined by this membrane, which continues over the developing viscera so that they are suspended and supported by the reflected peritoneum, principally from the dorsal body wall, but also from the ventral body wall in the region of the liver. *See* EPITHELIUM; FETAL MEMBRANE. [T.S.P.]

Peritonitis Inflammation of the peritoneum. The condition may be caused by infectious organisms or foreign substances introduced into the abdominal cavity. The small amount of serous fluid normally present as a lubricant acts as an excellent culture medium for bacterial growth and also as a means of spreading invading materials. The source of such substances or organisms is commonly a gastrointestinal inflammation, especially if perforation has occurred. Appendicitis, peptic ulcer, cancer of the bowel, gallbladder disease, and dysentery are common sources of infection that may produce peritonitis, as well as blood-borne forms of tuberculosis and pneumonia. *See* PERITONEUM. [E.G.St./N.K.M.]

Peritrichia A specialized subclass of the class Ciliatea composed of a large group of unusual-looking ciliate protozoans. Many are sessile and stalked, while some form colonies which may reach a large size. A number are attached as ectocommensals to a variety of animals and plants. A free-swimming stage in the life cycle, indispensable for distribution, is known as the telotroch. It is a small, mouthless form equipped with a single girdle of posteriorly located locomotor cilia. This is quite unlike the morphology of the mature, sedentary form, which is an inverted bell form atop a long stalk.

Vorticella (illustration *a*) and *Epistylis* are probably the best-

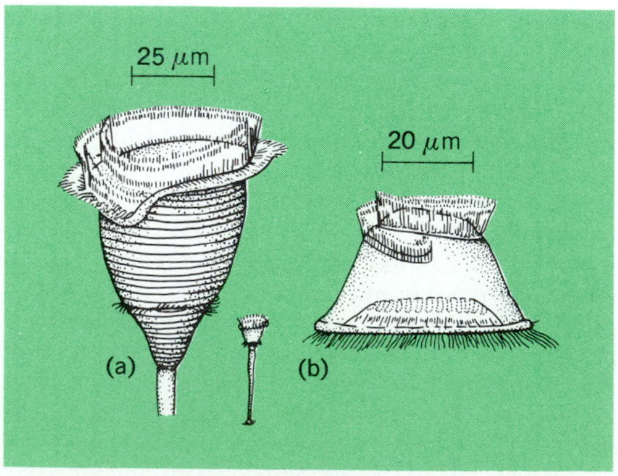

Peritrichida. (*a*) *Vorticella*, a stalked peritrich, (*b*) a mobile peritrich.

known stalked forms. The former is a solitary ciliate, the latter a colony builder. *Trichodina* (illustration *b*) belongs to the group of mobile peritrichs. *See* CILIATEA; HYMENOSTOMATIDA; THIGMOTRICHIDA. [J.O.C.]

Perlite A natural glass with abundant spherical or convolute cracks that cause it to break into small pearllike masses or "pebbles," usually less than a centimeter across. It is commonly gray or green with a pearly luster due to reflections from the thin air films formed along the perlitic fractures.

The water content of perlite is commonly 3–4% by weight. Under heat treatment (about 1500–2000°F or 820–1100°C) the contained moisture forms tiny steam bubbles in the softened glass, and the perlite is "popped" or exploded to roughly 15 or 20 times its original volume. Thus, the material is excellent for lightweight aggregate, insulation, fillers, and filters. Notable deposits are worked in California and New Mexico. *See* IGNEOUS ROCKS; VOLCANIC GLASS. [C.A.C.]

Permafrost Perennially frozen ground, occurring wherever the temperature remains below 32°F (0°C) for several years, whether the ground is actually consolidated by ice or not and regardless of the nature of the rock and soil particles of which the earth is composed. Perhaps 25% of the total land area of the Earth contains permafrost; it is continuous in the polar regions and becomes discontinuous and sporadic toward the Equator. During glacial times permafrost extended hundreds of miles south of its present limits in the Northern Hemisphere.

Temperature of permafrost at the depth of no annual change, about 30–100 ft (10–30 m), crudely approximates mean annual air temperature. It is below 23°F (−5°C) in the continuous zone, between 23–30°F (−5 and −1°C) in the discontinuous zone, and above 30°F (−1°C) in the sporadic zone. Temperature gradients vary horizontally and vertically from place to place and from time to time.

Ice is one of the most important components of permafrost, being especially important where it exceeds pore space. Physical properties of permafrost vary widely from those of ice to those of normal rock types and soil. The cold reserve, that is, the number of calories required to bring the material to the melting point and melt the contained ice, is determined largely by moisture content.

Permafrost develops today where the net heat balance of the surface of the Earth is negative for several years. Much permafrost was formed thousands of years ago but remains in equilibrium with present climates. Permafrost eliminates most groundwater movement, preserves organic remains, restricts or inhibits plant growth, and aids frost action. It is one of the primary factors in engineering and transportation in the polar regions. [R.F.B.]

Permanent-magnet movable-coil ammeter An ammeter in which a small coil rotates in the air gap of a permanent magnet. The magnet produces a uniform radial magnetic field across the air gap. This type of ammeter is used universally for the measurement of direct currents. The illustration shows the general arrangement of the mechanism.

It is important that the magnetic system be of such materials and proportions that a constant flux density is maintained in the air gap. The coil is usually wound on a metal form and swung on highly polished steel or hard-alloy pivots between V-cup jewels or bearings of sapphire or hard glass. Bronze springs serve to carry the current in and out of the coil, as well as to oppose the electrical torque, resulting in a deflection proportional to the coil current.

The lower limit of useful torque is that which will overcome the residual bearing friction by a factor of several times the expected accuracy in terms of deflection. This sets a definite

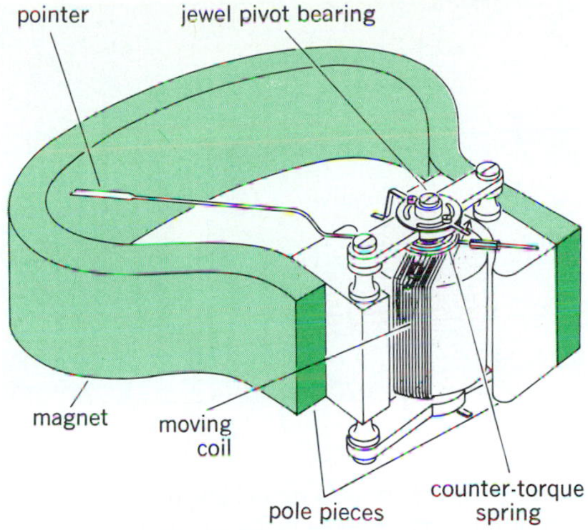

Permanent-magnet movable-coil mechanism. (*Weston Instruments, Inc.*)

lower limit to a pivoted instrument, which is usually in the order of 10–20 microamperes full scale in a coil of several thousand ohms' resistance.

The high limit of current through the coil is set by thermal limits; about 0.03 ampere is the maximum current in a moving coil. For higher values, the instrument is shunted; that is, the bulk of the current is bypassed around the moving coil through a shunt, with only a definite fraction of the current passing through the moving coil itself. *See* AMMETER. [J.H.M./J.Be.]

Permanganate The deep purple anion, MnO_4^-, which is derived from permanganic acid, $HMnO_4$. Although the parent acid is stable only in dilute aqueous solution, the salts are well characterized. The permanganates resemble the perchlorates in their oxidizing properties and solubility of both the heavy metal and alkaline-earth metal salts. Permanganates are used as disinfectants, oxidizing agents, wood preservatives, and bleaching agents. *See* OXIDIZING AGENT; PERCHLORATE. [E.E.W.]

Permeance The reciprocal of reluctance in a magnetic circuit. It is the analog of conductance (the reciprocal of resistance) in an electric circuit, and is given by Eq. (1), where **B** is

$$P_m = \frac{\text{magnetic flux}}{\text{magnetomotive force}} = \frac{\iint \mathbf{B} \cdot d\mathbf{S}}{\oint \mathbf{H} \cdot d\mathbf{l}} \qquad (1)$$

the magnetic flux density, **H** is the magnetic field strength, and the integrals are respectively over a cross section of the circuit and around a path within it. *See* CONDUCTANCE.

From Eq. (1), it can be shown that Eq. (2) is valid, where A

$$P_m = \mu A/l \qquad (2)$$

is the cross-sectional area of the magnetic circuit, l its length, and μ the permeability. If the material is ferromagnetic, as is often the case, then μ is not constant but varies with the flux density and the complete magnetization curve of B against H may have to be used to determine the permeance. *See* MAGNETIC CIRCUITS; MAGNETIC FIELD; MAGNETIC MATERIALS; MAGNETIC PERMEABILITY; MAGNETOMOTIVE FORCE; RELUCTANCE. [A.E.Ba.]

Permian The name applied to the last period of geologic time in the Paleozoic Era and to the corresponding system of rock formations that originated during that period. The Permian Period commenced approximately 280,000,000 years ago and ceased about 220,000,000 years ago. The system of rocks that originated during this interval of time is widely distributed on all the continents of the world. The Permian

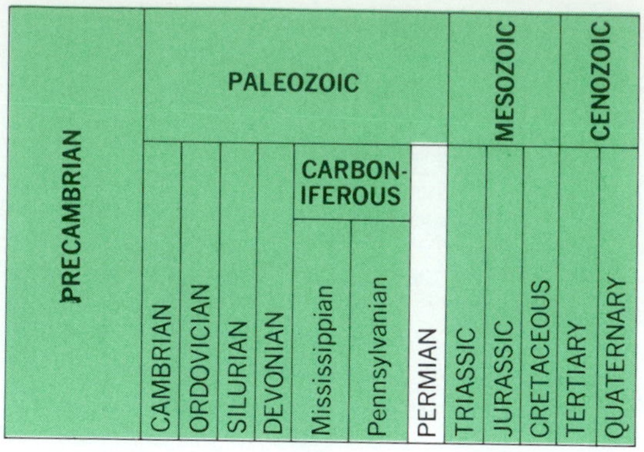

Period was a time of variable and changing climates, and during much of this time latitudinal climatic belts were well developed. During the latter half of Permian time, many long-established lineages of marine invertebrates became extinct and were not immediately replaced by new fossil-forming lineages. Rocks of Permian age contain many resources, including petroleum, coal, salts, and metallic ores.

During the Permian Period, several important changes took place in the paleogeography of the world. The joining of Gondwana to western Laurasia, which had started during the Carboniferous, was completed during Wolfcampian time (earliest Permian). The process of adding eastern Laurasia (Angara) to the eastern edge of western Laurasia was completed during Artinskian time (middle to latest Early Permian), and the supercontinent of Pangaea was formed. One very large world ocean, Panthalassa, occupied the remaining 75% of the Earth's surface, with a few small cratonic blocks, island arcs, and atolls. *See* CONTINENTAL DRIFT.

Most marine invertebrates of the Early Permian were continuations of well-established phylogenetic lines of Middle and Late Carboniferous ancestry. During Early Permian time, these faunas, dominated by brachiopods, bryozoans, corals, fusulinaceans, and ammonoids, gradually evolved into a number of specialized lineages. The tropical shallow-water faunas of southwestern North America evolved almost in isolation from those of the Tethyan region, because faunal exchanges had to cross either the cooler waters of a temperate shelf or deep waters of Panthalassa. Fluctuating climates permitted rare dispersals of some faunas between these two tropical faunal realms during the early part of Early Permian time. The closure of the Uralian seaway during Artinskian time, however, extended that dispersal path around Angara and into cold boreal waters through which the tropical species could not disperse. With the extension of this dispersal path, the faunas of the two tropical realms evolved independently, with only extremely rare dispersals between them.

Terrestrial faunas included insects which showed great advances over those of the Coal Measures. Several modern orders emerged, among them the Mecoptera, Odonata, Hemiptera, Trichoptera, Hymenoptera, and Coleoptera.

Land plants changed during the Permian from those adapted to the moist conditions of low-lying coal swamps, such as lepidodendrons and cordaites, to those adapted to the drier, well-drained conditions of mountains and alluvial plains, particularly conifers.

Of the vertebrates, the labyrinthodont amphibians were common and varied; however, the reptiles showed the most significant advances. Of several orders, the most significant were the Theriodonta, or mammallike reptiles, that evolved in the Triassic into mammals. Most of the known theriodonts are from South Africa and the Soviet Union. *See* PALEOZOIC.

[C.A.R.; J.R.P.R.]

Permittivity The permittivity ϵ of a material medium is related to the permittivity ϵ_0 of empty space by the equation $\epsilon = k\epsilon_0$, where k is the relative dielectric constant of the medium. The permittivity of empty space ϵ_0 has the value 8.85×10^{-12} coulomb^2/newton-meter^2. *See* CAPACITOR; COULOMB'S LAW; DIELECTRIC CONSTANT; ELECTRICAL UNITS AND STANDARDS.

[R.P.Wi.]

Perovskite A natural mineral and a structure type which includes no less than 150 synthetic compounds. The mineral perovskite is ideally $Ca[TiO_3]$ but extensive substitutions occur. The perovskite structure type is of great technical interest, since slight distortions away from cubic symmetry result in noncentrosymmetric (polar) arrangements which may have ferroelectric and antiferroelectric properties.

The mineral occurs as rounded cubes modified by the octahedral and dodecahedral forms. Hardness is $5\frac{1}{2}$ (Mohs scale), specific gravity 4.0, luster subadamantine to submetallic. The color ranges from yellow, brownish yellow, reddish, and dark brown to black. *See* BARIUM TITANATE.

[P.B.M.]

Peroxide A chemical compound which contains the peroxy (—O—O—) group, which may be considered to be a derivative of hydrogen peroxide (HOOH). An organic (or inorganic) peroxide is one in which some organic (or inorganic) substituent has replaced one or both hydrogens. Peroxides are used in such diverse reactions as oxidation, synthesis, polymerization, and oxygen generation. Inorganic peroxides include persulfates, hydrogen peroxide (H_2O_2), sodium peroxide, bivalent metal peroxides, and H_2O_2 addition compounds. Organic peroxides include peroxyacetic acid, dibenzoyl peroxide, and cumene peroxide. *See* HYDROGEN PEROXIDE; OXIDIZING AGENT; OXYGEN.

[S.S.N.]

Perpetual motion The expression perpetual motion, or perpetuum mobile, arose historically in connection with the quest for a mechanism which, once set in motion, would continue to do useful work without an external source of energy or which would produce more energy than it absorbed in a cycle of operation. This type of motion, now called perpetual motion of the first kind, involves only one of the three distinct concepts presently associated with the idea of perpetual motion.

Perpetual motion of the first kind refers to a mechanism whose efficiency exceeds 100%. Clearly such a mechanism violates the now firmly established principle of conservation of energy, in particular that statement of the principle of conservation of energy embodied in the first law of thermodynamics. (Indeed, the first law of thermodynamics is sometimes stated as "A perpetuum mobile of the first kind cannot exist."). *See* CONSERVATION OF ENERGY.

Perpetual motion of the second kind refers to a device that extracts heat from a source and then converts this heat completely into other forms of energy, a process which satisfies the principle of conservation of energy. A dramatic scheme of this type would be an ocean liner, which extracts heat from the nearly limitless oceanic source and then uses this heat for propulsion. This type of perpetual motion is, however, precluded by the second law of thermodynamics which is sometimes stated as "A perpetuum mobile of the second kind cannot exist."

The third type of perpetual motion is, in contrast to the two types described above wherein useful output was the goal, merely a device which can continue moving forever. It could result in actual systems if all mechanisms by which energy is dissipated could be eliminated. Since experience indicates that dissipative effects in mechanical systems can be reduced, by lubrication in the case of friction, for example, but not eliminated, mechanical perpetual motion of the third kind can be approximated but never achieved. An example of a genuine case of this kind occurs in a superconductor. If a direct current is caused to flow in a superconducting ring, this current will continue to flow undiminished in time without application of any external force. *See* SUPERCONDUCTIVITY; THERMODYNAMIC PRINCIPLES. [K.L.K.]

Perseus A compact circumpolar constellation of the northern sky, like its neighbor, Cassiopeia, on the east. Both constellations lie in a brilliant part of the Milky Way. The prominent stars in Perseus form the capital script letter *A* (see illustration). This group is represented by the figure of the hero

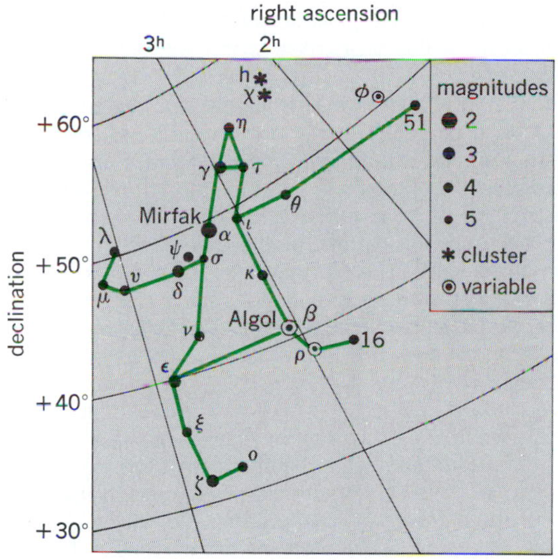

Line pattern of the constellation Perseus. The grid lines in the chart represent the coordinates of the sky. The apparent brightness, or magnitude, of the stars is shown by the size of the dots, which are graded by appropriate numbers as indicated.

Perseus. Mirfak, a navigational star, lies in the right shoulder. The constellation is noted for its clusters of stars. Just above the head are the famous double clusters *h* and χ in Perseus. Algol, the Demon Star, which is an eclipsing variable, is located in this constellation. *See* CASSIOPEIA; CONSTELLATION. [C.-S.Y.]

Persian melon A long-season cultivar of muskmelon, *Cucumis melo*, of the gourd family, Cucurbitaceae. The fruit is round and without sutures; it has dark-green skin and thin, abundant netting. The flesh is deep orange, very thick and firm, and distinctly sweet in flavor. The flesh is very rich in potassium, vitamin A, and vitamin C. *See* MUSKMELON; VIOLALES. [O.A.L.]

Persimmon A deciduous fruit tree species, *Diospyros kaki*. Persimmons originated in the subtropical regions of China but were cultivated more extensively in Japan, the current leader in world production. Persimmons have been introduced to a number of temperate zone countries; however, they have not attained substantial popularity, and only small commercial plantings exist in the United States (primarily in California), Italy, Brazil, and Israel. Persimmon cultivars adapt to a wide climatic range. Although they have a low chilling requirement, they can tolerate temperatures as low as 5°F (−15°C) if they are dormant. However, areas with late spring and early fall temperatures below 27°F (−3°C) should be avoided, as young growth and maturing fruit will be damaged. Mean annual temperatures averaging 57–59°F (14–15°C) are required for good growth and quality. Multiple, selective harvests are done in late fall to ensure that the fruits achieve the desirable deep orange-red color.

Persimmons are clonally propagated by budding onto seedling rootstocks. Seed extracted from mature fruits of *D. lotus*, *D. kaki*, and *D. virgiana* are germinated in the greenhouse in the fall. Seedlings are transplanted in outdoor nursery rows when temperatures are over 55°F (13°C). For horticultural purposes, persimmons are classified as astringent or nonastringent. Astringent persimmons have water-soluble tannins in the flesh that decrease as the fruit softens to ripeness. They are conical in shape. Nonastringent persimmons are firm when ripe, as their soluble tannins decrease with pollination. They have an oblate shape. Astringent cultivars produce best in cool climates, whereas nonastringent types require hot, dry climates for good quality. *See* FRUIT, TREE. [L.Fe.]

Personality theory A branch of psychology concerned with developing a scientifically defensible model or view of human nature—in the modern parlance, a general theory of behavior.

Most personality theories can be classified in terms of two broad categories, depending on their underlying assumptions about human nature. On the one hand, there are a group of theories that see human nature as fixed, unchanging, deeply perverse, and self-defeating. These theories emphasize self-understanding and resignation; in the cases of Freudian psychoanalysis and existentialism, they also reflect a distinctly tragic view of life—the sources of human misery are so various that the best that can be hoped for is to control some of the causes of suffering. On the other hand, there are a group of theories that see human nature as plastic, flexible, and always capable of growth, change, and development. Human nature is basically benevolent; therefore bad societies are the source of personal misery. Social reform will produce human happiness if not actual perfection. These theories emphasize self-expression and self-actualization—in the cases of Carl Rogers and Abraham Maslow, they reflect a distinctly optimistic and romantic view of life. [R.Hog.]

Persulfate A group of compounds more correctly known as peroxysulfates. Persulfates are salts of the two peroxy acids of sulfur, peroxymonosulfuric acid (H_2SO_5) and peroxydisulfuric acid ($H_2S_2O_8$). All these compounds contain sulfur in an oxidation state of 6+, the same as in sulfates. The unusual thing about their structure is the presence of the peroxy group, —O—O—.

Ammonium peroxydisulfate is used as an oxidizing agent and a bleach, and is converted to other salts, such as potassium peroxydisulfate which is used as a polymerization promoter. *See* PEROXIDE; SULFUR. [E.E.W.]

PERT An acronym for program evaluation and review technique; a planning, scheduling, and control procedure based upon the use of time-oriented networks which reflect the interrelationships and dependencies among the project tasks (activities). The major objectives of PERT are to give management improved ability to develop a project plan and to properly allocate resources within overall program time and cost limitations, to control the time and cost performance of the project, and to replan when significant departures from budget occur.

The basic requirements of PERT, in its time or schedule form of application, are:

1. All individual tasks required to complete a given program must be visualized in a clear enough manner to be put down in a network composed of events and activities. An event denotes a specified program accomplishment at a particular instant in

time. An activity represents the time and resources that are necessary to progress from one event to the next.

2. Events and activities must be sequenced on the network under a logical set of ground rules.

3. Time estimates can be made for each activity of the network on a three-way basis. Optimistic (minimum), most likely (modal), and pessimistic (maximum) performance time figures are estimated by the person or persons most familiar with the activity involved. The three-time estimates are used as a measure of uncertainty of the eventual activity duration.

4. Finally, critical path and slack times are computed. The critical path is that sequence of activities and events on the network that will require the greatest expected time to accomplish. Slack time is the difference between the earliest time that an activity may start (or finish) and its latest allowable start (or finish) time, as required to complete the project on schedule. *See* CRITICAL PATH METHOD (CPM).

5. The difference between the pessimistic (b) and optimistic (a) activity performance times is used to compute the standard deviation ($\hat{\sigma}$) of the hypothetical distribution of activity performance times [$\hat{\sigma} = (b - a)/6$]. The PERT procedure employs these expected times and standard deviations (σ^2 is called variance) to compute the probability that an event will be on schedule, that is, will occur on or before its scheduled occurrence time.

In the actual utilization of PERT, review and action by responsible managers is required, generally on a biweekly basis, concentrating on important critical path activities. A major advantage of PERT is the kind of planning required to create an initial network. Network development and critical path analysis reveal interdependencies and problem areas before the program begins that are often not obvious or well defined by conventional planning methods. [J.J.M.]

Perthite A parallel-to-subparallel intergrowth of potassium and sodium feldspar. Usually the intergrowth occurs in a lamellar fashion. In most cases the lamellae are approximately parallel. With increasing size of the intergrowths, cryptoperthites, microperthites, and macroperthites may be distinguished.

It is generally agreed that most—if not all—perthites are formed by unmixing of originally monoclinic $(K,Na)AlSi_3O_8$ into potassium- and sodium-rich feldspar domains. The unmixing starts with the formation of cryptoperthite, at any temperature below approximately 600°C (1100°F). Depending on the cooling rate, different unmixing textures can be expected. With decreasing temperature or with time, or both, the early formed domains grow continuously until they may become visible with a microscope or even to the naked eye. *See* FELDSPAR; MICROCLINE; ORTHOCLASE. [F.L.]

Perturbation (astronomy) Departure of a celestial body from the trajectory it would follow if moving only under the action of a single central force. Perturbations may be caused by either gravitational or nongravitational forces.

Planetary orbits are subject to two classes of disturbances: secular, or long-term, perturbations; and periodic, or relatively short-term, perturbations. Secular perturbations, so called because they are either progressive or have excessively long periods, arise because of the relative orientation of the orbits in space. They cause slow oscillatory changes of eccentricities and inclinations about their mean values with accompanying changes in the motions of the nodes and perihelia. Periodic perturbations arise from the relative positions of the planets in their orbits. When the disturbed and disturbing planets are aligned on the same side of the Sun, the perturbation reaches a maximum, and reduces to minimum when alignment is reached on opposite sides of the Sun.

The motions of planetary satellites, natural and artificial, reflect both gravitational and nongravitational perturbations. The centrifugal force arising from the rotation of a planet causes a deformation or oblateness of figure. In such a case the central mass does not attract as if it were concentrated at its center. For a close satellite the principal perturbation arises from the attraction of this equatorial bulge. [R.L.Du.]

Perturbation (mathematics) A modification in the mathematical structure of a problem changing the problem from one that can be solved exactly, the unperturbed problem, to one, the perturbed problem, for which it is usually possible to obtain only an approximate solution. The methods employed for this purpose form perturbation theory. These methods attempt to express the solution of the perturbed problem in terms of the properties of the solutions of the unperturbed problem.

Examples of perturbation problems can be found in nearly every branch of mathematics and physics, and in astronomy. The simplest case occurs in ordinary algebra. Suppose that the roots of the equation $f(x) = 0$ are known (the unperturbed problem), and that the roots of the equation $f(x) + \epsilon g(x) = 0$ are to be found (the perturbed problem). The parameter ϵ measures the size of the perturbation. Another set of examples occurs in linear differential equations and in particle dynamics. Possible perturbations include changes in the forces considered to be acting on the particle as well as changes in initial conditions. *See* PERTURBATION (ASTRONOMY).

Several examples occur in partial differential equations. One physical realization occurs in the theory of wave propagation where the perturbations can be changes in the index of refraction, changes in initial conditions, or changes in the nature or shape of the surfaces encountered by the waves. All of these changes can occur separately or concurrently. The first of these changes is called a volume perturbation, the second a perturbation of initial conditions, and the third a perturbation of boundary conditions. Similar examples can be taken from quantum mechanics, where the volume perturbation corresponds to a change in the hamiltonian, and perturbation of initial conditions to quantum mechanical time-dependent perturbation theory. Other partial differential equations of physics, such as the Laplace equation, the diffusion equation, and the equations of hydrodynamics, furnish further examples. *See* PERTURBATION (QUANTUM MECHANICS).

All of these problems are linear and can therefore be cast into an equation of the form $A\psi = \lambda\psi$, where ψ is the unknown quantity, λ is a constant, and A is an operator involving among other possibilities differentiation and integration. The quantity σ may be a scalar, a vector, or more generally a matrix quantity. When solutions can be obtained for only special values of λ, the eigenvalues, the equation is called the eigenvalue equation, and the associated problem is called the eigenvalue problem. The operator A contains the perturbation: that is, A equals $A_0 + \epsilon A_1$, where A_0 is the unperturbed operator and ϵA_1, the perturbing term. [H.F.]

Perturbation (quantum mechanics) An expansion technique useful for solving complicated quantum-mechanical problems in terms of solutions for simple problems. Perturbation theory in quantum mechanics provides an approximation scheme whereby the physical properties of a system, modeled mathematically by a quantum-mechanical description, can be estimated to a required degree of accuracy. Such a scheme is useful because very few problems occurring in quantum mechanics can be solved analytically. Consequently an approximation technique must be employed in order to give an approximate analytic solution or to provide suitable algorithms for a numerical solution. Even for problems which admit an exact analytic solution, the exact solution may be of such mathematical complexity that its physical interpretation is not apparent. For these situations, perturbation techniques are also desirable.

Here the discussion of the application of perturbation techniques to quantum mechanics is limited to the domain of nonrelativistic quantum theory. Applications of a similar but mathematically more intricate nature have also been made in quantum electrodynamics and quantum field theory. *See* QUANTUM ELECTRODYNAMICS; QUANTUM FIELD THEORY; QUANTUM MECHANICS.

Perturbation theory is applied to the Schrödinger equation, $H\Psi = (H_0 + \lambda V)\Psi = i\hbar(\partial/\partial t)\Psi$ [where $\hbar$ is Planck's constant h divided by 2π, and $(\partial/\partial t)$ represents partial differentiation with respect to the time variable t], for which the exact hamiltonian H is split into two parts: the approximate (unperturbed) time-independent hamiltonian H_0 whose solutions of the corresponding Schrödinger equation are known analytically, and the perturbing potential λV. The basic idea is to expand the exact solution Ψ in terms of the solution set of the unperturbed hamiltonian H_0 by means of a power series in the coupling constant λ. Such a procedure is expected to be successful if the system characterized by the unperturbed hamiltonian closely resembles that characterized by the exact hamiltonian. Supposedly the differences are not singular in character, but change as a continuous function of the parameter λ.

Perturbation theory is used in two contexts to provide information about the state of the system, which in quantum mechanics is determined by the wave function Ψ. If λV is time-independent, an objective may be to find the stationary states of the system Ψ_n whose time dependence is given by exp $(-iE_n t/\hbar)$, where $i = \sqrt{-1}$ and E_n represents the energy of the stationary state labeled by n. If λV is either time-independent or time-dependent, an objective may be to find the time evolution of a state which at some specified time was a stationary state of the unperturbed hamiltonian. The perturbing potential is then considered as causing transitions from the original state to other states of the unperturbed hamiltonian, and application of time-dependent perturbation theory provides the probability of such transitions. *See* PERTURBATION (MATHEMATICS). [D.M.Fr.]

Pesticide A material useful for the mitigation, control, or elimination of plants or animals detrimental to human health or economy. Algicides, defoliants, desiccants, herbicides, plant growth regulators, and fungicides are used to regulate populations of undesirable plants which compete with or parasitize crop or ornamental plants. Attractants, insecticides, miticides, acaricides, molluscicides, nematocides, repellants, and rodenticides are used principally to reduce parasitism and disease transmission in domestic animals, the loss of crop plants, the destruction of processed food, textile, and wood products, and parasitism and disease transmission in humans.

Materials used to control or alleviate disease conditions produced in humans and animals by plants or by animal pests are usually designated as drugs. For example, herbicides are used to control the ragweed plant, while drugs are used to alleviate the symptoms of hay fever produced in humans by ragweed pollen. Similarly, insecticides are used to control malaria mosquitoes, while drugs are used to control the malaria parasites transmitted to humans by the mosquito.

Some pesticides are obtained from plants and minerals. Examples include the insecticides cryolite, a mineral, and nicotine, rotenone, and the pyrethrins which are extracted from plants. A few pesticides are obtained by the mass culture of microorganisms. Most pesticides, however, are products of chemical manufacture. Two outstanding examples are the insecticide DDT and the herbicide 2,4-D.

By the mid-1960s, it became apparent that the benefits stemming from the ability of the new pesticides, particularly DDT, to control insect pests could be counterbalanced by adverse effects on other elements of the environment. Detailed reviews of the properties, stability, persistence, and impact upon all facets of the environment were carried out not only with DDT, but with other chlorinated, organic insecticides as well. Concern over the undesirable effects of pesticides culminated in the amendment of the Federal Insecticide, Fungicide, and Rodenticide Act (FIFRA) by Public Law 92-516, the Federal Environmental Pesticide Control Act (FEPCA) of 1972. The purpose behind this strengthening of earlier laws was to prevent exposure of either humans or the environment to unreasonable hazard from pesticides.

Pesticides must be selected and applied with care. Recommendations as to the product and method of choice for control of any pest problem—weed, insect, or varmint—are best obtained from county or state agricultural extension specialists. *See* AGRICULTURAL CHEMISTRY; FUNGICIDE; HERBICIDE; INSECTICIDE; RODENTICIDE. [G.F.L.]

Petalite A rare pegmatitic mineral with composition $LiAlSi_4O_{10}$. Its economic significance is markedly disproportionate to the number of its occurrences. It is the only basic raw material suitable for production of a group of materials known as crystallized glass ceramics (melt-formed ceramics). These extremely fine-grained substances are based on a keatite-type structure (stuffed silica derivative). Among the desirable properties of such submicroscopic aggregates are their exceedingly low thermal expansion and high strength, making them suitable for use in cooking utensils and telescopic mirror blanks.

The color is white, pink, pale green, or gray to black. Hardness is 6½ on Mohs scale. Petalite is mined from a large lithium-rich pegmatite at Bikita, Zimbabwe, the only major world source. [E.W.H.]

Petrifaction A mechanism by which the remains of extinct organisms are preserved in the fossil record. In petrifactions (though chiefly in plants rather than animals) the original shape and topography of the tissues, and occasionally even minute cytological details, are retained relatively undeformed.

The term petrifaction was adopted as a scientific term before knowledge existed of the geochemical mechanism or processes involved. It was formerly widely believed that in the formation of a petrifaction the organic matter of the organism or tissue was replaced molecule by molecule with mineral material entering in solution in percolating groundwater. It is now evident that what actually happens is that the mineral fills cell lumena and the intermicellar interstices of cell walls with insoluble salts depositing from solution. Petrifaction is hence a form of mineral emplacement or embedding, by which the organic residues are filled with solid substance which infiltrates in solution. The most common substances involved in petrifactions are silica, SiO_2, and calcium carbonate, $CaCO_3$ (calcite). Occasionally phosphate minerals, pyrite, hematite, and other less common minerals make up all or part of the petrifaction matrix. *See* FOSSIL; PALEOBOTANY; PETRIFIED FORESTS. [E.S.B.]

Petrified forests Exposures containing appreciable numbers of petrified tree trunks, either standing upright or lying prostrate in the enclosing sedimentary rocks; sometimes called fossil forests. The best-known examples are the Petrified Forest of Arizona, and the fossil forests near Cairo, Egypt; near Calistoga, California; near Vantage Bridge, Washington; and in Yellowstone National Park, Wyoming. *See* PALEOBOTANY; PETRIFACTION. [E.Do.]

Petrochemical One of a large number of substantially pure chemical substances produced commercially from petroleum or natural gas. Ordinarily the term does not include hydrocarbon fuels and lubricants nor chemicals produced by other than the processor handling the petroleum raw material. Organic compounds make up the great bulk, as well as number, of petrochemicals, but several inorganic compounds (ammonia, carbon black, sulfur, and hydrogen peroxide) also are produced in large amounts. *See* PETROLEUM PRODUCTS.

Petrochemicals should not be regarded as a particular type or class of chemical, since all of them have been, and many still are, made from other raw materials. However, a majority of the chemicals once made from other raw materials came to be made entirely, or almost entirely, from petroleum or natural gas because of the abundance of supplies and relatively low cost. It is anticipated that processes using other raw materials will become increasingly important. *See* ALKYLATION; CATALYTIC REFORMING; CRACKING; NATURAL GAS; ORGANIC CHEMICAL SYNTHESIS; PETROLEUM. [J.D.He.; W.W.R.]

Petrofabric analysis The systematic study of the fabrics of rocks, generally involving statistical study of the orientations and distribution of large numbers of fabric elements. The term fabric denotes collectively all the structural or spatial characteristics of a rock mass. The fabric elements are classified into two groups: (1) megascopic features, including bedding, schistosity, foliation, cleavage, faults, joints, folds, and mineral lineations; and (2) microscopic features, including the shapes, orientations, and mutual arrangement of the constituent mineral crystals (texture) and of internal structures (twin lamellae, deformation bands, and so on) inside the crystals.

The aim of fabric analysis is to obtain as complete and accurate a description as possible of the structural makeup of the rock mass with a view to elucidating its kinematic history. The fabric of a sedimentary rock, for example, may retain evidence of the mode of transport, deposition, and compaction of the sediment in the size, shape, and disposition of the particles; similarly, that of an igneous rock may reflect the nature of the flow or of gravitational segregation of crystals and melt during crystallization. The fabrics of deformed metamorphic rocks (tectonites) have been most extensively studied by petrofabric techniques with the objective of determining the details of the history of deformation and recrystallization. *See* STRUCTURAL GEOLOGY; STRUCTURAL PETROLOGY. [J.M.Ch.]

Petrographic province A region in which the igneous rocks, formed during a limited period, show a certain community of character (chemical, mineralogical, or petrological) that distinguishes them from other rocks in the area. Rocks of a petrographic province are consanguineous in that they are, theoretically at least, derived from a common parental magma. A classic province is represented by the highly potassic rocks around Rome and Naples.

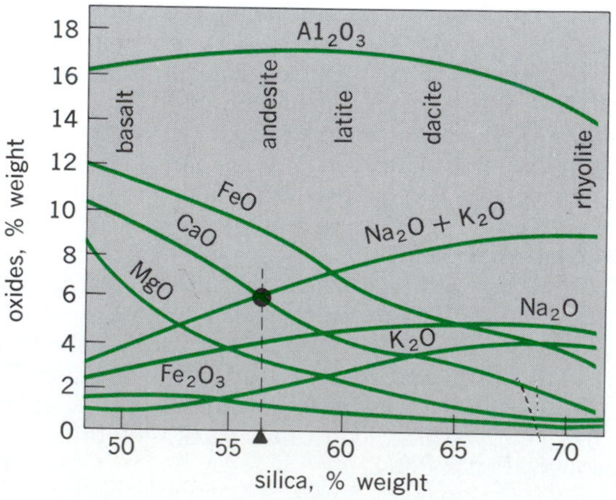

Igneous rock variation diagram of calc-alkali series of volcanic rocks from San Francisco Mountains, Arizona.

If the chemical analyses of rocks representing a genetically related series are plotted on a variation diagram, many characteristics of the series may be brought out. The illustration shows the percentages of oxides in each rock analysis as plotted against the silica content of that analysis.

All points lie on or close to smooth curves which represent the variation in composition with silica content (or roughly with time, because younger rocks tend to be more silicic). Noteworthy are the positive slopes for potassium oxide (K_2O) and sodium oxide (Na_2O), the arched curve for aluminum oxide (Al_2O_3), and the negative slopes for magnesium oxide (MgO), calcium oxide (CaO), ferrous oxide (FeO), and ferric oxide (Fe_2O_3). The combined alkali ($Na_2O + K_2O$) curve is seen to cross the CaO curve at 56.5% silica. This silica value is known as the lime-alkali index for the rock series. On the basis of lime-alkali indices, the numerous rock series are arbitrarily divided into groups as follows: alkalic < 51, alkali-calcic $> 51 < 56$, calc-alkalic $> 56 < 61$, and calcic > 61.

Certain rock series appear related to certain types of geological environments. In general, the more alkalic types occur in regions subjected to tension and vertical movement (faulting and subsidence), whereas more calcic series are found in compressional regions (fold-mountain belts). [C.A.C.]

Petrography The description of rocks with goals of classification and interpretation of origin. Most schemes for the classification of rocks are based on the size of grains and the proportions of various minerals. Interpretations of origin rely on field relations, structure, texture, and chemical composition as well as sizes and proportions of different kinds of grains. The names of rocks are based on the sizes and relative proportions of different minerals; boundaries between the names are arbitrary. The conditions of formation of a rock can be estimated from the types and textures of its constituent minerals.

The description of rocks begins in the field with observation of the shape and structure of bodies of rock at the scale of centimeters to kilometers. The geometrical relations between and structures within mappable rock units are generally the domain of field geology, but are simply rock descriptions at a reduced scale.

A petrographer can correctly name most rocks in which most crystals are larger than about 0.04 in. (1 mm) simply by examining the rock with a 10-power magnifying lens. Rocks with smaller grains require either microscopical examination or chemical analysis for proper classification.

Sizes and shapes of grains and voids are the most important features of a rock relevant to its origin. The same features also affect density, porosity, permeability, strength, and magnetic behavior. It is also essential to know the identity, abundance, and compositions of minerals constituting the grains in order to name a rock and infer its conditions of formation. *See* MINERALOGY; PETROLOGY. [A.T.A.]

Petrolatum A smooth, semisolid blend of mineral oil with waxes crystallized from the residual type of petroleum lubricating oil. The wax molecules are microneedles and hold a large amount of oil in a gel. Petrolatums are useful because they cling, lubricate, and resist both moisture and oxidation. They serve as lubricants in baking and candymaking; as carriers in polishes, cosmetics, and ointments; as rust preventives; as waterproofing agents for paper; and in other uses calling for an inert greaselike material. [J.K.R.]

Petroleum A naturally occurring, oily, flammable liquid composed principally of hydrocarbons, and occasionally found in springs or pools but usually obtained from beneath the Earth's surface by drilling wells. Formerly called rock oil, unrefined petroleum is now usually termed crude oil.

Composition. Petroleum is separated by distillation into fractions designed as (1) straight-run gasoline, boiling at up to about 390°F (200°C); (2) middle distillate, boiling at about 365–653°F (185–345°C), from which are obtained kerosene, heating oils, and diesel, jet, rocket, and gas turbine fuels; (3) wide-cut gas oil, which boils at about 653–1000°F (345–540°C), and from which are obtained waxes, lubricating oils, and feed stock for catalytic cracking to gasoline; and (4) residual oil, which may be asphaltic.

The physical properties and chemical composition of petroleum vary markedly, depending on its source. As it comes from the earth, it ranges from an occasional nearly colorless liquid consisting chiefly of gasoline to a heavy black tarry material high in asphalt content. Most crudes are black; many are amber, red, or brown by transmitted light and show a greenish fluorescence by reflected light and have a specific gravity in the range about 0.82–0.95.

Hydrocarbons constitute 50–98% of petroleum, and the remainder is composed chiefly of organic compounds containing oxygen, nitrogen, or sulfur, and trace amounts of organometallic compounds. The hydrocarbon types found in petroleum are paraffins (alkanes), cycloparaffins (naphthenes or cycloalkanes), and aromatics. Olefins (alkenes) and other unsaturated hydrocarbons are usually absent. *See* ALICYCLIC HYDROCARBON; ALKANE; AROMATIC HYDROCARBON.

The number of carbon atoms in hydrocarbons of a given boiling range depends on the hydrocarbon type. In general, gasoline will include hydrocarbons having 4–12 carbon atoms; kerosine, 10–14; middle distillate, 12–20; and wide-cut gas oil, 20–36. Five main classes of compounds are present in the gasoline fraction: straight-chain paraffins, branched-chain paraffins, alkylcyclopentanes, alkylcyclohexanes, and alkylbenzenes.

Asphalt is a dark-brown to black solid or semisolid consisting of carbon, hydrogen, oxygen, sulfur, and sometimes nitrogen. It is made up of three components: (1) asphaltene, a hard, friable, infusible powder; (2) resin, a semisolid to solid ductile and adhesive material; and (3) oil, which is structurally similar to the lubricating oil fraction from which it is derived. *See* ASPHALT AND ASPHALTITE; PETROCHEMICAL; PETROLEUM ENGINEERING; PETROLEUM PROCESSING; PETROLEUM PRODUCTS. [L.S.]

Origin. Accumulation of petroleum is believed to involve three steps: generation of oil, primary migration (the movement of oil from source to reservoir rock), and secondary migration (the redistribution of oil within the reservoir rock to form a pool).

Oil is generated in sedimentary basins. These basins are shallow depressions on the continents that have intermittently been covered with sea water, or offshore basins on continental shelves. They are hundreds of square kilometers in area and contain sediments of three types: (1) rock particles varying from sands to clay muds, which were eroded from hills and mountains and were carried to the basins by streams; (2) biochemical and chemical precipitates such as limestone, gypsum, anhydrite, and chert; and (3) organic matter from the plants and animals that lived in the sea or were carried in by rivers. The third type of sediment, the organic matter, is the source of petroleum.

It is believed that oil is generated from organic matter in two ways. A small amount, probably less than 10%, comes directly from the hydrocarbons that marine organisms form as part of their living cells. The second process, by which about 90% of the oil is formed, involves the formation of hydrocarbons from the decay and alteration of buried organic matter. Nearly all of the hydrocarbons containing up to 10 carbon atoms and probably over half of those of higher molecular weight are formed in this manner.

Primary migration of petroleum from source to reservoir, the second step in accumulation, is caused by the movement of water, which carries oil or its precursor out of compacting sediments. When the source muds are deposited, they contain 70–80% water. The remainder is solids, mostly clay minerals or carbonate particles; as they build up to great thicknesses in a sedimentary basin, water is squeezed out by the weight of the overlying sediments. The water and oil move in the direction of least resistance (lowest hydrostatic pressure) at rates of 0.8–3.9 in. (2–10 cm) per year. Early in compaction, the direction of fluid movement is straight up. As compaction progresses, there is lateral as well as vertical movement of the fluids.

In secondary migration, the last stage in the origin of an oil accumulation, the oil droplets are moved about within the reservoir to form the pool. Secondary migration includes in some instances a second step, during which crustal movements of the earth shift the position of the pool within the reservoir rock.

The composition and character of the oil ultimately formed is controlled by all phases of its origin and migration. The organic structures from which the oil originated, the increasing formation of light hydrocarbons with depth, the selectivity of the migration phase in carrying only part of the hydrocarbons from the source rock, and the alteration of the oil within the reservoir by subsurface water and ground movements, are believed important in determining the composition of petroleum. *See* PETROLEUM GEOLOGY. [J.M.H.]

Petroleum engineering An eclectic discipline comprising the technologies used for the exploitation of crude oil and natural gas reservoirs. It is usually subdivided into the branches of petrophysical, geological, reservoir drilling, production, and construction engineering. After an oil or gas accumulation is discovered, technical supervision of the reservoir is transferred to the petroleum engineering group, although in the exploration phase the drilling and petrophysical engineers have played a role in the completion and evaluation of the discovery.

By the use of down-hole logging tools and of laboratory analysis of cores made during the drilling operation, the petrophysical engineer estimates the porosity, permeability, and oil content of the reservoir rock that has been sampled at the drill site. *See* WELL LOGGING.

The geological engineer, using the petrophysical data, the seismic surveys conducted during the exploration operations, and an analysis of the regional and environmental geology, develops inferences concerning the lateral continuity and extent of the reservoir. *See* PETROLEUM GEOLOGY.

The reservoir engineer, using the initial studies of the petrophysicist and geological engineers together with the early performance of the wells drilled into the reservoir, attempts to assess the producing rates (barrels of oil or millions of cubic feet of gas per day) that individual wells and the entire reservoir are capable of sustaining. One of the major assignments of the reservoir engineer is to estimate the ultimate production that can be anticipated from both primary and enhanced recovery from the reservoir. *See* PETROLEUM ENHANCED RECOVERY; PETROLEUM RESERVES; PETROLEUM RESERVOIR ENGINEERING; PETROLEUM RESERVOIR MODELS.

The drilling engineer has the responsibility for the efficient penetration of the earth by a well bore, and for cementing of the steel casing from the surface to a depth usually just above the target reservoir. The drilling engineer or another specialist, the mud engineer, is in charge of the fluid that is continuously circulated through the drill pipe and back up to surface in the annulus between the drill pipe and the bore hole. *See* BOREHOLE LOGGING; ROTARY TOOL DRILL.

The production engineer, upon consultation with the petrophysical and reservoir engineers, plans the completion procedure for the well. This involves a choice of setting a liner across the formation or perforating a casing that has been extended and cemented across the reservoir, selecting appropriate pumping techniques, and choosing the surface collection, dehydration, and storage facilities. *See* OIL AND GAS WELL COMPLETION.

Major construction projects, such as the design and erection of offshore platforms, require the addition of civil engineers to the staff of petroleum engineering departments, and the design and implementation of natural gasoline and gas processing plants require the addition of chemical engineers. *See* OIL AND GAS, OFFSHORE; PETROLEUM. [T.M.D.]

Petroleum enhanced recovery Novel technology designed to enhance the fraction of the original oil in place in a reservoir. Interest in developing enhanced recovery technology accelerated as it became more certain that over two-thirds of the oil discovered in the United States, and a still greater percentage in the rest of the world, would remain unrecovered through the application of conventional primary and secondary (waterflood) operations. *See* OIL AND GAS FIELD EXPLOITATION; PETROLEUM RESERVES.

The problems encountered in developing such technology are very great because of the nature of fluid flow within a subsurface reservoir, and because of the inability of engineers to exercise any intimate degree of control over the flow and distribution of fluids within the reservoir. The only points of contact with the reservoir are at the surface of the producing and injection wells that lead down to the reservoir sands.

A subsurface reservoir comprises the interconnected pores of sandstone or porous carbonate rocks. There is a great tendency for oil being displaced by encroaching water or an expanding gas to be bypassed by the displacing phase. The tendency to bypass is exacerbated by the differences in wettability of the reservoir minerals by water and oil. If the encroaching water succeeds in bypassing oil, then pressure gradients, above an attainable level, will be required to effect displacement of the trapped ganglia. *See* PETROLEUM GEOLOGY.

There are certain shallow reservoirs at a low temperature and a relatively low pressure which have a high porosity that is saturated with medium- to high-viscosity oils. The viscosity is so high and the pressure so low that production rates are virtually uneconomic. Because of the much lower mobility of the viscous oils, injected water will readily finger through the reservoir without displacing a significant quantity of oil. Thermal technology has been developed by which the viscosity of the crude within the reservoir can be lowered and a suitable pressure gradient applied to achieve significant recovery of crude from such reservoirs. The methods are the steam soak, or cyclic stimulation, in which oil is produced from the same well into which steam had earlier been injected; the steam drive, which is much like a waterflood but with the substitution of steam for injected water; and wet, or quenched, in-situ combustion in which a steam drive is generated in situ by the combustion of a limited fraction of the oil in place. *See* OIL SAND; OIL SHALE. [T.M.D.]

Petroleum geology The application of geological concepts to finding and producing petroleum. Petroleum deposits are usually found in sedimentary rocks. With a few notable exceptions, economically important hydrocarbon deposits are found in the subsurface at depths ranging from several feet to several miles. *See* PETROLEUM.

Subsurface deposits. Petroleum occurs in porous rocks such as sandstones, limestones, and dolomites. In the subsurface, porosity, the space between the particles, is usually filled with water, but under certain conditions the fluid is a mixture of water and hydrocarbons, either oil or gas or both. The conditions necessary for the accumulation of hydrocarbons are a source and a trap to collect them. There are many circumstances which provide the right conditions to form an oil or gas field.

There are two basic types of hydrocarbon traps, stratigraphic and structural, but these often occur in combination with each other. The lens and the unconformity are examples of stratigraphic traps. The anticline, fault, and salt dome are examples of structural traps.

The quality of a reservoir rock is controlled primarily by porosity and permeability. Porosity is measured as a percentage of total rock volume. Permeability, the ability of the reservoir rock to transmit or allow movement of fluids, is provided by the interconnection of pores in the rock and allows oil and gas to flow from the reservoir into wells for production.

The voids or pore space in all reservoir rocks are filled with fluid, which may be water or various combinations of water, oil, and gas. In traps that contain oil, water, and free gas, the fluids occur in distinct zones: gas, the lightest fluid, occurring at the top of the trap, followed by oil, and then water. Where there is no oil, the gas is immediately above the water.

Surface occurrences. Petroleum deposits encountered on the surface are often outcrops of former subsurface deposits which have been exposed by erosion; others are leakages of hydrocarbons from subsurface accumulations. In some instances, particles of hydrocarbons (that is, heavy oil, asphalt, or tar) have been mixed with sediments and laid down as a primary part of the deposit. The hydrocarbons in surface deposits are normally heavy tars and asphalts which are residues from normal subsurface oils that have been altered by the escape of volatile fractions and oxidation. *See* ASPHALT AND ASPHALTITE.

Most oil or tar sand deposits are former subsurface deposits which have been exhumed by erosion. A few are primary deposits in which particles of tar or asphalt have been deposited along with conventional sediments. Only a few surface deposits are economically important. *See* OIL SAND.

Vast quantities of oil shale are exposed on the surface, with still greater quantities present at only modest depths of burial. *See* OIL AND GAS FIELD EXPLOITATION; SEDIMENTARY ROCKS. [F.E.Fo.]

Petroleum processing The recovery and processing of various usable fractions from the complex crude oils. The usable fractions include gasoline, jet fuel, kerosine, fuel oil, asphalt, lubricating oils, and many others.

Crude oil is a mixture of many different hydrocarbon compounds of the paraffin type (wax compounds) and of the naphthene type (asphalt compounds), making the chemistry of petroleum refining extremely complex. The refining processes can be grouped under three main headings: (1) separating (by distillation, treating, and blending) the crude oils to isolate the desired products; (2) breaking the remaining large chemical compounds into smaller chemical compounds by cracking; (3) building desired product properties by chemical reactions, such as reforming, alkylation, and isomerization. *See* ALKYLATION (PETROLEUM); CATALYTIC REFORMING; CRACKING; DISTILLATION; ISOMERIZATION; PETROLEUM PRODUCTS. [J.J.McK.; H.L.Ho.]

Petroleum products Crude petroleum is the starting material not only for the fuels used for transportation and energy production but also for petrochemical feedstocks, solvents, lubricants, asphalts, and many other specialties. The more complex the family of petroleum products, the more energy is needed to refine the crude.

Liquid products such as gasoline, jet fuel, diesel fuel, and marine fuel, which together account for more than half the volume of output, serve the transportation industry. Hydrocarbons from butane to C_{12} appear in gasoline blends; higher-boiling-fraction C_8 to C_{15} hydrocarbons are the blend stocks for jet fuels for the air-transport industry. Most specialized energy-consuming process units in refineries, such as catalytic reformers and cracking units, exist to convert crude fractions to gasoline of a quality that meets the needs of the modern automobile engine. Added to the usual engine requirements for antiknock performance and volatility of fuel are environmental demands for emission standards in exhaust. *See* ANTIKNOCK AGENT; CRACKING; GASOLINE; JET FUEL.

Diesel fuels are blended from both distilled and cracked fractions—up to C_{22}-type hydrocarbons. They range in volatility from kerosine for lightweight automotive diesel engines to gas oil for large, lower-speed industrial or marine diesel engines. Marine fuels are made from the highest-boiling crude fraction—the residuum from crude distillation blended with heavy cracked gas oils. *See* DIESEL FUEL; KEROSINE.

Petroleum fuels heat homes, factories, and offices and also generate electric power. Home-heating oils are similar to diesel fuels in boiling range but must exhibit storage stability and good atomization and combustion performance in small oil burners. *See* FUEL OIL.

From each of the fractions distilled from crude or its cracked or processed products, valuable and indispensable nonfuel materials are made. Some of the major products are: ethylene, propylene, butylene, and other reactive gases for the petrochemical industry; liquefied petroleum gases for heating, cooking, and drying; solvents for the paint and dry-cleaning industries and as vehicles for aerosol products; lubricants of all types; specialty oils for hydraulic fluids, transformers, emulsions, cutting fluids, pharmaceuticals, inks, preservatives, and so on; and wax for a host of applications. *See* LIQUEFIED PETROLEUM GAS (LPG); LUBRICANT; PETROCHEMICAL; SOLVENT.

The residues from processing crude can also yield such familiar products as asphalt, the major road-building material, and coke, burned as fuel or made into electrodes. *See* ASPHALT AND ASPHALTITE; COKE; PETROLEUM; PETROLEUM PROCESSING. [W.G.D.]

Petroleum reserves Proved reserves are the estimated quantities of crude oil liquids which with reasonable certainty can be recovered in future years from delineated reservoirs under existing economic and operating conditions. Thus, estimates of crude oil reserves do not include synthetic liquids which at some time in the future may be produced by converting coal or oil shale, nor do reserves include fluids which may be recovered following the future implementation of a supplementary or enhanced recovery scheme.

Indicated reserves are those quantities of petroleum which are believed to be recoverable by already implemented but unproved enhanced oil recovery processes or by the application of enhanced recovery processes to reservoirs similar to those in which such recovery processes have been proved to increase recovery.

Thus, crude oil reserves can be called upon in the future with a high degree of certainty, subject of course to the limitations placed on production rate by fluid flow within the reservoir and the capacity of the individual producing wells and surface facilities to handle the produced fluids. It is important to bear in the mind the distinction between resources and reserves. The former term refers to the total amount of oil that has been discovered in the subsurface, whereas the latter refers to the amount of oil that can be economically recovered in the future. The ratio of the ultimate recovery (the sum of currently proved reserves and past production) to the resource or original oil in place is the anticipated recovery efficiency. *See* NATURAL GAS; PETROLEUM; PETROLEUM ENHANCED RECOVERY. [T.M.D.]

Petroleum reservoir engineering The technology concerned with the prediction of the optimum economic recovery of oil or gas from hydrocarbon-bearing reservoirs. It is an eclectic technology requiring coordinated application of many disciplines: physics, chemistry, mathematics, geology, and chemical engineering.

Originally, the role of reservoir engineering was exclusively that of counting oil and natural gas reserves. The scope of reservoir engineering has broadened to include the analysis of optimum ways for recovering oil and natural gas, and the study and implementation of enhanced recovery techniques for increasing the recovery above that which can be expected from the use of conventional technology. *See* PETROLEUM RESERVES.

The overall recovery of crude oil from a reservoir is a function of the production mechanism, the reservoir and fluid parameters, and the implementation of supplementary recovery techniques. In general, recovery efficiency is not dependent upon the rate of production except for those reservoirs where gravity segregation is sufficient to permit segregation of the gas, oil, and water. Where gravity drainage is the producing mechanism, which occurs when the oil column in the reservoir is quite thick and the vertical permeability is high and a gas cap is initially present or is developed on producing, the reservoir will also show a significant effect of rate on the production efficiency. *See* PETROLEUM ENHANCED RECOVERY. [T.M.D.]

Petroleum reservoir models Physical and computational systems whose behavior is designed to resemble that of actual reservoirs. Early in the petroleum industry's history, there was a great deal of effort invested in the development of very complex physical (laboratory) models, while the computational models were quite simple. Since the explosive evolution of computers, the computational models have become more and more complex, and there has been a diminishing emphasis on the physical models. Today the emphasis is on combining the best features of the physical and computational models to permit evaluation of the complex behavior of enhanced oil recovery processes.

There are three basic requirements for a useful model: (1) a quantitative description of the relevant properties of a reservoir; (2) a means of describing the mechanics of fluids movements in a reservoir; and (3) a means of combining 1 and 2 by constructing analogous physical systems or by constructing mathematical systems which can be solved by computational techniques.

The purpose of a petroleum reservoir model is to estimate field performance (for instance, oil recovery) under a variety of operating strategies. Whereas the reservoir can be produced only once—and at considerable expense—a model can be produced many times. If this production can be accomplished at low expense and over a short period of time, then observation of the model's performance can give great insight into how to optimally produce a reservoir. Petroleum reservoir models can range in complexity from the intuition and judgment of an engineer to complex mathematical or physical systems describing enhanced oil recovery processes. *See* OIL FIELD MODEL; PETROLEUM RESERVOIR ENGINEERING. [H.S.P.]

Petrology The study of rocks, their occurrence, composition, and origin. Petrography is concerned primarily with the detailed description and classification of rocks, whereas petrology deals primarily with rock formation, or petrogenesis. Experimental petrology reproduces in the laboratory the conditions of high pressure and temperature which existed at various depths in the Earth where minerals and rocks were formed. A petrological description includes definition of the unit in which the rock occurs, its attitude and structure, its mineralogy and chemical composition, and conclusions regarding its origin. For a discussion of mineral identification, petrographic analysis, and the classification of rocks *see* MINERALOGY; PETROGRAPHY; ROCK. *See also* IGNEOUS ROCKS; METAMORPHIC ROCKS; SEDIMENTARY ROCKS. [W.I.R.]

Petromyzontida The order comprising the lampreys, sometimes called Petromyzontiformes, which are eellike, jawless vertebrates (class Agnatha) often grouped with the Myxinoidea as Cyclostomata. They differ from Myxinoidea as follows: The single nasal opening is on the dorsal side of the head and ends internally as a blind sac; the mouth is surrounded by a circular oral disk and provided with a rasping tongue (both disk and tongue are set with horny teeth); seven pairs of gill pouches open separately to the exterior; adults have well-developed eyes; dorsal and caudal fins are separate, and both are supported by fin rays; there are separate sexes and a distinct larval stage. *See* AGNATHA; CYCLOSTOMATA (CHORDATA); MYXINIFORMES; OSTEOSTRACI. [R.H.De.]

Pewter A tarnish-resistant alloy of lead and tin always containing appreciably more than 63% tin. Other metals are sometimes used with or in place of the lead; among them are copper, antimony, and zinc. Pewter is commonly worked by spinning and it polishes to a characteristic luster. Because pewter work hardens only slightly, pewter products can be finished without intermediate annealing. Early pewter, with high lead content, darkened with age. With less than 35% lead, pewter was used for decanters, mugs, tankards, bowls, dishes, candlesticks, and canisters. The lead remained in solid solution with the tin so that the alloy was resistant to the weak acids in foods. [F.H.R.]

pH A term used to describe the hydrogen-ion activity of a system. It is defined by the expression $pH = -\log a_{H^+}$. Here a_{H^+} is the activity of the hydrogen ion. In dilute solutions, the activity is essentially equal to the concentration, and the pH may be approximately defined as $pH = -\log_{10}[H^+]$, where $[H^+]$ is the concentration of the hydrogen ion in moles per liter. The use of the pH makes negative exponents unnecessary in describing the hydrogen-ion activity. In a system in which the hydrogen-ion activity is 10^{-3} mole/liter, the pH is 3. *See* ACID AND BASE; HYDROGEN ION. [F.J.J.]

pH regulation (biology) The processes operating in living organisms to preserve a viable acid-base state. In higher animals, much of the body substance (60–70%) consists of complex solutions of inorganic and organic solutes. For convenience, these body fluids can be subdivided into the cellular fluid (some two-thirds of the total) and the extracellular fluid. The latter includes blood plasma and interstitial fluid, the film of fluid that bathes all the cells of the body. For normal function, the distinctive compositions of these various fluids are maintained within narrow limits by a process called homeostasis. A crucial characteristic of these solutions is pH, an expression representing the concentration (or preferably the activity) of hydrogen ions, $[H^+]$, in solution. The pH is defined as $-\log [H^+]$, so that in the usual physiological pH range of 7 to 8, $[H^+]$ is exceedingly low, between 10^{-7} and 10^{-8} *M*. Organisms use a variety of means to keep pH under careful control, because even small deviations from normal pH can disrupt living processes. *See* HOMEOSTASIS; PH.

The most accessible and commonly studied body fluid is blood, and it, therefore, provides the most information on pH regulation. Blood pH in humans, and in mammals generally, is about 7.4. This value indicates that blood is slightly alkaline, because neutrality, the condition in which the concentration of hydrogen ions ($[H^+]$) equals the concentration of hydroxyl ions ($[OH^-]$), is pH 6.8 at mammalian body temperature of 98°F (37°C). The pH within cells, including the red blood cells, is typically lower by 0.2–0.6 unit, and is thus close to neutrality. In most animals other than warm-blooded mammals, blood pH deviates from the familiar value of 7.4. The major reason is that body temperature has an important influence on pH regulation. Consequently, animals that experience significant changes in body temperature have no single normal pH at which they regulate, but rather a series of values depending on body temperature.

Blood pH regulation is necessary because metabolic and ingestive processes add acidic or basic substances to the body and can displace pH from its proper value. For true regulation, active physiological mechanisms are required that can alter the acid-base composition of the blood in a controlled fashion. It is conventional to identify these control mechanisms with their effects on the principal buffer of the extracellular fluid, carbon dioxide (CO_2). Carbon dioxide is produced by cellular metabolism and distributes readily throughout the body because of its high solubility and rapid diffusion. In solution, CO_2 is hydrated to carbonic acid (H_2CO_3) which dissociates almost completely to H^+ and bicarbonate ions (HCO_3^-). Dissolved CO_2 can be identified with a particular partial pressure of CO_2 (PCO_2). To regulate blood pH, organisms have mechanisms to independently control PCO_2 and $[HCO_3^-]$.

The cells, while benefitting from the stability afforded by whole body mechanisms (respiratory and ionic), also have local means for their own pH regulations. An acute acid load on a cell, whether from its intrinsic metabolism or from an external source, is dealt with first by the cell's chemical buffering capacity, a capacity that exceeds that of the blood by severalfold. Other cellular mechanisms include the conversion of organic acids to neutral compounds through metabolic transformations, and the transfer of acid equivalents from the primary cell fluid, the cytoplasm, into cellular organelles. *See* BIOPOTENTIALS AND IONIC CURRENTS; CHEMIOSMOSIS; ENZYME. [D.C.J.]

Phaeophyceae A class of plants, commonly called brown algae, in the chlorophyll *a-c* phyletic line (Chromophycota). Brown algae occur almost exclusively in marine or brackish water, where they are attached to rocks, wood, sea grasses, or other algae. Approximately 265 genera and 1500 species are recognized, arranged in about 15 orders. *See* ALGAE; CHROMOPHYCOTA.

Phaeophyceae are characterized primarily by biochemical and ultrastructural features. The cells are typically uninucleate and contain one or more chloroplasts with or without pyrenoids. Photosynthetic pigments include chlorophyll *a* and *c*, ß-carotene, and several xanthophylls, principally fucoxanthin. The simplest thallus exhibited by Phaeophyceae is generally considered to be an erect, unbranched or branched, uniseriate filament arising from a prostrate filamentous base, Many Phaeophyceae are crustose or bladelike. Complex thalli differentiated into macroscopic organs are produced by kelps and rockweeds.

The geographic distribution of Phaeophyceae is bimodal. Kelps are most abundant and diverse on surf-swept rocky shores of the North Pacific, but they form an ecologically important vegetation belt in the lower intertidal and upper subtidal zones on all cold-water shores except Antarctica. Rockweeds are similarly abundant on cold-water shores, forming conspicuous belts in the upper intertidal zone. They also form extensive stands in salt marshes in the Northern Hemisphere. Tropical waters support a diverse array of Dictyotales and members of the fucalean family Sargassaceae. [P.C.Si.; R.L.Moe]

Phagocytosis A mechanism by which single cells of the animal kingdom, such as smaller protozoa, engulf and carry particles into the cytoplasm. It differs from endocytosis primarily in the size of the particle rather than in the mechanism; as particles approach the dimensions and solubility of macromolecules, cells take them up by the process of endocytosis.

Cells such as the free-living amebas or the wandering cells of the metazoa often can "sense" the direction of a potential food source and move toward it (chemotaxis). If, when the cell contacts the particle, the particle has the appropriate chemical composition, or surface charge, it adheres to the cell. The cell responds by forming a hollow, conelike cytoplasmic process around the particle, eventually surrounding it completely. Although the particle is internalized by this sequence of events, it is still enclosed in a portion of the cell's surface membrane and thus isolated from the cell's cytoplasm. The combined particle and membrane package is referred to as a food or phagocytic vacuole. *See* VACUOLE.

Ameboid cells of the metazoa also selectively remove foreign particles, bacteria, and other pathogens by phagocytosis. After the foreign particle or microorganism is trapped in a vacuole inside the macrophage, it is usually digested. To accomplish this, small packets (lysosomes) of lytic proenzymes are introduced into the phagocytic vacuole, where the enzymes are then dissolved and activated. *See* LYSOSOME. [P.W.B.]

Pharetronidia A subclass of sponges of the class Calcarea in which the skeleton is formed of quadriradite spicules cemented together into a compact network, or includes an aspiculous massive basal skeleton formed of irregular calcitic spherulites. Spicules shaped like tuning forks are characteristically present, and free triradiates, quadriradiates, and diactinal spicules may occur. *See* CALCAREA. [W.D.H.]

Pharmaceutical chemistry The chemistry of drugs and of medicinal and pharmaceutical products. The important aspects of pharmaceutical chemistry are as follows:

1. Isolation, purification, and characterization of medicinally active agents and materials from natural sources (mineral, vegetable, microbiological, or animal) used in treatment of disease and in compounding prescriptions.
2. Synthesis of medicinal agents not known from natural sources, or the synthetic duplication, for reasons of economy, purity, or adequate supply, of substances first known from natural sources.
3. Semisynthesis of drugs, whereby natural substances are transformed by means of comparatively simple steps into products which possess more favorable therapeutic or pharmaceutical properties.
4. Determination of the derivative or form of a medicinal agent which exhibits optimum medicinal activity and at the same time lends itself to stable formulation and elegant dispensing.
5. Determination of incompatibilities, chemical and biological, between the various ingredients of a prescription.
6. Establishment of safe and practical standards, with respect to both dosage and quality, to assure uniform and therapeutically reliable forms for all medication.
7. Improvement and promotion of the use of chemical agents for prevention of illness, alleviation of pain, cure of disease, and search for new therapeutic agents, particularly where no satisfactory remedy now exists. *See* PHARMACOGNOSY; PHARMACOLOGY; PHARMACY. [W.H.H.]

Pharmacognosy The general biology, biochemistry, and economics of nonfood natural products of value in medicine, pharmacy, and other health professions. The products studied are of biologic origin, either plant or animal. Pharmacognosy literally means knowledge of drugs, as do pharmacology and pharmacy. The center of interest in pharmacology, however, is on the mode of action of drugs. In pharmacy major attention is directed toward provision of suitable dosage forms, their production and distribution. Pharmacognosy is restricted to natural products with attention centered on sources of drugs, plant and animal, and on the biosynthesis and identity of their pharmacodynamic constituents.

Organs, or occasionally entire plants or animals, are dried or frozen for preservation and are termed crude drugs. They may be used medicinally in essentially this form or as sources of mixtures or of chemicals obtained by processes of extraction. Mixtures obtained by exudation from living plants include such drugs as opium, turpentine, and acacia. Processes of extraction are required to obtain such mixtures as peppermint oil (steam distillation), podophyllum resin (percolation), and parathyroid extract (solution).

Pure chemicals may be extracted from a crude drug (for example, the glycoside digitoxin from digitalis or the hormone insulin from pancreas), from a mixture obtained by exudation (for example, the alkaloid morphine from opium), or from an extracted mixture (for example, the terpene menthol from peppermint oil).

Vitamins as a class of natural products are within the scope of pharmacognosy, although many are obtained commercially by laboratory synthesis. Included also are antibiotics and biologicals (serums, vaccines, and diagnostic biological products).

Development of synthetic drugs related chemically to the active constituent of a natural product has frequently followed investigation of native use of the natural product as drug or poison. The objective of such development is usually to produce a drug having fewer undesirable side effects while retaining the useful therapeutic action. *See* PHARMACEUTICAL CHEMISTRY; PHARMACOLOGY; PHARMACY. [R.A.De.]

Pharmacology The science of detection and measurement of the effects of drugs or other chemicals on biological systems. The effect of chemicals may be beneficial (therapeutic) or harmful (toxic). The pure chemicals or mixtures may be of natural origin (plant, animal, or mineral) or may be synthetic compounds.

The broad area covered may be conveniently divided into a number of categories: chemotherapy, the use of chemicals to destroy invading organisms such as bacteria and molds in or on the host; pharmacotherapy, the use of drugs to restore or replace normal function in various tissue cells, organs, or integrated units; pharmacodynamics, studies on the mechanism of action of drugs which may utilize physiological, biochemical, or electrical techniques; toxicology, the study of the poisonous effects of chemicals; psychopharmacology, the study of the effects of chemicals on the behavior of humans or animals; biochemical pharmacology, the effects of chemicals on biochemical reactions in living systems, and the effects of these systems on the chemicals, that is, their metabolism; structure-activity relationship, relationship of biological activity to chemical structure and molecular properties; and clinical pharmacology, the study and evaluation of the effects of drugs in humans. *See* CHEMOTHERAPY; PATHOLOGY; TOXICOLOGY. [C.J.K.]

Pharmacy The health profession concerned with the discovery, development, production, and distribution of drugs. Drugs are substances (other than devices) used to diagnose, prevent, cure, or relieve the symptoms of disease. For relations to closely allied fields *see* MEDICINE; PHARMACEUTICAL CHEMISTRY; PHARMACOGNOSY; PHARMACOLOGY.

General practice is carried on in exclusive prescription pharmacies, semiprofessional pharmacies, and drug stores. It consists of compounding and dispensing drugs on order of the physician, dentist, or veterinarian; serving as consultant on drugs to the health professions and to the public; and selling other health supplies such as antiseptics, bandages, and home remedies.

A hospital pharmacy includes special administrative features, provision of drugs for nursing stations, manufacturing of pharmaceutical preparations, teaching of nurses and medical and pharmacy interns, service to the hospital committee on pharmacy and therapeutics, preparation and revision of a hospital formulary, and monitoring the drug regimen of the individual patient (clinical pharmacy). The pharmacist may have charge of investigational drugs, radioactive pharmaceuticals, medical and surgical sterile supplies, and gaseous drugs for inhalation therapy. [R.A.De.]

Pharynx A chamber at the oral end of the vertebrate alimentary canal, leading to the esophagus. In adult humans it is divided anteriorly by the soft palate into a nasopharynx and an oropharynx, lying behind the tongue but anterior to the epiglottis; there is also a retropharyngeal compartment, posterior to both epiglottis and soft palate. The nasopharynx receives the nasal passages and communicates with the two middle ears through auditory tubes. The retropharynx leads to the esophagus and to the larynx, and the paths of breathing and swallowing cross within it. *See* ESOPHAGUS; LARYNX; PALATE. [W.W.B.]

Phase (astronomy) In astronomy, the changing appearance of the Moon, inner planets, and Mars due to the angular difference between the incident light from the Sun and the viewing direction of the observer. Phases of the Moon are a familiar sight. During a lunar month (29.53 Earth days), the Moon completes a cycle of appearances or phases: dark of the moon, or new moon, during which the Moon is nearer the Sun than is the Earth; crescent until first quarter (about a week after new moon); half an illuminated disk, the Moon continues to wax, being gibbous, until it is full; it then wanes through third quarter and completes the cycle as illustrated.

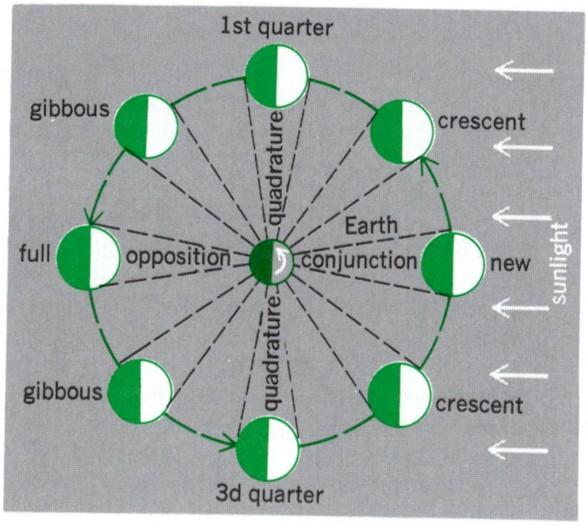

Viewed from Earth, Moon appears differently illuminated as it travels around Earth.

Mercury and Venus show phases like those of the Moon. Mars varies in appearance from full to gibbous. *See* MOON. [G.P.K.]

Phase (periodic phenomena) The fractional part of a period through which the time variable of a periodic quantity (alternating electric current, vibration) has moved, as measured at any point in time from an arbitrary time origin. In the case of a sinusoidally varying quantity, the time origin is usually assumed to be the last point at which the quantity passed through a zero position from a negative to a positive direction.

In comparing the phase relationships at a given instant between two time-varying quantities, the phase of one is usually assumed to be zero, and the phase of the other is described, with respect to the first, as the fractional part of a period through which the second quantity must vary to achieve a zero of its own (see illustration). In this case, the fractional part of the period is usually expressed in terms of angular measure, with one period being equal to 360° or 2π radians. *See* PHASE-ANGLE MEASUREMENT; SINE WAVE. [W.J.G]

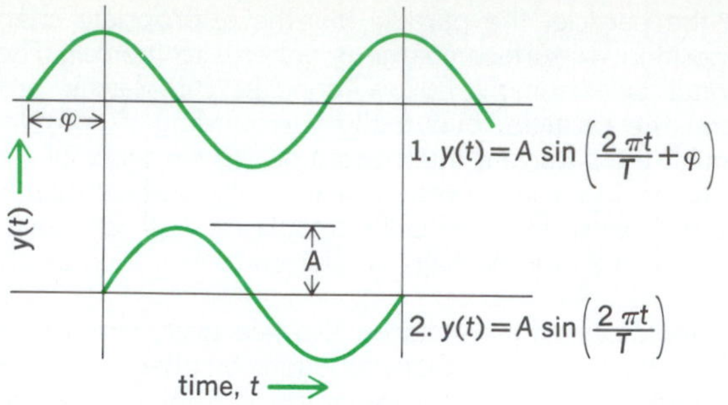

An illustration of the meaning of phase for a sinusoidal wave. The difference in phase between waves 1 and 2 is φ and is called the phase angle. For each wave, A is the amplitude and T is the period.

Phase-angle measurement The determination of the relative times at which alternating currents and voltages in a circuit take on zero values. If two voltages v_1 and v_2 are zero at the same instant, they are in phase, with zero phase difference (or out of phase with 180° difference). If one voltage v_1 passes through zero 1/8 cycle before a second voltage v_2, it leads by 360°/8 or 45°. The common phase meter, a commercial device for determining the angle between current and voltage, can be used when its presence will not disturb the circuits under measurement. When phase angles to be measured are in high-impedance or low-power circuits, this device is unsatisfactory and other measurement methods must be used. *See* PHASE (PERIODIC PHENOMENA); PHASE METER. [H.So./E.C.St.]

Phase-contrast microscope A microscope used for making visible differences in phase or optical path in transparent or reflecting specimens. It is one of the most important instruments available for studying living cells and is widely used in biological and medical research.

The essential features of a phase-contrast microscope are shown in the illustration. The practical problem is to find some way of separating the incident or direct light from that diffracted by the object. This is done by placing a diaphragm D of easily recognizable shape, such as an annulus, at the front focal plane of the substage condenser C. Light from each point of the focal plane passes as a parallel pencil of rays through the specimen S and is brought to a focus at the rear focal plane P of the objective O. Thus, on removing the eyepiece, an image of the annulus will be seen at the back of the objective lens. This image corresponds to the incident light. In addition, when a specimen is present, some light is diffracted by it and spreads out to fill the whole of the back lens of the objective. Thus, apart from the small area of overlap over the image of the annulus, the direct and diffracted waves are essentially separated at the plane P. A phase plate is now inserted at this level. This can be a transparent disk with an annular groove of such dimensions that it coincides exactly with the image of the diaphragm D. All the direct light now passes through the groove in the phase plate, whereas the diffracted light passes mainly outside the groove. Since the diffracted light has to pass through a greater thickness of transparent material than the direct light, a phase difference, depending on the refractive index of the phase-plate material and on the thickness of the groove, is introduced between them. If this phase difference is about one-quarter of a wavelength, the basic conditions for phase contrast will have been achieved. If the phase plate is made to retard the incident wave by a quarter of a wavelength, the crests and troughs of the two waves will coincide, giving a

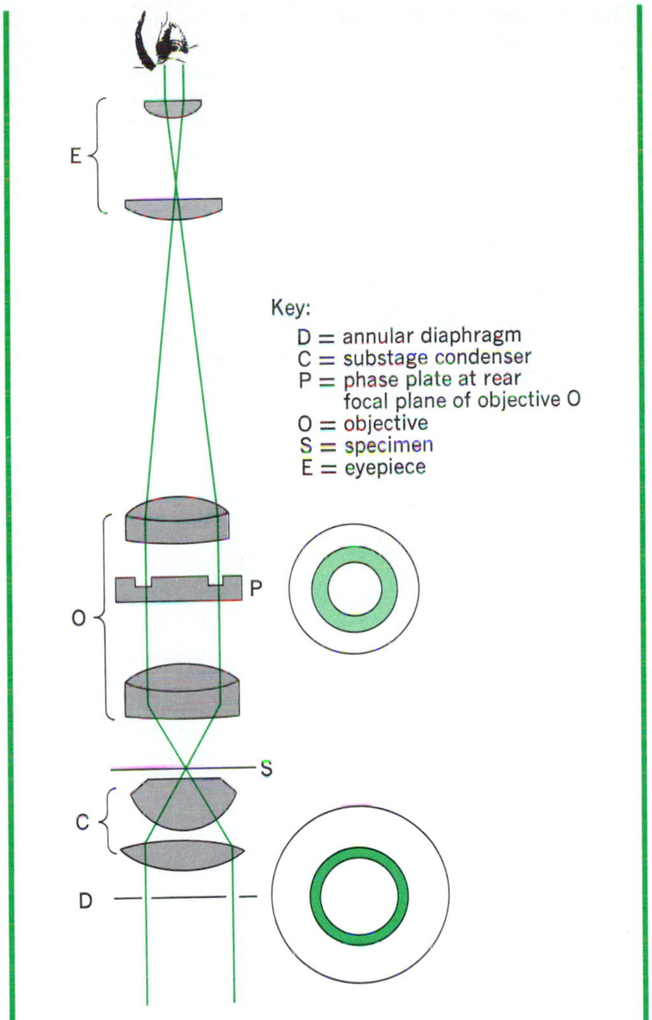

Diagram of a phase-contrast microscope.

resultant of greater amplitude. Refractile details will appear bright (negative contrast) instead of dark (positive contrast).

The phase-contrast microscope is the routine instrument for the examination of living cells because it is possible to study the cell structure under excellent optical conditions and with no loss in resolving power. The method is also useful for the study of unstained tissue sections and has found considerable use for the comparison of material in the electron and optical microscopes. *See* ELECTRON MICROSCOPE; MICROSCOPE; OPTICAL MICROSCOPE. [R.Ba.]

Phase equilibrium A general field of physical chemistry dealing with the various situations in which two or more phases (or states of aggregation) can coexist in thermodynamic equilibrium with each other, with the nature of the transitions between phases, and with the effects of temperature and pressure upon these equilibria. Many superficial aspects of the subject are largely qualitative, for example, the empirical classification of types of phase diagrams; but the basic problems always are susceptible to quantitative thermodynamic treatment, and in many cases, statistical thermodynamic methods can be applied to simple molecular models.

Thermodynamics requires that when two phases, α and β, are free to exchange heat, mechanical work, and matter (chemical species), the temperature T, the pressure P, and the chemical potential (partial molar free energy) μ_i of each particular component i must be equal in both phases at equilibrium. Algebraically, equilibrium exists when $T_\alpha = T_\beta$, $P_\alpha = P_\beta$, $\mu_{i,\alpha} = \mu_{i,\beta}$, and $\mu_{j,\alpha} = \mu_{j,\beta}$.

These conditions of thermal, mechanical, and material equilibrium need not all be present if the equilibrium between phases is subject to inhibiting restrictions. Thus, for a solution of a nonvolatile solute in equilibrium with the solvent vapor, the condition of equality of solute chemical potentials $\mu_{2,\alpha} = \mu_{2,\beta}$ need not apply, since there can be no solute molecules in the vapor phase. Similarly, in osmotic equilibria, in which solvent molecules can pass through a semipermeable membrane, whereas solute molecules cannot, $\mu_{1,\alpha} = \mu_{1,\beta}$ and $T_{1,\alpha} = T_{2,\beta}$, but the solute chemical potentials μ_2 are unequal, as are the pressures on opposite sides of the membrane. *See* OSMOSIS; SOLUTION.

If a system consists of P phases and C distinguishable components, there are $C + 2$ thermodynamic variables (C chemical potentials μ_i, plus the temperature and pressure) which are interrelated by an equation for each phase. Since there are P independent equations relating the $C + 2$ variables, only $F = C + 2 - P$ variables need be fixed to define completely the state of the system at equilibrium; the other variables are then beyond control. This relation for the number of degrees of freedom F, or variance, is called the phase rule. It has proved to be a powerful tool in interpreting and classifying types of phase equilibria.

When chemical changes may occur in the system, the number of components C is the number of independent components whose amounts can be varied by the experimenter; this is equal to the total number of chemical species present less the number of independent chemical equilibria between them.

An invariant system has no degrees of freedom ($F = 0$), for which the number of phases $P = C + 2$. For a one-component system, such an invariant point is a triple point at which three phases coexist at a single temperature and pressure only; for a two-component system, a quadruple point (four phases) would be invariant. *See* TRIPLE POINT. [R.L.S.]

Phase inverter A circuit having the primary function of changing the phase of a signal by 180°. The phase inverter is most commonly employed as the input stage for a push-pull amplifier. Therefore, the phase inverter must supply two voltages of equal magnitude and 180° phase difference. A variety of circuits are available for the phase inversion. *See* PUSH-PULL AMPLIFIER.

Overall fidelity of a phase inverter and push-pull amplifier can be adversely affected by improper design of the phase inverter. The principal design requirement is that frequency response of one input channel to the push-pull amplifier be identical to the frequency response of the other channel.

The simplest form of phase-inverter circuit is a transformer with a center-tapped secondary. Careful design of the transformer assures that the secondary voltages are equal. The transformer forms a good inverter when the inverter must sup-

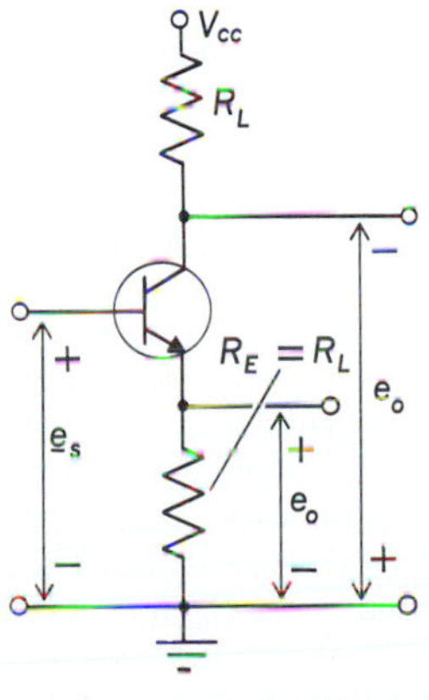

Single-transistor inverter. e_s = signal voltage; e_o = output voltage; V_{cc} = collector supply voltage; R_L = load resistance; R_E = emitter resistance.

ply power to the input of the push-pull amplifier. However, it usually costs more, occupies more space, and weighs more than a transistor or vacuum-tube circuit, and its frequency response may not be as uniform as that which can be obtained from solid-state or vacuum-tube circuits.

An amplifier that provides two equal output signals 180° out of phase is called a paraphase amplifier. If coupling capacitors can be omitted, the simplest paraphase amplifier is shown in the illustration. Approximately the same current flows through R_L and RE, and therefore if R_L and R_E are equal, the ac output voltages from the collector and from the emitter are equal in magnitude and 180° out of phase. *See* PHASE (PERIODIC PHENOMENA); PHASE-ANGLE MEASUREMENT. [H.F.K.]

Phase-locked loops Electronic circuits for locking an oscillator in phase with an arbitrary input signal. A phase-locked loop (PLL) is used in two fundamentally different ways: (1) as a demodulator, where it is employed to follow (and demodulate) frequency or phase modulation, and (2) to track a carrier or synchronizing signal which may vary in frequency with time. When operating as a demodulator, the PLL may be thought of as a matched filter operating as a coherent detector. When used to track a carrier, it may be thought of as a narrow-band filter for removing noise from the signal and regenerating a clean replica of the signal. *See* DEMODULATOR; ELECTRIC FILTER.

The basic components of a phase-locked loop are shown in the illustration. The input signal is a sine or square wave of

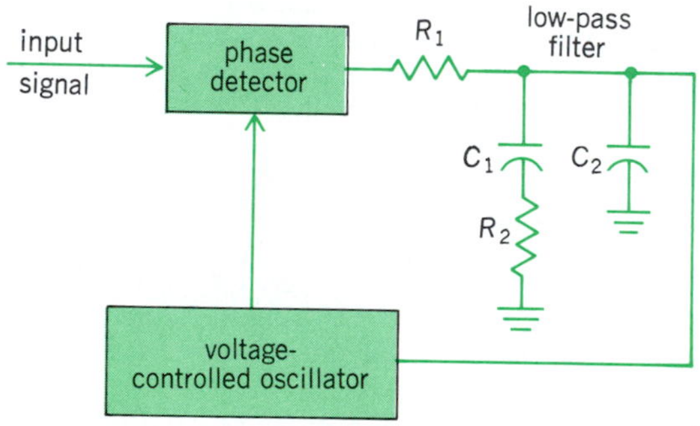

Phase-locked loop. R_1 and R_2 are resistors; C_1 and C_2 are capacitors.

arbitrary frequency. The voltage-controlled oscillator (VCO) output signal is a sine or square wave of the same frequency as the input, but the phase angle between the two is arbitrary. The output of the phase detector consists of a direct-current (dc) term, and components of the input frequency and its harmonics. The low-pass filter removes all alternating-current (ac) components, leaving the dc component, the magnitude of which is a function of the phase angle between the VCO signal and the input signal. If the frequency of the input signal changes, a change in phase angle between these signals will produce a change in the dc control voltage in such a manner as to vary the frequency of the VCO to track the frequency of the input signal.

The most widespread use of phase-locked loops is undoubtedly in television receivers. Synchronization of the horizontal oscillator to the transmitted sync pulses is universally accomplished with a PLL. The color reference oscillator is often synchronized with a phase-locked loop. Phase-locked loops are also used as frequency demodulators. They have been applied to stereo decoders made on silicon monolithic integrated circuits. High-performance amplitude demodulators may be built using phase-lock techniques. *See* AMPLITUDE-MODULATION DETECTOR. [T.B.M.]

Phase meter An instrument for the measurement of electrical phase angles. It is sometimes called the crossed-coil meter or the Tuma phase meter and is the basic element of power-factor meters and synchroscopes. When used for power-factor indication, it contains two movable coils A and B on a common shaft, as shown in the illustration. The two coils

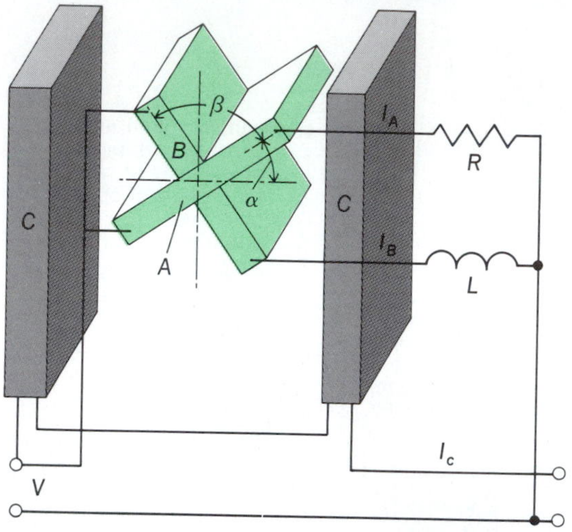

Diagram of the Tuma phase meter. I_A, I_B, and I_C are the currents flowing through coils *A*, *B*, and *C* respectively; *R* is a resistance; *L* is an inductance; *V* is a voltage.

move as a unit, the angle ß between them remaining fixed. There is no restraining spring acting on the shaft and the system will turn as long as currents produce an average nonzero torque.

The meter contains a fixed coil C, which carries the load current. The fixed coil is made in two sections so that its magnetic field will be nearly uniform in the neighborhood of the movable coils.

Suppose the current in coil B leads the current in fixed coil C by the angle θ, and the current in coil B lags the current in coil A by the angle ϕ. It can be shown that if ß equals ϕ, then the shaft will turn (α will vary) until α equals θ and the phase meter indicates directly the phase angle between the sources supplying coil C and the crossed coils. A change of 5° in phase angle produces a deflection of 5° on the phase meter. *See* PHASE-ANGLE MEASUREMENT; POWER-FACTOR METER. [H.So./E.C.St.]

Phase modulation A special kind of angle modulation in which the linearly increasing angle of a sine-wave carrier has added to it a phase angle that is proportional to the instantaneous value of the modulating wave (message to be communicated). Phase modulation (PM) is a scheme for impressing the message to be communicated upon a high-frequency sine-wave carrier. There is a direct proportionality between the message to be communicated and the phase variations imparted to the modulated wave propagated to the receiver. For basic concepts, technical terms, and supplementary information *see* MODULATION. *See also* ANGLE MODULATION.

Like other forms of angle modulation, PM reduces noise at the cost of extra bandwidth occupancy, transmits constant average signal power, transmits constant peak power equal to twice the average signal power, and has a channel-grabbing property whereby, if two signals reach the PM detector, the larger signal is accepted to the near exclusion of the smaller.

Important applications include certain types of telegraph, telemetering, and data-processing systems. PM is used in certain microwave radio relay systems, some of which carry telephone conversations and television programs simultaneously. PM techniques are used in many types of measuring and control systems.

The sharp limitation of range (channel grabbing) is especially important for some services. Typical examples using PM include mobile and fixed radio systems for police, airway, and military applications. *See* PHASE-MODULATION DETECTOR; PHASE MODULATOR. [H.S.Bl.]

Phase-modulation detector A device which recovers or detects the modulating signal from a phase-modulated carder. Any frequency-modulation (FM) detector with minor modifications will detect phase-modulated waves. *See* FREQUENCY-MODULATION DETECTOR; PHASE MODULATION.

The only difference between FM and phase modulation (PM) is the manner in which the modulation index varies with the modulating frequency. The modulation index is independent of the modulating frequency in PM but is inversely proportional to the modulating frequency in FM. Therefore an FM detector, when used to detect a phase-modulated wave, produces an output voltage which is proportional to the modulating frequency, assuming the original modulating signal to be of constant amplitude. Consequently, a low-pass filter with a single reactive element, such as an *RC* (resistance-capacitance) filter, is needed in the output of the FM detector which is used to detect a phase-modulated wave. [C.L.A.]

Phase modulator An electronic circuit that causes the phase angle of the modulated wave to vary (with respect to the unmodulated carrier) in accordance with the modulating signal. Since frequency is the rate of change of phase, a phase modulator will produce the characteristics of frequency modulation (FM) if the frequency characteristics of the modulating signal are so altered that the modulating voltage is inversely proportional to frequency. Commercial FM transmitters normally employ a phase modulator because a crystal-controlled oscillator can then be used to meet the strict carrier-frequency control requirements of the Federal Communications Commission. The chief disadvantage of phase modulators is that they generally produce insufficient frequency-deviation ratios, or modulation index, for satisfactory noise suppression. Frequency multiplication can be used, however, to increase the modulation index to the desired value, since the frequency deviation is multiplied along with the carrier frequency. *See* FREQUENCY MODULATION; PHASE MODULATION; PHASE-MODULATION DETECTOR.

Many types of phase modulators have been devised. A simple modulator is shown in the illustration. In this circuit the modulating voltage changes the capacitance of the varactor diode. The phase shift depends upon the relative magnitudes of the capacitive reactance of the varactor diode and the load resistance *R*. Therefore the phase shift varies with the modulating voltage and phase modulation (PM) is accomplished. However, the phase shift is not linearly related to the capacitance and the capacitance of the varactor diode is not linearly related to the modulating voltage. *See* VARACTOR.

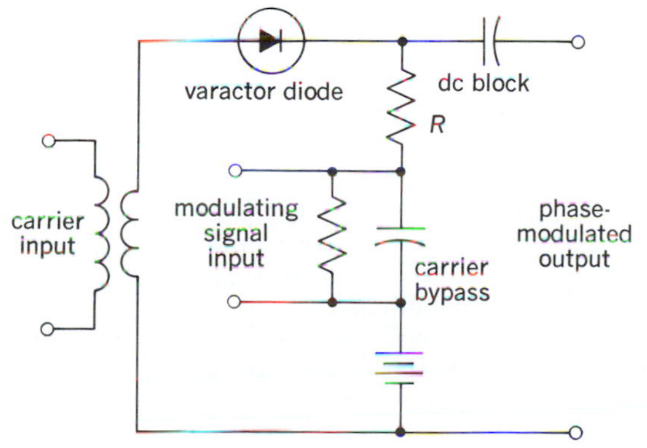

A simple phase modulator.

In one system of phase modulation the phase shift of the carrier is obtained by means of a special electron tube, the phasitron. The principal difficulty with normal phase modulators is a result of the relatively small amount of phase shift that can be produced without introducing nonlinearities and the resultant large amount of frequency multiplication which is usually needed. This difficulty is alleviated to a large degree by means of the phasitron in which a phase swing of as much as ± 200° can be obtained with low distortion. [C.L.A.]

Phase rule A relationship used to determine the number of state variables *F*, usually chosen from among temperature, pressure, and species compositions in each phase, which must be specified to fix the thermodynamic state of a system in equilibrium. It was derived by J. Willard Gibbs. The phase rule (in the absence of electric, magnetic, and gravitational phenomena) is given by the equation below, where *C* is the number of

$$F = C - P - M + 2$$

chemical species present at equilibrium, *P* is the number of phases, and *M* is the number of independent chemical reactions. Here phase is used to indicate a homogeneous, mechanically separable portion of the system, and the term independent reactions refers to the smallest number of chemical reactions which, upon forming various linear combinations, includes all reactions which occur among the species present. The number of independent state variables *F* is referred to as the degrees of freedom or variance of the system. *See* CHEMICAL EQUILIBRIUM; CHEMICAL THERMODYNAMICS; PHASE EQUILIBRIUM; THERMODYNAMIC PROCESSES. [S.I.S.]

Phase-transfer catalysis A process in which the rate of a reaction occurring in a two-phase organic-water system is enhanced by addition of a compound that helps transfer the water soluble reactant across the interface to the organic phase.

An important factor, which contributes to the slowness of many organic reactions, is the lack of homogeneity of the reaction mixture. This is particularly the case with nucleophilic substitution reactions such as the reaction below, where RX is

$$RX + Nu^- \rightarrow RNu + X^-$$

an organic reagent and Nu^+ is the nucleophilic reagent, The nucleophilic reagent is frequently an inorganic anion, which is soluble in water in which the organic substrate is insoluble, but is insoluble in the organic phase. The encounter rate between Nu^- and RX is consequently low, as they can only meet at the interface of the heterogeneous system. The water-soluble anion is also frequently highly solvated by water molecules, which stabilize the anion and thus reduce its nucleophilic reactivity. These problems have been overcome in the past by the use of polar aprotic solvents, which will dissolve both the organic and inorganic reagents, or by the use of homogeneous mixed-solvent systems, such as water:ethanol or water:dioxan. *See* ELECTROPHILIC AND NUCLEOPHILIC REAGENTS.

Phase-transfer catalysis involves the transportation of the inorganic anion, Nu^-, from the aqueous phase into the organic phase by the formation of a nonsolvated ion-pair with a cationic phase-transfer catalyst, Q^+. With highly lipophilic catalysts, the reactive ion pair $[Q^+Nu^-]$ is formed at the interface between the aqueous and organic phases, followed by rapid transportation into the bulk of the organic phase. The rate of the reaction is enhanced, as the encounter rate of the nucleophile, Nu^-, with the organic reagent, RX, in the single phase will be significantly higher than at the interface, Moreover, as the anion is transferred without water of solvation, its nucleophilic reactivity can be considerably higher in the organic phase than in the aqueous phase. Rate enhancements of greater than 10^7 have thus been observed. *See* Catalysis; Crown ethers; Heterogeneous catalysis; Homogeneous catalysis; Quaternary ammonium salts; Stereochemistry. [R.A.J.]

Phase transitions Changes of state brought about by a change in an intensive variable (for example, temperature or pressure) of a system. Some familiar examples of phase transitions are the gas-liquid transition (condensation), the liquid-solid transition (freezing), the normal-to-superconducting transition in electrical conductors, the paramagnet-to-ferromagnet transition in magnetic materials, and the superfluid transition in liquid helium. Further examples include transitions involving amorphous or glassy structures, spin glasses, charge-density waves, and spin-density waves. *See* Amorphous solid; Charge-density wave; Metallic glasses; Spin-density wave; Spin glass; Superconductivity; Superfluidity.

Typically the phase transition is brought about by a change in the temperature of the system. The temperature at which the change of state occurs is called transition temperature (usually denoted by T_c). For example, the liquid-solid transition occurs at the freezing point.

The two phases above and below the phase transition can be distinguished from each other in terms of some ordering that takes place in the phase below the transition temperature. For example, in the liquid-solid transition, the molecules of the liquid get "ordered" in space when they form the solid phase. In a paramagnet, the magnetic moments on the individual atoms can point in any direction (in the absence of an internal magnetic field), but in the ferromagnetic phase the moments are lined up along a particular direction, which is then the direction of ordering. Thus in the phase above the transition, the degree of ordering is smaller than in the phase below the transition. One measure of the amount of disorder in a system is its entropy, which is the negative of the first derivative of the thermodynamic free energy with respect to temperature. When a system possesses more order, the entropy is lower. Thus at the transition temperature the entropy of the system changes from a higher value above the transition to some lower value below the transition. *See* Entropy; Ferromagnetism; Paramagnetism.

This change in entropy can be continuous or discontinuous at the transition temperature. In other words, the development of order in the system at the transition temperature can be gradual or abrupt. This leads to a convenient classification of phase transitions into two types, namely, discontinuous and continuous.

Discontinuous transitions involve a discontinuous change in the entropy at the transition temperature. A familiar example of this type of transition is the freezing of water into ice. As water reaches the freezing point, order develops without any change in temperature. Thus there is a discontinuous decrease in the entropy at the freezing point. This is characterized by the amount of latent heat that must be extracted from the water for it to be "ordered" into the solid phase (ice). Discontinuous transitions are also called first-order transitions.

In a continuous transition, entropy changes continuously, and hence the growth of order below T_c is also continuous. There is no latent heat involved in a continuous transition. Continuous transitions are also called second-order transitions. The paramagnet-to-ferromagnet transition in magnetic materials is an example of such a transition.

The degree of ordering in a system undergoing a phase transition can be made quantitative in terms of an order parameter. At temperatures above the transition temperature the order parameter has a value zero, and below the transition it acquires some nonzero value. For example, in a ferromagnet the order parameter is the magnetic moment per unit volume (in the absence of an externally applied magnetic field). It is zero in the paramagnetic state since the individual magnetic moments in the solid may point in any random direction. Below the transition temperature, however, there exists a preferred direction of ordering, and as the temperature is lowered below T_c, more and more individual magnetic moments start to align along the preferred direction of ordering, leading to a continuous growth of the magnetization or the macroscopic magnetic moment per unit volume in the ferromagnetic state. Thus the order parameter changes continuously from zero above to some nonzero value below the transition temperature. In a first-order transition, the order parameter would change discontinuously at the transition temperature. [D.J.S.; S.J.]

Phase velocity The velocity of propagation of a pure sine wave of infinite extent. In one dimension, for example, the form of the disturbance for such a wave is given by $y(x,t) = A \sin [2\pi(x/\lambda - t/T)]$. Here x is the position at which the disturbance $y(x,t)$ exists at time t, λ is the wavelength, T is the period which is related to the wave frequency by $T = 1/f$, and A is the disturbance amplitude. The argument of the sine function is called the phase. The phase velocity is the speed with which a point of constant phase can be said to move. Thus $x/\lambda - ft$ = constant, so the phase velocity υ_p is given by $dx/dt = \upsilon_p = \lambda f$. This is the basic relationship connecting phase velocity, wavelength, and frequency. *See* Phase (periodic phenomena); Sine wave; Wave motion.

The phase velocity for waves in a medium is determined in part by intrinsic properties of the medium. For all mechanical waves in elastic media, the square of the phase velocity is proportional to the ratio of the appropriate elastic property of the medium to the appropriate inertia property. The phase velocity of electromagnetic waves depends upon the medium as well. In vacuum, the phase velocity c is given by $c^2 = 1/\epsilon_0\mu_0 \approx 9 \times 10^{16}$ m^2/s^2, where ϵ_0 and μ_0 are respectively the permittivity and permeability of the vacuum. Phase velocity may also depend upon the mode of wave propagation—in general, upon the frequency of the wave. Waves of different frequencies will travel at different speeds, resulting in a phenomenon called dispersion. *See* Electromagnetic radiation; Light; Wave equation; Young's modulus. [S.A.Wi.]

Phenanthrene Colorless, crystalline hydrocarbon ($C_{12}H_{10}$) which melts at about 100°C (212°F) and boils at 332°C (630°F). Phenanthrene (I) is usually obtained from coal tar, but it may also be produced by the hydrogenation of coal.

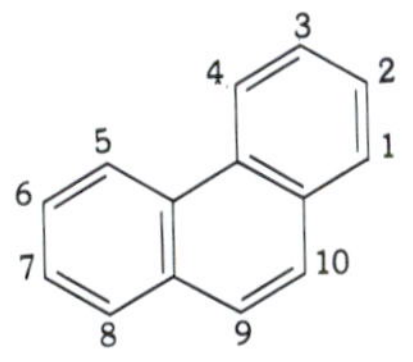

(I)

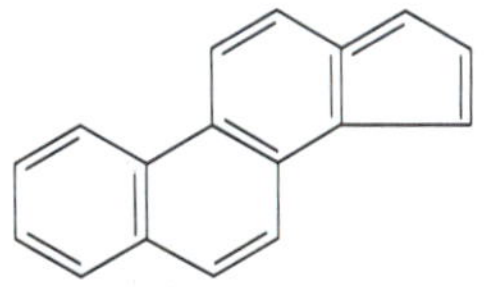

(II)

Phenanthrene has little commercial importance, except that it may be hydrogenated to symmetrical octahydrophenanthrene, which in turn may be rearranged and dehydrogenated to anthracene. Phenanthrene is of interest because the nucleus is found in some resin acids and is produced by the degradation of certain alkaloids. Reduction products of 1,2-cyclopentenophenanthrene (II) may be regarded as forming the skeleton of the steroids. *See* Aromatic hydrocarbon; Polynuclear hydrocarbon; Sterol. [C.K.B.]

Phenocryst A relatively large crystal embedded in a finer-grained or glassy igneous rock. The presence of phenocrysts gives the rock a porphyritic texture. Phenocrysts are represented most commonly by feldspar, quartz, biotite, hornblende, pyroxene, and olivine. Strictly speaking, phenocrysts crystallize from molten rock material (lava or magma). They commonly represent an earlier and slower stage of crystallization than does the matrix in which they are embedded. *See* IGNEOUS ROCKS; PORPHYROBLAST. [C.A.C.]

Phenol The simplest member of a class of organic compounds possessing a hydroxyl group attached to a benzene ring or to a more complex aromatic ring system. Phenol itself, C_6H_5OH, may also be called hydroxybenzene or carbolic acid. Pure phenol is a colorless solid melting at 42°C (108°F), moderately soluble in water, and weakly acidic (pK 9.9).

Phenol has broad biocidal properties, and dilute aqueous solutions have long been used as an antiseptic. At higher concentrations phenol causes severe skin burns; it is a violent systemic poison. *See* ANTISEPTIC.

Phenol has the structure shown. Simple substituted phenols, such as the three isomeric chlorophenols, are named as indicated, using the ortho (*o*), meta (*m*), and para (*p*) prefixes. In more highly substituted phenols the positions of substitution are indicated by numbers (as in 2,4-dichlorophenol). Compounds with more than one hydroxyl group per aromatic ring are known as polyhydric phenols, and include catechol, resorcinol, hydroquinone, phloroglucinol, and pyrogallol.

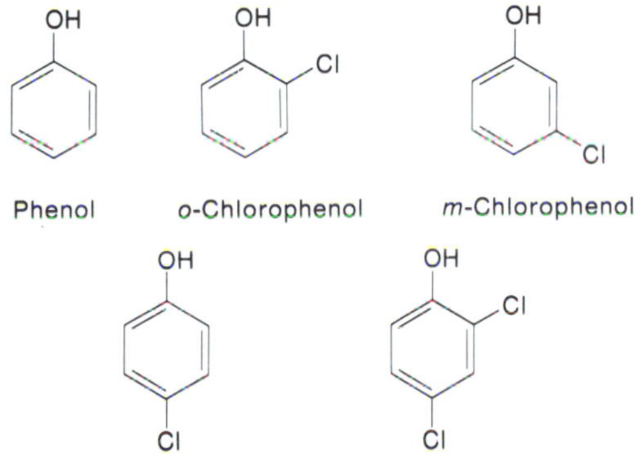

Phenol, *o*-Chlorophenol, *m*-Chlorophenol, *p*-Chlorophenol, 2,4,-Dichlorophenol

Until World War I phenol was essentially a natural coal tar product. However, synthetic methods have replaced extraction from natural sources. There are many possible syntheses.

Phenol is one of the most versatile and important industrial organic chemicals. It is the starting point for many diverse products used in the home and industry. A partial list includes: nylon, epoxy resins, surface active agents, synthetic detergents, plasticizers, antioxidants, lube oil additives, phenolic resins (with formaldehyde, furfural, and so on), polyurethanes, aspirin, dyes, wood preservatives, herbicides, drugs, fungicides, gasoline additives, inhibitors, explosives, and pesticides. *See* CATECHOL; CRESOL; HYDROQUINONE; PHENOLIC RESIN; PICRIC ACID; RESORCINOL. [R.I.S.; M.St.]

Phenolic resin One of the condensation products of phenols or phenolic derivatives with aldehydes such as formaldehyde and furfural. The phenol-formaldehyde resins, developed commercially between 1905 and 1910, were the first truly synthetic polymers and have found wide usage. They are characterized by low cost, dimensional stability, high strength, and resistance to aging.

Phenolic resins can be cast from syrupy intermediates or molded from B-stage solid resins. Laminated products can be produced by impregnating fiber, cloth, wood, and other materials with the resin. An important type of phenolic resin product is rigid foam. Cured phenolic plastics are rigid, hard, and resistant to chemicals (except strong alkali) and to heat.

Some of the uses for phenolic resins are for making precisely molded articles, such as telephone parts, for manufacturing strong and durable laminated boards, or for impregnating fabrics, wood, or paper. Phenolic resins are also widely used as adhesives, as the binder for grinding wheels, as thermal insulation panels, as ion-exchange resins, and in paints and varnishes. *See* ADHESIVE; EPOXY RESIN; ION EXCHANGE; PHENOL; PLASTICS PROCESSING; POLYMERIZATION. [J.A.M.]

Phenylalanine An amino acid considered essential for normal growth of animals. Dietary phenylalanine is the source of tyrosine in animal tissues. Phenylalanine originates biosynthetically from phosphoenolpyruvic acid and D-erythrose-4-phosphate by way of shikimic acid and prephenic acid. *See* AMINO ACIDS; TYROSINE. [E.A.Ad.]

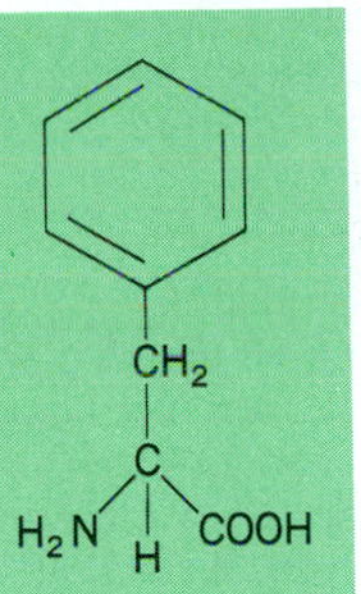

Phenylalanine

Phenylketonuria An inborn error of metabolism in which affected individuals lack the liver enzyme needed to metabolize phenylalanine, an amino acid essential for normal growth and development. If untreated, affected individuals may become severely mentally retarded, become microcephalic, have behavioral problems, develop epilepsy, or show other signs of neurological impairment. Phenylketonuria (PKU) is inherited as an autosomal recessive trait and is found in all ethnic groups but most frequently in individuals of northern European descent. Its incidence is about 1 per 14,000 births in the United States. *See* PHENYLALANINE.

Newborn screening programs for phenylketonuria have been successful in identifying most cases within a few weeks of birth. A low phenylalanine diet with restriction of proteins, if initiated early in infancy, can prevent the development of severe intellectual and neurological handicaps that would otherwise occur in virtually all untreated cases. *See* MENTAL DEFICIENCY; PROTEIN METABOLISM. [F.delaC.]

Pheromone A substance that acts as a molecular messenger, transmitting information from one member of a species to another member of the same species. A distinction is made between releaser pheromones, which elicit a rapid, behavioral response, and primer pheromones, which elicit a slower, developmental response and may pave the way for a future behavior.

Communication via pheromones is common throughout nature, including some eukaryotic microorganisms such as fungi that exchange vital chemical signals. The cellular slime molds form large aggregations of amebas which unite to form a sorocarp made up of a long, slender stalk that supports a spore-containing fruiting body. A pheromone is responsible for the aggregation. In several species of algae, relatively simple hydrocarbons act as sperm attractants.

By far the largest number of characterized pheromones come from insect species. In social insects, such as termites and ants, there may be as many as a dozen different types of messages that are used to coordinate the complex activities which must be carried out to maintain a healthy colony. These activities might require specialized pheromones such as trail pheromones (to lead to a food source), alarm pheromones (recruiting soldiers to the site of an enemy attack), or pheromones connected with reproductive behavior. Much less is known about mammalian pheromones because mammalian behavior is more difficult to study. There are, however, a small number of well-characterized mammalian pheromones from pigs, dogs, hamsters, mice, and marmosets.

There is great potential for controlling the behavior of a given species by manipulating its natural chemical signals. For example, pheromones have been used to disrupt the reproduction of certain insect pests. This approach can lead to reduced use of pesticides as well as advances in the control of both agricultural pests and disease vectors. *See* CHEMICAL ECOLOGY; CHEMORECEPTION; INSECT CONTROL, BIOLOGICAL; SOCIAL INSECTS. [J.Mein.]

Phlebitis An inflammatory process in the wall of a vein. Infection may come from a distance and gain a foothold in the vessel wall, or probably more commonly, veins become infected secondarily from continuous inflamed tissues. When a larger vein is infected, thrombosis or intravascular clotting is likely to occur which results in partial or complete occlusion of the vessel. Inflamed veins predispose to thrombosis because the damaged walls lose their smooth lining and become roughened, a condition which allows the deposition of clot material from the blood. Thrombi either become organized and recanalized or are dislodged to form emboli. Dislodged venous thrombi or emboli pass through the right side of the heart and on to the lungs. Small emboli plug pulmonary arteries, cause anoxia, and may result in necrosis or infarction of a region of the lung. Large emboli may result in massive pulmonary embolism with obstruction of several major pulmonary arteries, causing sudden death. *See* EMBOLUS; THROMBOSIS. [F.A.C.]

Phlebotomus fever A mild, insect-borne virus disease of humans occurring commonly in Mediterranean countries and in the Soviet Union, China, and India. It is also known as sandfly fever.

After the bite of an infected female sandfly (in the case of the Middle East members of the group), the person develops headache, malaise, conjunctivitis, nausea, pain, and stiffness. [J.L.Me.]

Phloem The principal food-conducting tissue in vascular plants. Its conducting cells are known as sieve elements, but phloem may also include companion cells, parenchyma cells, fibers, sclereids, rays, and certain other cells. As a vascular tissue, phloem is spatially associated with xylem, and the two together form the vascular system. *See* XYLEM.

Sieve elements differ from phloem parenchyma cells in the structure of their walls and to some extent in the character of their protoplasts. Sieve areas, distinctive structures in sieve element walls, are specialized primary pit fields in which there may be numerous modified plasmodesmata. Plasmodesmata are strands of cytoplasm connecting the protoplasts of two contiguous cells. These strands are often surrounded by callose, a carbohydrate material, that appears to form rapidly in plants when they are placed under stress.

Companion cells are specialized parenchyma cells that occur in close ontogenetic and physiologic association with sieve tube members. Some sieve-tube members lack companion cells. The precise functional relationship between these two kinds of cells is unknown.

Parenchyma cells in the phloem occur singly or in strands of two or more cells. They store starch, frequently contain tannins or crystals, commonly enlarge as the sieve elements become obliterated, or may be transformed into sclereids or cork cambium cells.

Phloem fibers vary greatly in length (from less than 0.04 in. or 1 mm in some plants to 20 in. or 50 cm in the ramie plant). The secondary walls are commonly thick and typically have simple pits, but may or may not be lignified. [M.A.W.]

Phlogopite A mineral of the mica group, also called bronze mica. Its composition is $K_2[Mg,Fe(II)]_6(Si_6,Al_2)O_{20}$-$(OH)_4$, including minor amounts of sodium, Na, that substitute for potassium, K, and containing small amounts of Mn, Fe(III), and Ti. With an increase in Fe(II), it grades into biotite from which there is no sharp distinction. It is the Mg-rich half of the biotite-phlogopite series known to the mica industry as amber mica. Phlogopite is widely used as an electrical insulator.

It occurs in disseminated flakes, foliated masses, or large crystals. The specific gravity is 2.8–3.0, and hardness is 2.5–3.0 on Mohs scale. Thin sheets are transparent in shades of light brown and green. Phlogopite occurs chiefly in certain peridotites (kimberlites), in carbonatites, in serpentinized peridotites, in marbles derived from impure dolomitic limestones, and as very large crystals of commercial importance in coarse-grained plagioclase-apatite-calcite-pyroxene rocks of pegmatitic affinity. It alters to vermiculite. *See* MICA; SILICATE MINERALS. [E.W.H.]

Phobic reaction A type of neurosis comprising intense fear whose irrationality the individual may realize without being able to dispel the fear itself. The specific forms which phobic reactions may take are varied. These have often been labeled to render them into a medically acceptable terminology—for example, nyctophobia, which is morbid fear of darkness; ocholophobia, fear of crowds; zoophobia, fear of animals; claustrophobia, fear of being in a confined space; agoraphobia, fear of open places; and hydrophobia, fear of water. These specific fears may be considered neurotic symptoms rather than relatively discrete patterns related to particular psychodynamic foundations and may occur within the context of broader patterns of maladaptive reaction. The origins of phobic reactions also tend to be quite varied; sometimes they originate in a generalized fear of a class of objects, one member of which initially caused a pain reaction. For example, when bitten by a dog, a child may learn morbidly to fear all dogs. Phobias sometimes result in a more complex, symbolic, and distorted displacement, as when a fear of sexual penetration leads to a fear of all sharp objects. *See* NEUROTIC DISORDERS. [J.S.Br.; W.Mis.]

Phoenicopteriformes The flamingos, a small monotypic order of wading birds that includes the family Phoenicopteridae, which has six species found worldwide in tropical marine and fresh waters; one species lives in the high Andes. The flamingos were formally included in the Ciconiiformes and are still placed there by many researchers; others have advocated a close relationship with either the Anseriformes or the Charadriiformes. None of these classifications, however, are strongly supported by available evidence, and so it is best to place these bizarre birds in a unique order. *See* ANSERIFORMES; CHARADRIIFORMES; CICONIIFORMES.

The earliest definite flamingo fossil is *Juncitarsus*, from the middle Eocene of Wyoming; it possesses several primitive traits for the family. Beginning in the late Oligocene, modern flamingos are found in the fossil records from all areas of the world.

Flamingos are long-legged, long-necked wading birds with a thick bill which is sharply bent downward at the midpoint. The three anterior toes are webbed, possibly for walking on soft substrates. The long and broad wings enable the bird to fly well. The adult plumage is pink to light red, with black flight feathers. Flamingos are gregarious, often congregative in flocks in excess of 1 million. They breed monogamously in large colonies, and both parents incubate the one or two eggs in a nest formed of a stout pillar of mud 1 ft (0.3 m) high on a mud flat. *See* AVES. [W.J.B.]

Pholidophoriformes An extinct actinopterygian group composed of mostly small fusiform fishes of an advanced holostean level that are found in both marine and fresh-water deposits and range from the Middle Triassic to the Lower Cretaceous. The best-known representative is *Pholidophorus bechei* (see illustration), which exhibits holostean features in its

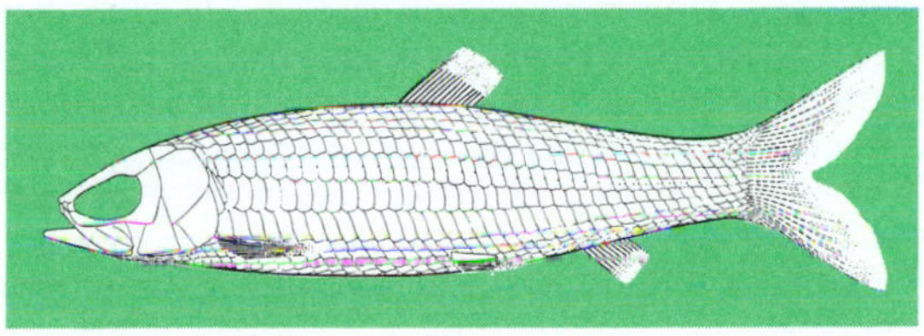

Pholidophorus bechei **Lower Jurassic of England; length to 8 in. (20 cm). (*After D. Rayner, The structure of certain Jurassic holostean fishes with special reference to their neurocrania, Phil. Trans. Roy. Soc. London, Ser. B, no. 601, Cambridge University Press, 1948*)**

enamel-covered ganoid scales, fin rays, and head bones; in the fulcra bordering all the fins with a strong series on the upper caudal lobe; and in the structure of its caudal skeleton. *See* ACTINOPTERYGII; HOLOSTEI; TELEOSTEI. [T.M.C.]

Pholidota An order of mammals comprising the living pangolins, or scaly anteaters, and their poorly known fossil predecessors. All living pangolins are assigned to the genus *Manis*. They are found in Africa south of the Sahara and in southeastern Asia, including certain islands of the East Indies.

Pangolins feed principally on termites and ants. The elongate tubular skull without teeth, long protrusive tongue, small eyes with heavy eyelids, thick skin, strong legs, five-toed feet with large claws, and large tail enable these unique animals to rip open ant nests and termite dens and devour the animals therein. The greatest peculiarity of animals in the genus *Manis* is a covering of all but the undersides of the body by an armor of large imbricating dermal horny scales. Living pangolins are frequently characterized as being animated pine cones. The position and number of hairs in relation to the scales are peculiar to each modern species. *See* MAMMALIA. [D.E.S.]

Phonetics The science that deals with the production, transmission, and perception of spoken language. At each level, phonetics overlaps with some other sciences, such as anatomy, physiology, acoustics, psychology, and linguistics. In each case, phonetics focuses on phenomena relevant to the study of spoken language.

Speech is normally produced by exhaling air from the lungs through the vocal tract. The vocal tract extends from the larynx through the pharynx and the oral cavity to the lips. If the velum (soft palate) is not raised, the air also passes through the nasal cavities. The shape and size of the oral cavity can be varied by the movement of active articulators: tongue, lips, and velum. *See* PALATE.

Phoneticians usually describe speech sounds with reference to their point (or place) of articulation and their manner of articulation. The point of articulation of a sound is the place of maximum constriction within the vocal tract. The great majority of sounds are produced by moving some part of the tongue toward some region on the roof of the mouth. Exceptions are articulations involving lips and those sounds in which the vocal folds serve as articulators.

At most of these points of articulation, sounds can be produced with several manners of articulation. One way to classify manners of articulation refers to the degree of stricture employed in producing the sound. Sounds produced with complete constriction of the vocal tract are stops, or plosives. If the closure is incomplete, but the articulators are brought close enough so that the air passing between them is set into turbulent motion, the resultant sounds are fricatives or spirants. If the articulators are approximated, but the constriction remains large enough so that air can pass through without friction, the sounds are called approximants—vowellike sounds functioning as consonants. Most of these consonant sounds can be voiced or voiceless; vowels are normally voiced. The terms "voiced" and "voiceless" refer to the presence and the absence of vocal fold vibration.

Acoustic phonetics deals with the manner in which the spoken message is encoded in the sound waves. According to the generally accepted source-filter theory of speech acoustics, sound is generated at a source (which for phonated speech is constituted by the vibrating vocal folds) and passed through the vocal tract. The opening and closing of the vocal folds create a succession of condensations and rarefactions of air molecules—variations in air pressure—and transform kinetic energy into acoustic energy. The sound wave generated at the glottis can be considered, for practical purposes, a complex periodic wave, and as such it contains energy at frequencies that are multiples of the fundamental frequency (harmonics).

The vocal tract acts as a filter, transmitting more energy at those frequencies that correspond to the resonances of the vocal tract than at other frequencies. Energy concentrations at the resonance frequencies of the vocal tract are referred to as formants.

In principle, the source and filter are independent of each other; consider the fact that the same vowel can be sung at different fundamental frequencies (pitches), and different vowels can be produced at the same pitch. The sound wave can be described by specifying its fundamental frequency, amplitude, and spectrum.

The subject matter of phonetics is not limited to the production and perception of vowels and consonants; of equal importance are such prosodic and suprasegmental aspects of spoken language as duration, fundamental frequency, and intensity, as they determine such linguistically relevant phenomena as tone and intonation, stress and emphasis, and the signaling of various boundaries—boundaries of morphemes and words, phrases, clauses, and sentences. *See* SPEECH. [I.Le.]

Phonocardiography The science of graphic visualization and interpretation of sound vibrations associated with each heartbeat. The phonocardiogram is the record obtained, the phonocardiograph the recording instrument. The latter generally consists of an audio amplifier coupled to a carbon crystal or capacitance microphone which is placed in turn over several areas of the chest of interest to the observer.

The technique is used in general to verify auscultatory findings obtained by the use of the stethoscope. However, since the various microphones in use do not resemble the frequency characteristics of the human ear, the records obtained are not

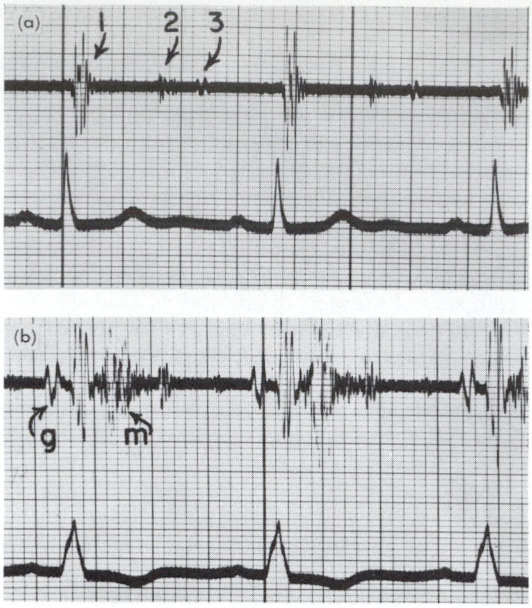

Phonocardiograms. (*a*) Normal subject, record from cardiac apex: First (1) and second (2) heart sounds appear split; a faint third (3) heart sound is also present. (*b*) Abnormal phonocardiogram: First and second heart sounds as in *a*. The first sound is preceded by an atrial gallop sound, at g, coinciding with atrial activity, as seen from the electrocardiogram. Between the first and second sound, rapid irregular vibrations, at m, occur which are characteristic of a systolic murmur. Simultaneous electrocardiograms are shown below each sound record. Time lines represent 0.04 s.

accurate representations of what is being heard. But phonocardiography serves well to time auscultatory events, particularly if the heart rate is rapid or irregular.

For timing purposes phonocardiograms are generally recorded simultaneously with a pulse record, or an electrocardiogram (see illustration) . For the purpose of a more detailed analysis, phonocardiograms have been obtained from other areas, such as the esophagus, and from the cavities of the heart and the large blood vessels by means of special phonocatheters advanced into the heart through the venous system (intracardiac phonocardiography). The miniature microphones placed on the tip of the catheter are related to hydrophones used in solar application. Spectral phonocardiography is an adaptation of the Potter sound spectrograph whereby frequency, time, and loudness of heart noises can be displayed in graphic form. *See* ELECTROCARDIOGRAPHY. [H.H.He.]

Phonolite A light-colored, aphanitic (not visibly crystalline) rock of volcanic origin, composed largely of alkali feldspar, feldspathoids (nepheline, leucite, sodalite), and smaller amounts of dark-colored (mafic) minerals (biotite, soda amphibole, and soda pyroxene). Phonolite is chemically the effusive equivalent of nepheline syenite and similar rocks. Rocks in which plagioclase (oligoclase or andesine) exceeds alkali feldspar are rare and may be called feldspathoidal latite. *See* FELDSPATHOID; MAGMA.

Phonolites are rare and highly variable rocks. They occur as volcanic flows and tuffs and as small intrusive bodies (dikes and sills). They are associated with trachytes and a wide variety of feldspathoidal rocks. *See* IGNEOUS ROCKS; TRACHYTE. [C.A.C.]

Phonon A sound quantum. The energy of a phonon is $h\nu$, where h is Planck's constant and ν the frequency of vibration of the sound wave. The phonon is thus analogous to the photon, a light quantum.

The concept of a phonon as a packet of sound waves, the wave packet having particlelike aspects, is particularly convenient in the theory of the thermal conductivity of insulators, where one may speak of a phonon gas, collisions between phonons, and a phonon mean free path. In the theory of the properties of superfluid helium, the quanta of longitudinal sound waves in the liquid helium are called phonons. *See* CONDUCTION (HEAT). [J.De.L.]

Phonoreception The perception of sound by animals through specialized sense organs. A sense of hearing is possessed by animals belonging to two divisions of the animal kingdom: the vertebrates and the insects. The sense is mediated by the ear, a specialized organ for the reception of vibratory stimuli. Such an organ is found in all except the most primitive vertebrates, but only in some of the many species of insects. The vertebrate and insect types of ear differ in evolutionary origin and in their modes of operation, but both have attained high levels of performance in the reception and discrimination of sounds. *See* SOUND.

Vertebrates. The vertebrate ear is a part of the labyrinth, located deep in the bone or cartilage of the head, one ear on either side of the brain. A complex assembly of tubes and chambers contains a membranous structure which bears within it a number of sensory endings of different kinds. *See* EAR.

The membranous labyrinth is shown in a generalized schematic form in Fig. 1. It is convenient to recognize two divisions: a superior division, which includes the three semicircular canals and the utricle, and an inferior division, which includes the saccule and its appendages, the lagena and the cochlea.

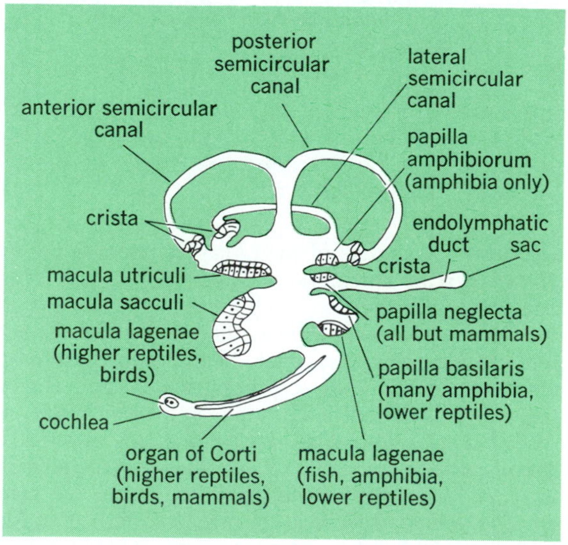

Fig. 1. Generalized sketch of the vertebrate labyrinth. The three cristae, macula ultriculi, and macula sacculi are always present in vertebrates, and the other endings appear as indicated, with a few exceptions.

The superior division is remarkably uniform in character from the higher fishes upward, but the inferior division shows many variations. The saccule is always present. The lagena is present in all classes except the mammals, although it is missing in occasional species. The cochlea is found in reptiles, birds, and mammals.

The sensory endings within these parts of the labyrinth also vary in the vertebrate series. Again there is uniformity for the superior division. There is a crista in each ampulla of the three semicircular canals and a utricular macula. In all but the mammals (with a few individual exceptions), there is a macula

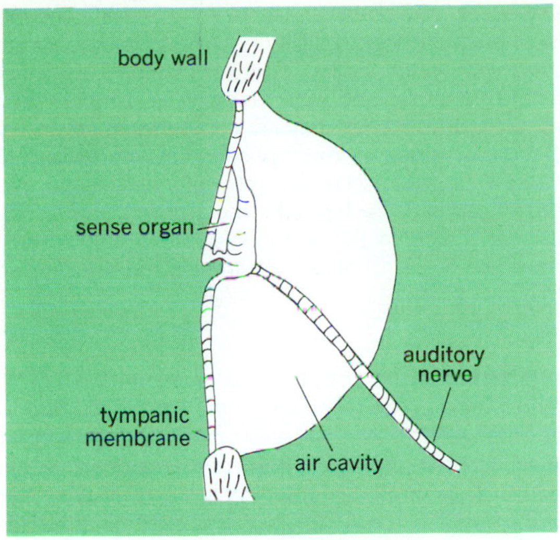

Fig. 2. Ear of a grasshopper.

neglecta, usually located on the floor of the utricle or close to the junction of utricle and saccule. All forms have a saccular macula. All those with a lagena (in general, all except the mammals) have a lagenar macula. All the amphibians have a papilla amphibiorum, but it is found in no other forms. A basilar papilla appears in certain amphibians, is continued in the reptiles, and then is developed in a more elaborate form as the cochlea of higher reptiles, birds, and mammals.

These endings contain ciliated cells (hair cells) which are supplied by fibers of the eighth cranial (auditory) nerve. In the cristae the cilia of the hair cells are particularly long and are embedded in a gelatinous substance that forms a cap or cupola. In the maculae the cilia are surmounted by a flat plate of gelatinous material in which numerous granules of calcium carbonate (otoliths) are usually embedded. The ciliated cells in the papillae lie on a movable membrane (basilar membrane) and have a membranous covering, the tectorial membrane.

The superior part of the labyrinth generally serves for bodily posture and equilibrium, whereas the saccule and its appendages (lagena and cochlea) serve for hearing. However, there are exceptions to this rule, the most important of which is that in the higher vertebrates, including mammals and probably birds and reptiles, the saccule serves only for equilibrium.

Beginning with the amphibians, which are the earliest vertebrates to spend a considerable portion of their lives on land, there appears a special mechanism, the middle ear, whose function is the transmission of aerial vibrations to the endings of the inner ear. All the vertebrates above the fishes, and certain of the fishes as well, have some type of sound-facilitative mechanism. *See* EAR; HEARING (HUMAN).

Invertebrates. The group of invertebrates which has received the most attention has been the insects. Other arthropods, such as certain crustaceans and spiders, have also been found to be sensitive to sound waves.

The insect ear consists of a superficial membrane of thin chitin with an associated group of sensilla called scolophores. Such an apparatus is shown in simplified form in Fig. 2. These ears are found in most species of katydids, crickets, grasshoppers, cicadas, waterboatmen, mosquitoes, and nocturnal and spinner moths. The occur in different places in the body: on the antennae of mosquitoes, on the forelegs of katydids and crickets, on the metathorax of cicadas and waterboatmen, and on the abdomen of grasshoppers. Probably these differently situated organs represent separate evolutionary developments, through the association of a thinned-out region of the body wall with sensilla that are found extensively in the bodies of insects and that by themselves seem to serve for movement perception.

The insects mentioned above are noted for their production of stridulatory sounds made by rubbing the edges of the wings together, or a leg against a wing, or by other means. These sounds are produced by the males and serve for enticing the females in mating. A most striking adaptation is that shown by mosquitoes: The ear of the male mosquito is sensitive only to a narrow range of frequencies around 380 Hz, and this frequency is the one which is produced by the wings of the female in flight. If the ear of the male mosquito is made nonfunctional, the mosquito fails to find a mate. [E.G.W.]

Phoresy A relationship between two different species of organisms in which the larger, or host, organism transports a smaller organism, the guest. It is regarded as a type of commensalism in which the relationship is limited to transportation of the guest. *See* ECOLOGICAL INTERACTIONS. [C.B.C.]

Phoronida A small, relatively homogeneous group of animals now generally considered to constitute a separate animal phylum. Two genera, *Phoronis* and *Phoronopsis*, and about 16 species are recognized.

Phoronids may occur in vertical tubes placed just below the surface in intertidal or subtidal mud flats, or as feltlike masses of intertwined tubes attached to rocks, pilings, or old logs in shallow water. In both cases the tubes, composed basically of a secreted, parchmentlike material, are encrusted with small particles of sand or shell. A third living habit concerns those phoronids found inside channels, probably self-made, in limestone rock or the shells of dead pelecypod mollusks.

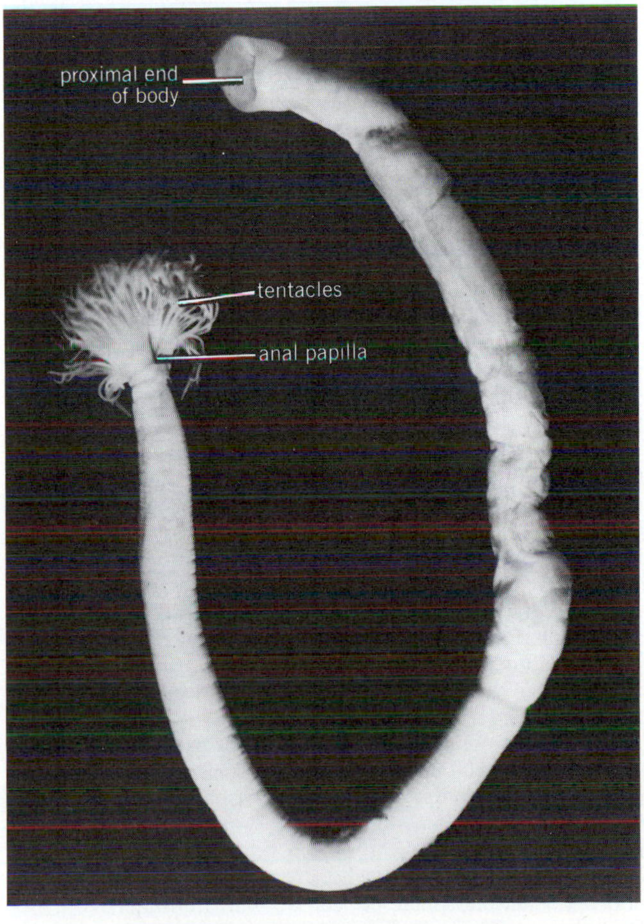

Phoronopsis harmeri removed from its tube.

The geographical distribution of phoronids appears to be worldwide in temperate and tropical seas. There are no records of phoronids from the polar regions.

The body is more or less elongate, ranging in length from about 1.6 to 8 in. (4 to 20 cm), and bears a crown of tentacles arranged in a double row surrounding the mouth which is usually crescent-shaped (see illustration). The anus occurs at the level of the mouth and is borne on a papilla immediately outside the double row of tentacles. The digestive tract is therefore U-shaped, the mouth and anus opening close together at one end of the animal. The tentacles rest on a connective tissue base known as the lophophore. Associated with the mouth is a ciliated flap of tissue known as the epistome. *See* LOPHOPHORE.

The phylum includes both dioecious animals and hermaphrodites. All phoronids may reproduce sexually, and in most cases the life history includes the pelagic actinotroch larva. Some species reproduce asexually by transverse fission. [J.R.M.]

Phosphate A negative ion having the formula PO_4^{3-}. Phosphates are derived from phosphoric acid, H_3PO_4.

The term phosphate is a broad term which encompasses all anions derived from acids containing phosphorus in the 5+ oxidation state, as indicated in the following list (all of those listed are obtained from P_4O_{10} and water):

$(HPO_3)_n$	Metaphosphoric acid
$H_5P_3O_{10}$	Triphosphoric or tripolyphosphoric acid
$H_4P_2O_7$	Pyrophosphoric acid
H_3PO_4	Orthophosphoric acid

Phosphates are important ingredients in commercial fertilizers. Certain organic phosphates have been used as insecticides and nerve gases. *See* FERTILIZER; ORGANOPHOSPHORUS COMPOUND; PHOSPHORUS. [E.E.W.]

Phosphate metabolism Organic phosphate compounds are present in the structural units of every animal cell, and inorganic phosphate is associated with calcium in bone and teeth. The total phosphorus in the adult human body is about 1.2% by weight, with only 0.14% by weight present in the soft tissues and the remainder in mineralized tissue in the form of apatite crystals. Blood phosphate plays an important role in regulating neutrality, and it is in equilibrium with both bone and cellular organic phosphates. The blood level is held relatively constant by regulating phosphate excretion by the kidney. This control is primarily mediated by action of parathyroid hormone. Vitamin D enhances the entry of phosphate into bone. Phosphate plays an important role in absorption of sugar from the intestine and reabsorption of glucose from the kidney. *See* PARATHYROID HORMONE; VITAMIN D.

The central role of phosphates in life processes is indicated by their occurrence in ribonucleic acid (RNA) and deoxyribonucleic acid (DNA). Through the formation of lecithins, phosphates are involved in fat metabolism. Phosphates play a major role in the conservation and transfer of energy, particularly of the energy produced in the tricarboxylic acid cycle (Krebs cycle), in glycolysis, and in the pentose shunt. They do so by participating in many phosphorylation and transphosphorylation reactions involving sugars and other organic compounds. *See* CARBOHYDRATE METABOLISM; CHROMOSOME; KREBS CYCLE; LIPID METABOLISM; NUCLEIC ACID.

Phosphorus-containing coenzyme systems include the pyridine (nicotinamide) and the riboflavin nucleotide systems concerned with oxidation-reduction reactions; coenzyme A, the functional form of pantothenic acid, concerned with transacetylation, acylation, and condensation reactions; the diphosphothiamine system concerned with decarboxylation; and pyridoxal phosphate concerned with transamination. *See* BIOCHEMISTRY; COENZYME; ENERGY METABOLISM. [M.K.S.]

Phosphate minerals Any naturally occurring inorganic salts of phosphoric acid, $H_3[PO_4]$. All known phosphate minerals are orthophosphates. There are over 150 species of phosphate minerals, and their crystal chemistry is often very complicated. Phosphate mineral paragenesis can be divided into three categories: primary phosphates (crystallized directly from a melt or fluid), secondary phosphates (derived from the primary phosphates by hydrothermal activity), and rock phosphates (derived from the action of water upon buried bone material, skeletons of small organisms, and so forth). *See* MINERAL; PHOSPHATE. [P.B.M.]

Phosphatide A complex lipid containing phosphorus. The phosphatides, also known as phospholipids, are usually divided into groups on the basis of compounds from which they are derived. For example, glycerophosphatides are derived from glycerophosphoric acid (with the structure shown, where $R_1 = R_2 = R_3 = H$), sphingophosphatides are derived from sphingosine phosphate, and inositol phosphatides are derived from inositol phosphates.

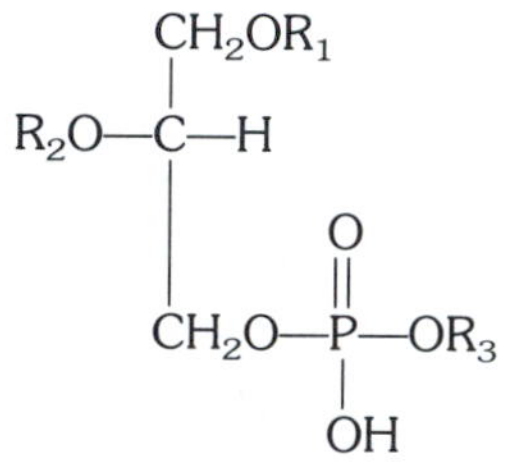

Phosphatidyl ethanolamine, lecithin, phosphatidyl inositol, and the plasmalogens are present in both plant and animal tissues; phytoglycolipids have been found only in plants; sphingomyelin has been found only in animal tissues. The phosphatides are important components of biological membranes.

Since an individual phosphatide may contain a variety of fatty acid residues, it may be described as pure only with this limitation in mind. Most of the highly unsaturated fatty acids of animal tissue lipids occur in the phosphatides. Phosphatides can act as protective colloids, as wetting and emulsifying agents, and as antioxidants, and are therefore used considerably in the food and petroleum industries. The chief source of commercial phosphatides is soybean. *See* LIPID. [H.E.Ca.; R.H.G.]

Phosphorescence A delayed luminescence, that is, a luminescence that persists after removal of the exciting source. It is sometimes called afterglow.

This original definition is rather imprecise, because the properties of the detector used will determine whether or not there is an observable persistence. There is no generally accepted rigorous definition or uniform usage of the term phosphorescence. In the literature of inorganic luminescent systems, some authors define phosphorescence as delayed luminescence whose persistence time decreases with increasing temperature. According to this usage, luminescence whose persistence time is independent of temperature is called fluorescence regardless of the length of the afterglow; a temperature-independent afterglow of long duration is called simply a slow fluorescence, which implies that the atomic or molecular transition involved is forbidden to a greater or lesser degree by the spectroscopic selection rules. The most common mechanism of phosphorescence in photoconductive inorganic systems, however, occurs when electrons or holes, set free by the excitation process and trapped at lattice defects, are expelled from their traps by the thermal energy in the system and recombine with oppositely charged carriers with the emission of light. *See* HOLE STATES IN SOLIDS; SELECTION RULES (PHYSICS).

In the organic literature the term phosphorescence is reserved for the forbidden luminescent transition from a metastable energy state M to the ground state G, while the afterglow corresponding to the M→E→G process (where E is a higher energy state) is called delayed fluorescence. *See* FLUORESCENCE; LIGHT; LUMINESCENCE. [C.C.K.; J.H.S.]

Phosphorus A chemical element, P, atomic number 15, atomic weight 30.9738. Phosphorus forms the basis of a very large number of compounds, the most important class of which are the phosphates. For every form of life, phosphates play an essential role in all energy-transfer processes such as metabolism, photosynthesis, nerve function, and muscle action. The nucleic acids which among other things make up the hereditary material (the chromosomes) are phosphates, as are a number of coenzymes. Animal skeletons consist of a calcium phosphate. *See* PHOSPHATE.

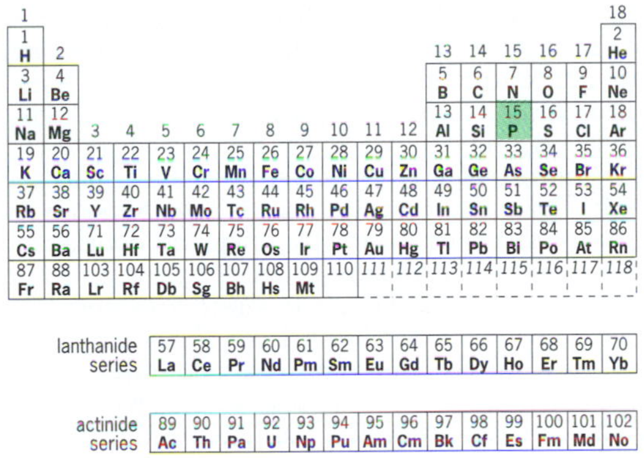

About three-quarters of the total phosphorus (in all of its chemical forms) used in the United States goes into fertilizers. Other important uses are as builders for detergents, nutrient supplements for animal feeds, water softeners, additives for foods and pharmaceuticals, coating agents for metal-surface treatment, additives in metallurgy, plasticizers, insecticides, and additives for petroleum products. *See* DETERGENT; WATER SOFTENING.

Of the nearly 200 different phosphate minerals, only one, fluorapatite, $Ca_5F(PO_4)_3$, is mined chiefly from large secondary deposits originating from the bones of dead creatures deposited on the bottom of prehistoric seas and from bird droppings on ancient rookeries. *See* PHOSPHATE MINERALS.

Research in phosphorus chemistry indicates that there may be as many compounds based on phosphorus as on carbon. In organic chemistry it has been customary to group the various chemical compounds based on carbon into families which are called homologous series. This can also be done in the chemistry of phosphorus compounds, even though many phosphorus-based families are incomplete. The best known of the families of compounds based on phosphorus is the group of chain phosphates. Phosphate salts consist of cations, such as sodium, along with chain anions, such as $(P_nO_{3n+1})^{(n+2)-}$, which may have 1–1,000,000 phosphorus atoms per anion.

The phosphates are based on phosphorus atoms tetrahedrally surrounded by oxygen atoms, with the lowest member of the series being the simple PO_4^{3-} anion (the orthophosphate ion). The family of chain phosphates is based on a row of alternating phosphorus and oxygen atoms in which each phosphorus atom remains in the center of a tetrahedron of four oxygen atoms. There is also a closely related family of ring phosphates, a member of which, the trimetaphosphate, is shown in Fig. 1.

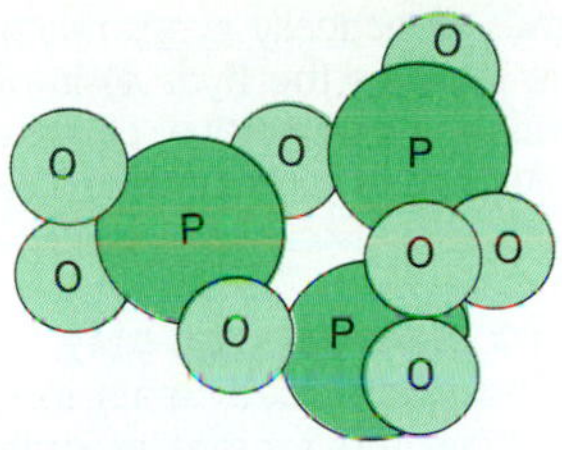

Fig. 1. Ring phosphate anion, $(P_3O_9)^{3-}$.

An interesting structural characteristic of many known phosphorus compounds is the formation of cagelike structures. Such cagelike molecules are exemplified by white phosphorus, P_4, and one of the phosphorus pentoxides, P_4O_{10} (Fig. 2). Network structures are also common; for example, black phosphorus crystals in which the atoms are bonded together in the form of vast, corrugated planes (Fig. 3).

In the majority of its compounds, phosphorus is chemically bonded to four neighboring atoms. There is a large number of compounds in which one of the four neighboring atoms is absent, and in which its place is taken by an unshared pair of electrons. There are also a few compounds in which there are five or six neighboring atoms bonded to the phosphorus. These compounds are very reactive and tend to be unstable.

During the 1960s and 1970s a large number of organic-phosphorus compounds were prepared. Most of these chemical structures involve three or four neighboring atoms bonded to the phosphorus, but stable structures having two, five, or six neighboring atoms per phosphorus are also known. *See* ORGANOPHOSPHORUS COMPOUND; PHOSPHATE.

Essentially all of the phosphorus used in commerce is in the form of phosphates. The majority of phosphatic fertilizers consist of highly impure monocalcium or dicalcium orthophosphate, $Ca(H_2PO_4)_2$ and $CaHPO_4$. These phosphates are salts of orthophosphoric acid. *See* FERTILIZER.

The phosphorus compound of major biological importance is adenosine triphosphate (ATP), which is an ester of sodium tripolyphosphate, widely employed in detergents and water-

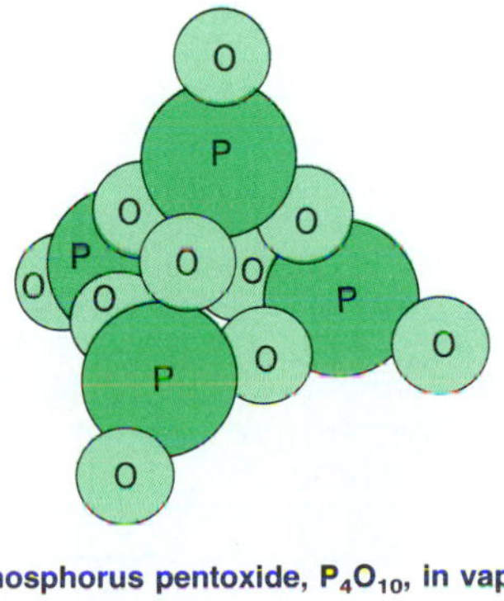

Fig. 2. Phosphorus pentoxide, P_4O_{10}, in vapor state.

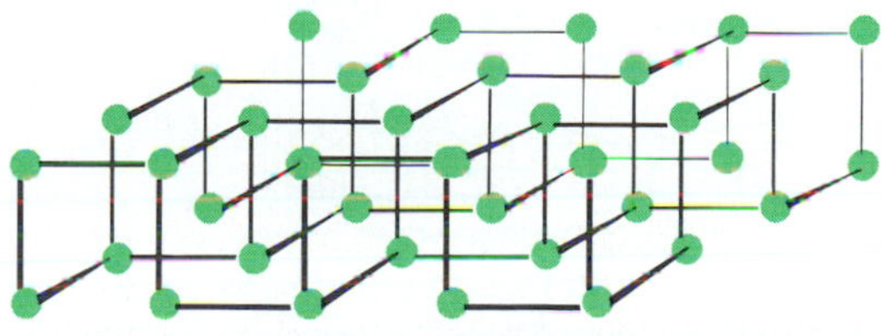

Fig. 3. Black phosphorus, P_n.

softening compounds. Practically every reaction in metabolism and photosynthesis involves the hydrolysis of this tripolyphosphate to its pyrophosphate derivative, called adenosine diphosphate (ADP). *See* ADENOSINE TRIPHOSPHATE (ATP). [J.R.V.W.]

Photoacoustic spectroscopy A technique for measuring small absorption coefficients in gaseous and condensed media, involving the sensing of optical absorption by detection of sound. It is frequently called optoacoustic spectroscopy.

During the transmission of optical radiation through a sample (gas, liquid, or solid), the absorption of radiation by the sample can be measured by several techniques. The straightforward detection technique requires a measurement of the optical radiation level with and without the sample in the optical path. The transmitted power P_{out} and the incident power P_{in} are related through the equation below, where α is the

$$P_{out} = P_{in}e^{-\alpha l}$$

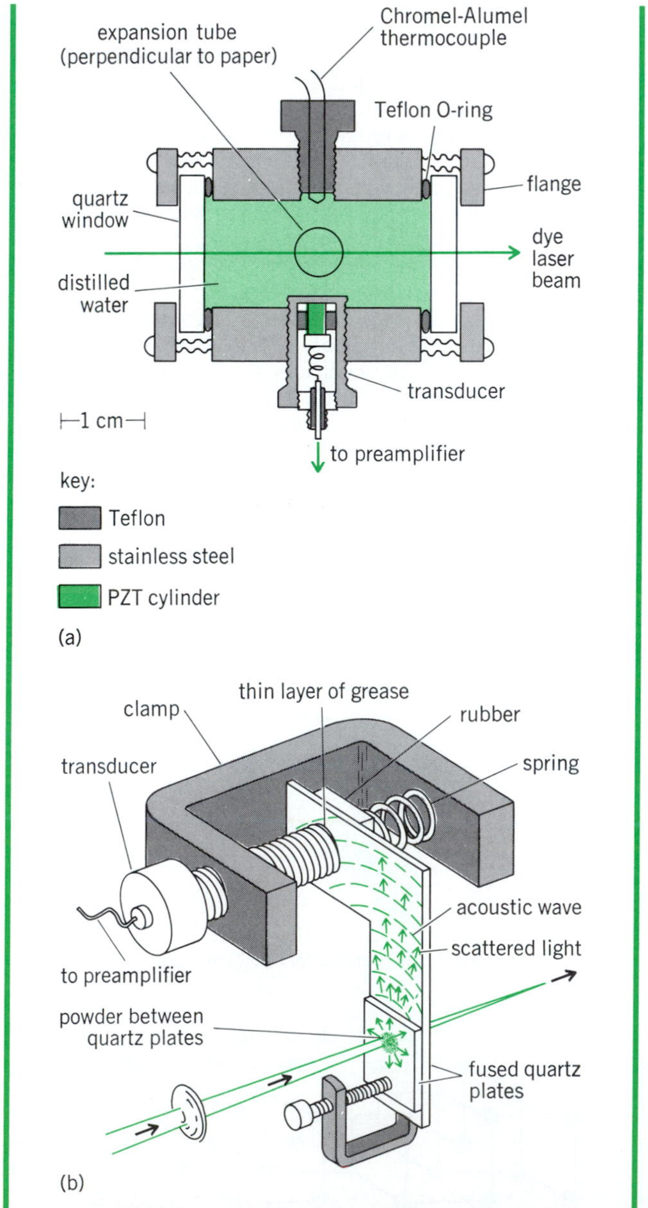

Arrangement for pulsed-laser (*a*) immersed and (*b*) contacted piezoelectric transducer optoacoustic spectroscopy.

absorption coefficient and l is the length of the absorber. With this technique, the minimum measurable value of αl is of the order of 10^{-4} unless special precautions have been taken to stabilize the source of radiation.

Optoacoustic detection is a calorimetric method where no direct detection of optical radiation is carried out but, instead, a measurement is made of the power absorbed by the medium from the incident radiation. The optoacoustic signal is proportional to the incident power and the absorption-length product αl. Thus, for given sources of noise from the detection transducers, the signal-to-noise ratio improves as the incident energy is increased.

If optical radiation is amplitude-modulated at an audio frequency, the absorption of such radiation by a gaseous medium that has been confined in a cell with appropriate optical windows for the entrance and exit of the radiation, and nonradiative relaxation of the medium, will cause a periodic variation in the temperature of the column of the irradiated gas. Such a periodic rise and fall in temperature gives rise to a corresponding periodic variation in the gas pressure at the audio frequency. The audio-frequency pressure fluctuations (that is, sound) are efficiently detected using a sensitive gas-phase microphone.

The capability of measuring extremely small absorption coefficients and correspondingly small concentrations of the absorption gases has many applications, including high-resolution spectroscopy of isotopically substituted gases, excited states of molecules and forbidden transitions, and pollution detection. The pollution measurements have demonstrated that the optoacoustic spectroscopy technique in conjunction with tunable lasers can be routinely used for on-line real-time in-place detection of undesirable gaseous constituents at sub-parts-per-billion levels. *See* LASER; LASER SPECTROSCOPY.

A very sensitive calorimetric spectroscopic technique has been developed for the study of weak absorption in liquids and solids. This technique uses a pulsed tunable laser for excitation and a submerged piezoelectric transducer, in the case of a liquid, or a contacted piezoelectric transducer, in the case of a solid, for the detection of the ultrasonic signal generated due to the absorption of the radiation and its subsequent conversion into a transient ultrasonic signal (see illustration). Because of the capability of measuring very small fractional absorptions, the technique is clearly applicable to the area of monitoring water pollution, impurity detection in thin semiconductor wafers, transmission studies of ultrapure glasses (used in optical fibers for optical communications), and so forth. *See* ABSORPTION. [C.K.N.P.]

Photochemistry The branch of chemistry concerned with reactions of excited molecules produced by the absorption of light. Chemical reactions involve the breaking (and formation) of chemical bonds which requires energies in the range of 200–600 kilojoules/mole; this corresponds to the energy of light quanta in the ultraviolet (100–400 nanometers), visible (400–700 nm), and near-infrared (700–1000 nm) regions of the electromagnetic spectrum. Light of shorter wavelengths (x-rays, γ-rays) has sufficient energy to ionize and to dissociate molecules. The effects of this light constitute the field of radiation chemistry, while light of longer wavelengths (infrared) is not sufficiently energetic to produce electronic excitation in single quanta excitations. *See* INORGANIC PHOTOCHEMISTRY; LASER PHOTOCHEMISTRY; RADIATION CHEMISTRY.

Electronic excitation of molecules reduces the ionization potential and increases the electron affinity of a molecule by an increment equal to the excitation energy, and therefore promotes electron transfer reactions. The change in electronic configuration produced by light absorption leads to dramatic changes in chemical properties. Since the Gibbs free energy is also increased by light absorption, electronically excited molecules may undergo spontaneous reactions to products which are thermodynamically inaccessible from the ground state. This is exemplified by the photosynthetic reaction in which the conversion of carbon dioxide (CO_2) and water (H_2O) to carbohydrate $(CH_2O)_n$ and oxygen (O_2) is mediated by excited chlorophyll molecules. The reverse process of respiration leads to a decrease in Gibbs free energy and takes place spontaneously in the absence of light. *See* PHOTOSYNTHESIS.

Quantum yield. The rate of a photochemical reaction is proportional to the rate at which electronically excited molecules M* are produced; since each absorbed photon (energy $h\nu$) of light produces one excited molecule, this excitation rate is equal to the rate of light absorption or the intensity of light absorbed, I_a. The reaction rate is therefore proportional to I_a, or rate = γI_a, where the proportionality constant, γ = rate/I_a, known as the quantum yield (or efficiency), is characteristic of the reaction under the conditions of examination. The quantum yield of the primary process (that involving the excited molecule) cannot exceed unity and is usually much lower than

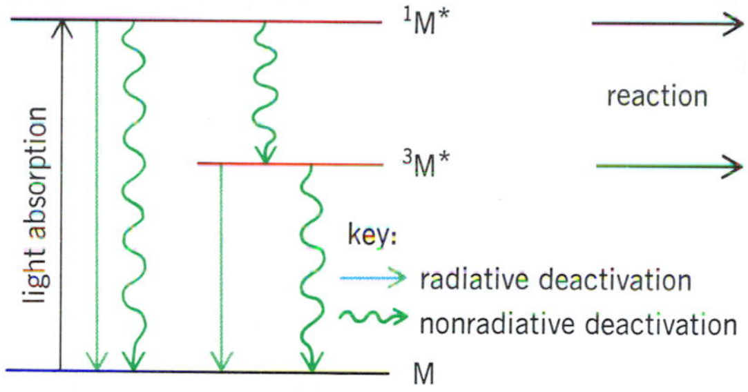

Radiative and nonradiative deactivation of electronically excited molecule compete with reaction of singlet state $^1M^*$ and of lower-energy triplet state $^3M^*$.

this since electronically excited molecules revert to the ground state (with or without the emission of luminescence) within a period of $\sim 10^{-3}$ s (triplet state) to 10^{-8} s (singlet states). These so-called photophysical processes, summarized in the illustration, complete with photochemical reactions, effectively reducing the quantum yield. Additionally, the excited molecule may be quenched by other molecules Q in a process of electron transfer [reaction (1)] or energy transfer [reaction (2)],

$$M^* + Q \rightarrow (M^+Q^-) \text{ or } (M^-Q^+) \rightarrow M + Q \quad (1)$$

$$M^* + Q \rightarrow M + Q^* \quad (2)$$

which leave the molecule M chemically unchanged.

On the other hand, if the primary photochemical products are atoms or free radicals, these may undergo secondary nonphotochemical chain reactions which amplify the primary process and lead to overall photochemical quantum yields much greater than unity. An example is reaction (3*a*), in which the primary process (3*b*) is followed by the chain propagating steps (3*c* and 3*d*) to produce $\sim 10^6$ molecules of HCl for each

$$H_2 + Cl_2 \xrightarrow{\text{light}} 2HCl \quad (3a)$$

$$Cl_2 + \text{light} \rightarrow 2Cl \quad (3b)$$

$$Cl + H_2 \rightarrow H + HCl \quad (3c)$$

$$H + Cl_2 \rightarrow Cl + HCl \quad (3d)$$

light photon absorbed by Cl_2, that is, an overall quantum yield of 10^6. *See* CHAIN REACTION (CHEMISTRY).

Mechanisms. Photochemical reaction mechanisms are deduced from measurements of the dependence of quantum yields on such different reaction variables as reactant concentration and the concentration of added substances. Primarily it is necessary to identify the nature of the electronically excited state responsible for the reaction, which is most often singlet ($^1M^*$) or triplet ($^3M^*$) following which a theoretical examination of the orbital transformations along the reaction coordinate may be attempted.

If the quantum yield is reduced by a selective triplet-state quencher (Q) which introduces the competing process [reaction (4)], then the triplet state $^3M^*$ may be regarded as the reactive

$$^3M^* + Q \rightarrow M + {}^3Q^* \quad (4)$$

state; otherwise the singlet state $^1M^*$ is assigned the role of reactive intermediate. The rate constant of the primary process may then be obtained by monitoring the decay of intermediate $^{1,3}M^*$ in absorption or emission following flash or laser pulse excitation. *See* EXCITED STATE; FREE RADICAL; QUANTUM CHEMISTRY; TRIPLET STATE.

Photochemical reactions are classified as unimolecular, in which the excited molecule itself undergoes chemical change, or bimolecular if the excited molecule reacts with another molecule present. *See* CHEMICAL DYNAMICS. [B.St.]

Photoclinometer A term applied to directional surveying instruments which record photographically the direction and magnitude of well deviations from the vertical. Two instruments of this type are in wide use, the Schlumberger photoclinometer and the Surwell clinograph. Both instruments record a series of deviation measurements on one trip into and out of the well. From this series of data it is possible to plot quite accurately the course of the well.

In the Schlumberger photoclinometer (see illustration) the deviation from the vertical is indicated by a small metal ball which rolls in a transparent glass bowl graduated in circular degrees. The direction of the deviation in azimuth is indicated by a magnetic compass. With the instrument suspended by an electrical cable, the positions of the compass and steel ball are photographed on a 35-mm film by operation of electrical controls at the surface. Correlation of the pictures with the depths at which they are taken yields a measure of the magnitude and direction of deviation of the hole as a function of depth.

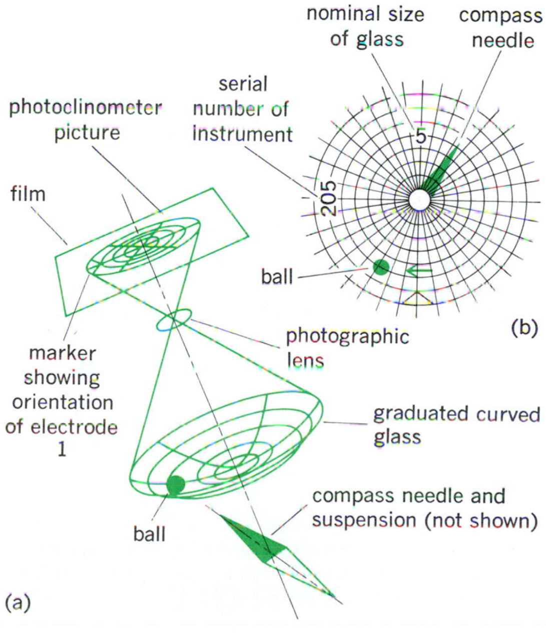

Diagram of a Schlumberger photoclinometer. (*a*) Principal features. (*b*) Type of record obtained. (*Schlumberger Well Surveying Corp.*)

The Surwell clinograph also operates electrically but is powered by batteries contained in the instrument. The deviation from the vertical is indicated by a box level gage and the direction in azimuth by a gyroscopic compass, permitting its use inside steel pipe. *See* SURVEYING. [H.G.Bo.]

Photoconductive cell A device for detecting electromagnetic radiation (photons) by variation of the electrical conductivity of a substance (a photoconductor) upon absorption of the radiation by this substance. During operation the cell is connected in series with an electrical source and current-sensitive meter, or in series with an electrical source and resistor. Current in the cell, as indicated by the meter, is a measure of the photon intensity, as is the voltage drop across the series resistor. Photoconductive cells are made from a variety of semiconducting materials in the single-crystal or polycrystalline form. *See* PHOTOCONDUCTIVITY; PHOTOELECTRIC DEVICES. [S.R.B.]

Photoconductivity The increase in electrical conductivity caused by the excitation of additional free charge carriers by light of sufficiently high energy in semiconductors and insulators. Effectively a radiation-controlled electrical resistance, a photoconductor can be used for a variety of light- and particle-detection applications, as well as a light-controlled switch. Other major applications in which photoconductivity plays a central role are television cameras (vidicons), normal silver halide emulsion photography, and the very large field of electrophotographic reproduction. *See* OPTICAL DETECTORS; OPTICAL MODULATORS; PARTICLE DETECTOR; PHOTOGRAPHY; TELEVISION CAMERA TUBE.

Although all insulators and semiconductors may be said to be photoconductive, that is, they show some increase in electrical conductivity when illuminated by light of sufficiently high energy to create free carriers, only a few materials show a large enough change, that is, show a large enough photosensitivity, to be practically useful in applications of photoconductors.

Since the electrical conductivity σ of a material is given by the product of the carrier density, its charge, and its mobility, an increase in the conductivity can be formally due to either an increase in carrier density or an increase in mobility. Although cases are found in which both types of effects are observable, photoconductivity in single-crystal materials is due primarily to an increase in carrier density. In polycrystalline materials, on the other hand, where transport may be limited by potential barriers between the crystalline grains, an increase in mobility due to photoexcitation effects on these intergrain barriers may dominate the photoconductivity.

The variation of photoconductivity with photon energy is called the spectral response of the photoconductor. Spectral response curves typically show a fairly well-defined maximum at a photon energy close to that of the bandgap of the material, that is, the minimum energy required to excite an electron from a bond in the material into a higher-lying conduction band where it is free to contribute to the conductivity. This energy ranges from 3.7 eV, in the ultraviolet, for zinc sulfide (ZnS) to 0.2 eV, in the infrared, for cooled lead selenide (PbSe).

Another major characteristic of a photoconductor of practical concern is the rate at which the conductivity changes with changes in photoexcitation intensity. If a steady photoexcitation is turned off at some time, for example, the length of time required for the current to decrease to $1/e$ of its initial value is called the decay time of photoconductivity, t_d. The magnitude of the decay time is determined by the lifetime π and by the density of carriers trapped in imperfections as a result of the previous photoexcitation, which must now also be released in order to return to the thermal equilibrium situation. *See* PHOTOCONDUCTIVE CELL. [R.H.Bu.]

Photocopying processes Those means by which a copy is created on a sensitized surface (generally paper, film, or metal plate) by the action of radiant energy. The term is generally applied only to documentary reproduction. The document to be photocopied must already have been prepared by other applications of photography, by manuscript, or by typewriter. A document in this case is classed as either a line drawing or a continuous tone illustration or a combination. Some photocopying processes do not handle tone satisfactorily. Photocopying offers practical printing methods for the production of a single copy or a limited number of copies or for the production of a stencil or master from which to run off larger numbers by use of diazo paper or offset lithography. *See* PHOTOGRAPHIC MATERIALS; PHOTOGRAPHY.

Advantages of photocopying are its photographic accuracy except for occasional problems with color; reduction and enlargement ability of some processes; speed in most instances; space saving in the case of microfilm; economy of labor and materials over any other short-run copying process; convenience of handling the thin flexible material as opposed to letterpress type or electrolytic plates; simplicity of machine operation by untrained staff for certain processes; and flexibility as achieved by the combination of photocopying processes with other printing processes.

Photocopying processes may be somewhat arbitrarily divided into seven classes: silver halide photocopying, transfer processes, thermography, plan copying, electrostatic processes, the electrolytic process, and microfilming. [D.C.W.]

Photodegradation Reduction in the useful properties of materials because of chemical changes resulting from the absorption of light. The chemical changes can include bond scission (especially of the molecular backbone), color formation, cross-linking, and chemical rearrangements. All organic materials can photodegrade, but the process has greatest practical relevance for polymers where scission of the polymer backbone is particularly important. Photodegradations of polymers in the absence of oxygen (photolysis) or using wavelengths shorter (more energetic) than those at the Earth's surface (< 280 nanometers) have been studied extensively, but only the more practical situation of polymers exposed to terrestrial sunlight (or its equivalent) in air is discussed in this article.

Although all organic polymers can be degraded by light, the rate of degradation varies enormously from polymer to polymer, and is also dependent on the incident wavelengths. Light containing ultraviolet (uv; shorter-wavelength) components is much more destructive than visible light, so that polymers exposed indoors, behind window glass (transmitting > 330 nm), will degrade much more slowly than samples exposed outdoors.

For many aromatic polymers, such as polyester and the aramids, in which the polymer itself is the chromophore (light-absorbing group), backbone scission results predominantly from this direct absorption of light energy. For many other polymers, including polyolefins, and polyvinyl chloride where only impurities absorb energy from sunlight, scission of a chemical bond by light to give free radicals is followed by reaction of these highly reactive free radicals with atmospheric oxygen. *See* FREE RADICAL.

Although numerous organic materials will undergo photodegradation, hydrocarbon polymers are particularly vulnerable because their useful properties depend entirely on their high molecular weights, in the tens or hundreds of thousands. Anything that reduces the molecular weight of polymeric systems will alter the characteristics of these systems and limit their service life. In fact, the scission of as few as one carbon-carbon bond in a thousand in a polymer molecule can completely destroy its useful physical properties. This sensitivity is not observed in lower-molecular-weight substances such as liquid hydrocarbons.

A general approach to reducing the rates of photodegradation for all types of polymers is the use of low levels of additives. These additives, known as photostabilizers or uv stabilizers, are effective at fractions of a weight percent. *See* PHOTOCHEMISTRY; POLYMER; STABILIZER (CHEMISTRY). [D.M.W.; D.J.Ca.]

Photodiode A semiconductor two-terminal component with electrical characteristics that are light-sensitive. All semiconductor diodes are light-sensitive to some degree, unless enclosed in opaque packages, but only those designed specifically to enhance the light sensitivity are called photodiodes.

Most photodiodes consist of semiconductor *pn* junctions housed in a container designed to collect and focus the ambient light close to the junction. They are normally biased in the reverse, or blocking, direction; the current therefore is quite small in the dark. When they are illuminated, the current is proportional to the amount of light falling on the photodiode. *See* JUNCTION DIODE.

Photodiodes are used both to detect the presence of light and to measure light intensity. *See* PHOTOELECTRIC DEVICES. [W.R.Si.]

Photodissociation laser A device which uses intense pulses of ultraviolet light to dissociate molecules in such a way as to leave one constituent in an excited state capable of sustaining laser action. The most notable examples are iodine compounds, which have given peak 1.3-micrometer pulse powers above 10^9 W from the excited iodide atoms. *See* LASER. [S.F.J.; A.L.S.]

Photoelasticity An experimental technique for the measurement of stresses and strains in material objects by means of the phenomenon of mechanical birefringence. Photoelasticity is especially useful for the study of objects with irregular boundaries and stress concentrations, such as pieces of machinery with notches or curves, structural components with slits or holes, and materials with cracks. The method provides a visual means of observing overall stress characteristics of an object by means of light patterns projected on a screen or photographic film. Photoelasticity is generally used to study objects stressed in two planar directions (biaxial), but with refinements it can be used for objects stressed in three spatial directions (triaxial). *See* BIREFRINGENCE.

When a stressed model is subjected to monochromatic polarized light in a polariscope the birefringence of the model causes the light to emerge refracted into two orthogonal planes. Because the velocities of light propagation are different in each direction, there occurs a phase shifting of the light waves.

Isochromatic fringe pattern for plate with hole. (*From M. M. Frocht, Photoelasticity, vol. 2, copyright © 1948 by John Wiley and Sons, Inc.; used with permission*)

When the waves are recombined with the polariscope, regions of stress where the wave phases cancel appear black, and regions of stress where the wave phases combine appear light. Therefore, in models of complex stress distribution, light and dark fringe patterns (isochromatic fringes) are projected from the model (see illustration). These fringes are related to the stresses. *See* POLARIZED LIGHT.

When white light is used in place of monochromatic light, the relative retardation of the model causes the fringes to appear in colors of the spectrum. White light is often used for demonstration, and monochromatic light is used for precise measurements. *See* STRESS AND STRAIN. [W.Z.]

Photoelectric devices Devices which give an electrical signal in response to visible, infrared, or ultraviolet radiation. They are often used in systems which sense objects or encoded data by a change in transmitted or reflected light. Photoelectric devices which generate a voltage can be used as solar cells to produce useful electric power. The operation of photoelectric devices is based on any of the several photoelectric effects in which the absorption of light quanta liberates electrons in or from the absorbing material. *See* PHOTOVOLTAIC EFFECT; SOLAR CELL.

Photoconductive devices are photoelectric devices which utilize the photo-induced change in electrical conductivity to provide an electrical signal. Photoemissive systems have also been used in photoelectric applications. These vacuum-tube devices utilize the photoemission of electrons from a photocathode and collection at an anode. *See* PHOTOCONDUCTIVE CELL; PHOTOEMISSION.

Many photoelectric systems now utilize silicon photodiodes or phototransistors. These devices utilize the photovoltaic effect, which generates a voltage due to the photoabsorption of light quanta near a *pn* junction. Modern solid-state integrated-circuit fabrication techniques can be used to create arrays of photodiodes which can be used to read printed information. *See* PHOTODIODE; PHOTOELECTRICITY; PHOTOTRANSISTOR. [R.A.C.]

Photoelectricity The process by which electromagnetic radiation incident on a solid, liquid, or gas liberates electrical charge, which is detectable in an electric field. The process is strictly quantum in nature.

The quantum nature of electromagnetic radiation also manifests itself in the liberation of electrons and positive holes within the interior of a solid, giving rise to photoconductive and photovoltaic effects. These phenomena are readily observable in the class of solids known as semiconductors. *See* PHOTOCONDUCTIVITY; PHOTOELECTRIC DEVICES; PHOTOEMISSION; PHOTOVOLTAIC EFFECT. [M.A.K.]

Photoemission The ejection of electrons from a solid (or less commonly, a liquid) by incident electromagnetic radiation. Photoemission is also called the external photoelectric effect. The visible and ultraviolet regions of the electromagnetic spectrum are most often involved, although the infrared and x-ray regions are also of interest. For important practical applications of photoemission *see* PHOTOTUBE; TELEVISION CAMERA TUBE. *See also* PHOTOELECTRICITY.

The salient experimental features of photoemission are the following: (1) There is no detectable time lag between irradiation of an emitter and the ejection of photoelectrons. (2) At a given frequency the number of photoelectrons ejected per second is proportional to the intensity of the incident radiation. (3) The photoelectrons have kinetic energies ranging from zero up to a well-defined maximum, which is proportional to the frequency of the incident radiation and independent of the intensity.

In 1905 Albert Einstein made the clarifying assumption that electromagnetic radiation had characteristics like those of particles when it delivered energy to electrons in the emitter. In Einstein's approach the light beam behaves like a stream of photons, each of energy $h\nu$, where h is Planck's constant, and ν is the frequency of the photon. The energy required to eject an electron from the emitter has a well-defined minimum value ϕ called the photoelectric threshold energy. When a photon interacts with an electron, the latter absorbs the entire photon energy. *See* PHOTON.

For $h\nu$ values below the threshold, photoelectrons are not ejected. Even though the electrons absorb photon energy, they do not receive enough to surmount the potential barrier at the surface, which normally holds the electrons in the solid. For photon energies above ϕ, the kinetic energies of photoelectrons range from zero up to a maximum value, $E = h\nu - \phi$. This is the Einstein photoelectric law, and E is commonly termed the Einstein maximum energy. *See* HEAT RADIATION; SCHOTTKY EFFECT. [L.Ap.]

Photoferroelectric imaging The process of storing an image in a ferroelectric material by utilizing either the intrinsic or extrinsic photosensitivity in conjunction with the ferroelectric properties of the material. Specifically, photoferroelectric (PFE) imaging is a process of storing photographic images or other optical information in transparent lead lanthanum zirconate titanate (PLZT) ceramics. The photoferroelectric imaging device consists simply of a thin flat plate (about 0.01 inch or 0.2–0.3 mm thick) of optically polished PLZT ceramic with transparent conductive indium–tin oxide (ITO) electrodes sputter-deposited on the two major surfaces. The image to be stored is exposed onto one of the ITO electroded surfaces by using near-ultraviolet illumination in the intrinsic photosensitivity region (corresponding to a bandgap energy of approximately 3.35 eV) of the PLZT. Simultaneously, a voltage pulse is applied across the electrodes to switch the ferroelectric polarization from one stable remanent state to another. Images are stored both as spatial distributions of light-scattering centers in the bulk of the PLZT and as surface deformation strains which form a relief pattern of the image on the exposed surface. Both the light scattering and surface strains are related to spatial distributions of ferroelectic domain orientations introduced during the image-storage process. These spatial distributions correspond to brightness variations in the image to which the PLZT is exposed. The stored image may be viewed directly or it may be projected onto a screen by using either transmitted or reflected light.

Important potential applications of photoferroelectric imaging are temporary image storage and display. Various types of image processing, including image contrast enhancement, are also offered by the capability of switching from a positive to a negative stored image in discrete steps. *See* ELECTRONIC DISPLAY; FERROELECTRICS; PHOTOCONDUCTIVITY. [C.E.L.]

Photogrammetry The practice of obtaining surveys by means of photography. The camera commonly is airborne with its axis vertical, but oblique and horizontal (ground-based) photographs also are applicable. Data reduction is accomplished by stereoscopic line-of-sight geometry with use of both analytical and analog methods. *See* AERIAL PHOTOGRAPH; REMOTE SENSING; SURVEYING.

In vertical aerial surveys adjacent photos are overlapped. The two images of the same terrain are then superimposed for three-dimensional viewing by human operators or automated sensors.

In a widely used analog procedure the two photos are placed in the projectors of a stereoplotting instrument. With the aid of visible ground-control points the photos are oriented to the relative positions they had at the instants of exposure. In a typical automated stereoplotting system (see illustration) scanning devices substitute for human eyes to sense model-surface slope

Typical automated stereoplotting system. (*Lockwood, Kessler, and Bartlett, Inc.*)

and thus to control servomechanisms that raise and lower the plotting table and translate it along a succession of closely spaced parallel horizontal dimensions, or ground profiles. [R.H.D.]

Photographic materials The common sensitive materials of photography, that is, plates, films, and papers. They consist of a support of glass, plastic sheet, or paper, respectively, coated with an emulsion, which usually is a suspension of silver halide crystals in gelatin and provides the light-sensitive layer in which the picture will be formed (see illustration). *See* PHOTOLYSIS.

Spectral sensitivity. The silver halides are normally sensitive only to the ultraviolet, violet and blue wavelengths, but

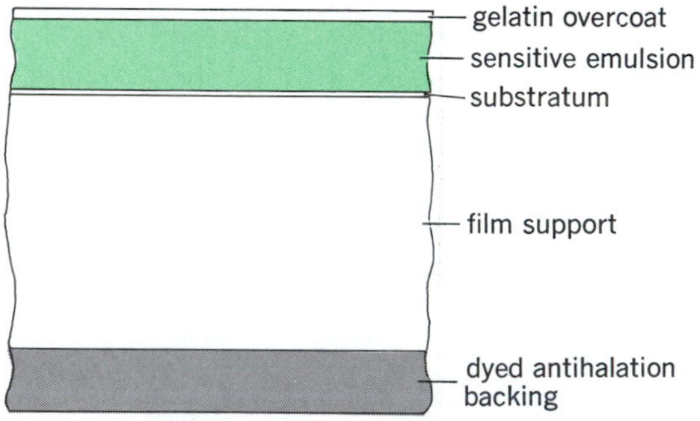

Diagrammatic cross section of film.

they can be made sensitive to longer wavelengths by adding special dyes to the emulsion; these dyes are usually of the carbocyanine and merocyanine class, and the process of adding them to the emulsion is known as optical sensitizing. To reproduce tone values as seen by the eye, emulsions must respond to wavelengths to which the eye is sensitive, that is, approximately 400–700 nanometers, or from violet to red.

Nonsensitized emulsions are also termed blue-sensitive, color-blind, or ordinary. Emulsions that have been treated with dyes to extend the sensitivity through the green are known as orthochromatic; when the sensitivity is extended through the red they are known as panchromatic. With ordinary emulsions, blues are reproduced light and greens and reds dark. Orthochromatic emulsions reproduce blues and greens as light and reds as dark, and they are used extensively in portraiture, commercial and industrial photography, and the graphic arts. Panchromatic emulsions give reasonably good reproduction in black and white of the tone values of colored subjects and are particularly useful with incandescent light sources. The furthest extension of sensitizing by dyes is to about 1300 nm, which is in the near-infrared region. Ultraviolet sensitivity is limited by the strong optical absorption by gelatin below 28 nm, but emulsions sensitive to much shorter wavelengths can be prepared with the gelatin greatly reduced.

Photographic products. Thousands of types and sizes of plate, film, and paper are available for a wide variety of applications. Generally, each field of use requires special properties. Amateur, professional, commercial, motion-picture, and industrial products for camera use range from an exposure index of less than 10 to over 1000 and from very fine grain and optimum sharpness to fairly coarse grain with corresponding loss in definition. They can be obtained in a wide range of contrasts and spectral sensitivity, especially for scientific and industrial photography. Films of extremely high contrast and density are used in graphic reproduction in industry and the printing trade to give high-contrast line and halftone negatives. Blue-sensitive emulsions are used for copying and making duplicate negatives. Many types of x-ray film are made, some coated on both sides and used with or without intensifying screens. Special multiple coatings are used for color photography.

Nonsilver processes. There are many special-purpose photographic processes that use compounds other than silver salts. In general, they have neither the sensitivity to light possible with silver compounds nor the amplification factor of development. They are largely used in the so-called reprography methods. They include inorganic compounds which are photosensitive, especially iron salts and dichromates in colloid layers such as gelatin, glue, albumen, shellac, polyvinyl alcohol, and other synthetic resins which become insoluble on exposure. Many important processes are based on light-sensitive organic compounds, especially diazo compounds; unsaturated compounds such as cinnamic acid derivatives, which are insolubilized by cross-linking on exposure; systems in which polymerization occurs as a result of free-radical formation on exposure; dyes which are formed or destroyed directly by light action; and thermographic systems, in which a thermal pattern produces an image by means of melting a composition or initiating a chemical reaction, or by producing a deformation in an electrostatically charged layer. Electrophotographic systems rely on the elimination of a charge by exposure of a charged layer and the subsequent development of an electrostatic image by charged powder, or by electrolytic deposition of metal in an image-sensitive photoconducting system. [W.Cl.]

Photography

The process of forming visible images directly or indirectly by the action of light or other forms of radiation on sensitive surfaces. In the traditional sense, photography utilizes the action of light to bring about changes in silver halides. These changes may be invisible, necessitating a developer to reveal the image, or they may be a directly visible darkening (print out). Most photography is of the first kind and the function of the developer is to convert the exposed silver halide to silver. The bright parts of the subject give more exposure than the dark parts so that a negative results; that is, the brighter parts of the subject correspond to the darker parts of the reproduction. A positive, in which the relation between light and dark areas corresponds to that of the subject, is obtained when a negative is printed onto a sheet of similar material so that the negative tones are reversed. In the reversal process, direct production of the positive occurs if the developed negative silver is removed chemically and the remaining silver halide is then redeveloped; direct positive images can also be obtained directly by using special materials. *See* PHOTOGRAPHIC MATERIALS.

The major branches of photography include infrared, ultraviolet, high-speed, and stereoscopic photography, photographic photometry, nuclear-particle recording, and microphotography (including microfilming). *See* AERIAL PHOTOGRAPH; ASTRONOMICAL PHOTOGRAPHY; CAMERA; CINEMATOGRAPHY; LENS (OPTICS); MICRORADIOGRAPHY; OPTICAL MICROSCOPE; OSCILLOSCOPE; PHOTOELASTICITY; PHOTOGRAMMETRY; RADIOGRAPHY; SCHLIEREN PHOTOGRAPHY; SPECTROGRAPHY; STROBOSCOPIC PHOTOGRAPHY; UNDERWATER PHOTOGRAPHY.

Infrared photography. Emulsions with special sensitizing dyes can respond to radiation up to about 1300 nanometers. Photographs can thus be made of subjects associated with radiation in the near-infrared, such as spectra, stars, and hot objects. Subjects which selectively transmit, emit, or reflect near-infrared radiation, especially in a manner different from visible radiation, can also be photographed. Infrared photographs taken from long distances or from high altitudes show improved clarity of detail because the atmosphere may selectively transmit the near-infrared and also because the contrast of ground objects may be higher as a result of their different reflectivities in the near infrared. Grass and foliage appear white because chlorophyll is transparent to the near-infrared.

Infrared photography is used to record the distribution of temperature at the surface of heated objects; for photography in total darkness (the object being illuminated only by infrared); in criminology, for deciphering altered or deteriorated documents and other objects; and for photographing textiles where dark dyes interfere with visual examination. It is used in medicine because the skin is somewhat transparent to infrared and the subcutaneous veins may be revealed for diagnosis. *See* INFRARED RADIATION.

Ultraviolet photography. This is used in the printing field, in spectrography, and in photomicrography and by mineralogists, police investigators, museums, and art galleries. The two methods of ultraviolet photography are (1) the fluorescence method, in which the subject is illuminated by ultraviolet and a filter is used on the camera to absorb the reflected ultraviolet and permit only the visible fluorescence to reach the film; and (2) the reflected ultraviolet method, in which an ultraviolet source is used and the camera is provided with a filter which permits only ultraviolet to reach the film. *See* ULTRAVIOLET RADIATION.

High-speed photography. High-speed photography deals with photography at exposure times shorter than those given by normal shutters, and at picture (frame) frequencies faster than those given by motion-picture cameras with intermittent film movement. It serves a great variety of purposes in technological studies.

The best normal still-camera mechanical shutters give exposures not much shorter than 1/1000 s. Very short exposures are given by magnetooptical shutters (Faraday effect), electrooptical shutters (Kerr effect), and pulsed electron image tubes. So-called pulsed light sources are used to give very intense illumination of short duration. They include the xenon gaseous-discharge tubes, an electric spark in air, exploding wires, the argon flash bomb, high-voltage flying-spot cathode-ray tubes, and lasers. High-speed single x-ray pictures have been taken by discharging a short-duration high potential through the x-ray tube. High-intensity repetitive flashes can be obtained with xenon flash tubes for so-called stroboscopic photography with frequencies up to 10^5 per second.

The classical camera for serial frame separation is the motion-picture camera. Normal cameras with intermittent movement of film are usually limited to 128 frames/s (standard rates are 16 or 24). For higher rates (up to 10,000 frames/s or more), continuous film movement is used with optical compensation for image motion, such as a rotating plane-parallel glass block or a ring of mirrors or lenses.

There is a mechanical limit to the rate at which film can be moved. Higher speeds of movement are obtained by using stationary film and moving the image by means of a rapidly rotating mirror.

Photomicrography. The apparatus for photography of the image formed by the microscope consists essentially of an optical bench on which are aligned an illuminating system, a compound microscope, and a lensless camera connected to the microscope eyepiece by a light-tight extension. Good photomicrographs can be made by attaching a simple camera to a microscope. Resolving power is more important than mere magnification. Maximum resolution is given by the electron microscope. *See* METALLOGRAPHY; MICROSCOPE.

Stereoscopic photography. Stereoscopic photography is possible with a single camera if two separate photographs are made, one after the other, from viewpoints separated by the interocular or other appropriate distance, by displacing the lens of a single camera for the two exposures, or by rotating the object or the camera so as to give a pair of exposures on one film. Most stereoscopic photography is done either by the simultaneous method, in which two photographs are made at the same time with two separate cameras; with a stereoscopic camera, essentially two cameras in one body with matched optical systems and coupled focusing movements; or with single cameras using beam splitters to give two photographs side by side on the film. *See* STEREOSCOPY.

Photographic photometry. The intensity of radiation, or the spectral distribution of intensity, can be measured by photography. The radiation whose intensity is to be measured is compared with that from a standard source by matching the photographic densities produced by both. The method is capable of high precision if the characteristics of photographic materials are accurately known and the results are interpreted intelligently. *See* PHOTOMETRY.

Microphotography. The process of making photographs on a greatly reduced scale is termed microphotography. Microfilming is the special technique of copying documents to reduced size on film, usually 16- and 35-mm film, but 70- and 102-mm widths are used in rolls, strips, and assemblages of strips, as well as sheet film. The negatives, or contact positive films or paper prints made from them, can be read in enlarging readers. Special projection equipment for reading the films may give an enlargement onto a table or a diffusing rear-projection screen, with provision for rapidly winding the film and framing any desired page. In reader-printers the image may be viewed and, if desired, a paper print may be made in the reader by stabilization processing or electrophotographic means. *See* COLOR PHOTOGRAPHY. [W.Cl.]

Photoionization

The ejection of one or more electrons from an atom, molecule, or positive ion following the absorption of one or more photons. The process of electron ejection from matter following the absorption of electromagnetic radiation has been under investigation for over a century. The earliest measurements involved the ultraviolet irradiation of metal surfaces. The theoretical interpretation of this phenomenon, known as the photoelectric effect, played an important role in establishing quantum mechanics. It was shown that, contrary to classical ideas, energy exchanges between radiation and matter are mediated by integral numbers of photons. In the gas phase the photoeffect is called either photoionization (atoms, molecules, and their positive ions) or photodetachment (atomic and molecular negative ions). *See* PHOTOEMISSION.

Photoionization involves a radiative bound-free transition from an initial state consisting of *n* photons and an atom, molecule, or ion in a bound state to a final continuum state consisting of a residual ion (or an atom in the case of photodetachment) and *m* free electrons: that is,

$$nh\nu + X \rightarrow X^{m+} + me^-$$

In the simplest atomic photoionization process a single electron is ejected from an atom following the absorption of a single photon. Each mode of fragmentation defines a final-state channel that is characterized by the energy and angular momentum of the outgoing electron as well as the excitation state of the residual ion. Since the photoionization process is endoergic, each channel has a well-defined threshold energy below which the channel is energetically closed. The threshold photon energy for a particular channel is equal to the binding energy of the electron that is to be ejected plus the excitation energy, if any, of the residual ion.

Above threshold, the energy carried off by the outgoing electron represents the balance between the energy supplied by the photon and the binding energy of the electron plus the excitation energy of the residual ion (neglecting the small recoil of the heavy ion). A photoelectron spectrum is characterized by a discrete set of peaks, each peak being associated with a particular state of the residual ion. Information on the excitation state of the ion following photoionization can also be obtained by monitoring the fluorescence emitted in the subsequent radiative decay of the state. One of the earliest applications of photoionization measurements was the investigation of the structure of atoms by determining the binding energies of both outer- and inner-shell electrons by means of photoelectron spectroscopy. *See* ATOMIC STRUCTURE AND SPECTRA; ELECTRON SPECTROSCOPY. [D.J.Pe.]

Photoluminescence

A luminescence excited in a body by some form of electromagnetic radiation incident on the body. The term photoluminescence is generally limited to cases in which the incident radiation is in the ultraviolet, visible, or infrared regions of the electromagnetic spectrum.

Photoluminescence may be either a fluorescence or a phosphorescence, or both. Energy can be stored in certain luminescent materials by subjecting them to light or some other exciting agent, and can be released by subsequent illumination of the material with light of certain wavelengths. This type of photoluminescence is called stimulated photoluminescence. *See* FLUORESCENCE; LUMINESCENCE; PHOSPHORESCENCE. [C.C.K.; J.H.S.]

Photolysis

Chemical decomposition by the action of radiant electromagnetic energy, especially light. Photolysis occurs in certain crystals, notably the silver halides, when they are exposed to radiation. When this occurs, the effect of the radiation is to produce a definite chemical change resulting in the separation of photolytic silver. Photolysis occurs in many other materials, such as the lead and thallium halides, zinc oxide, the metallic azides, and in organic compounds such as the oxalates, styphnates, and fulminates.

A photographic emulsion consists of microcrystalline grains of silver bromide, AgBr, or silver chloride, AgCl, embedded in gelatin. Upon prolonged exposure to light, so-called print-out specks of silver form within and on the surface of the grains. Much shorter exposures produce a latent image which can be made visible by the process of development. Experiments have shown that the latent image consists of only a very few atoms of silver in each grain. The high sensitivity of the photographic system comes about because of the enormous gain (10^9) which can be achieved by reduction of each exposed grain that occurs during development.

The Gurney-Mott theory of the photographic process proposes a two-stage mechanism. In the first stage a light quantum is absorbed at a point within the silver halide grain, releasing a mobile electron and a positive hole. These mobile defects diffuse to trapping sites (sensitivity centers) within the volume or on the surface of the grain. In the second stage, the trapped (negatively charged) electron is neutralized by an interstitial (positively charged) silver ion, which combines with the electron to form a silver atom. The silver atom at the sensitivity center is capable of trapping a second electron, after which the process repeats itself, causing the silver speck to grow. The positive holes are assumed to diffuse to the surface without recombining with electrons, where they escape or react with the gelatin.

The early Gurney-Mott theory has been criticized, especially because of the assumed lack of electron-hole recombination. The essential idea of an electronic process linking the initial absorption of light quanta with the formation of the image speck is nevertheless well founded. *See* PHOTOCHEMISTRY; PHOTOGRAPHIC MATERIALS. [F.C.Br.]

Photometer An instrument used for making measurements of light, or electromagnetic radiation, in the visible range. In general, photometers may be divided into two classifications: laboratory photometers, which are usually fixed in position and yield results of high accuracy; and portable photometers, which are used in the field or outside the laboratory and yield results of lower accuracy. Each class may be subdivided into visual (subjective) photometers and photoelectric (objective or physical) photometers. These in turn may be grouped according to function, such as photometers to measure luminous intensity (candelas or candlepower), luminous flux, illumination (illuminance), luminance (photometric brightness), light distribution, light reflectance and transmittance, color, spectral distribution, and visibility. Visual photometric methods have largely been supplanted commercially by physical methods, but because of their simplicity, visual methods are still used in educational laboratories to demonstrate photometric principles. *See* ILLUMINANCE; LUMINANCE; LUMINOUS FLUX; LUMINOUS INTENSITY. [G.A.Ho.]

Photometry That branch of science which deals with the calculation and measurement of light or of its time rate of flow. The term light is usually restricted to electromagnetic radiations of wavelengths that are capable of affecting the human eye. However, in some cases photometry has come to mean also the measurement of radiations in the nearby ultraviolet and infrared regions.

Photometry is usually concerned with measurements of luminous intensity, luminous flux, luminous flux density, luminance, light distribution, color, and the reflectance and transmittance of light; by extension it may even include visibility measurements. Photometric tests are made of light sources, lighting fixtures (luminaires), lighting materials, and lighting installations. *See* ILLUMINANCE; LUMINANCE; LUMINOUS FLUX; LUMINOUS INTENSITY.

Since light is defined as radiant energy that is capable of producing visual sensation, photometric measurements either are made by the human eye or are based upon its visual responses. Because the spectral response of the human eye differs with the individual, the International Commission on Illumination (ICI) has adopted a standard spectral luminous efficiency curve which has been accepted as being that of photopic vision of the normal eye. *See* VISION.

Modernization of photometric measurements using photoelectric solid-state cells rather than visual observers has removed the problem of the visual individuality of the observer. Such cells generate electric currents when irradiated and cover the range of the visual spectrum. Although their spectral response is not the same as the standard ICI spectral luminous efficiency curve, their response can be modified by filters to match the ICI curve. Also, geometric corrections can be accomplished by properly shaped translucent shields to correct for deviations from Lambert's cosine law of incidence caused by shadows or reflections. *See* PHOTOCONDUCTIVE CELL.

Although visual photometry has been largely supplanted by physical photometry, photometric measurements in both cases are eventually based on the primary standard of luminous intensity, the candela, cd. In October 1979 the General Conference on Weights and Measures redefined the base SI unit candela as the luminous intensity, in a given direction, of a source that emits monochromatic radiation of frequency 540×10^{12} hertz and of which the radiant intensity in that direction is 1/683 watt per steradian. The specified frequency corresponds to a wavelength of 555 nanometers in free space, which is at the maximum visual response of the human eye. *See* ILLUMINATION. [W.B.Bo.]

Photomorphogenesis The regulatory effect of light on plant form, involving growth, development, and differentiation of cells, tissues, and organs. Morphogenic influences of light on plant form are quite different from light effects that nourish the plant through photosynthesis, since the former usually occur at much lower energy levels than are necessary for photosynthesis. Light serves as a trigger in photomorphogenesis, frequently resulting in energy expenditure orders of magnitude larger than the amount required to induce a given response. Photomorphogenic processes determine the nature and direction of a plant's growth and thus play a key role in its ecological adaptations to various environmental changes. *See* PHOTOSYNTHESIS.

Morphogenically active radiation is known to control seed and spore germination, growth and development of stems and leaves, lateral root initiation, opening of the hypocotyl or epicotyl hook in seedlings, differentiation of the epidermis, formation of epidermal hairs, onset of flowering, formation of tracheary elements in the stem, and form changes in the gametophytic phase of ferns, to mention but a few of such known phenomena. Many nonmorphogenic processes in plants are also basically controlled by light independent of photosynthesis. Among these are chloroplast movement, biochemical reactions involved in the synthesis of flavonoids, anthocyanins, chlorophyll, and carotenoids, and leaf movements in certain legumes. [W.R.Br.]

Photomultiplier A very sensitive vacuum-tube detector of light or radiant flux containing a photocathode which converts the light to photoelectrons; one or more secondary-electron-emitting electrodes or dynodes which amplify the number of photoelectrons; and an output electrode or anode which collects the secondary electrons and provides the electrical output signal. It is also known as a multiplier phototube. Because of the very large amplification provided by the secondary-emission mechanism, and the very short time variation associated with the passage of the electrons within the device, the photomultiplier is applied to the detection and measurement of very low light levels, especially if very high speed of response is required.

The illustration is a schematic of a typical photomultiplier and shows its operation. Light incident on a semitransparent photocathode located inside an evacuated envelope causes photoelectron emission from the opposite side of the photocathode. The efficiency of the photoemission process is called the quantum efficiency, and is the ratio of emitted photoelec-

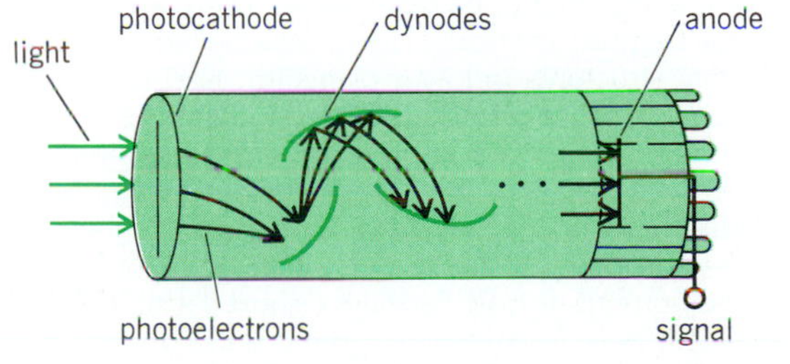

Schematic of a photomultiplier. (*After P. W. Engstrom, Photomultipliers—then and now, RCA Eng., 24(1):18–26, June–July 1978*)

trons to incident photons (light particles). Photoelectrons are directed by an accelerating electric field to the first dynode, where from 3 to 30 secondary electrons are emitted for each incedent electron, depending upon the dynode material and the applied voltage. These secondaries are directed to the second dynode, where the process is repeated and so on until the multiplied electrons from the last dynode are collected by the anode.

A typical photomultiplier may have 10 stages of secondary emission and may be operated with an overall applied voltage of 2000 V. In most photomultipliers the focusing of the electron streams is done by electrostatic fields shaped by the design of the electrodes. Some special photomultipliers designed for very high speed utilize crossed electrostatic and magnetic fields which direct the electrons in approximate cycloidal paths between electrodes.

Ever since their invention, photomultipliers have been found useful in low-level photometry and spectrometry. The most important applications of photomultipliers are related to scintillation counting, and these form the basis of a whole science of tracer chemistry that has been applied to agriculture, medicine, and industrial problems. In medicine, photomultiplier-scintillator combinations are used in the gamma-ray camera, computerized tomography, and the positron scanner. *See* COMPUTERIZED TOMOGRAPHY; GAMMA-RAY DETECTORS; NUCLEAR MEDICINE; RADIOLOGY; SCINTILLATION COUNTER. [R.W.E.]

Photon A quantum of a single mode (that is, single wavelength, direction, and polarization) of the electromagnetic field. There are also two other definitions of photon in use, not entirely consistent with the first definition or each other: an elementary light particle or "fuzzy ball," and an informal unit of light energy. The fuzzy-ball definition emphasizes a particle character of light suggested, for example, by momentum exhibited in the Compton effect and light levitation phenomena. However, the fuzzy-ball picture lacks a rigorous foundation and is not required for the explanation of any fundamental phenomenon. As an informal unit of energy, the photon equals $h\nu$, where h is Planck's constant ($= 6.626 \times 10^{-34}$ joule-second), and ν is the frequency of the light in hertz. *See* COMPTON EFFECT; PHOTOEMISSION; QUANTUM (PHYSICS). [M.Sa.]

Photoperiodism The growth, development, or other responses of organisms to the length of night or day or both. Photoperiodism has been observed in plants and animals, but not in bacteria (prokaryotic organisms), other single-celled organisms, or fungi.

A true photoperiodism response is a response to the changing day or night. Some species respond to increasing day lengths and decreasing night lengths (for example, by forming flowers or developing larger gonads); this is called a long-day response. Other species may exhibit the same response, or the same species may respond in some different way, to decreasing days and increasing nights; this is a short-day response. Sometimes a response is independent or nearly independent of day length, and is said to be day-neutral. There are many plant responses to photoperiod. These include development of reproductive structures in lower plants (mosses) and in flowering plants; rate of flower and fruit development; stem elongation in many herbaceous species as well as coniferous and deciduous trees (usually a long-day response and possibly the most widespread photoperiodism response in higher plants); autumn leaf drop and formation of winter dormant buds (short days); development of frost hardiness (short days); formation of roots on cuttings; formation of many underground storage organs such as bulbs (onions, long days), tubers (potato, short days), and storage roots (radish, short days); runner development (strawberry, short day); balance of male to female flowers or flower parts (especially in cucumbers); aging of leaves and other plant parts; and even such obscure responses as the formation of foliar plantlets (such as the minute plants formed on edges of *Bryophyllum* leaves), and the quality and quantity of essential oils (such as those produced by jasmine plants). Note that a single plant, for example, the strawberry, might be a short-day plant for one response and a long-day plant for another response.

Animal responses. There are also many responses to photoperiod in animals, including control of several stages in the life cycle of insects (for example, diapause) and the long-day promotion in birds of molting, development of gonads, deposition of body fat, and migratory behavior. Even feather color may be influenced by photoperiod (as in the ptarmigan). In several mammals the induction of estrus and spermatogenic activity is controlled by photoperiod (sheep, goat, snowshoe hare), as is fur color in certain species (snowshoe hare). Growth of antlers in American elk and deer can be controlled by controlling day length. Increasing day length causes antlers to grow, whereas decreasing day length causes them to fall off. By changing day lengths rapidly, a cycle of antler growth can be completed in as little as 4 months; slow changes can extend the cycle to as long as 2 years. When attempts are made to shorten or extend these limits even more, the cycle slips out of photoperiodic control and reverts to a 10–12-month cycle, apparently controlled by an internal annual "clock."

Seasonal responses. Response to photoperiod means that a given manifestation will occur at some specific time during the year. Response to long days (shortening nights) normally occurs during the spring, and response to short days (lengthening nights) usually occurs in late summer or autumn. Since day length is accurately determined by the Earth's rotation on its tilted axis as it revolves in its orbit around the Sun, detection of day length provides an extremely accurate means of determining the season at a given latitude. Such other environmental factors as temperature and light levels also vary with the seasons but are clearly much less dependable from year to year.

Mechanisms. It has long been the goal of researchers on photoperiodism to understand the plant or animal mechanisms that account for the responses. Light must be detected, the duration of light or darkness must be measured, and this time measurement must be metabolically translated into the observed response: flowering, stem elongation, gonad development, fur color, and so forth. In spite of many years of intensive research, relatively little is known about these mechanisms. Enough is known, however, to suggest that the basic mechanisms differ not only between plants and animals but among different species as well. The roles (synchronization, anticipation, and so on) are similar in all organisms that exhibit photoperiodism, but the mechanisms through which these roles are achieved are apparently quite varied.

Strongest inhibition of flowering in short-day plants comes when the light interruption occurs around the time of the critical night (about 7–9 h for cocklebur plants), but actual effectiveness also depends on the length of the dark period. With short-day cockleburs, the shorter the night, the less the flowering and the longer the time that light inhibits flowering.

Orange-red wavelengths used as a night interruption are by far the most effective part of the spectrum in inhibition of short-day responses and promotion of long-day responses (flowering in most studies), and effects of orange-red light can be completely reversed by subsequent exposure of plants to light of somewhat longer wavelengths, called far-red light. These observations led in the early 1950s to discovery of the phytochrome pigment system, which is apparently the molecular machinery that detects the light effective in photoperiodism of higher plants. *See* PHYTOCHROME.

In photoperiodism of short-day plants, an optimum response is usually obtained when phytochrome is in the far-red receptive form during the day and the red-receptive form during the night. Although normal daylight contains a balance of red and far-red wavelengths, the red-receptive form is most sensitive, so the pigment under normal daylight conditions is driven mostly to the far-red receptive form. At dusk this form is changed metabolically, and the red-receptive form builds up. It is apparently this shift in the form of phytochrome that initiates measurement of the dark period. This is how a plant "sees": when the far-red-sensitive form of the pigment is abundant, the plant "knows" it is in the light; the red-sensitive form (or lack of far-red form) indicates to the plant's biochemistry that it is in the dark.

The measurement of time—the durations of the day or night—is the very essence of photoperiodism. The discovery of a biological clock in living organisms was made in the late 1920s. It was shown that the movement of leaves on a bean plant (from horizontal at noon to vertical at midnight) continued uninterruptedly for several days, even when plants were placed in total darkness and at a constant temperature, and that the time between given points in the cycle (such as the most vertical leaf position) was almost but not exactly 24 h. In the case of bean leaves, it was about 25.4 h. Many other cycles have now been found with similar characteristics in virtually all groups of plants and animals. There is strong evidence that the clocks are internal and not driven by some daily change in the environment. Such rhythms are called circadian.

Circadian rhythms usually have period lengths that are remarkably temperature-insensitive, which is also true of time measurement in photoperiodism. Furthermore, the rhythms are normally highly sensitive to light, which may shift the cycle to some extent. Thus, daily rhythms in nature are normally synchronized with the daily cycle as the Sun rises and sets each day. Their circadian nature appears only when they are allowed to manifest themselves under constant conditions of light (or darkness) and temperature, so that their free-running periods can appear. [F.B.S.]

Photophore gland A highly modified integumentary gland which arises from an epithelial invagination into the dermis. It becomes cut off from its site of origin and develops into a luminous organ composed of a lens and a light-emitting gland, at the back of which is a pigmented reflector of probably dermal-cell origin. These luminous bodies occur in deep-sea teleosts and elasmobranchs which live in areas of total darkness. *See* EPITHELIUM; GLAND. [O.E.N.]

Photoreception The process of absorption of light energy by plants and animals and its utilization for biologically important purposes. In plants photoreception plays an essential role in photosynthesis and an important role in orientation. Photoreception in animals is the initial process in vision. *See* PHOTOSYNTHESIS; TAXIS; VISION.

The photoreceptors of animals are highly specialized cells or cell groups which are light-sensitive because they contain pigments which are unstable in the presence of light of appropriate wavelengths. These light-sensitive receptor pigments absorb radiant energy and then undergo physicochemical changes, which lead to the initiation of nerve impulses that are conducted to the central nervous system. *See* EYE (INVERTEBRATE); EYE (VERTEBRATE). [V.J.W.]

Photorespiration Plant respiration (especially CO_2 evolution) during exposure to light; it differs biochemically from normal respiration occurring in the dark and is specifically associated with the oxidation of compounds produced during photosynthesis. The rate of photorespiratory CO_2 release is often three to five times greater than the rate of dark respiration, and the biochemical reactions responsible for photorespiration are distinct from those of dark respiration. Strong evidence suggests that by blocking photorespiration, the net photosynthetic incorporation of CO_2 in many species could be increased by at least 50%. *See* PHOTOSYNTHESIS.

The rate of glycolic acid synthesis, which occurs largely in leaf chloroplasts, is probably the most important factor controlling photorespiration. A number of the biochemical reactions responsible for the synthesis of glycolic acid in any given photosynthetic tissue are known. At least several of these may occur simultaneously. The CO_2 produced during photorespiration arises from one or more reactions of the glycolate pathway. Photorespiratory CO_2 is released from the oxidative decarboxylation of glyoxylic acid and during the conversion of glycine to serine. *See* CELL PLASTIDS.

It has often been suggested that a function of photorespiration is to protect the chloroplast against photooxidative destruction in an environment of high irradiance and low CO_2 concentration. This hypothesis implies that photorespiration consumes toxic oxidants, thereby converting compounds to CO_2 and protecting the photochemical apparatus. Photosynthetic tissues possess a number of mechanisms for destroying toxic oxidants, however, and the mesophyll chloroplasts of C_4 species are protected against photooxidative damage without having an active photorespiratory system. The hypothesis also fails to consider that species occur that are intermediate between C_3 and C_4 species in their photorespiration. Some C_3 plants also have lower-than-usual rates of photorespiration. *See* PLANT RESPIRATION. [I.Z.]

Photosphere The visible surface of the Sun or other stars. The photosphere of the Sun is a gaseous layer a few hundred kilometers thick with an average effective temperature of 5780 K (9940°F), determined from the total radiation per square centimeter. *See* SUN. [J.W.E.]

Photosynthesis The light-dependent conversion of radiant energy into chemical energy as adenosine triphosphate (ATP) and reduced nicotinamide adenine dinucleotide phosphate (NADP) serves as a prelude to the utilization of these compounds for the reductive fixation of CO_2 into organic molecules. Such molecules, broadly designated as photosynthates, are usually but not invariably in the form of carbohydrates such as glucose, and form the base for the nutrition of all living things, as well as serving as the starting materials for fuel, fiber, animal feed, oil, and other compounds used by people. Collectively, the biochemical processes by which CO_2 is assimilated into organic molecules are known as the photosynthetic dark reactions, not because they must occur in darkness, but because light—in contrast to the photosynthetic light reactions—is not required.

C_3 photosynthesis. The essential details of C_3 (the reductive pentose cycle) photosynthesis can be seen in Fig. 1. Three molecules of CO_2 combine with three molecules of the five-carbon compound ribulose bisphosphate (RuBP) in a reaction catalyzed by RuBP carboxylase to form three molecules of an enzyme-bound six-carbon compound. These are hydrolyzed into six molecules of the three-carbon compound phosphoglyceric acid (PGA), which are phosphorylated by the conversion of six molecules of ATP (releasing adenosine diphosphate, ADP, for photophosphorylation via the light reactions). The resulting compounds are reduced by the NADP formed in photosynthetic light reactions to form six molecules of the three-carbon compound phosphoglyceraldehyde (PGAL). One molecule of PGAL is made available for combination with another PGAL (requiring a second "turn" of the Calvin-cycle wheel) to form a six-carbon sugar, and the other five PGAL molecules, through a complex series of enzymatic reactions, are rearranged into three molecules of RuBP, which can again be carboxylated with CO_2 to start the cycle turning again. The resulting sugar is usually

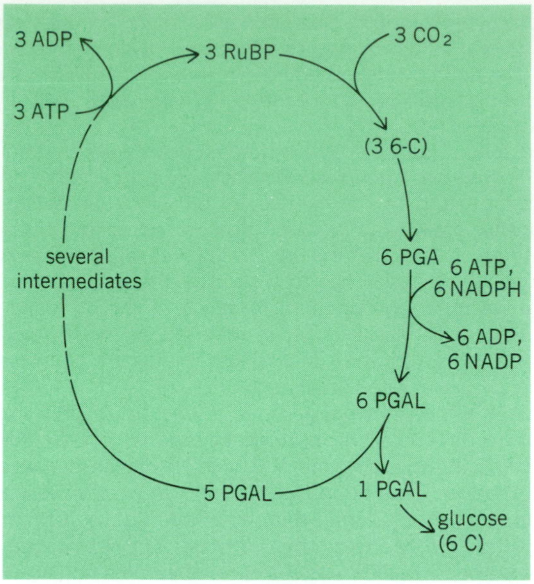

Fig. 1. **Schematic outline of the Calvin (C_3) carbon dioxide assimilation cycle.**

glucose-6-phosphate, which can be enzymatically converted into sucrose or other compounds that may remain in the leaf or may be transported from the leaf to other cells.

C_4 photosynthesis. Initially, the C_3 cycle was thought to be the only route for CO_2 assimilation, although it was recognized by plant anatomists that some rapidly growing plants (such as maize, sugarcane, and sorghum) possessed an unusual organization of the photosynthetic tissues in their leaves (Kranz morphology). Plants having the Kranz anatomy utilize an additional CO_2 assimilation route now known as the C_4–dicarboxylic acid pathway. The essential features of C_4 photosynthesis are seen in Fig. 2. Carbon dioxide enters a mesophyll cell, where it combines with the three-carbon compound phosphoenolpyruvate (PEP) to form a four-carbon acid, oxaloacetic acid, which is reduced to malic acid, or transmitted to aspartic acid. The malic acid moves into bundle sheath cells where the acid is decarboxylated, the CO_2 is assimilated via the C_3 cycle, and the resulting three-carbon compound, pyruvic acid, moves back into the mesophyll cell and is transformed into PEP, which can be carboxylated again. The two cell types, mesophyll and bundle sheath, are not necessarily adjacent (sedges are an example), but in all documented cases of C_4 photosynthesis the organism had two distinct types of green cells. C_4 metabolism is classified into three types, depending on the decarboxylation reaction used with the four-carbon acid in the bundle sheath cells.

1. NADP-ME type (sorghum):

$$\text{NADP}^+ + \text{malic acid} \xrightarrow{\text{NADP–malic enzyme}} \text{pyruvic acid} + CO_2 + \text{NADPH}$$

2. NAD-ME type (*Atriplex* species):

$$\text{NAD}^+ + \text{malic acid} \xrightarrow{\text{NAD–malic enzyme}} \text{pyruvic acid} + CO_2 + \text{NADH}$$

3. PCK type (*Panicum* species):

$$\text{Oxaloacetic acid} + \text{ATP} \xrightarrow[\text{carboxykinase}]{\text{Phosphoenolpyruvate}} \text{PEP} + CO_2 + \text{ADP}$$

CAM photosynthesis. Under arid and desert conditions, where soil water is in short supply, transpiration during the day when temperatures are high and humidity is low may rapidly deplete the plant of water, leading to desiccation and death. By keeping stomates closed during the day, water can be conserved, but the uptake of CO_2, which occurs entirely through the stomata, is prevented. Desert plants in the Crassulaceae, Cactaceae, Euphorbiaceae, and 15 other families evolved, apparently independently of C_4 plants, an almost identical strategy of assimilating CO_2 by which the CO_2 is taken in at night when the stomata open; water loss is low because of the reduced temperatures and correspondingly higher humidities. First studied in plants of the Crassulaceae, the process has been called crassulacean acid metabolism (CAM).

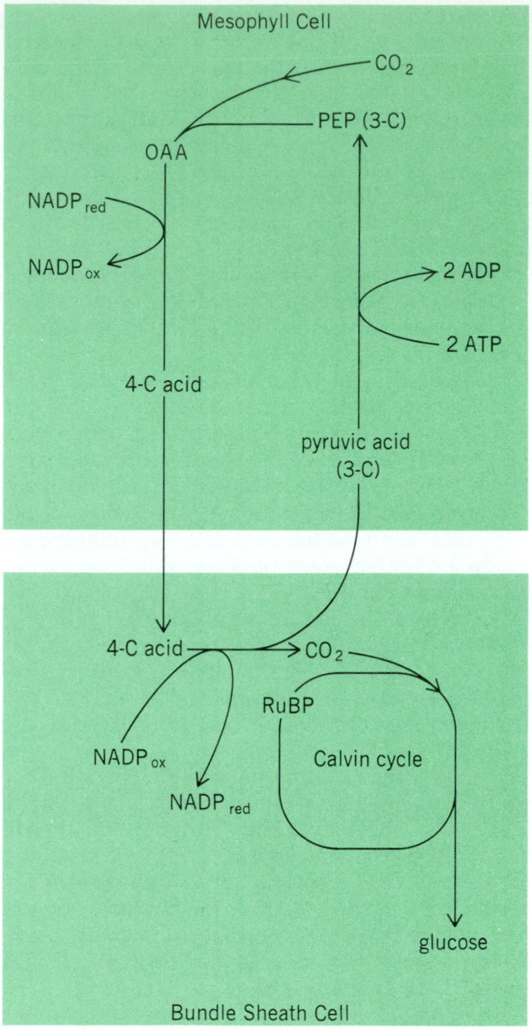

Fig. 2. **Schematic outline of the Hatch-Slack (C_4) carbon dioxide assimilation route in two cell types.**

In contrast to C_4, where two cell types cooperate, the entire process occurs within an individual cell. At night, CO_2 combines with PEP through the action of PEP carboxylase, resulting in the formation of oxaloacetic acid and its conversion into malic acid. The PEP is formed from starch or sugar via the glycolytic route of respiration. Thus, there is a daily reciprocal relationship between starch (a storage product of C_3 photosynthesis) and the accumulation of malic acid (the terminal product of nighttime CO_2 assimilation). [M.G.; Y.W.K.; C.C.B.]

Phototransistor A semiconductor device with electrical characteristics that are light-sensitive. Phototransistors differ from photodiodes in that the primary photoelectric current is multiplied internally in the device, thus increasing the sensitivity to light. *See* PHOTODIODE; TRANSISTOR.

Some types of phototransistors are supplied with a third, or base, lead. This lead enables the phototransistor to be used as a switching, or bistable, device. The application of a small amount of light causes the device to switch from a low current to a high current condition. *See* PHOTOELECTRIC DEVICES. [W.R.Si.]

Phototrophic bacteria Green and purple bacteria which thrive under anaerobic conditions and carry out photosynthesis without oxygen production, or anoxygenic photosynthesis. For the photoassimilation of CO_2, molecular nitrogen, and simple organic substrates, phototrophic bacteria require external hydrogen donors of a lower redox potential than water, such as hydrogen sulfide, elemental sulfur, thiosulfate, molecular hydrogen, or highly reduced organic compounds. The photosynthetic pigments of these bacteria are different from those of all other phototrophic organisms; they consist of the bacteriochlorophylls *a*, *b*, *c*, *d*, and *e*. In addition, a great variety of carotenoids are present which give phototrophic bacteria their characteristic colors. Most species multiply by binary fission. Under anaerobic conditions in the dark all species perform a fermentative metabolism.

The green and purple bacteria share their basic prokaryotic cellular organization with a third group of phototrophic bacteria, the cyanobacteria or blue-green algae. These bacteria perform the same type of oxygen-evolving (oxygenic) photosynthesis as the eukaryotic algae and higher plants; they are, therefore, conventionally classified with these groups of organisms. *See* PHOTOSYNTHESIS.

Two physiological-ecological groups are distinguished among both the green and the purple bacteria and taxonomically established in four families:

Phototrophic green bacteria (Chlorobiineae)
1. Chlorobiaceae, mesophilic green sulfur bacteria
2. Chloroflexaceae, thermophilic gliding green bacteria

Phototrophic purple bacteria (Rhodospirillineae)
3. Chromatiaceae, purple sulfur bacteria
4. Rhodospirillaceae, purple nonsulfur bacteria

The Chlorobiaceae constitute a small and physiologically uniform group of nonmotile bacteria. All species are strictly anaerobic and depend on both hydrogen sulfide and light for their development. In nature the Chlorobiaceae thrive in the anaerobic, sulfide-containing parts of all kinds of aquatic environments exposed to light, for example, ditches, ponds, lakes, rivers, sulfur springs, estuaries, and muddy marine habitats.

The only species of the family Chloroflexaceae is *Chloroflexus aurantiacus*. All strains are thermophilic and grow best between 122 and 140°F (50 and 60°C). The cells occur in long filaments and exhibit gliding motility on solid surfaces. *Chloroflexus* grows under anaerobic conditions with simple organic compounds; the color of the cells is dull green. The natural habitats of *Chloroflexus* are hot springs all over the world.

Like the Chlorobiaceae, the Chromatiaceae are strictly anaerobic and obligately phototrophic; their CO_2 assimilation is linked to the oxidation of sulfide and sulfur to sulfate. In nature, blooms are often visible as pink to purple-red layers on sulfide-containing mud in ditches, ponds, sulfur springs, estuaries, and marine tide pools. They also form "red water" in stratified lakes and sewage lagoons.

The Rhodospirillaceae are the group of phototrophic bacteria with the greatest structural and metabolic diversity. Most species have the capacity to thrive either as anaerobic phototrophs or as aerobic respiring chemoorganotrophs. In nature, the Rhodospirillaceae occur in all kinds of aquatic habitats, particularly where the decomposition of organic matter by chemoorganotrophic bacteria results in anaerobic conditions and the liberation of low-molecular-weight breakdown products; they regularly flourish in sewage lagoons. [N.P.]

Phototube An electron tube containing a photocathode from which electrons are emitted when it is exposed to light or other electromagnetic radiation. An elementary vacuum phototube consists of a photocathode, an anode or electron collector, and an evacuated envelope through which radiation is transmitted to the photocathode. A gas phototube contains, in addition, an inert gas which may be ionized by electron current from the photocathode. For a description of a phototube in which the electron current is amplified by means of a secondary-emission electron multiplier *see* PHOTOMULTIPLIER.

Phototubes serve as sensing elements in the detection and measurement of light and ultraviolet or infrared radiation. Phototubes also convert variations in intensity of incident radiation into corresponding variations in electron output current, as in light-controlled relay circuits, and in the conversion of sound modulation of photographic film into audio-frequency currents, as in the sound tracks on motion-picture film.

The fundamental characteristics of a phototube are its spectral sensitivity characteristic, or output current, expressed as a function of the wavelength of incident radiation at constant anode voltage, and its anode-current characteristics, which show the dependence of anode current on applied voltage and radiant flux input.

The anode current of a vacuum phototube is directly proportional to the intensity of radiation incident on its photocathode. The anode is normally connected to a positive potential of at least 20 volts relative to the photocathode in order that the anode current be limited by photoelectric emission rather than by space charge or electron emission velocity.

In a gas phototube the photoelectric current is amplified by partial ionization of a gas contained in the tube at low pressure. An inert gas such as neon or argon is used because photocathodes react chemically with other gases. At low anode voltages the anode current of a gas phototube is emission-limited like that of a vacuum phototube. At anode potentials greater than 25 volts, electrons emitted from the photocathode acquire sufficient energy to ionize some of the gas atoms. The total current is then the sum of the free-electron current, the positive-ion current, and the current due to secondary electrons which are produced by ion bombardment of the photocathode. The ratio of anode current, at an anode voltage sufficient to cause ionization, to the emission-limited current measured at a lower voltage is the gas amplification factor of a gas phototube. This factor ranges between 3 and 10 in commercial gas phototubes. *See* ELECTRICAL CONDUCTION IN GASES.

Photocathodes of practical importance contain one or more of the alkali metals lithium, sodium, potassium, rubidium, or cesium in complex combination with other metals or with oxides of certain metals. Because of their high reflectivity and conductivity, the pure alkali metals have lower quantum efficiencies than do the more complex photocathodes, which invariably are semiconductors. [J.L.We.]

Photovoltaic effect A term most commonly used to mean the production of a voltage in a nonhomogeneous semiconductor, such as silicon, by the absorption of light or other electromagnetic radiation. In its simplest form, the photovoltaic effect occurs in the common photovoltaic cell, used, for example, in solar batteries and exposure meters. The photovoltaic cell consists of an *np* junction between two different semiconductors (see illustration), an *n*-type material in which conduction is due to electrons, and a *p*-type material in which conduction is due to positive holes. When light is absorbed near such a junction, new mobile electrons and holes are released, as in photoconduction. An additional feature of a photovoltaic cell, however, is that there is an electric field in the junction region between the two semiconductor types. The released charge moves in this field. This current

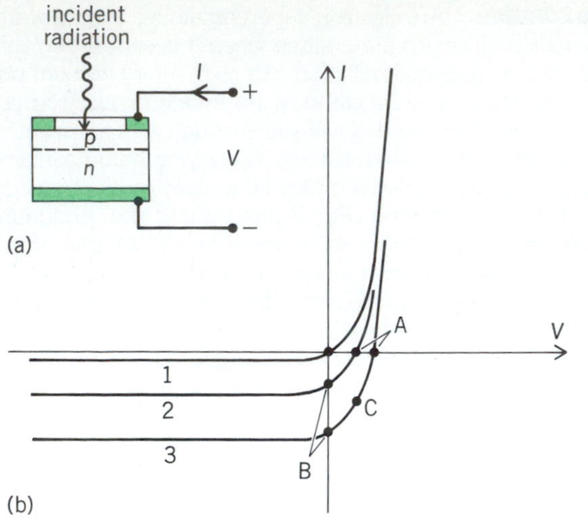

The *p-n* junction photovoltaic cell. (*a*) Cross section. (*b*) Current-voltage characteristics. Curve 1 is for no incident radiation, and curves 2 and 3 for increasing incident radiation. Under open-circuit conditions (current *I* = 0) the terminal voltage increases with increasing light intensity (points A), and under short-circuit conditions (voltage *V* = 0) the magnitude of the current increases with increasing light intensity (points B). When the current is negative and the voltage is positive (point C, for example), the photovoltaic cell delivers to the external circuit.

flows in an external circuit without the need for a battery as required in photoconduction. *See* EXPOSURE METER; PHOTOCONDUCTIVITY; PHOTOELECTRICITY; SEMICONDUCTOR; SOLAR CELL. [L.Ap.]

Phreatoicoidea A suborder of the Isopoda and class Crustacea. The body is subcylindrical, appearing laterally compressed, mainly because of the downward development of the pleura of the pleon (see illustration). The first and occasionally the second thoracic segment is fused with the head. Antennules are shorter than the antennae. The eyes may be large, small, or absent. Mouthparts are primitive.

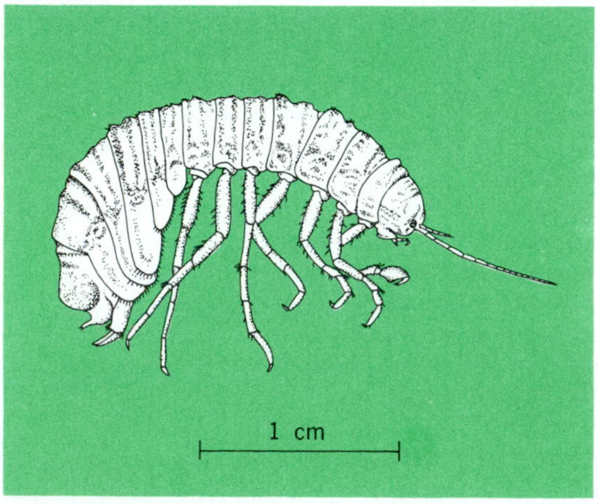

***Onchotelson brevicaudatus*, adult male.**

The suborder is divided into two families, the Amphisopidae, and the Phreatoicidae. The suborder is an ancient one and includes a fossil, *Protamphisopus wianamattensis*, from the Triassic beds of New South Wales. Three extant species are recorded from Australia, Tasmania, New Zealand, and South Africa and one that is subterranean from India. Most species occur in fresh water. Several are blind, subterranean forms and one occurs in hot water from deep artesian bores. A few are semiterrestrial, burrowing forms. [E.M.S.]

Phrynophiurida An order of Ophiuroidea in which the vertebrae usually articulate by means of hourglass-shaped surfaces and the arms are able to coil upward or downward in the vertical plane. There is usually a leathery integument, in which calcareous granules or platelets are embedded. Most species are found in deep water. Of the families, the Gorgonocephalidae often have branched arms, the Asteronychidae have a large disk and slender arms, and the Asteroschematidae have a small disk and stout arms.

These families share a number of characteristics and are grouped in one suborder, Euryalina. One remaining family, the Ophiomyxidae, differs in having a soft, unprotected integument. *See* OPHIUROIDEA. [H.B.F.]

Phthalic acid One of the three benzenedicarboxylic acids of formula $C_6H_4(COOH)_2$. Phthalic (or *o*-phthalic) acid, melting point 191°C (376°F), is the 1,2 isomer; isophthalic (or *m*-phthalic) acid, melting point 347–348°C (657–658°F), is the 1,3 isomer; terephthalic acid, melting point 425°C (797°F), is the 1,4 isomer. Phthalic acids are the most important of the three commercially. Phthalic acids are used in chemical analysis, in preparation of esters, and in preparation of phthaloyl chlorides. Phthalic anhydride, $C_6H_4(CO)_2O$, can be decarboxylated to benzoic acid and can be used in the synthesis of indigo and derivatives of anthraquinone. *See* ACID ANHYDRIDE; CARBOXYLIC ACID; PHTHALIMIDE; POLYESTER RESINS; XYLENE. [E.B.R.]

Phthalimide The imide of *o*-phthalic acid, also called 1,3-isoindoledione. The melting point of phthalimide is 238°C (460°F), and it is only slightly soluble in water. It is a weak acid. In the form of its sodium or potassium salt it is widely used in the synthesis of both primary amines and amino acids. It is combined with the malonic ester in synthesis of complex amino acids. Phthalimide is used as starting material in the synthesis of methyl anthranilate, the active principle in jasmine and orange oils. *See* AMIDE; AMINE; PHTHALIC ACID. [E.B.R.]

Phycobilin A protein-bound pigment which occurs in some groups of algae, namely Cyanophyceae, Rhodophyta, and Cryptophyceae. In conjugated proteins (generic name, biliproteins), phycobilins act as light absorbers in photosynthesis. There are two types of algal biliproteins, the red phycoerythrins and the blue phycocyanins. The phycobilins are open-chain tetrapyrroles and are chemically related to the mammalian bile pigments such as biliverdin and bilirubin. They are attached to proteins by covalent bonds which may be cleaved by acid, alkali, or methanol. Two phycobilins have been characterized, phycocyanobilin and phycoerythrobilin; and there is spectral evidence for at least one more, termed phycourobilin.

Phycoerythrins and phycocyanins serve as accessory pigments to chlorophyll *a* in photosynthesis. The light energy they absorb is passed on to chlorphyll *a* with an efficiency approaching 100% and is preferentially used in that system of photosynthesis in which cytochrome reduction and oxygen production from water take place. In addition, phycobilin-absorbed light is essential for efficient photosynthetic utilization of the light energy absorbed by chlorophyll a itself. *See* CHLOROPHYLL; PHOTOSYNTHESIS. [C.O.hE.]

Phylactolaemata A class of ectoproct bryozoans. Phylactolaemates have lophophores which usually are markedly U-shaped (or rarely nearly circular but still kidney-shaped) in basal outline, and relatively short, wide zooecia; these animals dwell only in fresh waters. *See* BRYOZOA; LOPHOPHORE.

Phylactolaemate colonies are either encrusting threadlike networks of relatively isolated zooecia with solid chitinous uncalcified walls, or small to large masses of gelatinous material in which the individual zooids are embedded side-by-side without definite separating zooecial walls. Stolons are not present.

Only a few (about 50) phylactolaemate species exist, all classified in a single order, the Plumatellida (or Plumatellina). Exclusively fresh-water, the phylactolaemates may have evolved relatively recently from ctenostomes, although some workers have suggested that phylactolaemates might be very primitive ectoprocts surviving as evolutionary relics. *See* CTENOSTOMATA. [R.J.Cu.]

Phyllite A large group of regional metamorphic rocks derived from argillaceous sediments and recrystallized in the greenschist facies (low degree of metamorphism), essentially composed of white mica and quartz. Phyllites are fine-grained, strongly schistose rocks, and the schistosity surfaces exhibit a glittery sheen given off by mica. With very low metamorphism the phyllites pass into slates, and with increasing metamorphism they pass into mica schists. *See* METAMORPHIC ROCKS; SCHIST; SLATE.

Phyllite has a wide distribution in the crystalline mountain ranges of the world: the Highlands of Scotland, mountains of Norway, Harz Mountains and mountains of Saxony, Alps, Appalachians, and Great Lakes district of the United States. [T.F.W.B.]

Phyllocarida A subclass of the crustacean class Malacostraca containing the extant order Leptostraca and the fossil order Archiostraca. The Phyllocarida has a long fossil record, and many early fossil taxa were referred to this subclass. However, studies of presumed phyllocarids from the Burgess Shale have shown that only the archiostracans agree with the definition of the Phyllocarida.

Phyllocarids are distinct from other malacostracan crustaceans because of two other characteristics considered to reflect the primitive condition, which strengthen the hypothesis of early separation from the main evolutionary line. The first is the presence of a bivalve carapace. The second is an abdomen consisting of seven fully formed somites and terminating in a telson that bears caudal rami. *See* CRUSTACEA; LEPTOSTRACA; MALACOSTRACA. [P.A.McL.]

Phyllonite A metamorphic rock—the name is a combination of phyllite and mylonite; the phyllonites occupy a position between the rock types representing the component parts of the name. There are two distinct stages of their development. In the first stage the original rock is granulated by extreme deformation and pulverized to a mylonite. In the second stage, but frequently overlapping the first stage in time, new minerals recrystallize and grow (this is called crystalloblastesis). *See* METAMORPHIC ROCKS; MYLONITE; PHYLLITE. [T.F.W.B.]

Phylogeny The history of living organisms. Although the concepts of organic evolution and phylogeny are closely interlocked, the terms are not synonyms as frequently given in dictionaries and used by biologists. Considerable confusion in theoretical discussions has resulted from misuse of "phylogeny" and "evolution." Evolution is change in organisms with respect to time. The study of evolution, or of evolutionary principles, is the investigation of the mechanisms of evolutionary modifications in organisms, be they large or small changes. Phylogeny is the history of living organisms—their lineages resulting from their evolution. Organisms evolve and leave a trail of phylogeny behind them. The study of phylogeny is the description of lineages of animals and plants from the origin of life to the present, the events being placed in a proper time scale. Phylogenetic studies may be purely descriptive or they may be explanatory, using evolutionary and other pertinent principles. Clearly, the study of phylogeny is a historical science and must be studied with the approaches used in any historical study. But the study of evolution is not a historical science, even though some evolutionary mechanisms, for example, speciation, require long periods of time for their completion.

Phylogenetic study began with the acceptance of the theory of organic evolution, and its origins can be traced to the publication of Charles Darwin's *Origin of Species* (1859). Because phylogeny is defined in terms of organic evolution, all phylogenetic theory and methodology must be based on evolutionary theory. Evolutionary mechanisms can be summarized into two basic ones that are fundamental to phylogenetic study. These are phyletic evolution or change in a single phyletic lineage with respect to time. A single phyletic lineage is a species reproducing generation after generation in time. The second is speciation or the multiplication of phyletic lineages, giving rise to two or more species from a single ancestral one. Phyletic evolution can occur without speciation, but speciation must involve phyletic evolution. A third evolutionary mechanism—extinction—is important to phylogenetic studies because of the resulting gaps produced between groups. The consequence of these evolutionary mechanisms is that an adaptive radiation of species will evolve from a single common ancestor. The goal of phylogenetic studies is to start with the species of this radiation and to ascertain what their phylogenetic history has been, what evolutionary mechanisms were involved, and finally to arrange the taxa into a classification and a phylogenetic diagram (dendrogram) that express this history.

It must be emphasized that the same methods of phylogenetic analysis must be used by neontologists studying recent species and by paleontologists studying fossils. Contrary to common belief, paleontologists cannot observe phylogeny directly, but must offer hypotheses of phylogenetic relationships and then test them with empirical evidence as must neontologists.

The basic method in phylogenetic study is determination of homologous features. Homology is defined in terms of phylogeny, usually in the following form: Homologous features or conditions in two or more organisms are ones that can be traced phylogenetically back (or stem phylogenetically from) the same feature or condition in the immediate common ancestor of these organisms. Homology is defined in terms of phylogeny, and phylogeny is defined in terms of evolution; hence these definitions are strictly noncircular.

After hierarchical series of homologies are established and the homologies are analyzed further to ascertain their relative taxonomic value, or to distinguish between pleisomorphs (primitive homologies), convergent apomorphs (advanced homologies), and synapomorphs (uniquely derived features), phylogenies can be constructed using this information. Thus the procedure of working methods is first to recognize homologs using similarity as the recognizing criterion, and second to recognize phylogenies using homologs.

Determination of phylogenies and the establishment of classifications are associated, but are not the same. Classifications are based on phylogenetic studies and attempt to arrange organisms into taxa which reflect the entire evolutionary history of the group. Concepts of the best classification vary considerably, but most workers accept the idea that classifications should reflect two major aspects of evolutionary history, namely: (1) the degree of genetical similarity as ascertained by the degree of phenotypical similarity; and (2) the phylogenetic series of events, including branching of lineages, in the history of the group. *See* ANIMAL SYSTEMATICS; EXTINCTION (BIOLOGY); SPECIATION; TAXONOMIC CATEGORIES. [W.J.B.]

Phymosomatoida An order of regular sea urchins, class Echinoidea, characterized by imperforate tubercles, complex ambulacral compounding in which one or more elements are occluded from the perradial suture, and a stirodont lantern with unfused epiphyses. They comprise two families. They first appeared in the Lower Jurassic and are probably paraphyletic, since they include the ancestors of camarodonts. The two extant genera are each known from a single species: *Glyptocidaris*, from a depth of 30–490 ft (10–150 m) around northern Japan, and *Stomechinus*, a common inhabitant of rocky shores around the Indo-West Pacific. They are epifaunal grazers. *See* ECHINODERMATA; ECHINOIDEA. [A.Sm.]

Physical anthropology The science which treats the biological aspects of humans and their relation to their historical or cultural aspects. Psychology and the medical sciences deal intensively with particular areas of human nature; physical anthropology integrates these and other approaches into the broadest view possible. This view regards humans as a species of animal, encompassing evolution, the functioning of various organ systems, humans' adaptive qualities in relation to environment, genetics, and variation in individuals, populations, or racial groups. *See* ANATOMY, REGIONAL; GENETICS; MEDICINE; PSYCHOLOGY, PHYSIOLOGICAL AND EXPERIMENTAL.

The main distinction of the evolution of humans from the higher primates (monkeys and apes) is the emergence of upright walking and an enlarged brain. Speech and the making of tools are often cited as characteristics of humans, and may be regarded as the functional equivalent of the anatomically expanding cerebral cortex. *See* FOSSIL HUMAN; PRIMATES.

The modern reason for the study of human regional variation, which one recognizes as race, lies not in classification but in the fact that this is the most important zoological and evolutionary information available for the human species. The local variation and intergrading is seen, not as the result of the hybridizing of "pure" races, but as the result of the basic processes of small-scale evolution, for example, adaptation, selection, genetic drift, and continuous gene exchange. Adaptation may be expressed in body size (especially weight), which is correlated inversely with mean annual temperature according to Bergmann's rule; that is, body bulk tends to be greater, for greater heat conservation, in colder areas. Thus some bodily traits are not essentially racial, but are environmental in origin.

Individual variation has been at all times the central field of research in anthropology. Early studies in racial classification were concerned more with the existence than with the explanation of racial differences, and with testing such differences by standard measurements, means, and probability statistics. Later work was addressed to the traits themselves, that is, to organization of body form or of physiological response. This called for the addition, to traditional methods, of hypotheses based on theoretical considerations, experimental design (difficult in the case of humans), selection of nonstandard measures, a variety of techniques, and more refined statistics dealing with variation and correlation. *See* HUMAN GENETICS; POPULATION GENETICS; SEROLOGY. [W.W.Ho.]

Physical chemistry The branch of chemistry that deals with the interpretation of chemical phenomena and properties in terms of the underlying physical processes, and with the development of techniques for their investigation. The term chemical physics is often employed to denote a branch of physical chemistry where the emphasis is on the interpretation and analysis of the physical properties of individual molecules and bulk systems, instead of their reactions. Theoretical chemistry is another major branch, where the emphasis is on the calculation of the properties of molecules and systems, and which used the techniques of quantum mechanics and statistical thermodynamics. It is convenient to regard physical chemistry as dealing with three aspects of matter: its equilibrium properties, structure, and ability to change.

Equilibrium properties. The study of matter in a state of equilibrium constitutes the field of chemical thermodynamics. In particular, chemical thermodynamics provides a technique for discussing the response of a system to a change in the external conditions (such as the shift in the boiling and freezing point of either a pure substance or a mixture when the applied pressure is changed, or when the composition of the mixture is modified), and for rationalizing the energy changes that occur in the course of a chemical reaction. The branch of thermodynamics dealing with the latter is called thermochemistry. Chemical thermodynamics also provides a framework for the determination of the maximum amount of work that may be generated by a system undergoing a specified change, and it therefore provides a way of establishing bounds for the efficiencies of a variety of devices, including engines, refrigerators, and electrochemical cells. Thermodynamics is used in chemistry to assess the position of equilibrium of a chemical reaction (that is, how far it will proceed), and to determine what conditions are necessary in order to optimize the yield of a particular product. The branch of chemical thermodynamics dealing with ionic reactions occurring in the presence of electrodes constitutes the field of equilibrium electrochemistry. *See* CHEMICAL EQUILIBRIUM; CHEMICAL THERMODYNAMICS; ELECTROCHEMISTRY; ENTHALPY; ENTROPY; FREE ENERGY; GIBBS FUNCTION; THERMOCHEMISTRY.

Structure. The principal role of quantum mechanics in chemistry is in the discussion of atomic and molecular structure, and in the interpretation of spectroscopic data. In the branch of physical chemistry known as computational quantum chemistry, interest centers on the numerical solution of the Schrödinger equation in order to obtain wave functions and geometries of molecules. Computational quantum chemistry is so developed that it is capable of being used to map the changes in the structures of molecules while they are in the course of reaction, when atoms and groups of atoms are being transferred from one molecule to another. *See* QUANTUM CHEMISTRY; SCHRÖDINGER'S WAVE EQUATION.

Spectroscopic techniques are used not only to identify molecules present in a sample, but also to determine their shape, size, and electron distribution. The techniques fall into four categories: absorption spectroscopy, emission spectroscopy, Raman spectroscopy, and resonance techniques. *See* ELECTRON PARAMAGNETIC RESONANCE (EPR) SPECTROSCOPY; ELECTRON SPECTROSCOPY; MOLECULAR STRUCTURE AND SPECTRA; MÖSSBAUER EFFECT; NUCLEAR MAGNETIC RESONANCE (NMR); PHOTOCHEMISTRY; RAMAN EFFECT; SPECTROSCOPY.

Techniques for the investigation of molecular structure based on diffraction depend on the observation of the direction through which radiation and particles are scattered when they impinge on a sample. Other techniques for investigating structure include the electric and magnetic properties of molecules, in particular, the determination of electric polarizabilities and dipole moments, magnetic properties, and the properties based on optical birefringence, such as optical activity and the Faraday effect. *See* X-RAY DIFFRACTION.

Structural properties and thermodynamic properties are brought together by statistical thermodynamics. This major theoretical procedure gives a way of predicting the thermodynamic properties of assemblies of molecules in terms of their individual energy levels.

Physical and chemical change. The third major branch of physical chemistry is concerned with change: physical change and chemical change. In particular, it is concerned with the rate of change. Physical change includes the diffusion of one substance into another, or the migration of ions in an electrode solution. The application of thermodynamics to change in general constitutes the field of nonequilibrium thermodynamics. *See* DIFFUSION IN GASES AND LIQUIDS; GAS; TRANSPORT PROCESSES.

Chemical change may be studied at a variety of levels. Empirical chemical kinetics is the study of reactions in order to determine how their rates depend on the concentrations of the participants in the reaction and on the conditions, mainly the temperature. Investigation of the time dependence of reactions yields a detailed picture of the sequence of molecular transformations involved in a complex chemical reaction. *See* CHEMICAL DYNAMICS; SHOCK TUBE; ULTRAFAST MOLECULAR PROCESSES.

An important extension of chemical kinetics is to the reactions that occur on surfaces; these are the processes involved in heterogeneous catalysis. A special application of surface chemistry is to the stability of colloidal suspensions of species in fluids, and another is to the processes that occur at the interface between an electrode and the solution in which it is immersed. *See* ADSORPTION; COLLOID; HETEROGENEOUS CATALYSIS; SURFACE PHYSICS. [P.W.A.]

Physical geography The description, analysis, classification, and genetic interpretation of the natural features and phenomena of the Earth's surface. Numerous specialized but overlapping disciplines have developed within the broad field of physical geography. Among these are geomorphology, hydrology, meteorology and climatology, and biogeography. *See* BIOGEOGRAPHY; CLIMATOLOGY; GEOMORPHOLOGY; HYDROLOGY; METEOROLOGY.

Further considerations of physical geography appear in various topical and regional articles. For examples of principal topical discussions *see* CARTOGRAPHY; CONTINENT; HILL AND MOUNTAIN TERRAIN; MATHEMATICAL GEOGRAPHY; PLAINS; TERRAIN AREAS. For physical geography of the continents and major island areas *see* AFRICA; ANTARCTICA; ARCTIC AND SUBARCTIC ISLANDS; ASIA; AUSTRALIA; EAST INDIES; EUROPE; NEW ZEALAND; NORTH AMERICA; PACIFIC ISLANDS; SOUTH AMERICA; WEST INDIES. [D.L.J.]

Physical law A physical phenomenon is said to be controlled or governed by a physical law when the phenomenon is one of a broad class of phenomena such that it is possible to formulate some regularity which applies to all members of the class. If the phenomenon is one which can be described in terms of numerical measurement, the law is often formulated in terms of mathematical relations between the numbers obtained by measurement, as in the inverse square law of universal gravitation. However, a mathematical formulation of a natural law is by no means necessary. [P.W.Br./G.Ho.]

Physical measurement Quantitative information on physical conditions, properties, or relations essential for coordination of activities, efficiency of communication, and understanding of the nature of things in science and engineering and in much of everyday life. Time, distance, mass, temperature, force, power, and all other physical quantities (or parameters or variables), as well as the properties of matter, materials, and devices, must be described and measured in terms which have the same meaning for everyone. The measuring device or instrument is calibrated (that is, the functional relationship between its indication and the magnitude of the measured quantity is determined) by direct or indirect comparison with a standard which embodies, possesses, or generates a fixed or reproducible magnitude of the physical quantity which is taken as the unit or some multiple or fraction of the unit. Any measured quantity may thus be expressed by a number (the magnitude ratio) and the name of the unit, for example, a length of 1.54 meters. The general area of scientific activity relating to standards and units and the accuracy of measurement is called metrology. *See* UNITS OF MEASUREMENT.

Standards. In 1793 the French government adopted the decimal metric system wherein the basic unit of length was defined as one ten-millionth of the Earth's polar quadrant (as determined from latitude surveys), to be called the meter. The basic unit for mass was defined as the mass of a cubic decimeter of water, to be called the kilogram. For working standards, a platinum bar was marked with fine lines a meter apart, and a platinum iridium cylinder was constructed equal in mass to a cubic decimeter of water. When later refinements in measurement showed that neither of these standards exactly realized the units as originally defined, the discrepancies were eliminated by redefining the units in terms of the material standards which had been constructed. *See* METRIC SYSTEM.

In 1975, the United States adopted the Metric Conversion Act, declaring that "the policy of the U.S. shall be to coordinate and plan the increasing use of the metric system in the United States," and established the U.S. Metric Board "to coordinate the voluntary conversion to the metric system." However, English units have become almost universal in some worldwide industries—for example, dimensions of oil-drilling equipment, or altitude measurement in aviation. Thus it is likely that there will always be exceptions to uniformity, requiring special knowledge of special units for at least some people even if the whole world "goes metric" in principle.

At present the International System of Units (abbreviated SI, from the French Système International d'Unités) is constructed from seven base units for independent quantities plus two supplementary units for plane angle and solid angle (Table 1). Units for all other quantities are derived from these nine units. In Table 2 are listed 19 SI derived units with special names. These units are derived from the base and supplementary units in a coherent manner, which means they are expressed as products and quotients of the nine base and supplementary units without numerical factors. All other SI derived units are similarly derived in a coherent manner from the 28 base, supplementary, and special-name SI units. For use with the SI units there is a set of 16 prefixes (Table 3) to form multiples and submultiples of these units. For mass, the prefixes are to be applied to the gram instead of to the SI unit, the kilogram.

The SI units together with the SI prefixes provide a logical and interconnected framework for measurements in science, industry, and commerce.

Units. In some cases, quantities are commonly expressed in terms of fundamental constants of nature, and use of these constants or "natural units" is acceptable. Typical examples of natural units, with their symbols, are:

elementary charge	e
electron mass	m_e
proton mass	m_p
Bohr radius	a_o
electron radius	r_e
Compton wavelength of electron	λ_c
Bohr magneton	μ_B
nuclear magneton	μ_N
speed of light	c
Planck constant	h

Table 1. SI base and supplementary units

Quantity*	Unit name	Unit symbol
SI base units		
Length	meter	m
Mass	kilogram	kg
Time	second	s
Electric current	ampere	A
Thermodynamic temperature	kelvin	K
Amount of substance	mole	mol
Luminous intensity	candela	cd
SI supplementary units		
Plane angle	radian	rad
Solid angle	steradian	sr

*Quantity here and in Table 2 means a measurable attribute.

Table 2. SI derived units with special names

Quantity	SI unit: Name	Symbol	Expression in terms of other units	Expression in terms of SI base units
Frequency	hertz	Hz		s^{-1}
Force	newton	N		$m \cdot kg \cdot s^{-2}$
Pressure, stress	pascal	Pa	N/M^2	$m^{-1} \cdot kg \cdot s^{-2}$
Energy, work, quantity of heat	joule	J	N·m	$m^2 \cdot kg \cdot s^{-2}$
Power, radiant flux	watt	W	J/s	$m^2 \cdot kg \cdot s^{-3}$
Quantity of electricity, electric charge	coulomb	C	A·s	$s \cdot A$
Electric potential, potential difference, electromotive force	volt	V	W/A	$m^2 \cdot kg \cdot s^{-3} \cdot A^{-1}$
Capacitance	farad	F	C/V	$m^{-2} \cdot kg^{-1} \cdot s^4 \cdot A^2$
Electric resistance	ohm	Ω	V/A	$m^2 \cdot kg \cdot s^{-3} \cdot A^{-2}$
Conductance	siemens	S	A/V	$m^{-2} \cdot kg^{-1} \cdot s^3 \cdot A^2$
Magnetic flux	weber	Wb	V·s	$m^2 \cdot kg \cdot s^{-2} \cdot A^{-1}$
Magnetic flux density	tesla	T	Wb/m^2	$kg \cdot s^{-2} \cdot A^{-1}$
Inductance	henry	H	Wb/A	$m^2 \cdot kg \cdot s^{-2} \cdot A^{-2}$
Celsius temperature	degree Celsius	°C		K
Luminous flux	lumen	lm		cd·sr*
Illuminance	lux	lx	lm/m^2	$m^{-2} \cdot cd \cdot sr$*
Activity (of a radionuclide)	becquerel	Bq		s^{-1}
Absorbed dose, specific energy imparted, kerma, absorbed dose index	gray	Gy	J/kg	$m^2 \cdot s^{-2}$
Dose equivalent, dose equivalent index	sievert	Sv	J/kg	$m^2 \cdot s^{-2}$

*In this expression the steradian (sr) is treated as a base unit.

Table 3. SI prefixes

Factor	Prefix	Symbol	Factor	Prefix	Symbol
10^{18}	exa	E	10^{-1}	deci	d
10^{15}	peta	P	10^{-2}	centi	c
10^{12}	tera	T	10^{-3}	milli	m
10^{9}	giga	G	10^{-6}	micro	μ
10^{6}	mega	M	10^{-9}	nano	n
10^{3}	kilo	k	10^{-12}	pico	p
10^{2}	hecto	h	10^{-15}	femto	f
10^{1}	deka	da	10^{-18}	atto	a

Certain units which are not part of the SI are used so widely that it is impractical to abandon them. The units that are accepted for continued use with the International System are listed in Table 4. It is likewise necessary to recognize, outside the International System, the following units which are used in specialized fields:

electronvolt	eV
unified atomic mass unit	u
astronomical unit	AU
parsec	pc

Logarithmic measures such as pH, dB (decibel), and Np (neper) are acceptable.

Table 4. Units in use with the International System

Name	Symbol	Value in SI unit
Minute	min	1 min = 60 s
Hour	h	1 h = 60 min = 3,600 s
Day	d	1 d = 24 h = 86,400 s
Degree	°	1° = (π/180) rad
Minute	′	1′ = (1/60)° = (π/10,800) rad
Second	″	1″ = (1/60)′ = (π/648,000) rad
Liter	L*	1 L = 1 dm^3 = 10^{-3} m^3
Metric ton	t	1 t = 10^3 kg
Hectare	ha	1 ha = 10^4 m^2

*An alternative symbol for liter is "l." Since "l" can be easily confused with the numeral 1, the symbol "L" is recommended for United States use.

The internationally accepted definitions for the seven base units follow.

Mass. The kilogram (kg) is equal to the mass of the International Prototype Kilogram. The International Prototype is a platinum-iridium cylinder preserved at the International Bureau of Weights and Measures at Sèvres, France. Mass is the only one of the base quantities for which the standard is an arbitrarily defined object. No basic property of matter involving mass can be measured with more precision than is possible in comparing kilogram masses by weighing, about 1 part in 10^8. *See* BALANCE; MASS; WEIGHT MEASUREMENT.

Length. The meter (m) is the length equal to 1 650 763.73 wavelengths in vacuum of the radiation corresponding to the transition between the levels $2p_{10}$ and $5d_5$ of the krypton-86 atom (an orange-red line).

Two monochromatic radiations, one in the visible, the other in the infrared spectral region, produced by helium-neon lasers stabilized on a saturated absorption line of iodine or of methane, are recommended as wavelength standards: iodine 127, R(127), band 11–5, component i [wavelength in a vacuum, 632 991.399 × 10^{-12} m]; methane, P(7), band ν_3 [wavelength in a vacuum 3 393 231.40 × 10^{-12} m]. These lines are reproducible with an uncertainty of the order of 1 in 10^{10} the value of their wavelength in meters is subject to the uncertainty of the standard (the wavelength of the ^{86}Kr line) estimated to be 4 in 10^9.

The wavelength of the methane line mentioned above multiplied by its frequency (measured by comparison with the ^{133}Cs transition of the definition of the second) yields the speed of propagation of electromagnetic waves in vacuum c = 299 792 458 m/s. This value of c will probably be kept unaltered in the future and will permit the meter to be defined as the distance electromagnetic radiation travels in a vacuum in 1/299 792 458 s. *See* ELECTROMAGNETIC RADIATION.

Time interval. The second (s) is the duration of 9 192 631 770 periods of the radiation corresponding to the transition between the two hyperfine levels of the ground state of the cesium-133 atom. In the best equipments the stability and accuracy of the cesium frequency generator correspond to an uncertainty of 1 in 10^{12} or even 1 in 10^{13}.

The second was long defined, for physical measurements as well as for civil affairs, as 1/86,400 of the time required for an average complete rotation of the Earth on its axis with respect to the Sun. Because of the slight slowing of the Earth's rotation rate, the universal second thus defined is not a constant. A time scale called Coordinated Universal Time (UTC) recommended by the General Conference of Weights and Measures (CGPM) in 1975 is defined in such a manner that it differs from international atomic time (TAI) by an exact whole number of seconds. This difference is adjusted occasionally by the use of a positive or negative leap second at the end of certain months to keep UTC in agreement with the time defined by the rotation of the Earth with an approximation better than 9/10 second. *See* ATOMIC CLOCK; ATOMIC TIME; DYNAMICAL TIME; TIME.

Temperature. The kelvin (K), the unit of thermodynamic temperature, is the fraction 1/273.16 of the thermodynamic temperature of the triple point of water. The unit kelvin and its symbol K should also be used to express an interval or differences of temperature.

To provide convenient and adequately accurate means for practical realization and measurement of temperature, the International Practical Temperature Scale is used, based on the assigned values of the temperatures of a number of reproducible equilibrium states (defining fixed points) and on standard instruments calibrated at those temperatures. Interpolation between the fixed-point temperatures is provided by formulas used to establish the relation between indications of the standard instruments and values of International Practical Temperature.

Electric current. The ampere (A) is that constant current which, if maintained in two straight parallel conductors of infinite length and of negligible circular sections, and placed 1 meter apart in a vacuum, would produce between these conductors a force equal to 2×10^{-7} newton per meter of length. *See* ELECTRIC CURRENT.

Luminous intensity. The CGPM, in 1979, redefined the base SI unit candela as the luminous intensity, in a given direction, of a source that emits monochromatic radiation of frequency 540×10^{12} hertz and of which the radiant intensity in that direction is 1/683 watt per steradian. *See* ILLUMINATION; LIGHT; LUMINOUS EFFICACY; LUMINOUS EFFICIENCY; LUMINOUS INTENSITY; PHOTOMETRY; RADIOMETRY.

Amount of substance. The mole is the amount of substance of a system which contains as many elementary entities as there are atoms in 0.012 kilogram of carbon-12. When the mole is used, the elementary entities must be specified, and may be atoms, molecules, ions, electrons, other particles, or specified groups of such particles. *See* ATOMIC MASS UNIT; ATOMIC WEIGHT; GRAM-MOLECULAR WEIGHT; MOLECULAR WEIGHT; RELATIVE ATOMIC MASS; RELATIVE MOLECULAR MASS. [W.A.Wi.]

Physical optics The study of the interaction of electromagnetic waves in the optical range with material systems. The optical range of wavelengths may be taken as the range from about 1 nanometer to about 1 millimeter.

The explanation of the absorption, reflection, scattering, polarization, and dispersion of light by a material medium in terms of the properties of the atoms and molecules making up the medium is the objective of physical optics. In the course of seeking this objective, physicists have found that optical investigations are powerful methods of determining the structures of atoms and molecules and of large systems composed thereof. *See* ABSORPTION; ATOMIC STRUCTURE AND SPECTRA; CRYSTAL OPTICS; DIFFRACTION; DISPERSION (RADIATION); ELECTROMAGNETIC RADIATION; ELECTROOPTICS; FARADAY EFFECT; FLUORESCENCE; INTERFERENCE OF WAVES; LASER; LIGHT; MAGNETOOPTICS; MOLECULAR STRUCTURE AND SPECTRA; POLARIZED LIGHT; REFLECTION OF ELECTROMAGNETIC RADIATION; REFRACTION OF WAVES; SCATTERING OF ELECTROMAGNETIC RADIATION; SPECTROSCOPY. [R.C.L.]

Physical organic chemistry A branch of science concerned with the scope and limitations of the various rules, effects, and generalizations in use in organic chemistry by means of physical and mathematical methods. It includes, but is not limited to, the dynamics and energetics of organic chemical transformations, transient intermediates in these reactions, rate comparisons between families of reactions, dynamic stereochemistry, conservation of orbital symmetry, the least-motion principle, the isomer number for a given elemental composition, conformational analysis, nonexistent compounds, aromaticity, tautomerism, strain and steric hindrance, and the double-bond rule. Spectroscopy is the main tool employed, with nuclear magnetic resonance being the most widely used spectroscopic technique. With the advent of modern fast computers, computational chemistry has also become an important tool. *See* NUCLEAR MAGNETIC RESONANCE (NMR); SPECTROSCOPY.

Physical organic chemistry is traditionally distinguished from, yet totally intertwined with, synthetic organic chemistry, which deals with the question of how to obtain desired products from available compounds. This distinction can be illustrated with a diagram (see illustration) showing how the energy might vary during a chemical reaction in which the reactant R yields a product P (R → P). Whereas the synthetic organic chemist will be interested primarily in the practical problem of how to convert R into P, the physical organic chemist studies the curve or curves connecting R and P as well as the structure and physical properties at all extrema, including R and P. However,the demarcation between synthetic organic chemistry and physical organic chemistry is not sharp. Physical organic chemists have contributed greatly to the understanding of the chemistry of hydrocarbons and their derivatives and have enhanced the repertoire of the synthetic organic chemists. In turn, synthetic organic chemists have made possible the construction of the custom-made, often intricate molecules that physical organic chemists use for their studies. The efforts of both groups, moreover, have made possible the birth of such new fields as molecular biochemistry and computational chemistry. *See* COMPUTATIONAL CHEMISTRY; MOLECULAR BIOLOGY; ORGANIC CHEMICAL SYNTHESIS.

Chemical reaction mechanisms. The diagram shown in the illustration is also useful in discussions of the dynamics of chemical reactions. It is an attempt to portray how the atoms in the reactant molecule R may move in space to their final positions in the product molecule P, and how the potential energy of the system would vary as a function of these positions. A complete correlation would be multidimensional; what is normally shown is a cross section in which the maximum potential energy is in fact a minimum (saddle point). While an essentially infinite number of pathways between R and P can be imagined and followed, the vast majority of the molecules will in practice use the one that makes the least demand on energy to reach the next maximum; this pathway is known as the reaction mechanism. The maxima (T terms in the illustration) are known as transition states, and the minima (I terms in the illustration) as intermediates.

If a reaction has a single transition state (and, hence, no intermediate), it is known as concerted; alternatively, it is stepwise. A stepwise reaction is simply a succession of concerted steps in which the intermediates are not isolated. *See* FREE RADICAL; ORGANIC REACTION MECHANISM; REACTIVE INTERMEDIATES; PHOTOCHEMISTRY.

Chemical kinetics. The most important route to quantitative information about a reaction is the study of its kinetics. This must begin with an experimental determination of the rate law: the expression that shows how the rate of formation of product $d[P]/dt$ (or loss of reactant: $-d[R]/dt$) depends on the concentration of all species involved in the reaction other than the solvent. The differential might be found to equal $k[R_1]$,

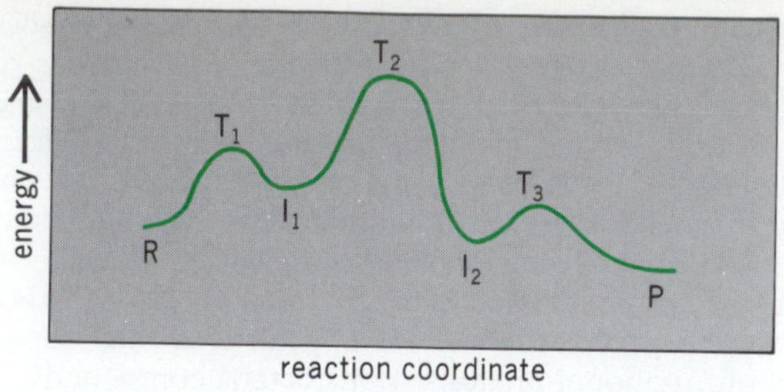

Energy profile of an organic reaction R → P. T terms indicate transition states and I terms indicate intermediates.

$k[R_1]\,[R_2]$, $k[R_1]^2$, and so on; k is known as the rate constant, and the reaction is described as first-order, second-order, and so forth, depending on the total number of concentration terms. One important feature is that the reaction order equals the sum of all molecules that have participated in the formation of the transition state (T_2 in the illustration). Thus, for a reaction to be concerted, it is necessary that the order equal the sum of all reactant molecules involved in the stoichiometry. *See* CATALYSIS; CHEMICAL DYNAMICS; PERICYCLIC REACTION.

Stereochemistry. This is also a powerful tool in physical organic chemistry. Experimental work has demonstrated that when a chiral compound such as $(-)HCR_1R_2X$ is converted into optically active HCR_1R_2Y by direct displacement with Y, the product obtained has an optical rotation that is opposite to that exhibited by the same material if it is produced in two steps, via initial displacement by A to give intermediate HCR_1R_2A as shown in the following reaction scheme. This

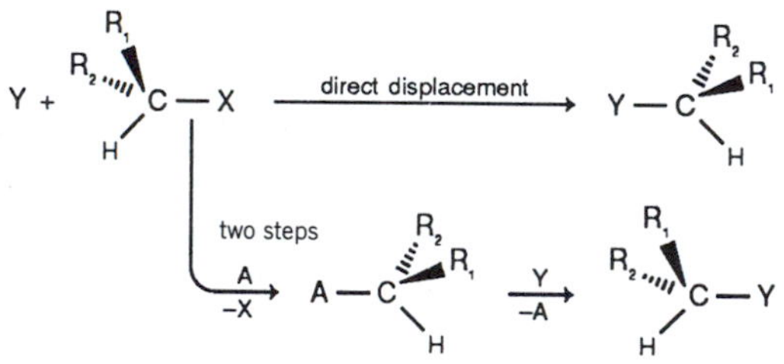

result demonstrates that displacement reactions occur with inversion; the reagent approaches in front, and the leaving group departs in the back. *See* OPTICAL ACTIVITY; STEREOCHEMISTRY.

Isomers. The isomer number has perhaps been the most important organizing principle in organic chemistry since its inception. Simply put, it means that for every elemental composition and molecular mass the isomer number can be predicted by writing all possible sequences of the atoms present, obeying the valence numbers of the atoms: four for carbon, three for nitrogen, two for oxygen, one for hydrogen and the halogens, and so forth. Once reliable atomic weights became available so that elemental compositions could be reported with confidence, this simple rule proved remarkably successful. *See* CHEMICAL BONDING; ORGANIC CHEMISTRY; VALENCE.

The need to specify molecular mass has proved more troublesome: it requires a definition of the concept of a molecule. Such definitions usually refer to covalent bonds as the entities that hold the atoms together, to rule out ionic species such as sodium chloride as candidates.

Among the extra compounds, none have affected organic chemistry more drastically than the stereoisomers. It turns out that for all but the simplest compounds a given sequence of the atoms may represent two, more than two, or even many more isomers. *See* CONFORMATIONAL ANALYSIS; MOLECULAR ISOMERISM.

Many compounds that are considered to be nonexistent, even though they are allowed by the simple rules of isomer numbers, in fact are transient intermediates in various reactions. Sometimes they can be detected spectroscopically, but cannot be isolated. There are instances in which neither of two isomers can be isolated, but mixtures of the two can. In other words, the barrier between the two is low, and the equilibrium constant is close to unity. An example is acetoacetic ester, which normally contains about 15% of the enol isomer. Such isomers are known as tautomers. *See* TAUTOMERISM. [W.J.LeN.]

Physical science The fields of inquiry to which the general designation science may be appropriately applied are broadly divided into social science and natural science. The latter is further subdivided into biology and physical science. Physical science is generally considered to include astronomy, chemistry, geology, mineralogy, meteorology, and physics. These overlap more or less, as illustrated by astrophysics, chemical physics, physical chemistry, and geophysics. There is overlap, likewise, between the physical and biological sciences, as seen in biochemistry, biophysics, virology, and the close relation between geology and paleontology. The boundaries implied in all such classifications are artificial and consist of regions where one field shades into another. *See* ASTRONOMY; BIOLOGY; CHEMISTRY; GEOLOGY; METEOROLOGY; MINERALOGY; PHYSICS; SCIENCE. [J.H.Hi.]

Physical theory A physical theory usually involves the attempt to explain a certain class of physical phenomena by deducing them as necessary consequences of other phenomena regarded as more primitive and less in need of explanation. The value of a theory depends on both the success with which it coordinates a wide range of presently known facts and its fertility in suggesting places to look for presently unknown phenomena. [P.W.Br./G.Ho.]

Physics Formerly called natural philosophy, physics is concerned with those aspects of nature which can be understood in a fundamental way in terms of elementary principles and laws. In the course of time, various specialized sciences broke away from physics to form autonomous fields of investigation. In this process physics retained its original aim of understanding the structure of the natural world and explaining natural phenomena.

The most basic parts of physics are mechanics and field theory. Mechanics is concerned with the motion of particles or bodies under the action of given forces. The physics of fields is concerned with the origin, nature, and properties of gravitational, electromagnetic, nuclear, and other force fields. Taken together, mechanics and field theory constitute the most fundamental approach to an understanding of natural phenomena which science offers. The ultimate aim is to understand all natural phenomena in these terms. *See* CLASSICAL FIELD THEORY; MECHANICS; QUANTUM FIELD THEORY.

The older, or classical, divisions of physics were based on certain general classes of natural phenomena to which the methods of physics had been found particularly applicable. The divisions are all still current, but many of them tend more and more to designate branches of applied physics or technology, and less and less inherent divisions in physics itself. The divisions or branches, of modern physics are made in accordance with particular types of structures in nature with which each branch is concerned.

In every area physics is characterized not so much by its subject-matter content as by the precision and depth of understanding which it seeks. The aim of physics is the construction of a unified theoretical scheme in mathematical terms whose structure and behavior duplicates that of the whole natural world in the most comprehensive manner possible. Where other sciences are content to describe and relate phenomena in terms of restricted concepts peculiar to their own disciplines, physics always seeks to understand the same phenomena as a special manifestation of the underlying uniform structure of nature as a whole. In line with this objective, physics is characterized by accurate instrumentation, precision of measurement, and the expression of its results in mathematical terms.

For the major areas of physics and for additional listings of articles in physics *see* ACOUSTICS; ASTROPHYSICS; ATOMIC PHYSICS; BIOPHYSICS; CLASSICAL MECHANICS; ELECTRICITY; ELECTROMAGNETISM; HEAT; LOW-TEMPERATURE PHYSICS; MOLECULAR PHYSICS; NUCLEAR PHYSICS; OPTICS; SOLID-STATE PHYSICS; THEORETICAL PHYSICS. [W.G.P.]

Physiological acoustics The study of specific responses that may occur in the ear or elsewhere along the central auditory pathways, following presentation of an appropriate stimulus at any level of the auditory system. Such responses may be recorded with the aid of various techniques which may be mechanical, electrical, optical, and so forth. The specific stimulus for the ear is acoustic energy. Experimentally, signals with well-defined parameters are used. The approach employed by physiological acoustics thus is purely analytical. This is in contrast to the holistic approach employed by psychoacoustics, which lends itself well to experiments on human subjects. Systematic physiological experiments can be performed only in animals, but differences between humans and other mammals are mainly in degree, not in principle. *See* EAR; HEARING (HUMAN); PSYCHOACOUSTICS. [J.T.]

Physiological ecology (animal) The study of biophysical, biochemical, and physiological processes used by animals to cope with factors of their physical environment, or employed during ecological interactions with other organisms. Loosely speaking, animal physiological ecology represents the interface between the fundamental physiological question of how the whole animal works and the fundamental ecological question of how and why the animal lives where it does. The discipline of animal physiological ecology is thus very broadly based, but for the most part it is characterized by highly empirical research that emphasizes physiological function and evolutionary adaptation under natural conditions, and is couched within the framework of intact individual organisms.

The methods used by animal physiological ecologists are diverse, but are usually not unique compared to those of general physiology. Physiological ecologists interpret their data, however, in terms of adaptation to the environment and hence ultimately in terms of natural selection and evolution. Since the questions it attempts to answer exist in nature, much of physiological ecology is predicated on studying animals under field conditions or on quantification of the biotic and abiotic environmental factors experienced by animals in their natural habitats. Hypotheses based on results from field work are often tested in the laboratory under controlled conditions that duplicate the field environment as much as possible. The reverse process is also true—insights and inferences gained in the laboratory may be examined and tested in the field. *See* METABOLISM.

Animals need to acquire nutrients from their environments for use as building blocks for maintenance, growth, and reproduction and to provide chemical energy for metabolism. Physiological ecologists are concerned with the kinds and amounts of nutrients animals require, the sources of the nutrients in the foods available in the animal's habitat, the roles and functional characteristics of specialized digestive glands and organs, and the rates and efficiencies of food processing and nutrient extraction in the gut. *See* NUTRITION.

Physiological ecologists are interested in the rates and routes of water and solute exchange, the precision and metabolic costs of osmotic regulation, the ability to tolerate osmotic stresses, and the relationship of these factors to the osmotic environments the animal must face in its natural habitat. *See* BEHAVIORAL ECOLOGY; ECOLOGY; HOMEOSTASIS; PHYSIOLOGICAL ECOLOGY (PLANT). [M.A.C.]

Physiological ecology (plant) The study of growth, biological processes, and reproduction within plant populations in natural or controlled environments. The five principal growth forms are unicellular, thallose, and herbaceous forms, and shrubs and trees. Life cycle processes include (1) germination, (2) growth, (3) absorption and loss of water, gases, and mineral nutrients, (4) photosynthesis and respiration, and (5) production of reproductive structures (flowers, cones, fruits, seeds, and spores). Each process is the result of a complex series of physical and chemical reactions within and between cells. All cellular processes are within the context of the whole plant as governed by the interaction of its genes and its environment through space and time.

Plant physiological ecology is part of the attempt to answer the question: Why does a plant grow where it does? An answer requires a knowledge of the genetic structure of the population, the population's physiological processes and their rates, its growth forms, the populational demography including seed sources, and the population's niche in the operation and structure of the ecosystem.

Actual and potential geographic ranges of populations or species seldom coincide. Environments change. Populations evolve, migrate, and become extinct in response to environmental changes. The potential range of a species or a local population depends upon the degree and nature of genetic variation among its individuals, the range of physiological and morphological adaptability within each individual (acclimatization and phenotypic plasticity), and the frequency, extent of occurrence, availability, and stability of suitable environments. These complex variables, operating together, determine the kinds and rates of physiological processes, and thus the degree of vegetative and reproductive success of a population within an ecosystem or in the biosphere. *See* BIOSPHERE; ECOSYSTEM. [W.D.B.]

Phytamastigophorea A class of the subphylum Sarcomastigophora, also known as the Phytomastigina. These are the plant flagellates which contain chlorophyll and other pigments, but colorless forms are also included. Encystment is frequent among phytoflagellates, cyst composition being one method of determining relationships for some colorless species.

The Phytamastigophorea include 10 orders: Chrysomonadida, Silicoflagellida, Coccolithophora, Heterochlorida, Cryptomonadida, Dinoflagellida, Ebriida, Euglenida, Chloromonadida, and Volvocida. See articles on these groups. *See* SARCOMASTIGOPHORA. [J.B.L.]

Phytoalexin Any antibiotic produced by plants in response to microorganisms. Plants use physical and chemical barriers as a first line of defense. When these barriers are breached, however, the plant must actively protect itself by employing a variety of strategies. Plant cell walls are strengthened, and special cell layers are produced to block further penetration of the pathogen. These defenses can permanently stop a pathogen when fully implemented, but the pathogen must be slowed to gain time.

The rapid defenses available to plants include phytoalexin accumulation, which takes a few hours, and the hypersensitive reaction, which can occur in minutes. The hypersensitive reaction is the rapid death of plant cells in the immediate vicinity of the pathogen. Death of these cells is thought to create a toxic environment of released plant components that may in themselves interfere with pathogen growth, but more importantly, damaged cells probably release signals to surrounding cells and trigger a more comprehensive defense effort. Thus, phytoalexin accumulation is just one part of an integrated series of plant responses leading from early detection to eventual neutralization of a potentially lethal invading microorganism.

The tremendous capacity of plants to produce complex chemical compounds is reflected in the structural diversity of phytoalexins. Each plant species produces one or several phytoalexins, and the types of phytoalexins produced are similar in related species. The diversity, complexity, and toxicity of phytoalexins may provide clues about their function. The diversity of phytoalexins may reflect a plant survival strategy. That is, if a plant produces different phytoalexins from its neighbors, it is less likely to be successfully attacked by pathogens adapted to its neighbor's phytoalexins. Diversity and complexity, therefore, may reflect the benefits of using different deterrents from those found in other plants. *See* PLANT PATHOLOGY. [A.R.A.]

Phytochrome A plant pigment which regulates plant growth and development in response to presence or absence of light, day length, or light quality. It is found either in a highly stable inactive form or a relatively less stable active form in plant tissues, depending upon their previous illumination history. Absorption of light by inactive phytochrome transforms it to the active form, in which conformation it may regulate phenomena such as stem elongation, leaf expansion, flowering, seed germination, the relative activity of different metabolic pathways, and a host of other physiological and biochemical processes.

The control of plant development by light, acting through phytochrome, enables a plant to respond most favorably to its own immediate environment. Thus it can be protected from squandering resources by expanding leaves in darkness; it can maximize elongation under conditions in which light for photosynthesis is absent or limiting, in order to reach illumination with highest efficiency; and it can regulate its flowering or dormancy in response to appropriate season by measuring the length of days and nights. Phytochrome thus provides a continuous monitor of the light environment.

Phytochrome consists of a protein bound to a group (chromophore) which absorbs visible light. The color and reactivity of phytochrome to visible light are strictly determined by the chromophore's properties. The chromophore has been identified as an open-chain tetrapyrrole closely related to the blue-green and red algal photosynthetic accessory pigments phycocyanin and phycoerythrin, which are also open-chain tetrapyrroles. It is somewhat less closely related to the chlorophylls, which are closed tetrapyrrole rings. *See* CHLOROPHYLL; PHYCOBILIN. [W.R.Br.]

Phytoplankton Mostly autotrophic microscopic algae which inhabit the illuminated surface waters of the sea, estuaries, lakes, and ponds. Many are motile. Some perform diel (diurnal) vertical migrations, others do not. Some nonmotile forms regulate their buoyancy. However, their locomotor abilities are limited, and they are largely transported by horizontal and vertical water motions.

A great variety of algae make up the phytoplankton. Diatoms (class Bacillariophyceae) are often conspicuous members of marine, estuarine, and fresh-water plankton. Dinoflagellates (class Dinophyceae) occur in both marine and fresh-water environments and are important primary producers in marine and estuarine environments. Coccolithophorids (class Haptophyceae) are also marine primary producers of some importance. They do not occur in fresh water.

Even though marine and fresh-water phytoplankton communities contain a number of algal classes in common, phytoplankton samples from these two environments will appear quite different. These habitats support different genera and species and groups of higher rank in these classes. Furthermore, fresh-water plankton contains algae belonging to additional algal classes either absent or rarely common in open ocean environments. These include the green algae (class Chlorophyceae), the euglenoid flagellates (class Euglenophyceae), and members of the Prasinophyceae.

The phytoplankton in aquatic environments which have not been too drastically affected by human activity exhibit rather regular and predictable seasonal cycles. Coastal upwelling and divergences, zones where deeper water rises to the surface, are examples of naturally occurring phenomena which enrich the mixed layer with needed nutrients and greatly increase phytoplankton production. In the ocean these are the sites of the world's most productive fisheries. *See* EUTROPHICATION. [R.W.H.]

Phytotronics Research using whole plants and conducted under controlled environmental conditions to determine responses to a single or known combination of environmental elements. Originally, the term phytotronics was used to identify research conducted specifically in phytotrons where controlled plant growth units are available for simultaneous use. The name phytotronics is also often applied to any research conducted with whole plants in a controlled environment plant growth chamber or room. [H.H.]

Piciformes An order of birds containing six families, which have in common some anatomical characters, including a peculiar arrangement of the tendons of the toes. The order is generally considered to be most closely related to the Passeriformes. All members of the Piciformes nest in holes, and young birds spend a long period in the nest.

The largest family is the Picidae, the woodpeckers, worldwide in distribution except for Madagascar, Australia, and most of Oceania. Three families are confined to the neotropical region: the jacamars (Galbulidae), many of which are highly iridescent and resemble giant hummingbirds; the puffbirds (Bucconidae), most of which have a peculiar notch at the tip of the bill; and the toucans (Ramphastidae), noted for large and often highly colorful bills. Barbets (Capitonidae) are pantropical, whereas honey guides (Indicatoridae) are confined to Africa and India. *See* AVES; PASSERIFORMES. [K.C.P.]

Picornaviridae A viral family made up of the small (18–30 nanometer) ether-sensitive viruses that lack an envelope and have a ribonucleic acid (RNA) genome. The name is derived from "pico" meaning very small, and RNA for the nucleic acid type. Picornaviruses of human origin include the following subgroups: enteroviruses (polioviruses, coxsackieviruses, and echoviruses) and rhinoviruses. There are also picornaviruses of lower animals (for example, bovine foot-and-mouth disease, a rhinovirus). *See* ANIMAL VIRUS; COXSACKIEVIRUS; ECHOVIRUS; ENTEROVIRUS; FOOT-AND-MOUTH DISEASE; POLIOMYELITIS; RHINOVIRUS. [J.L.Me.]

Picric acid A phenol in which nitro (NO_2) groups are present in the 2, 4, and 6 positions on the ring of carbon atoms. The formula for picric acid is shown. It approaches the mineral acids in acid strength (pK_a = 0.80).

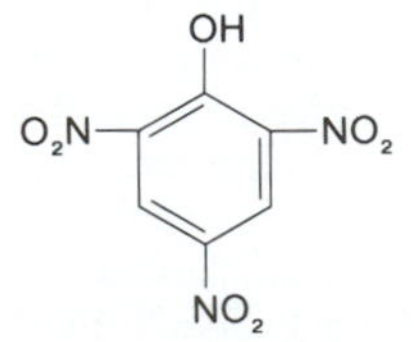

Picric acid was widely used as a yellow dye for silk before being replaced by modern dyes of greater fastness. It has also been used as an explosive, and can be detonated by both heat and shock. *See* PHENOL. [R.I.S./M.St.]

Picrite The term picrite has been used with several different meanings. It is generally considered to include certain medium- to fine-grained igneous rocks composed chiefly of olivine with smaller amounts of pyroxene, hornblende, and plagioclase feldspar (labradorite). Its feldspar content is slightly higher than that of peridotite and lower than that of gabbro. Certain analcite-bearing types, associated with teschenite, have also been included under the term picrite. Picrite is rare and is found in small intrusives (sills and dikes). *See* GABBRO; IGNEOUS ROCKS; PERIDOTITE. [C.A.C.]

Pictorial drawing A view of an object (actual or imagined) as it would be seen by an observer who looks at the object either in a chosen direction or from a selected point of view. Pictorial sketches often are more readily made and more clearly understood than are front, top, and side views of an object. Pictorial drawings, either sketched freehand or made with drawing instruments, are frequently used by engineers and architects to convey ideas to their assistants and clients. *See* DESCRIPTIVE GEOMETRY; ENGINEERING DRAWING.

In making a pictorial drawing, the viewing direction that shows the object and its details to the best advantage is chosen. The resultant drawing is orthographic if the viewing rays are considered as parallel, or perspective if the rays are considered as meeting at the eye of the observer. Perspective drawings provide the most realistic, and usually the most pleasing, likeness when compared with other types of pictorial views.

Several types of nonperspective pictorial views can be sketched, or drawn with instruments. In the isometric pictorial, the direction of its axes and all measurements along these axes are made with one scale (Fig. 1). Oblique pictorial drawings, while not true orthographic views, offer a convenient method for drawing circles and other curves in their true shape (Fig. 2).

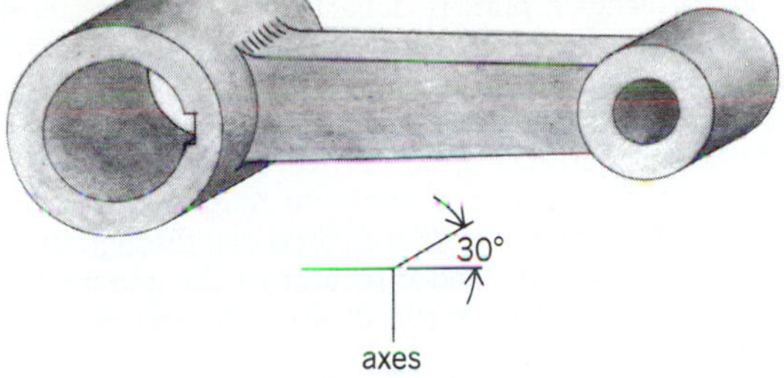

Fig. 2. Oblique pictorial drawing.

In order to reduce the distortion in an oblique drawing, measurements along the receding axis may be foreshortened. When they are halved, the method is called cabinet drawing. [C.J.B.]

Picture tube A cathode-ray tube used as a television picture reproducer. Modern television picture tubes usually have large glass envelopes (see illustration) on the inner face of which a light-emitting layer of luminescent materials is deposited. A modulated stream of high-velocity electrons is made to scan this layer of luminescent materials in a series of horizontal lines so that it recreates the picture elements (light and dark areas). *See* CATHODE-RAY TUBE.

The number of electrons in the stream at any instant of time is varied by electrical impulses corresponding to the signal sent out by the television transmitter. These electrical impulses (picture information) were originally generated by a studio television camera. The home television receiver picks up the signal, and after suitable amplification and detection the picture information is supplied to the picture tube so that it recreates the original picture. *See* TELEVISION.

Picture tubes are defined by the method used to focus the stream of electrons to a small spot. Electrostatic focus, accomplished by applying voltage differences to the electron-gun

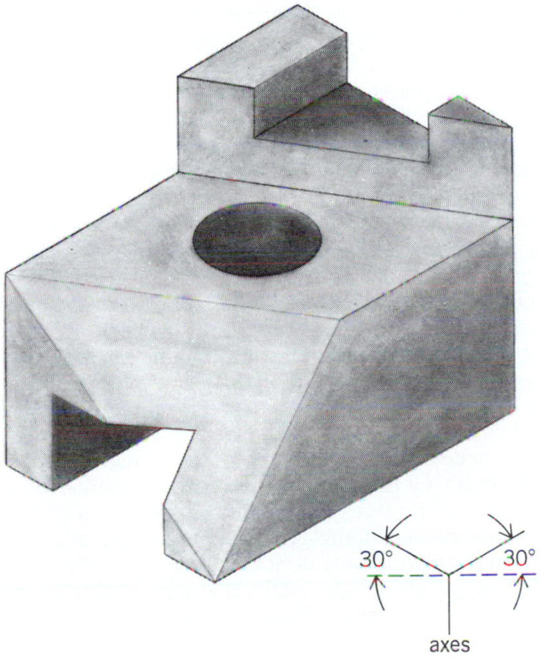

Fig. 1. Isometric drawing; measurements along each axis are made with the same scale.

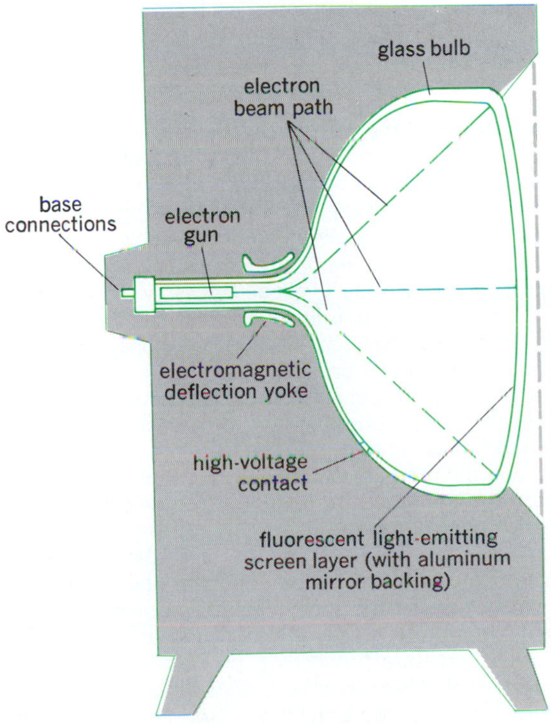

Black-and-white kinescope.

electrodes, has become universal. The electron beam, produced by an electron gun, is deflected electromagnetically to cause it to scan the picture area. This deflection is accomplished by a deflecting yoke, made up of two pairs of shaped coils which fit around the small neck of the picture tube. When pulsating electric currents of proper wave shape and phase are supplied to these coils, they generate magnetic fields which cause the electron beam to bend as it passes through them. By changing the magnitude and direction of the magnetic field, the electron beam can be made to arrive at any point on the face of the picture tube.

The glass envelope is made of a special composition to minimize optical defects and to provide electrical insulation for high voltages. The structural design of the glass bulb is made to withstand 3–6 times the force of atmospheric pressure.

The luminescent screen is made of a thin layer of phosphors. The phosphor materials are primarily zinc cadmium sulfide (emits yellow light) and zinc sulfide (emits blue light). The luminescent screen is aluminized by vacuum evaporation from a small molten aluminum pellet. In the operation of the completed tube, the high-velocity electron beam penetrates the aluminum film, and its energy is transferred primarily to the luminescent screen.

A color television picture tube is similar to the black-and-white picture tube but differs in two ways. (1) The light-emitting screen is made up of small elemental areas, each capable of emitting light in one of the three primary colors (red, green, or blue) laid in interlaced arrays. (2) Means are provided for selectively exciting any one or more of these phosphor primaries. *See* COLOR TELEVISION. [C.P.Sm.]

Pier A fingerlike structure projecting from shore and providing berths for ships to load and discharge passengers and cargo. Construction usually takes the form of a pile-supported platform using steel, timber, and concrete materials, thus allowing free water flow underneath. Sometimes bulkhead construction is used around the periphery with earth fill inside and pavement topping it. Piers may be housed over with sheds and may have special cargo-handling equipment. Uses of piers sometimes extend to recreational purposes, such as fishing piers, or to community purposes, such as car parking. Spaces between adjacent piers are called slips, *See* COASTAL ENGINEERING; WHARF. [E.J.Q.]

Piezoelectricity Electricity, or electric polarity, resulting from the application of mechanical pressure on a dielectric crystal. The application of a mechanical stress produces in certain dielectric (electrically nonconducting) crystals an electric polarization (electric dipole moment per cubic meter) which is proportional to this stress. If the crystal is isolated, this polarization manifests itself as a voltage across the crystal, and if the crystal is short-circuited, a flow of charge can be observed during loading. Conversely, application of a voltage between certain faces of the crystal produces a mechanical distortion of the material. This reciprocal relationship is referred to as the piezoelectric effect. The phenomenon of generation of a voltage under mechanical stress is referred to as the direct piezoelectric effect, and the mechanical strain produced in the crystal under electric stress is called the converse piezoelectric effect. *See* POLARIZATION OF DIELECTRICS.

The necessary condition for the piezoelectric effect is the absence of a center of symmetry in the crystal structure. Of the 32 crystal classes, 21 lack a center of symmetry, and with the exception of one class, all of these are piezoelectric. Hydrostatic pressure produces a piezoelectric polarization in the crystals of those 10 classes that show pyroelectricity in addition to piezoelectricity. *See* CRYSTALLOGRAPHY; PYROELECTRICITY.

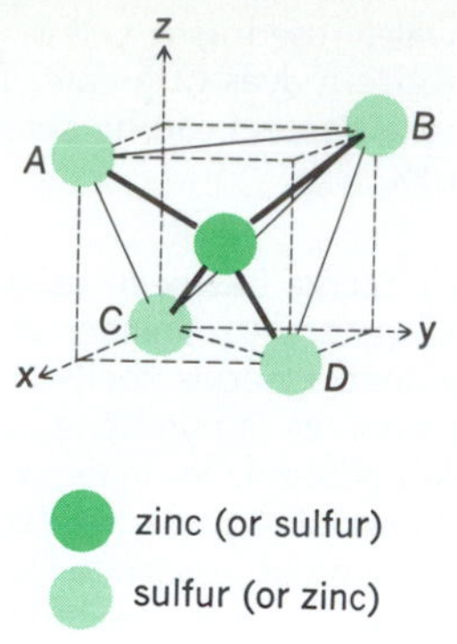

Tetrahedral structure of zincblende, ZnS. Only part of unit cell is shown. Size of circles has no relation to size of ions.

Molecular theory. Quantitative theories based on the detailed crystal structure are very involved. Qualitatively, however, the piezoelectric effect is readily understood for simple crystal structures. The illustration shows this for a particular cubic crystal, zincblende (ZnS). Every Zn ion is positively charged and is located in the center of a regular tetrahedron *ABCD*, the corners of which are the centers of sulfur ions, which are negatively charged. When this system is subjected to a shear stress in the xy plane, the edge *AB*, for example, is elongated, and the edge *CD* of the tetrahedron becomes shorter. Consequently, these edges are no longer equivalent, and the Zn ion will be displaced along the z axis, thus giving rise to an electric dipole moment. The dipole moments arising from different octahedrons sum up because they all have the same orientation with respect to the axes x, y, and z.

Applications. The sharp resonance curve of a piezoelectric resonator makes it useful in the stabilization of the frequency of radio oscillators. Quartz crystals are used almost exclusively in this application. In vacuum-tube oscillators, the crystal generally is part of the feedback circuit. Selective band-pass filters with low losses can be built by using piezoelectric resonators as circuit elements. A synthetic piezoelectric crystal which is often substituted for quartz in this application is ethylene diamine tartrate. *See* QUARTZ CLOCK.

Piezoelectric materials are used extensively in transducers for converting a mechanical strain into an electrical signal. Such devices include microphones, phonograph pickups, vibration-sensing elements, and the like. The converse effect, in which a mechanical output is derived from an electrical signal input, is also widely used in such devices as sonic and ultrasonic transducers, headphones, loudspeakers, and cutting heads for disk recording. *See* DISK RECORDING; MICROPHONE; ULTRASONICS; UNDERWATER TRANSDUCER. [H.G.]

Piezoelectric materials. The principal piezoelectric materials used commercially are crystalline quartz and rochelle salt, although the latter is being superseded by other materials, such as barium titanate. Quartz has the important qualities of being a completely oxidized compound (silicon dioxide), and is almost insoluble in water. Therefore, it is chemically stable against changes occurring with time. It also has low internal losses when used as a vibrator. Rochelle salt has a large piezoelectric effect, and is thus useful in acoustical and vibrational devices where sensitivity is necessary, but it decomposes at high temperatures (131°F or 55°C) and requires protection against moisture. Barium titanate provides lower sensitivity, but greater immunity to temperature and humidity effects. Other crystals that have been used for piezoelectric devices include tourmaline, ammonium dihydrogen phosphate (ADP), and ethylenediamine tartrate (EDT). *See* BARIUM TITANATE; QUARTZ; ROCHELLE SALT. [F.D.L.]

Pig A mammal which is a member of the superfamily Suoidea in the order Artiodactyla, the even-toed ungulates.

There are two families, the Suidae or Old World pigs (including wild and domestic species), and the Tayassuidae or peccaries of the Americas. The toes terminate in nails which are modified into hooves, the tail is short, and the body is covered sparsely with hair which is frequently bristlelike. These bristles are used in the manufacture of the highest-quality brushes. These animals are herbivorous. Tusklike canines are present in some species. *See* ARTIODACTYLA; TRICHINOSIS. [C.B.C.]

Pigeonite Monoclinic pyroxenes of the general formula (Mg,Fe)SiO_3 having some augite in solid solution. Pigeonite bears the same relation to the orthorhombic pyroxenes as augite does to the diopside-hedenbergite series. Pigeonite is the orthorhombic pyroxene equivalent in the volcanic rocks. *See* AUGITE; DIOPSIDE; ORTHORHOMBIC PYROXENE; PYROXENE. [G.W.DeV.]

Pigment A finely divided material which contributes to optical and other properties of paint, finishes, and coatings. Pigments are insoluble in the coating material, whereas dyes dissolve in and color the coating. Pigments are mechanically mixed with the coating and are deposited when the coating dries. Their physical properties generally are not changed by incorporation in and deposition from the vehicle. Pigments may be classified according to composition (inorganic or organic) or by source (natural or synthetic). However, the most useful classification is by color (white, transparent, or colored) and by function. Special pigments include anticorrosive, metallic, and luminous pigments. *See* DYE; LUMINOUS PAINT; PAINT. [C.R.Ma.; C.W.Si.]

Pigmentation The unique pigmentation of a given biological material is an expression of the quantity and quality of reflected incoming visible light. The light returning from the material is perceived as colored. The color is determined by the chemical composition and the physical properties of the light-exposed material. Consequently, a coloration may be due to pigments or may be of structural origin.

Pigments are essential constituents of the living world. Their contribution to the evolution and maintenance of life, and its manifold expressions, is most evident in the role of chlorophylls and the associated carotenoids of certain bacteria and most plants. These pigments harvest solar light energy for utilization in the photosynthesis of organic material from inorganic precursors. *See* CAROTENOID; CHLOROPHYLL; PHOTOSYNTHESIS.

The outermost structures on the animal skin are pigmented for many reasons, for example, to reduce the animal's visibility against a colored background, or to provide optical signals to the other sex or to other species. Conspicuously pigmented flowers attract pollinators, and colored fruits are easily found by animals, which eat them and then help disperse the seeds. *See* CHROMATOPHORE; PROTECTIVE COLORATION.

The role of pigments in communication depends on the ability of organisms to discriminate between different regions of the solar spectrum. In animals with eyes, this is accomplished by differently colored visual pigments contained in specialized receptor cells. Microorganisms, fungi, and plants also have special pigment systems that permit these organisms to move or grow toward, or away from, light (positive and negative phototaxis and phototropism, respectively). *See* PLANT MOVEMENTS.

Since organisms are totally dependent on light—at least indirectly—elaborate pigment systems have evolved which tune metabolic and activity patterns to the daily pattern of light and dark, and to the changes in the relative lengths of day and night occurring in the course of a year. The visual pigments of animals (the pigments of the extraretinal photoreceptor organs of many vertebrates and invertebrates) and the phytochromes of plants are typical representatives of pigments that correlate biological activity with light-dark cycles (photoperiodism). Colored coenzymes, such as flavins, may function as on-off switches of light-regulated metabolic processes. *See* COLOR VISION; PHOTOPERIODISM; PHOTORECEPTION; PHYTOCHROME.

In the examples listed above, pigments mediate, in various ways, the beneficial actions of light. Absorbed solar light energy may, however, also have detrimental effects by causing undesirable, or even destructive, reactions. The pigmentation of the human skin by melanin provides a light-absorbing layer which protects the tissue layers below from such potentially damaging radiation of the Sun. *See* INTEGUMENT; SKIN. [P.H.Ho.]

Pika Any member of the family Ochotonidae in the order Lagomorpha, which includes 14 species of small mammals restricted to northern Asia and northern North America. They may occur as high as 17,000 ft (5.2 km) in the Himalayas; all species inhabit rocky areas, dwelling deep under the rocks away from their natural enemies, hawks and foxes.

These mammals resemble rabbits, and have a vestigial tail; however, they have short, rounded ears (see illustration). Since they do not hibernate, they have evolved an interesting food-storage system to provide reserves for the winter. They cut and carry green herbage to the rocky areas they inhabit and spread it in the sun to dry. When the vegetation is dried, it is piled like a haystack. Common species include the Alaskan pika (*Ochotona collaris*) and the American pika (*O. princeps*). *See* LAGOMORPHA. [C.B.C.]

The pika, a rock-dwelling, diurnal mammal.

Pike Any of about five species of fish which compose the family Esocidae in the order Clupeiformes, known by a variety of names such as pickerel and muskellunge. These fish are voracious predators with an elongated beaklike snout and sharp teeth. The head is partly scaled and the body is covered with cycloid scales that have deeply scalloped edges. The body is cylindrical and compressed; thus, these fish are well adapted for rapid movements as they dart after prey. They prey upon each other, as well as other fish, amphibians, small aquatic birds and mammals, and rats. All species are edible but are considered second-rate game fish. *See* CLUPEIFORMES. [C.B.C.]

Pile foundation A part of a foundation system which transfers loads from a superstructure to the soil or rock below the surface by means of structural elements called piles. A pile foundation is used when the upper layers of soil are too soft and compressible to support the loads of the superstructure in a safe and economical manner by means of footings or mat

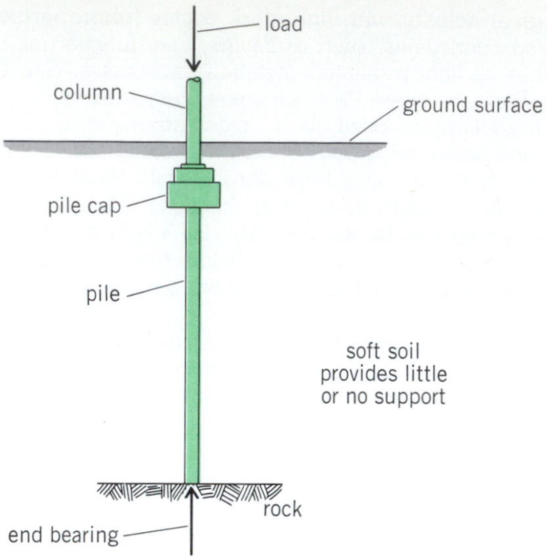

Fig. 1. End-bearing pile.

foundations. A pile foundation is often used to support structures over water or adjacent to water, in addition to supporting structures which are constructed over soft soils.

Piles, the structural elements which are installed into the soil or rock, may be wood, concrete, or steel; or concrete and wood or concrete and steel my be combined to form composite piles. Occasionally, wood piles may be reinforced with steel bands.

Piles obtain supporting capacity from the earth through end bearing or side friction or more generally through a combination of both (Figs. 1 and 2). Piles installed through very soft soils to rock or another firm stratum derive their capacity to support structural loads from the strength of the firm stratum on which the end of the pile is bearing. In such a situation the shaft of the pile acts as a column, and must have sufficient strength to transfer the imposed load from the top to the bottom of the pile.

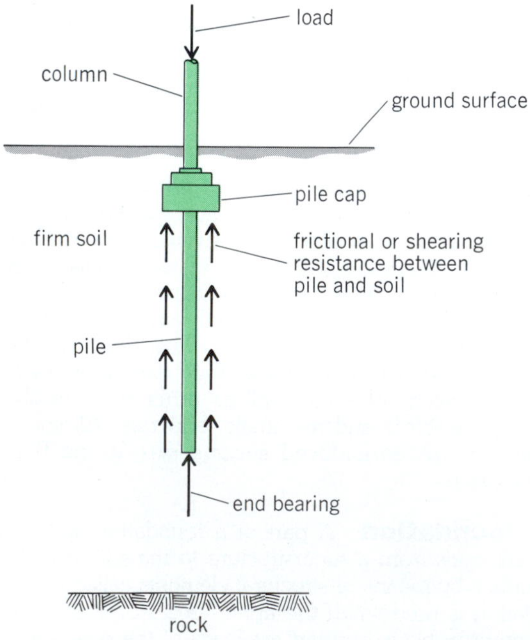

Fig. 2. Friction pile with some support from end bearing.

Soil is not always underlain by rock or another firm stratum within a depth which can be reached economically by piles. In this condition, the soil in which the pile is embedded and the soil below the pile tip provide the support for the pile. As the pile is subjected to load, it tends to move downward. The tendency to move downward is resisted by the friction between the sides of the pile and the soil in which the pile is embedded, or the shearing resistance of the soil immediately adjacent to the sides of the pile. In addition, as the tip of the pile tends to move downward, the soil below the tip resists this tendency due to its bearing strength. *See* Soil mechanics.

Most piles are installed by driving the piles into the soil. The equipment used consists of a steel-pile-driving hammer which operates between vertical leads. The leads are attached to a mobile crane. The simplest type of hammer is a steel weight raised above the top of the pile and allowed to fall by gravity. Another type of hammer used is a diesel hammer. In this type, the ram is lifted by an explosion of diesel fuel in a combustion chamber. When the ram reaches a predetermined height above the top of the pile, it drops by gravity; the next fuel explosion for lifting occurs at the moment of impact. A second type of diesel hammer is one which has an air chamber above the ram. As the ram is raised by the explosion in the combustion chamber, the air in the air chamber is compressed. As the ram drops, the compressed air adds downward force to the ram. Methods of installing piles other than using pile-driving equipment involve drilling a hole into the soil with an auger or a rotating bucket and filling the hole with concrete. *See* Caisson foundation; Cofferdam; Foundations; Retaining wall.

[G.M.R.]

Pilot production The production of a product, process, or piece of equipment on a simulated factory basis. In mass-production industries where complicated products, processes, or equipment are being developed, a pilot plan often leads to the presentation of a better product to the customer, lower development and manufacturing costs, more efficient factory operations, and earlier introduction of the product. Following the engineering development of a product, process, or complicated piece of equipment and its one-of-a-kind fabrication in the model shop, it becomes desirable and necessary to "prove out" the development on a simulated factory basis. *See* Industrial cost control; Mass production; Product design; Production engineering; Quality control.

[J.E.Wo.]

Pilotage One of four procedures used in navigating an aircraft. The other three are position fixing, homing, and dead reckoning. Pilotage is the procedure of using landmarks, such as cities, towns, rivers, railroads, and prominent highways, to guide an aircraft to a destination. The installation of lights at airports and prominent spots across the country enhanced the ability of the pilot to direct the aircraft. *See* Dead reckoning; Position fixing.

The introduction of radar brought a new dimension to pilotage. Airborne radar operating at microwaves produces very sharp maps of the terrain over which the aircraft is flying. Radar pilotage has also been adapted to missile guidance.

[P.C.S.]

Piloting The form of navigation in which position is determined relative to external reference points, usually fixed points on the Earth. It is the oldest form of navigation. With the development of electronic aids, piloting techniques were extended far from shore. However, the term "piloting" is generally associated with nearness of land, where tidal and other currents may be strong, shoals and other underwater obstructions may be in near proximity, and maneuvering room is limited when other vessels are encountered. Thus it is not unusual for ships to employ the services of a local expert, called a pilot, to assist in the navigation of the vessel while it enters or leaves port. Whether a pilot is used or navigation is performed by ship's officers, the period during which piloting is performed is a critical one, requiring frequent and urgent exercise of judgment and attention to detail.

A conspicuous object, structure, or light that serves as an indicator for establishing the position of a craft or otherwise assisting in its safe navigation is called a mark. To be useful, not only must a mark be identified, but its position must be known accurately. Artificial marks designed and erected specifically to assist the navigator are called aids to navigation and include beacons, both lighted and unlighted, lighthouses, buoys, both lighted and unlighted, and lightships. Unlighted aids are called daymarks. *See* BUOY; LIGHTHOUSE.

In addition to visible aids to navigation, bottom topography can be of assistance in locating the position of a vessel. Sound signals transmitted through water or air may be used for navigation. Electronic beacons and positioning systems have been established at a number of places to assist in navigation. *See* ELECTRONIC NAVIGATION SYSTEMS.

The measurements made for piloting purposes are of direction, distance, differential distance between two points, and distance to the bottom.

Bearings are usually measured (1) by noting when two objects are in range (directly in line); (2) by means of a suitable attachment to a compass or compass repeater; (3) by pelorus, a compasslike instrument without directive properties; or (4) electronically, by radio direction finder, radar, or by the indication of the receiver-indicator of an electronic system of navigation. Distance is generally measured by radar. The difference in distance from the ship to two points is usually measured electronically by means of the receiver-indicator of a hyperbolic navigation system. Depth measurement is usually made by an echo sounder. *See* DIRECTION-FINDING EQUIPMENT; ECHO SOUNDER; HYPERBOLIC NAVIGATION SYSTEM; RADAR.

Position is generally determined by means of lines of positions, each indicating a series of possible positions of the craft at the time of measurement. A measured bearing provides a straight line of position (actually part of a great circle) passing through the object sighted. A measured distance provides a circular line of position with the object as the center and the distance as the radius. A measured differential distance provides a hyperbolic line of position. A position, called a fix, is usually determined by crossing two or more lines of position taken simultaneously or nearly so. *See* CELESTIAL NAVIGATION; DEAD RECKONING; NAVIGATION; POLAR NAVIGATION. [A.B.M.]

Pimento A type of pepper, *Capsicum annuum*, grown for its thick, sweet-fleshed red fruit. A member of the plant order Polemoniales, pimento is of American origin, and gets its name from the Spanish word designating all sweet peppers. In the United States, however, the term pimento generally refers to the heart-shaped varieties (cultivars) grown for canning and used for stuffing olives and flavoring foods. Georgia is the only important pimento-producing state. *See* PEPPER; SOLANALES. [H.J.C.]

Piña A fiber, also known as pineapple fiber, obtained from the large leaves of the pineapple plant grown in tropical countries. This natural fiber is white and especially soft and lustrous. In the Philippine Islands, it is woven into piña cloth, which is soft, durable, and resistant to moisture. Piña is also used in making coarse grass cloth and for mats, bags, and clothing. *See* NATURAL FIBER; PINEAPPLE. [M.D.P.]

Pinales An order of the plant class Pinopsida, subdivision Pinicae, in the division Pinophyta (Gymnospermae), with more than 50 genera and over 540 species—the dominant forest-makers of the world. Here belong the pine, spruce, hemlock, cedar, fir, larch, juniper, cypress, yew, redwood, big tree, kauri, podocarpus, and many others. All are woody, being either shrubs or trees.

The conifers appeared in the Carboniferous, attained their zenith and widest distribution in the Jurassic, and then began to wane. However, the persistent species form some of the most extensive forests of modern times. Characteristically, they are mountain dwellers, ranging from the Arctic to the Antarctic circles, with the greatest display in the temperate zones and a good representation in the tropics.

The cones are unisexual. In some species, both male and female cones are borne on the same plant (monoecious); in others, on separate plants (dioecious). Usually, a year intervenes between pollination and fertilization, and another year between fertilization and embryo formation.

The conifers are used in landscaping and as the source of valuable lumber and such by-products as turpentine, tar, resin, and oils of various kinds. See separate articles on most of the major conifers by common names. *See* PINE; PINOPSIDA. [A.Cr.]

Pinch effect A name given to manifestations of the magnetic self-attraction of parallel electric currents having the same direction. The effect at modest current levels of a few amperes can usually be neglected, but when current levels approach a million amperes such as occur in electrochemistry, the effect can be damaging and must be taken into account by electrical engineers. The pinch effect in a gas discharge has been the subject of intensive study, since it presents a possible way of achieving the magnetic confinement of a hot plasma (a highly ionized gas) necessary for the successful operation of a thermonuclear or fusion reactor.

The law of attraction which describes the interaction between parallel electric currents was discovered by A. M. Ampère in 1820. For a cylindrical wire of radius r meters carrying a total surface current of I amperes, it manifests itself as an inward pressure on the surface (Fig. 1) given by $I^2/2 \times 10^7 \pi r^2$ pascals. For the electric currents of normal experience, this force is small and passes unnoticed, but it is significant that the pressure increases with the square of the current, I^2. For example, at 25,000 amperes the pressure amounts to about 1 atm (100 kilopascals) for a wire of 1-cm radius, but at 10^6 amperes the pressure is about 1600 atm or about 12 tons in.$^{-2}$ (160 megapascals).

There are a number of ways in which the magnetic field of a fusion reactor can be arranged around the plasma to hold it together, and one of these methods is the pinch effect. A fusion reactor using this type of confinement would ideally be a toroidal tube in which the confined plasma would carry a large electric current induced in it by magnetic induction from a transformer core passing through the major axis of the torus.

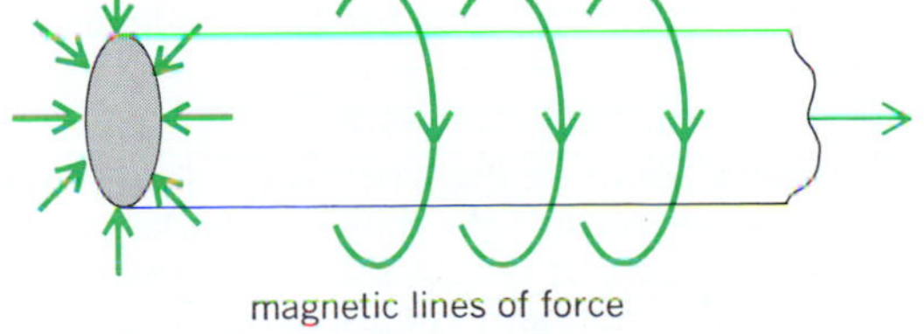

Fig. 1. Pinch pressure on a current-carrying conductor. Arrows at left show direction of pinch pressure.

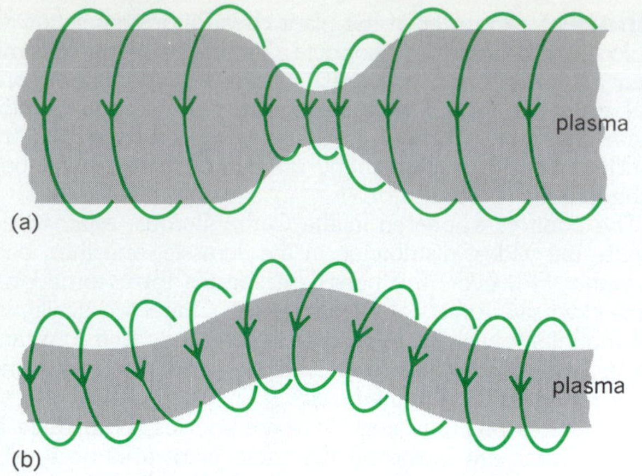

Fig. 2. Instability. (*a*) Sausage type. (*b*) Kink type.

The current would have the double function of ohmically heating the plasma and compressing the plasma toward the center of the tube.

Characteristically, as can be shown by high-speed photography, the pinch forms at the inner surface of a discharge tube wall and contracts radially inward, forming an intense line, the pinch, on the axis; the pinch rebounds slightly; the contracted discharge rapidly develops necks and kinks; and in a few microseconds all structure is lost in an apparently turbulent glowing gas which fills the tube. Thus, the pinch turns out to be unstable, and plasma confinement is soon lost by contact with the wall. The cause of the instability is easily seen qualitatively: The pinch confinement can be described as being caused by the magnetic field lines encircling the pinch which are stretched longitudinally but which are in compression transversely (Fig. 2). For a uniform cylindrical pinch, the magnetic pinch pressure is everywhere equal to the outward plasma pressure, but at a neck or on the inward side of a kink, the magnetic field lines crowd together, creating a higher magnetic pressure than the outward gas pressure. Consequently, the neck contracts still further, the kink cuts in on the concave side and bulges out on the convex side, and both perturbations grow. The instability has a disastrous effect on the confinement time.

The term theta pinch has come into wide usage to denote an important plasma confinement system which relies on the repulsion of oppositely directed currents and which is thus not in accord with the original definition of the pinch effect (self-attraction of currents in the same direction). Plasma confinement systems based on the original pinch effect are known as Z pinches.

Tokamak is essentially a low-density, slow Z pinch in a torus with a very strong longitudinal field. The helical magnetic field lines, resultant from the externally applied field and that of the pinch, do not close, that is, complete one revolution of the minor axis in going around the major axis of the torus once. This is known theoretically to prevent the growth of certain helical distortions of the plasma.

The performance of tokamak experiments has raised the possibility of achieving a net power balance; construction of several large tokamak installations has been undertaken. *See* NUCLEAR FUSION. [J.A.Ph.]

Pine The genus *Pinus*, of the pine family, characterized by evergreen leaves, usually in tight clusters (fascicles) of two to five, rarely single. There are about 80 known species distributed throughout the Northern Hemisphere. Botanically the leaves are of two kinds: (1) a scalelike form, the primary leaf, which subtends a much shortened and eventually deciduous shoot bearing (2) the secondary leaves or needles. The wood of pines is easily recognized by the numerous resin ducts and by the characteristic resinous odor. *See* PINALES; PINE NUT; PINE OIL. [A.H.G./K.P.D.]

Pine nut The edible seed of more than a dozen species of evergreen cone-bearing trees in the genus *Pinus*, native to the temperate zone of the Northern Hemisphere. The important nut-producing species are the stone pine (*P. pinea*) of southern Europe; the Swiss stone pine (*P. cembra*), native to the Swiss Alps and eastward through Siberia to Mongolia; and the piñon pine (*P. cembroides* var. *edulis*) of the arid regions of the southwestern United States. The seeds or nuts, variable in size according to species, are borne in cones which take 3–4 years to develop. *See* PINE. [L.H.MacD.]

Pine oil A material fractioned from oils recovered from longleaf pine wood. The pine is either destructively distilled or solvent-extracted to yield turpentine, rosin, charcoal, and other useful commercial products. The recovered oils are further refined, and pine oil results from a definite cut in the fractionation. Good grades of pine oil consist largely of terpineol. *See* WOOD CHEMICALS.

The textile industry uses pine oil as a penetrant, dispersing agent, wetting agent, and inhibitor of bacterial growth in practically all wet processing of cotton, silk, rayon, and woolen goods. In the paper industry it is used as a wetting and leveling agent in coating and as a preservative of casein. It also finds extensive use for the preparation of high-grade disinfectants, liniments, odorant blocks, dog soaps, cattle sprays, and other insecticidal preparations. *See* ESSENTIAL OILS. [W.Mos.]

Pine terpene A major component of the essential oils obtained from various *Pinus* species. The principal terpenes of the oil of southern pines [longleaf pine (*P. palustris*) and slash pine (*P. caribaea*)] are α- and ß-pinene, whose structures are shown below. *See* ESSENTIAL OIL; PINE.

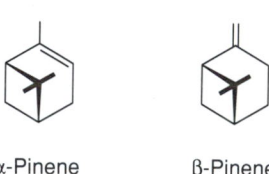

α-Pinene β-Pinene

Gum turpentine (gum spirits) is the volatile fraction of the oleoresin that exudes from cuts made in the trunks of live trees. The resin is collected and distilled by a process that yields about 20% turpentine, mainly α- and ß-pinene, and 70% rosin; it was the basis of the original naval stores industry.

Wood turpentine is obtained by steam distillation from stumps and other logging residues. The volatile material in this case consists of about 50% turpentine and 30–40% of higher-boiling-point alcohols; the latter fraction is known as pine oil. The bulk of the wood turpentine and pine oil produced by modern industrial processes is a by-product of the sulfate wood-pulping process (sulfate turpentine).

Important uses of turpentine or the purified pinenes derived from turpentine are in terpene resins, as a thinner in paints and varnishes, and as a starting material in the synthesis of other commercially valuable terpenes.

Pine oil is a mixture of monoterpene alcohols, mainly α-terpineol, obtained in large amounts mixed with wood turpentine or sulfate turpentine. The term pine oil is also used to designate the essential oil of various species of pine.

Much of the pine oil of commerce is prepared synthetically by acid-catalyzed hydration of α-pinene. This process involves a complex series of reactions that occur via cationic intermediates.

The composition of industrial-grade pine oil is approximately 65% α-terpineol, 20–25% of other monoterpene alcohols, and 10–15% hydrocarbons. Pine oil has surfactant and emulsifying properties and is also a disinfectant. Most of the pine oil manufactured is used in the manufacture of cleansers and textile penetrants. *See* SURFACTANT; TERPENE; WOOD CHEMICALS. [J.A.Mo.]

Pineal body The pineal body, or epiphysis as it is sometimes named, is an outgrowth from the roof of the diencephalon of the brain. It is widespread throughout the vertebrate animal kingdom. In lampreys and some lower bony fish it serves as a third eye. In other vertebrates it serves as an endocrine organ that is regulated by light entering the body via the paired eyes. Its role in humans remains to be clarified.

The pineal organ is an unpaired, elongate, club-shaped, knoblike, or occasionally threadlike organ attached by a stalk to the roof of the forebrain. When it serves as a light receptor, it projects upward through the skull to lie under the skin; in other vertebrates it is buried beneath the roof of the skull. It is small in birds and some mammals, and relatively large in cocks, echidnas, marsupials, rodents, ungulates, and humans. Nervous elements predominate in lower vertebrates, whereas glandular cells predominate in higher ones. [G.C.K.]

One of the most important contributions to the understanding of pineal physiology was the demonstration of an indole compound in the gland. This active principle is known as melatonin. In addition to melatonin, the pineal gland has been shown to contain physiologically active substances such as histamine, serotonin, and catecholamines in fairly high concentrations. In addition, some enzymes involved in the formation and metabolism of these animals are also present in this organ. The melatonin enzyme was found to be present in the pineal gland of many animal species, including the cat, rat, cow, monkey, bird, and human. Studies of indolalkylamine metabolism indicate that the pineal may not function continuously and that its function may vary in a circadian fashion. *See* CIRCADIAN RHYTHMS.

The physiological significance of these amines and indoles and their cyclic variations is not precisely known. Much attention has been given to the possible extrapineal function of melatonin because it appears to alter the state of the reproductive systems, the estrous cycle, and the brain and causes a change in the pigment cells in the integument of amphibians. Furthermore, removal of the gland produces changes in the reproductive system which are in contrast to those resulting from melatonin treatment in addition to affecting the pituitary, adrenal, and thyroid glands. [H.J.Cl.]

Pineapple A low-growing perennial plant, indigenous to the Americas. The cultivated varieties (cultivars) belong to the species *Ananas sativus* of the plant order Bromeliales.

The edible portion of the pineapple develops from a mass of ovaries on a fleshy flower stock having persistent bracts (see illustration). On the cultivated types, the flowers are usually abortive. The leaves are long and swordlike and usually rough-edged. Commercial plantings bear fruit at the age of 12–20 months, and may continue to be productive for as much as 8–10 years. *See* BROMELIALES.

The major producing area is Hawaii, where special methods of culture and harvesting have been developed. Pineapples are also grown in the West Indies and other tropical areas, and to a limited extent in southern Florida.

Pineapple (*Ananas sativus*), fruit and leaves. (*USDA*)

Pineapples are consumed fresh in considerable quantity, but because of distance from markets and the problems of transporting fresh fruit, most of the crop is canned as sliced pineapple or as juice. *See* FRUIT. [J.H.C.]

Pinene One of the most important terpenes. α-Pinene (I) is the chief constituent of turpentine and is a component of many essential oils. ß-Pinene (II) often accompanies it and is usually present in smaller quantities. Both exist as dextro, levo, and racemic forms of optical isomers. Physical properties are listed in the table. *See* TERPENE.

Physical properties of pinene

Property	α-Pinene *d*-, *l*-, and *dl*-	β-Pinene *d*-, *l*-
Boiling point, 760 mm	156°C	164–166°C
Density (20°C)	0.858–0.860	0.87
Refractive index, 20°/D	1.466	1.46–1.48
Specific rotation	*d*=+51° *l*= −51°	*l*=22°

Both pinenes are colorless oils which resinify on exposure to air. They undergo polymerization in the presence of anhydrous aluminum chloride or boron trifluoride. The polymers are used in the formulation of rubbers, paints, and varnishes.

The chemistry of the pinenes is intricate and complicated by their easy transformation into compounds with different ring structures. These molecular rearrangements permit the

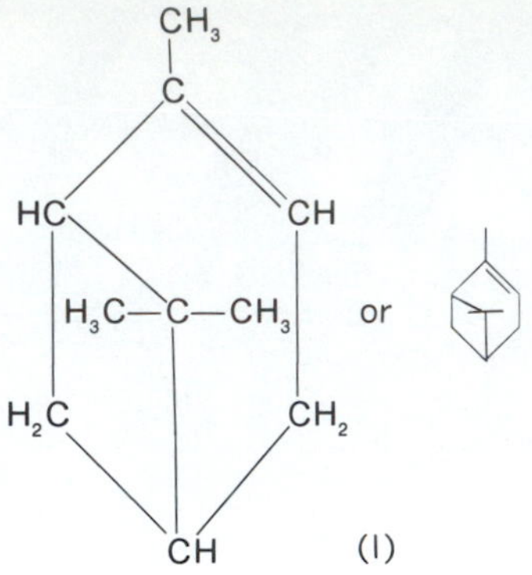

(I)

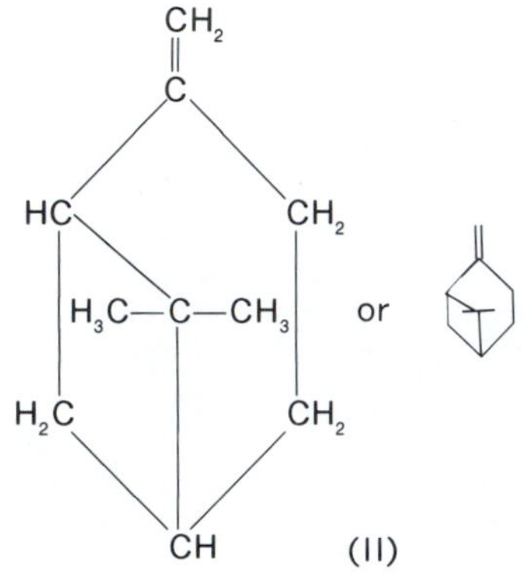

(II)

pinenes to be used as a convenient starting point for the synthesis of other important chemicals. *See* TURPENTINE. [F.W.]

Pinicae The largest and most important subdivision of the division Pinophyta (Gymnospermae), the other two subdivisions being the Cycadicae and Gneticae. The Pinicae are all woody plants, most of them trees with a central trunk and excurrent branches. They have simple, alternate or less often opposite or whorled leaves which are mostly rather small, often needlelike or scalelike. The wood lacks vessels and usually has resin canals. *See* CYCADICAE; GINKGOALES; GNETICAE; PINALES; PINOPHYTA; TAXALES. [A.Cr.]

Pinkeye An infection of the inner aspect of the eyelids caused by the Koch-Weeks bacillus. It is also known as epidemic conjunctivitis. The organism is regarded as identical with the nontypable form of *Hemophilus influenzae*. Pinkeye is widespread, occurs in epidemic outbreaks, and is said to be especially prevalent in Egypt. *See* CONJUNCTIVITIS. [W.L.Br.]

Pinnipeds Carnivorous mammals of the suborder Pinnipedia, which includes 32 species of seals, sea lions, and walrus in 3 families. All species of the order are found along coastal areas, from the Antarctic to the Arctic regions. Although many species have a restricted distribution, the group as a whole has worldwide distribution.

Pinnipeds are less modified both anatomically and behaviorally for marine life than are other marine mammals: Each year they return to land to breed, and they have retained their hindlimbs. They are primarily carnivorous mammals, with fish supplying the basic diet. They also eat crustaceans, mollusks, and, in some instances, sea birds. Their body is covered with a heavy coat of fur, the limbs are modified as flippers, the eyes are large, the external ear is small or lacking, and the tail is absent or very short.

The single species of the family Odobenidae is the walrus (*Odobenus rosmarus*); it grows to 10 ft (3 m) and weighs 3000 lb (1350 kg). It has no external ears but does have a distinct neck region. The upper canines of both sexes are prolonged as tusks which can be used defensively. The walrus feeds on marine invertebrates and fish.

The family Otariidae, the eared seals, includes the sea lions and fur seals, which are characterized by an external ear. The neck is longer and more clearly defined than that of true seals, and the digits lack nails. The California sea lion (*Zalophus californianus*) occurs along the Pacific coast, and is commonly seen in zoos and performing in circuses. The southern sea lion (*Otaria byronia*) is found around the Galapagos Islands and along the South American coast. Stella's sea lion (*Eumetopias jubatus*) is a large species found along the Pacific coast. Minor species are the Australian sea lion (*Phocarctos hookeri*).

Phocidae, the largest family of pinnipeds, comprises the true seals. It includes the monk seals, elephant seals, common seals, and other less well-known forms. The family is unique in that the digits have nails, the soles and palms are covered with hair, and the necks are very short. Most species live in marine habitats; however, the Caspian seal (*Pusa caspica*) lives in brackish water. The only fresh-water species is the Atlantic gray seal (*Halichoreus grypus*), a species found in the North Atlantic along the coasts of Europe, Iceland, and Greenland. *See* CARNIVORA; MAMMALIA.

Pinophyta A division of seed plants consisting of some 600 or 700 species, widely distributed on all continents, the most common and familiar ones being the cone-bearing evergreen trees of the order Pinales. Because the ovules (young seeds) are exposed directly to the air at the time of pollination, the Pinophyta are commonly called gymnosperms, in contrast to the angiosperms (division Magnoliophyta), which have the ovules enclosed in an ovary. The division Pinophyta consists of three subdivisions, the Pinicae, the Cycadicae, and the Gneticae.

The gymnosperms may be characterized as vascular plants with roots, stems, and leaves, and with seeds that are not enclosed in an ovary. The seeds are borne on cone scales in simple or compound cones, or exposed at the end of a stalk.

The gymnosperms evidently originated from Paleozoic vascular cryptogams at a time when the ferns (division Polypodiophyta) had not become fully differentiated from the ancestral rhyniophytes. Which of these two groups should be considered ancestral to the gymnosperms is a matter of convenience and arbitrary definition. The gymnosperms proliferated during the Carboniferous Period and became the dominant land plants by the beginning of the Mesozoic Era. During the Cretaceous Period they were progressively replaced by the angiosperms, and they now have the appearance of a relict group. Only the order Pinales now has fairly numerous and diversified species represented by many individuals over large areas. *See* CYCADICAE; GNETICAE; MAGNOLIOPHYTA; PINALES; PINICAE; POLYPODIOPHYTA. [A.Cr.]

Pinopsida One of the two classes (the other being the Ginkgoopsida) which collectively make up the subdivision Pinicae of the division Pinophyta (Gymnospermae). There are two living orders, the Pinales and Taxales, plus the fossil order Cordaitales. The Pinopsida are characterized by entire-margined or slightly toothed leaves that are usually narrow and never fan-shaped, in contrast to the fan-shaped, often bilobed leaves of the Ginkgoopsida. All modern species of Pinopsida have naked sperms, in contrast to the flagellated, swimming sperms of the Ginkgoopsida. *See* CORDAITALES; GINKGOALES; PINALES; PINICAE; TAXALES. [A.Cr.]

Pinta One of the treponematoses, which include yaws and syphilis, occurring principally in certain rural tropical areas of Central and South America: The infectious agent is a spirochete, *Treponema carateum*. The early lesions and the general evolution of pinta are similar to, but generally milder than, those of syphilis and yaws. Generalized lesions develop a reddish-purplish hue which is followed by atrophy and spotty achromia of the skin. Treponemes are present in the skin lesions. Serious late complications seem not to occur. Transmission is presumably by direct contact. *See* SYPHILIS; YAWS. [T.B.T.]

Pinworm infection An infection, also known as enterobiasis, of the human intestinal tract by the pinworm (*Enterobius vermicularis*). Both male and female nematodes inhabit the large intestine but the vagina may also be invaded. The female produces several thousand eggs which accumulate in its uteri. Pregnant worms, 2–4 weeks old, are expelled in the feces, or crawl out of the anus to lay the eggs and die. Their movements cause itching, and nervous disorders may ensue. The disease is diagnosed by observing the worm upon extrusion from the anus and by microscopical detection of eggs in anal swabs. *See* MEDICAL PARASITOLOGY. [J.F.M.]

Pionium An exotic atom, also called the pi-mu atom, which is similar in structure to the hydrogen atom but with the proton replaced by a pion and the electron replaced by a muon. Pionium is unique among atoms that have been observed in the laboratory in that all of its constituents are unstable particles not found in ordinary matter. Pionium is formed during the decay of a certain heavier particle called the neutral kaon. A kaon has many modes of decay, one of which results in the formation of a pion, a muon, and a neutrino. *See* ELEMENTARY PARTICLE.

The nomenclature of exotic atoms is not well established, and the name pionium may also refer to the pion-electron atom. [P.A.S.]

Pipeline A line of piping and the associated pumps, valves, and equipment necessary for the transportation of a fluid. Major uses of pipelines are for the transportation of petroleum, water (including sewage), chemicals, foodstuffs, pulverized coal, and gases such as natural gas, steam, and compressed air. Pipelines must be leakproof and must permit the application of whatever pressure is required to force conveyed substances through the lines. Pipe is made of a variety of materials and in diameters from a fraction of an inch up to 30 ft (9 m). Principal materials are steel, wrought and cast iron, concrete, clay products, aluminum, copper, brass, cement and asbestos (called cement-asbestos), plastics, and wood.

Pipe is described as pressure and nonpressure pipe. In many pressure lines, such as long oil and gas lines, pumps force substances through the pipelines at required velocities. Pressure may be developed also by gravity head, as for example in city water mains fed from elevated tanks or reservoirs.

Nonpressure pipe is used for gravity flow where the gradient is nominal and without major irregularities, as in sewer lines, culverts, and certain types of irrigation distribution systems.

Design of pipelines considers such factors as required capacity, internal and external pressures, water- or airtightness, expansion characteristics of the pipe material, chemical activity of the liquid or gas being conveyed, and corrosion. [L.N.McC.]

Piperales An order of flowering plants, division Magnoliophyta (Angiospermae) in the subclass Magnoliidae of the class Magnoliopsida (dicotyledons). The order includes only 3 families and about 1500 species, more than 1400 of them belonging to the family Piperaceae. Within its subclass the order is marked by ethereal oil cells (like those of the Magnoliales), a usually more or less herbaceous habit, and especially by reduced, crowded flowers with orthotropous ovules (that is, the ovules are straight, with the micropyle and funicules at opposite ends). *See* BETEL NUT; BLACK PEPPER; CUBEB; MAGNOLIIDAE; MAGNOLIOPSIDA. [A.Cr.]

Pipet A tube, usually made of glass or plastic, used almost exclusively to deliver accurately known volumes of liquids or solutions. There are two general categories of pipets: volumetric or transfer pipets and the graduated measuring type (see illustration). With volumetric pipets the liquid is sucked up

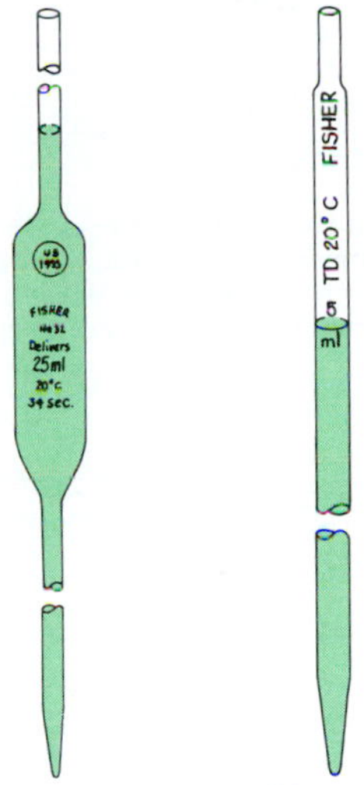

(a) (b)

Pipets. (*a*) Volumetric type. (*b*) Graduated type.

above the mark on the stem above the bulb and the upper end is quickly closed with the index finger; the level of liquid in the stem is allowed to fall slowly, by regulating the pressure on the finger, until the bottom of the meniscus is tangent to the mark.

A graduated measuring pipet is used in the same general way except that the volume of liquid delivered can be varied by allowing the liquid to drain from one calibration mark to another. These pipets are usually not as accurate in delivering volumes of liquid as are the volumetric type. *See* TITRATION; VOLUMETRIC ANALYSIS. [C.E.B.]

Pirani gage A type of instrument used to measure vacuum by utilizing a resistance change due to a temperature change in a filament. This fine-wire filament, one of the four electrical resistances forming a Wheatstone bridge circuit, is exposed to the vacuum to be measured. Electric current heats the wire; the surrounding gas (in the vacuum) conducts heat away from the wire. At a stable vacuum, the wire quickly reaches equilibrium temperature. If the pressure rises, the gas carries away more heat, and the temperature of the wire decreases. Since the resistance of the filament is a function of temperature, the electrical balance of the Wheatstone bridge is changed. Pressure measurement range of this type of gage is usually 1 to 10^{-4} torr (10^2 to 10^{-2} pascals). *See* VACUUM MEASUREMENT. [R.C.]

Pisces (constellation) The Fishes, in astronomy, a zodiacal constellation appearing in the autumn evening sky. Pisces is the twelfth and last sign of the zodiac. It is inconspicuous, having no star brighter than the fourth magnitude. But it is an important constellation because the vernal equinox,

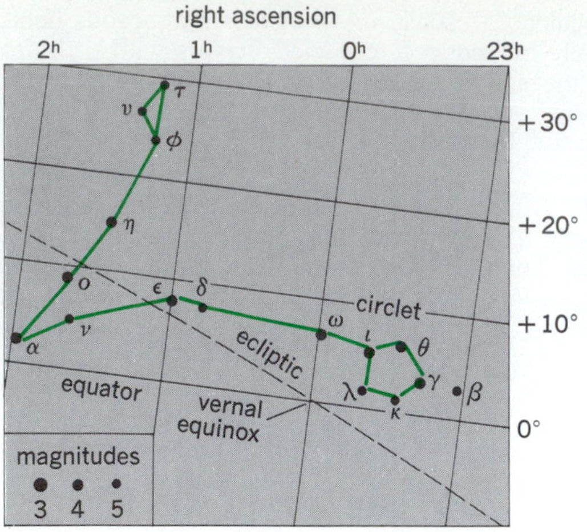

Line pattern in the constellation Pisces. The grid lines represent the coordinates of the sky. The apparent brightness, or magnitude, of the various stars is shown by the sizes of the dots, which are graded by appropriate numbers as indicated.

which marks the beginning of the astronomical year, is now located in it (see illustration). *See* CONSTELLATION. [C.-S.Y.]

Pisces (zoology) A term that embraces all fishes and fishlike vertebrates. The Pisces include four well-defined groups that merit recognition as classes: the Agnatha or jawless fishes, the most primitive; the Placodermi or armored fishes, known only as Paleozoic fossils; the Chondrichthyes or cartilaginous fishes; and the Osteichthyes or bony fishes. *See* AGNATHA; CHONDRICHTHYES; OSTEICHTHYES; PLACODERMI. [R.M.B.]

Pistachio A small spreading evergreen tree, *Pistacia vera*, native to the dry parts of Asia and Asia Minor and now grown in the Mediterranean region, Turkey, Iran, and Afghanistan as a nut crop. The green or reddish oval fruits are borne in clusters along the smaller branches. Classified botanically as a drupe, the fruit consists of an outer fleshy husk within which is a thin tough shell enclosing the usually green kernel which is covered with brown seed coats. The husk splits at maturity and is separated easily from the shell.

Pistachio nuts are used in the baking, ice cream and confectionary trades, and are an important food in central Asia, where much of the crop is used locally. *See* SAPINDALES. [L.H.MacD.]

Pitch That psychological property of sound characterized by highness or lowness. Pitch varies most directly with the frequency of sound waves (see illustration), and pitch discriminations can be made throughout the frequency range of normal hearing, from about 16 to 20,000 hertz (Hz).

A numerical scale of pitch has been constructed by the method of fractionation, selecting pitches that appear to be half as high as reference tones, and the method of bisection, selecting pitches that appear to fall halfway between two reference tones. The pitch unit was named the mel, and a value of 1000 mels was arbitrarily assigned to a tone of 1000 Hz.

The critical features of pitch perception are the detection and analysis of specific tonal values within a broad spectrum of frequency change. The ability to focus attention on specific tone patterns within larger patterns of sound is an outcome of several levels of receptor control including: neural feedback from the primary afferent centers of the brain; neuromotor control of frequency input by the middle ear muscles; and possibly dynamic tuning of the basilar membrane through middle ear tension on the spiral ligament. *See* EAR; HEARING (HUMAN). [K.U.S.]

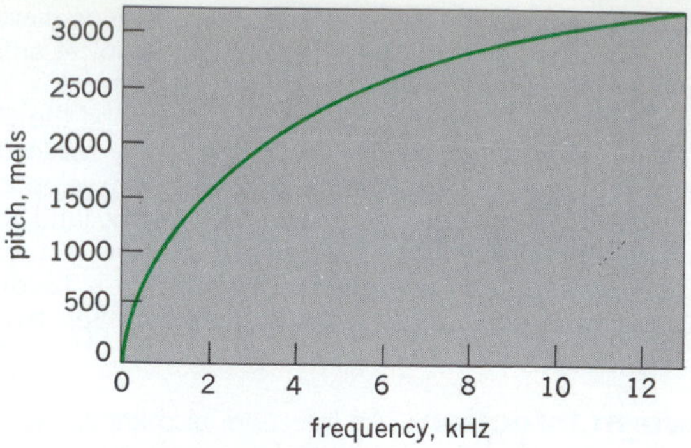

Pitch level of sounds of different frequency, in mels. A tone of 1000 Hz at an intensity level of 40 dB above absolute threshold has a pitch level of 1000 mels. (*After S. S. Stevens and J. Volkman, The relation of pitch to frequency, Amer. J. Psychol., 53(3):329–353, 1940*)

Pitcher plant Any member of the families Sarraceniaceae and Nepenthaceae. In these insectivorous plants the leaves form deep cups or pitchers in which water collects. Visiting insects, falling into this water, are drowned and digested by the action of enzymes secreted by cells located in the walls of the pitcherlike structures of these plants. Often these plants climb by tendrils. The end of a tendril may develop into a pitcher, which captures and digests insects. *See* INSECTIVOROUS PLANTS; NEPENTHALES. [P.D.St./E.L.C.]

Pitchstone A natural glass with dull or pitchy luster and generally brown, green or gray color. It is extremely rich in microscopic, embryonic crystal growths (crystallites) which may cause its dull appearance. The water content of pitchstone is high and generally ranges from 4 to 10% by weight. Pitchstone is formed by rapid cooling of molten rock material (lava or magma) and occurs most commonly as small dikes or as marginal portions of larger dikes. *See* IGNEOUS ROCKS; VOLCANIC GLASS. [C.A.C.]

Pith The central zone of tissue of an axis in which the vascular tissue is arranged as a hollow cylinder. Pith is present in most stems and in some roots. Stems without pith rarely occur in angiosperms but are characteristic of psilopsids, lycopsids, *Sphenophyllum*, and some ferns. Roots of some ferns, many monocotyledons, and some dicotyledons include a pith although most roots have xylem tissue in the center.

Pith is composed usually of parenchyma cells often arranged in longitudinal files. This arrangement results from predominantly transverse division of pith mother cells near the apical meristem. *See* PARENCHYMA; ROOT (BOTANY); STEM. [H.W.Bl.]

Pitot tube An instrument that measures the stagnation pressure of a flowing fluid; also called an impact tube. Stagnation (also called impact or total) pressure is the pressure that would be obtained if the fluid were brought to rest isentropically. When total pressure is being measured, the system consists of a primary sensing element mounted on a suitable support, pressure connecting lines, and a pressure-indicating device. Normally, the connecting lines and indicating devices are considered secondary elements and are not treated as part of the pitot tube. *See* MANOMETER.

The pitot tube is used primarily to obtain fluid velocity, total energy as measured being composed of impact or stagnation pressure P_2, static pressure P_1, and velocity v_1 of the fluid of density ρ. Then for incompressible (low-speed) flow, the equation shown below holds. For incompressible flow, the total en-

$$\frac{P_2}{\rho} = \frac{P_1}{\rho} + \frac{v_1^2}{2}$$

ergy is expressed by Bernoulli's theorem. For compressible flow, total energy can be expressed in terms of impact pressure, static pressure, and Mach number, which is related to velocity. *See* BERNOULLI'S THEOREM; FLOW MEASUREMENT. [L.N.Kr.]

Pituitary gland The most important single endocrine organ, because its secretions regulate many essential metabolic processes. The gland, also known as the hypophysis, is present in all vertebrates and is intimately related to the hypothalamus, a portion of the brain from which its main innervation is received. The human gland is lodged in the sella turcica, a deep depression in the sphenoid bone (see illustration).

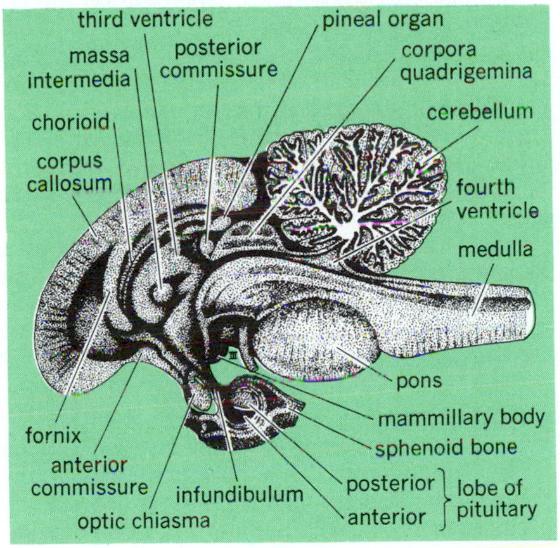

Relation of the human pituitary gland to the brain and sphenoid bone. (*After H. W. Rand, The Chordates, Blakiston, 1950*)

Although considerable anatomical variation is encountered among the vertebrate groups, the gland is roughly divisible into anterior, intermediate, and posterior lobes. The anterior and intermediate lobes collectively are referred to as the adenohypophysis. The posterior lobe, the pituitary stalk, and the median eminence of the floor of the brain are referred to collectively as the neurohypophysis. [G.C.Ke.]

The pituitary gland secretes several distinct hormones which may be grouped into three general classifications according to their composition: (1) hormones consisting of a large number of amino acids linked together, called protein hormones; (2) hormones consisting of a large number of amino acids linked together and associated with various types of sugar molecules, called glycoprotein hormones; and (3) hormones consisting of only a relatively few amino acids linked together, called peptide hormones.

The protein hormones secreted by the pituitary gland (growth hormone and prolactin) are all elaborated from the anterior lobe, called the adenohypophysis. The pituitary glycoprotein hormones—thyroid-stimulating hormone, follicle-stimulating hormones, and luteinizing hormone—are also secreted by the anterior lobe. Peptide hormones are secreted by the anterior lobe, as adrenocorticotrophin (ACTH) and by the posterior lobe, as oxytocin and vasopressin (or the antidiuretic hormone, ADH). Another peptide hormone, the melanocyte-stimulating hormone (MSH), is secreted by the pars intermedia of the hypophysis. *See* NEUROHYPOPHYSIS HORMONE.

Among the effects of removal of the pituitary gland (hypophysectomy), if performed on an immature individual, are: (1) a severe retardation of somatal (body) growth; (2) failure of sex organs to achieve mature development, with a consequential disruption of normal reproductive function in the adult; (3) atrophy of the adrenal cortex, with a concomitant suppression of the many biological processes dependent on that gland's steroid hormone production; and (4) emaciation of the thyroid gland, with accompanying depression of the body's metabolic rate, a bleaching effect on the skin (in some species), and a general adverse effect on body chemistry, leading to a variety of metabolic disorders. These phenomena are all attributable to deficiencies of anterior pituitary (adenohypophysial) hormones. Among the biological alterations associated with loss of posterior lobe function, the most serious is diabetes insipidus, or excessive loss of water through the kidney (diuresis). The hormones found in the posterior pituitary (neurohypophysis) are believed to be synthesized in the hypothalamus, and then transported to the posterior lobe through association with a carrier protein. Thus, it is probable that even excising the posterior lobe would not completely abolish the elaboration of the posterior lobe hormones. [L.E.R.]

Pituitary gland disorders Disorders resulting from deranged secretion of hormones by the pituitary gland cells.

Anterior lobe malfunction. Growth hormone, unlike other anterior pituitary hormones, does not have a specific target organ but affects the tissues and organs of the body in general. Hypersecretion of growth hormone usually results from a tumorous overgrowth of the cells. The hyperfunction is usually manifest by increased bone growth. When this arises before puberty, it results in abnormal tallness, or gigantism. When the hyperfunction has its onset after the age of 20, the bone growth is manifest by a general increase in the thickness of bones, a condition known as acromegaly.

Cushing's disease is a clinical disorder of anterior pituitary cells associated with enlargement of the cortex of the adrenal gland. This disease is characterized by obesity of the trunk of the body, high blood pressure, fatigability, weakness, hirsutism, edema, thinning of the bones, and in women, amenorrhea. *See* ADRENAL GLAND.

Simmond's disease is produced by extreme hypofunction of the anterior lobe, usually as a result of a vascular or tumorous destructive process. The vascular changes are due to infarction of the gland which can be associated with intrapartum or postpartum hemorrhage. The neoplasms that can cause Simmond's disease may be primary in the pituitary gland or metastatic from other sites in the body. With this disease there is a marked weight loss, weakness, and signs of hypothyroidism and hypoadrenalism.

Intermediate region. The intermediate region of the pituitary gland is the probable site of production of the melanocyte-stimulating hormone. This hormone is associated with the increase of pigmentation of the skin.

Posterior lobe malfunction. Diabetes inspididus is produced by the hypofunction of the posterior lobe of the pituitary gland. This disease is manifested by increased thirst and increased urine production. Deficiency of the antidiuretic hormone causes the tubules of the kidneys to lose their capacity to concentrate urine. The basic lesion is often in the posterior lobe of the pituitary or in the hypothalamus. Although oxytocin is highly important in the reproductive process, the identification of the specific syndromes or disorders associated with excess or insufficient oxytocin have not been described. *See* DIABETES; LACTATION; PITUITARY GLAND. [N.K.M.)

pK The logarithm (to the base 10) of the reciprocal of the equilibrium constant for a specified reaction under specified conditions (for example, solvent and temperature). The pK values are often more convenient to tabulate and use than the equilibrium constants themselves. The value of K for the dissociation of the HSO_4^- ion in aqueous solution at 25°C (77°F) is 0.0102 mole/liter. The logarithm is $0.008_6 - 2 = -1.991_4$ The pK is therefore $+ 1.991_4$. The choice of algebraic sign, although arbitrary, results in positive values for most dissociation constants applicable to aqueous solutions. The concept of pK is especially valuable in the study of solutions. *See* CHEMICAL EQUILIBRIUM; IONIC EQUILIBRIUM; pH. [T.F.Y.]

Placentation The intimate association or fusion of a tissue or organ of the embryonic stage of an animal to its parent for physiological exchange to promote the growth and development of the young. It enables the young, retained within the body or tissues of the mother, to respire, acquire nourishment, and eliminate wastes by bringing the bloodstreams of mother and young into close association but never in direct connection (see illustration). Placentation characterizes the early development of all mammals except the egg-laying duckbill platypus and spiny anteater. It occurs in some species of all other orders of vertebrates except the birds. In fact, in certain sharks and reptiles it is almost as well developed as in mammals. A few examples are also known among invertebrates (*Peripatus*, certain tunicates, and insects). *See* FETAL MEMBRANE. [H.W.Mo.]

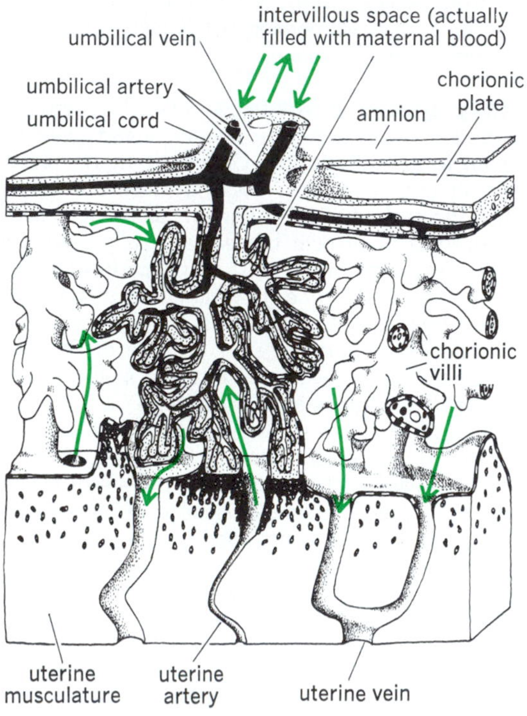

Block removed from center of human placenta.

Placer mining The working of deposits of sand, gravel, and other alluvium and eluvium containing concentrations of metals or minerals of economic importance. For many years gold has been the most important product obtained, although considerable platinum, cassiterite (tin mineral), phosphate, monazite, columbite, ilmenite, zircon, diamond, sapphire, and other gems have been produced. Other valuables recovered include native bismuth, native copper, native silver, cinnabar, and other heavy weather-resistant metals or minerals. A form of sluice box is the type of unit most often used to separate and recover the valuable metals or minerals.

A number of mining methods exist, but all have elements of preparing the ground, excavating the pay material, separating and recovering the valuable products, and stacking the tailings. A general classification of present placer mining methods is small-scale hand methods, hydraulicking, mechanical methods, dredging, and drift mining.

Small-scale hand methods include the use of pans, rockers, long toms, shoveling into boxes, ground sluicing, booming, and the use of dip boxes, puddling boxes, dry washers, surf washers, and small-scale mechanical placer machines.

Hydraulic mining, used if water supplies are adequate and streams not objectionably polluted, utilizes water under pressure, forced through nozzles, to break and transport the placer gravel to the recovery plant (often sluice box), where it is washed. The valuable material is separated and retained in the sluice, and the tailings pass through and are stacked by water under pressure.

Mechanical methods include the use of equipment such as bulldozers, draglines, front-end loaders, pumps, pipelines, hydraulic elevators, and recovery units which are used in a number of different combinations depending on the physical characteristics of placer deposits.

Large flat-lying areas are best suited for dredging because huge volumes of gravel can be handled fairly cheaply. Prior to dredging a deposit, the ground must be prepared. This preparation may consist of removing trees and other vegetation and overburden such as the frozen loess (muck) found in many parts of Alaska by use of equipment and water under pressure (stripping).

Because of relatively high costs, drift mining of placer deposits is not used to a large extent. It consists of sinking a shaft to bedrock, driving drifts up- and downstream in the valley deposits for 250–300 ft (75–90 m), and then extending crosscuts the width of the pay streak. [E.H.B.]

Placodermi A class of fishes (Pisces) known only from the Devonian Period, except for a few that survived into the base of the Carboniferous. They were true fishes or gnathostomes and can be distinguished from other fishes by the following characters: the gill chamber extends far under the cranium and is covered laterally by opercula; there is a neck joint between the cranium and the fused anterior vertebrae; the head and shoulder girdle are more or less covered with dermal bones composed typically of cellular bone and superficially of semidentine instead of true dentine; the bones are commonly ornamented with tubercles or ridges; the endoskeleton is cartilage and is sometimes calcified in a globular fashion or perichondrally ossified; the notochord is persistent, and the vertebrae consist only of neural and hemal arches; the tail is diphycercal or slightly heterocercal, and an anal fin is lacking.

Most placoderms were bottom-dwelling fishes, with the head and trunk dorsoventrally depressed; only one group had laterally compressed and deepened bodies, suggesting a more nectic manner of life. Most of them were small or moderate-sized, but a few were, for their time, gigantic, reaching a length of as much as 20 ft (6 m). They were the dominant fishes of the Devonian, and are found in both marine and fresh-water deposits.

The Placodermi are classified in nine orders, each with its own distinct specializations. The evolution and interrelations of these orders are matters of disagreement. *See* ANTIARCHA. [R.H.De.]

Placodontia A small but interesting order of marine reptiles of the subclass Euryapsida that are known only from deposits of Triassic age of Europe and the Near East (Israel). As the name implies, they are reptiles with grossly specialized dentitions—flat-crowned teeth are located in both the upper and lower jaws and on the palate—that probably functioned as crushing devices for hard-shelled prey.

The modification of the dentition had its effect on the entire skull, which became massive in relation to the body; and the coronoid region of the mandible, to which most of the masticatory muscles were attached, became greatly extended upward, giving it a superficial similarity to a mammalian jaw. The postcranial skeleton shows a number of aquatic adaptations. The thorax was box-shaped and the digits of hand and foot were probably webbed; the joint surfaces of the limb bones suggest that the animals did little if any walking on land. The advanced placodonts resemble sea turtles in overall appearance and in the fact that their body is encased in an armor of dermal bones similar to that of the dermochelyid sea turtle *Psephophorus*. The more generalized genera lack such an armor and possess a long tail. *See* EURYAPSIDA; REPTILIA. [R.Z.]

Plague An infectious disease of rodents and humans, occurring in bubonic or pneumonic form and caused by the bacillus *Pasteurella pestis*. This deadly malady has been notorious since the 3d century for its pandemics, most notably in 1332–1371 (the Black Death), 1630 (Milan), 1665 (London), and 1720–1721 (Marseilles and Provence). Domestic plague is transmitted from rat to rat and from rat to human by the efficient vector flea *Xenopsylla cheopis*. However, wild rodents serve as true reservoirs of plague and are frequently directly implicated as a source of human infection (sylvatic plague).

In warm climates plague is usually bubonic, so called because of the characteristic swollen lymph nodes or buboes. Bubonic plague is usually transmitted to humans by rat fleas. In cold climates it is more likely to take the much more fatal and contagious pneumonic, or tonsillar, form. Pneumonic plague spreads from patients with primary pneumonic plague or from patients with bubonic plague and secondary pneumonic infection.

Human infection is rare in the United States and arises from exposure to fleas of wild rodents. Plague was often a serious health problem in cities in India, Burma, and Indonesia, although it is presently in decline. It persists there, however, in the wild rodents. It is fairly common in Vietnam. [K.F.M.]

Plains The relatively smooth sections of the continental surfaces, occupied largely by gentle rather than steep slopes and exhibiting only small local differences in elevation. Because of their smoothness, plains lands, if other conditions are favorable, are especially amenable to many human activities. Thus it is not surprising that the majority of the world's principal agricultural regions, close-meshed transportation networks, and concentrations of population are found on plains. Large parts of the Earth's plains, however, are hindered for human use by dryness, shortness of frost-free season, infertile soils, or poor drainage. Because of the absence of major differences in elevation or exposure or of obstacles to the free movement of air masses, extensive plains usually exhibit broad uniformity or gradual transition of climatic characteristics.

Somewhat more than one-third of the Earth's land area is occupied by plains. With the exception of ice-sheathed Antarctica, each continent contains at least one major expanse of smooth land in addition to numerous smaller areas. The largest plains of North America, South America, and Eurasia lie in the continental interiors, with broad extensions reaching to the Atlantic (and Arctic) Coast. The most extensive plains of Africa occupy much of the Sahara and reach south into the Congo and Kalahari basins. Much of Australia is smooth, with only the eastern margin lacking extensive plains. *See* TERRAIN AREAS.

Surfaces that approach true flatness, while not rare, constitute a minor portion of the world's plains. Most commonly they occur along low-lying coastal margins, the lower sections of major river systems, or the floors of inland basins. Nearly all are the products of extensive deposition by streams or in lakes or shallow seas. The majority of plains, however, are distinctly irregular in surface form, as a result of valley-cutting by streams or of irregular erosion and deposition by continental glaciers. [E.H.Ha.]

Planaria A flatworm of the turbellarian order Tricladida. The majority of species are free-living, fresh-water forms. One commonly studied species is *Dugesia tigrina*, formerly known as *Planaria maculata*, a typical example of the order.

Dugesia reaches a length of 0.6–0.7 in. (15–18 mm). The animal is broad and blunt anteriorly with lateral projections (see illustration), the auricles, which have a tactile function. A pair of eyespots, present on the anterior dorsal surface, function as photoreceptors. The ventral surface of *Dugesia* is ciliated. A muscular pharynx may be protruded from the posterioventrally located mouth. The genital pore is located posterior to the mouth. Circulatory and respiratory systems are lacking but the muscular system is well developed with circular, longitudinal, and oblique fibers. There is no body cavity or coelom, and consequently these animals are categorized as acoelomates. The digestive system is incomplete; that is, there is no anal opening, so the mouth serves both for ingestion of food and egestion of solid waste.

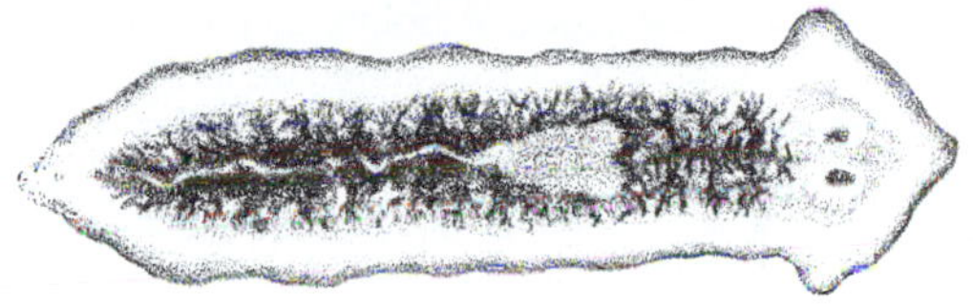

The planaria, a bilaterally symmetrical, dorsoventrally flattened, triploblastic organism.

When cut in half transversely, planarians regenerate a tail on the posterior of the anterior portion and a head on the anterior of the posterior part. If a transverse section is removed from the middle of the animal, the anterior of the section develops a head and the posterior, a tail. In addition, parts of one animal have been successfully grafted onto another animal. *See* REGENERATION (BIOLOGY).

One of the more unusual experimental observations on planaria was reported during behavioral studies in which ribonucleic acid (RNA), in the form of minced bodies of planarians that were trained to respond to certain stimuli, was fed to untrained planarians. The untrained planarians responded to that stimulus in a manner comparable to the response of the trained group. *See* TRICLADIDA. [C.B.C.]

Planck's constant A fundamental physical constant which represents the elementary quantum of action, action being defined as energy multiplied by time. Introduced by Max Planck in 1900, it has the value $h = 6.6261 \times 10^{-27}$ erg-second or 6.6261×10^{-34} joule-second. The symbol $\hbar$, sometimes called the Dirac h, is often used for convenience in physics to denote the quantity $h/2\pi$, where $\pi = 3.1416\ldots$.

As used by Planck in deriving his radiation law, h multiplied by the frequency of radiation represented a bundle of energy, that is, a quantum of energy. Radiant energy at any wavelength can occur only as multiples of this energy; thus energy is quantized. *See* COMPTON EFFECT; FUNDAMENTAL CONSTANTS; HEAT RADIATION; QUANTUM MECHANICS. [H.G.S./P.J.W.]

Planck's radiation law A law of physics which gives the spectral energy distribution of the heat radiation emitted from a so-called blackbody at any temperature. Discovered by Max Planck, this law laid the foundation for the advent of the quantum theory because it was the first physical law to postulate that electromagnetic energy exists in discrete bundles, or quanta. *See* HEAT RADIATION; QUANTUM MECHANICS. [H.G.S.; P.J.W.]

Plane In the euclidean definition, a surface is "that which has length and breadth only," and a plane surface is "a surface which lies evenly with the straight lines on itself." A definition of a plane that avoids this somewhat loosely defined concept of a surface is as follows: "A plane is a set of at least three points, not all collinear, such that if any two points *A*, *B* of the set are given, all points of the line *AB* are in the set, and such that if any four points *A*, *B*, *C*, *D* of the set are given, then at least one of the pairs of lines *AB* and *CD*, or *AC* and *BD*, or *AD* and *BC* have a common point." (This last restriction prevents a plane from occupying all of space.) Still another definition is this: "A plane is the locus of points each equidistant from two points in space." Points or lines in the same plane are called coplanar. Three or more planes that have a line in common are called coaxial planes. *See* ANALYTIC GEOMETRY; EUCLIDEAN GEOMETRY; LINE; POINT. [J.S.F.]

Plane curve The locus of points in the euclidean plane that satisfy some geometric or algebraic definition. Not all sets of points deserve to be called a curve, but the distinction is somewhat arbitrary. For most of this article, a curve is considered to be the locus of a set of points that satisfy an algebraic or transcendental equation in two variables.

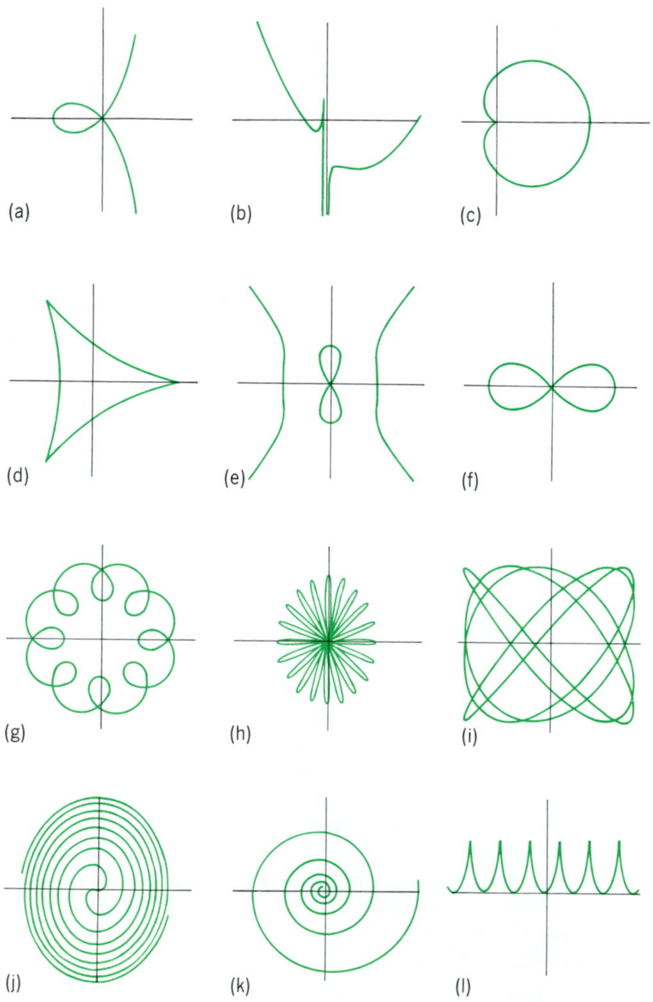

Plane curves. (*a*) Right strophoid. (*b*) Trident of Newton. (*c*) Cardioid. (*d*) Deltoid. (*e*) Devil on two sticks. (*f*) Lemniscate of Bernoulli. (*g*) Epitrochoid. (*h*) Rhodona. (*i*) Bowditch curve. (*j*) Fermat's spiral. (*k*) Logarithmic spiral. (*l*) Cycloid.

The most interesting geometric properties are those preserved by linear transformations, especially translations, rotations, reflections, and magnifications. Useful geometric properties include the number of branches into which the curve is divided; the number and degree of nodes, cusps, isolated points, and flex points; the number of loops; symmetries; branches that go to infinity; and asymptotes.

These terms can be defined informally as follows. A branch is a maximal smooth continuous portion of the curve. A multiple point is a point in the plane that lies on two or more branches; its degree is the number of branches involved. A node is a multiple point where the branches cross. A cusp is a multiple point where the branches meet but do not pass; that is, each of the branches ends at that point. An isolated point is a point of the curve through which no branches pass. Multiple and isolated points are collectively termed singular points. Any point that is not singular is termed ordinary. A flex (or point of inflection) is a point on the curve whose tangent cuts the curve. A smooth closed branch forms a loop. A curve is symmetric about a line *L* if every line perpendicular to *L* intersects the curve at equal distances from *L* on opposite sides of *L*; that is, portions of the curve form mirror images about *L*. The curve is symmetric about a point *P* if every line through *P* intersects the curve at equal distances from *P* in opposite directions. An asymptote is a line toward which a branch approaches as it moves to infinity from the origin; the curve and line are said to intersect at infinity.

In addition to these geometric properties, the form of the defining equation is of interest. This can be an algebraic (polynomial in *x* and *y*) or transcendental equation. In the former case, quadratic, cubic, and quartic equations are of special interest.

Once a coordinate system has been chosen, and the defining equation is known (in any of the forms, though the parametric form is usually the most useful), various properties of the curve can be defined in terms of the equation. These include the locations of *x*- and *y*-intercepts, local maxima and minima, flexes, nodes, and cusps. *See* ANALYTIC GEOMETRY.

The illustration shows some of the many plane curves of sufficient historical interest to have received names. *See* CARDIOID; CYCLOID; LEMNISCATE OF BERNOULLI; LISSAJOUS FIGURES; ROSE CURVE. [J.D.L.]

Plane geometry The branch of mathematics that deals with geometric figures, that is, collections of points that all lie in the same plane (coplanar). Although the words "point" and "plane" are undefined concepts, for elementary applications the intuitive meanings will serve: a point is a location, and a plane is a flat surface. For similar definitions, together with a discussion of the postulates and axioms (assumed truths) used in plane geometry, *see* EUCLIDEAN GEOMETRY.

Dimensions and measures. There are three spatial dimensions; geometric figures are classified as being zero, one, two, or three dimensional. Plane geometry deals only with geometric figures having fewer than three dimensions. Three-dimensional geometric figures, called solids, are dealt with in another branch of euclidean geometry. *See* SOLID (GEOMETRY).

A dimension is any measurement associated with a geometric figure that has units of length. The measure of a geometric figure is a number multiplied by a power of a length, with the result giving information about the size of the figure. The power of the length will be 1, 2, or 3, depending on whether the figure is one, two, or three dimensional. *See* UNITS OF MEASUREMENT.

Lines, line segments, and rays. Exactly one line passes through two given points. The part of a line between (and including) two points is called a line segment, with the two defining points being the end points of the segment. That part of a line that lies on one side of a point (together with that point) is called a ray.

Angles. An angle is the geometric figure formed by joining two rays having a common end point. Each ray is a side of the angle; the common end point is the vertex of the angle.

The concept of the measure of an angle may be understood by imagining that one ray of an angle is held fixed but the other ray is hinged at the vertex and allowed to rotate in the plane. A measure of the angle is a number, together with some unit of angular measure, that tells how much the hinged ray would need to be rotated so that it would overlie the fixed ray. If a ray were to be rotated exactly one revolution, it would return to its original position. The measure of an angle can be what fraction of one revolution would enable one side of the angle to become coincident with the other.

The revolution is a convenient unit of angular measure for many applications. However, a more commonly used unit is the degree, which is defined by 360° = 1 revolution. In the modern-day use of calculators and computers, fractions of degrees are most conveniently expressed by using decimal fractions; however, another subdivision of degrees is firmly entrenched in many applications: 1 degree = 60 minutes (1° = 60′), and 1 minute = 60 seconds (1′ = 60″).

An angle of measure 1/4 revolution, or 90°, is called a right angle. Two lines (or rays or segments) that intersect so as to form a right angle are said to be perpendicular. Two angles are complementary if their measures have a sum of 90°. An angle of measure 1/2 revolution, or 180°, is called a straight angle. Two angles are supplementary if their measures have a sum of 180°.

Polygons. A polygon is the geometric figure formed when line segments are joined end to end so as to enclose a region of the plane. The polygon having the least number of sides (three) is the triangle. An angle whose sides are two of those segments and whose arc lies inside the triangle is an interior angle of the triangle; an exterior angle of the triangle is any angle that is adjacent and supplementary to an interior angle. The interior angles of any triangle have measures that sum to 180°. This fact allows the determination of the measures of all angles formed by the intersections of three line if the measures of only two angles with different vertices are known. *See* POLYGON.

Congruent and similar geometric figures. Two geometric figures are congruent (≃) if they have exactly the same shape and size. If two geometric figures are congruent, one figure could be made to overlie the other by a combination of these types of motion: translation (sliding), rotation (twisting), and reflection about a line (flipping over). The parts of the geometric figures that would then coincide are called corresponding parts, where a part of a geometric figure is any set of points associated with that figure.

Two geometric figures are similar (∼) if they have the same shape but (perhaps) have different sizes. If two geometric figures are similar, one figure could be made to overlie the other by a combination of translation, rotation, reflection, and either expanding or shrinking. The parts that would then coincide are corresponding.

Circles. A circle is a collection of points in the plane, all of which are the same distance from another point, called the center. The region bounded by a circle sometimes is called a disk. Circumference means either the circle that is the boundary curve of a disk or the distance around that circle. A radius of a circle is any line segment that joins the center and a point of a circle. A chord is any line segment whose end points lie on the circle. A diameter is any chord that contains the center. A secant is a line that intersects a circle in two points. A tangent is a line that intersects a circle in only one point, called the point of tangency.

Associated with a circle are several important dimensions, including the lengths of a radius and a diameter, usually denoted, respectively, by r and d. These symbols appear in almost all formulas involving the length of the circumference C or the area A enclosed by a circle. Relationships between these variables for any circle are given by the equations

$$d = 2r \qquad C = \pi d \qquad A = \pi r^2$$

In these formulas, π (the lowercase Greek letter pi) represents the irrational number (value 3.141592 . . .) that is usually defined as the ratio of the circumference to the diameter of any circle. *See* CIRCLE. [H.L.Ba.]

Plane table A tripod-mounted drawing board on which topographic survey details are compiled in the field. The table and affixed manuscript sheet are set up and oriented over a ground point represented on the sheet. Stadia sightings with an alidade locate additional points to be plotted by direction-distance observations. Points also are located by intersections of the sightings from two different plane-table setups. Sufficient additional points are observed to permit sketching planimetric map details. Elevation differences also are observed for sighted points to allow for contour plotting. *See* ALIDADE; STADIA; TOPOGRAPHIC SURVEYING AND MAPPING. [B.A.B.]

Planer A machine for the shaping of long, flat, or flat contoured surfaces by reciprocating the workpiece under a stationary single-point tool or tools. Usually the workpiece is too large to be handled on a shaper.

Planers are built in two general types, open-side or double-housing. The former is constructed with one upright or housing to support the crossrail and tools. The double-housing type has an upright on either side of the reciprocating table connected by an arch at the top. *See* SHAPER; WOODWORKING. [A.H.T.]

Planet A relatively small, solid celestial body moving in orbit around a star, in particular the Sun. Besides Earth, the eight known planets of the solar system are Mercury, Venus, Mars, Jupiter, Saturn, Uranus, Neptune, and Pluto; in addition,

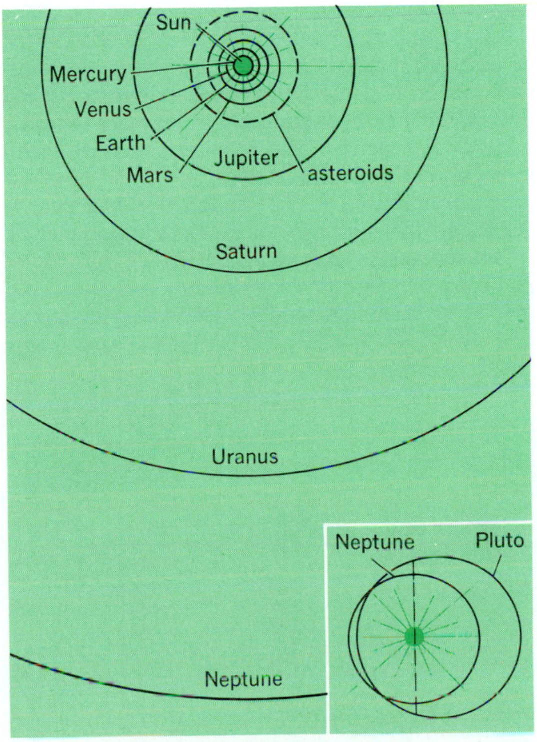

Plan of the solar system. (*After L. Rudaux and G. de Vaucouleurs, Astronomie, Larousse, 1948*)

over 2000 minor planets, or asteroids, mostly located between the orbits of Mars and Jupiter, are known (see illustration).

There are two basic groups of planets: the small, dense, terrestrial planets—Mercury, Venus, Earth, Mars, and Pluto—and the giant or Jovian planets—Jupiter, Saturn, Uranus, and Neptune. With the exception of Pluto, the terrestrial planets are all located within the inner solar system. The low-density Jovian planets extend outward from Jupiter to the remote outer reaches of the solar system. This distribution is not accidental, but is related to the fractionation of rocky, icy, and gaseous materials during the early stages of formation of the solar system.

The planets may also be divided into inferior planets, Mercury and Venus, located inside the Earth's orbit, and superior planets, from Mars to Pluto, circulating outside the Earth's orbit.

The motions of the planets in their orbits around the Sun are governed by Kepler's laws. Kepler's laws are true only when the mutual perturbations of the motions of the planets by the others are neglected. Table 1 gives the mean distance of the planets from the Sun and the sidereal period of revolution.

In the course of their motions around the Sun, Earth and other planets occupy a variety of relative positions or configurations. The inferior planets are in conjunction with the Sun when closest to the Earth-Sun direction, either between the Earth and the Sun (inferior conjunction) or beyond the Sun (superior conjunction). Between conjunctions, the geocentric angular distance from the planet to the Sun, or the elongation, varies up to a maximum value. The superior planets are not so limited, and their elongations can reach up to 180° when they are in opposition with the Sun.

The combinations of the orbital motions of Earth and of any other planet give rise to complicated apparent motions of the planets as seen from the Earth. Because the orbits of the main planets are, except for Pluto, only slightly inclined to the plane of the orbit of Earth, the apparent paths of the planets (except Pluto) are restricted to the zodiac, a belt 16° wide centered on the ecliptic. The ecliptic is the path in the sky traced out by the Sun in its apparent annual journey as the Earth revolves around it.

The apparent motions with respect to the celestial sphere, that is, to the fixed stars, appear for the inferior planets as oscillations back and forth about the position of the Sun steadily moving eastward among the stars. For the superior planets, the apparent motion is generally eastward or direct, but for short periods near the time of opposition it is westward or retrograde.

The size, mass, density, and rotation period of each of the planets is given in Table 2.

Minute perturbations in the motion of some nearby stars have indicated the existence of minor components of small mass. The masses of these satellite bodies, although larger than planetary masses in the solar system, are considered to be too small to be self-luminous; they are consequently more like planets than dwarf stars. From this evidence it is inferred that planetary systems are not as uncommon in the universe as was previously believed, and studies of multiple-star systems suggest that one star in every three may be attended by planetary companions. *See* ASTEROID; JUPITER; MARS; MERCURY (PLANET); NEPTUNE; PLANETARY PHYSICS; PLUTO; SATURN; SOLAR SYSTEM; URANUS; VENUS. [J.K.B.]

Table 1. Planetary orbits

Planet	Mean distance (semimajor axis) AU	Mean distance 10^6 km	Sidereal period of revolution Years	Sidereal period of revolution Days
Mercury	0.387	57.9	0.241	87.97
Venus	0.723	108.2	0.615	224.70
Earth	1.000	149.6	1.000	365.26
Mars	1.524	227.9	1.881	686.96
Jupiter	5.203	778.3	11.862	4332.59
Saturn	9.539	1427.	29.458	10759.
Uranus	19.18	2870.	84.014	30685.
Neptune	30.06	4497.	164.79	60189.
Pluto	39.44	5900.	247.7	90465.

Table 2. Physical elements of planets

Planet	Equatorial radius, r_e (Earth = 1)	Mass (Earth = 1)	Density, g/cm^3	Rotation period
Mercury	0.38	0.053	5.6	58.65d
Venus	0.95	0.815	5.2	243d
Earth	1.00	1.000	5.52	23h56m22.7s
Mars	0.53	0.107	3.95	24h37m22.6s
Jupiter	11.20	317.89	1.33	9h50m30s*
Saturn	9.47	95.15	0.69	10h14m†
Uranus	4.06	14.54	1.20	10h49m
Neptune	3.87	17.23	1.64	18h24m
Pluto	0.5?	0.002	0.8?	6.3d

*Latitude < 12° (system I); 9h55m40.6s, latitude > 12° (system II).
†Near Equator; 10h38m at intermediate latitudes.

Planetarium A projection device which accurately portrays the stars and planets at any time in the past, present, or future from any point on the Earth or the near region of space. The modern planetarium instrument is a mechanical-electrical analog of space.

Planetarium projectors are basically designed around three major components: a star lamp that has the intrinsic brilliance of stars in nature; a planet-projection system that allows for freedom of motion on or off the Earth; and a computer that solves all of the mechanical and mathematical problems incurred with modern planetarium programming. The planetarium dome is constructed of perforated aluminum sheets. With the large number of perforations, the planetarium dome can function as a theatrical scrim. This allows the instrument to project the stars on the surface of the dome and also allows for projection through the dome.

The star lamp is a high-pressure xenon arc which produces a tiny, intense-point source of light. This light is focused through thousands of individual lenses and pinholes and is projected to the dome. When the star projector is put in motion, the sky appears to move. There are two star balls, one for the northern hemisphere and one for the southern. [D.M.L.]

Planetary gear train An assembly of meshed gears consisting of a central gear, a coaxial internal or ring gear, and one or more intermediate pinions supported on a revolving carrier. Sometimes the term planetary gear train is used broadly as a synonym for epicyclic gear train, or narrowly to indicate that the ring gear is the fixed member. In a simple planetary gear train the pinions mesh simultaneously with the two coaxial gears. With the central gear fixed, a pinion rotates about it as a planet rotates about its sun, and the gears are named accordingly: The central gear is the sun, the pinions the planets. The illustration shows a construction typical in industrial planetary trains.

In operation, input power drives one member of a planetary gear train, the second member is driven to provide the output, and the third member is fixed. If the third member is not fixed, no power is delivered. This characteristic provides a convenient clutch action. A brake band about the intermediate member and fixed to the gear housing serves to lock or free the third

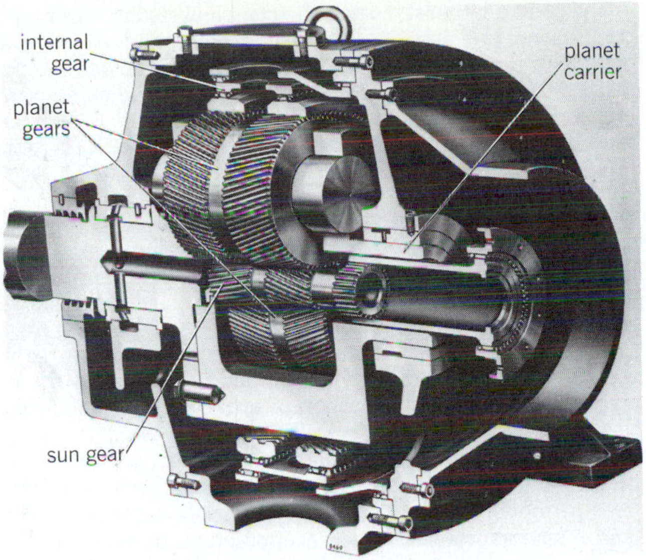

Stoecktícht planetary gear train. (*DeLaval Steam Turbine Co.*)

member; the band itself does not enter into the power path. *See* GEAR TRAIN. [J.R.Z.]

Planetary nebula A gaseous shell thrown off by a star during the later stages of its evolution. Planetary nebulae are so called because many display small greenish disks, akin to those of the planets Uranus and Neptune in small telescopes. They have nothing to do with planets; rather they are associated with late stages of stellar evolution, a planetary nebula being the shell thrown off by a dying star just before it settles down to become a degenerate white dwarf. Well over a thousand have been cataloged, and the total number in the Galaxy must be about 20,000.

Although a planetary nebula is often symmetrical about the central star, which is usually seen, a great variety of forms appear. Delicate filamentary wisps and knotlike concentrations are the rule, indicating that the ejected shells must be usually very inhomogeneous. Spectra of planetary nebulae are dominated by emission lines of abundant elements, in various ionization stages, all excited because of the rich ultraviolet radiation emitted by the central star.

Possibly most stars with masses between one and five times that of the Sun produce planetary nebulae. When hydrogen becomes exhausted in the core of a solar-type main-sequence star, it evolves into a red giant whose outer envelope is eventually ejected. As the shell density declines, the shell is exposed to energetic photons of the hot residual core. Hence, the characteristic bright-line nebular spectrum is produced. As the nebula expands further, this spectrum gradually fades away. The hot, blue-white core, deprived of all nuclear fuel, slowly evolves to a white dwarf. *See* STELLAR EVOLUTION; WHITE DWARF STAR.

Planetary nebulae appear to be important suppliers of dust and gas to the interstellar medium. *See* INTERSTELLAR MATTER; NEBULA. [L.H.Al.]

Planetary physics The study of the structure, composition, and physical and chemical properties of the planets of the solar system, including their atmospheres and their immediate cosmic environment.

Planetary scientists attempt to synthesize their information about the structure and properties of each of the planets by constructing models of them. The most obvious gross properties of the planet are its mass and its radius. In constructing a model, it is required that the model be in hydrostatic equilibrium. This means that at any interior point in the model, the pressure must be great enough to sustain the weight of the overlying mass of material. Thus, given the mass and the radius, estimating the interior pressures through the principles of hydrostatic equilibrium, and knowing something about the compressibilities of materials, it is generally possible to place constraints upon the interior composition of the planets. Knowledge of the equatorial bulge and the rate of spin provides additional information about the distribution of mass in the interior of the planet.

Classes of chemical composition. The planets in the solar system have an extremely wide range of properties. This distribution of characteristics can be understood in part from a knowledge of the more abundant elements in nature and their volatility properties. Approximately 98% of matter in the Sun, and therefore also presumably in the matter from which the Sun and the solar system were formed, consists of the gases hydrogen and helium. Most of the remaining material consists of carbon, nitrogen, and oxygen, which in the presence of very large amounts of hydrogen tends to form methane, ammonia, and water. These substances are collectively called ices, and they evaporate at relatively low temperatures. Both the light gases hydrogen and helium and the ices are of quite low abundance on the Earth and the other inner planets in the solar system. What comprises the bulk of the material in these planets is the rocky material, constituting only about 3 parts in 1000 of the solar mix of elements.

The differences in the volatilities of these materials, which correlate with the properties of the planetary bodies in the solar system, give information about the properties of the environment in which the planets formed in the solar system. The inner planets, composed predominantly of rocks, evidently formed in a rather hot environment, so that the volatile gases and ices were not condensed and did not collect along with the rocky material, which presumably was condensed. The comets, residing at very large distances from the Sun in the solar system, appear to be mixtures of rocky materials and of the ices. The outer giant planets, Uranus and Neptune, appear to be primarily composed of materials heavier than hydrogen and helium, probably mixtures of rocky and icy materials. The two largest planets in the solar system, Jupiter and Saturn, are much closer in composition to that of the Sun itself, although studies of their interior structures tend to indicate that there is some degree of enrichment in the heavier elements. These differences in composition thus indicate that the tendency to collect hydrogen and helium depends upon the size of the body which has formed, the larger bodies being more successful in gravitationally capturing the elusive hydrogen and helium. *See* COMET.

These compositional classes provide a natural means for dividing the planetary objects within the solar system into separate groups.

Giant planets. The giant planets are Jupiter, Saturn, Uranus, and Neptune. Nowhere in the interiors of the giant planets can anything resembling a solid surface be expected. The temperatures in the interiors tend to be thousands to tens of thousands of degrees Celsius. The pressures range up to tens of millions of bars (10^7 bars = 10^{12} pascals) and higher. Under these circumstances all materials behave like fluids. There may be a certain amount of compositional stratification, with denser fluids underlying lighter ones.

Terrestrial planets. The terrestrial planets include Mercury, Venus, Earth, and Mars. The Earth's Moon may also be considered a terrestrial planet. The Earth consists of a thin upper crust composed of rocks of relatively low density and low melting points, overlying a much thicker mantle composed predominantly of metallic silicates and oxides, which in turn overlies a substantial core, which is composed of much denser materials, believed predominantly to be iron with other elements, either alloyed or in solution.

It is conjectured that the interior of Venus is probably much like that of the Earth, with a core, a mantle, and a crust. Major structural features on the surface of the planet suggest extensive tectonic activity. The extent to which the crust of Venus is subject to extensive continental drift motions is quite unknown. *See* VENUS.

The mass of Mars is approximately one-tenth that of the Earth, and hence significant differences in the internal structure are to be expected. There appears to be less of a density contrast between the core of Mars and that of its mantle. Because the planet is smaller, the temperature increases less rapidly with depth than in the case of the Earth, and hence Mars should have a somewhat more rigid outer mantle and crust than the Earth. There is no indication that large amounts of continental drift have taken place on Mars. On the other hand, tectonic activity has clearly played a large role in the history of Mars. *See* MARS.

Mercury has only about half the mass of Mars. The mean density of Mercury is very high, indicating that Mercury probably has an abnormally large core predominantly composed of metallic iron. There is much evidence of extensive tectonic activity, although, like Mars, the increase of temperature below the surface of Mercury probably occurs sufficiently slowly so that the crust and upper mantle are relatively rigid, and nothing resembling continental drift has probably taken place. *See* MERCURY (PLANET).

The Moon has a history which includes extensive episodes of melting and differentiation. The upper layers of the Moon, which is only just over 1% of the mass of the Earth, are quite rigid, and there is no evidence for extensive horizontal motions of the structural units. The Moon is unique in the solar system in having a relatively low density among the inner planets, and at best a very small core, indicating that the planet is practically devoid of metallic iron. *See* MOON.

Galilean satellites. The four Galilean satellites of Jupiter—Io, Europa, Ganymede, and Callisto—have masses which are all roughly comparable to the mass of the Earth's Moon. The Galilean satellites appear to represent a composition class which is slightly more volatile-rich than the pure rocky materials characteristic of the inner solar system. *See* JUPITER.

Atmospheres. The temperature at the surface of a planetary body depends in a complex manner on the properties of the overlying atmosphere, as well as upon the distance of the planet from the Sun. The atmosphere of Venus is very much hotter relative to the Earth than would be expected purely on the basis of the relative distances from the Sun. The difference appears to arise from the extensive operation of the greenhouse effect within the very thick atmosphere of Venus.

The only terrestrial planets with atmospheres are Venus, Earth, and Mars. Both Mars and Venus have atmospheres composed predominantly of carbon dioxide. The element of next greatest abundance in the atmospheres of Mars and Venus is nitrogen, which is also the predominant element in the atmosphere of the Earth. The next most abundant element in the terrestrial atmosphere is oxygen, which is maintained there predominantly as the result of the operation of life.

Magnetospheres. Some of the planets contain substantial magnetic fields; others do not. Within the inner solar system, the Earth possesses a relatively strong field, Mercury a relatively weak one, and if Venus and Mars contain significant intrinsic fields, they are sufficiently weak that they have not been confirmed. On the other hand, Jupiter and Saturn have very strong magnetic fields. The generation of planetary magnetic fields appears to depend upon a combination of planetary rotation with an inner convecting layer having significant electrical conductivity. *See* MAGNETOSPHERE; PLANET; VAN ALLEN RADIATION.

[A.G.W.C.]

Plant anatomy

Plant anatomy The area of plant science concerned with the internal structure of plants. It deals both with mature structures and with their origin and development.

The plant anatomist dissects the plant and studies it from different planes and at various levels of magnification. At the level of the cell, anatomy overlaps plant cytology, which deals exclusively with the cell and its contents. Sometimes the name plant histology is applied to the area of plant anatomy directed toward the study of cellular details of tissues. *See* PLANT CELL; PLANT ORGANS.

[K.E.]

Plant-animal interactions

Plant-animal interactions The examination of the ecology of interacting plants and animals by using an evolutionary, holistic perspective. For example, the chemistry of defensive compounds of a plant species may have been altered by natural-selection pressures resulting from the long-term impacts of herbivores. Also, the physiology of modern herbivores may be modified from that of thousands of years ago as adaptations for the detoxification or avoidance of plant defensive chemicals have arisen.

The application of the theories based on an understanding of plant-animal interactions provides an understanding of problems in modern agricultural ecosystems. In addition, plant-animal interactions have practical applications in medicine. For example, a number of plant chemicals, such as digitalin from the foxglove plant, that evolved as herbivore-defensive compounds have useful therapeutic effects on humans. *See* AGROECOSYSTEM.

The evolutionary consequences of plant-animal interactions vary, depending on the effects on each participant. Interaction types range from mutualisms, that is, relationships which are beneficial to both participating species, to antagonisms, in which the interaction benefits only one of the participating species and negatively impacts the other. Interaction types are defined on the basis of whether the impacts of the interaction are beneficial, harmful, or neutral for each interacting species (see table).

[W.G.A.]

Effects of interaction types for each species*

Interaction	Effect on species A	Effect on species B
Mutualism	+	+
Commensalism	+	0
Antagonism	+	−
Competition	−	−
Amensalism	0	−
Neutralism	0	0

*+ = beneficial, − = harmful, 0 = neutral.

Fossil record. Plants and animals interact in a variety of ways within modern ecosystems. These interactions may range from simple examples of herbivory (animals eating plants) to more complex interactions such as pollination or seed and fruit dispersal. Animals also rely on plants for food and shelter. The complex interactions between these organisms over geologic time not only have resulted in an abundance and diversity of organisms in time and space but also have contributed to many of the evolutionary adaptations found in the biological world.

Paleobiologists have attempted to decipher some of the interrelationships that existed between plants and animals throughout geologic time. The ecological setting in which the organisms lived in the geologic past is being analyzed in association with the fossils. Thus, as paleobiologists have increased their understanding of certain fossil organisms, it has become possible to consider some aspects of the ecosystems in which they lived, and in turn, how various types of organisms interacted.

Herbivory. Perhaps the most widespread interaction between plants and animals is herbivory, in which plants are utilized as food. One method of determining the extent of herbivory in the fossil record is by analyzing the plant material that has passed through the digestive gut of the herbivore.

The stems of some fossil plants show tissue disruption similar to various types of wounds occurring in plant parts that have been pierced by animal feeding structures. As plants developed defense systems in the form of fibrous layers covering inner, succulent tissues, some animals evolved piercing mouthparts that allowed them to penetrate these thick-walled layers. In some fossil plants, it is also possible to see evidence of wound tissue that has grown over these penetration sites. *See* COAL BALLS; HERBIVORY.

Mimicry. Another example of the interactions between plants and animals that can be determined from the fossil record is mimicry. Certain fossil insects have wings that are morphologically identical to plant leaves, thus providing camouflage from predators as the insect rested on a seed fern frond.

Pollination. The transfer of pollen from the pollen sacs to the receptive stigma in angiosperms or to the seed in gymnosperms is an example of an ancient interaction between plants and animals. It has been suggested that pollination in some groups initially occurred as a result of indiscriminate foraging behavior by certain animals, and later evolved specifically as a method to effect pollination. The size, shape, and organization of fossil pollen grains provide insight into potential pollination vectors. *See* FOSSIL; PALEOBOTANY; POLLINATION. [T.N.T.]

Plant cell The basic unit of structure and function in nearly all plants. Although plant cells are variously modified in structure and function, they have many common features. The most distinctive feature of all plant cells is the rigid cell wall, which is absent in animal cells. The range of specialization and the character of association of plant cells is very wide. In the simplest plant forms a single cell constitutes a whole organism and carries out all the life functions. In just slightly more complex forms, cells are associated structurally, but each cell appears to carry out the fundamental life functions, although certain ones may be specialized for participation in reproductive processes. In the most advanced plants, cells are associated in functionally specialized tissues, and associated tissues make up organs such as the leaves, stem, and root. *See* CELL WALLS (PLANT).

Plant and animal cells are composed of the same fundamental constituents—nucleic acids, proteins, carbohydrates, lipids, and various inorganic substances—and are organized in the same fundamental manner. A characteristic of their organization is the presence of unit membranes composed of phospholipids and associated proteins and in some instances nucleic acids.

Perhaps the most conspicuous and certainly the most studied of the features peculiar to plant cells is the presence of plastids. Among the organisms generally classified as plants, only the bacteria, the blue-green algae, and some fungi lack plastids. The plastids are membrane-bound organelles with an inner membrane system. Chlorophylls and other pigments are associated with the inner membrane system. *See* CELL (BIOLOGY); CELL PLASTIDS; CHLOROPHYLL. [W.G.W.]

Plant communication Movement of signals or cues, presumably chemical, among individual plants or plant parts. These chemical cues are a consequence of damage to plant tissues and stimulate physiological changes in the undamaged "receiving" plant or tissue. There are very few studies of this phenomenon, and so theories of its action and significance are fairly speculative.

Plants produce a wealth of secondary metabolites that do not function in the main, or primary, metabolism of the plant, which includes photosynthesis, nutrient acquisition, and growth. Since many of these chemicals have very specific negative effects on animals or pathogens, ecologists speculate that they may be produced by plants as defenses. Plant chemical defenses either may be present all of the time (constitutive) or may be stimulated in response to attack (induced). Those produced in response to attack by pathogens are called phytoalexins. In order to demonstrate the presence of an induced defense, the chemistry of plant tissues or their suitability to some "enemy" (via a bioassay) must be compared before and after real or simulated attack. Changes found in the chemistry and suitability of control or unattacked plants when nearby experimental plants are damaged imply that some signal or cue has passed from damaged to undamaged plants. Controlled studies have shown that responses in undamaged plants are related to the proximity of a damaged neighbor. *See* ALLELOPATHY; PHYTOALEXIN; PLANT METABOLISM. [J.C.Sc.]

Plant evolution That phase of evolution dealing with the origin and development of the plant kingdom.

Evolutionary processes. The processes of plant evolution—mutation, genetic recombination, natural selection, and reproductive isolation—are the same as in all forms of life, but profound differences between animals and plants in their mode of life, basic structure, and individual development bring about corresponding modifications in the mode of action and relative importance of these four processes. The developmental cycle of any particular organ in plants is shorter, less complex, and less well integrated with other organs than in animals. Consequently, plant organs can be modified much more profoundly by both environmental and genetic effects than can animal structures without producing inviability.

Because of their general lack of motility, plants evolved special devices for the dissemination of propagules, such as spores and seeds. In addition, flowering plants evolved elaborate structures which promote cross-pollination through the aid of animals. Together with the structures which protect the spores and seeds, the progressive elaboration of these methods of cross-pollination and dispersal of spores and seeds forms the principal thread around which the evolution of land plants is centered. A secondary thread is formed by the vascular system, which serves for conduction and support. On the other hand, the modifications of the outward form of leaves and stems as well as of the inner physiological conditions of their cells which adapt plants to different habitats and modes of life are of such a general nature that they are repeated in an almost parallel fashion many times in unrelated groups of plants. They are consequently less reliable as signposts for pathways of evolution than are corresponding structures in animals. *See* POPULATION DISPERSAL.

In addition, the dioecious condition (that is, individuals distinct as to sex—male or female) is far less widespread in plants than in animals. In some plants, such as mosses, monoecious types (having individuals both male and female) were derived secondarily from dioecious ancestors.

All of the above-mentioned characteristics of plants facilitated evolution through recombination of genetic characteristics derived from different adaptive systems. Hybridization between subspecies and species is particularly common in plants, and the hybrid derivatives are often vigorous and well adapted to new habitats. They can become stabilized in three different ways: by segregation of relatively homozygous intermediate types, by introgression, and by polyploidy. In the ancestry of most present-day plant species one of these processes has occurred.

Evolutionary history. In the evolutionary history of the plant kingdom, five levels of advancement are recognized, each based upon an adaptive system offering new opportunities for adaptive radiation. These five stages include Procaryota, Eucaryota, Embryobionta, vascular plants, and seed plants. Each level was first reached by a group with generalized or primitive characteristics. Through adaptive radiation, this generalized group gave rise in each case to a number of evolutionary lines or phylads, the modern representatives of which are very different from one another. See PLANT KINGDOM. [G.L.St.]

Plant geography For botanists, the major subdivision concerned with all aspects of the spatial distribution of plants. This subdivision is also known as phytogeography, phytochorology, geographical botany, and geobotany, and is here restricted to terrestrial landscapes. By custom, it often involves some aspects of the distribution of plants in time, called historical plant geography, and is thus allied to paleoecology and paleobotany.

The function of plant geography is to record the observed empirical facts of plant distribution, and also to understand and interpret these facts. Attention is directed at species, population, community, and ecosystem levels of integration. Where possible, the study includes the prediction and control of distributional phenomena, especially as these relate to plant pests, parasites, and diseases, and to the introduction and spread of desirable species and vegetation types. All such knowledge is pertinent to forestry, agriculture, range and pasture management, wildlife habitat management, horticulture, and soil and water conservation. Except mainly for genetical research on speciation and subsequent dispersal and survival into island and continental areas, plant geography is not an experimental or quantitative science, and in general does not involve laboratory procedures and technologic equipment.

There are two major subdivisions of plant geography, focused upon the two most commonly employed levels of integration of plant life. Floristic plant geography embraces the spatial distribution of flora, while vegetational plant geography is the spatial distribution of vegetation.

Flora is a scientific term with no common usage. The flora of an area or period of time is the totality of all species within that geographical unit, independent of their relative abundances and their relationships to one another. The technical term "population," in this connection, refers collectively to all individuals of any one species within a locality.

Vegetation is a term of popular origin and refers to the mosaic of plant life that forms the natural or seminatural landscape. The vegetation of a region is the tapestry or carpet of plant life, developed by differential and varying combinations and growths of the numerous elements of the local flora. Technically, it is an organized and integrated whole, at a higher level of integration than the separate species, composed of those species and their populations. Vegetation possesses emergent properties not necessarily found in the species themselves, and is thus a holistic system in its own right. In turn, vegetation is a component which together with all factors of the environment becomes the ecosystem. See ECOSYSTEM. [F.E.E.]

Plant growth An irreversible increase in the size of the plant. As plants, like other organisms, are made up of cells, growth involves an increase in cell numbers by cell division and an increase in cell size. Cell division itself is not growth, as each new cell is exactly half the size of the cell from which it was formed. Only when it grows to the same size as its progenitor has growth been realized. Nonetheless, as each cell has a maximum size, cell division is considered as providing the potential for growth. See CELL (BIOLOGY); CELL DIVISION.

While growth in plants consists of an increase in both cell number and cell size, animal growth is almost wholly the result of an increase in cell numbers. Another important difference in growth between plants and animals is that animals are determinate in growth and reach a final size before they are mature and start to reproduce. Plants have indeterminate growth and, as long as they live, continue to add new organs and tissues. In a plant new cells are produced all the time, and some parts such as leaves and flowers may die, while the main body of the plant persists and continues to grow. The basic processes of cell division are similar in plants and animals, though the presence of a cell wall and vacuole in plant cells means that there are certain important differences. This is particularly true in plant cell enlargement, as plant cells, being restrained in size by a cellulose cell wall, cannot grow without an increase in the wall. Plant cell growth is thus largely a property of the cell wall. See CELL WALLS (PLANT).

Sites of cell division. Cell division in plants takes place in discrete zones called meristems. The stem and root apical meristems produce all the primary (or initial) tissues of the stem and root. The cylindrical vascular cambium produces more conducting cells at the time when secondary thickening (the acquisition of a woody nature) begins. The vascular cambium is a sheet of elongated cells which divide to produce xylem or water-conducting cells on the inside, and phloem or sugar-conducting cells on the outside. Unlike the apical meristems whose cell division eventually leads to an increase in length of the stem and root, divisions of the vascular cambium occur when that part of the plant has reached a fixed length, and lead only to an increase in girth, not in length. The final meristematic zone, the cork cambium, is another cylindrical sheet of cells on the outer edge of older stems and roots of woody plants. It produces new outer cells only, and these cells differentiate into the corky layers of the bark so that new protective layers are produced as the tree increases in circumference. See APICAL MERISTEM; BUD; LATERAL MERISTEM; PERIDERM; ROOT (BOTANY); STEM.

Controls. Plant growth is affected by internal and external factors. The internal controls are all the product of the genetic instructions carried in the plant. These influence the extent and timing of growth and are mediated by signals of various types transmitted within the cell, between cells, or all around the plant. Intercellular communication in plants may take place via hormones (or chemical messengers) or by other forms of communication not well understood. There are several hormones (or groups of hormones), each of which may be produced in a different location, that have a different target tissue and act in a different manner. See ABSCISIC ACID; AUXIN; CYTOKININS; GIBBERELLIN; PLANT HORMONES.

The external environments of the root and shoot place constraints on the extent to which the internal controls can permit the plant to grow and develop. Prime among these are the water and nutrient supply available in the soil. Because cell expansion is controlled by cell turgor, which depends on water, any deficit in the water supply of the plant reduces cell turgor and limits cell elongation, resulting in a smaller plant. See PLANT-WATER RELATIONS.

Mineral nutrients are needed for the biochemical processes of the plant. When these are in insufficient supply, growth will be less vigorous, or in extreme cases it will cease altogether. See PLANT MINERAL NUTRITION.

An optimal temperature is needed for plant growth. The actual temperature range depends on the species. In general, metabolic reactions and growth increase with temperature, though high temperature becomes damaging. Most plants grow slowly at low temperatures 32–50°F (0–10°C), and some tropical plants are damaged or even killed at low but above-freezing temperatures.

Light is important in the control of plant growth. It drives the process of photosynthesis which produces the carbohydrates that are needed to osmotically retain water in the cell for growth. See PHOTOSYNTHESIS.

Fruits and seeds. Fruits and seeds are rich sources of hormones. Initial hormone production starts upon pollination and is further promoted by ovule fertilization. These hormones promote the growth of both seed and fruit tissue. Fruits grow initially by cell division, then by cell enlargement, and finally sometimes by an increase in air spaces.

The growth of a seed starts at fertilization. A small undifferentiated cell mass is produced from the single-celled zygote. This proceeds to form a small embryo consisting of a stem tip bearing two or more leaf primordia at one end and a root primordium at the other. Either the endosperm or the cotyledons enlarge as a food store. *See* FRUIT; SEED.

Flowering. At a certain time a vegetative plant ceases producing leaves and instead produces flowers. This often occurs at a particular season of the year. The determining factor for this event is day length (or photoperiod). Different species of plants respond to different photoperiods. *See* FLOWER; PHOTOPERIODISM.

The light signal for flowering is received by the leaves, but it is the stem apex that responds. Exposing even a single leaf to the correct photoperiod can induce flowering. Clearly, then, a signal must travel from the leaf to the apex. Grafting a plant that has been photoinduced to flower to one not so induced can cause the noninduced plant to flower. It has been proposed that a flower-inducing hormone travels from the leaf to the stem apex and there induces changes in the development of the cells such that the floral morphology results.

Dormancy. At certain stages of the life cycle, most perennial plants cease growth and become dormant. Plants may cease growth at any time if the environmental conditions are unfavorable. When dormant, however, a plant will not grow even if the conditions are favorable. *See* DORMANCY.

Leaf abscission. As a perennial plant grows, new leaves are continuously or seasonally produced. At the same time the older leaves are shed because newer leaves are metabolically more efficient in the production of photosynthates. A total shedding of tender leaves may enable the plant to withstand a cold period or drought. In temperate deciduous trees, leaf abscission is brought about by declining photoperiods and temperatures. *See* ABSCISSION; PLANT MORPHOGENESIS.
[P.J.D.]

Plant hormones A plant hormone, or phytohormone, is an organic compound that is synthesized in minute quantities in one part of a plant and translocated to another part, where it influences a specific physiological process.

Plants contain many chemically and physiologically distinct types of plant-regulatory substances, some of which properly can be called hormones. For example, auxins, gibberellins, cytokinins, and growth inhibitors are classes of plant regulatory chemicals, some of which satisfy the criterion of being hormonal in nature.

Other plant regulatory substances, for example, abscisic acid, ethylene, and vitamins, can also be classified as hormones; and still other substances, for example, growth retardants, may prove to be hormonal in nature for some plants. *See* ABSCISIC ACID; AUXIN; CYTOKININS; GIBBERELLIN; PLANT GROWTH; PLANT METABOLISM; PLANT MOVEMENTS; PLANT PHYSIOLOGY. [R.H.B.]

Plant kingdom The worldwide array of plant life, including plants that have roots in the soil, plants that live on or within other plants and animals, plants that float on or swim about in water, and plants that are carried on dust particles in the air. There is great variation in body form and size, ranging from microscopic unicells such as bacteria and certain algae to the giant sequoia trees of California. There are more than 300,000 known species of plants, and discoveries add hundreds of "new" species every year.

This large number of forms necessitates some systematic method of classification. The species are therefore grouped into successively larger and more inclusive categories to make a format hierarchy of classification. The white oak, for example, may be classified as follows:

- Kingdom Plantae
  - Subkingdom Embryobionta
    - Division Magnoliophyta
      - Class Magnoliopsida
        - Subclass Hamamelidae
          - Order Fagales
            - Family Fagaceae
              - Genus *Quercus*
                - Species *alba*

One system of classification of the entire plant kingdom is shown in the table (the asterisk indicates groups known only as fossils). In the subkingdom Thallobionta the scheme is shown only to the level of the class. Common names are inserted in parentheses after some of the scientific names. It should also be noted that the distinction between the plant and animal kingdom is not absolutely sharp among the unicellular flagellates; some of the groups, such as the euglenoids, that are here

A system of classification for the plant kingdom

KINGDOM PLANTAE

- **Subkingdom Prokaryotae**
  - Division: Bacteria†
    - Cyanophyceae (blue-green algae)†
    - Prochlorophyceae
- **Subkingdom Thallobionta (thallophytes)†**
  - Division Rhodophycota (red algae)
    - Class Rhodophyceae
  - Division Chromophycota
    - Class: Chrysophyceae (golden or golden-brown algae)†
    - Prymnesiophyceae
    - Xanthophyceae (yellow-green algae)
    - Eustigmatophyceae
    - Bacillariophyceae (diatoms)
    - Dinophyceae (dinoflagellates)
    - Phaeophyceae (brown algae)
    - Raphidophyceae (chloromonads)
    - Cryptophyceae (cryptomonads)
  - Divison Euglenophycota (euglenoids)
    - Class Euglenophyceae
  - Division Chlorophycota (green algae)
    - Class: Chlorophyceae†
    - Charophyceae†
    - Prasinophyceae
  - Division Fungi†
    - Class: Myxomycetes (slime molds)
    - Oomycetes
    - Chytridiomycetes (chytrids)
    - Hyphochrytridiomycetes
    - Zygomycetes
    - Ascomycotina (sac fungi)†
    - Basidiomycotina (club fungi)†
- **Subkingdom Embryobionta (embryophytes)†**
  - Division Rhyniophyta*†
    - Class Rhyniopsida†
  - Division Bryophyta†
    - Class Hepaticopsida (liverworts)
      - Subclass Jungermanniidae
        - Order: Takakiales
        - Calobryales
        - Jungermanniales
        - Metzgeriales
      - Subclass Marchantiidae
        - Order: Sphaerocarpales
        - Monocleales
        - Marchantiales
    - Class: Anthocerotopsida (hornworts)
    - Sphagnopsida (peatmosses)
    - Andreaeopsida (granite mosses)
    - Bryopsida (mosses)†
      - Subclass: Archidiidae
      - Bryidae
        - Order: Fissidentales
        - Bryoxiphiales
        - Schistostegales
        - Dicranales†

(continued)

A system of classification for the plant kingdom (*cont.*)

Order: Pottiales†
Grimmiales†
Seligeriales
Encalyptales
Funariales†
Splachnales
Bryales
Mitteniales
Orthotrichales
Isobryales†
Hookeriales†
Hypnales
Subclass: Buxbaumiidae†
Tetraphididae
Dawsoniidae†
Polytrichidae
Division Psiolotophyta†
Class Psilotopsida
Division Lycopodiophyta†
Class: Lycopodiopsida (club mosses)†
Isoetopsida†
Division Equisetophyta†
Class: Hyeniopsida*
Sphenophyllopsida*
Equisetopsida (horsetails)†
Division Polypodiophyta (ferns)†
Class Polypodiopsida†
Order: Protopteridales*
Archaeopteridales*†
Ophioglossales
Noeggerathiales*†
Marattiales†
Polypodiales†
Marsileales†
Salviniales†
Division Pinophyta†
Class Ginkgoopsida
Order: Calamopityales*
Callistophytales*
Peltaspermales*
Ginkgoales†
Leptostrobales*
Caytoniales†
Arberiales*
Pentoxylales*
Ephedrales
Class Cycadopsida
Order: Lagenostomales
(= Lyginopteridales)*
Trigonocarpales
(= Medullosdales)*
Cycadales†
Order: Bennettitales (= Cycadeoidales†)*
Gnetales
Welwitschiales
Class Pinopsida†
Order: Cordaitales*†
Pinales†
Division Magnoliophyta (angiosperms;
flowering plants)†
Class Magnoliopsida (dicotyledons)†
Subclass Magnoliidae†
Order: Magnoliales†
Laurales
Piperales†
Aristolochiales†
Illiciales
Nymphaeales†
Ranunculales†
Papaverales†
Subclass: Hamamelidae†
Order: Trochodendrales†
Hamamelidales†
Daphniphyllales
Didymelales
Eucommiales†
Urticales†
Leitneriales†
Juglandales†
Myricales†
Fagales†
Casuarinales†
Subclass Caryophyllidae†
Order: Caryophyllales†
Polygonales†
Plumbaginales†
Subclass Dilleniidae†
Order: Dilleniales†
Theales†
Malvales†
Lecythidales†
Nepenthales
Violales†
Salicales†
Capparales†
Batales†
Ericales†
Diapensiales†
Ebenales†
Primulales
Subclass Rosidae†
Order: Rosales†
Fabales
Proteales†
Order: Podostemales†
Haloragales†
Myrtales†
Rhizophorales
Cornales†
Santalales†
Rafflesiales†
Celastrales†
Euphorbiales†
Rhamnales†
Linales†
Polygalales†
Sapindales†
Geraniales†
Apiales
Subclass Asteridae†
Order: Gentianales†
Solanales
Lamiales†
Callitrichales
Plantaginales†
Scrophulariales†
Campanulales†
Rubiales†
Dipsacales†
Calycerales
Asterales†
Class Liliopsida (monocotyledons)†
Subclass Alismatidae†
Order: Alismatales†
Hydrocharitales†
Najadales†
Triuridales†
Subclass Arecidae†
Order: Arecales†
Cyclanthales†
Pandanales†
Arales†
Subclass Commelinidae†
Order: Commelinales†
Eriocaulales†
Restionales†
Juncales†
Cyperales†
Hydatellales
Typhales†
Subclass: Zingiberidae
Bromeliales†
Zingiberales†
Subclass Liliidae†
Order: Liliales†
Orchidales†

classified as plants are treated as protozoan animals by many zoologists. *See* separate articles on names marked by daggers. *See* PLANT TAXONOMY. [A.Cr.]

Plant metabolism The complex of physical and chemical events of photosynthesis, respiration, and the synthesis and degradation of organic compounds. Photosynthesis produces the substrates for respiration and the starting organic compounds used as building blocks for subsequent biosyntheses of nucleic acids, amino acids, and proteins, carbohydrates and organic acids, lipids, and natural products. *See* PHOTORESPIRATION; PHOTOSYNTHESIS; PLANT RESPIRATION. [I.P.T.]

Plant mineral nutrition The relationship between plants and all chemical elements other than carbon, hydrogen, and oxygen in the environment. Plants obtain most of their mineral nutrients by extracting them from solution in the soil or the aquatic environment. Mineral nutrients are so called because most of them have been derived from the weathering of minerals of the Earth's crust. Nitrogen is exceptional in that little occurs in minerals; its primary source is gaseous nitrogen of the atmosphere.

Some of the mineral nutrients are essential for plant growth; others are toxic, and some absorbed by plants may play no role in metabolism. Many are also essential or toxic for the health and growth of animals using plants as food. Six basic facts concerning nutrients have been established: plants do not need any of the solid materials in the soil—they cannot even take them up; plants do not need soil microorganisms; plant roots must have a supply of oxygen; all plants require at least 13 mineral nutrients; all of the essential mineral nutrients may be supplied to plants as simple ions of inorganic salts in solution; and all of the essential nutrients must be supplied in adequate but nontoxic quantities.

A mineral nutrient is regarded as essential if, in its absence, a plant cannot complete its life cycle. The essential nutrients are: nitrogen, sulfur, phosphorus, calcium, potassium, magnesium, boron, manganese, zinc, copper, molybdenum, and chlorine. In addition, sodium and silicon had been shown to be essential for some plants, beneficial to some, and possibly of no benefit to others. Cobalt was also shown to be essential for the growth of legumes when relying upon fixating atmospheric nitrogen.

Increasing salt concentrations progressively depress the water potential of solutions, thus making it more difficult for plants to absorb water. Plants can offset this effect by accumulating organic molecules or salts within their own cells. The ability of plants to accumulate soluble salts in their leaves without damage varies widely, from most common species which can accumulate only 1–2% of their dry weight, to some desert plants growing in saline soils which accumulate 20–50%. *See* PLANT-WATER RELATIONS.

A number of nutrients, sometimes at quite low concentrations, interfere directly with other aspects of plant metabolism. The ions of the heavy metals cobalt, nickel, chromium, manganese, copper, and zinc are particularly toxic in low concentrations. So is the aluminum ion, although this can exist only in significant quantities in quite acid solutions. The toxicities of these ions may result from their multivalent properties which permit them to interfere with the function of proteins and other large molecules by forming cross-linkages through adsorption to negatively charged sites.

Plants are particularly sensitive to aluminum and heavy-metal toxicities when the concentration of calcium in solution is low: increasing calcium increases the plant's tolerance. Tolerance to heavy metals also varies with plant species and cultivars. In some cases, tolerance may be achieved by excluding the toxic element from the plant. In other cases, tolerance is accompanied by an unusually high accumulation of the normally toxic element in plant tops, possibly by synthesis of specific chelators to remove it from cell metabolism.

The nutrient composition of plants is important to the health and productivity of animals which graze them. With the exception of boron, all elements which are essential for plant growth are also essential for herbivorous mammals. Animals also require sodium, iodine, and selenium and, in the case of ruminant herbivores, cobalt.

Animal requirements for mineral nutrients may also differ quantitatively from plants. For example, plants require large amounts of potassium and little or no sodium, whereas animals require large quantities of sodium and little potassium.

Plants and animals differ also in their tolerance of high levels of nutrients, sometimes with unfortunate results for the grazing animals. For example, high levels of selenium in some species do not harm the plant but may kill animals which graze them. [J.F.L.]

Plant morphogenesis A biological science concerned with the origin and development of the form and structure of plants. It deals with those phenomena, both internal and external, by which form is determined. Form is the result of orderly growth in various dimensions. The form of the body and of such structures as leaves, flowers, hairs, and cells are examples of this controlled development. Form and structure in living things are visible expressions of the organizing character of life at every level.

In almost all organisms an axis appears during development and its two ends become different, as root and shoot in higher plants. This polarity may be seen very early, even in the fertilized egg itself. It is evident in regeneration and also appears in physiological processes. The arrangement of lateral structures around the main polar axis is not irregular, but shows patterns of symmetry of various sorts. Most plant axes are vertical, and the distribution of leaves around them (phyllotaxy) is radial and precise. The points of attachment of leaves may be opposite, whorled, or dispersed in a spiral (alternate). Such spirals fall into definite classes which have simple mathematical relationships to one another.

A conspicuous fact of development is that during its course the parts of the organism become different from one another, a process called differentiation. An example of such differentiation is that between the two ends of the polar axis. Development in higher plants is accompanied by the progressive appearance of distinct organs—roots, stems, leaves, flowers, fruits, and seeds—the outward sign of division of labor within the plant. Growth and differentiation generally proceed together but there are cases where one occurs without the other. In the course of the life cycle of an organism, differential changes occur in a definite series of stages. Internal differences that arise among tissues and among cells during development are as marked as is external form.

Physiological differences between parts of the plant are common. Cells of the root are unable to synthesize sugar and certain other necessary substances and must get them from the shoots. Many localized physiological differences are maintained because of the impermeability of the cell membranes to substances produced in the cells. [R.H.Goo.]

Plant movements The wide range of movements that allow plants to reorient themselves in relation to changed surroundings, to facilitate spore or seed dispersal, or, in the case of small free-floating aquatic plants, to migrate to regions optimal for their activities. There are two types of plant movement: abiogenic movements, which arise purely from the physical properties of the cells and therefore take place in nonliving tissues or organs; and biogenic movements, which occur in living cells or organs and require an energy input from metabolism.

Abiogenic movements. Drying or moistening of certain structures causes differential contractions or expansions on the two sides of cells and hence causes movements of curvature. Such movements are called hygroscopic and are usually associated with seed and spore liberation and dispersal. Examples of such movement occur in the "parachute" hairs of the fruit of dandelion (*Taraxacum officinale*), which are closed when damp but open when the air is dry to induce release from the heads and give buoyancy for wind dispersal.

Another type of abiogenic movement is due to changes in volume of dead water-containing cells. In the absence of a gas phase, water will adhere to lignocellulose cell walls. As water is lost by evaporation from the surface of these cells, considerable tensions can build up inside, causing them to decrease in volume while remaining full of water. The effect is most commonly seen in some grasses of dry habitats, such as sand dunes, where longitudinal rows of cells on one side of the leaf act as spring hinges, contracting in a dry atmosphere and causing the leaf to roll up into a tight cylinder, thus minimizing water loss by transpiration.

Biogenic movements. There are two types of biogenic movement. One of these is locomotion of the whole organism and is thus confined to small, simply organized units in an aqueous environment. The other involves the change in shape and orientation of whole organs of complex plants, usually in response to specific stimuli.

Locomotion. In most live plant cells the cytoplasm can move by a streaming process known as cyclosis. Energy for cyclosis is derived from the respiratory metabolism of the cell. The mechanism probably involves contractile proteins very similar to the actomyosin of animal muscles.

Cell locomotion is a characteristic of many simple plants and of the gametes of more highly organized ones. Motility in such cells is produced by cilia anchored in the peripheral layers of the cell and projecting into the surrounding medium. *See* CILIA AND FLAGELLA.

Cell locomotion is usually not random but is directed by some environmental gradient. Thus locomotion may be in response to specific chemicals, in which case it is called chemotaxis. Light gradients induce phototaxis; temperature gradients induce thermotaxis; and gravity induces geotaxis. One or more of these environmental factors may operate to control movement to optimal living conditions.

Movement of organs. In higher plants, organs may change shape and position in relation to the plant body. When bending or twisting of the organ is evoked spontaneously by some internal stimulus, it is termed autonomous movement. The most common movements, however, are those initiated by external stimuli such as light and the force of gravity. Of these there are two kinds. In nastic movements (nasties), the stimulus usually has no directional qualities (such as a change in temperature), and the movement is therefore not related to the direction from which the stimulus comes. In tropisms, the stimulus has a direction (for instance, gravitational pull), and the plant movement direction is related to it.

The most common autonomous movement is circumnutation, a slow, circular, sometimes waving movement of the tips of shoots, roots, and tendrils as they grow; one complete cycle usually takes from 1 to 3 h. These movements are due to differential growth, but some may be caused by turgor changes in the cells of special hinge organs and are thus reversible.

1. *Nastic movements.* There are two kinds of nastic movements, due either to differential growth or to differential changes in the turgidity of cells. They can be triggered by a wide variety of external stimuli.

Photonastic (light/dark trigger) movements are characteristic of many flowers and inflorescences, which usually open in the light and close in the dark. Thermonasty (temperature-change trigger) is seen in the tulip and crocus flowers, which open in a warm room and close again when cooled. The most striking nastic movements are seen in the sensitive plant (*Mimosa pudica*). Its multipinnate leaves are very sensitive to touch or slight injury. Leaflets fold together, pinnae collapse downward, and the whole leaf sinks to hang limply.

Epinasty and hyponasty occur in leaves as upward and downward curvatures respectively. They arise either spontaneously or as the result of an external stimulus, such as exposure to the gas ethylene in the case of epinasty; they are not induced by gravity.

2. *Tropisms.* Of these the most universal and important are geotropism (or more properly gravitropism) and phototropism; others include thigmotropism and chemotropism.

In geotropism, the stimulus is gravity. The main axes of most plants grow in the direction of the plumb line with shoots upward (negative geotropism) and roots downward (positive geotropism).

In phototropism the stimulus is a light gradient, and unilateral light induces similar curvatures; those toward the source are positively phototropic; those away from the source are negatively phototropic. Main axes of shoots are usually positively phototropic, while the vast majority of roots are insensitive.

In thigmotropism (sometimes called haptotropism), the stimulus is touch; it occurs in climbing organs and is responsible for tendrils curling around a support. In many tendrils the response may spread from the contact area, causing the tight coiling of the basal part of the tendril into an elaborate and elastic spring.

Chemotropism is induced by a chemical substance. Examples are the incurling of the stalked digestive glands of the insectivorous plant *Drosera* and incurling of the whole leaf of *Pinguicula* in response to the nitrogenous compounds in the insect prey. A special case of chemotropism concerns response to moisture gradients; for example, under artificial conditions in air, the primary roots of some plants will curve toward and grow along a moist surface. This is called hydrotropism and may be of importance under natural soil conditions in directing roots toward water sources. *See* PLANT HORMONES; PLANT PHYSIOLOGY. [L.J.A.]

Plant organs Plant parts having rather distinct form, structure, and function. Organs, however, are interrelated through both evolution and development and are similar in many ways.

Roots, stems, and leaves are vegetative, or asexual, plant organs. They do not produce sex cells or play a direct role in sexual reproduction. In many species, nevertheless, these organs or parts of them (cuttings), may produce new plants asexually (vegetative reproduction). Sex organs are formed during the reproductive stage of plant development. In flowering plants, sex cells are produced in certain floral organs. The flower as a whole is sometimes called an organ, although it is more appropriate to consider it an assemblage of organs. *See* FLOWER; FRUIT; LEAF; REPRODUCTION (PLANT); ROOT (BOTANY); STEM. [K.E.]

Plant pathology The study of disease in plants; it is an integration of many biological disciplines and bridges the basic and applied sciences.

Kinds of plant diseases. Diseases were first classified on the basis of symptoms. Three major categories of symptoms were recognized long before the causes of disease were known; necroses, destruction of cell protoplasts (rots, spots, wilts); hypoplases, failure in plant development (chlorosis, stunting); and hyperplases, overdevelopment in cell number and size (witches'-brooms, galls). This scheme remains useful for recognition and diagnosis.

When fungi, and then bacteria, nematodes, and viruses, were recognized as causes of disease, it became convenient to classify diseases according to the responsible agent. If the agents were infectious (biotic), the diseases were classified as being "caused by bacteria," "caused by nematodes," or "caused by viruses." To this list were added phanerogams and protozoans, and later mollicutes (mycoplasmas, spiroplasmas), rickettsias, and viroids. In a second group were those diseases caused by such noninfectious (abiotic) agents as air pollutants, inadequate oxygen, and nutrient excesses and deficiencies.

Symptoms of plant diseases. Symptoms are expressions of pathological activity in plants. They are visible manifestations of changes in color, form, and structure: leaves may become spotted, turn yellow, and die; fruits may rot on the plants or in storage; cankers may form on stems; and plants may blight and wilt.

Those symptoms that are external and readily visible are considered morphological. Others are internal and primarily histological, for example, vascular discoloration of the xylem of wilting plants. Microscopic examination of diseased plants may reveal additional symptoms at the cytological level, such as the formation of tyloses (extrusion of living parenchyma cells of the xylem of wilted tissues into vessel elements).

It is important to make a distinction between the visible expression of the diseased condition in the plant, the symptom, and the visible manifestation of the agent which is responsible for that condition, the sign. The sign is the structure of the pathogen, and when present it is most helpful in diagnosis of the disease.

All symptoms may be conveniently classified into three major types because of the manner in which pathogens affect plants. Most pathogens produce dead and dying tissues, and the symptoms expressed are categorized as necroses. Early stages of necrosis are evident in such conditions as hydrosis, wilting, and yellowing. As cells and tissues die, the appearance of the plant or plant part is changed, and is recognizable in such common conditions as blight, canker, rot, and spot.

Many pathogens do not cause necrosis, but interfere with cell growth or development. Plants thus affected may eventually become necrotic, but the activity of the pathogen is primarily inhibitory or stimulatory. If there is decrease in cell number or size, the expressions of pathological activity are classified as hypoplases; if cell number or size is increased, the symptoms are grouped as hyperplases. These activities are very specific and most helpful in diagnosis. In the former group are such symptoms as mosaic, rosetting, and stunting, with obvious reduction in plant color, structure, and size. In the latter group are gall, scab, and witches'-broom, all visible evidence of stimulation of growth and development of plant tissues. *See* CROWN GALL. [C.W.B.]

Causative agents of disease. The primary agents of plant disease are fungi, bacteria, viruses and viroids, nematodes, parasitic seed plants, and a variety of noninfectious agents.

Plants with symptoms caused by noninfectious agents cannot serve as sources of further spread of the same disorder. Such noninfectious agents may be deficiencies or excesses of nutrients, anthropogenic pollutants, or biological effects by organisms external to the affected plants.

Fungi. More plant diseases are caused by fungi than by any other agent. The fungi that cause plant disease derive their food from the plant (host) and are called parasites. Those that can live and grow only in association with living plant tissues are obligate parasites. Some fungi obtain their food from dead organic matter and are known as saprobes or saprophytes. Still others can utilize food from either dead organic matter or from living plant cells, and are referred to as either facultative parasites or facultative saprophytes. *See* FUNGI. [C.W.E.]

Bacteria. Over 100 species of bacteria mainly in five genera cause disease in hundreds of different species of flowering plants. Destructive bacterial diseases affect the major cereal, vegetable, and fruit crops. None of the bacterial pathogens of plants causes serious diseases of humans or animals, and certain groups of green plants (mosses, ferns, conifers, and hardwood trees) have few or no major bacterial diseases. *See* BACTERIA.

Each species of bacterium produces a distinctive pattern of symptoms on those hosts that it attacks. With a few exceptions, most of the bacteria that cause disease in plants are non-spore-forming, rod-shaped, gram-negative cells. Some species give a positive reaction with the Gram stain; these include species of *Streptomyces* and *Corynebacterium*, a few species of *Bacillus* that cause soft rots, and those species of *Clostridium* that can decay plant storage organs under anaerobic conditions. Rickettsialike bacteria have been observed in diseased plants; unlike other plant pathogenic bacteria, these have been extremely difficult to grow in culture or have not been isolated. A number of yellows-type diseases previously considered to be caused by viruses have now been shown to be caused by mycoplasmalike organisms (MPLO), members of the class Mollicutes. Pleomorphic MPLOs lacking cell walls have been observed in vascular tissues of yellows-affected plants; however, some have not as yet been characterized through isolation and cultivation techniques. [A.K.]

Viruses and viroids. Viruses and viroids are the simplest of the various causative agents of plant disease. The essential element of each of these two pathogens is an infective nucleic acid. The nucleic acid of viruses is covered by an exterior shell (coat) of protein, but that of viroids is not. *See* PLANT VIRUSES AND VIROIDS.

The control or prevention of virus diseases involves breeding for resistance, propagation of virus-free plants, use of virus-free seed, practices designed to reduce the spread by vectors and, in some cases, the deliberate inoculation of plants with mild strains of a virus to protect them from the deleterious effects of severe strains. [R.I.H.]

Nematodes. Nematodes (roundworms) are one of the most common groups of invertebrate animals. All soils that support plant life contain nematodes living in the water films that surround soil particles. Only a relatively small group of nematodes parasitize plants. *See* NEMATA.

Plant injury is of three general types and is related to feeding habits. Migratory endoparasites destroy tissues as they feed, producing necrotic lesions in the root cortex. Other migratory endoparasites invade leaf tissues and produce extensive brown spots. Sedentary endoparasites do not kill host cells, but induce changes in host tissues, which lead to an elaborate feeding site or gall. The third general type of symptom is produced by certain migratory ectoparasites, where root tips are devitalized and cease to grow without any associated swelling or necrosis.

In addition to the plant injury that they cause directly, nematodes are important factors in disease complexes. Lesions and galls provide entrance courts for soil fungi and bacteria, and many diseases caused by soil-borne pathogens are more severe when nematodes are present. Plant varieties resistant to soil-borne fungi sometimes lose this resistance when previously parasitized by nematodes. [R.A.R.]

Seed plants. Many parasitic seed plants (estimated at nearly 3000) attack other higher plants. They occur in at least 15 dif-

Many small plants of a dwarf mistletoe (*Arceuthobium vaginatum*) parasitizing a ponderosa pine branch. This is the most damaging disease of ponderosa pine in many parts of the West.

ferent families. In some families (for example, the mistletoes Loranthaceae and Viscaceae) all members are parasitic; in others, only a single genus is parasitic in an otherwise autotrophic family (for example, the dodder, genus *Cuscuta*, of the morning glory family Convolvulaceae).

Most parasitic plants are terrestrial; that is, the parasitic connection with the host plant is through the roots. Other parasitic plants grow on the aboveground parts of the host. Some plants are classed as semiparasites because they can live in the soil as independent plants for a time, but are not vigorous or may not flower if they do not become attached to a suitable host.

The nutritional status of parasitic plants ranges from total parasites with no chlorophyll (for example, the broomrapes, Orobanchaceae) to plants that are well supplied with chlorophyll and obtain primarily water and minerals from their hosts (many mistletoes; see illustration). [F.G.H.]

Plant phylogeny The determination of probable phylogeny among plants is based partly on the study of fossils and partly on detailed comparisons among modern groups. In general, the fossil record is less useful to plant than to animal phylogenists because the critically important parts of plants (especially the reproductive organs) are seldom adequately preserved, and there are great gaps in the record. The morphological integration which permits animal taxonomists to make numerous sound deductions from fragmentary remains is much less evident in plants, and a given type of vegetative structure may be compatible with several different sorts of reproductive structures.

Despite these difficulties, a broad outline of plant phylogeny is becoming reasonably clear. At the level of families and orders, on the other hand, there are still many major points of disagreement. Although there is a general consensus about evolution among the angiosperms in terms of many individual characteristics and general trends, an agreement on the outline of the phylogeny of families and orders is only beginning to be evident. *See* EMBRYOBIONTA; PALEOBOTANY; PLANT EVOLUTION; PLANT KINGDOM; THALLOBIONTA. [A.Cr.]

Plant physiology That branch of plant sciences that aims to understand how plants live and function. Its ultimate objective is to explain all life processes of plants by a minimal number of comprehensive principles founded in chemistry, physics, and mathematics.

Plant physiology seeks to understand all the aspects and manifestations of plant life. In agreement with the major characteristics of organisms, it is usually divided into three major parts: (1) the physiology of nutrition and metabolism, which deals with the uptake, transformations, and release of materials, and also their movement within and between the cells and organs of the plant; (2) the physiology of growth, development, and reproduction, which is concerned with these aspects of plant function; and (3) environmental physiology, which seeks to understand the manifold responses of plants to the environment. The part of environmental physiology which deals with effects of and adaptations to adverse conditions—and which is receiving increasing attention—is called stress physiology.

Plant physiological research is carried out at various levels of organization and by using various methods. The main organizational levels are the molecular or subcellular, the cellular, the organismal or whole-plant, and the population level. Work at the molecular level is aimed at understanding metabolic processes and their regulation, and also the localization of molecules in particular structures of the cell but with little if any consideration of other processes and other structures of the same cell. Work at the cellular level often deals with the same processes but is concerned with their integration in the cell as a whole. Research at the organismal level is concerned with the function of the plant as a whole and its different organs, and with the relationships between the latter.

Research at the population level, which merges with experimental ecology, deals with physiological phenomena in plant associations which may consist either of one dominant species (like a field of corn) or of numerous diverse species (like a forest). Work at the organismal and to some extent the population level is carried out in facilities permitting maintenance of controlled environmental conditions (light, temperature, water and nutrient supply, and so on). *See* PLANT METABOLISM; PLANT RESPIRATION; PHYSIOLOGICAL ECOLOGY (PLANT); PHYTOTRONICS. [A.L.]

Plant propagation The deliberate, directed reproduction of plants using plant cells, tissues, or organs.

Asexual propagation, also called vegetative propagation, is accomplished by taking cuttings, by grafting or budding, by layering, by division of plants, or by separation of specialized structures such as tubers, rhizomes, or bulbs. This method of propagation is used in agriculture, in scientific research, and in professional and recreational gardening. It has a number of advantages over seed propagation: it retains the genetic constitution of the plant type almost completely; it is faster than seed propagation; it may allow elimination of the nonfruiting, juvenile phase of the plant's life; it preserves unique, especially productive, or esthetically desirable plant forms; and it allows plants with roots well adapted for growth on poor soils to be combined with tops that produce superior fruits, nuts, or other products. *See* BREEDING (PLANT); REPRODUCTION (PLANT). [C.E.LaM.]

Tissue cultures and protoplast cultures are among the techniques that have been investigated for plant propagation; the success of a specific technique depends on a number of factors. Practical applications of such methods include the clonal propagation of desirable phenotypes and the commercial production of virus-free plants.

Plant tissue cultures are initiated by excising tissue containing nucleated cells and placing it on an enriched sterile culture medium. The response of a plant tissue to a culture medium depends on a number of factors: plant species, source of tissue, chronological age and physiological state of the tissue, ingredients of the culture medium, and physical culturing conditions, such as temperature, photoperiod, and aeration.

Though technically more demanding, successful culture of plant protoplasts involves the same basic principles as plant tissue culture. Empirical methods are used to determine detailed techniques for individual species; such factors as plant species, tissue source, age, culture medium, and physical culture conditions have to be considered. *See* PLANT CELL; TISSUE CULTURE. [K.G.F.]

Plant respiration A biochemical process whereby specific substrates are oxidized with a subsequent release of carbon dioxide, CO_2. There is usually conservation of energy accompanying the oxidation which is coupled to the synthesis of energy-rich compounds, such as adenosine triphosphate (ATP), whose free energy is then used to drive otherwise unfavorable reactions that are essential for physiological processes such as growth. Respiration is carried out by specific proteins, called enzymes, and it is necessary for the synthesis of essential metabolites, including carbohydrates, amino acids, and fatty acids, and for the transport of minerals and other solutes between cells. Thus respiration is an essential characteristic of life itself in plants as well as in other organisms.

Overall aerobic respiration is the end result of a sequence of many biochemical reactions that ultimately lead to O_2 uptake and CO_2 evolution. In the absence of O_2, as may occur in

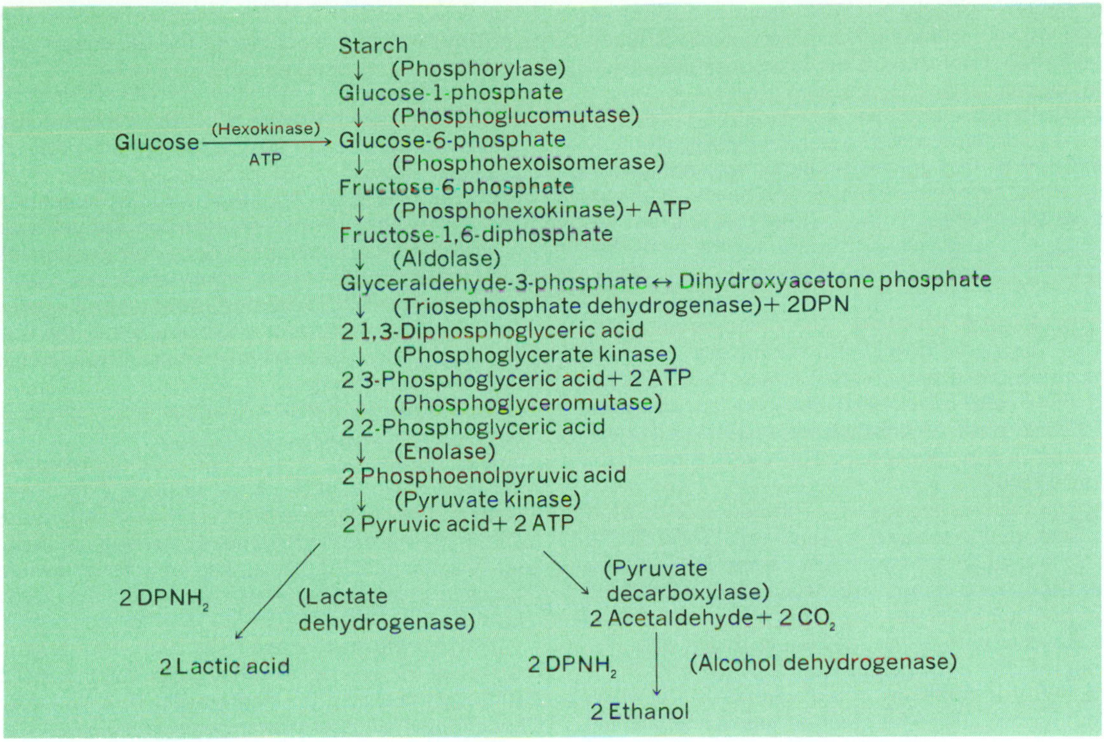

Reaction sequence for anaerobic glycolysis. The soluble enzymes are shown in parentheses.

bulky plant tissues such as the potato tuber and carrot root and in submerged plants such as germinating rice seedlings, the breakdown of hexose does not go to completion. The end products are either lactic acid or ethanol, which are produced by anaerobic glycolysis or fermentation.

The sequence of reactions of anaerobic glycolysis or fermentation is shown in the illustration. The enzymes associated with anerobic glycolysis have been isolated from many plant tissues, but more often ethanol and not lactic acid is the final product.

In aerobic tissues, pyruvic acid produced during glycolysis is completely oxidized with the accompanying synthesis of much more ATP than in anaerobic glycolysis. Pyruvic acid oxidation takes place in the mitochondria by means of a cyclic sequence of reactions, the Krebs cycle, which begins when the first product of pyruvate oxidation, acetyl coenzyme A, reacts with oxaloacetic acid to produce citric acid. Oxaloacetic acid is eventually regenerated. Thus the cycle (also known as the citric acid cycle) can be repeated. In terms of conservation of chemical energy, the Krebs cycle is about 12 times more efficient than anaerobic glycolysis per mole of glucose oxidized. *See* KREBS CYCLE.

In addition to anaerobic glycolysis and the Krebs cycle, there are two other sequences of biochemical reactions related to respiration that are important in plant tissues: (1) The pentose phosphate pathway permits an alternate mechanism for converting hexose phosphate to pyruvate, and (2) in germinating fatty seeds the reactions of the Krebs cycle are modified so that acetyl coenzyme A is converted to succinic acid and then to hexose by a pathway called the glyoxylate cycle. *See* PHOTORESPIRATION; PHOTOSYNTHESIS; PLANT GROWTH; PLANT METABOLISM. [I.Z.]

Plant taxonomy A study aimed at producing a system of classification of plants which best reflects the totality of their similarities and differences. The system is hierarchical, with the basic units being grouped into progressively larger and more inclusive units.

Most taxonomists take the species to be the basic unit of classification. In general, a species may be described as a self-perpetuating natural population which is permanently (in terms of human time) distinct and distinguishable from other such populations. In a typical species all individuals are capable of interbreeding to produce fertile offspring either directly or through other individuals of the same group, whereas such interbreeding between members of different species is restricted or prevented entirely by internal or external natural factors. *See* SPECIES CONCEPT.

A group of similar species is called a genus, a group of genera a family, a group of families an order, a group of orders a class, and a group of classes a division. Taxonomists sometimes find it useful to introduce intermediate categories, such as subgenus or subfamily, but these are optional groupings not required in the formal hierarchy. Conversely, it is sometimes useful to have an order with a single family or a family with a single genus.

The taxonomic system is a general-purpose system which seeks to balance the needs of all users by putting together those things that are most alike in all respects. This truly natural system is the ultimate but never fully attainable goal of taxonomists. Even if all the information that might be desired were available, taxonomists would often differ about how best to devise a natural system, but the range of choice would be limited and small. Artificial systems, on the other hand, based on a few arbitrarily chosen characters, are easily produced. There can be as many artificial systems as there are systematists and ends to be served. Modern taxonomists always prefer the natural system, falling back on an artificial system only when they feel unable to produce a reasonably natural one.

In an effort to assure the naturalness of the classification, taxonomists commonly rely heavily on the phylogenetic concept. An attempt is made to set up the system in such a way that every taxon (taxonomic group) has a unified ancestry. Such a group can be expected to be more homogeneous in features as yet unstudied than one which originated from diverse ancestors. The fossil record of plants is too fragmentary, however, and the critically important features are too seldom preserved to permit the kind of phylogenetic certitude that might be desired. In practice, if the taxonomist can devise a scheme that permits a logically possible phylogeny, without internal contradictions, the phylogenetic criterion is considered to be satisfied. The development of a natural classification and a phylogenetic scheme thus go hand in hand, each influencing the other. For many taxonomists the determination of probable phylogenetic relationships by this indirect method provides the chief stimulus and intellectual satisfaction for the work. *See* PLANT EVOLUTION; PLANT KINGDOM; PLANT PHYLOGENY. [A.Cr.]

Plant viruses and viroids Plant viruses are pathogens which are composed mainly of a nucleic acid (genome) normally surrounded by a protein shell (coat); they replicate only in compatible cells, usually with the induction of symptoms in the affected plant. Viroids are the smallest pathogens known, consisting of a single-stranded RNA with a molecular weight of about 120,000.

About 400 plant viruses are known. The genomes of most plant viruses, such as the tobacco mosaic virus (TMV) or the potato virus Y, are single-stranded RNAs; some RNA viruses (for instance, the wound tumor virus) have double-stranded RNA genomes. Cauliflower mosaic virus and bean golden mosaic virus are examples of viruses having double-stranded and single-stranded DNA. The genome of many plant viruses is a single piece and it is contained in a single particle whereas those of brome mosaic and some other viruses are segmented and distributed between several particles. There are also several low-molecular-weight RNAs (satellite RNAs) which depend on helper viruses for their replication. *See* RIBONUCLEIC ACID (RNA).

Less than a dozen viroids are known, but they cause very serious diseases in such diverse plants as chrysanthemum, citrus, coconut, and potato. Their replication is poorly understood; the available evidence does not allow discrimination between synthesis on an RNA (that is, viroid) template or synthesis on a DNA (that is, host genome) template. *See* PLANT PATHOLOGY; VIROID; VIRUS. [R.I.H.]

Plant-water relations Water is the most abundant constituent of all physiologically active plant cells. Leaves, for example, have water contents which lie mostly within a range of 55–85% of their fresh weight. Other relatively succulent parts of plants contain approximately the same proportion of water, and even such largely nonliving tissues as wood may be 30–60% water on a fresh-weight basis. The smallest water contents in living parts of plants occur mostly in dormant structures, such as mature seeds and spores. The great bulk of the water in any plant constitutes a unit system. This water is not in a static condition. Rather it is part of a hydrodynamic system, which in terrestrial plants involves absorption of water from the soil, its translocation throughout the plant, and its loss to the environment, principally in the process known as transpiration.

Cellular water relations. The typical mature, vacuolate plant cell constitutes a tiny osmotic system, and this idea is central to any concept of cellular water dynamics. Although the cell walls of most living plant cells are quite freely permeable to water and solutes, the cytoplasmic layer that lines the cell wall is more permeable to some substances than to others.

If a plant cell in a flaccid condition—one in which the cell sap exerts no pressure against the encompassing cytoplasm and cell wall—is immersed in pure water, inward osmosis of water into the cell sap ensues. This gain of water results in the exertion of a turgor pressure against the protoplasm, which in turn is transmitted to the cell wall. This pressure also prevails throughout the mass of solution within the cell. If the cell wall is elastic, some expansion in the volume of the cell occurs as a result of this pressure, although in many kinds of cells this is relatively small.

If a turgid or partially turgid plant cell is immersed in a solution with a greater osmotic pressure than the cell sap, a gradual shrinkage in the volume of the cell ensues; the amount of shrinkage depends upon the kind of cell and its initial degree of turgidity. When the lower limit of cell wall elasticity is reached and there is continued loss of water from the cell sap, the protoplasmic layer begins to recede from the inner surface of the cell wall. Retreat of the protoplasm from the cell wall often continues until it has shrunk toward the center of the cell, the space between the protoplasm and the cell wall becoming occupied by the bathing solution. This phenomenon is called plasmolysis. *See* OSMOREGULATORY MECHANISMS.

In some kinds of plant cells movement of water occurs principally by the process of imbibition rather than osmosis. The swelling of dry seeds when immersed in water is a familiar example of this process.

Stomatal mechanism. Various gases diffuse into and out of physiologically active plants. Those gases of greatest physiological significance are carbon dioxide, oxygen, and water vapor. The great bulk of the gaseous exchanges between a plant and its environment occurs through tiny pores in the epidermis that are called stomates. Although stomates occur on many aerial parts of plants, they are most characteristic of, and occur in greatest abundance in, leaves. *See* EPIDERMIS (PLANT); LEAF.

Transpiration process. The term transpiration is used to designate the process whereby water vapor is lost from plants. Although basically an evaporation process, transpiration is complicated by other physical and physiological conditions prevailing in the plant. Whereas loss of water vapor can occur from any part of the plant which is exposed to the atmosphere, the great bulk of all transpiration occurs from the leaves. There are two kinds of foliar transpiration: (1) stomatal transpiration, in which water vapor loss occurs through the stomates, and (2) cuticular transpiration, which occurs directly from the outside surface of epidermal walls through the cuticle. In most species 90% or more of all foliar transpiration is of the stomatal type. *See* EVAPOTRANSPIRATION.

Transpiration is a necessary consequence of the relation of water to the anatomy of the plant, and especially to the anatomy of the leaves. Terrestrial green plants are dependent upon atmospheric carbon dioxide for their survival. In terrestrial vascular plants the principal carbon dioxide–absorbing surfaces are the moist mesophyll cells walls which bound the intercellular spaces in leaves. Ingress of carbon dioxide into these spaces occurs mostly by diffusion through open stomates. When the stomates are open, outward diffusion of water vapor unavoidably occurs, and such stomatal transpiration accounts for most of the water vapor loss from plants. Although transpiration is thus, in effect, an incidental phenomenon, it frequently has marked indirect effects on other physiological processes which occur in the plant because of its effects on the internal water relations of the plant.

Water translocation. In terrestrial rooted plants practically all of the water which enters a plant is absorbed from the soil by the roots. The water thus absorbed is translocated to all parts of the plant. The mechanism of the "ascent of sap" (all translocated water contains at least traces of solutes) in plants, especially tall trees, was one of the first processes to excite the interest of plant physiologists.

The upward movement of water in plants occurs in the xylem, which, in the larger roots, trunks, and branches of trees and shrubs, is identical with the wood. In the trunks or larger branches of most kinds of trees, however, sap movement is restricted to a few of the outermost annual layers of wood. *See* XYLEM.

Root pressure is generally considered to be one of the mechanisms of upward transport of water in plants. While it is undoubtedly true that root pressure does account for some upward movement of water in certain species of plants at some seasons, various considerations indicate that it can be only a secondary mechanism of water transport.

Upward translocation of water (actually a very dilute sap) is engendered by an increase in the negativity of water potential in the cells of apical organs of plants. Such increases in the negativity of water potentials occur most commonly in the mesophyll cells of leaves as a result of transpiration.

Water absorption. The successively smaller branches of the root system of any plant terminate ultimately in the root tips, of which there may be thousands and often millions on a single plant. Most absorption of water occurs in the root tip regions, and especially in the root hair zone. Older portions of most roots become covered with cutinized or suberized layers through which only very limited quantities of water can pass. *See* Root (Botany).

Whenever the water potential in the peripheral root cells is less than that of the soil water, movement of water from the soil into the root cells occurs. There is some evidence that, under conditions of marked internal water stress, the tension generated in the xylem ducts will be propagated across the root to the peripheral cells. If this occurs, water potentials of greater negativity could develop in peripheral root cells than would otherwise be possible. The absorption mechanism would operate in fundamentally the same way whether or not the water in the root cells passed into a state of tension. The process just described, often called passive absorption, accounts for most of the absorption of water by terrestrial plants.

The phenomenon of root pressure represents another mechanism of the absorption of water. This mechanism is localized in the roots and is often called active absorption. Water absorption of this type only occurs when the rate of transpiration is low and the soil is relatively moist. Although the xylem sap is a relatively dilute solution, its osmotic pressure is usually great enough to engender a more negative water potential than usually exists in the soil water when the soil is relatively moist. A gradient of water potentials can thus be established, increasing in negativity across the epidermis, cortex, and other root tissues, along which the water can move laterally from the soil to the xylem. *See* Plant mineral nutrition. [B.S.M.]

Plantaginales An order of flowering plants, division Magnoliophyta (Angiospermae), in the subclass Asteridae of the class Magnoliopsida (dicotyledons). The order consists of only the family Plantaginaceae, with about 250 species. Within its subclass the order is marked by its small, chiefly wind-pollinated flowers that have a persistent regular corolla. The perianth and stamens of the flowers are attached directly to the receptacle (hypogynous) and there are typically four petals. The plants are herbs or seldom half-shrubs with mostly basal, alternate leaves. The common plantain (*Plantago major*) is a lawn weed of this order. *See* Asteridae; Magnoliopsida. [A.Cr.]

Plasma chromatography A trace analysis technique in which trace molecules in a gas are detected by subjecting them to an ion-molecule reaction and measuring the ion-molecules produced. It is capable of detecting and identifying a substance present in concentrations far below the parts-per-billion range. The instrumentation has been applied to a wide range of fundamental and analytical studies. Upper atmospheric ion-molecule reactions have been studied as well as many aspects of chemical analysis, such as the identification of specific organic molecules, biomedical analysis of large molecules, and pesticide analysis.

The great sensitivity of the method to trace compounds arises because the ion-molecule reaction occurs at atmospheric pressure, where millions of collisions of reactant ions with trace molecules are possible. This all occurs in the plasma chromatograph, where ion-molecules are created and then separated by drifting them through an electrical field toward a detector. The detector output of ion current versus time presents a plasmagram of the reactant ion and the different ion-molecules formed. Plasmagrams, both positive and negative, resemble chromatograms with a millisecond time scale. This technique uses a shutter grid to inject a discrete ion-molecule pulse into the drift region and a gating grid to produce a recordable scan of the plasmagram (see illustration). *See* Chromatography; Gas chromatography; Mass spectrometry; Trace analysis. [F.W.K.]

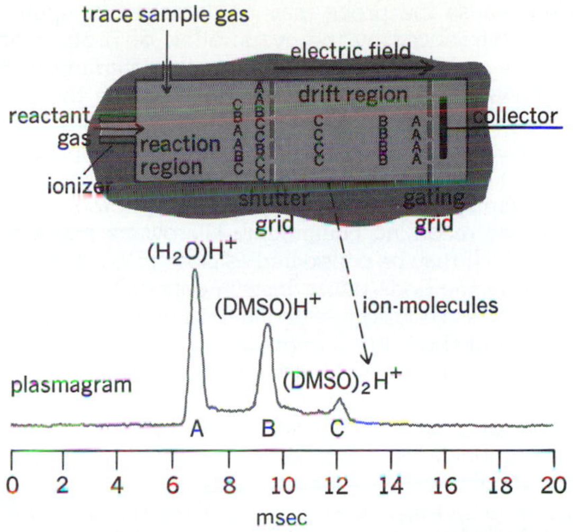

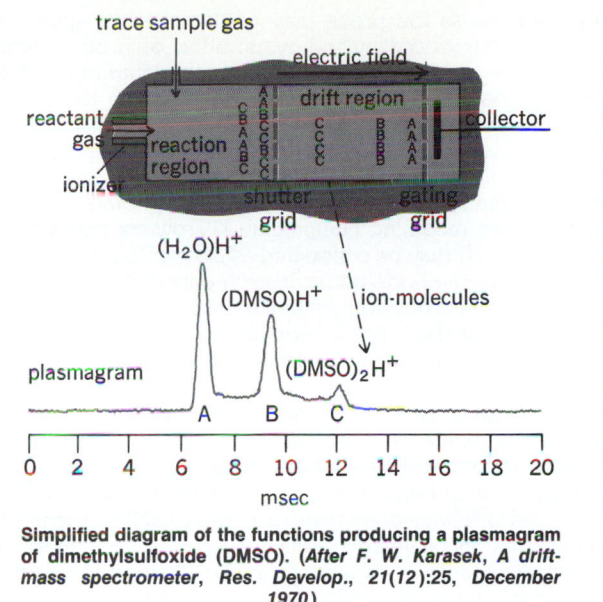

Simplified diagram of the functions producing a plasmagram of dimethylsulfoxide (DMSO). ***(After F. W. Karasek, A drift-mass spectrometer, Res. Develop., 21(12):25, December 1970)***

Plasma diagnostics The branch of science dealing with the experimental investigation of plasmas. Plasmas may appear in a stationary or in a flowing state, the latter being either in a steady or an unsteady motion.

Diagnostics deals with the determination of various physical properties, which are subdivided into microscopic properties, such as collision frequency, and macroscopic properties. The macroscopic properties are either static (such as density, pressure, and temperature); or transport (such as viscosity, diffusivity, heat conductivity, and electrical conductivity). *See* Plasma physics.

In addition, there are many other variables, for example, electrical (current, voltage, and power) and dynamic (mass flow, velocity, thrust, and others).

Very few methods of measurement used in the other branches of fluid dynamics, for instance, aerodynamics, can be used in plasmas. *See* Aerodynamics.

The two basic types of diagnostic methods are those measuring global quantities and those measuring local ones. The local quantities are measured as functions of both space and time. This is not always easy and sometimes mean values may be obtained as a compromise.

Methods of measurement lie between two extremes, namely, those interfering with the flow and those not interfering with it. The former group employs solid probes; the latter uses microwaves, optical spectroscopy, particle beams, and lasers. Methods for measuring global electrical quantities are well known from electrical engineering. *See* Electrical measurements.

Many diverse methods can be used in plasma diagnostics, including probe, optical, spectroscopic, microwave, laser, and energetic particle beam methods.

Probe methods have a limited applicability, since they usually disturb the very quantities intended to be measured. The various types of probes represent different degrees of disturbances to the flow and also possess different ranges of operation. Probes constitute the most common method used to obtain detailed information of the internal structure of plasmas. Sometimes the introduction of probes into a plasma is not permissible because the probe may be damaged by sputtering. Impurities introduced by the evaporation of probe material contaminate and cool the flow. If the probe interferes with the plasma flow, sometimes it may be used outside the plasma, that is, between the flow and the solid wall.

Spectroscopic methods may be applied in the optical, ultraviolet, and x-ray regions of the spectrum. These methods are based on emission spectra. They employ photographic and photoelectric recording equipment. Microwave methods use radiation which may be considered as close to equilibrium radiation. These methods utilize instruments called microwave interferometers. Thomson scattering of laser light may serve as a diagnostic method. In concept, the idea is similar to radar. *See* INTERFEROMETRY; LASER; MICROWAVE SPECTROSCOPY. [D.G.S.]

Plasma physics That field of physics which relates to the study of highly ionized gases. A gas which is composed of a nearly equal number of positive and negative free charges (positive ions and electrons) is called a plasma. Because it is composed of charged particles, a plasma exhibits many phenomena not encountered in ordinary gases. In addition to their importance in many new areas of applied science, these effects are evident in astrophysical phenomena (most of the matter in this universe exists in the plasma state), both in stellar atmospheres and in interstellar space. Plasma phenomena have also been observed in the tenuous ionized gases of the Earth's outer atmosphere. *See* COSMIC ELECTRODYNAMICS.

Practical interest in plasma physics arises from various applications of gas discharges, from the study of electron beams in electron tubes, and from the new research fields of ultra-high-temperature processes and controlled fusion. These latter experiments require millions or even hundreds of millions of degrees kinetic temperature (to be defined later). To obtain a comprehensive physical picture of plasma it is necessary to consider its behavior from two different aspects:

1. The microscopic picture, relating to its particlelike properties, such as the effects of interparticle collisions in producing diffusion and other transport phenomena, ionization, x-radiation, and other particulate processes. In this picture a plasma exhibits properties some of which are much like those of any gas.
2. The macroscopic picture, where the collective or fluidlike properties are most evident. These properties include conduction of electricity, propagation of various kinds of waves, and ability to support classes of unstable and turbulent behavior peculiar to conducting fluids. Many of the macroscopic behavioral properties of plasma are related to the general field of magnetohydrodynamics. *See* MAGNETOHYDRODYNAMICS.

The charged particles of a plasma interact with each other through the electrostatic or Coulomb field with which each is surrounded. On the microscopic scale, these electrostatic fields give rise to localized attractive or repulsive forces between the particles as they pass near to each other, resulting in mutual deflection. On the macroscopic scale, the summation of the many infinitesimal electrostatic and magnetic fields produced by the moving plasma particles results in a smeared-out or averaged electromagnetic field. The plasma then reacts collectively, that is, as a conducting fluid, to the total electromagnetic field in which it is immersed. This field consists of the combination of the plasma electromagnetic field and any externally imposed fields. The coupled nature of the plasma motion and the electromagnetic field in which it moves causes most of the complexity of plasma behavior. *See* NUCLEAR FUSION; PINCH EFFECT. [R.F.P.]

Plasma propulsion The imparting of thrust to a spacecraft through the acceleration of a plasma (ionized gas). A plasma can be accelerated by electrical means to exhaust velocities considerably higher than those attained by chemical rockets. The higher exhaust velocities (specific impulses) of plasma thrusters usually imply that for a particular mission the spacecraft would use less propellant than the amount required by conventional chemical rockets. This means that for the same amount of propellant a spacecraft propelled by a plasma rocket can increase in velocity over a set distance by an increment larger than that possible with a chemical propulsion system. Plasma propulsion is one of three major classes of electric propulsion, the others being electrothermal propulsion and ion (or electrostatic) propulsion. *See* ELECTROTHERMAL PROPULSION; ION PROPULSION; SPACECRAFT PROPULSION; SPECIFIC IMPULSE.

Pure electromagnetic acceleration. The most promising and thoroughly studied electromagnetic plasma accelerator is the magnetoplasmadyamic (MPD) thruster. In this device the plasma is both created and accelerated by a high-current discharge. The discharge is due to the breakdown of the gas as it is injected in the interelectrode region. The acceleration process can be described as being due to a body force acting on the plasma. This body force is the Lorentz force created by the interaction between the current conducted through the plasma and the magnetic field. The latter could either be externally applied by a magnet or self-induced by the discharge, if the current is sufficiently high. *See* MAGNETIC FIELD; MAGNETOHYDRODYNAMICS.

Microscopically, the acceleration process can be described as the momentum transfer from the electrons, which carry the current, to the heavy particles through collisions. Such collisions are responsible for the creation of the plasma (ionization) and its acceleration and heating (Joule heating).

Hybrid acceleration. The collision processes invariably heat the plasma. If the gas particles are exhausted hot, they are dissipating energy in kinetic modes useless to propulsion since their thermal motion is random. Moreover, if the exhausted atoms are in an excited or ionized state, the fraction of the internal energy tied in these internal modes is also not available for propulsion. If a fraction of these translational and internal modes is somehow recovered, the plasma acceleration is called hybrid (electromagnetic-electrothermal). Hybrid acceleration is an active area of research and is the most promising alternative for surpassing the 40% efficiency level of magnetoplasmadynamic thrusters.

Flight tests. Few tests of plasma thrusters in space are known publicly outside the Soviet Union. Most of the flown plasma propulsion systems are of the pulsed solid-fed (Teflon-ablative) type launched in the 1970s for satellite attitude control. Active research aims at flying plasma-propelled vehicles by the mid-1990s. [E.Y.C.]

Plasmal Those aldehydic components of lipids which give positive color tests with reagents used for detecting aldehydes in tissues. The Feulgen test for plasmalogens depends on the liberation of plasmal, principally derivatives of palmitaldehyde and stearaldehyde, from these lipids by the action of mercuric chloride and acetic acid. *See* LIPID; PHOSPHATIDE. [R.H.G.; H.E.Ca.]

Plasmid A circular extrachromosomal genetic element that is ubiquitous in prokaryotes and has also been identified in a number of eukaryotes. In general, bacterial plasmids can be classified into two groups on the basis of the number of genes and functions they carry. The larger plasmids are deoxyribonucleic acid (DNA) molecules of around 100 kilobase (kb) pairs, which is sufficient to code for approximately 100 genes. There is usually a small number of copies of these plasmids per host chromosome, so that their replication must be precisely coordinated with the cell division cycle. The plasmids in the second group are smaller in size, about 6–10 kb. These plasmids may harbor 6–10 genes and are usually present in multiple copies (10–20 per chromosome). *See* GENE.

Plasmids have been identified in a large number of bacterial genera. Some bacterial species harbor plasmids with no known functions (cryptic plasmids) which have been identified as small circular molecules present in the bacterial DNA. The host range of a particular plasmid is usually limited to closely related genera. Some plasmids, however, are much more promiscuous and have a much broader host range.

The functions specified by different bacterial plasmids are usually quite specialized in nature. Moreover, they are not essential for cell growth since the host bacteria are viable without a plasmid when the cells are cultured under conditions that do not select for plasmid-specified gene products. Plasmids thus introduce specialized functions to host cells which provide versatility and adaptability for growth and survival. Plasmids which confer antibiotic resistance (R plasmids) have been extensively characterized because of their medical importance. Plasmids have played a seminal role in the spectacular advances in the area of genetic engineering. Individual genes can be inserted into specific sites on plasmids in cell cultures and the recombinant plasmid thus formed introduced into a living cell by the process of bacterial transformation. *See* GENETIC ENGINEERING. [R.H.R.]

Plasmin A proteolytic enzyme which can digest many proteins through the process of hydrolysis. Plasmin (fibrinolysin) is found in plasma in the form of the inert precursor, or zymogen, plasminogen (profibrinolysin); its site of synthesis in the body is unknown. Plasma also contains several inhibitors which limit the action of plasmin.

Plasma itself has the potentiality of activating plasminogen, perhaps because it contains intrinsic activators. Plasmin activator is produced in blood vessel walls, from which it is released following vascular injury. Further, the process of blood coagulation may foster the activation of plasminogen. Moreover, the presence of a fibrin clot enhances activation of plasminogen by many agents, perhaps because plasminogen is adsorbed to the fibrin and thus separated from its inhibitors in the plasma.

Plasmin can act on many protein substrates. It liquefies coagulated blood by digesting fibrin (fibrinolysis), the insoluble meshwork of the clot; it also digests fibrinogen, the precursor of fibrin, rendering it incoagulable. The digestion products of fibrinogen and fibrin are anticoagulant substances which further interfere with the clotting process. Plasmin also inactivates several other protein procoagulant factors, particularly proaccelerin (factor V) and antihemophilic factor (factor VIII). Other substrates attacked by plasmin include gamma globulin and several protein or polypeptide hormones found in plasma.

The action of plasmin may be related to certain body defenses. It can convert the first component of complement to a proteolytic enzyme. It can also liberate polypeptide kinins (for example, bradykinin) from plasma precursors. The kinins can reproduce such elements of the inflammatory process as pain, dilatation and increased permeability of small blood vessels, and the migration of leukocytes. Besides these physiological actions, plasmin also hydrolyzes casein, gelatin, denatured hemoglobin, and certain synthetic esters of arginine and lysine, for example, *p*-toluenesulfonylarginine methyl ester (TAME). *See* BLOOD; COMPLEMENT; FIBRINOGEN; IMMUNOLOGY; INFLAMMATION. [O.D.R.]

Plasmodiophorida Protozoa composing an order of Mycetozoia. They are endoparasitic in plants, primarily causing, for example, club root of cabbage, powdery scab of potatoes, and galls in *Ruppia*. Underground portions of host plants are invaded by young parasites, often a flagellate which sometimes arises from a freshly excysted ameba. Becoming intracellular, the young parasite grows and develops into a plasmodium. At maturity, the plasmodium produces uninucleate cysts (spores) which are released upon degeneration of the damaged cell. Under favorable conditions, the released spores hatch into uninucleate stages which become the infective forms. In the invaded area, the host's tissue commonly undergoes hypertrophy to form a gall. [R.P.H.]

Plasmodiophoromycetes A class of slime molds (Myxomycota) with one order, Plasmodiophorales, and one family, Plasmodiophoraceae. These fungi are microscopic and occur in aquatic or semiaquatic habitats. They are obligate parasites of algae, vascular plants, and aquatic fungi. They are characterized by the formation of a naked multinucleate mass of protoplasm called a plasmodium. The plasmodium can migrate by amoeboid movement from cell to cell in the host, increasing its volume as nuclear division occurs. At a certain size it cleaves into several, usually multinucleate parts. Each part forms a membrane-bound reproductive structure (sporangium). When the sporangium is mature, motile spores (zoospores) are formed within it. Subsequently zoospores are discharged outside the host through pores in the host cell that are formed by the fungus. The zoospores have two flagella of unequal length, both inserted anteriorly and of the whiplash type. The flagella and an amoeboid motion carry the zoospores to the host plant, where they encyst. Subsequently the encysted zoospore produces a germ tube that penetrates a host cell, forming a plasmodium within it. Sexual reproduction is known in some species.

Serious diseases caused by members of the class are club root of crucifers and powdery scab of potato. All members of the class cause enlargement of the infected host tissue. Their distribution is worldwide. *See* MYXOMYCOTA. [R.A.P.]

Plasmon The quanta of waves produced by collective effects of large numbers of electrons in matter when the electrons are disturbed from equilibrium. Metals provide the best evidence of plasmons, because they have a high density of electrons free to move. The results of plasmon stimulation are seen in the illustration. The graph shows the probability of energy losses by fast electrons transmitted through a thin aluminum foil. The number of detected electrons in a beam is plotted against their energy loss during transit through the foil. Each energy-loss peak corresponds to excitation of one or more plasmons. Within experimental error, the peaks occur at integral multiples of a fundamental loss. Further evidence is the fact that the areas under the peaks (a measure of the energy-loss probability) follow a Poisson distribution. *See* DISTRIBUTION (PROBABILITY).

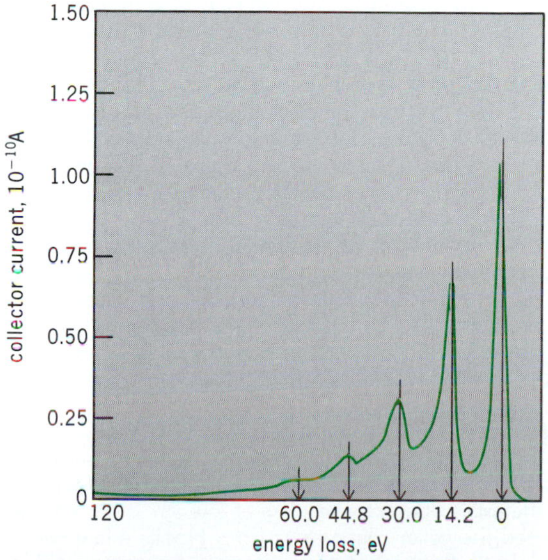

Number of detected electrons in a beam versus their energy loss during transit through a thin aluminum foil. (Number of electrons is expressed as a current; 10^{-14} A = 6.7×10^{6} electrons per second.) Peaks at approximate multiples of 14.2 eV correspond to energy donated to plasmons in the aluminum. (*After T. L. Ferrell, T. A. Callcott, and R. J. Warmack Plasmons and surfaces, Amer. Sci., 70:044-353, 1985*)

The name plasmon derives from the physical plasma as a state of matter in which the atoms are ionized. At the lowest densities this means an ionized gas, or classical plasma; but densities are much higher in a metal, or quantum plasma, the atoms of a solid metal being in the form of ions. In both types of physical plasma, the frequency of plasma-wave oscillation is determined by the electronic density. In a quantum plasma the energy of the plasmon is its frequency multiplied by Planck's constant, a basic relationship of quantum mechanics. *See* FREE-ELECTRON THEORY OF METALS; PLASMA PHYSICS; QUANTUM MECHANICS.

The plasmon energy for most metals corresponds to that of an ultraviolet photon. However, for silver, gold, the alkali metals, and a few other materials, the plasmon energy is sufficiently low to correspond to that of a visible or near-ultraviolet photon. This means there is a possibility of exciting plasmons by light. If plasmons are confined upon a surface, optical effects can be easily observed. In this case, the quanta are called surface plasmons, and they have the bulk plasmon energy as an upper energy limit.

Surface plasmons were first proposed to explain energy losses by electrons reflected from metal surfaces. Since then, numerous experiments have involved coupling photons to surface plasmons. Potential applications extend to new light sources, solar cells, holography, Raman spectroscopy, and microscopy. [T.L.F.]

Plaster A plastic mixture of solids and water which sets to a hard, coherent solid and which is used to line the interiors of buildings. A similar material of different composition, used to line the exteriors of buildings, is known as stucco. The term plaster is also used in the industry to designate plaster of paris.

Plaster is usually applied in one or more base (rough or scratch) coats up to $\frac{3}{4}$ in. (1.9 cm) thick, and also in a smooth, white, finish coat about $\frac{1}{16}$ in. (0.16 cm) thick. The solids in the base coats are hydrated (or slaked) lime, sand, fiber or hair (for bonding), and portland cement (the last may be omitted in some plasters). The finish coat consists of hydrated lime and gypsum plaster (in addition to the water). *See* LIME (INDUSTRY); MORTAR; PLASTER OF PARIS. [J.F.McM.]

Plaster of paris The hemihydrate of calcium sulfate, composition $CaSO_4 \cdot \frac{1}{2}H_2O$, made by calcining the mineral gypsum, composition $CaSO_4 \cdot 2H_2O$, at temperatures up to 480°F (250°C). It is used for making plasters, molds, and models.

When the powdered hemihydrate is mixed with water to form a paste or slurry, the calcining reaction is reversed and a solid mass of interlocking gypsum crystals with moderate strength is formed. Upon setting there is very little (a slight contraction) dimensional change, making the material suitable for accurate molds and models.

Diverse types of plaster, varying in the time taken to set, the amount of water needed to make a pourable slip, and the final hardness, are made for different applications. These characteristics are controlled by the calcination conditions (temperature and pressure) and by additions to the plaster. *See* GYPSUM; PLASTER. [J.F.McM.]

Plasticity The property of a solid body whereby it undergoes a permanent change in shape or size when subjected to a stress exceeding a particular value, called the yield value. Many solid materials obey Hooke's law at low stress. As the stress is increased, however, departures from Hooke's law occur, and some plastic flow takes place; that is, the material does not completely recover its original shape or size when the stress is released.

Plastic behavior is often accompanied by time-dependent effects such as creep (the increase in strain with time at constant stress), stress relaxation (the decay of stress with time at constant strain), and elastic aftereffect or recovery (the gradual decrease to a limiting permanent strain when the stress is removed). The study of these phenomena in all their manifestations is the science of rheology. For the influence of dislocations on plastic flow in crystals *see* CRYSTAL. *See also* CREEP (MATERIALS); ELASTICITY; HOOKE'S LAW; RHEOLOGY; STRESS AND STRAIN. [R.F.S.H.]

Plastics processing Those methods and techniques used to convert plastics materials in the form of pellets, granules, powders, sheets, fluids, or preforms into formed shapes or parts. The plastic materials may contain a variety of additives which influence the properties as well as the processability of the plastics. After forming, the part may be subjected to a variety of ancillary operations such as welding, adhesive bonding, machining, and surface decorating (painting, metallizing).

Injection moldings. Injection molding consists of heating and homogenizing plastics granules in a cylinder until they are sufficiently fluid to allow for pressure injection into a relatively cold mold where they solidify and take the shape of the mold cavity. Solid particles, in the form of pellets or granules, constitute the main feed for injection moldable plastics. The major advantages of the injection-molding process are the speed of production, minimal requirements for postmolding operations, and simultaneous multipart molding.

Extrusion. In this process, plastic pellets or granules are fluidized, homogenized, and continuously formed. Products made this way include tubing, pipe, sheet, wire and substrate coatings, and profile shapes. The process is used to form very long shapes or a large number of small shapes which can be cut from the long shapes. Pellets used for other processing methods, such as injection molding, are made by chopping long filaments of extruded plastic. *See* EXTRUSION.

Blow molding. This process consists of forming a tube (called a parison) and introducing air or other gas to cause the tube to expand into a free-blown hollow object or against a mold for forming into a hollow object with a definite size and shape. The parison is traditionally made by extrusion, although injection-molded tubes have been introduced.

Thermoforming. Thermoforming is the forming of plastics sheets into parts through the application of heat and pressure. Tooling for this process is the most inexpensive compared to other plastics processes. However, it can accommodate very large parts as well as small parts.

Rotational molding. Finely ground powders are heated in a rotating mold until melting or fusion occurs. If liquid materials are used, the process is often called slush molding. The melted or fused resin uniformly coats the inner surface of the mold. When cooled, a hollow finished part is removed.

Compression and transfer molding. Compression molding consists of charging a plastics powder or preformed plug into a mold cavity, closing a mating mold half, and applying pressure to compress, heat, and cause flow of the plastic to conform to the cavity shape.

Foam processes. Foamed plastics materials have achieved a high degree of importance in the plastics industry. Foams can be made soft and flexible to hard and rigid. There are three types of cellular plastics: blown (expanded matrix, such as a natural sponge), syntactic (the encapsulation of hollow organic or inorganic microspheres in the matrix), and structural (dense outer skin surrounding a foamed core).

There are seven basic processes used to generate plastics foams. They include: the incorporation of a chemical blowing agent that generates gas (through thermal decomposition) in the polymer liquid or melt; gas injection into the melt which expands during pressure relief; generation of gas as a byproduct of a chemical condensation reaction during cross-linking; volatilization of a low-boiling liquid (for example, Freon) through the exothermic heat of reaction; mechanical dispersion of air by mechanical means (whipped cream); incorporation of nonchemical gas-liberating agents (adsorbed gas on finely divided carbon) into the resin mix which is released by heating; and expansion of small beads of thermoplastic resin containing a blowing agent through the external application of heat.

Casting and encapsulation. Casting is a low-pressure process requiring nothing more than a container in the shape of the desired part. For thermoplastics, liquid monomer is poured into the mold and, with heat, allowed to polymerize in place to a solid mass. For vinyl plastisols, the liquid is fused with heat. Thermosets are poured into a heated mold wherein the cross-linking reaction completes the conversion to a solid. Encapsulation and potting are terms for casting processes in which a unit or assembly is encased or impregnated, respectively, with a liquid plastic which is subsequently hardened by fusion or chemical reaction.

Calendering. In the calendering process, a plastic is masticated between two rolls that squeeze it out into a film which then passes around one or more additional rolls before being stripped off as a continuous film. Fabric or paper may be fed through the latter rolls so that they become impregnated with the plastic. *See* POLYMER. [S.H.G.]

Plate girder A beam built up of steel plates and shapes which may be welded or bolted together to form a deep beam larger than can be produced by a rolling mill (see illustration). As such, it is capable of supporting greater loads on longer spans. The typical welded plate girder consists of flange plates welded to a deep web plate. A bolted configuration consists of flanges built of angles and cover plates bolted to the web plate. Both types may have vertical stiffeners connected to the web plate, and both may have additional cover plates on the flanges to increase the load capacity of the member. Box girders consist of common flanges connected to two web plates, forming a closed section.

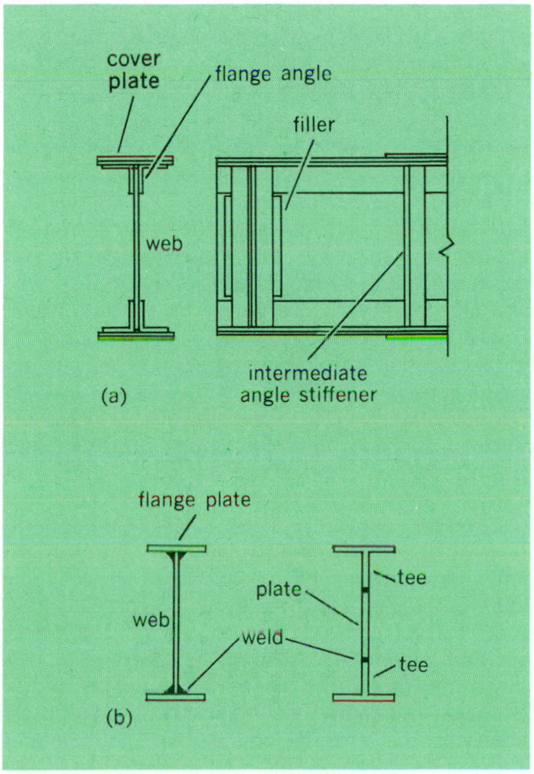

Plate girder configurations: (*a*) bolted type and (*b*) welded type.

In general, the depth of plate girders is one-tenth to one-twelfth of the span length, varying slightly for heavier or lighter loads. On occasion, the depth may be controlled by architectural considerations.

Stiffeners, plates or angles, may be attached to the girder web by welding or bolting to increase the buckling resistance of the web. Stiffeners are also required to transfer the concentrated forces of applied loads and reactions to the web without producing local buckling.

Splices are required for webs and flanges when full lengths of plates are not available from the mills or when shorter lengths are more readily fabricated. Splices provide the necessary continuity required in the web and flanges. [J.B.S.]

Plate tectonics Plate tectonics theory provides an explanation for the present-day tectonic behavior of the Earth, particularly the global distribution of mountain building, earthquake activity, and volcanism in a series of linear belts. Numerous other geological phenomena such as lateral variations in surface heat flow, the physiography and geology of ocean basins, and various associations of igneous, metamorphic, and sedimentary rocks can also be logically related by plate tectonics theory.

The theory is based on a simple model of the Earth in which a rigid outer shell 30–90 mi (50–150 km) thick, the lithosphere, consisting of both oceanic and continental crust as well as the upper mantle, is considered to lie above a hotter, weaker semiplastic asthenosphere. The asthenosphere, or low-velocity zone, extends from the base of the lithosphere to a depth of about 400 mi (700 km). The brittle lithosphere is broken into a mosaic of internally rigid plates which move horizontally across the Earth's surface relative to one another. Only a small number of major lithospheric plates exist, which grind and scrape against each other as they move independently like rafts of ice on water. Most dynamic activity such as seismicity, deformation, and the generation of magma occur only along plate boundaries, and it is on the basis of the global distribution of such tectonic phenomena that plates are delineated. *See* ASTHENOSPHERE; LITHOSPHERE.

The plate tectonics model for the Earth is consistent with the occurrence of sea-floor spreading and continental drift. Convincing evidence exists that both these processes have been occurring for at least the last 6×10^8 years. This evidence includes the magnetic anomaly patterns of the sea floor, the paucity and youthful age of marine sediment in the ocean basins, the topographic features of the sea floor, and the indications of shifts in the position of continental blocks which can be inferred from paleomagnetic data on paleopole positions, paleontological and paleoclimatological observations, the match-up of continental margins and geological provinces across present-day oceans, and the structural style and rock types found in ancient mountain belts.

Geological observations, geophysical data, and theoretical considerations support the existence of three fundamentally distinct types of plate boundaries, named and classified on the basis of whether immediately adjacent plates move apart from one another (divergent plate margins), toward one another (convergent plate margins), or slip past one another in a direction parallel to their common boundary (transform plate margins). The boundaries of plates can, but need not, coincide with the contact between continental and oceanic crust. The velocity at which plates move varies from plate to plate and within portions of the same plate, ranging between 0.8 and 8 in. (2 and 20 cm) per year.

Not only does plate tectonics theory explain the present-day distribution of seismic and volcanic activity around the globe and physiographic features of the ocean basins such as trenches and mid-oceanic rises, but most Mesozoic and Cenozoic mountain belts appear to be related to the convergence of lithospheric plates. Two different varieties of modern mobile belts have been recognized, cordilleran type and collision type. The Cordilleran range, which forms the western rim of North and South America (the Rocky Mountains, Pacific Coast ranges, and the Andes) have for the most part been created by the underthrusting of an ocean lithospheric plate beneath a continental plate. Underthrusting along the Pacific margin of South America is causing the continued formation of the Andes. The Alpine-Himalayan belt, formed where the collision of continental blocks buckled intervening volcanic belts and sedimentary strata into tight folds and faults, is an analog of the present tectonic situation in the Mediterranean, where the collision of Africa and Europe has begun.

Plate tectonics is considered to have been operative as far back as 2.5×10^9 years. Prior to that interval, evidence suggests that plate tectonics may have occurred, although in a markedly different manner, with higher rates of global heat flow producing smaller convective cells or more densely distributed mantle plumes which fragmented the Earth's surface into numerous small, rapidly moving plates. *See* CONTINENTAL DRIFT; OROGENY. [W.C.P.]

Plateau Any elevated area of relatively smooth land. Usually the term is used more specifically to denote an upland of subdued relief that on at least one side drops off abruptly to adjacent lower lands. In most instances the upland is cut by deep but widely separated valleys or canyons. Small plateaus that stand above their surroundings on all sides are often called tables, tablelands, or mesas. The abrupt edge of a plateau is an escarpment or, especially in the western United States, a rim. [E.H.Ha.]

Platinum A chemical element, Pt, atomic number 78, and atomic weight 195.09. Platinum is a soft, ductile, white noble metal. The platinum-group metals—platinum, palladium, iridium, rhodium, osmium, and ruthenium—are found widely distributed over the Earth. Their extreme dilution, however, precludes their recovery, except in special circumstances.

1	2	3	4	5	6	7	8	9	10	11	12	13	14	15	16	17	18
1 H																	2 He
3 Li	4 Be											5 B	6 C	7 N	8 O	9 F	10 Ne
11 Na	12 Mg											13 Al	14 Si	15 P	16 S	17 Cl	18 Ar
19 K	20 Ca	21 Sc	22 Ti	23 V	24 Cr	25 Mn	26 Fe	27 Co	28 Ni	29 Cu	30 Zn	31 Ga	32 Ge	33 As	34 Se	35 Br	36 Kr
37 Rb	38 Sr	39 Y	40 Zr	41 Nb	42 Mo	43 Tc	44 Ru	45 Rh	46 Pd	47 Ag	48 Cd	49 In	50 Sn	51 Sb	52 Te	53 I	54 Xe
55 Cs	56 Ba	71 Lu	72 Hf	73 Ta	74 W	75 Re	76 Os	77 Ir	78 Pt	79 Au	80 Hg	81 Tl	82 Pb	83 Bi	84 Po	85 At	86 Rn
87 Fr	88 Ra	103 Lr	104 Rf	105 Db	106 Sg	107 Bh	108 Hs	109 Mt	110	*111*	*112*	*113*	*114*	*115*	*116*	*117*	*118*

lanthanide series	57 La	58 Ce	59 Pr	60 Nd	61 Pm	62 Sm	63 Eu	64 Gd	65 Tb	66 Dy	67 Ho	68 Er	69 Tm	70 Yb
actinide series	89 Ac	90 Th	91 Pa	92 U	93 Np	94 Pu	95 Am	96 Cm	97 Bk	98 Cf	99 Es	100 Fm	101 Md	102 No

The platinum-group metals have wide chemical use because of their catalytic activity and chemical inertness. As a catalyst, platinum is used in hydrogenation, dehydrogenation, isomerization, cyclization, dehydration, dehalogenation, and oxidation reactions. *See* CATALYSIS; ELECTROCHEMICAL PROCESS.

Platinum is not affected by atmospheric exposure, even in sulfur-bearing industrial atmospheres. Platinum remains bright and does not visually exhibit an oxide film when heated, although a thin, adherent film forms below 842°F (450°C). Hydrogen or other reducing atmospheres are not directly harmful to platinum at elevated temperatures. Platinum may be worked to fine wire and thin sheet, and by special processes, to extremely fine wire. Important physical properties are given in the table.

Physical properties of platinum

Properties	Value
Atomic weight (^{12}C = 12.00000)	195.09
Crystal structure	Face-centered cubic
Common chemical valence	2.4
Density at 25°C, g/cm^3	21.46
Melting point, °C	1772
Boiling point, °C	3800
Specific heat at 0°C, cal/g	0.0314

Platinum can be made into a spongy form by thermally decomposing ammonium chloroplatinate or by reducing it from an aqueous solution. In this form it exhibits a high absorptive power for gases, especially oxygen, hydrogen, and carbon monoxide. The high catalytic activity of platinum is related directly to this property. Platinum strongly tends to form coordination compounds. *See* CRACKING; HYDROGENATION.

Platinum dioxide, PtO_2, is a dark-brown insoluble compound, commonly known as Adams catalyst. Platinum(II) chloride, $PtCl_2$, is an olive-green water-insoluble solid. Chloroplatinic acid, H_2PtCl_6, is the most important platinum compound. *See* IRIDIUM; OSMIUM; PALLADIUM; RHODIUM; RUTHENIUM. [H.J.A.]

Platyasterida A Paleozoic (Ordovician and Devonian) order of sea stars of the class Asteroidea; only two genera have been recognized. Platyasteridans have relatively elongate flat arms; broad, transversely aligned ventral arm plates; and paxilliform aboral plates. Fundamental differences in ambulacral system construction have been found between modern asteroids and all of the well-known early Paleozoic asteroids, including the platyasteridans. Similarities between platyasteridans and modern asteroids are attributable to evolutionary convergence. *See* ASTEROIDEA; ECHINODERMATA. [D.B.B.]

Platycopida An order of the Podocopa (Ostracoda) containing a single family, Cytherellidae, and two extinct superorders. Platycopids are small marine or brackish-water ostracods with an asymmetrical shell that is oblong, nearly rectangular, and laterally compressed; their six pairs of appendages are adapted for burrowing and filter feeding.

Species are dimorphic. The female carapace is wider near the rear and longer than that of the male, providing a brood chamber behind the body. All living species are benthic, and fossil forms probably were also. Some species live at great depths in the Atlantic Ocean, but others have been found in the Arabian, Caribbean, and Mediterranean seas in shallower water. *See* CRUSTACEA; OSTRACODA; PODOCOPA. [P.A.McL.]

Platyctenida An order of the phylum Ctenophora comprising four families (Ctenoplanidae, Coeloplanidae, Tjalfiellidae, Savangiidae) and six genera. All species are highly modified from the planktonic ctenophores. The platyctenes are fairly small (1–6 cm or 0.4–2.4 in.) and brightly colored. They have adopted a variety of swimming, creeping, and sessile habits, with concomitant morphological changes and loss of typical ctenophoran characteristics. The body is compressed in the oral-aboral axis, and the oral part of the stomodeum is everted to form a creeping sole. Sexual reproduction involves internal fertilization in many species, with retention of the developing cydippid larvae in brood pouches. Some species also reproduce asexually by fission. Most platyctenids are found in tropical coastal waters, where many are ectocommensals on benthic organisms. *See* CTENOPHORA. [L.P.M.]

Platyhelminthes A phylum of the invertebrates, commonly called the flatworms. They are bilaterally symmetrical, nonsegmented, dorsoventrally flattened worms characterized by lack of coelom, anus, circulatory and respiratory systems, and exo- or endoskeleton. They possess a protonephridial excretory system, a complicated hermaphroditic reproductive system, and a solid mesenchyme which fills the interior of the body (see illustration). Three classes occur in the phylum: (1)

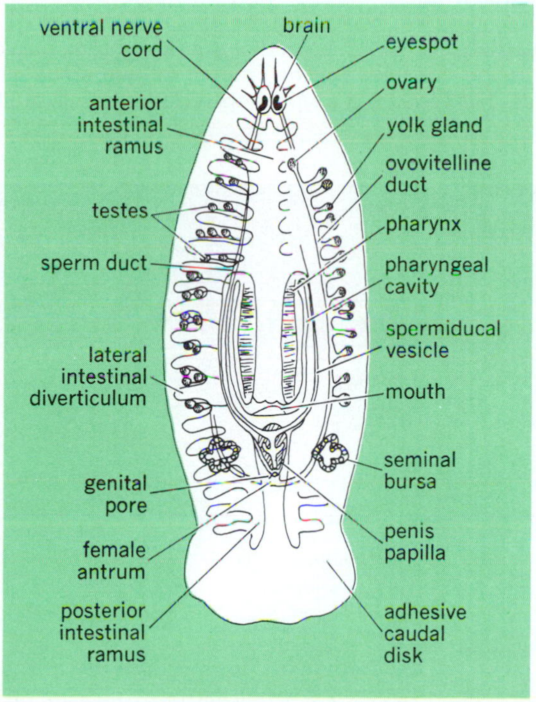

***Bdelloura candida* (Tricladida), ectocommensal on the king crab, *Limulus*. Complete digestive and male systems shown on left, female systems on right.**

the Turbellaria, mainly free-living, predacious worms; (2) the Trematoda, or flukes, holozoic ecto- or endoparasites; and (3) the Cestoda, or tapeworms, saprozoic endoparasites in the enteron of vertebrates, whose larvae are found in the tissues of invertebrates or vertebrates.

Turbellaria are widespread in fresh water and the littoral zones of the sea, while one group of triclads occurs on land in moist habitats. Adult trematodes occur on, or in, practically all tissues and cavities of the vertebrates on which they feed. They are responsible for troublesome diseases in humans and animals. Larval flukes are frequent in mollusks, mainly gastropods, and occasionally occur in pelecypods. Vector hosts, such as insects and fish, are often interpolated between mollusk and vertebrate. Adult tapeworms, living in the enteron or the biliary ducts, compete with the host for food and accessory food factors such as vitamins. Larval tapeworms reside chiefly in arthropods, but larvae of one group, the Cyclophyllidea, develop in mammals, which may be severely impaired, or even killed, by the infection. *See* CESTODA; TAPEWORM DISEASE; TREMATODA; TURBELLARIA. [C.G.G.]

Platypus The single species *Ornithorhynchus anatinus* of the family Ornithorhynchidae in the order Monotremata, which occurs in eastern Australia and Tasmania. This mammal

The platypus, a primitive aquatic mammal, showing the webbed digits which aid in swimming.

is a monotreme, which lays eggs and incubates them in a manner similar to birds. The platypus is also known as the duckbill, duckmole, and water mole. One of the most primitive mammals, it retains some reptilian characteristics. The female lacks a marsupium; however, marsupial or epipubic bones are well developed.

Ornithorhynchus is well adapted to aquatic life. It ranges from the tropical, sea-level streams to cold lakes at altitudes of 6000 ft (1800 m). It is covered with short dense fur, typical of many aquatic mammals; the external ears are lacking; and the tail is broad and flattened, resembling that of the beaver (see illustration). The rubbery bill is covered with a highly sensitive skin. There are no teeth, but horny ridges are used for crushing food and grubbing on the bottom of rivers. *See* MONOTREMATA. [C.B.C.]

Playa The low, essentially flat part of a basin or other undrained area in an arid region. In heavy rains the playa may be temporarily covered with a shallow sheet of water and is then a playa lake.

Five principal types of playa have been recognized. The dry playa develops where the water table is below capillary reach of surface; its surface is hard, flat, and smooth, composed of silt and clay. The moist playa or salina occurs where the water table is within capillary reach of surface; this type may be further subdivided into (1) salt-encrusted playa, where the water table is at the surface or so near it that salt water evaporating on the surface leaves a salt crust (see illustration); and (2) clay-encrusted playa, where the water table is near the limit of capillary action and salt brought to the surface is mixed with silt and clay, forming a puffy surface (self-rising ground). The crystal-body playa is essentially a massive body of crystalline salt at

Light-colored saline deposits encrust lowest parts of the basin or bolson in Death Valley, California.

or very close to the surface. The compound playa results from a water table at different levels in different parts; these exhibit characteristics of dry playa in one part and moist playa in another. The lime-pan playa is formed in basins receiving drainage from limestone terrain and has a floor of hard travertine. *See* SALINE EVAPORITE. [T.C.]

Plecoptera An order of primitive insects known as the stoneflies. Except for wings and tracheal gills, there are relatively slight differences between aquatic immature stages and adult mature ones. The soft and somewhat flattened body, strong legs, paired tarsal claws, chewing mouthparts, and rusty blacks, dull yellows, and browns are characteristic of both immature and adult stages (see illustration).

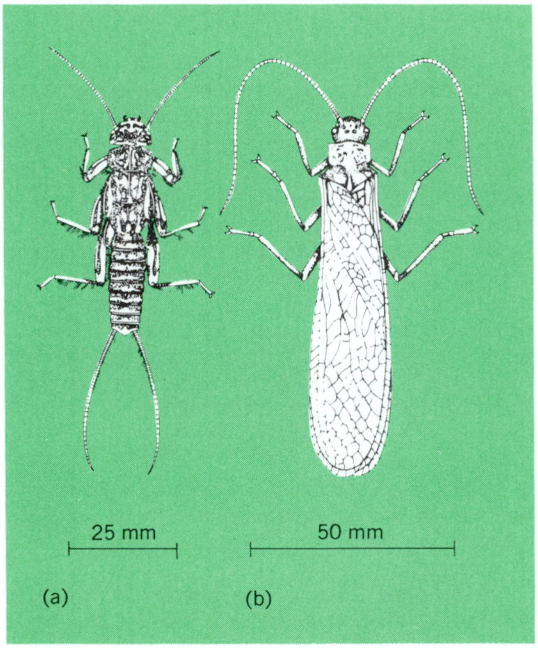

Plecoptera. (*a*) Nymph of *Perla*. (*b*) Adult of *Pteronarcys*. (*After A. H. Morgan, Field Book of Ponds and Streams, Putnam, 1930*)

Immature stoneflies live in the rapid, stony parts of clean, swift streams, although a few species occur along the rocky shores of large temperate lakes.

Adult stoneflies hold the wings close to the backs when at rest or walking. The hindwings are pleated and hidden. The name Plecoptera means pleated wings.

In various parts of North America stonefly adults emerge in every month of the year. Successions of species reach their peaks of abundance from November to March, with some extending to August. Nymphs and adults of some species are plant feeders; others are carnivorous. *See* INSECTA. [L.Ber.]

Plectomycetes A class of the subdivision Ascomycotina. The Plectomycetes produce a well-developed mycelium on which both sexual (asci) and asexual (conidia) states occur. The asci form in association with small ascocarps of varying complexity. These ascocarps arise from special coiled or intertwined hyphal branches that are sometimes differentiated into male (antheridium) and female (ascogonium) structures. The ascocarps may consist of a loose grouping of hyphae around the asci forming a gymnothecium, or may be spherical, with a definite wall or peridium that lacks an opening, forming a cleistothecium. The asci are usually spherical, are scattered throughout the ascocarp, and have thin walls that dissolve at maturity. The ascospores are small, colorless or sometimes pigmented, round in face view, and somewhat flattened in side view, and may be ornamented or flanged.

The two recognized orders, Onygenales and Eurotiales, are differentiated by the structure of the ascocarp and the type of asexual conidia produced. In the Onygenales, the ascocarp usually is not well developed, but many species form characteristic peridial hyphae. These fungi occur in soil, on dung, and on plant debris. In the Eurotiales, the ascocarps have a definite wall (peridium) enclosing the asci, and they are often brightly colored. The Eurotiales are common in soil and on plant materials. Some plectomycetes are pathogenic. *See* ASCOMYCOTINA; EUMYCOTA; FUNGI. [R.T.Ha.]

Pleiades A beautiful group of stars resembling a little dipper, in the constellation of Taurus, known since earliest records. The Pleiades is a typical open cluster (see illustration).

The Pleiades. (*Lick Observatory photograph*)

Its distance is 420 light-years (1 ly = 9.46×10^{12} km), its linear diameter about 15 light-years, its age 5×10^7 years. The cluster is permeated with diffuse nebulosity. Though early accounts refer to the Pleiades in terms of seven stars, only six are now conspicuous to the unaided eye, which raises a theory that one, the lost Pleiad, has faded. [H.S.H.]

Pleistocene A time-stratigraphic term referring to the sequence of geologic deposits (the Pleistocene Series) and a geologic-time term (the Pleistocene Epoch of the Quaternary Period) for the interval of time during which these sediments were deposited.

The lack of an agreed-upon definition has led to a concerted international effort to establish a stratotype section with a fixed lower boundary based on biostratigraphical data. The section will be located in the area of marine sediments in the province of Calabria in southern Italy.

The Pleistocene was characterized by numerous (at least 17), cyclic, worldwide changes of climate. These changes had a

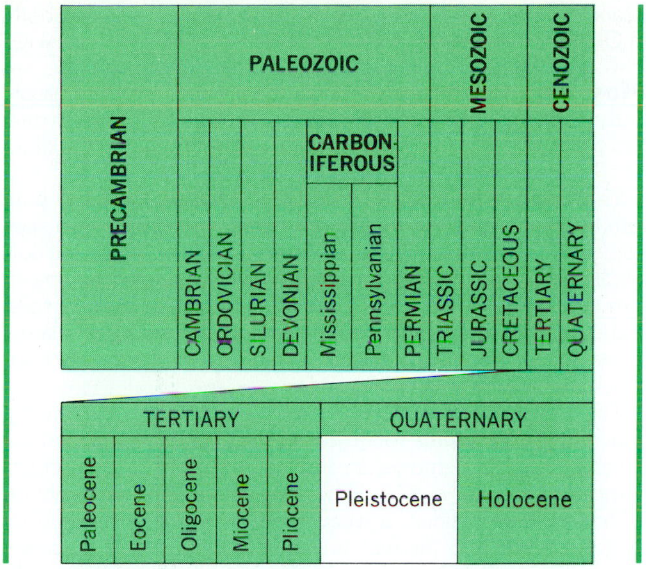

profound effect on the fauna and flora of the time, and on surficial geologic processes. Surficial sediments and landforms in many areas of the world date primarily from this interval of time. *See* GLACIAL EPOCH; QUATERNARY. [W.H.J.]

Pleochroic halos Small halos of color or color differences that are sometimes observed around inclusions in minerals. If halos occur in doubly refracting substances, they may show pleochroic discoloration, and the term pleochroic halos is loosely applied to such small colored halos generally. *See* PLEOCHROISM.

The halos are found only around minute inclusions of certain minerals, especially zircon, allanite, monazite, and others known to contain minor amounts of uranium or thorium. Pleochroic halos are most widespread in minerals such as biotite, which is never known in the metamict state. Halos have been reported in many rock-making minerals, including amphiboles, pyroxenes, and micas. *See* RADIOACTIVE MINERALS.

The halos are usually spheroidal (circular, as seen in microscopic section), sometimes consisting of several concentric rings, and are fairly sharply bounded. [P.M.H.]

Pleochroism In some colored transparent crystals, the effect wherein the color is quite different in different directions through the crystals. In such a crystal the absorption of light is different for different polarization directions. In colored transparent tourmaline the effect may be so strong that one polarized component of a light beam is wholly absorbed, and the crystal can be used as a polarizer. *See* DICHROISM; TRICHROISM. [B.H.Bi.]

Plesiosauria An order of extinct aquatic diapsid reptiles within the infraclass Sauropterygia, common throughout the world during the Jurassic and Cretaceous periods. These large carnivores are characterized by long, paddle-shaped limbs and short, dorsoventrally compressed bodies. Unlike the nothosaurs (more primitive members of the Sauropterygia), plesiosaurs have greatly expanded ventral portions of both pectoral and pelvic girdles to provide large areas for the attachment of muscles to move the limbs anteriorly and posteriorly. Lateral undulation of the trunk was severely restricted by elaboration of the ventral scales.

Approximately 40 genera of plesiosaurs are recognized, divided into two large groups: The plesiosauroids include the most primitive genera and others that have small heads and very long necks. Among the plesiosauroids, the elasmosaurids had as many as 76 cervical vertebrae. The pliosauroids had shorter necks but larger skulls. Both elasmosaurs and pliosaurs persisted until the end of the Mesozoic. *See* DIAPSIDA; NOTHOSAURIA; REPTILIA; SAUROPTERYGIA. [R.L.C.]

Pleural diseases Primary pleural diseases are relatively infrequent. However, involvement of the pleura following or in conjunction with some other underlying disease, most often pulmonary, is common. Pleural adhesions may be found in as many as two-thirds of autopsies in a general hospital. Types of pleural disorders are inflammatory conditions, noninflammatory conditions, and tumors.

Pleuritis, or pleurisy, will have different characteristics depending on the underlying cause and the stage of the disease. Most instances of pleuritis accompany or follow inflammatory diseases of the lungs, including tuberculosis, bacterial pneumonia, lung infarcts, septicemia, rheumatic fever, and uremia. An important reaction is the appearance of fluid in the pleural cavity, but it is important to differentiate between the fluid secondary to true inflammation of the pleura (exudate) and that following cardiac failure (transudate).

Hemorrhagic pleuritis is characterized by the presence of bloody fluid and may be encountered in cases of hemorrhagic diseases or tumor involvement.

Hydrothorax is the accumulation of serous fluid in the pleural cavity, often bilaterally. Most cases are related to heart failure. Hemothorax is the accumulation of bloody fluid in the pleural cavity. Whether caused by traumatic injury or spontaneous rupture of an aneurysm, this is a serious emergency. Chylothorax is the accumulation of milky fluid, usually of lymphatic origin. The condition may be due to trauma, tumor obstruction of lymphatics, or congenital anomalies. Pneumothorax is the accumulation of air or gas in the pleural cavity. Spontaneous pneumothorax may follow the rupture of alveoli or bullae and is usually associated with emphysema, asthma, or tuberculosis. Traumatic pneumothorax is a feared complication of penetrating chest wounds. [V.E.G.]

Pleuronectiformes One of the most distinctive orders of actinopterygian fishes, also called Heterosomata, composed of the flatfishes: halibut, plaice, flounders, soles, tonguesoles, and their allies. The striking feature of the group is the loss of bilateral symmetry (see illustration), a characteristic of almost all vertebrates. Like the ancestors of pleuronectiforms, young flatfishes are symmetrical and swim upright; early in life, however, one eye migrates across the top of the skull to lie on the same side as the other eye. This transformation is associated with deformation of the skull bones and nerves, a change in position so that the fish lies on one (the blind) side, partial or complete depigmentation of the blind surface, and sometimes modification and development of asymmetry in paired fins, dentition, squamation, visceral anatomy, and other structures. *See* ACTINOPTERYGII; FLATFISH. [R.M.B.]

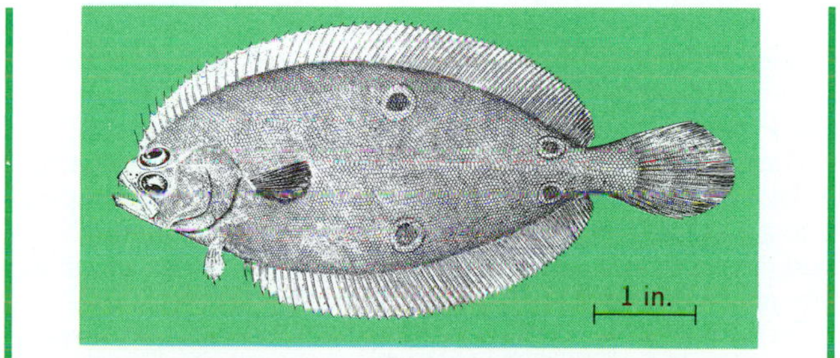

Fourspot flounder (*Paralichthys oblongus*), of the Pleuronectiformes. 1 in. = 2.5 cm. (*After G. B. Goode, Fishery Industries of the United States, sect. 1, 1884*)

Pleuropneumonia-like organism (PPLO) The nature of these organisms and their classification are uncertain. They may represent a special growth form of bacteria, which they resemble in many respects. At present they are classified in the separate order Mycoplasmatales in the class Mollicutes. *See* MOLLICUTES. [L.L.D.]

Pliocene The youngest of the five major worldwide divisions (epochs) of the Tertiary Period (Cenozoic Era), extending from the end of the Miocene to the beginning of the Pleistocene (Quaternary). Pliocene also denotes the series of rocks and their fossils formed during this epoch. Pliocene rocks

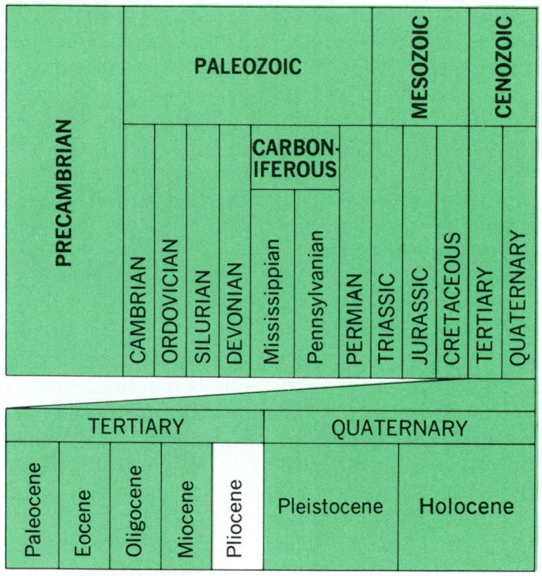

include most normal sedimentary types. Those of a mechanical (clastic) rather than chemical or biogenetic origin are the more common. High surface relief was characteristic of Pliocene times. Marine deposits are patchy though often thick. *See* CENOZOIC; MIOCENE; PLEISTOCENE; TERTIARY.

In its broadest usage, Pliocene time encompassed two intervals of widespread orogenic movements separated by an interval of general marine expansion. During Pliocene time the ocean withdrew from eastern Europe and Turkestan, never to return. Remnants were left such as the Aral, Caspian, and Black seas. A later Pliocene North Sea retreated progressively northward.

Through the Pliocene worldwide temperatures dropped steadily. By mid-Pliocene time world climates had attained temperatures essentially those of today though generally more arid. Since then climates have become colder (peaks of Ice Age glaciation), then again warmer than those prevalent during historic time.

Pliocene marine shellfish look modern. Equatorward expansion of bivalves, gastropods, and foraminiferids (especially cassidulinids) having cold-water environmental optima was general. Pliocene mammals were larger than those in preceding epochs. Grasslands were widespread. Mio-Pliocene land emergences permitted intercontinental interchanges of mammals before endemic evolution on land set in. Afro-Asian Pliocene seaways (Zanzibar, Red Sea, Persia, Mekran coast) semi-isolated much of the earlier Pliocene fauna in Africa. Late Pliocene and early Pleistocene land reconnections permitted further intercontinental mammal interchanges; northern forms went south, others migrated to the newly reconnected South America, while many distinctive South American mammals died out. Much of the climacteric Plio-Pleistocene mammal fauna remained in the modern Oriental Region, but north and northwest of the Himalayas selective scouring during the subsequent Ice Age was rigorous, and this fauna was modified both in Old World and New. [R.M.Kl.]

Plowing Cutting and turning a furrow with a plow is usually the first and most effective operation of tillage. There are two types of plow, the moldboard (see illustration) and the disk. *See* TILLAGE.

The working part of the moldboard plow is the plow bottom. The shape of the moldboard may be described as a section cut from a warped cylinder, on the inside of which the soil spirals. Moldboard plows have largely been adapted to power farming by the selection of bottoms suitable for certain soils and combination of these bottoms into gangs sufficiently large to use the power of the tractor at the approximate speed for which the bottoms were designed.

The effectiveness of plowing may be materially increased by attachments. A jointer, or a jointer and coulter combination, which cuts a small furrow in front of the plow shin, permits complete coverage of sod. Where there are large amounts of rubbish or crop residue, a weed chain or wire can be used to drag the debris into the furrow and hold it there until covered. A furrow wheel may reduce land-side friction, and a depth gage on the beam helps secure uniform depth. A modified form of plow bottom, called a lister, is in effect a right and a left moldboard joined at the shin so as to throw soil both to the right and to the left. This produces a furrow or trough, called a list, in which seed is planted. Because it concentrates rainfall in the furrow, this method is used largely in areas of light rainfall.

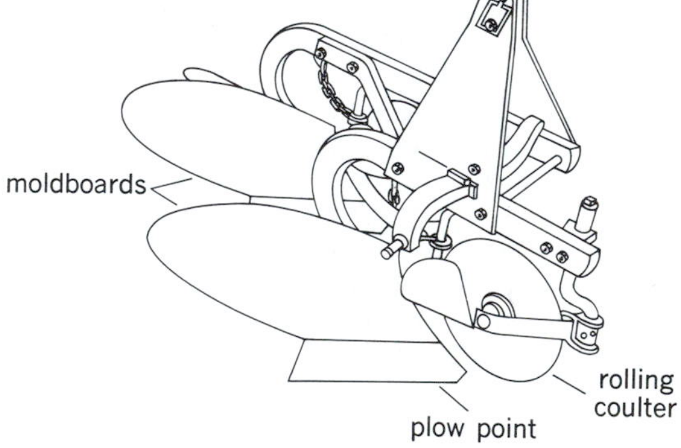

One type of moldboard plow. ***(Tractor and Implement Operations—North America, Ford Motor Co.)***

The disk plow consists of a number of disk blades attached to one axle or gang bolt. This plow is used for rapid, shallow plowing. In fields where numerous rocks and roots are present, the disk plow, which rolls over obstacles, is substituted for the moldboard. The disk is also used for sticky soils that will not scour on a moldboard. [M.L.N.]

Plug gage A cylinder of steel whose diameter is ground, then honed or lapped to gage tolerance, a plug gage is used for checking the diameter of a cylindrical cavity such as a bolt hole or bearing bore. A single plug gage controls one limit of tolerance of an inside diameter. [R.A.Bo.]

Plum Any of the smooth-skinned stone fruits grown on shrubs or small trees. Plums are widely distributed in all land areas of the North Temperate Zone, where many species and varieties are adapted to different climatic and soil conditions.

There are four principal groups: (1) Domestica (*Prunus domestica*) of European or Southwest Asian origin, (2) Japanese or Salicina (*P. salicina*) of Chinese origin, (3) Insititia or Damson (*P. insititia*) of Eurasian origin, and (4) American (*P. americana* and *P. hortulana*). The Domesticas are large, meaty, prune-type plums. A prune is a plum which dries without spoiling. [R.P.L.]

Plumbaginales An order of flowering plants, division Magnoliophyta (Angiospermae), in the subclass Caryophyllidae of the class Magnoliopsida (dicotyledons). The order consists of only the family Plumbaginaceae, with about 350 species. The plants are herbs or less often shrubs. The flowers are strictly pentamerous; that is, each floral whorl has five members; the petals are fused and all alike in shape and size. The family contains a few garden ornamentals, such as species of *Armeria*, known as thrift or sea-pink. See CARYOPHYLLIDAE; MAGNOLIOPSIDA. [A.Cr.]

Pluto The most distant known planet in the solar system. Pluto was discovered in 1930.

Pluto is visible only through fairly large telescopes. Periodic variations in its brightness demonstrate that the surface of Pluto is covered with bright and dark markings and indicate a period of rotation of 6.3 days. This long period and the high inclination and eccentricity of Pluto's orbit around the Sun have led several scientists to propose that Pluto was originally a satellite of Neptune.

Pluto has a small satellite, provisionally named Charon, discovered in 1978, in a synchronous orbit; that is, the satellite's period is identical to the rotational period of the planet. This discovery made possible a determination of Pluto's mass, which was found to be about one-seventh that of the Earth's Moon. With a diameter in the range of 1900–2500 mi (3000–4000 km) Pluto must have a mean density close to 1 g/cm^3. Pluto must be composed largely of ices, a conclusion that gains support from the detection of an infrared absorption band of solid methane in its spectrum. The average temperature of Pluto must be close to 50 K (−370°F), which is consistent with the presence of solid methane on the planet's surface, in equilibrium with a small amount of methane gas. See PLANETARY PHYSICS. [T.C.O.]

Pluton A rock body formed by the consolidation of magma (molten rock) below the surface of the Earth, but now exposed due to erosion. A rather elaborate terminology based on the supposed three-dimensional shape of plutons has evolved. But because of the difficulty in determining the real shape, the modern tendency is to use the general term pluton, except for relatively small bodies. See BATHOLITH; LACCOLITH. [M.P.B.]

Plutonium A chemical element, Pu, atomic number 94. Plutonium is a reactive, silvery metal in the actinide series of elements. The principal isotope of chemical interest is ^{239}Pu, with a half-life of 24,131 years. It is formed in nuclear reactors

1	2	3	4	5	6	7	8	9	10	11	12	13	14	15	16	17	18
1 H																	2 He
3 Li	4 Be											5 B	6 C	7 N	8 O	9 F	10 Ne
11 Na	12 Mg											13 Al	14 Si	15 P	16 S	17 Cl	18 Ar
19 K	20 Ca	21 Sc	22 Ti	23 V	24 Cr	25 Mn	26 Fe	27 Co	28 Ni	29 Cu	30 Zn	31 Ga	32 Ge	33 As	34 Se	35 Br	36 Kr
37 Rb	38 Sr	39 Y	40 Zr	41 Nb	42 Mo	43 Tc	44 Ru	45 Rh	46 Pd	47 Ag	48 Cd	49 In	50 Sn	51 Sb	52 Te	53 I	54 Xe
55 Cs	56 Ba	71 Lu	72 Hf	73 Ta	74 W	75 Re	76 Os	77 Ir	78 Pt	79 Au	80 Hg	81 Tl	82 Pb	83 Bi	84 Po	85 At	86 Rn
87 Fr	88 Ra	103 Lr	104 Rf	105 Db	106 Sg	107 Bh	108 Hs	109 Mt	110	111	112	113	114	115	116	117	118

lanthanide series	57 La	58 Ce	59 Pr	60 Nd	61 Pm	62 Sm	63 Eu	64 Gd	65 Tb	66 Dy	67 Ho	68 Er	69 Tm	70 Yb
actinide series	89 Ac	90 Th	91 Pa	92 U	93 Np	94 Pu	95 Am	96 Cm	97 Bk	98 Cf	99 Es	100 Fm	101 Md	102 No

by the process shown in the following reaction. Plutonium-239 is fissionable, but may also capture neutrons to form higher plutonium isotopes.

$$^{238}\mathrm{U} + n \longrightarrow {}^{239}\mathrm{U} \xrightarrow[23.5\ \mathrm{min}]{\beta^-} {}^{239}\mathrm{Np} \xrightarrow[2.33\ \mathrm{days}]{\beta^-} {}^{239}\mathrm{Pu}$$

Plutonium-238, with a half-life of 87.7 years, is utilized in heat sources for space application, and has been used for heart pacemakers. Plutonium-239 is used as a nuclear fuel, in the production of radioactive isotopes for research, and as the fissile agent in nuclear weapons. See NUCLEAR FUELS; NUCLEAR REACTOR.

Plutonium exhibits a variety of valence states in solution and in the solid state. Plutonium metal is highly electropositive. Numerous alloys of plutonium have been prepared, and a large number of intermetallic compounds have been characterized.

Reaction of the metal with hydrogen yields two hydrides. The hydrides are formed at temperatures as low as 150°C (300°F). Their decomposition above 750°C (1400°F) may be used to prepare reactive plutonium powder. The most common oxide is PuO_2, which is formed by ignition of hydroxides, oxalates, peroxides, and nitrates of any oxidation state in air of 870–1200°C (1600–2200°F). A very important class of plutonium compounds are the halides and oxyhalides. Plutonium hexafluoride, the most volatile plutonium compound known, is a strong fluorinating agent. A number of other binary compounds are known. Among these are the carbides, silicides, sulfides, and selenides, which are of particular interest because of their refractory nature.

Because of its radiotoxicity, plutonium and its compounds require special handling techniques to prevent ingestion or inhalation. Therefore, all work with plutonium and its compounds must be carried out inside glove boxes. For work with plutonium and its alloys, which are attacked by moisture and by atmospheric gases, these boxes may be filled with helium or argon. See ACTINIDE ELEMENTS; NEPTUNIUM; NUCLEAR CHEMISTRY; TRANSURANIUM ELEMENTS; URANIUM. [F.We.]

Plywood A wood product in which thin sheets of wood are glued together, grains of adjacent sheets being at right angles to each other in the principal plane. Because of this cross-grained orientation, mechanical properties of plywood are less directional than are those of natural lumber. Originally developed to replace wide wood boards that had been sawed directly from large logs, plywood was manufactured from logs not suitable for other purposes. It has since found such widespread use that tree farms are now cultivated specifically to yield logs suitable for processing into sheets for plywood.

Plywood with waterproof glue is designated exterior type; it is also used interiorly where moisture is present. Plywood with moisture-resistant glue is designated interior type; it can withstand an occasional soaking but neither repeated soakings nor continuous high humidity.

Most commonly used plywoods are ¼-in. (6.35 mm) sanded interior paneling or ½-in. (12.70 mm) exterior grade plywood sheeting. Other standard thicknesses extend to 1 in. (25.40 mm) for interior types and to 1⅛ in. (28.58 mm) for exterior types. The most common panel size is 4 × 8 ft (122 × 244 mm); larger sizes are manufactured for such special purposes as boat hulls. See WOOD PRODUCTS. [F.H.R.]

Pneumatolysis The alteration of rocks by the action of magmatic gases. The gases accompany magmatic intrusions and permeate the intruded rocks along fissures and other lines of least resistance. See METAMORPHISM; METASOMATISM.

Primary magmatic gases are acid and therefore react readily with limestone to form skarn rocks. Appreciable amounts of heavy metals (usually present as chlorides, fluorides, or sulfides) are associated with the magmatic gases. These are captured by the limestone and retained as deposits of skarn rocks or ores. In other places the heavy metals are deposited in granite or schist, causing the formation of greisen. *See* GREISEN; SKARN. [T.F.W.B.]

Pneumococcus The major causative microorganism (*Streptococcus pneumoniae*) of lobar pneumonia. Pneumococci occur singly or as pairs or short chains of oval or lancet-shaped cocci, flattened at proximal sides and pointed at distal ends. A capsule of polysaccharide envelops each cell or pair of cells. The organism is nonmotile and stains gram-positive unless degenerating.

Pneumococci have been isolated from the upper respiratory tract of healthy humans, monkeys, calves, horses, and dogs. Epizootics of pneumococcal infection have been described in monkeys, guinea pigs, and rats but are not the source of human infection. In humans, pneumococci may be found in the upper respiratory tract of nearly all individuals at one time or another. Following damage to the epithelium lining the respiratory tract, they may invade the lungs. They are the principal cause of lobar pneumonia in humans and may cause also pleural empyema, pericarditis, endocarditis, meningitis, arthritis, peritonitis, and infection of the middle ear. Approximately one of four cases of pneumococcal pneumonia is accompanied by invasion of the bloodstream by pneumococci, producing bacteremia. The mortality of untreated pneumococcal infection is high but has been reduced significantly by the use of antibiotics. A polyvalent vaccine of 14 pneumococcal capsular polysaccharides is available for the prevention of pneumococcal infection caused by the more virulent pneumococcal types in those segments of the population at high risk of fatal infection. [R.Au.]

Pneumocystosis A disease resulting from infection with protozoa of the genus *Pneumocystis*. The organisms grow in the lungs and form cysts. Pneumocystosis became important when outbreaks of a disease were observed in premature infants who developed cough, progressive difficulty in breathing, and x-ray changes in both lung fields.

The organism appears to be transmitted from person to person. With refinements in management of patients, the number of cases has declined. [J.K.F.]

Pneumonia An acute or chronic inflammation of the lung tissues, occurring in humans and in many animals. Pneumonia in humans may be caused by numerous microbial, immunological, physical, or chemical agents. It may also be associated with many systemic diseases. Any or all parts of the lung may be involved, with the inflammatory exudate filling the alveolar air spaces of one or more lobes, as in lobar pneumonia, or the smaller segments, as in lobular pneumonia. It may be disposed in and around the bronchi, as in bronchopneumonia, or may be limited predominantly to the interalveolar areas, as in interstitial pneumonia. Pneumonia may be a primary disease or a secondary event in other diseases.

The bacterial pneumonias may sometimes cause destruction of parts of the lung by abscess formation or may involve the pleura in an empyema or sterile effusions. Inflammation and dilation of bronchi (bronchiectasis) and spread of infection to remote organs is not unusual. Fortunately, however, most common pneumonias heal by resolution, leaving lung structure and function little changed. Viral pneumonias may have superimposed bacterial infection. [M.Fi.]

Pneumonitis An acute infectious disease of the respiratory tract in which localized air-sac infiltration of varying degree with fluid and a diversity of cells is a prominent feature. The most common causative agent in humans is a *Mycoplasma* pneumonia—the so-called Eaton agent of pneumonia. The disease has come to be known as primary atypical pneumonia. *See* PNEUMONIA. [K.F.M.]

Podicipediformes The order of birds (sometimes Podicipitiformes) containing the single family Podicipedidae, the grebes. Their legs are set far back on the body, with compressed, bladelike tarsi. The toes of grebes are not webbed, however, but are individually broadened and lobed. The plumage is dense and silky, and the tail is rudimentary. The larger species feed principally on fish, while the smaller ones eat a high proportion of aquatic invertebrates. Many grebes have elaborate and often spectacular aquatic courtship displays. The nests are built in water, and may be built up from the bottom, anchored to emergent vegetation, or even floating. The family as a whole is virtually cosmopolitan. *See* AVES. [K.C.P.]

Podocopa One of two Recent subclasses of Ostracoda. The two may share many characteristics but differ significantly in others. Shared characteristics include a carapace (shell) with a straight-to-concave ventral margin and somewhat convex dorsal margin; adductor muscle-scar patterns varying from numerous to few; the absence of an anterior notch in the bivalve shell; and the absence of lateral eyes, heart, and frontal organ (Bellonci organ). Podocopid ostracods are small, rarely exceeding 0.25 in. (7 mm). The carapace valves are unequal, weakly or strongly calcified, and may be heavily ornamented. The hinge varies from simple to complex. Six or seven pairs of appendages are present; the seventh is adapted for walking, for cleaning, or as a clasping leg in males. The subclass is represented by the two extant orders, Platycopida and Podocopida.

Podocopid ostracods are found in fresh-water, marine, and brackish-water environments, and a few are terrestrial. The majority are burrowing or crawling inhabitants of the benthos and are filter or detritus feeders or herbivores. Some are capable of swimming, but none appear to ever lead a planktonic existence. *See* OSTRACODA. [P.A.McL.]

Podocopida An order of the subclass Podocopa, class Ostracoda, that consists of the extant suborder Podocopina and the extinct Metacopina. There is agreement on the assignment of the superfamilies Cypridoidea, Bairdioidea, Cytheroidea, and Darwinuloidea to the Podocopida; however, some researchers also include the Sigillioidea, and others consider the Terrestrichtheroidea a podocopid taxon of equivalent rank. In all podocopids the two valves fit firmly, hermetically sealing the animal inside when closed.

Appendages include antennules, antennae, mandibles, maxillules, and three pairs of thoracic legs. Antennules and antennae are both used in locomotion; swimming forms are provided with long, feathered setae, whereas crawling forms have only short, stout setae. Mandibles are strongly constructed and, with rare exceptions, used for mastication. The maxilla has setiferous endites and bears a large branchial plate to circulate water alongside the body. The furca is variously developed but never lamelliform. If eyes are present, the two lateral eyes and the median eye are joined into one central structure. No heart is developed. *See* CRUSTACEA; PODOCOPA. [P.A.McL.]

Podostemales An order of flowering plants, division Magnoliophyta (Angiospermae), in the subclass Rosidae of the class Magnoliopsida (dicotyledons). The order consists of only the family Podostemaceae, with about 200 species, the greatest number occurring in tropical America. They are submerged aquatics with modified, branching, often thalluslike shoots, and small, perfect flowers with a much reduced perianth. *See* MAGNOLIOPSIDA; ROSIDAE. [A.Cr.]

Poecilosclerida An order of sponges of the class Demospongiae in which the skeleton includes two or more types of megascleres, each localized in a particular part of the sponge colony. Frequently one type of megasclere is restricted to the dermis and another type occurs in the interior of the sponge. Sometimes one category is embedded in spongin fibers while a second category, usually spinose, protrudes from the fibers at right angles. Spongin is always present but varies in amount from species to species. Microscleres are usually present; often several types occur in one species. A wide variety of microsclere categories is found in the order but asters are never present. *See* DEMOSPONGIAE. [W.D.H.]

Poecilostomatoida One of two major orders of parasitic Copepoda that were previously included in the Cyclopoida. The classification of parasitic copepods has been established on the basis of the structure of the mouth. In poecilostomatoids the mouth is represented by a transverse slit, partially covered by the overhanging labrum, which resembles an upper lip. Although there is variability in the form of the mandible among poecilostomatoids, it can be generalized as being falcate. Body segmentation is typically podoplean, having prosome-urosome articulation between the fifth and sixth thoracic somites; however, this segmentation is often lost with the molt to adulthood. The antennules frequently are reduced in size and the antennae modified to terminate in small hooks or claws that are used in attachment to host organisms.

Most poecilostomatoid copepods are ectoparasites of marine fishes or invertebrates, usually attaching to the external surface of the host or in the branchial cavity on the walls or gill surfaces. Representatives of one family, however, have successfully made the transition to fresh-water habitats, and a second family has evolved an endoparasitic mode of life. *See* COPEPODA; CRUSTACEA; CYCLOPOIDA. [P.A.McL.]

Pogonophora The beard worms—a phylum of sedentary marine worms living in cool waters of all the world's oceans, generally at depths between 330 and 13,200 ft (100 and 4600 m), shallower at higher latitudes and deeper in trenches. Often described as "the twentieth-century phylum," they were first dredged early in the century and thoroughly investigated only in the 1950s.

Pogonophorans construct a tube, and are the only nonparasitic metazoans to have no trace of a mouth, gut, or anus. These are long, slender worms, the diameter in most being less than 0.04 in. (1 mm) and the length being over 100 times the diameter. Superficially the tubes remind one of corn silk or coarse thread, but most have a characteristic banding pattern of brown or yellow pigments. The larger tubes are sometimes rigid, thicker, and dark-colored.

These worms absorb their nutrients through the pinnules and microvilli of tentacles without the aid of digestive enzymes. This is accomplished against concentration gradients and may be supplemented by limited pinocytosis and phagocytosis by tentacular surfaces in a few species.

The sexes are separate, with the gonopore location being the only sexual dimorphism. Fertilization has not been observed but must occur within the maternal tube, as fertilized eggs and developing larvae have been found there. [E.B.Cu.]

Point In axiomatic geometry, usually a completely undefined (primitive) element, although there are axiomatizations of geometry in which those properties of "point" that are desired are given by postulates. To illustrate: In a geometry in which line is a primitive element, points may be defined as classes of lines that conform to certain requirements (postulates). In n-dimensional metric analytic point geometry, a point may be defined as an ordered n-tuple of numbers. *See* ANALYTIC GEOMETRY; EUCLIDEAN GEOMETRY. [L.M.Bl.]

Point-contact diode A semiconductor rectifier using the barrier formed between a specially prepared semiconductor surface and a metal point to produce the rectifying action. The contact is usually maintained by mechanical pressure, but in some instances it may be welded or bonded. The rectifying action implies that the resistance of the contact is significantly greater for one direction of applied voltage (reverse direction) than for the other (forward direction). *See* SEMICONDUCTOR RECTIFIER.

Point-contact diodes have been widely used in radio and television, and most notably in computers, microwave detectors, and ultrahigh-frequency mixers. Today their use is very small, having been usurped by junction diodes and Schottky barrier diodes. [L.P.H.]

Point-contact transistor A transistor in which the emitter and collector consist of metal point contacts closely spaced on the surface of a block of semiconductor. The usual configuration is with both points on the same surface and about 2 mils (50 mm) apart, although good devices have also been made with the points on opposite sides of a thin wafer of semiconductor. This type of transistor was the first transistor-like device invented. Today point-contact transistors are almost unknown in practical use. [L.P.H.]

Point source A source having definite position but no extension in space. In discussing radiation, it is convenient to define the concept of point source. If the radiation propagates in radially straight lines (or, which is the same thing, in spherical waves) from the point source, conservation of energy demands that the intensity of the radiation decrease in any direction inversely as the square of the distance from the source. No physical source is actually a mathematical point, but for distances sufficiently large compared to dimensions of the source, the inverse-square law may be a good approximation. *See* INVERSE-SQUARE LAW. [McA.H.H.]

Poison A substance which by chemical action and at low dosage can kill or injure living organisms. Broadly defined, poisons include chemicals toxic for any living form: microbes, plants, or animals. For example, antibiotics such as penicillin, although nontoxic for mammals, are poisons for bacteria. In common usage the word is limited to substances toxic for humans and mammals, particularly where toxicity is a substance's major property of medical interest. Because of their diversity in origin, chemistry, and toxic action, poisons defy any simple classification. Almost all chemicals with recognized physiological effects are toxic at sufficient dosage.

Origin and chemistry. Many poisons are of natural origin. Some bacteria secrete toxic proteins (for example, botulinus, diphtheria, and tetanus toxins), among the most poisonous compounds known. Lower plants notorious for poisonous properties are ergot (*Claviceps purpurea*) and a variety of toxic mushrooms. *See* ERGOT.

Higher plants, which constitute the major natural source of drugs, contain a great variety of poisonous substances. Many of the plant alkaloids double as drugs or poisons, depending on dose. These include curare, quinine, atropine, mescaline, morphine, nicotine, cocaine, picrotoxin, strychnine, lysergic acid, and many others. *See* ATROPINE; COCAINE; MORPHINE; QUININE.

Poisons of animal origin (venoms) are similarly diverse. Toxic marine animals alone include examples of every phylum. Insects and snakes represent the best-known venomous land animals, but on land, too, all phyla include poison-producing species. Among mammalian examples are certain shrews with poison-producing salivary glands.

Poisons of nonliving origin vary in chemical complexity from the toxic elements, for example, the heavy metals, to complex synthetic organic molecules. Most of the heavy metals (gold, silver, mercury, arsenic, and lead) are poisons of high potency in the form of their soluble salts. Strong acids or bases are toxic largely because of corrosive local tissue injury.

The chemically reactive gases hydrogen sulfide, hydrocyanic acid, chlorine, bromine, and ammonia are also toxic, even at low concentration, both because of their corrosiveness and because of more subtle chemical interaction with enzymes or other cell constituents.

Many organic substances of synthetic origin are highly toxic and represent a major source of industrial hazard. Most organic solvents are more or less toxic on ingestion or inhalation. Many alcohols, such as methanol, are toxic. Many solvents (for example, carbon tetrachloride, tetrachloroethane, dioxane, and ethylene glycol) produce severe chemical injury to the liver and other viscera, sometimes from rather low dosage.

Actions. The action of poisons is generally described by the physiological or biochemical changes which they produce. For most poisons, a descriptive account can be given which indicates what organic system (for example, heart, kidney, liver, brain, and bone marrow) appears to be most critically involved and contributes most to seriously disordered body function or death. In many cases, however, organ effects are multiple, or functional derangements so generalized that a cause of death cannot be localized.

More precise understanding of the mechanism of poisons requires detailed knowledge of their action in chemical terms. Information of this kind is available for only a few compounds, and then in only fragmentary detail. Poisons that inhibit acetylcholinesterase have toxic actions traceable to a single blocked enzyme reaction, hydrolysis of normally secreted acetylcholine. Understanding of the mechanism of chemical inhibition of cholinesterase is not complete, but allows some prediction of chemical structures likely to act as inhibitors.

Carbon monoxide toxicity is also partly understood in chemical terms, since formation of carboxyhemoglobin, a form incapable of oxygen transport, is sufficient to explain the anoxic features of toxicity.

Heavy metal poisoning in many cases is thought to involve inhibition of enzymes by formation of metal mercaptides with enzyme sulfhydryl groups, the unsubstituted form of which is necessary for enzyme action. This is a general reaction that may occur with a variety of sulfhydryl-containing enzymes in the body. Specific susceptible enzymes whose inhibition explains toxicity have not yet been well documented.

Metabolic antagonists active as poisons function by competitive blocking of normal metabolic reactions. Some antagonists may act directly as enzyme inhibitors, others may be enzymatically altered to form derivatives which are even more potent inhibitors at a later metabolic step. *See* ENZYME INHIBITION.

Where poison mechanisms are relatively well understood it has sometimes been possible to employ rationally selected antidotes.

Health aspects. The environment of civilized people abounds in toxic chemicals: drugs, industrial products, fuels, pesticides, and many common household products such as cleaning agents and paint products. Acute and chronic poisoning represent an appreciable public health hazard.

The multiplication of toxic chemicals in everyday use led, in 1953, to the establishment of information centers known as Poison Control Centers, which provide physicians with emergency telephone information concerning toxicity and antidotes for common poisons as well as for brand-name chemical formulations. *See* TOXICOLOGY. [E.Ad.]

Poison gland The specialized gland of certain fishes, as well as the granular and some mucous glands of many aquatic and terrestrial Amphibia. The poison glands of fishes are simple or slightly branched acinous structures which use the holocrine method of secreting a mucuslike substance. The poison glands of snakes are modified oral or salivary glands. Amphibian glands are simple, acinous, holocrine, with granular secretion. In some cases these amphibian poison glands produce mucus by a merocrine method of secretion. These glands function as protective devices. *See* GLAND. [O.E.N.]

Poisonous plants Almost 400 species of vascular plants, representing 70 families, have been listed as poisonous plants occurring in the United States. About 100 of these species cause dermatitis. Among these are such common plants as poison ivy (*Rhus radicans*), poison sumac (*R. vernix*), and viper's bugloss (*Echium vulgare*). Plants representing nearly a score of species possess spines, thorns, or fruits having sharp awns, which often cause injury to animals. In addition to these are many other species more appropriately cataloged as poisonous plants because they contain toxic principles which initiate pathological conditions in animals. [P.D.St./E.L.C.]

Polar meteorology A regional branch of atmospheric science, taking special account of circumstances and phenomena prevailing in the northern and southern polar regions. The atmosphere in these regions has a distinctive character brought about principally by low solar elevation, as well as the given geographical configuration of oceans and land masses. [N.U.]

Polar molecule A molecule possessing a permanent electric dipole moment. Molecules containing atoms of more than one element are polar except where forbidden by symmetry; molecules formed from atoms of a single element are nonpolar (except ozone). The dipole moments of polar molecules result in stronger intermolecular attraction, increased viscosities, higher melting and boiling points, and greater solubility in polar solvents than in nonpolar molecules. [R.D.W.]

Polar navigation The complex of navigational techniques modified from those used in other areas to suit the distinctive regional character of polar areas. Although polar navigation has become routine to a rising number of navigators operating in and through such high-latitude parts of the world, their success continues to be based on a sound grasp of the regional differences and the developing adaptations of navigational principles and aids to suit these peculiar area needs. For example, in polar regions the meridians radiate outward from the poles, and parallels are concentric circles. Thus the rectangular coordinates familiar to the navigator accustomed to using the Mercator projection are replaced by polar coordinates.

Limitations. Piloting in polar regions is strongly affected by the absence of any great number of aids to navigation. Also, natural landmarks may not be shown on the chart, or may be difficult to identify. The appearance of some landmarks changes markedly under different ice conditions. When snow covers both the land and a wide ice foot attached to the shore and extending for miles seaward, even the shore line is difficult to locate. *See* PILOTING.

Charts of polar regions are less reliable than those of other regions, because relatively little surveying has been done in the polar areas. Because relatively few soundings are shown on charts, ships entering harbors often send small boats ahead to determine the depth of water available.

The availability of electronic aids in polar regions is somewhat limited. Radar is useful, but experience in interpretation of the scope in polar regions is essential for reliable results. This is particularly true in aircraft, where the relative appearance of water and land areas often reverses in winter and summer. A radio direction finder is useful, when radio signals are available. The use of electronics in polar regions is further restricted by magnetic storms, which are particularly severe in the auroral zones. *See* ELECTRONIC NAVIGATION SYSTEMS; RADAR.

Reliable dead reckoning depends upon the availability of accurate measurement of direction and distance (or speed and time). There are difficulties in meeting these requirements in polar regions. Direction is measured largely by a compass. The magnetic compass becomes unreliable in the vicinity of the magnetic poles of the Earth, and the north-seeking gyrocompass becomes unreliable in the vicinity of the geographical poles of the Earth. The directional gyro is widely used in aircraft operating in polar regions but is not in general use aboard ship. Distance or speed measurement in polar regions presents no problems in aircraft. When ships operate in ice, however, the sensing element in the water may be adversely affected or damaged by the ice. At best, dead reckoning is difficult aboard a ship operating in the ice, not only because of the difficulty encountered in measuring course and speed, but also because neither of these may be constant for very long. *See* AIRCRAFT COMPASS SYSTEM; DEAD RECKONING; GYROSCOPE.

Celestial navigation is of great importance in polar regions, sometimes providing the only means of determining position accurately, or establishing a directional reference. When operating in lower latitudes, navigators generally avoid observation of bodies near the horizon because of the uncertainty of the refraction correction there. In polar regions, even though refraction is more uncertain, navigators often have no choice. Near the equinoxes the Sun may be the only body available for several weeks, and it remains close to the horizon. During the polar summer the Sun is often the only celestial body available. When only one body is available, it is observed at frequent intervals, perhaps hourly, and a series of running fixes is plotted. *See* CELESTIAL NAVIGATION.

Modern technology. Advances in technology have contributed significantly to the safety and reliability of navigation in polar regions. The availability of modern echo sounders has made possible a continuous plot of the bottom profile beneath a vessel while under way. Sonar indicates the approach of a vessel to an underwater obstruction. Reliable inertial navigators are available to make aircraft navigation in polar regions almost routine, providing both positional and directional data. Ship inertial navigation systems are available and have made practicable submarine operations under sea ice. *See* INERTIAL GUIDANCE SYSTEM; SONAR.

The U.S. Navy Navigation Satellite System (NNSS) provides means for intermittent position determination for craft equipped with suitable instrumentation. The NAVSTAR Global Positioning System (GPS) will virtually eliminate the limitations for aircraft of the Navy Navigation Satellite System. *See* SATELLITE NAVIGATION SYSTEMS. [A.B.M.]

Polar triangle The poles of a great circle on a sphere are the points of the sphere where a perpendicular to the plane of the great circle at its center cuts the sphere. To obtain the polar triangle of a spherical triangle ABC, construct on its sphere three great circles having respective poles A, B, and C. Two great circles, one having B and the other C as poles, intersect in two points on opposite hemispheres defined by the great circle through BC. Denote by A' the point lying on the same hemisphere as A. Locate B' and C' by a like procedure. The spherical triangle $A'B'C'$ is the polar triangle of ABC. In geometry it is shown that, if one spherical triangle is the polar of a second, then the second is the polar of the first. *See* TRIGONOMETRY. [L.M.K.]

Polarimetric analysis A method of chemical analysis based on the optical activity of the substance being determined. Optically active materials are asymmetric; that is, their molecules or crystals have no plane or center of symmetry. Asymmetric substances possess the power of rotating the plane of polarization of plane-polarized light. Measurement of the extent of this rotation, polarimetry, is performed by an instrument known as a polarimeter. Polarimetry is applied to both organic and inorganic materials. *See* OPTICAL ACTIVITY; POLARIZED LIGHT.

Polarimetry may be used for either qualitative or quantitative analytical work. In qualitative applications, the presence of an optically active material is shown, and then a calculation of specific rotation often leads to the identification of the unknown. In quantitative work, the concentration of a given optically active material is determined either from a calibration curve of percentage of the constituent versus angular rotation or from the specific rotation, assuming the angular rotation to be a linear function of concentration.

Polarimetry is widely used in carbohydrate chemistry, especially in the analysis of sugar solutions. Also, since there is great difference between the biological activities of the different optical forms of organic compounds, polarimetry is widely used in biochemical research to identify the molecular configurations.

Optical rotary dispersion is the measurement of the specific rotation as a function of wavelength. The information obtained by this method has shown that minor changes in configuration of a molecule have a marked effect on its dispersion properties. By using the properties of compounds of known configuration, it has been possible to determine the absolute configurations of many other molecules and to identify various isomers. *See* COTTON EFFECT; OPTICAL METHODS OF CHEMICAL ANALYSIS; OPTICAL ROTATORY DISPERSION. [R.F.G.; J.N.L.]

Polarimetry The science of determining the polarization state of electromagnetic radiation (x-rays, light or radio waves). Radiation is said to be linearly polarized when the electric vector oscillates in only one plane. It is circularly polarized when the x-plane component of the electric vector oscillates 90° out of phase with the y-plane component.

To completely specify the polarization state, it is necessary to make six intensity measurements of the light passed by a quarter-wave retarder and a rotatable linear polarizer, such as a Polaroid or a Nicol prism. The retarder converts circular light into linear light.

Most starlight is unpolarized. However, atoms in the presence of a magnetic field align themselves at fixed, quantized angles to the field direction. Then the spectral lines they emit are circularly polarized when the magnetic field is parallel to the line of sight, and linearly polarized when the field is perpendicular. The light from sunspots is polarized because the magnetic fields impose some direction in the emitting gas. Other phenomena also remove isotropy and produce polarization. *See* SOLAR MAGNETIC FIELD; ZEEMAN EFFECT.

Electrooptical devices are rapidly replacing rotating polarizers and fixed retarders. The magnetograph consists of a spectrograph to isolate the atomic spectral line for study; a Pockels cell, an electrooptic crystal whose retardance depends on an applied voltage; a polarizing prism to isolate the polarization state passed by the retarder; a pair of photocells to detect the transmitted light; and a scanning mechanism to sweep the solar image across the spectrograph entrance slit. Two photocells are needed to simultaneously measure left- and right-circular polarization. *See* SPECTROGRAPH.

A magnetograph can be made sensitive to linear polarization, but the signal levels are about 100 times weaker for the inferred transverse fields than for longitudinal fields of comparable strength. To improve signal-to-noise levels, the spectrograph can be replaced with an optical filter having a narrow passband, and the photocells can be replaced with an array of photosensitive picture elements (pixels). [D.M.R.]

Polaris Alpha Ursae Minoris, the North Star, or polestar. Polaris is located approximately 1° of arc from the north celestial pole, the point where the Earth's axis of rotation now pierces the celestial sphere. Two bright stars at the end of the bowl of the Big Dipper point toward Polaris. Polaris has a faint, main-sequence companion. *See* STAR. [J.L.Gr.]

Polarity (electricity) The designation of positive and negative terminals in an electric circuit carrying a current. The choice of positive charge, and thereby of positive potential, is a purely arbitrary one which had been made long before the discovery of the electron. It is now known that the primary current carriers in most circuit elements are electrons, which possess negative charge.

For a seat of electromotive force, such as a battery or a generator, the positive terminal is by convention the one at which electrons enter from the external circuit; the negative terminal is the one from which they leave. For a "passive" element, such as an ammeter or a resistor, the positive terminal is the one through which the electrons leave the element; the negative terminal is the one through which they enter from the circuit. *See* POTENTIALS. [J.W.St.]

Polarization of dielectrics A vector quantity representing the electric dipole moment per unit volume of a dielectric material. *See* DIELECTRIC MATERIALS.

Dielectric polarization arises from the electrical response of individual molecules of a medium and may be classified as electronic, atomic, orientation, and space-charge or interfacial polarization, according to the mechanism involved.

Electronic polarization represents the distortion of the electron distribution or motion about the nuclei in an electric field.

Atomic polarization arises from the change in dipole moment accompanying the stretching of chemical bonds between unlike atoms in molecules. *See* MOLECULAR STRUCTURE AND SPECTRA.

Orientation polarization is caused by the partial alignment of polar molecules, that is, molecules possessing permanent dipole moments, in an electric field. This mechanism leads to a temperature-dependent component of polarization at lower frequencies.

Space-charge or interfacial polarization occurs when charge carriers are present which can migrate an appreciable distance through a dielectric but which become trapped or cannot discharge at an electrode. This process always results in a distortion of the macroscopic field and is important only at low frequencies. *See* DIELECTRIC CONSTANT; ELECTRIC FIELD; ELECTRIC SUSCEPTIBILITY. [R.D.W.]

Polarization of waves Polarization is the phenomenon which is exhibited when a transverse wave is polarized. The term polarization is also used to describe the process of polarizing a wave.

In an unpolarized wave, the vibrations in a plane perpendicular to the ray appear to be oriented in all directions with equal probability. In a polarized wave the displacement direction of the vibrations is completely predictable. For certain disturbances, such as the transverse acoustic wave produced when a steel bar is struck, the polarization is complete. Electromagnetic radiation is normally unpolarized if it is generated by atomic processes. Thus ultraviolet, visible, and infrared radiations produced by heated bodies or electrical discharges are generally unpolarized. Radiation generated by vacuum-tube oscillators or transistor oscillators is always polarized. The probability waves (matter waves) associated with atomic or nuclear particles are generally unpolarized. *See* ELECTROMAGNETIC RADIATION; POLARIZED LIGHT; QUANTUM MECHANICS. [B.H.Bi.]

Polarized light Light which has its electric vector oriented in a predictable fashion with respect to the propagation direction. In unpolarized light, the vector is oriented in a random, unpredictable fashion. Even in short time intervals, it appears to be oriented in all directions with equal probability. Most light sources seem to be partially polarized so that some fraction of the light is polarized and the remainder unpolarized.

According to all available theoretical and experimental evidence, it is the electric vector rather than the magnetic vector of a light wave that is responsible for all the effects of polarization and other observed phenomena associated with light. Therefore, the electric vector of a light wave, for all practical purposes, can be identified as the light vector. *See* CRYSTAL OPTICS; ELECTROMAGNETIC RADIATION; LIGHT; POLARIZATION OF WAVES.

One of the simplest ways of producing linearly polarized light is by reflection from a dielectric surface. At a particular angle of incidence, known as Brewster's angle, the reflectivity for light whose electric vector is in the plane of incidence becomes zero. The reflected light is thus linearly polarized at right angles to the plane of incidence.

Linear polarizing devices. The first polarizers were glass plates inclined so that the incident light was at Brewster's angle. Such polarizers are quite inefficient since only a small percentage of the incident light is reflected as polarized light.

Certain natural materials absorb linearly polarized light of one vibration direction much more strongly than light vibrating at right angles. Such materials are termed dichroic. Tourmaline is one of the best-known dichroic crystals, and tourmaline plates were used as polarizers for many years. *See* DICHROISM.

Other natural materials exist in which the velocity of light depends on the vibration direction. These materials are called birefringent. One of the best-known of these birefringent crystals is transparent calcite (Iceland spar). The Nicol prism is made of two pieces of calcite cemented together (Fig. 1). The cement is Canada balsam, in which the wave velocity is intermediate between the velocity in calcite for the fast and the slow ray. The angle at which the light strikes the boundary is such that for one ray the angle of incidence is greater than the critical angle for total reflection. Thus the rhomb is transparent for only one polarization direction. *See* BIREFRINGENCE; CRYSTAL; OPTICS.

A different type of polarizer, made of quartz, is shown in Fig. 2. Here the vibration directions are different in the two pieces so that the two rays are deviated as they pass through the material. The incoming light beam is thus separated into two oppositely linearly polarized beams which have an angular separation between them, and it is possible to select either beam.

A third mechanism for obtaining polarized light is the Polaroid sheet polarizer, of which there are three types. The first is a microcrystalline polarizer in which small crystals of a dichroic material are oriented parallel to each other in a plastic medium. The second type depends for its dichroism on a property of an iodine-in-water solution. The iodine appears to form a linear high polymer. If the iodine is put on a transparent oriented sheet of material such as polyvinyl alcohol (PVA), the iodine chains apparently line themselves parallel to the PVA molecules and the resulting dyed sheet is strongly dichroic. A third type of sheet polarizer depends for its dichroism directly on the molecules of the plastic itself. This plastic consists of oriented polyvinylene.

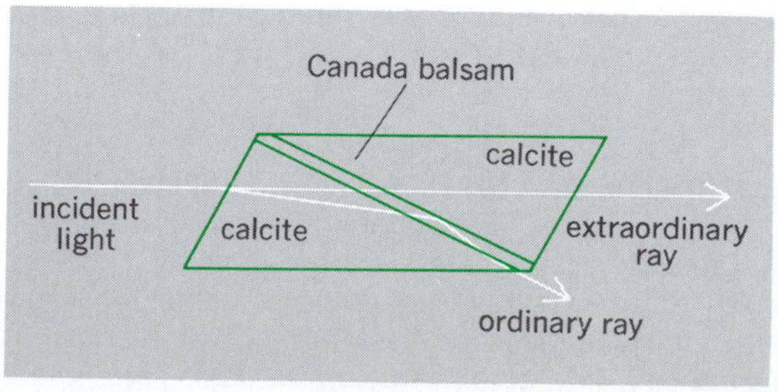

Fig. 1. Nicol prism. The ray for which Snell's law holds is called the ordinary ray.

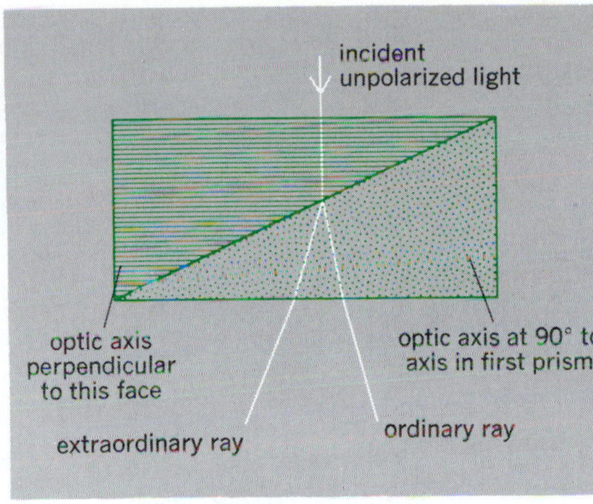

Fig. 2. Wollaston prism.

Polarization by scattering. When an unpolarized light beam is scattered by molecules or small particles, the light observed at right angles to the original beam is polarized. The best-known example of polarization by scattering is the light of the north sky. *See* SCATTERING OF ELECTROMAGNETIC RADIATION.

Types. Polarized light is classified according to the orientation of the electric vector. In linearly polarized light, the electric vector remains in a plane containing the propagation direction. For monochromatic light, the amplitude of the vector changes sinusoidally with time. In circularly polarized light, the tip of the electric vector describes a circular helix about the propagation direction. The amplitude of the vector is constant. The frequency of rotation is equal to the frequency of the light. In elliptically polarized light, the vector also rotates about the propagation direction, but the amplitude of the vector changes so that the projection of the vector on a plane at right angles to the propagation direction describes an ellipse.

Circular and elliptical polarizing devices. Circularly and elliptically polarized light are normally produced by combining a linear polarizer with a wave plate. A Fresnel rhomb can be used to produce circularly polarized light.

A plate of material (quartz, calcite, or other birefringent crystals) which is linearly birefringent is called a wave plate or retardation sheet. Wave plates have a pair of orthogonal axes which are designated fast and slow. Polarized light with its electric vector parallel to the fast axis travels faster than light polarized parallel to the slow axis. The thickness of the material can be chosen so that for light traversing the plate, there is a definite phase shift between the fast component and the slow component. A plate with a 90° phase shift is termed a quarter-wave plate.

If linearly polarized light is incident normally on a quarter-wave plate and oriented at 45° to the fast axis, the transmitted light will be circularly polarized. If the linearly polarized light is at an angle other than 45° to the fast axis, the transmitted radiation will be elliptically polarized.

Analyzing devices. Polarized light is one of the most useful tools for studying the characteristics of materials. The absorption constant and refractive index of a metal can be calculated by measuring the effect of the metal on polarized light reflected from its surface. *See* REFLECTION OF ELECTROMAGNETIC RADIATION.

The analysis of polarized light can be performed with a variety of different devices. If the light is linearly polarized, it can be extinguished by a linear polarizer and the direction of polarization of the light determined directly from the orientation of the polarizer. If the light is elliptically polarized, it can be analyzed with the combination of a quarter-wave plate and a linear polarizer. Any such combination of polarizer and analyzer is called a polariscope. [B.H.Bi.]

Polarized-vane ammeter

Polarized-vane ammeter An ammeter in which the current to be measured passes through a small coil, distorting the field of a circular permanent magnet, and an iron vane aligns itself with the axis of the distorted field, the deflection being roughly proportional to the current. Reversed electrical polarity reverses the direction of motion; the instrument thus indicates current flowing into or out of a battery. This instrument is of only moderate accuracy. It is used in large numbers in battery chargers and automobiles because of its low cost. *See* AMMETER. [J.H.M./J.Be.]

Polarographic analysis

Polarographic analysis An electrochemical technique used in analytical chemistry. Polarography involves measurements of current-voltage curves, obtained when voltage is applied to electrodes (usually two) immersed in the solution being investigated. One of these electrodes is a reference electrode: its potential remains constant during the measurement. The second electrode is an indicator, and its potential varies in the course of measurement of the current-voltage curve, because of the change of the applied voltage. In the simplest version, so-called dc polarography, the indicator electrode is a dropping mercury electrode, consisting of a mercury drop hanging at the orifice of a fine-bore glass capillary. The capillary is connected to a mercury reservoir so that mercury flows through the capillary at the rate of a few milligrams per second. The outflowing mercury forms a drop at the orifice, which grows until it falls off. The lifetime of each drop is several seconds (usually 2 to 5). Each drop forms a new electrode; its surface is practically unaffected by processes taking place on the previous drop. Hence each drop represents a well-reproducible electrode with fresh, clean surface. *See* ELECTROCHEMICAL TECHNIQUES; ELECTRODE.

The dropping-mercury electrode is immersed in the solution to be investigated and placed in a cell containing the reference electrode. Polarographic current-voltage curves can be recorded with a simple instrument consisting of a potentiometer or another source of voltage and a current-measuring device (such as a sensitive galvanometer). The voltage can be varied by manually changing the applied voltage in finite increments,

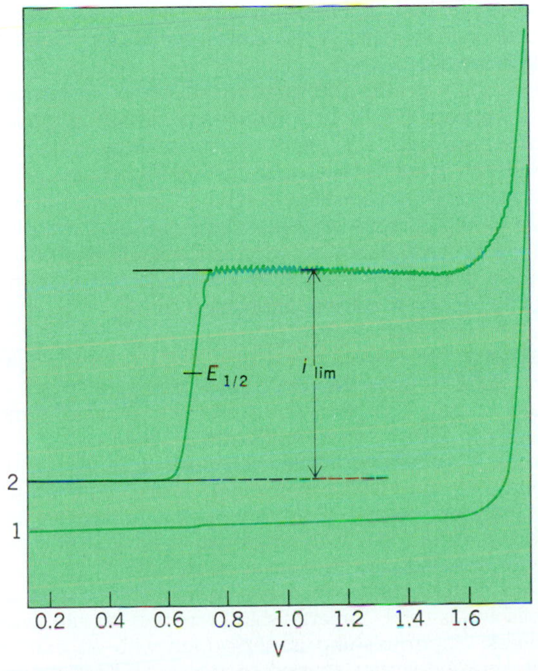

Polarographic current-voltage curves: 1 = supporting electrolyte; 2 = curve in the presence of an electroactive (reducible) compound; $E_{1/2}$ = half-wave potential; i_{lim} = limiting current (wave height).

measuring current at each, and plotting current as a function of the voltage. Alternatively, commercial instruments are available in which voltage is increased linearly with time (a voltage ramp), and current variations are recorded automatically (see illustration).

Polarographic studies can be applied to investigation of electrochemical problems, to elucidation of some fundamental problems of inorganic and organic chemistry, and to solution of practical problems. In electrochemistry polarography allows measurement of potentials, and yields information about the rate of the electrode process, adsorption-desorption phenomena, and fast chemical reactions accompanying the electron transfer. In fundamental applications, polarography allows determination of the form and charge of the species (for example, inorganic complex or organic ion) in the solution. Polarography also permits the study of equilibria (complex formation, acid-base, tautomeric), rates, and mechanisms. [P.Z.]

Polaron The object that results when an electron in the conduction band of a crystalline insulator or semiconductor polarizes or otherwise deforms the lattice in its vicinity. The polaron comprises the electron plus its surrounding lattice deformation. (Polarons can also be formed from holes in the valence band.) If the deformation extends over many lattice sites, the polaron is "large," and the lattice can be treated as a continuum. Charge carriers inducing strongly localized lattice distortions form "small" polarons. *See* BAND THEORY OF SOLIDS; HOLE STATES IN SOLIDS; SEMICONDUCTOR. [D.M.L.]

Poliomyelitis An acute infectious viral disease which in its serious form affects the central nervous system and, by destruction of motor neurons in the spinal cord, produces flaccid paralysis. However, about 99% of infections are either inapparent or very mild. *See* CENTRAL NERVOUS SYSTEM; VIRUS.

The virus probably enters the body through the mouth; primary multiplication occurs in the throat and intestine. The blood seems to be the most likely route to the central nervous system. The severity of the infection may range from a completely inapparent through minor influenzalike illness, or an aseptic meningitis syndrome (nonparalytic poliomyelitis) with stiff and painful back and neck, to the severe forms of paralytic and bulbar poliomyelitis. In all clinical types virus is regularly present in the enteric tract.

Poliomyelitis occurs throughout the world. In temperate zones it appears chiefly in summer and fall although winter outbreaks have been known. It occurs in all age groups, but less frequently in adults because of their acquired immunity.

Salk vaccine (formalin-killed), prepared from virus grown in monkey kidney cultures, was first developed and used in the United States. Oral poliovirus vaccine (Sabin) is also now generally used throughout the world as an effective immunizing agent. The vaccine is a living, attenuated virus. [J.L.Me.]

Polishing The smoothing of a surface by the cutting action of abrasive grit either glued to or impregnated in a flexible wheel or belt. Polishing, not a precision process, removes stock until the desired surface condition is obtained.

When a considerable amount of stock must be removed, operations may start with a coarse grit. Then a finer-grit wheel may be used for finishing. Common polishing machines usually are lathe-type machines with wheels at either end of a power spindle, or one end may drive an abrasive belt.

For a mirrorlike surface, a precision abrading process called superfinishing is used. Superfinishing removes minute flaws or inequalities. Superfinishing is performed with an extremely fine-grit abrasive stone, shaped to match and cover a large portion of the work surface. *See* LAPPING. [A.H.T.]

Pollen The small male reproductive bodies produced in the pollen sacs of seed plants (gymnosperms and angiosperms).

Fig. 1. Tetrad of cattail (*Typha latifolia*), grains cohering, one-pored, exine reticulate. (*Scanning electron micrograph by C. M. Drew, U.S. Naval Weapons Center, Cina Lake, California*).

On maturation in the pollen sac, a pollen grain may reach 0.00007 mg as in spruce, or less than 1/20 of this weight. A grain usually has two waxy, durable outer walls, the exine, and an inner fragile wall, the intine. These walls surround the contents with their nuclei and reserves of starch and oil.

Pollen identification depends on interpretation of morphological features. Exine and aperture patterns are especially varied in the more highly evolved dicots, so that recognition at family, genus, or even species level may be possible despite the small surface area available on a grain (Figs. 1 and 2). Since the morphological characters are conservative in the extreme, usually changing very slowly through geologic time, studies of fine detail serve to establish the lineal descent of many plants living today. *See* PALYNOLOGY.

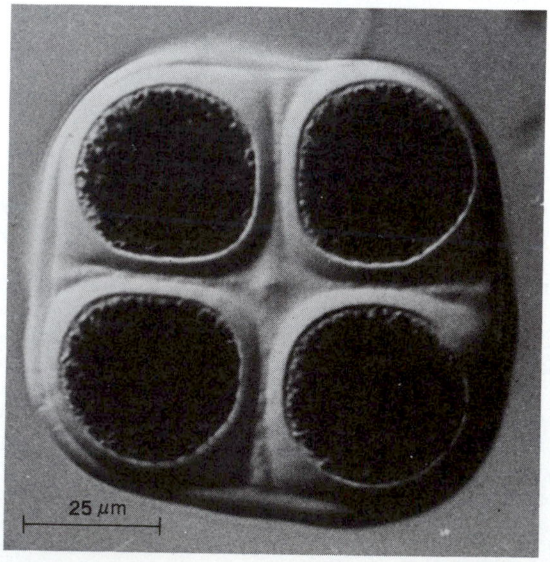

Fig. 2. Young tetrad of *Lavatera* (mallow family), grains free in callose of pollen mother cell. (*Photomicrograph by Luc Waterkeyn*)

Extreme variations in size may occur within a family, but pollen grains range mainly from 24 to 50 micrometers, with the dicot range being from 2 μm in *Myosotis* to 250 μm in *Mirabilis*; the monocots range from about 15 to 150 μm or more in the ginger family, with eelgrass (*Zostera*) having pollen measuring 2550 × 3.7 μm in a class of its own; living gymnosperms range from 15 μm in *Gnetum* to about 180 μm in *Abies* (including sacs), while fossil types range from about 11 μm (one-furrowed) to 300 μm.

Most grains are free (monads) though often loosely grouped because of the spines or sticky oils and viscin threads. Compound grains (polyads) are richly developed in some angiosperm families, commonly occurring in four (tetrads), or in multiples of four up to 64 or more.

In most microspores (pollen and spore) the polar axis runs from the inner (proximal) face to the outer (distal) face, as oriented during tetrad formation. The equator crosses it at right angles. Bilateral grains dominate in the gymnosperms and monocots, the polar axis usually being the shorter one, with the single aperture on the distal side. On the other hand, almost all dicot pollen is symmetrical around the polar axis (usually the long axis), with shapes ranging mainly from spheroidal to ellipsoidal, with rounded equatorial outlines, and sometimes "waisted" as in the Umbelliferae. Three-pored grains often have strikingly triangular outlines in polar view.

Rarely lacking, the flexible membranes of apertures are sometimes covered only by endexine. They allow for sudden volume changes, as for the emergence of the germ tube. They are classified as furrows, with elongate outlines, and pores usually more or less circular in shape. Short slitlike intermediate forms occur. A few families have no apertures, some as a result of reduction of exine as an adaptation to wind- or water-pollination. Gymnosperms may also lack openings, or have one small papilla; most have one furrow, or a long weak area.

Few pollen grains are completely smooth (psilate) at ordinary magnifications; most have sculpture both on the surface and the structure below it. *See* FLOWER; POLLINATION; REPRODUCTION (PLANT). [L.M.C.]

Pollination The transport of pollen grains from the plant parts that produce them to the ovule-bearing organs, or to the ovules (seed precursors) themselves. In gymnosperms, the pollen, usually dispersed by the wind, is simply caught by a drop of fluid excreted by each freely exposed ovule. In angiosperms, where the ovules are contained in the pistil, the pollen is deposited on the pistil's receptive end (the stigma), where it germinates. *See* FLOWER.

Without pollination, there would be no fertilization; it is thus of crucial importance for the production of fruit crops and seed crops. Pollination also plays an important part in plant breeding experiments aimed at increasing crop production through the creation of genetically superior types. *See* BREEDING (PLANT); REPRODUCTION (PLANT).

Self- and cross-pollination. In most plants, self-pollination is difficult or impossible, and there are various mechanisms which are responsible. For example, in dichogamous flowers, the pistils and stamens reach maturity at different times; in protogyny, the pistils mature first, and in protandry, the stamens mature before the pistils. Selfing is also impossible in dioecious species, where some plants bear flowers that have only pistils (pistillate or female flowers), while other individuals have flowers that produce only pollen (staminate or male flowers). In monoecious species, where pistillate and staminate flowers are found in the same plant, self-breeding is at least reduced. Heterostyly is another device that promotes outbreeding. Here some flowers (pins) possess a long pistil and short stamens, while others (thrums) exhibit the reverse condition; each plant individual bears only pins or only thrums.

Flower attractants. As immobile organisms, plants normally need external agents for pollen transport. These can be insects, wind, birds, mammals, or water, roughly in that order of importance. In some plants the pollinators are simply trapped; in the large majority of cases, however, the flowers offer one or more rewards, such as sugary nectar, oil, solid food bodies, perfume, sex, an opportunity to breed, a place to sleep, or some of the pollen itself. For the attraction of pollinators, flowers provide either visual or olfactory signals. Color includes ultraviolet, which is perceived as a color by most insects and at least some hummingbird species. Fragrance is characteristic of flowers pollinated by bees, butterflies, or hawkmoths, while carrion or dung odors are produced by flowers catering to certain beetles and flies. A few orchids, using a combination of olfactory and visual signals, mimic the females of certain bees or wasps so successfully that the corresponding male insects will try to mate with them, thus achieving pollination (pseudocopulation).

While some flowers are "generalists," catering to a whole array of different animals, others are highly specialized, being pollinated by a single species of insect only. Extreme pollinator specificity is an important factor in maintaining the purity of plant species in the field, even in those cases where hybridization can easily be achieved artificially in a greenhouse or laboratory, as in most orchids. The almost incredible mutual adaptation between pollinating animal and flower which can frequently be observed exemplifies the idea of coevolution. *See* POLLEN. [B.J.D.M.]

Pollucite A mineral tectosilicate having the composition $Cs_4Al_4Si_9O_{26} \cdot H_2O$. Well-formed crystals are usually cubic, but are rare; it is most commonly massive. Hardness is $6\frac{1}{2}$ on Mohs scale, and specific gravity is 2.90. The mineral is colorless to white with a vitreous luster. Pollucite is rare. It has been found in large masses in the Varuträsk pegmatite in Sweden and in the lithium-rich pegmatite at Bakita, Rhodesia. In the United States it was at one time worked as an ore of cesium at Newry, Maine, and in the Black Hills of South Dakota. *See* SILICATE MINERALS. [C.S.Hu.]

Polonium A chemical element, Po, atomic number 84. Marie Curie discovered the radioisotope ^{210}Po in pitchblende. This isotope is the penultimate member of the radium decay series. All polonium isotopes are radioactive, and all are short-lived except the three α-emitters, artificially produced ^{208}Po (2.9 years) and ^{209}Po (100 years), and natural ^{210}Po (138.4 days).

Polonium (^{210}Po) is used mainly for the production of neutron sources. It can also be used in static eliminators and, when incorporated in the electrode alloy of spark plugs, is said to improve the cold-starting properties of internal combustion engines.

Most of the chemistry of polonium has been determined using ^{210}Po, 1 curie of which weighs 222.2 micrograms; work

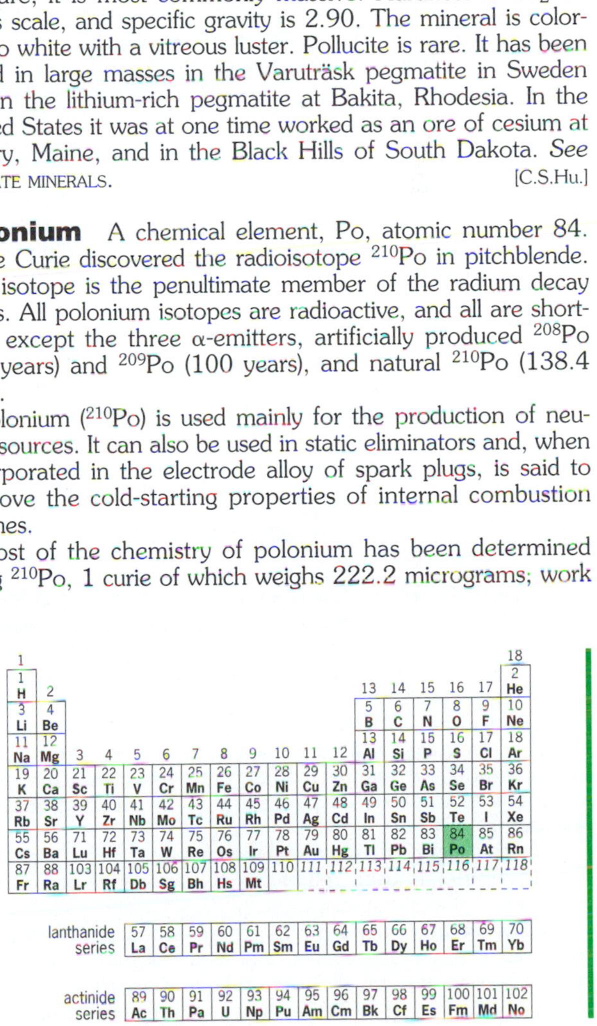
1 H; 2 He
3 Li; 4 Be; 5 B; 6 C; 7 N; 8 O; 9 F; 10 Ne
11 Na; 12 Mg; 13 Al; 14 Si; 15 P; 16 S; 17 Cl; 18 Ar
19 K; 20 Ca; 21 Sc; 22 Ti; 23 V; 24 Cr; 25 Mn; 26 Fe; 27 Co; 28 Ni; 29 Cu; 30 Zn; 31 Ga; 32 Ge; 33 As; 34 Se; 35 Br; 36 Kr
37 Rb; 38 Sr; 39 Y; 40 Zr; 41 Nb; 42 Mo; 43 Tc; 44 Ru; 45 Rh; 46 Pd; 47 Ag; 48 Cd; 49 In; 50 Sn; 51 Sb; 52 Te; 53 I; 54 Xe
55 Cs; 56 Ba; 71 Lu; 72 Hf; 73 Ta; 74 W; 75 Re; 76 Os; 77 Ir; 78 Pt; 79 Au; 80 Hg; 81 Tl; 82 Pb; 83 Bi; 84 Po; 85 At; 86 Rn
87 Fr; 88 Ra; 103 Lr; 104 Rf; 105 Db; 106 Sg; 107 Bh; 108 Hs; 109 Mt; 110; 111; 112; 113; 114; 115; 116; 117; 118
lanthanide series
57 La; 58 Ce; 59 Pr; 60 Nd; 61 Pm; 62 Sm; 63 Eu; 64 Gd; 65 Tb; 66 Dy; 67 Ho; 68 Er; 69 Tm; 70 Yb
actinide series
89 Ac; 90 Th; 91 Pa; 92 U; 93 Np; 94 Pu; 95 Am; 96 Cm; 97 Bk; 98 Cf; 99 Es; 100 Fm; 101 Md; 102 No

with weighable amounts is hazardous, requiring special techniques. Polonium is more metallic than its lower homolog, tellurium. The metal is chemically similar to tellurium, forming the bright red compounds $SPoO_3$ and $SePoO_3$. The metal is soft, and its physical properties resemble those of thallium, lead, and bismuth. Valences of 2 and 4 are well established; there is some evidence of hexavalency. Polonium is positioned between silver and tellurium in the electrochemical series.

Two forms of the dioxide are known: low-temperature, yellow, face-centered cubic (UO_2 type), and high-temperature, red, tetragonal. The halides are covalent, volatile compounds, resembling their tellurium analogs. *See* RADIOACTIVITY; TELLURIUM. [K.W.B.]

Polyacetal A polyether derived from aldehydes ($RC{=}H_2$ or CHR′) or ketones [R═C(R′R″)] and containing —O—R—O— groups in the main chain. Of the many possible polyacetals, the most common is a polymer or copolymer of formaldehyde, polyoxymethylene $(\text{—O—CH}_2\text{—})_n$. While the substance paraformaldehyde contains oligomers or low-molecular-weight polyoxymethylenes (*n* very small), high-molecular-weight, crystalline polyoxymethylenes constitute an important class of engineering plastics that, in commerce, is often simply referred to as polyacetal. Cellulose and its derivatives also have a polyacetal structure. For discussion of other polyethers and polymers containing acetal side groups *see* POLYETHER RESINS; POLYVINYL RESINS. *See also* ACETAL; CELLULOSE; FORMALDEHYDE.

Polyacetals are typically strong and tough, resistant to fatigue, creep, organic chemicals (but not strong acids or bases), and have low coefficients of friction. Electrical properties are also good. Improved properties for particular applications may be attained by reinforcement with fibers of glass or polytetrafluoroethylene and by incorporation of an elastomeric toughening phase. The combination of properties has led to many uses such as plumbing fittings, pump and valve components, bearings and gears, computer hardware, automobile body parts, and appliance housings. *See* POLYMER; POLYMERIZATION. [J.A.M.]

Polyacrylate resin Polymers obtained from a variety of acrylic monomers, such as acrylic and methacrylic acids, their salts, esters, and amides, and the corresponding nitriles. The most important monomers are shown here.

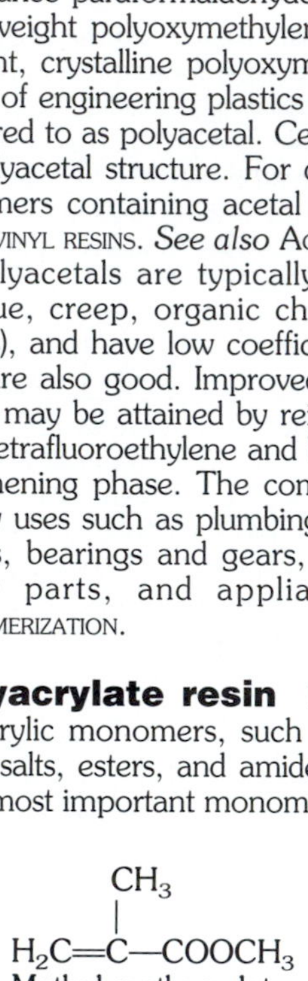

Methyl methacrylate

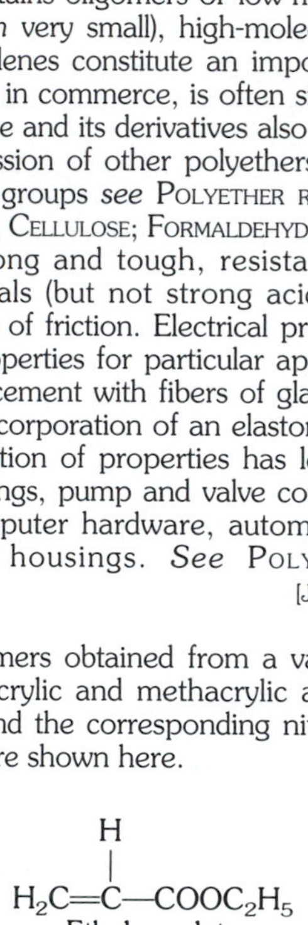

Ethyl acrylate

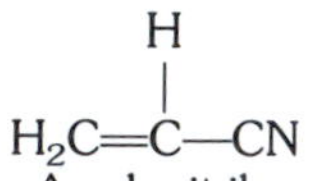

Acrylonitrile

Polymethyl methacrylate is a hard, transparent polymer with high optical clarity, high refractive index, and good resistance to the effects of light and aging. It and its copolymers are useful for lenses, signs, indirect lighting fixtures, transparent domes and skylights, dentures, and protective coatings.

Solutions of polymethyl methacrylate and its copolymers are useful as lacquers. Aqueous latexes formed by the emulsion polymerization of methyl methacrylate with other monomers are useful as water-based paints and in the treating of textiles and leather.

Polyethyl acrylate is a tough, somewhat rubbery product. The monomer is used mainly as a plasticizing or softening component of copolymers.

Methyl methacrylate is of interest as a polymerizable binder for sand or other aggregates, and as a polymerizable impregnant for concrete; usually a cross-linking acrylic monomer is also incorporated. The binder systems (polymer concrete) are used as overlays for bridge decks as well as for castings, while impregnation is used to restore concrete structures and protect bridge decks against corrosion by deicing salts. *See* ACRYLONITRILE; PLASTICS PROCESSING; POLYACRYLONITRILE RESINS; POLYMERIZATION. [J.A.M.]

Polyacrylonitrile resins Hard, horny, relatively insoluble, and high-melting materials produced by the polymerization of acrylonitrile. Polyacrylonitrile (polyvinyl cyanide) is used almost entirely in copolymers. The copolymers fall into three groups: fibers, plastics, and rubbers. The presence of acrylonitrile in a polymeric composition tends to increase its resistance to temperature, chemicals, impact, and flexing.

The major use of acrylonitrile is in the form of fibers. By definition an acrylic fiber must contain at least 85% acrylonitrile; a modacrylic fiber may contain less, from 35 to 85% acrylonitrile. The high strength, high softening temperature, resistance to aging, chemicals, water, and cleaning solvents, and the soft woollike feel of fabrics have made the product popular for many uses such as sails, cordage, blankets, and various types of clothing.

Copolymers of vinylidene chloride with small proportions of acrylonitrile are useful as tough, impermeable, and heat-sealable packaging films.

Use has been made of copolymers of acrylonitrile with butadiene, often called NBR (formerly Buna N) rubbers, which contain 15–40% acrylonitrile. The NBR rubbers resist hydrocarbon solvents such as gasoline, abrasion, and in some cases show high flexibility at low temperatures.

The development of blends and interpolymers of acrylonitrile-containing resins and rubbers represented a significant advance in polymer technology. The products, usually called ABS resins, typically are made by blending acrylonitrile-styrene copolymers with a butadiene-acrylonitrile rubber, or by interpolymerizing polybutadiene with styrene and acrylonitrile.

The combination of low cost, good mechanical properties, and ease of fabrication led to the rapid development of various uses for ABS resins. Applications include products requiring high impact strength, such as pipe, and sheets for structural uses, such as industrial duct work and components of automobile bodies. *See* ACRYLONITRILE; PLASTICS PROCESSING; RUBBER; STYRENE. [J.A.M.]

Polyamide resins Horny, whitish, translucent, high-melting polymers. They are used for fibers, bristles, bearings, gears, molded objects, coatings, and adhesives. The term nylon formerly referred specifically to synthetic polyamides as a class. Because of many applications in mechanical engineering, nylons are considered engineering plastics.

The most common commercial aliphatic polamides are nylons-6,6; -6; -6,10; -11; and -12. Nylon-6,6, nylon-6,10, nylon-6,12, and nylon-6 are the most commonly used polyamides for general applications as molded or extruded parts; nylon-6,6 and nylon-6 find general application as fibers.

As a group, nylons are strong and tough. Mechanical properties depend in detail on the degree and distribution of crystallinity, and may be varied by appropriate thermal treatment or by nucleation techniques. Because of their generally good mechanical properties and adaptability to both molding and extrusion, certain nylons are often used for gears, bearings, and electrical mountings. Nylon bearings and gears perform quietly and need little or no lubrication. Nylon resins are also used extensively as filaments, bristles, wire insulation, appliance parts, and film. Properties can also be modified by copolymerization. Reinforcement of nylons with glass fibers results in increased stiffness, lower creep and improved resistance to elevated temperatures. *See* HETEROCYCLIC POLYMER; PLASTICS PROCESSING; POLYETHER RESINS; POLYMERIZATION. [J.A.M.]

Polychaeta The largest class of the phylum Annelida, containing 68–70 families. About 1600 genera and 10,000

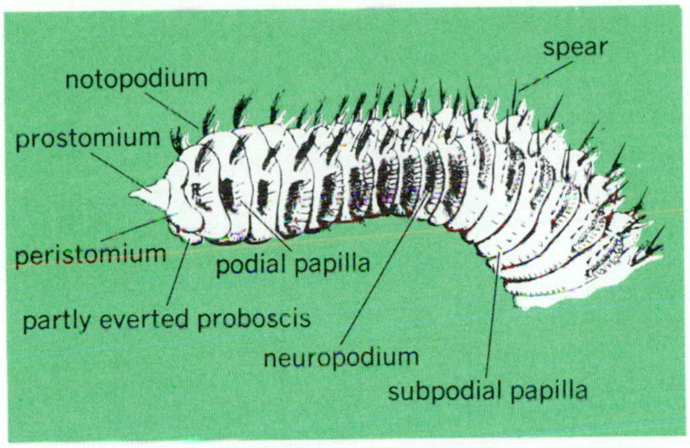

Fig. 1. Terminology of the anterior parts of the body, based on *Phylo* (Orbiniidae).

species have been named from worldwide areas; about one-fourth of this number may be synonymous. Polychaeta (meaning "many setae") is conveniently though not clearly divisible into the Errantia, or free-moving annelids, and Sedentaria, or tubicolous families. *See* ERRANTIA; SEDENTARIA.

The body may be long, cylindrical, and multisegmented, or short and compact, with a limited number of segments. It consists of prostomium (Fig. 1), or head; peristomium, or first segment around the mouth; trunk, or body proper; and tail region, or pygidium. Most segments have highly diagnostic paired, lateral fleshy appendages called parapodia. These are provided with secreted supporting rods and spreading fascicles of setae, or hooks, which display remarkable specificity.

The anterior end, or prostomium, may be a simple lobe derived from the larval trochophore, modified as a pseudoannulated cone, or covered by peristomial structures so as to be invisible. Oral tentacles for food gathering may be eversible from the buccal cavity; they may be long, slender, or thick and their surface smooth or papillated.

The anterior, preoral end may be developed as a thick, fleshy papillated, nonretractile proboscis (Fig. 2), or the prostomium may be completely retractile into the first several segments and protected by a cage formed of setae directed forward, or concealed by a compact operculum formed of setae of the first several segments. The anterior end of the alimentary tract is muscular or epithelial; it may be covered with soft papillae or hard structures. These structures function for secretion, food gathering, and maintaining traction; they are named for their form or function.

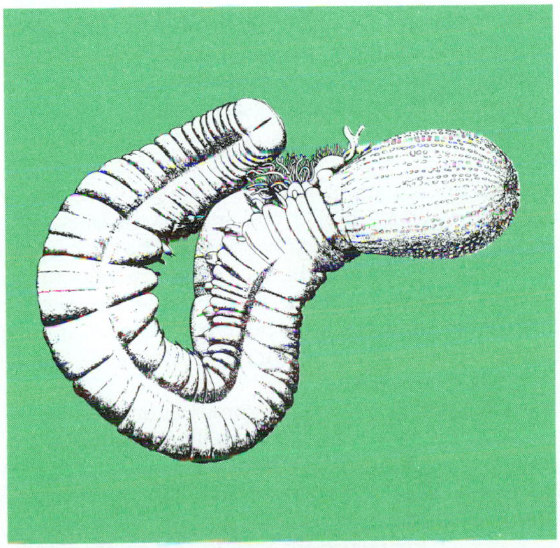

Fig. 2. Nonretractile proboscis organ preceding prostomium in *Artacama* (Terebellidae).

The trunk is the main body region and is composed of metameres numbering few to many. They may be similar to one another (homonomous) as in Errantia, or different (heteronomous) resulting in anterior thoracic and posterior abdominal regions.

Reproduction is highly evolved and diversified; it can be sexual or asexual. Sexual reproduction is usually dioecious, with the two sexes similar. In rare cases it is dimorphic.

Polychaetes range in length from a fraction of 0.04 in. (1 mm) to more than 144 in. (360 cm). Colors and patterns are varied and specific, due to pigment and refraction of light. Littoral, warm-water species may be brilliant and multicolored, whereas polar and deep-water species tend to be drab or sometimes melanistic to almost black.

Most polychaetes are free-living; some of the remaining members are commensal with another animal for attachment surface, for food, or for protection. Polychaetes are distributed in all marine habitats and show remarkable specificity according to latitude, depth, and kinds of substrata. Most of the families tend to be represented in any major geographic area, although taxa may differ with place. *See* ANNELIDA. [O.H.]

Polychlorinated biphenyls A generic term covering a family of partially or wholly chlorinated isomers of biphenyl. The commercial mixtures generally contain 40–60% chlorine with as many as 50 different detectable isomers present (see illustration). The polychlorinated biphenyl (PCB) mixture is a colorless, viscous fluid, relatively insoluble in water, that can withstand very high temperatures without degradation. PCBs

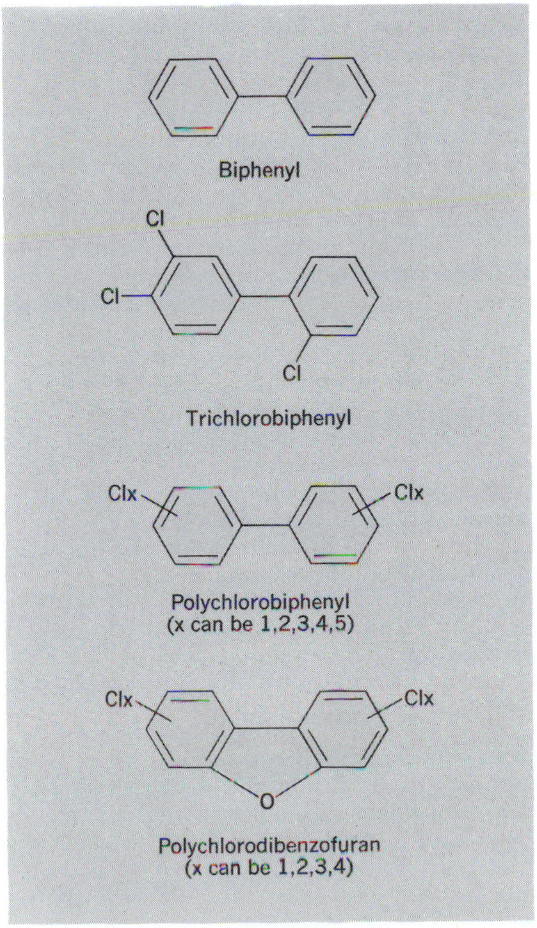

Structural formulas of chlorinated biphenyls and related compounds.

do not conduct electricity and the more highly chlorinated isomers are not readily degraded in the environment.

The major uses of PCBs can be grouped in three major categories: open uses, partially closed systems, and closed systems. Examples of uses in the first category are in paints, inks, plastics, and paper coatings. The PCBs in all these products are in direct contact with the environment and can be leached out by water or vaporize into the atmosphere. The so-called carbonless carbon paper contains PCBs in the encapsulated ink and has been claimed to be responsible for the PCBs found extensively in recycled paper. They have also been used as plasticizers in polyvinyl chloride (PVC) and chlorinated rubbers. Uses of PCBs regarded as partially closed systems include the working fluid in heat exchangers and hydraulic systems.

It is not known exactly how PCBs are released into the environment or in what quantities. However, sewage outfalls, industrial and municipal disposal, leaking from dumps, and burning of refuse are certainly important sources. PCBs are found universally in the sewage of major cities, although the atmosphere is the major pathway of global transport.

The U.S. Food and Drug Administration has placed strict limits on PCBs in food products. Studies of the toxic properties of PCBs have revealed some serious effects on humans. There is evidence that one class, the polybrominated biphenyls (PBBs), have caused abnormalities in the blood of exposed humans that indicate damage to the immune system.   [G.H.]

Polycladida   A class of marine Turbellaria which are several millimeters to several centimeters in length and whose leaflike bodies have a central intestine with radiating branches. Most species live in the littoral zone on the bottom, on seaweed or on other objects, or as commensals in the shells of mollusks and hermit crabs. None are parasitic. Except in warm waters, they are seldom brightly colored. *See* TURBELLARIA.   [E.R.J.]

Polycythemia   Any disorder characterized by an increase in the numbers of circulating red cells and usually by a concomitant increase in hemoglobin. Relative polycythemia is really the result of hemoconcentration in which there is fluid loss from the blood, with corresponding increases in proportions of cellular elements. Such polycythemia is seen in excessive vomiting, diarrhea, dehydration, and similar conditions. *See* ERYTHROCYTOSIS.

Secondary polycythemias result from a compensatory increase in the formation of red cells following hypoxia of the bone marrow. This condition is most often associated with chronic conditions, such as heart or lung diseases and prolonged exposure to high altitudes.

Primary polycythemia or polycythemia vera is a chronic and ultimately fatal disease of middle and old age in which there is a gradual increase in the number of red cells and usually an increase in the number of platelets and leukocytes. The viscosity of the blood is increased, which tends to slow the rate of blood flow; this factor, together with the high platelet count, predisposes to the thromboses commonly seen in the condition. Paradoxically there is also a hemorrhagic tendency. [C.Ho.]

Polyene   Any hydrocarbon containing two or more double bonds. The double bonds may be either alternating with single bonds, as in conjugated polyenes (I), or contiguous, as

$$-CH=CH-CH=CH- \qquad C=C=C$$
$$(I) \hspace{4cm} (II)$$

in cumulenes (II). Compounds with separated double bonds act much like simple olefins. *See* ALKENE; CHEMICAL BONDING.

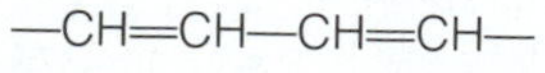

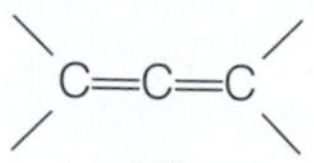

Interaction between conjugated double bonds is demonstrated by the fact that the single bond in polyenes is actually intermediate in length between a single bond and a double bond. This indicates an appreciable amount of double-bond character attributed to the polyene single bond. Thus, butadiene, one of the most important polyenes, can be considered to be a resonance hybrid (III).

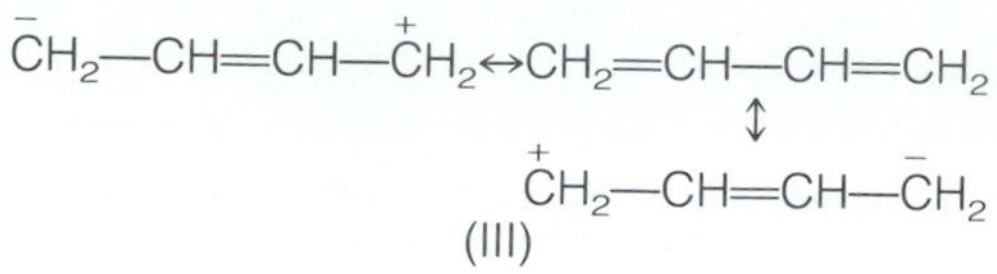

$$\overset{-}{C}H_2-CH=CH-\overset{+}{C}H_2 \leftrightarrow CH_2=CH-CH=CH_2$$
$$\updownarrow$$
$$\overset{+}{C}H_2-CH=CH-\overset{-}{C}H_2$$
$$(III)$$

The lowest member of the polyene series is allene (IV), which is found in petroleum refining in combination with its isomer, propyne, or methylacetylene (V). In allene the two terminal groups are in perpendicular planes; therefore, if the groups are different, optical isomers can be separated. Even numbered cumulenes, however, show only cis-trans isomerism, since substituents are in the same plane. Cumulenes are very reactive species; a few allenes occur in nature. *See* ALIPHATIC HYDROCARBON; MOLECULAR ISOMERISM; PETROLEUM PROCESSING; RESONANCE (MOLECULAR STRUCTURE).   [D.S.Br.]

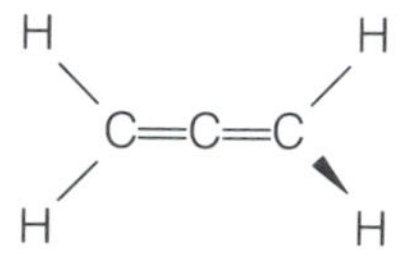

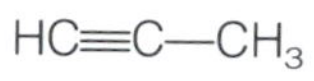

Polyester resins   Polymeric materials in which ester groups are in the main chains. The aliphatic polyesters tend to be relatively soft, and the aromatic derivatives are usually hard and brittle, or tough. The properties of either group may be modified by cross linking, crystallization, plasticizers, or fillers. *See* ESTER.

The commercial products are: alkyds which are used in paints, enamels, and molding compounds; unsaturated polyesters or unsaturated alkyds which are used extensively with fiber glass for boat hulls and panels; aliphatic saturated polyesters; aromatic polyesters, such as polyethylene terephthalate which is used in the form of fibers and films; and the aromatic polycarbonates. The polydiallyl esters, while frequently listed with the polyesters, are not true polyesters as defined above. For a discussion of polyvinyl esters *see* POLYVINYL RESINS.

The alkyds are commonly used as coatings. Combinations of conventional vegetable drying oils and alkyd resins represent the basis of most of the oil-soluble paints. The drying oil–alkyd may be further modified by the inclusion of a vinyl monomer, such as styrene. Some of the styrene polymerizes, probably as a graft polymer, and the remainder polymerizes and copolymerizes in the final drying or curing of the paint. *See* DRYING OIL; PAINT; POLYMERIZATION.

The unsaturated polyesters, in combination with glass fiber, have found applications as panels, roofing, radar domes, boat hulls, and protective armor for soldiers. The compositions are distinguished by ease of fabrication and high impact resistance. *See* POLYMERIC COMPOSITE.

Saturated aliphatic polyesters have long been frequently used as intermediates in the preparation of prepolymers for making segmented polyurethanes. Lactone rings can also be opened to yield linear polyesters.

The aromatic polyesters which have achieved general importance are the polyethylene terephthalates, which yield very strong and chemically resistant fibers and films. Polyethylene terephthalate is the principal ingredient of polyester fibers. Other aromatic polyesters include polybutylene terephthalate, which may be molded or extruded to yield materials that can replace metals or thermoset resins in some automotive, electrical, and specialty applications, especially when reinforced with glass fibers or mineral fillers.

Aromatic polycarbonates are a strong, tough group of thermoplastic polymers formed most frequently from bisphenol A and phosgene. The products, polycarbonates, are noted for high softening temperatures, and high impact resistance, clarity, and resistance to creep. Polycarbonate is usually available as a molding compound. Because of its high strength, toughness, and softening point, the resin, both by itself and as a glass-reinforced material, has found many electrical domestic and engineering applications. It is often used to replace glass and metals. Examples include bottles, unbreakable windows, appliance parts, electrical housings, marine propellers, and shotgun shells. Flame-retardant grades are of interest because of low toxicity and smoke emission on burning.

Polydiallyl esters are polymers of diallyl esters. Thermosetting molding compounds may be produced by careful limitation of the initial polymerization to yield a product which is fusible. Major applications are in electronic components, sealants, coatings, and glass-fiber composites. *See* PLASTICS PROCESSING. [J.A.M.]

Polyether resins Thermoplastic or thermosetting materials which contain ether-oxygen linkages, —C—O—C—, in the polymer chain. Depending upon the nature of the reactants and reaction conditions, a large number of polyethers with a wide range of properties may be prepared. The main groups of polyethers in use are epoxy resins, phenoxy resins, polyethylene oxide and polypropylene oxide resins, polyoxymethylene, and polyphenylene oxides.

The epoxy resins form an important and versatile class of cross-linked polyethers characterized by excellent chemical resistance, adhesion to glass and metals, electrical insulating properties, and ease and precision of fabrication. Various fillers such as calcium carbonate, metal fibers and powders, and glass fibers are commonly used in epoxy formulations in order to improve such properties as the strength and resistance to abrasion and high temperatures. Some reactive plasticizers act as curing agents, become permanently bound to the epoxy groups, and are usually called flexibilizers. Rubbery polymers are added to improve toughness and impact strength. Epoxies are commonly used in protective coatings. They are used as potting or encapsulating compositions for the protection of delicate electronic assemblies from the thermal and mechanical shock of rocket flight, and as dies for stamping metal forms.

Polyethylene oxide and polypropylene oxide are thermoplastic products whose properties are greatly influenced by molecular weight. Low-to-moderate-molecular-weight polyethylene oxides vary in form from oils to waxlike solids. They are relatively nonvolatile, are soluble in a variety of solvents, and have found many uses as thickening agents, plasticizers, lubricants for textile fibers, and components of various sizing, coating, and cosmetic preparations. The polypropylene oxides of similar molecular weight have somewhat similar properties, but tend to be more oil-soluble (hydrophobic) and less water-soluble (hydrophilic). While polyalkylene oxides are not of interest as such in structural materials, polypropylene oxides are used extensively in the preparation of polyurethane foams.

Phenoxy resins are transparent, strong, ductile, and resistant to creep, and, in general, resemble polycarbonates in their behavior. The major application is as a component in protective coatings, especially in metal primers. *See* POLYESTER RESINS.

Polyphenylene oxide (PPO) is the basis for an engineering plastic characterized by chemical, thermal, and dimensional stability. Polyphenylene oxide is outstanding in its resistance to water. Uses include medical instruments, pump parts, and insulation.

Polyoxymethylene, or polyacetal, resins are polymers of formaldehyde. Having high molecular weights and high degrees of crystallinity, they are strong and tough and are established in the general class of engineering thermoplastics. Polyacetals are typically resistant to fatigue, creep, organic chemicals (but not strong acids or bases), and have low coefficients of friction. Electrical properties are also good. The combination of properties has led to many uses such as plumbing fittings, pump and valve components, bearings and gears, computer hardware, automobile body parts, and appliance housings. *See* EPOXIDATION; PLASTICS PROCESSING; POLYMERIZATION. [J.A.M.]

Polyethylene glycol Any of a series of water-soluble polymers with the general formula

$$HO—(CH_2—CH_2—O)_n—H$$

These colorless, odorless compounds range in appearance from viscous liquids to waxy solids.

The low-molecular-weight members, diethylene glycol (n = 2) through tetraethylene glycol (n = 4), are produced as pure compounds and find use as humectants, dehydrating solvents for natural gas, textile lubricants, heat-transfer fluids, solvents for aromatic hydrocarbon extractions, and intermediates for polyester resins and plasticizers.

The intermediate members of the series with average molecular weights of 200 to 20,000 are used commercially in ceramic, metal-forming, and rubber-processing operations; as drug suppository bases and in cosmetic creams, lotions, and deodorants; as lubricants; as dispersants for casein, gelatins, and inks; and as antistatic agents.

The highest members of the series have molecular weights from 100,000 to 10,000,000. They are of interest because of their ability at very low concentrations to reduce friction of flowing water. *See* ETHYLENE OXIDE; GLYCOL. [R.K.Ba.]

Polyfluoroolefin resins Resins distinguished by their resistance to heat and chemicals and by the ability to crystallize to a high degree. Several main products are based on tetrafluoroethylene, $F_2C{=}CF_2$ (TFE), hexafluoropropylene, $F_2C{=}CFCF_3$ (HFP), and monochlorotrifluoroethylene, $FClC{=}CF_2$ (CTFE). *See* POLYVINYL RESINS.

The polymer polytetrafluoroethylene is insoluble, resistant to heat and chemical attack, and in addition, has the lowest coefficient of friction of any solid. Aqueous suspensions of the polymer can also be used for coating various articles. However, special surface treatments are required to ensure adhesion because polytetrafluoroethylene does not adhere well to anything. Polytetrafluoroethylene (TFE resin) is used for bearings, valve seats, packings, gaskets, coatings, and tubing, and can withstand relatively severe conditions. Because of its excellent electrical properties, polytetrafluoroethylene is useful when a dielectric material is required for service at a high temperature. The nonadhesive quality is turned to advantage in use of polytetrafluoroethylene to coat articles such as rolls and cookware to which materials might otherwise adhere.

The properties of polymonochlorotrifluoroethylene (CTFE resin) are generally similar to those of polytetrafluoroethylene; however, the presence of the chlorine atoms in the former causes the polymer to be a little less resistant to heat and to chemicals. Thus the applications of polychlorotrifluoroethylene are in general similar to those for polytetrafluoroethylene.

Polyvinylidene fluoride has properties generally similar to those of other fluorinated resins: relative inertness, low dielectric constant, and thermal stability (up to about 150°C or 300°F). The resins (PVF_2 resins) are, however, stronger and less susceptible to creep and abrasion than TFE and CTFE resins. Applications of polyvinylidene fluoride are mainly as electrical insulation, piping, process equipment, and as a protective coating in the form of a liquid dispersion.

Several types of fluorinated, noncrystallizing elastomers have been developed in order to meet needs (usually military) for rubbers which possess good low-temperature behavior with a high degree of resistance to oils and to heat, radiation, and weathering. *See* HALOGENATED HYDROCARBON; PLASTICS PROCESSING; POLYMERIZATION. [J.A.M.]

Polygalales An order of flowering plants, division Magnoliophyta (Angiospermae), in the subclass Rosidae of the class Magnoliopsida (dicotyledons). The order consists of 7 families and nearly 1900 species, mostly of tropical and subtropical regions. The vast majority of the species belong to only 3 families, the Malpighiaceae (about 800 species), Polygalaceae (about 750 species), and Vochysiaceae (about 200 species). Within its subclass the order is distinguished by its simple leaves and usually irregular flowers, which have the perianth and stamens attached directly to the receptacle (hypogynous), and often have the anthers opening by terminal pores instead of longitudinal slits. *See* MAGNOLIOPSIDA; ROSIDAE. [A.Cr.]

Polygon A polygon of n sides ($n < 2$) is a figure formed by joining an ordered set of n points called vertices by line segments called sides that connect each point to its immediate successor (where the first point is considered the successor of the last). Euclid considered a polygon to be this figure and also the plane region bounded by these vertices and sides. Modern usage considers the polygon to be any set of line segments and vertices forming a closed broken line. A polygon is simple if each side intersects only its two neighbors. It is plane or skew according to whether its sides do or do not lie in a plane. A simple plane polygon is convex if the polygonal region that it bounds lies wholly on one side of the (extended) line of each side. A polygon of 3, 4, 5, 6, 7, 8, 10, or 12 sides is called a triangle (3), quadrilateral (4), pentagon (5), hexagon (6), heptagon (7), octagon (8), decagon (10), or dodecagon (12). *See* HEXAGON; OCTAGON; PENTAGON; QUADRILATERAL; REGULAR POLYTOPES; SQUARE; TRAPEZOID; TRIANGLE. [J.S.F.]

Polygonales An order of flowering plants, division Magnoliophyta (Angiospermae), in the subclass Caryophyllidae of the class Magnoliopsida (dicotyledons). The order consists only of the family Polygonaceae, with about 800 species, most abundant in north temperate regions. Within its subclass the order is characterized by its well-developed endosperm; an ovary with but one chamber (unilocular), mostly tricarpellate with a single basal ovule which is usually straight and has the micropyle at the opposite end from the stalk (orthotropous); the flowers are often trimerous, that is, with floral parts in sets of three, and usually two or three sets of stamens. Rhubarb (*Rheum rhaponticum*) and buckwheat (*Fagopyrum esculentum*) are familiar members of the Polygonales. *See* BUCKWHEAT; CARYOPHYLLIDAE; MAGNOLIOPSIDA; RHUBARB. [A.Cr.]

Polyhedron A solid, all of whose boundary points lie in a finite number of planes (at least four). The bounding sections formed by these planes are plane polygonal regions called faces of the polyhedron, whose sides are called edges of the polyhedron, and whose vertices are called vertices of the polyhedron. According to the number of faces, a polyhedron is called a tetrahedron (4 faces), pentahedron (5), hexahedron (6), octahedron (8), dodecahedron (12), or icosahedron (20). For any convex polyhedron (but not for all polyhedrons) the numbers of vertices V, edges E, and faces F are related by the equation $V - E + F = 2$.

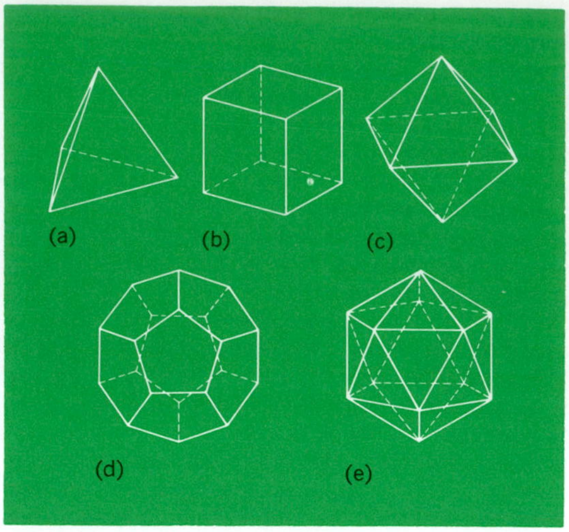

The five regular polyhedrons. (*a*) Tetrahedron. (*b*) Cube. (*c*) Octahedron. (*d*) Dodecahedron. (*e*) Icosahedron.

Regular polyhedrons (see illustration) are convex polyhedrons whose faces are congruent regular polygons, forming equal dihedral angles at each edge. There are only five such regular solids, called platonic solids. *See* CUBE; EUCLIDEAN GEOMETRY; OCTAHEDRON; PARALLELEPIPED; PRISM; PRISMATOID AND PRISMOID; PYRAMID AND FRUSTUM; REGULAR POLYTOPES; SOLID (GEOMETRY); TETRAHEDRON. [J.S.F.]

Polymer The terms polymer, high polymer, macromolecule, and giant molecule are used to designate high-molecular-weight materials of either synthetic or natural origin. Plastics are relatively stiff at room temperature, rubbers or elastomers are flexible and retract quickly after stretching, and fibers are especially strong filamentary materials. Coatings are generally either plastics or rubbers which have been applied as a thin layer on a substrate. In practice, plastics, rubbers, fibers, and coatings are used as formulations of the polymers with other ingredients such as fillers, pigments, plasticizers, flow improvers, and stabilizers against aging and degradation.

For a discussion of the general methods of preparation *see* POLYMERIZATION. For descriptions of typical synthetic polymers of the condensation type *see* CELLULOSE; PHENOLIC RESIN; POLYAMIDE RESINS; POLYESTER RESINS; POLYETHER RESINS; POLYSULFIDE RESINS; POLYURETHANE RESINS; SILICONE RESINS; UREA-FORMALDEHYDE-TYPE RESINS. For descriptions of some important members of the addition-type synthetic polymers *see* HYDROCARBON RESIN; POLYACRYLATE RESIN; POLYACRYLONITRILE RESINS; POLYFLUOROOLEFIN RESINS; POLYOLEFIN RESINS; POLYSTYRENE RESIN; POLYVINYL RESINS.

Natural products, rubbers, fibers, and paints are treated in other articles. Biological polymers, as in enzymes and living tissues, are treated in the appropriate articles. *See* MANUFACTURED FIBER; NATURAL FIBER; PROTEIN; RUBBER; SURFACE COATING; TERPENE.

Properties. The properties of polymeric materials are determined by the molecular properties of the macromolecules, the morphology, and the type of formulation involving plasticizers or fillers. The morphology in turn depends on the conditions of fabrication in which molecular orientation or crystallization may be induced. Properties also depend on the temperature and elapsed time of the measurement.

Molecular properties include molecular size and weight, molecular structure or architecture, molecular-weight distribution, polarity, and flexibility of the polymeric chains (or chain segments between cross-links in cured or vulcanized polymers). Taken together, molecular properties determine the attractive forces between the molecules, the morphology or arrangement of masses of molecules, and the general behavior of the polymer.

The desirable properties of high polymers (strength and resistance to solvents) increase rapidly with increasing molecular weight in the low ranges of molecular weight, and more slowly in the high ranges. On the other hand, the melt viscosity increases with molecular weight in the opposite manner, slowly at first and rapidly in the higher-molecular-weight range. The ease of fabrication (molding, extrusion, and shaping) of polymeric compositions varies inversely with the melt viscosity; that is, the materials become increasingly difficult to mold or extrude at very high values of molecular weight. The optimum molecular weight of a polymer frequently varies for different applications and different methods of fabrication.

The atoms in the chains are held together by primary valence bonds. If it were possible to apply a force of tension only to the primary valence bonds, then a tensile strength of more than 2,000,000 psi (1.4×10^{10} pascals) would be observed. In reality, however, under tension the molecules slip past one another and the resistance to that slippage is due to the effective attractive forces (van der Waals forces) between the molecules. Also, flaws reduce strength by initiating cracks.

The flexibility of linear polymer chains and of the segments between cross-links in cured products is decreased by the presence of polar groups and regularity in the molecular structure. A nonpolar, irregular chain should be the most flexible. Products containing highly flexible chains are rubbery, soft, and nonbrittle, with relatively high resistance to impact and to tear.

When a close fit between chains is possible, crystallization can take place spontaneously or upon drawing or cooling. The presence of polar attracting groups enhances the tendency to crystallize. It is also possible to form extended-chain crystallites in which molecules are packed together longitudinally in fibrillar arrays. This may be achieved by the use of special extrusion or spinning techniques. Such highly anisotropic crystallites confer high levels of strength and stiffness in the chain direction.

The illustration shows responses of typical rubbers, plastics, and fibers to large stresses. A reinforced or crystallizable rubber exhibits a relatively low value of breaking stress, but high elongation. A ductile plastic such as polyethylene exhibits yielding, drawing, and, at high elongations, some strengthening due to orientation. A brittle plastic such as polystyrene does not yield much, and breaks at a low elongation; a fiber exhibits the highest strength, high stiffness, and low elongation. In all cases, the shape and position of the stress-strain curve are determined by the balance struck between cohesion and segmental mobility at the temperature and testing rate concerned.

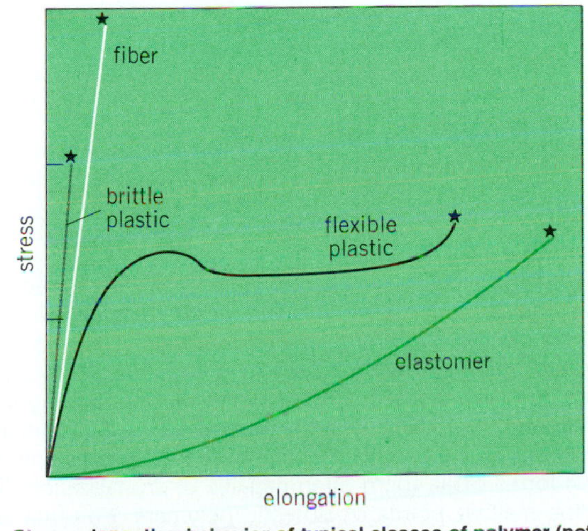

Stress-elongation behavior of typical classes of polymer (not drawn to scale). The star represents the point at which failure occurs.

As plastics are used extensively in engineering applications, the toughness or the ability to resist fracture assumes considerable importance. Much research has been directed to the understanding of fracture processes, on one hand, and to the improvement of toughness, on the other. In general, toughness appears to require some combination of high modulus with the ability to dissipate applied energy by relaxation of molecular segments or of added materials.

Compounding. Plastic masses, rubber formulations, coatings, and other polymeric compositions may contain age inhibitors, strengthening and coloring pigments, flow improvers, and plasticizing or softening agents.

The addition of plasticizers to a polymeric material causes it to be softer and more rubbery in character. Plasticizers are held in association with the polymer chains by secondary valence forces. They separate the molecules, thus reducing the effective intermolecular attractive forces. Finely divided polar substances may act as strengthening fillers for plastics. Age inhibitors, such as free-radical inhibitors and light-masking agents, are almost always incorporated in polymeric compositions.

Polyblends are produced by the addition of small amounts of a rubbery polymer to a polymeric glass. The impact strength of the glass thus is substantially increased. The rubbery polymer is not truly compatible and exists as a finely dispersed separate phase. Interpenetrating polymer networks, in which two different networks are intertwined, constitute an interesting special case of polyblends. The phenomenon has somewhat similar counterparts in inorganic glass technology and in physical metallurgy. *See* PLASTICS PROCESSING; POLYMERIC COMPOSITE. [J.A.M.]

Polymer-supported reaction An organic chemical reaction where one of the species, such as the substrate, the reagent, or a catalyst, is bound to a cross-linked, and therefore insoluble, polymer support. A major attraction is that at the end of the reaction period the polymer-supported species can be separated cleanly and easily, usually by filtration, from the soluble species. This easy separation can greatly simplify product isolation procedures, and it may even allow the polymer-supported reactions to be automated. Because it is possible to reuse or recycle polymer-supported reactants and because they are insoluble, involatile, easily handled, and easily recovered, polymer-supported reactants are attractive from an environmental viewpoint. *See* ASYMMETRIC SYNTHESIS; CATALYSIS.

Polymer-supported reactants are usually prepared in the form of beads of about 50–100 micrometers' diameter. Such beads can have a practically useful loading only when their interiors are functionalized, that is, carry functional groups. With a typical polymer-supported reactant, more than 99% of the reactive groups are inside the beads. An important consequence is that for soluble species to react with polymer-supported species the former must be able to diffuse freely into the polymer beads. *See* GEL; POLYMER.

In a typical application of reactions involving polymer-supported substrates, a substrate is first attached to an appropriately functionalized polymer support. The synthetic reactions of interest are then carried out on the supported species. Finally, the product is detached from the polymer and recovered. Often the polymer-supported species has served as a protecting group that is also a physical "handle" to facilitate separation.

Reactions involving polymer-supported reagents are generally more useful than those involving polymer-supported substrates, because no attachment or detachment reactions are needed, and it is not necessary for all the polymer-supported species to react in high yield. Indeed, polymer-supported reagents are often used in excess to drive reactions to high conversions. [P.H.]

Polymeric composite Any of the combinations or compositions that comprise two or more materials as separate phases, at least one of which is a polymer. By combining a polymer with another material, such as glass, carbon, or another polymer, it is often possible to obtain unique combinations or levels of properties. Typical examples of synthetic polymeric composites include glass-, carbon-, or polymer-fiber-reinforced thermoplastic or thermosetting resins, carbon-reinforced rubber, polymer blends, silica- or mica-reinforced resins, and polymer-bonded or mica-reinforced resins, and polymer-bonded or -impregnated concrete or wood. It is also often useful to consider as composites such materials as coatings (pigment-binder combinations) and crystalline polymers (crystallites in a polymer matrix). Typical naturally occurring composites include wood (cellulosic fibers bonded with lignin) and bone (minerals bonded with collagen). On the other hand, polymeric compositions compounded with a plasticizer or very low proportions of pigments or processing aids are not ordinarily considered as composites.

Typically, the goal is to improve strength, stiffness, or toughness, or dimensional stability by embedding particles or fibers in a matrix or binding phase. A second goal is to use inexpensive, readily available fillers to extend a more expensive or scarce resin. Still other applications include the use of some fillers such as glass spheres to improve processability, the incorporation of dry-lubricant particles such as molybdenum sulfide to make a self-lubricating bearing, and the use of fillers to reduce permeability.

For discussion of common thermosetting matrixes *see* POLYESTER RESINS; POLYETHER RESINS. For details on glass and carbon fibers *see* GLASS; GRAPHITE. For information on common thermoplastic matrixes, polyblends, polymer concretes, and reinforced rubber *see* POLYACRYLATE RESIN; POLYAMIDE RESINS; POLYMER; POLYSTYRENE RESIN; POLYVINYL RESINS; RUBBER. *See also* COMPOSITE MATERIAL. [J.A.M.]

Polymerization The linking of small molecules (monomers) to make larger molecules. Polymerization requires that each small molecule have at least two reaction points or functional groups. There are two distinct major types of polymerization processes, condensation polymerization, in which the chain growth is accompanied by elimination of small molecules such as H_2O or CH_3OH, and addition polymerization, in which the polymer is formed without the loss of other materials. There are many variants and subclasses of polymerization reactions.

An example of the condensation process is the reaction (1)

$$n\mathrm{H_2N{-}(CH_2)_5{-}\overset{\overset{\large O}{\|}}{C}{-}OH} \xrightarrow{\text{Catalyst}}$$

ε-Aminocaproic acid

$$\mathrm{H}\left[\mathrm{-\overset{\overset{\large H}{|}}{N}-(CH_2)_5-\overset{\overset{\large O}{\|}}{C}-}\right]_n\mathrm{OH} + (n-1)\mathrm{H_2O} \qquad (1)$$

Polyamide, nylon-6

of ε-aminocaproic acid in the presence of a catalyst to form the polyamide, nylon-6. The repeating structural unit is equivalent to the starting material minus H and OH, the elements of water. The molecules formed are linear because the total functionality of the reaction system (functional groups per molecule) is always two. However, if a trifunctional material, such as a tricarboxylic acid, were added to the nylon-6,6 polymerizing mixture, a branched polymeric structure would result, because two of the carboxylic groups would participate in one polymer chain, and the third carboxylic group would start the growth of another. Under appropriate conditions, these chains can become bridges between linear chains and the polymer becomes cross-linked. The arrangements of the chains are shown in Fig. 1. *See* CONDENSATION REACTION.

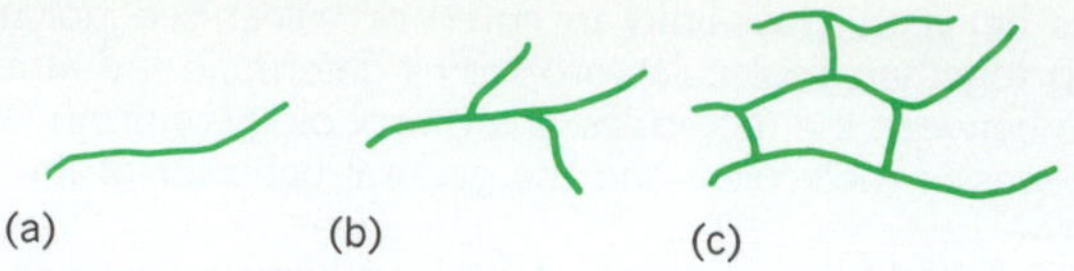

Fig. 1. Polymer chains. (*a*) Linear polymer chain. (*b*) Branched polymer chain. (*c*) Cross-linked polymer chain.

An example of addition polymerization is reaction (2). The

$$n\mathrm{H_2C{=}O} \xrightarrow[\text{Initiator (I)}]{\text{Catalyst or}} [\mathrm{-CH_2-O-}]_n \qquad (2)$$

Formaldehyde → Polyoxymethylene

structure of the repeating unit is the difunctional monomeric unit, or "mer." In the presence of catalysts or initiators, the monomer yields a polymer by the joining together of n mers. If n is a small number, 2–10, the products are dimers, trimers, tetramers, or oligomers, and the materials are usually gases, liquids, oils, or brittle solids. In most solid polymers, n has values ranging from a few score to several hundred thousand, and the corresponding molecular weights range from a few thousand to several million. The end groups of this example of addition polymers are shown to be fragments of the initiator.

If only one monomer is polymerized, the product is called a homopolymer. The polymerization of a mixture of two monomers of suitable reactivity leads to the formation of a copolymer, a polymer in which the two types of mer units have entered the chain in a more or less random fashion. If chains of one homopolymer are chemically joined to chains of another, the product is called a block or graft copolymer:

A—B—A—A—B—B—B
Random copolymer

A—A—A—A—B—B—B—B—B
Block copolymer

A—A—A—A—A
|
B
|
B
|
B
Graft polymer

Isotactic and syndiotactic (stereoregular) polymers are formed in the presence of complex catalysts, or by changing polymerization conditions, for example, by lowering the temperature. The groups attached to the chain in a stereoregular polymer are in a spatially ordered arrangement. The configuration of these ordered polymers and the disordered, atactic form is shown in Fig. 2. The regular structures of the isotactic and syndiotactic forms make them often capable of crystallization. The crystalline melting points of isotactic polymers are often substantially higher than the softening points of the atactic product.

In Fig. 2 each carbon atom to which a phenyl group is attached is asymmetrically substituted. For illustration, the

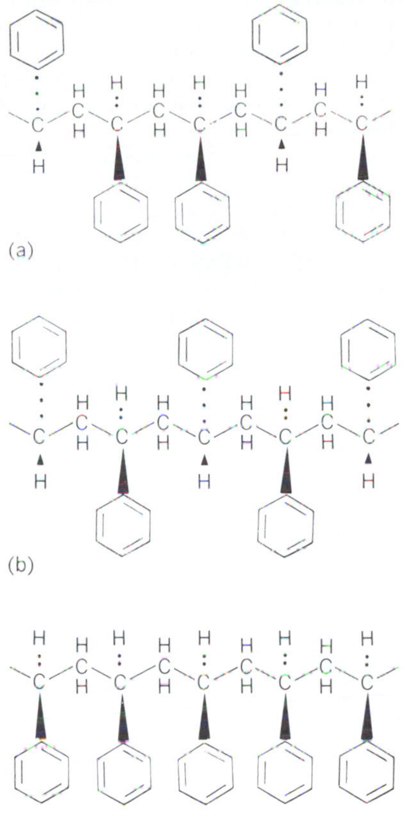

Fig. 2. Spatially oriented polymers. (*a*) Atactic (random; *dlldl* or *lddld*, and so on). (*b*) Syndiotactic (alternating; *dldl*, and so on). (*c*) Isotactic (right- or left-handed; *dddd*, or *llll*, and so on).

heavily marked bonds are assumed to project up from the paper, and the dotted bonds down. Thus in a fully syndiotactic polymer, asymmetric carbons alternate in their left- or right-handedness (alternating *d*, *l* configurations), while in an isotactic polymer, successive carbons have the same steric configuration (*d* or *l*).

Among the several kinds of polymerization catalysis, free-radical initiation has been most thoroughly studied and is most widely employed. Atactic polymers are readily formed by free-radical polymerization, at moderate temperatures, of vinyl and diene monomers and some of their derivatives. *See* CATALYSIS; FREE RADICAL.

Some polymerizations can be initiated by materials, often called ionic catalysts, that contain highly polar reactive sites or complexes. The term heterogeneous catalyst is often applicable to these materials because many of the catalyst systems are insoluble in monomers and other solvents. These polymerizations are usually carried out in solution from which the polymer can be obtained by evaporation of the solvent or by precipitation on the addition of a nonsolvent. A distinguishing feature of complex catalysts is the ability of some representatives of each type to initiate stereoregular polymerization at ordinary temperatures or to cause the formation of polymers which can be crystallized. *See* CHAIN REACTION (CHEMISTRY); CHEMICAL DYNAMICS; HETEROGENEOUS CATALYSIS; INHIBITOR (CHEMISTRY); INORGANIC POLYMER; ORGANIC REACTION MECHANISM; PLASTICS PROCESSING; POLYMER. [J.A.M.]

Polymorphism (crystallography) The property of crystallizing in two or more forms. The term is applied to crystals of the same substance having a different structure. Substances such as $CaCO_3$ which exist in two crystal forms are said to be dimorphous, while substances such as TiO_2 which appear in three forms are termed trimorphous. The modifications of these substances are stable at normal temperature and in a comparable temperature range. Other polymorphic forms, such as those of SiO_2 (quartz, cristoballite, and trydimite), have a specific nonoverlapping stability range. Many polymorphic modifications are only stable under high pressure. The polymorphic forms of the elements are called allotropic modifications. *See* CRYSTAL. [W.C.D.]

Polymorphism (genetics) A form of genetic variation, specifically a discontinuous variation, occurring within plant and animal species in which distinct forms exist together in the same population, even the rarest of them being too common to be maintained solely by mutation. Thus the human blood groups are examples of polymorphism, while geographical races are not; nor is the diversity of height among humans because height is "continuous," and does not fall into distinct tall, medium, and short types. *See* BLOOD GROUPS; MUTATION.

Distinct forms must be controlled by some switch which can produce one form or the other without intermediates such as those arising from environmental differences. This clear-cut control is provided by the recombination of the genes. Each gene may have numerous effects and, in consequence, all genes are nearly always of importance to the organism by possessing an overall advantage or disadvantage. They are very seldom of neutral survival value as minor individual variations in appearance often are. Thus a minute extra spot on the hindwings of a tiger moth is in itself unlikely to be of importance to the survival of the insect, but the gene controlling this spot is far from negligible since it also affects fertility. *See* RECOMBINATION (GENETICS).

Genes having considerable and discontinuous effects tend to be eliminated if harmful, and each gene of this kind is therefore rare. On the other hand, those that are advantageous and retain their advantage spread through the population so that the population becomes uniform with respect to these genes. Evidently, neither of these types of genes can provide the switch mechanism necessary to maintain a polymorphism. That can only be achieved by a gene which has an advantage when rare, yet loses that advantage as it becomes commoner.

Occasionally there is an environmental need for diversity within a species, as in butterfly mimicry. Mimicry is the resemblance of different species to one another for protective purposes, chiefly to avoid predation by birds. Sexual dimorphism also falls within the definition of genetic polymorphism. In any species, males and females are balanced at optimum proportions which are generally near equality. Any tendency for one sex to increase relative to the other would be opposed by selection. Batesian mimicry and sexual dimorphism, however, are rather special instances. *See* GENETICS. [E.B.F.]

Polymyxin A basic polypeptide antibiotic produced as a fermentation product by certain strains of the bacterium *Bacillus polymyxa*. The antibacterial action of the polymyxins is uniquely directed toward gram-negative bacteria. Cautious therapeutic applications have saved many lives in the treatment of urinary tract infections, bacterial meningitis, bacteremias, and burn and wound infections. The polymyxins are strongly bactericidal, and resistance toward them is not readily developed by susceptible microorganisms. *See* ANTIBIOTIC. [R.G.B.]

Polynomial systems of equations Systems of mathematical equations of the form of system (1). Each

$$\begin{aligned} f_1(x_1,x_2,\ldots,x_n) &= 0 \\ f_2(x_1,x_2,\ldots,x_n) &= 0 \\ &\ldots\ldots\ldots\ldots \\ f_m(x_1,x_2,\ldots,x_n) &= 0 \end{aligned} \qquad (1)$$

$f_i(x_1,x_2,\ldots,x_n)$, $i = 1, 2, \ldots, m$, is a sum of terms of the form shown as expression (2), where the coefficient $a_{i_1 i_2 \cdots i_n}$

$$a_{i_1 i_2 \ldots i_n} x_1^{i_1} x_2^{i_2} \ldots x_n^{i_n} \quad (2)$$

is a constant, or fixed number, and the exponent i_j of the variable x_j is a nonnegative whole number. An example of such a system in two variables is system (3). The expressions

$$\begin{aligned} x^2 - xy + y^2 - 1 &= 0 \\ x^2 + xy - 3y^2 - 2x + 2y + 1 &= 0 \end{aligned} \quad (3)$$

$f_i(x_1,x_2,\ldots,x_n)$ are called polynomials in several variables. The problem posed by system (1) is to find necessary and sufficient conditions that there exist values of the variables $x_1 = a_1$, $x_2 = a_2, \ldots, x_n = a_n$ which simultaneously satisfy each equation of the system, and to find all such sets of values, which are called solutions of the system. In example (3), a complete set of solutions is given by $x = 1$, $y = 0$; $x = 0$, $y = 1$; $x = 1$; $y = 1$; and $x = -1$, $y = -1$. *See* EQUATIONS, THEORY OF; LINEAR SYSTEMS OF EQUATIONS. [R.A.Be.]

Polynuclear hydrocarbon One of a class of hydrocarbons possessing more than one ring. The aromatic polynuclear hydrocarbons may be divided into two groups. In the first, the rings are fused, which means that at least two carbon atoms are shared between adjacent rings. Examples are naphthalene (I), which has two six-membered rings, and acenaphthene (II), which has two six-membered rings and one five-membered ring.

In the second group of polynuclear hydrocarbons, the aromatic rings are joined either directly, as in the case of biphenyl (III), or through a chain of one or more carbon atoms, as in 1,2-diphenylethane (IV).

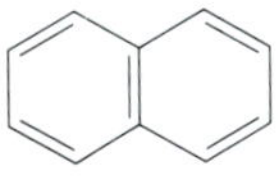

(I)

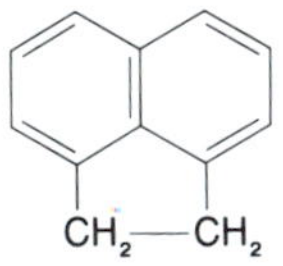

(II)

(III)

$C_6H_5CH_2CH_2C_6H_5$

(IV)

The higher-boiling polynuclear hydrocarbons found in coal tar or in tars produced by the pyrolysis or incomplete combustion of carbon compounds are frequently fused-ring hydrocarbons, some of which may be carcinogenic. *See* ANTHRACENE; AROMATIC HYDROCARBON; BIPHENYL; PHENANTHRENE; STEROID. [C.K.B.]

Polyol A compound containing more than one hydroxyl group (—OH). Each hydroxyl is attached to separate carbon atoms of an aliphatic skeleton. This group includes glycols, glycerol, and pentaerythritol and also such products as trimethylolethane, trimethylolpropane, 1,2,6-hexanetriol, sorbitol, inositol, and polyvinyl alcohol. Polyols are obtained from many plant and animal sources and are synthesized by a variety of methods.

Polyols such as glycerol, pentaerythritol, trimethylolethane, and trimethylolpropane are used in making alkyd resins for decorative and protective coatings. Glycols, glycerol, 1,2,6-hexanetriol, and sorbitol find application as humectants and plasticizers for gelatin, glue, and cork. Explosives are made by the nitration of glycols, glycerol, and pentaerythritol.

The polymeric polyols used in manufacture of the urethane foams represent a series of synthetic polyols. These polyols are generally polyoxyethylene or polyoxypropylene adducts of di- to octahydric alcohols. *See* GLYCEROL; GLYCOL. [P.C.J.]

Polyolefin resins Polymers derived from unsaturated hydrocarbons containing the ethylene or diene groups. Broadly, polyolefin resins may include virtually all addition polymers; however, the term polyolefin is specifically used for polymers of ethylene, the alkyl derivatives of ethylene (the α-olefins), and the dienes.

The polyvinyl resins, the fluorocarbon polymers, and other addition polymers are covered in other articles. *See* POLYFLUOROOLEFIN RESINS; POLYMERIZATION; POLYVINYL RESINS.

Polyethylene is a whitish, translucent polymer of moderate strength and high toughness. The physical properties vary markedly with the degree of crystallinity and with the size and distribution of the crystalline regions. With increasing crystallinity or density, the products generally become stiffer and stronger, and have higher softening temperatures and higher resistance to penetration by liquids and gases; at the same time, they lose some of their resistance to tear, impact, and stress cracking, and higher temperatures and pressures are needed for molding. The major uses of polyethylene are as packaging films, containers and bottles, molded articles, electrical insulation, coatings, and pipe. *See* ETHYLENE.

High-molecular-weight, isotactic, highly crystalline polypropylene is generally similar in properties to high-density polyethylene. In comparison with the latter, isotactic polypropylene is harder and stronger. The use of both unoriented and oriented films of polypropylene homopolymers and copolymers has grown rapidly in such applications as packaging for textiles and food, liners for bags, and heat-shrinkable wraps for records and other articles. *See* PROPYLENE.

The polyisobutylene polymers vary in properties from low-molecular-weight oils to high-molecular-weight rubbery solids. The main use of polyisobutylene is in the form of the copolymer with isoprene. The copolymer, known as butyl rubber, is distinguished by its impermeability to gases and its resistance to aging; it is used in automobile tires and tubes.

The discovery of the stereospecific catalysts has made possible the formation of high-molecular-weight, isotactic, crystalline polymers of other α-olefins, such as α-butene, α-octene, and α-dodecene. Typical melting temperatures of the crystalline, isotactic polymers of some of the α-olefins are relatively high. The high-melting polymers are of considerable interest because relatively few thermoplastic polymers are available which have high softening temperatures and at the same time can be easily fabricated.

Butadiene, isoprene, and 2-chlorobutadiene have gained the widest use of dienes employed in the polymer field. Butadiene and 2-chlorobutadiene have long been used in the production of synthetic rubbers by free-radical catalysis. The isoprene structure has long been recognized to correspond to the repeating unit in natural rubber. *See* POLYACRYLONITRILE RESINS; POLYMER; POLYSTYRENE RESIN. [J.A.M.]

Polyoma virus A virus which is capable of producing a number of tumors in newborn mice, rats, guinea pigs, hamsters, and rabbits. The virus grows readily in mouse embryo tissue culture and destroys the cells with the production of a hemagglutinin. It transforms hamster cells growing in tissue culture, which are then capable of producing sarcomas in hamsters. During the transformation process the infectious virus disappears, although an antigen to the virus can be demonstrated in the tumors. *See* ANIMAL VIRUS; TISSUE CULTURE. [A.E.Mo.]

Polyplacophora A class of the phylum Mollusca, sometimes termed Loricata. Commonly called chitons, polyplacophorans form a relatively stereotyped class of nearly 600 marine species placed in 43 genera, almost all living on wave-swept rocks of the lower littoral and upper sublittoral. The class is divided into two orders, the Lepidopleurida and Chitonida. *See* CHITON.

Adhering by a sucker foot, as in gastropod limpets, the flattened elliptical chiton body bears eight articulated shell plates. The detailed structure of the plates, and of the mantle tissues which secrete them, is consistent with shell-mantle anatomy throughout the other molluscan classes. *See* MOLLUSCA. [W.D.R.-H.]

Polyploidy The occurrence of related forms possessing chromosome numbers which are multiples of a basic number (n), the haploid number. Forms having $3n$ chromosomes are triploids; $4n$, tetraploids; $5n$, pentaploids, and so on. Autopolyploids are forms derived by the multiplication of chromosomes from a single diploid organism. As a result the homologous chromosomes come from the same source. These are distinguished from allopolyploids, which are forms derived from a hybrid between two diploid organisms. As a result, the homologous chromosomes come from different sources.

About one-third of the species of vascular plants have originated at least partly by polyploidy, and as many more appear to have ancestries which involve ancient occurrences of polyploidy. The condition can be induced artificially with the drug colchicine.

In animals, most examples of polyploidy occur in groups which are parthenogenetic, such as crustaceans of the genus *Artemia*, certain earthworms, weevils of the family Curculionidae, moths of the genus *Solenobia*, sawflies of the genus *Diprion*, lizards of the genus *Cnemidophorus*, and fishes of the genera *Carassius* and *Poecilopsis*; or in species which reproduce asexually by fission, such as the flatworm *Dendrocoelum infernale*. Sexual polyploids have been found in the frog genera *Odontophrynus*, *Ceratophrys*, and *Phyllomedusa*. *See* CHROMOSOME; GENE; GENETICS. [G.L.St.]

Polypodiales The largest order of modern ferns, commonly called the true ferns, with approximately 250 genera and 9000 species; also known as Filicales. Although well represented in the temperate regions, they reach their greatest development in the moist tropics. They vary in habit from small filmy structures to large treelike plants. Many are epiphytic (live perched on other plants) and a number are climbing species. A few are aquatic. Perhaps the most striking species are the tropical tree ferns with their upright, unbranched stems and terminal clusters of large graceful leaves.

The Polypodiales differ from the other fern orders in being leptosporangiate—that is, their sporangium, or spore sac, arises from a single surface cell—and in having small sporangia with a definite number of spores. The wall of the sporangium is almost encircled with a ring of cells having unevenly thickened walls. This ring is called the annulus. When the sporangium is mature, the annulus, acting as a spring, causes the sporangium wall to rupture, thus discharging the spores. These plants are valued for their beauty and for the clues they give to the evolutionary history of the Polypodiales which extends back through the coal measures of the Paleozoic. *See* PALEOBOTANY.

The sporophyte is the conspicuous phase of the true ferns, and like other vascular plants it has true roots, stems, and leaves (Fig. 1). In most ferns, especially those of the temperate regions, the mature stem is usually a creeping rhizome (underground stem) without aerial branches. However, in several species the stems are branched, and in some they are erect. Whereas in the tropics the leaves are usually persistent and evergreen, in temperate regions the leaves of most species die back each year and are replaced by new ones the next growing season. Characteristic of this

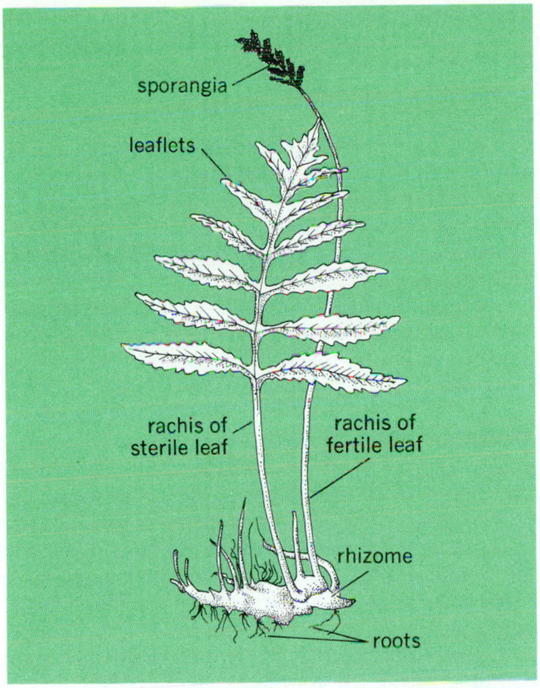

Fig. 1. The sensitive fern (*Onoclea sensibilis*), a representative of the Polypodiopsida. (*After W. W. Robbins, T. E. Weier, and C. R. Stocking, Botany: An Introduction to Plant Science, 3d ed., John Wiley and Sons, Inc., 1964*)

order is the apparent uncoiling of the leaves from the base toward the apex.

The internal structure of the blade of the leaf and of the root is very similar to that of these organs in the seed plants. The main difference is the presence of large intercellular spaces in the fern leaf and the frequent lack of apparent distinction between the spongy and palisade cells of the mesophyll, possibly because most ferns grow in the shade.

The life cycle of the fern consists of two independent (self-sustaining) alternating generations. The common leafy fern plant is the sporophytic (spore-producing) generation. When the mature spores are discharged and reach a suitable substrate, they germinate and produce a small, flat, green, heart-shaped structure known as the prothallium or gametophytic

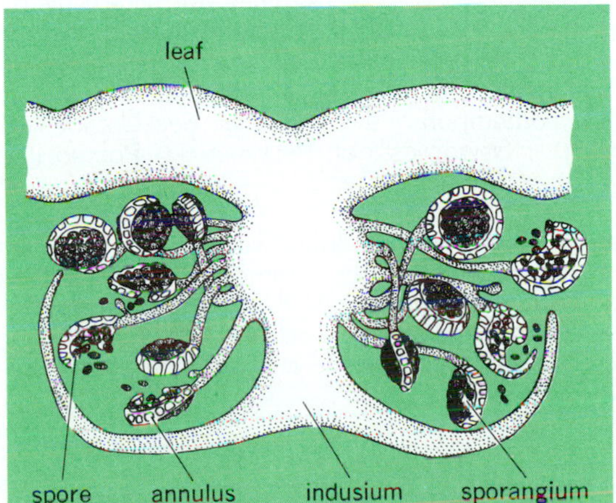

Fig. 2. Diagram of section through a fern leaf, showing details of a sorus. (*After W. W. Robbins, T. E. Weier, and C. R. Stocking, Botany: An Introduction to Plant Science, 3d ed., John Wiley and Sons, Inc., 1964*)

(gamete-producing) generation. The gametophyte produces the sex organs antheridia (male) and archegonia (female). The gametes (sperm and egg) unite in fertilization and the resultant cell, or zygote, develops into the spore-bearing (sporophytic) fern plant.

In all ferns, the spores are produced in special multicellular organs known as sporangia. Except for a few genera, the sporangia are arranged in groups or clusters called sori (Fig. 2). These are on the lower surface of the leaves or fertile fronds, either along the midrib of the pinnae, near the leaf margins, or scattered. Usually each sorus is covered by a flaplike structure called the indusium, which may be of various shapes and sizes. However, a few ferns have naked sori, and others have a false indusium formed by the folding or inrolling of the leaf margin. *See* POLYPODIOPHYTA; PSILOTOPHYTA. [P.A.V.]

Polypodiophyta A division of the plant kingdom, commonly called the ferns, which is widely distributed throughout the world but is most abundant and varied in moist, tropical regions, The Polypodiophyta are sometimes treated as a class Polypodiopsida of a broadly defined division Tracheophyta (vascular plants). The group consists of five living orders (Ophioglossales, Marattiales, Polypodiales, Marsileales, and Salviniales), plus several orders represented only by Paleozoic fossils. The vast majority of the nearly 10,000 species belong to the single order Polypodiales, sometimes also called Filicales. *See* POLYPODIOPSIDA.

The Polypodiophyta ordinarily have well-developed roots, stems, and leaves that contain xylem and phloem as conducting tissues. The central cylinder of vascular tissue in the stem usually has well-defined parenchymatous leaf gaps where the leaf traces depart from it. The leaves are spirally arranged on the stem and are usually relatively large, with an evidently branching vascular system. In most kinds of ferns the leaves, called fronds, are compound or dissected. *See* LEAF; PHLOEM; ROOT (BOTANY); STEM; XYLEM.

The Polypodiophyta show a well-developed alternation of generations, both the sporophyte and the gametophyte generation being detached and physiologically independent of each other at maturity. The sporophyte is much the more conspicuous, and is the generally recognized fern plant. On some or all of its leaves it produces tiny sporangia which in turn contain spores. *See* COENOPTERIDALES; MARATTIALES; OPHIOGLOSSIDAE; POLYPODIALES; REPRODUCTION (PLANT). [A.Cr.]

Polypodiopsida A class (also known as Filicineae) of the plant division Polypodiophyta containing a large group of plants commonly called ferns. They are widely distributed throughout the world with their greatest development in the moist tropics. Polypodiopsida is an old class with a good representation in the Paleozoic flora. Some of the plant fossils resemble contemporary living species, *See* COENOPTERIDALES; MARATTIALES; OPHIOGLOSSIDAE; POLYPODIALES; POLYPODIOPHYTA. [P.A.V.]

Polypteriformes A distinctive and apparently ancient order of actinopterygian fishes, also called the Cladistia or the bichirs. Their characters include thick, rhombic, ganoid scales with an enamellike covering; a slitlike spiracle behind the eye; a well-ossified internal skeleton; a symmetrical caudal fin, basically heterocercal, with the upper part continuous with the dorsal fin (see illustration); a dorsal series of free, spinelike finlets, each supported by a radial bone; a distinctive pectoral fin base with three enlarged radial bones; and paired ventral lungs.

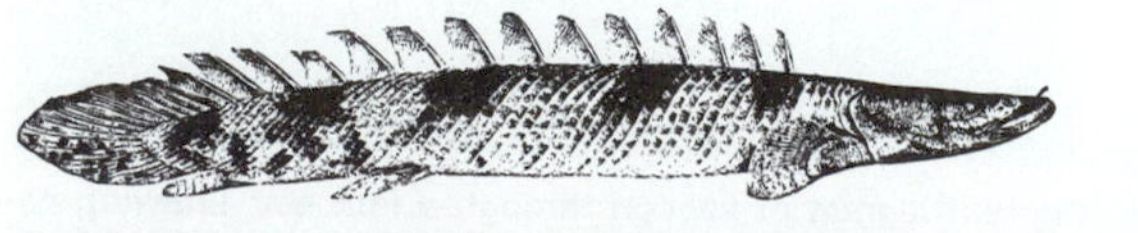

Bichir (*Polypterus endlicheri*), length to 3 ft (90 cm). (*After G. A. Boulenger, Catalogue of the Fresh Water Fishes of Africa in the British Museum, vol. 1, 1909*)

This order consists of a single family, the Polypteridae, that is known from the Eocene. The two Recent genera, *Polypterus*, with about 10 species, and *Erpetoichthys*, with 1 species, are confined to fresh waters of tropical Africa. *See* ACTINOPTERYGII; CHONDROSTEI. [R.M.B.]

Poly-p-xylylene resins Linear, crystallizable resins based on an unusual polymerization of *p*-xylene and derivatives. The polymers are tough and chemically resistant, and may be deposited as adherent coatings by a vacuum process. The vapor deposition process makes it possible to coat small microelectronic parts with a thin layer of the polymer. *See* XYLENE. [J.A.M.]

Polysaccharide A class of high-molecular-weight carbohydrates, colloidal complexes, which break down on hydrolysis to monosaccharides containing five or six carbon atoms. The polysaccharides are considered to be polymers in which monosaccharides have been glycosidically joined with the elimination of water. A polysaccharide consisting of hexose monosaccharide units may be represented by the reaction below.

$$nC_6H_{12}O_6 \rightarrow (C_6H_{10}O_5)_n + (n - 1)H_2O$$

The term polysaccharide is limited to those polymers which contain 10 or more monosaccharide residues. Polysaccharides such as starch, glycogen, and dextran consist of several thousand D-glucose units. Polymers of relatively low molecular weight, consisting of two to nine monosaccharide residues, are referred to as oligosaccharides. *See* DEXTRAN; GLUCOSE; GLYCOGEN; MONOSACCHARIDE; STARCH.

Polysaccharides are often classified on the basis of the number of monosaccharide types present in the molecule. Polysaccharides, such as cellulose or starch, that produce only one monosaccharide type (D-glucose) on complete hydrolysis are termed homopolysaccharides. On the other hand, polysaccharides, such as hyaluronic acid, which produce on hydrolysis more than one monosaccharide type (*N*-acetylglucosamine and D-glucuronic acid) are named heteropolysaccharides. *See* CARBOHYDRATE. [W.Z.H.]

Polystyrene resin A hard, transparent, glasslike thermoplastic resin. Polystyrene is characterized by excellent electrical insulation properties, relatively high resistance to water, high refractive index, clarity, and low softening temperature.

Besides the many applications of styrene in combination with other materials as in rubber and paints, large quantities of the homopolymer and the polyblends are employed in the injection molding of toys, panels, and novelty items, and also in the extrusion of sheets. The sheets are used for panels, or they may be further shaped by vacuum forming for uses such as liners for refrigerator doors. *See* ACRYLONITRILE; POLYMER; POLYMERIZATION; RUBBER; STYRENE. [J.A.M.]

Polysulfide resins Resins that vary in properties from viscous liquids to rubberlike solids. Organic polysulfide resins are prepared by the condensation of organic dihalides with a polysulfide.

Compounding and fabrication of the rubbery polymers can be handled on conventional rubber machinery. The polysulfide rubbers are distinguished by their resistance to solvents such as gasoline, and to oxygen and ozone. The polymers are relatively impermeable to gases. The products are used to form coatings which are chemically resistant and special rubber articles, such as gasoline bags. The polysulfide rubbers were among the first polymers to be used in solid-fuel compositions for rockets. *See* ORGANOSULFUR COMPOUND; POLYMERIZATION; RUBBER. [J.A.M.]

Polysulfone resins Engineering plastics whose molecules contain sulfone groups in the main chain. Polysulfones are characterized by high strength, stiffness, toughness, and thermal and oxidation stability. The aromatic structure and the presence of the sulfone groups are responsible for the resistance to heat and oxidation. The polymers can be used over a wide temperature range, and in the presence of acids, alkalies, aqueous media, and nonpolar organic solvents; resistance to creep and dimensional changes due to aging is high. Uses include electronic and automotive parts, hardware and houseware items, and plumbing parts. *See* SULFONE. [J.A.M.]

Polytrichidae A subclass of the mosses (class Bryopsida), consisting of one family and 19 genera. The plants are acrocarpous and perennial, grow on soil, often in dry habitats, and are usually unbranched; annual increments often grow up through the male inflorescences, but do not give the appearance of forked stems. The plants are nearly always dioecious, with both male and female inflorescences terminal. The stems generally have well-developed vascular tissues. The long, narrow leaves are sheathed at the base; above the base the midrib often occupies most of the leaf. Stomata are usually present at or near the base of the capsule. *See* BRYOPHYTA; BRYOPSIDA; DAWSONIIDAE. [H.Cr.]

Polytropic process A process which occurs with an interchange of both heat and work between the system and its surroundings. The nonadiabatic expansion or compression of a fluid is an example of a polytropic process. The interrelationships between the pressure (P) and volume (V) and pressure and temperature (T) for a gas undergoing a polytropic process

$$PV^a = \text{constant} \tag{1}$$

$$P^b/T = \text{constant} \tag{2}$$

are given by Eqs. (1) and (2), where a and b are the polytropic constants for the process of interest. These constants, which are usually determined from experiment, depend upon the equation of state of the gas, the amount of heat transferred, and the extent of irreversibility in the process. *See* GAS; ISENTROPIC PROCESS; ISOTHERMAL PROCESS; THERMODYNAMIC PROCESSES. [S.I.S.]

Polyurethane resins Resins that can be produced in forms varying from hard, glossy, solvent-resistant coatings, to abrasion- and solvent-resistant rubbers, fibers, and flexible-to-rigid foams. The foams have found the widest use. The more flexible foams are employed as upholstery material for furniture, as rug backing, insulation, and crash pads. The more rigid foams are employed as the core in structural and insulating laminates and as insulation in refrigerated appliances and vehicles. Polyurethanes are also used as adhesives, for example, in the bonding of rubber and of nylon.

Polyurethane (or polyisocyanate) resins are produced by the reaction of a diisocyanate with a compound containing at least two active hydrogen atoms, such as a diol or diamine. Linear, fiber-forming polymers are formed by the addition of diisocyanates to diols, while cross-linking is made possible by the use of polyols or isocyanates having more than two functional groups. Much urethane technology is based on the use of prepolymers in which a low-molecular-weight polyester or polyether is prepared. Some of these have terminal hydroxyl groups, while those reacted with an excess of diisocyanate have terminal isocyanate groups. Thus segments of a rubbery block alternate with segments of rigid urethane units. This growth process is called chain extension. *See* PLASTICS PROCESSING; POLYESTER RESINS; POLYETHER RESINS; POLYMERIZATION; URETHANE. [J.A.M.]

Polyvinyl resins Polymeric materials generally considered to include polymers derived from monomers having the structure

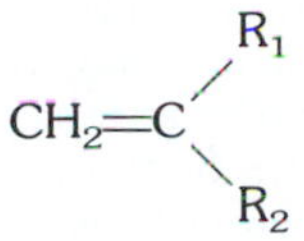

in which R_1 and R_2 represent hydrogen, alkyl, halogen, or other groups. This article refers to polymers whose names include the term vinyl. For discussions of other vinyl-type polymers *see* POLYACRYLATE RESIN; POLYACRYLONITRILE RESINS; POLYFLUOROOLEFIN RESINS; POLYOLEFIN RESINS; POLYSTYRENE RESIN.

Many of the monomers can be prepared by addition of the appropriate compound to acetylene. For example, vinyl chloride, vinyl fluoride, vinyl acetate, and vinyl methyl ether may be formed by the reactions of acetylene with HCl, HF, CH_3OOH, and CH_3OH, respectively. Processes based on ethylene as a raw material have also become common for the preparation of vinyl chloride and vinyl acetate.

The polyvinyl resins may be characterized as a group of thermoplastics which, in many cases, are inexpensive and capable of being handled by solution, dispersion, injection molding, and extrusion techniques. The properties vary with chemical structure, crystallinity, and molecular weight.

Polyvinyl acetals are relatively soft, water-insoluble thermoplastic products obtained by the reaction of polyvinyl alcohol with aldehydes. Properties depend on the extent to which alcohol groups are reacted. Polyvinyl butyral is rubbery and tough and is used primarily in plasticized form as the inner layer and binder for safety glass. Polyvinyl formal is the hardest of the group; it is used mainly in adhesive, primer, and wire-coating formulations, especially when blended with a phenolic resin.

Polyvinyl acetate is a leathery, colorless thermoplastic material which softens at relatively low temperatures and which is relatively stable to light and oxygen. The polymers are clear and noncrystalline. The chief applications are as adhesives and binders for water-based or emulsion paints. *See* PAINT.

Polyvinyl alcohol is a tough, whitish polymer which can be formed into strong films, tubes and fibers that are highly resistant to hydrocarbon solvents. Although polyvinyl alcohol is one of the few water-soluble polymers, it can be rendered insoluble in water by drawing or by the use of cross-linking agents. Two groups of products are available, those formed by the essentially complete hydrolysis of polyvinyl acetate, and those formed by incomplete hydrolysis.

The former may be plasticized with water or glycols and molded or extruded into films, tubes, and filaments which are resistant to hydrocarbons. These products are used for liners in gasoline hoses, for grease-resistant coating and paper adhesives, for treating paper and textiles, and as emulsifiers and thickeners.

Polyvinyl carbazole is a tough, glassy thermoplastic with excellent electrical properties and a relatively high softening temperature. Uses of the product has been limited to small-scale electrical applications requiring resistance to high temperatures.

Polyvinyl chloride (PVC) is a tough, strong thermoplastic material which has an excellent combination of physical and electrical properties. The products are usually characterized as plasticized or rigid types. Polyvinyl chloride (and copolymers) is the second most commonly used polyvinyl resin and one of the most versatile plastics. The plasticized types are somewhat elastic materials which are familiar in the form of shower curtains, floor coverings, raincoats, dishpans, dolls, bottle-top sealers, prosthetic forms, wire insulation, and films, among others. Rigid products, which may consist of the homopolymer, copolymer, or polyblends, are commonly used in the manufacture of phonograph records, pipe, chemically resistant liners for chemical-reaction vessels, and siding and window sashes.

Polyvinylidene chloride is a tough, hornlike thermoplastic with properties generally similar to those of polyvinyl chloride. Because of its relatively low solubility and decomposition temperature, the material is most widely used in the form of copolymers with other vinyl monomers, such as vinyl chloride. The copolymers are employed as packaging film, rigid pipe, and as filaments for upholstery and window screens.

Polyvinyl ethers exist in several forms varying from soft, balsamlike semisolids to tough, rubbery masses, all of which are readily soluble in organic solvents. Polymers of the alkyl vinyl ethers are used in adhesive formulations and as softening or flexibilizing agents for other polymers.

Polyvinyl fluoride is a tough, partially crystalline thermoplastic material which has a higher softening temperature than polyvinyl chloride. Films and sheets are characterized by high resistance to impact and cracking caused by flexing and temperature and by resistance to weathering.

Polyvinyl pyrrolidone is a water-soluble polymer of basic nature which has film-forming properties, strong absorptive or complexing qualities for various reagents, and the ability to form water-soluble salts which are polyelectrolytes. The main uses are as a water-solubilizing agent for medicinal agents such as iodine, and as a semipermanent setting agent in hair sprays. Certain synthetic textile fibers containing small amounts of vinylpyrrolidone as a copolymer have improved affinity for dyes. *See* PLASTICS PROCESSING; POLYMER; POLYMERIZATION. [J.A.M.]

Pomegranate A small deciduous tree, *Punica granatum*, belonging to the plant order Myrtales. Pomegranate is grown as an ornamental as well as for its fruit. The pomegranate is a native of Asia. It was originally known for its medicinal qualities, and cures for various ills were attributed to the fruit juice, the rind, and the bark of the roots. The fruit is a reddish, pomelike berry, containing numerous seeds imbedded in crimson pulp, from which an acid, reddish juice may be obtained. Limited quantities are grown in California and the Gulf states. *See* FRUIT, TREE; MYRTALES. [J.H.Cl.]

Poplar Any tree of the genus *Populus*, family Salicaceae, marked by simple, alternate leaves which are usually broader than those of the willow, the other American representative of this family. Poplars have scaly buds, bitter bark, flowers and fruit in catkins, and a five-angled pith. *See* SALICALES; WILLOW.

Some species are commonly called cottonwood because of the cottony hairs attached to the seeds. Other species, called aspens, have weak, flattened leaf stalks which cause the leaves to flutter in the slightest breeze. One of the important species in the United States is the quaking, or trembling, aspen (*P. tremuloides*). The soft wood of this species is used for paper pulp. The European aspen (*P. nigra*), which is similar to the quaking aspen, is sometimes planted, and its variety, *italica*, the Lombardy poplar of erect columnar habit, is used in landscape planting. The black cottonwood (*P. trichocarpa*) is the largest American poplar and is also the largest broad-leaved tree in the forests of the Pacific Northwest. The cottonwood or necklace poplar (*P. deltoides*) is native in the eastern half of the United States. In the balsam or tacamahac poplar (*P. balsamifera*), the resin is used in medicine as an expectorant. The wood is used for veneer, boxes, crates, furniture, paper pulp, and excelsior. [A.H.G./K.P.D.]

Poppy A plant, *Papaver somniferum* (Papaveraceae), which is probably a native of Asia Minor. It is cultivated extensively in China, India, and elsewhere. This plant is the source of opium, obtained by cutting into the fruits (capsules) soon after the petals have fallen. The white latex (juice) flows from the cuts and hardens when exposed to the air. This solidified latex is collected, shaped into balls or wafers, and often wrapped in the flower petals. This is the crude opium, which contains at least 20 alkaloids, including morphine and codeine. *See* DRUG ADDICTION; MORPHINE; PAPAVERALES. [P.D.St./E.L.C.]

Population dispersal The process by which groups of living organisms expand the space or range within which they live. Because of their reproductive capacity, all populations have a natural tendency to expand. As increased area supports more individuals, dispersal and reproduction are intimately correlated.

Distinction should be made between dispersal and seasonal migration. Birds, butterflies, salmon, and others migrate regularly without necessarily expanding their geographic range, since they usually return to their original areas or die out. *See* MIGRATORY BEHAVIOR.

Dispersal consists of several phases: (1) the production of units, that is, of individuals or parts of individuals (disseminules) fit or adapted for dispersal; (2) the transportation of individuals or disseminules to the new habitat; and (3) ecesis, the process of becoming established through germination, rooting, or physiological and psychological adjustment.

Certain disseminules (propagules or diaspores) may represent various stages of the life cycle of the individual. Many free-living animals do not produce special dispersal structures but rely upon the ability of the entire organism to move about (vagility). Organisms attached to a substratum, such as most plants and certain animals, produce disseminules adapted to certain agents of dispersal (see table).

Individuals or disseminules are transported in five general ways: self-dispersal (autochory), water dispersal (hydrochory), wind dispersal (anemochory), animal dispersal (zoochory), and dispersal by humans (anthropochory).

Success in population dispersal depends upon three factors: fitness of the new habitat, fitness of the migrating individuals, and the chance juxtaposition of these two which, in the long

Examples of dispersal stages in life cycle of plants and animals

Environment	Organism	Disseminule	Dispersal by
Sea	Kelp	Zoospore	Currents
	Coral	Planula	Currents
	Sea worm	Trochophore	Currents
	Clam	Trochophore	Currents
	Barnacle	Adult	Driftwood, ships
	Crab	Zoea	Currents
	Sea urchin	Pluteus	Currents
	Fish	Adult	Autochory
	Lamprey	Adult	Fish
Terrestrial	Mushroom	Spore	Wind
	Fern	Spore	Wind
	Pine	Seed	Wind
	Blueberry	Fruit	Birds
	Tumbleweed	Entire plant	Wind
	Insect	Adult	Autochory, wind
	Spider	Young animal	Wind
	Reptiles, birds, mammals	Adult	Autochory
Parasitic	Bacteria	Entire cell	Water, food, air
	Intestinal ameba	Cyst	Water, food, human
	Malaria parasite	Gamete, sporozoite	Mosquito
	Tapeworm	Egg	Pig
	Blood fluke	Egg, cercaria	Water, snail

run, depends on the number of individuals invading the new habitat. The probability for a new habitat to be favorable is greatest close to the parent population. Spores blown over great distances have less chance of landing in spots suited for germination than have seeds failing close to the parent plant. In wide-range dispersal larger numbers of disseminules are usually necessary than at close range, to ensure ecesis. *See* PARTHENOGENESIS; SPECIATION. [K.Le./H.G.Ba.]

Population genetics The study of both experimental and theoretical consequences of mendelian heredity on the population level, in contradistinction to classical genetics which deals with the offspring of specified parents on the familial level. The genetics of populations studies the frequencies of genes, genotypes, and phenotypes, and the mating systems. It also studies the forces that may alter the genetic composition of a population in time, such as recurrent mutation, migration, and intermixture between groups, selection resulting from genotypic differential fertility, and the random changes incurred by the sampling process in reproduction from generation to generation. This type of study contributes to an understanding of the elementary step in biological evolution. The principles of population genetics may be applied to plants and to other animals as well as humans. *See* GENETICS; MENDELISM. [C.C.L.]

Porcelain A high-grade ceramic ware characterized by high strength, a white color (under the glaze), very low absorption, good translucency, and a hard glaze. Equivalent terms are European porcelain, hard porcelain, true porcelain, and hard paste porcelain. *See* GLAZING; POTTERY.

Porcelain is distinguished from other fine ceramic ware, such as china, by the fact that the firing of the unglazed ware (the bisque firing) is done at a lower temperature (1800–2200°F or 1000–1200°C) than the final or glost firing, which may be as high as 2700°F (1500°C). In other words, the ware reaches its final state of maturity at the maturing temperatures of the glaze.

The white color is obtained by using very pure white-firing kaolin or china clay and other pure materials, the low absorption results from the high firing temperature, and the translucency results from the glass phase. *See* CERAMICS. [J.F.McM.]

Porcupine Any of about 26 species of rodents which have spines or quills in addition to regular hair. These mammals are included in two families, the Hystricidae, or Old World porcupines, and the Erethizontidae, or New World porcupines. The spines are sharply pointed, erectile hairs which serve the animal as defensive structures and are controlled by powerful skin muscles. These animals have short limbs which terminate in five functional digits in the Old World species and four in the New World species. *See* MAMMALIA; RODENTIA. [C.B.C.]

Porifera The sponges, a phylum of the animal kingdom which includes about 5000 described species. The body plan of sponges is unique among animals. Currents of water are drawn through small pores or ostia in the sponge body and leave by way of larger openings called oscula. The beating of flagella on collar cells or choanocytes, localized in chambers in the interior of the sponge, maintains the water current. Support for the sponge tissues is provided by calcareous or siliceous spicules, or by organic fibers, or by a combination of organic fibers and siliceous spicules. Some species have a compound skeleton of organic fibers, siliceous spicules, and a basal mass of aragonite or calcite. The skeletons of species with supporting networks of organic fibers have long been used for bathing and cleaning purposes. Because of their primitive organization, sponges are of interest to zoologists as an aid in understanding the origin of multicellular animals. *See* ANIMAL KINGDOM; PARAZOA.

The Porifera are divided into the classes Hexactinellida, Calcarea, Demospongiae, and Sclerospongiae on the basis of the skeletal structure. A taxonomic scheme of the Porifera follows.

Class Hexactinellida
  Subclass Amphidiscophora
    Order: Amphidiscosa
      Hemidiscosa
  Subclass Hexasterophora
    Order: Hexactinosa
      Lychniscosa
      Lyssacinosa
      Reticulosa
Class Calcarea
  Subclass Calcinea
    Order: Clathrinida
      Leucettida
  Subclass Calcaronea
    Order: Leucosoleniida
      Sycettida
  Subclass Pharetronidia
Class Demospongiae
  Subclass Tetractinomorpha*
    Order Homosclerophorida
      Choristida
      Spirophorida
      Hadromerida
      Axinellida
  Subclass Ceractinomorpha*
    Order Dendroceratida
      Dictyoceratida
      Halichondrida
      Haplosclerida
      Poecilosclerida
Class Sclerospongiae

See separate articles on each group, except those with an asterisk.

The Porifera have a fossil record extending from the Cambrian to Recent times. More than 1000 genera of fossil sponges have been described from the Paleozoic, Mesozoic, and Cenozoic eras. [W.D.H.]

Porocephalida One of two orders in the class Pentastomida of the phylum Arthropoda. In this order the larvae have four legs. The hooks on the adult articulate on a chitinous base or fulcrum and are arranged in a flattened trapezoidal pattern, or in a curved line or straight line. There are no podial or parapodial lobes. The two suborders are Porocephaloidea and Linguatuloidea.

The Porocephaloidea comprise five families—Porocephalidae, Sebekidae, Armilliferidae, Sambonidae, and Subtriquetridae—and include most of the pentastomid species. These animals are cylindrical, with rounded ends. The hooks are sessile; the mouth is anterior between the hooks. Adults occur in the respiratory passages of snakes, lizards, turtles, and crocodiles; the larvae are encysted in fishes, snakes, lizards, crocodiles, and mammals.

Linguatuloidea contains a single family, with a single genus, *Linguatula*. Adults live in the nasal cavities of carnivorous mammals; the larvae are encysted in herbivorous mammals. The body is elongate, flattened ventrally, annulate, and attenuated posteriorly. *See* PENTASTOMIDA. [H.W.S.]

Porphyrin A class of red-pigmented compounds with a cyclic tetrapyrrolic structure in which the four pyrrole rings are joined through their α-carbon atoms by four methene bridges (=CH—). The pyrrole formula is shown here.

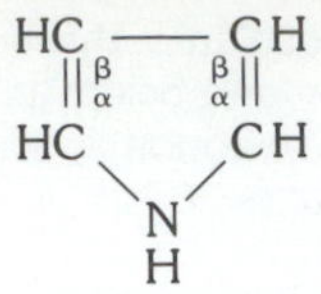

The porphyrins form part of the active nucleus of chlorophylls *a* and *b*, hemoglobin, myohemoglobin, cytochromes, and the enzymes catalase and peroxidase. The parent substance is synthetic porphin in which the hydrogen (H) atoms in the eight β-positions on the pyrrole rings are unsubstituted. Naturally occurring porphyrins differ from porphin and from each other by various side chains in these eight β-positions. *See* CHLOROPHYLL; CYTOCHROME; ENZYME; HEMOGLOBIN; SPECTROPHOTOMETRIC ANALYSIS. [R.S.]

Porphyroblast A relatively large crystal formed in a metamorphic rock. The presence of abundant porphyroblasts gives the rock a porphyroblastic texture. Minerals found commonly as porphyroblasts include biotite, garnet, chloritoid, staurolite, kyanite, sillimanite, andalusite, cordierite, and feldspar. Porphyroblasts are generally a few millimeters or centimeters across, but some attain a diameter of over 1 ft (30 cm). They may be bounded by well-defined crystal faces, or their outlines may be highly irregular or ragged. Very commonly they are crowded with tiny grains of other minerals that occur in the rock.

Most commonly, porphyroblasts develop in schist and gneiss during the late stages of recrystallization. As the rock becomes reconstituted, certain components migrate to favored sites and combine there to develop the large crystals. *See* GNEISS; METAMORPHIC ROCKS; SCHIST. [C.A.C.]

Porphyry An igneous rock characterized by porphyritic texture, in which large crystals (phenocrysts) are enclosed in a matrix of very fine-grained to aphanitic (not visibly crystalline) material. Porphyries are generally distinguished from other porphyritic rocks by their abundance of phenocrysts and by their occurrence in small intrusive bodies (dikes and sills) formed at shallow depth within the earth. In this sense porphyries are hypabyssal rocks. *See* IGNEOUS ROCKS; PHENOCRYST.

Porphyries occur as marginal phases of medium-sized igneous bodies (stocks, laccoliths) or as apophyses (offshoots) projecting from such bodies into the surrounding rocks. They are also abundant as dikes cutting compositionally equivalent plutonic rock, or as dikes, sills, and laccoliths injected into the adjacent older rocks. *See* LACCOLITH. [W.I.R.]

Porpoise Any one of several species of marine mammal of the family Phocaenidae found in tropical and temperate oceans of the world and in several large river systems in South America and Asia. The common porpoise (*Phocaena phocaena*) is the smallest member of the order Cetacea and is found in the North Atlantic Ocean, the North Sea, and the Baltic Sea.

The porpoise is an excellent swimmer, although the flippers are relatively small. They are social animals with a well developed brain and use an elaborate system of behavior and sounds to communicate between individuals. Moreover, they are able to echolocate with a highly developed sonar system. Hair is lacking and there is usually an insulating layer of fat beneath the smooth thick skin. Porpoises must surface periodically to breathe, but they are otherwise well adapted for their marine environment. During the summer, after a gestation period of 1 year, a single young is born.

The Pacific harbor porpoise (*P. vomerina*) is a common sight in waters along the West Coast of the United States. The black finless porpoise (*Neomeris phocaenoides*) of the Pacific and Indian oceans lacks a dorsal fin, but is similar to the common porpoise in its habits and has been found as far as 1000 mi (1600 km) up the Yellow River of China. The river porpoise of South America is blind. *See* CETACEA; DOLPHIN. [C.B.C.]

Porulosida An order of Radiolaria. The central capsule shows many pores, more or less uniformly distributed over the surface of the thick and usually globular layer. The development of skeletal structures ranges from none at all through scattered spines to well-developed latticework shells. Colonial aggregates, in which the capsules of a number of organisms are embedded in a matrix, occur in several genera. *See* RADIOLARIA. [R.P.H.]

Position fixing The process of determining a craft's present position without reference to a previous position. Two technologies are commonly used for applying this procedure. One is radio (electronic) and the other is astronomical (celestial). Position fixing by the use of radio technology is sometimes called radio location. Some experts consider celestial navigation a procedure apart from any other. *See* CELESTIAL NAVIGATION.

Many radio position-fixing systems were developed during World War II. Some of these employed continuous-wave technology, while others used pulse technology. Hyperbolic, radial, and polar geometry were all used. Four, hyperbolic systems were developed: loran, Decca, loran C, and Omega. *See* DECCA; LORAN; OMEGA.

A special system developed for aircraft was vhf omnidirectional radio range (VOR), but this system gave only radial guidance. DME, a distance-measuring system, was combined with VOR to form a circular/radial position system. Tacan, a military radial/distance system, supplied distance to the VOR to form VORTAC. *See* RADIO RANGE; RHO-THETA SYSTEMS; TACAN; VORTAC.

Since 1972 the Navy Navigation Satellite system (formerly called Transit) has been in use by the military and private industry. It is basically a hyperbolic system, with various positions of the satellite serving as the equivalent of hyperbolic radio transmitting stations. The accuracy of the system can be of the order of feet. [P.C.S.]

Positive-displacement flowmeter A quantity meter in which the flowing stream is separated into discrete quantities. These meters capture definite volumes one after another and pass them downstream.

In the nutating-disk meter a circular disk attached to a spherical center is restrained from rotation by a vertical partition through a slot in the disk (illustration *a*). It is allowed to nutate so that its shaft is in contact with a conical guide. The liquid entering the inlet port moves the disk in a nutating motion until the liquid discharges from the outlet port. The double conical shape of the measuring chamber makes a seal with the disk as it goes through the nutating motion. Normally a mechanical counter indicates the number of cycles as the fluid flows through smoothly with no pulsations.

In the rotary-vane meter an eccentrically mounted drum carries radial spring-loaded vanes. These vanes slide in and out, and seal against the meter casing to form pockets that carry a measured amount of fluid on each cycle.

In the oval-shaped gear meter two carefully formed gears provide their own synchronization, and with close-fitting teeth to the outer circular-shaped chambers they deliver a fixed quantity of liquid for each rotation. The lobed impeller is similar to the oval gear except that the synchronization is done by external circular gears and therefore the rotation is smooth.

In the rotary-abutment meter two displacement rotating vanes interleave with cavities on an abutment rotor (illustration *b*). The three elements are geared together, and the meter

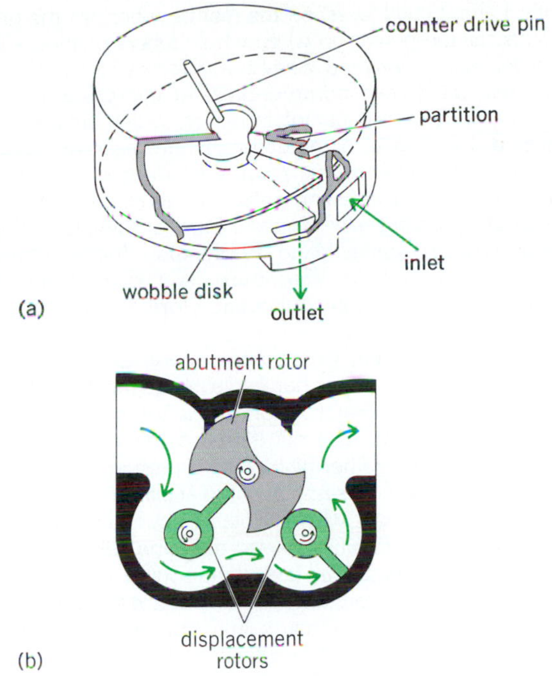

Positive-displacement flowmeters. (*a*) Nutating disk. (*b*) Rotary abutment.

depends upon close clearances rather than rubbing surfaces to provide the seals.

In the liquid-sealed drum-type gas meter a cylindrical chamber is filled more than half full with water and divided into four rotating compartments formed by trailing vanes. Gas entering through the center shaft from one compartment to another forces rotation that allows the gas then to exhaust out the top as it is displaced by the water. *See* FLOW MEASUREMENT; QUANTITY FLOWMETER. [M.Br.; L.P.E.]

Positron An elementary particle with mass equal to that of the electron, and positive charge equal in magnitude to the electron's negative charge. The positron is thus the antiparticle (charge-conjugate particle) to the electron. The positron has the same spin and statistics as the electron. Positrons, like electrons, appear as decay products of many heavier particles; electron-positron pairs are produced by high-energy photons in matter. *See* ELECTRON; ELECTRON-POSITRON PAIR PRODUCTION; ELEMENTARY PARTICLE.

A positron is, in itself, stable, but cannot exist indefinitely in the presence of matter, for it will ultimately collide with an electron. The two particles will be annihilated as a result of this collision, and photons will be created. However, a positron can first become bound to an electron to form a short-lived "atom" termed positronium. *See* POSITRONIUM. [C.J.G.]

Positronium The bound state of an electron and a positron. Positronium was discovered by studies of the so-called annihilation radiation from positrons stopped in gases. It is formed in a collision between a positron and a gas atom which results in the capture of an atomic electron by the positron. *See* POSITRON.

Positronium is of particular interest because it is the two-body system to which quantum electrodynamics is applicable, and its study has served as an important confirmation of the theory of quantum electrodynamics. *See* QUANTUM ELECTRODYNAMICS.

No states of positronium other than the ground $n = 1$ state ($n = 1,2,3,\ldots$, being the principal quantum number) have been found. Studies of positron annihilation in solids and liquids indicate that a perturbed form of positronium exists under certain conditions.

Positronium is an unstable atom and annihilates with the emission of photons. [V.Hu.]

Posttraumatic stress disorder An anxiety disorder in some individuals who have experienced an event that poses a direct threat to the individual's or another person's life. The characteristic features of anxiety disorders are fear, particularly in the absence of a real-life threat to safety, and avoidance behavior.

Diagnosis. A diagnosis of posttraumatic stress disorder requires that four criteria be met. (1) The individual must have been exposed to an extremely stressful and traumatic event beyond the range of normal human experience. (2) The individual must periodically and persistently reexperience the event, for example through recurrent dreams and nightmares, an inability to stop thinking about the event, flashbacks, or auditory hallucinations. (3) There is persistent avoidance of events related to the trauma, and psychological numbing that was not present prior to the trauma. (4) Enduring symptoms of anxiety and arousal are present. These symptoms can be manifested in different forms, including anger, irritability, a very sensitive startle response, an inability to sleep well, and physiological evidence of fear when the individual is reexposed to a traumatic event.

Physiological changes. There are also important biological events that accompany stress. Physiological arousal responses in individuals with posttraumatic stress disorder include increases in heart rate, respiration rate, and skin conductivity upon reexposure to traumatic stimuli. Posttraumatic stress disorder may also be associated with structural and physiological changes in the brain. In addition, extreme and prolonged stress is associated with a variety of physical ailments, including heart attacks, ulcers, colitis, and decreases in immunological functioning. Posttraumatic stress disorder is rarely the only psychological problem that an affected individual manifests. It is common to also find significant depression, generalized anxiety, substance abuse and dependence, marital problems, and intense, almost debilitating anger.

Treatment. Posttraumatic stress disorder can be treated by pharmacological means with psychotherapy. Most psychological treatments for the disorder involve reexposure to the traumatic event. Failure to expose oneself to stimuli that evoke fear allows the fear response to remain strong. Increased exposure, however, diminishes the fear response. In short, many symptoms of posttraumatic stress disorder may be self-perpetuating. Events that occur after the cessation of the traumatic event also seem to be important in whether symptoms become severe and intractable. Along with specific behavior interventions, individuals with posttraumatic stress disorder should become involved in psychotherapeutic treatment for secondary problems. *See* NEUROBIOLOGY; NORADRENERGIC SYSTEM; PSYCHOPHARMACOLOGY; STRESS (PSYCHOLOGY). [R.W.But.]

Postulate In a formal deductive system, a proposition accepted without proof, from which other propositions are deduced by the conventional methods of formal logic. There is a certain arbitrariness as to which propositions are to be treated as postulates, because when certain proved propositions are treated as such, other propositions which were originally postulates often become proved propositions. In strict usage, the term postulate is nearly equivalent to axiom, although axiom is often loosely used to denote a truth supposed to be self-evident. *See* LOGIC. [P.W.Br./H.Ma.]

Potassium A chemical element, K, atomic number 19, and atomic weight 39.102. It stands in the middle of the alkali metal family, below sodium and above rubidium. This lightweight, soft, low-melting, reactive metal melts at 63.7°C (147°F), boils at 760°C (1400°F) and has a density of 0.8199 g/cm^3 (0.4739 oz/in.^{3}) at 100°C (212°F). It is very similar to sodium in its behavior in metallic forms. *See* ALKALI METALS; SODIUM.

Potassium chloride finds its main use in fertilizer mixtures. It also serves as the raw material for the manufacture of other potassium compounds. Potassium hydroxide is used in the manufacture of liquid soaps, and potassium carbonate in making soft soaps. Potassium carbonate is also an important raw material for the glass industry. Potassium nitrate is used in matches, in pyrotechnics, and in similar items which require an oxidizing agent.

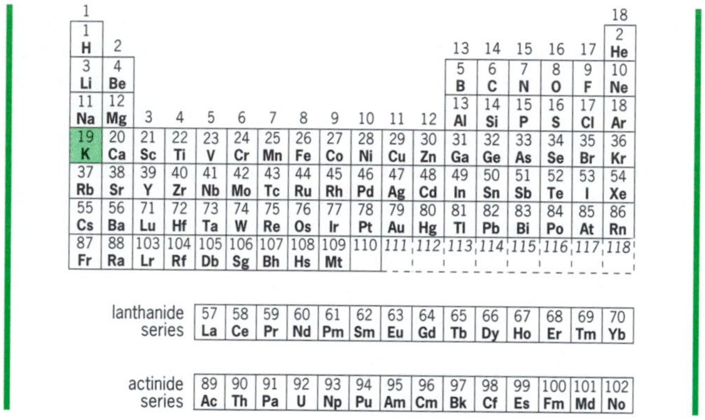

Potassium is a very abundant element, ranking seventh among all the elements in the Earth's crust, 2.59% of which is potassium in combined form. Sea water contains 380 parts per million, making potassium the sixth most plentiful element in solution.

Potassium is even more reactive than sodium. It reacts vigorously with the oxygen in air to form the monoxide, K_2O, and the peroxide, K_2O_2, In the presence of excess oxygen, it readily forms the superoxide, KO_2.

Potassium does not react with nitrogen to form a nitride, even at elevated temperatures. With hydrogen, potassium reacts slowly at 200°C (390°F) and rapidly at 350–400°C (660–752°F). It forms the least stable hydride of all the alkali metals.

The reaction between potassium and water or ice is violent, even at temperatures as low as −100°C (−148°F). The hydrogen evolved is usually ignited in reaction at room temperature. Reactions with aqueous acids are even more violent and verge on being explosive. [M.Si.]

Potato, Irish The plant *Solanum tuberosum* of the nightshade family, Solanaceae. The origin of the potato was in the South American continent. Many wild tuber-bearing species are found in Central and South America. Potatoes have been under cultivation for several thousand years in these areas, supplying a large portion of the food for the inhabitants. They were introduced into Europe early in the 16th century and into the United States from Ireland in 1719.

The potato is an annual, herbaceous, dicotyledonous plant. It is sometimes regarded as a potential perennial because of its ability to reproduce vegetatively by means of tubers. Tubers, which are shortened, thickened underground stems, form on the ends of lateral stems (stolons) underground. The principal areas in the mature tubers from the exterior inward are the periderm, cortex, vascular ring, perimedullary zone, and central pith. Lateral buds (eyes) on the mature tuber are the growing points for the new crop when whole tubers or pieces with at least one eye are planted. *See* SOLANALES.

Potatoes are grown commercially in every state in the United States. The number of farms on which potatoes are grown is decreasing at a rather rapid rate, accompanied by more potato-planted acres per farm. With more efficient methods of growing and harvesting potatoes, higher yields and better methods of storing have resulted. Most of the production is concentrated in 15 states: Idaho, Washington, Maine, Oregon, California, North Dakota, Wisconsin, New York, Minnesota, Colorado, Michigan, Pennsylvania, Florida, Nevada, and Texas.

Chemical composition varies with variety, area of growth, cultural practices, maturity at harvest, subsequent storage history, and other factors. Starch calorically is the most important nutritional component. The nutritional or biological value of protein in potatoes is rather high. Potato protein contains substantially more of all the essential amino acids, except histidine, than does that of whole wheat. Potatoes contain appreciable amounts of the B vitamins: niacin, thiamine, riboflavin, and pyridoxine. Potatoes are an excellent source of vitamin C. Inorganic constituents or minerals of potatoes are predominantly potassium, phosphorus, sulfur, chlorine, magnesium, and calcium. Iron is present in sufficient quantity for potatoes to be a good nutritional source of this mineral. [O.S.]

Potato, sweet The fleshy root of the plant *Ipomoea batatas*. The sweet potato was mentioned as being grown in Virginia as early as 1648. Louisiana is the leading state for commercial production of both the canned and fresh products.

There are two principal types of sweet potato, the kind erroneously called yam and the Jersey type. The chief difference between the two is that in cooking or baking the yam, much of the starch is broken down into simple sugars (glucose and fructose) and an intermediate product, dextrin. This gives it a moist, syrupy consistency somewhat sweeter than that of the dry (Jersey) type. On cooking, the sugar in the dry type remains as sucrose.

The true yam (*Dioscorea*) includes both edible and medicinal varieties. The latter are grown primarily for their cortisone, a steroid. Cortisone yams can be grown in the southern United States, particularly in Louisiana. *See* POTATO, IRISH; STEROID. [J.C.M.]

Potential barrier A region including a maximum of potential energy which prevents a particle on one side of the region from passing to the other side. According to classical physics, a particle must possess an energy exceeding the height of the potential barrier to surmount it. However, quantum mechanics shows that a particle with less energy has a finite probability for penetrating the barrier. The probability decreases rapidly as the particle energy decreases. Such barriers are exemplified by the negative field of the electrons surrounding an atom, by the positive charge in the nucleus which repels a positively charged particle, or by barriers to electrons or holes created in solid-state devices as a result of fixed charge distributions. *See* NUCLEAR STRUCTURE; SEMICONDUCTOR. [R.St.]

Potential energy The energy due to position or configuration. The electric field $\vec{E}$ and the magnetic field $\vec{B}$ in a region of space can be calculated from an electrostatic potential and a magnetic vector potential. Then the force on a charged particle can be calculated, and the work done on the particle along a specific path by the force can be obtained. Thus, macroscopic changes in potential energy that result from pushing an object from one place to another with a stick or pulling it with a string have their microscopic origin in the electromagnetic potential energy which holds the atoms in place in the stick or string together. However, there are macroscopic forces which produce different potential energy functions, such as the elastic spring.

Each potential energy function specifies the dependence of the force on position, and each one contains a constant, which is a reference potential energy. This constant part of the potential energy simply specifies a reference point and does not contribute to the force. *See* ENERGY; HOOKE'S LAW; POTENTIALS. [B.DeF.]

Potential flow Fluid flow which can be specified by a velocity potential. In contrast to creeping flow, it represents a condition where inertia is controlling and viscous forces are negligible. When the effect of viscosity of a fluid is negligible, in most cases the fluid flow will be irrotational at all times if it starts from rest. For irrotational flow, there exists a velocity potential such that the velocity vector is the gradient of the velocity potential. For an incompressible fluid, the velocity potential satisfies the Laplace equation, which has been investigated very thoroughly from a mathematical point of view. Potential flow applies to most subsonic flow problems of a fluid of small viscosity, such as air and water, far away from a solid wall but not in the boundary-layer flow near the wall. *See* BOUNDARY-LAYER FLOW; FLUID FLOW; LAPLACE'S IRROTATIONAL MOTION; POTENTIALS. [S.-I.P.]

Potentials Functions or sets of functions from whose first derivatives a vector can be formed. A vector is a quantity which has a magnitude and a direction, such as force.

A single function, the scalar potential, is used in gravitation theory, electricity and magnetism, fluid mechanics, and other areas. The vectors obtained from it by partial differentiation are in these cases the gravitational, electric and magnetic field strengths, and the velocity, respectively. The vector potential is a set of three functions whose first derivatives give the magnetic induction. *See* ELECTRIC FIELD; GRAVITATION; LAPLACE'S IRROTATIONAL MOTION: MAGNETIC FIELD; MAGNETIC INDUCTION. [F.Ro.]

Potentiometer (voltage meter) A device for the measurement of an electromotive force (emf) by comparison with a known potential difference. The known potential difference is established by the flow of a definite current through a known resistance, using a standard cell as a reference. The principal potentiometer circuits are (1) the constant-current dc potentiometer (historically known as Poggendorff's first method); (2) the Brooks deflectional dc potentiometer (a variant of the basic constant-current potentiometer); (3) the constant-resistance dc potentiometer (known as Poggendorff's second method); (4) the Drysdale ac potentiometer; and (5) the Tinsley-Gall ac potentiometer. *See* ELECTRICAL MEASUREMENTS; ELECTROMOTIVE FORCE (EMF); POTENTIALS; VOLTAGE MEASUREMENT; VOLTMETER. [A.J.Co.]

Pottery Vessels made entirely or partly of clay, and fired to a strong, hard product; occasionally, the term refers to just the lower grades of such ware. Pottery may be glazed or unglazed. *See* CERAMICS; GLAZING.

Grades of pottery (such as china, stoneware, earthenware, and other special types) are distinguished by their color, strength, absorption (the weight of water soaked up when the piece is submerged, expressed as a percentage of the original weight), and translucency (ability to transmit light). All these properties refer to the material or "body" under any glaze present. *See* PORCELAIN.

Absorption is due to the presence of open pores or voids in the fired material into which water can penetrate; in general, the higher the firing temperature, the lower the absorption. Body color is determined mainly by raw-material purity. Strength depends on the porosity and also on the amount and type of glass and crystals developed in the body on firing. Translucency is obtained in products in which there is low porosity and little differences in index of refraction between the glass and crystals in the body. *See* CLAY. [J.F.McM.]

Pottiales An order of mosses. The plants vary in height from minute (up to 0.8 in. or 5 mm) to large (up to 4 in. or 10 cm). The stems are erect, simple, or with dichotomous or fasciculate branching. The leaves are crowded, in several rows, and rarely three-ranked. They vary in shape, but the majority are lanceolate to broadly ovate or obovate. The costa is strong, mostly percurrent or excurrent. *See* BRYOPSIDA. [W.H.W.]

Poultry production Poultry production can be divided into two major categories, meat production and egg production. Heavy-breed chickens, turkeys, ducks, and geese are used for meat, while light-breed chickens are used primarily for egg production.

Trends in both broiler and egg production have been toward vertical integration. This means that the various aspects of production and marketing (production of hatching eggs, incubation, feeding, growing, processing, and marketing) are being merged within the framework of a single organization. [G.O.B.]

Pour point The lowest temperature at which an oil will pour when cooled under prescribed test conditions. Because petroleum oils are complex mixtures which become plastic solids when cooked, several solidification temperatures are used, each defined by a definite test procedure. The cloud point is the temperature at which a solid phase separates from solution. For a grease, the dropping point is the temperature at which the plastic solid becomes sufficiently fluid to flow through an orifice. For waxes, the melting point is the temperature at which the material becomes fluid enough to drop from the test thermometer; the congealing point is the temperature at which a sample wetting the thermometer appears to congeal and hence rotate with the thermometer. [M.Sou.]

Powder metallurgy A process for making a wide range of components and shapes from a variety of metals and alloys in the form of powder. The process is automated and uses pressure and heat to form precision metal parts into net or near-net shapes, requiring a minimum amount of secondary finishing.

The most common metals available in powder form are iron, fin, nickel, copper, aluminum, and titanium, and refractory metals such as tungsten, molybdenum, tantalum, and niobium (columbium). Prealloyed powders such as low-alloy steels, bronze, brass, and stainless steel are produced in which each particle is itself an alloy. Also available in powder form are nickel-cobalt-base superalloys and tool steels.

Although most powder metallurgy parts and products are pressed in mechanical or hydraulic compacting presses, other techniques such as cold and hot isostatic pressing, powder metallurgy forging, and direct powder rolling are used to make a variety of important end products.

Powder metallurgy has environmental advantages in terms of both materials and energy conservation. Instead of generating scrap, the process utilizes scrap from other metal processes and recycles it into useful powders. Energy savings as high as 50% can be realized using lower-density powder metallurgy parts in place of steel parts machined from bar stock that require extensive machining. Only the amount of raw material needed for the finished part is used. *See* METALLURGY. [P.K.J.]

Power The time rate of doing work. Like work, power is a scalar quantity, that is, a quantity which has magnitude but no direction. Some units often used for the measurement of power are the watt (1 joule of work per second) and the horsepower (550 foot-pounds of work per second). *See* WORK.

Power is a concept which can be used to describe the operation of any system or device in which a flow of energy occurs. In many problems of apparatus design, the power, rather than the total work to be done, determines the size of the component used. Any device can do a large amount of work by performing for a long time at a low rate of power, that is, by doing work slowly. However, if a large amount of work must be done rapidly, a high-power device is needed. High-power machines are usually larger, more complicated, and more expensive than equipment which need operate only at low power. A motor which must lift a certain weight will have to be larger and more powerful if it lifts the weight rapidly than if it raises it slowly. An electrical resistor must be large in size if it is to convert electrical energy into heat at a high rate without being damaged. [P.W.S.]

Power amplifier The final stage in multistage amplifiers, such as audio amplifiers and radio transmitters, designed to deliver appreciable power to the load. Power amplifiers may be called upon to supply power ranging from a few watts in an audio amplifier to many thousands of watts in a radio transmitter. In audio amplifiers the load is usually the dynamic impedance presented to the amplifier by a loudspeaker, and the problem is to maximize the power delivered to the load over a wide range of frequencies. The power amplifier in a radio transmitter operates over a relatively narrow band of frequencies with the load essentially a constant impedance. *See* AMPLIFIER. [H.F.K.]

Power factor The ratio of watts average (or active) power to the apparent power of an alternating-current circuit. By definition, and of general application, the equation below

$$\text{Power factor (pf)} = \frac{\text{watts average power}}{\text{rms volts} \times \text{rms amperes}}$$

holds, which is the ratio of instrument readings. A watt-meter indicates average power and electrodynamometer or iron-vane instruments show root-mean-square (rms) voltage and current. For the steady-state ac circuit under sinusoidal voltage and current, pf = cos θ, where θ is the phase angle between the voltage and current. This definition is restricted to sine waves of the same frequency. *See* ALTERNATING CURRENT; NONSINUSOIDAL WAVEFORM; POWER-FACTOR METER. [B.L.S.]

Power-factor meter An instrument used to indicate whether load currents and voltages are in time-phase with one another. *See* POWER FACTOR.

The single-phase meter contains a fixed coil that carries the load current, and crossed coils that are connected to the load voltage. There is no spring to restrain the moving system, which takes a position to indicate the angle between the current and voltage. The scale can be marked in degrees or in power factor. *See* PHASE METER.

The angle between the currents in the crossed coils is a function of frequency, and consequently each power-factor meter is designed for a single frequency and will be in error at all other frequencies. [H.So./E.C.St.]

Power integrated circuits Integrated circuits that are capable of driving a power load. The key feature of a power integrated circuit that differentiates it from other semiconductor technologies is its ability to handle high voltage, high current, or a combination of both.

In its simplest form, a power integrated circuit may consist of a level-shifting and drive circuit that translates logic-level input signals from a microprocessor to a voltage and current level sufficient to energize a load. For example, such a chip may be used to operate electronic display, where the load is usually capacitive in nature but requires drive voltages above 100 V, which is much greater than the operating voltage of digital logic circuits (typically 5 V). At the other extreme, the power integrated circuit may be required to perform load monitoring, diagnostic functions, self-protection, and information feedback to the microprocessor, in addition to handling large amounts of power to actuate the load. An example of this is an automotive multiplexed bus system with distributed power integrated circuits for control of lights, motors, air conditioning, and so forth. *See* AUTOMOTIVE ELECTRICAL SYSTEM; ELECTRONIC DISPLAY; MICROPROCESSOR.

Power integrated circuits are expected to have an impact on all areas in which power semiconductor devices are presently being used. In addition, they are expected to open up new applications based upon their added features. The wide spectrum of voltages and currents over which power semiconductor

Applications of power integrated circuits

Application	Power requirements: Voltage V	Power requirements: Current A	Operating frequency*	Circuit configuration
Display drives	100–400	0.01–0.1	High	dc
Computer power supplies	50	10–50	High	dc
Motor drives	300–600	10–50	Medium	dc
Factory automation	300–600	10–50	Low	ac
Telecommunications	100–600	1–10	High	ac
Appliance control	300–600	1–20	Low	ac
Consumer electronics	50	0.1–10	Medium	dc
Lighting	300–600	1–10	Medium	ac
Smart house	300–600	5–50	Low	ac
Aircraft electronics	100–600	10–50	Medium	dc
Automotive electronics	50–100	1–20	Low	dc

*Low to less than 1 kHz; medium is 1–50 kHz; high is above 50 kHz.

devices are utilized are summarized in the table. *See* INTEGRATED CIRCUITS. [B.J.Ba.]

Power plant A means for converting stored energy into work. Stationary power plants such as electric generating stations are located near sources of stored energy, such as coal fields or river dams, or are located near the places where the work is to be performed, as in cities or industrial sites. Mobile power plants for transportation service are located in vehicles, as the gasoline engines in automobiles and diesel locomotives for railroads. Most power plants convert part of the stored raw energy of fossil fuels into kinetic energy of a spinning shaft. Some power plants harness nuclear energy. Elevated water supply or run-of-the-river energy is used in hydroelectric power plants. For transportation, the plant may produce a propulsive jet, as in some aircraft, instead of the rotary motion of a shaft. Other sources of energy, such as winds, tides, waves, geothermal sources, ocean thermal, nuclear fusion, and solar radiation, have been of negligible commercial significance in the generation of power despite their magnitudes. *See* ENERGY SOURCES. [T.Ba.; L.M.Ol.; K.A.R.]

Power shovel A digging machine, usually self-propelled on crawler, rubber tire, or sometimes rail mountings. It is equipped with a shovel boom, dipper attached to the front end of a dipper stick, dipper trip mechanism, padlock (dipper sheave block), crowding mechanism, and cables all carried on a fully revolving superstructure. *See* BULK-HANDLING MACHINES; CONSTRUCTION EQUIPMENT. [A.M.P.]

Power steering A steering control system for a propelled vehicle in which an auxiliary power source assists the driver by providing the major force required to direct the road wheels. The term power steering commonly applies to any system with a variety of power booster arrangements. The principal components of the power booster are the control valve, the power actuator, and the power source. Dependent upon the type of system, these components normally operate in conjunction with a hand steering wheel, linkage system, and steered wheels. (Most systems provide for manual steering gear backup.) As a vehicle follows a driver-steered course, the valve senses any deviation between the course-set position of the steering wheel and the corresponding position of the steered wheels and releases power to the actuator until the error is corrected. *See* AUTOMOTIVE STEERING.

The power steering control system is a hydraulic servomechanism. As the driver's desired course is set at the steering wheel, the system (servoloop) compares the desired course with a measurement of the actual and "signals" a measure of the error. The error signal proportionately actuates the power to assist the steering maneuver. This servoloop essentially controls the power assist function of the steering system. Normally the actual servoloop is added to a manual steering arrangement; however, manual backup steering is not mandatory for off-highway vehicles. The main components of the system that maintain the servo function are the control valve and actuator, introduced in various arrangements (see illus.).

In each system an engine-driven pump with a reservoir acts as the power source. The pump delivers hydraulic fluid oil under a working pressure to the controller valve. The controller valve meters the working pressure or pump delivery to the actuator, a hydraulic cylinder. The controlling valve responds to a torque from the steering wheel and to reactive shock from the road wheels and directs oil as required to the actuator. In the stroking action the hydraulic cylinder repositions the steered wheels to maintain the vehicle course set by the driver. In order for the driver to "feel" the road at the steering wheel, the control valve is desensitized to low steering torque efforts, and then manual steer results. [J.L.R.]

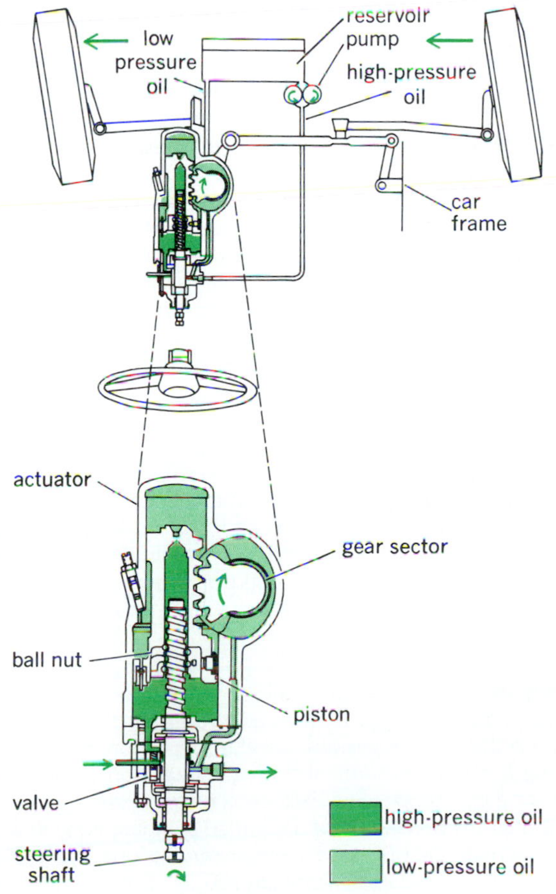

Integral power steering. (*TRW Inc.*)

Poynting's vector A vector, the outward normal component of which, when integrated over a closed surface in an electromagnetic field, represents the outward flow of energy through that surface. It is given by the equation below, where

$$\Pi = \mathbf{E} \times \mathbf{H} = \mu^{-1}\,\mathbf{E} \times \mathbf{B}$$

E is the electric field strength, **H** the magnetic field strength, **B** the magnetic flux density, and μ the permeability.

When an electromagnetic wave is incident on a conducting or absorbing surface, theory predicts that it should exert a force on the surface in the direction of the difference between the incident and the reflected Poynting's vector. *See* ELECTRIC FIELD; ELECTROMAGNETIC RADIATION; MAGNETIC FIELD; MAXWELL'S EQUATIONS; RADIATION PRESSURE. [W.R.Sm.]

Prairie dog A stout, fossorial rodent belonging to the family Sciuridae. Three species are recognized; the black-tailed prairie dog (*Cynomys ludovicianus*), once abundant on the prairies of the western United States, but now reduced in numbers; the white-tailed prairie dog (*C. gunnisoni*), discontinuously distributed in mountainous and high plains areas of Colorado, New Mexico, Utah, and Montana; and *C. mexicanus*, found in northern Mexico. All species have the same general features. The tail is short and flat, the ears are small, and the limbs are short and terminate in long claws which are used for burrowing. Prairie dogs are social animals and live in colonies. Their burrows are constructed with the main entrance surrounded by a mound of hard-packed dirt to protect against flooding. *See* RODENTIA; SOCIAL ANIMALS. [C.B.C.]

Praseodymium A chemical element, Pr, atomic number 59, and atomic weight 140.91. Praseodymium is a metal-

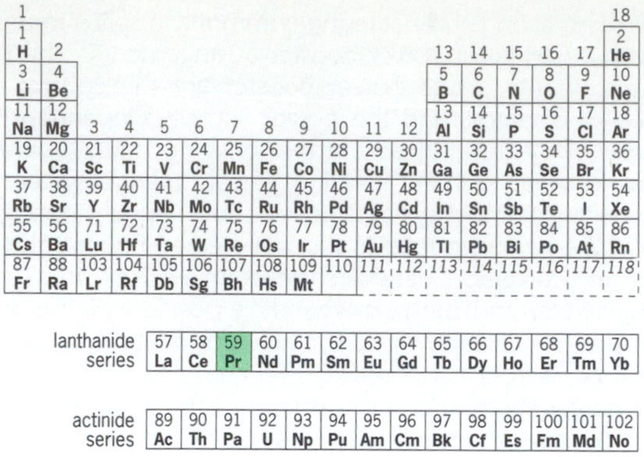
1 2 13 14 15 16 17 18
1 H
2 He
3 Li 4 Be 5 B 6 C 7 N 8 O 9 F 10 Ne
11 Na 12 Mg 3 4 5 6 7 8 9 10 11 12 13 Al 14 Si 15 P 16 S 17 Cl 18 Ar
19 K 20 Ca 21 Sc 22 Ti 23 V 24 Cr 25 Mn 26 Fe 27 Co 28 Ni 29 Cu 30 Zn 31 Ga 32 Ge 33 As 34 Se 35 Br 36 Kr
37 Rb 38 Sr 39 Y 40 Zr 41 Nb 42 Mo 43 Tc 44 Ru 45 Rh 46 Pd 47 Ag 48 Cd 49 In 50 Sn 51 Sb 52 Te 53 I 54 Xe
55 Cs 56 Ba 71 Lu 72 Hf 73 Ta 74 W 75 Re 76 Os 77 Ir 78 Pt 79 Au 80 Hg 81 Tl 82 Pb 83 Bi 84 Po 85 At 86 Rn
87 Fr 88 Ra 103 Lr 104 Rf 105 Db 106 Sg 107 Bh 108 Hs 109 Mt 110 111 112 113 114 115 116 117 118
lanthanide series 57 La 58 Ce 59 Pr 60 Nd 61 Pm 62 Sm 63 Eu 64 Gd 65 Tb 66 Dy 67 Ho 68 Er 69 Tm 70 Yb
actinide series 89 Ac 90 Th 91 Pa 92 U 93 Np 94 Pu 95 Am 96 Cm 97 Bk 98 Cf 99 Es 100 Fm 101 Md 102 No

lic element of the rare-earth group. The stable isotope 140.907 makes up 100% of the naturally occurring element. The oxide is a black powder, the composition of which varies according to the method of preparation. If oxidized under a high pressure of oxygen it can approach the composition PrO_2. The black oxide dissolves in acid with the liberation of oxygen to give green solutions or green salts which have found application in the ceramic industry for coloring glass and for glazes. For properties of the metal *see* RARE-EARTH ELEMENTS. [F.H.Sp.]

Prasinophyceae A small class of mostly motile, photosynthetic, unicellular algae in the chlorophyll *a-b* phyletic line (Chlorophycota). It is segregated from the Chlorophyceae primarily on the basis of ultrastructural characters, especially the possession of one or more layers of polysaccharide scales outside the plasmalemma. Prasinophytes are mainly members of the marine plankton, but they are also found in brackish- and fresh-water habitats. A few are benthic, with both coccoid and colonial forms known, while others live symbiotically within dinoflagellates, radiolarians, and turbellarian worms. Approximately 180 species are known in 13 genera. *See* ALGAE; CHLOROPHYCOTA; PHYTOPLANKTON. [P.C.Si.; R.L.Moe]

Preamplifier A voltage amplifier suitable for operation with a low-level input signal. It is intended to be connected to another amplifier with a higher input level. Preamplifiers are necessary when an audio amplifier is to be used with low-output transducers such as magnetic phonograph pickups. A preamplifier may incorporate frequency-correcting networks to compensate for the frequency characteristics of a given input transducer and to make the frequency response of the preamplifier-amplifier combination uniform. *See* AMPLIFIER; VOLTAGE AMPLIFIER. [H.F.K.]

Precambrian The term used to designate rocks and time older than the Cambrian, which is the oldest geologic period from which abundant fossils have been recovered. Age determinations by means of radioactive elements place the upper boundary of the Precambrian at about 600,000,000 years ago. The oldest dated rocks appear to be 3,800,000,000–4,000,000,000 years old. All evidence indicates that these rocks are not remnants of the original Earth crust, and it follows that the Precambrian interval of Earth history lasted well over 3,500,000,000 years. *See* CAMBRIAN; RADIOACTIVE MINERALS; ROCK AGE DETERMINATION.

The variety of rocks found in Precambrian terranes is great, but one continent differs little from another in this regard. They are represented by all kinds of sedimentary rocks (conglomerates, sandstones, shales, and limestones) and extensive lava flows. In general, the older rocks exhibit higher degrees of metamorphism, but this is not invariably true. Precambrian rocks underlie all other rocks of continents. They occur beneath a great blanket of flat or deformed sedimentary and other rocks, are near the surface covered by a thin veneer of strata or unconsolidated material, or are exposed.

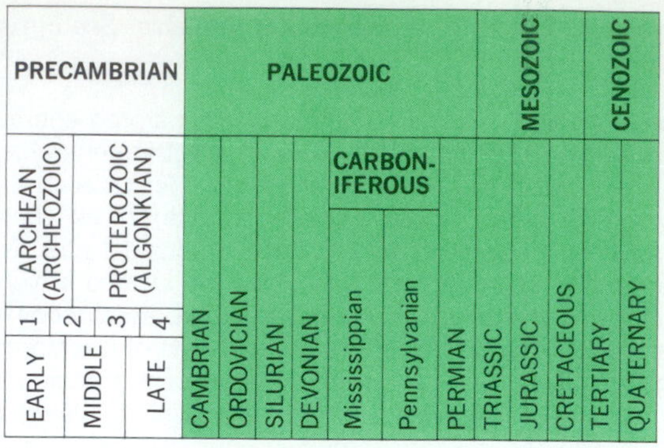
PRECAMBRIAN
PALEOZOIC
MESOZOIC
CENOZOIC
ARCHEAN (ARCHEOZOIC)
PROTEROZOIC (ALGONKIAN)
1
2
3
4
EARLY
MIDDLE
LATE
CAMBRIAN
ORDOVICIAN
SILURIAN
DEVONIAN
CARBON-IFEROUS
Mississippian
Pennsylvanian
PERMIAN
TRIASSIC
JURASSIC
CRETACEOUS
TERTIARY
QUATERNARY

Precambrian rocks are important sources of metalliferous ores, particularly iron, nickel, gold, uranium, and copper. Tremendous stores of iron ore have been mined and are still being mined from the Canadian shield, the Baltic shield, the Ukraine, and the Brazilian and Guianan shields. The ores are mostly sedimentary, but large masses of magnetite of probably igneous origin are also productive.

Life was probably abundant in the seas of late Precambrian time, since Early Cambrian formations contain abundant fossils, but forms with shells capable of being preserved were still few.

The atmosphere undoubtedly underwent great change in Precambrian time, but evidence from the rocks indicates that climatic variation from dry to wet and from hot to cold took place then as later. Glaciation was widespread. [J.P.F.]

Precast concrete Concrete that has been cast into a form which is later incorporated into a structure. A concrete structure may be constructed by casting the concrete in place on the site, by building it of components cast elsewhere, or by a combination of the two. Concrete cast in other than its final position is called precast.

In contrast with cast-in-place concrete construction, in which columns, beams, girders, and slabs are cast integrally or bonded together by successive pours, precast concrete requires field connections to tie the structure together. These connections can be a major design problem.

Precast units can be standardized. Savings can then result from repeated reuse of forms and assembly-line production. Furthermore, high quality can be maintained because of the controls that can be kept on production under plant conditions. However, there is always the possibility that transportation, handling, and erection costs for the precast units will offset the savings. *See* CONCRETE; PRESTRESSED CONCRETE. [F.S.M.]

Precession An angular velocity of the axis of spin of a spinning rigid body, which arises as a result of external torques acting on the body. Examples are the precession of a spinning top, the Earth (precession of the equinoxes), an airplane propeller, or a gyroscope. *See* PRECESSION OF EQUINOXES.

The uniform precession of a charged spinning body in a uniform magnetic field is called Larmor precession. This motion, similar to that of a rapidly spinning top, is of great importance in atomic physics.

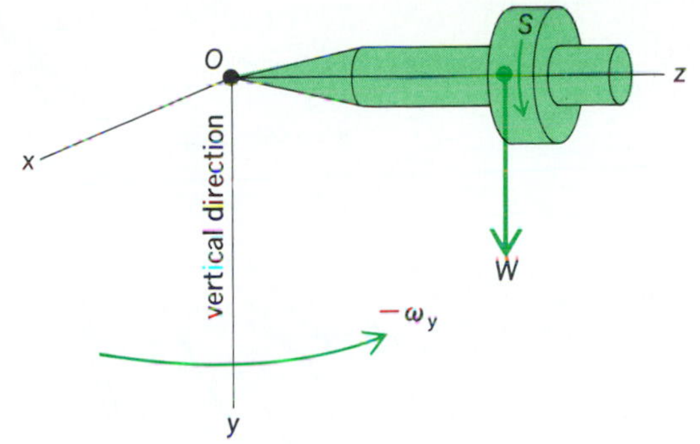

A fast-spinning top (spin velocity S) supported at point *O* and released in a horizontal plane processes about the vertical *y* axis with angular velocity $-\omega_y\omega$.

Consider a top spinning rapidly about its *z* axis of symmetry and placed horizontally on a point support at *O* (see illustration). The top does not fall as a result of the pull of gravity (its weight, *W*), as might be supposed; rather, the *z* axis of the top rotates slowly about the vertical *y* axis while maintaining its position in the horizontal *xz* plane. This rotation about the *y* axis is called precession. [R.E.Bo.]

Precession of equinoxes A slow change in the direction of orientation of Earth's axis of rotation which results in a gradual westward motion of the equinoxes. There are two types, known as lunisolar precession and planetary precession.

The lunisolar precession of the equinoxes is caused by the gravitational attraction of the Sun and the Moon which, as a result of the polar flattening of Earth and the inclination of Earth's axis, gives rise to a small turning moment, or torque, on the Earth in its orbit. As a result of this torque, Earth's axis describes a cone about the normal to the plane of its orbit (see illustration); the period of this precession is approximately 26,000 years in a direction opposite to that of Earth's rotation.

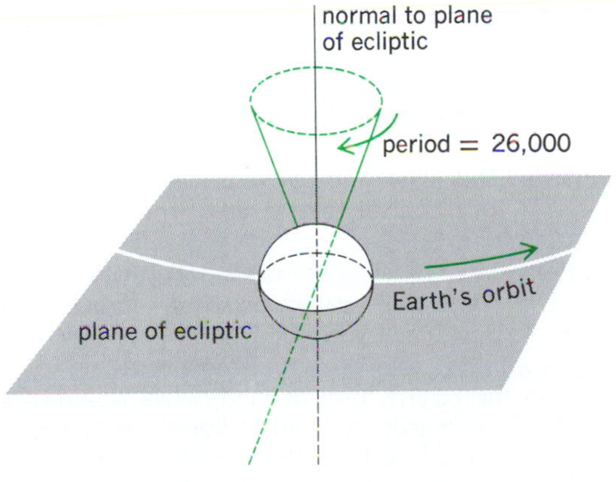

Conical motion of Earth's axis. One complete cycle of precession requires 26,000 years.

The planetary precession is a comparatively small eastward motion of the equinoxes caused by the action of the other planets in altering the plane of Earth's orbit.

As a result of precession, Polaris will not always be the pole star. Vega will be nearer to the north celestial pole in about 12,000 years; α Draconis was the pole star about 4600 years ago. Another effect of precession is that the signs of the zodiac no longer correspond to their respective constellations. *See* PRECESSION. [J.A.Hy.]

Precious stones The materials found in nature that are used frequently as gemstones, including amber, beryl (emerald and aquamarine), chrysoberyl (cat's-eye and alexandrite), coral, corundum (ruby and sapphire), diamond, feldspar (moonstone and amazonite), garnet (almandite, demantoid, and pyrope), jade (jadeite and nephrite), jet, lapis lazuli, malachite, opal, pearl, peridot, quartz (amethyst, citrine, and agate), spinel, spodumene (kunzite), topaz, tourmaline, turquois and zircon. *See* GEM.

The terms precious and semiprecious have been used to differentiate between gemstones on a basis of relative value. Because there is a continuous gradation of values from materials sold by the pound to those valued at many thousands of dollars per carat, and because the same mineral may furnish both, a division is essentially meaningless. [R.T.L.]

Precipitation (chemistry) The process of producing a separable solid phase within a liquid medium. In analytical chemistry, precipitation is widely used to effect the separation of a solid phase in an aqueous solution. For example, the addition of a water solution of silver nitrate to a water solution of sodium chloride results in the formation of insoluble silver chloride. Quite often, one of the components in the solution is thus virtually completely separated in a relatively pure form. It can then be isolated from the solution phase by filtration or centrifugation, and the substance determined by weighing. This procedure is known as gravimetric analysis. Precipitation may also be used merely to effect partial or complete separation of a substance for purposes other than that of gravimetric analysis. Such purposes might involve either the isolation of a relatively pure substance or the removal of undesirable components of the solution. *See* GRAVIMETRIC ANALYSIS.

The extent to which a component can be separated from solution can be determined from the solubility-product constant obtained by determining the quantity of dissolved substance present in a known amount of saturated solution. This value is known as the solubility. The solubility can be drastically altered merely by adding to the solution any of the ions that make up the precipitate, for example, by adding varying quantities of either silver nitrate or sodium chloride to a saturated solution of silver chloride. Although solubility can be altered over a wide range, the solubility product itself remains practically constant over this same range. *See* SOLUBILITY PRODUCT CONSTANT.

Various techniques may be employed in order to reduce contamination by foreign ions. Precipitation from dilute solution is often effective. Heating the reaction mixture speeds recrystallization processes by which incorporated foreign ions may be returned to the solution phase. Precipitation from homogeneous solution results in the slow formation of large crystals of small surface area and hence lessens coprecipitation. If all these methods fail to reduce adequately the quantity of foreign ions incorporated in the solid phase, the precipitate is dissolved and reprecipitated by the previous procedure. *See* CHEMICAL SEPARATION TECHNIQUES; CRYSTALLIZATION; NUCLEATION. [L.Go./R.W.Mu.]

Precipitation (meteorology) The fallout of water drops or frozen particles from the atmosphere. Liquid types are rain or drizzle, and frozen types are snow, hail, small hail, ice pellets (also called ice grains; in the United States, sleet), snow pellets (graupel, soft hail), snow grains, ice needles, and ice crystals. In England sleet is defined as a mixture of rain and snow, or melting snow. Deposits of dew, frost, or rime, and moisture collected from fog are occasionally also classed as precipitation. *See* HAIL; SNOW.

All precipitation types are called hydrometeors, of which additional forms are clouds, fog, wet haze, mist, blowing snow, and spray. Whenever rain or drizzle freezes on contact with the ground to form a solid coating of ice, it is called freezing rain, freezing drizzle, or glazed frost; it is also called an ice storm or a glaze storm, and sometimes is popularly known as silver thaw or erroneously as a sleet storm. *See* CLOUD; FOG.

Rain, snow, or ice pellets may fall steadily or in showers. Steady precipitation may be intermittent though lacking sudden bursts of intensity. Hail, small hail, and snow pellets occur only in showers; drizzle, snow grains, and ice crystals occur as steady precipitation. Showers originate from instability clouds of the cumulus family, whereas steady precipitation originates from stratiform clouds.

The amount of precipitation, often referred to as precipitation or simply as rainfall, is measured in a collection gage. It is the actual depth of liquid water which has fallen on the ground, after frozen forms have been melted, and is recorded in millimeters or inches and hundredths. A separate measurement is made of the depth of unmelted snow, hail, or other frozen forms. *See* PRECIPITATION GAGES; SNOW GAGE. For discussions of other topics related to precipitation *see* CLOUD PHYSICS; DEW; DEW POINT; HUMIDITY; HYDROLOGY; HYDROMETEOROLOGY; RAIN SHADOW; VAPOR PRESSURE; WEATHER MODIFICATION. [J.R.F.]

Precipitation gages Instruments used to measure the amount of rainfall or snowfall expressed as inches or millimeters of depth of water which falls on a level surface. When precipitation gages are equipped with a recorder, the time of occurrence and the rate or intensity are available as well. Other forms of precipitation, such as dew, frost, and moisture absorbed by the soil, are also of considerable interest, but they are not measured by the precipitation gages routinely used by the meteorological services.

The gage is basically an open container and funnel constructed to minimize any splashing out and mounted 1–3 ft (0.3–0.9 m) above the surface to prevent splashing into the container, yet not so high that its catch is affected by the wind which normally increases in velocity with height. From the volume of water calibrated and the area of the opening, the depth of water is easily obtained. A continuous type of recording rain gage uses a spring balance to record on a clock-driven chart the weight of the precipitation calibrated.

When a suitable wind screen is used, snowfall can be measured by melting the snow collected in a rain gage (the funnel must be removed). Fallen snow is also measured by taking sample cores. *See* SNOW GAGE. [V.E.S.]

Precipitin The visible result of the chemical interaction of antigen and antibody. Not all antibodies will result in precipitation, yet they may participate in agglutination reactions or add onto particulate antigens, and evidence for their occurrence together with precipitating antibody can be obtained for most sera. Precipitins may be noted qualitatively or be quantified by noting the end-point dilution (titer) of serum required to give a precipitate at the threshold of visibility, or the amount of antibody may be determined in milligrams or micrograms by analysis of the precipitate with correction for the antigen contained therein. *See* ANTIBODY; ANTIGEN. [D.R.]

Precision approach radar (PAR) A radar system located on an airfield for observation of the position of an aircraft with respect to an approach path and specifically intended to provide guidance to the aircraft during its approach to the field.

The PAR system consists of a ground radar equipment which is alternately connected to two antenna systems. One of these antenna systems sweeps a narrow beam over a 20° sector in the horizontal plane. The second antenna system sweeps a narrow beam over a 7° sector in the vertical plane. Since the reflected energy from the aircraft produces information on the location of the aircraft in terms of angles and distance, the two sweeps give the three-dimensional location of the aircraft (see illustration). *See* ELECTRONIC NAVIGATION SYSTEMS. [P.C.S.]

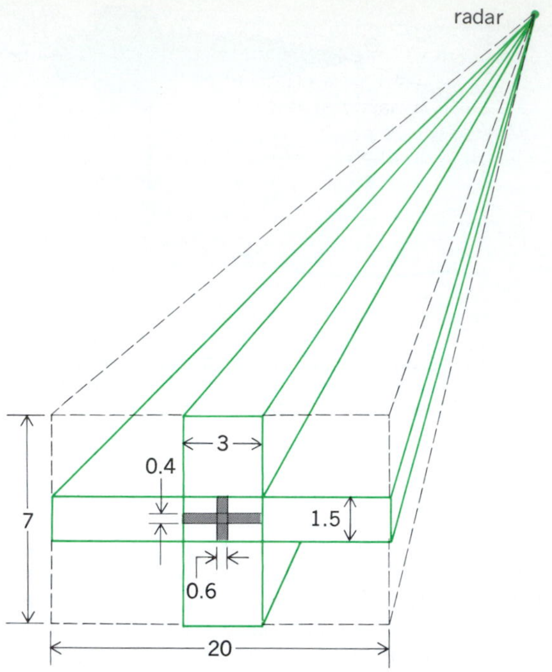

Field patterns of the antennas of the radar low-approach system. The dimensions shown are all in degrees. The broken lines indicate the total movement of the field patterns. The gray areas represent the dimension of the field patterns from the antennas when they are not sweeping.

Pregnancy In humans, the period of about 280 days required for the normal intrauterine development of a child. Changes in the mother are most marked in the reproductive organs. A slight enlargement and softening of the uterus may be detectable on examination by the end of the second month. With further enlargement, the muscular wall becomes hypertrophied and then stretched, so that it is somewhat flexible. During the fourth month the uterus ordinarily rises above the pelvis and becomes abdominal. By the end of pregnancy the upper portion of the uterus lies just beneath the anterior rib margins.

Changes in the breasts begin early in pregnancy with enlargement and tenderness being common. As term approaches, the nipples and areolae increase in size and deepen in pigmentation, reflecting the proliferation of the glandular elements of the breasts. A watery or slightly turbid discharge, called colostrum, may be expressed from the nipples, especially in women who have had previous pregnancies.

Changes in the vagina, ovaries, and cervix accompany the progression of pregnancy. Systemic alterations in renal function, cardiovascular adaptation, blood volume, and cell counts, as well as in chemical constituents of blood and tissue all reflect the complex physiologic changes required for pregnancy.

Pregnancies may proceed to term or may be concluded by natural or artificial means prior to normal delivery. The term abortion is used to indicate the passage of a fetus of less than 2 lb (1000 g), or before the twenty-eighth week of pregnancy. This may occur spontaneously as a result of any one of many causes, including induced methods. *See* MENSTRUATION; PREGNANCY DISORDERS. [E.G.St./N.K.M.]

Pregnancy disorders Pregnancy in humans leads to the relatively common minor complaints of nausea, vomiting, fatigue, and constipation but also to more serious disorders.

Hemorrhage in pregnancy may be due to many causes but

by far the most common is abortion; less common causes are ectopic pregnancy, premature labor, and premature separation of the placenta from the uterus.

In ectopic pregnancy, a not infrequent occurrence, a fertilized ovum becomes implanted in a tissue other than the uterine mucosa. Most often the site of implantation is some part of the Fallopian tube which becomes stretched and inflamed as growth proceeds; rupture is not uncommon. In other cases, the fertilized ovum may be implanted on some part of the peritoneum or mesentery and produce some symptoms if growth proceeds. There are occasional reports of the delivery of a full-term, normal infant by abdominal operation from such a site. In most cases, however, it is believed that the embryo does not survive and is gradually absorbed by the body.

Two important tumors of the placenta occur. They are the hydatidiform mole and choriocarcinoma. These growths are grouped together because in about one-half of the cases choriocarcinoma is preceded by a hydatidiform mole. They should be considered separate, however, because the mole in most cases is a benign lesion, whereas choriocarcinoma may be fatal.

It is estimated that of every 100 fertilizations, only 70% will develop into term infants. Of the remainder, one-third are abnormal so that implantation cannot occur, one-third result in abnormal embryos which are aborted, and one-third are lost by spontaneous abortion from other causes. *See* PLACENTATION; PREGNANCY. [N.K.M.]

Prehistoric technology The set of ideas that prescribe the manufacture and use of implements before written history. Technology is the principal means through which the human species has succeeded in occupying most of the world. The prehistoric record documents that for over 2 million years (m.y.) the human lineage has been making and using implements. Archeologists tend to use the term artifact for any material that was modified by ancient humans, whether this material was used or not, and the term tool for any material that was used by ancient humans, whether it was modified or not.

The archaeological evidence for prehistoric technologies may be biased because of the nature of the raw materials that were used and the conditions of burial and preservation of these materials. In general, artifacts of stone, bone, pottery, and metal preserve fairly well in many areas, whereas artifacts of wood and other vegetable materials, skin, and horn tend to decay fairly rapidly and are normally found only in exceptional prehistoric contexts.

In 1816, C. Thompsen, director of the Danish National Museum, began to chronologically order the museum's prehistoric collections into three major groups, based upon technology, and now called the Three Age System. The earliest was a Stone Age, followed by a Bronze Age, and finally an Iron Age. As time went on, the prehistory of Europe was further divided, based on regional sequences that could be documented through excavation. In other places, such as the Americas, Australia, Oceania, and sub-Sarahan Africa, different nomenclature was often used, since the regional technological sequences differed from that of western Europe.

New technological traits can be introduced into a society in a number of ways: (1) innovation or invention, which is the development of new ideas, including new technological characteristics; (2) diffusion, which involves the spread of ideas, including technological knowledge, from one group to another; and (3) migration, which is the spread of peoples, often with new technologies, into new areas. One of the goals of a prehistorian is to try to ascertain, based on archeological evidence, which factors best explain the technological changes seen in the archeological record. [N.To.]

Prehnite A mineral with the formula $Ca_2(Al,Fe^{3+})(OH)_2$ - $[Si_3AlO_{10}]$, with Al in parentheses in octahedral and Al in brackets in tetrahedral coordination by oxygens. The mineral usually occurs as stalactitic aggregates or as curved crystals, has a vitreous luster, and is yellowish green to pale green in color. Hardness is 6–6½ on Mohs scale; specific gravity 2.8–2.9. Common occurrences include vesicular basalts such as the Keweenaw basalts in the Upper Peninsula of Michigan, and the Watchung basalts in New Jersey. *See* SILICATE MINERALS. [P.B.M.]

Prenatal diagnosis The identification of disease before the birth of a fetus. It often implies genetic diagnosis, but identification of anatomical defects as well as assessment of fetal functions and maturity are also considered. Some of the relatively common diseases that can be diagnosed prenatally are Tay-Sachs disease, cystic fibrosis, Duchenne's muscular dystrophy, hemophilia A, congenital adrenal hyperplasia, thalassemia, and sickle cell anemia.

Ultrasonic data have vastly improved the understanding of normal growth and development, thus permitting earlier and more accurate diagnosis of fetal disease. Fetal movements can also be observed, allowing assessment of functional well-being. *See* MEDICAL IMAGING; MEDICAL ULTRASONIC TOMOGRAPHY.

Obstetricians typically review the health history of the pregnant woman, the father-to-be, and their families in order to identify any possible heritable disorders in the families. Certain risks related to the patient's age, race, and geographic origin may be noted. Based on these assessments, genetic testing of the unborn child may be discussed or recommended. Using cells from the fetus, a prenatal genetic diagnosis can be made for at least 10% of those disorders that are known or assumed to result from a gene mutation. Such diagnosis is based on deoxyribonucleic acid (DNA) analysis or on detection of an abnormal enzyme or other protein produced by the defective gene. *See* HUMAN GENETICS. [D.McN.]

Press fit A force fit that has negative allowance; that is, the bore in the fitted member is smaller than the shaft which is pressed into the bore. Tight fits have slight negative allowance so that light pressure is required to assemble the parts; they are used for gears, pulleys, cranks, and rocker arms. Medium force fits have somewhat greater negative allowance and require considerable pressure for assembly; they are used for fastening locomotive wheels, car wheels, and motor armatures. *See* ALLOWANCE; FORCE FIT; SHRINK FIT. [P.H.B.]

Pressure The ratio of force to area. The force per unit area at the interior of the Sun is estimated to be 3×10^{17} dynes/cm^2 (3×10^{16} Pa). In interstellar space, pressure approaches zero. Atmospheric pressure at the surface of Earth is in the vicinity of 14 lb/in.^{2} (100 kPa). Pressures in enclosed containers less than this value are spoken of as vacuum pressures. This is absolute pressure measured above zero pressure as a reference level. Inside a steam boiler, the pressure may be 800 lb/in.^{2} (5.5 MPa) or higher. Such pressure, measured above atmospheric pressure as a reference level, is gage pressure, designated psig. Differential pressure is the difference between any two pressures, neither of which is atmospheric. *See* PRESSURE MEASUREMENT. [F.H.R.]

Pressure altimeter A device that precisely measures the pressure of the air at the level an aircraft is flying and converts the pressure measurement to an indication of height above sea level according to a standard pressure-altitude relationship. In essence, a pressure altimeter is a highly refined aneroid barometer since it utilizes an evacuated capsule whose movement or force is directly related to the pressure on the outside of the capsule. Various methods are used to sense the capsule function and cause a display to respond such that the pilot sees the altitude level much as one looks at a watch.

Because altitude measured in this manner is also subject to changes in local barometric pressure, altimeters are provided with a barosetting that allows the pilot to compensate for these weather changes, the sea-level air pressure to which the altimeter is adjusted appearing in a window of the dial. Flights below 18,000 ft (5486 m) must constantly contact the nearest traffic center to keep the altimeters so updated. Flights above 18,000 ft and over international waters utilize a constant setting of 29.92 in. Hg, or 1013.2 millibars (101.32 kPa), so that all high-flying aircraft have the same reference and will be interrelated, thus providing an extra margin of safety. [J.W.A.]

Pressure measurement The determination of the magnitude of a fluid force applied to a unit area. Pressure measurements are generally classified as gage pressure, absolute pressure, or differential pressure. *See* PRESSURE.

Pressure gages generally fall in one of three categories, based on the principle of operation: liquid columns, expansible-element gages, and electrical pressure transducers.

Liquid-column gages include barometers and manometers. They consist of a U-shaped tube partly filled with a nonvolatile liquid. Water and mercury are the two most common liquids used in this type of gage. *See* BAROMETER; MANOMETER.

There are three classes of expansible metallic-element gages: bourdon, diaphragm, and bellows. Bourdon-spring gages, in which pressure acts on a shaped, flattened, elastic tube, are by far the most widely used type of instrument. These gages are simple, rugged, and inexpensive. In diaphragm-element gages, pressure applied to one or more contoured diaphragm disks acts against a spring or against the spring rate of the diaphragms, producing a measurable motion. In bellows-element gages, pressure in or around the bellows moves the end plate of the bellows against a calibrated spring, producing a measurable motion.

Electrical pressure transducers convert a pressure to an electrical signal which may be used to indicate a pressure or to control a process. Such devices as strain gages and resistive, magnetic, crystal, and capacitive pressure transducers are commonly used to convert the measured pressure to an electrical signal. *See* PRESSURE TRANSDUCER; STRAIN GAGE. [J.H.Z.]

Pressure seal A seal used to make pressure-proof the interface (contacting surfaces) between two parts that have frequent or continual relative rotational or translational motion; such seals are known as dynamic seals, as compared with static seals. While the pressure in seals is lower than that in gaskets, the motion hinders their effectiveness so that there are more types of seals than gaskets, each type attempting to serve its environment. The materials are leather, rubber, cotton, and flax, and for piston rings, cast iron. The forms of nonmetallic seals are rectangular, V-ring, and O-ring. Cartridge seals are available for rolling-contact bearings. Special seals include carbon ring and labyrinth seals for turbines and mechanical seals for pumps. *See* GASKET. [P.H.B.]

Pressure transducer An instrument component which detects a fluid pressure and produces an electrical, mechanical, or pneumatic signal related to the pressure. *See* TRANSDUCER.

In general, the complete instrument system comprises a pressure- sensing element such as a bourdon tube, bellows, or diaphragm element; a device which converts motion or force produced by the sensing element to a change of an electrical, mechanical, or pneumatic parameter; and an indicating or recording instrument. Frequently the instrument is used in an autocontrol loop to maintain a desired pressure. *See* PROCESS CONTROL.

Although pneumatic and mechanical transducers are commonly used, electrical measurement of pressure is often preferred because of a need for long-distance transmission, higher accuracy requirements, more favorable economics, or quicker response. Electrical pressure transducers may be classified by the operating principle as resistive transducers, strain gages, magnetic transducers, crystal transducers, capacitive transducers, and resonant transducers.

In resistive pressure transducers, pressure is measured by an element that changes its electrical resistance as a function of pressure. Many types of resistive pressure transducers use a movable contact, positioned by the pressure-sensing element. One form is a contact sliding along a continuous resistor, which may be straight-wire, wire- wound, or nonmetallic such as carbon.

Strain-gage pressure transducers might be considered to be resistive transducers, but are usually classified separately, They convert a physical displacement into an electrical signal. When a wire is placed in tension, its electrical resistance increases. The change in resistance is a measure of the displacement, hence of the pressure. Another variety of strain gage transducer uses integrated circuit technology. Resistors are diffused onto the surface of a silicon crystal within the boundaries of an area which is etched to form a thin diaphragm. *See* INTEGRATED CIRCUITS; STRAIN GAGE.

In magnetic pressure transducers, a change of pressure is converted into change of magnetic reluctance or inductance when one part of a magnetic circuit is moved by a pressure-sensing element—bourdon tube, bellows, or diaphragm.

Piezoelectric crystals produce an electric potential when placed under stress by a pressure-sensing element. Crystal transducers offer a high speed of response and are widely used for dynamic pressure measurements in such applications as ballistics and engine pressures. *See* PIEZOELECTRICITY.

Capacitive pressure transducers almost invariably sense pressure by means of a metallic diaphragm, which is also used as one plate of a capacitor.

The resonant transducer consists of a wire or tube fixed at one end and attached at the other (under tension) to a pressure-sensing element. The wire is placed in a magnetic field and allowed to oscillate. As the pressure is increased, the element increases the tension in the wire or tube, thus raising its resonant frequency. *See* PRESSURE MEASUREMENT. [J.H.Z.]

Pressure vessel A cylindrical or spherical metal container capable of withstanding pressures exerted by the material enclosed. Pressure vessels are important because many liquids and gases must be stored under high pressure. Special emphasis is placed upon the strength of the vessel to prevent explosions as a result of rupture. Codes for the safety of such vessels have been developed that specify the design of the container for specified conditions.

Most pressure vessels are required to carry only low pressures and thus are constructed of tubes and sheets rolled to form cylinders. Some pressure vessels must carry high pressures, however, and the thickness of the vessel walls must increase in order to provide adequate strength. Hydraulic and pneumatic cylinders are machine elements that are forms of pressure vessels. [J.J.R.]

Pressurized blast furnace A blast furnace operated under higher than normal pressure. The pressure is obtained by throttling the off-gas line, which permits a greater volume of air to be passed through the furnace at lower velocity and results in an increasing smelting rate. The process permits large increases in the weight of high-temperature air blown into the bottom of the furnace at lower gas velocities, thus increasing the rate of smelting and decreasing the rate of coke consumption, and also permitting smoother operation with less flue dust production through decreased pressure drop between bottom and top pressures. *See* FURNACE. [B.S.O.]

Prestressed concrete Concrete with stresses induced in it before use so as to counteract stresses that will be produced by loads. Prestress is most effective with concrete, which is weak in tension, when the stresses induced are compressive. One way to produce compressive prestress is to place a concrete member between two abutments, with jacks between its ends and the abutments, and to apply pressure with the jacks. The most common way is to stretch steel bars or wires, called tendons, and to anchor them to the concrete; when they try to regain their initial length, the concrete resists and is prestressed. The tendons may be stretched with jacks or by electrical heating.

Prestressed concrete is particularly advantageous for beams. It permits steel to be used at stresses several times larger than those permitted for reinforcing bars. It permits high-strength concrete to be used economically, for in designing a member with reinforced concrete, all concrete below the neutral axis is considered to be in tension and cracked, and therefore ineffective, whereas the full cross section of a prestressed concrete beam is effective in bending. *See* REINFORCED CONCRETE; STRESS AND STRAIN. [F.S.M.]

Priapulida One of the minor groups of wormlike marine animals, now regarded as a separate phylum of the animal kingdom with uncertain zoological affinities. The phylum is a small one with only two genera, *Priapulus* and *Halicryptus*.

Priapulida inhabit the colder waters of both hemispheres. They burrow in mud and sand of the sea floor, from the intertidal region to depths of 14,850 ft (4500 m).

Priapulids are small to medium-sized animals, the largest specimen attaining 6 in. (15 cm) in length. The body of *Priapulus* is made up of three distinct portions: proboscis, trunk, and caudal appendage (see illustration). Separated by a constriction from the trunk, the bulbous, introversible proboscis usually constitutes the anterior third of the body and is marked by 25 longitudinal ridges of papillae or spines. The mouth is located at the anterior end of the proboscis and is surrounded by concentric rows of teeth. The cylindrical trunk is annulated, but not segmented, and is often covered with irregularly dispersed spines and tubercles. At the posterior end of the trunk there are three openings: the anus and two urogenital apertures. [M.E.Ri.]

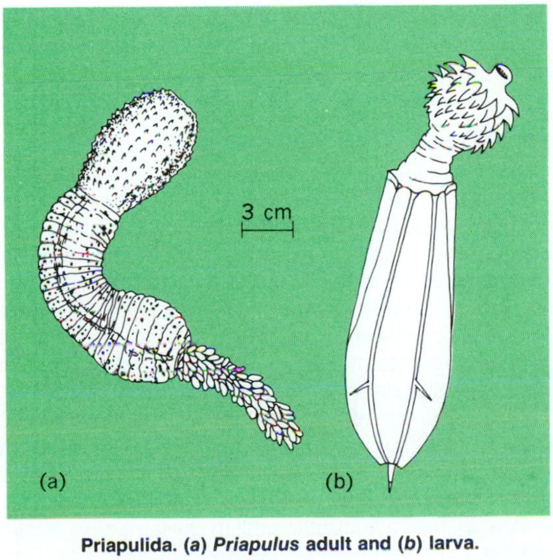

Priapulida. (*a*) *Priapulus* adult and (*b*) larva.

Prilling A combination spray-drying and crystallizing technique used to produce agglomerates (prills) of ammonium nitrate for fertilizer. In prilling, a hot concentrated solution of ammonium nitrate is sprayed into a tower in which it crystallizes as it descends through a rising current of atmospheric air. *See* CRYSTALLIZATION. [M.Ye.]

Primary battery An electric battery designed to deliver only one continuous or intermittent discharge. It cannot be recharged efficiently. Primary batteries are designed to deliver limited amounts of electric energy, determined by the materials used and the size of the cell. When the available energy drops to zero, the battery is usually discarded. Primary batteries may be classified by the type of electrolyte used.

Aqueous-electrolyte batteries use solutions of acids, bases, or salts in water as the electrolyte. These solutions have ionic conductivities and practically no electronic conductivity. Disadvantages of such cells include corrosion of the electrode materials by the electrolyte, a relatively high evaporation rate of water vapor which can cause cell failure, and the difficulties of preventing leakage. For examples of cells with aqueous electrolytes *see* BATTERY; DRY CELL; MERCURY BATTERY; RESERVE BATTERY.

Solid-electrolyte batteries use electrolytes of solid crystalline salts which have predominantly ionic conductivity. Solid electrolyte batteries may be classified in two broad categories: (1) cells with solid crystalline salt, such as silver iodide, as the electrolyte; (2) cells with ion-exchange membrane as the electrolyte. *See* ION-SELECTIVE MEMBRANES AND ELECTRODES.

Waxy-electrolyte batteries use waxy materials, such as polyethylene glycol, in which a small amount of a salt is dissolved in the molten wax. At room temperatures these materials are solid. The conductivity is small and the current output is limited.

Fused-electrolyte batteries use crystalline salts or bases which are solid at room temperature. In use, the cell is heated and maintained at a temperature above the melting point of the electrolyte. [J.D.]

Primary vascular system (plant) The arrangement of conducting elements which serves for two-way transportation of substances between different parts of a plant. The conducting elements are of two principal kinds: xylem, which is mainly responsible for the conduction of water together with dissolved inorganic substances upward from the root to other plant organs; and phloem, which is mainly responsible for the conduction of food materials (assimilates), a flow which may take place in either direction. In the shoot region of the plant, xylem and phloem are usually associated into vascular bundles. In the root, however, they usually alternate with one another on different radii. *See* PHLOEM; XYLEM. [W.R.P.]

Primates The order of mammals to which humans belong. The great diversity of niches occupied by primates within a mainly arboreal ecological zone renders the order difficult to define in terms of diagnostic morphological characteristics. Primates have been characterized in terms of a number of evolutionary trends, among which are retention of a generalized limb structure and dentition; increasing digital mobility; replacement of claws by flat nails; development of stereoscopic vision, with concomitant reduction in the olfactory apparatus, leading to shortening of the face; and progressive development of the brain, notably in the cerebral cortex. Not all these trends are observable in every primate line, but the list does serve to characterize the order as a whole.

Living forms. One of the most striking peculiarities of Primates is that the order contains a gradational series of living forms which reflect to some extent extinct stages in its evolution. Primates divide naturally into two suborders, Prosimii and

Anthropoidea, designated, respectively, the "lower" and "higher" primates. Each suborder contains six families, and these are individually discussed below.

Tupaiidae comprises the five genera of tree shrews; it is sometimes referred to the menotyphylan insectivores but is here considered, provisionally, the most primitive prosimian family. Whatever their correct classification, these small, slender, clawed, rather squirrellike creatures provide a model of the very earliest primates.

In the absence of competition, the prosimians of Madagascar were able to realize their full evolutionary potential. There are three living families, Lemuridae, Indriidae, and Daubentoniidae. Lemuridae contains six genera, all arboreal forest livers. The diurnal forms, *Lemur* and *Hapalemur*, are gregarious, while the nocturnal *Lepilemur*, *Cheirogaleus*, *Microcebus*, and *Phaner* tend to a solitary existence. All, apart from the vertical clinger and leaper *Lepilemur* and, to some extent, *Hapalemur*, are quadrupedal. Indriidae are all wholly arboreal vertical clingers and leapers. *Indri* and *Propithecus* are diurnal whereas *Avahi* is nocturnal, but all live in similar family groups of around two to five individuals. The single aberrant genus of Daubentoniidae is *Daubentonia*. With rodentlike anterior teeth, this solitary nocturnal animal is primarily insectivorous. *See* AYE-AYE.

Included in the family Lorisidae are the lorises of Asia and the galagos of Africa. Both Asian forms are nocturnal and probably solitary. Locomotion is by slow, deliberate climbing. All African forms are nocturnal.

Tarsiidae contains only *Tarsius* of Southeast Asia. The name *Tarsius* derives from the greatly elongated ankle region, a specialization associated with a vertical clinging and leaping way of life. Usually found solitary or in pairs, *Tarsius* is highly territorial. *See* TARSIER.

New World monkeys are characterized by the possession of three premolar teeth; there are two families, Callithricidae and Cebidae. Callithricidae includes the most primitive forms which make up the marmosets and the tamarins. These diurnal animals live in small family groups in the high forest canopy, where they adopt a springing form of locomotion. Except on the big toe, marmosets have claws, not nails. Cebidae is a very diverse family in which locomotion varies from arm-swinging quadrupedalism in forms such as *Ateles*, to quadrupedalism at all levels of the forest in *Cebus* and its relatives. All cebids are diurnal, apart from the large-eyed *Aotus*. The basic social unit is the small family group, although in forms such as the very vocal howler monkey, troops may be much larger. In four genera, *Ateles*, *Brachyteles*, *Lagothrix*, and *Alouatta*, the tail is prehensile and tactile, and is used as an extra "hand" in locomotion. *See* MARMOSET; MONKEY.

The family Cercopithecidae includes all of the Old World monkeys and is divided into two subfamilies, Cercopithecinae and Colobinae. Geographically, there are also two separate groups, African and Asian, with only *Macaca* common to both, but the geographical and taxonomic divisions do not exactly coincide. The Colobinae, for example, *Colobus*, the guereza, and *Presbytis*, the langur, are arboreal browsers, as are some cercopithecines, for example, *Cercocebus albigena*. *Macaca* is arboreal or terrestrial with equal facility. The baboons, for example, *Papio* and *Theropithecus*, are primarily terrestrial. Social organization ranges from loose groups among the langurs to highly organized troops in some species of *Papio*.

The superfamily of apes and humans, Hominoidea contains three families, Hylobatidae, Pongidae, and Hominidae. Hylobatidae includes *Hylobates* (the gibbon) and the larger *Symphalangus* (the siamang), both restricted to Southeast Asia, and both highly arboreal brachiators (arm swingers). *Pan* (the chimpanzee), *Gorilla* of Africa, and *Pongo* (the orangutan) of Sumatra and Borneo are members of the Pongidae. *Pan* is both a forest and woodland dweller; *Gorilla* and *Pongo* inhabit dense forests. *Pongo* is a deliberate climber, while *Gorilla* and *Pan*, though classed as semibrachiators, are secondarily adapted to quadrupedal knuckle walking. Hominidae contains one living species, *Homo sapiens*, a large, erect, omnivorous terrestrial biped. Humans' anterior dentition is greatly reduced and the brain greatly enlarged, features correlating with the highly developed manual dexterity. *See* CHIMPANZEE; GORILLA; ORANGUTAN. [I.M.T.]

Fossils. In evolution of primates, natural selection tended to favor those with larger brains, better sight, and more dexterous front feet that functioned as hands. The incisors are reduced to two in each of the jaws in the monkeys and apes.

Small prosimian primates were common in the middle latitudes of North America and Europe from the Paleocene until the middle Eocene. Apparently, faunal interchange between the two regions was greatly reduced or ceased during the middle Eocene. Then as aridity increased in the late Eocene and early Oligocene, the distribution of the primates shifted southward in both hemispheres. Some of these prosimians were extremely close to primitive stocks of other orders, such as the Insectivora and the Dermoptera. Even in the groups that were surely primates, infraordinal affinities are obscured by the presence of both lemuroid and tarsioid characters.

New World monkeys (Cebidae) appeared for the first time in the late Oligocene and early Miocene of Argentina. They were already typical ceboids at that time. Thus, no clue is provided concerning their ancestral relationships. Others have been found in the late Miocene of Columbia. It seems likely that the Cebidae arose from an early prosimian group from North America, probably the Omomyidae.

Old World monkeys (Cercopithecidae) and apes (Pongidae) seem to have descended from different groups of Eocene prosimians. The earliest cercopithecoid, *Moeripithecus*, occurs in the early Oligocene of Egypt. Other more advanced genera such as *Ankarapithecus*, *Mesopithecus*, and *Dolichopithecus* have been found in Pliocene faunas, mostly in central and southern Europe. Some of the living genera also extend back into the Pliocene. There are many divergent lineages of apes and apelike creatures, all restricted to the Eastern Hemisphere. *See* FOSSIL HUMAN. [M.C.McK.; R.A.St.]

Prime mover The component of a power plant that transforms energy from the thermal or the pressure form to the mechanical form. Mechanical energy may be in the form of a rotating or a reciprocating shaft, or a jet for thrust or propulsion. The prime mover is frequently called an engine or turbine and is represented by such machines as waterwheels, hydraulic turbines, steam engines, steam turbines, windmills, gas turbines, internal combustion engines, and jet engines. These prime movers operate by either of two principles: (1) balanced expansion, positive displacement, intermittent flow of a working fluid into and out of a piston and cylinder mechanism so that by pressure difference on the opposite sides of the piston, or its equivalent, there is relative motion of the machine parts; or (2) free continuous flow through a nozzle where fluid acceleration in a jet (and vane) mechanism gives relative motion to the machine parts by impulse, reaction, or both. *See* GAS TURBINE; HYDRAULIC TURBINE; IMPULSE TURBINE; INTERNAL COMBUSTION ENGINE; POWER PLANT; REACTION TURBINE; STEAM ENGINE; STEAM TURBINE; TURBINE. [T.Ba.]

Primer (explosive) An agent used with explosives, propellants, and pyrotechnics to produce the initial fire. The primer itself may be initiated by percussion, stab action, friction, or heat. The gases and hot particles from the primer serve in turn to ignite a large mass of material, for example, the gunpowder in a cartridge, the primary explosive in a detonator cap, or the powder train in a fuse. The term primer is sometimes loosely applied to a detonator.

Although primer compositions deflagrate rapidly, they do not detonate. In fact, a detonating action would be most undesirable because it would tend to blow away rather than to ignite the charge. *See* DETONATOR; MATCH; PYROTECHNICS. [W.E.Go.]

Primer (surface coating) A material used for the first coat of paint or as the prime coat in a protective coating system. Primers are designed to promote adhesion of the coating system to the substrate, to furnish a good base for further coatings, and to prevent attack on the substrate by air, water, or other materials.

Wood primers are formulated to give maximum adhesion and to provide adequate flexibility for adjustment to dimensional changes that occur when wood swells or shrinks because of changes in moisture content. Primers used as the undercoat for enamels must provide a smooth film and permit easy sanding.

Metal primers must provide excellent adhesion by mechanical anchorage or chemical bonding. Modern corrosion-resistant coating systems applied over steel contain inorganic or organic zinc dispersed in a suitable vehicle. Zinc is attacked by the electrochemical corrosion process preferentially before steel, and protects the steel substrate by this sacrificial property. Prime coat materials for nonferrous metallic substrates usually are generically related to materials used in subsequent coats.

Primers for porous surfaces such as concrete must seal the surface adequately to provide a uniform base for future coatings.

Primers are always pigmented. In clear finishes the coat which performs this function is described as a sealer, an undercoater, or a wash coat. *See* CORROSION; PAINT; PIGMENT; SURFACE COATING. [C.R.Ma.; C.W.Si.]

Primulales An order of flowering plants, division Magnoliophyta (Angiospermae), in the subclass Dilleniidae of the class Magnoliopsida (dicotyledons). The order consists of three families: the Myrsinaceae, with about 1000 species; the Primulaceae, with about 800 species; and the Theophrastaceae, with a little more than 100 species. These are plants with sympetalous flowers; that is, the petals are fused by their margins to form a corolla with a basal tube and terminal lobes. The functional stamens are opposite the corolla lobes, and there is a compound ovary that has a single style and two to numerous ovules. *See* DILLENIIDAE; MAGNOLIOPSIDA. [A.Cr.]

Printed circuit A generic term applied to circuits fabricated by any of several graphic art processes. Printed circuits greatly simplify mass production and increase equipment reliability. Their most important contribution, however, is the tremendous reduction achieved in size and weight of electronic devices.

Technology. Printed circuit technology may be divided into three basic parts: engineering, photography, and manufacturing.

The configuration in which circuit elements are located and the routing of conductor paths establish the precise circuit pattern. Using this pattern, an artwork master normally several times larger than the final size of the printed circuit is prepared. Scale depends on the type of circuit that is to be fabricated.

Enlarged masters are photographically reduced to the required size within the allowable tolerance, thereby reducing any dimensional inaccuracies which have occurred in the master. When a number of circuits are to be fabricated simultaneously, precision step-and-repeat equipment provides a composite master, with each circuit precisely positioned relative to key indexing locations, and all identical.

The circuit pattern masters are used to fabricate the screens and masks for the application of photoresistive materials in the actual formation of the required patterns on the finished parts. The masters are also used in the preparation of numerous types of tooling, for example, drill templates, tapes for operation of numerical-tape-controlled drilling equipment (especially common where high precision is required), routing templates and dicing fixtures for trimming printed circuits to final configuration, and laminating and holding fixtures. Numerous processes including etching, screening, plating, laminating, vacuum deposition, diffusion, and application of protective coatings, are used in combination to produce various types of printed circuits.

Applications. Printed circuits can be divided, in terms of application, into printed wiring, thick and thin films, hybrid circuits, and integrated circuits.

Printed wiring. The printed wiring board is a copper-clad dielectric material with conductors etched on one or both sides. Single-sided boards are commonly used in ground-support equipment in which space and weight are not critical.

Two-sided boards may use plated-through holes or eyelets to provide electrical continuity between the sides. They are used in those applications in which the maximum number of interconnections (conductors) in a given area are required for minimum cost.

Multilayer printed wiring boards capitalize on the reduced size of miniaturized and microelectronic parts. The surface area required for mounting of these subminiature parts and integrated circuit packages has decreased significantly while the number of interconnections has increased manyfold. In a multilayer board, conductors are located on several insulated internal layers. Common multilayer circuits consist of a number of two-sided etched copper foil boards, separated by an insulating layer (the prepreg) and laminated together under controlled temperature, pressure, and time.

Thick-film circuits. Thick-film circuits consist of such passive elements as resistors, capacitors, and inductors deposited on wafers or substrates of such dielectric materials as ceramic, glass, porcelain-coated metal, and the like. They are used commercially for mass fabrication of passive networks for use in linear microcircuits and large-signal digital modules. A large-scale electronic computer may contain millions of these circuits. Thick-film design and manufacture are usually based on film thickness of approximately 0.001 in. (25 micrometers).

Thin-film circuits. The deposition of thin films was the first application of printed circuit technology to microelectronics. Like thick-film circuits, thin-film circuits consist primarily of passive elements; however, such active elements as field-effect transistors and diode devices have become available. The most important operational difference between thick- and thin-film circuits is that thin films have greater precision and stability.

Hybrid circuits. Thick thin-film hybrid circuits consist of one or more substrates upon which the passive elements of the circuit (resistor networks and inductances) have been deposited and to which discrete components (such as diodes, transistors, integrated circuits, capacitors, and inductors) have been attached by split-tip series resistance welding or by thermocompression bonding, or by soldering (thick-film hybrids only). After components and external leads are attached, the circuit is normally encapsulated in a resin compound. Hybrid circuits allow advantageous use of thick- and thin-film fabrication techniques for such applications as complex switching circuits and differential amplifiers in the form of functional electronic building blocks.

Integrated circuits. An integrated circuit is a semiconductor device with both active and passive elements diffused into a silicon wafer to form a functional circuit. Many of the fabrication processes for integrated circuits such as photomasking, selective etching, and photoreduction are similar to those used in the manufacture of other printed circuits. *See* INTEGRATED CIRCUITS. [L.K.L.]

Printing A process in which an image is reproduced on a surface, such as paper. Although photocopying was not originally considered a printing process, the demarcation between the two has become blurred.

Printing processes. There are five general printing processes: relief, planographic, intaglio, screen printing, and electrostatic.

Relief printing is a process in which the printing element consists of a raised surface of type, lines, and points which can be inked and pressed onto the paper. The nonprinting areas are below the printing surface.

Planographic printing is a process in which the printing areas of the plate are on the same surface as the nonprinting areas, and printing is accomplished by using the principle that grease and water do not mix. Commercial lithography is a planographic process that uses thin metal plates made photomechanically and mounted on a press which prints indirectly by a process commonly called offset. The inked image on the plate is first transferred to an intermediate rubber-covered offset cylinder which then prints the image on the paper; thus the name offset. Relief and intaglio printing can also be printed by the offset principle.

Intaglio printing, also known as gravure printing, is accomplished by cutting or engraving and etching various sizes of minute cells (or wells) below the surface of a plate or cylinder to hold the ink. After the cells are flooded and loaded with ink, the surface of the plate is scraped of excess ink by a doctor blade. Therefore, the depth and size of each cell determine the amount of ink that is available to be transferred to the printed surface.

Screen printing, also called porous printing, is a process in which ink is brushed or squeegeed through a stencil image on a fine screen onto paper or other surface such as metal or glass. The screen holds the image area and may carry either pictorial or typographic material.

Electrostatic processes were once considered as copying rather than printing processes. However, copying equipment is becoming faster, and so-called printing equipment is being developed for much of the short- run printing required for offices and many in-plant operations. Xerography has been combined with electronic imaging to produce electronic printing systems. [M.H.B.]

Letterpress printing. The relief process of printing is the oldest of the major printing processes. The distinguishing characteristic of the relief process of printing is that the image areas of the image carrier stands out in relief from the body of the carrier itself (Fig. 1).

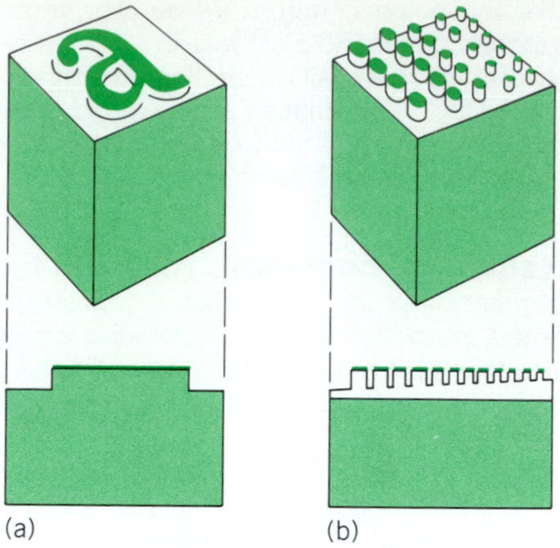

Fig. 1. Diagram of printing-image carriers for letterpress. (*a*) Type face. (*b*) Dots of a halftone. (*After V. Strauss, The Printing Industry, Printing Industries of America, 1967*)

By this definition, the relief printing process can be broken down further into two relief printing methods, each characterized by the nature of the image carrier. One method has been identified, by tradition, as letterpress printing, with the image carrier being rigid. The second relief printing method was originally identified as rubber-plate printing, whose image carrier is soft, flexible, and resilient. However, the application of this method has become so widespread, although highly specialized, that it has achieved recognition as a printing process in its own right, and is now identified as the flexographic process. However, the line between letterpress and flexographic printing has become somewhat blurred because of a development in flexible and resilient image carriers. Relief printing plates made of photopolymeric materials have been developed. These plates are not quite as flexible or resilient as rubber, but more importantly with respect to printing processes and methods, they are used in a letterpress printing application, namely, book presses. The distinguishing characteristics here are the nature of the ink used and the inking system of the book press.

A frequently heard term is wraparound. This is often incorrectly described as a process or method. Wraparound refers to a letterpress metal image carrier which is thin enough and flexible so that it can be mounted on (wrapped around) a cylinder. A press whose image-carrying segment is cylindrical is identified as a rotary press. With the exception of screen printing, all the major printing processes utilize rotary presses almost exclusively. The wraparound technique was developed in an attempt to provide lower-cost photoengraving techniques and to adapt the inherent speed advantage of a rotary press to job and commercial letterpress printing. The wraparound plate should be distinguished from the curved (rigid) plates used on letterpress rotary presses.

Letterpresses are generally classified by using the printing unit configuration as the criterion. The platen press (Fig. 2*a*) has a flat platen and bed. The flatbed cylinder press (Fig. 2*b*) has a flat bed, with an impression cylinder in place of a platen. The rotary press (Fig. 2*c*) has two cylindrical printing units, the flatbed being replaced by a plate cylinder, and both members are cylindrical.

The platen and flatbed cylinder presses of letterpress printing are inherently sheet-fed. For some special printing applications both have been adapted to roll feeding or, in the case of the platen press, to imprint continuous fan-folded business forms. Flatbed cylinder presses were built to print from rolls of paper to meet the needs of the so-called country newspaper.

Lithographic printing. Originally this was a term that signified printing from stone. Stone lithography has ceased to exist except as an art medium. The term offset is almost universally used as synonymous with lithography. Technically this is incor-

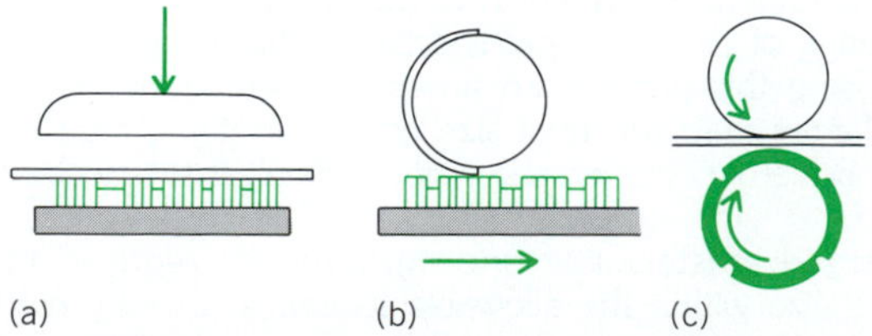

Fig. 2. Letterpress printing presses. (*a*) Platen press. (*b*) Flatbed-cylinder press. (*c*) Rotary press. (*After V. Strauss, The Printing Industry, Printing Industries of America, 1967*)

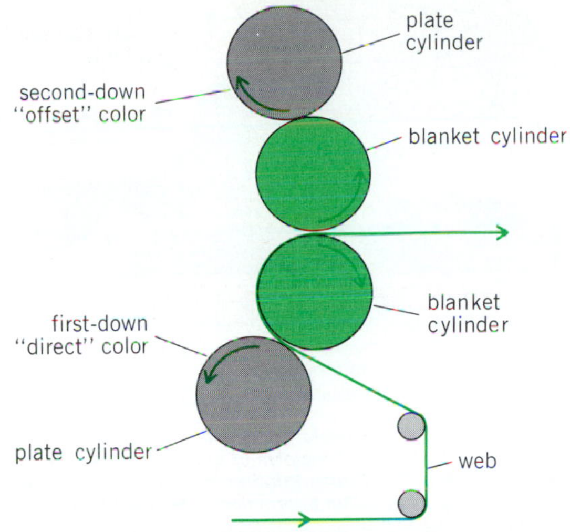

Fig. 3. Diagram of a blanket-to-blanket web for direct lithography. (*After E. J. Kelly, D. Crouse, and R. Supansic, Web Offset Press Operating, Graphic Arts Technical Foundation, 1974*)

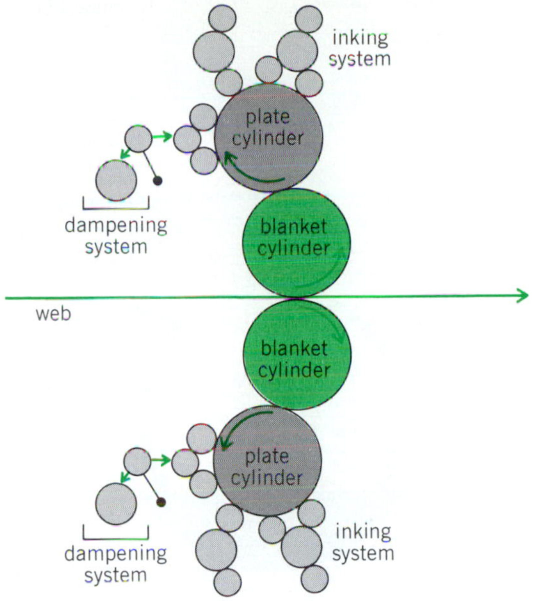

Fig. 5. Diagram of a typical blanket-to-blanket printing unit of the horizontal type. (*After E. J. Kelly, D. Crouse, and R. Supansic, Web Offset Press Operating, Graphic Arts Technical Foundation, 1974*)

rect because the offset technique has been applied to both gravure and letterpress printing. Direct lithography—pressing the inked lithographic plate directly against the substrate—predated modern offset lithography. While direct lithography virtually disappeared many years ago, there are applications in conjunction with offset lithography on blanket-to-blanket web (lithographic) offset presses (Fig. 3).

Lithographic presses, direct or indirect, are rotary presses—the plate is mounted on a cylinder which rotates during printing. For the sheet-fed configuration of a lithographic press (Fig. 4), the plate cylinder is provided with the means to rotate it independent of its gear drive. This feature is provided to make possible the desired front-to-back position of the image on the sheet. When the dampening system is disengaged and a shallow-etched flexible relief image carrier is mounted on the plate cylinder, printing is accomplished by indirect (offset) letterpress printing.

In offset printing, primarily lithographic, the basic configuration of rotary printing is universal. However, many adaptations of the rotary configuration have been designed. Offset makes it possible to print both sides (sheet or web) in the same printing nip (see Fig. 5). A printing unit in this configuration comprises two plate/blanket cylinder couples so assembled that the blanket cylinders run against each other. Each blanket cylinder transfers the ink for its image and, at the same time, acts as the impression cylinder for the opposing blanket cylinder. These presses are called blanket-to-blanket presses. *See* PRINTING PRESS.

Gravure printing. This process is used for printing and coating applications which fall into three major categories: publication, packaging, and specialties. The publication field includes magazines, mail-order catalogs, brochures, newspaper supplements, comics, and other types of commercial printing. Packaging consists of folding cartons and flexible packages. The specialty field represents items such as wall and floor coverings, decorated household paper products, cigarette filter tips, and a wide variety of other products. Gravure is also used for applying accurately metered quantities of coatings to paper and other substrates.

The gravure image carrier is a copper-plated cylinder or copper plate on which the image is in the form of cells or cups engraved in the surface. A printing unit, or a gravure press, consists of a gravure cylinder and an impression roll, with attached inking and doctoring mechanisms. A gravure press may consist of from one to eight or more units that are coupled together with paper-feeding and roll- or folder-delivery arrangements.

The basic gravure printing or coating unit consists of the engraved gravure image carrier cylinder at the bottom (Fig. 6) with a rubber-covered impression roll at the top. Gravure ink transfer is accomplished when the paper is pressed into contact with the liquid ink in the gravure cell. [H.F.G.; M.H.B.]

Flexographic printing. Flexography is a rotary letterpress printing process, traditionally using flexible rubber printed plates and fluid, fast-drying ink. Since the development of photopolymer flexo printing plates, the original definition of exclusively rubber flexo printing plates is no longer accurate. The flexo industry refers to its printing plates as being rubber or elastomeric in composition. In addition the inks used in flexography are no longer always limited to the original fluid, fast-drying variety. A typical flexographic printing ink distribution system is shown in Fig. 7.

As the result of the accuracy, versatility, speed, economy, and print quality of modern flexographic printing, it has become the major method for printing flexible packaging

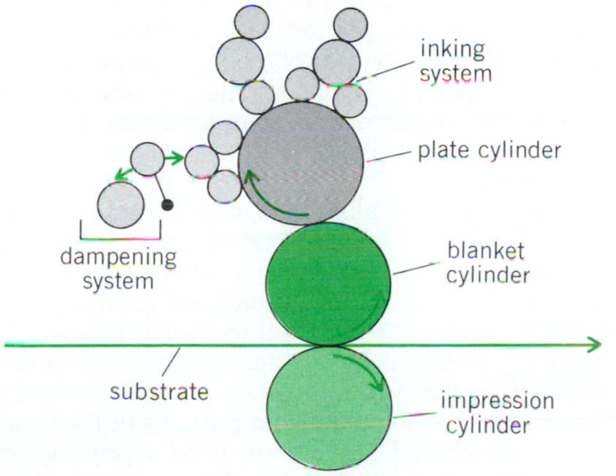

Fig. 4. Diagram of lithographic offset. (*After E. J. Kelly, D. Crouse, and R. Supansic, Web Offset Press Operating, Graphic Arts Technical Foundation, 1974*)

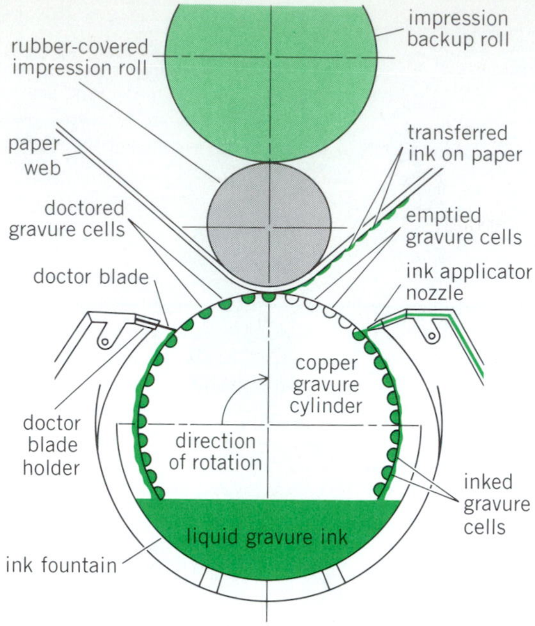

Fig. 6. Schematic diagram of a gravure printing unit.

materials, including paper and plastic films, in addition to multiwall bags, corrugated containers, envelopes, pressure-sensitive labels, and paperback books. [D.E.T.]

Screen printing. A printing process that differs markedly from the other printing methods in that the printing plate is actually a porous stencil, which is supported by a meshed screen of fabric or metal. Screen printing technology has, to a high degree, refined a basic stenciling process into a commercial and industrial imaging system that is virtually limitless in its application. The diverse range of substrates which may be printed in this way include paper, metal, plastics, textiles, glass, ceramic, and wood.

The basic screen printing method may be simply described as follows. A woven fabric generally of polyester, nylon, or stainless steel (although silk is still used occasionally) is stretched tightly upon a wood or aluminum rectangular frame and affixed. A stencil is then prepared either separately from the fabric (indirect method), or directly onto the mesh using a liquid, photosensitive emulsion (direct and direct/indirect methods).

The image, if photographic or consisting of fine, detailed lines, is developed within the stencil medium by using light-exposure techniques. An alternative method is to hand-cut the stencil by using paper or plastic film.

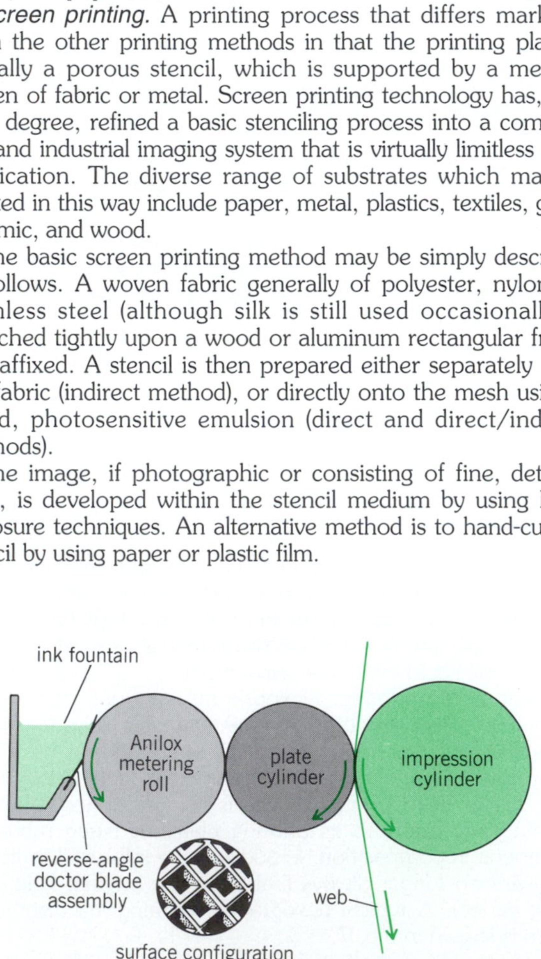

Fig. 7. Flexographic reverse-angle doctor blade ink distribution system.

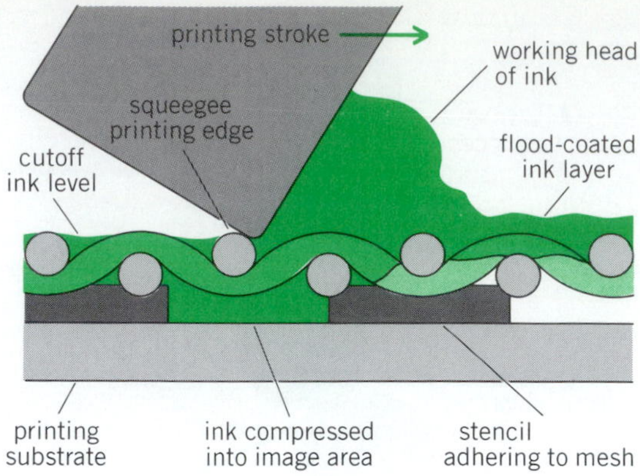

Fig. 8. Diagram of the basic factors in the screen printing action. (*After E. J. Kyle, Principles of quality control, Technical Guidebook of the Screen Printing Industry, vol. 4, Screen Printing Association International, March 1975*)

The fully prepared printing screen is mounted onto any of several automatic or semiautomatic types of press. A squeegee is attached in a position which permits a regular sweeping motion, and ink is introduced into the screen (Fig. 8). By means of a constant pressure exerted from one side of the screen to the other, the squeegee presses the ink through the mesh at every point which is not blocked off by the stencil. As the flexible screen is pressed downward, it contacts the substrate, and the fluid ink transfers, forming the printed image.

The foremost advantage of screen printing is in flexibility of application. Through the selection of different fabrics of varying mesh fineness, the amount of printing ink applied may range from a very thick wet film of 150 micrometers (0.006 in.) to a relatively thin dry ink deposit of 4 or 5 micrometers (0.00015 or 0.00020 in.).

The range of color reproduction by screen printing is actually limitless. In most instances, each color is applied singly and allowed to dry before the next color is printed.

Refinements such as the control of ink deposits, ultraviolet polymerization, and automatic substrate handling equipment have contributed greatly to significant increases in the speed of screen printing production and improved quality of the printed image. [T.McS.]

Screenless printing. Processes involving screenless printing are considered essentially specialty processes because of the difficulty of controlling them. These are collotype and screenless lithography. Other means are being developed to simulate the effects of screenless printing and these may well lift this type of printing out of the specialty area.

Collotype printing is known in different countries by different names, for example, photogelatin, phototype, heliotype, heliogravure, albertype, and artotype. The process, although not literally planographic, belongs in the planographic category. It has the distinguishing characteristics of using no screen and comes the closest of any process to simulating a true photographic gradation.

The plate, usually of aluminum when used on a rotary press (when flatbed presses are used, the base is heavy plate glass), is coated with a layer of light-sensitive (bichromated) gelatin, over which is laid an unscreened (continuous-tone) photographic negative of the copy to be reproduced. When the two are exposed to light, the gelatin is affected in varying degrees in proportion to the amount of light admitted to it through the dark, medium, and light portions of the negative. The plate is developed in water to yield a printing image which varies not only in thickness (in terms of thousandths of an inch) but also in moisture content and consequent ability to repel ink or hold it in varying amounts. The more hardened areas accept less moisture and swell less, thus accepting more ink.

Because of its screenlessness it has been used to great advantage in reproducing ancient manuscripts which either were falling to pieces from age or were likely to suffer from handling and exposure to the air. Of all processes, collotype is capable of giving the most faithful facsimile reproductions.

Screenless lithography is a means of printing lithographic plates made from continuous-tone positives or negatives on an offset press. Best results have been obtained with special photopolymer plates exposed from positives. Unlike collotype, the plates are run conventionally on the press with regular inks and fountain solutions. [M.H.B.]

Color printing. The basic principle of color reproduction through the media of modern color printing of pictures lies in the use of only three primary colors to create the illusion of seeing in the reproduction the hundreds or more colors the eye sees in the original.

Modern color printing, in which only three colored printing inks were used, came about by bringing together the invention of photography, the halftone screen, coal tar, and synthetic dyes to make transparent color inks and color separation filters. All of these technologies were coupled with a better understanding of color vision. *See* COLOR VISION.

Practically all commercial color printing also uses a fourth printing of black ink, and therefore could be considered a four-color process.

Color printing with inks is the controlled proportioning of red, green, and blue light, and this is accomplished in an indirect way called the subtractive process. Printing is done with yellow, magenta, and cyan inks because these colors ideally are complementary to blue, green, and red respectively, and each can individually shut off the reflectance of red, green, and blue light from the paper surface.

The first step in any color reproduction system is called color separation. In this process three individual photographic picture images are exposed separately through red, green, and blue filters. Each filtered separation image must also be exposed through a halftone screen to create different-size dots that control the hues and tone values. When the filtered color separation and halftone screen exposures are made simultaneously to the same photographic film, it is called the direct screen process. In the indirect process, the filtered color-separation negatives are normal continuous-tone photographic images, and the halftone dot screening is introduced in a second photographic step with a high-contrast film to create positive dot images.The direct screen method is more economical and thus more popular.

Electronic color scanning. The more sophisticated, contemporary electronic scanners are capable of producing, in one operation, completely color-corrected, enlarged or reduced, cropped, and properly screened sets of separations from original copy in a matter of minutes. In most cases, the quality of these separations exceeds that achieved by conventional photomechanical techniques. This capability, coupled with speed, makes electronic scanners highly cost-effective pieces of production equipment.

The primary function of an electronic color scanner is to produce the four black-and-white separations representing the yellow, magenta, cyan, and black images from an original full-color image. These final separation films may be either continuous-tone or halftone images. They must be produced in such a manner that the printing plates made from them are capable of producing a full-color image which is of optimum sharpness, color content, and contrast ratio.

The original copy is a transparent color-positive photograph, mounted on the transparent scanning drum. Prior to mounting the transparency, certain of its characteristics must be measured. Both the highlight (brightest) and shadow (darkest) areas of interest must be measured on an accurate densitometer in order to determine their optical densities. The neutral-colored area (true gray area) of the picture must also be determined and measured for its color content. These densities, along with information as to which printing process and inks will be used to print the final images, and information about the characteristics of the paper that the image is to be printed on, will be used by a trained scanner operator to set up the scanner's color computer.

After the operator has set up the color computer, the first piece of unexposed film is placed on the recording or output drum. The machine is now ready to scan and produce the first of the four output separations. The machine must make four consecutive runs in order to produce the four images. A fresh piece of film is loaded on the recording drum for each successive exposure.

The scanning light beam passes through the transparent color photo and into the optical color head. At any given moment, the scanning light beam emitted from the lamp, after being filtered through the transparent color photo, contains all of the colors contained in the transparency at the exact spot the beam is focused on. This filtered light beam is then broken down in the color head into the three additive primary colors of red, green, and blue.

Color masking is a process by which several deficiencies in the color printing process are corrected or compensated for. The biggest of these deficiencies is the result of imperfectly colored printing inks.

In addition to color masking, image enhancement can be performed. Enhancement is a general term for picture detail improvement which results in an apparent sharpening of the picture.

After all of this, the electronic signal representing the picture must be modified to compensate for the characteristics of the film being exposed on the recording drum. On most modern scanners, this is simply a matter of entering the output film's exposure characteristics into the scanner's computer.

The electronic output signal now represents the desired image to be exposed on the separation film. Next it is necessary to convert the electronic signal into light. This is done in the recording head. There are three commonly used light sources employed in recording heads: glow lamps, lasers, and xenon arc lamps.

Electronic prepress systems are capable of handling full-color images to make up full-color pages. These systems employ electronic scanners, color video terminals, and contemporary interactive computer technology.

There are systems that are totally interactive; that is, the operator can manipulate images on a computer-controlled color video terminal, in real time, changing position, cropping, and modifying colors. [B.M.C.]

Platemaking. Platemaking is the hub around which the conventional printing processes revolve. Printing plates or cylinders are made in any one or a combination of six different ways: (1) manually: by hand tools, engravers, knives, and so forth, to produce plates for the relief, intaglio, and stencil processes; greasy crayon or tusche can be used for making hand-drawn lithographic plates; (2) mechanically: by engraving and geometric lathes, ruling machines, pantographs, and so on, for the relief and intaglio processes; by hand transferring and benday machines for lithographic plates; (3) electrochemically: by electrodeposition of metals to produce longer-wearing images; (4) electronically: by the use of electronic engravers to make relief and intaglio image carriers; (5) electrostatically: by using xerographic or Electrofax principles to make relief and lithographic plates; (6) photomechanically: by combining the photographic process to produce images and a mechanical process to make plates. The photomechanical method is the most important and most widely used, as it has not only speeded up the reproduction of pictorial and text matter, but has improved their quality.

Photoengraving. This is the photomechanical process used to produce relief plates in line, halftone, and color for letterpress, letterset, and flexographic printing. There are two general types of engravings, line and halftone, which are divided into five groups: one-color line engravings; one-color halftone engravings; one-color combination line and halftone engravings; multicolor sets of engravings in line or halftone, or in line and half-tone combinations; and four-color process engravings. In photoengraving, the metal in the nonprinting areas is removed by chemical etching or mechanical routing.

Gravure platemaking and cylinder making. Gravure is printed principally on web presses from etched cylinders and is generally called rotogravure. In sheet-fed gravure, printing is done from a thin copper plate wrapped around the plate cylinder.

Lithographic platemaking. Before describing the lithographic platemaking process, the chemistry of the lithographic principle should be considered. A lithographic plate must have image areas that are ink-receptive and nonwettable by water, and the nonimage areas must be water-receptive and nonwettable by ink. The chemistry of the production of the images is related to the changes that occur to the photosensitive coatings on exposure to light and the development process. Some become insoluble and oleophilic (ink-receptive) and are used as the printing areas. Others serve as stencils, and other means of making the nonimage areas water-receptive and desensitized to ink.

Stencil process image carriers. There are two stencil processes in general use: screen printing and stencil duplicating.

Screen printing image carriers are produced manually or by photomechanical means. The printing screen is made by attaching the screen material to a rigid frame and stretching it so it is level and smooth. The stencil is applied to the bottom side of the screen which is in contact with the surface to be printed; and ink with a consistency similar to thick paint is transferred by rubbing it over the screen surface with a rubber squeegee.

Manual stencils are made by knife-cutting special film stencil materials which consist of two plastic layers, They can also be produced by artists drawing directly on the screens with special materials. [M.H.B.]

Printing press There are four major types of printing presses or printing systems: letterpress, offset lithography, intaglio, and screen printing. In letterpress the image to be reproduced is in relief. Ink is applied to the raised areas, and when a sheet of paper is pressed against them, the film of ink is split, the image from the plate being left on the paper surface. In offset lithography, a planographic process, the image is on the same plane as the nonimage areas. The printing or transfer of the image is from a flat surface which is chemically treated so that ink is accepted in the image areas and repelled in the nonimage areas of the plate. In the intaglio printing process the image to be reproduced is recessed below the surface of the printing plate or image carrier. Included in this category are engraving and gravure processes. In screen printing (serigraphy) the design to be reproduced is an open area in a finely woven screen, the nonprinting areas of the screen being blocked off; the ink is squeegeed through the open areas onto the surface which is to be printed.

Offset presses. There are several varieties of offset presses available: sheet-fed, single-color and multicolor, perfecting, web, and common impression cylinder presses.

Sheet-fed single-color offset presses have a thin, flexible printing plate wrapped around a plate cylinder. The plates may be made of zinc, aluminum, chrome-plated steel, plastic, or paper. Short- to medium-run work is usually done on a single-color press, and even two- to four-color may be printed on such a press by running the sheet through the required number of times. For longer runs and higher-quality work, two-color and multicolor presses are available.

In most web offset presses, two units print the top and bottom of the web simultaneously. The units are arranged in opposition to each other so that each blanket not only prints an image on the web of paper, but also acts as the impression cylinder for the opposing blanket cylinder.

The common impression cylinder press consists of a number of plate and blanket cylinders grouped around one large impression cylinder. The claimed chief advantage is better control of register, or accurate positioning of one color image over another.

Letterpress. Letterpress is the oldest and second in use of the printing processes. There are three basic designs of letterpresses: platen, flat-bed cylinder, and rotary.

A platen press is one in which the impression of the image is obtained by bringing two flat surfaces together. The form, or relief printing plate, is placed in a vertical position against the bed of the press, and by means of a clamshell or sliding action the form is pressed against the paper to be printed against a platen covered with a sheet of heavy oiled paper.

The flat-bed cylinder press utilizes a moving flat bed to hold the form or image carrier while a fixed cylinder provides the impression pressure. The paper which is held by grippers in the cylinder is rolled over the form, and the image is transferred as the bed passes under the cylinder. It prints while one revolution of the cylinder is made, and is raised during the second revolution of the cylinder to permit the bed to return to its original position.

The rotary letterpress is the fastest of the letterpress types and has the widest use. On rotary letterpresses, both the image-carrying (plate) surface and the impression surface are cylindrical. Any kind of plate that can be curved to fit the cylinder and has sufficient rigidity may be used, such as electrotypes, stereotypes, or wrap-around metal or plastic.

Intaglio printing processes. Intaglio processes include engraving, gravure, and rotogravure.

In the engraving process, the design is etched or mechanically cut in a copper or steel plate. An impression member presses the paper down on the inked plate with considerable pressure, producing very fine detail.

In sheet-fed gravure, the flat, flexible etched sheet of copper is clamped around the plate cylinder of a press. The sheets of paper are then pressed firmly against the plate by a rubber-covered impression roll. Sheet-fed gravure is used for short runs, for art reproductions and very high-quality printing.

A rotogravure unit consists of a printing or engraved cylinder, an impression cylinder, and an inking device. The image in the form of tiny cells etched into the copper-plated engraved cylinder has ink applied. The impression cylinder is covered with a resilient rubber composition that presses the paper into contact with the ink in tiny cells of the printing surface. Gravure inks are quite volatile and dry rapidly; however, a dryer after each printing unit dries the ink before the next color is applied. This permits color to be printed over dry color, resulting in the longest range of density of any process. Multicolor rotogravure presses have as many as twelve printing units, six colors on each side of the web, and modern presses have automatic register controls, which assure exact color register. Rotogravure is used for package printing, printing flexible films and foil, long-run mail-order catalogs, and high-quality magazines.

Screen printing. Although much screen printing is done by a hand method with very simple equipment consisting of a table, a frame, and a squeegee, most commercial screen printing is done on mechanical or power-operated presses. Screen printing is used to print both line and halftone on most any material. It has been used for art prints (where it is referred to as serigraphy), decalcomanias, posters, greeting cards, wallpaper, and many items which would be difficult to print by other processes. *See* INK; PRINTING. [W.A.R.]

Prion disease Transmissible spongiform encephalopathies in both humans and animals. Scrapie is the most common form in animals, while in humans the most prevalent form is Creutzfeldt-Jakob disease. This group of disorders is characterized at a neuropathological level by vacuolation of the brain's gray matter (spongiform change). They were initially considered to be examples of slow virus infections. Experimental work has consistently failed to demonstrate detectable nucleic acids—both ribonucleic acid (RNA) and deoxyribonucleic acid (DNA)—as constituting part of the infectious agent. Contemporary understanding suggests that the infectious particles are composed predominantly, or perhaps even solely, of protein, and from this concept was derived the acronym prion (proteinaceous infectious particles). Also of interest is the apparent paradox of how these disorders can be simultaneously infectious and yet inherited in an autosomal dominant fashion (from a gene on a chromosome other than a sex chromosome).

Disorders. Scrapie, which occurs naturally in sheep and goats, was the first of the spongiform encephalopathies to be described. An increasing range of animal species have been recognized as occasional natural hosts of this type of disease. Bovine spongioform encephalopathy, commonly known as mad cow disease, has been epidemic in British cattle. The first confirmed cases were reported in late 1986. By early 1995 it had been identified in almost 150,000 cattle and more than half of all British herds. Its exact origin is not known, but claims that it came from sheep are now discredited.

So far, animal models have indicated that only central nervous system tissue has been shown to transmit the disease after oral ingestion—a diverse range of other organs, including udder, skeletal muscle, lymph nodes, liver, and buffy coat of blood (white blood cells) proving noninfectious.

The currently recognized spectrum of human disorders encompasses kuru, Creutzfeldt-Jakob disease, Gerstmann-Straussler-Scheinker disease, and fatal familial insomnia. All, including familial cases, have been shown to be transmissible to animals and hence potentially infectious; all are invariably fatal with no effective treatments currently available.

Human-to-human transmission. A variety of mechanisms of human-to-human transmission have been described. Transmission is due in part to the ineffectiveness of conventional sterilization and disinfection procedures to control the infectivity of transmissible spongiform encephalopathies. Numerically, pituitary hormone–related Creutzfeldt-Jakob disease is the most important form of human-to-human transmission of disease. However, epidemiological evidence suggests that there is no increased risk of contracting Creutzfeldt-Jakob disease from exposure in the form of close personal contact during domestic and occupational activities. Incubation periods in cases involving human-to-human transmission appear to vary enormously, depending upon the mechanism of inoculation. Current evidence suggests that transmission of Creutzfeldt-Jakob disease from mother to child does not occur. Two important factors pertaining to transmissibility are the method of inoculation and the dose of infectious material administered. A high dose of infectious material administered by direct intracerebral inoculation is clearly the most effective method of transmissibility and generally provides the shortest incubation time. *See* BRAIN; MUTATION; NERVOUS SYSTEM DISORDERS; SCRAPIE; VIRUS INFECTION, LATENT, PERSISTENT, SLOW. [C.L.Mas.; S.J.Co.]

Prism A polyhedron of which two faces are congruent polygons in parallel planes, and the other faces are parallelograms (see illustration). The bases B are the congruent polygons; the lateral faces are the parallelograms; the lateral edges are the edges not lying in the bases; and the perpendicular distance between the bases is the altitude h. Sections parallel to the bases are congruent to the bases. A prism is a right prism if its lateral edges are perpendicular to the bases; an oblique prism otherwise. A prism is called a triangular prism if its bases are triangles; a pentagonal prism if its bases are pentagons; and a parallelepiped if its bases are parallelograms. The volume of any prism is equal to the area of its base times its altitude ($V = Bh$). *See* POLYHEDRON; PRISMATOID AND PRISMOID. [J.S.F.]

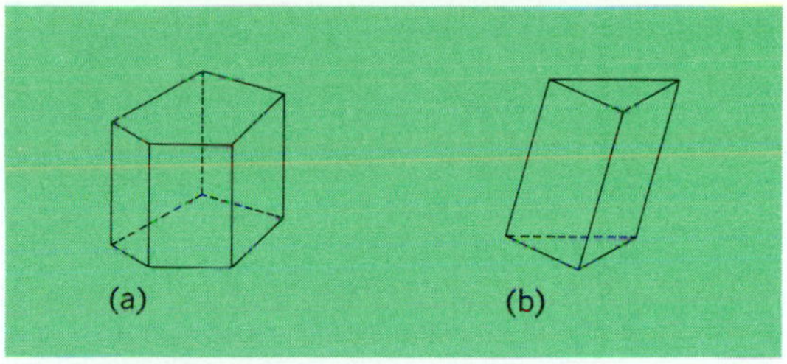

Prism configurations. (*a*) Right. (*b*) Oblique.

Prismatic astrolabe A surveying instrument used to make the celestial observations needed in establishing an astronomical position. The instrument (see illustration) consists of an accurate prism, a small pan of mercury to serve as an artificial horizon, an observing telescope with two eyepieces of different power, level bubbles and leveling screws, a magnetic compass and azimuth circle, adjusting screws, flashlight-battery power source, light, and a rheostat to control the intensity of illumination.

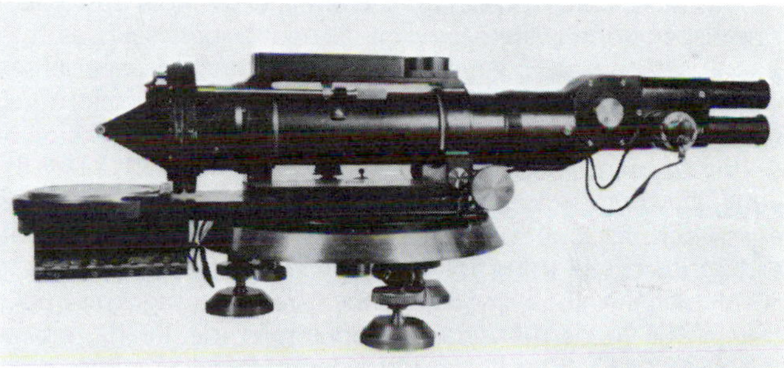

A prismatic astrolabe, used to make celestial observations. (*U.S. Naval Oceanographic Office*)

By using a fixed prism, the instrument measures a fixed altitude, usually 45°. As a rising star increases altitude past that for which the instrument was constructed, the direct image appears to move upward from the bottom of the field of vision to the top. The image reflected by the mercury horizon appears to move downward from top to bottom. At the established altitude the rays produce images at the center of the field of view. A fixed altitude is used to minimize error due to variations from standard atmospheric refraction. Each accurately timed observation provides one line of position. [A.B.M.]

Prismatoid and prismoid A prismatoid (see illustration) is a polyhedron all of whose vertices lie in two parallel base planes. If the two base polygons have the same number of sides, the prismatoid is called a prismoid. Among the prismatoids are included pyramids, frustums of pyramids, wedges, parallelepipeds, and other prisms. The area of a section of a prismatoid by a plane parallel to and between its bases is a

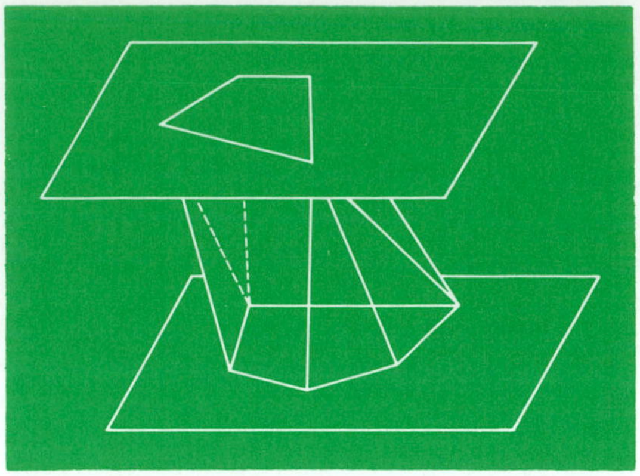

Diagram of a prismatoid.

quadratic function (possibly linear or constant) of the distance from the plane to one of the bases. A generalized prismoid is any solid for which the area of a section parallel to a fixed base is represented by a polynomial of degree less than or equal to 3 in terms of the distance from that base. Frustums of cones, segments of spheres, and many other solids satisfy this condition. *See* PARALLELEPIPED; POLYHEDRON; PRISM; PYRAMID AND FRUSTUM. [J.S.F.]

Private branch exchange A switching system, now always automatic, usually located on the business customer's premises to interconnect terminal equipment, such as telephone sets, and to connect terminal equipment to special-service equipment and to public and private communications networks. *See* TELEPHONE; TELEPHONE SERVICE.

Private branch telephone exchanges have evolved since the early 1900s from manual private branch exchanges in which all switching functions were performed by an attendant, through versions using rotary switch and relay technology, to an architecture that, since the early 1970s, strongly resembles that of the modern business computer. In the 1970s, private telephone exchanges began to be designed as stored-program-controlled (SPC) exchanges wherein the control of the system is provided by an integrated-circuit processor. A generic set of operating instructions used to control call-processing sequences is stored in the system memory. Additional memory is provided for a specific set of instructions related to the particular installation and the features to be provided for that installation. The stored program controls the entire system and directs the system to perform a number of tasks in a predetermined sequential order. When users of the terminal equipment dial address-signaling information into the system, the information is stored. When the stored program advances to the point where it is ready to process the call-handling request, various software program subroutines analyze the information, direct the system hardware to make the appropriate connections, and then maintain overall supervision of the call. The processor executes individual tasks so quickly that additional tasks, such as routine maintenance checks and audit routines, can be interleaved with call processing with no noticeable degradation in call-processing capacity. *See* DIGITAL COMPUTER; INTEGRATED CIRCUITS.

The generic term private branch exchange (PBX), in modern usage, encompasses those systems which not only provide voice communications but also support digital data terminals of various rates and a broad range of both voice and data information services and communications for the modern business community. A contemporary private branch exchange is sometimes called a business communications system. Contemporary private branch exchanges often use time-division switching in which voice signals are first converted to a digital format. Individual calls are then separated from one another on a timing basis. Because digital techniques were developed which permitted the transmission of data in the range of 1 million bits per second over the same twisted pair previously used for telephones, and because most buildings are already wired for telephones, private branch exchanges have made a virtually smooth evolution from applications restricted to voice and analog transmission to integrated voice and data communications. (Other names for contemporary private branch exchanges include electronic PBX or EPBX; private automatic branch exchange or PABX, and computerized branch exchange or CBX.) *See* ELECTRICAL COMMUNICATIONS.

Private branch exchanges connect integrated (voice and data) workstations, local-area networks (LANs), mainframe computers, applications processors, and centralized maintenance and administration facilities over dedicated data links as well as more common communications links. Applications processors, sometimes called outboard processors since they are not part of the basic call-processing function, besides enhancing standard private branch exchange features, allow a variety of business services to be provided. Private branch exchanges also serve as a local-area network for those applications where medium data rates are adequate and most of the connections between data devices are temporary. Lastly, private branch exchanges provide code, speed, and protocol conversion for data communications, and can act as gateways to public and private networks. *See* DATA COMMUNICATIONS; LOCAL-AREA NETWORK; MICROPROCESSOR; SWITCHING SYSTEMS (COMMUNICATIONS). [L.E.Th.]

Probability Although probability theory derives its notion and terminology from intuition, a vague statement such as "John will probably come" is as remote from it as the statement "John is forceful and energetic" is remote from mechanics. Probability theory constructs abstract models, mostly of a qualitative nature, and only experience can show whether these reasonably describe laws of nature or life. As always in mathematics, only logical relations and implications enter the theory, and the notion of probability is just as undefinable (and as intuitive) as are the notions of point, line, or mass.

The sample space. One speaks of probabilities only in connection with conceptual (not necessarily performable) experiments and must first define the possible outcomes. It is necessary to distinguish between elementary (indivisible) and compound outcomes or events. Each elementary outcome is called sample point; their aggregate is the sample space. The conceptual experiment is defined by the sample space, and it must be introduced and established at the outset.

Events. In examining a bridge hand, one may ask whether it contains an ace or satisfies some other condition. In principle each such event may be described by specifying the sample points which do satisfy the stipulated condition. Thus every compound event is represented by an aggregate of sample points, and in probability theory these terms are synonymous. The standard notations of set theory are used to describe relations among events. *See* SET THEORY.

Given an event A one may consider the case that A does not occur. This is the negation or complement of A, denoted by A'; it consists of those sample points that do not belong to A. Given two events A and B, the event C that either A or B or both occur is the union of A and B and denoted by $C = A \cup B$. In particular $A \cup A'$ is the whole sample space $\mathfrak{S}$ which therefore represents certainty. The event D, both A and B occur, is the intersection of A and B and written $D = A \cap B$. It consists of the points common to A and B.

Probabilities in finite spaces. If the sample space $\mathfrak{S}$ contains only N points $E_1, \ldots, E_N$ their probabilities may be any numbers such that $P\{E_j\} \geqq 0$ and $P\{E_1\} + \cdots + P\{E_N\} = 1$.

The probability $P\{A\}$ of an event A is the sum of the probabilities of all points contained in A; thus $P\{\mathfrak{S}\} = 1$.

Frequently considerations of symmetry lead one to consider all E_j as equally likely; that is, to set $P(E_j) = 1/N$. In this case $P(A) = n/N$ where n is the number of points in A; for a gambler betting on A, these represent the "favorable cases." For example, in throwing a pair of "perfect" dice, one naturally assumes that the 36 possible outcomes are equally likely. This model does not lose its justification or usefulness by the fact that actual dice do not live up to it. The assumption of perfect randomness in games, card shuffling, industrial quality control, or sampling is rarely realized, and the true usefulness of the model stems from the experience that noticeable departures from the ideal scheme lead to the detection of assignable causes and thus to theoretical or experimental improvements.

Conditional probability–independence. Suppose that a population of N people includes N_A color-blind persons and N_H females. To the event A "a randomly chosen person is color-blind" can be ascribed probability $P\{A\} = N_A/N$, and similarly for the event H that a person be female one has $P\{B\} = N_H/N$. If N_{AH} is the number of color-blind females, the ratio N_{AH}/N_H may be interpreted as probability that a randomly chosen female is color-blind; here the experiment "random choice in the population" is replaced by a selection from the female subpopulation. In the original experiment, N_{AH}/N is the probability of the simultaneous occurrence of both A and H, so that $N_{AH}/N_H = P\{A \cap H\}/P\{H\}$. Similar situations occur so frequently that it is convenient to define the conditional probability of the event A relative to H by Eq.(1).

$$P\{A \mid H\} = \frac{P\{A \cap H\}}{P\{H\}} \tag{1}$$

This concept is useful whenever it is desired to restrict the consideration to those cases where the event H occurs (or where the hypothesis H is fulfilled). Thus, in betting on an event A the knowledge that H occurred would induce one to replace $P\{A\}$ by $P\{A \mid H\}$.

Independent trials. The intuitive frequency interpretation of probability is based on the concept of experiments repeated under identical conditions; a theoretical model for this concept can be developed.

Consider an experiment described by a sample space $\mathfrak{S}$; for simplicity of language it can be assumed that $\mathfrak{S}$ consists of finitely many sample points $E_1, \ldots, E_N$. When the same experiment is performed twice in succession, the thinkable outcomes are the N^2 pairs of sample points (E_1, E_1), $(E_1, E_2), \ldots,$ (E_N, E_N), and these now constitute the new sample space. It is called the combinatorial product of $\mathfrak{S}$ by itself and denoted by $\mathfrak{S} \times \mathfrak{S}$.

Probabilities must be assigned to the events in $\mathfrak{S} \times \mathfrak{S}$. If the second trial is independent of the first, the probabilities in $\mathfrak{S} \times \mathfrak{S}$ follow the productive rule $P\{E_i, E_j\} = P\{E_i\}P\{E_j\}$.

In the case of n tossings of a coin, this rule leads to the probability 2^{-n} for each sample point in agreement with the requirement of equally likely cases. In the more general case of Bernoulli trials, each trial results in success S or failure F, and $P\{S\} = p$, $P\{F\} = q$ where $p + q = 1$. (This may be considered as the model of a skew coin.) A succession of n independent trials of this kind leads to the sample space of n-tuples ($SFFS \cdots FS$), and the probability of such a point is the product ($pqqp \cdots qp$) obtained on replacing each S by p and each F by q.

Markov chains. Markov chains represent an important scheme for dependent trials. Suppose that at each trial the possible outcomes are $E_1, \ldots, E_N$ and that whenever E_i occurs the conditional probability of E_j at the next trial is p_{ij}, independently of what happened at the preceding trials. Here, of course, $p_{ij} \geqq 0$ and $p_{i1} + p_{i2} + \cdots + p_{iN} = 1$ for each i. The p_{ij} are called transition probabilities. The whole process is now determined if the initial probabilities, π_i, at the first trial are known. For example, $P\{E_a E_b E_c\} = \pi_a p_{ab} p_{bc}$. The probability of the event "E_c at the third trial" is obtained by summation over all a and b, and so on. Markov chains, and their analog with continuous time, represent the simplest type of stochastic process. *See* STOCHASTIC PROCESS.

Random variables and their distributions. The theory of probability traces its origin to gambling, and the gambler's gain may still serve as the simplest example of a random variable. With every possible outcome (sample point) there is associated a number, namely, the corresponding gain. In other words, the gain is a function on the sample space, and such functions are called random variables. With the same experiment, one may associate many random variables.

Every random variable X has a distribution function $F(t) = P\{X \leqq t\}$. *If* X assumes only finitely many values, then $F(t)$ is a step function. The notion of independence carries over: Two random variables X and Y are independent if $p\{X \leqq x, Y \leqq t\} = P\{X \leqq s\} \cdot P\{Y \leqq t\}$.

Expectations. Given a random variable X one may interpret its distribution function $F(t)$ as describing the distribution of a unit mass along the real axis such that the interval $a < x \leqq b$ carries mass $F(b) - F(a)$. In the case of a discrete variable assuming the values $x_1, x_2, \ldots$ with probabilities $p_1, p_2, \ldots$ the entire mass is concentrated at the points x_i; if $F'(x) = f(x)$ exists, it represents the ordinary mass density as defined in mechanics. The center of gravity of this mass distribution is called the expectation of X; the usual symbol for it is $E(X)$, but physicists and engineers use notations such as $\langle X \rangle$, $\langle X \rangle_{Av}$, or X. In the cases mentioned, $E(X)$ is given by Eqs. (2) and (3).

$$E(X) = \Sigma p_i x_i \tag{2}$$

$$E(X) = \int_{-\infty}^{+\infty} x f(x)\, dx \tag{3}$$

Before discussing the significance of the new concept, a few frequently used definitions are appropriate. Put $m = E(X)$. Then $(X - m)^2$ is, of course, a random variable. In mechanics, its expectation represents the moment of inertia of the mass distribution. In probability, it is called variance of X, given by Eq. (4). Its positive root is the standard deviation.

$$\mathrm{Var}(X) = E(X - m)^2 = E(X^2) - m^2 \tag{4}$$

The variance is a measure of spread: It is zero only if the entire mass is concentrated at the point m, and it increases as the mass is moved away from m. In the case of two variables X_1 and X_2 with expectations m_1 and m_2 it is necessary to consider not only the two variances $s_i^2 = E[(X_i - m_i)^2]$ but also the covariance $\mathrm{Cov}(X_1, X_2) = E[(X - m_1)(X_2 - m_2)] = E(X_1 X_2) = m_1 m_2$. The covariance divided by $s_1 s_2$ is called the correlation coefficient of X_1, and X_2. If it vanishes, X_1 and X_2 are called uncorrelated. Every pair of independent variables is uncorrelated, but the converse is not true.

Laws of large numbers. To explain the meaning of the expectation and, at the same time, to justify the intuitive frequency interpretation of probability, consider a gambler who at each trial may gain the amounts $x_1, x_2, \ldots, x_n$ with probabilities $p_1, p_2, \ldots, p_n$. The gains at the first and second trials are independent random variables X_1, X_2 with the indicated distribution and the common expectation $m = \Sigma p_i x_i$. The event that an individual gain equals x_i as probability p_i, and the frequency interpretation of probability leads one to expect that in a large number n of trials this event should happen approximately np_i times. If this is true, the total gain $S_n = X_1 + X_2 + \ldots + X_n$ should be approximately nm; that is, the average gain $(1/n)S_n$ should be close to m. The law of large numbers in its simplest form asserts this to be true. *See* GAME THEORY; STATISTICS. [W.F.]

Problem solving (psychology) The voluntary ideational behavior sustained by an organism in order to attain a goal which is not immediately accessible. Nearly identical processes have been distinguished by different terms, depending upon the circumstances under which the problem solving takes place. Usually the term problem solving has been restricted to those studies which require the subject to manipulate a tool, assemble a device, or discover the significance of a detour. If, on the other hand, the solution calls for the recognition of a relationship, that is, the ideational classification of stimuli, the process has been called concept formation or discrimination learning, when observed in infrahuman animals. When the problem to be solved requires the correction of a malfunction of equipment, the process has been called troubleshooting. When the problem requires a choice between two or more alternative courses of action, the process has been called decision making.

Heuristics is defined as the study of the mental processes involved in problem solving. The various stages in the process of solving a problem for which there is no immediate solution appear to be as follows:

Initially the individual must comprehend the nature of the problem. The gestalt psychologists argue for the importance of this first stage since they believe that the manner in which a problem is "seen" will determine, at least in part, the likelihood of attaining a successful solution.

The second stage, sometimes called the preparation stage, is devoted to obtaining information which appears to be relevant to the final solution. This is also a crucial stage since if the individual establishes an inappropriate set or bias, he may reject as irrelevant information which is in fact relevant. A bias in favor of irrelevant information is almost as undesirable. During this stage the problem solver tries to examine all of the relevant information and its implications for his problem. Discovery of interrelationships is in itself additional information which may be relevant.

The third stage, sometimes called the incubation stage, is characterized by the termination of active seeking for the desired solution and the turning of attention to other matters. It is usually described as a passive period of hopeful waiting for insight. Successful problem solving may follow a period of apparent inattention to the problem because the problem solver has avoided inappropriate information-handling biases during this so-called incubation period.

The final stage, when the problem solver achieves the solution and comprehends its significance, is called insight.

Clearly, the process of problem solving depends upon memory of information as well as imaginative and creative handling of the information. *See* MEMORY. [E.J.A.]

Proboscidea An order of placental mammals derived from unknown Paleocene or Eocene condylarths of Africa and known from the late Eocene to the Recent. Although fossil representatives such as the deinotheres of the Old World and the mastodonts of both the Old and New worlds were once abundant, only the two species of living elephants survive.

As a rule, the order is characterized by predominantly herbivorous, large to gigantic animals. Either the upper or lower second incisors, or both, enlarge to become tusks. Each foot on the pillarlike legs has five toes which are bound together in a broad, rounded pad. Elephants are proboscis-bearing (hence the name of the order) and in the living species the trunks are sensitive and work together with the tusks during feeding. The skull, because of the effects of trunk development, has undergone profound morphological modifications in the course of elephant evolution.

Proboscideans were confined to Africa well into the Miocene, when different groups spread to Europe, Asia, and North America. From North America various proboscideans entered South America during the Pleistocene. *See* ELEPHANT; MAMMALIA. [F.S.S.]

Procellariiformes An order of oceanic birds characterized by tubelike nostril openings, webbed feet, dense plumage, compound horny sheath of the bill and, often, a peculiar musky odor. Included families are Diomedeidae (albatrosses), Procellariidae (petrels, fulmars, shearwaters), Hydrobatidae (storm petrels), and Pelecanoididae (diving petrels). The order encompasses extremes in size paralleled only by the Falconiformes, ranging from the wandering albatross (*Diomedea exulans*), with a wingspread of almost 12 ft (3.7 m), to the sparrow-sized least petrel (*Halocyptena microsoma*).

All of the families except the diving petrels are widely distributed; the latter are confined to the Southern Hemisphere, mostly in higher altitudes, although one species follows cold currents off western South America almost to the Equator. The diving petrels are convergent in anatomy and habits toward the auks (Alcidae) of the Northern Hemisphere. *See* AVES; OCEANIC BIRDS. [K.C.P.]

Process control A field of engineering dealing with ways and means by which conditions of processes are brought to and maintained at desired values, and undesirable conditions are avoided as much as possible. In general, a process is understood to mean any system where material and energy streams are made to interact and to transform each other. Examples are the generation of steam in a boiler; the separation of crude oil by fractional distillation into gas, gasoline, kerosine, gas-oil and residue; the sintering of iron ore particles into pellets; and the polymerization of propylene molecules for the manufacture of polypropylene. In the wide sense, process control also encompasses determining the desired values.

Process control in the wide sense includes a number of functions, which can be arranged in a hierarchy, as follows:

Scheduling
Mode setting
Quality control
Regulatory control/Sequence control
Coping with faults

Computerized instrumentation has revolutionized the interaction with plant personnel, in particular the process operators. Traditionally, the central control room was provided with long panels or consoles, on which alarm lights, indicators, and recorders were mounted. Costs were rather high, and surveyability was poor. In computerized instrumentation, visual display units can provide information in a concise and flexible way, adapted to human needs and capabilities. *See* AUTOMATION; CONTROL SYSTEMS. [J.E.R.]

Process engineering A branch of engineering in which a process effects chemical and mechanical transformations of matter, conducted continuously or repeatedly on a substantial scale. Process engineering constitutes the specification, optimization, realization, and adjustment of the process applied to manufacture of bulk products or discrete products. Bulk products are those which are homogeneous throughout and uniform in properties, are in gaseous, liquid, or solid form, and are made in separate batches or continuously. Examples of bulk product processes include petroleum refining, municipal water purification, the manufacture of penicillin by fermentation or synthesis, the forming of paper from wood pulp, and the manufacture of paint, whiskey, plastic resin, and so on. Discrete products are those which are separate and individual, although they may be identical or very nearly so. Examples of discrete product processes include the casting, molding, forging, shaping, forming, joining, and surface finishing of the component piece parts of end products or of the end products themselves. Processes are chemical when one or more essential steps involve chemical reaction. Almost no chemical process occurs without many accompanying mechanical steps such as pumping and conveying, size reduction of particles, classification of particles and their separation from fluid streams, evaporation and distillation with attendant boiling and condensation, absorption, extraction, membrane separations, and mixing. [E.F.L.]

Prochirality The property displayed by a prochiral molecule or a prochiral atom (prostereoisomerism). A molecule or atom is prochiral if it contains, or is bonded to, two constitutionally identical ligands (atoms or groups), replacement of one of which by a different ligand makes the molecule or atom chiral. Examples are shown in the illustration.

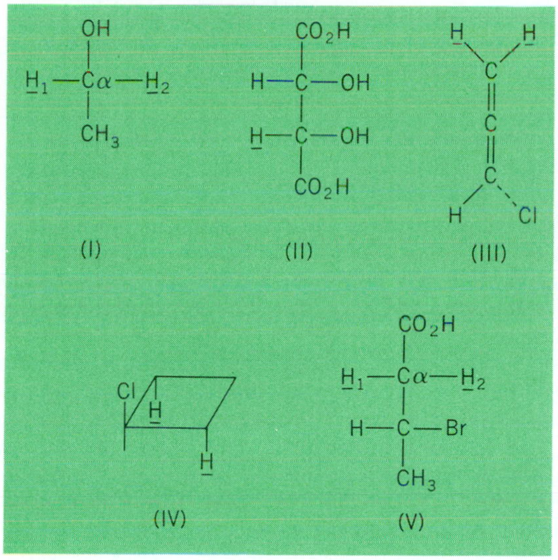

Prochiral structures.

None of molecules I–IV is chiral, but if one of the underlined pair of hydrogens is replaced, say, by deuterium, chirality results in all four cases. In compound I, ethanol, a prochiral atom or center can be discerned ($C\alpha$:CH_2); upon replacement of H by D, a chiral atom or center is generated, whose configuration depends on which of the two pertinent atoms (H_1 or H_2) is replaced. Molecule V is chiral to begin with, but separate replacement of H_1 and H_2 (say, by bromine) creates a new chiral atom at $C\alpha$ and thus gives rise to a pair of chiral diastereomers. No specific prochiral atom can be discerned in molecules II–IV, which are nevertheless prochiral (III has a prochiral axis).

Faces of double bonds may also be prochiral (and give rise to prochiral molecules), namely, when addition to one or other of the two faces of a double bond gives chiral products.

Although the term prochirality is widely used, especially by biochemists, a preferred term is prostereoisomerism. This is because replacement of one or other of the two corresponding ligands (called heterotopic ligands) or addition to the two heterotopic faces often gives rise to achiral diastereomers without generation of chirality. Thus not all compounds which display prostereoisomerism also display prochirality. *See* MOLECULAR ISOMERISM; STEREOCHEMISTRY. [E.L.E.]

Prochlorophyceae A class of prokaryotic organisms coextensive with the division Prochlorophycota in the kingdom Monera. Because prochlorophytes carry out oxygen-evolving photosynthesis, they may be considered algae. They are distinguished from Cyanophyceae, the only other prokaryotic algae, by the presence in their photosynthetic lamellae of chlorophyll *b* in addition to chlorophyll *a* and the absence of phycobilin pigments. Otherwise, they resemble Cyanophyceae biochemically and ultrastructurally. The class comprises a single genus, *Prochloron*, with one species. *See* ALGAE; CYANOPHYCEAE. [P.C.Si.; R.L.Moe]

Procyon Alpha Canis Minoris, a nearby bright star, 0.3 magnitude, of spectral type F5; Procyon is slightly brighter than a normal main-sequence F5 star. Thus it may be an old star that has begun to evolve off the main sequence. Its great age is indicated by a close faint companion of absolute magnitude +13, yellowish in color, that is, a white dwarf star, which is nearing the end of its evolution. *See* STAR. [J.L.Gr.]

Product design The determination and specification of the parts of a product and their interrelationship so that they become a unified whole. The design must satisfy a broad array of requirements in a condition of balanced effectiveness. A product is designed to perform a particular function or set of functions effectively and reliably, to be economically manufacturable, to be profitably salable, to suit the purposes and the attitudes of the consumer, and to be durable, safe, and economical to operate. For instance, the design must take into consideration the particular manufacturing facilities, available materials, know-how, and economic resources of the manufacturer. The product may need to be packaged; usually it will also need to be shipped so that it should be light in weight and sturdy of construction. The product should appear significant, effective, compatible with the culture, and appear to be worth more than the price. *See* PRODUCTION ENGINEERING; PRODUCTION PLANNING. [R.I.F.]

Product of inertia The product of inertia of area A relative to the indicated XY rectangular axes is $I_{XY} = \int xy\,dA$ (see illustration). The product of inertia of the mass contained in volume V relative to the XY axes is $I_{XY} = \int xy\rho\,dV$—similarly for I_{YZ} and I_{ZX}.

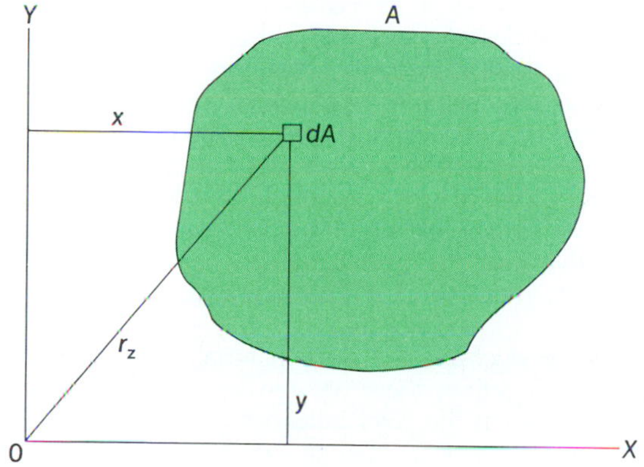

Product of inertia of an area.

Relative to principal axes of inertia, the product of inertia of a figure is zero. If a figure is mirror symmetrical about a YZ plane, $I_{ZX} = I_{XY} = 0$. *See* MOMENT OF INERTIA. [N.S.F.]

Production engineering The planning and control of the mechanical means of changing the shape, condition, and relationship of materials within industry toward greater effectiveness and value. Production engineering is a term applied to certain aspects of planning and control of manufacturing; it is a service function to the production department.

As industry and technology evolve to greater levels of sophistication, complexity, and specialization, the broad area of figuring out what to do becomes more involved and at the same time better understood. By this process some of what had been originally performed by either the production department or the industrial engineer becomes a separate activity with its own background of knowledge, principles, and techniques.

Production engineering as a planning activity takes place between product design and the planning of the overall manufacturing process. Overall manufacturing planning is usually considered within the profession of industrial engineering. The purpose of production engineering is to refine and adjust the design of the product to the problems involved in its proposed manufacture. Conversely it should solve certain problems, mainly mechanical, such as those involved in processing, tools, dies, and new or special equipment necessary to manufacture the product efficiently and according to the established specifications. *See* INDUSTRIAL ENGINEERING; PRODUCT DESIGN; PRODUCTION METHODS. [R.I.F.]

Production methods In product engineering, the processes that are used to obtain a given product. Basically all production processes and methods can be classified as one of two types. Analytic industrial processes break down a given material into several products, as in an ore-reducing plant. Synthetic processes create one product from several different materials, as in a blast furnace or an automobile assembly plant. Processes may be a combination analytic-synthetic (wood-furniture factory) or synthetic-analytic (feed mill).

Industry frequently classifies processes into those that (1) change the shape, called forming, including cutting, molding, bending, dissolving, and machining; (2) change the chemical or internal characteristics, called treating, including mixing, blending, heat treating, and refining; (3) change the external surface, called finishing, including rinsing, coating, drying, and painting; and (4) add other pieces, called assembling, including attaching, joining, packaging, fitting, and fastening.

Most industries use a combination of at least two basic processes. In fact, some processes can be placed logically in more than one class, for example, metal plating.

Processes have also been classified into continuous or process-type operations, as in an oil refinery, and intermittent (or repetitive) or manufacturing-type operations.

Almost all production processes or methods change the form or condition of some material, or add or deduct other materials, aided by workers or machinery, or both, with the end objective being a product which has greater utility by nature of its new form or characteristics than the initial material.

The end product and the start material are of fundamental importance. Together, these are termed the product-material factor; this factor is the chief influencing feature in the choice of production methods. A change in the end product or in the characteristics of the start material may cause or allow significant changes in the production processes or methods. *See* PROCESS ENGINEERING; PRODUCT DESIGN; PRODUCTION ENGINEERING. [R.Mu.]

Production planning The function of a manufacturing enterprise responsible for the efficient planning, scheduling, and coordination of all production activities. The planning phase involves forecasting demand and translating the demand forecast into a production plan that optimizes the company's objective, which is usually to maximize profit while in some way optimizing customer satisfaction. These twin objectives are not always synonymous. During the scheduling phase the production plan is translated into a detailed, usually day-by-day, schedule of products to be made. During the coordination phase actual product output is compared with scheduled product output, and this information is used to adjust production plans and production schedules. *See* INDUSTRIAL COST CONTROL; OPTIMIZATION.

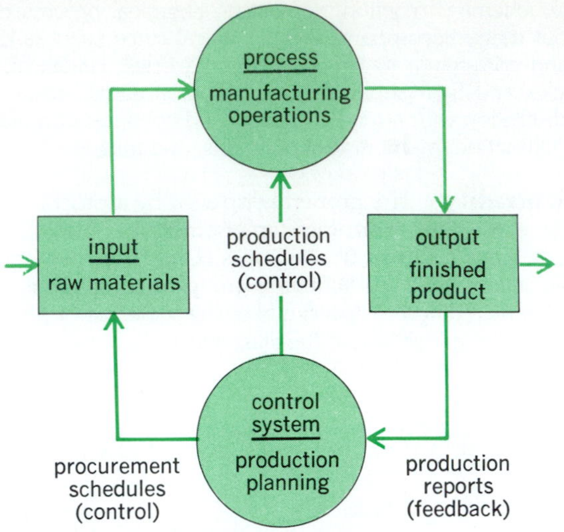

The production process as an input-output process.

If the production or manufacturing process is viewed as an input-output process, then the production planning function can be viewed as a control process with feedback (see illustration). The control is in the form of schedules and plans, while the feedback results from the comparison of the production reports with the production schedules. *See* CONTROL SYSTEMS; INVENTORY CONTROL; PRODUCTION ENGINEERING; PRODUCTION METHODS. [J.E.Bi.]

Productivity Output in relation to input. Financial measures such as profit-and-loss statements characteristically state the relation of money-valued outputs of an enterprise in relation to costs, which are money-valued inputs. However, productivity refers to measures of output and input that are cast primarily in physical units, for example, the number of units of product in relation to an input factor such as production worker labor-hours.

Productivity measures do not have universal validity. The units to be used for measuring output are characteristically a function of the problem which is being addressed. The output of a single product is of interest to the industrial engineer, whose task is to gage the change in efficiency that is traceable to the installation of a new piece of equipment or a new set of operations for a complete production system. By contrast, the economist is typically interested in output measures that require the calculation of an index number which is a composite representation of output change.

The problems in relation to gaging productivity are usually concentrated on the measure of output, owing to the requirement of measuring a "mix" of products that not only vary in physical characteristics at a given moment, but also vary in proportion through time. In that case it is necessary to "weight" the measure of output, thereby assigning differential importance to its components. *See* INDUSTRIAL COST CONTROL; PRODUCTION PLANNING. [S.M.]

Progesterone A steroid hormone produced in the corpus luteum and placenta. The hormone has an important physiological role in the luteal phase of the menstrual cycle and in the maintenance of pregnancy. In addition, progesterone produced in the testis and adrenals has a key role as an intermediate in the biosynthesis of androgens, estrogens, and the corticoids (adrenal cortex steroids). *See* ANDROGEN; CHOLESTEROL; ESTROGEN; MENSTRUATION; PREGNANCY; STEROID; STEROL. [R.I.D.]

Programmable controllers Control devices, normally used in industrial control applications, that employ the hardware architecture of a computer and a relay ladder diagram language. Programmable controllers (PCs), also referred to as programmable logic controllers (PLCs), were developed in the late 1960s for the automotive industry as replacements for large relay control panels. They are now widely accepted as an industrial control device in many industries, such as automotive, chemical, petroleum, pipelines, pulp and paper, food, rubber, off-road vehicles, power utilities, tobacco, mining, textiles, and machinery.

Architecture. The relations of the basic components of a programmable controller are diagrammed in the illustration. A programmer is temporarily connected to the programmable controller for the purpose of entering, examining, or editing the ladder diagram logic. Once the logic is established as desired by the control engineer, the programmer can be disconnected and used on another programmable controller. Thus one programmer can service many (10–50) programmable controllers of the same manufacturer.

A central processor unit (CPU) and its associated input/output (I/O) form the permanent part of the programmable controller. Each unit normally has its own power supply or supplies that accept ac power and produce a well-regulated dc voltage. The central processor unit includes a processor, a logic memory, and a storage memory. The processor operates upon instructions stored in the logic memory. *See* MICROPROCESSOR.

The logic memory stores the instructions that result from the ladder diagram display entered by the control engineer with the programmer. These instructions indicate how the processor is to react to various changes in inputs and what outputs to activate or deenergize. The storage memory is used to retain numerical values, such as those used with timing or counting or arithmetic operations. The storage memory may be a physically separate memory or a portion of the logic memory set aside specifically for numerical storage. *See* COMPUTER STORAGE TECHNOLOGY; DIGITAL COMPUTER.

The input/output section isolates the central processor unit from the outside world represented by customer-supplied field devices. This isolation protects the central processor unit from electrical noise and voltage variations that can occur in industrial environments. Without this protection the central processor unit would operate in an unreliable and erratic manner. The field devices are inputs providing data to the programmable controller, and outputs to be controlled by the programmable controller. The input/output structure can be local (up to 80 ft or 25 m) or remote up to 10,000 ft (3000 m) from the central processor unit. Four wires (central processor unit to input/output connection) replace hundreds of field device wires over long distances in large installations.

Programming. A relay ladder diagram format of programming was selected to be as compatible as possible with the previous control devices (relays, timers, and counters) that the programmable controller was designed to replace. Use of the relay ladder diagram eliminates the need to introduce a new language at this level of control and greatly enhances the acceptance of programmable controllers by all levels of industrial control personnel. *See* CONTROL SYSTEMS. [R.A.W.]

Programming languages Notations with which people can communicate algorithms to computers and to one another. To accomplish this, programmers specify a sequence of operations to be applied to data objects. *See* ALGORITHM.

Computers process a rather low-level machine language, which has simple instructions (load, add, and so forth) and data (for example, words or partial words including those containing instructions). The following example shows machine code that assigns the absolute value of B − C to A.

Location	*Machine code*	*Meaning*
040003	100020040001	load B into register A1
040004	250320040002	subtract C from register A1
040005	741020040007	GOTO 040007 if A1 positive
040006	110020000000	negate A1
040007	010020040000	store contents of A1 into A

Writing in machine code places a tremendous burden on the programmers. They must write in an unusual number base (for example, eight), and transfer data from memory to registers in order to perform operations. They must know the machine's operation codes for the operations. Finally, they must know the addresses of the operands (both registers and data words) and instructions (to be able to skip some operations with branches).

The introduction of assembler languages eased communication between programmers and machines by permitting symbolic rather than numeric or positional references for operation codes (opcodes) and addresses. Additional data and instructions could therefore be added to a program without changes to existing instructions. Thus, in the example below, which shows assembler code that assigns the absolute value of B − C to A, instructions could be added between the branch positive, BP, and its destination.

Label	*Opcode*	*Operands*	*Comments*
	L	A1,B	
	AN	A1,C	· A1 : = B − C
	BP	A1,NEXT	
	N	Al	· A1 : = − A1
NEXT	S	A1,A	· A : = |B − C|

Unfortunately, assembler language does not go very far in improving programmers' ability to communicate algorithms. A programmer must still be concerned with expressing algo-

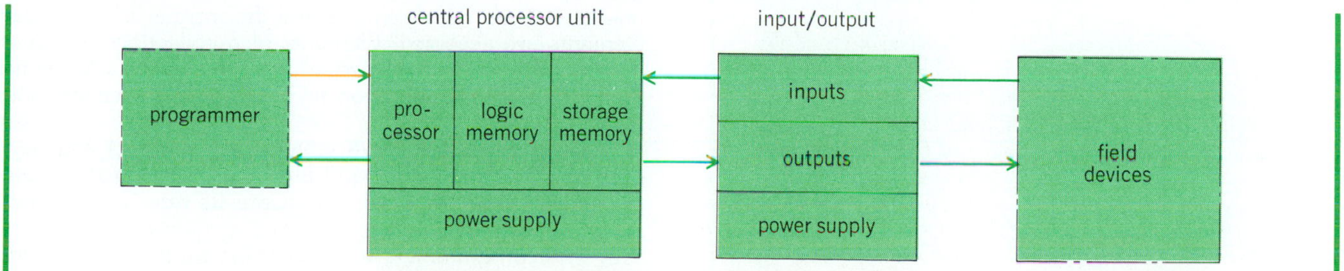

Generalized block diagram of a programmable controller.

rithms in a form suitable to particular machines. For instance, if an expression requires more registers than are available, the programmer must create temporary variables to hold partial results of expressions. High-level languages allow programmers to communicate more easily with one another by providing structures that hide the architecture of the machine. Expressions in high-level languages can be written almost as they are in algebra so that the programmer need not worry about which register contains a particular computation or which registers are available. *See* DIGITAL COMPUTER.

Structure of programming languages. The components of a program and their relationships are shown in the illustration. The basic executable unit of a programming language is a program. Programs contain declarations of data and procedures, and a sequence of statements to be executed. Programming languages manipulate objects or values like integers or files. Objects are referred to by their names, called identifiers. Constant objects, such as the number 2 (one of whose names is the literal "2"), do not change. Variable objects may contain either other objects or the names of other objects.

Expressions. Expressions combine objects and operators to yield new objects. They may be written in one of several forms. Infix operators appear between their operands; prefix operators precede their operands; and postfix operators follow their operands. Most languages adopt infix notation for binary operators and prefix notation for unary operators.

Statements. Important types of statements include assignment statements, compound statements, conditional statements, repetitive statements, and GOTO statements.

Assignment statements assign values to variables. In an assignment of form (1), the value in the object denoted by B is

A: = B (1)

assigned to the object denoted by A. The order of evaluation of the two sides of an assignment statement differs among languages—some insist on left-to-right evaluation, while others leave the order of evaluation undefined. Some languages permit the same value to be assigned to several variables in a single statement. Other languages treat assignment just like any other operator in an expression.

A compound statement (2) groups a sequence of statements

BEGIN stmt1; stmt2;...stmtn END (2)

into a single statement. Execution of statements within a sequence of statements is in the order in which the statements appear in the sequence.

The IF statement (3) allows a programmer to indicate that a

IF expression THEN statement (3)

statement is to be executed if a stated condition is true. If the expression is true, the statement following the THEN is executed followed by the statement after the IF statement. If the expression is false, the statement is skipped. Many languages permit execution of one of two alternate statements with the form of IF statement (4). If the expression is true, statement 1

IF EXPRESSION then statement1 ELSE statement2 (4)

is executed; otherwise statement 2 is executed. In either case, execution continues with the statement after the IF.

One of the advantages offered by computers is performing the same operations repetitively. Two statements in programming languages support this kind of execution: FOR and WHILE. The FOR statement (5) is generally used when a programmer knows how many times a statement is to be executed.

FOR variable-identifier: =
expression TO expression DO statement (5)

The WHILE statement (6) is used when continued execution depends on a condition that is true for an unknown number of repetitions. The statement is executed as long as the expres-

WHILE expression DO statement (6)

sion is true. When the expression becomes false, execution continues with the statement after the WHILE.

The lowest-level control statement in most languages is the GOTO statement (7). This statement provides for the direct

GOTO identifier (7)

transfer of control to a statement labeled with the identifier that is the argument of the GOTO statement.

Procedures. As problems grow more complex, they must be divided into subproblems that can be solved independently. This approach to problem solving is supported by procedure declarations, which group sequences of statements together and associate identifiers with the statements. A procedure identifier abstracts the details of the computation carried out by the statements of the procedure by encapsulating them. Procedures also help reduce the size of a program by permitting similar code to be written only once.

Data types. Data types are sets of values and the operations that can be performed on them. Data types offer programmers two principal advantages: abstraction and authentication. Data types hide representational details from programmers so they do not have to be concerned whether characters are represented by ASCII or EBCDIC codes, or whether they are left-aligned and blank-filled or right-aligned and zero-filled in machine words. The concept of data type also provides useful redundancy. The appearance of a variable as an operand in an expression implies a particular type for the variable (for example, A + B implies that A and B are either integer or real) that may be checked against the type of its current value.

Scalars are indivisible units of data; they are treated as single complete entities. Most languages contain several scalar types. The most common are integer, real, boolean, and character.

While variables usually hold values of objects, they may also hold the name of some object. Variables that can hold such values are called references or pointers.

Collections of related values can be represented by aggregate types, which can be characterized by the types of the objects that are their elements and the manner in which the elements are accessed. Elements of a collection can have homogeneous or heterogeneous types, and can be selected by their position in a collection or by a name associated with them.

Binding attributes to variables. Both constant and variable objects have values and attributes. Among the most important attributes of an object are its type, scope, and extent.

In most programming languages, the same identifier can name distinct objects in different parts of a program. The

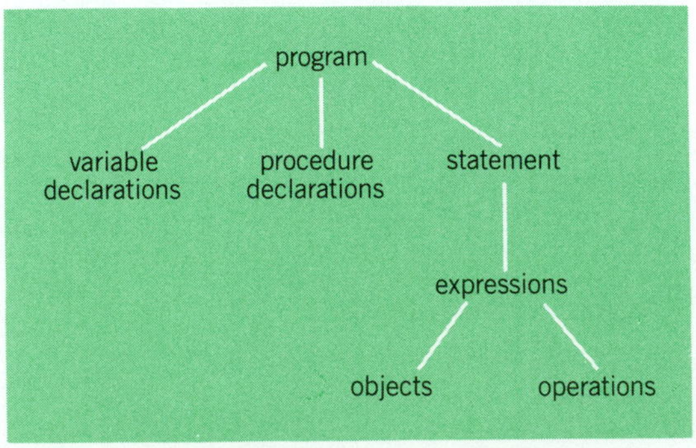

Components of a high-level language.

scope of an identifier is that part of a program in which the identifier has the same meaning. Generally that section of program is called a scope unit and corresponds to a procedure or compound statement. The set of objects that can be referenced in a scope unit is called the environment.

Data types may be associated with objects statically via declarations or dynamically via assignment statements. In statically typed languages, the type associated with an object remains the same throughout the scope of the object.

The extent or lifetime of a variable is the period of program execution during which the storage is allocated. Static storage is created at the beginning of the program execution and not deallocated until program termination. Local storage for variables is created on entry to the block or procedure containing the variable's declaration and destroyed on exit. Dynamic storage is created and destroyed by execution of special statements in the programming language.

Parameters. Parameters communicate information between the calling environment and the environment of the called procedure. Although this can be accomplished by referencing nonlocal variables, communicating with parameters permits a procedure to work on different variables each time it is called. Formal parameters are identifiers that are locally declared in a procedure definition; actual parameters are expressions in a procedure call statement.

Functions. Function declarations are similar to procedure declarations. But unlike a procedure, which communicates its results by assigning values to its parameters, a function communicates a distinguished value known as the result of the function. This result of the function is often indicated by assigning a value to the name of the function.

When a statement in a procedure or function alters the value of a nonlocal variable or reference or name parameter, the procedure is said to have a side effect. It may not be quite accurate to call such assignments in procedures side effects, because procedures must include such statements in order to communicate their results. Therefore, they are really main effects of procedures. Side effects in functions are less desirable because they make the results of expression evaluation unpredictable. Thus, it is good programming practice to avoid side effects in functions and to use procedures instead when side effects are desired.

Applicative and object-oriented languages. Most programming languages are imperative languages, because their programs largely consist of a series of commands to assign values to objects. ALGOL, PL/1, FORTRAN, Ada, COBOL and Pascal are imperative languages. Two other types of languages exist: applicative and object-oriented. No commonly used language is purely imperative, applicative, or object-oriented, so that every language has some characteristics of each type of language.

A purely applicative language would be one in which there are no statements, only expressions without side effects. The name "applicative" refers to the repeated application of functions to the results of other functions to specify an algorithm. LISP, APL, and VAL are three languages in which programs are largely applicative.

Programs written in object-oriented languages consist of sequences of commands directed at objects. An object receiving a command invokes an internal procedure to respond to the command. Assignment statements are replaced by commands to objects requesting that they change their values.

Defining new data types. Procedure and function declarations define new operations that operate on existing objects. Type declarations add new objects and operations to a programming language. Designing programs with user-defined types allows solutions to be stated in problem-oriented rather than machine-oriented terms. ALGOL 68- and Pascal-type declarations allow programmers to define the representation of new objects of the data types and ALGOL 68 permits the definition of operations.

Particular languages. The more widely used languages include FORTRAN, ALGOL, COBOL, BASIC, PL/I, APL, SNOBOL, LISP, Pascal, C, and Ada.

FORTRAN was designed in the mid-1950s for efficient scientific applications. To achieve this goal, the data and control components of the language are very simple, providing programmers with low-level hardware operations. FORTRAN control structures are particularly primitive; neither compound statements nor unbound repetitive statements are part of the language. Two conditional statements (arithmetic and logical IF's) and a bounded iteration statement (DO) provide the only relief from GOTO's and labels.

ALGOL was designed in the late 1950s and early 1960s as a language for scientific computation. The elegance of ALGOL's structure has made it the basis for several subsequent language design efforts, most notably ALGOL 68 and Pascal, and its control structures have even been adopted by PL/1 and later versions of FORTRAN. ALGOL was also the first language whose syntax was defined with a formal grammar (called Backus-Naur Form or BNF after two of the principal designers of ALGOL). ALGOL control statements are very much like statements (2)–(5). However, all repetitive statements are variants of the FOR statement, and compound statements are units of scope that may contain declarations of their own local variables.

COBOL was designed in 1959 and 1960 as a language to support business applications on computers. As was the case with FORTRAN, efficiency was a primary concern. Because of its intended application area, COBOL has relatively poor features for specifying computations (such as expressions, functions, and parameters), but strong features for data description and input/output. Two of the most noticeable features of COBOL programs are the English-like syntax and the division of programs into four parts.

BASIC was developed at Dartmouth College in the early 1960s to provide nonscience majors with a simple, interactive language. BASIC supports real and string data as well as one- or two-dimensional arrays of one of these primitive types. Dimension statements permit string lengths and array subscripts to be specified. Variable names are restricted to single letters, followed by an optional digit if the variable is numeric or $ if the variable is a string. Thus, types are statically associated with variables.

PL/1 was designed in the mid-1960s to meet the needs of a broad range of applications: scientific, business applications, and systems. PL/1 combines and extends many of the features of FORTRAN (parameter binding and formated input/ output), ALGOL (scope rules, variable extents, recursive procedures, and control structures), and COBOL (record structures and picture data types). By providing redundant features from different languages, the designers have made the language difficult to learn and encouraged users to stay in one of the FORTRAN, COBOL, or ALGOL subsets of PL/1.

APL, designed during the mid-1950s to the mid-1960s, is an interactive language whose operators accept and produce arrays with homogeneous elements of type number or character. The only control construct is a GOTO statement, but this is inconsequential since many APL programs have a strong applicative flavor. Although they often consist of an imperative sequence of assignments, APL programs generally operate on array objects to produce new array objects. Thus repetitive constructs are not needed as frequently as they are in ALGOL-like languages.

SNOBOL was developed during the 1960s as a string processing language. SNOBOL has no real control constructs. Instead, every statement either "succeeds" or "fails" and may optionally GOTO a new statement depending on success or failure. SNOBOL statements provide extremely powerful operators for string manipulation.

LISP, invented in the late 1950s, is a very widely used language in the artificial intelligence community. The basic type in LISP is the list, usually denoted by a sequence of items in parentheses. The power of LISP comes from the representation of everything by lists, which leads to very powerful environments of tools and extensions to LISP. LISP is a good language to write experimental programs, since its flexibility is useful for dealing with the unexpected. However, it executes slowly and uses a lot of memory. *See* ARTIFICIAL INTELLIGENCE.

Pascal was developed in 1969 to obtain a language which would be suitable to teach programming as a systematic activity with constructs that could be implemented reliably and efficiently. The control structures of Pascal are similar to statements (2)–(5) and (6). However, a CASE statement was added as an additional selection statement. This statement permits execution of one alternative among several statements based on the value of a selection expression.

C was designed in 1972 to implement the Unix operating system. Similar to Pascal in many ways, C has enumerated types, records (called structures), the CASE statement, and reference variables. The main difference between programming in C and other languages is the extensive use of reference variables, which are defined to be equivalent to arrays. *See* OPERATING SYSTEM.

Ada, designed in the mid-1970s, won a competition sponsored by the Department of Defense to design a new language to support development of embedded systems. Ada was designed using Pascal as a base, to meet the reliability and efficiency requirements imposed by these applications. [J.D.G.; M.D.W.]

Progression (mathematics) Ordered, countable sets of numbers, $x_1, x_2, x_3, \ldots$, not necessarily all different. In general such sets are called sequences, whereas the term progression is usually confined to the special types: the arithmetic, in which the difference $x_k - x_{k-1}$ between successive terms is constant; the geometric, in which the ratio x_k/x_{k-1} is constant; and the harmonic, in which the reciprocals of the terms are in arithmetic progression.

If the first term of an arithmetic progression is a and the common difference b, then the terms of the progression are given by Eqs. (1). The sum of the first n terms S_n is given by Eq. (2).

$$x_1 = a, x_2 = a + b, \quad x_3 = a + 2b, \ldots, \quad x_n = a + (n-1)\,b, \ldots \tag{1}$$

$$S_n = n\,\frac{x_1 + x_n}{2} = n\left(a + \frac{n-1}{2}\,b\right) \tag{2}$$

If the first term of a geometric progression is a and the common ratio r, then the terms of the progression are given by Eqs. (3). Excluding the case $r = 1$ (when all terms are the same), the sum S_n of the first n terms is given by Eq. (4).

$$x_1 = a, \quad x_2 = ar, \quad x_3 = ar^2, \ldots, x_n = ar^{n-1}, \ldots \tag{3}$$

$$S_n = a\,\frac{1 - r^n}{1 - r} \tag{4}$$

The arithmetic mean A and the geometric mean G of n positive numbers are defined by Eqs. (5) and (6).

$$A = \frac{x_1 + x_2 + \cdots + x_n}{n} \tag{5}$$

$$G = \sqrt{x_1 x_2 \cdots x_n} \tag{6}$$

The reciprocals of sequence (1) form a harmonic progression. There is no compact expression for the sum of n terms. If x_1, x_2, x_3 are in harmonic progression, then Eq. (7)

$$x_2 = \frac{2x_1x_3}{x_1 + x_3} \tag{7}$$

is called their harmonic mean. [L.B.]

Projective geometry A geometry that investigates those properties of figures that are unchanged (invariant) when the figures are projected from a point to a line or plane.

Two features of plane projective geometry are (1) introduction of an ideal line that each ordinary line g intersects (the intersection being common to all lines parallel to g), and (2) the principle of duality, according to which any statement that is obtained from a valid one (theorem) by substituting for each concept involved, its dual, is also valid. ("Line" and "point" are dual, "connecting two points by a line" is dual to "intersecting two lines," and so on.) The subject has been developed both synthetically (as a logical consequence of a set of postulates) and analytically (by the introduction of coordinates and the application of algebraic processes). *See* CONFORMAL MAPPING; EUCLIDEAN GEOMETRY. [L.M.Bl.]

Prokaryotae A group of predominantly unicellular microorganisms or infectious agents of cells (the viruses), lacking nuclei, and having asexual and chromonemal reproduction and unidirectional recombination. The Prokaryotae may be considered to include a kingdom for viruses, although such "organisms" are considered acellular or noncellular (even nonliving) by many authorities, and one for the typical moneran forms, the many kinds of bacteria plus the cyanobacteria (the blue-green algae) and the Prochlorophycota. *See* VIRUS.

Bacteria, viruses, and blue-green algae possess little in common, besides such superficial characters as microscopic (or ultramicroscopic) size and frequent involvement in causing diseases in other organisms, including human beings, and such negative characters as not being eukaryotic, not possessing any mouth opening, and not being multicellular (or, in the case of viruses, even cellular) in their organization. In the smallest dimension, prokaryotes measure from 0.2 to 10 micrometers; viruses show a diameter of 10 to 300 nanometers, the largest, therefore, just barely overlapping in width with the smallest bacterium.

Monerans (above the virus level) are generally solitary, unicellular forms; but some species are filamentous, colonial, or mycelial. Some are also motile, either by gliding or by the action of bacterial flagella containing the protein flagellin. Modes of nutrition are diverse: absorptive, chemosynthetic, photoheterotrophic, and photoautotrophic. Respiration is anaerobic or aerobic, or facultatively either one. *See* ALGAE; BACTERIA; CYANOBACTERIA; EUKARYOTAE; PROCHLOROPHYCEAE. [J.O.C.]

Proline A common amino acid of proteins occurring in varying amounts in essentially all individual proteins, and in mixtures of proteins at about 4% by weight. It is a major amino acid (10–15%) in the collagen proteins.

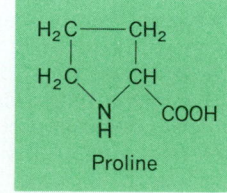

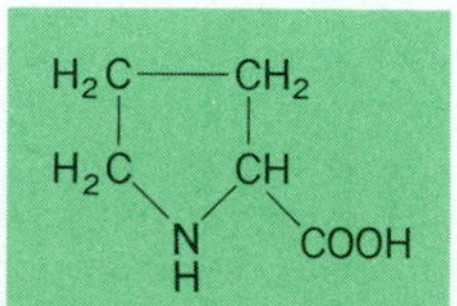

Proline

Nutritionally, proline is not required in the diet of humans or other mammals, since it is formed freely from glutamic acid, which in turn is derived from products of carbohydrate metabolism. *See* AMINO ACIDS; CARBOHYDRATE METABOLISM; COLLAGEN; GLUTAMIC ACID. [E.Ad.]

Promethium A chemical element, Pm, atomic number 61. Promethium is the "missing" element of the lanthanide rare-earth series. The atomic weight of the most abundant separated radioisotope is 147.

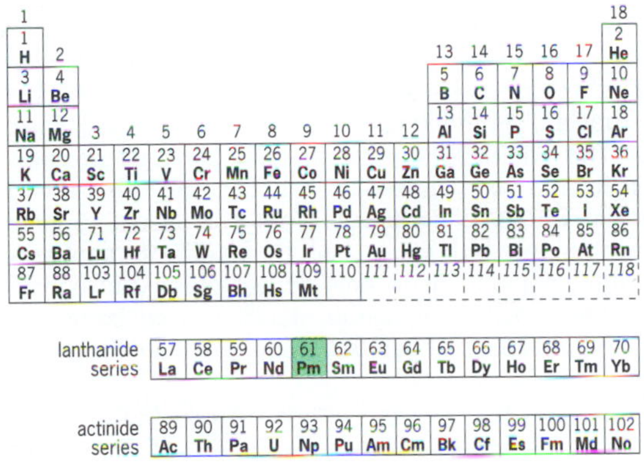

Although a number of scientists have claimed to have discovered this element in nature as a result of observing certain spectral lines, no one has succeeded in isolating element 61 from naturally occurring materials. It is produced artificially in nuclear reactors, since it is one of the products that results from the fission of uranium, thorium, and plutonium.

All the known isotopes are radioactive. Its principal uses are for research involving tracers. Its main application is in the phosphor industry. It has also been used to manufacture thickness gages and as a nuclear-powered battery in space applications. *See* RARE-EARTH ELEMENTS. [F.H.Sp.]

Pronghorn An antelopelike animal, *Antilocapra americana*, the sole representative of the family Antilocapridae, and of uncertain taxonomic affinities. This animal is reputed to be the fastest ungulate in North America and is the only hollow-horned ungulate with branched horns present in both sexes. Like the deer, the pronghorn sheds its horns each fall; the new growth is complete by midsummer.

The pronghorn live in small herds in rather wild, rocky desert country. They feed on cactus, sagebrush, and other vegetation. In late summer, a buck begins to accumulate a harem of about 10–15 does. Two young are usually born in the spring after a gestation period of 35 weeks. Average life-span for a pronghorn is about 8 years. *See* ARTIODACTYLA. [C.B.C.]

Prony brake An absorption dynamometer that applies a friction load to the output shaft by means of wood blocks, flexible band, or other friction surface. The prony brake (see illustration) provides means for measuring the torque T, developed as in the equation below, where L is the distance shown

$$T = L\,(W - W_0)$$

on the drawing, W is the scale weight with the brake operating, and W_0 is the scale weight with the brake free. Small prony brakes are air-cooled; the drum of larger ones may be

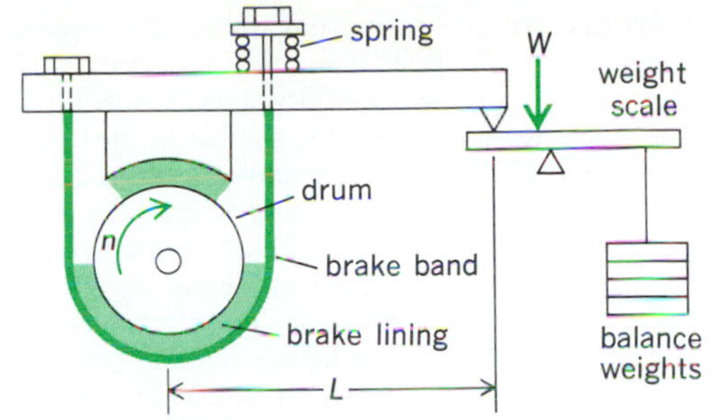

Diagram of a prony brake.

filled with water to absorb the heat during operation. *See* DYNAMOMETER. [F.H.R.]

Propagator (field theory) The probability amplitude for a particle to move or propagate to some new point of space and time when its amplitude at some point of origination is known. The propagator occurs as an important part of the probability in reactions and interactions in all branches of modern physics. Its properties are best described in the framework of quantum field theory for relativistic particles, where it is written in terms of energy and momentum. Concrete examples for electron-proton and proton-proton scattering are provided in the illustration. The amplitude for these processes contains the

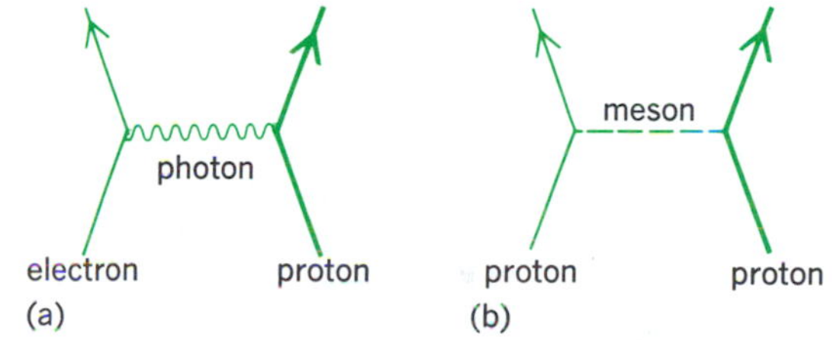

Feynman diagrams for scattering processes. (*a*) Electron-proton scattering via photon exchange. (*b*) Proton-proton scattering via meson exchange.

propagators for the exchanged proton and meson, which actually specify the dominant part of the probability of each process when the scattering occurs at small angles. In similar fashion, for any electromagnetic process, a propagator for each internal line of the Feynman diagram (each line not connected directly to the outside world) enters the probability amplitude. *See* FEYNMAN DIAGRAM; QUANTUM ELECTRODYNAMICS; QUANTUM FIELD THEORY; QUANTUM MECHANICS. [K.E.L.]

Propane A member of the alkane or paraffin series of hydrocarbons, formula $CH_3CH_2CH_3$. It makes up 3–18% of natural gas. It is readily liquefied (melting point, −187.7°C or −305.9°F; boiling point, −42.1°C or −43.8°F), and mixtures with liquefied butane are sold as liquefied petroleum gas (LPG) in cylinders under moderate pressure for domestic fuel. *See* LIQUEFIED PETROLEUM GAS (LPG).

At temperatures above about 650°C (1200°F), propane undergoes cracking to ethylene and methane. This reaction is the basis of an important commercial source of ethylene.

In the petroleum industry, propane is used as a combined solvent and refrigerant for the refining of lubricants and other products. *See* ALKANE; CRACKING; PETROLEUM PROCESSING. [L.S.]

Propanol One of the three-carbon saturated, aliphatic alcohols (alkanols). Normal propanol (also known as propyl alcohol, 1-propanol, or ethyl carbinol), $CH_3CH_2CH_2OH$, has chemical properties similar to other primary aliphatic alcohols. It is a colorless, mobile liquid of pungent odor and taste and has the following physical properties: molecular weight, 60.09; boiling point, 97.2°C (207°F); melting point, −126.2°C (−195.2°F); specific gravity, 0.8036 at 20°C (28°F). The compound is miscible with water and is soluble in most organic liquids. Normal propanol is toxic and flammable.

Isopropanol, $CH_3CHOHCH_3$ (also known as 2-propanol, isopropyl alcohol, and dimethyl carbinol), is the simplest of the secondary alcohols and is a major industrial organic chemical. It is a colorless, mobile, and toxic liquid (molecular weight, 60.09; boiling point, 82.3°C or 180°F; melting point, −88.5°C or −127.3°F; specific gravity, 0.786 at 20°C or 28°F) and has a pungent odor and taste. It is miscible with water and soluble in most organic liquids. Isopropanol is used primarily for the manufacture of acetone by catalytic dehydrogenation. It is also used as a solvent, an extractant, an antifreeze, and as a rubbing alcohol. *See* ALCOHOL. [P.D.S.]

Propellant Usually, a combustible substance that produces heat and supplies ejection particles, as in a rocket engine. A propellant is both a source of energy and a working substance; a fuel is chiefly a source of energy, and a working substance is chiefly a means for expending energy. Because the distinction is more decisive in rocket engines, the term propellant is used primarily to describe chemicals carried by rockets for propulsive purposes. *See* AIRCRAFT FUEL; ROCKET PROPULSION; THERMODYNAMIC CYCLE.

Propellants are classified as liquid or as solid. Even if a propellant is burned as a gas, it may be carried under pressure as a cryogenic liquid to save space. For example, liquid oxygen and liquid hydrogen are important high-energy liquid bipropellants.

Liquid propellants. A liquid propellant releases energy by chemical action to supply motive power for jet propulsion. The three principal types of propellants are monopropellant, bipropellant, and hybrid propellant. Monopropellants are single liquids, either compounds or solutions. Bipropellants consist of fuel and oxidizer carried separately in the vehicle and brought together in the engine. Hybrid propellants use a combination of liquid and solid materials to provide propulsion energy and working substance. Typical liquid propellants are listed in the table. *See* METAL-BASE FUEL.

The availability of large quantities and their high performance led to selection of liquefied gases such as oxygen for early liquid-propellant rocket vehicles. Liquids of higher density with low vapor pressure (see table) are advantageous for the practical requirements of rocket operation under ordinary handling conditions. Such liquids can be retained in rockets for long periods ready for use and are convenient for vehicles that are to be used several times. The high impulse of the cryogenic systems is desirable for rocket flights demanding maximum capabilities, however, such as space exploration or transportation of great weights for long distances. [S.Si.]

Physical properties of liquid propellants

Propellant	Boiling point, °F (°C)	Freezing point, °F (°C)	Density, g/ml	Specific impulse,* s
Monopropellants				
Acetylene	−119 (−84)	−115 (−82)	0.62	265
Hydrazine	236 (113)	35 (2)	1.01	194
Ethylene oxide	52 (11)	−168 (−111)	0.88	192
Hydrogen peroxide	288 (142)	13 (−11)	1.39	170
Bipropellants				
Hydrogen	−423 (−253)	−433 (−259)	0.07	
Hydrogen-fluorine	−306 (−188)	−360 (−218)	1.54	410
Hydrogen-oxygen	−297 (−183)	−362 (−219)	1.14	390
Nitrogen tetroxide	70 (21)	12 (−11)	1.49	—
Nitrogen-tetroxide-hydrazine	236 (113)	35 (2)	1.01	290
Red nitric acid	104 (40)	−80 (−62)	1.58	
Red fuming nitric acid–*uns*-dimethyl hydrazine	146 (63)	−71 (−57)	0.78	275

*Maximum theoretical specific impulse at 1000 psi (6.895 megapascals) chamber pressure expanded to atmospheric pressure.

Solid propellants. A solid propellant is a mixture of oxidizing and reducing materials that can coexist in the solid state at ordinary temperatures. When ignited, a propellant burns and generates hot gas. Although gun powders are sometimes called propellants, the term solid propellant ordinarily refers to materials used to furnish energy for rocket propulsion.

A solid propellant normally contains three essential components: oxidizer, fuel, and additives. Oxidizers commonly used in solid propellants are ammonium and potassium perchlorates, ammonium and potassium nitrates, and various organic nitrates, such as glyceryl trinitrate (nitroglycerin). Common fuels are hydrocarbons or hydrocarbon derivatives, such as synthetic rubbers, synthetic resins, and cellulose or cellulose derivatives. The additives, usually present in small amounts, are chosen from a wide variety of materials and serve a variety of purposes. Catalysts or suppressors are used to increase or decrease the rate of burning; ballistic modifiers may be used for a variety of reasons, as to provide less change in burning rate with pressure (platinizing agent); stabilizers may be used to slow down undesirable changes that may occur in tong-term storage.

Solid propellants are classified as composite or double base. The composite types consist of an oxidizer of inorganic salt in a matrix of organic fuels, such as ammonium perchlorate suspended in a synthetic rubber. The double-base types are usually high-strength, high-modulus gels of cellulose nitrate (guncotton) in glyceryl trinitrate or a similar solvent. [H.W.R.]

Propeller (aircraft) A hub-and-multiblade device for changing rotational power of an aircraft engine into thrust power for the purpose of propelling an aircraft through the air (see illustration). An air propeller operates in a relatively thin medium compared to a marine propeller, and is therefore characterized by a relatively large diameter and a fairly high rotational speed. It is usually mounted directly on the engine drive shaft in front of or behind the engine housing. *See* PROPELLER (MARINE CRAFT).

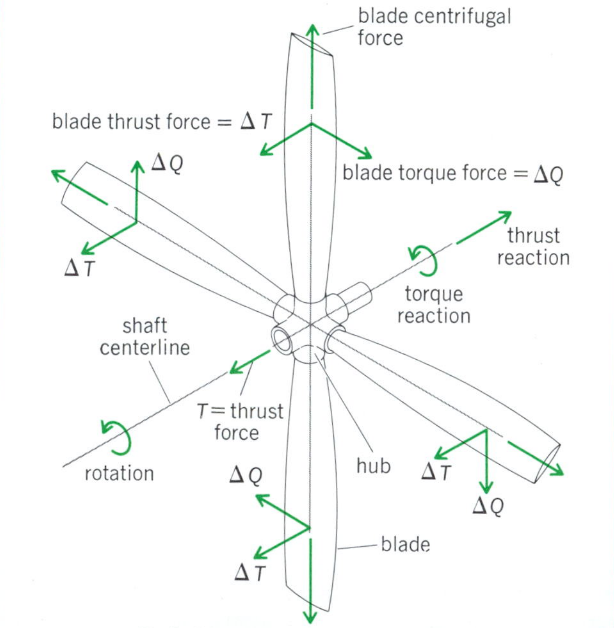

Typical four-bladed propeller system.

Usually propellers have two, three, or four blades; for high-speed or high-powered airplanes, six or more blades are used. In some cases these propellers have an equal number of opposite rotating blades on the same shaft, and are known as dual-rotation propellers.

A propeller blade advances through the air along an approximate helical path which is the result of its forward and rotational velocity components. This action is similar to a screw being turned in a solid surface, except that in the case of the propeller a slippage occurs because air is a fluid. Because of the similarity to the action of a screw, a propeller is also known as an airscrew. To rotate the propeller blade, the engine exerts a torque force. This force is reacted on by the blade in terms of lift and drag force components produced by the blade sections in the opposite direction. As a result of the rational forces reacting on the air, a rotational velocity remains in the propeller wake with the same rotational direction as the propeller. This rotational velocity times the mass of the air is proportional to the power input. The sum of all the lift and drag components of the blade sections in the direction of flight are equal to the thrust produced. These forces react on the air, giving an axial velocity component opposite to the direction of flight. By the momentum theory, this velocity times the mass of the air going through the propeller is equal to the thrust.

A propeller blade must be designed to withstand very high centrifugal forces. The blade also must withstand the thrust force produced plus any vibratory forces generated, such as those due to uneven flow fields. To withstand the high stresses due to rotation, propeller blades have been made from a number of materials, including wood, aluminum, hollow steel, and plastic composites. The most common material used has been solid aluminum. However, the composite blade constructions are being used for new turboprop installations because of their very light weight and high strength characteristics.

For a small, low-power airplane, very simple, fixed-pitch, single-piece, two-blade propellers are used. The rotational speed of these propellers depends directly on the power input and forward speed of the airplane. Because of the fixed-blade angle of this type of propeller, it operates near peak efficiency only at one condition. To overcome the limitations of the simple fixed-pitch propeller, configurations that provide for variable blade angles are used. The blades of these propellers are retained in their hub so that they can be rotated about their centerline while the propeller rotates. For the normal range of operation, the blade angle varies from the low blade angle needed for takeoff to the high blade angle needed for the maximum speed of the airplane. *See* AIRCRAFT PROPULSION; AIRPLANE; HELICOPTER. [H.V.B.]

Propeller (marine craft) A component of a ship-propulsion power plant which converts engine torque into propulsive force or thrust, thus overcoming a ship's resistance to forward motion by creating a sternward accelerated column of water. Since 1860 the screw propeller has been the only propeller type used in ocean transport, mainly because of the evolution of the marine engine toward higher rotative speed.

The advantages of a screw propeller include light weight, flexibility of application, good efficiency at high rotative speed, and relative insensitivity to ship motion. The fundamental theory of screw propellers is applicable to all forms of marine propellers. In its present form a screw propeller consists of a streamlined hub attached outboard to a rotating engine shaft, on which are mounted two to seven blades. The blades are either solid with the hub, detachable, or movable. The screw propeller has the characteristic motion of a screw; it revolves about the axis along which it advances. The screw blades are approximately elliptical in outline.

Stern view of *Great Land* in drydock, showing screw propeller. (*From E. Schorsch, R. T. Bicicchi, and J. W. Fu, Hull experiments on 24-knot RO/RO vessels directed toward fuel-saving application of copper-nickel, Soc. Nav. Archit. Mar. Eng. Trans., 86:254–276, 1978*)

One or more screw propellers are usually fitted as low as possible at the ship's stern to act as thrust-producing devices (see illustration). The low position of the propellers affords good protection and sufficient immersion during the pitching movements of the ship. The choice of the number of propellers to incorporate into a vessel design is based upon several factors. In general, a single-screw arrangement yields a higher propulsive efficiency than multiple screws, particularly when most of the propeller is operating in the boundary layer of the ship and can recover some of the energy loss. In addition, single-screw propulsion systems generally result in savings in machinery cost and weight in comparison to multiple-screw arrangements.

The formation and collapse of vapor-filled bubbles, or cavities, causes noise, vibration, and often rapid erosion of the propeller material, especially in fast, high-powered vessels. This phenomenon is known as cavitation. As long as the rotational and translational speeds of the propeller are not too high, the onset of cavitation can be delayed or limited to an acceptable amount by clever design of blade sections. *See* CAVITATION.

Supercavitating and superventilated propellers are designed to have fully developed blade cavities which spring from the leading edge of the blade, cover the entire back of the blade, and collapse well downstream of the blade trailing edge. The blade of such propellers has unique sections which usually are wedge-shaped with a sharp leading edge, blunt trailing edge, and concave face. Supercavitating propellers have cavities filled with water vapor and small amounts of gases dissolved in the fluid media. Superventilated propellers have cavities filled primarily with air from the water surface or gases other than water vapor from a gas supply system through the propeller shaft.

For ships which normally operate at widely varying speeds and propeller loadings (towboats, rescue vessels, trawlers, and ferryboats), the application of controllable-pitch (rotatable-blade) propellers permits the use of full engine power at rated rpm under all operational conditions, ensuring maximum thrust production, utmost flexibility, and maneuverability. Since these propellers are also reversible, they permit the use of nonreversible machinery (gas turbines). *See* MARINE ENGINE; MARINE MACHINERY. [J.B.H.]

Properdin A serum protein which acts only in conjunction with complement and Mg^{2+} and participates in diverse activities such as the destruction of bacteria, neutralization of viruses, and the lysis of certain red cells.

Properdin activity has been found in the serum of germ-free rats and humans with agammaglobulinemia, and its level varies little with a variety of disease states, although transient changes in the serum activity have been noted in pneumococcal pneumonia, in meningococcemia, and following total body irradiation. *See* SERUM. [D.R.]

Propionibacteriaceae A family of anaerobic bacteria. This family contains relatively anaerobic, nonmotile, nonsporulating, gram-positive, predominantly rod-shaped bacteria that ferment glucose and certain other substrates. Some of the Propionibacteria are involved in the production of Swiss cheese and others in the production of vitamin B_{12}. Two genera are included in the family, *Propionibacterium* and *Eubacterium*. [H.A.B.]

Propositional calculus A branch of automata theory that concerns itself with the manipulation of linguistic statements. Rules of symbolic logic are employed to reduce compound statements into simplified component statements. Of central interest is whether a particular statement is "true" or "false." For example, consider the following statements: "New York is a city" is true; "St. Louis is the mayor of New York" is false; "Cats are mammals" is true. Suppose now that the symbols X, Y, and Z are assigned to these statements, respectively. New statements can be made using the symbols. The new statement "Both X and Z" is true, since the first and last statements are both true. The statement "Both X and Y" is false, since the second statement is false. The statement "Either X or Y" is true, since at least one of them is true.

When the statements considered are limited to those that can unambiguously be said to be true (T) or false (F), or, more succinctly, when the sets A, S, and Z each have two members, $\{T,F\}$, then the language is referred to as switching algebra, switching theory, or boolean algebra. The study of switching functions is of particular importance in computing machines where binary sequences are used to represent numbers. *See* AUTOMATA THEORY; BOOLEAN ALGEBRA. [N.B.R.]

Proprioception The sense of position and movement of the limbs and the sense of muscular tension. The awareness of the orientation of the body in space and the direction, extent, and rate of movement of the limbs depend in part upon information derived from sensory receptors in the joints, tendons, and muscles. Information from these receptors, called proprioceptors, is normally integrated with that arising from vestibular receptors (which signal gravitational acceleration and changes in velocity of movements of the head), as well as from visual, auditory, and tactile receptors. Sensory information from certain proprioceptors, particularly those in muscles and tendons, need not reach consciousness, but can be used by the motor system as feedback to guide postural adjustments and control of well-practiced or semiautomatic movements such as those involved in walking.

Receptors for proprioception are the endings of peripheral nerve fibers within the capsule or ligaments of the joints or within muscle. These endings are associated with specialized end organs such as Pacinian corpuscles, Ruffini's cylinders, and Golgi organs (the latter resembling histologic Golgi structures in the skin), and muscle spindles. *See* CUTANEOUS SENSATION; SENSATION; SOMESTHESIS. [R.LaM.]

Propulsion The process of causing a body to move by exerting a force against it. Propulsion is based on the reaction principle, stated qualitatively in Newton's third law, that for every action there is an equal and opposite reaction. A quantitative description of the propulsive force exerted on a body is given by Newton's second law, which states that the force applied to any body is equal to the rate of change of momentum of that body, and is exerted in the same direction as the momentum change. *See* NEWTON'S LAWS OF MOTION.

In the case of a vehicle moving in a fluid medium, such as an airplane or a ship, the required change in momentum is generally produced by changing the velocity of the fluid (air or water) passing through the propulsive device or engine. In other cases, such as that of a rocket-propelled vehicle, the propulsion system must be capable of operating without the presence of a fluid medium; that is, it must be able to operate in the vacuum of space. The required momentum change is then produced by using up some of the propulsive device's own mass, which is called the propellant. *See* AERODYNAMIC FORCE; AIRFOIL; FLUID DYNAMICS; FLUID FLOW; FLUID MECHANICS; PROPELLANT.

The two terms most generally used to describe propulsion efficiency are thrust specific fuel consumption for engines using the ambient fluid (air or water), and specific impulse for engines which carry all propulsive media on board. *See* SPECIFIC FUEL CONSUMPTION; SPECIFIC IMPULSE.

The energy source for most propulsion devices is the heat generated by the combustion of exothermic chemical mixtures composed of a fuel and an oxidizer. An air-breathing chemical propulsion system generally uses a hydrocarbon such as coal, oil, gasoline, or kerosine as the fuel, and atmospheric air as the oxidizer. A non-air-breathing engine, such as a rocket, almost always utilizes propellants that also provide the energy source by their own combustion.

Where nuclear energy is the source of propulsive power, the heat developed by nuclear fission in a reactor is transferred to a working fluid, which either passes through a turbine to drive the propulsive element such as a propeller, or serves as the propellant itself. Nuclear-powered ships and submarines are accepted forms of transportation. Nuclear-powered airplanes and rockets are still in the developmental stages. *See* AIRPLANE; ELECTROMAGNETIC PROPULSION; JET PROPULSION; SHIP POWERING AND STEERING; TURBINE PROPULSION. [J.Gr.]

Propylene A gas, $CH_3{-}HC{=}CH_2$, with boiling point $-48°C$ ($-54°F$) and melting point $-185°C$ ($-301°F$). All processes for thermal or catalytic cracking of hydrocarbons yield propylene. Preparation outside of normal refinery operations is usually by catalytic dehydrogenation of propane.

Major uses for propylene include the production of isopropyl alcohol, from which acetone is obtained; polypropylene plastics; tripropylene for the alkylation of phenol to produce alkyl phenols, from which are derived nonionic detergents and lubricating oil additives; and propylene chlorohydrin for producing propylene oxide from which are obtained detergents, hydraulic fluids, and lubricants. *See* ALKENE; ETHYLENE; POLYOLEFIN RESINS. [C.A.Co.]

Prosobranchia The largest subclass in the class Gastropoda, containing most marine snails, the limpets and whelks, a limited number of nonpulmonate land snails, and nearly all of the nonpulmonate fresh-water snails. Three orders are commonly recognized, the Archaeogastropoda (Aspidobranchia), Mesogastropoda (Pectinibranchia), and Neogastropoda.

Respiration is usually by means of ctenidia, or gills, located in front of the heart. When gills are absent, respiration is carried on by means of a pallial outgrowth or a pulmonary chamber. An operculum is usually present, there is generally only one pair of tentacles, and the sexes are usually separate. The shells produced have evolved many shapes. Certain of the primitive forms have nacreous shells, such as *Haliotis*, the sea ear or abalone, and *Trochus*, the top shell. The shells are used for jewelry, inlay work, and buttons. *See* ARCHAEOGASTROPODA; GASTROPODA; MESOGASTROPODA; NEOGASTROPODA. [W.J.C.]

Prospecting The search for mineral deposits that can be worked at a profit. A prospect is an occurrence of minerals of potential value before that value has been determined by exploration and development. Mineral deposits include those containing metallic elements, such as gold, copper, lead, zinc, or iron; nonmetallic materials, such as asbestos, clay, phosphates, potash, or sulfur; and mineral fuels, such as coal or petroleum. Deposits worked for their aggregate of materials, such as sand, gravel, or dimension stone, are usually considered deposits of the rock itself.

Prospecting is generally done by groups of specialists. Such a group may consist of a geologist, a geophysicist, a mining engineer, and assistants as required for traversing and sampling. Specialists in photogeology and in geophysical and geochemical prospecting may be called in as needed.

Mineral prospecting normally proceeds from the general to the specific, from consideration of large regions to smaller favorable areas within the region, and finally to individual prospects. Following a preliminary investigation, including a library search and a study of available maps and reports, prospecting may often be concentrated on smaller areas immediately. Study of the geology will indicate what minerals might be found in a given area. Prospecting methods may be subdivided into direct and indirect methods.

Direct methods. A map of some sort is required to prospect large areas systematically. Published geologic and topographical maps are suitable bases for plotting mineral occurrence or guides to minerals. Aerial photographs are excellent for such purposes, though ground control is necessary to produce maps from them. Photogeologic studies combined with ground checking may indicate such guides as soil staining, alteration, or structures in deeply weathered areas that might otherwise be missed.

Ore guides are mostly associations of geologic and other regional factors. The ore guides that are significant depend on the characteristic rock associations and distribution of the ore in any given region or locality. For example, mineral stream gravel or placer deposits generally indicate veins or lodes in the country rock of the backland. Local ore guides include rock alteration and channel ways such as fractures, faults, contacts between dissimilar rocks, and breccia pipes. Ore shoots are likely to occur where mineralizing solutions encounter easily replaced rock such as limestone, where veins cross contacts, at intersections of veins and faults, at rolls in faults, and along veins wherever there is a change in physical or chemical environment.

Any departure from normal structure, topography, rock color, or vegetation may prove significant. Old mine workings, dumps, prospect pits, and burrows of animals may yield information on mineralization and on the size and trend of deposits. *See* ORE AND MINERAL DEPOSITS.

At the point where the mineral indications go beneath the surface, trenches or test pits must be dug or bulldozed. Ground sluicing or hydraulicking may be used to advantage where water is abundant, slopes are moderately steep, and stream pollution is unobjectionable. Drilling is sometimes necessary in prospecting areas covered by swamp, water, or rock or where deep overburden makes trenching or pitting unduly expensive.

Prospects are sampled to determine their composition. Methods for representative sampling depend on the size, character, and accessibility of the project.

Indirect methods. In a large proportion of indirect prospecting geophysical methods are used, involving the measurement of physical quantities associated with buried mineral deposits and geological structures. The measurements are interpreted in terms of geology. Plausible assumptions are made about the subsurface, and the physical effects of assumed structures or deposits are computed or estimated and compared with the geophysical measurements. The assumptions are modified until there is reasonable agreement between the computations and observed data.

Selection of a suitable geophysical method is facilitated by information on the geologic habits and environment of the deposits sought and on their physical properties compared with the surrounding rocks. Magnetic methods and the various electrical methods are commonly used in prospecting for metallic minerals, because these minerals usually have magnetic and electrical properties that contrast with those of the surrounding rocks. Seismic and gravitational methods have been less used because measurable contrasts in elasticity and density are less common, but they have been indirectly useful in locating buried structures. Little information on depth is theoretically obtainable with magnetic, gravitational, self-potential, and thermal methods, where physical effects are produced by the bodies or structures themselves. Information on depth can usually be obtained with resistivity or seismic methods, where effects are produced by transmitting electrical or seismic energy through the ground. The use of airborne magnetic, electromagnetic, spectral, and radioactivity methods makes it possible to cover large and otherwise inaccessible areas at comparatively low cost. All three types of surveys may be made simultaneously from the same aircraft, at little more than the cost of a survey by a single method. *See* GEOPHYSICAL EXPLORATION; MAGNETOMETER; REMOTE SENSING.

Geochemical and botanical methods. Geochemical prospecting involves the analysis of elements in the soils, rocks, surface and underground waters, organisms, and vegetation for the purpose of defining areas where the anomalous distribution of the elements indicates the presence of ore. Botanical prospecting involves analysis of elements found in deep-rooted plants or plants commonly associated with certain elements or mineral deposits and comparative study of morphological differences in plants growing in mineralized and nonmineralized areas. *See* GEOBOTANICAL INDICATORS; GEOCHEMICAL PROSPECTING. [A.F.B.]

Prostate gland A triangular body in men, the size and shape of a chestnut, that lies immediately in front of the bladder with its apex directed down and forward. It is found only in the male, having no female counterpart. The prostatic portion of the urethra extends through it, passing from the bladder to the penis. This organ contains 15–20 branched, tubular glands which form lobules. The gland ducts open into the urethra. Between the gland clusters, or alveoli, there is a dense, fibrous, connecting tissue, the stroma, which also forms a tough capsule around the gland, continuous with the bladder wall. Penetrating the prostate to empty into the urethra are the ejaculatory ducts from the seminal vesicles which are located above and behind the organ (see illustration). The prostatic gland

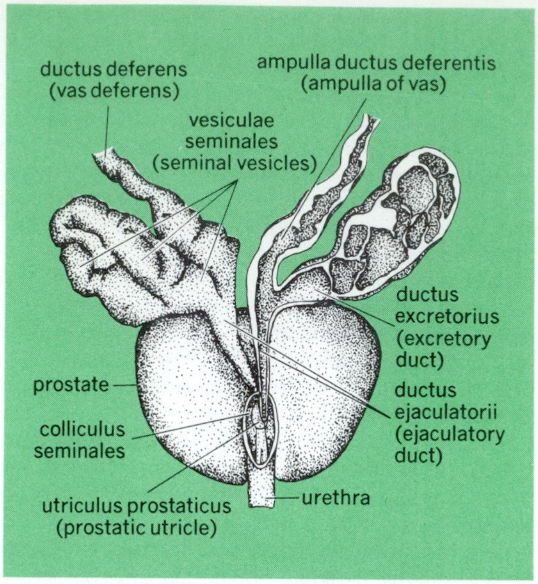

Prostate gland and seminal vesicles. (*After Eycleshymer and Jones, in W. A. N. Darland, ed., American Illustrated Medical Dictionary, 19th ed., Saunders, 1942*)

secretes a viscid, alkaline fluid which aids in sperm motility and in neutralizing the acidity of the vagina, thus enhancing fertilization.

After middle age, the prostate is sometimes subject to new tissue growth, usually benign, that may result in interference with urine flow through the compressed urethra. [W.J.B.]

Prosthesis A device designed and constructed to fill space on or within the living body, to restore function on or within the body, to achieve a cosmetic reconstruction of a superficial feature of the body, or combinations of these. Because of trauma or disease, it is frequently necessary or appropriate treatment to remove surgically portions of the body, such as a whole limb, a segment of long bone, interfaces of a joint, blood vessels, or a valve within the heart. In many cases such body features and components can be replaced by components or devices which restore substantially appearance and function. The variety of prostheses which are designed and successfully implanted increases each year. Prostheses of proved efficacy are manufactured in standard ranges of dimensions and also to custom dimensions. [W.R.]

Protactinium A chemical element, symbol Pa, atomic number 91. Isotopes of mass numbers 216, 217, and 222–238 are known, all of them radioactive. Only ^{231}Pa (the parent of actinium), ^{234}Pa, and ^{234m}Pa occur in nature. The most important of these is ^{231}Pa, an α-emitter with a half-life of 32,500 years. The artificial isotope, ^{233}Pa, is important as an intermediary in the production of fissile ^{233}U. Both ^{231}Pa and ^{233}Pa can be synthesized by neutron irradiation of thorium. *See* ACTINIUM; RADIOACTIVITY; URANIUM.

Protactinium is, formally, the third member of the actinide series of elements and the first in which a 5*f* electron appears, but its chemical behavior in aqueous solution resembles that of tantalum and niobium more closely than that of the other actinides. *See* NIOBIUM; TANTALUM.

Metallic protactinium is silver in color, malleable, and ductile. Samples exposed to air at room temperature show little or no tarnishing over a period of several months. The numerous compounds of protactinium that have been prepared and char-

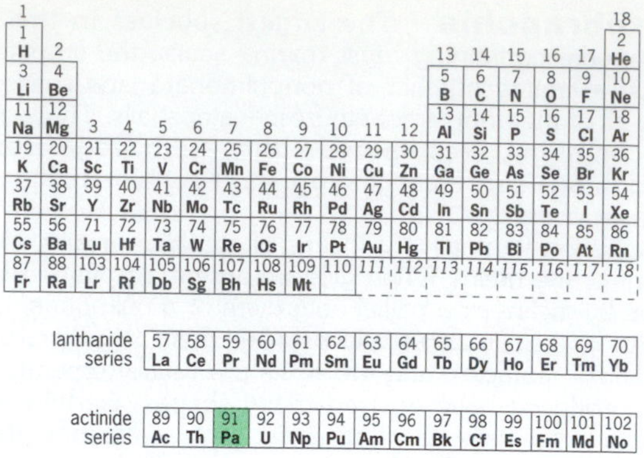

acterized include binary and polynary oxides, halides, oxyhalides, sulfates, oxysulfates, double sulfates, oxynitrates, selenates, carbides, organometallic compounds, and noble metal alloys. *See* ACTINIDE ELEMENTS. [H.W.Ki.]

Protandry That condition in which an animal is first a male and then becomes a female. It occurs in many groups, including oysters and cyclostomes. The reverse condition is protogyny. *See* PROTOGYNY. [T.I.S.]

Proteales An order of flowering plants, division Magnoliophyta (Angiospermae), in the subclass Rosidae of the class Magnoliopsida (dicotyledons). The order consists of only two families: the Proteaceae, with about 1400 species, and the Elaeagnaceae, with about 50. Within its subclass the order is marked by its flowers, which are strongly perigynous; that is, the perianth and stamens are united into a basal cup that is distinct from the ovary. The calyx is four-lobed and often corolla-like, and the true petals are much reduced or absent. *See* MAGNOLIOPSIDA; ROSIDAE. [A.Cr.]

Protective coloration When the color pattern of an animal increases the probability of its survival, the animal is said to be protectively colored. There are three types of protective coloration: cryptic or concealing coloration, which renders the animal inconspicuous; aposematic or warning coloration, which advertises the presence of an otherwise well-protected animal; and mimicry, whereby a species imitates an aposematic species. Batesian mimics are edible species which gain protection by mimicking genuine aposematics, whereas Müllerian mimicry involves the sharing of a pattern by two aposematic species, a type of double protection.

Cryptic coloration is widespread. The most common form is countershading, that is, dark pigmentation on the dorsal surface, light on the ventral surface. An example will explain its utility. When a small fish swims near the surface, a gull sees the dark dorsal surface blending into the dark waters, whereas a larger fish from below sees the light belly blending into the bright surface water. Background matching, however, may be much more precise, and correspondingly more difficult to interpret except as adaptation. It may involve morphology and behavior as well as color. Thus the walking stick insects of the family Phasmidae closely resemble twigs of the plants upon which they live, even to the angle at which they rest upon a branch. Numerous insects, when at rest, resemble the leaves of the plants upon which they feed. *Kallima*, the dead-leaf butterfly, exemplifies this (see illustration).

Aposematic coloration is displayed by animals that are distasteful to predators, such as some butterflies; those animals equipped with poison glands, such as many salamanders; and

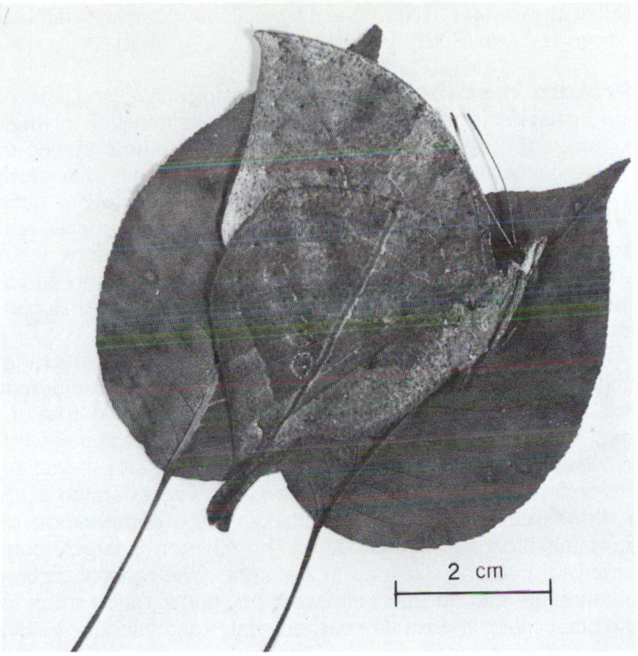

Kallima, known as the dead-leaf butterfly, photographed with two leaves from its habitat. (_General Biological Supply House, Inc._)

those animals, such as bees, which have stings or other formidable defenses. These may be in sharp contrast to patterns of related animals which are more subject to predation. *See* CHROMATOPHORE. [E.O.D.]

Protein A polymeric compound made up of various amino acids as the monomeric units. Proteins generally contain from 50 to 1000 amino acid residues per polypeptide chain. The amino acids are joined by peptide bonds between the α-carboxyl groups and the α-amino groups of adjacent amino acids in which the α-amino group of the first amino acid residue is usually free, as shown below. *See* AMINO ACIDS; PEPTIDE.

$$H_2N-\overset{\displaystyle R}{\overset{|}{CH}}-CO-NH-\overset{\displaystyle R'}{\overset{|}{CH}}-CO-NH-\overset{\displaystyle R''}{\overset{|}{CH}}-CO$$

Occurrence. Proteins are of importance in all biological systems, playing a wide variety of structural and functional roles. They form the primary organic basis of structures such as hair, tendons, muscle, skin, and cartilage. All of the enzymes, the catalysts in biochemical transformations, are protein in nature. Many hormones, such as insulin and growth hormone, are proteins. The substances responsible for oxygen and electron transport (hemoglobin and the cytochromes, respectively) are conjugated proteins that contain a metalloporphyrin as the prosthetic group. The chromosomes are highly complex nucleoproteins, that is, proteins conjugated with nucleic acid. The viruses are also nucleoprotein in nature. Thus, proteins play a fundamental role in the processes of life. *See* NUCLEOPROTEIN.

Specificity. The linear arrangement of the amino acid residues in a protein is termed its sequence (primary structure). The sequence in which the different amino acids are linked in any given protein is highly specific and characteristic for that particular protein.

This specificity of sequence is one of the most remarkable aspects of protein chemistry. The number of possible permutations of sequence in even so small a protein as insulin, of molecular weight 5732 and with 51 amino acid residues, is astronomic: 10^{51} permutations. Yet it has been established that the pancreatic cell of a given species has only one of these possible sequences. The elucidation of the mechanism conferring such a high degree of specificity on the biosynthetic reactions by which proteins are built up from free amino acids has been one of the key problems of modern biochemistry. *See* MOLECULAR BIOLOGY.

Structure. Finding the structure of proteins required the development of many new techniques in protein chemistry. The strategy in the sequencing of a protein has depended upon coupling of the free α-amino residue of the polypeptide with a chemical, particularly phenyl isothiocyanate, that will remove that residue under appropriate conditions and leave the remainder of the polypeptide chain intact. At this point in the process it is possible to identify the cleaved derivative of the first amino acid residue or the remaining amino acids by a variety of techniques such as chromatography. The stepwise process is then repeated in this manner as far as possible.

From the proteins that have been sequenced thus far, it is clear that proteins of similar function from different species and those with common genetic origins have similar sequences. When there are differences in the sequence, the changes are usually conservative, for example, the replacement of one aliphatic amino acid by a different aliphatic amino acid.

Proteins are not stretched polymers; rather, the polypeptide backbone of the molecule can fold in several ways by means of hydrogen bonds between the carbonyl oxygen and the amide nitrogen.

The long polypeptide chains of proteins, particularly those of the fibrous proteins, are held together in a rather well-defined configuration. The backbone is coiled in a regular fash-

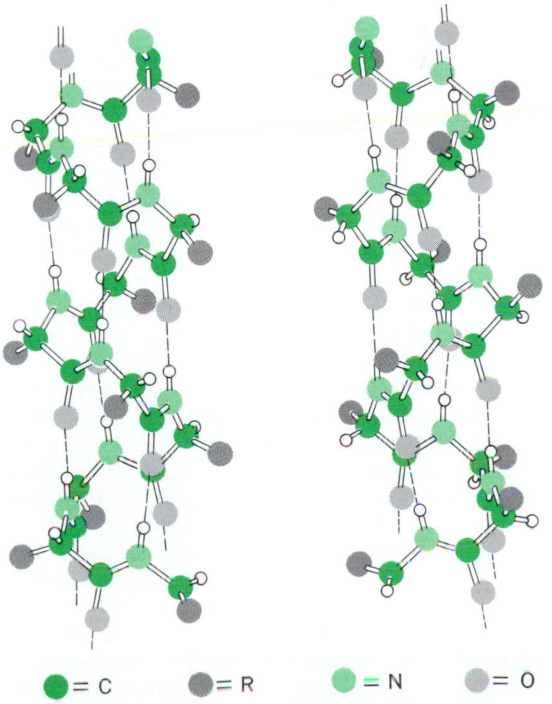

The α-helix proposed by L. Pauling and R. B. Corey. The repeating —N—C—C— units form the backbone which spirals up in a left-handed or a right-handed fashion. Hydrogen bonds are indicated by the broken lines. Note that the side chains (R) are all directed out from the helix. The pitch of the helix is about 0.54 nanometer, and there are 3.7 residues contained in one complete turn. (_After J. T. Edsall and J. Wyman, Biophysical Chemistry, vol. 1, Academic Press, 1958_)

ion, forming an extended helix. As a result of this coiling, peptide bonds separated from one another by several amino acid residues are brought into close spatial approximation. The stability of the helical configuration can be attributed to hydrogen bonds between these peptide bonds. The particular helical configuration which best fits the available x-ray data is that proposed by L. Pauling and R. B. Corey (see illustration). In this structure, each —NH group in peptide bond is hydrogen-bonded in the

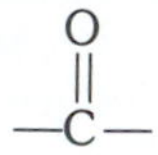

group three residues removed from it. *See* FIBROUS PROTEIN.

In addition to the α-helical sections of proteins, there are also segments that contain β-structures in which there are hydrogen bonds between two polypeptide chains that run in parallel or antiparallel fashion.

The third level of folding in a protein (that is, tertiary structure) comes about through various interactions between different parts of the molecule. The disulfide bridge formed between two cysteine residues at different linear locations in the molecule can stabilize parts of a three-dimensional structure by introducing a primary valence bond as a cross-link. Hydrogen bonds between different segments of the protein, hydrophobic bonds between nonpolar side chains of amino acids such as phenylalanine and leucine, and salt bridges such as those between positively charged lysyl side chains and negatively charged aspartyl side chains all contribute to the individual tertiary structure of a protein.

Finally, for those proteins that contain more than one polypeptide chain per molecule, there is usually a high degree of interaction between each subunit, for example, between the α- and β-polypeptide chains of hemoglobin. This feature of the protein structure is termed its quarternary structure.

Properties. The properties of proteins are determined in part by their amino acid composition. For example, the net charge on the macromolecule at any given hydrogen-ion concentration is largely a function of the relative number of basic (lysine, histidine, and arginine) and dicarboxylic amino acids (aspartic and glutamic acids). This net charge strongly influences the solubility of the protein at different pH values since the solubility depends in part on the proportion of polar groupings on the macromolecule. When the hydrogen-ion concentration is high (low pH), the net charge is positive; when the hydrogen-ion concentration is low (high pH), the net charge is negative. The pH at which the net charge of the protein is zero is defined as the isoelectric point (pI).

As macromolecules that contain many side chains that can be protonated and unprotonated depending upon the pH of the medium, proteins are excellent buffers. The fact that the pH of blood varies only very slightly in spite of the numerous metabolic processes in which it participates is due to the very large buffering capacity of the blood proteins.

As mentioned above, it is the interaction of the side chains of amino acids that provides some proteins (enzymes) with their ability to catalyze essential reactions in body metabolism. Likewise, the specific binding of small ions to the side chains of proteins permits the transport of essential metal ions into and out of cells, as in the action of the sodium pump within the membrane of some cells.

Biosynthesis. Although a few proteins such as collagen are stable indefinitely in adulthood, most body proteins are in a continual process of degradation and synthesis (turnover). Thus, the half-life of serum proteins in humans is about 10 days. The sequence of amino acid residues in a protein is controlled by the sequence of the DNA as expressed in a molecule called messenger RNA. *See* DEOXYRIBONUCLEIC ACID (DNA); RIBONUCLEIC ACID (RNA); RIBOSOMES. [G.E.Pe.; J.M.M.]

Protein metabolism The transformation and fate of food proteins from their ingestion to the elimination of their excretion products. Proteins are of exceptional importance to organisms because they are the chief constituents, aside from water, of all the soft tissue of the body. Special proteins have unique roles as structural and functional elements of cells and tissues. Examples are keratin of skin, collagen of tendons, actin and myosin of muscle, the blood proteins, enzymes in all tissues, and protein hormones of the hypophysis. *See* BLOOD; ENZYME; HORMONE; MUSCLE; PROTEIN.

Protein is digested to amino acids in the gastrointestinal tract. These are absorbed and distributed among the different tissues, where they form a series of amino acid pools that are kept equilibrated with each other through the medium of the circulating blood. The needs for protein synthesis of the different organs are supplied from these pools. Excess amino acids in the tissue pools lose their nitrogen by a combination of transamination and deamination. The nitrogen is largely converted to urea and excreted in the urine. The residual carbon products are then further metabolized by pathways common to the other major foodstuffs—carbohydrates and fats. *See* AMINO ACIDS; CARBOHYDRATE; LIPID.

The recommended daily protein intake is 0.016 oz/lb (1 g/kg) body weight for adults, for example, 2.45 oz (70 g) for a 165-lb (74-kg) man. This is increased to 0.056 oz/lb (3.5 g/kg) for infants up to 1 year, and from 0.024 to 0.032 oz/lb (1.5 to 2 g/kg) for growing children and for women during the latter half of pregnancy and during nursing.

Nutritive value. All proteins are not equally nutritious. Animal proteins are generally superior to vegetable proteins. Rarely, this may result from resistance to digestion, which usually is counteracted by cooking or heating. A well-known example is soybean meal. The nutritive value of its protein is improved by heating because this destroys a substance in the meal which inhibits digestion by trypsin. Overheating lowers the nutritional value of proteins by making lysine unavailable. This is a problem of some concern in connection with the manufacture of prepared breakfast cereals. The major cause of poor nutritional value, however, is a low content or unavailability of one or more of the indispensable amino acids. Vegetable proteins tend to be lacking in lysine and tryptophan.

Digestion. This occurs to a limited extent in the stomach and is completed in the duodenum of the small intestine. The main proteolytic enzyme of the stomach is pepsin, which is secreted in an inactive form, pepsinogen. Its transformation to the active pepsin is initiated by the acidity of the gastric juice and accelerated and completed by pepsin. Pepsin preferentially hydrolyzes peptide bonds containing an aromatic amino acid, and it requires an acid medium to function. *See* PEPSIN.

A second proteinase in the stomach, rennin, present only in infancy, is particularly adapted to the digestion of milk protein. Digestion is initiated by the well-known milk-clotting reaction used in cheese manufacture. Rennin requires less acid than pepsin to be active. In infancy, hydrochloric acid secretion by the stomach is not fully developed. *See* RENNIN.

The acid chyme is discharged from the stomach, containing partially degraded proteins, into a slightly alkaline fluid in the small intestine. This fluid is composed of pancreatic juice and succus entericus, the intestinal secretion. The pancreas secretes three known proteinases, trypsin, chymotrypsin, and carboxypeptidase. Trypsin and chymotrypsin are endopeptidases; that is, they cleave internal peptide bonds. The so-called peptidases are exopeptidases. They cleave terminal peptide bonds. Trypsin has a predilection for those containing the basic amino acid residues of lysine and arginine. These two proteinases perform the major share in hydrolyzing proteins to small peptides. Digestion to amino acids is completed by the exopeptidases. Carboxypeptidase acts on peptides from the free carboxyl end; aminopeptidases from the free amino end. Other peptidases act on di- or tripeptides, or peptides containing such special amino acids as proline.

The absorbed amino acids are carried by the portal blood system to the liver. From there, they are distributed to the rest of the body. [D.M.G.]

Proteocephaloidea An order of tapeworms of the subclass Cestoda. With one exception, these worms are intestinal parasites of fresh-water fishes, amphibians, and reptiles. The holdfast organ bears four suckers and, frequently, an apical organ which may be suckerlike. The segmental anatomy is very similar to that of the Tetraphyllidea. Most authorities recognize two families, the Proteocephalidae and the Monticellidae. *See* CESTODA. [C.P.R.]

Proteomyxidia A subclass of Actinopodea including protozoan organisms without protective coverings (tests) or skeletal elements. Pseudopodia are fine, branched, and interconnected (reticulopodia), or filamentous (filopodia). Many are invaders of algae and other plants. There is one order, Proteomyxida, which contains two families, Pseudosporidae and Vampyrellidae. *See* ACTINOPODEA. [R.P.H.]

Proterozoic A term, more or less synonymous with Algonkian, widely used for later Precambrian time in the twofold subdivision of the Precambrian. Many students of the Precambrian find this division unsuitable and no longer use it. The Geological Society of Canada has divided it into Lower, Middle, and Upper, and has endeavored to fit later Precambrian rocks and history into this framework. Many more recent investigators have used the term Proterozoic and this threefold subdivision for the youngest three divisions in the fourfold subdivision of the Precambrian. *See* PRECAMBRIAN. [J.P.F.]

Protista The kingdom comprising all single-celled forms of living organisms in both the five-kingdom and six-kingdom systems of classification. Kingdom Protista encompasses both Protozoa and Protophyta, allowing considerable integration in the classification of both these animallike and plantlike organisms, all of whose living functions as individuals are carried out within a single cell membrane. Among the kingdoms of cellular organisms, this definition can be used to distinguish the Protista from the Metazoa (sometimes named Animalia) for many-celled animals, or from the Fungi and from the Metaphyta (or Plantae) for many-celled green plants. *See* METAZOA.

The most significant biological distinction is that which separates the bacteria and certain other simply organized organisms, including blue-green algae (collectively, often designated Kingdom Monera), from both Protista and all many-celled organisms. The bacteria are described as prokaryotic; both the Protista and the cells of higher plants and animals are eukaryotic. Structurally, a distinguishing feature is the presence of a membrane, closely similar to the bounding cell membrane, surrounding the nuclear material in eukaryotic cells, but not in prokaryotic ones. *See* EUKARYOTAE; PROKARYOTAE; PROTOZOA.

The definition that can separate the Protista from many-celled animals is that the protistan body never has any specialized parts of the cytoplasm under the sole control of a nucleus. In some protozoa, there can be two, a few, or even many nuclei, rather than one, but no single nucleus ever has separate control over any part of the protistan cytoplasm which is specialized for a particular function. In contrast, in metazoans there are always many cases of nuclei, each in control of cells of specialized function.

Most authorities would agree that the higher plants, the Metazoa, and the Parazoa (or sponges) almost certainly evolved (each independently) from certain flagellate stocks of protistans. [W.D.R.-H.]

Protobranchia A subclass of bivalve mollusks characterized by a foot with a sole that is divided sagittally and longitudinally and has papillate margins. The Protobranchia are divided into two orders, Solemyoida and Nuculoida. The Solemyoida have an edentulous shell, and if teeth are present, they are not chevron-shaped. The palps are small, triangular, and without palp proboscides. This is in contrast to the Nuculoida, which have well-developed, chevron-shaped hinge teeth and large palps with palp proboscides. The Nuculoida are subdivided into two superfamilies, Nuculacea and Nuculanacea.

The Protobranchia are found throughout the seas of the world, and they are particularly common in the deep sea, where they may form up to 10% of the invertebrate infauna. The subclass has one of the longest geological records within the animal kingdom, dating from the Early Ordovician if not the Late Cambrian. *See* BIVALVIA; MOLLUSCA. [J.A.A.]

Protococcida An order of the protozoan subclass Coccidia in the class Telosporea, subphylum Sporozoa. This is a very small group containing perhaps six species in perhaps three genera. All are parasites of marine invertebrates. Only sexual reproduction is known. *See* COCCIDIA. [N.D.L.]

Protogyny A condition in hermaphroditic or dioecious animals in which the female reproductive structures mature before the male structures. It is of rare occurrence. Botanically, protogyny occurs in some plant species in which the stigma develops, withers, and dies before the anthers mature. *See* PROTANDRY. [T.I.S.]

Protolepidodendrales An extinct order of the class Lycopodiopsida (clubmosses) of the Devonian Period. Most of these clubmosses are poorly preserved and not well understood; the notable exception is the widespread *Leclercqia complexa*, which has been largely reconstructed from fragmentary fossils.

Leclercqia (Early to Middle Devonian) was a medium-sized herb; its horizontal rhizomes generated vertical axes up to 0.4 in. (1 cm) in diameter. The vascular tissue formed a star-shaped actinostele, with inward maturation of the metaxylem tracheids. Additional structural support was provided in the cortical cylinder by thickened hypodermal fibers. The microphyllous leaves were not routinely shed. Most branched into five unequal, needlelike lobes, each supplied by a lateral branch of the median vascular trace, and arranged three-dimensionally. Zones of sterile leaves alternated with zones of fertile sporophylls; both sterile and fertile leaves bore a small ligule on the upper surface. The stomata-bearing sporangia were located on the same surface of the sporophyll, somewhat closer to the stem than the ligule. The sporangia were elliptical and released spores of the same size; thus, *Leclercqia* is primitively homosporous.

Complex leaf morphologies are important characteristics in assigning genera to the Protolepidodendrales, even though it may be that they represent multiple independent evolutionary events. The presence of a ligule has been demonstrated only in *Leclercqia*; yet it is the combination of homosporous reproduction and the presence of a ligule that delimits the order and testifies to its evolutionary intermediacy between the eligulate homosporous Lycopodiales and the ligulate heterosporous Selaginellales. *See* LYCOPODIALES; LYCOPODIOPHYTA; LYCOPODIOPSIDA; SELAGINELLALES. [R.M.Ba.; W.A.DiM.]

Proton A particle that is the positively charged constituent of ordinary matter. Together with the neutron, the proton is the building stone of all atomic nuclei; a single proton constitutes the nucleus of the hydrogen atom. The most important properties that characterize the proton are its charge, which is identical in magnitude but of opposite sign to that of the electron (the negatively charged constituent of ordinary matter) and has the value of 4.8033×10^{-10} esu = 1.6022×10^{-19} coulomb; its mass, 1.6726×10^{-24} g = $1836.1m_e$ (m_e is the mass of the electron); its spin, $\frac{1}{2}\hbar = (\frac{1}{2})\ 1.0546 \times 10^{-27}$ erg-sec ($\hbar$ is Planck's constant h divided by 2π); its magnetic moment, 1.4106×10^{-23} erg/gauss; its lifetime, which, according to all available evidence, is infinite; and the fact that it obeys the Pauli exclusion principle, that is, it is a fermion (obeys Fermi-Dirac statistics). *See* ELEMENTARY PARTICLE; NUCLEON. [E.G.S.]

Proton-induced x-ray emission (PIXE) A highly sensitive analytic technique for determining the composition of elements in small samples. Proton-induced x-ray emission is a nondestructive method capable of analyzing many elements simultaneously at concentrations of parts per million in samples as small as nanograms. PIXE has gained acceptance in many disciplines; for example, it is the preferred technique for surveying the environment for trace quantities of such toxic elements as lead and arsenic. There has also been a rapid development in the use of focused proton beams for PIXE studies in order to produce two-dimensional maps of the elements at spatial resolutions of micrometers.

The typical PIXE apparatus uses a small Van de Graaff machine to accelerate the protons which are then guided to the sample. Nominal proton energies are between 1 and 4 MeV; too low an energy gives too little signal while too high an energy produces too high a background. The energetic protons ionize some of the atoms in the sample, and the subsequent filling of empty inner orbits results in the characteristic x-rays. These monoenergetic x-rays emitted by the sample are then efficiently counted in a high-resolution silicon (lithium) detector which is sensitive to the x-rays of all elements heavier than about sodium. *See* ACTIVATION ANALYSIS; JUNCTION DETECTOR; TRACE ANALYSIS; VAN DE GRAAFF GENERATOR; X-RAYS. [L.Gro.]

Proton-proton chain A group of nuclear reactions involving fusion of light nuclei that converts hydrogen into helium. It is believed to be the principal source of energy in main sequence stars of a little more than a solar mass and of less-massive stars. Completion of a chain results in the consumption of four protons, the synthesis of a helium nucleus and two neutrinos, and the release of 26.73 MeV of energy. The energy E reflects the mass difference between the four protons and the helium nucleus, and is calculated on the basis of the Einstein mass-energy equation $E = mc^2$, where m is the mass difference and c^2 is the square of the velocity of light. Because hydrogen is the fuel consumed in the process, it is referred to as hydrogen burning by means of the proton-proton chain. *See* NUCLEAR FUSION; NUCLEAR REACTION; STELLAR EVOLUTION. [G.R.C.]

Protorosauria An order of primitive diapsid reptiles in the infraclass Archosauria, known from the upper Permian to the end of the Triassic Series. Protorosaurs (also known as prolacertiforms) are the earliest and most primitive known reptiles that show modifications of the rear limb for a more upright posture and a fore-and-aft movement that culminated in bipedal dinosaurs and birds. Protorosaurs are characterized by elongate neck vertebrae, reduction of the lower temporal bar, and emargination of the quadrate to support a tympanum.

Protorosaurus, from the upper Permian of Europe, was up to 6 ft (2 m) in length. Both the neck and the limbs were long. *Protorosaurus* may have resembled a large varanid lizard in life, although the ankle bones were specialized in a much different way than those of lizards. The most distinctive protorosaur was *Tanystropheus* from the Middle Triassic Series of Europe. It reached 9 ft (3 m) in length, most of which consisted of a bizarrely elongated neck, at least three times the length of the trunk region. *Tanystropheus* was apparently an aquatic animal, as was the related North American genus *Tanytrachelus*. *See* ARCHOSAURIA; DIAPSIDA; REPTILIA. [R.L.C.]

Protostar A dense condensation of material that is still in the process of accreting matter to form a star. Protostars are expected in dense interstellar complexes of gas and dust. Since the gas is mostly in the form of molecules, especially molecular hydrogen (H_2), these complexes are called molecular clouds. *See* INTERSTELLAR MATTER; MOLECULAR CLOUD.

Molecular clouds contain clumps of material with relatively low temperatures, typically 10 to 50 K (−442 to −370°F), and densities significantly higher than in the surrounding medium, between 10,000 and 100,000 atoms per cubic centimeter (illus. *a*). Such condensations will collapse under the effect

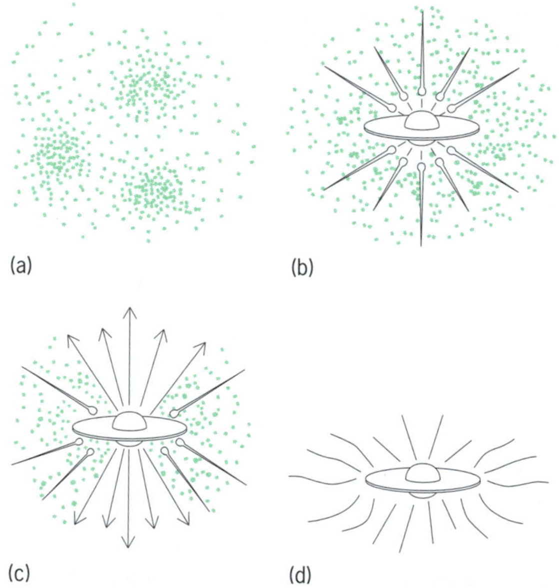

Stages of protostellar evolution. (*a*) Dense cores occur in a large molecular cloud. (*b*) A protostar surrounded by a disk has formed in a collapsing core. (*c*) A well-collimated wind has broken through at the opposite poles of the system. (*d*) A visible star with a circumstellar disk is revealed. (*After F. H. Shu, F. C. Adams, and S. Lizano, Star formation in molecular clouds: Observations and theory, Annu. Rev. Astron. Astrophys., 25:23–84, 1987*)

of gravity if they are more massive than a certain value, now called the Jeans mass. The protostar is the central core that forms as a cloud condensation contracts.

Because the collapsing core is also rotating, its outer parts are flattened into a disk (illus. *b*). At this stage, the protostar's mass is very small, perhaps only 1/1000 of the mass of the present Sun. Over a period of 10^4–10^6 years it continues to accrete material from the infalling envelope and eventually attains a stellarlike mass. Core collapse is checked, leaving a central protostar, when the gravitational heating produced by the contraction can no longer be radiated away as quickly as it is generated. Eventually, the core becomes sufficiently hot that the nuclear reactions that sustain stars can begin.

Much of this description of protostellar evolution is based on theoretical considerations. Although the velocity patterns of the gas around many objects have been measured, infalling material has not been detected. By contrast, in many cases (illus. *c*), gas is seen emanating from an embedded source in two well-collimated and oppositely directed outflows. The objects at the cores of these bipolar outflows are loosely referred to as protostars. *See* STELLAR EVOLUTION; T TAURI STAR. [A.I.S.]

Prototheria One of the four subclasses of the class Mammalia. Prototheria contains a single order, the Monotremata. No ancestral genera of fossil monotremes are known, and the structure of the living monotremes is so specialized that the affinities of the Prototheria are largely conjectural. Most mammalogists believe that the prototheres arose from a different stock of therapsid reptiles than the one that gave rise to the Theria.

No fossils earlier than the Pleistocene are known, and these come from Australia. The duckbilled platypus and several species of the echidnas are living representatives of this group. Everything indicates that the Prototheria represent a very small and relatively unsuccessful group that has miraculously survived in an isolated corner of the Earth. *See* MAMMALIA; MONOTREMATA; THERAPSIDA. [D.D.D./F.S.S.]

Prototype The term used at that stage in the design process when the component has been realized physically in a form that satisfies the functional, environmental, reliability, maintainability, packaging, and other requirements, but when the design does not necessarily reflect the techniques of manufacture by which it will be ultimately produced in quantity. Although considerations of the methods of manufacture play a role in the preliminary stages of the evolution of a component, early developmental versions, since they are produced in small quantities (one to several units), may not be designed to take advantage of the methods of fabrication peculiar to large-scale production. *See* PILOT PRODUCTION. [R.W.M.]

Protozoa A group of eukaryotic microorganisms traditionally classified in the animal kingdom. Although the name signifies primitive animals, some Protozoa (phytoflagellates and slime molds) show enough plantlike characteristics to justify claims that they are plants.

Protozoa are almost as widely distributed as bacteria. Free-living types occur in soil, wet sand, and in fresh, brackish, and salt waters. Protozoa of the soil and sand live in films of moisture on the particles. Habitats of endoparasites vary. Some are intracellular, such as malarial parasites in vertebrates, which are typical Coccidia in most of the cycle. Other parasites, such as *Entamoeba histolytica*, invade tissues but not individual cells. Most trypanosomes live in the blood plasma of vertebrate hosts. Many other parasites live in the lumen of the digestive tract or sometimes in coelomic cavities of invertebrates, as do certain gregarines. *See* COCCIDIA; GREGARINIA; TRYPANOSOMATIDAE.

Many Protozoa are uninucleate, others are binucleate or multinucleate, and the number of nuclei also may vary at different stages in a life cycle. Protozoa range in size from 1 to 10^6 micrometers. Colonies are known in flagellates, ciliates, and Sarcodina. Although marked differentiation of the reproductive and somatic zooids characterizes certain colonies, such as *Volvox*, Protozoa have not developed tissues and organs.

Morphology. A protozoan may be a plastic organism (ameboid type) but changes in form are often restricted by the pellicle. A protective layer is often secreted outside the pellicle, although the pellicle itself may be strengthened by incorporation of minerals. Secreted coverings may fit closely, for example, the cellulose-containing theca of Phytomonadida and Dinoflagellida, analogous to the cell wall in higher plants. The dinoflagellate theca (Fig. 1*a*) may be composed of plates arranged in a specific pattern. Tests, as seen in Rhizopodea (Arcellinida, Gromiida, Foraminiferida), may be composed mostly of inorganic material, although organic (chitinous) tests occur in certain species. Siliceous skeletons, often elaborate, characterize the Radiolaria (Figs. 1*d* and 2*c*). A vase-shaped lorica, from which the anterior part of the organism or its appendages may be extended, occurs in certain flagellates (Fig. 1*b*) and ciliates (Fig. 1*c*). Certain marine ciliates (Tintinnida) are actively swimming loricate forms.

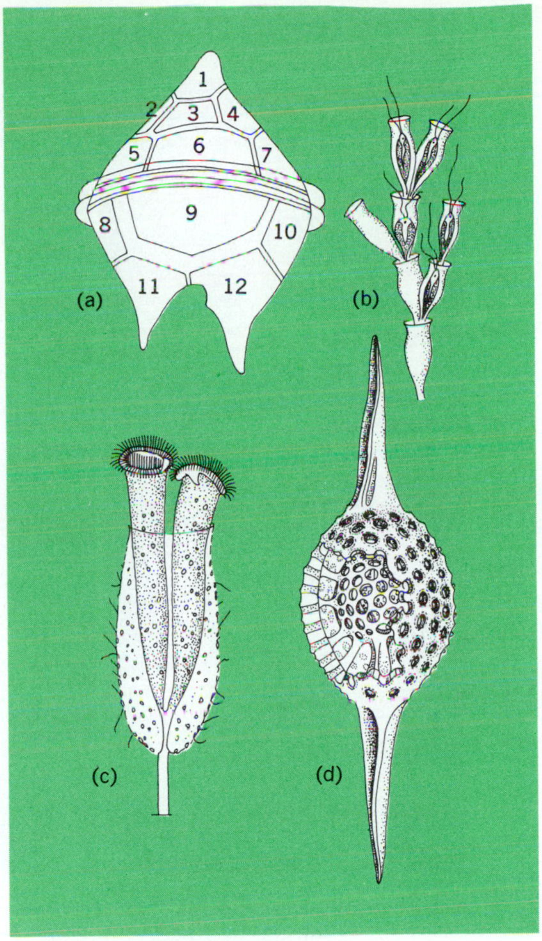

Fig. 1. External coverings of Protozoa. (*a*) Theca of dinoflagellate (*Peridinium*), showing separate plates. (*b*) Lorica of a colonial chrysomonad, *Dinobryon*. (*c*) Two zooids within a lorica of a peritrich, *Cothurnia*. (*d*) A radiolarian skeleton, siliceous type. (*After L. H. Hyman, The Invertebrates, vol. 1, McGraw-Hill, 1940*)

Flagella occur in active stages of Mastigophora and flagellated stages of certain Sarcodina and Sporozoa. A flagellum consists of a sheath enclosing a matrix in which an axoneme extends from the cytoplasm to the flagellar tip. In certain groups the sheath shows lateral fibrils (mastigonemes) which increase the surface area and also may modify direction of the thrust effecting locomotion. Although typically shorter than flagella, cilia are similar in structure. *See* CILIA AND FLAGELLA.

Two major types of pseudopodia have been described, the contraction-hydraulic and the two-way flow types. The first are lobopodia with rounded tips and ectoplasm denser than endoplasm. The larger ones commonly contain granular endoplasm and clear ectoplasm. Two-way flow pseudopodia include reticulopodia of Foraminiferida and related types, filoreticulopodia of Radiolaria, and axopodia of certain Heliozoia.

In addition to nuclei, food vacuoles (gastrioles) in phagotrophs, chromatophores and stigma in many phytoflagellates, water-elimination vesicles in many Protozoa, and sometimes other organelles, the cytoplasm may contain mitochondria, Golgi material, pinocytotic vacuoles, stored food materials, endoplasmic reticulum, and sometimes pigments of various kinds.

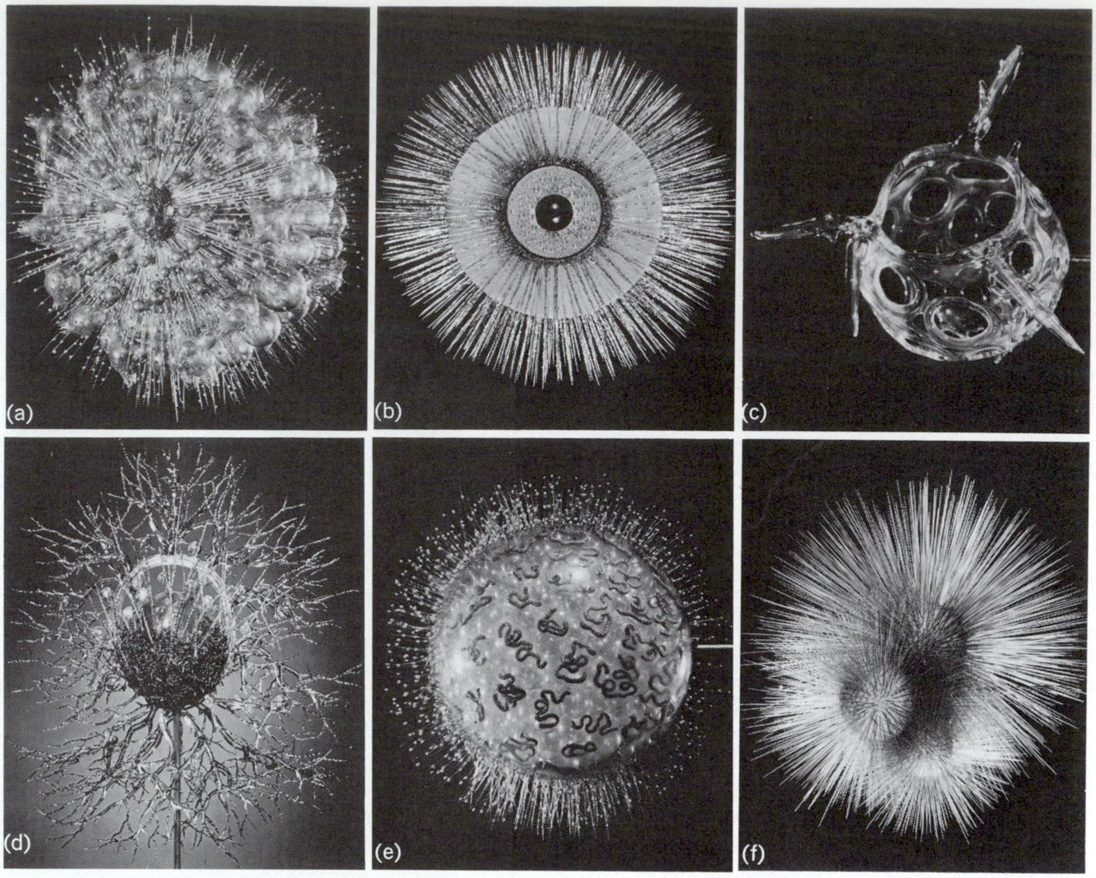

Fig. 2. Glass models of marine Protozoa. Radiolarian types: (*a*) *Trypanosphaera transformata* Haeckel (Indian Ocean); (*b*) *Actissa princeps* Haeckel (Indian and Pacific oceans); (*c*) *Peridium spinipes* Haeckel (Pacific Ocean); (*d*) *Lithocircus magnificus* Haeckel (Atlantic Ocean); (*e*) *Collozoum serpentinum* Haeckel (Atlantic Ocean). (*f*) Foraminiferan type: the pelagic *Globigerina bulloides* d'Orbigny (which is found in all seas). (*American Museum of Natural History*)

Nutrition. In protozoan feeding, either phagotrophic (holozoic) or saprozoic (osmotrophic) methods predominate in particular species. In addition, chlorophyll-bearing flagellates profit from photosynthesis; in fact, certain species have not been grown in darkness and may be obligate phototrophs.

Phagotrophic ingestion of food, followed by digestion in vacuoles, is characteristic of Sarcodina, ciliates, and many flagellates. Digestion follows synthesis of appropriate enzymes and their transportation to the food vacuole. Details of ingestion vary. Formation of food cups, or gulletlike invaginations to enclose prey, is common in more or less ameboid organisms, such as various Sarcodina, many flagellates, and at least a few Sporozoa. Entrapment in a sticky reticulopodial net occurs in Foraminiferida and certain other Sarcodina. A persistent cytostome and gullet are involved in phagotrophic ciliates and a few flagellates. Many ciliates have buccal organelles (membranes, membranelies, and closely set rows of cilia) arranged to drive particles to the cytostome. Particles pass through the cytostome into the cytopharynx (gullet), at the base of which food vacuoles (gastrioles) are formed. Digestion occurs in such vacuoles.

By definition saprozoic feeding involves passage of dissolved foods through the cortex. It is uncertain to what extent diffusion is responsible, but enzymatic activities presumably are involved in uptake of various simple sugars, acetate and butyrate. In addition, external factors, for example, the pH of the medium, may strongly influence uptake of fatty acids and phosphates.

Reproduction. Reproduction occurs after a period of growth which ranges, in different species, from less than half a day to several months (certain Foraminiferida). General methods include binary fission, budding, plasmotomy, and schizogony. Fission, involving nuclear division and replication of organelles, yields two organisms similar in size. Budding produces two organisms, one smaller than the other. In plasmotomy, a multinucleate organism divides into several, each containing a number of nuclei. Schizogony, characteristic of Sporozoa, follows repeated nuclear division, yielding many uninucleate buds.

Simple life cycles include a cyst and an active (trophic) stage undergoing growth and reproduction. In certain free-living and parasitic species, no cyst is developed. Dimorphic cycles show two active stages; polymorphic show several. The former include adult and larva (Suctoria); flagellate and ameba (certain Mastigophora and Sarcodina); flagellate and palmella (nonflagellated; certain Phytomonadida); and ameba and plasmodium (Mycetozoia especially).

Parasitic protozoa. Parasites occur in all major groups. Sporozoa are exclusively parasitic, as are some flagellate orders (Trichomonadida, Hypermastigida, and Oxymonadida), the Opalinata, Piroplasmea, and several ciliate orders (Apostomatida, Astomatida, and Entodiniomorphida). Various other groups contain both parasitic and free-living types. Protozoa also serve as hosts of other protozoa, certain bacteria, fungi, and algae.

Relatively few parasites are distinctly pathogenic, causing amebiasis, visceral leishmaniasis (kala azar), sleeping sickness, Chagas' disease, malaria, tick fever of cattle, dourine of horses, and other diseases. *See* AMEBIASIS; CHAGAS' DISEASE; CILIOPHORA; CNIDOSPORA; LEISHMANIASIS; MALARIA; SARCOMASTIGOPHORA; SLEEPING SICKNESS; SPOROZOA. [R.P.H.]

Protractor An instrument used to construct and measure angles formed by the lines of a plane (see illustration). In its

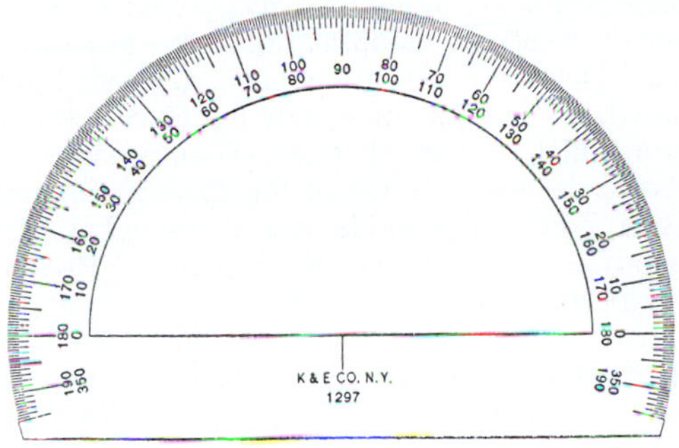

A protractor.

simplest form it consists of half a circular disk of metal or transparent material, with the bounding semicircle graduated in degrees, usually from 0 to 180°. The midpoint of the diameter of the semicircle is marked. It serves as the vertex of angles constructed or measured. *See* ANGLE. [L.M.Bl.]

Protura An order of ancestrally wingless insects (subclass Apterygota). The order is usually classified into four families in two suborders (Eosentomoidea and Acerentomoidea), and about 500 species are recognized.

The insects are under 0.08 in. (2 mm) in length and have an elongated body with a small head, relatively large abdomen, and three pairs of functional legs. The first pair of legs is normally held out in front of the head, antennalike. They serve to replace the sensory role of true antennae, which are absent. Paired pseudoculi (probably chemosensory) occupy eyelike positions on the head. The mouthparts include an elongate and styliform mandible, a maxilla, and a two-part labium. All parts are enclosed within a pocket in the head, so that only their tips are exposed. The thorax bears five-segmented legs, each with an apical claw. In the adult, the abdomen has 12 segments; the anus is terminal, and cerci (segmented sensory appendages on the last abdominal segment) are lacking. The sterna of the first, second, and third segments bear short styli, probably representing vestigial limbs. A large gland opens at the rear of the eighth dorsal plate, which sometimes has a comblike lid. The genitalia are enclosed within a ventral pouch between the eleventh and twelfth segments, and have paired protrusible stylets (slender, elongated appendages).

Protura live in forest soil, leaf litter, and similar places, and are thought to feed on mycorrhizal fungi. They are distributed globally but unevenly in temperate and tropical climates, and they sometimes occur in enormous numbers, as in some Oregon fir forests in the United States. *See* APTERYGOTA; INSECTA. [W.L.Bro.]

Proustite A mineral having composition Ag_3AsS_3. It occurs in prismatic crystals terminated by steep ditrigonal pyramids, but is more commonly massive or in disseminated grains. Hardness is 2–2.5 (Mohs scale) and specific gravity is 5.55. The luster is adamantine and the color ruby red. It is called light ruby silver in contrast to pyrargyrite, dark ruby silver. Proustite and pyrargyrite are found together in silver veins. Noted localities are at Chañiarcillo, Chile; Freiberg, Germany; Guanajuato, Mexico; and Cobalt, Ontario, Canada. *See* PYRARGYRITE. [C.S.Hu.]

Prymnesiophyceae A class of algae (also known as Haptophyceae) in the chlorophyll *a-c* phyletic line (Chromophycota). In protozoological classification these organisms constitute an order, Prymnesiida or Haptomonadida, in the class Phytamastigophora. Most of the approximately 300 species of prymnesiophytes are biflagellate monads. *See* CHROMOPHYCOTA.

This class has been segregated from the Chrysophyceae, with which it shares many biochemical and ultrastructural characters. Prymnesiophytes differ from chrysophytes, however, in several significant characters: (1) the typical monad bears a filiform organelle, the haptonema, between the two flagella; (2) except in the order Pavlovales, the flagella are of equal length and smooth; (3) organic scales, which may be calcified, cover most mottle cells.

Calcified scales (cellulosic scales impregnated with calcite) are characteristic of many prymnesiophytes. These scales, which are given the general term coccoliths, were discovered in marine sediment before they were observed on living cells. Coccoliths are classified into several morphological types (such as rhabdoliths, discoliths, zygoliths, and ceratoliths), which are of great diagnostic value in the taxonomy of the prymnesiophytes that bear them (coccolithophorids).

In most prymnesiophytes a nonmotile phase alternates with a motile phase. The nonmotile phase is a free-living unicell or a palmelloid or pseudofilamentous colony. In some cases the alternation is mediated by sexual reproduction. Usually, however, reproduction is effected by binary fission or the production of zoospores.

Prymnesiophytes are primary marine, with coccolithophorids constituting one of the three major components of phytoplankton (the others being diatoms and dinoflagellates). *See* ALGAE; COCCOLITHOPHORIDA; PHYTOPLANKTON; PROTOZOA. [P.C.Si.; R.L.Moe]

Pseudoborniales An order of fossil plants found in Middle and Upper Devonian rocks. The group is related to Sphenophyllales and includes a single family and two monotypic genera. *Pseudobornia ursina* is known from Bear Island (north of Norway), Alaska, and Germany. *Prosseria grandis* is found in New York State. Sphenopsid characters are more firmly established in this order than in Hyeniales. *See* HYENIALES; SPHENOPHYLLALES. [H.P.B.]

Pseudomonadaceae A family of gram-negative, aerobic nonsporeforming bacteria. There are four genera: *Pseudomonas*, *Xanthomonas*, *Zoogloea*, and *Gluconobacter*. A few species in the genus *Pseudomonas* are pathogenic for plants, animals, and humans; nearly all *Xanthomonas* species are pathogenic for plants; and species in the genus *Gluconobacter* are used in industrial microbiological processes for the production of vinegar, gluconic acid, and L-sorbose. Motile species invariably possess one or more polar flagella. Photosynthetic pigments are not found, but other types of pigments such as pyocyanin and fluorescein are common. [W.C.Ha.]

Pseudomonas A genus of gram-negative, nonsporeforming, rod-shaped bacteria. Motile species possess polar flagella. They are strictly aerobic, but some members do respire anaerobically in the presence of nitrate. Some species produce acids oxidatively from carbohydrates; none is fermentative and none photosynthetic.

Members of the genus *Pseudomonas* cause a variety of infective diseases; some species cause disease of plants. One species, *P. mallei*, is a mammalian parasite, and is the causative agent of glanders, an infectious disease of horses that occasionally is transmitted to humans by direct contact. *Pseudomonas aeruginosa* is the most significant cause of hospital-acquired infections, particularly in predisposed patients with metabolic, hematologic, and malignant diseases. The spectrum of clinical disease ranges from urinary tract infections to septicemia, pneumonia, meningitis, and infections of postsurgical and posttraumatic wounds. *See* GLANDERS; HOSPITAL INFECTIONS; MELIOIDOSIS; MENINGITIS; PNEUMONIA. [G.L.Gi.]

Pseudophyllidea An order of tapeworms of the subclass Cestoda, parasitic in the intestine of all classes of vertebrates. Typically, the head is simple in structure with two groovelike attachment organs, the bothria. Most pseudophyllideans are segmented and polyzoic with replication of the reproductive systems, although there are a number which do not show such replication and are monozoic.

Dibothriocephalus latus, the broad or fish tapeworm of humans and certain piscivorous mammals, is a pseudophyllidean. In humans, this worm sometimes precipitates a pernicious anemia by competing with the host for vitamin B_{12}. Larval pseudophyllideans are occasionally found as parasites in the extraintestinal tissues of humans, producing a condition known as sparganosis. *See* CESTODA; TAPEWORM DISEASE. [C.P.R.]

Pseudoscorpionida An order of terrestrial Arachnida having the general appearance of miniature scorpions without the postabdomen and sting. The body length is seldom greater than 0.2 in. (5.0 mm). Typically, each finger of the anterior appendages, or chelicerae, has a serrula composed of a row of ligulate plates. Ducts of silk glands open near the end of the movable finger, often in connection with a simple or branched spinneret. The second pair of appendages, or palpi, are large and conspicuous, usually with glands that discharge venom through a terminal tooth on one or both of the chelal fingers. The four pairs of legs are ambulatory.

Pseudoscorpions feed chiefly on small arthropods and, although frequently found on birds, mammals, and insects, are considered nonparasitic. Pseudoscorpions are common in the nests of mammals, birds, and social insects, in woody debris and forest litter, under stones, and in crevices in the bark of trees. About 2000 species have been described. *See* ARACHNIDA. [C.C.Ho.]

Pseudosphaeriales (lichenized) An order of the class Ascolichenes. The order is also called the Pleosporales. They resemble the typical pyrenomycetous lichens except for the structure of the ascocarp, which is not a true perithecium. It is flask-shaped and lined with a layer of interwoven, branched pseudoparaphyses. The asci, with bitunicate walls, are located in scattered locules.

There are two major families: the larger one, Arthophyreniaceae, is a widespread family with at least five genera; the Mycoporaceae is a small family with two well-known genera, *Dermatina* and *Mycoporellum*. All of the species in this order are crustose and many lack a well-defined thallus. [M.E.H.]

Pseudotuberculosis A disease of rodents and birds caused by the bacterium *Pasteurella pseudotuberculosis*. The disease is occasionally transmitted to humans.

This sporadic, or epizootic, plaguelike disease in rodents and animals causes small abscesses in the liver, spleen, and intestinal wall. In humans, the infection has been reported as an acute fatal septicemia in some cases. The organisms may localize in mesenteric lymph nodes of the ileocecal region and cause acute appendicitis and gastrointestinal symptoms. [K.F.M.]

Psilomelane A basic oxide of barium and manganese with the idealized composition $BaMnMn_8O_{16}(OH)_4$. The mineral is not well crystallized, and typically occurs as fine-grained masses and crusts. The color is iron black to dark steel gray. The hardness is about 5.5 on Mohs scale, and the specific gravity is 4.71. Psilomelane often occurs admixed with other manganese oxides, chiefly pyrolusite, and with clay and hydrated iron oxides. [C.Fr.]

Psilophytales A group long recognized as an order of fossil plants (subdivision Psilopsida) collected in rocks of Late Silurian and Devonian age. It has been subdivided into three categories whose descriptions include the chief kinds of plants formerly included in Psilophytales. They are given in this Encyclopedia as three classes of the division Rhyniophyta: Rhyniopsida, Zosterophyllopsida, and Trimerophytopsida. *See* EMBRYOBIONTA; PALEOBOTANY; RHYNIOPHYTA; RHYNIOPSIDA; TRIMEROPHYTOPSIDA; ZOSTEROPHYLLOPSIDA. [H.P.B.]

Psilotophyta A division of the plant kingdom consisting of only two genera with three living species, *Psilotum nudum*, *P. complanatum*, and *Tmesipteris tannensis*. *Psilotum* is widespread in tropical and subtropical regions of both hemispheres, but *Tmesipteris* is confined to Australia and some of the Pacific islands. The Psilotophyta have the typical life cycle of vascular cryptograms, with an alteration of sporophyte and gametophyte generations, the sporophyte being much the larger and more complex. The Psilotophyta have no economic importance, but they are interesting as possible remnants of an ancient (Silurian and Devonian) group of plants, the Rhyniophyta, which is regarded as ancestral to all other vascular plants. *See* LYCOPODIOPHYTA; POLYPODIOPHYTA; PSILOPHYTALES; RHYNIOPHYTA. [A.Cr.]

Psittaciformes An order of birds containing the single family Psittacidae, the parrots. This group, with over 300 species, is predominantly tropical, but has extended as far north as the central United States (the extinct Carolina parakeet, *Conuropsis carolinensis*) and as far south as Tierra del Fuego (the Austral parakeet, *Enicognathus ferrugineus*). Only one species, the monk parakeet (*Myiopsitta monachus*) of southern South America, is known to build its own nest of sticks. All other parrots nest in holes or cavities.

Parrots are highly popular as cage birds; many have brilliantly colored plumage, and some have exceptional powers of vocal mimicry. The order is a strongly differentiated one, relatively uniform anatomically in spite of superficial diversity in size and form. Its relationships to other orders are uncertain. *See* AVES; PARROT. [K.C.P.]

Psittacosis An infection of humans derived from contact with birds, and a frequent infection in psittacines caused by the bacterium *Chlamydia psittaci*. Of the 130 bird species involved, 71 are psittacine; parrots, parakeets, turkeys, ducks, and pigeons are the important sources. Inhalation of dust from infected droppings or handling of feathers, viscera, or waste contaminated by infected birds may cause human disease, which is usually respiratory and may range from fatal to subclinical. [J.S.]

Psocoptera An order of insects commonly referred to as book lice, bark lice, and psocids, the latter a general term for all members of the order. They are usually less than 0.25 in. (0.6 cm) long, though rarely some may reach about 0.5 in. (1.2 cm). Wings may be absent, and when present are of differing distinctive venational types.

Book lice are most common among old papers on dusty shelves, in cereals, or other domestic situations. They are usually pale, wingless types of insects. Many bark lice, the majority winged, occur on the bark or foliage of trees, and some are found under dead bark or beneath stones. Nymphs of a few species occur on tree trunks as clusters of gregarious individuals, but disperse when mature.

Psocoptera are worldwide, especially in warm countries, and some 1300 species are known. Current classification lists about 37 families for this group. *See* INSECTA. [A.B.Gu.]

Psoriasis An inflammatory skin disease of unknown cause, usually chronic or recurrent but often acute in nature.

The local lesions occur predominantly at certain sites, such as elbows, knees, and scalp. They consist of dull red, well-defined plaques or patches, usually covered by distinctive silvery scales which when removed disclose tiny capillary bleeding points. The lesions spread by peripheral extension and may involve huge areas of the body. The plaques are inconstant in size, shape, and location during the course of the disease, but they have a tendency to recur. Tension, fatigue, and overtreatment often produce flareups. [E.G.St./N.K.M.]

Psychoacoustics All of the psychological interactions between humans (and animals) and the world of sound. It encompasses all studies of the perception of sound, as well as the production of speech. *See* HEARING (HUMAN); SPEECH. [L.E.M.]

Psychoanalysis In the narrow sense, a form of psychotherapy invented and described by Sigmund Freud. In a broader (and incorrect) sense, psychoanalysis refers to almost any form of psychotherapy other than behavior modification, but especially those forms that emphasize the therapist-client relationship and the view that the client may be unaware of the reasons for his or her actions. Included in this second category are viewpoints (for example, existential psychoanalysis) that are radically inconsistent with the spirit of Freud's ideas. *See* PSYCHOTHERAPY. [R.Hog.]

Psycholinguistics An area of study which draws from linguistics and psychology and focuses upon the comprehension and production of language. Although psychologists have long been interested in language, and the field of linguistics is an older science than psychology, scientists in the two fields have had little contact until the work of Noam Chomsky was published in the late 1950s. Chomsky's writing had the effect of making psychologists acutely aware of their lack of knowledge about the structure of language, and the futility of focusing attention exclusively upon the surface structure of language. As a result, psycholinguists, who have a background of training in both linguistics and psychology, have been attempting since the early 1960s to gain a better understanding of how the abstract rules which determine human language are acquired and used to communicate appropriately created meaningful messages from one person to another via the vocal-auditory medium. Research has been directed to the evolutionary development of language, the biological bases of language, the nature of the sound system, the rules of syntax, the nature of meaning, and the process of language acquisition. [D.S.P.]

Psychology, physiological and experimental Experimental and physiological psychology are aspects of the discipline of psychology. That discipline comprises a large number of subareas which include the study of individual differences, developmental factors, learning, motivation, perception, and social influences, and applied areas such as clinical, counseling, and industrial psychology. The common core of psychology is its concern with the understanding and utilization of the systematic determinants of behavior. Because behavior is determined by the social world and by the biological nature of people, psychology is both a social science and a biological science.

Experimental psychology refers to both a method and a broad subject-matter area. As a method, experimental psychology uses certain procedures to obtain controlled observations which permit an accurate assessment of the relationships between the variables being studied. In the prototype design, a single variable (say, number of presentations) is varied, and the effect on a second variable (such as recall) is measured with all other conditions being held constant. More complex designs permit the control of two or more variables. Such experiments are most readily conducted under controlled laboratory conditions. However, the attempt to approximate the basic procedures of controlled observations underlie psychological research regardless of the setting or area.

Physiological psychology refers to the study of the interrelationships between the biology and the behavior of humans and animals. The predominant focus has been on the roles of sensory mechanisms, neurophysiology, and neurochemistry in mediating and modulating behavior. The procedures typically involve the modification of specific physiological systems and a determination of the effect on behaviors such as perception, learning, motivation, or performance. Because of the highly technical character of physiological psychology and its close relationship to other developed and developing disciplines such as biochemistry, neuroanatomy, pharmacology, and physiology, it is highly interdisciplinary. [W.B.W.]

Psychoneuroimmunology The study of the interactions among behavioral, neural and endocrine, and immune functions. This convergence of disciplines has evolved to achieve a more complete understanding of adaptive processes. At one time, the immune system was considered an independent agency of defense that protected the organism against foreign material (that is, proteins that were not part of one's "self"). Indeed, the immune system is capable of considerable self-regulation. However, converging data from the behavioral and brain sciences indicate that the brain plays a critical role in the regulation or modulation of immunity. This research indicates that the nervous and immune systems, the two most complex systems that have evolved for the maintenance of homeostasis, represent an integrated mechanism for the adaptation of the individual and the species. Thus, psychoneuroimmunology emphasizes the study of the functional significance of the relationship between these systems—not in place of, but in addition to, the more traditional analysis of the mechanisms governing the functions within a single system—and the significance of these interactions for health and disease. *See* NEUROIMMUNOLOGY.

Brain–immune system interactions. Evidence for nervous system–immune system interactions exists at several biological levels. Primary (thymus, bone marrow) and secondary (spleen, lymph nodes, gut-associated lymphoid tissues) lymphoid organs are innervated by the sympathetic nervous system, and lymphoid cells bear receptors for many hormones and neurotransmitters. These substances, secreted by the pituitary gland, are thus able to influence lymphocyte function. Moreover, lymphocytes themselves can produce neuropeptide substances. Thus, there are anatomical and neurochemical channels of communication that provide a structural foundation for the several observations of functional relationships between the nervous and immune systems.

Stress and immunity. The link between behavior and immune function is suggested by experimental and clinical observations of a relationship between psychosocial factors, including stress, and susceptibility to or progression of disease processes that involve immunologic mechanisms. Abundant data document an association between stressful life experiences and changes in immunologic reactivity. The death of a family member and other, less severe, stressful experiences (such as taking examinations) result in transient impairments in several parameters of immune function. *See* DISEASE; STRESS (PSYCHOLOGY).

In animals, a variety of stressors can influence a variety of immune responses. Since immune responses are themselves capable of altering levels of circulating hormones and neurotransmitters, these interactions probably include complex feedback and feedforward mechanisms. *See* ENDOCRINOLOGY.

The direction, magnitude, and duration of stress-induced alterations of immunity are influenced by (1) the quality and quantity of stressful stimulation; (2) the capacity of the individual to cope effectively with stressful events; (3) the quality and quantity of immunogenic stimulation; (4) the temporal relationship between stressful stimulation and immunogenic stimulation; (5) the sampling times and the particular aspect of immune function chosen for measurement; (6) the experiential history of the individual and the existing social and environmental conditions upon which stressful and immunogenic stimulation are superimposed; (7) a variety of host factors such as species, strain, age, sex, and nutritional state; and (8) interactions among these variables.

Conditioning. Central nervous system involvement in the modulation of immunity is dramatically illustrated by the classical (Pavlovian) conditioning of the acquisition and extinction of suppressed and enhanced antibody- and cell-mediated immune responses. In a one-trial taste-aversion conditioning situation, a distinctively flavored drinking solution (the conditioned stimulus) was paired with an injection of the immunosuppressive drug cyclophosphamide (the unconditioned stimulus). When subsequently immunized with sheep red blood cells, conditioned animals reexposed to the conditioned stimulus showed a reduced antibody response compared to nonconditioned animals and conditioned animals that were not reexposed to the conditioned stimulus. *See* CONDITIONED REFLEX.

The acquisition and the extinction (elimination of the conditioned response by exposures to the conditioned stimulus without the unconditioned stimulus) of the conditioned enhancement and suppression of both antibody- and cell-mediated immune responses—and nonimmunologically specific host defense responses as well—have been demonstrated under a variety of experimental conditions.

Prospects. An elaboration of the integrative nature of neural, endocrine, and immune processes and the mechanisms underlying behaviorally induced alterations of immune function is likely to have clinical and therapeutic implications that will not be fully appreciated until more is known about the extent of these interrelationships in normal and pathophysiological states. *See* ENDOCRINE SYSTEM (VERTEBRATE); IMMUNOLOGY; NERVOUS SYSTEM (VERTEBRATE). [R.Ad.]

Psychoneurosis Any of a number of seemingly unrelated mental illnesses, such as anxiety neurosis, phobic states, obsessive and compulsive states, and hysteria. Anxiety neurosis and neurocirculatory asthenia (its mild chronic form) are closely related to manic-depressive disease. Hysteria, which is essentially a condition found in adolescent girls and young women, is characterized by unconsciously motivated trances, seizurelike states, pseudoparalyses, and so forth. It is believed to have a particular type of psychopathology, as do the phobic and obsessive-compulsive states. *See* OBSESSIVE-COMPULSIVE DISORDER; PHOBIC REACTION. [R.D.A.]

Psychopathy The most disabling and protracted of the social maladjustments; also referred to as sociopathy. Individuals suffering from such disturbances exhibit obscure types of disorder of character and personality. Afflicted individuals find it impossible to curb their impulses and conform to the demands of school, employment, and society. Some of these psychopathies may be conditioned by the environment in which the child is brought up, and others probably represent delays in the maturation or occult diseases of the brain. Psychiatry and mental health workers have found no means of correcting or effectively coping with these disorders. Many tend to disappear once adulthood is reached. [R.D.A.]

Psychopharmacology The science of drugs that significantly influence behavior, the way a person thinks, feels, and acts. Psychopharmacologic drugs were first discovered accidentally through incidental behavioral effects noted in drug treatment of physical disease. They then were developed through chemical rearrangements of the parent compound and were tested on animals. Psychopharmacology, a science for the systematic study and synthesis of such drugs, was stimulated by the appearance of new drugs and their application to mental disorders, new laboratory methods, and extensive governmental financial support.

Psychopharmacologic drugs are used to treat mental and physical disorders by influencing behavioral excesses and imbalances. They are tools for research into social and biological processes which influence the organization of behavior. In humans, research is applied to normal or disturbed behavior, its subjective aspects (feeling, thought, and states of consciousness), objective aspects (sensorimotor and intellectual performance), and theoretical aspects (distribution of psychic energies, personality structure, and so on). Research on animals deals with drug effects on learning, maturation, drive (fear, hunger, sex, punishment, and reward), adaptation to stress and stimuli, motor activities, and associated neurochemistry.

Psychopharmacologic drugs are usually classified into seven groups: antipsychotics (major tranquilizers, neuroleptics); antidepressants (psychic energizers); anxiolytics (minor tranquilizers); sedative-hypnotics (sleeping medications); psychostimulants; psychodysleptics (psychotomimetics); and others (lithium, piracetam, and so on). *See* AMPHETAMINE; BARBITURATES; NARCOTIC; PSYCHOTOMIMETIC DRUG; TRANQUILIZER. [D.X.F.]

Psychosis A term that designates severe psychiatric disorders. Psychosis refers usually to psychiatric conditions which are characterized, among other symptoms, by serious impairment of judgment and power of willing or determining (volition). The rigidity of the symptoms and the marked incapacity for corrective learning are outstanding characteristics of the term. Although Sigmund Freud stressed that psychotic patients suffer from "a break with reality," neurotics suffer from an intrapsychic, unconscious conflict between instinctual and controlling forces. In practice there is no qualitative difference between psychosis and neurosis; it is rather one of degree. From a forensic (legal) viewpoint, the psychotic patient is not considered responsible for his acts or competent to enter legally binding agreements. *See* NEUROTIC DISORDERS; PARANOIA; PARANOID STATE; SCHIZOPHRENIA. [F.C.R.]

Psychosomatic disorders A type of neurosis. Perhaps the simplest and most diffuse form of neurotic symptomatology are the so-called neurasthenias in which the principal complaint is a general malaise involving a loss of energy and zest along with a vague sense of aches, pains, and fatigue. Neurasthenic symptoms tend to be centered about those activities in the person's life that have proved either notably unrewarding or internally stressful.

Hypochondriacal symptoms, in the form of exaggerated somatic complaints accompanied by a strong but ambivalent fear of failing ill and with much concern about diet, elimination, and health, represent another form of somatization. A common origin of hypochondriacal symptoms is a family background where parents were excessively concerned with the health of children and in which sickness was rewarded with attention. Feelings of inadequacy on the part of the patient later in life may then be translated into reward-seeking somatic complaints. [J.S.Br.; W.Mis.]

Psychosurgery Any surgical procedure on the brain, such as prefrontal lobotomy, leukotomy, transorbital lobotomy, and topectomy, that has been recommended as treatment for various psychoses (particularly schizophrenia and psychotic depression), severe neuroses, and chronic painful conditions. Most psychiatrists will not agree to the use of psychosurgery in neurosis and depressions.

The techniques vary according to the specific procedure. In most cases the operation consists of severing bilaterally the white projection fibers to the prefrontal areas. It is used sparingly, even for the treatment of psychoses, and has been largely replaced by tranquilizing drugs. *See* NEUROTIC DISORDERS; PSYCHOSIS; TRANQUILIZER. [F.C.R.]

Psychotherapy Any treatment or therapy that is psychological in nature. Although the term is popularly used as if there were one process or set of procedures that constitutes psychotherapy, this is true only in a very general way; in actuality, there are many (over 130) psychotherapies. In general, psychotherapeutic treatment has been perceived primarily as a treatment for the so-called psychoneurotic disorders. *See* NEUROTIC DISORDERS; PSYCHOSIS.

In addition to designating psychotherapies in terms of their theoretical or procedural differences, for example, psychoanalysis, client-centered therapy, or behavior therapy, the psychotherapies can also be classified in terms of the number or category of clients receiving therapy, or in terms of the length of psychotherapy. Thus therapies have been categorized as individual or group psychotherapy, family or marital therapy, child or adult psychotherapy, and brief or time-limited versus long-term psychotherapy.

Perhaps best known of all forms of psychotherapy is psychoanalysis. The goal of the therapy is to gradually overcome the patient's defenses and to help the patient attain insight into the repressed conflicts which are the source of difficulty. Apart from the historic departures of Alfred Adler and Carl Jung from the fold of psychoanalysis, and the subsequent theoretical modifications offered by Karen Horney and Harry Stack Sullivan, there has been the development of what is loosely termed psychoanalytic psychotherapy. *See* PSYCHOANALYSIS.

There are so many other psychotherapies that only a few can be mentioned. Client-centered therapy was developed by Carl Rogers during the 1940s. A major emphasis of this therapy is placed on the inherent capacity of the client to change as long as the therapist is emphatic and genuine and conveys nonpossessive warmth.

Another therapy which has received much attention is behavior therapy, which differs in noticeable ways from the psychoanalytic therapies. The focus is on ridding the patient of symptoms and in changing behavior. There is no concern with hypothesized inner conflicts or unconscious problems responsible for the symptoms; the therapist is generally more directive or active; and the length of therapy is much shorter.

Cognitive therapies have emerged which tend to emphasize faulty preceptions and cognitions (thoughts) as possible causes of personal difficulties. In these therapies attempts are made to change the client's cognitions so that they are more realistic and more positive. Also, there has been some integration of cognitive and behavioral emphases, so that some therapists refer to themselves as cognitive behaviorists. [S.L.G.]

Psychotomimetic drug A class of drugs reliably inducing temporary states of altered perception, often with symptoms similar to those of psychosis. The drug experiences are clearly perceived, vivid, and remembered. Sensory input is heightened and subjective experience is intensified, but control is diminished. The subject's feelings and momentary perceptions gain an independence from the normal corrections of logic; whatever stray item occupies the attention (a sensation or an unguarded and unevaluated memory or thought) becomes at the moment compellingly significant. Thinking and perceiving of this order coexist with the capacity for, but not an interest in, normal thought and function.

These drugs have been called psychedelic (or mind-manifesting) because of the convincing clarity with which a normally suppressed, more sensory, plastic, and primitive part of the mind is revealed to the drugged subject, contrasting sharply with the capacity to focus, discriminate, and judge.

The two drugs of chief interest, mescaline and LSD-25 (lysergic acid diethylamide), are of ancient lineage. Mescaline, used in highly structured Amerindian tribal religious rituals, is derived from peyote buttons, which are the dried tops of a cactus found in the southwestern United States. LSD has been synthesized with compounds derived from ergot, which infests such grasses as rye. (Some ergot derivatives are useful in the treatment of migraine and in obstetrics for the postpartum constriction of the uterus and blood vessels.) *See* ERGOT.

There are a number of other psychotomimetics (less potent than LSD but similar), such as those from mushrooms (psilocybin and psilocin) and those synthesized in the laboratory. Ring-substituted amphetamines resembling mescaline's structure (for example, "STP") have been produced in profusion. There are subtle differences among these newer compounds, some producing more or less euphoria and more or less dyscontrol and anxiety. [D.X.F.]

Psychrometer An instrument consisting of two thermometers which is used in the measurement of the moisture content of air or other gases. The bulb or sensing area of one of the thermometers either is covered by a thin piece of clean muslin cloth wetted uniformly with distilled water or is otherwise coated with a film of distilled water. The temperatures of both the bulb and the air contacting the bulb are lowered by the evaporation which takes place when unsaturated air moves past the wetted bulb. An equilibrium temperature, termed the wet-bulb temperature (T_W), will be reached; it closely approaches the lowest temperature to which air can be cooled by the evaporation of water into that air. The water-vapor content of the air surrounding the wet bulb can be determined from this wet-bulb temperature and from the air temperature measured by the thermometer with the dry bulb (T_D) by using an expression of the form $e = e_{SW} - aP(T_D - T_W)$. Here e is the water-vapor pressure of the air, e_{SW} is the saturation water-vapor pressure at the wet-bulb temperature, P is atmospheric pressure, and a is the psychrometric constant, which depends upon properties of air and water, as well as on speed of ventilation of air passing the wet bulb. *See* PSYCHROMETRICS. [R.M.Sch.]

Psychrometrics A study of the physical and thermodynamic properties of the atmosphere. The properties of primary concern in air conditioning are (1) dry-bulb temperature, (2) wet-bulb temperature, (3) dew-point temperature, (4) absolute humidity, (5) percent humidity, (6) sensible heat, (7) latent heat, (8) total heat, (9) density, and (10) pressure.

The dry-bulb temperature is the ambient temperature of the air and water vapor as measured by a thermometer or other temperature-measuring device in which the thermal element is dry and shielded from radiation. *See* AIR TEMPERATURE; TEMPERATURE.

If the bulb of a dry-bulb thermometer is covered with a silk or cotton wick saturated with distilled water and the air is drawn over it at a velocity not less than 1000 ft/min (5 m/s), the resultant temperature will be the wet-bulb temperature. Where the dry-bulb and wet-bulb temperatures are the same, the atmosphere is saturated.

The dew-point temperature is the temperature at which the water vapor in the atmosphere begins to condense. This is also the temperature of saturation at which the dry-bulb, wet-bulb, and dew-point temperatures are all the same. *See* DEW POINT.

The actual quantity of water vapor in the atmosphere is designated as the absolute humidity. Percentage or relative humidity is the ratio of the actual water vapor in the atmosphere to the quantity of water vapor the atmosphere could hold if it were saturated at the same temperature. *See* HUMIDITY.

Sensible heat, or enthalpy of dry air, is heat which manifests itself as a change in temperature. *See* ENTHALPY.

Latent heat, or enthalpy of vaporization, is the heat required to change a liquid into a vapor without change in temperature. Latent heat is sometimes referred to as the latent heat of vaporization and varies inversely as the pressure.

The total heat, or enthalpy, of the atmosphere is the sum of the sensible heat, latent heat, and superheat of the vapor above the saturation or dew-point temperature. Total heat is relatively constant for a constant wet-bulb temperature, deviating only about 1.5–2% low at relative humidities below 30%.

The density of the atmosphere varies with both altitude and percentage humidity. The higher the altitude the lower the density, and the higher the moisture content the lower the density. *See* DENSITY.

Atmospheric pressure is usually referred to as barometric pressure. Pressure varies inversely as elevation, as temperature, and as percentage saturation. *See* MOISTURE-CONTENT MEASUREMENT; PSYCHROMETER. [J.Ev.]

Pteraspidomorpha The subclass of Agnatha that includes the jawless vertebrates with paired nostrils. Pteraspidomorphi, sometimes called Diplorhina, includes only the extinct ostracoderm order Heterostraci and some or all of the problematical fossil genera now referred to the order Thelodonti. *See* AGNATHA; HETEROSTRACI. [R.H.De.]

Pteridospermae Seed ferns, extinct plants characterized by naked seeds borne on large fernlike fronds. Fossil evidence indicates that these plants reached their greatest abundance during the Pennsylvanian Period. Subsequently they declined, becoming extinct sometime during the Jurassic.

Reconstruction of a seed fern. Note appearance of adventitious roots on the lower portion of the stem.

Because some seed ferns had rather long, slender stems with relatively small amounts of supporting tissue, it has been surmised that they were vinelike and required the support of more sturdy neighboring plants. Others grew erect without the aid of other plants and in their general appearance they superficially resembled modern tree ferns. Stems of pteridosperms were mostly unbranched. Near the apex of their stems they produced a crown of large, spirally arranged fronds (see illustration). Both the fronds of tree ferns and the pteridosperm leaves were compound. Slender adventitious roots were borne on the lower parts of the stem. *See* PALEOBOTANY. [W.N.S.]

Pterobranchia A group of animals regarded as a class of the Hemichordata. They are all marine, widely distributed but not common, benthonic (chiefly deep-sea) inhabitants. All are small or microscopic, sessile, tube-dwelling (tubicolous) organisms. They may be colonial or pseudocolonial with a cuticular skeleton. A U-shaped gut is present. A lophophore of one or several pairs of arms arises from the collar or middle of the three body segments. *See* HEMICHORDATA. [T.H.B.]

Pteropsida A large group of vascular plants characterized by having parenchymatous leaf gaps in the stele and by having leaves which are thought to have originated in the distant past as branched stem systems. Some botanists regard the Pteropsida as a natural group which they recognize as a class, subdivision, or division. Others regard it as an artificial assemblage of plants that have undergone certain similar changes from a rhyniophyte ancestry. The various components of the Pteropsida are here treated as three separate divisions under the names Magnoliophyta, Pinophyta, and Polypodiophyta. *See* MAGNOLIOPHYTA; PINOPHYTA; POLYPODIOPHYTA. [A.CR.]

Pterosauria An extinct order of flying reptiles (Archosauria) of the Mesozoic Era. Impressions show that a thin membrane of skin extended from the side of the body and arm to the enormously elongated fourth finger of each hand and formed a wing analogous to that of a bat (see illustration).

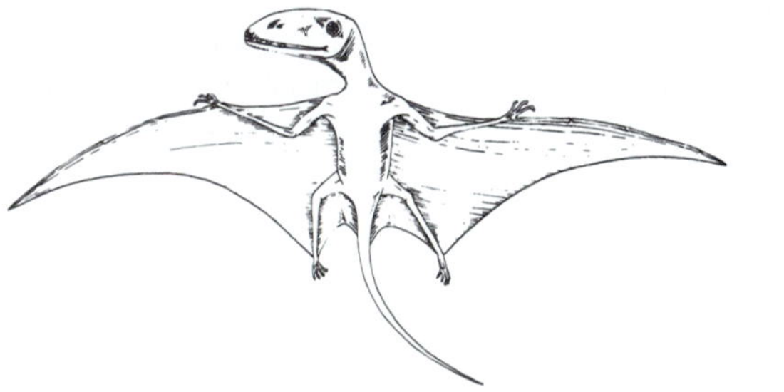

***Dimorphodon*. This drawing is a restoration of an Early Jurassic pterosaur.**

The first three fingers were short and clawed; the fifth was absent. A large keeled sternum supported strong wing muscles, and a slender coracoid bone braced the shoulder joint against the sternum in a fashion similar to that of birds. Pterosaurs had large, lightly constructed skulls; small bodies; light, hollow bones; and small, five-toed rear feet. *See* ARCHOSAURIA; REPTILIA. [J.T.G.]

Pterygota The larger of two subclasses of Insecta. All have wings in the adult stage or have been derived from winged ancestors; that is, if wingless, they are secondarily so. The primitive pterygotes make up the section Paleoptera; the mayflies, dragonflies, and damselflies are in this category. All other pterygotes constitute the section Neoptera. The more primitive Neoptera have exopterygote development, as in the grasshoppers, true bugs, and others, whereas the more specialized Neoptera have endopterygote development, as in the butterflies, moths, beetles, wasps, and so on. *See* APTERYGOTA; ENDOPTERYGOTA; EXOPTERYGOTA; INSECTA. [F.M.C.]

Ptychodactiaria An order of the zoantharian anthozoans of the phylum Coelenterata. This group of solitary sea anemones is known only from two genera, *Ptychodactis* and *Dactylanthus* from the Arctic and Antarctic. *See* ANTHOZOA; COELENTERATA. [C.H.]

Public health Organized social action for the maintenance and promotion of health. This broad concept has evolved from much more restricted definitions which limited public health to the actions of governmental health agencies directed at the prevention of infectious diseases. The advances of medical science and the growing complexity of society have made more imperative the need for extensions of organized social action for the preservation of the individual's health.

In broad terms, public health is concerned with determining the health status of the population (vital statistics), the genetic and bionomic factors in health and disease (epidemiology), and with building upon this knowledge the structure and function of organized health programs and services (public health administration). [E.M.C.]

Pulley A wheel with a flat, crowned, or grooved rim used with a flat belt, V-belt, or a rope to transmit motion and energy. Pulleys for use with V-belt and rope drives have grooved surfaces and are usually called sheaves. A combination of ropes, pulleys, and pulley blocks arranged to gain a mechanical advantage, as for hoisting a load, is referred to as block and tackle. *See* BELT DRIVE; BLOCK AND TACKLE.

Pulleys for flat belts are made of cast iron, fabricated steel, wood, and paper. A particular pulley design must be based on such considerations as the ability to resist shock, to conduct heat, and to resist corrosive environments. The face must be smooth enough to minimize belt wear; yet there must be adequate friction between belt and pulley face to carry the load.

The two common types of V-belt pulley are pressed-steel and cast-iron. The pressed-steel pulleys are suitable for single-belt drives. For multiple-belt drives, or in single-belt drives where pulley mass should be high to get a flywheel effect, cast-iron pulleys are used. [J.R.Z.]

Pulmonata A subclass of the molluscan class Gastropoda, containing about 23,500 species of snails that are grouped into three superorders: Systellommatophora, Basommatophora, and Stylommatophora. Pulmonates include most of the common snails found on land or in freshwaters, although a few prosobranchs have invaded both habitats. Certain pulmonates are intertidal to subtidal on rocky shores of tropical and temperate regions. *See* BASOMMATOPHORA; STYLOMMATOPHORA; SYSTELLOMMATOPHORA.

Fresh-water pulmonates belong to the superorder Basommatophora and are adapted to make use of accidental transport on birds or insects from one isolated body of water to another. Mostly they have short life-spans, and their numbers vary dramatically with the seasons.

Land pulmonates belong to the superorder Stylommatophora and are marginally terrestrial in that they can be active only when the humidity in their microhabitat is 90% or above. Thus many are cryptic or nocturnal in both feeding and reproduction. They can survive in deserts by estivating for up to 6 years. *See* GASTROPODA. [G.A.S.]

Pulsar A celestial radio source producing intense short bursts of radio emission. A total of about 330 pulsars have been found, and hundreds of thousands of pulsars must exist in the Milky Way Galaxy—most of them too distant to be detected with existing radio telescopes. *See* RADIO ASTRONOMY.

Pulsars are distinguished from all other types of celestial radio sources in that their emission, instead of being constant over time scales of years or longer, consists of periodic sequences of brief pulses. The interval between pulses, or pulse period, is nearly constant for a given pulsar, but for different sources ranges from approximately 1.5 ms to 4 s. The bursts of emission are generally confined to a window whose width is a few percent of the interpulse period.

Compelling evidence exists that pulsars are neutron stars, the collapsed cores left behind when certain types of moderate- to high-mass stars become unstable and undergo supernova explosions. A newly formed neutron star spins rapidly (perhaps many times per second) and has an extremely strong magnetic field strength. The neutron star's spin causes the beamed emission to sweep around the sky like a rotating searchlight, thereby causing the periodic pulses observed from a particular location such as the Earth.

Pulsars have provided astronomers with a unique set of probes for the investigation of the diffuse gas and magnetic fields in interstellar space. Measurement of absorption at 1420 MHz, the frequency of the hyperfine transition in ground-state neutral hydrogen atoms, gives information on the structure of gas clouds, and in many cases provides an estimate of the pulsar distance. The broadband, pulsed nature of pulsar signals makes them ideal for measurements of dispersion by ionized gas. [J.H.Ta.]

Pulse demodulator A device that recovers the modulating signal from a pulse-modulated wave. This recovery may or may not require a two-stage demodulation. Pulse-duration-modulated (PDM) or pulse-amplitude-modulated (PAM) signals may be completely recovered by a conventional detector in amplitude-modulation (AM) radio or a phase-quadrature-type demodulator in frequency-modulation (FM) radio. The reason is that the pulse area is proportional to the modulating-signal amplitude at the time of each pulse, and the low-pass demodulator output filter removes the pulse-frequency components as well as the carrier-frequency components and leaves only the original modulating-frequency components in the filter output. *See* AMPLITUDE-MODULATION DETECTOR; FREQUENCY-MODULATION DETECTOR.

Pulse-position-modulated (PPM) and pulse-code-modulated (PCM) signals require a two-stage demodulation because the pulse train must be processed before the original modulation can be recovered. The pulse-train modulation may be recovered from a radio-frequency (rf) carrier by any appropriate AM or FM detector, provided that the low-pass filter in the detector output passes the significant frequency components of the pulse. *See* PULSE MODULATION. [C.L.A.]

Pulse generator An electronic circuit capable of producing a waveform that rises abruptly, maintains a relatively flat top for an extremely short interval, and then rapidly falls to zero. A relaxation oscillator, such as a multivibrator, may be adjusted to generate a rectangular waveform having an extremely short duration, and as such it is referred to as a pulse generator. However, there is a class of circuits whose exclusive function is generating short-duration, rectangular waveforms. These circuits are usually specifically identified as pulse generators. An example of such a pulse generator is the triggered blocking oscillator, which is a single relaxation oscillator having transformer-coupled feedback from output to input. *See* BLOCKING OSCILLATOR; MULTIVIBRATOR; RELAXATION OSCILLATOR.

Pulse generators sometimes include, but are usually distinguished from, trigger circuits. Trigger circuits generate a short-duration, fast-rising waveform for initiating or triggering an event or a series of events in other circuits. In the pulse generator, the pulse duration and shape are of equal importance to the rise and fall times. *See* TRIGGER CIRCUIT.

The term pulse generator is often applied not only to an electronic circuit generating prescribed pulse sequences but to an electronic instrument designed to generate sequences of pulses with variable delays, pulse widths, and pulse train combinations, programmable in a predetermined manner, often microprocessor-controlled.

A network, formed in such a way as to simulate the delay characteristics of a lossless transmission line, and appropriate switching elements to control the duration of a pulse form the basis for a variety of types of pulse generators. Some delay-line-controlled pulse generators are capable of generating pulses containing considerable amounts of power for such applications as modulators in radar transmitters. *See* DELAY LINE; WAVE-SHAPING CIRCUITS. [G.M.G.]

Pulse jet A type of jet engine characterized by periodic surges of thrust. The pulse jet engine was widely known for its use during World War II on the German V-1 missile (see illustration). The basic engine cycle was invented in 1908. The inlet end of the engine is provided with a grid to which are attached flap valves. These valves are normally held by spring tension against the grid face and block the flow of air back out of the front of the engine. They can be sucked inward by a negative differential pressure to allow air to flow into the engine. Downstream from the flap valves is the combustion chamber. A fuel injection system is located at the entrance to the combustion chamber. The chamber is also fitted with a spark plug. Following the combustion chamber is a long exhaust duct which provides an inertial gas column.

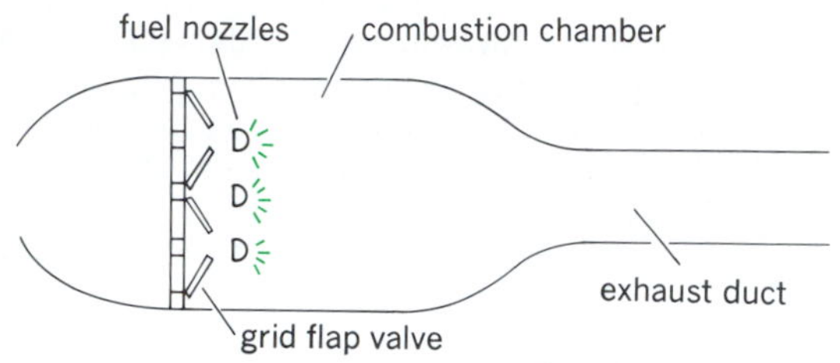

Diagram of a pulse jet.

Pulse jets have also been used to propel radio-controlled target drones and experimental helicopters. In the latter case, they were mounted on the blade tips for directly driving the rotor. The high fuel consumption, noise, and vibrations generated by the pulse jet limit its scope of applications. *See* PROPULSION. [B.Pi.]

Pulse modulation A system of modulation in which the amplitude, duration, position, or mere presence of discrete pulses may be so controlled as to represent the message to be communicated. These several forms of pulse modulation are commonly called, respectively, pulse-amplitude modulation (PAM), pulse-duration modulation (PDM), pulse-position modulation (PPM), and pulse-code modulation (PCM). *See* AMPLITUDE MODULATION; FREQUENCY MODULATION; MODULATION; PHASE MODULATION.

Ease of multiplexing channels by time division is one of the important economic advantages of all forms of pulse modulation. Circuits for accomplishing this are simple and low in cost, and because this low cost is shared by a number of channels, the cost per channel is even lower. When pulse modulation is applied to long-distance communications, these time-division techniques also permit substantial simplification at way stations and branching points. Because each information-bearing pulse keeps its individuality in journeying from the first transmitter to the last receiver, it is comparatively easy to drop and add message channels at various intermediate points along the way.

All forms of pulse modulation may be defined in terms of pulse carrier, and all involve sampling; in addition, PCM implies quantization and coding.

A pulse carrier (Fig. 1) consists of a series of regularly recurrent pulses. In general, the power associated with each pulse differs essentially from zero only during a limited interval of time, which is the pulse width. Sampling is a process of extracting successive portions of predetermined duration taken at regular intervals from a continuously varying magnitude-time wave.

Pulse-amplitude modulation. Pulse-amplitude modulation is amplitude modulation of a pulse carrier. The modulated wave (Fig. 1) is linearly proportional to equally spaced samples of the modulating wave. Chief interest in PAM lies in its application to time-division multiplexing. *See* MULTIPLEXING.

Pulse-duration modulation. Pulse-duration modulation is modulation of a pulse carrier wherein the value of each instantaneous sample of a modulating wave produces a pulse of proportional duration (Fig. 1) by varying the leading, trailing, or both edges of the pulse. PDM is also termed pulse-length modulation or pulse-width modulation. In contrast to PAM, PDM is able to trade extra bandwidth for noise reduction. Only the position of the modulated edge or edges conveys information, and the part of each PDM pulse that conveys no information represents wasted pulse power. When this wasted power is subtracted from PDM the result is PPM.

Pulse-position modulation. Pulse-position modulation is modulation of a pulse carrier wherein the value of each instantaneous sample of a modulating wave varies the position in time of a pulse relative to its unmodulated time of occurrence (Fig. 1). Channel pulses, one for each channel, are transmitted in turn and are preceded by a synchronizing pulse called a marker. This array of marker-plus-channel pulses repeats itself every 125 microseconds and is called a frame.

In practical applications, even though PPM is more efficient than PDM, both are highly inefficient when used for certain purposes, for example, when used for multiplexing ordinary telephone channels. Consequently, communication engineers have shown considerable interest in PCM, which not only is

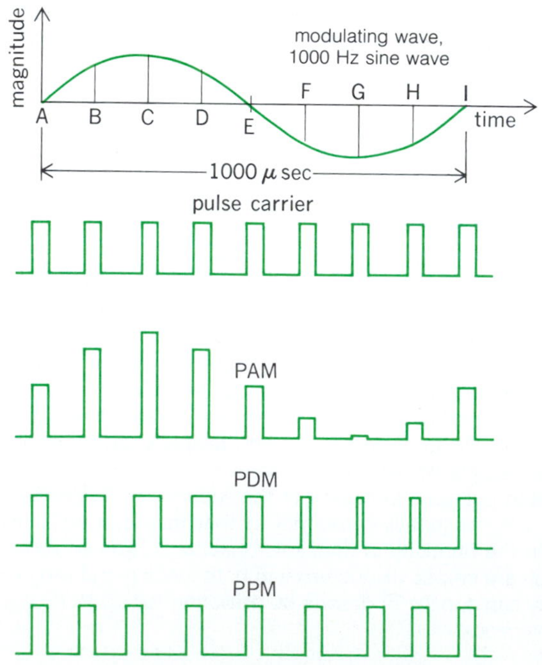

Fig. 1. Examples of pulse modulation. (*After H. S. Black, Modulation Theory, Van Nostrand, 1953*)

more efficient but also possesses many other very important advantages.

Pulse-code modulation. Pulse-code modulation is a method of transmitting continuously varying message waves in which (1) the message wave is sampled, (2) the value of each sample is replaced by the closest one of a finite set of permitted values, and (3) these permitted values are then each unambiguously represented by some one of the possible patterns of N on-or-off pulses. These three operations are known as sampling, quantizing, and coding, respectively.

Quantization changes a continuously varying signal to a stepped signal. Graphically quantization means that the straight line representing the relationship between input and output has been replaced by a stair-step function. Clearly, by using small enough steps, errors may be cut to an arbitrary minimum (Fig. 2).

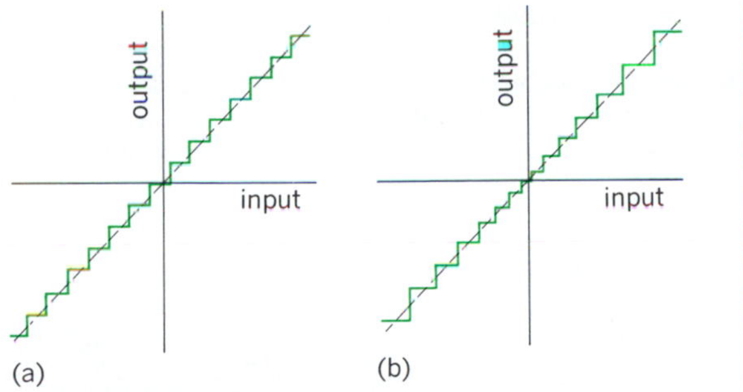

Fig. 2. Relation between input and quantized output. Quantization (*a*) uniform, (*b*) tapered. (*From H. S. Black, Modulation Theory, Van Nostrand, 1953*)

When speech samples are quantized, the steps are tapered as shown in Fig. 2*b*. By tapering most of the steps logarithmically, nearly uniform percentage precision is obtained over most of the range, and far fewer steps are required. To reproduce telephonic speech to a high degree of fidelity, 128 logarithmic steps will suffice.

In PCM these 128 values are represented by a 7-pulse code. Each voice sample is represented with adequate accuracy by some one of the possible patterns of 7 on-or-off pulses. Channels are multiplexed by time division.

Similarly, other forms of communication may also be represented by on-or-off pulses. In practice, a single PCM system may process many kinds of communications and may provide many channels of each kind.

Ordinarily, PCM systems are called upon to transmit large numbers of on-or-off pulses per second. For example, for each telephone conversation, PCM transmits 56,000 on-or-off pulses per second in each direction of transmission. A one-way television signal might be represented by 70,000,000 on-or-off pulses per second.

By having to make only on-or-off distinctions, PCM is able to deliver a high-quality signal even when noise and interference are so bad that it is barely possible to recognize the pulses. At each regenerative repeater, just so long as incoming pulses are correctly identified, a new pulse generator may be caused to generate new pulses of correct magnitude, waveform, and timing. [H.S.Bl.]

Pulse modulator

A device for the pulse modulation of a radio-frequency carrier signal. *See* PULSE MODULATION.

Pulse-amplitude scheme. Pulse-amplitude modulation (PAM) requires least complexity of any of the pulse-modulation schemes and is usually used in the initial stages of other pulse-modulation methods. The amplitude of the input signal is sampled periodically at a fixed frequency, producing a sequence of pulses whose amplitude varies in accordance with the modulating signal. In a common practical system the trains of pulses representing several input signals are interlaced to form a time-division multiplex system.

Illustration *a* is a block diagram of the essential parts of a four-channel time-division multiplex transmitter that uses PAM. Illustration *b* depicts some of the representative waveforms. The principal elements of this system are the distributor, or commutator, which can be either electromechanical or electronic; a source of square waves at the sampling frequency f_s; and a source of square waves of one-half the sampling frequency $f_s/2$ obtained from a bistable circuit arranged to provide both phases of the square wave.

In a practical PAM system, it is necessary to provide guard time between pulses to reduce interchannel crosstalk. Illustration *a* shows the distributor output pulses passing through a synchronous gate operating at twice the sampling frequency. This gate selects the center portion of each pulse in the train. The resultant output is a series of amplitude-modulated pulses having only one-half of their original width with blank time between each pulse. The waveform in the output line with signals due to channel A alone is shown in illustration *b*.

Pulse-duration scheme. Pulse-duration modulation (PDM) can be produced by converting a train of pulses from a PAM system. If the signals generated in a PAM system are added to constant-amplitude sampling-frequency pulses, a train of pulses of varying amplitude, but with a finite minimum value, is produced at the output of the adder. The resulting pulses can be

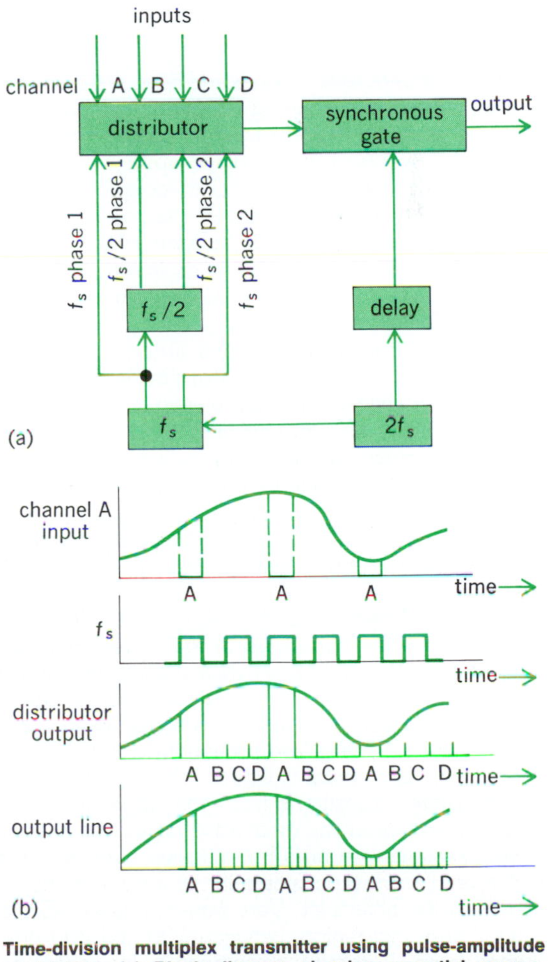

Time-division multiplex transmitter using pulse-amplitude modulation. (*a*) Block diagram showing essential components in four-channel system. (*b*) Waveforms associated with channel A.

differentiated and clipped, resulting in a series of reverse sawtooth shapes in which the width of each sawtooth becomes proportional to the amplitude of the particular pulse. When passed through a limiter, the desired sequence of pulses, having constant amplitude and varying in width in accordance with the signal, is generated.

Pulse-position scheme. Pulse-position modulation (PPM) can be derived from PDM. In one method, the variable-width pulses are first differentiated and then rectified to permit only the negative peaks to trigger a monostable multivibrator. In the latter, each negative trigger pulse causes a pulse of constant amplitude and width to be generated, whose time position is proportional to the amplitude of the modulating signal (input).

Pulse-code scheme. The sampling techniques needed for pulse-code modulation (PCM) can be similar to those employed in PAM. To quantize a signal, the amplitude sample is first matched with a series of amplitude standards and then converted to the one with the nearest standard value. One method of achieving this is to utilize a multiplicity of regenerative devices, each of which changes between its OFF and ON state within a very small input voltage range. By proper biasing, these devices can be set to operate at any given input voltage level. When the inputs and outputs of, for example, eight of these level selectors work in parallel, it is possible to convert input signals of varying amplitudes to outputs of eight discrete amplitudes. *See* ANALOG-TO-DIGITAL CONVERTER.

Coding can be accomplished with the aid of a special beam code tube, consisting of a cathode-ray tube with conventional gun and electrostatic deflection plates, a special aperture plate placed perpendicularly to the electron beam and an anode. The aperture plate has horizontal slots in eight vertical rows in accordance with the pulse code. [E.L.G./C.L.A.]

Pulse transformers

Iron-cored devices which are used in the transmission and shaping of low-power pulses whose widths range from a fraction of a microsecond to about 25 microseconds. Among the extensive applications of pulse transformers are the following: (1) to couple between the stages of pulse amplifiers; (2) to invert the polarity of a pulse; (3) to change the amplitude and impedance level of a pulse; (4) to differentiate a pulse; (5) to effect "dc isolation" between a source and a load; (6) to act as coupling element in certain pulse-generating circuits.

In many cases the functions listed above can be accomplished as well or better with transistor circuits. However, the pulse transformer, being a passive element, has none of the instability associated with active circuits.

A pulse transformer behaves as a reasonable approximation to a perfect transformer when used in connection with the fast waveforms it is intended to handle. The core of a pulse transformer is usually molded from a magnetic ceramic such as sintered manganese-zinc ferrite. *See* TRANSFORMER. [C.C.H.]

Pulsed gas laser

A device which uses pulsed gas discharges to achieve the population inversion required for laser action. Pulsed gas discharges permit a further departure from equilibrium than is possible in a continuous discharge. Thus pulsed laser action can be obtained in some additional gases which could not be made to lase continuously. In some of them the length of the laser pulse is limited when the lower state is filled by stimulated transitions from the upper level, and so introduces absorption at the laser wavelength. An example of such a self-terminating laser is the nitrogen laser, which gives pulses of several nanoseconds duration at a wavelength of 337 nanometers in the ultraviolet. Very powerful laser radiation in the vacuum ultraviolet region, between 100 and 200 nm, can be obtained from short-pulse discharges in hydrogen, and in rare gases such as xenon at high pressure. (High pressure leads to higher power.) When the gas pressure is too high to permit an electric discharge, excitation may be provided by an intense burst of fast electrons from a small accelerator, the so-called E beam.

Some gases, notably carbon dioxide, which can provide continuous laser action, also can be used to generate intense pulses of microsecond duration. For this purpose, gas pressures of about 1 atm (10^5 Pa) are used, and the electrical discharge takes place across the diameter of the laser column, hence the name transverse-electrical-atmospheric (TEA) laser. *See* GAS-DISCHARGE LASER; LASER. [S.F.J.; A.L.S.]

Pumice

A rock froth, formed by the extreme puffing up of liquid lava by expanding gases liberated from solution in the lava prior to and during solidification. Some varieties will float in water for many weeks before becoming waterlogged. Typical pumice is siliceous (rhyolite or dacite) in composition, but the lightest and most vesicular pumice (known also as reticulite and thread-lace scoria) is of basaltic composition. *See* LAVA; VOLCANIC GLASS. [G.A.M.]

Pump

A machine that draws a fluid into itself through an entrance port and forces the fluid out through an exhaust port (see illustration). A pump may serve to move liquid, as in a cross-country pipeline; to lift liquid, as from a well or to the top of a tall building; or to put fluid under pressure, as in a hydraulic brake. These applications depend predominantly upon the discharge characteristic of the pump. A pump may also serve to empty a container, as in a vacuum pump or a sump pump, in which case the application depends primarily on its intake characteristic. *See* CENTRIFUGAL PUMP; COMPRESSOR; DISPLACEMENT PUMP; FAN; FUEL PUMP; PUMPING MACHINERY; VACUUM PUMP. [E.F.W.]

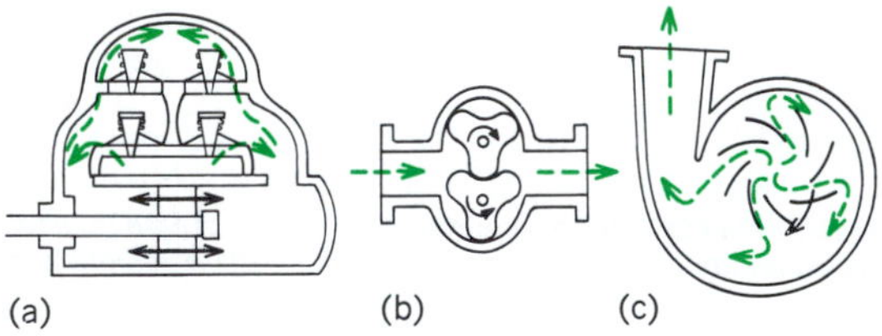

Pumps. (*a*) Reciprocating. (*b*) Rotary. (*c*) Centrifugal.

Pumped storage

A process, also known as hydroelectric storage, for converting large quantities of electrical energy to potential energy by pumping water to a higher elevation, where it can be stored indefinitely and then released to pass through hydraulic turbines and generate electrical energy. An indirect process is necessary because electrical energy cannot be stored effectively in large quantities. Storage is desirable, as the consumption of electricity is highly variable between day and night, between weekday and weekend, as well as among seasons. Consequently, much of the generating equipment needed to meet the greatest daytime load is unused or lightly loaded at night or on weekends. During those times the excess capability can be used to generate energy for pumping, hence the necessity for storage.

A typical pumped-storage development is composed of two reservoirs of essentially equal volume situated to maximize the difference in their levels. These reservoirs are connected by a system of waterways along which a pumping-generating station is located (see illustration). Under favorable geological conditions, the station will be located underground, otherwise it will be situated on the lower reservoir. The principal equipment of the station is the pumping-generating unit. In United States

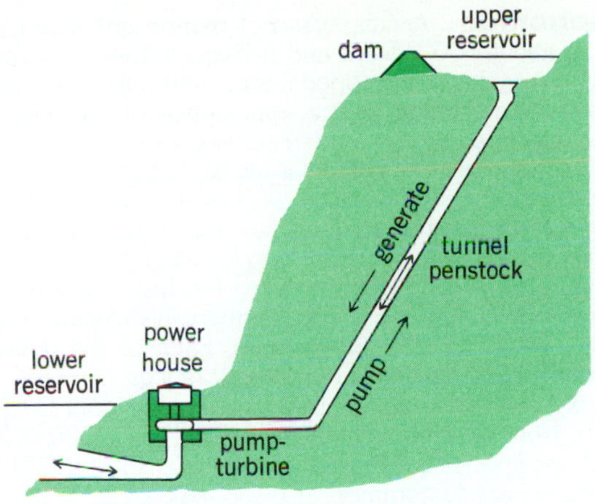

Schematic of a conventional pumped-storage development.

practice, the machinery is reversible and is used for both pumping and generating; it is designed to function as a motor and pump in one direction of rotation and as a turbine and generator in opposite rotation. *See* ELECTRIC POWER GENERATION; ENERGY STORAGE; HYDRAULIC TURBINE; HYDROELECTRIC GENERATOR; PUMPING MACHINERY; WATERPOWER. [D.L.G.]

Pumping machinery Devices which convey fluids, chiefly liquids, from a lower to a higher elevation or from a region of lower pressure to one of higher pressure. Pumping machinery may be broadly classified as mechanical or as electromagnetic.

In mechanical pumps the fluid is conveyed by direct contact with a moving part of the pumping machinery. The two basic types are (1) velocity machines, centrifugal or turbine pumps, which impart energy to the fluid primarily by increasing its velocity, then converting part of this energy into pressure or head, and (2) displacement machines with plungers, pistons, cams, or other confining forms which act directly on the fluid, forcing it to flow against a higher pressure. *See* CENTRIFUGAL PUMP; DISPLACEMENT PUMP.

Where direct contact between the fluid and the pumping machinery is undesirable, as in atomic energy power plants for circulating liquid metals used as reactor coolants or as solvents for reactor fuels, electromagnetic pumps are used. There are no moving parts in these pumps; no shaft seals are required. The liquid metal passing through the pump becomes, in effect, the rotor circuit of an electric motor. *See* ELECTROMAGNETIC PUMPS. [E.F.W.]

Pumpkin The term commonly applied to the larger, orange-colored fruit of the *Cucurbita* species, used when ripe as a table vegetable, in pies, or for autumn decoration. Although some taxonomists would restrict the term pumpkin to the species *Cucurbita pepper* and *C. moschata*, it is also used in referring to *C. mixta*. New Jersey, Illinois, and California are important producing states. *See* VIOLALES. [H.J.C.]

Puna An alpine biological community in the central portion of the Andes Mountains of South America. Sparsely vegetated, treeless stretches cover high plateau country (altiplano) and slopes of central and southern Peru, Bolivia, northern Chile, and northwestern Argentina. The poor vegetative cover and the puna animals are limited by short seasonal precipitation as well as by the low temperatures of high altitudes.

Like the paramos of the Northern Andes, punas occur above timberline, and extend upward, in modified form, to perpetual snow. Due to greater heights of the Central Andean peaks, aeolian regions, that is, regions supplied with airborne nutrients above the upper limit of vascular plants, generally the snowline, are more extensive here than above the paramos. *See* PARAMO. [H.G.B.]

Pure culture A culture that contains cells of one kind, all progeny of a single cell. The concept of a pure culture is an important one in microbiology because most considerations of microorganisms require dealing with only one kind of a cell at a time. Many techniques have been evolved for the isolation and maintenance of pure cultures. *See* MICROBIOLOGICAL METHODS. [J.W.Fo./R.E.K.]

Purine A heterocyclic organic compound containing fused pyrimidine and imidazole rings; the structural formula is shown below. A large number of substituted purine derivatives occur

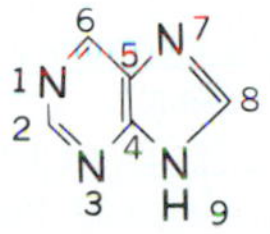

in nature, some of which, as components of nucleic acids and coenzymes, play vital roles in the genetic and metabolic processes of all living organisms. *See* COENZYME; NUCLEIC ACID.

The purine bases, adenine and guanine (see formulas),

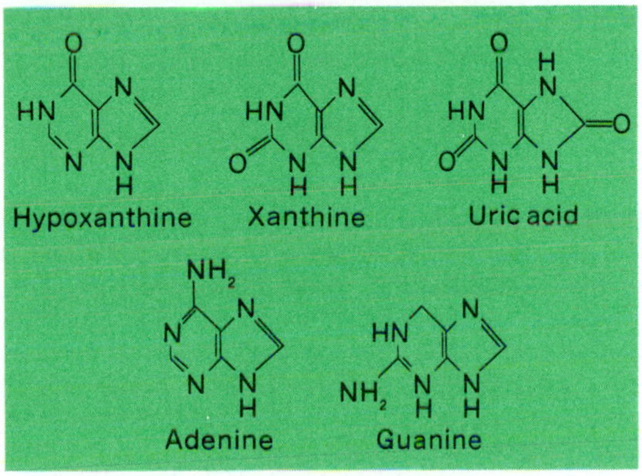

Structure of some important purines.

together with pyrimidines, are fundamental components of all nucleic acids. Certain methylated derivatives of adenine and guanine are also present in some nucleic acids in low amounts. In biological systems, hypoxanthine, adenine, and guanine occur mainly as their 9-glycosides, the sugar being either ribose or 2-deoxyribose, although other pentoses are found in rare cases. Such compounds are termed nucleosides generically, and inosine (hypoxanthine nucleoside), adenosine, or guanosine specifically. The name nucleotide refers to phosphorylated derivatives of nucleosides, the principal ones containing 5′-phosphate groups, as in guanosine 5′-phosphate and adenosine 5′-triphosphate (ATP). *See* ADENOSINETRIPHOSPHATE (ATP); PYRIMIDINE. [S.C.H.]

Push-pull amplifier A two-transistor or two-tube amplifier circuit often used as the power-output stage of a multistage amplifier. The power-output stage in an audio amplifier is normally expected to furnish from 5 to 50 watts or more. Use of one active device in this stage is not feasible, because the transistor or tube would have to operate class A with a low conversion efficiency, on the order of 10%. Use of two or more active devices in parallel does not improve efficiency. However, with two active devices operating in push-pull, it is possible to supply the required amount of power with conversion efficiency on the order of 50%. This higher efficiency

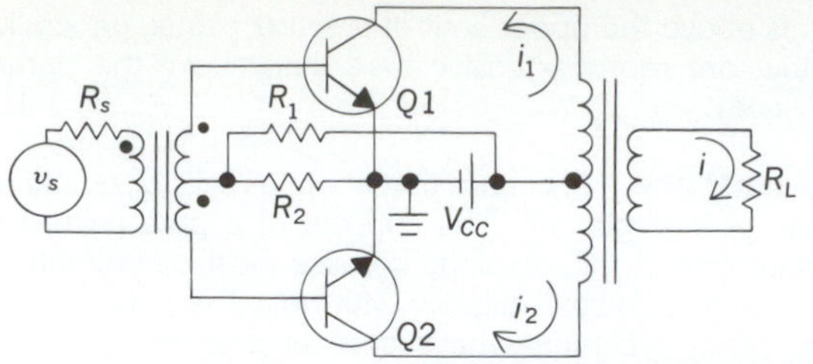

Two transistors in a push-pull arrangement. *Q1* and *Q2* are transistors; v_s is source voltage; R_s is source resistance; R_L is land resistance; V_{CC} *is collector supply voltage;* R_1 and R_2 are resistances; i, i_1 and i_2 are currents.

means that the amplifier does not require as much power from the power supply, because less power is dissipated as heat by the power amplifier active devices. *See* POWER AMPLIFIER.

A simplified circuit for a push-pull amplifier is shown in the illustration. The input signals must be of equal magnitude but 180° out of phase. The collector current in the transistor with the positive signal is increasing while the transistor current in the other transistor is decreasing. This relation between the collector currents gave rise to the name push-pull. Because of their phase relationship, the two currents flowing in the two halves of the primary winding of the transformer produce an output similar to that of a single source connected to the transformer.

Although the active devices are indicated as junction transistors in the illustration, other devices such as triodes, beam power tubes, or field-effect transistors (FET) can also be used in this push-pull arrangement. *See* AMPLIFIER. [H.F.K.]

Pycnodontiformes An order of very specialized deep-bodied fishes near the holostean level of organization that are known only from the fossil record. They were a widespread group that first appeared in the Upper Triassic of Europe, flourished during the Jurassic and Cretaceous, and persisted to the Upper Eocene. Pycnodontiforms are a closely interrelated group most commonly found preserved in marine limestone and associated with coraliferous facies.

They are characterized (see illustration) by a laterally compressed, disk-shaped body; long dorsal and anal fins, with each fin ray supported by its own endoskeletal element; an externally symmetrical tail; and an axial skeleton having greatly extended neural and hemal spines that are expanded in some forms, and neural and hemal arches which are well ossified—sometimes interdigitating around the notochord—but with no vertebral centra. [T.M.C.]

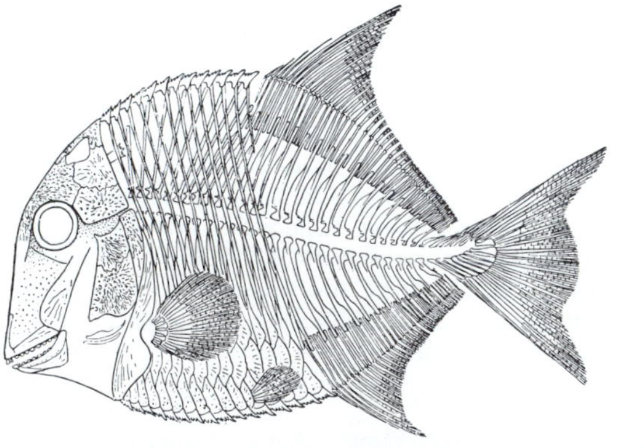

***Coelodus costue*, a pycnodont from Lower Cretaceous of Italy; length to 4 in. (10 cm). (*After A. S. Woodward*)**

Pycnogonida A subphylum of marine arthropods, consisting of about 600 Recent and perhaps 1 Devonian species. The Pycnogonida, or Pantopoda, are commonly called sea spiders. This name refers to the appearance of the adults and does not indicate affinity with terrestrial spiders.

The pycnogonids are characterized by reduction of the body to a series of cylindrical trunk somites supporting the appendages, a large specialized feeding apparatus called the proboscis, and a reduced abdomen. In many genera there are seven pairs of appendages, of which the first four are on the first or cephalic segment. This segment also bears a dorsal tubercle containing four simple eyes. Each of the remaining three trunk segments bears a single pair of legs.

Pycnogonids are found in all seas except the inner Baltic and Caspian, from intertidal regions to depths of 21,500 ft (6500 m), and one species is bathypelagic at about 3300 ft (1000 m). They are especially common in polar seas. Most of the intertidal species spend their lives in association with some coelenterate as encysted, parasitic larval and juvenile stages, or are ectoparasitic as adults, being attached to anemones and hydroids by their claws and proboscides. A few have been found riding in hydromedusae, and some occur in the mantle cavity of bivalves or on nudibranchs and holothurians. Most of the deep-sea species are known only as adults, and their mode of life is a mystery. *See* ARTHROPODA. [J.W.He.]

Pyelonephritis A bacterial infection of the kidney and the kidney pelvis which occurs in two forms, acute and chronic. It is very frequently associated with some disturbance of urine flow in the lower urinary tract. Examples of common causes of obstruction are stones in the ureter, cancer, or an enlarged prostate. The causative bacteria commonly are those that normally inhabit the large intestine. The symptoms are shaking chills, fever, pain in the flank, and frequent passage of small amounts of urine which has a burning quality. Antibiotic therapy brings prompt relief of symptoms and eradicates the infection. However, if the interference to urine flow is not relieved, reinfection often occurs. *See* KIDNEY; NEPHRITIS. [G.E.Str.]

Pygasteroida An order of Diadematacea which exhibits various stages in the backward migration of the anus out of the apical system. They have four genital pores (instead of five), noncrenulate tubercles, and simple ambulacral plates. All members are referred to a single family, the Pygasteridae. *See* DIADEMATACEA. [H.B.F.]

Pyramid and frustum A pyramid is a polyhedron of which one face is called the base, and the other faces (called lateral faces) are triangles having a common vertex which is called the vertex of the pyramid. The distance from the vertex to the base is called the altitude. The volume of a pyramid is one-third the product of its base times its altitude ($V = \frac{1}{3} Bh$). Sections of a pyramid formed by planes parallel to the base are similar to the base. A pyramid with a triangular base is a triangular pyramid, or tetrahedron. If the base of a pyramid is a regular polygon and the lateral edges are equal, the pyramid is called a regular pyramid; if the polygon is a square, the pyramid is called a square pyramid. A frustum of a pyramid is a segment of a pyramid included between two parallel planes. *See* POLYHEDRON; PRISMATOID AND PRISMOID. [J.S.F.]

Pyran One of a group of organic compounds containing a heterocyclic six-membered ring made up of one oxygen and five carbon atoms. Formulas (I) and (II) show pyrans of various kinds. *See* HETEROCYCLIC COMPOUNDS.

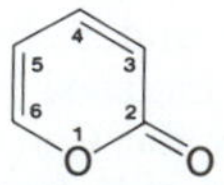

α-Pyrone (I)

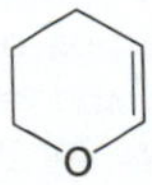

$\Delta^{2,3}$-Dihydropyran (II)

All pyranosidic carbohydrates are tetrahydropyran derivatives. Glycals, for example, glucal, are $\Delta^{2,3}$-dihydropyrans. The α-pyrone system is present in cardiac-active steroid derivatives, such as scillaren A, isolated from squill and from a variety of toad.

Formulas (III) and (IV) show some benzopyran types. Vitamin E, or α-tocopherol, is a chroman (III) derivative. Coumarin, or

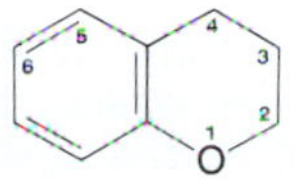

(III)

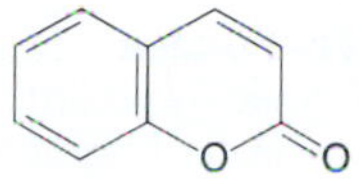

(IV)

α-chromone (IV), the fragrant body in clover, was formerly used in flavors and perfumes. [W.J.Ge.]

Pyrargyrite A mineral having composition Ag_3SbS_3. The mineral occurs as prismatic crystals and in massive form and in disseminated grains. The hardness is 2.5 on Mohs scale and specific gravity is 5.85. The luster is adamantine and the color a deep ruby red to black, giving it the name dark ruby silver. Pyrargyrite is an important silver ore when it is found in veins associated with proustite and other silver minerals. It has been mined as silver ore at Chañarcillo, Chile; Freiberg, Germany; Guanajuato, Mexico; and Cobalt, Ontario, Canada. *See* PROUSTITE. [C.S.Hu.]

Pyrazole One of a group of organic heterocyclic compounds with two nitrogen atoms occupying adjacent positions in a doubly unsaturated five-membered ring, as in the structure below. The pyrazole system is aromatic in character.

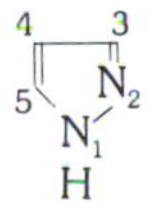

Electrophilic substitution occurs preferentially at the 4 position. The nucleus is resistant to disruption by oxidation. Certain drugs and dyes are pyrazole derivatives. *See* AZOLE; HETEROCYCLIC COMPOUNDS.

The parent compound shown above is a water-soluble, colorless solid, mp 70°C (158°F), bp 187°C (360°F), with an odor resembling that of pyridine. Pyrazole is both a weak base and a weak acid.

Antipyrine and aminopyrine are pyrazole derivatives of considerable value as analgesics and antipyretics, respectively. Azo coupling of 3-methyl-1-phenyl-5-pyrazolone or of related compounds gives 4-azopyrazolone derivatives, which are of interest as wool, food, and photographic dyes. *See* DYE. [W.J.Ge.]

Pyrenomycetes The largest class in the subdivision Ascomycotina. The distinguishing feature of these fungi is the ascocarp, or perithecium, that gives rise to the common name for the class, perithecial ascomycetes. Most ascocarps are brown or black, but some are blue, yellow, or red. Coiled branches form on the hyphae to initiate ascocarp formation, and they are quickly enveloped by layers of hyphae to form a wall around the coil, which subsequently differentiates into ascogenous hyphae that will form the asci. The internal tissues of the ascocarp vary between species. In most species, a neck with a canal forms through which the ascospores will be discharged. The asci are formed in a cluster or layer in the base of the ascocarp. Most have a single rigid wall (unitunicate), and the eight ascospores are forcibly discharged through a pore in the apex of the ascus, but in some the ascus wall dissolves. The ascospores are variable in color, size, shape, and number of cells. Most pyrenomycetes also form abundant asexual conidia.

Pyrenomycetes are widespread and occur primarily on living and dead plant materials. They may cause serious disease or act as agents of decomposition, producing a well-developed mycelium that may be superficial or immersed in the host or substrate. *See* ASCOMYCOTINA; EUMYCOTA; FUNGI. [R.T.Ha.]

Pyrenulales An order of the class Ascolichenes, also known as the Pyrenolichenes. The flask-shaped perithecia are uniformly immersed in the medulla of the thalli with a small ostiole opening at the surface. The asci and paraphyses arise from a blackened hypothecium and line the walls of the perithecium. The spores eventually burst the ascal walls and ooze out through the ostiole in a jelly matrix.

There are about 10 families, 50 genera, and more than 1500 species in the Pyrenulales. The major taxonomic criteria for separating genera and species are the septation and color of spores, since vegetative characters are so poorly developed. [M.E.H.]

Pyridine An organic heterocyclic compound containing a triunsaturated six-membered ring of five carbon atoms and one nitrogen atom. Pyridine (I) and pyridine homologs are obtained

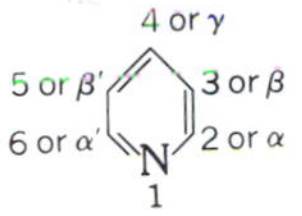

(I)

by extraction of coal tar or by synthesis. The pyridine system is found in natural products, for example, in nicotine (II) from tobacco, in ricinine (III) from castor bean, in pyridoxine or vitamin B_6 (IV), in nicotinamide or niacinamide or vitamin P (V), and in several groups of alkaloids. *See* HETEROCYCLIC COMPOUNDS.

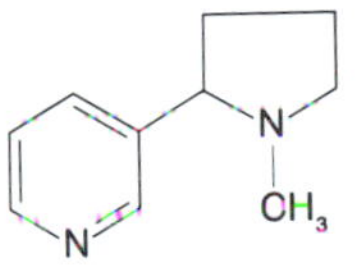

(II)

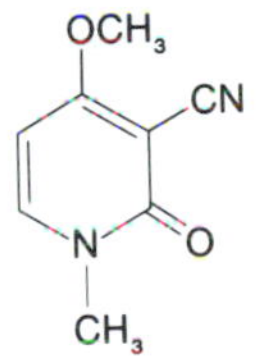

(III)

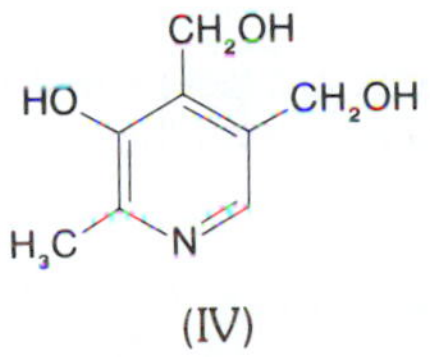

(IV)

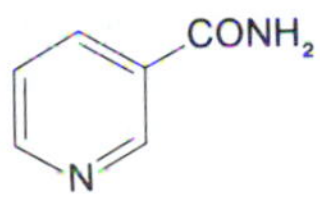

(V)

Pyridine (I) is a colorless, hygroscopic liquid with a pungent, unpleasant odor. When anhydrous it boils at 115.2–115.3°C (239.4–239.5°F). Pyridine is miscible with organic solvents as well as with water. The pyridine system is aromatic. It is stable to heat, to acid, and to alkali. Pyridine is used as a solvent for organic and inorganic compounds, as an acid binder, as a basic catalyst, and as a reaction intermediate.

Pyridine is an irritant to skin (eczema) and other tissues (conjunctivitis), and chronic exposure has been known to cause liver and kidney damage. Repeated exposure to atmospheric levels greater than 5 parts per million is considered hazardous. [W.J.Ge.]

Pyrimidine A heterocyclic organic compound (I) containing nitrogen atoms at positions 1 and 3. Naturally occurring

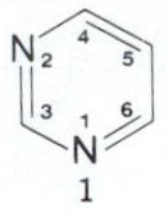

(I)

derivatives of the parent compound are of considerable biological importance as components of nucleic acids and coenzymes and, in addition, synthetic members of this group have found use as pharmaceuticals. *See* COENZYME; NUCLEIC ACID.

Pyrimidine compounds which are found universally in living organisms include uracil (II), cytosine (III), and thymine (IV).

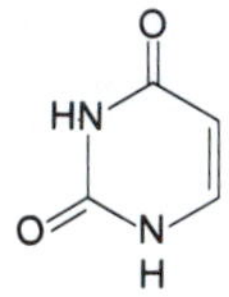

(II)

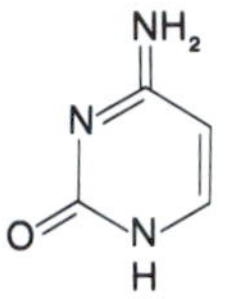

(III)

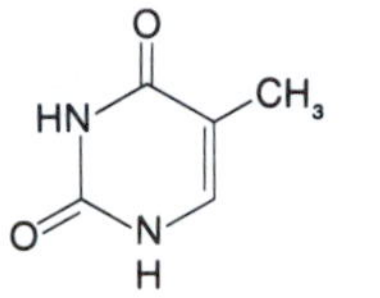

(IV)

Together with purines these substances make up the "bases" of nucleic acids, uracil and cytosine being found characteristically in ribonucleic acids, with thymine replacing uracil in deoxyribonucleic acids. A number of related pyrimidines also occur in lesser amounts in certain nucleic acids. Other pyrimidines of general natural occurrence are orotic acid and thiamine (vitamin B_1). *See* DEOXYRIBONUCLEIC ACID (DNA); PURINE; RIBONUCLEIC ACID (RNA).

Among the sulfa drugs, the pyrimidine derivatives, sulfadiazine, sulfamerazine, and sulfamethazine, have general formula (V). These agents are inhibitors of folic acid biosynthesis in

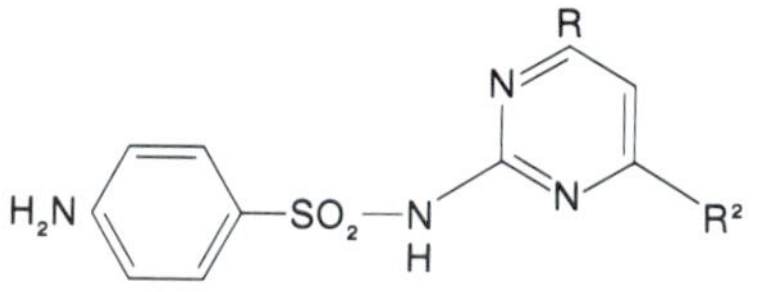

(V)

microorganisms. The barbiturates are pyrimidine derivatives which possess potent depressant action on the central nervous system. *See* BARBITURATES; SULFONAMIDE. [S.C.H.]

Pyrite A mineral having composition FeS_2. Pyrite has a Mohs hardness of 6–6.5 and a density of 5.02. The luster is metallic, the color brass yellow, and the streak greenish black or brownish black.

Pyrite, or iron pyrites, is the most common "fool's gold," but it is hard and brittle whereas gold is soft and sectile. Its hardness also distinguishes it from softer chalcopyrite. Marcasite was once thought to be polymorphous with pyrite, but precise analyses show that marcasite contains excess iron, whereas pyrite is stoichiometric FeS_2. *See* MARCASITE.

Pyrite is the most common and most widespread sulfide mineral. It forms under almost all known conditions of mineral deposition. Under oxidizing conditions, pyrite readily alters to iron sulfates and eventually to limonite, forming gossan, the surface expression of pyrite-rich mineral deposits. *See* LIMONITE.

Because of its high sulfur content (53.4%), pyrite has become a source of sulfur for the production of sulfuric acid. In some places, it is mined for sulfur alone. *See* SULFUR. [L.Gr.]

Pyroclastic rocks Rocks of extrusive (volcanic) origin, composed of rock fragments produced directly by explosive eruptions. Pyroclastic fragments may represent shattered and comminuted older rocks (volcanic, plutonic, sedimentary, or metamorphic) or solidified lava droplets formed by violent explosion. *See* TUFF; VOLCANO. [C.A.C.]

Pyroelectricity The property of certain crystals to produce a state of electric polarity by a change of temperature. Certain dielectric (electrically nonconducting) crystals develop an electric polarization (dipole moment per unit volume) when they are subjected to a uniform temperature change. This pyroelectric effect occurs only in crystals which lack a center of symmetry and also have polar directions (that is, a polar axis). These conditions are fulfilled for 10 of the 32 crystal classes. Typical examples of pyroelectric crystals are tourmaline, lithium sulfate monohydrate, cane sugar, and ferroelectric barium titanate.

Pyroelectric crystals can be regarded as having a built-in or permanent electric polarization. When the crystal is held at constant temperature, this polarization does not manifest itself because it is compensated by free charge carriers that have reached the surface of the crystal by conduction through the crystal and from the surroundings. However, when the temperature of the crystal is raised or lowered, the permanent polarization changes, and this change manifests itself as pyroelectricity.

The magnitude of the pyroelectric effect depends upon whether the thermal expansion of the crystal is prevented by clamping or whether the crystal is mechanically unconstrained. In the clamped crystal, the primary pyroelectric effect is observed, whereas in the free crystal, a secondary pyroelectric effect is superposed upon the primary effect. The secondary effect may be regarded as the piezoelectric polarization arising from thermal expansion, and is generally much larger than the primary effect. *See* PIEZOELECTRICITY. [H.Gr.]

Pyrolusite A mineral having composition MnO_2. Well-developed crystals (polianite) are rare; it is usually in radiating fibers or reniform coatings. The hardness is 1–2 on the Mohs scale (often soiling the fingers) and the specific gravity is 4.75. The luster is metallic and the color iron-black. It frequently forms pseudomorphs after other manganese minerals, notably manganite.

Pyrolusite is extensively mined as a manganese ore in many countries, chiefly in Russia, Ghana, India, the Republic of South Africa, Morocco, Brazil, and Cuba. *See* MANGANESE; MANGANITE. [C.S.Hu.]

Pyrolysis The chemical change of a substance by means of heat alone. Thermal rearrangements into isomers, thermal polymerizations, and thermal decompositions are all included in the term pyrolysis, but it does not include thermal changes that require catalysts or changes that are initiated by other forms of energy. A transformation promoted by ultraviolet radiation, for example, would be photolytic, not pyrolytic. *See* CATALYSIS; PHOTOCHEMISTRY.

An inorganic example of pyrolysis is shown in reaction (1).

$$\underset{\text{Sodium bicarbonate}}{2NaHCO_3} \rightarrow \underset{\text{Sodium carbonate}}{Na_2CO_3} + \underset{\text{Water}}{H_2O} + \underset{\text{Carbon dioxide}}{CO_2} \quad (1)$$

Prominent industrial examples include cracking of petroleum; destructive distillation of coal or wood; pyrolysis of methane into hydrogen and carbon black; and syntheses of biphenyl from benzene, or styrene from ethylbenzene, or ethylene from ethane. *See* CARBON BLACK; CHARCOAL; COAL CHEMICALS; COKE; CRACKING; DESTRUCTIVE DISTILLATION; LIME (INDUSTRY).

Many pyrolytic reactions involve the production of radicals as intermediates. Propane and butane are representative paraffins that have been carefully studied for their pyrolytic behavior. Propane changes at 500–600°C (930–1110°F) bidirectionally into methane and ethylene or into hydrogen and propylene. Butane gives rise to methane plus propylene, ethane plus ethylene, and to a lesser extent hydrogen plus butylene. In stable organic molecules each carbon is surrounded by eight electrons, as in the structures for methane and propane shown here, wherein each bond depicts an electron pair.

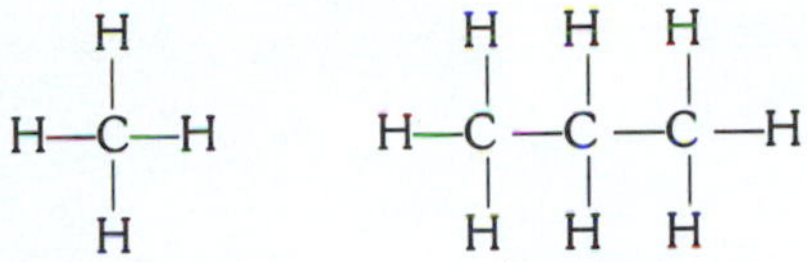

Methane Propane

The bond energies holding a molecule together become smaller with increasing temperature. If, eventually, one of the bonds comes to zero energy, the molecule falls apart into radicals. One of the carbons in a radical has only seven electrons around it, for example, the methyl radical,

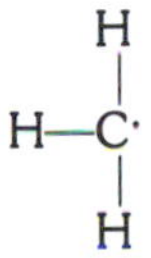

The dot represents one electron. Methylene represents a carbon with only six electrons around it, as in methylene itself,

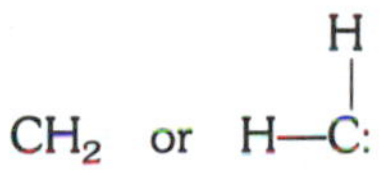

The products from propane or butane mentioned above are the result of initial thermal cleavage of the alkane into radicals, for example, methyl and ethyl from propane, as shown by reaction (2), followed by collision of these radicals (R·) with undecomposed hydrocarbon to form RH and create a new radical, as shown by reaction (3). The $C_3H_7\cdot$ radical then breaks into an unsaturated hydrocarbon and a smaller radical or atom, as seen in reaction (4). These new radicals take the place of the original R· and cause the breakdown to continue. This "chain reaction" is interrupted when the radicals (R· or H·) combine with each other to form R—R, R—H, or H—H. *See* CHAIN REACTION (CHEMISTRY); FREE RADICAL.

$$C_3H_8 \rightarrow CH_3\cdot + C_2H_5\cdot \qquad (2)$$

$$C_3H_8 + R\cdot \rightarrow RH + C_3H_7\cdot \qquad (3)$$

$$C_3H_7\cdot \begin{cases} \nearrow C_2H_4 \text{ (ethylene)} + CH_3\cdot \text{ (methyl radical)} \\ \searrow C_3H_6 \text{ (propylene} + H\cdot \text{ (hydrogen atom)} \end{cases} \qquad (4)$$

There are compounds which pyrolyze by mechanisms not involving bond rupture into radicals. Instead, pairs of electrons stay together and move as a unit, usually because of some structural feature within the molecule that makes it possible. Lower temperatures are usually observed for such processes than for the ones involving scission into radicals. For example, ethyl acetate decomposes at 500°C (930°F) into acetic acid and ethylene, isopropyl acetate at 450°C (1000°F) into acetic acid and propylene, and tertiary butyl acetate at 350°C (660°F) into acetic acid and isobutylene. It is thus evident that esters of primary alcohols are more stable than those of secondary alcohols which, in turn, are more stable than esters of tertiary alcohols. All three types pyrolyze by a mechanism involving a quasi six-membered ring in the transition state and realignment of electron pairs, as in reaction (5). The term R

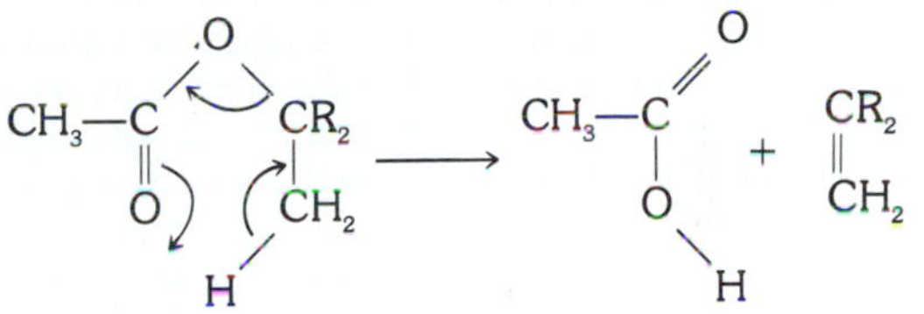

(5)

represents a hydrogen or a methyl group. This cyclic transition state mechanism has been adapted to countless other reactions of organic chemistry. *See* ORGANIC REACTION MECHANISM. [C.D.H.]

Pyrometallurgy Processes employing chemical reactions at elevated temperatures for the extraction of metals from ores and concentrates. Pyrometallurgy is the principal means of metal production.

The processes of pyrometallurgy may be divided into preparation processes which convert the raw material to a form suitable for further processing (for example, roasting to convert sulfides to oxides), reduction processes which reduce metallic compounds to metal (the blast furnace which reduces iron oxide to pig iron), and refining processes which remove impurities from crude metal (fractional distillation to remove iron, lead, and cadmium from crude zinc).

The complete production scheme, from ore to refined metal, may employ pyrometallurgical processes (steel, lead, tin, zinc), or only the primary extraction processes may be pyrometallurgical, with other methods used for refining (copper, nickel). In some cases (uranium, tungsten, molybdenum), isolated pyrometallurgical processes are used in a treatment scheme which is predominately nonpyrometallurgical. *See* IRON METALLURGY; METALLURGY. [H.H.K.]

Pyrometer A temperature-measuring device, originally an instrument that measures temperatures beyond the range of thermometers, but now in addition a device that measures thermal radiation in any temperature range. This article discusses radiation pyrometers; for other temperature-measuring devices *see* BOLOMETER; THERMISTOR; THERMOCOUPLE.

The illustration shows a very simple type of radiation pyrometer. Part of the thermal radiation emitted by a hot object is intercepted by a lens and focused onto a thermopile. The resultant heating of the thermopile causes it to generate an electrical signal (proportional to the thermal radiation) which can be displayed on a recorder.

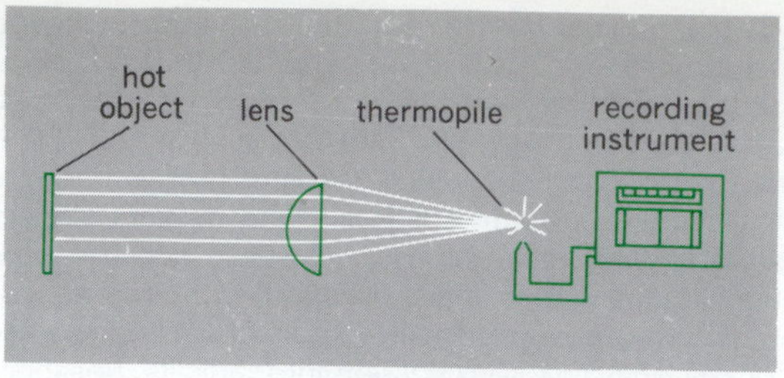

Elementary radiation pyrometer. (*After D. M. Considine and S. D. Ross, Process Instruments and Controls Handbook, McGraw-Hill, 1974*)

Unfortunately, the thermal radiation emitted by the object depends not only on its temperature but also on its surface characteristics. The radiation existing inside hot, opaque objects is so-called blackbody radiation, which is a unique function of temperature and wavelength and is the same for all opaque materials. However, such radiation, when it attempts to escape from the object, is partly reflected at the surface. In order to use the output of the pyrometer as a measure of target temperature, the effect of the surface characteristics must be eliminated. A cavity can be formed in an opaque material and the pyrometer sighted on a small opening extending from the cavity to the surface. The opening has no surface reflection, since the surface has been eliminated. Such a source is called a blackbody source, and is said to have an emittance of 1.00. By attaching thermocouples to the blackbody source, a curve of pyrometer output voltage versus blackbody temperature can be constructed. *See* BLACKBODY; HEAT RADIATION.

Pyrometers can be classified generally into types requiring that the field of view be filled, such as narrow-band and total-radiation pyrometers; and types not requiring that the field of view be filled, such as optical and ratio pyrometers. The latter depend upon making some sort of comparison between two or more signals.

The optical pyrometer should more strictly be called the disappearing-filament pyrometer. In operation, an image of the target is focused in the plane of a wire that can be heated electrically. A rheostat is used to adjust the current through the wire until the wire blends into the image of the target (equal brightness condition), and the temperature is then read from a calibrated dial on the rheostat.

The ratio, or "two-color," pyrometer makes measurements in two wavelength regions and electronically takes the ratio of these measurements. If the emittance is the same for both wavelengths, the emittance cancels out of the result, and the true temperature of the target is obtained. This so-called graybody assumption is sufficiently valid in some cases so that the "color temperature" measured by a ratio pyrometer is close to the true temperature. *See* TEMPERATURE MEASUREMENT; THERMOMETER. [T.P.M.]

Pyromorphite A mineral series in the apatite group, or in the larger grouping of phosphate, arsenate, and vanadate-type minerals. In this series lead (Pb) substitutes for calcium (Ca) of the apatite formula $Ca_5(PO_4)_3(F,OH,Cl)$, and little fluorine (F) or hydroxide (OH) is present. *See* APATITE.

The pyromorphite series crystallizes in the hexagonal system. Crystals are prismatic. Other forms are granular, globular, and botryoidal. Pyromorphite colors range through green, yellow, and brown; vanadinite occurs in shades of yellow, brown, and red.

Pyromorphites are widely distributed as secondary minerals in oxidized lead deposits. Pyromorphite is a minor ore of lead; vanadinite is a source of vanadium and minor ore of lead. *See* LEAD; VANADIUM. [W.R.Lo.]

Pyrophyllite A hydrated aluminum silicate with composition $Al_2Si_4O_{10}(OH)_2$. The mineral is commonly white, grayish, greenish, or brownish, with a pearly to waxy appearance and greasy feel. It occurs as compact masses, as radiating aggregates (see illustration), and as foliated masses. Pyrophyllite belongs to the layer silicate (phyllosilicate) group of minerals. The mineral is soft (hardness 1–1½ on the Mohs scale) and has easy cleavage parallel to the structural layers. The mineral is highly stable to acids.

Specimen of pyrophyllite. (*Pennsylvania State University*)

Pyrophyllite is used principally for refractory materials and in other ceramic applications. The main sources for pyrophyllite in the United States are in North Carolina. An unusual form from the Transvaal is called African wonderstone. *See* SILICATE MINERALS. [G.W.BR.]

Pyrotechnics Fireworks. The ingredients of fireworks are basically the familiar incendiary components. Chlorates and nitrates are the oxidizers, and sulfur, charcoal, antimony sulfide, and sometimes powdered metals are the fuels. Colors are produced by the metals present in the salts; sodium gives yellow; calcium, brick red; strontium, scarlet; barium, green; and copper, blue-green. Iron filings and aluminum powder give sparks and brilliant effusions.

Highway warning flares are usually made with strontium nitrate to impart a brilliant scarlet hue and employ a slow-burning fuel of wax and sawdust.

Colored smokes are used for military signaling and for daylight displays. They incorporate a variety of vaporizable organic dyes in the usual pyrotechnic compositions. *See* INCENDIARY. [W.E.Go.]

Pyrotheria An extinct order of primitive, mastodonlike, herbivorous, hoofed mammals restricted to the Eocene and Oligocene deposits of South America. There is only one family (Pyrotheriidae) in the order and four genera in the family.

The characters of this group superficially resembling those in early proboscideans are nasal openings over orbits indicating the presence of a trunk, strong neck musculature, and six upper and four lower bilophodont cheek teeth. Pyrotheres are distantly related to the members of the superorder Paenungulata, including Proboscidea, Xenungulata, and others. *See* PROBOSCIDEA; XENUNGULATA. [G.T.J.]

Pyroxene A family of diverse and important rock-forming minerals crystallizing over a wide range of temperature, pressure, and composition and occurring as constituents of a host of rock types. The table gives the formulas and space groups of the pyroxene family. The crystal chemistry of the pyroxene family is very complicated. Since the pyroxenes are so ubiquitous and since their compositions range extensively, research on these compounds reveals important information about the origin and thermal history of certain rocks.

The pyroxene family of minerals

Name	Formula	Space group
Orthorhombic pyroxenes		
Orthoenstatite-orthoferrosilite	$(Mg,Fe)_2(Si_2O_6)$	*Pbca*
Protoenstatite	$Mg_2(Si_2O_6)$	*Pbcn*
Monoclinic pyroxenes		
Diopside-hedenbergite	$Ca(Mg,Fe)\ (Si_2O_6)$	*C2/c*
Johannsenite	$CaMn(Si_2O_6)$	*C2/c*
Schefferite	$Ca(Mg,Mn)\ (Si_2O_6)$	*C2/c*
Augite	$(Ca,Mg,Fe,Al)_2((Si,Al)_2O_6)$	*C2/c*
Aegirine	$NaFe(Si_2O_6)$	*C2/c*
Jadeite	$NaAl(Si_2O_6)$	*C2/c*
Spodumene	$LiAl(Si_2O_6)$	*C2/c*
High clinoenstatite	$Mg_2(Si_2O_6)$	*C2/c*
Clinoferrosilite	$Fe_2(Si_2O_6)$	$P2_1/c$
Low clinoenstatite	$Mg_2(Si_2O_6)$	$P2_1/c$
Pigeonite	$(Mg,Fe,Ca)\ (Mg,Fe)\ (Si_2O_6)$	$P2_1/c$

Common pyroxenes range in color from white (pure diopside and enstatite) through shades of yellow and green to deep brown and greenish-black (ferrosilite, hedenbergite, and aegirine). Specific gravity ranges from 3.2 (enstatite) to 4.0 (ferrosilite). The hardness on Mohs scale ranges from $5\frac{1}{2}$ to 6.

Pyroxenes occur in most rock types of medium to high grade, particularly in basalts, pyroxenites, charnockites, and ultrabasic rocks derived from gravity settling of ferromagnesian minerals. As constituents in ultrabasic rocks, they occur often with olivine and spinel. Diopside and hedenbergite are typical metamorphic and skarn products. Augites occur as constituents of basalts and charnockites, and as common metamorphic products, often associated with hornblende amphibole. [P.B.M.]

Pyroxenite A heavy, dark-colored, phaneritic (visibly crystalline) igneous rock composed largely of pyroxene with smaller amounts of olivine or hornblende. Pyroxenite composed largely of orthopyroxene occurs with anorthosite and peridotite in large, banded gabbro bodies. Some of these pyroxenite masses are rich sources of chromium. Certain pyroxenites composed largely of clinopyroxene are also of magmatic origin, but many probably represent products of reaction between magma and limestone. Other pyroxene-rich rocks have formed through the processes of metamorphism and metasomatism. *See* GABBRO; IGNEOUS ROCKS; PERIDOTITE; PYROXENE. [C.A.C.]

Pyroxenoid A group of silicate minerals whose physical properties resemble those of pyroxenes. In contrast with the two-tetrahedra periodicity of pyroxene single silicate chains, the pyroxenoid crystal structures contain single chains of $(SiO_4)^{4-}$ silicate tetrahedra having repeat periodicities ranging from three to nine (see illus). The tetrahedron is a widely used geometric representation for the basic building block of most silicate minerals, in which all silicon cations (Si^{4+}) are bonded to four oxygen anions arranged as if they were at the corners of a tetrahedron. In pyroxenoids, as in other single-chain silicates, two of the four oxygen anions in each tetrahedron are shared between two Si^{4+} cations to form the single chains, and the other two oxygen anions of each tetrahedron are bonded to divalent cations, such as calcium (Ca^{2+}), iron (Fe^{2+}), or manganese (Mn^{2+}). These divalent cations bond to six (or sometimes seven or eight) oxygen anions, forming octahedral (or irregular seven- or eight-cornered) coordination polyhedra. *See* PYROXENE; SILICATE MINERALS.

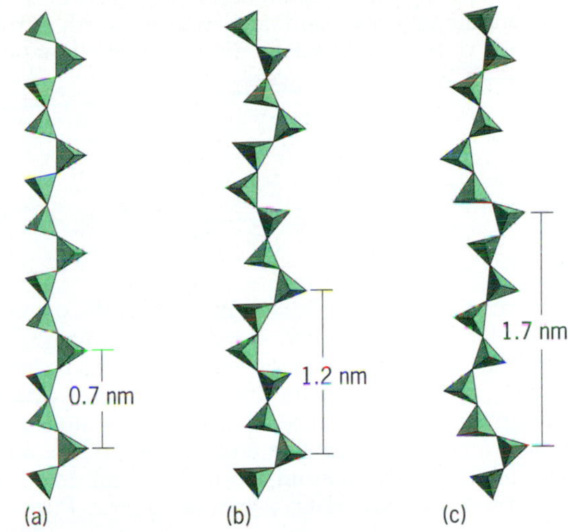

Tetrahedral silicate chains in three pyroxenoid structures: (*a*) wollastonite with three-tetrahedra periodicity, (*b*) rhodonite with five-tetrahedra periodicity, and (*c*) pyroxmangite-pyrox ferroite with seven-tetrahedra periodicity.

The pyroxenoid structures have composite structural units consisting of strips of octahedra (or larger polyhedra) two or more units wide formed by sharing of polyhedral edges, to which the silicate tetrahedral chains are attached on both top and bottom. The repeat periodicity of the octahedral strips is the same as that of the silicate chains to which they are attached. These composite units are cross-linked to form the three-dimensional crystal structures. The pyroxenoid minerals are triclinic, with either $C\bar{1}$ or $I\bar{1}$ space-group symmetry depending on the stacking of the composite units, and *c*-axis lengths ranging from about 0.71 nanometer for three-repeat silicate tetrahedral chains to about 2.3 nm for nine-repeat chains. *See* CRYSTAL STRUCTURE.

There are two series of pyroxenoid minerals, one anhydrous and one hydrous. The anhydrous pyroxenoids are significantly more abundant. A general formula for anhydrous pyroxenoids is $(Ca,Mn,Fe^{2+})SiO_3$. Silicate-chain repeat length is inversely proportional to mean divalent cation size. The hydrogen in hydrous pyroxenoids is hydrogen-bonded between two oxygen atoms; additional hydrogen in santaclaraite is bound as hydroxyl (OH) and as a water molecule (H_2O). [C.W.Bu.]

Pyrrhotite A mineral with composition $Fe_{1-x}S$ ($x = 0$ to 0.2). Eskebornite, $Fe_{1-x}Se$, is the selenium analog. The iron-deficient pyrrhotites are ferrimagnetic at room temperature, but at some higher temperature they become paramagnetic, presumably because of vacancy disorder.

The mineral occurs as rounded grains to large masses, more rarely as tabular pseudohexagonal crystals and rosettes. Color is brownish bronze-yellow with dark grayish-black streaks. Hardness is 4 on Mohs scale and specific gravity 4.6 (for the composition Fe_7S_8).

Pyrrhotite occurs in basic igneous rocks as a late-stage fractional differentiate, particularly in norites and gabbros, and sufficient quantities may constitute an ore of iron. Pyrrhotite also occurs with magnetite and chondrodite in contact metamorphic marbles, and in low-temperature veins with calcite and other sulfides and sulfosalts. [P.B.M.]

Pyrrole One of a group of organic compounds containing a doubly-unsaturated five-membered ring in which nitrogen occupies one of the ring positions. Pyrrole (I) is a representative compound.

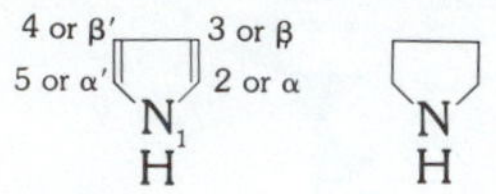

(I) (II)

The pyrrole system is found in the green leaf pigment, chlorophyll, in the red blood pigment, hemoglobin, and in the blue dye, indigo. Interest in these colored bodies has been largely responsible for the intensive study of pyrroles. Tetrahydropyrrole, or pyrrolidine (II), is part of the structures of two protein amino acids, proline and hydroxyproline, and of hygrine, which is an alkaloid from Peruvian coca. *See* HETEROCYCLIC COMPOUNDS; HYDROXYPROLINE; INDOLE; PORPHYRIN; PROLINE. [W.J.Ge.]

Pythagorean theorem This theorem states that in any right triangle the square on the hypotenuse is equal to the sum of the squares on the other two sides: $r^2 = x^2 + y^2$. More than 100 different proofs have been given for this extremely important theorem of euclidean plane geometry.

The three-dimensional Pythagorean theorem may be phrased "the square of the diagonal of a rectangular box is equal to the sum of the squares of three adjacent edges that meet at a vertex: $r^2 = x^2 + y^2 + z^2$." *See* SQUARE; TRIANGLE. [J.S.F.]

Python A reptile in the subfamily Pythoninae of the family Boidae. There are 9 species, all occurring in Africa, Southeast Asia, New Guinea, and Australia. The entire subfamily includes 7 genera and 21 species that are called pythons, but the genus *Python* is more commonly considered and will be discussed here. The pythons are distinguished from the closely related boas on the basis of skull structure and methods of reproduction. *See* BOA.

Pythons are large, muscular constrictors that have long, powerful teeth, but no poison apparatus. Pythons kill their prey by constriction and eventual suffocation. In this process, bones are rarely broken. Most pythons are semiarboreal, occurring in forest regions; however, some are found among shrubs and in grassy areas. *See* REPTILIA. [C.B.C.]

Q (electricity) Often called the quality factor of a circuit, Q is defined in various ways, depending upon the particular application. In the simple RL and RC series circuits, Q is the ratio of reactance to resistance, as in Eqs. (1), where X_L is the inductive reactance, X_C is the capacitive reactance, and R is the resistance. An important application lies in the dissipation factor or loss angle when the constants of a coil or capacitor are measured by means of the alternating-current bridge.

$$Q = X_L/R \quad Q = X_C/R \text{ (a numerical value)} \qquad (1)$$

Q has greater practical significance with respect to the resonant circuit, and a basic definition is given by Eq. (2), where Q_0 means evaluation at resonance. For certain circuits, such as cavity resonators, this is the only meaning Q can have.

$$Q_0 = 2\pi \frac{\text{max stored energy per cycle}}{\text{energy lost per cycle}} \qquad (2)$$

For the RLC series resonant circuit with resonant frequency f_0, Eq. (3) holds, where R is the total circuit resistance, L is the inductance, and C is the capacitance. Q_0 is the Q of the coil if it contains practically the total resistance R. The greater the value of Q_0, the sharper will be the resonance peak. In terms of the resonance curve, Eq. (4) holds, where f_0 is the frequency at resonance, and f_1 and f_2 are the frequencies at the half-power points. *See* RESONANCE (ALTERNATING-CURRENT CIRCUITS).

$$Q_0 = e\pi f_0 L/R = 1/2\pi f_0 CR \qquad (3)$$

$$Q_0 = f_0/(f_2 - f_1) \qquad (4)$$

[B.L.R.]

Q fever An acute, febrile, infectious disease in humans, characterized by sudden onset, patchy pneumonitis, absence of rash, and low mortality, and caused by a bacterialike microorganism, *Coxiella burneti*.

A primary cycle between certain small animals, such as bandicoots in Australia, and their tick parasites, occurs in nature. Secondary cycles of obscure relationship to the primary cycle occur in domestic animals. Placentas of domestic animals are highly infectious and become, on disintegration, the chief source of infection for humans and animals. Vaccination of valuable dairy herds gives some promise of reduction of incidence in humans and domestic animals. *See* RICKETTSIOSES.

[C.B.P.]

Q meter A direct-reading instrument widely used for measuring the Q of an electric circuit at radio frequencies. Originally designed to measure the Q of coils, the Q meter has been developed into a flexible, general-purpose instrument for determining many other quantities such as (1) the distributed capacity, effective inductance, and self-resonant frequency of coils; (2) the capacitance, Q or power factor, and self-resonant frequency of capacitors; (3) the effective resistance, inductance or capacitance, and the Q of resistors; (4) characteristics of intermediate- and radio-frequency transformers; and (5) the dielectric constant, dissipation factor, and power factor of insulating materials. *See* ELECTRICAL MEASUREMENTS; Q (ELECTRICITY).

[I.F.K./E.C.St.]

Quadraphonic sound system A sound reproduction system in which four discrete channels of program are fed to four loudspeakers located at various points around the listener. The listener's enjoyment of the reproduced sound is heightened through being immersed in the sound field. The usual arrangement of loudspeakers in the listening room is depicted in the illustration. The loudspeakers are situated near the corners of the room and are designated as left-front (L_F), left-back (L_B), right-back (R_B), and right-front (R_F).

The program producer may apportion segments of the total musical program among the four channels to achieve a variety of effects in reproduction. Thus, the sound heard in a concert hall may be simulated by feeding the direct sound from the orchestra to the L_F and R_F loudspeakers while feeding the concert hall reverberation sounds primarily to the L_B and R_B loudspeakers. Or the orchestral sounds may be distributed to all four loudspeakers, thereby giving a sense of intimacy as though the listener were seated among the musicians. Moreover, an illusion of moving sound sources may be achieved by sweeping the source gradually from one channel to another.

The variety of effects employed in quadraphonic sound reproduction are possible because of the psychoacoustic effect termed localization, that is, the ability of a listener to ascertain, through the sense of hearing, the location of a sound. In persons having normal hearing, the ability is highly developed for most sounds within the audible range of frequencies.

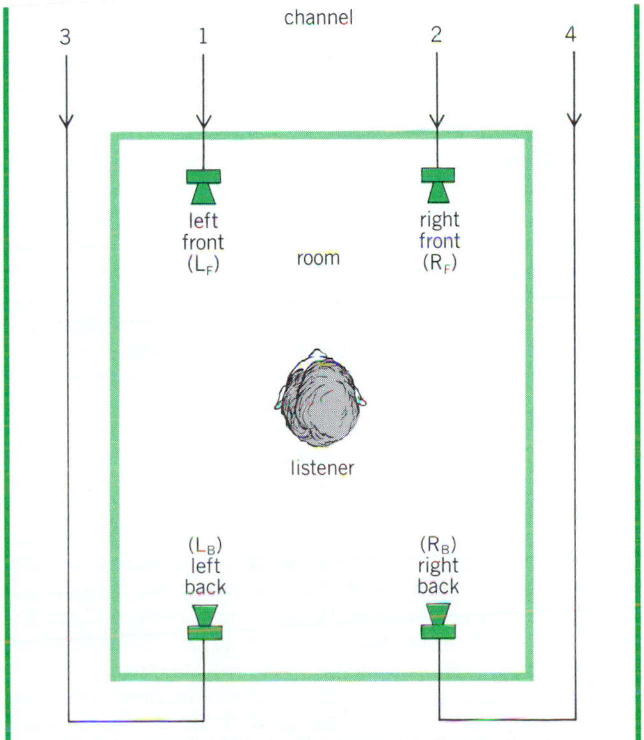

Diagram showing the placement of the four loudspeakers used in quadraphonic sound reproduction.

In 4-2-4 matrixing, the original four channels are combined, or matrixed, in electronic circuits to compress the signals into two channels that can be recorded or transmitted as conventional stereophonic signals. In a receiver or phonograph playback system, the two-channel signals are processed, or dematrixed, to yield four output signals that more or less resemble the four input signals. The term "matrix" comes from the mathematical terminology applied to an array of algebraic equations such as those used to describe the 4-2-4 encode-decode processes.

A quadraphonic sound reproduction system, needing twice as many loudspeakers and amplifiers, is clearly more complex and costly than a stereophonic system. However, the availability of relatively inexpensive and compact integrated circuits and solid-state amplifiers makes four-channel systems economically viable. Because of their ability to use existing two-channel records and FM transmitters, 4-2-4 quadraphonic systems entered the marketplace quickly. The technical problems initially encountered in true four-channel phonograph records and players have been solved, so that reliable equipment of this type is available. *See* AMPLIFIER; INTEGRATED CIRCUITS; SOUND-REPRODUCING SYSTEMS. [J.G.W.]

Quadrature The condition in which the phase angle between two alternating quantities is 90°, corresponding to one-quarter of an electrical cycle. The electric and magnetic fields of electromagnetic radiation are in space quadrature, which means that they are at right angles in space. *See* ELECTROMAGNETIC RADIATION.

The current and voltage of a perfect coil are in quadrature because the coil current lags behind the coil voltage by exactly 90°. The current and voltage of a perfect capacitor are also in quadrature, but here the current leads the voltage by 90°. In these last two cases the current and voltage are in time quadrature. [J. Mar.]

Quadric surface A surface defined analytically by an equation of the second degree in three variables. If these variables are x, y, z, such an equation has the form:

$$ax^2 + by^2 + cz^2 + 2exy + 2fxz + 2gyz + 2px + 2qy + 2rz + d = 0$$

Every plane section of such a surface is a conic. *See* CYLINDER; ELLIPSOID AND SPHEROID; HYPERBOLOID; PARABOLOID; SURFACE AND SOLID OF REVOLUTION. [J.S.F.]

Quadrilateral In the euclidean sense, the quadrilateral is a geometric figure bounded by four straight-line segments called sides, each of which intersects each of two adjacent sides in points called vertices, but fails to intersect the opposite side (considered as a finite segment). The four vertices are connected in pairs by the four sides and by two other segments called diagonals. If the latter intersect each other, the quadrilateral is a plane quadrilateral; if not, it is a skew quadrilateral. A plane quadrilateral is called a rhombus if its four sides are equal, or a kite if it has two pairs of equal adjacent sides. A quadrilateral with one pair of sides parallel is a trapezoid; if the other pair of sides are equal, the trapezoid is called isosceles. A quadrilateral with two pairs of parallel sides is a parallelogram. *See* PARALLELOGRAM; RECTANGLE; SQUARE; TRAPEZOID. [J.S.F.]

Qualitative chemical analysis The branch of chemistry concerned with identifying the elements and compounds present in a sample of matter. Inorganic qualitative analysis traditionally used classical "wet" methods to detect elements or groups of chemically similar elements, but instrumental methods have largely superseded the test-tube methods. Methods for the detection of organic compounds or classes of compounds have become increasingly available and important in organic, forensic, and clinical chemistry. Once it is known which elements and compounds are present, the role of quantitative analysis is to determine the composition of the sample. *See* ANALYTICAL CHEMISTRY; QUANTITATIVE CHEMICAL ANALYSIS.

Inorganic analysis. The operating principles of all systematic inorganic qualitative analysis schemes for the elements are similar: separation into groups by reagents producing a phase change; isolation of individual elements within a group by selective reactions; and confirmation of the presence of individual elements by specific tests.

Through usage and tradition, descriptive terms for sample sizes have been:

macro	0.1 gram or more
semimicro	0.01 to 0.1 gram
micro	1 milligram (1 mg or 10^{-3} g)
ultramicro	1 microgram (1 μg or 10^{-6} g)
submicrogram	less than 1 microgram

For defining the smallest amount of a substance that can be detected by a given method, the term "limit of identification" is used. Under favorable conditions, an extremely sensitive method can detect as little as 10^{-15}g.

Spot tests are selective or specific single qualitative chemical tests carried out on a spot plate (a glass or porcelain plate with small depressions in which drop-size reactions can be carried out), on paper, or on a microscope slide. On paper or specially prepared adsorbent surfaces, spot tests become one- or two-dimensional through the use of solvent migration and differential adsorption (thin-layer chromatography). Containing selected indicator dyes, pH indicator paper strips are widely used for pH estimation. Solid reagent monitoring devices and indicator tubes are used for the detection and estimation of pollutant gases in air. *See* CHROMATOGRAPHY; PH.

By use of the microscope, crystal size and habit can be used for qualitative identification. With the addition of polarized light, chemical microscopy becomes a versatile method. *See* CHEMICAL MICROSCOPY.

Any instrumental method of quantitative analysis can be adapted to qualitative analysis. Some, such as electrochemical methods, are not often used in qualitative analysis (an exception would be the ubiquitous pH meter). Others, such as column chromatography (gas and liquid) and mass spectrometry, are costly and uncomplicated but capable of providing unique results. Emission spectroscopy is important in establishing the presence or absence of a suspected element in forensic analysis. The simple qualitative flame test developed into flame photometry and subsequently into atomic absorption spectrometry, *See* ATOMIC SPECTROMETRY; FLAME PHOTOMETRY; SPECTROCHEMICAL ANALYSIS.

Bombardment of surfaces by x-rays, electrons, and positive ions has given rise to a number of methods useful in analytical chemistry. X-ray diffraction is used to determine crystal structure and to identify crystalline substances by means of their diffraction patterns. X-ray fluorescence analysis employs x-rays to excite emission of characteristic x-rays by elements. The electron microprobe uses electron bombardment in a similar manner to excite x-ray emission. *See* SECONDARY ION MASS SPECTROMETRY (SIMS); X-RAY DIFFRACTION; X-RAY FLUORESCENCE ANALYSIS.

Neutron activation analysis, wherein neutron bombardment induces radioactivity in isotopes of elements not naturally radioactive, involves measurement of the characteristic gamma radiation and modes of decay produced in the target atoms. *See* ACTIVATION ANALYSIS. [J.L.L.]

Organic analysis. Qualitative analysis of an organic compound is the process by which the characterization of its class and structure is determined. Due to the numerous classes of organic compounds and the complexity of their molecular structures, a systematic analytical procedure is often required.

A typical procedure entails an initial assignment of compound classification, followed by a complete identification of the molecular structure.

The initial step is an examination of the physical characteristics. Color, odor, and physical state can be valuable clues. The physical constants of an unknown compound provide pertinent data for the analyst. Constants such as melting point, boiling point, specific gravity, and refractive index are commonly measured.

The preliminary chemical tests which are applied are elemental analysis procedures. The elements generally associated with carbon, hydrogen, and oxygen are sulfur, nitrogen, and the halogens. The analyst is usually interested in the latter group. These elements are converted to water-soluble ionic compounds via sodium fusion. The resulting products are then detected with wet chemical tests. The solubility of a compound in various liquids provides information concerning the molecular weight and functional groups present in the compound. In order to indicate the presence or absence of a functional group, specific classification reactions are tested. The reactions are simple, are rapid, and require a small quantity of sample. No single test is conclusive evidence. A judicious choice of reactions can confirm or negate the presence of a functional group.

Instrumental methods are commonly applied for functional group determination and structure identification. Absorption spectroscopy and infrared absorption spectroscopy are among the most important techniques. Whenever a molecule is exposed to electromagnetic radiation, certain wavelengths cause vibrational, rotational, or electronic effects within the molecule. The radiation required to cause these effects is absorbed. The nature and configuration of the atoms determine which specific wavelengths are absorbed. Raman spectroscopy is slightly different from the other spectroscopic techniques. A sample is irradiated by a monochromatic source. Depending upon the vibrational and rotational energies of the functional groups in the sample, light is scattered from the sample in a way which is characteristic of the functional groups. Instrumental techniques are also valuable for structure identification. In addition to infrared spectroscopy, nuclear magnetic resonance and mass spectrometry are widely used. Other, less common techniques are electron spin resonance, x-ray diffraction, and nuclear quadrupole spectroscopy. *See* INFRARED SPECTROSCOPY; MASS SPECTROMETRY; NUCLEAR MAGNETIC RESONANCE (NMR); RAMAN EFFECT.

In addition to chemical structure, the physical structure of a sample can be important. Thermal analysis is a useful procedure for examining structural characteristics. Among the most popular methods are thermogravimetry, differential thermal analysis, differential scanning calorimetry, and thermal mechanical analysis. *See* CALORIMETRY.

Although the implementation of instrumental methods has greatly simplified qualitative analytical procedures, no individual instrument is capable of complete identification for all samples. A complement of instrumental and wet chemical techniques is generally required for adequate proof of identification. *See* SPECTROSCOPY. [S.S.; K.Lo.]

Quality control A system of inspection, analysis, and action applied to a manufacturing operation so that, by inspecting a small portion of the product currently produced, an estimate of the overall quality of the product can be made to determine what, if any, changes must be made in the operation in order to achieve or maintain the required level of quality.

The term is usually applied in the context of manufacturing operations. The manufacturer makes one or more items, each having a set of specifications for various characteristics, which must be met to satisfy the needs of the customer. The specifications may be set either by the customer in view of the intended use of the product of by the manufacturing organization in view of its understanding of the customer's intended use, or by both. Quality control is the set of activities in a manufacturing operation which undertakes to ensure that the specifications of the finished product are satisfied. More general terms such as quality assurance and total quality control embrace all the technical and management aspects of product quality and safety during the design, specification, development, manufacturing, and use stages. Periodic reviews of quality standards and quality operations are also undertaken to ensure that products remain satisfactory to the consumer and competitive in the market place. This can involve comparative testing of competitive products and the undertaking of quality improvement programs. Since most manufacturing organizations have a profit objective, quality control performs its tasks with the objective of minimizing costs, that is, with obtaining the specified product quality at the lowest possible cost. *See* INDUSTRIAL COST CONTROL; INSPECTION AND TESTING. [J.A.Cl.]

Quantitative chemical analysis That branch of analytical chemistry which deals with the determination of the relative proportions of constituents in a compound or of components in a mixture.

In an ultimate analysis, the amount of a single element or compound is determined. In a proximate analysis, certain constituents are determined as a group of indefinite relative composition, for example, the determination of ash in a sample of coal.

Methods of analysis may be direct or indirect. In a direct gravimetric method, the desired constituent is converted to a compound of definite, known composition and this compound is weighed. In a direct volumetric method, the desired constituent is determined by measuring the volume of reagent of known concentration required to react completely with the constituent.

In an indirect gravimetric method, a mixture of substances which includes the desired constituent is weighed and then wholly, or in part, converted to some other substance, or mixture of substances, of known composition and weighed. The amount of desired constituent can then be calculated. *See* GRAVIMETRIC ANALYSIS.

In an indirect volumetric method, a measured quantity of reagent is added in excess of the amount required to react with the desired constituent. The excess reactant is then determined by titration, and the amount of it which reacts with the desired constituent is determined by difference. *See* TITRATION; VOLUMETRIC ANALYSIS.

Methods of quantitative analysis vary with the amount of sample taken and of constituent being determined. A macroanalysis (decigram analysis) uses a sample of about 0.1–0.5 gram; a semimicroanalysis (centigram analysis) uses a sample of about 0.01–0.1 g; a microanalysis (milligram analysis) uses a sample of about 1–1 0 mg; and an ultramicroanalysis (microgram analysis) uses a sample of approximately 0.001 mg. *See* BALANCE.

Since the accuracy of a quantitative analysis depends in part on the nature of the material and on the nature and quantities of foreign constituents present, calibration of methods of unknown applicability is often necessary.

In the correction factor method, the procedure is applied to a sample containing all the constituents in the given sample except the desired one. Any numerical result obtained is applied as a correction factor to the value obtained on the unknown.

In the synthetic sample method, the procedure of analysis is applied in parallel to the unknown and to a sample containing constituents of the same nature and approximate amounts as those in the unknown, and with the desired constituent added in known amount. Any discrepancy in the value obtained on the synthetic sample is applied to correct the analysis of the unknown. *See* ANALYTICAL CHEMISTRY; SPECTROCHEMICAL ANALYSIS; SPECTROPHOTOMETRIC ANALYSIS; VOLUMETRIC ANALYSIS. [S.G.S.]

Quantity flowmeter A device that measures fluid flow by measuring or passing fixed amounts of fluid. By introducing a timing element, all quantity meters can be converted to volume flow rate meters. Quantity meters may be classified as positive-displacement meters or rotating impeller-type meters. *See* POSITIVE-DISPLACEMENT FLOWMETER; ROTATING IMPELLER FLOWMETER.

The accuracy of quantity meters is dependent to varying degrees on the physical properties of the fluid being measured. Changes from the reference (calibration) conditions of viscosity and density must be considered if accurate measurement is required. *See* FLOW MEASUREMENT. [M.Br.; L.P.E.]

Quantization A term applied to the transition from a description of a system of particles or fields in the classical approximation where canonically conjugate variables commute to a description where these variables are treated as noncommuting operators. In this transition, Poisson bracket relations among dynamical variables are replaced by commutation relations. So-called second quantization is the analogous treatment of a field, such as the electromagnetic field solution of Maxwell's equations, the wave-function solution of Schrödinger's equation, or the solution of Dirac's equation, with commutation relations imposed upon the field as the canonical coordinate and with the generalized momentum defined through the Lagrangian. The wave equation in the latter two cases resulted from the first quantization. *See* CANONICAL TRANSFORMATIONS; LAGRANGE'S EQUATIONS; QUANTUM ELECTRODYNAMICS; QUANTUM FIELD THEORY.

A result of treating a system through quantum mechanics is that some observable quantities become quantized, or assume discrete values. Independently of the above operator discussion, the process leading to discrete values for an observable is also called quantization. For example, energy and angular momentum historically were first treated this way through quantization postulates hypothesizing discrete eigenvalue spectra. *See* ATOMIC STRUCTURE AND SPECTRA; QUANTUM MECHANICS. [K.E.L.]

Quantized electronic structure (QUEST) A material that confines electrons in such a small space that their wavelike behavior becomes important and their properties are strongly modified by quantum-mechanical effects. Such structures occur in nature, as in the case of atoms, but can be synthesized artificially with great flexibility of design and applications. They have been fabricated most frequently with layered semiconductor materials. Generally, the confinement regions for electrons in these structures are 1–100 nanometers in size. The allowable energy levels, motion, and optical properties of the electrons are strongly affected by the quantum-mechanical effects. The structures are referred to as quantum wells, wires, and dots, depending on whether electrons are confined with respect to motion in one, two, or three dimensions (see illus.). Multiple closely spaced wells between which electrons can move by quantum-mechanical tunneling through intervening thin barrier-material layers are referred to as superlattices. *See* ARTIFICIALLY LAYERED STRUCTURES; QUANTUM MECHANICS; TUNNELING IN SOLIDS.

The most frequently used fabrication technique for quantized electronic structures is epitaxial growth of thin single-crystal semiconductor layers by molecular-beam epitaxy or by chemical vapor growth techniques. These artificially synthesized quantum structures find major application in high-performance transistors and solid-state lasers. They also have important scientific applications for the study of fundamental physics problems in which particles are confined so that they have free motion in only two, one, or zero directions.

The optical applications are based on the interactions between light and electrons in the quantum structures. The absorption of a photon by an electron in a quantum well raises the electron from occupied quantum states to unoccupied quantum states. Electrons and holes in quantum wells may also recombine, with the resultant emission of photons from the quantized electronic structure as the electron drops from a higher state to a lower state.

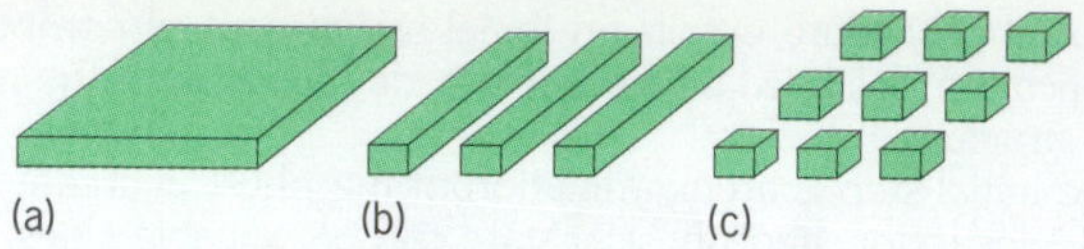

Geometric configurations of (*a*) quantum well, (*b*) quantum wires, and (*c*) quantum dots.

The photon emission is the basis for quantum-well semiconductor lasers, which have widespread applications in optical fiber communications and compact disk and laser disk optical recording. The photon absorption is the basis for quantum-well photodetectors and light modulators. *See* COMPACT DISK; LASER; OPTICAL DETECTORS; OPTICAL MODULATORS; OPTICAL RECORDING.

Modulation doping is a special way of introducing electrons into quantum wells for electrical applications. The electrons come from donor atoms lying in adjacent barrier layers. Modulation doping thus produces carriers in the quantum well without introducing impurity dopant atoms there. Since there are no impurity atoms to collide with in the well, electrons there are free to move with high mobility along the quantum-well layer. Resistance to electric current flow is thus much reduced. This enhances the low-noise and high-speed applications of quantum wells and is the basis of the high-electron-mobility transistor (HEMT), which is also known as the modulation-doped field-effect transistor (MODFET). HEMTs are widely used in microwave receivers for direct reception of satellite television broadcasts. *See* TRANSISTOR. [A.C.Go.]

Quantized vortices A type of flow pattern exhibited by superfluids, such as liquid ^{4}He below 2.17 K (−455.76°F). The term vortex designates the familiar whirlpool pattern where the fluid moves circularly around a central line and the velocity diminishes inversely proportionally to the distance from the center. The strength of a vortex is determined by the circulation, which is the line integral of the velocity around any path enclosing the central line. *See* VORTEX.

A superfluid is believed to be characterized by a macroscopic (that is, large-scale) quantum-mechanical wave function ψ. This wave function locks the superfluid into a coherent state. Since the velocity around the vortex increases without limit as the center is approached, the superfluid density and thus ψ must vanish at the center in order to avoid an infinite energy. Thus the central core of the vortex marks the zeros, or nodal lines, in the macroscopic wave function. *See* QUANTUM MECHANICS.

Quantized vortex lines are usually produced by rotating a vessel containing superfluid helium. At very low rotation speeds, no vortices exist: the superfluid remains at rest while the vessel rotates. At a certain speed the first vortex appears and corresponds to the first excited rotational state of the system. If the container continues to accelerate, additional quantized vortices will appear. At any given speed the vortices form a regular array which rotates with the vessel.

Quantized vortex lines were first detected in the mid-1950s by their influences on superfluid thermal waves traveling across the lines. In the late 1950s it was discovered that electrons in liquid helium form tiny charged bubbles which can become trapped on the vortex core but can move quite freely along the line. These electron bubbles (often referred to as ions) have been one of the most useful probes of quantized vortices. Researchers have been able to use ions to detect single quantized vortex lines. In one experiment the trapped ions are pulled out at the top of the vortex lines, accelerated, and focused onto a phosphor screen. The pattern of light thus pro-

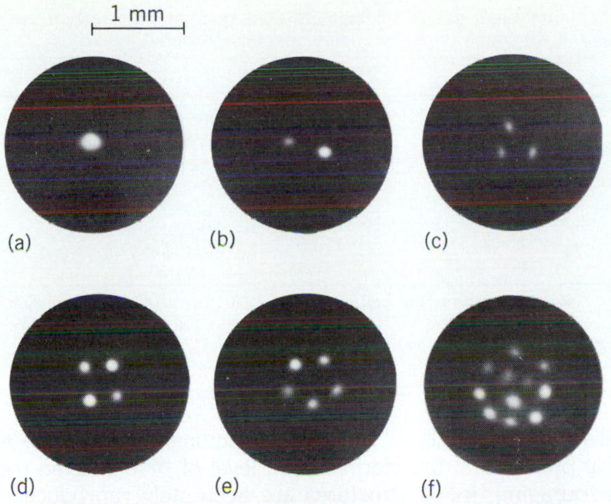

Stationary configurations of vortices which appear when a cylindrical container of superfluid ^{4}He is rotated about its axis. As the rotation speed increases from *a* to *f*, more vortices appear and the patterns become more complex. (*From E. J. Yarmchuk, M. J. V. Gordon, and R. E. Packard, Observation of stationary vortex arrays in rotating superfluid helium, Phys. Rev. Lett., 43:214–217, 1979*)

duced on the phosphor is a map of the position of the vortices where they contact the liquid meniscus (see illustration). *See* LIQUID HELIUM; SUPERFLUIDITY. [R.E.P.]

Quantum (physics) A term characterizing an excitation in a wave or field, connoting fundamental particlelike properties such as energy or mass, momentum, and angular momentum for this excitation. In general, any field or wave equation that is quantized, including systems already treated in quantum mechanics that are second-quantized, leads to a particle interpretation for the excitations which are called quanta of the field. This term historically was first applied to indivisible amounts of electromagnetic, or light, energy usually referred to as photons. The photon, or quantum of the electromagnetic field, is a massless particle, best interpreted as such by quantizing Maxwell's equations. Analogously, the electron can be said to be the quantum of the Dirac field through second quantization of the Dirac equation, which also leads to the prediction of the existence of the positron as another quantum of this field with the same mass but with a charge opposite to that of the electron. In similar fashion, quantization of the gravitational field equations suggests the existence of the graviton. The pi meson or pion was theoretically predicted as the quantum of the nuclear force field. Another quantum is the quantized lattice vibration, or phonon, which can be interpreted as a quantized sound wave since it travels through a quantum solid or fluid, or through nuclear matter, in the same manner as sound goes through air.

The use of quantum as an adjective (quantum mechanics, quantum electrodynamics) implies that the particular subject is to be treated according to the modern rules that have evolved for quantized systems. *See* ELEMENTARY PARTICLE; GRAVITATION; GRAVITON; MAXWELL'S EQUATIONS; MESON; PHONON; PHOTON; QUANTIZATION; QUANTUM ELECTRODYNAMICS; QUANTUM FIELD THEORY; QUANTUM MECHANICS. [K.E.L.]

Quantum acoustics The study of the properties of propagating sound waves that are directly attributable to the underlying quantum-mechanical nature of the medium. Effects of a quantum origin generally appear either at low temperatures or at short wavelengths or high frequency. At sufficiently high temperatures and long wavelengths, sound waves are well described by the usual or classical acoustics.

Comparison of light and sound. Absorption of discrete bundles of light by the electrons in a metal is responsible for the photoelectric effect wherein the photons kick the electrons out of a metal. Similarly, sound waves which are vibrations of ponderable matter such as liquids and solids are also absorbed in bundles called phonons. Although the phonon atomic effect is in nature and interpretation quite similar to the photoelectric effect, there are definite differences between light and sound. Light propagates in vacuum, whereas sound waves are oscillations of a ponderable medium which cannot be transmitted through vacuum. *See* LATTICE VIBRATIONS; PHONON; PHOTOEMISSION.

There are other significant differences between light and sound owing to the role of interactions between the waves. Light in vacuum is an extremely linear system. The huge intensifies and frequencies required for nonlinear response such as the scattering of light by light have not yet been achieved. Sound waves, however, are enormously nonlinear; shock waves and nonlinear beating are an everyday occurrence. When light propagates in a dielectric crystal, its properties can be modified so interaction of light by light can occur in the crystal. However, even for the most nonlinear crystals and strongest lasers these effects are many orders of magnitude weaker than similar effects for sound waves. Thus nonlinear quantum acoustics is important because it is related to aspects of nature not readily accessible in other systems. *See* NONLINEAR ACOUSTICS; NONLINEAR OPTICS.

Sound in dielectrics. In dielectric substances such as helium or quartz the electrons are not free to flow or conduct electricity. Thus the only motion possible is a vibration of the constituent ions or molecules. These oscillations can be thought of as sound waves (or phonons) bouncing around inside the medium. In a dielectric temperature is a measure of the intensity of the sound waves randomly propagating inside the medium in equilibrium.

Sound scattering. The basic quantity of interest in sound scattering is the attenuation coefficient of sound, which is the rate at which the energy in an impressed wave is absorbed by the medium. In the presence of nonlinearities a given wave will interact with other waves to create sum and difference frequencies of the original wave and the various components of the random thermal motion. This occurs because nonlinear effects lead to new sound waves whose strength is determined by the product of a given wave, the wave with which it interacts, and the coefficient of nonlinearity.

In addition to thermally excited vibrations, any oscillator has a zero-point motion which is due to the fundamental quantum of action h as required by the uncertainty principle. This random motion, which persists even at absolute zero (0 K), is responsible for the fact that helium remains a liquid, and is also capable of scattering sound. However, in applying nonlinear acoustics to calculate the attenuation of a sound wave due to this mechanism, those interactions which remove energy from the zero-temperature noise cannot be allowed to occur, as the energy at 0 K is a minimum according to the second law. The only nonlinear process which meets this criterion involves scattering of sound from zero-point noise at a lower frequency through the production of difference frequencies. The attenuation due to this effect represents a spontaneous decay of sound waves into a spectrum of subharmonic frequencies.

Parametric amplification of decay. When sound is produced by a coherent source, the amplitude can be quite large. In this situation the sound wave can act like a pump in a parametric amplifier and dramatically amplify one of its decay products. A large-amplitude sound wave is made up of an enormous number of phonons all with the same frequency. When just one of those phonons spontaneously decays, its decay products will be in the proper phase relation to the entire sound wave so as to be immediately amplified. In this language the decay products are the signal and idler waves in a parametric amplifier. Now the entire beam can rapidly decay into the signal and idler channels at a rate which is much faster than that of low-amplitude spontaneous decay. *See* PARMETRIC AMPLIFIER. [S.Pu.]

Quantum chemistry A branch of chemistry concerned with the application of quantum mechanics to chemical problems. More specifically, it is concerned with the electronic structure of molecules. Methods developed since 1960 permit the quantum chemist to obtain reliable approximate solutions to the nonrelativistic Schrödinger equation. The method which dominates the field of quantum chemistry is the Hartree-Fock or self-consistent-field approximation. *See* HAMILTON'S EQUATIONS OF MOTION; QUANTUM MECHANICS.

For closed-shell molecules, the form of the Hartree-Fock wave function is given by Eq. (1), in which $A(n)$, the antisymmetrizer for n electrons, has the effect of making a Slater determinant out of the orbital product on which it operates.

$$\psi_{HF} = A(n)\phi_1(1)\phi_2(2)\ldots\phi_n(n) \qquad (1)$$

The ϕ's are spin orbitals, products of a spatial orbital χ and a one-electron spin function α or β. For any given molecular system, there are an infinite number of wave functions of form (1), but the Hartree-Fock wave function is the one for which the orbitals ϕ have been varied to yield the lowest possible energy [Eq. (2)].

$$E = \int \psi \times_{HF} H \psi_{HF} d\tau \qquad (2)$$

The resulting Hartree-Fock equations are relatively tractable due to the simple form of the energy E for single determinant wave functions [Eq. (3)].

$$E_{HF} = \sum_{i} I(i \mid i) + \sum_{i}\sum_{j>i} [(ij \mid ij) - (ij \mid ji)] \qquad (3)$$

To solve the Hartree-Fock equations exactly, either the orbitals ϕ must be expanded in a complete set of analytic basis functions or strictly numerical (that is, tabulated) orbitals must be obtained. The former approach is impossible from a practical point of view for systems with more than two electrons, and the latter has been accomplished only for atoms and for a few diatomic molecules. Therefore, the exact solution of the Hartree-Fock equations is abandoned for polyatomic molecules. Instead an incomplete (but reasonable) set of analytic basis functions is adopted and solved for the best variational [that is, lowest energy given by Eq. (2)] wave function of form (1). Such a wave function is referred to as being of self-consistent-field (SCF) quality. For very large basis sets, then, it is reasonable to refer to the resulting SCF wave function as near-Hartree-Fock.

For large chemical systems, only minimum basis sets (MBS) can be used in ab initio theoretical studies. The term "large" includes molecular systems, with 100 or more electrons.

Ab initio theoretical methods have had the greatest impact on chemistry in the area of structural predictions. The most encouraging aspect of ab initio geometry predictions is their reliability. Essentially all molecular structures appear to be reliably predicted at the Hartree-Fock level of theory. Even more encouraging, many structures are accurately reproduced by using only minimum-basis-set self-consistent-field methods. A fairly typical example is methylenecyclopropane (see illustration), with its minimum-basis-set self-consistent-field structure compared with experiment in the table. Carbon-carbon bond distances differ typically by 0.002 nm from experiment, and angles are rarely in error by more than a few degrees. Thus, for many purposes, theory may be considered complementary to experiment in the area of structure prediction.

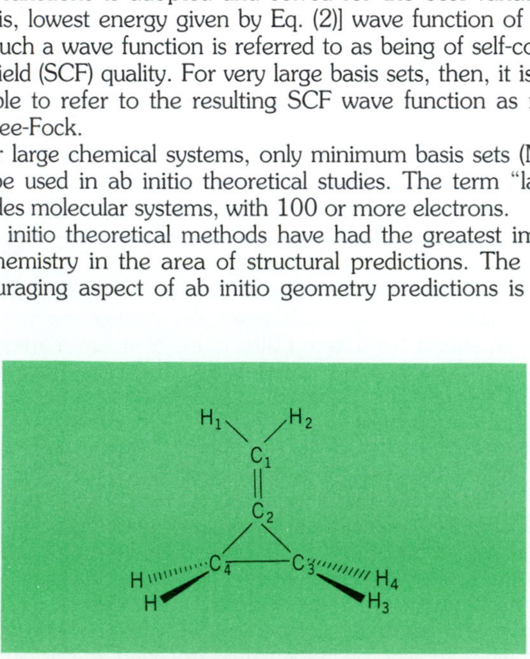

Methylenecyclopropane structure.

Minimum-basis-set self-consistent-field geometry prediction compared with experiment for methylenecyclopropane

Parameter*	Theory	Experiment
$r\,(C_1{=}C_2)$	0.1298 nm	0.1332 nm
$r\,(C_2{-}C_3)$	0.1474 nm	0.1457 nm
$r\,(C_3{-}C_4)$	0.1522 nm	0.1542 nm
$r\,(C_1{-}H_1)$	0.1083 nm	0.1088 nm
$r\,(C_3H_3)$	0.1083 nm	0.109 nm
$\theta(H_1C_1H_2)$	116.0°	114.3°
$\theta(H_3C_3H_4)$	113.6°	113.5°
$\theta(H_{34}C_3C_4)$	149.4°	150.8°

*Here r represents the carbon-carbon bond distance; θ represents the bond angle in degrees of H-C-H bonds; the numbers on C and H correspond to the numbered atoms in the illustration.

The Hartree-Fock formalism has very powerful predictive capabilities. However, large classes of chemical problems cannot be reasonably described, and a good deal of discretion is required on the part of the theoretician. Although state-of-the-art quantum chemistry has gone well beyond the Hartree-Fock model, the model is certain to remain the cornerstone of electronic structure theory for many years to come. *See* CHEMICAL BONDING; CHEMICAL STRUCTURES; MOLECULAR ORBITAL THEORY; MOLECULAR STRUCTURE AND SPECTRA; RESONANCE (MOLECULAR STRUCTURE). [H.F.S.]

Quantum chromodynamics A theory of the strong ("nuclear") interactions among quarks, which are regarded as fundamental constituents of matter. Quantum chromodynamics (QCD) seeks to explain why quarks combine in certain configurations to form the observed patterns of subnuclear (or "elementary") particles, such as the proton and pi meson. According to this picture, the strong interactions among quarks are mediated by a set of force particles known as gluons. Strong interactions among gluons may lead to new structures that correspond to as yet undiscovered particles. The long-studied nuclear force which binds protons and neutrons together in atomic nuclei is regarded as a collective effect of the elementary interactions among constituents of the composite protons and neutrons. *See* NUCLEAR STRUCTURE.

Quantum chromodynamics has not yet been subjected to rigorous experimental tests. Several qualitative predictions of quantum chromodynamics do seem to have been borne out. Part of the esthetic appeal of the theory is due to the fact that quantum chromodynamics is nearly identical in mathematical structure to quantum electrodynamics (QED) and to the unified theory of weak and electromagnetic interactions put forward by S. Weinberg and A. Salam. This resemblance encourages the hope that a unified description of the strong, weak, and electromagnetic interactions may be at hand. *See* QUANTUM ELECTRODYNAMICS; WEAK NUCLEAR INTERACTIONS.

Gauge theories. At the heart of current theories of the fundamental interactions is the idea of gauge invariance, which draws its name from some early investigations by H. Weyl into a possible connection between scale changes and the equations of electrodynamics. Weyl attempted to deduce electromagnetism from a symmetry principle, the conjectured invariance of physical laws under a change of length scale chosen independently at every position of space and time. This specific undertaking ran afoul of quantum mechanics, but the general strategy and the name have survived. It is widely believed that gauge theories constructed to embody various symmetry principles represent the correct quantum-mechanical descriptions of the strong, weak, and electromagnetic interactions. *See* SYMMETRY LAWS (PHYSICS).

Color. Although the quark model reproduces the properties of the observed hadron states, a theoretical inconsistency arises from the fact that the quarks must be spin-½ particles. According to the Pauli exclusion principle, identical spin-½ particles cannot occupy the same quantum state. The properties of several of the baryon resonances can be reproduced only by having three quarks of the same flavor in a configuration which is symmetric under the interchange of any pair of quarks, contrary to the exclusion principle. It was shown that the quark model could be made to conform with the Pauli principle if a new distinguishing property was imputed to the quarks. Each quark flavor is thus said to come in three colors, arbitrarily labeled red, blue, and green. Color may be regarded as the strong-interaction analog of the electric charge. Color cannot be created or destroyed by any of the known interactions. Like electric charge, it is said to be conserved. *See* COLOR (QUANTUM MECHANICS); EXCLUSION PRINCIPLE.

In the face of evidence that color could be regarded as the conserved charge of the strong interactions, it was natural to seek a gauge symmetry which would have color conservation as its consequence. An obvious candidate for the gauge symmetry group is the unitary group SU(3), now to be applied to color rather than flavor. The theory of strong interactions among quarks which is prescribed by local color gauge symmetry is known as quantum chromodynamics. The mediators of the strong interactions are eight massless spin-1 bosons, one for each generator of the symmetry group. These strong-force particles are named gluons because they make up the "glue" which binds quarks together into hadrons. Gluons also carry color, and hence have strong interactions among themselves.

Asymptotic freedom. The theoretical description of the strong interactions has historically been inhibited by the very strength of the interaction, which renders low-order perturbative calculations untrustworthy. In 1973 it was found that, in many gauge theories, the effective strength of the interaction becomes increasingly feeble at short distances. For quantum chromodynamics, this implies that the interaction between quarks becomes weak at small separations, a property known as asymptotic freedom. This discovery raises the hope that some aspects of the strong interactions might be treated by using familiar computational techniques that are predicated upon the smallness of the interaction strength.

Quarkonium. It was suggested in 1974 that the bound system of an extremely massive quark with its antiquark would be so small that the strong force would be extremely feeble. In this case, the binding between quark and antiquark is mediated by the exchange of a single massless gluon, and the spectrum of bound states resembles that of an exotic atom composed of an electron and an antielectron (positron) bound electromagnetically in a Coulomb potential generated by the exchange of a massless photon. Since the electron-positron atom is known as positronium, the heavy quark-antiquark atom has been called quarkonium. Two families of heavy quark-antiquark bound states, the ψ/J system composed of charmed quarks and the Y system made up of *b* quarks, have been discovered. Both have level schemes characteristic of atomic spectra, which have been analyzed by using tools of nonrelativistic quantum mechanics developed for ordinary atoms. *See* CHARM; ELEMENTARY PARTICLE; FUNDAMENTAL INTERACTIONS; GLUONS; J PARTICLE; POSITRONIUM; QUANTUM FIELD THEORY; QUARKS; STRONG NUCLEAR INTERACTIONS. [C.Q.]

Quantum electrodynamics The study of the properties of electromagnetic radiation and its interaction with electrically charged matter; in particular, with atoms and their constituent electrons. The fundamental equations governing quantum electrodynamics actually are believed to encompass all of atomic physics, chemistry, properties of bulk matter, and classical electromagnetic theory. Almost all the phenomena readily perceived by the senses are believed to be ultimately understandable in terms of the laws of quantum electrodynamics. In the scope of its application and implications, and in the simplicity of the underlying assumptions, quantum electrodynamics is rivaled only by gravitational theory. In practice, the term quantum electrodynamics refers to those phenomena specific to the quantum nature of electromagnetic radiation. This includes the study of emission and absorption of light by atoms and the basic interactions of light with electrons and other fundamental particles. *See* ELECTROMAGNETIC RADIATION; GRAVITATION; QUANTUM FIELD THEORY.

Quantum electrodynamics was formulated by P. A. M. Dirac, W. Heisenberg, and W. Pauli shortly after the foundations of quantum mechanics were laid down in 1925. This formulation gave a satisfactory interpretation of the wave-particle duality of light, in which it is possible for light to manifest itself both in ways appropriate to a particle (photon) description or a wavelike description, in accordance with the experimental foundations of quantum mechanics. About the same time, Dirac discovered an equation describing the motion of electrons which incorporated both the requirements of quantum theory and of relativity. However, introduction of electromagnetic interaction into Dirac's equation, while successfully describing the magnetic properties of the electron and the existence of its oppositely charged counterpart, or antiparticle, the positron, led to mathematical disasters which inhibited the further development of the theory until after World War II. Stimulated by experiments using microwave techniques developed during World War II, S. Tomonaga, R. P. Feynman, J. S. Schwinger, F. J. Dyson, and many others found a way to by-pass, but not solve, these difficulties. Since that time the theory has been confronted by very sophisticated experimental tests and has passed them satisfactorily. *See* QUANTUM MECHANICS. [J.D.B.]

Quantum electronics A loosely defined field concerned with the interaction of radiation and matter, particularly those interactions involving quantum energy levels and resonance phenomena, and especially those involving lasers and masers. Quantum electronics encompasses useful devices such as lasers and masers and their practical applications; related phenomena and techniques, such as nonlinear optics and light modulation and detection; and related scientific problems and applications, such as quantum noise processes, laser spectroscopy, picosecond spectroscopy, and laser-induced optical breakdown.

In one sense any electronic device, even one as thoroughly classical in nature as a vacuum tube, may be considered a quantum electronic device, since quantum theory is accepted to be the basic theory underlying all physical devices. In practice, however, quantum electronics is usually understood to refer to only those devices such as lasers and atomic clocks in which stimulated transitions between discrete quantum energy levels are important, together with related devices and physical phenomena which are excited or explored using lasers. Other devices such as transistors or superconducting devices which may be equally quantum-mechanical in nature are not usually included in the domain of quantum electronics. *See* ATOMIC CLOCK; LASER; LASER SPECTROSCOPY; MASER; NONLINEAR OPTICS; OPTICAL DETECTORS; OPTICAL MODULATORS; OPTICAL PULSES; QUANTUM MECHANICS. [A.E.S.]

Quantum field theory Quantum theory of physical systems possessing an infinite number of degrees of freedom, such as the electromagnetic field, gravitational field, or wave fields in a medium. Its major applications lie in the attempted description of fundamental particles and their associated wave fields under circumstances in which both the effects of quantum mechanics and of special relativity are important. Quantum field theory is also used in nonrelativistic quantum theory for systems of many particles such as electrons in a metal, or for sound waves in liquids and solids. *See* QUANTUM MECHANICS; RELATIVISTIC QUANTUM THEORY; RELATIVITY.

The classical electromagnetic field is a dynamical system which must be subjected to the rules of quantum mechanics. The consequence of this requirement is the emergence of a particle (photon) interpretation from the classical Maxwell equations, manifesting the particle-wave duality of quantum mechanics in its most perfect form. The experimental success of this theory for electrodynamics has led to its imitation for the more complex interactions of other fundamental particles. For example, in 1935 H. Yukawa tried to interpret the strong force between protons and neutrons responsible for the existence of nuclei in the same way. This force is only effective when the protons and neutrons are within a distance on the order of 10^{-13} cm (4×10^{-14} in.) of each other. Yukawa supposed the nuclear force was due to the exchange of a particle, called a meson, between the nucleons, just as the electrostatic force between charged particles is interpreted in terms of exchange of a photon, as illustrated. *See* ELEMENTARY PARTICLE; MESON; QUANTUM ELECTRODYNAMICS.

The short range of the force can be understood as a consequence of a finite rest mass m for the meson. When the nucleon emits the meson, the energy cost is approximately mc^2. According to the uncertainty principle, if the energy of the system is observed to the accuracy ΔE, the observation must be made over an interval of time Δt longer than $h/\Delta E$. Thus, an energy fluctuation of a magnitude of $\sim mc^2$ can only take place over a time interval $\Delta t < h/mc^2$, consistent with the uncertainty principle. In this time interval, the meson can travel no further than $c\Delta t < h/mc^2$, a distance identified with the range of the force. In this way Yukawa was led to predict the mass of the meson ($m \simeq 270\ m_{\text{electron}}$) intermediate between the electron and proton mass. To describe this Yukawa meson, now known as π-meson, one writes an analog of the Maxwell field equations and then passes to the quantum description in a way similar to that described for the electromagnetic field. A satisfactory particle description is thereby obtained, again embodying the characteristic wave-particle duality of the quantum theory. *See* UNCERTAINTY PRINCIPLE.

However, although the theory of isolated mesons is satisfactory, when interactions such as the couplings to the nucleons are introduced, chaos emerges. The field equations have not been solved. There are grave doubts as to whether solutions exist. There is, aside from the electromagnetic interactions, no small parameter in terms of which there might exist a series expansion.

Faced with this dilemma, theoretical physicists have concentrated their efforts on finding general properties of any relativistic quantum field theory, rather than detailed consequences of a specific set of equations. These efforts have been fruitful in three different areas: (1) the quantum theory of symmetries and their connection with conservation laws such as conservation of energy, momentum, and angular momentum; (2) the study of certain exact properties; in particular the mathematical property of analyticity of the functions used to describe collision processes; (3) the use of the diagrammatic method of R. P. Feynman, developed for quantum electrodynamics and appropriately modified for the study of more general interactions, to interpret qualitatively many of the features of collision processes between relativistic particles. *See* SYMMETRY LAWS (PHYSICS). [J.D.B.]

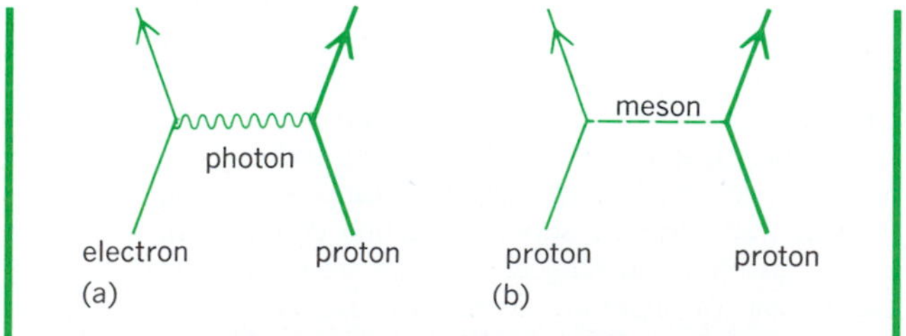

Analogy between (*a*) electromagnetic force via photon exchange and (*b*) nuclear force via meson exchange.

Quantum gravitation The quantum theory of the gravitational field; also, the study of quantum fields in a curved space-time. In classical general relativity, the gravitational field is represented by the metric tensor $g_{\mu\nu}$ of space-time. This tensor satisfies Einstein's field equation, with the energy-momentum tensor of matter and radiation as a source. On the other hand, the equations of motion for the matter and radiation fields also depend on the metric.

However, classical field theories such as Maxwell's electromagnetism or the classical description of particle dynamics are approximations valid only at the level of large-scale macroscopic observations. At a fundamental level, elementary interactions of particles and fields must be described by relativistic quantum mechanics, in terms of quantum fields. Because the geometry of space-time in general relativity is inextricably connected to the dynamics of matter and radiation, a consistent theory of the metric in interaction with quantum fields is only possible if the metric itself is quantized. *See* MAXWELL'S EQUATIONS; QUANTUM FIELD THEORY; RELATIVISTIC QUANTUM THEORY; RELATIVITY.

Under ordinary laboratory conditions the curvature of space-time is so extremely small that in most quantum experiments gravitational effects are completely negligible. Quantization in Minkowski space is then justified. Gravity is expected to play a significant role in quantum physics only at rather extreme conditions of strongly time-dependent fields, near or inside very dense matter. The scale of energies at which quantization of the metric itself becomes essential is given by $(\hbar > c^5/G)^{1/2} \approx 10^{-19}$ GeV, where G is the gravitational constant, $\hbar$ is Planck's constant divided by 2π, and c is the velocity of light. Energies that can be reached in the laboratory or found in cosmic radiation are far below this order of magnitude. Only in the very early stages of the universe, within a proper time of the order of $(G\hbar/c^5)^{1/2} \approx 10^{-43}$) s after the big bang, would such energies have been produced. *See* BIG BANG THEORY.

In most physical systems the metric is quasistationary over macroscopic distances so that its fluctuations can be ignored. A quantum description of fields in a curved space-time can then be given by treating the metric as a classical external field in interaction with the quantum fields. The most striking quantum effect in curved space-time is the emission of radiation by black holes, discovered by S. W. Hawking. *See* BLACK HOLE.

A black hole is an object that has undergone gravitational collapse. Classically this means that it becomes confined to a space-time region in which the metric has a singularity (the curvature becomes infinite). This region is bounded by a surface, called the horizon, such that any matter or radiation falling inside becomes trapped. Therefore, classically the mass of a black hole can only increase. However, this is no longer the case if quantum effects are taken into account. When, due to fluctuations of the quantum field, particle-antiparticle or photon pairs are created near the horizon of a black hole, one of the particles carrying negative energy may move toward the hole, being absorbed by it, while the other moves out with positive energy.

It is found that the total rate of emission is inversely proportional to the square of the mass. For stellar black holes whose masses are of the order of a solar mass, the emission rate is negligibly small and unobservable. Only primordial black holes, of mass less than 10^{13} kg, formed very early in the quantum era of the universe, would have been small enough to produce quantum effects that could play any significant role in astrophysics or in cosmology. *See* ASTROPHYSICS; COSMOLOGY.

There are basically two approaches to the quantization of the metric, the canonical and the covariant quantization. A third method, which can be derived from the first and is now most widely used, is based on the Feynman path integral representation for the vacuum-to-vacuum amplitude, which is the generator of Green's functions for the quantum theory. One important feature of this method is that since the topology of the manifold is not specified at the outset it is possible to include a sum over paths in different topologies. The outcome of this idea, suggested by J. A. Wheeler and developed by Hawking, is that the vacuum would, at the level of the Planck length, $(G\hbar/c^3)^{1/2} \approx 10^{-35}$ m, acquire a foamlike structure. *See* GREEN'S FUNCTION.

At present a complete, consistent theory of quantum gravity is still lacking. The formal theory fails to satisfy the power counting criterion for renormalizability. In every order of the perturbation expansion, new divergences appear which could only be canceled by counterterms that do not exist in the original lagrangian. This may not be just a technical problem but the reflection of a conceptual difficulty stemming from the dual role, geometric and dynamic, played by the metric. *See* FUNDAMENTAL INTERACTIONS; GRAVITATION; RENORMALIZATION; SUPERGRAVITY.

[S.W.MacD.]

Quantum mechanics The modern theory of matter, of electromagnetic radiation, and of the interaction between matter and radiation; also, the mechanics of phenomena to which this theory may be applied. Quantum mechanics, also termed wave mechanics, generalizes and supersedes the older classical mechanics and Maxwell's electromagnetic theory. Atomic and subatomic phenomena provide the most striking evidence for the correctness of quantum mechanics and best illustrate the differences between quantum mechanics and the older classical physical theories. Quantum mechanics is needed to explain many properties of bulk matter, for instance, the temperature dependence of the specific heats of solids.

The formalism of quantum mechanics is not the same in all domains of applicability. In approximate order of increasing conceptual difficulty, mathematical complexity, and likelihood of future fundamental revision, these domains are the following: (i) Nonrelativistic quantum mechanics, applicable to systems in which particles are neither created nor destroyed, and in which the particles are moving slowly compared to the velocity of light. Here a particle is defined as a material entity having mass, whose internal structure either does not change or is irrelevant to the description of the system. (ii) Relativistic quantum mechanics, applicable in practice to a single relativistic particle (one whose speed equals or nearly equals c); here the particle may have zero rest mass, in which event, its speed must equal c. (iii) Quantum field theory, applicable to systems in which particle creation and destruction can occur; the particles may have zero or nonzero rest mass. This article is concerned mainly with nonrelativistic quantum mechanics, which apparently applies to all atomic and molecular phenomena, with the exception of the finer details of atomic spectra. Nonrelativistic quantum mechanics also is well established in the realm of low-energy nuclear physics. *See* ATOMIC STRUCTURE AND SPECTRA; NUCLEAR PHYSICS; QUANTUM FIELD THEORY; RELATIVISTIC QUANTUM THEORY.

Planck's constant. The quantity 6.61×10^{-34} joule-second, first introduced into physical theory by Max Planck in 1901, is a basic ingredient of the formalism of quantum mechanics. Planck's constant commonly is denoted by the letter h; the notation $\hbar = h/2\pi$ also is standard.

Uncertainty principle. In classical physics the observables characterizing a given system are assumed to be simultaneously measurable (in principle) with arbitrarily small error. For instance, it is thought possible to observe the initial position and velocity of a particle and therewith, using Newton's laws, to predict exactly its future path in any assigned force field. According to the uncertainty principle, accurate measurement of an observable quantity necessarily produces uncertainties in one's knowledge of the values of other observables. In particular, for a single particle relation (1*a*) holds, where Δx represents

$$\Delta x \Delta p_x > \hbar \quad (1a)$$

$$\Delta t\ \Delta E \gtrsim \hbar \quad (1b)$$

the uncertainty (error) in the location of the x coordinate of the particle at any instant, and Δp_x is the simultaneous uncertainty in the x component of the particle momentum. Relation (1*a*) asserts that under the best circumstances, the product $\Delta x \Delta p_x$ of the uncertainties cannot be less than about 10^{-34} joule second.

The uncertainty relation (1*b*) is derived and interpreted somewhat differently than relation (1*a*); it asserts that for any system, an energy measurement with error ΔE must be performed in a time not less than $\Delta t \sim \hbar/\Delta E$. If a system endures for only Δt seconds, any measurement of its energy must be uncertain by at least $\Delta E \sim \hbar \Delta t$. *See* UNCERTAINTY PRINCIPLE.

Wave-particle duality. It is natural to identify such fundamental constituents of matter as protons and electrons with the mass points or particles of classical mechanics. According to quantum mechanics, however, these particles, in fact all material systems, necessarily have wavelike properties. Conversely, the propagation of light, which, by Maxwell's electromagnetic theory, is understood to be a wave phenomenon, is associated in quantum mechanics with massless energetic and momentum-transporting particles called photons. The quantum-mechanical synthesis of wave and particle concepts is embodied in the de Broglie relations, given by Eqs. (2*a*) and (2*b*).

$$\lambda = h/p \quad (2a)$$

$$f = E/h \quad (2b)$$

These give the wavelength λ and wave frequency f associated with a free particle (a particle moving freely under no forces) whose momentum is p and energy is E; the same relations give the photon momentum p and energy E associated with an electromagnetic wave in free space (that is, in a vacuum) whose wavelength is λ and frequency is f. *See* PHOTON.

The wave properties of matter have been demonstrated conclusively for beams of electrons, neutrons, atoms (hydrogen, H, and helium, He), and molecules (H_2). When incident upon crystals, these beams are reflected into certain directions, forming diffraction patterns. Diffraction patterns are difficult to explain on a particle picture; they are readily understood on a wave picture, in which wavelets scattered from regularly spaced atoms in the crystal lattice interfere constructively along certain directions only. *See* ELECTRON DIFFRACTION; NEUTRON DIFFRACTION.

The particle properties of light waves are observed in the photoelectric effect and the compton effect. *See* COMPTON EFFECT; PHOTOEMISSION.

Complementarity. Wave-particle duality and the uncertainty principle are thought to be examples of the more profound principle of complementarity, first enunciated by Niels Bohr (1928). According to the principle of complementarity, nature has "complementary" aspects; an experiment which illuminates one of these aspects necessarily simultaneously obscures the complementary aspect. To put it differently, each experiment or sequence of experiments yields only a limited amount of information about the system under investigation; as this information is gained, other equally interesting information (which could have been obtained from another sequence of experiments) is lost. Of course, the experimenter does not forget the results of previous experiments, but at any instant, only a limited amount of information is usable for predicting the future course of the system.

Quantization. In classical physics the possible numerical values of each observable, meaning the possible results of exact measurement of the observable, generally form a continuous set. For example, the x coordinate of the position of a particle may have any value between $-\infty$ and $+\infty$. In quantum mechanics the possible numerical values of an observable need not form a continuous set, however. For some observables, the possible results of exact measurement form a discrete set; for other observables, the possible numerical values are partly discrete, partly continuous; for example, the total energy of an electron in the field of a proton may have any positive value between 0 and $+\infty$, but may have only a discrete set of negative values, namely, -13.6, $-13.6/4$, $-13.6/9$, $-13.6/16$ eV,....Such observables are said to be quantized; often there are simple quantization rules determining the quantum numbers which specify the allowable discrete values. Spectroscopy, especially the study of atomic spectra, probably provides the most detailed quantitative confirmation of quantization. *See* QUANTIZATION.

Probability considerations. The uncertainty and complementarity principles, which limit the experimenter's ability to describe a physical system, must limit equally the experimenter's ability to predict the results of measurement on that system. Suppose, for instance, that a very careful measurement determines that the x coordinate of a particle is precisely $x = x_0$. This is permissible in nonrelativistic quantum mechanics. Then formally, the particle is known to be in the eigenstate corresponding to the eigenvalue $x = x_0$ of the x operator. Under these circumstances, an immediate repetition of the position measurement again will indicate that the particle lies at $x = x_0$. Knowing that the particle lies at $x = x_0$ makes the momentum p_x of the particle completely uncertain, however, according to relation (la). A measurement of p_x immediately after the particle is located at $x = x_0$ could yield any value of p_x from $-\infty$ to $+\infty$.

More generally, suppose the system is known to be in the eigenstate corresponding to the eigenvalue α of the observable A. Then for any observable B, which is to some extent complementary to A, that is, for which an uncertainty relation of the form of relations (1) limits the accuracy with which A and B can simultaneously be measured, it is not possible to predict which of the many possible values $B = \beta$ will be observed. However, it is possible to predict the relative probabilities $P_\alpha(\beta)$ of immediately thereafter finding the observable B equal to β, that is, of finding the system in the eigenstate corresponding to the eigenvalue $B = \beta$.

To the eigenvalues correspond eigenfunctions, in terms of which $P_\alpha \leqslant \beta)$ can be computed. In particular, when α is a discrete eigenvalue of A, and the operators depend only on x and p_x, the probability $P_\alpha(\beta)$ is postulated as in Eq. (3), where $u(x,\alpha)$

$$P\alpha(\beta) = \left| \int_{-\infty}^{\infty} dx\, v\times(x,\beta)u(x,\alpha) \right|^2 \tag{3}$$

is the eigenfunction corresponding to $A = \alpha$; $v(x,\beta)$ is the eigenfunction corresponding to $B = \beta$; and the asterisk denotes the complex conjugate. The integral in Eq. (3) is called the projection of $u(x,\alpha)$ on $v(x,\beta)$. The quantity $|u(x,\alpha)|^2\, dx$ is the probability that the system, known to be in the eigenstate $A = \alpha$, will be found in the interval x to $x + dx$. *See* EIGENVALUE (QUANTUM MECHANICS).

Wave function. When the system is known to be in the eigenstate corresponding to $A = \alpha$, the eigenfunction $u(x,\alpha)$ is the wave function; that is, it is the function whose projection on an eigenfunction $v(x,\beta)$ of any observable B gives the probability of measuring $B = \beta$. The wave function $\psi(x)$ may be known exactly; in other words, the state of the system may be known as exactly as possible (within the limitations of uncertainty and complementarity), even though $\psi(x)$ is not the eigenfunction of a known operator. This circumstance arises because the wave function obeys Schrödinger's wave equation. Knowing the value of $\psi(x)$ at time $t = 0$, the wave equation completely determines $\psi(x)$ at all future times. In general, however, if $\psi(x,0) = u(x,\alpha)$, that is, if $\psi(x,t)$ is an eigenfunction of A at $t = 0$, then $\psi(x,t)$ will not be an eigenfunction of A at later times $t > 0$.

A system described by a wave function is said to be in a pure state. Not all systems are described by wave functions, however. For example, a beam of hydrogen atoms streaming out of a small hole in a hydrogen discharge tube can be regarded as a statistical ensemble or mixture of pure states oriented with equal probability in all directions.

Schrödinger equation. Equation (4) describes a plane wave

$$\psi(x,t) = A(\lambda)\, exp\left[2\pi i\left(\frac{x}{\lambda} - ft\right)\right] \tag{4}$$

of frequency f, wavelength λ, and amplitude $A(\lambda)$, propagating in the positive x direction. The previous discussion concerning wave-particle duality suggests that this is the form of the wave function for a beam of free particles moving in the x direction with momentum $p = p_x$, with Eq. (2) specifying the connections between f, λ, and E, p. Differentiating Eq. (4), it is seen that Eqs. (5) hold. Since for a free particle $E = p^2/2m$, it follows also that Eq. (6) is valid. *See* WAVE MOTION.

$$p_x\psi = \frac{h}{\lambda}\psi = \frac{\hbar}{i}\frac{\partial\psi}{\partial x} \tag{5a}$$

$$E\psi = hf\psi = -\frac{\hbar}{i}\frac{\partial\psi}{\partial t} \tag{5b}$$

$$\frac{-\hbar^2}{2m}\frac{\partial^2\psi}{\partial x^2} = -\frac{\hbar}{i}\frac{\partial\psi}{\partial t} \tag{6}$$

Equation (6) holds for a plane wave of arbitrary λ, and therefore for any superposition of waves of arbitrary λ, that is, arbitrary p_x. Consequently, Eq. (6) should be the wave equation obeyed by the wave function of any particle moving under no forces, whatever the projections of the wave function on the eigenfunctions of p_x. Equations (5) and (6) further suggest that for a particle whose potential energy $V(x)$ changes, in other words, for a particle in a conservative force field, $\psi(x,t)$ obeys Eq. (7).

$$\frac{-\hbar^2}{2m}\frac{\partial^2\psi}{\partial x^2} + V(x)\psi = -\frac{\hbar}{i}\frac{\partial\psi}{\partial t} \tag{7}$$

Equation 7 is the time-dependent Schrödinger equation for a one-dimensional (along x), spinless particle. Noting Eq. (5*b*), and observing that Eq. (7) has a solution for the form of Eq. (8), it is inferred that $\psi(x)$ of Eq. (8) obeys the time-dependent Schrödinger equation, Eq. (9). *See* FORCE.

$$\psi(x,t) = \psi(x)\, exp\,(-iEt/\hbar) \tag{8}$$

$$\frac{-\hbar^2}{2m}\frac{\partial^2\psi}{\partial x^2} + V(x)\psi = E\psi \tag{9}$$

Equation (9) is solved subject to reasonable boundary conditions, for example, that ψ must be continuous and must not become infinite as x approaches $\pm\infty$. These boundary conditions restrict the values of E for which there exist acceptable solutions $\psi(x)$ to Eq. (9), the allowed values of E depending on $V(x)$. In this manner, the allowed energies of atomic hydrogen listed in the earlier discussion of quantization are obtained.

The forms of Eqs. (5*a*), (7), and (9) suggest that the classical observable p_x, must be replaced by the operator $(\hbar/i)\,(\partial/\partial x)$. With this replacement, Eq. (10) holds. In other words,

$$(xp_x - p_x x)\psi = i\hbar\psi \qquad (10)$$

whereas the classical canonically conjugate variables x and p_x are numbers, obeying the commutative law in Eq. (11*a*), the quantum-mechanical quantities x and p_x are noncommuting operators, obeying Eq. (11*b*).

$$xp_x - p_x x = 0 \qquad (11a)$$

$$xp_x - p_x x = i\hbar \qquad (11b)$$

Correspondence principle. Since classical mechanics and Maxwell's electromagnetic theory accurately describe macroscopic phenomena, quantum mechanics must have a classical limit in which it is equivalent to the older classical theories. Although there is no rigorous proof of this principle for arbitrarily complicated quantum-mechanical systems, its validity is well established by numerous illustrations. [E.G.]

Quantum mineralogy Quantum mechanics applied to mineralogical systems. A theoretical understanding of chemical bonding and the electronic structure of minerals is fundamental to understanding the behavior of minerals. The principal goal of quantum mineralogy is to calculate from first principles the properties and transformations of solid earth materials. One of the advantages of this theoretical tool is the ability to predict the behavior of materials at conditions which are not readily accessible to experiment. More importantly, though, experimental observations may be interpreted in terms of quantum-mechanical theory. *See* CHEMICAL BONDING; QUANTUM CHEMISTRY; QUANTUM MECHANICS.

The methods used in quantum mineralogy are no different from those used for other chemical systems. They range from simple conceptions of the chemical bond in terms of quantum-mechanical expressions to rigorous calculation of wave functions for model systems. Too often, the more exact the quantum-mechanical calculation, the less understandable are the results for the nonspecialist. Despite such limitations, quantum mineralogy is a powerful theoretical probe into the nature, of structure, bonding, and properties of mineral systems. The size and complexity of model systems that can be considered in rigorous calculations continue to grow rapidly with increasing efficiency of computer hardware and program algorithms. *See* CHEMICAL STRUCTURES; MINERALOGY; SOLID-STATE CHEMISTRY; SOLID-STATE PHYSICS. [B.C.C.]

Quantum numbers The quantities, usually discrete with integer or half-integer values, which are needed to characterize a physical system of one or more atomic or subatomic particles. Specification of the set of quantum numbers serves to define such a system or, in other words, to label the possible states the system may have. In general, quantum numbers are obtained from conserved quantities determinable by performing symmetry transformations consisting of arbitrary variations of the system which leave the system unchanged. For example, since the behavior of a set of particles should be independent of the location of the origin in space and time (that is, the symmetry operation is translation in space-time), it follows that momentum and energy are rigorously conserved. *See* SYMMETRY LAWS (PHYSICS).

In general, each physical system must be studied individually to find the symmetry transformations, and thus the conserved quantities and possible quantum numbers. The quantum numbers themselves, that is, the actual state labels, are usually the eigenvalues of the physical operators corresponding to the conserved quantities for the system in question. *See* EIGENVALUE (QUANTUM MECHANICS); ELEMENTARY PARTICLE; PARITY (QUANTUM MECHANICS).

It is not necessary that the conserved quantity be "quantized" in order to be regarded as a quantum number; for example, a free particle possesses energy and momentum, both of which can have values from a continuum but which are used to specify the state of the particle. [K.E.L.]

Quantum solids A class of solids whose atoms or molecules undergo large zero-point motion even in the quantum ground state (at temperature $T = 0$ K $= -459.67$°F) as a result of their small mass and the weak attractive part of their interaction potential. The most striking examples are the isotopes of helium, ^{3}He and ^{4}He, which have a root-mean-square displacement from their lattice sites of approximately 25%. Further examples are the molecular hydrogens, H_2, D_2, and HD, as well as some heavier molecular solids. *See* INTERMOLECULAR FORCES; QUANTUM MECHANICS.

These materials display quantum effects in their bulk properties when cooled to temperatures near absolute zero so that the chaotic thermal motion is reduced. Both of the helium isotopes remain liquid all the way to absolute zero, unless external pressure (~3 megapascals ≈ 30 atm) is applied. This is because the atoms are not at rest at 0 K; the zero-point motion acts as an internal pressure which must be overcome in order to bring the atoms close enough together for solidification. All other substances, including the hydrogens, freeze under their own vapor pressure above 10 K.

The two melting curves are quite different in detail because of the different types of quantum statistics which the particles obey. There is a pronounced minimum in the ^{3}He melting pressure which is unique, and can be understood by considering the entropies of the liquid and the solid. The pressure minimum leads to the bizarre situation of the addition of heat causing freezing. The inverse of this process, the adiabatic formation of the solid by compression, is an important process which has been used extensively to cool liquid and solid ^{3}He to temperatures of approximately 1 mK. *See* CRYOGENICS; LIQUID HELIUM. [E.D.A.]

Quantum statistics The statistical description of particles or systems of particles whose behavior must be described by quantum mechanics rather than by classical mechanics. As in classical, that is, Boltzmann statistics, the interest centers on the construction of appropriate distribution functions. However, whereas these distribution functions in classical statistical mechanics describe the number of particles in given (in fact, finite) momentum and positional ranges, in quantum statistics the distribution functions give the number of particles in a group of discrete energy levels. In an individual energy level there may be, according to quantum mechanics, either a single particle or any number of particles. This is determined by the symmetry character of the wave functions. For antisymmetric wave functions only one particle (without spin) may occupy a state; for symmetric wave functions, any number is possible. Based on this distinction, there are two separate distributions, the Fermi-Dirac distribution for systems described by antisymmetric wave functions and the Bose-Einstein distribution for systems described by symmetric wave functions. *See* BOLTZMANN STATISTICS; BOSE-EINSTEIN STATISTICS; EXCLUSION PRINCIPLE; FERMI-DIRAC STATISTICS; KINETIC THEORY OF MATTER; QUANTUM MECHANICS; STATISTICAL MECHANICS. [M.Dr.]

Quantum theory of matter The microscopic explanation of the properties of condensed matter, that is, solids and liquids, based on the fundamental laws of quantum mechanics. Without the quantum theory, some properties of matter such as magnetism and superconductivity have no explanation at all, while for others only a phenomenological description can be obtained. With the theory, it is at least possible to comprehend what is needed to approach a complete understanding.

The theoretical problem of condensed matter—large aggregates of elementary particles with mutual interactions—is the quantum-mechanical many-body problem: an enormous number, of order 10^{23}, of constituent particles in the presence of a heat bath and interacting with each other according to quantum-mechanical laws. What makes the quantum physics of matter different from the traditional quantum theory of elementary particles is that the fundamental constituents (electrons and ions) and their interactions (Coulomb interactions) are known but the solutions of the appropriate quantum-mechanical equations are not. This situation is not due to the lack of a sufficiently large computer, but is caused by the fact that totally new structures, such as crystals, magnets, ferroelectrics, superconductors, liquid crystals, and glasses, appear out of the complexity of the interactions among the many constituents. The consequence is that entirely new conceptual approaches are required to construct predictive theories of matter. The usual technique for approaching the quantum many-body problem for a condensed-matter system is to try to reduce the huge number of variables (degrees of freedom) to a number which is more manageable but still can describe the essential physics of the phenomena being studied. *See* CRYSTAL; FERROELECTRICS; GLASS; LIQUID CRYSTALS; MAGNETIC MATERIALS; QUANTUM MECHANICS; SOLID-STATE PHYSICS; SUPERCONDUCTIVITY. [E.A.]

Quantum theory of measurement The attempt to reconcile the counterintuitive features of quantum mechanics with the hypothesis that quantum mechanics is in principle a complete description of the physical world, even at the level of everyday objects. A paradox arises because, at the atomic level where the quantum formalism has been directly tested, the most natural interpretation implies that where two or more different outcomes are possible it is not necessarily true that one or the other is actually realized, whereas at the everyday level such a state of affairs seems to conflict with direct experience.

The resolution of this paradox that is probably most favored by practicing physicists proceeds in two stages. At stage 1, it is pointed out that, quite generically, whenever the quantum formalism appears to generate a superposition of macroscopically distinct states it is impossible to demonstrate the effects of interference between them. The reasons for this claim include the facts that the initial state of a macroscopic system is likely to be unknown in detail; the initial state has extreme sensitivity to random external noise; and most important, merely by virtue of its macroscopic nature any such system will rapidly have its quantum-mechanical state correlated (entangled) with that of its environment in such a way that no measurement on the system alone (without a simultaneous measurement of the complete state of the environment) can demonstrate any interference between the two states in question—a result often known as decoherence. Thus, it is argued, the outcome of any possible experiment on the ensemble of macroscopic systems prepared in this way will be indistinguishable from that expected if each system had actually realized one or the other of the two macroscopically distinct states in question. Stage 2 of the argument (often not stated explicitly) is to conclude that if this is indeed true, then it may be legitimately asserted that such realization of a definite macroscopic outcome has indeed taken place by this stage.

Most physicists agree with stage 1 of the argument. However, not all agree that the radical reinterpretation of the meaning of the quantum formalism which is implicit at stage 2 is legitimate; that is, an interpretation in terms of realization, by each indvidiual system, of one alternative or the other, forbidden at the atomic level by the observed phenomenon of interference, is allowed once, on going to the macroscopic level, the phenomenon disappears. Consequently, various alternative interpretations have been developed. *See* QUANTUM MECHANICS. [A.J.L.]

Quark-gluon plasma A predicted state of matter containing deconfined quarks and gluons. According to the theory of strong interactions, called quantum chromodynamics, hadrons such as mesons and nucleons (the generic name for protons and neutrons) are bound states of more fundamental objects called quarks. The quarks are confined within the individual hadrons by the exchange of particles called gluons. However, calculations indicate that at sufficiently high temperatures or densities, hadronic matter should evolve into a new phase of matter containing deconfined quarks and gluons, called a quark-gluon plasma or quark matter. Such a state of matter is thought to have existed briefly about 1 microsecond after the big bang, and might also exist inside the core of dense neutron stars. *See* BIG BANG THEORY; HADRON; NEUTRON STAR; QUANTUM CHROMODYNAMICS.

The study of such a new state of matter requires a means for producing it under controlled laboratory conditions. Theory suggests that if ordinary nuclear matter is sufficiently heated or compressed, a transition from the hadronic to the quark-gluon phase may result. Experimentally this requires collisions of beams of heavy ions such as nuclei of gold or uranium (although lighter nuclei can be used) with other heavy nuclei at high enough energies to produce the necessary extreme conditions of heat and compression. Quantum chromodynamics calculations using the lattice gauge model indicate that energy densities of at least 1–2 GeV/fm^3 (1 femtometer = 10^{-15} m), about 10 times that found in ordinary nuclear matter, must be produced in the collision of plasma formation to occur. *See* RELATIVISTIC HEAVY-ION COLLISIONS.

Early accelerator experiments suggest that interesting phenomena are being observed with energy densities on the order of 1 GeV/fm^3. Future experiments will provide critical tests of the theory of the strong interaction and illuminate the earliest moments of the universe. *See* ELEMENTARY PARTICLE; GLUONS; QUARKS. [L.S.S.]

Quarks The basic constituent particles, of which "elementary" particles are now believed to be composed. Theoretical models built on the quark concept have been very successful in understanding and predicting many phenomena in particle physics. However, the experimental observation of free quarks remains ambiguous.

The evolution in the late 1950s and early 1960s of hadron spectroscopy revealed an order and symmetry among the states of hadronic matter that could be interpreted in terms of representations of the SU(3) symmetry group. This in turn was interpreted by M. Gell-Mann, and independently by G. Zweig, as a consequence of the grouping of elementary constituents of fractional electric charge (christened quarks by Gell-Mann), in pairs and triplets to form the observed hadrons. The general features of the quark model of hadrons have withstood the tests of time, and many of the static properties of hadrons are consistent with predictions of this model. *See* SYMMETRY LAWS (PHYSICS); UNITARY SYMMETRY.

The deep inelastic scattering of electrons on protons revealed form factors corresponding to pointlike constituents of the proton. J. D. Bjorken referred to these proton constituents as partons, although from the beginning it was recognized the partons and quarks might be merely different manifestations of the same entities.

Physicists now believe that the proton and neutron are not fundamental constituents of matter, but that they are made of quarks, very much as the nuclei of ^3H and ^3He are made of protons and neutrons and as the molecules of NO_2 and N_2O are made of oxygen and nitrogen atoms.

Until 1974 only three flavors of quarks were known; two of very nearly equal mass, of which the proton, neutron, and pi mesons are composed, and a third, more massive quark which is a constituent of strange particles such as the K mesons and hyperons such as Λ^0. The names attached to these quarks are

Properties of quarks

Flavor:	u	d	c	s	t*	b
Mass (GeV/c^2)†	0.39	0.39	1.55	0.51	(>15)	4.72
Electric charge	+2/3	−1/3	+2/3	−1/3	(+2/3)	−1/3
Baryon number	1/3	1/3	1/3	1/3	(1/3)	1/3
Spin (in units of $\hbar$)	1/2	1/2	1/2	1/2	(1/2)	1/2
Isotopic spin	+1/2	−1/2	0	0	(0)	0
Strangeness	0	0	0	−1	(0)	0
Charm	0	0	+1	0	(0)	0

*The t quark has not yet been found. Listed are its predicted properties.
†Masses are uncertain. Values listed here are half the mass of the lowest-lying quark-antiquark vector meson.

the up quark (u), the down quark (d), and the strange quark (s). Baryons are presumed to be composed of three quarks, such as the proton (uud), neutron (udd), $\Lambda^0(uds)$, and $\Xi^-(dss)$. Mesons are composed of a quark-antiquark pair, such as the $\pi^+(u\bar{d})$, $\pi^-(\bar{u}d)$, $K^+(u\bar{s})$, and $K^-(\bar{u}s)$. Antiparticles such as the antiprotons are formed by the antiquarks of those forming the particle, for example, the antiproton $\bar{p}(\overline{uud})$. *See* BARYON; MESON.

The quantum numbers of quarks are simply added to give the quantum numbers of the elementary particle which they form on combination. The natural unit of electric charge of a quark is $+2/3$ or $-1/3$ of the charge on a proton (1.6×10^{-19} coulomb), and the baryon number of each quark is $+1/3$; the charge, baryon number, and so forth, of each antiquark is just the negative of that for each quark.

The properties of quarks, to the extent that they are now understood, are presented in the table, where additional quarks discussed below are also included.

In 1974 two very different experiments found evidence for unusual elementary particles. These particles (labeled the J or ψ) were the first evidence for the existence of a new quark, the c (charm) quark. The J/ψ meson is understood to be the vector meson composed of the $c\bar{c}$ quark pair, and its decay into mesons is strongly inhibited, so that its lifetime is about 1000 times typical heavy, unstable mesons. This new quark had been predicted earlier by S. Glashow. Subsequently mesons and baryons containing the c quark were found and studied. *See* CHARM; J PARTICLE.

In 1977 a new resonance at about 9.4 GeV was reported; subsequent detailed study confirmed that this is the lowest-lying state of a new quark system. The quarks which make up this Υ (upsilon) particle have been called b for "bottom" (or "beauty").

Theoretical physicists expect that there is a heavier quark, as yet undiscovered, which may have the same relationship to b that the charm quark does to the strange quark. This quark, together with its presumed quantum numbers, is indicated in the table as the t quark, where t stands for "top" or "truth." *See* UPSILON PARTICLES.

In spite of the dramatic progress in the understanding of elementary particles which has accompanied the quark concept and the corresponding revolution in the concept of matter at its most fundamental level, at least three major questions remain. First, it has not been determined whether or not there are free quarks. Second, the number of flavors (kinds) of quarks has not been determined. The predicted t quark has not been found, and it is not known whether there are an infinite number of quarks, or whether the flavor spectrum stops at five or six. Third, it is not known whether quarks are truly fundamental entities, or whether quarks in turn have their own internal substructure. *See* ELEMENTARY PARTICLE. [L.W.J.]

Quarrying The process of extracting stone for commercial use from natural rock deposits. The industry has two major branches: a dimension-stone branch, involving preparation of blocks of various sizes and shapes for use as building stone, monumental stone, paving stone, curbing, and flagging; and a crushed-stone branch, involving preparation of crushed and broken stone for use as a basic construction, chemical, and metallurgical raw material. [S.H.Bo.]

Quartz The most abundant and widespread of all minerals. Quartz is an important rock-forming mineral that occurs as a subordinate constituent of many igneous, metamorphic, and sedimentary rocks; it is the principal constituent of sandstone and quartzite and of unconsolidated sands and gravels.

Quartz does not vary much in its chemical composition (SiO_2, silicon dioxide). The hardness is arbitrarily designated as 7 on Mohs scale, and the specific gravity is 2.650. Ordinarily, quartz is colorless and transparent, and the luster is vitreous. Amethyst is a purple or bluish-violet gem variety, and citrine is an orange-brown gem variety. Rose quartz is a massive type found in pegmatites. Quartz also may have a smoky yellow to dark smoky brown color, varying to brownish-black and almost opaque. *See* AMETHYST.

Quartz is enantiomorphous; crystals are either right-handed or left-handed, and correspondingly rotate the plane of polarization of transmitted light. It also is piezoelectric. Twinning is common in quartz. Large flawless crystals of quartz are employed for the manufacture of quartz oscillator plates, as prisms in optical spectrographs, and as other optical devices. Quartz also is employed as an abrasive and in the manufacture of firebrick and other refractories and of glass. *See* CRYSTAL OPTICS; PIEZOELECTRICITY; POLARIZED LIGHT; SPECTROSCOPY. [C.Fr.]

Quartz clock A clock that uses the piezoelectric property of a quartz crystal. When a quartz crystal vibrates, a difference of electric potential is produced between two of its faces. The crystal has a natural frequency of vibration, depending on its size and shape, and if it is introduced into an oscillating electric circuit having nearly the same frequency as the crystal, two effects take place simultaneously: The crystal is caused to vibrate at its natural frequency, and the frequency of the entire circuit becomes the same as the natural frequency of the crystal.

In the clock the alternating current from the oscillating circuit is amplified, and the frequency is subdivided in steps such as from 100 kHz to 1 kHz. This setup finally drives a synchronous motor and gear train to display time by hands on a clock face. The best crystals run for a year with accumulated errors of less than 0.1 s. *See* CLOCK; OSCILLATOR. [G.M.C.]

Quartzite A metamorphic rock consisting largely or entirely of quartz. Most quartzites are formed by metamorphism of sandstone; but some have developed by metasomatic introduction of quartz, SiO_2, often accompanied by other chemical elements, for example, metals and sulfur (ore quartzites). *See* METAMORPHIC ROCKS; METASOMATISM; SANDSTONE.

Pure sandstones yield pure quartzites. Impure sandstones yield a variety of quartzite types. The cement of the original sandstone is in quartzite recrystallized into characteristic silicate minerals, whose composition often reflects the mode of development. Even the Precambrian quartzites correspond to types that are parallel to present-day deposits. *See* QUARTZ. [T.F.W.B.]

Quasar An astronomical object that appears starlike on a photographic plate but possesses many other characteristics, such as a large redshift, that prove that it is not a star. The name "quasar" is a contraction of the term "quasistellar object" (QSO) which was originally given to these objects for their photographic appearance. The objects appear starlike because their angular diameters are less than about 1 second of arc, which is the resolution limit of ground-based optical telescopes imposed by atmospheric effects.

The color of quasars is generally much bluer than that of most stars with the exception of white dwarf stars. The blueness of quasars as an identifying characteristic led to the discovery that many blue starlike objects have a large redshift and are therefore quasars. The quasistellar objects (QSOs) discovered this way turned out to emit little or no radio radiation and to be about 100 times more numerous than the radio-emitting quasistellar radio sources (QSSs). Why some should be strong radio emitters and most others not is unknown. The orbiting *Einstein Observatory* for x-ray study has found that most quasars also emit strongly at x-ray frequencies. The optical continuum emission is highly variable, with time scales ranging from fractions of a day to several years. The rapid fluctuations indicate that there are some components in quasars that have diameters less than a light-day or about 10^{10} km, the size of the solar system. The radiation mechanism responsible for the optical and x-ray continuum emission is not known, but the most likely possibilities are either synchrotron emission or inverse Compton radiation. *See* GALAXY, EXTERNAL.

The many similarities of the observed characteristic of quasars with radio galaxies, Seyfert galaxies, and BL Lacertae objects strongly suggest that quasars are also active nuclei of galaxies. There is good statistical evidence which shows that quasars with large red shifts are spatially much more numerous than those with small redshifts. Because high-redshift objects are very distant and emitted their radiation at an earlier epoch, quasars must have been much more common in the universe about 10^{10} years ago. Since this is about the same epoch when galaxies are thought to form, it is possible that quasars may be associated with the birth of some galaxies.

The prodigious amount of energy released by quasars over their lifetime must involve a quantity of matter more than 10^9 times the mass of the Sun, and contained within a volume similar in size to the solar system. The source of this energy and the mechanism by which particles are accelerated to velocities very near the speed of light are not known. *See* ASTRONOMICAL SPECTROSCOPY; BLACK HOLE; INFRARED ASTRONOMY; NEUTRON STAR; SUPERMASSIVE STARS. [W.D.]

Quasiatom A transient electronic structure, formed in an atom-atom or ion-atom collision, which closely approximates the characteristics usually identified with a stable atom whose atomic number equals the combined charge carried by the colliding nuclei. Such short-lived atomic states can be formed in energetic atomic collisions, when the distance of closest approach between nuclear charge centers shrinks beyond the mean classical orbiting radius of the most-bound electrons, so that even these innermost electrons see a nearly monopolar electric field generated by the single charge center momentarily formed by the two nuclear collision partners. Thus, for the short time of this close nuclear promixity, all the surrounding electrons behave as if their motion were governed by this single charge center, leading to the formation of an electronic energy-level structure and electronic shells associated with a united atom having atomic number $Z = Z_1 + Z_2$, where Z_1 and Z_2 are the individual atomic numbers of the colliding system. *See* ATOMIC STRUCTURE AND SPECTRA.

An especially important potential utilization of quasiatom formation was suggested in 1969 as a possible vehicle for investigating superheavy atoms, considerably beyond the heaviest stable systems available. Such atomic systems are of intrinsic interest in quantum field theory, and particularly in the quantum electrodynamics of the very strong fields which can be generated in atoms when an electron is bound by a charge center whose value Z exceeds $1/\alpha \cong 137$, where α is the fine-structure constant. Under the latter conditions the usual perturbative expansion of all quantum-electrodynamic effects in terms of the coupling constant $Z\alpha$ is of doubtful validity, and any new theoretical approaches require experimental verification. *See* QUANTUM ELECTRODYNAMICS; QUANTUM FIELD THEORY. [J.S.Gr.]

Quasicrystal A solid with conventional crystalline properties but exhibiting a point-group symmetry inconsistent with translational periodicity. Like crystals, quasicrystals display discrete diffraction patterns, crystallize into polyhedral forms, and have long-range orientational order, all of which indicate that their structure is not random. But the unusual symmetry and the finding that the discrete diffraction pattern does not fall on a reciprocal periodic lattice suggest a solid that is quasiperiodic. Their discovery in 1982 contradicted a long-held belief that all crystals would be periodic arrangements of atoms or molecules.

It is easily shown that in two and three dimensions the possible rotations that superimpose an infinitely repeating periodic structure on itself are limited to angles that are $360°/n$, where n can be only 1, 2, 3, 4, or 6. Various combinations of these rotations lead to only 32 point groups in three dimensions, and 230 space groups which are combinations of the 14 Bravais lattices that describe the periodic translations with the allowed rotations. Until the 1980s, all known crystals could be classified according to this limited set of symmetries allowed by periodicity. Periodic structures diffract only at discrete angles (Bragg's law) that can be described by a reciprocal lattice, in which the diffraction intensities fall on lattice points that, like all lattices, are by definition periodic, and which has a symmetry closely related to that of the structure. *See* CRYSTAL; CRYSTALLOGRAPHY; X-RAY CRYSTALLOGRAPHY; X-RAY DIFFRACTION.

Icosahedral quasicrystals were discovered in 1982 during a study of rapid solidification of molten alloys of aluminum with one or more transition elements, such as manganese, iron, and chromium. Since then, many different alloys of two or more metallic elements have led to quasicrystals with a variety of symmetries and structures. The illustration shows the external polyhedral form of an icosahedral aluminum-copper-iron alloy.

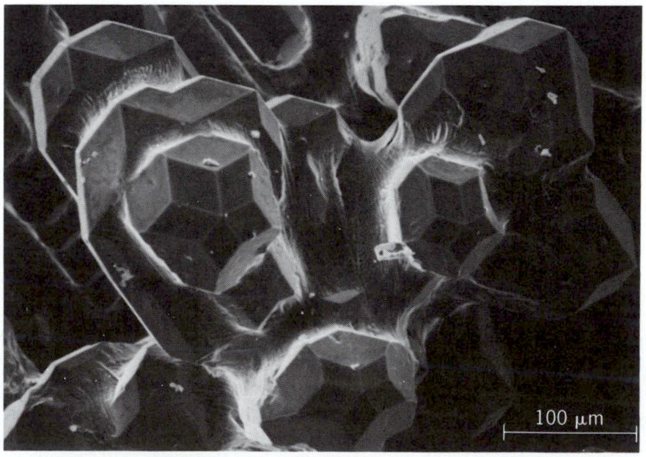

Quasicrystals of an alloy of aluminum, copper, and iron, displaying an external form consistent with their icosahedral symmetry.

The diffraction patterns of quasicrystals violate several predictions resulting from periodicity. Quasicrystals have been found in which the quantity n is 5, 8, 10, and 12. In addition, most quasicrystals exhibit icosahedral symmetry in which there are six intersecting fivefold rotation axes. Furthermore, in the electron diffraction pattern the diffraction spots do not fall on a (periodic) lattice but on what has been called a quasilattice. *See* ELECTRON DIFFRACTION. [J.W.C.; D.S.]

Quasielastic light scattering Small frequency shifts or broadening from the frequency of the incident radiation in the light scattered from a liquid, gas, or solid. The term quasielastic arises since the frequency changes are usually so small that, without instrumentation specifically designed for their detection, they would not be observed and the scattering process would appear to occur with no frequency changes at all, that is, elastically. The technique is used by chemists, biologists, and physicists to study the dynamics of molecules in fluids, mainly liquids and liquid solutions.

Several distinct experimental techniques are grouped under the heading of quasielastic light scattering (QLS). Intensity fluctuation spectroscopy (IFS) is the technique most often used to study such systems as macromolecules in solution and critical phenomena where the molecular motions to be studied are rather slow (see illustration). This technique, also called photon

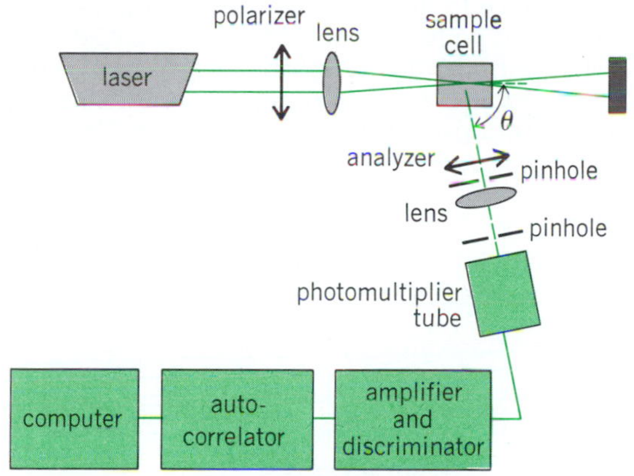

Schematic diagram of an intensity fluctuation spectroscopy apparatus.

correlation spectroscopy and, less frequently, optical mixing spectroscopy, is used to measure the dynamical constants of processes with relaxation time scales slower than about 10^{-6} s. For faster processes, dynamical constants are obtained by utilizing techniques known as filter methods, which obtain direct measurements of the frequency changes of the scattered light by utilizing a monochromator or filter much as in Raman spectroscopy. *See* INTERFEROMETRY; RAMAN EFFECT; SCATTERING OF ELECTROMAGNETIC RADIATION.

If light is scattered by a collection of scatterers, the scattered intensity at a point far from the scattering volume is the result of interference between the wavelets scattered from each of the scatterers and, consequently, will depend on the relative positions and orientations of the scatterers, the scattering angle θ, and the wavelength λ of the light used. The structure of scatterers in solution whose size is comparable to $(4\pi\lambda)\sin\theta/2$ ($\equiv q$) where q is the length of the scattering vector, may be studied by a technique variously called static light scattering, integrated intensity light scattering, or in the older literature simply light scattering. [R.P.]

Quaternary A period that encompasses at least the last 3,000,000 years of the Cenozoic Era, and is concerned with major worldwide glaciations and their effect on land and sea, on worldwide climate, and on the plants and animals that lived then. The Quaternary is divided into the Pleistocene Epoch and Holocene. The universal term Pleistocene is gradually re-placing Quaternary;

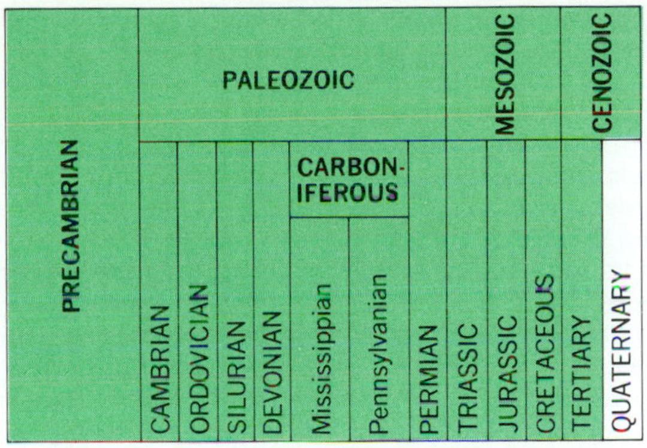

Holocene involves the last 7000 years since the Pleistocene. *See* CENOZOIC; GLACIAL EPOCH; HOLOCENE; PLEISTOCENE. [S.E.Wh.]

Quaternary ammonium salts Analogs of ammonium salts in which organic radicals have been substituted for all four hydrogens of the original ammonium cation. Substituents may be alkyl, aryl, or aralkyl, or the nitrogen may be part of a ring system. Such compounds are usually prepared by treatment of an amine with an alkylating reagent under suitable conditions. They are typically crystalline solids which are soluble in water and are strong electrolytes. Treatment of the salts with silver oxide, potassium hydroxide, or an ion-exchange resin converts them to quaternary ammonium hydroxides, which are very strong bases, as shown in the reaction below.

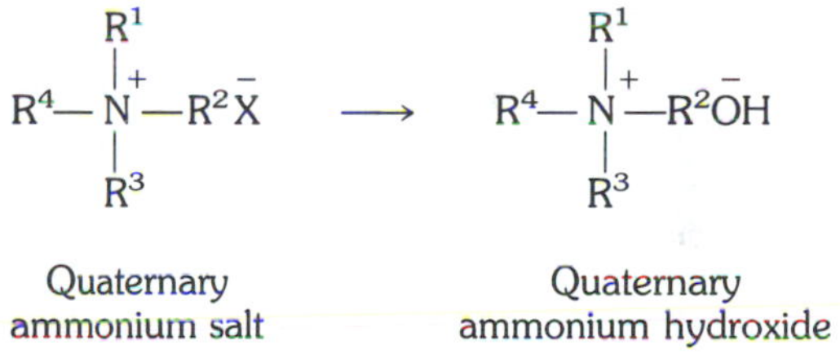

Some quaternary ammonium salts have found use as water repellents, fungicides, emulsifiers, paper softeners, antistatic agents, and corrosion inhibitors. *See* AMINE; AMMONIUM SALT; SURFACTANT. [P.E.F.]

Quaternions An associative, noncommutative algebra based on four linearly independent units or basal elements. Quaternions were originated in 1843, by W. R. Hamilton.

The four linearly independent units in quaternion algebra are commonly denoted by 1, i, j, k, where 1 commutes with i, j, k and is called the principal unit or modulus. These four units are assumed to have the following multiplication table:

$$1^2 = 1 \quad i^2 = j^2 = k^2 = ijk = -1$$

$$i(jk) = (ij)k = ijk$$

$$1i = i1 \qquad 1j = j1 \qquad 1k = k1$$

The i, j, k do not commute with each other in multiplication, that is, $ij \neq ji$, $jk \neq kj$, $ik \neq ki$, etc. But all real and complex numbers do commute with i, j, k, thus if c is a real number, then $ic = ci$, $jc = ci$, and $kc = ck$. On multiplying $ijk = -1$ on the left by i, so that $iijk = i(-1) = -i$, it is found, since $i^2 = -1$, that $jk = i$. Similarly $jjk = ji = -k$; when exhausted, this process leads to all the simple noncommutative relations for i, j, k, namely,

$$ij = -ji = k \quad jk = -kj = i \quad ki = -ik = j$$

More complicated products, for example, $jikjk = -kki = i$, are evaluated by substituting for any adjoined pair the value given in the preceding series of relations and then proceeding similarly to any other adjoined pair in the new product, and so on until the product is reduced to ± 1, $\pm i$, $\pm j$, or $\pm k$. Multiplication on the right is also permissible; thus from $ij = k$, one has $ijj = kj$, or $-i = kj$. Products such as jj and jjj may be written j^2 and j^3.

All the laws and operations of ordinary algebra are assumed to be valid in the definition of quaternion algebra, except the commutative law of multiplication for the units i, j, k. Thus the associative and distributive laws of addition and multiplication apply Without restriction throughout. Addition is also commutative, for example, $i + j = j + i$. [D.M.Y.]

Quebracho Any of a number of trees belonging to different genera but having similar qualities, all indigenous to South America and valuable for both wood and bark. The heartwood of one South American tree, *Schinopsis lorentzii* (family Anacardiaceae), is called quebracho (meaning ax-breaker) in reference to the exceedingly hard wood, one of the hardest known. Quebracho is the world's most important source of tannin. See SAPINDALES; TANNIN. [P.D.St./E.L.C.]

Queensland tick typhus A benign infectious disease found in rural northeastern Australia, caused by bacterialike microorganisms, *Rickettsia australis*, and presumed to be carried by the tick *Ixodes holocyclus*. The symptomatology, including eschar, is similar to rickettsialpox. See RICKETTSIOSES. [C.B.P.]

Quellung reaction Swelling of the capsule of a bacterial cell, caused by contact with serum containing antibodies capable of reacting with polysaccharide material in the capsule. The swelling is thought to be caused mainly by incorporation of fluid. The quellung phenomenon is sometimes employed as a method for rapid identification of bacteria seen in sputum or spinal fluid, being applicable to the *Pneumococcus*, *Klebsiella*, and *Hemophilus*. See HEMOPHILUS; IMMUNOLOGY; KLEBSIELLA; MEDICAL BACTERIOLOGY; PNEUMOCOCCUS. [P.B.B.]

Queueing theory The mathematical theory of the formation and behavior of queues or waiting lines. The name is also applied loosely to the mathematical study of a wide variety of problems connected with traffic congestion and storage systems. Uneven flow through a service point, with fluctuating arrivals and service times, constitutes a major topic of operations research. For the mathematician, queueing theory is particularly interesting because it is concerned with relatively simple stochastic processes, which are in general non-Markovian and possibly stationary. See OPERATIONS RESEARCH; STOCHASTIC PROCESS.

The principal pioneer of queueing theory, A. K. Erlang, began in 1908 to study problems of telephone congestion. It is of interest to study the waiting times of subscribers in a manually operated system—for example, the average waiting time and the chance that a subscriber will obtain service immediately without waiting—and to examine how much the waiting times will be affected if the number of operators is altered, or conditions are changed in any other way. If there are more operators or if service can be speeded up, subscribers will be pleased because waiting will be reduced, but the improved facility will be more expensive to maintain; therefore, a reasonable balance must be struck.

Related problems in the use of automatic telephone exchanges and of long-distance lines able to carry only a limited number of messages simultaneously have resulted in much mathematical study of telephone traffic problems. Similar problems arise in other contexts. In a factory a number of machines, such as looms, may be under the care of one or more repairer. If a machine breaks down, it must stand idle until a repairer is free from repairing other machines. Machines here correspond to telephone subscribers, breakdown corresponds to attempts to make a call, and repair corresponds to connection. Other examples of congestion situations are aircraft flying around in circles waiting to use an airport landing strip, automobiles lining up at a turnpike toll booth, and customers lining up at the counter of a retail shop, waiting for service.

What is most interesting to investigate varies with the circumstances. Sometimes it is the mean waiting time of customers, sometimes the frequency with which the queue length exceeds a given limit, sometimes the proportion of the servers' time that is idle, and sometimes the average duration of a period during which a server is continuously occupied. In the study of stocking a warehouse or retail shop, known generally as the theory of inventories, the frequency with which the stock will be exhausted is considered under various reordering policies. Similar considerations apply in the theory of dams and water storage. See LINEAR PROGRAMMING; SYSTEMS ENGINEERING. [F.J.An.]

Quince The deciduous tree *Cydonia oblonga*, originally from Asia, grown for its edible fruit. The fruit is a pear-shaped or apple-shaped pome, characteristically tomentose, aromatic, sour, astringent, and green, turning clear yellow at maturity. Used mostly for jam and jelly or as a stewed fruit, the fruit of the quince develops a pink color in cooking. See DECIDUOUS PLANTS; FRUIT; FRUIT, TREE; ROSALES. [H.B.T.]

Quinine The chief alkaloid of the bark of the cinchona tree, which is indigenous to certain regions of South America. The structure of quinine is shown below.

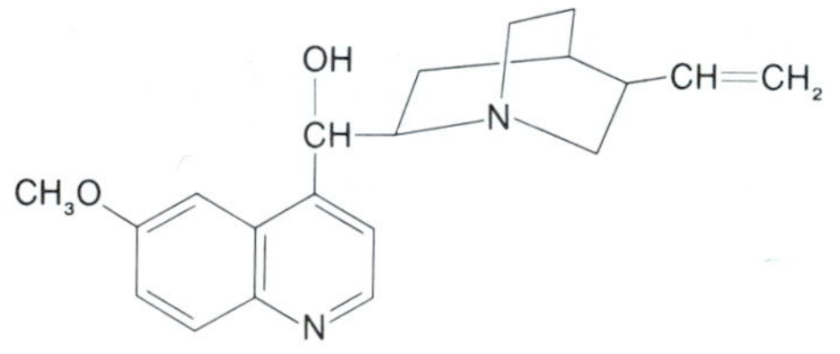

Until the 1920s quinine was the best chemotherapeutic agent for the treatment of malaria. However, clinical studies definitely established the superiority of the newer synthetic antimalarials such as primaquine, chloroquine, and chloroguanide. See ALKALOID; MALARIA. [S.M.K.]

Quinoa An annual herb, *Chenopodium quinoa* (family Chenopodiaceae), a native of Peru, and the staple food of many people in South America. These plants, grown at high altitudes, produce large quantities of highly nutritious seeds used whole in soups or ground into flour, which is made into bread or cakes. The seeds are also used as poultry feed, in medicine, and in making beer. In the United States the leaves are sometimes used as a substitute for spinach. See CARYOPHYLLALES. [P.D.St./E.L.C.]

Quinoline One of a group of organic compounds containing a benzene ring fused to the 2,3 positions of pyridine. Positions α, β, and γ are the same, respectively, as positions 2, 3, and 4. Quinoline and some of its homologs are obtained as coal tar extractives. The quinoline ring system appears in synthetic chemotherapeutic agents and in dyes. Quinine and other natural alkaloids also contain the quinoline ring. Quinoline itself is useful as a solvent, a source of nicotinic acid, an acid acceptor and dehydrohalogenating agent, and the starting point in organic syntheses. *See* HETEROCYCLIC COMPOUNDS; PYRIDINE.

Quinoline (1) is a colorless, steam-volatile liquid, with bp

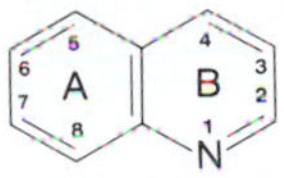

(I)

237.1°C (458.8°F), mp −15°C (59°F), specific gravity (15°C) 1.09771. Quinoline is a weakly basic tertiary amine that forms simple salts with acids, and quaternary salts with alkylating agents. The molecule is aromatic in character.

Quinoline is oxidized by nitric acid and other reagents to quinolinic acid (II). In contrast, quinolinium quaternary salts are

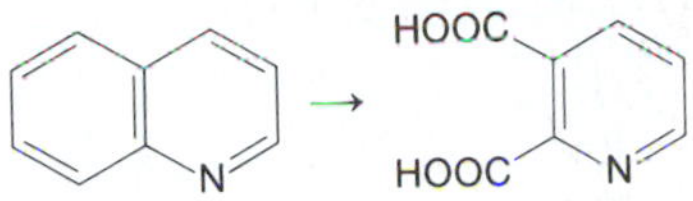

(II)

oxidized by alkaline permanganate to give products of pyridine ring disruption. *See* ISOQUINOLINE; QUININE. [W.J.Ge.; M.Sh.]

Quinone One of a class of aromatic diketones in which the carbon atoms of the carbonyl groups are part of the ring structure. The name quinone is applied to the whole group, but it is often used specifically to refer to *p*-benzoquinone (I). *o*-Benzoquinone (II) is also known but the meta isomer does not exist.

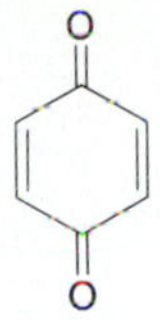

(I)

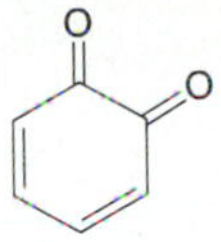

(II)

Quinones are prepared by oxidation of the corresponding aromatic ring systems containing amino ([bd]NH_2) or hydroxyl ([bd]OH) groups on one or both of the carbon atoms being converted to the carbonyl group.

Three of the several possible quinones derived from naphthalene are known: 1,4-naphthoquinone (III), 1,2-naphthoquinone, and 2,6-naphthoquinone (IV).

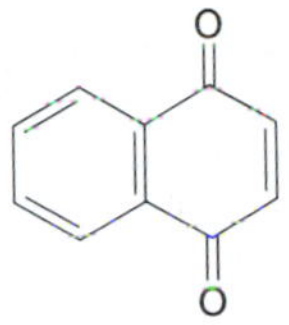

(III)

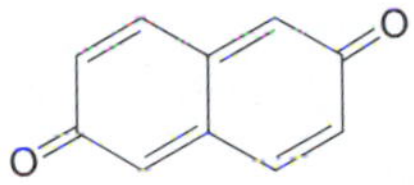

(IV)

Important naturally occurring naphthoquinones are vitamins K_1 and K_2 which are found in blood and are responsible for proper blood clotting reaction. A number of quinone pigments have been isolated from plants and animals. Illustrative of these are juglone found in unripe walnut shells and spinulosin from the mold *Penicillium spinulosum*. 9,10-Anthraquinone derivatives form an important class of dyes of which alizarin is the parent type. *p*-Benzoquinone is manufactured for use as a photographic developer. *See* ANTHRAQUINONE PIGMENTS; AROMATIC HYDROCARBON; DYE; HYDROQUINONE; KETONE; OXIDATION-REDUCTION; VITAMIN K. [D.A.S.]

R

Rabbit Any of the mammals which make up the family Leporidae in the order Lagomorpha. This family contains about 52 species of rabbits and hares, which are essentially grassland animals. They are found in all regions of the world, from intensely cold to very hot climates, and are absent only in Antarctica. In some regions, such as Australia and New Zealand, they are not native but have been introduced by humans and have become pests. There are 28 Old World species and 24 New World species, most of which fall into two genera, *Lepus* (hares) and *Sylvilagus*. All are short-tailed animals with long ears and elongated hindlimbs.

Unlike the hares, rabbits are social animals, and the wild European rabbit lives in large colonies. Rabbits use burrows but the hare rests in a depression in grass which is called a form. While both are active during the cool of the morning and early evening, the hare is more active nocturnally. These animals have many predatory enemies such as birds, foxes, weasels, and various species of cats. Their swiftness and their keen senses of hearing and smell afford them some protection. *See* LAGOMORPHA. [C.B.C.]

Rabies An acute, encephalitic viral infection. In humans it is almost invariably fatal. Human beings are infected from the bite of a rabid animal, usually a dog. Canine rabies can infect all warm-blooded animals, and death usually results. Some animals show chiefly paralytic signs, whereas others manifest encephalitic hyperexcitability and viciousness. However, the vampire bat may transmit virus for months while apparently not infected, and in other wild animals infection may take other courses than the fatal encephalitis.

The virus is believed to move from the saliva-infected wound through sensory nerves to the central nervous system, multiply there with destruction of brain cells, and thus produce encephalitis, with severe excitement, throat spasm upon swallowing (hence hydrophobia, or fear of water), convulsions, and death—with paralysis sometimes intervening before death.

Diagnosis in the human is made by observation of Negri bodies in brains of animals inoculated with the patient's saliva, or in the patient's brain after death. Diagnosis in dogs is essential for guidance concerning human vaccination. A dog which has bitten a person is isolated and watched for 7 days for signs of rabies; if none occur, rabies was absent. If signs do appear, the animal is killed and the brain examined for Negri bodies, or for rabies antigen by testing with fluorescent antibodies.

The wound is cleansed immediately and sometimes cauterized with nitric acid. The patient is vaccinated. Indiscriminate human vaccination is inadvisable, because there is risk of vaccination encephalitis. Vaccination is avoided if the biting animal is found not to be rabid. Human vaccination is done with a series of doses of fixed virus. The hope is that antibodies will develop in the patient as a result of vaccination before the infecting street virus has had sufficient time to multiply.

Rabies is controlled by compulsory vaccination of dogs and destruction of stray dogs; cattle may also be vaccinated. A living attenuated virus vaccine is available, and has been suggested for use in persons subjected to a high risk, for example, those who work with stray dogs in an area where canine vaccination is not compulsory. [J.L.Me.]

Raccoon Any of the carnivorous mammals which belong to the family Procyonidae, of which there are 16 species, found only in the New World. The robust body is covered with long gray-brown hair, the face is marked by a black mask across the eyes, and the bushy tail has five to seven black rings. The raccoon is plantigrade when standing but walks with its heels off the ground. The digits are long and terminate in nonretractile claws. This animal is found in woodland areas near forest edges, and is a good climber. Primarily terrestrial, it is also a good swimmer, although it cannot pursue prey underwater. A litter of five or six is born each year, and the animals live and travel in family groups. *See* CARNIVORA; MAMMALIA. [C.B.C.]

Racemization The formation of a racemate from a pure enantiomer. Alternatively stated, racemization is the conversion of one enantiomer in a 50:50 mixture of the two enantiomers (+ and −, or R and S) of a substance. Racemization is normally associated with the loss of optical activity over a period of time since 50:50 mixtures of enantiomers are optically inactive. *See* OPTICAL ACTIVITY.

Racemization is an energetically favored process since it reflects a change from a more ordered to a more random state. But the rate at which enantiomers racemize is typically quite slow unless a suitable mechanistic pathway is available, since racemization usually, but not always, requires that a chemical bond at the chiral center of an enantiomer be broken. Racemization of enantiomers possessing more than one chiral center requires that all chiral centers of half of the molecules invert their configurations. *See* ENTROPY.

The observation and study of racemization have important implications for the understanding of the mechanisms of chemical reactions and for the synthesis and analysis of chiral natural products such as peptides. Moreover, racemization is of economic importance since it provides a way of converting an unwanted enantiomer into a useful one. Synthetic medicinal agents are often produced industrially as racemates. After resolution and isolation of the desired enantiomer, half of the product would have to be discarded were it not for the possibility of racemizing the unwanted isomer and of recycling the resultant racemate. [S.H.W.]

Radar An acronym for radio detection and ranging, the original and still principal application of radar. The name is applied to both the technique and the equipment used.

Radar is a sensor; its purpose is to provide estimates of certain characteristics of its surroundings of interest to a user, most commonly the presence, position, and motion of such objects as aircraft, ships, or other vehicles in its vicinity. In other uses, radars provide information about the Earth's surface (or that of other astronomical bodies) or about meteorological conditions. To provide the user with a full range of sensor capability, radars are often used in combinations or with other elements of more complete systems.

Radar operates by transmitting electromagnetic energy into the surroundings and detecting energy reflected by objects. If a narrow beam of this energy is transmitted by the directive antenna, the direction from which reflections come and hence the bearing of the object may be estimated. The distance to the reflecting object is estimated by measuring the period between the transmission of the radar pulse and reception of the echo. In most radar applications this period will be very short since electromagnetic energy travels with the velocity of light.

Kinds of radar. The physical nature of radars varies greatly. Several radars are available for use on small boats as a safety and navigation aid, some so small as to be carried by an operator. Another radar seen in a hand-held form is that used by police to measure the speed of automobiles. *See* MARINE NAVIGATION.

Perhaps the largest radars are those covering acres of land, long arrays of antennas all operating together to monitor the flight of space vehicles or astronomical bodies. Other very large radars are designed to monitor flight activity at substantial distances. These are large mainly because they must use longer-than-usual radio wavelengths associated with ionospheric containment of the signal for over-the-horizon operations.

More common in size are those radars seen at airports, with rotating antennas 10–30 ft (3–9 m) wide. Radars intended for mobile use, particularly airborne radars, are quite compact. *See* AIRBORNE RADAR.

Radars intended principally to determine the presence and position of reflecting targets in a region around the radar are called search radars. Other radars examine further the targets detected: examples are height finders with antennas that scan vertically in the direction of an assigned target, and tracking radars that are aimed continuously at an assigned target to obtain great accuracy in estimating target motion. In some modern radars, these search and track functions are combined, usually with some computer control. Surveillance radar connotes operation of this sort, somewhat more than just search alone. There are also very complex and versatile radars with considerable computer control, with which many functions are performed and which are therefore called multifunction radars. Very accurate tracking radars intended for use at missile test sites or similar test ranges are called instrumentation radars. Radars designed to detect clouds and precipitation are called meteorological or weather radars. *See* RADAR METEOROLOGY; SURVEILLANCE RADAR.

Some radars have separate transmit and receive antennas sometimes located miles apart. These are called bistatic radars, the more conventional single-antenna radar being monostatic. Some useful systems have no transmitter at all and are equipped to measure, for radarlike purposes, signals from the targets themselves. Such systems are often called passive radars, but the terms radiometers or signal intercept systems are generally more appropriate. *See* PASSIVE RADAR.

The terms primary and secondary are used to describe, respectively, radars in which the signal received is reflected by the target and radars in which the transmission causes a transponder (transmitter-responder) carried aboard the target to transmit a signal back to the radar. *See* AIR-TRAFFIC CONTROL; ELECTRONIC NAVIGATION SYSTEMS.

Operation. It is convenient to consider radars composed of four principal parts: the transmitter, antenna, receiver, and display (see illustration).

The transmitter provides the rf signal in sufficient strength (power) for the radar sensitivity desired and sends it to the antenna, which causes the signal to be radiated into space in a desired direction. The signal propagates (radiates) in space, and some of it is intercepted by reflecting bodies. These reflections, in part at least, are radiated back to the antenna. The antenna collects them and routes all such received signals to the receiver, where they are amplified and detected. The presence of an echo of the transmitted signal in the received signal reveals the presence of a target. The echo is indicated by a sudden rise in the output of the detector, which produces a voltage (video) proportional to the sum of the rf signals being received and the rf noise inherent in the receiver itself. The time between the transmission and the receipt of the echo discloses the range to the target. The direction or bearing of the target is disclosed by the direction the antenna is pointing when an echo is received.

A duplexer permits the same antenna to be used on both transmit and receive, and is equipped with protective devices to block the very strong transmit signal from going to the sensitive receiver and damaging it. The antenna forms a beam, usually quite directive, and, in the search example, rotates throughout the region to be searched. *See* ANTENNA (ELECTROMAGNETISM).

The radar reflections are among the signals received by the antenna in the period between transmissions. Most search

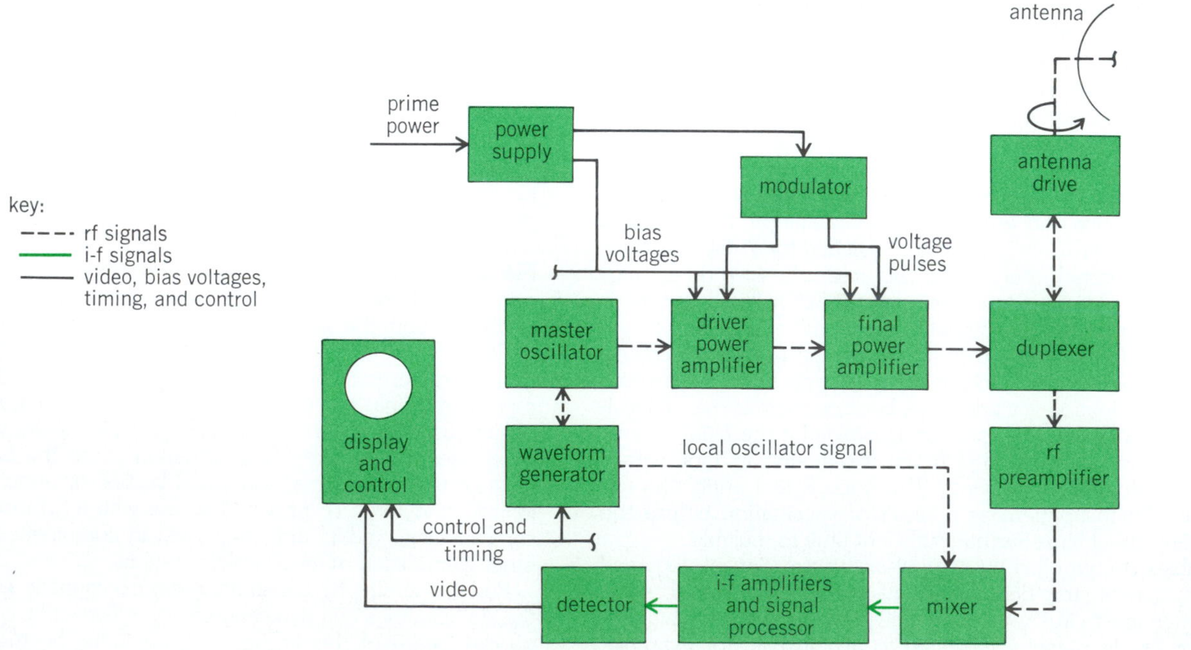

Block diagram of a pulse radar.

Radar carrier-frequency bands

Band designation	Nominal frequency range	Representative wavelength
HF	3–30 MHz	30 m at 10 MHz
VHF	30–300 MHz	3 m at 100 MHz
UHF	300–1000 MHz	1 m at 300 MHz
L	1000–2000 MHz	30 cm at 1000 MHz
S	2000–4000 MHz	10 cm at 3000 MHz
C	4000–8000 MHz	5 cm at 6000 MHz
X	8000–12,000 MHz	3 cm at 10,000 MHz
K_u	12.0–18 GHz	2 cm at 15 GHz
K	18–27 GHz	1.5 cm at 20 GHz
K_a	27–40 GHz	1 cm at 30 GHz
mm	40–300 GHz	0.3 cm at 100 GHz

radars have a pulse repetition frequency (prf), antenna beamwidth, and rotation rate such that several pulses are transmitted (perhaps 20 to 40) while the antenna scans past a target. This allows a buildup of the echo being received. Some radars are equipped with low-noise rf preamplifiers to improve sensitivity. Almost all radars use some form of mixing of the received signal with a local oscillator signal to produce an intermediate frequency (i-f) signal, commonly at 30 or 60 MHz, more conveniently amplified and processed. The local oscillator signal, offset from the transmitted rf by precisely the i-f, can be supplied by a waveform generator. The mixer is sometimes called the first detector. The i-f signal is then fed to a detector (sometimes called the second detector) which produces a video signal, a voltage proportional to the strength of the i-f at its imput. This video may be amplified and used in the cathode-ray tube (CRT) of the radar display to cause a bright spot to be formed on the face of the tube when an echo is received, in a manner indicating the range and bearing of the target. *See* CATHODE-RAY TUBE; ELECTRONIC DISPLAY; MIXER; RADIO RECEIVER.

Radar carrier frequencies are broadly identified by a nomenclature that originated in wartime secrecy and has since been found very convenient and widely accepted. The spectrum is divided into bands, the frequencies and wavelengths of which are given in the table. The charged layers of the ionosphere present a highly refractive shell at radio frequencies well below the microwave frequencies of most radars. Consequently, over-the-horizon radars have been built in the 10-MHz area to exploit this skip path. *See* CONTINUOUS-WAVE RADAR; MONOPULSE RADAR; RADIO SPECTRUM ALLOCATIONS. [R.T.H.]

Radar-absorbing materials Materials that are designed to reduce the reflection of electromagnetic radiation by a conducting surface in the frequency range from approximately 100 MHz to 100 GHz. The level of reduction achieved varies from a few decibels to greater than 50 dB, in percentage terms reducing the reflected energy by up to 99.999%.

Two methods have been widely adopted in order to produce practical absorbing materials. The first is to avoid a discrete change of impedance at the material surface by gradually varying the impedance, for example by the use of a thick profiled lossy layer. The removal of the discrete discontinuity at the surface allows the microwave energy to be transmitted into the absorbing medium without reflection. Tapered carbon-loaded pyramids which are used for the lining of anechoic chambers are typical of this type of absorber. To produce such absorbers, it is necessary in practice to taper the material over distances which are large compared with the wavelength, in the material, of the frequencies to be absorbed. While this type of absorber is capable of producing a very high degree of absorption over a broad bandwidth, it is at the same time a relatively thick material. *See* ANECHOIC CHAMBER.

The second method of absorber design has been developed to give much thinner absorbing layers which are capable of producing good absorption (≧25 dB) with restricted bandwidths. These materials consist of lossy layers where the absorption is produced by a destructive interference at the frequency for which the material is electrically a quarter wavelength. *See* INTERFERENCE OF WAVES.

Microwave-absorbing materials are widely used both within the electronics industry and for defense purposes. Their uses can be classified into three major areas: (1) for test purposes so that accurate measurements can be made on microwave equipment unaffected by spurious reflected signals (the most common example is the anechoic chamber); (2) to improve the performance of any practical microwave system by removing unwanted reflections which can occur if there is any conducting material in the radiation path; and (3) to camouflage a military target by reducing the reflected radar signal. *See* ELECTRONIC WARFARE; MICROWAVE; REFLECTION OF ELECTROMAGNETIC RADIATION. [R.J.L.]

Radar astronomy A powerful astronomical technique that uses radar echoes to furnish otherwise-unavailable information about bodies in the solar system. By comparing a radar echo to the transmitted signal, information can be obtained about the target's size, shape, topography, surface bulk density, spin vector, and orbital elements. While other astronomical techniques rely on passive measurement of reflected sunlight or naturally emitted radiation, the illumination used in radar astronomy is a coherent signal whose polarization and time modulation or frequency modulation are tailored to meet specific scientific objectives. Through measurements of the distribution of echo power in time delay or Doppler frequency, radar achieves spatial resolution of a planetary target despite the fact that the radar beam is typically much larger than the angular extent of the target. This capability is particularly valuable for asteroids and planetary satellites, which appear as unresolved point sources through optical telescopes. Moreover, the centimeter-to-meter wavelengths used in radar astronomy readily penetrate cometary comas and the optically opaque clouds that conceal Venus and Titan, and also permit determination of near-surface roughness (abundance of wavelength-scale rocks), bulk density, and metal concentration in planetary regoliths. *See* ASTEROID; SATELLITE (ASTRONOMY); SATURN; VENUS.

A radar telescope is esssentially a radio telescope equipped with a high-power transmitter (a klystron vacuum-tube amplifier) and specialized instrumentation that links the transmitter, low-noise maser receiver, high-speed data-acquisition computer, and antenna together in an integrated radar system. Planetary radars, which must detect echoes from targets at distances from about 10^6 km for closely approaching asteroids and comets to more than 10^9 km for Saturn's rings and satellites, are the largest and most sensitive radars on Earth. *See* KLYSTRON; MASER; RADAR; RADIO TELESCOPE. [S.J.O.]

Radar meteorology The study of the scattering of radar waves by all types of atmospheric phenomena and the use of radar for making weather observations and forecasts. In general, radar is useful in meteorology because it is capable of detecting water and ice particles, but it can also be used to observe lightning and regions of the atmosphere which have large gradients of temperature and water vapor. Important applications of radar treated in meteorology include use of radar for rainfall measurement; study of cloud and precipitation formation; observations of tornadoes and hurricanes; and radar echoes from targets other than water or ice particles. *See* DOPPLER RADAR; METEOROLOGY; RADAR; STORM DETECTION. [L.J.B.]

Radian measure A radian is the angle subtended at the center of a circle by an arc of the circle equal in length to its radius. It is proved in geometry that equal central angles of two circles subtend arcs proportional to their radii; and the con-

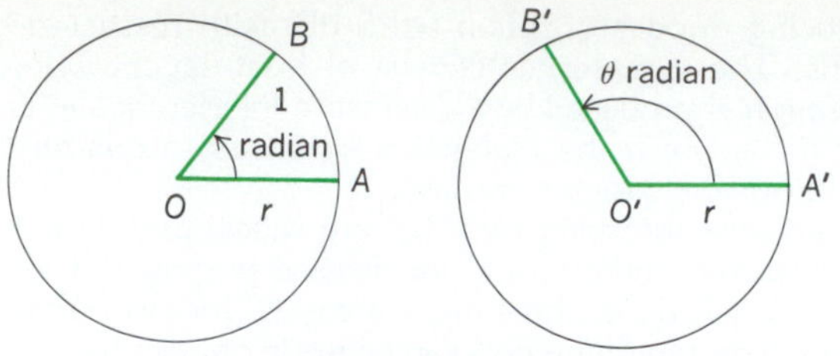

Diagrams showing radian measurement.

verse is true. Hence the radian is independent of the length of the radius. The illustration represents two circles of radius r. Arc AB of length r subtends 1 radian (rad) at the center O of the circle, and arc $A'B'$ of length s subtends θ rad at its center. Since arcs on equal circles are proportional to their subtended central angles, $s/r = \theta/1$, or the formula $s = r\theta$ holds. If $\theta = 2\pi$, $s = 2\pi r$, the circumference of the circle. Therefore 2π rad is the complete angle about a point or 360°, and 2π rad = 360°, 1 rad = $360°/2\pi = 57.2958\pi$.

The degree as a unit of angle has come down from antiquity. However, its use in various theories involves clumsy constants. The use of the radian avoids these constants. The radian is employed generally as a measure of angle in theoretical discussions; when no unit of angle is mentioned, the radian is understood. [L.M.K.]

Radiance The physical quantity that corresponds closely to the visual brightness of a surface. A simple radiometer for measuring the (average) radiance of an incident beam of optical radiation (light, including invisible infrared and ultraviolet radiation) consists of a cylindrical tube, with a hole in each end cap to define the beam cross section there, and with a photocell against one end to measure the total radiated power in the beam of all rays that reach it through both holes (see illustration). If A_1 and A_2 are the respective areas of the two holes, D

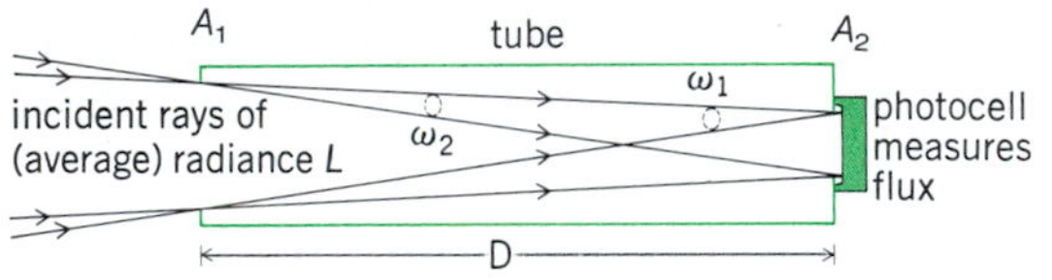

A simple radiometer.

is the length of the tube (distance between holes), and Φ is the radiant flux or power measured by the photocell, then the (average) radiance is approximately given by the equation $L = \Phi/(A_1 \cdot A_2/D^2)$ W · m^{-2} · sr^{-1}. [F.E.N.]

Radiant heating Any system of space heating in which the heat-producing means is a surface that emits heat to the surroundings by radiation rather than by conduction or convection. The surfaces may be radiators such as baseboard radiators or convectors, or they may be the panel surfaces of the space to be heated. *See* PANEL HEATING AND COOLING.

The heat derived from the Sun is radiant energy. Radiant rays pass through gases without warming them appreciably, but they increase the sensible temperature of liquid or solid objects upon which they impinge. The same principle applies to all forms of radiant-heating systems, except that convection currents are established in enclosed spaces and a portion of the space heating is produced by convection. Any radiant-heating system using a fluid heat conveyor may be employed as a cooling system by substituting cold water or other cold fluid. However, the technique is not practical on the scale required for comfort control of an occupied space. [E.L.W./R.Ko.]

Radiation The emission and propagation of energy; also, the emitted energy itself. The etymology of the word implies that the energy propagates rectilinearly, and in a limited sense, this holds for the many different types of radiation encountered.

The major types of radiation may be described as electromagnetic, acoustic, and particle, and within these major divisions there are many subdivisions. Electromagnetic radiation is classified roughly in order of decreasing wavelength as radio, microwave, visible, ultraviolet, x-rays, and γ-rays. Acoustic or sound radiation may be classified by frequency as infrasonic, sonic, or ultrasonic in order of increasing frequency, with sonic being between about 16 and 20,000 Hz. The traditional examples of particle radiation are the α- and β-rays of radioactivity. *See* ELECTROMAGNETIC RADIATION; RADIOACTIVITY; SOUND. [McA.H.H.]

Radiation biology A study of the influence of light or ionizing radiation, such as x-rays or fast particles, on living systems. Radiation biology ranges from a consideration of the effects of visible light on metabolism to the effects of cosmic rays on whole organisms. Because of the breadth of the subject, it is studied to a large extent in separate areas. The first division separates the fields of ionizing radiation effects and photon effects. The term photon effects means the action of ultraviolet, visible, and infrared light in the region where ionization does not occur. *See* RADIATION; X-RAYS.

Ionizing radiation. Ionizing radiation produces random energy releases of great size and generally of great disruptive effect. When such a disruptive effect is on functioning units of the organism, it is termed direct action. It is termed indirect action when it occurs on the water moiety through formation of active radicals, notably OH and H, which exert a gentler action, and which, because of radical diffusion, can act at some distance from the original ionization. To measure any effect of ionizing radiation the amount of radiation must be measured. Such measurement is called dosimetry; the amount given is the dose. *See* DOSIMETER.

A considerable variety of effects exists. A brief sampling follows.

1. Survival after whole-body irradiation. A whole animal, such as a mouse, which has been irradiated, shows no marked immediate effect. After a few days, however, a definite increase in mortality occurs if exposure exceeded 400 roentgens, with a mean lethal dose at 500 R and a rapid increase in mortality rising to 100% at 1000 R. In such a process death is due to effects on a variety of organs, the blood-forming organs and any rapidly dividing tissue being the most sensitive.

2. Production of mutations. Ionizing radiation produces mutations in any living organism, with the possible exception of viruses. The majority of such mutations are lethal. *See* MUTATION.

3. Production of chromosome breaks. AH kinds of chromosome abnormalities, including chromosome breaks, are produced by ionizing radiation. Chromosome breaks can restitute and probably the majority do so.

4. Delay of cell division. Ionizing radiation inhibits cell division to a marked degree. The degree of delay increases with the dose. *See* CELL DIVISION.

5. Formation of giant cells. Cells in which division is delayed may grow into giant cells many times the normal size in the case of mammalian cells, or into long filaments in the case of bacteria.

6. Reduction of survival. If means for studying the ability of

a cell to divide are available, it is found that animal, plant, and bacterial cells and viruses are reduced in survival.

7. Action on biological macromolecules. All important biological macromolecules—deoxyribonucleic acid (DNA), ribonucleic acid (RNA), enzymes, and antigens—are destructively affected by ionizing radiation. Roughly speaking, if one ionization occurs within the molecule, or if one active radical in the case of nucleic acid, or 1–10 radicals in the case of protein reach the molecule, it loses its function. *See* DEOXYRIBONUCLEIC ACID (DNA); RIBONUCLEIC ACID (RNA).

8. Action on metabolism and protein synthesis. Metabolism is reduced by ionizing radiation but much less than is division or the formation of lethal mutations. Also reduced are protein synthesis and formation of microsomal particles, or ribosomes, which are agents in protein synthesis. *See* RIBOSOMES.

9. There is evidence that ionizing radiation can precipitate the destructive action of enzymes already in the cell. One is a nuclease which rapidly degrades the DNA, drastically damaging the cell.

10. The phenomenon of recovery from the effects of ionizing radiation has been established. Cellular DNA which has become fragmented by ionizing radiation is found to be restored to much larger size if the cells in which damage has occurred are held for 20 min (bacteria) or several hours (mammalian cells).

11. Ionizing radiation, in rather high doses, acts to depress the immune system.

Ultraviolet light. Ultraviolet light differs from ionizing radiation in that it is less energetic and much more specific. Before any of its effects can occur, ultraviolet light must first be absorbed, a process which is wavelength-dependent. The most significant absorption of ultraviolet light is by nucleic acids. There is also absorption in protein, notably by aromatic amino acids and cystine.

Light of high intensity (such as sunlight), though not readily absorbed, may yet produce marked biological action, the high intensity compensating for the lack of absorption. Light in the near-ultraviolet range, where absorption by nucleic acids is inefficient, has been found to have significant actions.

Because it must be absorbed, ultraviolet light rarely exerts lethal action on whole animals. The effects produced by ultraviolet radiation are discussed below.

Burns. Ultraviolet light produces drastic action on the skin, familiar as sunburn. Prolonged exposure to ultraviolet light or sunlight can cause skin cancer, particularly in individuals with fair skin.

Chromosome breaks. Ultraviolet light produces chromosome breaks, though of a less drastic kind than ionizing radiation, being more commonly in one of two chromatids rather than the whole chromosome.

Delay of cell division and giant-cell formation. Cell division is inhibited by ultraviolet light, and such cells can grow to giant size.

Reduction of survival. Ultraviolet light reduces survival, though not so simply, in general, as ionizing radiation, there being a tendency for a fraction of the cells to show less sensitivity.

Photoreactivation. If cells or virus-infected cells are subjected to illumination by light in the blue, or near-ultraviolet, range, the degree of survival is markedly increased, a phenomenon known as photoreactivation.

Excision repair. There is also a process whereby dimers and cross-links in the DNA can be enzymatically removed and the DNA then accurately patched. This occurs in the dark. Mutant cells lacking one or more of the enzymes needed have been isolated.

Action on metabolism and protein synthesis. These are affected by ultraviolet light, though not in the same way as by ionizing radiation. More effect takes place on the DNA-synthetic process, which is less affected by ionizing radiation.

Recombination repair. A still only partly understood mechanism exists by which one intact genome can be assembled from two damaged genomes. This is related to genetic recombination, and the enzymes needed have not yet been characterized.

Error-prone repair. It has been found that in some cells the effect of radiation is to induce a new repair system which is less specific in its action than those described above.

Mutations. Ultraviolet light, including near ultraviolet, produces mutation in all cells, including bacteria. It does so by a relatively small chemical alteration in the DNA. [E.C.Pol.]

Radiation chemistry The study of the chemical effects of the absorption of high-energy radiation in matter. High-energy radiation includes the emanation associated with radioactive decay and fission (that is, α-particles, electrons, γ-rays, and neutrons), together with their related atom and fission recoils; and the artificial analogs of such emanations produced by accelerating electrons, protons, deuterons, and helium nuclei, as well as charged nuclei of higher atomic number and x-rays.

Sources of high-energy radiations in the laboratory and industry include radioactive nuclides (for example, ^{60}Co, ^{90}Sr, and ^{3}H) and instruments such as x-ray tubes, Van de Graaff generators, the betatron, the cyclotron, and the synchroton. The linear electron accelerator (linac) has proved particularly valuable for the study of transient species which have lifetimes as short as 16 picoseconds. A variety of fast reaction techniques are used to study the reactions of short-lived species produced by such short pulses. These include optical absorption, fluorescence, ionic conductivity, electron spin resonance, nuclear magnetic resonance, resonance Raman scattering, and polarography. *See* CERENKOV RADIATION; ELECTRON PARAMAGNETIC RESONANCE (EPR) SPECTROSCOPY; FLUORESCENCE; IONIC CRYSTALS; NUCLEAR MAGNETIC RESONANCE (NMR); POLAROGRAPHIC ANALYSIS; RAMAN EFFECT.

Energy is transmitted to irradiated material by momentum transfer and by excitation and ionization. The latter two always accompany the former. Momentum transfer is characteristic of processes involving neutrons; it is always involved to some extent in particle effects and is an important contribution to heavy-particle (for example, proton) effects. In a momentum transfer interaction, the usual effect is the ejection of a nucleus from its molecular or crystalline structure. In the case of solids, this is known as the Wigner effect. When crystalline material is involved, the process is called discomposition. Discomposition results not only from neutron impact, but also secondarily from impacts involving high-energy displaced nuclei.

For other than momentum transfer, the principal effects are due to fast-moving charged particles. A 1-MeV charged particle, unlike a parent photon which produces only one ionization (as in a Compton process), may produce a total of 10^5 ions (and electrons) and excited molecules. The distribution of such primarily produced entities is inhomogeneous, and is affected greatly by the nature of the radiation and by the state of aggregation of the material irradiated. It is this inhomogeneity of the distribution of activated species which is an important fact differentiating radiation chemistry from photochemistry and thermally induced chemistry. *See* PHOTOCHEMISTRY.

Chemical effects of high-energy radiation must be guarded against in nuclear reactors (where effects include radiation corrosion, water decomposition, and the Wigner effect) and in living systems (because of mutations, cancer production, and so on). Such effects may be deliberately employed to induce polymerization of special kinds: to cross-link, to graft, and to thermally stabilize polymers; to sterilize foods, medicinals, and surgical materials; to change the properties of catalysts; and to induce reactions not possible by other means or to induce them under unusual environmental conditions, such as under extremely low temperature, in very thick layers, and in heavy-walled (pressure) vessels. *See* RADIATION BIOLOGY; RADIATION DAMAGE TO MATERIALS; RADIOACTIVITY AND RADIATION APPLICATIONS. [S.G.]

Radiation damage to materials Harmful changes in the properties of liquids, gases, and solids, caused by interaction with nuclear radiations. For a discussion of radiation damage in minerals *see* METAMICT STATE. For a description of damage caused to biological systems by radiation *see* RADIATION BIOLOGY.

Radiation damage is usually associated with materials of construction that must function in an environment of intense high-energy radiation from a nuclear reactor. Materials that are an integral part of the fuel element or cladding and nearby structural components are subject to such intense nuclear radiation that a decrease in the useful lifetime of these components can result. Radiation damage will also be a factor in thermonuclear reactors. *See* NUCLEAR FUSION; NUCLEAR REACTOR.

Electronic components are extremely sensitive to even moderate radiation fields. Transistors malfunction because of defect trapping of charge carriers. Ferroelectrics such as $BaTiO_3$ fail because of induced isotropy; quartz oscillators change frequency and ultimately become amorphous. High-permeability magnetic materials deteriorate because of hardening. Plastics used for electrical insulation rapidly deteriorate.

There are several mechanisms that function on an atomic and nuclear scale to produce radiation damage in a material if the radiation is sufficiently energetic, whether it be electrons, protons, neutrons, x-rays, fission fragments, or other charged particles.

Electronic excitation and ionization is most severe in liquids and organic compounds and appears in a variety of forms such as gassing, decomposition, viscosity changes, and polymerization in liquids. Rapid deterioration of the mechanical properties of plastics takes place either by softening or by embrittlement, while rubber suffers severe elasticity changes at low fluxes. Cross-linking, scission, free-radical formation, and polymerization are the most important reactions. *See* RADIATION CHEMISTRY.

In an environment of neutrons, transmutation effects may be important. Even materials that have a low cross section such as aluminum can show an appreciable accumulation of impurity atoms from transmutations. The elements boron and europium have very large cross sections and are used in control rods. Damage to the rods is severe in boron-containing materials. *See* TRANSMUTATION.

Displaced atoms are the most important source of radiation damage in nuclear reactors outside the fuel element. It is a consequence of the ability of the energetic neutrons born in the fission process to knock atoms from their equilibrium position in their crystal lattice, displacing them many atomic distances away into interstitial positions and leaving behind vacant lattice sites.

Nuclear irradiations performed at low temperatures (4 K) result in the maximum retention of radiation-produced defects. As the temperature of irradiation is raised, many of the defects are mobile and some annihilation may take place. The increased mobility, particularly of vacancies and vacancy agglomerates, may lead to acceleration of solid-state reactions, such as precipitation, short- and long-range ordering, and phase changes. These reactions may lead to undesirable property changes.

The presence of small amounts of impurities may profoundly affect the behavior of engineering alloys in a radiation field. It has been observed that helium concentrations as low as 10^{-9} seriously reduce the high-temperature ductility of a stainless steel. Small amounts of copper, phosphorus, and nitrogen have a strong influence on the increase in the ductile-brittle transition temperature of pressure vessel steels under irradiation. Heat treatment prior to irradiation determines the retention of both major alloying components and impurities in solid solution in metastable alloys. It also affects the number and disposition of dislocations. Thus heat treatment is an important variable in determining subsequent radiation behavior. [D.S.B.]

Radiation hardening The protection of semiconductor electronic devices and electronic systems from the effects of high-energy radiation. Applications for such devices are in three major areas: (1) satellites, which are exposed to natural space radiation from the Van Allen belts, solar flares, and cosmic rays; (2) electronics, especially sensor and control electronics for commercial nuclear power-generating plants; and (3) equipment designed to survive the radiation from nuclear explosions.

Although most radiation effects have been explained and are well understood, much remains to be done. Among the better-understood phenomena are displacement damage from neutrons and photocurrent transients produced by ionizing-radiation pulses. Basic electromagnetic-pulse interactions are also well understood, although their effects on complex electronic systems are extremely difficult to predict. *See* ELECTROMAGNETIC PULSE (EMP).

The quasipermanent effects of exposure to ionizing radiation are least understood, and understanding the response of semiconductor devices to this radiation is probably the single most important remaining radiation-hardening problem. Hardening of metal-oxide-semiconductor (MOS) devices has been accomplished by lower-temperature processing, which probably reduces physical or crystalline defects, and by developing extremely clean processes, which probably reduce chemical defects. The most important electrical manifestations of ionizing dose damage in MOS devices are an increase in leakage current; a shift in threshold voltage; and a decrease in speed, transconductance, and channel conductance. *See* CRYSTAL.

Dose-rate effects are well understood. The generation of electron-hole pairs in semiconductors is proportional to the dose. Carriers generated in or near a *pn* junction result in a transient photocurrent proportional to dose rate and the effective volume of the junction. High dose rates can damage semiconductor devices through logic upset, latch-up, and burnout. *See* PHOTOVOLTAIC EFFECT.

Displacement damage effects are caused by neutrons, protons, electrons, and other high-energy particles. The production of lattice defects is proportional to the nonionizing energy absorbed by the lattice. The dominant effect in bipolar silicon devices is a reduction in common-emitter current gain.

Single-event upsets are caused at a very low rate in logic and memory circuits by cosmic rays. Rates are low enough (less than 10^{-3} upset per bit per day) that error-detection-and-correction (EDAC) software can be effectively used. Single-event phenomena have led to the development of fault-tolerant architectures to mitigate the effects of random digital upsets, and to the design modification of integrated circuits to prevent an upset even when the active volume is struck by a cosmic ray. *See* FAULT-TOLERANT SYSTEMS.

Hardening of electronic systems is accomplished by a combination of selecting hardened components, designing circuits more tolerant to radiation-induced degradations, and shielding. Shielding is effective against x-rays and electrons, which have short ranges in materials, and relatively ineffective against gamma radiation, neutrons, and cosmic rays, which have long ranges in materials. *See* INTEGRATED CIRCUITS; RADIATION DAMAGE TO MATERIALS; RADIATION SHIELDING; SEMICONDUCTOR. [G.C.M.]

Radiation pressure Pressure exerted by electromagnetic radiation on objects on which it impinges. This pressure is caused by the fact that electromagnetic radiation transmits energy and possesses momentum. These pressures are very small. The effect is conspicuous in the case of a comet near the Sun, where the radiation pressure from the Sun forces the lighter cometary constituents away from the Sun. *See* ELECTROMAGNETIC RADIATION. [W.R.Sm.]

Radiation shielding Physical barriers designed to provide protection from the effects of ionizing radiation; also, the technology of providing such protection. Major sources of radiation are nuclear reactors and associated facilities, medical and industrial x-ray and radioisotope facilities, charged-particle accelerators, and cosmic rays. Types of radiation are directly ionizing (charged particles) and indirectly ionizing (neutrons, gamma rays, and x-rays). In most instances, protection of human life is the goal of radiation shielding. In other instances, protection may be required for structural materials which would otherwise be exposed to high-intensity radiation, or for radiation-sensitive materials such as photographic film and certain electronic components.

Charged particles lose energy and are thus attenuated and stopped primarily as a result of coulombic interactions with electrons of the stopping medium. Gamma-ray and x-ray photons lose energy principally by three types of interactions: photoemission, Compton scattering, and pair production. Neutrons lose energy in shields by elastic or inelastic scattering. Elastic scattering is more effective with shield materials of low atomic mass, notably hydrogenous materials, but both processes are important, and an efficient neutron shield is made of materials of both high and low atomic mass. *See* COMPTON EFFECT; ELECTRON-POSITRON PAIR PRODUCTION; PHOTOEMISSION.

The most common criteria for selecting shielding materials are radiation attenuation, ease of heat removal, resistance to radiation damage, economy, and structural strength.

For neutron attenuation, the lightest shields are usually hydrogenous, and the thinnest shields contain a high proportion of iron or other dense material. For gamma-ray attenuation, the high-atomic-number elements are generally the best. For heat removal, particularly from the inner layers of a shield, there may be a requirement for external cooling with the attendant requirement for shielding the coolant to provide protection from induced radioactivity.

Metals are resistant to radiation damage, although there is some change in their mechanical properties. Concretes, frequently used because of their relatively low cost, hold up well; however, if heated they lose water of crystallization, becoming somewhat weaker and less effective in neutron attenuation.

If shielding cost is important, cost of materials must be balanced against the effect of shield size on other parts of the facility, for example, building size and support structure. If conditions warrant, concrete can be loaded with locally available material such as natural minerals (magnetite or barytes), scrap steel, water, or even earth.

Radiation shields vary with application. The overall thickness of material is chosen to reduce radiation intensities outside the shield to levels well within prescribed limits for occupational exposure or for exposure of the general public. The reactor shield is usually considered to consist of two regions, the biological shield and the thermal shield. The thermal shield, located next to the reactor core, is designed to absorb most of the energy of the escaping radiation and thus to protect the steel reactor vessel from radiation damage. It is often made of steel and is cooled by the primary coolant. The biological shield is added outside to reduce the external dose rate to a tolerable level. *See* RADIATION BIOLOGY; RADIATION DAMAGE TO MATERIALS. [R.E.F.]

Radiation therapy The use of ionizing radiation to treat disease; the method is also known as radiotherapy and therapeutic radiology. Radiotherapy was widely used in the past to treat diseases of the skin, lymph nodes, and other organs. However, because radiation can cause cancer and because alternative treatments for these diseases have been discovered, radiation therapy is now mainly limited to treating malignant tumors: the medical specialty is called radiation oncology. *See* ONCOLOGY.

The exact mechanisms by which radiation kills cells remain uncertain. Most likely, electrons dislodged from water or biological molecules disrupt the bonds between atoms of the nuclear deoxyribonucleic acid (DNA), resulting in double-strand breaks. Although usually not immediately fatal, such damage may cause the death of cells when they attempt to divide. Complex enzymatic mechanisms can repair some of the damage if given sufficient time.

The fraction of cells surviving after irradiation depends on many factors. Radiation affects both normal and cancerous cells in a similar manner qualitatively, but different cell lines vary greatly in their quantitative sensitivity. Rapidly dividing tissues—such as the skin, bone marrow, and gastrointestinal mucosa—usually display the greatest sensitivity experimentally and the most immediate side effects clinically.

The goal of radiation therapy is either to cure the disease permanently (radical treatment) or to reduce or eliminate symptons (palliative treatment) by destroying tumors without causing unacceptable injuries to normal tissues. For many cancers—such as cancers of the reproductive organs, lymphomas, and small head and neck tumors—a cure usually is possible, and the chance of functionally significant complications is small. Some tumors, however, contain too many cells to be entirely destroyed by tolerable doses. Also, some neoplasms, such as sarcomas and glioblastomas, are relatively resistant to irradiation and are difficult to eradicate even when only small numbers of cells are present. Such situations are best handled by using surgery to remove all visible tumor. *See* CANCER (MEDICINE); RADIATION BIOLOGY. [A.R.]

Radiative transfer The study of the propagation of energy by radiative processes; it is also called radiation transport. Radiation is one of the three mechanisms by which energy moves from one place to another, the other two being conduction and convection. *See* ELECTROMAGNETIC RADIATION; HEAT TRANSFER.

The kinds of problems requiring an understanding of radiative transfer can be characterized by looking at meteorology, astronomy, and nuclear reactor design. In meteorology, the energy budget of the atmosphere is determined in large part by energy gained and lost by radiation. In astronomy, almost all that is known about the abundance of elements in space and the structure of stars comes from modeling radiative transfer processes. Since neutrons moving in a reactor obey the same laws as radiation being scattered by atmospheric particles, radiative transfer plays an important part in nuclear reactor design.

Each of these three fields—meteorology, astronomy, and nuclear engineering—concentrates on a different aspect of radiative transfer. In meteorology, situations are studied in which scattering dominates the interaction between radiation and matter; in astronomy, there is more interest in the ways in which radiation and the distribution of electrons in atoms affect each other; and in nuclear engineering, problems relate to complicated, three-dimensional geometry.

Radiative transfer is a complicated process because matter interacts with the radiation. This interaction occurs when the photons that make up radiation exchange energy with matter. These processes can be understood by considering the transfer of visible light through a gas made up of atoms. Similar processes occur when radiation interacts with solid dust particles or when it is transmitted through solids or liquids. *See* PHOTON.

If a gas is hot, collisions between atoms can convert the kinetic energy of motion to potential energy by raising atoms to an excited state. Emission is the process which releases this energy in the form of photons and cools the gas by converting the kinetic energy of atoms to energy in the form of radiation. The reverse process, absorption, occurs when a photon raises an atom to an excited state, and the energy is converted to kinetic energy in a collision with another atom. Absorption heats the gas by converting energy from radiation to kinetic energy. Occasionally an atom will absorb a photon and reemit another photon of the same energy in a random direction. If the photon is reradiated before the atom undergoes a collision, the photon is said to be scattered. Scattering has no net effect on the temperature of the gas. *See* ABSORPTION; ATOMIC STRUCTURE AND SPECTRA; SCATTERING OF ELECTROMAGNETIC RADIATION. [A.H.K.]

Radiator Any of numerous devices, units, or surfaces that emit heat, mainly by radiation, to objects in the space in which they are installed. Because their heating is usually radiant, radiators are of necessity exposed to view. They often also heat by conduction to the adjacent thermally circulated air.

Radiators are usually classified as cast-iron (or steel) or nonferrous. They may be directly fired by wood, coal, charcoal, oil, or gas (such as stoves, ranges, and unit space heaters). The heating medium may be steam, derived from a steam boiler, or hot water, derived from a water heater, circulated through the heat-emitting units.

Electric heating elements may be substituted for fluid heating elements in all types of radiators, convectors, and unit ventilators. *See* HOT-WATER HEATING SYSTEM; RADIANT HEATING; STEAM HEATING. [E.L.W./R.Ko.]

Radio Communication between two or more points, employing electromagnetic waves as the transmission medium.

Radio waves transmitted continuously, with each cycle an exact duplicate of all others, indicate only that a carrier is present. The message must cause changes in the carrier which can be detected at a distant receiver. The method used for the transmission of the information is determined by the nature of the information which is to be transmitted as well as by the purpose of the communication system.

In code telegraphy the carrier is keyed on and off to form dots and dashes. The technique, often used in ship-to-shore and amateur communications, has been largely superseded in many other point-to-point services by more efficient methods.

In frequency-shift transmission the carrier frequency is shifted a fixed amount to correspond with telegraphic dots and dashes or with combinations of pulse signals identified with the characters on a typewriter. This technique is widely used in handling the large volume of public message traffic on long circuits, principally by the use of teletypewriters. *See* TELETYPEWRITER.

In amplitude modulation the amplitude of the carrier is made to fluctuate, to conform to the fluctuations of a sound wave. This technique is used in AM broadcasting, television picture transmission, and many other services. *See* AMPLITUDE-MODULATION RADIO.

In frequency modulation the frequency of the carrier is made to fluctuate around an average axis, to correspond to the fluctuations of the modulating wave. This technique is used in FM broadcasting, television sound transmission, and microwave relaying. *See* FREQUENCY-MODULATION RADIO.

In pulse transmission the carrier is transmitted in short pulses, which change in repetition rate, width, or amplitude, or in complex groups of pulses which vary from group to succeeding group in accordance with the message information. These forms of pulse transmission are identified as pulse-code, pulse-time, pulse-position, pulse-amplitude, pulse-width, or pulse-frequency modulation. Such techniques are complex and are employed principally in microwave relay systems. *See* PULSE MODULATION.

In radar the carrier is normally transmitted as short pulses in a narrow beam, similar to that of a searchlight When a wave pulse strikes an object, such as an aircraft, energy is reflected back to the station, which measures the round-trip time and converts it to distance. A radar can display varying reflections in a maplike presentation on a cathode-ray tube. *See* RADAR.

Hundreds of thousands of radio transmitters exist, each requiring a carrier at some radio frequency. To prevent interference, different carrier frequencies are used for stations whose service areas overlap and receivers are built to select only the carrier signal of the desired station. Resonant electric circuits in the receiver are adjusted, or tuned, to accept one frequency and reject others.

All nations have a sovereign right to use freely any or all parts of the radio spectrum. But a growing list of international agreements and treaties divides the spectrum and specifies sharing among nations for their mutual benefit and protection. Each nation designates its own regulatory agency. In the United States all nongovernmental radio communications are regulated by the Federal Communications Commission (FCC). *See* AMATEUR RADIO; RADIO BROADCASTING; RADIO BROADCASTING NETWORKS; RADIO SPECTRUM ALLOCATIONS. [J.D.Si.]

Radio altimeter A low-power radar which measures the distance of an aircraft (or other aerospace vehicle) above the ground. Radio altimeters are often used in aircraft during bad-weather landings. They are an essential part of many blind-landing and automatic navigation systems and are used over mountains to indicate terrain clearance. Special types are used in surveying for quick determination of profiles. Radio altimeters are used in bombs, missiles, and shells as proximity fuses to cause detonation or to initiate other functions at predetermined altitudes. Similar devices are used as landing aids on spacecraft.

Like other radar devices, the altimeter measures distance by determining the time required for a radio wave to travel to and from a target, in this case the ground. If the ground were a perfectly flat horizontal plane, specular (mirror-type) reflection would occur and the distance measured would be exactly the altitude. Actually, the ground is not smooth, and energy is scattered back to the radar from all parts of the ground illuminated by the transmitter. For the radar to measure distance to the ground accurately, it must distinguish between the energy from points near the vertical and that from more distant points. Most radio altimeters use either pulse or frequency modulation, the former being more popular for high altitudes, and the latter for low altitudes.

In a pulse altimeter the radio-frequency carrier is modulated with short pulses (<0.25 microsecond). The short pulse permits measurement, even at low altitudes, of the time delay between the leading edge of the transmitted pulse and that of the pulse returned from the ground. The simplest pulse altimeter displays the received signal on a cathode-ray tube with circular sweep, allowing the pilot to make a determination of the leading edge position for the echo signal.

In an fm altimeter, a continuous carrier is swept in frequency in some manner, usually to give a triangular or sinusoidal frequency-time curve. The difference in frequency between that received from the ground (but transmitted earlier) and that being transmitted is a measure of the time delay. [R.K.Mo.]

Radio astronomy The study of celestial objects by measurement and analysis of the electromagnetic radiation they emit in the wavelength range from 1 mm to 30 m (0.04 in. to 100 ft). Radio astronomers study an entire range of celestial objects, including the normal stars, planets, galaxies, and the exotic quasars, pulsars, and x-ray sources. The universe itself is also a source of microwave radiations. The cosmic microwave radiation and radio emission from pulsars and quasars are discussed in separate articles. *See* COSMIC BACKGROUND RADIATION; PULSAR; QUASARS.

Radio universe. Since the late 1940s, radio telescopes have been used to map the skies and determine the positions and intensities, or fluxes, of individual sources of radio emission. Such maps have been made with increasing sensitivity and angular resolution; the latter property enables astronomers to determine the position of the radio sources accurately. Knowing the position, astronomers can refer to optical photographs of the sky and establish precisely which object is emitting radio waves. This procedure has led to the identification of radio sources with many bright galaxies and even with the most distant objects in the universe, the quasistellar objects (or quasars). However, nearly one-fifth of all radio sources are unidentified, that is, excellent photographs taken at the radio source positions show no object at all from which the radiation could arise. One concludes from this that these unidentified sources are "normal" galaxies and quasars at such great distances that they cannot be seen optically. *See* RADIO TELESCOPE.

It appears that there are more sources at great distances per unit volume of space than there are nearby, a result that means the universe is expanding, and that those distant sources are, in general, stronger than ones nearby. These two conclusions are fundamentally linked, and together they mean that further work on cosmological problems must await a clearer understanding of the nature and evolution of individual radio sources. *See* COSMOLOGY.

Galaxies and radio galaxies. Spiral galaxies, such as the Milky Way Galaxy, are often radio sources, although most are quite weak. Elliptical galaxies are usually not radio sources. However, the few elliptical galaxies that are radio sources are very spectacular ones, being among the most energetic radio objects in the sky. These are known as the radio galaxies. The radio emission of these galaxies emanates from the whole volume of the galaxy and, in most cases, a huge volume outside the galaxy. Frequently there are two large regions on opposite sides of the galaxy from which most of the radio energy is emitted. The radio emission from galaxies and radio galaxies is almost certainly generated by the electron synchrotron process, in which relativistic electrons spiral around magnetic field lines and emit a continuous radiation spectrum throughout the band accessible to radio astronomers. *See* GALAXY, EXTERNAL.

Solar system astronomy. The Sun is an intense radio source, but only because it is so close to Earth. If it were at the distance of the nearest stars, its radio emission could not be detected. Solar radio emission tends to be intermittent: solar flares that produce cosmic rays and plasma streams that interact with Earth are visible as radio bursts; these are most frequent during the peak of the 11-year solar cycle. *See* COSMIC RAYS; SUN.

Jupiter is a much stronger radio source than had been expected from estimates of its surface temperature by optical astronomers. Most of its radio emission is caused by electron synchrotron emission in its very strong magnetic field. The very-long-wavelength emission of Jupiter is impulsive, and its strength depends upon the position of Io, one of Jupiter's moons. Similar impulsive long-wavelength bursts have also been discovered from Saturn. *See* JUPITER; SATURN.

Radio stars. Several nearby, apparently normal stars are detectable at radio wavelengths. Such stars as Algol, β Persei, and AR Lacerta are multiple star systems that have been found to be radio sources. The radio emission from these stars is dominated by radio bursts in which the radio fluxes may increase by a factor of 100 or more. It is clear that the radio bursts are initiated by or are a product of mass exchange processes going on between (at least) two closely bound stars.

An extreme example of radio emission from stars comes from stars that are also x-ray sources. Again, these objects are usually binary systems in which mass exchange plays a deciding role in their continuing evolution, but with x-ray sources one of the component stars appears to be a star that has exhausted its reservoir of nuclear fuel and is collapsing to its final state. *See* BINARY STAR; X-RAY ASTRONOMY.

Supernova remnants. During the formation of a supernova, the atmospheric envelope of a star is ejected and in this process becomes a rapidly expanding cloud of relativistic particles, magnetic field, and filaments of ionized gas. These conditions are precisely those necessary for the generation of radio emission through electron synchrotron radiation, and very intense radio sources indeed exist at the positions of old supernovae chronicled by ancient astronomers. One of the most interesting of the radio sources associated with a supernova remnant is the Crab Nebula. *See* CRAB NEBULA; SUPERNOVA.

HII regions. An HII region (a region of ionized hydrogen) is a large cloud of interstellar gas that has been ionized and heated by one or more bright, hot stars located within. These nebulae are sources of both continuum and line energy at radio and optical wavelengths. Since cosmic matter consists mostly of hydrogen, the ionized gas consists mainly of protons and electrons that emit continuum energy by the bremsstrahlung process. *See* NEBULA.

Hydrogen line. Study of the 21-cm line of neutral atomic hydrogen has been exceptionally rewarding in its contribution to the knowledge of galactic structure and of the physical characteristics of interstellar gas. Line intensity normally reflects the amount of gas in the line of sight; line wavelength and width indicate the line-of-sight velocity of the gas and the state of internal motion, just as with recombination lines in HII regions. If the gas overlies a strong radio source, the gas temperature can be inferred by observing the 21-cm line in absorption.

The structure of the Galaxy has been elucidated by the study of the amount and velocity of the hydrogen within it. A prime advantage of this method is that the very distant gas is just as visible as nearby gas, whereas optical studies of the whole Galaxy are impossible because the very distant stars are made invisible by intervening clouds of dust. The results of these radio studies indicate that the Milky Way Galaxy is a spiral. *See* MILKY WAY GALAXY.

Molecular lines. It has long been believed that simple molecules could not exist in the tenuous gas between stars, because the starlight radiation field would be sufficiently intense to break apart even the simplest molecular species. In spite of these arguments, by 1968 radio astronomers had found rotational transitions of three simple molecules, OH, H_2O, and NH_3. These molecules were found in dark clouds of gas and dust usually associated with HII regions; such dark clouds are believed to be the sites of recent and continuing star formation. Since 1968 dozens of molecular species have been detected in interstellar space. Observations of the molecular species may make it possible to establish the chemistry and thermodynamics in the interstellar clouds from which, ultimately, stars, planets, and life itself must form. *See* MOLECULAR STRUCTURE AND SPECTRA; RADIO TELESCOPE. [R.L.Bro.]

Radio broadcasting

The transmission of audible program material for direct reception by the general public. *See* RADIO.

Frequency allocations. By international agreement, certain portions of the radio-frequency (rf) spectrum are set aside for radio broadcasting. In the United States over 4400 stations occupy the standard broadcast band of 107 channels, extending from 535 to 1605 kHz, each channel being 10 kHz wide. In addition, there are over 3300 stations on 100 channels utilizing the frequency-modulation (FM) band from 88 to 108 MHz, each channel being 200 kHz wide. These stations are all licensed to provide domestic service. There are also some shortwave broadcast stations operating in high-frequency bands between 5950 and 26,100 kHz that provide international service. In Europe and Asia the low-frequency band 150–290 kHz is also used for broadcasting. All radio stations of any kind in the United States must be licensed by the Federal Communications Commission (FCC). *See* RADIO SPECTRUM ALLOCATIONS.

AM medium-frequency band. The power of the rf carrier is specified in the FCC construction permit and license. Nine power levels are recognized, ranging from 100 W for small stations serving local areas to 50,000 W for the largest stations serving large areas and long distances. In other countries much higher powers are sometimes employed.

It is often possible to avoid interfering with other stations by the use of a directional antenna, which radiates low power in the directions of other stations assigned to the same or adjacent channels and concentrates its power over the areas to be served. The majority of stations in the United States and Canada are directional.

Signals from standard broadcast stations arrive at the receiving points in two ways, by ground wave and by sky, or ionospheric, wave. In the daytime, only ground waves are received, but at night both ground and sky waves are effective. *See* RADIO-WAVE PROPAGATION.

In the United States the geographical area over which a standard broadcast station projects its signals is divided by the FCC into two categories: A primary service area is served by ground-wave propagation with sufficient field intensity to provide consistent day and night service. A secondary service area is serviced by ground waves of insufficient strength to provide consistent day and night service, or is served by sky waves at night.

Except for a limited number of channels reserved for service over large areas, each broadcast channel is used by a large number of geographically separated stations. All these stations do not have the same priority status. A dominant station is permitted to broadcast full-time. Some secondary stations are permitted to operate only between the hours of sunrise and sunset, so that they will not interfere with other stations having higher priority status on the channel.

VHF-FM band. FM broadcasting has full provision for high-fidelity reproduction, is ordinarily free of static, substantially suppresses noise and interference, and is relatively free of fading. The service range is ordinarily 40–80 mi (65–130 km) line of sight, but stations with high power and high antenna elevation are often receivable at distances up to 150 mi (250 km). *See* FREQUENCY MODULATION; FREQUENCY-MODULATION DETECTOR; FREQUENCY-MODULATION RADIO; FREQUENCY MODULATOR.

Receiving systems for AM and FM radio signals are quite different, but combination receivers and tuners are now commonplace.

In the United States and many countries, FM broadcasting is assigned frequencies between 88 and 108 MHz. The channel spacing is 200 kHz. This facilitates the broadcasting of high-fidelity broad-band stereo.

Shortwave broadcasting. Shortly after World War I, experimental transmissions on radio frequencies between about 4000 and 20,000 kHz proved that communication was possible over great distances by reflection from layers of ionized air, which vary in altitude from about 150 to 250 mi (250 to 400 km). Radio waves from the transmitting antennas follow a path at low angles above Earth until they arrive at the reflecting ionized layers of air where they are reflected back to Earth. A number of successive reflections between the layers of the ionosphere and Earth make transmission for many thousands of miles possible and, at times, completely around the world. *See* AMATEUR RADIO.

Satellite broadcasting. Intercontinental communications by satellite are now commonplace. The satellites in use are capable of handling several thousand narrow-channel telephone, telegraph, and teletype channels and many wide-channel television channels, all simultaneously.

Experimental broadcast and communications satellites, in geosynchronous orbit, are received throughout much of North America on small earth terminals in the 4- or 12-GHz bands, by broadcasters, cable television operators, hotels, apartment complexes, camps, and individuals in isolated areas. All of them provide a choice of television channels and some also offer FM radio. The proliferation of small TVRO (television receive only) earth stations will promote the global village concept by adding a new dimension to international broadcasting. *See* COMMUNICATIONS SATELLITE. [J.G.E.]

Radio broadcasting networks A group of broadcast stations interconnected by leased channels on wire, microwave, or satellite to one or more central feed points for the purpose of receiving and rebroadcasting program material of a timely nature. Networks make it possible to broadcast live programs simultaneously to the public through affiliated radio stations; they make national and regional markets available to advertisers and offer stations quality entertainment and public service programs.

For over 40 years, all domestic radio broadcast networks utilized telephone company facilities exclusively in the AT&T toll system for their program distribution. In 1971 the Federal Communications Commission (FCC) regulations allowed other common carriers to begin offering program channel service to radio networks in direct competition with AT&T.

In 1972 Microwave Communications Incorporated (MCI) began offering high-quality program channel service between New York City and Washington on its terrestrial microwave system. Other common carriers were quick to follow suit, picking up program distribution contracts in areas of the country where this service was not cost-effective for AT&T.

A satellite relay system is basically a microwave technology using high-power amplifiers at the transmit end coupled to high-gain antennas which beam a wide-band signal to domestic communications satellites in geostationary orbit over the Earth. These "fixed-position" satellites, operated and maintained in their orbits by the common carriers, repeat the signal on assigned frequencies, beaming a broad pattern, or footprint, which can cover an entire continent. The signal may be received at any location in the coverage area using high-technology, low-noise amplifiers (LNAs) coupled to standard microwave reflectors aimed toward the satellite. *See* COMMUNICATIONS SATELLITE.

Domestic radio networks first began leasing point-to-point microwave and satellite channels in the early 1970s to supplement their AT&T distribution networks. Satellite channels in particular offered a higher-quality audio service at lower rates due to the decreased number of radio systems and audio amplifiers in the chain from transmitting station to final receive point. Several radio networks have taken advantage of this new technology, and many new networks have entered into the field as a direct result of the availability of high-quality audio transmission systems via satellite. Developments in satellite technology have vastly improved the signal quality that networks can provide to stations, and have reopened programming possibilities which had been all but lost to radio. *See* RADIO; RADIO BROADCASTING. [W.J.Wi.]

Radio compass A popular term for an automatic radio direction finder used for navigation purposes on ships and aircraft. It is not strictly a compass, because it indicates direction with respect to the radio station to which it is tuned, rather than to the north magnetic pole. The modern radio compass uses a nondirectional antenna in combination with a bidirectional loop antenna to provide a unidirectional bearing indication. Navigators thus know at all times whether they are traveling toward or away from the radio station used as a reference. The antennas are used with a special radio receiver that provides a visual indication of direction on a meter or cathode-ray indicator. *See* DIRECTION-FINDING EQUIPMENT. [J.Mar.]

Radio-frequency amplifier A tuned amplifier that amplifies the high-frequency signals commonly used in radio communications. The frequency at which maximum gain occurs in a radio-frequency (rf) amplifier is made variable by changing either the capacitance or the inductance of the tuned circuit. A typical application is the amplification of the signal received from an antenna before it is mixed with a local oscillator signal in the first detector of a radio receiver. The amplifier that follows the first detector is a special type of rf amplifier known as an intermediate-frequency (i-f) amplifier. *See* AMPLIFIER; INTERMEDIATE-FREQUENCY AMPLIFIER.

An rf amplifier is distinguished by its ability to tune over the desired range of input frequencies. The shunt capacitance, which adversely affects the gain of a resistance-capacitance coupled amplifier, becomes a part of the tuning capacitance in the rf amplifier, thus permitting high gain at radio frequencies. The power gain of an rf amplifier is always limited at high radio frequencies, however.

Two typical rf amplifier circuits are shown in the illustration. The conventional bipolar transistor amplifier of illustration *a* uses tapped coils in the tuned circuits to provide optimum gain-bandwidth characteristics consistent with the desirable value of tuning capacitance. Inductive coupling provides the desired impedance transformation in the input and output circuits. The tuning capacitors are usually ganged so as to rotate on a single shaft, providing tuning by a single knob. Sometimes varactor diodes are used to tune the circuits, in which case the tuning control is a potentiometer that controls the diode voltage. Automatic gain control (AGC) is frequently used on the rf amplifier, as shown. AGC voltage controls the bias and hence the transconductance of the amplifier. In the field-effect transistor (FET) circuit (illustration *b*), tapped coils are not required because of the very high input and output resistances of the FET. Characteristics of the FET are similar to those of a pentode tube except for the lower drain voltage and higher capacitance between the input and the output. Thus the FET circuit is similar to a pentode circuit except that it is much simpler because it has no heater, screen grid, or suppressor grid equivalents. *See* AUTOMATIC GAIN CONTROL (AGC); SEMICONDUCTOR; TRANSISTOR. [C.L.A.]

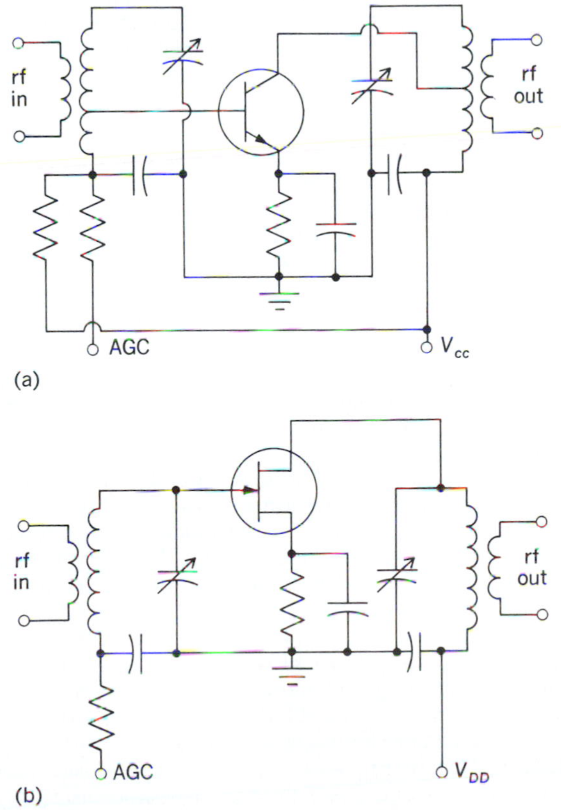

Typical rf amplifiers. Circuits with (*a*) bipolar transistor and (*b*) field-effect transistor. V_{cc} = collector supply voltage; V_{DD} = drain supply voltage.

Radio-frequency impedance measurements Measurements of electrical impedance at frequencies ranging from a few tens of kilohertz to about 1 gigahertz. In the electrical context, impedance is defined as the ratio of voltage to current (or electrical field strength to magnetic field strength), and it is measured in units of ohms (Ω). *See* ELECTRICAL IMPEDANCE.

At zero frequency, that is, when the current involved is a direct current, both voltage and current are expressible as real numbers. Their ratio, the resistance, is a scalar (real) number. However, at nonzero frequencies, the voltage is not necessarily in phase with the current, and both are represented by vectors, and therefore are conveniently described by using complex numbers. To distinguish between the scalar quantity of resistance at zero frequency and the vectorial quantity at nonzero frequencies, the word impedance is used for the complex ratio of voltage to current. *See* ALTERNATING CURRENT; DIRECT CURRENT; ELECTRICAL RESISTANCE.

The measurement of impedance at radio frequencies cannot always be performed directly by measuring an rf voltage and dividing it by the corresponding rf current, for the following reasons: (1) it may be difficult to measure rf voltages and currents without loading the circuit by the sensing probes; (2) the distributed parasitic reactances (stray capacitances to neighboring objects, and lead inductances) may be altered by the sensing probes; and (3) the spatial voltage and current distributions may prevent unambiguous measurements (in waveguides, for instance).

At low frequencies, impedance measurements are often carried out by measuring separately the resistive and reactive parts, using either Q-meter instruments (for resonance methods), or reconfigurable bridges, which are sometimes called universal LCR (inductance-capacitance-resistance) bridges. In one such bridge the resistive part of the impedance is measured at dc with a Wheatstone bridge. Capacitive reactance is measured with a series-resistance-capacitance bridge, and inductive reactance is measured with a Maxwell bridge, using alternating-current (ac) excitation and a standard capacitance. *See* WHEATSTONE BRIDGE.

Transformer bridges are capable of operating up to 100 MHz. The use of transformers offers the following advantages: (1) only two bridge arms are needed, the standard, and the unknown arms, and (2) both the detector and the source may be grounded at one of their terminals, minimizing ground-loop problems and leakage. *See* TRANSFORMER.

A coaxial line admittance bridge is usable from 20 MHz to 1.5 GHz. The currents flowing in three coaxial branch lines are driven from a common junction, and are sampled by three independently rotatable, electrostatically shielded loops, whose outputs are connected in parallel.

A quantity related to impedance is the complex (voltage) reflection coefficient, defined as the ratio of the reflected voltage to the incident voltage, when waves propagate along a uniform transmission line in both directions. Usually, uppercase gamma (Γ) or lowercase rho (ρ) is used to represent the reflection coefficient. When a transmission line of characteristic impedance Z_0 is terminated in impedance Z_T, the reflection coefficient at the load is given by Eq. (1), and the voltage standing-wave ratio (VSWR) is related to the magnitude of Γ by Eq. (2).

$$\Gamma = \frac{Z_T - Z_0}{Z_T + Z_0} \quad (1)$$

$$\text{VSWR} = \frac{1 + |\Gamma|}{1 - |\Gamma|} \quad (2)$$

See TRANSMISSION LINES.

When it is sufficient to measure only the voltage standing-wave ratio (VSWR), resistive bridges may be used. Resistive bridges employed as reflectometers use a matched source and detector, and therefore differ from the Wheatstone bridge, which aims to use a zero-impedance voltage source and an infinite-impedance detector.

Some specialized electronic instruments make use of the basic definition of impedance, and effectively measure voltage and current. One such instrument is called an rf vector impedance meter. Instead of measuring both the voltage and the current, it drives a constant current into the unknown impedance, and the resultant voltage is measured.

Vector voltmeters (VVM) are instruments with two (high-impedance) voltmeter probes, which display the voltages at either probe (relative to ground) as well as the phase difference between them. One type operates from 1 MHz to 1 GHz, and linearly converts to a 20-kHz intermediate frequency by sampling.

When the magnitude of the reactive part of the impedance is much greater than the resistive part at a given frequency, resonance methods may be employed to measure impedance. The most commonly used instrument for this purpose is the Q meter. *See* Q METER.

At the upper end of the rf range, microwave methods of impedance measurement may also be used, employing slotted lines and six-port junctions. *See* MICROWAVE MEASUREMENTS.

[P.I.S.]

Radio paging systems Systems, consisting of three basic elements—a personal paging receiver, radio transmitter, and an encoding device—whose primary purpose is to alert an individual, or group of individuals, and deliver a short message of a temporary or perishable nature. Characteristics that are used to define a specific paging system include distance covered, radio frequency, modulation type, paging code format, and message type.

On-site systems cover a single building or a small complex of buildings typically utilizing one low-power transmitter. Wide-area systems can cover an entire city or country and usually use multiple transmitters which simulcast the paging signals. Most paging systems now utilize the very high-frequency (VHF) or ultrahigh-frequency (UHF) radio spectrum using frequency modulation (FM). *See* FREQUENCY MODULATION; RADIO SPECTRUM ALLOCATIONS.

Paging receivers fall into four basic categories: tone alert, tone and voice, numeric, and alphanumeric. Tone pagers emit a "beep" when they are signaled. Some models silently alert the user with a vibration in place of a beep; other models use differing staccato beeps to provide the user with several alert messages. Tone and voice pagers allow the initiator of the page to transmit a simple voice message which will follow a pager's beep alert. Numeric pagers, sometimes called digital pagers, allow the initiator to convey numerical information. These messages are typically composed by using a tone telephone key pad. Alphanumeric pagers allow the initiator to send a complete textual message to the pager user. These messages are composed on word processors, personal computers, or dedicated terminals which can connect to a paging terminal. *See* MICROCOMPUTER; WORD PROCESSING.

[J.A.Wr.]

Radio range A radio facility emitting signals which, when received by appropriate companion equipment, provide a direct indication of the bearing of the facility from the vehicle.

The vhf omnidirectional radio range (VOR) operates at very high frequencies and provides radial lines of position in any direction. It is the most important type of radio range in civil air operations, and several hundred VOR ranges are in use throughout the world.

This facility operates in the 112–118-MHz range and employs a directional and a nondirectional antenna. The emission from the directional antenna, when combined with radiation from the nondirectional antenna, results in a cardioid space pattern. This space pattern is rotated at a speed of 30 Hz either physically or through the use of a goniometer. No other modulation appears on the rotating (variable phase) pattern. The rotation is synchronous with a frequency modulation which is imposed on a small amount of rf energy derived from the source that feeds the variable phase antenna and radiated by the nondirectional antenna. The output of the nondirectional antenna is therefore a carrier frequency that is amplitude-modulated with frequencies that vary cyclically from 9480 to 10,440 Hz at a 30-Hz rate. This signal appears everywhere in space within the coverage of the station. A special receiver is necessary to derive navigational information from the VOR. *See* DOPPLER VOR; ELECTRONIC NAVIGATION SYSTEMS.

[P.C.S.]

Radio receiver That part of a radio communication system which abstracts the desired information from the radio-frequency (rf) energy collected by the antenna. All radio receivers must perform three basic functions: selectivity, amplification, and detection. For basic discussion of radio principles *see* RADIO.

Functions. Many radio signals transmitted at the same time are available at the antenna. From these many signals the receiver must select the single one desired. This is done by tuning the receiver to the frequency of the desired carrier. The tuning circuit contains a combination of inductances L and capacitances C, one or more of which are variable. The frequency f selected is determined by the relation below, where f

$$f = 1/2\pi\sqrt{LC}$$

is in hertz (Hz), L is in henries, and C is in farads. Tuning the receiver, therefore, consists in changing the inductance or capacitance, usually the latter. When so tuned the inductance-capacitance circuit accepts the desired frequency and rejects other frequencies. By using several such circuits in series, a high degree of selectivity may be obtained.

Because the incoming signal may be weak and because a certain minimum energy is required to operate the loudspeaker, the headphones, or the television picture tube, considerable amplification must take place between the input of the receiver and its output. This is usually called the gain of the receiver. It may amount to 10,000,000 times in voltage or 140 decibels (dB). *See* AMPLIFIER.

The energy collected by the antenna and presented to the input of the receiver is in the form of rf waves which act as a carrier for the information to be transmitted. The purpose of the detector in a receiver is to remove the desired communication from this carrier and to convert it into a form that will actuate the output device, such as a loudspeaker. *See* AMPLITUDE-MODULATION DETECTOR; FREQUENCY-MODULATION DETECTOR.

Types. Two general types of receiver are in use today, the tuned-radio-frequency (TRF) and the superheterodyne. Both of these can be used for amplitude-modulated (AM) signals. Frequency-modulation (FM) receivers are almost always superheterodyne. *See* AMPLITUDE MODULATION; AMPLITUDE-MODULATION RADIO; FREQUENCY MODULATION; FREQUENCY-MODULATION RADIO.

In a TRF receiver all amplification up to the detector takes place at the frequency of the incoming signal. This usually requires several stages of tuned amplification. Each stage is

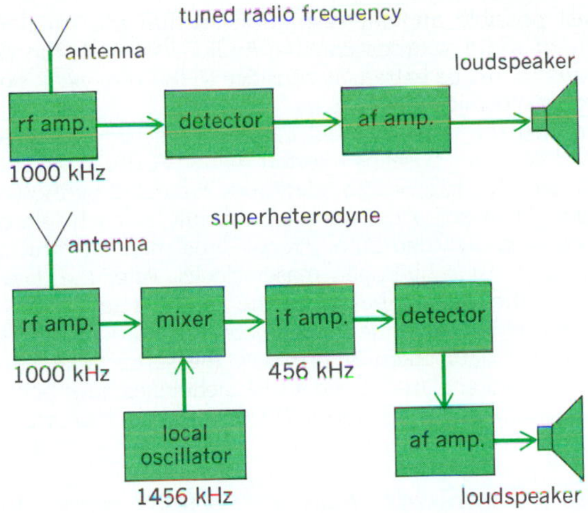

Block diagrams showing comparison of TRF and superheterodyne receivers.

tuned to the same frequency and the tuning elements are ganged together for convenience when tuning. TRF receivers are especially applicable to the low-frequency and very-low-frequency bands from about 10,000 to 300,000 Hz. The illustration shows a block diagram of a TRF receiver compared to a superheterodyne receiver.

With the increased use of higher frequencies for broadcasting and communications, a high degree of selectivity became necessary. The TRF receiver was inadequate in this respect. This selectivity was obtained by circuit techniques involving frequency changing. The superheterodyne circuit changes the frequency by heterodyning, or beating, two frequencies together to get a third. When two signals of small difference in frequency are superimposed on a nonlinear device, the output consists of energy at two major combination frequencies which are the sum and difference of the original frequencies. The sum of the two beating frequencies is usually eliminated by tuned circuits following the heterodyne process. The difference frequency, referred to as the intermediate frequency, is passed on for further amplification and, what is more important, through stages of high selectivity because selectivity at lower frequencies is more easily obtained than at high frequencies.

FM and SSB receivers. In an FM system the carrier is modulated by the desired message by varying the frequency of the carrier instead of varying its amplitude. The receiver for such a system must have some means of producing a varying voltage amplitude to correspond to the varying frequency. In other words, the FM signal must be converted to an AM signal. A limiter is employed to eliminate all amplitude variations of the carrier and to deliver to the final detector a signal which is free of noise. Several types of FM detectors, usually called discriminators, are employed.

The single-sideband (SSB) receiver is designed for the reception of but one sideband of intelligence on a reduced carrier. The independent-sideband (ISB) receiver is used for the reception of two channels of intelligence, one on an upper and one on a lower sideband. To effect this type of reception without undue distortion, the carrier is exalted or reinforced, and in some instances a local carrier synchronized by the incoming reduced carrier is employed for demodulation, sometimes referred to as product detection. To separate the sidebands and carrier, carefully designed filters are employed. *See* SINGLE SIDEBAND. [W.Lyo.]

Radio spectrum allocations The specification of the frequency bands of the radio spectrum which are made available for use by the various radio services. The radio spectrum is the part of the natural spectrum of electromagnetic radiation lying between the frequency limits of 10^4 and 3×10^{11} hertz. Since the characteristics of the radiation do not change abruptly, these limits are not sharply defined and may change, depending upon the demands for changes in service and upon changes in technology.

For purposes of identification the radio spectrum is divided into bands differing from adjacent bands by frequency ratios of 10. These bands are identified by the metric wavelength of the shortest waves in each band (kilometric band, and so forth) or by adjective or numerical designators (for example, very-low-frequency band or band 4). The wavelength λ in meters is related to the frequency f in hertz by the relationship $c = f\lambda$, where c, the velocity of propagation of radio waves in space, is about 3×10^8 m/s.

The major classes of radio services are provided with international allocations within the various frequency bands. Within the United States there is a national allocation which conforms in general to the international allocation. The Radio Regulations divide the Earth into three regions and contain a detailed table of allocations, with some differences between regions, together with exceptions to the listed allocations and conditions for frequency use. Regional and national allocations can be made within the worldwide framework, but any departures by signatories of the Regulations must be accomplished on the basis of noninterference with services operating in accordance with the agreed allocations.

The characteristics of radio waves, due to natural phenomena such as atmospherics, vary greatly with frequency. Attempts are made to take advantage of propagation characteristics when making allocations, and normally the frequencies allocated have characteristics satisfying the operational requirements of the service with regard to distance, bandwidth, and other technique factors. Allocations are sometimes made for a new service which has already begun to use a particular band. Since many services have varied and continually growing requirements, allocations to them will be found in several frequency bands. In making specific frequency assignments to individual radio stations, the separation between assignable frequencies must be such as to avoid harmful interference in view of the bandwidth requirements, the frequency stability of the transmitters, the frequency selectivity of the receivers, the transmitter power and its distance separation from receivers operating on adjacent channels, and other factors. Thus the separation between adjacent assigned frequencies varies widely between services and between different locations in the frequency spectrum. *See* BANDWIDTH REQUIREMENTS (COMMUNICATIONS).

In most of the frequency bands it has been found practical to make block allocations of frequencies to given services in each region. Individual frequency assignments can then be made to radio stations operating in that service without the need for coordination with frequency assignments to stations in other services. Interference problems are more easily resolved, as they occur between compatible users. Some of these allocations are exclusive to a particular service, and some are shared between specified services to varying degrees. Similarly, in some services the stations have individual frequency assignments which are protected from interference from other stations, whereas the stations in other services may share time or geographic coverage on frequencies common to several stations. These arrangements are based upon the nature and the needs of the particular services being provided. [L.M.P.; J.M.L.]

Radio telescope An instrument used in astronomical research to detect and measure the radio-frequency power coming from various directions in the sky. It consists of three complementary parts: the large reflecting surface that collects and focuses the incident radiation; the electronic receiver that amplifies and detects cosmic radio signals; and a data display device. From the ground, observations with radio telescopes must be made at wavelengths shorter than 100 ft (30 m), because of ionospheric attenuation, and longer than 0.04 in. (1 mm), because the very-short-wavelength radio radiation is absorbed by atmospheric H_2O, CO_2, and O_3.

The fundamental principle of a radio telescope is identical to that of a reflecting telescope used at visual wavelengths. The incoming waves (radio or optical) are intercepted by a precise mirror and reflected to a common focal point. The shape of the reflecting surface or "dish" is important: the radio waves must arrive "in phase" at the focal point following their reflection from the dish; that is, the path length from the point of reflection to the focus must be exactly the same for all points on the dish. This restriction can be most simply satisfied if the shape of the reflecting surface is made paraboloidal; consequently, most modern radio telescopes have this shape (see illustration). *See* OPTICAL TELESCOPE; TELESCOPE.

Once the radio waves are collected and brought together at the focal point of the telescope, they are in general still extremely weak. The incoming radio-frequency (rf) signals are first amplified at the focus 10 to 1000 times and then converted to a lower frequency, the intermediate frequency (i-f), that can be easily transmitted by cables from the focal point to the telescope-control building. There the i-f is further amplified, and the signal is detected and displayed in the manner the astronomer finds most suited to the particular investigation.

The types of astronomical objects that emit radio-frequency radiation and hence can be studied by radio astronomers are of such a diverse nature that a variety of radio telescopes and receiving equipment are necessary for a modern radio observatory. Two general astronomical considerations dictate what instruments are needed: first, radio telescopes should have the highest possible angular resolution so that the small-scale details of radio sources can be studied; second, the radio receivers should be extremely sensitive to the very weak signals emitted by cosmic radio sources.

The 328-ft-diameter (100-m) radio telescope operated by the Max Planck Institut für Radioastronomie at Effelsberg, Germany. (*Max Planck Insititut für Radioastronomie*)

The ultimate in angular resolution is achieved by the technique of very-long-baseline interferometry (VLBI), in which one simultaneously utilizes radio telescopes separated by thousands of miles. Data are acquired independently at each telescope and recorded on video tape. Precise time markings are made on the tape using hydrogen maser clocks. After the data are recorded, the video tapes from the separate telescopes are brought together; the time markings on the individual tapes are aligned, and data taken at precisely the same times can be compared and analyzed. Such VLBI techniques have achieved angular resolutions of about 0.0003 second of arc. *See* ASTRONOMY. [R.L.Bro.]

Radio transmitter A generator of high-frequency electric current whose characteristics of amplitude, frequency, or phase angle may be altered, or modulated, in accordance with the intelligence to be transmitted. A radio transmitter consists of several distinct major components to accomplish the objectives of a particular design for a particular requirement.

The power a transmitter delivers to the antenna may vary from a fraction of a watt to 1,000,000 watts. Lower powers are used mainly for portable or mobile services, while higher powers are required for broadcasting over large areas and in point-to-point communications. Transmitters may be classified by the type of modulation used.

Amplitude-modulation (AM) transmitters have two principal design types, either low-level modulated or high-level modulated. The low-level modulated transmitter is modulated at its low power stages, requiring little modulation power. The high-level modulated transmitter requires modulation power to be equal to about 50% of the carrier's power. In order to amplify faithfully and reproduce the modulation at low levels, the power amplifiers must be linear. These are usually class B linear amplifiers. Because it operates over a large band of frequencies, the high-power modulator must have linear power amplification to achieve low distortion. *See* AMPLIFIER.

Frequency modulation (FM) transmitters are basically similar to AM transmitters, except that the AM modulation amplifying system is dispensed with and the exciter must be a variable frequency source. One method of modulating the frequency at the exciter utilizes a reactance in the frequency-determining section of the oscillator in the exciter. The reactance value is changed electronically or electrically in accordance with the low-frequency modulating signal. The remainder of the transmitter, as in AM transmitters, is made up of frequency doubling and tripling stages of power amplification and a power output stage.

The use of single sideband/independent sideband (SSB/ISB) transmission for communications has become very popular because of several peculiar system characteristics. Its ability to transmit intelligence occupying only half the bandwidth, as compared to AM and FM modes, and its adaptability for modulating by many low-frequency subcarriers are principal advantages.

Typical equipment arrangement for SSB transmission comprises an SSB generator (exciter), which makes the SSB or ISB signal (using a system of balanced modulators usually) and filters to separate out the carrier for reinsertion in the desired amount, select the desired sideband, and reject the undesired sideband. The output, at low levels of several watts, is then amplified without frequency changing in linear class B amplifiers to the final power output for coupling to the antenna. This method is commonly referred to as linear SSB transmission. A principal design objective is the maximizing of power output in order to minimize spurious radiation. [W.Lyo.]

Radio-wave propagation The means by which radio signals are transported through space from a transmitting antenna to a receiving antenna. *See* RADIO.

The frequencies around 20 kHz can be received reliably at distances of thousands of miles but are limited to telegraph-type signals and require very large transmitting antennas. Higher frequencies are needed for voice, and still higher frequencies for television transmission. As the frequency increases, the transmission range tends to decrease. Frequencies above 100 MHz can transmit wide-band signals, but they are limited to approximately line-of-sight distances with the usual type of equipment. However, distances of 200 mi (320 km) or more are possible by the use of high power and large antennas to provide narrow "searchlight" beams.

Reflections from the ionosphere (ionized layers 50–250 mi or 80–400 km above the Earth's surface) provide a useful but variable long-distance service at frequencies less than about 30 MHz. These reflections account for the long-range broadcast coverage at night and for the shortwave intercontinental communication. *See* RADIO BROADCASTING.

The principal components of the received radio signal are shown symbolically in the illustration. The vector sum of the direct, reflected, and surface waves has been called the space wave, ground wave, or tropospheric transmission to differentiate it from the ionospheric reflections. The ionospheric and surface waves are the principal components at frequencies below 10–30 MHz. The direct and reflected rays are the principal factors at frequencies above 30–50 MHz. Although the ionospheric, direct, and ground-reflected waves can be easily visualized as rays, the surface wave is more difficult to understand; it originates at the air-Earth boundary because the Earth is not a perfect reflector.

Variations in signal level with time are caused by changing atmospheric conditions. The severity of the fading usually increases as either the frequency or path length increases. Most fading is temporary diversion of energy to some direction other than the intended location, associated with refraction or interference, but absorption effects are important in the microwave region.

The dielectric constant of the atmosphere normally decreases gradually with increasing altitude. The result is that on the average the radio ray is bent or refracted toward the Earth so that the distance to the radio horizon is slightly greater than to the optical horizon. The amount of refraction is variable, and exceptionally long-range transmission may occur occasionally.

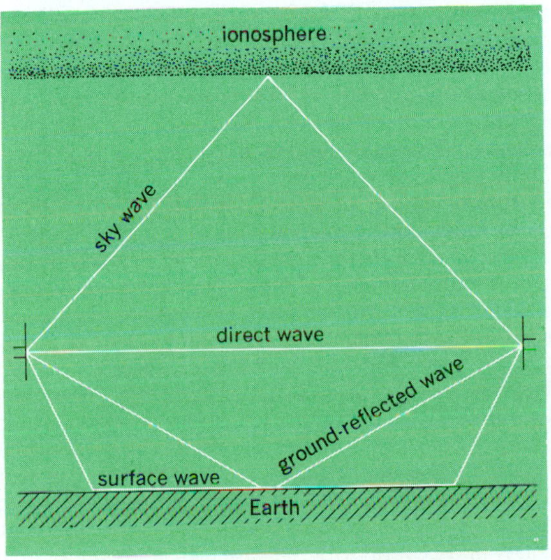

Possible transmission paths between antennas.

Conversely, when the radio energy is bent away from the Earth (upward bending), the transmission loss is increased. *See* REFRACTION OF WAVES. [K.B.]

Radioactive fallout The radioactive material which results from a nuclear explosion in the atmosphere. In particular the term radioactive fallout applies to the debris which is deposited on the ground, but common usage has extended its coverage to include airborne material as well.

In a nuclear explosion, the device itself and the material immediately surrounding it are vaporized into a fireball which then rises in the atmosphere. The altitude at which the fireball cools sufficiently to stabilize depends very much on the yield of the explosion. In general, an atomic explosion fireball stabilizes as a cloud high in the troposphere, while a thermonuclear fireball tends to break through into the stratosphere and stabilize at altitudes above 10 mi (16 km).

As the fireball cools, the vaporized materials condense to fine particles. If the explosion has taken place high above the Earth, the only material present is that of the device itself, and the fine particulates are carried by the winds and distributed over a wide area before they descend to Earth. Bursts at or near the surface carry large amounts of inert material up with the fireball. A large fraction of the radioactive debris condenses out onto the large inert particles of soil and other material, and many small radioactive particulates attach themselves to the larger inert particles. The larger particles settle rapidly by gravity, and large percentages of the radioactivity may be deposited in a few hours near the site of the explosion. In contrast, the radioactivity from a thermonuclear explosion at high altitude takes months or years to reach the surface.

Local fallout is the deposition of large radioactive particles near the site of the explosion. It may present a hazard near the test site or in the case of nuclear warfare. The smaller particulates, which are spread more widely, constitute the worldwide fallout. The radioactivity comes from the isotopes (radionuclides) of perhaps 60 different elements formed in fission, plus a few others formed by activation and any unfissioned uranium or plutonium. Each of the radioactive isotopes produced in the fission process goes through three or four successive radioactive decays before becoming a stable nuclide. The radioactive half-lives of these various fission products range from fractions of a second to about 100 years. Thus decay is very rapid at first and then slows down as the short-lived isotopes disappear.

The debris in the troposphere circles the Earth in about 2 weeks, while material injected into the stratosphere may travel much faster. The horizontal distribution in the north-south direction is much slower, and it may take several months for radioactivity to cover the Northern Hemisphere, for example, following a test in the tropics. Radioactive debris in the troposphere is brought down to the surface of the Earth mainly in precipitation. Radioactive material remains about 30 days in the troposphere.

Once the fallout has actually been deposited, its fate depends largely on the chemical nature of the radionuclides involved. Since the level of gamma radiation from worldwide fallout is negligible, attention has been paid to those nuclides which might possibly present some hazard when they enter the biosphere. Plants, for example, may be contaminated directly by foliar deposition or indirectly by deposition on the soil that is followed by root uptake. Animals may be contaminated by eating the plants, and humans by eating plant or animal foods.

The three radionuclides of greatest interest are strontium-90, cesium-137, and iodine-131. Strontium-90 is a beta emitter which follows calcium metabolically. and tends to deposit in the bone. The radiation emitted could result in bone cancer or leukemia if the levels were to become sufficiently high. Cesium-137 is distributed throughout soft tissues and emits both beta and gamma radiation. It is considered to be a possible genetic hazard but not as dangerous to the individual as strontium-90. Iodine-131 is relatively short-lived and probably is a hazard only as it appears in tropospheric fallout. *See* RADIATION BIOLOGY. [J.H.Ha.]

Radioactive minerals The so-called radioactive minerals loosely comprise species that contain uranium or thorium as an essential part of their chemical composition, together with minerals in which these elements are sometimes present in solid solution, usually in small and variable amounts. About 150 minerals fall into the first category, including many that are rare or that are imperfectly known. The principal uranium minerals from the point of view of their economic importance are the oxide uraninite and its variety pitchblende, the vanadates carnotite and tyuyamunite, the silicate coffinite, the phosphates autunite and torbernite, and the complex oxides brannerite and davidite. The chief thorium minerals of economic interest are the silicates thorite and thorogummite and the oxide thorianite. *See* RADIOACTIVITY; RADIUM; THORIUM; URANIUM. [C.Fr.]

Radioactive tracer A radioactive isotope which, when injected into a chemically similar substance or artificially attached to a biological or physical system, can be traced by radiation detection devices. Many problems in biology, medicine, and industrial engineering not amenable to other approaches can be solved by the use of these tracers. *See* RADIOACTIVITY; RADIOACTIVITY AND RADIATION APPLICATIONS; RADIOISOTOPE; RADIOISOTOPIC ASSAY.

The simplest radioactive tracer studies consist of the tagging of a biological entity with a radioactive isotope (radioisotope). The entity is then tracked by following the radiation from the isotope. The operation becomes more complex when a large number of biological particles are labeled, for example, in the tagging of red blood cells or bacteria. When the labeled substance is injected into an animal, it is impossible to follow the individual labeled particles, but their average movement can be tracked by observations of the radiation.

A radioisotope may also be used to tag its own element. Phosphorus-32 can be introduced into the soil where a plant is growing, and the amount of phosphorus absorbed and its distribution throughout the plant can be studied.

In most biological tracer experiments, the radioisotope is introduced into the system and its radiation subsequently measured with Geiger-Müller counters or scintillation detectors. Extremely soft (low-intensity) radiations may be detected by the use of photographic film. *See* AUTORADIOGRAPHY; GEIGER-MÜLLER COUNTER; SCINTILLATION COUNTER.

Many medical tracer techniques have achieved the status of important and widely used diagnostic tests. For example, radioiodine has found wide use in determining the status of the human thyroid gland.

There are various scanning techniques which enable one to locate a radioisotope within the body and consequently to delineate an organ or an abnormality within an organ. Scanners have been devised which can accurately fix the size of the thyroid gland and determine the presence of nodules or abnormalities in the shape of the gland. Other scanning techniques have been highly successful in locating tumors within the brain.

Radioactive tracers have become increasingly useful in industrial applications, such as the testing of wear and corrosion in mechanical components. On a larger scale, the entire operation of an industrial plant can be tested and controlled by tracer techniques. *See* NONDESTRUCTIVE TESTING. [G.L.B.]

Radioactive waste management The treatment and containment of radioactive wastes. Radioactive waste management is required to some degree in all operations associated with the use of nuclear energy for national defense or peaceful purposes. Liquid, solid, and gaseous radioactive wastes are produced in the mining of ore, production of reactor fuel materials, reactor operation, processing of irradiated reactor fuels, and numerous related operations. Wastes also result from the use of radioactive materials, for example, in research laboratories, industrial operations, and medical treatment.

Aspects of management. The chief aim in the safe handling and containment of radioactive wastes is the prevention of radiation damage to humans and the environment by controlling the dispersion of radioactive materials. Liquid or solid waste products containing significant quantities of the more toxic radioactive materials require isolation and permanent containment in media from which any potential escape into the human environment would be at tolerable levels. The radioactive materials of major concern are those that may be readily incorporated into the body and those that have relatively long half-lives, ranging from a few years to thousands of years. *See* RADIATION BIOLOGY; RADIATION DAMAGE TO MATERIALS.

Waste management is focused on those radioisotopes which originate in nuclear reactors. Here the fission products (chemical elements formed by nuclear fragmentation of actinide elements such as uranium or plutonium) accumulate in the nuclear fuel, along with plutonium and other transuranic nuclides. *See* NUCLEAR REACTOR; PLUTONIUM; TRANSURANIUM ELEMENTS; URANIUM.

If fuel discharged from the nuclear reactor is reprocessed, uranium and plutonium are recovered after chemical dissolution. During recovery, the favored treatment processes produce high-level waste in the form of an acidic aqueous stream. Regulations require that the high-level wastes from nuclear fuel reprocessing plants be solidified within 5 years after reprocessing and shipped to a Federal repository within 10 years after reprocessing. *See* NUCLEAR FUELS REPROCESSING.

Mined cavity disposal and sub-seabed geologic disposal are the chief options under investigation for final storage and disposal of radioactive waste. The basic requirement for acceptable final storage or disposal of radioactive waste is the capability to contain and isolate the waste safely until decay has reduced the radioactivity to nonhazardous levels—or at least to levels found in nature.

Exactly how long near-total containment and isolation of nuclear waste must be maintained is much debated. However, two characteristics of waste change radically in a few hundred years. First, the heat generation rate decreases an order of magnitude in the period of 10 to 100 years and another order of magnitude in the period of 100 to 1000 years. Second, the toxicity of the high-level waste decreases about three orders of magnitude in the first 300 to 400 years due to decay of short-lived fission products. It is then at levels comparable with those of an equal volume of average ores of common toxic elements. [A.M.Pl.]

Decommissioning of a nuclear facility. Because of the buildup of radioactive contamination in a nuclear facility such as a nuclear reactor or reprocessing plant, detailed plans must be made, beginning with the design and construction stages of the facility, to provide adequate protection from radiation exposure during the final decommissioning stage.

There are four principal methods of decommissioning in use: layaway, mothballing, dismantlement, and entombment. Layaway and mothballing are temporary measures to postpone final action until some of the radioactivity has decayed. They involve removing the easily removable high-level radioactive objects. Entombment, as the name implies, is making the radioactive contamination inaccessible by the use of demolition techniques and covering the residue with reinforced concrete. *See* DECONTAMINATION OF RADIOACTIVE MATERIALS; HEALTH PHYSICS; RADIOACTIVITY. [K.Z.M.]

Radioactivity A phenomenon resulting from an instability of the atomic nucleus in certain atoms whereby the nucleus experiences a spontaneous but measurably delayed nuclear transition or transformation with the resulting emission of radiation. The discovery of radioactivity by H. Becquerel in 1896 marked the birth of nuclear physics.

All chemical elements may be rendered radioactive by adding or by subtracting (except for hydrogen and helium) neutrons from the nucleus of the stable ones. The availability of this wide variety of radioactive isotopes has stimulated their use in science and technology in an enormous number of applications. For a discussion of some of the more important applications *see* RADIOACTIVITY AND RADIATION APPLICATIONS. *See also* ALPHA PARTICLES; BETA PARTICLES; GAMMA RAYS.

A particular radioactive transition may be delayed by less than a microsecond or by more than a billion years, but the existence of a measurable delay or lifetime distinguishes a radioactive nuclear transition from a so-called prompt nuclear transition, such as is involved in the emission of most gamma rays. The delay is expressed quantitatively by the radioactive decay constant, or by the mean life, or by the half-period for each type of radioactive atom, discussed below.

There are five types of radioactivity commonly found in nearly all elements, each characterized by the particular type of nuclear radiation which is emitted by the transforming parent nucleus: alpha-ray decay, negatron beta decay, positron beta decay, electron capture, and gamma-ray decay. In addition, there are several other decay modes that are observed more rarely in specific regions of the periodic table.

Transition rates and decay laws. The rate of radioactive transformation, or the activity, of a source equals the number A of identical radioactive atoms present in the source, multiplied by their characteristic radioactive decay constant λ. Thus Eq. (1) holds, where the decay constant λ has dimensions of

$$\text{Activity} = A\lambda \text{ disintegrations per second} \qquad (1)$$

second^{-1}. The numerical value of λ expresses the statistical probability of decay of each radioactive atom in a group of identical atoms, per unit time. For example, if $\lambda = 0.01\ s^{-1}$ for a particular radioactive species, then each atom has a chance of 0.01 (1%) of decaying in 1 s, and a chance of 0.99 (99%) of not decaying in any given 1-s interval. The constant λ is one of the most important characteristics of each radioactive nuclide: λ is essentially independent of all physical and chemical conditions such as temperature, pressure, concentration, chemical combination, or age of the radioactive atoms.

Many radioactive nuclides have two or more independent and alternative modes of decay. For example, ^{238}U can decay either by alpha-ray emission or by spontaneous fission. When two or more independent modes of decay are possible, the nuclide is said to exhibit dual decay. The competing modes of decay of any nuclide have independent partial decay constants given by the probabilities $\lambda_1, \lambda_2, \lambda_3, \ldots$, per second, and the total probability of decay is represented by the total decay constant λ, defined by Eq. (2).

$$\lambda = \lambda_1 + \lambda_2 + \lambda_3 + \cdots \qquad (2)$$

The actual life of any particular atom can have any value between zero and infinity. The average or mean life of a large number of identical radioactive atoms is, however, a definite and important quantity. The total L of the life-spans of all the A_0 atoms initially present is given by Eq. (3). Then the average

$$L = \frac{A_0}{\lambda} \qquad (3)$$

lifetime L/A_0, which is called the mean life τ, is given by Eq. (4).

$$\tau = 1/\lambda \qquad (4)$$

The time interval over which the chance of survival of a particular radioactive atom is exactly one-half is called half-period T. The half-period T is related to the total radioactive decay constant λ, and to the mean life τ, by Eq. (5). For mnemonic

$$T = 0.693/\lambda = 0.693\tau \qquad (5)$$

reasons, the half-period T is much more frequently employed than the total decay constant λ or the mean life τ.

Radioactive series decay. In a number of cases a radioactive nuclide *A* decays into a nuclide *B* which is also radioactive; the nuclide *B* decays into *C* which is also radioactive, and so on. For example, $^{232}_{90}$Th decays into a series of 10 successive radioactive nuclides. Substantially all the primary products of nuclear fission are negatron beta-ray emitters which decay through a chain or series of two to six successive beta-ray emitters before a stable nuclide is reached as an end product.

Alpha-ray decay. Alpha-ray decay is that type of radioactivity in which the parent nucleus expels an alpha ray (a helium nucleus). The atomic number, or nuclear charge, of the decay product is 2 units less than that of the parent, and the nuclear mass of the product is 4 atomic mass units less than that of the parent, because the emitted alpha particle carries away this amount of nuclear charge and mass. This decrease of 2 units of atomic number or nuclear charge between parent and product means that the decay product will be a different chemical element, displaced by 2 units to the left in a periodic table of the elements.

In the simplest case of alpha decay, every alpha ray would be emitted with exactly the same velocity and hence the same kinetic energy. However, in most cases there are two or more discrete energy groups called lines. For example, in the alpha decay of a large group of ^{238}U atoms, 77% of the alpha decays will be by emission of alpha rays whose kinetic energy is 4.20 MeV, while 23% will be by emission of 4.15-MeV alpha rays. When the 4.20-MeV alpha ray is emitted, the decay product nucleus is formed in its ground (lowest energy) level. When a 4.15-MeV alpha ray is emitted, the decay product is produced in an excited level, 0.05 MeV above the ground level. This nucleus promptly transforms to its ground level by the emission of a 0.05-MeV gamma ray or alternatively by the emission of the same amount of energy in the form of a conversion electron and the associated spectrum of characteristic x-rays. Thus in all alpha-ray spectra, the alpha rays are emitted in one or more discrete and homogeneous energy groups, and alpha-ray spectra are accompanied by gamma-ray and conversion electron spectra whenever there are two or more alpha-ray groups in the spectrum.

Among all the known alpha-ray emitters, most alpha-ray energy spectra lie in the domain of 4–6 MeV, although a few extend as low as 2 MeV and as high as 10 MeV. There is a systematic relationship between the kinetic energy of the emitted alpha rays and the half-period of the alpha emitter. The highest-energy alpha rays are emitted by short- lived nuclides, and the lowest-energy alpha rays are emitted by the very-long-lived alpha-ray emitters. H. Geiger and J. M. Nuttall showed that there is a linear relationship between log λ and the energy of the alpha ray.

The Geiger-Nuttall rule is inexplicable by classical physics, but emerges clearly from quantum, or wave, mechanics. In 1928 the hypothesis of transmission through nuclear potential barriers was shown to give a satisfactory account of the alpha-decay data, and it has been altered subsequently only in details. *See* POTENTIAL BARRIER.

Beta-ray decay. Beta-ray decay is a type of radioactivity in which the parent nucleus emits a beta ray. There are two types of beta decay established: in negatron beta decay (β^-) the emitted beta ray is a negatively charged electron (negatron); in positron beta decay (β^+) the emitted beta ray is a positively charged electron (positron). In beta decay the atomic number shifts by one unit of charge, while the mass number remains unchanged. In contrast to alpha decay, when beta decay takes place between two nuclei which have a definite energy difference, the beta rays from a large number of atoms will have a continuous distribution of energy.

For each beta-ray emitter, there is a definite maximum or upper limit to the energy spectrum of beta rays. This maximum energy, E_{max}, corresponds to the change in nuclear energy in the beta decay. As in the case of alpha decay, most beta-ray spectra include additional continuous spectra which have less maximum energy and which leave the product nucleus in an excited level from which gamma rays are then emitted.

For nuclei very far from stability, the energies of these excited states populated in beta decay are so large that the excited states may decay by proton, neutron, two-neutron or alpha emission, or spontaneous fission and, it is predicted theoretically, by two-proton emission.

The continuous spectrum of beta-ray energies implies the simultaneous emission of a second particle besides the beta ray, in order to conserve energy and angular momentum for each decaying nucleus. This particle is the neutrino. The neutrino has zero charge and presumably zero rest mass, travels at the same speed as light (3×10^{10} cm/s), and is emitted as a companion particle with each beta ray. *See* NEUTRINO.

Gamma-ray decay. Gamma-ray decay is a transition between two excited levels of a nucleus, or between an excited level and the ground level. A nucleus in its ground level cannot emit any gamma radiation. Therefore gamma-ray decay occurs only as a sequel of another radioactive decay process or of some other process whereby the product nucleus is left in an excited state. Such additional processes include the fusion of two nuclei, Coulomb excitation, and induced nuclear fission. *See* COULOMB EXCITATION; NUCLEAR FISSION; NUCLEAR FUSION.

A gamma ray is high-frequency electromagnetic radiation (a photon) in the same family with radio waves, visible light, and x- rays. The energy of a gamma ray is given by $h\upsilon$, where h is Planck's constant and υ is the frequency of oscillation of the wave in hertz. The gamma-ray or photon energy $h\upsilon$ lies between 0.05 and 3 MeV for the majority of known nuclear transitions. Higher- energy gamma rays are seen in neutron capture and some reactions. *See* ELECTROMAGNETIC RADIATION.

Gamma rays carry away energy, linear momentum, and angular momentum, and account for changes of angular momentum, parity, and energy between excited levels in a given nucleus. This leads to a set of gamma-ray selection rules for nuclear decay and a classification of gamma-ray transitions as "electric" or as "magnetic" multipole radiation of multipole order 2^l, where $l = 1$ is called dipole radiation, $l = 2$ is quadrupole radiation, and $l = 3$ is octupole, l being the vector change in nuclear angular momentum. The most common type of gamma-ray transition in nuclei is the electric quadrupole (E2). There are cases where several hundred gamma rays with different energies are emitted in the decays of atoms of only one isotope. *See* MULTIPOLE RADIATION.

Spontaneous fission. This involves the spontaneous breakup of a nucleus into two heavy fragments and neutrons. After the discovery of fission in 1939, it was subsequently discovered that isotopes like ^{238}U had very weak decay branches for spontaneous fission, with branching ratios on the order of 10^{-6}. New isotopes subsequently identified like ^{252}Cf have large (3.1%) spontaneous fission branching. [J.H.H.]

Radioactivity and radiation applications The field in which the subatomic fragments emitted in radioactive decay (alpha-, beta-, gamma-rays) or produced by high-voltage accelerators (electrons, protons, x-rays) are applied to the problems of science, engineering, industry, and medicine. The techniques are extraordinarily versatile and sensitive and are basically inexpensive. A disadvantage that limits the range and extent of these applications is the health hazard that may be involved. *See* RADIOACTIVITY; RADIOISOTOPE.

Tracer applications are based on two principles. First is the chemical similarity of radioactive atoms and other atoms of the same element. Periodically a few of the radioactive atoms decay, emitting some penetrating subatomic fragments that can be detected one by one, usually through their ability to cause ionization. Thus the movement of a particular element can be followed through various chemical, physical, and biological steps. The second principle involves the characteristic half-life and nature of the emitted fragments. This makes a radioactive species unique and thereby detectable above a background of radioactive emitters associated with elements. For discussions of radioisotope techniques relating to tracer methodology *see* ACTIVATION ANALYSIS; ISOTOPE DILUTION TECHNIQUES. *See also* NONDESTRUCTIVE TESTING; RADIOACTIVE TRACER; RADIOECOLOGY; RADIOISOTOPIC ASSAY; RADIOMETRIC ASSAY.

Penetration and scattering applications arise from the fact that subatomic fragments can penetrate a thick section of a material, and yet a small fraction of the incident particles can be backscattered by a relatively thin section. The oldest application of the penetrating properties of energetic ionizing photons is radiography. An extension of this technique is autoradiography. Since World War II the penetration and scattering properties of beta- and gamma-rays have been applied in industry in the form of thickness gages. *See* AUTORADIOGRAPHY; RADIOGRAPHY.

The absorption of small amounts of energy from ionizing particles and ionizing photons has chemical effects that have been the basis of several practical applications. The oldest application of this principle is radiation therapy. For example, in cancer therapy the local affected areas are irradiated by external beams of gammas from cobalt-60 or of radiation from accelerators. Radioactive sources have also been administered internally to induce beneficial biochemical reactions in patients afflicted with various ailments. *See* ISOTOPIC IRRADIATION; RADIOLOGY.

A related area is the radiation sterilization of biomedical supplies. The advantages to this method of biochemical destruction of microscopic life are that (1) unlike steam sterilization, it can be performed at low temperatures on plastics and other thermally unstable materials, and (2) unlike germicidal gases, ionizing radiation can reach every point in the treated product. Radiation-sterilized objects are not radioactive.

The radiation preservation of food is an area of considerable promise. Small doses can inhibit sprouting in potatoes, kill insects in wheat, and sterilize pork products but practical applications have been sharply limited due to a cautious role by regulatory authorities in approving such procedures.

Kinetic energy of emissions in radioactive decay can be converted to useful forms of light, heat, and electricity. *See* LUMINOUS PAINT; NUCLEAR BATTERY. [J.Sil.]

Radiocarbon dating A method of obtaining age estimates on organic materials which has been used to date samples as old as 70,000 years. A date is obtained based on radiocarbon (^{14}C) which gives a temporal index of when a specific sample was removed from its reservoir.

Radiocarbon determinations can be obtained on wood, charcoal, marine and fresh-water shell, bone and antler, peat and organic- bearing sediments, carbonate deposits such as tufa, caliche, and marl, and dissolved CO_2 and carbonates in ocean, lake, and groundwater sources. Each sample type has specific problems associated with its use for dating purposes, including contamination and special environmental effects. While the impact of ^{14}C dating has been most profound in archeological research and particularly in prehistoric studies, extremely significant contributions have also been made in hydrology and oceanography.

Carbon has three naturally occurring isotopes. Two (^{12}C and ^{13}C) are stable, but the third, ^{14}C, decays by very weak beta decay (electron emission) to ^{14}N with a half-life of approximately 5700 years. Naturally occurring ^{14}C is produced as a secondary effect of cosmic-ray bombardment of the upper atmosphere. As $^{14}CO_2$, it is distributed on a worldwide basis into various atmospheric, biospheric, and hydrospheric reservoirs on a time scale much less than its half-life. Metabolic processes in living organisms and relatively rapid turnover of carbonates in surface ocean waters maintain ^{14}C levels at approximately constant levels in most of the biosphere. *See* Isotope.

To the degree that ^{14}C production has proceeded long enough without significant variation to produce an equilibrium or steady- state condition, ^{14}C levels observed in contemporary materials may be used to characterize the original ^{14}C activity in the corresponding carbon reservoirs. Once a sample has been removed from exchange with its reservoir, as at the death of an organism, the amount of ^{14}C begins to decrease as a function of its half-life. A ^{14}C age determination is based on a measurement of the residual ^{14}C activity in a sample compared to the activity of a sample of assumed zero age (a contemporary standard) from the same reservoir.

A measurement of the ^{14}C content of an organic sample will provide an accurate determination of the sample's age if it is assumed that (1) the production of ^{14}C by cosmic rays has remained essentially constant long enough to establish a steady state in the $^{14}C/^{12}C$ ratio in the atmosphere; (2) there has been a complete and rapid mixing of ^{14}C throughout the various carbon reservoirs; (3) the carbon isotope ratio in the sample has not been altered except by ^{14}C decay; and (4) the total amount of carbon in any reservoir has not been altered. In addition, the half-life of ^{14}C must be known with sufficient accuracy, and it must be possible to measure natural levels of ^{14}C to appropriate levels of accuracy and precision. Studies have shown that the primary assumptions on which the method rests have been violated both systematically and to varying degrees for particular sample types. Methods for calibration and corrections of conventional ^{14}C values have been developed. [R.E.T.; R.A.Mu.]

Radiochemical laboratory An installation to provide a safe environment for handling and investigating radioactive materials. Common features of radiochemical laboratories include readily decontaminated surfaces, good ventilation, cleanliness, good lighting, special arrangements for waste disposal, and filtration of both supply and exhaust air. Specific features of these laboratories depend primarily on the type and amount of radioactivity handled. [S.L.]

Radiochemistry A subject which embraces all applications of radioactive isotopes to chemistry. It is not precisely defined and is closely linked to nuclear chemistry. *See* Nuclear chemistry; Radioisotope.

In general, radiochemical studies can be classified according to whether the use of isotopes represents a convenient or a unique solution to a problem. Convenient applications usually exploit the high sensitivity of tracer techniques because alternative analytical procedures are slower and often less accurate. The efficiencies of chemical separations are studied by labeling the desired compound and following the radioactivity during the separations. [D.R.S.]

Radioecology The study of the interaction of radioisotopes or ionizing radiation with populations, ecological communities, or ecosystems. Radioecology began with attempts to predict the impact of the developing nuclear industry on environmental systems: the fates of radioactive substances released into the environment, and the effects of ionizing radiation on ecological processes. The value of radioactive tracers for ecological studies emerged rapidly, as did the utility of ionizing radiation effects as a tool in ecological research. Radioecology now encompasses the use of radioactive tracers, as well as fates and effects of radioactive substances in the environment. *See* Radioactive tracer; Radioisotope. [R.G.W.]

Radiography The technique of producing a photographic image of an opaque specimen by the penetration of radiation such as gamma rays, x-rays, neutrons, or charged particles. When a beam of radiation is transmitted through any heterogeneous object, it is differentially absorbed, depending upon the varying thickness, density, and chemical composition of this object. The image registered by the emergent rays on a photographic film adjacent to the specimen under examination constitutes a shadowgraph or radiograph of its interior. Radiography is the general term applied to this nondestructive technique of testing the gross internal structure of any object, whether it be of the chest of a human patient for evidence of tuberculosis, silicosis, heart pathology, or embedded foreign objects; of bones in case of fractures, arthritis or other bone diseases; or of a weld in a pipe to observe cracks, inclusions, or voids.

The term roentgenography specifically applies to the use of x- radiation, and is derived from the name of its discoverer, W. C. Röntgen. Microradiography is simply the extension of radiography to small specimens and the photographic enlargement of these images (up to 500 ×). When the image is registered on a fluorescent screen and visually observed, the technique is called fluoroscopy, The photography of the fluorescent image, as in mass chest examinations, is called photofluoroscopy. The radiation-sensitive silver halide emulsion of photographic film may be replaced by the dry-plate electrostatically formed image of xerography, and the technique becomes xeroradiography. High-speed radiography, for instance, of fast-moving objects, can be achieved by specially designed x-ray apparatus with exposure times as short as 0.2×10^{-6} s. *See* Computerized tomography; Microradiography; Radiology.

A newer development in the field of radiography is the use of neutron beams. These neutrons can penetrate matter with relative ease since they are not electrically charged. There are low attenuations of thermal neutrons in most of the heavy elements; therefore, large thicknesses of heavy materials can be inspected by neutrons in less time than would be required by x-radiography or gamma radiography. On the other hand, the neutron absorption coefficients of elements with low atomic numbers are high; hydrogen, lithium, and boron are particularly attenuating. With neutrons, it is possible to visualize materials such as liquids, adhesives, rubber, plastic, or explosives even when they are in metal assemblies.

Proton radiography employs beams of protons. A monoenergetic charged particle has a well-defined distance it will travel in a given material before it is stopped; the distance is called the range. Since most of the attenuation of the charged particles occurs near the end of the range, a very small change in material thickness will result in a large change in radiation transmission. Therefore, the sensitivity of this method to small changes in object thickness is very great, if the total path for the radiation approximates the range. [G.H.T.; G.L.C; H.B.]

In heavy-ion radiography, accelerated beams of heavy charged particles are used to produce detailed radiographs that can be used for diagnostic applications to cancer research and therapy. These beams have special physical qualities that make them particularly well- suited to imaging research: they have an exact range-energy relationship, and they scatter less than electrons or x-rays. To make a radiograph, the beam passes through the tissue or organ to be imaged, and is collected in a stack of approximately 50 thin plastic sheets. If the particle passes through dense tissue, it stops in the front part of the detector stack; conversely, if the tissue is less dense, the particles travel farther to the back of the stack. The data can be viewed in oblique light and photographed, but for quantitative reconstruction a light beam scattered off the plastic sheets is digitized with a television-type vidicon camera and a computer, and the integrated data are stored on magnetic tapes or disks. Heavy-ion radiography thus provides a more precise resolution of density and composition of tissue than does conventional x-ray radiography. [C.A.T.]

Radioimmunoassay A general method employing the reaction of antigen with specific antibody, permitting measurement of the concentration of virtually any substance of biologic interest, often with unparalleled sensitivity. The basis of the method is summarized in the competing reactions shown in the illustration. The unknown concentration of the antigenic substance in a sample is obtained by comparing its inhibitory effect on the binding of radioactively labeled antigen to a limited amount of specific antibody with the inhibitory effect of known standards.

A typical radioimmunoassay is performed by the simultaneous preparation of a series of standard and unknown mixtures in test tubes, each containing identical concentrations of labeled antigen and specific antibody. After an appropriate reaction time the antibody- bound (B) and free (F) fractions of the labeled antigen are separated by one of a variety of techniques. The B/F ratios in the standards are plotted as a function of the concentration of unlabeled antigen (standard curve), and the unknown concentration of antigen is determined by comparing the observed B/F ratio with the standard curve.

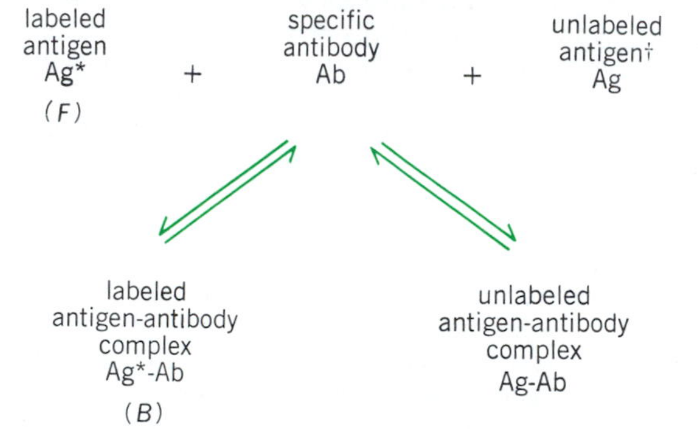

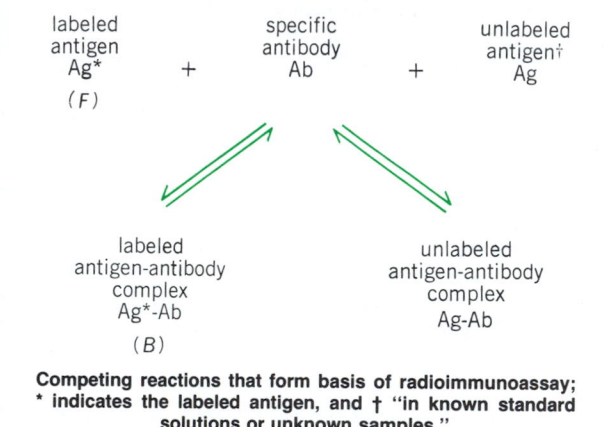

Competing reactions that form basis of radioimmunoassay; * indicates the labeled antigen, and † "in known standard solutions or unknown samples."

The radioimmunoassay principle has found wide application in the measurement of a large and diverse group of substances in a variety of problems of clinical and biological interest. It is therefore not unexpected that there are differences in the specific methods employed for the assay of a particular substance. The full potential of the method has yet to be exploited. It seems that virtually any substance of biologic interest can be measured, the method being modified according to the characteristics of the particular substance. *See* Antibody; Antigen; Immunology. [R.S.Y.]

Radioisotope A radioactive isotope (as distinguished from a stable isotope) of an element. Atomic nuclei are of two types, unstable and stable. Those in the former category are said to be radioactive and eventually are transformed, by radioactive decay, into the latter. One of the three types of radioactive ray (α-, β-, and γ-rays) is emitted during each stage of the decay.

The term radioisotope is also loosely used to refer to any radioactive atomic species. Whereas approximately a dozen radioisotopes are found in nature in appreciable amounts, hundreds of different radioisotopes have been artificially produced by bombarding stable nuclei with various atomic projectiles. *See* Isotope; Radioactivity. [H.E.D.]

Radioisotopic assay An analytical technique including procedures for separating and reproducibly measuring a radioactive tracer. Separations for the assay may be made carrier-free (without stable atoms added) or with a few milligrams of added carrier by techniques such as ion exchange, solvent extraction, distillation, precipitation, electrodeposition, and isotopic exchange. Once separated, the radioisotopes may be measured as gases, liquids, or solids in Geiger-Müller, proportional, or scintillation counters, or in ionization chambers.

Geiger and proportional counters are used primarily for measuring beta rays, whereas solid scintillation detectors are used for gamma rays. Liquid scintillation counters eliminate many self-absorption problems with weak beta emitters. *See* Activation analysis; Particle detector; Radioactive tracer; Radiochemistry; Radiometric assay. [W.W.M.]

Radiolaria A group of marine protozoa, regarded as a subclass of Actinopodea in older classifications, but not recognized as a natural group in some modern systems owing to its heterogeneity. In certain modern systems, the Radiolaria are subdivided into two classes, Polycystinea and Phaeodarea. Radiolarians occur almost exclusively in the open ocean as part of the plankton community, and are widely recognized for their ornate siliceous skeletons produced by most of the groups (illustration). Their skeletons occur abundantly in ocean sediments and are used in analyzing the layers of the sedimentary record (biostratigraphy).

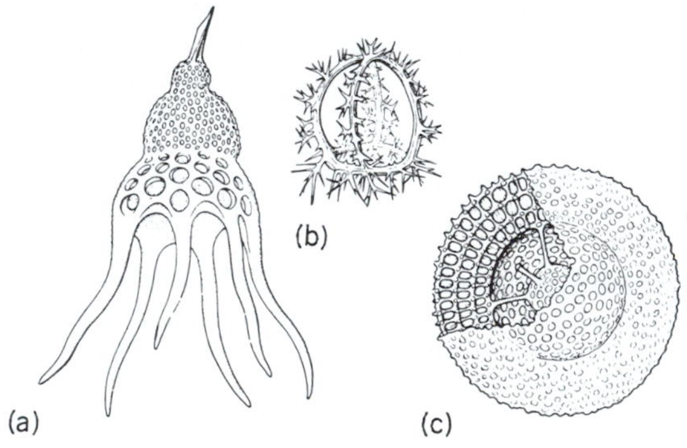

Radiolaria. Skeletons representing (*a, b*) certain Nasselarida (or Monopylina) and (*c*) certain Spumellarida (or Peripylina). (*After L. H. Hyman, The Invertebrates, vol. 1, McGraw-Hill, 1940*)

A characteristic feature of the group is the capsule, a central mass of cytoplasm bearing one or more nuclei, food reserves, and metabolic organelles. This is surrounded by a perforated wall and a frothy layer of cytoplasm known as the extracapsulum, where food digestion generally occurs and where numerous axopodia (stiffened strands of cytoplasm) and rhizopodia radiate toward the surrounding environment. Algal symbionts including dinoflagellates, green algae, and golden-brown pigmented algae occur profusely in the extracapsulum. The algal symbionts living within the protection of the extracapsulum provide photosynthetically derived food for the radiolarian host.

Radiolarians have a fossil record that extends back to early Paleozoic time, about a half billion years ago. Compared to other groups of shell-bearing marine microplankton, they are highly diverse, several hundred species having inhabited the oceans at any given time. Because they are planktonic and have undergone continuous evolutionary change, radiolarians are particularly useful for determining time equivalence (and geological ages) of marine sedimentary deposits at widely separated localities. The Cenozoic record of radiolarians in sediments, particularly on the deep-sea floor, is sufficiently complete to show the course of evolutionary change in considerable detail.

Assemblages of fossil radiolarians also provide clues to oceanic conditions during the geological past. Each of the major oceanic water masses has its characteristic radiolarian fauna, and so changes in the distribution or composition of these assemblages can be interpreted in terms of changes in the pattern of water masses, or in their oceanographic properties. *See* ACTINOPODA; OCULOSIDA; PROTOZOA; RADIOLARIAN EARTH. [O.R.A.; W.R.R.]

Radiolarian earth A porous, earthy, unconsolidated sediment formed from the opaline silica skeletal remains of Radiolaria. It is formed from radiolarian oozes that accumulate on the deep ocean floor. The indurated equivalent with pores filled by silica is radiolarite. The radiolarian earths and radiolarites are white and cream-colored. Radiolarites are known from Devonian to Tertiary age. *See* CHERT; RADIOLARIA; SEDIMENTARY ROCKS. [R.Si.]

Radiology The medical science concerned with x-rays, radioactive materials, and other ionizing radiations, and the application of the principles of this science to diagnosis and treatment of disease. Nonionizing radiations of infrared and ultrasound are also used for diagnosis.

Diagnostic radiology uses radiation, usually x-rays, to study the configuration of anatomical structures or the function of body organ systems. *See* RADIOGRAPHY.

Radioactive isotopes are used to obtain images of organ systems and functions. The accumulation of isotope in a tumor or an organ such as the thyroid is recorded by a suitable γ-ray detector attached to an electronic amplifier and recording equipment. The image of the radioactivity concentrated in an organ is viewed on a television-type screen and recorded on a photographic print. *See* RADIOACTIVE TRACER.

Sound waves of 1–10 MHz are transmitted from a crystal transducer, and after amplification are displayed on an oscilloscope and recorded on a photographic print. The ultrasound pulses demonstrate organ structures such as the heart, liver, and spleen. Although the resolution is less fine than that obtained with x-ray, there is an advantage in that the ultrasound is nonionizing radiation. Ultrasound is particularly useful, therefore, in determining the size and degree of development of the human fetus. *See* ULTRASONICS.

Infrared radiation from the human body is used to detect tumors such as breast tumors, which are near the body surface. The technique, thermography, is based on the idea that tumors are warmer than the surrounding normal tissue. This increase in temperature is detected by an infrared device, and the "hot spot" scan is displayed on a television-type screen, with permanent records kept on photographic prints.

Radiation therapy deals with the treatment of disease with ionizing radiation. The diseases most commonly treated are cancer and allied diseases. Radiation therapy has been found useful in the management of some diseases such as ringworm of the scalp and bursitis, but because of possible serious complications occurring many years later, the use of ionizing radiation is generally avoided if alternative methods of treatment are available.

In cancer therapy the objective is to destroy a tumor without causing irreparable radiation damage in normal body tissues that must of necessity be irradiated in the process of delivering a lethal dose to the tumor. This applies particularly to important normal structures in the vicinity of the tumor. The relative radiosensitivity of the tumor with respect to these normal structures is the chief factor determining the success of the treatment. [L.B.Lu.]

Radiometric assay An analytical technique that includes procedures for measuring, by tracer methods, elements which are not themselves radioactive. It is particularly useful for small quantities of inorganic ions or complex organic compounds.

In radiometric titrations a radioactive tracer is used to determine the equivalence point in a volumetric determination, where a highly insoluble compound of definite composition is formed. At least one of the solutions used in the titration must be radioactive, because progress of the titration is followed by plotting the radioactivity of the solution versus volume of reagent added. The equivalence point is determined by the intersection of two straight lines on this plot, because a definite change in solution activity appears after equivalent amounts of reagents have reacted. *See* RADIOACTIVE TRACER; RADIOCHEMISTRY; RADIOISOTOPIC ASSAY; TITRATION. [W.W.M.]

Radiometry The detection and measurement of radiant electromagnetic energy. Conventionally, radiometry is concerned with infrared radiation. Generally, the devices used in radiometry can be, and are, used with visible light. However, the use of devices applicable only to visible light is commonly excluded from the term radiometry. *See* ELECTROMAGNETIC RADIATION; INFRARED RADIATION; PHOTOMETRY.

Detectors such as the thermometer which respond to the increase in temperature resulting from the absorption or radiant energy are termed thermal detectors. Improvement in thermal detectors has been concerned with securing a large and rapid rise in temperature and high sensitivity in the detection of changes in temperature. To secure the largest rise for a given radiation, the detector should absorb it as completely as possible; the loss of heat must be as small as possible. The heat capacity of a thermal detector should be small, so that a rapid rise in temperature will occur. The area of the absorbing surface should be small to permit the measurement of narrow beams of radiation.

The thermocouple, one of the oldest and still one of the best radiation detectors, produces an electromotive force (emf) when heated. The sensitive portion of a bolometer undergoes a change in resistance when heated. *See* BOLOMETER; THERMOCOUPLE.

The photographic effect does not depend upon a rise in temperature but upon the freeing of a bound electron by the absorption of a single quantum of radiation. Detectors utilizing this principle are termed quantum detectors, or photodetectors. Photoemissive cells and photomultipliers are of interest for visible radiation, but have limited application in the near infrared to about 1 μm. Photoconductive cells using materials such as lead sulfide, lead selenide, and lead telluride are useful to 5 μm. Photoconductive cells of single crystals of gold-doped germanium, cooled with liquid helium, have sensitivity extending to much longer wavelengths. *See* PHOTOCONDUCTIVE CELL; PHOTODIODE; PHOTOMULTIPLIER; PHOTOTRANSISTOR; PHOTOTUBE. [H.W.Ru./G.R.H.]

Radish A cool-season annual or biennial crucifer, *Raphanus sativus*, of Chinese origin belonging to the plant order Capparales. The radish is grown for its thickened hypocotyl, which is eaten uncooked as a salad vegetable. Colors include red, yellow, white, black, pink, and red-white combinations. *See* CAPPARALES. [H.J.C.]

Radium A chemical element, Ra, with atomic number 88. Radium is a rare radioactive element found in uranium minerals to the extent of 1 part for about every 3,000,000 parts of uranium. Chemically, radium is an alkaline- earth metal having properties quite similar to those of barium. Biologically, radium concentrates in bones by replacing calcium and, as a result of prolonged irradiation, causes anemia and cancerous growths. Because radiations from radium and its decay products prefer-

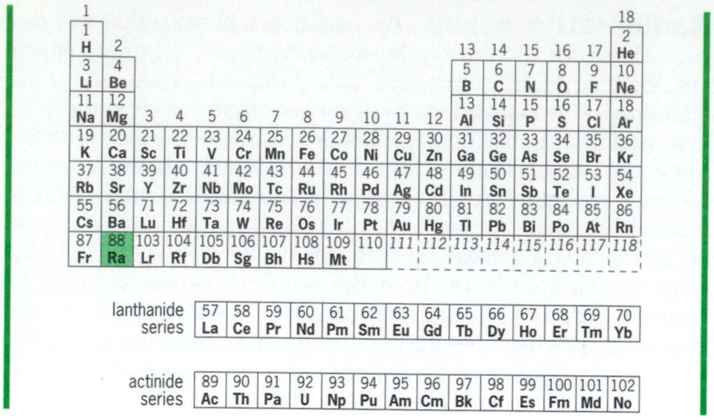

entially destroy malignant tissue, radium has been used to check the growth of cancer. For this use pure radium compounds are sealed in tubes or needles; or radon, the gaseous decay product of radium, is pumped into small tubes. The use of radium in luminescent paints for watch, clock, and meter dials, and signs visible in the dark depends on its α-radiation striking a scintillator such as zinc sulfide.

Thirteen isotopes of radium are known; all are radioactive; four occur naturally; the rest are produced synthetically. Only ^{226}Ra is technologically important. It is distributed widely in nature, usually in exceedingly small quantities. The most concentrated source is pitchblende.

When first prepared, nearly all radium compounds are white, but they discolor on standing because of intense radiation. Radium salts ionize the surrounding atmosphere, thereby appearing to emit a blue glow. Radium compounds will discharge an electroscope, fog a light-shielded photographic plate, and produce phosphorescence and fluorescence in certain inorganic compounds such as zinc sulfide. The emission spectrum of radium compounds is similar to those of the other alkaline earths; radium halide imparts a carmine-red color to a flame. *See* ALKALINE-EARTH METALS; RADIOACTIVITY; RADON. [M.L.S.]

Radius A line segment that joins the center of a circle with any one of its points. If r denotes the length of a radius, the length (circumference) of a circle is $2\pi r$, and the area enclosed is πr^2. The radius of a circle may be marked off on the circle six times as a chord, and a regular hexagon obtained. *See* CIRCLE. [L.M.Bl.]

Radius of gyration A relation of the area or mass of a figure to its moment of inertia. If I is the moment of inertia about a line of a figure whose area is A, the figure's radius of gyration with respect to that line is $k = +\sqrt{I/A}$. Accordingly, $I = k^2A$. For a figure of mass M, $k = +\sqrt{I/M}$; $I = k^2M$. In these equations, k is measured in length units such as feet. Geometrically similar figures have equal radii of gyration about corresponding centroidal axes. *See* MOMENT OF INERTIA. [N.S.F.]

Radome A strong, but electrically transparent, thin shell used to house a radar antenna, or a space-communications antenna of similar structure. The shell must be large enough not to interfere with the scanning motion of the antenna. In airborne radar the radome prevents the antenna from upsetting the aerodynamic characteristics of the airplane or missile and protects the antenna against aerodynamic forces. Shipboard radars frequently require radomes to protect them against wind and water damage and blast pressures from nearby guns. Large land-based radars are usually shielded by radomes, especially in severe climatic conditions. *See* ANTENNA (ELECTROMAGNETISM); RADAR. [R.I.B.]

Radon A chemical element, symbol Rn, atomic number 86. Radon is produced as a gaseous emanation from the radioactive decay of radium. The element is highly radioactive and decays by the emission of energetic alpha particles. It is the heaviest of the noble, or inert, gas group and thus is characterized by chemical inertness. All isotopes are radioactive with short half-lives. *See* INERT GASES; RADIOACTIVITY; RADIUM.

In addition to the three natural isotopes, 22 isotopes have been synthesized by nuclear reactions of artificial transmutation in cyclotrons and linear accelerators, but none of these is as long-lived as ^{222}Rn. *See* NUCLEAR REACTION.

Any surface exposed to ^{222}Rn becomes coated with an active deposit which consists of a group of short-lived daughter products. The radiations of this active deposit include energetic alpha, beta, and gamma rays.

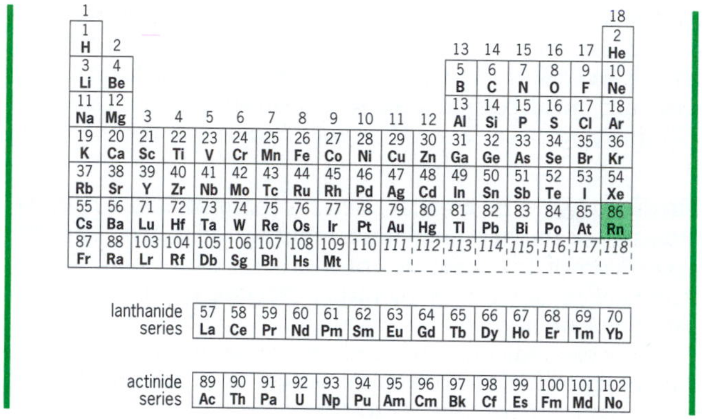

Radon possesses a particularly stable electronic configuration, which gives it the chemical properties characteristic of noble-gas elements. It has a boiling point of −65°C (−85°F) and a melting point variously reported as −71°C and −113°C (−96°F and −171°F). The spectrum of radon has been extensively studied, and resembles that of the other inert gases. [E.K.H.]

Raffinose The best-known trisaccharide (oligosaccharide), consisting of one melibiose and two sucrose units, widely distributed in higher plants. Sugarbeets contain about 0.5% of this sugar. The best-known sources are cottonseed meal and the manna of *Eucalyptus*. It is also known as melitose, melitri-ose, gossypose, and *O*-α-D- galactopyranosyl-(1→6)-*O*-α-D-glucopyranosyl-(1→2)-β-D- fructofuranoside. *See* OLIGOSAC-CHARIDE.

Raffinose was found to be enzymically synthesized in plants from uridine diphosphate D-galactose and sucrose by an enzyme which transfers the D-galactose moiety of this sugar nucleotide to sucrose, resulting in the formation of raffinose. [W.Z.H.]

Rafflesiales An order of flowering plants, division Magnoliophyta (Angiospermae), in the subclass Rosidae of the class Magnoliopsida (dicotyledons). The order consists of three families and fewer than a hundred species, all tropical or subtropical. The plants are highly specialized, nongreen, rootless parasites which grow from the roots of the host. They have few or solitary, rather large to very large flowers with numerous ovules and a single set of tepals that are commonly united into a conspicuous, corolloid calyx. *See* MAGNOLIOPSIDA; ROSIDAE. [A.Cr.]

Railroad engineering That part of transportation engi-neering involved in the planning, design, development, construction, maintenance, and use of facilities for the transportation of goods and people in wheeled units of rolling stock running on, and guided by, rails normally supported on cross ties and held to fixed alignment.

Freight handling systems. In the United States and in most other countries the part of railroad engineering concerned with fixed properties, including tracks, roadbeds, bridges, buildings, yards, and terminals, is considered to be a part of the broad discipline of civil engineering. In a like manner the part of railroad engineering concerned with rolling stock, including locomotives, cars, car ferries, and other such equipment, is considered to be a part of the broad discipline of mechanical engineering. In addition, the staffs of railroad organizations include many architects and chemical, electrical, and industrial engineers.

Probably the most important transformation that has taken place on railroads in the United States in recent times was the change from steam to diesel power. Electric locomotives are used on some 700 mi (1100 km) of Conrail lines, with only a scattering of such power at other locations. *See* Diesel engine; Locomotive.

Many different kinds of freight cars have been developed. Specially equipped long flat cars with tie-down devices are used for handling trailers or containers in piggyback service; similar cars are equipped to carry containers (see illustration). Other innovations include designs to carry double stacks of containers and low-profile piggyback cars which carry trailers in a well to allow clearance through eastern tunnels. Other long flat cars are equipped with fully enclosed bilevel and trilevel racks for handling assembled automobiles and trucks. Covered hopper cars, some of large capacity, are used for shipment of cement, grain, flour, and other dry bulk commodities. Large-capacity tank cars are in demand for liquid bulk shipments. Most refrigerator cars are equipped with gasoline engine–driven mechanical refrigeration units.

Main lines are equipped with signal systems to aid in the expedition and safety of train movements. Increasingly the signaling facilities are augmented by two-way communication with train crews.

Electronic computers are widely used by railroads for data processing and for storage and retrieval of information. A widely used automation device is the hotbox detector, which locates and records overheated car axle bearings in moving trains, thus permitting the defective car to be set out before complete journal failure and possible derailment. [W.J.He.]

Low-profile piggyback car.

Passenger systems. Experience has shown that, as a country develops, its need for a reliable and efficient railway-system grows. There are many distinct characteristics and functions of the guided rail transport system that cannot be achieved as efficiently, as economically, or as acceptably within environmental constraints by other modes of transport. This is particularly true of the passenger services.

Transportation systems in industrialized nations are subject to increasing pressures from population expansion, congestion and overcrowding, energy shortages, and rising costs.

The post–World War II Japanese experience is considered a model of development. The high-speed rail passenger system, Shinkansen, is considered one of the finest technological achievements of the 20th century. It is very profitable and reliable, operating comfortable trains that run frequently. It has an excellent safety record and has transported more than 1,600,000,000 passengers during its years of service without a single casualty.

TGV (Très Grande Vitesse), the French high-speed passenger system serving the Paris-Lyon route, has a roadbed that takes advantage of flat or rolling terrain by following the natural contours, with no tunnels. This roadbed has the general appearance of a slowly curving motorway. However, while it is less than half the width of a four-lane highway, it has three times the passenger- carrying capacity per hour.

After World War II the United States railroads emphasized freight services to the great detriment of passenger operations. With the availability of more powerful locomotives and the invention of the Janney coupler, along with the air brake, it became possible to make up trains of great length. As trains get longer and heavier, they must get slower. The average freight train speed in the United States (calculated by dividing total train miles by total train hours) is 20.1 mi/h (32.3 km/h). Such a train must have a flat track, which is unsuitable for passenger operations.

As passenger trains were forced to slow for flattened curves, they ran late consistently, and their schedules were lengthened as a consequence. Slower track carrying heavier freight became uncomfortable and unreliable. The passenger services were all but immobilized by the heavy freight, slow speed, and flat track. Orders for new passenger cars were discontinued.

In 1970 Congress passed the Rail Passenger Service Act, and in compliance with the provisions of this act the National Railroad Passenger Corporation (Amtrak) was created on May 1, 1971. In the early 1980s Amtrak's network served all but 4 of the top 50 metropolitan areas. However, in many of them the amount of service was not more than one train a day, with many scheduled times at inconvenient hours. With over 60% of its system running no more than once a day, Amtrak cannot even offer a reasonable round trip connection to a very large segment of its prospective passengers. [L.F.P.]

Rain shadow An area of diminished precipitation on the lee side of mountains. There are marked rain shadows, for example, east of the coastal ranges of Washington, Oregon, and California, and over a larger region, much of it arid, east of the Cascade Range and Sierra Nevadas. All mountains decrease precipitation on their lee; but rain shadows are sometimes not marked if moist air often comes from different directions, as in the Appalachian region.

The causes of rain shadow are (1) precipitation of much of the moisture when air is forced upward on the windward side of the mountains, (2) deflection or damming of moist air flow, and (3) downward flow on the lee slopes, which warms the air and lowers its relative humidity. [J.R.F.]

Rainbow Colored arcs seen in the skies when the Sun or Moon is illuminating large numbers of falling raindrops. Such arcs are centered around the antisolar point (180° from the Sun). Among the parallel rays striking a water droplet there is one ray which after one internal reflection leaves the drop at the smallest angle of deviation from the direction of the incident rays. All other rays emerge from the drop, after an internal reflection, at larger angles of deviation. The ray of minimum deviation, also called Descartes ray, has the greatest intensity and thus is more visible than the others. Since the angle of deviation for such a ray of minimum deviation is smaller for the red light and larger for the blue, the rays of minimum deviation from a large number of water drops of the same size are seen as arcs of different colors, red on the upper part, and blue on the lower part of the bow. The angular distance of these arcs from the antisolar point is greater than 42°, and the system of arcs is called the primary rainbow. The rays of minimum deviations, after two internal reflections, form similar colored arcs at the angular distance of 51° from the antisolar point to make the secondary rainbow. In this rainbow the color sequence is reversed, with the red arcs visible on the lower part, the blue on the upper part of the bow. [Z.S.]

Rainforest A term used loosely in plant geography for forests of broad-leaved (dicotyledonous), mainly evergreen trees found in continually moist climates in the tropics, subtropics, and some parts of the temperate zones. Sometimes the term is unjustifiably extended to include other very wet forests such as the Olympic Rain Forest of the state of Washington in which the trees are mostly conifers.

The tropical rainforest includes the vast Amazon forest as well as large areas in western and central Africa, Malaysia, Indonesia, and New Guinea. It is the home of an enormous number of plant and animal species. Tropical rainforests are usually mixed in composition, no one species forming a large proportion of the whole stand, but in some parts of the tropics there are rainforests dominated by a single species.

Certain structural features are characteristic of tropical rainforests, such as the thin, flangelike buttresses of the larger trees, and flowers and fruit that are produced, as in the cacao (*Theobroma cacao*), on the trunk (cauliflorous) instead of on the branches.

Orchids and other epiphytes and woody vines (lianes) are common. In a rainforest that has not been culled for timber or recently disturbed, the undergrowth is generally rather thin, and visibility is about 60 ft (18 m) or more on the ground.

In the rainforest, animals of most groups, like plants, show great species diversity. For example, there are several times as many species of birds as in an equal area of North American broad-leaved forest. Insects are far more numerous than other animals, in terms of species and sheet numbers. Interactions between plants and animals are extremely complex, and many very specialized relationships have evolved, such as those between certain types of ants and the trees in whose hollow twigs they live. In the tropical rainforest an unusually large variety of organisms live together in a state of balance, so that it is one of the most stable as well as most complex ecosystems on Earth. [P.W.R.]

Raman effect A phenomenon observed in the scattering of light as it passes through a material medium, whereby the light suffers a change in frequency and a random alteration in phase. Raman scattering differs in both these respects from Rayleigh and Tyndall scattering, in which the scattered light has the same frequency as the unscattered and bears a definite phase relation to it. The intensity of normal Raman scattering is roughly one-thousandth that of Rayleigh scattering in liquids and smaller still in gases. *See* SCATTERING OF ELECTROMAGNETIC RADIATION; TYNDALL EFFECT.

Because of its low intensity, the Raman effect was not discovered until 1928, although the scattering of light by transparent solids, liquids, and gases had been investigated for many years before. The development of the laser has led to a resurgence of interest in the Raman effect and to the discovery of a number of related phenomena. *See* LASER.

When the exciting radiation falls within the frequency range of a molecule's absorption band in the visible or ultraviolet spectrum, the radiation may be scattered by two different processes, resonance fluorescence or the resonance Raman effect. Both these processes give much more intense scattering than the normal nonresonant Raman effect. The absolute frequencies of the resonance Raman effect shift by exactly the amount of any shift in the exciting frequency, just as do those of the normal Raman effect. Thus the main characteristic of the resonance as compared to the normal Raman effect is its intensity, which may be greater by two or three orders of magnitude. *See* FLUORESCENCE.

Raman scattering is analyzed by spectroscopic means. The collection of new frequencies in the spectrum of monochromatic radiation scattered by a substance is characteristic of the substance and is called its Raman spectrum. Although the Raman effect can be made to occur in the scattering of radiation by atoms, it is of greatest interest in the spectroscopy of molecules and crystals. In a typical experiment monochromatic radiation from a laser impinges on the sample in an appropriate transparent cell. Raman scattering is approximately uniform in all directions and is usually studied at right angles. In this way the intense radiation of the laser beam interferes least with the observation of the weak scattered light.

Raman spectroscopy is of considerable value in determining molecular structure and in chemical analysis. Molecular rotational and vibrational frequencies can be determined directly, and from these frequencies it is sometimes possible to evaluate the molecular geometry, or at least to find the molecular symmetry. Even when a precise determination of structure is not possible, much can often be said about the arrangement of atoms in a molecule from empirical information about the characteristic Raman frequencies of groups of atoms. This kind of information is closely similar to that provided by infrared spectroscopy; in fact, Raman and infrared spectra often provide complementary data about molecular structure. Raman spectra also provide information for solid-state physicists, particularly with respect to lattice dynamics but also concerning the electronic structures of solids. *See* INFRARED SPECTROSCOPY; LATTICE VIBRATIONS; MOLECULAR STRUCTURE AND SPECTRA. [R.C.L.]

Ramapithecus The genus name given by G. E. Lewis in 1934 to a fragmentary upper jaw of an extinct hominoid from the early Pliocene of the Siwalik Hills, northern India. *Ramapithecus*, which includes at least two species, is now represented by jaws and teeth from the later Miocene of eastern Africa and eastern Europe and from the early Pliocene of Asia. The fossils exhibit characteristics which suggest that they may be the earliest known hominids, direct ancestors of modern humans. *See* FOSSIL HUMAN. [E.T.]

Ramie The plant *Boehmeria nivea*, a stingless member of the nettle family; the only member of the family used commercially for fiber. This herbaceous perennial is erect, usually nonbranching, and 3 to 9 ft (1 to 2 m) tall at maturity. The fiber, which comes from the inner bark, is exceptionally strong and has uses similar to those for fiber flax. Ramie is grown mostly in the tropics and subtropics of the Far East and Brazil. *See* NATURAL FIBER. [E.G.N.]

Ramjet The simplest of the air-breathing propulsion engines (see illustration). In flight, air enters the front of the diffuser at high velocity. The diffuser is shaped to reduce the airspeed and hence the kinetic energy of the air as it passes through. With an efficient diffuser, the reduction in kinetic energy results in a nearly equal increase in potential energy, in

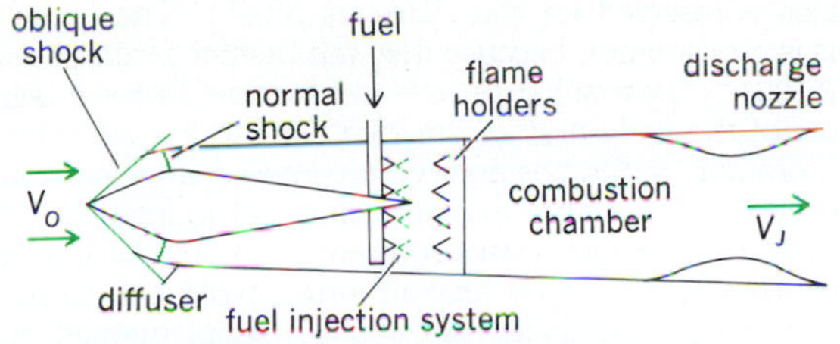

Diagram of a ramjet engine.

the form of an increase in air pressure. This higher-pressure air enters the combustion chamber, where fuel is continuously injected and burned. The hot gas is then ejected rearwardly through the discharge nozzle at velocity V_J greater than flight speed V_0. To a first approximation, thrust F is given in the equation below, where M is the mass of air per second flowing

$$F = M\,(V_J - V_0)$$

through the engine. *See* AIRCRAFT PROPULSION; JET PROPULSION.

The efficiency of a ramjet in converting the chemical energy in the fuel into kinetic energy of the airstream depends upon the ratio of the pressure in the combustion chamber to the ambient air pressure. This pressure ratio in turn depends upon the flight speed or, more exactly, upon flight Mach number. Ramjet engines are usually considered for applications in the range of flight Mach numbers between 2.5 and 8.

The scramjet (supersonic combustion ramjet) is essentially a ramjet engine intended for flight at hypersonic speeds (that is, above Mach 6), with the gases flowing through the combustion chamber and burning at supersonic speeds. By designing the engine inlet diffuser to provide supersonic speeds in the combustion chamber for flight at hypersonic speeds, the problem of extreme air temperature and dissociation in the combustion chamber, which limits the practical flight speed of ramjet engines, is avoided in the scramjet. Thus while ramjets cease to be practical at Mach numbers above 8, scramjets, on the other hand, are believed to be feasible up to Mach 25. [B.Pi.]

Rangefinder (optics) An optical instrument for measuring distance, usually from its position to a target point. Light from the target enters the optical system through two windows spaced apart, the distance between the windows being termed the base length of the rangefinder. The rangefinder operates as an angle-measuring device for solving the triangle comprising the rangefinder base length and the line from each window to the target point. Rangefinders can be classified in general as being of either the coincidence or the stereoscopic type.

In coincidence rangefinders, one-eyed viewing through a single eyepiece provides the basis for manipulation of the rangefinder adjustment to cause two images or parts of each to match or coincide. This type of device is used, in its simpler forms, in photographic cameras. The basic optical arrangement is shown in the illustration, where M_1 and M_2 are a semitransparent mirror and a reflecting mirror, respectively. When coincidence is obtained, that is, when the target T is seen in the same apparent position along either path, the rangefinder equation is satisfied.

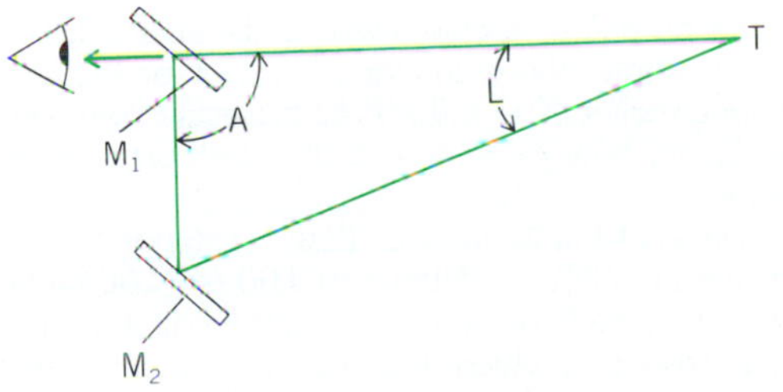

Simple coincidence rangefinder. *A* is a right angle; *L* = convergence angle at target *T*.

Stereoscopic rangefinders are entirely different, although externally they resemble coincidence rangefinders except for the fact that they possess two eyepieces. It is essentially a large stereobinocular fitted with special reticles which allow a skilled user to superimpose the stereo image formed by the pair of reticles over the images of the target seen in the eyepieces, so that the reticle marks appear to be suspended over the target and at the same apparent distance. [E.E.K.]

Rankine cycle A thermodynamic cycle used as an ideal standard for the comparative performance of heat-engine and heat-pump installations operating with a condensable vapor as the working fluid. Applied typically to a steam power plant, as shown in the illustration, the cycle has four phases: (1) heat addition *bcde* in a boiler at constant pressure p_1 changing water at b to superheated steam at e, (2) isentropic expansion *ef* in a prime mover from initial pressure p_1 to back pressure p_2, (3) heat rejection *fa* in a condenser at constant pressure p_2 with wet steam at f converted to saturated liquid at a, and (4) isentropic compression *ab* of water in a feed pump from pressure p_2 to pressure p_1.

This cycle more closely approximates the operations in a real steam power plant than does the Carnot cycle. Between given temperature limits it offers a lower ideal thermal efficiency for the conversion of heat into work than does the Carnot standard. Losses from irreversibility, in turn, make the conver-

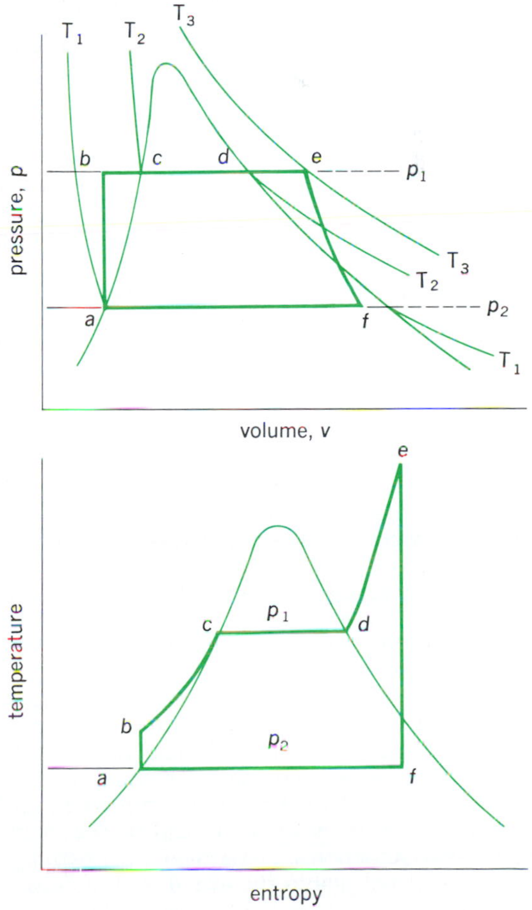

Rankine-cycle diagrams (pressure-volume and temperature-entropy) for a steam power plant using superheated steam. Pressure-volume diagrams shows curves for constant temperatures T_1, T_2, and T_3 (isothermals).

sion efficiency in an actual plant less than the Rankine cycle standard. *See* CARNOT CYCLE; REFRIGERATION CYCLE; THERMODYNAMIC CYCLE; VAPOR CYCLE. [T.Ba.]

Ranunculales An order of flowering plants, division Magnoliophyta (Angiospermae), in the subclass Magnoliidae of the class Magnoliopsida (dicotyledons). The order consists of 8 families and about 3200 species. The vast majority of the species belong to only 3 families, the Ranunculaceae (2000), Berberidaceae (650), and Menispermaceae (425). Within its subclass, the order is characterized by its mostly separate carpels, triaperturate pollen, herbaceous or only secondarily woody habit, frequently numerous stamens, generally more than two sepals, and lack of ethereal oil cells. Many members of the order contain isoquinoline alkaloids, most notably berberin. The barberry (*Berberis*), in the family Berberidaceae, and the buttercup (*Ranunculus*), columbine (*Aquilegia*), larkspur (*Delphinium*), and wind flower (*Anemone*), in the family Ranunculaceae, are familiar genera of the Ranunculales. *See* MAGNOLIIDAE; MAGNOLIOPSIDA. [A.Cr.]

Rapakivi granites Granites containing large, pink, usually ellipsoidal, alkali feldspars (orthoclase or microcline), which are rimmed by mantles of white plagioclase (albite or oligoclase). Typically these mantled feldspar phenocrysts occur in a finer-grained matrix of quartz, plagioclase, alkali feldspar, and biotite or hornblende. *See* GRANITE. [W.A.E.]

Rape Rape (*Brassica napus*) and turnip rape (*B. campestris*) plants are members of the Cruciferae family. The name is derived from the Latin *rapum*, meaning "turnip," to which these plants are closely related. The aerial portions of rape plants have been bred to produce oilseeds, fodder, and vegetable crops. Rape seed is small, round, and usually black, although varieties with yellow seed coats are also grown.

The seeds contain over 40% oil. Rapeseed contributes approximately 10–12% of the world's total edible vegetable oil supply and is one of the few edible oilseed crops that can be produced in northern Canada, Europe, and Asia, or, as a cool season crop, in subtropical areas. Rapeseed meal, the by-product of oilseed extraction, is high-quality protein feed supplement for livestock and poultry. In some Asian countries it is also used as a fertilizer and soil conditioner for specialty crops such as tobacco and citrus fruits. Rape is widely used for forage. In some countries the whole plant is cut and fed to cattle. *See* CAPPARALES; TURNIP. [R.K.D.]

Raphidophyceae A small class of poorly known biflagellate unicellular algae (raphidomonads) in the chlorophyll *a–c* phyletic line (Chromophycota). It is sometimes called Chloromonadophyceae, and its members called chloromonads, but this nomenclature is confusing because it calls to mind *Chloromonas*, a totally unrelated genus of green algae (Chlorophyceae). The alternative nomenclature used here is derived from Raphidomonas, a generic name within the class. *See* CHROMOPHYCOTA.

Two families are recognized. All photosynthetic raphidomonads are placed in the Vacuolariaceae. The Thaumatomastigaceae comprises a few colorless forms that bear pseudopodia. They are osmotrophic or phagotrophic or both.

Most genera in both families occur in fresh water (acidic to neutral), but there are brackish-water and marine forms that produce conspicuous blooms. *Heterosigma*, for example, is a frequent cause of red tide in the inland seas of Japan. *See* ALGAE. [P.C.Si.; R.L.Moe]

Rare-earth elements The group of 17 chemical elements with atomic numbers 21, 39, and 57–71; the name lanthanides is reserved for the elements 58–71. The name rare earths is a misnomer, because they are neither rare nor earths. *See* ACTINIDE ELEMENTS; LANTHANIDE CONTRACTION; PERIODIC TABLE.

Most of the early uses of the rare earths took advantage of their common properties and were centered principally in the glass, ceramic, lighting, and metallurgical industries. Today these applications use a very substantial amount of the mixed rare earths just as they are obtained from the minerals, although sometimes these mixtures are supplemented by the addition of extra cerium or have some of their lanthanum and cerium fractions removed.

The elements exhibit very complex spectra, and the mixed oxides, when heated, give off an intense white light which resembles sunlight, a property finding application in cored carbon arcs, such as those employed in the movie industry.

The rare-earth metals have a great affinity for the nonmetallic elements, as, for example, hydrogen, carbon, nitrogen, oxygen, sulfur, phosphorus, and the halides. Considerable amounts of the mixed rare earths are reduced to metals, such as misch metal, and these alloys are used in the metallurgical industry. Alloys made of cerium and the mixed rare earths are used in the manufacture of lighter flints. Rare earths are also used in the petroleum industry as catalysts. Yttrium aluminum garnets (YAG) are used in the jewelry trade as artificial diamonds.

Although the rare earths are widely distributed in nature, they generally occur in low concentrations. They are found in high concentrations as mixtures in a number of minerals. The relative abundance of the different rare earths in various rocks, geological formations, and the stars is of great interest to the geophysicist, astrophysicist, and cosmologist.

The rare-earth elements are metals possessing distinct individual properties. Many of the properties of the rare-earth metals and alloys are quite sensitive to temperature and pressure. They are also different when measured along different crystal axes of the metal; for example, electrical conductivity, elastic constants, and so on. The rare earths form organic salts with certain organic chelate compounds. These chelates, which have replaced some of the water around the ions, enhance the differences in properties among the individual rare earths. Advantage is taken of this technique in the modern ion- exchange methods of separation. *See* CHELATION; ION EXCHANGE; TRANSITION ELEMENTS. [F.H.Sp.]

Rarefied gas dynamics That branch of gas dynamics dealing with the flow of gases at such low density that the molecular mean free path is not negligibly small. Under these conditions the gas no longer behaves as a continuous fluid. Important modifications in flow phenomena occur which are ascribable to the discrete molecular structure of the gas. The subject is also called superaerodynamics. *See* GAS DYNAMICS.

It is convenient to divide rarefied gas dynamics into three flow regimes. These are called free molecule flow, transition flow, and slip flow, corresponding respectively to highly rarefied, moderately rarefied, and only slightly rarefied flow conditions. Phenomena in the three regimes are quite dissimilar, so the subdivision is useful. The term "rarefied" is relative, but can be made quantitative in terms of a dimensionless number, called the Knudsen number K, defined as the ratio of the mean free path λ divided by a characteristic dimension L of the flow field, for example, the diameter of a duct or the length of a high-altitude rocket ($K = \lambda/L$). Free molecule flow corresponds to $K \gg 1$, slip flow to $K \ll 1$, and transition to intermediate values of K.

The regime of highly rarefied flow corresponds in the aerodynamic case to flight at altitudes of 100 mi (160 km) or more. The mean free path is long compared to dimension L. It follows that molecules, which impinge on a surface and are then reemitted, travel very far before colliding with another molecule. The essential simplification of free molecule flow is that such collisions can be neglected. For the aerodynamic case of flow past a convex body the incident molecules are thus not disturbed by the presence of the body.

Slip flow is the flow regime of only slight rarefaction, corresponding in the aerodynamic case to flight at altitudes of the order 20–50 mi (32–80 km). Noncontinuum effects can be thought of in terms of small corrections to ordinary continuum flow. The term slip flow arises from the phenomenon of slip, according to which a rarefied gas adjacent to a surface does not adhere rigidly to it but rather has a finite velocity, known as the slip velocity, determined by the local stress and temperature gradients.

No simple formulation for the transition regime has yet been developed, except for the fundamental Maxwell-Boltzmann equation itself. Under some situations this equation can be solved for arbitrary values of λ/L, including those corresponding to transition flow conditions. These special solutions, together with a large number of experimental results, indicate that simple interpolation between slip flow on the one hand, and free molecule flow on the other hand, will usually suffice for the transition flow regime. [S.A.S.]

Raspberry The horticultural name for certain species of the genus *Rubus*, plant order Rosales. In these species the fruit, when ripe (unlike the blackberry), separates thimblelike from the receptacle. Raspberry plants are upright shrubs with perennial roots and prickly, biennial canes (stems). There are several species, both American and European, from which the cultivated raspberries have been developed. Varieties are grouped as to color of fruit—black, red, and purple, the last being hybrids between the red and black types. Leading states in commercial production are Michigan, Oregon, New York, Washington, Ohio, Pennsylvania, New Jersey, and Minnesota. The fruit is sold fresh for dessert purposes, is canned, and is made into jelly or jam, but quick freezing is the most important processing method. *See* FRUIT; ROSALES. [J.H.Cl.]

Rat The name for over 650 species of mammals in a number of different families of the order Rodentia. These animals are of great economic importance to humans. In addition to harboring many diseases transmissible to humans, such as bubonic plague, endemic typhus, rat-bite fever, and infectious jaundice, ingested rats can transmit trichinosis to swine. *See* MURINE TYPHUS FEVER; PLAGUE; RAT-BITE FEVER; TRICHINOSIS.

Rats are usually active at night, feeding on nearly every type of food. The teeth are modified for gnawing, with sharp and chisellike incisors that grow continually and have heavily enameled fronts. There is a large space between the one pair of incisors and the molars into which the flesh of the cheeks is drawn during gnawing. This prevents the cheek teeth from meeting and wearing down and also prevents the swallowing of dirt and debris. *See* RODENTIA. [C.B.C.]

Rat-bite fever Either of two diseases transmitted by the bite of a rat; each has a distinct etiology and symptomatology. One is due to a spiral organism, *Spirillum minus*, and the other to *Streptobacillus moniliformis*.

Following a bite and infection with *Spirillum*, the wound heals promptly, but after a period of 5–28 days there is sudden onset of symptoms, including a flare-up of the wound site which may subsequently ulcerate, regional lymphangitis and lymphadenitis, chills, relapsing type of fever, macular skin rash, malaise, and headache. Fever may continue for weeks in untreated cases.

Most cases of streptobacillary fever have followed a rat bite, although sporadic cases have been reported in which no direct contact with rats could be established. Following a bite, the local wound ordinarily heals, but occasionally an abscess develops. The onset of the disease after an incubation period of 1–5 days is abrupt, with chills, fever, vomiting, headache, severe pains in the back and joints, a maculopapular rash, and leukocytosis. Acute arthritis is one of the most prominent and persistent symptoms. [T.B.T.]

Ratchet A wheel, usually toothed, operating with a catch or a pawl so as to rotate in a single direction (see illustration). A ratchet and pawl mechanism locks a machine such as a hoisting winch so that it does not slip. The locking action may serve to produce rotation in a desired direction and to disengage in the undesired direction as in a drill brace. The catch or pawl may be of various shapes such as an eccentrically mounted disk or ball bearing. Gravity, a spring, or centrifugal force (with the catch mounted internal to the ratchet) are commonly used to hold the pawl against the ratchet. A ratchet and pawl provides an arresting action. *See* BRAKE; ESCAPEMENT; PAWL. [F.H.R.]

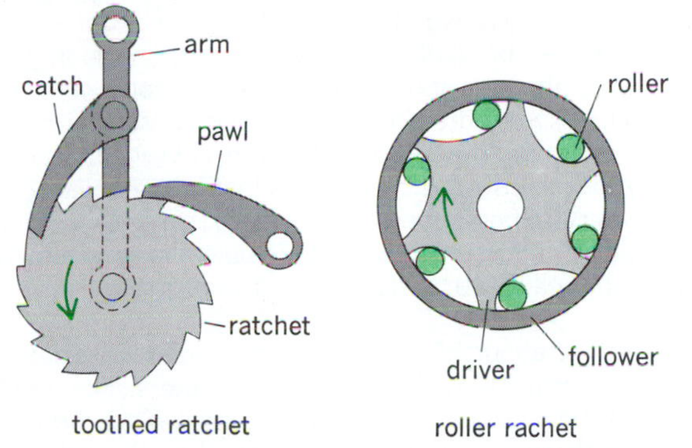

Ratchets. (*a*) Toothed ratchet is driven by catch when arm moves to left; pawl holds ratchet during return stroke of catch. (*b*) In roller ratchet, rollers become wedged between driver and follower when driver turns faster than follower in direction of arrow.

Rate-of-climb indicator A device used to indicate changes in the vertical position of an aircraft by comparing the actual outside air pressure to a reference volume that lags the outside pressure because a calibrated restrictor imposes a lag-time constant to the reference pressure volume. The faster the aircraft changes its height, the greater the pressure differential which will exist across a differential pressure capsule. As the capsule expands or contracts, a very delicate mechanism converts the motion to a readout usually graduated in feet per minute. [J.W.A.]

Ratites A group of flightless (except for tinamous), mostly large, running birds formerly segregated as a superorder of birds, the Palaeognathae, but whose interrelationships have been a long-standing controversy. The ratites represent two or three phyletic lines (the emus, cassowaries, moos and kiwis, ostriches and elephant birds, and rheas and tinamous—the last two groups may be very closely related) which evolved from a common ancestral stock, possibly much like the still volant tinamous of Central and South America. *See* AVES; NEOGNATHAE; STRUTHIONIFORMES. [W.J.B.]

Rauwolfia A genus of mostly poisonous, tropical trees and shrubs of the dogbane family (Apocynaceae). Certain species are the source of valuable emetics and cathartics. The species *Rauwolfia serpentina* has received special attention as the source of tranquilizing drugs. Among the purified alkaloids obtained from *R. serpentina*, reserpine is perhaps the one most used as a tranquilizing agent. *See* GENTIANALES; TRANQUILIZER. [P.D.St./E.L.C.]

Raw water Water obtained from natural sources such as streams, reservoirs, and wells. Natural water always contains impurities in the form of suspended or dissolved mineral or organic matter and as dissolved gases acquired from contact with earth and atmosphere. Industrial or municipal wastes may also contaminate raw water. *See* WATER POLLUTION.

If admitted to a steam-generating unit, such contaminations may corrode metals or form insulating deposits of sediments or scale on heat-transfer surfaces, with resultant overheating and possible failure of pressure parts.

Raw water can be treated to remove objectionable impurities or to convert them to forms that can be tolerated. For steam generation, suspended solids are removed by settling or filtration. Scale-forming hardness is diminished by chemical treatment to produce insoluble precipitates that are removable by filtration, or soluble compounds that do not form scale. Essentially complete purification is achieved by demineralizing treatment or evaporation. *See* WATER SOFTENING; WATER TREATMENT. [F.G.E.]

Ray Any of about 350 species of the order Batoidea, subclass Elasmobranchii, which also contains the skates, sawfishes, guitar fishes, electric rays, and sting rays. Their bodies are flattened with large pectoral fins attached to the side of the head, and ventral gill slits. Spiracles are located behind the eyes of the upper surface of the head. These openings lead to the pharynx and serve for the intake of water in respiration. The rays are characterized by a long spikelike tail. Some males have well-developed claspers, copulatory structures. The manta ray (*Manta birostris*) is a familiar example; it is noted for its ability to leap above the surface of the water to a height of 6 ft (1.8 m), using the pectoral fins for propulsion and balance. *See* BATOIDEA. [C.B.C.]

Reactance The opposition that inductance and capacitance offer to alternating current through the effect of frequency. Reactance alters the magnitude of current and also changes the circuit phase angle.

Inductive reactance X_L equals $2\pi fL$, where f is the frequency in hertz and L is the self-inductance in henrys. The voltage E across an inductance reaches its peak 90° before the current I reaches its peak, and $I = E/X_L$ amperes. Capacitive reactance X_C equals $1/(2\pi fC)$, where C is the capacitance in farads. The voltage E across a capacitance reaches its peak 90° after the current reaches its peak, and $I = E/X_C$ amperes. Reactances are components of impedance which, in general, includes resistance R and reactance. [B.L.R.]

Reaction turbine A power-generation prime mover utilizing the steady-flow principle of fluid acceleration, where nozzles are mounted on the moving element. The rotor is turned by the reaction of the issuing fluid jet and is utilized in varying degrees in steam, gas, and hydraulic turbines. All turbines contain nozzles; the distinction between the impulse and reaction principles rests in the fact that impulse turbines use only stationary nozzles, while reaction turbines must incorporate moving nozzles. *See* IMPULSE TURBINE; PRIME MOVER. [T.Ba.]

Reactive intermediates Unstable compounds which are formed as necessary intermediate stages during a chemical reaction. Thus, if a reaction in which A is converted to B requires that A first be converted to C, then C is an intermediate in the reaction (A → C → B). The term reactive further implies a certain degree of instability of the intermediate; reactive intermediates are typically isolable only under special conditions, and most of the information regarding the structure and properties of reactive intermediates comes from indirect experimental evidence.

In organic reactions the most common types of reactive intermediates are those arising from dissociative reactions, in which carbon has a decreased valence. Associative reactions can also give rise to some of the same intermediates, and to others in which carbon has an increased valence. *See* VALENCE.

Carbocations are compounds in which carbon bears a positive charge. Classical carbocations (also called carbenium ions) are trivalent, and have only six valence electrons. Nonclassical carbocations (also called carbonium ions) are tetra- or pentavalent, and have eight valence electrons. Examples are the methyl cation (classical) CH_3^+, and the methonium ion (nonclassical) CH_5^+.

Carbanions are compounds in which carbon bears a negative charge. A carbanion will always have a positive counterion in association with it; depending upon the particular cation and the stability of the carbanion, the association may be ionic, covalent, or some intermediate combination of ionic and covalent bonding, as shown below (M = metal). Carbanions are trivalent, with eight valence electrons.

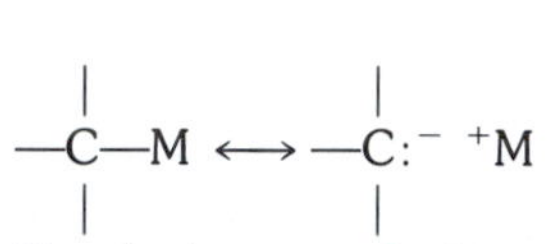

Covalent Ionic

Free radicals are neutral compounds having an odd number of electrons and therefore one unpaired electron. Carbon free radicals are trivalent, with seven valence electrons, and typically assume a planar structure. Free radicals are primarily electron-deficient species and are stabilized by structural features which donate electron density or delocalize the odd electron by resonance. *See* FREE RADICAL; RESONANCE (MOLECULAR STRUCTURE).

Radical ions are charged compounds with an unpaired electron, and are either radical cations (positively charged) or radical anions (negatively charged). In many cases a radical ion is derived from a stable neutral molecule by addition of one electron (radical anion) or removal of one electron (radical cation).

Carbenes are compounds which have a divalent carbon. The divalent carbon also has two nonbonded electrons, for a total of six valence electrons. The two nonbonded electrons may have either the same spin quantum number, which is a triplet state, or an opposite spin quantum number, which is a singlet state. Generation of carbenes is most commonly by photolysis or thermolysis of diazo compounds or ketenes, or by alpha-elimination reactions.

There are many other kinds of reactive intermediates which do not fit into the previous classifications. Some are simply compounds which are unstable for a variety of possible reasons, such as structural strain or an unusual oxidation state. *See* CHEMICAL DYNAMICS; ELECTROPHILIC AND NUCLEOPHILIC REAGENTS; MOLECULAR ORBITAL THEORY. [C.C.W.]

Reactor (electricity) A device for introducing an inductive reactance into a circuit. Inductive reactance x is a function of the product of frequency f and inductance L; thus, $x = 2\pi fL$. For this reason, a reactor is also called an inductor. Since a voltage drop across a reactor increases with frequency of applied currents, a reactor is sometimes called a choke. All three terms describe a coil of insulated wire. *See* CHOKE (ELECTRICITY); INDUCTOR.

According to their construction, reactors can be divided into those that employ iron cores and those where no magnetic material is used within the windings. The first type consists of a coil encircling a circuit of iron which usually contains an air gap or a series of air gaps. The air gaps are used to attenuate the effects of saturation of the iron core. The second type, called an air-core reactor, is a simple circular coil, wound around a cylinder constructed of nonmagnetic material for greater mechanical strength. This strength is necessary for the coil to withstand the electromagnetic forces acting on each conductor. These forces become very large with heavy current flow, and their direction tends to compress the coil into less space: radial forces tend to elongate internal conductors in the coil and to compress the external ones while the axial forces press the end sections toward the center of the coil.

Both iron-core and air-core reactors may be of the air-cooled dry type or immersed in oil or a similar cooling fluid. Both types of reactors are normally wound with stranded wire in order to reduce losses due to eddy currents and skin effect. In addition, it is important to avoid formation of short-circuited metal loops when building supporting structures for air-core reactors since these reactors usually produce large magnetic fields external to the coil. If these fields penetrate through closed-loop metal structures, induced currents will flow, causing both losses and heating of the structures. Which of these two reactor types should be used depends on the particular application. *See* EDDY CURRENT; SKIN EFFECT (ELECTRICITY). [V.R.S.]

Reactor physics The science of the interaction of the elementary particles and radiations characteristic of nuclear reactors with matter in bulk. These particles and radiations include neutrons, beta rays, and gamma rays of energies between zero and about 10^7 electronvolts. *See* BETA PARTICLES; GAMMA RAYS.

The study of the interaction beta and gamma radiations with matter is, within the field of reactor physics, undertaken primarily to understand the absorption and penetration of energy through reactor structures and shields. *See* RADIATION SHIELDING.

With this exception, reactor physics is the study of those processes pertinent to the chain reaction involving neutron-induced nuclear fission with consequent neutron generation. Reactor physics is differentiated from nuclear physics, which is concerned primarily with nuclear structure. Reactor physics makes direct use of the phenomenology of nuclear reactions. Neutron physics is concerned primarily with interactions between neutrons and individual nuclei or with the use of neutron beams as analytical devices, whereas reactor physics considers neutrons primarily as fission-producing agents. In the hierarchy of professional classification, neutron physics and reactor physics are both ranked as subfields of the more generalized area of nuclear physics. *See* CHAIN REACTION (PHYSICS); NEUTRON; NUCLEAR FISSION; NUCLEAR PHYSICS.

Concepts. Reactor physics borrows most of its basic concepts from other fields. From nuclear physics comes the concept of the nuclear cross section for neutron interaction, defined as the effective target area of a nucleus for interaction with a neutron beam. The total interaction is the sum of interactions by a number of potential processes, and the probability of each of them multiplied by the total cross section is designated as a partial cross section. An outgrowth of this is the definition of macroscopic cross section, which is the product of cross section (termed microscopic, for specificity) with atomic density of the nuclear species involved.

Cross sections vary with energy according to the laws of nuclear structure. In reactor physics this variation is accepted as input data to be assimilated into a description of neutron behavior. Common aspects of cross section dependence, such as variation of absorption cross section inversely as the square root of neutron energy, or the approximate regularity of resonance structure, form the basis of most simplified descriptions of reactor processes in terms of mathematical or logical models.

The concept of neutron flux is related to that of macroscopic cross section. This may be defined as the product of neutron density and neutron speed, or as the rate at which neutrons will traverse the outer surface of a sphere embedded in the medium, per unit of spherical cross-sectional area. The product of flux and macroscopic cross section yields the reaction rate per unit volume and time.

Criticality. The critical condition is what occurs when the arrangement of materials in a reactor allows, on the average, exactly one neutron of those liberated in one nuclear fission to cause one additional nuclear fission. If a reactor is critical, it will have fissions occurring in it at a steady rate. This desirable condition is achieved by balancing the probability of occurrence of three competing events: fission, neutron capture which does not cause fission, and leakage of neutrons from the system. If υ is the average number of neutrons liberated per fission, then criticality is the condition under which the probability of a neutron causing fission is $1/\upsilon$. Generally, the degree of approach to criticality is evaluated by computing k_{eff}, the ratio of fissions in successive links of the chain, as a product of probabilities of successive processes.

Reactivity. Reactivity is a measure of the deviation of a reactor from the critical state at any frozen instant of time. The term reactivity is qualitative, because several sets of units are in current use to describe it.

Reflectors. Reflectors are bodies of material placed beyond the chain-reacting zone of a reactor, whose function is to return to the active zone (or core) neutrons which might otherwise leak. Reflector worth can be crudely measured in terms of the albedo, or probability that a neutron passing from core to reflector will return again to the core.

Good reflectors are materials with high scattering cross sections and low absorption cross sections. The first requirement ensures that neutrons will not easily diffuse through the reflector, and the second, that they will not easily be captured in diffusing back to the core.

Beryllium is the outstanding reflector material in terms of neutronic performance. Water, graphite, D_2O, iron, lead, and ^{283}U are also good reflectors.

Reactor dynamics. Reactor dynamics is concerned with the temporal sequence of events when neutron flux, power, or reactivity varies. The inclusive term takes into account sequential events, not necessarily concerned with nuclear processes, which may affect these parameters. There are basically three ways in which a reactor may be affected so as to change reactivity. A control element, absorbing rod, or piece of fuel may be externally actuated to start up, shut down, or change reactivity or power level; depletion of fuel and poison, buildup of neutron-absorbing fission fragments, and production of new fissionable material from the fertile isotopes ^{232}Th, ^{234}U, ^{238}U, and ^{240}Pu make reactivity depend upon the irradiation history of the system; and changes in power level may produce temperature changes in the system, leading to thermal expansion, changes in neutron cross sections, and mechanical changes with consequent change of reactivity.

Reactor control physics. Reactor control physics is the study of the effect of control devices on reactivity and power level. As such, it includes a number of problems in reactor statics, because the primary question is to determine the absorption of the control elements in competition with the other neutronic processes. It is, however, a problem in dynamics, given the above information, to determine what motions of the control devices will lead to stable changes in reactor output.

Reactivity changes. Long-term reactivity changes may represent a limiting factor in the burning of nuclear fuel without costly reprocessing and refabrication. As the chain reaction proceeds, the original fissionable material is depleted, and the system would become subcritical if some form of slow addition of reactivity were not available. This is the function of shim rods in a typical reactor. The reactor is originally loaded with enough fuel to be critical with the rods completely inserted. As the fuel burns out, the rods are withdrawn to compensate. *See* NUCLEAR FUELS.

Reactor kinetics. This is the study of the short-term aspects of reactor dynamics with respect to stability, safety against power excursion, and design of the control system.

Control is possible because increases in reactor power often reduce reactivity to zero (the critical value) and also because there is a time lapse between successive fissions in a chain resulting from the finite velocity of the neutrons and the number of scattering and moderating events intervening, and because a fraction of the neutrons is delayed. *See* DELAYED NEUTRON; NUCLEAR REACTOR. [B.I.S.]

Reagent chemicals High-purity chemicals used for analytical reactions, for the testing of new reactions where the effects of impurities are unknown, and in general for chemical work where impurities must either be absent or at known concentrations.

Chemicals are purified by a variety of methods, the most common being recrystallization from solution. If the desired chemical is volatile and the impurities are not volatile, sublimation is an effective method of purification. For liquid chemicals, distillation is an effective procedure. Finally, the simplest procedure may be to synthesize the desired reagent from pure materials. *See* CHEMICAL SEPARATION TECHNIQUES; CRYSTALLIZATION; DISTILLATION.

Commercial chemicals are available at several levels of purity. Chemicals labeled "technical" or "commercial" are usually quite impure. The grade "USP" indicates only that the chemical meets the requirements of the United States Pharmacopeia. The term "CP" means only that the chemical is purer than "technical." Chemicals designated "reagent grade" or "analyzed reagent" are specially purified materials which usually have been analyzed to establish the levels of impurities. The American Chemical Society has established specifications and tests for purity for some chemicals. Materials which meet these specifications are labeled "Meets ACS Specifications." [K.G.S./C.Ru.]

Real-time systems Computer systems in which the computer is required to perform its tasks within the time restraints of some process or simultaneously with the system it is assisting. Usually the computer must operate faster than the system assisted in order to be ready to intervene appropriately.

Real-time computer systems and applications span a number of different types.

In real-time control and real-time process control the computer is required to process systems data (inputs) from sensors for the purpose of monitoring and computing system control parameters (outputs) required for the correct operation of a system or process. The type of monitoring and control functions provided by the computer for subsystem units ranges over a wide variety of tasks, such as turn-on and turn-off signals to switches; feedback signals to controllers (such as motors, servos, and potentiometers) to provide adjustments or corrections; steering signals; alarms; monitoring, evaluation, supervision, and management calculations; error detection, and out-of-tolerance and critical parameter detection operations; and processing of displays and outputs.

In real-time assistance the computer is required to do its work fast enough to keep up with a person interacting with it (usually at a computer terminal device of some sort, for example, a screen and keyboard). The computer supports the person or persons interacting with it and provides access, retrieval, and storage functions, usually through some sort of database management system, as well as data processing and computational power. System access allows the individual to intervene in the system's operation. The real-time computer also often provides monitoring or display information, or both. *See* DATABASE MANAGEMENT SYSTEMS; MULTIACCESS COMPUTER.

In real-time robotics the computer is a part of a robotic or self-contained machine. Often the computer is embedded in the machine, which then becomes a smart machine. If the smart machine also has access to, or has embedded within it, artificial intelligence functions (for example, a knowledge base and knowledge processing in an expert system fashion), it becomes an intelligent machine. *See* ARTIFICIAL INTELLIGENCE; COMPUTER; DIGITAL COMPUTER; EMBEDDED SYSTEMS; EXPERT SYSTEMS; ROBOTICS. [E.C.J.]

Real variable A variable whose range is a subset of the real numbers. By extension the term is also used to refer to the theory of functions of one or more real variables. This theory has to do with properties of broad classes of functions, such as continuity, types of discontinuities, differentiability of functions, oscillation and variation of functions, and the various kinds of integrals. *See* CALCULUS; INTEGRATION.

Real numbers are those commonly used in the geometric theory of measurement. The integers and fractions, also called rational numbers, are included among the real numbers. In practice an irrational number x is specified by telling which rational numbers are less than x and which are greater than x. Such a division of the rational numbers into two classes was used by J. W. R. Dedekind as the formal definition of a real number and is called a Dedekind cut. [L.M.G.]

Realgar A mineral having composition AsS and crystallizing in the monoclinic system. Realgar can occur in short, vertically striated crystals, but more frequently is granular and in crusts. The hardness is 1.5–2 (Mohs scale) and the specific gravity is 3.48. The luster is resinous and the color red to orange. Realgar is found in ores of lead, silver, and gold associated with orpiment and stibnite. It occurs with the silver and lead ores in Hungary, Czechoslovakia, and Germany. Good crystals have come from Binnenthal, Switzerland, and Allchar, Macedonia. In the United States it is found at Manhattan, Nevada; Mercer, Utah; and as deposits from geyser waters in Yellowstone National Park. *See* ARSENIC; ORPIMENT; STIBNITE. [C.S.Hu.]

Reamer A multiple-cutting-edge tool designed to enlarge or accurately size and finish an existing hole in solid material by removal of a small amount of stock. The cutting edges may be ground on the apexes between longitudinal flutes or grooves, or cutting may take place on chamfered edges at the end of the reamer. Reaming is performed either manually or by machine. [A.H.T.]

Rearrangement reaction A reaction in which an atom or bond moves or migrates, having been initially located at one site in a reactant molecule and ultimately located at a different site in a product molecule. A rearrangement reaction may involve several steps, but the key feature is that a bond shifts from one site of attachment to another. The simplest examples of rearrangement reactions are intramolecular, that is, reactions in which the product is simply a structural isomer of the reactant. *See* MOLECULAR ISOMERISM.

More complex rearrangement reactions occur when the rearrangement is accompanied by another reaction, for example, a substitution reaction. *See* CHEMICAL BONDING; ENANTIOMER; STEREOCHEMISTRY.

Rearrangement reactions are classified on the basis of the group that migrates and the initial and final location of the migrating bond. The initial bond location is designated as position 1, and the final location as position i, where the number of atoms is simply counted along the connection from 1 to i. Such a migration is called a $[1,i]$ rearrangement or $[1,i]$ shift. If the migrating group also reattaches itself at a different site from the one to which it had originally been attached, both shifts are indicated, as in $[i,i]$ shift.

In addition, classification can be based on how many electrons move with the migrating group. Of the two electrons in the initial bond that breaks, the migrating group may bring with it both electrons (nucleophilic or anionotropic), one electron (radical), or no electrons (electrophilic or cationotropic). If the rearrangement is a concerted reaction in which there is a cyclic delocalized transition state that results in shifts of pi bonds as well as sigma bonds, the reaction is called a sigmatropic rearrangement. *See* DELOCALIZATION; PERICYCLIC REACTION. [C.C.W.]

Receiving gage A fixed gage designed to inspect a number of dimensions and also their relation to each other. It is termed a receiving gage because of its similarity to the cavity which receives the part in actual service. The gage checks on "go" limits. A typical example is a chamber gage which resembles the chamber of a rifle but which is made to the largest limits of size of a cartridge. Thus any cartridge which fits the receiving gage will fit any rifle chamber of that caliber. [R.A.Bo.]

Reciprocating aircraft engine A fuel-burning internal combustion piston engine specially designed and built for minimum fuel consumption and light weight in proportion to developed shaft power. The rotating output shaft of the engine may be connected to a propeller, ducted fan, or helicopter rotor.

Reciprocating aircraft engines are used in about 86% of all powered aircraft flying in the United States. Most of the aircraft powered by these engines belong to the general aviation segment of the domestic aviation fleet. The reciprocating aircraft engine is used to power single-engine and multiengine airplanes, helicopters, and airships. It is the principal engine used in aircraft for air taxi, pilot training, business, personal, and sport flying as well as aerial application of seed, fertilizer, herbicides, and pesticides for farming. *See* AGRICULTURAL AIRCRAFT; GENERAL AVIATION.

Predominantly, reciprocating aircraft engines operate on a four-stroke cycle, where each piston travels from one end of its stroke to the other four times in two crankshaft revolutions to complete one cycle. The cycle is composed of four distinguishable events called intake, compression, expansion (or power), and exhaust, with ignition taking place late in the compression stroke and combustion of the fuel-air charge occurring early in the expansion stroke. These spark-ignition engines burn specially formulated aviation gasolines. *See* INTERNAL COMBUSTION ENGINE.

Most modern aircraft using engines with up to 336 kW (450 hp) output are powered by air-cooled, horizontally opposed, reciprocating engines. The trend in modern reciprocating engine development is toward lower engine weight and improved fuel economy rather than increased power. [K.J.S.]

Reciprocity principle In the scientific sense, a theory that expresses various reciprocal relations for the behavior of some physical systems. Reciprocity applies to a physical system whose input and output can be interchanged without altering the response of the system to a given excitation. Optical, acoustical, electrical, and mechanical devices that operate equally well in either direction are reciprocal systems, whereas unidirectional devices violate reciprocity. The theory of reciprocity facilitates the evaluation of the performance of a physical system. If a system must operate equally well in two directions, there is no need to consider any nonreciprocal components when designing it.

Some systems that obey the reciprocity principle are any electrical network composed of resistances, inductances, capacitances, and ideal transformers; systems of antennas, which obey certain restrictions; mechanical gear systems; and light sources, lenses, and reflectors.

Devices that violate the theory of reciprocity are transistors, vacuum tubes, gyrators, and gyroscopic couplers. Any system that contains the above devices as components must also violate the reciprocity theory. The gyrator differs from the transistor and vacuum tube in that it is linear and passive, as opposed to the active and nonlinear character of the other two devices. *See* GYRATOR; TRANSISTOR. [H.S.La.]

Recombination (genetics) The formation of new combinations of genes by the replacement of a portion of the genetic material from one cell lineage with its counterpart derived from another lineage. Genetic recombination is a normal consequence of sexual reproduction, and the increased adaptability conferred on organisms by recombination is believed to be an evolutionary advantage that is responsible for the widespread occurrence of sex in diverse forms of life. Genetic recombination is also known to occur by processes such as transduction in microorganisms that more nearly resemble infection than sexual union. Recombination has provided geneticists with their most powerful experimental tool for resolving the hereditary material into its genic elements and for investigating chromosome behavior and structure. *See* CHROMOSOME.

Two cells from separate lineages are brought together in sexual organisms by fertilization, followed by fusion of their haploid nuclei, each containing a single set of chromosomes, to form a single diploid nucleus which contains two sets. Reassortment of genes and chromosomes of the diploid into new combinations occurs when single sets of chromosomes are segregated into haploid eggs, sperm, or spores during nuclear divisions during meiosis. *See* MEIOSIS; SYNGAMY.

An inherited difference referable to corresponding places (loci) in homologous chromosomes is said to be due to homologous (allelic) genes. When an allelic difference is present the corresponding locus is said to be genetically marked. Genes at nonhomologous loci are termed nonallelic. Genes at different loci can recombine with one another.

Differences at different gene loci within a pair of homologous chromosomes do not ordinarily behave independently of one another during meiosis, but remain in their original combinations unless a mutual exchange of equivalent segments occurs between paired homologs through a process known as crossing-over. Genes are said to be linked, and therefore located in the same chromosome pair, when parental gene combinations occur more frequently than recombinations among the products of meiosis. *See* LINKAGE (GENETICS). [D.D.P.]

Recording Any process for preserving signals, sounds, data, or other information for future reference or reproduction. Recording systems in common use include magnetic tape and wire recording, disk recording as on phonograph records, photographic recording of waveforms, and facsimile recording of photographic and other material. The term recording is also used to denote the end product of a recording process, such as a magnetic tape, disk recording, or paper chart from a graphic-level recorder. *See* DATA-PROCESSING SYSTEMS; DISK RECORDING; FACSIMILE; GRAPHIC RECORDING INSTRUMENTS; MAGNETIC RECORDING; OPTICAL RECORDING; SOUND RECORDING. [H.F.O.]

Rectangle A parallelogram whose adjacent sides are perpendicular. A rhombus is a plane quadrilateral having four equal sides. A square is a rectangle with four equal sides and is therefore also a rhombus. The area of either the rectangle or the rhombus is equal to the product of the base times the altitude ($A = bh$). *See* QUADRILATERAL; SQUARE. [J.S.F.]

Rectifier A nonlinear circuit component that allows more current to flow in one direction than in the other. An ideal rectifier is one that allows current to flow in one (forward) direction unimpeded but allows no current to flow in the other (reverse) direction. Thus, ideal rectification might be thought of as a switching action, with the switch closed for current in one direction and open for current in the other direction. Rectifiers are used primarily for the conversion of alternating current (ac) to direct current (dc). *See* ELECTRONIC POWER SUPPLY.

A variety of rectifier elements are in use. The vacuum-tube rectifier can efficiently provide moderate power. Its resistance to current flow in the reverse direction is essentially infinite because the tube does not conduct when its plate is negative with respect to its cathode. In the forward direction, its resistance is small and almost constant. Gas tubes, used primarily for higher power requirements, also have a high resistance in the reverse direction. The semiconductor rectifier has the advantage of not requiring a filament or heater supply. This type of rectifier has approximately constant forward and reverse resistances, with the forward resistance being much smaller. Mechanical rectifiers can also be used. The most common is the vibrator, but other devices are also used. *See* GAS TUBE; MECHANICAL RECTIFIER; SEMICONDUCTOR RECTIFIER; VIBRATOR.

If the average current is subtracted from the current flowing in the rectifier, an alternating current results. This ripple current flowing through a load produces a ripple voltage which is often undesirable. Filter and regulator circuits are used to reduce it to as low a value as is required. *See* ELECTRIC FILTER; VOLTAGE REGULATOR.

A half-wave rectifier circuit is shown in Fig. 1. The rectifier, a diode, is practically ideal. The ac input is applied to the primary of the transformer; secondary voltage *e* supplies the rectifier and load resistor R_L. The rectifying action of the diode is shown in Fig. 2, in which the current *i* of the rectifier is plotted against the voltage e_d across the diode. The applied sinusoidal voltage from the transformer secondary is shown under the voltage axis; the resulting current *i* flowing through the diode is shown at the right to be half-sine loops.

A full-wave rectifier circuit uses two separate diodes. The resulting current wave shape is shown in Fig. 3. A more continuous flow of direct current is produced because the first diode conducts for the positive half-cycle and the second diode conducts for the negative half-cycle.

When high dc power is required by an electronic circuit, a polyphase rectifier circuit may be used. It is also desirable when

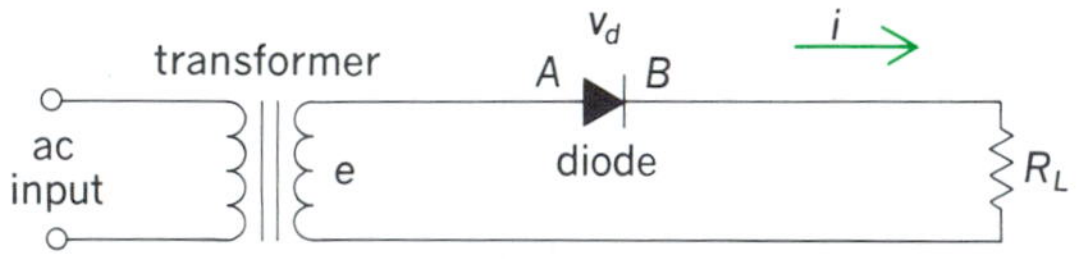

Fig. 1. Half-wave diode rectifier. V_d = voltage across diode. Ideal diode allows current *i* to flow only in forward direction from *A* to *B*.

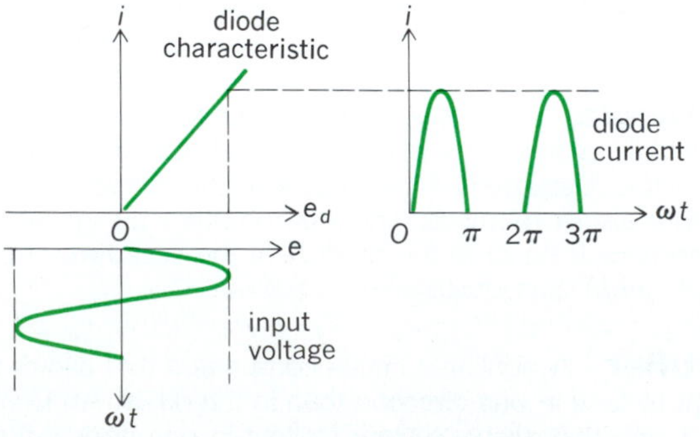

Fig. 2. Rectifying action of half-wave diode rectifier. *t* = time; ω = angular frequency of input voltage.

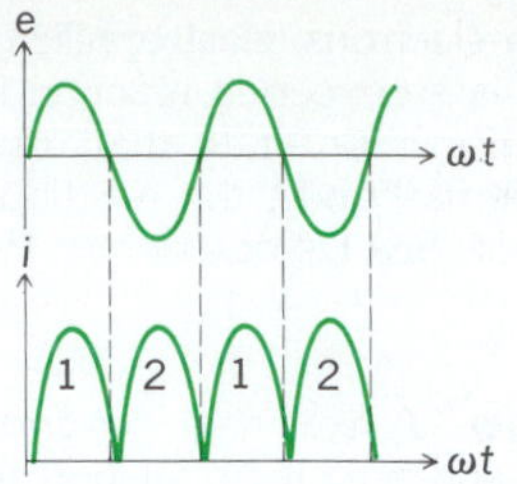

Fig. 3. Applied voltage and output current of full-wave rectifier.

expensive filters must be used. This is particularly true of power supplies for the final radio-frequency and audiofrequency stages of large radio and television transmitters. [D.L.W.]

Rectilinear motion Motion is defined as continuous change of position of a body. If the body moves so that every particle of the body follows a straight-line path, then the motion of the body is said to be rectilinear. *See* MOTION.

When a body moves from one position to another, the effect may be described in terms of motion of the center of mass of the body from a point *A* to a point *B* (see illustration). If the

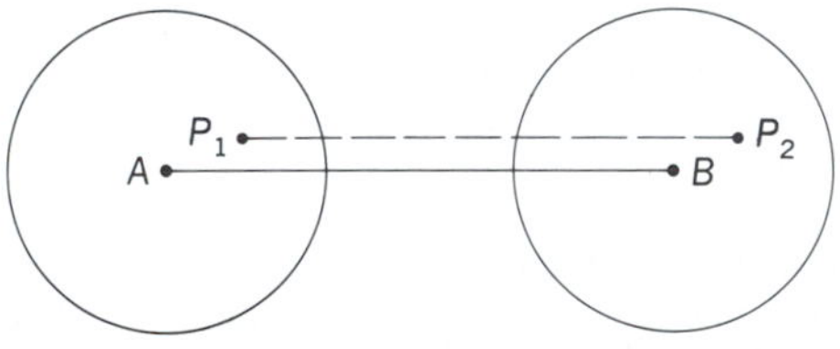

Rectilinear motion. All points move parallel to the center of mass.

center of mass of the body moves along a straight line connecting the points *A* and *B*, then the motion of the center of mass of the body is rectilinear. If the body as a whole does not rotate while it is moving, then the path of every particle of which the body is composed is a straight line parallel to or coinciding with the path of the center of mass, and the body as a whole executes rectilinear motion. This is shown by the straight line connecting points P_1 and P_2 in the illustration. *See* CENTER OF MASS.

Rectilinear motion is an idealized form of motion which rarely, if ever, occurs in actual experience, but it is the simplest imaginable type of motion and thus forms the basis for the analysis of more complicated motions. However, many actual motions are approximately rectilinear and may be treated as such without appreciable error. For example, a ball thrown directly upward may follow, for all practical purposes, a straight-line path. The motion of a high-speed rifle bullet fired horizontally may be essentially rectilinear for a short length of path, even though in its larger aspects the ideal path is a parabola. The motion of an automobile traveling over a straight section of roadway is essentially rectilinear if minor variations of path are neglected. The motion of a single wheel of the car is not rectilinear, although the motion of the center of mass of the wheel may be essentially so. *See* BALLISTICS. [R.D.Ru.]

Recursive function A function that maps natural numbers to natural numbers and is special in that it must be calculable by using a precisely specified algorithm. The mathematical definitions of partial recursive functions and recursive functions were developed to give a precise mathematical characterization of those functions or operations on the natural numbers which are computable by using effective procedures.

An effective procedure is a procedure or process determined by a finite list of precise instructions. There is no upper bound on the number of instructions in this list. These instructions can be carried out in a discrete one-step-at-a-time fashion. The only equipment needed is pencil and paper, and an unlimited supply is assumed to be available. No creativity is involved in applying these instructions, and no random devices such as flipping a coin are allowed in carrying them out.

An effective procedure may be used to calculate a function as follows: The calculator inputs to the effective procedure a natural number (0, 1, 2, 3, . . .) and allows the procedure to be carried out on this input. If an ouput is obtained, the function is said to be defined at this input and its value at this input is the output so obtained. Any function which can be computed in this way is called effectively computable. For example, addition and multiplication of natural numbers are effectively computable functions.

The notion of effective procedure is intuitive and vague. In the setting of an idealized computing device called a Turing machine, a more precise definition of certain effectively computable functions can be given. *See* AUTOMATA THEORY; GÖDEL'S THEOREM; LOGIC. [G.C.N.]

Recycling technology Methods for reducing solid waste by reusing discarded materials to make new products. The three integral phases of recycling are the collection of recyclable materials, manufacture or reprocessing of these materials into new products, and purchase of these products. Various techniques have been developed to recycle plastics, glass, metals, paper, and wood.

Plastics. Plastic discards represent an estimated 10% by weight and up to 26% by volume of the municipal solid waste in the United States discarded after materials recovery. About 2% by weight of discarded plastics is recovered. Approximately half of plastic waste consists of single-use convenience packaging and containers. Many manufacturers prefer plastic for packaging because it is lightweight, resists breakage and environmental deterioration, and can be processed to suit specific needs. Once plastics are discarded, these attractive physical properties become detriments.

The collected plastic waste is usually separated manually from the waste stream, and often it is cleaned to remove adhesives or other contaminants. It is sorted further, based on different resins. Mechanical separation techniques can be used to sort plastics based on unique physical or chemical properties.

A significant problem is the presence of contaminants such as dirt, glass, metals, chemicals from previous usage, toxicants from metallic-based pigments, and other materials that are part of or have adhered to the plastic products. Other constraints involve inconsistencies in the amount of different plastic resins in commingled plastic wastes used for recycling, and engineering aspects of recycled plastic products, such as lessened chemical and impact resistance, strength, and stiffness, and the need for additional chemicals to counteract other types of degradation for reprocessing. There may be limitations to the number of times that a particular plastic product can be effectively recycled as compared to steel, glass, or aluminum, which can be recycled many times with no loss of their properties and virtually no contamination. [R.L.Sw.; V.T.B.; M.L.Bo.]

Glass. Glass containers are a usual ingredient in community recycling programs; they are 100% recyclable and can be recycled indefinitely. In 1993 in the United States, glass containers, which constitute 6% of the solid-waste stream by weight, were recycled at a rate of 35%. Nearly one-third of the glass containers available for consumption in the United States were cycled back into glass containers and other useful items such as glasphalt, or were returned as refillable bottles.

The process of recycling glass is straightforward. Cullet (scrap glass) in the form of used glass bottles and jars is mixed with silica sand, soda ash, and limestone in a melting furnace at temperatures up to 2800°F (1540°C). The molten glass is poured into a forming machine, where it is blown or pressed into shape. The new containers are gradually cooled, inspected, and shipped to the customer. Before glass can be recycled, however, it must be furnace ready, that is, sorted by color and free of contaminants.

Cullet must meet a standard of quality similar to that of the raw material it replaces. Contamination from foreign material will result in the cullet being rejected by the plants, as it poses a serious threat to the integrity and purity of the glass packaging being produced. Contaminants include metal caps, lids, stones, dirt, and ceramics. Paper labels do not need to be removed for recycling, as they burn off at high furnace temperatures. [N.T.; N.U.R.]

Metals. Metals must be recycled to alleviate the need to mine more ore, to reduce energy consumption, to limit the dissemination of metals into the environment, and to reduce the cost of metals. In the United States a substantial portion of these needs are met by recycling metals.

The extensive recycling is important for three reasons. (1) The energy required to recycle a metal is considerably less in comparison to producing it from ore. (2) Extracting the metal from ore produces a tremendous amount of waste material. (3) Metals that are not recycled become dissipated throughout the environment; since many metals are toxic, this can result in the pollution of water and soil. *See* HAZARDOUS WASTE.

While it is beneficial to recycle, two important problems hinder recycling: collection and impurity buildup. When a metal becomes scrap and is a candidate for recycling, it must be collected at a cost that makes it attractive to recyclers. It is useful to divide scrap into three categories—home scrap, new or prompt scrap, and old or obsolete scrap—whose methods for collection differ significantly. Home scrap is waste produced during fabrication, and includes casting waste (for example, risers), shearings and trimmings, and rejected material. This scrap is usually recycled within the plant, and therefore it is not recognized as recycled material in recycling statistics. New or prompt scrap is waste generated by the user of semifinished material, that is, scrap from machining operations (such as turnings or borings), trimmings, and rejected material. This material is collected and sold to recyclers and, if properly labeled and segregated, it is easy to recycle and is valuable. Old or obsolete scrap is waste derived from products that have completed their life cycle, such as used beverage cans, old automobiles, and defunct batteries. The collection and the impurity buildup problems are most severe when considering old or obsolete scrap. [D.F.S.]

Paper. Paper and paperboard for recycling come from a variety of sources, including offices, retail businesses, coverters, printers, and households. Paper products that have been distributed, have been purchased, and have served their intended purposes are considered postconsumer waste. Other sources, such as scrap paper generated in the papermaking process (mill broke) or converting operations (such as trimmings from envelopes and boxes), are considered preconsumer waste.

Recycled paper fibers are used in the manufacture of many recycled-content paper products such as paperboard, corrugated containers, tissue products, newspapers, and printing and writing paper. They can also be used in other products such as insulation, packing materials, and molded egg cartons and flowerpots. *See* PRINTING.

Collection is the crucial first step in recycling. It occurs in curbside programs, in drop-off centers, in paper drives, and increasingly in commercial collection systems run side by side with waste collection for landfill or incineration.

Reprocessing begins by sorting waste papers by grade and level of cleanliness. Next, the waste paper (usually in bales) is mixed with water in a slusher or pulper to produce a fiber-and-water slurry. In this pulping stage the paper is agitated until broken down into fibers, and large-size contaminants (greater than about 5 mm or 0.2 in.) are removed when the pulper is emptied through the screen plate. Depending on the intended product, chemicals such as surfactants are added to the pulper to help remove undesirable materials from the fibers for separation in later operations.

The pulp is then pumped through several different-size slotted or perforated screens to separate medium-size contaminants (usually 5–0.2 mm or 0.2–0.05 in.) from the pulp. Screening is generally followed by centrifugal cleaning, where the pulp is subjected to a vortex in a tapered cone. Using specific designs, cleaners separate high-specific-gravity materials, such as dirt and sand, and low-specific-gravity materials, such as styrofoam and some plastics, from the pulp. *See* PAPER. [J.S.S.]

Wood. Waste is generated at every stage of the process by which a forest tree is turned into consumer and industrial products. Additional waste is generated in the disposal of those products. Wood waste is also produced by the homeowner and by large and small businesses and is generated from landscaping and agricultural operations such as pruning and tree removal. While these processes do not strictly return wood to the economy in its original form, they have the effect of diverting wood residues from the landfill, and thus they may be included under a broadly interpreted definition of recycling.

Wood recycling begins with wood separation from the waste stream. Recovered materials can be processed into various products, including fuel, raw material for particleboard or other wood-composite panel products, compost, landscaping mulch, animal bedding, landfill cover, amendments for municipal solid waste and sludge compost, artificial firewood, wood-plastic composite lumber and other composite products, charcoal, industrial oil absorbents, insulation, and specialty concrete.

Most of these products require that the wood be ground into small particles. A typical grinder is a hammermill, although a variety of grinders are used. The size of the particles is determined by the end use of the wood; sizes smaller than about 20 mesh are called wood flour (the particles passing the 20-mesh screen are usually less than 0.8 mm in size). Wood may then be passed over an electromagnet which removes items made of ferrous metals such as nails or staples. If additional processing is performed, it is typically to separate wood particles by size. This is accomplished in two ways: the particles can be passed through a series of screens of different mesh size and the various-sized particles can be collected from the screens; or the particles may be separated in a tower with air blown in the bottom and out the top; the particles distribute themselves in the tower, with the small, light particles on top and the large, dense ones at the bottom. *See* WOOD PRODUCTS. [J. Simo.]

Red dwarf star A red star of low luminosity. Dwarf stars are commonly those main-sequence stars fainter than an absolute magnitude of about +1. Red dwarfs are the faintest and coldest of the dwarfs. They are present among both old and young stars. Red dwarfs are the most numerous class of stars in space, although they are so faint that their presence in remote parts of the Galaxy must be inferred from their frequency near the Sun and their contribution to the total mass of the Galaxy. They have a low-energy output per unit mass and a long nuclear lifetime. *See* DWARF STAR; STAR. [J.L.Gr.]

Red Sea The Red Sea lies between Arabia and northeastern Africa. It is about 1200 mi (2000 km) long and 180 mi (300 km) wide, and has a maximum depth of about 7600 ft (2300 m). Details of its bathymetry remain incomplete.

Geophysical and geological findings have indicated that the Red Sea basin is a complex system of rifts formed during several periods of crustal movement and accompanying vulcanism. These movements have resulted in the separation of the ancient crystalline rock masses of Africa and Arabia. The Red Sea basin is a segment of the African rift system.

The water that enters the Red Sea basin is surface water from the Gulf of Aden, which flows through the straits over a sill. No permanent rivers discharge into the Red Sea, and the Suez Canal affords no appreciable interchange with the Mediterranean.

Because of a peculiar thermohaline situation set up by the cooling and evaporation of northward-flowing surface water, a shallow density-driven outflow is produced that carries saline water back to the straits, where it spills over the sill and into the Gulf of Aden.

The well-mixed water that fills the Red Sea basin from 660 ft (200 m) to over 6600 ft (2000 m) exhibits remarkably uniform properties. Its salinity is nearly constant at 40.6 parts per thousand and its temperature remains near 22°C (72°F). [A.C.N.; D.A.Ro.]

Redbeds Detrital sedimentary rocks pigmented by red ferric oxide that coats grains, fills pores as cement, or is dispersed in a muddy matrix. These conspicuously colored rocks commonly constitute thick sequences of nonmarine to shallow marine deposits. Detrital redbeds accumulated in many parts of the globe during the past 1,000,000,000 years of Earth history. Development of red pigment in most redbeds was complex and is difficult to decipher. Ferric oxides also pigment marine chert, limestone, and iron formations, but these chemical deposits are not usually included among redbeds.

Some redbeds contain abundant grains of sedimentary and low-grade metamorphic rocks and relatively few grains of iron-bearing minerals. Most of them, however, contain feldspar and relatively abundant grains of opaque black oxides derived from igneous and high-grade metamorphic source rocks. Clay minerals in redbeds, as in most other ancient detrital deposits, are predominantly illite and chlorite, thus providing no specific clue to the climate in the source area or at the place of deposition.

As one tectonic end member, extensive uniform redbed-evaporite sequences accumulated on stable cratons, whereas some thick wedges of desert redbeds with border conglomerates accumulated in rift valleys. On a global scale, paleomagnetic evidence of the distribution of redbeds relative to their pole position corroborates paleogeographic data, suggesting that most redbeds, evaporites, and eolian sandstones accumulated less than 30° north and south of a paleoequator where hot, dry climate generally prevailed. Moreover, continental drift reconstructions reveal that the most widespread redbeds in the geologic record developed near the Equator in late Paleozoic and early Mesozoic time when the continents were assembled in a great landmass, Pangea. *See* CONTINENTAL DRIFT; PALEOMAGNETISM; SEDIMENTARY ROCKS. [F.B.V.H.]

Redshift A systematic displacement toward longer wavelengths of lines in the spectra of distant galaxies, and also of the continuous part of the spectrum. First studied systematically by E. Hubble, redshift is central to observational cosmology, in which it provides the basis for the modern picture of an expanding universe.

There are two fundamental properties of redshifts. First, the fractional redshift $\Delta\lambda/\lambda$ is independent of wavelength. This rule has been verified from 21 cm (radio radiation from neutral hydrogen atoms) to about 6×10^{-5} cm (the visible region of the electromagnetic spectrum) and leads to the interpretation of redshift as resulting from a recession of distant galaxies. Though this interpretation has been questioned, no other mechanism is known that would explain the observed effect.

Second, redshift is correlated with apparent magnitude in such a way that when redshift is translated into recession speed and apparent magnitude into distance, the recession speed is found to be nearly proportional to the distance. This rule was formulated by Hubble in 1929, and the constant of proportionality bears his name. *See* HUBBLE CONSTANT; MAGNITUDE (ASTRONOMY). [D.La.]

Redtop grass One of the bent grasses, *Agrostis alba* and its relatives, which occur in cooler, more humid regions of the United States on a wide variety of soils. Redtop tolerates both wet and dry lands and acid and infertile soils, and it is used where other species of grasses do not thrive. Redtop is a perennial and makes a coarse, loose turf. The inflorescence is a reddish open panicle. Redtop is used for pasture and hay and is fairly nutritious if harvested promptly when heading occurs. It is effective in preventing erosion by holding banks of drainage ditches, waterways, and terrace channels. *See* CYPERALES. [H.B.S.]

Reduced-pressure distillation A technique used to distill high-molecular-weight substances. Materials that have large molecules require high temperatures for distillation. There is a limit to this temperature increase, for above 750°F (400°C), thermal decomposition begins. Therefore, to distill a material such as a motor oil whose distillation temperature is well over 750°F (400°C), some other distillation technique must be used. If a distillation vessel is put under a vacuum, thus removing atmospheric pressure from the surface of the boiling liquid, the vapor can escape more easily from the liquid surface and at an appreciably lower temperature.

Distillation at reduced pressure is widely used as a separation process. Equipment ranges in size from that which will handle micro quantities in the laboratory to plant installations that process thousands of barrels per day. One of the largest uses of this process has evolved in the petroleum industry. The same types of distillation equipment are used to purify and separate fatty acids, a major ingredient in soap and in lubricating greases. Another large user of vacuum-steam distillation is the edible fat and oil industry. The process is called deodorization, which removes objectionable taste and odor from cottonseed and soybean oils, which are the main constituents in shortenings, and in salad and cooking oils. *See* DISTILLATION. [E.S.Ba.]

Redwood A member of the pine family, *Sequoia sempervirens*, is the tallest tree in the Americas, attaining a height of 350 ft (107 m) and a diameter of 27 ft (8.2 m). Its present range is limited to a strip along the Pacific Coast, extending from southwest Oregon to south of San Francisco. The leaves are evergreen, sharply pointed, small, disposed in two vertical rows on short branches, and scalelike on the main stem. The cones are egg-shaped. The bark is a dull red-brown, on old trees sometimes 1 ft (0.3 m) thick, densely fibrous, and highly resistant to fire. The tree gets its common name from the color of the bark as well as that of the heartwood.

The wood holds paint well and is used for bridge timbers, tanks, flumes, silos, posts, shingles, paneling, doors, caskets, furniture, siding, and many other building purposes. *See* PINALES; PINE. [A.H.G./K.P.D.]

Reed-type frequency meter An instrument for measuring frequency using the principle of vibrating reeds. The reeds for such instruments, assembled into a so-called reed comb, are made of specially selected and properly tempered steel. They have bent tips which are enameled white for visibility. When the supply voltage is applied to the instrument, all of the reeds receive vibrational impulses; the effect is visible only in the reed or reeds which are in resonance. The reed in resonance behaves somewhat like a whip, the tip swinging through a readily visible arc, the natural period of vibration depending upon length and thickness. The reeds in the comb are vibrated by an electromagnet energized from the source of which the frequency is to be measured. *See* FREQUENCY MEASUREMENT. [E.B.C.]

Reef A mass or ridge of rock or rock-forming organisms in a water body, a rock trend on land or in a mine, or a rocky trend in soil. Usually the term reef means a rocky menace to navigation, within 6 fathoms (11 m) of the water surface. Various kinds of calcium carbonate–secreting animals and plants create biogenic, or organic, reefs throughout the warmer seas. Most biogenic reefs are made of corals and associated organisms, but some entire reefs and important parts of others consist mainly of lime-secreting algae, hydrozoans, annelids, oysters, or sponges. *See* ALGAE; SCLERACTINIA.

The term fringing reef refers to a coral or other biogenic reef that fringes the edge of the land. A barrier reef ordinarily made of corals or other organisms parallels the shore at the seaward side of a natural lagoon. An atoll is an annular coral reef that surrounds a lagoon. *See* ATOLL. [P.Cl.]

Reference electrode An electrode with an invariant potential. In electrochemical methods, where it is necessary to observe, measure, or control the potential of another electrode (denoted indicator, test, or working electrode), it is necessary to use a reference electrode, which maintains a potential that remains practically unchanged during the course of an electrochemical measurement. Potentials of indicator or working electrodes are measured or expressed relative to reference electrodes. *See* POTENTIALS.

One such electrode, the normal hydrogen electrode, has been chosen as a reference standard, relative to which potentials of other electrodes and those of oxidation-reduction couples are often expressed. By maintaining a constant pressure of hydrogen gas the potential of a hydrogen electrode can be used for determination of the activity of hydrogen ions in the tested solution. However, in practice the determination of the hydrogen-ion activity (pH) is performed by using a glass electrode. The hydrogen electrode itself is used only in fundamental studies and some nonaqueous solutions. The hydrogen electrode, however, remains important for providing a reference standard. *See* ACTIVITY (THERMODYNAMICS); HYDROGEN ELECTRODE; pH.

In practice, potentials are measured against reference electrodes that are easier to work with than the normal hydrogen electrode. Such electrodes are known as secondary reference electrodes; the most common are the calomel and silver–silver chloride electrodes. *See* CALOMEL ELECTRODE; ELECTRODE; SILVER CHLORIDE ELECTRODE; SOLVENT. [P.Z.]

Reflecting microscope A microscope whose objective is composed of two mirrors, one convex and the other concave (see illustration). The imaging properties are independent of the wavelength of light, and this freedom from chromatic aberration allows the objective to be used even for infrared and ultraviolet radiation. Although the reflecting microscope is simple in appearance, the construction tolerances are so small and so difficult to achieve that the system is used only when refracting objectives are unsuitable. The distance from the objective

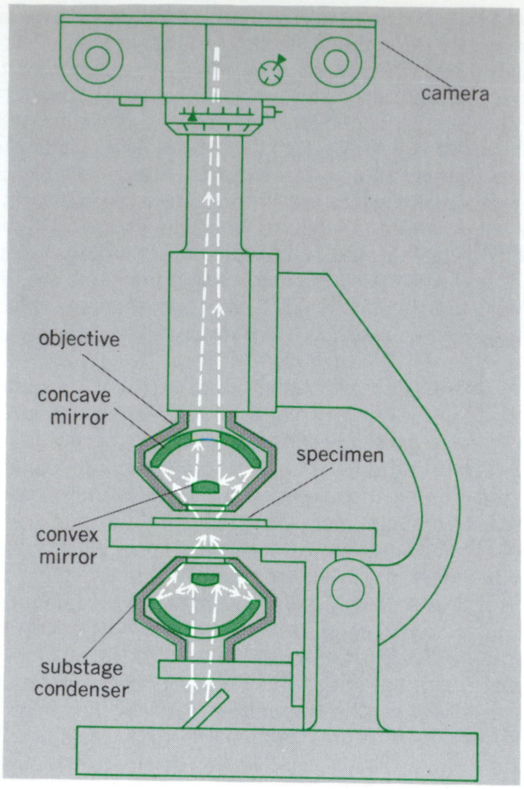

Reflecting microscope arranged for photomicrography.

to the specimen can be made very large; this large working distance is useful in special applications, such as examining objects situated within metallurgical furnaces. Reflecting microscopes have been mainly used for microspectrometry in the infrared and the ultraviolet, and for ultraviolet microphotography. *See* MICROSCOPE; OPTICAL MICROSCOPE. [D.S.G.]

Reflection of electromagnetic radiation The returning or throwing back of electromagnetic radiation such as light, ultraviolet rays, radio waves, or microwaves by a surface upon which the radiation is incident. In general, a reflecting surface is the boundary between two materials of different electromagnetic properties, such as the boundary between air and glass, air and water, or air and metal. Devices designed to reflect radiation are called reflectors or mirrors.

The simplest reflection laws are those that govern plane waves of radiation. The law of reflection concerns the incident and reflected rays (as in the case of a beam from a flashlight striking a mirror) or, more precisely, the wave normals of the incident and reflected waves. The law states that the incident and reflected rays and the normal to the reflecting surface all lie in one plane, called the plane of incidence, and that the reflection angle equals the angle of incidence as in Eq. (1) [*see* illustration]. The angles θ_{inc} and θ_{refl} are measured between

$$\theta_{refl} = \theta_{inc} \tag{1}$$

the surface normal and the incident and reflected rays, respectively. The surface (in the above example, that of the mirror) is assumed to be smooth, with surface irregularities small compared to the wavelength of the radiation. This results in so-called specular reflection. In contrast, when the surface is rough, the reflection is diffuse. An example of this is the diffuse scattering of light from a screen or from a white wall where light is returned through a whole range of different angles.

The reflectivity of a surface is a measure of the amount of reflected radiation. It is defined as the ratio of the intensities of

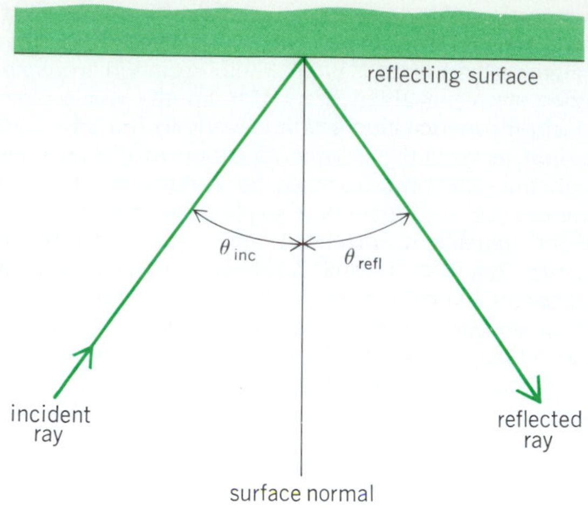

Reflection of electromagnetic radiation from smooth surface.

the reflected and incident radiation. The reflectivity depends on the angle of incidence, the polarization of the radiation, and the electromagnetic properties of the materials forming the boundary surface. These properties usually change with the wavelength of the radiation. Reflecting materials are divided into two groups: transparent materials also called dielectrics, and opaque conducting materials, usually metals.

The reflectivity of polished metal surfaces is usually quite high. Silver and aluminum, for example, reflect more than 90% of visible light. In ordinary mirrors the reflecting surface is the interface between metal and glass, which is thus protected from oxidation, dirt, and other forms of deterioration. When it is not permissible to use this protection for technical reasons, one uses "front-surface" mirrors, which are usually coated with evaporated aluminum.

The material property that determines the amount of radiation reflected from an interface between two dielectric media is the phase velocity v of the electromagnetic radiation in the two materials. In optics one uses as a measure for this velocity the refractive index n of the material, which is defined by Eq. (2) as

$$n = c/v \tag{2}$$

the ratio of the velocity of light c in vacuum and the phase velocity in the material. For visible light, for example, the refractive index of air is about $n = 1$, the index of water is about $n = 1.33$, and the index of glass is about $n = 1.5$. *See* PHASE VELOCITY; REFRACTION OF WAVES.

For normal incidence ($\theta_{inc} = 0$) the reflectivity R of the interface is given by Eq. (3), in which the material constants are

$$R = \left(\frac{v_1 - v_2}{v_1 + v_2}\right)^2 = \left(\frac{n_2 - n_1}{n_2 + n_1}\right)^2 \tag{3}$$

labeled 1 and 2, where the radiation is incident in material 1. The reflectivity of an air-water interface is about 2% ($R = 0.02$) and that of an air-glass interface about 4% ($R = 0.04$); the other 98% or 96% are transmitted through the water or glass, respectively. *See* ALBEDO; GEOMETRICAL OPTICS; MIRROR OPTICS; REFLECTION OF SOUND. [H.K.]

Reflection of sound The return of sound waves from surfaces on which they are incident. The geometrical laws for reflection of sound waves are the same as those for light waves. The apparent differences involve only questions of scale, because the average wavelength of sound is about 100,000 times that of light. For example, a mirror or lens used to produce a beam of sound waves must be enormously large compared to mirrors and lenses used in optical systems. *See* REFLECTION OF ELECTROMAGNETIC RADIATION.

A concave surface tends to concentrate the reflected sound waves. Convex reflectors tend to spread the reflected waves. Therefore, when placed at the boundaries of a room, they tend to diffuse the sound throughout the room. For this reason, some radio-broadcasting studios employ cylindrical convex panels as part of their wall construction to promote diffusion. *See* ARCHITECTURAL ACOUSTICS; ECHO; SOUND. [C.M.H.]

Reflex A simple, unlearned, yet specific behavioral response to a specific stimulus. Reflexes are exhibited by virtually all animals from protozoa to primates. Along with other, more complex stimulus-bound responses such as fixed action patterns, they constitute much of the behavioral repertoire of invertebrates. In higher animals, such as primates, where learned behavior dominates, reflexes nevertheless persist as an important component of total behavior.

The simplest known reflexes require only one neuron or, in the strictest sense, none. For example, ciliated protozoa, which are single cells and have no neurons, nevertheless exhibit apparently reflexive behaviors. However, most reflexes require activity in a large sequence of neurons. The neurons involved in most reflexes are connected by specific synapses to form functional units in the nervous system. Such a sequence begins with sensory neurons and ends with effector cells such as skeletal muscles, smooth muscles, and glands, which are controlled by motor neurons. The central neurons which are often interposed between the sensory and motor neurons are called interneurons. The sensory side of the reflex arc conveys specificity as to which reflex will be activated. The remainder of the reflex response is governed by the specific synaptic connections that lead to the effector neurons. A familiar reflex is the knee-jerk or stretch reflex. It involves the patellar (kneecap) tendon and a group of upper leg muscles. Other muscle groups show similar reflexes. [J.L.La.]

Reforming processes Methods for producing higher-octane gasolines. An early process known as thermal reforming is essentially an extension of thermal cracking. Cracking is used to convert the higher-molecular-weight (higher-boiling-range) feedstocks into gasoline; reforming processes convert (reform) gasoline into higher-octane gasoline (reformate). The equipment for thermal reforming is, essentially, the same as that used for thermal cracking but with the added impetus of higher temperature. Thermal cracking processes conventionally use temperatures on the order of 450–515°C (840–960°F), whereas reforming processes use temperatures on the order of 510–600°C (950–1110°F). *See* CRACKING; GASOLINE; OCTANE NUMBER.

Like thermal cracking and the related catalytic cracking, thermal reforming also has its catalytic counterpart: catalytic reforming. Catalytic reforming is also used to convert low-octane gasolines into higher-octane gasolines. However, many thermal processes will produce reformed gasolines with octane numbers in the range 65–80, but catalytic forming is capable of producing reformates with octane numbers on the order of 90–95. Catalytic reforming is conducted in the presence of hydrogen over hydrogenation-dehydrogenation catalysts, which may be supported on alumina or silica-alumina. Depending on the catalyst, a definite sequence of reactions takes place, involving structural changes in the charge stock. Furthermore, this process has actually rendered thermal reforming somewhat obsolescent. *See* CATALYSIS.

Upgrading, by reforming, may be accomplished, in part, by an increase in the volatility (reduction of molecular size) or chiefly by the conversion of *n*-paraffins to isoparaffins, olefins, and aromatics, and naphthenes to aromatics. The nature of the final product is, of course, influenced by the source and composition of the straight-run naphtha charge. In thermal reforming, the reactions resemble those taking place in the cracking of gas oils; molecular size is reduced, while olefins and some aromatics are synthesized.

The thermal reformate liquid has the potential for further reaction by virtue of the presence of the olefinic constituents, but still contains appreciable quantities of these unsaturated materials and, as such, is classed as being unstable. Further hydrotreatment is required for stabilization. *See* DEHYDROGENATION; HYDROCRACKING; ISOMERIZATION; PETROLEUM PROCESSING. [J.G.S.]

Refraction of waves The change of direction of propagation of any wave phenomenon which occurs when the wave velocity changes. The term is most frequently applied to visible light, but it also applies to all other electromagnetic waves, as well as to sound and water waves.

The physical basis for refraction can be readily understood with the aid of the illustration. Consider a succession of equally spaced wavefronts approaching a boundary surface obliquely. The direction of propagation is in ordinary cases perpendicular to the wavefronts. In the case shown, the velocity of propagation is less in medium 2 than in medium 1, so that the waves are slowed down as they enter the second medium. Thus, the direction of travel is bent toward the perpendicular to the boundary surface (that is, $\theta_2 < \theta_1$). If the waves enter a medium in which the velocity of propagation is faster than in their original medium, they are refracted away from the normal.

Snell's law. The simple mathematical relation governing refraction is known as Snell's law. If waves traveling through a medium at speed v_1 are incident on a boundary surface at angle θ_1 (with respect to the normal), and after refraction enter the second medium at angle θ_2 (with the normal) while traveling at speed v_2, then Eq. (1) holds. The index of refraction *n of*

$$\frac{v_1}{v_2} = \frac{\sin \theta_1}{\sin \theta_2} \tag{1}$$

a medium is defined as the ratio of the speed of waves in vacuum c to their speed in the medium. Thus $c = n_1 v_1 = n_2 v_2$, and therefore Eq. (2) holds. The refracted ray, the normal to the

$$n_1 \sin \theta_1 = n_2 \sin \theta_2 \tag{2}$$

surface, and the incident ray always lie in the same place.

The relative index of refraction of medium 2 with respect to that of medium 1 may be defined as $n = n_1/n_2$. Snell's law then becomes Eq. (3). For sound and other elastic waves which

$$\sin \theta_1 = n \sin \theta_2 \tag{3}$$

require a medium in which to propagate, only this last form has meaning. Equation (3) is frequently used for light when one medium is air, whose index of refraction is very nearly unity.

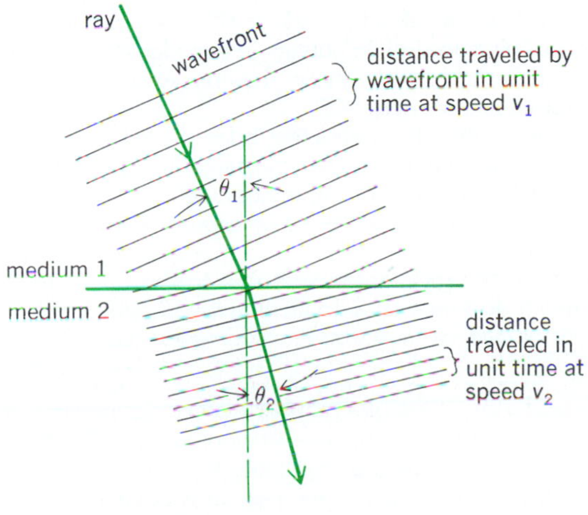

Physical basis for Snell's law.

When the wave travels from a region of low velocity (high index) to one of high velocity (low index), refraction occurs only if $(n_1/n_2) \sin \theta_1 \leqq 1$. If θ_1 is too large for this relation to hold, then $\sin \theta_2 > 1$, which is meaningless. In this case the waves are totally reflected from the surface back into the first medium. The largest value that θ_1 can have without total internal reflection taking place is known as the critical angle θ_c. Thus $\sin \theta_c = n_2/n_1$.

Atmospheric refraction. The index of refraction of the Earth's atmosphere increases continuously from 1.000000 at the edge of space to 1.000293 (yellow light) at 0°C and 760 mmHg (101.3 kilopascals) pressure. Thus celestial bodies as seen in the sky are actually nearer to the horizon than they appear to be. The effect decreases from a maximum of about 35 minutes of arc for an object on the horizon to zero at the zenith, where the light enters the atmosphere at perpendicular incidence.

Other manifestations of atmospheric refraction are the mirages and "looming" of distant objects which occur over oceans or deserts, where the vertical density gradient of the air is quite uniform over a large area. *See* MIRAGE.

Sound waves. The velocity of sound in a gas is proportional to the square root of the absolute temperature. Because of the vertical temperature gradients in the atmosphere, refraction of sound can be quite pronounced. As in mirage formation, to allow large-scale refraction the temperature at a given height must be uniform over a rather large horizontal area. *See* ATMOSPHERIC ACOUSTICS; SOUND.

Seismic waves. The velocity of elastic waves in a solid depends upon the modulus of elasticity and upon the density of the material. Waves propagating through solid earth are refracted by changes of material or changes of density. Worldwide observations of earthquake waves enable scientists to draw conclusions on the distribution of density within the Earth. *See* SEISMOLOGY.

Water waves. As the waves enter shallower water they travel more slowly. As a train of waves approaches a coastline obliquely, its direction of travel becomes more nearly perpendicular to the shore because of refraction. *See* WAVE MOTION.

[J.W.St.]

Refractometric analysis A method of chemical analysis based on the measurement of the index of refraction of a substance. The most common type of refractometer is the Abbe refractometer. It is simple to use, requiring but a drop or two of sample and allowing a measurement of refractive index to be made in 1–2 min, with a precision of 0.0001. More precise measurements of refractive indices may be made by using a dipping or immersion refractometer, the prism of which is completely immersed in the sample. The most precise measurements of the refractive indices of gases or solutions containing small traces of impurities are made with an interferometer. For measurements in flowing systems, differential refractometers are used.

The measurement of refractive index is used to identify compounds whose other physical constants are quite similar. Because minute amounts of impurities often cause a measurable change in the refractive index of a pure material, refractive index is often used as a criterion for purity. A measurement of refractive index gives information as to the gross amount of impurity; it does not serve to identify the impurity. *See* OPTICAL METHODS OF CHEMICAL ANALYSIS; REFRACTION OF WAVES. [R.F.G.; J.N.L.]

Refractory One of a number of ceramic materials for use in high-temperature structures or equipment. The term high temperatures is somewhat indefinite but usually means above about 1830°F (1000°C), or temperatures at which, because of melting or oxidation, the common metals cannot be used. In some special high-temperature applications, the so-called refractory metals such as tungsten, molybdenum, and tantalum are used. *See* CERAMICS.

The greatest use of refractories is in the steel industry, where they are used for construction of linings of equipment such as blast furnaces, hot stoves, and open-hearth furnaces. Other important uses of refractories are for cement kilns, glass tanks, nonferrous metallurgical furnaces, ceramic kilns, steam boilers, and paper plants. Special types of refractories are used in rockets, jets, and nuclear power plants. Many refractory materials, such as aluminum oxide and silicon carbide, are also very hard and are used as abrasives; some applications, for example, aircraft brake linings, make use of both characteristics.

Refractory materials are commonly grouped into (1) those containing mainly aluminosilicates; (2) those made predominantly of silica; (3) those made of magnesite, dolomite, or chrome ore, termed basic refractories (because of their chemical behavior); and (4) a miscellaneous category usually referred to as special refractories. [J.F.McM.]

Refrigeration The cooling of a space or substance below the environmental temperature. Mechanical refrigeration is primarily an application of thermodynamics wherein the cooling medium, or refrigerant, goes through a cycle so that it can be recovered for reuse. The commonly used basic cycles, in order of importance, are vapor-compression, absorption, steam-jet or steam-ejector, and air. Each cycle operates between two pressure levels, and all except the air cycle use a two-phase working medium which alternates cyclically between the liquid and vapor phases. *See* REFRIGERATION CYCLE.

Vapor-compression cycle. The vapor-compression cycle consists of an evaporator in which the liquid refrigerant boils at low temperature to produce cooling, a compressor to raise the pressure and temperature of the gaseous refrigerant, a condenser in which the refrigerant discharges its heat to the environment, usually a receiver for storing the liquid condensed in the condenser, and an expansion valve through which the liquid expands from the high-pressure level in the condenser to the low-pressure level in the evaporator. This cycle may also be used for heating if the useful energy is taken off at the condenser level instead of at the evaporator level. *See* HEAT PUMP.

Absorption cycle. The absorption cycle accomplishes compression by using a secondary fluid to absorb the refrigerant gas, which leaves the evaporator at low temperature and pressure. Heat is applied, by means such as steam or gas flame, to distill the refrigerant at high temperature and pressure. The most-used refrigerant in the basic cycle is ammonia; the secondary fluid is then water. The condenser, receiver, expansion valve, and evaporator are essentially the same as in any vapor-compression cycle. The compressor is replaced by an absorber, generator, pump, heat exchanger, and controlling-pressure reducing valve.

Steam-jet cycle. The steam-jet cycle uses water as the refrigerant. High-velocity steam jets provide a high vacuum in the evaporator, causing the water to boil at low temperature and at the same time compressing the flashed vapor up to the condenser pressure level. Its use is limited to air conditioning and other applications for temperatures above 32°F (0°C).

Air cycle. The air cycle, used primarily in airplane air conditioning, differs from the other cycles in that the working fluid, air, remains as a gas throughout the cycle. Air coolers replace the condenser, and the useful cooling effect is obtained by a refrigerator instead of by an evaporator. A compressor is used, but the expansion valve is replaced by an expansion engine or turbine which recovers the work of expansion. Systems may be open or closed. In the closed system, the refrigerant air is completely contained within the piping and components, and is continuously reused. In the open system, the refrigerator is replaced by the space to be cooled, the refrigerant air being expanded directly into the space rather than through a cooling coil.

Refrigerants. The working fluid in a two-phase refrigeration cycle is called a refrigerant. Ammonia and Freon-22 are most important for industrial refrigeration, Freon-11 and Freon-12 being used most often for commercial and air conditioning work in which nontoxic refrigerants are necessary. A secondary cooling liquid that does not change from the liquid phase is called a brine. Solutions of sodium chloride or calcium chloride in water are frequently used as circulating brines in refrigeration systems. *See* AIR CONDITIONING; AIR COOLING; AUTOMOTIVE AIR CONDITIONING; COOLING TOWER; DRY ICE; MARINE REFRIGERATION; REFRIGERATOR. [C.F.K.]

Refrigeration cycle A sequence of thermodynamic processes whereby heat is withdrawn from a cold body and expelled to a hot body. Theoretical thermodynamic cycles consist of nondissipative and frictionless processes. For this reason, a thermodynamic cycle can be operated in the forward direction to produce mechanical power from heat energy, or it can be operated in the reverse direction to produce heat energy from mechanical power.

One of the first cycles used for mechanical refrigeration was the reverse Brayton cycle. With the development of Freon and other condensable fluids for the vapor-compression cycle, the reverse Carnot cycle became more important. The reversed cycle is used primarily for the cooling effect that it produces during a portion of the cycle and so is called a refrigeration cycle. It may also be used for the heating effect, as in the comfort warming of space during the cold season of the year. *See* HEAT PUMP; THERMODYNAMIC PROCESSES.

In the refrigeration cycle a substance, called the refrigerant, is compressed, cooled, and then expanded. In expanding, the refrigerant absorbs heat from its surroundings to provide refrigeration. After the refrigerant absorbs heat from such a source, the cycle is repeated. Compression raises the temperature of the refrigerant above that of its natural surroundings so that it can give up its heat in a heat exchanger to a heat sink such as air or water. Expansion lowers the refrigerant temperature below the temperature that is to be produced inside the cold compartment or refrigerator. The sequence of processes performed by the refrigerant constitutes the refrigeration cycle. When the refrigerant is compressed mechanically, the refrigerative action is called mechanical refrigeration. *See* BRAYTON CYCLE; CARNOT CYCLE; REFRIGERATION; THERMODYNAMIC CYCLE. [T.Ba.]

Refrigerator An insulated, cooled compartment. If it is large enough for the entry of a person, it is termed a walk-in box; otherwise it is called a reach-in refrigerator. Cooling may be by mechanical or gas refrigeration, by water or dry ice, or by brine circulation. Temperatures maintained depend upon the requirements of the product stored, generally varying from 55°F (13°C) down to 0°F (−18°C), and sometimes lower.

A household or domestic refrigerator is a factory-built, self-contained cabinet. The range of storage capacities is wide and varies among manufacturers. Modem designs have a main compartment for holding food above freezing, a second compartment for storage below freezing, and trays for the freezing of ice cubes. Low-temperature household refrigerators, or home freezers, for the storage of frozen foods are manufactured in both the chest and the upright, or vertical, types.

A commercial refrigerator is any factory-built refrigerated fixture, cabinet, or room that can be assembled and disassembled readily. Commercial or built-in refrigerators are used in restaurants, markets, hospitals, hotels, and schools for the storage of food and other perishables. *See* REFRIGERATION. [C.F.K.]

Regeneration (biology) The replacement by an organism of parts of the body which have been lost or severely injured. The term is comprehensive and covers a wide range of restorative activities in a variety of organisms. Some authors prefer to use the term reconstitution rather than regeneration. The capacity for regeneration varies greatly among different groups of organisms. Among the invertebrates, many of the hydroids, flatworms, annelids, echinoderms, and arthropods can replace major portions of the body. In certain instances, particularly in sponges, a few cells or a small fragment of the original organism is capable of reconstituting a completely new individual. In the vertebrates, the highest capacity for regeneration is found in the Amphibia, of which many species can regenerate a complete limb, a tail, portions of the eye, the lower jaw, and a number of other highly organized structures.

As a rule, the structures formed as a result of regeneration are duplicates of the original structures and possess all their functional characteristics. Under some circumstances, however, the regenerate may be of a different type than the original structure. After removal, the eye of a crustacean may be replaced by an antenna. A head, instead of a tail, may be formed at the posterior end of a flatworm. An amphibian, through regenerative activities, may establish supernumerary limbs that were not previously present.

Many organisms, both invertebrate and vertebrate, although they may be incapable of regenerating complex organs, or major portions of the body, have the capacity for reconstituting various types of tissues. Examples are the continual or periodic replacement by various animals of skin, scales, feathers, teeth, antlers, the lining of the alimentary canal, and some components of the reproductive tract. Such activities, which represent a phase of the normal life cycle of an individual, are often referred to as repetitive or physiological regeneration. Wound healing and the repair of bone fractures can likewise be regarded as types of regenerative activity.

Among the major problems are the sources of the cells which enter into regenerative activities and the manner in which they achieve the requisite potentialities for forming specialized tissues, new organs, or discrete portions of the body. Also of primary importance are problems of polarity and the relationship of local and organismic factors in the establishment of growth patterns. [E.G.B.]

Regeneration (engineering) The process of feeding back a portion of the output signal of an amplifier to its input in such a way that the input signal is reinforced. The result is greatly increased amplification. The feedback must be positive; that is, the two signals must be in phase, and it must be limited in magnitude to prevent the circuit from going into oscillation. *See* FEEDBACK CIRCUIT. [J.Mar.]

Regenerative braking A system of dynamic braking in which the electric drive motors are used as generators and return the kinetic energy of the motor armature and load to the electric supply system. This method is employed when the load is losing a large amount of potential energy, as in the case of an electric train descending a long grade or a pumped storage reservoir being emptied. The potential energy of the load is first converted to kinetic energy in the motor armature and load and then to electrical energy. *See* DYNAMIC BRAKING. [A.R.E.]

Regolith The mantle or blanket of unconsolidated or loose rock material that overlies the intact bedrock and nearly everywhere forms the land surface. The regolith may be residual (weathered in place), or it may have been transported to its present site. The undisturbed residual regolith may grade from agricultural soil at the surface, through fresher and coarser weathering products, to solid bedrock several feet or more beneath the surface. The transported regolith includes the alluvium of rivers, sand dunes, glacial deposits, volcanic ash, coastal deposits, and the various mass-wasting deposits that occur on hillslopes. The lunar surface also has a regolith. This layer of fragmental debris is believed to derive from prolonged meteoritic and secondary fragment impact. *See* BASEMENT ROCK; SOIL; WEATHERING PROCESSES. [V.R.B.]

Regular polytopes The *n*-dimensional analogs of the regular polygons ($n = 2$) and platonic solids ($n = 3$). They are conveniently denoted by their Schläfli symbols $\{p,q\ldots\}$; for instance, the pentagon, hexagon, octagon, tetrahedron, octahedron are denoted by {5}, {6}, {8}, {3,3}, {3,4}. The cube is {4,3} because its faces are squares {4} and there are three of them at each vertex. The five platonic solids $\{p,q\}$, are determined by the inequality below. The numbers of vertices, edges,

$$(p - 2)(q - 2) < 4$$

faces (V,E,F) can be deduced from the obvious relations $pF = 2E = qV$ with the help of Euler's formula $V - E + F = 2$.

The general polytope (sometimes loosely called a "polyhedron" [see illustration] regardless of the number of dimensions)

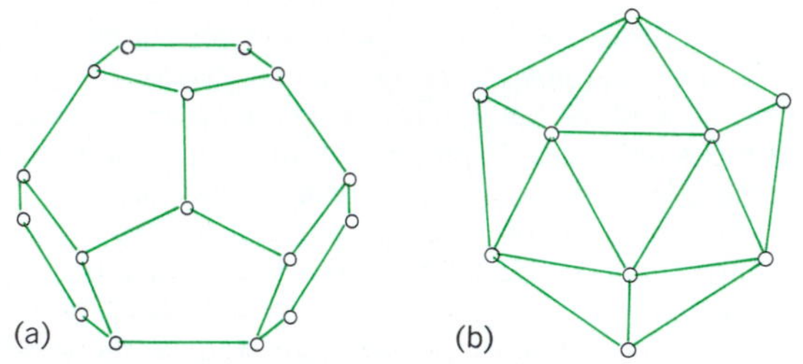

Two regular polyhedrons. (*a*) Dodecahedron {5,3}. (*b*) Icosahedron {3,5}.

is a finite region of *n*-dimensional space enclosed by a finite number of hyperplanes. When any redundant hyperplanes have been discarded, those that remain contain ($n - 1$)-dimensional polytopes called cells or "facets." For instance, the cells of a polygon are its sides, those of a polyhedron are its faces, and those of a four-dimensional polytope are solids. *See* ANALYTIC GEOMETRY; EUCLIDEAN GEOMETRY. [H.S.M.C.]

Regularia The name given by G. Cuvier in 1817 to an assemblage of echinoids in which the anus and periproct lie within the apical system. The test is globular and preponderantly radially symmetrical, and the ambulacral plates are commonly compound. The group included, in effect, all those echinoids which did not fall in the Irregularia. *See* ECHINOIDEA; IRREGULARIA. [H.B.F.]

Regulation The process of maintaining a quantity or condition essentially constant despite variations in such factors as line voltage and load. In an industrial process-control system, the speed, temperature, voltage, or position of a critical element can be kept constant by measuring the condition being regulated and feeding back into the system a signal representing the difference between the actual and the desired quantities. *See* CONTROL SYSTEMS; REGULATOR.

The term regulation is also used in the opposite sense to indicate the difference between the maximum and minimum voltages at the terminals of a tube, transformer, generator, or other device over the range of normal operating conditions. *See* SERVOMECHANISM; VOLTAGE REGULATOR. [F.H.R.]

Regulator A control device designed to maintain the value of some quantity substantially constant. The value to be maintained can usually be established at any value within the range of the regulator by making an appropriate setting. A regulated system is a feedback control system employing a regulator to maintain some quantity of the system at a constant value. *See* CONTROL SYSTEMS. [J.A.Hr.]

Reheating The addition of heat to steam of reduced pressure after the steam has given up some of its energy by expansion through the high-pressure stages of a turbine. The reheater tube banks are arranged within the setting of the steam-generating unit in such relation to the gas flow that the steam is restored to a high temperature. Under suitable conditions of initially high steam pressure and superheat, one or two stages of reheat can be advantageously employed to improve thermodynamic efficiency of the cycle. *See* STEAM-GENERATING UNIT; STEAM TURBINE; SUPERHEATER; VAPOR CYCLE. [R.A.M.]

Reindeer A ruminant, *Rangifer tarandus*, of the deer family, Cervidae. Reindeer inhabit the Arctic region and have a circumpolar distribution. They have been domesticated for centuries and are economically important to the Laplanders, who use them as draft animals as well as for their skins, flesh, and milk.

Both sexes have antlers, a characteristic peculiar to this deer species. Although their senses of sight and hearing are not sharp, their sense of smell is quite keen. The hooves are round and broad, giving the animal stable footing in snow and on ice. In the wild, they are the most migratory of all deer, traveling in large herds in search of food during the winter months. The rutting season begins during the autumn migrations; fawns are born during the spring. *See* ARTIODACTYLA; DEER. [C.B.C.]

Reinforced concrete Portland-cement concrete with steel embedded in it to assist in carrying loads. Steel in several of the following forms may be used: bars or rods, wire, pipe, and structural shapes such as wide-flange beams.

Loads are transferred between concrete and steel by the bond along the surface of intersection. The bond may be substantially improved mechanically by giving reinforcing bars raised surfaces, in which case the bars are called deformed bars. *See* CONCRETE. [F.S.M.]

Relapsing fever An acute infectious disease caused by various species of the genus *Borrelia*, a spirochete. The disease is characterized by episodes of fever which subside spontaneously and recur over a period of weeks. Epidemiologically, two types are recognized, a louse-borne type which often occurs in epidemics, and a tick-borne type which usually is endemic. The clinical characteristics of the two are similar. [T.B.T.]

Relative atomic mass The ratio of the average mass per atom of the natural nuclidic composition of an element to $\frac{1}{12}$ of the mass of an atom of nuclide ^{12}C. For example, $\mu(Cl) = 35.453$. Relative atomic mass replaces the concept of atomic weight. It is also known as relative nuclidic mass. *See* NUCLIDE. [T.C.W.]

Relative molecular mass The ratio of the average mass per formula unit of the natural nuclidic composition of a substance to $\frac{1}{12}$ of the mass of an atom of nuclide ^{12}C. For example, $\mu(KCl) = 74.555$. Relative molecular mass replaces the concept of molecular weight. *See* NUCLIDE. [T.C.W.]

Relative motion All motion is relative to some frame of reference. The simplest laboratory frame of reference is three mutually perpendicular axes at rest with respect to an observer. In terms of the frame of reference of an observer some distance from Earth, the laboratory frame of reference would be moving with Earth as it rotates on its axis and as it revolves about the Sun. What would be a simple form of motion in the laboratory frame of reference would appear to be a much more complicated motion in the frame of reference of the distant observer. *See* FRAME OF REFERENCE.

Motion means continuous change of position of an object with respect to an observer. To another observer in a different frame of reference the object may not be moving at all, or it may be moving in an entirely different manner. The motions of the planets were found in ancient times to appear quite complicated in the laboratory frame of reference of an observer on Earth. By transferring to the frame of reference of an imaginary observer on the Sun, Johannes Kepler showed that the relative motion of the planets could be simply described in terms of elliptical orbits. The validity of one description is no greater than the other, but the latter description is far more convenient. [R.D.Ru.]

Relativistic electrodynamics The study of the interaction between charged particles and electric and magnetic fields when the velocities of the particles approach that of light. Relativistic electrodynamics, which can be considered as an extension of the everyday laws of electricity and magnetism, is an important consideration in high-energy particle accelerators, in high-current, high-voltage vacuum tubes, and in electromagnetic radiation.

The laws which relate the electric and magnetic fields to the charges and currents which produce them are known as Maxwell's equations. Charged particles or current elements in such fields experience a force which is called a Lorentz force. The motion of these charged particles and current elements in such fields is then determined by Newton's laws, appropriately generalized for relativistic velocities. *See* MAXWELL'S EQUATIONS; RELATIVITY.

J. C. Maxwell's contribution to relativistic electrodynamics was to formulate his four equations and to introduce the concept of displacement current. Although each equation was originally deduced for static or other restrictive conditions, Maxwell implied the validity of each for fields which varied in arbitrary ways with time. *See* DISPLACEMENT CURRENT.

One direct consequence of Maxwell's equations comes from the fact that the fields **E** and **B**, in the absence of charges and currents, satisfy a wave equation with wave velocity given by the equation below. The fact that this is the same as the veloci-

$$c = \frac{1}{\sqrt{\epsilon_0 \mu_0}} \approx 3 \times 10^8 \text{ m/s}$$

ty of light led Maxwell to infer that light, like electricity, magnetism, and optics, is an electromagnetic wave phenomenon.

The property of the Maxwell equations that makes them applicable to problems in relativistic electrodynamics is their relativistic invariance. Specifically, this means that Maxwell's equations will seem correct to an observer traveling with a constant velocity as well as to an observer at rest. One must realize, however, that a magnetic field in the rest frame will appear to be both an electric and a magnetic field in the moving frame. *See* FRAME OF REFERENCE.

One interesting consequence of the way in which the fields transform is the diminution of the repulsive force between charges moving with high velocity in parallel paths. An important application of this effect is to relativistic beams of particles which might be expected to disperse as a result of space-charge forces. In linear electron accelerators, for example, this transverse divergence can be neglected. Another way of looking at the phenomenon is that in the rest system of the electrons, the accelerator appears extremely short because of the Lorentz contraction, and the space-charge forces have little time to spread the beam apart.

The tendency for relativistically charged beams not to diverge is of course not restricted to accelerators. It is important in consideration of electron optics in high-current, high-voltage vacuum tubes, and may even be important in controlled fusion devices.

Most electron linear accelerators produce extremely relativistic electrons whose longitudinal motion is also inhibited by their large relativistic mass. Indeed, the phase oscillations which are responsible for longitudinal stability become extremely slow; since all electrons are effectively traveling at the same velocity *c*, they neither fall behind nor overtake one another and are therefore less inclined to get out of phase.

For relativistic effects in the motion of charged particles in circular particle accelerators *see* SYNCHROTON. [R.L.G.]

Relativistic heavy-ion collisions Collisions of the heaviest available nuclei at very high energies. At these energies the nuclei move at relativistic velocities, close to the speed of light. Since 1985 a new field of physics has led to studies of the states of high density of particles over nuclear volumes, created by such collisions. These studies explore the nature of the forces holding protons together. Experiments on scattering of high-energy protons and other particles indicate that the proton comprises smaller units, quarks and gluons, but free quarks have never been observed. If they were simply bound together to form protons, sufficiently violent collisions would free them. The mechanism for combining quarks inside particles must be more subtle, and may lie in the interaction between a free quark and the physical vacuum, the ordinary empty space found in the laboratory. *See* GLUONS; PROTON; QUARKS.

This notion of the physical vacuum endowed with properties may seem novel, but is at the heart of many aspects of modern theoretical physics. The unique feature of the physics of high particle density is that it may be able to produce an actual change in the vacuum. The highly ordered state of the physical vacuum can be destroyed by a sufficient density of particles, reverting to a state better described as a simple vacuum, with quarks and gluons moving freely, except for their mutual scattering. *See* QUANTUM CHROMODYNAMICS; QUARK-GLUON PLASMA.

Programs have been undertaken with beams of heavy ions with energies of 15–200 GeV/per atomic mass unit. The projectile nuclei employed have ranged from oxygen to sulfur, impinging on heavy targets. The energies are high enough to produce a sufficient density of particles, and the regions may

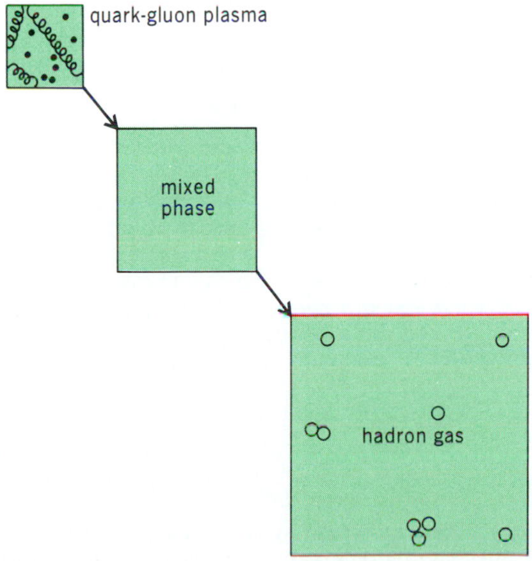

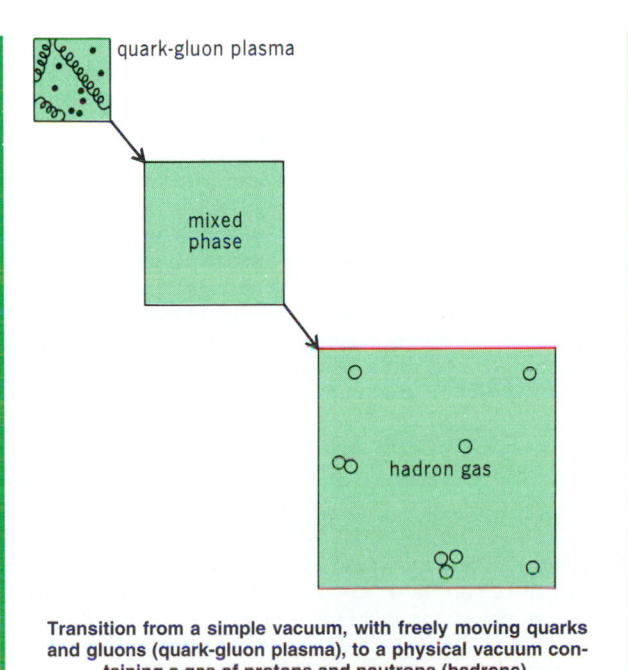

Transition from a simple vacuum, with freely moving quarks and gluons (quark-gluon plasma), to a physical vacuum containing a gas of protons and neutrons (hadrons).

be large enough for the sequence of states shown in the illustration to be observed in the tiny space of the nuclear volume. The spatial dimensions are a few times 10^{-13} cm and the times of the order of 10^{-22} s. The experiments must observe the debris that emerges from such collisions and deduce what happened during these very short times. *See* PARTICLE ACCELERATOR.

The first question is whether the ordered kinetic energy of two incident nuclei can be efficiently converted to a large number of particles in random motion, making it possible to speak of a high particle and energy density. The answer has been strongly affirmative. Even at the highest energy studies, a large fraction of the incident energy is converted to particles. When the two nuclei collide head-on at the highest energies, almost a thousand nucleons and pions emerge. This number of particles, distributed over the volume of the nuclei, is probably sufficient to break down the mechanism of confinement of quarks inside protons, although whether this state persists long enough to produce really interesting effects must be settled by more suitable measurements. *See* ELEMENTARY PARTICLE; NUCLEAR REACTION. [W.J.W.]

Relativistic quantum theory The quantum theory of particles which is consistent with the special theory of relativity, and thus can describe particles moving arbitrarily close to the speed of light. It is now realized that the only satisfactory relativistic quantum theory is quantum field theory; the attempt to relativize the Schrödinger equation for the wave function of a single particle fails, as shown below. However, with a change of interpretation, relativistic wave equations do correctly describe some aspects of the motions of particles in an electromagnetic field. *See* QUANTUM FIELD THEORY; QUANTUM MECHANICS; RELATIVITY.

The Schrödinger equation for the wave function $\psi(\mathbf{r},t)$ of a particle is Eq. (1), where E is the energy operator $i\hbar(\partial/\partial t)$, $\mathbf{p}$ is

$$E\psi = H(\mathbf{p},\mathbf{r})\psi \tag{1}$$

the momentum operator $-i\hbar\boldsymbol{\nabla}$, and $H(\mathbf{p},\mathbf{r})$ is the classical hamiltonian. For a nonrelativistic particle in free space, $H = \mathbf{p}^2/2m$. The naive way to relativize Eq. (1) would be to use the relativistic hamiltonian, Eq. (2). However, this equation is not relativistically invariant.

$$H = \sqrt{(mc^2)^2 + \mathbf{p}^2c^2} \tag{2}$$

The so-called Klein-Gordon equation, Eq. (3), is relativistical-

$$E^2\varphi = [(mc^2)^2 + \mathbf{p}^2c^2]\varphi \tag{3}$$

ly invariant. However, the only possible density of a conserved quantity formed from [φ is of the form shown in (4). But this

$$\rho \propto \varphi^* E\varphi - \varphi E\varphi^* \tag{4}$$

cannot be a probability density, because it is not positive definite (it changes sign when φ is replaced by φ^*).

But ρ, in (4), can be interpreted as a charge density (when multiplied by a unit charge e); φ is then to be interpreted as a matrix element of a field operator Φ of a quantized field whose quanta are particles with mass m and charge e and no spin.

P. A. M. Dirac found a relativized form of Eq. (1) which is both linear in $\mathbf{E}$ and has a positive definite density form given by Eq. (5), where β and α are constants which obey Eq. (6).

$$E\psi = [\beta mc^2 + \alpha \cdot pc]\psi \ (\rho \propto \psi^*\psi) \tag{5}$$

$$\alpha_i\alpha_j + \alpha_j\alpha_i = 0, \quad i \neq j \tag{6}$$

$$\alpha\beta + \beta\alpha = 0 \quad \alpha_i^2 = 1 \quad \beta^2 = 1 \quad i, j = 1, 2, 3$$

Obviously the four constants β and α_i cannot be numbers; however, they can be 4×4 matrices, and ψ is then a four-component object called a Dirac spinor. Taking Eq. (5) as an eigenequation for E, one finds four eigenstates (because H is a 4×4 matrix), two with $E = +\sqrt{(mc^2)^2 + p^2c^2}$ and two with $E = -\sqrt{(mc^2)^2 + p^2c^2}$. The interpretation of the two positive energy states is that they are the two spin states of a particle with spin ½[hsl]. But the two negative energy states are an embarrassment; even if one started with a particle in a positive energy state, it would quickly make radiative transitions down through the negative energy states. Dirac's solution was to observe that if the particle described by ψ obeyed the Pauli principle, then one can suppose that all the negative energy states are already filled with particles, thus excluding any more.

There are still four single-particle states for a given momentum $\mathbf{p}$: the two spin states of a particle with positive energy, and the two states obtained by subtracting a negative energy particle (of momentum $-\mathbf{p}$). These last states ("hole states") have positive energy and a charge opposite the charge of the particle. The hole is in fact the antiparticle; if the particle is an electron, the hole is a positron. However, with the filling up of the negative energy states, one no longer has a single-particle system, and ψ, just as in the Klein-Gordon case, no longer can be interpreted as a wave function but as a matrix element of a field operator Ψ. *See* POSITRON. [C.J.G.]

Relativity A general theory of physics, primarily conceived by Albert Einstein, which involves a profound analysis of time and space, leading to a generalization of physical laws, with far-reaching implications in important branches of physics and in cosmology. Historically, the theory developed in two stages. Einstein's initial formulation in 1905 (now known as the special, or restricted, theory of relativity) does not treat gravitation; and one of the two principles on which it is based, the principle of relativity (the other being the principle of the constancy of the speed of light), stipulates the form invariance of physical laws only for inertial reference systems. Both restrictions were removed by Einstein in his general theory of relativity developed in 1915, which exploits a deep-seated equivalence between inertial and gravitational effects, and leads to a successful "relativistic" generalization of Isaac Newton's theory of gravitation. Because of the extreme weakness of gravitation compared with all other known forces, the non-newtonian consequences of the general theory have been verified only in a few instances, and their applications are largely confined to cosmology, the structure of neutron stars and black holes, and the motion of bodies in the solar system. For the same reason, where it can be applied, the special theory suffices in the treatment of relativistic effects in atomic, nuclear, and high-energy physics, and its confirmation in numerous instances of this kind renders it among the most securely established of modern scientific theories.

Special (restricted) theory. The scientific developments that led to the birth of special relativity arose from a dilemma confronting physicists in the latter part of the 19th century in their attempts to reconcile Maxwell's electromagnetic equations with the principles of newtonian mechanics. The latter

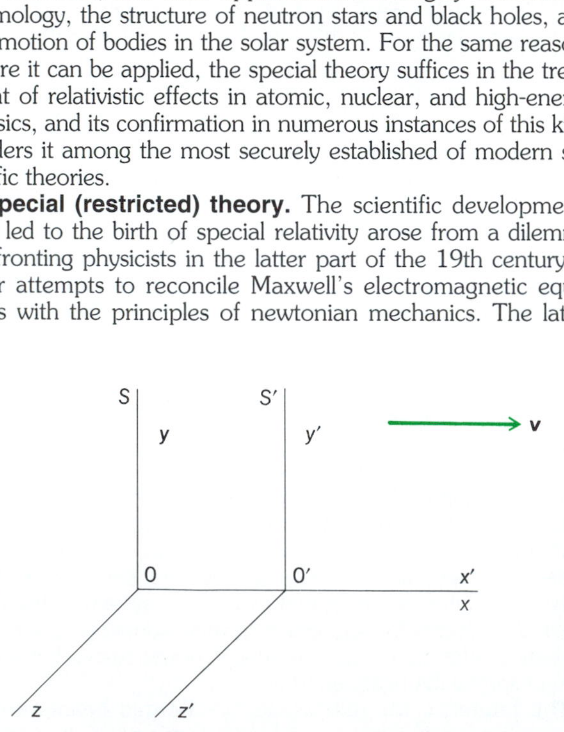

Fig. 1. Inertial reference frames *S*, *S′* with parallel coordinate axes; *v* is the velocity of *S′* relative to *S*.

were then widely held to have universal validity, but Maxwell's equations do not preserve their form under galilean transformations, such as that given by Eqs. (1) [see Fig. 1], that is, they

$$x' = x - vt \qquad y' = y \qquad z' = z \qquad t' = t \tag{1}$$

do not satisfy the galilean principle of relativity. Hence, it appeared necessary to assume that these equations apply only in one special inertial frame, one in which the electromagnetic, or luminiferous, ether is at rest, and which, it was also natural to suppose, served to define Newton's absolute space and absolute velocity. In particular, the absolute velocity of the Earth in its orbital motion around the Sun could be expected to reveal itself in appropriate electromagnetic or optical experiments. However, all such experiments yielded negative results. The most famous of these, the interferometer experiment of A. A. Michelson and E. W. Morley in 1881, presented the greatest challenge. *See* FRAME OF REFERENCE; MAXWELL'S EQUATIONS.

H. A. Lorentz was led in 1892 to postulate that a body whose absolute speed of translational motion is v is contracted in the direction of motion by the factor $\sqrt{1 - v^2/c^2}$. This hypothesis was independently suggested by G. F. FitzGerald, and is thus known as the Lorentz-FitzGerald contraction hypothesis. Other experiments, as well as critical methodological considerations, led Lorentz to arrive gradually, in 1899 and 1904, at a general solution of the problem posed by all the unexpected negative ether-drift results. This solution involved replacing galilean transformations (1) by space-time transformation equations (2), and introducing associated transformation equations for the electromagnetic field quantifies. Henri Poincaré corrected and completed Lorentz's theory in 1905 and 1906, and named Eqs. (2) Lorentz transformations. *See* LIGHT; LORENTZ TRANSFORMATIONS.

$$\begin{aligned} x' &= x - \frac{vt}{\sqrt{1 - v^2/c^2}} \\ y' &= y \\ z' &= z \\ t' &= t - \frac{vx/c^2}{\sqrt{1 - v^2/c^2}} \end{aligned} \tag{2}$$

Although the mathematical structure of the Lorentz-Poincaré theory is the same as that of the theory advanced independently by Einstein in 1905, there is nevertheless a vast difference in the underlying conceptual framework. Einstein started with the idea that the absoluteness of time and space is imposed not by nature but by custom, and that therefore, in particular, the notion of simultaneity must be reexamined critically.

Einstein's formulation. The theory is erected on two fundamental postulates and two definitions. The postulates, which Einstein called the principle of relativity and the principle of the constancy of the velocity of light, can be stated respectively as follows: (I) The analytical form of physical laws is the same in all inertial reference systems. (II) The speed of light in vacuum is a universal constant.

Postulate I is a generalization of the galilean principle of relativity to embrace all phenomena rather than only those of mechanics (excepting, however, gravitational phenomena). It is II that, taken together with I, clearly represents a radical breaking away from traditional thinking. The new formulation of simultaneity is also intimately connected with II. Two spatially separated localized occurrences (or events) are simultaneous when the readings of two identical clocks adjacent to the events are the same, and it is known that the clocks are synchronized. However, when the clocks are not near each other, their synchronism must be defined.

The definition given by Einstein is tied to II, and can be stated as follows: (A) Two identical clocks C, C', situated at two distant points P, P' fixed in a given inertial frame S, synchronize in S, when the respective C and C' times t_1, t'_1 of the sending of a light signal at P and its arrival at P' are connected by the formula $t'_1 - t_1 = t_2 - t'_1$ with the C time t_2 of its return to P after reflection at P' back to P.

Two events that are simultaneous in terms of clocks synchronized in S according to definition A are not simultaneous when referred to identical clocks synchronized according to A in a reference frame that is moving relative to S. In other words, simultaneity is a relative concept: it depends on the reference system under consideration. This relativity is at the heart of many relativistic phenomena.

Einstein's second definition clarifies the operational meaning of the length of a moving rod, and can be stated as follows: (B) The length, as measured in a given inertial frame S, of a rigid rod that is in uniform motion relative to S is the length between the instantaneous positions of the ends of the rod, where the instantaneousness and the length are determined respectively by clocks synchronized in S and by meter sticks at rest in S.

With the aid of these postulates and definitions, Einstein succeeded in deriving the Lorentz transformations and deduced the Lorentz-FitzGerald contraction.

Time dilation. Another remarkable result concerns the relativity of time. As measured in S', the time interval between a given pair of events which have the same coordinates in S' is smaller by the factor $\sqrt{1 - v^2/c^2}$ than it is as measured in S. In this respect, too, there is complete reciprocity, as there must be, between S and S': the time interval between two events is a relative concept. Einstein pointed out also the following corollary of this time dilation effect. If one of two identical clocks, initially synchronized and adjacently at rest in an inertial frame, makes a round trip, it will lag behind the other upon its return. Convincing verification of time dilation is provided by the measured lifetimes of fast mesons and by more precise experiments employing the Mössbauer effect. *See* CLOCK PARADOX; MÖSSBAUER EFFECT.

Minkowski's formulation. Hermann Minkowski's geometric approach, formulated in 1908, which was an important link in the development of general relativity, centers on absolute quantities in the special theory, and provides it with the efficient formalism of tensors in space-time.

Minkowski's space-time, his "world," is the continuum of all events. It is a four-dimensional, quasieuclidean space, with the line element ds^2 (the distance squared between two neighboring events) given by Eq. (3), defining the metric of the space

$$ds^2 = \sum_{\alpha,\beta=0}^{3} \eta_{\alpha\beta} dx^{\alpha dx}\beta \equiv (dx^0)^2 - \sum_{i=1}^{3} (dx_i)2 \tag{3}$$

$$x^0 = ct \qquad x^1 = x \qquad x^2 = y \qquad x^3 = z$$

$$\eta_{\alpha\beta} = 0 \ (if\ \alpha \neq \beta)$$

$$\eta_{00} = -\eta_{11} = -\eta_{22} = -\eta_{33} = 1$$

the "Minkowski metric." *See* SPACE-TIME.

The tensors of special relativity represent absolute quantities, which are determined by their components, relative quantities, obeying the appropriate transformation rules associated with the Lorentz group.

One may replace dx^α in Eqs. (3) by finite Δx^α, and according to whether Δs^2 is greater than 0, less than 0, or equal to 0, one has a timelike, spacelike, or null interval (displacement four-vector). The first and last cases correspond to possible displacements of a particle with $m_0 \neq 0$ or $m_0 = 0$. The integral $\int ds/c$ along a particle's space-time path (or "world line") gives the time as measured by a clock accompanying the particle: the "proper time" on the world line.

Applications. The branches of classical physics can usually be relativistically generalized by applying postulate I in Minkowski's formulation of expressing the physical laws in tensor form. Minkowski's method is successful in the case of a particle moving under the action of an external field of force, and important but limited results exist for systems of interacting particles. A dramatic consequence is Einstein's mass-energy equivalence, which has implications ranging from nuclear reactor energy production to atomic bombs. [H.M.S.]

General theory. One of the basic tenets of special relativity is that no physical effect can propagate with a velocity greater than the speed of light, and so *c* represents a universal speed limit. On the other hand, classical gravitational theory describes the gravitational field of a body throughout space as a function of its instantaneous position, which is equivalent to the assumption that gravitational effects propagate with an infinite velocity; that is, classical gravitational theory is an action-at-a-distance theory. Thus, special relativity and classical gravitational theory are inconsistent, and a modified theory of gravity is necessary. This is the theory developed by Einstein in 1905–1915, following his discovery of special relativity.

Principle of equivalence. It had long been considered a fundamental and puzzling question why bodies of different mass fall with the same acceleration in a gravitational field, or equivalently why the trajectory of a test body is independent of its mass. This situation was explained by Newton with the statement that both the gravitational force on a body and its inertial resistance to acceleration are proportional to its mass.

The explanation by Newton is not very profound and is more in the nature of an ad hoc description. A deeper and more natural explanation occurred to Einstein. In physics there are numerous examples of forces other than gravitation which are mass-proportional; these generally arise due to the use of accelerated coordinate systems to describe the motion. One well-known example is the centrifugal force encountered in a rotating coordinate system. Consider one observer in the gravitational field of the Earth and another in an accelerating elevator or rocket in free space (Fig. 2). If both drop a test body, they will observe it to accelerate relative to the floor. According to classical theory, the Earth-based observer would attribute this to a gravitational force and the elevator-based observer would attribute it to the accelerated floor overtaking the uniformly moving body. However, Einstein reasoned that the effects are identical and the theory of gravity should provide an equivalent description of the two systems. This is the famous principle of equivalence that Einstein made the physical cornerstone of general relativity; it states that on a local scale the physical effects of a gravitational field are indistinguishable from the physical effects of an accelerated coordinate system. From the point of view of the principle of equivalence, it is evident why the motion of a test body in a gravitational field is independent of its mass. *See* CENTRIFUGAL FORCE.

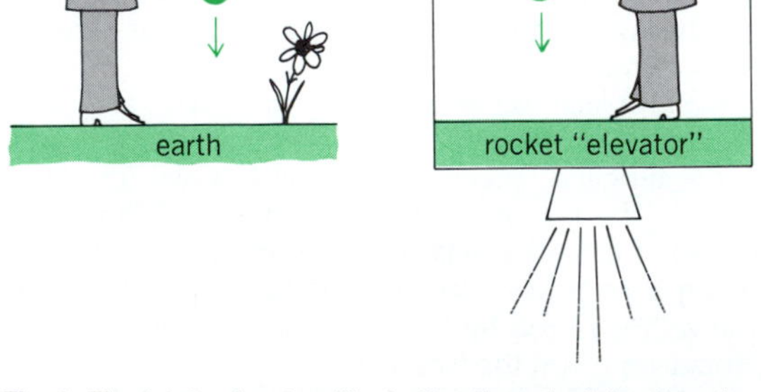

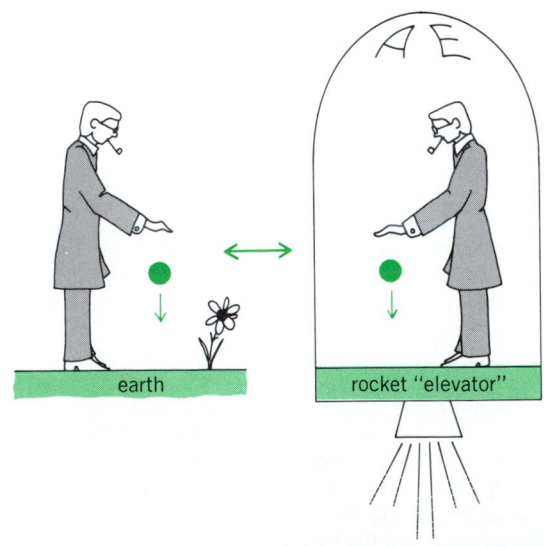

Fig. 2. Einstein's elevator, illustrating the principle of equivalence. The elevator acceleration is equal to *g*, the acceleration of gravity.

Tensor field equations. The close connection between gravity and accelerating coordinate systems suggests that the equations describing the gravitational field be cast in a form that is manifestly independent of the coordinate system in order to achieve the maximum simplicity and elegance. The study of systems of equations that are independent of coordinate systems was begun by K. Gauss in connection with his study of geometry on curved two-dimensional surfaces, and was carried to a high state of development by B. Riemann and T. Levi-Civita in the tensor calculus. Tensors are a basically simple generalization of vectors. *See* TENSOR ANALYSIS.

The space-time of relativity contains one covariant second-rank tensor of particularly great importance, called the metric tensor $g_{\mu\nu}$, which is a generalization of the Lorentz metric of special relativity $\eta_{\alpha\beta}$ introduced in Eqs. (3). Nearby points in space-time, known as events, which are separated by coordinate distances dx^{μ} have an invariant physical separation whose square, the line element, is given by Eq. (4). This quantity is a

$$ds^2 = g_{\mu\nu}\, dx^{\mu}\, dx^{\nu} \tag{4}$$

generalization of the line element in special relativity and has the same relation to the concept of proper time. *See* RIEMANNIAN GEOMETRY.

Tensor equations are equations in which one tensor of a given rank and type (contravariant or covariant) is set equal to another of the same type. The field equations of general relativity are tensor equations for the metric tensor, which completely describes the geometry of the space. The Riemann tensor (or curvature tensor), $R^{\alpha}_{\mu\beta\nu}$ plays a central role in the geometric structure of a space; if it is zero, the space is termed flat and has no gravitational field; if nonzero, the space is termed curved. In terms of the contracted Riemann tensor, a Riemann tensor summed over $\alpha = \beta$, the Einstein field equations for free space are given by Eq. (5). In a region of space

$$R^{\alpha}_{\mu\alpha\nu} \equiv R_{\mu\nu} = 0 \tag{5}$$

containing matter or energy, the zero on the right side of this equation is replaced by a tensor representing the energy content of space.

The field equations are a set of 10 second-order partial differential equations since the four-by-four symmetric tensor $R_{\mu\nu}$ has 10 independent components; they are to be solved for the metric tensor. A solution in a given coordinate system defines an Einstein space-time. The curvature of this space corresponds to the intrinsic presence of a gravitational field. That is, the concept of a field of mechanical force in classical gravitational theory is replaced by the geometric concept of curved space in relativity theory. *See* DIFFERENTIAL EQUATION.

Motion of test bodies. The path of a test body is a generalization of a straight line in euclidean space; it is the shortest "distance" (in terms of intervals *ds*) between points in space-time, known as a geodesic. General relativity theory possesses an extraordinary property; because the field equations are nonlinear, unlike newtonian theory, the motion of a test body in a gravitational field is not arbitrary since the body itself has mass and contributes to the field. Indeed, it turns out that the field equations are so restrictive that the geodesic equation of motion is a necessary consequence and need not be treated as a separate postulate.

Schwarzschild solution. A very important solution of the field equations, obtained by K. Schwarzschild in 1916, represents the field in free space around a spherically symmetric body such as the Sun. It is the basis for a relativistic description of the solar system and all of the experimental tests of general relativity which have been carried out.

Gravitational redshift. There are a number of ways to show that electromagnetic radiation of a given frequency emitted in a gravitational field will appear to an outside observer to have a lower frequency; that is, it will be redshifted. Accurate and dependable measurements involve the use of the Mössbauer effect in terrestrial experiments. With this technique the redshift has been measured by R. Pound and G. Rebka to be within about 1% of the value predicted by general relativity theory.

Perihelion shift. The equations of motion can be solved for a planet considered as a test body in the Schwarzschild field of the Sun. As should be expected, the orbits obtained are very similar to the ellipses of classical theory. Small differences occur, however, the most interesting of which is that the ellipse rotates very slowly in the plane of the orbit so that the perihelion, the point of closest approach of the planet to the Sun, is at a slightly different angular position on each orbit. This shift is extremely small. It is greatest in the case of the planet Mercury, whose perihelion advance is predicted to be only 43 seconds of arc in a century. This is in excellent agreement with the value for the discrepancy between classical theory and observation, which was well known and unaccountable for many years before the discovery of general relativity.

Deflection of star light. The principle of equivalence suggests an extraordinary phenomenon of gravity. Light or other electromagnetic radiation crossing the Einstein elevator horizontally will appear to be deflected downward in a parabolic arc because of the upward acceleration of the elevator. The same phenomenon must occur for light in the gravitational field of the Sun; it must be deflected toward the Sun. A calculation of this deflection gives 1.75 seconds for the net deflection of starlight grazing the edge of the Sun. Modern measurements are made by tracking quasars as they pass near or behind the Sun. With these techniques, the deflection has been measured to be within 1% of the value predicted by general relativity. *See* QUASARS; RADIO ASTRONOMY.

Radio time delay. In the curved space around the Sun the distance between points in space, for example between two planets, is not the same as it would be in flat space. In particular, the round-trip travel time of a radar signal sent between the Earth and Venus will be measurably increased by the curvature effect when the Earth, the Sun, and Venus are approximately lined up. Although the maximum time delay is only a few hundred microseconds, it has been accurately measured. Subsequent experiments using the Viking spacecraft were able to achieve an accuracy of about one-half of 1%, and agree with the predictions of general relativity.

For the applications of general relativity to the study of black holes, to cosmology, and to gravitational radiation *see* BLACK HOLE; COSMOLOGY; GRAVITATION; PULSAR. [R.J.A.]

Relaxation oscillator An electronic circuit which has two stable states, resulting in two distinct output levels, and which switches between the two states at a rate determined by the rate of rise or decay of voltage across the storage element in an *RC* or *RL* circuit. The output waveform is usually nonsinusoidal and may be approximately a square wave, a sawtooth wave, or a series of short repeating pulses. *See* WAVE-SHAPING CIRCUITS.

In one of the most widely used forms of relaxation oscillator, the astable multivibrator, two devices are connected so that they are alternately ON and OFF. The blocking oscillator is representative of a class of circuits in which a single device has two stable conditions. There are other relaxation oscillators using single transistors or transistorlike devices with a current gain greater than unity, such as the point-contact transistor and the unijunction transistor, sometimes called a double-base diode. Negative-resistance devices such as tunnel diodes can also be used as a relaxation oscillator. *See* BLOCKING OSCILLATOR; MULTIVIBRATOR; NEGATIVE-RESISTANCE CIRCUITS; TUNNEL DIODE. [G.M.G.]

Relaxation time of electrons The characteristic time for a distribution of electrons in a solid to approach or "relax" to equilibrium after a disturbance is removed. The most familiar example is a transport property such as electrical resistivity, in which the disturbance is the electric field, equilibrium is the state of zero current, and relaxation time is inversely proportional to the resistivity. Closely related is the concept of a lifetime, which is related to equilibrium properties rather than transport properties. The lifetime is the mean time that an electron will remain in a given quantum state before changing its state as a result of a collision. A related concept is the electron mean free path, the average distance traveled before a collision. Although characteristic collision times are quite short (of the order of 10^{-14} s at room temperature), mean free paths are surprisingly long, ranging from about 100 atom distances at room temperature to 1,000,000 atom distances in a pure metal near absolute zero. This is surprising because atoms in a solid are very densely packed. It corresponds classically to the unlikely event that a rifle bullet might travel for miles through a dense forest without hitting a tree. From a practical point of view, that is what makes metals useful as electrical conductors. *See* FREE-ELECTRON THEORY OF METALS.

The conduction process may be viewed as the steady-state balance between the accelerating effect of an electric field and the decelerating effect of collisions. This process is best viewed in terms of the probability distribution function for the electron gas, $f(\mathbf{k},\mathbf{r},t)$, where $\mathbf{k}$ specifies the momentum, $\mathbf{r}$ the position, and t the time. Viewed in $\mathbf{k}$-space, the entire distribution shifts from its equilibrium distribution f_0 under the influence of a perturbation such as an electric field (see illustration). A steady

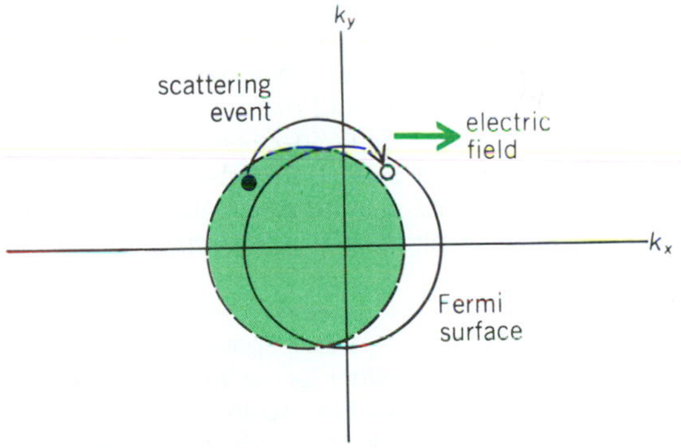

Effect of electric field on electron distribution in a solid, viewed in k-space. Shaded area indicates occupied states in distribution *f* which result when field is applied.

state is reached due to collisions which tend to restore the distribution to f_0. The relaxation time τ is most simply defined in terms of the rate at which the perturbed distribution approaches its equilibrium value once the perturbation is removed, as given in the equation below.

$$\left(\frac{\partial f}{\partial t}\right)_{coll.} = \frac{f - f_0}{\tau}$$

Electron mean free paths in solids are extremely long because the electron wave (viewed quantum-mechanically) readjusts in a perfectly periodic crystal lattice to avoid the atom cores and spend most of its time in the spaces between. In the analogy of the rifle in the forest, the bullet will not travel very far, but the sound of the gun can, because sound waves bend around the trees. In a perfectly periodic lattice, the scattering vanishes because the electron waves scattered from the ion cores interfere in a coherent and constructive way. Any disturbance to the lattice periodicity tends to destroy this wave pattern, resulting in less transmission of the electron wave. The conductivity of solids at room temperature, for example, is limited by the scattering produced by atomic vibrations. At lower temperatures, when the vibrations are reduced, the conductivity is limited by scattering from impurities and lattice imperfections. *See* CRYSTAL; ELECTRICAL RESISTIVITY; LATTICE VIBRATIONS. [R.J.H.]

Relay An electromechanical or solid-state device operated by variations in the input which, in turn, operate or control other devices connected to the output. They are used in a wide variety of applications throughout industry, such as in telephone exchanges, digital computers, motor and sequencing controls, and automation systems. Highly sophisticated relays are utilized to protect electric power systems against trouble and power blackouts as well as to regulate and control the generation and distribution of power. In the home, relays are used in refrigerators, automatic washers and dishwashers, and heat and air-conditioning controls. Although relays are generally associated with electrical circuitry, there are many other types, such as pneumatic and hydraulic. Input may be electrical and output directly mechanical, or vice-versa. *See* RELAY CONTROL SYSTEMS.

Relays using discrete solid-state components, operational amplifiers, or microprocessors can provide more sophisticated designs. Their use is increasing, particularly in applications where the relay and associated equipment are packaged together. *See* AMPLIFIER; MICROPROCESSOR. [J.L.Bl.]

Relay control systems Feedback systems in which the actuator input applied to the plant is either at its maximum or minimum allowed value as a result of a preceding relay. The use of relays is motivated by their low cost and high reliability as well as the fact that for many systems they represent the realization of an optimal control policy. For example, if it is required to transfer the state of a linear system from one point to another in a minimum amount of time, under the constraint that the input does not exceed specified bounds, the solution can be shown to consist of a control that uses only its minimum and maximum values and thus dictates a feedback structure that would be implemented using relays as the primary actuators.

Relay control systems are generally characterized by a set of regions in state space (the vector space of state vectors) in which the control assumes either its maximum or minimum values. The surfaces separating these regions are called switching surfaces. Each point on a switching surface may be identified as being normal, if a trajectory enters that point from one side of the surface and exists from the other, or as being an end point, if trajectories on both sides of the surface at that point are directed toward the surface. Trajectories entering a switching surface at an end point do not leave the surface but "chatter" along it (the relay switches back and forth infinitely fast) until a normal point is encountered and the trajectory exits from the switching surface. Depending on the controlled system, the existence of the chattering mode may or may not be acceptable. *See* CONTROL SYSTEMS; NONLINEAR CONTROL THEORY. [P.E.B.; C.T.C.]

Reliability (engineering) The probability that a component part, equipment, system, or software will satisfactorily perform its intended function under given circumstances, such as environmental conditions, limitations as to operating time, and frequency and thoroughness of maintenance.

Reliability is influenced by all aspects of an engineering effort; the ultimate reliability of a component or a system depends upon the quality of research involved in its conception, its design, the manner in which it is manufactured, the external influences on its operation, maintenance considerations, and other factors. *See* SYSTEMS ENGINEERING.

In specific circumstances reliability can mean a number of different things. Reliability can be considered as a function of time, the time that a piece of equipment operates satisfactorily compared with the total time over which it was expected to operate. A second specific definition of reliability is the average number of hours of maintenance for each hour of satisfactory performance. A third definition is the fraction of the total number of identical equipments which operate satisfactorily over a given period of time. A fourth measure of reliability compares the actual performance of a device with its ideal performance. This fourth definition of reliability permits a direct comparison between the performance of devices under idealized conditions and under actual conditions, thus introducing the influence of environment on performance, and it also considers the realization of a performance ratio rather than a good-bad or operate-nonoperate evaluation. [R.W.M.]

Reluctance A property of a magnetic circuit analogous to resistance in an electric circuit.

Every line of magnetic flux is a closed path. Whenever the flux is largely confined to a well-defined closed path, there is a magnetic circuit. That part of the flux that departs from the path is called flux leakage. *See* MAGNETIC CIRCUITS; MAGNETIC FLUX.

For any closed path of length l in a magnetic field H, the line integral of $H \cos \alpha \, dl$ around the path is the magnetomotive force (mmf) of the path, as in Eq. (1), where α is the angle

$$\text{mmf} = \oint H \cos \alpha \, dl \tag{1}$$

between H and the path. If the path encloses N conductors, each with current I, Eq. (2) holds. *See* MAGNETOMOTIVE FORCE.

$$\text{mmf} = \oint H \cos \alpha \, dl = NI \tag{2}$$

Consider the closely wound toroid shown in the illustration. For this arrangement of currents, the magnetic field is almost

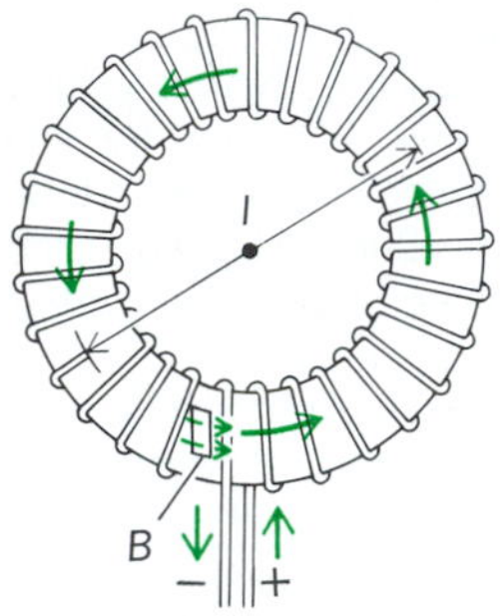

A toroidal coil.

entirely within the toroidal coil, and there the flux density or magnetic induction B is given by Eq. (3), where l is the mean

$$B = \mu \frac{NI}{l} \tag{3}$$

circumference of the toroid and μ is the permeability. The flux Φ within the toroid of cross-sectional area A is given by either form of Eqs. (4), which is similar in form to the equation for

$$\Phi = BA = \frac{\mu A}{l} NI$$

$$\Phi = \frac{NI}{l/\mu A} = \frac{\text{mmf}}{l/\mu A} = \frac{\text{mmf}}{\mathcal{R}} \tag{4}$$

the electric circuit, although nothing actually flows in the magnetic circuit. The factor $l/\mu A$ is called the reluctance $\mathcal{R}$ of the magnetic circuit. The reluctance is not constant because the permeability μ varies with changing flux density. *See* MAGNETIC PERMEABILITY. [K.V.M.]

Reluctance motor A synchronous motor which starts as an induction motor and, upon nearing full speed, locks into step with the rotating field and runs at synchronous speed. The stator and rotor windings are similar to those of an induction motor. The rotor is of squirrel-cage construction, to allow induction-motor starting, and has salient-pole projections which provide synchronous operation at full speed. The reluctance motor is built only in small sizes and for situations in which low cost and simplicity are mandatory and efficiency is of little concern. *See* INDUCTION MOTOR; SYNCHRONOUS MOTOR. [L.V.B.]

Remipedia A unique class of the Crustacea whose members, at first glance, resemble polychaete worms in overall body form. This impression comes from their multiple serial segments and lack of obvious body divisions. The lack of body organization suggests that remipeds are the most primitive crustaceans yet recognized. When compared with the other presumably primitive crustacean class, the Cephalocarida, remipeds are relatively large, with mature individuals ranging from 0.6 to 1.7 in. (15 to 42 mm) in body length. Although the trunk appendages are biramous and paddle-shaped in both classes, endites and epipods that are present in cephalocarids are lacking in remipeds. *See* CEPHALOCARIDA; CRUSTACEA.

As now constituted, the Remipedia comprise two orders, three families, and four genera; however, as more marine caves are explored additional species will be discovered. [P.A.McL.]

Remote-control system A system in which there is an appreciable distance separating the controlled quantity and the controlling quantity. There are three essential components in a remote-control system: a controlling quantity, a transmission medium, and a controlled quantity.

A familiar example of a remote-control system is the telephone system. A voice at one end of the telephone cable controls the motion of a sound-producing diaphragm at the other end of the cable. The voice is the controlling quantity, and the sound produced by the diaphragm is the controlled quantity. Three other examples of remote-control systems are a master-slave manipulator, a telemetering system, and television. *See* CONTROL SYSTEMS; REMOTE MANIPULATORS; TELEMETERING; TELEVISION. [J.G.Tr.]

Remote manipulators Mechanical, electromechanical, or hydromechanical devices which enable a person to perform manual operations while separated from the site of the work. Manipulators are often equipped with touch and vision sensors, and thus can extend human hands and eyes over distances from a few inches to many miles. Reasons which keep a person from performing the work directly include a hazardous work environment or one which is difficult or impossible to gain access to because it is blocked, too large, too small, or too far away.

Manipulators are controlled directly by a person who moves control handles or operates switches similar to those on a derrick, thereby causing the manipulator to move about on wheels or tracks, or to reach out and grasp objects with a mechanical arm. Manipulators are different from industrial robots, which operate without direct and continuous manual control. *See* ROBOTICS.

The table lists some typical manipulator applications, together with the typical distance between operator and work.

A manipulator usually has several links and joints arranged to make an arm. On the end is a gripper or tool which permits objects to be grasped, handles turned, force exerted, buttons pushed, doors opened, and so on. If the operator can be located nearby, he or she can operate the arm like a large puppet, but electric or hydraulic power is more common.

The typical elements of a manipulator system are shown pictorially in the illustration which is a radioactive hot lab situation in which the operator views the work through a lead-glass window. Manipulator systems have control stations, transmission links for command and sensing signals, an effector system at the remote site with various sensor systems, and display systems for helping the operator see and feel what the effector system is doing.

Typical manipulator applications

Application	Reason for separation between operator and work	Typical separation distance
Radioactive laboratory	Protect operator from radiation	6–10 ft (2–3 m)
Undersea research	Protect operator from deep ocean pressure	3–6 ft (1–2 m)
Undersea oil operations	Protect operator from deep ocean pressure	600–1300 ft (200–400 m)
Factory warehouses	Allow operator to lift very heavy loads	3–15 ft (1–5 m)
Factory operations	Protect operator from heat, noise, or shock	3–10 ft (1–3 m)
Microsurgery	Task is too small for customary surgical tools and surgeon's hands	A fraction of an inch to a few inches (a few millimeters to a few centimeters)

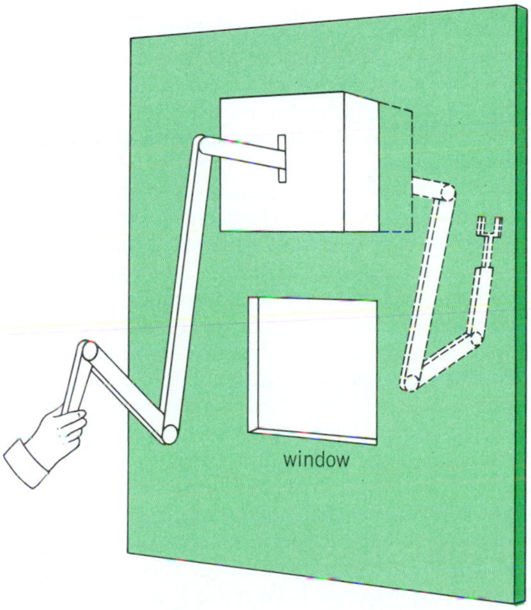

Typical elements of a manipulator system.

Visual sensing is provided by direct view or television. Large manipulators such as that on the space shuttle may have television cameras mounted at the shoulder, elbow, and wrist to give the operator a choice of views. Touch or force sensing is provided either by direct mechanical connection or by electric motors which drive the control handle back against the operator's hand so as to imitate the force being exerted by the distant gripper. *See* CONTROL SYSTEMS; HUMAN-MACHINE SYSTEMS. [D.E.W.]

Remote sensing The gathering and recording of information about terrain and ocean surfaces without actual contact with the object or area being investigated. Remote sensing uses the visual, infrared, and microwave portions of the electromagnetic spectrum. Remote sensing is generally conducted by means of remote sensors installed in aircraft and satellites.

Photography. Photography is probably the most useful remote sensing system. Much of the experience gained over the years from photographs of the terrain taken from aircraft is being drawn upon for use in space.

Multispectral photography isolates the reflected energy from a surface in a number of given wavelength bands and records each spectral band separately on film. This technique allows selection of the significant bandwidths in which a given area of terrain displays maximum tonal contrast and, hence, increases the effective spectral resolution of the system over conventional black-and-white or color systems. Because of its spectral selectivity capabilities, the multispectral approach provides a means of collecting a great amount of specific information.

Multispectral imagery. Multispectral scanning systems record the spectral reflectance by photoelectric means (rather than by photochemical means as in multispectral photography) simultaneously in several individual wavelengths within the visual and near-infrared portions of the electromagnetic spectrum.

In satellite applications, optical energy is sensed by an array of detectors simultaneously in four spectral bands from 0.47 to 1.1 micrometers. As the optical sensors for the various frequency bands sweep across the underlying terrain in a plane perpendicular to the flight direction of the satellite, they record energy from individual areas on the ground. The smallest individual area distinguished by the scanner is called a picture element or pixel, and a separate spectral reflectance is recorded in analog or digital form for each pixel. The spectral reflectance values for each pixel can be transmitted electronically to ground receiving stations in near-real time, or stored on magnetic tape in the satellite until it is over a receiving station. When the signal intensities are received on the ground, they can be reconstructed almost instantaneously into the virtual equivalent of conventional aerial photographs.

Infrared. Thermal infrared radiation is mapped by means of infrared scanners similar to multispectral scanners, but in this case radiated energy is recorded generally in the 8–14-μm portion of the electromagnetic spectrum. The imagery provided by an infrared scanning system gives information that is not available from ordinary photography or from multispectral scanners operating in the visual portion of the electromagnetic spectrum. *See* EMISSIVITY; INFRARED RADIATION.

In the past, thermal infrared images were generally recorded on photographic film. Videotape records are replacing film as the primary recording medium and permit better imagery to be produced and greater versatility in interpretation of data.

Thermal infrared mapping (thermography) from satellite altitudes is proving to be useful for a number of purposes, one of which is the mapping of thermal currents in the ocean. Thermal infrared mapping from aircraft and satellite altitudes has many other uses also, including the mapping of volcanic activity and geothermal sites, location of groundwater discharge into surface and marine waters, and regional pollution monitoring.

Microwave radar. This type of remote sensing utilizes both active and passive sensors. The active sensors such as radar supply their own illumination and record the reflected energy. The passive microwave sensors record the natural radiation. A variety of sensor types are involved. These include imaging radars, radar scatterometers and altimeters, and over-the-horizon radars using large ground-based antenna arrays, as well as passive microwave radiometers and imagers. One of the most significant advantages of these instruments is their all-weather capability, both day and night. *See* MICROWAVE; RADAR.

High-frequency (hf) radar. Such radars utilize frequencies in the 3–30-MHz portion of the electromagnetic spectrum (median wavelength of about 20 m) and are thus not within the microwave part of the spectrum. The energy is transmitted by ground-based antennas in either a sky-wave or surface-wave mode. In the sky-wave mode, the energy is refracted by the various ionospheric layers back down to the Earth's surface some 480–1800 mi or 800–3000 km (on a single-hop basis) away from the hf radar antenna site. The incident waves are reflected from such surface features as sea waves. [P.C.B.]

Renewable resources Agricultural materials used as feedstocks for industrial processes. For many centuries agricultural products were the main sources of raw material for the manufacturing of soap, paint, ink, lubricants, grease, paper, cloth, drugs, and a host of other nonfood products. During the early 1900s, the advances in organic synthesis in western Europe and the United States led to the use of coal as an alternative resource; in the 1940s, oil and natural gas were added as starting materials as a result of great advances in catalysis and polymer sciences. Since then the petrochemical industry has grown rapidly as the result of the abundance and low price of the starting materials as well as the development of new products. However, with the rapidly increasing economies of the nations of the world, these developments did not ever result in reduction in the utilization of agricultural products as industrial materials.

Animal fats, marine and vegetable oils, and their fatty acid derivatives have always played a major role in the manufacturing of many industrial products. Some of these commodities are produced solely for industrial end uses; examples are linseed, tung, castor (not counting minor amounts used for medicinal purposes), and sperm whale oils. Others, such as tallow and soybean oil, are used for both edible and industrial products. *See* DETERGENT; FAT AND OIL; LUBRICANT; SOAP.

Starch, cellulosics, and gums also have been used for many centuries as industrial materials, whereas sugar crops, such as sugarcane and sugarbeet, have mainly satisfied world food requirements. *See* CELLULOSE; GUM; STARCH.

Natural rubber and turpentine are excellent examples of plant-derived hydrocarbons. The development of synthetic rubbers during and after World War II has never threatened the demand for natural rubber; there is generally a world shortage. Turpentine is a product of the wood and paper pulp industry and is used as a solvent and thinner in paints and varnishes. *See* RUBBER; TURPENTINE.

The threat that industrial nations might be separated from part or all of their traditional sources of raw materials through political and economic upheavals or natural calamities has resulted in a renewed effort to develop additional crops for local agriculture. In the United States, research has provided a number of candidate species that either are now in commercial development or are ready for the time when circumstances warrant such development. Examples are jojoba (liquid wax ester to replace sperm whale oil), guayule (alternate source of natural rubber), kenaf (paper fiber with annual yields much higher than available from trees), and crambe and meadowfoam (long-chain fatty acids, since erucic acid is no longer available from rapeseed oil). There is also active research involving *Cuphea* species (alternate source of lauric and other medium-chain fatty acids, to augment coconut oil), *Vernonia* (source of epoxy oil), and several other promising plants. For example, the Chinese tallow tree has the potential of producing 2.2 tons per acre (5 metric tons per hectare) of seed oil that could be used for manufacturing fuel and other chemicals. [L.H.P.]

Renner-Teller effect The splitting, into two, of the potential function along the bending coordinate in degenerate electronic states of linear triatomic or polyatomic molecules. Most of the areas and methods of molecular physics and spectroscopy assume the validity of the Born-Oppenheimer approximation. The nuclei generally move much more slowly than the electrons, the frequencies associated with electronic transitions are much higher than vibrational frequencies, and one can consider separately the three types of molecular motion: electronic, vibrational, and rotational. These statements are no longer necessarily valid for electronic states which are degenerate or at least close to degeneracy, and the Born-Oppenheimer approximation breaks down.

Degenerate electronic states usually occur in molecules having a high degree of symmetry. The symmetric equilibrium geometry which causes the electronic degeneracy is, in general, lowered in the course of molecular vibrations, and this may lead to splitting of the potential. The molecular potential is usually expressed in terms of a polynomial expansion in displacements r, and, in nonlinear molecules, the linear terms may lead to coupling of the electronic and vibrational degrees of freedom. The resulting breakdown of the Born-Oppenheimer approximation is in this case known as the Jahn-Teller effect. In linear molecules the symmetry is lowered during bending vibrations. In the bending potential the linear (and other odd) terms are zero by symmetry. The first nonvanishing terms which can couple the degenerate electronic states are quadratic in the bending coordinate. The results of this coupling in linear molecules are referred to as the Renner-Teller effect, or simply the Renner effect. *See* JAHN-TELLER EFFECT; MOLECULAR STRUCTURE AND SPECTRA. [V.E.B; T.A.M.]

Rennin An enzymatic protein used for coagulating milk casein in cheese making. It is also related to the cheese-ripening process through its proteolytic activity. *See* CHEESE; ENZYME.

Rennin is secured from rennet In preparing rennet extract, only milk-fed calves are used. After butchering, the fourth stomach, or abomasum, is removed and freed of its food content. At the rennet factory, the stomachs are processed to extract the rennin. [P.H.T.]

Renormalization A program in quantum field theory consisting of a set of rules for calculating S-matrix amplitudes which are free of ultraviolet (or short-distance) divergences, order by order in perturbative calculations in an expansion with respect to coupling constants. *See* QUANTUM FIELD THEORY; SCATTERING MATRIX. [I.Ba.]

Reproduction (animal) The formation of new individuals, which may occur by asexual or sexual methods. In the asexual methods, which occur mainly among the lower animals, the offspring are derived from a single individual. Sexual methods are general throughout the animal kingdom, with offspring ordinarily derived from the paired union of special cells, the gametes, from two individuals. Basic to all processes of reproduction is the origin of the new individual from one or more living cells of the parent or parents.

Asexual reproduction. Asexual processes of reproduction include binary fission, multiple fission, fragmentation, budding, and polyembryony. Among the protozoa and lower metazoa, these are common methods of reproduction. However, the last-mentioned process can occur in mammals, including humans.

Binary fission involves an equal, or nearly equal, longitudinal or transverse splitting of the body of the parent into two parts, each of which grows to parental size and form. This method of reproduction occurs regularly among protozoans.

Multiple fission, schizogony, or sporulation produces several new individuals from a single parent. It is common among the Sporozoa, such as the malarial parasite, which form cystlike structures containing many cells, each of which gives rise to a new individual.

Fragmentation is a form of fission occurring in some metazoans, especially the Platyhelminthes, or flatworms; the Nemertinea, or ribbon worms; and the Annelida, or segmented worms. The parent worm breaks up into a number of parts, each of which regenerates missing structures to form a whole organism.

Budding is a form of asexual reproduction in which the new individual arises from a relatively small mass of cells that initially forms a growth or bud on the parental body. The bud may assume parental form either before separation from the body of the parent as in external budding, or afterward, as in internal budding. External budding is common among sponges, coelenterates, bryozoans, flatworms, and tunicates. Internal budding occurs among fresh-water sponges and bryozoans. In the sponges the internal buds, termed gemmules, consist of groups of primitive cells surrounded by a dense capsule formed by the body wall. If the parent animal dies as a result of desiccation or low temperature, the cells of the gemmules can later be released and form new sponges. In the bryozoans the similarly functioning buds are known as statoblasts.

Polyembryony is a form of asexual reproduction, occurring at an early developmental stage of a sexually produced embryo, in which two or more offspring are derived from a single egg. Examples are found scattered throughout the animal kingdom, including humans; in humans it is represented by identical twins, triplets, or quadruplets.

Sexual reproduction. Sexual reproduction in animals assumes various forms which may be classified under conjugation, autogamy, fertilization (syngamy), and parthenogenesis. Basically, the various processes all involve the occurrence of certain special nuclear changes, termed meiotic divisions, preliminary to the production of the new individual. *See* GAMETOGENESIS; MEIOSIS.

Conjugation occurs principally among the ciliate protozoans, such as *Paramecium*, and involves a temporary union of two individuals during which each is "fertilized" by a micronucleus from the other.

In autogamy the nuclear changes described for conjugation take place, but since there is no mating, there is no transfer of micronuclei. Instead, the prospective migratory micronucleus reunites with the stationary one. The process may be considered related to parthenogenesis.

Fertilization, or syngamy, comprises a series of events in which two cells, the gametes, fuse and their nuclei, which had previously undergone meiotic divisions, fuse. In metazoans, the gametes are of two morphologically distinct types: spermatozoa, or microgametes, and eggs, also called ova or macrogametes. These types are produced by male and female animals, respectively, but in some cases both may be produced by a single, hermaphroditic, individual. The nucleus of the spermatozoon has half the number of chromosomes characteristic of the ordinary (somatic) cells of the animal. The nucleus of the ripe egg in some animals, for instance, coelenterates and echinoderms, also has attained this haploid condition, but in most species of animals it is at an early stage of the meiotic divisions when ready for fertilization. In the latter situation, the meiotic divisions of the egg, characterized by formation of small, nonfunctional cells termed polar bodies, are completed after the sperm enters, whereupon the haploid egg nucleus fuses with the haploid sperm nucleus. Fertilization thus produces a zygote with the diploid chromosome number typical of the somatic cells of the species (23 pairs in humans) and this is maintained during the ensuing cell divisions.

Parthenogenesis is the development of the egg without fertilization by a spermatozoon. It is listed as a form of sexual reproduction because it involves development from a gamete. Rotifers, crustaceans, and insects are the principal groups in which it occurs naturally. It has also been induced (artificial parthenogenesis) in species from all the major phyla by various kinds of chemical or physical treatment of the unfertilized egg. Even in mammals, several adult rabbits have reportedly been thus produced. *See* ESTRUS; OOGENESIS; OVUM; SPERM CELL; SPERMATOGENESIS. [A.T./H.L.H.]

Reproduction (plant) Plants produce new individuals by either asexual or sexual means.

Asexual reproduction. In simple one-celled plants, asexual reproduction is accomplished by division of the vegetative cell and the subsequent growth of the two nearly equal daughter cells which form new individuals. This is known as simple (binary) fission. Another method, known as budding or gemmation, occurs in the unicellular yeast plant. In this plant a small protuberance, which later becomes detached by constriction, emerges from the cell. The new cell thus formed may develop directly into a new plant, or it may undergo division and give rise to more new individuals. In some liverworts and mosses, the plants produce small multicellular structures called gemmae, which, when detached, may grow, forming new plants. Still another method occurring in many kinds of plants is effected by means of spores, which are usually formed by multiple division of an ordinary vegetative cell or in a specialized organ (sporangium).

Asexual reproduction also takes place by fragmentation, in which a part of the vegetative plant becomes detached and develops into a new plant. Many of the higher plants have underground stems, such as rhizomes, tubers, bulbs, or corms, which are efficient agents in vegetative or asexual propagation. *See* BUDDING.

Sexual reproduction. Sexual reproduction appeared early in the evolution of the plant kingdom. In this kind of reproduction, specialized sex cells called gametes fuse, forming a zygote which has the potential to produce a new individual. In some lower plants such as the alga *Spirogyra*, two protoplasts of ordinary vegetative cells may serve as gametes and unite, but in others the protoplast undergoes multiple fission, producing many motile cells. If these are relatively large and few in number, they will function as spores, germinating and producing new plants asexually. However, if these motile cells are quite small and numerous, they will function as gametes, fusing in pairs to form zygotes. When these gametes are similar in size, appearance, and structure they are called isogametes; when morphologically different, they are heterogametes. Therefore, fusion of isogametes is known as isogamy, and that of heterogametes as heterogamy. In heterogamy, the large, nonmotile gamete (female) is the egg and the smaller, motile gamete (male) is the sperm. In thalloid plants, if the plant (thallus) produces both kinds of gametes, it is homothallic; if only one kind, it is heterothallic. *See* THALLOBIONTA.

Early in the course of evolution, certain cells of plants became specialized as sex organs to produce differentiated gametes. The male organ producing numerous motile gametes (sperms) is called the antheridium, and the female organ usually producing only one egg is the oogonium. In more highly organized plants, the one-celled oogonium is replaced by a more complicated, multicellular archegonium containing one egg. The plant bearing antheridia alone is the male gametophyte, and the one with archegonia is the female gametophyte. Eventually, the seed plants appeared at the top of the evolutionary scale, bearing flowers with stamens and pistils. The stamens bear anthers (microsporangia) which produce the pollen grains (microspores) comparable to the asexual spores found in lower plants. In the same flower, or in another, there is a pistil composed usually of an ovary, a style, and a stigma. The ovary contains one or more ovules (megasporangia), in which the megaspores are formed. One of these will develop into the embryo sac (female gametophyte). The pollen grain will germinate only on the stigma of a flower. The transfer of pollen to the stigma (pollination) is accomplished by various agencies; some of these agencies are gravitation, wind, insects, birds, and other animals. *See* EMBRYOBIONTA; FLOWER; POLLINATION.

After pollination the pollen tube penetrates the ovule and enters the embryo sac, where it swells and bursts, releasing the two sperms. The tube nucleus degenerates. One of the sperms unites with the fusion nucleus, forming the so-called triple fusion (endosperm) nucleus which divides and grows into the endosperm, a food-accumulating tissue. The other sperm unites with the egg, forming a zygote which may soon begin to grow at the expense of the endosperm, giving rise to the embryo.

Pollination and fertilization may affect structures outside the embryo and endosperm, resulting in abscission of the pistil, or they may affect chemical composition, color, and time of ripening of the fruit. Such influence is called ectogony. In triple fusion, genetic factors which are responsible for certain characteristics of the endosperm such as color, form, shape, and chemical composition may be introduced by the male gamete. Appearance of these characteristics is known as xenia.

As the embryo and endosperm are developing, the ovule enlarges, its integuments become modified into protective coverings (seed coats) and the seed is fully formed.

Seed development is accompanied by changes in the ovary, which enlarges and becomes the fruit. The fruits of plants display a wide range in form, size, texture, design, and chemical composition, but in all cases fruits aid in dissemination of seeds. *See* FRUIT; SEED. [P.D.St./E.L.C.]

Reproductive behavior The systems of behavior pattern by means of which, in different types of animals, the sperm is brought to the egg and the parental care of the resulting young is ensured. The mere existence of eggs and sperm and of reproductive organs is not enough to guarantee that the eggs will reach the sperm. There must also be, for each species, a set of behavior patterns through which the male and female approach each other and, by their mutual reactions to each other, facilitate the movement of the sperm toward the egg. Similarly, there must exist patterns of parental behavior, appropriate for each type of animal, by which the parent approaches the young and provides the necessary care. The various patterns of reproductive behavior thus constitute one of the ways in which animals are adapted to their environment. These patterns have evolved under the influence of natural selections just as have other aspects of structure, function, and behavior. [D.S.L.]

Reproductive system The structures concerned with the production of sex cells (gametes) and perpetuation of the species. The reproductive function constitutes the only vertebrate physiological function that necessitates the existence of two morphologically different kinds of individuals in each animal species, the males and the females (sexual dimorphism). The purpose of the reproductive function is fertilization, that is, the fusion of a male and a female sex cell produced by two distinct individuals.

Anatomy. Egg cells, or ova, and sperm cells, or spermatozoa, are formed in the primary reproductive organs, which are collectively known as gonads. Those of the male are called testes; those of the female are ovaries. The gonads are paired structures, although in some forms what appears to be an unpaired gonad is the result either of fusion of paired structures or of unilateral degeneration.

The reproductive elements formed in the gonads must be transported to the outside of the body. In most vertebrates, ducts are utilized for this purpose. These ducts, together with the structures that serve to bring the gametes of both sexes together, are known as accessory sex organs. The structures used to transport the reproductive cells in the male are known as deferent ducts and those of the female as oviducts. In a few forms no ducts are present in either sex, and eggs and sperm escape from the body cavity through genital or abdominal pores.

Oviducts, except in teleosts and a few other fishes, are modifications of Möllerian ducts formed during early embryonic development. In all mammals, each differentiates into an anterior, nondistensible Fallopian tube and a posterior, expanded uterus. In all mammals except monotremes the uterus leads to a terminal vagina which serves for the reception of the penis of the male during copulation. The lower part, or neck, of the uterus is usually telescoped into the vagina to a slight degree. This portion is referred to as the cervix.

In most vertebrates the reproductive ducts in both sexes open posteriorly into the cloaca. In some, modifications of the cloacal region occur and the ducts open separately to the outside or, in the male, join the excretory ducts to emerge by a common orifice. *See* COPULATORY ORGAN; OVARY; PENIS; REPRODUCTION (ANIMAL); TESTIS. [C.K.W.]

Physiology. The physiological process by which a living being gives rise to another of its kind is considered one of the outstanding characteristics of plants and animals. It is one of the two great drives of all animals; self-preservation and racial perpetuation.

Estrous and menstrual cycles. The cyclic changes of reproductive activities in mammalian females are known as estrous or menstrual cycles.

Most mammalian females accept males only at estrus (heat). Estrus in mammals can occur several times in one breeding season; the mare, ewe, and rat come to estrus every 21, 16, and 5 days respectively if breeding does not take place. This condition is called polyestrus. The bitch is monestrous; she has only one heat, or estrus, to the breeding season and if not served then, she does not come into heat again for a prolonged interval, 4–6 months according to different breeds. In monestrous and seasonally polyestrous species the period of sexual quiescence between seasons is called anestrus. *See* ESTRUS.

The reproductive cycle of the female in the primate and human is well marked by menstruation, the period of vaginal blood flow. Menstruation does not correspond to estrus but occurs between the periods of ovulation at the time the corpus luteum declines precipitously. *See* MENSTRUATION.

Mating. Mating, also called copulation or coitus, is the synchronized bodily activity of the two sexes which enables them to deposit their gametes in close contact. It is essential for successful fertilization because sperm and ovum have a very limited life span.

The logistics of sperm transport to the site of fertilization in the oviduct present many interesting features in mammals, but it is important to distinguish between passive transport of sperm cells in the female genital tract, and sperm migration, which clearly attributes significance to the intrinsic motility of the cell. Viable spermatozoa are actively motile, and although myometrial contractions play a major role in sperm transport through the uterus, progressive motility does contribute to migration into and within the oviducts. Even though a specific attractant substance for spermatozoa has not yet been demonstrated to be released from mammalian eggs or their investments, some form of chemotaxis may contribute to the final phase of sperm transport and orientation toward the egg surface.

Although in most mammalian species the oocyte is shed from the Graafian follicle in a condition suitable for fertilization, ejaculated spermatozoa must undergo some form of physiological change in the female reproductive tract before they can penetrate the egg membranes. The interval required for this change varies according to species, and the process is referred to as capacitation. The precise changes that constitute capacitation remain unknown, although there is strong evidence that they are—at least in part—membrane-associated phenomena, particularly in the region of the sperm head, that permit release of the lytic acrosomal enzymes with which the spermatozoon gains access to the vitelline surface of the egg.

Fertilization takes place in the oviducts of mammals and the fertilized eggs or embryos do not descend to the uterus for some 3 to 4 days in most species. During this interval, the embryo undergoes a series of mitotic divisions until it comprises a sphere of 8 or 16 cells and is termed a morula. Formation of a blastocyst occurs when the cells of the morula rearrange themselves around a central, fluid-filled cavity, the blastocoele. As the blactocyst develops within the uterine environment, it sheds its protective coat and undergoes further differentiation before developing an intimate association with the endometrium, which represents the commencement of implantation or nidation.

Association of the embryo with the uterine epithelium, either by superficial attachment or specific embedding in or beneath the endometrium, leads in due course to the formation of a placenta and complete dependence of the differentiating embryo upon metabolic support from the mother. Implantation and placentation exhibit a variety of forms, but in all instances the hormonal status of the mother is of great importance in determining whether or not implantation can proceed. *See* PREGNANCY.

Endocrine function. The endocrine glands secrete certain substances (hormones) which are necessary for growth, metabolism, reproduction, response to stress, and various other physiological processes. The endocrine glands most concerned with the process of reproduction are the pituitary and the gonads.

The posterior lobe of the pituitary gland secretes two neurohumoral agents, vasopressin and oxytocin. These are involved in reproduction only indirectly, through their effect on uterine contractility in labor and on the release of milk from the mammary gland when a suckling stimulus is applied. The anterior lobe secretes a variety of trophic hormones, including two gonadotrophic hormones, the follicle-stimulating hormone (FSH) and the luteinizing or interstitial-cell stimulating hormone (LH or ICSH). These hormones act directly on both ovaries and testes. *See* PITUITARY GLAND.

The gonadal (steroidal) hormones control the secretion of gonadotrophins by acting on the hypothalamus. It has been suggested that steroids act by means of a "negative feedback"; that is, high levels of circulating gonadal hormones stop further release of gonadotropins. However, although this is true for experiments involving pharmacological doses of such hormones, it may not be the case with endogenous physiological levels. It is certainly true that less steroid is required to inhibit pituitary function in the female than in the male. Under certain circumstances small doses of gonadal hormones can stimulate release of gonadotropic hormones from the pituitary. Estrogen can simulate the release of LH; hence the occurrence of ovulation in rats, rabbits, sheep, and women. Progesterone can also facilitate ovulation in persistently estrous rats, in chickens, and in estrous sheep and monkeys. *See* ESTROGEN; PROGESTERONE.

The formation of gametes (spermatogenesis and oogenesis) is controlled by anterior pituitary hormones. The differentiation of male and female reproductive tracts is influenced, and mating behavior and estrous cycles are controlled, by male or female hormones. The occurrence of the breeding season is mainly dependent upon the activity of the anterior lobe of the pituitary, which is influenced through the nervous system by external factors, such as light and temperature. The transportation of ova from the ovary to the Fallopian tube and their subsequent transportation, development, and implantation in the uterus are controlled by a balanced ratio between estrogen and progesterone. Furthermore, it is known that estrogens, androgens, and progesterone can all have the effect of inhibiting the production or the secretion, or both, of gonadotrophic hormones, permitting the cyclic changes of reproductive activity among different animals.

Mammary glands are essential for the nursing of young. Their growth, differentiation, and secretion of milk, and in fact the whole process of lactation, are controlled by pituitary hormones as well as by estrogen and progesterone. Other glands and physiological activities also influence lactation, although this is largely via the trophic support of other pituitary hormones. [M.C.C.; M.J.K.H.; R.H.F.H.]

Reproductive system disorders Those disorders which involve the structures of the female and male reproductive systems.

Female system. Failure of the ovaries to form normally results in short stature, sterility, lack of development of female secondary sex characteristics, such as breast growth, fat deposition in buttocks, thighs, and mons pubis, and female escutcheon. Destruction of the ovaries after puberty results in loss of fertility, cessation of menses, and the loss of secondary sex characteristics.

Neoplastic enlargement of the ovary can be cystic or solid, benign or malignant. Malignant ovarian tumors are commonly asymptomatic in their early stages, and may be quite widely spread in the pelvic and abdominal cavity before they are discovered. *See* OVARY.

Inflammation is the most common disorder of the Fallopian tubes, and if it is repeated or severe, destruction of the tubal lining with closure of the outer ends of the tubes can occur. This inflammation can be caused by several bacterial organisms. Sterility commonly results because the tube is permanently closed to passage of eggs and sperm.

The second most common problem involving the tube is pregnancy. The egg is fertilized in the outer portion of the tube and descends to implant in the uterus, but in some cases the passage is delayed and the conceptus attaches to the wall of the tube. The tube has a small lumen and thin wall, and the growing pregnancy quickly enlarges and grows through the tube, leading usually to rupture and hemorrhage into the peritoneal cavity.

The muscle (myometrium) and lining (endometrium) of the uterus are susceptible to various problems, including tumors, infections, and hormonal derangements. Benign tumors of the myometrium (fibroids) are a common disorder, producing irregular enlargement of the uterus and sometimes causing pain, obstruction of the urinary tract, and heavy vaginal bleeding.

Infection of the endocervical glands with gonococcus or chlamydia trachomatis agent can occur. This may be asymptomatic, except for producing a mucopurulent discharge, or it may cause pain when the cervix is manipulated, particularly during intercourse. The infection can ascend from this area into the internal genital organs and adjacent structures. *See* GONORRHEA.

Cancer of the surface tissue of the cervix is a relatively common, potentially lethal problem. Early malignant changes can be detected by examining properly stained exfoliated cells in the manner of G. N. Papanicolaou. This cancer produces few symptoms, except for irregular bleeding (usually postcoital).

Inflammation of the lining of the vagina can be due to several common organisms: *Monilia*, a fungus; *Trichomonas*, a protozoan; and *Haemophilus vaginalis*, a bacterium. Infection of the female external genitalia can be due to *Monilia*, bacterial agents, and viruses. This infection can be diffuse and widespread, or localized to hair follicles, Bartholin's glands, or lesions such as warts or ulcers. Inflammation can be caused by allergic reactions to soap, powders, semen, lubricant, or even clothing.

Male system. During fetal life the testicles form in the abdominal cavity near the kidneys and migrate down into the scrotum. Failure of descent permanently damages the sperm-producing cells causing infertility but allows the interstation cells which produce hormones to survive. Other causes of male infertility are organic problems which block passage of seminal fluid or interfere with sperm production.

The testes are sites of malignant tumors which carry a high mortality rate. Undescended testicles are more susceptible to malignant transformation, and for that reason should be removed when discovered, and appropriate hormonal replacement instituted. *See* TESTIS.

The major disorder of the epididymis is infection and inflammation, which can lead to scarring and permanent blockage of the ducts. The vas deferens and seminal vesicles are rarely afflicted by disease, although the network of veins surrounding the vas can become engorged and tortuous, and is called a variococoele. Inflammation is the chief disorder of the urethra causing dysuria and a discharge.

Inflammation and development of small calculi are relatively minor ailments of the prostate. Benign overgrowth of this gland is the most common and troublesome complaint. This enlargement results in constriction of the urethra, which obstructs urinary outflow and leads to an increasing residual of urine in the bladder. Cancer of the prostate is another relatively frequent problem in older males. This tumor spreads primarily to bones, where it is markedly painful.

Inflammation (balanitis) or narrowing (phimosis) of the foreskin which covers the glans penis can occur and may interfere with urination. These conditions are unknown in parts of the world where the foreskin is removed for social or medical reasons. Viral infections can produce warty growths (condyloma acuminata) or painful ulcers (herpetic lesions) of the glans or shaft of the penis, and a syphilitic infection can result in the firm, painless ulcer known as a chancre.

Disorders of erectile capability range from tumescence unaccompanied by sexual desire (Peyronie's disease, priapism) to impotence, both psychologic and secondary to old age. Carcinoma can arise from the surface epithelium of the penis and spread both locally and via the lymphatic system to the nodes of the groin. This disease is virtually unknown in populations where males are circumcised at birth. [G.M.Gu.]

Reproductive technology Any procedure undertaken to aid in conception, intrauterine development, and birth when natural processes do not function normally. The most common are in vitro fertilization and gamete intrafallopian transfer.

Infertility, which is the inability to conceive during at least 12 months of unprotected intercourse, is an increasingly common problem. Hormonal therapy and microsurgery are used to overcome many hormonal and mechanical forms of infertility, but many types of infertility do not respond to such treatment. *See* INFERTILITY.

In vitro fertilization bypasses the Fallopian tubes. Immediately prior to ovulation the mature oocyte is removed from the ovary and placed, together with prepared sperm, in a petri dish for 2–3 days. Fertilization takes place during this time and the fertilized oocyte develops into a two- to eight-cell embryo, which is then transferred into the uterus. In vitro fertilization is useful for females with absent or severely damaged Fallopian tubes; couples in which the female has endometriosis; when the male has severely reduced sperm counts; or when the couple has immunologic or unexplained infertility for a period of 2 or more years. The four principal steps of in vitro fertilization are induction and timing of ovulation, oocyte retrieval, fertilization, and embryo transfer.

Gamete intrafallopian transfer, or GIFT, is similar to in vitro fertilization with a few important distinctions. In gamete intrafallopian tube transfer, the spermatozoa and oocytes are placed into the fimbriated end of the Fallopian tube during the laparoscopy. It is unusual for more than two oocytes to be placed into either Fallopian tube. Indications for gamete intrafallopian transfer include unexplained infertility of two or more years' duration, cervical stenosis, immunologic infertility, oligospermia, and endometriosis. At least one Fallopian tube must appear normal; where there has been severe pelvic adhesions or distorted tubal anatomy from any cause, gamete intrafallopian transfer should not be considered. *See* PREGNANCY; REPRODUCTIVE SYSTEM DISORDERS. [M.M.S.]

Reprographics The generation of graphic images (characters of various alphabets, line drawings, photographs) which use any one or several of a broad spectrum of mechanical, electronic, and electromechanical devices. With the advent of computers into the field of graphics, a number of disciplines that were formerly separate, such as calligraphy, orthography, photography, lithography, planography, telegraphy, telephony, radiotelegraphy, television, and the skills of art and drafting, have been linked through the medium of electronic communications between people and machines to form an interactive communications network. In a basic reprographic system it is possible for a person to communicate with a machine by inputting a message via typewriter-optical character reader or magnetic disk. Voice-to-digital communication, and vice versa, is in various stages of development. Reprographics permits the composition of an entire page on a video display terminal with subsequent reproduction in multiple copies by electronically communicating the image to a bond copy output device. The hard copies can be generated within a facility, or the entire page can be transmitted over a network for viewing remotely as soft copy on a page layout terminal or a video display terminal, or for remote reproduction of hard copies.

Reprographics has even been extended to the ends of the solar system. The pictures of the planets, especially Jupiter and Saturn, sent back by the Voyagers and Pioneers and subsequently reproduced have added considerably to documentation of the solar system. *See* Character recognition; Microprocessor; Photocopying processes; Printing; Telegraphy; Teletypewriter; Word processing. [E.L.Ga.]

Reptilia A class of vertebrates composed of four living orders, the turtles or Chelonia, the tuatara or Rhynchocephalia, the lizards and snakes or Squamata, and the crocodilians or Crocodilia. Numerous extinct orders are also known. The group first appeared in the Carboniferous and underwent a culminating evolutionary radiation in the Mesozoic, often called the age of reptiles. Although the major portion of the class is not extinct, several Recent groups, particularly the Squamata, are very successful, and there are approximately 5000 living species of reptiles as compared to about 4000 living mammals. A classification for the class is given here; see separate articles on each of the groups listed.

Class Reptilia
- Subclass Anapsida
  - Orders: Cotylosauria
  - Mesosauria
  - Chelonia
- Subclass Ichthyopterygia
  - Order: Ichthyosauria
- Subclass Euryapsida
  - Orders: Sauropterygia
  - Placodontia
  - Araeoscelidia
- Subclass Lepidosauria
  - Orders: Eosuchia
  - Rhynchocephalia
  - Squamata
- Subclass Archosauria
  - Orders: Thecodontia
  - Crocodilia
  - Pterosauria
  - Saurischia
  - Ornithischia
- Subclass Synapsida
  - Orders: Pelycosauria
  - Theropsida
  - Ictidosauria

The reptiles are the most primitive of the completely terrestrial vertebrates and are consequently the first to exhibit amniote features. In the reptiles the eggs are covered by a complex series of protective layers, including a leathery or calcareous shell. A rich supply of food material in the form of yolk is deposited inside the ovum to furnish food for the developing embryo. A series of protective extraembryonic membranes, the serosa and amnion, appears later in embryogenesis to protect the embryo from water loss and shock. A third such membrane, the allantois, functions as a storage sac for nitrogenous wastes. The serosa and allantois usually fuse to form a respiratory structure. Gaseous exchanges take place across the shell and seroallantoic membrane between the outside air and the blood vessels of the allantois. All of these adaptations made it possible for the reptile egg to be deposited on land, undergo its development there, and hatch into a fully developed form without a gilled larval stage. Most reptilian eggs are buried in the soil or in rotting vegetation out of the direct sunlight. *See* Amniota.

If all intermediate fossil forms are taken into account, the above features of the life history are the only ones that serve to distinguish the reptiles from the amphibians. As a matter of fact, several fossil groups near the evolutionary transition between the two classes, known only from bony remains, cannot be satisfactorily placed in one or the other class because information on their life histories is not available. *See* Amphibia.

In recent reptiles, discrete dry horny scales cover the body; mucous glands are lacking in the skin; there is a single occipital condyle; mammary glands are lacking; the middle ear region has a single ossicle (stapes), if functional, transmitting vibrations to inner ear; the heart is composed of two auricles and a single partially divided ventricle (fully divided in Crocodilia); and pleural cavities and peritoneal cavities are continuous (separated in crocodilians). Although primarily oviparous, some lizards and snakes are ovoviviparous or viviparous.

The Mammalia are descended from one reptilian as is the class Aves. Both groups differ markedly from the Reptilia. [J.M.S.]

Repulsion motor An alternating-current (ac) commutator motor designed for single-phase operation. The chief distinction between the repulsion motor and the single-phase series motors is the way in which the armature receives its power. In the series motor the armature power is supplied by conduction from the line power supply. In the repulsion motor, however, armature power is supplied by induction (transformer action) from the field of the stator winding. For discussion of the ac series motor *see* Universal motor. *See also* Alternating-current motor.

The repulsion motor primary or stationary field winding is connected to the power supply. The secondary or armature

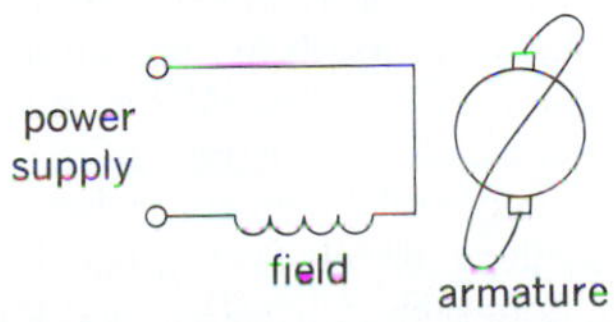

Schematic of a repulsion motor.

winding is mounted on the motor shaft and rotates with it. The terminals of the armature winding are short-circuited through a commutator and brushes. There is no electrical contact between the stationary field and rotating armature (see illustration). *See* Windings in electric machinery. [I.L.K.]

Reserve battery A battery which is inert until an operation is performed which brings all the cell components into the proper state and location to become active. *See* BATTERY.

There are several types of reserve batteries. In water-activated or electrolyte-activated batteries the water or electrolyte component is not present during storage. It is added just before the cell is put into use. In thermal batteries the electrolyte is a solid at room temperature and has very low conductivity. If the temperature is raised above the melting point, the conductivity of the electrolyte becomes excellent and the cell is capable of delivering significant power.

Practical, water-activated battery systems have been developed using magnesium anodes against silver chloride or cuprous chloride cathodes. The most important design factor is to ensure that the electrolyte is delivered as quickly as possible at the time of activation, at the same time avoiding chemical short-circuiting of the cells. The dry elements of water-activated batteries are stored in a sealed container and are capable of indefinite storage life.

Any cell can be made as a reserve-electrolyte cell. If the electrodes are in place, it is necessary only to add the electrolyte to make a complete cell. In practice, however, the separation of the electrolyte is done only when excessive deterioration would occur during wet storage prior to use.

The liquid-activated batteries have disadvantages which are difficult to overcome; for example, automatic electrolyte-charging equipment may cause intercell shorting. An alternative approach is to introduce a gas which reacts with the spacer material to form a conducting electrolyte. Such gas-activated batteries operate over a wide temperature range.

Thermal batteries are also known as heat-activated or fused-electrolyte batteries. Some compounds, such as sodium chloride and potassium hydroxide, show very low conductivity in the solid state of room temperature but very good conductivity in the molten state. For example, a mixture of sodium hydroxide and potassium hydroxide becomes an excellent ionic conductor when heated about 338°F (170°C). If a zinc anode and a silver oxide cathode are combined with solid pads of the eutectic mixture, all the elements of a cell are present. Thermal batteries are capable of operation at very low ambient temperatures, provided that a suitable heat source is available to melt the electrolyte. [J.D.]

Reservoir A pond or lake built for the storage of water, usually by the construction of a dam across a river. The size of reservoir needed is a function of the water demands, the natural flows of the river impounded, and the extent of droughts to be encountered. Water supplies may be taken directly from the reservoirs, or water may be released from the reservoirs to augment river flows past a water supply intake downstream. *See* DAM; WATER SUPPLY ENGINEERING. [R.H.]

Resin Originally a category of vegetable substances soluble in ethanol but insoluble in water, but generally in modern technology an organic polymer of indeterminate molecular weight. The class of flammable, amorphous secretions of conifers or legumes are considered true resins. Water- swellable secretions of various plants, especially the Burseraceae, are called gum resins. The natural vegetable resins are largely polyterpenes and their acid derivatives, which find application in the manufacture of lacquers, adhesives, varnishes, and inks.

The synthetic resins, originally viewed as substitutes for certain natural resins, have a large place of their own in industry and commerce. Phenol-formaldehyde, phenol-urea, and phenol-melamine resins are important commercially. Any unplasticized organic polymer is considered a resin, thus nearly any one of the common plastics may be viewed as a synthetic resin. Water-soluble resins are marketed chiefly as substitutes for vegetable gums and in their own right for highly specialized applications. Carboxymethylcellulose, hydroxyalkylated cellulose derivatives, modified starches, polyvinyl alcohol, polyvinylpyrrolindone, and polyacrylamides are widely used as thickening agents for foods, paints, and drilling muds, as fiber sizings, in various kinds of protective coatings, and as encapsulating substances. *See* POLYMER. [F.W.]

Resistance heating The generation of heat by electric conductors carrying current. The degree of heating for a given current is proportional to the electrical resistance of the conductor. If the resistance is high, a large amount of heat is generated, and the material is used as a resistor rather than as a conductor. *See* ELECTRICAL RESISTANCE.

In addition to having high resistivity, heating elements must be able to withstand high temperatures without deteriorating or sagging. Other desirable characteristics are low temperature coefficient of resistance, low cost, formability, and availability of materials. Most commercial resistance alloys contain chromium or aluminum or both, since a protective coating of chrome oxide or aluminum oxide forms on the surface upon heating and inhibits or retards further oxidation.

Since heat is transmitted by radiation, convection, or conduction or combinations of these, the form of element is designed for the major mode of transmission. The simplest form is the helix, using a round wire resistor, with the pitch of the helix approximately three wire diameters. This form is adapted to radiation and convection and is generally used for room or air heating. It is also used in industrial furnaces, utilizing forced convection up to about 1200°F (650°C). Such helixes are stretched over grooved high-alumina refractory insulators and are otherwise open and unrestricted.

The electrical resistance of molten salts between immersed electrodes can be used to generate heat. Limiting temperatures are dependent on decomposition or evaporization temperatures of the salt, Parts to be heated are immersed in the salt. Heating is rapid and, since there is no exposure to air, oxidation is largely prevented. Disadvantages are the personnel hazards and discomfort of working close to molten salts.

A major application of resistance heating is in electric home appliances, including electric ranges, clothes dryers, water heaters, coffee percolators, portable radiant heaters, and hair dryers. Resistance heating also has application in home or space heating.

If the resistor is located in a thermally insulated chamber, most of the heat generated is conserved and can be applied to a wide variety of heating processes. Such insulated chambers are called ovens or furnaces, depending on the temperature range and use. The term oven is generally applied to units which operate up to approximately 800°F (430°C). Typical uses are for baking or roasting foods, drying paints and organic enamels, baking foundry cores, and low-temperature treatments of metals. The term furnace generally applies to units operating above 1200°F (650°C). Typical uses of furnaces are for heat treatment or melting of metals, for vitrification and glazing of ceramic wares, for annealing of glass, and for roasting and calcining of ores. *See* ELECTRIC HEATING; FURNACE. [W.Ro.]

Resistance measurement The quantitative determination of that property of an electrically conductive material, component, or circuit called electrical resistance. The unit of measurement is called the ohm. Measurements for engineering applications usually range from 0.1 microhm (10^{-7} ohm) to 1000 teraohms (10^{15} ohms). Measurements often required for materials development, basic research, or experiments under extreme environmental conditions extend this span to include zero ohms (superconductors) and 10^{18} ohms. *See* ELECTRICAL RESISTANCE; ELECTRICAL RESISTIVITY; OHM'S LAW; SUPERCONDUCTIVITY.

Resistance may be measured by using either direct or alternating current (dc or ac). If dc is used, the true resistance of the conductor is measured. When the measurement is made with ac, the result is usually called the effective resistance. The effect of leakage can usually be eliminated by suitably guarding one terminal or portion of the resistance to be measured. Physically a guard consists of a low-resistance conductor, electrically insulated from the guarded terminal and located to intercept the leakage current.

Commonly used methods of measuring resistance may be classified as either deflection methods or comparison methods. As implied by the names, deflection methods utilize the deflection of ammeters or voltmeters, which may be calibrated in terms of resistance under specific operating conditions, while comparison methods are based upon the use of a calibrated resistor, which can be compared to an unknown resistor. The fundamental deflection method is known as the voltmeter-ammeter method. Basic comparison methods are by potential drop, Wheatstone bridge, and Kelvin bridge.

The voltmeter-ammeter method of measuring resistance is illustrated in Fig. 1. Simultaneous readings of the voltmeter and ammeter are taken, and the unknown resistance R_x is calculated from Ohm's law, as shown in Eq. (1). For the voltmeter

$$R_x = V/A \tag{1}$$

connection shown in Fig. 1*a*, the true resistance R_x is slightly larger than the calculated value, because the ammeter measures the sum of the currents in R_x and the voltmeter. For the connection of Fig. 1*b*, R_x will be slightly smaller than that calculated, since the voltmeter reading is larger than the potential drop across R_x. The circuit in Fig. 1*a* causes the least error when R_x is low; the circuit in Fig. 1*b* causes the least error when R_x is high. A practical and widely used simplification based on this method is used in the ohmmeter. *See* OHMMETER.

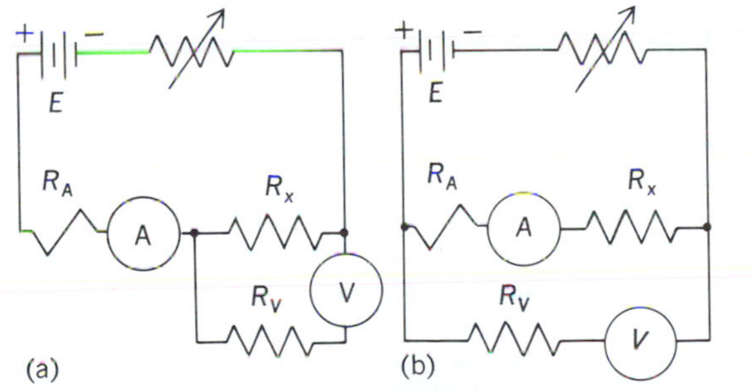

Fig. 1. Voltmeter-ammeter method which is used to measure resistance. (*a*) Least error when R_x is low. (*b*) Least error when R_x is high.

Comparison by potential drop, accomplished with the circuit of Fig. 2, is a logical development from the voltmeter-ammeter method. Only one meter is used for the measurement. The unknown resistance R_x is connected in series with a standard resistance R_s, which may be either fixed or adjustable. Since the same current flows in both resistors, by Ohm's law Eqs. (2) hold. This method is seldom used commercially, but it is capable of high accuracy for comparing like-value resistors, using a potentiometer for the potential difference measurement. *See* WHEATSTONE BRIDGE.

$$I = V_s/R_s = V_x/R_x$$
$$R_x = R_s V_x/V_s \tag{2}$$

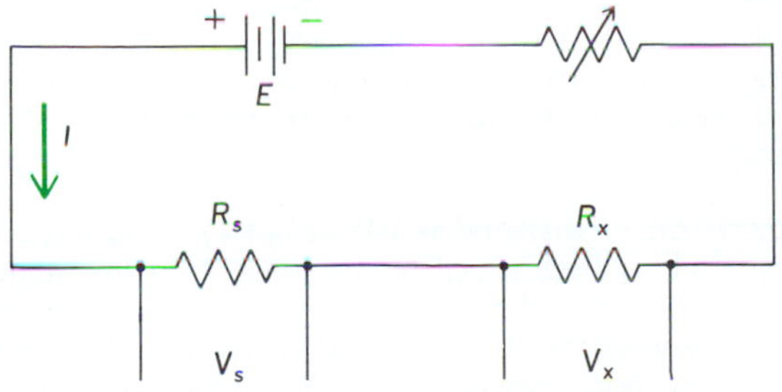

Fig. 2. Resistance comparison by potential drop.

High-quality standard resistors are used in comparison methods of resistance measurement. The unknown resistance is measured by comparing it to the accurately known value of the standard resistor. These standards are specially constructed and treated to achieve (1) constancy over long periods of time, (2) a low temperature coefficient, (3) a low internal thermal voltage, and (4) stability under varying humidity conditions. When intended for use in ac circuits, an additional requirement is to reduce the self-inductance, the distributed capacitance, and the skin effect of the resistor to the lowest practical values so that the standard will have a minimum frequency coefficient. *See* ELECTRICAL UNITS AND STANDARDS. [C.E.A.]

Resistance welding A process in which the heat for producing the weld is generated by the resistance to the flow of current through the parts to be joined. The application of external force is required; however, no fluxes, filler metals, or external heat sources are necessary. Most metals and their alloys can be successfully joined by resistance welding processes. Several methods are classified as resistance welding pro-cesses: spot, roll-spot, seam, projection, upset, flash, and percussion.

In resistance spot welding, coalescence at the faying surfaces is produced in one spot by the heat obtained from the resistance to electric current through the work parts held together under pressure by electrodes. The size and shape of the individually formed welds are limited primarily by the size and contour of the electrodes. *See* SPOT WELDING.

In roll resistance spot welding, separated resistance spot welds are made with one or more rotating circular electrodes. The rotation of the electrodes may or may not be stopped during the making of a weld.

In resistance seam welding, coalescence at the faying surfaces is produced by the heat obtained from resistance to electric current through the work parts held together under pressure by electrodes. The resulting weld is a series of overlapping resistance spot welds made progressively along a joint by rotating the electrodes.

In projection welding, coalescence is produced by the heat obtained from resistance to electric current through the work parts held together under pressure by electrodes. The resulting welds are localized at predetermined points by projections, embossments, or intersections.

In upset welding, coalescence is produced simultaneously over the entire area of abutting surfaces or progressively along a joint, by the heat obtained from resistance to electric current through the area of contact of those surfaces. Pressure is applied before heating is started and is maintained throughout the heating period.

In flash welding, coalescence is produced simultaneously over the entire area of abutting surfaces by the heat obtained from resistance to electric current between the two surfaces and by the application of pressure after heating is substantially completed. Flash and upsetting are accompanied by expulsion of the metal from the joint. *See* FLASH WELDING.

In percussion welding, coalescence is produced simultaneously over the entire abutting surfaces by the heat obtained from an arc produced by a rapid discharge of electrical energy with pressure percussively applied during or immediately following the electrical discharge.

Most metals and alloys can be resistance-welded to themselves and to each other. The weld properties are determined by the metal and by the resultant alloys which form during the welding process. Stronger metals and alloys require higher electrode forces, and poor electrical conductors require less current. Copper, silver, and gold, which are excellent electrical conductors, are very difficult to weld because they require high current densities to compensate for their low resistance. Medium- and high-carbon steels, which are hardened and embrittled during the normal welding process, must be tempered by multiple impulses. *See* WELDING AND CUTTING OF METALS. [E.F.N.]

Resistor A component of an electric circuit that produces heat while offering opposition, or resistance, to the flow of electric current. All conductors exhibit resistance in varying degrees; however, the term resistor is generally used only to describe a device specifically used to introduce resistance into an electric circuit. The unit of resistance measurement is the ohm. Resistors are described by stating their total resistance in ohms along with their safe power-dissipating ability in watts. A more detailed description would specify the residual inductance and stray shunt capacitance of the resistor. *See* ELECTRICAL RESISTANCE; ELECTRICAL RESISTIVITY.

Resistors may be classified according to the general field of engineering in which they are used. Power resistors range in size from about 5 watts to many kilowatts and may be cooled by air convection, air blast, or water. Direct-current (dc) ammeters employ resistors as meter shunts to bypass the major portion of the current around the low-current elements.

Voltmeters of both the dc and the ac types employ scale-multiplying resistors designed for accuracy and stability. By far the greatest number of resistors manufactured are intended for use in the electronics field.

Resistors are also classified according to their construction, which may be composition, film-type, or wire-wound. Further classification may be made according to whether the resistor has a fixed or adjustable resistance. Adjustable resistors may be further classified as adjustable-slide, rheostat, and potentiometer types. *See* RHEOSTAT.

The composition resistor is by far the most widely manufactured type because of its low cost, reliability, and small size. Basically it is a mixture of resistive materials, usually carbon, and a suitable binder molded into a cylinder. Copper wire leads are attached to the ends of the cylinder, and the entire resistor is molded into a plastic or ceramic jacket.

The film-type resistor is rapidly replacing the composition resistor in applications in which greater stability of resistance with voltage, temperature, and humidity is demanded. Basically this resistor consists of a conducting film of carbon, metal, or metal oxide deposited upon a ceramic cylinder. The value of the resistance is controlled by controlling the thickness and length of the film.

Wire remains the most stable form of resistance material available; therefore all high-precision instruments rely upon wire-wound resistors.

For discussion of nonlinear resistors *see* THERMISTOR; VARISTOR. [R.L.R.]

Resolving power (optics) A quantitative measure of the ability of an optical instrument to produce separable images. The images to be resolved may differ in position because they represent (1) different points on the object, as in telescopes and microscopes, or (2) images of the same object in light of two different wavelengths, as in prism and grating spectroscopes. For the former class of instruments, the resolving limit is usually quoted as the smallest angular or linear separation of two object points, and for the latter class, as the smallest difference in wavelength or wave number that will produce separate images. Since these quantities are inversely proportional to the power of the instrument to resolve, the term resolving power has generally fallen into disfavor. It is still commonly applied to spectroscopes, however, for which the term chromatic resolving power is used, signifying the ratio of the wavelength itself to the smallest wavelength interval resolved. The figure quoted as the resolving power or resolving limit of an instrument may be the theoretical value that would be obtained if all optical parts were perfect, or it may be the actual value found experimentally. Aberrations of lenses or defects in the ruling of gratings usually cause the actual resolution to fall below the theoretical value, which therefore represents the maximum that could be obtained with the given dimensions of the instrument in question. This maximum is fixed by the wave nature of light and may be calculated for given conditions by diffraction theory. *See* DIFFRACTION. [F.A.J./G.R.H.]

Resonance (acoustics and mechanics) When a mechanical or acoustical system is acted upon by an external periodic driving force whose frequency equals a natural free oscillation frequency of the system, the amplitude of oscillation becomes large and the system is said to be in a state of resonance.

A knowledge of both the resonance frequency and the sharpness of resonance is essential to any discussion of driven vibrating systems. When a vibrating system is sharply resonant, careful tuning is required to obtain the resonance condition. Mechanical standards of frequency must be sharply resonant so that their peak response can easily be determined. In other circumstances, resonance is undesirable. For example, in the faithful recording and reproduction of musical sounds, it is necessary either to have all vibrational resonances of the system outside the band of frequencies being reproduced or to employ heavily damped systems. *See* ACOUSTIC RESONATOR; SYMPATHETIC VIBRATION; VIBRATION. [L.E.K.]

Resonance (alternating-current circuits) A condition in a circuit characterized by relatively unimpeded oscillation of energy from a potential to a kinetic form. In an electrical network there is oscillation between the potential energy of charge on capacitance and the kinetic energy of current in inductance. This is analogous to the mechanical resonance seen in a pendulum.

Three kinds of resonant frequency in circuits are officially defined. Phase resonance is the frequency at which the phase angle between sinusoidal current entering a circuit and sinusoidal voltage applied to the terminals of the circuit is zero. Amplitude resonance is the frequency at which a given sinusoidal excitation (voltage or current) produces the maximum oscillation of electric charge in the resonant circuit. Natural resonance is the natural frequency of oscillation of the resonant circuit in the absence of any forcing excitation. These three frequencies are so nearly equal in low-loss circuits that they do not often have to be distinguished.

Resonance is of great importance in communications, permitting certain frequencies to be passed and others to be rejected. Thus a pair of telephone wires can carry many messages at the same time, each modulating a different carrier frequency, and each being separated from the others at the receiving end of the line by an appropriate arrangement of resonant filters. A radio or television receiver uses much the same principle to accept a desired signal and to reject all the undesired signals that arrive concurrently at its antenna; tuning a receiver means adjusting a circuit to be resonant at a desired frequency. [H.H.Sk.]

Resonance (molecular structure) A feature of the valence-bond method, which is a mathematical procedure to obtain approximate solutions to the Schrödinger equation for molecules. The valence-bond method is based on the theorem that if two or more solutions to the Schrödinger equation are available, certain linear combinations of them will also be solutions. It has this basis in common with its rival, the molecular orbital method. The valence-bond and molecular orbital approaches are both approximations and, if carried out to their logical and exact extremes, must yield identical results; nevertheless, both are often described as theories. In the valence-bond theory, combinations of solutions represent hypothetical structures of the molecule in question. These structures are said to be resonance (or contributing) structures, and the real molecule is said to be the resonance hybrid (or just simply the hybrid) of these structures. *See* MOLECULAR ORBITAL THEORY; SCHRÖDINGER'S WAVE EQUATION.

The resonance theory provided a solution for a molecule which had baffled and preoccupied chemists for a century—benzene. The principal use of resonance still lies in the qualitative description of molecules whose properties would otherwise be difficult to understand. *See* BENZENE.

Until the beginning of the 20th century, benzene posed a baffling challenge to organic chemists. In spite of its relatively simple formula, C_6H_6, they were unable to conceive of a suitable structure for it. While a great many structures were proposed, the properties of benzene corresponded to none of them.

In the early 1870s F. A. Kekulé proposed a revolutionary idea; benzene must be represented by two structures, (I) and (II),

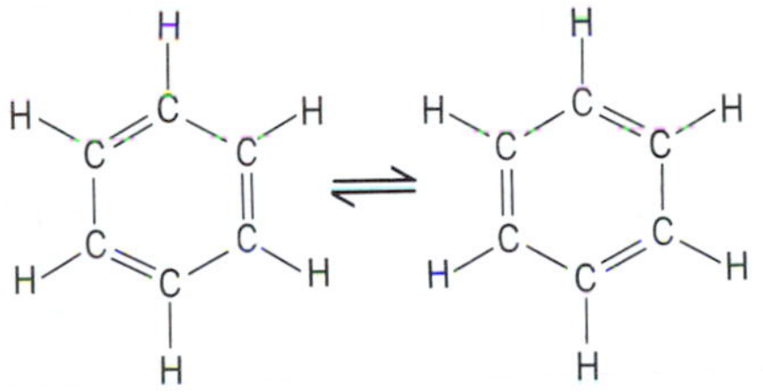

(I) (II)

rather than one, and all compounds containing the benzene skeleton must be subject to a rapid equilibration (oscillation) between the two. Kekulé's description of benzene was not completely satisfactory. While it accounted for the number of substituted benzene isomers, it did not explain why the compound failed to exhibit reactivity indicating the presence of multiple bonds. The problem was resolved with the advent of quantum mechanics in the early part of this century. In a sense, this solution is an expansion of Kekulé's oscillating pair; the so-called activation energy (the energy which must be imparted to a molecule in order to make it overcome the barrier that keeps it from being converted into another molecule) is negative in the case of benzene with respect to the oscillation, and this molecule therefore exists neither as (I) nor as (II) at any time, but it is an intermediate form (III) all the time. This intermediate struc-

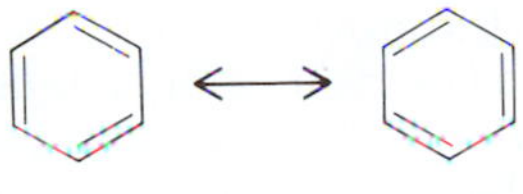

(III)

ture of benzene is described in terms of Kekulé's structures with the symbol ↔ between them; this is intended to signify that benzene has neither structure, but in fact is a hybrid of the two. The properties of benzene are thereby indicated to be those of neither (I) nor (II), but to be intermediate between the two.

The only property of the hybrid which is not intermediate between those of the hypothetical contributing structures is the energy: the energy of a resonance hybrid is by definition always at a minimum. This fact is responsible for the abnormal reluctance of benzene to undergo addition reactions; such reactions would lead to products that no longer have the resonance energy.

Although benzene is the classical example of resonance, the phenomenon is certainly not limited to it. Furthermore, the properties of all compounds are affected by resonance to some degree.

Although the molecular orbital approach has largely supplanted the valence-bond method, the resonance language remains so convenient that it is still used. *See* CHEMICAL BONDING; MOLECULAR STRUCTURE AND SPECTRA; QUANTUM CHEMISTRY. [W.J.LeN.]

Resonance (quantum mechanics) An enhanced coupling between quantum states with the same energy. The concept of resonance in quantum mechanics is closely related to resonances in classical physics. *See* RESONANCE (ACOUSTICS AND MECHANICS); RESONANCE (ALTERNATING-CURRENT CIRCUITS).

The matching of frequencies is central to the concept of resonance. An example is provided by waves, acoustic or electromagnetic, of a spectrum of frequencies propagating down a tube or waveguide. If a closed side tube is attached, its characteristic natural frequencies will couple and resonate with waves of those same frequencies propagating down the main tube. This simple illustration provides a description of all resonances, including those in quantum mechanics. The propagation of all quantum entities, whether electrons, nucleons, or other elementary particles, is represented through wave functions and thus is subject to resonant effects. *See* ACOUSTIC RESONATOR; CAVITY RESONATOR; HARMONIC (PERIODIC PHENOMENA); WAVEGUIDE.

An important allied element of quantum mechanics lies in its correspondence between frequency and energy. Instead of frequencies, differences between allowed energy levels of a system are considered. In the presence of degeneracy, that is, of different states of the system with the same energy, even the slightest influence results in the system resonating back and forth between the degenerate states. These states may differ in their internal motions or in divisions of the system into subsystems. The above example of wave flow suggests the terminology of channels, each channel being a family of energy levels similar in other respects. These energies are discretely distributed for a closed channel, whereas a continuum of energy levels occurs in open channels whose subsystems can separate to infinity. If all channels are closed, that is, within the realm of bound states, resonance between degenerate states leads to a theme of central importance to quantum chemistry, namely, stabilization by resonance and the resulting formation of resonant bonds. *See* DEGENERACY (QUANTUM MECHANICS); ENERGY LEVEL (QUANTUM MECHANICS); RESONANCE (MOLECULAR STRUCTURE).

Resonances occur in scattering when at least one channel is closed and one open. Typically, a system is divided into two parts: projectile + target, such as electron + atom or nucleon + nucleus. One channel consists of continuum states with their two parts separated to infinity. The other, closed channel consists of bound states. In the atomic example, a bound state of the full system would be a state of the negative ion and, in the nuclear example, a state of the larger nucleus formed by incorporating one extra nucleon in the target nucleus. *See* QUANTUM MECHANICS; SCATTERING EXPERIMENTS (ATOMS AND MOLECULES); SCATTERING EXPERIMENTS (NUCLEI). [A.R.P.R.]

Resonance ionization spectroscopy A form of atomic and molecular spectroscopy in which wavelength tunable light sources are used to remove electrons from (that is, ionize) a given kind of atom or molecule. Laser-based resonance ionization spectroscopy (RIS) techniques have been developed and used with ionization detectors such as the proportional counter to show that single atoms can be detected. Both RIS and one-atom detectors find a wide range of applications in physics, chemistry, and oceanography, and in the environmental sciences. [G.S.H.]

Resonance transformer An electrostatic particle accelerator in which the high-voltage terminal oscillates over the voltage range $\pm V$. These machines are used principally for acceleration of electrons. Electron current is allowed to pass only when the terminal voltage is near its peak value of $-V$. Thus the energy spread of the electrons can be held to a moderate value.

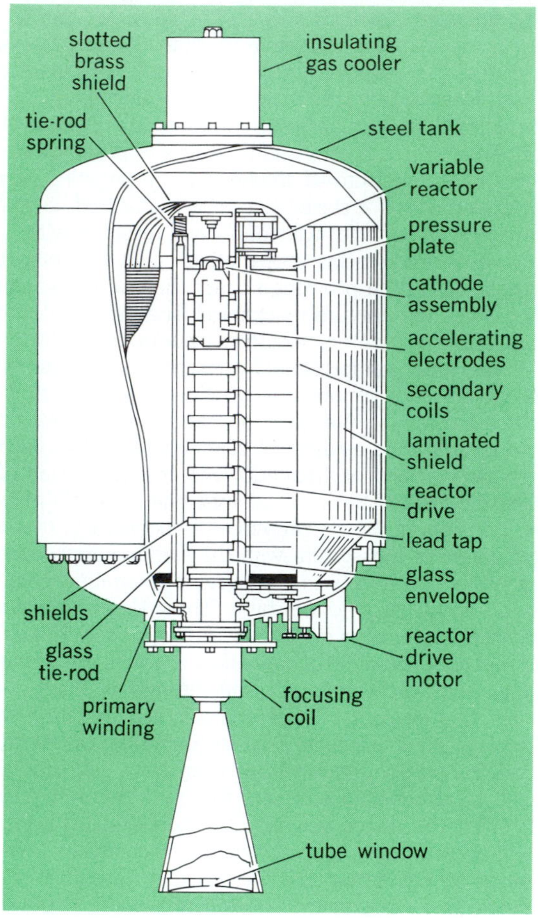

Schematic diagram of 1-MeV resonance transformer. (*General Electric Co.*)

The resonant transformer (see illustration) consists of a low-voltage primary winding which surrounds the lower end of a high-voltage coil stack. The stack is made up of a number of thin flat windings called pancake coils, in which the multisection vacuum tube is coaxially mounted. The inductance and capacitance of the high-voltage secondary have values such that the resonant frequency of this circuit is equal to the frequency at which primary power is supplied. *See* PARTICLE ACCELERATOR.

Resonance transformers are used for industrial radiography, for x-ray therapy, for food and drug sterilization, and for the processing of plastics. *See* COCKCROFT-WALTON ACCELERATOR.

[R.G.H.]

Resorcinol A dihydric phenol ($C_6H_6O_2$, also called *m*-dihydroxybenzene. Resorcinol, with the structure shown below, is a water-soluble, crystalline compound (mp 110°C or 230°F) that is colorless when pure. Aqueous solutions of resorcinol are

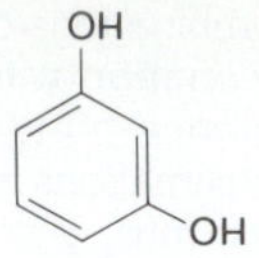

mildly acidic; (p*K* 9.2); in this and other respects its properties are similar to those of phenol. Resorcinol is irritating to the skin, particularly that of individuals who have become sensitized, and it is a severe systemic poison.

The largest industrial use for resorcinol is in the manufacture of adhesives and resins. Adhesives based on resorcinol are exceptionally water-resistant, and are used extensively in marine plywood and other products which are constantly exposed to water. *See* CATECHOL; CRESOL; HYDROQUINONE; PHENOL. [M.St.]

Respiration The various processes associated with the biochemical transformation of the energy available in the organic substrates derived from foodstuffs, to energy usable for synthetic and transport processes, external work, and, eventually, heat. This transformation, generally identified as metabolism, most commonly requires the presence of oxygen and involves the complete oxidation of organic substrates to carbon dioxide and water (aerobic respiration). If the oxidation is incomplete, resulting in organic compounds as end products, oxygen is typically not involved, and the process is then identified as anaerobic respiration. *See* METABOLISM.

The term "external respiration" is more appropriate for describing the exchange of O_2 and CO_2 between the organism and its environment. In most multicellular organisms, and nearly all vertebrates (with the exception of a few salamanders lacking both lungs and gills), external respiration takes place in specialized structures termed respiratory organs, such as gills and lungs. *See* LUNG; RESPIRATORY SYSTEM.

The ultimate physical process causing movement of gases across living tissues is simple passive diffusion. Respiratory gas exchange also depends on two convective fluid movements. The first is the bulk transport of the external medium, air or water, to and across the external respiratory exchange surfaces. The second is the transport of coelomic fluid or blood across the internal surfaces of the respiratory organ. These two convective transports are referred to as ventilation and circulation (or perfusion). They are active processes, powered by ciliary or muscular pumps.

In all vertebrates and many invertebrates, the circulating internal medium (coelomic fluid, hemolymph, or blood) contains a respiratory pigment, for example, hemocyanin or hemoglobin, which binds reversibly with O_2, CO_2, and protons. Respiratory pigments augment respiratory gas exchange, both by increasing the capacity for bulk transport of the gases, and by influencing gas partial pressure (concentration) gradients across tissue exchange surfaces. *See* BLOOD; HEMOGLOBIN; RESPIRATORY PIGMENTS (INVERTEBRATE).

The physiological adjustment of organisms to variations in their need for aerobic energy production involves regulated changes in the exchange and transport of respiratory gases. The adjustments are effected by rapid alterations in the ventilatory and circulatory pumps and by longer-term modifications in the respiratory properties of blood. [K.J.]

Respirator A device to protect the eyes and respiratory tract from noxious gases, vapors, and aerosols. Three general types of respirators are in common use, depending on the principle of operation utilized. The military protective mask (see illustration) and the commercial gas mask furnish clean air to the wearer by removing the contamination from the ambient atmosphere by means of a filter and a bed of adsorbent material. Rebreathing masks use closed systems which absorb

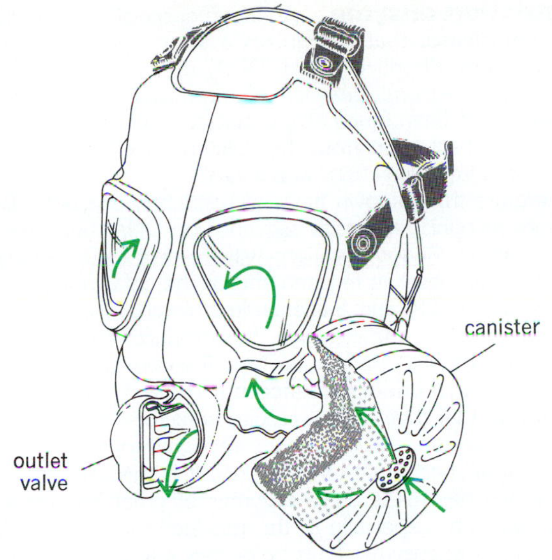

Protective mask. Arrows show airflow direction.

the carbon dioxide and water from exhaled air and generate oxygen for inhalation. Air-supply masks furnish clean air from a source independent of the surrounding atmosphere. [S.D.S.]

Respiratory pigments (invertebrate) Colored, metal-containing proteins which combine reversibly with oxygen, and which are found in the body fluids or tissues of invertebrate animals. The role of these pigments is primarily to aid in the transport of molecular oxygen. Thus they are distinguished from respiratory enzymes, which are concerned with the metabolic consumption of oxygen. Four distinctly colored groups of respiratory pigments exist among invertebrates: hemoglobins (purple, become orange-red with oxygen), chlorocruorins (green, become red with oxygen), hemocyanins (colorless, become blue with oxygen), and hemerythrins (colorless, become red with oxygen).

Each of the pigments is composed of two parts, a large protein molecule to which is bound one or more small moieties called prosthetic groups, each of which is or contains a metal. The metal binds the oxygen, and this binding imparts the characteristic color to the pigment. In hemoglobins the prosthetic group is an iron porphyrin compound called heme. Chlorocruorin contains a similar iron porphyrin which differs from heme only in that a vinyl group in the molecule is replaced by formyl. The prosthetic group of hemerythrin consists of two adjacent iron atoms which bind an oxygen molecule between them. The prosthetic group of hemocyanin is analogous and consists of two adjacent copper atoms. Pigments containing vanadium have been found in tunicates, but these substances do not combine reversibly with oxygen and so cannot be considered respiratory pigments.

The protein part of the pigment confers reversibility upon the combination of the metal with oxygen. In the absence of protein, the prosthetic groups lose their capacity to combine with oxygen reversibly. Instead, the metals are irreversibly oxidized: Electrons are transferred from metal to oxygen. The bonds between metal and protein so alter the electronic energy levels of the metal that this transfer, if it occurs, is reversible. For this reason, the combination of hemoglobin with oxygen is described as oxygenation rather than oxidation. The protein is also responsible for certain physiological adaptations of the pigment to the environment. Thus the affinity of the pigment for oxygen is often highest in animals which inhabit environments with the lowest oxygen content. *See* HEMOGLOBIN. [A.F.R.]

Respiratory syncytial virus A virus belonging to the Paramyxoviridae, genus *Pneumovirus*. This virus, although unrelated to any other known respiratory disease agent and differing from the parainfluenza viruses in a number of important characteristics, has been associated with, a large proportion of respiratory illnesses in very young children, particularly bronchiolitis and pneumonia. It appears to be one of the major causes of these serious illnesses of infants. It is the only respiratory virus that occurs with its greatest frequency in infants in their first 6 months of life. In older infants and children, a milder illness is produced.

The clinical disease in young infants may be the result of an antigen-antibody reaction that occurs when the infecting virus meets antibody transmitted from the mother. For this reason respiratory syncytial vaccines that stimulate production of antibodies in the serum, but not in the nasal secretions, may do more harm than good. *See* ANIMAL VIRUS; VIRUS CLASSIFICATION. [J.I.L.; M.E.Re.]

Respiratory system The system of organs involved in the acquisition of oxygen and the elimination of carbon dioxide by an organism. The lungs and gills are the two most important structures of vertebrates involved in the phase known as external respiration, or gaseous exchanges, between the blood and environment. Internal respiration refers to the gaseous exchanges which occur between the blood and cells. Certain other structures in some species of vertebrates serve as respiratory organs; among these are the integument or skin of fishes and amphibians. The moist, highly vascular skin of anuran amphibians is important in respiration. Certain species of fish have a vascular rectum which is utilized as a respiratory structure, water being taken in and ejected regularly by the animal. Saclike cloacal structures occur in some aquatic species of turtles. These are vascular and are intermittently filled with, and emptied of, water. It is thought that they may function in respiration. During embryonic life the yolk sac and allantois are important respiratory organs in certain vertebrates. *See* ALLANTOIS; GILL; YOLK SAC.

Structurally, respiratory organs usually present a vascular surface that is sufficiently extensive to provide an adequate area of absorption for gaseous exchange. This surface is moist and thin enough to allow for the passage of gases. [R.S.McE./T.S.P.]

The shape and volume of the lung, because of its pliability, conforms almost completely to that of its cavity. The lungs are conical; each has an apex and a base, two surfaces, two borders, and a hilum. The apex extends into the superior limit of the thoracic cavity. The base is the diaphragmatic surface. The coastal surface may show bulgings into the intercostal spaces. The medial surface has a part lying in the space beside the vertebral column and a part imprinted by the form of structures bulging outward beneath the mediastinal pleura. The cardiac impression is deeper on the left lung because of the position of the heart.

For convenience the lung may be divided into anatomical areas. The bronchial tree branches mainly by dichotomy. The ultimate generations, that is, the respiratory bronchioles, alveolar ducts, and alveoli constitute all of the respiratory portion of the lung. The trachea and extrapulmonary bronchi are kept open by C-shaped bars of hyaline cartilage. When in their branching the bronchi and bronchioles are reduced to a diameter of 1 mm or less, they are then free of cartilage and are called terminal bronchioles. One of the terminal bronchioles enters the apex of a secondary lobule of the lung. These secondary lobules are anatomic units of the lung, whose hexagonal bases rest on the pleura or next to a bronchiole or blood vessel. Finer lines divide the bases of the secondary lobules into smaller areas. These are the bases of primary lobules, each served by a respiratory bronchiole. *See* LUNG; RESPIRATION. [L.P.C.]

Respiratory system disorders Any disease of the respiratory system, airways or blood circulation is apt to reduce the respiratory function. It is well known, however, that less than one-third of the lung is needed to sustain life at rest. The other two-thirds more or less represent reserve volume which is used when more oxygen is needed, as when the body is working. Hence, slight reduction of the respiratory function manifests itself in shortness of breath, dyspnea, at work. Dyspnea at rest is a sign of severe disease of the respiratory system. In extreme conditions this is accompanied by incomplete oxygenation of the blood in the lung. The blood cannot get rid of all carbon dioxide and cannot take up enough oxygen. It becomes darker and has a bluish appearance where it shines through the skin, as on the lips or nails. This condition is called cyanosis. If the respiratory insufficiency continues to increase, basic functions of the organism are disturbed by the lack of oxygen. The patient loses consciousness, since the brain is most sensitive to lack of oxygen, and finally dies if no relief can be given.

Diseases of the respiratory system can be located in the lung tissue itself, interfering directly with the gas exchange; in the airways, thus reducing the air brought into the lung; or in the blood vessels, disturbing the blood circulation through the lung. Under rare conditions the respiratory musculature can be paralyzed, as in poliomyelitis, or injured by an accident. *See* BRONCHIAL DISORDERS; LUNG DISORDERS.

Among the diseases of pulmonary circulation, congenital malformations of the heart and pulmonary artery account for many cases of respiratory insufficiency of newborn and younger children, called blue babies because of the cyanosis. In adults, acquired heart diseases greatly influence the pulmonary circulation. *See* CARDIOVASCULAR SYSTEM; CIRCULATION DISORDERS.

[S.P.H.]

Response A quantitative expression of the manner in which a microphone, amplifier, loudspeaker, or other component or system performs its intended function. A linear re-sponse means that the output signal is exactly proportional to the input signal for the entire range of frequencies over which the device is intended to operate. A logarithmic response means that the output signal is a logarithmic function of the input signal. The frequency response of a device, often presented as a curve on a graph, is the deviation over the frequency range from the response at some selected frequency, such as 1000 Hz. *See* AMPLIFIER; CHARACTERISTIC CURVE; LOUDSPEAKER; MICROPHONE. [J.MAR.]

Rest mass A constant associated with a material body which determines its inertial properties and its internal energy content. It is sometimes called the inertial mass.

For a particle of rest mass m_0, Newton's second law of motion is given by Eq. (1), where $\mathbf{a} = d\mathbf{v}/dt$ is the acceleration

$$m_0 d\mathbf{v}/dt = \mathbf{F} \qquad (1)$$

of the particle and **F** is the force applied to it. For a particle moving with a speed near that of light Eq. (1) must be replaced by the relativistic equation of motion, Eq. (2), where c is the

$$d\backslash dt\left(\frac{m_0 \boldsymbol{v}}{\sqrt{1-(v^2/c^2)}}\right) = \mathbf{F} \qquad (2)$$

speed of light. These equations are used for the measurement of the rest masses of particles by deflection in suitable force fields. *See* INERTIA; RELATIVITY. [E.L.Hi.]

Restionales An order of flowering plants, division Magnoliophyta (Angiospermae), in the subclass Commelinidae of the class Liliopsida (monocotyledons). The order consists of 5 families and nearly 500 species, some 400 of them belonging to the Restionaceae. The vast majority of the species grow in temperate regions of the Southern Hemisphere. The Restionaies have reduced flowers and a single, pendulous, orthotropous ovule in each of the 1–3 locules of the ovary. *See* COMMELINIDAE; LILIOPSIDA. [A.Cr.]

Restriction enzyme An enzyme, specifically an endodeoxyribonuclease, that recognizes a short specific sequence within a deoxyribonucleic acid (DNA) molecule and then catalyzes double-strand cleavage of that molecule. Restriction enzymes have been found only in bacteria, where they serve to protect the bacterium from the deleterious effects of foreign DNA. *See* DEOXYRIBONUCLEIC ACID (DNA).

There are three known types of restriction enzymes. Type I enzymes recognize a specific sequence on DNA, but cleave the DNA chain at random locations with respect to this sequence. They have an absolute requirement for the cofactors adenosine triphosphate (ATP) and *S*-adenosylmethionine. Because of the random nature of the cleavage, the products are a heterogeneous array of DNA fragments. Type II enzymes also recognize a specific nucleotide sequence but differ from the type I enzymes in that they do not require cofactors and they cleave specifically within or close to the recognition sequence, thus generating a specific set of fragments. It is this exquisite specificity which has made these enzymes of great importance in DNA research, especially in the production of recombinant DNAs. Type III enzymes have properties intermediate between those of the type I and type II enzymes. They recognize a specific sequence and cleave specifically a short distance away from the recognition sequence. They have an absolute requirement for the ATP cofactor, but they do not hydrolyze it.

A key feature of the fragments produced by restriction enzymes is that when mixed in the presence of the enzyme DNA ligase, the fragments can be rejoined. Should the new fragment carry genetic information that can be interpreted by the bacterial cell containing the recombinant molecule, then the information will be expressed as a protein and the bacterial cell will serve as an ideal source from which to obtain that protein. For instance, if the DNA fragment carries the genetic information encoding the hormone insulin, the bacterial cell carrying that fragment will produce insulin. By using this method, the human gene for insulin has been cloned into bacterial cells and used for the commercial production of human insulin. The potential impact of this technology forms the basis of the genetic engineering industry. *See* ENZYME; GENETIC ENGINEERING. [R.Ro.]

Resultant of forces A system of at most a single force and a single couple whose external effects on a rigid body are identical with the effects of the several actual forces that act on the body. For analytic purposes, forces are grouped and replaced by their resultant. Forces can be added graphically (see illustration) or analytically. The sum of more than two vector forces can be found by extending the method of illustration c to a three-dimensional vector polygon in which one force is drawn from the tip of the previous one until all are laid out.

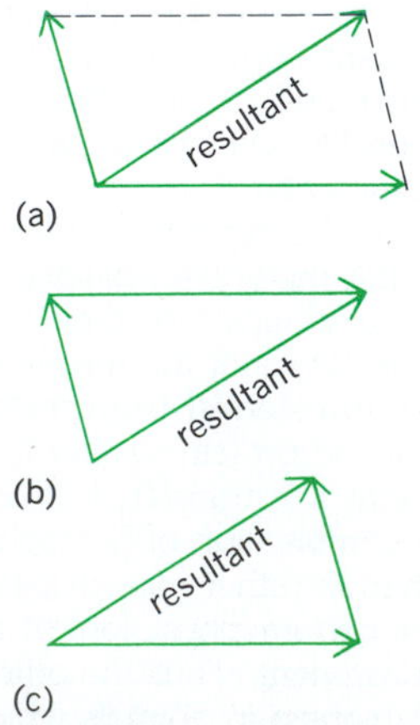

Resultant of two forces acting through a common center. (*a*) Diagonal of parallelogram. (*b*, *c*) Hypotenuse of triangle.

The resultant force is the force vector required to close the polygon directed from the tail of the first force vector to the tip of the last. A force system has a zero force resultant if its vector polygon closes. *See* CALCULUS OF VECTORS.

Two force systems are equivalent if their resultant forces, as described above, are equal and if their total vector moments about the same point are also equal. Vector moments are combined in the same manner as forces, that is, by parallelograms, triangles, or polygons. A resultant is the equivalent force system having the fewest possible forces and couples. *See* COUPLE; FORCE; STATICS. [N.S.F.]

Retaining wall A wall designed to maintain differences in ground elevations by holding back a bank of material. Sometimes a retaining wall also serves as a foundation wall. The illustration shows the major types of retaining wall.

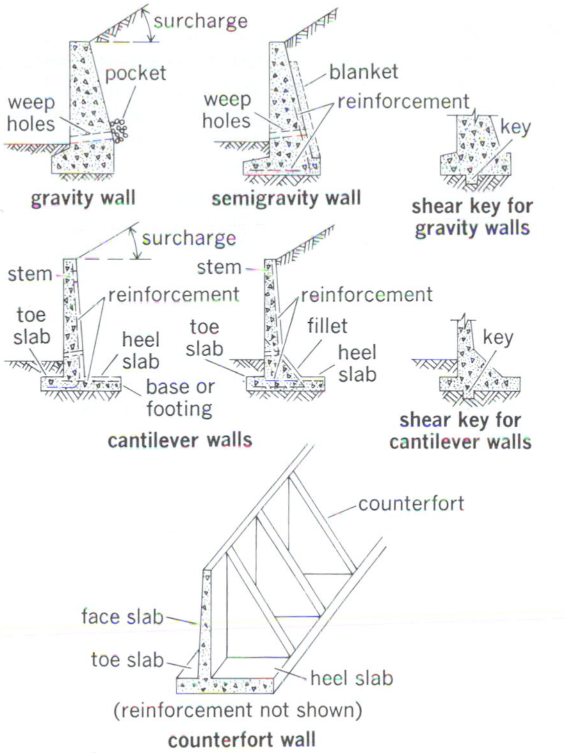

The major types of retaining wall.

Material that exerts pressure on the back of the wall is called backfill. It includes material taken from the excavation and placed behind the wall. The load applied on backfill above the top of the wall level is called surcharge.

Earth pressure against the back of a wall is known as active pressure and acts at an oblique angle; the precise angle depends upon the character of the backfill, slope of the ground surface, presence of surcharge, and groundwater level. Earth pressure against the front of a wall is known as passive pressure, although the soil must be somewhat compressed before this force develops.

External stability is achieved when a wall is proportioned so that it will neither rotate nor slide under all dead-load and applied forces. *See* FOUNDATIONS. [R.D.Che.]

Reticular formation Characteristic clusters of nerve cell bodies (gray matter) and their meshwork or reticulum, of fibers which are found in the brainstem and the diencephalon. The reticular formation is thought to be a complex, highly integrated mechanism which exerts both inhibition and facilitation on almost every type of activity of the central nervous system. *See* NERVOUS SYSTEM (VERTEBRATE). [D.B.W.]

Reticulosa An order of the subclass Hexasterophora in the class Hexactinellida. This is a group of Paleozoic hexactinellids with a branching form. Each branch is provided with dermal, parenchymal, and gastral spicule reticulations. *See* HEXASTEROPHORA. [W.D.H.]

Retinitis Any inflammatory condition involving the retina. In the majority of cases the adjacent middle layer of the eyeball (the choroid layer which contains the blood vessels of the eyeball) is also involved, in which case the condition is known as chorioretinitis.

Retinitis may be due to infectious agents (usually bacteria, less commonly fungi, and rarely viruses), mechanical injury of the eyeball (contusion), or intense light (photoretinitis). Chronic progressive damage to the retina (degenerative retinopathy) is frequently associated with conditions such as diabetes, high blood pressure, arteriosclerosis, senility, leukemia, and anemia. [W.R.A.]

Retortamonadida An order of parasitic flagellate protozoa belonging to the class Zoomastigophora. All retortamonads are medium to large in size and have a complicated blepharoplast-centrosome-axostyle apparatus (see illustration).

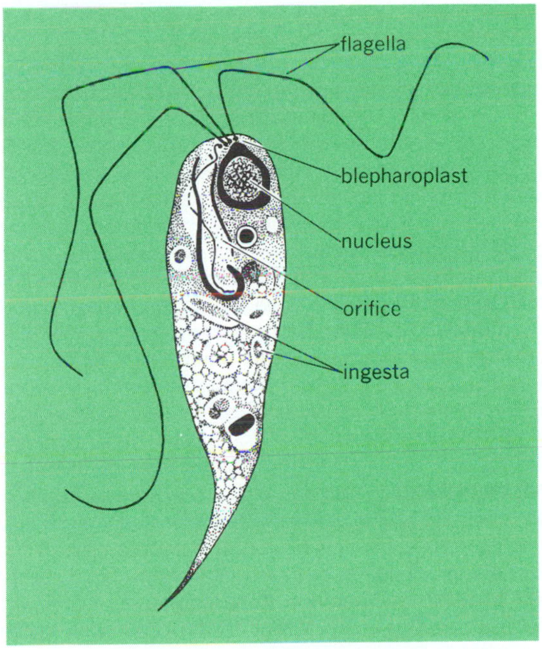

A rotortamonad, *Chilomastix aulastomi*.

Retortamonadida have two or four flagella, one turned ventrally into a cytostomal depression. The nucleus is located at the anterior tip. The body is twisted. These organisms are actually symbionts, ingesting bacteria in the digestive tracts of their hosts. *See* ZOOMASTIGOPHOREA. [J.B.L.]

Retrograde motion (astronomy) In astronomy, either an apparent east-to-west motion of a planet or comet with respect to the background stars or a real east-to-west orbital motion of a comet about the Sun or of a satellite about its primary. The majority of the objects in the solar system revolve from west to east about their primaries. However, near the time of closest approach of Earth and a superior planet, such as Jupiter, because of their relative motion, the superior planet appears to move from east to west with respect to the background stars. The same apparent motion occurs for an inferior planet, such as Venus, near the time of closest approach to Earth.

Actual, rather than apparent, retrograde motion occurs

among the satellites and comets; the eighth and ninth satellites of Jupiter and the ninth satellite of Saturn are examples. *See* ORBITAL MOTION. [R.L.Du.]

Retrovirus A family of viruses distinguished by three characteristics: (1) genetic information in ribonucleic acid (RNA); (2) virions possess the enzyme reverse transcriptase; and (3) virion morphology consists of two proteinaceous structures, a dense core and an envelope that surrounds the core. Some viruses outside the retrovirus family have some of these characteristics, but none has all three. Numerous retroviruses have been described; they are found in all families of vertebrates. *See* ANIMAL VIRUS; REVERSE TRANSCRIPTASE; RIBONUCLEIC ACID (RNA).

The genome is composed of two identical molecules of single- stranded RNA, which are similar in structure and function to cellular messenger RNA. Deoxyribonucleic acid (DNA) is not present in the virions of retroviruses. The reverse transcriptase in each virus makes a DNA copy of the RNA genome shortly after entry of the virus into the host cell. The discovery of this enzyme changed thinking in biology. Previously, the only known direction for the flow of genetic information was from DNA to RNA, yet retroviruses make DNA copies of their genome by using an RNA template. This reversal of genetic information was considered backward and hence the family name retrovirus, meaning backward virus.

Once the DNA copy of the RNA genome is made, it is inserted directly into one of the chromosomes of the host cell. This results in new genetic information being acquired by the host species. The study of reverse transcriptase has led to other discoveries of how retroviruses add a variety of new genetic information into the host. One such class of genes carried by retroviruses is oncogenes, meaning tumor genes. Retroviral oncogenes appear to be responsible for tumors in animals. *See* ONCOGENES; VIRUS CLASSIFICATION.

Two distinct retroviruses have been discovered in humans. One is human T-cell lymphotropic virus type 1 (HTLV-1), a type C-like virus associated with adult T-cell leukemia. The other is the human acquired immune deficiency syndrome (AIDS) virus, a type E lentivirus. *See* ACQUIRED IMMUNE DEFICIENCY SYNDROME (AIDS); LEUKEMIA. [P.A.Ma.]

Reverberation After sound has been produced in, or enters, an enclosed space, it is reflected repeatedly by the boundaries of the enclosure, even after the source ceases to emit sound. This prolongation of sound after the original source has stopped is called reverberation. A certain amount of reverberation adds a pleasing characteristic to the acoustical qualities of a room. However, excessive reverberation can ruin the acoustical properties of an otherwise well-designed room. A typical record representing the sound-pressure level at a given point in a room plotted against time, after a sound source has been turned off, is given in the decay curve shown in the illustration. The rate of sound decay is not uniform but fluctuates about an average slope. *See* ARCHITECTURAL ACOUSTICS; SOUND. [C.M.H.]

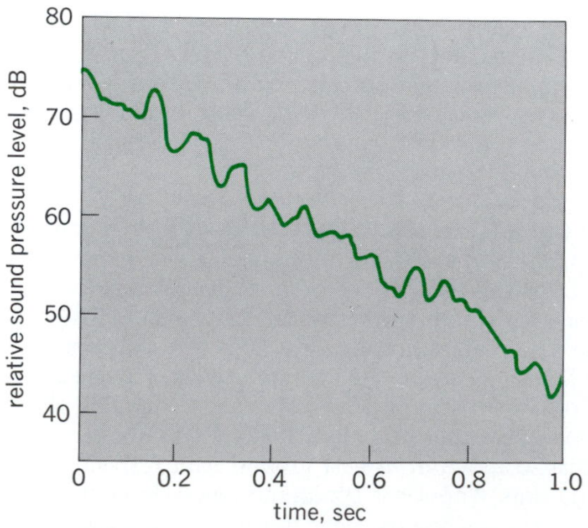

Typical decay curve illustrating reverberation.

Reverse transcriptase A viral deoxyribonucleic acid (DNA) polymerase capable of effecting reverse transcription, that is, DNA synthesis directed by a ribonucleic acid (RNA) template. Reverse transcription is the opposite of the usual transcription, that is, RNA synthesis using a DNA template. Reverse transcriptases are present in tumor viruses and are related to retroviruses. They are used in genetic engineering to obtain DNA copies from purified RNA. *See* DEOXYRIBONUCLEIC ACID (DNA); RIBONUCLEIC ACID (RNA). [H.M.T.]

Revetment A means of protecting river banks against erosion. In some cases the revetment must act as a solid barrier; however, as a basic principle, it should be designed to induce the high-velocity thread of the current to move away from the threatened area. This is most readily accomplished when (1) the revetment is initiated at a point upstream from the point of attack and at an angle of not more than about 15° to the current, and (2) the surface of the revetment is rough.

Revetments may consist of the following types: (1) rock paving with a heavy rock toe base; (2) rock, asphaltic, or concrete paving with brush, lumber, or concrete mattress for the underwater portions; and (3) pile dikes or other permeable fences. Rock paving has the advantage of flexibility in settling to heal breaches and permits easy repair.

Groins are short dikes, approximately perpendicular to the bank, which are sometimes used in lieu of revetment, particularly in areas where the bed is relatively stable. For other information on control of rivers *see* RIVER ENGINEERING. *See also* GROIN (ENGINEERING); RETAINING WALL. [W.E.J.]

Reynolds number In fluid mechanics, the ratio $\rho v d/\mu$, where ρ is fluid density, v is velocity, d is a characteristic length, and μ is fluid viscosity. The Reynolds number is significant in the design of a model of any system in which the effect of viscosity is important in controlling the velocities or the flow pattern. In the evaluation of drag on a body submerged in a fluid and moving with respect to the fluids, the Reynolds number is important.

The Reynolds number also serves as a criterion of type of fluid motion. In a pipe, for example, laminar flow normally exists at Reynolds numbers less than 2000, and turbulent flow at Reynolds numbers above about 3000. *See* DYNAMIC SIMILARITY; FLUID MECHANICS. [G.Mu.]

Rh incompatibility A condition in which red blood cells of the fetus become coated with an immunoglobulin (IgG) antibody [Rh (rhesus) antibody] of maternal origin which is directed against an antigen (Rh D antigen) of paternal origin that is present on fetal cells. *See* ANTIBODY; ANTIGEN; BLOOD GROUPS; IMMUNOGLOBULIN.

A pregnant woman may develop an antibody to a red blood cell antigen that the fetus has but that she does not possess. This occurs because the fetus has inherited the antigen from the father; fetal red cells pass into the maternal circulation and are recognized as foreign by the mother. The mother develops antibodies to the foreign protein; the antibodies pass across the placenta and attack the fetal red blood cells, producing a hemolytic anemia. This phenomenon is known as alloimmunization. Alloimmunization of the mother results in hemolytic disease of the newborn (also known as erythroblastosis fetalis). *See* ANEMIA; AUTOIMMUNITY.

The development of Rh incompatibility is determined by the Rh types of the parents. Some Rh-incompatible pregnancies do not result in sensitization. Factors influencing the occurrence of sensitization include the timing and extent of the transplacental hemorrhage, the degree and strength of antibody development in the mother, and the ABO status of mother and fetus. Most immunizations result from transplacental hemorrhage during placental separation at delivery.

The at-risk mother is monitored for the development and progression of Rh incompatibility with serial serum antibody titers to anti-D. The presence of anti-D and its titer do not necessarily correlate with the status of the fetus. To monitor the status, amniocentesis is performed. This procedure permits measurement of bile pigment concentration and assessment of lung maturity. These tests are done serially to decide when to deliver the fetus or whether to transfuse the fetus until it is mature enough to be delivered safely. *See* PRENATAL DIAGNOSIS.

Following delivery, the infant may require transfusion immediately or during the postnatal period. This may involve simple replacement or exchange transfusion, in which the infant's blood volume is replaced with compatible blood. The aim of exchange is to remove bilirubin and antibody-coated red blood cells and to correct the anemia while replacing the infant's blood with blood that is antigen negative for the offending antibody. Other medical treatment for jaundice may also be required. *See* TRANSFUSION.

Rh immune globulin (RhIG) is a concentrated solution of IgG anti-D obtained from pooled human plasma. Its action is to interfere with antigen recognition. RhIg should be administered to any Rh-negative mother of an Rh-positive infant within 72 h after delivery. Administration of RhIG at the twenty-eighth week of pregnancy followed by a second dose at birth has been very successful in all but eliminating Rh hemolytic disease. *See* PREGNANCY. [N.L.C.L.]

Rhabdiasoidea A group of parasitic nematodes set apart by their characteristic morphology and by their life cycles, which may include a parasitic generation alternating with one or more free-living generations. One species, *Strongyloides stercoralis*, causes an important diarrheal disease of humans; others parasitize domestic animals and cause great economic loss; still others are of great biological interest because of their varied methods of reproduction and complicated life cycles. *See* NEMATA. [J.A.S.]

Rhabditida An order of nematodes in which the number of labia varies from a full complement of six to three or two or none. The tubular stoma may be composed of five or more sections called rhabdions. The three-part esophagus always ends in a muscular bulb that is invariably valved. The excretory tube is cuticularly lined, and paired lateral collecting tubes generally run posteriorly from the excretory cell; some taxa have anterior tubules also. Females have one or two ovaries; when only one is present, the vulva shifts posteriorly. The cells of the intestine may be uninucleate, binucleate, or tetranucleate, and the hypodermal cells may also be multinucleate.

There are eight superfamilies in the order: Rhabditoidea, Allionematoidea; Bunonematoidea, Cephaloboidea, Panagrolaimoidea, Robertioidea, Chambersiellolidea, and Elaphonematoidea. The Rhabditoidea are one of the largest nematode superfamilies and contain many important parasites of humans and domestic animals. This superfamily is distinguished by the well-developed cylindrical stoma and three-part esophagus that ends in a valved terminal bulb. In parasitic species, adult stages and some larval stages lack the valved terminal bulb. Though most species of Rhabditoidea are free-living feeders on terrestrial bacteria, others are important in the biological control of insects or as parasites of mammals. *See* MEDICAL PARASITOLOGY. [A.R.M.]

Rhabdocoela Formerly considered an order of the Turbellaria, the group Rhabdocoela is now usually divided into three orders, Catenulida, Macrostomida, and Neorhabdocoela. The temocephalids are a group of rhabdocoeles closely related to the Dalyelloida. They are sometimes considered a distinct order but are usually classified under the Neorhabdocoela. They differ from other rhabdocoeles chiefly in the possession of tentacles and adhesive organs and in the absence of epidermal cilia in most species.

The rhabdocoeles have a simple, unbranched intestine, with little if any diverticulations. The pharynx is simple or bulbous, the gonads are usually compact, a cuticular apparatus is associated with the copulatory organ, and there are two main longitudinal nerves. This large group includes most small freshwater Turbellaria as well as many marine and some terrestrial species. *See* TURBELLARIA. [E.R.J.]

Rhamnales An order of flowering plants, division Magnoliophyta (Angiospermae), in the subclass Rosidae of the class Magnoliopsida (dicotyledons). The order consists of 3 families; the Rhamnaceae with about 900 species, the Vitaceae with about 700, and the Leeaceae with about 70. The Rhamnales have a single set of stamens, opposite the petals, and usually a well-developed intrastaminal disk. The ovary has two or more locules, with one or two erect, bitegmic ovules from the base in each locule. The seeds have a well-developed endosperm. *See* MAGNOLIOPSIDA; ROSIDAE. [A.Cr.]

L-Rhamnose A methyl pentose, known also as 6-deoxy-L-mannose and L-mannomethylose. The free sugar, with the structure shown, occurs in leaves and flowers of poison ivy

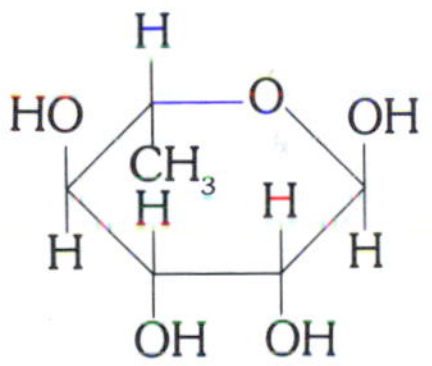

α-L-Rhamnopyranose

(*Rhus toxicodendron*). It is a constituent of many of the plant glycosides such as quercitrin and rutin. The type II *Pneumococcus* specific polysaccharides and a wide variety of gums and mucilages contain L-rhamnose, and there is evidence that this sugar is present in a number of bacterial polysaccharides. *See* MONOSACCHARIDE. [W.Z.H.]

Rhenium A chemical element, Re, with atomic number 75 and atomic weight 186.2. Rhenium is a transition element. It is a dense metal (21.04) with the very high melting point of 3440°C (6220°F).

Rhenium is similar to its homolog technetium in that it

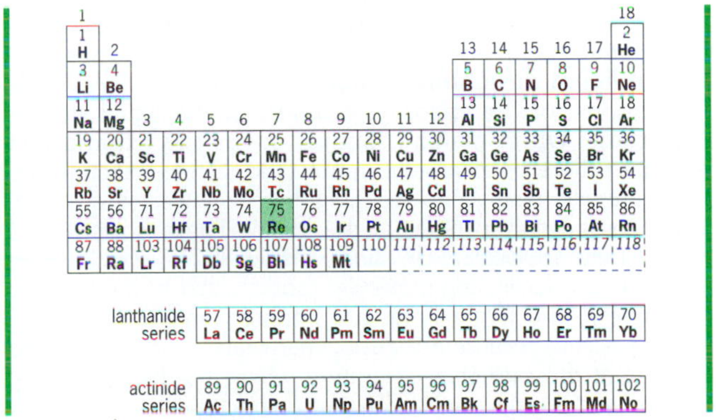

may be oxidized at elevated temperatures by oxygen to form the volatile heptoxide, Re_2O_7; this in turn may be reduced to a lower oxide, ReO_2. The compounds ReO_3, Re_2O_3, and Re_2O, are well known. Perrhenic acid, $HReO_4$, is a strong monobasic acid and is only a very weak oxidizing agent. Complex perrhenates, such as cobalt hexammine perrhenate, $[Co(NH_3)_6(ReO_4)_3]$, are also known.

The halogen compounds of rhenium are very complicated, and a large series of halides and oxyhalides have been reported. Rhenium forms two well-characterized sulfides, Re_2S_7 and ReS_2, as well as two selenides, Re_2Se_7 and $ReSe_2$. The sulfides have their counterparts in the technetium compounds, Tc_2S_7 and TcS_2. *See* TECHNETIUM; TRANSITION ELEMENTS. [S.F.]

Rheology The study of the deformation and flow of matter. The states of matter differ strikingly in their density and in the ease with which they can be deformed. The less dense the state of matter, the more easily deformable it ordinarily is. The viscosity of a gas arises from the crossing-over of molecules from a fast-moving layer into a neighboring more slowly moving layer, and vice versa. Because this crossing-over increases with temperature rise, the viscosity of a gas increases with temperature. Solids and liquids, however, become more fluid with temperature rise. *See* VISCOSITY.

Viscosity and structure. The extra fluidity of liquids, as compared with solids, arises from the great increase in the number of imperfections introduced with the 10% expansion in melting shown by normal liquids. Ice is an exception since it contracts by about 10% on melting. As a result, water has a viscosity of about 17 millipoises at the melting point, which is close to the normal value for ordinary liquids. (A poise is a unit of viscosity equal to 1 dyne s/cm^2.) A solid melts at that temperature at which the entropy from imperfections mutiplied by the temperature equals the heat of introducing them.

That the viscosity of a system closely reflects structure is exemplified by the fact that for nearly all simple substances the viscosity of the liquid at the melting point is approximately 2 centipoises. Also, the reciprocal of the viscosity, the fluidity, is a linear function of the molecular volume.

Because holes are necessary to permit viscous flow and since pressure tends to decrease the number of holes in a system, it follows that pressure normally increases viscosity. The pressure coefficient of viscosity indicates that the empty space a molecule requires in order to flow viscously is about one-seventh its normal volume. To provide such vacant sites, an activation energy of about one-third the heat of vaporization is required, as shown by the temperature coefficient of viscosity. The normal effect of pressure rise and temperature drop in increasing viscosity is modified in the rare cases where such changes promote significant modifications in structure.

When linear high polymers change their state, they introduce holes only in directions normal to the length of the chain. Thus a linear polymer melts in two directions but retains its solidlike properties along the chain length. The result is that high polymers progress by wriggling a segment at a time, as shown by the fact that they exhibit the expected pressure and temperature coefficients of a molecule the size of a segment.

Significant structures in liquids. To obtain quantitative expressions for the viscosity, a model of the liquid state is important. Upon melting, liquids ordinarily expand and become orders of magnitude more fluid. Since melting does not materially change nearest-neighbor distances, the expansion must be due to holes in the liquid. These holes should approximate molecular size, since smaller holes would not be as effective in facilitating molecular motion, that is, are wasteful of entropy, and larger holes are unduly wasteful of energy. Melting occurs because a thermodynamically more stable state appears at the melting temperature where the kinetic energy can support the fluidized vacancies, characteristic of the liquid state, against the potential energy of melting, which tends to collapse the holes. Such considerations lead to a model of the liquid in which fluidized vacancies in the liquid closely mirror the vapor molecules in their size, motion, and concentration.

High polymer viscosity. Liquid high polymers shear as a result of individual kinetic units jumping in a direction to reduce the stress. If the high polymer is long enough, there will be entanglements and a given molecule will be obligated to "slalom" around a neighbor with which it is entangled.

Non-newtonian viscosity. Frequently, systems are found which do not obey Newton's equation for viscosity; instead, a small increase in stress is found to disproportionately increase the rate of flow. *See* NON-NEWTONIAN FLUID. [H.E.]

Rheostat A variable resistor constructed so that its resistance value may be changed without interrupting the circuit to which it is connected. It is used to vary the current in a circuit. The resistive element of a rheostat may be a metal wire or ribbon, carbon disks, or a conducting liquid. *See* RESISTOR.

The metallic rheostat is the most common. The wire or ribbon is constructed in a coil or a grid, and taps are brought out from different sections of the element to a multicontact switch which can short-circuit any section of the resistor or switch it out of the circuit. For more continuous control, as is needed for laboratory rheostats, a sliding-contact finger bears directly on closely wound coils of resistive wire.

Typical applications of rheostats are for starting or controlling the speed of motors, for adjusting generator characteristics, for controlling storage-battery charging, for dimming lights, and for imposing artificial loads on electrical equipment during tests. [F.H.R.]

Rheostatic braking A system of dynamic braking in which direct-current drive motors are used as generators and convert the kinetic energy of the motor rotor and connected load to electrical energy, which in turn is dissipated as heat in a braking rheostat connected to the armature. This is accomplished by disconnecting the motor armature from the supply lines and connecting them to a braking rheostat, while the field circuit remains connected to the supply line. The polarity of the armature remains unchanged, but the direction of the armature current reverses, resulting in a negative torque. The torque can be controlled by means of the braking rheostat. *See* DIRECT-CURRENT MOTOR; DYNAMIC BRAKING. [A.R.E.]

Rheumatic fever A childhood illness which follows an infection by *Streptococcus hemolyticus*, group A, by about 3 weeks. It is characterized by arthritis (redness and swelling of joints) and carditis (inflammation of heart tissue). Typically the polyarthritis is migratory with involvement of two or more larger joints. Other symptoms include nosebleeds and chorea, also known as St. Vitus dance. Fever, rapid pulse, paleness, and indisposition are present. Rheumatic fever is important because it can cause heart disease. The likelihood of this occurring and of permanent heart damage are increased by rheumatic fever's natural tendency to recur. Heart damage can show up later, though the active illness was decidedly mild. [P.L.B.]

Rheumatism Any combination of muscle or joint pain, stiffness, or discomfort arising from nonspecific disorders. It is generally used as a lay expression to indicate a chronic or recurrent condition affecting a certain area and precipitated by cold, dampness, or emotional stress.

Lumbago, wryneck, charleyhorse and shinsplint are commonly used expressions included under the catchall category of rheumatism. [R.Se.]

Rhinoceros Odd-toed ungulates (order Perissodactyla) which are members of the family Rhinoceratidae. The bodies and limbs of these mammals are massive and thick-skinned.

Also characteristic of the family are the horns (epidermal derivatives), either one or two, which are composed of a solid mass of hairs attached to a bony prominence on the skull. The feet are tridactyl, with three hooves; the middle one is the most developed. There are five species, found in Asia and Africa.

Sight is poorly developed in these animals; however, hearing is quite good and the sense of smell is excellent. Most species are solitary and aggressive and are extremely dangerous since they are unpredictable in behavior. All species are vegetarians. *See* PERISSODACTYLA. [C.B.C.]

Rhinovirus A subgroup of the picornavirus group, including at least 70 antigenically distinct types. Like the enteroviruses, the rhinoviruses are small (17–30 nanometers), contain ribonucleic acid (RNA), and are not inactivated by ether. Unlike the enteroviruses, they are isolated from the nose and throat rather than from the enteric tract, and are unstable if kept under acid conditions. *See* ENTEROVIRUS.

Rhinoviruses have been recovered chiefly from adults with colds and only rarely from persons with more severe respiratory diseases. *See* COMMON COLD; PICORNAVIRIDAE. [J.L.Me.]

Rhizobiaceae A family of aerobic, rod-shaped, gram-negative, soil-inhabiting bacteria consisting of two genera, *Rhizobium* and *Agrobacterium*. The bacteria that make up the two genera are chemoorganotrophs; metabolism is respiratory and molecular oxygen is the terminal receptor. Carbohydrates are utilized by all strains without appreciable acid formation; that is, gas is not produced. [O.N.A.]

Rhizocephala An order of crustacean parasites related to the barnacles. Worldwide in distribution, they prey on other crustaceans, principally Decapoda, such as crabs, shrimp, and their allies. Rhizocephala produce modifications affecting the abdomen of the crab, making males resemble females and causing immature females to acquire precociously the adult form. These parasites have become so modified by their mode of life that, as adults, they are no longer recognizable as barnacles, or even as crustaceans. *See* PARASITIC CASTRATION.

An adult rhizocephalan is a thin-walled sac enclosing a visceral mass, composed chiefly of ovaries and testes. It shows no trace of segmentation, appendages, or sense organs. Even an alimentary tract is missing. Instead, it possesses a threadlike root system which penetrates the interior of the host in all directions and absorbs nourishment from the body fluids of the crab. [P.G.Re.]

Rhizomastigida An order of the class Zoomastigophorea, also known as the Rhizomastigina. All Rhizomastigida species are microscopic and ameboid, and have one or two flagella. Rhizomastigida are small, colorless, and rarely abundant. They generally occur at the mud–water interface and are holozoic, saprozoic, or parasitic. Life histories are poorly known, except for *Mastigella*. The group is small, but species occur in both fresh and salt water. *See* ZOOMASTIGOPHOREA. [J.B.L.]

Rhizophorales An order of flowering plants, division Magnoliophyta (Angiospermae), in the subclass Rosidae of the class Magnoliopsida (dicotyledons). The order contains a single family, Rhizophoraceae, with about 100 species widely distributed in the tropics. The plants are mostly tanniferous trees and shrubs with the leaves opposite, simple, and entire. The flowers are regular, mostly perfect, and variously perigynous or epigynous. The sepals are four or five and commonly fleshy or leathery; the petals are the same number as the sepals and likewise small and fleshy. The stamens are twice as many as the petals or sometimes more. The fruit is berrylike or rarely a capsule. Most members of the family are inland species, but the most conspicuous group are some 17 species of shoreline shrubs, the mangroves. *See* MAGNOLIOPHYTA; MAGNOLIOPSIDA; PLANT KINGDOM. [T.M.Ba.]

Rhizopodea A class of Sarcodina including both parasitic and free-living species found in fresh and salt water and the soil. No species forms true axopodia; instead, pseudopodia may be filopodia, lobopodia, or reticulopodia; or there may be no pseudopodia in some cases. Rhizopodea include five subclasses: Lobosia, Filosia, Granuloreticulosia, Mycetozoia, and Labyrinthulia. *See* FILOSIA; GRANULORETICULOSIA; LABYRINTHULIA; LOBOSIA; SARCODINA. [R.P.H.]

Rhizosphere The soil region subject to the influence of plant roots. It is characterized by a zone of increased microbiological activity and is an example of the relationship of soil microbes to higher plants.

A sharp boundary cannot be drawn between the rhizosphere and the soil unaffected by the plant (edaphosphere). At the root surface the rhizosphere effect is most intense, falling off sharply with increasing distance.

Growth of a plant markedly changes the microbial population of soil within its influence. In the rhizosphere there are more microorganisms than in soil distant from the plant. This increase is most pronounced with bacteria but is evident with other groups. The rhizosphere effect is seen in seedling plants; it increases with the age of the plant and usually reaches a maximum at the stage of greatest vegetative growth. Upon death of the plant the microbial population reverts to the level of the surrounding soil. Leguminous plants support higher rhizosphere populations than nonlegumes. The stimulation of microorganism growth in the rhizosphere results chiefly from the liberation of readily available organic substances by the growing plant. *See* SOIL MICROBIOLOGY. [A.G.L.]

Rhizostomeae An order of the class Scyphozoa with the most highly organized features of this class. The umbrella is generally higher than it is wide. The margin of the umbrella is divided into many lappets but is not provided with tentacles. Many radial canals, which are connected with each other to form a complicated network, issue from the cruciform stomach. The oral part is eight-sided, with many suctorial mouths surrounded by numerous small tentacles (see illustration). Usually there is no central mouth. No species of this group is

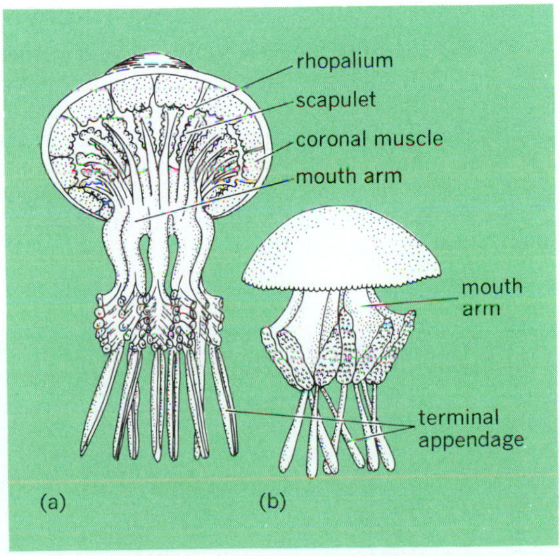

Rhizostomeae. (a) *Rhizostoma*. (b) *Mastigias*. (After L. Hyman, *The Invertebrates*, vol. 1, McGraw-Hill, 1940)

injurious to the human skin. Some large forms, such as *Rhopilema*, are used as food in China and Japan.

A fair number of fossils of this order were found in the strata of the Jurassic Period. *See* SCYPHOZOA. [T.U.]

Rho-theta systems Systems in which one or more signals are emitted from a facility (or colocated facilities) to produce simultaneous indication of bearing and distance. Since a bearing is a radial line of position and a distance is a circular line of position, the rho-theta system always ensures a position fix produced by the intersection of two lines of position which are at right angles to each other. Since the reference for both lines of position is at a common origin, computation for any course referred to this origin is simplified. *See* ELECTRONIC NAVIGATION SYSTEMS. For discussions of specific systems *see* DOPPLER VOR; RADIO RANGE; TACAN; VORTAC. [P.C.S.]

Rhodesian man The name frequently given to a group of archaic *Homo sapiens* found in eastern and southern Africa during the middle and the early upper Pleistocene, probably between 200,000 and 100,000 years ago. The best-known specimens are from Broken Hill, Northern Rhodesia, now Kabwe, Zambia. They include a nearly complete cranium, the upper jaw of another individual, a child's skull fragment, and limb bones of at least four adults. *See* FOSSIL HUMAN. [E.T.]

Rhodium A chemical element, Rh, atomic number 45, and atomic weight 102.905. Rhodium is a hard, white metal, considerably less ductile than either platinum or palladium, but much more ductile than any of the other platinum metals.

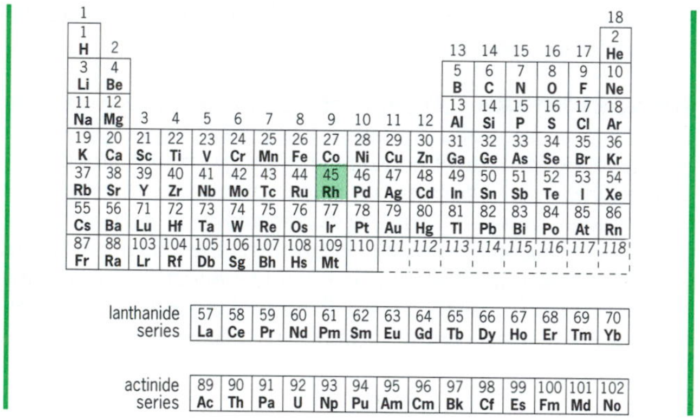

The principal use of rhodium is as an alloying element for platinum. Rhodium is an excellent hydrogenation catalyst and is active in catalytic reforming of hydrocarbons. Rhodium is also used in electrical contact applications. It is easily electroplated to form a hard, wear-resistant, permanently bright surface, used on both stationary and sliding electrical contacts, for mirrors and reflectors, and as a finish on jewelry. *See* ELECTROPLATING OF METALS; PLATINUM.

Physical properties of rhodium

Properties	Values
Atomic weight, $^{12}C = 12.00000$	102.905
Naturally occurring isotopes and percent abundance	103 (100)
Common chemical valence	3,4
Density at 25°C, g/cm^3	12.43
Melting point, °C	1963
Boiling point, °C	3700
Specific heat at 0°C, cal/g (J/kg)	0.0589 (246)

Rhodium is resistant to most common acids, including aqua regia, even at moderate temperatures. It is attacked by hot sulfuric acid, hot hydrobromic acid, sodium hypochlorite, and free halogens at 200–600°C (390–1110°F). The values of some important physical properties are given in the table.

Rhodium trichloride, $RhCl_3$, is a red compound which is insoluble in water. Rhodium trihydroxide is soluble in some acids and may be used to produce rhodium salts. Rhodium sulfate, $Rh_2(SO_4)_3 \cdot xH_2O$, is red or yellow and soluble in water. [H.J.A.]

Rhodochrosite The mineral form of manganese carbonate. Calcium, iron, magnesium, and zinc have all been reported to replace some of the manganese. The equilibrium replacement of manganese by calcium increases with the temperature of crystallization.

Rhodochrosite occurs more often in massive or columnar form than in distinct crystals. The color ranges from pale pink to brownish pink. Hardness is 3.5–4 on Mohs scale, and specific gravity is 3.70.

Well-known occurrences of rhodochrosite are in Europe, Asia, and South America. In the United States large quantities occur at Butte, Montana. As a source of manganese, rhodochrosite is also important at Chamberlain, South Dakota, and in Aroostook County, Maine. *See* CARBONATE MINERALS; MANGANESE. [R.I.Ha.]

Rhodonite A mineral inosilicate with composition $MnSiO_3$. Hardness is 5.5–6 on Mohs scale, and specific gravity is 3.4–3.7. The luster is vitreous and the color is rose red, pink, or brown. Rhodonite is similar in color to rhodochrosite, manganese carbonate, but it may be distinguished by its greater hardness and insolubility in hydrochloric acid. It has been found at Langban, Sweden; near Sverdlovsk in the Ural Mountains; and at Broken Hill, Australia. Fine crystals of a zinc-bearing variety, fowlerite, are found at Franklin, New Jersey. *See* SILICATE MINERALS. [C.S.Hu.]

Rhodophyceae A large class of plants, commonly called red algae, coextensive with the division Rhodophycota. Most red algae are found in the ocean, growing on rocks, wood, other plants, or animals in the intertidal zone and to depths limited by the availability of light. A few genera and species occur in fresh water, and these are usually found in rapidly flowing, well-aerated, cold streams. Some, however, grow in quiet warm water, while a few are subaerial. Most red algae are photosynthetic, but some grow on other algae with varying degrees of parasitism. Approximately 675 genera and 4100 species are recognized. *See* ALGAE.

Rhodophyceae are characterized by a unique combination of biochemical, reproductive, and ultrastructural features. The primary photosynthetic pigment is chlorophyll *a*. Water-soluble tetrapyrrolic compounds called phycobilins serve as accessory photosynthetic pigments. The chief food reserve, floridean starch, is a branched polymer of glucose similar to amylopectin of green plants. It occurs as granules in the cytoplasm. Multinucleate cells are common. Rhodophyceae are distinctive among eukaryotic algae in their lack of flagella, a feature shared among major groups only by the chlorophycean order Zygnematales. Some unicellular forms and many spores and male gametes of multicellular forms are capable of gliding or feeble ameboid motion. Unicellular red algae, which may form mucilaginous colonies, are considered primitive. Most red algae have multicellular thalli of microscopic or macroscopic size, including individual filaments, blades, and complex plants of distinctive form produced by the interplay of filamentous systems. *See* CORALLINALES; ZYGNEMATALES.

Two subclasses of Rhodophyceae are traditionally recognized: the Bangiophycidae and the Florideophycidae. Classification within the Bangiophycidae is based largely on vegetative and asexual reproductive features, while that of the Florideophycidae is based primarily on details of the development of the female reproductive system and carposporophyte, secondarily on vegetative features. Rhodophyceae seem most closely related to Cyanophyceae in their use of chlorophyll *a* and phycobilins as photosynthetic pigments and the absence of flagella. They probably did not evolve directly from Cyanophyceae, however, but from a colorless, nonflagellate, eukaryotic ancestor that acquired pigments from an endosymbiotic blue-green alga.

The most salient feature of red algae is their beauty, which has drawn admiration from generations of seaside visitors. Their greatest significance, however, is their role in the formation of coral reefs, the Corallinales being responsible for cementing together various animal and algal components. Of more apparent economic importance is their use as food, a centuries-old tradition of maritime peoples in many parts of the world. *See* AGAR; CARRAGEENAN; REEF. [P.C.Si.; R.L.Moe]

Rhombifera An extinct class of Echinodermata in which the thecal canals crossed the sutures at the edges of the plates, so that one-half of any canal lay in one plate and the other half on an adjoining plate. The canals sometimes occurred in lozenge-shaped clusters, called rhombs (see illustration). The theca was ovoid and sessile in earlier forms and comprised numerous irregular plates. The ambulacral grooves were trimerous and were restricted to the brachioles. In later forms the thecal plates became fewer and larger, and were arranged in five horizontal rows, termed cycles. A stem developed aborally and the ambulacra became pentamerous, traversing part of the theca before ascending the brachioles. Rhombifera probably died out in the Devonian. *See* ECHINODERMATA. [H.B.F.]

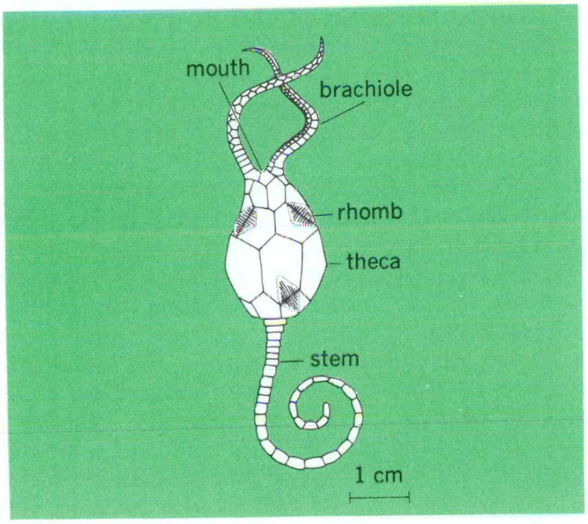

Ordovician *Pleurocystis*. (*Simplified after O. Jaekel*)

Rhubarb A herbaceous perennial, *Rheum rhaponticum*, of Mediterranean origin, belonging to the plant order Polygonales. Rhubarb is grown for its thick petioles which are used mainly as a cooked dessert; it is frequently called the pieplant. The leaves, which are high in oxalic acid content, are not commonly considered edible. Outdoor rhubarb is a common garden vegetable in most areas of the United States except the South. Michigan and Washington are important centers for forced or hothouse rhubarb. *See* POLYGONALES. [H.J.C.]

Rhynchobdellae An order of the class Hirudinea. These leeches possess an eversible proboscis and lack hemoglobin in the blood. They may be divided into two families, the Glossiphoniidae and the Ichthyobdellidae. Glossiphoniidae are flattened, mostly small leeches occurring chiefly in fresh water. Ichthyobdellidae typically have cylindrical bodies with conspicuous, powerful suckers used to attach themselves to passing fish. They frequently have lateral appendages which aid in respiration. *See* HIRUDINEA. [K.H.M.]

Rhynchocephalia One of the two surviving orders of lepidosaurian reptiles; they are represented today, however, by a single species, *Sphenodon punctatus*. They have a typical diapsid skull, and, also typically, the teeth are fused to the edges of the jaws; their only characteristic specialization is that the upper jaw forms an overhanging beak.

The Rhynchocephalia appear to have originated from the Eosuchia in the Lower Triassic. The fairly numerous fossil genera are classified into five families. *See* LEPIDOSAURIA; REPTILIA. [A.J.C.]

Rhynchocoela A phylum of bilaterally symmetrical, unsegmented, ribbonlike worms, frequently referred to as the Nemertinea. They have an eversible proboscis and a complete digestive tract with an anus. There is no coelom or body cavity, and the mesenchyme or parenchyma and the muscle fibers fill the area between the ciliated epidermis and the cellular lining of the digestive tract. Many species are brightly colored, sometimes having stripes or transverse bars.

The nemertineans are the simplest animals with a circulatory system. There are two lateral blood vessels and in some a third, unpaired dorsal vessel. The blood consists of a colorless fluid which may contain blood cells of several types. In species in which the blood is colored, the pigment is present in the cells. There is no heart, but the walls of the principal vessels may be contractile.

The nervous system has a pair of cerebral ganglia forming the brain as well as two longitudinal nerve cords and many smaller nerves. The ganglia and lateral cords may contain unusually large neurochord cells. In the epidermis there are scattered sensory nerve cells, probably tactile. A few to many simple eyes, or ocelli, may be present in front of the cerebral ganglia. There are no special respiratory organs; respiration occurs through the body surface. Nemertineans are usually either male or female, but a few individuals have both sex organs. Fertilization occurs outside the body in many species but may be internal in certain forms.

The nemertineans are mostly marine, bottom-dwelling worms, found in greatest numbers along the coasts of northern temperate regions. They live under stones, among the tangled masses of plants, in sand, mud, or gravel, and sometimes form mucus-lined tubes. A few are pelagic, fresh-water, or terrestrial. Certain species are commensal with other animals, but none can be regarded as parasitic in a strict sense.

That the Rhynchocoela represent the most highly organized acoelomate animals is indicated by the circulatory system, the presence of an anus, and the specialization of the epidermis. All groups of animals more complex than the nemertineans have some kind of cavity, a pseudocoele or coelom, between the body wall and the gut, instead of solid mesenchyme. *See* COELOM.

The phylum Rhynchocoela, containing about 550 known species, is divided into two classes, Anopla and Enopla. *See* ANOPLA; ENOPLA. [A.G.Hu.; J.B.J.]

Rhynchonellida An order of articulate brachiopods, representatives of which are first known in beds of Middle Ordovician age and which is still extant. *See* ARTICULATA (BRACHIOPODA).

Typical forms are dorsibiconvex, with a dorsal fold and ventral sulcus; the posterior margin is curved, the dorsal interarea absent, and the ventral one greatly reduced. The majority of forms are ornamented by ribs. The majority of the order are impunctate. *See* BRACHIOPODA. [A.J.R.]

Rhynchosauria An order of herbivorous diapsid reptiles in the infraclass Archosauria, limited to the Triassic System but with a worldwide distribution. Rhynchosaurs were pig- or sheep-sized quadrupedal reptiles that were common in the Middle and Upper Triassic Series of India, South America, Europe, and eastern Africa. Except for the earliest South African genus, *Mesosuchus*, they are characterized by multiple rows of teeth on both the upper and lower jaws that were fused into deep sockets. Teeth were not replaced as they were worn but were added posteriorly as the jaw grew. In most genera, the premaxillae are devoid of teeth and overhang the front of the lower jaw like a beak. The external nostril is median rather than paired.

Rhynchosaurs were elements of the early archosauromorph radiation that included the protorosaurs and primitive archosaurs. The structure of the ankle is nearly identical in the early members of these groups, but later rhynchosaurs enlarged the centrale which contributed to a simple hinge joint between the lower leg and the tarsals.

Rhynchosaurs were locally common in the Late Triassic but are unknown in the Jurassic. Their rapid extinction may have resulted from changes in the vegetation on which they fed, or from predation from the expanding community of predaceous archosaurs, including early dinosaurs. *See* ARCHOSAURIA; DIAPSIDA; PROTOROSAURIA; REPTILIA. [R.L.C.]

Rhyniophyta A division of the subkingdom Embryobionta. The bryophytes and vascular plants are included in this subkingdom. The category Rhyniophyta was devised for the relatively simple Silurian-Devonian vascular plants long held to be ancestral to other groups of vascular plants and usually referred to as Psilophytales. These plants have leafless stems and lack roots; their general morphological structure is not complex. The three classes of Rhyniophyta currently recognized are Rhyniopsida, Zosterophyllopsida, and Trimerophytopsida. *See* EMBRYOBIONTA; PSILOPHYTALES; RHYNIOPSIDA; TRIMEROPHYTOPSIDA; ZOSTEROPHYLLOPSIDA. [H.P.B.]

Rhyniopsida The earliest demonstrable vascular land plants, appearing in Silurian (mid- Ludlovian) time. Their small, leafless axes were usually branched dichotomously in three planes; adventitious and perhaps pseudomonopodial branching also occurred. Sporangia were usually terminal on the main axes. Some terminated in lateral branches, and some were subtended by adventitious branches. *See* RHYNIOPHYTA. [H.P.B.]

Rhyolite A light-colored, aphanitic (not visibly crystalline) or porphyritic rock of volcanic origin. If completely crystallized, rhyolite consists largely of alkali feldspar and quartz (or another polymorph of SiO_2), and subordinate oligoclase, and minor amounts of mafic minerals (biotite, hornblende, or pyroxene). *See* IGNEOUS ROCKS.

Rhyolites composed almost entirely of glass are called obsidian, pitchstone, perlite, or pumice. Typical rhyolites are mixtures of microcrystalline aggregates and glassy material, but many are entirely crystalline, and some grade into microgranite. Porphyritic varieties are those in which numerous large crystals (phenocrysts) are disseminated throughout an aphanitic matrix. *See* OBSIDIAN; PHENOCRYST; PORPHYRY; VOLCANIC GLASS.

Next to basalt and andesite, rhyolite is considered the most common volcanic rock. It occurs most frequently as a fragmented rock (tuff) which originated from a pyroclastic eruption, either as an ash flow or an air-fall deposit. The rock is widespread and occurs in large quantities associated with andesite, basalt, and small amounts of dacite and quartz latite. *See* MAGMA; PETROGRAPHIC PROVINCE; TUFF. [W.I.R.]

Riboflavin Also known as vitamin B_2, riboflavin is widely distributed in nature, and is found mostly in milk, egg white, liver, and leafy vegetables. It is a water-soluble yellow-orange fluorescent pigment.

Riboflavin deficiency results in poor growth and other pathologic changes in the skin, eyes, liver, and nerves. Riboflavin deficiency in humans is usually associated with a cracking at the corners of the mouth called cheilosis; inflammation of the tongue, which appears red and glistening (glossitis); corneal vascularization accompanied by itching; and a scaly, greasy dermatitis about the corners of the nose, eyes, and ears. *See* VITAMIN. [S.N.G.]

Ribonucleic acid (RNA) A class of long, unbranched nucleic acid macromolecules mainly involved in translating into proteins the genetic information that is carried in DNA. There are two main functions of RNAS: that of acting as tools in the synthesis of any protein and that of acting specifically in the syntheses of one protein. The major classes of RNA of the first kind include transfer RNA (tRNA) and ribosomal RNA (rRNA), and of the second kind messenger RNA (mRNA); viral RNA may belong to both classes. RNA is the intermediate of the pathway DNA $\rightarrow$ RNA $\rightarrow$ protein. The information in DNA is first transcribed into RNA which then directs the synthesis of specific proteins. *See* DEOXYRIBONUCLEIC ACID (DNA).

RNA is a long chain of repeating units called nucleotides. Each nucleotide is composed of three parts: a five-carbon sugar called ribose; a phosphate group; and most important, attached to each ribose, is one of four bases—adenine (A), cytosine (C), guanine (G), or uracil (U). Thus there are four different nucleotides, depending on which base is present. The structure and function of the RNA depend on the sequence and the number of nucleotides of which it is composed.

Transfer RNA is the smallest class of RNA known that has a biological function, which is to carry an amino acid to the ribosome during the process of protein synthesis. Transfer RNAs have a molecular weight of 25,000 daltons and contain from 73 to 90 nucleotides.

Ribosomes are particles made up of RNA and protein, which act as workbenches on which the process of linking amino acids together to form proteins occurs. Each ribosome consists of two subunits, a large one and a small one. Ribosomal RNAS, like tRNAs, are stable molecules, but, unlike tRNAs, they contain very few unusual bases. *See* RIBOSOMES.

The concept of a messenger (m) RNA that was involved in the expression of the genetic information in DNA was a big step forward in understanding the mechanisms involved in molecular genetics. Messenger RNA is synthesized by the enzyme RNA polymerase from one of the two strands of DNA. It is made up of a sequence of ribose nucleotides with bases complementary to those of that strand of DNA. *See* PROTEIN.

While most organisms carry their genetic information in the form of DNA, certain viruses have RNA as their genetic material. In RNA viruses, the RNA plays the role of DNA (since it contains the genetic information) and often acts as mRNA as well. There are several different types of viral RNA. In some RNA viruses the RNA is present as a long, single-stranded piece (such as in polio virus), while in other cases the RNA is in several small single-stranded pieces (such as in influenza virus, which contains about 10 different pieces of RNA in each virus). *See* VIRUS. [B.S.D.]

Ribose A water-soluble pentose, also known as D-ribose, which, together with 2-deoxy-D-ribose, makes up the carbohydrate constituents of nucleic acids, which are found in all living organisms. The structural formula is shown.

The universal occurrence of nucleic acids in all living cells makes this pentose highly interesting to biochemists and biologists. The type of nucleic acid that yields D-ribose is referred to as ribonucleic acid (RNA). D-Ribose is a constituent not only of

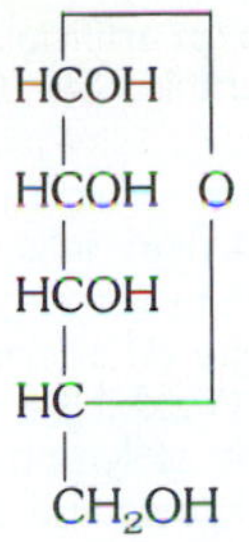

the nucleic acids, but also of several vitamins and coenzymes. As in the nucleic acids, this sugar occurs in the furanose configuration in these natural products. *See* Nucleic acid. [W.Z.H.]

Ribosomes Complex particles, about 25 nanometers in diameter, which synthesize proteins within the living cell. They are composed of many different proteins in addition to three molecules of ribonucleic acid (RNA). Each ribosome consists of two subunits, one approximately twice the size of the other. The small subunit contains one large molecule of RNA, whereas the large subunit has two RNA molecules, one twice as large as that in the small subunit, and the other is considerably smaller (see illustration). The ribosomal proteins make up the remainder of the ribosome particle. There are about 20 different proteins in the small subunit and 40 in the large one; as far as has been determined, each kind of protein molecule is present in just a single copy in the entire ribosome. The ribosomes in higher organisms (with a total mass of 4,000,000 daltons) are one-third to one-half larger than the ribosomes in bacteria and other microorganisms (which have a mass of about 2,700,000). A considerably greater mass of proteins accompanies the somewhat larger RNA molecules in higher organisms. Fully one-half of the mass is represented by proteins in these ribosomes.

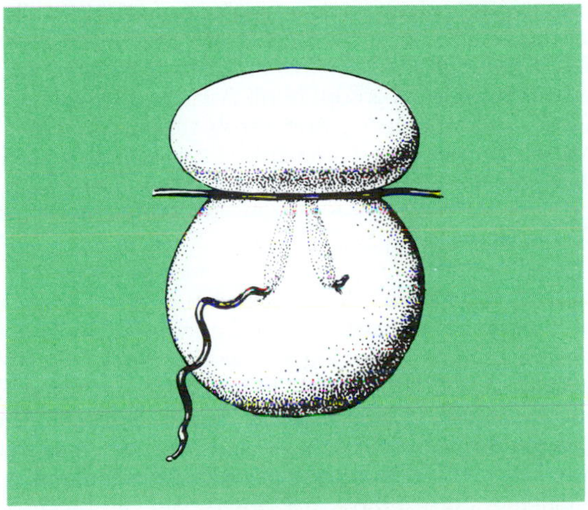

Large ribosome subunit (below) binding two tRNA molecules, one coupled to a nascent polypeptide chain (left) and the other to a single amino acid (right). mRNA lies in the groove between the subunits.

Ribosomes play an important part in protein synthesis. Part of the deoxyribonucleic acid (DNA) in the nucleus serves as a template for the synthesis of a messenger RNA (mRNA) molecule, whose sequence of nucleotides is complementary with that of the DNA. This mRNA strand then passes along a ribosome, where it mediates the production of a specific protein molecule as follows. When each successive set of three nucleotides arrive at the appropriate site on the ribosome, a specific amino acid is added to the polypeptide. The mechanism of this "translation" depends upon each kind of the 20 amino acids being coupled enzymatically with a particular transfer RNA (tRNA) molecule. If a particular set of three nucleotides (triplet) on this coupled tRNA is complementary with the mRNA triplet in position, the coupled amino acid is added to the growing polypeptide chain by the ribosome. *See* Genetic code; Ribonucleic acid (RNA).

Several ribosomes may be spaced along the length of the mRNA molecule, simultaneously utilizing different parts of its coded structure. This complex is called a polyribosome or polysome. Polysomes are not formed by a simple association between intact ribosomes and the mRNA strand. Instead, a union is first established between the small subunit and the messenger. A large subunit is then added to the assembly, resulting in the complete protein-synthesizing machine. Further ribosomes are joined together in this sequential manner at the leading end of the messenger, thus forming the polysome. In the cells of higher organisms, polysomes may exist free in the cytoplasm, or may be attached to membranes of the endoplasmic reticulum, making proteins to be contained within a membranous enclosure. *See* Endoplasmic reticulum. [B.E.B.]

Ribozyme A ribonucleic acid (RNA) molecule that, like a protein enzyme, can catalyze, or lower the activation energy for, specific biochemical reactions. *See* Enzyme; Protein; Ribonucleic acid (RNA).

Protein enzymes and ribozymes are similar in their catalytic efficiency. It is not surprising that ribozymes and protein enzymes are able to bind their substrates with similar affinities. Both rely on the same three types of interactions for folding and substrate binding: hydrogen bonding, hydrophobic interactions, and charge pairing interactions. These interactions allow formation of the precise three-dimensional shape and, furthermore, can be positioned in the active site for specific binding of the substrate. Proteins are synthesized from a pool of 20 different amino acids, whereas RNA molecules are polymers derived from a pool of only 4 nucleic acids. In this respect, proteins have more versatility in the reactions they catalyze and the substrates they bind. However, RNA catalysts have one capability that proteins do not have: they can form specific hydrogen bonds with a complementary sequence of nucleotides. Thus, RNA may be particularly well suited for catalysis of reactions that involve nucleotide or polynucleotide substrates. *See* Hydrogen bond. [B.B.]

Ribulose A pentose sugar, also known as D-riboketose and D-erythropetulose; it has never been prepared in crystalline form, and exists only as a syrup. The structural formula of ribulose is shown below.

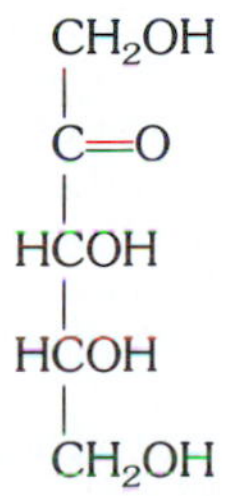

Ribulose-5-phosphate occurs in animal and plant tissues. It can be converted to ribulose-1,5-diphosphate by a phosphokinase enzyme acting in the presence of adenosine triphosphate (ATP). The ribulose-5-phosphate is also a significant intermediate in the carbohydrate metabolism through the pentose phosphate pathway. *See* Adenosine triphosphate (ATP); Carbohydrate metabolism; Monosaccharide. [W.Z.H.]

Rice The plant *Oryza sativa* is the major source of food for nearly one-half of the world's population. In China, Japan, Korea, the Philippines, India, and other countries of Asia, rice is far more important than wheat as a source of carbohydrates. The most important rice-producing countries are mainland China, India, and Indonesia, but in many smaller countries rice is the leading food crop. In the United States rice production is largely concentrated in selected areas of Arkansas, California, Louisiana, and Texas. Although this represents less than 1% of the world rice acreage, the United States often is the world's largest exporter of rice, since most of the world rice crop is consumed in the countries where it is produced.

Over 95% of the world rice crop is used for human food. Although most rice is boiled, a considerable amount is consumed as breakfast cereals. Rice starch also has many uses. Broken rice is used as a livestock feed and for the production of alcoholic beverages. The bran from polished rice is used for livestock feed; the hulls are used for fuel and cellulose. The straw is used for thatching roofs in the Orient and for making paper, mats, hats, and baskets. Cultivated rices are classified as upland and lowland. Upland types, which can be grown in high-rainfall areas without irrigation, produce relatively low yields. The lowland types, which are grown submerged in water for the greater part of the season, produce higher yields. In contrast to most plants, rice can thrive when submerged because oxygen is transported from the leaves to the roots. Rice varieties are also classified as long- or short-grain. Most long-grain rices have high amylose content and are dry or fluffy when cooked, while most short-grain rices have lower amylose content and are sticky when cooked. *See* CYPERALES. [J.A.Sh.]

Ricinulei An order of extremely rare arachnids, also known as the Podogona, with a body less than 1 in. (2.5 cm) in length. Superficially, they resemble ticks in general appearance and movement, and are found only in tropical Africa and in the Americas, from the Amazon to Texas. The two anterior pairs of appendages are chelate. Less than 25 modern species are known. The occurrence of several fossils from Carboniferous time suggests that the group was formerly more common. *See* ARACHNIDA. [C.C.Ho.]

Rickets A disorder of calcium and phosphorus metabolism, primarily affecting bony structures, due to vitamin D deficiency. Precursor substances from the diet are normally converted to vitamin D by the action of ultraviolet light (sunlight) on the skin. Therefore, infants are affected, especially in winter. Dark-skinned peoples require additional vitamin D as dietary supplement because their skin pigmentations interfere with natural production. *See* VITAMIN D.

In children, rickets consists of defective calcification and excessive production of cartilage at the ends of growing bones. Restlessness, constant movement often with hair loss from pillow contact, and defects of the ribs and long bones are typical symptoms. Bowlegs and pigeon breast may result, in addition to craniotabes, or softening of the flat skull bones, which later develop abnormally to form a square, boxlike skull. There is an increased susceptibility to fracture, delayed tooth eruption and enamel defects, and other indications of faulty mineral deposition. *See* TOOTH DISORDERS.

In adults, vitamin D deficiency produces osteomalacia, or demineralization of bones. In addition, low calcium levels in the blood may be present with resulting irritability, spasms, and convulsions, particularly of the hands, face, and larynx. [E.G.St./N.K.M.]

Rickettsiales An order of very small bacteria (0.2–0.5 micrometer in diameter) which are mostly obligate intracellular parasites of animals. The order is divided into three families, Rickettsiaceae, Bartonellaceae, and Anaplasmaceae. Only the species causing trench fever in humans, among mammalian parasites, has been grown on artificial, cell-free media. Another, the Q-fever agent, is filterable. *See* CARRION'S DISEASE; EPIDEMIC TYPHUS FEVER; Q FEVER. [C.B.P.]

Rickettsialpox A benign, infectious disease, similar to typhus, caused by bacterialike microorganisms, *Rickettsia akari*. It is transmitted by the mite *Allodermanyssus sanguineus*. In addition to a rash, patients exhibit a primary eschar, or ulcer, and often show a swelling of lymph glands similar to that in scrub typhus and fièvre boutonneuse. *See* RICKETTSIOSES. [C.B.P.]

Rickettsioses Infectious diseases affecting humans and some animals, caused by certain bacterial microorganisms; the human pathogens are members of the genera *Rickettsia*, *Rochalimaea*, and *Coxiella*, which are classified in the family Rickettsiaceae. These rickettsial agents make up a restricted group found in parasitic arthropods, such as lice, fleas, ticks, and mites, which act as vectors and often as the primary hosts.

Rickettsia species are the etiologic agents of such human diseases as epidemic and endemic, or murine, typhus, the Rocky Mountain spotted fever group of diseases, including fièvre boutonneuse of the Old World, and several tick typhuses, as well as the rickettsialpox of the Atlantic coast of the United States and of the former Soviet Union, and tsutsugamushi disease, or scrub typhus, of the Far East.

Clinically all are characterized by fever and a cutaneous eruption which is absent only in related Q fever. In addition, a primary ulcer, or eschar, at the site of vector attachment and swollen glands which drain this area are associated with the fièvre boutonneuse subgroup, north Asian and Queensland tick typhuses, rickettsialpox, and scrub typhus. Orchitis and swelling of the scrotum customarily occur in male guinea pigs infected with murine typhus and the spotted fever group. *See* EPIDEMIC TYPHUS FEVER; FIÈVRE BOUTONNEUSE; MURINE TYPHUS FEVER; Q FEVER; RICKETTSIALPOX; SCRUB TYPHUS. [C.B.P.]

Riemann surface A generalization of the complex plane that was originally conceived to make sense of mathematical expressions such as $\sqrt{z}$ or $\log z$. These expressions cannot be made single-valued and analytic in the punctured plane $\mathbf{C}\backslash\{0\}$ (that is, the complex plane with the point 0 removed). The difficulty is that for some closed paths the value of the expression when reaching the end of the path is not the same as it is at the beginning. For example, the closed path can be chosen to be the unit circle centered at $z = 0$ and followed counterclockwise from $z = 1$. If $\sqrt{z}$ is assigned the value $+1$ at $z = 1$, its value at the end of the circuit is -1. Similarly, if $\log z$ is assigned the value 0 at $z = 1$, at the end of the circuit, allowing the values to change continuously, the value is $2\pi i$. *See* COMPLEX NUMBERS.

The construction of an abstract surface for $\sqrt{z}$ on which $\sqrt{z}$ has a single value at each point and on which the values of $\sqrt{z}$ vary continuously around each closed path, always ending with the starting value, proceeds as follows. From the complex plane $\mathbf{C}$ the nonnegative half of the real axis $\{z = x + iy : y = 0,\ 0 \le x\}$ is removed. This slit is thought of as having two edges: an upper or + edge and a lower or − edge ($z = 0$ is not counted). Another copy of this slit plane is then placed over, and parallel to, the original, as the first and second floors in a building. Next, the two sheets are connected by attaching the − edge of the bottom sheet to the + edge of the top sheet, and the + edge of the bottom sheet to the − edge of the top sheet. The abstract surface thus constructed is not embedded in euclidean 3-space because the attachments cannot be done in 3-space without creating self-intersections. The expression $\sqrt{z}$ is defined at the point over z on the upper sheet to have positive imaginary part and on the lower to have negative part.

The abstract surface constructed is called the Riemann sur-

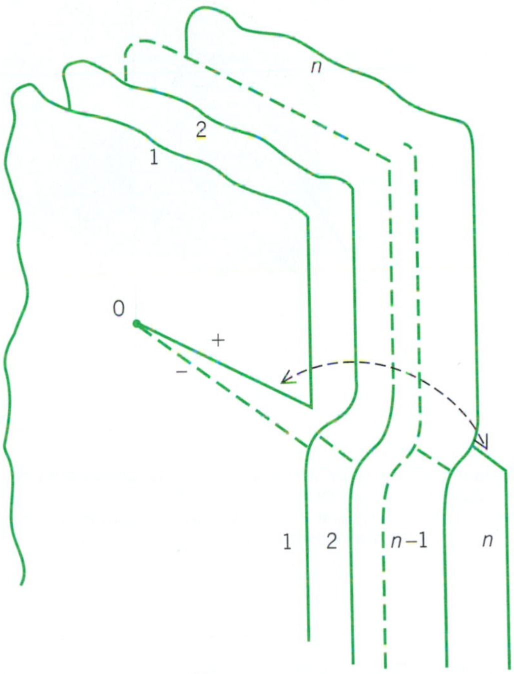

Riemann surface for $\sqrt[n]{z}$. *(After A. D. Sveshnikob and A. N. Tikhonov, The Theory of Functions of a Complex Variable, Mir Publishers, 1973)*

face for $\sqrt{z}$. It is two-sheeted over the z plane, except that only one point lies over $z = 0$, which is therefore called a branch point. The Riemann surface for $\sqrt[n]{z}$, which has n branches instead of 2, is shown in the illustration.

In the early 1920s, H. Weyl defined a Riemann surface in the context of the newly developing field of topology. This definition does not depend on having a particular function in hand, and does not require the use of such seemingly artificial devices as slits. A Riemann surface is a two-dimensional connected topological manifold (surface) with a complex structure. *See* MANIFOLD (MATHEMATICS); TOPOLOGY. [A.M.]

Riemannian geometry The geometry of an N-dimensional space which bears a coordinate system (x) and has associated with it and the given coordinate system a set of N^2 functions $g_{ab}(x)$ which are involved in the determination of certain fundamental geometric magnitudes, including lengths of arcs in the space, lengths of vectors, angles between vectors, the measures of parts of the space (areas, volumes, and hypervolumes), and curvatures. The arc length s, for example, of a parameterized arc $x^a = x^a(t)$ of class C' is given by the formula below.

$$\Delta s = s(t_2) - s(t_1) = \int_{t_1}^{t_2} \sqrt{g_{ab}x'^a x'^b dt}$$

Here, the prime (′) indicates differentiation with respect to t so that $x'^a = dx^a/dt$ and Einstein's summation convention is in force. The quantities g_{ab} are, as indicated, functions of the N coordinate variables $x^1, x^2, \ldots, x^N$ denoted briefly by x. The general subject may be classified into subspecies by means of the value of N and the conditions imposed on the fundamental quantities g_{ab} either directly or through derived quantities.

Riemannian geometry was initiated in 1854 by G. F. B. Riemann (1826–1866), with the quadratic forms $g_{ab}x'^a x'^b$ of course limited to the positive definite case. In later developments this restriction is removed.

The importance of this subject is due to several facts. First, it includes both euclidean geometry and the geometry of surfaces in euclidean space and, in addition, provides their most natural extension, a generalization of great richness and scope. Second, it provides a geometrical realization or application for certain of the major abstractions of tensor analysis and together with that discipline, with which it merges, forms the pattern and much of the motivation for general relativity. Third, most of the work on the incorporation of electricity and magnetism into the mold of general relativity (unified field theory) almost necessarily involves a riemannian geometry of one type or another (nonsymmetric g_{ab}'s symmetric g_{ab} with $N = 5$, g_{ab}'s in projective coordinates). Fourth, riemannian geometry provides the steppingstone to a variety of generalized differential geometries.

Finally, in the absence of contrary evidence, the success of the general theory of relativity suggests that actual physical space may correspond more closely to a noneuclidean riemannian geometry than to the euclidean variety. *See* RELATIVITY. [H.V.C.]

Rift valley An elongated, relatively narrow depression caused by the subsidence of a crustal block between two or more faults. The boundary faults are steeply inclined down toward the downthrown block, and where the direction of displacement has been ascertained, it indicates that the dislocations are gravity faults. Thus, rift valleys are the surface expression of large graben (see illustration). *See* FAULT AND FAULT STRUCTURES; GRABEN.

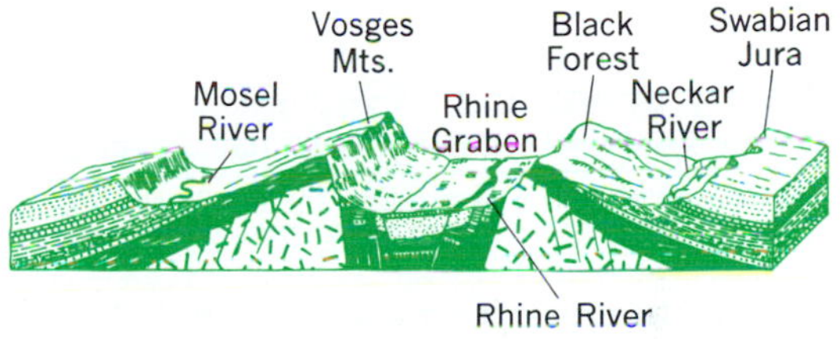

A generalized cross section of the rift valley of the Rhine and the adjoining block mountains of the Vosges and the Black Forest. *(After P. E. James, An Outline of Geography, 2d ed., copyright 1943 by Ginn and Co.)*

Rift valleys have been interpreted to be a part of the major tectonic pattern of the Earth and to be the result of deep-seated deforming pressures. They are associated with many mid-oceanic ridges, where they are interpreted to be graben in an uplifted segment of oceanic crust over a rising column of mantle material. These rift valleys occupy positions where the oceanic crust has split apart and new crust is forming. Smaller, less extensive rift valleys occur in regions underlain by continental crust. [P.H.O.]

Rift Valley fever An arthropod-borne (primarily mosquito), acute, febrile, viral disease of humans and numerous species of animals. Rift Valley fever is caused by a ribonucleic acid (RNA) virus in the genus *Phlebovirus* of the family Bunyaviridae. In sheep and cattle, it is also known as infectious enzootic hepatitis. First described in the Rift Valley of Africa, the disease presently occurs in west, east, and south Africa and has extended as far north as Egypt. Historically, outbreaks of Rift Valley fever have occurred at 10–15-year intervals in normally dry areas of Africa subsequent to a period of heavy rainfall.

In humans, clinical signs of Rift Valley fever are influenzalike, and include fever, headache, muscular pain, weakness, nausea, epigastric pain, and photophobia. Most people recover within 4–7 days, but some individuals may have impaired vision or blindness in one or both eyes; a small percentage of infected individuals develop a hemorrhagic syndrome and die.

Rift Valley fever should be suspected when high abortion rates, high mortality, or extensive liver lesions occur in newborn animals. The diagnosis is confirmed by isolating the virus from tissues of the infected animal or human. Control of the disease is best accomplished by widespread vaccination of susceptible animals to prevent amplification of the virus and, thus, infection of vectors. Any individual that works with infected animals or live virus in a laboratory should be vaccinated. *See* ANIMAL VIRUS; VACCINATION. [C.A.Me.]

Rigel Beta Orionis, a blue supergiant of spectral type B8. Although Rigel is one of the apparently brightest stars in the sky, it is too distant to have a measurable parallax or proper motion. It has a faint companion with which it is probably physically connected. Having a luminosity about 60,000 times that of the Sun means that Rigel is an exceptionally young star in rapid evolution, with a life span of only a few million years. *See* STAR. [J.L.Gr.]

Rigid body An idealized extended solid whose size and shape are definitely fixed and remain unaltered when forces are applied. Treatment of the motion of a rigid body in terms of Newton's laws of motion leads to an understanding of certain important aspects of the translational and rotational motion of real bodies without the necessity of considering the complications involved when changes in size and shape occur. Many of the principles used to treat the motion of rigid bodies apply in good approximation to the motion of real elastic solids. *See* RIGID-BODY DYNAMICS. [D.Wi.]

Rigid-body dynamics A rigid body is defined as an assemblage or system of mass particles that are located rigidly with respect to one another and therefore can have no motion relative to each other. Motion of a rigid body can occur by movement of all points of the body in a parallel direction through equal distances during a given interval of time, called translation, or by movement of all points in circles about a common axis with a common angular velocity, called rotation, or by combined translation and rotation.

Center of mass. The center of mass of a rigid body is a single point located within the body such that any force acting externally on the body along a line of action which passes through this point will result in pure translation, that is, no rotation. *See* CENTER OF MASS.

To calculate the position of the center of mass, consider a rigid body divided into elemental volumes (or particles) labeled dV in the illustration. Each such elemental volume is located from a fixed point O by means of a position vector $\mathbf{r}$ and has a mass density ρ.

The total mass m of the rigid body can then be expressed by the volume integral in Eq. (1).

$$m = \int_V \rho \, dV \tag{1}$$

The center of mass of the rigid body is then defined as the point within the rigid body whose position vector $\mathbf{r}_c$, is given by Eq. (2).

$$\mathbf{r}_c = \frac{1}{m}\int_V \mathbf{r}\rho \, dV \tag{2}$$

Translational motion. The motion of the center of mass of a rigid body can be derived directly from Newton's second law applied to each particle of the rigid body and summed for all such particles. All internal forces sum to zero because such forces between particles of the body occur in equal and opposite pairs (Newton's third law). The motion of the center of mass is the same as would be the case if all of the mass of the body were concentrated at this mass center and all external

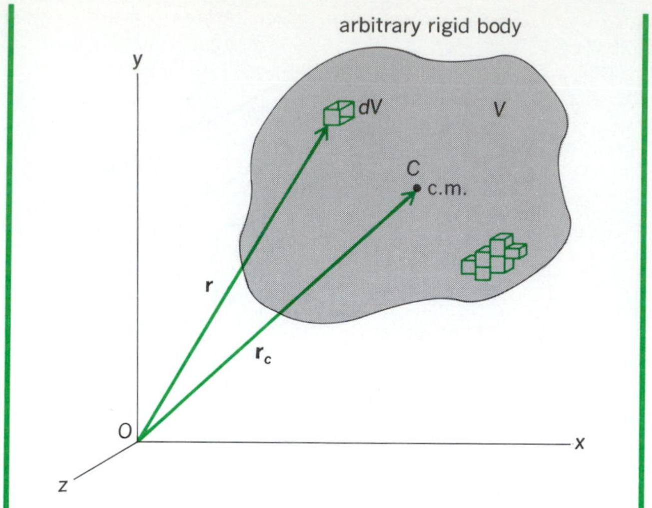

A rigid body of arbitrary shape may be considered to be composed of a continuous distribution of volume elements (or particles). A typical volume element *dV* and position vectors are shown. c.m. = center of mass.

forces were applied there. The motion of translation of the center of mass of a rigid body is therefore determined by the same methods as those used to determine the motion of a particle. The one significant difference from particle dynamics, however, is that the forces are those acting on the actual rigid body and as such often depend upon the shape of the body, its orientation in space, and sometimes upon the time rate of change of this orientation in space. Examples are the motion of an airplane or projectile through air or of a submarine through water.

The motion in translation may also be expressed in terms of a vector quantity called linear momentum. The quantity $\mathbf{P} = m\mathbf{v}_c$ where $\mathbf{v}_c$ is the velocity of the center of mass is defined as the linear momentum of the rigid body. The resultant external force F acting on a rigid body is equal to the time rate of change of the linear momentum of the body.

If no external forces are acting, the linear momentum of the body remains constant—a statement of the law of conservation of linear momentum. *See* CONSERVATION OF MOMENTUM; MOMENTUM.

Rotational motion. The equations that describe the rotational or angular motion of a rigid body are again derivable from Newton's second law. In the illustration, a rigid body of arbitrary shape is located and moving in an inertial system. This body may be subdivided into mass particles dV of mass m_i and Newton's law applied to each particle, resulting in Eq. (3).

$$m_i \frac{d^2\mathbf{r}}{dt^2} = \mathbf{F} \tag{3}$$

If both sides of Eq. (3) are multiplied by the position vector $\mathbf{r}$ (by means of a cross product), Eq. (4) is derived. Here the vec-

$$m_i\left(\mathbf{r} \times \frac{d^2\mathbf{r}}{dt^2}\right) = (\mathbf{r} \times \mathbf{F}) \tag{4}$$

tor product $(\mathbf{r} \times \mathbf{F})$ is defined as the moment $\mathbf{M}$ or torque about the point O of the resultant force $\mathbf{F}$ acting on dV. *See* TORQUE.

Equation (4) may be written as Eq. (5) where $\mathbf{r}$, $\mathbf{v}$, and m_i.

$$m_i\left(\mathbf{r} \times \frac{d^2\mathbf{r}}{dt^2}\right) = m_i \frac{d}{dt}\left(\mathbf{r} \times \frac{d\mathbf{r}}{dt}\right) = \frac{d}{dt}\left(\mathbf{r} \times m_i \frac{\mathbf{r}}{dt}\right)$$

$$= \frac{d}{dt}(\mathbf{r} \times m_i\mathbf{v}) \tag{5}$$

are the position vector, velocity, and mass, respectively, of the element dV. Equation (5) is valid for each elemental particle of the rigid body, and by summing the equation for all such particles of the body, the equation of angular motion is obtained from Eq. (6a), where m_i has been replaced by $\rho\, dV$ as before.

$$\sum \mathbf{M}_O = (\mathbf{M}_O)_{ext} = \frac{d}{dt}\int_V (\mathbf{r} \times \mathbf{v})\rho\, dV = \frac{d\mathbf{H}_O}{dt} \qquad (6a)$$

$$\int_V (\mathbf{r} \times m_i\mathbf{v}) = \int_V (\mathbf{r} \times \mathbf{v})\rho\, dV = \mathbf{H}_O \qquad (6b)$$

Equation (6b) expresses the moment of linear momentum or the angular momentum about a point O of a rigid body. On the left-hand side of Eq. (6a), the summation process results in zero for all equal and opposite collinear forces between particles (internal forces), and only the moment $\mathbf{M}_O$ of the external forces exists.

Equation (6a) expresses the principle of angular momentum, that the time rate of change of angular momentum of a rigid body about any point O fixed in an inertial system is equal to the resultant moment of external forces about the same point O. If the rigid body is not acted upon by any external moment, the angular momentum must remain constant. This is the principle of conservation of angular momentum. *See* ANGULAR MOMENTUM.

An important extension of Eq. (6a), the angular momentum equation, is as follows. Equation (6a) is precisely valid for the case in which the angular momenta and the moments of external forces are taken about the center of mass of a rigid body, even though the center of mass does not remain at rest in an inertial system. In this case the equation is that given as Eq. (7),

$$\mathbf{M}_c = \frac{d\mathbf{H}_c}{dt} \qquad (7)$$

where c denotes the center of mass. Equation (7) is useful, for example, in predicting the motion of a projectile moving through air under accelerated conditions. The fact that the angular equation can be written about the moving center of mass fixed within the projectile, instead of about some fixed point located exterior to the projectile, results in great simplification in the analysis of the motion. *See* BALLISTICS.

Work and energy. The change in the kinetic energy T of a rigid body in going from locations 1 to 2 in space is equal to the work done on the rigid body during this period. Or, the rate of change of kinetic energy of the body is equal to the rate at which work is done by all external forces acting on the body. If the forces acting are a function of the coordinates only and the work done is therefore independent of the path the body follows (gravitational forces, for example), the system is said to be conservative (frictionless, without energy dissipation). Then the work done on the body by the forces acting on it, when this body moves in a conservative force field from locations 1 to 2 in space, is called the potential energy of the body at 2 with respect to 1, or is equal to the negative of the change in potential energy V of the body in passage from 1 to 2. The sum of kinetic and potential energies therefore remains constant throughout the motion. This is the principle of conservation of energy for a conservative system. *See* CONSERVATION OF ENERGY; ENERGY; WORK. [R.E.Bo.]

Rinderpest An acute or subacute, contagious viral disease of ruminants and swine, manifested by high fever, lacrimal discharge, profuse diarrhea, erosion of the epithelium of the mouth and of the digestive tract, and high mortality.

Rinderpest (also known as cattle plague) is caused by a ribonucleic acid (RNA) virus classified in the genus *Morbillivirus* within the Paramyxoviridae family. This virus is closely associated with the viruses of human measles, peste des petits ruminants of sheep and goats, canine distemper, and phocine distemper, and with the dolphin morbillivirus. Although there are significant differences in virulence, only one serotype of rinderpest virus is known. The rinderpest virus is easily inactivated by heat and survives outside the host for a short time. Therefore, transmission of rinderpest is by direct contact between animals.

Although all cloven-hoofed animals are considered susceptible to rinderpest, clinical cases are mostly seen in cattle and water buffalo. Rinderpest is characterized by the development of high fever that lasts for several days until just prior to death. The morbidity in susceptible cattle and buffalo is greater than 90%, with death of almost all clinically affected animals. Rinderpest has been controlled in most African nations, but persists in regions of eastern Africa. *See* ANIMAL VIRUS; PARAMYXOVIRUS. [A.To.]

Ring A tie member or chain link. Tension or compression applied through the center of a ring produces bending moment, shear, and normal force on radial sections. Because shear stress is zero at the boundaries of the section where bending stress is maximum, it is usually neglected. [W.J.K./W.G.B.]

Ring gages Cylindrical rings of steel whose inside diameters are ground and then honed or lapped to gage tolerance; used for checking the diameter of a cylindrical object such as a shaft. The gaging surface is sometimes chrome-plated or made of carbide for wear resistance. A single gage controls one limit of tolerance of the diameter. The "go" gage will pass over the largest acceptable part but will not pass over any larger part. The "not go" gage will not pass over any part within tolerance but will pass over an undersize part. A ring gage is not preferred as a "not go" gage; an out-of-round or bent part could have many diameters undersize but still be "accepted" because the gage would "not go" if the part were bent or if even one diameter were up to the low limit. [R.A.Bo.]

Ring theory The mathematical term ring is used to designate a type of algebraic system with two compositions satisfying most but not all the properties of addition and multiplication in the system of integers, 0, ±1, ±2,.... In precise terms a ring is a set R with two binary compositions called addition and multiplication whose results on an ordered pair (a,b), a,b in R, are denoted by $a + b$ and ab, respectively.

These compositions must satisfy the following conditions:

C. $a + b$ and ab belong to R (closure).
A1. $a + b = b + a$ (commutative law).
A2. $(a + b) + c = a + (b + c)$ (associative law).
A3. There exists an element 0 (called zero) in R satisfying $a + 0 = a$ for every a in R.
A4. For each a in R there exists an element $-a$ (called the negative of a) in R such that $a + (-a) = 0$.
M1. $(ab)c = a(bc)$.
D. $a(b + c) = ab + ac$; $(b + c)a = ba + ca$ (distributive laws).

In the ring I of integers (addition and multiplication as usual) there are further conditions, for example, the commutative law of multiplication ($ab = ba$) and the cancellation law that if $a \neq 0$ and $ab = ac$, then $b = c$. *See* SET THEORY.

The importance of the concept of a ring stems from the fact that it embraces many special cases which are fundamental in all branches of mathematics. Thus it includes the ring I of integers, the ring R_0 of rational numbers, the ring $R^{\#}$ of real numbers, the ring C of complex numbers, various rings of functions, rings of matrices, and so on.

The conditions A1–A4 on the addition composition are exactly equivalent to the statement that any ring is a commutative group relative to its addition composition. This group is called the additive group of the ring. The algebraic system consisting of the set of elements of a ring together with its multiplication composition is called the multiplicative semigroup of the ring. *See* GROUP THEORY.

Various classes of rings are singled out by imposing conditions on the multiplicative semigroup. Thus integral domains are rings in which the product of nonzero elements is nonzero. Division rings are rings whose sets of nonzero elements are groups relative to the multiplication composition, and fields are division rings satisfying the commutative law of multiplication.

[N.J.]

Ripple tank A shallow tray containing a liquid and equipped with a means for generating surface waves. Ripple tanks are used to study a number of physical phenomena which can be described in terms of wave mechanics. Water, acoustic, light, and electromagnetic waves can be investigated with equal facility.

Control of water depth in the ripple tank by contouring the bottom permits the simulation of a variable or discontinuous medium for wave propagation. For example, a step change in the depth results in a wave diffraction in accord with Snell's law. Acoustical and optical wave diffractions have been studied in ripple tanks, and phase fronts near two-dimensional models of antenna structures and radomes have also been investigated. Ripple-tank models of large-scale water-wave motions in harbors and along seacoasts can be useful. *See* WAVE MOTION.

[D.R.F.H.]

Ripple voltage The total voltage across the load resistor of a rectifier minus the average voltage across the same resistor. The ripple can be expressed as a Fourier series. To reduce the ripple voltage, a low-pass filter is usually placed between the rectifier and the load. The filter is more efficient in reducing the ripple voltage if the fundamental frequency of the ripple voltage is high. *See* ELECTRIC FILTER; RECTIFIER. [D.L.W.]

Risk analysis The scientific study of risk, the potential realization of undesirable consequences from hazards arising from a possible event. For example, the probability of contracting lung cancer (unwanted consequence) is a risk caused by carcinogens (hazards) contained in second-hand tobacco smoke (event). Risk management is the term for the systematic analysis and control of risks. Risks are caused by exposure to hazards. Sudden hazards are referred to as acute (for example, a flash flood caused by heavy rains); prolonged hazards are referred to as chronic (for example, carcinogens in second-hand tobacco smoke and polluted air).

The definition of risk contains two components: the probability of an undesirable consequence of an event and the seriousness of that consequence. Sometimes the probability is expressed as a return period, which means, for instance, that a flood of a specified magnitude is expected to occur once every 100 years.

Most human activities involve risk. The risk of driving, for example, can be subdivided according to property damage, human injuries, fatalities, and harm to the environment. Even the stress and lack of exercise due to driving create health risks. Although risk pervades modern society and is widely acknowledged, it continues to cause unending controversy and debate.

Risk estimates are seldom accurate to even two orders of magnitude, and widely varying perceptions of risk by different interest groups can add confusion and conflict to the risk management process. As a result, risk analysis and assessment are often carried out conservatively with a safety factor as high as 1000. This, in turn, leads to nonoptimal decision making, because the risks are exaggerated.

The illustration shows risk management for persons, including the role of risk analysis and assessment. Thousands of natural and other hazards are subjected to the statistical analysis of mortality and morbidity data. Society selects a small number of risks to manage, but often some high risks (such as radon in houses), may not be managed, while some low risks (such as movement of dangerous goods), may be selected for management. Management alternatives include banning of the hazard (drugs), regulating the hazard (drivers' tests and licensing), controlling the release and exposure of hazardous materials, treatment after exposure, and penalties for damages. Each management alternative may be analyzed to estimate the impact on risk, which may be known only within a few orders of magnitude.

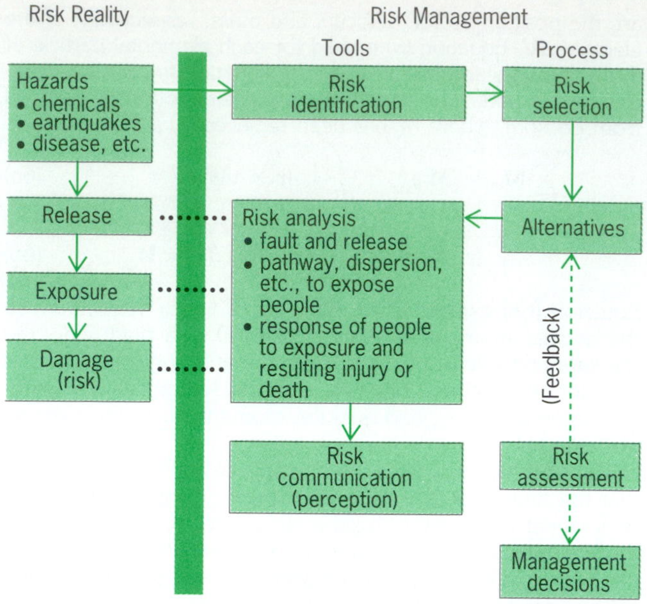

Risk analysis, assessment, and management.

Risk estimates are uncertain, are described in technical language, and are outside the general understanding or experience of most people. Perception plays a crucial role, tending to exaggerate the significance, for example, of risks that are involuntary, catastrophic, or newsworthy. Effective risk management therefore requires effective risk communication.

Risk assessment is the evaluation of the relative importance of an estimated risk with respect to other risks faced by the population, the benefits of the activity source of the risk, and the costs of managing the risk. *See* AIR POLLUTION; ENVIRONMENTAL ENGINEERING; NUCLEAR POWER; ONCOLOGY; RADIOACTIVITY WASTE MANAGEMENT; RADON. [K.W.H.; J.S.]

Ritz's combination principle The empirical rule, formulated by W. Ritz in 1905, that sums and differences of the frequencies of spectral lines often equal other observed frequencies. The rule is an immediate consequence of the quantum-mechanical formula $hf = E_i - E_f$ relating the energy hf of an emitted photon to the initial energy E_i and final energy E_f of the radiating system; h is Planck's constant and f is the frequency of the emitted light. *See* ATOMIC STRUCTURE AND SPECTRA; ENERGY LEVEL (QUANTUM MECHANICS); QUANTUM MECHANICS. [E.G.]

River A natural, fresh-water surface stream that has considerable volume compared with its smaller tributaries. The tributaries are known as brooks, creeks, branches, or forks. Rivers are usually the main stems and larger tributaries of the drainage systems that convey surface runoff from the land. Rivers flow from headwater areas of small tributaries to their mouths, where they may discharge into the ocean, a major lake, or a desert basin. *See* RIVER TIDES.

Rivers flowing to the ocean drain about 68% of the Earth's land surface. The remainder of the land either is covered by ice or drains to closed basins (common in desert regions). Regions draining to the sea are termed exoreic, while those draining to interior closed basins are endoreic. Areic regions are those which lack surface streams because of low rainfall or lithologic conditions.

Sixteen of the largest rivers account for nearly half of the total world river flow of water. The Amazon River alone carries nearly 20% of all the water annually discharged by the world's rivers. Rivers also carry large loads of sediment. The total sediment load for all the world's rivers averages about 22×10^9 tons (20×10^9 metric tons) brought to the sea each year. Sediment loads for individual rivers vary considerably. The Yellow River of northern China is the most prolific transporter of sediment. Draining an agricultural region of easily eroded loess, this river averages about 2.2×10^9 tons (2×10^9 metric tons) of sediment per year, one-tenth of the world average. *See* STREAM TRANSPORT AND DEPOSITION.

Perfectly straight rivers are uncommon. Meandering is the most common river pattern. Braided rivers have channels that are divided into anastomosing branches by alluvial islands and bars. Braided rivers often have steeper gradients, more variable discharges, coarser sediment loads, and lower sinuosity than meandering rivers.

River discharge varies over a broad range, depending on many climatic and geologic factors. The low flows of the river influence water supply and navigation. The high flows are a concern as threats to life and property. However, floods are also beneficial in providing new soil and moisture for crops.

Floods are but one attribute of rivers that affect human society. Many of the world's rivers are managed to conserve the natural flow for release at times required by human activity, to confine flood flows to the channel and to planned areas of floodwater storage, and to maintain water quality at optimum levels. *See* RIVER ENGINEERING. [V.R.B.]

River engineering A branch of civil engineering concerned with the improvement and stabilization of the channels of rivers (particularly channels in erodible alluvium) to better serve the needs of people. In many major river systems there is a need to improve channels in order to provide greater flood flow capacity, provide navigable waterways, improve water supplies, and stabilize the channels to permit development of adjacent valley ares. This has led to development of extensive and costly river control or management schemes. *See* RIVER.

Channel improvement and stabilization works have been integral parts of river control plans where meandering, shifting channels or inappropriate channel alignments and cross sections are problems. Large-scale channel improvement and stabilization plans have been carried out on many of the world's major rivers. In the United States one of the most extensive examples of river engineering is found in the Mississippi Basin. [H.B.Wi.]

River tides Tides that occur in rivers emptying directly into tidal seas. These fides show three characteristic modifications of ocean tides. (1) The speed at which the tide travels upstream depends on the depth of the channel. (2) The further upstream, the longer the duration of the falling tide and the shorter the duration of the rising tide. (3) The range of the tide decreases with distance upstream. *See* TIDE.

In a river the difference between the depths of water at high and low tides may be relatively large, leading to a marked difference between the speeds at which high and low tides move. The difference in depth between various points on the river also partially explains the second modification, or duration of fall and rise. In addition, the river flow, which may fluctuate widely, helps a failing tide but hinders a rising tide, increasing the difference in duration.

The third modification or decrease in tidal range upstream may be accounted for by loss of energy of the water through friction with the sides and bottom of the channel. Although friction always saps energy from the tide, if the channel becomes constricted within a short distance, the water may be forced into a smaller space, thus producing a larger tidal range. [B.K.]

Rivet A short rod with a head formed on one end. A rivet is inserted through aligned holes in two or more parts to be joined; then by pressing the protruding end, a second head is formed to hold the parts together permanently. The first head is called the manufactured head and the second one the point. In forming the point, a hold-on or dolly bar is used to back up the manufactured head and the rivet is driven, preferably by a machine riveter.

Blind rivets are special rivets that can be set without access to the point. They are available in many designs but are of three general types: screw, mandrel, and explosive (see illustration). In the mandrel type the rivet is set as the mandrel is pulled through. In the explosive type an explosive charge in the point is set off by a special hot iron; the explosion expands the point and sets the rivet. *See* RIVETED JOINT. [P.H.B.]

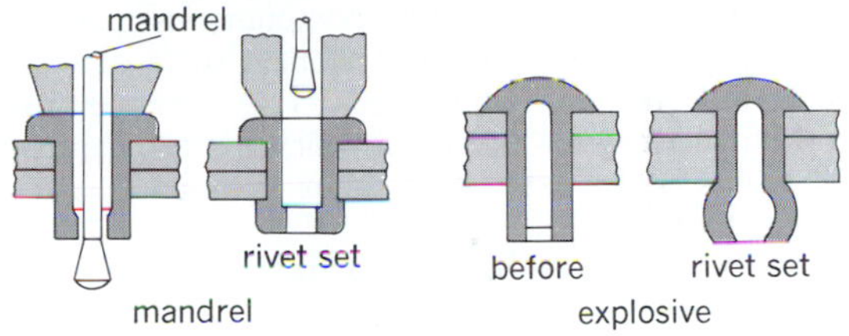

Two types of blind rivet.

Riveted joint The permanent joining of two or more machine or structural members, usually plates, by means of rivets. The plates may be lapped or butted. In the butt joint one or more cover plates must be used to accomplish the joining. One of these cover plates is often made wider than the other. The rivets in the joint may be disposed in several ways to form single or multiple rows in a regular or staggered arrangement. *See* RIVET. [L.S.L.]

Robotics The study of problems associated with the design, application, and control and sensory systems of robots. Historically, the term robot has been used loosely, and has been applied to almost any feedback-controlled mechanical system. While the exact usage of the term is a matter of preference, the computer-controlled mechanical arm used in industrial applications probably represents a reasonable middle ground for definition by example. Most concerns of practitioners of robotics are involved in such devices, and so much of the work in robotics relates directly to these devices.

Design. Some robots have very simple mechanical designs, involving only a few degrees of freedom of movement. However, the design of robot manipulators can also be quite complex. In a typical industrial robot arm, six degrees of freedom of movement (exclusive of gripper closure) are required to enable the gripper to approach an object from any orientation.

The problem of powering the robot's joints is made more difficult by the complex mechanical articulation. One approach is to place a prime mover (electrical, hydraulic, or pneumatic) at the joint itself. However, the weight and bulk of such motors and their associated gearing place constraints on the performance and mechanical design of the arm. A second approach is to place the prime movers in the immobile base of the robot and to transmit motion to the joints through mechanical linkages such as shafts, belts and cables, or gearing. This overcomes many of the problems associated with the first approach, but introduces a new set of problems in designing intricate, backlash-free mechanical linkages which can transmit power effectively through the complex articulations of the arm in all of its positions. No single approach has clearly dominated the field.

Control systems. A robot control system is the apparatus (usually electronic) which directs the activities of the mechanical parts. This may consist of only a sequencing device and a set of mechanical stops, so that the mechanism moves in a repetitive pattern between selected positions. However, more sophisticated systems employ servo-controlled positioning of the joints, and a measure of the actual joint position is obtained from a transducer, such as an optical shaft encoder, and is compared with the position specified for the joint. If the desired position and the measured position differ, the circuitry applies a correcting drive signal to the joint motor. *See* SERVOMECHANISM.

More sophisticated systems generate the robot's trajectory automatically by computer. Such computations may be based on mathematical descriptions of work objects or tasks contained in the computer's data base. The computer may also generate trajectories for the robot which are not fixed but vary with the state of the external world as reported by the robot's sensory system.

The most advanced robot control systems make use of hierarchical control. Each level of a hierarchy of control stages accepts, from its superior level, a statement of a goal to be achieved. Higher levels attempt to achieve their current goal by issuing sequences of commands (subgoals) to their subordinate levels. In selecting these subgoals, each level takes into account its own goal, its sensory input describing the state of the external world, and the status of its subordinate level. These systems are sensory-interactive and constitute a task-decomposition hierarchy similar to that of many functions in the human nervous system. They allow the robot to be instructed with very general commands at the highest level. *See* CONTROL SYSTEMS.

Sensory systems. The purpose of a robot sensory system is to gather specific information needed by the control system and, in more advanced systems, to maintain an internal, predictive model of the environment. Visual, proximity, tactile, acoustic, and force or torque senses are all used.

The most commonly used means of sensing objects at a distance is some form of visual sense. Usually this is done by computer analysis of an image from a television camera. Two important approaches are ambient light systems and structured light systems. Ambient light systems rely on normal sources of scene illumination, while structured light systems provide special patterns of illumination whose shape and orientation are known to the sensory system.

The most advanced robot sensory systems are multimodal (based on several senses) and are also hierarchically structured to generate successive levels of description, at increasing degrees of complexity and decreasing rates, suitable for use by the successive levels of a hierarchical control system. *See* ARTIFICIAL INTELLIGENCE; COMPUTER-AIDED DESIGN AND MANUFACTURING; DIGITAL CONTROL. [E.K.]

Roche limit The closest distance which a satellite, revolving around a parent body, can approach the parent without being pulled apart tidally. The simplest formal definition is that the Roche limit is the minimum distance at which a satellite can be in equilibrium under the influence of its own gravitation and that of the central mass about which it is describing a circular orbit. If the satellite is in a circular orbit and has negligible mass, the same density as the primary, and zero tensile strength, the Roche limit is 2.46 times the radius of the primary.

When a star has exhausted the supply of hydrogen in its core, its radius will increase by a factor of 10 to 100. A star in a binary system may then exceed its Roche limit, material will thus escape from that star, and its companion will receive the excess material. *See* BINARY STAR. [P.K.S.]

Rochelle salt The sodium potassium salt of the *d*-tartaric acid $NaKC_4H_4O \cdot 4H_2O$, also called Seignette salt, the first crystalline solid discovered to possess the properties of ferroelectricity. Such crystals have been widely used, for example, in microphones and phonograph pickup cartridges, because of their large piezoelectric effect. *See* FERROELECTRICS; PIEZOELECTRICITY. [H.Gr.]

Rock A relatively common aggregate of mineral grains. Some rocks consist essentially of but one mineral species (monomineralic, such as quartzite, composed of quartz); others consist of two or more minerals (polymineralic, such as granite, composed of quartz, feldspar, and biotite). Rock names are not given for those rare combinations of minerals that constitute ore deposits, such as quartz, pyrite, and gold. In the popular sense rock is considered also to denote a compact substance, one with some coherence; but geologically, friable volcanic ash also is a rock. A genetic classification of rocks is shown in the following list.

- Igneous
  - Intrusive
    - Plutonic (deep)
    - Hypabyssal (shallow)
  - Extrusive
    - Flow
    - Pyroclastic (explosive)
- Sedimentary
  - Clastic (mechanical or detrital)
  - Chemical (crystalline or precipitated)
  - Organic (biogenic)
- Metamorphic
  - Cataclastic
  - Contact metamorphic and pyrometasomatic
  - Regional metamorphic (dynamothermal)
- Hybrid
  - Metasomatic
  - Migmatitic

Exceptions to the requirement that rocks consist of minerals are obsidian, a volcanic rock consisting of glass; and coal, a sedimentary rock which is a mixture of organic compounds. *See* COAL; IGNEOUS ROCKS; METAMORPHIC ROCKS; OBSIDIAN; ROCK MECHANICS; SEDIMENTARY ROCKS; VOLCANIC GLASS. [E.W.H.]

Rock age determination Age determinations of rocks and minerals based on the radioactivity of certain chemical elements that occur within them. Measurements of rock ages have enabled geologists to reconstruct the geologic history of the Earth from the time of its formation to the present. Age determinations of rocks from the Moon have also contributed to knowledge of the history of the Earth. *See* RADIOACTIVITY.

Many rocks and minerals contain radioactive atoms that decay spontaneously to form stable atoms of other elements. Under certain conditions these radiogenic daughter atoms accumulate within the mineral crystals so that the ratio of the daughter atoms divided by the parent atoms increases with time. This ratio can be measured very accurately with a mass spectrometer, and is then used to calculate the age of the rock by means of an equation based on the law of radioactivity. The radioactive atoms used for dating rocks and minerals have very long half-lives, measured in billions of years. They occur in nature only because they decay very slowly. The pairs of par-

Parent-daughter pairs used for dating rocks and minerals

Parent	Daughter	Half-life, 10^9 years
Potassium-40	Argon-40	11.8
Potassium-40	Calcium-40	1.47
Rubidium-87	Strontium-87	48.8
Samarium-147	Neodymium-143	107
Rhenium-187	Osmium-187	43
Thorium-232	Lead-208	14.008
Uranium-235	Lead-207	0.7038
Uranium-238	Lead-206	4.468

ents and daughters used for dating are listed in the table. *See* DATING METHODS.

The rubidium-strontium (Rb-Sr) method is based on the nuclide rubidium-87, which decays to stable strontium-87 by emitting a beta particle from its nucleus. The abundance of the radiogenic strontium-87 therefore increases with time at a rate that is proportional to the Rb/Sr ratio of the rock or mineral. This method has been used to date not only igneous but also metamorphic rocks. The method is particularly well suited to the dating of very old rocks such as the ancient gneisses near Godthaab in Greenland, which are almost 3.8×10^9 years old.

The potassium-argon method is based on the assumption that all of the atoms of radiogenic argon-40 that form within a potassium-bearing mineral from the decay of potassium-40 accumulate within it. This assumption is satisfied only by a few minerals and rocks because argon is an inert gas that does not readily form bonds with other atoms. These conditions are satisfied only by the micas, sanidine feldspar, hornblende, and the clay minerals glauconite and illite. The K-Ar method of dating has been used to establish a chronology of mountain building events in North America beginning about 2.8×10^9 years and continuing to the present. In addition, the method has been used to date reversals of the polarity of the Earth's magnetic field during the past 1.3×10^7 years.

All of the atoms of uranium and thorium are radioactive and decay through a series of daughters to stable atoms of lead. Minerals that contain both elements can be dated by three separate methods based on the decay of uranium-238 to lead-206, uranium-235 to lead-207, and thorium-232 to lead-208.

In fission-track dating, the trail of damage from particles emitted by the decay of uranium-238 is enlarged by etching a polished surface of the solid with a suitable reagent. The tracks can then be counted visually by use of a microscope. The number of spontaneous fission tracks per square centimeter is proportional to the concentration of uranium and to the age of the sample. *See* FISSION TRACK DATING; RADIOCARBON DATING. [G.Fau.]

Rock burst A sudden and violent rock failure around a mining excavation on a sufficiently large scale to be considered a hazard endangering the existence of mine openings, equipment, and personnel. It has been estimated that the energy released in some big bursts was equivalent to that released in exploding 200 tons (180 metric tons) of TNT. Such bursts resemble small earthquakes and may be detected several hundred miles away.

Rock bursts are related to the fracture of rock in place and require two conditions for their occurrence: a stress in the rock mass sufficiently high to exceed its strength, and physical characteristics of the rock which enable it to store energy up to the threshold value for sudden rupture. Rocks which yield gradually in plastic strain when under load usually do not generate rock bursts. *See* ROCK MECHANICS. [S.H.B.]

Rock cleavage A secondary foliation produced in a rock by deformation. Since there exists no general agreement on the terminology or the mechanical significance of the various types of cleavage, it is described as a special type of foliation. Foliation is a descriptive term that includes all types of parallel or subparallel primary and secondary planar structures that pervade a rock. Such surfaces usually exist as planes of weakness along which a rock may split.

Rock cleavage is usually associated with folding, but may also occur in response to faulting. It is generally best developed in fine-grained rocks and in rocks with a high content of platy minerals such as mica.

Most rock cleavage is associated with dynamically metamorphosed rocks. The processes most commonly recognized in the generation of cleavage include shear, flow, extension and rotation, flattening, solution, and recrystallization. However, rock cleavage also occurs in deformed unmetamorphosed rocks. [P.J.R.]

Rock control (geomorphology) The influence of differences in earth materials on development of landforms. The materials out of which landforms are carved are not homogeneous; even in a small area they differ from one another in composition, physicochemical properties, mechanical behavior, and various other ways. These differences and variations are visibly reflected in landforms. Examples of the problems of rock control include such instances as cuestas, structural benches, knickpoints, dike ridges, mesas, hogbacks, structural plains, and karst and inversion of topography. [E.Y.]

Rock exfoliation The splitting off of concentric thin sheets or shells from the surface of massive rocks, a common weathering process in moist climates. Both above and beneath the ground surface, hydration of complex silicate minerals through the action of percolating water carrying minute amounts of carbon dioxide causes peripheral expansion in rocks. The process is particularly effective at edges and comers of joint blocks where penetration can be from several directions. Separation of successive curved shells tends to produce increasingly rounded forms. Chemical exfoliation is distinct from the spalling of rocks by fire or sheeting of rocks in response to release of load. *See* WEATHERING PROCESSES. [C.F.S.S.]

Rock magnetism The natural remanent magnetization (NRM) of igneous, metamorphic, and sedimentary rocks has enabled inferences to be made concerning changes in the direction and intensity of the geomagnetic field in past geological times and the relative movements of the poles and the continents.

The magnetic properties of rocks result from the iron oxide minerals present to a few percent in many rocks. Two groups are especially important: the magnetite (Fe_3O_4)–ulvospinel (Fe_2TiO_4) solid solution series and the hematite (Fe_2O_3)–ilmenite ($FeTiO_3$) solid solution series. Intergrowths between magnetic minerals are of frequent occurrence.

Magnetization induced by the present geomagnetic field is also important in the interpretation of air or ground geomagnetic surveys. Laboratory experiments, applying magnetic fields to rocks at different temperatures, are important not only in the interpretation of NRM but also as a petrological and mineralogical tool. *See* PALEOMAGNETISM. [S.K.R.]

Rock mechanics Application of the principles of mechanics and geology to quantify the response of rock when it is acted upon by environmental forces, particularly when human-induced factors alter the original ambient conditions. Rock mechanics is an interdisciplinary engineering science that requires interaction between physics, mathematics, and geology, and civil, petroleum, and mining engineering. The present state of knowledge permits only limited correlations between theoretical predictions and empirical results. Therefore, the most useful principles are based upon data obtained from laboratory and in-place measurements and from prototype behavior (behavior of the completed engineering works). Increasing emphasis is upon in-place measurements because rock properties are regarded as site-specific; that is, the properties of the rock system at one site probably will be significantly different from those at another site, even if geologic environments are similar. *See* ENGINEERING GEOLOGY; TECTONOPHYSICS.

Because of the interdisciplinary aspects, there is no standardization of rock mechanics terminology. However, the following terms and definitions are useful.

Environmental factors are the natural factors and human influences that require consideration in engineering problems in rock mechanics. The major natural factors are geology, ambient stresses, and hydrology. The human influences derive from the application of chemical, electrical, mechanical, or thermal energy during construction (or destruction) processes.

The ambient stress field is the distribution and numerical value of the stresses in the environment prior to its disturbance by humans.

The term rock system includes the complete environment that can influence the behavior of that portion of the Earth's crust that will become part of an engineering structure. Generally, all natural environmental factors are included.

A rock element is the coherent, intact piece of rock that is the basic constituent of the rock system and which has physical, mechanical, and petrographic properties that can be described or measured by laboratory tests on each such element. The concepts of rock system and rock element enable the concomitant engineering design to be optimized according to the principles of system engineering. *See* SYSTEMS ENGINEERING.

"Rock failure" occurs when a rock system or element no longer can perform its intended engineering function. Failure may be evidenced by fractures, distortion of shape, or reduction in strength. "Failure mechanism" includes the causes for the manner of rock failure. *See* ROCK BURST; SOIL MECHANICS; TUNNEL; UNDERGROUND MINING. [W.R.J.]

Rocket Either a propulsion system or a complete vehicle driven by such a propulsive engine. A rocket engine provides the means whereby chemical matter is burned to release the energy stored in it and the energy is expended, by ejection at high velocity of the products of combustion (the working fluid). The ejection imparts motion to the vehicle in a direction opposite to that of the ejected matter. A rocket vehicle is propelled by rocket reaction and includes all components necessary for such propulsion, and a payload such as an explosive charge, scientific instruments, or a crew. A rocket vehicle also includes guidance and control equipment mounted in a structural airframe or spaceframe. *See* ROCKET PROPULSION; ROCKET STAGING; SATELLITE (SPACECRAFT); SPACE FLIGHT; SPACE PROBE. [G.P.S.]

Rocket astronomy The discipline comprising measurements of the electromagnetic radiation from the Sun and other celestial bodies of wavelengths that are almost completely absorbed between the 150-mi (250-km) level reached by sounding rockets and 25-mi (40-km) level attained by balloons.

Rocket astronomy's principal subdiscipline is the Sun. The results from solar rocket astronomy have changed and refined the model of the solar atmosphere, have thrown new light on the role of magnetic fields in the Sun's atmosphere, and have led to a new understanding of solar flares. Rocket astronomy is more difficult for the study of celestial bodies other than the Sun because of the low intensity of their emissions.

Closely connected with and forming a part of rocket astronomy is the study of the ionosphere, the airglow, and the composition of the Earth's outer atmosphere by measuring the attenuation of solar extreme ultraviolet radiation with rocket-borne instrumentation. This work is often placed in the interdisciplinary fields of solar-terrestrial relationships, aeronomy, and ionospheric physics. *See* IONOSPHERE.

Astronomical research from space vehicles is conducted now to a greater degree from orbiting observations than from rockets. Nevertheless, rockets will continue to provide means for making new discoveries and for proving experiments for satellites at costs that are modest. *See* SATELLITE ASTRONOMY. [R.To.]

Rocket propulsion The process of imparting a force to a flying vehicle, such as a missile or a spacecraft, by the momentum of ejected matter. This matter, called propellant, is stored in the vehicle and ejected at high velocity. In chemical rockets the propellants are chemical compounds that undergo a chemical combustion reaction, releasing the energy for thermodynamically accelerating and ejecting the gaseous reaction products at high velocities. Chemical rocket propulsion is thus differentiated from other types of rocket propulsion, which use nuclear, solar, or electrical energy as their power source and which may use mechanisms other than the adiabatic expansion of a gas for achieving a high ejection velocity. Propulsion systems using liquid propellants (such as kerosine and liquid oxygen) have traditionally been called rocket engines, and those that use propellants in solid form have been called rocket motors. *See* ELECTROTHERMAL PROPULSION; INTERPLANETARY PROPULSION; ION PROPULSION; PLASMA PROPULSION; PROPULSION; SPACECRAFT PROPULSION.

Performance. The performance of a missile or space vehicle propelled by a rocket propulsion system is usually expressed in terms of such parameters as range, maximum velocity increase of flight, payload, maximum altitude, or time to reach a given target. Propulsion performance parameters (such as rocket exhaust velocity, specific impulse, thrust, or propulsion system weight) are used in computing these vehicle performance criteria. The table gives typical performance values. *See* SPECIFIC IMPULSE; THRUST.

Typical performance values of rocket propulsion systems*

Propulsion system parameter	Typical range of values
Specific impulse at sea level	180–390 s
Specific impulse at altitude	215–470 s
Exhaust velocity at sea level	5800–15,000 ft/s (1800-4500 m/s)
Combustion temperature	4000–7200 F°(2200–4000°C)
Chamber pressures	100–3000 lb/in.^{2} (0.7–20 MPa)
Ratio of thrust to propulsion system weight	20–150
Thrust	0.01–6.6 × 10^6 lb (0.05–2.9 × 10^7 n)[1]
Flight speeds	0–50,000 ft/s (0–15,000 m/s)

*Exact values depend on application, propulsion system design, and propellant selection.

†Maximum value applies to a cluster; for a single rocket motor it is 3.3 3 106 lb (14,7000 kN).

Applications. Rocket propulsion is used for different military missiles or space-flight missions. Each requires different thrust levels, operating durations, and other capabilities. In addition, rocket propulsion systems are used for rocket sleds, jet-assisted takeoff, principal power plants for experimental aircraft, or weather sounding rockets. For some space-flight applications, systems other than chemical rockets are used or are being investigated for possible future use. *See* GUIDED MISSILE; MISSILE; ROCKET-SLED TESTING; SATELLITE (SPACECRAFT); SPACE FLIGHT; SPACE PROBE.

Liquid-propellant rocket engines. These use liquid propellants stored in the vehicle for their chemical combustion energy. The principal hardware subsystems are one or more thrust chambers, a propellant feed system, which includes the propellant tanks in the vehicle, and a control system.

Bipropellants have a separate oxidizer liquid (such as liquefield oxygen or nitrogen tetroxide) and a separate fuel liquid (such as liquefied hydrogen or hydrazine). Monopropellants consist of a single liquid that contains both oxidizer and fuel ingredients. A catalyst is required to decompose the monopropellant into gaseous combustion products. Bipropellant combinations allow higher performance (higher specific impulse) than monopropellants. *See* PROPELLANT.

The three principal components of a thrust chamber are the combustion chamber, where rapid, high-temperature combustion takes place; the converging-diverging nozzle, where the hot reaction-product gases are accelerated to supersonic velocities; and an injector, which meters the flow of propellants in the desired mixture of fuel and oxidizer, introduces the propellants into the combustion chamber, and causes them to be atomized or broken up into small droplets. Some thrust chambers (such as the space shuttle's main engines and orbital maneuvering engines) are gimbaled or swiveled to allow a change in the direction of the thrust vector for vehicle flight motion control.

Solid-propellant rocket motors. In rocket motors the propellant is a solid material that feels like a soft plastic or soap. The solid propellant cake or body is known as the grain. It can have a complex internal geometry and is fully contained inside the solid motor case, to which a supersonic nozzle is attached.

The propellant contains all the chemicals necessary to maintain combustion. Once ignited, a grain will burn on all exposed surfaces until all the usable propellant is consumed; small unburned residual propellant slivers often remain in the chamber. As the grain surface recedes, a chemical reaction converts the solid propellant into hot gas. The hot gas then flows through internal passages within the grain to the nozzle, where it is accelerated to supersonic velocities. A pyrotechnic igniter provides the energy for starting the combustion.

The nozzle must be protected from excessive heat transfer, from high-velocity hot gases, from erosion by small solid or liquid particles in the gas (such as aluminum oxide), and from chemical reactions with aggressive rocket exhaust products. The highest heat transfer and the most severe erosion occur at the nozzle throat and immediately upstream from there. Special composite materials, called ablative materials, are used for heat protection, such as various types of graphite or reinforced plastics with fibers made of carbon or silica. The development of a new composite material, namely, woven carbon fibers in a carbon matrix, has allowed higher wall temperatures and higher strength at elevated temperatures; it is now used in nozzle throats, nozzle inlets, and exit cones. It is made by carbonizing (heating in a nonoxidizing atmosphere) organic materials, such as rayon or phenolics. Multiple layers of different heat-resistant and heat-insulating materials are often particularly effective. A three-dimensional pattern of fibers created by a process similar to weaving gives the nozzle extra strength. *See* NOZZLE.

Nozzles can have sophisticated thrust-vector control mechanisms. In one such system the nozzle forces are absorbed by a doughnut-shaped, confined, liquid-filled bag, in which the liquid moves as the nozzle is canted. The space shuttle solid rocket boosters have gimbaled nozzles for thrust-vector control, with actuators driven by auxiliary power units and hydraulic pumps.

Hybrid rocket propulsion. A hybrid uses a liquid propellant together with a solid propellant in the same rocket engine. The arrangement of the solid fuel is similar to that of the grain of a solid-propellant rocket; however, no burning takes place directly on the surface of the grain because it contains little or no oxidizer. Instead, the fuel on the grain surface is heated, decomposed, and vaporized, and the vapors burn with the oxidizer some distance away from the surface. The combustion is therefore inefficient.

Testing. Because flights of rocket-propelled vehicles are usually fairly expensive and because it is sometimes difficult to obtain sufficient and accurate data from fast-moving flight vehicles, it is accepted practice to test rocket propulsion systems and components extensively on the ground under simulated flight conditions. Components such as an igniter or a turbine are tested separately. Complete engines are tested in static engine test stands; the complete vehicle stage is also tested statically. In the latter two tests the engine and vehicle are adequately secured by suitable structures. Only in flight tests are they allowed to leave the ground. [G.P.S.]

Rocket-sled testing

Rocket-sled testing A method of subjecting structures and devices to high accelerations or decelerations and aerodynamic flow phenomena under controlled conditions. The test object is mounted on a sled chassis running on precision steel rails and accelerated by rockets and decelerated by water scoops. This captive track testing of full-scale aircraft and missile components makes possible the recovery of expensive test sections, facilitates the instrumentation of the test and the reception and recording of the data, and provides a degree of repeatability not normally achieved under conditions of free flight. The components tested may be the rocket engines themselves, sections of flight vehicles, or equipment intended to operate under conditions of high acceleration or other specific environments such as structural vibration, acoustic vibration, and related aerodynamic conditions.

The test vehicle takes many diversified forms, ranging from the highly faired, aerodynamically clean type used to explore the transonic and supersonic speed ranges to the relatively simple all-purpose utility type used for testing in the subsonic and low transonic ranges. All sleds are carried on shoes or slippers that grip the railhead in order to prevent derailing. They may be powered by either liquid- or solid-propellant rocket engines.

Tests conducted on the track provide initial design data, developmental evaluation, and performance measurements of completed components and equipments. [E.Co.]

Rocket staging

Rocket staging The use of successive rocket sections, each having its own engine or engines. One way to minimize the weight of large missiles, or space vehicles, is to use multiple stages. The first or initial stage is usually the heaviest and

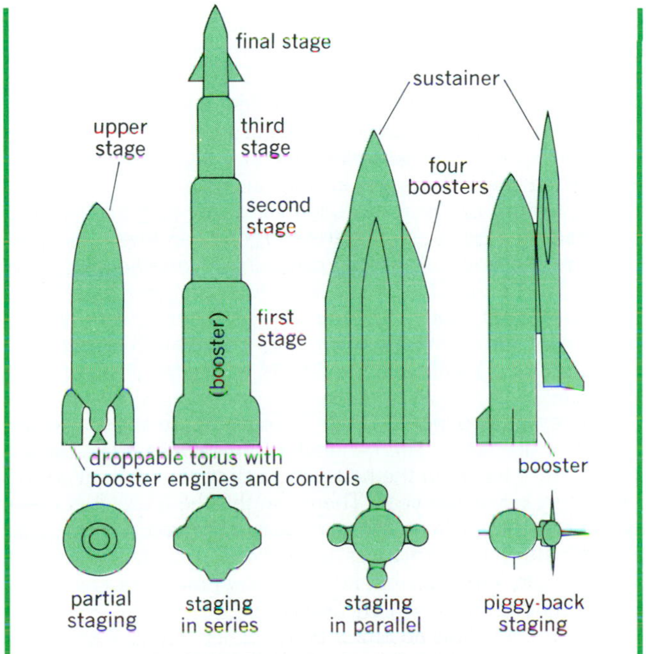

Typical schemes for staging missiles.

biggest and often called the booster; the next few stages are successively smaller and are generally called sustainers. Each stage is a complete vehicle in itself and carries its own propellant (either solid or liquid; both fuel and oxidizer), its own propulsion system, and has its own tankage and control system.

Once the propellant of a given stage is expended, the dead weight of that stage including empty tanks, rocket engine, and controls is no longer useful in contributing additional kinetic energy to the succeeding stages. By dropping off this useless weight, the mass that remains to be accelerated is made smaller; therefore it is possible to accelerate the payload to higher velocity than would be attainable if multiple staging were not used.

It is quite possible to employ different types of power plants, different types of propellants, and entirely different configurations in successive stages of any one multistage vehicle (see illustration). Because staging adds complications, it is impractical to have more than four to seven stages in any one vehicle. *See* ROCKET PROPULSION. [G.P.S.]

Rocky Mountain spotted fever An acute, infectious, typhuslike disease of humans caused by *Rickettsia rickettsi* and transmitted by species of ixodid, or hard-shelled, ticks. The disease is also called American spotted fever. The primary cycle in nature involves chiefly rodents, hares and rabbits, and dogs in some areas plus the opossum and cavy in Brazil together with the appropriate ticks which infest them. *See* ACARINA; RICKETTSIALES.

Incubation in areas with highly virulent strains may be as short as 2–3 days after tick attachment. Onset is sudden with headache, chills, high fever to 105°F (40.6°C), prostration, and appearance in 3–4 days of a measleslike rash. The rash, appearing initially on the forearms and ankles, becomes maculopapular over the entire body, including the palms and soles. The temperature may remain high for 2 weeks, falling gradually during the third week if death has not supervened. In milder disease, incubation may be 1–2 weeks and all symptoms are reduced in severity; ambulatory cases have been recognized. According to virulence and to age of the victim, mortality has varied in different areas from about 5 to 80%. Lasting immunity follows recovery. [C.B.P.]

Rodentia An order of mammals. Members are characterized by (1) a single pair of ever-growing upper and lower incisors, the enamel being limited to the front of the teeth so that wear produces and maintains a permanent chisel edge; (2) posterior cheek teeth that are reduced to a maximum of five upper and four lower teeth on each side; and (3) modification of the lower jaw articulation to permit free movement of the lower jaw in an anteroposterior direction, separating the cheek teeth when the incisors are in use, and vice versa. The most striking feature of the rodents is their ability to gnaw, as well illustrated by a beaver cutting trees for dams. Rodents range in size from the South American capybara (*Hydrochoerus*), with a skull 10–12 in. (25–31 cm) long and weighing more than 100 lb (45 kg), down to mouse-sized rodents, among the smallest of all mammals, weighing less than 3–4 oz (85–114 g). Most rodents are the size of a squirrel or a rat. *See* MAMMALIA.

The gestation period is short (20 days in the house mouse) and lifters are large. The generation time is also short (sexual maturity at 6 weeks in the house mouse), so that several generations may occur per year. There can thus be a rapid increase of numbers under favorable conditions. The resultant population densities lead to mass migrations, legendary among Scandinavian lemmings but known among other rodents, such as gray squirrels in New York State.

Some rodents (rats and mice of the family Muridae) associate closely with humans, living on stored food (especially grains) or refuse. Rodents serve as vectors for disease. Both wild and domestic rodents are an important source of furs. [A.E.Wo.]

Rodenticide One of several groups of materials commonly referred to as pesticides. Rodenticides are commonly thought of in terms of lethal chemical agents used to kill rodents whether applied as baits or dusts or used as fumigants. Rodenticides are frequently used to control other mammals that are not rodents, such as moles, rabbits, and hares.

Several common forms of rodenticides include: anticoagulants, which cause death by internal bleeding in rats and mice but are relatively safe for humans and pets; toxicants, or acute rodenticides, which cause death when a single dose is ingested and are usually restricted by law to use by trained individuals; lethal gases, used to kill rodents in their burrows or other confined areas; and dusts, which are used as contact poisons or which may adhere to the animal's body and be ingested during grooming. There are laws and regulations governing the sale and use of rodenticides, as there are for other types of pesticides. No rodenticide is universally effective; therefore, the proper rodenticide must be selected that is most satisfactory for a particular problem. *See* PESTICIDE; RODENTIA. [W.E.H.]

Roll mill A series of rolls operating at different speeds and used to grind paint or to mill flour. In paint grinding, a paste is fed between two low-speed rolls running toward each other at different speeds. Because the next roll in the mill is turning faster, it develops shear in the paste and draws the paste through the mill. The film is scraped from the last high-speed roll. For grinding flour, rolls are operated in pairs, rolls in each pair running toward each other at different speeds. Grooved rolls crush the grain; smooth rolls mill the flour to the desired fineness. *See* GRINDING MILL. [R.M.H.]

Rolling contact Contact between bodies such that the relative velocity of the two contacting surfaces at the point of contact is zero. Common applications of rolling contact are the friction gearing of phonograph turntables, speed changers, and wheels on roadways. Rolling contact mechanisms are, generally speaking, a special variety of cam mechanisms. The concepts of rolling contact are used in the study of antifriction bearings and in the study of the behavior of toothed gearing. *See* ANTIFRICTION BEARING; CAM MECHANISM; GEAR.

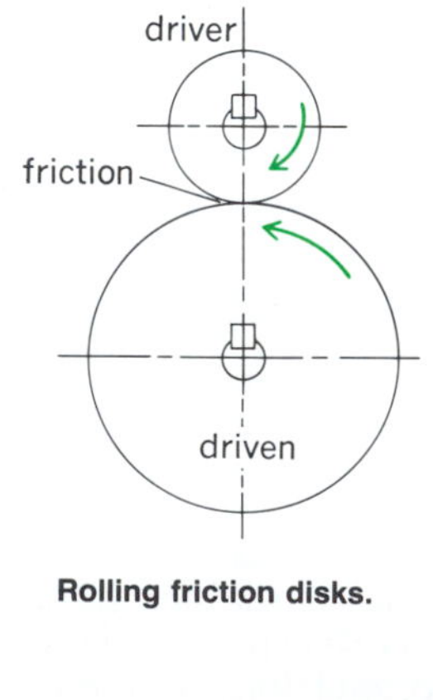

Rolling friction disks.

Pure rolling contact can exist between two cylinders rotating about their centers, with either external or internal contact. Two friction disks (see illustration) have external rolling contact if no slipping occurs between them. The rotational speeds of the disks are then inversely proportional to their radii. [J.R.Z.]

Roof construction The structural form used to provide the cover for homes, buildings, arenas, stadiums, and other

Intersecting barrel shell roof of the Administration Building, St. Louis Airport.

activity areas. Whatever form the roof construction takes, it must be capable of resisting the loads to which it will be subjected, carrying them satisfactorily to the walls or other supports, and providing protection from the elements. Principal roof loads are created by high winds and, in the northern climates, by snow and ice.

The most common type of roof is encountered in the construction of wood-framed structures used as dwelling units. This type of construction supports the roof surface, which consists of composition shingles or file placed on sheathing or plywood on rafters running from the peak of the roof to its eaves. Industrial buildings use essentially the same construction, except that trusses, longitudinal members, and roofing itself are generally fabricated from steel. *See* TRUSS.

Concrete, reinforced plastics, and steel are used in roofing large and small areas. Thin reinforced concrete sections have been used for shell roofs of cylindrical or spherical shape, or in the form of hyperbolic paraboloids (see illustration). Another form of equilibrium roof is a series of flat plates meeting at various angles to provide support one for the other.

A radically different type of structure has provided support by differential air pressure. These structures consist of a durable fabric or film supported by cables or, alternatively, make use of a double membrane which provides its own support. In either case, the interior of the structure is maintained at a slightly higher air pressure than the ambient atmospheric pressure, and in this way the shape of the structure is maintained. *See* BUILDINGS. [C.M.A.]

Root (botany) The absorbing and anchoring organ of vascular plants. Roots are simple axial organs that produce lateral roots, and sometimes buds, but bear neither leaves nor flowers. Elongation occurs in the root tip. The older portion of the root, behind the root tip, may thicken through cambial activity. Some roots, grass for example, scarcely thicken, but tree roots can become 4 in. (10 cm) or more in diameter near the stem. Roots may be very long. The longest maple (*Acer*) roots are usually as long as the tree is tall, but the majority of roots are only a few inches long. The longest roots may live for many years, while small roots may live for only a few weeks or months.

Root tips and the root hairs on their surface take up water and minerals from the soil. They also synthesize amino acids and growth regulators (gibberellins and cytokinins). These materials move up through the woody, basal portion of the root to the stem. The thickened, basal portion of the root anchors the plant in the soil. Thickened roots, such as carrots, can store food that is later used in stem growth. *See* CYTOKININ; GIBBERELLIN.

Roots usually grow in soil where: it is not too dense to stop root tip elongation; there is enough water and oxygen for root growth; and temperatures are high enough (above 39°F or 4°C) to permit root growth, but not so high that the roots are killed (above 104°F or 40°C). In temperate zones most roots are in the uppermost 4 in. (10 cm) of the soil; root numbers decrease so rapidly with increasing depth that few roots are found more than 6 ft (2 m) below the surface. Roots grow deeper in areas where the soil is hot and dry; roots from desert shrubs have been found in mines more than 230 ft (70 m) below the surface. In swamps with high water tables the lack of oxygen restricts roots to the uppermost soil layers. Roots may also grow in the air. Poison ivy vines form many small aerial roots that anchor them to bark or other surfaces.

The primary root originates in the seed as part of the embryo, normally being the first organ to grow. It grows downward into the soil and produces lateral second-order roots that emerge at right angles behind the root tip. Sometimes it persists and thickens to form a taproot. The second-order laterals produce third-order laterals and so on until there are millions of roots in a mature tree root system. In contrast to the primary root, most lateral roots grow horizontally or even upward. In many plants a few horizontal lateral roots thicken more than the primary, so no taproot is present in the mature root system.

Adventitious roots originate from stems or leaves rather than the embryo or other roots. They may form at the base of cut stems, as seen in the horticultural practice of rooting cuttings. [B.F.W.]

Root (mathematics) If a function $f(x)$ has the value 0 for $x = a$, a is a root of the equation $f(x) = 0$. The fundamental theorem of algebra states that any algebraic equation of the form $a_0x^n + a_1x^{n-1} + \cdots a_{n-1}x + a_n = 0$, where the a_k's are real numbers, has at least one root. From this it follows readily that such an equation has roots, real or complex, in number equal to the index (here n) of the highest power of x.

Furthermore, if $a + ib$ (where $i = \sqrt{-1}$) is a complex root of the given equation, so is $a - ib$, the conjugate of $a + ib$. Equations of degrees up to four may be solved algebraically. This statement means that the roots may be expressed as functions of the coefficients, the functions involving the elementary arithmetical processes of addition, multiplication, raising a number to a power, or extracting the root of a certain order of a given number. It was proved by H. Abel and by E. Galois that it is not possible to solve algebraically the *general* algebraic equation of degree higher than four. However, it is possible to determine the real roots of an algebraic equation to any desired degree of approximation. *See* CALCULUS; EQUATIONS, THEORY OF; NUMERICAL ANALYSIS. [A.N.L./S.Bo.]

Root-mean-square The square root of the arithmetic mean of the squares of a set of numbers is called their root-mean-square. If the numbers are $x_1, x_2, x_3, \ldots, x_n$, the root-mean-square is equal to

$$\sqrt{\frac{x_1^2 + x_2^2 + x_3^2 \cdots + x_n^2}{n}}$$

It is valuable as an average of the magnitudes of quantities, and it is not affected by the signs of the quantities.

Among applications of root-mean-square the most important is the standard deviation from the arithmetic mean. If the arithmetic mean $\bar{x}$ =

$$\frac{x_1 + x_2 + x_3 \cdots + x_n}{n}$$

then the standard deviation s =

$$\sqrt{\frac{(x_1 - \bar{x})^2 + (x_2 - \bar{x})^2 + (x_3 - \bar{x})^2 + \cdots (x_n - \bar{x})^2}{n}}$$

Thus standard deviation from the mean is the root-mean-square of the deviations from the mean. *See* STATISTICS. [H.R.C.]

Rope A long flexible structure consisting of many strands of wire, plastic, or vegetable fiber such as manila. Rope is classified as a flexible connector and is used generally for hoisting, conveying, or transporting loads; transmitting motion; and occasionally transmitting power. For flexibility and to reduce stresses as the rope bends over the sheave (pulley), a rope is made of many small strands. *See* PULLEY. [P.H.B.]

Rosales An order of flowering plants, division Magnoliophyta (Angiospermae), which gives its name to the subclass Rosidae in the class Magnoliopsida (dicotyledons). The order consists of 17 families and nearly 20,000 species; some 13,000 of these belong to the Leguminosae, 3000 to the Rosaceae, 1400 to the Crassulaceae, and 700 to the Saxifragaceae. *See* MAGNOLIOPSIDA.

The legume family (Leguminosae) generally has stipulate, compound leaves; 10 to many stamens which are often united (connate) by their filaments; and a single carpel which typically ripens into a dry fruit that opens at maturity down both sutures, releasing the nonendospermous seeds. Many or most members of the family harbor symbiotic nitrogen-fixing bacteria in their roots. *See* ALFALFA; BEAN; CLOVER; COWPEA; LESPEDEZA; LOCUST (FORESTRY); PEA; PEANUT; SOYBEAN; VETCH.

The ill-defined family Rosaceae typically has stipulate leaves and perigynous flowers; that is, the perianth and stamens are united into a basal cup distinct from the ovary. The important genus *Prunus*, includes cherries, plums, apricots, peaches, and almonds. *Pyrus* includes apples and pears. Some other familiar members of the Rosaceae are roses (*Rosa*), strawberries (*Fragaria*), raspberries and blackberries (*Rubus*), and hawthorn (*Crataegus*). *See* ALMOND; APPLE; APRICOT; BLACKBERRY; CHERRY; CRABAPPLE; CURRANT; GOOSEBERRY; NECTARINE; PEACH; PEAR; PLUM; QUINCE; RASPBERRY; STRAWBERRY; WILD CHERRY.

Members of the family Crassulaceae are notable for their very succulent leaves and resistance to desiccation. They have more or less separate carpets, which are often of the same number as the petals. Species of *Sedum*, *Kalanchoe*, and *Sempervivum* are often cultivated as ornamentals. *See* ROSIDAE. [A.Cr.]

Rose curve A type of plane curve that consists of loops (leaves, petals) emanating from a common point and that has a roselike appearance. Taking the common point O as the pole of a polar coordinate system (see illustration), these curves have equations of the form $\rho = a \cdot \sin n\theta$, where $a > 0$ and n is a positive integer (also $\rho = a \cdot \cos n\theta$, with a different choice of the initial line of the coordinate system). The curve is a circle of diameter a for $n = 1$. It has n or $2n$ leaves, according as n is an odd or even integer, respectively. The lemniscate is sometimes called a two-leaved rose, though its equation $\rho^2 = a^2 \cos 2\theta$ is not of the form given above. *See* LEMNISCATE OF BERNOULLI. [L.M.Bl.]

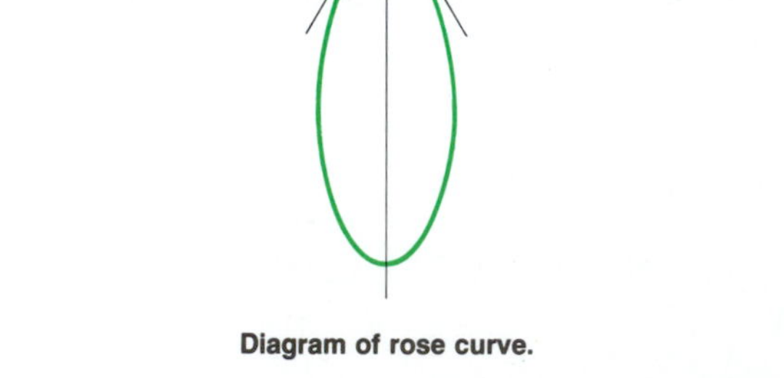

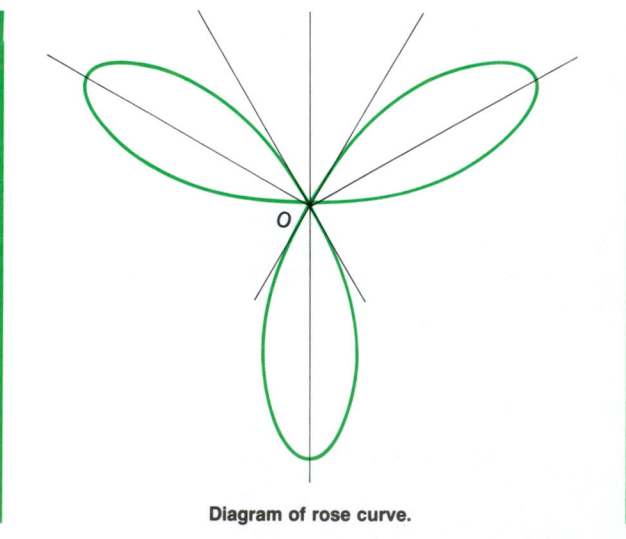

Diagram of rose curve.

Rosemary *Rosmarinus officinalis*, a member of the mint family, grown for its highly aromatic leaves and as an ornamental. Rosemary is an evergreen and perennial which can live as long as 20 years under favorable conditions. Although many varieties exist, *R. officinalis* is the only species. Most varieties are suitable for culinary use, although some (such as Pine Scented) contain high levels of terpenes and have a turpentine-like scent. *See* LAMIALES.

Rosemary is native to the Mediterranean area, and much of the world production is harvested from wild plants growing there. Though it is widely cultivated mainly in France, Spain, and California, it will grow in most temperate areas not subject to hard frosts.

Rosemary oil has been found to contain chemicals such as rosmanol which inhibit oxidation and bacterial growth. Rosemary is used in poultry seasoning and for flavoring soups and vegetables. *See* SPICE AND FLAVORING. [S.Kir.]

Roseola infantum A mild, sometimes epidemic viral disease with abrupt onset, high fever, and rash. It is also known as roseola infantum. It occurs almost exclusively in infants 6 months to 3 years old. Sometimes the disease may be confused with rubella, from which it may be distinguished by its abrupt onset; distinction from other exanthematous diseases is chiefly by abrupt termination of fever before the rash appears. *See* RUBELLA. [J.L.Me.]

Rosidae A large subclass of the class Magnoliopsida (dicotyledons) in the division Magnoliophyta (Angiospermae, the flowering plants), consisting of 16 orders, 108 families, and about 60,000 species. Most of them have a well-developed corolla with the petals separate from each other. A well-developed nectary disk is very often present. The largest orders in the subclass are the Rosales (20,000 species), Myrtales (9000 species), Euphorbiales (7500 species), and Sapindales (6700 species). *See* APIALES; CELASTRALES; CORNALES; EUPHORBIALES; FLOWER; GERANIALES; HALORAGALES; LINALES; MAGNOLIOPSIDA; MYRTALES; PODOSTEMALES; POLYGALALES; PROTEALES; RAFFLESIALES; RHAMNALES; ROSALES; SANTALALES; SAPINDALES. [A.Cr.]

Rosin A brittle resin ranging in color from dark brown to pale lemon yellow and derived from the oleoresin of pine trees. Rosin is insoluble in water, but soluble in most organic solvents. It softens at about 180–190°F (80–90°C). Rosin consists of about 90% resin acids and about 10% neutral materials such as anhydrides, sterols, and diterpene aldehydes and alcohols.

Rosin is obtained by wounding living trees and collecting the exudate (gum rosin), by extraction of pine stumps (wood rosin), and as a by-product from the kraft pulping process (sulfate or tall oil rosin).

The largest single use of rosin is in sizing paper to control water absorption, an application in which fortified rosin is important. Rosin soaps are used as emulsifying and tackifying agents in synthetic rubber manufacture. Other rosin uses include adhesives, printing inks, and chewing gum. [I.S.G.]

Rotary engine Internal combustion engine that duplicates in some fashion the intermittent cycle of the piston engine, consisting of the intake-compression-power-exhaust cycle, wherein the form of the power output is directly rotational.

Four general categories of rotary engines can be considered: (1) cat-and-mouse (or scissor) engines, which are analogs of the reciprocating piston engine, except that the pistons travel in a circular path; (2) eccentric-rotor engines, wherein motion is imparted to a shaft by a principal rotating part, or rotor, that is eccentric to the shaft; (3) multiple-rotor engines, which are based on simple rotary motion of two or more rotors; and (4) revolving-block engines, which combine reciprocating piston and rotary motion. *See* AUTOMOBILE; COMBUSTION CHAMBER; DIESEL CYCLE; DIESEL ENGINE; GAS TURBINE; INTERNAL COMBUSTION ENGINE; OTTO CYCLE. [W.Ch.]

Rotary tool drill A bit and shaft used for drilling wells. A turntable on the derrick floor rotates a string of hollow steel drill pipe at the bottom of which is a steel bit. The bit grinds the rock. A drilling fluid is pumped down through the drill pipe; the fluid flushes out the rock cuttings and returns up the space between drill string and hole side.

The drilling fluid may be air, water, or, most commonly, mud (a mixture of various clays and chemicals, each having a special function). The mud cools and lubricates the bit, removes cuttings from the hole, and cakes the wall of the hole to prevent caving before steel casing is set. The hydrostatic pressure exerted by the column of mud in the hole prevents blowouts which may result when the bit penetrates a high-pressure oil or gas zone. [A.L.P.]

Rotating impeller flowmeter A quantity meter that converts velocity to a proportional rotational speed. They include various types of turbine meters, propellers, helical impellers, and vortex cage meters.

The gas turbine meter is characterized by a large central hub that increases the velocity of the gas and also puts it through the tips of the rotor to increase the torque. Readout can be by a directly connected mechanical register or by a proximity sensor counting the blade tips as they pass by. Most of the available types of liquid-turbine meters are designed to cause a fluid flow or hydraulic lift to reduce end thrust, and several depend upon the flowing fluid to lubricate the bearings,

The elements in the propeller, helical impeller, and vortex cage meters (see illustration) rotate at a speed that is linear with fluid velocity. The useful rangeability is large and limited mostly by the bearings. *See* FLOW MEASUREMENT; QUANTITY FLOWMETER. [M.Br.; L.P.E.]

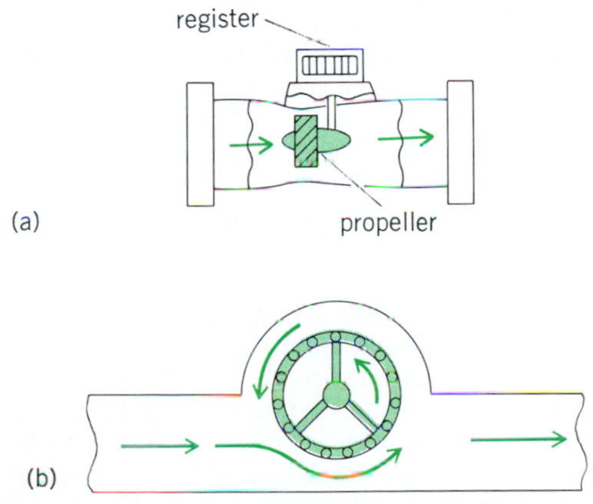

Types of rotating impeller flowmeters. (*a*) Propeller. (*b*) Vortex cage.

Rotating reference frame A frame of reference whose coordinate axes are rotating with respect to some inertial system. A point P fixed in the rotating frame moves uniformly about a circle in the inertial frame and is accelerated toward the axis of rotation. There is an additional acceleration (with respect to the inertial system) if P moves in the rotating system. For example, if the point moves away from the axis along a moving radius, it gains circular speed also. *See* CENTRIPETAL FORCE; CORIOLIS ACCELERATION.

Newton's second law can be modified to take account of these two accelerations that are not observed in the rotating system. Let A_{Ac}, A_C, A_{Co} be the observed, centripetal, and Coriolis accelerations, respectively. The total acceleration in the inertial frame is given by Eq. (1), where the accelerations

$$A = A_{Ac} + A_C + A_{Co} \quad (1)$$

must be added vectorially. Defining the centrifugal force, $FC = -mAC$, and the Coriolis force, $F_{Co} = -mA_{Co}$, permits Newton's second law to be written as Eq. (2). In rotating sys-

$$mA_{Ac} = F + F_C + F_{Co} \quad (2)$$

tems a consistent mechanical theory is obtained if the two additional forces are added to the impressed forces. Conversely, the presence of such forces will enable the observer to determine the motion of the reference frame. *See* FRAME OF REFERENCE; INERTIAL REFERENCE FRAME. [B.G.]

Rotational flow Fluid flow in which each minute particle of fluid rotates about its own axis. Rotational flow is possible only with a real or viscous fluid. A specific case is a forced vortex in which a fluid rotates like a solid body. Such a case might be obtained with a fluid in a cylindrical vessel rotating about its central axis, assuming no motion of the fluid relative to the container. A more common case is that in which the container is fitted with vanes, as the impeller of a centrifugal pump. As a small element rotates about the central axis, it also rotates about its own axis. *See* FLUID FLOW; VORTEX. [R.L.D.]

Rotational motion The motion of a rigid body which takes place in such a way that all of its particles move in circles about an axis with a common angular velocity; also, the rotation of a particle about a fixed point in space. Rotational motion is illustrated by (1) the fixed speed of rotation of the Earth about its axis; (2) the varying speed of rotation of the flywheel of a sewing machine; (3) the rotation of a satellite about a planet; (4) the motion of an ion in a cyclotron; and (5) the motion of a pendulum. Circular motion is a rotational motion in which each particle of the rotating body moves in a circular path about an axis. Such motion is exhibited by the first and second examples. For information concerning the other examples *see* HARMONIC MOTION; PARTICLE ACCELERATOR; PENDULUM.

The speed of rotation, or angular velocity, remains constant in uniform circular motion. In this case, the angular displacement θ experienced by the particle or rotating body in a time t is $\theta = \omega t$, where ω is the constant angular velocity.

A special case of circular motion occurs when the rotating body moves with constant angular acceleration. If a body is moving in a circle with an angular acceleration of α radians/s^2, and if at a certain instant it has an angular velocity ω_0, then at a time t seconds later, the angular velocity may be expressed as $\omega = \omega_0 + \alpha t$, and the angular displacement as $\theta = \omega_0 t + \frac{1}{2}\alpha t^2$. *See* ACCELERATION; VELOCITY.

A rotating body possesses kinetic energy of rotation which may be expressed as $T_{\text{rot.}} = \frac{1}{2}I\omega^2$, where ω is the magnitude of the angular velocity of the rotating body and I is the moment of inertia, which is a measure of the opposition of the body to angular acceleration. The moment of inertia of a body depends on the mass of a body and the distribution of the mass relative to the axis of rotation. For example, the moment of inertia of a solid cylinder of mass M and radius R about its axis of symmetry is $\frac{1}{2}MR^2$.

The action of a torque L is to produce an angular acceleration α according to the equation below, where $I\omega$, the product

$$L = I\alpha = I\frac{d\omega}{dt} = \frac{d}{dt}(I\omega)$$

of moment of inertia and angular velocity, is called the angular momentum of the rotating body. This equation points out that the angular momentum $I\omega$ of a rotating body, and hence its angular velocity ω, remains constant unless the rotating body is acted upon by a torque. Both L and $I\omega$ may be represented by vectors.

It is readily shown that the work done by the torque L acting through an angle θ on a rotating body originally at rest is exactly equal to the kinetic energy of rotation. *See* ANGULAR MOMENTUM; MOMENT OF INERTIA; RIGID-BODY DYNAMICS; TORQUE; WORK. [C.E.H./R.J.S.]

Rotational viscometers Instruments, such as the concentric cylinder device designed by M. M. Couette in 1890, for examining non-newtonian fluids. The basic concept is to rotate one cylinder while the shear stress (converted to torque) is by the deflection of a pointer attached to a torsion wire, but far more sophisticated systems have been developed including a unit designed for very low shear stress measurements. In this design, friction is reduced by means of an air bearing, and the torque is measured with an electronic null-balance system. The average shear rate is readily controlled by means of a variable-speed drive.

In addition to variable shear rate, the Couette design has the advantage of a small dispersion of shear rates within the sample if the annular gap is small compared to the radii. [H.L.G.]

Rotifera A phylum of pseudocoelomate, microscopic, mainly free-living aquatic animals, characterized by an anterior ciliary apparatus, the corona, whose cilia when in motion have the appearance of a pair of rapidly rotating wheels. This structure is implicit in the phyletic name (literally "wheel bearers") and the older popular name wheel animalcules.

The Rotifera show considerable diversity in form and structure, but all are bilaterally symmetrical, pseudocoelomate animals possessing complete digestive, excretory, nervous, and reproductive systems, but lacking separate respiratory and circulatory systems. They possess two features unique to their phylum: the corona, which is a retractile trochal disk, and the mastax, which is a gizzardlike structure derived from the modified pharynx.

Rotifers are dioecious and sexually dimorphic; females are commoner than males, some of which are degenerate organisms lacking a mouth and digestive organs. Males, when produced in the life cycle, are short-lived and survive for only hours or at the most a few days.

The three major subdivisions of the Rotifera, now given class status, are the Seisonacea, Bdelloidea, and Monogononta. *See* BDELLOIDEA; MONOGONONTA; SEISONACEA. [J.B.J.]

Rous sarcoma The first filterable agent (virus) known to cause a solid tumor in chickens. It was discovered in 1911 by P. F. Rous, who won the Nobel prize in 1967 for his discovery. Certain strains of the virus cause tumors in hamsters, rabbits, monkeys, and other species. The Rous virus is known as a "defective" virus in that it is incapable of producing tumors by itself. It is a ribonucleic acid (RNA) virus and belongs to the avian leukosis group. *See* ANIMAL VIRUS; AVIAN LEUKOSIS. [A.E.Mo.]

Rubber Originally, a natural or tree rubber, which is a hydrocarbon polymer of isoprene units. With the development of synthetic rubbers having some rubbery characteristics but differing in chemical structure as well as properties, a more general designation was needed to cover both natural and synthetic rubbers. The term elastomer, a contraction of the words elastic and polymer, was introduced, and defined as a substance that can be stretched at room temperature to at least twice its original length and, after having been stretched and the stress removed, returns with force to approximately its original length in a short time.

Three requirements must be met for rubbery properties to be present in both natural and synthetic rubbers: long threadlike molecules, flexibility in the molecular chain to allow flexing and coiling, and some mechanical or chemical bonds between molecules.

Natural rubber and most synthetic rubbers are also commercially available in the form of latex, a colloidal suspension of polymers in an aqueous medium. Natural rubber comes from trees in this form; many synthetic rubbers are polymerized in this form; some other solid polymers can be dispersed in water. *See* POLYMER.

Latexes are the basis for a technology and production methods completely different from the conventional methods used with solid rubbers.

In the crude state, natural and synthetic rubbers possess certain physical properties which must be modified to obtain useful end products. The raw or unmodified forms are weak and adhesive. They lose their elasticity with use, change markedly in physical properties with temperature, and are degraded by air and sunlight. Consequently, it is necessary to transform the crude rubbers by compounding and vulcanization procedures into products which can better fulfill a specific function.

Smoke sheet and pale crepe represent the forms in which the major portion of natural rubber is commercially available. The smoked sheet is obtained by acid coagulation of the tree latex, sheeting the coagulum, and then drying and smoking the sheets of rubber. The chemicals in the smoke preserve the rubber from molds and other organisms. Pale crepe is obtained by treating the latex, before or after coagulation, with sodium bisulfite and then washing, drying, and sheeting. Synthetic rubbers are usually available as bales, pellets, or flakes made from the polymerization process.

Although natural rubber may be obtained from hundreds of different plant species, the most important source is the rubber tree (*Hevea brasiliensis*). Natural rubber is *cis*-1,4-polyisoprene, containing approximately 5000 isoprene units in the average polymer chain. *See* RUBBER TREE.

Butadiene-styrene rubber (SBR) is the most important synthetic rubber and the most widely used rubber in the entire world. Formerly designated GR-S, SBRs are obtained by the emulsion polymerization of butadiene and styrene in varying ratios. However, in the most commonly used type, the ratio of butadiene to styrene is approximately 78:22. Unlike natural rubber, SBR does not crystallize on stretching and thus has low tensile strength unless reinforced. The major use for SBR is in tires and tire products. Other uses include belting, hose, wire and cable coatings, flooring, shoe products, sponge, insulation, and molded goods.

Butyl rubber is essentially a polyisobutylene except for the presence of diolefin, usually isoprene, to provide the unsaturation necessary for vulcanization. Butyl rubbers have excellent resistance to oxygen, ozone, and weathering. In addition, these rubbers exhibit good electrical properties and high impermeability to gases. The high impermeability to gases results in use of butyl as an inner liner in tubeless tires. Other widespread uses are for wire and cable products, injection-molded and extruded products, hose, gaskets, and sealants, and where good damping characteristics are needed.

Ethylene-propylene polymers are produced by the copolymerization of ethylene and propylene. These copolymers exhibit outstanding resistance to heat, oxygen, ozone, and other aging and degrading agents. Abrasion resistance in tire treads is excellent. The mechanical properties of their vulcanizates are generally approximately equivalent to those of SBR.

One of the first synthetic rubbers used commercially in the rubber industry is neoprene, a polymer of chloroprene, 2-chlorobutadiene-1,3. The neoprenes have exceptional resistance to weather, sun, ozone, and abrasion. They are good in resilience, gas impermeability, and resistance to heat, oil, and flame. They are fairly good in low temperature and electrical properties. This versatility makes them useful in many applications requiring oil, weather, abrasion, or electrical resistance or combinations of these properties, such as wire and cable, hose, belts, molded and extruded goods, soles and heels, and adhesives.

Nitrile-type rubbers are copolymers of acrylonitrile and a diene, usually butadiene. The nitrile rubbers can be blended with natural rubber, polysulfide rubbers, and various resins to provide characteristics such as increased tensile strength, better solvent resistance, and improved weathering resistance.

There are other specialty rubbers available which are important because of specific properties, but in the aggregate they make up only approximately 2% of the world production of synthetic rubbers. One of these, silicone rubber, is a linear condensation polymer based on dimethyl siloxane.

In general, the silicone rubbers have relatively poor physical properties and are difficult to process. They are unaffected by ozone, are resistant to hot oils, and have excellent electrical properties. *See* SILICON; SILICONE RESINS.

Proper choice of catalyst and order of procedure in polymerization have led to development of thermoplastic elastomers. The leading commercial types are styrene block copolymers having a structure which consists of polystyrene segments or blocks connected by rubbery polymers such as polybutadiene, polyisoprene, or ethylene-butylene polymer. Thermoplastic elastomers are very useful in providing a fast and economical method of producing a variety of products, including molded goods, toys, sporting goods, and footwear. One of the disadvantages for many applications is the low softening point of the thermoplastic elastomers. *See* POLYSULFIDE RESINS; POLYURETHANE RESINS. [E.G.P.]

Rubber tree *Hevea brasiliensis*, a member of the spurge family (Euphorbiaceae) and a native of the Amazon valley. It is the natural source of commercial rubber.

It has been introduced into all the tropical countries supporting the rainforest type of vegetation, and is grown extensively in established plantations, especially in Malaysia. The latex from the trees is collected and coagulated. The coagulated latex is treated in different ways to produce the kind of rubber desired. Rubber is made from the latex of a number of other plants, but *Hevea* is the rubber plant of major importance. *See* EUPHORBIALES; RUBBER. [P.D.St./E.L.C.]

Rubella A benign, infectious virus disease of humans characterized by coldlike symptoms and transient, generalized rash. This disease, also known as German measles, is primarily a disease of childhood. However, maternal infection during early pregnancy may result in infection of the fetus, giving rise to serious abnormalities and malformations. The congenital infection persists in the infant, who harbors and sheds virus for many months after birth.

In rubella infection acquired by ordinary person-to-person contact, the virus is believed to enter the body through respiratory pathways. Antibodies against the virus develop as the rash fades, increase rapidly over a 2–3 week period, and then fall during the following months to levels that are maintained for life. One attack confers life-long immunity, since only one antigenic type of the virus exists. Immune mothers transfer antibodies to their offspring, who are then protected for approximately 4–6 months after birth. *See* IMMUNITY.

Gamma globulin has been used for pregnant women exposed to rubella, but there is no convincing evidence that it offers any protection to the mother or to the fetus. A number of live, attenuated strains have been developed for use in vaccines; several are commercially available. [J.L.Me.]

Rubellite The red to red-violet variety of the gem mineral tourmaline. Perhaps the most sought-for of the many colors in which tourmaline occurs, it was named for its resemblance to ruby. The color is thought to be caused by the presence of lithium. It has a hardness of 7–7.5 on Mohs scale, a specific gravity near 3.04, and refractive indices of 1.624 and 1.644. Fine gem-quality material is found in Brazil, Madagascar, Maine, southern California, the Ural Mountains, and elsewhere. *See* GEM; TOURMALINE. [R.T.L.]

Rubiales An order of flowering plants, division Magnoliophyta (Angiospermae), in the subclass Asteridae of the class Magnoliopsida (dicotyledons). The order consists of the single family Rubiaceae, with about 6500 species. The Rubiales are marked by their inferior ovary; regular or nearly regular corolla with the petals grown together by their margins; stamens (equal in number to the petals) which are attached to the corolla tube alternate with the lobes; and opposite leaves with interpetiolar stipules or whorled leaves without stipules. *See* COFFEE; MAGNOLIOPSIDA. [A.Cr.]

Rubidium A chemical element, Rb, atomic number 37, and atomic weight 85.47. Rubidium is an alkali metal. It is a light, low-melting, reactive metal.

Most uses of rubidium metal and rubidium compounds are the same as those of cesium and its compounds. The metal is used in the manufacture of electron tubes, and the salts in glass and ceramic production.

Rubidium is a fairly abundant element in the Earth's crust, being present to the extent of 310 parts per million (ppm). This places it just below carbon and chlorine and just above fluorine and strontium in abundance. Sea water contains 0.2 ppm of rubidium, which (although low) is twice the concentration of lithium. Rubidium is like lithium and cesium in that it is tied up in complex minerals; it is not available in nature as simple halide salts as are sodium and potassium.

Rubidium has a density of 1.53 g/cm^3 (95.5 lb/ft^3, a melting point of 39°C (102°F), and a boiling point of 688°C (1270°F).

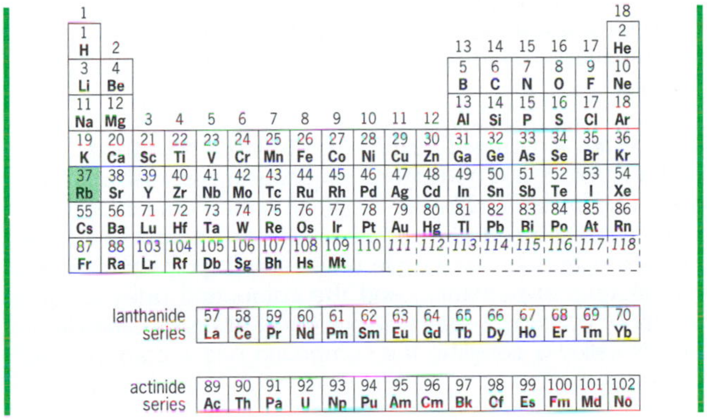

Rubidium is so reactive with oxygen that it will ignite spontaneously in pure oxygen. The metal tarnishes very rapidly in air to form an oxide coating, and it may ignite. The oxides formed are a mixture of Rb_2O, Rb_2O_2, and RbO_2. The molten metal is spontaneously flammable in air.

Rubidium reacts violently with water or ice at temperatures down to −100°C (−148°F). It reacts with hydrogen to form a hydride which is one of the least stable of the alkali hydrides. Rubidium does not react with nitrogen. With bromine or chlorine, rubidium reacts vigorously with flame formation. Organorubidium compounds can be prepared by techniques similar to those used for sodium and potassium. *See* ALKALI METALS; CESIUM. [M.Si.]

Rubidium-gas cell clock A type of atomic clock which makes use of microwave transitions of rubidium atoms that are mixed with a buffer gas in a glass cell. In an atmosphere consisting of an inert buffer gas mixture of suitable pressure, atoms can be confined by collisions with the buffer gas atoms which are not perturbing and thus can have a long interaction time with the exciting radiation, leading to narrow lines. This principle is used in passive and active rubidium-gas cell clocks. A small amount of rubidium and the buffer gas are placed in a glass cell inside a microwave cavity and kept at the proper temperature for optimum rubidium vapor pressure. Light from a rubidium lamp is passed through a filter and then through the cell through windows in the cavity. In the passive clock, under proper conditions the light transmitted through the cell is reduced in intensity, if the exciting radiation in the cavity is at the hyperfine resonance frequency (6834 megahertz). This decrease in intensity is detected with a photocell and is used to control the frequency of the exciting radiation. In an active clock, the light causes a population excess in the upper hyperfine level of the rubidium atoms, and maser oscillator action takes place. In both rubidium clocks the collisions with the buffer gas atoms change the resonance frequency by a nonnegligible amount, and therefore the clocks must be calibrated initially and are subject to drifts in rate. *See* ATOMIC CLOCK; MASER. [L.S.C.]

Ruby The red variety of the mineral corundum, in its finest quality the most valuable of gemstones. Only medium to dark tones of red to slightly violet-red or very slightly orange-red are called ruby; light reds, purples, and other colors are properly called sapphires. In its pure form the mineral corundum, with composition Al_2O_3, is colorless. The rich red of fine-quality ruby is the result of the presence of a minute amount of chromic oxide. The chromium presence permits rubies to be used for lasers producing red light. *See* CORUNDUM; LASER; SAPPHIRE.

The finest ruby is the transparent type with a medium tone and a high intensity of slightly violet-red, which has been likened to the color of pigeon's blood. Star rubies do not command comparable prices, but they, too, are in great demand. The ruby was among the first of the gemstones to be duplicated synthetically and the first to be used extensively in jewelry. *See* GEM. [R.T.L.]

Rugosa An order of extinct corals which flourished during the Paleozoic Era. The Rugosa, or Tetracorallia, first appeared in the Ordovician and became extinct in the Permian. Nothing is known of the soft parts. Both simple and compound skeletons are common, the simple ones having typically a curved, horn-shaped appearance, and the compound ones forming groups of cylindrical stems or polygonal columns. The simple form is called a corallite, the compound one a corallum. *See* COELENTERATA. [E.C.Stu.]

Runge vector The Runge vector describes certain unchanging features of a nonrelativistic two-body interaction for which the potential energy is inversely proportional to the distance r between the bodies or, alternatively, in which each body exerts a force on the other that is directed along the line between them and proportional to r^{-2}. Two basic interactions in nature are of this type: the gravitational interaction between two masses (called the classical Kepler problem), and the Coulomb interaction between like or unlike charges (as in the hydrogen atom). Both at the classical level and the quantum-mechanical level, the existence of a Runge vector is a reflection of the symmetry inherent in the interaction. *See* COULOMB'S LAW; QUANTUM MECHANICS; SYMMETRY LAWS (PHYSICS). [D.M.Fr.]

Running birds Any of 10 living and a number of recently extinct species of flightless, mostly large, heavy birds which are included in six orders, Struthioniformes, Rheiformes, Casuariiformes, Aepyornithiformes, Dinornithiformes, and Apterygiformes, usually categorized as the Ratites. They are characterized by a flat, keelless sternum and reduced wings. *See* RATITES; STRUTHIONIFORMES. [C.B.C.]

Running fit The intentional difference in dimensions of mating mechanical parts that permits them to move relative to each other. A free running fit has liberal allowance; it is used on high-speed rotating journals or shafts. A medium fit has less allowance; it is used on low-speed rotating shafts and for sliding parts. Running fits are affected markedly by their surface finish and the effectiveness of lubrication. *See* ALLOWANCE. [P.H.B.]

Runway A strip of land surfaced for the takeoff and landing of aircraft. The design and construction of runways for airports have engaged the attention of engineers for a long time. Not only is it necessary to provide a level, obstruction-free landing and approach area, it is also necessary to select the type of pavement, location, and construction of runways and taxiways which permits maximum use of the landing area. Further, it is necessary to develop durable surfaces which require minimum maintenance and long life. In the design of runways, consideration must be given to the type of lighting which will be used and how it can be installed so as to minimize possible interference with aircraft movements. *See* AIRPORT ENGINEERING.

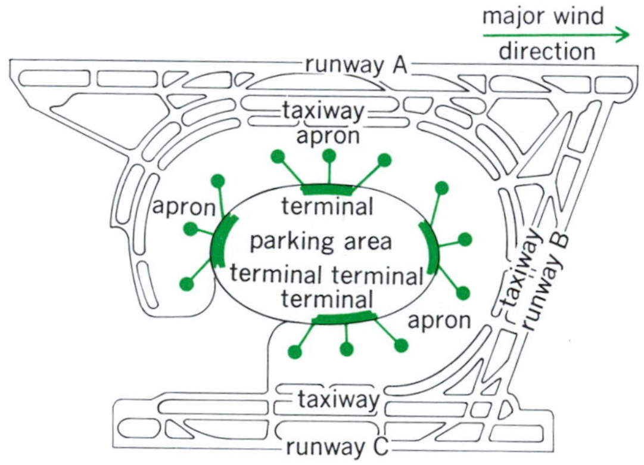

Typical configuration of the parallel runway concept.

A runway design for a given location is based on the traffic requirements, the type of aircraft using the field, and the relative positions of the terminals, taxiways, and runways. For most major airports, an alignment is sought which gives minimum noise interference with adjacent communities. All these elements must be compatible with runway orientation, which ideally provides that the major or instrument runway is in the direction of maximum wind conditions (see illustration). [J.M.Ky.]

Rust (microbiology) Fungi of the order Uredinales, or diseases caused by these fungi. Rusts are important plant pathogens of wheat and flax: other rusts, such as those of oats, barley, rye, white pines, apple, and snapdragon are also of economic importance. Rusts vary in specificity of parasitism.

Microcyclic rusts, such as *Puccinia malvacearum* (hollyhock rust), produce only teleutospores and basidiospores. Macrocyclic rusts produce other types of spores in addition. Autoecious rusts, such as *P. asparagi* (asparagus rust), complete their life cycle on one species of host. Heteroecious rusts require two species of host.

Control of rusts by dusting with fungicides is practical for ornamentals, but breeding resistant varieties of hosts offers the greatest promise for control of all rusts. Eradication of alternate hosts is effective in controlling white pine blister rust. [R.M.P.]

Rutabaga The plant *Brassica napobrassica*, a cool-season, hardy biennial crucifer of European origin, belonging to the order Capparales and probably resulting from the natural crossing of cabbage and turnip. The fleshy roots are cooked and usually eaten mashed as a vegetable. Rutabagas have been widely grown as a livestock feed in northern Europe and eastern Canada. Commercial production is limited to Canada and the northern part of the United States. *See* CAPPARALES; TURNIP. [H.J.C.]

Ruthenium A chemical element, Ru, atomic number 44, and atomic weight 101.07. Ruthenium is a hard, white metal, workable only at high temperatures and then only with difficulty.

Ruthenium is an excellent catalyst, useful in reactions involving hydrogenation, isomerization, oxidation, and reforming. Uses of pure ruthenium metal are minor. Ruthenium is an effective hardener for platinum and palladium. Alloys with large percentages (30–70%) of ruthenium containing other precious or base metals have been used for electrical contacts and applications, where extreme wear resistance and corrosion resistance are required, as in fountain pen nibs and instrument pivots.

Ruthenium is resistant to the common acids, including aqua regia, at temperatures up to 100°C (212°F), and up to 300°C (570°F) in the case of sulfuric acid. It also resists hydrofluoric and phosphoric acids at 100°C (212°F).

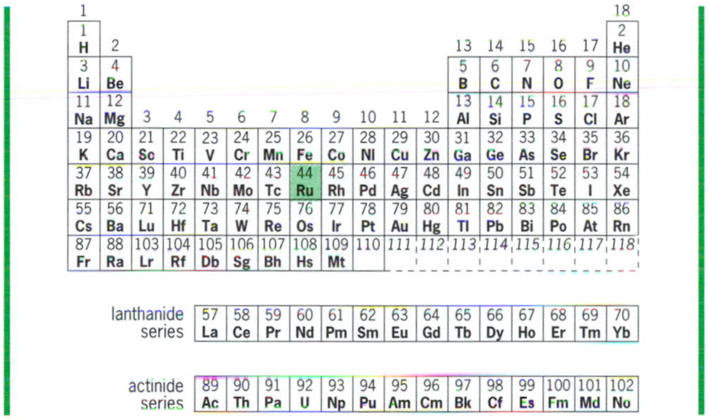

Potassium ruthenate, $KRuO_2 \cdot H_2O$, is soluble in water and is useful in purifying ruthenium. Ruthenium trichloride, $RuCl_3$, is soluble in water but decomposes in hot water. Ruthenium tetroxide is highly volatile and poisonous. [H.J.A.]

Rutherfordium A chemical element, symbol Rf, atomic number 104. Rutherfordium is the first element beyond the actinide series. In 1964 G. N. Flerov and coworkers at the Dubna Laboratories in Russia claimed the first identification of rutherfordium. A. Ghiorso and coworkers made a definitive identification at the Lawrence Radiation Laboratory, Berkeley, University of California, in 1969.

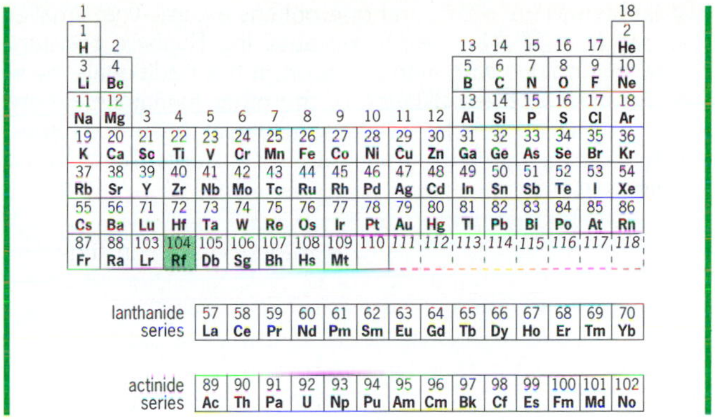

The Dubna group claimed the preparation of rutherfordium, mass number 260, by irradiating plutonium-242 with neon-22 ions in the heavy-ion cyclotron. The postulated nuclear reaction was $^{242}Pu + ^{22}Ne \rightarrow ^{260}Rf + 4$ neutrons. By 1969 the Berkeley group had succeeded in discovering two alpha-emitting isotopes of rutherfordium with mass numbers 257 and 259 by bombarding ^{249}Cf with ^{12}C and ^{13}C projectiles from the Berkeley heavy-ion linear accelerator (HILAC).

A number of years after the discovery at Berkeley, a team at Oak Ridge National Laboratory confirmed discovery of the isotope ^{257}Rf by detecting the characteristic nobelium x-rays following alpha decay. *See* ACTINIDE ELEMENTS; NOBELIUM; TRANSURANIUM ELEMENTS. [A.Gh.]

Rutile The most frequent of the three polymorphs of titania, TiO_2; the two other polymorphs are brookite and anatase.

The mineral occurs as striated tetragonal prisms and needles, commonly repeatedly twinned. The color is deep blood red, reddish brown, to black, rarely violet or yellow. Specific gravity is 4.2, and hardness 6.5 on Mohs scale. Melting point is 1825°C (3317°F).

Rutile occurs as an accessory in many rock types, ranging from plutonic to metamorphic rocks, and even as detrital material in sediments and placers because of its resistance to weathering. Large crystals have been found in some granite pegmatites. Other rocks include contact metamorphic limestones, alpine gneisses and schists, and pyrophyllite-lazulite schists. Rutile is commonly associated with apatite in high-temperature veins. In sufficient quantities, it is marketed as an ore of titanium. *See* TITANIUM. [P.B.M.]

Rydberg atom An atom which possesses one valence electron orbiting about an atomic nucleus within an electron shell well outside all the other electrons in the atom. Such an atom approximates the hydrogen atom in that a single electron is interacting with a positively charged core. *See* ELECTRON CONFIGURATION. [J.E.B.]

Rydberg constant An atomic constant describing the binding energy between electron and atomic nucleus. It is connected with the other universal physical constants by the relation, in cgs electrostatic units, $R_\infty = 2\pi^2 me^4/ch^3$g. Here, m and e are the mass and charge of the electron, c the velocity of light, and h is Planck's constant. (In SI units it is given by $R_\infty = \mu_0{}^2 me^4c^3/8h^3$ where $\mu_0 = 4\pi \times 10^{-7}$ henry/meter is the permeability of vacuum.) The Rydberg constant is determined from measurements of the wavelength of spectral lines of hydrogenlike atoms, which permit accurate theoretical calculations. The relationship between the wavelength and the Rydberg constant is given by Bohr's formula, corrected for finite nuclear mass, relativistic effects, and Lamb shifts. The symbol R_∞ indicates that the constant refers to a hypothetical atom of infinite mass. To correct for the finite nuclear mass M of a real atom, one replaces the electron mass m in the preceding formula by the reduced mass of electron and nucleus, that is, by $mM/(m + M)$, and one obtains a somewhat smaller Rydberg $R = R_\infty/(1 + mM)$. Because the Rydberg constant can be determined with high precision, it has traditionally been a cornerstone in the evaluation of the other fundamental constants; it gives a precise relation between these constants, which must be satisfied by any set of constants that may be adopted. [T.W.Ha.]

Rye A winter-hardy and drought-resistant cereal plant, *Secale cereale*, in the grass family (Graminae). Rye grain is used for animal feed, human food, and production of spirits. Ground rye is mixed with other feeds for livestock. It is often fall-sown to provide soil cover and important pasturage for livestock. Next to wheat, rye has the most desirable gluten for breadmaking. [J.A.Sh.]

Sable A carnivore, *Martes zibellina*, of the family Mustelidae, found in northern Asia from the Urals to Japan; it is a valuable fur-bearing animal and quite similar to the American marten. The sable obtains food by preying upon other arboreal animals, principally birds and small mammals. It is well adapted to these activities, with its slender, supple body and short limbs with five short toes on each foot that terminate in sharp curved claws. The long bushy tail serves as a balancing organ. Mating occurs in summer, and three or four young are born the following spring. *See* CARNIVORA; MARTEN. [C.B.C.]

Saccharin An organosulfur compound, also called *o*-sulfobenzoic imide. The material used as a sweetening agent (about 500–700 times as sweet as cane sugar) is the sodium salt, which passes largely unchanged through the body and is excreted in the urine. The slightly bitter aftertaste of sodium saccharin is mainly that of impurities from the conventional synthesis and can be avoided by syntheses starting from anthranilic acid or benzothiophene. Saccharin is used in food preparation for low-caloric diets and in diabetes therapy, where normal sugars cannot be tolerated. *See* ORGANOSULFUR COMPOUND. [N.K.]

Sacoglossa An order of the gastropod subclass Opisthobranchia containing a thousand living species of herbivorous sea slugs; sometimes called Ascoglossa. They occur in all the oceans, at shallow depths, reaching their greatest size and diversity in tropical seas.

Members of this abundant order have two common features: the herbivorous habit (except for *Olea* and a species of *Stiliger* which eat the eggs of other mollusks) and the possession of a uniseriate radula in which the oldest, often broken, teeth are usually retained within the body, not discarded.

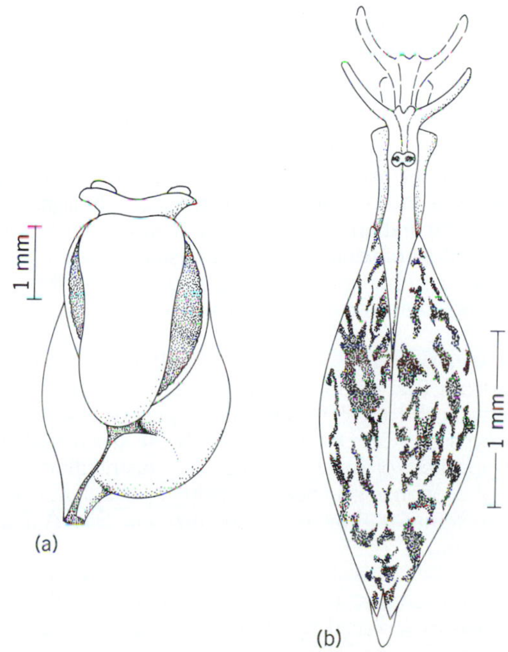

Representative sacoglossans: (a) *Volvatella*; (b) *Berthelinia*.

Sacoglossans possess varied shells; there may be a single shell, capacious and coiled (*Volvatella*, illustration *a*; *Cylindrobulla*), a flattened open dorsal shell (*Lobiger*), or two lateral shells (the bivalved gastropods, *Berthelinia*, illustration *b*). In the highest sacoglossans (*Elysia*, *Limapontia*) the true shell is completely lost after larval metamorphosis. *See* OPISTHOBRANCHIA. [T.E.T.]

Safety factor An empirical number by which the strength of a material is divided to obtain a conservative design stress. A safety factor is used because of uncertainties in operating conditions that may be encountered, nonuniformities in materials, simplifying design assumptions, effects of aging such as corrosion, and strains introduced inadvertently during fabrication and transportation and because of the seriousness of failure. Safety factors vary widely depending on the material, consequences of failure, and operating conditions. For ductile materials, safety factors applied to yield strength are often 1.5–4. For brittle materials that fracture with no prior evidence of incipient failure, factors of 5–8 may be appropriate. *See* STRESS AND STRAIN. [W.J.K./W.G.B.]

Safety glass A unitary structure formed of two or more sheets of glass between each of which is interposed a sheet of plastic, usually polyvinyl butyral. In usual manufacture, two clean and dry sheets of plate glass and a sheet of plastic are preliminarily assembled as a sandwich under slight pressure to produce a void-free bond. The laminate is then pressed under heat long enough to unite. For use in surface vehicles the finished laminated glass is approximately $\frac{1}{4}$ in. (6 mm) thick; for aircraft it is thicker. *See* GLASS. [F.H.R.]

Safety valve A relief valve set to open at a pressure safely below the bursting pressure of a container, such as a boiler or compressed air receiver. Typically, a disk is held against a seat by a spring; excessive pressure forces the disk open (see illustration). Construction is such that when the valve opens slight-

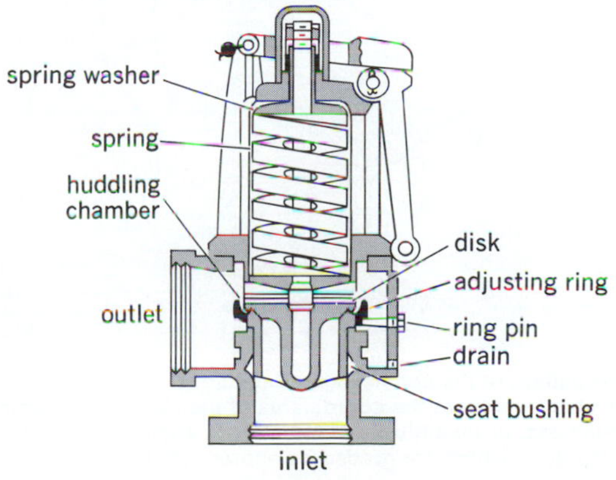

Diagram of a typical safety valve.

ly, the opening force builds up to open it fully and to hold the valve open until the pressure drops a predetermined amount. This differential or blow-down pressure and the initial relieving pressure are adjustable. *See* VALVE. [T.Ba.]

Safflower An oilseed crop (*Carthamus tinctorius*) that is a member of the thistle (Compositae) family and produces its seed in heads. Flowers vary in color from white through shades of yellow and orange to red. The seed is shaped like a small sunflower seed, and is covered with a hull that may be white with a smooth surface or off-white to dark gray with a ridged surface. Depending on hull thickness, the oil content varies from 25 to 45%. Safflower is grown commercially in Mexico, Australia, and California. *See* ASTERALES.

There are two types of safflower oil, both with 6–8% palmitic acid and 1–2% stearic acid. One type, the standard or polyunsaturated type, has 76–79% linoleic acid and 11–17% oleic acid. This high-linoleic type has been used in soft margarines, in salad oils, and in the manufacture of paints and varnishes. It has had limited use for frying foods, because heat causes it to polymerize and form a tough film on the cooking vessel. The second type of oil, called the high–oleic-acid or monounsaturated type, has 76–79% oleic acid and 11–17% linoleic acid. Its fatty-acid composition is similar to that of olive oil, but the flavor is bland. High-oleic safflower oil is a premium frying oil. *See* FAT AND OIL (FOOD). [P.F.K.]

Saffron The plant *Crocus sativus*, a member of the iris family (Iridaceae). A native of Greece and Asia Minor, it is now cultivated in various parts of Europe, India, and China. This crocus is the source of a potent yellow dye used for coloring foods and medicine. The dye is extracted from the styles and stigmas of the flowers. *See* LILIALES. [P.D.St./E.L.C.]

Sage The plant *Salvia officinalis*, a member of the mint family (Labiatae), the leaves of which yield a spice and an aromatic oil. It is a half-shrub native to the Mediterranean region but is now widely cultivated. It is much used as a flavoring in stuffing for fowl and in meats, especially sausage. Oil of sage is used in making perfumes. *See* LAMIALES. [P.D.St./E.L.C.]

Sagittarius The Archer, in astronomy, a zodiacal and summer constellation, the major portion of which lies directly in the Milky Way. Sagittarius is the ninth sign and the southernmost constellation of the zodiac (see illustration). The Milky Way in Sagittarius is very bright, containing rich star fields and clusters, because its direction lies in the center of the Milky Way stellar system. *See* CONSTELLATION. [C.-S.Y.]

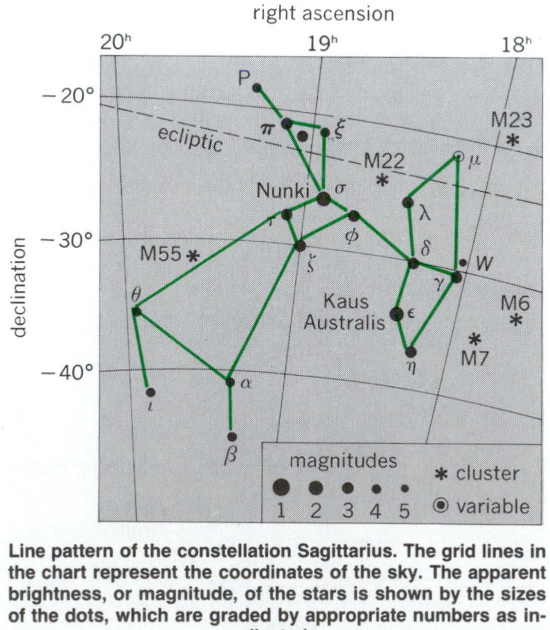

Line pattern of the constellation Sagittarius. The grid lines in the chart represent the coordinates of the sky. The apparent brightness, or magnitude, of the stars is shown by the sizes of the dots, which are graded by appropriate numbers as indicated.

Sailplane A high-performance glider. Sailplane construction traditionally has been of wood and plywood, although the use of aluminum alloy has become common. The use of fiber glass as primary structure has also come into prominence, since it is possible to produce the external shapes in accurate molds with greater precision, resulting in improved performance. *See* GLIDER. [E.S.]

Saint Elmo's fire A type of corona discharge observed on ships under conditions approaching those of an electrical storm. The charge in the atmosphere induces a charge on the masts and other elevated structures. The result of this is a corona discharge which causes a spectacular glow around these points. *See* CORONA DISCHARGE. [G.H.M.]

Salenioida An order of Echinacea in which the apical system includes one or several large angular plates covering the periproct, with the other characters similar to those of the hemicidaroid urchins. There are two families: the Acrosaleniidae, an extinct group confined to the Jurassic and Cretaceous; and the Saleniidae, which ranged from the Jurassic onward, with two surviving deep-sea genera. *See* ECHINACEA. [H.B.F.]

Salicales An order of flowering plants, division Magnoliophyta (Angiospermae), in the subclass Dilleniidae of the class Magnoliopsida (dicotyledons). The order consists of the single family Salicaceae, with about 350 species. There are only two genera, *Salix* (willow) and *Populus* (poplar and cottonwood), the former by far the larger. The Salicales are dioecious, woody plants, with alternate, simple, stipulate leaves and much reduced flowers that are aggregated into catkins. The mature seeds are plumose-hairy and are distributed by the wind. *See* DILLENIIDAE; MAGNOLIOPSIDA; POPLAR; WILLOW. [A.Cr.]

Salicylate A salt or ester of salicylic acid having the general formula shown below and formed by replacing the carboxylic

$$(o)\text{-}C_6H_4(OH)C(=O)\text{—}O\text{—}M \quad \text{or} \quad \text{—}R$$

hydrogen of the acid by a metal (M) to give a salt or by an organic radical (R) to give an ester. Alkali-metal salts are water-soluble; the others, insoluble. Sodium salicylate is used in medicines as an antirheumatic and antiseptic, in the manufacture of dyes, and as a preservative (illegal in foods). Salicylic acid is used in the preparation of aspirin. The methyl ester is the chief component of oil of wintergreen. This ester is used in pharmaceuticals as a component of rubbing liniment. It is also used as a flavoring agent and an odorant. *See* ASPIRIN. [E.H.H.]

Saline evaporite A sedimentary deposit of soluble salts resulting from the evaporation of a standing body of water. Quantitatively the most important evaporites are anhydrite, $CaSO_4$; gypsum, $CaSO_4 \cdot 2H_2O$; and halite (rock salt), NaCl. Other evaporites, of much more limited distribution and volume but of economic significance, include potassium chlorides, sodium carbonates, borates, and nitrates. *See* ANHYDRITE; GYPSUM; HALITE.

Evaporites, being soluble, are rarely exposed at the surface except in arid regions. Gypsum is common near the surface but is invariably replaced by anhydrite at greater depths, the latter mineral being vastly more abundant. Other evaporite minerals are not related to depth of burial, except for the effects of near-surface solution. [L.L.S.]

Saline water reclamation The partial or almost complete demineralization of sea and brackish waters, geothermal brines, wastewaters, and industrial effluents to make fresh water suitable for human or animal consumption, diverse industrial uses, irrigation, recreation, or aquifer recharge—broadly referred to as desalination.

A diversity of approaches are used for the separation of water from saline solutions. Thermal processes effect separation by means of phase changes and include distillation and freezing processes. In the membrane processes, one or more suitably designed organic membranes accomplish the separation process. In a single membrane design, reverse osmosis pressure forces the fresh water through the membrane. In electrodialysis, a system using multiple membranes, direct current leads to formation of pure water and brine streams and drives the salt ions toward the electrodes through charge-selective membranes. There are also chemical processes. In the ion exchange process, substances are added to exchange the ions in the solution or to precipitate the salts. In the solvent extraction process, chemicals with greater affinity for water can remove the wastes from solutions. *See* ION EXCHANGE; OSMOSIS; WATER TREATMENT.

There is no optimum or universal process for demineralization of all water streams. Each so-called contaminated or modified water stream must be evaluated for its peculiarities and an optimal process for purification for a specific end use recommended. Among the criteria to be considered are: salinity, identity of the ions present, presence of other contaminants, nature of the water, availability, energy costs, quantity, and quality. [R.B.]

Salmonella A genus of the bacterial family Enterobacteriaceae, causing enteric infections with or without blood invasion (septicemia). Most species are motile, utilize citrate, decarboxylate ornithine, form gas from glucose, and produce hydrogen sulfide; but notable exceptions exist, for example, *S. typhi* and *S. paratyphi A*. Salmonellae do not ferment lactose, produce indole, or split urea; the Voges-Proskauer reaction is negative. *See* ENTEROBACTERIACEAE.

The habitats of the salmonellae are many animal species and humans—which may be carriers or may suffer from *Salmonella* infection. Salmonellae are frequently found in chickens, ducks, turkeys, pigs, rodents, and turtles; they are not uncommon in cattle. Animal infections are often septicemic. Although the present nomenclature suggests otherwise, there is no animal specificity for a serotype. Some serotypes, however, are found in humans only (for example, *S. typhi* and *S. paratyphi A*). *See* SALMONELLOSES. [A.V.G.]

Salmonelloses Diseases caused by *Salmonella*. These include enteritis and septicemia with or without enteritis (5–10%). *S. typhi*, *S. paratyphi A*, *B*, and *C*, and occasionally *S. cholerae suis*, cause particular types of septicemia called typhoid and paratyphoid fever, respectively; while all other types may cause enteritis or septicemia, or both together. *See* SALMONELLA.

Typhoid fever has an incubation period of 5–14 days. It is typified by a slow onset with initial bronchitis, diarrhea or constipation, a characteristic fever pattern (increase for 1 week, plateau for 2 weeks, and decrease for 2–3 weeks), a slow pulse rate, development of rose spots, swelling of the spleen, and often an altered consciousness; complications include perforation of the bowel and osteomyelitis. Typhoid fever leaves the individual with a high degree of immunity. Paratyphoid fever has a shorter course and is generally less severe than typhoid fever.

Enteric fevers, that is, septicemias due to types of *Salmonella* other than those previously mentioned, are more frequent in the United States than typhoid and paratyphoid fever but much less frequent than *Salmonella* enteritis. In children and in previously healthy adults, enteric fevers are most often combined with enteritis and have a favorable outlook. The organisms involved are the same as those causing *Salmonella* enteritis.

Inflammation of the small bowel, enteritis, due to *Salmonella* is one of the most important bacterial zoonoses. The most frequent agents are *S. typhimurium*, *S. enteritidis*, *S. newport*, *S. heidelberg*, *S. infantis*, and *S. derby*. The incubation period varies from 6 h to several days. Diarrhea and fever are the main symptoms; the intestinal epithelium is invaded, and early bacteremia is probable. [A.W.C.V.G.]

Salmoniformes An order of soft-rayed fishes comprising salmon and their allies. They make up the stem group from which most higher teleostean fishes have evolved.

Salmoniformes are generalized fishes characterized by soft or articulated fin rays; adipose dorsal fin usually present; cycloid (smooth) scales; pelvic fin in an abdominal position (well back on the trunk), usually with more than six rays, and with pelvic girdle free from pectoral girdle; pectoral fin placed low on the side and more or less horizontal; upper jaw usually bordered by premaxillae and maxillae; premaxillae not protractile; no weberian apparatus connecting the swim bladder with the inner ear, but a duct joining the bladder with the gut; and branchiostegal rays variable in number and arrangement.

Salmoniform fishes have an imperfect fossil record; they were present in the Cretaceous and likely earlier. The order consists of 8 suborders, 37 families, some 212 genera, and about 950 species. Most are marine fishes, and they are common in all seas. A few are adapted to pelagic or shore waters, most live in the bathypelagic zone, some tolerate arctic seas, and many inhabit warm or cold fresh water, or are anadromous; that is, they enter fresh water to reproduce.

The suborder Salmonoidei includes many of the most important and best known of all fishes, especially in the family Salmonidae, or salmon, trouts, whitefishes, and graylings. The Argentinoidei (argentines, deep-sea smelts, and related fishes), Alepocephaloidei (slickheads), and Bathylaconoidei are found chiefly in deep marine waters. The Galaxioidei are mostly small, fresh-water fishes of south temperate lands, especially Australia and New Zealand. Esocoidei are a small group, the pikes and mudminnows of northern fresh waters. The Stomiatoidei, lightfishes and allies, inhabit the middle depths of the oceans. They are of small size and often grotesque form, and are equipped with photophores. The Myctophoidei is the largest of the suborders. All species are marine; some, including lizardfishes, inhabit shore waters, but the majority are oceanic. Those that live at moderate to considerable depths, such as the lantern fishes, commonly have photophores on the head and body. *See* PHOTOPHORE GLAND; TELEOSTEI. [R.M.B.]

Salt (chemistry) A compound formed when one or more of the hydrogen atoms of an acid are replaced by one or more cations of the base. The common example is sodium chloride in which the hydrogen ions of hydrochloric acid are replaced by the sodium ions (cations) of sodium hydroxide. There is a great variety of salts because of the large number of acids and bases which has become known.

Salts are classified in several ways. One method—normal, acid, and basic salts—depends upon whether all the hydrogen ions of the acid or all the hydroxide ions of the base have been replaced:

Class	*Examples*
Normal salts	$NaCl$, NH_4Cl, Na_2SO_4, Na_2CO_3, Na_3PO_4, $Ca_3(PO_4)_2$
Acid salts	$NaHCO_3$, NaH_2PO_4, Na_2HPO_4, $NaHSO_4$
Basic salts	$Pb(OH)Cl$, $Sn(OH)Cl$

The other method—simple salts, double salts (including alums), and complex salts—depends upon the character of completeness of the ionization:

Class	*Examples*
Simple salts	$NaCl$, $NaHCO_3$, $Pb(OH)Cl$
Double salts	$KCl{\cdot}MgCl_2$
Alums	$KAl(SO_4)_2$, $NaFe(SO_4)_2$, $NH_4Cr(SO_4)_2$
Complex salts	$K_3Fe(CN)_6$, $Cu(NH_3)_4Cl_2$, $K_2Cr_2O_7$

See ACID AND BASE; CHEMICAL BONDING. [A.B.G.]

Salt (food) The chemical compound sodium chloride. It is used extensively in the food industry as a preservative and flavoring, as well as in the chemical industry to make chlorine and sodium. *See* CHLORINE; FOOD PRESERVATION; SALT (CHEMISTRY); SODIUM.

Salt was originally made by evaporating sea water (solar salt). This method is still in common usage; however, impurities in solar salt make it unsatisfactory for most commercial uses, and these impurities also lead to clumping. Salt, freshly produced from sea-water evaporation ponds, may contain large numbers of halophilic (salt-loving) microorganisms. In the United States refined salt is obtained from underground mines located in Michigan and Louisiana. Salt is usually handled during the refining processes as brine.

Salt is liable to clumping during periods of high humidity, so preventives are added. Materials used include magnesium carbonate and certain silicates. Iodides are added in those areas where iodine deficiencies exist. [R.E.M.]

Salt dome An intrusive body of rock salt which has penetrated large thicknesses of overlying sedimentary rock. Salt domes are distinguished from other geological deformations involving salt in being roughly circular or elliptical in cross section and in having horizontal dimensions of the same order of magnitude or less than their vertical dimensions. *See* SALINE EVAPORITE.

Salt domes are best known along the Gulf Coast of the United States, both onshore and offshore, where they are economically important because of their association with deposits of oil and sulfur. [L.L.N.]

Salt-effect distillation A process of extractive distillation in which a salt that is soluble in the liquid phase of the system being separated is used in place of the normal liquid additive introduced to the extractive distillation column in order to effect the separation.

Extractive distillation is a process used to separate azeotrope-containing systems or systems in which relative volatility is excessively low. An additive, or separating agent, that is capable of raising relative volatility and eliminating azeotropes in the system being distilled is supplied to the column, where it mixes with the feed components and exerts its effect. The agent is subsequently recovered from one or both product streams by a separate process and recycled for reuse. *See* AZEOTROPIC MIXTURE.

In salt-effect distillation, the process is essentially the same as for a liquid agent, although the subsequent process used to recover the agent for recycling is different; that is, evaporation is used rather than distillation. The salt is added to the system by being dissolved in the reentering reflux stream at the top of the column. Being nonvolatile, it will reside in the liquid phase, flowing down the column and out in the bottom product stream.

The major commercial use of salt-effect distillation is in the concentration of aqueous nitric acid, using the salt magnesium nitrate as the separating agent. Other commercial applications include acetone-methanol separation using calcium chloride and isopropanol-water separation using the same salt. *See* AZEOTROPIC DISTILLATION; DISTILLATION. [W.F.F.]

Salt gland A specialized gland located around the eyes and nasal passages in marine turtles, snakes, and lizards, and in birds such as the petrels, gulls, and albatrosses, which spend much time at sea. In the marine turtle it is an accessory lacrimal gland which opens into the conjunctival sac. In seagoing birds and in marine lizards it opens into the nasal passageway. Salt glands copiously secrete a watery fluid containing a high percentage of salt, higher than the salt content of urine in these species. As a consequence, these animals are able to drink salt-laden sea water without experiencing the dehydration necessary to eliminate the excess salt via the kidney route. *See* GLAND. [O.E.N.]

Saltmarsh Generally, a maritime habitat characterized by special plant communities, which occur primarily in the temperate regions of the world; however, typical saltmarsh communities can be formed in association with mangrove swamps in the tropics and subtropics. In inland areas where saline springs emerge or where there are salt lakes, typical saltmarsh communities can be found, dominated sometimes by the same species that occur on maritime saltmarsh. The extensive areas of inland salt desert, while exhibiting some features of similarity with saltmarsh, are nevertheless best regarded as a separate entity. Excess sodium chloride is the predominant environmental feature of saltmarsh (maritime or inland). In the case of salt deserts sodium chloride is only one of the alkali salts that may occur in excess.

Saltmarsh can form on mud, muddy sand, or sandy mud, but not on pure sand because mobile sand does not provide a sufficiently stable substrate. Typical physiographic features associated with saltmarsh are the creeks, which serve as drainage channels, and pools, which are known as pans.

The plants that grow on saltmarshes must tolerate the excess sodium chloride and are termed halophytes. These plants possess features associated with the halophytic environment, for example, development of succulence, waxy cuticle, and salt-excreting glands; overall there is a tendency for the vegetation to exhibit a drab grayness. The predominant life forms on saltmarshes are the hemicryptophytes and therophytes. There are also characteristic algae associated with saltmarsh vegetation. Particular communities of green, red, and blue-green algae are common, and in the Atlantic–North Sea area there may be extensive communities of free-living brown fucoids. [V.J.C.]

Salviniales A small order of heterosporous, leptosporangiate ferns (division Polypodiophyta) which float on the surface of the water. The delicate, branching stem is provided with small, simple to bifid or more or less dissected leaves. The sporangia are enclosed in specialized appendages of the leaves, called sporocarps. The order contains only a single family, with two widely distributed genera, *Salvinia* and *Azolla*, with only about 20 species in all. *See* POLYPODIOPHYTA. [A.Cr.]

Samarium A chemical element, Sm, atomic number 62, belonging to the rare-earth group. Its atomic weight is 150.35,

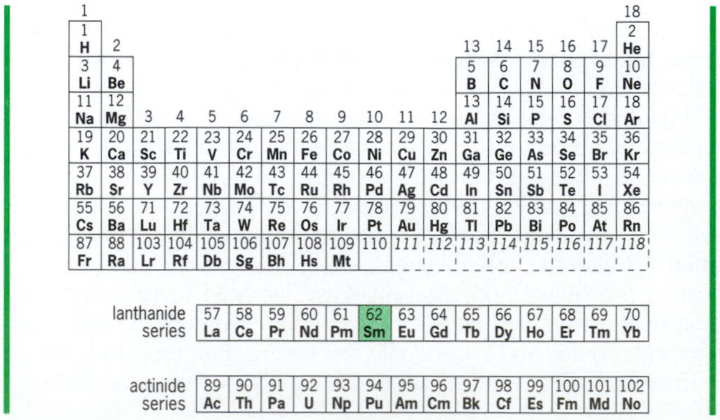
1 H 2 He
3 Li 4 Be 5 B 6 C 7 N 8 O 9 F 10 Ne
11 Na 12 Mg 13 Al 14 Si 15 P 16 S 17 Cl 18 Ar
19 K 20 Ca 21 Sc 22 Ti 23 V 24 Cr 25 Mn 26 Fe 27 Co 28 Ni 29 Cu 30 Zn 31 Ga 32 Ge 33 As 34 Se 35 Br 36 Kr
37 Rb 38 Sr 39 Y 40 Zr 41 Nb 42 Mo 43 Tc 44 Ru 45 Rh 46 Pd 47 Ag 48 Cd 49 In 50 Sn 51 Sb 52 Te 53 I 54 Xe
55 Cs 56 Ba 71 Lu 72 Hf 73 Ta 74 W 75 Re 76 Os 77 Ir 78 Pt 79 Au 80 Hg 81 Tl 82 Pb 83 Bi 84 Po 85 At 86 Rn
87 Fr 88 Ra 103 Lr 104 Rf 105 Db 106 Sg 107 Bh 108 Hs 109 Mt 110 111 112 113 114 115 116 117 118
lanthanide series 57 La 58 Ce 59 Pr 60 Nd 61 Pm 62 Sm 63 Eu 64 Gd 65 Tb 66 Dy 67 Ho 68 Er 69 Tm 70 Yb
actinide series 89 Ac 90 Th 91 Pa 92 U 93 Np 94 Pu 95 Am 96 Cm 97 Bk 98 Cf 99 Es 100 Fm 101 Md 102 No

and there are 7 naturally occurring isotopes; ^{147}Sm, ^{148}Sm, and ^{149}Sm are radioactive and emit α particles.

Samarium oxide is pale yellow, is readily soluble in most acids, and gives topaz-yellow salts in solutions. Samarium has found rather limited use in the ceramic industry, and it is used as a catalyst for certain organic reactions. One of its isotopes has a very high cross section for the capture of neutrons, and therefore there has been some interest in samarium in the atomic industry for use as control rods and nuclear poisons. *See* LANTHANUM; RARE-EARTH ELEMENTS. [F.H.Sp.]

Sampled-data control system An information system for controlling a process, or for transforming information to meet desired requirements, in which the signals or data in one or more components of the system appear in the form of a sequence of pulses or codes. A control system which incorporates sampling and usually a digital controller or digital processor is a sampled-data control system. Sampled-data controls find a wide range of both military and civilian applications, such as missile guidance, ground-controlled interception, satellite launch, energy processing, machine tool control, assembly-line automation, and laboratory instrumentation. With the advent of inexpensive micro- and minicomputers, most control systems designed today are various versions of sampled-data control systems.

The operation which converts continuous data into discrete and intermittent form is often referred to as sampling. Usually the sampled information is expressed in digital form for digital control computation. In this case the sampling takes the form of analog-to-digital conversion. Such operation may be performed by an electronic scanning device, a digital data link, or a mechanical switching device. The digital controller can be realized as a computer program describing a control algorithm if a digital computer is available for on-line operation, or a microprocessor implementing a control procedure if a microcomputer is incorporated into the control system. The output of the digital controller is a digital signal which is converted to a continuous control signal by a digital-to-analog converter. In some applications the digital-to-analog converter is an integral part of the process. The main purpose of the digital controller is to generate data or codes for controlling the process so that the performance requirements of the control system may be fulfilled. *See* ANALOG-TO-DIGITAL CONVERTER; DIGITAL COMPUTER; DIGITAL-TO-ANALOG CONVERTER; MICROCOMPUTER; MICROPROCESSOR. [J.T.T.]

Sand A loose material consisting of small mineral particles, or rock and mineral particles, distinguishable by the naked eye. Most sands are formed by natural agencies. Many deposits of such sands contain clay and silt in varying amounts; some deposits contain pebbles. The mineral composition of sands varies as does the size of the grains composing them. Sands are widely distributed and have many industrial uses.

The term sand also is applied, especially commercially, to small mineral or rock particles produced by crushing larger materials; for example, limestone sand made by crushing limestone, slag sand from slag, or sand made by crushing quartzite. Various granular materials, not necessarily of inorganic composition, likewise may be called sand because they consist of sand-size particles.

Natural sands result primarily from the disintegration of rocks by weathering or erosion. Streams and the waves and currents of lakes and oceans are major agencies eroding rock into sand. The grinding action of glaciers is another important sand-producing agency.

Some sand deposits are formed in place by the weathering of rocks, such as sandstone. Others are the result of the sorting out and concentration of sand from particulate material (composed of particles of various sizes) by the running water of streams or by the waves and currents of lakes and seas. Wind also concentrates sand from certain materials. Deposits of sand accumulate as bars in rivers and streams, in river deltas, as beaches and bars of lakes or seas, and as dunes built up by wind. *See* DUNE. [J.E.L.]

Sand dollar An echinoderm belonging to the order Clypeasteroida in the class Echinoidea. Sand dollars have a flat, disk-shaped body, with the mouth in a mid-ventral position and with the anus also on the ventral surface. There are several species.

Sand dollars live in sand, on the surface or partly buried, from the low-tide mark to depths of 4800 ft (1460 m). Burrowing and locomotion are assisted by the short spines which cover the body. Sand dollars ingest sand grains covered with diatoms or other algae. *See* CLYPEASTEROIDA. [C.B.C.]

Sandalwood The name applied to any species of the genus *Santalum* of the sandalwood family (Santalaceae). However, the true sandalwood is the hard, close-grained, aromatic heartwood of a parasitic tree, *S. album*, of the Indo-Malayan region. This fragrant wood is used in ornamental carving, cabinet work, and as a source of certain perfumes. The odor of the wood is an insect repellent, and on this account the wood is much used in making boxes and chests. The fragrant wood of a number of species in other families bears the same name, but none of these is the real sandalwood. *See* SANTALALES. [P.D.St./E.L.C.]

Sandstone A variety of sedimentary rocks. There are three major groups of sandstone: terrigenous, carbonate, and pyroclastic. Terrigenous sandstones consist of clasts derived from the erosion of preexisting rocks located outside the depositional basin. They are transported by moving currents of wind and water and are deposited in a wide variety of environments. Carbonate sandstones consist mainly of fragments of calcium carbonate in the form of skeletal debris, oolites or other layered carbonate pellets, and locally derived (intraformational) carbonate rock debris (intraclasts). Most carbonate sands are deposited in marine environments and are conventionally classified and analyzed as limestones. Pyroclastic sandstones are composed of rock fragments which were produced directly by explosive volcanic activity. They are volumetrically the least important sandstone variety. *See* LIMESTONE.

Sandstones consist of two major components: the framework fraction composed of all sand-sized clasts, regardless of composition, and the interstitial areas between the grains. In both modern sands and ancient sandstones, these interstices may be empty voids. However, in most ancient sandstones the interstices are commonly filled with either of two essentially mutually exclusive materials: crystalline chemical cement (usually silica or calcium carbonate) or fine-grained matrix. This dichotomy produces the two principal sandstone groups: arenites and wackes.

Sandstones represent between 10 and 20% of the total volume of sedimentary rocks in the Earth's crust. Most sandstone is found within the continental blocks rather than in ocean basins, and most sandstone in continental blocks occurs within the mobile mountain belts rather than in the stable cratons. *See* SEDIMENTARY ROCKS. [F.L.S.]

Sanitary engineering A specialty field generally developed in civil engineering but not limited to that branch. The sanitary engineer has the responsibility (1) to conceive, design, appraise, direct and manage engineering works and projects developed, as a whole or in part, for the protection and promotion of the public health, particularly as it relates to improvement of the environment, and (2) to investigate and correct engineering works and other projects that are capable of injury to the public health by being or becoming faulty in conception, design, direction, or management.

Sanitary engineering practice includes surveys, reports, designs, reviews, management, operation and investigation of works or programs for (1) water supply, treatment and distribution; (2) sewage collection, treatment and disposal; (3) control of pollution in surface and underground waters; (4) collection, treatment, and disposal of refuse; (5) sanitary handling of milk and food; (6) housing and institutional sanitation; (7) rodent and insect control; (8) recreational place sanitation; (9) control of atmospheric pollution and air quality in both the general air of communities and in industrial work spaces; (10) control of radiation hazards exposure; and (11) other environmental factors affecting health, comfort, safety, and well-being of people. *See* AIR POLLUTION; RADIATION BIOLOGY; SEWAGE; SEWAGE COLLECTION SYSTEMS; SEWAGE DISPOSAL; SEWAGE TREATMENT; WATER POLLUTION; WATER SUPPLY ENGINEERING. [W.T.I.]

Santalales An order of flowering plants, division Magnoliophyta (Angiospermae), in the subclass Rosidae of the class Magnoliopsida (dicotyledons). The order consists of 10 families and about 2200 species. The largest families are the Loranthaceae (about 1400 species), Santalaceae (about 400 species), Olacaceae (about 230 species), and Balanophoraceae (about 100 species). A few of the more primitive members of the Santalales are autotrophic, but otherwise the order is characterized by progressive adaptation to parasitism, accompanied by progressive simplification of the ovules. Some members of the Santalales, such as sandalwood, are rooted in the ground and produce small branch roots which invade and parasitize the roots of other plants. Others, such as mistletoe (*Viscum* and other genera of the Loranthaceae), grow on trees, well above the ground. *See* MAGNOLIOPSIDA; MISTLETOE; ROSIDAE; SANDALWOOD. [A.Cr.]

Sapindales An order of flowering plants, division Magnoliophyta (Angiospermae), in the subclass Rosidae of the class Magnoliopsida (dicotyledons). The order consists of 17 families and about 6700 species. The largest families are the Rutaceae (about 1600 species), Sapindaceae (1500 species), Meliaceae (about 1400 species), Anacardiaceae (about 600 species), and Burseraceae (about 600 species). Most of the Sapindales are woody plants, with compound or lobed leaves and polypetalous, hypogynous to perigynous flowers with one or two sets of stamens and only one or two ovules in each locule of the ovary. *See* BUCKEYE; CASHEW; CITRON; GRAPEFRUIT; KUMQUAT; LEMON; LIGNUMVITAE; LIME (BOTANY); MAGNOLIOPSIDA; MAHOGANY; MANDARIN; MAPLE; ORANGE; PISTACHIO; POISONOUS PLANTS; QUEBRACHO; ROSIDAE; TANGERINE. [A.Cr.]

Sapphire The name given to all gem varieties of the mineral corundum, except those that have medium to dark tones of red that characterize ruby. Although the name sapphire is most commonly associated with the blue variety, there are many other colors of gem corundum to which sapphire is applied correctly; these include yellow, brown, green, pink, orange, purple, colorless, and black. Sapphire has a hardness of 9, a specific gravity near 4.00, and refractive indices of 1.76–1.77. Asterism, the star effect, is the result of reflections from tiny, lustrous, needlelike inclusions of the mineral rutile, plus a domed form of cutting. *See* CORUNDUM; GEM; RUBY; RUTILE. [R.T.L.]

Sapropel A mud, slime, or ooze deposited in more or less open water. Sapropel may vary widely in composition depending upon relative contributions from decomposing substances derived from plants and animals. Hydrogen sulfide, produced during the initial biochemical degradation of these substances, promotes preservation of the more resistant parts of the organisms. Most marine sapropelic deposits contain no appreciable contribution from humic substances of terrestrial origin, but certain carbonaceous marine shales appear to contain some humic matter that was part of the original sapropelic deposit. Metamorphism of a sapropelic deposit leads to such products as torbanite, oil shales, and asphaltites. *See* ASPHALT AND ASPHALTITE; OIL SHALE; TORBANITE. [I.A.B.]

Sarcodina A superclass of Protozoa (in the subphylum Sarcomastigophora) in which movement involves protoplasmic flow, often with recognizable pseudopodia. Gametes may be flagellated, as in certain Foraminiferida. Most species are floating or creeping; a few are sessile. The pellicle is relatively thin, and the body is apt to be plastic unless restrained by skeletal structures. Sarcodina live in fresh, brackish, or salt waters; soil or sand; and as endoparasites in animals and plants. A group may be limited to a specific habitat, but many have a rather wide range. Sarcodina include two major classes: Rhizopodea and Actinopodea. *See* ACTINOPODEA; RHIZOPODEA; SARCOMASTIGOPHORA. [R.P.H.]

Sarcomastigophora A subphylum of Protozoa, including those forms that possess flagella or pseudopodia or both. Organisms have a single type of nucleus, except the developmental stages of some Foraminiferida. Sexuality, if present, is syngamy, the fusion of two gametes. Spores typically are not formed. Flagella may be permanent or transient or confined to a certain stage in the life history; this is true also of pseudopodia. Both flagella and pseudopodia may be present at the same time.

Three superclasses are included: (1) Mastigophora, commonly flagellates, contains 19 orders. (2) Opalinata includes 1 order; these organisms were once considered as ciliates, but further research has indicated flagellate kinships. (3) Sarcodina comprises organisms which normally possess pseudopodia and are flagellated only in the developmental stages; 13 orders possess irregularly distributed lobose or filose and branching pseudopodia, while 7 orders have radially distributed axopodia, often with axial filaments. *See* MASTIGOPHORA; OPALINATA; SARCODINA.

Most of the plant flagellates will live in either fresh water or in both fresh and salt water. The zooflagellates are small and are not sufficiently abundant to enter markedly into the food chain. But the parasites and symbionts are of considerable interest economically and theoretically, for example, the trypanosomes and the peculiar xylophagous (wood-eating) symbionts of termites. In the termites these parasites actually digest the wood eaten by the host. Conspicuous in the ecology of marine waters are the dinoflagellates, radiolarians, and acantharians, especially in tropical waters of otherwise low productivity. *See* PROTOZOA. [J.B.L.]

Sarcopterygii A name often employed to unite the lobefin, or crossopterygian, fishes and the lungfishes as a subclass of the Osteichthyes. The older names Amphibioidei and Choanichthyes are equivalent to Sarcopterygii. The structural differences between lobefin fishes and lungfishes are great, and their common ancestry, if any, lies in the Lower Devonian, long antedating the origin of the earliest tetrapods which have since diverged into four classes. It seems best to rank the Crossopterygii (lobefins) and Dipnoi (lungfishes) each equivalent in rank to the subclass Actinopterygii (rayfin fishes), in which the vast majority of fishes are classified. *See* ACTINOPTERYGII; CROSSOPTERYGII; DIPNOI; OSTEICHTHYES. [R.M.B.]

Sarcoptiformes A suborder of the order Acarina. These are minute (0.01–0.06 in. or 0.3–1.5 mm long) globular mites without stigmata, but there may be a tracheal system. The legs may be simple, or enlarged in the male, and may terminate in suckers, claws, or in a modification of both. The chelicerae are normally pincerlike. In normal development these creatures pass through four stages (larva and first, second, and third nymphal stages) before becoming adult. Under certain conditions a dispersal form, the hypopus, may develop between the first and second nymphal stages.

This group can be separated rather easily into the pale, weakly sclerotized Acaridiae and the dark, heavily sclerotized Oribatei. In the Acaridiae are found some of the most serious pests of stored food products; these destroy large quantities of wheat and other grains. The sarcoptids are skin parasites of warm-blooded vertebrates. *Sarcoptes scabiei* and its varieties burrow in the skin of humans, cattle, sheep, goats, camels, horses, and dogs under crowded winter conditions and produce an itching that can lead to secondary infections.

The oribatid mites were at one time regarded as free-living soil or compost organisms of no economic significance. Investigations, however, have shown that they play an important part in the breakdown of organic matter, and that they act as an intermediate host of tapeworms. [H.H.J.N.]

Sarcosporida An order of Protozoa of the class Haplosporea which comprises parasites in skeletal and cardiac muscle of vertebrates. The organisms have a very wide distribution both geographically and in host species, infecting reptiles, birds, and mammals (including marsupials). Humans are occasionally infected by the parasite referred to as *Sarcocystis lindemanni*. It is doubtful whether Sarcosporida found in different hosts are themselves always different, though this has often been assumed. Host specificity is known not to be strict; however, it is unlikely that there is only one species—*S. miescheriana*—as proposed by some investigators.

The life cycle of Sarcocystis is quite typical of the Sporozoa, since sexual stages are lacking and schizogony (multiple fission), though sometimes claimed, apparently does not occur. Instead, reproduction is by binary fission, with the eventual development of cysts (Miescher's tubules) in the muscles. These cysts are relatively large structures, easily visible with the unaided eye, and contain myriads of the minute crescentic spores. Infection of a new host is believed to be by the ingestion of these spores with food and water contaminated by feces from an infected animal.

Cyst morphology is probably more stable than anything else about the parasite. In general, the cyst wall is said to have one or two layers, the outer one being either smooth or provided with spines or villi. The genesis of the wall is in some dispute; some think it is formed by the host as a reaction to the parasite, but the majority opinion is that the parasite itself forms the cyst. The cyst is divided internally into compartments and the outer wall is smooth or rough, depending on the species. The spores are crescent-shaped. *See* HAPLOSPOREA; PROTOZOA; SPOROZOA. [R.D.M.]

Sargasso Sea A region of the west-central North Atlantic Ocean surrounded entirely by water. The western and northern boundaries of the Sargasso are formed by the Gulf Stream. The eastern and southern boundaries are more vague, but generally extend east to about 30°W and south to about 20°N (see illustration). The Sargasso is a clockwise gyre; that is, the currents revolve anticyclonically about its center. The Sargasso supplies much of the deep circulation of the Gulf Stream. The Bermuda Islands rise in the center of the Sargasso. *See* GULF STREAM.

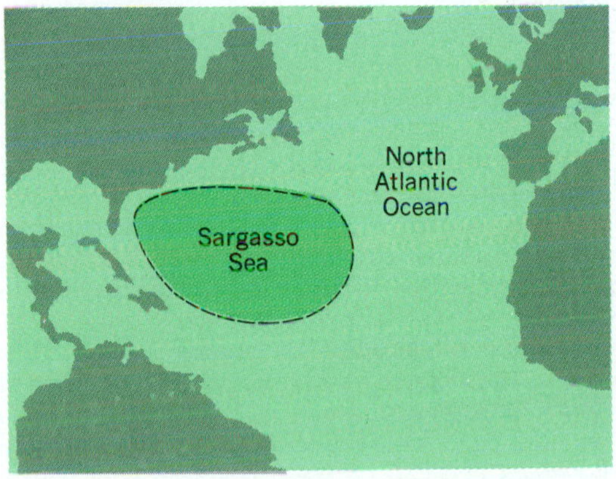

Location of Sargasso Sea.

The Sargasso gets its name from yellow-brown sargassum weed, which collects in the center of the gyre supported by translucent amber floats the size of small grapes scattered over the plants. The color contrast between weeds and water is emphasized by the electric blue color of the water caused by the extremely low plankton productivity. Sargassum is a brown alga which grows attached in shallow water in the Caribbean. Some breaks loose and is transported north to the Sargasso, where it is caught in the gyre, and this is presumably the original source for the floating weed. [J.M.T.]

Sassafras A medium-sized tree, *Sassafras albidum*, of the eastern United States, extending north as far as southern Maine. Sometimes it is only a shrub in the north, but from Pennsylvania southward heights of 90 ft (27 m) or more with diameters of 4–7 ft (1.2–2.1 m) have been reported for this plant. Sassafras is said to live from 700 to 1000 years. It can be recognized by the bright-green color and aromatic odor of the twigs and leaves. The leaves are simple or mitten-shaped (hence a common name "mitten-tree"), or they may have lobes on both sides of the leaf blade. *See* MAGNOLIALES. [A.H.G./K.P.D.]

Satellite (astronomy) A relatively small celestial body moving in orbit about a planet. The Moon is the only natural satellite of Earth. In all, the solar system contains 47 known satellites distributed among the nine planets as follows: Jupiter, 15; Saturn, 21; Uranus, 5; Mars, 2; Neptune, 2; Earth, 1; and (probably) Pluto, 1. Venus and Mercury have no satellites.

The largest satellites reach planetary dimensions, with four (Ganymede and Callisto of Jupiter, Titan of Saturn, and possibly Triton of Neptune) having diameters greater than 3000 mi (5000 km), that is, exceeding that of the planet Mercury. The smallest known satellites, found around Jupiter, Saturn, and Mars, are less than 30 mi (50 km) in diameter, and some are as small as 6 mi (10 km).

The surface gravity of most satellites is too feeble to retain an atmosphere having a density comparable to Earth's. An exception is Titan, with an atmosphere composed primarily of hydrogen and methane and having a density about equal to Earth's. *See* JUPITER; MARS; MOON; NEPTUNE; PLANET; PLUTO; SATURN; URANUS. [J.K.B.]

Satellite (spacecraft) A spacecraft that is in orbit about a planet (usually the Earth). Spacecraft are devices intended for observation, research, or communications in space. Even those spacecraft which are destined to probe the outer reaches of the solar system usually complete at least a fractional revolution around the Earth before being placed into the transfer orbit. Devices such as military missiles which usually follow ballistic paths between the launching site and the target do not orbit the Earth and thus are not referred to as satellites.

The path of rotation of a satellite might be either a circle or an ellipse, but the center of the Earth is at its center or focal point. Unless the satellite is disturbed by the gravitational pull of another body, it remains in a flat plane known as the orbit plane. The Earth revolves through this plane with a period of approximately 24 h. The orbit plane also rotates very slowly. *See* ORBITAL MOTION.

Satellites in low Earth orbits are located several hundred miles above the surface of the Earth. They have typical revolution times (also called periods) of about 90 min. These orbits have fairly short lifetimes (of the order of days, weeks, or sometimes months). Fairly large masses (weights) can be placed into these orbits by simple launch vehicles. Satellites in these orbits are primarily used for observation and military functions. Additional uses for these orbits include aeronomy, as well as the study of radiation belts and other natural phenomena. *See* MILITARY SATELLITES; SCIENTIFIC SATELLITES.

With the proper choice of orbit parameters, the Sun-synchronous (or heliosynchronous) orbit plane requires 1 entire Earth year to make a full revolution. During this time the Earth makes a full revolution about the Sun. As observed from a position on the Sun, the satellite orbit plane remains in the same apparent orientation throughout the year. Since the shadows on the Earth and the illumination of clouds remain more or less constant throughout the year (varying with the normal seasonal changes), this is very valuable in terms of making weather forecasts or observations. *See* METEOROLOGICAL SATELLITES.

Earth-synchronous or geosynchronous orbits have a period that exactly matches the Earth's rotation. However, they may have an inclination, in which case the orbit plane does not lie in the Earth's equatorial plane. While the satellite appears to hover at a fixed longitude over the Earth, it moves up and down along a latitude line in a narrow figure-eight motion.

A geostationary satellite orbit is a special type of geosynchronous orbit in which the inclination is 0°. Since the satellite plane lies in the equatorial plane of the Earth, a perfectly positioned satellite appears to hover directly over the Equator at one particular longitude. An observer positioned on a far-distant planet and looking back at the Earth would see the geostationary satellite rotating with the Earth through space. The geosynchronous and geostationary satellites are typically 22,238 mi (35,788 km) above the surface of the planet or approximately 26,000 mi (42,000 km) from its center. They are used primarily for communications and observation of weather systems. *See* COMMUNICATIONS SATELLITE.

In addition to the research, communications, observation, meteorological (weather), and military satellites, there are radio navigation and Earth survey satellites. The radio navigation satellites provide position information to ships, airplanes, land vehicles, and even backpack receivers. Earth resources satellites provide observation information on potential mineral and petroleum reserves on the Earth. They also help to monitor land usage and pollution. *See* APPLICATIONS SATELLITES; SATELLITE NAVIGATION SYSTEMS. [W.L.M.]

Satellite astronomy The study of astronomical objects using detectors mounted on Earth-orbiting satellites or deep-space probes so that observations unobstructed by the Earth's atmosphere can be made. Many astronomical satellites carrying detectors to record electromagnetic radiation at wavelengths shorter than visible light (ultraviolet, x-, and gamma rays) have been launched. In general, one or two wavelength regions have been observed from each satellite. Successive satellites incorporated significant developments in operating modes and improvements in detector sensitivity and resolution such that new discoveries were possible.

Ultraviolet astronomy. Since the Earth's atmosphere is opaque to ultraviolet light with wavelengths shorter than about 310 nm, ultraviolet astronomy had to await the space age. The *Orbiting Solar Observatory* (*OSO*) satellites were the first to be devoted primarily to astronomical ultraviolet and x-ray observations. Eight satellites (*OSO 1–8*) were launched, over a 12-year period beginning in 1963. The solar ultraviolet experiments included both low- and high-spectral-resolution spectrometers. The resulting detailed maps of the Sun greatly increased understanding of the temperature versus height profile of the solar atmosphere. Time-resolved spectra and images were also recorded so that the first ultraviolet observations of solar flares were conducted.

Extensive ultraviolet observations of stars and nebulae in the Galaxy but outside the solar system were first conducted with ultraviolet telescopes on the *Orbiting Astronomical Observatory* (*OAO*) satellites. Two of these spacecraft were successfully launched.

The *International Ultraviolet Explorer* (*IUE*), launched in 1978 into a geosynchronous orbit, carried a 45-cm-aperture telescope. *IUE* has been used for a great many observations of objects within the Galaxy, such as galactic x-ray binaries, hot stars, and globular clusters. However, its greater sensitivity has also permitted for the first time ultraviolet observations of extragalactic objects. Active galaxies and nearby bright quasars such as 3C273 have been some of the primary targets. *See* GALAXY, EXTERNAL; QUASAR; ULTRAVIOLET ASTRONOMY.

X-ray astronomy. X-ray astronomy, which can only be done from above the Earth's atmosphere, really came of age with the launch of the first *Small Astronomy Satellite* (*SAS 1*) in 1970, designated *Uhuru*. It carried two proportional-counter x-ray detectors in which a pulse of electric charge proportional to the energy of the incident x-ray photon is detected. *Uhuru* enabled the key discovery of x-ray binary systems in which a collapsed object (neutron star or black hole) accretes gas from the atmosphere of a "normal" companion star.

The *OSO* satellites also carried cosmic x-ray detectors (proportional counters similar to *Uhuru*) and contributed much to the detailed understanding of individual sources. Qualitatively different cosmic x-ray satellite experiments, however, were launched in 1974 and 1975 with the *Astronomical Netherlands Satellite* (*ANS*) and *SAS 3* satellite respectively. *ANS* was the first x-ray observatory: it was a pointed instrument carrying a variety of x-ray detectors that could be operated by an on-board computer in a variety of modes. The most significant discovery made with *ANS* was that of the x-ray burst sources. The *SAS 3* satellite contained a variety of x-ray detectors. It obtained precise source positions which allowed many x-ray sources to be identified optically for the first time.

A major increase in sensitivity for x-ray astronomy was achieved with the *High Energy Astronomical Observatory* (*HEAO*) satellite. The total number of x-ray sources known was quadrupled to some 1500 as a result of *HEAO 1*, and high-quality continuum spectra were obtained for many sources. The increased sensitivity meant that fainter objects (primarily extragalactic) such as active galaxies and quasars could be well observed for the first time. The *HEAO 2* satellite, or *Einstein Observatory*, launched in 1978, provided the first x-ray images of celestial objects and detected thousands of new sources. Perhaps foremost of its discoveries is the finding that even the most distant quasars known (and many previously unknown) are detected as strong x-ray sources and that collectively these may contribute most of the mysterious cosmic x-ray background.

Gamma rays, more energetic than x-rays, still do not penetrate the Earth's atmosphere. The *SAS 2*, launched in 1972, established the existence of gamma-ray point sources (such as the Crab and Vela pulsars) at energies of about 100 MeV. The detector used was a digitized spark chamber in which gamma rays are detected by the secondary electron pair they produce upon interacting in the detector. A similar but much longer-lived (and slightly more sensitive) spark chamber gamma-ray telescope was launched in 1975 by the Europeans as the *COS-B* mission. *See* GAMMA-RAY ASTRONOMY. [J.E.Gr.]

Hubble Space Telescope. The Hubble Space Telescope is a large telescope operated in orbit above the disturbing effects of the Earth's atmosphere. It was placed in orbit by the space shuttle on April 24, 1990. The telescope provides broad wavelength coverage and can be used to collect light with wavelengths from 115 nanometers in the ultraviolet, through the optical (or visible) portion of the spectrum, and on to wavelengths of 1 millimeter in the far infrared. With its 8-ft (2.4-m) primary mirror diameter, it is the largest astronomical telescope ever placed above the atmosphere. The Hubble Space Telescope is a long-lived mission. Periodic visits by the shuttle allow in-orbit replacement of failed or obsolete components.

The telescope optics are of the Ritchey-Chrétien type. Light enters the telescope through the aperture door, strikes the 8-ft-diameter (2.4-m) primary mirror, and is reflected to the secondary mirror where it is reflected again, back through a hole, to a focus behind the primary mirror. At the focal plane, the light is shared among at most eight instruments, and only four may be operating simultaneously. The initial set of scientific instruments includes two cameras, two spectrographs, and a photometer, thus forming a powerful complement of instruments capable of carrying out most types of observations performed by ground-based observatories.

Above the atmosphere, the Hubble Space Telescope can observe at wavelengths where the atmosphere is opaque or only partially transparent, as in the ultraviolet and the infrared. But by far the most compelling reason to place a telescope above the atmosphere is the greatly improved angular resolution that can be achieved. Under good conditions, images with a diameter of about 1 arc-second may be formed with ground-based telescopes. At visible wavelengths, the Space Telescope forms images with a diameter of 0.07 arc-second. When the Space Telescope observes a faint star, the area of the image occupied by the star is 50 times smaller and includes 100 times less background light (because the sky is twice as dark in space) than if the star is observed by the same-sized ground-based telescope. This means that the Space Telescope can detect stars 10 times fainter than the faintest stars that can be detected by a similar ground-based telescope. *See* Optical telescope; Scientific satellites; Telescope. [E.J.G.]

Satellite meteorology That branch of meteorological science where use of sensing elements on meteorological satellites plays a role in definition of the past and present state of the atmosphere. Meteorological satellites can measure radiation in a wide spectrum of electromagnetic wavelengths, providing the meteorologist with a supplemental source of data including cloud imagery from visual sensors and ground surface, cloud, and atmospheric layer temperatures, and water vapor concentration from infrared sensors. *See* Meteorological satellites.

Through the use of sensors, satellite data provide measurements of phenomena from the largest-scale global heat and energy budgets down to details of individual thunderstorms. Having both polar orbiting and geosynchronous satellites maintains coverage over a specific location on Earth at time intervals from 5 min to 12 h. Satellites can gather accurate data on the fundamental source of energy of the Earth's atmosphere as defined by the radiation budget. In addition, satellites can closely monitor land surface and sea temperatures and allow recognition of anomalies. Satellite monitoring of polar ice packs helps identify important seasonal variations. Ice caps and areas where vegetation is decreasing (owing to drought or human exploitation) modify the albedo (fraction of incident radiation reflected) and influence the effective solar radiation reaching the Earth. Superimposed on the great atmospheric circulations are the transient eddies or cyclones which ultimately maintain the heat balance of the Earth by moving warm air northward and cold air southward. Satellites play a great role in monitoring these eddies, which are the major weather-producing phenomena over most of the globe. *See* Front; Jet stream.

The greatest gain made with the introduction of weather satellites was in early detection, positioning, and monitoring of the strength of tropical storms (hurricanes, typhoons, and so forth). Lack of conventional meteorological data over the tropics (particularly the oceanic areas) makes satellite data indispensable for this task. *See* Hurricane; Tropical meteorology.

On a regional scale, most significant weather events experienced by a population—heavy rain or snow, severe thunderstorms, or high winds—are organized by systems that have horizontal dimensions of about 60 mi (100 km). These weather systems often fall between stations of conventional observing networks. Hence, the meteorologist might miss them were it not for satellite sensing. *See* Storm; Thunderstorm.

Although weather radar is the major source of small scale data, individual clouds can be closely watched with respect to structure, height, and location (and hence movement) by frequent satellite scans. Large thunderstorms with tornadic potential often exhibit a characteristic appearance which can be seen from satellites prior to tornado development. Rainfall estimates derived from satellite imagery, although crude, can guide the forecaster in issuing quantitative precipitation forecasts. Empirical quantitative methods under development show it may be possible to estimate a rainfall rate. *See* Applications satellites; Hail; Meteorology; Precipitation (meteorology); Tornado; Weather forecasting and prediction. [J.A.MacG.]

Satellite navigation systems Electronic navigation systems employing artificial satellites as signal sources, signal relays, or position references. Earth satellites can provide the advantages of rectilinear (straight-line) propagation between distant points on the surface of the Earth. The satellites' position in space can be determined so accurately that they may be used as position references in systems for precision navigation. *See* Electronic navigation systems.

NNSS. There is only one satellite navigation system now in operation, the U.S. Navy's Navigation Satellite System (NNSS). Developed initially for ballistic missile submarines for use in updating their inertial navigation system, NNSS has been extended for use by ships. The Navy's system provides high-accuracy fixes several times a day at every point on the Earth. It is passive and uses low-orbit satellites and the Doppler technique. *See* Doppler effect.

NNSS satellites are at an altitude of about 500 nautical miles (900 km), where they circle the world in 100 min. Their orbit planes are inclined 70° from the Earth's equatorial plane, so that worldwide service is available. Accuracy of NNSS is approximately 0.1 nautical mile (0.2 km) when ionospheric corrections are applied. Accuracy is reduced for fast-moving vehicles such as aircraft unless their velocity is known accurately, because their own motion introduces Doppler shift on the received signals.

Global Positioning System. The Global Positioning System (GPS) when fully implemented will be a universal positioning or navigation system that will provide for the first time three-dimensional position accuracies to 10 m (30 ft), velocity to an accuracy of 0. 03 m/s (0.1 ft/s), and time to atomic clock accuracy. This information will be available anywhere on Earth. A total of 18–24 satellites, called NAVSTAR, will be placed in orbit in the fully implemented system. Master stations and satellite monitor stations control the system and transmit data to the satellites. It is anticipated that user equipment will be installed on airplanes, ships, and ground vehicles and will also be available as a portable navigation set.

The basis for the GPS accurate position (and time) determination is the precise measurement of the transit time of the radiated signals from 4 satellites from the constellation. Since accurate time is essential, the satellites carry atomic clocks. To use the GPS and obtain the highest accuracy, great care must be taken to model and correct any errors in the received time delay such as clock drift and propagation delay. The satellites are continuously monitored, and error corrections are transmitted to the individual NAVSTAR satellites for inclusion in the satellite-generated signal so that the receiver can automatically correct for position and time.

The GPS is designed to provide precision velocity measurements to the user by measuring the Doppler shift in the satellite-transmitted carrier frequency. The accuracy of this measure is primarily influenced by the dynamics of the host vehicle of the receiver.

Satellites will be in 12 h orbits. The orbital inclination is 63°, and the satellites will be divided into three groups of eight, each group in a separate orbital plane, staggered by 120°. This arrangement permits continuous four-satellite coverage of the entire Earth. New satellites will be added as older satellites wear out after 5–7 years of service life. [L.F.]

Saturable reactor An iron-core inductor in which the effective inductance is changed by varying the permeability of the core. Saturable-core reactors are used to control large alternating currents where rheostats are impractical. Theater light dimmers often employ saturable reactors.

In the illustration of two types of saturable-core reactors, illustration *a* shows two separate cores, while in illustration *b* a three-legged core is formed by placing two two-legged cores together. The load winding, connected in series with the load, carries the alternating current and acts as an inductive element. The control winding carries a direct current of adjustable magnitude, which can saturate the magnetic core.

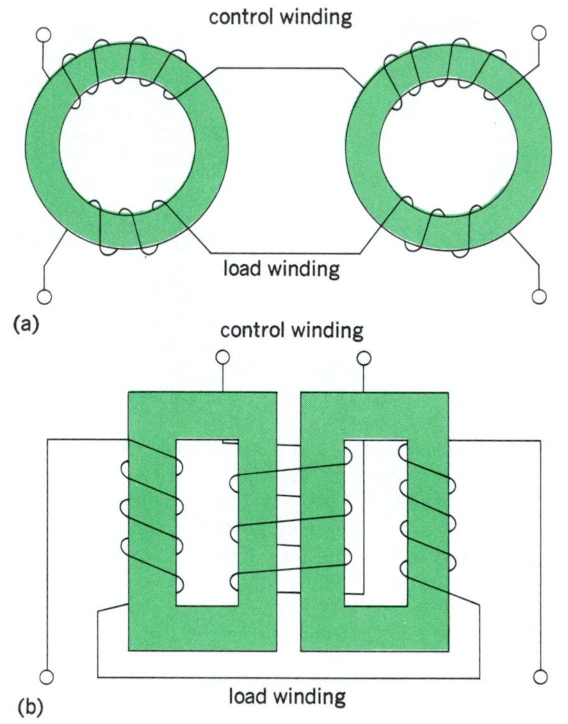

Typical construction of saturable-core reactors. (*a*) Two separate cores. (*b*) Three-legged cores.

Reducing the magnitude of the control current reduces the intensity of saturation. This increases the reactance of the load winding. As the reactance increases, the voltage drop in the load winding increases and causes a reduction in the magnitude of the voltage applied to the load. See INDUCTANCE; MAGNETIC PERMEABILITY; MAGNETIZATION. [W.S.P.]

Saturation The condition in which, after a sufficient increase in a causal force, further increase in the force produces no additional increase in the resultant effect. Many natural phenomena display saturation. For example, after a magnetizing force becomes sufficiently strong, further increase in the force produces no additional magnetization in a magnetic circuit; all the magnetic domains have been aligned, and the magnetic material is saturated. *See* MAGNETIC MATERIALS.

After a sponge has absorbed all the liquid it can hold, it is saturated. In thermionic vacuum tubes thermal saturation is reached when further increase in cathode temperature produces no (or negligible) increase in cathode current; anode saturation is reached when further increase in plate voltage produces substantially no increase in anode current. *See* DISTORTION (ELECTRONIC CIRCUITS): SATURATION CURRENT; VACUUM TUBE.

In colorimetry the purer a color is, the higher its saturation. Radiation from a color of low saturation contains frequencies throughout much of the visible spectrum. *See* COLORIMETRY. [F.H.R.]

Saturation current A term having a variety of specific applications but generally meaning the maximum current which can be obtained under certain conditions.

In a simple two-element vacuum tube, it refers to either the space-charge-limited current on one hand or the temperature-limited current on the other. In the first case, further increase in filament temperature produces no significant increase in anode current, whereas in the latter a further increase in voltage produces only a relatively small increase in current. *See* VACUUM TUBE.

In a gaseous-discharge device, the saturation current is the maximum current which can be obtained for a given mode of discharge. Attempts to increase the current result in a different type of discharge. *See* ELECTRICAL CONDUCTION IN GASES.

A third case is that of a semiconductor. Here again the saturation current is that maximum current which just precedes a change in conduction mode. *See* SEMICONDUCTOR. [G.H.M.]

Saturn The second-largest planet in the solar system and the sixth in order of distance to the Sun. The outermost planet known prior to 1781, Saturn is surrounded by a beautiful system of rings. Saturn is also the only planet that has a satellite (Titan) with a dense atmosphere. This distant planetary system has been visited by three NASA spacecraft: a preliminary survey by *Pioneer 11* in September 1979, and a more sophisticated reconnaissance by *Voyager 1* in November 1980 and *Voyager 2* in August 1981.

Saturn makes one revolution about the Sun in 29.46 years. The equatorial diameter of Saturn is about 75,000 mi (120,800 km), and the polar diameter about 67,700 mi (109,000 km). The volume is 769 (Earth = 1) with a few percent uncertainty. The mass is about 95.2 (Earth = 1) or 1/3500 (Sun = 1). The mean density is 0.68 g/cm^3, the lowest mean density of all the planets. The rotation axis of both the planet and the rings is inclined 26°45′ to the perpendicular to the orbital plane. The disk of Saturn is much more homogeneous than that of Jupiter. There is no feature comparable to the Great Red Spot, and the contrast of the features that are visible is very low (Fig. 1).

The optical spectrum of Saturn is characterized by strong absorption bands of methane (CH_4) and by much weaker bands of ammonia (NH_3). Absorption lines of molecular hydrogen (H_2) have also been detected.

The temperature the planet should assume in response to solar heating is calculated to be about −323°F (76 K), somewhat lower than the measured value of −294°F (92 K). This suggests that Saturn has an internal heat source of roughly the same magnitude as that on Jupiter. As in the case of Jupiter, a thermal inversion exists in the upper atmosphere. The inversion region is well above the main cloud layer, which is thought to consist primarily of frozen ammonia crystals, with an admix-

Fig. 1. Saturn, viewed from *Voyager 1*. The soft, velvety appearance of the low-contrast banded structure is due to scattering by a haze layer above the planet's cloud deck. (*NASA*)

ture of some other substances to provide the yellowish color sometimes observed in the equatorial zone.

The theoretical models for the internal structure of Saturn are similar to those for Jupiter, that is, a dense core surrounded by hydrogen compressed to a metallic state which gradually merges into an extremely deep atmosphere. The fact that the two planets radiate comparable amounts of energy despite their difference in size means that smaller Saturn must have some additional energy source besides gravitational contraction. The existence of a magnetic field and belts of trapped electrons has been deduced from observations of nonthermal radiation and mapped out in detail by the Pioneer and Voyager spacecraft.

Jupiter and Saturn are relatively similar bodies. Both seem to have compositions virtually identical with that of the Sun and the other stars—rich in hydrogen and helium. In that sense, they may represent the primitive material from which the entire solar system was formed, whereas the other planets

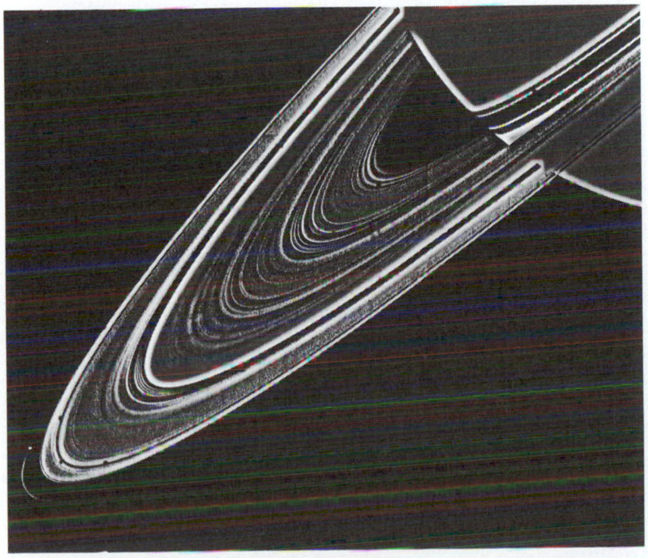

Fig. 2. Saturn's rings, viewed from *Voyager 1*. Approximately 95 individual concentric features are visible. One of the satellites discovered by *Voyager 1* is visible just inside the narrow F ring. (*NASA*)

have undergone fractionation processes resulting in the loss of most of the light gases. *See* PLANETARY PHYSICS.

The most remarkable feature associated with Saturn is the complex ring system that surrounds the planet (Fig. 2). The system is divided into six main regions, designated A through F. Both theory and observations prove that the ring system is made up of myriad separate particles that move independently in flat, mostly circular orbits in Saturn's equatorial plane. Periodic perturbations by the major satellites are responsible, in part, for the main divisions of Saturn's rings.

Saturn has 21 known satellites. The largest and brightest, Titan, was discovered by C. Huygens in 1655 and is visible with small telescopes; the other satellites are much fainter. Titan's mean apparent diameter corresponds to a linear diameter of approximately 3600 mi (5800 km). But this diameter refers to the satellite's atmosphere, which is filled with a dense aerosol produced photochemically by incident sunlight. The solid surface of Titan has a diameter of 3182 mi (5120 km), making this satellite larger than Mercury but smaller than Jupiter's giant Ganymede. This object contains a large fraction of icy material and is thus quite different from the Moon or the inner planets in composition. Furthermore, it is large and cold enough to retain an atmosphere of gases with relatively high molecular weights. The main constituent of this atmosphere is molecular nitrogen. The surface of Titan is so cold (−292°F or 93 K) that methane can liquefy. Hence, this substance may play the same role on Titan as water does on Earth. *See* PLANET; SATELLITE (ASTRONOMY). [T.C.O.]

Saurischia One of two orders of extinct reptiles popularly known as dinosaurs, the other being the Ornithischia. The order is distinguished by the triradiate shape of the pelvis. The order originated from thecodont reptiles by Middle Triassic times and is classified in the subclass Archosauria. Saurischian bones have been found on all continents except Antarctica, and in rock strata ranging from Middle Triassic to latest Cretaceous in age. *See* REPTILIA.

Most recent classifications recognize two suborders, the Sauropodomorpha and the Theropoda. The latter includes all carnivorous dinosaurs, and the former consists of the *Brontosaurus*-like animals (sauropods), together with their ancestral stock, the prosauropods. The Prosauropoda includes the oldest known saurischians, ranging from Middle Triassic time to the close of the Triassic Period. Like their ancestral thecodont stock, many show pronounced anatomical tendencies toward bipedal posture. Most, like *Plateosaurus*, appear to have been herbivores, but fragmentary evidence indicates that some may have been carnivorous. In general, prosauropods were of moderate size, up to 15–20 ft (4.5–6 m) long, and less than a ton (1 short ton = 0.9 metric ton) in weight.

The saurischians became extinct at the end of the Cretaceous, due to unknown causes. Most saurischians died out, leaving no descendants, but studies indicate that birds, via *Archaeopteryx*, probably are direct descendants of a small theropod dinosaur. *See* ARCHAEORNITHES; ARCHOSAURIA; DINOSAUR; ORNITHISCHIA; THECODONTIA. [J.H.O.]

Sauropterygia An order of Mesozoic reptiles (subclass Euryapsida) that are, without exception, adapted to the marine environment. The order includes the closely related nothosaurs and plesiosaurs. These reptiles, along with the ichthyosaurs, played a significant role as predators within the marine animal community of the Mesozoic Era. *See* REPTILIA.

The Nothosauria are the relatively generalized stem group from which the plesiosaurs evolved. With the exception of a single New World species and a record from Japan, nothosaurs are known primarily from Europe and the Near East (Israel) in rocks of Triassic age. The nothosaurs are notably diverse in the mode and degree of secondary aquatic modification. The directions of aquatic specialization involve shortening, or more often lengthening, of the neck; enlargement of the orbits or the temporal fenestrae; and reduction, or more commonly increase, in the number of phalanges in manus (hand), pes (foot), or both. A feature of considerable evolutionary significance in the light of plesiosaurian differentiation is the great individual variability in the number of presacral vertebrae (32–42) in *Pachypleurosaurus edwardsi*. *See* SKELETAL SYSTEM.

The Plesiosauria are the successful, compact, and highly specialized offshoot of the nothosaurs that attained worldwide distribution. Two major trends of plesiosaur evolution may be discerned since the Early Jurassic: In the one group there was a tendency toward a shortening of the neck from 27 to 13 vertebrae and an increase in skull size: in the other group the opposite trend led to forms of bizarre body proportions, for example, *Elasmosaurus*, with a neck containing 76 vertebrae. The plesiosaurs were carnivorous. *See* EURYAPSIDA; REPTILIA.

[R.Z.]

Savanna The term savanna was originally used to describe a tropical grassland with more or less scattered dense tree areas. This vegetation type is very abundant in tropical and subtropical areas, primarily because of climatic factors. The modern definition of savanna includes a variety of physiognomically or environmentally similar vegetation types in tropical and extratropical regions. The physiognomically savannalike extratropical vegetation types (forest tundra, forest steppe, and everglades) differ greatly in environment and species composition.

In the widest sense savanna includes a range of vegetation zones from tropical savannas with vegetation types such as the savanna woodlands to tropical grassland and thornbush. In the extratropical regions it includes the "temperate" and "cold savanna" vegetation types known under such names as taiga, forest tundra, or glades. *See* GRASSLAND ECOSYSTEM; TAIGA; TUNDRA.

[H.Li.]

Savory A herb of the mint family in the genus *Satureja*. There are more than 100 species, but only *S. hortensis* (summer savory) and *S. montana* (winter savory) are grown for flavoring purposes. *See* LAMIALES.

Summer savory, an annual herb, is characterized by long thin wiry stems with long internodes between small leaves. Winter savory is a perennial in most climates and, unlike summer savory, will tolerate some freezing weather. It can become woody after one or two growing seasons.

Savory is indigenous to areas surrounding the Mediterranean Sea. *Satureja montana* occurs wild from North Africa to as far north as Russia but is little cultivated. *Satureja hortensis*, which is widely cultivated, is native to Europe. Both types of savory are harvested or cut two or three times a year, after which the leaves are dehydrated and separated from the stems to be used as a spice in various foods, including poultry seasoning and beans. *See* SPICE AND FLAVORING.

[S.Kir.]

Sawing The parting of material by using metal disks, blades, bands, or abrasive disks as the cutting tools. Sawing a piece from stock for further machining is called cutoff sawing, while shaping or forming a piece is referred to as contour sawing.

Machine sawing of metal is performed by five types of saws or processes: hacksawing, band sawing, cold sawing, friction sawing, and abrasive sawing. Hacksaws are used principally as cutoff tools. The toothed blade, held in tension, is reciprocated across the workpiece. Band saws cut rapidly and are suited for either cutoff or contour sawing. They are basically a flexible endless band of steel running over pulleys or wheels. Cold sawing is principally a cutoff operation. The blade is a circular disk with cutting teeth on its periphery. Friction sawing is a rapid process used to cut steel as well as certain plastics. Cutting is done as the high-speed blade wipes the metal from the kerf after softening it with frictional heat. Abrasive sawing is a cutoff process using thin rubber or bakelite bonded abrasive disks. In addition to steel, other materials such as nonferrous metals, ceramics, glass, certain plastics, and hard rubber are cut by this method. *See* WOODWORKING.

[A.H.T.]

Sawtooth-wave generator A device which generates a current or voltage waveform whose magnitude is a continuously increasing function of time for a fixed interval, and then repeats the sequence periodically. The most widely used sawtooth waveform is ideally a linear function of time during the forward or rising interval and appears as shown in Fig. 1*a*, with the total period T made up of the active forward interval T_f, a retrace interval T_r, and an inactive interval T_i. Often it is desirable to make the two intervals T_r and T_i as small as possible, with T*i* often being zero.

Electronic circuits can generate only an approximation of the idealized waveform which, as a result, tends to have deviations which are often of the type shown in Fig. 1*b*.

Sawtooth waveforms are used as time-base elements or sweep generators and in time-delay and time-measuring equipment. *See* SWEEP GENERATOR.

The complexity of electronic circuitry required to generate linear sawtooth voltage waves depends upon the accuracy to which such generation is specified. An approximate linear sawtooth may be generated by a dc voltage source, a series *RC* circuit, and a switch which has a small but nonzero resistance when closed. This elementary sawtooth generator is shown schematically in Fig. 2.

A periodic sawtooth waveform can be generated by using an astable relaxation oscillator as a switch in the above circuit. When used as switches in sawtooth generators, relaxation oscillators may be synchronized with external pulses to maintain an accurately controlled period. *See* RELAXATION OSCILLATOR.

If the switch of Fig. 2 is the switch in a synchronous or keyed clamp each interval of the total period of the waveform may be controlled directly from an external source of pulses. *See* CLAMPING CIRCUIT.

Either a positive-going or a negative-going sawtooth may be generated by making V_B either positive or negative and replac-

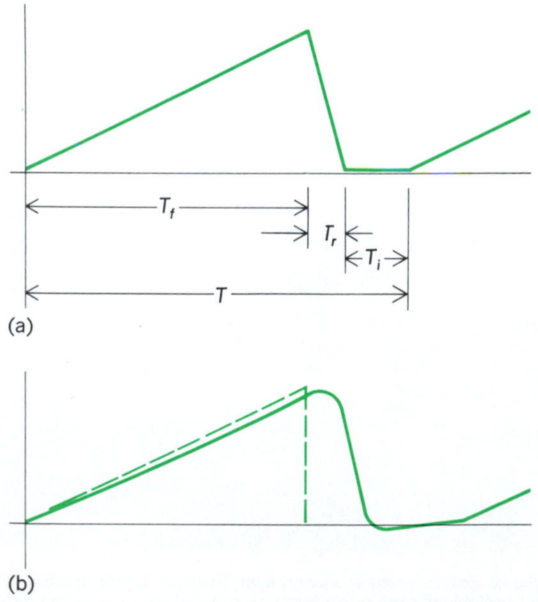

Fig. 1. Sawtooth wave. (*a*) An ideal linear sawtooth. (*b*) Approximate sawtooth generated by actual circuits.

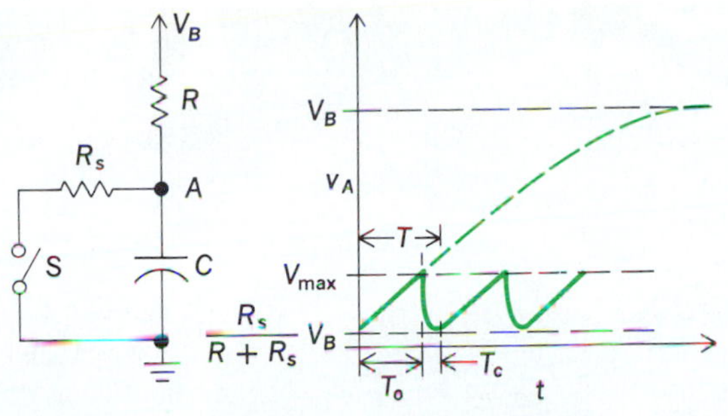

Fig. 2. Elements of sawtooth sweep circuit. V_B = voltage above ground at the point indicated; R and R_s are resistances; C is a capacitor. Graph plots voltage V_A at point A versus time T when switch S is opened for time interval T_0 and closed for time interval T_c. Rise of voltage toward V_B is interrupted at value V_{max} by closing of switch.

ing the clamp with a bidirectional clamp. If the potential V_B is variable at a rate which is slow compared to the periodicity of the waveform, a succession of sawtooth waveforms of varying amplitude will be generated. One particular application of such a circuit would be to generate one of the components of the rotating radial sweep if V_B were to be made to vary sinusoidally with the desired angular modulation. [G.M.G.]

Scabies A severe skin condition caused by sarcoptic mites. *Sarcoptes scabiei* infects a wide range of mammals, including humans, horses, hogs, sheep, dogs, cattle, foxes, and goats. Infection may begin with a few mites and may rapidly increase to a heavy infection within a few weeks. As the mite population becomes large, a hypersensitive condition develops, called dermatitis, which results in intense itching. [R.Su.]

Scalar A term synonymous in mathematics with "real" in real number or real function. The magnitude of a vector two units in length is the real number of scalar 2. The dot or scalar product of two vectors is the product of three real numbers associated with them (the magnitude of the first vector times the magnitude of the second, times the cosine of the angle between them) and is therefore a scalar. If in the functional relationships $S = S(s,y,z)$, $\mathbf{F} = \mathbf{F}(x,y,z)$, S is a real number and $\mathbf{F}$ is a vector, then $S(s,y,z)$ is a scalar function but $\mathbf{F}(x,y,z)$ is a vector function. *See* CALCULUS OF VECTORS. [H.V.C.]

Scale (music) A series of notes arranged from low to high by a specified scheme of intervals, suitable for musical purposes. A musical scale is an arrangement of intervals evolved in the making of music. When a given note of a musical scale is repeated, it need not be given exactly the same frequency; the frequency may be modified significantly in accordance with the artistic requirements of the musical context. Thus it is difficult to be very specific about the "specified scheme of intervals."

The number of musical scales and the intervals they contain are myriad, but most of them have in common the interval called the octave. The musical import of two notes an octave apart is so often the same that the two notes are given the same letter name. *See* OCTAVE.

A diatonic scale is one in which the octave is divided into intervals of two different sizes, five of one and two of the other. If the size of the smaller interval (often called a half-step) is taken as one unit, the sequence of intervals in the major diatonic scale is 2, 2, 1, 2, 2, 2, 1; in the minor diatonic scale (melodic ascending), the sequence is 2, 1, 2, 2, 2, 2, 1. The relative sizes shown by these numbers are approximate; only in equal temperament are the sizes exactly as indicated here. [R.W.Y.]

Scale (zoology) The fundamental unit of the primary scaled integument of vertebrates, of which the epidermal or dermal component may be the more conspicuous or elaborated. Although there is great diversity in detailed structure among vertebrate groups, scaled integuments may be interpreted as the most effective way of producing a physically strong external body surface, with maximum flexibility necessitated by the fundamental pattern of vertebrate locomotion relying on lateral sinusoidal movements of the body.

In the majority of extant fish the epidermis is extremely thin. The term "scale" when used in an ichthyological context usually refers to the relatively large, prominent, dermal ossification, on the outer surface of which lie so-called dental tissues. The latter—enamel or enamellike mineralizations and dentine—are therefore present at the dermoepidermal boundary. The scaled integument of fish shows a definite pattern which is governed by the orientation of the underlying myomeres and by changes in body depth and length. The general evolutionary tendency toward thinning, or size reduction, of integumentary sclerifications seen in various fish lineages is probably associated with weight reduction and increased locomotor efficiency.

Scaled integuments are absent from modern amphibians, although nonoverlapping dermal ossifications are seen in some apodans. It seems probable that the absence of scales in other modern species is associated with the secondary utilization of the integument as a respiratory surface. In most reptilian lineages, there is a ubiquity of scaled integuments, with or without dermal ossifications, but always with elaboration of epidermal tissues. The distribution of keratinaceous protein types varies in different groups of reptiles, and in this respect modern crocodilian scales exactly resemble the leg scales of birds. Scaled integuments prompted the subclass name for lepidosaurs—the tuatara, lizards, and snakes, the last two constituting the order Squamata. Lepidosaurian scales may or may not possess dermal ossifications. Among extant reptiles the extraordinary development of the dermal skeleton in turtles forms the characteristic carapace. [P.F.M.]

Scandium A chemical element, Sc, atomic number 21, and atomic weight 44.956. It is the first transition element of the first long period. Some physical properties of scandium are listed in the table.

Isotopes of scandium include ^{40}Sc and ^{51}Sc and one corresponding to every intermediate value. Except for ^{45}Sc, which occurs naturally, the isotopes occur only when generated by nuclear reactions.

The oxide and other compounds of scandium are used as catalysts in the conversion of acetic acid to acetone, in making propanol, and in converting dicarboxylic acids to ketones and cyclic compounds. Treatment with scandium sulfate solution is an economical means of improving the germination of the seeds of many species of plants.

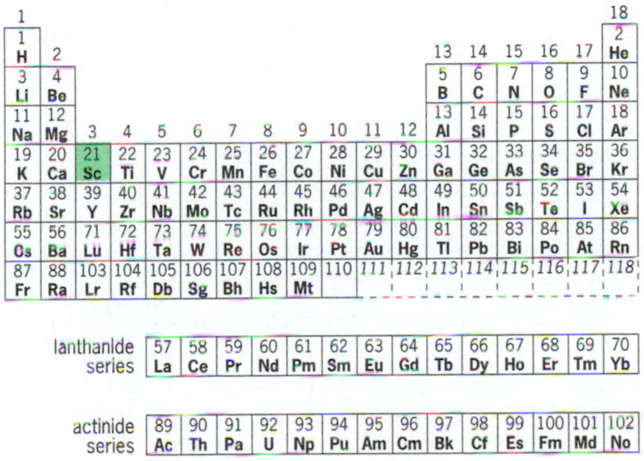

1 18
1 H 2 13 14 15 16 17 2 He
3 Li 4 Be 5 B 6 C 7 N 8 O 9 F 10 Ne
11 Na 12 Mg 3 4 5 6 7 8 9 10 11 12 13 Al 14 Si 15 P 16 S 17 Cl 18 Ar
19 K 20 Ca 21 Sc 22 Ti 23 V 24 Cr 25 Mn 26 Fe 27 Co 28 Ni 29 Cu 30 Zn 31 Ga 32 Ge 33 As 34 Se 35 Br 36 Kr
37 Rb 38 Sr 39 Y 40 Zr 41 Nb 42 Mo 43 Tc 44 Ru 45 Rh 46 Pd 47 Ag 48 Cd 49 In 50 Sn 51 Sb 52 Te 53 I 54 Xe
55 Cs 56 Ba 71 Lu 72 Hf 73 Ta 74 W 75 Re 76 Os 77 Ir 78 Pt 79 Au 80 Hg 81 Tl 82 Pb 83 Bi 84 Po 85 At 86 Rn
87 Fr 88 Ra 103 Lr 104 Rf 105 Db 106 Sg 107 Bh 108 Hs 109 Mt 110 111 112 113 114 115 116 117 118
lanthanide series 57 La 58 Ce 59 Pr 60 Nd 61 Pm 62 Sm 63 Eu 64 Gd 65 Tb 66 Dy 67 Ho 68 Er 69 Tm 70 Yb
actinide series 89 Ac 90 Th 91 Pa 92 U 93 Np 94 Pu 95 Am 96 Cm 97 Bk 98 Cf 99 Es 100 Fm 101 Md 102 No

Physical properties of scandium

Property	Value
Natural atomic weight	44.956
Outer valence electrons	$3d^1 4s^2$
Density	3.0 g/cm^3
Melting point	1538°C (2800°F)
Boiling point	2870°C (5198°F)

Scandium-47 has a satisfactory half-life for use as a tracer, and it can be prepared carrier-free. The presence of 2.5–25 at. % of scandium in the anode increases the voltage, the voltage stability, and the service life of a nickel alkaline storage battery.

The chief scandium ore mineral is thortveitite, which is found in granite pegmatites, and in some ores of tin, of tungsten, and of the rare earths. Scandium is widely distributed in many parts of the world, *See* RARE-EARTH ELEMENTS; TRANSITION ELEMENTS. [W.D.Wi.]

Scanning acoustic microscope A type of microscope based on a sapphire pellet with a thin-film piezoelectric transducer on the back side. This serves to convert electromagnetic energy into acoustic energy in the form of a collimated beam which propagates through the pellet to the front surface. There a spherical cavity filled with the coupling fluid forms an acoustic lens. This lens is ideal—it does not suffer from aberrations—and it brings the acoustic beam into a diffraction-limited focus near the center of curvature of the spherical surface. The wave velocity decreases by a factor of 7.5 in passing through this solid-liquid interface, The strong refraction at the lens surface reduces the spherical aberration below the diffraction limit. The object under examination is placed near the focus. It is then scanned by mechanical techniques, point by point and line by line, in a raster pattern familiar in television display. The cathode-ray tube used to monitor the image has a scan synchronized with the mechanical motion of the object. Thus each point on the monitor screen corresponds to a given point on the object. An electronic signal proportional to the acoustic power reflected from or transmitted through the object is used to modulate the intensity of the electron beam in the monitor. The image on screen which reproduces the "image of acoustic reflectivity" is recorded photographically. *See* ACOUSTIC MICROSCOPE. [C.F.Q.]

Scanning electron microscope An electron microscope that builds up its image as a time sequence of points in a manner similar to that employed in television.

The imaging method of the scanning electron microscope (SEM) allows separation of the two functions of a microscope, localization and information transfer. The SEM utilizes a very fine probing beam of electrons which sweeps over the specimen to emit a variety of radiations. The signal, which is proportional to the amount of radiation leaving an individual point of the specimen at any instant, can be used to modulate the brightness of the beam of the display cathode-ray tube as it rests on the corresponding point of the image. In practice, the points follow one another with great rapidity so that the image of each point becomes an image of a line, and the line in turn can move down the screen so rapidly that the human eye sees a complete image as in television. The image can also be recorded in its entirety by allowing the point-by-point information to build up in sequence on a photographic film (see illustration).

As with all microscopy research, it is very often the preparative methodology that determines the success or failure of the research. A specific requirement of scanning electron microscope preparation is that the material be dried. This preparative step must be done very carefully to minimize surface tension effects. The two methods most often used—critical-point drying and freeze drying—have yielded excellent results even for very fragile biological tissue. Alternatively, the drying step can be avoided completely by observing the specimen while it is still frozen. Interior structure can be revealed by using techniques in which the specimen is broken at a low temperature. *See* ELECTRON MICROSCOPE; MICROTECHNIQUE. [T.L.H.]

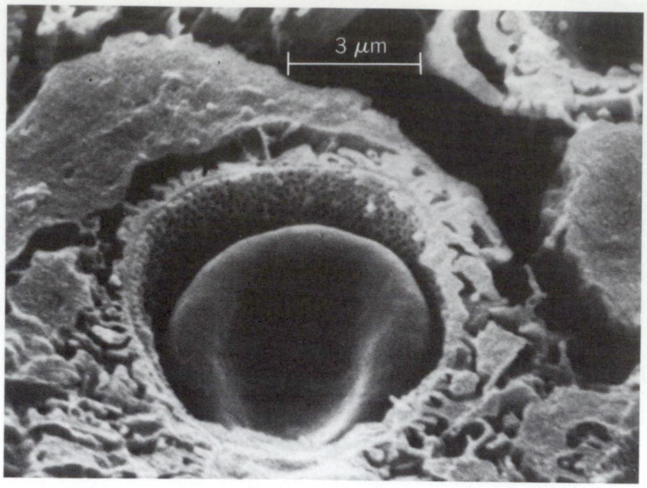

Red blood cell in a capillary of the kidney. Ethanol cryofracture technique was followed by critical-point drying. (*From W. J. Humphreys, B. O. Spurdock, and J. S. Johnson, Critical point drying of ethanol-infiltrated, cryofractured biological specimens for scanning electron microscopy, Proceedings of the Scanning Electron Microscopy Symposium, pp. 275–282, 1974*)

Scanning tunneling microscope A device for imaging the surface of conductors and semiconductors with atomic resolution. Due to its relatively simple construction and operation and its ability to achieve atomic resolution with relative ease, the scanning tunneling microscope (STM) has gained worldwide acceptance by scientists studying surface phenomena.

The operating principles of the scanning tunneling microscope are shown in the illustration. A wire made of tungsten, gold, or some other noble metal is etched electrochemically at one end down to a fine tip whose radius is 0.1 micrometer, or less; it is then attached onto a piezoelectrically scanned (x,y,z) stage. Under a powerful microscope, such as a transmission electron microscope, the tip of the wire typically looks as illustrated, with one atom always protruding farther from the surface. The piezoelectric tripod can be scanned in x, y, or z with a resolution far better than 0.1 nanometer (approximately one-half the diameter of typical atoms) by applying a voltage across its x, y, or z electrode, respectively. A voltage V is applied between the wire tip and the conducting sample to be imaged. As the end atom on the tip approaches to within a few atomic diameters of a sample atom, electrons can tunnel from the tip to the sample (or vice versa) through the potential barrier imposed by the work functions of the tip and sample, thereby generating a current. A constant tunnel current is maintained as the tip is raster-scanned in x and y to record an image. The tunneling image is usually recorded on a computer as variations in the z position (measured through the z-piezoelectric voltage) corresponding to each (x,y) coordinate of the tip. *See* COMPUTER GRAPHICS; ELECTRON MICROSCOPE; PIEZOELECTRICITY; TUNNELING IN SOLIDS; WORK FUNCTION (THERMODYNAMICS).

One application of the scanning tunneling microscope is in the mapping of the atomic reconstructions of crystal surfaces. Other areas of application of the scanning tunneling micro-

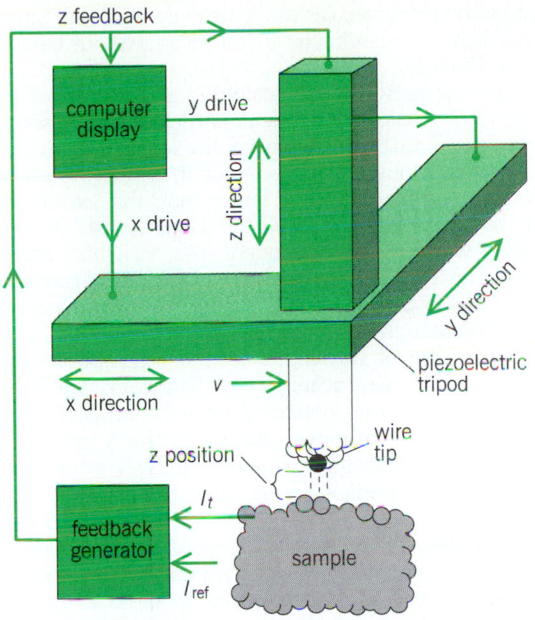

Schematic diagram of the scanning tunneling microscope showing the configuration of the piezoelectric tripod, wire tip, and sample. I_t = tunnel current; I_{ref} = constant reference current.

scope are in the study of electrochemical processes and biological imaging. It is also being used for surface modifications on an atomic scale. [H.K.Wi.]

Scaphopoda A minor class in the phylum Mollusca, comprising two families, Dentaliidae and Siphonodentaliidae, each with few and little-known genera. It is exclusively marine and widely distributed at moderate depths in temperate and tropical seas.

Dentalium, the "elephant's tusk shell," is the best-known example. The tubular cylindrical shell is slightly concave on its dorsal side; it tapers gently posteriorward. The animal lives embedded almost vertically in sand, shell gravel, or gravel, the posterior tip of the shell protruding slightly.

Dentalium burrows through the substratum with its muscular, protrusible foot, the firm pointed tip easily penetrating the deposits. The extended foot becomes anchored in the bottom by anterolateral folds which are inflated by blood pressure. Contraction of the foot then draws the animal forward, and these actions are associated with pulses of water expelled through the posterior mantle orifice. There is a spacious hemocoel, but no obvious heart and no pericardium. The sexes are separate. *See* MOLLUSCA. [R.D.P.]

Scapolite An aluminosilicate mineral. It is commonly found as light-colored, translucent tetragonal prisms. Scapolite is normally white, but many other colors are known, including some used as semiprecious gems resembling amethyst and citrine. The mineral has a Mohs hardness of 5–6. The formula of scapolite is $(Na,Ca)_4(Al,Si)_6SI_6O_{24}(Cl,CO_3,SO_4)$. In nature significant amounts of K and SO_4 substitute for Na and CO_3.

Scapolite is a common mineral in metamorphic rocks, particularly in those which contain calcite. It is found in marbles, gneisses, skarns, and schists. Scapolite probably forms about 0.1% of the Earth's upper crust. It is commonly found as inclusions in igneous rocks derived from deep within the Earth's crust, and probably makes up several percent of the lower crust. *See* SILICATE MINERALS. [D.E.E.]

Scarlet fever An acute contagious disease which is the classic example of infection by the microorganism *Streptococcus hemolyticus*. Fever, sore throat, rash, headache, and vomiting follow 2–7 days after contact with a carrier, whose illness was probably not scarlet fever. The bright-red rash over the body is papular, or rough, with sandpaperlike quality. It blanches on pressure to leave a transient image, as in sunburn. A deeply flushed face with circumoral pallor, or paleness around the mouth, completes the patient's outward appearance. The tongue is aptly described as "strawberry," sometimes "raspberry." There is a firey redness of the throat, and the swollen tonsils are covered with patches of white exudate. The rash fades in a few days to be followed by desquamation, or peeling, while symptoms abate.

Early complications such as visibly swollen neck glands and draining ears mean bacterial spread from the nose and throat. Late complications, such as deafness, rheumatic fever, and nephritis, or kidney disease, appear during the convalescence. *See* RHEUMATIC FEVER. [P.L.B.]

Scattering experiments (atoms and molecules) Experiments in which an incident particle or system of particles, such as an electron, atom, or molecule, is deflected by collision with an atom or a molecule. Such experiments are useful for many reasons: They provide checks on the theory of scattering and yield information on the nature of atomic and molecular forces. They are also important since the experiments can be designed to simulate conditions in planetary atmospheres, to provide information on electric discharges, to assist in the development of gaseous lasers, and to aid in the theory of stellar absorption and of planetary nebulae. *See* ELECTRICAL CONDUCTION IN GASES; LASER; PLANETARY NEBULA; STAR. [H.W.M.]

Scattering experiments (nuclei) Experiments in which beams of particles such as electrons, nucleons, alpha particles and other atomic nuclei, and mesons are deflected by elastic collisions with atomic nuclei. Much is learned from such experiments about the nature of the scattered particle, the scattering center, and the forces acting between them. Scattering experiments, made possible by the construction of high-energy particle accelerators and the development of specialized techniques for detecting the scattered particles, are one of the main sources of information regarding the structure of matter. *See* NUCLEAR STRUCTURE; PARTICLE ACCELERATOR; PARTICLE DETECTOR; SCATTERING MATRIX. [K.A.Er.]

Scattering layer A layer of organisms in the sea which causes sound to scatter and returns echoes. Recordings by sonic devices of echoes from sound scatterers indicate that the scattering organisms are arranged in approximately horizontal layers in the water, usually well above the bottom. The layers are found in both shallow and deep water. [J.B.H.]

Scattering matrix A matrix which expresses the initial state in a scattering experiment in terms of the possible final states, and hence enters the calculation of the probabilities that certain reactions will occur in a collision of two or more particles. The problem is compounded in the relativistic domain, in which particles may be created or destroyed in a collision and the forces between elementary particles are known only approximately. It was therefore suggested by W. Heisenberg in 1943 that the *S* matrix should play a fundamental rather than a subsidiary role in relativistic quantum mechanics. *See* ELEMENTARY PARTICLE; MATRIX MECHANICS; NUCLEAR REACTION; QUANTUM MECHANICS; SCATTERING EXPERIMENTS (NUCLEI).

A Lorentz covariant perturbation theory for the direction calculation of *S* on the basis of quantum field theory was developed in 1948–1949 by J. Schwinger, R. P. Feynman, and

F. J. Dyson. The covariant perturbation techniques have been remarkably successful in quantum electrodynamics; the fine structure constant is small, and provides a natural expansion parameter for the perturbation series. In contrast, the strength of the specifically nuclear forces between elementary particles prevents a meaningful perturbation expansion of S for particle reactions. Much emphasis has consequently been given to Heisenberg's original conjecture that Lorentz invariance and unitarity could be used in part to determine S. It has in fact been possible to develop at least a partial dynamical theory of the S matrix by supplementing the requirements of Lorentz invariance and unitarity by analyticity conditions derived from perturbation theory. Although this dispersion-relation, or S-matrix, theory of strong interactions has been quite successful in some instances, particularly in the study of low-energy baryon-meson scattering, the theoretical basis of the approach remains obscure. *See* DISPERSION RELATIONS; PERTURBATION (QUANTUM MECHANICS); QUANTUM ELECTRODYNAMICS; QUANTUM FIELD THEORY. [L.D.]

Scattering of electromagnetic radiation The process in which energy is removed from a beam of electromagnetic radiation and reemitted with a change in direction, phase, or wavelength. All electromagnetic radiation is subject to scattering by the medium (gas, liquid, or solid) through which it passes.

It has been known since the work of J. Maxwell in the nineteenth century that accelerating electric charges radiate energy and, conversely, that electromagnetic radiation consists of fields which accelerate charged particles. Light in the visible, infrared, or ultraviolet region interacts primarily with the electrons in gases, liquids, and solids—not the nuclei. The scattering process in these wavelength regions consists of acceleration of the electrons by the incident beam, followed by reradiation from the accelerating charges. *See* ELECTROMAGNETIC RADIATION.

Scattering processes may be divided according to the time between the absorption of energy from the incident beam and the subsequent reradiation. True "scattering" refers only to those processes which are essentially instantaneous. Mechanisms in which there is a measurable delay between absorption and reemission are usually termed luminescence. *See* LUMINESCENCE.

Instantaneous scattering processes may be further categorized according to the wavelength shifts involved. Some scattering is "elastic"; there is no wavelength change, only a phase shift. In 1928 C. V. Raman discovered the process in which light was inelastically scattered and its energy was shifted by an amount equal to the vibrational energy of a molecule or crystal.

In liquids or gases two distinct processes generate inelastic scattering with small wavelength shifts. The first is Brillouin scattering from pressure waves. When a sound wave propagates through a medium, it produces alternate regions of high compression (high density) and low compression (or rarefaction). Brillouin scattering of light to higher (or lower) frequencies occurs because the medium is moving toward (or away from) the light source. This is an optical Doppler effect. *See* DOPPLER EFFECT.

The second kind of inelastic scattering studied in fluids is due to entropy and temperature fluctuations, and is known as Rayleigh scattering. These entropy fluctuations produce a broadening in the scattered radiation centered about the exciting wavelength, rather than sharp, well-defined wavelength shifts. Under the assumption that the scattering in fluids is from particles much smaller than the wavelength of the exciting light, Lord Rayleigh derived in 1871 an equation, shown below, for such scattering. The dependence of scattering inten-

$$r^2 I(\theta)/I_0 = \pi d\lambda^{-4} v^2 (1 + \cos^2 \theta)(n - 1)^2$$

sity upon the inverse fourth power of the wavelength given in Rayleigh's equation is responsible for the fact that daytime sky looks blue and sunsets red: blue light is scattered out of the sunlight by the air molecules more strongly than red; at sunset, more red light passes directly to the eyes without being scattered. *See* ENTROPY.

Rayleigh's derivation of his scattering equation relies on the assumption of small, independent particles. Under some circumstances of interest, both of these assumptions fail. Colloidal suspensions provide systems in which the scattering particles are comparable to or larger than the exciting wavelengths. Such scattering is called the Tyndall effect and results in a nearly wavelength-independent (that is, white) scattering spectrum. The Tyndall effect is the reason clouds are white (the water droplets become larger than the wavelengths of visible light). *See* TYNDALL EFFECT.

The breakdown of Rayleigh's second assumption—that of independent particles—occurs in all liquids. There is strong correlation between the motion of neighboring particles. This leads to fixed phase relations and destructive interference for most of the scattered light. The remaining scattering arises from fluctuations in particle density discussed above. [J.F.S.]

Scent gland A specialized skin gland of the tubuloalveolar or acinous variety found in many mammals. These glands produce substances having peculiar odors. In some instances they are large, in others small. Examples of large glands are the civet gland in the civet cat, the musk gland in the musk deer, and the castoreum gland in the beaver. The civet gland is an anal gland, whereas the musk and castoreum are preputial. Examples of small scent glands are the preputial or Tyson's glands in the human male which secrete the smegma, and the vulval glands in the female. The secretions in all of the above glands are sebaceous. *See* GLAND. [O.E.N.]

Scheelite A mineral consisting of calcium tungstate, $CaWO_4$. Scheelite occurs in colorless to white, tetragonal crystals; it may also be massive and granular. Its fracture is uneven, and its luster is vitreous to adamantine. Scheelite has a hardness of 4.5–5 on Mohs scale and a specific gravity of 6.1. Its streak is white. The mineral is transparent and fluoresces bright bluish-white under ultraviolet light.

Scheelite is an important tungsten mineral and occurs in small amounts in vein deposits. The most important scheelite deposit in the United States is near Mill City, Nevada. *See* TUNGSTEN. [E.C.T.C.]

Schematic drawing Concise, graphical symbolism whereby the engineer communicates to others the functional relationship of the parts in a component and, in turn, of the com-

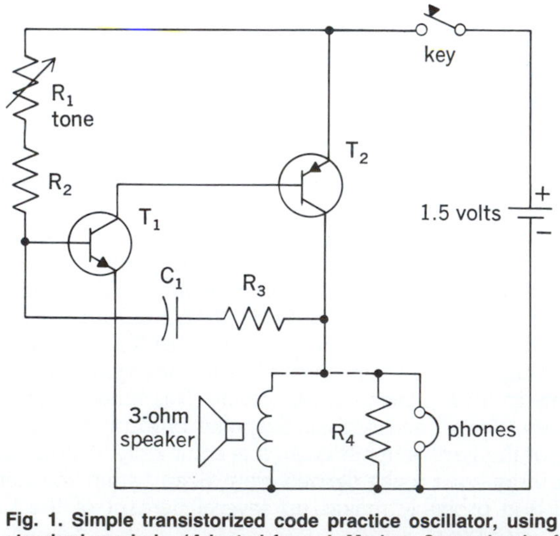

Fig. 1. Simple transistorized code practice oscillator, using standard symbols. (*Adapted from J. Markus, Sourcebook of Electronic Circuits, McGraw-Hill, 1968*)

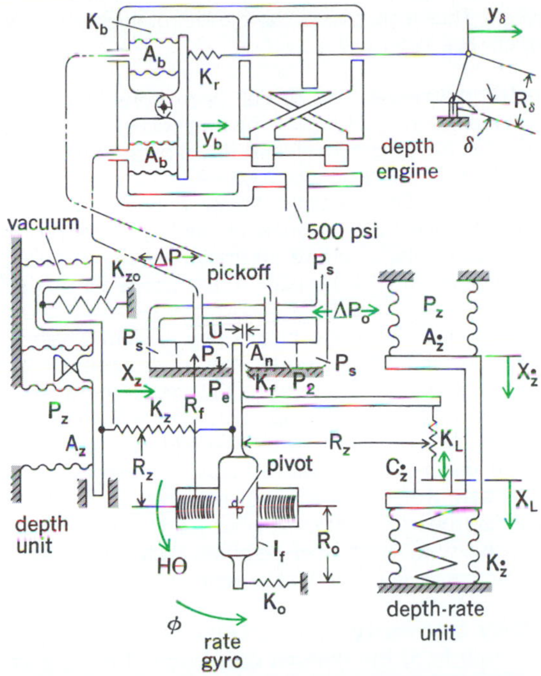

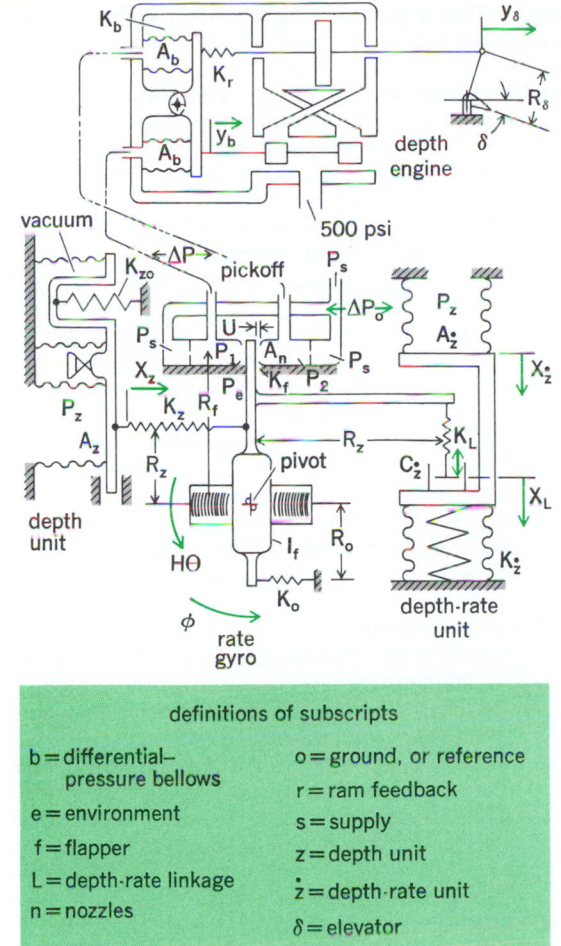

definitions of subscripts

- b = differential–pressure bellows
- e = environment
- f = flapper
- L = depth-rate linkage
- n = nozzles
- o = ground, or reference
- r = ram feedback
- s = supply
- z = depth unit
- $\dot{z}$ = depth-rate unit
- δ = elevator

Fig. 2. **Mechanical schematic of the depth-control mechanism of a torpedo.**

ponents in a system. The symbols do not attempt to describe in complete detail the characteristics or physical form of the elements, but they do suggest the functional form which the ensemble of elements will take in satisfying the functional requirements of the component. They are different from a block diagram in that schematics describe more specifically the physical process by which the functional specifications of a block diagram are satisfied.

An electrical schematic is a functional schematic which defines the interrelationship of the electrical elements in a circuit, equipment, or system. The symbols describing the electrical elements are stylized, simplified, and standardized to the point of universal acceptance (Fig. 1).

In a mechanical schematic, the graphical descriptions of elements of a mechanical system are more complex and more intimately interrelated than the symbolism of an electrical system and so the graphical characterizations are not nearly as well standardized or simplified (Fig. 2). However, a mechanical schematic illustrates such features as components, acceleration, velocity, position force sensing, and viscous damping devices. *See* DRAFTING; ENGINEERING DRAWING. [R.W.M.]

Schist A large group of rocks which by deformation during regional metamorphism have acquired a schistosity; that is, they show a more or less perfect cleavage along which they easily split up in flaggy slabs. There are orthoschists derived from igneous rocks and paraschists derived from sedimentary rocks. Crystalline schist is a common designation for all rocks which have recrystallized during regional metamorphism. *See* METAMORPHISM; SCHISTOSITY.

The chemical composition of a schist is similar to that of most silicate rocks, but mineralogically all schists have one feature in common, namely, that one or more of the major mineral constituents are flaky or fibrous (having the crystal structure of phyllosilicates or inosilicates). [T.F.W.B.]

Schistosity A type of cleavage characteristic of metamorphic rocks, notably schists and phyllites. The rocks tend to split along parallel planes defined by the distribution and parallel arrangement (preferred orientation) of platy mineral crystals such as muscovite, biotite, chlorite, talc, and graphite or, less commonly, rodlike crystals such as tremolite, actinolite, and hornblende. The preferred orientation of these minerals is produced during deformation and recrystallization (metamorphism) of the rocks. The schistosity may have a consistent attitude over large areas but local folding is also common. *See* PHYLLITE; ROCK CLEAVAGE; SCHIST. [J.M.Ch.]

Schistosomiasis A disease in which humans are parasitized by any of three species of blood flukes: *Schistosoma mansoni*, *S. haematobium*, and *S. japonicum*. Adult *S. mansoni* prefer the veins of the hemorrhoidal plexus, *S. haematobium* those of the vesical plexus, and *S. japonicum* those of the small intestine. The disease is also known as bilharziasis. *See* DIGENEA.

An embryonated egg passed in feces or urine hatches in fresh water, liberating a miracidium larva which penetrates into specific gastropod snails. The larval cycle in the snail lasts for about 1 month. The cercaria emerges from the mollusk, swims in the water, and penetrates the skin of the final host upon coming in contact with it.

Schistosomiasis is an agricultural hazard for all ages in irrigated lands or swamps. Elsewhere fluvial waters are the main source of infection, in which case incidence is marked in human beings who are less than 15 years old and is higher among boys than among girls. [J.F.M.]

Schistostegales An order of the true mosses (subclass Bryidae), consisting of a single species, *Schistostega pennata*, the cave moss, which is especially characterized by its leaf arrangement and the form of its protonema. The plants grow in dimly lit, cavelike places, such as the undersides of upturned tree roots. The cave moss is widely distributed in north temperate regions. An unrelated moss found in Australia and New Zealand, *Mittenia plumula*, has an identical protonema and habitat.

The leafy plants are erect, unbranched, and dimorphous: sterile shoots have leaves wide-spreading in two rows and confluent because of broad basal decurrencies. Fertile plants have leaves in five rows and erect-spreading in a terminal tuft. All leaves are ecostate, entire, and unbordered. The inflorescences are terminal, the setae elongate, and the capsules subglobose, with a flat operculum but no peristome. *See* BRYIDAE; BRYOPHYTA; BRYOPSIDA. [H.Cr.]

Schizogoniales A small, unique order of the Chlorophyta containing algae that are submicroscopic filaments or macroscopic ribbons and sheets a few centimeters wide. They are attached by rhizoids to rocks in salt or fresh water, usually occurring in alpine streams, on highly nitrogenous soil, or on bones, especially in the Arctic. There is one stellate chloroplast and a central pyrenoid in subrectangular cells; in *Prasiola* the cells are often grouped in fours. [G.W.P.]

Schizomida An order of Arachnida with two families, Protoschizomidae (*Agastoschizomus*, *Protoschizomus*) and Schizomida (*Schizomus*, *Trithyreus*, *Megaschizomus*), including about 130 described species, mainly in the genus *Schizomus*. Schizomids occur in most tropical and warm temperate regions, but each species has a restricted distribution.

Many undescribed species probably occur in tropical leaf litter and caves. Schizomids are most closely related to Uropygi, with which they share several evolutionary novelties. *See* UROPYGI.

Schizomids are 0.12–0.44 in. (3–11 mm) in size, are white to tan, and lack eyes, although some retain vestigial "eye-spots." The carapace is distinctively tripartite and the abdomen has a short, terminal flagellum.

Because schizomids are vulnerable to desiccation, they inhabit only moist places such as leaf litter, soil, caves, or beneath stones and logs. They are strict carnivores and fast runners. *See* ARACHNIDA. [J.A.Co.]

Schizophrenia A group of mental disorders which are characterized by withdrawal from reality and by disturbances in thinking and feeling. It is also called dementia praecox. In many cases it leads to disorganization of the personality but not necessarily to mental deterioration. One of the outstanding symptoms of schizophrenia is a lack of rapport and serious misjudgment of reality processes. Certain types of schizophrenics are refractory to influence and suggestion; this behavior in its extreme form is called negativism.

The main symptom of schizophrenia is a rather severe dissociation in thinking and feeling. The patients think and speak incoherently and illogically. There is a blocking of the thinking process; ideas become very vague. Emotions become flattened and also extremely ambivalent or contradictory at the same time.

The course of schizophrenia is often progressive, leading to a terminal state of deterioration; however, there are many cases which do not reach such a terminal state and many even recover spontaneously. *See* PSYCHOSIS. [F.C.R.]

Schlieren photography An optical technique that detects density gradients occurring in a gas flow. The schlieren system is used particularly in supersonic wind tunnels because it clearly shows the density gradients created by the shock and expansion waves of the airflow around the wind tunnel model.

A simple schlieren system operates as shown in the illustration. A source of light is shielded so that only a small rectangular slit emits light. A lens is placed at its focal distance from the slit so that the light is bent into a parallel beam. A second lens collects the parallel beam into an image of the slit and forms an inverted image on the screen or photographic plate. If a knife-edge is moved into the light stream near the slit image, the image at the screen darkens uniformly. Consider the system just described to be oriented so that the parallel light beam crosses a wind tunnel section. A light ray *a*, bent from the parallel path by density gradients in the test section, cannot be brought to focus at the slit image and is interrupted by the knife-edge so that a dark spot occurs at *a′* on the screen. Light ray *b*, deflected the opposite way by a different density gradient, escapes the knife-edge and appears as a light spot at *b′* on the screen. Thus a picture of the density gradients appears on the screen. *See* WIND TUNNEL. [D.P.An.]

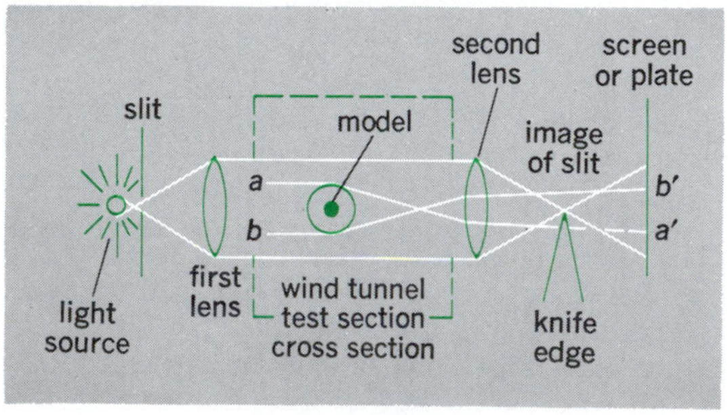

Basic schlieren optical system.

Schmidt camera An optical system consisting of a spherical mirror with a corrector plate near its focus. Cameras of the Schmidt type are of great value in direct astronomical photography and also as parts of spectroscopes.

The corrector plate is a figured plate with one plane surface, its second surface being aspherical, slightly deviating from a plane, such that the aperture aberration of the mirror is balanced. The curvature of the plate at the axis is zero. *See* OPTICAL SURFACES.

The Schmidt camera is a catadioptric system which is free from aperture aberration, asymmetry, and chromatic aberrations. It can therefore be used at the very large aperture of $f/2.0$ or more. The image, although sharp, has curvature of field, but the field can be flattened by adding a second mirror to produce what could be called a Schmidt-Cassegrain system.

The success of the Schmidt camera, which at first was used mostly for astronomical purposes, has led to numerous designs of catadioptric systems for other purposes, such as microscope optics and camera lenses. *See* ASTRONOMICAL PHOTOGRAPHY. [M.J.H.]

Schottky anomaly A contribution to the heat capacity of a solid arising from the thermal population of discrete energy levels as the temperature is raised. The effect is particularly prominent at low temperatures, where other contributions to the heat capacity are generally small. *See* SPECIFIC HEAT.

Discrete energy levels may arise from a variety of causes, including the removal of orbital or spin degeneracy by magnetic fields, crystalline electric fields, and spin orbit coupling, or from the magnetic hyperfine interaction. Such effects commonly occur in paramagnetic ions. *See* LOW-TEMPERATURE THERMOMETRY.

Corresponding to the Schottky heat capacity, there is a contribution to the entropy. This can act as a barrier to the attainment of low temperatures if the substance is to be cooled either by adiabatic demagnetization or by contact with another cooled substance. Conversely, a substance with a Schottky anomaly can be used as a heat sink in experiments at low temperatures (generally below 1 K or −457.9°F) to reduce temperature changes resulting from the influx or generation of heat. *See* ADIABATIC DEMAGNETIZATION; LOW-TEMPERATURE PHYSICS. [W.P.W.]

Schottky effect The enhancement of the thermionic emission of a conductor resulting from an electric field at the

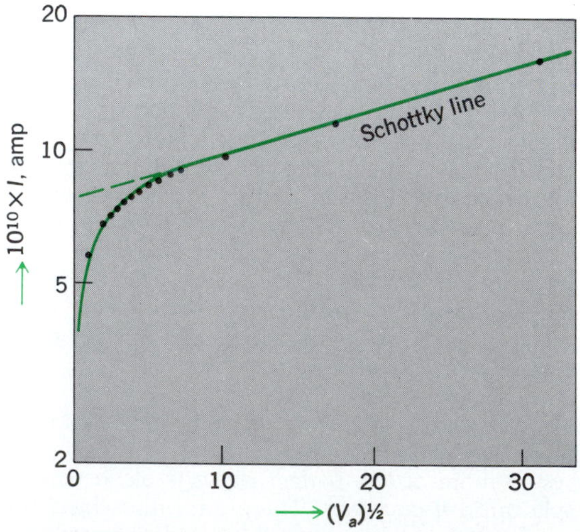

Logarithm of thermionic emission current *I* of tungsten as function of square root of anode voltage V_a. (*After W. B. Nottingham, Phys. Rev., 58:927–928, 1940*)

conductor surface. Since the thermionic emission current is given by the Richardson formula, an increase in the current at a given temperature implies a reduction in the work function of the emitter. The reduction in work function can be calculated by considering the effect of a constant externally applied field on the potential energy of an electron near the conductor surface, and is found to be proportional to the square root of the field. *See* THERMIONIC EMISSION; WORK FUNCTION (ELECTRONICS).

A plot of the logarithm of the current versus the square root of the anode voltage should yield a straight line. An example is given in the illustration for tungsten; the deviation from the straight line for low anode voltages is due to space-charge effects. *See* SPACE CHARGE. [A.J.D.]

Schrödinger's wave equation The differential equation of quantum mechanics (first proposed by Erwin Schrödinger in 1926) whose solution determines the average result, also termed expectation value, of every conceivable experiment on the physical system under examination. When solved, the Schrödinger equation yields the wave function; from the wave function, expectation values are computed. The term wave equation comes from the resemblance of the Schrödinger equation to those differential equations describing acoustic and electromagnetic waves. Also, its consequences and mode of derivation are consistent with the tenet that electrons and other particulate constituents of matter have wave-like properties. *See* QUANTUM MECHANICS; WAVE EQUATION; WAVE MOTION. [E.G.]

Schuler pendulum Any apparatus which swings, because of gravity, with a natural period of 84.4 min, that is, with the same period as a hypothetical simple pendulum whose length is the Earth's radius. In 1923 Max Schuler showed that such an apparatus has the unique property that the pendulum arm will remain vertical despite any motions of its pivot. It is therefore useful as a base for navigational instruments. Schuler also showed how gyroscopes can be used to increase the period of a physical pendulum to the desired 84.4 min.

Gyrocompasses employ the Schuler principle to avoid errors due to ship accelerations. The principle has become the foundation of the science of inertial navigation. *See* GYROCOMPASS; INERTIAL GUIDANCE SYSTEM. [R.H.C.]

Sciatica Pain in the lower extremities, hips, and back caused by irritation of the sciatic nerves. The location of the specific pain and the causes producing it are quite varied.

Sciatica may result from mechanical pressure on the nerves or their roots in the cord. Trauma, herniated intervertebral disks, pregnancy, inflammation, or tumors may cause compression. Toxic or metabolic disorders, such as lead poisoning, diabetes mellitus, alcoholism, and vitamin-B deficiency may induce sciatic pain by producing changes in the nerves. Inflammations, both local and systemic, may also cause temporary or permanent nerve injury. Certain viral diseases, syphilis, and local infections act in this manner. *See* SYPHILIS; VIRUS; VITAMIN.

In addition, lesions in the anal region and the prostate may induce sciatica through reflex stimulation. Joint diseases, pelvic strain, and injury most often precipitate an attack. The pain of sciatica may begin suddenly and violently or gradually, as a nagging discomfort. The pain is usually along the leg, with later extension to the thigh and back. Numbness of the outside of the foot may occur. *See* PAIN. [E.G.St./N.K.M.]

Science In common usage the word science is applied to a variety of disciplines or intellectual activities which have certain features in common. Usually a science is characterized by the possibility of making precise statements which are susceptible of some sort of check or proof. This often implies that the situations with which the special science is concerned can be made to recur in order to submit themselves to check, although this is by no means always the case. There are observational sciences such as astronomy or geology in which repetition of a situation at will is intrinsically impossible, and the possible precision is limited to precision of description.

A common method of classifying sciences is to refer to them as either exact sciences or descriptive sciences. Examples of the former are physics and, to a lesser degree, chemistry; and of the latter, taxonomical botany or zoology. The exact sciences are in general characterized by the possibility of exact measurement. One of the most important tasks of a descriptive science is to develop a method of description or classification that will permit precision of reference to the subject matter. *See* PHYSICAL SCIENCE. [P.W.Br./G.Ho.]

Scientific satellites Satellites used to gain scientific knowledge. They may be satellites of Earth or of other planets. Satellites used to apply space knowledge and techniques to practical purposes are called applications satellites. Many spacecraft are used for multiple purposes, combining space exploration, science, and applications in various ways. Scientific satellites are often called research satellites, and applications satellites are commonly designated by the application field, for example, navigation satellites. *See* SATELLITE (SPACECRAFT).

The space environment was one of the first areas investigated in scientific satellites. In this environment are cosmic rays, dust, magnetic fields, and various radiations from the Sun and galaxies. The Van Allen radiation belts were discovered by the first United States satellites. Soon thereafter satellites and space probes confirmed the existence of the solar wind. The continued investigation of these phenomena by satellites and space probes has led to the knowledge that a planet with a strong magnetic field, like Earth, is surrounded by a complex region, called a magnetosphere.

Numerous investigations are made of the effect of the space environment on materials and processes, including effects of radiation damage, meteoric erosion processes, and extremely high vacuum for very long periods. The influence of prolonged weightlessness on welding, alloying metals, growing crystals, and biological processes constitutes a promising field for future practical applications.

Scientific satellites, sounding rockets, and space probes are having a profound effect on earth science. The Earth and its atmosphere can now be studied in comparison with the other planets and their atmospheres, providing greater insight into the formation and evolution of the solar system. Satellites also make possible precision measurements of the size and shape of Earth and its gravitational field, and even measurements of the slow drifting of continents relative to each other.

Satellites are also having a great influence on astronomy by making it possible to observe celestial objects in all the wavelengths that reach the vicinity of Earth, whereas the ionosphere and atmosphere prevent most of these radiations from reaching telescopes on the ground. Moreover, even in the visible wavelengths, the small-scale turbulence and continuously varying refraction in the lower atmosphere distorts the image of a stellar object viewed at the ground, so that there is a limit to the improvement that can be achieved by increasing the size of a ground-based telescope. *See* SATELLITE ASTRONOMY.

The applications satellites emphasize the continuing day-to-day practical utilization of the satellite. They are operational in nature, although some of them, either directly or as a by-product of their operational output, contribute significantly to research, for example, with meteorological or Earth survey satellites. An applied research program generally precedes establishment of an applications satellite system. Thus, Tiros and Nimbus laid the groundwork for operational meteorological satellites. Likewise, Syncom satellites preceded the Intelsat communications system. *See* Applications satellites; Communications satellite; Meteorological satellites; Military satellites; Satellite navigation systems; Space flight. [H.E.N.]

Scintillation counter A particle or radiation detector which operates through emission of light flashes that are detected by a photosensitive device, usually a photomultiplier. The scintillation counter not only can detect the presence of a particle, gamma ray, or x-ray, but can measure the energy, or the energy loss, of the particle or radiation in the scintillating medium. The sensitive medium may be solid, liquid, or gaseous, but is usually one of the first two. The scintillation counter is one of the most versatile particle detectors, and is widely used in industry, scientific research, and radiation monitoring, as well as in exploration for petroleum and radioactive minerals that emit gamma rays. Many low-level radioactivity measurements are made with scintillation counters. *See* Low-level counting; Photomultiplier.

Scintillation counters are made of transparent crystalline materials or liquids or plastics. In order to be an efficient detector, the bulk scintillating medium must be transparent to its own luminescent radiation, and since most detectors are quite extensive, covering meters in length, the transparency must be of a high order. One face of the scintillator is placed in optical contact with the photosensitive surface of the photomultiplier, as shown in the illustration. In order to direct as much as possible of the light flash to the photosensitive surface, reflecting material is placed between the scintillator and the inside surface of the container.

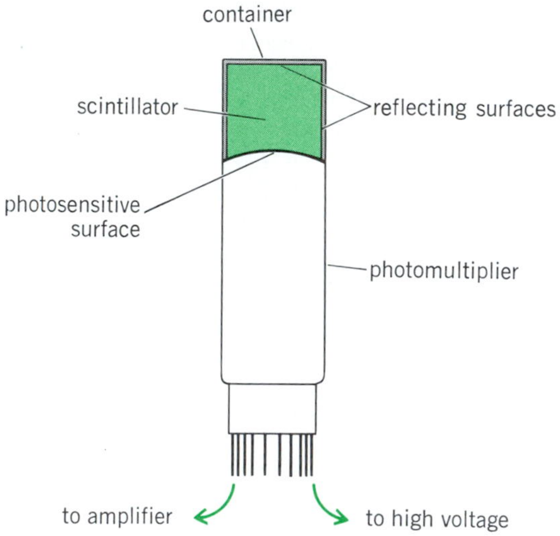

Diagram of a scintillation counter.

Scintillation counters are often used in temporal coincidence with other particle or radiation detectors. The coincidence technique eliminates or reduces the problem of local background or false events. *See* Particle detector. [R.Ho.]

Scleractinia An order of the subclass Zoantharia which comprises the true or stony corals (see illustration). These are solitary or colonial anthozoans which attach to a firm substrate. They are profuse in tropical and subtropical waters and contribute to the formation of coral reefs or islands. Some species are free and unattached.

Most of the polyp is impregnated with a hard calcareous skeleton. The polyps increase rapidly by budding, and the skeletons of polyps which settle in groups may fuse to form a colony. The pyriform, ciliated planula larva swims with its aboral extremity, which is composed of an ectodermal sensory layer, directed anteriorly. Planulation occurs periodically in conformity with lunar phases in many tropical species. *See* Zoantharia. [K.At.]

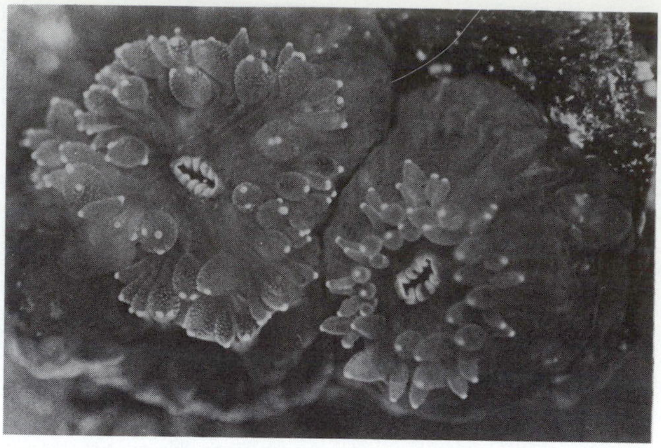

Solitary coral polyps, *Oulangia* sp.

Sclerenchyma Single cells or aggregates of cells whose principal function is thought to be mechanical support of plants or plant parts. Sclerenchyma cells have thick secondary walls and may or may not remain alive when mature. They vary greatly in form and are of widespread occurrence in vascular plants. Two general types, sclereids and fibers, are widely recognized, but since these intergrade, the distinction is sometimes arbitrary. [N.H.B.]

Sclerospongiae A class of sponges that lay down a compound skeleton including an external, basal mass of calcium carbonate, either aragonite or calcite, as well as internal siliceous spicules and protein fibers. The living tissue forms a thin layer over the basal calcareous skeleton and extends into surface depressions of the skeleton; the organization of the tissue is similar to that of encrusting demosponges. Sclerosponges are common inhabitants of cryptic habitats on coral reefs in both the Caribbean and Indo-Pacific biogeographic regions. *See* Demospongiae. [W.D.H.]

Scorpiones An order of the Arachnida which have chelate pedipalps and chelicera, a terminal caudal sting, and abdominal pectines (see illustration). The body is divided into a

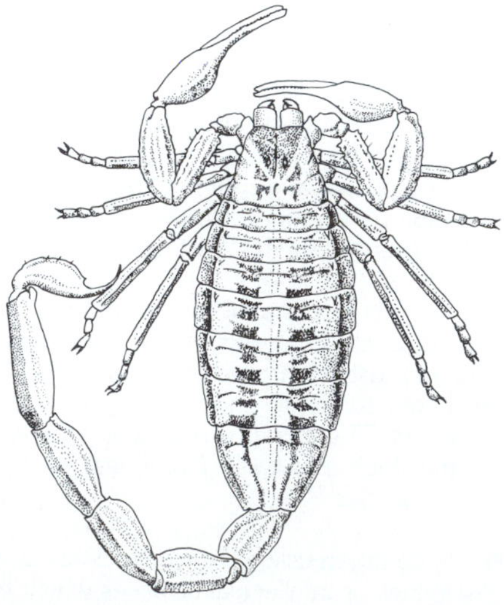

***Centrurus* sp. (After E. L. Palmer and H. S. Fowler, *Fieldbook of Natural History*, McGraw-Hill, 1975)**

cephalothorax, which is covered by the unsegmented carapace, and a segmented abdomen. This latter division is differentiated into an anterior preabdomen and a postabdomen which, plus the terminal sting, constitutes the "tail" or cauda. The cephalothorax bears the chelicera, the pedipalps, and four pairs of walking legs. The preabdomen contains seven segments, while the postabdomen has five. The carapace bears three groups of small, simple eyes, a median pair and two groups of anterolateral eyes. True cave scorpions lack eyes.

Approximately 1000 species are known. Scorpions are widely distributed throughout the tropical zone and the warmer areas of the temperate zone. About 56 species are found in the United States, 22 of which are in Arizona. California and Florida also have a rich scorpion fauna. Scorpions are found over three-fourths of the United States; they are absent from the New England states, Iowa, and areas immediately surrounding the Great Lakes.

The only species in the United States that are known to be lethal are *Centruroides sculpturatus* and *C. gertschi*, found in southern and central Arizona and the adjacent areas of California, New Mexico, and Mexico. The venom of these scorpions has proved fatal to healthy children up to 16 years of age and to adults suffering from hypertension and general debility. *See* ARACHNIDA. [H.L.St.]

Scorpius The Scorpion, in astronomy, one of the most beautiful and vivid constellations in the sky. Scorpius is the eighth sign of the zodiac (see illustration). The bright red star

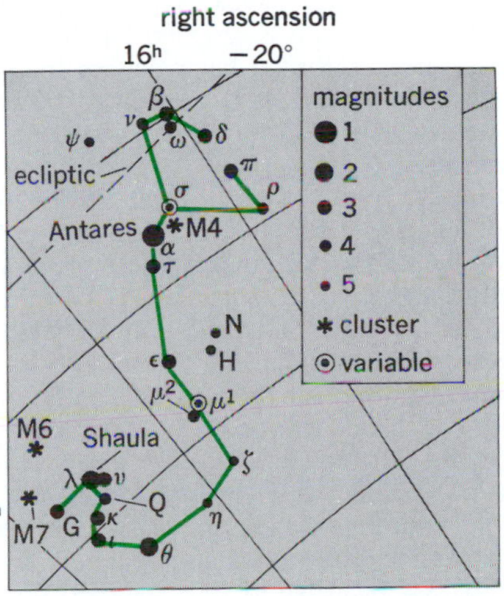

Line pattern of the constellation Scorpius. The grid lines represent the coordinates of the sky. The apparent brightness, or magnitudes, of the stars is shown by the sizes of the dots, which are graded by appropriate numbers as indicated.

Antares is situated at the heart. Antares is one of the largest stars known, having a diameter over 450 times that of the Sun. As in Sagittarius, the Milky Way in Scorpius is bright and rich in star clouds and clusters. *See* CONSTELLATION. [C.-S.Y.]

Scouring A process that removes natural encrustants which are of a resistant nature from fibers. In natural fibers such as cotton or linen, very resistant and water-repellent waxes are the main concern. These waxes require relatively severe treatment with caustic soda and other auxiliaries at high temperatures to form a wax soap (saponification). Thus the waxes are emulsified prior to being rinsed or washed out. With protein fibers such as wool, the repellent is a natural grease. This also must be removed by emulsification with a slightly alkalinized soap or detergent, but with much lower temperature and alkaline content. Synthetic fibers, including rayons and synthetics, contain no natural encrustants but have so-called producer oils added to them as protective agents. These are self-emulsifiable oils and are removed concurrently with any sizing agent used in the desizing process, thus eliminating the need for a scour. *See* DESIZING. [J.J.McD.]

Scrapie A transmissible, usually fatal disease of adult sheep characterized by degeneration of the central nervous system. (It is classed as a slow virus disease.) The disease is known in Great Britain, France, Belgium, Iceland, the United States, Canada, and northern India. Scrapie has certain similarities with kuru, a human disease in New Guinea, and mink encephalopathy.

Scrapie affects both sexes and is insidious in its onset, starting with hyperexcitability and progressive itch. Later, loss of wool occurs when the animal rubs against fixed objects or bites and nibbles its skin. Some animals do not rub but are either nervous and tremble when approached or appear sleepy. Incoordination of gait is constant and usually more evident in the hindquarters. In the final stages the sheep, being unable to stand, lie down, become emaciated, and die. *See* VIRUS INFECTION, LATENT, PERSISTENT, SLOW. [I.Zl.]

Screening A mechanical method of separating a mixture of solid particles into fractions by size. The mixture to be separated, called the feed, is passed over a screen surface containing openings of definite size. Particles smaller than the openings fall through the screen and are collected as undersize. Particles larger than the openings slide off the screen and are caught as oversize. A single screen separates the feed into only two fractions. Two or more screens may be operated in series to give additional fractions. Screening occasionally is done wet, but most commonly it is done dry.

Industrial screens may be constructed of metal bars, perforated or slotted metal plates, woven wire cloth, or bolting cloth. The openings are usually square but may be circular or rectangular. *See* MECHANICAL CLASSIFICATION; MECHANICAL SEPARATION TECHNIQUES; SEDIMENTATION (INDUSTRY). [W.L.McC.]

Screw A cylindrical body with a helical groove cut into its surface. For practical purposes a screw may be considered to be a wedge wound in the form of a helix so that the input motion is a rotation while the output remains translation. The screw is to the wedge much the same as the wheel and axle is to the lever in that it permits the exertion of force through a greatly increased distance.

The screw is by far the most useful form of inclined plane or wedge and finds application in the bolts and nuts used to fasten parts together; in lead and feed screws used to advance cutting tools or parts in machine tools; in screw jacks used to lift such objects as automobiles, houses, and heavy machinery; in screw-type conveyors used to move bulk materials; and in propellers for airplanes and ships. *See* PROPELLER (AIRCRAFT); PROPELLER (MARINE CRAFT); SCREW FASTENER; SCREW JACK; SCREW THREADS; SIMPLE MACHINE. [R.M.Ph.]

Screw fastener A threaded machine part used to join parts of a machine or structure. Screw fasteners are used when a connection that can be disassembled and reconnected and that must resist tension and shear is required. A nut and bolt is a common screw fastener. Bolt material is chosen to have an extended stress-strain characteristic free from a pronounced yield point. Nut material is chosen for slight plastic flow.

The nut is tightened on the bolt to produce a preload tension in the bolt, as illustrated. This preload has several advantageous effects. It places the bolt under sufficient tension so that during vibration the relative stress change is slight with consequent improved fatigue resistance and locking of the nut. Preloading also increases the friction between bearing surfaces of the joined members so that shear loads are carried by the

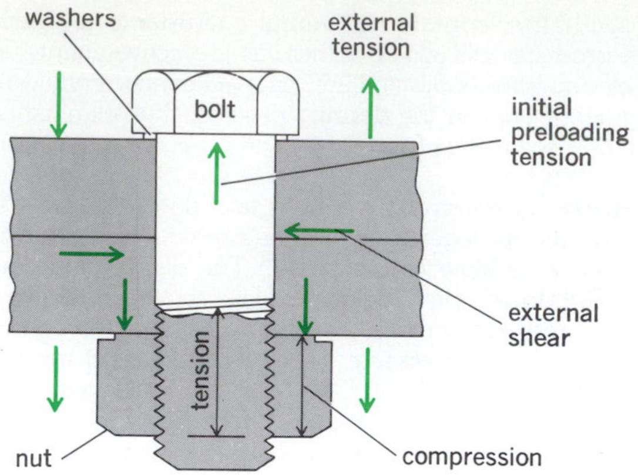

Forces on a bolt fastener under preloading.

friction forces rather than by the bolt. *See* BOLT; JOINT (STRUCTURES); NUT (ENGINEERING); SCREW. [F.H.R.]

Screw jack A mechanism for lifting and supporting loads, usually of large size. A screw jack mechanism consists of a thrust collar and a nut which rides on a bolt; the threads between the nut and bolt normally have a square shape. A standard form of screw jack has a heavy metal base with a central threaded hole into which fits a bolt capable of rotation under a collar thrusting against the load. *See* SCREW; SIMPLE MACHINE. [J.J.R.]

Screw-thread gage This type of gage may be used to inspect the pitch diameter, major diameter, minor diameter, lead, straightness, and thread angle of a screw thread (see illustration). Screw-thread gages may be plug gages, ring gages,

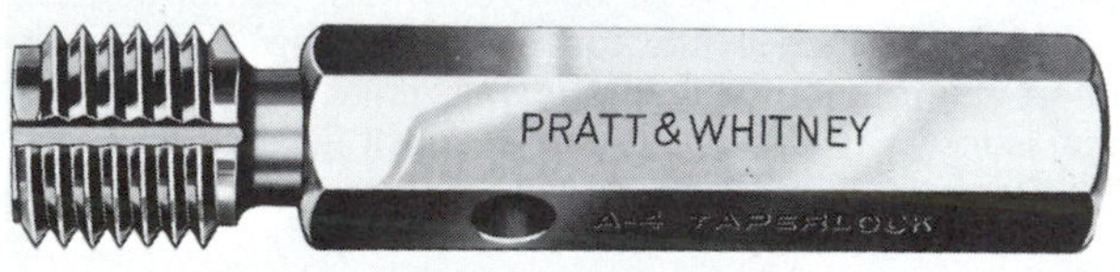

Thread plug gage. (*Pratt and Whitney Co.*)

snap gages, or indicating gages. A "go" plug or ring gage inspects all these characteristics. A "not go" plug or ring gage checks only pitch diameter and lead. *See* PLUG GAGE; RING GAGES; SNAP GAGE. [R.A.Bo.]

Screw threads Continuous helical ribs on a cylindrical shank. Screw threads are used principally for fastening, adjusting, and transmitting power. To perform these specific functions, various thread forms have been developed. A thread on the outside of a cylinder or cone is an external (male) thread; a thread on the inside of a member is an internal (female) thread (Fig. 1). *See* SCREW.

Types of thread. A thread may be either right-hand or left-hand. A right-hand thread on an external member advances into an internal thread when turned clockwise; a left-hand thread advances when turned counterclockwise. If a single helical groove is cut or formed on a cylinder, it is called a single-thread screw. Should the helix angle be increased sufficiently for a second thread to be cut between the grooves of the first thread, a double thread will be formed on the screw. Double, triple, and even quadruple threads are used whenever a rapid advance is desired, as on valves.

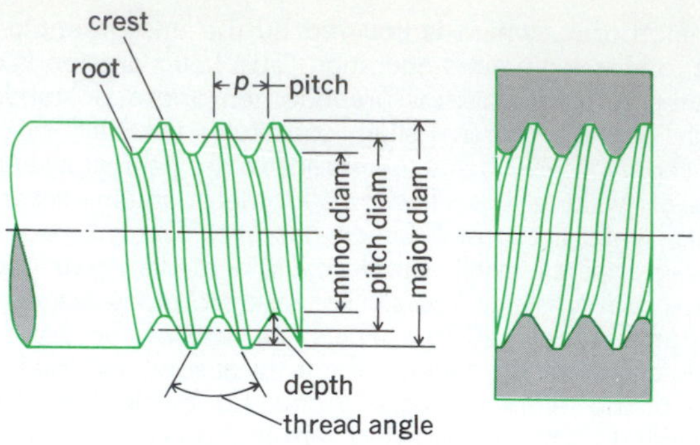

Fig. 1. Screw thread nomenclature.

Pitch and major diameter designate a thread. Lead is the distance advanced parallel to the axis when the screw is turned one revolution. For a single thread, lead is equal to the pitch; for a double thread, lead is twice the pitch. For a straight thread, the pitch diameter is the diameter of an imaginary coaxial cylinder that would cut the thread forms at a height where the width of the thread and groove would be equal.

Thread forms have been developed to satisfy particular requirements. Where strength is required for the transmission of power and motion, a thread having faces that are more nearly perpendicular to the axis is preferred. These threads, with their strong thread sections, transmit power nearly parallel to the axis of the screw.

Thread fastener. Most threaded fasteners are a threaded cylindrical rod with some form of head on one end. Of the many forms available, five types meet most requirements for threaded fasteners and are used for the bulk of production work: bolt, stud, cap screw, machine screw, and set screw (Fig. 2). Bolts and screws can be obtained with varied heads and points.

A bolt is generally used for drawing two parts together. A stud is a rod threaded on both ends. Studs are used for parts that must be removed frequently and for applications where bolts would be impractical.

Cap screws (plated or unplated) are widely used in machine tools and for assembling parts in automotive and aeronautical equipment. They are available in four standard heads: hexagon, flat, round, and fillister. Flathead and roundhead screws

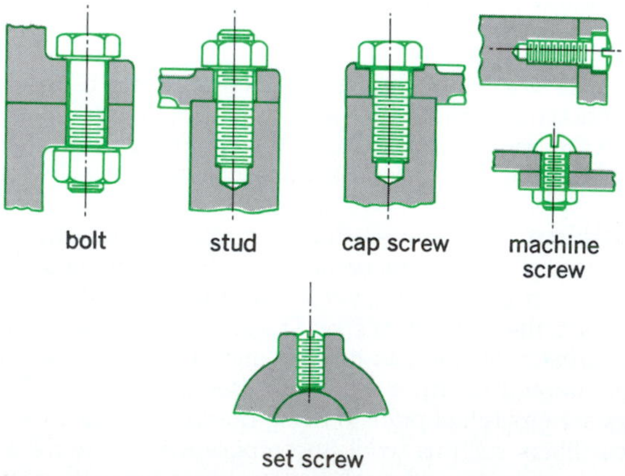

Fig. 2. Common types of fasteners. (*After T. E. French and C. J. Vierck, Engineering Drawing and Graphic Technology, 12th ed., McGraw-Hill, 1978*)

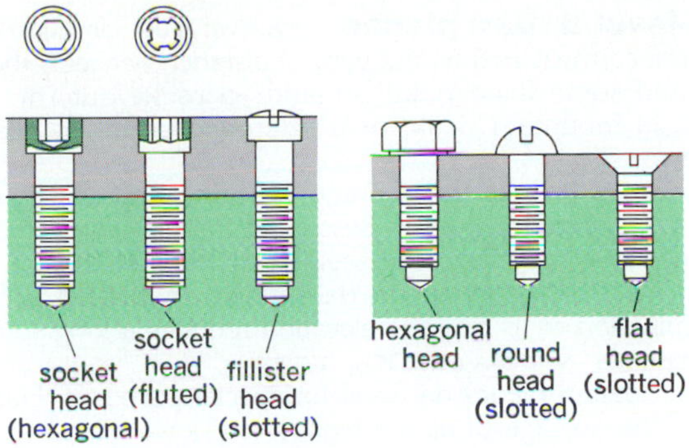

Fig. 3. Cap screws. (*After W. J. Luzadder, Fundamentals of Engineering Drawing, 8th ed., Prentice-Hall, 1981*)

have slotted heads (Fig. 3). Machine screws are similar to cap screws and fulfill the same purpose, being employed principally on small work. Set screws made of hardened steel are used to hold parts in a position relative to one another. Wood screws, lag screws, and hanger bolts are used in wood. [W.J.L.]

Scrophulariales An order of flowering plants, division Magnoliophyta (Angiospermae), in the subclass Asteridae of the class Magnoliopsida (dicotyledons). The order consists of 12 families and nearly 10,000 species. The largest families are the Scrophulariaceae (about 2700 species), Acanthaceae (about 2600 species), Gesneriaceae (about 1800 species), Bignoniaceae (about 800 species), and Oleaceae (about 600 species). The Scrophulariales usually have a superior ovary and generally either an irregular corolla or one with fewer stamens than corolla lobes, or commonly both. They uniformly lack stipules. *See* Asteridae; Magnoliopsida; Olive. [A.Cr.]

Scrub typhus An acute mite-borne infection due to the bacterialike microorganism *Rickettsia tsutsugamushi*. It was first called tsutsugamushi disease in Japan. It is characterized in humans by high fever, a macular rash, usually a primary eschar or ulcer, and swelling of the regional lymph glands. The vectors are larval trombiculid mites whose species vary according to location and season, but in most areas are chiefly *Trombicula akamushi* and *T. deliensis*. The primary disease cycle is between these mites, which are parasitic on vertebrates only in the larval stage, and field mice and rats. Transmission to humans occurs from the next generation of larval mites. *See* Acari; Rickettsioses. [C.B.P.]

Sculpin Any of several species of small fish of the family Cottidae, known as sea scorpions in Europe. There are 24 species in North America, many being common throughout the fresh waters of the United States, and most marine species occurring along the Pacific Coast. These fish have bony plates under the eyes, spiny pelvic and anal fins, and sometimes spines on the head. All the fresh-water species are small, and many spawn under rocks or logs. These fish are not of economic importance since many, especially most of the large species, are coarse and tasteless; they are used as bait. *See* Perciformes. [C.B.C.]

Scurvy An acute or chronic disease due to vitamin-C (ascorbic acid) deficiency. In adults scurvy is rarely due to dietary deficiency in normally fed populations, since 3–12 months of severe deficiency are required to produce symptoms. These include weakness, weight loss, irritability, and vague pains. Clinical signs include such things as delayed wound healing, tooth changes, gingivitis, petechial hemorrhages, and nosebleeds. Stress, infection, and hemorrhage may aggravate the situation. The disease is often related to chronic or severe gastrointestinal disease or to food idiosyncrasies. Any type of healing requires increased vitamin C supply, and infection similarly increases body demand. *See* Ascorbic acid. [N.K.M.]

Scyphozoa A class of the phylum Coelenterata, containing five living orders—Stauromedusae, Cubomedusae, Coronatae, Semaeostomeae, and Rhizostomeae—and a fossil order, Stromatoporoidea. They are all marine and usually take two forms, the polyp, or scyphopolyp, and the medusa, or scyphomedusa. However, some are polyplike and sessile throughout their lives, while others are always pelagic and lack the sessile polyp stage. Among the Coelenterata, the Scyphozoa are characterized by having well-developed medusae of large size and fairly well-organized polyps of small size.

The scyphomedusae are generally found near the coast. Exceptions are certain pelagic forms and the Coronatae, which are abyssal. Most of the Stauromedusae have a circumpolar distribution, while the Cubomedusae and Rhizostomeae are found mostly in warm and tropical seas. Scyphomedusae are carnivorous, except the Rhizostomeae which are plankton eaters. Some Semaeostomeae and Cubomedusae are injurious to humans because of their nematocysts. On the other hand, some Rhizostomeae are used as food in the Orient. *See* Coelenterata. [T.U.]

Sea anemone Any polyp of the nearly 1000 anthozoan coelenterates belonging to the order Actiniaria. They occur intertidally and subtidally in marine and estuarine habitats, attached to solid substrates or burrowing into soft sediments. No fresh-water or truly planktonic species are known. Anemones may be very small, a fraction of an inch long, to large individuals more than 3 ft (90 cm) in length or diameter. *See* Actiniaria.

Sea anemones enter into a number of interesting symbiotic partnerships. Some are host to single-celled marine algae, which grow within the cells of the anemone. The algae provide organic materials which aid in the nutrition of the anemone. Other anemones live on gastropod shells inhabited by hermit crabs. *See* Anthozoa; Coelenterata. [C.H.]

Sea breeze A diurnal, thermally driven circulation in which a surface convergence zone often exists between airstreams having over-water versus over-land histories. The sea breeze is one of the most frequently occurring small-scale (mesoscale) weather systems. It results from the unequal sensible heat flux of the lower atmosphere over adjacent solar-heated land and water masses. Because of the large thermal inertia of a water body, during daytime the air temperature changes little over the water while over land the air mass warms. Occurring during periods of fair skies and generally weak large-scale winds, the sea breeze is recognizable by a wind shift to onshore, generally several hours after sunrise. On many tropical coastlines the sea breeze is an almost daily occurrence. It also occurs with regularity during the warm season along midlatitude coastlines and even occasionally on Arctic shores. Especially during periods of very light winds, similar though sometimes weaker wind systems occur over the shores of large lakes and even wide rivers and estuaries (lake breezes, river breezes). At night, colder air from the land often will move offshore as a land breeze. Typically the land breeze circulation is much weaker and shallower than its daytime counterpart. *See* Atmospheric general circulation; Mesometeorolgy; Meteorology.

The occurrence and strength of the sea breeze is controlled by a variety of factors, including land-sea surface temperature differences; latitude and day of the year; the synoptic wind and its orientation with respect to the shoreline; the thermal stability of the lower atmosphere; surface solar radiation as affected by haze, smoke, and stratiform and convective cloudiness; and the geometry of the shoreline and the complexity of the surrounding terrain. *See* WIND. [W.A.Ly.]

Sea horse Any of the 50 species of marine fishes which occur principally in tropical and subtropical waters and are members of the genus *Hippocampus* in the order Gasterosteiformes. These fishes, together with the pipefishes, make up the family Syngnathidae. The sea horses vary in length from about 2 in. to 1 ft (5 to 30 cm). As with other members of the order, they have an elongate, tubelike snout.

The short compressed body is encompassed by bony rings which afford rigidity (see illustration). The head is bent ventrally, and the mouth is located at the end of the tubiform snout. Small dorsal and pectoral fins are present, but a caudal fin is lacking. These fishes have a prehensile tapering tail, which is used for clinging to seaweed. *See* GASTEROSTEIFORMES. [C.B.C.]

The sea horse *Hippocampus hudsonius*, which has gill tufts and 11 trunk rings.

Sea ice Ice formed by the freezing of seawater. Ice in the sea includes sea ice, river ice, and land ice. Land ice is principally icebergs. River ice is carried into the sea during spring breakup and is important only near river mouths. The greatest part, probably 99% of ice in the sea, is sea ice. *See* ICEBERG.

The freezing point temperature and the temperature of maximum density of seawater vary with salinity. When freezing occurs, small flat plates of pure ice freeze out of solution to form a network which entraps brine in layers of cells. As the temperature decreases more water freezes out of the brine cells, further concentrating the remaining brine so that the freezing point of the brine equals the temperature of the surrounding pure ice structure. The brine is a complex solution of many ions.

The brine cells migrate and change size with changes in temperature and pressure. The general downward migration of brine cells through the ice sheet leads to freshening of the top layers to near zero salinity by late summer. During winter the top surface temperature closely follows the air temperature, whereas the temperature of the underside remains at freezing point, corresponding to the salinity of water in contact.

The sea ice in any locality is commonly a mixture of recently formed ice and old ice which has survived one or more summers. Except in sheltered bays, sea ice is continually in motion because of wind and current. [W.Ly.]

Sea-level datum planes Sea level is the elevation of the sea surface measured as the vertical distance between the surface and some fixed point on land—a rocky outcrop on a beach, a mountain peak, or a reference point installed by humans.

Mean sea level is the average elevation and is frequently used as a reference level in describing the elevation of points on land, or of depths in the sea. The surface of the sea is by no means a stationary spheroidal surface, so that the accurate determination of its average elevation requires a long series of observations. *See* SEA-LEVEL FLUCTUATIONS.

Other datum planes are used for special purposes. Half-tide level is the average of all the highest and lowest readings during the period (called high and low waters). Mean low water and mean high water are the averages of all of the low waters and high waters, respectively; half-tide level lies midway between these two means. *See* TIDE. [J.G.P.]

Sea-level fluctuations The height of sea level is constantly changing and its average value (mean sea level) will depend critically on the length of a series of observations and the period in time during which they are made. Furthermore, the sea surface is not even level. *See* SEA-LEVEL DATUM PLANES.

Sea level stands about 10 in. (25 cm) higher along the coast of Maine than it does off Florida, and the steepest slope occurs off Cape Hatteras, North Carolina. This occurs because the water in the sea is always in motion. The motion of any fluid over the surface of the Earth is influenced by the rotation of the Earth itself; fluids are deflected to the right (looking downstream) in the Northern Hemisphere, to the left in the Southern. This effect results from the Coriolis acceleration. *See* CORIOLIS ACCELERATION; OCEAN CIRCULATION.

Sea level also deviates from the mean under the influence of differential atmospheric pressure. If pressure is high over one area of the ocean and low somewhere else, the water will tend to flow toward the low-pressure area. Thus the sea surface behaves as an inverted barometer; it stands high where air pressure is low, and low where air pressure is high. *See* AIR PRESSURE.

A third effect can be observed where parts of the sea surface are separated from each other by a land barrier. Precise leveling across the continent of North America shows that sea level is about 20 in. (50 cm) higher on the Pacific than on the Atlantic side of the continent. Currents and winds probably cause part of this difference in level, but another important factor is the difference in the density of the seawater. [J.G.P.]

Sea of Okhotsk A semienclosed basin adjacent to the North Pacific Ocean, bounded on the north, east, and west by continental Russia, the Kamchatka Peninsula, and northern Japan. On its southeast side the Sea of Okhotsk is connected to the North Pacific via a number of straits and passages through the Kurile Islands. The sea covers approximately 590,000 mi^2 (1,500,000 km^2), or about 1% of the total area of the Pacific, and has a maximum depth of over 9000 ft (3000 m). The mean depth of the Sea of Okhotsk is about 2500 ft (830 m). Because of the cold, wintertime arctic winds that blow to the southeast, from Russia toward the North Pacific Ocean, the Sea of Okhotsk is nearly completely covered with ice during the winter months, from November through March. *See* BASIN; PACIFIC OCEAN. [S.C.Ri.]

Sea squirt A sessile tunicate of the class Ascidiacea (phylum Chordata), so named because water is squirted from two openings when the animal is touched. Sea squirts may be solitary, colonial, or compound, but always have a permanent enclosing structure, the test or tunic, composed of polysaccharide material structurally similar to cellulose. In colonies, individuals are loosely attached together; in compound ascidians the individual zooids are embedded in a common test.

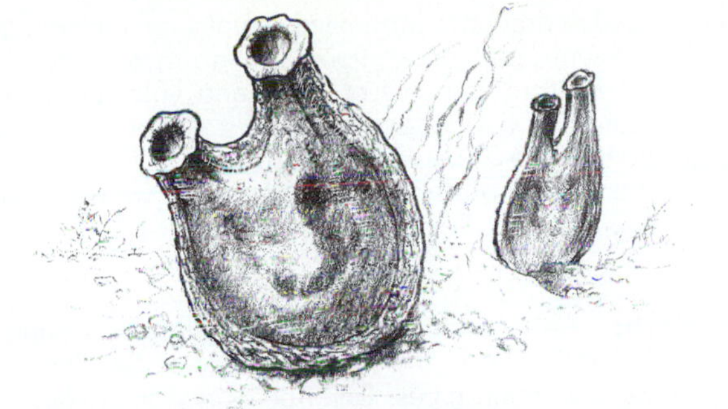

Sea squirts; these saclike tunicates serve as scavengers and as food for higher forms.

The individual is roughly cylindrical in shape and attached at the base (see illustration). The mantle encloses a space or atrial cavity. Suspended within the atrial cavity is a pharynx or branchial sac perforated by numerous oval slits (stigmata), each one bearing a ring of marginal cilia. Constant beating of the cilia draws in water through the inhalant or oral aperture into the branchial sac. Water passes through the stigmata into the atrial cavity and is discharged at the exhalant or atrial aperture. In the branchial sac, food particles from the water are filtered on a mucous sheet which is then rolled up and passed to the esophagus. Sea squirts are dioecious, and fertilization may be internal or external.

Sea squirts occur in all seas and at all depths in the ocean. Most require a hard substratum on which to settle, but a few live in sand or mud. They are particularly common in shallow-water environments such as rocky shores, piers, and boat hulls, and in mangrove lagoons. *See* ASCIDIACEA; TUNICATA. [I.Go.]

Sea state The description of the ocean surface or state of the sea surface with regard to wave action. Wind waves in the sea are of two types: Those still growing under the force of the wind are called sea: those no longer under the influence of the wind that produced them are called swell. Differences between the two types are important in forecasting ocean wave conditions.

Those waves which are still growing under the force of the wind have irregular, chaotic, and unpredictable forms. The unconnected wave crests are only two to three times as long as the distance between crests and commonly appear to be traveling in different directions. As the waves grow, they form regular series of connected troughs and crests with wave lengths commonly ranging from 12 to 35 times the wave heights. The appearance of the sea surface is termed state of the sea (Table 1). The height of a sea is dependent on the strength of the wind, the duration of time the wind has blown, and the fetch (distance of sea surface over which the wind has blown). *See* OCEAN WAVES.

Table 1. Sea height code*

Code	Height, ft†	Description of sea surface
0	0	Calm, with mirror-smooth surface
1	0–1	Smooth, with small wavelets or ripples with appearance of scales but without crests
2	1–3	Slight, with short pronounced waves or small rollers; crests have glassy appearance
3	3–5	Moderate, with waves or large rollers; scattered whitecaps on wave crests
4	5–8	Rough, with waves with frequent whitecaps; chance of some spray
5	8–12	Very rough, with waves tending to heap up; continuous whitecapping; foam from white caps occasionally blown along by wind
6	12–20	High, with waves showing visible increase in height, with extensive whitecaps from which foam is blown in dense streaks
7	20–40	Very high, with waves heaping up with long frothy crests that are breaking continuously; amount of foam being blown from the crests causes sea surface to take on white appearance and may affect visibility
8	40+	Mountainous, with waves so high that ships close by are lost from view in the wave troughs for a time; wind carries off crests of all waves, and sea is entirely covered with dense streaks of foam; air so filled with foam and spray as to affect visibility seriously
9		Confused, with waves crossing each other from many and unpredictable directions, developing complicated interference pattern that is difficult to describe; applicable to conditions 5–8

*Modified from *Instruction Manual for Oceanographic Observations*, H.O. Publ. no. 607, 2d ed., U.S. Navy Hydrographic Office, 1955.
†1 ft = 0.3 m.

Table 2. Swell-condition code*

Code	Description	Height, ft†	Length, ft†
0	No swell	0	0
	Low swell	1–6	
1	Short or average		0–600
2	Long		600+
	Moderate swell	6–12	
3	Short		0–300
4	Average		300–600
5	Long		600+
	High swell	12+	
6	Short		0–300
7	Average		300–600
8	Long		600+
9	Confused		

**Instruction Manual for Oceanographic Observations*, H.O. Publ. no. 607, 2d ed., U.S. Navy Hydrographic Office, 1955.
†1 ft = 0.3 m.

As sea waves move out of the generating area into a region of weaker winds, a calm, or opposing winds, their height decreases as they advance, their crests become rounded, and their surface is smoothed. These waves are more regular and more predictable than sea waves and, in a series, tend to show the same form or the same trend in characteristics. Wave lengths generally range from 35 to 200 times wave heights. A descriptive classification of swell waves is given in Table 2. [N.A.B.]

Sea urchin A marine echinoderm of the class Echinoidea. These invertebrates are found commonly in shallow waters. The soft internal organs are enclosed in and protected by a test or shell (see illustration) consisting of a number of plates which

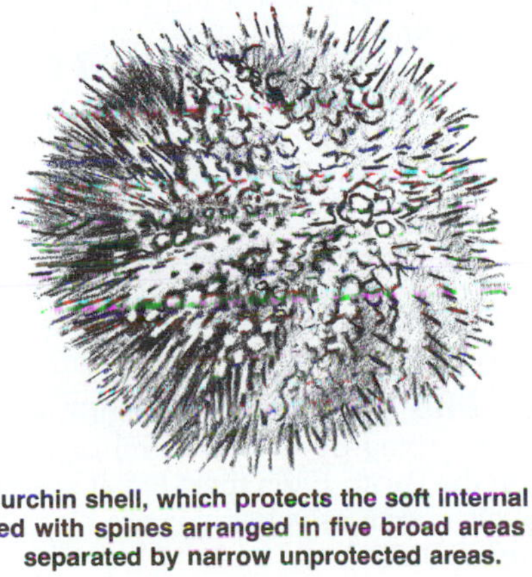

The sea urchin shell, which protects the soft internal organs, is covered with spines arranged in five broad areas that are separated by narrow unprotected areas.

fit closely together and are located under the skin. The oral surface is in contact with the substratum. Five teeth, located in the mouth, form part of Aristotle's lantern, a complex chewing structure. Like many other species of echinoderms, sea urchins use tube feet for locomotion and pincerlike structures called pedicellariae to keep the shell clean. Sea urchins feed on most available animal and vegetable materials.

Many species burrow into the sand, while others move into rock crevices to protect themselves from severe tidal action. *See* ECHINOIDEA. [C.B.C.]

Sea wall A structure at the water's edge to resist encroachment of the sea and to retain the natural soil or deposited fill behind the wall. A usual type of sea wall has a sloped or inclined face toward the sea to deflect and dissipate the wave forces. Sometimes it resembles a breakwater. Timber, steel, and concrete are used in a variety of forms. *See* COASTAL ENGINEERING. [E.J.Q.]

Seaborgium A chemical element, symbol Sg, atomic number 106. Seaborgium has chemical properties similar to tungsten. It was synthesized and identified in 1974. This discovery of seaborgium took place nearly simultaneously in two

1	2	3	4	5	6	7	8	9	10	11	12	13	14	15	16	17	18
1 H																	2 He
3 Li	4 Be											5 B	6 C	7 N	8 O	9 F	10 Ne
11 Na	12 Mg											13 Al	14 Si	15 P	16 S	17 Cl	18 Ar
19 K	20 Ca	21 Sc	22 Ti	23 V	24 Cr	25 Mn	26 Fe	27 Co	28 Ni	29 Cu	30 Zn	31 Ga	32 Ge	33 As	34 Se	35 Br	36 Kr
37 Rb	38 Sr	39 Y	40 Zr	41 Nb	42 Mo	43 Tc	44 Ru	45 Rh	46 Pd	47 Ag	48 Cd	49 In	50 Sn	51 Sb	52 Te	53 I	54 Xe
55 Cs	56 Ba	71 Lu	72 Hf	73 Ta	74 W	75 Re	76 Os	77 Ir	78 Pt	79 Au	80 Hg	81 Tl	82 Pb	83 Bi	84 Po	85 At	86 Rn
87 Fr	88 Ra	103 Lr	104 Rf	105 Db	106 Sg	107 Bh	108 Hs	109 Mt	110	*111*	*112*	*113*	*114*	*115*	*116*	*117*	*118*

lanthanide series	57 La	58 Ce	59 Pr	60 Nd	61 Pm	62 Sm	63 Eu	64 Gd	65 Tb	66 Dy	67 Ho	68 Er	69 Tm	70 Yb
actinide series	89 Ac	90 Th	91 Pa	92 U	93 Np	94 Pu	95 Am	96 Cm	97 Bk	98 Cf	99 Es	100 Fm	101 Md	102 No

nuclear laboratories, the Lawrence Berkeley Laboratory at the University of California and the Joint Institute for Nuclear Research at Dubna in Russia.

The Berkeley group, under the leadership of A. Ghiorso, used as its source of heavy ions the Super-Heavy Ion Linear Accelerator (SuperHILAC). The production and positive identification of the isotope of seaborgium with the mass number 263 decays with a half-life of 0.9 + 0.2 s by the emission of alpha particles of principal energy 9.06 + 0.04 MeV. This isotope is produced in the reaction in which four neutrons are emitted: ^{249}Cf(^{18}O,4*n*).

The Dubna group, under the leadership of G. N. Flerov and Y. T. Oganessian, produced its heavy ions with a heavy-ion cyclotron. They found a product that decays by the spontaneous fission mechanism with the very short half-life of 7 ms. They assigned it to the isotope ^{259}Sg, suggesting reactions in which two or three neutrons are emitted: ^{207}Pb(^{54}Cr,2*n*) and ^{208}Pb(Cr^{54},3*n*). *See* NOBELIUM; NUCLEAR CHEMISTRY; PERIODIC TABLE; TRANSURANIUM ELEMENTS. [G.T.S.]

Seamount and guyot A seamount is an isolated submarine mountain rising 3000 ft (900 m) or more above the ocean floor. In the Pacific Basin there are at least 10,000 such mountains, which occur as volcanic peaks on ridges, or rises, or as individual peaks.

Flat-topped seamounts are called guyots, or tablemounts. They are present on all ocean floors but are most common in the Pacific. Bottom samples of coral and volcanic erosion debris indicate that the flattops were once at sea level though they are now 1000–7000 ft (300–2100 m) below the ocean surface. Thus guyots are ancient islands which were truncated to sea level by erosion. *See* OCEANIC ISLANDS. [E.L.H.; H.W.Me.]

Seaplane An airplane capable of navigating on, taking off from, and alighting upon the surface of water. Seaplanes are grouped into two main types: flying boats and float planes. In the flying boat, the hull, which provides buoyancy and planing area, is an integral part of the airframe, a specially designed fuselage which supports the wings and tail surfaces and houses the crew, equipment, and cargo. Although multihull flying boats have been built, modern use is confined to single-hull boats with lateral stability on the water provided by small floats or pontoons attached to the wings. The float plane is a standard landplane made capable of water operation by the addition of floats which are attached to the airframe by struts. In practice the twin float is used exclusively, lateral stability on the water being provided by the separation of the two identical floats (see illustration). A seaplane with retracting wheels which

A single-engine float plane.

permit either land or water operation is known as an amphibian. *See* AIRPLANE. [R.A.Ho.]

Search coil A device used for measuring the flux density in a small region of a magnetic field. The apparatus consists of a small coil connected by flexible leads to a ballistic galvanometer. If the coil is placed with its plane perpendicular to the magnetic field, the flux threading the coil is $\Phi = BA$, where A is the area of the coil and B the magnetic induction. *See* MAGNETIC FLUX; MAGNETIC INDUCTION.

When the coil is quickly turned through a quarter turn or is withdrawn from the field to a place where B is zero, the flux through the coil is changed from BA to zero. During this change, there is an electromotive force (emf) and a current in the closed circuit whose instantaneous values are proportional to the rate of change of flux through the coil. The deflection of the ballistic galvanometer is proportional to the charge q, which is the integral of the current over time, and is thus proportional to the total flux change BA. However, if the instrument is properly calibrated, the value of B can be determined. *See* FLUXMETER; GALVANOMETER. [K.V.M.]

Seawater Water is most often found in nature as seawater ≅ 98%). Seawater is an aqueous solution of salts of a rather constant composition of elements. Its presence determines the climate and makes life possible on the Earth. The boundaries of seawater are the boundaries of the oceans, the mediterranean seas, and their embayments. *See* HYDROSPHERE.

Physical properties. Seawater is basically a concentrated electrolyte solution containing many dissolved salts. The ratio of water molecules to salt molecules is about 100 to 1. Since nearly all the salt exists as electrically conducting ions, the ratio of water molecules to ions is about 50 to 1. In the neighborhood of ions, extremely high electric fields exist and water molecules near them become aligned; water molecules that remain in the vicinity of the ions for a long time constitute a hydration shell, and the ions are said to be solvated. *See* SOLVATION.

As a consequence of the salts in seawater, its physical properties differ from those of pure water, the difference being closely proportional to the concentration of the salts or the salinity. Salinity measurements, along with pressure and temperature data, are used to differentiate water masses. In studying the movement of water masses in the oceans and their small and large-scale circulation patterns, including geostrophic flow, properties such as density, compressibility, thermal expansion coefficients, and specific heats need to be known as functions of temperature, pressure, and salinity.

The large value of the osmotic pressure of seawater is of great significance to biology and desalination by reverse osmosis. Related to osmotic pressure and very important to the formation of ice is the reversal of the freezing point and temperature of maximum density of seawater compared to pure water: the freezing point temperature is lowered to 28.6°F (−1.9°C), and the temperature of maximum density is decreased from just below 39°F (4°C) for pure water to about 25.7°F (−3.5°C) for 35% salinity seawater.

Since the greatest ocean depths exceed 33,000 ft (10,000 m) and more than 54% of the oceans' area is at pressure above 400 bars (40 megapascals), it is necessary to consider the effect of pressure, as well as temperature, on the physical properties of seawater. The pressure corresponding to the maximum ocean depth is about 1100 bars (110 MPa). The usual pressure dependence of viscosity is a consequence of the open structure of water which is altered by pressure, temperature, and solutes.

Because electromagnetic radiation can propagate in the ocean for only limited distances, sound waves are the principal means of communication in this medium. *See* UNDERWATER SOUND. [F.H.F.]

Composition of seawater. The concentrations of the various components of seawater are regulated by numerous chemical, physical, and biochemical reactions. The present-day compositions of seawaters (see table) are controlled both by the makeup of the ultimate source materials and by the large number of reactions, of chemical and physical natures, occurring in the oceans.

Only calcium, among the major cations of seawater, is present in a state of saturation, and such a situation generally occurs only in surface waters. Here its concentration is governed by the solubility of calcium carbonate. Barium concentrations in deep waters can be limited by the precipitation of barium sulfate. The noble gases and dissolved gaseous nitrogen have their marine concentrations determined by the temperature at which their water mass was in contact with the atmosphere and are in states of saturation or very nearly so.

The formation and alteration of minerals on the sea floor apparently are responsible for controlling the concentrations of the major cations Na, K, Mg, and Ca. Such clay minerals as illite, chlorite, and montmorillonite are presumably synthesized from these dissolved species and the river-transported weathered solids (aluminosilicates, such as kaolinite). The SiO_2 is introduced in part as diatom frustules.

Chemical abundances in the marine hydrosphere

Element	Concentration, mg/liter	Element	Concentration, mg/liter
H	108,000	Ag	0.0003
He	0.000007	Cd	0.00011
Li	0.17	In	0.000004
Be	0.0000006	Sn	0.0008
B	4.6	Sb	0.0003
C	28	Te	—
N	15	I	0.06
O	857,000	Xe	0.00005
F	1.2	Cs	0.0003
Ne	0.0001	Ba	0.03
Na	10,500	La	1.2×10^{-5}
Mg	1350	Ce	5.2×10^{-6}
Al	0.01	Pr	2.6×10^{-6}
Si	3.0	Nd	9.2×10^{-6}
P	0.07	Pm	—
S	885	Sm	1.7×10^{-6}
Cl	19,000	Eu	4.6×10^{-7}
A	0.45	Gd	2.4×10^{-6}
K	380	Tb	—
Ca	400	Dy	2.9×10^{-6}
Sc	<0.00004	Ho	8.8×10^{-7}
Ti	0.001	Er	2.4×10^{-6}
V	0.002	Tm	5.2×10^{-7}
Cr	0.00005	Yb	2.0×10^{-6}
Mn	0.002	Lu	4.8×10^{-7}
Fe	0.01	Hf	<0.000008
Co	0.0004	Ta	<0.000003
Ni	0.007	W	0.0001
Cu	0.003		0.0000084
Zn	0.01	Os	—
Ga	0.00003	Ir	—
Ge	0.00006	Pt	—
As	0.003	Au	0.00001
Se	0.00009	Hg	0.0002
Br	65	Ti	<0.00001
Kr	0.0002	Pb	0.00003
Rb	0.12	Bi	0.00002
Sr	8.0	Po	—
Y	0.00001	At	—
Zr	0.00002	Rn	0.6×10^{-15}
Nb	0.00001	Fr	—
Mo	0.01	Ra	1.0×10^{-10}
Tc	—	Ac	—
Ru	0.0000007	Th	0.000001
Rh	—	Pa	2.0×10^{-9}
Pd	—	U	0.003

The formation of ferromanganese minerals, sea-floor precipitates of iron and manganese oxides, may govern the concentrations of a suite of trace metals, including zinc, manganese, copper, nickel, and cobalt. These elements are in highly undersaturated states in seawater but are highly enriched in these marine ores, the so-called manganese nodules. *See* AUTHIGENIC MINERALS.

Cation-exchange reactions between positively charged species in seawater and such minerals as the marine clays and zeolites appear to regulate, at least in part, the amounts of sodium, potassium, and magnesium, as well as other members of the alkali and alkaline-earth metals which are not major participants in mineral formations.

Superimposed upon the chemical processes are changes in the chemical makeup of seawater by the melting of ice, evaporation, mixing with runoff waters from the continents, and upwelling of deeper waters. The net effects of the first three processes are changes in the absolute concentrations of all of the elements but with no major changes in the relative amounts of the dissolved species. [E.D.G.]

In the open sea, all the organic matter is produced by the photosynthesis and growth of unicellular planktonic forms. During this growth, all the elements essential for living matter are obtained from the seawater. Some elements are present in great excess, such as the carbon of CO_2, the potassium, and the sulfur (as sulfate). Other elements—for example, phosphorus, nitrogen, and silicon—are present in small enough quantities so that plant growth removes virtually all of the supply from the water. During photosynthesis, as these elements are being removed from the water, oxygen is released. *See* PHOTOSYNTHESIS.

The organic matter formed by photosynthesis and growth of the unicellular plants may be largely eaten by the zooplankton, and these in turn form the food for larger organisms. At each step of the food chain a large proportion of the eaten material is digested and excreted, and this, along with dead organisms, is decomposed by bacterial action. The decomposition process removes oxygen from the water and returns to the water those elements previously absorbed by the phytoplankton.

The distribution of oxygen and essential nutrient elements in the sea is modified by the spatial separation of these biological processes. Photosynthesis is limited to the surface layers of the ocean, generally no more than 330 ft (100 m) or so of depth, but the decomposition of organic material may take place at any depth. Reflecting this separation of processes, the concentration of nutrient elements in the surface is low, rises to maximum values at intermediate depths 990–2600 ft (300–800 m), and decreases slightly to fairly constant values which extend nearly to the bottom. *See* SEAWATER FERTILITY. [B.H.K.]

Seasons The four divisions of the year based upon variations of sunlight intensity (solar energy per unit area at the Earth's surface) at local solar noon (noontime) and daylight period. The variations in noontime intensity and daylight period are the result of the Earth's rotational axis being tilted 23°.5 from the perpendicular to the plane of the Earth's orbit around the Sun. The direction of the Earth's axis with respect to the stars remains fixed as the Earth orbits the Sun. If the Earth's axis were not tilted from the perpendicular, there would be no variation in noontime sunlight intensity or daylight period and no seasons. The Earth-Sun distance does not influence the seasons because it varies only slightly, and it is overwhelmed by the effects of variations in sunlight intensity and daylight period due to the alignment of the Earth's axis.

Sunlight intensity at a location depends upon the angle from the horizon to the Sun at local solar noon; this angle in turn depends upon the location's latitude and the position of the Earth in its orbit. At increased angles, a given amount of sunlight is spread over smaller surface areas, resulting in a greater concentration of solar energy, which produces increased surface heating. Intensity is a more important factor than the number of daylight hours in determining the heating effect at the Earth's surface. *See* EARTH ROTATION AND ORBITAL MOTION. [H.P.C.]

Seawater fertility A measure of the potential ability of seawater to support life. Fertility is distinguished from productivity, which is the actual production of living material by various trophic levels of the food web. Fertility is a broader and more general description of the biological activity of a region of the sea, while primary production, secondary production, and so on, is a quantitative description of the biological growth at a specified time and place by a certain trophic level. *See* BIOLOGICAL PRODUCTIVITY.

The potential of the sea to support growth of living organisms is determined by the fertilizer elements that marine plants need for growth. Fertilizers, or inorganic nutrients as they are called in oceanography, are required only by the first trophic level in the food web, the primary producers; but the supply of inorganic nutrients is a fertility-regulating process whose effect reaches throughout the food web. When there is an abundant supply to the surface layer of the ocean that is taken up by marine plants and converted into organic matter through photosynthesis, the entire food web is enriched, including zooplankton, fish, birds, whales, benthic invertebrates, protozoa, and bacteria.

The elements needed by marine plants for growth are divided into two categories depending on the quantities required: The major nutrient elements that appear to determine variations in ocean fertility are nitrogen, phosphorus, and silicon. The micronutrients are elements required in extremely small, or trace, quantities including essential metals such as iron, manganese, zinc, cobalt, magnesium, and copper, as well as vitamins and specific organic growth factors such as chelators. The overall pattern of ocean fertility is set by the major fertilizer elements—nitrogen, phosphorus, and silicon—and not by micronutrients.

The regions of the world's oceans differ dramatically in overall fertility. Extremes exist because the overall pattern of fertility is determined by the processes that transport nutrients to the sunlit upper layer of the ocean where there is energy for photosynthesis. *See* SEAWATER. [R.T.B.]

Sebaceous gland A gland which produces and liberates sebum, a mixture composed of fat, cellular debris, and keratin. When the gland arises in association with a hair follicle, it forms a thickened outpushing from the side of the developing follicle near the epidermis. Central cells in these sebaceous glands form oil droplets within the cytoplasm. These cells disintegrate to liberate the sebaceous substance and are therefore of the holocrine type. The Meibomian or tarsal glands, within the tarsus or supporting plate at the edge of the eyelids, are sebaceous and complex tubuloacinous structures. The numerous separate glands open along the entire edge of the upper and lower lids. Retained secretions of the tarsal glands produce a chalozion or Meibomian cyst. *See* GLAND. [O.E.N.]

Second sound A type of wave propagated in the superfluid phase of liquid helium (helium II) and in certain other substances under special conditions. The name is misleading since second sound is not in any sense a sound wave, but a temperature or entropy wave. In ordinary or first sound, pressure and density variations propagate with very small accompanying variations in temperature; in second sound, temperature variations propagate with no appreciable variation in density or pressure. *See* LIQUID HELIUM; SUPERFLUIDITY.

The two-fluid model of helium II provides further insight into the nature of second sound. In this model the liquid can be described as consisting of superfluid and normal components of densities ρ_s and ρ_n, respectively, such that the total density $\rho = \rho_s + \rho_n$. The superfluid component is frictionless and devoid of entropy; the normal component has a normal viscosity and contains the entropy and thermal energy of the system. In a temperature or second-sound wave, the normal and superfluid flows are oppositely directed so that $\rho_s \mathbf{V}_s + \rho_n \mathbf{V}_n = 0$, where $\mathbf{V}_s$ and $\mathbf{V}_n$ are the superfluid and normal flow velocities. Thus a variation in relative densities of the two components, and hence a temperature fluctuation, propagates with no change in total density or pressure. In a first-sound wave, the two components move in phase, that is, $\mathbf{V}_n \cong \mathbf{V}_s$.

Theoretical predictions that second sound should exist in certain solid dielectric crystals under suitable conditions have been confirmed experimentally for solid helium single crystals at temperatures between 0.4 and 1.0 K (−459.0 and −457.9°F). *See* DIELECTRIC MATERIALS.

Another quite different class of materials can exhibit second sound. In smectic A liquid crystals, when the wave vector is oblique with respect to the layers of these ordered structures, a modulation of the interlayer spacing can propagate at nearly constant density. *See* LIQUID CRYSTALS. [H.A.F.]

Secondary emission The emission of electrons from the surface of a solid into vacuum caused by bombardment with charged particles, in particular with electrons. The mechanism of secondary emission under ion bombardment is quite different from that under electron bombardment; it is only in the latter case that the term secondary emission is generally used.

The bombarding electrons and the emitted electrons are referred to, respectively, as primaries and secondaries. Secondary emission has important practical applications because the secondary yield, that is, the number of secondaries emitted per incident primary, may exceed unity. Thus, secondary emitters are used in electron multipliers, especially in photomultipliers, and in other electronic devices such as television pickup tubes, storage tubes for electronic computers, and so on.

The emission of secondary electrons can be described as the result of three processes: (1) excitation of electrons in the solid into high-energy states by the impact of high-energy primary electrons, (2) transport of these secondary electrons to the solid-vacuum interface, and (3) escape of the electrons over the surface barrier into the vacuum. The efficiency of each of these three processes, and hence the magnitude of the secondary emission yield δ, varies greatly for different materials.

Most of the materials used in practical devices are semiconductors or insulators whose band-gap energies are much larger than their electron affinities. Examples are magnesium oxide (MgO), beryllium oxide (BeO), cesium antimonide (Cs_3Sb), and potassium chloride (KCl). Maximum δ values in the 8–15 range are typically obtained at primary energies of several hundred volts.

In certain semiconductors the bands are bent downward to such an extent that the vacuum level lies below the bottom of the conduction band in the bulk. A material with this characteristic is said to have negative effective electron affinity. The most important material in this category is cesium-activated gallium phosphide, GaP(Cs). The illustration shows the curve of yield δ versus primary energy E_p for GaP(Cs) by comparison with MgO. Values of δ exceeding 100 are readily obtained, with maximum yields at energies in the 5–10-kV region. *See* BAND THEORY OF SOLIDS; SEMICONDUCTOR. [A.H.So.]

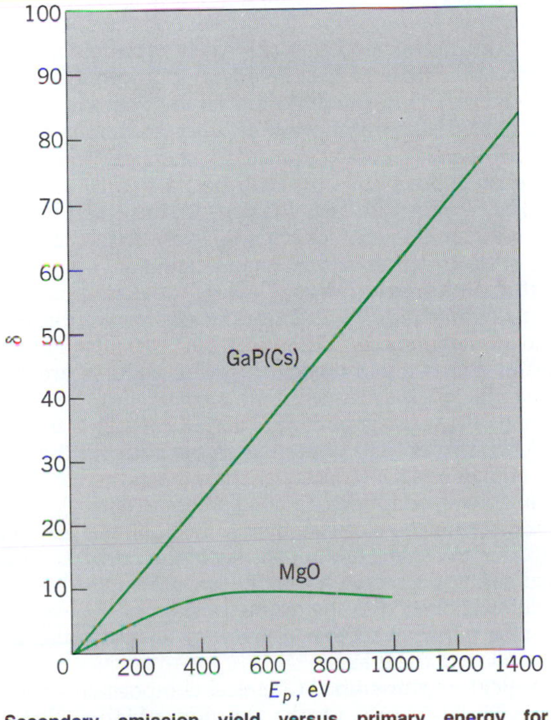

Secondary emission yield versus primary energy for GaP(Cs). The curve for MgO is shown for comparison.

Secondary ion mass spectrometry (SIMS) A technique for microchemical analysis that permits the visualization of a three-dimensional chemical image of the solid state using primary ions. It is also referred to as ion probing.

The ion probing of a material is accomplished by bombarding the sample with a 5–20-keV beam of primary ions. This process sputters off the surface layers of the sample, producing a variety of secondary species including neutral atoms and molecules, secondary electrons, photons, and positive and negative ions. Mass spectrometric analysis of the positive and negative sputtered secondary ions can provide microchemical characterization with lateral resolution better than that of the electron probe and can produce sampling depths similar to those of the ion scattering and Auger electron spectrometry techniques. In addition, mass spectrometric analysis provides full elemental coverage from hydrogen to uranium, including the capability of isotopic characterization. This combination of ion-bombardment excitation and mass spectrometric detection can provide detectabilities in the range of 10^{-15} to 10^{-19} g. *See* ELECTRON SPECTROSCOPY; MASS SPECTROMETRY; MASS SPECTROSCOPE.

Like the electron microprobe, the ion probe is used to analyze samples on a microscale. Applications include the analysis of small mineral samples or patches in geological and lunar materials and the analysis of airborne particulates and semiconductor devices. The ion probe provides much better elemental sensitivity and broader elemental coverage than does the electron probe in these applications. In addition, the ion probe provides isotopic detection, which is unavailable with the electron probe. Another feature of the microanalytical capability of the ion probe is the ability to acquire secondary ion images of the elements present in the sample under study. *See* ANALYTICAL CHEMISTRY. [C.A.E.]

Secretory structures (plant) Cells or organizations of cells which produce a variety of secretions. The secreted substance may remain deposited within the secretory cell itself or may be excreted, that is, released from the cell. Substances may be excreted to the surface of the plant or into intercellular cavities or canals. Some of the many substances contained in the secretions are not further utilized by the plant (resins, rubber, tannins, and various crystals), while others take part in the functions of the plant (enzymes and hormones). Secretory structures range from single cells scattered among other kinds of cells to complex structures involving many cells; the latter are often called glands.

Epidermal hairs of many plants are secretory or glandular. Such hairs commonly have a head composed of one or more secretory cells borne on a stalk. The hair of a stinging needle is bulbous below and extends into a long, fine process above. If one touches the hair, its tip breaks off, the sharp edge penetrates the skin, and the poisonous secretion is released.

Glands secreting a sugary liquid—the nectar—in flowers pollinated by insects are called nectaries. Nectaries may occur on the floral stalk or on any floral organ: sepal, petal, stamen, or ovary.

The hydathode structures discharge water—a phenomenon called guttation—through openings in margins or tips of leaves. The water flows through the xylem to its endings in the leaf and then through the intercellular spaces of the hydathode tissue toward the openings in the epidermis. Strictly speaking, such hydathodes are not glands because they are passive with regard to the flow of water.

Some carnivorous plants have glands that produce secretions capable of digesting insects and small animals. These glands occur on leaf parts modified as insect-trapping structures. In the sundews (*Drosera*) the traps bear stalked glands, called tentacles. When an insect lights on the leaf, the tentacles bend down and cover the victim with a mucilaginous secretion, the enzymes of which digest the insect. *See* INSECTIVOROUS PLANTS; VENUS' FLYTRAP.

Resin ducts are canals lined with secretory cells that release resins into the canal. Resin ducts are common in gymnosperms and occur in various tissues of roots, stems, leaves, and reproductive structures.

Gum ducts are similar to resin ducts and may contain resins, oils, and gums. Usually, the term gum duct is used with reference to the dicotyledons, although gum ducts also may occur in the gymnosperms.

Oil ducts are intercellular canals whose secretory cells produce oils or similar substances. Such ducts may be seen, for example, in various parts of the plant of the carrot family (Umbelliferae).

Laticifers are cells or systems of cells containing latex, a milky or clear, colored or colorless liquid. Latex occurs under pressure and exudes from the plant when the latter is cut. [K.E.]

Sedative A medication capable of producing a mild state of inhibition of the central nervous system (CNS) associated with reduced awareness of external stimuli. Numerous pharmacologic agents can induce different degrees of sedation, depending on the following variables: dosage; route of administration; absorption, metabolism, and excretion rates of the compound; specific receptor sites in the central nervous system that are affected by the agent; environmental setting; and state of the patient. *See* CENTRAL NERVOUS SYSTEM.

Ethanol was probably the first sedative compound and was widely used for its analgesic and hypnotic properties, as well as for its ability to decrease inhibitory anxiety with resultant relaxation and occasional euphoria. Subsequent sedative compounds include the barbiturates (for example, phenobarbital and secobarbital) and the benzodiazepines (for example, diazepam and alprazolam). Although these sedative compounds possess properties of tolerance and habituation, they vary in their addictive potential according to specific receptor sites and the particular type of patient. *See* ADDICTIVE DISORDERS; BARBITURATES.

Other classes of chemical agents that are used as sedatives include the antihistamines as well as some antidepressant drugs which possess sedative side effects in addition to their primary pharmacologic properties. Since these classes of compounds are not addicting, they can be safely used as hypnotics. *See* ANXIETY STATES; TRANQUILIZER. [D.M.Ga.]

Sedentaria A group of 28 or more families of polychaete annelids in which the anterior, or cephalic, region is more or less completely concealed by overhanging peristomial structures, or the body is divided into an anterior thoracic and a posterior abdominal region; the pharynx or proboscis is usually soft and epithelial, lacking hard jaws or paragnaths. *See* POLYCHAETA. [O.H.]

Sedimentary rocks One of the three major groups of rocks that make up the crust of the Earth, the other two being igneous and metamorphic. Most sedimentary rocks are layered, and, as is implied by the name, have originated by the sedimentation, or settling, of particles. Thus layering, or stratification, is the most important single characteristic of sediments and sedimentary rocks, even though there are some igneous and metamorphic rocks that show some kind of stratification or pseudostratification. *See* IGNEOUS ROCKS; METAMORPHIC ROCKS; SEDIMENTOLOGY.

The distinction between sedimentary and other rocks is understood best by considering their origin. Sediments are formed at or very near the surface of the Earth as a result of processes operating at the surface, at normal surface temperatures and pressures. Most igneous and metamorphic rocks, on the other hand, are formed as a result of conditions deep in the crust of the Earth, where temperatures and pressures may be very high. Some overlapping between the three rock families exists; for example, it is difficult to classify rocks that originate as volcanic ash falls (igneous) but are then transported and become interlayered with normal sediments.

Though sedimentary rocks account for only 5% of the Earth's outer crust, they make up 75% of the exposed rocks at the surface. In general, sediments are distributed as a rather thin layer at the surface. The thickness of this thin layer may vary greatly from place to place; the thickness of the total sedimentary volume may vary from a few tens of feet to well over 30,000 ft (9000 m) in some places where the crust is subsiding rapidly. Sedimentary rocks are important as the source of many major mineral resources, notably coal, oil and gas, iron ores, and limestone.

The three most abundant kinds of sedimentary rocks, shale, sandstone, and limestone, together account for over 95% of all sediments.

Origin. Sedimentary rocks originate primarily as the result of the fragmentation and destruction of preexisting rocks. As rocks are weathered by the action of water, wind, frost, and organic decay, large masses become mechanically broken into finer sizes and some of the constituents dissolve in rain or soil water. The solid fragments, ions in solution, and colloids in suspension, are transported, primarily by running water and secondarily by wind and groundwater, from the site of weathering, the source area, to this site of deposition. Transportation of detritus may be temporarily interrupted by sedimentation in streams or lakes, resulting in river bars, alluvial fans, or lake deltas. Eventually, however, most of the material reaches the site of the lowest gravitational potential energy on the Earth's surface, the bottom of the sea. After final deposition the soft, water-saturated muds, silts, and sands become buried under successive layers of later sediment, the water is squeezed out, the sediments become compacted, and chemical changes result in cementing the original unconsolidated material to a rock. *See* DIAGENESIS.

Sedimentary petrologists commonly divide the sedimentary rocks into two large groups, the detrital and the chemical. The detrital (sometimes called clastic) rocks are formed by the sedimentation of mineral or rock fragments that were derived from the mechanical disintegration of preexisting rocks in the source area and have been transported, as solids, to the site of deposition. The chemical rocks originate as chemical precipitates at the site of deposition; they may be inorganic precipitates formed from supersaturated solutions or they may be formed by the biochemical action of organisms, as are the calcium carbonate shells of mollusks. *See* PETROLOGY; ROCK.

Textural characteristics. Textural analysis has led to a fuller understanding of the genetic factors involved in the formation of sediment by settling of particles through a fluid (water) or gaseous (air) medium, for the textures are directly related to the hydrodynamics of the medium.

Perhaps the most important textural property is the size of individual particles. Size of particles is the basis for the division of the detrital rocks into lutite, or shale (fine); arenite, or sandstone (medium); and rudite, or conglomerate (coarse).

In addition to the size distribution, textural analysis includes the study of the shape and roundness of the particles. Shape is defined as the degree of sphericity, or approach to a sphere. Roundness is defined as the degree of sharpness of corners or edges of a particle. Another element of texture is the surface configuration of the grains.

Chemical composition. Chemical composition of rocks is expressed in terms of the oxides of the elements. Determinations are made normally by wet chemical analysis and the assumption is made that the elements are present as oxides, the oxygen not being determined directly. The average sedimentary rock contains about 58% silicon dioxide (SiO_2), 13% aluminum oxide (Al_2O_3), 6% calcium oxide (CaO), 5% ferrous oxide (FeO) and ferric oxide (Fe_2O_3), 5% carbon dioxide (CO_2), and smaller amounts of magnesium oxide (MgO), potassium oxide (K_2O), sodium oxide (Na_2O), and many trace elements. This average sediment may be thought of as a combination of sandstone, shale, and limestone.

Sedimentary structures. These are the larger textural features of sediments that were formed during or shortly after deposition, as distinct from the still larger elements of structure, folds and faults, which were produced much later than deposition. Sedimentary structures include the mechanical (such as bedding and ripple marks), made by the currents that transport sediment; the chemical (such as oolites, concretions, and nodules), produced by inhomogeneous precipitation; and the organic (such as fossils), produced by organisms living in the environment. *See* STRUCTURAL GEOLOGY.

Classification and nomenclature. Sedimentary rocks have been classified on two bases, the purely descriptive, and the genetic. The disadvantage of the purely descriptive approach is that rocks of widely divergent origin might be lumped together. The disadvantage of the genetic approach is that the origin of the rock must be known before it can be classified.

The more fruitful approach has been to combine the two into a classification based on objective descriptive characteristics that have some genetic significance. Using this as a basis, geologists subdivide the sedimentary rocks into two broad groups, the clastic or detrital, and the nonclastic or chemically precipitated rocks. The clastic rocks are characterized by individual grains, often of heterogeneous composition, derived from the erosion of preexisting rocks; the grains may be more or less rounded by abrasion during transport from the erosional area to the site of deposition. The chemical precipitates are normally fairly homogeneous in composition and are characterized by an interlocking crystalline texture, where the crystal sizes may vary greatly. *See* ARENACEOUS ROCKS; ARGILLACEOUS ROCKS; CHERT; CONGLOMERATE; LIMESTONE; SANDSTONE; SHALE; WEATHERING PROCESSES. [R.Si.]

Sedimentation (industry) A process based on the settling of solid particles through a liquid. Sedimentation is used in several ways in industrial processes. It may be used to obtain a concentrated slurry from a dilute suspension of a solid in a liquid. This is called thickening. It may also be used to remove solid particles from a liquid to obtain a clear supernatant liquid. This is called clarification. The driving force for the process is the difference in density between the solid and liquid. Ordinarily, sedimentation is accomplished by the force of gravity, and the liquid is water or an aqueous solution. When the particles are fine, or when the density difference is small, gravity settling may be too slow to be practicable; then centrifugal force can be used in place of gravity. When centrifugal force is not adequate, the more positive method of filtration may be used. All these methods of treating solids and liquids fall into the generic group of mechanical separations. *See* CENTRIFUGATION.

Particles too small to be settled at practicable rates are often amenable to flocculation, accomplished by adding agents such as sodium silicate, alum, lime, and alumina. The particles agglomerate into coarse flocs, which act like single large particles, settle at a practicable speed, and leave a clear supernatant liquid behind. *See* CLARIFICATION; FLOCCULATION; MECHANICAL SEPARATION TECHNIQUES; THICKENING. [W.L.McC.]

Sedimentology The study of natural sediments, both lithified (sedimentary rocks) and unlithified, and of the processes by which they are formed. Sedimentology includes all those processes that give rise to sediment or modify it after deposition: weathering, which breaks up or dissolves preexisting rocks so that sediment may form from them; mechanical transportation; deposition; and diagenesis, which modifies sediment after deposition and burial within a sedimentary basin and converts it into sedimentary rock. Sediments deposited by mechanical processes (gravels, sands, muds) are known as clastic sediments, and those deposited predominantly by chemical or biological processes (limestones, dolomites, rock salt, chert) are known as chemical sediments. *See* WEATHERING PROCESSES.

The raw materials of sedimentation are the products of weathering of previously formed igneous, metamorphic, or sedimentary rocks. In the present geological era, 66% of the continents and almost all of the ocean basins are covered by sedimentary rocks. Therefore, most of the sediment now forming has been derived by recycling previously formed sediment. Identification of the oldest rocks in the Earth's crust, formed more than 3×10^9 years ago, has shown that this process has been going on at least since then. Old sedimentary rocks tend to be eroded away or converted into metamorphic rocks, so that very ancient sedimentary rocks are seen at only a few places on Earth. *See* IGNEOUS ROCKS; METAMORPHIC ROCKS.

Major controls. The major controls on the sedimentary cycle are tectonics, climate, worldwide (eustatic) changes in sea level, the evolution of environments with geological time, and the effect of rare events.

Tectonics are the large-scale motions (both horizontal and vertical) of the Earth's crust; they are driven by forces within the interior of the Earth but have a large effect on sedimentation. These crustal movements largely determine which areas of the Earth's crust undergo uplift and erosion, thus acting as sources of sediment, and which areas undergo prolonged subsidence, thus acting as sedimentary basins. Rates of uplift may be very high (over 10 m or 33 ft per 1000 years) locally, but probably such rates prevail only for short periods of time. Over millions of years, uplift even in mountainous regions is about 1 m (3.3 ft) per 1000 years, and it is closely balanced by rates of erosion. Rates of erosion, estimated from measured rates of sediment transport in rivers and from various other techniques, range from a few meters per 1000 years in mountainous areas to a few millimeters per 1000 years averaged over entire continents. *See* BASIN; PLATE TECTONICS.

Climate plays a secondary but important role in controlling the rate of weathering and sediment production. The more humid the climate, the higher these rates are. A combination of hot, humid climate and low relief permits extensive chemical weathering, so that a larger percentage of source rocks goes into solution, and the clastic sediment produced consists mainly of those minerals that are chemically inert (such as quartz) or that are produced by weathering itself (clays). Cold climates and high relief favor physical over chemical processes. *See* CLIMATOLOGY.

Tectonics and climate together control the relative level of the sea. Changes in sea level, whether local or worldwide, strongly influence sedimentation in shallow seas and along coastlines; sea-level changes also affect sedimentation in rivers by changing the base level below which a stream cannot erode its bed. *See* SEA-LEVEL FLUCTUATIONS.

One of the major conclusions from the study of ancient sediments has been that the general nature and rates of sedimentation have been essentially unchanged during the last billion years of geological history. However, this conclusion, uniformitarianism, must be qualified to take into account progressive changes in the Earth's environment through geological time, and the operation of rare but locally or even globally important catastrophic events. The most important progressive changes have been in tectonics and atmospheric chemistry early in Precambrian times, and in the nature of life on the Earth, particularly since the beginning of the Cambrian. *See* CAMBRIAN.

Sediment is moved either by gravity acting on the sediment particles or by the motions of fluids (air, water, flowing ice), which are themselves produced by gravity. Deposition takes place when the rate of sediment movement decreases in the direction of sediment movement; deposition may be so abrupt that an entire moving mass of sediment and fluid comes to a halt (mass deposition, for example, by a debris flow), or so slow that the moving fluid (which may contain only a few parts per thousand of sediment) leaves only a few grains of sediment behind.

A fundamental dynamic property of sediment grains is their settling velocity. This property is the constant velocity of fall that a grain attains in a fluid at rest, when the force of gravity is balanced by the forces of buoyancy and fluid resistance. The settling velocity depends on the density and viscosity of the fluid, as well as on the size, shape, and density of the grains. *See* STREAM TRANSPORT AND DEPOSITION; TURBULENT FLOW.

Sedimentary environments. Sedimentary rocks preserve the main direct evidence about the nature of the surface environments of the ancient Earth and the way they have changed through geological time. Thus, besides trying to understand the basic principles of sedimentation, sedimentologists have studied modern and ancient sediments as records of ancient environments. For this purpose, fossils and primary sedimentary structures are the best guide. These structures are those formed at the time of deposition, as opposed to those formed after deposition by diagenesis, or by deformation. In describing sequences of sedimentary rocks in the field (stratigraphic sections), sedimentologists recognize compositional, structural, and organic aspects of rocks that can be used to distinguish one unit of rocks from another. Such units (sedimentary facies) can generally be interpreted as having formed in different environments of deposition. *See* FACIES (GEOLOGY); TRACE FOSSILS.

Basin analysis involves studies of the larger-scale aspects of sediment accumulation, such as the different types of sedimentary basins, and how they form, and the types of source areas and how they are linked to particular types of basins. Clastic sediments derived from preexisting rocks (terrigenous sediments) have textures and mineralogy that preserve clues about the nature of the source from which they were derived (their provenance). Sedimentary structures and textures can be used to indicate the direction of sediment transport at the place of deposition (paleocurrents). By carefully dating the deposits, or at least determining which deposits are the same age (stratigraphic correlation), by studying their provenance and paleocurrents, and by reconstructing their depositional environments, it is possible to reconstruct how a sedimentary basin developed and where the sediment came from. *See* SEDIMENTARY ROCKS. [G.V.M.]

Seebeck effect A thermoelectric phenomenon discovered in 1821 by T. J. Seebeck. He found that near a closed circuit composed of two linear conductors of different metals, a magnetic needle would be deflected if, and only if, the two junctions were at different temperatures; and, further, that if the cooler junction were to become the warmer, the direction of deflection would be reversed.

The Seebeck electromotive force which gives rise to the current is directly measurable with a voltmeter or potentiometer. It depends only upon the material and temperatures. [J.W.St.]

Seed A fertilized ovule containing an embryo which forms a new plant upon germination. Seed-bearing characterizes the higher plants—the gymnosperms (conifers and allies) and the angiosperms (flowering plants). Gymnosperm (naked) seeds arise on the surface of a structure, as on a seed scale of a pine cone. Angiosperm (covered) seeds develop within a fruit, as the peas in a pod. *See* FLOWER; FRUIT.

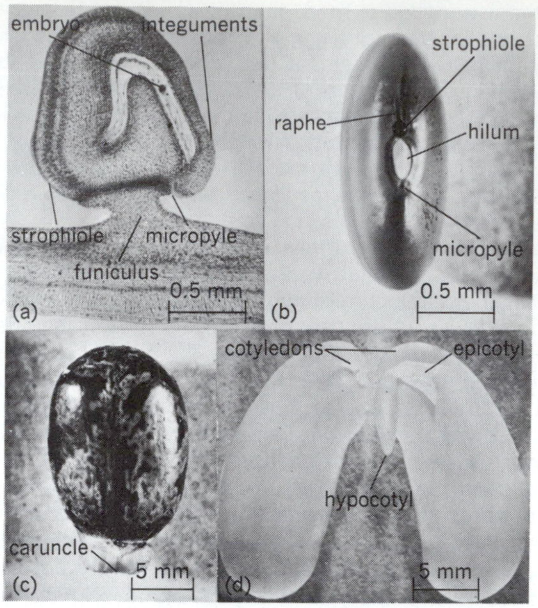

Seed structures. (*a*) Median longitudinal section of pea ovule shortly after fertilization, showing attachment to pod tissues. (*b*) Mature kidney bean. (*c*) Mature castor bean. (*d*) Opened embryo of mature kidney bean.

Structure. One or two tissue envelopes, or integuments, form the seed coat which encloses the seed except for a tiny pore, the micropyle (see illustration). The micropyle is near the funiculus (seed stalk) in angiosperm seeds. The hilum is the scar left when the seed is detached from the funiculus. The embryo consists of an axis and attached cotyledons (seed leaves). The part of the axis above the cotyledons is the epicotyl (plumule); that below, the hypocotyl, the lower end of which bears a more or less developed primordium of the root (radicle). The epicotyl, essentially a terminal bud, possesses an apical meristem (growing point) and, sometimes, leaf primordia. The seedling stem develops from the epicotyl. An apical meristem of the radicle produces the primary root of the seedling, and transition between root and stem occurs in the hypocotyl. *See* APICAL MERISTEM; MAGNOLIOPHYTA; PINOPHYTA; ROOT (BOTANY); STEM.

Two to many cotyledons occur in different gymnosperms. The angiosperms are divided into two major groups according to number of cotyledons: the monocotyledons and the dicotyledons. Mature gymnosperm seeds contain an endosperm which surrounds the embryo. In some mature dicotyledon seeds the endosperm persists, the cotyledons are flat and leaflike, and the epicotyl is simply an apical meristem. In other seeds, such as the bean, the growing embryo absorbs the endosperm, and food reserve for germination is stored in fleshy cotyledons. The endosperm persists in common monocotyledons, for example, corn and wheat, and the cotyledon, known as the scutellum, functions as an absorbing organ during germination. Grain embryos also possess a coleoptile and a coleorhiza sheathing the epicotyl and the radicle, respectively. The apical meristems of lateral seed roots also may be differentiated in the embryonic axis near the scutellum of some grains.

Dispersal. Mechanisms for seed dispersal include parts of both fruit and seed. Some dry fruits have membranous air sacs which aid in seed dispersal by water. In cockelburrs and beggarticks the seeds are enclosed in spiny, barbed fruits which are readily carried in the fur of animals. Seed dispersal by wind is aided by hair tufts of dandelion achenes, wings of elm and maple fruits and of conifer seeds, and by seed-coat hairs of willow, cottonwood, and milkweed seeds. Cotton fibers consist of greatly elongated epidermal hairs of the seed coat as do also the kapok fibers of the silk-cotton tree (*Ceiba pentandra*). [R.M.R.]

Seiche A short-period oscillation in an enclosed or semienclosed body of water, analogous to the free oscillation of water in a dish. The initial displacement of water from a level surface can arise from a variety of causes, and the restoring force is gravity, which always tends to maintain a level surface. Once formed, the oscillations are characteristic only of the geometry of the basin itself and may persist for many cycles before decaying under the influence of friction. *See* WAVE MOTION.

The most straightforward example of a seiche is that occurring in a long, narrow lake of uniform depth; in this case the water surface elevations have a simple physical analogy with the vibrations of a plucked string. Seiches are also a common feature of semienclosed basins such as bays, gulfs, and harbors, the natural period for an open-end rectangular basin of uniform depth being twice that of the same sized basin closed at both ends. A problem of great practical significance is the oscillation of water in harbors, which can generate strong and unpredictable currents affecting the safety of ships entering or leaving the harbor; under extreme conditions these can also cause damage to moored vessels.

A seiche can be generated when the water is subject to changes in wind or atmospheric pressure gradients or, in the case of semienclosed basins, by the oscillation of adjacent connected water bodies having a periodicity close to that of the seiche or of one of its harmonics. Other, less frequent causes of seiches include heavy precipitation over a portion of the lake, flood discharge from rivers, seismic disturbances, and submarine mudslides or slumps. [D.M.F.]

Seismic risk The probability that social or economic consequences of earthquakes will equal or exceed specified values at a site, at several sites, or in an area, during a specified exposure time.

Although the term seismic risk is occasionally used in a general sense to mean the potential for both the occurrence of natural phenomena and the economic and life loss associated with earthquakes, it is useful to differentiate between the concepts of seismic hazard and seismic risk. Seismic hazard may be defined as any physical phenomena that result either from surface faulting during shallow earthquakes or from the ground shaking resulting from an earthquake and that may produce adverse effects on human activities.

The exposure time is the time period of interest for seismic hazard or risk calculations. In practical applications, the exposure time may be considered to be the design lifetime of a building or the length of time over which the numbers of casualties will be estimated. [S.T.A.]

Seismic stratigraphy Determination of the nature of sedimentary rocks and their fluid content from analysis of seismic data. Seismic stratigraphy is divided into seismic-sequence (facies) analysis and reflection-character analysis.

In seismic-sequence analysis the first step is to separate seismic-sequence units, also called seismic-facies units. This is usually done by mapping unconformities where they are shown by angularity. Angularity below an unconformity may be produced by erosion at an angle across the former bedding surfaces or by toplap (offlap), and angularity above an unconformity may be produced by onlap or downlap, the latter distinction being based on geometry. The unconformities are then followed along reflections from the points where they cannot be so identified, advantage being taken of the fact that the unconformity reflection is often relatively strong. The procedure often followed is to mark angularities in reflections by small arrows before drawing in the boundaries. *See* UNCONFORMITY.

Reflection-character analysis may be based on information from boreholes which suggests that a particular interval may change nearby in a manner which increases its likelihood to contain hydrocarbon accumulations. Lateral changes in the wave shape of individual reflection events may suggest where the stratigraphic changes or hydrocarbon accumulations may be located.

Where sufficient information is available to develop a reliable model, expected changes are postulated and their effects are calculated and compared to observed seismic data. The procedure is called synthetic seismogram manufacture; it usually involves calculating seismic data based on sonic and density logs from boreholes, sometimes based on a model derived in some other way. The sonic and density data are then changed in the manner expected for a postulated stratigraphic change, and if the synthetic seismogram matches the actual seismic data sufficiently well, it implies that the changes in earth layering are similar to those in the model. *See* GEOPHYSICAL EXPLORATION; SEDIMENTARY ROCKS; SEISMOGRAPHIC INSTRUMENTATION; SEISMOLOGY; STRATIGRAPHY. [R.E.Sh.]

Seismographic instrumentation A variety of devices or systems of devices for measuring the motion of the Earth, generally as a result of the passing of seismic waves, but also as a result of earth tides or deformation in response to tectonic forces. Typically consisting of a sensor (seismometer), a signal conditioning element or elements (galvanometer, mechanical or electronic amplifier, filters, analog-to-digital conversion circuitry, telemetry modules, and so on), and a recording system (analog visible or direct, FM, or digital magnetic tape or disk), seismographs find a wide range of application, from industrial vibration measurement through petroleum exploration and earthquake studies to investigations of the Earth's gravity field.

The seismograph records a measure of ground motion as a function of time. The seismoscope is a simpler type of device which indicates only the occurrence of relatively strong ground shaking, not its time of occurrence or duration.

The seismometer is the basic sensing element in seismographic instruments, and is found in one of two fundamentally different types, termed inertial and strain. The inertial seismometer (Fig. 1) generates an output signal proportional to the relative motion between its frame (usually attached to the ground or the point of interest) and an internal inertial reference mass. The strain seismometer (Fig. 2), or, better, the linear extensometer, is an instrument capable of providing a continuous precise measurement of the distance between two points, spaced typically 3 ft to 0.6 mi (1 m to 1 km) apart.

A variation on the linear extensometer is the tiltmeter. Here change in the relative elevations of two points is monitored, usually with respect to a liquid-level surface. Horizontal distance between the reference points may be as little as a fraction of an inch or as large as several hundred feet. Tiltmeters are used primarily in monitoring long-term deformation of the

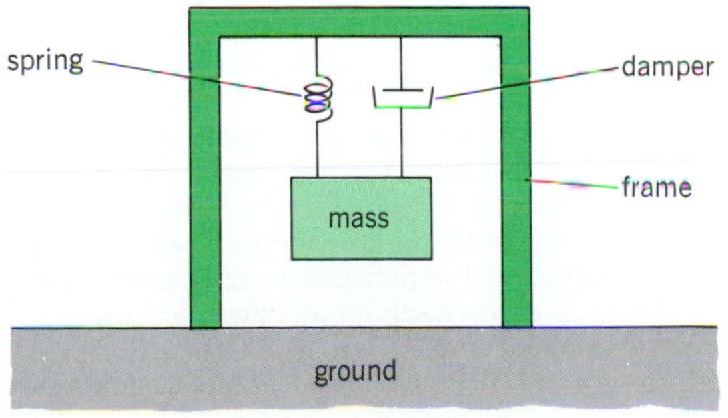

Fig. 1. Principle of vertical component inertial seismometer.

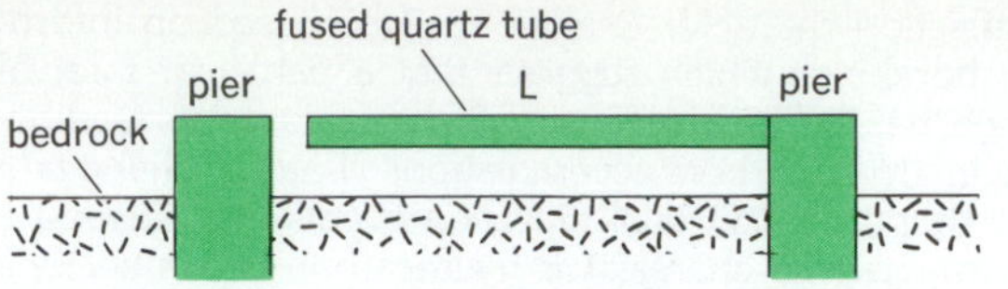

Fig. 2. Principle of linear extensometer in strain seismograph. Change in reference length *L* is measured.

Earth's surface due to geological processes, with particular application to investigations of possible precursory phenomena to earthquakes and volcanic eruptions.

The gravimeter is no more than a vertical-component accelerometer, that is, a pendulum sensing vertical ground motion and equipped with a displacement transducer, analogous to the above inertial tiltmeter. Gravimeters are widely used in geophysical exploration, in the study of earth tides, and in the recording of very-low-frequency (0.003 to 0.01 Hz) seismic waves from earthquakes.

The complete seismograph produces a record of the properly conditioned signal from the seismometer, along with appropriate timing information. The recording system may be as simple as a mechanical stylus scratching a line on a smoke-covered chart in a portable microearthquake seismograph, or as complex as a high-speed multichannel computer-controlled system handling 25,000 18-bit digital words per second. *See* EARTHQUAKE; GEOPHYSICAL EXPLORATION; SEISMOLOGY. [T.V.McE.]

Seismology The science of strain-wave propagation in the planets and their naturally occurring satellites. Strain waves propagate in the interior from a source and are recorded by seismographs distributed over the surface. From wave analysis it is possible to infer the subsurface structure and sometimes the mechanism of the source. *See* SEISMOGRAPHIC INSTRUMENTATION.

Earthquake. An earthquake is the abrupt release of strain waves through the action of natural forces within the Earth. The depth of occurrence of the earthquake focus (that is, the region of the Earth at which the earthquake originates) provides a basis for classification. Shallow-focus earthquakes occur at depths down to 36 mi (60 km), intermediate focus at 36–180 mi (60–300 km), and deep focus at 180–420 mi (300–700 km). Most earthquakes have shallow foci.

The size of an earthquake can be viewed in two ways, either subjectively by considering the intensity of shaking at a given location, or objectively by considering the magnitude of the earthquake source. A commonly used intensity scale is that of Medvedev, Sponheuer, and Kárník (msk scale). Observed shaking levels are thereby characterized numerically from I (not felt, but instrumentally observed) to XII (landscape permanently changed, and almost all human-made structures totally destroyed).

The magnitude scale is instrumental and characterizes the earthquake magnitude quantitatively, independently of individual sensations, and in a manner roughly related to the total energy output from the earthquake focus. Its development was originally the work of C. F. Richter in 1935, and it is still usually associated with his name despite modifications and extensions by others which have taken place since. *See* EARTHQUAKE.

Propagation theory. Understanding of seismic phenomena in the Earth relies heavily on the theory of strain propagation in solids. Observational experience indicates that the stress-strain relationship of the Earth for infinitesimal strains and seismic frequencies is highly elastic (with the possible exception of a surficial weathered layer less than 180 mi or 300 km thick) so that Hooke's law is applicable for seismic propagation theory to a high degree of approximation. *See* HOOKE'S LAW.

In a large solid body made up of regions which are individually homogeneous and isotropic, there can occur irrotational waves (sometimes called dilational or compressional waves) and equivoluminal waves. Since the velocity of the irrotational waves is always greater than that of the equivoluminal waves, irrotational waves arrive first and are designated by the symbol *P* (for primus), while equivoluminal waves coming in second are designed *S* (secundus). These are the basic body-wave types. Particle displacement in *P* is orthogonal to the wavefront whereas in *S* it is parallel (see illustration).

Considerations so far have been limited to the interior of an elastic solid. For seismic purposes the boundary of greatest interest is the free surface of the Earth. Special wave types called Rayleigh waves are characterized by particle trajectories which are retrograde ellipses; that is, when the particle is at the top of its ellipse, it is moving in a direction opposite to that of the propagation of the wave (see illustration). There is no component of particle motion which is both transverse to the ray path and parallel to the free surface of the solid supporting the Rayleigh-wave propagation. The speed of these waves is usually about 0.9 of the shear-wave velocity in the medium.

If the boundary value problem under consideration is further complicated by the introduction of a layer over the half space, an additional wave type will be found to propagate within the upper layer but with its particle motion horizontally polarized with respect to the bedding planes (see illustration). The Love wave is dispersive with phase velocities lying between the shear velocities for the two media involved.

Elastic waves incident upon an interface between two media having differing elastic properties undergo reflection and refraction analogously to electromagnetic and acoustic waves. In the elastic case the phenomena are more complicated because of the occurrence of wave conversions from *P* to *S* at the interface. *See* REFRACTION OF WAVES.

With the application of digital computers to seismic problems it has become feasible to study wave propagation in multiple horizontally layered structures with a large number of layers. Using available computational programs, it is possible to predict the seismogram which would result from an Earth model consisting of multiple horizontal layers or from an unlayered sphere, either liquid or solid.

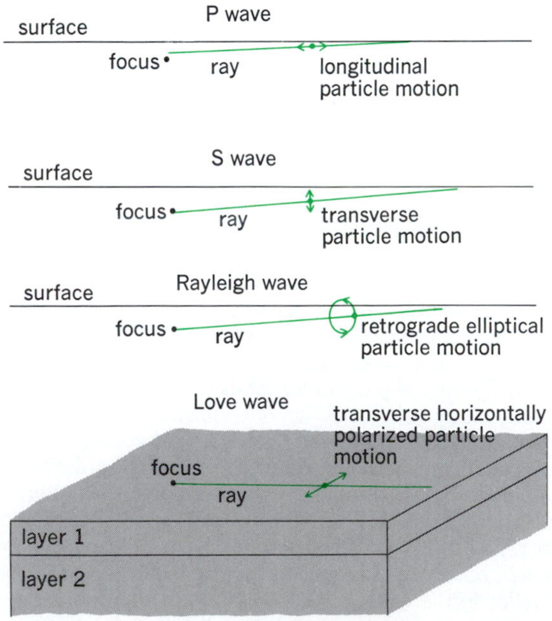

Fundamental seismic waves.

Earth's interior. The primary application of seismology has been to determine the internal structure of the Earth. In seismic prospecting the Earth is excited by explosives, dropping weights, eccentric wheels, and so forth, and the reflections and refractions of the seismic energy produced are observed and analyzed in terms of the subsurface structure. The motivation is usually the determination of sedimentary formations favorable for the occurrence of natural deposits of commercially valuable materials (usually petroleum). *See* GEOPHYSICAL EXPLORATION.

The Earth behaves like a low-pass filter to seismic waves so that at longer recording times, corresponding roughly to deeper penetration of the waves into the Earth, the waves will be longer than at shorter observation times, corresponding to closer proximity to the source and less penetration into the Earth. As a consequence, the structural detail available about the Earth from seismology is somewhat greater at shallower depths than nearer the Earth's center.

Monitoring nuclear explosions. An important research goal of seismology has been the development of adequate techniques for distinguishing underground nuclear explosions from earthquakes. The problem of remote monitoring of underground nuclear explosions by the detection and analysis of the seismic signals they produce has been reduced essentially to the questions of how weak a signal can be detected with confidence and whether the discrimination criteria for separating explosions from earthquakes, which have been established for moderate and strong events, are applicable to those yielding low levels of radiated energy. Seismic monitoring consists of three steps: detection of the signal from an event, location of the source, and identification of the source as an earthquake or an explosion. [K.C.T.]

Seisonacea A class of the phylum Rotifera which comprises a group of little-known marine animals. They form a single family with about seven species and are found only in Europe. The Seisonacea are epizoic or possibly ectoparasitic on crustacea. They have a very elongated jointed body with a small head; a long, slender neck region; a thicker, fusiform trunk; and an elongated foot, terminating in a perforated disk (see illustration). The Seisonacea are larger than other rotifers, attaining sizes up to 0. 12 in. (3 mm) in length. *See* ROTIFERA. [E.H.A.]

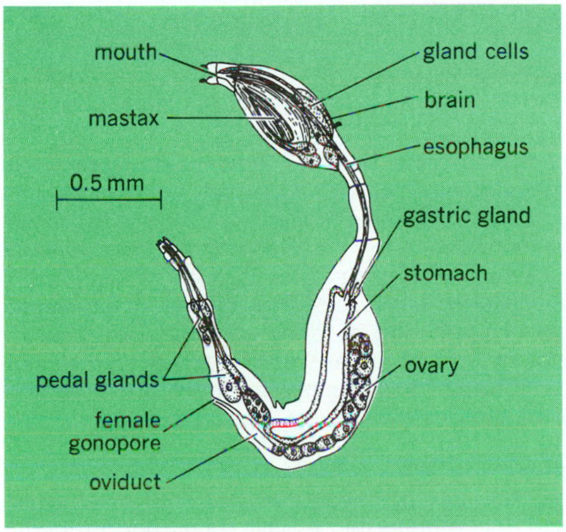

***Seison*, a rotifer. (*After Plate, 1887*)**

Seizure disorders Disorders characterized primarily by seizures involving the whole body or part of it, often—though not always or necessarily—associated with loss of consciousness. Seizures are defined as episodes of involuntary, generally violent muscular contractions, rhythmically alternated with periods of relaxation. These are part of the symptomatology of a large number of systemic and neurological diseases in which the brain is affected by various pathological processes (inflammatory, tumoral, degenerative, traumatic, and so forth), or they can result from various types of intoxications. They are the most characteristic, though not exclusive, manifestation of epilepsy and are commonly referred to as epileptic seizures. On the other hand, their occurrence, especially if in the form of isolated episodes, does not justify the diagnosis of epilepsy. *See* EPILEPSY. [C.A.M.]

Selachii An order that includes all fossil and Recent sharks, with the possible exception of Devonian cladoselachians. The selachian endoskeleton is entirely cartilaginous, but it is frequently calcified superficially. The body is covered in minute scales, which in Recent sharks are simple and nongrowing, but which in earlier sharks often fused together and became enlarged during life and may have been more regularly arranged over the body.

Sharks periodically shed their teeth, sometimes replacing several thousand in a lifetime, and their teeth are common in the fossil record. It is therefore rather surprising that the earliest records of sharks are from the Late Devonian, when they were already highly diversified.

There are 5 groups of fossil and living sharks: cladoselachians, symmoriids, ctenacanthids, hybodontids, and neoselachians. *See* CHONDRICHTHYES; SHARK. [J.G.Ma.]

Selaginellales The plant order of small club mosses, containing only one living genus, *Selaginella*. Of the approximately 700 species, most live in the tropics, although a few species give the genus a worldwide distribution in moist, shady habitats. One species, the "resurrection plant" (*S. lepidophylla*), is tolerant of xeric (dry) conditions, reacting to the absence of water by contracting into a ball-shaped mass which unrolls when moisture becomes available. A few species are cultivated for their ornamental value.

The order has many characteristics of other members of the phylum Lycopsida. It differs mainly from the order Lycopodiales in having two sizes of spores and a special, small membranous outgrowth called a ligule which is borne on the upper or adaxial base of the leaf. The sex organs always develop on separate gametophytes, and the sporangia always occur on clusters of sporophylls collectively called strobili. Arising at the base of each leafless branch is a rootlike organ called a rhizophore which bears adventitious roots at its tip. *See* LYCOPODIALES; ROOT (BOTANY). [P.A.V.]

Selection rules (physics) General rules concerning the transitions which may occur between the states of a quantum-mechanical physical system. They derive in almost all cases from the symmetry properties of the states and of the interaction which gives rise to the transitions. The system may have a classical (nonquantum) counterpart, and in this case the selection rules may often be related to the classical conserved quantities. A first use of selection rules is in determining the symmetry classes of the states; but in a great variety of ways they may yield other information about the system and the conservation laws. *See* QUANTUM MECHANICS; SYMMETRY LAWS (PHYSICS).

For an isolated system the total angular momentum is a conserved quantity; this fact derives from a fundamental fact of nature, namely, that space is isotropic. Each state is then classifiable by angular momentum J and its z component M ($= -J, -J+1, \ldots, +J$). Angular momenta combine in a vectorial fashion. Thus, if the system makes a particle-emitting transition $J_1,M_1 \rightarrow J_2,M_2$, the emitted particles must carry away angular momentum (j, μ), where $\mathbf{j} = \mathbf{J}_1 - \mathbf{J}_2$. This implies that $\mu = M_1 - M_2$ and that j takes on values $J_1 - J_2, J_1 - J_2+1, \ldots,$

(J_1+J_2). Thus in transitions ($J = 4$ [mim] $J = 2$) the possible j values comprise only 2, 3, 4, 5, 6, and, if it is also specified that $M_1 - M_2 = \pm 4$, only 4, 5, 6. Observe that J_z is additive. *See* ANGULAR MOMENTUM; QUANTUM NUMBERS.

Another fundamental symmetry, the parity, which determines the behavior of a system (or of its description) under inversion of the coordinate axes, is conserved by the strong and electromagnetic interactions, and gives a classification of systems as even ($\pi = +1$) or odd ($\pi = -1$). Under combination the parity combines multiplicatively. Thus, if the transition above is $4^{\pm} \rightarrow 2^{\pm}$, it follows that $j^{\pi} = 2^-, \ldots, 6^-$, while $4^{\pm} \rightarrow 2^{\pm}$ would give $j^{\pi} = 2^+, \ldots, 6^+$. The angular momentum **j** may be a combination of intrinsic spin **s** and orbital angular momentum **l**. Scalar, pseudoscalar, vector, and pseudovector particles are respectively characterized by $s^{\pi s} = 0^+, 0^-, 1^-, 1^+$, where π_s is the "intrinsic" parity, while l always carries $\pi_l = (-1)^l$. *See* PARITY (QUANTUM MECHANICS); SPIN (QUANTUM MECHANICS).

The isospin symmetry of the elementary particles is almost conserved, being broken by electromagnetic and weak interactions. It is described by the group SU(2), of unimodular unitary transformations in two dimensions. Since the SU(2) algebra is identical with that of the angular momentum SO(3), isospin behaves like angular momentum with its three generators **T** replacing **J**. *See* ISOBARIC SPIN.

The isospin group is a subgroup of SU(3) which defines a more complex fundamental symmetry of the elementary particles. Two of its eight generators commute, giving two additive quantum numbers, T_z and strangeness S' (or, equivalently, charge and hypercharge). The strangeness is conserved ($\Delta S' = 0$) for strong and electromagnetic, but not for weak, interactions. The selection rules and combination laws for SU(3) and its many extensions, and the quark-structure underlying them, correlate an enormous amount of information and make many predictions about the elementary particles. *See* BARYON; ELEMENTARY PARTICLE; MESON; QUARKS; UNITARY SYMMETRY.

A great variety of other groups have been introduced to define relevant symmetries for atoms, molecules, nuclei, and elementary particles. They all have their own selection rules, representing one aspect of the symmetries of nature. [J.B.Fr.]

Selectivity The ability of a radio receiver to separate a desired signal frequency from other signal frequencies, some of which may differ only slightly from the desired value. Selectivity is achieved by using tuned circuits that are sharply peaked and by increasing the number of tuned circuits. With a sharply peaked circuit, the output voltage falls off rapidly for frequencies increasingly lower or higher than that to which the circuit is tuned. *See* Q (ELECTRICITY); RADIO RECEIVER; RESONANCE (ALTERNATING-CURRENT CIRCUITS). [J.Mar.]

Selenium A chemical element, Se, atomic number 34, atomic weight 78.96. The properties of this element are similar to those of tellurium. *See* TELLURIUM.

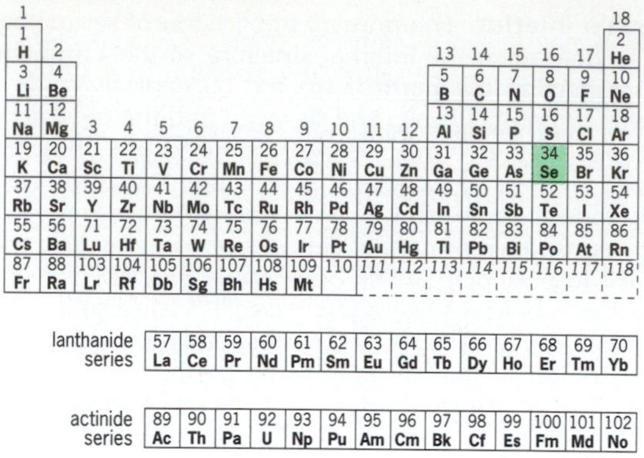

The abundance of this widely distributed element in the Earth's crust is estimated to be about $7 \times 10^{-5}\%$ by weight, occurring as the selenides of heavy elements and to a limited extent as the free element in association with elementary sulfur. Selenium minerals do not occur in sufficient quantity, however, to be useful as commercial sources of the element; instead, selenium-bearing copper sulfide ores represent the primary sources.

Major uses of selenium include the photocopying process of xerography, the decolorization of glasses tinted by the presence of iron compounds, and use as a pigment in plastics, paints, enamels, glass, ceramics, and inks. The use of selenium in rectifiers has declined with increased use of silicon and germanium for this purpose. Selenium is also employed in photographic exposure meters and as a metallurgical additive to improve the machinability of certain steels.

Selenium burns in air with a blue flame to give selenium dioxide, SeO_2. The element also reacts directly with a variety of metals and nonmetals, including hydrogen and the halogens. Nonoxidizing acids fail to react with selenium, but nitric acid, concentrated sulfuric acid, and strong alkali hydroxides dissolve the element.

The only important compound of selenium with hydrogen is hydrogen selenide, H_2Se, a colorless flammable gas possessing a distinctly unpleasant odor, and a toxicity greater and a thermal stability less than that of hydrogen sulfide. Dissolved in water, hydrogen selenide can precipitate many heavy-metal ions as very slightly soluble selenides. Organic compounds containing C-Se bonds are numerous and vary from the simple selenols, RSeH, selenenic acids, RSeOH, organyl selenium halides, RSeX, and diorganyl selenides and diselenides, R_2Se and R_2Se_2, to those molecules exhibiting biological activity such as selenoamino acids and selenopeptides. [J.W.G.]

Seligeriales An order of the true mosses (Bryidae), consisting of one family and five or six genera. The plants grow on rocks, and may be exceedingly small and gregarious to moderate in size and tufted. The stems are erect and simple or forked. The linear to lance-subulate leaves have a single costa which often fills the subula. The operculate capsules are terminal, and the peristome is usually present. The calyptra is generally cucullate. *See* BRYIDAE; BRYOPHYTA; BRYOPSIDA. [H.Cr.]

Semaeostomeae An order of the class Scyphozoa including most of the common medusae. The umbrella of these medusae is more flat than high and is usually domelike. The margin of the umbrella is divided into many lappets. Sensory organs are situated between the lappets. The tentacles are generally well developed and very long except in a few forms.

The life history of this group shows the typical alternation of generations, with forms passing through several larval stages. The Semaeostomeae are distributed mostly in temperate zones and are generally coastal forms with a few exceptions. Some of them are known to have violent poison on their tentacles. Several fossils of this group were found in the strata of the Jurassic period. *See* SCYPHOZOA. [T.U.]

Semiconductor A solid crystalline material whose electrical conductivity is intermediate between that of a metal and an insulator. Semiconductors exhibit conduction properties that may be temperature-dependent, permitting their use as thermistors, or voltage-dependent, as in varistors. By making suitable contacts to a semiconductor or by making the material suitably inhomogeneous, electrical rectification and amplification can be obtained. Semiconductor devices, rectifiers, and transistors have replaced vacuum tubes almost completely in low-power electronics, making it possible to save volume and power consumption by orders of magnitude. In the form of integrated circuits, they are vital for complicated systems. The optical properties of a semiconductor are important for the understanding and the application of the material. Photodiodes, photoconductive detectors of radiation, injection lasers, light-emitting diodes, solar-energy conversion cells, and so forth are examples of the wide variety of optoelectronic devices. *See* INTEGRATED CIRCUITS; LASER; LIGHT-EMITTING DIODE; PHOTODIODE; PHOTOELECTRIC DEVICES; SEMICONDUCTOR RECTIFIER; THERMISTOR; TRANSISTOR; VARISTOR.

Conduction in semiconductors. The electrical conductivity of semiconductors ranges from about 10^3 to 10^{-9} ohm^{-1} cm^{-1}, as compared with a maximum conductivity of 10^7 for good conductors and a minimum conductivity of 10^{-17} ohm^{-1} cm^{-1} for good insulators. *See* ELECTRIC INSULATOR.

The electric current is usually due only to the motion of electrons, although under some conditions, such as very high temperatures, the motion of ions may be important. The basic distinction between conduction in metals and in semiconductors is made by considering the energy bands occupied by the conduction electrons. *See* BAND THEORY OF SOLIDS; IONIC CRYSTALS.

At absolute zero temperature, the electrons occupy the lowest possible energy levels, with the restriction that at most two electrons may be in the same energy level. In semiconductors and insulators, there are just enough electrons to fill completely a number of energy bands, leaving the rest of the energy bands empty. The highest filled energy band is called the valence band. The next higher band, which is empty at absolute zero temperature, is called the conduction band. The conduction band is separated from the valence band by an energy gap which is an important characteristic of the semiconductor. In metals, the highest energy band that is occupied by the electrons is only partially filled. This condition exists either because the number of electrons is not just right to fill an integral number of energy bands or because the highest occupied energy band overlaps the next higher band without an intervening energy gap. The electrons in a partially filled band may acquire a small amount of energy from an applied electric field by going to the higher levels in the same band. The electrons are accelerated in a direction opposite to the field and thereby constitute an electric current. In semiconductors and insulators, the electrons are found only in completely filled bands, at low temperatures. In order to increase the energy of the electrons, it is necessary to raise electrons from the valence band to the conduction band across the energy gap. The electric fields normally encountered are not large enough to accomplish this with appreciable probability. At sufficiently high temperatures, depending on the magnitude of the energy gap, a significant number of valence electrons gain enough energy thermally to be raised to the conduction band. These electrons in an unfilled band can easily participate in conduction. Furthermore, there is now a corresponding number of vacancies in the electron population of the valence band. These vacancies, or holes as they are called, have the effect of carriers of positive charge, by means of which the valence band makes a contribution to the conduction of the crystal.

The type of charge carrier, electron or hole, that is in largest concentration in a material is sometimes called the majority carrier and the type in smallest concentration the minority carrier. The majority carriers are primarily responsible for the conduction properties of the material. Although the minority carriers play a minor role in electrical conductivity, they can be important in rectification and transistor actions in a semiconductor.

Intrinsic semiconductors. A semiconductor in which the concentration of charge carriers is characteristic of the material itself rather than of the content of impurities and structural defects of the crystal is called an intrinsic semiconductor. Electrons in the conduction band and holes in the valence band are created by thermal excitation of electrons from the valence to the conduction band. Thus an intrinsic semiconductor has equal concentrations of electrons and holes. The carrier concentration, and hence the conductivity, is very sensitive to temperature and depends strongly on the energy gap. The energy gap ranges from a fraction of 1 eV to several electron volts. A material must have a large energy gap to be an insulator.

Impurity semiconductors. Typical semiconductor crystals such as germanium and silicon are formed by an ordered bonding of the individual atoms to form the crystal structure. The bonding is attributed to the valence electrons which pair up with valence electrons of adjacent atoms to form so-called shared pair or covalent bonds. These materials are all of the quadrivalent type; that is, each atom contains four valence electrons, all of which are used in forming the crystal bonds. *See* CRYSTAL.

Atoms having a valence of +3 or +5 can be added to a pure or intrinsic semiconductor material with the result that the +3 atoms will give rise to an unsatisfied bond with one of the valence electrons of the semiconductor atoms, and +5 atoms will result in an extra or free electron that is not required in the bond structure. Electrically, the +3 impurities add holes and the +5 impurities add electrons. They are called acceptor and donor impurities, respectively. Typical valence +3 impurities used are boron, aluminum, indium, and gallium. Valence +5 impurities used are arsenic, antimony, and phosphorus.

Semiconductor material "doped" or "poisoned" by valence +3 acceptor impurities is termed *p*-type, whereas material doped by valence + 5 donor material is termed *n*-type. The names are derived from the fact that the holes introduced are considered to carry positive charges and the electrons negative charges. When the carrier concentration is predominantly determined by the impurity content, the conduction of the material is said to be extrinsic.

Materials. The group of chemical elements which are semiconductors includes germanium, silicon, gray (crystalline) tin, selenium, tellurium, and boron. Germanium, silicon, and gray tin belong to group 14 of the periodic table and have crystal structures similar to that of diamond. Germanium and silicon are two of the best-known semiconductors. They are used extensively in devices such as rectifiers and transistors.

A large number of compounds are known to be semiconductors. A group of semiconducting compounds of the simple type AB consists of elements from columns symmetrically placed with respect to column 14 of the periodic table. Indium antimonide (InSb), cadmium telluride (CdTe), and silver iodide (AgI) are examples of 13–15, 2–13, and 1–16 compounds, respectively. Many practical applications have been found for the various 13–15 compounds. These compounds have zincblende crystal structure which is geometrically similar to the diamond structure possessed by the elemental semiconductors, germanium and silicon, of column 14, except that the four nearest neighbors of each atom are atoms of the other kind.

The properties of semiconductors are extremely sensitive to the presence of impurities. It is therefore desirable to start with the purest available materials and to introduce a controlled amount of the desired impurity. The zone refining method is often used for further purification of obtainable materials. *See* ZONE REFINING.

For basic studies as well as for many practical applications, it is desirable to use single crystals. Various methods are used for growing crystals of different materials. For many semiconductors the Czochralski method is commonly used. The method of condensation from the vapor phase is used to grow crystals of a number of semiconductors.

The introduction of impurities, or doping, can be accomplished by simply adding the desired quantity to the melt from which the crystal is grown. When the amount to be added is very small, a preliminary ingot is often made with a larger content of the doping agent; a small slice of the ingot is then used to dope the next melt accurately. Impurities which have large diffusion constants in the material can be introduced directly by holding the solid material at an elevated temperature while this material is in contact with the doping agent in the solid or the vapor phase.

A doping technique, ion implantation, has been used extensively. The impurity is introduced into a layer of semiconductor by causing a controlled dose of highly accelerated impurity ions to impinge on the semiconductor. *See* ION IMPLANTATION.

A growing subject of scientific and technological interest is amorphous semiconductors. The representative amorphous semiconductors are selenium, germanium, and silicon in their amorphous states, and arsenic and germanium chalcogenides. *See* AMORPHOUS SOLID.

Rectification in semiconductors. In semiconductors, narrow layers can be produced which have abnormally high resistances. The resistance of such a layer is nonohmic; it may depend on the direction of current, thus giving rise to rectification. Rectification can also be obtained by putting a thin layer of semiconductor or insulator material between two conductors of different material.

A narrow region in a semiconductor which has an abnormally high resistance is called a barrier layer. A barrier may exist at the contact of the semiconductor with another material, at a crystal boundary in the semiconductor, or at a free surface of the semiconductor. In the bulk of a semiconductor, even in a single crystal, barriers may be found as the result of a nonuniform distribution of impurities. The thickness of a barrier layer is small, usually 10^{-3} to 10^{-5} cm (4×10^{-4} to 4×10^{-6} in.).

A barrier is usually associated with the existence of a space charge. In an intrinsic semiconductor, a region is electrically neutral if the concentration n of conduction electrons is equal to the concentration p of holes. Any deviation in the balance gives a space charge equal to $e(p - n)$, where e is the charge on an electron. In an extrinsic semiconductor, ionized donor atoms give a positive space charge and ionized acceptor atoms give a negative space charge. [H.Y.F.]

Semiconductor diode A two-terminal electronic device that utilizes the properties of the semiconductor from which it is constructed. In a semiconductor diode without a *pn* junction, the bulk properties of the semiconductor itself are used to make a device whose characteristics may be sensitive to light, temperature, or electric field. In a diode with a *pn* junction, the properties of the *pn* junction are used. The most important property of a *pn* junction is that, under ordinary conditions, it will allow electric current to flow in only one direction. Under the proper circumstances, however, a *pn* junction may also be used as a voltage-variable capacitance, a switch, a light source, a voltage regulator, or a means to convert light into electrical power. *See* SEMICONDUCTOR.

The conductivity of a semiconductor is proportional to the number of electrical carriers (electrons and holes) it contains. In a temperature-compensating diode, or thermistor, the number of carriers changes with temperature. *See* THERMISTOR.

In a photoconductor the semiconductor is packaged so that it may be exposed to light. Light photons whose energies are greater than the band gap can excite electrons from the valence band to the conduction band, increasing the number of electrical carriers in the semiconductor. *See* PHOTOCONDUCTIVITY.

In some semiconductors the conduction band has more than one minimum. This results in a region of negative differential conductivity, and a device operated in this region is unstable. The current pulsates at microwave frequencies, and the device, a Gunn diode, may be used as a microwave power source. *See* MICROWAVE SOLID-STATE DEVICES.

A rectifying junction is formed whenever two materials of different conductivity types are brought into contact. Most commonly, the two materials are an *n*-type and a *p*-type semiconductor, and the device is called a junction diode. However,

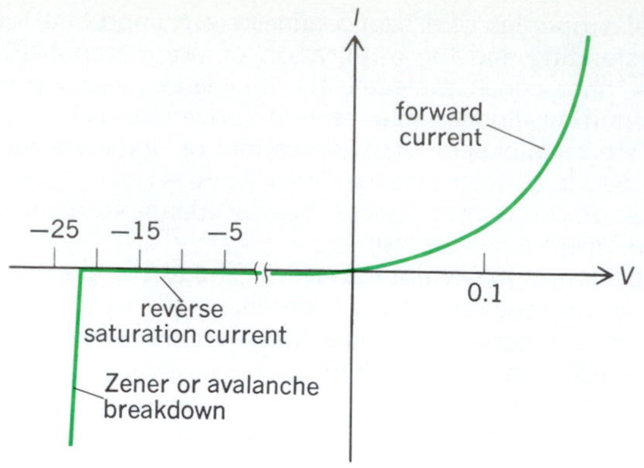

Current-voltage characteristic of a *pn* junction.

rectifying action also occurs at a boundary between a metal and a semiconductor of either type. If the metal contacts a large area of semiconductor, the device is known as a Schottky barrier diode; if the contact is a metal point, a point-contact diode is formed. *See* POINT-CONTACT DIODE; SCHOTTKY EFFECT.

The contact potential between the two materials in a diode creates a potential barrier which tends to keep electrons on the *n* side of the junction and holes on the *p* side. When the *p* side is made positive with respect to the *n* side by an applied field, the barrier height is lowered and the diode is forward biased. Majority electrons from the *n* side may flow easily to the *p* side, and majority holes from the *p* side may flow easily to the *n* side. When the *p* side is made negative, the barrier height is increased and the diode is reverse-biased. Then, only a small leakage current flows: Minority electrons from the *p* side flow into the *n* side, and minority holes from the *n* side flow into the *p* side. The current-voltage characteristic of a typical diode is shown in the illustration. Rectifying diodes can be made in a variety of sizes, and much practical use can be made of the fact that such a diode allows current to flow in essentially one direction only. *See* JUNCTION DIODE; SEMICONDUCTOR RECTIFIER; TUNNEL DIODE. [S.N.]

Semiconductor heterostructures Structures consisting of two different semiconductor materials in junction contact, with unique electrical or electrooptical characteristics. A heterojunction is a junction in a single crystal between two dissimilar semiconductors. The most important differences between the two semiconductors are generally in the energy gap and the refractive index. In semiconductor heterostructures, differences in energy gap permit spatial confinement of injected electrons and holes, while the differences in refractive index can be used to form optical waveguides. Semiconductor heterostructures have been used for diode lasers, light-emitting diodes, optical detector diodes, and solar cells. In fact, heterostructures must be used to obtain continuous operation of diode lasers at room temperature. Heterostructures also exhibit other interesting properties such as the quantization of confined carrier motion in ultrathin heterostructures and enhanced carrier mobility in modulation-doped heterostructures. Structures of current interest utilize III–V and IV–VI compounds having similar crystal structures and closely matched lattice constants. *See* BAND THEORY OF SOLIDS; LASER; LIGHT-EMITTING DIODE; OPTICAL DETECTORS; REFRACTION OF WAVES; SOLAR CELL.

The most intensively studied and thoroughly documented materials for heterostructures are GaAs and $Al_xGa_{1-x}As$. Several other III–V and IV–VI systems also are used for semiconductor heterostructures. A close lattice match is necessary in heterostructures in order to obtain high-quality crystal layers by epitaxial growth and thereby to prevent excessive carrier recombination at the heterojunction interface.

When the narrow energy gap layer in heterostructures becomes a few tens of nanometers or less in thickness, new effects that are associated with the quantization of confined carriers are observed. These ultrathin heterostructures are referred to as superlattices or quantum well structures, and they consist of alternating layers of GaAs and $Al_xGa_{1-x}As$. These structures are generally prepared by molecular-beam epitaxy. Each layer is 5 to 40 nanometers thick.

In the GaAs layers, the motion of the carriers is restricted in the direction perpendicular to the heterojunction interfaces, while they are free to move in the other two directions. The carriers can therefore be considered as a two-dimensional gas. The Schrödinger wave equation shows that the carriers moving in the confining direction can have only discrete bound states. *See* QUANTUM MECHANICS.

Another property of semiconductor heterostructures is illustrated by a modulation doping technique that spatially separates conduction electrons in the GaAs layer and their parent donor impurity atoms in the $Al_xGa_{1-x}As$ layer. Since the carrier mobility in semiconductors is decreased by the presence of ionized and neutral impurities, the carrier mobility in the modulation-doped GaAs is larger than for a GaAs layer doped with impurities to give the same free electron concentration. Higher carrier mobilities should permit preparation of devices that operate at higher frequencies than are possible with doped layers. *See* SEMICONDUCTOR. [H.C.C.]

Semiconductor laser A device in which laser action takes place through the stimulated recombination of free electrons in the conduction band of a direct-gap conductor with holes in the valence band. In recombining, the electrons give up energy corresponding nearly to the band gap. This energy is radiated as a light quantum. In direct-gap semiconductors, such as gallium arsenide, recombination occurs directly without the emission or absorption of a quantum of lattice vibrations. A flat junction between *p*-type and *n*-type material may be used. When a current is passed through this junction in the forward direction, a large number of holes and electrons are brought together. A light wave passing along the plane of the junction can be amplified by stimulating such recombination of electrons and holes. The ends of the semiconducting crystal provide the mirrors to complete the laser structure. *See* LASER; SEMICONDUCTOR. [S.F.J.; A.L.S.]

Semiconductor memories Devices for storing digital information that are fabricated by using integrated circuit technology. Semiconductor memories are widely used to store programs and data in almost every digital system. Initially developed as a replacement for magnetic core memories, which were used as the main computer storage memory, semiconductor memories started to appear in the early 1970s and have almost totally replaced core memories as the main computer memory elements.

Many different types of semiconductor memories are used in computer systems to perform various functions—bulk data storage, program storage, temporary storage, and cache (or intermediate) storage. Almost all of the memories are a form of random-access memory (RAM), where any storage location can be accessed in the same amount of time. However, there are many different types of RAMs, the most frequently used of which is the writable and readable memory that is simply referred to as a RAM.

Although the RAM acronym indicates the random-access capability, it is a misnomer since almost all semiconductor memories except for a few specialty types can be randomly accessed. A more appropriate name for the memory would be a read/write RAM to indicate that data can be written into the memory as well as be read out of it.

There are many other forms of semiconductor memories in use—mask-programmable read-only memories (ROMs), fuse-programmable read-only memories (PROMs), ultraviolet-erasable programmable read-only memories (UV EPROMs), electrically alterable read-only memories (EAROMs), electrically erasable programmable read-only memories (EE PROMs), and nonvolatile static RAMs (NV RAMs). All of these memory types are also randomly accessible, but their main distinguishing feature is that once information is loaded into the storage cells, it stays there even if the power is shut off.

The first of these memory types, the ROM, is programmed by the memory manufacturer during the actual device fabrication. Here, though, there are two types of ROMs; one is called last-mask or contact-mask programmable, and the other is often called a ground-up design.

As an alternative to the mask-programmable memories, semiconductor manufacturers developed all the other programmable memory types to permit the users to program the memories themselves. The first of the user-programmable devices was the fuse PROM. The fuse-programmable memories are one-time programmable memories—once the information is programmed in, it cannot be altered.

The birth of the microprocessor in the early 1970s brought with it new memory types that offered a feature never before available—reusability. Information stored in the memory could be erased—in the case of the UV EPROM, by an ultraviolet light, and in the case of the EAROM, EE PROM, or NV RAM, by an electrical signal—and then the circuit could be reprogrammed with new information that could be retained indefinitely. All of these memory types are starting to approach the ideal memory element for the computer, an element that combines the flexibility of the RAM with the permanence of the ROM when power is removed. *See* COMPUTER STORAGE TECHNOLOGY; INTEGRATED CIRCUITS; LOGIC CIRCUITS; MICROPROCESSOR. [D.Bur.]

Semiconductor rectifier An electrical component which conducts current preferentially in one direction and inhibits the flow of current in the other direction by utilizing the properties of a semiconductor material such as silicon. A major use of the semiconductor rectifier is as a component in electrical equipment designed to convert electrical power from alternating current (ac) to direct current (dc). Such total equipment, consisting of not only the semiconductor components but also circuit breakers, bus work, fuses, sharing reactors, control circuits, and so forth, is often referred to as a semiconductor rectifier equipment. *See* CIRCUIT BREAKER; CONTROLLED RECTIFIER; RECTIFIER; SEMICONDUCTOR.

In 1957 the first commercial silicon-controlled rectifier (SCR) was developed in the United States. This electrical semiconductor component has, besides the anode and cathode of the rectifier diode, a third control electrode, or gate, which allows initiating conduction through the device by means of a small trigger pulse signal from a suitable control circuit. By the late 1960s the SCR had become the dominant power-control component. Its application ranges from adjustable speed control of fractional horsepower motors, to lighting and heating controls, to control of the largest dc motors.

Silicon rectifier diodes. The electrical heart of the rectifier diode is the junction between a *p*-type region and an *n*-type region of conductivity within a monocrystalline slice of silicon semiconductor material called the pellet or the chip. The processing required to form such a *pn* junction is accomplished in commercially available silicon devices by either a diffusion or an alloying process. *See* SEMICONDUCTOR.

Once the *pn* junction is formed, it exhibits its rectifying characteristic by allowing current to flow in the conventional sense from anode (*p*-type) to cathode (*n*-type) under a very small forward bias voltage; when the voltage polarity is reversed, the *pn* junction blocks the flow of current and only a very small reverse blocking or leakage current will flow under high-reverse-voltage bias.

Single-pellet silicon diodes are commercially available in current ratings from under 1 A to the order of 5000 A. Diodes may be connected in parallel for greater current capability. Voltage ratings of single-pellet silicon rectifier diodes commercially available cover a range from 50 to 5000 V. Individual diodes can be connected in series for greater voltage capability and diodes can also be obtained prepackaged in series strings for high-voltage applications.

Schottky diodes. Unlike the *pn* junction of a regular silicon rectifier diode, the Schottky diode makes use of the rectification effect of a metal-to-silicon barrier. The Schottky diode, sometimes called Schottky barrier diode, overcomes a major limitation of the *pn* junction diode; being a majority-carrier device, it has both higher forward conductivity (lower forward voltage drop) and faster switching speeds than its minority-carrier *pn* junction counterpart. However, other factors confine its use to low-voltage power supply applications—chiefly its limited reverse blocking voltage, typically 45 V, with a maximum of 80 V available. Secondary shortcomings are high reverse blocking current and restricted temperature of operation, with commercial devices providing a maximum of 347°F (175°C), compared with 392°F (200°C) for *pn* junction diodes.

Integrated circuits used in computer and instrument systems commonly require voltages less than 15 V and frequently as low as 5 V. Thus, the advantage of low forward voltage drop and faster switching favors the Schottky rectifier diode, due to its inherent efficiency. This is particularly true for the high-frequency-switching regulator supply applications where power at switching frequencies of 20–50 kHz has to be rectified. The higher reverse blocking current of the Schottky diode is more than compensated for by the Schottky's superior switching speeds.

Silicon-controlled rectifier. The SCR is a triode reverse-blocking thyristor. Whereas diodes use two alternate layers of *pn*-type semiconductor material and transistors use three such layers, thyristor devices utilize four layers forming three or more junctions within a slice of silicon semiconductor material.

The most widely used of all thyristor devices for power control is the SCR. For specialized ac-switching power control, such as in lamp dimmers and heating controls, the bidirectional triode thyristor, popularly called the triac, has also come into widespread usage.

The illustration shows a typical arrangement of alternate *p* and *n* layers in an SCR structure. The thickness is exaggerated in proportion for clarity of illustration. With positive voltage on its cathode with respect to its anode, the SCR blocks the flow of reverse current in a manner similar to that of a conventional silicon rectifier diode. When the voltage is reversed, the SCR blocks forward current flow until a low-power trigger signal is applied between the gate terminal and the cathode, whereupon the SCR switches into a highly conductive state with a voltage drop of approximately 1 V between anode and cathode similar to that of the rectifier diode. Once in conduction, the SCR continues to conduct even after the gate signal is removed, provided the anode (or load) current remains above the holding current level, typically in the order of milliamperes. If anode current momentarily drops below the holding current level or if the anode voltage is momentarily reversed, the SCR reverts to its blocking state and the gate terminal regains control.

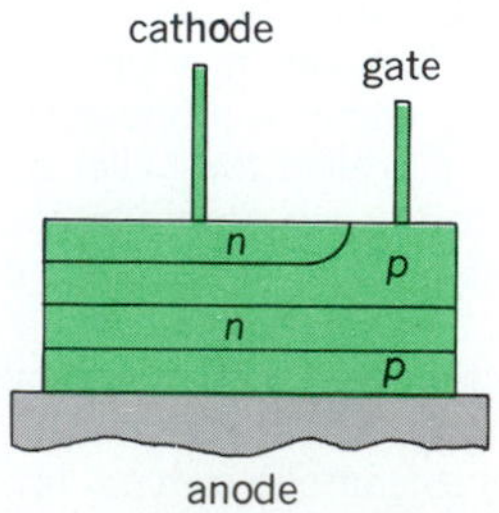

Diagrammatic view of typical SCR structure.

ASCRs, RCTs, and GTOs. These devices are all in the thyristor family and are mainly used in place of SCRs in power circuits requiring operation from a dc source. The ASCR, asymmetrical silicon-controlled rectifier, and RCT, reverse-conducting thyristor, have the advantage of faster turnoff time than the SCR, and thus require a less costly auxiliary circuit to effect turnoff. The RCT has an added circuit advantage, as it has a built-in reverse rectifier diode in parallel with the device. Along with faster turnoff times, the ASCR and RCT devices have lower forward-voltage drops for comparable forward-blocking voltage ratings and silicon area, thus increasing the device's efficiency.

The GTO, gate turnoff device, is also a thyristor. Like the SCR, it is a symmetrical reverse-blocking triode thyristor (unlike the ASCR and RCT, which cannot block reverse voltages), but it has the added advantage of being able to turn off current when a negative signal is applied to the gate. Thus the GTO does not require an auxiliary circuit to commutate it off as do the SCR, ASCR, and RCT devices. The added complexity of gate turnoff makes the GTO higher priced than similarly rated SCRS. [F.B.G.]

Semionotiformes An order of actinopterygian fishes which appeared first in the upper Permian, reached maximum development in the Triassic and Jurassic, and persists in the Recent fauna as the gars.

In the Semionotiformes the body is encased in a heavy armor of interlocking ganoid scales, which are thick, are more or less rhomboidal, and have an enamellike surface. Modern forms have an elongate body and bony jaws provided with enlarged conical teeth (see illustration).

Spotted gar (*Lepisosteus oculatus*). (*After G. B. Goode, Fishery Industries of the United States, 1884*)

The single Recent family, Lepisosteidae, contains one genus, *Lepisosteus*, with seven species restricted to lowland fresh and brackish waters of North and Central America. *See* ACTINOPTERYGII. [R.M.B.]

Sendai virus A member of the viruses in the type species Parainfluenza 1, genus *Paramyxovirus*, family Paramyxoviridae; it is also called hemagglutinating virus of Japan (HVJ). Sendai virus was originally recovered in Sendai, Japan, from mice inoculated with autopsy specimens from newborns who died of fatal pneumonitis in an epidemic in 1952. Subsequent attempts to isolate this virus from humans were, however, mostly unsuccessful, although mice are commonly infected with Sendai virus along with rats, guinea pigs, hamsters, and pigs. It is believed that the natural host of Sendai virus is the mouse and that the virus is usually nonpathogenic for humans. *See* ANIMAL VIRUS; PARAINFLUENZA VIRUS. [N.I.]

Senescence The deteriorative biological changes relating to aging in all living forms, including humans. Maturation and development during infancy and childhood are also age-related but are usually set apart from senescence.

From the moment the ovum is fertilized and a new individual is created until death in old age, the processes of aging are at work. As organs are formed and molded into their normal shapes, degeneration of excesses of cells takes place. In other words, the processes of cell multiplication, organ formation, and growth overlap the degenerative processes of aging at every moment in the life cycle. However, it is only after the attainment of full maturation and growth in early adulthood that the deteriorative effects of aging predominate over those of maturation and growth.

It is known that there are human structural and physiological changes that appear to be due to the aging process in every organ system. The impaired efficiency of the cardiovascular, pulmonary, renal, neuromuscular, and other systems from the fourth to the eighth decades of human life have been documented in many scientific studies. The declines are as follows: brain weight, 10%; blood flow to brain, 20%; cardia output at rest, 30%; number of glomeruli in kidneys, 44%; glomerular filtration rate, 31%; number of fibers in nerves, 37%; nerve conduction velocities, 10%; number of taste buds, 64%; maximum pulmonary ventilation, 44%; basal metabolic rate, 16%.

The structural changes in all the various organs consist of: reduction in the size of the constituent cells of the various organs; loss of cells; increase in connective, supporting (interstitial) tissues; accumulation of lipofuscin (wear-and-tear pigment) in the cytoplasm of cells; and reduction in the activity of many enzymes. Collagen, the substance of which connective tissues are composed, and elastin undergo structural and chemical alterations, accounting for the age changes in skin and the stiffening of blood vessels. Fibroblasts, the cells that produce collagen, are no longer as capable of mitotic division as in earlier life, and this causes a slowness in healing of wounds.

Senescence, then, appears to be a diverse and complex process with functional manifestations referable to all structures of the human organism. Of course the declining forces of vitality result ultimately in death. *See* AGING; DEATH. [R.D.A.]

Sensation A term commonly used to refer to the subjective experience resulting from stimulation of a sense organ, for instance, a sensation of warm, sour, or green. As a general scientific category, the study of sensation is the study of the operation of the senses. Sense receptors are the means by which information presented as one form of energy, for example, light, is converted to information in the form used by the nervous system, that is, impulses traveling along nerve fibers. *See* SENSE ORGAN.

Each sense has mechanisms and characteristics peculiar to itself, but all display the phenomena of absolute threshold, differential threshold, and adaptation. Not until sufficient stimulation impinges on a receptor can the presence of a stimulus be detected. The quantity of stimulation required is known as the absolute threshold. Not until a sufficient change occurs in some aspect of a stimulus can the change be detected. The magnitude of the change required is called the differential threshold. Under steady stimulation there is a decrease in sensitivity of the corresponding sense, as indicated by a shift in the absolute threshold and in the magnitude of sensation. After the stimulation ceases, sensitivity increases. An obvious example of visual adaptation occurs when one goes from bright to dim surroundings or vice versa.

With fairly good accuracy humans can localize visual objects, sounds, and cutaneous contacts and can discriminate the spatial orientation of the body and its members. With rather poor accuracy humans can localize many of the stimuli originating within the body.

With the exception of hearing, in which sense localization depends on differences in the acoustic stimuli reaching the two ears, there appears to be a common principle involved in giving spatially separated receptors their different local signs. Stimulation at different points on the receptive surface results in peaks of electrical activity at different loci in the brain. In no sense is there anything like a private wire from each sensory cell to a corresponding point in the brain. In fact, there are so many opportunities for a signal to go astray on its way from the receptor to the brain that it is surprising that spatial discrimination is as good as it is. Nevertheless, there is clear evidence that, by a combination of anatomical and functional arrangements, spatial differences at the receptor level are translated into topologically similar spatial differences in brain activity. *See* HEARING (HUMAN).

The nerve fibers between receptor and brain do not serve merely as transmitters of sensory information. Their interconnections enable them to influence one another's sensitivity and to perform logical operations like those carried out inside computers. As a result the information arriving in the sensory areas of the brain is not merely a more or less faithful replica of that presented to the receptors but in addition has had certain aspects of the information selected for special signaling. *See* CHEMICAL SENSES; PAIN; PARESTHESIA; SOMESTHESIS; TASTE; TOUCH; VISION. [J.F.H.]

Sense amplifier A device at the heart of a dynamic random-access memory (DRAM) that is meant to sense extremely small voltage differences between two DRAM bit lines and then latch-in the correct data state. In the simplest sense it can be likened to a cross-coupled latch or a static random-access memory (SRAM) cell. The major difference is that latches and SRAM cells are meant to be hard-driven to one state or another. The sensitivity of a sense amplifier can be varied in its design, but for DRAM applications it is usually set to sense a voltage difference from as low as 30 mV to as high as 70 mV. Sense amplifiers are often fabricated in complementary metal-oxide semiconductor (CMOS) technology, the most advanced current technology in memory design, but sense amplifiers have been designed in older technologies like *n*-channel MOS (NMOS).

The key element in the design of a sense amplifier is the consideration of circuit noise during sensing and latching of the data. When trying to sense 50 mV, any upset or noise on the nodes can cause the improper data state to be latched. The parasitic resistances and capicitances must be carefully considered in the design. These are also reflected in the design time delay from onset of sensing. *See* ELECTRICAL NOISE; SEMICONDUCTOR MEMORIES. [A.Hy.]

Sense organ A structure which is a receptor for external or internal stimulation. A sense organ is often referred to as a receptor organ. External stimuli affect the sensory structures which make up the general cutaneous surface of the body, the exteroceptive area, and the tissues of the body wall or the proprioceptive area. These somatic area receptors are known under the general term of exteroceptors. Internal stimuli which originate in various visceral organs such as the intestinal tract or heart affect the visceral sense organs or interoceptors. A receptor structure is not necessarily an organ; in many unicellular animals it is a specialized structure within the organism. Receptors are named on the basis of the stimulus which affects them, permitting the organism to be sensitive to changes in its environment.

Photoreceptors are structures which are sensitive to light and in some instances are also capable of perceiving form, that is, of forming images. Light-sensitive structures include the stigma of phytomonads, photoreceptor cells of some annelids, pigment cup ocelli and retinal cells in certain asteroids, the eyespot in many turbellarians, and the ocelli of arthropods. The compound eye of arthropods, mollusks, and chordates is capable of image formation and is also photosensitive. See PHOTORECEPTION.

Phonoreceptors are structures which are capable of detecting vibratory motion or sound waves in the environment. The most common phonoreceptor is the ear, which in the vertebrates has other functions in addition to sound perception. See EAR; PHONORECEPTION.

Statoreceptors are structures concerned primarily with equilibration, such as the statocysts found throughout the various phyla of invertebrates and the inner ear or membranous labyrinth filled with fluid.

The sense of smell is dependent upon the presence of olfactory neurons, called olfactoreceptors, in the olfactory epithelium of the nasal passages among the vertebrates. See OLFACTION.

The sense of taste is mediated by the taste buds, or gustatoreceptors. In most vertebrates these taste buds occur in the oral cavity, on the tongue, pharynx, and lining of the mouth; however, among certain species of fish, the body surface is supplied with taste buds as are the barbels of the catfish. See TASTE.

The surface skin of vertebrates contains numerous varied receptors associated with sensations of touch, pain, heat, and cold. See CHEMICAL SENSES; CUTANEOUS SENSATION; SENSATION.

[C.B.C.]

Sensible temperature The temperature at which air with some standard humidity, motion, and radiation would provide the same sensation of human comfort as existing atmospheric conditions. Of the many sensible temperature formulas thus far proposed, none is completely satisfactory or generally accepted. Some are purely empirical, modifying the actual temperature according to the humidity; others are theoretical, and express estimated heat loss rather than an equivalent temperature. See AIR TEMPERATURE; HUMIDITY.

Most used is effective temperature, represented by lines of equal comfort on a chart of dry-bulb versus wet-bulb temperature; it relates existing comfort to that in motionless, saturated air. Effective temperature is approximated by the temperature-humidity index (THI), given in °F by Eq. (1), where t is the sum

$$\text{THI} = 15 + 0.4t \qquad (1)$$

of the dry-bulb and wet-bulb temperatures. The THI is tabulated at major Weather Bureau stations, and accumulated into so-called cooling degree days. See TEMPERATURE-HUMIDITY INDEX.

Under cold conditions, atmospheric moisture is negligible and wind becomes important in heat removal. Wind chill, given by Eq. (2), does not estimate sensible temperature as

$$\text{Wind chill} = (10.45 + 10\sqrt{v} - v)\,(33 - t) \qquad (2)$$

such, but heat loss in kcal/m^2 h, for wind speed v in m/s and air temperature t in °C. It is used by the U.S. Army, government agencies, construction contractors, and others in cold areas.

[A.Cou.]

Sensitivity The ability of the output of a device or system to respond to an input stimulus. Mathematically, sensitivity is expressed as the ratio of the response or change induced in the output to a stimulus or change in the input. If the sensitivity varies with the level of the input signal, then sensitivity is usually expressed in terms of the derivative of the output with respect to the input at a specified input level.

The reciprocal of sensitivity is called the scale factor or figure of merit, and it represents the conversion factor by which the output indicator or scale reading must be multiplied to obtain the magnitude of the input. Occasionally the scale factor is called the sensitivity through loose usage.

[J.Mar.]

Sensory learning Learning in which an organism is trained to respond to changes in, or differences between, some aspects of the physical stimulus presented to one of the sense organs. Included are discrimination learning and perceptual learning, both of which are distinct from motor learning. Discrimination learning is a form of sensory learning in which the organism is required to respond to a difference between stimuli; reward or punishment or both are presented to provide differential reinforcement. Simple pavlovian conditioned-response formation is excluded from sensory learning. This involves the learning of a given muscular or glandular response as it becomes reliably elicited when a particular stimulus or sensory event occurs, usually following a series of trials with an effective eliciting stimulus.

Sensory preconditioning is a particular experimental approach to the study of sensory learning in which two or more stimuli, such as tones, lights, or skin contacts, are paired a few times. One of these stimuli is then made a conditioned stimulus to elicit a certain response. The test for preconditioning is to present the other element or elements from the original combination in order to determine its power to elicit the response. A positive outcome is accepted if the original pairing has been effective, that is, if it has resulted in a stronger or more reliable response than would be the case without the original pairing of stimuli. Simple stimulus generalization is ruled out by testing control subjects. See CONDITIONED REFLEX.

Examples of sensory learning are numerous and familiar. Learning a melody or development of familiarity with a face are two of the most common. Cross-modality associations are also formed in everyday experiences. A face recalls a voice, and vice versa. A particular landmark recalls another that is nearby but not simultaneously present. Much of knowledge (cognition) is derived from sequential experience that is the essence of sensory learning. See COGNITION; MEMORY. [A.H.R.]

Sepioidea An order of the class Cephalopoda (subclass Coleoidea) including the cuttlefishes (*Sepia*), the bobtail squids (*Sepiola*), and the ram's-horn squid (*Spirula*). The group is characterized by an internal shell that is calcareous and broad with closely packed laminate chambers (the cuttlebones of cuttlefishes). The mouth is surrounded by ten appendages (eight arms and two longer tentacles) that bear suckers with chitinous rings. The tentacles are contractile and retractile into pockets at their bases (see illustration).

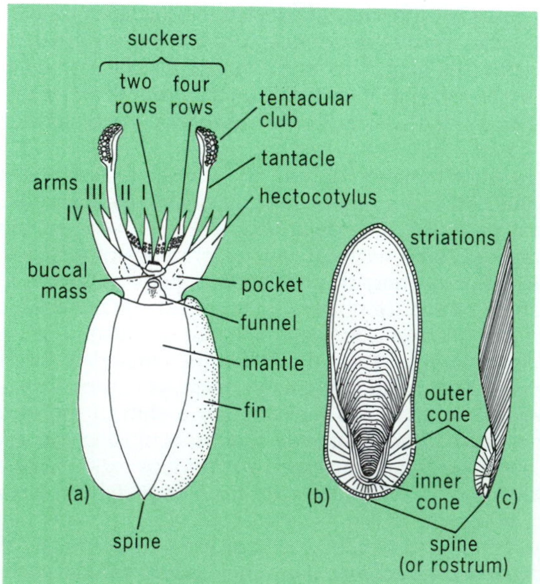

Diagram of some features of (*a*) a cuttlefish and of (*b*) a cuttlebone shown in ventral view and (*c*) in cross section.

Cuttlefishes are common benthic or epibenthic (living on or just above the bottom, respectively) animals that occur in the warm and temperate waters of the nearshore and continental shelf zones of the Old World, but they are excluded from the Western Hemisphere (North and South America). They prey on shrimps, crabs, small fishes, and other cuttlefishes. The sexes are separate, and during mating, which follows a colorful ritualistic courting behavior, sperm is transferred to the female in cylindrical packets (spermatophores) by a modified arm (the hectocotylus) of the male.

The cuttlebones are used to control the buoyancy of the cuttlefishes as fluid is pumped into and out of the laminar chambers. Dried cuttlebones are a source of calcium for cage birds and are used for fine jewelry molds, dentifrices, and cosmetics. The cuttlefishes eject an attention-getting blob of brownish-black ink when threatened by predators, then change color, become transparent, and jet-swim away, leaving the predator to attack the false body (pseudomorph) of ink. Artists have used the ink, called sepia, for centuries. Cuttlefishes are important in world fisheries; about 200,000 metric tons are caught each year for human food. *See* CEPHALOPODA; COLEOIDEA. [C.F.E.R.]

Sepiolite A complex hydrated magnesium silicate mineral named for its resemblance to cuttlefish bone, alternately named meerschaum (sea foam). The ideal composition, $Mg_8(H_2O)_4(OH)_4Si_{12}O_{30}$, is modified by some additional water of hydration, but is otherwise quite representative. Interlaced disoriented fibers aggregate into a massive stone so porous that it floats on water. These stones are easily carved, take a high polish with wax, and harden when warmed. *See* CLAY MINERALS; SILICATE MINERALS. [W.F.B.]

Septibranchia An order containing three marine genera in two families, Poromyidae and Cuspidariidae. Septibranchia show an extreme modification toward a scavenging habit. Small crustaceans that are trapped are ingested by the wide esophagus and crushed in the muscular stomach. Septibranchs can therefore be considered as predacious carnivores, unlike the majority of class Bivalvia which are microphagous filter feeders. *See* BIVALVIA; MOLLUSCA. [R.D.P.]

Septic tank A single-story settling tank in which settled sludge is in immediate contact with sewage flowing through the tank while solids are being decomposed by anaerobic bacterial action. Such tanks have limited use in municipal treatment, but are the primary resource for the treatment of sewage from individual residences which are not serviced by a sewage collection system.

Septic tanks have a capacity of approximately 1 day's flow. Since sludge is collected in the same unit, additional capacity is provided for sludge. About one-half of a 500-gal (1900-liter) tank is occupied by sludge in 5 years in an ordinary household installation. The majority of states require a minimum capacity of 500 gal (1900 liters) in a single tank. Home disposal units are buried in the ground and are not serviced until the system gives trouble from clogging or overflow.

Municipal and institutional units are designed to hold 12–14 hours' flow, with additional sludge capacity provided. Provision is made for sludge withdrawal about once a year.

Septic tank effluent is dangerous and odorous. It will contain pathogenic bacteria and sewage solids. Particles of sludge and scum are trapped in the flow and will cause nuisance at the point of discharge unless properly handled. *See* SEWAGE TREATMENT. [W.T.I.]

Septicemia A condition in which infection is disseminated through the body in the bloodstream; this has also been referred to as blood poisoning. Bacteremia and viremia are related terms used when the presence of bacteria or viruses in the blood has been demonstrated or is postulated. Usually in these states there is an identifiable local focus of infection from which the infecting agents are discharged into the blood. The principal clinical manifestations are fever, rigors, generalized aching, prostration, and local signs of infection in the lung, kidney, joints, and so on. Without treatment the death rate is high in septicemia, but many lives can be saved by administration of appropriate antimicrobial drugs. [P.B.B.]

Sequence stratigraphy The study of depositional relationships among sedimentary units (facies), which are distinguished by observable attributes in a time-stratigraphic framework. Sequence stratigraphy combines detailed analysis of sedimentary facies with depositional geometries to construct a hierarchy of time-stratigraphic units that stack into progressively larger-scale cycles. All elements of each cycle are produced in response to changes in accommodation space for accumulation of sediment within a depositional setting. The sequence stratigraphic mapping of the preserved depositional facies within each discrete time interval results in the construction of paleogeographic maps suitable for predicting the distribution of rock types either unexposed or removed by erosion. *See* FACIES (GEOLOGY); PALEOGEOGRAPHY.

Sequence stratigraphy can be divided into three primary areas of application: an interpretation methodology, a set of predictive depositional models, and a global correlation hypothesis and cycle chart. Sequence stratigraphy methodology is driven by observation and incorporates traditional stratigraphic practices. The methodology developed from a recognition that seismic reflection profiles imaged stratigraphic surfaces that paralleled relative time lines rather than depositional facies (formations). This observation has been reinforced by detailed physical and biostratigraphic correlation of rock units.

The primary unit of sequence stratigraphy is the depositional sequence which consists of genetically related strata bounded by surfaces of erosion or nondeposition and their correlative conformities. The term "genetically related" refers to the depositional sequence being deposited during one complete accommodation cycle, often known as one cycle of relative water-level change, consisting of four phases: rising, highstand, falling, and lowstand.

Each depositional sequence is composed of systems tracts defined by their position within the sequence and by the stacking pattern of smaller depositional units known as parasequences. Systems tracts consist of all sediments deposited during one phase of a depositional cycle. *See* SEDIMENTOLOGY; STRATIGRAPHIC NOMENCLATURE.

Sequence stratigraphic interpretation begins with the assembly of observations of stratigraphic geometries from outcrop studies, from correlation of wireline logs from well to well, or from interpretation of seismic reflection profiles. The bounding surfaces of each stratigraphic unit are defined and correlated through the study area. *See* WELL LOGGING.

The distribution of each depositional systems tract, calibrated as to specific depositional environments by using outcrop or well data, is used to construct maps that facilitate inference of depositional environments between calibration points, or extrapolation beyond wells or outcrops by using the regional depositional geometries defined by seismic record sections.

By using well data to calibrate seismic facies within each of the mapped systems tracts, the distribution of depositional environments can be inferred and then interpolated or extrapolated within the context of the mapped pattern. It is the systematic assembly of data from many sources into the time-stratigraphic framework of sequence stratigraphic depositional phases that forms the basis for predicting depositional patterns and sediment types beyond local control.

The fact remains that the cycles, rapid changes in both lake and sea levels (depositional cyclicity), are recorded in the stratigraphic record, and they are the consequence of primary forcing factors, including rate of sediment supply, variations in water-level cycles, climate change, and differing tectonic styles. These forcing factors have the potential to construct depositional geometries at several scales, including parasequence, parasequence set, sequence, and megasequence. The distribution of specific rock types and sedimentary facies will vary between depositional settings within each of these scales of depositional geometries. It is therefore important to clearly define the scale of observed patterns and to consider those scales in selecting the most appropriate interpretation of depositional setting and processes, and the inference of primary forcing factors. *See* STRATIGRAPHY. [J.M.A.]

Sequoia The giant sequoia or big tree (*Sequoia gigantea*) occupies a limited area in California and is said to be the oldest and most massive of all living things. The leaves are evergreen, scalelike, and overlapping on the branches. In height sequoia is a close second to the redwood (300–330 ft or 90–100 m) but the trunk is more massive. Sequoia trees may be 27–30 ft (8–9 m) in diameter 10 ft (3 m) from the ground. The stump of one tree showed 3400 annual rings. The red-brown bark is 1–2 ft (0.3–0.6 m) thick and spongy. Vertical grooves in the trunk give it a fluted appearance. The heartwood is dull purplish-brown and lighter and more brittle than that of the redwood. The wood and bark contain much tannin, which is probably the cause of the great resistance to insect and fungus attack. The most magnificent trees are within the General Grant and Sequoia National Parks. *See* PINALES; REDWOOD. [A.H.G./K.P.D.]

Series The indicated sum of a succession of numbers or terms. Series are used to obtain approximate values of infinite repeating decimals, to solve transcendental equations, to obtain values of logarithms or trigonometric functions, to evaluate integrals, and to solve boundary value problems.

For a finite series, with only a limited number of terms, the sum is found by addition. For an infinite series, with an unlimited number of terms, a sum or value can be assigned only by some limiting process. When the simplest such process yields a value, the infinite series is convergent. Many tests for convergence enable one to learn whether a sum can be found without actually finding it.

If each term of an infinite series involves a variable x and the series converges for each value of x in a certain range, the sum will be a function of x. Often the sum is a given function of x, $f(x)$, for which a series having terms of some given form is desired. Thus the Taylor's series expansion

$$f(x) = \sum_{n=0}^{\infty} f^{(n)}(a)\,\frac{(x-a)^n}{n!}$$

can be found for a large class of functions, the analytic functions, and represent such functions for sufficiently small values of $|x - a|$. For a much less restricted type of function on the interval $-\pi < x < \pi$, a Fourier series expansion of the form

$$\tfrac{1}{2}A_0 + \sum_{n=1}^{\infty} (A_n \cos nx + B_n \sin nx)$$

can be found.

Finite series. Here the problem of interest is to determine the sum of the first n terms,

$$S_n = u_0 + u_1 + u_2 + \cdots + u_{n-1}$$

when u_n is a given function of n. Examples are the arithmetic series, with $u_n = a + nd$ and $S_n = (n/2)[2a+(n-1)d]$, and the geometric series, with $u_n = ar^n$ and $S_n = a(1 - r^n)/(1 - r)$. *See* PROGRESSION (MATHEMATICS).

Convergence and divergence. An infinite series is the indicated sum of an unlimited number of terms

$$u_0+u_1+u_2+\cdots u_n \cdots$$

or more briefly

$$\sum_{n=0}^{\infty} u_k$$

or simply Σu_n, read "sigma of u_n." The sum S_n of the first n terms is known as the nth partial sum. Thus S_n is the finite sum

$$\sum_{k=0}^{n-1} u_k$$

If, as n increases indefinitely or becomes infinite, the partial sum S_n approaches a limit S, then the infinite series Σu_n is convergent. S denotes the sum or value of the series. For example, if $|r| < 1$,

$$S = \sum ar^n = \frac{a}{1-r} \text{ since } S_n = a\,\frac{1-r^n}{1-r}$$

If, as n becomes infinite, the partial sum S_n does not approach a finite limit, then the infinite series Σu_n is divergent. For example, $\Sigma 1$ diverges, since here $S_n = n$ becomes infinite with n. Also $\Sigma(-1)^n$ diverges, since here $S_n = \tfrac{1}{2}[1 - (-1)^{n+1}]$ which is alternately one and zero.

Positive series are series each of whose terms is a positive number or zero. For such series, the partial sum S_n increases as n increases. If for some fixed number A no sum S_n ever exceeds A, the sums are bounded and admit A as an upper bound. In this case, S_n must approach a limit, and the series is convergent. If every fixed number is exceeded by some S_n, the sums are unbounded. In this case, S_n must become positively infinite and the series is divergent. The tests for convergence of positive series are tests for boundedness, and this is shown by a comparison of S_n with the partial sums of another series or with an integral.

For any series Σu_n, which may have both positive and negative terms, the series of absolute values, $\Sigma |u_n|$, is a positive series whose convergence may be proved by one of the tests for positive series. If $\Sigma |u_n|$ converges, then Σu_n necessarily converges and is said to converge absolutely. The sum of an absolutely convergent series is independent of the order of the terms.

A series which converges but which does not converge absolutely is said to be conditionally convergent. For such a series, a change in the order of the terms may change the sum or cause divergence.

Power series. There are series with $u_n = a_n x^n$. For such a series, it may happen that

$$\lim_{n\to\infty} \left| \frac{a_{n+1}}{a_n} \right| = A$$

If $A = 0$, the series converges for all values of x. If $A \neq 0$, the series converges for all x of the interval $-1/A < x < 1/A$. It will diverge for all x with $|x| > 1/A$. For any power series, the interval of convergence is related in this way to a number A, which, however, in the general case has to be given by the superior limit of

$$\sqrt{|a_n|}$$

Similar remarks apply to the series with $u_n = a_n(x - c)^n$. Here the interval of convergence is $|x - c| < 1/A$.

One of the most important power series is the binomial series:

$$1 + mx + \frac{m(m-1)}{1\cdot 2}x^2 + \cdots + \frac{m(m-1)(m-2)\cdots(m-n+1)}{n!}x^n + \cdots$$

When m is a positive integer, this is a finite sum of $m + 1$ terms which equals $(1 + x)^m$ by the binomial theorem. When m is not a positive integer, the interval of convergence is $-1 < x$

< 1, and for x in this interval, the sum of the series is $(1 + x)^m$. *See* BINOMIAL THEOREM.

Let the power series $\Sigma a_n(x - c)^n$ have the sum function $f(x)$. Then $a_0 = f(c)$, $a_n = f^{(n)}(c)/n!$, and the series is the Taylor series of $f(x)$ at $x = c$. Thus, every power series whose interval of convergence has positive length can be put in the form

$$f(x) = f(c) + f'(c)\,\frac{x - c}{1!} + \cdots + f^{(n)}(c)\,\frac{(x - c)^n}{n!} + \cdots$$

where $f(x)$ is the sum function.

The Maclaurin series is the special case of Taylor series with $c = 0$:

$$f(x) = f(0) + f'(0)\,\frac{x}{1!} + \cdots + f^{(n)}(0)\,\frac{x^n}{n!} + \cdots$$

Uniform convergence. Let each term of a series be a function of z, $u_n = g_n(z)$. Let S_n be the sum of the first n terms, and S the sum to which the series converges for a particular value of z. Then $R_n = S - S_n$ is the remainder after n terms, and for the particular value of z, $\lim R_n$ must equal zero. If, for a given range of z, it is possible to make $R_n(z)$ arbitrarily small for sufficiently large n without specifying which z in the range is under consideration, the series converges uniformly. *See* FOURIER SERIES. [P.F./S.Bo.]

Series sometimes appear in disguised form in arithmetic. Thus the approximation of a rational number by an infinite repeating decimal is really a geometric series.

Series circuit An electric circuit in which the principal circuit elements have their terminals joined in sequence so that a common current flows through all the elements. The circuit may consist of any number of passive and active elements, such as resistors, inductors, capacitors, electron tubes, and transistors. *See* CIRCUIT (ELECTRICITY). [R.L.R.]

Serine An amino acid. Serine reacts with periodate to yield glyoxylate, ammonia, and formaldehyde. Serine is a biosynthetic precursor of several important metabolites: glycine,

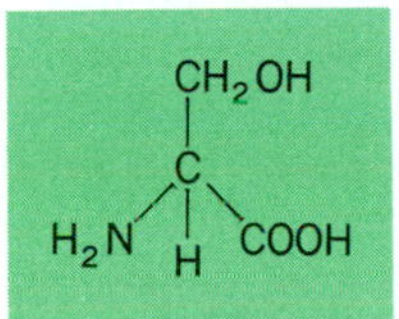

Serine

cysteine, choline (and hence betaine), and the side chain of tryptophan. *See* AMINO ACIDS; CYSTEINE; TRYPTOPHAN. [E.A.Ad.]

Serology The division of biological science concerned with antigen-antibody reactions in serum. It properly encompasses any of these reactions, but is often used in a limited sense to denote laboratory diagnostic tests, especially for syphilis. The techniques of blood grouping have come from the study of antigen-antibody reactions in serum, as have techniques for identification of genetic polymorphism and quantitation of numerous serum proteins. With these advances came the means for developing transfusion therapy with cells and plasma. In addition, these techniques led to identification of antibodies involved in incompatibility reactions, such as in erythroblastosis fetalis, and the development of effective measures to prevent their occurrence. Further, extension of these techniques to identification of antigens on white cells led to effective methods of histocompatibility typing, facilitating organ transplantation. *See* TRANSPLANTATION BIOLOGY. [D.R.]

Serotonin A compound, also known as 5-hydroxytryptamine, derived from tryptophan, an indole containing amino acid. It is widely distributed in the animal and vegetable kingdoms. In mammals it is found in gastrointestinal enterochromaffin cells, in blood platelets, and in brain and nerve tissue.

Serotonin is a local vasoconstrictor, plays a role in brain and nerve function and in regulation of gastric secretion and intestinal peristalsis, and has pharmacologic properties. Administration of serotonin produces a type of sedation and other depressant conditions of the nervous system. These facts suggest that in the brain serotonin may be a neurohormone.

The interaction of serotonin with alkaloids such as reserpine is of clinical importance. These drugs, which are used as tranquilizing agents in treatment of nervous and mental disorders, release serotonin from the brain. The hallucinogen D-lysergic acid diethylamide (LSD) may act by antagonism to serotonin. *See* TRANQUILIZER. [M.K.S.]

Serpentine The name applied to the hydrous magnesium silicate mineral assemblage of the rock serpentinite. Serpentinization alters host rocks to compositions approximating $3MgO \cdot 2SiO_2 \cdot 2H_2O$, and the names serpentine, chrysotile, antigorite, and others are applied with regard to degree of crystallinity achieved locally in given serpentinized groundmasses. *See* SERPENTINITE; SILICATE MINERALS.

The most important serpentine mineral is the fibrous variety, chrysotile, which accounts for as much as 95% of the asbestos of commerce. Although chrysotile is not a crystal in the sense of having plane-bounding faces, it nevertheless has a crystal structure. *See* ASBESTOS.

Serpentine is easily cut and polished for ornamental stone. It may be brackish-green through leek green to nearly white, or brownish or yellowish. Serpentinized carbonate rocks sometimes show an attractive clouded green color and are then called verde antique, or serpentine marble. [W.F.B.]

Serpentinite An abundantly occurring, fine-grained massive rock, generally considered to have been derived by pneumatolytic or hydrothermal processes from preexisting basic and ultrabasic igneous intrusions. The inferred processes are called serpentinization and the product serpentine. *See* METAMORPHIC ROCKS; PNEUMATOLYSIS; SERPENTINE.

Serpentine differs from the more conventional rocks—those composed of separable grains of various minerals—for serpentine has no idiomorphic crystal grains of its own. When grains are apparent, they are relicts from the preexisting crystalline rock which was altered to serpentine while retaining the old grain boundaries (called pseudomorphs after the older grain). [T.F.W.B.]

Serum The liquid portion that remains when blood is allowed to clot spontaneously and is then centrifuged to remove the blood cells and clotting elements. It is approximately of the same volume (55%) as plasma and differs from it only by the absence of fibrinogen which, after conversion into fibrin, is incorporated in the separated clot. *See* FIBRINOGEN.

Blood serum contains 6–8% solids, including proteins, such as albumin and globulins; hormones; antibodies; and enzymes as well as certain nutritive organic materials in small amount, such as amino acids, glucose, and fats. Somewhat less than 1% of the serum consists of inorganic substances. Small amounts of waste material such as urea, uric acid, xanthine, creatinine, creatine, bile pigments, and ammonia are also present. Trace amounts of gases such as oxygen and carbon dioxide are dissolved in the serum. *See* BLOOD.

Certain types of sera, both human and animal, are used in clinical medicine. Immune serum and hyperimmune serum are developed either by naturally occurring disease or are deliberately prepared by repeated injection of antigens to increase antibody titer for either diagnostic tests or the treatment of active disease. These sera are referred to as antisera, since they have a specific antagonistic action against specific antigens. *See* SEROLOGY.

By custom, the clear portion of any liquid material of animal origin separated from its solid or cellular elements is also referred to as sera, for example, the fluid in a blister and the fluids present in certain natural cavities of the body, such as the peritoneal, pleural, or pericardial sacs. These are more properly referred to, however, as effusions. [R.Str.]

Servomechanism An automatic feedback control system for mechanical motion. The name servomechanism, or "servo" for short, is often used interchangeably with feedback control system, but strictly speaking, it applies only to those feedback control systems in which the controlled quantity or output is mechanical position or one of its derivatives (velocity and acceleration). *See* CLOSED-LOOP CONTROL SYSTEM; CONTROL SYSTEMS.

The purpose of a servomechanism is to provide one or more of the following objectives: (1) accurate control of the motion of a shaft or rod without the need for human attendants (automatic control); (2) maintenance of accuracy with mechanical load variations, changes in the environment, power supply fluctuations, aging and deterioration of components (regulation and self-calibration); (3) control of a high-power load from a low-power command signal (power amplification); (4) control of an output shaft from a remotely located input shaft, without the use of mechanical linkages (remote control, shaft repeater).

Many of these objectives can be attained by a nonfeedback system if accurately calibrated components are available. The salient advantage of the feedback method of control is that it permits economical design and production of accurate systems using individually inaccurate components. A typical servomechanism requires only a few of its components to have high accuracy.

Historically, the first applications of servomechanisms were in speed governing of engines, automatic steering of ships, automatic control of guns, and electromechanical analog computers. Today servomechanisms are employed in almost every industrial field. Among the most significant applications are cutting tools for discrete parts manufacturing, rollers in sheet and web processes (such as steel rolling, paper calendering, high-speed printing), elevator control and vehicle drives, robots and remote manipulators in nuclear and other hazardous environments, positioning of cameras, telescopes, and antennas, stabilization and guidance of spacecraft, recording tape and disk drives, and mechanical knee and arm prostheses. For these and other applications *see* AUTOMATION; FLIGHT CONTROLS; GOVERNOR; GUIDANCE SYSTEMS; HUMAN-FACTORS ENGINEERING; HUMAN-MACHINE SYSTEMS; MEDICAL CONTROL SYSTEMS; PROCESS CONTROL; REGULATOR; REMOTE-CONTROL SYSTEM; ROBOTICS. [G.W.]

Set theory A mathematical term referring to the study of collections or sets. Consider a collection of objects (such as points, dishes, equations, chemicals, numbers, or curves). This set may be denoted by some symbol, such as X. It is useful to know properties that the set X has, irrespective of what the elements of X are. The cardinality of X is such a property.

Two sets A and B are said to have the same cardinal written $C(A) = C(B)$, provided there is a one-to-one correspondence between the elements of A and the elements of B. For finite sets this notion coincides with the phrase "A has the same number of elements as B." However, for infinite sets the above definition yields some interesting consequences. For example, let A denote the set of integers and B the set of odd integers. The function $f(n) = 2n - 1$ shows that $C(A) = C(B)$. Hence, an infinite set may have the same cardinal as a part or subset of itself.

A is called a subset of B if each element of A is an element of B, and it is expressed as $A \subset B$. The collection of odd integers is a subset of itself.

One approved method of forming a set is to consider a property P possessed by certain elements of a given set X. The set of elements of X having property P may be considered as a set Y. The expression $p \in X$ is used to denote the fact that p is an element of X. Then $Y = \{p \mid p \in X$ and p has property $P\}$. Another approved method is to consider the set Z of all subsets of a given set X. Paradoxically, it is not permissible to regard the collection of all sets as a set.

In set theory, one is interested not only in the properties of sets but also in operations involving sets: addition, subtraction, multiplication, and mapping. The sum of A and B ($A + B$ or $A \cup B$) is the set of all elements in either A or B; that is, $A + B = \{p \mid p \in A$ or $p \in B\}$. The intersection of A and B ($A \cdot B$, $A \cap B$, or AB) is the set of all elements in both A and B; that is, $A \cdot B = \{p \mid p \in A$ and $p \in B\}$. If there is no element which is in both A and B, one says that A does not intersect B and writes $A \cdot B = 0$. The expression $A - B$ is used to denote the collection of elements of A that do not belong to B; that is $A - B = \{p \mid p \in A$ and $p \notin B\}$. [R.H.Bi.]

Sewage A combination of (1) liquid wastes conducted away from residences, institutions, and business buildings and (2) the liquid wastes from industrial establishments with (3) such surface, ground, and storm water as may find its way or be admitted into the sewers. Category 1 is known as sanitary or domestic sewage; 2 is usually referred to as industrial waste; 3 is known as storm sewage.

Sewage is the waste water reaching the sewer after use; hence it is related in quantity and in flow fluctuation to water use. The quantity of sewage is generally less than the water consumption since some portion of water used for fire fighting, lawn irrigation, street washing, industrial processing, and leakage does not reach the sewer. These losses are compensated for partly by the addition of water from private wells, groundwater infiltration, and illegal connections from roof drains.

Sewage is actually water with a small amount of impurity in it. Examination of sewage is required to know the effects of these impurities. Various tests are used to aid in determining the characteristics, composition, and condition of sewage. These include physical examination, solids determination, tests for determining the oxygen requirement of organic matter, chemical and bacteriological tests, and examination under the microscope. *See* SEWAGE COLLECTION SYSTEMS; SEWAGE DISPOSAL; SEWAGE SOLIDS; SEWAGE TREATMENT. [W.T.I.]

Sewage collection systems Systems of pipes and conduits, together with control devices, pumping stations, and appurtenances used for the collection and transfer of waste waters. *See* SEWAGE.

Sewer pipe is manufactured from a number of materials, such as vitrified clay, concrete, asbestos cement, corrugated iron, cast iron, and steel. Plastics, bituminous wood fiber, and wood stave pipe also are used. All sewer pipes may be surcharged or filled at some time and must be capable of withstanding some hydraulic pressure. Sewers are laid underground and must be able to withstand external pressures such as those caused by soil, water, and the extra weight of traffic. Large-diameter reinforced concrete sewers or conduits may be constructed in place.

Sewer appurtenances are access holes, inlets, regulators, inverted siphons, and outfalls. An access hole is an opening at the ground or street surface permitting a worker to enter to make examinations and repairs. An inlet provides an opening for storm water and is usually placed at the curb line of the street. The structure below the inlet is called a catch basin. A regulator is a device designed to divert sewage flow from one sewer to another channel. It is most frequently used in combined sewer systems to regulate the amount of storm water permitted to flow to a sewage treatment plant. An inverted siphon is a length of sewer set below hydraulic grade line so that it flows under pressure between an inlet chamber and an outlet chamber. The outfall is a structure designed to admit treated or untreated sewage to a receiving body of water.

Major factors affecting the quantity and flow patterns of sanitary sewage are (1) population and population increase; (2) population density and density change; (3) water use, demand, and consumption; (4) industrial requirements, present and future; (5) commercial requirements, present and future; (6) expansion of service geographically; (7) groundwater geology of area; and (8) topography of area. *See* SEWAGE DISPOSAL. [W.T.I.]

Sewage disposal The discharge of waste waters into surface-water or groundwater courses, which constitute the natural drainage of an area. Most waste waters contain offensive and potentially dangerous substances, which can cause pollution and contamination of the receiving water bodies. Contamination is defined as the impairment of water quality to a degree that creates a hazard to public health. Pollution refers to the adverse effects on water quality that interfere with its proper and beneficial use. The principal sources of pollution are domestic sewage and industrial wastes.

Water-quality criteria deal with the physical, chemical, and biological parameters of pollution. The most common standards are concerned with physical appearance, odor production, dissolved-oxygen concentration, pathogenic contamination, and potentially toxic or harmful chemicals. The allowable quantity and concentration of these characteristics and substances vary with the water usage. *See* SEWAGE TREATMENT; WATER ANALYSIS; WATER POLLUTION; WATER PURIFICATION. [D.J.O.]

Sewage solids A semiliquid mass, called sludge, removed from the liquid flow of sewage. Solids treatment depends on the source of the solid and its characteristics. Solids are removed as screenings, grit, primary sludge, secondary sludge, and scum. Sewage solids must be disposed of without nuisance or hazard to health. Burial, incineration, and drying for use as fertilizer are means of final disposal. Lagooning may be used; in seacoast cities sludge may be taken out to sea. Disposal of sewage solids on land has limited application. It has certain public-health dangers and must be closely supervised. [W.T.I.]

Sewage treatment Any process to which sewage is subjected in order to remove or alter its objectionable constituents and thus render it less offensive or dangerous. These processes may be classified as preliminary, primary, secondary, or complete, depending on the degree of treatment accomplished. Preliminary treatment may be the conditioning of industrial waste prior to discharge to remove or to neutralize substances injurious to sewers and treatment processes, or it may be unit operations which prepare the water for major treatment. Primary treatment is the first and sometimes the only treatment of sewage. It is the removal of floating solids and coarse and fine suspended solids. Secondary treatment utilizes biological methods, that is, oxidation processes following primary treatment by sedimentation. Complete treatment removes much of the suspended, colloidal, and organic matter.

Septic tanks and Imhoff tanks are considered secondary treatment methods because sedimentation is combined with biological digestion of the sludge. *See* IMHOFF TANK; SEPTIC TANK. [W.T.I.]

Sewing machine A mechanism that stitches cloth, leather, book pages, and other material by means of a double-pointed needle or any eye-pointed needle. In ordinary two-threaded machines, a lock stitch is formed (see illustration). A

Lock stitch; upper or needle thread is dark color, and under or bobbin thread is light color.

presser foot held against the material with a yielding spring adjusts itself automatically to variations in thickness of material and allows the operator to turn the material as it feeds through the machine, A cluster of cams, any one of which can be selected to guide the needle arm, makes possible a variety of stitch patterns. *See* CAM MECHANISM. [F.H.R.]

Sex-influenced inheritance That part of the inheritance pattern on which sex differences operate to promote character differences. These characters are expressed differently in the two sexes, even when their genotypes are identical. Some confusion exists between this term and sex-limited inheritance. Such characteristics as baldness in humans and some eye colors in *Drosophila* all show influence of sex. The effects frequently are attributed to hormones secreted by the sex cells, but often hormones cannot be demonstrated. The differences found within the individual body cells of the sexes may be the responsible factors. Sex-limited inheritance differs from sex-influenced in that the inheritance finds expression only in one sex. [J.W.Go.]

Sex-linked inheritance The transmission to successive generations of differences that are due to genes located in the sex chromosomes. Because of their ready identification and straightforward expression, sex-linked genes have frequently provided the basis for crucial observations and experiments in genetics. The best-known sex-linked traits in humans concern color vision and hemophilia. *See* COLOR VISION.

Sex-linked genes are distinguished as such not by the traits they control but by characteristic patterns of transmission with respect to the sex difference, paralleling in all respects the distribution of the sex-determining X or Y chromosomes. Except for the sex-determining factors themselves, sex-linked genes are no more likely than nonsex-linked genes to be concerned with differences in the primary or secondary sexual traits that differentiate male from female.

The following tabulation is an example of sex-linked inheritance in *Drosophila*. In this example X is the X chromosome, Y the Y chromosome, and *W*, *w* are sex-linked genes concerned with eye pigment:

Parents:	X^wX^w White-eyed female	×	X^WY Red-eyed males	
	Eggs: All X^w		Sperm: $\frac{1}{2}X^W$, $\frac{1}{2}Y$	
Offspring:	$\frac{1}{2}X^WX^w$ Red-eyed females		$\frac{1}{2}X^wY$ White-eyed males	
Parents:	X^WX^w Red-eyed female	×	X^wY White-eyed male	
Eggs:	$\frac{1}{2}X^W$, $\frac{1}{2}X^w$		Sperm: $\frac{1}{2}X^w$, $\frac{1}{2}Y$	
Offspring:				
$\frac{1}{4}X^WX^w$ Red-eyed females	$\frac{1}{4}X^wX^w$ White-eyed females		$\frac{1}{4}X^WY$ Red-eyed males	$\frac{1}{4}X^wY$ White-eyed males

Sex linkage is termed complete in the case of genes located in the differential segment of the X chromosome which is not represented by a homologous counterpart in the Y. Sex linkage is termed incomplete or partial if genes are located in homologous X- and Y-chromosome segments that pair and exchange parts during meiosis. Cases of completely Y-linked genes are rare or doubtful, and evidence for partially sex-linked genes in humans is inconclusive. *See* GENETICS; HUMAN GENETICS; SEX-INFLUENCED INHERITANCE. [D.D.P.]

Sextant A navigation instrument used for measuring angles, primarily altitudes of celestial bodies. Originally, the sextant had an arc of 60°, or 1/6 of a circle, from which the instrument derived its name. Because of the double-reflecting principle used, such an instrument could measure angles as large as 120°. In modern practice, the name sextant is commonly applied to all instruments of this type regardless of the length of the arc, which is seldom exactly 60°. The optical principles of the sextant are similar to those of the prismatic astrolabe. *See* PRISMATIC ASTROLABE.

Modern sextants may be grouped into two classes, marine and air. The marine sextant (see illustration) is designed for use

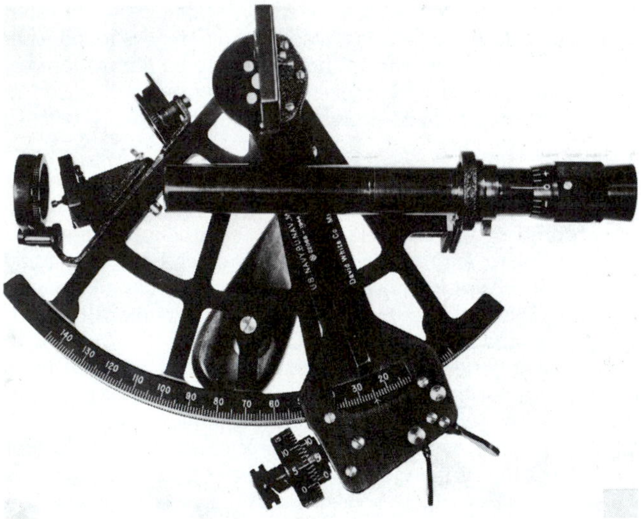

A marine sextant.

by mariners. It utilizes the visible sea horizon as the horizontal reference. An instrument designed for use in aircraft is called an air sextant. Such sextants have built-in artificial horizons. Most modern air sextants are periscopic to permit observation of celestial bodies without need of an astrodome in the aircraft. *See* CELESTIAL NAVIGATION. [A.B.M.]

Sexual dimorphism A fundamental phenomenon of life is bisexuality. In some organisms the functional separation of the sexes is not marked by obvious morphological differences. Commonly, however, the sexes can be diagnosed by differences of form or pattern. Such diagnostic differences between the sexes make up the phenomenon of sexual dimorphism. Sexual dimorphism among animals ranges from differences detectable only with care to differences so extreme that careful study has been required to assign the sexes to the same species (see illustration). In the great majority of flowering plants, a single flower includes both stamens and pistils. Just as in hermaphroditic animals, sexual dimorphism is out of the question here. However, staminate and pistillate flowers are separate in some species, and thus sexual dimorphism can occur.

Another type of sexual dimorphism, polymorphism, is exemplified by termites. In this group it appears to be one aspect of a complex polymorphism based upon the caste society of these insects. *See* SOCIAL INSECTS. [E.O.D.]

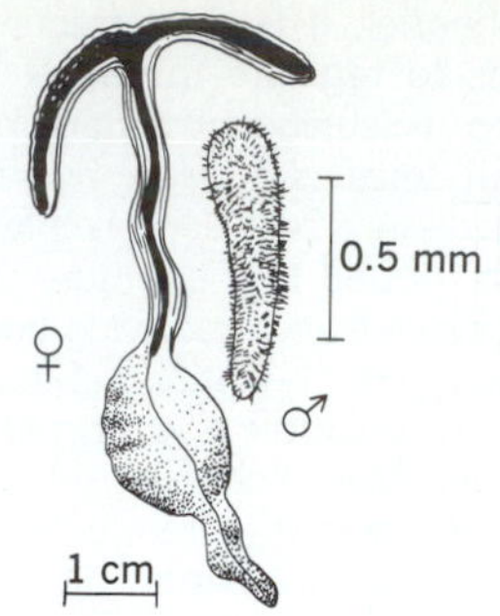

Bonellia **female (left) and male (right). (*After E. O. Dodson, Genetics, Saunders, 1956*)**

Sexually transmitted diseases A group of infectious diseases usually transmitted by sexual contact. Gonorrhea, syphilis, chancroid (soft chancre), and others less common occur in the United States. *See* ACQUIRED IMMUNE DEFICIENCY SYNDROME (AIDS); GONORRHEA; GRANULOMA INGUINALE; LYMPHOGRANULOMA VENEREUM; SOFT CHANCRE; SYPHILIS. [E.G.St./N.K.M.]

Sferics The electromagnetic radiations originating in atmospheric electrical discharges. Thunderstorms are the principal sources of these discharges. These radiations are also known as atmospherics, but the contracted form sferics is more common. *See* ATMOSPHERIC ELECTRICITY; ELECTROMAGNETIC RADIATION; THUNDERSTORM.

Sferics are the major cause of static heard in amplitude-modulated radio receivers. This same noise can, however, be put to good use in the detection and tracking of severe storms such as squall lines, tornadoes, snowstorms, and hailstorms. Sferics are important in two major scientific areas: (1) the determination of the characteristics of the source of the radiations; and (2) the study of the propagation of these radiations within and beyond the Earth's atmosphere. *See* RADIO-WAVE PROPAGATION; STORM DETECTION. [C.G.St.]

Shadow A region of darkness caused by the presence of an opaque object interposed between such a region and a source of light. A shadow can be totally dark only in that part called the umbra, in which all parts of the source are screened off. With a point source, the entire shadow consists of an umbra, since there can be no region in which only part of the source is eclipsed. If the source has an appreciable extent, however, there exists a transition surrounding the umbra, called the penumbra, which is illuminated by only part of the source. Depending on what fraction of the source is exposed, the illumination in the penumbra varies from zero at the edge of the full shadow to the maximum where the entire source is exposed. The edge of the umbra is not perfectly sharp, even with an ideal point source, because of the wave character of light. *See* DIFFRACTION; ECLIPSE. [F.A.J./W.W.W.]

Shadowgraph A simple method of making visible the disturbances that occur in fluid flow at high velocity. Light passing through a flow field of varying density is retarded differently through the field, resulting in a turning of the wavefronts, that is, a refraction of the rays. This is the basis for shadowgraph flow visualization.

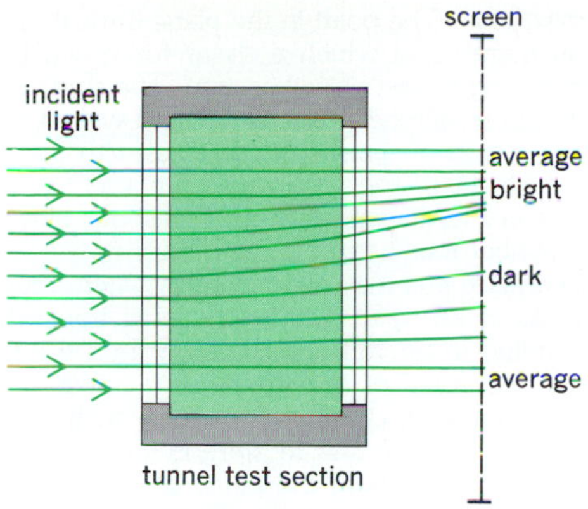

Shadowgraph produced by light passing through fluid in test section indicates flow pattern.

The application of the shadowgraph to a wind tunnel is shown in the illustration. The rays enter normal to the sidewall through the window at the left. As they pass through the test section, they encounter a change in density of the fluid in the tunnel and are deflected. The light rays then fall on a screen, where they are observed or photographed. Where the rays have crowded together, the screen is brightly illuminated, and where the rays diverge, the screen is dark. Where the spacing is unchanged, the illumination is normal, even though there has been a change of density along the path of the ray.

The light need not be parallel when it enters the test section, so a slit source and lens system are not required. This simplicity makes the shadowgraph system considerably less expensive than the other principal methods of flow visualization, that is, schlieren and interferometer. It is often used where the finer resolution of the other systems is not required or desirable. *See* INTERFEROMETRY; SCHLIEREN PHOTOGRAPHY. [R.F.C.]

Shaft balancing The process (often referred to as rotor balancing) of redistributing the mass attached to a rotating body in order to reduce vibrations arising from centrifugal force.

A rotating shaft supported by coaxial bearings (for example, ball bearings together with any attached mass, such as a turbine disk or motor armature) is called a rotor. If the center of mass of a rotor is not located exactly on the bearing axis, a centrifugal force will be transmitted via the bearings to the foundation. The horizontal and vertical components of this force are periodic shaking forces that can travel through the foundation to create serious vibration problems in neighboring components.

Any rigid shaft may be dynamically balanced by adding or subtracting a definite amount of mass at any convenient radius in each of two arbitrary transverse cross sections of the rotor. The so-called balancing planes selected for this purpose are usually located near the ends of the rotor, where suitable shoulders or balancing rings have been machined to permit the convenient addition of mass (lead weights, calibrated bolts, and so on) or the removal of mass (by drilling or grinding). Long rotors, running at high speeds, may undergo appreciable elastic deformations. For such flexible rotors it is necessary to utilize more than two balancing planes. *See* MECHANICAL VIBRATION. [B.P.]

Shafting The machine element that supports a roller and wheel so that they can perform their basic functions of rotation. Shafting, made from round metal bars of various lengths and machined to dimension the surface, is used in a great variety of shapes and applications. Because shafts carry loads and transmit power, they are subject to the stresses and strains of operating machine parts. Most shafting is rigid and carries bending loads without appreciable deflection. Some shafting is highly flexible; it is used to transmit motion around corners.

Shafts used in special ways are given specific names, although fundamentally all applications involve transmission of torque. The primary shafting connection between a wheel and a housing is an axle. A short shaft is a spindle. A short stub shaft mounted as part of a motor or engine or extending directly therefrom is a head shaft. A secondary shaft that is driven by a main shaft and from which power is supplied to a machine part is called a countershaft. *See* BELT DRIVE; PULLEY. [J.J.R.]

Shale A fine-grained laminated or fissile sedimentary rock made up of silt- and clay-sized particles. The average shale consists of about one-third quartz, one-third clay minerals, and one-third miscellaneous minerals, including carbonates, iron oxides, feldspars, and organic matter. The tendency for shales to split along bedding planes is called fissility.

No general agreement on a classification scheme for shales has been reached. On the basis of the composition of the silt fraction, shales have been divided into groups roughly paralleling sandstone types: quartzose shale (orthoquartzite), feldspathic shale (arkose), chloritic shale (graywacke), and micaceous shale (subgraywacke). Shales may also be subdivided on the basis of origin into residual (from reworked soils), transported, and hybrid. The hybrid shales are mixtures of clastic and nonclastic materials (carbonate, organic matter, iron oxides). In this group belong the black shales. *See* BLACK SHALE; SEDIMENTARY ROCKS. [R.Si.]

Shape memory alloys Special alloys which, after being deformed, can recover their original shape when heated. If an ordinary metal is strained beyond its elastic limit, permanent deformation of the material is produced. Therefore, it would be unusual for a crumpled wire to straighten out spontaneously when heated. However, this is precisely how certain shape memory alloys behave. If one of these alloys is deformed (when below a critical temperature and with a strain limit of about 10%), it can recover its original, undeformed shape when heated. The heating in effect "reminds" the alloy that it prefers a different crystal structure, and thus an associated shape, at higher temperature. This unusual behavior is called shape memory. *See* STRESS AND STRAIN.

Shape memory behavior is inherently connected with that type of solid-state phase transformation known as martensitic transformation. This is a type of microstructural change wherein the initial crystal structure transforms to a new crystal structure by a mechanism that does not involve diffusional mixing of atoms but rather is dominated by coordinated shear displacements of sections of crystal. This mode of transformation is best known and technologically very important in steel; however, martensitic transformation in steel cannot lead to shape memory behavior, because there is an additional requirement on the transformation that the martensitic transformation be thermoelastic, and this is missing in steel.

The main practical requirement for shape memory behavior is probably an appropriate alloy, since only certain martensitic alloys exhibit the effects described. Some of the best-known and most-developed shape memory alloys come from the Cu-Zn-Al, Cu-Al-Ni, and Ni-Ti alloys systems. Applications of shape memory take advantage of two inherent parameters: the shape change on heating through a certain temperature range, and the internal stress developed if this reverse transformation is opposed (such as by physically blocking the movement or adding an external load). These two parameters, the reversion strain and reversion stress respectively, have been combined to invent a wide variety of heat-activated fasteners, switches, couplings, controls, actuators, deployment devices, force-applying members, and even heat engines (by combining the strain and stress parameters to do work). There also have been some medical and dental applications, including teeth braces, artificial muscles, orthopedic devices, and blood clot filters. *See* ALLOY. [J.Pe.]

Shaper A machine tool for cutting flat or flat, contoured surfaces by reciprocating a single-point tool across the workpiece. A shaper is usually used for small pieces requiring a short cutter stroke rather than for large pieces requiring longer strokes provided by a planer. Shapers are classified as either horizontal or vertical depending on the plane of motion of the reciprocating ram. [A.H.T.]

Shark Any of about 225 species of elasmobranchs, cartilaginous fishes which occur in all oceans but are most abundant in tropical and subtropical regions. Some species frequent shore waters while others occur near the surface in deep waters; some deep-sea forms may be luminescent. All species are carnivorous; *Carcharhinus nicaraguensis*, a fresh-water species, will eat humans.

Frilled and cow sharks are the most primitive modern sharks. The cow shark (*Hexanchus griseum*) is a European species with six gill slits, eyes that lack a nictitating membrane, primitive teeth, and a single dorsal fin. *Heptranchias maculatus* is one of several species known as the seven-gill sharks. The oil shark (*H. deani*) is found in Japanese waters.

Cat sharks and their allies, characterized by five gill slits, an anal fin, and two dorsal fins, are the most highly specialized sharks. The lesser spotted dogfish (*Schyliorhinus caniculus*) and the greater spotted dogfish (*S. stellaris*) are common European species. The most ferocious shark is the human-eating mackerel shark (*Lamnia nasus*). The largest sharks are the basking sharks, which may reach a length of 40 ft (12 m).

Angel and dogfish sharks are small, a foot or more in length. *Squalus acanthias* may reach a length of 3 ft (0.9 m). It is common on both sides of the Atlantic, the North Sea, and the Mediterranean. In front of each dorsal fin, there is a spine with attached poison glands that can cause painful wounds. *See* ELASMOBRANCHII; SELACHII. [C.B.C.]

Shear A straining action wherein applied forces produce a sliding or skewing type of deformation. A shearing force acts parallel to a plane as distinguished from tensile or compressive forces, which act normal to a plane. Examples of force systems producing shearing action are forces transmitted from one plate to another by a rivet that tend to shear the rivet, forces in a beam that tend to displace adjacent segments by transverse shear, and forces acting on the cross section of a bar that tend to twist it by torsional shear (see illustration). Shear forces are usually accompanied by normal forces produced by tension, thrust, or bending. Shearing stress is the intensity of distributed force expressed as force per unit area. *See* STRESS AND STRAIN. [J.B.S.]

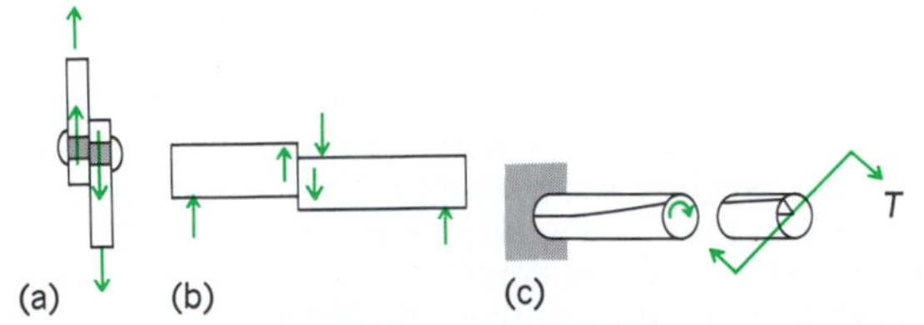

Shearing actions. (*a*) Single shear on rivet. (*b*) Transverse shear in beam. (*c*) Torsion.

Shear center The point in the plane through a section of a structural member at which a shear force can be applied without producing a twist of that section. The shear center of a section through a member, such as a rolled wide-flange beam, that has two planes of symmetry coincides with the geometric center or centroid of the section. When such a member is loaded transversely, it bends without twisting. However, the structural member may be unsymmetrical or may have but one plane of symmetry and may be loaded elsewhere than through that plane, as in an open section made of thin material for resisting bending in aircraft construction; then the load force and the reaction force of the structural member constitute a couple that, in general, causes the member to twist. The load can be applied through a unique point in the plane of the section so that the moments in the plane of the section are balanced and the beam will not twist. This unique point is called the shear center. In the illustration, in *a* the channel bends when the load is along the axis of symmetry; in *b* the channel twists, even with the load through the centroid if the load is not along the axis of symmetry; in *c* the channel resists twisting when the load is through the shear center. *See* LOADS, TRANSVERSE; SHEAR. [J.B.S.]

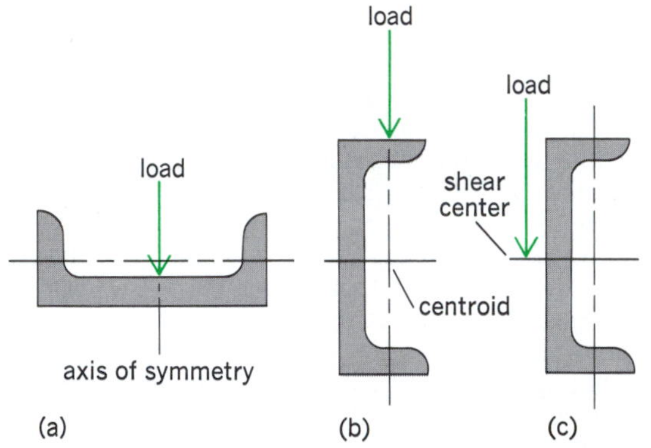

Three conditions of channel section under load. Channel (*a*) resists bending, (*b*) twists, and (*c*) resists twisting.

Sheathed bacteria Bacteria consisting of chains of cells, usually rod-shaped, enclosed in a hyaline envelope or sheath. Seven genera are recognized: *Sphaerotilus*, *Leptothrix*, *Streptothrix*, *Lieskeella*, *Phragmadiothrix*, *Crenothrix*, and *Clonothrix*. *Sphaerotilus natans* and *S. discophorus* oxidize ferrous and manganous ions and deposit the oxides on the sheaths in large amounts. Thus they play an important role in the cycles of iron and manganese in nature and may be responsible in part for the accumulation of such metal oxides. Also, *S. natans* is involved in water pollution. Massive growths of *Sphaerotilus* frequently develop in streams and rivers polluted by domestic sewage and industrial wastes, especially those from paper pulp mills. The extensive, white, cottony growths destroy the natural beauty of rivers and beaches and cause considerable financial losses to commercial fishermen by fouling lines, clogging and destroying nets, and even coating the fish. [J.L.St.]

Sheep Sheep are members of the family Bovidae in the order Artiodactyla, the even-toed, hoofed mammals. Sheep were possibly the first animals to be domesticated about 11,000 years ago in southwestern Asia. *See* ARTIODACTYLA.

Wild sheep range over the mountainous areas of the Northern Hemisphere. They have long, curved, massive horns and mixed coats with a hairy outercoat and a woolly undercoat. Although wild sheep vary in size, they are usually larger than domestic sheep and have shorter tails.

Sheep are called lambs until about 12 months of age, from which time to 24 months they are referred to as yearlings, and thereafter as two-year-olds, and so on. The female sheep is called a ewe and the male, a ram or buck; the castrated male is called a wether. Sheep meat is called lamb or mutton depending on the age at slaughter.

A precise definition is difficult since sheep are so variable. Horns, if present, tend to curl in a spiral. If horns occur in both sexes, they are usually larger for the ram. In some breeds only the rams have horns but many breeds are hornless. The hairy outercoat common to wild sheep was eliminated by breeding in most domestic sheep. Wool may cover the entire sheep, or the face, head, legs, and part of the underside may be bare; or wool may be absent as in some breeds (bred for meat) with a short hairy coat or a coat which constantly sheds. Although domestic sheep are mostly white, shades of brown, gray, and black occur, sometimes with spotting or patterns of color. *See* Wool.

Hundreds of breeds of sheep of all types, sizes, and colors are found over the world. Wool-type breeds, mostly of Merino origin, are important in the Southern Hemisphere, but both fine- and long-wool types are distributed all over the world. Sheep with fat tails or fat rumps are common in the desert areas of Africa and Asia. These usually produce carpet wool. Milk breeds are found mostly in central and southern Europe. Meat breeds from the British Isles are common over the world. [C.E.T.]

Sheet-metal forming The shaping of thin sheets of metal (usually less than ¼ in. or 6 mm) by applying pressure through male or female dies or both. Parts formed of sheet metal have such diverse geometries that it is difficult to classify them. Sheet forming is accomplished basically by processes such as stretching, bending, deep drawing, embossing, bulging, flanging, roll forming, and spinning. In most of these operations there are no intentional major changes in the thickness of the sheet metal. *See* Metal forming.

Stretch forming is a process in which the sheet metal is clamped between jaws and stretched over a form block. The process is used primarily in the aerospace industry to form large panels with varying curvatures.

Bending is one of the most common processes in sheet forming. The part may be bent not only along a straight line, but also along a curved path (stretching, flanging). In addition to male and female dies used in most bending operations, the female die can be replaced by a rubber pad (Fig. 1). The roll-forming process replaces the vertical motion of the dies by the rotary motion of rolls with various profiles. Each successive roll bends the strip a little further than the preceding roll.

While many sheet-forming processes are carried out in a press with male and female dies usually made of metal, there are some processes which utilize rubber to replace one of the dies. The simplest of these processes is the Guerin process (Fig. 2).

A great variety of parts are formed by the deep-drawing process (Fig. 3), the successful operation of which requires a careful control of factors such as blank-holder pressure, lubrication, clearance, material properties, and die geometry.

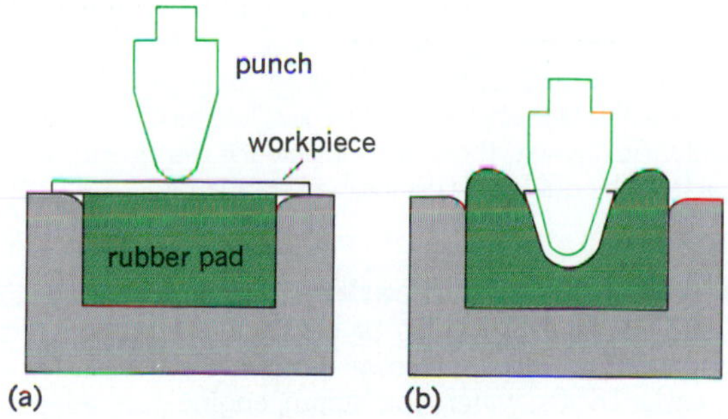

Fig. 1. Bending process with a rubber pad. (*a*) Before forming. (*b*) After forming.

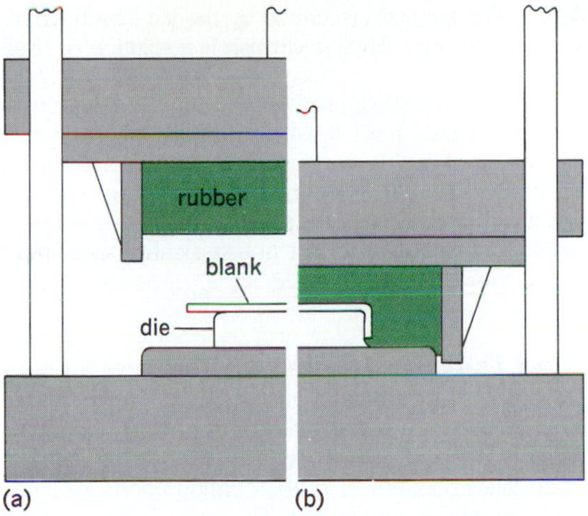

Fig. 2. The Guerin process, the simplest rubber forming process. (*a*) Before forming. (*b*) After forming.

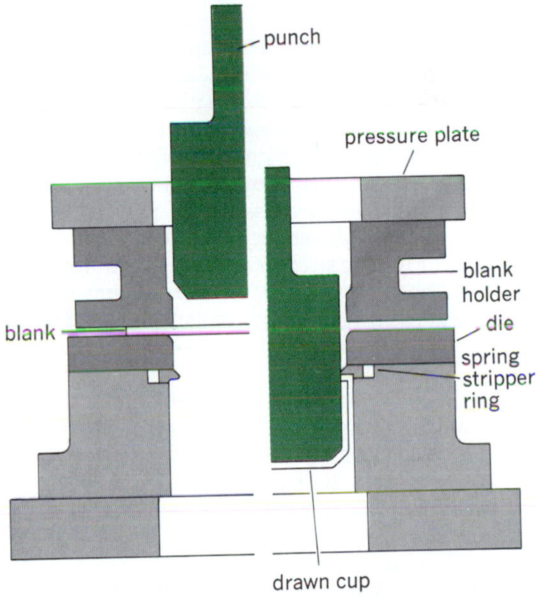

Fig. 3. The deep-drawing process.

Many parts require one or more additional processes. Embossing consists of forming a pattern on the sheet by shallow drawing. Coining consists of putting impressions on the surface by a process that is essentially forging, the best example being the two faces of a coin. Shearing is separation of the material by the cutting action of a pair of sharp tools, similar to a pair of scissors. *See* Coining.

The spinning process forms parts with rotational symmetry over a mandrel with the use of a tool or roller. There are two basic types of spinning: conventional or manual spinning, and shear spinning. The conventional spinning process forms the material over a rotating mandrel with little or no change in the thickness of the original blank. In shear spinning (hydrospinning, floturning) the deformation is carried out with a roller in such a manner that the diameter of the original blank does not change but the thickness of the part decreases by an amount dependent on the mandrel angle. Shear spinning produces parts with various shapes (conical, curvilinear, and also tubular by tube spinning on a cylindrical mandrel) with good surface finish, close tolerances, and improved mechanical properties. *See* Metal coatings; Spinning (metals). [S.Ka.]

Shellac The lac resin (secreted by the lac insect) when used in flake (or shell) form. Shellac varnish is a solution of shellac in denatured alcohol.

Shellac varnish is widely used in wood finishing where a fast-drying, light-colored, hard finish is desired. Shellac varnish is not water-resistant and is not suitable for exterior coatings, but is widely used for finishing floors and for sealing materials (for example, asphalt, coal tar, oil stains, and resinous knots) which are soluble in mineral spirits but not in alcohol. *See* WOOD FINISHING. [C.R.Ma.; C.W.Si.]

Shielded thermocouple An instrument for obtaining the total temperature inside a pitot tube using a thermocouple rather than a pressure-sensing device. The thermocouple must be shielded to minimize losses due to radiation and conduction. Calibration takes account of probe configurations, wall conductivity, and gas stream conditions. The same instrument may be used for subsonic and supersonic flows. *See* AIR-VELOCITY MEASUREMENT; THERMOCOUPLE. [H.Fo.]

Ship design The design of a ship involves a selection of the features of form, size, proportions, and other factors which are open to choice, in combination with those features which are imposed by circumstances beyond the control of the design naval architect. As ships have increased in size and complexity, plans for building them have become more detailed and more varied. The intensive research since the period just prior to World War II has brought about many technical advances in the design of ships. These changes have been brought about principally by the development of new welding techniques, developments in main propulsion plants, advances in electronics, and changes in materials and methods of construction. *See* SHIPBUILDING.

Ships are designed, built, and operated to fulfill the requirements and limitations specified by the operator and owner. These owner's requirements denote the essential considerations which are to form the basis for the design. They may be generally stated as (1) a specified minimum deadweight carrying capacity, (2) a specified measurement tonnage limit, (3) a selected speed at sea, or a maximum speed on trial, and (4) maximum draft combined with other draft limitations. Apart from these requirements, the ship owner expects the designer to provide a thoroughly efficient ship.

All ships have many requirements which are common to all types, whether they are naval, merchant, or special-purpose ships. The first of such requirements is that the ship must be capable of floating when carrying the load for which it was designed. One of the first things which a designer must do is to determine the weight and size of the ship and decide upon a suitable hull form to provide the necessary buoyancy to support the weight that has been chosen. *See* BUOYANCY.

From the specified requirements, an approach is made to the selection of the dimensions, weight, and displacement of the new design. This is a detailed operation, but some rather direct approximations can be made to start the design process. This is usually done by analyzing data available from an existing ship which is closely similar. Length and speed are related to the hull form, the propulsion machinery, and the propeller design. Speed has an important influence on the length selected for the ship. The speed of the ship is related to the length in terms of the ratio $V/\sqrt{L}$, where V is the speed in knots and L is the effective waterline length of the ship. As the speed-length ratio increases, the resistance of the ship increases.

A ship must be stable under all normal conditions of loading and performance at sea. This means that when the ship is inclined from the vertical by some external force, it must return to the vertical when the external force is removed. Stability may be considered in the transverse or in the longitudinal direction. In surface ships, longitudinal stability is of much less concern than transverse stability. Submarines, however, are concerned with longitudinal stability in the submerged condition. The transverse stability of a surface ship must be considered in two ways, first at small angles of inclination, called initial stability, and second at large angles of inclination. Initial stability depends upon two factors, (1) the height of the center of gravity of the ship above the base line and (2) the underwater form of the ship.

A ship at sea is subjected to many forces because of the action of the waves, the motion of the ship, and the cargo and other weights which are distributed throughout the length of the ship. These forces produce stresses in the structure, and the structure must be of suitable strength to withstand the action. The determination of the minimum amount of material required for adequate strength is essential to attaining the minimum weight of the hull. The structure as a whole bends in a longitudinal plane, with the maximum bending stresses being found in the bottom and top of the hull girder. Therefore, depth is important because as it is increased, less material is required in the deck and bottom shell. However, there are limits which control the maximum depth in terms of practical arrangement and efficiency of design.

The desired hydrostatic and hydrodynamic characteristics which the new ship form should have are formulated by considering resistance and propulsion, stability, volumetric requirements, steering and maneuverability, performance in rough weather, and the freeboard necessary for seaworthiness and safety. After these factors have been decided, the work on the drawing of the hull form is started. The geometry of the ship is usually delineated on a drawing which is commonly referred to as the lines drawing of the ship, or simply the lines.

After the lines have been completed, they are usually sent to a model basin (test facility) for the construction of a scale model and for testing in a basin. These tests are generally carried out in still water. The power determined from these experiments represents the power required to drive the ship under ideal sea conditions, and allowances must be made for many departures from these ideals, such as probable rough water performance and increased resistance as a result of fouling.

After having established the principal dimensions, form, and general arrangement of the ship, the designer undertakes the problem of providing a structure capable of withstanding the forces which may be imposed upon it. The hull of a steel merchant ship is a complex structure, unique in the field of engineering structures in that it is primarily a plate structure, depending for its major overall strength on the plating of the shell, decks, and in most cases, also on the inner bottom and longitudinal bulkheads. The framing members, each of which has its own function to perform, are designed primarily to maintain the plate membranes to the planned contours and their positions relative to each other when subjected to the external forces of water pressure and breaking seas, as well as to the internal forces caused by the services for which the ship is designed. Unlike most other large engineering structures, the forces supporting the ship's hull as well as the loads which may be imposed upon it vary considerably, and in many cases, cannot be determined accurately. As a result, those responsible for the structural design of ships must be guided by established standards. [J.C.N.]

Ship powering and steering The earliest propulsion of boats and vessels by paddles or oars was succeeded or augmented by the use of sails. Power propulsion essentially began in the early 1800s, when the steam engine was adapted to marine use. The first screw-propelled steamer made the Atlantic crossing in 1845. The great advantages of the screw propeller have made it almost the only form of propulsion of marine craft since that time. There has been some use of pump jets for propulsion of small craft and hydrofoil boats. *See* PROPELLER (MARINE CRAFT).

Steam reciprocating engines were the only form of power plant used until about 1890, when both steam turbines and diesel engines were developed. These latter plants have largely supplanted steam reciprocating engines. Nuclear steam turbine units are now used for submarine and some surface ship propulsion, mainly aboard naval ships. The relatively high cost of nuclear power precludes its widespread use for merchant ships. Another power plant which has been used in ships is the gas turbine. *See* MARINE ENGINE; SHIP PROPULSION REACTOR.

For steering and maneuvering, ships are normally fitted with one or more rudders at the stern and in most cases located directly aft of the propeller or propellers. Multiple rudders are commonly fitted to warships where a high degree of maneuverability is required and where twin or quadruple screws are employed. Rudders are controlled from the bridge and other locations through an electrical or hydraulic system which actuates large hydraulic steering engines.

Because the gyrocompass is more sensitive in detecting initial deviations from the course than is the human eye, automatic steering devices offer an improvement over hand steering. The improvement may be greater than sometimes realized, since reduction of rudder drag reduces the fuel used, reduction of rudder angle required reduces the work of the steering engine, and reduction of deviations from course reduces the distance traveled. *See* GYROCOMPASS.

By international law, oceangoing ships for years have been required to have sufficient power for astern operation to secure proper control of the ship in all normal circumstances. In addition, maximum safe speed under conditions of reduced visibility is an important operational consideration and is closely related to stopping ability. The problem varies with type of power plant. A system finding great favor with ship operation is the controllable and reversible pitch propeller which avoids the necessity of reversing machinery. [R.B.C.]

Ship propulsion reactor A nuclear reactor used to supply motive power to a ship. Nuclear reactors for shipboard propulsion can, in theory, be of any type used for the production of useful power. *See* NUCLEAR REACTOR.

In all the shipboard nuclear power plants that have been built, energy conversion is based on the steam-turbine cycle, and that portion of the plant is more or less conventional. Gas-turbine applications are also possible, and these have been studied. *See* MARINE ENGINE.

There are four radical differences between shipboard reactors and similar installations ashore, involving (1) problems of weight and space limitations, (2) problems of plant reliability and on-board maintenance, (3) problems in plant safety, and (4) problems inherent in location on a moving platform.

Only two types of reactors—the pressurized-water reactor and the sodium-cooled reactor—have actually been applied to operating vessels. The pressurized-water plant has been the favorite because it can be more easily shielded and maintained in working order.

The U.S. Navy has in operation and under construction a large number of nuclear-powered submarines and surface vessels. The first nonnaval marine installations of nuclear power were on the Soviet icebreaker *Lenin* (1957) and on the United States demonstration merchant ship NS *Savannah* (1959), both of which used pressurized-water reactors. [L.H.R.; M.G.P.]

Ship routing The selection of the most favorable route for a ship's transit, based upon the environment conditions expected during the voyage, and the effect of these conditions upon the ship's performance. Ship routing is also referred to as weather routing, optimum track ship routing, and meteorological navigation.

Routes can be computed to minimize transit time, avoid waves over a specific height, or, more typically, provide a compromise between travel time and the roughness of the seaway to be encountered. Routing offices compute the initial route based on the long-range weather forecast, but maintain a daily plot of both the ship and the storm positions. If conditions warrant, an amended route is sent to the ship master based on the latest weather changes. *See* OCEAN WAVES; WEATHER FORECASTING AND PREDICTION.

Although the effect of waves is the predominant factor considered, the ship router also utilizes knowledge of ocean currents, wind, and hazards to navigation such as ice fields or fog banks. On United States east coast runs, the currents are more important than the waves. Most major oil companies have their tankers follow the Gulf Stream on the northbound runs and avoid it on the return trip. *See* OCEAN CIRCULATION. [R.W.J.]

Ship salvage Rendering aid to distressed property from a peril at sea or while on other navigable waters. The property in distress could be a vessel of any size or character, from a rowboat to a cruise ship, or the cargo aboard a ship.

Typical salvage situations involve, in order of frequency, rescue tourings, strandings, and sunken vessels.

Ships frequently are disabled at sea as a result of breakdown of engines, fire, collision, loss of propeller, or loss of steering control. The shipowner arranges for a salvage tug to tow the disabled vessel to a port of repair.

Crew error, failure or malfunction of a navigational aid, and storm conditions are often the cause of a vessel's running aground on a submerged reef and occasionally right up on a shoreline. If the vessel is not too hard aground, it can frequently be refloated by pulling of tugs on high tide. More serious strandings may require the employment of beach-gear ground tackles.

When a vessel is sunk with the main deck still out of water, a combination of patching any holes or openings and pumping out the water is all that is necessary to refloat. When the vessel's main deck is under water, more sophisticated techniques are required. If the vessel's main deck is not deeply submerged, a waterproof wall may be built around the main deck to above the water surface and the vessel then patched and pumped. This wall or barrier is known as a cofferdam. Plastics technology has given birth to the so-called foam-in-salvage technique, whereby chemicals are combined by a mixing gun held by a diver in the sunken ship to form urethane foam. The foam hardens quickly, forcing the water out of the vessel and making it buoyant. Another similar technique entails pumping of particles of expanded plastic in bead or disk form into the sunken vessel to displace the water. A third technique involves the pumping of hollow polystyrene balls the size of basketballs into the sunken vessel.

One of the quickest and most convenient methods of raising sunken vessels, particularly smaller ones, is the use of heavy lift cranes. Divers attach slings to the sunken vessel, and floating cranes lift the vessel to the surface where it can be pumped out.

A similar technique involves the use of heavy lift craft. Wire cables are passed under the sunken vessel. The lift craft take on ballast water to increase their draft. At low tide, the wires are pulled tight on the deck of the lift craft. On the rising tide, the salvage crews pump out the lift craft to raise the wreck off the bottom.

Occasionally, sunken barges and smaller vessels can be refloated by use of pontoons. Pontoons can be metal tanks or large rubber bags. The pontoon is flooded to sink it alongside the sunken vessel. A diver then attaches the pontoon to the wreck with chain or wire cable. The water in the pontoon is forced out by pumping in compressed air, and this gives the necessary buoyancy to lift the vessel. [P.S.B.]

Shipbuilding The construction of large vessels which travel over seas, lakes, or rivers. Many different approaches have been used in the construction of ships. Sometimes a ship must be custom-built to suit the particular requirements of a low-volume trade route with unique cargo characteristics. On the other hand, there are many instances where a significant number of similar ships are constructed, providing an opportunity to employ procedures which take advantage of repetitive processes.

The building of a ship can be divided into seven phases: design, construction planning, work prior to keel laying, ship erection, launching, final outfitting, and sea trials. *See* SHIP DESIGN.

The construction planning process establishes the construction techniques to be used and the schedules which all of the shipbuilding activities must follow. Construction planners generally start with an erection diagram on which the ship is shown broken down into erection zones and units. To facilitate the fabrication of steel, insofar as possible, the erection units are designed to be identical. The size (or weight) of the erection units selected is usually limited by the amount of crane capacity available. Once the construction planners have established the manner in which the ship is to be erected and the sequence of construction, the schedules for construction can be developed. Working backward from the time an erection unit is required in the dock, with allowances made for the many processes involved, a schedule of working plans and for procurement of purchased equipment is prepared.

Before the keel of a ship is laid (or when the first erection unit is placed in position) a great deal of work must have been accomplished for work to proceed efficiently. The working drawings prepared by ship designers completely define a ship, but often not in a manner that can be used by the construction trades people. Structural drawings prescribe the geometry of the steel plates used in construction, but they cannot be used, in the form prepared, to cut steel plates. Instead, the detailed structural drawings must be translated into cutting sketches, or numerical-control cutting tapes, which are used to fabricate steel. Several organizations have developed sophisticated computer programs which readily translate detailed structural drawings into machine-sensible tapes which can be used to drive cutting torches.

If all of the preceding work has been accomplished properly and on schedule, the erection of a ship can proceed rapidly; however, problem areas invariably arise. When erecting a ship one plate at a time, there are no serious fitting problems; but when 900-metric-ton erection units do not fit (or align) properly, there are serious problems which tend to offset some of the advantages for this practice.

A ship is launched as soon as the hull structure is sufficiently complete to withstand the strain. Ships may be launched endwise, sidewise, or by in-place flotation (for example, graving docks). The use of a graving dock requires a greater investment in facilities than either of the other two methods, but in some cases there may be an overall advantage due to the improved access to the ship and the simplified launch procedure. *See* DRYDOCKING.

The final outfitting of a ship is the construction phase during which checks are made to ensure that all of the previous work has been accomplished in a satisfactory manner; and last-minute details, such as deck coverings and the top coat of paint, are completed. It is considered good practice to subject as much of the ship as possible to an intensive series of tests while at the dock, where corrections and final adjustments are more easily made than when at sea. As a part of this test program, the main propulsion machinery is subjected to a dock trial, during which the ship is secured to the dock and the main propulsion machinery is operated up to the highest power level permissible.

When a comprehensive program of dockside tests have been completed, the only capabilities which have not been demonstrated are the operation of the steering gear during rated-power conditions and the operation of the main propulsion machinery at rated power; these capabilities must be demonstrated during trials at sea. [R.L.Ha.]

Shipping fever A severe inflammation of the lungs (pneumonia) commonly seen in North American cattle after experiencing the stress of transport. This disease occurs mainly in 6–9-month-old beef calves transported to feedlots. The characteristic pneumonia is caused primarily by the bacterium *Pasteurella haemolytica* serotype A1; thus a synonym for shipping fever is bovine pneumonic pasteurellosis.

Pasteurella haemolytica replicates rapidly in the upper respiratory tract and is inhaled into the lungs, where pneumonia develops in the deepest region of the lower respiratory tract (the pulmonary alveoli). Viruses can also concurrently damage pulmonary alveoli and enhance the bacterial pneumonia.

Symptoms initially include reduced appetite, high fever, rapid and shallow respiration, depression, and a moist cough. During the later stages of disease, cattle lose weight and have labored breathing. The lesion causing these symptoms is a severe pneumonia accompanied by inflammation of the lining of the chest cavity, and is called a fibrinous pleuropneumonia. Without vigorous treatment, shipping fever can cause death in 5–30% of affected cattle.

Treatment is aimed at eliminating bacteria by using antibiotics, reducing inflammation by using nonsteroidal anti-inflammatory drugs, and separating affected cattle. Vaccines can stimulate immunity to the viruses and bacteria associated with shipping fever. *See* ANIMAL VIRUS; INFLAMMATION; PNEUMONIA; VACCINATION. [A.W.Co.]

Shipworm A wood-boring mollusk of the family Teredinidae. These borers, also called pileworms, are found around the world in tropical to boreal and marine to nearly freshwater habitats, from the intertidal to depths of about 2000 ft (600 m). It was not until humans became navigators that shipworms became pests. Before the advent of steel ships, they were greatly feared, and even today hundreds of millions of dollars are spent annually on prevention and repair due to their activity. *See* BIVALVIA. [R.D.T.]

Shock absorber Effectively a spring, a dashpot, or a combination of the two, arranged to minimize the acceleration of the mass of a mechanism or portion thereof with respect to its frame or support.

The spring type of shock absorber is generally used to protect delicate mechanisms, such as instruments, from direct impact or instantaneously applied loads. Such springs are often made of rubber or similar elastic material. *See* SHOCK ISOLATION.

An example of the dashpot type of shock absorber is the direct-acting shock absorber in an automotive spring suspension system (see illustration). Here the device is used to dampen and control a spring movement. The energy of the mass in motion is converted to heat by forcing a fluid through a restriction, and the heat is dissipated by radiation and conduction from the shock absorber. *See* VIBRATION DAMPING.

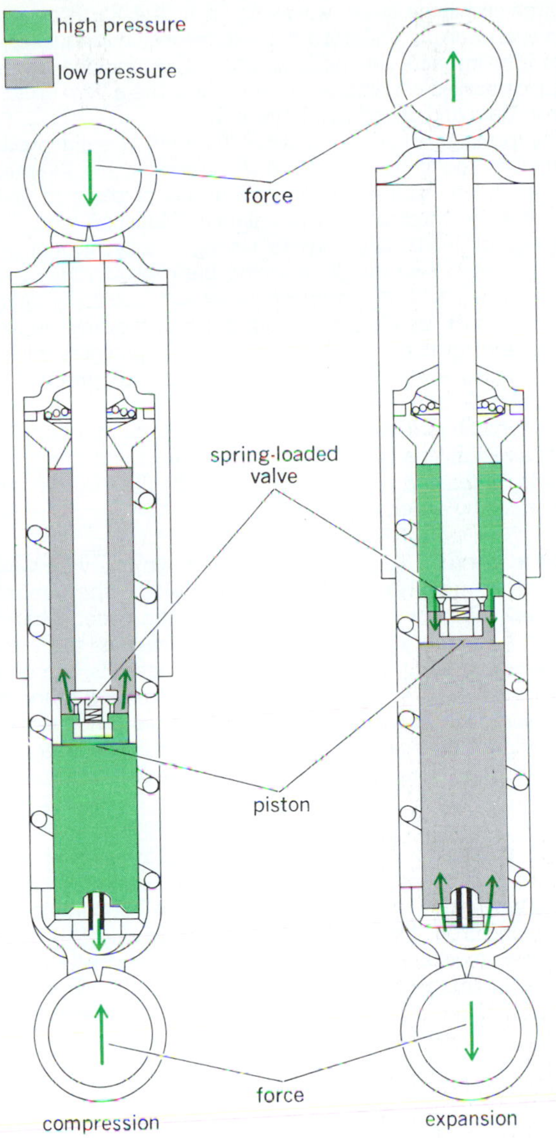

A dashpot-type shock absorber. (*Plymouth Division, Chrysler Corp.*)

There are also devices available which combine springs and viscous damping (dashpots) in the same unit. They use elastic solids (such as rubber or metal), compressed gas (usually nitrogen), or both for the spring. A flat-viscosity hydraulic fluid is used for the viscous damping. [L.S.L.]

Shock isolation The application of isolators to alleviate the effects of shock on a mechanical device or system. The term shock generally denotes suddenness, either in the application of a force or in the inception of a motion. *See* SHOCK WAVE.

Shock isolation is accomplished by storing energy in a resilient medium (isolator, cushion, and so on) and releasing it at a slower rate. The effectiveness of an isolator depends upon the duration of the shock impact.

Rubber is the most common material used in commercial shock isolators. Rubber isolators are generally used where the shock forces are created through small displacements. For larger displacement shock forces, such as those experienced by shipping containers in rough handling conditions, thick cushions of felt, rubberized hair, sponge rubber, cork, or foam plastics are used.

The shock load must be divided between the case, the shock cushion, and the equipment. The case, since it must withstand effects of rough handling such as sliding and dropping, is by necessity rigid. The more rigid the case the closer to a 1:1 ratio will be the transfer of the shock from outside to inside. The absorption of the shock is primarily between the cushion and the equipment. *See* DAMPING; SHOCK ABSORBER; SPRING (MACHINES); VIBRATION DAMPING. [K.W.J.]

Shock syndrome A state characterized by prostration and hypotension due to diminished circulating blood volume, with reduced capillary blood flow and tissue hypoxia.

The five major varieties of shock include neurogenic, oligemic, vasogenic or endotoxic, cardiogenic, and anaphylactic shock. The most common types of shock are neurogenic and oligemic. Neurogenic shock is synonymous with fainting and is produced by nervous or psychic stimuli such as sudden pain, fright, or fear. These stimuli result in a sudden outflow of autonomic impulses to the arterioles, producing vasodilatation and rapid pooling of blood in the splanchnic bed and voluntary muscles. Cerebral blood flow is thus reduced, and fainting occurs. Neurogenic shock is usually self-limiting, and rest in a recumbent position is usually sufficient to restore cerebral blood flow.

Oligemic shock is synonymous with hypovolemic or hemorrhagic shock. It is caused by such things as massive hemorrhage, severe crushing injuries, and burns. There is a true reduction in blood volume which may lead to hypotension and irreversible damage to vital centers.

"Shock" is also used, sometimes inaccurately, to denote superficially similar states, such as emotional, insulin, and electrical shock. Pallor, apprehension, cold skin, sweating, low blood pressure, and rapid, shallow pulse and breathing are typical findings. Any form of severe stress may induce this state; the most common causes are hemorrhage, physical injury with severe pain, and burns. [N.K.M.]

Shock tube A laboratory device for rapidly raising confined samples of gases to preselected high temperatures and densities. This is accomplished by a shock wave, generated by breaking a fragile diaphragm, which separates the low-pressure section (test) from the high-pressure section (driver) of the tube. Shock tubes can be circular, square, or rectangular in cross section.

Samples which attained thermodynamic equilibrium after passage through the shock front have thus been prepared under homogeneous conditions at high temperatures. Their optical and spectroscopic properties can be measured. Conversely, when the indices of refraction or the spectroscopic parameters of the various species suspected of being present in the shock-heated gases are known (or can be reliably estimated), their measurement provides information on the exact composition and hence of the thermodynamic state of the gas. *See* CHEMICAL THERMODYNAMICS; SPECTROSCOPY. [S.H.B.]

Shock wave A fully developed compression wave of large amplitude. Shock waves arise from sharp and violent disturbances generated from a lightning stroke, bomb blast, or other form of intense explosion, and from steady supersonic flow over bodies. Measurements of fluid density, pressure, and temperature across the surfaces show that these quantities always increase along the direction of flow, and that the rates of change are usually so rapid as to be beyond the spatial resolution of most instruments. These surfaces of abrupt change in fluid properties are called shock waves or shock fronts. [S.-C.Li.]

Display. Shock waves can be made visible for purposes of photography. Shock-wave display techniques are used most extensively in aerodynamic test facilities. The most widely used techniques are the schlieren and shadowgraph techniques;

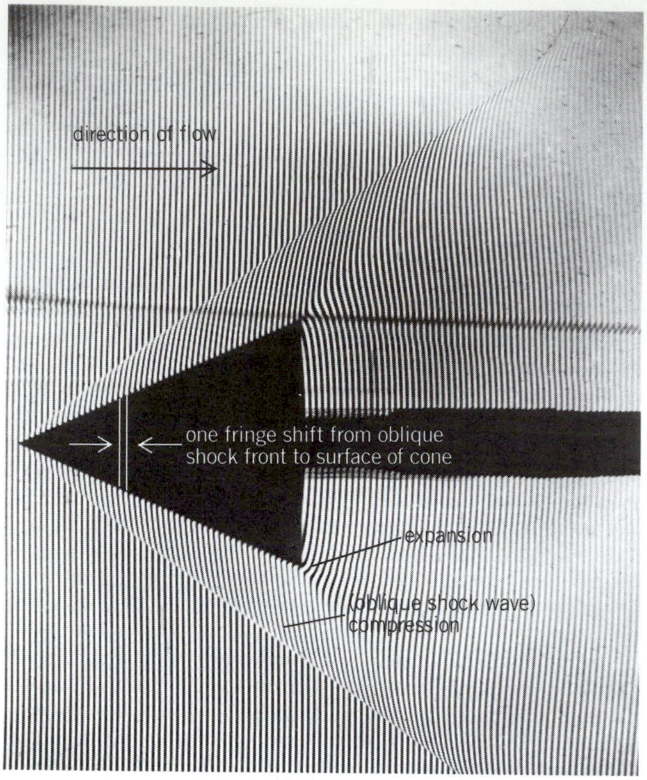

Interferogram showing density change across shock wave formed around 45° cone at Mach number approximately 4. (*U.S. Naval Ordnance Laboratory*)

optical interferometry is also used. All three methods are based on the variation of the index of refraction with density in the medium traversed by light. The schlieren and shadowgraph methods have the advantage of simplicity of experimental setup, and are usually preferred because they readily display even weak disturbances such as nearly sonic shocks. *See* INTERFEROMETRY; SCHLIEREN PHOTOGRAPHY; SHADOWGRAPH.

The interferometric technique permits a direct measurement of the density change across the shock wave. The change is presented as displacement of interference fringes (see illustration). The method is best suited for obtaining quantitative information on shock waves and flow fields.

For the visualization of shock-wave phenomena in gases of low density, the above methods are too insensitive. Under such conditions, techniques utilizing the absorption of radiation or corpuscular rays provide better means for detection of shock waves. Best suited for the radiation absorption techniques are the spectral regions of strong continuous absorption, which for most gases are in the ultraviolet, or the region of soft x-rays. Corpuscular-ray absorption techniques use monoenergetic particles such as electrons, protons, or α particles.

Two other methods for visualization of especially low-density flows are the glow discharge and afterglow techniques. In these techniques the gas, prior to entering the test chamber, is subjected to a strong electrical discharge which excites it to luminescence. The light emission accompanying the discharge is used in the first method; the glow persisting after the exciting discharge has been cut off is used in the second method. Shock waves become visible because density changes affect the luminous intensity as well as the spectral distribution of the glow. [E.M.Wi.]

Solids. The term shock wave, with reference to solids, generally means a large-amplitude compressional stress wave or pressure pulse that produces an almost instantaneous increase in the value of stress or pressure in a material.

Large-amplitude stress waves occur in the Earth as a result of events such as underground explosions, earthquakes, and meteorite impacts. In the laboratory they can be produced under controlled conditions and have become a very useful tool in solid-state and geophysical research.

The passage of a shock wave through a solid produces changes in the physical state of the material. For example, it can redistribute atoms, change the levels of electron energy, and alter the internal energy balance. Measurements of the behavior of solids subjected to strong compressional pulses have provided new insight into the basic properties of solids and have led to the development of commercially valuable processes such as explosive cladding and welding, explosive cutting and forming, shock compaction of powders into solid parts, explosive excavation of the Earth, and production of new materials or those with special properties.

Experimenters usually produce shock waves in a material by the detonation of a high explosive in contact with the material or by the impact of a high-speed projectile fired from a gun or accelerated by a high explosive. [R.K.L.]

Shore birds A general term, synonymous with waders, applied to members of the 19 families in the suborders Charadrii, Pterocii, Lari, and Alcae of the order Charadriiformes. These include such common species as the plovers, curlews, jacanas, phalaropes, woodcocks, avocets, snipes, and oyster catchers. Most of these birds are always found near water, although their habitats are quite varied. Wide variations are noted for their feet, which may be webbed and may terminate in three or four long or short toes. *See* CHARADRIIFORMES. [C.B.C.]

Short circuit A term commonly used to describe an electrical connection of negligible impedance connected across a pair of terminals. There is implication of an undesirable condition arising from electrical insulation failure due to improper operation, mechanical damage, or damage by natural causes such as lightning and rain. Protection against such short circuits is a major field in electrical power engineering. *See* ELECTRIC PROTECTIVE DEVICES.

Short circuits are often used to advantage on high-frequency transmission lines. For instance, a stub transmission line, one-quarter wavelength long and short-circuited at one end, acts as an insulator at the opposite end and therefore is used as a support for a high-frequency transmission line. *See* CIRCUIT (ELECTRICITY); TRANSMISSION LINES. [R.L.R.]

Short takeoff and landing (STOL) The term applied to heavier-than-air craft that cannot take off and land vertically, but can operate within areas substantially more confined than those normally required by aircraft of the same size. A pure STOL aircraft is a fixed-wing vehicle that derives lift primarily from free-stream airflow over the wing and its high lift system, sometimes with significant augmentation from the propulsion system. Although all vertical takeoff and landing (VTOL) machines, including helicopters, can lift greater loads by developing forward speed on the ground before liftoff, they are still regarded as VTOL (or V/STOL craft), operating in the STOL mode. *See* VERTICAL TAKEOFF AND LANDING (VTOL).

It has been customary to define STOL capability in terms of the runway length required to take off or land over a 50-ft (15-m) obstacle, the concept of "short" length being variously defined as from 500 to 2000 ft (150 to 600 m), depending on the high-lift concept employed and on the mission of the aircraft. In addition to being able to operate from short runways, STOL aircraft are usually expected to be able to maneuver in confined airspace so as to minimize the required size of the terminal area. Such aircraft must therefore have unusually good slow-flight stability and control characteristics, especially in turbulence and under instrument flight conditions. *See* AIRPLANE. [R.E.K.]

Shrew An insectivorous mammal of the family Soricidae found in Asia, Africa, Europe, and North America. The family includes 265 species grouped into three subfamilies, the red-toothed, white-toothed, and armored shrews. Shrews are small, savage animals with extremely sharp teeth. The tail is moderately long, the eyes are minute, the snout is sharp-pointed, and the ears are small (see illustration).

The Eurasian common shrew (*Sorex araneus*).

Shrews are solitary individuals that establish distinct territories. Couples are found together only during mating. The female builds the nest, a hole in the ground lined with leaves and moss, where she deposits her litter of 5–10 young, born after a gestation period of between 2–3 weeks. The female mates at the age of 1 year; however, since the life-span of the shrew is only 15 months, the females always die during the fall after the litter is born. *See* INSECTIVORA. [C.B.C.]

Shrimp The common name applied to a number of crustaceans which are members of various orders or suborders; however, the edible and economically important species are decapod crustaceans of the suborder Natantia. The various ostracod orders, including small to microscopic forms, are referred to as the seed shrimp. The Stomatopoda are known as the mantis shrimp, while the opossum shrimp are members of the order Mysidaceae. Mantis shrimps are marine crustaceans which inhabit burrows on the bottom of the sea. The body is elongate with a broad abdomen; the second thoracic limb, used to capture smaller crustaceans, is quite large and resembles the front leg of a praying mantis. The head is unique in that there are two movable segments bearing the antennules and eyes. Largest species is about 1 ft (30 cm) long. *See* DECAPODA (CRUSTACEA); OSTRACODA; STOMATOPODA. [C.B.C.]

Shrink fit A fit that has considerable negative allowance so that the diameter of a hole is less than the diameter of a shaft that is to pass through the hole, also called a heavy force fit. Shrink fits are used for permanent assembly of steel external members, as on locomotive wheels. The difference between a shrink fit and a force fit is in method of assembly. In shrink fits, the outer member is heated, or the inner part is cooled, or both, as required. The parts are then assembled and returned to the same temperature. *See* ALLOWANCE; FORCE FIT. [P.H.B.]

Shunting The act of connecting one device to the terminals of another so that the current is divided between the two devices in proportion to their respective admittances. Shunting is widely used in ammeters, galvanometers, and other current-measuring instruments to bypass part of the current around the instrument so as to change the measuring range. Resistors are frequently shunted across tuned circuits to broaden the tuning characteristics. [J.Mar.]

Siberian tick typhus A relatively benign, rash- and eschar-producing, spotted feverlike disease in northern Asia, caused by the bacterialike microorganism *Rickettsia sibirica*. The disease is transmitted by four species of *Dermacentor* and two of *Haemaphysalis*. Clinically, Siberian tick typhus resembles most closely fièvre boutonneuse, including low mortality, but epidemiologically, it is like American spotted fever, with the primary natural cycle between ground squirrels and other small rodents, and their immature tick parasites. *See* FIÈVRE BOUTONNEUSE; RICKETTSIALES; RICKETTSIOSES; ROCKY MOUNTAIN SPOTTED FEVER. [C.B.P.]

Sickle cell disease A disease caused by the presence of an abnormal hemoglobin, sickle cell hemoglobin (hemoglobin S, HbS). The genetic defect in sickle cell disease is a single nucleotide change from GAG to GTG at the sixth codon of the ß-globin gene. The nucleotide change results in a single amino acid change at position six of the ß-globin protein chain with the substitution of a valine for a glutamic acid. The sickle ß-globin ($ß^s$) chains are then incorporated into intact hemoglobin tetramers. The normal major adult hemoglobin (hemoglobin A, HbA) contains two α and two ß chains to form an $\alpha_2ß_2$ tetramer. Sickle cell hemoglobin is $\alpha_2ß_2{}^s$. *See* HEMOGLOBIN.

In individuals with sickle cell anemia, the aggregation and polymerization of HbS has two major effects: stasis, or slowing of blood flow, and anemia, This in turn leads to increasing local deoxygenation within the small blood vessels. Deoxygenation itself is created by this slowed blood flow and leads to further aggregation of sickle hemoglobin and increasing sickling. This process produces a vicious cycle, of stasis—deoxygenation—more stasis, and is responsible for the painful episodes which accompany sickle cell anemia, so-called painful crises. In addition, the slowed and decreased blood flow to vital tissues can result in tissue damage by ischemia and infarction of organs such as the liver, kidney, and heart.

Sickle cell anemia can be distinguished from sickle cell trait and other sickle syndromes by hemoglobin electrophoresis. The presence of large amounts of HbA indicates the presence of sickle cell trait. In addition, a variety of tests based on the decreased solubility of sickle hemoglobin as compared with normal hemoglobin are available for screening for the presence of some sickle cell hemoglobin; however, these tests do not distinguish between sickle cell trait, sickle cell anemia, or its variants. Gene diagnosis of the sickle syndromes is also available. *See* HUMAN GENETICS. [A.Ba.]

Sideband The frequency band located either above or below the carrier frequency within which fall the frequency components of the wave produced by the process of modulation. Apart from the carrier, all components of an amplitude-modulated sinusoidal carrier, when taken together, form a pair of sidebands extending on either side of the carrier frequency in mirror symmetry and containing all the frequency components of the modulating wave. The sidebands above and below the carrier frequency are called upper sideband and lower sideband, respectively. *See* AMPLITUDE MODULATION; CARRIER; MODULATION; SINGLE SIDEBAND. [H.S.Bl.]

Sidereal time One of several kinds of time scales used in astronomy, whose primary application is as part of the coordinate system to locate objects in the sky. It is also the basis for determining the solar time used in everyday living.

The common measurements of time are based on the motions of the Earth that most affect everyday life: Earth's rotation on its axis, and revolution in orbit around the Sun. Objects in the sky reflect these motions and appear to move westward, crossing the meridian each day. A particular object or point is chosen as a marker, and the interval between its successive crossings of the local meridian is defined to be a day, divided into 24 equal parts called hours. The actual length of the day for comparison between systems depends on the reference object chosen. The time of day is reckoned by the angular distance around the sky that the reference object has moved westward since it last crossed the meridian. In fact, the angular distance west of the meridian is called the hour angle. *See* MERIDIAN.

The reference point for marking sidereal time is the vernal equinox, one of the two points where the planes of the Earth's Equator and orbit appear to intersect on the celestial sphere. The sidereal day is the interval of time required for the hour angle of the equinox to increase by 360°. One rotation of the Earth with respect to the Sun is a little longer, because the Earth has moved in its orbit as it rotates and hence must turn approximately 361° to complete a solar day. A sidereal year is the time required for the mean longitude of the Sun to increase 360°, or for the Sun to make one circuit around the sky with respect to a fixed reference point. *See* EARTH ROTATION AND ORBITAL MOTION; TIME. [A.D.F.]

Siderite The mineral form of ferrous carbonate, often containing appreciable amounts of magnesium and manganese substituting for iron.

Siderite has hexagonal (rhombohedral) symmetry and the same structure as calcite. Individual crystals are often rhombohedral in shape, sometimes with curved faces. Massive varieties also occur. Siderite is often brownish and sometimes gray or greenish. The specific gravity is 3.9 and the hardness is 4 on Mohs scale. *See* CALCITE; CARBONATE MINERALS. [R.I.Ha.]

Siderocapsaceae A family of chemolithotrophic bacteria found in iron-bearing waters. It is a heterogeneous group of gram-negative organisms, possessing in common the ability to deposit iron or manganese compounds around the cells. Cell morphology may be apparent only after treatment with dilute acid. The species are organized into four genera. *See* CHEMOLITHOTROPHIC BACTERIA. [R.S.W.]

Signal detection theory A theory in psychology which characterizes not only the acuity of an individual's discrimination but also the psychological factors that bias the individual's judgments. Failure to separate these two aspects of discrimination had tempered the success of theories based upon the classical concept of a sensory threshold. The theory provides a modern and more complete account of the process whereby an individual makes fine discriminations.

The theory of signal detection has two parts of quite different origins. The first comes from mathematical statistics and is a translation of the theory of statistical decisions. The major contribution of this part of the theory is that it permits a determination of the individual's discriminative capacity, or sensitivity, that is independent of the judgmental bias or decision criterion the individual may have had when the discrimination was made. The second part of the theory comes from the study of electronic communications. It provides a means of calculating for simple signals, such as tones and lights, the best discrimination that can be attained. The prediction is based upon physical measurements of the signals and their interfering noise.

This opportunity to compare the sensitivity of human observers with the sensitivity of an "ideal observer" for a variety of signals is of considerable usefulness, and of growing interest, in sensory psychology. Signal detection theory has been applied to several topics in experimental psychology in which separation of intrinsic discriminability from decision factors is desirable. Included are attention, imagery, learning, conceptual judgment, personality, reaction time, manual control, and speech.

The analytical apparatus of the theory has been of value in the evaluation of the performance of systems that make decisions based on uncertain information. Such systems may involve only people, or people and machines together, or only machines, Examples come from medical diagnosis, where clinicians may base diagnostic decisions on a physical examination, or on an x-ray image, or where machines make diagnoses, perhaps by counting blood cells of various types. [J.A.Sw.]

Signal generator A piece of electronic test equipment that delivers a sinusoidal output of accurately calibrated frequency. The frequency may be anywhere from audio to microwave, depending upon the intended use of the instrument. The frequency and the amplitude are adjustable over a wide range. The oscillator must have excellent frequency stability, and its amplitude must remain constant over the tuning range.

The Wien-bridge oscillator is commonly used for frequencies up to about 200 kHz. For a radio-frequency signal generator up to about 200 MHz, a resonant circuit oscillator is used (such as a tuned-plate tuned-grid, Hartley, or Colpitts). Beyond this range vhf and microwave oscillators are used. *See* OSCILLATOR. [J.Mi.]

Signal-to-noise ratio The ratio, at some location, of some measure of the desired signal to the same measure of the total noise, abbreviated S/N. This is a primary consideration in the design of any communication system, and is a measure of the efficiency of the system. Distortion of the signal by noise causes errors. The objective of engineering is the maintenance of error-free communications over the smallest possible S/N value. *See* ELECTRICAL NOISE. [W.Lyo.]

Significant figures The numbers considered here are real numbers in decimal form. All numbers are considered as positive although the theory applies also to negative numbers. Every digit in a number is significant unless its sole purpose is to fix the position of the decimal point. Thus the significant figures in 31.50 are 3, 1, 5, and 0; in 0.0315 only 3, 1, and 5 are significant; and in 3005 both zeros, 3, and 5 are significant.

A number having n significant figures has n-figure accuracy. Thus 31.50 and 3005 have four-figure accuracy but 0.00315 has three-figure accuracy. Evidently the number 3150 may have four-figure or three-figure accuracy. The so-called powers-of-ten notation is used to remove ambiguity. In the powers-of-ten notation a number is written in the form $a \times 10^k$ where k is an integer and a equals 1 or lies between 1 and 10. In this notation all figures in a are significant. Thus 3150 would be written 3.150×10^3 to indicate four-figure accuracy and written 3.15×10^3 to indicate three-figure accuracy. [L.M.K.]

Silica minerals Silica (SiO_2) occurs naturally in at least nine different varieties (polymorphs), which include tridymite, cristobalite, coesite, and stishovite, in addition to high (ß) and low (α) quartz. These forms are characterized by distinctive crystallography, optical characteristics, physical properties, pressure-temperature stability ranges, and occurrences.

The crystal structures of all silica polymorphs except stishovite contain silicon atoms surrounded by four oxygens, thus producing tetrahedral coordination polyhedra. Each oxygen is bonded to two silicons, creating an electrically neutral framework. Stishovite differs from the other silica minerals in having silicon atoms surrounded by six oxygens (octahedral coordination.) Ideal high tridymite is composed of sheets of SiO_4 tetrahedra oriented perpendicular to the c crystallographic axis (Fig. 1) with adjacent tetrahedra in these sheets pointing in opposite directions. High cristobalite, like tridymite, is composed of parallel sheets of SiO_4 tetrahedra with neighboring tetrahedra pointing in opposite directions. However, the hexagonal rings are distorted and adjacent sheets are rotated

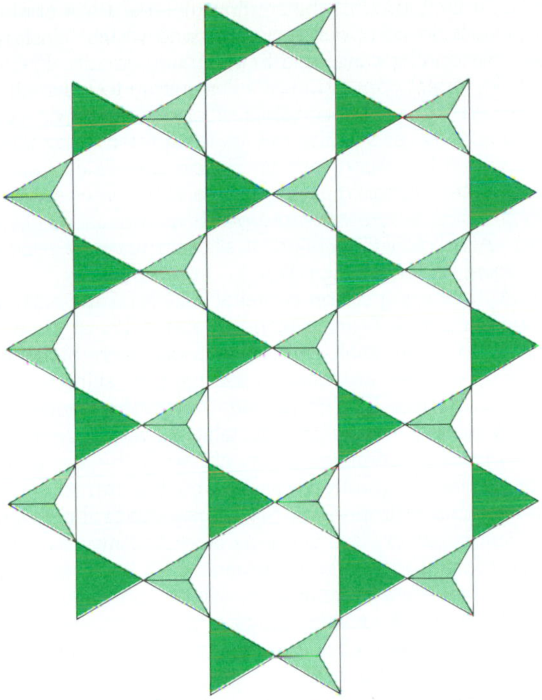

Fig. 1. Portion of an idealized sheet of tetrahedrally coordinated silicon atoms similar to that found in tridymite and cristobalite. Sharing of apical oxygens (which point in alternate directions) between silicons in adjacent sheets generates a continuous framework. (*After J. J. Papike and M. Cameron, Crystal chemistry of silicate minerals of geophysical interest, Rev. Geophys. Space Phys., 14:37–80*; copyright © *1976 by American Geophysical Union*)

60° with respect to one another, resulting in the geometry shown in Fig. 2. Coesite also contains silicon atoms tetrahedrally coordinated by oxygen. These polyhedra share corners to form chains composed of four-membered rings. Silicon in stishovite is octahedrally coordinated by oxygen. These coordination polyhedra share edges and corners to form chains of octahedra parallel to the *c* crystallographic axis.

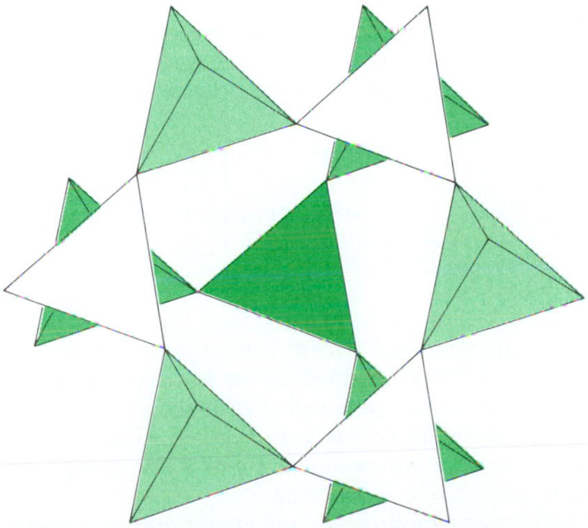

Fig. 2. Portion of cubic high cristobalite illustrating the distortion of tetrahedral sheets, with are oriented parallel to (111), and the 60° rotation of adjacent sheets. (*After J. J. Papike and M. Cameron, Crystal chemistry of silicate minerals of geophysical interest, Rev. Geophys. Space Phys., 14:37–80*; copyright © *1976 by American Geophysical Union*)

Chemically, all silica polymorphs are ideally 100% SiO_2. However, unlike quartz which commonly contains few impurities, the compositions of tridymite and cristobalite generally deviate significantly from pure silica. This usually occurs because of a coupled substitution in which a trivalent ion such as Al^{3+} or Fe^{3+} substitutes for Si^{4+}, with electrical neutrality being maintained by monovalent or divalent cations occupying interstices In the relatively open structures of these two minerals. *See* COESITE; QUARTZ. [J.C.D.]

Silicate minerals All silicates are built of a fundamental structural unit, the so-called SiO_4 tetrahedron. The crystal structure may be based on isolated SiO_4 groups or, since each of the four oxygen ions can bond to either one or two silicon (Si) ions, on SiO_4 groups shared in such a way as to form complex isolated groups or indefinitely extending chains, sheets, or three-dimensional networks. Mixed structures in which more than one type of shared tetrahedra are present also are known. *See* SILICON.

Silicates are classified according to the nature of the sharing mechanism, as revealed by x-ray diffraction study. The sharing mechanism gives rise to a characteristic ratio of Si to O, but it is possible for oxygen ions that are not bonded to Si to be present in the structure, and sometimes some or all of any aluminum present must be counted as equivalent to Si.

The detailed crystallographic and physical properties of the various silicates are broadly related to the type of silicate framework that they possess. Thus, the phyllosilicates as a group typically have a platy crystal habit, with a cleavage parallel to the plane of layering of the structure, and are optically negative with rather high birefringence. The inosilicates, based on an extended one-dimensional rather than two-dimensional linkage of the SiO_4 tetrahedra, generally form crystals of prismatic habit; if cleavage is present, it will be parallel to the direction of elongation. The tectosilicates commonly are equant in habit, without marked preference for cleavage direction, and tend to have a relatively low birefringence.

Silicate minerals make up the bulk of the outer crust of the Earth and form in a wide range of geologic environments. Many silicates are of economic importance. For discussions of certain silicate mineral groups *see* AMPHIBOLE; ANDALUSITE; CHLORITE; CHLORITOID; EPIDOTE; FELDSPAR; FELDSPATHOID; GARNET; HUMITE; MICA; OLIVINE; PYROXENE; SCAPOLITE; SERPENTINE; ZEOLITE. [C.Fr.]

Silicate polymers Crystalline and amorphous inorganic polymers of silicates. Examples include the naturally occurring fiberlike asbestos and sheetlike mica. The industrially important water-soluble alkali metal silicates can give highly viscous polymeric solutions. Borosilicate glasses form another important group of silicate polymers. The Pyrex type is well known for its resistance to thermal shock; the leached Vycor type is porous and can be used for filtering bacteria and viruses. Asbestos occurs as ladder polymers, of which crocidolite is the most important, and as layer polymers exemplified by chrysotile. The zeolites, many of which have been found naturally or have been synthesized, are three-dimensional network polymers. Their uses as molecular sieves are well known. *See* ASBESTOS; INORGANIC POLYMER; MOLECULAR SIEVE; SILICATE MINERALS; ZEOLITE. [R.A.S.]

Siliceous sinter A porous silica deposit formed around hot springs. It is white to light gray and sometimes friable. Geyserite is a variety of siliceous sinter formed around geysers. The siliceous sinters are deposited as the hot subterranean waters cool after issuing at the surface and become supersaturated with silica that was picked up at depth. The sinters are frequently deposited on algae that live in the pools around the hot springs. *See* GEYSER. [R.Si.]

Silicoflagellata A class of unicellular flagellate microorganisms of the plant division Chrysophyta, which are a part of the marine plankton. Their exoskeletons, siliceous and coarsely perforate (see illustration), resemble those of the Radiolaria, with which they have been grouped. They are usually subpyramidal or hemispherical in shape, and delicately filigreed. Two families and 11 genera of silicoflagellates have been described from siliceous sedimentary rocks ranging in age from Upper Cretaceous to Recent, in association with abundant diatoms and siliceous sponge spicules. *See* MICROPALEONTOLOGY; PHYTOPLANKTON. [D.J.J.]

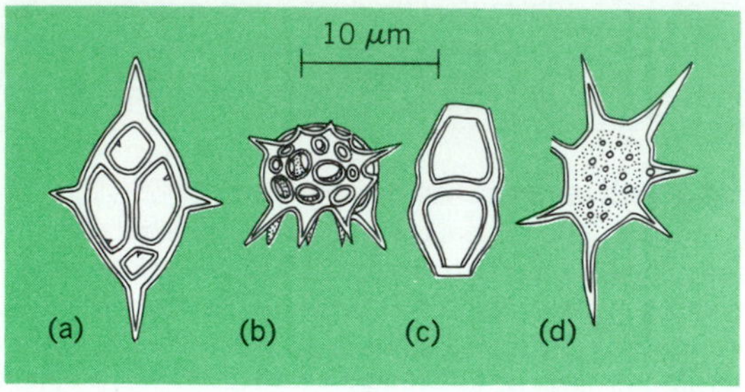

Examples of fossil and modern Silicoflagellata. (*a*) *Dictyocha*, Cretaceous to Recent; (*b*) *Cannopilus*, Miocene; (*c*) *Naviculopsis*, Eocene to Miocene; and (*d*) *Vallacerta*, Upper Cretaceous.

Silicoflagellida An order of the phylum Protozoa, class Phytamastigophorea. These organisms are marine flagellates which have an internal, siliceous, tubular skeleton; numerous small, discoid, yellow chromatophores; and a single flagellum. At times the organisms also put forth, from the ends of their skeletal tubes, long, rather threadlike pseudopodia. The skeleton forms a basket within which the moiety of protoplasm lies, but skeletal elements always have at least a thin covering. There is a single genus, *Dictyocha*, with four species. *See* PHYTAMASTIGOPHOREA. [J.B.L.]

Silicon A chemical element, Si, atomic number 14, and atomic weight 28.086. Silicon is the most abundant electropositive element in the Earth's crust. The element is a metalloid with a decided metallic luster; it is quite brittle. It has a specific gravity of 2.42 at 20°C (68°F), melts at 1420°C (2588°F), and boils at 3280°C (5936°F). The element is usually tetravalent in its compounds, although sometimes divalent, and is decidedly electropositive in its chemical behavior. In addition, pentacoordinate and hexacoordinate compounds of silicon are known. *See* METALLOID.

Crude elementary silicon and its intermetallic compounds are used in alloying constituents to strengthen aluminum, magnesium, copper, and other metals. Metallurgical silicon of

1	2	3	4	5	6	7	8	9	10	11	12	13	14	15	16	17	18
1 H																	2 He
3 Li	4 Be											5 B	6 C	7 N	8 O	9 F	10 Ne
11 Na	12 Mg											13 Al	14 Si	15 P	16 S	17 Cl	18 Ar
19 K	20 Ca	21 Sc	22 Ti	23 V	24 Cr	25 Mn	26 Fe	27 Co	28 Ni	29 Cu	30 Zn	31 Ga	32 Ge	33 As	34 Se	35 Br	36 Kr
37 Rb	38 Sr	39 Y	40 Zr	41 Nb	42 Mo	43 Tc	44 Ru	45 Rh	46 Pd	47 Ag	48 Cd	49 In	50 Sn	51 Sb	52 Te	53 I	54 Xe
55 Cs	56 Ba	71 Lu	72 Hf	73 Ta	74 W	75 Re	76 Os	77 Ir	78 Pt	79 Au	80 Hg	81 Tl	82 Pb	83 Bi	84 Po	85 At	86 Rn
87 Fr	88 Ra	103 Lr	104 Rf	105 Db	106 Sg	107 Bh	108 Hs	109 Mt	110	*111*	*112*	*113*	*114*	*115*	*116*	*117*	*118*

lanthanide series	57 La	58 Ce	59 Pr	60 Nd	61 Pm	62 Sm	63 Eu	64 Gd	65 Tb	66 Dy	67 Ho	68 Er	69 Tm	70 Yb
actinide series	89 Ac	90 Th	91 Pa	92 U	93 Np	94 Pu	95 Am	96 Cm	97 Bk	98 Cf	99 Es	100 Fm	101 Md	102 No

98–99% purity is used as the starting material for manufacturing organosilicon compounds and silicone resins, elastomers, and oils. Silicon chips are used in integrated circuits. Photovoltaic cells for direct conversion of solar energy to electricity use wafers sliced from single crystals of electronic-grade silicon. Silicon dioxide is used as the raw material for making elementary silicon and for silicon carbide. Sizable crystals of it are used for piezoelectric crystals. Fused quartz sand becomes silica glass, used in chemical laboratories and plants as well as an electrical insulator. A colloidal dispersion of silica in water is used as a coating agent and as an ingredient in certain polishes.

Naturally occurring silicon contains 92.2% of the isotope of mass number 28, 4.7% of silicon-29, and 3.1% of silicon-30. In addition to these stable, natural isotopes, several artificially radioactive isotopes are known. Elementary silicon has the physical properties of a metalloid, resembling germanium below it in group IV of the periodic table. In very pure form silicon is an intrinsic semiconductor, although the extent of its semiconduction is greatly increased by the introduction of minute amounts of impurities. Silicon resembles the metals in its chemical behavior. It is about as electropositive as tin, and decidedly more positive than germanium or lead. In keeping with this rather metallic character, silicon forms tetrapositive ions and a variety of covalent compounds; it appears as a negative ion in only a few silicides and as a positive constituent of oxy acid or complex anions.

Several series of hydrides are formed, a variety of halides (some of which contain silicon-to-silicon bonds), and also many series of oxygen-containing compounds which may be either ironic or covalent in their properties.

Silicon occurs in many forms of the dioxide and as almost numberless variations of the natural silicates. For a discussion of the structures and compositions of the representative classes *see* SILICATE MINERALS.

In abundance, silicon exceeds by far every other element except oxygen. It constitutes 27.72% of the solid crust of the Earth, whereas oxygen constitutes 46.6%, and the next element after silicon, aluminum, accounts for 8.13%.

Silicon is reported to form compounds with 64 of the 96 stable elements, and it probably forms silicides with 18 other elements. Besides the metal silicides, used in large quantities in metallurgy, silicon forms useful and important compounds with hydrogen, carbon, the halogen elements, nitrogen, oxygen, and sulfur. In addition, useful organosilicon derivatives have been prepared. [E.P.P.]

Silicone resins Polymers composed of alternating atoms of silicon and oxygen with organic substituents attached to the silicon atoms, as shown in the formula below.

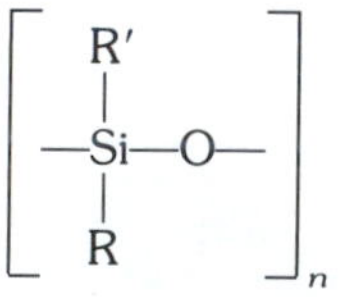

Silicones are obtained by the condensation of hydroxy organosilicon compounds formed by the hydrolysis of organosilicon halides. Silicones, also called organopolysiloxanes, may exist as liquids, greases, resins, or rubbers. Silicone polymers have good resistance to water and oxidation, stability at high and low temperatures, and lubricity.

The wide range of structural variations makes it possible to tailor compositions for many kinds of applications. Low-molecular-weight silanes containing amino or other functional groups are employed as treating or coupling agents for glass fiber and other reinforcements in order to cause unsaturated polyesters and other resins to adhere more firmly.

The liquids, generally dimethyl silicones of relatively low molecular weight, have low surface tension, great wetting power and lubricity for metals, and very small change in viscosity with temperature. They are used as hydraulic fluids, as antifoaming agents, as treating and waterproofing agents for leather, textiles, and masonry, and in cosmetic preparations.

Silicone resins are frequently selected for coating applications in which thermal stability in the range 300–500°C is required.

Silicone rubbers are compositions containing high-molecular-weight dimethyl silicone linear polymer, finely divided silicon dioxide as the filler, and a peroxidic curing agent. The silicone rubbers remain flexible at very low temperatures and stable at high temperatures. An application of increasing importance is as implants in the body for cosmetic or prosthetic purposes. *See* Inorganic polymer; Rubber; Silicon. [J.A.M.]

Silk The lustrous fiber produced by the larvae of silkworms; also the thread or cloth made from such fiber.

The cocoons of the silkworm are delivered to a factory, called a filature, where the silk is unwound from the cocoons and the strands are collected into skeins. The process of unwinding the filament from the cocoon is called reeling. As the filament of a single cocoon is too fine for commercial use, 3–10 strands are usually reeled at a time to produce the desired diameter of raw silk thread. The usable length of the reeled filament is from 1000 to 2000 ft (300 to 600 m). The remaining part of the filament is valuable raw material for the manufacture of spun silk.

The term reeled silk is applied to the raw silk strand that is formed by combining several filaments from separate cocoons. It is reeled into skeins, which are packed in small bundles called books. From the filature, the books of reeled silk go to the throwster where they are transformed into silk yarn, also called silk thread, by a process known as throwing. Silk throwing is analogous to the spinning process that changes cotton, linen, or wool fibers into yarn. The manufacture of those fibers, however, unlike that of silk yarm, does not include carding, combing, and drawing out, the usual processes for producing a continuous yarn. [M.D.P.]

Sillimanite A nesosilicate mineral with the composition $Al_2O[SiO_4]$, crystallizing in the orthorhombic system. It commonly occurs in slender crystals or parallel groups, and is frequently fibrous, hence the synonym fibrolite. There is one perfect cleavage, luster is vitreous, color is brown, pale green, or white, hardness is 6–7 on Mohs scale, and the specific gravity is 3.23. Sillimanite, andalusite, and kyanite are polymorphs of $Al_2O[SiO_4]$. *See* Andalusite; Kyanite; Silicate minerals. [P.B.M.]

Silurian A geologic period of time during which a system of rocks, the Silurian System, was deposited. The strata are identified in terms of the principal forms of life extant at that time as determined by means of the fossils. Silurian rocks are present in many parts of the world, but three of the best-known sections are in Great Britain, eastern United States, and central Bohemia. *See* Paleozoic.

Silurian deposits are represented by both terrestrial and marine strata, but the latter are predominant. The marine strata are composed largely of carbonate facies, either limestone or dolomite, but there are sandstones and shales, including dark graptolite shales. The marine strata commonly contain a large invertebrate fossil fauna.

In Late Silurian times the seaways of eastern North America began to dry up, leading to the formation of salt beds. Lower and Middle Silurian strata bear extensive iron deposits, mostly in the form of hematite.

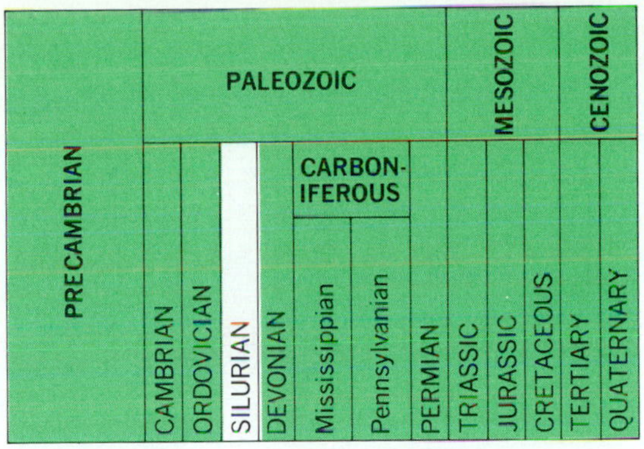

Silurian rocks contain a prolific invertebrate fossil assemblage. Among the best-represented groups were the brachiopods belonging to the class Articulata (Pygocaulia). The coral faunas also were prolific. Crinoids (echinoderms) became common for the first time. Primitive fish (Agnatha and Placodermi) are the only known Silurian vertebrates. [T.W.A.]

Siluriformes The catfishes, a highly distinctive order (also called Nematognathi) of actinopterygian fishes. In the Siluriformes the Weberian apparatus is more complex than in the related Cypriniformes. The body is usually naked or the scales are enlarged and modified into large, overlapping bony plates. There are one to four pairs of barbels, and both the pectoral and dorsal fins usually have a strong spine (see illustration). This is a large group, including about 26 families and perhaps 2000 species, of which nearly 1200 live in South America. Catfishes are known from Eocene times. *See* Actinopterygii; Cypriniformes. [R.M.B.]

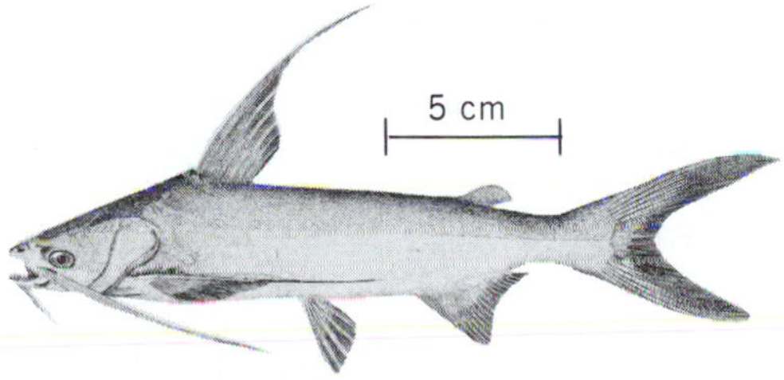

Gafftopsail catfish (*Bagre marinus*). (*After G. B. Goode, Fishery Industries of the United States, 1884*)

Silver A chemical element, Ag, atomic number 47, atomic mass 107.868. It is a gray-white, lustrous metal. Chemically it is one of the heavy metals and one of the noble metals; commercially it is a precious metal. There are twenty-five isotopes of silver. Their atomic masses range from 102 to 117.

In most of its uses, silver is alloyed with one or more other metals. Silver, with the highest thermal and electrical conductivities of all the metals, is used for electrical and electronic contact points. Silver also has well-known uses in jewelry and silverware. Alloys in which silver is an ingredient include dental amalgam and metals for engine pistons and bearings. *See* Amalgam; Silver alloys.

Silver is a rather rare element. Sometimes it occurs in nature as the free element (native silver) or alloyed with other metals. For the most part, however, silver is found in ores containing silver compounds. The principal silver ores are argentite, cerargyrite or horn silver, and several minerals in which silver sulfide is combined with sulfides of other metals. About three-fourths

1	2	3	4	5	6	7	8	9	10	11	12	13	14	15	16	17	18
1 H																	2 He
3 Li	4 Be											5 B	6 C	7 N	8 O	9 F	10 Ne
11 Na	12 Mg											13 Al	14 Si	15 P	16 S	17 Cl	18 Ar
19 K	20 Ca	21 Sc	22 Ti	23 V	24 Cr	25 Mn	26 Fe	27 Co	28 Ni	29 Cu	30 Zn	31 Ga	32 Ge	33 As	34 Se	35 Br	36 Kr
37 Rb	38 Sr	39 Y	40 Zr	41 Nb	42 Mo	43 Tc	44 Ru	45 Rh	46 Pd	47 Ag	48 Cd	49 In	50 Sn	51 Sb	52 Te	53 I	54 Xe
55 Cs	56 Ba	71 Lu	72 Hf	73 Ta	74 W	75 Re	76 Os	77 Ir	78 Pt	79 Au	80 Hg	81 Tl	82 Pb	83 Bi	84 Po	85 At	86 Rn
87 Fr	88 Ra	103 Lr	104 Rf	105 Db	106 Sg	107 Bh	108 Hs	109 Mt	110	111	112	113	114	115	116	117	118

lanthanide series	57 La	58 Ce	59 Pr	60 Nd	61 Pm	62 Sm	63 Eu	64 Gd	65 Tb	66 Dy	67 Ho	68 Er	69 Tm	70 Yb
actinide series	89 Ac	90 Th	91 Pa	92 U	93 Np	94 Pu	95 Am	96 Cm	97 Bk	98 Cf	99 Es	100 Fm	101 Md	102 No

of the silver produced is a by-product of the extraction of other metals, copper and lead in particular. *See* SILVER METALLURGY.

Pure silver is a white, moderately soft metal (2.5–3 on Mohs hardness scale), somewhat harder than gold. When polished, it has a brilliant luster and reflects 95% of the light falling on it. Its density is 10.5 times that of water. The quality of silver, its fineness, is expressed as parts of pure silver per 1000 parts of total metal. Commercial silver is usually 999 fine.

Although silver is the most active chemically of the noble metals, it is not very active in comparison with most other elements. It does not oxidize at all readily (as iron does when it rusts), but it reacts with sulfur or hydrogen sulfide to form the familiar silver tarnish. Electroplating silver "with rhodium will prevent this discoloration. Silver itself does not react with dilute nonoxidizing acids (hydrochloric or sulfuric acids) or strong bases (sodium hydroxide). However, oxidizing acids (nitric or concentrated sulfuric acids) dissolve it by reaction to form the unipositive silver ion, Ag^+. This ion, which is present in solutions of all simple, soluble compounds of silver, is rather easily reduced to the free metal, as in the deposition of silver mirrors by organic reducing agents.

Silver is almost always monovalent in its compounds, but an oxide, fluoride, and sulfide of divalent silver are known. Some coordination compounds of silver, also called silver complexes, contain divalent and trivalent silver. Although silver does not oxidize when heated, it can be oxidized chemically or electrolytically to form silver oxide or peroxide, a strong oxidizing agent. Because of this activity, silver finds considerable use as an oxidation catalyst in the production of certain organic materials. *See* PHOTOGRAPHIC MATERIALS.

Soluble silver salts, especially $AgNO_3$, have proved lethal in doses as small as 0.070 oz (2 g). Silver compounds may be slowly absorbed by the body tissues, with a resulting bluish or blackish pigmentation of the skin (argyria). [W.E.C.]

Silver alloys Combinations of silver with one or more other metals. Pure silver is very soft and ductile but can be hardened by alloying. Copper is the favorite hardener and normally is employed in the production of sterling silver, which must contain a minimum of 92.5% silver, and also in the production of coin silver.

Silver-copper eutectic and modifications containing other elements such as zinc, tin, cadmium, phosphorus, or lithium are widely used for brazing purposes, where strong joints having relatively good corrosion resistance are required. Where higher strengths at elevated temperature are required, silver-copper-palladium alloys and other silver-palladium alloys are suitable. The addition of a small amount of silver to copper raises the recrystallizing temperature without adverse effect upon the electrical conductivity.

Silver may be alloyed with gold or palladium in any ratio, producing soft and ductile alloys; certain of these intermediate alloys are useful for electrical contacts, where resistance to sulfide formation must be achieved.

Silver has proved to be a useful component for high-duty bearings in aircraft engines, where it may be overlaid with a thin layer of lead and finally with a minute coating of indium. Specially developed alloys of silver with tin, plus small percentages of copper and zinc in the form of moderately fine powder, can be mixed with mercury to yield a mass which is plastic for a time and then hardens, developing relatively high strength despite the fact that it contains about 50% mercury. This material was developed specifically for dental use and is generally known as amalgam, although the term amalgam actually includes all the alloys of mercury with other metals. *See* AMALGAM; SILVER. [E.M.Wis.]

Silver chloride electrode An electrode made of silver, covered or intimately mixed with silver chloride. One method of preparation consists of coating a silver wire or a silver-plated noble metal with silver chloride by electrolysis as an anode in a chloride solution. A second method consists of three steps: (1) pasting silver oxide or oxalate on a platinum helix, (2) reducing the oxide or axalate to silver by heating to 500°C (930°F), and (3) chlorodizing part of the silver to silver chloride. A third method consists of pasting an intimate mixture of silver oxide and silver chlorate on a platinum helix and heating to 500°C (930°F). Although these systems are frequently referred to as silver chloride electrodes, they are, in reality, not electrodes until immersed in a chloride solution. The standard potential of the silver chloride electrode is −0.2224 volt relative to the normal hydrogen electrode at 25°C (77°F). *See* CALOMEL ELECTRODE; ELECTRODE POTENTIAL; HYDROGEN ELECTRODE. [W.J.H.]

Silver metallurgy The art and science of extracting silver metal economically from various ores, and the reclamation of silver from the myriad types of industrial processes or scrap produced therefrom. It includes all processes of separating silver from its ores, alloys, and solutions, as well as the smelting, refining, and working of the metal and its alloys and compounds. It deals with the technical application of the chemical and physical properties of silver to its concentration, extraction, purification, alloying, working, and compounding to meet the requirements of technical needs.

Chloridization is an extractive process in which silver is precipitated from aqueous solutions. In chloridization methods the silver contained in ores or metallurgical products is first converted by means of a chloridizing roast into a compound which is soluble in water or in certain aqueous solutions. The silver is then precipitated as an insoluble compound by suitable reagents and the precipitate worked for the metal.

The cyanidation process differs only in minor details, such as strength of the solution and time of treatment, from the cyanide process used in the recovery of gold. No preliminary metallurgical treatment other than fine grinding is required. *See* GOLD METALLURGY.

The Parkes process is based on the greater affinity of silver for zinc than for lead. Slab zinc is added to an argentiferous lead bath, the temperature of which has been raised higher than the melting point of the zinc. When the zinc has been melted and thoroughly mixed into the lead bath, the bath temperature is lowered and a silver-zinc alloy separates and floats on the top of the kettle. This zinc crust, as it is now called, is pressed off to remove the excess lead. The resultant high silver-lead retort bullion is then cupeled to recover the silver and gold.

Cupellation is the oldest and most widely known method of separating and recovering gold and silver from lead. The silver-lead bullion is charged to a reverberatory-type cupellation furnace. After the charge has melted, air is blown across the top of the molten bath, causing oxidation of the lead and other impurities which separate as in impure litharge slag, leaving behind the silver and gold as a doré alloy.

By far the most important method of silver production results from the smelting and subsequent treatment of the silver-rich slimes resulting from the electrolytic refining of copper. Slimes which have been leached with sulfuric acid or air are filtered and charged to a reverberatory-type furnace with appropriate fluxes. The resulting melt slag contains most of the base-metal impurities. Upon completion of this step, the material remaining in the furnace consists of gold, silver, selenium, tellurium, and residual base metals, and is called matte. Treatment of the matte eventually yields doré which is then cast into anodes for subsequent electrolytic refining. *See* SILVER. [R.D.Mu.]

Silverfish An insect of the order Thysanura, also known as a bristle tail. There are over 350 cosmopolitan species, all considered among the most primitive insects. They are small, wingless insects with biting mouthparts (see illustration).

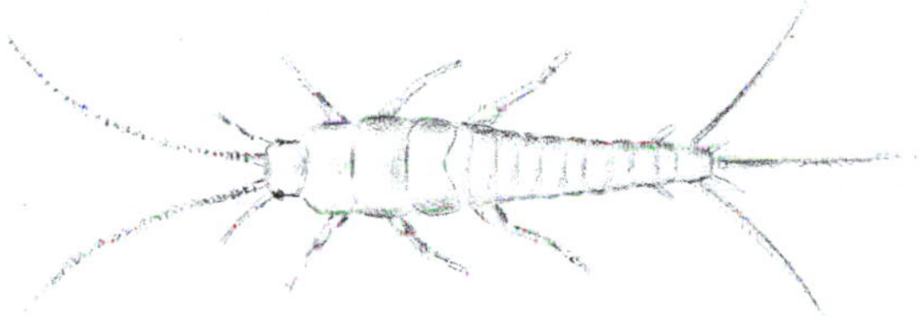

The silverfish, a primitive insect pest.

Silverfish belong to two families, the Machilidae and Lepismatidae. These insects are active and can run rapidly. They are fond of paste and glue, frequently being destructive to book bindings, and also feed on starched fabrics and rayon. They can be serious pests in libraries. *See* THYSANURA. [C.B.C.]

Silviculture The theory and practice of controlling the establishment, composition, and growth of stands of trees for any of the goods and benefits that they may be called upon to produce. The techniques proceed on the assumption that the natural vegetation of any site normally tends to extend itself to occupy all available growing space, making the fullest possible use of the available growth factors, such as light and moisture. The forester attempts, usually by cutting in the act of harvesting useful wood products, to create vacancies in the forest vegetation that will provide environments that are favorable either to the establishment of new, desirable trees (reproduction cuttings) or to the enhanced growth of those that remain (intermediate cuttings).

Integrated schedules of treatment for stands are called silvicultural systems. They cover both intermediate and reproduction cuttings but are classified and named in terms of the general method of reproduction cutting contemplated. Such programs are evolved for particular situations and kinds of stands with due regard for all the significant biological and economic considerations involved. These considerations include the desired uses of the land, kind of wood products sought, prospective costs and returns of the enterprise represented by management of the stand, funds available for long-term investment in stand treatments, harvesting techniques and equipment employed, reduction of losses from damaging agencies, and the natural requirements that must be met in reproducing the stand and fostering its growth. *See* FOREST AND FORESTRY. [D.M.S.]

Similitude The use in scientific studies and engineering designs of the corresponding behavior between large and small objects of similar nature. Two structures behave similarly if they are geometrically, kinematically, and dynamically similar. For geometric similarity, the ratios of critical dimensions, such as ratios of diameters to lengths, must be equal. For kinematic similarity, corresponding velocities and velocity gradients must be in the same ratios at corresponding locations. For dynamic similarity, ratios of forces acting within the two structures, such as viscosity and inertia, must be equal. *See* DIMENSIONAL ANALYSIS; DYNAMIC SIMILARITY.

Model tests in wind and water tunnels, towing tanks, dynamometers, antenna test ranges, and plasma reacting with magnetic fields are predicated on similitude relations between the model being tested and the full-sized object being studied. The use of small models greatly increases the speed with which design changes can be explored, and considerable economies are achieved. *See* TOWING TANK; WIND TUNNEL. [F.H.R.]

Simple machine Any of several elementary machines, one or more of which is found in practically every machine. The group of simple machines usually includes only the lever, wheel and axle, pulley (or block and tackle), inclined plane, wedge, and screw. However, the gear drive and hydraulic press may also be considered as simple machines. The principles of operation and typical applications of simple machines depend on several closely related concepts. *See* EFFICIENCY; FRICTION; MECHANICAL ADVANTAGE; POWER; WORK.

Two conditions for static equilibrium are used in analyzing the action of a simple machine. The first condition is that the sum of forces in any direction through their common point of action is zero. The second condition is that the summation of torques about a common axis of rotation is zero. Corresponding to these two conditions are two ways of measuring work. In machines with translation, work is the product of force and distance. In machines with rotation, work is the product of torque and angle of rotation. *See* BLOCK AND TACKLE; GEAR DRIVE; HYDRAULIC PRESS; INCLINED PLANE; LEVER; SCREW; TORQUE; WEDGE; WHEEL AND AXLE. [R.M.Ph.]

Simulation The development and use of computer models for the study of actual or postulated dynamic systems. The essential characteristic of simulation is the use of models for study and experimentation rather than the actual system modeled. In practice, it has come to mean the use of computer models because modern electronic computers are so much superior for most kinds of simulation that computer modeling dominates the field. "Systems," as used in the definition, refers to an interrelated set of elements, components, or subsystems. "Dynamic systems" are specified because the study of static systems seldom justifies the sophistication inherent in computer simulation. *See* ANALOG COMPUTER; COMPUTER; DIGITAL COMPUTER.

"Postulated" systems as well as "actual" ones are included in the definition because of the importance of simulation for testing hypotheses, as well as designs of systems not yet in existence. The "development" as well as "use" of models is included because, in the empirical approach to system simulation, a simplified simulation of a hypothesized model is used to check educated guesses, and thus to develop a more sophisticated and more realistic simulation of the simuland. The simuland is that which is simulated, whether real or postulated.

Among the systems which have been simulated are aerospace systems; chemical and other industrial processes; structural dynamics; physiological and biological systems; automobile, ship, and submarine dynamics; social, ecological, political, and economic systems; corporations and small businesses; traffic and transportation systems; electrical, electronic, optical, and acoustic systems; electrical energy systems and other energy related systems; and learning, thinking, and problem-solving systems. *See* CHEMICAL ENGINEERING; ECOLOGY; INDUSTRIAL ENGINEERING; OPTIMIZATION; PROBLEM SOLVING (PSYCHOLOGY); PROCESS ENGINEERING; PRODUCTION ENGINEERING; STRUCTURAL ANALYSIS.

Mathematical modeling is a recognized and valuable adjunct, and usually a precursor, of computer simulation. Mathematical modeling does not necessarily precede simulation, however; sometimes the simuland is not well enough understood to permit rigorous mathematical description. In such cases it is often possible to postulate a functional relationship of the elements of the simuland without specifying mathematically what that relationship is. This is the building-block approach, for which analog computers are particularly well suited. Parameters related to the function of the blocks can be adjusted until some functional criteria are met. Thus the mathematical model can be developed as the result of, rather than as a requirement for, simulation.

Analog computers, in which signals are continuous and are processed in parallel, were originally the most popular for simulation. Their modular design made it natural to retain the simulation-simuland correspondence, and their parallel operation gave them the speed required for real-time operation. The result was unsurpassed human-machine rapport. However, block-oriented digital simulation languages were developed which allow a pseudo–simulation-simuland correspondence, and digital computer speeds have increased to a degree that allows real-time simulation of all but very fast or very complex systems. *See* DATA-PROCESSING SYSTEMS.

Hybrid simulation, in which both continuous and discrete signals are processed, both in parallel and serially, is the result of a desire to combine the speed and human-machine rapport of the analog computer with the precision, logic capability, and memory capacity of the digital computer. Hybrid simulation now makes possible the simulation of a new array of systems which require combinations of computer characteristics unavailable in either all-analog or all-digital computers. [J.H.McL.]

Sine wave A wave having a form which, if plotted, would be the same as that of a trigonometric sine or cosine function. The sine wave may be thought of as the projection on a plane of the path of a point moving around a circle at uniform speed. It is characteristic of one-dimensional vibrations and one-dimensional waves having no dissipation. *See* HARMONIC MOTION.

The sine wave is the basic function employed in harmonic analysis. It can be shown that any complex motion in a one-dimensional system can be described as the superposition of sine waves having certain amplitude and phase relationships. The technique for determining these relationships is known as Fourier analysis. *See* FOURIER SERIES; WAVE EQUATION; WAVE MOTION; WAVEFORM. [W.J.G.]

Single crystal In crystalline solids the atoms or molecules are stacked in a regular manner, forming a three-dimensional pattern which may be obtained by a three-dimensional repetition of a certain pattern unit called a unit cell. When the periodicity of the pattern extends throughout a certain piece of material, one speaks of a single crystal, A single crystal is formed by the growth of a crystal nucleus without secondary nucleation or impingement on other crystals. *See* CRYSTAL; CRYSTALLOGRAPHY.

Ideally, single crystals are free from internal boundaries. They give rise to a characteristic x-ray diffraction pattern. Many types of single crystal exhibit anisotropy, that is, a variation of some of their physical properties according to the direction along which they are measured. For example, the electrical resistivity of a randomly oriented aggregate of graphite crystallites is the same in all directions. [D.T.]

Single sideband An electronic signal-processing technique in which a spectrum of intelligence is translated from a zero reference frequency to a higher frequency without a change of frequency relationships within the translated spectrum. Single-sideband (SSB) signals have no appreciable carrier.

Amplitude-modulated (AM) signals have identical upper and lower sidebands symmetrically located on each side of the translation frequency, which is often called the carrier. The SSB spectrum differs from the AM spectrum in having little or no carrier and only one sideband. *See* AMPLITUDE MODULATION.

In the SSB signal-processing action, the intelligence spectrum to be translated is applied to the signal input port of a balanced modulator. A higher-frequency sinusoidal signal, often called a carrier, is applied to the other input port of this circuit. Its function is to translate the zero reference spectrum to the carrier frequency and to produce the upper and lower sidebands, which are symmetrically located on each side of the carrier. The carrier frequency power is suppressed to a negligible value by the balanced operation of the modulator and does not appear at the output. Generally, the balanced modulator operates at an intermediate frequency which is lower than the frequency of transmission. Following the balanced modulator is a sideband filter which is designed to remove the unwanted sideband signal power and to allow only the desired intelligence spectrum to pass. *See* AMPLITUDE MODULATOR; MODULATOR.

There are many advantages in the use of SSB techniques for communication systems. The two primary advantages are the reduction of transmission bandwidth and transmission power. The bandwidth required is not greater than the intelligence bandwidth and is one-half that used by amplitude modulation. The output power required to give equal energy in the intelligence bandwidth is one-sixth that of amplitude modulation.

Propagation of radio energy via ionospheric refraction provides the possibility for multiple paths of differing path length which can cause a selective cancellation of frequency components at regular frequency spacings. This produces in amplitude modulation a severe distortion of the intelligence because of the critically dependent carrier-to-sideband amplitude and phase relationships. SSB is much less affected under these conditions. [D.M.H.]

Sink flow A point in three-dimensional flow, into which fluid is presumed to flow uniformly from all directions. The strength of a sink is defined as the volume per unit time flowing into the point. A sink may also be defined as a negative source. *See* SOURCE FLOW.

In two-dimensional flow, in which all flow occurs in parallel planes that have identical flow patterns, a sink is a straight line into which fluid flows uniformly from all directions at right angles to the line. It appears as a point on the customary two-dimensional flow diagram. *See* FLUID FLOW. [V.L.S.]

Sintering The welding together and growth of contact area between two or more initially distinct particles at temperatures below the melting point, but above one-half of the melting point in kelvins. Since the rate of sintering is greater with smaller than with larger particles, the process is most important with powders, as in powder metallurgy and in firing of ceramic oxides.

Although sintering does occur in loose powders, it is greatly enhanced by compacting the powder, and most commercial sintering is done on compacts. Compacting is generally done at room temperature, and the resulting compact is subsequently sintered at elevated temperature without application of pressure. For special applications, the powders may be compacted at elevated temperatures and therefore simultaneously pressed and sintered. This is called hot pressing or sintering under pressure.

Certain compacts from a mixture of different component powders may be sintered under conditions where a limited amount of liquid is formed at the sintering temperature. This is called liquid-phase sintering, important in certain powder-metallurgy and ceramic applications. *See* CERAMICS; POWDER METALLURGY. [F.V.L.]

Sinus Any space in an organ, tissue, or bone, but usually referring to the paranasal sinuses of the face. In humans, four such sinuses, lined with ciliated, mucus-producing epithelium, communicate with each nasal passage through small apertures. The ethmoid and sphenoid sinuses are located centrally between and behind the eyes. The frontal sinuses lie above the nasal bridge, and the maxillary sinuses are contained in the upper jaw beneath the orbits. The mastoid portion of the temporal bone contains air cells lined with similar epithelium. [T.S.P.]

Siphonaptera An order of insects commonly known as the fleas. These animals are of importance because they are bloodsucking pests of humans and animals and transmit serious diseases from animals to humans.

Fleas in the adult stage are recognized with ease (see illustration). They are small, dark brown in color, laterally flattened, and with three pairs of legs modified for jumping. The body is more or less oval in shape and armed with spines and setae, adapting the flea for living among the hairs of animals. The head is provided with mouthparts modified for sucking blood.

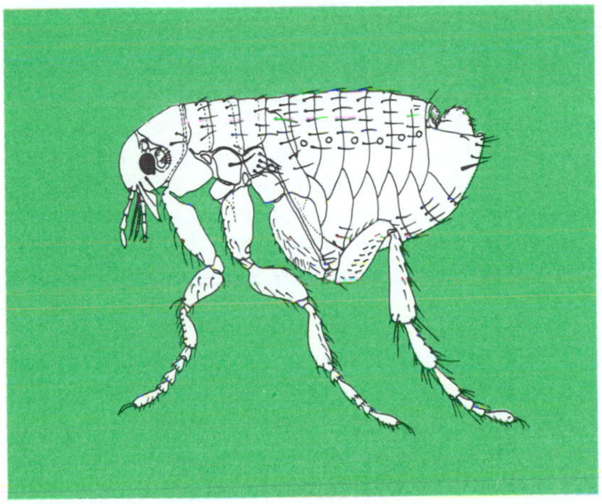

Human flea, *Pulex irritans*. (*After E. O. Essig, College Entomology, Macmillan, 1942*)

Fleas have a complete metamorphosis; that is, the life history involves four distinct phases: the egg, larva, pupa, and adult. Each adult female flea lays a number of eggs, 400 or more in some species, over a long period of time.

The great majority of species are obscure ectoparasites of mammals, and to a lesser extent birds, and do not affect humans. Host specificity is exhibited in that, ordinarily, a certain taxonomic group of fleas will parasitize a specific taxonomic group of hosts. Nevertheless, some species may leave their favored hosts and attack humans, and these species are the pests and disease carriers. Bubonic plague, which was responsible for millions of deaths in India during the first two decades of this century, is a disease of rats. Murine or endemic typhus fever is another disease transmitted from domestic rats to humans by rat fleas. Flea allergy is the term applied to the severe reactions which sometimes result from flea bites. *See* INSECTA. [I.F.]

Siphonocladales An order of green algae in the division Chlorophytia. Four families are recognized in this order, all marine and mostly tropical: Valoniaceae, Siphonocladaceae, Boodleaceae, and Anadyomenaceae.

In the Valoniaceae the plants are essentially unicellular, coenocytic vesicles, and spherical or clavate (see illustration).

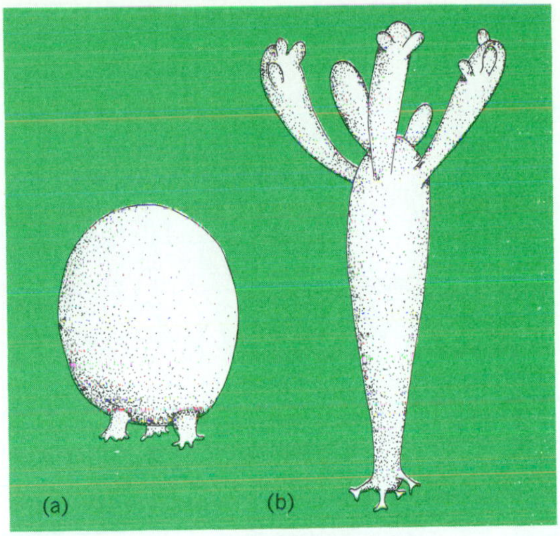

***Valonia*. (*a*) Spherical form. (*b*) Clavate form.**

The other families exhibit erect tufts of branched filaments, uni- or multiseriate, or expanded blades, some of which have the branch elements forming reticulations and anastomosing networks. Chloroplasts are usually close or loose networks, with pyrenoids. Starch accumulates as food. [G.W.P.]

Siphonophora An order of the class Hydrozoa of the phylum Coelenterata, characterized by an extremely complex organization of components of several different types, some having the basic structure of a jellyfish, others of a polyp. The components may be connected by a stemlike region or may be more closely united into a compact organism.

Most siphonophores possess a float and are animals of the open seas. Best known is the Portuguese man-of-war, *Physalia*, with a float as much as 16 in. (40 cm) long and tentacles which extend downward for many feet. These animals may be swept shoreward and may make swimming not only unpleasant but dangerous. *See* COELENTERATA; HYDROZOA. [S.Cr.]

Siphonostomatoida An order of Copepoda; all members are parasites of marine and fresh-water fishes or a variety of invertebrate hosts. In fact, it is estimated that 67% of copepod parasites of fishes belong to the Siphonostomatoida. Their obvious success in the parasitic mode of life appears at least in part to be a result of two morphologic adaptations. The first is the modification of the buccal apparatus (mouth and appendages) resulting in a mouth cone that has a small opening near the base through which the mandibles are free to enter. The maxilliae are subchelate or brachiform and serve as the appendage for attachment to the host. The second adaptation is the development of a frontal filament, which is a larval organ of attachment. In those siphonostomatoids possessing a frontal filament, the brachiform second maxilla may be used to manipulate the frontal organ, and in some species fusion of those two structures forms the attachment structure. *See* COPEPODA; CRUSTACEA; POECILOSTOMATOIDA. [P.A.McL.]

Sipunculida A phylum of marine worms which dwell in burrows, secreted tubes, or adopted shells. The body is greatly elongated in the dorsoventral axis (see illustration). As a result, both mouth and anus occur close together at one end of the body, and the alimentary tract is a long, twisted loop which extends ventrally and shows little differentiation along its course. The jawless mouth, surrounded by tentacles, is situated in an eversible proboscis, or introvert, which is projected by hydrostatic pressure generated by the strongly muscular body wall. The proboscis can be withdrawn by special retractor muscles which are anchored in the trunk wall and extend to its tip.

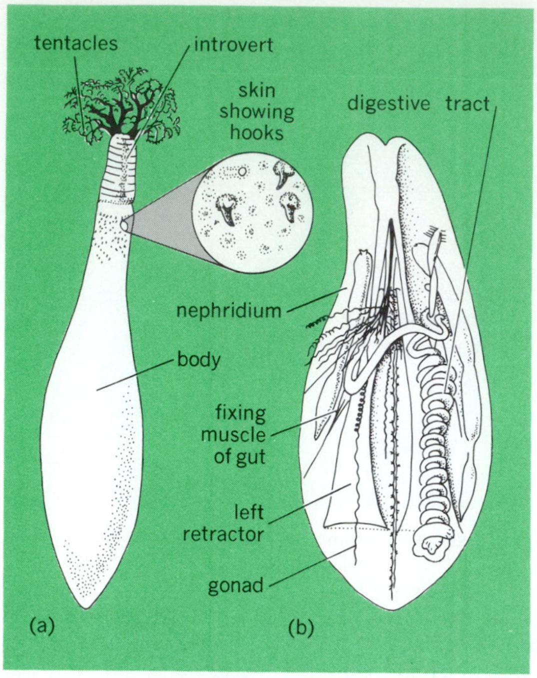

Two examples of Sipunculida. (*a*) *Dendrostoma petraeum*. (*b*) *Dendrostomum pyroides* (after Chamberlin).

The body cavity is an extensive schizocoele which shows no evidence of segmentation but is filled with fluid and traversed by fibers. A pair of giant metanephridia whose ventral pores open to the exterior serve in excretion. The nervous system includes a dorsal brain that is continuous, by right and left connectives, with a nonsegmented ventral cord which extends to the posterior tip of the worm. The sexes are separate.

There are about a dozen genera with probably more than 200 species. Sipunculids are widely distributed. All are bottom dwellers, feeding by means of tentacles and cilia on the organic matter in the ingested mud and sand. [F.A.B.]

Siren (acoustics) An apparatus for generating sound by the mechanical interruption of the flow of fluid (usually air) by a perforated rotating disk or cylinder. The disk may be so shaped that the fluid under pressure causes the disk to rotate, or conversely, the rotation of the disk may cause the flow of the fluid. The resulting frequency is the product of the speed of rotation of the disk and the number of perforations.

Before the advent of the jet engine, the siren was the most powerful of human-made steady sound sources, and as such it found application in fog signaling and as an emergency warning signal. *See* SOUND. [R.W.Y.]

Sirenia An order of thoroughly aquatic placental mammals which include the living manatees and dugongs, the extinct (since the 18th century) Steller's sea cow, and their numerous fossil relatives. Included also is the peculiar, late Eocene–early Oligocene *Moeritherium*. Living sirenians have torpedo-shaped bodies ending in a horizontal tail fluke. These mammals are nearly hairless, thick-skinned, and without hindlimbs and with paddlelike forelimbs. The manus of living forms, although pentadactyl, shows no digits externally. The anterior parts of the palate and mandible are covered with rough, horny plates. The large head sits on a short neck, and the nostrils are on the dorsal surface of the snout. These animals have a small mouth with lips bearing hairs, the ears have no pinnae, and the small tongue is rugose and fixed. *See* MAMMALIA. [F.S.S.]

Sirius Alpha Canis Majoris, the brightest star visible from Earth. From the Northern Hemisphere, Sirius appears in the south in midwinter. It is the sixth nearest star to the Sun. Absolute magnitude +1.4 and spectral type A1 are representative of early type A stars, with an effective temperature near 10,000 K (17,500°F), strong hydrogen lines, and large Balmer continuous absorption. Special interest attaches to the system of Sirius because of the close companion, of the ninth magnitude (more than 10^4 times fainter)—the white dwarf α CMaB. The period of the binary system is 50 years. *See* BINARY STAR; STAR; WHITE DWARF STAR. [J.L.Gr.]

Sirocco A southerly or southeasterly wind current from the Sahara or from the deserts of Saudi Arabia which occurs in advance of cyclones moving eastward through the Mediterranean Sea. The sirocco is most pronounced in the spring, when the deserts are hot and the Mediterranean cyclones are vigorous. It is observed along the southern and eastern coasts of the Mediterranean Sea from Morocco to Syria as a hot, dry wind capable of carrying sand and dust great distances from the desert source. The sirocco is cooled and moistened in crossing the Mediterranean and produces an oppressive, muggy atmosphere when it extends to the southern coast of Europe. *See* AIR MASS; WIND. [F.S.]

Sisal A fiber obtained from the leaves of *Agave sisalana*, produced in Brazil, Haiti, and several African countries, including Tanzania, Kenya, Angola, and Mozambique.

Sisal is used mainly for twine and rope, but some of the lower grades are used for upholstery padding and paper. The greatest quantity goes into farm twines, followed by industrial tying twine and rope. Most sisal-fiber ropes made in the United States are small to medium in size, intended for light duty. Sisal is sometimes used for marine cordage in Europe. *See* NATURAL FIBER. [E.G.N.]

Skarn A rock term generally reserved for rocks composed entirely, or almost so, of lime-bearing silicates, and derived from nearly pure limestones and dolomites into which large amounts of silicon, aluminum, iron, and magnesium have been introduced. *See* METAMORPHIC ROCKS; PNEUMATOLYSIS.

Skarn is an old term in mining. Now it is applied to a coarse-grained rock or association of minerals formed by reaction between hot silica-rich solutions or acid gases and a limestone or dolomite. The reaction rock is made up of coarse masses of minerals of various kinds, depending upon the temperature and the composition of the reacting gases or solutions. The most important skarn minerals are andradite-garnet, hedenbergite-diopside, iron-rich hornblende, and actinolite-tremolite. [T.F.W.B.]

Skeletal system A structure composed of bone or cartilage or a combination of both which provides a framework for the animal body and serve as attachment for muscles. Relative movements of bony parts are permitted by the joints (articulations) between them. Although the skeleton of vertebrates is characteristically internal (endoskeleton), many species also possess external or exoskeletal structures of bony or horny material to provide support or protection, as the shell of the turtle. Such exoskeletal structures and their derivatives make up the dermal skeleton. Invertebrates which possess skeletal structures have an exoskeleton composed of chitin. *See* BONE; CARTILAGE.

The vertebrate skeleton is primarily an endoskeleton of cartilage and bone. There are two types of bone: cartilage (endochondral, or replacing bone) and dermal (membrane, or investing bone). The first type is preformed in cartilage and ossifies later, while the second type ossifies directly from membrane without any cartilaginous predecessor. Otherwise they are simi-

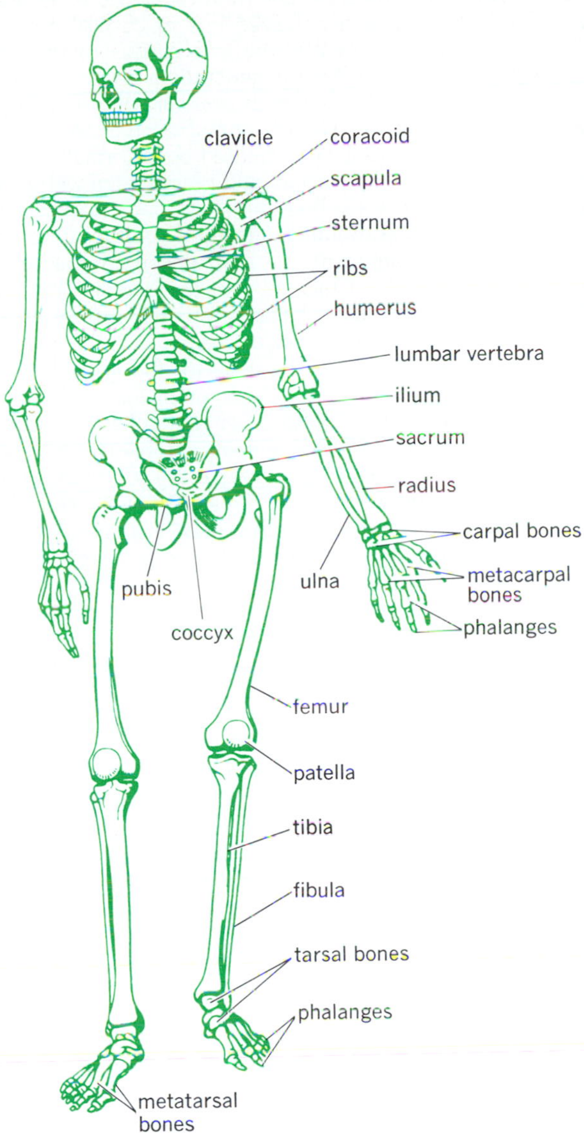

Fig. 1. Human skeleton, anterior view.

lar. Dermal bone occurs only in the skull and shoulder region. An exoskeleton of hair, claws, nails, hoofs, horns, antlers, feathers, scales, and bony ossicles in the skin may also be present.

The skeleton consists of the axial skeleton, comprising the skull, vertebral column, and associated structures, and the appendicular skeleton or appendages (Fig. 1). The skull is subdivided into the neurocranium and the branchiocranium, also known as the splanchnocranium or viscerocranium.

Axial skeleton. Skeletal organization is quite uniform except for variations in the number, size, and arrangement of the bones.

Skull. The neurocranium consists of the braincase and three pairs of capsules associated with the organs of smell, sight, equilibrium, and hearing. The olfactory capsules develop around the olfactory lobes, the optic capsules around the eyeballs, and the otic or auditory capsules around the ear. The olfactory and otic capsules unite with the rest of the neurocranium, but the optic capsules remain free to permit eye movements. The neurocranium has openings (foramina) for blood vessels and nerves.

The Chondrichthyes, that is, the sharks, rays, and chimeras, have a cartilaginous skull, but adult Osteichthyes, or bony fishes, and tetrapods have an ossified skull. Some bony fishes have more than 150 separate skull bones, but during evolution the number was reduced to approximately 27 bones in the human skull.

In the living amphibians, all specialized, the number of skull bones is reduced from 40 in *Eryops* to 18 in some of the legless Apoda (*Ichthyophis*) and the bullfrog, *Rana*, and to 15 in the mudpuppy, *Necturus*. Much of the reduction in number of skull bones is due to fusion.

The skull of some Permian cotylosaurs, the most primitive reptiles, is very similar to that of labyrinthodont amphibians. The skull of the living lizard *Iguana* has most of the neurocranial bones found in *Eryops*.

The adult bird skull is fused into four complexes: (1) a thin, rigid, continuous neurocranium; (2) a vertically hinged upper beak supported on a basically reptilian upper jaw and palate; (3) a mandible of fused reptilian elements; and (4) a jointed hyobranchial chain.

Adult mammalian skulls typically lack paired prefrontal, postfrontal, postorbital, quadratojugal, and septomaxillary bones. The reduction in total number of bones is due to fusion. Paired frontal, jugal, lacrimal, nasal, parietal, periotic, and squamosal bones plus unpaired basioccipital, basisphenoid, dermosupraoccipital, ethmoid, presphenoid, and vomer form the neurocranium. The paired premaxillary, maxillary, pterygoid and palatine bones form the upper jaw, and the dentary, the lower jaw.

In the adult human skull (Fig. 2) the pre- and basisphenoids fuse to form the sphenoid; the basi-, ex-, and supraoccipitals to form the occipital; the periotic, squamosal, and tympanic to form the temporal; and the premaxillary and maxillary to form the maxilla. The upper and middle turbinates, or conchae, fuse with the ethmoid. The basi-, cerato-, and thyrohyals form the hyoid bone.

Vertebrae. The term vertebrate refers to the vertebral column composed of vertebrae. A typical vertebra consists of a body, or centrum, and a neural arch surrounding the spinal cord lying dorsal to the centrum. The neural arch has spinal and transverse processes for the attachment of muscles and ligaments and articular processes for articulation with adjacent vertebrae. Fish have little regional differentiation in the vertebral column, but tetrapods typically have cervical, thoracic, lumbar, sacral, and caudal vertebrae, Legless forms lack sacral vertebrae. The first two cervical vertebrae, the atlas and axis, of amphibian fossils and amniotes are specialized for support and movements of the head. Humans and some other tetrapods have fibrocartilaginous intervertebral disks between

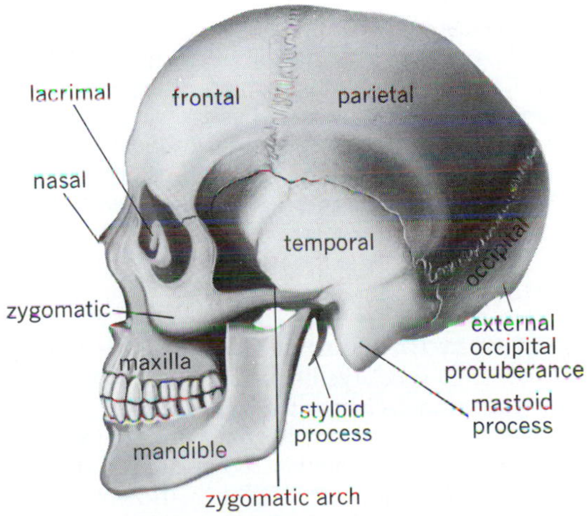

Fig. 2. Adult human skull. (*After F. G. Evans, Atlas of Human Anatomy, Littlefield, Adams, 1957*)

the bodies of most of the movable vertebrae. There are seven cervical vertebrae in most mammals, including humans and giraffes.

Two types of ribs, dorsal and ventral, occur in vertebrates. Dorsal ribs lie in the septum dividing the trunk musculature into dorsal and ventral masses; ventral or pleural ribs lie in the septa dividing the trunk musculature into segments. Many fish have both types of ribs, but only dorsal ribs occur in tetrapods. The earliest tetrapods had ribs on all of the presacral vertebrae, but in living forms, especially birds and mammals, movable ribs are usually restricted to the thoracic region. In tetrapods with functional hindlegs, one or more sacral ribs articulate with the pelvis. Cervical ribs, when present, are usually short, whereas thoracic ribs are typically long and curved. In turtles the trunk vertebrae and ribs are immovably fused to the carapace or upper shell. In certain mammals (living armadilloes and fossil glyptodonts) there is considerable fusion of the trunk vertebrae.

Most tetrapods have a sternum, which in amniotes is connected with the ventral ends of the ribs via costal cartilages. The sternum in living amphibians is mostly cartilaginous and incorporated with the pectoral girdle; it does not connect with the ribs, which are short and straight. The sternum is absent in the legless living amphibians, turtles, legless lizards, and snakes. The bird sternum is an ossified quadrangular bone with a prominent median keel in flying forms and penguins. The keel is absent in ostriches and related forms. The mammalian sternum is an ossified median structure of several parts which may remain separate or fuse.

Appendicular skeleton. This consists of the skeleton of the pectoral (shoulder) and the pelvic (hip) girdles and of the paired appendages. The girdles provide attachment and support for the appendages, which may be fins, flippers, legs, wings, or arms. The pectoral girdle differs from the pelvic one in consisting of both dermal and cartilage bones and, except in skates and pterodactyls, in having no cartilaginous or bony connection with the vertebral column. The pectoral girdle is located close to the head; in bony fishes (and some early tetrapods) it is actually attached to the skull, while the pelvic girdle is associated with the outlets of the digestive and reproductive systems. However, in some bony fishes the pelvic fins are in the throat region anterior to the pectoral fins.

Birds have a swordlike scapula, an enlarged coracoid, and clavicles joined by an interclavicular element, the furcula, to form the wishbone. The avian pelvis consists of an expanded ilium, articulating with several sacral vertebrae, and a slender pubis paralleling the expanded ischium. The pubic symphysis is typically lacking, but ostriches and rheas have both a pubic and an ischiatic symphysis.

Mammals typically have a clavicle, a scapula, and no coracoid except in monotremes (egg-laying mammals). Horses have only the scapula. The three pelvic bones are usually fused into a single pelvic bone. Monotremes, marsupials, insectivores, many rodents, hoofed mammals, and carnivores have both a pubic and an ischiatic symphysis.

The skeleton of the paired appendages (fins) of fish consists of cartilaginous or bony fin rays. Tetrapod limbs are all modifications of a single basic pattern. Most variation occurs in the number and shape of hand and foot bones. The forearm and leg bones fuse in frogs, and the ulna and fibula are small splints in hoofed mammals. [F.G.Ev.]

Skeletal system disorders

Skeletal system disorders The skeletal system, because of its highly specialized structure and function, reacts to injury and disease in distinctive ways.

Fractures. Trauma-induced fractures are the most common bone disorder. Repair of traumatic bone injury is similar to that observed in other connective tissues; however, the cells involved are highly specialized. The repair of a fractured bone can be so complete that a few years after fracture an x-ray may not reveal the presence of any residual change. *See* BONE.

Osteoporosis. The most common metabolic bone disease is osteoporosis. It is a generalized loss of cortical bone substance associated with metabolic changes in the body. Generally it is demineralized degeneration of the bones that accompanies aging. Although both sexes are affected, it occurs earlier and is more rapidly progressive in females. Osteoporosis is the result of decreased bone formation, rather than increased bone destruction, and is associated with a negative calcium balance in the body.

Rickets and osteomalacia. Vitamin D deficiency results in a bone metabolic abnormality called rickets in children and osteomalacia in adults. Children have an unusually high need for vitamin D and calcium because of their rapidly growing bone structure. The precise role of vitamin D in this process is not known. It appears to be essential for the normal mineralization of the bone matrix. *See* CALCIUM METABOLISM; VITAMIN D.

Osteitis deformans. This disorder is of unknown cause and is characterized by continuous destruction of bone and its simultaneous replacement by an abnormally soft, poorly mineralized bone. As the normal bone is resorbed, it is replaced by a light, bulky, porous osteoid matrix that is unusually rich in blood vessels. Although thicker than the normal bone, it is softer and prone to fracture.

Fibrous dysplasia. This condition is characterized by focal areas of replacement of bone by fibrous connective tissue. The cause is unknown; however, the mechanism of change suggests that it is an abnormality in the process of remodeling bone.

Neoplasms. Although metastatic neoplasms to bone are common, primary osteogenic tumors derived from bone-forming tissues are very rare. The three principal osteogenic tumors are osteoma, osteoid osteoma, and osteogenic sarcoma. Osteogenic sarcoma is the most important because of its high rate of malignancy. Osteoma and osteoid osteoma are benign excessive proliferations or scarring of the bone tissue similar to that which occurs following fracture. They result in an excessive production of bone in a localized region of the skeleton and do not undergo malignant change.

A rare malignant neoplasm arising in the soft tissues within bones is Ewing's sarcoma. The neoplasm is apparently derived from the endothelial cells of blood vessels within marrow. It is most frequent in adolescents and young adults, usually in long tubular bones or the pelvic bones, or both. *See* ONCOLOGY.

Osteomyelitis. This is an inflammatory disease of bone in which the infectious organism enters the bone substance either by way of the bloodstream or through direct penetration due to trauma. The organisms are usually pyogenic (staphylococci, streptococci, pneumococci, or gonococci), although other organisms such as *Hemophilus influenzae*, coliform bacilli, and tubercle bacilli may rarely be involved. Necrosis and spread of the infection may ensue, with extensive bone destruction and pus formation.

Arthritis. Inflammatory joint lesions are termed arthritis. When caused by a specific organism, it is termed septic arthritis. *See* ARTHRITIS; DYSCHONDROPLASIA. [N.K.M.]

Skin

Skin The external covering, or integument, of the body. There are two layers, the outer epidermis, which is composed of flattened, ectodermal, epithelial cells, and the inner corium or dermis, which is formed by fibrous yet elastic mesodermal connective tissue containing a rich supply of blood vessels, nerves, and accessory structures.

The outer portion of the epidermis of higher vertebrates consists of many layers of dead cornified (horny) cells; the inner portion is usually a single layer of dividing cells, the stratum germinativum, which is in contact with the underlying dermis. In more primitive forms there are fewer layers of dead cells and little cornification.

The integument has many important functions: protection; pliability and extensibility, which allow movement; regulation of body fluids; excretion of wastes; regulation of body temperature; development of sense organs in the skin; and the formation of specialized accessory structures. *See* INTEGUMENT. [T.S.P.]

Skin disorders The skin is subject to localized and generalized disorders, as well as those of primary occurrence in the skin and those secondary to involvement of other tissue.

There are many rather common skin disorders that so far have offered little evidence of their causes. Common examples include psoriasis, parapsoriasis, and lichen planus. The skin is the major recipient of damage from trauma of all kinds. Sunburn, frostbite, lacerations, hematomas, thermal burns, and chemical irritations and burns are commonplace.

Skin conditions occurring in regions which have a large number of oil, or sebaceous, glands usually produce an increase in secretion so that an oily skin is characteristic. Acne is the most common example and probably results from endocrinologic imbalance and other factors. *See* ACNE.

The skin is the site of direct and indirect involvement in many infections. The most common diseases produced by the pyogenic, or pus-forming, bacteria include erysipelas, impetigo, folliculitis, and furunculosis, or boils.

Viral infections account for many of the infectious exanthemas such as measles, chickenpox, and smallpox, and also for other disorders. They are responsible for the formation of warts and may be implicated in certain other skin tumors. Viruses are also the agents of herpes simplex (fever blisters) and herpes zoster (shingles). *See* CHICKENPOX AND SHINGLES; MEASLES; SMALLPOX.

Fungus infections include the superficial types, such as tinea or ringworm, moniliasis or athlete's foot, and similar disorders. The infrequent but potentially dangerous systemic fungus infections such as actinomycosis, blastomycosis, and histoplasmosis often are marked by skin lesions. *See* ACTINOMYCOSIS; BLASTOMYCOSIS; CANDIDIASIS; HISTOPLASMOSIS.

The term eczema is generally used to include any noninfectious, inflammatory lesion in which there are papules, blisters, and oozing of serum from the affected parts. The erythemas are disorders in which the prominent feature is involvement of the vasculature of the affected skin. Such involvement often produces a characteristic swelling, or edema, in these disorders and includes urticatia, or hives, which results from any of the allergic agents. *See* ECZEMA.

Skin lesions are common in many psychic disorders. These vary from simple pruritus (itching) to widespread, compound lesions of great severity.

Primary tumors of the skin are common and most often are of the benign varieties. Skin cancer, however, has one of the highest rates of incidence of any malignancy. Its superficial location and relative ease of treatment also produce one of the highest rates of cure, if diagnosis is made early. *See* SKIN. [E.G.St./N.K.M.]

Skin effect (electricity) The crowding of high-frequency electric current into a thin surface layer of a conductor.

For a steady unidirectional flow of electricity, the current is uniformly distributed over the cross section of a uniform conductor. For an alternating current, there is no longer this simple uniformity, but the current density is greater near the outer surface than at the center. The magnitude of the nonuniformity increases as the frequency rises. For low frequencies, the effect is very small, but at frequencies for which the wavelength within the conducting material is comparable with the dimensions of the conductor, or smaller, the entire current may be considered to be within a relatively thin surface layer.

The skin effect is largely due to self-induced electromotive forces (emf's) which are different for different paths within the conductor. These emf's increase with frequency, since they depend upon the rate of change of flux. The skin effect results in a resistance for a conductor that is greater for an alternating current than for a direct current, since the effective cross section of the conductor is decreased. [K.V.M.]

Skin friction A type of friction force which exists at the surface, or skin, of a solid body immersed in a much larger volume of fluid which is in motion with velocity u_1 relative to the body, as illustrated. The magnitude of skin friction per unit surface area, the shear stress τ_w, was equated by Isaac Newton to the rate of deformation of an adjacent fluid element $(\partial u/\partial y)_w$ times a transport property of the fluid called the absolute or dynamic viscosity coefficient μ. Such flow distortion is significant only in a thin boundary layer, which may be laminar or turbulent and outside of which, with y greater than $\delta(x)$, the motion is essentially inviscid. *See* VISCOSITY.

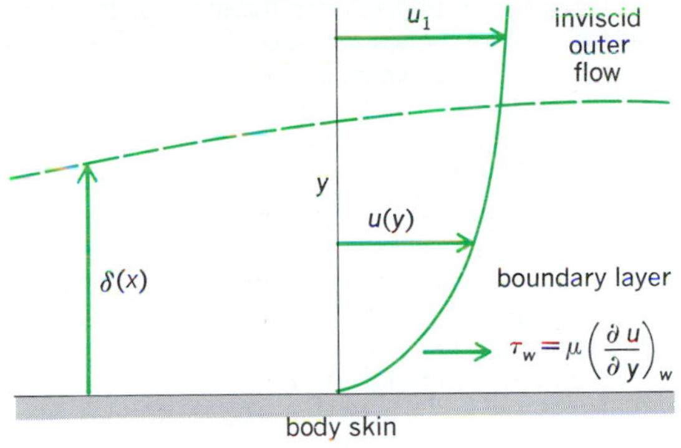

Boundary-layer velocity profile.

The skin friction force contributes directly to the drag of the body; it also contributes indirectly because, by its action, the inviscid outer flow may be modified with effect upon the pressure distribution. *See* BOUNDARY-LAYER FLOW. [B.M.L.]

Skin test A procedure for evaluating the immunity status, involving the introduction of a reagent into or under the skin. Certain toxic antigens applied in minute doses will give a visible, but readily tolerated, lesion if less than a threshold amount of antibody is present in the skin. Examples are the Schick test for diphtheria antitoxin and the Dick test for scarlet fever antitoxin. If sufficient antibody is present, the toxin will be largely or completely neutralized, and the lesion will be minimal or absent. False positive reactions occur in allergies to the reagent toxin, but these can be controlled with inactivated toxins. Other tests involve substances that are not reactive with normal individuals, but give a visible skin reaction in the presence of antibodies acquired as a result of hypersensitivity or allergy to an infecting organism. A positive reaction is thus presumptive for previous contact with a specific infectious agent, for example, the tuberculin and Mantoux tests for tuberculosis, and the Mallein test for glanders. In a few instances, antibody may be injected as a reagent and the neutralization of toxins present in the skin observed, as in the Schultz-Charlton blanching test in scarlet fever. [H.P.T.]

Skink A cosmopolitan reptile belonging to the family Scincidae. They are small- to medium-sized, mostly terrestrial or fossorial lizards with a cylindrical body and short legs; in some instances the legs are vestigial. The body scales have cores of bone. In many species the eyelids are transparent. Skinks exhibit pleurodont dentition; that is, the teeth are attached on the side of the jaw. Most species are ovoviviparous. Three genera, *Neoseps*, *Lygosoma*, and *Eumeces*, occur in the United States. A number of species of *Tiliqua*, largest of the skinks, inhabit Malaysia and Australia. *See* REPTILIA; SQUAMATA. [C.B.C.]

Skull The bones or cartilages of the head which form the cranium and the face. The skull surrounds and protects the brain and major sense organs and includes the jaws.

The skull is formed from a large number of individual pieces, In humans the eight cranial bones which form a hollow, protective braincase are the occipital, sphenoid, ethmoid, and frontal, as well as the paired temporal and parietal bones. There are 14 facial bones: the vomer, mandible (lower jaw), and the paired nasal, lacrimal, zygomatic (cheek), palatine, maxillary (upper jaw), and inferior turbinate of the nasal passage. Several of these bones represent fusions of what are several bones in most other animals. Skull articulations are generally fixed, serrated sutures. The skull base contains openings, or foramina, for blood vessels and nerves. Certain parts of the skull bones are hollow and form air sinuses such as frontal, maxillary, and mastoid. *See* Sinus; Skeletal system. [T.S.P.]

Skunk Any one of a group of carnivores in the family Mustelidae, which also includes otters, weasels, badgers, and martens. Skunks are found only in the New World and range from North America to South America, with six species in the United States. They are characterized by their glossy black and white coat and two musk glands at the base of the tail.

The common striped skunk (*Mephitis mephitis*) has a white stripe on each side running into the tail. These animals are terrestrial and somewhat fossorial (burrowing). They are carnivores primarily but eat vegetation such as seeds, leaves, and nuts during the winter. During the summer they feed on insects, fruit, eggs, and rodents. The spotted skunks (*Spilogale*) range from the eastern United States to Colorado and Wyoming. *See* Carnivora; Mammalia; Marten; Otter; Weasel. [C.B.C.]

Skutterudite A mineral with composition $(Co,Ni)AS_3$, an ore of cobalt and nickel. Commonly the mineral is massive with metallic luster and tin-white color. The hardness is $5\frac{1}{2}$–6 on Mohs scale and the specific gravity 6.6. Skutterudite is found at Freiberg, Annaberg, and Schneeberg in Germany, and at Cobalt, Ontario. *See* Cobalt; Nickel. [C.S.Hu.]

Slate A group name for various very fine-grained rocks derived from mudstones, siltstones, and other clayey sediments as a result of low-degree regional metamorphism. Highly characteristic of slates is the perfect fissility or slaty cleaveage which is a regular and perfect planar schistosity, the slates themselves thus grading into phyllites. The chief minerals of slates are muscovite (as sericite), chlorite, and quartz. Common accessories are tourmaline, rutile, epidote, sphene, hematite, and ilmenite. *See* Metamorphic rocks; Phyllite; Rock cleavage; Schistosity.

Slates are widely used for roofing purposes. They are easily prepared and fixed, are weatherproof and durable, and in many areas are cheaper and better than any other thatching materials. The making of slates is still performed by hand using chisel and mallet. Big slabs are split into separate slates, the thickness of which varies with the size required and the quality of the rock. The slates are afterward trimmed to size either by hand or by machine-driven rotating knives. [T.F.W.B.]

Sleep and dreaming There has been extensive research relative to sleep and dreaming including studies of sleep patterns, sleep structure and physiology, as well as the cognitive aspects of sleep, which include dreaming. More recently, emphasis was placed on sleep disorders.

Patterns and structure of sleep. Patterns of sleep refer to the occurrence or absence of sleep, usually described in a 24-hour context, and involve such indices as amount of sleep, number of sleep episodes, and placement of sleep, for example, nocturnal or diurnal sleep. The typical adult pattern is a single, nocturnal episode of about 7 to 8 h.

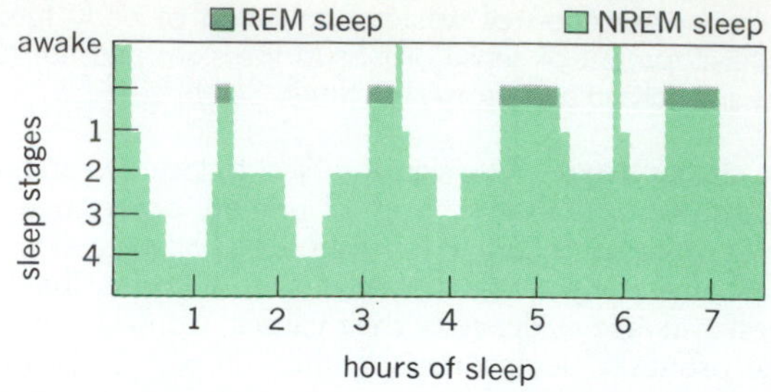

Typical sleep cycle of a young adult showing early preponderance of stages 3 and 4. (*After A. Kales, Sleep and dreams: Recent research on clinical aspects, UCLA symposium, Ann. Intern. Med., 68:1078–1104, 1968*)

Sleep structures refer to the characteristics of the ongoing process of sleep and are generally indexed by the electroencephalogram (EEG). These measures are taken by pasting electrode cups to the head. Some electrodes pick up "brain waves," others eye movements, and others muscle tonus. Frequently, heart rate, respiration, and temperature are also measured.

The EEG-defined sleep structure has been found to display five distinct patterns or "stages" of sleep in a night of sleep in human subjects. These have been designated stages 1–4 and stage 1-REM, which shows the presence of stage 1 (an EEG essentially identical with waking) plus a number of phasic occurrences, particularly rapid eye movements (REM).

Many studies of normal subjects have revealed consistent overall characteristics of adult human sleep. A typical night of sleep for a young adult is shown in the illustration. Summary aspects of the figure typify the overall structure of human sleep; about 50% of the night is spent in stage 2 sleep, about 20% in slow-wave sleep (stages 3 and 4), about 25% in REM sleep, and about 5% in stage 1 (light sleep). Most of the slow-wave sleep occurs early in the sleep period, and REM sleep occurs episodically about every 90 to 100 min, with REM periods increasing in duration across the night.

Experimental variations. Sleep may be viewed as a biological rhythm—an endogenously determined tendency of a biological event to repeat itself on a time schedule. As such, it may be expected that temporal variations significantly influence sleep. Sleep schedules may be varied in three, often interacting ways: time awake prior to sleep may be short or long, sleep length may be reduced, and the placement of sleep may be varied. Sleep latency and prior wakefulness are inversely related, reaching an asymptote at about 60 h. Increased prior wakefulness increases the amount of slow-wave sleep and decreases the time spent in light stages of sleep (stage 0 after sleep onset and stage 1). Prior wakefulness has only limited effects on 1-REM. The time of day of sleep occurrence sharply affects the latency of stage 1-REM. Shortened sleep amounts down to about 3–4 h result in disproportionately less 1-REM sleep, but have little effect on stage 4. Subsequent sleep shows a sleep debt, evidenced by extended amounts of sleep. Displaced sleep, which results from such activities as shiftwork schedules or jet travel, is characterized by temporal variations in the sleep structure such as reduced 1-REM latencies and by increased awakening tendencies.

Deprivation is a major variable in sleep. The most certain finding of sleep deprivation studies is increased sleepiness, and many of the performance variations associated with sleep deprivation may be attributable to this factor. Under extreme deprivation, such as 10 days, biochemical changes are best described as subtle. The established neurological aftereffects of sleep deprivation are decreased pain thresholds, mild muscular tremors, and double vision. Reflexes and sensory systems seem unaffected. There is also some evidence for a short-term memory deficit. However, short and highly motivating tasks are remarkably resistant to sleep deprivation.

The sleep structure variable which has received the most intensive experimental attention is stage 1-REM deprivation. Clearly 1-REM sleep is strongly resistant to deprivation conditions, as evidenced by increased difficulty in maintaining deprivation and a "rebound" of 1-REM amounts after deprivation. However, its effects on waking behavior are less clear. Gross personality effects and simple learning, verbal memory, or performance effects are not present. Indeed, stage 1-REM deprivation appears facilitative in patients showing clinical depression. The most positive evidence suggests that 1-REM plays a role in long-term memory storage of material which requires extensive adaptational changes (unexpected learning).

Sleep disorders. Sleep disorders are deviations from normative sleep with associated clinical aspects. These include the typical childhood disorders of sleepwalking, night terrors, and enuresis, and the more general anomalies such as narcolepsy, hypersomnia, and insomnias. Also included are the secondary sleep disorders associated with other pathological conditions, for example, those typically present in depressions or associated with central nervous system pathology such as steeping sickness or encephalitis lethargica.

While "insomnia" can be taken as a general term for all sleep disorders, it now specifically refers to a behavioral syndrome in which the person complains of an inability to initiate sleep or maintain sleep (awakening during sleep or early awakening). Because it is thought of as a symptom (rather than a disorder per se), it, like body temperature, may reflect a wide range of causes. The amount of insomnia clearly depends on definition of degree, chronicity, and the sex and age of the subject. It is likely that all persons have transient periods of disturbed sleep.

Because insomnia is a syndrome, the insomnia expressed by a particular person may require a detailed diagnosis and very different responses. A few "types" of insomnia are illustrative:

Situational insomnias result from transient life stress such as work requirements, significant psychological changes such as loss of a loved one, location of occupational changes, or financial stress. They are typically transient, arising from situational changes and disappearing when adaptation has occurred.

In pseudo-insomnia, the individual may believe that he or she is getting less than is needed, and attributes feelings of inadequacy or stress to lack of sleep.

In rhythmic insomnia, the person's sleep period may be "entrained" by habit or circumstances to late hours of retiring; and because of work requirements the individual must be awakened early, resulting in inadequate amounts of sleep.

Psychopathological insomnias occur because sleep is sensitive to chronic psychological disturbances, such as anxiety and particularly depression. The insomnia may be reflecting this underlying psychological circumstance.

Aging insomnia is the type experienced by many individuals as nighttime awakening becomes prevalent as a part of the aging process.

All of these examples of insomnia require different approaches and treatment (or nontreatment). In the past there has been a tendency for a single mode of treatment, the sleeping pill. A great deal of sleep research has been focused on its role, and has added to knowledge of another insomnia, the drug-related insomnia, in which chronic use of sleeping pills may not yield a cure, but in fact may cause disturbed sleep.

Dreams. The discovery in the 1950s of the relationship between the electrophysiological index of an "active" phase of sleep and the index's very close association with the subjective state of dreaming joined these essentially independent interests. Simply, the identification of stage 1-REM displayed a unique and significant part of sleep as an active and exciting biological state. This stage's close indexing of dreams indicated their presence across the night far beyond the morning recall to which dreams had been previously confined.

The REM stage of sleep occurs in bursts of 90-min cycles and occupies some 20–25% of each night of all normal human sleep. The initial report that some 85% of awakenings from this state yielding a recall readily judged to be dreaming emphasized the large amount of sleep time occupied by dreaming. Detailed studies of REM awakenings have shown that reported materials from REM periods across the night become progressively longer, more vivid, and more unrealistic with successive REM.

Awakenings from non-REM periods clearly established that all cognitive aspects of sleep are not confined to the REM period. While judges can distinguish between reports of REM awakenings and NREM awakenings with high accuracy, on the basis of those having "dreamlike" quality (REM reports) and those having "thoughtlike" characteristics (NREM reports), as many as 30–50% of the NREM awakenings may contain some cognitive aspects.

A primary question which goes beyond the issue of dreams being reflective is whether dreams have a functional role. Many roles have been ascribed: At an affectional level, gratification of impulses, catharsis, synthesis, rehearsal, and self-education have been suggested. At a cognitive level, information processing and consolidation of acquired material have been attributed to the dream state. Few of the affective notions have received experimental tests, though clinical testimony had often been asserted. The more cognitive roles have been explored by REM deprivation experiments with limited success, although these experiments have seldom considered the dream per se as a critical variable.

In general, the modern era of sleep research has provided much information on the biological basis of the dream state and the substantial place and character of that state in the animal kingdom and human existence. [W.B.W.]

Sleep disorders Sleep is a reversible state during which the individual's voluntary functions are suspended but the involuntary functions, such as circulation and respiration, are uninterrupted; the sleeping subject assumes a characteristic posture with relative immobility and decreased responses to external stimuli.

The sleep state can be divided into nonrapid eye movement (NREM) and rapid eye movement (REM) sleep. NREM sleep is divided into four stages based on electroencephalographic criteria. The first REM in a normal adult occurs 60–90 min after sleep onset, and there are usually four or five NREM-REM cycles, each lasting for 90–110 min. Most dreams occur during REM sleep. The sleep-wake pattern for humans follows a circadian rhythm. An average adult needs approximately 7–8 hours of sleep, but elderly people have frequent awakenings.

Clinical disorders. Sleep disorders are classified as dyssomnias (intrinsic, extrinsic, or circadian rhythm disorders), parasomnias, and disorders associated with medical, psychiatric, and neurological illnesses. Obstructive sleep apnea syndrome, narcolepsy, psychophysiological and idiopathic insomnia, and restless legs syndrome are examples of intrinsic sleep disorders. Extrinsic sleep disorders include inadequate sleep hygiene, insufficient sleep syndrome, and hypnotic-, stimulant-, and alcohol-dependent sleep disorders. Circadian-rhythm disorders include jet lag and shift-work sleep disorders.

Parasomnias are characterized by abnormal movements or behavior intruding into sleep (for example, sleep walking, sleep terrors, sleep talking, nightmares, sleep paralysis, tooth grinding, and bed wetting).

Excessive daytime sleepiness, insomnia (sleeplessness), and abnormal movements and behavior during sleep are the major sleep complaints. An individual with insomnia may have difficulty initiating or maintaining sleep; repeated awakenings or early morning awakenings; or daytime fatigue and impairment of performance. Individuals with hypersomnia complain of excessive daytime sleepiness. Circadian-rhythm sleep disorders may be associated with either insomnia or hypersomnia.

There are two forms of sleep apnea. In central apnea, both the airflow at the mouth and nose and the effort by the diaphragm decrease. In obstructive apnea, the airflow stops but the effort by the diaphragm continues. Obstructive sleep apnea syndrome is common in middle-aged and elderly obese men.

Narcoleptic sleep attacks usually begin in individuals between the ages of 15 and 25. Narcolepsy is characterized by an irresistible desire to sleep, and the attacks may last 15–30 min. Other symptoms may include fearful dreams or feeling of loss of power at sleep onset or offset. Narcolepsy cannot be cured.

The restless legs syndrome occurs during middle age and is characterized by intense disagreeable feelings in the legs at rest and repose with compulsion to move the legs to get relief. Most individuals with this problem also have periodic limb movements in sleep.

The most common cause of insomnia is psychiatric or psychophysiologic disorder (for example, depression, anxiety, or stress), but other causes include medical disorders or pain. Early morning awakening is characteristic of depression.

Diagnosis and treatment. Overnight polysomnography involves recording of brain waves, muscle activities, eye movements, heart activity, airflow at the nose and mouth, respiratory effort, and oxygen saturation. Polysomnography is needed for individuals with excessive daytime sleepiness, narcolepsy, parasomnias, restless legs–period limb movements, and nocturnal seizures. The multiple sleep latency test is important in documenting pathologic sleepiness and diagnosing narcolepsy; it records brain waves, muscle activities, eye movements, and heart activity.

Treatment of obstructive sleep apnea syndrome consists of avoidance of sedatives or alcohol consumption, weight loss, and use of continuous positive airway pressure which will open the upper airway passage. Narcolepsy is treated by stimulants and short naps. Cataplexy is treated with tricyclics. Chronic insomnia can best be treated by sleep hygiene and behavioral modification. Most parasomnias do not require special treatment; however, psychotherapy may be helpful in some cases. Circadian-rhythm disorders may be treated by bright light and chronotherapy. *See* BIOLOGICAL CLOCKS; SLEEP AND DREAMING.

[S.Cho.]

Sleeping sickness

Sleeping sickness A specific disease of humans confined to tropical Africa. It is caused by *Trypanosoma gambiense* or *T. rhodesiense*, protozoa which are transmitted by tsetse flies of the genus *Glossina*. The disease is also known as African trypanosomiasis, maladie du sommeil, and Schlafkrankheit. *See* TRYPANOSOMATIDAE.

The incubation period is 2–3 weeks. The first signs are a local reaction at the site of the bite, fever, enlargement of adjacent lymph nodes, skin rash, edema, and cardiovascular disturbances. Asthenia develops early and is often very pronounced. The meningoencephalitic stage is accompanied by increasing lassitude and general indifference to surroundings. Somnolence, indifference to food, and emaciation characterize the late phase of the disease.

Diagnosis can be made from the clinical signs of the primary trypanosome chancre, the irregular fever, glandular enlargement, particularly in the neck, coupled with the demonstration of trypanosomes in stained thick films and smears from gland puncture, peripheral blood, and lumbar puncture.

The organic arsenical compounds tryparsamide and melarsen BAL are effective, especially when administered early in the disease. Antrypol (Bayer 205) and pentamidine have also been used in treatment but are more valuable as prophylactic drugs. Control against tsetse flies is the main objective. *See* MEDICAL PARASITOLOGY.

[M.Y.]

Slide rule

Slide rule A mechanical analog computing aid which is used extensively for multiplication and division and to a lesser degree for looking up functions. In its most common form a slide rule consists of a body formed from two parallel members rigidly fastened together, a slide which can be moved left or right between the body members, and a transparent indicator which carries a hairline and can be moved left or right over the face of the body and the slide. Scales are provided on the body and the slide as shown in the illustration.

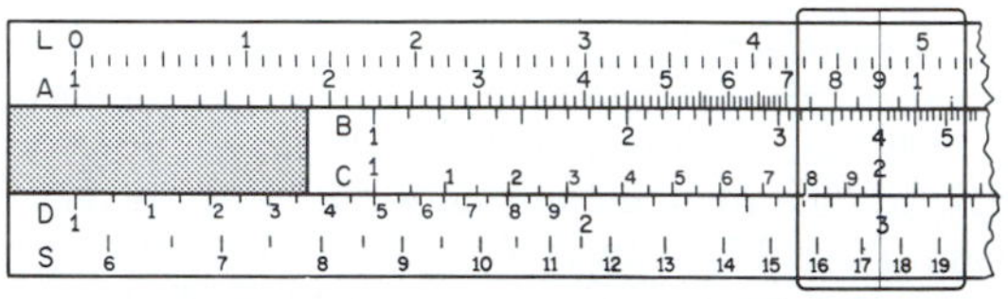

Left end of slide rule, showing multiplication of 1.5 by 2.

The C and D scales, used for multiplication and division, are graduated from 1 on the left to 10 on the right, with intermediate numbers distributed logarithmically. Multiplication on a slide rule is based upon the fact that the product of two numbers can be obtained by adding the logarithms of the two numbers and then taking the antilogarithm. Division, which is accomplished by subtraction of the logarithm of the divisor from that of the dividend, is performed in reverse sequence to multiplication. *See* LOGARITHM.

Many slide rules also carry scales from which sines, cosines, tangents, natural logarithms, logarithms to the base 10, squares, and cubes can be read.

[W.W.S.]

Slider-crank mechanism

Slider-crank mechanism A four-bar linkage with output crank and ground member of infinite length. A slider crank (see illustration) is most widely used to convert reciprocating to rotary motion (as in an engine) or to convert rotary to reciprocating motion (as in pumps), but it has numerous other applications. Positions at which slider motion reverses are called dead centers. When crank and connecting rod are extended in a straight line and the slider is at its maximum distance from the axis of the crankshaft, the position is top dead center (TDC); when the slider is at its minimum distance from the axis of the crankshaft, the position is bottom dead center (BDC).

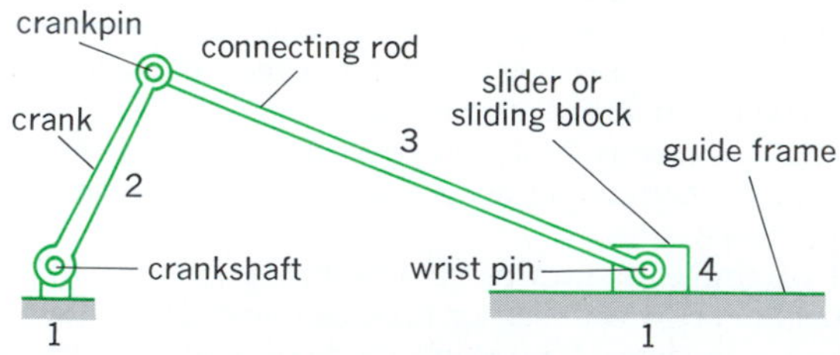

Principal parts of slider-crank mechanism.

The conventional internal combustion engine employs a piston arrangement in which the piston becomes the slider of the slider-crank mechanism. Radial engines for aircraft employ a single master connecting rod to reduce the length of the crankshaft. The master rod, which is connected to the wrist pin in a piston, is part of a conventional slider-crank mechanism. The other pistons are joined by their connecting rods to pins on the master connecting rod.

To convert rotary motion into reciprocating motion, the slider crank is part of a wide range of machines, typically pumps and compressors. Another use of the slider crank is in toggle mechanisms, also called knuckle joints. The driving force is applied at the crankpin so that, at TDC, a much larger force is developed at the slider. See FOUR-BAR LINKAGE. [D.P.Ad.]

Slip The difference between the operating speed of an induction motor and its synchronous speed (the speed of the rotating field). Slip s is usually expressed as a decimal fraction of synchronous speed n_s as in the equation below, where n is

$$s = \frac{n_s - n}{n_s}$$

the rotor, or operating, speed. See INDUCTION MOTOR. [W.W.S.]

Slip rings Electromechanical components which, in combination with brushes, provide a continuous electrical connection between rotating and stationary conductors. Typical applications of slip rings are in electric rotating machinery, synchros, and gyroscopes. Slip rings are also employed in large assemblies where a number of circuits must be established between a rotating device, such as a radar antenna, and stationary equipment.

Slip rings are usually constructed of steel with the cylindrical outer surface concentric with the axis of rotation. Insulated mountings insulate the rings from the shaft and from each other. Conducting brushes are arranged about the circumference of the slip rings and held in contact with the surface of the rings by spring tension. See ELECTRIC ROTATING MACHINERY; MOTOR. [A.R.E.]

Slope The trigonometric tangent of the angle α that a line makes with the x axis. In the illustration the slope of a plane curve C at a point P of C is the slope of the line that is tangent

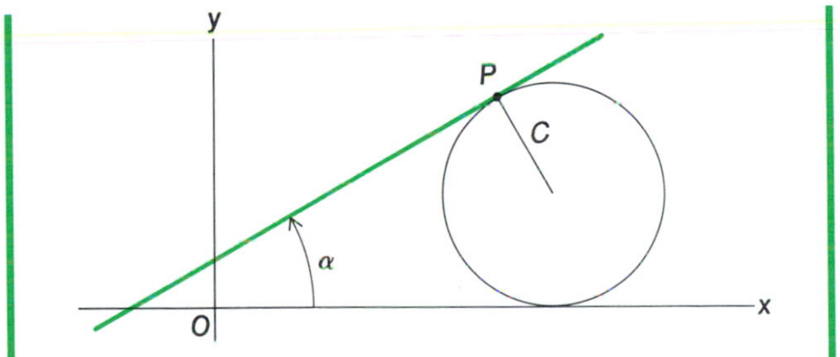

Slope of a curve.

to C at P. If $y = f(x)$ is an equation in rectangular coordinates of curve C, the slope of C at $P(x_0, y_0)$ is the value of the derivative $dy/dx = f'(x)$ at P, denoted by $f'(x_0)$, and hence an equation of the nonvertical tangent to C at P is $y - y_0 = f'(x_0)(x - x_0)$. See ANALYTIC GEOMETRY; CALCULUS. [L.M.Bl.]

Sloth A mammal, found exclusively in Central and South America, of the family Bradypodidae in the order Edentata. These animals are all slow-moving, arboreal species that cling to branches upside down by means of long, curved claws, and are difficult if not impossible to dislodge. The long hairs of the fur, which are a mixture of white, black, brown, and yellow or green, are unusual in that they are grooved. Algae inhabit the grooves, being green when conditions are damp and turning yellow in times of drought. The sloth has a short stumpy tail, round eyes, a small mouth, and horny lips which assist in eating leaves, a preferred food. The small pinnae of the ears are hidden in the fur. See EDENTATA. [C.B.C.]

Slug (zoology) A terrestrial pulmonate mollusk in which the shell is absent or reduced to a small internal or external rudiment. The slug form has evolved independently several times. The incorporation into the muscular foot region of the body organs (which are contained within the shell in other mollusks) results in a streamlined body shape (see illustration), enabling the animal to enter small holes or crevices.

***Limax maximus*. There are two pairs of tentacles on the head, and the opening to the lung is clearly visible.**

Lung respiration occurs as in other pulmonates, but skin respiration is probably at least as important.

A few slugs are carnivorous, such as *Testacella* which eats earthworms, but the majority are herbivores and may become serious horticultural and agricultural pests. See PULMONATA. [N.Ru.]

Smallpox An acute, infectious, viral disease characterized by severe systemic involvement and a single crop of skin lesions which proceeds through macular, papular, vesicular, and pustular stages. Smallpox is also known as variola major.

The virus enters through mucous membranes of the upper respiratory tract and propagates in lymphoid or other tissues. After a 12-day incubation period, the disease begins with 1–5 days of fever and malaise. During this period of time the virus multiplies, circulates in the bloodstream, localizes in the epidermis, and produces skin lesions. The lesions form crusts which fall off 2–4 weeks afterward and leave pink scars which fade slowly. Permanent scarring occurs because there is involvement of all skin layers, including necrosis of the corium.

Smallpox has had worldwide distribution, but with vaccination it has been virtually eliminated from some countries. It is transmitted by contact with oral or nasal secretions of infected persons or with material from skin lesions. Because the virus survives drying, it also has been transmitted through infected bedclothes or on cotton imported from distant countries.

All human beings are susceptible. Children are born with maternal antibodies, but these are soon lost. Control measures in addition to vaccination include rigorous quarantine regulation of travelers and of those in contact with known or suspected cases, and strict isolation procedures with patients. [J.L.Me.]

Smithsonite The natural form of zinc carbonate. Smithsonite may contain small amounts of iron, calcium, cobalt, copper, manganese, cadmium, and magnesium. It occurs chiefly as a secondary mineral formed by the oxidation of primary zinc ores, such as sphalerite.

Smithsonite is rarely found in well-formed crystals and is more often in compact or porous masses with a cauliflowerlike appearance. Its color is often dirty white, but a variety of tints, including greens, blues, and browns, may be found, depending on the impurities present. For pure smithsonite the hardness is about 4 or 4½ on Mohs scale and the specific gravity is 4.43. See CARBONATE MINERALS; ZINC. [R.I.Ha.]

Smog The word smog can be defined in several ways. It was originally used to refer to a combination of coal smoke and fog, the particles of the smoke often serving as nuclei on which the water vapor condensed. Later, the term was used to refer to any dirty urban atmosphere. A unique type was recognized in the late 1940s and early 1950s and came to be known as photochemical smog, defined as that type of air pollution which owes many of its properties to the products of photochemical reactions involving the vapor of various organic substances, especially hydrocarbons, oxides of nitrogen, and atmospheric oxygen. Occasionally fog may also be involved in the formation of such smog.

Photochemical smog is particularly apparent in cities where little coal is burned, there is little industrial pollution, and there are large concentrations of automobiles. The intensity of photochemical smog at any time and place, like that of any type of air pollution, is markedly influenced by meteorological conditions and the geographical features of the region.

Usually the most obvious and immediately annoying aspect of the urban pollution is the murkiness of the air which it produces. This visibility decrease is mainly caused by the suspended particles. The median particle size in photochemical smog is so small that the particles preferentially scatter blue light, and depending upon the nature and angle of illumination, the air may appear blue or brown. The brown appearance is accentuated by the presence of nitrogen dioxide (NO_2), which is itself brown in color and results from various combustion operations, including driving automobiles.

Another especially annoying feature of photochemical smog is the eye irritation and lacrimation it produces. This irritation seems to result from the combined effects of several gases, especially formaldehyde, acrolein, peroxyacetyl nitrate (PAN), and possibly peroxybenzoyl nitrate. *See* AIR POLLUTION. [R.D.C.]

Smoke Fumes and smoke are dispersions of finely divided solids or liquids in a gaseous medium. The particle-size range is 0.01–5.0 micrometers. Typical dispersions are smokes from incomplete combustion of organic matter such as tobacco, wood, and coal; soot or carbon black; oil-vapor mists; chemical fumes such as sulfur trioxide (SO_3) and phosphorus pentoxide (P_2O_5) mists, ammonium chloride (NH_4Cl), and metal oxides; and the products of hydrolysis of metal chlorides by moist air. Oil-vapor and P_2O_5 mists (formed by burning phosphorus in moist air) have been extensively used in military operations to produce screening smokes. *See* AIR POLLUTION. [H.H.St./A.J.T.]

Smooth manifold A manifold with an appropriate additional structure that allows one to speak of differentiability or smoothness of real-valued functions. Such manifolds are also called differentiable manifolds. In this setting one can define the notion of a smooth mapping between smooth manifolds and can introduce the differential of a smooth mapping, which is a generalization of derivative. *See* MANIFOLD (MATHEMATICS). [A.W.K.]

Smut (microbiology) A term applied both to fungi of the order Ustilaginales of the subclass Heterobasidiomycetidae and to the diseases which they cause. Plants of several families are subject to smuts, but smuts of cereals are of greatest economic importance. The control measure most effective against all types of smuts is the development and planting of resistant varieties of grain. Bunt of wheat is also controlled by treating seed with organic mercurial fungicides to kill spores adhering to the seed coat. *See* BASIDIOMYCOTINA. [R.M.P.]

Snail Any of the approximately 74,000 species in the class Gastropoda of the phylum Mollusca or, alternatively, any of the 12 or so species of land pulmonate gastropods used as human food.

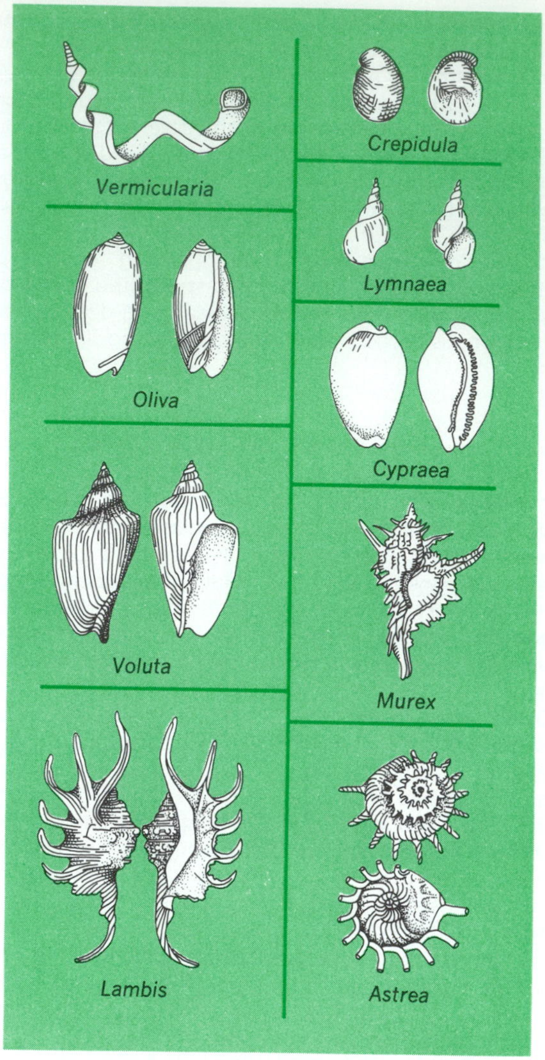

Diversity of snail shells. ***(After R. R. Shrock and W. H. Twenhofel, Principles of Invertebrate Paleontology, 2d ed., McGraw-Hill, 1953)***

The shell of snails is in one piece and typically turbinate (see illustration), but may be planospiral or limpet-shaped, or may be secondarily lost (as in land slugs and marine nudibranchs). In development, gastropods have undergone torsion (the visceral mass and the mantle-shell covering it have become twisted through 180° in relation to the head and foot) so that the mantle cavity is placed anteriorly above the head.

In feeding, all snails use a characteristic rasping tongue or radula. This is a chitinous ribbon bearing teeth which is moved over a supporting protrusible "tongue" with a to-and-fro action. The radular apparatus has a twofold function: it serves both for rasping off food material (mechanically like an inverted version of the upper incisor teeth of a beaver) and for transporting the food back into the gut like a conveyor belt.

Both fresh-water and terrestrial snail species serve as vectors (first or sole intermediate hosts) in the transmission of flukes (Trematoda) infecting humans or domestic animals. For further discussion of these parasitic diseases *see* CLONORCHIASIS; DIGENEA; FASCIOLIASIS; SCHISTOSOMIASIS. *See also* GASTROPODA; PULMONATA. [W.D.R.-H.]

Snake Any of the 3000 species of reptiles which belong to the 13 living families of the suborder Serpentes in the order Squamata. Except for Antarctica, they have a worldwide distribution but are most abundant in warm climates. The more primitive pythons and boas have external vestiges of hindlimbs, but most other snakes are limbless. The elongate body is covered with scales, external ear openings are lacking, and the eyes lack mobility. The eyelids are fused together to form a transparent membrane over the eyes which is cast off during the periodic shedding of the skin.

The kinetic skull, mandible-quadrate articulation, and ligamentous connection between the anterior ends of the two mandibular rami permit extreme opening of the mouth, thus enabling snakes to swallow animals several times their size. Poisonous snakes with canaliculate fangs are called proteroglyphous. These teeth are erectile and are used only for stabbing.

There are oviparous species, which produce shell-covered eggs shortly after being fertilized, as well as viviparous species, which give birth to well-developed young. The eggs are usually deposited in a hole in the ground where there is sufficient humidity and temperature to permit development. *See* Boa; Python; Reptilia; Squamata. [C.B.C.]

Snap gage A gage with two surfaces that are flat and parallel; these are spaced to control one limit of tolerance of an outside diameter or a length. Adjustable snap gages have adjustable gaging surfaces so that the distance between gaging surfaces can be set with gage blocks and used to gage any dimension falling within the limits of the frame size. Snap gaging is faster than ring gaging, but it checks only one diameter or length at a time. *See* Ring gages. [R.A.Bo.]

Snap ring A form of spring used principally as a fastener. Piston rings are a form of snap ring used as seals. The ring is elastically deformed, put in place, and allowed to snap back toward its unstressed position into a groove or recess. The snap ring may be used externally to provide a shoulder that retains a wheel or a bearing race on a shaft, or it may be used internally to provide a construction that confines a bearing race in the bore of a machine frame, as illustrated. [L.S.L.]

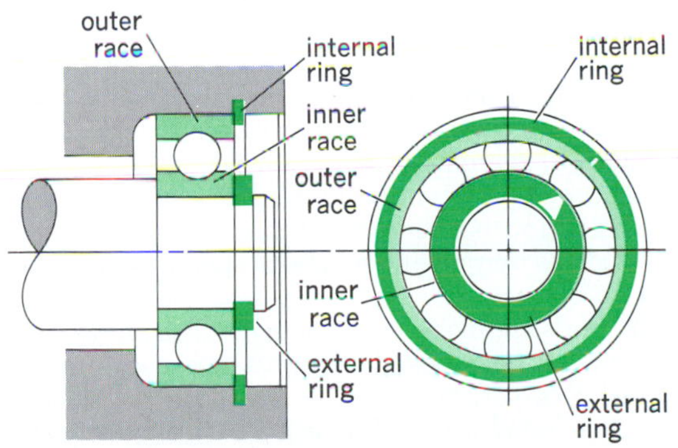

Snap rings hold ball-bearing race in place. Internal ring supports axial thrust of axle. External ring aligns inner race against shoulder of shaft.

Snow The most common form of frozen precipitation, usually flakes of starlike crystals, matted ice needles, or combinations, often rime-coated; also transitions between these forms, or combinations of them with ice columns, plate crystals, snow pellets, snow grains, or ice pellets. *See* Precipitation (meteorology).

When newly fallen, snow has a gravity ratio commonly 1:10 but varying greatly. Wet snow is heavier, but dry, fluffy snow may have a specific gravity of 1:30 or less. The density of snow cover is increased by packing, melting, and refreezing, and also because snow changes to ice granules through transfer of water by sublimation from small to larger ice particles. Deep in glaciers entrapped air is compressed and density approaches that of pure ice.

Snow, especially when new, is a good insulator because air is entrapped by the flakes. Thus it often protects low vegetation, such as grains, from freezing. Air temperatures tend to be low over snow because it reflects sunshine, is a good radiator at night, and interferes with conduction of heat from the ground. [J.R.F.]

Snow gage An instrument for measuring the amount of water equivalent in snow; more commonly known as a snow sampler. Frequently snow samplers are made of a lightweight seamless aluminum tube consisting of easily coupled lengths. Other snow samplers have been developed from material such as fiber glass and plastic for use in shallow snow, deep dense snow, and so forth.

To obtain a measurement the sampler is pushed vertically through the snow to the ground surface. The sampler, together with its snow core, is withdrawn and weighed. The water equivalent of the snow layer is obtained by subtracting the weight of the sampler from the total. In addition to any error introduced by the scale, there is usually a 6–8% error in the weight of samples taken in this manner.

Automatic devices that permit the remote observation of the water equivalent of the snow have been developed. These devices also permit telemetering the data to a central location, eliminating the need for travel to the snowfields. *See* Snow; Snow surveying. [R.T.Be.]

Snow line A term generally used to refer to the elevation of the lower edge of a snow field. In mountainous areas, it is not truly a line but rather an irregular, commonly patchy border zone, the position of which in any one sector has been determined by the amount of snowfall and ablation. These factors may vary considerably from one part to another. On glacier surfaces the snow line is sometimes referred to as the glacier snow line or névé line (the outer limit of retained winter snow cover on a glacier).

Year-to-year variation in the position of the orographical snow line is great. The mean position over many decades, however, is important as a factor in the development of nivation hollows and protalus ramparts in deglaciated cirque beds. *See* Glaciology; Snowfield and névé. [M.M.Mi.]

Snow surveying An inventory of the winter's accumulation of snow. A snow survey is made at the same location each year, and this site is referred to as the snow course. A snow course consists of a series of sampling points, usually 10–20, to give a dependable sampling average. These points are situated at a measured spacing, 50–100 ft (15–30 m) apart, along a permanently marked and mapped route. The significant information obtained by means of a snow survey is the average water equivalent of all of the samples taken at the snow course. The snow courses may be sampled several times a year, year after year. *See* Snow gage.

While snow surveys were initially made to forecast water supplies for irrigation, they are also used for water-management purposes, flood forecasting, power requirements, and municipal water supplies. *See* Surface water. [R.T.Be.]

Snowfield and névé The term snowfield is usually applied to mountain and glacial regions to refer to an area of snow-covered terrain with definable geographic margins. Where the connotation is very general and without regard to geographical limits, the term snow cover is more appropriate; but glaciology requires more precise terms with respect to snowfield areas.

These terms differentiate according to the physical character and age of the snow cover. Technically, a snowfield can embrace only new or old snow (material from the current accumulation year). Anything older is categorized as firn or ice. The term névé is a descriptive phrase used to refer to consolidated granular snow not yet changed to glacier ice. Because of the need for simple terms, however, it has become acceptable to use the term névé when specifically referring to a geographical area of snowfields on mountain slopes or glaciers (that is, an area covered with perennial "snow" and embracing the entire zone of annually retained accumulation). *See* GLACIOLOGY.

[M.M.Mi.]

Soap A particular type of detergent in which the water-solubilizing group of the 10–18-carbon chain is carboxylate, $—COO^-$, and the positive ion is usually sodium, Na^+, or potassium, K^+. All similar compounds with hydrophilic groups other than carboxylate are synthetic detergents or surfactants. *See* DETERGENT.

Because of the marked imbalance of polarity within the molecules of soaps, they have unusual surface and solubility characteristics. The feature of their molecular structure which causes these properties is the location of the hydrophilic function at or near the end of a long hydrocarbon (hydrophobic) chain. Thus one part of the molecule is water-seeking while the other portion is oil-seeking. This explains why these molecules concentrate and orient themselves at interfaces, such as an oil-solution interface where the hydrophobic portion enters the oil and the hydrophilic portion stays in the water.

Soap was formerly manufactured by an alkaline hydrolysis reaction called saponification, according to the reaction below

$$\underset{\text{Fat}}{C_3H_5(OOCR)_3} + \underset{\text{Caustic soda}}{3NaOH} \rightarrow \underset{\text{Soap}}{3NaOOCR} + \underset{\text{Glycerin}}{C_3H_5(OH)_3}$$

(R represents the hydrocarbon chain). An important modern process for making soap is the direct hydrolysis of fats by water at high temperatures. This permits fractionation of the fatty acids, which are neutralized to soap in a continuous process. Advantages for this process are close control of the soap concentration, the preparation of soaps of certain chain lengths for specific purposes, and easy recovery of the by-product glycerin. After the soap is recovered, it is pumped into giant mixers called crutchers where builders, perfumes, and other desired ingredients are added. (Builders are alkaline compounds which improve the cleaning performance of the detergent.) Finally, the soap is rolled into flakes, cast or milled into bars, or spray-dried into soap powder.

There is one serious disadvantage to the general use of soap which has caused its replacement by synthetic detergents for home laundering. The problem is that carboxylate ions of the soap react with the calcium and magnesium ions in natural hard water to form insoluble materials called lime soap. In areas having a hard water supply, this is the floating curd which appears in the washbowl and as bathtub ring. The greatest market remaining for soap is as bar soap for personal bathing.

Metallic soaps are alkaline-earth or heavy-metal long-chain carboxylates which are insoluble in water but soluble in nonaqueous solvents. They are used as additives to lubricating oils, in greases, as rust inhibitors, and in jellied fuels. [R.C.M.]

Soapstone A soft talc-rich rock. Soapstones are rocks composed of serpentine, talc, and carbonates (magnesite, dolomite, or calcite). They represent original periodites which were altered at low temperatures by hydrothermal solutions containing silicon dioxide, carbon dioxide, and other dissolved materials (products of low-grade metasomatism). The whole group of rocks may loosely be referred to as soapstones because of their soft, soapy consistency. [T.F.W.B.]

Social animals Those animals which exhibit social behavior. This may be defined as any behavior stimulated by or acting upon another animal of the same species. In this broad sense, almost any animal which is capable of behavior is to some degree social. Even those animals which are completely sedentary, such as adult sponges and sea squirts, have a tendency to live in colonies and are social to that extent. Social reactions are occasionally given by other species than the animals's own, like the relations between domestic animals and humans. *See* SOCIAL INSECTS.

There are three general types of social animals: those which are found only in temporary aggregations, including all the lower invertebrates; those which live in permanent groups bound together by care-giving behavior, such as the insect societies; and those whose permanent groups are organized through mutual imitation and agonistic behavior (often in addition to care giving), such as many vertebrate societies. Humans belong to a fourth type in which a highly developed communication system has been added to the above social processes.

Fish. The behavioral capacities of fish reflect their watery habitat. As to psychological capacities, fish are capable of rapid conditioning. Trout in hatcheries quickly learn to rise for food at any splashing noise. Fish can learn simple mazes and are able to find their way back to their home ranges quickly under natural conditions. On the other hand, much of their social behavior, particularly that connected with mating and nest building, is largely organized by instinct.

Compared with birds and mammals, fish lack some of the basic types of social behavior. As water-living animals they have no problem of waste disposal and consequently have no special patterns of eliminative behavior. Fish never feed their young after hatching, with the exception of certain species whose newly hatched young feed on mucus from the adult's body, and their care-giving behavior is limited to offering shelter and driving off predators. The connection between offspring and parents is brief at best, and the young show little care-soliciting behavior.

Much of the social life of fishes is concentrated around the school, in which mutual imitation or allelomimetic behavior coordinates the group and provides mutual safety. Other social life centers around mating and nest building. The school is perhaps the most typical part of the social life of fishes. It is usually an association between newly hatched young, but many species such as herring and mackerel spend most of their lives in such a group.

Birds. Birds are highly social animals. They move so rapidly and freely that their social behavior is not always apparent to the casual observer, but when their actions are studied more carefully it is found that birds develop a great variety and complexity of social organization. Their biological differentiation of labor is simpler than that of insects since there are only three principal types of individuals: males, females, and young. However, they exhibit all the basic types of social behavior in well-developed forms and are capable of considerable differentiation of behavior on the basis of learning and experience.

With their great mobility, birds frequently encounter other birds of different species, which creates a problem of separation between different societies. The bright colors and plumage changes found in many birds are closely related to social behavior and serve to differentiate between species, the sexes, and adults and young. Display behavior, emphasizing the plumage, is a prominent part of sexual behavior in many species.

Display is a type of communication. In addition, birds have a wide variety of vocal signals; best known of these is the song of the perching birds, now shown to be a territorial signal. In some species the song is in part learned; in others it is developed even by birds reared in isolation. Certain species, such as crows and parrots, have powers of vocal imitation superior to those of mammals, but even when human words are learned, there is no evidence that the birds do more than learn a complicated trick. Such parroting is found in nature in the songs of mockingbirds. Vocal and visual signals are important mechanisms in the social organization of birds but are not used as a true language.

Social behavior is affected still more directly by aerial life. One major adaptation is rapid development by the young. This means that in many birds there is little opportunity for a parent-offspring relationship to develop. The most prominent social relationship of most birds is consequently the male-female relationship or pair bond.

Primary socialization or imprinting takes place very rapidly in precocious birds such as ducks and geese, but is spread over a longer period in birds which are hatched in an immature state. The parasitic cowbirds and cuckoos do not become imprinted by their foster parents but join their own kind after leaving the nest.

Another fundamental social process is that of becoming attached to particular places or home sites, or localization. Many species of birds have two homes, one during the breeding season and another at other times. Migrations between the two may cover thousands of miles and require extraordinary capacities for orientation, the basic nature of which is the subject of much research. *See* MIGRATORY BEHAVIOR.

Flocking behavior is typical of birds. In the smaller flying birds extraordinarily complex and rapid maneuvers based on allelomimetic behavior are seen; in some larger birds there appears considerable leadership. In contrast, the hawks and owls are almost entirely solitary in their movements. Territory is another important feature of bird life, ranging from the large breeding territories of the perching birds to the small nest area in gulls and other colonial sea birds. In still others there are no definite territories.

Mammals. The postnatal development of each species of mammal is closely related to the social organization typical of the adults. Every highly social animal has a short period early in life when it readily forms attachments to any animal with which it has prolonged contact. This critical period for primary socialization can be demonstrated by removing young from their natural parents and rearing them by hand. Experience before this period has no effect; during the period the young animal transfers all its native social behavior to the human; thereafter this process becomes increasingly difficult, although captivity and hand feeding have some effect on adults.

The process of socialization begins almost immediately after birth in ungulates like the sheep, and the primary relationship is formed with the mother. In dogs and wolves the process does not begin until about 3 weeks of age, at a time when the mother is beginning to leave the pups. Consequently the strongest relationships are formed with litter mates, thus forming the foundation of a pack. Many rodents stay in the nest long after birth; primary relationships are therefore formed with nest mates. Young primates are typically surrounded by a group of their own kind, but because they are carried for long periods the first strong relationship tends to be with the mother.

Mammals may develop all types of social behavior to a high degree, but not necessarily in every species. Mammals have great capacities for learning and adaptation, which means that social relationships are often highly developed on the basis of learning and habit formation as well as on the basis of heredity and biological differences. The resulting societies tend to be malleable and variable within the same species and to show considerable evidence of cultural inheritance from one generation to the next. Mammalian societies have been completely described in relatively few forms, and new discoveries will probably reveal the existence of a greater variety of social organization.

Basic social organization and behavior of humans obviously differs from that of all other primates, although there is some relationship. At the same time, the range of variability of human societies as seen in the nuclear family does not approach that in mammals as a whole. Human societies are characterized by the presence of all fundamental types of social behavior and social relationships rather than by extreme specialization. [J.P.S.]

Social hierarchy Social hierarchy was first adequately described in flocks of domestic hens. The fighting behavior of each individual hen is organized so that in any given pair hen A always pecks hen B and never the reverse. In this way an entire flock can be described in terms of social rank, or a "peck order." This type of social organization is an important general phenomenon among higher animals.

A social hierarchy is composed of many dominance-subordination relationships. When two hens are brought together for the first time, they usually fight. The next time they come together the fighting has a shorter duration. One hen forms a habit of winning and the other of losing. Eventually the relationship is reduced to a threat or peck by the dominant hen and avoidance or submission to the peck by the subordinate one. The hens thus form a social relationship based on agonistic behavior, which is defined as any behavior connected with conflict between two individuals and including fighting, escape, and submissive behavior.

There are several possible arrangements of dominance hierarchies in a group of three animals, assuming that each relationship shows either a complete dominance or complete peacefulness. These groupings are: (1) straight-line dominance order, commonly found in flocks of hens in aggressive breeds; (2) triangular order, less common; (3) one dominant and two subordinate, common in mouse groups; (4) two dominant animals and one subordinate; (5) all animals peaceful, with no dominance order, frequently found in young animals reared together.

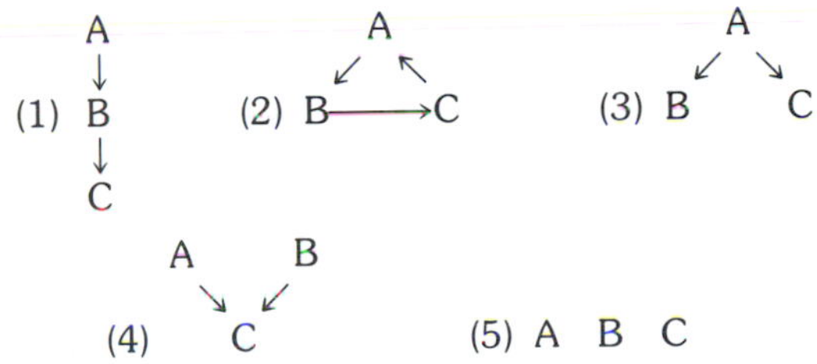

Since most of the lower invertebrates do not exhibit agonistic behavior, social hierarchies have so far been found only in arthropods and vertebrates. In arthropods, the best examples are found in crayfish and other Crustacea. While fighting occurs in some insects, it is doubtful whether true dominance-subordination relationships are formed. In the higher social insects, such as ants, bees, wasps, or termites, no fighting takes place within the colony. On the other hand, dominance-subordination relationships have been described in all classes of vertebrates, although they have little importance among amphibia, which are poorly equipped for fighting behavior. The social hierarchy is thus a typical attribute of vertebrate societies, reaching its highest development in birds and mammals.

The dominance hierarchy can be either extremely important or completely nonexistent, depending on the species of animal involved. Well-developed dominance relationships tend to be relatively permanent, persisting over periods of years in long-lived animals. Such social organization may have important effects on the control of fighting within a species; on the division of food, territory, and mates; and upon the selection of surviving individuals. At the same time it should be remembered that this is only one aspect of social organization and that important relationships which do not involve fighting, such as care-dependency and leader-follower relationships, also exist in animal societies. *See* SOCIAL ANIMALS; SOCIAL INSECTS. [J.P.S.]

Social insects Social organization has evolved independently several times in unrelated groups of the class Insecta. The best examples of these social insects are the termites of the order Isoptera and the social wasps, bees, and ants of the order Hymenoptera. Characteristically, social insects are differentiated and specialized in structure, function, and behavior into castes, which live in colonies and exhibit group integration and division of labor. The principal castes are the reproductives, which include the king and queen, and the steriles, the workers and soldiers. The reproductive caste performs the fundamental functions of reproduction, distribution of the species accompanied by swarming or colonizing flight, choosing the site for the new colony, excavation of the first galleries, and feeding and care of the first young of the new colony.

The termite sterile castes consist of workers and soldiers represented by both sexes; but among social wasps, bees, and ponerine ants (Ponerinae), the workers are the only sterile caste, consisting of nonreproductive females. The function of the workers, which form a majority in the population, is that of tending the eggs, young nymphs, or larvae, feeding and cleaning other castes, constructing and repairing the nest, and gathering food. The soldiers are structurally modified to defend the community against predators. *See* HYMENOPTERA; ISOPTERA. [K.K.]

Sociobiology A scientific discipline that applies evolutionary biology to the study of animal (including human) social behavior. It is a synthesis of ethology, ecology, and evolution, in which social behavior is viewed as the result of natural selection and other biological processes.

The two basic principles that underlie sociobiology are the interaction principle and the central principle of fitness maximization. The interaction principle simply acknowledges that all characteristics (including behavior) are the result of an interaction between genotype and experience. Thus, the ancient nature-nurture debate has not proven useful. Sociobiologists emphasize that neither nature (genes) nor nurture (environment), acting alone, determines behavior. Rather, all behavior is influenced by varying combinations of both interacting factors. Although this does not in any way deny a role for experience in influencing behavior, it makes room for examination of the role of natural selection. The central principle of fitness maximization holds that living things will behave in a manner that maximizes their fitness, their success in projecting copies of their genes into succeeding generations. In fact, living things can be defined as structures produced by successful genes, whose function is to project genes into the future.

Biologists have long been puzzled by the seeming paradox that living things sometimes behave in a way that reduces their personal reproductive success while increasing that of another individual. For example, ground squirrels and many birds give warning calls that alert other individuals to the presence of a predator, while seemingly rendering themselves more conspicuous. Female honeybees do not reproduce; instead, they labor for the success of the queen. With the important concept of inclusive fitness, sociobiologists have identified the adaptive significance of such behavior. Thus, a "gene's-eye" view of evolution recognizes that Darwinian fitness (the success of an individual in producing offspring) is only a special case of the more general process of inclusive fitness, by which genes strive to project copies of themselves into the future. Thus, whereas genes are contained in offspring, they are also copied in other relatives, although the more distant the relative, the lower the chance that such a copy is present. It has frequently been found that when animals seem to behave altruistically as individuals, they are at the same time being genetically selfish, since their behavior enhances their inclusive fitness: the success of the underlying genes, contained in the bodies of relatives.

In addition, sociobiologists have found that altruistic behavior could occur even among unrelated individuals as a result of reciprocity: if a recipient of such an act is likely to repay it, both individuals (and their genes) ultimately benefit. It is also possible that altruism could result from selection acting at the level of groups, rather than individuals or genes, if groups containing altruists were at a sufficient advantage compared to groups composed of selfish individuals. However, selection acting on individuals within such groups will tend to favor selfishness over altrium. In any event, much of the cogency of sociobiology has resulted from the recognition that selection seems to be most powerful when it acts at lower levels—notably, genes and individuals rather than groups or species.

Although most of the research in sociobiology has involved animals, there has been controversy, most of it centering on *Homo sapiens*. Some people have objected to what they see as political or ethical implications of evolutionary biology applied to the human species. Much of this concern is well founded, given the past abuses of Social Darwinism and racism based on misunderstandings of genetics. However, a sociobiologic perspective on the human species would emphasize the existence of cross-cultural universals, characteristics that all human beings share by virtue of their biology and independent of their diverse cultures; as such, sociobiology could be a potent antidote to racism. Sociobiology may well shed considerable light on why human beings do the things they do. *See* ECOLOGY; ETHOLOGY; GENETICS; ORGANIC EVOLUTION. [D.P.B.]

Soda niter A nitrate mineral having chemical composition $NaNO_3$ (sodium nitrate). Soda niter is by far the most abundant of the nitrate minerals. It is usually massive granular. Its hardness is 1.5 to 2 on Mohs scale, and its specific gravity is 2.266. It has a vitreous luster and is transparent. It is colorless to white, but when tinted by impurities, it is reddish brown, gray, or lemon yellow. *See* NITRATE MINERALS.

Soda niter is a water-soluble salt found principally as a surface efflorescence in arid regions, or in sheltered places in wetter climates. The only large-scale commercial deposits of soda niter in the world occur in a belt along the eastern slope of the coast ranges in the Atacama, Tarapaca, and Antofagasta deserts of northern Chile. Chilean nitrate had a monopoly of the world's fertilizer market for many years, but now occupies a subordinate position owing to the development of synthetic processes for nitrogen fixation which permit the production of nitrogen from the air. This has led to the commercial production of artificial nitrates and has reduced the need to import Chilean nitrates for use in the manufacture of fertilizers and explosives. *See* FERTILIZER; NITRATE. [G.Sw.]

Sodalite A mineral of the feldspathoid group with chemical composition $Na_4Al_3Si_3O_{12}Cl$. Sodalite is usually massive or granular with poor cleavage. The Mohs hardness is 5.5–6.0, and the density 2.3. The luster is vitreous, and the color is usually blue but may also be white, gray, or green. Notable occurrences are at Mount Vesuvius; Bancroft, Ontario; and on the Kola Peninsula of Russia. *See* FELDSPATHOID; SILICATE MINERALS. [L.Gr.]

Sodium A chemical element, Na, atomic number 11, and atomic weight 22.9898. Sodium is between lithium and potassium in the periodic table. The element is a soft, reactive, low-

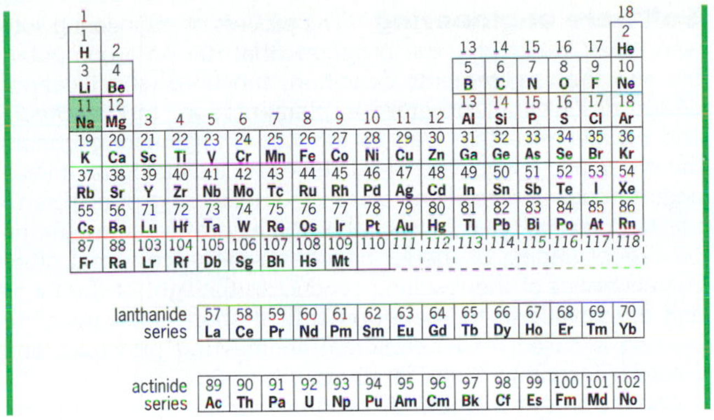

melting metal with a specific gravity of 0.97 at 20°C (68°F). Sodium is commercially the most important alkali metal. *See* ALKALI METALS.

Sodium ranks sixth in abundance among all the elements in the Earth's crust, which contains 2.83% sodium in combined form. Sodium is, after chlorine, the second most abundant element in solution in sea water. The important sodium salts found in nature include sodium chloride (rock salt), sodium carbonate (soda and trona), sodium borate (borax), sodium nitrate (Chile saltpeter), and sodium sulfate. Sodium salts are found in sea water, salt lakes, alkaline lakes, and mineral springs.

Sodium reacts rapidly with water, and even with snow and ice, to give sodium hydroxide and hydrogen. When exposed to air, freshly cut sodium metal loses its silvery appearance and becomes dull gray because of the formation of a coating of sodium oxide. Sodium does not react with nitrogen, even at very high temperatures, but will react with ammonia to form sodium amide. Sodium and hydrogen react above about 200°C (390°F) to form sodium hydride. Sodium reacts with carbon with difficulty, if at all, but will react with the halogens. Sodium also reacts with various metal halides to give the metal plus sodium chloride.

Sodium does not react with paraffin hydrocarbons but does form addition compounds with naphthalene and other polycyclic aromatic compounds and with arylated alkenes. The reaction of sodium with alcohols is similar to, but less rapid than, the reaction of sodium with water. There are two general reactions with organic halides. One of these involves condensation of two organic, halogen-bearing compounds by removal of the halogen. The second type of reaction involves replacement of the halogen by sodium, giving an organosodium compound. [M.Si.]

Sodium-vapor lamp A low-pressure electric discharge lamp of monochromatic yellow light. The arc tube contains a small amount of sodium and some neon gas to facilitate starting. Sodium radiation is confined to a single band in the yellow region of the spectrum. This results in high luminous efficiency, but extremely poor color rendition limits low-pressure sodium lamps to tasks where color is unimportant.

The high-pressure sodium lamp that is used in the United States utilizes an amalgam of sodium and mercury inside an arc tube surrounded by an evacuated glass jacket. Operation of the arc at a higher temperature and pressure produces a continuous radiation pattern rather than the line pattern of yellow light associated with the earlier low-pressure sodium lamp. Its high luminous efficiency (over 100 lumens per watt), combined with good color rendition, has brought increasing use of the high-pressure sodium lamp in many general lighting applications. *See* ILLUMINATION; LAMP; VAPOR LAMP. [V.G.McC.]

Sofar The name associated with explosive signals which follow transmission paths through deep ocean layers. Sounds originating moderately deep in the ocean are refracted so that they propagate to large distances with small losses. For example, a small explosive charge set off at a 3000- or 4000-ft (900- or 1200-m) depth is detectable many hundreds of miles away. If such an explosive signal is received at two or more points, the geographical position of the explosion can be calculated. This signaling technique has been used to locate aviators who have been forced down in the open ocean, and explains the origin of the word sofar, which stands for sound fixing and ranging. *See* UNDERWATER SOUND. [R.W.Mo.]

Soft chancre An ulcerative venereal disease of the genitals or adjacent tissues of humans caused by the Ducrey bacillus (*Hemophilus ducreyi*). It can be distinguished from a syphilitic lesion by cultural methods.

The disease begins as a macule or pustule upon the prepuce or glans in the male or upon the labia minora or vestibule in the female a few hours after sexual intercourse. The pustule soon forms a painful, ragged, soft ulcer with undermined edges. The regional lymph nodes become swollen and tender, and lesions may become multiple by autoinoculation. *See* HEMOPHILUS; SEXUALLY TRANSMITTED DISEASES. [W.L.Br.]

Soft-iron ammeter An ammeter in which current in a coil causes two pieces of magnetic material within the coil, one fixed and one attached to a pointer, to become similarly magnetized and to repel each other, moving the pointer. These ammeters are widely used for alternating-current measurements. The illustration shows a repulsion-vane type with pointer, control spring, and damping vane on the moving element.

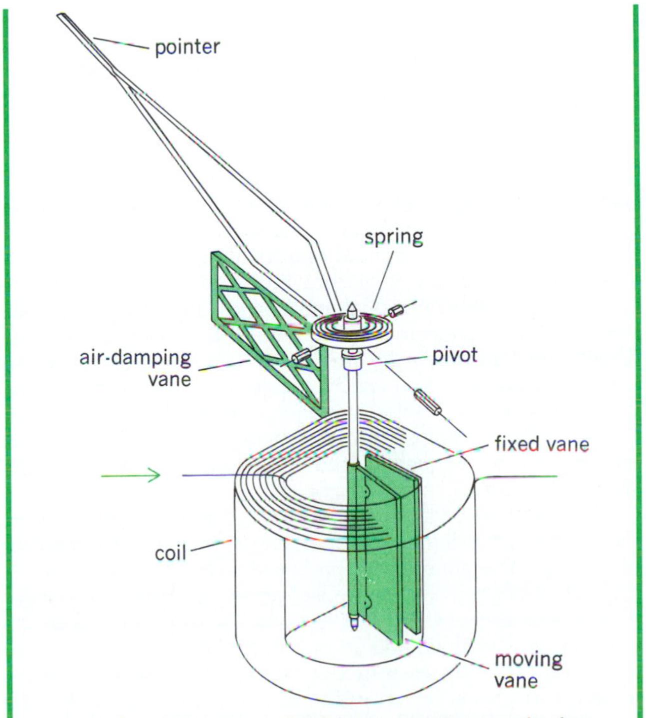

Repulsion vane on a book-type ammeter mechanism. *I* = current. (*Weston Instruments, Inc.*)

Requiring about 300 ampere-turn for full-scale deflection in the larger sizes, the common 5-ampere range will have a 60-turn coil of 0.06-in.-diameter (1.5-mm) copper wire, ample to carry even considerable overloads. The iron vanes tend to saturate magnetically on overload and thereby act as further overload protection. Actuating forces are high, and only the limitations in the iron vanes in responding exactly to the coil current limit the accuracy. The most accurate instruments of this type are accurate to within 0.25%. *See* AMMETER. [J.H.M./J.Be.]

Software Programs, procedures, and routines that are usable on a particular kind of computer and enable the computer system to function. A computer comprises two major parts: the hardware, which consists of components that a person can see and touch, and the software, which makes the computer actually work.

A computer can be made to perform many different kinds of tasks by providing it with different programs, which are detailed sets of instructions telling the computer exactly what to do. Unfortunately, the kinds of instructions that the hardware of a computer can execute are very primitive (or low level); a vast number of these instructions are needed to get a "raw" computer to accomplish even very simple tasks. A modern computer, which can run many programs at the same time and has a mouse-based windowing system, laser printer, large hard disks, network connection, and multimedia capabilities, requires many millions of low-level instructions simply to provide these facilities, even before starting to run an application such as a word processor, computer mail, or video game.

A common set of software that implements support for network interfaces, windowing systems, allocation of computer memory to programs, printer and disk interfaces, and so forth is known as systems software or operating systems. The extensive use of computers and the need for standardization have reduced the number of such systems to only a handful in wide use. *See* OPERATING SYSTEM.

Programs that are written to run on top of an operating system, such as mail readers, word processors, spreadsheets, and games, are termed applications software. An application is a program that is intended to achieve a particular task, whereas operating systems software is intended to provide generic support to applications software. A third class of software is the application programs that are used to help write other programs, such as compilers, debuggers, and program editors.

The computer's hardware is effectively hidden within a layer (the operating-systems or systems-software layer) that provides the illusion of a machine that is more powerful and easy to use. The operating-systems layer is usually not used directly by users, who run applications software which gives the illusion of a collection of highly specialized and powerful machines.

Commonly used applications include word processors, spreadsheets, graphics or drawing tools, tools to manage accounting for both personal and business use, electronic mail, personal information managers, games, and packages for the development of new software. Almost any application written for a windowing system can be expected to have a menu bar, which gives access to the operations provided by the application by selecting items from the menu; and a toolbar, which makes the most frequently used operations immediately accessible to the user.

When writing applications software, the programmer uses the facilities provided by the operating system through a family of libraries. The libraries are packaged collections of functions, each of which provides controlled access to some component of the operating system. For example, if a program needs to create a new window, add an option to a menu, or be notified when the user presses a mouse button, the operating system provides functions which will accomplish these tasks for the program. *See* DIGITAL COMPUTER; PROGRAMMING LANGUAGES. [S.M.Ka.]

Software engineering The process of engineering software, that is, not only the programs that run on a computer, but also the requirements definition, functional specification, design description, program implementation, test methods, and maintenance that lead to this code. Engineering implies the systematic application of scientific and technological knowledge through the medium of sound engineering principles to reach a practical goal. This goal becomes practical through the process of trading off performance, reliability, cost, and other characteristics of the resulting product in the light of the funds and time available for development. In short, software engineering is based on fundamental engineering principles and guided by economic considerations.

Development stages. A programmer tends to program, that is, to write a series of instructions in a programming language, and to give less thought to other stages of what is really, considered as a whole, a design process. On the other hand, in terms of software engineering, the problem-solving steps that have been learned in the other branches of engineering may be applied also here.

In general, these steps include formulating the problem, searching for and developing solutions, evaluating the solutions in terms of the goals, and refining and verifying the solution selected. In each branch of engineering these general steps have been adapted to the needs of the branch. In software engineering there is now considerable agreement on six stages: requirements definition, functional specification, design, implementation, test, and maintenance. As work proceeds through this series of stages, the findings of later stages flow back to influence earlier stages.

Requirements definition. In the first stage, requirements definition, the basic question is what exactly the user expects the software to do. This problem-oriented definition provides the guide to later development stages and a checklist upon which the using and the development organizations can achieve agreement on what must be done. It provides a base against which the resulting software can eventually be compared for acceptance by the user.

Functional specification. At the functional specification stage, the user requirements are converted into a specification of the functions to be performed by the proposed system. While the requirements definition may have set values for some performance objectives that were closely related to the problem circumstances, the functional specification carries the goal-setting process further. Under constraints set by the current capabilities of computer technology and the user's resources, the specification sets target numbers for performance, reliability, compatibility, and other product objectives.

Design. Software design is the process of going from what a system is to do to how it is to do it. In this stage the designer transforms the functional specification into a form from which the programmer can more easily code the problem represented by the specification into a list of programming instructions. While small, simple problems can sometimes be coded directly by a programmer, experience shows that most real problems cannot be coded directly and that an intervening design step is needed.

Design includes the task of partitioning (or decomposing) the entire problem area into subsystems, routines, subroutines, modules, and so forth, which individually become easier to understand than the whole. Together these elements may be thought of as a hierarchy, like an organization chart, or alternatively, a flow diagram, like a strip road map. In either case, a key task of the design process is determining the manner in which these elements relate to one another—their interconnections. Developing the best possible arrangement—best in terms of the constraints imposed by the product specified and development resources—makes design an engineering activity.

Implementation. In its essentials the implementation stage transforms the output documents of the design stage into program instructions. These instructions are converted by a software support program called a compiler or an interpreter into machine code, that is, the binary digits that constitute the internal language of a computer.

In practice, however, if the program is of more than trivial size, the probability of making errors in programming is high. One of the most effective ways of finding these errors is the program review, conducted before the program is compiled.

Errors can occur in the earlier stages, too. In fact, studies of the development process indicate that errors tend to be more frequent in the early stages than in the later ones, and more expensive to correct in later stages than in earlier ones. Thus it is important to review the results at each stage and to eliminate errors as early as possible.

Testing. After the program is compiled, it is tested, that is, a set of values is assigned to the inputs, the program is run, and the program outputs are compared to the correct outputs. Since the test is being done by computer, a large number of input sets may be run. Moreover, the task of comparing the test output with the known correct output can also be performed by computer. When this match fails, the circumstances are reported back to the programmer to find and correct the errors. Even at computer speeds, the test of all combinations and paths might run up into years of test time. So it becomes the task of test personnel to devise test procedures that exercise the program adequately in a reasonable time and at affordable cost.

Maintenance. The term maintenance, adopted from hardware terminology, has a different meaning in software. Software maintenance takes place in response to finding a previously unknown error, or to a change in the environment in which the system functions, or to an opportunity to improve performance. The "repair" does not return the software to its original state, as hardware repair does. It advances the system to a new and sometimes poorly understood state. One result is often the introduction of further errors, leading to the need for further maintenance as soon as they are discovered, and so on. *See* DIGITAL COMPUTER.

Software management. The methods of software engineering find application in business, industry, and government. The problems are often very large. The resulting projects employ many people and substantial resources over long time periods. Personnel, money, and time are the factors with which management works, so that software development becomes subject to the methods of management: planning, organizing, scheduling, directing, monitoring, controlling, and so on. But because software engineering is relatively new, managers lack working experience in it, and software engineers have not had time to grow into management. Even when some of them do, the pace of oncoming technology may outmode them.

If management could confine itself to management and engineers to engineering, the problems would be fewer and less severe. In software, however, there is reason to believe that management decisions, that is, decisions properly within the management sphere, can influence design matters, and vice versa. [W.My.]

Soil Finely divided rock-derived material containing an admixture of organic matter and capable of supporting vegetation. Soils are independent natural bodies, each with a unique morphology resulting from a particular combination of climate, living plants and animals, parent rock materials, relief, the groundwaters, and age. Soils support plants, occupy large portions of the Earth's surface, and have shape, area, breadth, width, and depth. Soil, as used here, differs in meaning from the term as used by engineers, where the meaning is unconsolidated rock material. *See* PEDOLOGY.

Origin and classification. Soil covers most of the land surface as a continuum. Each soil grades into the rock material below and into other soils at its margins, where changes occur in relief, groundwater, vegetation, kinds of rock, or other factors which influence the development of soils. Soils have horizons, or layers, more or less parallel to the surface and differing from those above and below in one or more properties, such as color, texture, structure, consistency, porosity, and reaction (see illustration). The succession of horizons is called the soil profile.

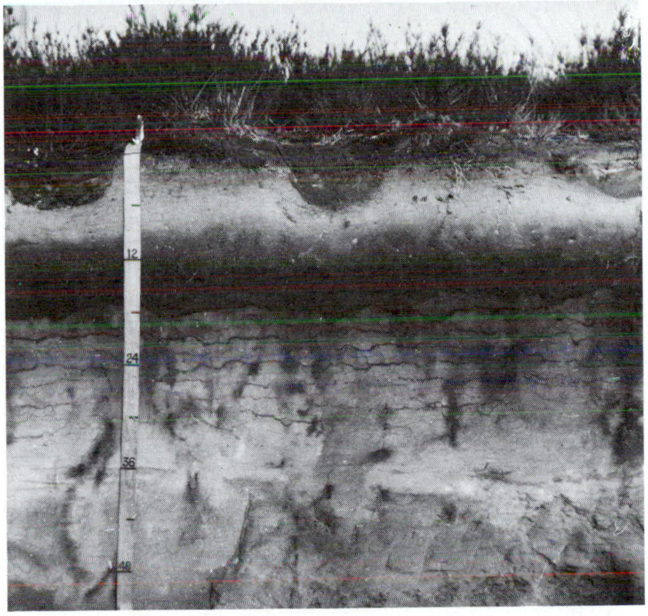

Photograph of a soil profile showing horizons. The dark crescent-shaped spots at the soil surface are the result of plowing. The dark horizon is the principal horizon of accumulation of organic matter that has been washed down from the surface. The thin wavy lines were formed in the same manner. 1 in. = 2.5 cm.

Soil formation proceeds in stages, but these stages may grade indistinctly from one into another. The first stage is the accumulation of unconsolidated rock fragments, the parent material. Parent material may be accumulated by deposition of rock fragments moved by glaciers, wind, gravity, or water, or it may accumulate more or less in place from physical and chemical weathering of hard rocks. The second stage is the formation of horizons. This stage may follow or go on simultaneously with the accumulation of parent material. Soil horizons are a result of dominance of one or more processes over others, producing a layer which differs from the layers above and below. *See* WEATHERING PROCESSES.

Systems of soil classification are influenced by concepts prevalent at the time a system is developed. The earliest classifications were based on relative suitability for different crops, such as rice soils, wheat soils, and vineyard soils. Over the years, many systems of classification have been attempted but none has been found markedly superior. Two bases for classification have been tried. One basis has been the presumed genesis of the soil; climate and native vegetation were given major emphasis. The other basis has been the observable or measurable properties of the soil.

The Soil Survey staff of the U.S. Department of Agriculture and the land-grant colleges adopted the current classification scheme in 1965. This system differs from earlier systems in that it may be applied to either cultivated or virgin soils. Pre-

Soil orders

Order	Formative element in name	General nature of soils
Alfisols	alf	Gray to brown surface horizons, medium to high base supply, with horizons of clay accumulation; usually moist, but may be dry during summer
Aridisols	id	Pedogenic horizons, low in organic matter, and usually dry
Entisols	ent	Pedogenic horizons lacking
Histosols	ist	Organic (peats and mucks)
Inceptisols	ept	Usually moist, with pedogenic horizons of alteration of parent materials but not of illuviation
Mollisols	oll	Nearly black organic-rich surface horizons and high base supply
Oxisols	ox	Residual accumulations of inactive clays, free oxides, kaolin, and quartz; mostly tropical
Spodosols	od	Accumulations of amorphous materials in subsurface horizons
Ultisols	ult	Usually moist, with horizons of clay accumulation and a low supply of bases
Vertisols	ert	High content of swelling clays and wide deep cracks during some season

vious systems have been based on virgin profiles, and cultivated soils were classified on the presumed characteristics or genesis of the virgin soils. The new system has six categories, based on both physical and chemical properties. These categories are the order, suborder, great group, subgroup, family, and series, in decreasing rank. The orders and the general nature of the included soils are given in the table. The suborder narrows the ranges in soil moisture and temperature regimes, kinds of horizons, and composition, according to which of these is most important. The taxa (classes) in the great group category group soils that have the same kinds of horizons in the same sequence and have similar moisture and temperature regimes. The great groups are divided into subgroups that show the central properties of the great group, intergrade subgroups that show properties of more than one great group, and other subgroups for soils with atypical properties that are not characteristic of any great group. The families are defined largely on the basis of physical and mineralogic properties of importance to plant growth. The soil series is a group of soils having horizons similar in differentiating characteristics and arrangement in the soil profile, except for texture of the surface portion, and developed in a particular type of parent material.

Surveys. Soil surveys include those researches necessary (1) to determine the important characteristics of soils, (2) to classify them into defined series and other units, (3) to establish and map the boundaries between kinds of soil, and (4) to correlate and predict adaptability of soils to various crops, grasses, and trees; behavior and productivity of soils under different management systems; and yields of adapted crops on soils under defined sets of management practices. Although the primary purpose of soil surveys has been to aid in agricultural interpretations, many other purposes have become important, ranging from suburban planning, rural zoning, and highway location, to tax assessment and location of pipelines and radio transmitters. This has happened because the soil properties important to the growth of plants are also important to its engineering uses.

Two kinds of soil maps are made. The common map is a detailed soil map, on which soil boundaries are plotted from direct observations throughout the surveyed area. Reconnaissance soil maps are made by plotting soil boundaries from observations made at intervals. The maps show soil and other differences that are of significance for present or foreseeable uses. [G.D.S.]

Physical properties. Physical properties of soil have critical importance to growth of plants and to the stability of cultural structures such as roads and buildings. Such properties commonly are considered to be: size and size distribution of primary particles and of secondary particles, or aggregates, and the consequent size, distribution, quantity, and continuity of pores; the relative stability of the soil matrix against disruptive forces, both natural and cultural; color and textural properties, which affect absorption and radiation of energy; and the conductivity of the soil for water, gases, and heat. These usually would be considered as fixed properties of the soil matrix, but actually some are not fixed because of influence of water content. The additional property, water content—and its inverse, gas content—ordinarily is transient and is not thought of as a property in the same way as the others. However, water is an important constituent, despite its transient nature, and the degree to which it occupies the pore space generally dominates the dynamic properties of soil. Additionally, the properties listed above suggest a macroscopic homogeneity for soil which it may not necessarily have. In a broad sense, a soil may consist of layers or horizons of roughly homogeneous soil materials of various types that impart dynamic properties which are highly dependent upon the nature of the layering. Thus, a discussion of dynamic soil properties must include a description of the intrinsic properties of small increments as well as properties it imparts to the system.

From a physical point of view it is primarily the dynamic properties of soil which affect plant growth and the strength of soil beneath roads and buildings. While these depend upon the chemical and mineralogical properties of particles, particle coatings, and other factors discussed above, water content usually is the dominant factor. Water content depends upon flow and retention properties, so that the relationship between water content and retentive forces associated with the matrix becomes a key physical property of a soil. *See* EROSION; GROUNDWATER HYDROLOGY; SOIL MECHANICS. [W.H.G.]

Soil chemistry The study of the composition and chemical properties of soil. Soil chemistry involves the detailed investigation of the nature of the solid matter from which soil is constituted and of the chemical processes that occur as a result of the action of hydrological, geological, and biological agents on the solid matter. Because of the broad diversity among soil components and the complexity of soil chemical processes, the application of a wide variety of concepts and methods employed in the chemistry of aqueous solutions, of amorphous and crystalline solids, and of solid surfaces is required.

Elemental composition. The elemental composition of soil varies over a wide range, permitting only a few general statements to be made. Those soils that contain less than 12–20% organic carbon are termed mineral. All other soils are termed organic. Carbon, oxygen, hydrogen, nitrogen, phosphorus, and sulfur are the most important constituents of organic soils and of soil organic matter in general. Carbon, oxygen, and hydrogen are most abundant; the content of nitrogen is often about one-tenth that of carbon, while the content of phosphorus or sulfur is usually less than one-fifth that of nitrogen (Table 1).

Table 1. Average percentages of total carbon, total nitrogen, and organic phosphorus in selected soils

Soil	% C	% N	% P
Sand	2.5	.23	.04
Fine sandy loam	3.3	.23	.06
Medium loam	2.3	.22	.05
Clay loam, well drained	4.6	.36	.10
Clay loam, poorly drained	8.0	.43	.05
Peat	46.1	1.32	.03

Table 2. Average amounts of trace elements commonly found in soils and crustal rocks

Trace element	Soil, ppm*	Crustal rocks, ppm
As	6	1.8
B	10	10
Cd	.06	.2
Co	8	25
Cr	100	100
Cu	20	55
Mo	2	1.5
Ni	40	75
Pb	10	13
Se	.2	.05
V	100	135
Zn	50	70

*ppm = parts per million.

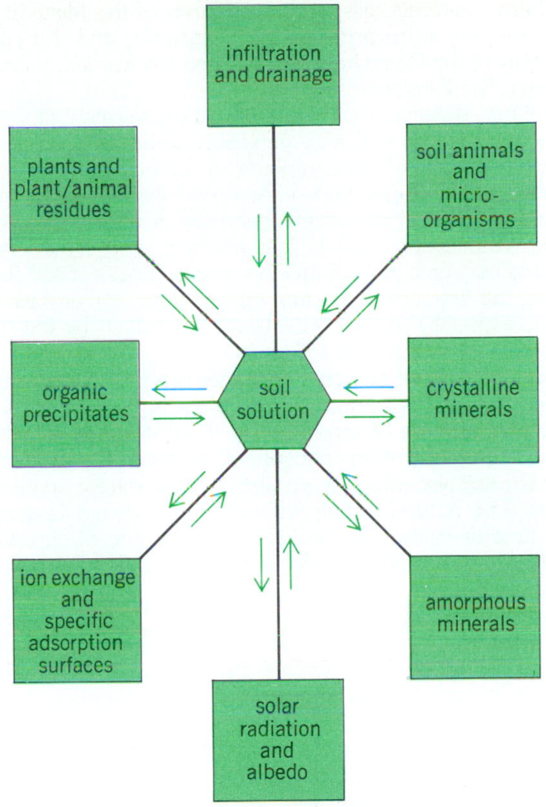

Factors influencing the chemistry of soil solution. (*Modified from J. F. Hodgson, Chemistry of the micronutrients in soil, Adv. Agron., 15:141, 1963*)

Besides oxygen, the most abundant elements found in mineral soils are silicon, aluminum, and iron. The distribution of chemical elements will vary considerably from soil to soil and, in general, will be different in a specific soil from the distribution of elements in the crustal rocks of the Earth. The most important micro or trace elements in soil are boron, copper, manganese, molybdenum, and zinc, since these elements are essential in the nutrition of green plants. Also important are cobalt, selenium, cadmium, and nickel. The average distribution of trace elements in soil is not greatly different from that in crustal rocks (Table 2).

The elemental composition of soil varies with depth below the surface because of pedochemical weathering. The principal processes of this type that result in the removal of chemical elements from a given soil horizon are: (1) soluviation (ordinary dissolution in water), (2) cheluviation (complexation by organic or inorganic ligands), (3) reduction, and (4) suspension. The principal effect of these four processes is the appearance of alluvial horizons in which compounds such as aluminum and iron oxides, aluminosilicates, or calcium carbonate have been precipitated from solution or deposited from suspension. *See* WEATHERING PROCESSES.

Minerals. The minerals in soils are the products of physical, geochemical, and pedochemical weathering. Soil minerals may be either amorphous or crystalline. They may be classified further, approximately, as primary or secondary minerals, depending on whether they are inherited from parent rock or are produced by chemical weathering, respectively.

The bulk of the primary minerals that occur in soil are found in the silicate minerals. Chemical weathering of the silicate minerals is responsible for producing the most important secondary minerals in soil. These are found in the clay fraction and include aluminum and iron hydrous oxides (usually in the form of coatings on other minerals), carbonates, and aluminosilicates. *See* CLAY MINERALS; SILICATE MINERALS.

Ion exchange. A portion of the chemical elements in soil is in the form of cations that are not components of inorganic salts but that can be replaced reversibly by the cations of leaching salt solutions or acids. These cations are said to be exchangeable, and their total quantity is termed the cation exchange capacity (CEC) of the soil. The CEC of a soil generally will vary directly with the amounts of clay and organic matter present and with the distribution of clay minerals.

The stoichiometric exchange of the anions in soil for those in a leaching salt solution is a phenomenon of relatively small importance in the general scheme of anion reactions with soils. Under acid conditions ($pH < 5$) the exposed hydroxyl groups at the edges of the structural sheets or on the surfaces of clay-sized particles become protonated and thereby acquire a positive charge. The degree of protonation is a sensitive function of pH, the ionic strength of the leaching solution, and the nature of the clay-sized particle.

Soil solution. The solution in the pore space of soil acquires its chemical properties through time-varying inputs and outputs of matter and energy that are mediated by the several parts of the hydrologic cycle and by processes originating in the biosphere (see illustration). The soil solution thus is a dynamic and open natural water system whose composition reflects the many reactions that can occur simultaneously between an aqueous solution and an assembly of mineral and organic solid phases that varies with both time and space. *See* SOIL. [G.Sp.]

Soil conservation The practice of arresting and minimizing artificially accelerated soil deterioration. Its importance has grown because cultivation of soils for agricultural production, deforestation and forest cutting, grazing of natural range, and other disturbances of the natural cover and position of the soil have increased greatly in the last 100 years.

Erosion and deterioration. The exact extent of accelerated soil erosion in the world today is not known, particularly as far as the rate of soil movement is concerned. However, it may be said that nearly every semiarid area with cultivation or long-continued grazing, every hill land with moderate to dense settlement in humid temperate and subtropical climates, and all cultivated or grazed hill lands in the Mediterranean climate areas suffer to some degree from such erosion. Recognized problems of erosion are found in such culturally diverse areas as southern China, the Indian plateau, south Australia, the South African native reserves, Russia, Spain, the southeastern and midwestern United States, and Central America.

Within the United States the most critical areas have been the hill lands of the Piedmont and the interior Southeast, the Great Plains, the Palouse area hills of the Pacific Northwest, southern California hills, and slope lands of the Midwest. The high-intensity rainstorms of the Southeast, and the cyclical droughts of the Plains have predisposed the two larger areas to erosion. *See* EROSION.

Soil may deteriorate either by physical movement of soil particles from a given site or by depletion of the water-soluble elements in the soil which contribute to the nourishment of crop plants, grasses, trees, and other economically usable vegetation. The physical movement generally is referred to as erosion. Wind, water, glacial ice, animals, and tools used by humans may be agents of erosion. For purposes of soil conservation, the two most important agents of erosion are wind and water, especially as their effects are intensified by the disturbance of natural cover or soil position.

Depletion of soil nutrients obviously is a part of soil erosion. However, such depletion may take place in the absence of any noticeable amount of erosion. The disappearance of naturally stored nitrogen, potash, phosphate, and some trace elements from the soil also affects the usability of the soil for human purposes. The natural fertility of virgin soils always is depleted over time as cultivation continues, but the rate of depletion is highly dependent on management practices. *See* PLANT MINERAL NUTRITION; SOIL.

Accelerated erosion may be induced by any land use practice which denudes the soil surfaces of vegetative cover. For example, cultivation of any row crop on a slope without soil-conserving practices is an invitation to accelerated erosion. Cultivation of other crops, like the small grains, also may induce accelerated erosion, especially where fields are kept bare between crops to store moisture. Forest cutting, overgrazing, grading for highway use, urban land use, or preparation for other large-scale engineering works also may speed the erosion of soil.

Causes of soil mismanagement. One of the chief causes of erosion-inducing agricultural practices in the United States has been ignorance of their consequences. The cultivation methods of the settlers of western European stock who set the pattern of land use in this country came from a physical environment which was far less susceptible to erosion than North America, because of the mild nature of rainstorms and the prevailing soil textures in Europe. Corn, cotton, and tobacco, moreover, were crops unfamiliar to European agriculture. In eastern North America the combination of European cultivation methods and American interfilled crops resulted in generations of soil mismanagement.

On the Plains and in other susceptible western areas, small grain monoculture, particularly of wheat, encouraged the exposure of the uncovered soil surface so much of the time that water and wind inevitably took their toll. On rangelands, lack of knowledge as to the precipitation cycle and range capacity, and the urge to maximize profits every year contributed to a slower, but equally sure denudation of cover.

Finally, the United States has experienced extensive erosion in mountain areas because of forest mismanagement. Clear-cutting of steep slopes, forest burning for grazing purposes, inadequate fire protection, and shifting cultivation of forest lands have allowed vast quantities of soil to wash out of the slope sites where they could have produced timber and other forest values indefinitely.

Effects on other resources. Accelerated erosion may have consequences which reach far beyond the lands on which the erosion takes place and the community associated with them. During periods of heavy wind erosion, for example, the dust fall may be of economic importance over a wide area beyond that from which the soil cover has been removed. The most pervasive and widespread effects, however, are those associated with water erosion. Removal of upstream cover changes the regimen of streams below the eroding area.

A long chain of other effects also ensues. Because of the extremes of low water in denuded areas during dry seasons, water transportation is made difficult or impossible without regulation, fish and wildlife support is endangered or disappears, the capacity of streams to carry sewage and other wastes safely may be seriously reduced, recreational values are destroyed, and run-of-the-river hydroelectric generation reaches a very low level. *See* WATER CONSERVATION. [E.A.A.; D.J.P.]

Soil degradation Loss in the quality or productivity of soil as a result of human activities. Degradation is attributed to changes in soil nutrient status, loss of soil organic matter, deterioration of soil structure, and toxicity due to accumulations of naturally occurring or anthropogenic materials. The effects of soil degradation include loss of agricultural productivity, negative impacts on the environment and economic stability, and increased clearing of virgin forest and exploitation of marginally suitable land. *See* SOIL.

Soil degradation may impair the function of soil organisms and thus result in further problems. The soil microbial community is essential to cycling of nutrients and decomposition of wastes; thus hindrance of microbial activities may have serious ecological results. Degradative forces such as erosion that result in loss of organic matter may also result in long-term losses of microbial activity. *See* EROSION; SOIL ECOLOGY.

Soil degradation often affects productivity through depletion of plant nutrients. Excessive leaching of cations involved in buffering soil pH may result in soil acidification, changing the solubility, and thus availability, of certain nutrients to plants. Misuse can also result in concentrations of chemicals that are toxic to plants. *See* SOIL CHEMISTRY; SOIL FERTILITY.

Historically, mining of coal and various metal ores has been among the most devastating forms of land use. Mining often exposes large amounts of reduced (decreased oxidation state) minerals in the form of mine tailings and rubble. The acidic mine tailings are extremely difficult to revegetate, and without intervention they remain exposed, continuing to produce acidic drainage until oxidizable material is depleted. Left untreated, surface (strip) mines may require 50–150 years to recover. *See* AUTOXIDATION; SURFACE MINING.

In some cases, compounds are toxic to soil organisms when present at significant concentrations, and thus produce large shifts in microbial community structure. Eventually, microorganisms present in the soil degrade most organic contaminants, thus alleviating toxic effects.

Unlike organic contaminants, which are degraded to nontoxic forms, toxic metals usually become essentially permanent features of soils. Fortunately, adsorption of metals to soil colloids decreases the availability of the metals for movement in the environment or uptake by plants, animals, and microorganisms. *See* AGRICULTURAL CHEMISTRY. [G.K.S.]

Soil ecology The study of the interactions among soil organisms, and between biotic and abiotic aspects of the soil environment. Soil is made up of a multitude of physical, chemical, and biological entities, with many interactions occurring among them. Soil is a variable mixture of broken and weathered minerals and decaying organic matter. Together with the proper amounts of air and water, it supplies, in part, sustenance for plants as well as mechanical support. *See* SOIL.

Abiotic and biotic factors lead to certain chemical changes in the top few decimeters (8–10 in.) of soil. The work of the soil ecologist is made easier by the fact that the surface 10–15 cm (4–6 in.) of the A horizon has the majority of plant roots, microorganisms, and fauna. A majority of the biological-chemical activities occur in this surface layer.

The biological aspects of soil range from major organic inputs, decomposition by primary decomposers (bacteria, fungi, and actinomycetes), and interactions between microorganisms and fauna (secondary decomposers) which feed on them. The detritus decomposition pathway occurs on or within the soil after plant materials (litter, roots, sloughed cells, and soluble compounds) become available through death or senescence. Plant products are used by microorganisms (primary decomposers). These are eaten by the fauna which thus affect flows of nutrients, particularly nitrogen, phosphorus, and sulfur. The immobilization of nutrients into plants or microorganisms and their subsequent mineralization are critical pathways. The labile inorganic pool is the principal one that permits subsequent microorganism and plant existence. Scarcity of some nutrient often limits production. Most importantly, it is the rates of flux into and out of these labile inorganic pools which enable ecosystems to successfully function. *See* ECOLOGICAL INTERACTIONS; ECOLOGY; ECOSYSTEM; GUILD; SYSTEMS ECOLOGY. [D.C.C.]

Soil fertility The ability of a soil to supply plant nutrients. Sixteen chemical elements are required for the growth of all plants: carbon, oxygen, and hydrogen (these three are obtained from carbon dioxide and water), plus the elements nitrogen, phosphorus, potassium, calcium, magnesium, sulfur, iron, manganese, zinc, copper, boron, molybdenum, and chlorine. Some plant species also require one or more of the elements cobalt, sodium, vanadium, and silicon. *See* PLANT MINERAL NUTRITION.

While carbon and oxygen are supplied to plants from carbon dioxide in the air, the other essential elements are supplied primarily by the soil. Of the latter, all except hydrogen from water are called mineral nutrients. Only part of the 13 essential mineral nutrients in soil are in a chemical form that can be immediately used by plants. The unusable (unavailable) parts, which eventually do become available to plants, are of two kind: they may be in organic combination (such as nitrogen in soil humus) or in solid inorganic soil particles (such as potassium in soil clays). The time for complete decomposition and dissolution of these compounds varies widely, from days to hundreds of years.

Soils exhibit a variable ability to supply the mineral nutrients needed by plants. This characteristic allows soils to be classified according to their level of fertility. This can vary from a deficiency to a sufficiency, or even toxicity (too much), of one or more nutrients. A serious deficiency of only one essential nutrient can still greatly reduce crop yields. Several soil properties are important in determining a soil's inherent fertility. One property is the adsorption and storage of nutrients on the surfaces of soil particles. Such adsorption of a number of nutrients is caused by an attraction of positively charged nutrients to negatively charged soil particles. This adsorption is called cation exchange (adsorbed cations can be exchanged with other cations in solution), and the quantity of nutrient cations a soil can adsorb is called its cation-exchange capacity. *See* ADSORPTION; ION EXCHANGE; SOIL CHEMISTRY.

The negative charge in soils is associated with clay particles, but some of the soil's cation-exchange capacity may arise from organic matter (humus) in the soil. Negative charges of organic matter arise largely from carboxylic and phenolic acid functional groups. Since these functional groups are weak acids, the negative charge from organic matter increases as the soil pH increases. The negative charges of soil organic matter can adsorb the same cations as described for the soil clays. The proportion of cation-exchange capacity arising from mineral clays and from organic matter depends on the proportions of each in the soil and on the kinds of clays. In most mineral soils, the soil clays comprise the greater proportion of cation-exchange capacity. Within the class of mineral soils, those soils with more clay and less sand and silt have the greatest cation-exchange capacity.

The amount and kind of acids on the cation exchange sites can have a substantial influence on a soil's perceived fertility. Two factors cause soils to become acid: when crops are harvested, exchangeable bases are removed as part of the crop; and exchangeable bases move with drainage water below the crop's root zone (leaching). Since much of the nitrogen fertilizer supplied to crops contains ammonium nitrogen (this is true of both manure and chemical fertilizers), the addition of high rates of nitrogen also enhances soil acidification. *See* FERTILIZER; NITROGEN CYCLE; SOIL. [D.E.Ki.]

Soil mechanics The study of the response of masses composed of soil, water, and air to imposed loads. Because both water and air are able to move through the soil pores, the discipline also involves the prediction of these transport processes. Soil mechanics provides the analytical tools required for foundation engineering, retaining wall design, highway and railway subbase design, tunneling, earth dam design, mine excavation design, and so on. Because the discipline relates to rock as well as soils, it is also known as geotechnical engineering. *See* ENGINEERING GEOLOGY.

Soil consists of a multiphase aggregation of solid particles, water, and air. This fundamental composition gives rise to unique engineering properties, and the description of the mechanical behavior of soils requires some of the most sophisticated principles of engineering mechanics. The terms multiphase and aggregation both imply unique properties. As a multiphase material, soil exhibits mechanical properties that show the combined attributes of solids, liquids, and gases. Individual soil particles behave as solids, and show relatively little deformation when subjected to either normal or shearing stresses. Water behaves as a liquid, exhibiting little deformation under normal stresses, but deforming greatly when subjected to shear. Being a viscous liquid, however, water exhibits a shear strain rate that is proportional to the shearing stress. Air in the soil behaves as a gas, showing appreciable deformation under both normal and shear stresses. When the three phases are combined to form a soil mass, characteristics that are an outgrowth of the interaction of the phases are manifest. Moreover, the particulate nature of the solid particles contributes other unique attributes. *See* VISCOSITY.

When dry soil is subjected to a compressive normal stress, the volume decreases nonlinearly; that is, the more the soil is compressed, the less compressible the mass becomes. Thus, the more tightly packed the particulate mass becomes, the more it resists compression. The process, however, is only partially reversible, and when the compressive stress is removed the soil does not expand back to its initial state.

When this dry particulate mass is subjected to shear stress, an especially interesting behavior owing to the particulate nature of the soil solids results. If the soil is initially dense (tightly packed), the mass will expand because the particles must roll up and over each other in order for shear deformation to occur. Conversely, if the mass is initially loose, it will compress when subjected to a shear stress. Clearly, there must also exist a specific initial density (the critical density) at which the material will display zero volume change when subjected to shear stress. The term dilatancy has been applied to the relationship between shear stress and volume change in particulate materials. Soil is capable of resisting shear stress up to a certain maximum value. Beyond this value, however, the material undergoes large, uncontrolled shear deformation. *See* SHEAR.

The other limiting case is saturated soil, that is, a soil whose voids are entirely filled with water. When such a mass is initially loose and is subjected to compressive normal stress, it tends to decrease in volume; however, in order for this volume decrease to occur, water must be squeezed from the soil pores. Because water exhibits a viscous resistance to flow in the microscopic pores of fine-grained soils, this process can require considerable time, during which the pore water is under increased pressure. This excess pore pressure is at a minimum near the drainage face of the soil mass and at a maximum near the center of the soil sample. It is this gradient (or change in pore water pressure with change in position within the soil mass) that causes the outflow of water and the corresponding decrease in volume of the soil mass.

Conversely, if an initially dense soil mass is subjected to shear stress, it tends to expand. The expansion, however, may be time-dependent because of the viscous resistance to water being drawn into the soil pores. During this time the pore water will be under decreased pressure. Thus, in saturated soil masses, changes in pore water pressure and time-dependent volume change can be induced by either changes in normal stress or by changes in shear stress. *See* ROCK MECHANICS; SOIL. [C.A.Mo.]

Soil microbiology A study of the microorganisms in soil, their functions, and the effect of their activities on the character of soil and the growth and health of plant life, particularly cultivated crops. It embraces the biology of microorganisms—their morphology, physiology, and taxonomy—as well as their biochemistry. It is related to soil chemistry and physics and, in its application, to agronomy, plant physiology, and plant pathology.

The soil microorganisms are viruses, myxomycetes (slime fungi), protozoa, algae, yeasts, fungi, actinomycetes, and bacteria. Soil is distinguished from subsoil chiefly by the presence of organic matter in the soil. The organic matter is composed of dead plant and animal tissues and the products of their decomposition by microorganisms. The microorganisms derive their energy by oxidizing plant and animal residues. Plants depend upon nutrients made available by microorganisms which form an essential link in the cycle of food in nature.

The soil is the greatest natural reservoir of microorganisms. These take part in many reactions in the soil. Some microorganisms are specific, taking part only in certain reactions (*Nitrosomonas*, which oxidize ammonia to nitrite). Some are more general in their metabolism (heterotrophic types, which depend on organic material for energy source) and take part in many reactions. One group of organisms may use the products of another, for example, during the course of decomposition of complex nitrogenous material to nitrate. In addition to forming simpler degradation products, microorganisms synthesize many complex substances. These are bound up in microbial protoplasm, while excess amounts may be liberated to provide essential substances for other groups of microorganisms. In contrast to this associative action are antagonisms caused by competition for the same food or by the ability of many organisms to produce antibiotics.

Microorganisms are not uniformly distributed throughout the soil, because soil is not homogeneous but comprises a variety of microenvironments. The organisms congregate in colloidal films about the surface of soil particles and are particularly abundant around fragments of decaying plant and animal debris.

The majority of soil microorganisms are most active at pH 6.0–6.8. Some sulfur-oxidizing bacteria tolerate a pH as low as 2.0 and other organisms a pH as high as 10. A supply of free oxygen is important to most organisms. With abundant oxygen, decomposition proceeds rapidly to completion. However, in close-textured and wet soils where anaerobic conditions (molecular oxygen not available) prevail, organic matter is decomposed slowly and incompletely. The number of microorganisms is highest in spring and fall. Proximity to growing plants markedly increases the number of soil organisms. [A.G.L.]

Sol-gel process A chemical synthesis technique for preparing gels, glasses, and ceramic powders. The sol-gel process generally involves the use of metal alkoxides, which undergo hydrolysis and condensation polymerization reactions to give gels.

The sol-gel process comprises solution, gelation, drying, and densification. The preparation of silica glass begins with an appropriate alkoxide, which is mixed with water and a mutual solvent to form a solution. As the hydrolysis and condensation polymerization reactions continue, viscosity increases until the solution ceases to flow. This sol-to-gel transition is irreversible, and there is little if any change in volume. The resultant gels are generally subjected to an aging process spanning hours to days. The gels are kept in sealed containers, and very little solvent loss or shrinkage occurs.

The drying process involves the removal of the liquid phase. As evaporation occurs, drying stresses can cause catastrophic cracking of bulk materials. The ability to dry without cracking serves to limit the sizes of monolithic pieces produced by the sol-gel method. The gel network becomes stiffer, shrinkage stops, and drying stresses arise because of spatial variability in strain rate; that is, gel shrinkage is faster at the exterior surface than in the interior. It is at this point that maximum stress is developed at the surface of the gel and cracking is prone to occur.

The final stage of the sol-gel process is densification. Here the gel-to-glass conversion occurs and the gel achieves the properties of the glass. As the temperature increases, several processes occur, including elimination of residual water and organic substances, relaxation of the gel structure, and ultimately densification. *See* GEL; GLASS.

The production of glasses by the sol-gel method has important scientific and technological implications. For example, the sol-gel approach permits preparation of glasses at far lower temperatures than is possible by using conventional melting. It also makes possible synthesis of compositions that are difficult to obtain by conventional means because of problems associated with volatilization, high melting temperatures, or crystallization. The sol-gel approach is a high-purity process that leads to excellent homogeneity. Finally, the sol-gel approach is adaptable to producing films and fibers as well as bulk pieces, that is, monoliths (solid materials of macroscopic dimensions, at least a few millimeters on a side). *See* MATERIALS SCIENCE AND ENGINEERING; POLYMERIZATION. [B.Du.; M.Ya.]

Solanales An order of flowering plants, division Magnoliophyta (Angiospermae), in the subclass Asteridae of the class Magnoliopsida (dicotyledons). The order (also known as Polemoniales) consists of 8 families with about 5000 species. The Solanaceae, having about 2300 species, are the largest family of the order, followed by the Convolvulaceae (about 1500 species), Hydrophyllaceae and Polemoniaceae (about 300 species each), and Cuscutaceae (about 150 species). In general, the flowers of the Polemoniales are sympetalous, that is, the petals are fused by their margins; they have a regular or nearly regular, usually five-lobed corolla; and the stamens, being equal in number with the petals or corolla lobes, alternate with the lobes. The fruit is usually capsular or softly fleshy, only seldom drupaceous. Except for many members of the Polemoniaceae, the leaves are nearly always alternate.

Petunias (in the Solanaceae), potatoes (*Solanum tuberosum*), sweet potatoes (*Ipomoea batatas*, in the Convolvulaceae), morning glory (*Ipomoea pupurea*), dodder (*Cuscuta*, a nongreen, twining parasite), and phlox (family Polemoniaceae) are familiar members of the Polemoniales. *See* ASTERIDAE; EGGPLANT; FRUIT; MAGNOLIOPSIDA; PAPRIKA; PIMENTO; POTATO, IRISH; POTATO, SWEET; TOBACCO; TOMATO. [A.Cr.]

Solar cell A semiconductor electrical junction device which absorbs and converts the radiant energy of sunlight directly and efficiently into electrical energy. Solar cells may be used individually as light detectors, for example in cameras, or connected in series and parallel to obtain the required values of current and voltage for electric power generation.

Most solar cells are made from single-crystal silicon and have been very expensive for generating electricity, but have found application in space satellites and remote areas where low-cost conventional power sources have been unavailable.

The conversion of sunlight into electrical energy in a solar cell involves three major processes: absorption of the sunlight in the semiconductor material; generation and separation of free positive and negative charges to different regions of the solar cell, creating a voltage in the solar cell; and transfer of these separated charges through electrical terminals to the outside application in the form of electric current.

When light is absorbed in the semiconductor, a negatively charged electron and positively charged hole are created. The heart of the solar cell is the electrical junction which separates these electrons and holes from one another after they are created by the light. An electrical junction may be formed by the contact of: a metal to a semiconductor (this junction is called a Schottky barrier); a liquid to a semiconductor to form a photoelectrochemical cell; or two semiconductor regions (called a *pn* junction).

The fundamental principles of the electrical junction can be illustrated with the silicon *pn* junction. Pure silicon to which a trace amount of a group V element (in the periodic table) such as phosphorus has been added is an *n*-type semiconductor, where electric current is carried by free electrons. Each phosphorus atom contributes one free electron, leaving behind the phosphorus atom bound to the crystal structure with a unit positive charge. Similarly, pure silicon to which a trace amount of a group III element such as boron has been added is a *p*-type semiconductor, where the electric current is carried by free holes. The interface between the *p*- and *n*-type silicon is called the *pn* junction. The fixed charges at the interface due to the bound boron and phosphorus atoms create a permanent dipole charge layer with a high electric field. When photons of light energy from the Sun produce electron-hole pairs near the junction, the built-in electric field forces the holes to the *p* side and the electrons to the *n* side. This displacement of free charges results in a voltage difference between the two regions of the crystal. When a load is connected at the terminals, an electron current flows and useful electrical power is available at the load. *See* SEMICONDUCTOR; SOLAR-ELECTRIC POWER GENERATION; SOLAR ENERGY. [D.G.Sc.]

Solar constant The rate at which energy is received from the Sun just outside Earth's atmosphere. At Earth's mean distance from the Sun, the solar constant is 1.36×10^3 watts per square meter. Depending on the Sun's distance from the zenith, up to a third of this energy may be scattered in Earth's atmosphere. From the measured solar constant, the total radiation of the Sun is 3.86×10^{26} watts. *See* SUN. [J.W.E.]

Solar corona The outermost extension of the atmosphere of the Sun which is normally visible only at a total solar eclipse. The corona consists of gas at a density of about 10^{-15} g/cm^3 (or 10^{-12} atmospheric density) and a normal kinetic temperature between 1 and 2×10^6 K (between 2 and 4×10^6 °F). It is easily observable at an eclipse to a height of 2–3 solar diameters. The outer envelope often consists of beautiful radial streamers composed of electrons and protons ejected by the Sun. *See* SUN. [J.W.E.]

Solar-electric power generation For solar-electric power plants, the annual sunshine hours should be as large as possible, and the humidity, which causes absorption and scattering, should be low. Four methods can be used to generate solar-electric power: the solar-thermal distributed receiver system; the solar-thermal central receiver system; the photovoltaic system; and photoelectrochemical conversion. In each case, the overall system may serve as backup, that is, operating only when the Sun shines, equipped only with a minor energy storage capacity (for example, for 1 h full output) to bridge temporary cloud coverage; alternatively, the overall system may include a conventional fuel system to replace solar energy at night or during cloudy days; and finally, the independent solar-electric system includes a storage system to ensure continuous power-generating capacity based on solar energy only.

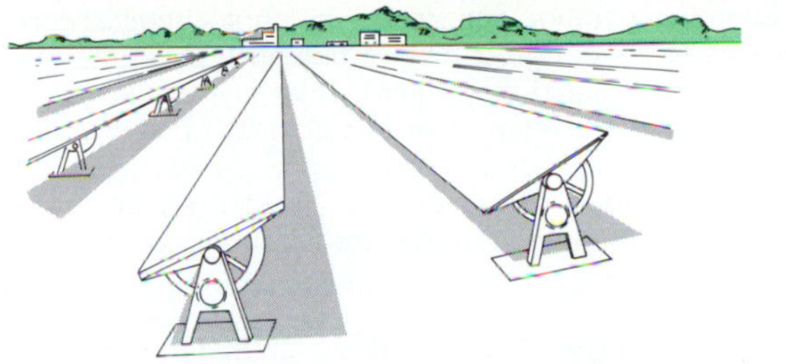

Fig. 1. Solar thermal distributed power station system.

In the solar-thermal distributed system, solar radiation is absorbed over a large area covered with flat-plate (nonconcentrating) collectors or parabolic-trough concentrators, which focus the sunlight on a heat pipe carrying the working fluid (Fig. 1). The flat-plate collector operates at turbine inlet temperatures of 250–500°F (121–260°C). With the parabolic trough, temperatures of 550–1000°F (288–538°C) at turbine inlet are attainable. The higher the temperature, the higher the efficiency and the smaller the land area needed for a given power level, but the more expensive is the system.

In the solar-thermal central receiver system, sunlight is concentrated on a receiver by a large number of mirrors designed to follow the Sun (heliostats). The receiver is a heater located atop a tower (Fig. 2) which is served by a certain collector area. A given solar power plant may consist of an arbitrary number of these modules.

In the photovoltaic systems, electricity is produced by means of solar cells. The photovoltaic system offers the advantage that its performance and, to a large extent, its economy are independent of size over a wide range. *See* SOLAR CELL.

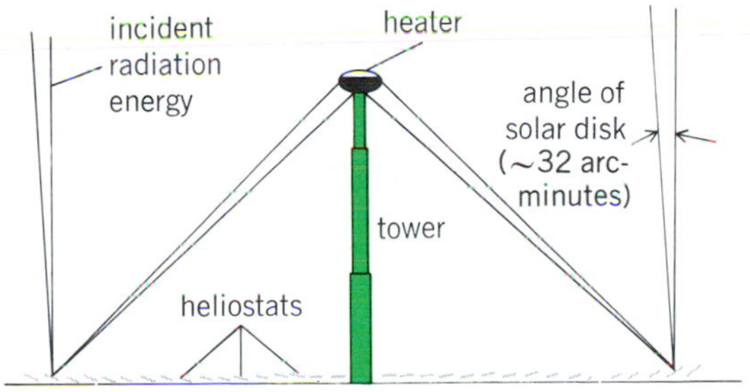

Fig. 2. Solar-thermal central-receiver system. 32 ft = 9.75 m.

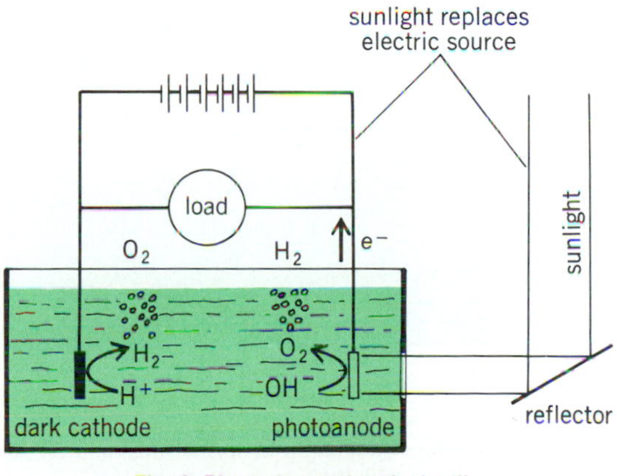

Fig. 3. Photoelectrochemical cell.

Energy generation and storage by photochemical conversion appears to have a great deal of potential. By immersing semiconductors in liquid electrolytes, solar energy can be changed by the same converter either to electricity or to chemical energy by generating chemical fuels, especially hydrogen. In a photoelectrochemical converter (Fig. 3), light replaces a conventional power source in driving the electrolytical decomposition of water. [K.A.E.]

Solar energy The energy transmitted from the Sun. This energy is in the form of electromagnetic radiation. The Earth receives about one-half of one-billionth of the total solar energy output. In 1971, based on radiation measurements in space, the National Aeronautics and Space Administration proposed a new space solar constant of 1353 W/m^2, and a standard spectral irradiance in W/m^2 over a small range of wavelengths (bandwidth) centered at the wavelength (in micrometers) shown in the illustration. Accordingly, the solar radiation energy in the ultraviolet is 105.8 W/m^2 (7.82% of the solar constant), in the visible 640.4 W/m^2 (47.33%), and in the infrared 606.8 W/m^2 (44.85%). The solar radiation energy output is essentially constant. However, because of the ellipticity of the Earth's orbit, the solar constant varies between 1398 W/m^2 about January 3 and 1308 W/m^2 about July 4, or 3% about the mean value. Based on its cross-sectional area, the rotating Earth receives therefore 751 $\times$ 10^{15} kWh annually.

Actually, passage through the atmosphere splits the radiation reaching the surface in a direct and a diffuse component, and reduces the total energy through selective absorption by dry air molecules, dust, water molecules, and thin cloud layers, while heavy cloud coverage eliminates all but the diffuse radiation.

On a global basis, about 50% of the total annual incident radiation is reflected back into space by clouds, 15% by the surface, and about 5.3% is absorbed by bare soil. Of the remaining 29.7%, only about 1.7% is absorbed by marine vegetation and 0.2% by land vegetation. By far the largest portion is used to evaporate water and lift it into the atmosphere. The evaporation energy is radiated into space by vapor condensation to clouds. The solar energy spent to lift the water can be partly recovered in the form of waterpower (hydraulic energy).

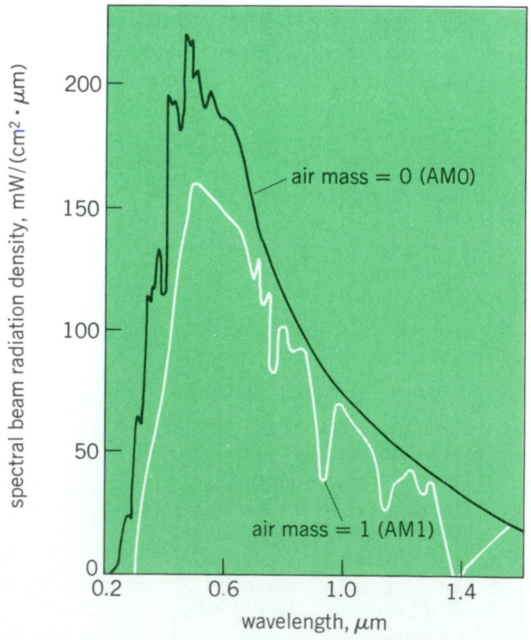

Spectral distribution of sunlight in space (AM0) and on surface of Earth with the Sun at the zenith (AM1).

Solar energy can be utilized in the form of heat, organic chemical energy through photosynthesis, and wind power, and also in the form of photovoltaic power (generating electricity by means of solar cells). The two greatest problems in utilizing solar energy are its low concentration and its irregular availability due to the diurnal cycle and to seasonal and climatic variations. *See* SOLAR-ELECTRIC POWER GENERATION; SOLAR HEATING AND COOLING.

Since solar energy is more time-variant than its use, it is necessary to supply either nonsolar makeup energy or to store solar energy. Solar energy storage systems may be divided into two groups: controlled energy storage and environmental energy storage. Controlled energy storage systems store the energy in as concentrated form as possible in a variety of forms—as thermal, chemical, mechanical, or potential (gravitational) energy; environmental energy storage systems utilize the thermal energy stored by solar irradiation in the ground, air, and water. In contrast to thermal controlled energy storage, environmental energy storage generally is thermal energy storage at temperatures below the user level. *See* ENERGY STORAGE. [K.A.E.]

Solar heating and cooling The use of solar energy to produce heating or cooling for technological purposes. Beneficial uses include distillation of sea water to produce salt or potable water; heating of swimming pools; space heating; heating of water for domestic, commercial, and industrial purposes; cooling by absorption or compression refrigeration; and cooking. *See* SOLAR ENERGY.

Distillation. Production of potable water from sea water by solar distillation is accomplished in several parts of the world by use of glass-roofed solar stills (see illustration). Production of salt from the sea has been accomplished for hundreds of years by trapping ocean water in shallow ponds at high tide and simply allowing the water to evaporate under the influence of the Sun. *See* DISTILLATION.

Swimming pool heating. Swimming pool heating is a moderate-temperature application which, under suitable weather conditions, can be accomplished with a simple unglazed and uninsulated collector. For applications where a significant temperature difference exists between the fluid within the collector passages and the ambient air, both glazing and insulation are essential.

Space heating. Space heating can be carried out by active systems which use separate collection, distribution, and storage subsystems, or by passive designs which use components of a building to admit, store, and distribute the heat resulting from absorbing the incoming solar radiation within the building itself.

Passive systems can be classified as direct-gain when they admit solar radiant energy directly into the structure through large south-facing windows, or as indirect-gain when a wall or a roof absorbs the solar radiation, stores the resulting heat, and then transfers it into the building. Passive systems are generally effective where the number of hours of sunshine during the

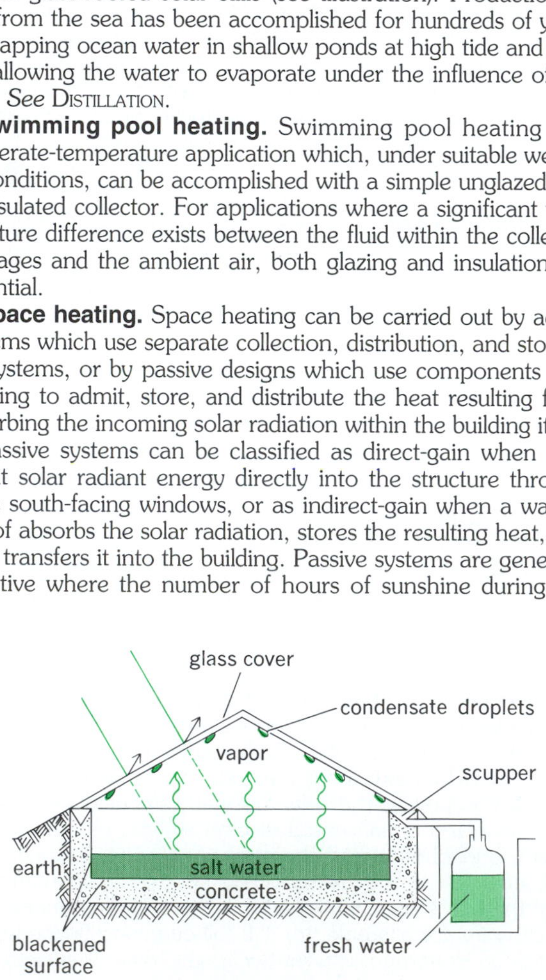

Roof-type solar still.

winter months is relatively high, where moderate indoor temperature fluctuations can be tolerated, and where the need for summer cooling and dehumidification is moderate or nonexistent.

Active systems may use either water or air to transport heat from roof-mounted south-facing collectors to storage in rock beds or water tanks. The stored heat may be withdrawn and used directly when air is the transfer fluid. When the heat is collected and stored as hot water, fan-coil units are generally used to transfer the heat to air which is then circulated through the warmed space. Standby energy sources are included in designs for active systems, since some method of providing warmth must be included for use when the Sun's radiant energy is inadequate for long periods of time. The standby heater may be something as simple as a wood-burning stove or fireplace, or as complex as an electrically powered heat pump. *See* COMFORT HEATING; HEAT PUMP.

Service water heating. Solar water heating for domestic, commercial, or industrial purposes is an old and successful application of solar-thermal technology. The most widely used water heater, and one that is suitable for use in relatively warm climates where freezing is a minor problem, is the thermosiphon type. A flat-plate collector is generally used with a storage tank which is mounted above the collector. A source of water is connected near the bottom of the tank, and the hot water outlet is connected to its top. A downcomer pipe leads from the bottom of the tank to the inlet of the collector, and an insulated return line runs from the top of the collector to the upper part of the storage tank which is also insulated.

The system is filled with water, and when the Sun shines on the collector, the water in the tubes is heated. It then becomes less dense than the water in the downcomer, and the heated water rises by thermosiphon action into the storage tank. It is replaced by cool water from the bottom of the tank, and this action continues as long as the Sun shines on the collector with adequate intensity.

For applications where the elevated storage tank is undesirable or where very large quantities of hot water are needed, the tank is placed at ground level. A small pump circulates the water in response to a signal from a controller which senses the temperatures of the collector and the water near the bottom of the tank. Heat exchangers may also be used with water at operating pressure within the tubes of the exchanger and the collector water outside to eliminate the necessity of using high-pressure collectors. *See* HOT-WATER HEATING SYSTEM.

Cooling. Cooling can be provided by both active and passive systems.

The two feasible types of active cooling systems are Rankine cycle and absorption. The Rankine cycle system uses solar collectors to produce a vapor (steam or one of the fluorocarbons generally known as Freon) to drive an engine or turbine. A condenser must be used to condense the spent vapor so it can be pumped back through the vaporizer. The engine or turbine drives a conventional refrigeration compressor which produces cooling in the usual manner. *See* RANKINE CYCLE; REFRIGERATION.

Passive cooling systems make use of three natural processes: convection cooling with night air; radiative cooling by heat rejection to the sky on clear nights; and evaporative cooling from water surfaces exposed to the atmosphere. The effectiveness of each of these processes depends upon local climatic conditions. *See* ENERGY STORAGE; SOLAR CELL. [J.I.Y.]

Solar magnetic field The magnetic field that pervades the ionized and highly conducting gas composing the Sun. The field is observable by optical means at the Sun's surface; subsurface fields are inferred from the laws of magnetohydrodynamics and from the behavior of sunspots; the effects of magnetic fields external to the Sun are seen in solar prominences, flares, and coronal streamers, and in relation to the solar wind, an outflow of tenuous plasma into interplanetary space. *See* SOLAR WIND; SUN.

Direct measurement of the strength and polarity of the complex patterns of magnetic fields at the Sun's surface is accomplished by means of an instrument known as the solar magnetograph that utilizes the Zeeman effect. *See* ZEEMAN EFFECT. [H.W.B.]

Solar neutrinos Neutrinos produced in nuclear reactions inside the Sun. The first direct test of how the Sun produces its luminosity (observed most conspicuously on Earth as sunlight) was carried out by observing these particles. The results of this experiment were in disagreement with theoretical predictions based upon the supposedly well-established theory of stellar evolution. This discrepancy between theory and observation suggests that either the process by which the Sun shines is not understood as well as previously believed, or some modification in the classical physical theory of neutrinos is required. *See* NEUTRINO; NUCLEAR REACTION; PROTON-PROTON CHAIN. [J.N.B.]

Solar radiation The electromagnetic radiation and particles (electrons, protons, and rarer heavy atomic nuclei) emitted by the Sun. Electromagnetic energy has been observed over the whole spectrum with wavelengths varying from 0.01 nanometer to 30 kilometers. The bulk of the energy is in the spectrum of visible light (400–800 nm). The solar spectrum doubtless extends far beyond the observed limits in both directions. *See* SOLAR ENERGY.

The Sun also emits a continuous stream of electrons with shorter bursts of electron and proton showers sufficiently intense to affect the ionization of the upper terrestrial atmosphere. *See* IONOSPHERE; SOLAR WIND; SUN. [J.W.E.]

Solar system The Sun and the bodies moving about it. There are more than 10^8 stars in the Galaxy, but only in the case of the solar system is a star with an accompanying planetary system known. The other stars are all so far away that there has been no means of detecting any possible accompanying planets. The Sun is about 3×10^7 light-years (3×10^{20} km) from the center of the Galaxy, part of the flattened distribution of stars forming the disk of the Galaxy. The Galaxy also has a substantial spheroidal content of stars which are believed to have been the first to form when the Galaxy was formed. The disk stars, such as the Sun, are believed to have formed later, and many examples of the process of star formation are observed continuing within the galactic disk today. By observing how such star formation occurs, it may be possible to gain clues as to the way in which the Sun, and possibly its accompanying planetary system, was formed. *See* MILKY WAY GALAXY; STELLAR EVOLUTION; SUN.

The planetary system contains an inner group of four planets, Mercury, Venus, Earth, and Mars, which are primarily composed of rocks and iron. A large number of very small bodies, typically only a few kilometers in radius or less, called asteroids, occur in orbits beyond that of Mars. These too are primarily rocky in composition. *See* ASTEROID; EARTH; MARS; MERCURY (PLANET); VENUS.

Beyond the asteroids lies the largest planet in the solar system, Jupiter. Jupiter is primarily composed of the light gases hydrogen and helium. Saturn, the second most massive planet in the solar system, lying beyond Jupiter, is basically similar in composition and structure. *See* JUPITER; SATURN.

Beyond Saturn lie two intermediate-mass planets, Uranus and Neptune. While these planets contain substantial amounts of hydrogen and helium, the majority of their mass appears to be composed of heavier elements, probably some combination of rocky materials plus "ices," water, ammonia, and methane. *See* NEPTUNE; URANUS.

Still farther out is a very small planet, Pluto, considerably less massive than the Earth and apparently composed primarily of rocks and frozen ices. *See* PLANET; PLUTO.

Well beyond the region of the planets lies a huge region of space, out to about 100,000 astronomical units (1.5×10^3 km) from the Sun, which appears to contain a very large number of small, frozen bodies called comets. The space within the solar system is also filled with much smaller bodies. Some of these weigh many kilograms. These are the meteoroids, which become bright meteors or fireballs after entering the Earth's atmosphere, and are called meteorites when they are found on the ground. *See* COMET; METEOR; METEORITE.

There is a long history of speculation about the origin of the solar system. These theories can be classed as monistic, in which usually the Sun and planets are thought to have formed as an isolated system, and dualistic, in which the planets are generally thought to have been formed as a result of an interaction between two stars. It is now generally believed that some form of monistic theory is required to understand the origin of the solar system. [A.G.W.C.]

Solar wind The continuous outward flow of ionized solar gas and a "frozen-in" remnant of the solar magnetic field through the solar system. This flow arises from strong outward pressure in the solar corona, becomes supersonic at a few solar radii (1 solar radius = 6.96×10^5 km or 4.32×10^5 mi) above the visible surface of the Sun (the photosphere), and attains speeds in the range 250–750 km/s (155–465 mi/s) in interplanetary space. The solar wind is believed to remain supersonic out to a distance from the Sun of 50–100 astronomical units (AU), where it is slowed by interaction with the interstellar gas and magnetic field. *See* INTERSTELLAR MATTER; SOLAR MAGNETIC FIELD.

In 1962, sampling instruments on the *Mariner 2* spacecraft revealed an important pattern in the variations of solar wind speed with time. The observed speed rose systematically from low values to high values in 1 or 2 days and then returned to low values during the next 3 to 5 days. Each of these high-speed streams tended to be seen at approximately 27-day intervals or to recur with the rotation period of the Sun as viewed from the spacecraft. A similar recurrence tendency had been noted in geomagnetic activity, and was widely interpreted as the effect of localized, long-lived streams of particles emitted from the Sun and swept past the Earth once during each solar rotation. The high-speed solar wind streams have been linked to recurrent geomagnetic activity and found to be prominent features of the solar wind much of the time since 1962. *See* GEOMAGNETISM; MAGNETOSPHERE.

X-ray and ultraviolet images of the Sun have brought attention to the features known as coronal holes. These are regions of abnormally low corona] density that appear dim in any radiation emitted (x-rays and the ultraviolet) or scattered (the white-light corona observed at eclipse) from the corona. Detailed examination of these images suggests that the holes are regions where the solar magnetic field is open, with field lines reaching from their visible roots in the photosphere out into interplanetary space. In contrast, the structure seen in the bright or dense corona suggests closed magnetic fields, with field lines connecting separated locations in the photosphere. *See* SOLAR CORONA.

The largest and longest-lived coronal holes have been observed to occur in the polar regions of the Sun (as defined by its rotation axis); conspicuous polar holes existed for at least 8 years during the last 11-year sunspot cycle. These polar holes are related to the weak, dipolelike, general magnetic field of the Sun and are thus of opposite polarity in the two hemispheres. *See* SUN. [A.J.Hu.]

Soldering The joining of metals by causing a lower-melting-point metal to wet or alloy with the joint surfaces and then freeze in place. A solder is defined as a joining material that melts below 427°C (800°F); brazing alloys melt above this temperature. Solders are used to establish reliable electrical connections, to make a liquid- or gas-tight joint, and to hold parts together physically. The usual practice is to rivet, crimp, or otherwise support the load and to seal the space with solder.

The most commonly used solders are alloys of tin and lead that melt below the melting point of tin. Antimony, bismuth, cadmium, silver, and arsenic are sometimes added to improve strength, wetting qualities, or grain size, or to produce alloys having desired melting ranges.

In order that surfaces will accept solder readily, they and the solder must be free from oxide or other obstructing films. When necessary, parts are cleaned chemically or by abrasion. Also, readily solderable coatings such as gold, silver, or tin may be applied. Even so, fluxes are usually used to aid the soldering process. Rosin, a pine product, is an effective and practically harmless flux used widely for electrical connections. When less stringent requirements exist and when less carefully prepared surfaces are to be soldered, rosin is mixed with chemically active agents that aid materially in soldering. The rosin-type fluxes may be incorporated as the core of wire solders or dissolved in various solvents for direct application to joints prior to soldering. Inorganic salts, such as zinc chloride and ammonium chloride, are widely used where stronger fluxes are needed.

In applying solders, joints are heated by soldering irons, torches, induction heaters, or furnaces, or by immersion in molten solder. In the first two instances, solder is fed to the joints by hand. In the third and fourth, preformed shapes are placed close to the joints before fluxing and heating. *See* BRAZING. [G.M.B.]

Solenoid (electricity) An electrically energized coil of insulated wire which produces a magnetic field within the coil. If the magnetic field produced by the coil is used to magnetize and thus attract a plunger or armature to a position within the coil, the device may be considered to be a special form of electromagnet and in this sense the words solenoid and electromagnet are synonymous. In a wider scientific sense the sole-

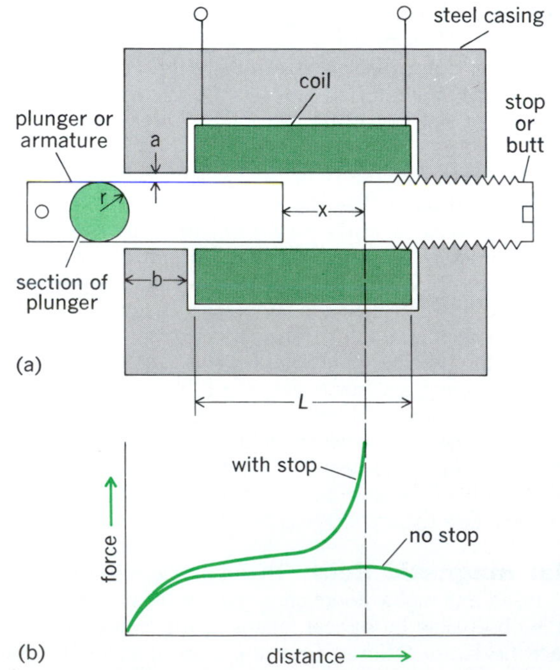

Steel-clad solenoid. (*a*) Cross-sectional view. (*b*) Relation of the force acting on the armature to the displacement of the armature.

noid may be used to produce a uniform magnetic field for various investigations. So long as the length of the coil is much greater than its diameter (20 or more times), the magnetic field at the center of the coil is sensibly uniform, and the field intensity is almost exactly that given by the equation for a solenoid of infinite length.

When used as an electromagnet of the plunger type, the solenoid usually has an iron or steel casing. The casing increases the mechanical force on the plunger and also serves to constrain the magnetic field. The addition of a butt or stop at one end of the solenoid greatly increases the force on the plunger when the distance between the plunger and the stop is small. The illustration shows a steel-clad solenoid with plunger and plunger stop. The relation of force versus distance with and without the stop is also shown. *See* ELECTROMAGNET. [J.Mei.]

Solenoid (meteorology) In meteorological usage, solenoids are hypothetical tubes formed in space by the intersection of a set of surfaces of constant pressure (isobaric surfaces) and a set of surfaces of constant specific volume of air (esoteric surfaces). The isobaric and isosteric surfaces are such that the values of pressure and specific volume, respectively, change by one unit from one surface to the next. The state of the atmosphere is said to be barotropic when there are no solenoids, that is, when isobaric and isosteric surfaces coincide. The number of solenoids cut by any plane surface element of unit area is a measure of the torque exerted by the pressure gradient force, tending to accelerate the circulation of air around the boundary of the area. *See* BAROCLINIC FIELD; BAROTROPIC FIELD; ISOPYCNIC. [F.S.]

Solid (geometry) A geometric solid, usually called a solid, is a set of points forming a finite portion of three-dimensional space, continuously joined together, and separated from the rest of space by a set of points called its boundary. An interior point of the solid is a point *I* such that all points of space within a sufficiently small positive distance *d* from *I* are points of the solid. A boundary point is a point *B* such that, no matter how small a positive distance *d* is chosen, there are points of the solid and points not of the solid within distance *d* of *B*. A plane section of the solid consists of the points common to the solid and to a given plane that intersects the solid. A bounding section contains boundary points but no interior points of the solid. Any two bounding sections cut by parallel planes may be called bases of the solid. The distance between these planes is then called the altitude, and the section by a plane halfway between the bases is called a midsection of the solid. A solid is called convex if every line segment joining a pair of its points belongs to the solid. *See* POLYHEDRON; SURFACE AND SOLID OF REVOLUTION. [J.S.F.]

Solid solution The crystalline state, where at least one atomic position may accommodate more than one atomic species. In mineralogy and metallurgy, solid solution is a frequent and important phenomenon. In metallurgy, examples are the systems Au-Ag-Hg and Fe-Ni. In mineralogy, the most important examples are the systems $CaO—MgO—CO_2$, $Na_2O—CaO—Al_2O_3—SiO_2$, and $MgO—Al_2O_3—SiO_2$.

It is easiest to conceptualize the phenomenon by defining structure type, mineral species, and end-member composition. The structure type is a design of distinct-symmetry independent atomic positions possessing an asymmetric unit of atomic positions which must be specified along with the unit cell properties, including axial parameters, crystal system, and space group. Mineral species is defined on the basis of structure type and end-member composition, where each end member in principle is a distinct species. *See* CRYSTAL; CRYSTALLOGRAPHY; MINERALOGY. [P.B.M.]

Solid-state battery A battery in which both the electrodes and the electrolyte are solid-state materials. Solid-state batteries have been developed as a logical application for a relatively new class of materials, the solid electrolytes, otherwise known as superionic conductors. These materials are very good conductors for ions, but they maintain a very high resistance toward electronic conduction. These properties are necessary for any electrolyte; the high ionic conductivity minimizes the internal resistance of the battery, while the high electronic resistance minimizes its self-discharge rate. Typical materials are β-Al_2O_3 for Na^+ conduction, Ag_4RbI_5 for Ag^+ conduction, and Li-β-Al_2O_3 for Li^+ conduction. The polycrystalline electrolytes typically employed in batteries have ionic conductivities about one-fifth that of the corresponding single crystals.

Solid-state batteries generally fall into the low-energy-density category. This limitation arises from the polarization, at high current densities, at the solid-solid interfaces. However, in certain applications the inherent advantages of solid-state batteries are more important than the net energy content or the current output. From a manufacturing point of view, they are easier to miniaturize, and there is no problem with the electrolyte leaking out of the container. They have long shelf life and are resistant to shock and vibration. They have a wider operating temperature range, and there is no abrupt decrease in performance such as that which occurs when a liquid electrolyte freezes or boils. *See* BATTERY; LITHIUM PRIMARY CELL. [G.G.B.]

Solid-state chemistry The science of the elementary, atomic compositions of solids and the transformations that occur in and between solids and between solids and other phases to produce solids. Solid-state chemistry deals primarily with those microscopic features which are uniquely characteristic of solids and which are the causes for the macroscopic chemical properties and the chemical reactions of solids. As with other branches of the physical sciences, solid-state chemistry also includes related areas that furnish concepts and knowledge essential to understandings and explanations of those phenomena which are more characteristic of the subject itself.

The overlap of solid-state chemistry and solid-state physics is extensive. In general, solid-state physics treats properties, such as energy and entropy, which are continuously variable in the solid, whereas solid-state chemistry concerns those properties which are discontinuous because of chemical reactions; in a sense, by definition the properties are chemical because they are discontinuous. In another perspective, solid-state chemistry tends to be based on structure in configuration space, whereas solid-state physics tends to be based on momentum space. *See* SOLID-STATE PHYSICS. [R.J.T.]

Solid-state physics The study of the physical properties of solids, such as electrical, dielectric, elastic, and thermal properties, and their understanding in terms of fundamental physical laws. Most problems in solid-state physics would be called solid-state chemistry if studied by scientists with chemical training, and vice versa. Solid-state physics emphasizes the properties common to large classes of compounds rather than the dependence of properties upon compositions, the latter receiving greater emphasis in solid-state chemistry. In addition, solid-state chemistry tends to be more descriptive, while solid-state physics focuses upon quantitative relationships between properties and the underlying electronic structure. *See* SOLID-STATE CHEMISTRY.

Many of the scientists who study the physics of liquids identify with solid-state physics, and the term "condensed-matter physics" has been used by some researchers to replace "solid-state physics" as a division of physics. It includes noncrystalline solids such as glass as well as crystalline solids. *See* AMORPHOUS SOLID; GLASS.

In solid-state physics it is generally assumed that the electronic states can be described as wavelike. The individual electronic states, called Bloch states, have energies which depend upon the wave number (a vector equal to the momentum divided by $\hbar$, which is Planck's constant divided by 2π), and the wave number is restricted to a domain called the Brillouin zone. This energy given as a function of the wave number is called the band structure. There are several curves, called bands, for each line in the Brillouin zone. *See* BRILLOUIN ZONE.

The total energy of a solid includes a sum of the energies of the occupied electronic states. Since the energy bands depend upon the positions of the atoms, so does the total energy, and the stable crystal structure is that which minimizes this energy. The theory has not proved adequate to really predict the crystal structure of various solids, but it is possible to predict the changes in energy under various distortions of the lattice. There are in fact three times as many independent distortions, called normal modes, as there are atoms in the solid. Each has a wave number, and the frequencies of the normal vibrational modes, as a function of wave number in the Brillouin zone, form vibrational bands in direct analogy with the electronic energy bands. These can be directly calculated from quantum theory or measured by using neutron or x-ray diffraction. *See* CRYSTAL; LATTICE VIBRATIONS; NEUTRON DIFFRACTION; X-RAY DIFFRACTION. [W.A.H.]

Solids pump A device used to move solids upward through a chamber or conduit. It is able to overcome the large dynamic forces at the base of a solids bed and cause the entire bed to move upward.

Solids pumps are used to cause motion of solids in process-type equipment in which treatment of solids under special conditions of temperature, oxidation, and reduction can be combined with upward motion and discharge of the spent solids overhead from the reacting vessel. The solids pump has found its principal application in the operation of oil-shale retorts.

Solids pumps are inherently of the positive displacement type. One practical method uses a reciprocating piston mounted on a trunnion permitting it to swing into an inclined position for filling and then to swing back into vertical position for discharge. The illustration shows a mechanically driven solids pump in four positions through its cycle of operation. *See* BULK-HANDLING MACHINES. [C.Be.]

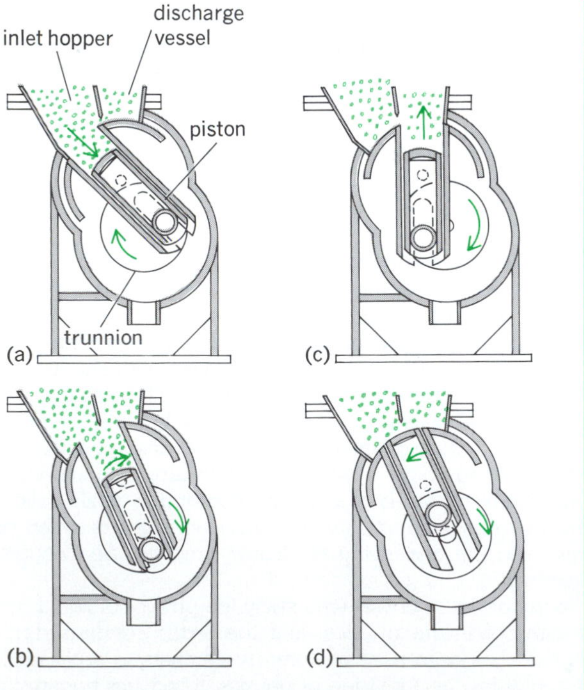

Operation cycle of mechanically driven solids pump. (*a*) Filling with solids from inlet hopper. (*b*) Piston rotating on a trunnion toward its discharge position. (c) The discharge position, with piston pushing charge of solids upward. (*d*) Piston rotating back toward original filling position.

Solifugae An order of nonvenomous, spiderlike predatory arachnids found chiefly in arid and semiarid, tropical, and warm-temperate regions. They are also known as sun spiders. The relatively large anterior appendages, or chelicerae, are used for holding and crushing prey. The sun spiders are agile and usually stalk their prey during the night. A fossil form is known from Pennsylvanian time. *See* ARACHNIDA. [C.C.Ho.]

Soliton An isolated wave which propagates without dispersing its energy over larger and larger regions of space. In most of the scientific literature, the requirement that two solitons emerge unchanged from a collision is also added to the definition; otherwise one speaks of a solitary wave.

There are many equations of mathematical physics which have solutions of the soliton type. Correspondingly, the phenomena which they describe, be it the motion of waves in shallow water or in an ionized plasma, exhibit solitions. The first observation of this kind of wave was made in 1834 by John Scott Russell, who followed on horseback a soliton propagating in the windings of a channel. In 1895, D. J. Korteweg and H. de Vries proposed an equation for the motion of waves in shallow waters which possesses soliton solutions, and thus established a mathematical basis for the study of the phenomenon. Interest in the subject, however, lay dormant for many years, and the major body of investigations began only in the 1950s. Researches done by analytical methods and by numerical methods made possible with the advent of computers gradually led to a complete understanding of solitons. *See* WAVE MOTION.

Eventually, the fact that solitons exhibit particlelike properties, because the energy is at any instant confined to a limited region of space, received attention, and solitons were proposed as models for elementary particles. However, it is difficult to account for all of the properties of known particles in terms of solitons. It has been realized that some of the quantum fields which are used to describe particles and their interactions also have solutions of the soliton type. The solitons would then appear as additional particles, and may have escaped experimental detection because their masses are much larger than those of known particles. In this context the requirement that solitons emerge unchanged from a collision has been found too restrictive, and particle theorists have used the term soliton where traditionally one would speak of a solitary wave. *See* ELEMENTARY PARTICLE; QUANTUM FIELD THEORY. [C.R.]

Solstice The two days during the year when the Earth is so located in its orbit that the inclination (about $23^1/_2°$, or 23.45°) of the polar axis is toward the Sun. This occurs on

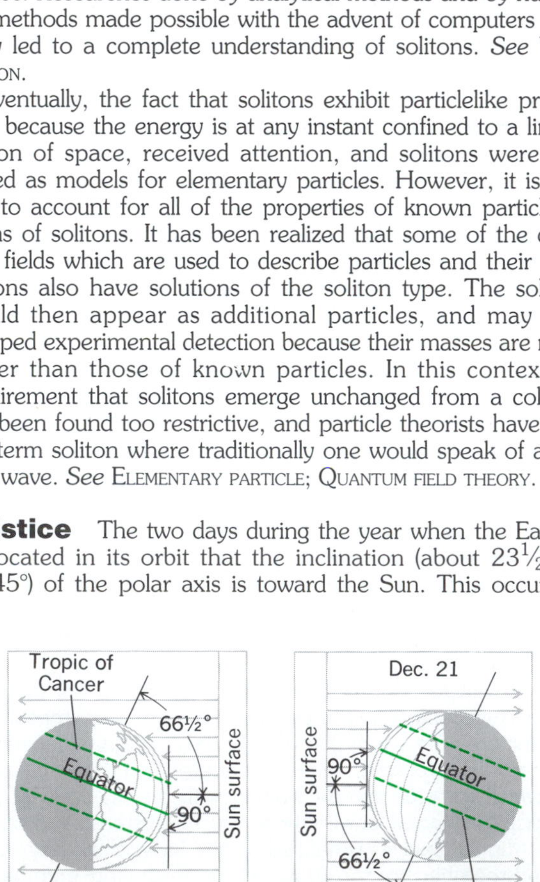

The Earth at the time of the summer and winter solstices. The dates may vary because of the extra one-fourth day in the year.

June 21, called the summer solstice, when the North Pole is tilted toward the Sun; and on December 22, called the winter solstice, when the South Pole is tilted toward the Sun (see illustration). The adjectives summer and winter, used above, refer to the Northern Hemisphere; seasons are reversed in the Southern Hemisphere.

At the time of the summer solstice every place north of the Arctic Circle will have 24 h of sunlight and the length of day at all places north of the Equator will be more than 12 h, increasing in length with increasing latitude. Identical conditions are found in the Southern Hemisphere at the time of the Northern Hemisphere's winter solstice. *See* MATHEMATICAL GEOGRAPHY.

[V.H.E.]

Solubility product constant A special type of simplified equilibrium constant (symbol K_{sp}) defined for, and useful for, equilibria between solids (s) and their respective ions in solution, for example, reaction (1). For this relatively simple equilibrium, Eqs. (2) and (3) apply.

$$AgCl(s) \rightleftharpoons Ag^+ + Cl^- \quad (1)$$

$$[Ag^+][Cl^-] \cong K_{sp} \quad (2)$$

$$(Ag^+)(Cl^-)/(AgCl) = K_{sp} \quad (3)$$

It can be demonstrated experimentally that a small increase in the molar concentration of chloride ion $[Cl^-]$ causes a reduction in the concentration of silver present as Ag^+. Similarly, an increase in $[Ag^+]$ reduces $[Cl^-]$. The product of the two concentrations is approximately constant as indicated by Eq. (2) and equal to the K_{sp} of Eq. (3). Equation (3) is exact since the variables are activities instead of concentrations. In accordance with the choice of standard state usually made for a solid, the activity of solid AgCl is unity, hence Eq. (4) holds.

$$(Ag^+)(Cl^-) = K_{sp} = 1.8 \times 10^{-10}\ \text{mole}^2\ \text{liter}^{-2} \quad (4)$$

In practice, various complications arise: addition of too much of either ion produces more complicated ions and hence actually increases the apparent concentration of the other ion. Addition of a salt without a common ion (that is, a salt supplying neither Ag^+ nor Cl^-) either may react with Ag^+ or Cl^- or may merely increase the concentration of both ions by a lowering of the mean ionic activity coefficient. *See* IONIC EQUILIBRIUM; PRECIPITATION (CHEMISTRY).

[T.F.Y.]

Solubilizing of samples The process by which samples that do not dissolve easily are converted into different chemical compounds which are soluble. The sample may be heated in air to evolve volatile components or to oxidize a component to a volatile higher oxidation state with the formation of an acid-soluble form, as in the roasting of a sulfide to form the oxide and sulfur dioxide. Most frequently, the sample is treated with a solvent which reacts with one or more constituents of the sample. The choice of solvent is determined by the chemical reactions which are required. Reactions used include solvation, neutralization, complex formation, metathesis, displacement, oxidation-reduction, or combinations of these.

Many substances do not dissolve at temperatures obtainable in the presence of liquid water. Fused salt reactions employing temperatures of 750–2030°F (400–1100°C) are necessary for the attack and decomposition of many types of samples. The material used as the solvent is called a flux, and the process of melting the mixture of dry, solid flux with the sample is called a fusion.

[C.L.R.]

Solution A homogeneous mixture of two or more components whose properties vary continuously with varying proportions of the components. A liquid solution can be distinguished experimentally from a pure liquid by the fact that during transfers into other single phases at equilibrium (freezing and vaporizing at constant pressure) the temperature and other properties vary continuously, whereas those of a pure liquid remain constant. For an apparent exception *see* AZEOTROPIC MIXTURE. *See also* SOLVENT.

Gases, unless highly compressed, are mutually soluble in all proportions.

A solid solution is, similarly, a single phase whose composition and other properties vary continuously with changing composition of the liquid phase with which it is in equilibrium. *See* SOLID SOLUTION.

The extent to which substances can form solutions depends upon the kind and strength of the attractive forces between the several molecular species involved. It is necessary to consider the attractive forces exerted by molecules of the following types: (1) nonpolar molecules; (2) polar molecules, that is, those containing electric dipoles; (3) ions; and (4) metallic atoms.

Actual solutions may be considered in terms of their departure from a simple idealized model—a mixture of components having the same attractive fields, which mix without change in volume or heat content. This is analogous to an ideal gas mixture, which is formed with no heat of mixing and in which the total pressure is the sum of the partial pressures. In such a solution the escaping tendency of the individual molecules is the same, whether they are surrounded by similar or by different molecules.

[J.H.Hi.]

Solution mining The extraction of soluble mineral values from a variety of ores using appropriate solvents or chemicals; also known as borehole, nonshaft, or chemical mining. Originally the term solution mining applied to a dissolution, washing, or "brining" of soluble salts such as halite (common table salt), potash, or trona. After the mid-1970s the definition was expanded to include heap-and-dump leaching, in-place (insitu) leaching, removal of oil from tar sands, extraction of sulfur from deep wells by the Frasch process, and recovery of metalliferous solutions from geothermal sources.

Solution mining frequently involves chemical changes; for example, copper usually goes into solution as copper sulfate, gold and silver as a cyanide compound, and uranium as a sulfate or uranyl tricarbonate. The valuable constituent is removed from the solid body by mass transfer to a usually aqueous solution. Solution mining, especially heap-and-dump leaching, has become an attractive metals extraction technique for gold, silver, copper, and uranium.

Solution mining includes the recovery of values from mine-waste dumps, old mine stopes, caved areas, and peripheries of abandoned open-pit mines. It also includes the extraction of minerals from deposits that are too low-grade to be successfully treated by conventional mining and milling.

A successful leaching operation involves three main steps: preparation, dissolution, and recovery. The preparation step may include size reduction of the ore, pelletizing, or roasting to enhance availability of the mineral to be recovered. Dissolution involves the solubilization of the desired material, sometimes at elevated temperatures. Then in the recovery steps the valuable liquids are separated from the leached residues and the metal is recovered from the liquid phase. Recovery may include techniques such as ion-exchange solvent extraction, adsorption or charcoal or resins, chemical precipitation, or electrolysis. *See* ADSORPTION; ELECTROCHEMICAL TECHNIQUES; ION EXCHANGE; PRECIPITATION (CHEMISTRY); SOLVENT EXTRACTION.

Solution mining offers several advantages: (1) The operation requires relatively low capital and start-up costs. (2) The lead time is shorter than in other types of mining, and there is usually a greater flexibility in start-up, shut-down, and plant expansion. (3) Fewer complications are encountered. (4) Finally, personnel requirements are usually less than with conventional operations, energy consumption is lower, and work-related accidents are substantially decreased.

Solution mining also has disadvantage and limitations: (1) Not all ores are amenable to leaching. (2) Conditions may not be favorable for the solution control that must be maintained. (3) Generally speaking, leaching of coarsely fractured material in mine dumps or heaps or in place is a relatively slow process, and the amount extracted is usually less than that attained by agitation leaching or flotation. (4) Leaching is more restrictive than flotation. (5) Leaching requires extra safeguards, because leaching solutions usually contain corrosive or hazardous chemicals. [D.D.Ra.]

Solvation The association or combination of a solute unit (ionic, molecular, or particulate) with solvent molecules. This association may involve chemical or physical forces, or both, and may vary in degree from a loose, indefinite complex to the formation of a distinct chemical compound. Such a compound contains a definite number of solvent molecules per solute molecule. Solvation occurring in aqueous solutions is referred to as hydration. In certain colloidal suspensions, solvation is, to a large extent, responsible for the stability of the sol. *See* COLLOID; HYDRATION; SOLUTION; SOLVENT. [F.J.J.]

Solvent By convention, the component present in the greatest proportion in a homogeneous mixture of pure substances (solutions). Components of mixtures present in minor proportions are called solutes. Thus, technically, homogeneous mixtures are possible with liquids, solids, or gases dissolved in liquids; solids in solids; and gases in gases. In common practice this terminology is applied mostly to liquid mixtures for which the solvent is a liquid and the solute can be a liquid, solid, or gas. *See* SOLUTION.

Three broad classes of solvents are recognized—aqueous, nonaqueous, and organic. Formalistically, the nonaqueous and organic classifications are both not aqueous, but the term organic solvents is generally applied to a large body of carbon-based compounds that find use industrially and as media for chemical synthesis. Organic solvents are generally classified by the functional groups that are present in the molecule, for example, alcohols, halogenated hydrocarbons, or hydrocarbons; such groups give an indication of the types of physical or chemical interactions that can occur between solute and solvent. Nonaqueous solvents are generally taken to be inorganic substances and a few of the lower-molecular-weight, carbon-containing substances such as acetic acid, methanol, and dimethylsulfoxide. Nonaqueous solvents can be solids (for example, fused LiI), liquids (H_2SO_4), or gases (NH_3) at ambient conditions; the solvent properties of fused sodium iodide (NaI) are manifested in the molten state, whereas hydrogen sulfate (H_2SO_4) and ammonia (NH_3) must be liquefied to act as solvents. [J.J.L.]

Solvent extraction A technique, also called liquid extraction, for separating the components of a liquid solution. This technique depends upon the selective dissolving of one or more constituents of the solution into a suitable immiscible liquid solvent. It is particularly useful industrially for separation of the constituents of a mixture according to chemical type, especially when methods that depend upon different physical properties, such as the separation by distillation of substances of different vapor pressures, either fail entirely or become too expensive.

Industrial plants using solvent extraction require equipment for carrying out the extraction itself (extractor) and for essentially complete recovery of the solvent for reuse, usually by distillation. *See* DISTILLATION.

The petroleum refining industry is the largest user of extraction. In refining virtually all automobile lubricating oil, the undesirable constituents such as aromatic hydrocarbons are extracted from the more desirable paraffinic and naphthenic hydrocarbons. By suitable catalytic treatment of lower boiling distillates, naphthas rich in aromatic hydrocarbons such as benzene, toluene, and the xylenes may be produced. The latter are separated from paraffinic hydrocarbons with suitable solvents to produce high-purity aromatic hydrocarbons and high-octane gasoline. Other industrial applications include so-called sweetening of gasoline by extraction of sulfur-containing compounds; separation of vegetable oils into relatively saturated and unsaturated glyceride esters; recovery of valuable chemicals in by-product coke oven plants; pharmaceutical refining processes; and purifying of uranium.

Solvent extraction is carried out regularly in the laboratory by the chemist as a commonplace purification procedure in organic synthesis, and in analytical separations in which the extraordinary ability of certain solvents preferentially to remove one or more constituents from a solution quantitatively is exploited. Batch extractions of this sort, on a small scale, are usually done in separatory funnels, where the mechanical agitation is supplied by handshaking of the funnel. *See* EXTRACTION. [R.E.Tr.]

Somatic cell genetics There are two approaches to somatic cell genetics: (1) the genetics of somatic cells, especially in culture, of higher organisms; and (2) the genetics of higher organisms by means of their somatic cells, especially in culture.

The first approach, pioneered by T. T. Puck, handles somatic cells in culture as cultures of unicellular organisms for the purpose of genetic research. It involves the same sorts of operations used, for example, in the genetics of bacteria, yeast, unicellular algae, and protozoa. It includes studies on colonies or cultures derived from single cells (clones) in respect of mutation; control of differentiation; metabolic cooperation; gene complementation; reactions to physical, chemical, or biological agents; cell fusion; uptake and integration of foreign genetic material, especially viral; and so on.

The second approach started with Curt Stern's discovery in 1936 of somatic gene segregation in the fruit fly *Drosophila melanogaster*. In the mid-1950s the "parasexual cycle" in the fungus *Aspergillus nidulans* (later, in many others) was discovered and harnessed to genetic analysis. This made it clear that mitotically multiplying cells in which loss of one or more chromosomes occurs or can be induced could provide an alternative to the germ cells for genetic analysis. How to obtain chromosome loss in mammalian cells came unexpectedly in 1967 from the new avenue of somatic cell hybridization. [G.Po.]

Somatic cell hybridization, the experimental fusion of somatic cells, has provided an alternative to sex as a way of studying genetics in higher organisms. Particularly striking has been the impact of somatic cell hybridization in the field of human genetics, especially for human gene mapping. *See* GENETIC MAPPING; HUMAN GENETICS.

In somatic cells grown in tissue culture, fusion occurs spontaneously as a rare event, and the frequency with which this happens can be greatly increased by treating the cells with Sendai virus or polyethylene glycol. In a small proportion of cells, the nuclei also fuse. By making a suitable choice of parental cells, and growing the cells in culture media which will support only the growth of cells having some particular phenotype, it is possible to select mononuclear cell hybrids which contain chromosomes from each parent. Intraspecific (human-human) or interspecific (human-mouse) hybrids can be obtained.

An exciting use of intraspecific hybrids was first described by C. Milstein and coworkers. They immunized a mouse against a particular antigen and then fused spleen cells from this mouse with a cell line from a mouse myeloma (a tumor which is capable of producing immunoglobulin). A small proportion of the hybrid clones which were selected and grown were making a single specific antibody against the original antigen. Similar hybrids, usually known as hybridomas, have proved an invaluable source of highly specific monoclonal antibodies for use in research and in diagnosis. *See* GENETIC ENGINEERING. [S.P.]

Somatostatin A naturally occurring peptide that acts primarily to inhibit the release of growth hormone from the anterior pituitary. The peptide consists of 14 amino acids; the two cysteine residues are joined by a disulfide bond so that the peptide forms a ring structure.

Somatostatin was first detected and isolated from the hypothalamus, a region of the brain known to control the release of hormones from the anterior pituitary gland. Many tissues other than the hypothalamus have been found to contain somatostatin, indicating that this peptide has a variety of different functions. Somatostatin is present in widespread areas of the nervous system, including the brain, brainstem, spinal cord, and spinal sensory ganglia. There is evidence that it acts as neurotransmitter or neuromodulator, providing chemical communication between nerve cells involved in the perception of sensory stimuli. Somatostatin is also present in high concentrations in the D cells of the pancreas, and has the ability to inhibit the release of both pancreatic insulin and glucagon. It is widely distributed in the gastrointestinal tract, where it inhibits the secretion of several gut hormones, gastric acid, and pancreatic electrolytes and digestive enzymes; it also inhibits the contraction of the stomach, intestine, and gallbladder, and reduces absorption of nutrients.

Synthetic analogs of somatostatin have been synthesized which are more potent, longer-acting, and more specific in their biological effects than natural somatostatin. Clinical applications which have been studied in animals and humans include treatment of (1) diabetes mellitus; (2) acromegaly and prolactin-secreting pituitary tumors; (3) pancreatic islet cell tumors; and (4) gastrointestinal disorders. *See* ENDOCRINE SYSTEM (VERTEBRATE); HORMONE; NEUROSECRETION; PITUITARY GLAND. [W.H.Si.]

Somesthesis A general term for the somatic sensibilities aroused by stimulation of bodily tissues such as the skin, muscles, tendons, joints, and the viscera. Six primary qualities of somatic sensation are commonly recognized: touch-pressure (including temporal variations such as vibration), warmth, coolness, pain, itch, and the position and movement of the joints. These basic sensory qualities exist because each is served by a different set of sensory receptors (the sensory endings of certain peripheral nerve fibers) which differ not only in their sensitivities to different types of stimuli, but also in their connections to structures within the central nervous system.

The somatic sensory pathways are dual in nature. One major part, the lemniscal system, receives input from large-diameter myelinated peripheral nerve fibers (for example, those serving the sense of touch-pressure). The second major somatic pathway is called the anterolateral system. It receives input from small-diameter myelinated and unmyelinated peripheral nerve fibers carrying pain and temperature information. *See* CUTANEOUS SENSATION; PAIN; PARESTHESIA; PROPRIOCEPTION; TEMPERATURE SENSES. [R.LaM.]

Sonar A term that refers both to the application of underwater sound to the detection and location of objects in the sea and to the apparatus used in such applications. The word is derived from "sound navigation and ranging"; the British use the word asdic. Since electromagnetic radiations, such as visible light or radar, do not penetrate the sea significantly, sonar is the most successful method of underwater detection. Except for underwater telephones, sonar is also the only means by which a fully submerged submarine can apprise itself of what is happening around it. Thus sonar is used by submarines to locate surface vessels and other submarines and to navigate through a mine field or under the ice in the Arctic Ocean. *See* UNDERWATER TELEPHONE.

There are two general types of sonar methods, active and passive. In an active sonar system a pulse of sound is generated by the searcher and projected into the water. This sound is reflected back from the target and detected by the searcher as an echo. Since the speed of sound in sea water is known, the range and also the bearing of the target can be determined. This method is also called echo-ranging. In a passive system the searcher detects the noises emitted by the target. Unless more than one listening station is involved, passive sonar provides information only as to the existence of a noise source and its beating from the searcher. Because the noise radiated from a ship is often characteristic of the type of ship, passive sonar, in contrast to active sonar, may help in classifying the target. *See* UNDERWATER SOUND.

Since active sonar systems involve the deliberate emission of sound, which discloses the searcher to easy detection by other vessels, they are used sparingly by submarines, which rely on concealment for their safety. The active systems are therefore used mainly by surface antisubmarine vessels, such as destroyers, since they generally operate in an environment which is too noisy for successful passive detection of a quiet submarine.

Sonar may also be used by aircraft for submarine detection by dropping buoys (called sonobuoys) in the water, by mine sweepers in locating mines, or by acoustic torpedoes. Sonar may be used for detection of schools of fish or for the location of submerged wrecks. An echo sounder, or depth indicator, is essentially an active sonar system which sends sound pulses to the sea bottom. *See* ACOUSTIC TORPEDO; ECHO SOUNDER; SONOBUOY. [R.W.Mo.]

Sonic barrier An aeronautical term coined during World War II to symbolize the technical difficulties for crewed aircraft in accelerating through the speed of sound. In designing and testing subsonic airplanes to fly at increasingly higher speeds, aeronautical engineers found the following adverse effects as the speed of the airplane approached the speed of sound: a rapid rate of increase in drag, breakdown of lift, and loss of control and maneuverability of the airplane. Such experience suggested that these violent and uncontrollable aerodynamic phenomena might be inevitable in the transonic range of speeds (that is, in the immediate vicinity of the speed of sound). These adverse effects are collectively and loosely termed the sonic barrier.

As a result of intensive research the source of most of the difficulties has been traced to the formation of local shock waves, which in turn interact with the boundary layer and induce flow separation from the aerodynamic surfaces. By careful choice of geometrical shapes, such as use of thin wings and slim bodies, it has been found possible to design airplanes with relatively smooth flow characteristics through the transonic speed range. *See* AIRPLANE; TRANSONIC FLIGHT. [S.-C.Li.]

Sonic boom Strong pressure waves (shock waves) generated by aircraft in supersonic flight and heard along the ground as explosivelike sounds called booms or bangs. Contrary to popular belief, a sonic boom does not occur at only one location at the instant the aircraft exceeds the local speed of sound. Instead, aircraft flying faster than the speed of sound (approximately 1100 ft/s or 330 m/s) generate shock waves that radiate away from and behind the aircraft and are dragged with it as long as it is flying supersonic. Where this trailing shock wave intercepts the ground, it is heard as an impulsive type of sound. In some situations, depending on the shock-wave pattern, an observer may hear two distinct booms or a double boom rather than hear the single sound. *See* SHOCK WAVE; SUPERSONIC FLIGHT.

Some factors affecting the sonic boom signal heard on the ground, such as speed, altitude, and route of the aircraft, can be reasonably well controlled. Others, such as meteorological conditions, topography, and ground-level air turbulence, cannot be modified. Consequently, the extent to which sonic booms can be predicted and controlled for known flight profiles and flying conditions is limited by those environmental factors beyond human control.

Aircraft altitude is the primary contributor to the magnitude of the sonic boom and is one of the factors most accessible to control measures. Current U.S. Air Force aircraft are restricted from supersonic flight over inhabited areas below an altitude of 30,000 ft (9 km) except for special missions or in the event of emergency. It appears that there is no possibility to significantly reduce or eliminate the sonic boom from current and next-generation aircraft which fly at supersonic speeds. The operational controls and flight regulations necessary to ensure acceptable levels of sonic booms on the ground have not been achieved for present-day aircraft flying over land. Commercial transports are prohibited by law in the United States from supersonic operations over land or over water where the sonic booms may be propagated to impact on land. [H.E.V.G.; C.W.Ni.]

Sonobuoy A miniature broadcasting station placed in the sea to receive underwater sounds and transmit them to an aircraft or other remote point. Its original and still most common application is to provide an "ear in the water" to permit an aircraft to detect typical sounds made by submarines.

Aircraft have unique advantages in antisubmarine warfare because of their mobility and relative invulnerability to submarine counterattack. Without the ability to launch sonobuoys, however, the aircraft's capability against submerged submarines would be very limited, since underwater acoustics is probably the most effective means of detection. *See* UNDERWATER SOUND.

While the most common type of sonobuoy merely listens for underwater sounds arriving from all directions, there are designs which even operate as miniature sonars. Such active sonobuoys are needed if the submarine is masked by natural sea background. *See* SONAR.

Sonobuoys have been used as geophysical instruments, providing a convenient means of picking up underwater acoustic returns in ocean bottom profiling. [I.H.G.]

Sonochemistry The study of the chemical changes which occur in the presence of sound or ultrasound. While there are many industrial applications of ultrasound, until the mid-1980s application of ultrasound to chemical processes had been sporadic. With the advent of inexpensive and reliable sources of ultrasound, there was a resurgence of interest in its chemical applications.

Ultrasound spans the frequencies of roughly 20 kHz to 10 MHz (human hearing has an upper limit of less than 18 kHz). Since the velocity of sound in liquids is approximately 5000 ft/s (1500 m/s), ultrasound has acoustic wavelengths of roughly 7.5 to 0.015 cm. Clearly, no direct coupling of the acoustic field with chemical species on a molecular level can account for sonochemistry. Instead, the chemical effects of ultrasound derive from several different physical mechanisms, depending on the nature of the system.

The most important of these mechanisms is cavitation, the formation of gas bubbles in a liquid which occurs when the pressure within the liquid drops significantly below the vapor pressure of liquid. When sound passes through a liquid, it consists of expansion (negative-pressure) waves and compression (positive-pressure) waves. This can cause the formation, growth, and rapid recompression of vapor bubbles in the liquid. The implosive bubble collapse generates localized heating and associated high-energy chemistry. In homogeneous media the generally accepted sonochemical mechanism involves intense localized heating which is produced by the implosive collapse of a bubble during cavitation. *See* CAVITATION.

When a liquid-solid interface is subjected to ultrasound, cavitation still occurs, but with major changes in the nature of the bubble collapse. No longer do cavities implode spherically. Instead, a markedly asymmetric collapse occurs, which generates a jet of liquid directed at the surface. The impingement of this jet can create a localized erosion (and even melting), responsible for surface pitting and ultrasonic cleaning. Enhanced chemical reactivity of solid surfaces is associated with these processes.

Since ultrasound can induce ligand dissociation from organometallic complexes, the initiation of homogeneous catalysis by ultrasound becomes practical. The potential advantages of such sonocatalysis include (1) the use of low ambient temperatures to preserve thermally sensitive substrates and to enhance selectivity; (2) the ability to generate high-energy species unobtainable from photolysis or simple pyrolysis; (3) the mimicry, on a microscopic scale, of bomb reaction conditions; and (4) the potential ease of scale-up. *See* HOMOGENEOUS CATALYSIS.

The effects of ultrasound on liquid-solid heterogeneous organometallic reactions has been a matter of intense investigation. The first use of ultrasound involved preparation of organometallic complexes of the main-group metals (for example, lithium, magnesium, and aluminum) from organic halides. Much interest was initiated by a 1980 report of the use of an ultrasonic cleaner to accelerate lithiation reactions. Various research groups have dealt with extremely reactive metals, such as lithium (Li), magnesium (Mg), or zinc (Zn), as stoichiometric reagents for a variety of common transformations. The activation of less reactive metals continues to attract major efforts in heterogeneous catalysis, metal-vapor chemistry, and synthetic organometallic efforts. *See* CATALYSIS; HETEROGENEOUS CATALYSIS; ULTRASONICS. [K.S.S.]

Sonoscan A type of acoustic microscope that uses unfocused beams with a wavelength of 15 micrometers. It does not require mechanical scanning. The acoustic beam deforms a liquid-solid interface in a pattern that represents the image. A laser beam reflected from this deformed surface is modulated in a way that corresponds to the image imprinted on the surface. The information on the laser beam is extracted with suitable electronics and presented for display on the face of a television monitor. *See* ACOUSTIC MICROSCOPE. [C.F.Q.]

Sorghum Sorghum includes many widely cultivated grasses having a variety of names. Sorghum is known as guinea corn in West Africa, Kafir corn in South Africa, mtama in East Africa, and durra in the Sudan. In India sorghums are called juar, jowar, or cholam; in China, kaoliang; and in America, milo. Cultivated sorghums in the United States are classified as a single species, *Sorghum bicolor*, although there are many varieties and hybrids. The two major types of sorghum are the grain, or nonsaccharine, type, cultivated primarily for grain production and to a lesser extent for forage, and the sweet, or saccharine, type, used for forage production and for making syrup and sugar.

Grain sorghum is grown in the United States chiefly in the Southwest and the Great Plains. It is a warm-season crop which withstands heat and moisture stress better than most other crops, but extremely high temperatures and extended drought may reduce yields. It is most extensively grown in Texas, Kansas, Nebraska, Oklahoma, Missouri, Colorado, and South Dakota. This grain production is fed to cattle, poultry, swine, and sheep primarily. Sorghum is considered nearly equal to corn in feed value. [F.R.M.]

Sound Subjectively, what is heard with the ears; objectively, a mechanical disturbance from equilibrium in an elastic material medium, that is, a material medium which will return to its original equilibrium state after the disturbing influence has been removed. Most of the solids, liquids, and gases of common experience possess such elasticity to a certain degree and can therefore act as sources, transmitters, and receivers of sound. *See* ACOUSTICS.

Originally sound referred entirely to what is heard. Later, however, interest developed in sounds which human beings cannot hear, either because the frequency is too low (infrasound) or too high (ultrasound). Audible sound is usually described as audio or sonic. The characterization of sound by frequency (associated with pitch in hearing—the higher pitch corresponding to higher frequency) is well known, since most sounds of interest are produced by the mechanical vibration of material media, for which the frequency means the number of complete vibrations or cycles per second (hertz, Hz). Another important quantity characterizing sound is the sound pressure, that is, the excess pressure in the medium being traversed by the sound over the normal equilibrium pressure in the undisturbed medium. *See* PITCH; SOUND PRESSURE.

Nature of sound propagation. Sound is propagated through a medium such as air by means of wave motion. In wave propagation the medium as a whole does not move, but only the disturbance, for example, the hump on the water surface in the case of a water wave. For a sound wave in a fluid like air, the "hump" is the excess pressure p in the medium constituting the disturbance. Such an excess pressure produced at one point does not remain there but moves through the fluid with velocity c, which is called the velocity of sound; in the case of air at 20°C this is 1129 ft/s (344 m/s). *See* WAVE (PHYSICS); WAVE MOTION.

Harmonic waves. The most important type of sound wave in practice is that called harmonic. For such a wave in the x direction with the velocity c, the excess pressure is given by Eq. (1). The quantity p_0 is called the amplitude of the wave; it is the

$$p = p_0 \cos\left[2\pi\nu\left(t - \frac{x}{c}\right)\right] \quad (1)$$

maximum value of the excess pressure as the wave travels along. The minimum value is of course $-p_0$. The quantity ν is the frequency of the wave. At any given point the value of the excess pressure repeats itself ν times a second or in the time $1/\nu$, which is called the period of the wave, designated by T. Another important quantity connected with the wave is the wavelength, designated by λ. At any given time the excess pressure repeats itself after a space interval given by Eq. (2).

$$\lambda = \frac{c}{\nu} = cT \quad (2)$$

This relation between wave length, velocity, and frequency holds for harmonic waves of all kinds and in all material media. Low frequencies are always associated with long wavelengths, and vice versa. For an audible sound wave in air with frequency 10^3 Hz, the wavelength is about 13 in. (34 cm). On the other hand, for an ultrasonic wave of frequency 10^6 Hz, the wavelength is about 1.3×10^{-2} in. (3.4×10^{-2} cm).

Associated with the excess pressure p of the sound wave, there is also an actual displacement of the medium from equilibrium in the x direction. This is usually called the particle displacement.

A very important quantity descriptive of a sound wave is the intensity I, qualitatively a measure of what might be called its strength but precisely defined as the time average rate of flow of energy per unit time through unit area of the medium through which the wave is passing, with the understanding that the area in question is perpendicular to the direction of wave propagation. This is the time average power transported by the wave. From the principles of mechanics it is the time average of the product of the particle velocity u and the acoustic pressure p.

A very important unit of intensity in practical sound problems is the decibel (dB). A sound of intensity I is said to be a number of decibels above or below a standard reference intensity I_0, given by expression (3). Thus, if $I = 2I_0$, I is about 3.01

$$10 \log_{10} I/I_0 \quad (3)$$

dB above I_0, or if $I = I_0/2$, I is about 3.01 dB below I_0. For many problems in audio acoustics I_0 is taken as the threshold of hearing (the minimum audible intensity), which for the average person with so-called normal hearing is approximately 10^{-16} W/cm². In terms of this reference, normal conversational speech directly in front of the mouth has an intensity of about 65 dB, whereas next to an aircraft jet engine the intensity is about 160 dB. For a normal human ear the intensity at which sound ceases to be heard but produces pain in the ear and incipient damage is about 120 dB. *See* DECIBEL; LOUDNESS.

The harmonic sound wave considered so far is called a plane wave since the particle displacement and pressure are uniform over any plane surface perpendicular to the direction of propagation. Such a plane is called a wavefront.

A type of sound wave of even greater importance than the plane wave is the spherical wave. Such a wave spreads equally in all directions from a point source; it can be realized ideally by a small sphere that pulsates radially. The radiated sound then moves out so that the particle velocity u and the sound pressure p are uniform over a sphere or radius r with center at the source and are, of course, dependent on r.

Velocity of sound. The general theoretical expression for the velocity of sound in a compressible fluid medium in which the instantaneous pressure and density are respectively p and ρ is given by Eq. (4), where the ratio $dp/d\rho$ must be taken under

$$c = \sqrt{\frac{dp}{d\rho}} \quad (4)$$

adiabatic conditions, that is, such that no heat enters or leaves the element of the medium as the sound wave passes through it. For a gas, Eq. (4) gives Eq. (5), where γ is the ratio of the

$$c = \sqrt{\frac{\gamma p}{\rho}} \quad (5)$$

specific heat at constant pressure to that at constant volume. For air, $\gamma = 1.4$. From the general gas equation of an ideal gas, Eq. (4) takes the form of Eq. (6), where R is the gas con-

$$c = \sqrt{\gamma R T} \quad (6)$$

stant per unit mass, and T is the absolute temperature. Denoting the velocity of sound in the gas at 0°C by c_0 permits Eq. (6) to be put in the form of Eq. (7), where Θ is temperature

$$c = c_0 \sqrt{1 + \frac{\Theta}{273}} \quad (7)$$

in Celsius degrees. *See* GAS.

Sound is propagated through the bulk of an elastic solid by means of a longitudinal compressional wave. The word longitudinal means that the displacement of the medium through which the sound is passing lies in the direction of propagation. In addition to longitudinal waves, solids can also propagate pure shear waves which are transverse in character, in which the displacement of the medium is at right angles to the direction of propagation. Though not strictly a sound wave in the usually accepted sense, this type of wave is considered acoustical in character; it is exemplified as a component of seismological or earthquake waves.

The velocity of sound varies with the temperature [Eq. (7) for gases], usually increasing as the temperature increases, and hence can have different values in different parts of a medium. This is of particular importance in sound propagation through the atmosphere. The same is true for sound transmitted through the ocean. Even in an isothermal medium the velocity will change with the frequency. This is known as sound dispersion. It is a very small effect at low frequencies and becomes perceptible only at very high or ultrasonic frequencies. *See* ULTRASONICS; UNDERWATER SOUND.

Sound rays. Though sound, like light, is propagated by waves, under certain conditions its transmission may be described adequately by means of rays, which are lines perpendicular to the wavefront in each case, giving the direction of sound propagation. The necessary conditions for the use of rays are: the wavefront must not change its direction materially over a distance of one wavelength; the amplitude of the sound pressure must not change materially over the same distance; finally the wavelength itself (in the case of transmission through a nonhomogeneous medium) must change very little over a distance of its own order of magnitude.

Reflection and refraction. When a sound wave strikes a surface separating media of different density and elasticity, reflection and refraction can take place. The laws governing these phenomena are the same as those for light. *See* REFLECTION OF SOUND; REFRACTION OF WAVES.

Diffraction and scattering. Like light waves, sound waves bend around obstacles, that is, experience diffraction. Indeed, sound can be heard around a corner on account of diffraction, which in the case of audible sound is a much more obvious phenomenon than the diffraction of visible light because the wavelength of audible sound is much greater than that of visible light. The bending of sound waves due to diffraction becomes less marked as the wavelength decreases relative to the spatial dimensions of the diffracting obstacles. With ultrasonic radiation it is possible to produce sound shadows. Another common name for sound diffraction is scattering, particularly if many obstacles are involved. *See* DIFFRACTION; SCATTERING OF ELECTROMAGNETIC RADIATION.

Transducers. Sound sources are referred to as transducers, since they transfer energy of various forms into sound. The most important source of sound in human experience is the human voice, which in ordinary conversation and singing produces power outputs of from 30 to 60 microwatts. Other important sources are mechanically vibrating solids, liquids, and gases. Solid vibrators include strings, rods, plates, and membranes (as in musical instruments). The organ pipe is an example of a sound source based on vibrating air. The turbulent flow of air in the exhaust of a jet plane produces intense sound. The break of the surf on the shore is a liquid source of sound. *See* ACOUSTIC NOISE; VIBRATION.

More sophisticated sound sources depend on the ability to produce mechanical vibrations by means of magnetic and electrical effects. The ordinary telephone receiver and the radio loudspeaker are illustrations. Still more elaborate and valuable for many applications are piezoelectric and magnetostrictive oscillators. These electroacoustic transducers have found wide use as both sources and receivers, for example, in underwater sound, in sound recording and reproduction, in radio and television, and in the industrial applications of sound. *See* DISK RECORDING; EARPHONES; LOUDSPEAKER; MAGNETIC RECORDING; MAGNETOSTRICTION; MICROPHONE; PIEZOELECTRICITY; SONAR; SOUND RECORDING; SOUND-REPRODUCING SYSTEMS; STEREOPHONIC SOUND; TELEPHONE; TRANSDUCER; UNDERWATER TRANSDUCER. *See also* ATMOSPHERIC ACOUSTICS; BIOACOUSTICS.

Attenuation of sound. When a sound wave from a given source passes through any actual physical medium, its intensity decreases with distance from the source. This is known as attenuation. Part of this is due to the geometrical spreading of the wavefront, unless it is confined in a tube or duct. The rest of the attenuation is due to the influence of the medium and the surfaces over which or through which the sound passes. Reflection, refraction, and scattering can all serve to increase the effective attenuation of sound propagation in a given direction. The medium itself produces attenuation by changing a part of the acoustic energy into heat, a kind of attenuation known as sound absorption. Mechanisms responsible for this are viscosity, heat conduction, and molecular relaxation processes. *See* SOUND ABSORPTION. [R.B.L.]

Sound absorption

The process by which the intensity of sound is diminished due to the conversion of the energy of the sound wave into heat. The absorption of sound is a special but important case of sound attenuation. *See* SOUND.

In actual material media the geometrical attenuation is supplemented by absorption due to the interaction between the sound wave and the physical properties of the medium. This dissipates the sound energy by transforming it into heat and hence decreases the intensity of the wave. In all practical cases the attenuation due to such absorption is exponential in character.

The four so-called classical origins of dissipative sound absorption in material media are: shear viscosity, heat conduction, heat radiation, and diffusion. In the flow of a fluid, viscosity represents the tendency for the fluid flowing in one layer parallel to the direction of motion to retard the motion in adjoining layers. The role of heat conduction can be qualitatively understood as follows. The compression associated with the propagation of a sound wave in a fluid produces a local increase in temperature, which sets up a temperature gradient through which heat will flow by conduction before the succeeding rarefaction cools the gas. This is an irreversible phenomenon, leading to energy dissipation into heat. For ordinary gases in which the components differ little in molecular weight, the effect of diffusion on absorption is negligible compared with that of viscosity and heat conduction. [R.B.L.]

Sound pressure

The incremental variation in the static pressure of a medium when a sound wave is propagated through it. This incremental variation is also known as the excess pressure.

The unit of pressure commonly used in acoustics is the microbar (= 0.1 newton/m^2 = 1 dyne/cm^2). One microbar is approximately 10^{-6} times the normal atmospheric pressure.

Instantaneous sound pressure at a point is the incremental change from the static pressure of the medium at a given instant caused by the presence of a sound wave.

Peak sound pressure for any specified time interval is the maximum absolute value of the instantaneous sound pressure in that interval.

Effective sound pressure at a point is the root-mean-square (rms) value of the instantaneous sound pressure over a time interval at that point. The time interval should be an integral number of periods if the sound pressure is periodic. In the case of nonperiodic sound pressures, the interval should be long enough to make the value obtained essentially independent of small changes in the length of the interval. [W.J.G.]

Sound recording

The technique of entering sound, especially music, on a storage medium for playback at a subsequent time. The storage medium most widely used is magnetic tape. *See* MAGNETIC RECORDING.

Monophonic. In the simplest form of sound recording, a single microphone picks up the composite sound of a musical ensemble, and the microphone's output is recorded on a reel of ¼-in. (6.4-mm) magnetic tape. This single-track, or monophonic, recording suffers from a lack of dimension, since in playback the entire ensemble will be heard from a single point source: the loudspeaker.

Stereophonic. An improved recording, with left-to-right perspective and an illusion of depth, may be realized by using two microphones which are spaced or angled appropriately and whose outputs are routed to separate tracks on the tape recorder. These two tracks are played back over two loudspeakers, one each on the listener's left and right. Under ideal conditions this stereophonic system will produce an impressive simulation of the actual ensemble. *See* STEREOPHONIC SOUND.

Binaural. In this system, two microphones are placed on either side of a small acoustic baffle in an effort to duplicate the human listening condition. The recording is played back over headphones, so that the microphones are, in effect, an extension of the listener's hearing mechanism.

Multitrack. To give the recording engineer more technical control over the recording medium, many recordings are now made using a multiple-microphone technique. In place of the stereo microphone, one or more microphones are located close to each instrument or group of instruments. *See* MICROPHONE.

In the control room the engineer mixes the outputs of all microphones to achieve the desired musical balance. As a logical extension of this technique, the microphone outputs may not be mixed at the time of the recording, but may be routed to 16 or more tracks on a tape recorder, for mixing at a later date.

When many microphones are so used, each instrument in the ensemble is recorded—for all practical purposes—simultaneously. The complex time delay and acoustic relationships within the room are lost in the recording process, and the listener hears the entire ensemble from one perspective only, as though he or she were as close to each instrument as is each microphone. Electronic signal processing devices may not be entirely successful in restoring the missing information, and the listener hears a recording that may be technically well executed, yet lacking the apparent depth and musical cohesiveness of the original. However, the multitrack technique becomes advantageous when it is impractical or impossible to record the entire ensemble at once stereophonically. *See* SOUND-REPRODUCING SYSTEMS. [J.M.Wor.]

Sound-reinforcement system A sound-reinforcement system (public address system) is an electronic means for augmenting the sound output of a speaker, singer, or musical instrument. The use of a sound-reinforcement system makes it possible to render speech intelligible where it was either too weak to be heard above the general noise or too reverberant. In certain instances, for example, in large halls or outdoors, the volume range of the orchestra at the listener's location is inadequate for full artistic appeal and the use of a sound-reinforcement system is required to augment the original sound. The basic elements of a sound-reinforcement system are microphones, amplifiers, volume controls, and loudspeakers. *See* AMPLIFIER; LOUDSPEAKER; MICROPHONE.

Sound-reinforcement systems for auditorium-sized rooms may use a single horn loudspeaker located at the ceiling above a stage; direct-radiator loudspeakers placed on the walls; or two column loudspeakers at each side of the stage. The directivity pattern of the microphone should be such that the region of least response is directed toward the loudspeakers and the audience. Sound-reinforcement systems for outdoor theaters may employ a single loudspeaker station located above the stage and directed into the audience area, or a number of loudspeakers, each supplying a portion of the audience area with sound. [H.F.O.]

Sound-reproducing systems Units which pick up sound at one point and reproduce it at the same point or some other point at the same time or some subsequent time. Sound-reproducing systems may be classified as monaural, binaural, monophonic, stereophonic, or quadraphonic.

The purpose of the several types of sound-reproducing systems is to provide esthetic and pragmatic communication—namely, music and speech. A major requirement is that the sound-reproducing system have the capabilities for controlling and processing the audio signal so as to supply the sound characteristics that the listener feels is appropriate for the program. The impact of the music thus depends both upon engineering and subjective factors. Engineering factors include, in approximate order of importance, distortion that is low or appropriate; smoothness of response with frequency; full frequency range of response; low system internal noise; ability to produce sound levels at least 50 decibels (dB) above the background noise in the listening space; proper amount of auditory perspective; and appropriately coordinated room acoustics in originating and listening spaces.

Monaural system. A monaural sound-reproducing system utilizes one or more microphones connected to a single transducing channel which, in turn, is coupled to one or two earphones worn by the listener. The most common example is the telephone in which there is, in general, a single source of sound, one microphone, a transducer, and one earphone coupled to the ear of the listener. *See* EARPHONES; MICROPHONE; TELEPHONE.

Binaural system. A binaural sound-reproducing system is a closed-circuit type of sound-reproducing system in which two microphones, used to pick up the original sound, are each connected to two independent corresponding transducing channels which, in turn, are coupled to two independent corresponding earphones worn by the listener. Use of the binaural sound-reproducing system is limited to specific applications, as in conference or court recording. The microphones are mounted in a dummy simulating the human head in shape and dimensions and at the locations corresponding to the ears of the human head. *See* BINAURAL SOUND.

Monophonic system. A monophonic sound-reproducing system employs one or more microphones, used to pick up the original sound, which are coupled to a single transducing channel which, in turn, is coupled to one or more loudspeakers in reproduction. The monophonic sound-reproducing system is the most widely employed of all sound-reproducing systems. Examples are the disk phonograph, radio, sound motion picture, television, and magnetic tape reproducer sound systems. The monophonic sound-reproducing system is of the field type, in which the sound is picked up by a microphone in a field and reproduced by means of a loudspeaker into a field. *See* LOUDSPEAKER.

Stereophonic system. A stereophonic sound-reproducing system is a field-type sound-reproducing system in which two or more microphones, used to pick up the original sound, are each coupled to a corresponding number of independent transducing channels which, in turn, are each coupled to a corresponding number of loudspeakers arranged in substantial geometrical correspondence to that of the microphones. Each ear of the listener receives sound from all sources. The relative amounts depend on the acoustics of the space of origination, the microphone array used, the mix-down process, the acoustics of the listening space, the placement and directionality of the loudspeakers, and the location and orientation of the listener. *See* STEREOPHONIC SOUND.

Quadraphonic system. A quadraphonic sound-reproducing system is a field-type system in which each of the four microphones used to pick up the original sound is coupled to four transducing channels which, in turn, are coupled to four loudspeakers arranged in substantial geometrical correspondence to the arrangement of the microphones. *See* QUADRAPHONIC SOUND SYSTEM.

High-quality sound-reproducing systems can reproduce the sounds of speech and music without frequency discrimination. Furthermore, these sound-reproducing systems cover the frequency range of normal hearing. These considerations show that high-quality sound-reproducing systems provide the capabilities of reproducing the entire audio frequency range. High-grade sound-reproducing systems also provide the capabilities of covering the amplitude range of sound reproduction in the home and can be built so that harmonic distortion is not perceptible.

The sounds of speech and music vary continuously in frequency and amplitude. Therefore, in order to reproduce these sounds the transient response of the reproducing system must exhibit a high order of excellence. The response of a part or the whole of a sound-reproducing system to a sudden change in the amplitude or frequency is a measure of the transient response. In general, transient response is good if system frequency response (including source and listening spaces) is smooth, with no significant, sudden peaks or dips. These deviations from a flat response are usually due to resonances, often in the loudspeaker. The effects are most marked at the extremes of the frequency range. [V.S.]

Source flow A source, in three-dimensional flow, is a point from which fluid issues at a uniform rate in all directions. The strength of a source is defined as the volume per unit time issuing from the point. Because the flow is outward and is uniform in all directions, velocity v_r at distance r from the source is the strength m divided by the area of sphere through the point with center at the source, or $v_r = m/4\pi r^2$.

A sink is a negative source, or a point into which fluid flows uniformly in all directions. A well point is the physical equivalent of a sink. It is a relatively small region in a porous medium into which fluid flows uniformly from all directions. Similarly, if fluid were injected into the porous medium through the well point, it would be equivalent to a source flow.

In two-dimensional flow, all flow occurs in a set of parallel planes, with no flow normal to them. The flow is identical in each of these parallel planes.

A source, in two-dimensional flow, is a line normal to the planes of flow, from which fluid is imagined to flow uniformly in all directions at right angles to the line. The source appears as a point on the customary two-dimensional flow diagram.

Sources and sinks are used as flow elements in conjunction with doublets, vortices, and uniform flow to develop complex flow situations. The combination of a source, an equal sink, and a uniform flow, properly placed, results in flow about a closed body in three-dimensional flow and about a cylinder in two-dimensional flow. The source and sink concepts are generally useful for frictionless flow situations, and have little meaning in frictional situations, such as flow in a pipe. *See* SINK FLOW; STREAMLINE FLOW. [V.L.S.]

Sourwood A deciduous tree, *Oxydendrum arboreum*, of the heath family, indigenous to the southeastern section of the United States, and found from Pennsylvania to Florida and west to Indiana and Louisiana. It is usually a small or medium-sized tree. The wood is not used commercially. Sourwood is also known as sorrel tree, and it is widely planted as an ornamental. *See* ERICALES. [A.H.G./K.P.D.]

South African tick-bite fever An infectious tick-borne disease of humans which is similar to fièvre boutonneuse but has some biological and ecological differences. A similar, or the same, disease occurs in East Africa, where it is known as Kenya tick typhus. The chief vector in Kenya is the dog tick (*Haemaphysalis leachi*), and in South Africa it is this vector plus *Rhipicephalus evertsi*. *See* FIÈVRE BOUTONNEUSE; RICKETTSIOSES. [C.B.P.]

South America The southernmost of the New World or Western Hemisphere continents, three-fourths of which lies within the tropics. South America is approximately 4500 mi (7200 km) long and at its greatest expanse 3000 mi (4800 km) wide. Its total area is estimated to be about 7,000,000 mi^2 (18,000,000 km^2), more than two-thirds of which consists of plains. The land portions stand upon a continental platform somewhat larger than the land above sea level so that varying widths of shallow sea extend over the continental shelf to the steep scarp of the submarine continental slopes. *See* CONTINENT.

At least three-fourths of the South American coastline is composed of highland masses; consequently, the areas of maximum elevation on the continent generally lie not far from the ocean. This topography has an important influence on the continental climate and causes drainage of streams to be toward the interior more often than is true of any other continent. Moreover, there is a paucity of large independent rivers, for they have little chance of developing when the coasts throughout the well-watered parts are guarded by mountains and plateaus that overlook the seas, forcing the bulk of the drainage to find an ultimate outlet at a small number of well-marked gaps. Thus the Andes, the Brazilian and Guiana Highlands, and to some extent the Patagonian Plateau have determined the present form and configuration of the entire South American land mass.

Because of the vast extent of the Andes, a greater proportion of South America than of any other continent lies above 10,000 ft (3000 m). Throughout the Andean mountain mass there are high plateaus and deep longitudinal valleys, especially in the middle and northern parts. Although in general the Andes resulted from folding and faulting, volcanic formations are common in southern Colombia, Ecuador, central and southern Peru, and western Bolivia. Possibly nowhere else on Earth is there a more spectacular series of volcanic peaks than in western Bolivia and on each side of the structural depression in Ecuador.

The broadest lowland routes to the interior are gapways between the three great barriers separating coastal from interior South America: the Andes cordilleran belt, the Guiana Highlands, and the Brazilian Highlands. The breach between the Andes and the Guiana Highlands is some 300 mi (480 km) wide. The western part, near the Andes, is composed of rolling and irregular plains. The climate of the gap is warm to hot at all times. Rainforest vegetation covers the southern part, and tropical savanna and semideciduous forest predominate in sections of the northern region. The low-relief plain occurs between the hilly margins of the Brazilian Highlands and the Guiana Highlands. Through it courses the Amazon River. The climate is monotonously warm to hot. Vegetation is evergreen tropical rainforest along rivers and in low sections; semideciduous forest and areas of savanna are found on interstream uplands and the bordering hillier country.

There are three great river systems in South America and a number of important rivers which are not a part of these systems. First of the river systems in size is the Amazon which, with its many tributaries, drains a basin covering 2,700,000 mi^2 (7,000,000 km^2), or about 40% of the continent. Next is the system composed of the Paraguay, Paraná, and La Plata rivers, the last being a huge estuary. The third river complex, located in southern Venezuela, is the Orinoco, which drains water from 365,000 mi^2 (945,000 km^2) of land, emptying into the Atlantic Ocean along the northeast edge of the continent. The major independent rivers are the São Francisco, arising on the Brazilian plateau, and the Magdalena-Cauca, which empties into the Caribbean.

South America is unique in having a west-coast desert that extends almost to the Equator, an east-coast desert poleward from latitude 40° (the Patagonian), and a desert probably receiving less rain than any on Earth (the Atacama). Three major islands or island groups lie off the South American coast. Trinidad, in the north, contains a range of mountains exceeding at points 3000 ft (900 m) elevation. The Falklands stand on the shallow continental shelf off the rocky coast of

Patagonia, near the southern tip of the continent. On the Equator and about 550 mi (885 km) west of Ecuador are the Galápagos, a group of 13 volcanic islands. [C.L.W.]

South Pole That end of the Earth's rotational axis opposite the North Pole. It is the southernmost point on the Earth and one of the two points through which all meridians pass (the North Pole being the other point). This is the geographic pole and is not the same as the south magnetic pole. The South Pole lies inland from the Ross Sea, within the land mass of Antarctica, at an elevation of about 9200 ft (2800 m).

There is no natural way to determine local Sun time because there is no noon position of the Sun, and shadows point north at all times, there being no other direction from the South Pole. *See* MATHEMATICAL GEOGRAPHY; NORTH POLE. [V.H.E.]

Southeast Asian waters All the seas between Asia and Australia and the Pacific and the Indian oceans. They form a geographical and oceanographical unit because of their special structure and position, and make up an area of 3,450,000 mi^2 (8,940,000 km^2), or about 2.5% of the surface of all oceans.

The surface circulation is completely reversed twice a year by the changing monsoon winds. The subsurface circulation carries chiefly the outrunners of the intermediate waters of the Pacific Ocean into these seas. The tides are mostly of the mixed type. Diurnal tides are found in the Java Sea, in the Gulf of Tonkin, and in the Gulf of Thailand. Semidiurnal tides with high amplitudes occur in the Malacca Straits. *See* INDIAN OCEAN; PACIFIC OCEAN. [K.W.]

The Southeast Asian seas are characterized by the presence of numerous major plate boundaries. In Southeast Asia the plate boundaries are identified, respectively, by young, small ocean basins with their spreading systems and associated high heat flow; deep-sea trenches and their associated earthquake zones and volcanic chains; and major strike-slip faults such as the Philippine Fault (similar to the San Andreas Fault in California). The Southeast Asian seas are thus composed of a mosaic of about 10 small ocean basins whose boundaries are defined mainly by trenches and volcanic arcs. The dimensions of these basins are much smaller than the basins of the major oceans. The major topographic features of the region are believed to represent the surface expression of plate interactions, the scars left behind on the sea floor. *See* PLATE TECTONICS. [D.E.H.]

Soybean *Glycine max*, a legume native to China that has become a major source of vegetable protein and oil for human and animal consumption and for industrial usage. The valued portion of the plant is the seed, which contains about 40% protein and 21% oil. Illinois, Iowa, Arkansas, Missouri, Indiana, Mississippi, Minnesota, Ohio, Louisiana, and Tennessee are the major soybean producers in the United States. *See* FAT AND OIL (FOOD). [W.R.F.]

Space Physically, space is that property of the universe associated with extension in three mutually perpendicular directions. Space, from a newtonian point of view, may contain matter, but space exists apart from matter. Through usage, the term space has come to mean generally outer space or the region beyond Earth. Geophysically, space is that portion of the universe beyond the immediate influence of Earth and its atmosphere. From the point of view of flight, space is that region in which a vehicle cannot obtain oxygen for its engines or rely upon an atmospheric gas for support (either by buoyancy or by aerodynamic effects). Astronomically, space is a part of the space-time continuum by which all events are uniquely located. Relativistically, the space and time variables of uniformly moving (inertial) reference systems are connected by the Lorentz transformations. Gravitationally, one characteristic of space is that all bodies undergo the same acceleration in a gravitational field and therefore that inertial forces are equivalent to gravitational forces. Perceptually, space is sensed indirectly by the objects and events within it. Thus, a survey of space is more a survey of its contents. *See* EUCLIDEAN GEOMETRY; LORENTZ TRANSFORMATIONS; RELATIVITY; SPACE-TIME. [S.F.S.]

Space biology The inclusive term for the various biological sciences concerned with the study of living things in the space environment. Medical studies that are intrinsically linked to the purposes of biology are termed biomedical; and exobiology is the search for and study of extraterrestrial life. Areas of space biology can also be identified by the inherent physical factors of space and space travel and their effects on living organisms: (1) biophysics—factors due to the environment of space, such as temperature, pressure, radiation, and weightlessness; (2) biodynamics—factors due to the flight dynamics of the spacecraft, including acceleration, vibration, and noise; and (3) aerospace medicine—factors created by the involvement of human beings in the artificial environment of the spacecraft, including breathing gases, nutrition, toxicology, and isolation. *See* AEROSPACE MEDICINE; EXOBIOLOGY. [T.W.H.]

Space charge The net electric charge within a given volume. If both positive and negative charges are present, the space charge represents the excess of the total positive charge diffused through the volume in question over the total negative charge. [E.G.R.]

Space communications Communication between a vehicle in outer space and the Earth, using high-frequency electromagnetic radiation. Provision for such communication is an essential requirement in any space mission. The total communication system ordinarily includes (1) transmission of command signals to the spacecraft; (2) radio equipment for tracking by ground stations; and (3) transmission of scientific and applications data from the spacecraft to the Earth, sometimes singled out as telemetry. For space probes and satellites generally, telemetry is a key element for the success of the mission. A somewhat special communication system is the communications satellite system in which the spacecraft serves solely as a relay station between remote points on the Earth, and is here excluded from consideration. *See* COMMUNICATIONS SATELLITE; SATELLITE (SPACECRAFT); SPACE FLIGHT; SPACE PROBE.

Certain characteristic constraints distinguish space communication systems from their terrestrial counterparts. Although only line-of-sight propagation is required, both the transmitter and receiver are in motion. The movement of spacecraft in relation to the Earth, for instance, makes it necessary to provide geographically scattered stations for adequate continuous coverage.

Because enormously greater distances are involved (millions of kilometers in the case of space probes to Mars, Venus, and beyond), the signal received on Earth is so small that local interference, both artificial and natural, has to be drastically reduced. For this purpose, the transmitting frequency has to be made sufficiently high, and highly sensitive receivers (large antennas and special low-noise amplifiers such as masers) are used which eliminate all but the unavoidable emanations from space (galactic noise). For deep-space missions, large steerable antennas, such as at Goldstone, California (see illustration), are located as far from civilization as practical. Sophisticated data processing is required, and the ground-receiver complex includes several large, high-speed digital computers and associated processing equipment. *See* MASER; RADIO RECEIVER; RADIO TELESCOPE.

The spacecraft installations for communication are constrained by weight and space considerations, since there are practical limitations to these spacecraft parameters. An additional concern is reliability, since the equipment must operate,

Goldstone 210-ft (64-m) antenna. (*Jet Propulsion Laboratories*)

sometimes for years, unattended under the extremes of the space environment. To increase reliability, highly reliable components and equipment are developed, and redundancy is used, where practical, to eliminate single-point system failures. In some applications the spacecraft antenna must point toward Earth stations for acceptable transmission of data; hence angular pointing accuracies of 1° or less are often required. Spacecraft power is always at a premium, and techniques must be employed to minimize the consumption by the communications system. The transmitter is usually a major consumer of power, and efforts must be made to increase the transmitter efficiency to reduce total power requirements. All aspects of data transmission must meet much closer tolerances than are common on terrestrial systems.

The Tracking and Data Relay Satellite System (TDRSS) consists of two geostationary relay satellites 130° apart in longitude, and a ground terminal located at White Sands, New Mexico. The purpose of TDRSS is to provide telecommunication services between low-Earth-orbiting user spacecraft and user control centers. A principal advantage of the system is the elimination of many of the worldwide ground stations for tracking low-orbiting spacecraft. [D.S.He.]

Space flight The penetration by humans into the reaches of the universe above the terrestrial atmosphere and investigation of these regions by automated, remote-controlled, and crewed vehicles. The purpose of space flight is to provide significant contributions to the physical and mental needs of humanity on a national and global basis. Such contributions fall specifically in the areas of (1) earth resources of food, forestry, atmospheric environment, energy, minerals, water, and marine life; (2) earth and space sciences for research; (3) commercial materials processing, manufacturing in space, and public services. More general goals of space flight include expansion of knowledge; exploration of the unknown, providing a driving force for technology advancement and, hence, improved Earth-based productivity; development and occupation of new frontiers with access to extraterrestrial resources and unlimited energy; strengthening of national prestige, self-esteem, and security; and providing opportunity for international cooperation and understanding.

To conduct crewed space flight, the two leading spacefaring nations, the United States and Russia, formerly the Soviet Union, developed spacecraft systems and the ground facilities, research and development base, operational know-how, planning experience, and management skills. In the United States, crewed space programs are conducted by the National Aeronautics and Space Administration (NASA), a federal agency established in 1958 for the peaceful exploration of space. In Russia, crewed space flights were under the auspices of the U.S.S.R. Academy of Sciences; they are now the responsibility of the Russian Space Agency (RKA). The first crewed spacecraft, the *Vostok 1*, piloted by Yuri A. Gagarin, was launched on April 12, 1961, and returned after completing one revolution of the Earth. The first American crewed space flight took place 3 weeks later, when NASA launched Alan B. Shepard for a 15-min suborbital test flight. Since then, a large number of crewed spacecraft have been put into orbit around the Earth by both nations, and nine have been launched toward the Moon by the United States (see table).

Crewed spacecraft. A crewed spacecraft is a vehicle capable of sustaining humans above the terrestrial atmosphere. In a more limited sense, the term "crewed spacecraft" is usually understood to apply to vehicles for transporting and sustaining human crews in space for time periods limited by prestored onboard supplies, as distinct from orbital space stations which support theoretically unlimited habitation of humans in space by autonomous systems, crewed maintenance, and periodic resupply. *See* SPACE STATION.

The basic requirements of crewed spacecraft are quite different from those of uncrewed space probes and satellites. The presence of humans on board necessitates a life-sustaining environment and the means to return safely to Earth. The major common feature of all crewed spacecraft, therefore, is the atmospheric-return element or reentry module. It consists basically of a pressure-tight cabin for the protection, comfort, and assistance of the crew, and surrounded by heat-resistant material, the thermal protection system, to cope with the high frictional and radiative heating accompanying the energy dissipation phase of the atmospheric-return flight to Earth.

The main system that distinguishes crewed spacecraft from other spacecraft is the environmental control and life support system. It provides the crew with a controlled atmosphere. Because all atmospheric supplies must be carried into space, it is essential to recirculate and purify the spacecraft atmosphere to keep the total vehicle weight within reasonable limits.

Space suits or pressure suits are mobile spacecraft or chambers that house the astronauts and protect them from the hostile environment of space. They provide atmosphere for breathing, pressurization, and thermal control; protect astronauts from heat, cold, glare, radiation, and micrometeorites; contain a communication link and hygiene equipment; and must have adequate mobility. The suit is worn during launch, docking, and other critical flight phases.

Because of the potential hazards of space flight to personnel, crewed spacecraft must meet stringent requirements of safety and hardware reliability. Reliability is associated with the probability that the spacecraft systems will operate properly for the required length of time and under the specified conditions. To assure survival and return in all foreseeable emergencies, the design of a "crew-rated" spacecraft includes standby systems (double or triple redundancy) and allows for launch escape, alternate and "degraded" modes of operation, contingency plans, emergency procedures, and abort trajectories to provide maximum probability of mission success. The priority order of this reliability requirement is: (1) crew safety, (2) minimum achievable mission fulfillment, (3) mission data return, and (4) minimal degradation.

Soviet/Russian programs. After the first *Sputnik* launch on October 4, 1957, developments of Soviet crewed space flight capability followed in quick succession, leading from the first-generation *Vostok* (East) to the second-generation *Voskhod* (Ascent) and to the third-generation *Soyuz* (Union) spacecraft.

The *Vostok*, a single-seater for short-duration missions and ballistic reentry from Earth orbits, consisted of a near-spherical cabin, having three small viewports and external radio antennas. The approximately 7-ft-diameter (2-m) sphere of the cabin was attached to a service module. The second-generation *Voskhod*, essentially a greatly modified *Vostok*, was a short-duration multiperson craft and was designed to permit EVA or spacewalking by one of the crew.

The *Soyuz* is much heavier, larger, and more advanced in its orbital systems, permitting extended orbital stay times. It consists of three main sections or modules: a descent vehicle, an orbital module, and an instrument-assembly module. The three elements are joined together, with the descent vehicle in the middle. Shortly before atmospheric reentry, the two outer modules are jettisoned. *Soyuz* spacecraft carried originally up to three cosmonauts. However, when an accidental explosive decompression of the descent cabin during reentry caused the death of the *Soyuz 11* crew in 1971, the third seat was removed to make room for the necessary additional life support equipment, and the *Soyuz* thereafter carried only two cosmonauts. An improved version, the *Soyuz T* (for "Transport") was introduced in 1980. A new version, the *Soyuz-TM*, with extended mission duration and new subsystems, replaced this vehicle in February 1987, after a crewless test flight in May 1986.

Between 1971 and 1981, five *Salyut* space stations operated in low Earth orbit successively. The most successful of them, *Salyut 6*, was the first of a "second generation" of such stations. Unlike its predecessors, it had two docking ports, instead of only one. This enabled it to receive visiting crews and resupply ships. Together with other new on-board systems, this feature was the key to missions of considerably extended duration. Its successor, *Salyut 7*, was launched on April 19, 1982.

On February 19, 1986, the Soviet Union launched the core vehicle in its new *Mir* (Peace) space station complex series. This complex represents a new-generation space station which evolved from the *Salyut* and *Kosmos* series vehicles. The core, an advanced version of *Salyut*, has six docking ports and consists of four sections. Connected to various docking ports are four laboratory modules, which were launched separately between 1989 and 1996. These modules are dedicated to different scientific and technical disciplines or functions, including technological production with a shop, astrophysics, biological research, and medical research.

Phase 1 of the International Space Station program, begun in 1995, involves dockings of *Mir* with the United States space shuttle, ferrying of cosmonauts and supplies on the space shuttle to *Mir*, and the participation of United States astronauts as members of *Mir* crews. However, in 1997 the performance of *Mir* was adversely affected by a series of malfunctions and mishaps aboard the aging space station.

With the successful test flight of the powerful expendable heavy-lift launcher *Energia* on May 15, 1987, the Soviet Union gained a tremendous new launch capability. With its second, and so far last, flight on November 15, 1988, the *Energia* launched the Soviet space shuttle *Buran* on its only (crewless) orbital test flight.

United States programs. During the first decade after its inception in 1961, the United States crewed space program was conducted in three major phases—Mercury, Gemini, and Apollo.

Mercury was, in its basic characteristics, similar to the Soviet *Vostok*, but it weighed only about a third as much, as necessitated by the smaller missiles of the United States at that time (the Redstone and Atlas). The one-person *Mercury* capsules used ballistic reentry and were designed to answer the basic questions about humans in space: how they were affected by weightlessness, how they withstood the gravitational forces of boost and entry, how well they could perform in space.

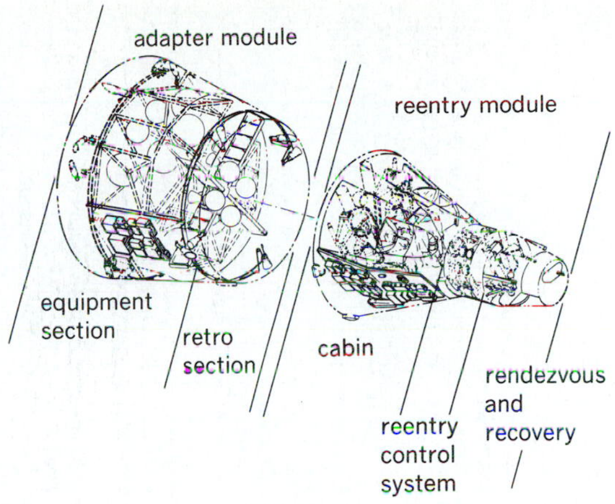

Fig. 1. **Gemini spacecraft. (NASA)**

The second United States step into space was the Gemini Program. With the two-person *Gemini* capsule (Fig. 1), for the first time a crewed spacecraft had been given operational maneuverability in space. Its reentry module flew a lifting reentry trajectory for precise landing point control. In addition, its design permitted EVA by one of the crew.

The third-generation spacecraft, *Apollo*, had five distinct parts: the command module (CM), the service module (SM), the lunar module (LM), the launch escape system (LES), and the spacecraft/lunar module adapter (SLA). The three modules made up the basic spacecraft; the LES and SLA were jettisoned early in the mission after they had outlived their function. The CM served as the control center for the spacecraft and provide living and working quarters for the three-member crew for the entire flight, except for the period when two persons entered the LM for the descent to the Moon and return. The CM was the only part of the spacecraft that returned to Earth, flying a lifting trajectory with computer-steered maneuvers. The LM carried two astronauts from the orbiting command/service module (CSM) down to the surface of the Moon, provided a base of operations there, and returned the two astronauts to a rendezvous with the CSM in orbit. On the last three lunar landings, *Apollo 15–17*, lunar exploration was supported by the lunar roving vehicle. *See* MOON.

A more powerful booster was required to lift the *Apollo* spacecraft to Earth orbit and thence to the Moon. At the Army Ballistic Missile Agency (ABMA) in 1958, a team of engineers under Wernher von Braun set out to prove that vastly more powerful space rockets could be built from existing hardware by clustering engines and tanks. This project evolved into the Saturn Program of NASA. The Saturn 5, which became the Apollo lunar launch vehicle, had the capability of lift 250,000 lb (113 metric tons) into low Earth orbit and to send 100,000 lb (45 metric tons) to the Moon.

Skylab. The experimental space station *Skylab* (Fig. 2) was the largest object ever placed in space, and the first crewed project in the U.S. Space Program with the specific purpose of developing the utility of space flight in order to expand and enhance humanity's well-being on Earth. To that end, *Skylab's* equipment included Earth resources remote sensing instruments and the first crewed solar telescopes in space. A total of three crews of three astronauts each carried out experiments and observations on *Skylab*. *Skylab* was launched uncrewed by Saturn 5 in 1973. Subsequently visited by three astronaut crews the space station underwent repair in orbit, a first for the space program.

Space Transportation System. After the joint American-Soviet space mission in 1975, attention began to focus on the routine application of newly acquired know-how, systems, and

Crewed space flights, 1961–1996*

Dates launched and recovered	Designation	Crew	Weight, lb	Revolutions; max. distance from Earth, mi	Duration	Remarks
1961						
Apr. 12	*Vostok 1*	Yuri A. Gagarin	10,419	1; 203	1 h 48 min	First person in space, conducted radio and TV communication with Earth. Spacecraft contained life-support systems, telemetry, ejection seat, and recovery system. Cosmonaut and spacecraft landed at preselected area. Automatic control with cosmonaut available as backup. Two-gas air supply at sea-level pressure.
May 5	*Freedom 7*	Alan B. Shepard, Jr.	2,855	—; 116	15 min	*Mercury* spacecraft launched in ballistic suborbital trajectory by Redstone booster (*Mercury-Redstone 3*). Downrange distance 302 mi. First United States person in space; weightless for 5 min with no ill effects. Astronaut exercised manual control of spacecraft.
July 2–21	*Liberty Bell 7*	Virgil I. Grissom	2,836	—; 118	15 min	*Mercury* spacecraft launched into ballistic trajectory by Redstone booster (*Mercury-Redstone 4*). Downrange distance 303 mi. Weightless for 5 min with no ill effects. Hatch opened prematurely during recovery, spacecraft filled with water and sank in 2500 fathoms of water in Atlantic Ocean. Astronaut was recovered.
Aug. 6–7	*Vostok 2*	Gherman S. Titov	10,432	16; 152	25 h 11 min	First test of prolonged weightlessness; cosmonaut ate, worked, and slept in space. Monitored by TV and radio. Vestibular disturbances produced motion sickness but apparently no significant aftereffects.
1962						
Feb. 20	*Friendship 7*	John H. Glenn, Jr.	2,987	3; 162	4 h 55 min	*Mercury* spacecraft launched into orbit by Atlas booster (*Mercury-Atlas 6*). First United States crewed orbital flight, achieving original Project Mercury objectives of placing a human in orbit, observing reactions to space environment, and making a safe recovery. No adverse physiological effects from the 4.5 h of weightlessness.
May 24	*Aurora 7*	M. Scott Carpenter	2,975	3; 167	4 h 56 min	*Mercury* spacecraft launched into orbit by Atlas booster (*Mercury-Atlas 7*). Flight plan similar to *Friendship 7*. Manual attitude control exercised during retrofire. Landed 250 mi beyond predicted impact point.
Aug. 11–15	*Vostok 3*	Andrian G. Nikolayev	10,412	60; 146	94 h 22 min	Launched as first half of tandem mission with *Vostok 4*. Radiation measurements taken and effects of biological specimens studied. Extensive physiological measurements recorded during prolonged exposure to space environment with no adverse effects noted. TV monitoring, radio and TV contact with *Vostok 4*. Ccsmonaut floated free of restraint for a total of 3.5 h, was ejected to land separately from spacecraft.
Aug. 12–15	*Vostok 4*	Pavel R. Popovich	10,425	45; 147	70 h 57 min	Launched from some facility as *Vostok 3* into tandem flight. Passed to within 4 mi of *Vostok 3* at closest point. Extensive radiation and physiological experiments conducted. TV monitoring, radio and TV contact with *Vostok 3*. Cosmonaut floated free of restraint for a total of about 3 h, was ejected to land separately from spacecraft.
Oct. 3	*Sigma 7*	Walter M. Schirra, Jr.	3,029	5; 176	9 h 13 min	*Mercury* spacecraft launched into orbit by Atlas booster (*Mercury-Atlas 8*). Automatic reentry mode utilized, landed 9000 yd from prime recovery ship for first recovery in Pacific Ocean. Experiments included camera, radiation sensors, and ablative samples on the neck of spacecraft.
1963						
May 15–16	*Faith 7*	L. Gordon Cooper, Jr.	3,033	21; 166	34 h 20 min	*Mercury* spacecraft launched into orbit by Atlas booster (*Mercury-Atlas 9*). Long-endurance mission with no adverse physiological effects as astronaut ate, slept, and worked in the space environment. Reentry sequence was with astronaut orienting spacecraft and firing retrorockets manually. A very accurate landing was made in the Pacific Ocean 7000 yd from prime recovery ship. Experiments included infrared and standard photography, radiation measurement, ejection of flashing light (sighted in fifth and sixth orbits), and observation of lights on the ground.
June 14–19	*Vostok 5*	Valery F. Bykovsky	10,408	76; 146	119 h 6 min	New time and distance space-flight records set, last 45 revolutions in tandem with *Vostok 6*. Radio contact with *Vostok 6*. Direct TV broadcast to Earth in real time. Extensive medical and biological experiments conducted.

June 16–19	*Vostok 6*	Valentina V. Tereshkova	10,392	45; 143	70 h 50 min	Orbited first woman and first person not trained as a pilot. Passed within proximity of *Vostok 5* on first orbit, in different orbital plane. Radio contact with *Vostok 5*. Direct TV broadcast to Earth in real time. Parachuted separately from spacecraft on landing.
1964						
Oct. 12–13	*Voskhod 1*	Vladimir M. Komarov, Konstantin P. Feoktistov, Boris B. Yegorov	11,731	15; 254	24 h 17 min	First multicrewed space flight; first "shirt-sleeve" environment in space flight, no pressure suits; record altitude for crewed space flight. Feoktistov, scientist, and Yegorov, physician, as experimenters and observers in addition to Komarov as pilot. Both Feoktistov and Yegorov experienced motion sickness and disorientation illusions. Crew worked, ate, drank, and slept unrestrained. Crew remained in spacecraft through landing.
1965						
Mar. 18–19	*Voskhod 2*	Pavel I. Belyayev, Aleksei A. Leonov	12,529	16; 308	26 h 2 min	New record altitude for crewed space flight. During second orbit Leonov took first space walk with autonomous life-support system, moved 17 ft from spacecraft on tether, 23 min in space environment, including 11 min in air lock through which he exited from spacecraft. While outside, Leonov was observed through the spacecraft TV camera. First Soviet landing sequence using manual system in lieu of automatic. Crew remained in spacecraft through landing. Planned landing point was overshot: landed in deep snow.
Mar. 23	*Gemini 3*	Virgil I. Grissom, John W. Young	7,111	3; 140	4 h 53 min	*Gemini* spacecraft launched into orbit by modified Titan 2 booster (*Gemini-Titan 3*). First United States two-person space flight, Grissom first person in space for second time. One orbital maneuver was conducted in each of the three orbits. Manual control was exercised throughout the reentry phase, using the limited lifting characteristics of the spacecraft to steer toward touchdown.
June 3–7	*Gemini 4*	James A. McDivitt, Edward H. White II	7,879	62; 184	97 h 56 min	*Gemini* spacecraft launched into orbit by modified Titan 2 booster (*Gemini-Titan 4*). Endurance mission of 4 days to test crew and spacecraft in buildup to longer missions. Eleven scientific and engineering experiments, in addition to basic medical checks on 4 days of exposure to the space environment. White spent 22 min outside the spacecraft in EVA, propelling himself with an oxygen-jet gun. During space walk he was attached to the spacecraft with a 25-ft tether-and-oxygen line; he carried emergency oxygen and camera. Spacecraft computer malfunction necessitated a ballistic reentry sequence.
Aug. 21–29	*Gemini 5*	L. Gordon Cooper, Jr., Charles Conrad, Jr.	7,947	120; 217	190 h 55 min	*Gemini* spacecraft launched into orbit by modified Titan 2 booster (*Gemini-Titan 5*). Endurance mission of 8 days confirmed the physiological feasibility of Apollo lunar landing mission. First flight of fuel-cell electrical power system. On Aug. 23 a simulated rendezvous was conducted with a series of four maneuvers through two orbits which raised the perigee to 124 mi and apogee to 192.6 mi. Five medical experiments measured physiological effects; extensive weather and terrain photography was conducted. Voice communication was conducted with *Sea Lab 2* under the Pacific Ocean.
Dec. 4–18	*Gemini 7*	Frank Borman, James A. Lovell, Jr.	8,076	206; 204	330 h 35 min	*Gemini* spacecraft boosted into orbit by modified Titan 2 booster (*Gemini-Titan 7*). Endurance mission of 14 days, longest scheduled flight in Gemini program. Station-keeping conducted with booster second stage after separation; spacecraft maneuvered to serve as rendezvous target for *Gemini 6*. Accurate controlled reentry was made, landing 7.6 mi from target. One astronaut at a time doffed his pressure suit to enhance comfort. No adverse physiological effects from 2 weeks in the space environment. Twenty experiments were incorporated into the mission, ranging from detailed medical measurements and sleep analysis to Earth photography and radiometry.

*1 lb = 0.45 kg. 1 mi = 1.6 km. 1 ft = 0.3 m. 1 yd = 0.9 m. 1 fathom = 1.8 m. 1 lb/in.^{2} = 6.9 kPa.

Crewed space flights, 1961–1996* (cont.)

Dates launched and recovered	Designation	Crew	Weight, lb	Revolutions; max. distance from Earth, mi	Duration	Remarks
1965 (cont.)						
Dec. 15–16	*Gemini 6*	Walter M. Schirra, Jr., Thomas P. Stafford	7,817	16; 193	25 h 51 min	*Gemini* spacecraft launched into orbit by modified Titan 2 booster (*Gemini-Titan 6*). Rendezvous with *Gemini 7* as target. A series of posigrade maneuvers began the rendezvous chase and shortened the initial 1200 mi separation distance to 730 mi. During second revolution the distance was reduced to 300 mi with another thruster firing, followed by a plane change to match *Gemini 7's* orbit. Orbit was adjusted to near circular at beginning of third revolution; as fourth revolution began, the terminal phase of rendezvous was initiated, resulting in approach to within 1 ft of *Gemini 7*. Station-keeping and fly-around maneuvers were conducted for 5 h 19 min. Controlled reentry to landing within 13 mi of prime recovery ship.
1966						
Mar. 16	*Gemini 8*	Neil A. Armstrong, David R. Scott	8,351	6½; 186	10 h 41 min	*Gemini* spacecraft launched into orbit by modified Titan 2 booster, 1 h 41 min after Gemini-Agena target vehicle (GATV)-Atlas booster combination. Rendezvous with the GATV was accomplished in fourth orbit, 6 h after launch, and first docking of two vehicles in space was achieved 33 min later. After 28 min of flight in the docked configuration, a thruster in the orbital-attitude maneuvering system stuck open and the spacecraft began to spin. *Gemini* spacecraft was undocked from GATV and, after 25 min, was stabilized by use of reentry control system. Mission terminated early with landing in secondary recovery area in western Pacific.
June 3–6	*Gemini 9a*	Thomas P. Stafford, Eugene A. Cernan	8,268	45; 194	72 h 21 min	Mission utilizing augmented target docking adapter (ATDA) as the target was placed into orbit on June 1 with an Atlas booster. The *Gemini* was launched into orbit on June 3 by its modified Titan 2 booster. Rendezvous with ATDA achieved on third orbit. Crew confirmed that shroud covering the docking collar had failed to separate, and docking was canceled. Two additional rendezvous were performed as planned, one using visual techniques, the last from above the ATDA. After 1 h of EVA, Cernan's visor accumulated fog, and communications between Stafford and Cernan were poor. Planned use of the self-contained life-support and propulsion systems to maneuver was canceled. Total EVA was 2 h 7 min. Controlled reentry to 0.42 mi of target.
July 18–21	*Gemini 10*	John W. Young, Michael Collins	8,248	43; 476	70 h 47 min	GATV-Atlas booster combination launched 100 min before *Gemini* spacecraft launched into orbit by modified Titan 2 booster. Rendezvous and docking with GATV 10 accomplished in fourth orbit. GATV engine used to propel the docked combination to record altitude, then proper orbit to rendezvous with GATV 8. After separation from GATV 10, *Gemini 10* effected rendezvous with the passive target. First period of standup EVA was terminated at 50 min when both crew members suffered eye irritation. Second EVA period, after rendezvous with passive GATV, consisted of Collins on umbilical moving to GATV and recovering micrometeorite experiment package.
Sept. 12–15	*Gemini 11*	Charles Conrad, Jr., Richard F. Gordon, Jr.	8,509	44; 853	71 h 17 min	GATV-Atlas booster combination launched 1 h 37 min before *Gemini* spacecraft was launched into orbit by modified Titan 2 booster within 2-s launch window. Rendezvous accomplished on first orbit using on-board information exclusively. Docking at 1 h 34 min after launch. Docking accomplished twice by each crew member. During umbilical EVA, Gordon removed the nuclear-emulsion experiment package and attached tether between *Gemini* spacecraft and GATV. Excessive fatigue from these activities caused early termination of EVA at 33 min. Using GATV primary propulsion system, docked configuration was propelled to an altitude of 853 mi. Hatch opened third time for 2 h 10 min standup EVA, during which scheduled photography was accomplished. The two space-

						craft undocked, and station-keeping by use of a 100-ft nylon tether was exercised with a slow spin rate imparted to the tethered combination. After separation of the tether, a second rendezvous with GATV was conducted from 25 mi separation. Reentry sequence was fully automatic, with impact 2.9 mi from the aiming point.
Nov. 11–15	*Gemini 12*	James A. Lovell, Jr., Edwin E. Aldrin, Jr.	8,297	59; 187	94 h 35 min	GATV-Atlas booster combination launched 1 h 38 min before *Gemini* spacecraft was launched into orbit by modified Titan 2 booster. Rendezvous was accomplished in third orbit with docking 4 h 14 min after *Gemini* launch. Two standup EVA periods, on first and third days, totaled 3 h 34 min. On second day Aldrin conducted 2 h 6 min umbilical EVA, testing restraint devices to overcome body-positioning problems experienced on prior flights. Resting frequently, Aldrin used portable hand rails, foot restraints, and various tethers. He completed 19 tasks, demonstrating that useful EVA work was feasible with proper planning and restraining devices. *Gemini* and GATV spacecraft undocked, remaining joined by 100-ft tether for station-keeping and gravity-gradient stabilization experiment. Fourteen experiments were conducted. Automatic reentry sequence was used for second time.
1967						
Apr. 23–24	*Soyuz 1*	Vladimir M. Komarov	14,000	18; 139	24 h 45 min	After a 2-year hiatus, Soviet Union crewed space flight resumed with *Soyuz 1*. During reentry on its eighteenth orbit, after 25.2 h of flight, *Soyuz 1*'s parachutes fouled and the 14,000-lb spacecraft crashed, killing Komarov.
1968						
Oct. 11–22	*Apollo 7*	Walter M. Schirra, Jr., Donn Eisele, Walter Cunningham	87,382	163; 244	260 h 9 min 3 s	First crewed flight of *Apollo* command and service modules. Launched by a Saturn 1B, this was an engineering test flight and was a complete success. Rendezvous and simulated docking exercises with the Saturn upper stage were practiced. A TV camera was carried, and a series of in-flight programs were staged by the astronauts, showing their activities within the command module. Transmissions were relayed live to home receivers all over the United States.
Oct. 26–30	*Soyuz 3*	Georgi T. Beregovoi	14,000	64; 151	94 h 51 min	First successful crewed flight of the *Soyuz* spacecraft. Accomplished orbital rendezvous with the uncrewed *Soyuz 2*, which was orbited several days previously. Docking was not attempted. Automatic control was used to bring the two craft within 650 ft of each other. At the point the pilot assumed manual control, and he made at least two simulated dockings. Television pictures of the inside of the spacecraft were relayed to home receivers in the Soviet Union and Europe.
Dec. 21–27	*Apollo 8*	Frank Borman, James A. Lovell, Jr., William Anders	87,000	10; 231,000 (lunar	146 h 59 min 49 s	First crewed circumlunar orbital mission. Also first use of Saturn 5 launch vehicle for a crewed flight. Remained in an orbit about 69 mi above the lunar surface for approximately 20 h (10 revolutions). The ability to use lunar landmarks for navigation to the lunar landing sites was successfully verified. Communications with Earth were excellent, and TV pictures of good quality, of both the lunar surface and the Earth from lunar and translunar distances, were transmitted to the Earth from circumlunar reentry speed. Landing was within 3 mi of the recovery ship.
1969						
Jan. 14–17	*Soyuz 4*	Vladimir A. Shatalov	14,000	48; 158	71 h 14 min	Rendezvoused and docked with *Soyuz 5*, to form the "first experimental space station." Cosmonaut Shatalov in *Soyuz 5* performed the actual docking. The two spacecraft remained docked for 4 h 35 min, during which time cosmonauts Khrunov and Yeliseyev transferred from *Soyuz 4* to *Soyuz 5*, spending about 1 h outside the spacecraft in the process.
Jan 15–18	*Soyuz 5*	Boris V. Volynov, Yevgeny V. Khrunov, Aleksei S. Yeliseyev	14,000	50; 158	72 h 46 min	After docking with *Soyuz 4*, cosmonauts Khrunov and Yeliseyev transferred to *Soyuz 5* and cosmonaut Volynov remained in the spacecraft. Details of the transfer operation were transmitted over TV.
Feb. 28–Mar. 13	*Apollo 9*	James A. McDivitt, David R. Scott, Russell L. Schweikart	95,231	151; 240	241 h 1 min	Earth orbital flight designed to give the lunar module its first crewed flight test. This vehicle was crewed for 28 h by astronauts Schweikart and McDivitt. During this period the hatches of both command module and the lunar module were opened. Schweikart got out onto the "front porch" of the lunar module and thus was able to exercise the suit and backpack

*1 lb = 0.45 kg. 1 mi = 1.6 km. 1 ft = 0.3 m. 1 yd = 0.9 m. 1 fathom = 1.8 m. 1 lb/in.^{2} = 6.9 kPa.

Crewed space flights, 1961–1996* (cont.)

Dates launched and recovered	Designation	Crew	Weight, lb	Revolutions; max. distance from Earth, mi	Duration	Remarks
1969 (cont.)						designed for operation on the lunar surface. Some curtailment of this activity was deemed advisable because of Schweikart's previous nausea. Subsequently, the lunar module was separated from the command module, and after 6 h 24 min was rendezvoused and docked with the command module. The astronauts in the lunar module reentered the command module and then jettisoned the lunar module. Numerous navigation and control exercises and Earth photography experiments were accomplished. Because of weather conditions in the prime recovery areas, the landing area was moved 500 mi south. Splashdown occurred within 3 mi of the recovery ship.
May 18–26	*Apollo 10*	Thomas P. Stafford, John W. Young, Eugene A. Cernan	96,000	31; 251,000 (lunar)	192 h 3 min 23 s	Lunar module tested in lunar orbit. Astronauts Stafford and Cernan entered the lunar module, which separated from the command module, descended to within 50,000 ft of the lunar surface, and then rendezvoused and docked with the command module, which had remained in an orbit about 69 mi above the lunar surface. Color TV pictures were transmitted to the Earth of the lunar surface, of the Earth from lunar and translunar distances, and of the activities of the astronauts inside the command module. This flight successfully verified that the *Apollo* spacecraft was ready to attempt a lunar landing on the next flight.
July 16–24	*Apollo 11*	Neil A. Armstrong, Michael Collins, Edwin E. Aldrin, Jr.	96,000	31; 250,000 (lunar CM)	195 h 17 min 12 s	The objective of the Apollo program (a crewed lunar landing) was achieved in this flight. Astronaut Armstrong emerged from the lunar module on July 20, 1969, followed shortly afterward by Aldrin. This event was recorded by a TV camera and relayed to home receivers in many parts of the world by satellite. The astronauts gathered samples of the lunar surface material for return to Earth. After deploying a United States flag, they installed a passive seismic instrument and a laser reflector on the lunar surface so that data of scientific interest could be obtained after their return to Earth. The astronauts found that mobility on the lunar surface in the one-sixth gravity environment presented no unanticipated problems, and walking was relatively easy. During the whole flight all the systems functioned correctly and the other parts of the flight followed *Apollo 10* pattern closely.
Oct. 11–16	*Soyuz 6*	Georgi S. Shonin, Valeri N. Kubasov	14,000	79; 144	118 h 42 min	Tested effectiveness of various welding techniques in space. Along with *Soyuz 7* and *8*, tested multiple launch capacity of Soviet Union and performed various scientific investigations.
Oct. 12–17	*Soyuz 7*	Anatoly V. Filipchenko, Vladislav N. Volkov, Viktor V. Gorbatko	14,000	—; 141	118 h 41 min	Along with *Soyuz 6* and *8*, tested capacity of Soviet Union for multiple launch and rendezvous; also performed navigation experiments and various scientific investigations.
Oct. 13–18	*Soyuz 8*	Vladimir A. Shatalov, Aleksei S. Yeliseyev	14,000	—; 140	119 h	Along with *Soyuz 6* and *7*, tested capacity of Soviet Union for multiple launch and rendezvous; also tracked other ships' maneuvers and performed various scientific investigations.
Nov. 14–22	*Apollo 12*	Charles Conrad, Jr., Richard F. Gordon, Jr., Alan L. Bean	96,000	45; 250,000 (lunar CM)	142 h 52 min	More lengthy exploration of the Moon's surface.
1970						
Apr. 11–17	*Apollo 13*	James A. Lovell, Jr., Fred W. Haise, Jr., John L. Swigert, Jr.	97,000	—; 250,000	142 h 52 min	Rupture of a fuel-cell oxygen tank in service module forced cancellation of the third crewed trip to the Moon. The rupture was caused by high pressure which developed when the failure of two thermal switches responsible for regulating the tank heaters led to abnormally high temperatures within the tank, produced by the combustion of Teflon in the supercritical (cryogenic liquid) oxygen. Lunar module life-support system enabled crew to return to Earth.
June 1–19	*Soyuz 9*	Andrian G. Nikolayev, Vitali I. Sevastyanov	14,000	300; 167	424 h 59 min	Predominantly biomedical flight, testing human ability to live and work for prolonged periods of time in spacecraft environment. Earth resources applications experiments were also performed.

1971						
Jan. 31–Feb. 9	*Apollo 14*	Alan B. Shepard, Jr., Stuart A. Roosa, Edgar D. Mitchell	98,000	34; 250,000 (lunar CM)	216 h 2 min	Third lunar landing. Precision touchdown accomplished by Shepard and Mitchell in "Antares." On two EVAs, crew deployed Apollo lunar surface scientific experiment package (ALSEP), completed other scientific tasks. First extraterrestrial use of a two-wheeled cart, modular equipment transporter (MET). New record surface time, 9 h 24 min. Surface samples returned to Earth, 94 lb.
Apr. 19	*Salyut 1*	Crewless	40,000	—; 176	175 days	First Soviet space station. Prototype. Launched crewless. Reentered Oct. 11, 1971.
Apr. 23–24	*Soyuz 10*	Vladimir A. Shatalov, Aleksey S. Yeliseyev, Nikolai Rukavishnikov	14,500	32; 165	47 h 46 min	Three-person spacecraft, docked with *Salyut 1* space station. Remained docked for 5.5 h, but no crew transfer took place, probably because of mechanical difficulties.
June 6–30	*Soyuz 11*	Georgi T. Dobrovolskiy, Vladislav N. Volkov, Viktor I. Patsayev	14,300	385; 140	570 h 22 min	Second docking with *Salyut 1*. Crew boarded *Salyut* and remained for 23 days 17 h 40 min, breaking previous *Soyuz 9* record. After extensive on-board activities, crew transferred to *Soyuz* for Earth return and was killed during reentry by pulmonary embolisms, caused by rapid decompression of cabin through leaking hatch valve.
July 26–Aug. 7	*Apollo 15*	David R. Scott, Alfred M. Worden, James B. Irwin	107,000	74; 250,000 (lunar CM)	295 h	Perfect landing by Scott and Irwin in "Falcon" at Hadley Rill. First use of lunar roving vehicle (LRV), driven a total of 17.5 mi during three surface EVAs. Brought back more than 170 lb of samples. First subsatellite, carrying particles and fields (P&F) experiment, launched in lunar orbit by crewed spacecraft. First inflight EVA performed by Worden in deep space on return leg.
1972						
Apr. 16–27	*Apollo 16*	John W. Young, Thomas K. Mattingly II, Charles M. Duke	103,000	64; 250,000 (lunar CM)	265 h 51 min 5 s	Landing by Young and Duke in "Orion" at Descartes. Second use of LRV, for a total of 16.6 driven miles. Three EVAs totaling 20 h 14 min yielded 209 lb of samples. Fourth ALSEP station was deployed on surface and second P&F subsatellite launched in lunar orbit. On return leg, Mattingly undertook second deep-space EVA to retrieve film from external bay.
Dec. 7–19	*Apollo 17*	Eugene A. Cernan, Ronald E. Evans, Harrison H. Schmitt	103,000	75; 250,000 (lunar CM)	301 h 51 min	Last *Apollo* lunar landing, by Cernan and Schmitt, a geologist, in "Challenger" in mountainous region at Taurus-Littrow. Third LRV was driven 21 mi; fifth ALSEP deployed. Longest lunar stay time (74 h 59 min 38 s), with three EVAs setting new record of 22 h 5 min 6 s. Most samples returned to Earth, 250 lb. Third deep-space film retrieval EVA, by Evans.
1973						
Apr. 3	*Salyut 2*	Crewless	40,000	—	—	Second Soviet space station. Failed in orbit.
May 14–July 11, 1979	*Skylab 1*	Crewless	170,000 (without payload shroud)	—; 276	—	First United States (experimental) space station, launched crewless on two-stage Saturn 5. At 60 s after launch, the workshop meteoroid/thermal shield tore loose. Primary solar array wing was lost, the other jammed closed by debris. As a result, inside temperatures became too high and power output too low for long-term human habitation. Station was saved through determined effort by NASA engineers on ground and astronauts in space, to become successful beyond expectations. Station reentered July 11, 1979.
May 25–June 22	*Skylab 2*	Charles Conrad, Jr., Joseph P. Kerwin, Paul J. Weitz	31,000	404; 276	28 days 49 min	Crew launch on Saturn 1B postponed from May 15 to analyze problems. After entering *Skylab*, crew deployed parasollike thermal blanket on workshop outside through small air lock. On June 7, Conrad and Kerwin performed hazardous EVA to repair jammed solar wing. Three EVAs totaled 5 h 51 min.
July 28–Sept. 25	*Skylab 3*	Alan L. Bean, Jack R. Lousma, Owen Garriott	31,000	858; 270	59 days 1 h 9 min	Launch vehicle, Saturn 1B. Crew erected new twin-pole sun shield on EVA, installed new rate gyros. Extended solar observations with Apollo telescope mount (ATM), and 39 Earth observation passes. Three EVAs totaled 13 h 44 min. Space manufacturing and maneuvering unit tests added to mission schedule.
Sept. 27–29	*Soyuz 12*	Vasily G. Lazarev, Oleg G. Makarov	14,460	32; 216	Approx. 48 h	First flight after *Soyuz 11* tragedy. Short-duration test of a modified *Soyuz* with new hatch/valve design, improved maneuvering and manual control capabilities. Tested new space suit design and cabin suit loop.
Nov. 16–Feb. 8, 1974	*Skylab 4*	Gerald P. Carr, William R. Pogue, Edward Gibson	31,000	1214; 270	84 days 1 h 16 min	Launch vehicle, Saturn 1B. Last *Skylab* mission with 4 EVAs for film exchange and mechanical repairs, totaling 22 h 21 min. Mission included first exoatmospheric observation of a comet, Kohoutek.

*1 lb = 0.45 kg. 1 mi = 1.6 km. 1 ft = 0.3 m. 1 yd = 0.9 m. 1 fathom = 1.8 m. 1 lb/in.^{2} = 6.9 kPa.

Crewed space flights, 1961–1996* (cont.)

Dates launched and recovered	Designation	Crew	Weight, lb	Revolutions; max. distance from Earth, mi	Duration	Remarks
1973 (cont.)						
Dec. 18–26	*Soyuz 13*	Pyotr I. Klimuk, Valentin V. Lebedev	14,400	—; 169	7 days 20 h 55 min	Similar to *Soyuz 12*, with addition of *Orion 2* telescope system for solar, stellar, and cometary studies. Also included were biomedical studies, such as Levka brain-bloodflow device.
1974						
June 25	*Salyut 3*	Crewless	40,000	—; 172	214 days	Third Soviet space station. Military type. New solar panel and other systems.
July 3–19	*Soyuz 14*	Pavel R. Popovich, Yuri P. Artukhin	14,400	—; 169	15 days 17 h 39 min	Crew docked and transferred to *Salyut 3* space station. Purpose was to test the new space station. Stay on board was 14½ days.
Aug. 26–28	*Soyuz 15*	Gennady V. Sarafanov, Lev S. Demin	14,400	32; 183	48 h 12 min	Intended as second crew transfer to *Salyut 3*, but unable to dock because remotely controlled docking system failed.
Dec. 2–8	*Soyuz 16*	Anatoly V. Filipchenko, Nikolai Rukavishmikov	14,400	—; 140	5 days 22 h 24 min	Soviet test and simulation of ASTP mission, including new compatible docking system and reduction of spacecraft pressure from 14.7 to 10 lb/in.^{2} in flight. To approximate the United States docking system for docking tests, a dummy structural ring was taken along and separated in orbit.
Dec. 26	*Salyut 4*	Crewless	40,000	—; 226	770 days	Fourth Soviet space station. Civilian type.
1975						
Jan. 11–Feb. 9	*Soyuz 17*	Aleksei Gubarev, Georgiy Grechko	14,400	—; 225	29 days 13 h 20 min	Crew docked and transferred to *Salyut 4* space station.
Apr. 5	*Soyuz 18A*	Vasiley Lazarev, Oleg Makarov	14,400	—; —	Approx. 20 min	Suborbital flight. Mission aborted after launch due to vehicle failure after approximately 20 min of flight.
May 24–July 26	*Soyuz 18*	Pyotr I. Klimuk, Vitali I. Sevastyanov	14,400	—; 220	62 days 23 h 20 min	Second crew transfer to *Salyut 4*, which had completed 3352 revolutions at the time. Flew part of the time simultaneously with ASTP.
July 15–24	Apollo-Soyuz Test Project (ASTP)	United States: Thomas P. Stafford, Vance D. Brand, Donald K. Slayton	32,500	138; 143	9 days 1 h 28 min	First international link-up in space and last flight of an *Apollo* spacecraft. *Apollo* launched on Saturn 1B 7½ h after *Soyuz*. First handshake between Stafford and Leonov occurred 3:19 P.M. EDT on July 17. Crews performed interspacecraft transfer four times and conducted joint activities. A total of 31 experiments were performed by American crew, some of them jointly with Soviets, and docking tests with *Soyuz* were practiced. After final undocking, each spacecraft continued on own mission. *Soyuz* returned July 21. Overall Soyuz-Apollo mission duration 9 days 8 h 58 min.
		Soviet Union: Alekseiy A. Leonov, Valerity N. Kubasov	15,000	96; —	5 days 22 h 31 min	
				31 (docked)	1 day 23 h 17 min (docked)	
1976						
June 22	*Salyut 5*	Crewless	40,000	—; 169	412 days	Fifth Soviet space station. Civilian type.
July 6–Aug. 24	*Soyuz 21*	Vitaly Zholobov, Boris V. Volynov	14,400	—; 169	49 days 6 h 24 min	Crew docked on July 7 and transferred to *Salyut 5* space station, launched June 22. On-board activities included earth resources photography, weather observations; formation of crystals and metals-melting in zero *g*; plant and insect growth for genetic studies. Crew reportedly suffered sensory deprivation effects. Flight had been preceded by uncrewed *Soyuz 20* (launched Oct. 17, 1975, docked to *Salyut 4* on Oct. 19).
Sept. 15–23	*Soyuz 22*	Valery F. Bykovsky, Vladimir Aksenov	14,400	127; 175	7 days 21 h 54 min	Earth surveillance mission. Test of a special East German camera system for use on space stations.
Oct. 14–16	*Soyuz 23*	Vyacheslav Zudov, Valery Rozhdestvenskiy	14,400	32; 172	2 days 6 min	Attempt to dock with *Salyut 5* space station failed. Crew returned to Earth.
1977						
Feb. 7–25	*Soyuz 24*	Viktor V. Gorbatko, Yuri Glazkov	14,400	285; 176	17 days 17 h 23 min	Crew transferred to *Salyut 5* on Feb. 8. Space station had mostly military mission with special photo reconnaissance runs.
Sept. 29	*Salyut 6*	Uncrewed	42,000	—; 221	—	Uncrewed launch of new Soviet space station. Reentered July 29, 1982.
Oct. 9–11	*Soyuz 25*	Vladimir Kovalyonok, Valery Ryumin	14,400	32; 221	2 days 46 min	Attempt to dock with *Salyut 6* failed. Crew returned to Earth.
Dec. 10–Jan. 16	*Soyuz 26*	Yuri Romanenko, Georgiy Grechko	14,400	590; 221	37 days 10 h 6 min Crew duration: 96 days 10 h	Crew transferred to *Salyut 6* for a 96-day mission, breaking United States record of 84 days on Mar. 4, 1978. Crew returned on Mar. 16, 1978, with *Soyuz 27*.

1978						
Jan. 10–16	*Soyuz 27*	Vladimir Dzhanibekov, Oleg G. Makarov	14,400	1024; 219	64 days 22 h 53 min Crew duration: 5 days 22 h 59 min	Crew transferred to *Salyut 6* to join *Soyuz 26* crew. Resupplied by first uncrewed logistics flight, *Progress 1* (launched Jan. 20). On return to Earth, crew used the "older" *Soyuz 26* spacecraft, leaving *Soyuz 27* for the long-duration crew (returned Mar. 16).
Mar. 2–10	*Soyuz 28*	Aleksei Gubarev, Vladimir Remek	14,400	125; 221	7 days 22 h 17 min	Crew transferred to *Salyut 6* on Mar. 4 to join *Soyuz 26* crew. First non-Soviet cosmonaut: Remek (Czechoslovakia).
June 16–Sept. 3	*Soyuz 29*	Vladimir Kovalyonok, Alexander Ivanchenko	14,400	125; 230	79 days 15 h 23 min Crew duration: 140 days	Crew transferred to *Salyut 6* on June 17 to establish new duration record (140 days). Returned on Nov. 2 with *Soyuz 31* spacecraft. Resupplied by: *Progress 2* (launched July 7), *Progress 3* (launched Aug. 7), *Progress 4* (launched Oct. 3).
June 27–July 5	*Soyuz 30*	Pyotr I. Klimuk, Miroslav Hermaszewski	14,400	125; 225	7 days 22 h 4 min	Crew transferred to *Salyut 6* to join *Soyuz 29* crew. Second non-Soviet cosmonaut: Hermaszewski (Poland).
Aug. 26–Nov. 2	*Soyuz 31*	Valery F. Bykovsky, Sigmund Jaehn	14,400	1070; 221	67 days 20 h 14 min Crew duration: 7 days 20 h 49 min	Crew transferred to *Salyut 6* to join *Soyuz 29* crew. Third non-Soviet cosmonaut: Jaehn (East Germany). Crew returned with *Soyuz 29* spacecraft.
1979						
Feb. 25–June 13	*Soyuz 32*	Vladimir Lyakhov, Valery Ryumin	14,400	1711; 232	108 days 4 h 24 min Crew duration: 175 days	Crew transferred to *Salyut 6* to establish new endurance record (175 days). Returned on Aug. 19 with *Soyuz 34*. *Soyuz 32* returned uncrewed. Resupplied by *Progress 5* (launched Mar. 12) and *Progress 6* (launched May 13).
Apr. 10–12	*Soyuz 33*	Nikolai Rukavishnikov, Georgi Ivanov	14,400	31; 221	1 day 23 h 1 min	Docking with *Salyut 6* failed. Crew returned to Earth. Fourth non-Soviet cosmonaut: Ivanov (Bulgaria).
June 6–Aug. 19	*Soyuz 34*		14,400	1152; 257	73 days 18 h 17 min	Launched uncrewed and docked to *Salyut 6*. Returned *Soyuz 32* crew. Resupplied by *Progress 7* (launched June 28). New type of uncrewed Soyuz (*Soyuz T1*) launched Dec. 16 on flight test to dock with *Salyut 6*.
1980						
Apr. 9–June 3	*Soyuz 35*	Leonid Popov, Valery Ryumin	14,400	869; 220	55 days 22 min Crew duration: 184 days 20 h 12 min	Crew transferred to *Salyut 6* for record-duration mission (184 days). Returned Oct. 11 with *Soyuz 37* spacecraft. Resupplied by *Progress 8* (launched Mar. 27), *Progress 9* (launched Apr. 27), *Progress 10* (launched June 29), *Progress 11* (launched Sept. 28).
May 26–July 31	*Soyuz 36*	Valery Kubasov, Bertalan Farkas	14,400	1040; 220	65 days 20 h 54 min	Crew transferred to *Salyut 6* to join with *Soyuz 35* crew. Returned June 3 with *Soyuz 35* spacecraft. Fifth non-Soviet cosmonaut: Farkas (Hungary).
June 5–9	*Soyuz T2*	Yuriy Malyshev, Vladimir Aksenov	14,400	62; 220	3 days 22 h 41 min	First crewed test of new type spacecraft. Crew transferred to *Salyut 6* on June 6. Tenth crewed visit to *Salyut 6*.
July 23–Oct. 11	*Soyuz 37*	Viktor V. Gorbatko, Pham Tuan	14,400	1258; 220	79 days 15 h 17 min Crew duration: 7 days 20 h 42 min	Crew transferred to *Salyut 6* on July 25 to join with *Soyuz 35* crew. Returned July 31 with *Soyuz 36* spacecraft. Sixth non-Soviet cosmonaut: Pham Tuan (Vietnam).
Sept. 18–26	*Soyuz 38*	Yuri Romanenko, Arnaldo T. Mendez	14,400	124; 220	7 days 20 h 43 min	Crew transferred to *Salyut 6* to join with *Soyuz 35* crew. Returned Sept. 26 with *Soyuz 38*. Seventh non-Soviet cosmonaut: Mendez (Cuba).
Nov. 27–Dec. 10	*Soyuz T3*	Leonid Kizim, Oleg G. Makarov, Gennady Strekalov	14,400	202; 220	12 days 19 h 8 min	Crew transferred to *Salyut 6*. Thirteenth crewed visit to *Salyut 6*. First three-person mission since ill-fated *Soyuz 11* in 1971. Returned Dec. 10. *Salyut 6* resupplied by *Progress 12* (launched Jan. 24).
1981						
Mar. 12–May 26	*Soyuz T4*	Vladimir Kovalyonok, Viktor Savinykh	14,400	1183; 220 (estimated)	75 days	Crew transferred to *Salyut 6* on Mar. 14. Savinykh became 100th human to go into space. To date: 43 U.S. astronauts, 57 East-bloc cosmonauts.
Mar. 22–30	*Soyuz 39*	Vladimir Dzhanibekov, Jugderdemidiyn Gurragcha	14,400	125; 220	7 days 20 h 43 min	Crew transferred to *Salyut 6* on Mar. 23 to join with *Soyuz T4* crew. Returned on Mar. 30 with *Soyuz 39*. Eighth non-Soviet cosmonaut: Gurragcha (Mongolia).
Apr. 12–14	*STS 1*	John Young, Robert Crippen	205,000	36; 172	2 days 6 h 20 min 59 s	First crewed orbital test flight of the U.S. space shuttle *Columbia*, for a full, successful orbital check of the revolutionary reusable spacecraft and its many subsystems. Return to Earth, featuring atmospheric entry at Mach 25, hypersonic/supersonic maneuvering flight and subsonic unpowered landing at Edwards AFB, California, was celebrated by millions around the world. Time of landing on Apr. 14: 1:20 P.M. EST.
May 14–22	*Soyuz 40*	Leonid Popov, Dumitru Prunariu	14,400	124; 220	7 days 20 h 41 min	Crew transferred to *Salyut 6* on May 15 to join with *Soyuz T4* crew. Returned May 22 with *Soyuz 40*. Ninth non-Soviet cosmonaut: Prunariu (Romania).

*1 lb = 0.45 kg. 1 mi = 1.6 km. 1 ft = 0.3 m. 1 yd = 0.9 m. 1 fathom = 1.8 m. 1 lb/in.^{2} = 6.9 kPa.

Crewed space flights, 1961–1996* (cont.)

Dates launched and recovered	Designation	Crew	Weight, lb	Revolutions; max. distance from Earth, mi	Duration	Remarks
1981 (cont.)						
Nov. 12–14	*STS 2*	Joe Engle, Richard Truly	210,000	36; 166	2 days 6 h 13 min	Second crewed orbital test flight of the U.S. space shuttle *Columbia.* Launch at Cape Canaveral after two postponements and a liftoff delay due to hardware problems. Mission duration reduced from the planned 124 h to 54 h due to malfunction of on-board fuel cell. Despite shortened mission, crew achieved all objectives successfully. Landing on Rogers Dry Lake, Edwards AFB.
1982						
Mar. 22–30	*STS 3*	Jack R. Lousma, C. Gordon Fullerton	210,000	130; 150	8 days 4 min	Third crewed orbital test of space shuttle *Columbia*. Launch at Cape Canaveral after 1 h delay. Fully successful. Flight extended 1 day to enable landing at Northrup Strip, New Mexico, after rains rendered Rogers Dry Lake, California, too soft.
Apr. 19	*Salyut 7*	Crewless	42,000	—; 163	—	Crewless launch of new Soviet space station, replacing *Salyut 6* (deorbited July 29, 1982). Eleventh anniversary of launch of *Salyut 1. Salyut 7* and an attached module *Cosmos 1686* remained in orbit for almost 9 years, reentering on Feb. 6, 1991, after about 50,200 orbits of the Earth.
May 13–Dec. 10	*Soyuz T5*	Anatoliy Berezovoy, Valentin Lebedev	14,400	Approx. 3380 (crew); 216	106 days Crew duration: 211 days	Crew transferred to *Salyut 7* on May 14 and went on to establish new duration record of 211 days. Was host to two visiting crews, *T6* and *T7*, and returned with *Soyuz T7* spacecraft.
June 25–July 2	*Soyuz T6*	Vladimir Dzhanibekov, Aleksandr Ivanchenkov, Jean-Loup Chretien	14,400	125; 194	7 days 21 h 51 min	Crew transferred to *Salyut 7* on June 25 to join with *T5* crew. Returned July 2 with *T6* spacecraft. Tenth non-Soviet cosmonaut: Jean-Loup Chrétien (France).
June 27–July 4	*STS 4*	Thomas K. Mattingly, Henry W. Hartsfield	210,000	112; 150	7 days 1 h 10 min	Fourth and last orbital test flight of *Columbia,* after 74 workdays of refurbishment. Mission accomplished and flight test program concluded with great success. Landing at Edwards AFB.
Aug. 19–27	*Soyuz T7*	Leonid Popov, Aleksandr Serebrov, Svetlana Savitskaya	14,400	125 (crew); 194	113 days Crew duration: 7 days 21 h 52 min	Crew transferred to *Salyut 7* on Aug. 20 to join *T5* crew. Returned Aug. 27 with *T5* spacecraft. Second woman in space: Svetlana Savitskaya.
Nov. 11–16	*STS 5*	Vance D. Brand, Robert Overmyer, Joseph Allen, William Lenoir	210,000	81; 150	5 days 2 h 14 min	First operational mission of shuttle *Columbia* successfully delivers reimbursable cargo of 2 commercial communications satellites (comsats) to orbit. First four-person crew. After landing at Edwards, *Columbia* returns to manufacturer for changes necessitated by *Spacelab* payload *(STS 9).*
1983						
Mar. 10–Aug. 14	*Kosmos 1443*	Crewless	42,000	—; 214	157 days	*Salyut 7* temporarily enlarged by docking of uncrewed "Kosmodul" *Kosmos 1443* resupply/reentry vehicle.
Apr. 4–9	*STS 6*	Paul Weitz, Karol Bobko, Donald Peterson, Story Musgrave	250,000	80; 183	5 days 24 min	First flight of space shuttle *Challenger* featuring first EVA by shuttle astronauts (Musgrave and Peterson) after 9 years. Cargo of 2-ton *Tracking and Data Relay Satellite (TDRS)* deploys successfully but attains geostationary orbit (GEO) later as planned after major rescue program by NASA (58 days) involving 39 maneuvers by mini-thrust-control jets.
Apr. 20–22	*Soyuz T8*	Vladimir Titov, Gennady Strekalov, Alexandr Serebrov	14,400	32; 194	48 h 18 min	Mission aborted. Crew returned after failure of link-up with *Salyut 7* space station when rendezvous radar deployed incorrectly.
June 18–24	*STS 7*	Robert Crippen, Frederick Hauck, John Fabian, Sally Ride, Norman Thagard	250,000	97; 187	6 days 2 h 24 min	Shuttle *Challenger* with first five-person crew including first American woman in space (Ride). Successful deployment of two commercial comsats *(Anik C-2, Palapa B-1).* First deployment/retrieval of free-flying experiment-carrying platform, *Shuttle Palette Satellite (SPAS 01)* of West Germany, with Canadian-built robot-arm remote manipulator system.
June 27–Nov. 24	*Soyuz T9*	Vladimir Liakov, Aleksandr Aleksandrov	14,400	2390; 214	149 days 9 h 46 min	Crew transferred to *Salyut 7/Kosmos 1443* on June 28, conducted close to 300 experiments during their stay—which was extended unexpectedly when a planned follow-on mission did not come because of explosion of carrier rocket during the launch attempt on Sept. 26. Its crew, Titov and Strekalov, already ill-fated on *T8*, escaped serious injury. *T9* crew added solar arrays, delivered by *Kosmos 1443*, to *Salyut 7*, and returned on Nov. 24 with *T9* spacecraft.

Aug. 30–Sept. 5	*STS 8*	Richard Truly, Daniel Brandenstein, Dale Gardner, William Thornton, Guion Bluford	250,000	97; 193	6 days 1 h 8 min	Third flight of *Challenger*, featuring first night launch (Canaveral) and first night landing (Edwards). First black NASA astronaut in space (Bluford). Successful delivery of Indian comsat *Insat 1-B*, and test of robot-arm remote manipulation system with 4-ton proof mass. First inflight investigations into space motion sickness phenomenon (space adaptation syndrome; SAS) by physician Thornton.
Nov. 28–Dec. 8	*STS 9/SL 1*	John Young, Brewster Shaw, Byron Lichtenberg, Ulf Merbold, Owen Garriott, Robert Parker	250,000 (*Spacelab*: 26,000)	166; 156	10 days 7 h 47 min	First flight of European-built (ESA) *Spacelab* aboard shuttle *Columbia*. First six-person crew. First foreign shuttle astronaut: Merbold (West Germany). Work aboard *Spacelab* on more than 70 experiments in many scientific-technical disciplines went so well that mission was extended by 1 day. Landing at Edwards, after mission had conclusively proved the practical worth of humans in space.
1984						
Feb. 3–11	*STS 41-B*[†] (10)	Vance D. Brand, R. Gibson, Bruce McCandless, Robert Stewart, Ron McNair	250,000	127; 185	7 days 23 h 16 min	Tenth flight of shuttle and fourth flight of *Challenger*, carrying two reimbursable commercial comsats (*Westar 6*, *Palapa B-2*). Satellites deployed successfully, but payload assist module (PAM) upper stage failed in both cases. Comsats retrieved later by mission 51-A. First free (untethered) EVA performed by McCandless, wearing manned maneuvering unit (MMU) back-pack system, to 300 ft from orbiter. Second black astronaut (McNair). First landing at Kennedy Space Center, Florida.
Feb. 8–Oct. 2	*Soyuz T10*	Leonid Kizim, Vladimir Solovyov, Oleg Atkov	14,400	3790 (crew); 200	63 days Crew duration: 236 days 21 h 50 min	Crew transferred to *Salyut 7* on Feb. 9 and went on to set new endurance record of 237 days. During the 34-week stay, Kizim and Solovyov made six EVAs and spent almost 23 h outside the station, mainly for major repair work on *Salyut 7*. Crew took more than 25,000 photographs of Earth and was joined twice by visiting crews (*T11* and *T12*). Resupplied by *Progress 19* on Feb. 23, *Progress 20* on Apr. 17, *Progress 21* on May 8, *Progress 22* on May 30, and *Progress 23* on Aug. 15. Returned on Oct. 2 with *Soyuz T11* spacecraft.
Apr. 2–11	*Soyuz T11*	Juri Malishev, Grennady Strekalov, Rakesh Sharma	14,400	125 (crew); 200	183 days Crew duration: 7 days 21 h 42 min	Crew transferred to *Salyut 7* to join *T10* crew. Returned with more than 1000 photographs of India and other Earth areas with *Soyuz T10* spacecraft. Eleventh non-Soviet cosmonaut: Sharma (India).
Apr. 6–13	*STS 41-C* (11)	Robert Crippen, Richard Scobee, George Nelson, Terry Hart, James van Hoften	250,000 (LDEF: 21,000)	107; 313	6 days 23 h 41 min	Fifth flight of *Challenger*. Flew new "direct" ascent to higher-than-usual altitude for rendezvous with solar telescope satellite *Solar Maximum Mission* (*SMM*), inoperative since end of 1980. Crew deployed large Long Duration Exposure Facility (LDEF), then captured and repaired *SMM* with full success, setting new EVA record of 7 h 18 min.
July 17–19	*Soyuz T12*	Vladimir Dzhanibekov, Sventlana Savitskaya, Igor Volk	14,400	196; 221	12 days 7 h 14 min	Crew transferred to *Salyut 7* on July 18 to join *T10* crew. Second space mission of Savitskaya, who also became world's first female spacewalker (3 h 35 min), performing experiments with electron-beam tool for space assembly and repair. Crew returned with *T12* spacecraft.
Aug. 30–Sept. 5	*STS 41-D* (12)	Henry W. Hartsfield, Michael Coats, Richard Mullane, Stephen Hawley, Judith Resnik, Charles Walker	250,000	96; 185	6 days 56 min	First flight of space shuttle *Discovery*. Cargo: three reimbursable comsats (*SBS 4*, *Telstar 3-C*, *LEASAT 1*) successfully deployed. Also: OAST-1 experiment pallet. Second U.S. woman in space (Resnik). First "passenger" from industry (Walker, of McDonnell Douglas).
Oct. 5–13	*STS 41-G* (13)	Robert Crippen, Dave Walker, Kathryn Sullivan, Sally Ride, David Leestma, Marc Garneau, Paul Scully-Power	250,000	132; 220	8 days 5 h 24 min	Sixth flight of *Challenger*. First seven-person crew, including two women, one Canadian (Garneau), one Australian (Scully-Power). Payload: *Earth Radiation Budget Satellite* (*ERBS*), successfully deployed; *OSTA 3* experiments. First U.S. woman to walk in space (Sullivan). Second landing at Kennedy Space Center.

*1 lb = 0.45 kg. 1 mi = 1.6 km. 1 ft = 0.3 m. 1 yd = 0.9 m. 1 fathom = 1.8 m. 1 lb/in.^{2} = 6.9 kPa.

†Beginning with 41-B, shuttle flights were no longer numbered consecutively. In the new system, the first digit indicates the last digit of the current fiscal year (begins Oct. 1 of previous calendar year), that is, 4 = 1984. The second digit indicates the launch site: 1 = Kennedy Space Center; 2 = Vandenberg Air Force Base. The letter indicates which serial flight in the fiscal year; for example, B = second flight.

Crewed space flights, 1961–1996* (cont.)

Dates launched and recovered	Designation	Crew	Weight, lb	Revolutions; max. distance from Earth, mi	Duration	Remarks
1984 (cont.)						
Nov. 8–16	*STS 51-A* (14)	Frederick Hauck, Dave Walker, Joseph Allen, Anna Fisher, Dale Gardner	250,000	126; 220	7 days 23 h 45 min	Second flight of *Discovery*. Two reimbursable comsats *(Telesat-H, LEASAT/Syncom IV-1)* successfully deployed. Two others, *Palapa B-1* and *Westar 6* from mission 41-B, successfully retrieved and returned to Earth for repair. Fifth U.S. woman in space (Fisher). Third landing at Kennedy Space Center.
1985						
Jan. 24–27	*STS 51-C* (15)	Thomas Mattingly, Loren Shriver, Gary Payton, James Buchli, Ellison Onizuka	—	47;—	3 days 1 h 34 min	Third flight of *Discovery*. First dedicated Department of Defense mission, classified secret. Fourth landing at Kennedy Space Center.
Apr. 12–19	*STS 51-D* (16)	Karol Bobko, Donald Williams, Margaret Rhea Seddon, Jeffrey Hoffman, David Griggs, Charles Walker, E.J. "Jake" Garn	250,000	109; 283	6 days 23 h 55 min	Fourth flight of *Discovery*. Payload of two reimbursable comsats *(Telesat-I/Anik, Syncom IV-3)* successfully deployed, but one *(Syncom)* failed to ignite its upper stage despite heroic attempts by crew with remote manipulator system. First flight of a U.S. politician (Senator Garn, Utah). Second flight of industry representative Walker. Sixth U.S. woman in space (Seddon). Landing at Kennedy Space Center resulted in damaged tires.
Apr. 29–May 6	*STS 51-B/ Spacelab 3* (17)	Robert Overmyer, Frederick Gregory, Don Lind, Norman Thagard, William Thornton, Lodewijk van den Berg, Taylor Wang	250,000	110; 220	7 days 8 min	Second flight of *Spacelab* pressurized module. Crew conducted over a dozen major experiments and tested research-animal holding facility with rats and two squirrel monkeys. Mission accomplished with great success. Landing of *Challenger* at Edwards AFB.
June 6–Sept. 26	*Soyuz T13*	Vladimir Dzhanibekov, Viktor Savinykh	14,400	—; 200	112 days/ Dzhanibekov; 168 days 3 h 51 min/Savinykh	Crew transferred to *Salyut 7* space station on June 8, after testing new flight control systems in *Soyuz* spacecraft. Performed extensive repair/maintenance of power system on station. Resupplied by *Progress 24* (launched June 21), *Cosmos 1669*, a new type of free-flying, uncrewed platform (docked July 21, separated Aug. 29). On Sept. 26, Dzhanibekov returned with "visiting" Georgi Grechko in *Soyuz T13*, leaving Savinykh with Vasyutin and Volkov in *Salyut* until Nov. 21.
June 17–24	*STS 51-G* (18)	Daniel Brandenstein, John Creighton, Shannon Lucid, Steven Nagel, John Fabian, Patrick Baudrey, Sultan Salman Al-Saud	250,000	111; 240	7 days 1 h 40 min	Fifth flight of *Discovery*, with cargo of three reimbursable comsats *(Morelos-A/Mexico; Arabsat 1B/Arabia; Telstar 3D/AT&T)*. Crew also deployed and later retrieved freeflying instrument platform SPARTAN 1 (Shuttle Pointed Autonomous Research Tool for Astronomy). Payload included six scientific-technological experiments from U.S., France, and Germany. Two foreign astronauts: Prince Al-Saud (Saudi Arabia), Baudry (France). Seventh U.S. woman (Lucid).
July 29–Aug. 6	*STS 51-F/ Spacelab 2* (19)	Charles G. Fullerton, Roy D. Bridges, Jr., Karl G. Henize, Anthony W. England, F. Story Musgrave, Loren W. Acton	252,800	126; 197	7 days 22 h 45 min	Eighth flight of *Challenger*, first flight of *Spacelab* pallets-only version. Liftoff after 1 h 37 min delay due to computer problem. One of three orbiter engines shut down prematurely (after 5 min 45 s) but not early enough to cause mission abort. Flight extended 1 day to help recover from onboard experiment problems. Mission accomplished. Landing at Edwards AFB.
Aug. 27–Sept. 3	*STS 51-I* (20)	Joe H. Engle, Richard O. Covey, James D. A. van Hoften, John M. Lounge, William F. Fisher	337,000	111; 280	7 days 2 h 17 min	Sixth flight of *Discovery*. Liftoff after two launch delays, one (Aug. 24) caused by weather, the other (Aug. 25) by backup computer problem. Payload of three reimbursable comsats [*ASC 1, AUSSAT 1* (Australia), *Leasat (Syncom) IV-4*] successfully deployed—the first two on the same day—a first. Crew performed rendezvous with *Leasat (Syncom) IV-3*, deployed Apr. 13 by *STS 51-D* and drifting lifeless, and successfully conducted salvage and repair operation in space on Aug. 31 and Sept. 1, redeploying the comsat for normal operation. Landing at Edwards AFB.

Sept. 17–Nov. 21	*Soyuz T14*	Vladimir Vasyutin, Georgy Grechko, Alexander Volkov,	14,400	—; 200	9 days/Grechko; 64 days 21 h 52 min/Vasyutin & Volkov; 168 days 3 h 51 min/Savinykh	Crew transferred to *Salyut 7*/*Soyuz T13* on Sept. 18. Veteran cosmonaut Grechko returned with Dzhanibekov on Sept. 26 in *Soyuz T13*, making this the first successful space station crew rotation. Vasyutin, Volkov, and Savinykh returned suddenly on Nov. 21 with *Soyuz T14* because of illness of Vasyutin, requiring hospitalization.
Oct. 3–7	*STS 51-J* (21)	Karol Bobko, Ronald J. Grabe, Robert Stewart, William A. Pailes, David C. Hilmers		63; 320	4 days 1 h 45 min	First flight of fourth shuttle orbiter *Atlantis*, setting new altitude record (after *41-C*). Payload: military (classified). Landing at Edwards AFB.
Oct. 30–Nov. 6	*STS 61-A*/ *Spacelab D1* (22)	Henry W. Hartsfield, Steven R. Nagel, Guion Bluford, Bonnie J. Dunbar, James F. Buchli, Reinhard Furrer, Ernst Messerschmid, Wubbo Ockels	250,000 (*Spacelab*: 26,000)	111; 202	7 days 32 min	Ninth flight of *Challenger* with the West German *D-1* mission, the first foreign-chartered *Spacelab* flight. Crew included three foreign astronauts: Furrer, Messerschmid (both W. Germany), and Ockels (Netherlands). Eighth U.S. woman (Dunbar). First eight-member crew. Mission concluded on Nov. 6 with great success. Landing at Edwards AFB with test of new nose wheel steering.
Nov. 26–Dec 3.	*STS 61-B* (23)	Brewster H. Shaw, Bryan D. O'Connor, Sherwood C. Spring, Mary L. Cleave, Jerry L. Ross, Rudolfo Neri Vela, Charles D. Walker	261,455	109; 235	6 days 22 h 54 min	Second flight of *Atlantis*, carrying three commercial comsats: *Morelos-B* (Mexico), *Aussat 2* (Australia), and *Satcom KU-2*. Ross and Spring conducted major outboard activities demonstrating space-station-type assembly and handling of large structures in space (payload designations: EASE/ACCESS). One foreign payload specialist: Neri (Mexico). Third flight of industry representative Walker. Ninth U.S. woman (Cleave). First flight of *PAM-D2* upper stage. Heaviest PAM payload (Satcom). Landing at Edwards AFB after full mission accomplishment.
1986						
Jan. 12–18	*STS 61-C* (24)	Robert L. Gibson, Charles F. Bolden, Jr., George D. Nelson, Steven A. Hawley, Franklin E. Chang-Diaz, Robert J. Cenker, C. William Nelson	250,000	96; 201	6 days 2 h 4 min 9 s	First flight of *Columbia* after major modifications since its last flight (*STS 9*) in Nov. 1983. Originally scheduled for Dec. 18, 1985, launch was delayed seven times due to systems problems and inclement weather. Crew included first Hispanic American in space (Chang-Diaz, Costa Rica), a payload specialist from industry (Cenker, RCA), and the second U.S. congressman in space (Florida U.S. Rep. Bill Nelson). Crew deployed advanced *Satcom K-1* communications satellite, took photographs of Comet Halley, and performed numerous on-board experiments. Landing at Edwards AFB after two postponements due to bad weather at Kennedy Space Center in Florida. Second night landing.
Jan. 28	*STS 51-L* (25)	Francis R. Scobee, Michael J. Smith, Judith A. Resnik, Ellison S. Onizuka, Ronald E. McNair, Gregory B. Jarvis, Christa McAuliffe	250,000	—	1 min 13 s	*Challenger* was launched at 11:38 a.m. EST after a night of below-freezing temperatures. Its right solid-rocket booster developed a catastrophic leak due to malfunction of a joint seal between two motor elements. The shuttle was destroyed 73 s after liftoff, killing its crew of seven and bringing the United States space program to a halt for 32 months.
Feb. 19	*Mir*	Uncrewed	235,000	210; —	—	Crewless launch of new Soviet space station *Mir* (Peace), similar in size and design to *Salyut 7*, with estimated length of 57 ft. Represents a new generation station, with six docking ports for future buildup with add-on modules and docking of *Soyuz* and *Progress* logistics flights.
March 13–July 16	*Soyuz T15*	Leonid Kizim, Vladimir Solovyev	14,400	2000 (est.); 210	125 days	Crew docked at *Mir* on Mar. 15 about 49 h after liftoff and transferred to the space station. First resupply flight followed on Mar. 19 with launch of crewless *Progress 25* which docked to *Mir* on Mar. 21. *Progress 26* was launched on Apr. 23, and the tanker-transport docked with *Mir* on Apr. 26. On May 5, Kizim and Solovyev transferred from *Mir* to *Salyut 7* using *Soyuz T15*. Their stay in the *Salyut 7*/*Soyuz T15*/*Cosmos 1686* complex, trailing *Mir* in orbit by 15 min, included 3 h 50 min spacewalk on May 28 to demonstrate the ability to work with a 15-m-long triangular lattice. An crewless *Soyuz TM*, launched on May 21, docked with *Mir* on May 23 and returned to Earth on May 30—a test of a modified version of the

*1 lb = 0.45 kg. 1 mi = 1.6 km. 1 ft = 0.3 m. 1 yd = 0.9 m. 1 fathom = 1.8 m. 1 lb/in.^{2} = 6.9 kPa.

Crewed space flights, 1961–1996* (cont.)

Dates launched and recovered	Designation	Crew	Weight, lb	Revolutions; max. distance from Earth, mi	Duration	Remarks
1986 (cont.)						*Soyuz* with improved computers, rendezvous-docking systems, and landing parachute. A second EVA on May 31, 5 h long, also demonstrated structural deployment. *Progress 26* undocked on June 22 and deorbited on June 23. On June 26, the crew returned to *Mir* with *Soyuz T15*. On July 16, the crew returned in *Soyuz T15* to Earth. Kizim had set a new record of 373 days of cumulative time in space by a human. Landing was 34 mi northeast of Arkalyk in Kazakhstan.
1987						
Feb. 6–July 30	*Soyuz TM-2*	Yuri Romanenko, Alexander Laveikin	14,400	Approx. 145; 230	174 days 1 h 22 min/ Laveikin; 8 days 20 h 1 min/Faris and Viktorenko	First flight to *Mir* in 1987 was the crewless *Progress 27* tanker/transport, launched on Jan. 16, and docked to *Mir* on Jan. 18, to provision the station for the subsequent crew visit. After liftoff in a new, more advanced *Soyuz* version, crew docked at *Mir* on Feb. 8 for a long-duration attempt. After an initial zero-g adaptation problem experienced by Laveikin, crew settled down to conduct materials processing work. On Mar. 5, *Progress 28*, a new tanker/transport, docked with *Mir's* rear port and was unloaded. On Mar. 9, its engines boosted *Mir* to increase its orbit altitude.
March 31	*Kvant*	Crewless	43,000	—; 250		An astrophysics module, *Kvant*, launched on Mar. 31, failed to dock to the *Mir* core on Apr. 5. After a second fruitless attempt on Apr. 9, the crew corrected the problem–a piece of cloth left by the ground crew–in a 3 h 40 min EVA on Apr. 12, and docking was accomplished. *Kvant's* propulsion module was jettisoned Apr. 13, making the module's rear docking port accessible. Resupply ship *Progress 29*, launched Apr. 21, docked to this port on Apr. 23. After unloading and taking on stored waste, it burned up in reentry on May 11. On May 13, *Progress 30* was launched with fresh supplies and mail. In a second and third EVA, on June 12 and 16, the crew installed a third solar power array on the station.
July 21–Dec. 29	*Soyuz TM-3*	Alexander Viktorenko, Alexander P. Alexandrov, Mohammed Faris	14,400	Approx. 2576; 230	161 days/Alexandrov; 326 days/ Romanenko; 8 days/Levchenko	After launch from Baikonur, the crew, which included the first Syrian cosmonaut (Faris), docked with *Mir* on July 24. On July 29, Viktorenko and Faris, accompanied by Laveikin who had shown an abnormal electrocardiogram reading, undocked *Soyuz TM-2* and landed in Kazakhstan early on July 30. Romanenko and Alexandrov continued the mission on *Mir*. Resupply: *Progress 31*, docked Aug. 7, separated Sept. 22; *Progress 32*, launched Sept. 22, docked Sept. 26. On Sept. 30, Romanenko broke the current record of 237 days. After separation of *Progress 32*, resupply ship *Progress 33* docked on Nov. 23. When Yuri Romanenko returned Dec. 29, 1987, in *TM-3* with Alexandrov and Anatoliy Levchenko (of *TM-4*), he had set a new record for the longest crewed spaceflight to date—326 days.
Dec. 21, 1987– June 17, 1988	*Soyuz TM-4*	Vladimir Titov, Musa Manarov, Anatoliy Levchenko	14,400	—; 230	366 days 22 h 39 min Titov and Manarov; 7 days 21 h 2 min Levchenko	Docked on Dec. 23, after 33 revolutions, to *Mir*. When, on Dec. 29, Levchenko—a shuttle test pilot, Alexandrov, and Romanenko returned to Earth in *TM-3*, landing at 12:16 p.m. about 50 mi from Arkalyk in Kazakhstan, Romanenko had set a new world record of 326 days in space. Titov and Manarov went on for a long-duration stay in *Mir*, slightly over 1 year. Resupply ship *Progress 34* was launched Jan. 21, 1988, and docked early on Jan. 23, *Progress 35* followed on Mar. 24 (undocked May 5), and *Progress 36* on May 15.

1988						
June 7–Sept. 7	*Soyuz TM-5*	Anatoly Solovyov, Viktor Savinykh, Alexander Alexandrov	14,400	162; 225	10 days 8 h 5 min (crew)	Docked to *Mir* on June 9. The 10-day mission was the 63d crewed flight since Gagarin. Alexandrov was the second Bulgarian (after Georgi Ivanov) and the 13th foreign national flown by the Soviets since 1978. Countdown and launch were widely televised. Visitor crew returned to Earth on June 17 in *TM-4*, leaving the "fresh" *TM-5* docked to *Mir.* On June 18, Titov and Manarov transferred the *TM-5* from the aft to the forward docking port. During a 5-h EVA on June 30, Titov and Manarov did not succeed in replacing a faulty detector unit on a British/Dutch x-ray telescope outside *Mir*, because of a broken wrench. *Progress 37*, the next resupply ship, was launched July 19.
Aug. 29–Dec. 21	*Soyuz TM-6*	Vladimir Lyakhov, Valery Polyakov, Abdullah Ahad Mohmand	14,400	138; 221	8 days 20 h 27 min/ Lyakov and Mohmand; 241 days/ Polyakov	*TM-6* was launched for a trip to *Mir*, with a crew including a physician, Polyakov, who was to stay aboard *Mir* to observe Titov and Manarov, and an Afghan cosmonaut, Mohmand. Docking occurred on Aug. 31. Since Lyakhov had received special training as a rescuer-cosmonaut, he was able to handle *TM-6* alone. After undocking from *Mir* in *Soyuz TM-5* of the previous mission, Lyakhov and Mohmand prepared for reentry at 3 a.m. but became stranded in orbit due to malfunction of a guidance system sensor, followed by computer input errors. With oxygen supply threatening to run out after 48 h, the twice-failed retroburn was achieved 24 h later after preventive measures. *TM-5* landed on Sept. 7, 100 mi from Dzhezkazgan in Kazakhstan. On Sept. 12, the crewless resupply ship *Progress 38* docked with *Mir* after *Soyuz TM-6* had been moved, on Sept. 8, from the *Kvant* astrophysics module to the airlock of the station. *Progress 39* was launched Sept. 9, carrying French equipment for an upcoming visit by a French cosmonaut. In a 4-h EVA on Oct. 20, Titov and Manarov successfully repaired a telescope in the *Kvant* module, finishing an earlier attempt of June 30.
Sept. 29–Oct. 3	*STS 26*† (26)	Frederick H. Hauck, Richard O. Covey, John M. (Mike) Lounge, David C. Hilmers, George D. Nelson,	253,700	645; 188	4 days 1 h 57 min	The United States' return to space. After a 32-month hiatus, NASA astronauts were back in space on the orbiter *Discovery*, after a flawless launch at 11:37 a.m. EDT. At 6 h 13 min into the mission Nelson deployed the prime payload, NASA's second *Tracking and Data Relay Satellite* (*TDRS*) for transfer to geostationary orbit. Secondary payloads were 11 experiments, including 2 by students. It was the seventh flight of *Discovery*, with over 200 engineering changes since the *Challenger* loss. Touchdown occurred at 12:37 p.m. EDT at Edwards, with a rollout distance of 7451 ft.
Nov. 15	"*Buran*"	Crewless	224,000	2; 156	3 h 25 min	First flight of Soviet reusable space shuttle "*Buran*" (Blizzard). Launched by second flight of Energia launch vehicle at 6 a.m. local time, landed two orbits later on runway at Baikonur, 7.5 mi from launch pad.
Nov. 26, 1988–Apr. 27, 1989	*Soyuz TM-7*	Alexandrovitch Volkov, Sergei Krikalev, Jean-Loup Chrétien	14,400	—; 230	25 days/ Chrétien; 152 days/Volkov and Krikalev	Second flight of French cosmonaut Chrétien to a Soviet space station. Mission, named "Aragatz," docked on *Mir* on Nov. 28. Chrétien became first French citizen to perform an EVA when he installed the Dec. deployable antenna structure ERA in space on Dec. 9. On Dec. 21, Chrétien returned in *Soyuz TM-6* with the long-duration crew, Titov and Manarov, who on this day completed 366 days on *Mir*. Volkov and Krikalev stayed in space with Polyakov until Apr. 27, 1989, when they returned in *TM-7*, leaving *Mir* crewless for the first time in 2 years. During its occupied period, the space station had been resupplied by a total of 17 *Progress* flights with about 40 tons of goods.
Dec. 2–6	*STS 27* (27)	Robert L. Gibson, Guy S. Gardner, Mike Mullane, Jerry L. Ross, William M. Shepherd	—	70(?); —	4 days 9 h 6 min 19s	Launched as second flight after the *Challenger* accident, *Atlantis* carried a radar imaging reconnaissance satellite on a classified Department of Defense mission into a high-inclination orbit. Returned to Edwards after highly successful mission.

*1 lb = 0.45 kg. 1 mi = 1.6 km. 1 ft = 0.3 m. 1 yd = 0.9 m. 1 fathom = 1.8 m. 1 lb/in.^{2} = 6.9 kPa.

†After the *Challenger* accident, shuttle mission designations returned to consecutive numbering. Unavoidable changes in flight assignments, however, result in numbered flights not necessarily being flown in sequence.

Crewed space flights, 1961–1996* (cont.)

Dates launched and recovered	Designation	Crew	Weight, lb	Revolutions; max. distance from Earth, mi	Duration	Remarks
1989						
March 13–18	*STS 29* (28)	Michael L. Coats, John E. Blaha, James F. Buchli, Robert C. Springer, James P. Bagian	263,290	79; 206	4 days 23 h 39 min	*Discovery*, on this third flight after the *Challenger* disaster, deployed its main payload, the third *Tracking and Data Relay Satellite, TDRS-D*, completing the constellation of NASA's new communications system. After conducting a number of physical, biological, medical, and materials experiment, *Discovery* returned to Edwards AFB.
May 4–8	*STS 30* (29)	David M. Walker, Ronald J. Grabe Norman E. Thagard, Mary L. Cleave, Mark C. Lee,	217,513	64; 185	4 days 56 min 38 s	Originally scheduled for Apr. 28, the launch of *Atlantis* was aborted at T-31 due to failure of the hydrogen recirculation pump of a main engine and rescheduled for May 4. On that day, *Atlantis* lifted off and, 6½ h later, placed the Venus radar mapper probe *Magellan* on its trajectory to the planet Venus, the first interplanetary probe launched by the United States in nearly 11 years. *Atlantis* landed at Edwards after a highly successful mission.
Aug. 8–13	*STS 28* (30)	Brewster H. Shaw, Jr., Richard N. Richards, James C. Adamson, Mark N. Brown, David C. Leestma	—	80; 192	5 days 1 h 56 s	Fifth flight after the *Challenger* accident. The extensively refurbished *Columbia*, on its first trip after over 3 years of inactivity, carried a classified Department of Defense payload into a 57° inclination orbit. Landing at Edwards took place after a successful mission.
Sept. 6, 1989– Feb. 19, 1990	*Soyuz TM-8*	Alexander Viktorenko, Alexander Serebrov	14,400	Approx. 2662; 230	166 days 7 h 58 min	After a pause of almost 5 months in its crewed program, the Soviet Union launched *TM-8* to *Mir*, the 67th crewed flight of Soviet Union, carrying *Mir*'s 20th and 21st occupants. Docking on *Mir* occurred on Sept. 8, executed manually after a last-minute technical problem with the automatic rendezvous/docking system of the *TM-8*. The arrival of the crew had been preceded by the docking of *Progress M1*, the first of a new series of crewless resupply craft, able to carry 5952 lb of cargo (882 lb more than the earlier version), and equipped with a recoverable entry capsule (not used on this flight).
Oct. 18–23	*STS 34* (31)	Donald E. Williams, Michael J. McCulley, Shannon W. Lucid, Franklin R. Chang-Diaz, Ellen S. Baker	264,775	79; 206	4 days 23 h 40 min	Launch of *Atlantis* with the *Galileo* Jupiter probe as primary payload occurred after two launch postponements, one from Oct. 12 due to a main engine controller problem, the other from Oct. 17 due to inclement weather. At 7:15 p.m., the 38,500-lb payload was successfully deployed; at 8:15 p.m. the Inertial Upper Stage (IUS) first stage was fired, followed by second-stage ignition at 8:20. The 2-ton *Galileo* separated from the IUS at 9:05 with deployed booms, was spun up to 2.8 revolutions per min at 9:30 p.m., and began its 6-year voyage to Jupiter, which includes one Venus and two Earth flybys for gravity assist. After conducting in-space experiments on material processing, plant life, ozone monitoring, ice crystals, and human physiology, *Atlantis* returned to Earth two orbits early due to adverse wind at the landing site.
Nov. 22–27	*STS 33* (32)	Frederick D. Gregory, John E. Blaha, Manley L. (Sonny) Carter, Kathryn C. Thornton, F. Story Musgrave		78 (?); —	5 days 7 min	On the seventh shuttle flight after the *Challenger* accident, the *Discovery* lifted off in darkness, carrying a classified military payload into a low-inclination orbit. After a successful mission, the vehicle landed at Edwards. it was the third night launch of the shuttle program.
Nov. 26	*Kvant 2*	Crewless	43,000	—; 250		*Kvant 2*, the second of four large building blocks of the Soviet *Mir* complex, was launched on a *Proton*. It docked to *Mir* on Dec. 6, after a lengthy delay caused by a multistep trouble-shooting program to deploy a stuck solar array and subsequent rendezvous problems. After the successful docking, *Kvant 2*, which carried systems and equipment to expand *Mir*'s operational capability and a crewed maneuvering unit for extravehicular activity, generated 3.5 kW of power from each of its two solar arrays.

1990						
Jan. 9–20	*STS 32* (33)	Daniel C. Brandenstein, James D. Wetherbee, Bonnie J. Dunbar, Marsha S. Ivins, G. David Low	229,848	172; 224	10 days 21 h 52 s	Longest shuttle mission to date. Originally scheduled for Dec. 18, 1989, but delayed due to unfinished launch pad work and one weather-caused slip (on Jan. 8). Launched on its mission to deploy the Navy communications satellite *Syncom IV-F5* (on Jan. 10) and retrieve the Long Duration Exposure Facility (LDEF), a 12-sided, open-grid aluminum satellite, 30 ft long, 14 ft in diameter, and 10.5 tons in mass, carrying 57 experiments. LDEF was originally orbited by *Challenger* (*STS 41C*) on Apr. 6, 1984, but its planned retrieval 1 year later was delayed by schedule slips and the *Challenger* accident. After *Columbia*'s rendezvous with LDEF on Jan. 12, Mission Specialist Dunbar used Remote Manipulator System arm to grapple the research facility, 2093 days and 32,419 orbits after LDEF's launch. *Columbia* landed at Edwards, delayed 2 days by ground fog. Heaviest landing weight (by 5 tons) to date.
Feb. 11–Aug. 9	*Soyuz TM-9*	Anatoliy Solovyev, Alexander Balandin	14,400	230	179 days	Docked with *Mir* on Feb. 13 to relieve Viktorenko and Serebrov. Earlier in February, they had conducted several spacewalks (EVA) to test a crewed maneuvering unit named *Ikar* (Icarus): Serebrov on four flights on Feb. 1 during a 5-h EVA, Viktorenko several times on a 3-h 45-min EVA on Feb. 5. The cosmonauts also tested a modified space suit and portable life-support system. The Soviet crewed maneuvering unit weighs 440 lb and is propelled by 36 compressed-air thrusters. Solovyev and Balantin became the 22d and 23d occupants of *Mir*. Viktorenko and Serebrov returned to Earth on Feb. 19, landing in *Soyuz TM-8* near Arkalyk, 205 mi northeast of Baikonur, in good health after 166 days in space, including five spacewalks and dozens of experiment programs.
Feb. 28–Mar. 4	*STS 36* (34)	John O. Creighton, John H. Casper, Michael Mullane, David C. Hilmers, Pierre J. Thuot	—	71; —	4 days 10 h 19 min	Launched as ninth flight since *Challenger* loss, after five delays due to head cold of Commander, bad weather, and a ground computer glitch, *Atlantis* deployed a secret intelligence-gathering satellite on a Department of Defense mission in a highly inclined orbit (62°). Returned to Edwards AFB after completing a fully successful mission.
Apr. 24–29	*STS 31* (35)	Loren J. Shriver, Charles F. Bolden, Steven A. Hawley, Bruce McCandless, Kathryn D. Sullivan	259,229	76; 381	5 days 1 h 15 min	The long-awaited launch of the Hubble Space Telescope, 20 years in planning and development, took place after an earlier launch attempt, on Apr. 10, was scrubbed due to malfunction of a valve in one of *Discovery*'s auxiliary power units. The 12.5-ton spacecraft, 42.5 × 14 ft, was placed into space by the manipulator arm on Apr. 25 at 8:45 a.m. EDT and released, after deployment of its solar arrays and a checkout, at 3:38 p.m. For the remainder of the mission, the crew carried out medical and engineering tests in the shuttle middeck. *Discovery* landed at Edwards AFB.
May 31	*Kristall*	Crewless	43,000	250		As third major addition to *Mir*, the technology module *Kristall* docked to the Soviet space station on June 10. Nearly identical to the previous module, *Kvant 2*, with length of 45 ft, width of 14.3 ft, and volume of 2000 ft^3, *Kristall* forms the second of four spokes of the *Mir* complex. Equipped with furnaces for materials experiments in space, it also has a docking port for the *Buran* space shuttle. Docking took two attempts, using backup system to a failed maneuvering jet. On June 11, *Kristall* was moved by the *Ljappa* pivot arm from forward docking port to a lateral port opposite the *Kvant 2* module. With core, *Kvant 1*, *Kvant 2*, *Kristall*, and *Soyuz-TM*, *Mir* now weighed 183,000 lb. *Kristall* also carried equipment to repair loose insulation panels on *Soyuz TM-9*. Repair took place on July 17, but Solovyev and Balandin encountered difficulties in closing the *Kvant 2* outer airlock door, damaged on exit, and exceeded the life-support capability of their spacesuits. Crew used emergency entry method (depressurizing *Kvant 2*), conducted 3-h EVA to repair hatch on July 26, and returned to Earth on Aug. 9, one week after arrival of *TM-10*, landing at Arkalyk in Kazakhstan with 286 lb of material.

*1 lb = 0.45 kg. 1 mi = 1.6 km. 1 ft = 0.3 m. 1 yd = 0.9 m. 1 fathom = 1.8 m. 1 lb/in.^{2} = 6.9 kPa.

Crewed space flights, 1961–1996* (cont.)

Dates launched and recovered	Designation	Crew	Weight, lb	Revolutions; max. distance from Earth, mi	Duration	Remarks
1990 (cont.)						
Aug. 1–Dec. 10	*Soyuz TM-10*	Gennaidy Manakov, Gennaidy Strekalov	14,400	Approx. 2096; 235	131 days	Launched to replace *TM-9*, it docked with *Mir* on Aug. 3 to continue work of previous crew (which had completed 506 of 520 planned science experiments), concentrating mostly on materials manufacturing with the *Kristall* module and switching solar panels from *Kristall* to *Kvant 1*. Crewless *Progress M-4* was launched Aug. 15 and docked Aug. 17, to bring supplies and boost *Mir* again up to 248 mi altitude. *Progress M-4*, was used to check out a new feature, a returnable capsule for returning material to Earth from *Mir*, prior to its use on *Progress M-5*. Manakov and Strekalov returned on Dec. 10.
Oct. 6–10	*STS 41* (36)	Richard N. Richards, Robert D. Cabana, Bruce E. Meinick, William M. Shepherd, Thomas D. Akers	256,330	65; 206	4 days 2 h 10 min	Replacing the postponed *STS 35 (Columbia)*, *Discovery* was launched with the European-built out-of-the-ecliptic solar probe *Ulysses* plus additional scientific experiments on board. *Ulysses* was deployed during the fifth orbit and launched by its own IUS/PAM rocket system. The fastest artificial object, it started its 5-year mission via a gravity-assist slingshot maneuver around Jupiter in Feb. 1992. It passed out of the Sun in summer 1994 and spring 1995, respectively.
Nov. 15–20	*STS 38* (37)	Richard O. Covey, Frank L. Culbertson, Carl J. Meade, Robert C. Springer, Charles D. ("Sam") Gemar		79; 130	4 days 21 h 54 min 32 s	On this fifth night launch, *Atlantis*, with a classified DOD payload, lifted off at 6:48 P.M. EST on it seventh flight. The planned landing at Edwards on Nov. 19 was cancelled due to high winds and took place 1 day later at Kennedy at about 4:43 P.M. EST. It was the sixth Kennedy landing, $5^1/_2$ years after the last one (*Discovery*, on Apr. 5, 1985)
Dec. 2–11	*STS 35* (38)	Vance D. Brand, Guy S. Gardner, John M. Lounge, Robert A. R. Parker, Ronald A. Parise, Jeffrey A. Hoffman, Samuel T. Durrance	267,513	143; 220	8 days 23 h 6 min 5 s	The first shuttle mission in 7 years dedicated to a single science discipline: astrophysics. After 6 months delay due to a faulty cooling system valve, hydrogen leaks, and payload electronics repair, *Columbia*'s tenth flight took off with ASTRO 1 (two Spacelab pallets with three ultraviolet telescopes on a pointing system, and one x-ray telescope) to explore the hottest, albeit invisible parts of the universe from the Orbiter cargo bay. The flight featured more than 80 h of observations (about 394) of about 135 science targets in the universe, starting 23 h after orbit insertion and ending 4 h before deorbit burn. Also, an experiment of communicating with amateur radio stations within line-of-sight of *Columbia* in voice and data mode (SAREX) was conducted by Parise. During orbit 39 on Dec. 4, *Columbia* and *Mir* were only 31 mi apart, and *Mir* was sighted twice from the shuttle. *STS 35* landed at Edwards one day earlier than planned because of bad weather that was forecast for later, after a mission troubled by computer, pointing, and plumbing problems.
Dec. 2–May 26	*Soyuz TM-11*	Viktor Afanasyev, Musa Manarov, Toyohiro Akiyama	14,400	Approx. 2740; 230	8 days/Akiyama; 131 days/ Manakov and Strekalov	Lifting off with the first Japanese in space on board, a journalist, *Soyuz TM-11* docked at *Mir* 2 days later, making up a record number of 12 humans in space at the same time (5 in *Mir*, 7 in *STS 35*). Cosmonauts Manakov, Strekalov, and Akiyama returned on Dec. 10, in *TM-10* with Afanasyev and Manarov taking over on *Mir*. Resupply ship *Progress M7*, lauched March 19, failed twice to dock to *Mir*'s *Kvant* module. On the second attempt on March 23, ground controllers barely avoided a collision between *Mir* and *M7*. On March 26, the crew relocated *Soyuz TM-11* from the main port to the *Kvant 1* module and docked *M7* successfully to that port on March 28.

1991						
Apr. 5–11	*STS 37* (39)	Steven Nagel, Kenneth D. Cameron, Jerry L. Ross, Jerome Apt, Linda M. Godwin	240,809	92; 281	5 days 23 h 32 min	*Atlantis* lifted off with the 35,000 lb Gamma Ray Observatory (GRO), the heaviest NASA science satellite ever carried by a shuttle. Equipped with four instruments to observe celestial high-energy sources with 10 times the sensitivity of prior experiments, GRO was set free on April 7 after an unplanned emergency EVA by Ross and Apt to deploy GRO's stuck high-gain antenna, the first United States spacewalk since Dec. 1985. On April 8, Ross and Apt performed a scheduled 6-h EVA to test various crew equipment translation aids such as carts on a track, for use on the exterior of the space station.
Apr. 28–May 6	*STS 39* (40)	Michael L. Coats, L. Blaine Hammond, Jr., Gregory J. Harbaugh, Donald R. McMonagle, Guion S. Bluford, C. Lacy Veach, Richard J. Hieb	236,623	133; 162	8 days 7 h 22 min	*Discovery* made the first unclassified DOD-dedicated shuttle mission. The mission was highlighted by around-the-clock observations of the atmosphere, gas release from subsatellites, shuttle engine firings, and the shuttle's orbital environment in wavelengths ranging from infared to the far-ultraviolet, including the aurora australis.
May 18–Oct. 10	*Soyuz TM-12*	Anatoly Artsebarsky, Sergei Krikalev, Helen Sharman	14,400	Approx. 2265; 230	7 days 19 h 5 min/ Sharman; 175 days 2 h 48 min/Afanasyev and Manarov	With the first British space traveler, food technologist Helen Sharman, as part of Britain's commercial project Juno, *Soyuz* docked at *Mir* on May 20, the ninth crew to visit the station. Earlier, on Apr. 25, the *Mir* crew, during a 5-h EVA, inspected the automatic docking system antenna that had nearly caused the collision with the *Progress M7* automated transport on Mar. 23. *M7* was discarded to burn up in the atmosphere. Sharman, Afanasyov, and Manarov returned on May 26 in *TM-11*, landing in Kazakhstan. In the first 3 months of their stay, Artsebarsky and Krikalev made six spacewalks totalling a record 32 h, for such tasks as repairing a docking antenna and a television camera, deploying instruments, and assembling a 46-ft-tall towerlike structure called *Sofora*, to carry reaction jets, on *Mir*. On Aug. 16, *Progress M8* was separated from *Mir*, and on Aug. 20 new supplies were brought up on *M9* in a night launch. *Soyuz TM-12* returned on Oct. 10 with Artsebarsky, Aubakirov, and Viehboeck.
June 5–14	*STS 40* (41)	Bryan D. O'Connor, Sidney M. Gutierrez, Tamara E. Jernigan, M. Rhea Seddon, James P. Bagian, F. Drew Gaffney, Millie Huges-Fulford	241,175	145; 185	9 days 2 h 15 min	*Columbia* carried Spacelab Life Sciences 1 (SLS 1), with three medical doctors, one biochemistry researcher, and one physicist on board, to perform 18 major experiments on the cardiovascular, renal/endocrine, blood, immune, musculoskeletal, and neurovestibular systems. Test objects included white rats and jellyfish.
Aug. 2–11	*STS 43* (42)	John E. Blaha, Michael A. Baker, Shannon W. Lucid, G. David Low, James C. Adamson	247,965	140; 207	8 days 21 h 21 min	The *Atlantis* flight put NASA's fourth Tracking and Data Relay Satellite (TDRS-E) into space, boosted successfully to a geosynchronous orbit by an Inertial Upper Stage (IUS). Other payloads included an atmospheric ultraviolet calibration instrument to measure the Earth's ozone layer, a heat pipe experiment, and an optical communications experiment. *Atlantis* landed at Kennedy.
Sept. 12–18	*STS 48* (43)	John O. Creighton, Kenneth S. Reightler, Mark N. Brown, James F. Buchli, Charles D. Gemar	224,141	81; 356	5 days 8 h 27 min	*Discovery* carried the giant Upper Atmosphere Research Satellite (UARS) into a 57°-inclination orbit, the first major flight element of NASA's Mission to Planet Earth program. The 14,400-lb satellite carried nine complementary scientific instruments to provide a more complete understanding of the chemistry, dynamics, and energy input of the upper atmo-

*1 lb = 0.45 kg. 1 mi = 1.6 km. 1 ft = 0.3 m. 1 yd = 0.9 m. 1 fathom = 1.8 m. 1 lb/in.² = 6.9 kPa.

Crewed space flights, 1961–1996* (cont.)

Dates launched and recovered	Designation	Crew	Weight, lb	Revolutions; max. distance from Earth, mi	Duration	Remarks
1991 (cont.)						sphere, including the ozone layer, plus a radiometer measuring the Sun's energy output. Other onboard work included microgravity research on crystal growth, fluids and structures behavior, life sciences, and radiation effects.
Oct. 2–Mar. 25	*Soyuz TM-13*	Alexander Volkov, Toktar Aubakirov, Franz Viehboeck	14,400	Approx. 2740; 230	8 days/Viehboeck and Aubakirov; 144 days 19 h/ Artsebarsky	*Soyuz* docked with *Mir* on Oct. 4. Cosmonauts Artsebarsky, Viehboeck (first space traveller from Austria), and Aubakirov (first Kazakh crew member), returned on Oct. 10 in *TM 12*, leaving Volkov and Krikalev in control of the space station.
Nov. 24–Dec. 1	*STS 44* (44)	Frederick D. Gregory, Terence T. Henricks, Story Musgrave, Mario Runco, Jr., James S. Voss, Thomas J. Hennen	247,087	108; 195	6 days 22 h 51 min	*Atlantis*, on a DOD-dedicated mission, carried a Defense Support Program (DSP) satellite for detection, from geosynchronous orbit, of nuclear detonations, and missile and space launches. After its deployment on the first day of the mission, the crew worked with a variety of secondary payloads, including naked-eye Earth observations, radiation measurements, and medical investigations. Due to malfunction of one of three Inertial Measurement Units (IMUs), the flight had to be shortened by 3 days.
1992						
Jan. 22–30	*STS 42* (45)	Ronald J. Grabe, Stephen S. Oswald, Norman E. Thagard, William F. Readdy, David C. Hilmers, Roberta L. Bondar, Ulf D. Merbold	231,497	128; 188	8 days 1 h 16 min	*Discovery* carried the International Microgravity Laboratory 1 (IML 1), dedicated to worldwide research in the behavior of materials and life in weightlessness. 225 scientists from 15 countries, experimenting in 42 research disciplines, cooperated in this fight. Along with IML 1, the payload consisted of 12 Get Away Special (GAS) containers, the large-format IMAX camera, and two student experiments. The crew included foreign astronauts Bondar (Canada) and Merbold (Germany).
Mar. 17–Aug. 10	*Soyuz TM-14*	Alexander Viktorenko, Alexander Kaleri, Klaus-Dietrich Flade	14,400	Approx. 2280; 245	312 days 6 h 1 min/Krikalev; 175 days 3 h 52 min/Volkov; 7 days 21 h 57 min/Flade	The crew, including the German guest cosmonaut Flade, docked with *Mir* on Mar. 19. Flade conducted a crowded science program and returned with former *Mir* occupants Volkov and Krikalev on Mar. 25, to land in *Soyuz TM-13* in Kazakhstan, leaving Viktorenko and Kaleri in *Mir*. In space since May 18, 1991, Krikalev was originally slated to return in late 1991, but the schedule was revised, leaving him in orbit for five extra months. On Feb. 20-21 he conducted his seventh spacewalk, fetching experiments left outside during previous EVAs, and setting a new record.
Mar. 24–Apr. 2	*STS 45* (46)	Charles Bolden, Brian Duffy, C. Michael Foale, David Leestma, Kathryn Sullivan, Dirk Frimout, Byron Lichtenberg	222,086	142; 182	8 days 22 h 9 min 20 s	The multinational flight of *Atlantis*, including a Belgian payload specialist (Frimout), carried the first ATLAS (Atmospheric Laboratory for Applications and Science) associated with NASA's Mission-to-Planet-Earth initiative. The ATLAS payload of 14 instruments remained attached to the Orbiter, most of them mounted on two Spacelab pallets. A Far Ultraviolet Space Telescope (FAUST) was also employed.
May 7–16	*STS 49* (47)	Daniel C. Brandenstein, Kevin P. Chilton, Richard J. Hieb, Bruce E. Melnick, Pierre J. Thuout, Kathryn C. Thornton, Thomas D. Akers	246,008	140; 181	8 days 21 h 18 min 36 s	This was the first mission featuring four EVAs, and the maiden voyage of the orbiter *Endeavour*. The main mission was reboosting the 11.7-ft diameter, 17.5-ft-high communications satellite *INTELSAT VI* (F-3) launched 2 years earlier by a Titan rocket, but stranded in a useless orbit when the Titan's second stage failed to separate. After rendezvous with the satellite, Thuot and Hieb, on two spacewalks on flight days 4 and 5, failed to capture the 8960-lb satellite. Joined by Akers on May 13, however, they succeeded in grasping it with their hands, berthing it in the shuttle cargo bay with the remote manipulator arm, attaching it to a new perigee kick motor, and redeploying it. This first three-person EVA in history lasted 8 h 36 min, the longest spacewalk ever conducted. The

						solid-fuel motor was fired remotely from the ground on May 14, taking *INTELSAT VI* to its final destination in a geosynchronous orbit. In a fourth EVA on May 14, Thornton and Akers evaluated space-station assembly equipment and techniques as well as several prototype crew self-rescue devices.
June 25–July 9	*STS 50* (48)	Richard N. Richards, Kenneth D. Bowersox, Bonnie J. Dunbar, Carl J. Meade, Ellen S. Baker, Lawence J. DeLucas, Eugene H. Trinh	245,902	220;185	13 days 19 h 30 min 30 s	The first extended-duration (EDO) shuttle flight. *Columbia*, specially modified for 2-week missions, carried the first U.S. Microgravity Laboratory 1 (USML 1), a Spacelab module equipped with 31 experiments for advanced research in zero *g*. The longest shuttle flight to date, it supported investigations of the effects of weightlessness on plants, humans, and materials. The mission also carried investigations into polymer membrane manufacturing, the Space Shuttle Amateur Radio Experiment II (SAREX-II), and numerous other cargo bay, middeck, and secondary payloads.
July 27–Feb. 1	*Soyuz TM-15*	Anatoliy Solovyev, Sergei Avdeyev, Michel Tognini	14,400	2946; 245	146 days/Kaleri and Viktorenko 14/Tognini	The third Russian-French mission since 1982, *Soyuz* carried French researcher Tognini, to conduct 10 biomedical, fluid and materials science and technological experiments on board *Mir*. Tognini returned on Aug. 10 in *Soyuz TM-14* along with Viktorenko and Kaleri. At touchdown in Kazakhstan, the spherical *TM 14* rolled over, briefly trapping but not injuring the crew. Solovyev and Avdeyev remained in *Mir* with numerous tasks such as installing new gyrodynes and rocket thrusters for attitude control. *Progress M-13*, was launched on Aug. 14, carrying supplies and medical gear to *Mir*.
July 31–Aug. 8	*STS 46* (49)	Loren J. Shriver, Andrew M. Allen, Jeffrey A. Hoffman, Franklin R. Chang-Diaz, Marsha S. Ivins, Claude Nicollier, Franco Malerba	241,797	126; 266	7 days 23 h 14 min 12 s	*Atlantis* successfully deployed (on Aug. 2) the first European Retrievable Carrier (EURECA) platform, carrying seeds, spores, crystals, and other experimental materials for research in zero *g*, to be retrieved about 9 months later. Due to data errors, EURECA's orbital self-transfer to its final altitude of 313 mi occurred later than planned but was accomplished on Aug. 7. Deployment of the main payload, the 1142-lb Italian Tethered Satellite System (TSS 1) on a 12.5-mi-long cable from the *Atlantis*, however, was terminated when a breakdown in the reel mechanism halted the unreeling of the tether after only 850 ft. The crew included two foreign astronauts, Malerba (Italy) and Nicollier (Switzerland) for ESA.
Sept. 12–20	*STS 47* (50)	Robert L. Gibson, Curtis L. Brown, Jr., Mark C. Lee, Jay Apt, N. Jan Davis, May C. Jemison, Mamoru Mohri	232,661	126; 192	7 days 22 h 30 min	*Endeavor* carried the Japan-chartered Spacelab SL-J with 44 materials and life sciences experiments and nine Get Away Special (GAS) payload canisters. The cargo also included the Israeli Space Agency Investigation About Hornets (ISAIAH) experiment and another flight of the Shuttle Amateur Radio Experiment (SAREX). The crew included the first Japanese shuttle astronaut (Mohri).
Oct. 22–Nov. 1	*STS 52* (51)	James D. Wetherbee, Michael A. Baker, C. Lacy Veach, William M. Shepherd, Tamara E. Jernigan, Steven G. MacLean	239,178	158; 189	9 days 20 h 56 min	*Columbia* carried the U.S. Microgravity Payload (USMP 1) as main cargo. The crew, with the third Canadian astronaut (MacLean), performed materials and science experiments in micro-*g* with the U.S. Microgravity Payload 2 (USMP 2), deployed the second Italian/U.S. Laser Geodynamic Satellite (LAGEOS 2), and conducted other physical experiments including the Canadian Experiment #2 (CANEX 2), and space-station related investigations.
Dec. 2–9	*STS 53* (52)	David M. Walker, Robert D. Cabana, Guion S. Bluford, James S. Voss, Michael Richard Clifford	230,731	116; 237	7 days 7 h 20 min	As the last dedicated DOD mission, *Discovery* carried the satellite DOD 1 as primary and two major experiments, the Glow/Cryogenic Heat Pipe Experiment (GCP) and Orbital Debris Radar Calibration Spheres (ODERACS), as secondary payloads. With GCP, remotely commanded special sensors observed orbiter surfaces, contamination events, and airglow, and measured the zero-*g* performance of liquid oxygen heat pipes. ODERACS had to be canceled when its container in the cargo bay failed to open.

*1 lb = 0.45 kg. 1 mi = 1.6 km. 1 ft = 0.3 m. 1 yd = 0.9 m. 1 fathom = 1.8 m. 1 lb/in.^{2} = 6.9 kPa.

Crewed space flights, 1961–1996* (cont.)

Dates launched and recovered	Designation	Crew	Weight, lb	Revolutions; max. distance from Earth, mi	Duration	Remarks
1993						
Jan. 13–19	*STS 54* (53)	John H. Casper, Donald R. McMonagle, Mario Runco, Jr., Susan J. Helms, Gregory J. Harbaugh	248,338	188; 95	5 days 23 h 39 min	*Endeavor* carried the sixth Tracking and Data Relay Satellite (TDRS), its Inertial Upper Stage (IUS), and the Diffuse X-Ray Spectrometer (DXS), consisting of two instruments attached to the sides of the cargo bay. After successful deployment of TDRS-F, crew activities included several microgravity experiments on materials and rodents, and the first in a series of test spacewalks leading up to space-station construction. Harbaugh and Runco, assisted by Helms from the cabin, tested EVA training procedures for the space station and the Hubble repair mission (*STS 61*). Using children's toys, the crew responded to students from various elementary schools asking questions about physics in zero-*g*.
Jan. 24–July 22	*Soyuz TM-16*	Gennadiy Manakov, Alexander Polishchuk	14,400	Approx. 2790; 242	189 days/Solovyev and Avdeyev	Equipped with the new androgynous docking system APAS-89, to be tested for use in the newly planned *Mir*/space shuttle docking mission, *Soyuz* docked with *Mir*, for the first time, at the *Kristall* module's docking port. The previous crew, Solovyev and Avdeyev, returned on Feb. 1 in *Soyuz TM-15* after 189 days in space. On Feb. 4, *Progress M-15* was undocked from *Mir*'s aft port and used to deploy, by centrifugal force, a 66-ft-diameter solar sail made from superthin aluminum foil. The purpose of the test, which reflected a visible light beam to Earth, was to assess the force of photons from the Sun for propulsive application.
Apr. 8–17	*STS 56* (54)	Kenneth D. Cameron, Stephen S. Oswald, Michael Foale, Kenneth D. Cockrell, Ellen Ochoa	225,597	147; 185	9 days 6 h 8 min	*Discovery* carried ATLAS 2, the second in NASA's series of Atmospheric Laboratory for Applications and Science missions. The remote sensing instruments of ATLAS studied the Sun's energy output and Earth's middle-atmosphere chemical makeup, particularly with regard to ozone depletion. The crew on Apr. 11 deployed Spartan-201, a free-flying platform carrying instruments to study the Sun's corona and the solar wind, and recovered the platform on Apr. 13.
Apr. 26–May 6	*STS 55* (55)	Steven R. Nagel, Terence T. Henricks, Jerry L. Ross, Charles J. Precourt, Bernard A. Harris, Ulrich Walter, Hans Wihelm Schlegel	244,156	158; 187	9 days 23 h 40 min	*Columbia* carried the Spacelab module on its eleventh flight, chartered for the second time by the German space agency DARA, with two German payload specialists (Walter, Schlegel) among its crew. The Spacelab D2 mission was devoted to zero-*g*-relevant investigations in a wide range of science disciplines, including materials research; life sciences; technology experiments in optics, robotics, automation, and medical applications; Earth observations; atmospheric physics; and astronomy, for a total of about 90 different projects. Many investigations were aimed at developing, testing, and verifying future operations/utilization in permanent space-station laboratories.
June 21–July 1	*STS 57* (56)	Ronald Grabe, Brian Duffy, David Low, Nancy Sherlock, Peter (Jeff) J. K. Wisoff, Janice Voss	239, 319	158; 289	9 days 23 h 46 min	Highlights of the *Endeavour*'s flight were the first flight of the privately developed middeck module SPACEHAB and the retrieval of the European space platform EURECA, in orbit since its deployment from *STS 46* on Aug. 1, 1992. SPACEHAB experiments included 13 commercial payloads, ranging from drug improvement to feeding plants, cell splitting, the first soldering experiment in space by United States astronauts, and high-temperature melting of metals; and six NASA experiments involving biotechnology, human factors, and water recycling for the future space station. Other payloads included SHOOT (Superfluid Helium On-Orbit Transfer) to demonstrate the resupply of liquid helium containers in space. A 5½-h EVA by Low and Wisoff on June 25 served to latch two EURECA antennas and to refine procedures for the planned Hubble repair mission.

July 1–Jan. 14	*Soyus TM-17*	Vasili Tsibliyev, Alexander Serebrov, Jean-Pierre Haignere	14,400	Approx. 2790; 242	179 days/Polishchuk and Manakov 21 days/Haignere	Carrying the fourth French cosmonaut since 1982, *Soyuz* docked with *Mir* on July 3. Resupply craft *Progress M-18*, docked on May 24, was undocked on July 4 and reentered destructively, after separating a small ballistic *Raduga* ("Rainbow") reentry capsule carrying experiments and recordings for a soft landing in Russia. During his stay at the station, Haignere conducted 11 major Altair experiments, then returned to Earth on July 22 with the previous crew, Polishchuk and Manakov, in *TM 16*.
Sept. 12–22	*STS 51*	Frank Culbertson, William Readdy, Jim Newman, Dan Bursch, Carl Walz	250,023	157; 296	9 days 20 h 11 min	*Discovery* carried two payloads. An Advanced Communication Technology Satellite (ACTS), a testbed for technology leading to a new generation of comsats, was deployed and injected by a Transfer Orbit Stage (TOS) toward its geosynchronous location. The German Orbiting and Retrievable Far and Extreme Ultraviolet Spectrometer–Shuttle Pallet Satellite (ORFEUS–SPAS) was set free by the robot arm and was retrieved 6 days later, the first of a series of such missions. Other mission tasks included a 7-h EVA by Walz and Newman on Sept. 16 as part of planning for the Hubble repair mission.
Oct. 18–Nov. 1	*STS 58* (58)	John E. Blaha, Richard A. Searfoss, M. Rhea Seddon, William S. McArthur, Jr., David A. Wolf, Shannon W. Lucid, Martin J. Fettman	244,860	224; 179	14 days 13 min 40 s	*Columbia* carried SLS-2, the second Spacelab dedicated to life sciences research. NASA's longest shuttle mission to date conducted, in a 39° inclination orbit, a series of experiments to gain more knowledge on how the human body adapts to the weightless environment of space. The 14 experiments were performed on the crew, which included two medical doctors (Seddon, Wolf), a physicist/biochemist (Lucid), and a veterinarian (Fettman), and on laboratory animals (48 rats).
Dec. 2–13	*STS 61* (59)	Richard O. Covey, Kenneth D. Bowersox, Tom Akers, Jeffrey A. Hoffman, F. Story Musgrave, Claude Nicollier, Kathryn Thornton	236,670	162; 370	10 days 19 h 58 min 35 s	*Endeavour* made the first in a series of planned visits to the orbiting Hubble Space Telescope (HST). The crew included four astronauts with considerable EVA experience (Hoffman, Musgrave, Thornton, Akers) and Swiss astronaut Nicollier (ESA). On Dec. 4, Nicollier grappled the HST with the shuttle's robot arm, berthing it in the cargo bay. In five spacewalks during the following 5 days, the astronauts installed the Wide Field/Planetary Camera-II and the Corrective Optics Space Telescope Axial Replacement (COSTAR) mechanism, both designed to compensate for the erroneous spherical aberration of the primary mirror. They also replaced the spacecraft's solar arrays and installed new gyroscopes, additional computer memory, and other systems to increase HST reliability. Total EVA time accumulated by all crews: 35 h 28 min. HST was set free on Dec. 10.
1994 Jan. 8–July 9	*Soyuz TM-18*	Viktor Afanasyev, Yuri Usachov, Valery Polyakov	14,400	3072; 242	197 days 3 h 2 min/ Tsibliyev and Serebrov 437 days 17 h 59 min/ Polyakov	This was the fifteenth mission to the *Mir* station. Docking occurred on Jan. 10. Polyakov, a physician cosmonaut who had spent 241 days in space in 1988 (*TM 6*), intended to conduct the longest human space flight in history to date to test the effects of long periods of weightlessness on the human body. He returned on *TM 20* on Mar. 22, 1995, after 14.6 months in space. During the mission, three *Progress* supply ships were to be sent to *Mir*, which was repaired and maintained by the crews. The previous crew, Vasily Tsibliyev and Alexander Serebrov, returned to Earth on Jan. 14 after 6 months, despite a last-minute slight collision of their *Soyuz TM-17* with *Mir*.
Feb. 3–11	*STS 60* (60)	Charles F. Bolden, Kenneth S. Reightler, Jr., N. Jan Davis, Ronald M. Sega, Franklin R. Chang-Diaz, Sergei K. Krikalev	233,290	130; 220	8 days 7 h 9 min	This *Discovery* mission included the first Russian cosmonaut (Krikalev). The mission also carried the commercial SPACEHAB facility (on its second flight) with 12 biotechnology and materials processing payloads, and the free-flying Wake Shield Facility, a 12-ft-diameter disk designed to generate an ultravacuum environment in space in its wake, within which to grow thin semiconductor crystal films of gallium arsenide. WSF 1 deployment, however, did not take place due to problems with a horizon sensor on the satellite; crystals were grown with WSF attached to the robot arm. The mission also included deployment of the German satellite BREMSAT.

*1 lb = 0.45 kg. 1 mi = 1.6 km. 1 ft = 0.3 m. 1 yd = 0.9 m. 1 fathom = 1.8 m. 1 lb/in.^{2} = 6.9 kPa.

Crewed space flights, 1961–1996* (cont.)

Dates launched and recovered	Designation	Crew	Weight, lb	Revolutions; max. distance from Earth, mi	Duration	Remarks
1994 (cont.)						
Mar. 4–18	*STS 62* (61)	John H. Casper, Andrew M. Allen, Charles D. Gemar, Marsha S. Ivins, Pierre J. Thuot	245,457	223; 188	13 days 23 h 17.5 min	This *Columbia* mission was dedicated to microgravity research with the U.S. Microgravity Payload-2 (USMP 2) and the Office of Aeronautics and Space Technology-2 (OAST 2) payloads. The four experiments on USMP 2 were designed to study solidification techniques, crystal growth, and gas behavior in zero *g*. OAST 2 carried five experiments for various other effects of materials and environmental phenomena in space. The Orbiter also carried special acceleration measurement systems, and the Shuttle Solar Backscatter Ultraviolet (SSBUV) instrument.
Apr. 9–20	*STS 59* (62)	Sidney M. Gutierrez, Kevin P. Chilton, Jerome Apt, Michael R. Clifford, Linda M. Godwin, Thomas D. Jones	237,048	181; 139	11 days 5 h 50 min	*Endeavour,* in a 57° orbit, carried the first Space Radar Laboratory (SRL 1) for internationally conducted studies of the Earth's global environment. SRL consisted of the Spaceborne Imaging Radar-C (SIR-C) and the X-Band Synthetic Aperture Radar (X-SAR). Other payloads included microgravity experiments (CONCAP-IV) and medical investigations.
July 1–Nov. 4	*Soyuz TM-19*	Yuri Malenchenko, Talgat Musabayev	14,400	1960; 242	125 days 20 h 53 min 182 days/ Afanasyev and Usachov	This mission docked with *Mir* on July 3, carrying a Kazakhstan cosmonaut (Musabayev) besides a Russian. The two crew members replaced Afanasyev and Usachev, who returned to Earth in *Soyuz TM-18* on July 9, after 6 months in space. Valery Polyakov continued his orbital stay. Malenchenko and Musabayev returned in *TM 19* on Nov. 4, 1994 (with Ulf Merbold of *TM 20*).
July 8–23	*STS 65* (63)	Robert D. Cabana, James Donald Halsell, Richard J. Hieb, Carl E. Walz, Leroy Chiao, Donald A. Thomas, Chiaki Naito-Mukai	228,640 (landing)	235; 188	14 days 17 h 55 min	*Columbia* carried the second mission of the International Microgravity Laboratory (IML 2) with furnaces and other facilities to study the behavior of materials and life in the weightless environment, and to produce a variety of material structures, from crystals to metal alloys. More than 200 scientists contributed 82 investigations, of which 81 were successfully completed in 19 IML 2 facilities, plus other middeck experiments. The crew included the first Japanese woman in space (Naito-Mukai).
Sept. 9–20	*STS 64* (64)	Richard N. Richards, L. Blaine Hammond, Jerry Linenger, Susan J. Helms, Carl J. Meade, Mark C. Lee	210,916 (landing)	176; 162	10 days 22 h 50 min	*Discovery* was launched for a mission of atmospheric research, robotic atmospheric research, robotic operations, and an EVA. LITE (Lidar In-Space Technology Experiment) was flown and used in space. Operating from the shuttle payload bay, the neodymium YAG laser, an optical radar, probed the Earth's atmosphere to collect measurements on clouds and airborne dust. Other payloads were the Robot-Operated Processing System (ROMPS), a Simplified Aid for EVA Rescue (SAFER) backpack used experimentally, and the astronomy satellite SPARTAN 201. deployed and retrieved after 40 h of free flight.
Sept. 30–Oct. 11	*STS 68* (65)	Michael A. Baker, Terrence W. Wilcutt, Thomas D. Jones, Steven L. Smith, Daniel W. Bursch, Peter (Jeff) J. K. Wisoff	223,040 (landing)	182; 139	11 days 5 h 46 min	*Endeavour* carried the Space Radar Laboratory into space for the second time. An international team used its instruments to study how the Earth's global environment is changing.

Oct. 4–Mar. 22	*Soyuz TM-20*	Alexander Viktorenko, Elena Kondakova, Ulf Merbold	14,400	2637; 247	31 days 11 h 36 min/Merbold 169 days 5 h 22 min/Viktorenko and Kondakova	This mission docked at *Mir* on Oct. 6, its crew including the third Russian woman (Kondakova) since inception of the Soviet/Russian space program, and ESA-astronaut Merbold on his third space trip. *TM 20* was the first of two ESA Euro-Mir missions, and its launch had depended on successful docking of *Progress M 24,* accomplished by *Mir* commander Malenchenko on Sept. 2 manually. Malenchenko and Musabayev then performed two EVAs, on Sept. 9 and 13, to inspect the (undamaged) orbital station where it had been struck by *M 24* on Aug. 30 and by *Soyuz TM-17* after its undocking in January, as well as *Mir*'s *Sofora* control mast. On Oct. 11, the crew had to cope with a serious power and attitude control failure of the station's core module. After performing experiments in life sciences, materials processing, and technology development, though limited by a failure of the CSK 1 materials furnace, Merbold, with Malenchenko and Musabayev, returned to *TM 19* on Nov. 14, leaving Kondakova and Viktorenko in space with Polyakov (to return on Mar. 22, 1995).
Nov. 3–14	*STS 66* (66)	Donald R. McMonagle, Curtis L. Brown, Ellen Ochoa, Joseph R. Tanner, Jean-François Clervoy, Scott E. Parazynski	209,857 (landing)	174; 193	10 days 22 h 35 min	*Atlantis* carried the Atmospheric Laboratory for Applications and Science (ATLAS) on its third flight for Mission to Planet Earth. The crew, with a French astronaut (Clervoy) for ESA, used the remote-sensing laboratory for studying the Sun's energy output, the middle atmosphere's chemical makeup, and the effects of these factors on global ozone levels. On Nov. 4, they deployed the German-built reusable satellite CRISTA-SPAS, carrying cryogenic infrared spectrometers, telescopes, and a high-resolution spectrograph for atmosphere research. They recovered it on Nov. 12, using a new rendezvous method in preparation for later rendezvous/docking flights to the Russian *Mir* station.
1995 Feb. 3–11	*STS 63* (67)	James D. Wetherbee, Eileen M. Collins, Bernard A. Harris, C. Michael Foale, Janice Voss, Vladimir G. Titov	211,318 (landing)	129; 250	8 days 6 h 28 min 15 s	*Discovery* lifted off with the second Russian cosmonaut (Titov) and the first woman shuttle pilot (Collins) within a narrow launch window of 5 min, as required by its mission; rendezvous and fly-around of the Russian space station *Mir* at 51.6° inclination. The rendezvous, a dress rehearsal of docking missions to follow, took place on Feb. 6, validating new flight techniques required for those missions, though no docking was performed as yet. Also on board: Spacehab on its third flight and SPARTAN 204, a satellite carrying various science experiments. It was deployed by Titov on Feb. 7 and retrieved with the robot arm on Feb. 9. Foale and Harris performed a 4.5-h EVA on Feb. 9.
Mar. 2–18	*STS 67* (68)	Stephen S Oswald, William G. Gregory, Tamara E. Jernigan, John M. Grunsfeld, Wendy B. Lawrence, Samuel T. Durrance, Ronald A. Parise	217,683 (landing)	262; 220	16 days 15 h 9 min	Setting a shuttle duration record of about 16½ days in orbit, *Endeavour* carried the second flight of the ASTRO observation platform with three unique telescopes for ultraviolet astronomy. Its tasks included mapping areas of space still largely uncharted in ultraviolet and measuring intergalactic gas abundance. Ultraviolet spectra were taken of about 300 celestial objects, including Jupiter and the Moon. Other experiments concerned large-structure control dynamics and materials processing.
Mar. 14–Sept. 11	*Soyuz TM-21*	Vladimir Dezhurov, Gennady Strekalov, Norman Thagard	14,400	2820; 246	115 days 9 h 44 min	This eighteenth crew visit to *Mir* carried the first American (Thagard) on a Russian spacecraft. Rendezvous and docking took place on Mar. 16. With the current *Mir* crew of 3 cosmonauts and the 7 astronauts on *STS 67/Endeavour*, the number of humans simultaneously in space was 13, a record (until the *STS 67* landing on Mar. 18). *Mir 17* crew Viktorenko, Kondakova, and Polyakov returned on Mar. 22 in *TM 20*, with a new long-duration record of 437 days 17 h 59 min set by Polyakov and longest space flight by a woman (Kondakova) of 169 days 5 h 22 min. After a 3-month stay, Thagard, Dezhurov, and Strekalov were to be returned to Earth by the shuttle *Atlantis* (*STS 71*). Known as *Mir 18*, Thagard's mission was part of Phase One of joint activities leading to the International Space Station. At the end of his mission, Thagard had established a new United States record for time spent in space of 115 days 9 h 44 min.

*1 lb = 0.45 kg. 1 mi = 1.6 km. 1 ft = 0.3 m. 1 yd = 0.9 m. 1 fathom = 1.8 m. 1 lb/in.^{2} = 6.9 kPa.

Crewed space flights, 1961–1996* (cont.)

Dates launched and recovered	Designation	Crew	Weight, lb	Revolutions; max. distance from Earth, mi	Duration	Remarks
1995 (cont.)						
May 20	*Spektr*	Crewless	43,000	—; 257		Russia's new four-solar-array research module, launched on a heavy-lift Proton, carried extensive atmospheric science instrumentation and about 1760 lb of United States equipment for cardiovascular and other biomedical research by astronaut Thagard, later by his replacement, Bonnie Dunbar. *Spektr* docked automatically to *Mir* on June 1, requiring five spacewalks by Dezhurov and Strekalov between May 12 and June 2 and the repositioning of the older module, *Kristall*. This prepared the station for the arrival and docking of the shuttle *Atlantis* (*STS 71*).
June 27–July 7	*STS 71* (69)	Robert L. (Hoot) Gibson, Charles J. Precourt, Ellen S. Baker, Gregory J. Harbaugh, Bonnie Dunbar, Anatoly Y. Solovyev (up), Nikolai M. Budarin (up), Vladimir Dezhurov (down), Gennady Strekalov (down), Norman Thagard (down)	214,700 (landing)	153; 246	9 days 19 h 22 min 17 s; 75 days 11 h 19 min/Solovyev and Budarin	*Atlantis* opened Phase One of the International Space Station program as the first of a series of shuttle–*Mir* linkups, 20 years after the joint United States–Russian ASTP mission. Docking took place on June 29, forming the largest human-made structure in space. Joint scientific (mostly medical) investigations were carried out inside the Spacelab module in the *Atlantis* cargo bay. Solovyev and Budarin then took over as new occupants of *Mir* (*Mir 19*), while the *Mir* 18 crew of Dezhurov, Strekalov, and Thagard returned to Earth with *Atlantis*. Undocking was on July 4, 4 days 21 h 35 min after linkup.
July 13–22	*STS 70* (70)	Terrence (Tom) Henricks, Kevin Kregel, Donald Thomas, Nancy Currie, Mary Ellen Weber	195,195 (landing)	142; 206	8 days 22 h 21 min 2 s	*Discovery* carried the seventh Tracking and Data Relay Satellite (TDRS 7). It was released out of the payload bay and boosted into geosynchronous orbit with its Inertial Upper Stage (IUS). Other payloads included a Commercial Protein Crystal Growth (CPCG) facility, a Bioreactor Demonstrator, and various biological experiments.
Sept. 3–Feb. 29	*Soyuz TM-22*	Yuri Gidzenko, Sergei Avdeyev, Thomas Reiter	14,400	2789; 246		EUROMIR'95 docked to *Mir* on Sept. 5, after *Progress M-28* was undocked and jettisoned. ESA astronaut Reiter (Germany) served as flight engineer, marking the first time for Europe in Russia's space program. On Oct. 20, he (with Avdeyev) installed a European experiment on the outside of *Spektr*. Two cassettes exposed to natural and human-made dust and debris were returned by Reiter and Gidzenko during another EVA (the third) on Feb. 8, 1996. The *Mir 19* crew (Solovyev and Budarin) returned in *TM-21* on Sept. 11. On-board activities included experiments in life sciences and micro-*g* materials processing. Most of ESA's equipment had been brought to *Mir* by *Progress M-28* in July and by *Spektr*. Another load arrived on *Progress M-29* (launch, Oct. 8, docking, Oct. 10, at aft end of *Kvant1*), including the Munich Space Chair, set up Oct. 11 in *Spektr*. In November, the crew hosted the visiting shuttle *Atlantis/STS 74*. The crew returned 6 days after arrival of *TM-23* with the *Mir 21* crew.
Sept. 7–18	*STS 69* (71)	David M.Walker, Kenneth D.Cockrell, James S.Voss, Jim Newman, Michael L.Gernhardt	219,71 (landing)	170;232	10 days 20 h 29 min	*Endeavour* carried the SPARTAN 210 free-flyer on its third flight and the Wake Shield Facility (WSF 2) on its second flight as main payloads. SPARTAN, released on Sept. 8 and retrieved 2 days later, observed the solar wind and the Sun's outer atmosphere in conjunction with the *Ulysses* probe while it passed over the Sun's north polar region. The WSF, deployed by the robot arm on Sept. 11 and retrieved on Sept. 14, was partially successful in growing thin semiconductor films in the ultravacuum created in its wake. Other payloads included the International Extreme

						Ultraviolet Hitchhiker (IEH 1) and the Capillary Pumped Loop 2/Gas Bridge Assembly (CAPL 2/GBA) for microgravity studies of fluid systems. Voss and Gernhardt on an EVA tested construction tools, an arm-sleeve strap-on computer, and suit modifications for space-station assembly.
Oct. 20–Nov. 5	*STS 73* (72)	Kenneth D. Bowersox, Kent Rominger, Kathryn Thornton, Catherine Coleman, Michael Lopez-Alegria, Fred Leslie, Albert Sacco	230,158 (landing)	255; 174	15 days 21 h 53 min	*Columbia* carried the second U.S. Microgravity Laboratory (USML 2) Spacelab with science and technology experiments in areas such as fluid physics, materials science, biotechnology, combustion science, and commercial space processing technologies. Students interacted with the crew, discussing and comparing on board microgravity experiments with similar ground-based experiments.
Nov. 12–20	*STS 74* (73)	Kenneth Cameron, James Halsell, Chris Hadfield, Jerry Ross, William McArthur	205,000 (landing)	128; 185	8 days 4 h 31 min	*Atlantis* carried a Russian-built Docking Module (DM), two new solar arrays, and supplies for *Mir* on the second shuttle-*Mir* docking mission. After attaching the DM to *Mir*'s *Kristal* module with the robot arm and linking the shuttle's airlock to it on Nov. 15, the *Atlantis* crew, including a Canadian (Hadfield), was welcomed by Gidzenko, Avdeyev, and Reiter. During 3 days of joint operation, the crews transferred the United States biomedical and microgravity science samples and data collected by the *Mir 18, 19,* and *20* resident crews from the station to the shuttle, and fresh supplies to *Mir.* On Nov. 18, *Atlantis* undocked, leaving the DM attached to the station for future linkups.
1996						
Jan. 11–20	*STS 72* (74)	Brian Duffy, Brent W. Jett, Jr., Leroy Chiao, Winston E. Scott, Koichi Wakata, Daniel T. Barry	217,000 (landing)	141; 290	8 days 22 h 1 min	*Endeavour* retrieved, on Jan. 13, the Japanese satellite SFU (Space Flyer Unit) launched by an H-2 rocket in March 95. On Jan. 14, the Japanese crew member (Wakata) released the OAST-Flyer, a free-flying platform with four autonomous experiments. On Jan. 16, the satellite was retrieved. On Jan. 14, Chiao and Barry performed a spacewalk to evaluate a portable work platform. A second EVA, by Chiao and Scott, took place on Jan. 17 for further testing of spacesuit modifications and space station assembly tools and techniques.
Feb. 21–Sept. 2	*Soyuz TM-23*	Yuri Onufrienko, Yuri Usachev	14,400	3027; 247	193 days 19h 11 min	The *Mir 21* crew docked to the Mir station on Feb. 23, joining the three crew members of *Mir 20,* who returned in TM 22 on Feb. 29. In the first month, on-board activities focused on national research, followed by the United States science program NASA 2 when astronaut Shannon Lucid joined the crew on Mar. 23. On May 21 and 24, Onufrienko and Usachev performed their second and third spacewalk to install and unfurl the Mir Cooperative Solar Array brought up by *Atlantis* (*STS 74*) and a fourth on May 30 to install a European earth resources camera on the *Priroda* module. On June 13, they installed a truss structure called *Rapana* to *Kvant-1*, taking the place of the similar *Strela*. Later, they worked on the French research program Cassiopeia when *Mir 22* arrived with French astronaut Andre-Deshays. Onufrienko and Usachev returned with her on Sept. 2.
Feb. 22–Mar. 9	*STS 75* (75)	Andrew Allen, Scott Horowitz, Franklin Chang-Diaz, Jeffrey Hoffman, Claude Nicollier, Maurizio Cheli, Umberto Guidoni	229,031 (landing)	250; 185	15 days 17 h 39 min	*Columbia* carried three foreign astronauts among its crew (Cheli/ESA-Italy, Nicollier/ESA-Switzerland, Guidoni/ASI-Italy). On Feb. 25, the Tethered Satellite System, on a reflight from *STS 46* in July 1992, was deployed on the end of its thin 12.5-mi-long tether line. Experiments could not be completed, however, since, with 12 mi of tether deployed, the cable broke close to the shuttle, and the satellite was lost in orbit. On-board activities then focused on the operation of the third U.S. Microgravity Payload (USMP 3), continuing research aimed at improving basic knowledge of materials under zero-*g* conditions.

*1 lb = 0.45 kg. 1 mi = 1.6 km. 1 ft = 0.3 m. 1 yd = 0.9 m. 1 fathom = 1.8 m. 1 lb/in.^{2} = 6.9 kPa.

Crewed space flights, 1961–1996* (cont.)

Dates launched and recovered	Designation	Crew	Weight, lb	Revolutions; max. distance from Earth, mi	Duration	Remarks
1996 (cont.)						
Mar. 22–31	*STS 76* (76)	Kevin Chilton, Richard Searfoss, Ronald Sega, M. Richard Clifford, Linda Godwin, Shannon Lucid	246,335 (landing)	144; 246	9 days 5 h 15 min 53 s	*Atlantis* lifted off for the third linkup with Russia's *Mir* station as part of Phase 1 of the International Space Station program. After docking on Mar. 23 and hatch opening, the two crews of eight conducted joint operations and transferred 5000 lb of supplies and water to *Mir.* The first spacewalk by United States astronauts (Godwin, Clifford) with a shuttle attached to the Russian station took place on Mar. 27, to mount four experiments to its outside for assessing its environment. After concluding science and technology experiments in the Spacehab module in the cargo bay, *Atlantis* undocked on Mar. 28, leaving Lucid, the first United States astronaut to be ferried to *Mir* by shuttle, aboard the station to join Onufrienko and Usachev for the next 188 days.
Apr. 23	*Priroda*	Crewless	43,000	—; 246		The *Priroda* module of *Mir* was launched on a Proton rocket. Docking occurred on the first attempt on Apr. 26, despite the failure of two of the module's batteries on Apr. 25. On Apr. 27, *Priroda* was rotated from an axial to a radial docking port. The module carries a set of remote-sensing instruments and some microgravity experiments.
May 19–29	*STS 77* (77)	John H. Casper, Curtis L. Brown, Andrew S. W. Thomas, Daniel W. Bursch, Mario Runco, Jr., Marc Garneau	254,879 (landing)	161; 177	10 days 39 min 18 s	*Endeavour* carried into orbit the commercial Spacehab module (fourth flight), with nearly 3000 lb of experiments and support equipment, an Inflatable Antenna Experiment (IAE) on the free-flying Spartan 207 platform, and four technology experiments called TEAMS. The crew included a Canadian (Garneau). The Spartan was released on May 20, and the 50-ft-diameter IAE inflated on its three 92-ft-long struts, the most complex and precise inflatable space structure ever. It was separated from Spartan later and burned up in the atmosphere. The satellite was retrieved on May 21, followed by further experiments which included rendezvous with the passive aerodynamically stabilized, magnetically damped satellite PAMS/STU (Satellite Test Unit).
June 20–July 7	*STS 78* (78)	Terence Henricks, Kevin Kregel, Richard M. Linnehan, Susan J. Helms, Charles E. Brady, Jr. Jean-Jacques Favier, Robert Brent Thirsk	256,170 (landing)	271; 177	16 days 21 h 48 m	*Columbia* carried the Life and Microgravity Sciences (LMS) Spacelab, equipped to focus on both living organisms in the low-gravity environment and the processing of various substances when gravity effects are greatly reduced. Crew included a French and a Canadian payload specialist (Favier, Thirsk). Mission set a shuttle duration record.
Aug. 17–	*Soyuz TM-24*	Valerly Korzun, Aleksandr Kaleri, Claude André-Deshays	14,400	243	32 days 18 h 27 m/ André-Deshays	The *Mir 22* crew included a French female researcher-cosmonaut (Deshays). The 3 cosmonauts joined the *Mir 21* crew on board *Mir* on Aug. 19. Prior to their arrival, on Aug. 1, the uncrewed cargo ship *Progress M-31* was undocked and deorbited over the Pacific; on Aug. 2, *Progress-32* arrived. It was undocked on Aug.16 to free up a docking port, and linked up again with *Mir* on Sept. 3.
Sept. 16–26	*STS 79* (79)	William F. Readdy, Terrence W. Wilcutt, Jay Apt, Tom Akers, Carl E. Watz, John E. Blaha(up), Shannon Lucid(down)	249,352 (landing)	159; 239	10 days 3 h 18 min; 188 days 5 h/Lucid	*Atlantis* took off for its 4th docking with Russia's *Mir,* the second flight to carry a commercial SPACEHAB module in its cargo bay, and the first to carry a double version of it. Main mission purposes: to replace Lucid on *Mir* with Blaha for a 4-month stay; to carry 4600 lb logistics, including food, clothing, experiment supplies, and spare equipment, to the space station; and to return 2200 lb of science samples and hardware to Earth. Five different experiments were turned off on the shuttle and rapidly transferred to *Mir,* the first such transfer of powered scientific apparatus. The two vehicles docked on Sept. 18, and separated on Sept. 23. The mission began a new era of permanent space occupancy by U. S. astronauts. Lucid set records for longest space flight, both for women and for U. S. astronauts (188 days 5 hours).

*1 lb = 0.45 kg. 1 mi = 1.6 km. 1 ft = 0.3 m. 1 yd = 0.9 m. 1 fathom = 1.8 m. 1 lb/in.^{2} = 6.9 kPa.

Fig. 2. Space station Skylab in orbit. The sunshade rigged by crew repair is visible. (NASA)

experience, specifically in form of the emerging Space Transportation System (STS), with the ability to transport inexpensively a variety of useful payloads to orbit, as the mainstay and "work horse" of the United States space program in the 1980s and 1990s. The two major components of the STS are the space shuttle and the Spacelab.

The space shuttle is a reusable system, It has three major elements: the orbiter, an external tank (ET) containing the liquid propellants to be used by the orbiter main engines for ascent, and two solid-propellant rocket boosters (SRBs) "strapped on" to the external tank. The orbiter and SRB casings are reusable; the ET is expanded on each launch. *See* SPACE SHUTTLE.

With the launch of the twenty-fifth shuttle mission, *51-L*, on January 28, 1986, tragedy struck the American space program. At approximately 11:40 A.M. EST, 72 s after liftoff, the flight of *Challenger*, on its tenth mission, abruptly ended in an explosion triggered by a leak in the right solid rocket booster, killing the seven-member crew. Continuation of the United States crewed space program was suspended pending a thorough reassessment of flight safety issues and implementation of necessary improvements. All shuttle systems, operations, and its entire management were subjected to meticulous reviews and modifications, overviewed by outside expert committees. Shuttle flight operations resumed on September 29, 1988.

The Spacelab is a major adjunct of the shuttle. Developed and funded by 10 member nations of the European Space Agency (ESA), the large pressurized Spacelab module with an external equipment pallet is designed to be the most important payload carrier during the space shuttle era. With its large transport capacity, the Spacelab is intended to support a wide spectrum of missions in science, applications, and technology by providing versatile and economical laboratory and observation facilities in space for many users. Another objective is to reduce significantly the time from experiment concept to experiment result, compared with previous space practice, and able to reduce the cost of space experimentation. It allows the direct participation of qualified scientists and engineers to operate their own equipment in orbit.

As the next logical step after the development of a space transportation system, NASA has turned to the development of permanent human presence in space. Focal point of the orbital infrastructure required for this future program will be the International Space Station. In addition to living quarters, the facility will provide utilities, work space, and a docking hub to allow tending by the space shuttle. This will permit crew rotation and resupply. *See* SPACE TECHNOLOGY; SPACECRAFT STRUCTURE. [J.V.P.]

Space navigation and guidance The determination of the position and velocity of a space probe with respect to a target body or a known reference body such as the Earth (navigation) and, based upon this determination, the application of propulsive maneuvers to alter the subsequent path of the probe (guidance).

Space navigation can be viewed as determining the current position and velocity of the probe and then using that determination as the basis of predicting future motion. This determination is made by taking a series of measurements relating to the probe's motion and combining these measurements in such a manner as to make the most accurate estimate of the probe's current position and velocity, taking into account possible small errors or inaccuracies in the measurements themselves. One of the most powerful (and accurate) measurements which can be made is the relative velocity between an Earth tracking station and the space probe itself. This is accomplished by broadcasting an electromagnetic signal which consists of a single tone having a stable frequency to the probe. The probe will receive this signal shifted in frequency in exact proportion to the relative station-probe velocity. This frequency shift is known as the Doppler effect. It is also possible to measure the station-probe distance (or range) using electromagnetic signals.

For missions to the outer planets (Jupiter and beyond) or for a mission to a body such as a comet or asteroid, the position of the target may be sufficiently uncertain as to make a strictly Earth-relative navigation scheme inadequate. Here it is necessary to make measurements that directly involve the target. Typical of these is to obtain optical measurements of the location of the target relative to a star as seen from the probe itself.

In its simplest form, guidance consists of comparing the predicted future motion of the probe against that which is desired, and if these are sufficiently different, executing a propulsive maneuver to modify that future position. Typically, the probe will contain a small rocket motor which can be fired in any desired direction by first rotating the spacecraft away from its cruise orientation and then holding the new attitude fixed while the rocket motor is firing.

Multiplanet missions use an accurate delivery to the first target not only to scientifically explore that planet, but also to use the planet's gravitational attraction to change the course of velocity of the probe advantageously for the next leg of the mission toward the second and subsequent targets. The encounter with a planet becomes in itself a type of guidance correction as the planet changes the path of the probe. *See* GUIDANCE SYSTEMS; SPACE PROBE. [D.W.C.]

Space power systems On-board assemblages of equipment to generate, store, and distribute electrical energy on satellites and spacecraft. Present-generation spacecraft fly power systems from tens of watts to several kilowatts.

With few exceptions, the power systems for United States satellites have used photovoltaic generation for power, batteries for energy storage, and a host of electrical equipment for appropriate regulation, conversion, and distribution. Fuel cells have been limited primarily to use in the crewed space program, supplying power for *Gemini*, *Apollo*, and the space shuttle. Radioisotope thermoelectric generators (RTGs) have powered, or augmented solar power on, a few planetary missions and probes, powered lunar instruments left on the Moon by the *Apollo* missions, and augmented the solar-array battery power system on at least one Earth-orbiting spacecraft. [F.E.F.]

Space probe An automated, crewless vehicle, the payload of a rocket-launching system, designed for flight missions to other planets, to the Moon, and into interplanetary space, as distinguished from Earth-orbiting satellites (see table).

The space probe is used primarily for scientific purposes, which are stated as the mission objectives. Missions generally fall into three categories according to destination: those to the rocky bodies in the inner solar system, including the Earth-like planets, asteroids, and comets; those to the giant gaseous planets in the outer solar system; and those designed to study solar physics and the properties of interplanetary space. Most spacecraft launched to a planet or other body also study the environment of charged particles and electromagnetic fields in interplanetary space during their cruise phase en route to the destination.

Missions may also be categorized by complexity. The simplest are flyby spacecraft, which study their target body during a relatively brief encounter period from a distance of hundreds to thousands of miles as they continue past. Next are orbiters, which circle a planet or other body for extended study; some may carry atmospheric descent probes. Even more complex are lander missions, which touch down on a planet or other body for the collection of on-site data; some may bear exploration rovers designed to range beyond the immediate landing site. Finally, the most complex space probes envisaged are sample-return missions, which would collect specimen material from a target body and return it to Earth for detailed study.

Spacecraft subsystems. In the broadest terms, a space probe may be considered a vehicle that transports a payload of sensing instruments to the vicinity of a target body. Thus, the spacecraft must include a number of subsystems to provide power, to communicate with Earth, to maintain and modify attitude and perform maneuvers, to maintain acceptable onboard temperature, and to manage the spacecraft overall. *See* SPACE TECHNOLOGY.

Scientific instruments. The scientific payload may be divided

Important space probes

Name	Launch date	Comments
Luna 1	Jan. 2, 1959	Lunar probe; now in solar orbit; passed within 3278 mi (5275 km) of the Moon
Pioneer 4	Mar. 3, 1959	Cosmic rays; passed 37,300 mi (60,000 km) from Moon
Luna 2	Sept. 2, 1959	Impacted Moon
Luna 3	Oct. 4, 1959	Photographed far side of Moon
Pioneer 5	Mar. 11, 1960	First deep-space probe; magnetic fields and cosmic rays
Mariner 2	Aug. 26, 1962	First planetary flyby; Venus probe
Ranger 7	July 28, 1964	Lunar impact and approach photography
Mariner 4	Nov. 28, 1964	Mars encounter; photography, magnetic fields, cosmic rays
Ranger 8	Feb. 17, 1965	Lunar impact and approach photographs
Ranger 9	Mar. 21, 1965	Lunar impact and Alphonsus; approach photography
Zond 3	July 18, 1965	Photographs from lunar encounter
Pioneer 6	Dec. 16, 1965	Solar orbit
Luna 9	Jan. 31, 1966	First photographs from lunar surface
Luna 10	Mar. 31, 1966	Lunar and interplanetary data
Surveyor 1	May 30, 1966	Soft landing on Moon: environmental data and photography
Lunar Orbiter 1	Aug. 10, 1966	Lunar photographs
Pioneer 7	Aug. 17, 1966	Solar orbit
Luna 11	Aug. 24, 1966	Lunar data
Luna 12	Oct. 22, 1966	Lunar orbital photography and other data
Lunar Orbiter 2	Nov. 6, 1966	Lunar orbital photography
Luna 13	Dec. 21, 1966	Lunar surface photography and soil information
Lunar Orbiter 3	Feb. 5, 1967	Lunar orbital photography
Surveyor 3	Apr. 17, 1967	Lunar surface photography and surface properties
Lunar Orbiter 4	May 4, 1967	Lunar orbital photography
Venera 4	June 12, 1967	Analysis of Venus atmosphere; first instrumented landing on another planet
Mariner 5	June 14, 1967	Venus probe; atmospheric and magnetospheric data
Lunar Orbiter 5	Aug. 1, 1967	Lunar orbital photography
Surveyor 5	Sept. 8, 1967	Lunar surface photography and surface properties, including elemental analysis of surface
Surveyor 6	Nov. 7, 1967	Same as *Surveyor 5*; landing in Sinus Medii
Pioneer 8	Dec. 13, 1967	Solar orbit
Surveyor 7	Jan. 7, 1968	Same as *Surveyor 5*
Zond 5	Sept. 14, 1968	Circled Moon; recovered Sept. 21, 1968
Pioneer 9	Nov. 8, 1968	Solar orbit
Zond 6	Nov. 10, 1968	Circled Moon; recovered Nov. 17, 1968
Venera 5	Jan. 5, 1969	Same as *Venera 4*
Venera 6	Jan. 10, 1969	Same as *Venera 4*
Mariner 6	Feb. 25, 1969	Photography and analysis of surface and atmosphere of Mars
Mariner 7	Mar. 27, 1969	Same as *Mariner 6*
Luna 15	July 14, 1969	Lunar reconnaissance (crashed during attempted lunar landing)
Zond 7	Aug. 8, 1969	Reentered Aug. 14, 1969; third uncrewed circumlunar flight; recovered in the Soviet Union
Venera 7	Aug. 17, 1970	Lander capsule transmitted 23 min from surface of Venus, Dec. 15, 1970
Luna 16	Sept. 12, 1970	Reentered Sept. 24, 1970; uncrewed Moon lander touched down on Sea of Fertility Sept. 20, 1970; returned lunar soil samples
Zond 8	Oct. 20, 1970	Circled Moon; recovered Oct. 27, 1970
Luna 17	Nov. 10, 1970	Landed on Moon Nov. 17, 1970; uncrewed Moon rover
Mars 2	May 19, 1971	First Soviet Mars landing
Mars 3	May 28, 1971	Mars probe
Mariner 9	May 30, 1971	Mars probe
Luna 18	Sept. 2, 1971	Impacted Moon Sept. 11, 1971
Luna 19	Sept. 28, 1971	Lunar photography mission
Luna 20	Feb. 14, 1972	Recovered Feb. 25, 1972; returned lunar sample

into remote-sensing instruments, such as cameras, and direct-sensing instruments, such as magnetometers or dust detectors. They may be classified as passive instruments, which detect radiance given off by a target body, or active ones, which emit energy such as radar pulses to characterize a target body. See REMOTE SENSING.

Power subsystem. Electrical power is required for all spacecraft functions. The total required power ranges from about 300 to 2500 W for current missions, depending on the complexity of the spacecraft. The power subsystem must generate, store, and distribute electrical power. All space probes launched so far have generated power either via solar panels or via radioisotope thermoelectric generators (RTGs). See NUCLEAR BATTERY; SOLAR CELL; SPACE POWER SYSTEMS.

Telecommunications subsystem. In order to accomplish its mission, the spacecraft must maintain communications with Earth, such as receiving commands sent from ground controllers, and transmitting scientific data and routine engineering "housekeeping" data. All of these transmissions are made in various segments of the microwave spectrum. The design of the telecommunications subsystem takes into account the volume of data to be transmitted and the distance from Earth at which the spacecraft will operate, dictating such considerations as the size of antennas and the power of the on-board transmitters. See MICROWAVE; TELEMETERING.

Advanced planetary probes have carried a dish-shaped high-gain antenna which is the chief antenna used to both transit and receive. These antennas typically consist of a large parabolic reflector, with a subreflector mounted at the main reflector's focus in a Cassegrain-type configuration. In the interest of redundancy and in the event that Earth pointing is lost, spacecraft virtually always carry other on-board antennas. These may be low-gain antennas, which typically offer nearly omnidirectional coverage except for blind spots shadowed by the spacecraft body, or medium-gain antennas, which provide a beam width of perhaps 20–30°. See ANTENNA (ELECTROMAGNETISM).

Attitude-control subsystem. It would be impossible to navigate the spacecraft successfully or point its scientific instruments or antennas without closely controlling its orientation in space, or attitude. Some spacecraft, particularly earlier ones, have been spin-stabilized; during or shortly after launch, the spacecraft is set spinning at a rate on the order of a few revolutions per minute. Much like a rotating toy top, the spacecraft's orientation is stabilized by the gyroscopic action of its spinning

Important space probes (cont.)

Name	Launch date	Comments
Pioneer 10	Mar. 2, 1972	Jupiter encounter; transjovian interplanetary probe
Venera 8	Mar. 27, 1972	Venus landing July 22, 1972
Luna 21	Jan. 8, 1972	Moon landing Jan. 16, 1972, with *Lunikhod* rover
Pioneer 11	Apr. 5, 1973	Jupiter encounter and transjovian interplanetary probe; also Saturn encounter
Mars 4	July 21, 1973	Mars orbiter
Mars 5	July 25, 1973	Mars orbiter
Mars 6	Aug. 5, 1973	Mars lander
Mars 7	Aug. 9, 1973	Mars lander
Mariner 10	Nov. 3, 1973	Venus and Mercury encounter
Luna 22	May 29, 1974	Lunar probe
Helios 1	Dec. 10, 1974	Inner solar system, solar wind exploration
Venera 9	June 8, 1975	Venus probe
Venera 10	June 14, 1975	Venus probe
Viking 1	Aug. 20, 1975	Mars lander and orbiter
Viking 2	Sept. 9, 1975	Mars lander and orbiter
Helios 2	Jan. 15, 1976	Interplanetary; similar objectives to those of *Helios 1*
Luna 24	Aug. 9, 1976	Recovered Aug. 25, 1976; returned lunar sample
Voyager 2	Aug. 20, 1977	Jupiter, Saturn, Uranus, and Neptune encounters; also satellites and ring systems
Voyager 1	Sept. 5, 1977	Same objectives as *Voyager 2* with some orbital differences giving differing encounter trajectories
Pioneer Venus Orbiter	May 20, 1978	Returning atmospheric, surface, and particle and field information
Pioneer Venus Multi-Probe Bus	Aug. 8, 1978	Penetration of Venus atmosphere by four probes; returned atmospheric data
Venera 11	Sept. 8, 1978	Venus lander; returned information on surface properties; detection of lightning and thunderlike sounds
Venera 12	Sept. 14, 1978	Similar mission to *Venera 11*
Venera 13	Oct. 30, 1981	Venus lander
Venera 14	Nov. 4, 1981	Venus lander
Venera 15	June 2, 1983	Venus lander; surface topography
Venera 16	June 7, 1983	Similar mission to *Venera 15*
International Cometary Explorer (ICE)	—	Originally *International Sun-Earth Explorer 3 (ISEE 3)* Earth satellite, redirected using a lunar swingby on Dec. 22, 1983, to encounter with Comet Giacobini-Zinner; plasma and magnetic field
Vega 1	Dec. 15, 1984	Venus probe–Halley intercept
Vega 2	Dec. 21, 1984	Venus probe–Halley intercept
Sakigake	Jan. 8, 1985	Halley intercept; precursor to *Suisei*, upgraded to full mission
Giotto	July 2, 1985	Halley intercept
Suisei	Aug. 19, 1985	Halley intercept; plasma and magnetic field measurements
Phobos 1	July 7, 1988	Mars/Phobos probe, lost by command error
Phobos 2	July 12, 1988	Mars/Phobos probe; some data but communications lost
Magellan	May 4, 1989	Venus radar mapper
Galileo	Oct. 18, 1989	Jupiter orbiter and atmospheric probe
Muses	Jan. 24, 1990	Moon orbiter and relay probe; orbiter transmitter malfunctioned
Ulysses	Oct. 6, 1990	Solar polar orbiter
Mars Observer	Sept. 25, 1992	Contact lost 3 days before Mars arrival
Clementine	Jan. 25, 1994	Orbited Moon; thruster malfunction prevented asteroid flyby
Near Earth Asteroid Rendezvous (NEAR)	Feb. 17, 1996	Asteroid Eros orbiter
Mars Global Surveyor (MGS)	Nov. 7, 1996	Mars orbiter
Mars 96	Nov. 17, 1996	Mars orbiter and lander; failed to leave Earth orbit
Mars Pathfinder	Dec. 4, 1996	Mars lander, with *Sojourner* rover
Cassini	Oct. 15, 1997	Saturn orbiter, with *Huygens* Titan probe

mass. Most planetary spacecraft, however, are three-axis-stabilized, meaning that their attitude is fixed in relation to space. The spacecraft's attitude is maintained and changed via on-board thruster jets or reaction wheels, or a combination of both.

Propulsion subsystem. Most spacecraft are outfitted with a series of thruster jets, each of which produces approximately 0.2–2 pounds-force (1–10 newtons) of thrust. Thrusters are usually fueled with a monopropellant, hydrazine, which decomposes explosively when it contacts an electrically heated metallic catalyst within the trajectory-correction maneuvers. In addition to maintaining the spacecraft's attitude, on-board thrusters are used for trajectory-correction maneuvers. Spacecraft designed to orbit a planet or similar target body must carry a larger propulsion element capable of decelerating the spacecraft into orbit upon arrival. *See* SPACECRAFT PROPULSION.

Thermal control subsystem. In order to minimize the impact of temperature variations on the electronics on board, spacecraft nearly always incorporate some form of thermal control. Mechanical louvers, controlled by bimetallic strips similar to those in terrestrial thermostats, are often used to selectively radiate heat from the interior of the spacecraft into space. Other thermal strategies include painting exterior surfaces. In some cases, spacecraft may also carry one or more active forms of heating to maintain temperature at required minimums.

Command and data subsystem. This designation is given to the main computer that oversees management of spacecraft functions and handling of collected data. Blocks of commands transmitted from Earth are stored in memory in the command and data subsystem and are executed at prescribed times. This subsystem also contains the spacecraft clock in order to accurately pace its activities, as well as all the activities of the spacecraft.

Structure subsystem. The spacecraft's physical structure is considered a subsystem itself for the purposes of planning and design. Usually the heart of this structure is a spacecraft bus, often consisting of a number of bays, which houses the spacecraft's main subsystems. *See* SPACECRAFT STRUCTURE.

Significant missions. Soon after the Soviet Union launched the Earth-orbiting *Sputnik* in 1957 and the United States the *Explorer 1* in early 1958, both countries turned to developing probes that would escape Earth orbit and travel to the Moon and beyond.

Luna. The first successful mission to escape Earth orbit was the Soviet Union's *Luna 1,* which executed a 3728-mi (5275-km) lunar flyby after launch on January 2, 1959. The Soviet Union later that year launched *Luna 2*, which made the first lunar impact, and *Luna 3*, which studied the far side of the Moon. *Luna 9* made the first lunar soft landing on February 2, 1966, and returned photographs from the Moon's surface. Other successful missions in the series, which concluded in 1976, were lunar orbiters or landers. *Luna 16*, *20*, and *24* were ambitious automated missions which returned lunar samples to Earth. *Luna 17* and *21* were lunar landers that deployed rover vehicles. *See* MOON.

Pioneer. This family of largely dissimilar United States spacecraft was used mainly for interplanetary physics, although some later missions in the series achieved success as planetary probes. *Pioneer 1*, *2*, and *3* were failed lunar probes in 1958. *Pioneer 4* passed within 37,300 mi (60,000 km) of the Moon in 1959, followed by *Pioneer 5* (Fig. 1) in 1960, which relayed solar system data to a distance of 22.5×10^6 mi (36.2×10^6 km). *Pioneer 6*, *7*, *8*, and *9*, launched in 1965–1968, studied the space environment in solar orbit.

The series took a very different direction with *Pioneer 10* and *11*, the first space probes sent into the outer solar system, in 1972 and 1973, respectively. *Pioneer 10* executed a flyby of Jupiter in December 1973, while *Pioneer 11* flew by Jupiter in December 1974 and Saturn in 1979. In 1978 the series took yet another departure with *Pioneer Venus*, which used two launches to send an orbiter and multiple atmospheric probes to Venus. *See* JUPITER; SATURN; VENUS.

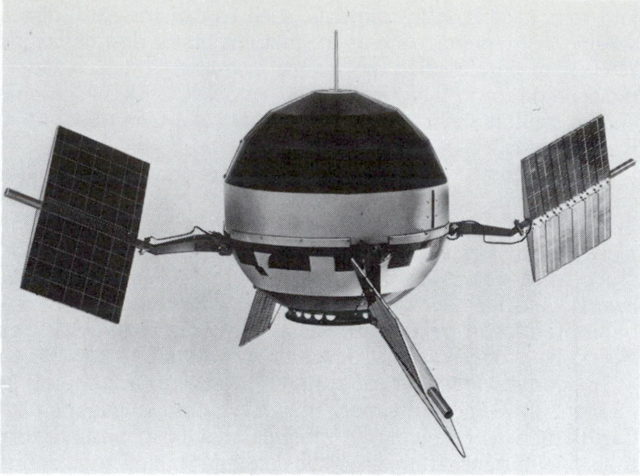

Fig. 1. ***Pioneer* 5 in flight configuration. Solar panels are erected as they would be in space. (*TRW Systems*)**

Ranger. In the early 1960s the United States carried out this series of missions to relay photographs of the Moon as the spacecraft descended to a crash landing. *Ranger 7* was the first success, returning 4308 photos during its lunar descent in July 1964. In 1965 *Ranger 8* and *9* sent back 7137 and 5814 photos, respectively.

Surveyor. These United States missions (Fig. 2) soft-landed robot spacecraft on the Moon as precursors to the crewed Apollo missions. *Surveyor 1* landed on the Moon on June 2, 1966, and returned 11,150 photos as it operated for 6 weeks. *Surveyor 3*, *5*, *6*, and *7* soft-landed in 1967 and 1968, relaying photos and conducting soil analyses. *Surveyor 6* executed the first takeoff from the Moon during its 1967 mission.

Lunar Orbiter. These spacecraft, launched during the same

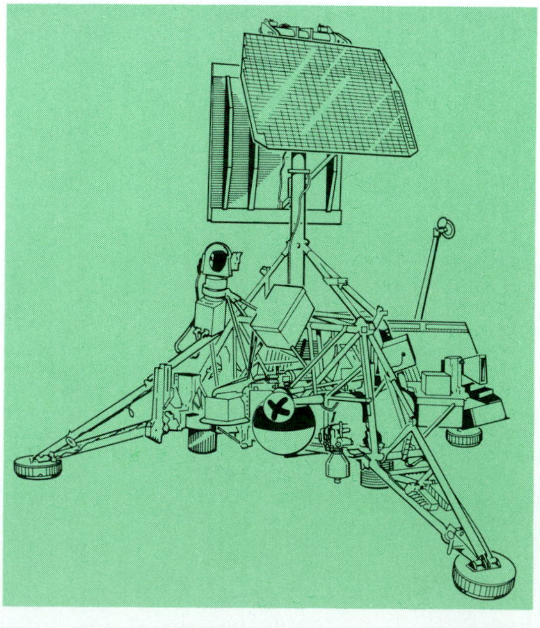

Fig. 2. Surveyor full-scale science training model which is used in operational simulation of the instrument complement. (*NASA*)

era as the *Surveyors*, were the first attempt to make detailed maps of the Moon. *Lunar Orbiter 1* was launched August 10, 1966, and relayed photos for the next 20 days. *Lunar Orbiter 2* sent 205 photographic frames later that year; *Lunar Orbiter 3* sent 182 photos in 1967, followed by *Lunar Orbiter 4* and *5*.

Mariner. This United States early planetary probes chiefly studied Venus and Mars, although the final in the series also visited Mercury. *Mariner 2* became the first spacecraft to encounter another planet when it flew by Venus in 1962. *Mariner 4* became the first spacecraft to encounter Mars in 1964. *Mariner 5* executed a Venus flyby in 1967. *Mariner 6* and *7* again targeted Mars, sending back 75 and 126 television pictures, respectively, during flybys in 1969. *Mariner 8* and *9* were Mars orbiters, sending back detailed pictures after their launches in 1971. *Mariner 10* was the first spacecraft to use a gravity-assist trajectory, encountering Venus before going on to two flybys of Mercury in 1973 and 1974. *See* MARS; MERCURY (PLANET).

Venera, Mars, and Zond. These missions were the Soviet Union's equivalent of the *Mariners*. The Soviets had the greatest success with *Venera*, which lofted a number of spacecraft to Venus over more than two decades. *Venera 4* transmitted during its 94-min descent to Venus's extremely hot surface. *Venera 7* made the first soft landing on Venus in 1970, and transmitted for 23 min, which was remarkable considering the planet's 900°F (500°C) surface temperature. *Venera 15* and *16*, launched in 1983, were orbiters that used synthetic-aperture radar to pierce Venus's opaque atmosphere and map the planet's surface.

The Soviets' Mars program achieved more mixed results. *Mars 2* and *3* in 1971 and four spacecraft in 1973 all reached Mars or its vicinity, but none returned useful data.

Zond was the designation given to several Soviet missions with various targets from 1964 through 1970. *Zond 3* returned photos from a lunar flyby in 1965. *Zond 5*, *6*, *7*, and *8*, in 1968–1970, flew around the Moon and returned to Earth.

Viking. One of the most ambitious planetary explorations ever mounted was the United States' Viking program. Two launches of combined orbiter-landers, *Viking 1* and *2* (Fig. 3), took place in August and September 1975, with arrival at Mars in June and August 1976. *Viking's* chief purpose was the search for biological activity on the Martian surface. Both orbiters carried out a global mapping program, while the landers sent back the first clear, detailed pictures from the surface of another planet. The spacecraft failed to detect any sign of life, although the possibility of lifeforms in the planet's past has not been ruled out.

Fig. 3. Viking spacecraft in final test configuration. The upper ovoid is the sterile shield surrounding the lander. (*Langley/NASA*)

Helios. Because of its close approach to the Sun (0.31 astronomical unit), this German-built solar system probe is called a solar probe. *Helios 1* was launched in 1974, and *Helios 2* in 1976. The experiments, directed toward phenomena taking place in the solar wind, were chiefly magnetometers and particle detectors. *See* SOLAR WIND.

Voyager. The launch in 1977 of *Voyager 1* and *2* began a decade-long "grand tour" of the outer solar system. Although *Voyager 2* was launched first, *Voyager 1* took a shorter trajectory to Jupiter and arrived in 1979; it went on to encounter Saturn in 1980. *Voyager 1*'s trajectory at Saturn was optimized to study the planet's major satellite, Titan. *Voyager 2* visited Jupiter in 1979 and Saturn in 1981, then continued on to flybys of Uranus in 1986 and Neptune in 1989. During the 1990s, both spacecraft continued to relay back fields and particles data as they exited the solar system. *See* NEPTUNE; URANUS.

Comet missions. The passage of Comet Halley into the inner solar system in the mid-1980s prompted a great deal of international interest in sending spacecraft to study it. The Soviet Union and Japan each dispatched two spacecraft, and the European Space Agency a single craft. The efforts were notable for a high degree of international cooperation, including the use of data obtained by the NASA Deep Space Network from the early-arriving Soviet craft to target more precisely the flyby of the European craft. *See* COMET; HALLEY'S COMET.

United States mission planners developed an ingenious scheme to redirect the *International Sun-Earth Explorer 3* (*ISEE 3*) satellite, renamed the *International Comet Explorer* (*ICE*), to conduct a flyby of the tail of Comet Giacobini-Zinner on September 11, 1985, about 6 months before the Comet Halley flybys.

The *Suisei* and *Sakigake* spacecraft represented the first solar system exploration mission launched by the Japanese Institute of Space and Astronautical Science (ISAS). *Sakigake* flew within a few million miles of Comet Halley on March 11, 1986, while the *Suisei* craft, also called *Planet-A*, approached within about 90,000 mi (150,000 km) of Comet Halley on March 9.

The enterprising *VEGA* mission by the Soviet Union sent a pair of spacecraft on flybys of Venus, where each released a lander and an atmospheric balloon, before continuing on to encounter Comet Halley. *VEGA 1*'s Halley encounter took place March 6, 1986, and *VEGA 2*'s followed 3 days later. The *VEGA*s achieved a closer flyby distance from Halley than did the Japanese probes, encountering the comet at distances of 5523 and 4990 mi (8889 and 8030 km).

The *Giotto* spacecraft was the European Space Agency's representative in the Comet Halley fleet. At its closest approach on March 14, 1986, *Giotto* came within about 360 mi (600 km) of the comet. Because the environment near the comet's nucleus was known to be very dusty, *Giotto* was designed to endure a high number of particle impacts. The last usable image was taken at a distance of about 1220 mi (1963 km). In addition, particle impacts caused the craft to wobble, resulting in a loss of radio signal 14 s before closest approach. The wobble was overcome, however, about 32 min later, and *Giotto* remained a highly functional probe years after the encounter.

Phobos. Phobos 1 and *2* were launched from the Soviet

Union in July 1988. The two nearly identical craft were to arrive at Mars in January 1989 and begin orbiting the planet. A series of maneuvers would have taken the craft near the satellite Phobos, where they would approach the rocky object's surface and deploy a long-duration lander.

A series of commands set to *Phobos 1* in August 1988 apparently contained a coding error that instructed the spacecraft to turn off its attitude-control system, and all contact was permanently lost. *Phobos 2* survived the transit to Mars, began its scientific surveys of the red planet, and was positioned near Phobos, capturing high-detail images from a distance of several hundred miles. On March 27, 1989, however, a failure in the attitude-control system resulted in loss of communications with the craft.

Magellan. The first craft to perform radar imaging at Venus, *Pioneer Venus* of the United States, was followed by the Soviet *Venera 15* and *16* probes, which were able to improve resolution by about a factor of 10. They were followed by the United States' *Magellan* mission, which in turn achieved a factor-of-10 improvement over the resolution delivered by the Soviet missions. Launched in May 1989, *Magellan* was also the first of several solar system probes to be lofted by the United States space shuttle. *See* SPACE SHUTTLE.

Magellan entered Venus orbit in August 1990, initiating a highly successful 4-year study of the planet. It produced radar maps of 98% of the Venusian surface with a resolution of 1000 ft (300 m) before compiling a high-resolution, comprehensive gravity field map for 95% of the planet. With various components gradually failing, mission controllers decided to plunge the spacecraft into Venus's atmosphere to study the craft's behavior and atmospheric characteristics. Contact was lost with *Magellan* on October 12, 1994.

Galileo. After the initial reconnaissance of the solar system performed by the *Mariners* and *Voyagers* from the 1960s through the 1980s, mission planners began to consider the next step in planetary exploration. Their plans called for a series of orbiters that would return to each of the planets visited by flyby spacecraft to conduct detailed studies. The first such approved mission was *Galileo*, which would orbit the solar system's largest planet, Jupiter. The orbiter also carried a descent probe designed to relay data as it dived into the giant planet's atmosphere.

En route to Jupiter, *Galileo* performed a flyby of Venus, two flybys of Earth, and the first-ever flybys of asteroids when it encountered the rocky body Gaspra in October 1991 and a second asteroid, Ida, in August 1993. *See* ASTEROID.

The failure of *Galileo*'s umbrellalike high-gain antenna to open fully prompted an extensive redesign of planned observations at Jupiter. After arrival in December 1995, the mission's highest priority was to relay data from the atmospheric probe during its descent. *Galileo* then began over 2 years in orbit, sending selected images and other scientific data that had been compressed in on-board computers to make the best use of the craft's available low-gain antennas.

Ulysses. By the 1980s the Sun had been studied for many years, but all observing equipment was limited to the two-dimensional surface of the ecliptic, the plane in which most planets orbit the Sun. *Ulysses* was designed to extend that point of view by sending a spacecraft out of the ecliptic and into the unexplored regions above and below the Sun's poles. After collaborative work between NASA and the European Space Agency, a project emerged wherein the European agency built and provided the single *Ulysses* spacecraft. NASA provided launch by a United States space shuttle.

Launched in October 1990, *Ulysses* flew first to Jupiter, where the giant planet's gravity sent the craft out of the ecliptic in February 1992. The encounter of the Sun's south pole took place in 1994, with an encounter of the north pole in 1995. *Ulysses*' observations continued, and it will make another passage over the Sun's polar regions under conditions of high solar activity in 2000 and 2001.

Mars Observer. This mission was to be the United States' first return to the red planet since the *Viking* effort of the 1970s. A polar orbiter, *Mars Observer* carried a high-resolution camera and other instruments to study Mars's geology and atmosphere. The mission proceeded according to plan until 3 days before Mars arrival in August 1993, when all contact was lost with the craft. A failure probably occurred in the propulsion system that either ruptured a fuel line or caused an explosion.

Clementine. Unlike most United States missions, carried out by the National Aeronautics and Space Administration (NASA), this novel effort was the work of the Department of Defense. *Clementine*'s flight plan called for it to orbit Earth for several days before departing for the Moon, where it would spend more than 2 months in orbit capturing high-resolution images of the lunar surface. The spacecraft was then to leave once more on a trajectory that would take it within 60 mi (100 km) of the asteroid 1620 Geographos.

Clementine arrived in lunar orbit in February 1994 and carried out systematic mapping. Two days after it left lunar orbit in early May, a malfunction forced the cancellation of the asteroid flyby. Although it returned a wealth of lunar imagery, the chief objective of *Clementine* was testing advanced technologies of interest to the military.

Mars initiatives. Following the failure of *Mars Observer*, NASA quickly approved *Mars Global Surveyor*, an orbiter that would duplicate most of the scientific mission of the lost spacecraft. *Mars Global Surveyor* was launched in November 1996 during a time window shared by a second mission to the red planet, *Mars Pathfinder*. This latter effort placed a lander on the planet with a small robot rover to range around the landing site as a technology demonstration. *Mars Pathfinder* landed on Mars in July 1997, successfully deployed the rover vehicle, and began carrying out meteorological and geological observations.

Following uncertainties in Russia, plans for a *Mars 94* mission were delayed until the 1996 window. However, the *Mars 96* probe failed to leave Earth orbit when booster rockets misfired, and reentered the Earth's atmosphere.

Cassini. Considered the last of the large United States planetary exploration missions. *Cassini*, launched in October 1997, will study the ringed planet Saturn and its major satellite, Titan. The *Cassini* orbiter will carry an instrumented probe, *Huygens*, for descent to the surface of Titan. Like *Galileo*, the craft will make use of multiplanet gravity assists, flying by Venus, Earth, and Jupiter on the way to its destination.

Discovery programs. Further *Cassini*-class missions are unlikely in the foreseeable future. Instead, there is a major effort to develop innovative mission designs that accomplish more with smaller, simpler, lightweight probes.

To encourage such developments, NASA initiated an effort called the Discovery program to develop small-to-moderate, inexpensive mission designs. The first two missions approved under the program were the *Mars Pathfinder* (mentioned above) and the *Near Earth Asteroid Rendezvous* (*NEAR*). The *NEAR* probe was launched in February 1996 to rendezvous with and orbit the asteroid Eros, one of the small, rocky bodies that passes relatively near Earth. [F.OD.]

Space processing The carrying out of various processes on materials aboard orbiting spacecraft. The low-gravity, high-vacuum environment associated with orbiting space vehicles offers unique opportunities to Investigate various processes in ways that cannot be duplicated on Earth. Gravity-driven convective flows, sedimentation, and hydrostatic pressure are virtually eliminated. Materials can be melted, shaped, and solidified in the absence of a container. Processes that require an extremely high vacuum in conjunction with large heat loads or the evolution of large quantifies of gas may be performed in the wake of an orbiting vehicle. Because orbital velocity Is much greater than the molecular speed of the residual atmosphere, the wake of an orbiting vehicle is an almost perfect vacuum chamber with virtually unlimited pumping speed and a 4 K heat sink.

The absence of sedimentation allows materials with large density differences to be maintained in stable dispersion for long periods of time under diffusion-controlled conditions. This offers an excellent opportunity to study processes such as Ostwald ripening, bubble coalescence and dissolution, the formation and stability of flocs, and solidification of alloys and composites.

The absence of hydrostatic pressure eliminates the tendency for materials to sag and deform when melted. This offers new dimensions in the important process of float zone crystal growth, not only in the length of the zone that can be maintained (which provides better thermal control), but also in the extension of the float zone process to materials with such low surface tension that float zone processing is unfeasible in Earth's gravity. *See* CRYSTAL.

Containerless processing eliminates problems such as container contamination and nucleation. Thermal physical measurements can be made at temperatures well above the melting points of any known material or on materials that are extremely reactive in the melt. Unique glasses and nonequilibrium metastable phases may be formed by allowing materials to cool well below their normal freezing points by denying them nucleation sites, Containerless melting or distillation in the high vacuum in the wake of an orbiting vehicle could produce metal samples of unprecedented purity. [R.J.N.]

Space shuttle A crewed reusable orbital spacecraft. The space shuttle along with expendable launch vehicles make up the United States' Space Transportation System (STS). The shuttle provides the unique capability for in-flight repair of satellites and their retrieval and return to Earth.

Shuttle missions are generally of two categories: attached payloads and satellites. An example of an attached payload is the Spacelab module, developed by the European Space Agency (ESA), which carries a wide variety of experiments that take advantage of the environment with very low–to–zero gravity. The position of the experiments in space allows them to view the Earth on a very broad scale and to view the celestial sphere without the encumbrance of the Earth's atmosphere. The other category is made up of two types of satellites. One type is deployed from the shuttle in low-Earth orbit with its own propulsion stage to boost it to higher orbit or to start it on a journey through the solar system. The other type is intended either to stay in low-Earth orbit or be retrieved at a later time by using the remote manipulation system. *See* REMOTE MANIPULATORS; SATELLITE (SPACECRAFT).

The space shuttle flight system (see illustration) consists of the orbiter, three space shuttle main engines, an external tank, and two solid-fuel rocket boosters. These components are reusable, except for the external tank, which is expended on each launch. The orbiter carries a crew of four to seven and payloads. It can remain in orbit for up to 10 days. *See* ROCKET PROPULSION.

The space shuttle is launched with the main engines and solid-fuel rocket boosters burning. Solid-fuel rocket booster separation occurs 2 min after lift-off. Main engine cutoff takes place a little over 8 min after lift-off. External tank separation occurs soon after main engine cutoff. The orbital maneuvering system engines provide the additional thrust to insert the orbiter into its final orbit.

To start the final phase of flight (entry), the orbiter is placed tail-first in its orbit, and its velocity is reduced by firing the

Shuttle flight system lifting off from Kennedy Space Center.

orbital maneuvering system engines. This causes the orbit to decay in a prescribed manner so that the landing may occur on the desired runway at Kennedy Space Center in Florida or Edwards Air Force Base in California. During entry, many small reaction control engines fire as needed to keep the orbiter on its flight path. As the orbiter descends into the atmosphere, its aerodynamic control surfaces become effective and take over the control function from the reaction control engines.

The thermal protection system, which is not redundant, presented a major technological and developmental challenge to the program. This system, unlike those of previous single-use spacecraft, has a design requirement for reuse for 100 missions.

Silica ceramic tiles, some 24,000, cover most of the orbiter's surface. Black borosilicate glass-coated high-temperature tiles are used on the undersurface. White-coated low-temperature tiles were originally used on the upper surface. Better understanding of the thermal environment for the upper surface and continued development of new materials led to most of the upper-surface tiles being replaced with a silica blanket made of two outer layers of woven fabric and an insulating battinglike center layer stitched together in a quiltlike pattern. *See* ATMOSPHERIC ENTRY.

The Orbital Flight Test Program commenced on April 12, 1981, with the launch of the orbiter *Columbia* for mission STS 1. STS 5, launched on November 11, 1982, started the Operational flight phase. Each flight brought new achievements to the space program. STS 5 accomplished the first commercial satellite deployments. STS 41-C (April 1984) accomplished the world's first in-orbit satellite repair by replacing faulty elements in the *Solar Maximum Mission* (*SMM*) satellite's control system and a primary experiment sensor. The crew of STS 51-A (November 1984) recovered and returned to Earth the inoperative *Palapa* and *Westar* satellites.

The launches from the shuttle of the *Magellan* spacecraft to Venus (STS 30, May 1989), *Galileo* to Jupiter (STS 34, October 1989), and *Ulysses* to investigate the Sun (STS 41, October 1990) opened a new era of space exploration. The Hubble Space Telescope was launched on STS 31 on April 26, 1990. *See* SATELLITE ASTRONOMY; SPACE PROBE.

A wealth of data has been accumulated by using the space shuttle. *Spacelab 1* conducted 70 different experiments during its 10-day mission. *Spacelab 3* collected over 250 million bits of data from 15 experiments. Experiments that have been conducted in biosciences and materials-processing sciences are precursors to the development of space-based microgravity production facilities. *See* SPACE PROCESSING.

Mission STS 51-L, the twenty-fifth launch of the space shuttle, lifted off at 11:38 A.M. EST, January 28, 1986. Suddenly, at 72 s into the flight, the external tank exploded. The orbiter *Challenger* was destroyed, and all seven crew members perished. NASA announced the suspension of further flight operations, and a Presidential Commission was established to review the circumstances surrounding the accident.

The Presidential Commission report, submitted on June 6, 1986, contained recommendations to be implemented prior to returning the space shuttle to flight status. The cause of the accident was determined to be inadequate design of the solid rocket motor field joint. Deflection of the joint with deformation of the seal at cold temperature allowed hot gases to pass by the O-ring seal, resulting in erosion and the eventual failure of the O-ring. The new joint has an overlapping tang that helps prevent separation during the start-up phase of motor operation when the case is responding dynamically to the internal pressure buildup. A third O-ring seal was also included in the new design along with the ability to verify the seal's integrity after installation.

Many activities in addition to those that resulted from the Commission's recommendations were accomplished by NASA during the return-to-flight period. A total of 254 modifications were made to the orbiter, 30 to the solid rocket booster, 13 to the external tank, and 24 to the space shuttle main engine.

On September 29, 1988, the space shuttle *Discovery* (STS 26) achieved orbit, and since then the space shuttle has flown many times. On May 7, 1992, a replacement for *Challenger*, named *Endeavor*, was launched on its first flight. [R.A.C.]

Space station An autonomous facility in space for the conduct of scientific and technological research, Earth-oriented applications, and support of other space activities. A space station provides the facilities, centralized computers and data management, and crew support to conduct a wide variety of space activities.

Initially, the space station would be placed in a low orbit a few hundred miles above Earth. Future possibilities, however, include stations in 24-h synchronous orbits at 23,000 mi (37,000 km) altitude, orbits of the Moon, and even orbits of other planets.

Skylab, the first United States space station, was launched on May 14, 1973, and boarded by a three-man crew on May 25, 1973. *Skylab* was made up primarily from space systems developed for the Apollo lunar program and modified as necessary for a space station–type mission. *Skylab* had three separate flight crews for a total of 171 days in orbit. *Skylab* conducted a multidiscipline program of research and applications, including Earth observations, astronomy, and manufacturing in space.

The National Aeronautics and Space Administration (NASA) envisions the future space station as an assembly of shuttle transportable modules. Individual modules would be provided for living quarters, utilities such as power and stabilization, and various types of payload activities, for example, space manufacturing or astronomy.

The space station and the reusable space shuttle provide major new opportunities for cost-effective space operations. The reusable space shuttle not only will provide a lower-cost means of placing a spacecraft in orbit, but will also materially reduce the cost of each space mission by providing a capability to maintain or repair spacecraft in orbit and by returning the spacecraft to Earth for use in subsequent missions. The space station, as the companion element to the shuttle, will provide the long-duration, space-maintainable, crew-supported and -operated facility which will further reduce the cost of space operations. *See* SPACE SHUTTLE.

With the space station logistically supported by the low-cost reusable space shuttle, activities in Earth orbit will assume a routine character like those in laboratories and test facilities on Earth. Personnel will include scientists and others without specialized space flight training. [T.H.]

Space technology The systematic application of engineering and scientific disciplines to the exploration and utilization of outer space. Space technology developed so that spacecraft and humans could function in an environment quite different from the Earth's surface. In outer space there is no atmosphere, no apparent gravity, no means of mechanical support, no simple source of power, and no easy access for repair. The lack of air pressure means that there is no natural convection cooling of electronic components, that there is no oxygen for burning of fuel or breathing, and that liquids quickly evaporate. For each mission and each requirement, space technology provides an answer, always with the goal of greatest reliability and lowest spacecraft mass.

Some technology is fundamental to the functioning of almost all spacecraft: structure and materials, thermal control, tracking and positioning, attitude control, and electrical power. The spacecraft structure must be designed to survive the forces encountered during testing, ground handling, launching, and any deployment of appendages. It is usually made of a metal (aluminum, magnesium, beryllium) or a composite (boron/epoxy or graphite/epoxy). It not only must support all the spacecraft parts, but must also be shaped to fit inside the available launch vehicle volume. *See* SPACECRAFT STRUCTURE.

To maintain temperatures at acceptable levels, various active and passive devices are used: surfaces with special absorptivities and emissivities, numerous types of thermal insulation, mechanical louvers to vary the heat radiated to space, heat pipes, and electrical heaters.

The location of a spacecraft can be measured by determining its distance from the transit time of radio signals or by measuring the direction of received radio signals, or by both. *See* SPACE NAVIGATION AND GUIDANCE; SPACECRAFT PROPULSION.

Most spacecraft either have no appreciable rotation (three-axis body-stabilized) or spin around a well-defined axis (spinner). A body-stabilized spacecraft is simpler in overall concept than the spinner, and may be necessary in many missions, but often requires more hardware than a spinning spacecraft. The orientation of a spacecraft is measured with Sun sensors (the simplest method), star trackers (the most accurate), and horizon or radio-frequency (rf) sensors (usually to determine the direction toward Earth). Attitude corrections are made by small thrusters or by reaction or momentum wheels. *See* STAR TRACKER.

Primary electrical power is most often provided by solar cells made from a thin section of crystalline silicon, protected by a thin glass cover. When sunlight is interrupted by an eclipse, power is often obtained from nickel-cadmium batteries that serve as a secondary power source. *See* SOLAR CELL; SPACE POWER SYSTEMS.

The status and condition of a spacecraft are determined by telemetry. *See* SPACE COMMUNICATIONS; TELEMETERING.

Many spacecraft missions have special requirements and hence necessitate special equipment. Satellites that leave the Earth's gravitational field to travel around the Sun and visit other planets have special requirements. Spacecraft that return to Earth require special protection. In some missions one spacecraft must find, approach, and make contact with another spacecraft. *See* SATELLITE (SPACECRAFT); SPACE PROBE. [G.D.Go.]

Space-time A term used to denote the geometry of the physical universe as suggested by the theory of relativity. It is also called space-time continuum. Whereas in Newtonian physics space and time had been considered quite separate entities, A. Einstein and H. Minkowski showed that they are actually intimately intertwined.

Einstein showed that in general two observers, each using the same techniques of observation but being in motion relative to each other, will disagree concerning the simultaneity of distant events. But if they do disagree, they are also unable to compare unequivocally the rates of clocks moving in different ways, or the lengths of scales and measuring rods. Instead, clock rates and scale lengths of different observers and different frames of reference must be established so as to assure the principal observed fact. Each observer, using his or her own clocks and scales, must measure the same speed of propagation of light. This requirement leads to a set of relationships known as the Lorentz transformations. *See* LORENTZ TRANSFORMATIONS.

In accordance with the Lorentz transformations, both the time interval and the spatial distance between two events are relative quantities, depending on the state of motion of the observer who carries out the measurements. There is, however, a new absolute quantity that takes the place of the two former quantities. It is known as the invariant, or proper, space-time interval τ and is defined by Eq. (1), where T is the ordi-

$$\tau^2 = T^2 - \frac{1}{c^2} R^2 \qquad (1)$$

nary time interval, R the distance between the two events, and c the speed of light in empty space. Whereas T and R are different for different observers, τ has the same value. In the event that Eq. (1) would render τ imaginary, its place may be taken by σ, defined by Eq. (2). If both τ and σ are zero, then a

$$\sigma^2 = R^2 - c^2T^2 (2)$$

light signal leaving the location of one event while it is taking place will reach the location of the other event precisely at the instant the signal from the latter is coming forth.

The existence of a single invariant interval led the mathematician Minkowski to conceive of the totality of space and time as a single four-dimensional continuum, which is often referred to as the Minkowski universe. In this universe, the history of a single space point in the course of time must be considered as a curve (or line), whereas an event, limited both in space and time, represents a point. So that these geometric concepts in the Minkowski universe may be distinguished from their analogs in ordinary three-dimensional space, they are referred to as world curves (world lines) and world points, respectively. *See* GRAVITATION; RELATIVITY. [P.G.B.]

Spacecraft ground instrumentation Instrumentation located on the Earth for monitoring, tracking, and communicating with crewed spacecraft, satellites, and space probes. Radars, communication antennas, and optical instruments are classified as ground instrumentation. They are deployed in networks and, to a lesser extent, in ranges. Ranges are relatively narrow chains of ground instruments used to follow the flights of missiles, sounding rockets, and spacecraft ascending to orbit. Some ranges are a few miles long; others, such as the U.S. Air Force's Eastern Test Range, stretch for thousands of miles. Networks, in contrast, are dispersed over wide geographical areas so that their instruments can follow satellites in orbit as the Earth rotates under them at 15° per hour, or space probes on their flights through deep space.

Networks are of two basic kinds: networks supporting satellites in Earth orbit, and networks supporting spacecraft in deep space far from Earth. A third concept will be added in the 1980s, when a series of Telecommunications and Data Relay Satellites (TDRS) will be operational to replace most of the network of the Earth-orbit stations. The TDRSs will be placed in geosynchronous orbits to relay signals to and from other orbiting spacecraft during more than 85% of each orbit, to and from a single ground station.

Ranges and networks have various technical functions:

1. Tracking: determination of the positions and velocities of space probes and satellites through radio and optical means.
2. Telemetry and acquisition: reception of telemetered signals from scientific instruments and spacecraft housekeeping functions. *See* TELEMETERING.
3. Voice reception and transmission: provision for communication with the crew of a spacecraft, such as the space shuttle.
4. Command: transmission of coded commands to spacecraft equipment, including scientific instruments.
5. Television reception: provision for observation of the crew, spacecraft environment, and so on.
6. Ground communication: telemetry, voice, television, command, tracking data, and spacecraft acquisition data transmission between network sites and the central mission control center.
7. Computing: calculation of orbital elements and radar acquisition data prior to transmission to a central point; also, computation of the signals that drive visual displays at a mission control center. *See* SPACE COMMUNICATIONS; SPACE NAVIGATION AND GUIDANCE. [T.R.]

Spacecraft propulsion A system that provides control of location and attitude of spacecraft by using rocket engines to generate motion. Spacecraft propulsion systems come in various forms depending on the specific mission requirements. Each exhibits considerable variation in such parameters as thrust, specific impulse, propellant mass and type, pressurization schemes, cost, and materials. All of these variables must be considered in deciding which propulsion system is best suited to a given mission. Typical spacecraft applications include communications satellites, science and technology spacecraft, and Earth-monitoring missions such as weather satellites. Orbital environments range from low-Earth to geosynchronous to interplanetary. *See* ASTRONAUTICAL ENGINEERING; ROCKET PROPULSION; SATELLITE (SPACECRAFT); SPACE FLIGHT; SPACE PROBE; SPECIFIC IMPULSE; THRUST.

The two fundamental variables that define the design of spacecraft propulsion systems are the total velocity change to be imparted to the spacecraft for translational purposes, and the impulse necessary to counteract the various external torques imposed on the spacecraft body. From these, the required quantity of a given propellant combination can be specified. Propellant accounts for almost 60% of the lift-off mass of a communications satellite.

The specific impulse has a significant effect on the total propellant load that a spacecraft must carry to perform its assigned mission. Since a massive satellite must be boosted into space by the use of expensive launch vehicles, such as the space shuttle and Ariane, significant cost savings may be gained if smaller, less expensive launch vehicles may be used. The size of the required launch vehicle is directly proportional to the mass of the payload. Since most of the other components that make up spacecraft are relatively fixed in weight, it is critical to utilize propellant combinations that maximize specific impulse.

For modern spacecraft the choices are either bipropellants, which utilize a liquid oxidizer and a separate liquid fuel; solid propellants, which consist of oxidizer and fuel mixed together; or monopropellants, which are liquid fuels that are easily dissociated by a catalyst into hot, gaseous reaction products. High specific impulse is offered by bipropellants, followed by solid propellants and monopropellants.

Spacecraft attitude control schemes play an important role in defining the detailed characteristics of spacecraft propulsion systems. Essentially, there are three methods for stabilizing a spacecraft: three-axis control, spin control, and gravity gradient. In three-axis systems the body axes are inertially stabilized with reference to the Sun and stars, and utilize rocket engines for control in all six degrees of freedom. Spin-stabilized spacecraft use the inertial properties of a gyroscope to permanently align one of the axes by rotating a major portion of the spacecraft body about this axis. This approach significantly reduces the number of thrusters needed for control. Gravity gradient control is a nonactive technique that relies on the Earth's tidal forces to permanently point a preferred body axis toward the Earth's center. *See* GYROSCOPE; INERTIAL GUIDANCE SYSTEM; SPACECRAFT STRUCTURE.

Translation of a spacecraft, independent of its control technique, requires thrusters aligned parallel to the desired translational axis. Usually, all three axes require translational capability. Combining the two requirements for attitude control and translation results in the minimum number of rocket engines required to perform the mission. These are supplemented with additional thrusters to allow for failures without degrading the performance of the propulsion system. Simplistically, it would be reasonable to assume that the propellant-engine combination with the highest specific impulse would be the preferable choice. However, the ultimate requirement is the lowest possible mass for the entire propulsion system. The complexity of the system is greatly influenced by, and is roughly proportional to, the specific impulse, since bipropellants require more tanks, valves, and so forth than either solid systems or monopropellant systems. This is primarily due to the differences in density between liquids and solids, and the fact that bipropellants require high-pressure gas sources to expel the fluid from the tanks and into the rocket engine chamber. For communications satellites in the lift-off weight range of 3000 lbm (1360 kg), the trade-off between specific impulse and system mass dictates the use of a solid rocket motor for the main-orbit circularizing burn and a monopropellant propulsion system for on-orbit attitude control and translation. Spacecraft launch masses above about 5000 lbm (2268 kg) require the use of all-bipropellant systems. [K.D.]

Spacecraft structure The supporting structure for systems capable of leaving the Earth and its atmosphere, performing a useful mission in space or around the Moon or the Earth, Sun, and other planets, sometimes returning to the Earth and sometimes landing on the bodies. Among the principal technologies that enter into the design of spacecraft structures are aerodynamics, aerothermodynamics, heat transfer, structural mechanics, dynamics, materials technology, and systems analysis.

The structural aspects of space flight can be divided into six broad regions or phases: (1) transportation, handling, and storage; (2) testing; (3) boosting; (4) Earth-orbiting; (5) reentry, landing, and recovery on the Earth; (6) interplanetary flight, with orbiting of or landing on other planets. Each phase has its own structural design criteria based upon the specific environment, the configuration existing during the phase, the mission objectives, and whether the craft is crewed or uncrewed. These design criteria require detailed consideration of heat, static loads, dynamic loads, rigidity, vacuum effects, radiation, meteoroids, acoustical loads, atmospheric pressure loads, foreign atmospheric composition, solar pressure, fabrication techniques, magnetic forces, sterilization requirements, accessibility for repair, and interrelation of one effect with the others.

The basic spacecraft structural design considerations apply equally well to both crewed and uncrewed spacecraft. The degree of reliability of the design required is, however, much greater for crewed missions. Also, the spacecraft structures in the case of crewed missions must include life-support systems, and reentry and recovery provisions. In the case of lunar or planetary missions where landing and leaving the foreign body is required, additional provisions for propulsion, guidance, control, spacecraft sterilization, and life-support systems must be realized and the structure must be designed to accommodate them.

Testing. To ensure that spacecraft structures will meet mission requirements criteria in general requires testing to levels above the expected environmental conditions by a specific value. The test level must be set to provide for variations in materials, manufacture, and anticipated loads. In cases where structures are required to perform dynamic functions repeatedly, life testing is required to ensure proper operation over a given number of cycles of operation. *See* INSPECTION AND TESTING.

Boost. The purpose of the boost phase is to lift the vehicle above the sensible atmosphere, to accelerate the vehicle to the velocity required, and to place the spacecraft at a point in space, heading in the direction required for the accomplishment of its mission. For space missions, the required velocities range from 26,000 ft/s (8 km/s) for nearly circular orbits to 36,000 ft/s (11 km/s) for lunar or interplanetary missions. Achievement of these velocities requires boosters many times the size of the spacecraft itself. Generally, this boosting is accomplished by a chemically powered rocket propulsion system using liquid or solid propellants. Multiple stages are required to reach the velocities for space missions. Vertical launching from a surface stand is the method generally employed to boost a spacecraft on the mission trajectory.

Space phase. The space phase begins after the boost phase and continues until reentry. In contrast with the boost phase, effects are repeated numerous times in various combinations. The spacecraft structure must be designed for the most severe combination of effects to be encountered during the entire mission.

The primary function of the structure in a spacecraft is to provide an enclosure containing an environment satisfactorily conditioned for the proper operation of the equipment and crew. This environment generally consists of active or passive temperature control, energetic particle radiation protection, an artificial atmosphere and other life-support systems, and, for the future spacecraft and space stations, an artificial gravity. *See* SPACE FLIGHT; SPACE STATION.

Design considerations. Spacecraft structures must accommodate numerous design considerations. The most significant loads that must be considered are propulsion, control system, pressure vessel loads, and reentry loads. Other subtle loads to be considered are gravity gradient loads, atmospheric drag in orbit, solar radiation pressure, meteoric particles, energetic particles, thermal extremes and thermal temperature gradients, inertial loads, sloshing, and acceleration loads. *See* SATELLITE (SPACECRAFT); SPACE PROBE.

Propulsion loads vary widely among spacecraft, however, the general loads to be considered are the same. These are shock, acceleration, deceleration, vibration, torsion, sloshing, pressure vibrations, and temperature.

The spacecraft control system imparts inertial loads throughout the structure. In the zero gravity environment every change in loading or orientation must be reacted through the structure.

Spacecraft structural design usually requires that part of the principal structure be a pressure vessel. Efficient pressure vessel design is therefore imperative. An important material property, especially in pressure vessel design, is notch sensitivity, referring to the material's apparent brittleness under biaxial strain. This apparent brittleness contributed to premature failure of some early boosters. *See* PRESSURE VESSEL.

A principal problem in the design of spacecraft arises from the presence of meteoric particles in space. These particles may have extremely high velocities relative to the spacecraft (up to 225,000 ft/s or 68 km/s). Probability of collision with larger particles is extremely low, but small-particle collisions will be frequent. Hence a protective structure of some type is required to prevent rupturing of pressurized compartments, damage to equipment, or erosion of surfaces needed for thermal protection in the space or reentry phases by collision with such particles. *See* METEOR.

Radiation shielding may be required for some vehicles, particularly those operating for extended times within the Earth's magnetically trapped radiation belts or during times of high-spot activity. The shielding may or may not be an integral part of the structure.

Temperature extremes in the structure and the enclosed environment are controlled by several techniques. Passive control is accomplished by surface coatings which maintain a balance between absorbed and emitted thermal radiation. Another passive technique is multilayer thermal blankets which control the radiation transfer from the spacecraft to space and vice versa. Other techniques used to actively control spacecraft temperatures are thermal louvers and heat pipes.

Thermal gradients must be considered in spacecraft design, especially when the spacecraft has one surface facing the Sun continuously. In some cases it is desirable to slowly rotate the spacecraft to eliminate such gradients.

Reentry phase. Although the atmospheric layer of the Earth is relatively thin, it is responsible for the reduction of vehicle velocity and the resulting deceleration loads, as well as for the severe heating experienced by reentering vehicles. A body entering the Earth's atmosphere possesses a large amount of energy. This energy must be dissipated in a manner which allows the reentering vehicle to survive. Most of the vehicle's original energy can be transformed into thermal energy in the air surrounding the vehicle, and only part of the original energy is retained in the vehicle as heat. The fraction that appears as heat in the vehicle depends upon the characteristics of the flow around the vehicle. In turn, the flow around the vehicle is a function of its geometry, attitude, velocity, and altitude. Several different approaches are used, such as the zero-lift, blunt, high-drag body (as exemplified by the intercontinental ballistic missiles and the human-in-space programs), and high lift-to-drag ratio glide vehicles. *See* ATMOSPHERIC ENTRY.

A third approach is to combine the rather compact ballistic shape with some of the lift capability of the glide vehicles. Figure 1 shows typical examples of the three shapes.

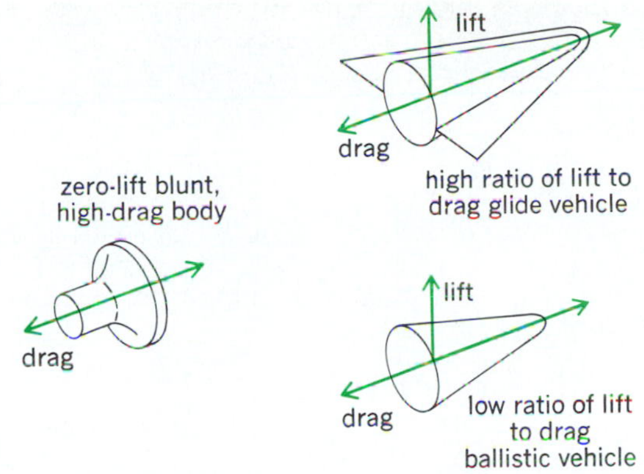

Fig. 1. Typical reentry shapes.

In crewed applications, vehicles employing aerodynamic lift during reentry have several advantages over zero-lift ballistic bodies. (1) The use of lift allows a more gradual descent, thus reducing the deceleration forces on both vehicle and occupants. (2) The vehicle's ability to glide and maneuver within the atmosphere gives it greater accuracy in either hitting a target or landing at a predetermined spot. (3) It can accommodate greater errors of guidance systems because for a given deceleration it can tolerate a greater range of entry angles. (4) Greater temperature control is afforded because aerodynamic lift may be varied to control altitude with velocity.

Spacecraft structures. Spacecraft structures may be divided into two broad categories, erectable and rigid.

It is sometimes required to confine a spacecraft within a protective nose fairing which is part of the boost vehicle system. This fairing provides an efficient aerodynamic shape for the front end of the booster and also provides protection to the enclosed spacecraft from aerodynamic and other effects. This fairing, however, constrains the shape of the spacecraft to something usually resembling a cylinder. In order to meet some mission objectives, however, it is necessary to design erectable structures.

Erectable structures take many and varied shapes. They are sometimes relatively simple hinged booms, while on other occasions they become quite large and massive.

An example of a large rigid deployable spacecraft structure is the 3300-lb (1500-kg) *Pegasus* spacecraft with its 2000 ft^2 (185 m^2) of aluminum target-sheet capacitor detectors.

Another type of unique erectable structure is the antennas on the radio astronomy *Explorer* spacecraft. The four 750-ft (229-m) V antennas were erected in flight as were two 300-ft (91-m) damper tubes and two 60-ft (18-m) dipole antennas. Each of these antenna booms was stored as a 0.002-in.-thick (50-micrometer) beryllium copper alloy tape approximately 2 in. (5 cm) wide in a mechanism within the spacecraft. Upon commands from the Earth the booms were deployed from the mechanism to form ½-in.-diameter (13-mm) tubes. In addition to serving the function of radio astronomy antennas, the long booms provided gravity stabilization so that one axis of the spacecraft always faced the Earth.

Another unique type of erectable structure is the large inflatable sphere, such as the *Echo* and *Pegeos* spacecraft. This type of structure provides a relatively large surface area for relatively little weight.

A spacecraft with a relatively large variety of elements to be erected was the *Orbiting Geophysical Observatory* (*OGO*). Figure 2 shows an artist's conception of the *OGO* in orbit with the booms fully extended.

Many more spacecraft structures are rigid than erectable or inflatable. Some examples of rigid spacecraft structures are the *Orbiting Astronomical Observatory* (*OAO*), the Nimbus weather spacecraft, the Telstar communications spacecraft, the Syncom communications spacecraft, the *Orbiting Solar Observatory* (*OSO*), and the lunar and command service modules.

In cases such as the *Orbiting Astronomical Observatory*, where the least distortion of the structure affects the alignment of the astronomical telescope, great care must be taken to eliminate thermal gradients. The passive temperature control techniques previously described are utilized to keep the thermal gradients to a minimum.

The *Nimbus* spacecraft structure illustrates yet another unique concept in design—the basic wheel section connected by trusses to the attitude control system housing. The solar array tracks the Sun during the sunlight portion of the spacecraft orbit.

Fig. 2. Orbiting Geophysical Observatory. (*NASA*)

The *Telstar* communications spacecraft is a classic example of the basic spherical-shaped rigid spacecraft. The solar cells on the surface supply power to the electronics. The white coating provides passive thermal control.

The *Syncom* spacecraft (Fig. 3) was one of the relatively early spacecraft to incorporate its own propulsion system. Its cylindrical sides are covered with solar cells.

The *Orbiting Solar Observatory* (*OSO 7*) spacecraft (Fig. 4) features a sail which tracks the Sun while the main spacecraft body rotates.

An example of one of the largest rigid uncrewed spacecraft is the *Applications Technology Satellite* (*ATS 7*). This spacecraft—over 50 ft (15 m) across—embodies structural trusses, a flexible erectable parabolic reflector 30 ft (9 m) in diameter, active solar array, active controls, active thermal system, and a propulsion system.

Prospects. The small uncrewed spacecraft have become more and more sophisticated electronically and have incorporated lightweight three-axis control systems. Graphite-fiber-reinforced plastics are universally used to reduce the structural weight factor, allowing a greater percentage of the available weight to go to useful payload.

Fig. 3. Syncom communications satellite. (*NASA*)

Fig. 4. *Orbiting Solar Observatory 7*. (*NASA*)

The Space Transportation System (STS), whose first orbital flight was launched in April 1981, heralds a new era in spacecraft and structural design. Spacecraft can now rely on reaching orbit by using the space shuttle. The shuttle can place payloads up to 29,484 kg (65,000 lb) into orbit and return payloads up to 14,515 kilograms (32,000 lb).

Crewed missions will require that complex spacecraft structural assemblies be made in space. This will require unique designs, built on modular concepts that can be transported by the space shuttle and assembled in space. These large space stations will be the stepping stones for the interplanetary travel of the distant future. *See* SPACE SHUTTLE. [E.W.T.]

Spallation reaction A nuclear reaction in which the interacting nuclei are disintegrated into a large number of the constituent protons, neutrons, and other light particles. Compared to more conventional reactions involving the simple transfer of nucleons between the colliding nuclei, spallation is a violent process occurring at high incident energy. In extreme cases the nuclei are apparently completely shattered into individual nucleons. Spallation can be induced by lighter particles, such as high-energy protons; however, these reactions are usually less violent and produce only a few particles (typically 3 to 10). Although the most abundant final products are protons, neutrons, and alpha particles, the spallation of larger fragments such as lithium and carbon also occurs.

The spallation reaction plays an important role in determining the abundance of elements in the universe. In particular, ^{6}Li, ^{9}Be, ^{10}B, and ^{11}B are created from spallation of interstellar gas by galactic cosmic rays. *See* NUCLEAR REACTION. [D.K.S.]

Spark chamber A triggered electronic particle-detecting device whose purpose is to make visible and to locate accurately in space the tracks of charged particles. It is generally classified as wide-gap or narrow-gap; the most common form is a narrow-gap array of parallel-plate condensers, the plates of which are spaced about 3⁄8 in. (10 mm) apart, filled with a mixture of helium and neon at atmospheric pressure. A spark-chamber system requires an external initiating signal from an auxiliary particle-detection system, which triggers the application of a short-duration high-voltage pulse to the array of plates. The high-voltage pulse produces a spark discharge that follows or marks the path of the ionizing particles. If several tracks are present they are all visible, up to at least 20 (although the chamber will not resolve tracks less than about 0.04 in. or 1 mm apart). Stereo cameras usually provide accurate track location; electronic and digital methods of track location are also used. Spark chambers are characterized by somewhat lower resolution than bubble chambers, a high rate of data collection, and a unique triggering capability that selects only those events that have passed logical selection tests to be photographed. *See* BUBBLE CHAMBER; PARTICLE DETECTOR; SPARK COUNTER. [A.R.]

Spark counter A particle detector which uses the ionization produced in a gas by high-speed charged particles to trigger a spark between two electrodes. Spark counters react in a very short time (about 10^{-9} s) to the particle, and thus can be used for fast timing, The spark is visible and can be photographed. *See* PARTICLE DETECTOR.

The principal components of a spark counter are two plane, parallel metallic electrodes, with a gas between the electrodes consisting of a mixture of argon and an organic gas such as xylene. A potential difference of about 2000 volts is placed across the electrodes, which are spaced about 0.08 in. or 2 mm apart. *See* SPARK CHAMBER. [W.B.Fr.]

Spark gap The region between two electrodes in which a disruptive electrical spark may take place. The gap should be taken to mean the electrodes as well as the intervening space. Such devices may have many uses. The ignition system in a gasoline engine furnishes a very important example. Another important case is the use of a spark gap as a protective device in electrical equipment. Here, surges in potential may be made to break down such a gap so that expensive equipment will not be damaged. *See* BREAKDOWN POTENTIAL; ELECTRIC SPARK. [G.H.M.]

Spark knock In spark-ignition engines, the sound and other effects associated with ignition and rapid combustion of the last part of the charge to burn, before the flame front reaches it. *See* COMBUSTION CHAMBER.

As the flame travels from the spark plug toward the far end of the combustion space, the gas behind the flame front, heated by combustion, expands and rapidly compresses the unburned mixture ahead of the flame. In this way the temperature of the unburned mixture is raised above its self-ignition temperature. Time-consuming chemical reactions must take place in this gas, however, before actual ignition takes place. If the flame front travels rapidly enough to consume the unburned mixture before these reactions are completed, normal combustion without knock takes place. If the reactions proceed too rapidly or if the flame front travels too slowly, the entire last part of the charge may burn almost instantaneously, producing a strong pressure wave, which is reflected back and forth across the combustion space at the speed of sound. The motion of the cylinder head and walls, under the action of this shock wave, causes the high-pitched sound known as spark knock.

The results of spark knock or engine detonation include undesirable noise and overheating or mechanical damage to engine parts such as pistons and spark plugs. Most of the power loss commonly attributed to spark knock is probably an indirect effect, the result of overheated sparkplug electrodes, which ignite the mixture far in advance of the normal ignition spark.

A fuel of high octane number resists spark knock principally because it has a longer self-ignition delay than other fuels under a given set of operating conditions. High-octane gasoline burns in the same manner and with the same flame velocity as low-octane gasoline. Spark knock can thus be eliminated by using high-octane fuel, without loss of power or efficiency. The long delay of high-octane fuels results either from its natural chemical makeup or from the addition of an antiknock material. One of the most effective antiknock additives is tetraethyllead, $Pb(C_2H_5)_4$, which is added to almost all gasoline sold for automobile or aircraft use. *See* ANTIKNOCK AGENT; OCTANE NUMBER; TETRAETHYLLEAD. [A.R.R.]

Spark plug A device that screws into the cylinder of an internal combustion engine to provide a pair of electrodes between which an electrical discharge is passed to ignite the combustible mixture. It consists of an outer steel casing that is electrically grounded to the engine and a ceramic insulator, sealed into the casing, through which a central electrode passes (see illustration). The high-tension current jumps the gap between

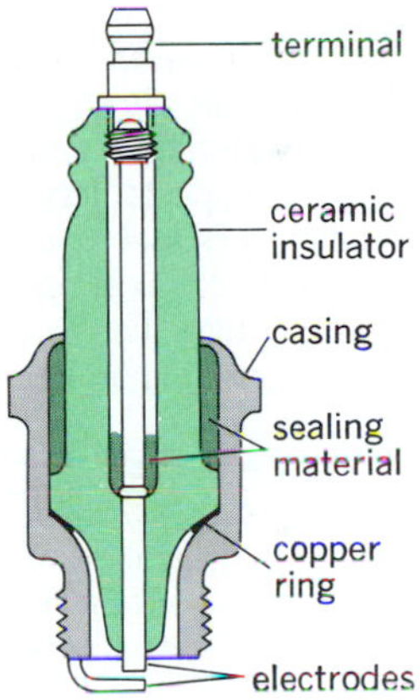

Cross section of a typical spark plug.

this electrode and a similar one fixed to the outer casing. The electrodes are made of alloys that resist electrical and chemical erosion. The parts exposed to the combustion gases are designed to operate at temperatures hot enough to prevent electrically conducting deposits, but cool enough to avoid ignition of the mixture before the spark occurs. *See* IGNITION SYSTEM. [A.R.R.]

Spatangoida An order of exocyclic Euechinoidea in which the posterior ambulacral plates form a shield-shaped area behind the mouth. The plates are arranged in two similar parallel longitudinal series. The apical system is compact. The families are defined mainly by reference to the fascioles, which are ribbonlike bands of minute, close-set uniform ciliated spinules on various parts of the test. The order reached its maximum in the mid-Tertiary, but is still richly represented today in most seas. *See* EUECHINOIDEA. [H.B.F.]

Spearmint Either of two vegetatively propagated, clonal cultivar species (*Mentha spicata* and *M. longifolia*) of mints of the family Lamiaceae (Labiatae). They are grown primarily in Idaho, Indiana, Michigan, Washington, and Wisconsin as a source of essential oil of spearmint.

Principal uses of the oil are in flavoring gum, toothpaste, and candy. Chopped fresh leaves of *M. spicata* preserved in vinegar are used as a condiment served with lamb, especially in England, and dried or freeze-dried leaves of several strains are used in flavoring soups, stews, tea, or sauces. Sprigs of the decorative curly mint *M. crispa* (or *M. spicata* var. *crispata*) are often used in mixed drinks such as mint juleps. *See* LAMIALES. [M.J.M.]

Special functions Functions which occur often enough to acquire a name. Some of these, such as the exponential, logarithmic, and the various trigonometric functions, are extensively taught in school and occur so frequently that routines for calculating them are built into many pocket calculators. *See* DIFFERENTIATION; LOGARITHM; TRIGONOMETRY.

The more complicated special functions, or higher transcendental functions as they are often called, have been extensively studied by many mathematicians because they arose in the problems which were being studied. Among the more useful functions are the following: the gamma function defined by Eq. (1), which generalizes the factorial; the related beta function defined by Eq. (2), which generalizes the binomial coefficient;

$$\Gamma(x) = \int_0^\infty t^{x-1}e^{-t}dt \qquad (1)$$

$$B(x,y) = \int_0^1 t^{x-1}(1-t)^{y-1}dt \qquad (2)$$

and elliptic integrals, which arose when mathematicians tried to determine the arc length of an ellipse, and their inverses, the elliptic functions. The hypergeometric function and its generalizations includes many of the special functions which occur in mathematical physics, such as Bessel functions, Legendre functions, error functions, and the classical orthogonal polynomials of Jacobi, Laguerre, and Hermite. The zeta function defined by Eq. (3) has many applications in number theory, and it also

$$\zeta(s) = \sum_{n=1}^{\infty} n^{-s} \qquad (3)$$

arises in M. Planck's work on radiation. *See* BESSEL FUNCTIONS; ELLIPTIC FUNCTION AND INTEGRAL; GAMMA FUNCTION; HEAT RADIATION; HYPERGEOMETRIC FUNCTIONS; LEGENDRE FUNCTIONS; NUMBER THEORY; ORTHOGONAL POLYNOMIALS. [R.A.]

Speciation The process by which new species of organisms evolve from preexisting species. It is part of the whole process of organic evolution, and the modern period of its study began with the publication in 1858 of Charles Darwin's and Alfred Russel Wallace's theory of evolution by natural selection, and in 1859 of Darwin's *On the Origin of Species*.

There was no problem of speciation during the period when it was believed that species were divinely created and immutable. Belief in the fixity of species was almost universal before the middle of the 19th century. Then it was gradually realized that all species continuously change, or evolve; however, the causative mechanism remained to be discovered.

Darwin proposed a mechanism. He argued that: (1) Within any species population there is always some heritable variation; the individuals differ among themselves in structure, physiology, and behavior. (2) Natural selection acts upon this variation by eliminating the less fit. Thus if two members of an animal population differ from each other in their ability to find a mate, obtain food, escape from predators, resist the ravages of parasites and pathogens, or survive the rigors of the climate, the more successful will be more likely than the less successful to leave descendants. The more successful is said to have greater fitness, to be better adapted, or to be selectively favored. Likewise among plants: one plant individual is fitter than another if its heritable characteristics make it more successful than the other in obtaining light, water, and nutrients, in protecting itself from herbivores and disease organisms, or in surviving adverse climatic conditions. Over the course of time, as the fitter members of a population leave more descendants than the less fit, their characteristics become commoner. *See* SPECIES CONCEPT.

This is the process of natural selection, which tends to preserve the well adapted at the expense of the ill adapted in a variable population. The genetic variability that must exist if natural selection is to act is generated by genetic mutations in the broad sense, including chromosomal rearrangements together with point mutations. *See* GENETICS; MUTATION.

If two separate populations of a species live in separate regions, exposed to different environments, natural selection will cause each population to accumulate characters adapting it to its own environment. The two populations will thus diverge from each other and, given time, will become so different that they are no longer interfertile. At this point, speciation has occurred: two species have come into existence in the place of one. This mode of speciation, speciation by splitting, is probably the most common mode. [E.C.P.]

Species concept The idea that the diversity of nature is divisible into a finite number of definable kinds, or species. In biology the species concept is of threefold importance: as the basic unit of classification, as a step of irreversibility in evolution (origin of species), and as the unit of ecological diversity (niche utilization and so on). The idea of species is fundamental to all biological research.

The members of a species form: (1) a reproductive community in which the individuals of a species (of animals) recognize each other as potential mates and seek each other for the purpose of reproduction—a multitude of devices (isolating mechanisms) ensures intraspecific reproduction in all organisms; (2) an ecological unit which, regardless of the individuals composing it, interacts as a unit with other species with which it shares the environment—each species fills a particular ecological niche; and (3) a genetic unit consisting of a large, intercommunicating gene pool, whereas the individual is merely a temporary vessel holding a small portion of the contents of the gene pool for a short period of time. These three properties raise the species above the typological interpretation of a "class of objects."

The species is a relational (rather than intrinsic) concept, because a species is better characterized by its property of reproductive isolation from other species (its relation of noninterbreeding) than by intrinsic properties. Derived from the observation of the reproductive gap between species that coexist at a given locality, species are most meaningful where the relation of two coexisting populations is described, that is, without the complications of the dimensions of area (longitude and latitude) and time. Such species are nondimensional for all practical purposes. Two or more species are called sympatric when their geographical ranges overlap. Species that occupy mutually exclusive geographical areas are allopatric. *See* SPECIATION. [W.J.B.]

Specific charge The ratio of charge to mass expressed as e/m, of a particle. The acceleration of a particle in electromagnetic fields is proportional to its specific charge. Specific charge can be determined by measuring the velocity which the particle acquires in falling through an electric potential; by measuring the frequency of revolution in a magnetic field; or by observing the orbit of the particles in combined electric and magnetic fields. *See* ELEMENTARY PARTICLE. [C.J.G.]

Specific fuel consumption Weight flow rate of fuel required to produce a unit of power or of thrust, often abbreviated as SFC. In reciprocating piston engines, SFC is usually expressed as pounds of fuel per hour required to deliver 1 shaft horsepower. For ram and turbojet engines it is given as pounds per hour for one pound of thrust. For turboprops, where both shaft and jet power are taken from the engine, SFC is given as fuel rate for equivalent shaft horsepower. *See* RAMJET; TURBOJET. [R.R.H.]

Specific gravity The specific gravity of a material is defined as the ratio of its density to the density of some standard material, such as water at a specified temperature, for example, 60°F (15°C), or (for gases) air at standard conditions of temperature and pressure. Specific gravity is a convenient concept because it is usually easier to measure than density, and its value is the same in all systems of units. *See* DENSITY. [L.N.]

Specific heat The ratio of the amount of heat required to raise unit mass of a material 1 degree in temperature to the amount of heat required to raise the same mass of a reference substance 1 degree in temperature. Both measurements are made at a reference temperature and in nearly all cases at either constant volume or constant pressure. Water is usually the reference substance. Because the heat capacity of water is nearly unity, the value of specific heat for a material is nearly equal to its heat capacity. Specific heat, as defined here, is a ratio without units, although it is often defined differently. For clarity it is recommended that thermodynamic discussion be carried out in terms of heat capacity instead of specific heat. Also, it is desirable to define heat units in electrical terms. *See* CHEMICAL THERMODYNAMICS; HEAT CAPACITY. [H.C.W.]

Specific impulse The quantity used to define the amount of thrust potentially available from rocket propellant combinations; also called specific thrust. Specific impulse is the thrust that theoretically can be obtained when unit weight of the propellant reacts in unit time, that is, pound thrust per pound per second of propellant flow. This ratio for specific impulse has the dimension of time and can be expressed as seconds. For example, a propellant having an impulse of 300 will theoretically deliver 300 lb of thrust when the propellant is consumed at the rate of 1 lb/s. Alternatively, 1 lb of propellant will deliver 1 lb of thrust for 300 s. *See* JET FUEL; PROPULSION; THRUST. [R.R.H.]

Speckle The generation of a random intensity distribution, called a speckle pattern, when light from a highly coherent source, such as a laser, is scattered by a rough surface or inhomogeneous medium. Although the speckle phenomenon has been known since the time of Isaac Newton, the development of the laser is responsible for the present-day interest in speckle. Speckle has proved to be a universal nuisance as far as most laser applications are concerned, and it was only in the mid-1970s that investigators turned from the unwanted aspects of speckle toward the uses of speckle patterns, in a wide variety of applications. The principle applications for speckle patterns fall into two areas: metrology, and stellar speckle interferometry. *See* LASER.

Photograph of speckle pattern generated by illuminating rough surface with laser radiation.

Objects viewed in coherent light acquire a granular appearance as shown in the illustration. The surfaces of most materials are extremely rough on the scale of an optical wavelength. When nearly monochromatic light is reflected from such a surface, the optical wave resulting at any moderately distant point consists of many coherent wavelets, each arising from a different microscopic element of the surface. Since the distances traveled by these various wavelets may differ by several wavelengths if the surface is truly rough, the interference of the wavelets of various phases results in the granular pattern of intensity called speckle. [J.C.Wy.]

Spectral type An indicator of the physical and chemical characteristics of a star, based on study of the star's spectrum. Stars possess a remarkable variety of spectra, some simple, others complex. To understand the natures of the stars, it was first necessary to bring order to the subject and to classify the spectra.

The modern system of classification was initiated about 1890. The spectra were ordered by letter, A through O, largely on the basis of the strengths of the hydrogen lines. Several letters were found to be unnecessary or redundant, and on the basis of continuity of lines other than hydrogen, it was found that B preceded A and O preceded B. The result is the classical spectral sequence, OBAFGKM. The classes were decimalized, setting up the sequence O5, . . . , O9, B0, . . . , B9, A0, and so forth. (Not all the numbers are used.) The modern standard sequence, called the Harvard sequence after the observatory where it was formulated, runs from O3 to M8.

Class A has the strongest hydrogen lines, B is characterized principally by neutral helium (with weaker hydrogen), and O by ionized helium. Hydrogen weakens notably through F and G, but the metal lines, particularly those of ionized calcium, strengthen. In K, hydrogen becomes quite weak, while the neutral metals grow stronger. The M stars effectively exhibit no hydrogen lines at all but are dominated by molecules, particularly titanium oxide (TiO). At G, the sequence branches downward into R and N, whose stars are rich in carbon molecules. In class S, the titanium oxide molecular bands of class M are replaced by zirconium oxide (ZrO).

At first appearance, the different spectral types seem to reflect differences in stellar composition. However, within the standard sequence, OBAFGKM, the elemental abundances are roughly similar. The dramatic variations in spectra are strictly the result of changes in temperature. The different spectra of the R, N, and S stars, however, are caused by true and dramatic variations in the chemical composition, the result of internal thermonuclear processing and convection. *See* STELLAR EVOLUTION.

In the 1940s, W. W. Morgan, P. C. Keenan, and E. Kellman expanded the Harvard sequence to include luminosity. A system of roman numerals is appended to the Harvard class to indicate position on the Hertzsprung-Russell diagram: I for supergiant, II for bright giant, III for giant, IV for subgiant, and V for dwarf or main sequence. *See* ASTRONOMICAL CATALOGS; HERTZSPRUNG-RUSSELL DIAGRAM. [J.B.Ka.]

Spectrochemical analysis Any of a number of techniques for the determination of the presence or the concentration of elemental or molecular constituents in a sample through the use of spectrometric measurements. Spectrometric measurements entail the monitoring of electromagnetic radiation as it is caused to emanate from, or interact with, the sample of interest.

The interaction of an analyte with electromagnetic radiation is based on changes in the level of some characteristic energy state of the analyte, for example, the oscillatory motion of a chemical bond, or the orbital location of a valence electron, or the rotational motion of the magnetic vector of an atomic nucleus. All of these types of characteristic states are quantized in energy by the principles of quantum mechanics. The change in energetic state of an analyte can be caused by the absorption of emission of energy of an amount exactly equal to the difference in energies of the two states. *See* QUANTUM MECHANICS.

In the basic spectrometric experiment which is performed for spectrochemical analysis, the sample is put into a state in which it will interact in the manner desired with the radiation of choice (see illustration). For absorption and luminescence measurements, a source of the radiation of choice is focused onto the sample so that the radiation may be absorbed. For emission measurements the sample is supplied with external energy which is not radiative, for example, the heat from a flame, or an applied voltage, or momentum from mechanical motion. The sample then emits the gained energy as photons (packets, quanta) of radiation.

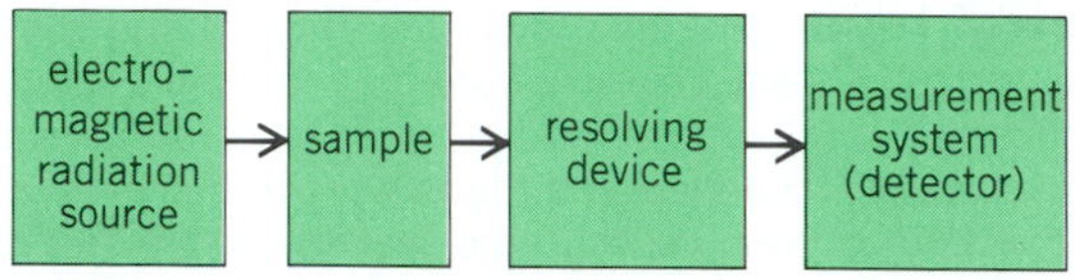

Schematic diagram of the basic spectrometric experiment.

The next stage of the experiment separates all the wavelengths of radiation present so that they may be measured independently of all others. Finally, there is a device for detecting the amount of radiation present (emission or luminescence) or absent (absorption). The identity of the wavelengths gives qualitative information. The amount of radiation gives quantitative information. *See* GAMMA-RAY DETECTORS; GEIGER-MÜLLER COUNTER; PHOTOTUBE; SPECTROSCOPY; TRANSDUCER. [A.T.Z.]

Spectrograph An optical instrument that consists of an entrance slit, collimator, disperser, camera, and detector and that produces and records a spectrum. A spectrograph is used to extract a variety of information about the conditions that exist where light originates and along the paths of light. It reveals the details that are stored in the light's spectral distribution, whether this light is from a source in the laboratory or a quasistellar object a billion light-years away.

Spectrograph design takes into account the type of light source to be measured, and the circumstances under which these measurements will be made. Since observational astronomy presents unusual problems in these areas, the design of astronomical spectrographs may also be unique.

Astronomical spectrographs have the same general features as laboratory spectrographs. The width of the entrance slit influences both spectral resolution and the amount of light entering the spectrograph, two of the most important variables in spectroscopy. The collimator makes this light parallel so that the disperser (a grating or prism) may properly disperse it. The camera then focuses the dispersed spectrum onto a detector, which records it for further study.

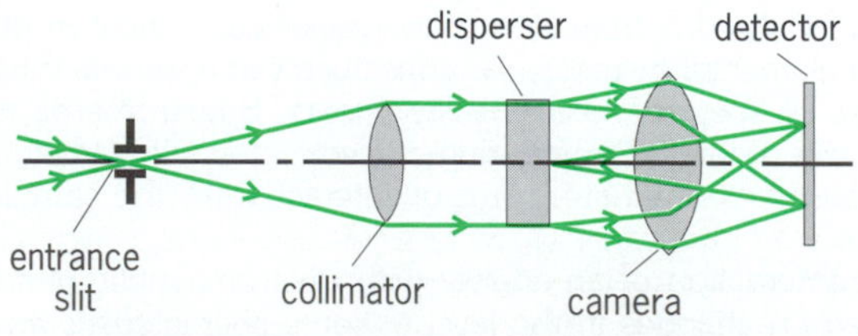

Basic optical components of a spectrograph.

Laboratory spectrographs usually function properly only in a fixed orientation under controlled environmental conditions. By contrast, most astronomical spectrographs are used on a moving telescope operating at local temperature. Thus, their structures must be mechanically and optically insensitive to orientation and temperature.

The brightness, spectral characteristics, and geometry of laboratory sources may be tailored to experimental requirements and to the capabilities of a spectrograph. Astronomical sources, in the form of images at the focus of a telescope, cannot be manipulated, and their faintness and spectral diversity make unusual and difficult demands on spectrograph performance.

Typical laboratory spectrographs use either concave gratings, which effectively combine the functions of collimator, grating, and camera in one optical element, or plane reflection gratings with spherical reflectors for collimators and cameras. *See* ASTRONOMICAL SPECTROSCOPY. [R.Hi.]

Spectrography The use of photography to record the electromagnetic spectrum displayed in a spectroscope. The technique is used mainly in atomic and molecular physics, in analysis of the chemical composition of materials, and in astronomical photography. *See* ASTRONOMICAL PHOTOGRAPHY; SPECTROCHEMICAL ANALYSIS; SPECTROSCOPY. [W.Cl.]

Spectroheliosope An instrument for the monochromatic visual observation of the Sun. A telescope projects an image of the Sun on the first slit of a powerful spectroscope. The resulting spectrum is imaged in the plane of a second slit which permits only a single line element of the spectrum to emerge from the instrument. The emergent line element is a monochromatic image of that part of the Sun that falls on the first slit. When the two slits are vibrated synchronously at high frequency, persistence of vision permits monochromatic observation of an area of the solar surface. The slits may also be moved at a slow rate and the image recorded photographically. This modification of the spectrohelioscope is a simple form of the spectroheliograph. [R.R.McM./J.W.E.]

Spectrophotometric analysis A method of chemical analysis based on the absorption or attenuation by matter of electromagnetic radiation of a specified wavelength or frequency. The region of the electromagnetic spectrum most useful for chemical analysis is that between 200 nanometers and 300 micrometers. Since the sample being analyzed absorbs the radiation, spectrophotometric analysis is sometimes referred to as absorptimetric analysis.

The instruments used in this work are referred to as spectrophotometers. A simple spectrophotometer consists of a source of radiation, such as a light bulb; a monochromator containing a prism or grating which disperses the light so that only a limited wavelength, or frequency, range is allowed to irradiate the sample; the sample itself; and a detector, such as a photocell, which measures the amount of light transmitted by the sample.

In most quantitative analytical work, a calibration or standard curve is prepared by measuring the absorption of known amounts of the absorbing material at the wavelength at which it strongly absorbs. Such a calibration curve is shown in Fig. 1 for the absorbing material whose absorption spectrum is shown in Fig. 2. The absorbance of the sample is read directly from the measuring circuit of the spectrophotometer.

When the transmittance or the absorbance of a sample is measured and plotted as a function of wavelength, an absorption spectrum is obtained. The spectrum shown in Fig. 2 indicates that the sample transmits the least light at 410 nm and transmits the most around 700 nm.

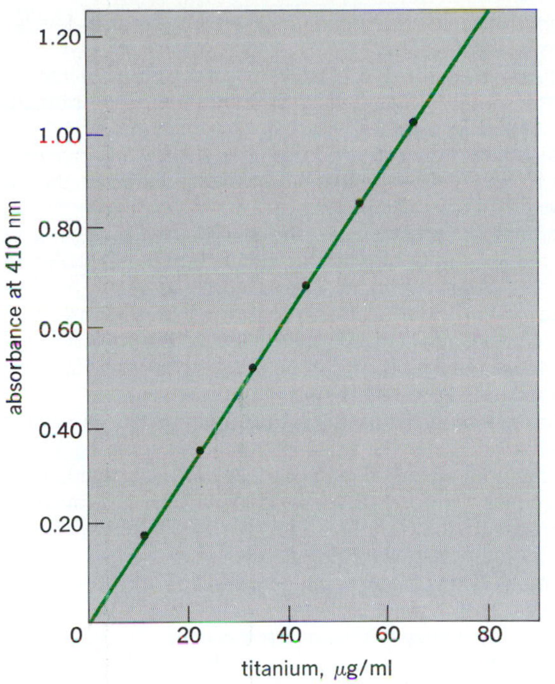

Fig. 1. Calibration curve for determination of titanium by its color formed with hydrogen peroxide.

Infrared spectrophotometry. The interaction with matter of electromagnetic radiation of wavelength between 1 and 300 μm (infrared region) induces either rotational or vibrational energy level transitions, or both, within the molecules involved. The frequencies of infrared radiation absorbed by a molecule are determined by its rotational energy levels and by the force constants of the bonds in the molecule. Since these energy levels and force constants are usually unique for each molecule, so also the infrared spectrum of each molecule is usually unique. Because of their individuality, infrared spectra of organic compounds are considered equivalent to, or superior to, the preparation of chemical derivatives for the identification of species in organic chemistry. For this reason the infrared portion of the spectrum is often called the fingerprint region.

Attenuated total reflectance. This is a technique in which the infrared spectrum of a surface can be obtained without any chemical treatment of the sample; its principle is based on the phenomenon of energy reflection at the interface of two media which have different refractive indexes and are in optical contact with each other. The spectrum obtained is similar to that of a transmission spectrum of a material. This technique is used extensively for analyzing coatings, pastes, paints, fibers, and fabrics.

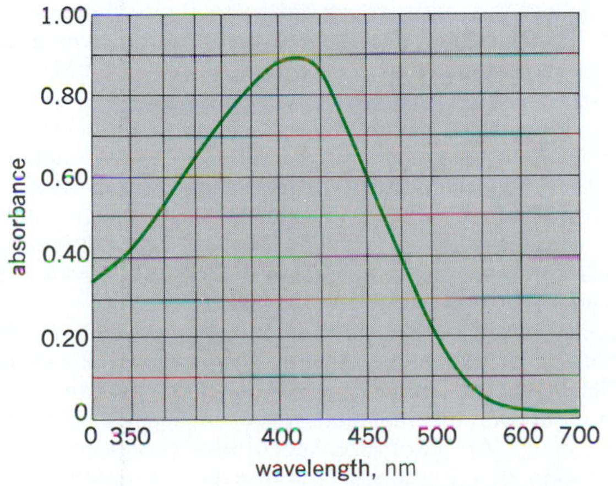

Fig. 2. Absorption spectrum of the peroxytitanate complex in region 340–700 nm.

Near-infrared spectrophotometry. This designates work carried out between 0.78 and 3 μm. The absorption bands in this region are mainly overtones (harmonics) of bands in the infrared region. These bands are quite sharp and are of great value in quantitative analysis for various functional groups, particularly those containing hydrogen atoms.

Visible spectrophotometry. The visible region of the spectrum covers the narrow range from about 380 to 780 nm. The spectrophotometers for this region use tungsten lamps as light sources, glass or quartz prisms or gratings in the monochromators, and photomultiplier cells as detectors. Within this narrow portion of the electromagnetic spectrum, a majority of the spectrophotometric analyses are made. Figure 2 shows a typical visible absorption spectrum. The substance, peroxytitanate ion, absorbs light in the region below 500 nm; that is, it absorbs violet, blue, and green light and transmits red, orange, and yellow. For analytical work, the wavelength of maximum absorption is usually used, in this case 410 nm.

Ultraviolet spectrophotometry. The spectral region from 200 to 400 nm, called the near ultraviolet, is commonly used in chemical analysis. Ultraviolet spectrophotometers usually have a hydrogen lamp as a radiation source; a quartz prism, or a grating in the monochromator; and a photomultiplier tube as a detector. Simple inorganic ions and their complexes as well as organic molecules can be detected and determined in this region.

Filter photometry. In filter photometry, the monochromator of the spectrophotometer is replaced by a filter. This filter passes a band of light of a much wider range of wavelengths than those passed by even the poorest monochromator. The filter is chosen so as to transmit best the light which the sample absorbs most. Filter photometers are generally much less expensive than spectrophotometers. By careful use of calibration curves, filter photometers can give sufficiently accurate and precise results for a wide variety of applications. *See* ANALYTICAL CHEMISTRY; MOLECULAR STRUCTURE AND SPECTRA; OPTICAL METHODS OF CHEMICAL ANALYSIS. [J.N.L.]

Spectroscopy An analytic technique concerned with the measurement of the interaction (usually the absorption or the emission) of radiant energy with matter, with the instruments necessary to make such measurements, and with the interpretation of the interaction both at the fundamental level and for practical analysis.

A display of such data is called a spectrum, that is, a plot of the intensity of emitted or transmitted radiant energy (or some function of the intensity) versus the energy of that light. Spectra due to the emission of radiant energy are produced as energy is emitted from matter, after some form of excitation, then collimated by passage through a slit, then separated into components of different energy by transmission through a prism (refraction) or by reflection from a ruled grating or a crystalline solid (diffraction), and finally detected. Spectra due to the absorption of radiant energy are produced when radiant energy from a stable source, collimated and separated into its components in a monochromator, passes through the sample whose absorption spectrum is to be measured, and is detected. Instruments which produce spectra are variously called spectroscopes, spectrometers, spectrographs, and spectrophotometers. *See* SPECTRUM.

Interpretation of spectra provides fundamental information on atomic and molecular energy levels, the distribution of species within those levels, the nature of processes involving change from one level to another, molecular geometries, chemical bonding, and interaction of molecules in solution. At the practical level, comparisons of spectra provide a basis for the determination of qualitative chemical composition and chemical structure, and for quantitative chemical analysis.

Origin of spectra. Atoms, ions, and molecules emit or absorb characteristically; only certain energies of these species are possible; the energy of the photon (quantum of radiant energy) emitted or absorbed corresponds to the difference between two permitted values of the energy of the species, or energy levels. (If the flux of photons incident upon the species is great enough, simultaneous absorption of two or more photons may occur.) Thus the energy levels may be studied by observing the differences between them. The absorption of radiant energy is accompanied by the promotion of the species from a lower to a higher energy level; the emission of radiant energy is accompanied by falling from a higher to a lower state; and if both processes occur together, the condition is called resonance.

Instruments. Spectroscopic methods involve a number of instruments designed for specialized applications.

An optical instrument consisting of a slit, collimator lens, prism or grating, and a telescope or objective lens which produces a spectrum for visual observation is called a spectroscope.

If a spectroscope is provided with a photographic camera or other device for recording the spectrum, the instrument is called a spectrograph. *See* SPECTROGRAPHY.

A spectroscope that is provided with a calibrated scale either for measurement of wavelength or for measurement of refractive indices of transparent prism materials is called a spectrometer.

A spectrophotometer consists basically of a radiant-energy source, monochromator, sample holder, and detector. It is used for measurement of radiant flux as a function of wavelength and for measurement of absorption spectra.

An interferometer is an optical device that measures differences of geometric path when two beams travel in the same medium, or the difference of refractive index when the geometric paths are equal. Interferometers are employed for high-resolution measurements and for precise determination of relative wavelengths. *See* INTERFEROMETRY.

Methods and applications. Since the early methods of spectroscopy there has been a proliferation of techniques, often incorporating sophisticated technology.

Astronomical spectroscopy involves the study of radiant energy emitted by celestial objects by combined spectroscopic and telescopic techniques to obtain information about their chemical composition, temperature, pressure, density, magnetic fields, electric forces, and radial velocity. *See* ASTRONOMICAL SPECTROSCOPY.

Atomic absorption and fluorescence spectroscopy is a branch of electronic spectroscopy that uses line spectra from atomized samples to give quantitative analysis for selected elements at levels down to parts per million, on the average. *See* FIRE ASSAYING; TRACE ANALYSIS.

Attenuated total reflectance spectroscopy is the study of spectra of substances in thin films or on surfaces obtained by the technique of attenuated total reflectance or by a closely related technique called frustrated multiple internal reflection. In either method the sample is penetrated by a radiant-energy beam one or more times. The technique is employed primarily in infrared spectroscopy for qualitative analysis of coatings and of opaque liquids.

Electron paramagnetic spectroscopy is a microwave technique, based on the splitting of electronic energy levels in a magnetic field and is used to establish structures of species containing unpaired electrons. *See* ELECTRON PARAMAGNETIC RESONANCE (EPR) SPECTROSCOPY.

Electron spectroscopy includes a number of subdivisions, all of which are associated with electronic energy levels. *See* ELECTRON SPECTROSCOPY.

Fourier transform spectroscopy is a family of techniques consisting of a fundamentally different method of irradiation of the sample, in which all pertinent wavelengths simultaneously irradiate it for a short period of time and the absorption spectrum is obtained by mathematical manipulation of the cyclical power pattern so obtained.

Gamma-ray spectroscopy includes the techniques of activation analysis and Mössbauer spectroscopy. *See* ACTIVATION ANALYSIS; MÖSSBAUER EFFECT; NEUTRON SPECTROMETRY.

Information on processes which occur on a picosecond time scale can be obtained by making use of the coherent properties of laser radiation, as in coherent anti-Stokes Raman spectroscopy. *See* LASER SPECTROSCOPY.

The source of a mass spectrometer produces ions, often from a gas, but also in some instruments from a liquid, a solid, or a material adsorbed on a surface. The dispersive unit provides either temporal or spatial dispersion of ions according to their mass-to-charge ratio. *See* MASS SPECTROMETRY; SECONDARY ION MASS SPECTROMETRY (SIMS).

In multiplex or frequency-modulated spectroscopy each optical wavelength exiting the spectrometer output is encoded with an audio frequency that contains the optical wavelength information. Use of a wavelength analyzer then allows recovery of the original optical spectrum.

Raman spectroscopy, in which Raman scattering of light by the molecules under study is analyzed by spectroscopic means, provides information of use in structural chemistry. *See* RAMAN EFFECT.

X-ray spectroscopy involves the excitation of inner electrons in atoms which is manifested as x-ray absorption; emission of a photon as an electron falls from a higher level into the vacancy thus created is x-ray fluorescence. The techniques are used for chemical analysis. *See* X-RAY FLUORESCENCE ANALYSIS; X-RAY SPECTROMETRY. [M.M.Bu.]

Spectrum The term spectrum is applied to any class of similar entities or properties strictly arrayed in order of increasing or decreasing magnitude. In general, a spectrum is a display or plot of intensity of radiation (particles, photons, or acoustic radiation) as a function of mass, momentum, wavelength, frequency, or some other related quantity. For example, a β-ray spectrum represents the distribution in energy or momentum of negative electrons emitted spontaneously by certain radioactive nuclides, and when radionuclides emit α-particles, they produce an α-particle spectrum of one or more characteristic energies. A mass spectrum is produced when charged particles (ionized atoms or molecules) are passed through a mass spectrograph in which electric and magnetic fields deflect the particles according to their charge-to-mass ratios. The distribution of sound-wave energy over a given range of frequencies is also called a spectrum. *See* MASS SPECTROSCOPE; SOUND.

In the domain of electromagnetic radiation, a spectrum is a series of radiant energies arranged in order of wavelength or of frequency. The entire range of frequencies is subdivided into wide intervals in which the waves have some common characteristic of generation or detection, such as the radiofrequency spectrum, infrared spectrum, visible spectrum, ultraviolet spectrum, and x-ray spectrum.

Spectra are also classified according to their origin or mechanism of excitation, as emission, absorption, continuous, line, and band spectra. An emission spectrum is produced whenever the radiations from an excited light source are dispersed. An absorption spectrum is produced against a background of continuous radiation by interposing matter that reduces the intensity of radiation at certain wavelengths or spectral regions. The energies removed from the continuous spectrum by the interposed absorbing medium are precisely those that would be emitted by the medium if properly excited. A continuous spectrum contains an unbroken sequence of waves or frequencies over a long range. Line spectra are discontinuous spectra characteristic of excited atoms and ions, whereas band spectra are characteristic of molecular gases or chemical compounds. *See* ATOMIC STRUCTURE AND SPECTRA; BAND SPECTRUM; ELECTROMAGNETIC RADIATION; LINE SPECTRUM; MOLECULAR STRUCTURE AND SPECTRA; SPECTROSCOPY. [W.F.M./W.W.W.]

Spectrum analyzer A device which sweeps over a portion of the radio-frequency spectrum, responds to signals whose frequencies lie within the swept band, and displays them in relative magnitude and frequency on a cathode-ray-tube screen. In essence, it is a superheterodyne receiver having a local oscillator whose frequency is varied cyclically, usually at the power-line frequency. *See* RADIO RECEIVER.

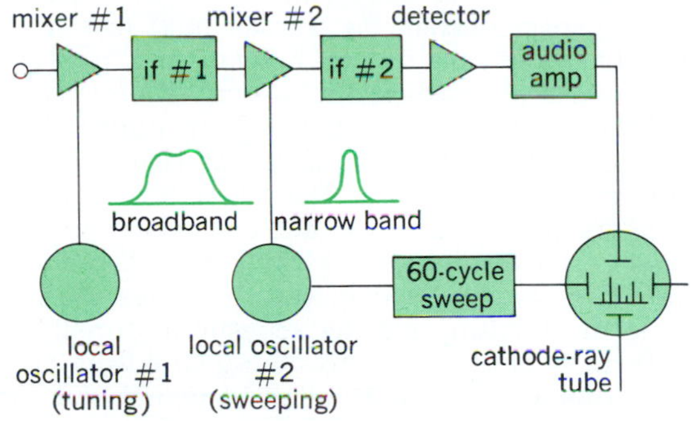

Block diagram of typical spectrum analyzer.

Spectrum analyzers are used specifically to study the spectra of pulsed transmitters, such as radar, to make sure that they are operating properly, without spurious emissions. Spectrum analyzers are often built into test equipment and military identification receivers. The output display shown in the illustration can be used to monitor radio traffic in the frequency band being covered; transmitter activity which might be overlooked by a standard receiver groping about in its tuning range will show up on the panorama display with its carrier frequency identified. *See* ELECTRIC POWER MEASUREMENT; FREQUENCY MEASUREMENT. [M.G.F.]

Speech A set of audible sounds produced by disturbing the air through the integrated movements of certain groups of anatomical structures. Humans attach symbolic values to these sounds for communication. There are many approaches to the study of speech.

The physiology of speech production may be described in terms of respiration, phonation, and articulation. These interacting processes are activated, coordinated, and monitored by acoustical and kinesthetic feedback through the nervous system.

Respiration. Most speech sounds are formed during exhalation. Consequently, during speech the period of exhalation is generally much longer than that of inhalation. The aerodynamics of the breath stream influence the rate and mode of the vibration of the vocal folds. This involves interactions between the pressures initiated by thoracic movements and the position and tension of the vocal folds.

The pressure pattern of the sublaryngeal air is closely related to the loudness of the voice and appears to be correlated with the perception of stress. For example, the word "permit" may be either a noun or a verb, depending on the placement of stress. Various attempts have been made to correlate units such as the syllable and the phrase to the contractions of specific respiratory muscles.

Phonation. The phonatory and articulatory mechanisms of speech may be regarded as an acoustical system whose properties are comparable to those of a tube of varying cross-sectional dimensions. At the lower end of the tube, or the vocal tract, is the larynx. The upper end of the vocal tract may terminate at the lips, at the nose, or both. The length of the vocal tract averages 6.4 in. (16 cm) in men and may be increased by either pursing the lips or lowering the larynx.

The larynx is the primary mechanism for phonation, that is, the generation of the glottal tone. The vocal folds attach anteriorly to the thyroid cartilage and posteriorly to the vocal processes of the arytenoid cartilages. The vibrating edge of the vocal folds measures about 0.92–1.08 in. (23–27 mm) in men and considerably less in women. The aperture between the vocal folds is called the glottis. The tension and position of the vocal folds are adjusted by the intrinsic laryngeal muscles, primarily through movement of the two arytenoid cartilages. By contraction of groups of muscles, the arytenoid cartilages may be pulled either anterior-posteriorly or laterally in opposite directions. They may also be rotated about the vertical axis. During whispering and unvoiced sounds, the glottis assumes the shape of a triangle, with the apex directly behind the thyroid cartilage. *See* LARYNX.

When the vocal folds are brought together and there is a balanced air pressure to drive them, they vibrate laterally in opposite directions. During phonation, the vocal folds control the energy by regulating the frequency and amount of air passing through the glottis. Their rate and mode of opening and closing are dependent upon the position and tension of the folds and the pressure and velocity of airflow. The tones are produced by the recurrent puffs of air passing through the glottis and striking into the supralaryngeal cavities.

As the air passes through the glottis with increasing velocity, the pressure perpendicular to the direction of air flow is reduced. This allows the tension of the folds to draw them together, either partly or completely, until sufficient pressure is built up to drive them apart again.

Speech sounds produced during phonation are called voiced. Almost all of the vowel sounds of the major languages and some of the consonants are voiced. In English, voiced consonants may be illustrated by the initial and final sounds in the following words: bathe, dog, man, jail. The speech sounds produced when the vocal folds are apart and are not vibrating are called unvoiced; examples are the consonants in the words hat, cap, sash, faith. During whispering all the sounds are unvoiced.

The rate of vibration of the vocal folds is the fundamental frequency of the voice and correlates well with the perception of pitch. The frequency increases when the vocal folds are made taut. Relative differences in the fundamental frequency of the voice are utilized in almost all languages to signal some aspects of linguistic information.

Articulation. The activity of the structures above and including the larynx in forming speech sound is called articulation. It involves some muscles of the pharynx, palate, tongue, and face and of mastication.

The primary types of speech sounds of the major languages may be classified as vowels, nasals, plosives, and fricatives. They may be described in terms of degree and place of constriction along the vocal tract. *See* PHONETICS.

The only source of excitation for vowels is at the glottis. During vowel production the vocal tract is relatively open and the air flows over the center of the tongue, causing a minimum of turbulence. The phonetic value of the vowel is determined by the resonances of the vocal tract, which are in turn determined by the shape and position of the tongue and lips.

The nasal cavities can be coupled onto the resonance system of the vocal tract by lowering the velum and permitting airflow through the nose. Vowels produced with the addition of nasal resonances are called nasalized vowels. Nasalization may be used to distinguish meanings of words made up of otherwise identical sounds, such as *bas* and *banc* in French. If the oral passage is completely constricted and air flows only through the nose, the resulting sounds are nasal consonants.

Plosives are characterized by the complete interception of airflow at one or more places along the vocal tract. The pressure which is built up behind the intercepting mechanism may not be immediately released or may be released through the oral or nasal orifice. The places of constriction and the manner of the release are the primary determinants of the phonetic properties of the plosives. The words par, bar, tar, car begin with plosives.

Fricatives are produced by partial constriction along the vocal tract which results in turbulence. Their properties are determined by the place or places of constriction and the shape of the modifying cavities. The fricatives in English may be illustrated by the initial and final consonants in the words vase, this, faith, hash. *See* SPEECH DISORDERS. [R.S.T.; W.S.-Y.W.]

Speech disorders Speech disorders may be classified according to their causes or symptoms. The major causes are organic, imitative environmental, and psychogenic. Organic disorders may result from disease, impairment, or absence of the organs of speech. Imitative disorders occur when the child imitates defective speech. A speech disorder has a psychogenic origin when there is a psychological basis for its presence. A classification includes disorders of articulation, rhythm, voice, and symbolization.

Disorders of articulation may be so severe that the resultant speech is unintelligible. Specific sounds or groups of sounds may be omitted, added, substituted, or distorted. The following are some examples of this type of disorder: Lalling, involving misarticulated r, l, t, and d sounds, may be caused by poor control of the tongue tip. In lisping, misarticulated sibilant sounds, particularly s and z, are often substituted by the th sound. Delayed speech, involving the absence of many consonants and poor intelligibility, is often caused by slow physical or psychological maturation. Dysarthria, generalized sound substitutions and distortions, is caused by lesions in the peripheral or central nervous system.

Disorders of rhythm are characterized by disruptions of the normal rate of speech. Two common disorders of rhythm are stuttering or stammering and cluttering.

Voice disorders are usually described as defects of pitch, loudness, and voice quality. Improper use of the voice may cause an injury to the vocal folds and intensify an existing voice disorder. Disorders of pitch are characterized as too high, too low, monotonous, and repeated pitch patterns. A high-pitched voice is most often caused by psychological tension. The muscles of the larynx are contracted so that the pitch is raised beyond its normal range. Voice quality disorders are usually described as hoarseness, nasality, denasality, and similar conditions. Hoarseness may be caused by pharyngeal or laryngeal pathologies. Cleft palate speech is a striking example of nasality. Nasal speech may also follow a paralysis of the palatal muscles. The nasal speech results because the palate is unable to aid in achieving nasopharyngeal closure.

Symbolization disorders involve an impairment of language formulation and expression. They may occur without a concomitant impairment of speech production. Common types of these disorders are aphasia and delayed speech. *See* SPEECH. [R.S.T.; W.S.-Y.W.]

Speech perception The process by which a human listener transforms a speech waveform into the linguistically relevant sounds of a natural language. Although the study of spoken language processing is concerned with the mapping of meaning onto sounds, speech perception is concerned more specifically with the study of how humans perceive and produce the sounds of their language. In this regard, speech perception focuses on the most basic level of language processing—the individual sounds, syllables, and words.

The study of speech perception is concerned with answering questions regarding different stages in the speech chain, such as what stages of perceptual analysis occur between the excitation of the listener's hearing mechanism and the understanding of the speaker's message; what information in the acoustic waveform is used to cue specific speech sounds; or what perceptual mechanisms are used in the speech perception process.

A convenient way to view the speech perception process is in the context of how a speaker transmits a spoken message to a listener. In this chain of events, which is often called the speech chain, the speaker must first choose the words that reflect his or her thoughts and organize them into grammatical sentences that express the intended meaning. This linguistic information is then transmitted along motor nerves from the brain to the articulators so that the utterance can be produced. The neural impulses move the articulators (the tongue, lips, jaw, and velum) appropriately to produce pressure changes in the surrounding air (a sound wave). The resulting sound wave activates the listener's hearing mechanism, and the acoustic information in the waveform is converted into neural impulses along the acoustic nerve. The combination of neural and cognitive activity in the listener's brain results in the speaker's message being understood.

Research has demonstrated that the perception of speech does not rely exclusively on the analysis and recognition of segmental acoustic features, but that listeners also use context and other knowledge sources to understand speech. Early studies in speech perception showed that words presented in noise were easier to recognize in a sentence context than when presented in isolation. These findings were obtained despite the fact that words produced in isolation are acoustically better defined than words produced in continuous speech.

In the past, variability due to different contexts has been viewed as a source of noise in attempting to extract phonetic segments from an acoustic signal. Researchers have begun, however, to recognize the importance of context effects and to view the speech waveform as a reliable source of acoustic-phonetic information. In addition to responding to the acoustic information in the signal, listeners have knowledge about their language and the world available to them. Phonological, lexical, and syntactic knowledge as well as pragmatic real-world constraints are combined with the acoustic information in the signal to support the speech perception process. *See* HEARING (HUMAN); PERCEPTION; PSYCHOACOUSTICS; SPEECH. [L.M.S.; D.B.P.]

Speech recognition The process of analyzing an acoustic speech signal to identify the linguistic message that was intended, so that a machine can correctly respond to spoken commands.

Speaking commands to a computer is the ultimate in natural human-to-machine communications. Natural, spontaneous interactions are possible, with little or no user training, permitting direct access by the physically handicapped or those who cannot type well. Speech allows rapid data entry and frees the user's hands and eyes for other tasks.

Fluent conversation with a machine is difficult to achieve

because of intrinsic variabilities and complexities of speech, as well as limitations in current practical capabilities. Recognition difficulty increases with vocabulary size, confusability of words, reduced frequency bandwidth, noise, frequency distortions, the population of speakers that must be understood, and the form of speech to be processed. To limit the problem, speech recognizers have been primarily confined to small vocabularies of words with large differences in sound structure, and isolated words, with pauses before and after each word.

Isolated word recognizers are thus the easiest to develop, followed in turn by recognizers of digit strings and strictly formatted word sequences. Sometimes the identification of every spoken word is not necessary, and detecting key words in context is enough to determine the topic of a conversation. Such a word-spotting system usually deals only with the important words, which are stressed and well articulated, easing the recognition, despite arbitrary contexts. Full understanding of the total sentence meaning and intended machine response involves cooperative use of many "knowledge sources" like acoustics, phonetics, prosodics, restricted syntax, semantics, and task-dictated discourse constraints. The term speech-understanding system is frequently used to describe such a sentence-understanding device.

Speech recognition requires the transformation of the continuous-speech signal into discrete representations which may be assigned proper meanings, and which, when comprehended, may be used to effect responsive behavior. Only some of the information in the speech signal is related to selecting the correct machine response, so that a critical task is to extract all, and only, those parts that convey the message. Recognizers primarily differ on how they reduce the data to message-distinguishing features, and how they classify utterances from that reduced data.

Since most devices must be trained to the specific pronunciation of each talker, a precompiling stage involves the user speaking sample pronunciations for each allowable utterance (word, phrase, or sentence), while identifying each with its typewritten form or a vocabulary item number. Later, when an unknown utterance is spoken, it is compared with all the lexicon of expected pronunciations to find which training sample it most closely resembles. For complex utterances, other linguistic and situational information can be used to help guide and confirm hypothesized word sequences. [W.A.Le.]

Speed The time rate of change of position of a body without regard to direction. It is the numerical magnitude only of a velocity and hence is a scalar quantity. Linear speed is commonly measured in such units as meters per second, miles per hour, or feet per second.

Average linear speed is the ratio of the length of the path traversed by a body to the elapsed time during which the body moved through that path. Instantaneous speed is the limiting value of the foregoing ratio as the elapsed time approaches zero. *See* VELOCITY. [R.D.Ru.]

Speed regulation The change in steady-state speed of a machine, expressed in percent of rated speed, when the machine load is reduced from rated load to zero. The definition of regulation is usually taken to mean the net change in a steady-state characteristic, and does not include any transient deviation or oscillation that may occur prior to reaching the new operation point. This same definition is used for stating the speed regulation of electric motors as well as for certain drive systems, such as steam turbines. [P.M.A.]

Speedometer A device for indicating the speed of a vehicle. There are three types of speedometers in general use: mechanical analog, quartz electric analog, and digital microprocessor.

The mechanical analog speedometer is driven by a cable housed in a casing and connected to a gear at the transmission. This gear is designed for the particular vehicle model, considering the vehicle's tire size and rear axle ratio. In most cases, the speedometer is designed to convert 1001 revolutions of the drive cable into registering 1 mi on the odometer, which records distance traveled by the vehicle. The speed-indicating portion of the speedometer operates on the magnetic principle. In the speedometer head, the drive cable attaches to a revolving permanent magnet that rotates at the same speed as the cable. Floating on bearings between the upper frame and the revolving permanent magnet is a nonmagnetic movable speed cup. The magnet revolves within the speed cup, producing a rotating magnetic field. The magnetic field is constant, and the amount of speed cup movement is at all times in proportion to the speed of the magnet rotation. A pointer, attached to the speed cup spindle, indicates the speed on the speedometer dial. *See* MAGNETIC FIELD.

The quartz speedometer utilizes an accurate clock signal supplied by a quartz crystal, along with integrated electronic circuitry to process an electrical speed signal. This signal is generated by a permanent-magnet generator mounted in the transmission. This permanent-magnet generator, designed to be used with both quartz and digital speedometers, provides a sinusoidal speed signal that is proportional to vehicle speed at the rate of 4004 pulses per mile (2503 per kilometer).

In the digital microprocessor speedometer, the vehicle speed is monitored by the permanent speed sensor mounted in the transmission. The signal is transmitted to the microprocessor where the counter converts the speed signal to a digital signal and stores it in memory. The timing circuit has the capacity to handle the counter and memory storage in less than 0.25 s. Memory circuit signals are sent to the electronic display circuit, which selects the display numerals representing the vehicle's speed, according to the number of pulses received from the speed sensor. *See* AUTOMOTIVE TRANSMISSION; ELECTRONIC DISPLAY. [R.A.Gr.]

Spelaeogriphacea A crustacean order within the Malacostraca: Peracarida, erected for a single living species. *Spelaeogriphus lepidops*. This shrimplike animal lives in a stream inside a cave on Table Mountain, South Africa. *Spelaeogriphus* has a slender, flexible body, seven pairs of walking legs, five pairs of swimmerets, and a very long flagellum on each second antenna. A short carapace is fused to the first thoracic segment, covers the second, and forms ventrolateral branchial chambers, within which are cuplike gills, attached to the first thoracic segment. The animals creep over the stream bottom with their walking legs and swim by undulations of the whole trunk. *See* PERACARIDA. [J.H.L.]

Sperm cell The male gamete. The typical sperm of most animals has a head containing the nucleus and acrosome, a middle piece with the mitochondria, and a tail (see illustration). Sperm, as well as the acrosome shape, varies with the species. The nucleus consists of condensed chromatin (deoxyribonucleic acid, DNA) and histone proteins. Each human sperm contains 22 autosomes and one sex chromosome, either an X or Y. Thus in many animals the sperm determines the genetic sex of the embryo. The acrosome, derived from the Golgi complex, contains hydrolytic enzymes, that is, hyaluronidase capable of lysing the egg coats at fertilization. Actin molecules which aid in sperm-egg interaction are found in the area between the acrosome and nucleus. The mitochondria in the middle piece apparently provide the energy necessary for the motility created by the tail. The tail has a central core, or axial filament, made up of

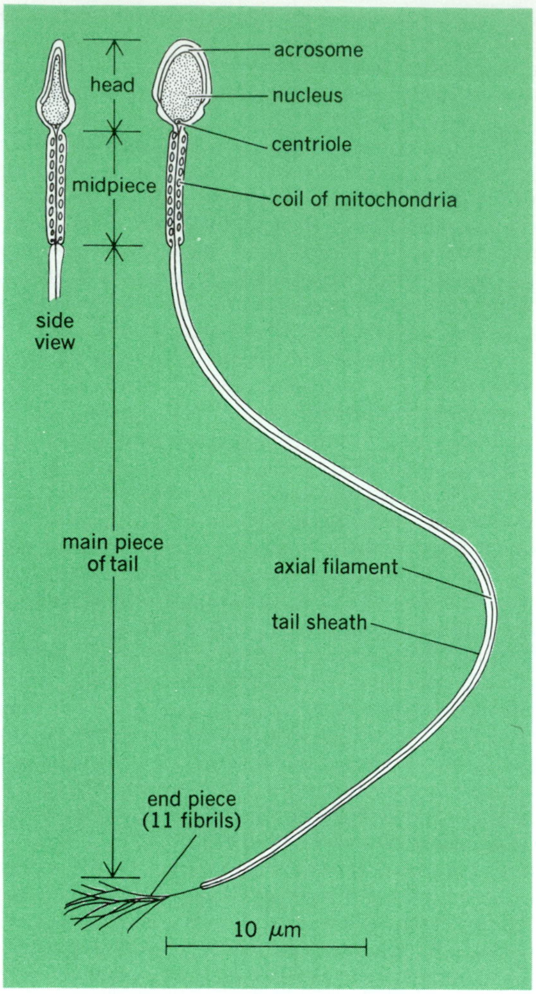

A diagram of a human spermatozoon.

nine double tubules and two central tubules (9 + 2 microtubule pattern). *See* Cilia and flagella; Spermatogenesis. [G.W.H.]

Spermatogenesis The process by which spermatogonia become sperm through meiosis. Spermatocytes arise by mitotic divisions of spermatogonia. The spermatocytes undergo meiotic divisions to produce the spermatids with *N* (haploid) number of chromosomes. *See* Gametogenesis.

Postmeiotically the spermatids undergo spermiogenesis, a transformation of nuclear and cytoplasmic constituents. The DNA within the nucleus condenses and assumes the shape characteristic of that species. Vesicles derived from the Golgi complex fuse to form the acrosomal vesicle anterior to the nucleus. The mitochondria fuse with each other. The distal centriole located posterior to the nucleus forms the point at which the nine doublet and two central microtubules of the axial filament in the flagellum arise. At its proximal end, the axial filament passes through the center of the mitochondrial helix. With maturation, all the other cytoplasmic structures are set aside in the cytoplasmic body, which is later lost. *See* Meiosis; Sperm cell. [G.W.H.]

Sperrylite A mineral with composition $PtAs_2$. Crystals are usually cubes or cubo-octahedrons. Its hardness is 6–7 on Mohs scale, and its specific gravity 10.59. The luster is metallic, and the color tin-white. Sperrylite is a rare mineral, found originally in the Sudbury district of Ontario, Canada, where it is mined as an ore of platinum. It has also been found in the Bushveld Igneous Complex. Transvaal, South Africa, and in gravels of the Timpton River, eastern Siberia. *See* Platinum. [C.S.Hu.]

Sphaeractinoidea An extinct group of fossil marine sponges closely related to modern Sclerospongiae and fossil Stromatoporoidea. They may have appeared as early as Late Carboniferous (Pennsylvania Period) of the Late Paleozoic Era. They were definitely present during the Permian Period at the close of the Paleozoic and flourished during the Mesozoic Era, especially during the Jurassic and Cretaceous, declining rapidly during the late Cretaceous, and may have persisted until Eocene time, during the Cenozoic Era.

Like the stromatoporoids, they have a radial canal system; in vertical thin section the skeleton is calcareous and characterized by a latticelike pattern. The vertical structures are round, short to long rodlike pillars. However, unlike the stromatoporoids, the horizontal component of the reticulum is trabecular or rodlike rather than separated parallel or concentric plates, or they may be thin, flat, or broadly arched cyst plates. *See* Sclerospongiae; Stromatoporoidea. [J.St.J.]

Sphaerioidaceae A family of fungi of the order Sphaeropsidales containing many plant pathogens (also called Sphaeropsidaceae or Phomaceae). There are 300 genera and 5000 species recognized by some authorities.

The pycnidia, or fruit bodies, containing the asexual spores (conidia) are either black or dark colored. They are flask-, cone-, or lens-shaped structures with thin walls and a round, relatively small pore. The conidia are discharged from the pycnidia in tendrils or cirrhi.

The genera are usually arranged into spore groups, depending on the number of cells, shape of spore cluster, and whether the spores are bright or dark. [N.F.Bu.]

Sphaerocarpales An order of liverworts in the subclass Marchantiidae, consisting of two families, the terrestrial Sphaerocarpaceae (*Sphaerocarpos* and *Geothallus*) and the aquatic Riellaceae (*Riella*). The plants are characterized by envelopes surrounding each antheridium and archegonium, absence of elaters, poor development of seta, and absence of thickenings in the unilayered wall of an indehiscent capsule.

The rhizoids are smooth, and air chambers and pores are lacking. Oil bodies may be present. The antheridia are separately contained within receptacles on the dorsal surface or at the margins, while the archegonia are separately produced in receptacles dorsally or behind the growing point. The sporophyte consists of a bulbous foot, short or obsolescent seta, and globose indehiscent capsule with a unilayered wall. *See* Bryophyta. [H.C.]

Sphaerularoida A superfamily of parasitic nematodes in the order Tylenchida. Adult females are hemocoel parasites of insects and mites; a few taxa contain both plant and insect parasites. In general, nematodes belonging to this group have three distinct phases in their life cycles: two free-living and one parasitic. In the free-living phase the female gonad is single, anteriorly directed, and with few developing oocytes and a prominent uterus filled with sperm. In the parasitic phase either the body becomes grossly enlarged and degenerates to a reproductive sac, or the uterus prelapses and gonadal development takes place outside the body. The males are always free-living and not infective. The most interesting taxon is *Sphaerularia bombi*, which prelapses the uterus. When totally prelapsed, the gonad becomes the parasite and the original female a useless appendage. It is not unusual for the prolapsed gonad to attain a volume 15,000 times that of the original female. *See* NEMATA. [A.R.M.]

Sphalerite A mineral, β-ZnS, also called blende. It is the low-temperature form and more common polymorph of ZnS. Pure β-ZnS on heating inverts to wurtzite, α-ZnS, at 1868°F (1020°C).

The mineral is most commonly in coarse to fine, granular, cleavable masses. The luster is resinous to submetallic; the color is white when pure, but is commonly yellow, brown, or black, darkening with increased percentage of iron. There is perfect dodecahedral cleavage; the hardness is 3½ on Mohs scale; specific gravity is 4.1 for pure sphalerite.

Sphalerite is a common and widely distributed mineral. It occurs both in veins and in replacement deposits in limestones. As the chief ore mineral of zinc, sphalerite is mined on every continent. The United States is the largest producer, followed by Canada, Mexico, Russia, Australia, Peru, the Congo River area, and Poland. *See* WURTZITE; ZINC. [C.S.Hu.; P.B.M.]

Sphene A nesosilicate mineral composition $CaTiO(SiO_4)$, also known as titanite. Sphene usually occurs as well-formed crystals with a characteristic wedge shape. There is prismatic cleavage and often a well-developed parting. Hardness is 5–5½ on Mohs scale; specific gravity 3.4–3.5. Luster is resinous to adamantine and the color brown, green, yellow, gray, or black. Fine crystals are found at Binnental and St. Gothard, Switzerland; Arendal, Norway; and in Ontario and Quebec. Canada. Sphene, associated with nepheline and apatite, occurs in huge masses on the Kola Peninsula, Russia, where it is mined as a source of titanium. *See* SILICATE MINERALS. [C.S.Hu.; P.B.M.]

Sphenisciformes The penguins, a small monotypic order of flightless, marine swimming birds found in the southern oceans. Classification schemes that hypothesize a link between penguins and loons have no support in fact.

Penguins are medium-sized to large birds. They are completely flightless, their wings having been modified into stiff, flattened flippers. They stand upright on legs that are far posterior and that terminate in four toes, the anterior three of which are webbed. Penguins swim and dive well, using only their wings for propulsion: their feet are used only for steering. Terrestrial locomotion is by walking, hopping, or sliding on the belly while pushing with the wings. The plumage consists of dense, scalelike feathers that are black dorsally and white ventrally. A distinctive pattern or crest, often yellow, occurs on the head. Penguins are gregarious, breeding in large colonies along the coast. The males and females, which are identical, form strong pair bonds and share in the incubation and care of the downy nestlings. The older young of some species are kept in large groups, or creches. The emperor penguin (*Aptenodytes forsteri*) breeds on the ice pack along the Antarctic coasts during the fall. Incubation of the one or two eggs is the responsibility of the male, which remains on the nest for over 2 months in the winter without eating.

Penguins are found only in the cold southern oceans, on the Antarctic continent and its surrounding islands and northward to Australia, New Zealand, South America, and Africa. One species, the Galápagos penguins (*Spheniscus mendiculus*), is found in the Galápagos Islands, which are on the Equator but are surrounded by the cold Humboldt Current. *See* AVES. [W.J.B.]

Sphenophyllales An extinct group of articulate land plants, common during Late Pennsylvanian and Early Permian times. They are typified by *Sphenophyllum*, a small, branching plant, probably of trailing habit. The long, jointed stems had superposed, longitudinal, surficial ribs between nodes. The vascular system contained a solid xylem core with triangular primary wood. The leaves were wedge-shaped and had toothed, notched, or rounded distal margins. Long, terminal cones, when found detached, contained sporangia and spores. Most species were homosporous (produced spores of a single type). *See* PALEOBOTANY. [S.H.M.]

Sphere Both in euclidean solid geometry and in common usage the word sphere denotes a solid of revolution obtained by revolving a semicircle of radius r about its diameter. Its total volume is $V = \frac{4}{3}\pi r^3$.

However, in analytic geometry, and more generally in modern mathematics, the word sphere denotes a spherical surface that bounds a solid sphere. In this sense a sphere is the locus of all points P in three-dimensional space whose distance from a fixed point O (called the center) is equal to a given number. The word radius may refer either to one of the segments OP, or to their common length r. A plane that intersects a sphere in just one point is called a tangent plane and is perpendicular to the radius drawn from the center of the sphere to that point. A plane that intersects a sphere in more than one point intersects it in a circle. The circle is called a great circle or a small circle of the sphere according to whether the plane does or does not pass through the center of the sphere. If two parallel planes intersect a sphere, the spherical surface between them is called a zone.

Any great circle of a sphere divides it into two hemispheres. A second great circle cuts a hemisphere into two lunes. A third great circle cuts each lune into two spherical triangles. *See* SURFACE AND SOLID OF REVOLUTION. [J.S.F.]

Spherical aberration An optical aberration which has symmetry of rotation. Such aberrations arise from the fact that the rays of different aperture generally do not come to the same focus, and are also known as aperture aberrations. An axis point of a system with symmetry of rotation has only aperture errors. *See* MIRROR OPTICS.

The aperture errors change with wavelength, and in a system designed for a wide range of wavelengths, the spherical and the chromatic aberrations have to be balanced against each other. A system in which the aperture aberrations do not change with wavelength is said to be spherochromatically corrected.

For an off-axis point, the asymmetry deformation errors must be separated from the aperture errors. Image analysis enables the aperture errors to be isolated and compared for different field angles. This procedure, again, enables the designer to balance out aperture aberrations as functions of the field angle by eventually introducing small aberrations of opposite order at the axis. *See* ABERRATION (OPTICS). [M.J.H.]

Spherical harmonics A spherical harmonic or solid spherical harmonic of degree n is a homogeneous function, $R_n(x,y,z)$, of degree n which satisfies Laplace's equation below.

$$\Delta R \equiv \frac{{}^2R}{\partial y^2} + \frac{\partial^2 R}{\partial y^2} + \frac{\partial^2 R}{\partial z^2} \equiv 0$$

Here n is any number, and $(x^2 + y^2 + z^2)^{(-n-1)/2}$. $R_n(x,y,z)$ is a spherical harmonic of degree $-n - 1$. There are analogous definitions for spaces of any number of dimensions. In the present article, n is a nonnegative integer and R_n, a polynomial in x, y, z (polynomial spherical harmonic). In terms of spherical coordinates r, θ, ϕ, $R_n(x,y,z) = r^n S_n(\theta,\phi)$, where S_n, a polynomial in $\cos\theta$, $\sin\theta$, $\cos\phi$, $\sin\phi$, is a spherical surface harmonic of degree n. There are $2n + 1$ linearly independent spherical surface harmonics of degree n; any spherical surface harmonic of degree n is a linear combination of these, and conversely any linear combination of spherical surface harmonics of degree n is again a spherical surface harmonic of degree n.

Spherical harmonics occur in potential theory. They occur in connection with Laplace's equation not only in spherical coordinates but also in spheroidal coordinates (spheroidal harmonics) and confocal coordinates (ellipsoidal surface harmonics). In spherical coordinates, spherical surface harmonics occur in connection with Laplace's and Poisson's equations, the wave equation, and the Schrödinger equation. In mathematical physics, spherical harmonics appear in the theories of gravitation, electricity and magnetism, hydrodynamics, and in other fields.

[A.Er.]

Sphingolipid A complex lipid which contains the amino alcohols sphingosine (I) and dihydrosphingosine or phytosphingosine (II).

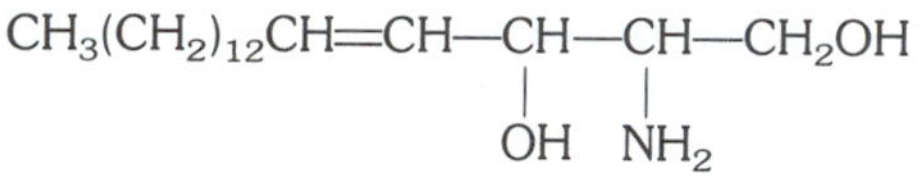

$$CH_3(CH_2)_{12}CH\!=\!CH\!-\!CH\!-\!CH\!-\!CH_2OH \quad (I)$$
$$\underset{\textstyle OH}{|} \quad \underset{\textstyle NH_2}{|}$$

$$CH_3(CH_2)_{13}CH\!-\!CH\!-\!CH\!-\!CH_2OH \quad (II)$$
$$\underset{\textstyle OH}{|} \quad \underset{\textstyle OH}{|} \quad \underset{\textstyle NH_2}{|}$$

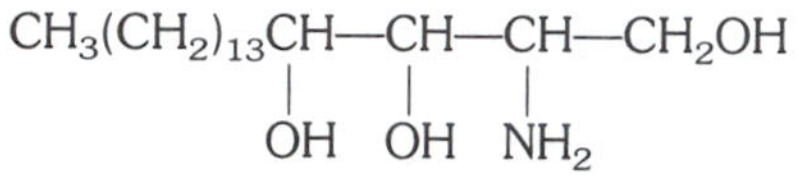

Ceramides are fatty acid amides of sphingosines. Ceramides of sphingosine are the basic components of cerebrosides (glycosides of ceramides) and sphingomyelins (phosphoryl choline esters of ceramides). Oligosaccharide phosphate esters of the ceramides of phytosphingosine (phytoglycolipids) are present in plant seeds. See GLYCOLIPID; LIPID; PHOSPHATIDE. [R.H.G.; H.E.Ca.]

Spica Alpha Virginis, one of the brightest and nearest of the hot, main-sequence stars, spectral type B1. Spica is 1500 times brighter than the Sun and has a temperature of nearly 20,000 K. Spica is a spectroscopic binary of 4-day period consisting of two nearly identical stars. Rapidly rotating, and relatively close, they are nearly unstable and are connected by streams of matter. See STAR. [J.L.Gr.]

Spice and flavoring Ingredients added to food to provide all or a part of the flavor. Spices are pungent or aromatic substances of vegetable origin used in foods at levels that yield no significant nutritive value. Flavor is the perception of those characteristics of a substance taken orally that affect the senses of taste and olfaction. The term flavoring refers to a substance which may be a single chemical species or a blend of natural or synthetic chemicals whose primary purpose is to provide all or part of the particular flavor effect to any food or other product taken orally. Flavorings are categorized by source: animal, vegetable, mineral, and synthetic. See OLFACTION; TASTE.

Other than by source, flavorings can be divided into two groups; one group affects primarily the sense of taste, and the other affects primarily the sense of olfaction. The members of the first group are called seasonings, the members of the second group are called flavors. The same terms can be used to divide flavorings in another way. The term flavors can be applied to those products which provide a characterizing flavor to a food or beverage. The term seasoning can then be applied to those products which modify or enhance the flavor of a food or beverage—spices added to meats, blends added to potato chips, lemon added to apple pie.

In the United States most flavorings are processed. Therefore another classification is by method of manufacture. Chemicals are produced by chemical synthesis or physical isolation and include such substances as vanillin, salt, monosodium glutamate, citric acid, and menthol. Concentrates are powders manufactured by dehydrating vegetables, such as onions and garlic, or concentrated fruit juices. Condiments are single ingredients or blends of flavorful foods, spices, and seasonings, some of which may have been derived by fermentation, enzyme action, roasting, or heating. They are usually designed to be added to prepared food at the table (for example, chut-

ney, vinegar, soy sauce, prepared mustard). Other manufactured flavorings include essential oils, hydrolyzed plant proteins, process flavors, and compounded or blended flavorings. See ESSENTIAL OILS.

Spices are natural substances that have been dehydrated and consist of whole or ground aromatic and pungent parts of plants, for example, anise, cinnamon, dill, nutmeg, and pepper. Such spices are also classified as herbs if grown in temperate climates.

Extracts are natural substances produced by extraction from solutions of the sapid constituents of spices and other botanicals in food-grade solvents. They are also available as synthetics. [E.J.M.]

Spilite An aphanitic (microscopically crystalline) to very-fine-grained igneous rock, with more or less altered appearance, resembling basalt but composed of albite or oligoclase, chlorite, epidote, calcite, and actinolite.

In spite of the highly sodic plagioclase, spilites are generally classed with basalts because of the low silica content (about 50%). They also retain many textural and structural features characteristic of basalt.

Spilites are found most frequently as lava flows and more rarely as small dikes and sills. Spilitic lavas typically show pillow structure, in which the rock appears composed of closely packed, elongated, pillow-shaped masses up to a few feet across. Pillows are typical of subaqueous lava flows. Vesicles, commonly filled with various minerals, may give the rock an amygdaloidal structure. See AMYGDULE; BASALT; IGNEOUS ROCKS.

[W.I.R.]

Spin (quantum mechanics) The intrinsic angular momentum of a particle. It is that part of the angular momentum of a particle which exists even when the particle is at rest, as distinguished from the orbital angular momentum. The total angular momentum of a particle is the sum of its spin and its orbital angular momentum resulting from its translational motion. The general properties of angular momentum in quantum mechanics imply that spin is quantized in half integral multiples of $\hbar$ ($= h/2\pi$, where h is Planck's constant); orbital angular momentum is restricted to half even integral multiples of $\hbar$. A particle is said to have spin $\frac{3}{2}$, meaning that its spin angular momentum is $\frac{3}{2}$. See ANGULAR MOMENTUM.

A nucleus, atom, or molecule in a particular energy level, or a particular elementary particle, has a definite spin. The spin is an intrinsic or internal characteristic of a particle, along with its mass, charge, and isotopic spin. See ISOBARIC SPIN; QUANTUM MECHANICS; SYMMETRY LAWS (PHYSICS). [C.J.G.]

Spin-density wave The ground state of a metal in which the conduction-electron spin density has a sinusoidal variation in space. The periodicity of the wave is unrelated to the lattice periodicity. Instead it is determined by the dimensions of the conduction-electron Fermi surface in momentum space.

[A.W.O.]

Spin glass One of a wide variety of materials which contain interacting atomic magnetic moments and also possess some form of disorder, in which the temperature variation of the magnetic susceptibility undergoes an abrupt change in slope, that is, a cusp, at a temperature generally referred to as the freezing temperature. At lower temperatures the spins have no long-range magnetic order, but instead are found to have static or quasistatic orientations which vary randomly over macroscopic distances. The latter state is referred to as spin-glass magnetic order. Spin-glass ordering is usually detected by means of magnetic susceptibility measurements, although additional data are required to demonstrate the absence of long-range order. Closely related susceptibility cusps can also be observed by using neutron diffraction. It is not generally agreed whether spin glasses undergo a phase transition or not. *See* MAGNETIC SUSCEPTIBILITY; NEUTRON DIFFRACTION; PHASE TRANSITIONS. [R.E.W.]

Spin label A molecule which contains an unpaired electron spin which can be detected with electron spin resonance (ESR) spectroscopy. Molecules are labeled when an atom or group of atoms which exhibit some unique physical property is chemically bonded to a molecule of interest. Groups containing unpaired electrons include organic free radicals and a variety of types of transition-metal complexes (such as vanadium, copper, iron, and manganese). Through analysis of ESR spectra, rates of molecular motion and relative orientations of spin-labeled molecules whose motion is restrained by surrounding molecules can be determined. Measurement of rates of molecular motion and molecular orientation has proved to be very important in the study of a variety of types of biologically important problems. *See* ELECTRON PARAMAGNETIC RESONANCE (EPR) SPECTROSCOPY. [R.Kr.]

Spinach A cool-season annual of Asiatic origin. *Spinacia oleracea*, belonging to the plant order Caryophyllales. It is grown for its foliage and served as a cooked vegetable or as a salad. New Zealand spinach (*Tetrogonia expansa*) and Mountain spinach (*Atriplex hortense*) are also called spinach but are less commonly grown. Spinach plants are usually dioecious. *See* CARYOPHYLLALES. [H.J.C.]

Spinal cord The portion of the central nervous system within the spinal canal of the vertebral column, that is, the entire central nervous system except the brain. The spinal cord extends from the foramen magnum at the base of the skull to a variable level of the spinal canal; it terminates at the lumbar level in humans and extends well into the caudal region in fishes.

The outer portion of the spinal cord is made up of nerve fibers most of which are oriented longitudinally and carry information between parts of the spinal cord, between spinal cord and brain, and between brain and spinal cord. The outer white matter is divided into dorsal, lateral, and ventral columns. The interior of the spinal cord consists of gray matter and is divided into a dorsal sensory horn (or column) and a ventral motor horn (or column). In the thoracic and lumbar regions of the cord there is also a small lateral horn (or column) which contains preganglionic sympathetic neurons. In the very center of the bilaterally symmetrical spinal cord is a small central canal, containing cerebrospinal fluid. *See* SYMPATHETIC NERVOUS SYSTEM.

Paired spinal nerves enter the spinal canal between each pair of vertebrae and connect with the spinal cord. The number of spinal nerves varies widely in vertebrates; in humans there are 31 pairs (8 cervical, 12 thoracic, 5 lumbar, 5 sacral, and 1 coccygeal). Each spinal nerve divides into a dorsal sensory root and a ventral motor root before entering the spinal cord. The motor neurons of the ventral horn, in addition to receiving synapses from dorsal root axons, also receive synaptic endings from neurons in other parts of the spinal cord and from long axons coming from the brain. The axons of the ventral horn neurons leave the cord through the ventral root of the spinal nerve and run with peripheral nerves to innervate the muscles of the body. With this complex synaptic and fiber organization, the spinal cord can act as the integrating center for spinal reflexes (such as the knee jerk reflex), send sensory information from the brain, and receive information from the brain to initiate or inhibit muscular activity. *See* MOTOR SYSTEMS; NERVOUS SYSTEM (VERTEBRATE); SENSATION. [D.B.W.]

Spinal cord disorders In addition to those disorders common to the brain, the spinal cord is subject to certain lesions because of its position or structure. A few of the more important disorders are mentioned.

Spinal cord injury results from dislocation, fracture, or compression in many cases, but a special form, called spinal shock, may result from a severe blow without actual distortion of adjacent tissue. In this case, there is a temporary paralysis which gradually clears. In direct damage, the cord may be slightly, partially, or completely damaged at one or more levels. Typical motor and sensory losses follow, with a poor prognosis for recovery if the nerve tissue is severely injured.

A fairly common type of potential cord injury is seen in a number of cases of slipped disks, in which the inner, soft part of the vertebral column extrudes into the spinal canal. If this compresses the cord, functional loss of temporary or permanent degree can follow; more often, pressure is exerted on spinal roots so that pain, numbness, and some type of muscle weakness intervene.

Inflammations may result from known or unknown agents and in meningitis may involve primarily the coverings; in myelitis, the cord itself. The meningococcus, pneumococcus, streptococcus, tubercle bacillus, and other microorganisms frequently cause meningitis. The most widely known cause of myelitis is the poliomyelitis virus group. *See* MENINGITIS; SPINAL CORD. [E.G.St./N.K.M.]

Spine The backbone, or vertebral column. The spine is one of the most characteristic structures of all vertebrates, but wide

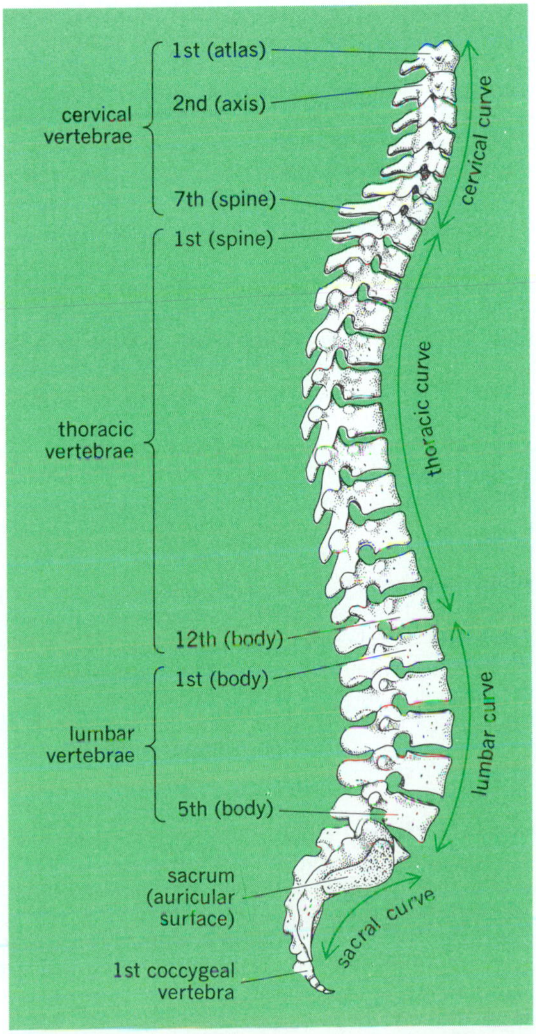

The human spine in lateral view. (*After W. J. Hamilton et al., Textbook of Human Anatomy, Macmillan, 1956*)

variations in numbers of vertebrae and development of particular regions are common.

Humans have 33 vertebrae (see illustration): 7 cervical (neck), 12 thoracic, 5 lumbar, 5 sacral, and usually 4 coccygeal (caudal). The cervical segments curve forward, while the thoracic segments curve backward and also articulate with the 12 pairs of ribs. The heavy lumbar vertebrae of the loin are followed by the fused sacral bones which form the dorsal wall of the pelvis. The coccyx consists of a small, curved group that forms a semiflexible "tail" at the base of the spine. Despite common general features, the vertebrae vary in shape and size. Their thick ventral bodies, or centra, form the main supporting portion of the vertebral column and are separated by cartilaginous intervertebral disks. Dorsal to the centra, neural arches cover and protect the spinal cord. *See* SKELETAL SYSTEM; SPINAL CORD. [T.S.P.]

Spinel The mineral of ideal composition $Al_2^{VI}[Mg^{IV}O_4]$. The spinel structure type is one of the most important and most comprehensively studied of all structure types. Its general formula can be written $M_2^{VI}[T^{IV}O_4]$, where the Roman numerals indicate the coordination numbers: the M atoms are octahedrally coordinated and the T atoms are tetrahedrally coordinated. The oxygen atoms are in cubic closest-packed arrangement, making the structure quite dense.

No less than 200 different spinels have been synthesized, usually by fusion of the component oxides. Synthetic spinels have many uses. The ferrimagnetic properties of some spinels are of great importance in the solid-state industry.

Mineralogically, the spinels are divided into three series: the spinel (Al^{3+}) series, the magnetite (Fe^{3+}) series, and the chromite (Cr^{3+}) series. Some members of these series, such as magnetite and chromite, are valuable ores of their principal metals. The Al_2MgO_4 spinel frequently occurs as rounded grains. The hardness is 8 on Mohs scale and the specific gravity 3.5, but extensive substitutions can significantly alter these values. Thus spinel may be blue, gray, green, yellow, red, brown, black, or colorless. *See* CHROMITE; MAGNETITE. [P.B.M.]

Spinicaudata An order of fresh-water branchiopod crustaceans formerly included in the order Conchostraca. The entire body and its appendages are covered by a bivalve carapace up to about 17 mm (0.7 in.) long, hence the name clam shrimp. The carapace is often bilaterally compressed and usually displays several lines of growth. The head cannot be protruded. Reproduction is usually bisexual. All species frequent temporary pools in many parts of the world, but *Cyclestheria* also occurs in permanent waters. *See* BRANCHIOPODA. [G.Fr.]

Spinning (metals) A production technique for shaping and finishing metal. In the spinning of metal, a sheet is rotated and worked by a round-ended tool. The sheet is formed over a mandrel. Spinning may serve to smooth wrinkles in drawn parts, provide a fine finish, or complete a forming operation as in curling an edge of a deep-drawn part. Spun products range from precision reflectors and nose cones to kitchen utensils. *See* SHEET-METAL FORMING. [R.L.Fr.]

Spinning (textiles) The making of yarn either from fiber or from a spinning solution. In the drawing out, twisting, and winding of either natural or synthetic fibers into a continuous thread or yarn, several operations, collectively called spinning, are performed. The fiber is drawn in or drafted from an initial length per unit weight as it comes from the carding operation into a longer length per unit weight. The fiber is also twisted to the required number of turns per unit length, forming yarn. Finally, the yarn is wound as a cop on a paper tube or is delivered to a bobbin or spool.

In the production of synthetic fiber, the dope, melt, or other liquid from which the artificial filament is produced is filtered and pumped through small holes in a spinneret. Liquid synthetic material emerges from the spinneret into conditions necessary to convert it to a pliant solid: an acid bath, an evaporation column, or a cooling chamber. Where the filaments emerge, they are stretched, and then twisted and taken up on a bobbin. *See* TEXTILE. [F.H.R.]

Spinulosida An order of Asteroidea in which pedicellariae rarely occur, are never of the crossed type, and do not form rosettes around the spines. The marginal plates are small, and the mouth plates are conspicuous but unkeeled. The spines normally occur in groups; in the deep-water Pterasteridae they are webbed. The tube feet have suckers and usually lie in two longitudinal rows along each ambulacrum. The order is generally considered to include 13 families, and is represented at all depths except the ultraabyssal. Best known are the sun stars (Solasteridae), the starlets (Asterinidae), and the genus *Acanthaster* (Acanthasteridae), notorious for damaging coral reefs. *See* ASTEROIDEA. [A.C.C.]

Spiral A term used generically to describe any geometrical entity that winds about a central point or axis while also receding from it. Spiral staircases, helices, and nonplanar loxodromes (curves that intersect those of a given class at a constant angle, for example, rhumb lines, in case the curves are on a sphere whose meridians form the given class) are examples of spirals whose windings do not lie in a plane. *See* HELIX. [L.M.Bl.]

Spiriferida An order of fossil articulate brachiopods whose earliest members are found in rocks of Middle Ordovician age and whose latest representatives are known in Lower Jurassic sediments. *See* ARTICULATA (BRACHIOPODA).

Four quite distinct groups are regarded as suborders of the order: Athyrididina, Atrypidina, Spiriferidina, and Retziidina (see illustration). The principal feature shared by these groups is the spiralium, a pair of spirally coiled ribbons of calcite supported by the crura. The direction of the axes of coiling and the nature of any connection between the two spirals varies considerably within the order. *See* BRACHIOPODA. [A.J.R.]

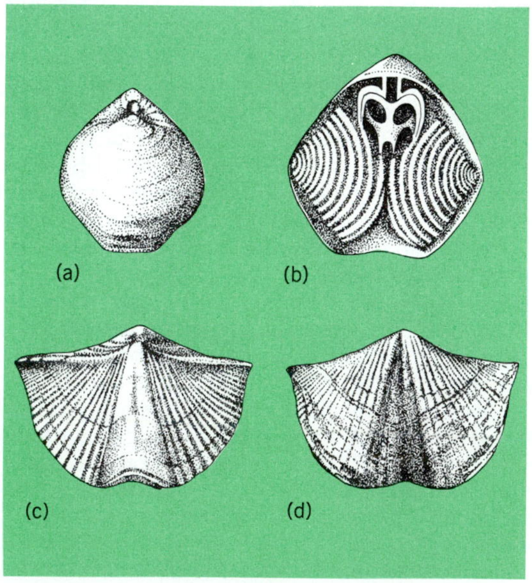

Spiriferida. Athyrididina: (*a*) Dorsal view of *Composita* and (*b*) ventral view with pedicle valve cut away to show spire. Spiriferidina: (c) Dorsal view of *Spinocyrtia* and (*d*) ventral view. (After R. C. Moore, ed., *Treatise on Invertebrate Paleontology, pt. H*, Geological Society of America, Inc., and University of Kansas Press, 1965)

Spirillaceae A family of bacteria containing gram-negative, non-spore-forming, helically curved, relatively nonflexible rods which, if motile, have polar flagella. Members possess a respiratory type of metabolism (they obtain energy by oxidation of nutrients via an electron transport system, with molecular oxygen as the terminal electron acceptor). Organic compounds, usually amino acids or salts of organic acids, serve as carbon and energy sources. Two genera are assigned to this family, *Spirillum* and *Campylobacter*. [N.R.K.]

Spirochaetales The spirochetes, an order of bacteria characterized by elongate cells twisted three-dimensionally into a spiral shape. Some members of the order are pathogens, causing syphilis (in humans), relapsing fever (in humans), and Weil's disease (in humans and other animals), while others are parasitic or free-living forms in water or sediments. The pitch of the spiral varies so that different species may appear as tightly coiled, almost closed springs; as regular, open-coiled forms; or as irregularly twisted cells. The spirochetes are unusually flexible, being able to elongate and contract or to superimpose secondary waves on the primary coils of the spiral. *See* RELAPSING FEVER; SYPHILIS.

All spirochetes are motile, usually swimming rapidly with a motion that involves both forward progress and spinning around the long axis. Flagella are absent in this group. Electron microscopy reveals axial filaments or bundles of axial filaments anchored near each pole of the cell, which are wound around the cell proper within an outer membrane and overlap in the center of the cell. It is presumed, without direct evidence, that the axial filaments are the organelle of motility.

There is a single family, Spirochaetaceae, with five genera (*Spirochaeta*, *Cristispira*, *Treponema*, *Borrelia*, and *Leptospira*). [S.C.R.]

Spirophorida An order of sponges of the class Demospongiae, subclass Tetractinomorpha, with a globular shape and a skeleton of oxeas and triaenes as megascleres, and microspined, contorted, sigmalike microscleres. Members of this order range down to depths of at least 5900 ft (1800 m). See DEMOSPONGIAE. [W.D.H.]

Spirotrichia A major subclass of the class Ciliatea which contains those ciliate Protozoa that are typified by conspicuous, compound ciliary structures. The structures are the cirri, which occur on the ventral surface of the body, and the buccal organelles, which include the undulating membrane and the adoral zone of membranelles. This group of ciliates is classified into six orders. Separate articles appear on each of the six orders listed below:

Subclass Spirotrichia
Order: Heterotrichida
Odontostomatida
Oligotrichida
Tintinnida
Entodiniomorphida
Hypotrichida

The orders in this subclass contain those organisms that are considered to be the most highly evolved ciliates. [J.O.C.]

Spirurida An order of nematodes in which the labial region is usually provided with two lateral labia or pseudolabia; in some taxa there are four or more lips; rarely lips are absent. Because of the variability in lip number, there is variation in the shape of the oral opening, which may be surrounded by teeth. The amphids are most often laterally located; however, in some taxa they may be located immediately posterior to the labia or pseudolabia. The stoma may be cylindrical and elongate or rudimentary. The esophagus is generally divisible into an anterior muscular portion and an elongate swollen posterior glandular region, where the multinucleate glands are located. Eclosion larvae are usually provided with a cephalic spine or hook and a porelike phasmid on the tail.

All known spirurid nematodes utilize an invertebrate in their life cycle; the definitive hosts are mammals, birds, reptiles, and rarely amphibians. The order contains four superfamilies: Spiruroidea, Physalopteroidea, Filarioidea, and Drilonematoidea.

Spiruroidea. The Spiruroidea comprise parasitic nematodes whose life cycle always requires an intermediate host for larvae to the third stage. The definitive hosts are mammals, birds, fishes, reptiles, and rarely amphibians; and spiruroids may be located in the host's digestive tract, eye, or nasal cavity or in the female reproductive system. Morphologically, the lip region is variable in Spiruroidea, ranging from four lips to none. When lips are present, the lateral lips are well developed and are referred to as pseudolabia. The cephalic and cervical region may be ornamented with cordons, collarettes, or cuticular rings. The stoma is always well developed and is often provided with teeth just inside the oral opening. In birds the nematodes are often associated with the gizzard, and the damage caused results in death, generally by starvation. When the muscles of the gizzard are destroyed, seeds pass intact and cannot be digested.

This superfamily contains the largest of all known nematodes, *Placentonema gigantissima*, parasitic in the placenta of sperm whales. Mature females attain a length of 26 ft (8 m) and a diameter of 1 in. (2.5 cm). The adult female has 32 ovaries, which produce great numbers of eggs. [A.R.M.]

Filarioidea. The Filarioidea contain highly specialized parasites of most groups of vertebrates. They are particularly common in amphibians, birds, and mammals. While they cannot be classified as completely harmless, most of the many hundreds of known species are not associated with any recognized disease. A limited number of species produce serious diseases in humans, and a few others produce serious diseases in domestic or wild animals. The filarial parasites of humans are found almost exclusively in the tropics, with some extension into the subtropics.

There are no conspicuous divisions into distinct body regions. Sexual dimorphism is the rule; in common with other nematodes, the female filarioid is at least twice as long as the male, and often the difference is much greater. The adult worms are found in a wide variety of places in the body of the vertebrate host, but each species has its preferred host and preferred location within that host.

All the known filariae require a bloodsucking arthropod intermediate host, usually an insect and commonly a dipteron, in which to complete embryonation. The microfilariae are ingested as the arthropod feeds. After embryonation is completed, the resulting infective larvae gain entrance into the definitive vertebrate in association with the next feeding of the arthropod. *See* NEMATA.

Filariasis. Filariasis is a disease caused by Filarioidea in humans or lower animals. The term is loosely used to indicate mere infection by such organisms. In human medicine, filariasis commonly refers to the disease caused by, or to infection with, one of the mosquito-borne, elephantoid-producing filarioids—most frequently *Wuchereria bancrofti*, less frequently *Brugia malayi*, and more recently *B. timori*. The only specific laboratory aid to diagnosis is the detection and identification of the microfilariae.

Onchocerciasis. This disease is caused by *Onchocerca volvulus* in the subcutaneous lymphatics. It is characterized by subcutaneous nodules which are most conspicuous where the skin lies close over bony structures, via cranium, pelvic girdle, joints, and shoulder blades. When they are on the head, the microfilariae reach the eyes. Ocular disturbances vary from mild transient bleary vision to total and permanent blindness.

Loa loa. The African eye worm, *Loa loa*, is the filarioidan worm most commonly acquired by Caucasian immigrants, including missionaries, in Africa. Transmission is by daytime-feeding sylvan deer-flies, genus *Chrysops*. The only preventive measures are protective clothing, including head nets. Repellents have some value. Fortunately, serious damage is rare even when the worm gets into the eye. The areas of pitting edema known as calabar swellings are painful and diagnostic. They commonly occur on the wrists, hands, arms, or orbital tissues. [G.F.O.]

Splachnales An order of the true mosses (subclass Bryidae), whose members are remarkable perennial plants that grow mainly on nitrogenous substrates, such as dung, and show considerable differentiation of neck tissue below the spore-bearing part of the capsule. The order consists of two families with about eight genera.

The plants are gregarious or dense-tufted, erect, and often forked. The leaves are soft, lanceolate or obovate, sometimes bordered at the margins, and commonly toothed. The single costa may be excurrent or terminate at or below the apex. The setae are elongate and the capsules erect with a noticeably differentiated neck. Stomata are very numerous in the neck, and have two guard cells. A single peristome is usually present, entire or forked with 16 teeth sometimes paired or joined in twos and fours. *See* BRYIDAE; BRYOPHYTA; BRYOPSIDA. [H.Cr.]

Spleen An organ of the circulatory system present in most vertebrates, lying in the abdominal cavity usually in close proximity to the left border of the stomach.

In humans the spleen normally measures about 1 by 3 by 5 in. (2.5 × 7.5 × 12.5 cm) and weighs less than ½ lb (230 g). It is a firm organ with an oval shape and is indented on its inner surface to form the hilum, or stalk of attachment to the peritoneum. This mesentery fold also carries the splenic artery and vein to the organ.

The spleen is an important part of the blood-forming, or hematopoietic, system; it is also one of the largest lymphoid organs in the body and as such is involved in the defenses against disease attributed to the reticuloendothelial system. Although the chief functions of the spleen appear to be the production of lymphocytes, the probable formation of antibodies, and the destruction of worn-out red blood cells, other less well-understood activities are known. For example, in some animals it may act as a reservoir for red blood cells, contracting from time to time to return these cells to the bloodstream as they are needed. In the fetus and sometimes in later life, the spleen may be a primary center for the formation of red blood cells. Another function of the spleen is its role in biligenesis. Because the spleen destroys erythrocytes, it is one of the sites where extrahepatic bilirubin is formed. *See* BILIRUBIN; SPLEEN DISORDERS. [W.J.B.]

Spleen disorders The spleen is rarely the site of primary disorders except those of vascular origin, but it is frequently involved in systemic inflammations, metabolic diseases, and generalized blood disorders.

Among vascular disturbances, acute and chronic congestion are prominent, particularly chronic congestion caused by cardiac failure, cirrhosis of the liver, and obstruction of the blood flow from the spleen by thrombi, scarring, or tumor tissue. Obstruction of the splenic artery or its branches by thrombi may result in an infarct caused by either cardiac or blood disease.

Inflammations include acute and chronic forms. The characteristic engorgement of blood often causes a marked enlargement of the organ. Bacteremias frequently produce this enlargement, or splenomegaly, and inflammation, but any severe infectious disease such as diphtheria or pneumonia may do so.

The leukemias, especially when of the lymphocytic or neutrophilic varieties cause some of the most prominent cases of splenomegaly as well as other changes.

Tumors originating in the spleen are rare and usually limited to such benign growths as hemangiomas, lymphangiomas, and fibromas, but malignant lymphomas and lymphosarcomas also occur. Secondary tumors, which originate elsewhere and metastasize to the spleen, are not uncommon, particularly the lymphoma group. *See* SPLEEN. [E.G.St./N.K.M.]

Splines A series of projection and slots used instead of a key to prevent relative rotation of cylindrically fitted machine parts. Splines are several projections machined on the shaft; the shaft fits into a mating bore called a spline fitting. Splines are made in two forms, square and involute, as illustrated.

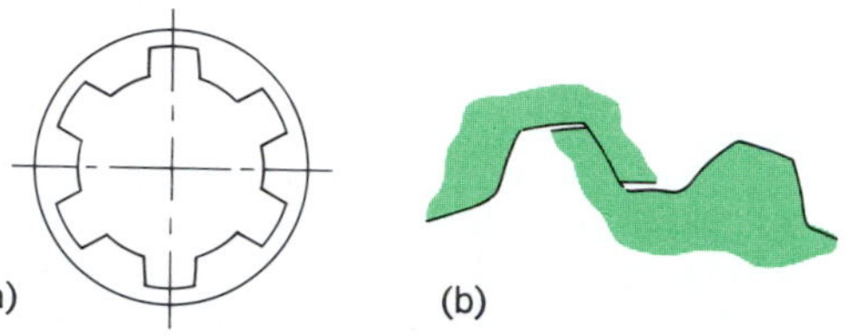

Diagrams of (*a*) square spline and (*b*) involute spline profile.

Since there are several projections (integral keys) to share the force in transmitting power, the splines can be shallow, thereby not weakening the shaft as much as a standard key.

Three classes of fits are used for square splines: sliding (as for gear shifting) under load, sliding when not loaded, and permanent fit. Square splines have been used extensively for machine parts. In the automotive industry, square splines have been replaced generally by involute splines.

Involute splines are used to prevent relative rotation of cylindrically fitted machine parts and have the same functional characteristics as square splines. The involute spline, however, is like an involute gear, and the spline fitting (internal part) is like a mating internal gear. *See* MACHINE KEY. [P.H.B.]

Spodumene The name given to the monoclinic lithium pyroxene $LiAl(SiO_3)_2$. Spodumene commonly occurs as white to yellowish prismatic crystals, often with a "woody" appearance.

Spodumene is usually found as a constituent in certain granitic pegmatites. The emerald-green variety, hiddenite, and a lilac variety, kunzite, are used as precious stones. Spodumene from pegmatites is used as an ore for lithium. *See* LITHIUM; PYROXENE. [G.W.DeV.]

Sponge The name for all members of the phylum Porifera. The majority of these organisms are marine; however, one family of sponges, the Spongillidae, occurs in fresh water. Sponges of commercial value are collected in the Mediterranean, in the Gulf of Mexico, and near the West Indies. *See* PORIFERA. [C.B.C.]

Spongiomorphida A small extinct order of Triassic and Jurassic marine invertebrate fossils of uncertain biologic affinities. The spongiomorphs had calcareous skeletons composed of small vertical rods called pillars and interconnecting horizontal rods called trabeculae or synapticulae. The pillars consist of upwardly inflected cone-in-cone layers of fibrous calcareous deposits. The trabeculae are formed from thickened zones at the base of the flare of the cones in the pillars. Spongiomorphs were sessile, benthonic, shallow marine organisms associated with calcareous deposits, especially reefs. *See* HYDROZOA. [J.St.J]

Spontaneous combustion Combustion that occurs when certain materials are stored in bulk. The oxidizing action of microorganisms often produces the initial heat. As the temperature increases, the air trapped in the material takes over the oxidation process, liberating more heat. Because the heat cannot be dissipated to the surroundings, the temperature of the material rises still more and the rate of oxidation increases. Eventually the material reaches an ignition point and bursts into flame. For example, coal is subject to spontaneous combustion and is generally stored in shallow piles to allow the heat of oxidation to dissipate. *See* COMBUSTION.

For gases at ordinary temperatures, molecular collisions do not usually cause combustion. At elevated temperatures the collisions of the thermally agitated molecules are more frequent. More important as a cause of chemical reaction is the greater energy involved in the collisions. Moreover, there is very little combustion attributable to direct reaction between the molecules. Instead, a high-energy collision dissociates a molecule into atoms, or free radicals. These molecular fragments react with greater ease, and the combustion process proceeds generally by a chain reaction involving these fragments. *See* CHAIN REACTION (CHEMISTRY).

Under certain conditions where the rate of chain branching equals or exceeds the rate at which chains are terminated, the combustion process speeds up to explosive proportions; because of the rapidity of molecular events, a large number of chains are formed in a short time so that essentially all of the gas undergoes reaction at the same time; that is, an explosion results. Another cause of explosion in gaseous combustion arises when the rate at which heat is liberated in the reaction is greater than the rate at which the heat dissipates to the surroundings. *See* CHEMICAL DYNAMICS; FLAME. [B.L.]

Sporotrichosis A mycotic infection of humans caused by *Sporothrix schenckii*, a fungus that is worldwide in distribution. This organism has been isolated from timber, plants, and soil. The disease is most frequently seen in farmers and horticulturists. The primary lesion develops within 2–8 weeks following introduction of the fungus into the tissue, and appears as a hard, pink nodule. Gradually, the lesion becomes darker in color and undergoes necrosis with discharge of purulent material. As the infection progresses, numerous nodules develop along the lymphatic chain of the arm. Rarely does the disease progress beyond this form. Although the primary lesion may heal, the secondary nodules may persist for months to years if untreated. [L.D.H.]

Sporozoa A subphylum of Protozoa, typically with spores. The spores are simple and have no polar filaments. There is a single type of nucleus. There are no cilia or flagella except for flagellated microgametes in some groups. In most Sporozoa there is an alternation of sexual and asexual stages in the life cycle. In the sexual stage, fertilization is by syngamy, that is, the union of male and female gametes. All Sporozoa are parasitic. The subphylum is divided into three classes—Telosporea, Toxoplasmea, and Haplosporea. *See* HAPLOSPOREA; PROTOZOA; TELOSPOREA; TOXOPLASMEA. [N.D.L.]

Sports medicine A branch of medicine concerned with the effects of exercise and sports on the human body, including treatment of injuries. Sports medicine can be divided into three general areas: clinical sports medicine, sports surgery, and the physiology of exercise. Clinical sports medicine includes the prevention and treatment of athletic injuries and the design of exercise and nutrition programs for maintaining peak physical performance. Sports surgery is also concerned with the treatment of injuries from contact (human or object) sports. Exercise physiology, a growing field of sports medicine, involves the study of the body's response to physical stress. It comprises the science of fitness, the preservation of fitness, and the role of fitness in the prevention and treatment of disease. [O.A.]

Spot welding A resistance-welding process in which coalescence is produced by the flow of electric current through the resistance of metals held together under pressure. Usually the upper electrode moves and applies the clamping force. Pressure must be maintained at all times during the heating cycle to prevent flashing at the electrode faces. Electrodes are water-cooled and are made of copper alloys because pure copper is soft and deforms under pressure. The electric current flows through at least seven resistances connected in series for any one weld (see illustration). After the metals have been fused together, the electrodes usually remain in place sufficiently long to cool the weld. *See* RESISTANCE WELDING; WELDING AND CUTTING OF METALS. [E.J.L.]

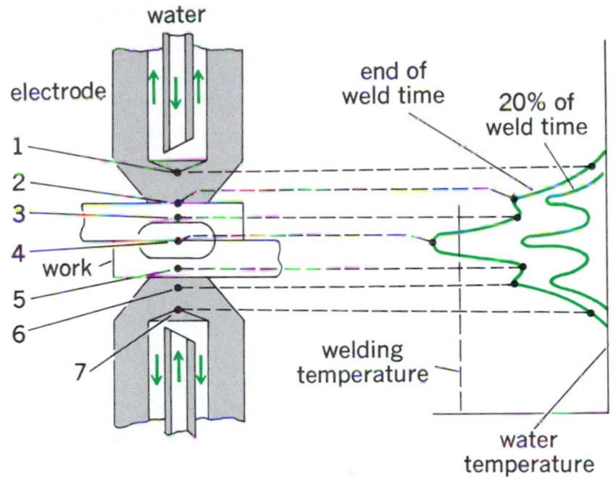

Distribution of temperature in local (numbered) elements of a spot-welding operation.

Spotted fever group A term that includes several disease entities in humans in various parts of the world, caused by the Rickettsiae. The diseases have in common transmission to humans by various species of ticks or certain parasitic mites. Included in the group are Rocky Mountain spotted fever, fièvre boutonneuse, several tick typhuses, and rickettsialpox. *See* FIÈVRE BOUTONNEUSE; QUEENSLAND TICK TYPHUS; RICKETTSIALES; RICKETTSIALPOX; ROCKY MOUNTAIN SPOTTED FEVER; SIBERIAN TICK TYPHUS; SOUTH AFRICAN TICK-BITE FEVER. [C.B.P.]

Spread spectrum communication A means of communicating by purposely spreading the spectrum (frequency extent or bandwidth) of the communication signal well beyond the bandwidth of the unspread signal. Spread spectrum signals are typically transmitted by electromagnetic waves in free space with usage in both nonmilitary and military systems.

Motivation for using spread spectrum signals is based on the following facts: (1) These systems have the ability to reject intentional and unintentional jamming by interfering signals so that information can be communicated. (2) Spread spectrum signals have a low probability of being intercepted or detected since the power in the transmitted wave is "spread" over a large bandwidth or frequency extent. (3) Since these signals cannot be readily demodulated without knowing the code and its precise timing, message privacy is obtained. (4) The wide bandwidth of the spread spectrum signals provides tolerance to multipath (reflected waves that take longer to arrive at the receiver than the direct, desired signal). (5) A high degree of precision in ranging (distance measuring) can be obtained by using one type of spread spectrum signal, with applications to navigation. (6) Multiple access, or the ability to send many independent signals over the same frequency band, is an important feature in spread spectrum signals. *See* ELECTRONIC WARFARE; RADIO-WAVE PROPAGATION.

There are four generic types of spread spectrum signals: direct sequence (DS) or pseudonoise (PN), frequency hopping (FH), time hopping (TH), and linear frequency modulation (chirp). The latter two types are not of as much interest as the first two types; frequency hopping systems are the most commonly used.

Direct sequence systems were once the most prevalent method of communicating in spread spectrum communications. Direct sequence modulation is characterized by phase-modulating a sine wave by an unending string of pseudonoise code chips (symbols of much smaller duration than a bit). This unending string is typically based on pseudonoise code that generates an apparently random sequence of code chips that repeats only after the pseudonoise code period. Digital data representing the information to be transmitted is biphase-modulated onto the carrier, and then the pseudonoise code generator biphase-modulates the carrier. *See* PHASE MODULATION.

In a direct sequence system, the phase of the carrier changes pseudorandomly with the pseudonoise code. In a frequency-hopping system, the frequency of the carrier changes according to a pseudonoise code with a consecutive group of pseudonoise code chips defining a particular frequency. Typically either multiple frequency shift keying (MFSK) or differential phase-shift keying (DPSK) is used. Multiple frequency shift keying is a modulation scheme in which one of a number of tones (2, 4, 8, and so forth) are transmitted at a given time according to a group of consecutive data bits (n bits produce 2^n tones). In conjunction with frequency hopping, multiple frequency shift keying would imply, at each instant of time, a given carrier frequency that depends on the hop-pseudonoise code sequence and the consecutive group of the most recent n bits. Differential phase-shift keying is similar to phase shift keying except that only the differences of the phases are encoded and noncoherent techniques (involving no carrier loop) can be employed at the receiver. *See* FREQUENCY MODULATION.

The actual frequency selection is achieved by a device called a frequency synthesizer. For example, a 12-bit segment of the pseudonoise code may correspond to a particular frequency, so that every 12 bits a new frequency is produced in the frequency synthesizer. Synthesizers are used for both the transmitter and the receiver. *See* MODULATION. [J.K.H.]

Spring (hydrology) A place where groundwater discharges upon the land surface because the natural flow of groundwater to the place exceeds the flow from it. Springs are ephemeral, discharging intermittently, or permanent, discharging constantly. Springs are usually at mean annual air temperatures. The less the discharge, the more the temperature reflects seasonal temperatures. Spring water usually originates as rain or snow (meteoric water).

Hot-spring water may differ in composition from meteoric water through exchange between the water and rocks. Common minerals consist of component oxides. Oxygen of minerals has more ^{18}O than meteoric water. Upon exchange, the water is enriched in ^{18}O. Most minerals contain little deuterium, so that slight deuterium changes occur. Some hot-spring waters are acid from the oxidation of hydrogen sulfide to sulfate.

Mineral spring waters have high concentrations of solutes and wide ranges in chemistry and temperatures; hot mineral springs may be classified as hot springs as well as mineral springs. Most mineral springs are high either in sodium chloride or sodium bicarbonate (soda springs) or both; other compositions are found, such as a high percentage of calcium sulfate from the solution of gypsum.

The chemical compositions of spring waters are seldom in chemical equilibrium with the air. Groundwaters whose recharge is through grasslands may contain a thousand times as much CO_2 as would be in equilibrium with air, and those whose recharge is through forests may contain a hundred times as much as would be in equilibrium with air. Sulfate in groundwater may be reduced in the presence of organic matter to H_2S, giving some springs the odor of rotten eggs. *See* GEYSER; GROUNDWATER HYDROLOGY. [I.B.]

Spring (machines) A machine element for storing energy as a function of displacement. Force applied to a spring member causes it to deflect through a certain displacement, thus absorbing energy.

A spring may have any shape and may be made from any elastic material. Even fluids can behave as compression springs and do so in fluid pressure systems. Most mechanical springs take on specific and familiar shapes such as helix, flat, or leaf springs. All mechanical elements behave to some extent as springs because of the elastic properties of engineering materials.

The most frequent use of springs is to supply motive power in a mechanism. Common examples are clock and watch springs, toy motors, and valve springs in auto engines. A special case of the spring as a source of motive power is its use for returning displaced mechanisms to their original positions, as

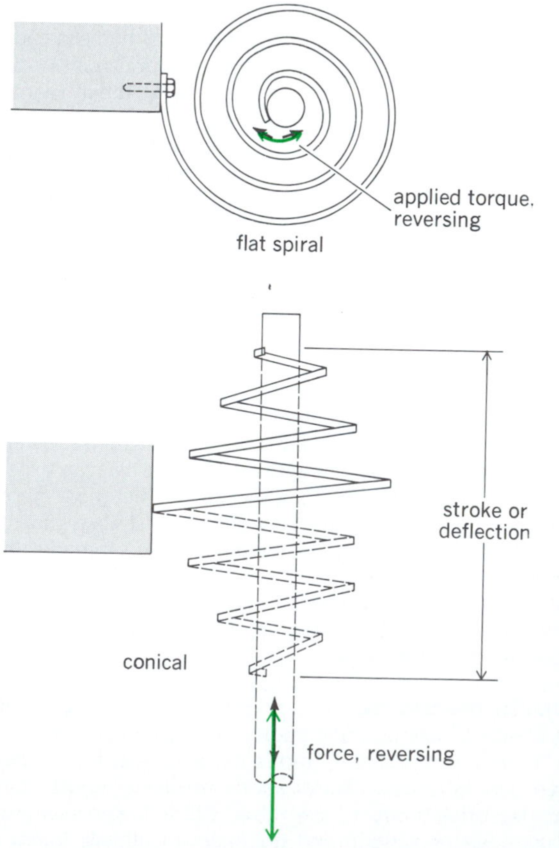

Spiral spring is unique in responding to torsional or translation forces.

in the door-closing device, the spring on the cam follower for an open cam, and the spring as a counterbalance. Frequently a spring in the form of a block of very elastic material such as rubber absorbs shock in a mechanism. Springs also serve an important function in vibration control. *See* Shock absorber; Shock isolation.

Springs may be classified into six major types according to their shape. These are flat or leaf, helical, spiral, torsion bar, disk, and constant force springs. A leaf spring is a beam of cantilever design with a deliberately large deflection under a load. The helical spring consists essentially of a bar or wire or uniform cross section wound into a helix. In a spiral spring, the spring bar or wire is wound in an Archimedes spiral in a plane. A spiral spring is unique in that it may be deflected in one of two ways or a combination of both of them (see illustration). A torsion bar spring consists essentially of a shaft or bar of uniform section. The disk spring consists essentially of a disk or washer supported at the outer periphery by one force and an opposing force on the center or hub of the disk. A constant force spring is used when a constant force must be applied regardless of displacement. [L.S.L.]

Spruce Evergreen tree belonging to the genus *Picea* of the pine family. The needles are single, usually four-sided, and borne on little peglike projections; the cones are pendulous. Resin ducts in the wood may be seen with a magnifying lens, but they are fewer than in *Pinus*.

The white spruce (*P. glauca*), ranging from northern New England to the Lake States and Montana and northward into Alaska, is distinguished by the somewhat bluish cast of its needles, small cylindrical cones, and gray or pale-brown twigs without pubescence (hairs). Red spruce (*P. rubens*) is a similar tree but with greener foliage; smaller, more oval cones; and more or less pubescent twigs. Occurring naturally with white spruce in the northeastern United States and adjacent Canada, red spruce extends southward along the Appalachians into North Carolina. Black spruce (*P. mariana*) ranges from northern New England and Newfoundland to Alaska. However, it occurs sparingly in the Appalachians to West Virginia. The cones are smaller than in the white and red species and are egg-shaped or nearly spherical and persistent. The twigs are pubescent.

Blue spruce (*P. pungens*), also known as Colorado blue spruce, is probably the best known of the western species because of its wide use as an ornamental tree. The twigs are glabrous (without pubescence). Engelmann spruce (*P. engelmanni*) has needles usually of a deep blue-green color, sometimes much like those of the blue spruce but the young twigs are slightly hairy. The cones, although cylindrical, are smaller than in blue spruce. This species is also a Rocky Mountain tree like the blue spruce, but it is more widely distributed from British Columbia to Arizona and also in the mountains of Oregon and Washington. Sitka spruce (*P. sitchensis*) is the largest spruce in the Northern Hemisphere. The leaves have a pungent odor, are considerably flattened, and stand out from the twig in all directions. It ranges from Alaska to northern California. The Norway spruce (*P. abies*), the common spruce of Europe, is much planted in the United States for timber, as well as for ornamental purposes. It can be recognized by the dark-green color of the leaves; glabrous, pendent, short branchlets; and large cones, usually near the top of the tree. *See* Pinales. [A.H.G./K.P.D.]

Spur gears Gears that transmit power between parallel shafts and have straight teeth parallel to the gear axis (Fig. 1). It is common practice, however, to group helical gears that have parallel shafts under the heading of spur gears. Spur gears are classified as external, internal, and rack and pinion.

Fig. 1. External spur gear and pinion, the commonest type of spur gear.

External spur gears, the most common, have teeth which point outward from the center of the gear. Internal or annular gears have teeth pointing inward toward the gear axis. A rack may be considered as a gear having an infinite pitch circle radius. Thus its pitch surface is a plane. A rack and pinion running together transform rectilinear motion into rotary motion, or vice versa.

Helical gears. Gears running on parallel axes and with teeth twisted oblique to the gear axis are essentially spur gears (Fig. 2). Because of the twist, contact is progressive across the tooth surface, starting at one edge and proceeding across the face of the tooth. The action results in reduced impact and quieter operation, particularly at high speed. Herringbone gears are equivalent to two helical gears of opposite hand placed side by side. They are especially suited for high-speed operation and eliminate the axial thrust produced by single helical gears.

Crossed helical gears. Where shafts cross obliquely, motion is transmitted by crossed helical gears. The teeth are helical but differ from the teeth of worm gears in that no one tooth (thread) makes a complete turn on the pitch circle. Pitch surfaces of crossed helical gears are cylindrical as with spur gears. However, with crossed shafts, the oblique teeth have point contact rather than the line contact that occurs with parallel shafts. Helical gears are referred to as right- or left-hand in

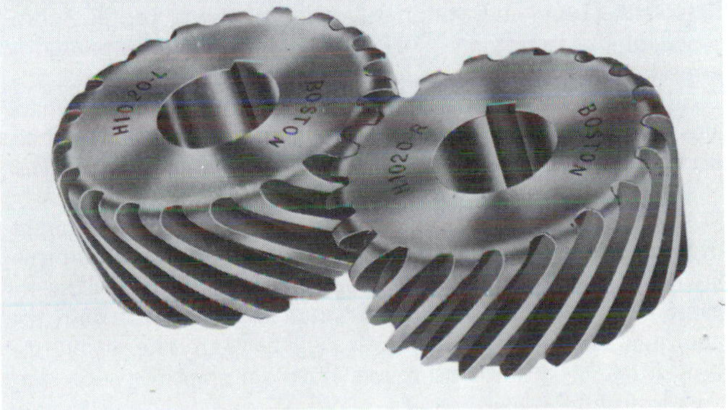

Fig. 2. Single helical gear-set. (*Boston Gear Works*)

the same manner as screw threads, a right-hand gear being one on which the teeth twist clockwise as they recede from an observer looking along the axis. *See* WORM GEARS. [J.R.Z.]

Sputtering The process by which atoms or groups of atoms are ejected from a metal surface as the result of heavy-ion impact. It generally takes place at the cathode of a self-maintained gaseous discharge, indicating that the important agent is the positive ion. Although sputtering is useful for certain processes, such as the generation of a clean surface, it is usually harmful. In the case of an oxide-coated thermionic cathode, sputtering by positive-ion bombardment may destroy the surface completely. [G.H.M.]

Squall A strong wind with sudden onset and more gradual decline, lasting for several minutes. Wind speeds in squalls commonly reach 30–60 mi/h (13–27 m/s), with a succession of brief gusts of 80–100 mi/h (36–45 m/s) in the more violent squalls. Squalls may be local in nature, as with isolated thunderstorms, or may occur over a wide area in the vicinity of a well-developed cyclone, where the squalls locally reinforce already strong winds.

The most common type of squall is the thundersquall or rain-squall associated with heavy convective clouds, frequently of the cumulonimbus type. It is formed when cold air, descending in the core of the thunderstorm rain area, reaches the Earth's surface and spreads out. Particularly in desert areas, the thunderstorm rain may largely or wholly evaporate before reaching the ground, and the squall may be dry, often associated with dust storms. *See* DUST STORM; SQUALL LINE; THUNDERSTORM.

Squalls of a different type result from cold air drainage down steep slopes. The force of the squall is derived from gravity and depends on the descending air which is colder and more dense than the air it replaces. So-called fall winds of this kind are common on mountainous coasts of high latitudes, where cold air forms on elevated plateaus and drains down fiords or deep valleys. [C.W.N.]

Squall line A line of thunderstorms, near whose advancing edge squalls occur along an extensive front. The thundery region, 12–30 mi (20–50 km) wide and up to 1200 mi (2000 km) long, moves at a typical speed of 30 knots (15 m/s) for 6–12 h or more and sweeps a broad area. In the United States, severe squall lines are most common in spring and early summer when northward incursions of maritime tropical air east of the Rockies interact with polar front cyclones. Ranking next to hurricanes in casualities and damage caused, squall lines also supply most of the beneficial rainfall in some regions. *See* FRONT; SQUALL; THUNDERSTORM. [C.W.N.]

Squamata The dominant order of living reptiles composed of the lizards and snakes. The group first appeared in Jurassic times and today is found in all but the coldest regions. Various forms are adapted for arboreal, burrowing, or aquatic lives, but most squamates are fundamentally terrestrial. There are about 4700 Recent species: 2200 lizards and 2500 snakes.

The order is readily distinguished from all known reptiles by its highly modified skull; an enlarged and movable quadrate; and a temporal opening that is lost or reduced in many forms. No other reptiles show these modifications, which allow for great kinesis in the lower jaw since it articulates with the quadrate. In addition, the order is distinct from other living reptile groups because its members have no shells or secondary palates and the males possess paired penes.

Traditionally the Squamata have been divided into two major subgroups, the lizards, suborder Sauria, and the snakes, suborder Serpentes. The latter group is basically a series of limbless lizards, and it is certain that snakes are derived from some saurian ancestor. There are many different legless lizards, and it has been suggested that more than one line has evolved to produce those species currently grouped together as snakes.

Sauria. The majority of saurians are insectivorous, but a few feed on plants while others, notably the Varanidae and allies, feed on larger prey including birds and mammals. The largest living lizard is the Komodo dragon (*Varanus komodoensis*).

The majority of lizards are quadrupedal in locomotion and are usually ambulatory scamperers or scansorial. Some forms are bipedal, at least when in haste. Perhaps the most famous of the bipedal forms is the Jesus Cristo lizard, *Basiliscus*, of Middle America which runs for some distance across the surface of small streams when frightened.

The coloration of each species of lizard is characteristic. Most forms exhibit marked differences in coloration between the sexes, at least during the breeding season, and frequently the young are markedly different from the parents. Color changes occur in rapid fashion among some species, and all are capable of metachrosis or changing color to a certain extent. Contrary to popular legend, the changes are not made to match the color of the background; they seem, rather, to be under neurohormonal control and to fluctuate with temperature and light and with the activities of the lizard. *See* CROMATOPHORE; SEXUAL DIMORPHISM.

There are but two species of venomous lizards, both members of the genus *Heloderma*, in the family Helodermatidae. The Gila monster (*H. suspectum*) is found in western New Mexico, Arizona, extreme southern Nevada, southwestern Utah, eastern California, and northwestern Mexico. The beaded lizard (*H. horridum*) is a Mexican species. Venom is produced in glands along the lower jaw and penetrates wounds inflicted by the anterior recurved teeth. No mechanism for injection of the venom into the wound is present, contrary to the situation in most snakes.

The following list indicates the major evolutionary lines and families of lizards. Families indicated by an asterisk contain limbless, snakelike species. All members of the families Pygopodidae, Anelytropsidae, Dibamidae, Amphisbaenidae, Feylinidae, and Anniellidae are snakelike lizards.

Iguania line
  Family: Iguanidae
    Agamidae
    Chamaeleontidae
Gekkota line
  Family: Gekkonidae
    Pygopodidae
Scincomorpha line
  Family: Xantusiidae
    Teiidae
    Lacertidae
    Gerrhosauridae*
    Scincidae*
    Anelytropsidae*
    Dibamidae*
Annulata line
  Family: Amphisbaenidae*
Anguimorpha line
  Family: Anguidae*
    Anniellidae*
    Feyliniidae*
    Xenosauridae
    Helodermatidae
    Varanidae
    Lanthonotidae

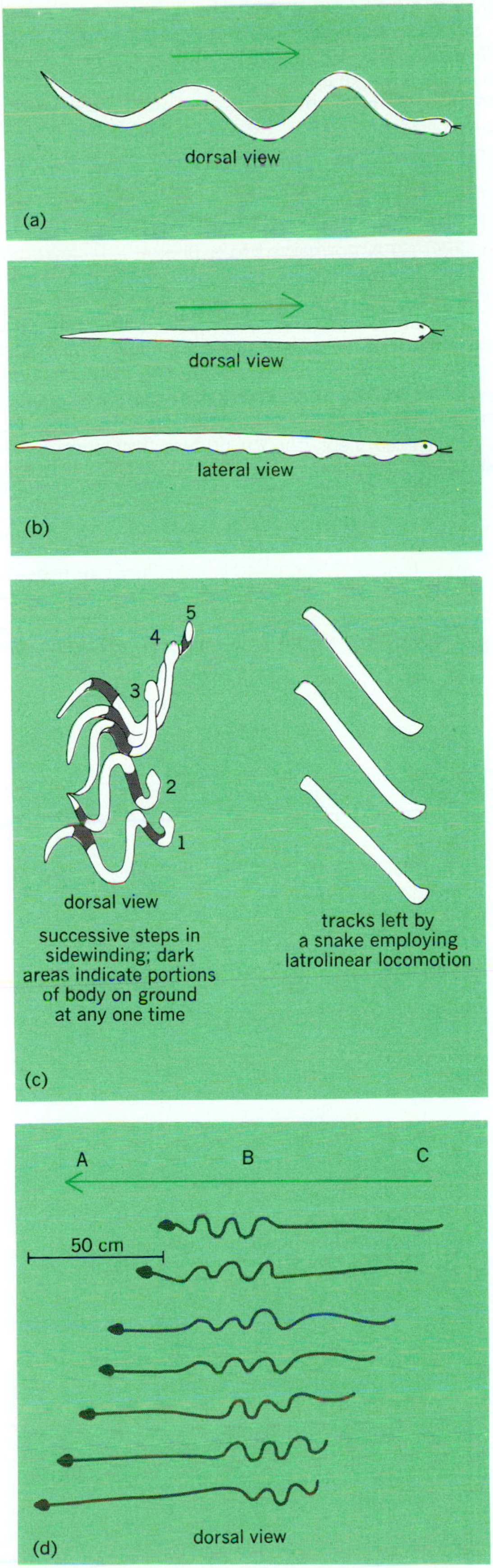

Serpentine locomotion. (*a*) Curvilinear. (*b*) Rectilinear. (*c*) Laterolinear. (*d*) Concertina; successive stages showing extension of the head and neck followed by pulling forward of the body and tail; the coils serve to anchor the body.

Serpentes. Snakes are basically specialized, limbless lizards which were probably evolved from burrowing forms but have now returned from subterranean habitats to occupy terrestrial, arboreal, and aquatic situations. The following characteristics are typical of all serpents. There is no temporal arch so that the lower jaw and quadrate are very loosely attached to the skull. This gives the jaw even greater motility than is the case in lizards. The body is elongate with 100–200 or more vertebrae, and the internal organs are elongate and reduced. A spectacle covers the eye.

The largest living snake is the Indian python (*Python reticulatus*), which reaches 30 ft (9 m) in length and a weight of 250 lb (113 kg). The largest venomous snake is the king cobra (*Ophiophagus hannah*), of southern Asia, which is known to attain a length of 18 ft (5.5 m).

The senses of snakes are fundamentally similar to those of all terrestrial vertebrates. Great dependence is placed upon olfaction and the Jacobson's organs (olfactory canals in the nasal mucosa). The tongue of all snakes is elongate and deeply bifurcated. When not in use it can be retracted into a sheath located just anterior to the glottis, but it is protrusible and is constantly being projected to pick up samples for the Jacobson's organs from the surrounding environment. Snakes are deaf to airborne sounds and receive auditory stimuli only through the substratum via the bones of the head. The eyes are greatly modified from those in lizards, and there is no color vision. Some groups are totally blind and have vestigial eyes covered by scales or skin.

Four basic patterns of locomotion are found in snakes, and several may be used by a particular individual at different times (see illustration). The most familiar type is curvilinear. In this pattern the snake moves forward by throwing out lateral undulations of the body and pushing them against any irregularity in the surface (illustration *a*). Snakes using rectilinear locomotion move forward in a straight line, without any lateral undulations, by producing wavelike movements in the belly plates (illustration *b*). Laterolinear locomotion, or sidewinding, is used primarily on smooth or yielding surfaces and is very complex. In essence, the snake anchors a portion of the body in the substratum and lifts the rest out laterally to a new position. When the lifted part is anchored again, the portions of the body left behind are lifted to the new position (illustration *c*). By a constant lifting and anchoring of alternate parts the snake moves in a lateral direction. Concertina locomotion movement resembles the expansion and contraction of that musical instrument (illustration *d*).

The vast majority of living snakes are harmless to humans, although a number are capable of inflicting serious injury with their venomous bites. The venom apparatus has evolved principally as a method of obtaining food, but it is also advantageous as a defense against attackers. Fangs are teeth modified for the injection of venom into the victim, and the venom glands are modified salivary glands connected to the grooved fangs by a duct. Special muscles are present in all proglyphous snakes to force the venom into the wound. The venom itself is a complex substance containing a number of enzymes. Certain of these enzymes attack the blood, others in the nervous system, and some are spreaders.

The following list indicates the major groups of living snakes.

Family: Typhlopidae
Leptotyphlopidae
Aniliidae
Boidae
Colubridae
Elapidae
Hydrophiida
Viperidae
Crotalidae

[J.M.S.]

Square A portion of a plane bounded by four equal line segments, or sides, each two of which are either parallel or mutually perpendicular. The intersecting sides determine four points, vertices of the square. The diagonals join pairs of vertices that are not on a side (see illustration). They are equal and mutually perpendicular. If a is the length of a side of a square, its area is $a \times a$ or a^2, and this is the motivation for calling a^2 "a squared." *See* QUADRILATERAL; RECTANGLE. [L.M.Bl.]

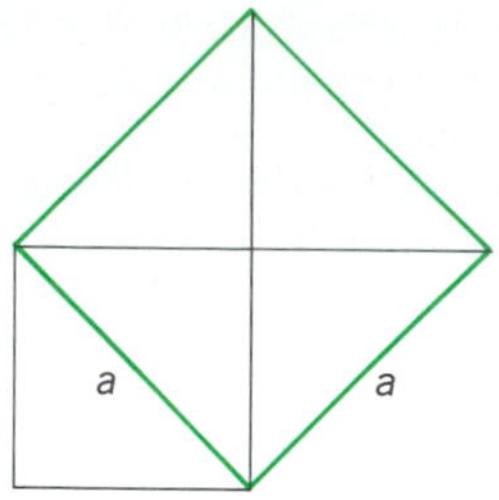

Square with side *a*.

Squash The common name for edible fruits of several species of the genus *Cucurbita: C. pepo, C. moschata, C. maxima*, and *C. mixta*. Those species originated in the Americas but are now grown in most countries around the world. Within squash there is tremendous variation in size, shape, color, and usage.

The most clearly defined group is summer squash, fruit of any species of *Cucurbita* eaten as a vegetable when immature. It is most commonly *C. pepo*. Fruit color may be white, yellow, or light or dark green, and the green may be solid or striped. Shapes may be flattened disks as in Pattypan, cylindrical as in Zucchini and Cocozelle, or with necks as in the straightneck and crookneck types. Summer squash has mild flavor, high water content, and relatively low nutritional value.

Winter squash is fruit of *Cucurbita* eaten when mature and derives its name from its ability to be stored for several weeks or months before consumption. Varieties of winter squash are found in all four species. The Table Queen group, synonymous with Acorn, is *C. pepo*, Butternut belongs to *C. moschata*, Green-striped Cushaw is *C. mixta*, while *C. maxima* has the widest range of types, including Buttercup, Hubbards, and Delicious of various colors, Banana, and Boston Marrow. Flesh color varies from light yellow to dark orange, and the edible portion ranges from thin to very thick. [H.M.M.]

Squeezed quantum states Quantum states for which certain variables can be measured more accurately than is normally possible.

All matter and radiation fluctuate. Much random fluctuation derives from environmental influence, but even if all these influences are removed, there remains the intrinsic uncertainty prescribed by the laws of quantum physics. The position and momentum of a particle, or the electric and magnetic components of an electromagnetic field, are conjugate variables that cannot simultaneously possess definite values (Heisenberg uncertainty principle). It is possible, however, to have the position of a particle more and more accurately specified at the expense of increasing momentum uncertainty; the same applies to electromagnetic field amplitudes. This freedom underlies the phenomenon of squeezing or the possibility of having squeezed quantum states. With squeezed states, the inherent quantum fluctuation may be partly circumvented by focusing on the less noisy variable, thus permitting more precise measurement or information transfer than is otherwise possible. *See* UNCERTAINTY PRINCIPLE.

According to quantum electrodynamics, the vacuum is filled with a free electromagnetic field in its ground state that consists of fluctuating field components with significant noise energy. If ϕ is a phase angle and $a(\phi)$ and $a[\phi + (\pi/2)]$ are two quadrature components of the field (for example, the electric and magnetic field amplitudes), the vacuum mean-square field fluctuation is given by Eq. (1), independently of the phase angle.

$$\langle \Delta a^2(\phi) \rangle = \frac{1}{4} \tag{1}$$

Equation (1) is normalized to a photon; the corresponding equivalent noise temperature at optical frequencies is thousands of kelvins. Equation (1) also gives the general fluctuation of an arbitrary coherent state, which is the quantum state of ordinary lasers. Further environment-induced randomness is introduced in addition to Eq. (1) for other conventional light sources, including light-emitting diodes. *See* COHERENCE; ELECTRICAL NOISE; LASER; QUANTUM ELECTRODYNAMICS.

In a squeezed state, the quadrature fluctuation is reduced below Eq. (1) for some ϕ, as given in Eq. (2). At that point,

$$\langle \Delta a^2(\phi) \rangle < \frac{1}{4} \tag{2}$$

squeezing, that is, reduction of field fluctuation below the coherent state level, occurs. The fluctuation of the conjugate quadrature is correspondingly increased to preserve the uncertainty relation, Eq. (3). In a two-photon coherent state, or

$$\langle \Delta a^2(\phi) \rangle \left\langle \Delta a^2\left(\phi + \frac{\pi}{2}\right) \right\rangle \geq \frac{1}{16} \tag{3}$$

squeezed state in the narrow sense, Eq. (3) is satisfied with equality. As seen in the illustration, the designation "squeezed state" is partly derived from the fact that the noise circle of Eq. (1) is squeezed to an ellipse when Eq. (3) is satisfied with equality.

Squeezed light can be generated by a variety of processes, especially nonlinear optical processes. The first successful experimental demonstration of squeezing, in 1985, involved a four-wave mixing process in an atomic beam of sodium atoms.

Squeezing was first studied in connection with optical communication, although it is evident that reduced quantum fluctu-

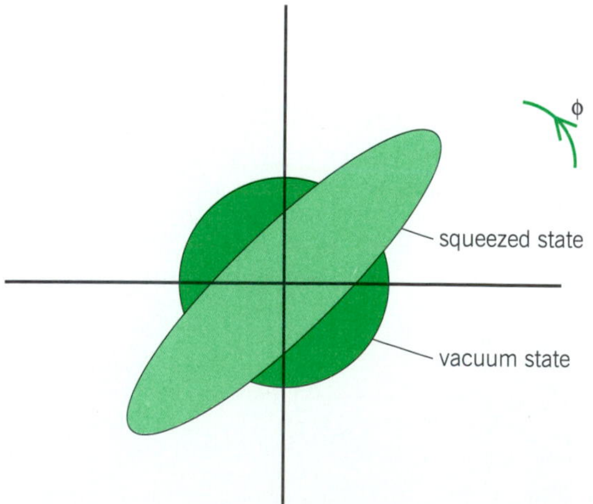

Field-amplitude fluctuation $\langle \Delta a^3(\phi) \rangle$ as a function of phase angle ϕ. The noise circle of the vacuum state, Eq. (1) in the text, is squeezed to an ellipse according to Eq. (3).

ation might find applications in precision measurements. *See* NONLINEAR OPTICS; OPTICAL COMMUNICATIONS; QUANTUM-MECHANICS. [H.P.Yu.]

Squid The name for a number of marine cephalopod mollusks characterized by a reduced internal shell, either calcareous or chitinous in composition. There are 10 tentacles; one pair is elongate with suckers on the terminal end, while the other four pairs have suckers along their entire length. The ink sac, together with the protective coloration provided by chromatophores, serves to protect these animals. The nervous system and sensory organs are well developed. Squid feed on small fish, crustaceans, and other marine organisms. They in turn are preyed upon by large fish, whales, and oceanic birds. The sexes are separate. The giant squid (*Architeuthis principles*) is the largest invertebrate, with a body length that may exceed 12 ft (3.7 m). *See* CEPHALOPODA. [C.B.C.]

SQUID A device which, in its original form, consists of two Josephon tunnel junctions connected in parallel on a superconducting loop (see illustration). The term is an acronym for

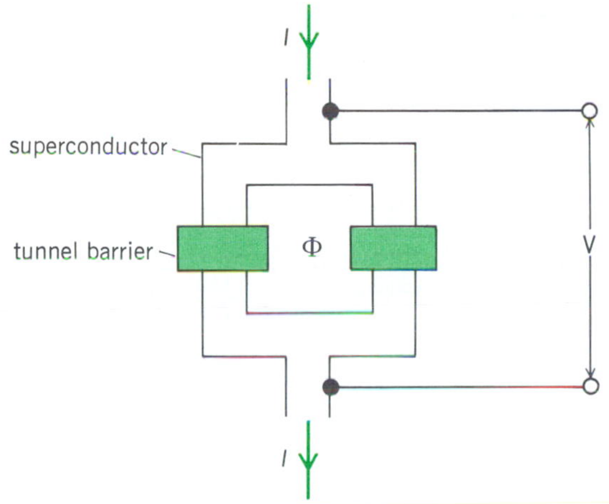

Direct-current SQUID with enclosed magentic flux Φ.

superconducting quantum interference device. A small applied current *I* flows through the junctions as a supercurrent, without developing a voltage, by means of Cooper pairs tunneling through the barriers. However, when the applied current exceeds a certain critical value a voltage *V* is generated. *See* INTERFERENCE OF WAVES; JOSEPHSON EFFECT; SUPERCONDUCTIVITY.

The SQUID has important device applications. When each Josephson tunnel junction is shunted with an external resistance to eliminate hysteresis on the current-voltage charactertistic and the SQUID is biased with a constant current greater than the critical value, the voltage across the SQUID is also an oscillatory function of the magnetic flux. If one measures the change in voltage produced by the application of a flux equivalent to a small fraction of one flux quantum, one has a very sensitive magnetometer. Another important potential application is as a logic element or memory cell in high-speed computers. *See* COMPUTER STORAGE TECHNOLOGY; LOGIC CIRCUITS; MAGNETOMETER.

The rf SQUID, which consists of a single junction interrupting a superconducting loop, is in fact misnamed, since no interference takes place. *See* SUPERCONDUCTING DEVICE. [J.Cl.]

SS 433 A remarkable stellar object with unique properties: it shows evidence of ejection of two narrow streams of cool gas traveling in oppositely directed beams from a central object at a velocity of almost one-quarter the speed of light—the beams executing a repeating, rotating pattern about the central object once every 164 days.

One peculiar characteristic of SS 443 is that the spectrum possesses not only a set of emission lines due to hydrogen and helium, but two further sets of lines, one displaced to longer (redder) wavelengths from the familiar lines, and the second displaced to shorter (bluer) wavelengths. These displacements can be understood in terms of the Doppler effect. In addition, the wavelengths of the lines change every night in a smoothly progressing pattern, indicating that the velocity of the emitting regions is also changing.

Especially intriguing is the observation that on a given night the average velocity of the approaching and receding beams is not zero, but rather a large positive value, about 7500 mi/s (12,000 km/s), despite the fact that SS 433 is approximately stationary with respect to the Earth. This proves to be a direct consequence of Einstein's special theory of relativity. An outside observer perceives a change in measured times and lengths of a system moving at very large velocity. The beam velocity in SS 433 is large enough that special relativity is important. *See* RELATIVITY.

There has been much speculation as to the type of star present in SS 443, with many astronomers now agreeing that the enormous velocities in the beams require a highly collapsed, compact star with a strong gravitational field. A neutron star, the same end point of stellar evolution responsible for pulsars, or possibly a black hole could satisfy this requirement. *See* BINARY STAR; BLACK HOLE; NEUTRON STAR; PULSAR. [B.M.]

Stability augmentation Modification of inherent aircraft handling qualifies by automatic control devices which supplement the pilot's control inputs. These automatic control devices were first known as stabilizers, then stability augmentation systems (SAS), and subsequently command augmentation systems (CAS) in certain applications.

Stability augmentation systems use instruments that measure the aircraft motion (sensors) and automatically control the aerodynamic control surfaces to modify the motion in a desired manner. This feedback control is generally in series with the pilot's input so that the cockpit controllers do not move and the pilot is unaware of the feedback control inputs. On the other hand, fully automatic guidance and control systems, such as autopilots and automatic landing systems, are mechanized in parallel with the pilot's inputs. The stability augmentation is typically an inner loop of the fully automatic control systems. *See* AUTOMATIC LANDING SYSTEM; AUTOPILOT; CONTROL SYSTEMS. [H.A.R.]

Stabilizer (aircraft) The horizontal or vertical aerodynamic wing surfaces that provide aircraft stability and longitudinal balance in flight. Horizontal and vertical stabilizers (fins) are

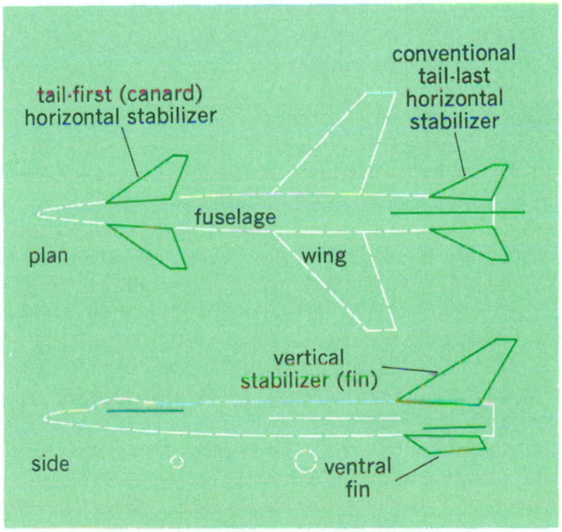

Various stabilizer arrangements.

similar to the aircraft wing in structural design and function of providing lift at angle of attack to the wind. However, stabilizers are not required to supply lift to overcome aircraft weight during flight, and when the wing-fuselage center of pressure is behind the aircraft center of gravity, the aerodynamic load on the horizontal stabilizer may be downward. Various stabilizer arrangements are shown in the illustration. *See* FLIGHT CONTROLS. [M.J.A.]

Stabilizer (chemistry) Any substance that tends to maintain the physical and chemical properties of a material. Degradation, that is, irreversible changes in chemical composition or structure, is responsible for the premature failure of materials. Stabilizers are used to extend the useful life of materials as well as to maintain their critical properties above the design specifications. There are three general approaches to stabilization: surface modification, modification of chemical composition, and use of additives.

Wood and metals are protected against destructive weathering and corrosion by covering with paints or other surface finishes. Laminated structures are also used to minimize degradation. For example, moisture-resistant polymers have been laminated over metals to provide corrosion resistance. Polymers have also been stabilized by crosslinking at the surface to restrict penetration of reactants into the bulk of the material. The protection of food products by plastic wrappings is an extension of the approach to stabilization by surface modification. *See* PAINT.

Stainless steels are stabilized steels; the structure of steel is modified by alloying with chromium or nickel corrosive chemical reactions. Both natural and synthetic polymers degrade during processing and long-term exposure. Different polymers exhibit considerable variation in the level of stability to hostile environments. There is evidence that some synthetic polymers have structural irregularities that could serve as sites for degradation to begin. Reduction of such irregularities during synthesis or replacement with less labile groups could also be used as an approach to stabilization. *See* POLYMER.

A wide variety of additives have been developed to stabilize polymers against degradation. Stabilizers are also available that inhibit thermal oxidation, burning, photodegradation, and ozone deterioration of elastomers. Stabilization of hydrocarbon polymers can be accomplished with preventative or chain-breaking antioxidants. Preventative antioxidants stabilize by reducing the number of radicals formed in the initiation stage. Stabilization of polymers against photooxidation, the principal component of outdoor weathering, is accomplished by addition of light screens, ultraviolet absorbers, or radical scavengers. Carbon black is the most effective of all light screens, functioning by preventing the penetration of ultraviolet radiation into the polymer. *See* ANTIOXIDANT; CARBON BLACK; INHIBITOR (CHEMISTRY); PHOTODEGRADATION; POLYMER. [W.L.H.]

Stadia A surveying distance-measuring method in which the interval intercepted on a vertically held rod by two cross hairs in the alidade or transit telescope is convertible to the horizontal distance from instrument to rod. The rod is called a stadia rod; the cross hairs are called stadia hairs and are fixed equidistant above and below the horizontal cross hair. Their distance apart, i, is a fixed ratio (conveniently 1:100, or 0.3:100 in levels) of the telescope's focal length. Hence the distance D from the focal point to an object is basically proportional to the interval R observed on the rod.

For horizontal stadia sighting, the small distance from the focal point to the instrument center is added to the distance obtained by reading the rod intercept as shown in the illustration. The distance is about 1 ft (30 cm) for externally focusing instruments; it is negligible for internally focusing telescopes. In all of stadia instruments, the figures and graduations on the stadia rod are read through the instrument telescope. *See* SURVEYING. [B.A.B.]

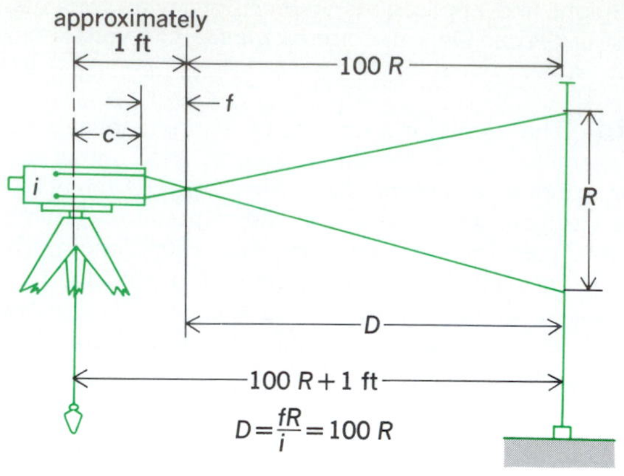

Stadia measurement of distance. 1 ft = 30 cm.

Stain (microbiology) Any colored, organic compound, usually called dye, used to stain tissues, cells and components, or cell contents. The dye may be natural or synthetic. The object stained is called the substrate. The small size and transparency of microorganisms make them difficult to see even with the aid of a high-power microscope. Staining facilitates the observation of a substrate by introducing differences in optical density or in light absorption between the substrate and its surroundings or between different parts of the same substrate. In electron microscopy, and sometimes in light microscopy (as in the silver impregnation technique of staining flagella or capsules), staining is accomplished by depositing on the substrate ultraphotoscopic particles of a metal such as chromium or gold (the so-called shadowing process); or staining is done by treating the substrate with solutions of metallic compounds such as uranyl acetate or phosphotungstic acid. *See* ACID-FAST STAIN; GRAM'S STAIN. [G.K.]

Stained glass Glass colored by any of several means. It is frequently assembled to produce a varicolored mosaic or representation for decorative uses. The glass is colored by one of three means: fusing with metallic oxides, enameling, or painting. The addition of small percentages of metal compounds, usually oxides, to molten glass or pot metal produces throughout the glass a color characteristic of the compound. If the percentage of metal compound is low, the glass is tinted; if the percentage is high, the glass is saturated. *See* SATURATION.

The glass can also be coated with transparent metallic oxides and fired to bond the enameling to the glass base. Transparent pigments can be painted onto the glass in solid colors, or in several colors to render a picture; the pigments are then baked or burned onto the glass. With a hot iron, the glazier cuts the glass to shape, joins it into the design with channeled lead strips, and solders the joints. Either the tracery of the window or iron frames support the finished glaziery. *See* GLASS. [F.H.R.]

Stainless steel The generic name commonly used for that entire group of iron-base alloys which exhibit phenomenal resistance to rusting and corrosion because of chromium (Cr) content. Contents of Cr exceeding 10%, with carbon (C) held suitably low, make iron effectively rustproof.

Other alloy elements, notably nickel (Ni) and molybdenum (Mo), can also be added to the basic stainless composition to produce both variety and improvement of properties. Over 100 different stainless steels are produced commercially, about half as standardized grades. Some are more properly classed as stainless irons since they do not harden as steel; others are true steels to which corrosion resistance becomes an added feature. Still others that are neither properly steels nor irons introduce totally new classes of materials, from both mechanical and chemical standpoints. *See* ALLOY; STEEL. [C.A.Z.]

Stalactites and stalagmites Stalactites, stalagmites, dripstone, and flowstone are travertine deposits in limestone caverns, formed by the evaporation of waters bearing calcium carbonate. Stalactites grow down from the roofs of caves and tend to be long and thin, with hollow cores. The water moves down the core and precipitates at the bottom, slowly extending the length while keeping the core open for more water to move down.

Stalagmites grow from the floor up and are commonly found beneath stalactites; they are formed from the evaporation of the same drip of water that forms the stalactite. Stalagmites are thicker and shorter than stalactites and have no central hollow core. *See* CAVE; LIMESTONE. [R.Si.]

Stall-warning indicator A device that determines the critical angle of attack for a given aircraft, at which point the drag coefficient overpowers the lift coefficient and the aircraft will no longer sustain itself in steady-state condition (level flight or climb/descent). The indicator usually operates from vane sensors, airflow pressure sensors, tabs on the leading edge of the wings, and computing devices which include accelerometers, airspeed detectors, and vertical gyros. [J.W.A.]

Standard model The most comprehensive theory of elementary particles that has been confirmed experimentally. It describes the first three of the four fundamental interactions (or forces), which are the strong, electromagnetic, weak, and gravitational interactions, in decreasing order of strength (see table). These interactions are mediated by vector mesons (of spin 1 in units of $\hbar$, Planck's constant divided by 2π). The vector mesons produce forces between the matter particles. The matter particles are fermions (named after E. Fermi, who studied the statistics of particles that obey the exclusion principle and have antisymmetric wave functions) of spin ½, in contrast to the vector mesons, which are bosons. These particles obey the spin statistics theorem that particles of integer spin must be bosons and particles of odd half-integer spin must be fermions. *See* BOSE-EINSTEIN STATISTICS; EXCLUSION PRINCIPLE; FERMI-DIRAC STATISTICS; FUNDAMENTAL INTERACTIONS.

Gauge principle. The three interactions of the standard model follow from the gauge principle, which states that the fundamental laws of nature must be invariant under independent symmetry transformations at each space-time point. The local character of this invariance contrasts with the weaker global invariance, in which a theory is invariant only under constant symmetry transformations. Each gauge theory is characterized by its symmetry group, coupling, constants, matter fields, and by whether or not the symmetry group is broken. See GAUGE THEORY, SYMMETRY LAWS (PHYSICS).

Quantum electrodynamics. The simplest gauge theory is quantum electrodynamics (QED), the theory of the interaction of the electron-positron field with the Maxwell field of electromagnetism. For quantum electrodynamics the symmetry group is the group of complex numbers of modulus 1, and the single generator corresponds to the electric charge in units of the positron charge. See QUANTUM ELECTRODYNAMICS.

Yang-Mills theories. Gauge theories were generalized to nonabelian theories, whose generators do not commute, by C. N. Yang and R. L. Mills. Yang and Mills studied the isospin symmetry (which has three generators) of the pion-nucleon system. The quanta of the gauge fields should behave as massless spin-1 particles. Since no such particles connected with the pion-nucleon system are known, this theory remained a curiosity for some time. See ISOBARIC SPIN.

Gauge theory of the strong interaction. The successful uses of the gauge principle in elementary particle physics required new ingredients. For the strong interaction, M. Gell-Mann and, independently, G. Zweig introduced fractionally charged particles called quarks as fundamental constituents of pions, nucleons, and other hadrons (that is, strongly interacting particles). O. W. Greenberg, and M. Y. Han and Y. Nambu introduced a new three-valued degree of freedom, which later was colloquially called color. When the color model is gauged, that is, the symmetry among the three colors is taken as the symmetry group of a Yang-Mills gauge theory and the quarks are taken as the fundamental triplet of the color group, the resulting theory is quantum chromodynamics (QCD), the fundamental theory of the strong interaction and one of the building blocks of the standard model. Two unexpected properties of quantum chromodynamics, confinement and asymptotic freedom, allow the theory to reconcile seemingly contradictory physical phenomena. See COLOR (QUANTUM MECHANICS); QUANTUM CHROMODYNAMICS; QUARKS.

Spontaneously broken symmetry. The application of Yang-Mills theory to the electromagnetic and weak interactions uses spontaneous symmetry breaking. This is the phenomenon in which the symmetry is broken by the vacuum state, although the lagrangian and equations of motion of the theory obey the symmetry. P. W. Higgs and T. W. B. Kibble, and others, showed that spontaneous symmetry breaking in a local gauge theory gives rise to a mass for those gauge particles associated with the generators for which the symmetry is broken. *See* ELECTROWEAK INTERACTION; SYMMETRY BREAKING.

Open questions. The standard model accounts for all elementary particle physics at energies below about 100 GeV. This is a remarkable consolidation of knowledge of the microworld. However, there are still important questions. It is not known what determines the gauge group of the model, why there are three generations of quarks and leptons, what produces the masses and mixings of the quarks and leptons, what accounts for the wide range of masses (ranging from the electron neutrino whose mass is known to be less than 18 eV, and could be zero, to the *Z* boson whose mass is 92 GeV, a

Properties of the fundamental interactions

Interaction	Strength*	Range	Symmetries that are obeyed
Strong	$\alpha s = 0.14$	Infinite	All
Electromagnetic	$\alpha = 1/137.036$	Infinite	All
Weak	$GFM2pc/\hbar 3 = 1.03 \times 10 - 5$	$2.44 \times 10 - 18$ m	CP almost exact, TCP
Gravitational	$GNM2p/\hbar c = 5.90 \times 10 - 39$	Infinite	All

*The strength $\alpha = e2/(4\pi \epsilon 0 \hbar c)$ is scale-dependent. The value given is at electron mass me = 0.511 MeV. Here, e is the electric charge of the proton, $\epsilon 0$ is the permittivity of the vacuum, $\hbar$ is Planck's constant divided by 2π, and c is the speed of light. The strength $\alpha s = g2/(4\pi \epsilon 0 \hbar c)$ is also scale-dependent. The value given is at 34 GeV. Here, g is the color (quantum chromodynamics) coupling constant. The color coupling constant should not be confused with the pion-nucleon coupling constant, $g\pi N$, sometimes also called g, for which $g2\pi N/(4\pi \epsilon 0 \hbar c)$ has a value of about 15. GF is the Fermi constant, GN is Newton's gravitational constant, and Mp is the proton mass.

range of 5 × 10^9), or how gravity can be incorporated into a still more fundamental theory.

Because of these questions, there has been an extensive search for phenomena that contradict the standard model. So far, no such contradictions have been found. Experiments at higher-energy accelerators seem essential to provide the insight necessary for further progress in elementary particle physics. *See* ELEMENTARY PARTICLE; PARTICLE ACCELERATOR. [O.W.G.]

Standing wave A disturbance which is oscillatory in time and which has an amplitude that varies in space between zero and a maximum value. Standing waves may be formed, for example, near an ideal boundary by the interaction of incident and perfectly reflected traveling waves. The points in space where the amplitudes are zero are called nodes. Antinodes are points where amplitude is a maximum. Standing waves are a limiting case of stationary waves. *See* WAVE (PHYSICS); WAVE MOTION. [W.J.G.]

Standing-wave detector An electric indicating instrument used for detecting standing waves along a transmission line or in a waveguide and measuring the resulting standing-wave ratio. It can also be used to measure the wavelength and hence the frequency of an electromagnetic wave in a line. The detecting device is usually a bolometer, thermocouple, or crystal, connected to an indicating meter directly or through an amplifier. The detecting device is moved along the line while observing the meter indication; the positions along the line at which maximum and minimum readings are obtained correspond to the nodes and antinodes of the standing wave. *See* STANDING WAVE; WAVEGUIDE; WAVELENGTH; WAVEMETER. [J.Mar.]

Staphylococcus A genus of the bacterial family Micrococcaceae. The staphylococci are parasites of humans, their usual habitat being the nasopharynx and skin. They may cause boils, osteomyelitis, wound infections, and some cases of septicemia, pneumonia, and kidney infection. They are gram-positive, spherical, pathogenic bacteria, and typically occur in grapelike clusters. Staphylococci grow readily on the usual laboratory culture media; meat-infusion agar, containing human or animal blood, is frequently employed for their isolation. Staphylococci are relatively resistant to adverse physical and chemical agents. Antibiotic-resistant strains of staphylococci have become increasingly numerous. Infections due to these resistant strains have raised serious problems of treatment and control.

Staphylococci produce several serologically distinct toxins and enzymes which are assumed to contribute to their capacity to produce disease. Among these substances, exotoxin produces intense destruction of the local tissues when injected into the skin of experimental animals and is rapidly fatal when injected intravenously. Three hemolysins are produced which dissolve red blood cells.

Enterotoxin is distinct from exotoxin and other staphylococcal products and is responsible for the acute gastrointestinal symptoms of staphylococcal food poisoning. Leukocidin is a toxin which destroys the white blood cells, thereby aiding the staphylococci in their attempt to counteract an important defense mechanism of the host. The production of the enzyme coagulase is confined essentially to pathogenic staphylococci. In the test tube, coagulase produces a clot in the blood plasma of some animal species. Hyaluronidase, an enzyme produced by most pathogenic staphylococci, breaks down certain constituents of the intracellular ground substance of many body tissues. [J.E.Bl.]

Star A celestial body, consisting of a large, self-luminous mass of hot gas held together by its own gravity. The composition by weight of the average star is about 70% hydrogen, 28% helium, 1.5% carbon, nitrogen, oxygen, and neon, and 0.5% iron group and heavier elements.

The stars contain by far the largest fraction of the mass of the universe. Stars are born, produce nuclear energy, evolve, and eventually die. Their life-spans range from 10^6 years, for a star of high luminosity, to 10^{10} for the Sun, and up to 10^{13} years for the faintest main-sequence stars. The oldest known in the Milky Way Galaxy are over 10^{10} years old.

The nomenclature for the identification of stars given their general location and brightness. The sky was subdivided into constellations and the brighter stars were named; Greek alphabet letters generally describe them, with α representing the brightest star, visually, in a given constellation. Thus Betelgeuse, or α Orionis, is the visually brightest star in Orion. Fundamentally a star is defined by its coordinates on the celestial sphere, right ascension and declination, and its brightness, or apparent magnitude. Because of the precession of the equinoxes, celestial coordinates must be specified for a given epoch. About 6000 stars are visible to the naked eye, but over 10^{12} exist in the Milky Way Galaxy.

Many stars have a total space motion of from 100 to 500 km/s (60 to 300 mi/s) with respect to the Sun. Such objects are called high-velocity stars and belong, according to the nomenclature of Walter Baade, to population II, that is, stars found in the spheroidal halo of a galaxy. The more slowly moving stars are younger, and may be members of population I like the Sun; such stars are found in the spiral regions or flattened disk of a galaxy.

Figure 1, a Hertzsprung-Russell or H-R diagram, is a means for plotting the relation between the luminosity and surface temperature or other temperature-dependent parameter, such as color or spectral type of the stars. Along the main sequence the stars are distinguished chiefly by mass; luminosity is a steep function of the mass. Data for stars out to 20 parsecs (6.2 × 10^{17} m or 65 light-years) are plotted in Fig. 1. The use of photoelectric color, like that of spectral type, is essentially an

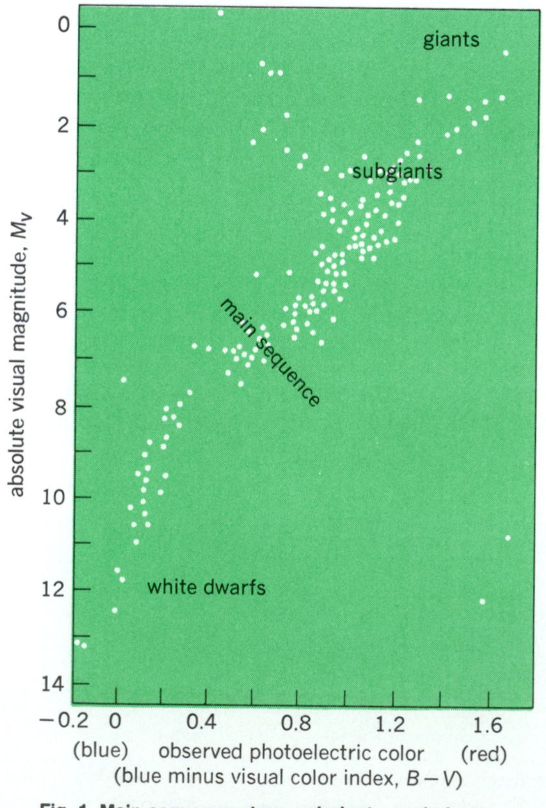

Fig. 1. Main-sequence stars, subgiants, and giants.

arrangement by surface temperature and is convenient and accurate. *See* HERTZSPRUNG-RUSSELL DIAGRAM; PARALLAX (ASTRONOMY); PARSEC.

Stellar spectra. The spectrum of a star in a large majority of cases shows absorption lines superposed on a continuous background. The interior of the star is at high temperature and pressure; its spectrum is nearly that of a blackbody. The star shades off into space through a reversing layer or stellar atmosphere in which the continuous spectrum and absorption lines are formed. In the Sun this reversing layer is about 150 mi (250 km) thick; its base is about 9900°F (5750 K), and its outer layer is almost 8100°F (4200 K). Outside the normal reversing layer in many stars there may be a temperature inversion in the low-density chromosphere and corona. The temperature of the corona eventually reaches 2,700,000°F (1,500,000 K). The emission lines of these outer layers affect only slightly the spectrum of the integrated light of a star and carry about one-millionth of the total energy. The x-ray radiation of the Sun is highly variable. Some stars have been found by their strong (and variable) x-ray emission. *See* SUN; X-RAY ASTRONOMY; X-RAY STAR.

Spectral classification. The spectral classification of a star gives in a simple symbolic form the essential features of its complex spectrum. By inspection, many features are found to vary in a smooth way from one star to another. This variation is correlated with the colors of the stars. As a result, spectral classification includes a vast majority of the stars and represents a sequence of decreasing reversing-layer temperature, from the left to the right (Fig. 2). There are several side sequences whose temperatures approximately correspond with those of normal stars, in which apparent abundance differences exist. Both dwarf and giant stars exist over a wide range of spectral type. Decimal subdivisions are used; in more refined analyses prefixes and suffixes are added.

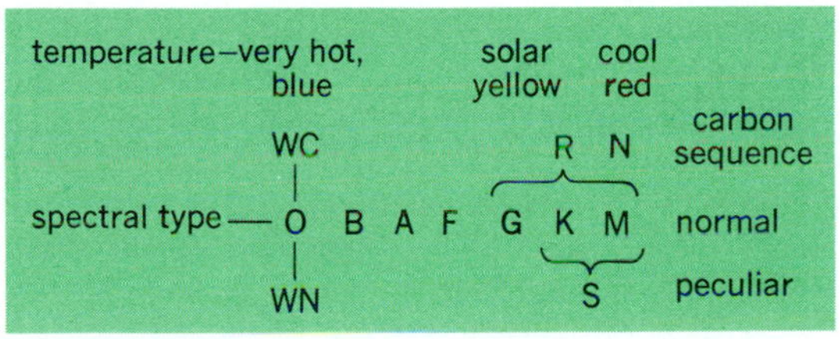

Fig. 2. Spectral classification from Draper catalog.

A more refined system of spectral classification has been widely used. In this system, a two-parameter set of criteria provides by inspection both spectral type and estimated luminosity. The luminosity is indicated by a suffix ranging from Ia, extremely bright supergiant, to III, normal giant, to V, main-sequence or dwarf star. Thus the designation G2 V represents a star like the Sun, G2 III a giant of nearly the same temperature, with certain luminosity-sensitive features enhanced.

Temperature and luminosity. Most differences in appearance of stellar spectra are caused by changes in the surface temperature. The degree of ionization and excitation of the atoms at a given temperature dictates whether an element will have an appreciable concentration of atoms in the lower atomic level that produces the absorption line.

Absolute magnitude effects occur because the ionization equation depends on pressure, the electron concentration setting the recombination rate. Consequently, stars of low surface gravity, which have lower pressures, will show a given percentage of ionization at a temperature about 900°F (500 K) lower than those of the main sequence.

Age, evolution, and mass. A group of color-magnitude curves for clusters, open (population I) and globular (very old population II), shows that, although the fainter end of the main sequence is essentially the same in all groups of stars, the brighter ends vary from one group to another in accordance with differences in age and composition. Stellar evolution causes such variations.

These H-R diagrams are most significant for the study of the ages, nuclear energy sources, and evolution of the stars. The location of a star in an H-R diagram is completely determined by its mass and chemical composition, if the latter is specified in detail throughout the star. A point in an H-R diagram, specified by L and R, does not uniquely determine the mass, however, because of the possible variation of composition with depth. (Such variation is caused by the consumption of hydrogen and its conversion to helium by thermonuclear processes.)

The masses of the stars are determined for stars which are members of a double or multiple system, either visual or spectroscopic binaries. Application of Newton's laws provides the mass, often with considerable uncertainty. *See* BINARY STAR.

The stars of the main sequence, for which stellar evolution has not yet been a substantial factor in displacement from their normal positions in an H-R diagram, obey a fairly well-established mass-luminosity relationship. However, because of stellar evolution, stars move off the main sequence. If they belong to population I, they essentially move horizontally in an H-R diagram, so that they still obey the mass-luminosity relationship. Population II stars with low metal abundances brighten by factors up to a hundred in the subgiant and red-giant stage. *See* MASS-LUMINOSITY RELATION; STELLAR REVOLUTION.

Stellar rotation. Another important property of stars is their rotation on their axes. The prevalence of wide double stars and of close spectroscopic binaries indicates that a large amount of angular momentum is often contained in the material that condensed to form the stars. In close double systems, revolution and rotation are often synchronous. In single stars, especially those of early spectral type, rotation is rapid.

Composition. The chemical composition of a star can be deduced from the spectrum of its atmosphere or from the theory of its internal structure. There is evidence that stars need not be chemically homogeneous. Products of nuclear reactions may concentrate in the center so that the ratio of helium to hydrogen increases inward. Very slow mixing, however, counteracts the tendency toward complete diffusive separation by gravity of heavy from light elements.

Studies of the composition of stars can be made only in their atmospheres. The ratio of hydrogen to heavy elements is about 8000–12,000 to 1, by number of atoms, for the Sun and young population I stars. For extreme high-velocity stars, old objects of population II, the abundance of the metals may be up to 400 times lower.

Motions. Galactic dynamics and kinematics are an important part of the subject of stellar statistics and are intimately connected with the distribution of the stars in space. The Milky Way galaxy has a mass of about 6×10^{11} Suns, of which only a small fraction is visible from Earth. These stars move in orbits around their common center of gravitation, located about 8500 parsecs (2.6×10^{20} m or 1.6×10^{17} m) from Earth in the constellation Sagittarius. The galaxy is highly flattened by its systematic rotation, the linear rotational velocity being approximately 220 km/s (137 mi/s) at the Sun's distance from the galactic center; this corresponds to a rotation period of 200,000,000 years. *See* GALAXY, EXTERNAL; MILKY WAY GALAXY.

Superposed on the systematic rotation of the galaxy are individual motions of the stars. Each star moves in a somewhat elliptical orbit and therefore shows a peculiar velocity with respect to the local standard of rest, the standard moving in a circular orbit around the galactic center. The Sun has an orbit of small ellipticity and inclination, so that solar motion with respect to the mean of neighboring stars can be detected by analysis of either radial velocities or proper motions of nearby stars.

Spatial distribution. The distribution of stars in space is the subject of studies of stellar statistics and galactic structure. The Milky Way is the dominant feature of the galaxy, even to the eye, and represents the mean place in which the stars are concentrated. It is nearly a great circle, indicating that the Sun lies near the galactic plane.

The number of stars at a given apparent magnitude is a function of galactic latitude and longitude; in the galactic plane the complex structure of interstellar clouds of dust that absorb light and the irregularities of spiral structure produce a somewhat irregular, patchy appearance, with a maximum number of faint stars in the direction of the galactic center in Sagittarius, with subsidiary maxima in Cygnus and Carina. These maxima are probably caused by spiral arms seen lengthwise. [J.L.Gr.]

Star clouds Aggregations of thousands or millions of stars spread over hundreds or thousands of light-years in space. The Milky Way is composed of such star clouds, the heaviest clouds being in the richest parts, such as Cygnus, Sagittarius, Carina, and Scutum. The stars in such clouds may appear unevenly distributed because of the presence of obscuring interstellar dust and gas. *See* MILKY WAY GALAXY. [H.S.H.]

Star clusters Groups of stars held together by gravitational attraction. The two chief types are open clusters, formerly called galactic clusters, containing from a dozen up to many hundreds of stars, and globular clusters, composed of thousands to hundreds of thousands of stars. A relative of the star cluster is the stellar association, a group of dozens or hundreds of relatively young stars spread loosely over a large volume of space. Star clusters are important in outlining the shape and extent of the Milky Way Galaxy and in deriving theories of stellar evolution on the assumption that stars of a given cluster were formed at the same time.

Open clusters lie along the backbone of the Milky Way Galaxy, strongly concentrated to the central plane of the Milky Way. A dozen are visible to the unaided eye, over 1000 are cataloged, and many more must exist. Most open clusters have an asymmetrical appearance. Most stars in open clusters fall along the highly populated branch of the spectrum-luminosity diagram known as the main sequence. The point where this sequence starts furnishes a criterion of the age of the cluster. At one extreme are young clusters, formed in recent geologic times, like the Orion Nebula cluster with an age of about 10^6 years. At the other extreme is NGC 188, with an age of 5×10^9 years, comparable with that of old systems like globular clusters.

Groups of thousands to hundreds of thousands of stars in globular symmetry constitute globular clusters. Though they are scattered widely in galactic latitude, their strong concentration toward the region of Sagittarius-Scorpio led Harlow Shapley in 1917 to postulate this as the center of the Milky Way Galaxy. Several are visible to the unaided eye, like Messier 13, the great cluster in Hercules (see illustration). A total of 131 have been cataloged in the Milky Way Galaxy, including several so far distant that they are really intergalactic. A few dozen more may be undetected. Color-magnitude diagrams differ appreciably from those of open clusters. The spectra show low metal content (ratio of iron to hydrogen), but the ratio varies from cluster to cluster. Many globular clusters probably formed in various parts of the Galaxy during its first 10^9 years. Some globular cluster stars may be the oldest stars in the Galaxy. Their ages, based on low-metallicity evolutionary models, can be as great as 1.41×10^{11} years. *See* MILKY WAY GALAXY; STAR; STELLAR EVOLUTION. [H.S.H.]

The great globular cluster in Hercules, Messier 13. Photographed with 200-in. (5.08-m) telescope. (*California Institute of Technology/Palomar Observatory*)

Star tracker A device for automatically following a star or a planet to measure the angles between specified reference directions and the direction from the celestial body's apparent place to the observer's position. With this data it is then possible by completion of geometric computations, using the difference between the measured angles and the expected angles, to update the observer's position.

Star trackers, sometimes referred to as astrotrackers, are used for celestial navigation of aircraft and spacecraft, for sensing of a space vehicle's angular attitude in flight, and for aiming large telescopes or tracking antennae at stars, planets, and other bodies of scientific interest. A star tracker is an essential part of any automatic celestial navigation system. Any tracker system solves the same celestial problem as a human navigator with a hand-held sextant. *See* CELESTIAL NAVIGATION.

The basic operation of one form of star tracker is shown in the illustration. Starlight enters the objective lens system to form an image at the focal plane. Vibrating reeds are located

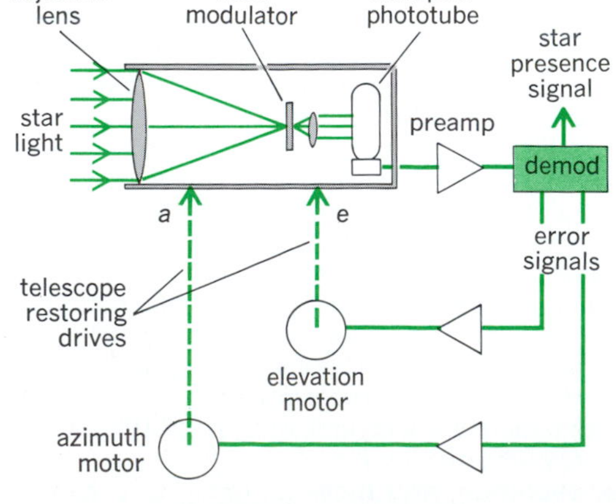

Mechanically driven star tracker.

at the focal plane so as to periodically modulate the image before the light reaches a photodetector. The detector converts the modulated light into an electric voltage or current which is amplified and demodulated to develop the pointing error. In addition, the signal electronics develop a star presence signal to confirm that a star is present and to control the gain of the tracking servoloops. The pointing-error signal indicates the direction and magnitude of telescope error. Each signal (up-down and left-right) actuates a motor as part of a servoloop to redirect the optics to the two respective angles, the elevation angle e and the azimuth angle a; this step centers the star image and nulls the error signal. The photodetector commonly used for this form of star tracker is the photomultiplier. *See* PHOTOMULTIPLIER.

Space-borne star-sensing and -tracking systems have been developed as an important tool in the astrionics of rocket, satellite, and deep-space vehicles. Used primarily as a pointing and attitude reference for the vehicle or its instruments and for the generation of navigation and guidance data, these devices in the United States have been given a vital role in many missions of the National Aeronautics and Space Administration. *See* SPACE NAVIGATION AND GUIDANCE. [E.P.E.]

Starburst galaxy A galaxy that is observed to be undergoing an unusually high rate of formation of stars. It is often defined as a galaxy that, if it continues to form stars at the currently observed rate, will exhaust its supply of star-forming material, the interstellar gas and dust, in a time period that is very short compared to the age of the universe. For a typical starburst galaxy, this gas exhaustion time scale is less than 10^8 years, that is, less than 1% of the age of the universe. Such galaxies must be undergoing a passing burst of star formation.

The term starburst usually also implies that the burst of star formation is occurring in the nuclear regions of the galaxy, because the term was coined to describe a sample of luminous spiral galaxies with bright, pointlike nuclei. However, related objects meet the definition of short gas-exhaustion time scale but exhibit more widespread star formation. These include the nearest, best-studied starburst galaxy, M82.

There are several theories as to why starbursts occur. One likely cause is the interaction between two galaxies as they pass close to, or collide with, one another. The tidal forces generated result in shock-wave compression of the interstellar material and star formation in the compressed clouds. Many starburst galaxies show evidence of interactions, including distorted appearance and long, wispy tails of material. Not all starbursts can be due to interactions, however, since some display no evidence for any recent disturbance. Other mechanisms that are thought responsible for high star-formation rates in galaxies are very strong spiral density waves and central bar instabilities. *See* GALAXY, EXTERNAL. [C.J.L.]

Starch The reserve carbohydrate stored usually in the seeds, roots, or stems of plants. It is second in abundance only to cellulose as a source of carbohydrates. Although starch is widespread in plants, only a few sources are sufficiently abundant to make extraction commercially feasible. These sources are corn, tapioca, potato, sago, waxy maize, wheat, sorghum, rice, and arrowroot. *See* CARBOHYDRATE.

There are two basic types of starch molecules: the linear starch polymer and the branched starch polymer. Amylose molecules vary in size from about 100 to over 1000 glucose units. Amylopectin molecules are three or more times larger. The sizes of the molecules and the amounts of each type present in the starch granule determine the specific properties of the individual starch.

Natural starches find a variety of uses as thickeners in foods and industrial processes. Many times, however, these starches are processed further to obtain viscosity or texture differences of even greater use. [T.A.W.]

Stark effect The effect of an electric field on spectrum lines. The electric field may be externally applied; but in many cases it is an internal field caused by the presence of neighboring ions or atoms in a gas, liquid, or solid. Discovered in 1913 by J. Stark, the effect is most easily studied in the spectra of hydrogen and helium, by observing the light from the cathode dark space of an electric discharge. Because of the large potential drop across this region, the lines are split into several components. For observation perpendicular to the field, the light of these components is linearly polarized.

The linear Stark effect exhibits large, nearly symmetrical patterns. The interpretation of the linear Stark effect was one of the first successes of the quantum theory. According to this theory, the effect of the electric field on the electron orbit is to split each energy level of the principal quantum number n into $2n - 1$ equidistant levels, of separation proportional to the field strength. *See* ATOMIC STRUCTURE AND SPECTRA.

The quadratic Stark effect occurs in lines resulting from the lower energy states of many-electron atoms. The quadratic Stark effect is basic to the explanation of the formation of molecules from atoms, of dielectric constants, and of the broadening of spectral lines.

The intermolecular Stark effect is produced by the action of the electric field from surrounding atoms or ions on the emitting atom. The intermolecular effect causes a shifting and broadening of spectrum lines. The molecules being in motion, these fields are inhomogeneous in space and also in time. Hence the line is not split into resolved components but is merely widened. [F.A.J./W.W.W.]

The quantum-confined Stark effect is the Stark effect observed in structures in which the hydrogenic system is confined in a layer of thickness much less than its normal diameter. This is not practical with atoms, but the effect is observed with excitons in semiconductor quantum-well heterostructures. It is important that quantum-confined Stark shifts can be much larger than the binding energy of the hydrogenic system. The resulting shifts of the exciton optical absorption lines can be used to make optical beam modulators and self-electrooptic-effect optical switching devices. *See* ARTIFICIALLY LAYERED STRUCTURES; ELECTROOPTICS; EXCITON; OPTICAL MODULATORS; SEMICONDUCTOR HETEROSTRUCTURES. [D.A.B.M.]

Static A hissing, crackling, or other sudden sharp noise that tends to interfere with the reception, utilization, or enjoyment of desired signals or sounds. Perhaps the commonest form of static is that heard in ordinary broadcast receivers during electrical storms. Interference in radio receivers caused by improperly operating electric devices in the vicinity is sometimes also called static. *See* SFERICS.

The crackling sounds heard when long-playing plastic phonograph records are played are also called static. These sounds are caused by sudden deflection of the phonograph needle by dust particles, which are attracted by the grooves of the record by surface electric charges caused by friction on dry days. Static appears as momentary white specks in a television picture. *See* ELECTRICAL INTERFERENCE; ELECTRICAL NOISE. [J.Mar.]

Static electricity The study of electric charges at rest and the fields they produce. The fundamental fact of static electricity (or electrostatics) is that similarly electrified bodies repel each other, whereas oppositely electrified bodies attract each other. Coulomb's law of force is the basic quantitative law of electrostatics. *See* COULOMB'S LAW; ELECTRIC CHARGE; ELECTROSTATICS. [R.P.Wi.]

Static var compensator A thyristor-controlled (hence static) generator of reactive power, either lagging or leading, or both. The word var stands for volt ampere reactive, or reactive power. The device is also called a static reactive compensator.

Need for reactive compensation. Reactive power is the product of voltage times current where the voltage and current are 90° out of phase with one another. Thus, reactive power flows one way for one-quarter of a cycle, the other way for the next quarter of a cycle, and so on (in contrast to the real power, or active power, which flows in one direction only). This back-and-forth flow results in no net power being delivered by the generator to the load. However, current associated with reactive power does flow through the conductor and creates extra losses. *See* ALTERNATING CURRENT; ELECTRIC POWER MEASUREMENT; VOLT-AMPERE.

Most loads draw lagging reactive power, which causes electric power system voltage to sag. On the other hand, under light loads, the capacitance of high-voltage lines can create excessive leading reactive power, causing the voltage at some locations to rise above the nominal value. Finally, it is prudent to keep reactive power flows to a minimum in order to allow the lines to carry more active power.

Mechanical versus static compensation. Utilities frequently install capacitors connected from line to ground to compensate for lagging reactive power and reactors connected from line to ground to compensate for leading reactive power. These reactors and capacitors are switched in and out with mechanical switches based on the level of line loading as it varies throughout the day. However, frequent operation of these mechanical switches may reduce their reliability. *See* CAPACITOR; REACTOR (ELECTRICITY).

It is desirable to have a controllable source of reactive power (leading or lagging); and the static var compensator, controlled with static switches, called thyristors, for higher reliability, fulfills this function. It is more expensive than mechanically switched capacitors and reactors (due to the cost of thyristor valves and associated equipment), and hence its use is based on an economic trade-off of benefits versus cost. *See* SEMICONDUCTOR RECTIFIER. [N.G.H.]

Statics The branch of mechanics that describes bodies which are acted upon by balanced forces and torques so that they remain at rest or in uniform motion. This includes point particles, rigid bodies, fluids, and deformable solids in general. Static point particles, however, are not very interesting, and special branches of mechanics are devoted to fluids and deformable solids. For example, hydrostatics is the study of static fluids, and elasticity and plasticity are two branches devoted to deformable bodies. Therefore this article will be limited to the discussion of the statics of rigid bodies in two- and three-space dimensions. *See* BUOYANCY; ELASTICITY; HYDROSTATICS; MECHANICS.

In statics the bodies being studied are in equilibrium. The equilibrium conditions are very similar in the planar, or two-dimensional, and the three-dimensional rigid body statics. These are that the vector sum of all forces acting upon the body must be zero; and the resultant of all torques about any point must be zero. Thus it is necessary to understand the vector sums of forces and torques.

In studying statics problems, two principles, superposition and transmissibility, are used repeatedly on force vectors. They are applicable to all vectors, but specifically to forces and torques (first moments of forces). The principle of superposition of vectors is that the sum of any two vectors is another vector. The principle of transmissibility of a force applied to a rigid body is that the same mechanical effect is produced by any shift of the application of the force along its line of action. To use the superposition principle to add two vectors, the principle of transmissibility is used to move some vectors along their line of action in order to add to their components.

The moment of a force about a directed line is a signed number whose value can be obtained by applying these two rules: (1) The moment of a force about a line parallel to the force is zero. (2) The moment of a force about a line normal to a plane containing the force is the product of the magnitude of the force and the least distance from the line to the line of the force. *See* EQUILIBRIUM OF FORCES; FORCE; TORQUE. [B.DeF.]

Stationary state In quantum mechanics, an energy state for which the probability of any observation is independent of time, that is, stationary. Because stationary states endure in time, they conveniently characterize physical systems. For example, at any instant, it is customary and useful to think that each atom in a discharge tube is in one of its stationary states, with most atoms in the ground state, and the radiation from the tube a measure of the populations of the excited states. *See* ENERGY LEVEL (QUANTUM MECHANICS); EXCITED STATE; GROUND STATE; QUANTUM MECHANICS. [E.G.]

Statistical mechanics That branch of physics which endeavors to explain the macroscopic properties of a system on the basis of the properties of the microscopic constituents of the system. Usually the number of constituents is very large. All the characteristics of the constituents and their interactions are presumed known; it is the task of statistical mechanics (often called statistical physics) to deduce from this information the behavior of the system as a whole.

Scope. Elements of statistical mechanical methods are present in many widely separated areas in physics. For instance, in the classical Boltzmann problem one attempts to explain the thermodynamic behavior of gases on the basis of classical mechanics applied to the system of molecules.

Statistical mechanics gives more than an explanation of already known phenomena. By using statistical methods, it often becomes possible to obtain expressions for empirically observed parameters, such as viscosity coefficients, heat conduction coefficients, and virial coefficients, in terms of the forces between molecules. Statistical considerations also play a significant role in the description of the electric and magnetic properties of materials. *See* BOLTZMANN STATISTICS; INTERMOLECULAR FORCES; KINETIC THEORY OF MATTER.

If the problem of molecular structure is attacked by statistical methods, the contributions of internal rotation and vibration to thermodynamic properties, such as heat capacity and entropy, can be calculated for models of various proposed structures. Comparison with the known properties often permits the selection of the correct molecular structure.

Perhaps the most dramatic examples of phenomena requiring statistical treatment are the cooperative phenomena or phase transitions. In these processes, such as the condensation of a gas, the transition from a paramagnetic to a ferromagnetic state, or the change from one crystallographic form to another, a sudden and marked change of the whole system takes place. *See* PHASE TRANSITIONS.

Statistical considerations of quite a different kind occur in the discussion of problems such as the diffusion of neutrons through matter. In this case, one knows the probability of the various events which affect the neutron, such as the capture probability and scattering cross section. The problem here is to describe the physical situation after a large number of these individual events. The procedures used in the solution of these problems are very similar to, and in some instances taken over from, kinetic considerations. Similar problems occur in the theory of cosmic-ray showers.

It happens in both low-energy and high-energy nuclear physics that a considerable amount of energy is suddenly liberated. An incident particle may be captured by a nucleus, or a high-energy proton may collide with another proton. In either case, there is a large number of ways (a large number of degrees of freedom) in which this energy may be utilized. To survey the resulting processes, one can again invoke statistical considerations. *See* SCATTERING EXPERIMENTS (NUCLEI).

Of considerable importance in statistical physics are the random processes, also called stochastic processes or sometimes fluctuation phenomena. The Brownian motion, the motion of a particle moving in an irregular manner under the influence of molecular bombardment, affords a typical example. The stochastic processes are in a sense intermediate between purely statistical processes, where the existence of fluctuations may safely be neglected, and the purely atomistic phenomena, where each particle requires its individual description. *See* BROWNIAN MOVEMENT; STOCHASTIC PROCESS.

All statistical considerations involve, directly or indirectly, ideas from the theory of probability of widely different levels of sophistication. The use of probability notions is, in fact, the distinguishing feature of all statistical considerations. *See* PROBABILITY; STATISTICS.

Methods. Consider a system of N particles, each of the mass m, contained in a volume V. Call the positions of the particles $x_1, y_1, z_1, \ldots, x_N, y_N, z_N$, their cartesian velocities $v_{x1}, \ldots, v_{zN}$, and their momenta $P_{x1}, \ldots, P_{zN}$. This simplest statistical description concentrates on a discussion of the distribution function $f(x,y,z;v_x,v_y,v_z;t)$. The quantity $f(x,y,z;v_x,v_y,v_z;t).\ (dxdydzdv_xdv_ydv_z)$ gives the (probable) number of particles of the system in those positional and velocity ranges where x lies between x and $x + dx$; v_x between v_x and $v_x + dv_x$, and so on. These ranges are finite.

Observations made on a system always require a finite time; during this time the microscopic details of the system will generally change considerably as the phase point moves. The result of a measurement of a quantity Q will therefore yield the time average, as in Eq. (1). The integral is along the trajectory

$$\overline{Q}_t = \frac{1}{t} \int_0^t Q\,dt \tag{1}$$

in phase space; Q depends on the variables $x_1, \ldots, P_{zN}$, *and t.* to evaluate the integral, one must know the trajectory, which requires the solution of the complete mechanical problem.

Ensembles. J. Willard Gibbs first suggested that instead of calculating a time average for a single dynamical system, one should instead consider a collection of systems, all similar to the original one. Such an ensemble of systems is to be constructed in harmony with the available knowledge of the single system, and may be represented by an assembly of points in the phase space, each point representing a single system. If, for example, one knows the energy of a system precisely, but nothing else, the appropriate representative example would be a uniform distribution of ensemble points over the energy surface, and no ensemble points elsewhere. An ensemble is characterized by a density function $\rho(x_1,\ldots,z_N;\ p_{x1},\ldots,p_{zN};t) \equiv p(x,p,t)$. The significance of this function is that the number of ensemble systems dN_e contained in the volume element $dx_1 \cdots dz_N;\ dp_x \cdots dp_{zN}$ of the phase space (this volume element will be called $d\Gamma$) at time t is as given in Eq. (2).

$$\rho(x,p,t)\,d\Gamma = dN_e \tag{2}$$

The ensemble average of any quantity Q is given by Eq. (3).

$$\overline{Q}_{\text{ens}} = \frac{\int Q\rho\,d\Gamma}{\int \rho\,d\Gamma}$$

The basic idea now is to replace the time average of an individual system by the ensemble average, at a fixed time, of the representative ensemble. Stated formally, one identifies $\overline{Q}_t$ defined by Eq. (1), in which no statistics is involved, with $\overline{Q}_{\text{ens}}$ defined by Eq. (3), in which probability assumptions are explicitly made.

Relation to thermodynamics. It is certainly reasonable to assume that the appropriate ensemble for a thermodynamic equilibrium state must be described by a density function which is independent of the time, since all the macroscopic averages which are to be computed as ensemble averages are time-independent.

The so-called microcanonical ensemble is defined by Eq. (4*a*), where c is a constant, for the energy E between E_0 and $E_0 + \Delta E$; for other energies Eq. (4*b*) holds. By using Eq. (3), one

$$\rho(p,x) = c \tag{4a}$$

$$\rho(p,x) = 0 \tag{4b}$$

may calculate any microcanonical average. The calculations, which involve integrations over volumes bounded by two energy surfaces, are not trivial. Still, one may obtain in this way many of the results of classical Boltzmann statistics. For applications and for the interpretation of thermodynamics, the canonical ensembles is much more preferable. In this case one describes a system which is not isolated but which is in thermal contact with a heat reservoir.

There is yet another ensemble which is extremely useful and which is particularly suitable for quantum-mechanical applications. Much work in statistical mechanics is based on the use of this so-called grand canonical ensemble. In the great ensemble one has a collection of systems; the number of particles in each system is no longer the same, but varies from system to system. The density function $p(N,p,x)\ d\Gamma_N$ gives the probability that there will be in the ensemble a system having N particles, and that this system, in its $6N$-dimensional phase space Γ_N, will be in the region of phase space $d\Gamma_N$. [M.Dr.]

Statistics The field of knowledge concerned with collecting, analyzing, and presenting data. Not only workers in the physical, biological, and social sciences, but also engineers, business managers, government officials, market analysts, and many others regularly use statistical methods in their work. The methods range from simple counting to complex mathematical systems designed to extract the maximum amount of information from very extensive data.

In an important sense statistics may be regarded as a field of application of probability theory. The common problem faced by any worker who must collect, analyze, and present data is that of random variation which prevents repetition of exactly the same result when a measurement is repeated. Statistical methods are employed to assess the magnitude of random variation, to minimize it, to balance it out, to remove it by calculation procedures, and to analyze it by suitably arranged patterns of observation. The theory of probability is concerned with the properties of random variables and hence furnishes the basis for developing techniques for controlling them. *See* PROBABILITY.

Viewing statistics from another direction, it is the science of deriving information about populations by observing only samples of those populations. A population is any well-specified collection of elements. Thus, one may refer to the population of adults in the continental United States viewing television screens at 8:14 P.M. on August 6, 1970. Populations may be finite or infinite. An element of a univariate population is characterized by the value of a random variable which measures some single attribute of interest in the population. Thus, one may be interested in whether or not individuals of the television audience were or were not viewing program A; with each individual one may associate a random variable; let it be X, which takes on the value of 1 if the individual is watching A and 0 if he or she is not. If one were interested in a second characteristic of the elements of the television audience (such as age), one would be dealing with a bivariate population; a third character-

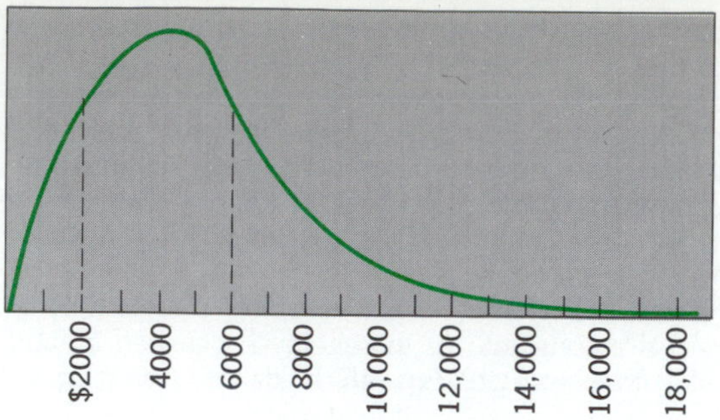

Fig. 1. Distribution of incomes.

istic (such as economic status) would make it a trivariate or, less specifically, a multivariate population.

Distributions. In a univariate population, the population distribution is a curve (function of the random variable which characterizes the elements of the population) from which one can determine the proportion of the population which has elements in a certain range of the random variable. For example, the curve of Fig. 1 provides the distribution of annual incomes of family units in the United States in 1954. The total area under the curve is 1. The area under the curve between any two vertical lines gives the proportion of the families having annual incomes between the two values marked on the horizontal scale by the two vertical lines.

The distribution is also referred to as the distribution function, the density function, the frequency function, or the probability density.

The total area under the distribution curve to the left of each point can also be plotted to give a curve which starts at zero and reaches unity as the variable becomes large; the resulting curve is sometimes called the cumulative distribution function, the probability distribution, or simply the distribution. The cumulative form of the curve of Fig. 1 is shown in Fig. 2; the height of the curve at any point on the horizontal scale equals the area to the left of that point under the curve of Fig. 1 and is the proportion of the population having incomes less than the value at that point. The distribution (in either frequency or cumulative form) gives complete information about the way the characterizing variable is spread through the population.

Population parameters. Populations (or population distributions) are often specified incompletely by certain population parameters. Some of these parameters are location parameters or measures of central tendency; a second class of important parameters consists of measures of dispersion or scale parameters.

The most widely used location parameters are the mean, the median, and the mode. The mean is the average over all the population of the values of the random variable. It is often represented by the Greek letter μ. In mathematical terms, if x is the random variable, $f(x)$ the frequency function for a given population, and $F(x)$ its cumulative form, then the mean is as shown in Eq. (1).

$$\mu = \int_{-\infty}^{\infty} xf(x)\,dx = \int_{-\infty}^{\infty} x\,dF(x) \tag{1}$$

The median, often designated by Med, M, or $X_{.50}$, is a number such that, at most, one-half the values of the variable associated with the elements of the population fall above or below it. The mode is the most frequent value of the random variable; if the frequency function has a unique maximum value, the mode is the value of the random variable at which the frequency function reaches its maximum. Location parameters are numbers near the center of the range over which the random variable of the population varies.

The extent to which a population is scattered on either side of its center is roughly indicated by measures of dispersion such as the standard deviation, the mean deviation, the interquartile range, the range, and sometimes others. The standard deviation is the square root of the mean square of the deviations from the mean; it is usually denoted by the Greek letter σ; σ^2 is called the variance and is expressed by Eq. (2).

$$\sigma^2 = \int_{-\infty}^{\infty} (x-\mu)^2 f(x)\,dx$$

$$= \int_{-\infty}^{\infty} (x-\mu)^2\,dF(x) \tag{2}$$

The mean deviation, shown in Eq. (3), is the average over the

$$\text{Mean deviation} = \int_{-\infty}^{\infty} |x-\mu|\, f(x)\,dx$$

$$= \int_{-\infty}^{\infty} |x-\mu|\, dF(x) \tag{3}$$

population of the deviations from the mean, all taken to be positive. The interquartile range (often denoted by Q) is the difference $X_{.75} - X_{.25}$, where $X_{.75}$ is the value of the random variable such that one-quarter of the population has values larger than $X_{.75}$, and $X_{.25}$ is the number such that one-quarter of the population has values smaller than $X_{.25}$. The three numbers, $X_{.25}$, $X_{.50}$, $X_{.75}$, are called quartiles; these divide the population into quarters. The range is the difference between the largest and the smallest of the population elements.

Sampling. If one examines every element of a population and records the value of the random variable for each, complete information is obtained about the distribution of the random variable in the population, and there is no statistical problem. It is usually impossible or uneconomical to make a complete enumeration (or census) of a population, and one must therefore be content to examine only a part or sample of the population. On the basis of the sample, one draws conclusions about the entire population; the conclusions thus drawn are not certain in the sense that they would likely have been somewhat different if a different sample of the population had been examined. The problem of drawing valid conclusions from samples and of specifying their range of uncertainty is known as the problem of statistical inference.

Simple random sampling is a method of selecting a sample of n elements out of a population of N elements so that all such samples have an equal probability of being drawn. This may be done by selecting a first element at random from the population, then a second element at random from the remaining population, and so on until the n elements are selected.

The observations of a sample, besides providing estimates of population parameters, can also be used to obtain an estimate of the population's frequency function. This estimate is determined by dividing the range of the sample observations into several intervals of equal length L and counting the number of

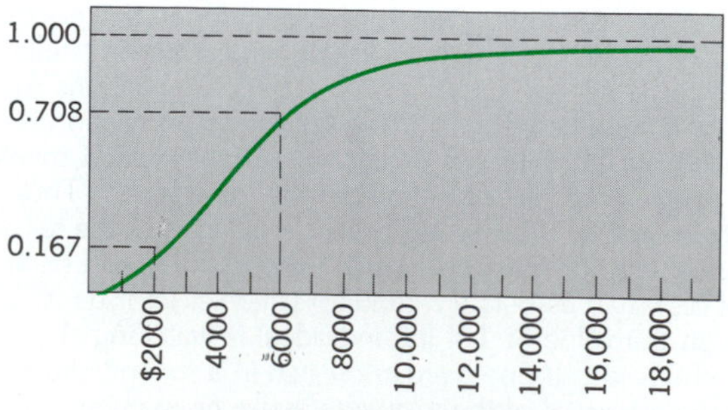

Fig. 2. Cumulative distribution of incomes.

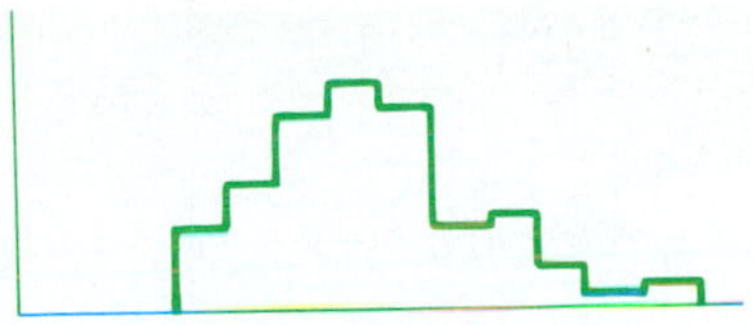

Fig. 3. Histogram.

observations occurring in each interval; these numbers are then divided by nL to determine fractions giving the relative density of the sample occurring in each interval; then on a sheet of graph paper one lays out the intervals on a horizontal axis and plots horizontal lines above each interval at a height equal to the fraction corresponding to the interval; finally the successive plotted horizontal lines are connected by vertical lines to form a broken line curve known as a histogram (Fig. 3).

When a population can be regarded as being made up of several nonoverlapping subpopulations, one may draw a sample from it by drawing a simple random sample from each subpopulation (or stratum); this procedure is called stratified random sampling. Systematic sampling may be regarded as a sampling of units at regular intervals in the population, for example, every tenth unit. When the elements of a population are in mutually exclusive groups called the primary units of the population, then a sample of these primary units might be made, and then, from those selected, a sample of the individual elements would be made. This procedure is called two-stage sampling or subsampling. Multistage sampling can involve more than two stages of sampling.

Many important sampling distributions are derived for random samples drawn from a normal or Gaussian distribution, which is a bell-shaped symmetrical distribution centered at its mean μ.

Estimation. In making an estimate of the value of a parameter of a population from a sample, a function (called the estimator) of the observations is used. For example, for estimating the mean of a normal population the mean of the sample observations is usually taken as the estimator. Another estimator is the average of the two most extreme observations. In fact, there is an infinitude of estimators. The problem of estimation is to find a "good" estimator.

A good estimator may be regarded as one which results in a distribution of estimates concentrated near the true value of the parameter and which can be applied without excessive effort.

Tests of hypotheses. Besides estimation of parameters, another major area of statistical inference is the testing of hypotheses. A hypothesis is merely an assertion that a population has a specific property. The test consists of drawing a sample from the population and determining whether or not it is consistent with the assertion. Very often the hypothesis is a statement about the mean of a population; that it has a given value, that it is the same as that of another population, that it exceeds that of another population by at least 10 units, and the like. Thus, one may be comparing a new blend of gasoline with a current blend, a new drug with a standard one, a new manufacturing process with an existing one.

Experimental design. An experiment is performed to obtain information about the relations between several variables. For example, one may study the effect of storage temperature and duration of storage on the flavor of a frozen food. Three variables (flavor, temperature, and duration) are involved; one (flavor) is called the subject of the experiment; the other two are called factors which influence the subject. Sometimes the factors have intrinsic value in themselves; sometimes they are merely nuisance variables which must be taken into account because it is impossible to perform the experiment without them.

There exist in the statistical literature great numbers of specific experimental designs. These are patterns for making experimental observations; the actual construction of the designs requires quite advanced mathematics based on group theory, finite geometries, and combinatorial analysis. The mathematical problem is to find a pattern from which it is possible to extract the desired information and yet minimize the number of observations.

Regression and correlation. The regression problem is that of estimating certain unknown constants or parameters occurring in a function which relates several variables; the variables may be random or not. By far the most easily handled cases are those in which the function is linear in the unknown parameters, and it is worth considerable effort to transform the function to that form if at all possible. *See* ANALYSIS OF VARIANCE; DISTRIBUTION (PROBABILITY). [A.M.M.]

Staurolite A nesosilicate mineral $(Al,Mg)_{18}(Fe,Al)_4(Si,Al)_8O_{48}H_{\sim 4}$. Staurolite is frequently in crystals, usually a combination of the vertical prism with the basal and side pinacoid. The hardness is 7–7½ on Mohs scale; specific gravity is 3.7. The luster is resinous to vitreous but may be dull when the mineral is impure or altered; the color is reddish brown to black. At St. Gothard, Switzerland, staurolite is found on kyanite in parallel orientation. In the United States it is found in schists in many states, notably New Hampshire, Massachusetts, North Carolina, Georgia, Virginia, and New Mexico. *See* SILICATE MINERALS. [C.S.Hu.; P.B.M.]

Stauromedusae An order of the class Scyphozoa, usually found in circumpolar regions. The egg develops into a planula which can only creep since it lacks cilia. The planula changes into a polyp that metamorphoses directly into a combined polyp and medusa form. The medusa is composed of a cuplike bell called a calyx (medusan part) and a stem or stalk (polyp part) which terminates in a pedal disk (see illustration). The calyx is eight-sided and has eight groups of short, capped ten-

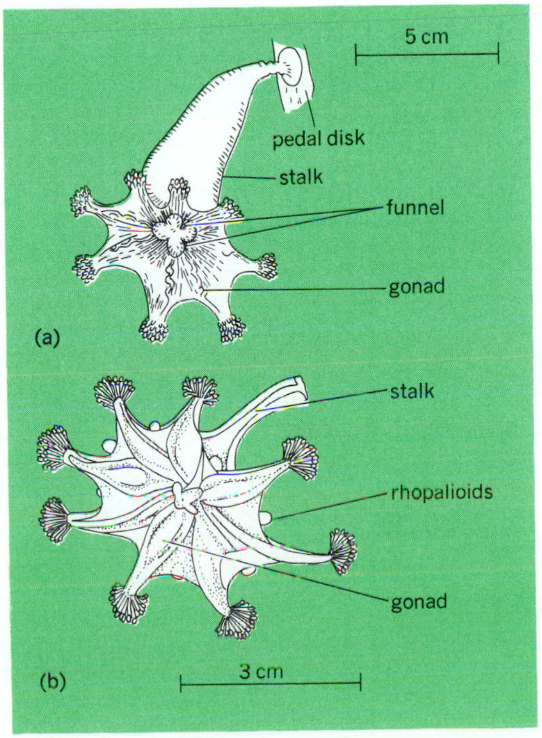

Stauromedusae. (a) *Lucernaria*. (b) *Haliclystus*. (After L. H. Hyman, *The Invertebrates*, vol. 1, McGraw-Hill, 1940)

tacles and eight sensory bodies, called anchors, on its margin. The mouth, situated at the center of the calyx, has four thin lips and leads to the stomach in which gastral filaments are arranged in a row on either side of each interradius. Though sessile, the medusa can move in a leechlike fashion by alternate attachment and release of the pedal disk, using the substratum as an anchor. *See* SCYPHOZOA. [T.U.]

Steady flow Fluid flow in which all conditions at any one point are constant with respect to time. True steady flow is found only with laminar flow. In turbulent flow there are continual fluctuations in velocity and pressure at every point. However, if the values fluctuate on both sides of an average value that is constant, then the flow is mean steady flow. *See* FLUID FLOW; LAMINAR FLOW. [R.L.D.]

Steam Water vapor, or water in its gaseous state. Steam is the most widely used working fluid in external combustion engine cycles, where it will utilize practically any source of heat, that is, coal, oil, gas, nuclear fuel (uranium and thorium), waste fuel, and waste heat. It is also extensively used as a thermal transport fluid in the process industries and in the comfort heating and cooling of space. The universality of its availability and its highly acceptable, well-defined physical and chemical properties also contribute to the usefulness of steam. *See* BOILER; STEAM CONDENSER; STEAM ENGINE; STEAM-GENERATING UNIT; STEAM HEATING; STEAM TURBINE; THERMODYNAMIC CYCLE; THERMODYNAMIC PRINCIPLES; WATER. [T.Ba.]

Steam condenser A heat-transfer device used for condensing steam to water by removal of the latent heat of steam and its subsequent absorption in a heat-receiving fluid, usually water, but on occasion air or a process fluid. Steam condensers may be classified as contact or surface condensers.

In the contact condenser, the condensing takes place in a chamber in which the steam and cooling water mix. The direct contact surface is provided by sprays, baffles, or void-effecting fill. In the surface condenser, the condensing takes place separated from the cooling water or other heat-receiving fluid (or heat sink). A metal wall, or walls, provides the means for separation and forms the condensing surface.

Both contact and surface condensers are used for process systems and for power generation serving engines and turbines. Modern practice has confined the use of contact condensers almost entirely to such process systems as those involving vacuum pans, evaporators, or dryers, and to condensing and dehumidification processes inherent in vacuum-producing equipment such as steam jet ejectors and vacuum pumps. The steam surface condenser is used chiefly in power generation but is also used in process systems, especially in those in which condensate recovery is important. Air-cooled surface condensers are used in process systems and in power generation when the availability of cooling water is limited. *See* STEAM; STEAM TURBINE; VAPOR CONDENSER. [J.F.Se.]

Steam electric generator An electric generator driven by a steam turbine. Such machines are also called turboalternators. To achieve the economy of scale that is the chief advantage of large central-station power plants, turboalternators are designed for higher continuous power ratings than any other class of electrical machines, approaching 2000 MW. The illustration shows the installation of a 618-MW unit with three major equipment components: collector house, generator, and dual steam turbine. The person in the photograph is standing approximately at the point at which the generator shaft connects to the turbine. The collector house contains equipment to supply direct current to the magnetizing windings on the rotor of the generator. *See* STEAM TURBINE.

The special characteristics of turboalternators result from their huge size and the fact that steam turbines are more efficient at high speeds. Turboalternators are synchronous machines; that is, their speed has a fixed relationship to the frequency of the alternating voltage at their electrical terminals. Most steam-driven generators have two poles and run at 3600 rev/min when connected to a 60-Hz system. *See* ALTERNATING CURRENT GENERATOR.

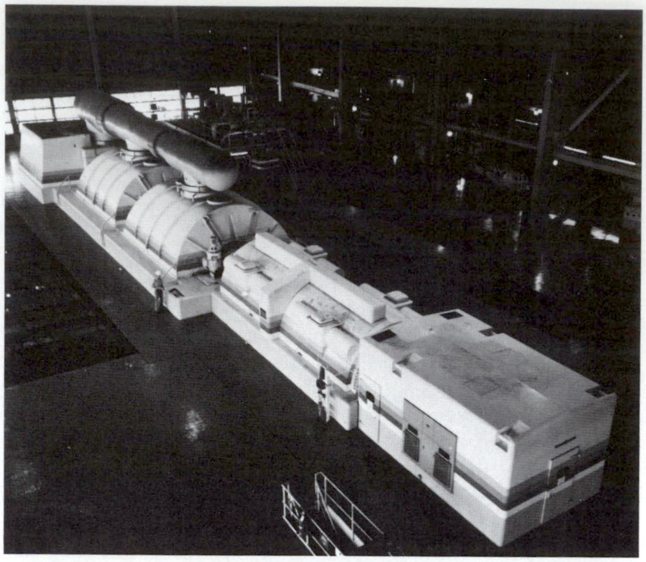

A 618-MW steam turbine-generator power plant showing the three major equipment components, the collector house (foreground), generator, and steam turbine. (*General Electric Co., Public Service Company of Indiana*)

Rotors with protruding or salient poles lack the mechanical strength to withstand the torques and centrifugal forces that prevail in machines operating at such high power levels and at these speeds. Consequently turboalternators have cylindrical rotors. The coils of the magnetizing winding are embedded in deep lots cut into the rotor surface, running parallel to the rotor axis. *See* WINDINGS IN ELECTRIC MACHINERY.

Cooling is critical in such large machines, and passages are usually provided in the winding conductors of the armature for the circulation of highly purified cooling water in direct contact with the conductors. Some machines are designed for water cooling of the rotor windings as well. A substantial amount of heat is removed by circulating hydrogen gas through the gap between the rotor and stator and through passages in the rotor and stator magnetic cores. *See* ELECTRIC ROTATING MACHINERY; GENERATOR. [G.McP.]

Steam engine A machine for converting the heat energy in steam to mechanical energy of a moving mechanism, for example, a shaft. The steam engine can utilize any source of heat in the form of steam from a boiler. Most modern machine elements had their origin in the steam engine: cylinders, pistons, piston rings, valves and valve gear crossheads, wrist pins, connecting rods, crankshafts, governors, and reversing gears. *See* BOILER; STEAM.

The 20th century saw the practical end of the steam engine. The steam turbine replaced the steam engine as the major prime mover for electric generating stations. The internal combustion engine, especially the high-speed automotive types which burn volatile (gasoline) or nonvolatile (diesel) liquid fuel, has completely displaced the steam locomotive with the diesel locomotive and marine steam engines with the motorship and motorboat. Because of the steam engine's weight and speed limitations, it was also excluded from the aviation field. *See* DIESEL ENGINE; GAS TURBINE; INTERNAL COMBUSTION ENGINE; STEAM TURBINE.

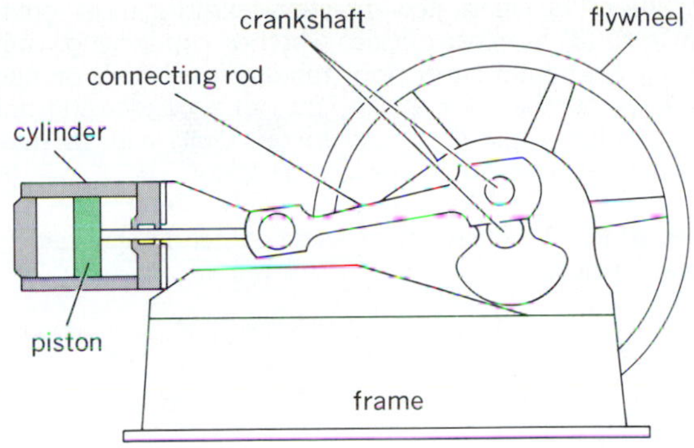

Fig. 1. Principal parts of horizontal steam engine.

A typical steam reciprocating engine consists of a cylinder fitted with a piston (Fig. 1). A connecting rod and crankshaft convert the piston's to-and-fro motion into rotary motion. A flywheel tends to maintain a constant-output angular velocity in the presence of the cyclically changing steam pressure on the piston face. A D slide valve admits high-pressure steam to the cylinder and allows the spent steam to escape (Fig. 2). The power developed by the engine depends upon the pressure and quantity of steam admitted per unit time to the cylinder.

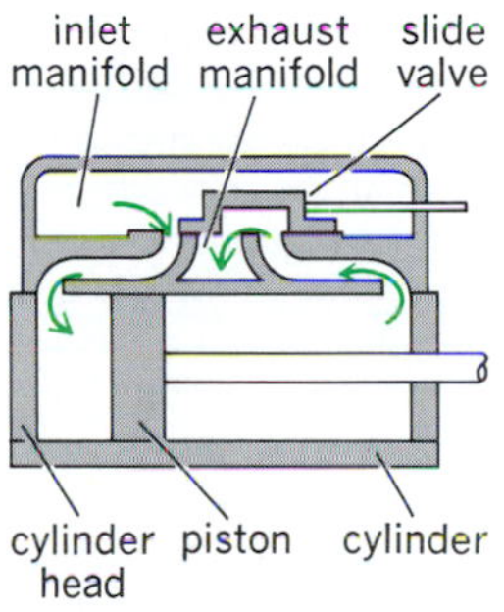

Fig. 2. Single-ported slide valve on counterflow double-acting cylinder.

Engines are classified as single- or double-acting, and as horizontal (Fig. 1) or vertical depending on the direction of piston motion. If the steam does not fully expand in one cylinder, it can be exhausted into a second, larger cylinder to expand further and give up a greater part of its initial energy. Thus, an engine can be compounded for double or triple expansion.

Steam engines can also be classed by functions, and are built to optimize the characteristics most desired in each application. Stationary engines drive electric generators, in which constant speed is important, or pumps and compressors, in which constant torque is important. [T.Ba.]

Steam-generating furnace An enclosed space provided for the combustion of fuel to generate steam. The closure confines the products of combustion and is capable of withstanding the high temperatures developed and the pressures used. Its dimensions and geometry are adapted to the rate of heat release, to the type of fuel, and to the method of firing so as to promote complete burning of the combustible and suitable disposal of ash. In water-cooled furnaces the heat absorbed materially affects the temperature of gases at the furnace outlet and contributes directly to the generation of steam. *See* FURNACE, STEAM-GENERATING UNIT. [G.W.K.]

Steam-generating unit The wide diversity of parts, appurtenances, and functions needed to release and utilize a source of heat for the practical production of steam at pressures to 5000 psi (34 megapascals) and temperatures to 1100°F (600°C); often referred to as a steam boiler for brevity.

The essential steps of the steam-generating process include (1) a furnace for the combustion of fuel, or a nuclear reactor for the release of heat by fission, or a waste heat system; (2) a pressure vessel in which feedwater is raised to the boiling temperature evaporated into steam, and generally superheated beyond the saturation temperature; and (3) in many modern central station units, a reheat section or sections for resuperheating steam after it has been partially expanded in a turbine. This aggregation of functions requires a wide assortment of components, which may be variously employed in the interests, primarily, of capacity and efficiency in the steam-production process. Their selection, design, operation, and maintenance constitute a complex process which, in the limit, calls for the highest technical skill. *See* BOILER; SUPERHEATER; WATER-TUBE BOILER. [T.Ba.]

Steam heating A heating system that uses steam generated from a boiler. The steam heating system conveys steam through pipes to heat exchangers, such as radiators, convectors, baseboard units, radiant panels, or fan-driven heaters, and returns the resulting condensed water to the boiler. Such systems normally operate at pressure not exceeding 15 pounds per square inch gage (psig) or 103 kilopascals gage, and in many designs the condensed steam returns to the boiler by gravity because of the static head of water in the return piping.

In a one-pipe steam heating system, a single main serves the dual purpose of supplying steam to the heat exchanger and conveying condensate from it. Ordinarily, there is but one connection to the radiator or heat exchanger, and this connection serves as both the supply and return. A two-pipe system is provided with two connections from each heat exchanger, and in this system steam and condensate flow in separate mains and branches. When the steam condensate cannot be returned by gravity to the boiler in a two-pipe system, an alternating return lifting trap, condensate return pump, or vacuum return pump must be used to force the condensate back into the boiler. *See* BOILER; COMFORT HEATING. [J.W.J.]

Steam jet ejector A steam-actuated device for pumping compressible fluids, usually from subatmospheric suction pressure to atmospheric discharge pressure. A steam jet ejector is most frequently used for maintaining vacuum in process equipment in which evaporation or condensation takes place. Because of its simplicity, compactness, reliability, and generally low first cost, it is often preferred to a mechanical vacuum pump for removing air from condensers serving steam turbines, especially for marine service.

Two or more stages may be arranged in series, depending upon the total compression ratio required. Two or more sets of series stages may be arranged in parallel to accommodate variations in capacity. Vapor condensers are usually interposed between the compression stages of multistage steam jet ejectors to condense and remove a significant portion of the motive steam and other condensable vapors. [J.F.Se.]

Steam separator A device for separating a mixture of the liquid and vapor phases of water. Steam separators are used in most boilers and may also be used in saturated steam lines to separate and remove the moisture formed because of heat loss.

Steam separators have many forms and may be as fundamental as a simple baffle that utilizes inertia caused by a change of direction. Most modern, high-capacity boilers use a combination of cyclone separators and steam dryers. Cyclone separators slice the steam-water mixture into thin streams so that the steam bubbles need travel only short distances through the mixture to become disengaged, and then they whirl the mixture in a circular path, creating a centrifugal force many times greater than the force of gravity. Steam dryers remove small droplets of water from the steam by providing a series of changes in direction and a large surface area to intercept the droplets. *See* BOILER FEEDWATER; STEAM; STEAM TURBINE. [E.E.Co.]

Steam turbine A machine for generating mechanical power in rotary motion from the energy of steam at temperature and pressure above that of an available sink. By far the most widely used and most powerful turbines are those driven by steam. Until the 1960s essentially all steam used in turbine cycles was raised in boilers burning fossil fuels (coal, oil, and gas) or, in minor quantities, certain waste products. However, modern turbine technology includes nuclear steam plants as well as production of steam supplies from other sources. *See* NUCLEAR REACTOR.

The illustration shows a small, simple mechanical-drive turbine of a few horsepower. It illustrates the essential parts for all steam

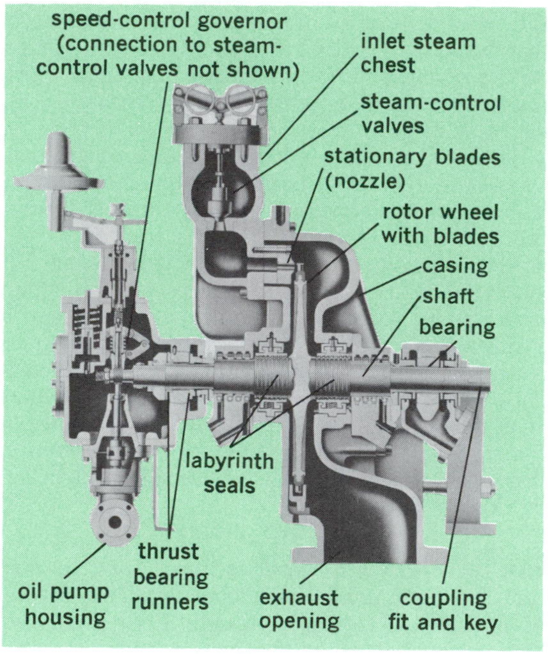

Cutaway of small, single-stage steam turbine. (*General Electric Co.*)

turbines regardless of rating or complexity: (1) a casing, or shell, usually divided at the horizontal center line, with the halves bolted together for ease of assembly and disassembly; it contains the stationary blade system; (2) a rotor carrying the moving buckets (blades or vanes) either on wheels or drums, with bearing journals on the ends of the rotor; (3) a set of bearings attached to the casing to support the shaft; (4) a governor and valve system for regulating the speed and power of the turbine by controlling the steam flow, and an oil system for lubrication of the bearings and, on all but the smallest machines, for operating the control valves by a relay system connected with the governor; (5) a coupling to connect with the driven machine; and (6) pipe connections to the steam supply at the inlet and to an exhaust system at the outlet of the casing or shell.

Steam turbines are ideal prime movers for driving machines requiring rotational mechanical input power. They can deliver constant or variable speed and are capable of close speed control. Drive applications include centrifugal pumps, compressors, ship propellers, and, most important, electric generators.

Steam turbines are classified (1) by mechanical arrangement, as single-casing, cross-compound (more than one shaft side by side), or tandem-compound (more than one casing with a single shaft); (2) by steam flow direction (axial for most, but radial for a few); (3) by steam cycle, whether condensing, noncondensing, automatic extraction, reheat, fossil fuel, or nuclear; and (4) by number of exhaust flows of a condensing unit, as single, double, triple flow, and so on. Units with as many as eight exhaust flows are in use. *See* TURBINE. [F.G.B.]

Stearate A salt (soap) or ester of stearic acid having the general formula

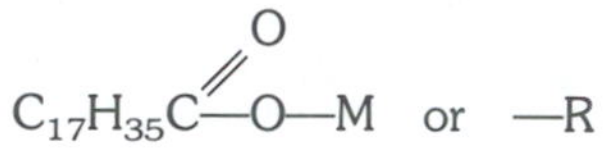

and formed by replacing the carboxylic hydrogen by a metal (M) to give a salt, or by an organic radical (R) to give an ester. Stearates occur in substantial amounts in animal and vegetable fats. The esters of long-chain alcohols are known as waxes. Other esters of monohydric and polyhydric alcohols are used in cosmetics, lacquers, nonionic surface-active agents, and plasticizers. Alkali-metal salts are major components of toilet and laundry soaps. Other metal salts are used in paints, waterproofing, pharmaceuticals, cosmetics, fungicides, and lubricant and grease additives. *See* FAT AND OIL; FAT AND OIL (FOOD); SOAP; WAX, ANIMAL AND VEGETABLE. [E.H.H.]

Steel Any of a great number of alloys that contain the element iron as the major component and small amounts of carbon as the major alloying element. These alloys are more properly referred to as carbon steels, and make up well over 90% of the tonnage of steels produced throughout the world. Small amounts, generally on the order of a few percent, of other elements such as manganese, silicon, chromium, molybdenum, and nickel may also be present in carbon steels. However, when large amounts of alloying elements are added to iron to achieve special properties, other designations are used to describe the alloys. For example, large amounts of chromium, over 12%, are added to produce the important groups of alloys known as stainless steels. *See* STAINLESS STEEL.

Low-carbon steels, sometimes referred to as mild steels, usually contain less than 0.25% carbon. These steels are easily hot-worked and are produced in large tonnages for beams and other structural applications. The medium-carbon steels contain between 0.25 and 0.70% carbon, and are most frequently used for machine components that require high strength and good fatigue resistance. High-carbon steels contain more than 0.7% carbon and are in a special category because of their high hardness and low toughness. This combination of properties makes the high-carbon steels ideal for bearing applications where wear resistance is important and the compressive loading minimizes brittle fracture that might develop on tensile loading. *See* FERROALLOY; HEAT TREATMENT (METALLURGY); IRON; IRON ALLOYS; STEEL MANUFACTURE. [G.Kr.]

Steel manufacture The sequence through which blast-furnace iron, scrap iron, and alloying additions are processed to combine and to separate them into desired compositions of steel and residual slag. The manufacture of steel is a tremendous industry with extensive ramifications resulting from the complex interrelation of technical and economic considerations.

Chemically, all steelmaking processes may be classified as acid or basic, depending upon the refractory and slag combination. Acid processes use silica (SiO_2) refractories throughout and are able to accommodate slags that become saturated with this component under operating conditions. These acid systems can be used to eliminate C, Mn, and Si from the charge, but require select raw materials within final steel specifications for P and S. Basic processes use magnesite (MgO) or equivalent refractories in the portions of the furnace that contact molten slag and metal, and accommodate lower silica slags with compensating amounts of lime (CaO). These systems can eliminate C, Mn, and Si as effectively as can the acid systems; they will also eliminate P and appreciable amounts of S.

Technologically, processes can be grouped into three main types: (1) pneumatic, (2) open-hearth, and (3) electric. In pneumatic processes all heat is derived from the initial heat content of the charge materials, principally molten, and the thermochemical balance of the refining reactions; the selective oxidation of the refining is accomplished by blowing air or commercial oxygen. In open-hearth processes the major source of heat is the combustion of fuel (usually gas or oil); this in turn depends for its success on the regenerative principle of preheating air to attain steelmaking temperatures efficiently. In electric processes the major source of heat is electric current (arc or resistance or both). Because this heat can be produced in the presence or absence of oxygen, electric furnaces can operate in a neutral or nonoxidizing atmosphere or vacuum, and thus are either preferred or required for alloys with significant amounts of easily oxidized elements and for other special grades where improved control of gases is required.

To use available raw materials and heat sources effectively for particular grades of steel, steelmaking processes of the pneumatic, open-hearth, and electric types exist, and any of these may be either acid or basic in their chemistry. In many cases, there is an overlapping or combination of the above principles in a single process. *See* STEEL. [G.D.]

Stellar evolution The large-scale, systematic, and irreversible changes with time of the structure and composition of a star. Stars are born, age, and die much like living beings, but the time scales involved are incomparably longer: for stars like the Sun, the lifetimes are on the order of 10^{10} years, and even for the most massive stars, whose life expectancy is much shorter, the life span is still on the order of 10^7 or 10^6 years. As a consequence, stellar evolution cannot be observed directly, except for rare cases of objects "caught" at stages of unusually fast structural changes.

Evidence on stellar evolution is therefore mainly indirect. Many stars are observed at various stages of evolution; the task is to identify these stages. It is useful to plot two of the most important stellar characteristics: luminosity (the total amount of energy emitted per second) and effective temperature (which represents the stars' surface conditions) against each other in a Hertzsprung-Russell diagram, as is done schematically in the illustration. Most stars will fall on the main sequence, which runs diagonally across the diagram from hot and luminous stars in the upper left corner to cool and faint stars in the lower right corner. In addition, there are two fairly distinct groups: the rather cool but very luminous giants (and even more luminous supergiants) and the fairly hot but quite faint white dwarfs.

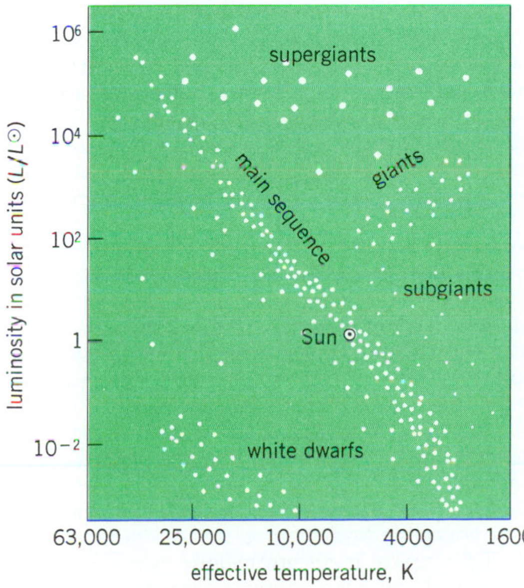

Hertzsprung-Russell diagram.

Stars are believed to be formed from condensations in the interstellar medium. Once a reasonably massive initial condensation forms, it attracts surrounding particles by its force of gravitation. Although other forces may also assist, gravitation appears to be the force that builds stars. However, once they have formed, gravitation tends to destroy them. *See* GRAVITATION.

A star's life is a perpetual struggle against self-destruction, namely collapse under the effect of self-gravitation. At the outset, the object (called a protostar) is little more than a huge contracting cloud of cold material. Its contraction accelerates all the time, and outright would collapse into a black hole if forces were not generated to counteract the gravitational contraction. Such a force is the thermal pressure of the gas. As the particles fall closer together, their potential energy is converted into kinetic energy of motion, and redistributed in numerous collisions; the resulting random motions are perceived as rising temperature of the gas, and associated with it is the pressure which eventually begins to balance gravitation.

A temperature gradient is established in the star in the sense that temperature steadily decreases from the center to the surface. According to the second law of thermodynamics, energy must flow down the temperature gradient. A star must therefore shine simply because it is hotter inside, even if it does not possess adequate energy sources. Because of the stars' enormous masses, gravitational potential energy is a large source of radiative energy, and the Sun could shine at its present luminosity for about 10^8 years without nuclear sources.

When the central temperature of a contracting star reaches about 5×10^6 to 1×10^7 K (9×10^6 to 1.8×10^7°F), thermonuclear reactions are ignited in which hydrogen is converted into helium. The onset of thermonuclear reactions provides sufficient heat to keep the pressure sufficiently high to halt the star's contraction. The star enters a long and quiet period of its life, and is now a main-sequence star. Eventually, however, all the available hydrogen in its interior will be consumed. As the energy source at the star's center vanishes, gravitation prevails again, and contraction resumes. Contraction heats up the star everywhere, but most important is the rise in temperature in the layers immediately surrounding the helium-rich core in which hydrogen has been exhausted. In a relatively thin shell, which still contains hydrogen, the temperature will rise again high enough to ignite the thermonuclear conversion of hydrogen into helium. Instead of a nuclear-burning core, the star now has a nuclear-burning shell. This shell-burning stage is short, since the energetically inactive core cannot maintain a sufficiently high temperature to support the weight of the outer layers by gas pressure, and begins to contract again. However, this time the star is no longer a homogeneous body: the nuclear burning shell acts as a "watershed" for internal motions, which are roughly speaking, reflected on it. Thus, while the core is contracting, the envelope outside the nuclear-burning shell begins to expand. The size of the star begins to increase, but at the same time the outer layers cool off as they expand, there is a drop in the surface temperature, and the color changes from white to yellow to red: the star evolves rapidly into a red giant. *See* GIANT STAR; THERMONUCLEAR REACTION.

If massive enough, the star reaches, in its contracting core, a temperature of or above 10^8 K (1.8×10^8°F). Then a new thermonuclear reaction starts in the core, namely, a transformation of three helium nuclei into one nucleus of carbon. This new energy source again temporarily stabilizes the star and halts core contraction and envelope expansion, and a second quiet stage ensues. As nuclear fuel, helium is much inferior to hydrogen, and is therefore consumed much faster. Another contraction of the core follows, accompanied eventually by an enormous expansion of the outer layers, and the star becomes a red supergiant, with a radius of 500 solar radii or so.

Sooner or later the star runs out of energy, and it may happen so abruptly that the inevitable gravitational contraction changes to a real collapse. In a matter of hours, the cool material of the envelope falls quite close to the center and heats up enormously, and thermonuclear reactions ignite instantaneously. The implosion suddenly changes into an explosion, which releases a large amount of luminous energy but also shafters the star to pieces. A supernova is observed. *See* GRAVITATIONAL COLLAPSE; SUPERNOVA.

The final evolutionary stage of a star of small mass is one of complete electron degeneracy. The star does not generate energy, and shines only by gradually releasing the heat energy accumulated at earlier stages in the random motions of the particles. At the early stages it is still hot enough to maintain a fairly high surface temperature, and it is observed as a white dwarf. The low luminosity, combined with high effective temperature, implies that white dwarfs are very small stars. Only when a solar mass is squeezed into the size of the Earth does the complete electron degeneracy set in. The radius of a degenerate dwarf is smaller for a larger mass of the dwarf, and theoretical calculations show that the radius tends to zero when the mass approaches about 1.4 solar masses (this value is known as the Chandrasekhar limit). Beyond that limit no stable configuration is possible for a degenerate dwarf, since the pressure of the degenerate electrons is not strong enough to counteract self-gravitation. By the sheer weight of the material, electrons are squeezed into atomic nuclei and combine with protons to form neutrons. This permits further drastic contraction of the star, which continues until the neutrons are squeezed to such a small volume per particle that the uncertainty principle begins to operate again, this time on the degenerate neutron gas, and generates a new strong pressure opposing further contraction. A star can stabilize again, and exist for a long time as a neutron star. At least some neutron stars are observed as pulsars. *See* NEUTRON STAR; PULSAR; WHITE DWARF STAR.

One way of producing neutron stars is the supernova explosion. The supernova observed in A.D. 1054 by Chinese astronomers in Taurus left behind a nebulosity (Crab Nebula) surrounding a pulsar. It is not clear if all supernovae leave a stellar remnant. If the mass of the remnant is larger than about 3 solar masses, not even the pressure of the degenerate neutrons can stop its collapse. Therefore it has been concluded that such an object would collapse without a limit and become a black hole. *See* BLACK HOLE; STAR. [M.Pl.]

Stellar population One of the categories into which stars may be classified, based on their place in the evolution of the galaxy containing them. The stellar component of the Milky Way Galaxy consists of three populations: the thin disk, the thick disk, and the halo.

The thin disk, originally referred to as population I, is the youngest component of the galactic stellar population. Still actively forming massive stars from molecular clouds, it is confined to within about 0.35 kiloparsec of the plane. (1 kpc equals 3300 light-years or 3.1×10^{16} km or 1.9×10^{16} mi.) All of the stars have metallicities lying between about one-fifth and twice the solar value, and star formation appears to have remained constant in this population for about the past 8×10^9 years. One reason for the relatively small thickness of the disk is the low velocity dispersion of the component stars; their motion is completely dominated by the differential rotation of the disk. These stars are found associated with H II regions and OB associations as well as open clusters. *See* INTERSTELLAR MATTER; MOLECULAR CLOUD; SUPERNOVA.

The thick disk is an older population, approximately $9–10 \times 10^9$ years, roughly corresponding to the range between what was once called population II and population I. Its metallicity lies between about one-tenth and one-third of the solar value. The stars in this population are distributed over greater distances from the plane, up to 1.5 kpc, and have correspondingly larger velocity dispersion. This population also includes globular clusters and subdwarfs that overlap at the lowest end of the abundances with the properties of the halo globulars, although the system of old disk globulars is distributed differently than those of the halo.

Lying around the disk and the nuclear spheroidal bulge, there is a halo, roughly corresponding to the original population II, that extends to considerable distances from the plane, some as distant as 30 kpc. This population has an age of order $10–15 \times 10^9$ years. The stars in this region have very large velocity dispersions and do not appear to participate in the differential rotation as much as other stars. Their metallicities are all lower than about one-twentieth that of the Sun and may extend down to 10^{-3} of the solar value. The most metal-poor globular clusters belong to this population. This stellar halo is not the same as the dark matter halo, but is probably embedded within it. *See* MILKY WAY GALAXY; STAR. [S.N.S.]

Stem The organ of vascular plants that usually develops branches and bears leaves and flowers. On woody stems a branch that is the current season's growth from a bud is called a twig. The stems of some species produce adventitious roots. *See* ROOT (BOTANY).

General characteristics. While most stems are erect, aerial structures, some remain underground, others creep over or lie prostrate on the surface of the ground, and still others are so short and inconspicuous that the plants are said to be stemless, or acaulescent. When stems lie flattened immediately above but not on the ground, with tips curved upward, they are said to be decumbent, as in juniper. If stems lie flat on the ground but do not root at the nodes (joints), the stem is called procumbent or prostrate, as in purslane. If a stem creeps along the ground, rooting at the nodes, it is said to be repent or creeping, as in ground ivy.

Most stems are cylindrical and tapering, appearing circular in cross section; others may be quadrangular or triangular.

Herbaceous stems (annuals and herbaceous perennials) die to the ground after blooming or at the end of the growing season. They usually contain little woody tissue. Woody stems (perennials) have considerable woody supporting tissue and live from year to year. A woody plant with no main stem or trunk, but usually with several stems developed from a common base at or near the ground, is known as a shrub. [N.A.]

External features. A shoot or branch usually consists of a stem, or axis, and leafy appendages. Stems have several distinguishing features. They arise either from the epicotyl of the embryo in a seed or from buds. The stem bears both leaves and buds at nodes, which are separated by leafless regions or internodes, and sometimes roots and flowers (see illustration).

The nodes are the regions of the primary stem where leaves and buds arise. The number of leaves at a node is usually specific for each plant species. In deciduous plants which are leafless during winter, the place of former attachment of a leaf is marked by the leaf scar. The scar is formed in part by the abscission zone formed at the base of the leaf petiole. The stem regions between nodes are called internodes. Internode length varies greatly among species, in different parts of the same stem, and under different growing conditions.

Lenticels are small, slightly raised or ridged regions of the stem surface that are composed of loosely arranged masses of cells in the bark. Their intercellular spaces are continuous with those in the interior of the stem, therefore permitting gas exchange similar to the stomata that are present before bark initiation.

There are three major types of stem branching: dichotomous, monopodial, and sympodial. Dichotomy occurs by a division of the apical meristem to form two new axes. If the terminal bud of an axis continues to grow and lateral buds grow out as branches, the branching is called monopodial. If

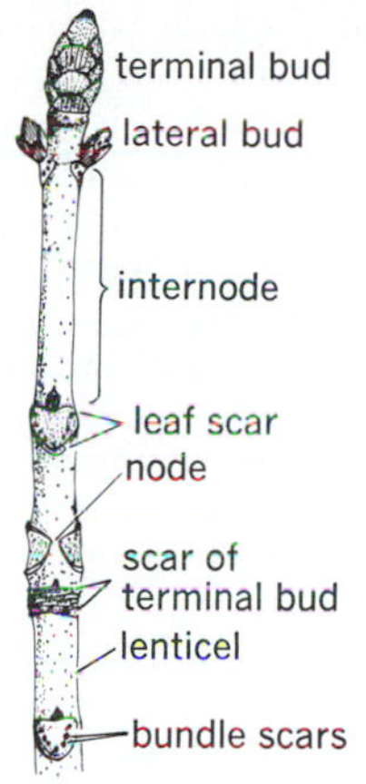

Winter woody twig (horse chestnut) showing apical dominance. (*After E. W. Sinnott and K. S. Wilson, Botany: Principles and Problems, 5th ed., McGraw-Hill, 1955*)

the apical bud terminates growth in a flower or dies back and one or more axillary buds grow out, the branching is called sympodial. Often only one bud develops so that what appears to be single axis is in fact composed of a series of lateral branches arranged in linear sequence.

The large and conspicuous stems of trees and shrubs assume a wide variety of distinctive forms. Columnar stems are basically unbranched and form a terminal leaf cluster, as in palms, or lack obvious leaves, as in cacti. Branching stems have been classified either as excurrent, when there is a central trunk and a conical leaf crown, as in firs and other conifers, or as decurrent (or deliquescent), when the trunk quickly divides up into many separate axes so that the crown lacks a central trunk, as in elm. *See* TREE. [J.B.F.]

Internal features. The stem is composed of the three fundamental tissue systems that are found also in all other plant organs: the dermal (skin) system, consisting of epidermis in young stems and peridem in older stems of many species; the vascular (conducting) system, consisting of xylem (water conduction) and phloem (food conduction); and the fundamental or ground tissue system, consisting of parenchyma and sclerenchyma tissues in which the vascular tissues are embedded. The arrangement of the vascular tissues varies in stems of different groups of plants, but frequently these tissues form a hollow cylinder enclosing a region of ground tissue called pith and separated from the dermal tissue by another region of ground tissue called cortex. *See* CORTEX (PLANT); EPIDERMIS (PLANT); PARENCHYMA; PHLOEM; PITH; SCLERENCHYMA; XYLEM.

Part of the growth of the stem results from the activity of the apical meristem located at the tip of the shoot. The derivatives of this meristem are the primary tissues; epidermis, primary vascular tissues, and the ground tissues of the cortex and pith. In many species, especially those having woody stems, secondary tissues are added to the primary. These tissues are derived from the lateral meristems, oriented parallel with the sides of the stem: cork cambium (phellogen), which gives rise to the secondary protective tissue periderm, which consists of phellum (cork), phellogen (cork cambium), and phelloderm (secondary cortex) and which replaces the epidermis; and vascular cambium, which is inserted between the primary xylem and phloem and forms secondary xylem (wood) and phloem. *See* APICAL MERISTEM; LATERAL MERISTEM.

The vascular tissues and the closely associated ground tissues—pericycle (on the outer boundary of vascular region), interfascicular regions (medullary or pith rays), and frequently also the pith—may be treated as a unit called the stele. The variations in the arrangement of the vascular tissues serve as a basis for distinguishing the stelar types. The word stele means column and thus characterizes the system of vascular and associated ground tissues as a column. This column is enclosed within the cortex, which is not part of the stele. *See* PERICYCLE. [J.E.Gu.]

Stenolaemata A class of bryozoans. Entirely marine, stenolaemates have lophophores which are circular in basal outline, and zooecia which are tubular or prismatic, gradually tapering to their proximal ends, and are short and fat to long and slender. *See* BRYOZOA.

Stenolaemate colonies vary greatly in size and shape, some being small and delicate and others large and massive. Many colonies are regionated or divisible into distinct endozone and exozone portions. The individual zooids within a stenolaemate colony may be relatively isolated, loosely bundled, or compactly fused.

The Stenolaemata include several thousand species ranged among six orders, the Fenestrata, Cryptostomata, Rhabdomesonata, Trepostomata, Cystoporata, and Cyclostomata. First appearing late in the Early Ordovician, stenolaemates expanded quickly to dominate Paleozoic bryozoan assemblages from the mid-Ordovician through the Permian; much reduced since then, they were locally important during the Jurassic and Cretaceous, and some cyclostomes persist today as evolutionary remnants of this formerly dominant class. *See* CRYPTOSTOMATA; CYCLOSTOMATA (BRYOZOA); CYSTOPORATA; TREPOSTOMATA. [R.J.Cu.]

Stenurida A Paleozoic order of the subphylum Asterozoa, order Ophiuroidea, comprising the brittle stars. Stenurids have a double row of plates (ambulacra) that abut across the arm axis either directly opposite one another or slightly offset. In contrast, modern ophiuroids have a single series of axial arm plates termed vertebrae. In stenurids, as in modern ophiuroids, lateral plates are present at the sides of ambulacrals, and prominent lateral spines are typical. Stenurids lack the dorsal and ventral arm shields that are found in most ophiuroids. The content of the order is poorly established, and fewer than 10 genera are known.

The relationship among ophiuroids, asteroids, and all other echinoderms provide an enduring problem in invertebrate evolution. Developmental and other studies based on modern organisms imply that asteroids and ophiuroids are not closely related within the echinoderms. Stenurid morphology, in contrast suggests a close common ancestry for the two. *See* ECHINODERMATA; OPHIUROIDEA. [D.B.B.]

Stepping motor A device which operates from electrical power to produce mechanical rotation and which, unlike a conventional motor, does not produce a speed dependent on the signals applied to its inputs but steps from one angular position to another as determined by which of its inputs are energized. This means that the stepping motor can serve as a link between any digital device and a world characterized by analog processes. The stepping motor has found extensive use as an interface between computers and mechanical devices. They are used extensively in equipment which couples to computers, such as printers, plotters, and magnetic memory disks. The main advantage of the stepping motor lies in the fact that it can provide very accurate mechanical positioning without any feedback. This means that, in general, a stepping motor and amplifier can be built more economically than other devices to perform the job of converting digital signals to mechanical rotation. For example, in addition to computer interfaces, stepping motors can be used in automobile control systems, home appliances, and even watches. *See* COMPUTER GRAPHICS; DIGITAL-TO-ANALOG CONVERTER.

There are three major types of stepping motors, the permanent-magnet stepping motor, the variable-reluctance motor, and a cam-operated motor. This last type relies on magnetic forces to move the output to mechanical detent positions. The first two types have the highest performance and are most commonly used. *See* MOTOR. [C.K.T.]

Stereochemistry The study of the three-dimensional arrangement of atoms or groups within molecules and the properties which follow from such arrangement. Molecules that have identical molecular structures but differ in the relative spatial arrangement of component parts are stereoisomers. Inorganic and organic compounds exhibit stereoisomerism. Examples are structures (I)–(VIII).

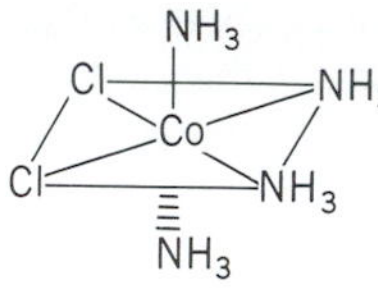
(I)

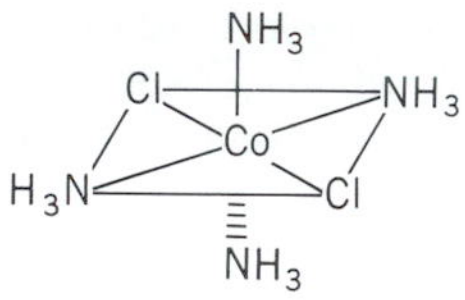
(II)

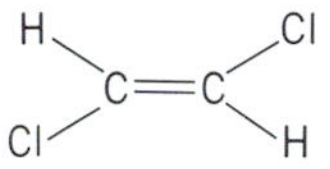
(III)

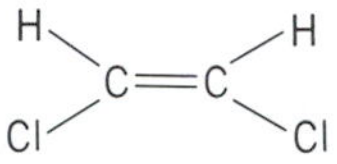
(IV)

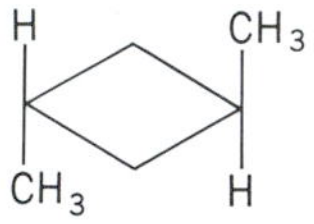
(V)

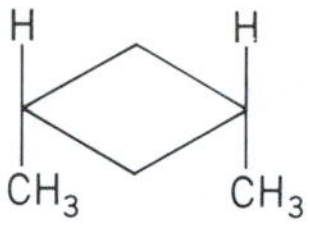
(VI)

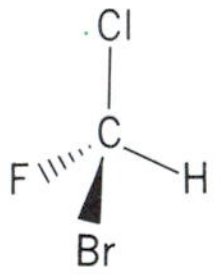
(VII)

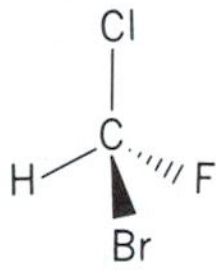
(VIII)

The nature of the stereochemistry of a molecule is determined by its symmetry. The symmetry elements to be considered are: planes of symmetry, axes of symmetry, centers of symmetry, and reflection or mirror symmetry. Two types of stereoisomers are known. Those such as (VII) and (VIII), which are devoid of reflection symmetry—which cannot be superimposed on their image in a mirror—are called enantiomers. All other stereoisomers, such as the pairs (I)–(II), (III–(IV), and (V)–(VI), are called diastereomers. The configuration of a stereoisomer designates the relative position of the atoms associated with a specific structure. The structures of stereoisomers (I) and (II) differ only in configuration. The same is true for (III) and (IV), (V) and (VI), and (VII) and (VIII). *See* ENANTIOMER.

[S.H.W.]

Stereoisomer One of a set of compounds whose molecular formulas and atomic arrangement are identical but which differ in their spatial relationships. Stereoisomers are divided into two types: enantiomers, which are mirror images of each other, and diastereomers, which are not mirror images of each other.

Enantiomers are pairs of isomers whose molecules are nonsuperimposable mirror images. They are also known as enantiomorphs. One molecular configuration of such dissymmetric substances is capable of rotating plane-polarized light to the right, dextro or (+) form, while the mirror image rotates the light equally to the left, levo or (−) form. Each member of such an enantiomeric pair (optical isomers) possesses identical chemical and physical properties except for interaction with other dissymmetric systems, that is, other optically active substances or plane-polarized or circularly polarized light. With the dextro form of another dissymmetric system, a given (+) form behaves exactly as the corresponding (−) form does with the levo form. *See* OPTICAL ACTIVITY.

[W.R.V.]

Diastereomers are pairs of stereoisomers that are not related as object and mirror image; they are also known as diastereoisomers. The stereoisomers need not be optically active. For example, *cis*- and *trans*-2-butene are geometrical isomers related as diastereomers. A compound that contains two or more chiral centers may be converted into a diastereomer by changing the configuration at one chiral center. In this case, the various stereoisomers are composed of pairs of enantiomers (which are related as mirror images), and each pair is related to the other parts as diastereomers. When a compound contains chiral centers, the maximum number of stereoisomeric forms is 2^n, where n is equal to the number of chiral centers. If a compound containing two or more chiral centers contains a mirror plan, the compound will not be optically active. Such compounds are called meso isomers.

Diastereomers can differ in chemical and physical properties and thus may be physically separated. This property may be used to separate enantiomers. A racemic mixture of enantiomers A(+) and A(−) is treated with an optically active reagent [B(+), for example] to provide diastereomers A(+)B(+) and A(−)B(+). The diastereomers are then separated and the reaction reversed to provide the resolved optically active enantiomers A(+) and A(−). The process is called resolution. *See* ASYMMETRIC SYNTHESIS; MOLECULAR ISOMERISM; RACEMIZATION; STEREOCHEMISTRY.

[M.M.Mid.]

Stereophonic radio transmission The standard method for stereo broadcast over frequency-modulation (FM) transmitters, utilizing means to transmit a signal which can be received without impairment, by the usual monaural FM receiver as well. *See* STEREOPHONIC SOUND.

In order for a stereo FM receiver to separate the sounds between the right and left microphones, the output from the right microphone is inverted in phase and then added to that of the left microphone. The voltage thus obtained is the difference voltage between the two microphones and is used to amplitude-modulate a supersonic subcarrier (38 kHz) in double-sideband suppressed carrier mode. This audio signal is inserted together with the sum signal to modulate the transmission. *See* SINGLE SIDEBAND.

The final process in the receiver for retrieval of the tones from the left and right microphones is to again take the sum, this time of the left plus right and the difference left minus right, and the difference between these pairs. *See* FREQUENCY MODULATION; FREQUENCY-MODULATION RADIO; RADIO BROADCASTING.

[W.Lyo.]

Stereophonic sound A system of sound recording or transmission in which two or more microphone channels are connected to corresponding loudspeakers in the listening room in order to create a sound field at the listener's ears which is perceived to be very much like that in the original space. Binaural is a special form of stereophonic sound in which two microphones replace the ears of a dummy head and are connected through an amplifier/recorder system to a pair of headphones worn by the listener. Quadraphonic is a special case of stereo in which four channels and loudspeakers are employed. *See* BINAURAL SOUND; QUADRAPHONIC SOUND SYSTEM.

There are three basic microphone techniques for stereophonic pickup: coincident, spaced apart, and individual instrument (also called close milking). The techniques are sometimes combined.

The coincident technique employs two microphones located very close together, often enclosed in the same case. Usually the two microphones will have cardioid pickup patterns, or one will be a cardioid and the other a figure-eight. The coincident technique relies only on differences in sound intensity to determine position of the phantom images. *See* MICROPHONE.

By placing the microphones several feet apart, it is possible to provide a good pickup of a large group without moving the microphones as far into the reverberant field. The spaced-apart technique provides both time and intensity cues.

Beginning in the late 1960s it became common for most popular music to be recorded in an acoustically dead studio using a large number of microphones, each located close to one instrument or small group of instruments. By proportionally mixing the microphone outputs between the two stereo channels, it is possible to provide the intensity cues necessary for perceiving phantom images. No time cues are produced. *See* SOUND RECORDING; SOUND-REPRODUCING SYSTEMS. [R.Bro.]

Stereoscopy The phenomenon of simultaneous vision with two eyes, producing a visual experience of the third dimension, that is, a vivid perception of the relative distances of objects in space. In this experience the observer seems to see the space between the objects located at different distances from the eyes.

Stereopsis, or stereoscopic vision, is believed to have an innate origin in the anatomic and physiologic structures of the retinas of the eyes and the visual cortex. It is present in normal binocular vision because the two eyes view objects in space from two points, so that the retinal image patterns of the same object points in space are slightly different in the two eyes. The stereoscope, with which different pictures can be presented to each eye, demonstrates the fundamental difference between stereoscopic perception of depth and the conception of depth and distance from the monocular view. *See* VISION. [K.N.O.]

Stereospecific catalyst Stereospecific polymerization catalysts lead to the formation of stereoregular (tactic) polymers, that is, polymers where the centers of steric isomerism in the main chain are arranged in a regular fashion with respect to their configurations. Three factors determine the tacticity of a polymeric chain during its formation: (1) the kind of monomer approach to the growing chain end, (2) the kind of attack of the growing chain end on the double bond (cis or trans opening), and (3) the configuration in the initiation step. In addition to a regular head-to-tail configuration and absence of branching, reactions have to be assumed. The kind of monomer approach is strongly affected by electrical and stereochemical forces, and therefore changes in monomer structure and environment greatly influence polymer tacticity. The intermediate radical or ion can be assumed to have a planar or near-planar structure, and in an uncomplexed form it should be able to rotate freely around its axis with cis and trans addition being equally possible. Upon addition of the next monomer this carbon changes into a tetrahedral structure, thereby creating two isomeric forms, isotactic or syndiotactic. *See* ASYMMETRIC SYNTHESIS; CATALYSIS; POLYMERIZATION; STEREOCHEMISTRY. [A.Sc.]

Steric effect (chemistry) The influence of the spatial configuration of reacting substances upon the rate, nature, and extent of reaction. The sizes and shapes of atoms and molecules, the electrical charge distribution, and the geometry of bond angles influence the courses of chemical reactions. The steric course of organochemical reactions is greatly dependent on the mode of bond cleavage and formation, the environment of the reaction site, and the nature of the reaction conditions (reagents, reaction time, and temperature). The effect of steric factors is best understood in ionic reactions in solution. *See* STEREOCHEMISTRY. [E.W.]

Sterilization An act of destroying all forms of life on and in an object. A substance is sterile, from a microbiological point of view, when it is free of all living microorganisms. Sterilization is used principally to prevent spoilage of food and other substances and to prevent the transmission of diseases by destroying microbes that may cause them in humans and animals. Microorganisms can be killed either by physical agents, such as heat and irradiation, or by chemical substances.

Heat sterilization is the most common method of sterilizing bacteriological media, foods, hospital supplies, and many other substances. Either moist heat (hot water or steam) or dry heat can be employed, depending upon the nature of the substance to be sterilized. Moist heat is also used in pasteurization, which is not considered a true sterilization technique because all microorganisms are not killed; only certain pathogenic organisms and other undesirable bacteria are destroyed. *See* PASTEURIZATION.

Many kinds of radiations are lethal, not only to microorganisms but to other forms of life. These radiations include both high-energy particles as well as portions of the electromagnetic spectrum. *See* RADIATION BIOLOGY.

Filtration sterilization is the physical removal of microorganisms from liquids by filtering through materials having relatively small pores. Sterilization by filtration is employed with liquid that may be destroyed by heat, such as blood serum, enzyme solutions, antibiotics, and some bacteriological media and medium constituents. Examples of such filters are the Berkefeld filter (diatomaceous earth), Pasteur-Chamberland filter (porcelain), Seitz filter (asbestos pad), and the sintered glass filter.

Chemicals are used to sterilize solutions, air, or the surfaces of solids. Such chemicals are called bactericidal substances. In lower concentrations they become bacteriostatic rather than bactericidal; that is, they prevent the growth of bacteria but may not kill them. Other terms having similar meanings are employed. A disinfectant is a chemical that kills the vegetative cells of pathogenic microorganisms but not necessarily the endospores of spore-forming pathogens. An antiseptic is a chemical applied to living tissue that prevents or retards the growth of microorganisms, especially pathogenic bacteria, but which does not necessarily kill them.

The desirable features sought in a chemical sterilizer are toxicity to microorganisms but nontoxicity to humans and animals, stability, solubility, inability to react with extraneous organic materials, penetrative capacity, detergent capacity, noncorrosiveness, and minimal undesirable staining effects. Rarely does one chemical combine all these desirable features. Among chemicals that have been found useful as sterilizing agents are the phenols, alcohols, chlorine compounds, iodine, heavy metals and metal complexes, dyes, and synthetic detergents, including the quaternary ammonium compounds. [C.F.N.]

Steroid A compound, principally of plant or animal origin, composed of a series of four carbon rings joined together to form a structural unit called cyclopentenoperhydrophenanthrene (see illustration). There are six centers of asymmetry so that 64 compounds (stereoisomers) with this structure are possible. Natural steroids are made by enzymes in living cells and are limited in number. The most abundant members of the

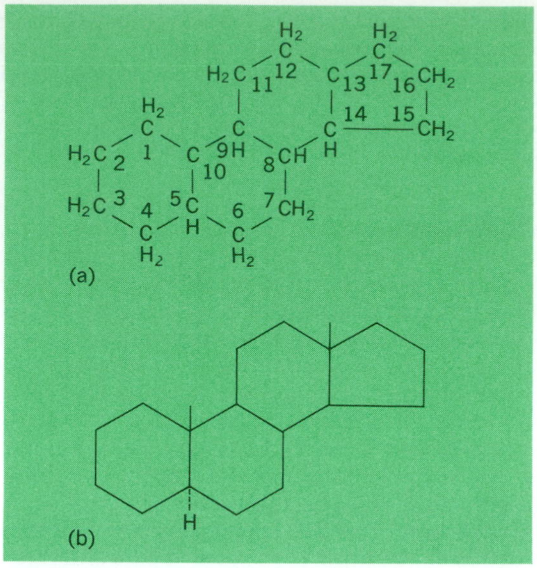

Steroid skeleton. ***(a)*** **Structure and numbering.** ***(b)*** **Shorthand formulation; the lines attached to the rings represent methyl groups.**

class are the sterols; male and female sex hormones are steroids. *See* HORMONE; STEROL. [T.F.G.]

Sterol Any one of the natural products derived from the steroid nucleus. Sterols are components of virtually all plant and animal cells with the notable exception of bacteria, where they are either totally absent or present in small amounts. They are waxy, colorless solids soluble in most organic solvents and virtually insoluble in water. They contain one alcohol functional group and may be either saturated (plant sterols; see illustration) or unsaturated (all animal sterols and many plant sterols).

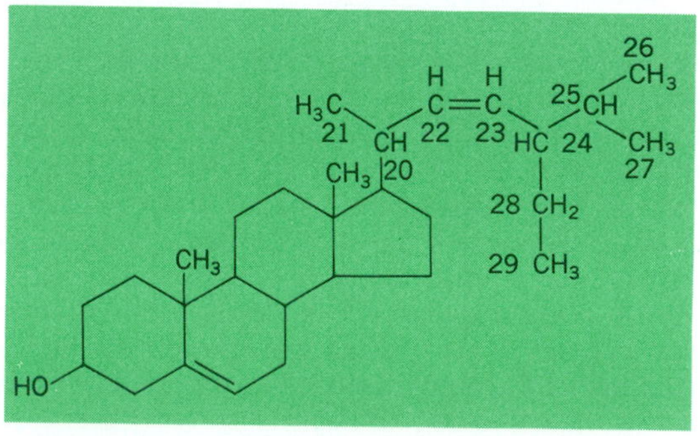

Stigmasterol, a plant sterol. It has 29 carbon atoms, in contrast with a typical animal sterol, cholesterol, which has 27 carbon atoms.

In nature they are often esterified with higher fatty acids. In living cells they often serve as part of cell membranes. In vertebrates cholesterol is the principal although not the sole sterol. *See* CHOLESTEROL; STEROID. [T.F.G.]

Stibnite A mineral with composition Sb_2S_3 (antimony trisulfide), the chief ore of antimony. It crystallizes in slender, prismatic, vertically striated crystals which may be curved or bent. It is often in bladed, granular, or massive aggregates. The hardness is 2 on Mohs scale and the specific gravity 4.5–4.6. The luster is metallic and the color lead-gray to black. It is one of few minerals that fuses easily in the match flame (525°C or 977°F).

Stibnite has been found in various mining districts in Germany, Rumania, France, Bolivia, Peru, and Mexico. In the United States the Yellow Pine mine at Stibnite, Idaho, is the largest producer. Other deposits are in Nevada and California. The finest crystals have come from the island of Shikoku, Japan. *See* ANTIMONY. [C.S.Hu.]

Stichosomida An order of nematodes composed of taxa that are parasitic in either vertebrates or invertebrates. The most distinguishing characteristic is the modification of the posterior esophagus into a stichosome, a series of glands exterior to the esophagus proper. The stichosome may be in one or two rows. The early larval stages possess a protrusible stylet that is absent in the adults. Amphids are postlabial.

There are three stichosomid superfamilies: Trichocephaloidea parasitize vertebrates. In Mermithoidea, adult worms are free living, but their larvae are parasitic in a variety of insects, arachnids, and pulmonate snails. Echinomermelloidea parasitize marine invertebrates; the superfamily comprises three families: Echinomermellidae, Marimemithidae, and Benthimermithidae. *See* NEMATA. [A.R.M.]

Stickleback Any fish which is a member of the family Gasterosteidae in the order Gasterosteiformes. They have a variable number of free spines in front of the dorsal fin. These small, fresh-water and marine fishes are found in cold and temperate waters of the Northern Hemisphere. All species are of some economic importance since they feed largely on mosquito larvae. *See* GASTEROSTEIFORMES. [C.B.C.]

Stilbite A mineral belonging to the zeolite family of silicates. It crystallizes in sheaflike aggregates of thin tabular crystals. There is perfect cleavage parallel to the side pinacoid, and here the mineral has a pearly luster; elsewhere the luster is vitreous. The color is usually white but may be brown, red, or yellow. Hardness is 3½–4 on Mohs scale; specific gravity is 2.1–2.2. *See* ZEOLITE.

Stilbite is found in Iceland, India, Scotland, Nova Scotia, and in the United States at Bergen Hill, New Jersey, and the Lake Superior copper district in Michigan. [C.S.Hu.]

Stirling engine An engine in which work is performed by the expansion of a gas at high temperature to which heat is supplied through a wall. Like the internal combustion engine, a Stirling engine provides work by means of a cycle in which a

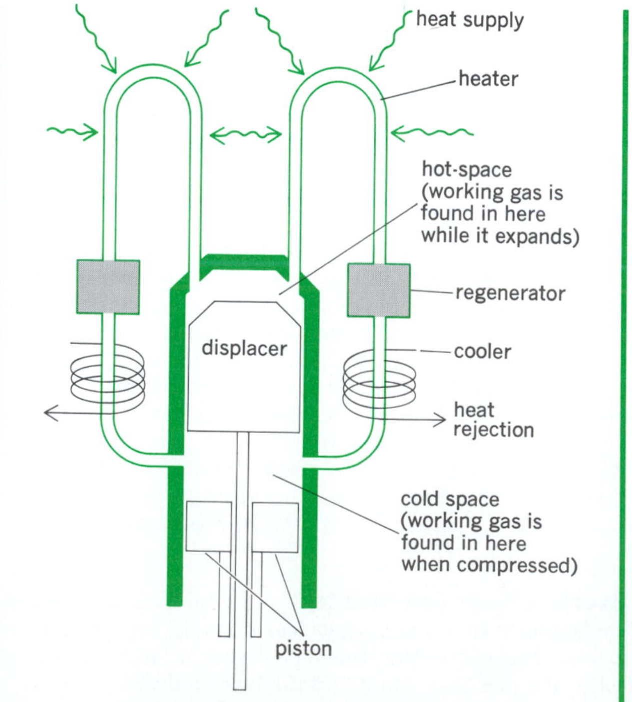

Principle of Stirling engine, displacer type.

piston compresses gas at a low temperature and allows it to expand at a high temperature. In the former case the heat is provided by the internal combustion of fuel in the cylinder, but in the Stirling engine the heat (obtained from externally burning fuel) is supplied to the gas through the wall of the cylinder (see illustration).

The rapid changes desired in the gas temperature are achieved by means of a second piston in the cylinder, called a displacer, which in moving up and down transfers the gas back and forth between two spaces, one at a fixed high temperature and the other at a fixed low temperature. When the displacer is raised, the gas will flow from the hot space via the heater and cooler tubes into the cold space. When it is moved downward, the gas will return to the hot space along the same path. During the first transfer stroke the gas has to yield up a large amount of heat to the cooler; an equal quantity of heat has to be taken up from the heater during the second stroke. *See* INTERNAL COMBUSTION ENGINE. [R.J.M.]

Stirling numbers Numbers which arise in enumerating the ways of distributing the elements of a set S into r indistinguishable cells so that no cell remains empty. There results a set-partition or partition of S into r blocks, and these are counted by the Stirling number of the second kind $S(n,r)$. The Stirling numbers satisfy the boundary conditions $S(n,1) = S(n,n) = S(n,r) = 1$, and the recusion $S(n-1,r-1) + rS(n-1,r)$ (which may be established by distinguishing within the family of partitions between the distributions for which the first $n-1$ elements of S leave a cell unoccupied which must be filled by the last element and those for which the first $n-1$ elements occupy every cell). In addition, they satisfy formula (1), where $(x)_r$ stands for the failing factorial $x(x-1)(x-2)\cdots(x-r+1)$. The Bell number, given by Eq. (2), is the total number of partitions of an n-element set. These numbers also give the number of rhyme schemes for a poem of n lines. Thus, there are $B(4) = 15$ ways to rhyme a quatrain; *abcd*, *aabc*, *abac*,. . ., *aabb*,. . ., *abbb*, *aaaa*. The Stirling numbers of the first kind, $s(n,r)$ are defined as the coefficients of the generating function of $(x)_n$. *See* COMBINATORIAL THEORY; ENUMERATION OF ARRANGEMENTS. [T.Br.]

$$x_n = \sum_{r}^{n} = 1\, S(n,r)\,(x)r \qquad (1)$$

$$B(n) = \sum_{r=1}^{n} S(n,r) \qquad (2)$$

Stochastic control theory A branch of control theory which aims at predicting and minimizing the magnitudes and limits of the random deviations of a control system through optimizing the design of the controller. Such deviations occur when random noise and disturbance processes are present in a control system, so that the system does not follow its prescribed course but deviates from the latter by a randomly varying amount.

In contrast to deterministic signals, random signals cannot be described as given functions of time such as a step, a ramp, or a sine wave. The exact function is unknown to the system designer; only some of its average properties are known.

A random signal may be generated by one of nature's processes, for instance, radar noise and wind- or wave-induced forces and moments on a radar antenna or a ship. Alternatively, it may be generated by human intelligence, for instance, the bearing of a zigzagging aircraft, or the contour to be followed by a duplicating machine.

One outstanding experimental fact about nature's random processes is that these signals are very closely Gaussian. The word "Gaussian" is a mathematical concept which describes one or more signals, $i_1, i_2, \ldots, i_n$ having the following properties:

1. The amplitude of each signal is normally distributed.

2. The joint distribution function of any number of signals at the same or different times taken from the set is a multivariate normal distribution. This experimental fact is not surprising in view of the fact that a random process of nature is usually the sum total of the effects of a large number of independent contributing factors. For instance, thermal noise is due to the thermal motions of billions of electrons and atoms. An ocean-wave height at any particular time and place is the sum of wind-generated waves at previous times over a large area. *See* DISTRIBUTION (PROBABILITY); ELECTRICAL NOISE; STOCHASTIC PROCESS.

The underlying mechanism that generates a random process can usually be described in physical or mathematical terms. For instance, the underlying mechanism that generates shot-effect noise is thermionic emission. If the generating mechanism does not change with time, any measured average property of the random process is independent of the time of measurement aside from statistical fluctuations, and the random process is called stationary. If the generating mechanism does change, the random process is called nonstationary. *See* CONTROL SYSTEMS. [S.S.L.C.]

Stochastic process A physical stochastic process is any process governed by probabilistic laws. Examples are (1) development of a population as controlled by Mendelian genetics; (2) Brownian motion of microscopic particles subjected to molecular impacts or, on a different scale, the motion of stars in space; (3) succession of plays in a gambling house; and (4) passage of cars by a specified highway point.

In each case, a probabilistic system is evolving; that is, its state is changing with time. Thus the state at time t depends on chance: It is a random variable $x(t)$. The parameter set of values of t involved is usually (and will always be in this article) either an interval (continuous parameter stochastic process) or a set of integers (discrete parameter stochastic process). Some authors, however, apply the term stochastic process only to the continuous parameter case.

If the state of the system is described by a single number, $x(t)$ is numerical-valued. In other cases, $x(t)$ may be vector-valued or even more complicated. For the numerical case, as the state changes, its values determine a function of time, the sample function, and the probability laws governing the process determine the probabilities assigned to the various possible properties of sample functions.

A mathematical stochastic process is a mathematical structure inspired by the concept of a physical stochastic process, and studied because it is a mathematical model of a physical stochastic process or because of its intrinsic mathematical interest and its applications both in and outside the field of probability. The mathematical stochastic process is defined simply as a family of random variables. That is, a parameter set is specified, and to each parameter point t a random variable $x(t)$ is specified. If one recalls that a random variable is itself a function, if one denotes a point of the domain of the random variable $x(t)$ by ω, and if one denotes the value of this random variable at ω by $x(t,\omega)$, it results that the stochastic process is completely specified by the function of the pair (t,ω) just defined, together with the assignment of probabilities. If t is fixed, this function of two variables defines a function of ω, namely, the random variable denoted by $x(t)$. If ω is fixed, this function of two variables defines a function of t, a sample function of the process.

Probabilities are ordinarily assigned to a stochastic process by assigning joint probability distributions to its random variables. These joint distributions, together with the probabilities derived from them, can be interpreted as probabilities of properties of sample functions. For example, if t_0 is a parameter value, the probability that a sample function is positive at time t_0 is the probability that the random variable $x(t_0)$ has a positive value. The fundamental theorem at this level is that, to any self-consistent assignment of joint probability distributions, there corresponds a stochastic process.

Stationary processes are the stochastic processes for which the joint distribution of any finite number of the random variables is unaffected by translations of the parameter; that is, the distribution of $x(t_1 + h),\ldots, x(t_n + h)$ does not depend on h. *See* PROBABILITY.

A Markov process is a process for which, if the present is given, the future and past are independent of each other. More precisely, if $t_1 < \cdots < t_n$ are parameter values, and $1 < j < n$, then the sets of random variables $[x(t_1),\ldots,x(t_{j-1})]$ and $[x(t_{j+1}),\ldots,x(t_n)]$ are mutually independent for given $x(t_j)$. Equivalently, the conditioned probability distribution of $x(t_n)$ for given $x(t_1),\ldots,x(t_{n-1})$ depends only on the specified value of $x(t_{n-1})$ and is in fact the conditional probability distribution of $x(t_n)$, given $x(t_{n-1})$. An important and simple example is the Markov chain, in which the number of states is finite or denumerably infinite.

A martingale is a stochastic process with the property that, if $t_1 < \cdots < t_n$ are parameter values, the expected value of $x(t_n)$ for given $x(t_1),\ldots,x(t_{n-1})$ is equal to $x(t_{n-1})$. That is, the expected future value, given present and past values, is equal to the present value. The interpretation that a martingale can be thought of as the fortune of a player after the successive plays of a fair gambling game is obvious.

A process with independent increments is a continuous-parameter process with the property that, if $t_1, < \cdots < t_n$ are parameter values, the successive increments in $x(t_2) - x(t_1),\ldots,x(t_n) - x(t_{n-1})$ are mutually independent. If $y(t) = x(t) - x(t_0)$, where t_0 is fixed, the $y(t)$ process is then a Markov process. *See* GAME THEORY; INFORMATION THEORY; LINEAR PROGRAMMING; OPERATIONS RESEARCH. [J.L.D.]

Stoichiometry The interpretation of mass and energy relationships indicated by chemical reaction equations. It is also defined as those calculations used to find the quantities of reactants and products in a chemical reaction.

Such calculations fall into four groups. The first group includes the law of conservation of matter, the law of chemical combining weights, and the law of combining proportions. The second group is based on the law of conservation of energy and includes heat of reaction, heat added, heat lost, and other energy effects associated with the reaction. The third group comprises the equilibrium relationships of the system and includes not only chemical equilibria but also physical equilibria such as gas-liquid systems. The fourth group includes the rate of reaction relationships of both chemical changes and physical changes.

In the chemical laboratory only the first group is usually considered. The second and fourth groups are neglected, or the reactions are forced to completion artificially. The third group is utilized in a limited number of cases. The units used are grams, moles, milliliters, or liters. The types of calculations include weight-weight, weight-volume (either gases or liquids), and thermal relationships. For any calculation a knowledge of the reaction under consideration and a balanced reaction equation are necessary.

All four groups of principles must be used in industrial stoichiometry. Flowing systems require the use of reaction rates and detailed considerations of heat losses, purity of materials, effects of catalysts, and side reactions. [K.G.S.; C.L.R.]

Stoker A mechanical means for feeding coal into, and for burning coal in, a furnace. There are three basic types of stokers. Chain or traveling-grate stokers have a moving grate on which the coal burns; they carry the coal from a hopper into the furnace and move the ash out (see illustration). Spreader stokers mechanically or pneumatically distribute the coal from a hopper at the furnace front wall and move it onto the grate which usually moves continuously to dispose of the ash after the coal is burned. Underfeed stokers are arranged to force fresh coal from the hopper to the bottom of the burning coal bed, usually by means of a screw conveyor. The ash is forced off the edges of the retort peripherally to the ashpit or is removed by hand. [G.W.K.]

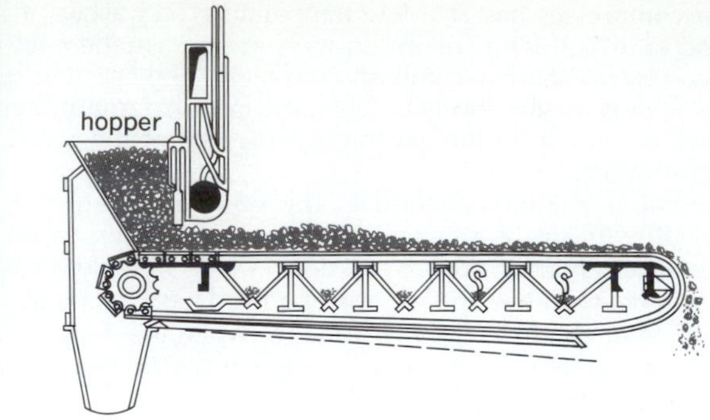

A chain grate stoker.

Stokes stream function A degenerate (one-component) vector potential used in analyzing and describing axially symmetric fluid-flow fields. In a steady axially symmetric flow, the rotation of a streamline about the axis of symmetry generates a stream surface. A certain mass rate of flow exists inside this stream surface which is the same at every axial station because, by definition, there is no flow through the stream surface. The value of the Stokes stream function ψ at a point in the flow is equal to $1/2\pi$ times mass rate of flow inside the stream surface passing through that point. If the fluid motion is slow enough to neglect compressibility, ψ also measures the volume rate of flow inside a stream surface. *See* FLUID FLOW. [A.E.Br.]

Stokes' theorem The assertion that under certain light restrictions the surface integral of $(\nabla \mathfrak{F} \times \mathfrak{F}) \cdot \mathbf{v}$ over a surface patch S is equal to the line integral of $\mathfrak{F} \cdot \tau$ taken around C, the boundary curve of S, provided the sense of transcription of C is right-handed relative to $\mathbf{v}$. This can be expresed as

$$\iint (\nabla \times \mathfrak{F}) \cdot \mathbf{v}\, dS = \int \mathfrak{F} \cdot \tau\, ds$$

Here $\mathfrak{F}$ is a vector function, $\mathbf{v}$ is one of the two unit normals to the two-sided surface S, s is arc length measured positively in the sense which is right-handed relative to $\mathbf{v}$, and τ is the unit tangent vector to C in the sense of increasing s. *See* CALCULUS OF VECTORS. [H.V.C.]

Stolonifera An order of the Alcyonaria which lacks coenenchyme. They form either simple or rather complex colonies. The polyp has a cylindrical body with a retractile oral portion which can withdraw into a solid anthostele or calyx protected by many calcareous spicules. The base of the mature polyp is attached to a creeping stolon which is a ribbonlike network or thin flat mat from which daughter polyps arise. Daughter polyps never bud from the wall of the primary polyp. Each polyp is connected by solenial tubes of the stolons, or by transverse platforms. *See* ALCYONARIA. [K.At.]

Stomach The stomach is a pouchlike dilatation of the gut immediately beyond the esophagus, in which food is held for a time and subjected to vigorous digestive action.

The stomach of fishes is generally simple and saccular in contour. Some are more or less straight; others are curved or J-shaped. Entrance into the stomach is controlled by a cardiac sphincter, which also serves to prevent undue entrance of water.

Although reptiles display great variety in body form, they generally show no remarkable variations in stomach configuration other than those required to fit it within the available space. In crocodiles the anterior part of the stomach is enlarged and very muscular and has a structure resembling the gizzard of birds.

The stomach of birds is divided into two chambers, the relatively thin-walled, glandular proventriculus and the very muscular gizzard. Within the proventriculus the food is mixed with mucus and digestive juices produced by the glandular mucous membrane. The softened and partially digested mass is then passed on into the gizzard, where it is vigorously ground and comminuted as digestion continues. Grinding action of the gizzard, highly developed in seed-eating birds, is augmented by the presence of gizzard stones which these birds ingest.

The most elaborate mammalian stomachs are found among the cud-chewing artiodactyl (even-toed) ungulates, such as the cow, sheep, deer, goat, giraffe, and camel. Stomachs of the ruminants are not only complex but large; that of an ox has a capacity of up to 60 gal (228 liters). Typically the stomach is divided into four chambers: rumen (paunch) reticulum (honeycomb), omasum (psalterium or manyplies), and abomasum. The first three are generally considered to be expansions of the lower end of the esophagus. Only the fourth has a glandular wall and corresponds to a true stomach. This arrangement makes it possible for these animals to ingest large amounts of food in a relatively short period of time, and then rework and digest it.

Carnivores have a stomach with general features typical of the majority of mammals, including humans (see illustration).

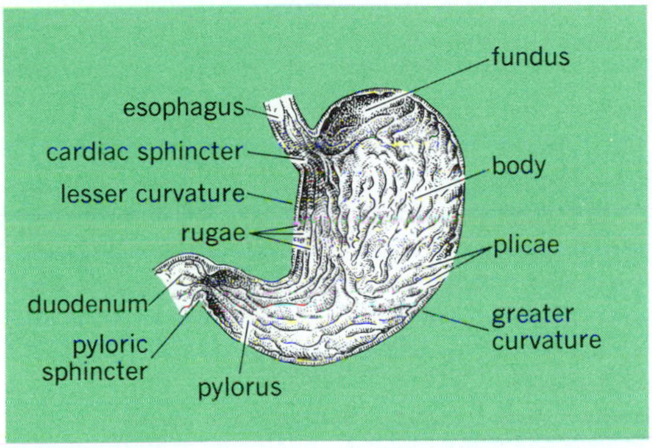

Longitudinal section of the human stomach showing gross appearance of interior.

The region of the stomach adjacent to the esophageal opening is known as the cardia, and a portion of the smooth muscle at this junction forms the relatively weak cardiac sphincter. The expansible muscular fundus and body curve downward and to the right, gradually narrowing to form the pyloric antrum and the somewhat more muscular pyloric canal. The distal part of the encircling muscle constitutes the pyloric sphincter, which controls the passage of material from stomach to duodenum. *See* DIGESTIVE SYSTEM; INTESTINE [V.M.E.]

Stomach disorders Disorders of the stomach include congenital defects, inflammations, tumors, and reaction to toxic materials and foreign bodies.

One of the most common congenital lesions is hypertrophic pyloric stenosis, which is heralded by regurgitation or vomiting of food. The male infant is affected most frequently; however, the mode of genetic inheritance has not been fully established.

Acute gastritis, an inflammatory process, is an extremely frequent lesion, the result of irritation to the mucosa of the stomach by a variety of substances, including caffeine, alcohol, salicylates, other drugs, and acid and alkaline solutions. The mucosa responds with a wide variety of changes, ranging from edema to erosion and ulceration. Chronic gastritis, another inflammatory process, leads to a variety of changes in the stomach. Different areas of the stomach can be affected, depending on the type of injury.

Stress can cause changes in the mucosa of the stomach. Peptic ulcers are found in 10% of the population of North America. The stomach is the second most frequent site of involvement after the duodenum. *See* PEPTIC ULCER.

Foreign bodies in the stomach are quite varied in their number and type. In most cases, small objects pass through into the bowel. Larger objects however are retained and may, if sharp, lacerate the stomach wall and cause perforation. Bezoars are accumulations of foreign material, often hair, which may reach immense size.

Tumors of the stomach are both benign and malignant. Of the two most frequent benign tumors, one is derived from the epithelium of the mucosa, an adenoma, and the other is derived from the smooth muscle of the underlying layers, a leiomyoma. Hemorrhage is frequent in persons with leiomyoma. The most frequent malignant tumor in the stomach arises from the epithelium of the mucosa, an adenocarcinoma. Extension and metastasis are common in this disease even when treated early in its course. *See* STOMACH. [H.T.N.]

Stomatopoda The only order of the superorder Hoplocarida belonging to the subclass Malacostraca of the class Crustacea. This order of the mantis shrimps contains four families: Lysioquillidae, Bathysquillidae, Squillidae, and Gonodactylidae. Stomatopoda are among the larger Crustacea; the size of an adult ranges from about 0.5 in. (12 mm) to over 12 in. (300 mm).

The bodies of stomatopods are narrow and elongate, almost eruciform. Only part of the cephalic and thoracic somites are fused and covered by the dorsal shield, the carapace. Those cephalic somites which bear the antennulae and the eyes are free and visible anterior to the carapace, and the last four thoracic somites are similarly exposed. The tail fan consists of a well-developed, sometimes peculiarly sculptured or deformed, median plate, the telson, and the two uropods. Anteriorly, the carapace bears a flattened movable plate, the rostrum. The eyes are large, stalked, and movable. Of eight pairs of thoracic appendages the first is narrow, slender, and hairy, probably being used for cleaning purposes. The second thoracic leg is very strong and heavy. It has become a large raptorial claw that shows a great resemblance to that of the praying mantis, and for this reason the Stomatopoda are often given the name mantis shrimps.

The Stomatopoda are marine animals rarely found in brackish water. Most species are confined to tropical and subtropical areas, though some occur in the boreal and antiboreal regions. The majority of stomatopods live in the littoral and sublittoral zones, but a few species have been found in greater depths, down to 2500 ft (760 m). *See* MALACOSTRACA. [L.B.Ho.]

Stone and stone products The term stone is applied to rock that is cut, shaped, broken, crushed, and otherwise physically modified for commercial use. The two main divisions are dimension stone and crushed stone. Other descriptive terms may be used, for example, building stone, roofing stone, or precious stone. *See* GEM; ROCK.

The term dimension stone is applied to blocks that are cut and milled to specified sizes, shapes, and surface finishes. The principal uses are for building and ornamental applications. Granites, limestones, sandstones, and marbles are widely used; basalts, diabases, and other dark igneous rocks are used less extensively. Soapstone is used to some extent. Rock suitable for use as dimension stone must be obtainable in large, sound blocks, free from incipient cracks, seams, and blemishes, and must be without mineral grains that might cause stains as a result of weathering. It must have an attractive color, and generally a uniform texture.

Slate differs from other dimension stone because it can be split into thin sheets of any desired thickness. Commercial slate must be uniform in quality and texture and reasonably free from knots, streaks, or other imperfections, and have good splitting properties. Roofing slates are important products of most slate quarries. However, the roofing-slate industry has declined considerably because of competition from other types of roofing. Slate is also used for milled products such as blackboards, electrical panels, window and door sills and caps, baseboards, stair treads, and floor tile. *See* SLATE.

Nearly all the principal types of stone—granite, diabase, basalt, limestone, dolomite, sandstone, and marble—may be used as sources of commercial crushed stone; limestone is by far the most important. Crushed stone is made from sound, hard stone, free from surface alteration by weathering. Stone that breaks in chunky, more or less cubical fragments is preferred. Commercial stone should be free from certain deleterious impurities, such as opalescent quartz, and free from clay or silt. Crushed stone is used principally as concrete aggregate, as road stone, or as railway ballast. Other uses for limestone are as a fluxing material to remove impurities from ores smelted in metallurgical furnaces, in the manufacture of alkali chemicals, calcium carbide, glass, paper, paint, and sugar, and for filter beds and for making mineral wool. [R.L.B.]

Storage battery An assembly of identical voltaic cells in which the electrochemical action is reversible so that the battery may be recharged by passing a current through the cells in the opposite direction to that of discharge. While many nonstorage batteries have a reversible process, only those that are economically rechargeable are classified as storage batteries. Storage batteries, sometimes known as electric accumulators or secondary batteries, have two general classifications: lead-acid and alkaline. *See* BATTERY; PRIMARY BATTERY.

Some of the important uses of storage batteries are to start gasoline and diesel engines; to operate communications circuits; switch tripping and closing in power-generating and handling systems; emergency lighting; emergency power both with and without conversion to alternating current; railway car lighting and air conditioning; rapid transit car controls; marine power systems; power for underwater exploratory vehicles and submarines; to activate photographic and portable sound systems as well as portable TV and radio; and various military applications.

Lead-acid battery. The lead-acid type of storage battery is so classified because the electrolyte is an acid and the plates are largely lead. The positive active material is lead peroxide and the negative active material is lead sponge. The active materials are supported by grids made of lead alloys.

A great many types of lead-acid cells are produced, but all have certain features in common. One is the open-circuit cell electromotive force (emf), which exists between a positive lead peroxide (PbO_2) electrode and a negative sponge lead (Pb) electrode when the two are immersed in sulfuric acid electrolyte ($H_2SO_4 + H_2O$). The emf value is independent of the quantities of lead peroxide, lead, or electrolyte present but does vary with temperature and sulfuric acid concentration. The relatively small variation with temperature is given in millivolts/°C over a range 0–40°C, as 0.30 for 1.200 sp gr electrolyte, 0.22 for 1.250 sp gr, 0.19 for 1.280 sp gr, and 0.18 for 1.300 sp gr.

Reaction (1) represents the cell reactions insofar as beginning and end materials are concerned.

$$PbO_2 + Pb + 2H_2SO_4 \underset{\text{charge}}{\overset{\text{discharge}}{\rightleftharpoons}} 2PbSO_4 + 2H_2O \qquad (1)$$

On discharge the overall effect is a reduction of PbO_2 at the positive electrode and an oxidation of Pb at the negative electrode, accompanied by sulfation in both cases. In charging, a counter voltage is imposed on the cell terminals, and current is forced through the cell in a direction opposite to that in which the cell discharges. This reverses the ionic movements in relation to the electrodes and, in effect, reverses the cell reactions. On discharge the electrolyte specific gravity decreases, and on charge it increases. Specific gravity serves as a measure of the sulfuric acid concentration and thus as an index of state of charge.

Alkaline-type battery. The alkaline-type storage battery is so classified because the electric energy is obtained from chemical action of an alkaline solution. One type of battery has positive plates of some nickel compound and negative plates of iron. Another type uses a nickel compound and cadmium. A third uses silver oxide and zinc. The nickel-iron cell will serve as an example, as described below.

The nickel-iron alkaline battery is composed of cells having a hydrated nickel oxide and iron in an alkaline solution. The positive active material in this cell is a higher oxide or hydroxide of nickel. The negative material is fine iron powder. The electrolyte is potassium hydroxide, to which a little lithium hydroxide is sometimes added.

The chemical behavior of the nickel-iron cell is shown in reaction (2). The KOH electrolyte supplies ions for conductivity

$$2NiOOH \cdot H_2O + Fe \underset{\text{charge}}{\overset{\text{discharge}}{\rightleftharpoons}} 2Ni(OH)_2 + Fe(OH) \qquad (2)$$

but unlike the lead-acid battery, the concentration of electrolyte does not undergo any net change in the chemical action of the cell. As a consequence, the specific gravity of the electrolyte does not change and cannot be used to indicate the state of charge of an alkaline battery, as in the case of the lead-acid battery. However, it also means that the gravity stays up all times, and the battery is much less susceptible to accidental damage from freezing than lead-acid. [W.W.Sm.]

Storage rings Annular vacuum chambers in which beams of high-energy charged particles can be stored and caused to collide nearly head-on. The motivation for using this technique is found in the following kinematic considerations. Unlike fixed-target systems (cyclotrons, synchrotrons, linear accelerators) in which a beam of accelerated particles traverses a fixed target, colliding-beam systems use the full energy of each particle to produce reactions.

A positron-electron (e^+e^-) storage ring consists of a ring of bending and focusing magnets enclosing a doughnut-shaped vacuum chamber in which counterrotating beams of e^+ and e^- are stored for periods of several hours. The two beams are made to collide with each other about 10^6 times per second in straight interaction sections of the vacuum chamber. These are surrounded with detectors for the observation of collision products. An alternate design consists of two separate but intersecting rings, one to store e^-, the other to store e^- or e^+ with collisions taking place at the intersection points.

Collisions between antiparticle and particle such as e^+e^- are of particular interest since most of the resulting processes proceed through annihilation which forms a particularly simple final state with the quantum numbers of the photon. This is one of the main reasons why e^+e^- collisions have proved to be particularly fruitful in deepening understanding of the fundamental structure of matter.

Because they have a small mass, e^+ and e^- in circular orbits radiate a substantial amount of energy in the form of synchrotron radiation, just as they do in electron synchrotrons. As in electron synchrotrons, the radiation itself constitutes a very intense source of x-rays which has proved very useful in various branches of biology, chemistry, and physics. *See* SYNCHROTRON; SYNCHROTRON RADIATION. [K.St.]

Proton storage rings consist of a pair of annular vacuum chambers in which high-energy protons can be stored and caused to collide nearly head-on. The motivation for using this technique is the same as for electron-positron storage rings. Namely, the colliding-beam storage ring permits studying reactions at energies that cannot be achieved with economically feasible conventional accelerators. See PARTICLE ACCELERATOR. [M.B.]

Storage tank A container for storing liquids or gases. A tank may be constructed of ferrous or nonferrous metals or alloys, reinforced concrete, wood, or filament-wound plastics, depending upon its use. Tanks resting on the ground have flat bottoms; those supported on towers have either flat or curved bottoms. Standpipes, which are usually cylindrical shells of steel or reinforced concrete resting on the ground, are frequently of great height and comparatively small diameter. They are built to contain water for a distribution system, and height is required to maintain pressure in the system. Tanks for other liquids and for gases, where storage is more important than pressure, are generally lower and of greater diameter. [C.M.A.]

Storage tube An electron tube into which information can be introduced and then extracted at a later time; also called a memory tube. See ELECTRON TUBE.

The process of introducing information into a storage tube is known as writing, and that of extracting useful information is called reading. The deliberate removal of information from the storage surface is called erasing. In some tubes, reading automatically effects erasing; in others, a separate operation is required. A charge-storage tube is one in which information is retained on a surface in the form of electric charges.

The characteristics of a storage tube are largely governed by the nature of the storage surface. This surface is usually a deposit or sheet of insulating or semiconducting material. The point-by-point potential of this surface is varied, in a controlled way, by the dielectric charging processes associated with electron bombardment of an insulator. The low conductivity of the storage-surface material ensures that the charge (or potential) pattern will not be dissipated before the reading operation is initiated and completed.

Storage tubes utilize electron guns to address the storage surface and effect the operations of reading, writing, and erasing. The electron gun will produce a fine pencil beam of electrons which may be deflected, as in a cathode-ray tube, to any desired location on the storage surface, or it will produce a flood beam which uniformly and simultaneously floods the entire storage surface with electrons. See CATHODE-RAY TUBE. [N.W.P.]

Storm An atmospheric disturbance involving perturbations of the prevailing pressure and wind fields on scales ranging from tornadoes (0.6 mi or 1 km across) to extratropical cyclones (1.2–1900 mi or 2–3000 km across); also, the associated weather (rain storm, blizzard, and the like). Storms influence human activity in such matters as agriculture, transportation, building construction, water impoundment and flood control, and the generation, transmission, and consumption of electric energy. See WIND.

The form assumed by a storm depends on the nature of its environment, especially the large-scale flow patterns and the horizontal and vertical variation of temperature; thus the storms most characteristic of a given region vary according to latitude, physiographic features, and season. Extratropical cyclones and anticyclones are the chief disturbances over roughly half the Earth's surface. Their circulations control the embedded smaller-scale storms. Large-scale disturbances of the tropics differ fundamentally from those of extratropical latitudes. See HURRICANE; SQUALL; THUNDERSTORM; TORNADO; TROPICAL METEOROLOGY.

Cyclones form mainly in close proximity to the jet stream, that is, in strongly baroclinic regions where there is a large increase of wind with height. Weather patterns in cyclones are highly variable, depending on moisture content and thermodynamic stability of air masses drawn into their circulations. Warm and occluded fronts, east of and extending into the cyclone center, are regions of gradual upgliding motions, with widespread cloud and precipitation but usually no pronounced concentration of stormy conditions. Extensive cloudiness also is often present in the warm sector. Passage of the cold front is marked by a sudden wind shift, often with the onset of gusty conditions, with a pronounced tendency for clearing because of general subsidence behind the front. Showers may be present in the cold air if it is moist and unstable because of heating from the surface. Thunderstorms, with accompanying squalls and heavy rain, are often set off by sudden lifting of warm, moist air at or near the cold front, and these frequently move eastward into the warm sector. See CYCLONE; JET STREAM; WEATHER.

Extratropical cyclones alternate with high-pressure systems or anticyclones, whose circulation is generally opposite to that of the cyclone. The circulations of highs are not so intense as in well-developed cyclones, and winds are weak near their centers. In low levels the air spirals outward from a high; descent in upper levels results in warming and drying aloft. Anticyclones fall into two main categories, the warm "subtropical" and the cold "polar" highs.

Between the scales of ordinary air turbulence and of cyclones, there exist a variety of circulations over a middle-scale or mesoscale range, loosely defined as from about one-half up to a few hundred miles. Alternatively, these are sometimes referred to as subsynoptic-scale disturbances because their dimensions are so small that they elude adequate description by the ordinary synoptic network of surface weather stations. Thus their detection often depends upon observation by indirect sensing systems. See METEOROLOGICAL SATELLITES; RADAR METEOROLOGY; STORM DETECTION. [C.W.N.]

Storm detection Microbarographs, radars, satellite-bome instruments, and sferics detectors (radio receivers) are used to detect storms and to assess their potential for destruction. See METEOROLOGY.

Certain severe thunderstorms emit an identifiable kind of infrasound, or ultralow-frequency acoustic wave. Traveling at the speed of sound and ducted between the ground and high-temperature layers in the upper atmosphere, these waves are often so powerful that they can be detected by sensitive pressure detectors, called microbarographs, more than 900 mi (1500 km) from the emitting storm. Acoustical detection is not used for storm warning, and its main value lies in storm research. [T.M.G.]

Radar can penetrate storm clouds to provide a three-dimensional, inside view of the storm. Advanced weather radars provide accurate images of both precipitation intensity inside a storm's shield of clouds and precipitation velocity. Contoured reflectivity maps are routinely displayed by radars of the U.S. National Weather Service on widely used plan position indicator scopes giving range and azimuth to precipitation targets whose reflectivity (intensity) is indicated by a stepped brightness scale. While a storm's reflectivity image is valuable for rainfall assessment and severe weather warnings, it is not a highly reliable tornado indicator. Highest reflectivity areas often signify hail. Radar warnings are primarily based on reflectivity values, on storm top heights, and sometimes on circulatory features of hook echoes seen in the patterns of reflectivity. See RADAR. [R.J.Do.]

Weather satellites provide photographs of nearly an entire hemisphere to users throughout the United States within 30 min of scan time. Scans are made routinely every 30 min. Intensity and development of weather systems are monitored by combining sequential pictures into motion picture loops. In addition, infrared data allow display of cloud systems at night, and indications of both temperature and water vapor by both night and day. Satellite photos reveal components of storm systems heretofore inaccessible. For example, frontal boundaries can be located precisely in most cases, since sharp cloud lines usually accompany fronts. The location and configuration of jet streams can often be identified through analysis of cirrus cloud streaks. Low-level moisture can be tracked via low clouds moving from the Gulf of Mexico into the central plains of the United States in advance of severe thunderstorm and tornado outbreaks. Developing low-pressure systems are identifiable as soon as they generate even the smallest of cloud bands. Two of the most destructive weather phenomena, hurricane and the severe thunderstorm, are efficiently monitored by using satellite technology. [J.W.]

Lightning discharges produce a wide spectrum of electromagnetic signals that are detected by radio receivers and provide effective means for locating and tracking thunderstorms. Both the electric and the magnetic components of lightning-produced radio signals have been used to locate and track thunderstorms through triangulation methods from the directions of arrival at two or more spaced stations. *See* HURRICANE; LIGHTNING; SFERICS; STORM; THUNDERSTORM. [W.L.T.]

Storm electricity Processes responsible for the separation of positive and negative electric charges in the atmosphere during storms, including the spectacular manifestation of this charge separation: lightning discharges. Cloud electrification is almost invariably associated with convective activity and with the formation of precipitation in the form of liquid water (rain) and ice particles (graupel and hail). The most vigorous convection and active lightning occurs in the summertime, when the energy source for convection, water vapor, is most prevalent. Winter snowstorms can also be strongly electrified, but they produce far less lightning than summer storms. Electrified storm clouds occasionally occur in complete isolation; more commonly they are found in convective clusters or in lines that may extend horizontally for hundreds of kilometers. *See* PRECIPITATION (METEOROLOGY).

Measurements of electric field at the ground and from instrumented balloons within thunderclouds have disclosed an electrostatic structure that appears to be fairly systematic throughout the world. The measurements show that the principal variations in charge occur in the vertical and are affected by the temperature of the cloud. The charge structure within a thundercloud is tripolar, with a region of dominant negative charge sandwiched between an upper region of positive charge and a subsidiary lower region of positive charge. In addition to the charge accumulations within the cloud, electrical measurements disclose the existence of charge-screening layers at the upper cloud boundary and a layer of positive charge near the Earth's surface beneath the cloud.

Large differences of electrical potential are associated with the distribution of charge maintained by active thunderclouds. These large differences in potential are maintained by charging currents that result from the motions of air and particles. The charging currents range from milliamperes in small clouds that are not producing lightning to several amperes for large storms with high rates of lightning. *See* CLOUD PHYSICS.

In response to charge separation within a thundercloud, the electric field increases to a value of approximately 10^6 V/m (300,000 V/ft) at which point dielectric breakdown occurs and lightning is initiated. The acoustic disturbance caused by the sudden heating of the atmosphere by lightning is responsible for thunder. *See* LIGHTNING; THUNDER.

During the 1980s, several networks were assembled for lightning detection and location (with 1–10 km or 0.6–6 mi accuracy) over areas as large as the United States. Research is investigating the relationships between lightning characteristics and the meteorological evolution of different types of storms. *See* ATMOSPHERIC ELECTRICITY; STORM DETECTION; THUNDERSTORM. [E.Wi.]

Storm surge A transient localized disturbance at sea level, resulting from the action of a tropical cyclone, an extratropical cyclone, or a squall over the sea. Storm surges, or storm tides, are not to be confused with tsunamis, or tidal waves, which result from seismic or molar disturbances of the Earth. *See* TSUNAMI.

The time history of the surge at a given location at shore is represented by the surge hydrograph. This is a time sequence of the difference between the measured tide and the predicted periodic tide. Maximum surge elevations of 15 ft (4.6 m) above predicted tide are not uncommon. In the case of hurricane-induced surges, the peak water level seems to depend primarily upon the atmospheric pressure at the hurricane center. However, the horizontal scale, the direction and speed of propagation of the hurricane, and the coastal geometry and bottom topography are important influencing factors in the storm surge behavior.

A storm surge is essentially a forced inertiogravitational wave of great wavelength. This implies that the duration or speed of the storm determines the dynamic augmentation of the water level at shore above that which would occur if the storm were stationary. Also, the inertial character of surges can explain quasi-periodic resurgences that often follow the primary forced surge. [R.O.Re.]

Straight-line mechanism A mechanism that produces a straight-line (or nearly so) output motion from an input element that rotates, oscillates, or moves in a straight line. Common machine elements, such as linkages, gears, and cams, are often used in ingenious ways to produce the required controlled motion. The more elegant designs use the properties of special points on one of the links of a four-bar linkage. *See* CAM MECHANISM; FOUR-BAR LINKAGE; GEAR; SLIDER-CRANK MECHANISM. [J.A.Sm.]

Strain Deformation or change in shape of a material as a consequence of applied forces. Strain is directly measurable, and from such measurement, within the elastic limit, the internal stress that accompanied the strain can be determined. The strain-producing action sets up stresses in the material, and these stresses cause a deformation or strain; that is, an initial strain is always accompanied by a stress. For this reason, the two phenomena are usually dealt with together. *See* ELASTIC LIMIT; STRAIN GAGE; STRESS AND STRAIN. [F.H.R.]

Strain gage A device which measures mechanical deformation (strain). Normally it is attached to a structural element, and uses the change of electrical resistance of a wire or semiconductor under tension. Capacity, inductance, and reluctance are also used.

The strain gage converts a small mechanical motion to an electrical signal by virtue of the fact that when a metal (wire or foil) or semiconductor is stretched, its resistance is increased. The change in resistance is a measure of the mechanical motion. In addition to their use in strain measurement, these gages are used in sensors for measuring the load on a mechanical member, forces due to acceleration on a mass, or stress on a diaphragm or bellows. *See* STRAIN. [J.H.Z.]

Strake The slender forward extension of the inboard region of the wing of a combat aircraft used to provide increased lift in the high angle-of-attack maneuvering condition. In contrast to the normal attached-flow design principles, these strakes are built to allow the flow to separate along the leading edge in the high angle-of-attack range and to roll up into strong leading-edge vortices. The illustration shows a typical strake installation and the vortices generated. These are highly stable, and their strong swirling motion creates a lower pressure area on the strake upper surface, resulting in a large incremental increase in lift force known as vortex lift. This vortex lift increases rapidly in the high angle-of-attack range and is less susceptible to

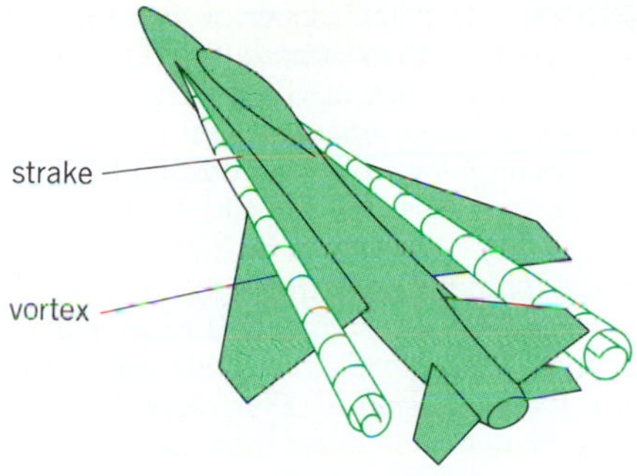

Strake and resulting vortex flow.

the normal stalling characteristics encountered with conventional lifting surfaces. *See* Airfoil; Subsonic flight; Vortex; Wing. [E.C.Po.]

Strand line The line which marks the boundary between land and a lake, sea, or ocean. The strand line is quite dynamic. It moves with the tides, it may move with the seasons, and it also moves as the sea level changes. Many varieties of organisms respond to shifts in the strand line, as to textural parameters of sediments and the coastal morphology.

Geologists are often concerned with locating the position of ancient strand lines in the rock record. The shifting of strand line positions through time as recorded in the stratigraphic record may be determined by analysis of fossils, rock textures, and structures. This enables the reconstruction of environments and helps to establish a paleogeography. *See* Facies (geology); Paleogeography; Sea-level fluctuations. [R.A.D.]

Strange particles Particles possessing the attribute of strangeness, a quantum number associated with one of the several quarks which are thought to constitute their structure. The nomenclature arose from the fact that the production rates for strange particles, which are compatible with strong interaction times of the order of 10^{-23} s, appeared to be inconsistent with their relatively long lifetimes (in the range 10^{-8} to 10^{-10} s). This inconsistency was resolved by experimental observations that strange particles are invariably produced in pairs (associated production) of equal and opposite strangeness, while their decay channels are restricted to those in which the strangeness quantum number is not conserved. This nonconservation of strangeness greatly suppresses the decay rates relative to the production rates. *See* Elementary particle; Quarks. [H.D.T.]

Strangles A highly contagious disease of the upper respiratory tract of horses and other members of the family Equidae, characterized by inflammation of the pharynx and abscess formation in lymph nodes. It occurs in horses of all ages throughout the world.

The causative agent is *Streptococcus equi* ssp. *equi*. The organism is usually found in intact abscesses in lymph nodes, where it may be observed as long chains of gram-positive cocci in smears. It is an obligate parasite of horses, donkeys, and mules, which it requires for survival.

Strangles is most common and most severe in young horses and is therefore very prevalent on breeding farms. Survival in the environment is of moderate duration. The causative agent has been reported to survive for 7 weeks in pus. Transmission is either direct by nose or mouth contact or aerosol, or indirect by flies, drinking buckets, pasture, and feed. The mean incubation period is about 10 days, with a range of 3–14 days. The animal becomes quieter, has a fever of 39.0–40.5°C (102–105°F), nasal discharge, loss of appetite, and swelling of one or more lymph nodes of the mouth. The disease is highly contagious under conditions of crowding, exposure to severe climatic conditions such as rain and cold, and prolonged transportation.

Carrier animals, although of rare occurrence, are critical in maintenance of the streptococcus and in initiation of new outbreaks. Wherever possible, horses should be left in small groups, and weanlings should be separated from yearlings and mares. During an outbreak, affected horses and those known or suspected to be infected should be isolated immediately. Because the infected horse is usually the means by which *S. equi* is transported from one location to another, animals coming onto a farm should be held in an isolation area for 14–21 days and carefully monitored. Procaine penicillin G is the antibiotic of choice and quickly relieves enlargement of lymph nodes and fever. *See* Antibody; Penicillin; Streptococcus. [J.F.T.]

Stratigraphic nomenclature A nomenclatural system for classifying the rock record of Earth's geological history. This record consists of layered or stratified rocks (the stratigraphic record) that were laid down as sediments in the ocean (marine sediments), in lakes (lacustrine sediments), and by rivers in floodplains (fluvial sediments). Igneous rocks can also be layered (lava and tuff), but the stratigraphic record is largely concerned with sediments and their fossils. Sedimentary rocks can vary greatly, both vertically (in time) and horizontally (in sediment facies). *See* Sedimentary rocks.

Locally, a succession of sedimentary strata can be correlated to another succession through their physical relationship, if continuity of layering exists between them. To correlate rocks over long distances, it is necessary to know their ages. The age of the rock can be determined directly through radiometric dating or biostratigraphy. *See* Dating methods; Radiocarbon dating; Rock age determination.

In general, in the European tradition the stratigraphic column has been preferably subdivided by using the fossil content (and thus, by implication, chronostratigraphy) of the sediments. In North America, geologists have often emphasized the physical characteristics of sediments and their lateral facies equivalence (lithostratigraphy) to do the same. A hierarchy of lithostratigraphic units has been designated, but these do not imply equivalence with the geological time and time-stratigraphic units. *See* Facies (geology); Geological time scale.

Modern usage recognizes two kinds of stratigraphic units: material units (dealing with rocks and their fossil content without time implication) and temporal units (time-related units). The material units include lithostratigraphic (or pure rock) units. The basic unit of lithostratigraphy is the formation, which is defined as a mappable body of rock with distinctive lithologic characteristics and stratigraphic position. Formations are generally named after their geographic location.

Another material unit is the basic unit of biostratigraphy, the biozone, which is defined by its distinctive fossil content. The magnetic polarity (reversals of Earth's magnetic field) unit, polarity zone, lacking time implication, is also considered a material unit. *See* Rock magnetism.

Temporal units include the geochronologic (time) units, and the chronostratigraphic (time-rock) units. The geochronologic units are subdivisions of geological time, whereas the chronostratigraphic are their equivalents and represent rocks deposited during each interval of time.

Another type of chronostratigraphic unit, the synthem, is defined as an unconformity-bounded regional body of sediments that represents a cycle of sedimentation in response to changes in relative sea level or tectonics. *See* Sea-level fluctuations.

Stratigraphic nomenclature is applicable mainly to the rocks deposited since the appearance of common microscopic and

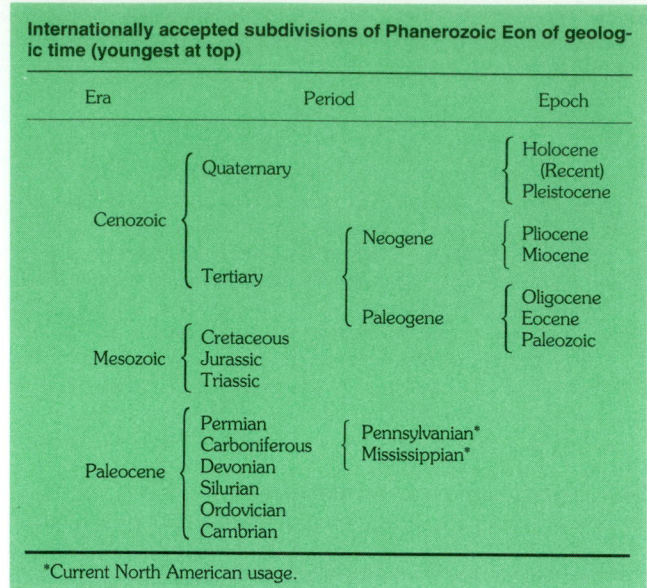

Internationally accepted subdivisions of Phanerozoic Eon of geologic time (youngest at top)

Era	Period		Epoch
Cenozoic	Quaternary		Holocene (Recent)
			Pleistocene
	Tertiary	Neogene	Pliocene
			Miocene
		Paleogene	Oligocene
			Eocene
			Paleozoic
Mesozoic	Cretaceous		
	Jurassic		
	Triassic		
Paleocene	Permian		
	Carboniferous	Pennsylvanian*	
		Mississippian*	
	Devonian		
	Silurian		
	Ordovician		
	Cambrian		

*Current North American usage.

correlatable fossils in the Earth history (the Phanerozoic Eon, or the last 5.9×10^8 years). Sediments of the pre-Phanerozoic (the Precambrian or Cryptozoic Eon; see table) contain only traces of life, and thus their subdivision is based largely on radiometric dating of volcanic rocks.

Sequence stratigraphy is that branch of stratigraphy that subdivides the sedimentary record along continental margins and in interior basins into a succession of depositional sequences as regional and interregional correlative units. Modern concepts of sequence stratigraphy were largely developed by geologists working in the petroleum exploration industry. *See* Basin; Continental margin; Seismic stratigraphy; Well logging.

The basic lithologic unit of sequence stratigraphy is the depositional sequence; a sequence is the body of rock deposited during a complete cycle of sea-level change, starting with a sea-level fall, followed by a rise, to the subsequent fall. *See* Stratigraphy. [B.U.H.]

Stratigraphy The branch of geology that studies layered or stratified rocks. Chiefly it concerns sedimentary rocks, but its principles may also be applied to layered igneous rocks, such as lavas, and tuffs, and to metamorphic rocks that were formed from sedimentary or volcanic rocks. It deals with the observed interrelations of the layers of such rocks and with the historical conclusions that can be inferred from those interrelations. Other aspects of sedimentary geology are sedimentary petrography (the study of the materials composing sedimentary rocks) and sedimentation (the study of the processes by which sediments are formed at present). Stratigraphy is of great practical value with important application to petroleum exploration. *See* Sedimentary rocks; Sedimentology.

The first task of stratigraphy is the description of local sequences of strata; from these descriptions local geologic history can be inferred. The second task of stratigraphy is the correlation of these local sequences, that is, the determination of their mutual time relations and the integration of the local histories into a regional or worldwide chronologic framework. The third task of stratigraphy is interpretation of the geologic history of the earth from the scattered data of local sequences and criteria of correlation. The conclusions of sedimentation, sedimentary petrology, and plant and animal ecology provide clues for this interpretation. *See* Rock age determination; Stratigraphic nomenclature. [J.R.]

Stratosphere The stable layer of the atmosphere extending from the top of the troposphere (6–10 mi or 10–16 km) to the temperature maximum at 30–34 mi (50–55 km). In the lower stratosphere below about 19 mi (30 km), temperature slowly decreases (in winter high latitudes), remains constant (in mid-latitudes), or increases with height (in low latitudes). In the upper stratosphere (= 19–30 mi or 30–50 km) there is a general increase of temperature with height to above 62°F (290 K) over the summer pole and −9°F (250 K) over the winter pole. Winds in summer are steady moderate easterlies, but large-scale systems with a westerly "polar night" jet stream form in winter. In late winter and spring these systems also affect the lower stratosphere, causing large temperature changes called "sudden warmings" and also large increases in the tracer (ozone and radioactivity) content. *See* Jet stream.

In composition the stratosphere is notable for its photochemically formed ozone layer and its low water vapor content. Its thermal structure can be broadly explained in terms of the radiation balance due to its minor constituents of ozone, carbon dioxide, and water vapor. *See* Atmosphere; Tropopause; Troposphere. [R.J.Mu.]

Strawberry Low-growing perennials, spreading by stolons, with fruit consisting of a fleshy receptacle, and "seeds" in pits or nearly superficial on the receptacle. The strawberry in the United States is derived from two species: *Fragaria chiloensis*, which grows along the Pacific Coast of North and South America, and *F. virginiana*, the eastern meadow strawberry, both members of the order Rosales. *See* Rosales.

The strawberry is the most universally grown of the small fruits, both in the home garden and in commercial plantings. Home garden production is possible in nearly all of the states, provided water can be supplied where rainfall is insufficient. Commercial production is important in probably three-fourths of the states. The following states are large producers: Oregon, California, Tennessee, Michigan, Louisiana, Washington, Arkansas, Kentucky, and New York. *See* Fruit. [J.H.C.]

Stream function In fluid mechanics, a mathematical idea which satisfies identically, and therefore eliminates completely, the equation of mass conservation. If the flow field consists of only two space coordinates, for example, x and y, a single and very useful stream function $\psi(x,y)$ will arise. If there are three space coordinates, such as (x,y,z), multiple stream functions are needed, and the idea becomes much less useful and is much less widely employed.

The stream function not only is mathematically useful but also has a vivid physical meaning. Lines of constant ψ are streamlines of the flow; that is, they are everywhere parallel to the local velocity vector. No flow can exist normal to a streamline; thus, selected ψ lines can be interpreted as solid boundaries of the flow.

Further, ψ is also quantitatively useful. In plane flow, for any two points in the flow field, the difference in their stream function values represents the volume flow between the points. *See* Creeping flow; Fluid flow. [F.M.W.]

Stream gaging The measurement of streamflow. A stream-gaging station is a site on a river where a record of streamflow is obtained. It usually consists of an instrument installation to record the fluctuating water level and also consists of facilities for making current-meter measurements of discharge. The discharge of a river is measured in the field by a direct observation of velocity and of the cross-sectional area through which this velocity is applicable. By definition, discharge is the product of mean velocity times cross-sectional area. Direct measurement of velocity is made by a current meter, an instrument with vanes, blades, or rotor cups that turn a shaft at a rate depending on the speed of the water (see illus-

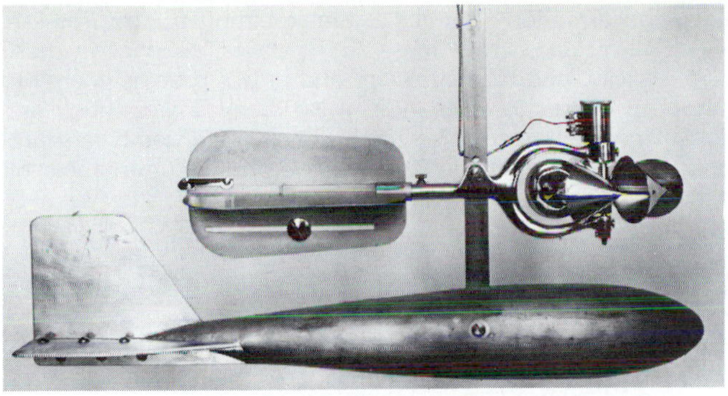

Price current meter and sounding weight for making periodic streamflow measurements. (*USGS*)

tration). Each rotation of the shaft causes a click that is audible in an earphone worn by the observer. The number of clicks in a clocked period of time can be translated into velocity by a calibration table. Indirect measurements are made by a water-stage recorder, which makes a graphical record of water level in a well that is connected by an open pipe to the natural stream. Fluctuations of water surface are followed by a float connected by a wire to the recorder. *See* FLOW MEASUREMENT. [L.B.L.]

Stream pollution Biological, or bacteriological, pollution in a stream indicated by the presence of the coliform group of organisms. While nonpathogenic itself, this group is a measure of the potential presence of contaminating organisms. Because of temperature, food supply, and predators, the environment provided by natural bodies of water is not favorable to the growth of pathogenic and coliform organisms. Physical factors, such as flocculation and sedimentation, also help remove bacteria. Any combination of these factors provides the basis for the biological self-purification capacity of natural water bodies.

Nonpolluted natural waters are usually saturated with dissolved oxygen. They may even be supersaturated because of the oxygen released by green water plants under the influence of sunlight. When an organic waste is discharged into a stream, the dissolved oxygen is utilized by the bacteria in their metabolic processes to oxidize the organic matter. The oxygen is replaced by reaeration through the water surface exposed to the atmosphere. This replenishment permits the bacteria to continue the oxidative process in an aerobic environment. In this state, reasonably clean appearance, freedom from odors, and normal animal and plant life are maintained.

An increase in the concentration of organic matter stimulates the growth of bacteria and increases the rates of oxidation and oxygen utilization. If the concentration of the organic pollutant is so great that the bacteria use oxygen more rapidly than it can be replaced, only anaerobic bacteria can survive and the stabilization of organic matter is accomplished in the absence of oxygen. Under these conditions, the water becomes unsightly and malodorous, and the normal flora and fauna are destroyed. Furthermore, anaerobic decomposition proceeds at a slower rate than aerobic. For maintenance of satisfactory conditions, minimal dissolved oxygen concentrations in receiving streams are of primary importance.

Municipal sewage and industrial wastes affect the oxygen content of a stream. Cooling water, used in some industrial processes, is characterized by high temperatures, which reduce the capacity of water to hold oxygen in solution. Municipal sewage requires oxygen for its stabilization by bacteria. Oxygen is utilized more rapidly than it is replaced by reaeration, resulting in the death of the normal aquatic life. Further downstream, as the oxygen demands are satisfied, reaeration replenishes the oxygen supply.

Polluted waters are deprived of oxygen by the exertion of the biochemical oxygen demand, which is defined as the quantity of oxygen required by the bacteria to oxidize the organic matter. Factors such as the turbulence of the stream flow, biological growths on the stream bed, insufficient nutrients, and inadequate bacteria in the river water influence the rate of oxidation in the stream as well as the removal of organic matter.

When a significant portion of the waste is in the suspended state, settling of the solids in a slow-moving stream is probable. The organic fraction of the sludge deposits decomposes anaerobically, except for the thin surface layer which is subjected to aerobic decomposition due to the dissolved oxygen in the overlying waters. In warm weather, when the anaerobic decomposition proceeds at a more rapid rate, gaseous end products, usually carbon dioxide and methane, rise through the supernatant waters. The evolution of the gas bubbles may raise sludge particles to the water surface. Although this phenomenon may occur while the water contains some dissolved oxygen, the more intense action during the summer usually results in depletion of dissolved oxygen.

Water may absorb oxygen from the atmosphere when the oxygen in solution falls below saturation. Dissolved oxygen for receiving waters is also derived from two other sources: that in the receiving water and the waste flow at the point of discharge, and that given off by green plants.

Unpolluted water maintains in solution the maximum quantity of dissolved oxygen. The saturation value is a function of temperature and the concentration of dissolved substances, such as chlorides. When oxygen is removed from solution, the deficiency is made up by the atmospheric oxygen, which is absorbed at the water surface and passes into solution. The oxygen balance in a stream is determined by the concentration of organic matter and its rate of oxidation, and by the dissolved oxygen concentration and the rate of reaeration. *See* ESTUARINE OCEANOGRAPHY; FRESH-WATER ECOSYSTEM; SEWAGE DISPOSAL; WATER POLLUTION. [D.J.O.]

Stream transport and deposition The sediment debris load of streams is a natural corollary to the degradation of the landscape by weathering and erosion. Eroded material reaches stream channels through rills and minor tributaries, being carried by the transporting power of running water and by mass movement, that is, by slippage, slides, or creep. The size represented may vary from clay to boulders. At any place in the stream system the material furnished from places upstream either is carried away or, if there is insufficient transporting ability, is accumulated as a depositional feature. The accumulation of deposited debris tends toward increased ease of movement, and this tends eventually to bring into balance the transporting ability of the stream and the debris load to be transported. [L.B.L.]

Streaming potential The potential which is produced when a liquid is forced to flow through a capillary or a porous solid. The streaming potential is one of four related electrokinetic phenomena which depend upon the presence of an electrical double layer at a solid-liquid interface. This electrical double layer is made up of ions of one charge type which are fixed to the surface of the solid and an equal number of mobile ions of the opposite charge which are distributed through the neighboring region of the liquid phase. In such a system the movement of liquid over the surface of the solid produces an electric current, because the flow of liquid causes a displacement of the mobile counterions with respect to the fixed charges on the solid surface. The applied potential necessary to reduce the net flow of electricity to zero is the streaming potential. [Q.V.W.]

Streamline flow A condition of fluid flow characterized by the absence of turbulence. Other designations employed are laminar flow or viscous flow.

Fluid-flow particles in streamline flow follow well-defined continuous paths or streamlines. At a fixed point in streamline flow, the flow velocity either remains constant (steady flow) or varies in a regular fashion with time (unsteady flow). In some instances, streamline flow can best be depicted as formed from thin layers of fluid which slip past each other (lamellar flow). As an illustration, streamline flow in a straight pipe might be considered as formed from layers in the shape of concentric annuli. If the flow can properly be represented by thin, plane layers (laminae) sliding past each other, the flow is commonly referred to as laminar flow, although this term is often used to designate streamline flow in general.

The persistence of streamline flow in a given system is largely dependent on the value of a nondimensional parameter called the Reynolds number. When the Reynolds number of a particular flow exceeds a certain value (the critical Reynolds number), transition from streamline flow to turbulent flow generally takes place. *See* BOUNDARY-LAYER FLOW; FLUID MECHANICS; LAMINAR FLOW; REYNOLDS NUMBER; STREAMLINING; TURBULENT FLOW.

[A.G.H.]

Streamlining The contouring of a body to reduce its resistance to motion through a fluid. The resistance to motion is referred to as the drag of the body.

Streamlining of a body must take into account the nature of drag forces. Drag forces are of four types: induced drag, wave drag, pressure (or form) drag, and skin friction. Induced drag is usually important for finite lifting surfaces such as wings; wave drag is important for bodies moving at supersonic speeds. Bodies moving at subsonic speeds are influenced by pressure drag and skin friction. Pressure drag is caused by inequalities in pressure forces acting in the direction of motion of the body. The action of viscosity tends to cause the flow to separate from the surface of the body with the consequent formation of a region of swirling or eddy flow termed the body wake. This eddy formation leads to a reduction in the downstream pressure on the body and hence gives rise to a force opposite to the body motion.

In general, the streamlining of a body in subsonic flow is the contouring of the body in such a manner that the wake is reduced to a minimum. Drag is then mainly the result of skin friction. Because one of the main causes of flow separation, and hence wake formation, is rapid deceleration of flow along the body, the contouring must make deceleration gradual. These considerations lead to the following general rules for streamlining: (1) The forward portion of the body should be well rounded, and (2) the body should curve back gradually from the forward section to a tapering after-section with the avoidance of sharp corners along the body surface. These conditions are well illustrated by teardrop shapes, as in the illustration. *See* STREAMLINE FLOW.

At supersonic speeds the airflow can accommodate sudden changes in direction by being compressed or expanded. Where this change in direction first occurs, a compression shock wave is created. The flow changes direction again at the midpoint of the body, creating an expansion shock wave. At the tail of the body the direction changes again, creating a compression shock wave. At each of these shock waves, changes in pressure, density, and velocity occur, and in this process energy is lost. This energy loss results in wave drag. Bodies which are streamlined for supersonic speeds are characterized by a sharp pointed nose, a sharp pointed tail, and a minimum number of direction changes between the nose and the tail. *See* AIRPLANE; SHOCK WAVE.

[A.G.H.]

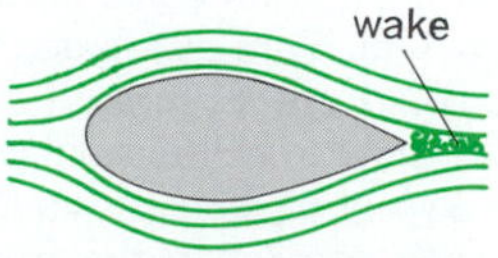

Flow indicated about a streamlined body which is traveling at subsonic speed.

Strength of materials A branch of applied mechanics concerned with the behavior of materials under load, relationships between externally applied loads and internal resisting forces, and associated deformations.

A material offers resistance to external load only insofar as the component elements can furnish cohesive strength, resistance to compaction, and resistance to sliding. The relations developed in strength of materials analysis evaluate the tensile, compressive, and shear stresses that a material is called upon to resist. The most important factors in determining the suitability of a structural or machine element for a particular application are strength and stiffness.

[J.B.S.]

Strepsiptera An order of twisted-wing insects that spend most of their life cycle as internal parasitoids of other insects. The adult male is only a few millimeters in wing span, is free living, and has only one pair of full flight wings (the posterior, or metathoracic, pair). The front (mesothoracic) wings are reduced to narrow, clublike organs that may function as halteres, or flight balancers. The male eyes are coarsely faceted and berrylike, and the antennae have four to seven segments, with some segments having finger- or bladelike extensions. Most adult females are immobile, blind, and larviform, and live inside the insect host. Rarely, the female is free living, with legs and eyes but no wings. More than 600 species of Strepsiptera are known, many of them not yet formally described and named. The order's relationship to other insect orders remains uncertain.

Both male and female begin larval life as mobile first-stage organisms with eyes and three pairs of functional legs. These triunguloids eventually attack the immature or adult forms of the host and enter their bodies, where they molt to an apodous instar. Two subsequent molts result in the divergence of the sexes and the differentiation of a hardened forebody that eventually protrudes through the host integument; the molted larval integuments are not shed but are incorporated in the general puparial wall and the neotenous adult female capsule. Males (and some adult females) emerge from the puparium and begin the search for a mate, probably through the mediation of airborne pheromones.

Strepsiptera are found worldwide in temperate and tropical habitats. They parasitize insects, mostly of orders Hemiptera and Hymenoptera, but also some cockroaches, mantids, orthopterans, flies, and silverfish. Their effect on the host is variable, ranging from reproductive failure to death. *See* ENDOPTERYGOTA; PHEROMONE.

[W.L.Bro.]

Streptococcus A large genus of spherical or ovoid bacteria that are characteristically arranged in pairs or in chains resembling strings of beads. Many of the streptococci that constitute part of the normal flora of the mouth, throat, intestine, and skin are harmless commensal forms; other streptococci are highly pathogenic. The cells are gram-positive and can grow either anaerobically or aerobically, although they cannot utilize oxygen for metabolic reactions. Glucose and other carbohydrates serve as sources of carbon and energy for growth. All members of the genus lack the enzyme catalase. Streptococci can be isolated from humans and other animals.

Streptococcus pyogenes is well known for its participation in many serious infections. It is a common cause of throat infection, which may be followed by more serious complications such as rheumatic fever, glomerulonephritis, and scarlet fever. Other beta-hemolytic streptococci participate in similar types of infection, but they are usually not associated with rheumatic fever and glomerulonephritis. Group B streptococci, which are usually beta-hemolytic, cause serious infections in newborns (such as meningitis) as well as in adults. Among the alpha-hemolytic and nonhemolytic streptococci, *S. pneumoniae* is an important cause of pneumonia and other respiratory infections. Vaccines that protect against infection by the most prevalent capsular serotypes are available. The viridans streptococci comprise a number of species commonly isolated from the mouth and throat. Although normally of low virulence, these streptococci are capable of causing serious infections (endocarditis, abcesses). *See* ENDOCARDITIS; NEPHRITIS; PNEUMONIA; RHEUMATIC FEVER, SCARLET FEVER. [K.Ru.]

Streptomycetaceae A family of bacteria comprising aerobic actinomycetes that usually produce a typical vegetative or substrate and aerial mycelium. They occur abundantly in soil and in other natural substrates. They produce a variety of enzymes (proteolytic, diastatic, and oxidative), vitamins (B_{12}), and antibiotics. Most of the important antibiotics isolated since the discovery of penicillin are produced by members of this family. It is sufficient to mention streptomycin, chloramphenicol, the tetracyclines, neomycin, erythromycin, novobiocin, nystatin, candicidin, and oleandomycin. *See* ANTIBIOTIC; ENZYME. [S.A.W.]

Streptomycin A colorless antibiotic substance produced by certain species of *Streptomyces*, mainly *S. griseus*. Streptomycin is reduced chemically to dihydrostreptomycin, giving a preparation that is frequently used in clinical practice in preference to streptomycin itself. Streptomycin and dihydrostreptomycin are active against gram-positive, gram-negative, and acid-fast bacteria, such as mycobacteria. These compounds have been used widely to treat infections of humans, animals, and plants. [S.A.W.]

Streptothricosis An acute or chronic infection of the epidermis, caused by the bacterium *Dermatophilus congolensis*, which results in an oozing dermatitis with scab formation. Streptothricosis includes dermatophilosis, mycotic dermatitis, lumpy wool, strawberry foot-rot, and cutaneous streptothricosis—diseases having a worldwide distribution and affecting a wide variety of species, including humans.

The infectious form of the bacterium is a coccoid, motile zoospore that is released when the skin becomes wet. Thus, the disease is closely associated with rainy seasons and wet summers. Zoospores lodge on the skin of susceptible animals and germinate by producing filaments which penetrate to the living epidermis, where the organism proliferates by branching mycelial growth.

Early cutaneous lesion of dermatophilosis in cattle reveals small vesicles, papules, and pus formation under hair plaques. An oozing dermatitis then appears as the disease progresses and exudates coalesce to form scabs, which change to hard crusts firmly adherent to the skin. The crusts enlarge and harden, and are often devoid of hair. Lesions occur on most areas of sheep, but the characteristic lesions in the wooled areas occur as numerous hard masses of crust or scab scattered irregularly over the back, flanks, and upper surface of the neck. Lesion resolution has been found to correlate with the presence of immunoglobulin A–containing plasma cells in the dermis and with the antibody levels to *D. congolensis* at the skin surface of infected sheep and cattle.

No single treatment is considered specific for dermatophilosis. Some observers claim that topical agents are successful for sheep and cattle. A number of systemic antibiotics are effective in treating the disease. A combination of streptomycin and penicillin has given good therapeutic results in both bovine and equine infections. [S.S.S.]

Stress (psychology) A term used to designate both environmental events and the body's response to such events. Of particular concern has been the relationship between stress and disease. Body responses has been thought to extend from behavioral aberrations to a host of other bodily problems referred to in some circles as psychosomatic disorders, that is, diseases resulting from an individual's interaction with the environment.

Designation of stressor. A specific event can be considered a stressor only if it evokes some aberrant bodily response. For example, if homeostasis of the electrolyte balance or body temperature is disrupted, there are immediate and serious consequences. However, most diseases that are believed to be stressor-related involve regulatory processes of the body, such as cardiovascular and gastrointestinal functioning. In these cases it is not as apparent when one is dealing with an aberrant stress response, particularly in the early phases of the disease process. An approach to the problem is suggested by the following, dealing specifically with cardiovascular functioning. *See* HOMEOSTASIS.

The cardiovascular system is a dynamic one, capable of altering its function over a wide range of values as metabolic demands vary. For example, the amount of blood pumped by the heart can increase four- or fivefold as one goes from sleep to strenuous exercise in order to meet the tissues' increasing needs for oxygen, nutrients, and the elimination of metabolic waste products. Since exercise in the healthy individual is not generally considered to be a potential disease-causing event, change in bodily functioning in itself cannot be considered to indicate the presence of a stressor. Rather, what must be determined is whether a bodily response to this event deviates in some manner from the immediate metabolic demands of the situation in such a way that the response is excessive and inefficient.

This point is illustrated by the following two observations. With exercise as the reference point, the cardiac output has been observed to increase disproportionately more to a threatening environmental event than the resultant increase in O_2 consumption. For example, exercising on a treadmill increased the cardiac output by approximately 100% and the O_2 consumption by 160%. On the other hand, exposure to the threatening event, while having a similar effect on the cardiac output, increases the O_2 consumption only half as much. A similar impairment of kidney function has also been reported under these circumstances. Although the kidney is not part of the cardiovascular system, it is intimately associated with cardiovascular functioning because it can control the blood pressure through the regulation of sodium excretion. Sodium excretion was observed to be impeded (that is, sodium was retained in the plasma) when a particularly threatening event was encountered, but not in the absence of the event. The net result is an acute elevation of the blood pressure, which may in the long run result in hypertension and its various sequelae, such as heart disease and stroke.

Designating the individual at risk. This is a particularly difficult research problem because it takes long periods of time to ascertain whether a given bodily response to environmental events is the precursor of an eventual disease state. The problem can be illustrated by research dealing with the causes of high blood pressure, a condition present in about 20% of the population by the age 60. The issue is to determine how to identify in early adulthood those individuals who will eventually develop high blood pressure, and then to try to prevent it or at least impede its course of development. One approach is suggested by the observation of appreciable individual differences in cardiovascular reactivity, such as heart rate, cardiac force, and blood pressure, to a standardized laboratory task inducing some pain or fright in young adult college men. For example, when measuring heart rate, some individuals do not react at all, some react modestly with increases of 10–25 beats per minute, while others react appreciably, in excess of 50 bpm.

Based on this observation, it may be asked whether those who react most are more apt to become hypertensive later in life. While this question cannot be answered with absolute certainty, there are other observations suggestive of this likelihood. First, the stress response of high heart rate appears to be a stable characteristic of some individuals. That is, individuals who are particularly reactive to one challenging event usually remain so to the other quite different challenges. Similarly, the less reactive subjects remain so under these varying circumstances. Second, particularly reactive subjects are more apt to have a hypertensive parent than the less reactive. Together, these data suggest that there is some sort of predisposition to overreact cardiovascularly to environmental events. These observations reinforce the point that events which are stressors for some are not so for others, although there are other bodily responses and possible disease states which may be influenced by these events in the cardiovascularly nonreactive individuals.

Other views. There are viewpoints about the nature of stress which differ somewhat from those discussed above. A most popular one is that of H. Selye, who viewed any environmental event as a stressor if it increases the demand on the body to readjust itself, with the intensity of the demand being the critical quality, not the nature of the demand. Another approach is illustrated by J. W. Mason, whose research has detailed the involvement of various psychoendocrines (hormones) to stressors. One facet of his work suggests that these various hormones prepare the body for action so that the individual can cope with the stress. This is very similar to Cannon's viewpoint in regard to epinephrine, the hormone released by the adrenal gland which not only stimulates the cardiovascular system, but as importantly, facilitates metabolism. Finally, a position offered by S. R. Burchfield looks at the stress response as maladaptive only when the individual uses coping styles which after numerous exposures to stress fail to conserve bodily resources and promote homeostasis. This view emphasizes the importance of behavioral adjustments to stressors. *See* EMOTION; HORMONE. [P.A.O.]

Stress and strain Related terms defining the intensity of internal reactive forces in a deformed body and associated unit changes of dimension, shape, or volume caused by externally applied forces. Stress is a measure of the internal reaction between elementary particles of a material in resisting separation, compaction, or sliding that tend to be induced by external forces. Total internal resisting forces are resultants of continuously distributed normal and parallel forces that are of varying magnitude and direction and are acting on elementary areas throughout the material. These forces may be distributed uniformly or nonuniformly. Stresses are identified as tensile, compressive, or shearing, according to the straining action.

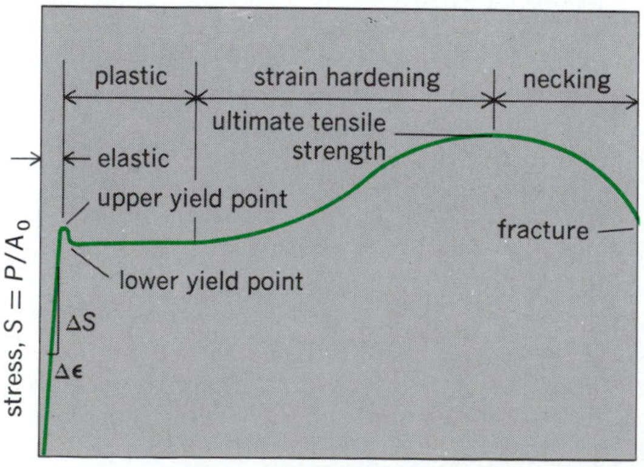

Stress-strain diagram for a low-carbon steel. ΔS = change in stress; $\Delta\epsilon$ = change in strain; P = force; A_0 = area of cross section.

Strain is a measure of deformation such as (1) linear strain, the change of length per unit of linear dimensions; (2) shear strain, the angular rotation in radians of an element undergoing change of shape by shearing forces; or (3) volumetric strain, the change of volume per unit of volume. The strains associated with stress are characteristic of the material. Strains completely recoverable on removal of stress are called elastic strains. Above a critical stress, both elastic and plastic strains exist, and that part remaining after unloading represents plastic deformation called inelastic strain. Inelastic strain reflects internal changes in the crystalline structure of the metal. Increase of resistance to continued plastic deformation due to more favorable rearrangement of the atomic structure is strain hardening.

A stress-strain diagram is a graphical representation of simultaneous values of stress and strain observed in tests and indicates material properties associated with both elastic and inelastic behavior (see illustration). It indicates significant values of stress-accompanying changes produced in the internal structure. *See* ELASTICITY; STRENGTH OF MATERIALS. [J.B.S.]

Stress concentration A condition in which a stress distribution has high localized stresses. A stress concentration is usually induced by an abrupt change in shape of a member. In the vicinity of notches, holes, changes in diameter of a shaft, or application points of concentrated loads, maximum stress is several times greater than where there is no geometrical discontinuity. Local stress disturbance is rapidly dissipated and effectively disappears at distances from the discontinuity equal to the major dimension of the section. *See* STRESS AND STRAIN. [J.B.S.]

Strigiformes The owls, an order of nocturnal, worldwide predacious birds that are probably most closely related to the goatsuckers. The strigiforms are arranged in four families: Ogygoptyngidae (fossil), Protostrigidae (fossil), Tytonidae (barn owls; 11 species), and Strigidae (typical owls; 135 species). The bay owls (*Phodilus*), occurring in southern Asia and Africa, are somewhat intermediate between the Tytonidae and the Strigidae and are sometimes placed in a separate family; here they are treated as a subfamily, Phodilinae, of the Tytonidae.

Owls are small to medium in size, with soft plumage of somber colors. The large head has a facial disk that covers the feathered parabolic reflectors of the bird's acute directional hearing system. The eyes are large and capable of sight in very dim light, and the bill is strong and hooked. The wings are long and rounded, and the flight feathers are fringed for silent flight. The ulna has a unique bony arch on the shaft. Their strong legs are short to medium in length and terminate in strong feet. Owls are excellent fliers but walk poorly. Of the four toes on each foot, two point forward and two backward and bear strong claws. Prey is detected by acute night vision or by directional hearing; owls can locate and catch their prey in total darkness. Owls are generally nonmigratory and solitary, but some species live in small flocks. Courtship takes place at night with a male hooting to a female, which answers. A strong pair bond exists between the monogamous male and female, which usually build their nest in a tree cavity. Some species nest on the ground, on cliff ledges, or in abandoned crow or hawk nests. The clutch of up to seven eggs is incubated by both sexes, and after hatching, the young stay in the nest and are cared for by both parents. *See* AVES. [W.J.B.]

Stripping The removal of volatile component from a liquid by vaporization. The stripping operation is an important step in many industrial processes which employ absorption to purify gases and to recover valuable components from the vapor phase. In such processes, the rich solution from the absorption step must be stripped in order to permit recovery of the absorbed solute and recycle of the solvent. *See* GAS ABSORPTION OPERATIONS.

Stripping may be accomplished by pressure reduction, the application of heat, or the use of an inert gas (stripping vapor). Many processes employ a combination of all three; that is, after absorption at elevated pressure, the solvent is flashed to atmospheric pressure, heated, and admitted into a stripping column which is provided with a bottom heater (reboiler). Solvent vapor generated in the rebolier or inert gas injected at the bottom of the column serves as stripping vapor which rises countercurrently to the downflowing solvent. When steam is used as stripping vapor for a system not miscible with water, the process is called steam stripping.

In addition to its use in conjunction with gas absorption, the term stripping is also used quite generally in technical fields to denote the removal of one or more components from a mixed system. Such usage covers (1) the distillation operation which takes place in a distilling column in the zone below the feed point, (2) the extraction of one or more components from a liquid by contact with a solvent liquid, (3) the removal of organic or metal coatings from solid surfaces, and (4) the removal of color from dyed fabrics. *See* DISTILLATION; ELECTROPLATING OF METALS; SOLVENT EXTRACTION. [A.L.K.]

Stroboscope An instrument for observing moving bodies by making them visible intermittently and thereby giving them the optical illusion of being stationary. A stroboscope may operate by illuminating the object with brilliant flashes of light or by imposing an intermittent shutter between the viewer and the object.

Stroboscopes are used to measure the speed of rotation or frequency of vibration of a mechanical part or system. They have the advantage over other instruments of not loading or disturbing the equipment under test. Mechanical equipment may be observed under actual operating conditions with the aid of stroboscopes. Parasitic oscillations, flaws, and unwanted distortion at high speeds are readily detected.

The flashing-light stroboscopes employ gas discharge tubes to provide a brilliant light source of very short duration. Tubes may vary from neon glow lamps, when very little light output is required, to special stroboscope tubes capable of producing flashes of several hundred thousand candlepower with a duration of only a few millionths of a second. *See* NEON GLOW LAMP; VAPOR LAMP. [A.R.E.]

Stroboscopic photography Stroboscopic or "strobe" photography generally refers to pictures of both single and multiple exposure taken by flashes of light from electrical discharges. Originally the term referred to multiple-exposed photographs made with a stroboscopic disk as a shutter. One essential feature of modern stroboscopic photography is a short exposure time, usually much shorter than can be obtained by a mechanical shutter.

High-speed photography with stroboscopic light has proved to be one of the most powerful research tools for observing fast motions in engineering and in science. Likewise, the electrical system of producing flashes of light in xenon-filled flash lamps is of great utility for studio, candid, and press photography. *See* PHOTOGRAPHY; STROBOSCOPE. [H.E.E.]

Stromatolite Any of the minute finely laminated structures mostly preserved in carbonate rock. Stromatolites may be planar, but are usually undulose or built up into finely layered domes or hemispherical masses. Such structures form today in marine intertidal zones and on certain lake shores by the growth of abundant mucilaginous filamentous blue-green algae which trap carbonate muds or clay. The very fine layers result from periodic influx of fine-gain sediment carried over shoreward flats by rising tides or winds. The filamentous algae glue this material in place, and during periods of nondeposition grow up through it, preparing the surface for the next ingress of sediment.

Algally induced stromatolites occur throughout the geologic column. Many distinctive and very large forms are known from the North American and Siberian middle Precambrian through Cambrian strata—the earliest organic structures known in the geologic record. The prevalence size, and diversity of form seen in the Precambrian and Cambrian have been proposed to result from the lack of the ubiquitous hordes of browsing gastropods seen today living and devouring the algal scum responsible for the mats. *See* BIOHERM; BIOSTROME. [J.L.Wi.]

Stromatoporoidea An extinct order thought to be sponges because of its close similarity to the class Sclerospongiae. Generally accepted stromatoporoids first appeared at the beginning of the Middle Ordovician and died out at the end of the Devonian. The Stromatoporoidea are worldwide in distribution, most commonly associated with bedded limestones and fossil reefs.

The stromatoporoid skeleton is called a coenosteum, which begins development as a thin, encrusting layer of irregular rods and plates called the peritheca. Most coenostea are hemispherical, or thick laminar or lens-shaped structures. However, some stromatoporoids are thin and irregularly undulatory and others are cylindrical, cone-shaped, branching, or anastomosing. *See* SCLEROSPONGIAE. [J.St.J.]

Strong nuclear interactions One of the fundamental physical interactions, which acts between a pair of hadrons. Hadrons include the nucleons, that is, neutrons and protons; the strange baryons, such as lambda (Λ) and sigma (Σ); the mesons, such as pion (π) and rho (ρ); and the strange meson, kaon (K). The nature of the interaction is determined principally through observation of the collision of a hadron pair. From this one learns that the interaction has a short range of about 10^{-15} m and is by far the dominant force within this range, being much larger than the electromagnetic interaction, which is next in magnitude. The strong interaction conserves parity and is time-reversal-invariant. *See* BARYON; HADRON; MESON; NUCLEON; PARITY (QUANTUM MECHANICS); STRANGE PARTICLES; SYMMETRY LAWS (PHYSICS).

The interaction between the baryons, including the nucleons and the strange baryons, is thought to arise from the exchange of mesons. The interaction for relatively large distances between nucleons is generated by the exchange of single pions (illustration *a*). At shorter separation distances the exchange of two-pion systems, such as the ρ (illustration *b*), dominates. An important contribution is furnished by the exchange of two separate pions with (illustration *c*) or without (illustration *d*) the formation of an excited state of the nucleon. The interaction between the strange baryons, and between the strange baryons and the nucleons, also arises in part from the exchange of pions, but the exchange of kaons can be equally important.

The range of the interaction generated by the exchanges can be calculated by using the simple formula below, where m

$$\text{Range} = \hbar/mc$$

is the mass of the exchange particles, $\hbar$ is Planck's constant divided by 2π, and c is the speed of light. According to the above equation, the range of the interaction developed when a single pion is exchanged (illustration *a*) is equal to 1.4×10^{-15} m, while that due to illustration *c* is 0.7×10^{-15} m.

The internal structure and finite size of the baryons (they consist of three quarks enclosed in a "bag," the quarks interacting

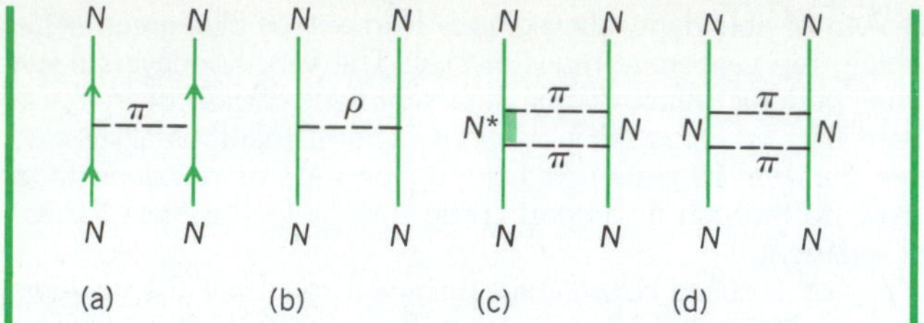

Interaction between nucleons (*a*) from exchange of single pion, (*b*) from exchange of ρ-meson, a two-ion system, (*c*) from exchange of two separate pions with formation of excited state of nucleon, and (*d*) without formation of excited state.

through the exchange of gluons) must be considered when the baryons are very close together, that is, for separation distances less than approximately 0.7×10^{-15} m. The consequences for nuclear forces are being actively investigated. *See* ELEMENTARY PARTICLE; FUNDAMENTAL INTERACTIONS; GLUONS; QUARKS. [H.F.]

Strongylida An order of nematodes in which the cephalic region may be adorned with three or six labia or the labia may be replaced by a corona radiata. All strongylid nematodes are parasitic. The order embraces eight superfamilies: Strongyloidea, Diaphanocephaloidea, Ancylostomatoidea, Trichostrongyloidea, Metastrongyloidea, Cosmocercoidea, Oxyuroidea, and Heterakoidea.

Strongyloidea. The Strongyloidea contain important parasites of reptiles, birds, and mammals. The early larval stages may be free-living microbivores, but the adults are always parasitic. Three species are important parasites of horses, *Strongylus vulgaris*, *S. equinus*, and *S. edentatus*. All three undergo direct life cycles; that is, infestations are acquired by ingestion of contaminated food.

Trichostrongyloidea. The Trichostrongyloidea comprise obligate parasites of all vertebrates but fishes. Normally they are intestinal parasites, but some are found in the lungs. The species are important parasites of sheep, cattle, and goats. The adult females lay eggs in the intestinal tract, which are passed out with the feces. In the presence of oxygen the eggs hatch in a few days. When the larvae are ingested by an appropriate host, their protective sheath is lost, and they proceed through the fourth larval stage to adulthood in the intestinal tract, where they may enter the mucosa. No migration takes place outside the gastrointestinal tract.

Metastrongyloidea. The Metastrongyloidea comprise obligate parasites of terrestrial and marine mammals, found commonly in the respiratory tract. In their life cycle they utilize both paratenic and intermediate hosts, among them a variety of invertebrates, including earthworms and mollusks. Two important species are *Metastrongylus apri* (swine lungworm) and *Angiostrongylus cantonensis* (rodent lungworm).

Heterakoidea. The Heterakoidea are capable of parasitizing almost any warm-blooded vertebrate as well as reptiles and amphibians. The species *Ascaridia galli* is the largest known nematode parasite of poultry; males are 2–3 in. (50–76 mm) long, and females 3–4.5 in. (75–116 mm). [A.R.M.]

Oxyuroidea. The Oxyuroidea constitute a large group of the phylum Nemata. Hosts include terrestrial mammals, birds, reptiles, amphibians, fishes, insects, and other arthropods.

The species are small to medium sized and thin bodied. With one exception, known life cycles are direct. Typically the eggs pass out of the host's alimentary tract onto the ground, where they become fully embryonated and infective. Normally the infective egg does not hatch until a susceptible animal ingests it. The cecum and colon of the host are the typical locations of these parasites. Larvae in all stages of development and adults occur in the gut.

The human pinworm, *Enterobius vermicularis*, is probably the most contagious of all helminthic diseases. It is estimated that 10% of the world's population suffer from this parasite, the majority being children. Indeed, incidence among schoolchildren in the cool regions of the world often approaches 100%. Infection occurs when eggs are inhaled or ingested. The most common method of transmission is from anus to mouth. Because of the aerial transmission, this disease is highly contagious. Though the infection is seldom serious, the behavioral symptoms are disturbing: nail biting, teeth grinding, anal scratching, insomnia, nightmares, and even convulsions. Several medical treatments are available, but there is often the danger of reinfestation from contaminated objects within the household or institution. *See* NEMATA. [J.T.L.]

Strontianite The mineral form of strontium carbonate, usually with some calcium replacing strontium. It characteristically occurs in veins with barite or celestite or as masses in certain sedimentary rocks. Strontianite is normally prismatic, but it may also be massive. It may be colorless or gray with yellow, green, or brownish tints. The hardness is 3½ on Mohs scale, and the specific gravity of 3.76. It occurs at Strontian, Scotland, and in Germany, Austria, Mexico, and India and, in the United States, in the Strontium Hills of Calfornia. *See* CARBONATE MINERALS; STRONTIUM. [R.I.Ha.]

Strontium A chemical element, Sr, atomic number 38, and atomic weight 87.62. Strontium is the least abundant of the alkaline-earth metals. The crust of the Earth is 0.042%

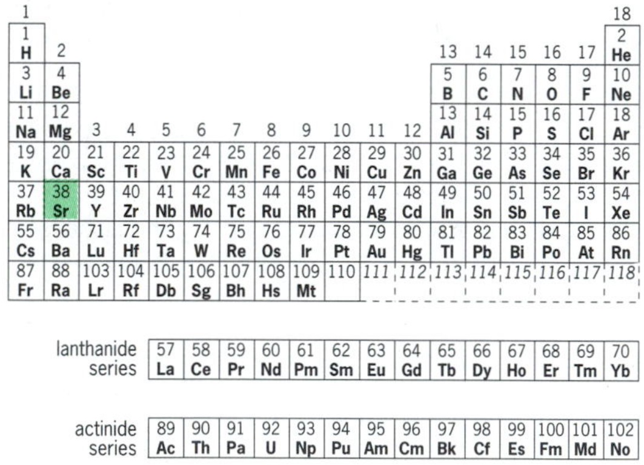

strontium, making this element as abundant as chlorine and sulfur. The main ores are celestite, $SrSO_4$, and strontianite, $SrCO_3$. *See* ALKALINE-EARTH METALS; STRONTIANITE.

Strontium nitrate is used in pyrotechnics, railroad flares, and tracer bullet formulations. Strontium hydroxide forms soaps and greases with a number of organic acids which are struc-

Properties of strontium

Property	Value
Atomic number	38
Atomic weight	87.62
Isotopes (stable)	84,86,87,88,90
Boiling point, °C	1638(?)
Melting point, °C	704(?)
Density, g/cm^3 at 20°C	2.6

turally stable, resistant to oxidation and breakdown over a wide temperature range.

Strontium is divalent in all its compounds which are, aside from the hydroxide, fluoride, and sulfate, quite soluble. Strontium is a weaker complex former than calcium, giving a few weak oxy complexes with tartrates, citrates, and so on. Some physical properties of the element are given in the table. [R.F.R.]

Strophomenida The largest order of articulate brachiopods which first appeared in Lower Ordovician times and ultimately became extinct in the Late Triassic. The order is diverse and shows an extraordinary degree of morphological variation. Many of the more bizarre forms adopted an attached mode of life, either anchored by spines or cemented by part of the pedicle valve; such a change is accompanied or preceded by atrophy of the pedicle. Four suborders are recognized: Strophomenidina, Chonetidina, Productidina, and Oldhaminidina. *See* ARTICULATA (BRACHIOPODA); BRACHIOPODA. [A.J.R.]

Structural analysis The determination of stresses and strains in a given structure. All structures must be designed to carry loads without danger of overall collapse or failure of their components. One way to assure structural safety is to ascertain that stresses and strains produced by loads are less than those allowed by established design codes. This determination of stresses and strains in proposed new structures is a primary objective of structural analysis. *See* STRESS AND STRAIN.

Generally, analysis begins with a check of the overall stability of a structure. This involves first the determination of the forces exerted by the structure against its supports. If the supports are adequate to withstand these forces, they in turn react with equal but opposite forces against the structure.

If computation shows that the reactions balance the loads (weight of structure, occupants, stored materials, vehicles, wind, and earthquake forces), the structure is in static equilibrium. *See* STATICS.

The next step is determination of internal forces and unit stresses in the components of the structure. Finally, if necessary, the deformation of the structure as a whole and of its components may be calculated. Many of these tools are based on the assumption that the structure is elastic under loads; that is, stress is proportional to strain. *See* STRUCTURAL CONNECTIONS. [J.B.S.]

Structural connections Methods of joining the individual members of a structure to form a complete assembly. The connections furnish supporting reactions and transfer loads from one member to another. Loads are transferred by fasteners (rivets, bolts) or welding supplemented by suitable arrangements of plates, angles, or other structural shapes. When the end of a member must be free to rotate, a pinned connection is used.

The suitability of a connection depends on its deformational characteristics as well as its strength. Rotational flexibility or complete rigidity must be provided according to the degree of end restraint assumed in the design. A rigid connection maintains the original angles between connected members virtually unchanged after loading. Flexible or nonrestraining connections permit rotation approximately equal to that at the ends of a simply supported beam. Intermediate degrees of restraint are called semirigid.

A commonly used form of connection for rolled-beam sections, called a web connection, consists of two angles which are attached to opposite sides of a member and which are in turn connected to the web of a supporting beam, girder, column, or framing at right angles. A shelf angle may be added to facilitate erection (Fig. 1).

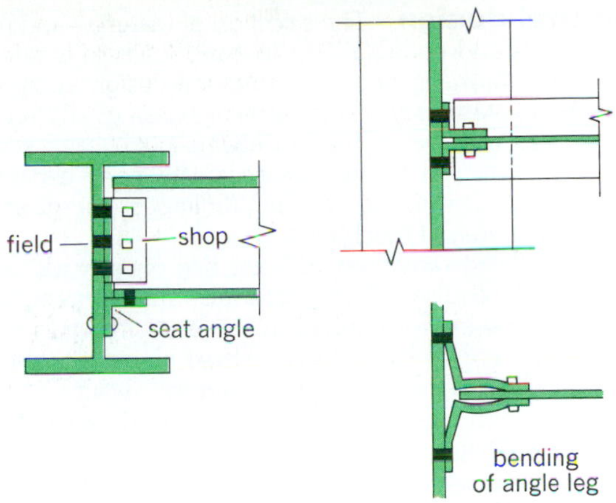

Fig. 1. Riveted or bolted web connections.

A bracket or seat on which the end of the beam rests is a seat connection; it is intended to furnish the end reaction of the supported beam. Two general types are used: The unstiffened seat provides bearing for the beam by a projecting plate or angle leg which offers resistance only by its own flexural strength (Fig. 2); the stiffened seat is supported by a vertical plate or angle which transfers the reaction force to the supporting member without flexural distortion of the outstanding seat.

When the action line of a transferred force does not pass through the centroid of the connecting fastener group or welds, the connection is subjected to rotational moment which produces additional shearing stresses in the connectors. The load transmitted by diagonal bracing to a supporting column flange through a gusset plate is eccentric with reference to the connecting fastener group.

In beam-to-column connections and stiffened seat connections or when members transfer loads to columns by a gusset plate or a bracket, the fasteners are subjected to tension forces caused by the eccentric connection. Although there are initial tensions in the fasteners, the final tension is not appreciably greater than the initial tension.

Rigidity and moment resistance are necessary at the ends of beams forming part of a continuous framework which must resist lateral and vertical loads. Wind pressures tend to distort a building frame, producing bending in the beams and columns which must be suitably connected to transfer moment and shear. The resisting moment can be furnished by various forms of angle T for fasteners or welded or bracket connections.

Where appreciably angular change between members is expected, and in special cases where a hinge support without moment resistance is desired, connections are pinned. Many bridge trusses and large girder spans have pin supports. *See* BOLT; JOINT (STRUCTURES); RIVET. [J.B.S.]

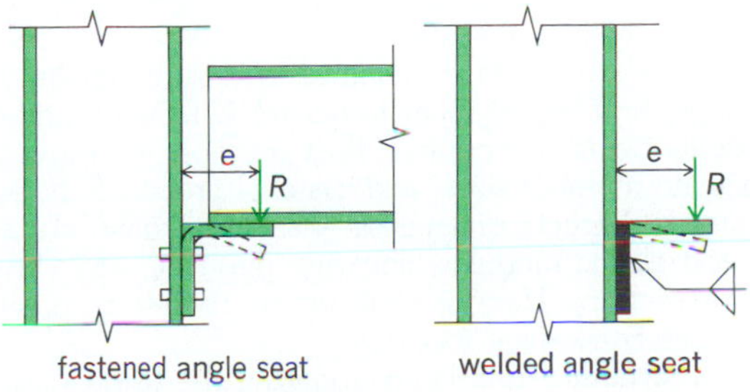

Fig. 2. Unstiffened seat connections. R = reaction load; e = distance to reaction from column face.

Structural design The selection of materials and member type, size, and configuration to carry loads in a safe and serviceable fashion. In general, structural design implies the engineering of stationary objects such as buildings and bridges, or objects that may be mobile but have a rigid shape such as ship hulls and aircraft frames. Devices with parts planned to move with relation to each other (linkages) are generally assigned to the area of mechanical design.

Structural design involves at least five distinct phases of work: project requirements, materials, structural scheme, analysis, and design. For unusual structures or materials a sixth phase, testing, should be included. These phases do not proceed in a rigid progression, since different materials can be most effective in different schemes, testing can result in changes to a design, and a final design is often reached by starting with a rough estimated design, then looping through several cycles of analysis and redesign. Often, several alternative designs will prove quite close in cost, strength, and serviceability. The structural engineer, owner, or end user would then make a selection based on other considerations.

Before starting design, the structural engineer must determine the criteria for acceptable performance. The loads or forces to be resisted must be provided. For specialized structures this may be given directly, as when supporting a known piece of machinery, or a crane of known capacity. For conventional buildings, building codes adopted on a municipal, county, or state level provide minimum design requirements for live loads (occupants and furnishings, snow on roofs, and so on). The engineer will calculate dead loads (structure and known, permanent intallations) during the design process. For the structure to be serviceable or useful, deflections must also be kept within limits, since it is possible for safe structures to be uncomfortably "bouncy." Very tight deflection limits are set on supports for machinery, since beam sag can cause driveshafts to bend, bearings to burn out, parts to misalign, and overhead cranes to stall. Beam stiffness also affects floor "bounciness," which can be annoying if not controlled. In addition, lateral deflection, sway, or drift of tall buildings is often held within approximately height/500 (1/500 of the building height) to minimize the likelihood of motion discomfort in occupants of upper floors on windy days. *See* Loads, dynamic; Loads, transverse.

Technological advances have created many novel materials such as carbon fiber- and boron fiber-reinforced composites, which have excellent strength, stiffness, and strenth-to-weight properties. However, because of the high cost and difficult or unusual fabrication techniques required, glass-reinforced composites such as fiberglass are more common, but are limited to lightly loaded applications. The main materials used in structural design are more prosaic and include steel, aluminum, reinforced concrete, wood, and masonry. *See* Composite material; Masonry; Precast concrete; Prestressed concrete; Reinforced concrete; Structural materials.

In an actual structure, various forces are experienced by structural members, including tension, compression, flexure (bending), shear, and torsion (twist). However, the structural scheme selected will influence which of these forces occurs most frequently, and this will influence the process of material selection. *See* Shear; Torsion.

Analysis of structures is required to ensure stability (static equilibrium), find the member forces to be resisted, and determine deflections. It requires that member configuration, approximate member sizes, and material properties be known or assumed. Aspects of analysis include: equilibrium; stress, strain, and elastic modulus; linearity; plasticity; and curvature and plane sections. Various methods are used to complete the analysis. *See* Structural analyses.

Once a structure has been analyzed (by using geometry alone if the analysis is determinate, or geometry plus assumed member sizes and materials if indeterminate), final design can proceed. Deflections and allowable stresses or ultimate strength must be checked against criteria provided either by the owner or by the governing building codes. Safety at working loads must be calculated. Several methods are available, and the choice depends on the types of materials that will be used. Once a satisfactory scheme has been analyzed and designed to be within project criteria, the information must be presented for fabrication and construction. This is commonly done through drawings, which indicate all basic dimensions, materials, member sizes, the anticipated loads used in design, and anticipated forces to be carried through connections.

[R.L.T.; L.M.J.]

Structural geology The branch of geology which, in the broadest sense, deals with the description and analysis of the forms and interrelations of rock bodies. Structural features that came into being during the formation of the rocks are termed primary; those that were imposed upon the rocks after their formation, secondary. Although it also considers primary structures such as bedding and igneous features, the major role of structural geology is the description and analysis of secondary structures. Tectonics is synonymous with structural geology. Investigations of tectonic features involve their descriptions and interrelations, the analysis of the mechanics by which they were formed, and finally the synthesis of all structural data. *See* Fault and fault structures; Geology; Tectonophysics.

[P.H.O.]

Structural materials Construction materials which, because of their ability to withstand external forces, are considered in the design of a structural framework. Materials used primarily for decoration, insulation, or other than structural purposes are not included in this group.

Clay products include the solid masonry units such as brick and the hollow masonry units such as clay tile or terra-cotta. Building stones generally used are limestone, sandstone, granite, and marble. Until the advent of steel and concrete, stone was the most important building material. Concrete is a mixture of cement, mineral aggregate, and water, which, if combined in proper proportions, form a plastic mixture capable of being placed in forms and of hardening through the hydration of the cement. Wood is unique among the common structural materials. Its tensile strength is generally greater than its compressive strength. The principal structural metals are the structural steels, steel castings, aluminum alloys, magnesium alloys, and cast and wrought iron. Composite materials have made important contributions to the design of naval, air, and space vehicles and structures. The most familiar composite is formed using filaments or woven cloth of fiberglass embedded in a polyester or epoxy resin base. *See* Brick; Composite material; Concrete; Steel; Stone and stone products; Structural steel; Terra-cotta; Tile.

[C.M.A.]

Structural petrology The study of the structural aspects of rocks, as distinct from the purely chemical and mineralogical studies that are generally emphasized in other branches of petrology. The term was originally used synonymously with petrofabric analysis, but is sometimes restricted to denote the analysis of only microscopic structural and textural features. *See* Petrofabric analysis; Petrography.

[J.M.Ch.]

Structural plate A flat plate or slab which is supported along the edges, such as the bottom or cover of a tank, cylinder head, bulkhead, or floor panel. A structural plate bends when subjected to forces applied normal to its surface. The bending produces curvature in all planes normal to the plate; this dishing action differs from bending of a beam where curvature occurs in a single plane under symmetrical loads. The analysis of stresses and deflections in a plate, while based on assumptions essentially the same as those used for a beam, is more complicated because of simultaneous bending in all planes. Simplified approximate analyses predict stresses and deflections useful in design. *See* BEAM. [J.B.S.]

Structural steel Steel used in engineering structures, usually manufactured by either the open-hearth or the electric-furnace process. The exception is carbon-steel plates and shapes whose thickness is $^7/_{16}$ in. (11 mm) or less and which are used in structures subject to static loads only. These products may be made from acid-Bessemer steel. The physical properties and chemical composition are governed by standard specifications of the American Society for Testing and Materials (ASTM). Structural steel can be fabricated into numerous shapes for various construction purposes. *See* STEEL. [W.G.B.]

Structures (engineering) Definite arrangements of related elements or members joined together to enable given sets of loads to be supported in a prescribed position. The principal structures designed by civil engineers are bridges, buildings, dams, docks, retaining walls, tanks and bins for storage, transmission towers, radio and television towers, highway pavements, and aircraft landing strips. As exploration of space and of the ocean depths is expanded, engineering structures differing in type and function from those now in use will be necessary. *See* AIRPORT ENGINEERING; BRIDGE; BUILDINGS; COASTAL ENGINEERING; DAM; FOUNDATIONS; HIGHWAY ENGINEERING; PAVEMENT; STORAGE TANK; TOWER.

A structure should be useful, safe, economical, and esthetically attractive. Primary emphasis must be placed upon safety, and attention once given to economy as the secondary design consideration is giving way to concern with structural appearance and esthetics. [C.M.A.]

Struthioniformes A small order of weak-flying, partridge-like birds and giant flightless ratite birds found in the southern continents. Their relationship to other birds is unknown. The struthioniforms are characterized by a palaeognathous palate which provides freedom of movement between segments of the palate and the jaw. For this reason they are frequently placed in a separate superorder, the Palaeognathae. However, that distinction places too much emphasis on their separation from other birds. Contrary to common opinion, no evidence supports the concept that the palaeognathous birds are primitive among living birds.

The order Struthioniformes can be divided into three suborders, Lithornithi, Tinami, and Struthioni, each of which includes both fossil and extant representatives.

The struthioniforms are medium-sized to giant birds. The ostriches are the largest extant birds, but some of the moas, elephant birds, and dromornithids were even larger. The head is small; a medium flattened bill is a common feature except in the kiwis, which have a long bill used for probing into the ground for worms. The wings are reduced in all forms except the lithionithids and tinamous, which can still fly, and the plumage is soft. Their legs are strong, and all forms run well. The struthioniforms eat a variety of foods, especially large fruits and other large food items. Breeding is polygamous, with two or more females laying eggs in a single nest. The males are responsible for incubation, and they assume the major or sole role in caring for the downy young, which leave the nest after hatching. *See* AVES; RATITES. [W.J.B.]

Strychnine The principal alkaloid present in nux vomica, the seeds of a tree native to India. *Strychnos nux-vomica*. The complex structure is shown here. In the medical practice of an early day strychnine had a reputation as a cardiovascular stimulant, respiratory stimulant, and bitter tonic. The modern view is

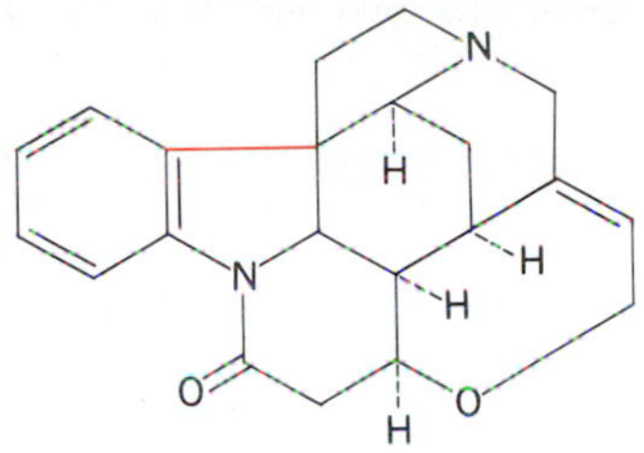

that the therapeutically desirable effects are obtainable only with doses bordering on the toxic. Pharmacological studies have shown that many of the therapeutic applications of strychnine have little or no rationale. *See* ALKALOID. [S.M.K.]

Sturgeon Any of 10 species of large fish which comprise the family Acipenseridae in the order Acipenseriformes. These fish are found in North Temperate Zone waters, where they are almost exclusively bottom-living and feed on organisms such as mollusks, worms, and larvae. The body has five rows of bony plates, of which one is situated dorsally, two laterally, and two ventrally. The snout is elongate and there are four barbels on its lower surface; the mouth is ventrally located, and in the adult the jaws lack teeth (see illustration). The skeleton is mainly cartilaginous.

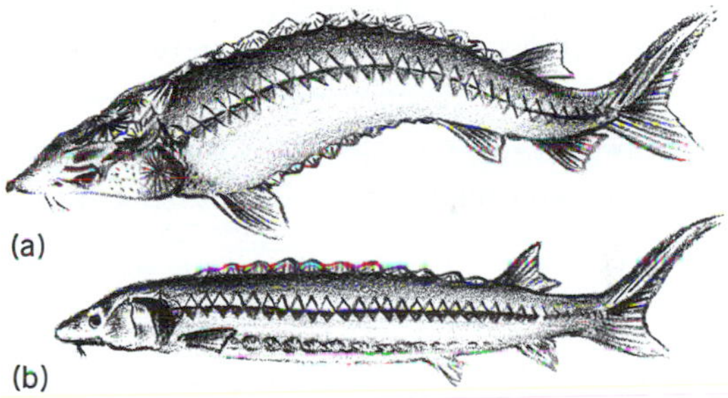

Two species of sturgeons found in United States coastal waters. *(a)* Atlantic sturgeon (*Acipenser oxyrhynchus*). *(b)* Short-nosed sturgeon (*A. brevirostrus*).

Sturgeons are known primarily for the roe, which is processed as caviar. A single female can produce millions of eggs, which may be removed from dead fish or stripped from living fish. The smoked flesh of the sturgeon is a delicacy. *See* ACIPENSERIFORMES. [C.B.C.]

Sty An acute inflammation of one of the glands of the eyelids. A painful swelling appears at the root of an eyelash, usually an infection caused by a staphylococcus organism. The swelling results from pus formation, local edema, and obstruction of the gland duct. Drainage is usually spontaneous. In certain cases, the inner glands may become chronically obstructed and the inflammation will change to a small hard scar or nodule called a chalazion. [E.G.St./N.K.M.]

Stylasterina An order of the class Hydrozoa of the phylum Coelenterata, including several brightly colored branching or encrusting "corals" of warm seas. (True corals belong to a different class, the Anthozoa). The calcareous skeleton is covered by living tissue and is penetrated by ramifying tubes. Nutritive polyps, the gastrozooids, lie in cups on one surface or along certain edges of the skeletal substance. A spine, or style, at the base of each cup gives the order its name.

Some authorities combine the Stylasterina with the Milleporina in a single order, the Hydrocorallina. *See* ANTHOZOA; HYDROZOA. [S.Cr.]

Stylolites Irregular surfaces occurring in certain rocks, mostly parallel to bedding planes, in which small toothlike projections on one side of the surface fit into cavities of like shape on the other side (see illustration). Stylolites are most common in limestones and dolomites but are also present in many other kinds of rock, including sandstones, gypsum beds, and cherts. *See* DOLOMITE; LIMESTONE; SEDIMENTARY ROCKS. [R.Si.]

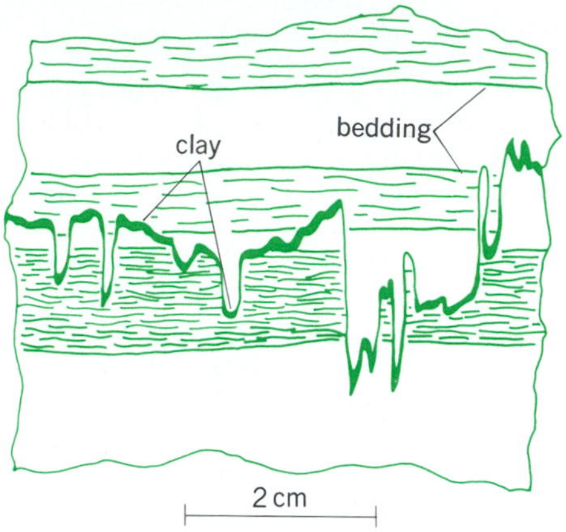

Stylolite in limestone.

Stylommatophora A superorder of the molluscan subclass Pulmonata containing about 20,500 species that are grouped into 56 families. Nearly all land snails without an operculum are stylommatophorans. They have eyespots on the tips of a pair of retractile tentacles, hermaphroditic reproduction with partial fusion of the male and female system, and, normally, a second pair of retractile tentacles that function as chemoreceptors. All are air-breathing and most are truly terrestrial. Three orders, based on excretory structures, are recognized. The more primitive Orthurethra and Mesurethra have less efficient water conservation devices than do the more specialized members of the order Sigmurethra, in which a closed ureter functions in water conservation. The latter specialization apparently is a prerequisite for evolution toward a sluglike structure, since all 16 families with slugs or sluglike taxa belong to the Sigmurethra. *See* PULMONATA. [G.A.S.]

Styrene A colorless, liquid hydrocarbon with the formula $C_6H_5CH{=}CH_2$. It boils at 145.2°C (293.4°F) and freezes at −30.6°C (−23.1°F). The ethylenic linkage of styrene readily undergoes addition reactions and under the influence of light, heat, or catalysts undergoes self-addition or polymerization to yield polystyrene.

The majority of the styrene used is converted into polystyrene, but other thermoplastic or even thermosetting resins are prepared from styrene by copolymerization with suitable comonomers. A smaller quantity of styrene goes into the manufacture of elastomers or synthetic rubbers.

Styrene is a skin irritant. Prolonged breathing of air containing more than 400 ppm of styrene vapor may be injurious to health. *See* POLYMERIZATION; POLYSTYRENE RESIN. [C.K.B.]

Subduction zones Elongate areas on the Earth defined by deep trenches, lines of volcanoes parallel to the trenches, and zones of large earthquakes dipping from the trenches landward beneath the volcanoes. Subduction zones mark areas where one of the Earth's plates, usually a dense oceanic plate, descends beneath another plate. The subducted plate (or lithosphere), comprising the crust and part of the upper mantle, is mixed back into the Earth's deeper mantle. This process balances the creation of lithosphere that occurs at the mid-ocean ridge system. The convergence of two plates occurs at rates of 1–10 km (0.6–6 mi) per million years. Stress and phase changes in the upper part of the cold descending plate produce large earthquakes in a narrow band called the Benioff-Wadati zone, which is confined to the upper portion of the plate and can extend as deep as 600 km (360 mi). The plate is heated as it descends, and the resulting release of water leads to melting of the overlying mantle. This melt rises to produce the linear volcanic chains that are one of the most striking features of subduction zones. *See* EARTHQUAKE; LITHOSPHERE; PLATE TECTONICS.

The forearcs of subduction zones, between the trench and the volcanic arc, are where much of the mechanical interaction between the overriding and subducting plates occurs. Forearcs are classified as accretionary or nonaccretionary. In accretionary forearcs, layers of sediment carried into the trench are scraped off the subducting plate and added to the upper plate as deformed, fault-bounded slices. The addition of successive slices of sediment causes the overriding plate to build outward and the older accreted material to be uplifted and rotated landward to form accretionary prisms.

Trenches that lack a thick sequence of turbidite sands usually have nonaccretionary forearcs. The sediments that reach these trenches are entirely subducted, to be either underplated to the upper plate at deeper levels or carried down and mixed back into the mantle. Some of these forearcs are actually eroded by the downgoing plate.

Some subduction zones are characterized by strong coupling between the two plates; in effect, the upper plate is moving outward over the lower plate. These subduction zones typically have shallow dips and thick accretionary prisms and are characterized by high earthquake activity; their back-arc regions develop foreland fold and thrust belts, in which seaward-dipping faults stack up slices of crustal material.

Other subduction zones show a very different type of interaction between the upper and lower plates. These typically have steeply dipping slabs; these steep dips result from the vertical sinking of the slabs. Old oceanic lithosphere is denser than the asthenosphere on which it sits and, once the lithosphere begins to subduct, tends to sink vertically in addition to moving downdip. A consequence of this motion is that the slab rolls back; that is, the hinge at which it bends moves away from the volcanic arc. [S.H.Bl.]

Subgiant star A member of the family of stars intermediate between giants and the main sequence in the Hertzsprung-Russell (H-R) diagram. The mean luminosity in a subgiant is about 10 times the Sun; the surface temperature lies between 4000 and 7000 K (6700 and 12,100°F). The masses are about 1.4 times that of the Sun. *See* HERTZSPRUNG-RUSSELL DIAGRAM; STAR.

The subgiants are of particular importance in current theories of stellar evolution. If a main-sequence star has exhausted about 12% of its mass of hydrogen, the star begins to evolve, expanding, cooling at the surface, and brightening. Old stars of population II, of masses about 1.35 times the Sun, are now evolving into the subgiant region of the H-R diagram. The age of the oldest known stellar systems can be determined from the luminosities of the subgiants to be between 5,000,000,000 and 12,000,000,000 years. *See* STELLAR EVOLUTION. [J.L.Gr.]

Subgraywacke An argillaceous sandstone with a composition intermediate between graywacke and orthoquartzite.

Precise definitions of the boundaries of this group vary among sedimentary petrologists, but there is general agreement that these rocks contain moderate to large amounts of rock fragments, some clay matrix, and at least a small amount of feldspar.

The rock fragments in subgraywackes may be dominated by chert and other sedimentary species rather than metamorphic or igneous rocks. The pore spaces are filled with a combination of clay matrix and mineral cement, usually quartz and carbonate. The clay matrix is mainly muscovite (illite) with smaller amounts of kaolinite and, in some few cases, biotite and chlorite.

Subgraywackes are probably the most abundant sandstone type and are found in deposits of all ages. They occur in moderately thick stratigraphic sections and are of wide lateral extent. *See* GRAYWACKE; SANDSTONE; SEDIMENTARY ROCKS. [R.Si.]

Sublimation The process by which solids are transformed directly to the vapor state without passing through the liquid phase. Sublimation is of considerable importance in the purification of certain substances such as iodine, naphthalene, and sulfur.

Sublimation is a universal phenomenon exhibited by all solids at temperatures below their triple points. For example, it is a common experience to observe the disappearance of snow from the ground even though the temperature is below the freezing point and liquid water is never present. The rate of disappearance is low, of course, because the vapor pressure of ice is low below its triple point. Sublimation is a scientifically and technically useful phenomenon, therefore, only when the vapor pressure of the solid phase is high enough for the rate of vaporization to be rapid. *See* PHASE EQUILIBRIUM; TRIPLE POINT; VAPOR PRESSURE. [N.H.N.]

Sublitharenite A sandstone that contains between 5 and 25% rock fragments and in which rock fragments are more abundant than feldspar grains. Sublitharenites are intermediate in composition between more quartz-rich sandstone (quartzarenites) and more rock-fragment-rich sandstones (litharenites). The terms litharenite and sublitharenite refer to these two members of the lithic sandstone clan. Such sandstones are called graywacke, subgraywacke, or lithic or sublithic arenite in other classifications.

Like litharenites, sublitharenites are found in diverse depositional environments and in a wide spectrum of sedimentary basins. The texture and composition of the rock fragments in sublitharenite are used to interpret the history of sandstones in the same manner as for litharenites. Sublitharenites are the most common type of terrigenous sandstone in the Earth's crust. *See* ARKOSE; GRAYWACKE; LITHARENITE; SANDSTONE; SUBGRAYWACKE. [E.F.McB.]

Submarine A ship which can operate completely submerged in the water. The term formerly was applied to any ship or vessel capable of operating completely underwater, but is now usually restricted to a ship built for military purposes. The term submersible usually is applied to smaller underwater vehicles built for commercial work or pleasure.

Throughout World Wars I and II, submarines operated primarily on the surface and submerged only in the final stage of an attack or to evade detection. Since high surface speed was essential in this type of naval warfare, submarine hulls were designed for minimum surface resistance, sacrificing submerged performance. The increasing effectiveness of radar and air patrols in locating and attacking surfaced submarines eventually forced them to remain submerged for longer periods. In the late stages of World War II the snorkel, a breathing tube permitting the submarine to get air for its diesel engines without surfacing, was developed. However, even a snorkeling submarine may be detected visually, especially by aircraft, because it is submerged only slightly, and the snorkel tube leaves a wake. *See* NAVAL SHIP.

The nuclear reactor, developed after World War II, does not require air for its fuel, and enhanced the undersea capabilities of the submarine. As the submarine began to operate more extensively beneath the sea, its hull was modified in successive designs to give minimum submerged resistance, which necessitated sacrifice of surface performance. *See* SHIP POWERING AND STEERING; SHIP PROPULSION REACTOR.

Submarines can be classified by their primary military missions. Attack submarines are fast, long-range ships used primarily with torpedoes, but they may also carry mines which can be laid from the torpedo tubes. They carry sensitive underwater sound receivers and transmitters (sonar) in order to fulfill their mission. They may be armed with acoustic homing torpedoes and cruise missiles.

Ballistic-missile submarines carry long-range, solid-fueled missiles that can be launched while submerged and may be fitted with nuclear warheads. The submarine can remain submerged for many days and, undetected, launch missiles on any target within range. The missiles are stowed in and launched from vertical tubes. *See* MISSILE.

Experimental submarines are occasionally built to test new designs of hull shape, deeper depth capability, power plants, or controls.

Submersibles are usually small, deep-diving vehicles of limited operation. Their primary use is for exploration and oceanographic study of the ocean depths, development of equipment, or commercial work. Some designs take advantage of the natural forces of gravity and buoyancy for vertical motion. Other designs use vertically oriented propellers to propel the craft up and down. Movement is restricted to short distances and slow speed because of small size and small battery capacity. Submersibles are usually carried or towed to their sites of operation. *See* UNDERSEA VEHICLES.

There are many similarities between a surface ship and a submarine. However, the submarine has some unique features that enable it to submerge and resist great sea pressure. Most submarines have two hulls, a pressure hull and a nonpressure hull. The pressure hull is the watertight, pressureproof envelope in which equipment operates and the officers and crew live. The structure of this pressure hull is very strong to resist the forces created by the weight of water above the ship when it is submerged. In certain areas of the submarine there is a nonpressure hull of lighter structure, forming the main ballast tanks. A nonwatertight superstructure provides a smooth, fair envelope to cover pipes, valves, and fittings on top of the hull. Above the superstructure the fairwater similarly encloses the bridge, the periscope, mast supports, and, if one is provided, the conning tower. [M.S.F.; J.H.We.]

Submarine cable A cable, mainly for communications usage, laid on the ocean floor to provide overseas links. This term is sometimes applied to power cables in water, but these are not usually of great length.

The first successful submarine telegraph cable was laid across the North Atlantic in 1866. The technology of submarine telephone cables was developed over a period of many years, partly in response to the limitations of overseas radio telephone service. Submarine telephone cables over large distances require periodically spaced amplifiers, called repeaters, to restore transmission losses in the cables. In May 1950, the first two repeatered submarine telephone cables (SA type) were laid between Key West and Havana. These were the prototype for the first transatlantic telephone cable (TAT-1, SB type) which was laid in 1955–1956. *See* TELEPHONE SERVICE.

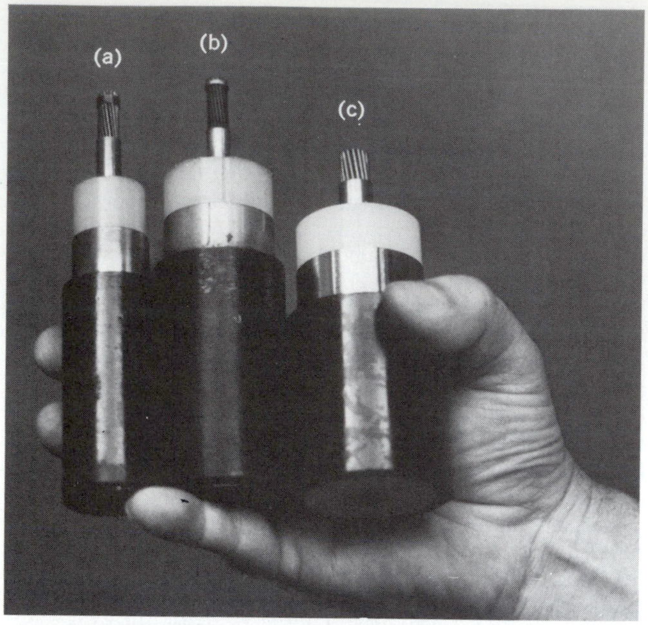

Submarine cables (*a*) SD, (*b*) SF, and (*c*) SG. (*American Telephone and Telegraph Company*)

The SA and SB systems used twin directional armored cables with long, flexible "sausage-like" type repeaters having the same appearance as the cable except for larger diameter. In 1954, before TAT-1 went in service, traffic volume studies showed submarine cable systems of greater capacity would be required to handle future growth. To obtain wider-band systems, a new approach to cable and repeater design was necessary. In the Bell System's SD and subsequent systems, flexible repeaters were replaced with repeaters in rigid beryllium-copper cylinders, and instead of armored deep-sea cable the strength members were placed in the center of the coaxial structure (see illustration). Coaxial cable used in submarine cable systems differs from that used on land in that the ocean type uses only a single coaxial tube rather than multiple tubes for transmission. *See* COAXIAL CABLE; COMMUNICATION CABLES; TRANSMISSION LINES.

Undersea telecommunication cables offer an efficient use for optical fiber technology since the potential cost-per-channel advantage over present systems is very high. Special cable design is required for light-wave communication. Mechanically the cable must protect the optical fibers from excessive microbending and tensile stress during laying and recovery. It must also be heavy enough to sink to the bottom, and provide a suitable power conductor to transmit the required system current and voltage. *See* OPTICAL COMMUNICATIONS. [C.H.Sp.]

Submarine canyon A relatively narrow, deep depression in the sea floor with steep slopes, the bottom of which grades continuously downward. Great submarine canyons are cut into the continental slopes off most coasts of the world. Many of these have rocky walls thousands of feet high. They have narrow inner gorges, winding courses, numerous tributaries and are, in fact, quite comparable to the great canyons of the land (see illustration). Some of the canyons are direct continuations of land canyons, and others occur off large rivers which flow through broad flat-floored valleys on land. The submarine canyons extend outward down the slope virtually to the deep ocean floor. The outer portions of these sea valleys are of modest dimensions and extend across broad, gently sloping fans, comparable to the piedmont fans along the fronts of

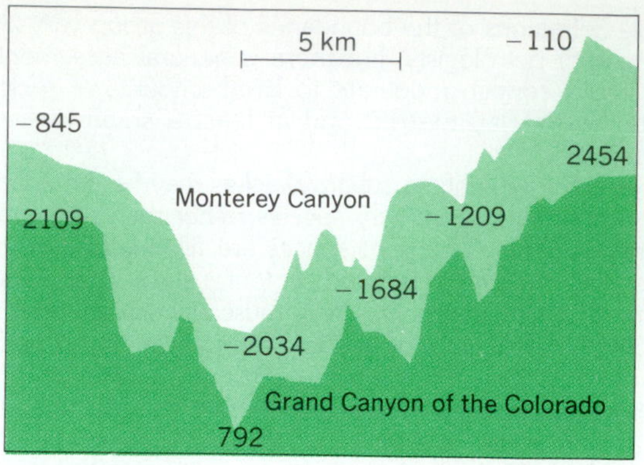

Transverse profile of the submarine canyon in Monterey Bay compared to a profile of the Grand Canyon of the Colorado River in Arizona. Elevations are in meters (1 m = 3.3 ft), and the vertical is magnified five times.

mountain ranges in arid regions. *See* CONTINENTAL MARGIN; MARINE GEOLOGY. [F.P.S.]

Submarine fan A thick sequence of land- and shelf-derived sediments that accumulates at the base of the continental slope and along the continental rise in the deep-marine environment. Submarine fans commonly develop seaward of submarine canyons. Fully developed submarine fans are composed of channel and channel-mouth deposits. They have been recognized both in modern deep-marine environments and in ancient deep-marine sequences preserved in the rock record. Fanlike features may also develop in deep lakes.

By definition, submarine fans are fan- or cone-shaped deposits that are more or less analogous to alluvial fans on land. In reality, however, outlines of both modern and ancient fans show great variability in shape. In fact, very few modern fans are actually cone-shaped. Fan-shaped deposits are more likely to develop in large open basins than in small restricted basins. Most fans are elongate, some are trapezoidal, and some are complex in shape. Submarine fans also vary greatly in size. *See* BASIN; SUBMARINE CANYON.

Turbidity currents are the most dominant process responsible for transporting and depositing sediment in the deep sea. Major portions of submarine-fan sequences are composed of deposits of turbidity currents. These currents flow downslope on the ocean floor because they possess a higher density than the surrounding seawater. Turbidity currents are capable of transporting large volumes (tens to hundreds of cubic kilometers) of sediment over long distances (hundreds of kilometers) at great velocities in deep-marine environments; coarse particles can be suspended and transported by these currents for long distances. This mechanism of sediment transport is responsible for the presence of gravels and coarse sand in the deep-sea environment. *See* TURBIDITY CURRENT.

A single depositional event from a turbidity current generates a turbidite bed. Although submarine fans consist predominantly of turbidites, turbidites, can occur also as thin sheet deposits unrelated to typical fan growth. In addition to turbidites, deposits of debris flows and slumps may also be present in submarine fans. *See* TURBIDITE.

Submarine fans can be classified into four types based on tectonic settings: immature passive margin, mature passive margin, active margin, and mixed setting. In all tectonic settings, sediment thickness of fans commonly ranges from 1000 to 5000 m (3300 to 16,500 ft). Fans deposited in some trenches and remnant ocean basins may reach a total thickness of up to 10,000 m (33,000 ft).

The immature passive-margin setting (North Sea type) represents an early stage of basin evolution in a divergent margin. Fans developed in this setting are small, rich in sand, and possess well-developed lobes.

The mature passive-margin setting (Atlantic type) represents an advanced stage of basin evolution in a divergent margin. Distal source, wide coastal plains and shelves, low gradients, low sand/mud ratios, and large basins are diagnostic features of this setting. Fans developed in this setting are large, rich in mud, and lack well-developed lobes.

The active-margin (Pacific type) represents convergent, transform, and collision margins. Proximal source, narrow coastal plains and shelves, high gradients, high sand/mud ratios, and small basins are diagnostic features of this setting. Fans developed in active-margin setting are small, rich in sand, and exhibit well-developed lobes.

The mixed-setting category is necessary to accommodate fans that cannot be classified readily into one of the other three types. For example, the Bengal Fan (Bay of Bengal), bounded by both passive and active margins, is considered to occupy an elongate remnant ocean basin. This type of setting refers to ocean basins that exist just prior to crustal collision in areas of plate consumption. As collision proceeds, orogenic highlands shed great quantities of sediments that are eventually deposited as submarine fans in elongate remnant ocean basins. *See* OROGENY; PLATE TECTONICS.

Global lowering of seal level is a primary factor in the development of submarine fans in passive continental margins.

Many hydrocarbon-bearing deposits in submarine canyons and fans that occur at both active- and passive-margin settings correlate with periods of low sea level. *See* DELTA; GLACIAL EPOCH; SEA-LEVEL FLUCTUATIONS.

Submarine-fan sandstones comprise economically important oil and gas reservoirs throughout the world. Mature passive-margin fans are mud-rich systems, and their lenticular sand bodies within meandering channels may not develop economically important reservoirs. Active-margin fans, on the other hand, are sand-rich systems, and their sheetlike sand bodies of braided channels and depositional lobes usually constitute important hydrocarbon reservoirs. *See* CONTINENTAL MARGIN; PETROLEUM GEOLOGY; SANDSTONE. [G.Sh.]

Submillimeter astronomy Astronomical observations carried out in the region of the electromagnetic spectrum with wavelengths from approximately 0.3 to 1.0 millimeter. Submillimeter astronomy is one of the last windows of the electromagnetic spectrum to be investigated, because of the effect of the Earth's atmosphere. Molecules in the atmosphere, primarily water vapor, absorb (dim) infrared and submillimeter light. This can be circumvented by placing telescopes on high dry mountaintops, flying them on airplanes, or orbiting them above the Earth's atmosphere. The *Infrared Astronomical Satellite* (*IRAS*), which operated in 1983, was such an experiment. Since 1974, observations have been carried out with a 36-in (0.9-m) telescope on the Kuiper Airborne Observatory, an airplane that flies as high as 45,000 ft (13.7 km).

However, in observational astronomy it is desirable to achieve great sensitivity, and this requires a telescope with maximum collecting area, hence of large diameter, which is difficult to mount on a satellite or airplane. Also, it is desirable to make maps with the greatest amount of resolved detail.

To achieve these objectives, large reflectors at dry mountain sites are now used for submillimeter astronomy. The James Clerk Maxwell Telescope (see illus.), the Swedish-ESO Submillimeter Telescope, and the Caltech Submillimeter Telescope are not constructed like conventional optical telescopes, with a single aluminized glass mirror. Instead, their parabolic surfaces are formed from a large number of carefully contoured aluminum panels. *See* SUBMILLIMETER-WAVE TECHNOLOGY; TELESCOPE.

From the Earth's surface it is not possible to observe all submillimeter wavelengths. At wavelengths outside three atmospheric windows at 0.35 to 0.45, and 0.8 mm, it is necessary to fly or orbit telescopes to make submillimeter observations.

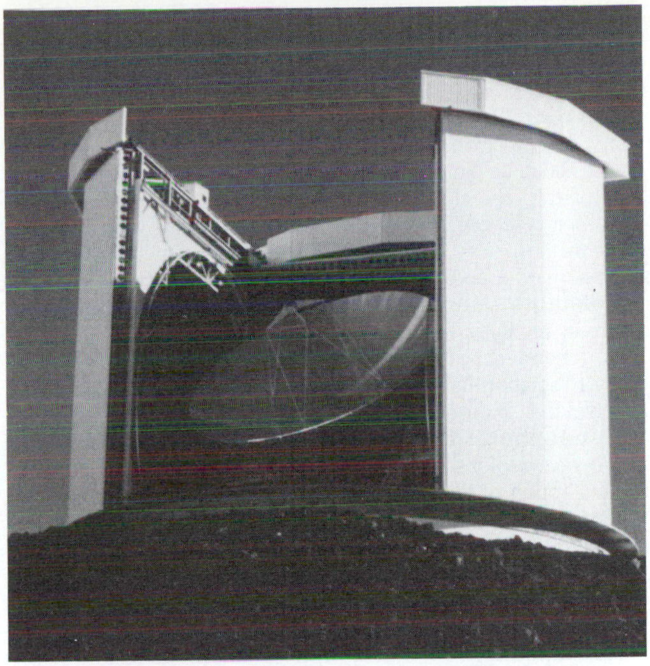

James Clerk Maxwell Telescope, 49-ft-diameter (15-m) submillimeter telescope situated at Mauna Kea, Hawaii.

Submillimeter observations carried out so far and planned for the future involve interstellar chemistry, protostars, interstellar dust, and cosmology. *See* COSMOCHEMISTRY; COSMOLOGY; INTERSTELLAR MATTER. [K.Kri.]

Submillimeter-wave technology The generation, detection, and application of radiation in the submillimeter-wave region of the electromagnetic spectrum. In practical terms, this region corresponds to wavelengths between about 30 micrometers and 3 millimeters (see illus.). At its extremes the region overlaps with the microwave and infrared. *See* ELECTROMAGNETIC RADIATION; INFRARED RADIATION; MICROWAVE.

Widespread use of the submillimeter-wave region did not occur until two revolutionary developments in the late 1950s and early 1960s: (1) the practical realization of Fourier transform spectroscopy, which led to significant improvements in the quality of broad band submillimeter-wave spectroscopy; and (2) the discovery of two reasonably intense submillimeter-wave laser sources, the hydrogen cyanide and the water-vapor electrically excited lasers. *See* INFRARED SPECTROSCOPY; LASER.

Radiation sources. The main application of incoherent radiation sources is as broad band sources in Fourier transform

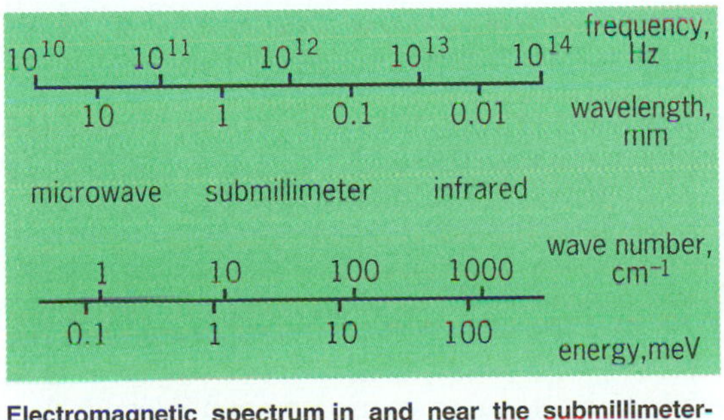

Electromagnetic spectrum in and near the submillimeter-wave region, showing its frequency, wavelength, wave number, and photon energy equivalents.

spectrometers. The two most commonly used sources are the mercury-vapor lamp and the globar. *See* MERCURY-VAPOR LAMP.

There are several categories of coherent submillimeter-wave radiation sources: (1) nonrelativistic electron-beam tubes such as klystrons; (2) relativistic electron-beam devices such as gyrotrons, free-electron lasers, and synchrotrons; and (3) solid-state devices such as Gunn and IMPATT oscillators. With the exception of the free-electron laser, these are long-wave devices. The final category, gas lasers, covers the entire submillimeter-wave region. *See* KLYSTRON; MICROWAVE SOLID-STATE DEVICES; SYNCHROTRON RADIATION.

Detectors. Most submillimeter-wave detectors are either thermal, rectifying, or photon detectors. The first respond to heat generated by absorbed radiation, the second by rectification of radiation-field-induced currents, the third by absorption of individual photons.

Atmospheric transparency. Between the visible and microwave regions the atmosphere is effectively opaque, with the exception of two windows in the near-infrared and three in the submillimeter-wave region, at 94, 140, and 220 GHz. The significant feature of the submillimeter-wave windows is that the levels of attenuation involved are not as seriously increased by adverse atmospheric conditions as those of the near-infrared.

Applications. In the past, applications of radiation in the submillimeter-wave region were distributed evenly over the region and largely concerned research activities. Now, however, as a result of developments toward integrated, inexpensive transmitter and receiver systems, applications include many radar- and telecommunications-related activities centered on the 94-GHz atmospheric window. The telecommunications-related applications are mainly military, but spin-offs into the civil area are expected. [J.R.B.]

Subsonic flight Movement of a vehicle through the atmosphere at a speed appreciably below that of sound waves. Subsonic flight extends from zero (hovering) to a speed approximately 85% of sonic speed corresponding to the ambient temperature. At higher vehicle velocities the local velocity of air passing over the vehicle surface may exceed sonic speed, and true subsonic flight no longer exists.

Vehicle type may range from a small helicopter, which operates at all times in the lower range of the velocity scale, to an intercontinental ballistic missile, which is operative throughout this and other velocity regimes, but is in subsonic flight for only a few seconds. The design of each is affected by the same principles of subsonic aerodynamics. Subsonic flow of a fluid such as air may be subdivided into a range of velocities in which the flow may be considered incompressible without appreciable error (below a velocity of approximately 300 mi/h or 135 m/s), and a higher range in which the compressible nature of the fluid becomes significant. In both cases the viscosity of the fluid is important. The theories which apply to compressible, inviscid fluids may be used almost without modification in some low-subsonic problems, and in other cases the results offered by these theories may be modified to account for the effects of viscosity and compressibility. *See* COMPRESSIBLE FLOW; TRANSONIC FLIGHT; VISCOSITY.

A typical subsonic wing cross section (airfoil) has a rounded front portion (leading edge) and a sharp rear portion (trailing edge). Air approaching the leading edge comes to rest at some point on the leading edge, with flow above this point proceeding around the upper airfoil surface to the trailing edge, and flow below passing along the lower surface to the same point, where the flow again theoretically has zero velocity. The two points of zero local velocity are known as stagnation points. If the path from front to rear stagnation point is longer along the upper surface than along the lower surface, the mean velocity of flow along the upper surface must be greater than that along the lower surface. Thus, in accordance with the principle of conservation of energy, the mean static pressure must be less on the upper surface than on the lower surface. This pressure difference, applied to the surface area with proper regard to force direction, gives a net lifting force. Lift is defined as a force perpendicular to the direction of fluid flow relative to the body, or more clearly, perpendicular to the free-stream velocity vector. *See* AIRFOIL; BERNOULLI'S THEOREM.

The wing, as the lifting device, whether fixed, as in the airplane, or rotating, as in the helicopter, is probably the most important aerodynamic part of an aircraft. However, stability and control characteristics of the subsonic airplane depend on the complete structure. Control is the ability of the airplane to rotate about any of the three mutually perpendicular axes meeting at its center of gravity. Static stability is the tendency of the airplane to return to its original flight attitude when disturbed by a moment about any of the axes. [J.E.Ma.]

Substitution reaction One of a class of chemical reactions in which one atom or group (of atoms) replaces another atom or group in the structure of a molecule or ion. Usually, the new group takes the same structural position that was occupied by the group replaced.

Substitution reactions involve the attack of a reagent, which is the source of the new atom or group, on the substrate, the molecule or ion in which the replacement occurs. They involve the formation of a new bond and the breaking of an old bond. Substitution reactions are classified according to the nature of the reagent (electrophilic, nucleophilic, or radical) and according to the nature of the site of substitution (saturated carbon atom or aromatic carbon atom). *See* ELECTROPHLIC AND NUCLEOPHILIC REAGENTS.

Systematic names for substitution reactions are composed of the parts: name of group introduced + de + name of group replaced + ation, with suitable elision or change of vowels for euphony. Thus, the replacement of bromine by a methoxy group is called methoxydebromination. *See* ORGANIC REACTION MECHANISM. [J.F.B.]

Subsynchronous resonance An electrical resonant frequency in an electric power system, less than the line frequency, that results from the insertion of series capacitors in power transmission lines. These series capacitors are inserted in some alternating-current transmission lines to cancel out a part of the line and system reactance and thereby increase the maximum power that can be transmitted over the line.

One serious consequence that must be analyzed and avoided or dealt with by countermeasures is the torsional interaction of mechanical oscillations in turbine generators with these subsynchronous electrical oscillations. This can produce self-excitation or negative damping, resulting in a buildup of dangerous shaft torques, which in one case resulted in machine failure.

For cases where the negative damping is a problem, several countermeasures have been used. One is simply to avoid a degree of compensation that can give a resonant frequency at any existing mechanical frequency. Usually this strategy is backed up by a subsynchronous relay to trip the generating unit if any unusual line configuration causes it to be resonant. *See* RELAY.

Another scheme is a dynamic stabilizer consisting of a thyristor-controlled reactor and a fast, sensitive detector of generator oscillations used to inject a modulated current to cancel, in effect, the subsynchronous oscillating current. It is necessary that the phase relation be held very accurately or a residual negative damping will exist. *See* ELECTRIC POWER SYSTEMS; SEMICONDUCTOR RECTIFIER; STATIC VAR COMPENSATOR. [L.A.K.]

Subtraction One of the four fundamental operations of arithmetic and algebra. Subtraction is often regarded as an operation inverse to addition, that is, if a and b are numbers, the number $a - b$ is defined as that number which added to b gives a. The more modern viewpoint eliminates subtraction completely by considering the number $a - b$ as the sum of a and that number (denoted by $-b$) which added to b gives 0. The number symbolized by $-b$ is called the inverse of b (with respect to addition). Every real number has a unique inverse (the number 0 is its own inverse). In this sense "subtraction" may be performed on objects of many different kinds, and the original numerical operation greatly extended. *See* ADDITION; ALGEBRA; DIVISION; MULTIPLICATION; NUMBER THEORY. [L.M.Bl.]

Subway engineering A branch of transportation relating to planning, general layout, detailed design, construction, and operation of subways. These underground systems are a major element in mass transportation. As metropolitan areas increase in size and population and vehicular traffic becomes more congested, subways are being given increased consideration in planning urban mass-transit systems.

The most modern subway trains are designed for high-speed travel with maximum safety in comfortable air-conditioned cars. Train movements are controlled by automatic equipment. Since construction of a subway involves putting a railway of special design underground, most of the engineering specialties relating to railways are required in building a subway. However, in subway work there is an emphasis on extensive and costly tunnel construction, with tunnels often being at sufficient depth to pass under bodies of water; underground passenger terminals; elaborate ventilation facilities; lengthy underground electric power distribution systems; lighting facilities; escalators for transporting passengers to and from street level; noise control; and safe and reliable signaling facilities.

Often difficult to install, subways are constructed by either the open-cut method, the tunneling method, or both. With the open-cut method, a deep open trench is excavated. Open-trench procedures limit the depth at which work can be performed. Tunneling permits subways to be installed at great depths. However, the operation can be costly due to the type of soil or foundation material encountered, unstable conditions, and extensive flows of water.

A factor which must be considered is nearby buildings and other structures, especially if these are of sufficient weight to cause heavy loads on the underlying foundation material and large pressures in the area of the proposed subway. Frequently, heavy underpinning must be installed to support adjacent buildings during subway construction. This underpinning may consist of steel pilings and heavy horizontal members on which buildings can be supported. Large concrete columns reinforced with steel also are used as supports.

Determination of the best location for a subway requires much careful study and consideration of various general plans, including underpinning. In some metropolitan areas the subway is built underground in the central business area and aboveground at other locations. *See* RAILROAD ENGINEERING; TRANSPORTATION ENGINEERING. [A.N.C.]

Sucrose An oligosaccharide, α-D-glucopyranosyl-β-D-fructofuranoside, also known as saccharose, cane sugar, or beet sugar. The structure is shown below. Sucrose is very soluble in

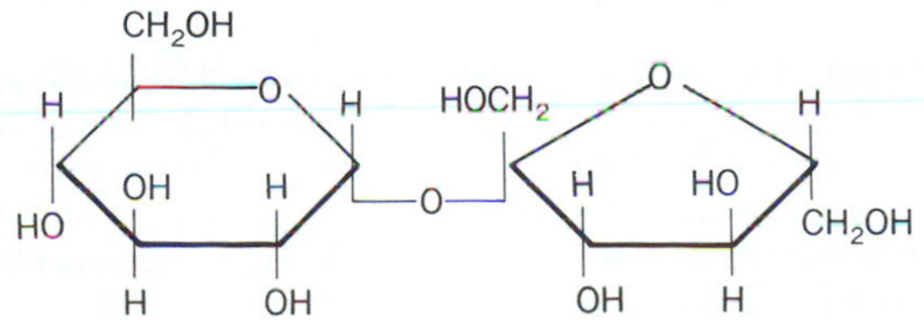

water and crystallizes from the medium in the anhydrous form. The sugar occurs universally throughout the plant kingdom in fruits, seeds, flowers, and roots of plants. Honey consists principally of sucrose and its hydrolysis products. Sugarcane and sugarbeets are the chief sources for the preparation of sucrose on a large scale. Another source of commercial interest is the sap of maple trees. *See* OLIGOSACCHARIDE; SUGARBEET; SUGARCANE. [W.Z.H.]

Suctoria A small specialized subclass of the protozoan class Ciliatea whose members were long considered entirely separate from the "true" ciliates. The sole order of this subclass is Suctorida. These forms show a number of highly specialized features. Most conspicuous are their tentacles, often numerous, which serve as mouths. These multiple organelles of ingestion fasten to the pellicle of prey organisms, generally passing ciliates. By forces not entirely understood, the tentacles are used to suck out the prey's protoplasm to provide sustenance for the suctorian. Nearly all species are stalked, and the sedentary, mature forms are devoid of any external ciliature. Young larval forms are produced by both endogenous and exogenous budding. These forms bear locomotor cilia and

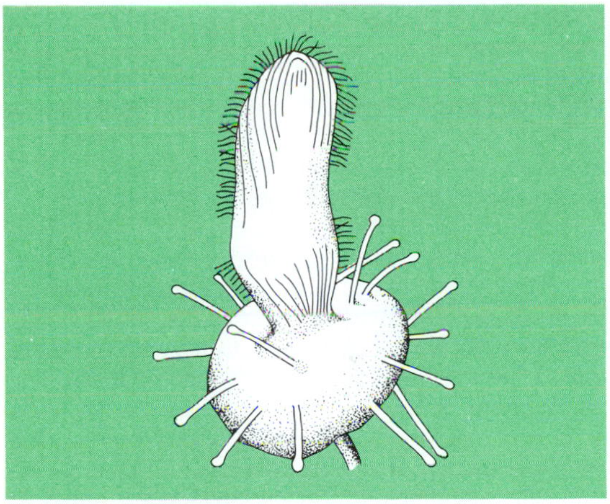

Endogenous budding in the suctorian *Podophrya*, a species which measure 10–28 micrometers.

serve, as in the case of species of the Peritrichia, for dissemination (see illustration). *See* CILIATEA; PERITRICHIA. [J.O.C.]

Sudangrass An annual, warm-season grass of tropical origin said to have been grown in Egypt since early times, though its value was first recognized in Sudan only in 1909. In that same year it was introduced into the United States as a replacement for johnsongrass, which had become a noxious weed in many southern states. Sudangrass (*Sorghum bicolor* var. *sudanense*, also called *S. sudanense* and *S. vulgare* var. *sudanense*) and its hybrids are commonly used as pasture, greenchop, silage, or hay. They fill an important need in many regions of the United States, because they produce high-quality forage for cattle and sheep during the summer, when other pasture is in short supply or of low quality. Many of the varieties and hybrids produce forage until frost. *See* GRASS. [M.R.G.]

Sudden infant death syndrome (SIDS) The sudden and unexpected death of an apparently normal infant that remains unexplained after the performance of an adequate autopsy. The incidence of SIDS in the United States is about 2.0 cases per 1000 live births, which makes SIDS a leading cause of postnatal death in the first year of life. Few SIDS cases occur in the first weeks of life; most occur at 2–4 months of age. SIDS is more common in males, prematurely born infants, multiple births, and the economically disadvantaged. It also occurs more frequently during winter months. The rate among Native Americans is greater than among Blacks, which is greater than among Caucasians; Asians have the lowest rate. The extent to which the cause of SIDS may be genetic is unresolved.

The physiologic and postmortem findings have been used in support of the apnea (respiratory failure) hypothesis, proposing that SIDS is due to a chronic abnormality in the mechanisms that control respiration. There are several such hypotheses, which propose prolonged central apnea, airway obstruction, or failure of arousal as the main effects of the unknown underlying abnormality. The available evidence seems to suggest that SIDS is due to a chronic congenital defect in the control of basic body functions. *See* CONGENITAL ANOMALIES; HUMAN GENETICS. [A.St.; K.Wi.]

Sugarbeet The plant *Beta vulgaris*, developed in modern times to fill the need for a sugar crop that could be grown in temperate climates. The sugarbeet was the source of only 5% of the world's commercial sugar in 1840, but by 1890 it supplied almost 50%, and since about 1920 this has dropped to about 40% with 60% derived from sugarcane. European countries produce most of their sugar from sugarbeet, with some countries having surplus sugar for export. *See* SUGARCANE.

The sugarbeet, fodder beet, garden beet, and leaf beet (Swiss chard) are cultivars of *B. vulgaris* and are genetically related. In addition to *B. vulgaris,* which includes all cultivated forms of beet, there are 12 species of wild beet in the Mediterranean and Middle Eastern regions. These wild beets are of interest in sugarbeet improvement as sources of disease resistance. Most varieties of sugarbeet in the United States are hybrid. [D.Ste.]

Sugarcane *Saccharum officinarum*, a member of the grass family. This crop originated in New Guinea about 15,000–8000 B.C., and was later moved by primitive peoples westward into Southeast Asia and India and eastward into Polynesia. Most current commercial varieties are interspecific hybrids involving primarily two or more of the following species: *S. officinarum, S. robustum,* and *S. spontaneum.*

Sugar is generally removed from sugarcane in large factories by either milling or diffusion. Milling crushes the stalks as they pass between a series of large metal rollers separating the fiber (bagasse) from the juice laden with sugar. The diffusion process separates sugar from finely cut stalks by dissolving it in hot water or hot juice. Processing of the juice is completed by clarification to remove nonsugar components, by evaporation to remove water, and by removal of molasses in high-speed centrifuges to produce centrifugal sugar. Bagasse and molasses are the principal by-products from processing sugarcane. [R.E.Co.]

Sulfate A negative ion having the formula SO_4^{2-} and derived from sulfuric acid, H_2SO_4. Because a molecule of sulfuric acid contains two hydrogen atoms, both normal sulfates and bisulfates are known. Of the normal sulfates, most are quite soluble in water, with the exception of those of silver, mercury(I), lead, strontium, barium, and calcium. Sulfate is determined both qualitatively and quantitatively by precipitation as barium sulfate, $BaSO_4$, which is the most insoluble sulfate. *See* SULFUR; SULFURIC ACID. [E.E.W.]

Sulfide A negative ion having the formula S^{2-} and derived from hydrogen sulfide or hydrosulfuric acid, H_2S. Because hydrosulfuric is a dibasic acid, sulfides, M_2S, and bisulfides, MHS, are formed. Although most of the metal sulfides are insoluble in water, they dissolve in acids with the evolution of H_2S gas. This escaping gas causes paper moistened with a solution of lead acetate to turn black, a test that can be used to detect the presence of sulfide in the original solution. *See* SULFUR. [E.E.W.]

Sulfite A negative ion having the formula SO_3^{2-} and derived from unstable sulfurous acid, H_2SO_3.

Because the sulfur in the sulfite ion has a 4+ oxidation state, it undergoes oxidation to the sulfate ion or reduction to free sulfur or to the sulfide ion. The sulfite ion can be identified by the fact that SO_2 gas is liberated when solutions are acidified, as represented by the reaction below.

$$2H^+ + SO_3^{2-} \rightleftharpoons [H_2SO_3] \rightleftharpoons H_2O + SO_2$$

Because sulfurous acid is dibasic, two series of salts, the normal M_2SO_3 and acid $MHSO_3$, are formed. Most of these salts, with the exception of the alkali metal and ammonium salts, are only slightly soluble. *See* SULFUR. [E.E.W.]

Sulfonamide One of a group of organosulfur compounds, RSO_2NH_2, that are readily prepared by the reaction of sulfonyl chlorides and ammonia, as in the reaction below. There is

$$RSO_2Cl + 2NH_3 \rightarrow RSO_2NH_2 + NH_4Cl$$

great interest in these substances because of the therapeutic sulfa drugs, but they are also of chemical value. *See* AMINE; ORGANOSULFUR COMPOUND. [N.K.]

Sulfonation and sulfation Sulfonation is a chemical reaction in which a sulfonic acid group, —SO_3H, is introduced into the structure of a molecule or ion in place of a hydrogen atom. Sulfation involves the attachment of the —OSO_2OH group to carbon, yielding an acid sulfate, $ROSO_2OH$, or of the —SO_4— group between two carbons, forming the sulfate, $ROSO_2OR$.

Sulfonation of aromatic compounds is the most important type of sulfonation. This is accomplished by treating the aromatic compound with sulfuric acid, as in the reaction below. The product of sulfonation is a sulfonic acid.

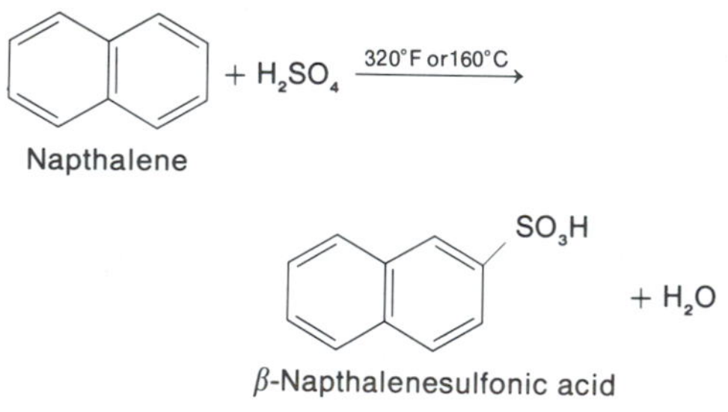

Sulfonation may also be defined as any chemical process by which the sulfonic acid group, —SO_2OH, or the corresponding salt or sulfonyl halide group, for example, —SO_2Cl, is introduced into an organic compound. These groups may be bonded to either a carbon or a nitrogen atom. The latter compounds are designated *N*-sulfonates or sulfamates.

Most sulfonates are employed as such in acid or salt form for applications where the strongly polar hydrophilic —SO_2OH group confers needed properties on a comparatively hydrophobic nonpolar organic molecule. Some sulfonates, such as methanesulfonic and toluenesulfonic acids, are used as catalysts. The major quantity of sulfonates and sulfates is both marketed and used in salt form. This category includes detergents; emulsifying, demulsifying, wetting, and solubilizing agents; lubricant additives; and rust inhibitors. *See* ORGANOSULFUR COMPOUND; SUBSTITUTION REACTION. [P.H.G./R.S.K.]

Sulfone One of a group of well-known, stable, and generally crystalline organosulfur compounds, RSO_2R'. Sulfones are best prepared by the oxidation of sulfides, and also by the reaction of sulfonyl chlorides. RSO_2Cl, with aromatic compounds in the presence of Friedel-Crafts catalysts, or by reaction of sodium sulfinates and alkyl bromides. Sulfones resist further oxidation, but can be reduced, under special conditions, to sulfides. Pyrolysis of sulfones frequently involves elimination of sulfur dioxide in a synthetically useful way. For example, alkenyl aryl sulfones smoothly yield alkenes. The sulfone group exerts a strong inductive effect and increases the acidity of α-hydrogen atoms. In $(CH_3SO_2)_3CH$, the acidity becomes very high. The —SO_2— group also deactivates the benzene ring to attack by electrophilic reagents. *See* ORGANOSULFUR COMPOUND; SULFIDE. [N.K.]

Sulfur A chemical element, S, atomic number 16, and atomic weight 32.064. The known stable isotopes and approximate percent abundances in natural sulfur are ^{32}S (95.1%); ^{33}S (0.74%); ^{34}S (4.2%); and ^{36}S (0.016%). The abundance of sulfur in the Earth's crust is 0.03–0.1%. It is often found as the free element near volcanic regions (impure deposits).

1	2	3	4	5	6	7	8	9	10	11	12	13	14	15	16	17	18
1 H																	2 He
3 Li	4 Be											5 B	6 C	7 N	8 O	9 F	10 Ne
11 Na	12 Mg											13 Al	14 Si	15 P	16 S	17 Cl	18 Ar
19 K	20 Ca	21 Sc	22 Ti	23 V	24 Cr	25 Mn	26 Fe	27 Co	28 Ni	29 Cu	30 Zn	31 Ga	32 Ge	33 As	34 Se	35 Br	36 Kr
37 Rb	38 Sr	39 Y	40 Zr	41 Nb	42 Mo	43 Tc	44 Ru	45 Rh	46 Pd	47 Ag	48 Cd	49 In	50 Sn	51 Sb	52 Te	53 I	54 Xe
55 Cs	56 Ba	71 Lu	72 Hf	73 Ta	74 W	75 Re	76 Os	77 Ir	78 Pt	79 Au	80 Hg	81 Tl	82 Pb	83 Bi	84 Po	85 At	86 Rn
87 Fr	88 Ra	103 Lr	104 Rf	105 Db	106 Sg	107 Bh	108 Hs	109 Mt	110	111	112	113	114	115	116	117	118

lanthanide series	57 La	58 Ce	59 Pr	60 Nd	61 Pm	62 Sm	63 Eu	64 Gd	65 Tb	66 Dy	67 Ho	68 Er	69 Tm	70 Yb
actinide series	89 Ac	90 Th	91 Pa	92 U	93 Np	94 Pu	95 Am	96 Cm	97 Bk	98 Cf	99 Es	100 Fm	101 Md	102 No

Properties. The allotropes (different crystalline forms) of sulfur have been studied intensively, but as yet the many modifications which exist for every state (gas, liquid, and solid) of elemental sulfur are not fully understood.

Rhombic sulfur, also called brimstone and alpha sulfur (α-sulfur), is the stable modification of the element below 95.5°C (204°F, the transition point), and most of the other forms revert to this modification if allowed to stand below this temperature. Rhombic sulfur is lemon-yellow, insoluble in water, slightly soluble in ethyl alcohol, diethyl ether, and benzene, and very soluble in carbon disulfide. Its density is 2.06 g/cm^3 (1.19 oz/in.^{3}) and its hardness is 2.5 on the Mohs scale. Its molecular formula is S_8.

Monoclinic sulfur, also called prismatic sulfur and beta sulfur (β-sulfur), is the stable modification of the element above the transition temperature and below the melting point. It crystallizes from molten sulfur in needlelike prisms which are almost colorless. It has a density of 1.96 g/cm^3 (1.13 oz/in.^{3}) and a melting point of 119.3°C (246.7°F). Its molecular formula is also S_8.

Plastic sulfur, also called gamma sulfur (γ-sulfur), is formed when molten sulfur at or near its normal boiling point is quenched to the solid state (such as by pouring it into cold water). This form of sulfur is amorphous and is only partially soluble in carbon disulfide.

Liquid sulfur exhibits the remarkable property of increasing in viscosity as its temperature is raised. Its color becomes dark reddish-black as its viscosity increases, and both color darkening and viscosity reach a maximum at 200°C (392°F). Above this temperature, the color lightens and the viscosity diminishes.

At the normal boiling point of the element (444.60°C or 832.28°F) gaseous sulfur is orange-yellow in color. As the temperature is raised, the color becomes deep red and then becomes lighter until, at 650°C (1202°F), it is straw-yellow.

Sulfur is an active element which combines directly with most of the known elements. It can exist in both positive and negative oxidation states, and can form ionic as well as covalent and coordinate covalent compounds. The uses of sulfur are limited primarily to the manufacture of sulfur compounds. However, large quantities of elemental sulfur are used in the vulcanization of rubber, in lime-sulfur sprays to destroy plant parasites, in the manufacture of artificial fertilizer and certain types of cements and electric insulators, in certain ointments and medicinals, and in the manufacture of gunpowder and matches. Sulfur compounds are used in the manufacture of chemicals, textiles, soaps, fertilizers, leather, plastics, refrigerants, bleaching agents, drugs, dyes, paints, paper, and other products.

Principal compounds. Hydrogen sulfide (H_2S) is the most important compound containing only hydrogen and sulfur. It is a colorless gas having a foul odor (similar to that of rotten eggs) and is considerably more poisonous than carbon monoxide, but warning of its presence (odor) is usually given before its concentration in the atmosphere is considered dangerous.

Metal sulfides can be classified into three categories: acid sulfides (hydrosulfides, MHS, where M = a univalent metal ion), normal sulfides (M_2S), and polysulfides (M_2S_3). Other sulfides include carbon-sulfur compounds and compounds containing the carbon-sulfur bond. Some important compounds are: carbon disulfide, CS_2, a liquid which is an excellent solvent for elemental sulfur and phosphorus; carbon monosulfide, CS, an unstable gas formed by passing an electric discharge through carbon disulfide, and carbon oxysulfide, SCO, formed from carbon monoxide and free sulfur at an elevated temperature.

Nitrogen-sulfur compounds which have been characterized are sulfur nitride, N_4S_4 (also called tetranitrogen tetrasulfide), nitrogen disulfide, NS_2, and nitrogen pentasulfide, N_2S_5. They should properly be referred to as nitrides because of the greater electro-negativity of nitrogen, although in the literature they are usually called sulfides.

Phosphorous-sulfur compounds which have been characterized are P_4S_3, P_4S_5 P_4S_7, and P_4S_{10}. All four are yellow crystalline materials that are used in converting organic oxygen compounds (for example, alcohols) into the corresponding sulfur analogs.

The oxides of sulfur which have been characterized have the formulas SO, S_2O_3, SO_2, SO_3, S_2O_7, and SO_4. Sulfur dioxide, SO_2, and sulfur trioxide, SO_3, are of far greater importance than the others. Sulfur dioxide can act as an oxidizing agent and as a reducing agent. It reacts with water giving an acidic solution (often called sulfurous acid) and bisulfite (HSO_3^-) and sulfite (SO_3^{2-}) ions. The dioxide is used as a refrigerant gas, as a disinfectant and preservative, as a bleaching agent, and in the refining of petroleum products. Its major use, however, is in the manufacture of sulfur trioxide and sulfuric acid. Sulfur trioxide is used primarily in the preparation of sulfuric acid and sulfonic acids.

Although salts (or esters) of all the oxy acids are known, in many cases the free acids themselves have not been isolated because of their instability. Sulfurous acid is not actually known as a pure substance. Sulfuric acid (H_2SO_4) is a colorless, viscous liquid with melting point 10.31°C (50.56°F). It is a strong acid in water and reacts with most metals in either the dilute or concentrated form. The concentrated acid is a strong oxidizing agent, especially at elevated temperatures. Pyrosulfuric acid ($H_2S_2O_7$) is an excellent sulfonating agent and loses sulfur trioxide on being heated. It also reacts vigorously with water, liberating a considerable quantity of heat. The persulfuric acids (peroxymonosulfuric acid, H_2SO_5, called Caro's acid, and peroxydisulfuric acid, $H_2S_2O_8$, called Marshall's acid) are known as the acids and salts. Sulfenic acids are known as the esters and halides. Sulfinic acids are formed by the reduction of the chlorides of sulfonic acids with zinc or by the reaction of Grignard reagents on sulfur dioxide in ether solution. Sulfonic acids (alkyl) are prepared by oxidizing mercaptans (RSH) or alkyl sulfides with concentrated nitric acid, by treatment of sulfites with alkyl halides, or by the oxidation of sulfinic acids. Other important organic oxygen-sulfur-containing compounds include the sulfoxides, R_2SO (which may be considered as being derived from sulfurous acid), and the sulfones, R_2SO_2 (from sulfuric acid).

Important halogen derivatives of sulfuric acid are the organic sulfonyl halides and the halosulfonic acids. Halogen-sulfur compounds which have been well characterized are S_2F_2 (sulfur monofluoride), SF_2, SF_4, SF_6, S_2F_{10}, S_2Cl_2 (sulfur monochloride), SCl_2, SCl_4, and S_2Br_2 (sulfur monobromide). The sulfur chlorides are used in the commercial manufacture of rubber; and the monochloride, which is a liquid at room temperature, is also used as solvent for organic compounds, sulfur, iodine, and certain metal compounds. [S.Ki.]

Sulfuric acid A strong mineral acid with the chemical formula H_2SO_4. It is a colorless, oily liquid, sometimes called oil of vitriol or vitriolic acid. The pure acid has a density of 1.834 at 25°C (77°F) and freezes at 10.5°C (50.90°F). It is an important industrial commodity, used extensively in petroleum refining and in the manufacture of fertilizers, paints, pigments, dyes, and explosives.

Sulfuric acid is produced on a large scale by two commercial processes, the contact process and the lead-chamber process. In the contact process, sulfur dioxide, SO_2, is converted to sulfur trioxide, SO_3, by reaction with oxygen in the presence of a catalyst. Sulfuric acid is produced by the reaction of the sulfur trioxide with water. The lead-chamber process depends upon the oxidation of sulfur dioxide by nitric acid in the presence of water, the reaction being carried out in large lead rooms.

Sulfuric acid reacts vigorously with water to form several hydrates. The concentrated acid, therefore, acts as an efficient drying agent, taking up moisture from the air and even abstracting the elements of water from such compounds as sugar and starch. The concentrated acid also acts as a strong oxidizing agent. It reacts with most metals upon heating to produce sulfur dioxide. *See* SULFUR. [F.J.J.]

Sum rules Formulas in quantum mechanics for transitions between energy levels, in which the sum of the transition strengths is expressed in a simple form. Sum rules are used to describe the properties of many physical systems, including solids, atoms, atomic nuclei, and nuclear constituents such as protons and neutrons. The sum rules are derived from quite general principles, and are useful in situations where the behavior of individual energy levels is too complex to describe by a precise quantum-mechanical theory. *See* ENERGY LEVEL (QUANTUM MECHANICS).

In general, sum rules are derived by using Heisenberg's quantum-mechanical algebra to construct operator equalities, which are then applied to particles or the energy levels of a system. *See* QUANTUM MECHANICS. [G.F.Be.]

Sumatra The largest western island of Indonesia: about 1600 mi (2575 km) long and 250 mi (400 km) across at the widest portion. It has an area of 166,800 mi^2 (432,000 km^2) and extends from 6°N to 6°S latitude. The island is made up of three distinct physiographic sections. Tertiary fold mountains on the west rise against the Sunda Platform Peaks of these mountains reach from 3000 to 8000 ft (900 to 2400 m). East of these mountains there is a belt of low hills of folded Tertiary formations. The eastern half of the island is composed of extensive, swampy, alluvial lowlands, which form part of the Sunda Platform. Important petroleum deposits are found near Palembang and Meda. Small islands to the east of Sumatra—Lingga, Bangka, and Billiton—contain major tin deposits associated with granites of the Sunda Platform.

Most of Sumatra has an annual average rainfall of over 150 in. (3.8 m). The western slopes of the mountains may have as much as 30 in. (75 cm) more, but the intermontane basins receive less. This results in a tropical rainforest in the uplands and a tropical swamp forest in the lowlands. *See* EAST INDIES. [M.G.A.-C.]

Sun The star around which the Earth revolves, and the planet's source of light and heat. The Sun is a globe of gas, 8.65×10^5 mi (1.4×10^6 km) in diameter, held together by its own gravity. Because of the weight of the outer layers, the density and temperature increase inward, until a central temperature of over 27,000,000°F (15,000,000 K) and density more than 90 times that of water is reached. At these great temperatures and densities, thermonuclear reactions converting hydrogen into helium take place, releasing the energy which streams outward.

The surface temperature of the Sun is about 10,000°F (6000 K); since solids and liquids do not exist at these temperatures, the Sun is entirely gaseous. Almost all the gas is in atomic form, although a few molecules exist in the coolest regions at the surface.

The Sun is a typical member of a numerous class of stars, the spectral type dG2 (d indicates dwarf).

Solar structure. The interior of the Sun can be studied only by inference from the observed properties of the entire star. A number of theoretical models using different assumptions have led to more or less similar results. A central density of near 3.25 lb/in.^{3} (90 g/cm^3) has been found, decreasing to 4×10^{-9} lb/in.^{3} 10^{-7} g/cm^3) at the surface. The central temperature is about 27,000,000°F (15,000,000 K), decreasing to 8500°F (5000 K) at the surface. Since this takes place over 430,000 mi (700,000 km), the temperature gradient is only 60°F per mile (20 K per kilometer). The radiation produced at the center by nuclear interactions flows outward rapidly.

The energy of the Sun is produced by the conversion of hydrogen into helium. For each hydrogen atom converted, one neutrino is produced. These neutrinos cannot be detected; only higher-energy neutrinos produced by subordinate processes can be observed. These have been detected at the Earth, but in smaller quantities than expected. However, the neutrino emission theory has sufficient uncertainties so that the theory of nuclear burning is still generally accepted. *See* SOLAR NEUTRINOS.

Although the material at the center of the Sun is dense, the photons created by nuclear reactions are continually absorbed and reemitted and thus make their way to the surface. The atoms in the center of the Sun are entirely stripped of their electrons by the high temperatures, and most of the absorption is by continuous processes, such as scattering of light by electrons.

In the outer regions of the solar interior, the temperature is low enough for ions and even neutral atoms to form and, as a result, atomic absorption becomes very important. The high opacity makes it very difficult for the radiation to continue outward; steep temperature gradients are established which result in convective currents. Most of the outer envelope of the Sun is in such convective equilibrium. These large-scale mass motions produce many interesting phenomena at the surface, including sunspots and solar activity.

Radiation. Electromagnetic energy is produced by the Sun in essentially all wavelengths. However, more than 95% of the energy is concentrated in the relatively narrow band between 290 and 2500 nanometers and is accessible to routine observation from ground stations on Earth. The maximum radiation is in the green region, and the eyes of human beings have naturally evolved to be sensitive to this range of the spectrum. The total radiation received from the Sun is termed the solar constant. *See* ELECTROMAGNETIC RADIATION; SOLAR RADIATION.

Atmosphere. Although the Sun is gaseous, it is seen as a discrete surface from which practically all the heat and light are radiated. One sees through the gas to the point at which the density is so high that the material becomes opaque. This

Fig. 1. The solar corona observed about 2 s after second contact during the eclipse of November 12, 1966, at Pulacayo, Bolivia (altitude 13,000 ft or 4000 m). (*G. A. Newkirk, National Center for Atmospheric Research*)

layer, the visible surface of the Sun, is termed the photosphere. Light from farther down reaches the Earth by repeated absorption and emission by the atoms, but the deepest layers cannot be seen directly. The surface is actually not sharp; the density drops off by about a factor of 3 for each 90 mi (150 km). However, the Sun is so far away that the smallest distance that can be resolved with the best telescope is about 300 km, and thus the edge appears sharp. *See* SOLAR CORONA; PHOTOSPHERE.

Looking at the Sun in isolated wavelengths absorbed by its atmospheric gases, one can no longer see down to the photosphere, but instead one sees the higher levels of the atmosphere, known as the chromosphere. This name results from the rosy color seen in this region at a solar eclipse. The chromosphere is a rapidly fluctuating region of jets and waves coming up from the surface. When all the convected energy coming up from below reaches the surface, it is concentrated in the thin material and produces considerable activity. As a result, this outer region is considerably hotter than the photosphere, with temperatures up to 54,000°F (30,000 K). *See* CHROMOSPHERE.

When the Moon obscures the Sun at a total solar eclipse, the vast extended atmosphere of the Sun called the corona can be seen (Fig. 1). The corona is transparent and is visible only when seen against the dark sky of an eclipse. Its density is low, but its temperature is high. The hot gas evaporating out from the corona flows steadily to the Earth and farther in what is called the solar wind. *See* SOLAR WIND.

The photospheric features of the Sun (sunspots and faculae) are readily observed with a small telescope by projection of the solar image through the eyepiece onto a shaded white card. This is the only safe method for observing the Sun without a specially designed solar eyepiece. Observations of chromospheric phenomena require special filters to exclude all light except that emitted by the atmospheric gases. Observations of the corona require either a total eclipse or a coronagraph.

Coronal holes. Early coronal observations showed that the corona was occasionally not visible over certain regions. In particular, most of the time it was quite weak over the poles. X-ray pictures revealed great bands of the solar surface essentially devoid of corona for many months. These proved to be regions where the local magnetic fields were connected to quite distant places, so the fields actually reached out to heights from which the solar wind could sweep the gas outward. Analysis of solar wind data showed that equatorial coronal holes were associated with high-velocity streams in the solar wind, and recurrent geomagnetic storms were associated with the return of these holes. Thus the great intensity of the corona over sunspot regions is partly due to their strong, closed magnetic fields which trap the coronal gas. *See* SOLAR MAGNETIC FIELD.

Solar activity. The transient phenomena, known collectively as solar activity, are all connected with sunspots or their remnants, and wax and wane in a remarkable cycle of activity. The sunspot cycle consists of variations in the sizes, numbers, and positions of the sunspots, expressed quantitatively as the sunspot number. The number of sunspots peaks soon after the beginning of each cycle and decays to a minimum in 11 years. The magnetic polarity of the sunspot groups reverses in each successive cycle so that the complete cycle lasts 22 years. The first spots of a cycle always occur at higher latitudes, between 20 and 35°, and as the spots increase in size they occur closer to the equator. Almost no spots are observed outside the latitude range of 5–35°. The average duration of the cycle from the first appearance of the last low-latitude spots is nearly 14 years. The difference between these two figures is the result of overlap of consecutive cycles.

Sunspots. Sunspots are the most conspicuous features of solar activity and are easily seen through a small telescope. They are dark areas on the Sun produced by the most intense magnetic fields (see Fig. 2). The magnetic fields produce a cooling of the surface, most likely by suppression of the normal convection which transports energy from the lower levels. Deprived of this supply of heat, the area cools by radiation and becomes a dark sunspot. Curiously enough, the region close to the cool sunspot is the scene of the hottest and most intense activity because of the great magnetic energy present in the sunspot. *See* SUNSPOT.

Prominences. Although prominences appear dark against the disk, they appear bright against the dark sky. They occur only in regions of horizontal magnetic fields, because these fields support them against the solar gravity. Thus filaments on the disk are good markers of the transition from one magnetic

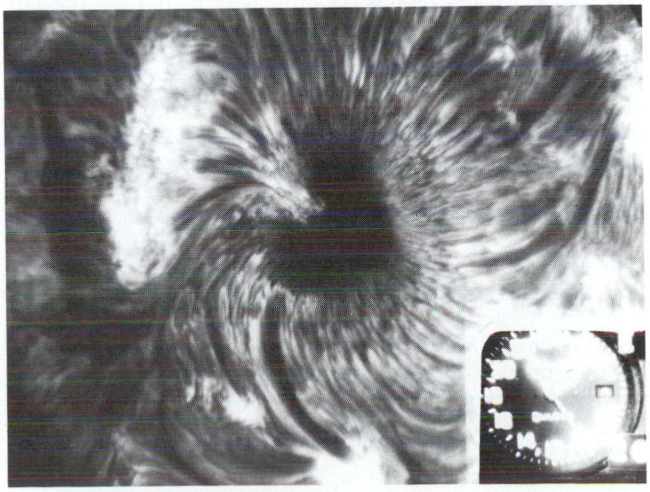

Fig. 2. A large symmetric sunspot photographed in H-α light. Clock indicates time of photograph. (*Big Bear Solar Observatory*)

Fig. 3. The great "sea horse" flare of August 7, 1972, late in the flare, photographed in the blue wing of the H-α line. The neutral line between two bright strands is crossed by an arcade of bright loop prominences raining down from the corona. (*Big Bear Solar Observatory*)

polarity to the opposite. Prominences are among the most beautiful of solar phenomena.

Plages. Just as prominences occur when the magnetic field changes from one sign to the other, plages occur whenever the magnetic field is vertical and relatively strong. They are bright regions visible in any strong spectrum line. It is now believed that the field is, in fact, stable only when it exceeds 1000 gauss (0.1 tesla), and the difference between apparently weaker field regions, such as the chromospheric network, and plages is only the density of clumps of strong field. Plages are normally associated with sunspots; a typical active region will have the preceding magnetic field clumped in a sunspot and the following field spread out in a plage.

Flares. The most spectacular activity associated with sunspots is the solar flare (Fig. 3). A flare is defined as an abrupt increase in the H-α emission from the sunspot region. The brightness of the flare may be five times that of the associated plage: the rise time is seldom longer than a few minutes, sometimes only 10 s. Of course, the hydrogen brightening is only a symptom; observations by other means show that there is a tremendous energy release, of as much as 10^{33} ergs (10^{26} J) in a large flare. An active sunspot group will show many small flares the size of a large sunspot. Once or twice in the lifetime of a large group, a great flare covering the entire region occurs. The energy released in the three or four great flares each year equals that in all the small flares put together.

Solar flares occur only as the result of sunspot activity. Practically all sunspots produce some flares, but certain spots are prodigiously active. The source of the flare energy is the magnetic fields surrounding the sunspots. Every magnetic field has a certain minimum energy form, normally that where no currents flow, a so-called potential field. But if the fields have been strongly twisted, currents are set up which stabilize the twisted fields until they become unstable and jump to lower energy states, producing flares. [H.Zi.]

Sundew Any plant of the genus *Drosera* (90 species) of the family Droseraceae. Sundews are small, herbaceous, insectivorous plants that grow on all the continents, especially Australia. Numerous glandular hairs (tentacles) on the leaf secrete a viscous fluid which traps a visiting insect. The tentacles then bend inward about the victim, bringing it into contact with the surface of the leaf where it is digested. The droplets secreted by the glands on the leaves glitter like dewdrops in the morning sunlight; hence, the name, sundew. *See* INSECTIVOROUS PLANTS; NEPENTHALES; SECRETORY STRUCTURES (PLANT). [P.D.St./E.I.C.]

Sundial An instrument for telling time by the Sun. It is composed of a style that casts a shadow and a dial plate, which is the surface upon which hour lines are marked and upon which the shadow falls. The style lies parallel to Earth's axis. The construction of the hour lines is based on the assumption that the apparent motion of the Sun is always on the celestial equator. The most widely used form is the horizontal dial that indicates local apparent time (Sun time). Other forms of the sundial indicate local mean time, and standard time. *See* TIME. [R.N.M.]

Sundog One of two bright spots (or parhelia) which are seen on both sides of the Sun or Moon, usually with red coloring on the part closer to the Sun. They are produced by prismatic refraction on the alternate sides of the hexagonal ice crystals floating in the air with vertical axes of symmetry. For the Sun at the horizon they are situated in the inner halo; with higher Sun they approach the outer halo. The parhelia are the second most frequently observed halo phenomena. *See* HALO. [Z.S.]

Sunfish A name given to certain members of the freshwater family Centrarchidae as well as to members of the marine family Molidae. There are two species of oceanic sunfishes in the genus *Mola;* both are large, stout, deep-bodied fishes with rough skin and are found in warm seas, where they are frequently seen near the surface. These fishes appear to be tailless.

The fresh-water sunfishes are grouped together with the crappies. The warmouth sunfish (*Chaenobryttus gulosus*) is the only species found in the United States, where it occurs in the Mississippi River drainage and in streams along the Atlantic coast. It differs from other sunfishes in having teeth on the tongue.

The crappies are game fish of considerable importance. Crappies are a favorite species for pond culture. They can be readily transplanted and, under favorable conditions, multiply prodigiously. *See* PERCIFORMES. [C.B.C.]

Sunflower *Helianthus annuus,* the most widely distributed of the 50 native North American species of this genus of the family Compositae. It is an extremely variable species, with two main divisions. The first involves wild weedy plants found along roadways and other recently disturbed areas; the second, domesticated plants grown in fields and gardens. *See* ASTERALES.

Within the domestic type there are two categories of plants: the ornamental, which has a few branches with larger heads than the wild, and the crop type, which has only a single stem and the largest head of all sunflowers (see illustration). Crop types are either oil or nonoil. Plant breeders have modified the plant for adaptation to modern, large-scale farming and have increased the oil content of the seeds. The present worldwide interest in growing sunflowers as a crop is due to the increased yield of the new commercially available oilseed hybrids.

Sunflowers are grown on all the continents and in many countries throughout the world. Russia is the major producer, followed by Argentina, the United States, and Canada. Sunflower oil (sunoil) is the second most important vegetable crop oil. It is a high-quality oil and is high in linoleic fatty acid.

Maturing sunflower. (*U.S. Department of Agriculture*)

The oil is used in cooking, salad dressing, mayonnaise, margarine, and soap. [B.H.B.]

Sunlamp A special form of mercury arc discharge lamp designed to produce ultraviolet radiation. These lamps also produce some radiant energy in the visible region of the spectrum, thus having a light output as well as an ultraviolet output. The lamps are principally used for producing a skin tan on the human body. The less common uses include therapeutically producing vitamin D in the body for the treatment of rickets and causing fluorescence or photochemical reactions.

Sunlamps have ultraviolet radiation at wavelengths above 280 nanometers. The lower limit is set by the fact that the quartz and high-silica glass used for lamp envelopes do not transmit below 280 nm. The two lamps designed for tanning are the FS-40 and the RS. The FS-40 tubular fluorescent lamp (illustration *a*) operates with a low-pressure mercury arc that causes a special chemical phosphor coating inside the tube to radiate ultraviolet energy. The RS sunlamp (illustration *b*) is a reflector unit containing a high-pressure mercury arc tube for generating ultraviolet energy plus a tungsten filament (similar to incandescent lamp filaments) in series with the arc tube to serve as a ballast. *See* MERCURY-VAPOR LAMP; ULTRAVIOLET LAMP. [G.R.P.]

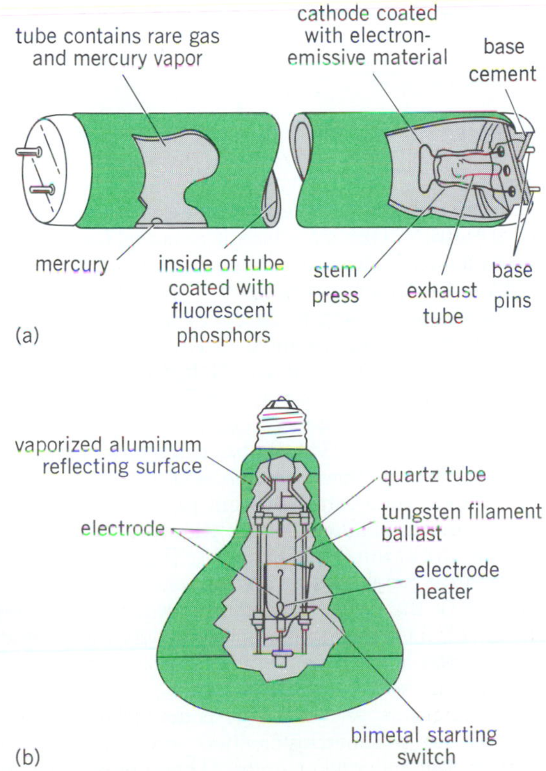

Structure of sunlamps. (*a*) FS-40. (*b*) RS.

Sunspot A dark area in the photosphere of the Sun caused by a lowered surface temperature. The temperature at the center of a spot is about 4000 K (7200°F), and the surface brightness is one-fifth that of the normal photosphere. The sizes and numbers of sunspots vary in the 11-year sunspot cycle, which is shared by all other forms of solar activity. *See* SUN. [J.W.E.]

Superacid An acid which has an extremely great proton-donating ability. It has proved convenient to define a superacid somewhat arbitrarily as an acid, or more generally, an acidic medium, which has a proton-donating ability equal to or greater than that of anhydrous (100%) sulfuric acid.

Superacids belong to the general class of proton or Brönsted acids. A proton acid is defined as any species which can act as a source of protons and which will therefore protonate a suitable base, as in reaction (1).

$$HA + B \rightleftharpoons BH^+ + A^- \qquad (1)$$

The strengths of acids are often compared by measuring the extent of their ionization in water, that is, the extent to which they can protonate the base water, as in reaction (2).

$$HA + H_2O \rightleftharpoons H_3O^+ + A^- \qquad (2)$$

However, all strong acids are fully ionized in dilute aqueous solution, and they therefore appear to have the same strength. Their strengths are said to be reduced or leveled to that of the hydronium ion (H_3O^+), which is the most highly acidic species that can exist in water. In any case, many of the superacids react with and are destroyed by water. For these reasons, the strengths of superacids cannot be measured by the conventional means of utilizing their aqueous solutions. The acidities of superacids can, however, be conveniently measured in terms of the Hammett acidity function. *See* ACID AND BASE.

The Hammett acidity function is a method of measuring acidity based on the determination of the ionization ratios of suitable weak bases (indicators), usually by means of the change in absorption spectrum that occurs on protonation of the base, although the nuclear magnetic resonance (NMR) spectrum has also been used. The Hammett acidity function (H_0) is defined by Eq. (3), where K_{BH+} is the dissociation con-

$$H_0 = pK_{BH+} - \log \frac{[BH^+]}{[B]} \qquad (3)$$

stant of the acid form of the indicator and $[BH^+]/[B]$ is the ionization ratio of the indicator. Hammett acidity function (H_0) values for a number of superacids are given in the table. In each case the value refers to the 100% (anhydrous) acid. Each of the

Hammett acidity function values for several superacids

Superacid	Formula	$-H_0$
Sulfuric acid	H_2SO_4	11.9
Chlorosulfuric acid	HSO_3Cl	13.8
Trifluoromethane sulfonic acid	HSO_3CF_3	14.0
Disulfuric acid	$H_2S_2O_7$	14.4
Fluorosulfuric acid	HSO_3F	15.1
Hydrogen fluoride	HF	15.1

superacids in the table is a liquid at room temperature, and each forms the basis of a solvent system. *See* IONIC EQUILIBRIUM; SOLUTION. [R.J.Gi.]

Supercavitating propeller A propeller designed to overcome the limitation on ship speed caused by loss of thrust when cavitation is severe. In this design, cavitation on the backs of the blades (forward side) is induced by special blade sections at relatively low forward speed so that, when revolutions and engine power are increased, the whole back of each blade becomes enveloped by a sheet of cavitation. When this is completed, further increase in thrust at still higher engine power and rpm is obtained by the increase in positive pressure on the blade face (rear side); erosion is avoided because the collapse of the cavitation bubbles occurs some distance behind the trailing edges of the blades. A photograph of such a supercavitating propeller in action is shown in the illustration. The supercavitating propeller is no replacement for the conventional propeller, being suitable only for very high ship and engine speeds. *See* CAVITATION; PROPELLER (MARINE CRAFT). [K.E.Sch.]

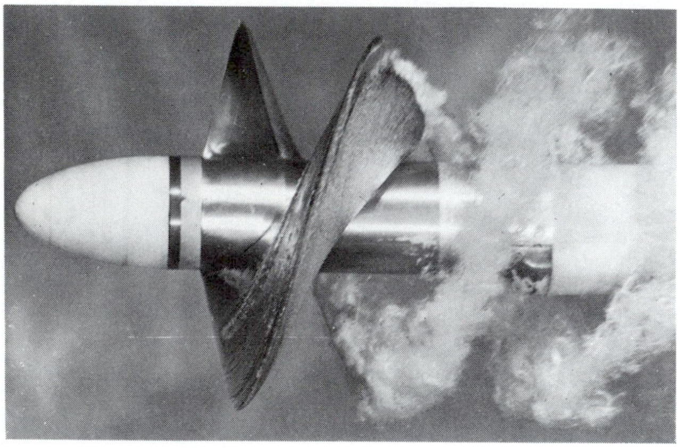

Supercavitating propeller in a water tunnel. (*U.S. Navy photograph*)

Supercharger An air pump or blower in the intake system of an internal combustion engine. Its purpose is to increase the air charge weight and power output from a given engine size. In an aircraft engine, the supercharger counteracts the power loss resulting from decreasing atmospheric pressure with increase of altitude. *See* INTERNAL COMBUSTION ENGINE.

In the case of the diesel engine, supercharging makes ignition of the fuel easier, and it does not require a fuel of better quality, as in the spark-ignited engine. Two-stroke-cycle engines always require some form of blower or air compressor to assist in charging the cylinder. However, only if the air pressure in the cylinder at the beginning of compression is substantially above atmospheric is the engine said to be supercharged. Large diesel engines all employ the two-stroke cycle, and are supercharged in order to be commercially competitive. *See* DIESEL ENGINE.

To enable a piston-type aircraft engine to develop its rated sea-level horsepower at altitude, a supercharger must be used to increase the pressure and weight of the intake air charge. Centrifugal compressors are used for this duty because of their relatively small size for a given capacity, and are driven either by a gear drive from the crankshaft or by a gas turbine powered from the engine exhaust. *See* RECIPROCATING AIRCRAFT ENGINE. [F.C.M./J.A.B.]

Superclusters Associations of galaxy clusters and groups, typically composed of a few rich clusters and many poorer groups and isolated galaxies. Most investigators find superclusters to be arranged in broad sheets and chains, and some believe that they are connected, forming cells which enclose nearly empty volumes of space. If so, the large-scale structure of luminous matter in the universe is therefore apparently best visualized as spongelike, at least on scales up to about 10^9 light-years (9.46×10^{24} m), which is about as far as currently available data reliably extend.

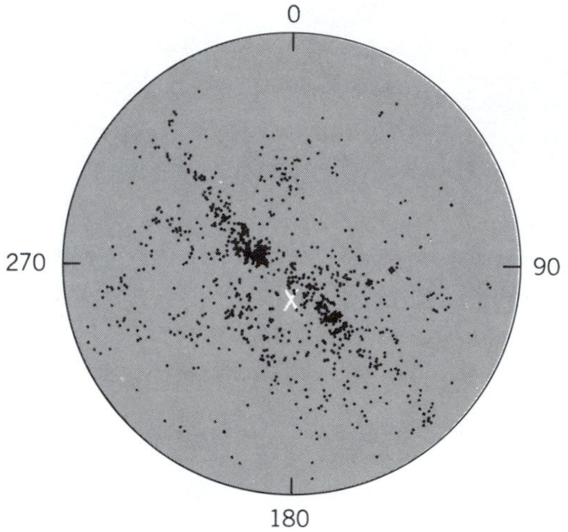

Local Supercluster as shown by the apparent positions of all galaxies in the northern galactic hemisphere with known radial velocities less than 2000 km/s (1243 mi/s). Numbers are galactic longitudes. (*After R. B. Tully and J. R. Fisher, A picture of the Supercluster, Bull. Amer. Astron. Soc., 8:555, 1976*)

In 1784 William Herschel noted that most of the "nebulae" he had discovered seemed to be concentrated in a wide belt across the northern sky, oriented roughly perpendicularly to the Milky Way. However, it was 1953 before Gérard de Vaucouleurs realized that observers on Earth are simply seeing from the inside a great flattened system of groups and clusters of galaxies, with a few individual objects scattered between. De Vaucouleurs called this the Local Supercluster.

The illustration shows the apparent positions of all galaxies in the northern galactic hemisphere with known radial velocities less than 2000 km/s (1243 mi/s). These objects are the Galaxy's nearest neighbors in extragalactic space, and most are members of the Local Supercluster, which shows well as the band of objects running diagonally across the illustration. The Virgo Cluster is the concentration of points in this band to the upper left of center. The cross shows the point toward which de Vaucouleurs' analysis suggests the Galaxy is moving.

Besides the Local Supercluster, there are perhaps 50 other superclusters that have been noted. These have been found primarily through examination of the various surveys of faint galaxies and of clusters of galaxies. *See* GALAXY, EXTERNAL. [H.G.C.]

Supercomputer A computer which, among existing computers at any given time, is superlative, often in several senses: highest computation rate, largest memory, or highest cost. Predominantly, the term refers to the fastest "number crunchers," that is, machines designed to perform numerical calculations at the highest speed that the latest electronic device technology and the state of the art of computer architecture allow. In the 1980s the term supercomputer also began to be applied to predicted future machines designed to perform symbolic, or artificial intelligence-oriented, computations. *See* ARTIFICIAL INTELLIGENCE.

Scientific applications. The demand for the ability to execute arithmetic operations at the highest possible rate originated in computer applications areas collectively referred to as scientific computing. Reseachers in fields such as physics, structural mechanics, meterorology, and aerodynamics often need to do large-scale numerical simulations of physical processes. A technique common to several of these disciplines is to compute an approximate numerical solution to a set of partial differential equations which mathematically describe the physical process of interest, but are too complex to be solved by formal mathematical methods. The approximate numerical solution is obtained by first superimposing a grid on a region of space, with a set of numerical values attached to each grid point. Each of these values represents a physical quantity hypothetically measurable at that point in space, such as temperature, pressure, or velocity. Then, the physical process is computationally "simulated" by repetitively updating the values at each grid point according to a system of algebraic equations, derived from the defining partial differential equations, which specify the relationships between the values. Each repeated step might represent a span of time; that is, the simulated time is divided into a series of discrete steps, just as space is divided into a set of grid points. A more accurate solution requires a finer spatial grid and shorter time intervals. Large problems in computational physics and aerodynamics sometimes require 100 h of computer time on current supercomputers. *See* NUMERICAL ANALYSES; SIMULATION.

Electronics technology. Supercomputers generally employ a variant of silicon-based integrated circuitry known as emitter-coupled logic (ECL). It is less dense than other types of integrated circuits, offering at most a few thousand transistors per chip, but providing switching speeds in the range of a few nanoseconds. The next generation of supercomputers is expected to employ faster versions of emitter-coupled logic, complementary metallic oxide semiconductors (CMOS), or switching devices which are based on gallium arsenide (GaAs) semiconductor technologies instead of the silicon-based materials currently employed. *See* COMPUTER STORAGE TECHNOLOGY; INTEGRATED CIRCUITS; LOGIC CIRCUITS; SEMICONDUCTOR MEMORIES.

Advances in computer architecture. Increases in computing speed which are purely due to the architectural structure of a computer can largely be attributed to the introduction of some form of parallelism into the machine's design: two or more operations which were performed one after the other in previous computers can now be performed simultaneously. *See* COMPUTER SYSTEMS ARCHITECTURE.

Pipelining is a technique which allows several operations to be in progress in the central processing unit at once. It is analogous to an assembly line, in the sense that operations which take several steps pass through a sequence of stages, each stage represents one step closer to completion of the entire operation, and already-passed stages can be simultaneously active, performing their respective steps of the operation on separate sets of operands.

The central processing unit nearly always has a much faster cycle time than the memory. This implies that the central processing unit is capable of processing data items faster than a memory unit can provide them. Interleaved memory is an organization of memory units which at least partially relieves this problem. Consecutive memory addresses are located in different memory units, allowing the central processing unit to initiate fetches from several units in sequence whenever the instructions of a program call for the fetching of the contents of each successive word from a contiguous block of memory words.

Parallelism within arithmetic and logical circuitry has been introduced in several ways. Adders, multipliers, and dividers now operate in bit-parallel mode, while the earliest machines performed bit-serial arithmetic.

Array processing is a form of parallelism in which the instruction execution portion of a central processing unit is replicated several times and connected to its own memory device as well as to a common instruction interpretation and control unit. In this way, a single instruction can be executed at the same time on each of several execution units, each on a different set of operands.

Vector processing is the term applied to a form of pipelined arithmetic units which are specialized for performing arithmetic operations on vectors, which are uniform, linear arrays of data values.

Multiprocessing is a form of parallelism that has complete central processing units operating in parallel, each fetching and executing instructions independently from the others. *See* CONCURRENT PROCESSING; DATA FLOW SYSTEMS; DIGITAL COMPUTER; MULTIPROCESSING. [D.W.M.]

Superconducting computers High-performance computers whose circuits employ superconductivity and the Josephson effect to reduce cycle time. Presently, high-performance, synchronous digital computers have operating cycle times in the range of 12 to 60 nanoseconds. Josephson technology and superconductivity are being explored as ways of reducing the cycle time further; initially the goal is to reduce the cycle time below 4 ns, and ultimately it is thought that a subnanosecond cycle time will be possible.

Cycle time. There are principally two components to the cycle time of a computer: the circuit delay and the package delay. Usually, for a given technology, higher circuit speed can be obtained by increasing dissipation or reducing circuit dimensions. However, there are two principal limitations to available circuit speed, both of which are related to power requirements. *See* INTEGRATED CIRCUITS.

The first is associated with the wasted power: the power dissipated in the form of heat. This has to be removed by a cooling medium, and the level of heat removal from circuit chips is limited on the one hand by the cost and complexity of the heat removal structures, and on the other hand by the allowable rise in temperature that the circuits can sustain while still operating correctly. Heat removal structures usually also limit the volumetric efficiency of packing circuit chips.

The second limitation to the power comes from the circuit engineers' ability to provide regulated power to the circuit chips. A large fraction of the chip input/output connections is devoted to power input—to carry the current, but even more importantly to reduce the power voltage swings caused by varying current drain on the chip as the circuits change their binary-encoded state under program control. This inductive variation of chip voltage becomes an increasingly difficult problem to solve as the circuit chip performance is improved. Of the two principal power limitations to circuit performance, this one appears the most demanding and severe.

The package delay component of the cycle time depends on the way the integrated circuit chips are arranged. The ultimate goal of high-performance computer packaging schemes is to achieve the highest possible density of circuits in a given volume, and this not only means high circuit density on each chip, but also high-density packing of chips. It is necessary to approach as closely as possible the ideal of a three-dimensional arrangement, and a good approximation is the card-on-board package, where chips are attached to both sides of the card, and the cards in turn are plugged closely together into the board.

Josephson technology. The Josephson effect was quickly identified as being the basis of a very promising computer technology. Initially, the very high switching speed and low power dissipation of this device were the main attractions, but it soon developed that many other aspects of the physics and engineering were very favorably inclined toward superconducting computer technology. Principal among these was the essentially lossless-transmission-line characteristics of all superconducting circuit interconnection lines. *See* JOSEPHSON EFFECT.

The power removal problem is no longer a concern for such circuits. This has enabled particularly attractive and novel techniques to be applied to power supply and regulation concerns faced by the circuit designer; active power supply and regulation circuits can be incorporated into the on-chip design, thereby removing the problem of inductive variation of chip-voltage.

These advantages naturally accrue from the basic physics of the superconducting state and the Josephson effect. The energy gap that characterizes these devices has shrunk from the hundreds of meV of the semiconductor electronic devices to just a few meV with the superconducting Josephson devices.

With little or no cooling constraint, simple attachment of high-performance chips to the package is feasible; there are no heat removal structures required, and the package can be fabricated in a high-density three-dimensional card-on-board arrangement. Josephson technology also makes possible the high-density superconducting interconnection lines and the ability, for the first time, to make a monolithic package in the sense that the package components are all made from the same substrate material: silicon. This has significant importance in that it allows the engineer to work with little concern for differential thermal expansion in the package design.

Much detailed engineering and materials work remains to be done before a superconducting computer becomes a reality. Josephson devices are not simple extensions of transistor circuits. They involve a much more significant break with the past than there was in going from the triode vacuum tube to the transistor. The biggest challenge is perhaps the sheer magnitude of the tasks involved in introducing a fundamentally new technology: chip technology, circuit design, package technology, design techniques, testing, and so forth.

There are also obvious difficulties associated with the use of liquid helium. Cooling the computer from 300 to 4.2 K (80 to −452°F) will undoubtedly produce stresses. These have to be understood and taken care of; they affect the choice of materials both for the devices and for the package. *See* SUPERCONDUCTING DEVICES. [D.J.H.]

Superconducting devices

Devices that perform functions in the superconducting state that would be difficult or impossible at room temperature, or that contain components which perform such functions. *See* SUPERCONDUCTIVITY.

Small-scale devices. The unique properties of superconductors have led to the development of tiny measuring and computing devices that are superior in performance to their nonsuperconducting counterparts. Most of the devices operate at or below 4.2 K (−452.0°F) and involve Josephson tunneling. The initial development of these devices took place in the second half of the 1960s. Fabrication of the devices has increasingly involved photolithography and electron-beam lithography, and superconducting junctions and circuits are now at the forefront of electronic ultraminiaturization.

There are two types of superconducting quantum interference device (SQUID) for detecting changes in magnetic flux: the dc SQUID and the rf SQUID. The dc SQUID, which operates with a dc bias current, consists of two Josephson junctions incorporated into a superconducting loop. The rf SQUID consists of a single Josephson junction incorporated into a superconducting loop and operates with a rf bias. Although rf SQUIDs have been used more widely than dc SQUIDS, their sensitivity is considerably lower, and it seems unlikely that their performance will be able to match that of the dc SQUID. *See* SQUID.

Typical high-speed digital computers, using large-scale integration of semiconductor devices, have cycle times of 30 to 50 nanoseconds. To achieve the goal of further reducing this time to 1 ns, both the device switching time and the time required to transmit a signal between different parts of the computer must be less than 1 ns. Because the signal propagation speed is typically 10^8 m · s^{-1}, the second requirement implies that the largest dimension of the computer must be no greater than 0.1 m. Although semiconducting devices with switching times substantially less than 1 ns are available, they dissipate considerable amounts of power. Because of the difficulty of extracting this power from a small volume, it does not appear feasible to reduce the physical size of a main-frame computer sufficiently to achieve a cycle time of 1 ns. On the other hand, preliminary studies of Josephson junction computer elements established that these devices had the combined requirements of very high switching speed and very low dissipation. *See* INTEGRATED CIRCUITS.

The Josephson effects have an established role in standards laboratories. The most important applications are the measurement of the fundamental constant ratio e/h, and maintaining the standard volt. When a Josephson junction is irradiated with microwaves of frequency f, constant-voltage steps are induced on the *I-V* characteristic at voltages $nhf/2e$, where n is an integer. Precise measurements of the voltages at which the steps are induced by a known frequency have led to the most accurate determination available of e/h. Furthermore, the standard volt at the United States National Bureau of Standards and at a number of other national laboratories is maintained (but not defined) by these voltage steps. *See* ELECTRICAL MEASUREMENTS; FUNDAMENTAL CONSTANTS.

The superconductor-insulator-superconductor (SIS) quasiparticle mixer is a superconducting detector of microwave radiation. The device consists of a small-area tunnel junction, fabricated from lead alloys, with the *I-V* characteristic shown in the illustration. The mixer is operated near the sharp onset in the current, where the characteristics are highly nonlinear. The nonlinearity is used to mix the signal frequency with the local oscillator frequency to produce an intermediate frequency that is coupled out of the junction into a low-noise preamplifier. Such receivers are likely to have a major impact on radio astronomy, and could be of great importance in such applica-

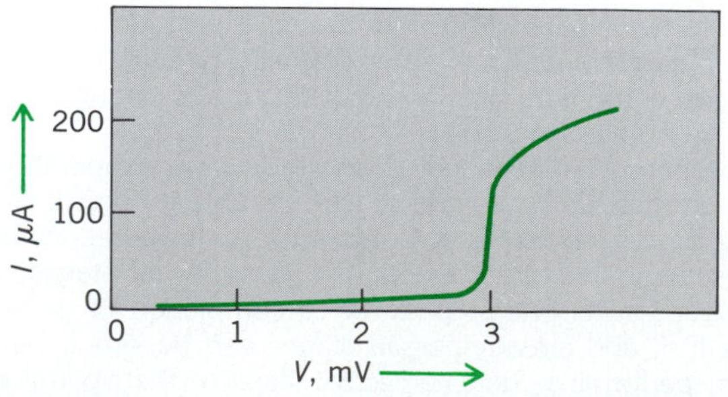

Current-voltage (*I-V*) characteristic of superconductor-insulator-superconductor (SIS) quasiparticle junction.

tions as space communications. *See* RADIO ASTRONOMY; SPACE COMMUNICATIONS. [J.H.Cla.]

Large-scale devices. Large-scale superconducting devices comprise magnets, motors, generators, and cables using zero resistance superconducting windings. Typical values of current densities used in superconducting coils are 10^9 A/m^2 or more—about two orders of magnitude higher than values used in normal conductors, resulting in a large reduction in size and weight as well as in power consumption.

The most commonly used superconducting alloy is the niobium-titanium alloy NbTi. This consists of titanium impurities in solution in niobium, the effect of the titanium impurities being to transform the niobium into a strong superconductor. Commerical niobium-titanium wires are composed of thin niobium-titanium filaments, typically 25 micrometers in diameter, embedded in a copper matrix. The purpose of the copper matrix is to stabilize the superconducting wire: it provides a parallel low-resistance conducting path for the case where a normal spot would appear accidentally along one of the niobium-titanium filaments, thus avoiding an avalanche effect that could quench the whole superconducting coil.

Magnetic fields are used extensively in high-energy physics to accelerate, bend, focus, and store particle beams and to detect and identify elementary particles. Superconducting magnets allow the utilization of higher fields in larger volumes at a lower cost in capital investment and energy expenses. As an example, the bubble chamber magnet at CERN would require a power of 70 MW with conventional coils, while the superconducting version consumes less than 1 MW.

Electric power generation has emerged as one of the most promising areas of application of superconductivity. Because of the losses occurring in a superconductor submitted to a high-intensity ac field, superconducting ac generators have a superconducting rotor and a normal armature. The advantages of the superconducting machine over conventional generators derive from the absence of electrical losses in the rotor and from the higher magnetic field intensity provided by the superconducting coils. *See* ALTERNATING-CURRENT GENERATOR; GENERATOR.

Short sections of flexible niobium-based superconducting ac cables have been successfully tested. However, due to large refrigeration costs, it is only at ratings of the order of 5 GVA that they could compete with conventional forced-cooling cables. *See* DIRECT-CURRENT TRANSMISSION; TRANSMISSION LINES.

Very large superconducting magnets with stored energies of the order of 10 gigajoules or more have several areas of application, including MHD power plants, controlled fusion reactors and energy storage.

Among the most promising applications for superconducting magnets for the short term are motors for marine propulsion. The much smaller weight of the superconducting motor makes electrical propulsion practical.

One of the most spectacular applications of superconducting magnets is for levitated trains for high-speed transportation. Their principle of operation is quite simple: When a magnet moves over an electrically conducting medium (such as a metallic sheet), the excited eddy currents are equivalent to an image magnet of opposite polarity located beneath the conducting sheet. Repulsion between the magnet and its image results in levitation of the magnet, provided the field intensity and the speed of motion of the magnet are high enough. [G.De.]

Superconductivity A phenomenon occurring in many electric conductors, in which the electrons responsible for conduction undergo a collective transition to an ordered state with many unique and remarkable properties. These include the vanishing of resistance to the flow of electric current, the appearance of a large diamagnetism and other unusual magnetic effects, substantial alteration of many thermal properties, and the occurrence of quantum effects otherwise observable only at the atomic and subatomic level.

Superconductivity was discovered by H. Kamerlingh Onnes in 1911, while studying the variation with temperature of the electrical resistance of mercury within a few degrees of absolute zero. He observed that the resistance dropped sharply to an unmeasurably small value at a temperature of 4.2 K (−452.0°F). The temperature at which the transition occurs is called the transition or critical temperature T_c.

In 1933 W. Meissner and R. Ochsenfeld discovered that a metal cooled into the superconducting state in a not-too-large magnetic field expels the field from its interior. This discovery demonstrated that superconductivity involves more than simply very high or infinite electrical conductivity, remarkable as that alone is. *See* MEISSNER EFFECT.

In 1957, J. Bardeen, L. N. Cooper, and J. R. Schrieffer reported the first successful microscopic theory of superconductivity. It described how and why the electrons in a conductor form the ordered superconducting state. The Bardeen-Cooper-Schrieffer (BCS) theory still stands as the basic explanation of superconductivity, even though extensive theoretical work has embellished it.

Transition temperatures. It was realized from the start that practical applications of superconductivity could become much more widespread if a high-temperature superconductor, that is, one with a high transition temperature T_c, could be found. Between 1911 and 1986, several thousand superconducting alloys and compounds were found, but the record for the highest transition temperature was only 23 K (−418°F), held by a specially prepared alloy of niobium and germanium. Then in 1986 the discovery of transition temperatures possibly as high as 30 K (−406°F) was reported in a compound containing barium, lanthanum, copper, and oxygen. In 1987 a compound of yttrium, barium, copper, and oxygen was shown to be superconducting above 90 K (−298°F). In 1988 researchers showed that a bismuth, strontium, calcium, copper, and oxygen compound was superconducting below 110 K (−262°F), and transition temperatures as high as 139 K (−209°F) have been found in a mercury, thallium, barium, calcium, copper, and oxygen compound.

Occurrence. Some 29 of the metallic elements are known to be superconductors in their normal forms, and another 17 become superconducting under pressure or when prepared in the form of highly disordered thin films (Fig. 1). The number of

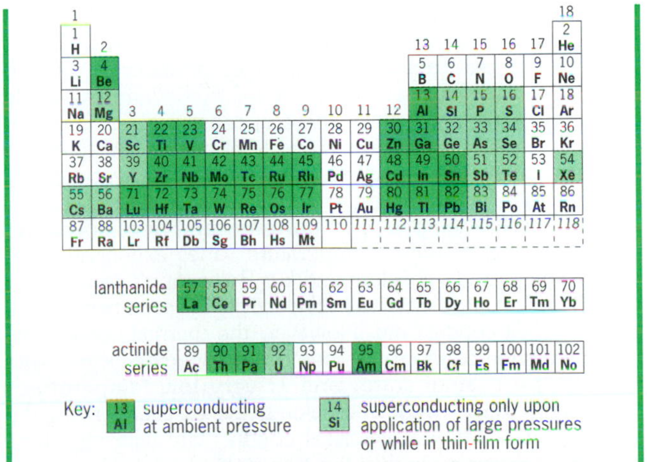

Fig. 1. Superconducting elements in the periodic table. (*After N. W. Ashcroft and N. D. Mermin, Solid State Physics, Holt, Rinehart and Winston, 1976*)

known superconducting compounds and alloys runs into the thousands. Superconductivity is thus a rather common characteristic of metallic conductors.

Despite the existence of a successful microscopic theory of superconductivity, there are no completely reliable rules for predicting whether a metal will be a superconductor. Certain trends and correlations are apparent among the known superconductors, however—some with obvious bases in the theory—and these provide empirical guidelines in the search for new superconductors.

The ordered superconducting state appears to be incompatible with any long-range-ordered magnetic state: None of the ferromagnetic or antiferromagnetic metals are also superconducting. The presence of nonmagnetic impurities in a superconductor usually has very little effect on the superconductivity, but the presence of impurity atoms which have localized magnetic moments can markedly depress the transition temperature even in concentrations as low as a few parts per million.

There are three classes of superconductors, all of which share the common feature that they contain one or more conducting planes of copper and oxygen atoms.

Magnetic properties. The existence of the Meissner-Ochsenfeld effect, the exclusion of a magnetic field from the interior of a superconductor, is direct evidence that the superconducting state is not simply one of infinite electrical conductivity. Instead, the Meissner-Ochsenfeld effect implies that the superconducting state is a true thermodynamic equilibrium state, a new phase which has lower free energy than the normal state of temperatures below the transition temperature and which somehow requires the absence of magnetic flux.

The exclusion of magnetic flux by a superconductor costs some magnetic energy. So long as this cost is less than the condensation energy gained by going from the normal to the superconducting phase, the superconductor will remain completely superconducting in an applied magnetic field. If the applied field becomes too large, the cost in magnetic energy will outweigh the gain in condensation energy, and the superconductor will become partially or totally normal. The manner in which this occurs depends on the geometry and the material of the superconductor. The geometry which produces the simplest behavior is a very long cylinder with field applied parallel to its axis. Two distinct types of behavior may then occur, depending on the type of superconductor.

Thermal properties. The appearance of the superconducting state is accompanied by quite drastic changes in both the thermodynamic equilibrium and thermal transport properties of a superconductor.

The heat capacity of a superconducting material is quite different in the normal and superconducting states (Fig. 2). In the normal state (produced at temperatures below the transition temperature by applying a magnetic field greater than the critical field), the heat capacity is determined primarily by the normal electrons and is nearly proportional to the temperature. In zero applied magnetic field, there appears a discontinuity in the heat capacity at the transition temperature. At temperatures just below the transition temperature, the heat capacity is larger than in the normal state. It decreases more rapidly with decreasing temperature, however, and at temperatures well below the transition temperature varies exponentially as $e^{-\Delta/kT}$, where Δ is a constant and k is Boltzmann's constant.

Ordinarily a large electrical conductivity is accompanied by a large thermal conductivity. However, the thermal conductivity of a pure superconductor is less in the superconducting state than in the normal state, and at very low temperatures approaches zero. Crudely speaking, the explanation for the association of infinite electrical conductivity with vanishing thermal conductivity is that the transport of heat requires the transport of disorder (entropy). The superconducting state is one of perfect order (zero entropy), and so there is no disorder to transport and therefore no thermal conductivity. *See* ENTROPY; THERMAL CONDUCTION IN SOLIDS.

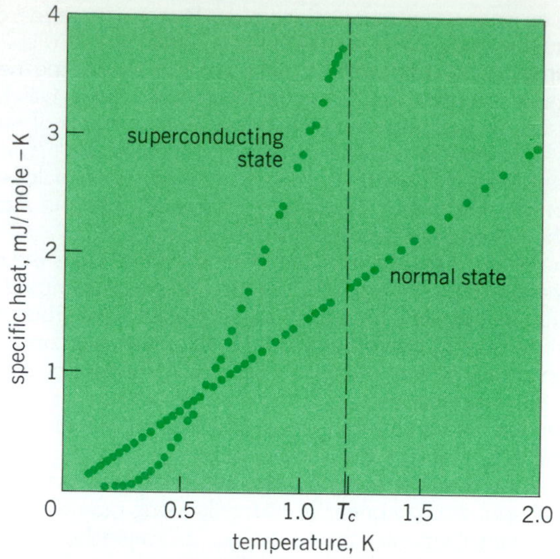

Fig. 2. Low-temperature specific heat of normal and superconducting aluminum. (*After N. E. Phillips, Heat Capacity of Aluminum between 0.1 K and 4 K, Phys. Rev., 114:676–685, 1959*)

Two-fluid model. C. J. Gorter and H. B. G. Casimir introduced in 1934 a phenomenological theory of superconductivity based on the assumption that in the superconducting state there are two components of the conduction electron "fluid" (hence the name given this theory, the "two-fluid model"). One, called the superfluid component, is an ordered condensed state with zero entropy, hence is incapable of transporting heat. It does not interact with the background crystal lattice, its imperfections, or the other conduction electron component and exhibits no resistance to flow. The other component, the normal component, is composed of electrons which behave exactly as they do in the normal state. It is further assumed that the superconducting transition is a reversible thermodynamic phase transition between two thermodynamically stable phases, the normal state and the superconducting state, similar to the transition between the liquid and vapor phases of any substance. The validity of this assumption is strongly supported by the existence of the Meissner-Ochsenfeld effect and by other experimental evidence. This assumption permits the application of all the powerful and general machinery of the theory of equilibrium thermodynamics. The results tie together the observed thermodynamic properties of superconductors in a very satisfying way.

Microscopic (BCS) theory. The key to the basic interaction between electrons which gives rise to superconductivity was provided by the isotope effect. It is an interaction mediated by the background crystal lattice and can crudely be pictured as follows: An electron tends to create a slight distortion of the elastic lattice as it moves, because of the Coulomb attraction between the negatively charged electron and the positively charged lattice. If the distortion persists for a brief time (the lattice may ring like a struck bell), a second passing electron will see the distortion and be affected by it. Under certain circumstances, this can give rise to a weak indirect attractive interaction between the two electrons which may more than compensate their Coulomb repulsion.

The first step toward a theory based on the interaction was taken by Cooper in 1956, when he showed that two electrons wi "bound pair" (often called a Cooper pair) if they are in the presence of a high-density fluid of other electrons, no matter how weak the interaction is. The two partners of a Cooper pair have opposite momenta and spin angular momenta. Then, in 1957, Bardeen, Cooper, and Schrieffer showed how to construct a wave function in which all of the electrons (at least, all of the important ones) are paired. Once this wave function is adjusted to minimize the free energy, it can be used as the basis for a complete microscopic theory of superconductivity.

The successes of the BCS theory and its subsequent elaborations are manifold. One of its key features is the prediction of an energy gap. Excitations called quasiparticles (which are something like normal electrons) can be created out of the superconducting ground state by breaking up pairs, but only at the expense of a minimum energy of Δ per excitation; Δ is called the gap parameter. The original BCS theory predicted that Δ is related to T_c by $\Delta = 1.76\ kT_c$ at $T = 0$ for all superconductors. This turns out to be nearly true, and where deviations occur they are understood in terms of modification of the BCS theory. The manifestations of the energy gap in the low-temperature heat capacity and in electromagnetic absorption provide strong confirmation of the theory. [D.N.L.; R.J.So.]

Supercritical fields Static fields that are strong enough to cause the normal vacuum, which is devoid of real particles, to break down into a new vacuum in which real particles exist. This phenomenon has been observed for electric fields, and is predicted for other fields such as gravitational fields and the gluon field of quantum chromodynamics.

Vacuum decay in quantum electrodynamics. The decay of the vacuum in strong electrostatic fields is a phenomenon in quantum electrodynamics that can be studied only through low-energy heavy-ion collisions. The best starting point for discussing this concept is to consider the binding energy of atomic electrons as the charge Z of a heavy nucleus is increased. If the nucleus is assumed to be a point charge, the total energy E of the $1s_{1/2}$ level drops to 0 when $Z = 137$. This so-called $Z = 137$ catastrophe had been well known, but it was argued loosely that it disappears when the finite size of the nucleus is taken into account. However, in 1969 it was shown that the problem is not removed but merely postponed, and reappears around $Z = 173$. Any level $E(nj)$ can be traced down to a binding energy of twice the electronic rest mass if the nuclear charge is further increased. At the corresponding charge number, called Z_{cr}, the state dives into the negative-energy continuum of the Dirac equation (the so-called Dirac sea). The overcritical state acquires a width and is spread over the continuum. *See* ANTIMATTER; RELATIVISTIC QUANTUM THEORY.

When Z exceeds Z_{cr}, a K-shell electron is bound by more than twice its rest mass, so that it becomes energetically favorable to create an electron-positron pair. The electron becomes bound in the $1s_{1/2}$ orbital and the positron escapes. The overcritical vacuum is therefore said to be charged. *See* POSITRON.

Clearly, the charged vacuum is a new ground state of space and matter. The normal, undercritical, electrically neutral vacuum is no longer stable in overcritical fields: it decays spontaneously into the new stable but charged vacuum. Thus the standard definition of the vacuum, as a region of space without real particles, is no longer valid in very strong external fields. The vacuum is more accurately defined as the energetically deepest and most stable state that a region of space can have while being penetrated by certain fields.

Superheavy quasimolecules. Inasmuch as the formation of a superheavy atom of $Z > 173$ is very unlikely, a new idea is necessary to test these predictions experimentally. That idea, based on the concept of nuclear molecules, was put forward in 1969: a superheavy quasimolecule forms temporarily during the slow collision of two heavy ions. It is sufficient to form the quasimolecule for a very short instant of time, comparable to the time scale for atomic processes to evolve in a heavy atom, which is typically of the order 10^{-18} to 10^{-20} s. Suppose a uranium ion is shot at another uranium ion at an energy corresponding to their Coulomb barrier, and the two, moving slowly (compared to the K-shell electron velocity) on Rutherford hyperbolic trajectories, are close to each other (compared to the K-shell electron orbit radius). Then the atomic electrons move in the combined Coulomb potential of the two nuclei, thereby experiencing a field corresponding to their combined charge of 184. This happens because the ionic velocity (of the order of $c/10$) is much smaller than the orbital electron velocity (of the order of c), so that there is time for the electronic molecular orbits to be established, that is, to adjust to the varying distance between the charge centers, while the two ions are in the vicinity of each other. *See* QUASIATOM.

Giant nuclear systems. The energy spectrum for positrons created in, for example, a uranium-curium collision consists of three components: the induced, the direct, and the spontaneous, which add up to a smooth spectrum. The presence of the spontaneous component leads only to 5–10% deviations for normal nuclear collisions along Rutherford trajectories. This situation raises the question as to whether there is any way to get a clear qualitative signature for spontaneous positron production. Suppose that the two colliding ions, when they come close to each other, stick together for a certain time Δt before separating again. The longer the sticking, the better is the static approximation. For Δt very long, a very sharp line should be observed in the positron spectrum with a width corresponding to the natural lifetime of the resonant positron-emitting state. The observation of such a sharp line will indicate not only the spontaneous decay of the vacuum but also the formation of giant nuclear systems ($Z > 180$). *See* LINEWDITH; NUCLEAR MOLECULE.

Observation of spontaneous positron emission. Positron spectra from uranium-238 and curium-248 colliding at an energy close to that of the Coulomb barrier display a well-defined peak centered at an energy of about 320 keV. Analysis indicates that the intrinsic width of the peak is less than 20 keV. It has been suggested that the observation of spontaneous positron emission as a sharp line necessarily implies that, at bombarding energies close to that of the Coulomb barrier, metastable giant nuclear composite systems are formed with a rather long lifetime. Widths of 20 keV or less correspond to lifetimes for the nuclear molecular system longer than about 1000 times the Rutherford scattering collision time, during which the $1s\sigma$ state is overcritically bound. Such a lifetime could be supplied by the formation of a rather cold intermediate giant nuclear complex as the nuclei barely touch in overcoming the Coulomb barrier.

Other field theories. The idea of overcriticality also has applications in other field theories, such as those of pion fields, gluon fields (quantum chromodynamics), and gravitational fields (general relativity).

A model of a heavy meson consists of a heavy quark Q and antiquark $\overline{Q}$ located in the foci of an ellipsoidal bag. The color-electric or glue-electric field lines do not penetrate the bag surface. The Dirac equation may be solved for light quarks q in this field of force. In the spherical case the potential is zero, and the solutions with different charges degenerate. As the source charges Q and $\overline{Q}$ are pulled apart, the wave functions start to localize. At a critical deformation of the bag, positive and negative energy states cross; that is, overcriticality is reached and the color field is strong enough that the so-called perturbative vacuum inside the bag rearranges so that the wave functions are pulled to opposite sides and the color charges of the heavy quarks are completely shielded. Hence two new mesons of type $\overline{Q}q$ and $Q\overline{q}$ appear; the original meson fissions. *See* GLUONS; QUANTUM CHROMODYNAMICS; QUARKS.

An example of superstrong gravitational fields is given by the solutions of the Dirac equation in the field of a gravitating mass shell. The energy gap vanishes completely when the radius of the shell equals the Schwarzschild radius. This is where overcriticality is reached and pair production will set in. This phenomenon is connected with the so-called Hawking radiation. *See* BLACK HOLE. [W.Gr.]

Supercritical fluid Any fluid at a temperature and a pressure above its critical point; also, a fluid above its critical temperature regardless of pressure. Below the critical point the fluid can coexist in both gas and liquid phases, but above the critical point there can be only one phase. Supercritical fluids are of interest because their properties are intermediate between those of gases and liquids, and are readily adjustable. *See* CRITICAL PHENOMENA; PHASE EQUILIBRIUM.

In a given supercritical fluid the thermodynamic and transport properties are a function of density, which depends strongly on the fluid's pressure and temperature. The density may be adjusted from a gaslike value of 0.1 g/ml to a liquidlike value as high as 1.2 g/ml. Furthermore, as conditions approach the critical point, the effect of temperature and pressure on density becomes much more significant. Increasing the density of supercritical carbon dioxide from 0.2 to 0.5 g/ml, for example, requires raising the pressure from 85 to 140 atm (8.6 to 14.2 megapascals) at 158°F (70°C), but at 95°F (35°C) the required change is only from 65 to 80 atm (6.6 to 8.1 MPa).

For a given fluid, the logarithm of the solubility of a solute is approximately proportional to the solvent density at constant temperature. Therefore, a small increase in pressure, which causes a large increase in the density, can raise the solubility a few orders of magnitude. While almost all of a supercritical fluid's properties vary with density, some of these properties are more like those of a liquid while others are more like those of a gas. *See* SUPERCRITICAL-FLUID CHROMATOGRAPHY.

In most supercritical-fluid applications the fluid's critical temperature is less than 392°F (200°C) and its critical pressure is less than 80 atm (8.1 MPa). High critical temperatures require operating temperatures that can damage the desired product, while high critical pressures result in excessive compression costs. In addition to these pure fluids, mixed solvents can be used to improve the solvent strength. [K.Jo.; R.Len.]

Spercritical-fluid chromatography Any separation technique in which a supercritical fluid is used as the mobile phase. For any fluid, a phase diagram can be constructed to show the regions of temperature and pressure at which gases and liquids, gases and solids, and liquids and solids can coexist. For the gas–liquid equilibrium, there is a certain temperature and pressure, known as the critical temperature and pressure, below which a gas and a liquid can coexist but above which only a single phase (known as a supercritical fluid) can form.

For example, the density of supercritical fluids is usually between 0.25 and 1.2 g/ml and is strongly pressure-dependent. Their solvent strength increases with density, so that molecules that are retained on the column can often be eluted simply by increasing the pressure under which the fluid is compressed.

Diffusion coefficients of solutes in supercritical fluids are tenfold greater than the corresponding values in liquid solvents (although about three orders of magnitude less than the corresponding values in gases). The high diffusivity of solutes in supercritical fluids decreases their resistance to mass transfer in a chromatographic column and hence allows separations to be made either very quickly or at high resolution. Molecules can be separated in a few minutes or less by using packed columns of the type used for high-performance liquid chromatography (HPLC). *See* DIFFUSION IN GASES AND LIQUIDS.

Despite the relatively large number of separations made using packed-column supercritical-fluid chromatography that have been described since 1962, supercritical-fluid chromatography became popular only in the late 1980s with the commercial introduction of capillary (open-tubular) supercritical-fluid chromatographs. While supercritical-fluid chromatography may never be as widely used as gas chromatography or high-performance liquid chromatography, it provides a useful complement to these chromatographies. *See* CHEMICAL SEPARATION TECHNIQUES; CHROMATOGRAPHY; FLUIDS; GAS CHROMATOGRAPHY; LIQUID CHROMATOGRAPHY. [P.R.G.]

Supercritical wing A wing with special streamwise sections, or airfoils, which provide substantial delays in the onset of the adverse aerodynamic effects which usually occur at high subsonic flight speeds.

When the speed of an aircraft approaches the speed of sound, the local airflow about the airplane, particularly above the upper surface of the wing, may exceed the speed of sound. Such a condition is called supercritical flow. On previous aircraft, this supercritical flow resulted in the onset of a strong local shock wave above the upper surface of the wing (illustration *a*). This local wave caused an abrupt increase in the pressure on the surface of the wing, which may cause the surface boundary-layer flow to separate from the surface, with a resulting severe increase in the turbulence of the flow. The increased turbulence leads to a severe increase in drag and loss in lift, with a resulting decrease in flight efficiency. The severe turbulence also caused buffet or shaking of the aircraft and substantially changed its stability or flying qualities. *See* AERODYNAMIC FORCE; AERODYNAMIC WAVE DRAG; LOW-PRESSURE GAS FLOW; SHOCK WAVE; TRANSONIC FLIGHT.

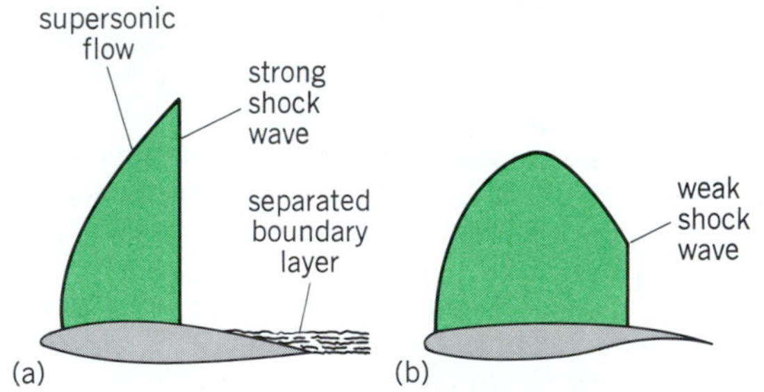

Comparison of airflow about airfoils. (*a*) Conventional airfoils, Mach number = 0.7. (*b*) Supercritical airfoils. Mach number = 0.8.

Supercritical airfoils are shaped to substantially reduce the strength of the shock wave and to delay the associated boundary-layer separation (illustration *b*). Since the airfoil shape allows efficient flight at supercritical flight speeds, a wing of such design is called a supercritical wing. *See* AIRPLANE; WING. [R.T.Wh.]

Superfluidity The frictionless flow of liquid helium at low temperature; also, the flow of electric current without resistance in certain solids at low temperature (superconductivity).

Both helium isotopes have a superfluid transition, but the detailed properties of their superfluid states differ considerably because they obey different statistics. ^{4}He, with an intrinsic spin of 0, is subject to Bose-Einstein statistics, and ^{3}He, with a spin of ½, to Fermi-Dirac statistics. There are two distinct superfluid states in ^{3}He called A and B.

The term "superfluidity" usually implies He II or the A and B phases of ^{3}He, but the basic similarity between these and the "fluid" consisting of pairs of electrons in superconductors is sufficiently strong to designate the latter as a charged superfluid. Besides flow without resistance, superfluid helium and superconducting electrons display quantized circulating flow patterns in the form of microscopic vortices. *See* BOSE-EINSTEIN STATISTICS; LIQUID HELIUM; QUANTIZED VORTICES; SECOND SOUND; SUPERCONDUCTIVITY. [L.J.C.]

Supergiant star A member of the family containing the intrinsically brightest stars, populating the top of the Hertzsprung-Russell diagram. Supergiant stars occur at all temperatures, from 54,000 to 5000°F (30,000 to 3000 K), and have luminosities ranging from 10^4 to 10^6 times that of the Sun. The hot supergiants have radii 20 times that of the Sun. The cool supergiants are the largest known stars, reaching several thousand solar radii. Among the bright supergiants are Deneb (α-Cygni), Rigel (β-Orionis), and Betelgeuse (α-Orionis). Because few have masses exceeding 20–40 Suns, the ratio of luminosity to mass is high. Thermonuclear energy sources will therefore be rapidly exhausted, hydrogen being consumed in a time scale of only 10^6 to 10^7 years. The supergiants are thus young but rapidly evolving massive stars. Nearly all are slightly variable in light and radial velocity, have highly turbulent and extended atmospheres, and show broadened spectral lines. *See* HERTZSPRUNG-RUSSELL DIAGRAM; STAR; STELLAR EVOLUTION. [J.L.Gr.]

Supergranulation A system of convective cells, with typical diameters of 30,000 km (19,000 mi), covering the Sun's surface, but invisible in ordinary photographs.

Relation to granulation. High-resolution photographs of the Sun's visible surface reveal the granulation, a closely packed cellular grid with bright (hot) centers surrounded by dark (cool) lanes. The granules have lifetimes of 10–30 min and average diameters of 1000 km (600 mi). In the 1930s, L. Biermann suggested that, in the outer 25–30% of the Sun, heat from the interior is transported to the surface by convection, similar to a pot of boiling soup heated from below. The granules are the surface manifestation of this process, corresponding to the boiling bubbles (plumes). A mixing-length theory proposed that the bubble sizes are approximately equal to the local scale height H (the distance in which density or pressure changes by a factor $e \approx 2.7$). Since H varies from about 1000 km (600 mi) at the surface to 100,000 km (60,000 mi) at the base of the convection zone, a very large range of plume sizes was hypothesized.

Discovery and description. In 1959, R. Leighton modified the spectroheliograph to image the Sun by using the Doppler and Zeeman effects. Images of line-of-sight (approaching and receding) gas motions and of magnetic fields could be obtained. Immediately obvious on the Doppler photographs was a new cell structure with an area 900 times that of the granulation, having a mainly horizontal flow pattern. Lifetimes of supergranules are 1–2 days. The cells form an irregular polygonal structure, with most of the downflow occurring at the polygon vertices. The supergranules fit neatly within an essentially identical network (grid) structure seen in the magnetic photographs. The kinetic energy of the supergranular gas flow at the Sun's surface exceeds the magnetic field's ability to resist such motions, and thus the magnetic field is dragged forcibly to the supergranule boundaries until the two patterns coincide. This magnetic field, in turn, causes local heating of the upper solar atmosphere (chromosphere), producing a similar chromospheric network pattern seen in high-temperature spectral lines. *See* DOPPLER EFFECT; FRAUNHOFER LINES; SOLAR MAGNETIC FIELD; SPECTROGRAPH; SPECTROHELIOSCOPE; SUN; ZEEMAN EFFECT. [G.W.Si.]

Supergravity A theory that attempts to unify gravitation with the other fundamental interactions. The notion of a unification theory of all interactions (forces) responsible for natural phenomena has fascinated physicists for over a century. The first, and only, completely successful unification theory was constructed by James Clerk Maxwell, in which the up-to-then unrelated electric and magnetic phenomena were unified in his electrodynamics. *See* FUNDAMENTAL INTERACTIONS; GRAVITATION; MAXWELL'S EQUATIONS.

Electroweak theory. The second stage of unification concerns the unification of electromagnetic and weak interactions, using Maxwell's theory as a guide. This was accomplished by making use of the non-Abelian gauge theories invented by C. N. Yang and R. L. Mills, and of spontaneous symmetry breaking. The symmetry of Maxwell's theory is very similar to spatial rotations about an axis, rotating the vector potentials while leaving the electric and magnetic fields unchanged. It is a local invariance because the rotations about a fixed axis can be made by different amounts at different points in space-time. Thus, Maxwell's theory is invariant under a one-parameter group of transformations U(1). In Yang-Mills theory this local invariance was generalized to theories with large symmetry groups such as the three-dimensional rotation group SO(3) ≃ SU(2) which has three parameters. The number of parameters of the local symmetry (gauge) group is also equal to the number of 4-vector potentials in the gauge theory based on that group. A detailed analysis of weak and electromagnetic forces shows that their description requires four 4-vector potentials (gauge fields), so that the gauge group must be a four-parameter group. In fact, it is the product SU(2) · U(1). *See* GAUGE THEORY; GROUP THEORY; SYMMETRY BREAKING; WEAK NUCLEAR INTERACTIONS; WEINBERG-SALAM MODEL.

Grand unification theories. In the third stage of unification, electroweak and strong forces are regarded as different components of a more general force which mediates the interactions of particles in a grand unification model. Strong forces are responsible for the interactions of hadrons and for keeping quarks confined inside hadrons. They are described by eight massless 4-vector potentials (gluons), the corresponding eight-parameter group being SU(3). This local symmetry is called color, and the corresponding theory quantum chromodynamics (QCD). Thus the gauge group of a grand unification theory must include (SU(3) · SU(2) · U(1) as a subsymmetry. The most dramatic prediction of these theories is the decay of proton. *See* GLUONS; GRAND UNIFICATION THEORIES; PROTON; QUANTUM CHROMODYNAMICS; QUARKS.

Superunification theories and supersymmetry. The last, and the most ambitious, stage of unification deals with the possibility of combining grand unification and gravitation theories into a superunification theory, also known as supergravity. To achieve this, use is made of the dual role played by local symmetry groups. On the one hand, they describe the behavior of forces. On the other hand, they classify the elementary particles (fields) of the theory into multiplets: spin-zero fields in one multiplet, spin-½ fields in another multiplet, and so forth, but never fermions and bosons in one irreducible multiplet. This last restriction used to be a major obstacle on the way to superunification. This is because, of all the elementary particles, only the quanta of gravitational field (gravitons) have spin 2, so that a multiplet of elementary particles including the graviton must of necessity involve particles of different spin. But then by an internal symmetry transformation, which is by definition distinct from space-time (Lorentz) transformations, one can "rotate" particles of different spin into one another, thus altering their space-time transformation properties. This apparent paradox can be circumvented if both the internal symmetry and Lorentz transformations are part of a larger group which also includes the spin-changing transformations. This is how supersymmetry makes its appearance in supergravity theories. *See* GRAVITON; LORENTZ TRANSFORMATIONS; RELATIVITY; SUPERSYMMETRY. [F.M.]

Superheater A component of a steam-generating unit in which steam, after it has left the boiler drum, is heated above its saturation temperature. The amount of superheat added to the steam is influenced by the location, arrangement, and amount of superheater surface installed, as well as the rating of the boiler. The superheater may consist of one or more stages of tube banks arranged to effectively transfer heat from the products of combustion. *See* STEAM-GENERATING UNIT. [G.W.K.]

Superheterodyne receiver A receiver that uses the heterodyne principle to convert the incoming modulated radio-frequency signal to a predetermined lower carrier frequency, the intermediate-frequency value. This is done by using a local oscillator tuned simultaneously with the input stage of the receiver, so that the oscillator frequency always differs from that of the incoming carrier by the i-f value.

With a fixed and favorably chosen i-f value, the i-f amplifier can efficiently provide the major portion of the amplifcation and selectivity required by the receiver. After amplification, the i-f signal is demodulated to the second detector to obtain the desired audio output signal. A similar circuit arrangement is used in television and radar receivers to obtain the desired video output signal. *See* HETRODYNE PRINCIPLE; RADIO RECEIVER. [J.Mar.]

Superinfection Either of two types of infection: a new infection occurring as a consequence of treatment of another infection, often because of the use of excessive numbers or amounts of antibiotics; or infection arising in individuals with specific underlying mechanisms promoting infections but not related to treatment of a preexisting infection. In most instances the superinfection is much more difficult to treat than the original infection.

An example of the first type is an individual who acquires virus influenza and may develop a complicating pneumococcal pneumonia. Then, if treated with two antibiotics, staphylococcal pneumonia may supervene as a superinfection.

The second type of superinfection can be subdivided into five categories: (1) Infection may arise solely as a consequence of an underlying disease. (2) Infections may arise as a consequence of the underlying disease plus therapy. (3) Infections may be due solely to treatment or procedures. (4) Infections may be increased in severity in compromised hosts, but not in frequency. (5) There are also certain societally related infections, including infections consequent to use of illicit drugs. *See* HOSPITAL INFECTIONS; INFECTION. [D.Lo.]

Superluminal motion Apparent proper motion exceeding the velocity of light, *c,* in an astronomical object. Many quasars exhibit systematic changes in images of the radio emission from their nuclei over periods of months to years. In some cases, features in the image appear to separate at more than 10 times the speed of light, given the great distance of the quasars from Earth.

Superluminal motion was one of the most exciting discoveries to emerge from a technique in radio astronomy called very long baseline interferometry (VLBI). This method involves the tape recording of radio signals from large antennas at up to 10–15 locations across the Earth, and the combination of these signals in a computer to form a radio image of the quasar at extremely high resolution (less than 0.001 second of arc). *See* QUASAR; RADIO ASTRONOMY; RADIO TELESCOPE.

Superluminal motion is seen mostly in quasars but also in some other active galactic nuclei. This rapid motion is confined to within a few tens of parsecs of the nucleus, whose power source is believed to be a massive black hole. Over 30 examples of superluminal motion are now known, with apparent proper motions up to 10*c*. *See* GALAXY, EXTERNAL.

Novae and supernovae in the Milky Way Galaxy can also show superluminal motion. An example is the elliptical ring of light centered on, and illuminated by, the supernova SN1987A in the Large Magellanic Cloud. *See* NOVA; SUPERNOVA.

Announcement of the discovery of superluminal motion caused widespread concern because of the apparent violation of A. Einstein's special theory of relativity. Many explanations were proposed (besides Einstein's theory being incorrect), but only the relativistic jet model has stood the test of time. Superluminal motion is explained in this model as mainly a geometric effect. A feature moves away from the nucleus of the quasar at high (relativistic) speed (but less than *c*) at a small angle to the line of sight to the Earth. Radio waves from the moving feature arrive only slightly later than waves from the nucleus, whereas the feature took a much longer time to reach its current position; the motion appears superluminal because the former measure of time is used rather than the latter. As the speed approaches *c* and the angle to the line of sight decreases, the apparent speed can be arbitrarily large. [S.C.U.]

Supermassive stars Hypothetical objects with masses exceeding 60 solar masses, the mass of the largest known ordinary stars (1 solar mass = 2×10^{30} kg). The term is most often used in connection with objects larger than 10^4 solar masses that might be the energy source in quasars and active galaxies. Since these objects have not been observed and are hypothetical at present, very little is known about their nature and behavior. About the only thing known for certain about supermassive stars is that they are unstable. [H.L.Sh.]

Supermultiplet A generalization of the concept of a multiplet. A multiplet is a set of quantum-mechanical states each of which has the same value of some fundamental quantum numbers and differs from the other members of the set by other quantum numbers, which take values from a range of numbers dictated by the fundamental quantum numbers. There are as many states in the set as there are different values of the varying quantum numbers. The number of states in the set is the multiplicity or dimension of the multiplet. For example, there are $2S+1$ orientations or projections of spin S, and these are labeled by M_s with values from $-S$ to $+S$ in integer steps. Each of the $2S+1$ members of this multiplet, which has the dimension $2S+1$, is labeled by S and M_s, where S labels the multiplet, and M_s a state within the multiplet. Suppose that T denotes another spinlike quantity but refers to a different physical property. Then the $2T+1$ states also constitute a multiplet labeled by T with members labeled by, say T_3. If sets of these multiplets are grouped together, then one speaks of a supermultiplet. For example, one supermultiplet might be denoted by a set of fundamental quantum numbers $(\lambda_1\lambda_2\lambda_3)$, and S and T would range over a set of values dictated by the values of λ_1, λ_2, and λ_3. For a fixed value of S, M_s would have its usual $2S+1$ values, and for fixed value of T, T_3 would have $2T+1$ different values, ranging from $-T$ to $+T$. The dimension of a supermultiplet in this case would be the sum over all values of S and T of the product $(2S+1)(2T+1)$. A supermultiplet then is a multiplet of multiplets. *See* QUANTUM MECHANICS; QUANTUM NUMBERS.

The term supermultiplet was first used by Eugene P. Wigner in a nuclear theory in which protons and neutrons are treated equivalently, and the dependence of the nuclear force on their differences and intrinsic spins is ignored. When used in this connection, one speaks of Wigner supermultiplets. *See* SYMMETRY LAWS (PHYSICS). [S.A.Wi.]

Supernova The sudden brightening of a star, by a factor of up to 10^{10}, so that it may briefly outshine the entire galaxy in which it is located. Fading occurs slowly, over several years or more. The supernova phenomenon was first distinguished from that of ordinary novae by Walter Baade and Fritz Zwicky in 1934. They also suggested that the most likely source of the enormous energy required was the formation of a neutron star, and that a certain star in the Crab Nebula, remnant of the 1054 supernova, was probably such a neutron star. Recent observations and calculations have essentially confirmed their suggestions. *See* CRAB NEBULA; NEUTRON STAR; NOVA; PULSAR.

A large galaxy like the Milky Way probably has about one supernova explosion every 30 years. Because dust blocks the view of much of the Milky Way, only six have been recorded in the last 1000 years (in 1006, 1054, 1572, 1604, and probably 1181 and 1680). More than 100 supernovae have been found in other galaxies since 1885.

In all cases the spectra show large amounts of gas being thrown off at speeds of thousands of miles per second. The blown-off gas travels rapidly out into space, sweeping up more (interstellar) gas as it goes, and remains visible as a bright nebula for thousands of years. Such supernova remnants are also characteristically copious sources of radio and x-rays (indicating the presence of magnetic fields, relativistic electrons, and shock waves). Some dozens of them have thereby been identified in the galaxy. *See* ASTRONOMICAL SPECTROSCOPY; RADIO ASTRONOMY; X-RAY ASTRONOMY. [V.T.]

Superoxide chemistry A branch of chemistry that deals with the reactivity of the superoxide ion (O_2^-), a one-electron (e^-) adduct of molecular oxygen (dioxygen; O_2) formed by the combination of O_2 and e^-. Because 1–15% of the O_2 that is respired by mammals goes through the O_2^- oxidation state, the biochemistry and reaction chemistry of the species are important to those concerned with oxygen toxicity, carcinogenesis, and aging. Although the name superoxide has prompted many to assume an exceptional degree of reactivity for O_2^-, the use of the prefix in fact was chosen to indicate stoichiometry. Superoxide was the name given in 1934 to the newly synthesized potassium salt (KO_2) to differentiate its two-oxygens-per-metal stoichiometry from that of most other metal–oxygen compounds (NA_2O, Na_2O_2, NaOH, Fe_2O_3).

Ionic salts of superoxide (yellow-to-orange solids), which form from the reaction of dioxygen with metals such as potassium, rubidium, or cesium, are paramagnetic, with one unpaired electron per two oxygen atoms.

In 1969, by means of electron spin resonance (ESR) spectroscopy, superoxide ion was detected as a respiratory intermediate, and metalloproteins were discovered that catalyze the disproportionation of superoxide, that is, superoxide dismutases (SODs), as shown in the reaction below.

$$2O_2^- + 2H^+ \xrightarrow{\text{SOD}} O_2 + H_2O_2$$

The biological function of superoxide dismutases is believed to be the protection of living cells against the toxic effects of superoxide. The possibility that superoxide might be an important intermediate in aerobic life provided an impetus to the study of superoxide reactivity.

The most general and universal property of O_2^- is its tendency to act as a strong Brønsted base. Its strong proton affinity manifests itself in any media. Another characteristic of O_2^- is its ability to act as a moderate one-electron reducing agent. *See* ACID AND BASE; OXIDATION-REDUCTION.

In general, superoxide ion chemistry does not appear to be sufficiently robust to make superoxide ion a toxin. However, it can interact with protons, halogenated carbons, and carbonyl compounds to yield peroxy radicals that are toxic. *See* OXYGEN TOXICITY; REACTIVE INTERMEDIATES.

Superoxide does not appear to have exceptional reactivity. Nevertheless, superoxide will continue to be an interesting species for study because of the multiplicity of its chemical reactions and because of its importance as an intermediate in reactions that involve dioxygen and hydrogen peroxide. *See* BIOINORGANIC CHEMISTRY; OXYGEN. [D.T.S.]

Superplastic forming A process for shaping superplastic materials, a unique class of crystalline materials that exhibit exceptionally high tensile ductility. Superplastic materials may be stretched in tension to elongations typically in excess of 200% and more commonly in the range of 400–2000%. There are rare reports of higher tensile elongations reaching as much as 8000%. The high ductility is obtained only for superplastic materials and requires both the temperature and rate of deformation (strain rate) to be within a limited range. The temperature and strain rate required depend on the specific material. A variety of forming processes can be used to shape these materials; most of the processes involve the use of gas pressure to induce the deformation under isothermal conditions at the suitable elevated temperature. The tools and dies used, as well as the superplastic material, are usually heated to the forming temperature. The forming capability and complexity of configurations producible by the processing methods of superplastic forming greatly exceed those possible with conventional sheet forming methods, in which the materials typically exhibit 10–50% tensile elongation. *See* SUPERPLASTICITY.

There are a number of commercial applications of superplastic forming and combined superplastic forming and diffusion bonding, including aerospace, architectural, and ground transportation uses. Examples are wing access panels in the Airbus A310 and A320, bathroom sinks in the Boeing 737, turbo-fan-engine cooling-duct components, external window frames in the space shuttle, front covers of slot machines, and architectural siding for buildings. *See* METAL FORMING. [C.H.Ha.]

Superplasticity The unusual ability of some metals and alloys to elongate uniformly thousands of percent at elevated temperatures, much like hot polymers or glasses. Under normal creep conditions, conventional alloys do not stretch uniformly, but form a necked-down region and then fracture after elongations of only 100% or less. The most important requirements for obtaining superplastic behavior include a very small metal grain size, a well-rounded (equiaxed) grain shape, a deformation temperature greater than one-half the melting point, and a slow deformation rate. *See* ALLOY; CREEP (MATERIALS); EUTECTICS.

Superplasticity is important to technology primarily because large amounts of deformation can be produced under low loads. Thus, conventional metal-shaping processes (for example, rolling, forging, and extrusion) can be conducted with smaller, and cheaper equipment. Nonconventional forming methods can also be used; for instance, vacuum-forming techniques, borrowed from the plastics industry, have been applied to sheet metal to form car panels, refrigerator door linings, and TV chassis parts. *See* METAL FORMING; PLASTICITY. [E.E.U.]

Superposition principle In classical wave theories (optics, acoustics), as in all theories characterized by linear homogeneous differential equations, the sum of any number of solutions to the equations is another solution. Thus (assuming one-dimensional waves for simplicity), the amplitude of the resultant wave at a point x and time t is the linear superposition of the amplitudes of all waves reaching x at time t. This fact, known as the principle of superposition, often also is taken to mean that if each of $\psi_1(x)$ and $\psi_2(x)$ are possible waveforms at $t = 0$, then any linear combination $\psi(x) = c_1\psi_1 + c_2\psi_2$ is a possible waveform at $t = 0$. This latter version is not a consequence of the linearity of the differential equations, but rather is an affirmation of the belief that the waveform can be chosen arbitrarily at any initial instant. *See* WAVE MOTION.

The preceding paragraph holds equally well for quantum theory, where the wave function $\psi(x,t)$ obeys the linear homogenous Schrödinger equation. In quantum theory, however, because the wave function represents physical states, the principle of superposition has a profound significance. In particular, if the observable A is certain to have value α_1 in the state represented by u_1, and is certain to have value α_2 in the state represented by u_2, then $c_1u_1 + c_2u_2$ represents a state in which measurement of the observable A is certain to yield either precisely α_1, or else precisely α_2. Thus, because probabilities are proportional to $|\psi|^2$, superposition accounts for phenomena which are difficult to understand from a classical viewpoint. *See* QUANTUM MECHANICS. [E.G.]

Superposition theorem (electric networks) Essentially, that it is permissible, if there are two or more sources of electromotive force in a linear electrical network, to compute at any element of the network the response of voltage or of current that results from one source alone, and then the response resulting from another source alone, and so on for all sources, and finally to compute the total response to all sources acting together by adding these individual responses.

Thus, if a load of constant resistance is supplied with electrical energy from a linear network containing two batteries, two generators, or one battery and one generator, it would be correct to find the current that would be supplied to the load by one source (the other being reduced to zero), then to find the current that would be supplied to the load by the second source (the first source now being reduced to zero), and finally to add the two currents so computed to find the total current that would be produced in the load by the two sources acting simultaneously.

By means of the principle of superposition, effects are added instead of causes. This principle seems so intuitively valid that there is far greater danger of applying superposition where it is incorrect than of failing to apply it where it is correct. It must be recognized that for superposition to be correct the relation between cause and effect must be linear. [H.H.Sk.]

Superpressure balloon A nonextensible (constant-volume) free balloon which converts its free lift into superpressure by the time that float level is reached. As long as the balloon is superpressurized, it will continue to float at a constant density level. The main advantage of the superpressure balloon is the possibility of long-duration flights which can circumnavigate the globe. Small superpressure balloons 4 m (13 ft) in diameter have been released in the southern hemisphere, and some have remained aloft for over a year. Balloons having diameters of over 30 m (100 ft) have been flown with limited results. These ballooons are relatively new, and the objective of the research has been to produce a reliable superpressure balloon capable of carrying 300 kg (660 lb) to an altitude of 42 km (26 mi). [W.R.N.]

Supersaturation A solution is at the saturation point when dissolved solute in its crystallizes from it at the same rate at which it dissolves. Under prescribed experimental conditions of temperature and pressure, a solution can contain at saturation only one fixed amount of dissolved solute. However, it is possible to prepare relatively stable solutions which contain a quantity of a dissolved solute greater than that of the saturation value provided solute phase is absent. Such solutions are said to be supersaturated. They can be prepared by changing the experimental conditions of a system so that greater solubility is obtained, perhaps by heating the solution, and then carefully returning the system to or near its original state. The addition of solute phase will immediately relieve supersaturation. Solutions in which there is no spontaneous formation of solute phase for extended periods of time are said to be metastable. There is no sharp line of demarcation between an unstable and metastable solution. The process whereby initial aggregates within a supersaturated solution develop spontaneously into particles of new stable phase is known as nucleation. The greater the degree of supersaturation, the greater will be the number of nuclei formed. *See* NUCLEATION; PHASE EQUILIBRIUM; PRECIPITATION (CHEMISTRY). [L.Go./R.W.Mu.]

Supersonic diffuser A passive compressor (or shaped duct) in which gas enters at a velocity greater than the speed of sound, is decelerated in a contracting section, and reaches sonic speed at a throat.

Supersonic compression systems can be categorized to three basic types (see illustration). External-compression inlets have

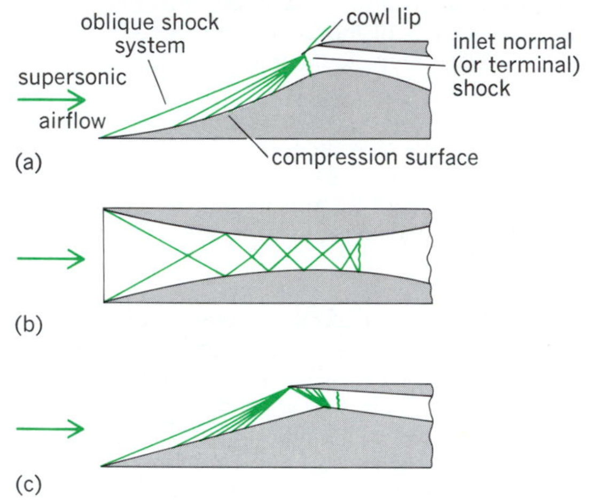

Basic supersonic compression systems. (*a*) External compression. (*b*) Internal compression. (*c*) Combination of external and internal compression.

the supersonic diffusion taking place at or ahead of the cowl lip (or throat station) and generally employ one or more oblique waves ahead of the normal shock. Internal-compression inlets accomplish supersonic diffusion internally downstream of the cowl lip. Deceleration of the flow is produced by a number of weak reflecting waves in a gradually convergent channel. The third system is a combination of external and internal compression and appears to represent an effective compromise. [J.F.C.; L.J.O.]

Supersonic flight Relative motion of a solid body and a gas at a velocity greater than that of sound propagation under the same conditions. The general characteristics of supersonic flight can be understood by considering the laws of propagation of a disturbance or pressure impulse, in a compressible fluid.

If the fluid is at rest, the pressure impulse propagates uniformly with the velocity of sound in all directions, the effect always acting along an ever-increasing spherical surface. If, however, the source of the impulse is placed in a uniform stream, the impulse will be carried by the stream simultaneously with its propagation at sonic velocity relative to the stream. Hence the resulting propagation is faster in the direction of the stream and slower against the stream. If the velocity of the stream past the source of disturbance is supersonic, the effect of the impulse is restricted to a cone whose vertex is the source of the impulse and whose vertex angle decreases from 90° (corresponding to Mach number equal to 1) to smaller and smaller values as the Mach number of the stream increases (see illustration). If the source of the pressure impulse travels through the air at rest, the conditions are analogous. *See* MACH NUMBER.

Consider the supersonic motion of a wing moving into air at rest. Because signals cannot propagate ahead of the wing, the

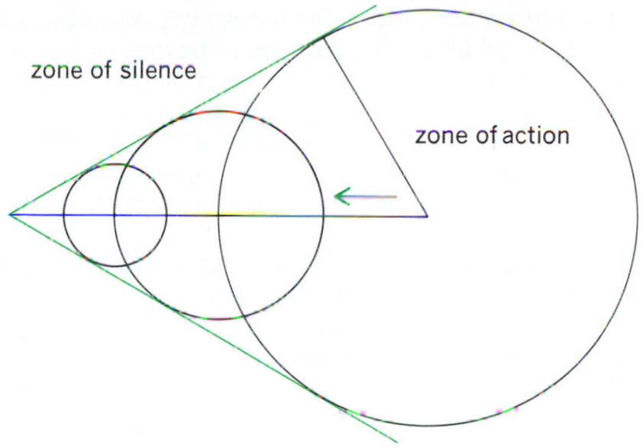

Generation of Mach wave by body at supersonic velocity; zones of action and silence are separated.

presence of the wing has no effect on the undisturbed air until the wing passes through it. Hence there must be an abrupt change in the properties of the undisturbed air as it begins to flow over the wing. This abrupt change takes place in a shock wave which is attached to the leading edge of the wing, provided that the leading edge is sharp and the flight Mach number is sufficiently large. As the air passes through the shock wave, its pressure, temperature, and density are markedly increased.

Further aft of the leading edge, the pressure of the air is decreased as the air expands over the surface of the wing. Hence the pressure acting on the front part of the wing is higher than the ambient pressure, and the pressure acting on the rear part of the wing is lower than the ambient pressure. The pressure difference between front and rear parts produces a drag, even in the absence of skin friction and flow separation. The wing produces a system of compression and expansion waves which move with it. This phenomenon is similar to that of a speedboat moving with a velocity greater than the velocity of the surface waves. Because of this analogy, supersonic drag is called wave drag. It is peculiar to supersonic flight, and it may represent the major portion of the total drag of a body. *See* HYPERSONIC FLIGHT; SUBSONIC FLIGHT; TRANSONIC FLIGHT. [J.E.Sc.]

Supersonic flow

Fluid motion in which the Mach number M, defined as the speed of the fluid relative to the sonic speed in the same medium, is more than unity. It is, however, common to call the flow transonic when $0.8 < M < 1.4$, and hypersonic when $M > 5$. *See* MACH NUMBER.

Mach waves. A particle moving in a compressible medium, such as air, emits acoustic disturbances in the form of spherical waves. These waves propagate at the speed of sound ($M = 1$). If the particle moves at a supersonic speed, the generated waves cannot propagate upstream of the particle. The spherical waves are enveloped in a circular cone called the Mach cone. The generators of the Mach cone are called Mach lines or Mach waves.

Shock waves. When a fluid at a supersonic speed approaches an airfoil (or a high-pressure region), no information is communicated ahead of the airfoil, and the flow adjusts to the downstream conditions through a shock wave. Shock waves propagate faster than Mach waves, and the flow speed changes abruptly from supersonic to less supersonic or subsonic across the wave. Similarly, other properties change discontinuously across the wave. A Mach wave is a shock wave of minimum strength. A normal shock is a plane shock normal to the direction of flow, and an oblique shock is inclined at an angle to the direction of flow. The velocity upstream of a shock wave is always supersonic. Downstream of an oblique shock, the velocity may be subsonic resulting in a strong shock, or

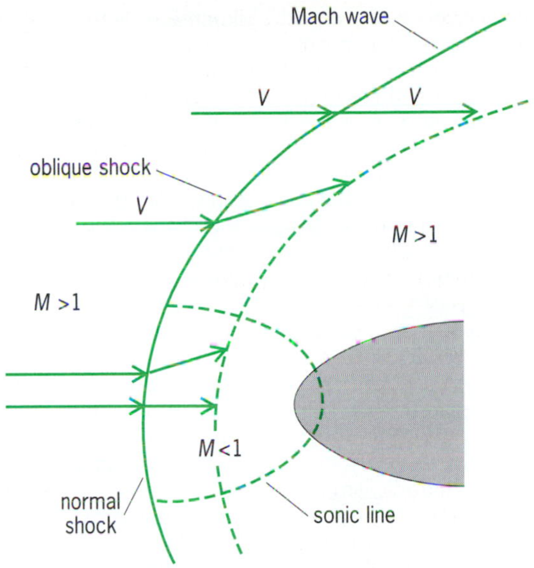

Typical normal shock, oblique shock, and Mach wave pattern in supersonic flow past a blunt body. *M* is the Mach number and *V* is the particle speed. The curved line parallel to normal and oblique shock waves indicates the end of the velocity vectors.

supersonic resulting in a weak shock. The downstream velocity component normal to any shock wave is always subsonic. There is no change in the tangential velocity component across the shock.

In a two-dimensional supersonic flow around a blunt body (see illus.), a normal shock is formed directly in front of the body, and extends around the body as a curved oblique shock. At a sufficient distance away, the flow field is unaffected by the presence of the body, and no discontinuity in velocity occurs. The shock then reduces to a Mach wave. *See* COMPRESSIBLE FLOW; FLUID FLOW; SUPERSONIC FLIGHT. [M.A.S.]

Superstring theory

A proposal for a unified theory of all interactions, including gravity. At present, there is a model that successfully describes the strong, weak, and electromagnetic interactions. This model, known as the standard model, correctly describes experiments up to the highest energies performed so far, and gives a complete description of the elementary particles and their interactions down to distances of the order of 10^{-18} m. Nevertheless, it has serious limitations, including the existence in the model of many parameters whose values are unexplained. Attempts to overcome these problems and to unify the forces of nature have been only partly successful. Moreover, these attempts, including the grand unification theories and supersymmetry, have left standing fundamental difficulties in reconciling gravitation and the laws of quantum mechanics. Superstring theory represents an ambitious program to unify all of the interactions observed in nature, including gravitation, in a theory with no unexplained parameters. In other words, this theory, if successful, should be able to account for all of the particles observed in nature and their interactions. *See* ELEMENTARY PARTICLE; FUNDAMENTAL INTERACTIONS; GRAND UNIFICATION THEORIES; GRAVITATION; STANDARD MODEL; SUPERSYMMETRY.

String concept. In string theory, the fundamental objects are not point particles, as in standard theories of elementary particles, but one-dimensional extended objects, the open and closed strings. In such a theory, what are usually called the elementary particles are simply particular quantum states of the string. In superstring theories, space-time is ten-dimensional (space is nine-dimensional). If such theories are to describe nature, six dimensions must be "curled up" or "compact." The main consequence of such extra dimensions is the existence of certain very massive particles.

The essential features of string theories can be understood by analogy with the strings of a musical instrument. Such strings vibrate as a characteristic frequency, as well as any integer multiple of that frequency. In classical physics, the amplitudes of vibration of each mode can take on a continuum of values. If they were a string of atomic dimensions, subject to the laws of quantum mechanics, the amplitudes of this quantum string could take on only discrete values, corresponding to particular quantum states. *See* QUANTUM MECHANICS; VIBRATION.

The strings of superstring theory are quite similar. The main difference is that they obey Einstein's principles of special relativity. As a result, since each quantum state has a particular energy, it has a definite mass. Thus, each state of the string behaves as a particle of definite mass. Because it is possible, in principle, to pump an arbitrarily large amount of energy into the string, the theory contains an infinity of different types of particles of arbitrarily large mass. The interactions of these particles are governed by the ways in which the strings themselves interact. To be consistent with the principles of relativity, a string can interact only by splitting into two strings or by joining together with another string to form a third string. As a result, the interactions of strings are nearly unique. This geometric picture of string interactions translates into a precise set of rules for calculating the interaction of individual string states, that is, particles. *See* RELATIVITY.

Classicial solutions. By various techniques, an enormous number of solutions of the classical string equations have been found. These include states in which space-time has any dimension between one and ten, and states with many bizarre symmetries and spectra. Each of these solutions then corresponds to a possible ground state of the system. The theories built around some of the states look very much like the real world. Not only are four dimensions flat while six are compact, but they possess gauge symmetries close to that of the standard model. Some have three or four generations of quarks and leptons, as well as light Higgs particles, which are of crucial importance in the standard model. Many of these solutions possess space-time supersymmetry. *See* HIGGS BOSON.

Problems. While the construction of these solutions, with their many desirable properties, represents a striking success, there are many problems that must be solved before string theory can be viewed as a successful theory of nature. The huge number of potential string ground states is one of the most serious of these problems. If the theory describes nature, it must have some mechanism that chooses one of these possible ground states from among all of the possible states. Because the masses and couplings of the elementary particles depend only on the choice of ground state, determining this true ground state will yield a set of predictions for these quantities. If string theory is a correct theory, these predictions must agree with the experimental values. [M.Di.]

Supersymmetry A symmetry that relates bosons to fermions, and vice versa. Since the 1960s there has been an unprecedented expansion in the use of symmetry considerations in the study of problems in physics. Supersymmetry is the latest and most elaborate application. Although supersymmetry was introduced in the early 1970s for studying problems in elementary particle physics, its use and applications have spread to other fields of physics. Experimental examples of symmetry have been discovered in the spectra of atomic nuclei.

Physicists believe that all constituents of matter have a property called angular momentum. According to quantum theory, angular momentum can be either an integer or half-integer multiple of Planck's constant divided by 2π, $\hbar$. Particles for which the angular momentum is an integer multiple of $\hbar$ are called bosons; those for which the angular momentum is a half-integer multiple of $\hbar$ are called fermions. Up to the early 1970s, when applying symmetry considerations to the study of phenomena in physics, physicists always considered symmetry transformations relating either bosons to bosons or fermions to fermions. In the early 1970s it was suggested that there may exist symmetries relating bosons to fermions, and vice versa. Because of their remarkable properties, and in order to distinguish them from the usual symmetries, these symmetries were called supersymmetries. *See* ANGULAR MOMENTUM; QUANTUM STATISTICS.

Supersymmetries have been used in attempts to construct a unified theory of all known interactions in physics: the gravitational, the weak, the electromagnetic, and the strong interactions. Particularly important here is the attempt to include the gravitational interaction, since there is at present no workable theory of gravitation consistent with the principles of quantum mechanics. Theories of gravitation that make use of supersymmetry have been called supergravity theories. See ELEMENTARY PARTICLE; FUNDAMENTAL INTERACTIONS; SUPERGRAVITY.

In medium-mass and heavy nuclei, two protons or two neutrons appear to bind into pairs. The pairs have angular momenta that are integer multiples of $\hbar$ and thus are bosons. In nuclei with an even number of protons and neutrons (even-even nuclei), all particles are paired. However, in nuclei with an even number of protons and an odd number of neutrons, or vice versa (even-odd nuclei), complete pairing is impossible and in addition to pairs (bosons) there are unpaired neutrons or protons. These unpaired particles are fermions. If the theory describing these nuclei is required to have supersymmetry, a set of relations is obtained linking together properties of even-even and even-odd nuclei. The most important relation is that linking the excitation spectra of atomic nuclei, that is, the set of their allowed quantum-mechanical energy levels. Several examples of supersymmetry in nuclei have been found. A general feature of the experimental examples found so far is that supersymmetry in nuclei appears to be somewhat broken. This is not unexpected, since supersymmetry is a rather complex type of symmetry requiring very stringent conditions on the interactions between the constituent particles. Despite the partial breaking, the fact that the experimental situation can be described, at least approximately, by invoking purely symmetry concepts is a remarkable result. This result has been exploited systematically in the study of complex atomic nuclei. *See* NUCLEAR STRUCTURE; SYMMETRY LAWS (PHYSICS). [F.I.]

Supertransuranics A group of predicted elements beyond the present periodic table of known elements, with atomic numbers around 114, expected to possess half-lives of the order of a year or longer. These are also referred to as the superheavy elements. Although they have not been discovered experimentally, it is generally believed that the difficulty lies in the synthesis of these elements, rather than in their stability once they are made.

The underlying physics responsible for the limited extent of the periodic table is the competition between attractive "nuclear forces" among the nucleons (that is, protons and neutrons) and the repulsive electrostatic forces among all the positively charged protons. The limit of the periodic table at an atomic number Z of approximately 109 is then set by the process of nuclear fission, which takes place when the disruptive effect of electrostatic forces overcomes the cohesive effect of the nuclear forces. *See* NUCLEAR FISSION.

The picture began to change when, in 1964, a clearer understanding was reached of the relation between fission half-lives and a well-known property of the nucleus called the magic numbers. If the number of protons (or neutrons) in a nucleus is equal to the proton magic number (or neutron magic number), then such a nucleus displays some special features; one example is that it is spherical in shape. Such a nucleus also displays an extra stability against fission. This means that this nucleus would have a longer half-life than would be expected otherwise. The superheavy elements are expected to occur with the center of the island at the proton magic number 114 and the neutron magic number 184. *See* MAGIC NUMBERS; NUCLEAR STRUCTURE; RADIOACTIVITY. [C.F.T.]

Suppression The elimination of some undesired component of a signal. In automobile radio installations, suppression techniques are essential to prevent ignition interference from reaching the radio circuits. Radio receivers themselves often contain special noise-suppression circuits and devices; this is particularly true for communication receivers that operate in crowded and noisy shortwave bands. In many radar installations, suppression circuits and techniques are used to reduce clutter caused by the ground and by fixed objects close to the antenna. *See* ELECTRIC FILTER; ELECTRICAL INTERFERENCE; ELECTRICAL NOISE; SUPPRESSOR. [J.Mar.]

Suppressor A device used to reduce or eliminate noise or other signals that interfere with the operation of a communication system. The term may be applied to a noise filter in a radio receiver, but it is more frequently used to describe a device applied at the noise source, such as a resistor used in series with spark plugs of a gasoline engine, or a capacitor across the terminals of a commutator motor or other sparking device that acts as a noise generator. The term suppressor may also be applied to a filter used in power leads of an electronic device to eliminate unwanted signals of noise. *See* ELECTRIC FILTER; ELECTRICAL INTERFERENCE; ELECTRICAL NOISE; SUPPRESSION. [W.R.LeP.]

Supramolecular chemistry The chemistry of assemblies with sizes between those of simple molecules and those of entities as large as biological cells. Organization is a basic characteristic of physical matter. In contrast to the chemistry of individual molecules, which focuses on the intramolecular interactions that hold atoms together in well-defined structures, supramolecular chemistry focuses on intermolecular interactions. The latter include hydrogen bonding, electrostatic attraction, charge transfer, and van der Waals forces, all of which are relatively weak; therefore, in order for stable structures to form, there must be a multiplicity of contact points between the two or more molecules composing the supramolecule. *See* HYDROGEN BOND.

The simplest supramolecule (also known as supermolecule) contains two entities, often specified as the receptor (host) and the substrate (guest). Naturally occurring receptor-substrate interactions are involved in numerous biological processes such as enzyme catalysis, immunological responses, translation and transcription of the genetic code, and signal induction by neurotransmitters. Chemists in the laboratory seek to mimic these phenomena of molecular recognition by constructing analogs of naturally occurring supramolecules. *See* ENZYME; MOLECULAR RECOGNITION.

While supramolecular chemistry begins with the relatively simple assemblies, the literal interpretation of the term also embraces more elaborate assemblies, all the way to those as complex as a living cell. Assemblies of bonded atoms with dimension in the range 1–100 nanometers have been characterized as nanostructures, and their synthesis, particularly at the upper end of this scale, presents one of the major challenges to modern chemistry.

Supramolecular chemistry has utilitarian promise in addition to its fascinating intellectual interest. Nanostructures, synthesized by processes directed by molecular recognition, make solid-state engineering possible at the molecular level. For example, chemionics, the chemistry of molecular components and devices that operate on photons, electrons, and ions, carries miniaturization to its ultimate level. *See* NANOCHEMISTRY; NANOSTRUCTURE. [C.D.Gu.]

Surface (geometry) A two-dimensional geometric figure (a collection of points) in three-dimensional space. The simplest example is a plane—a flat surface. Some other common surfaces are spheres, cylinders, and cones, the names of which are also used to describe the three-dimensional geometric figures that are enclosed (or partially enclosed) by those surfaces. In a similar way, cubes, parallelepipeds, and other polyhedra are surfaces. *See* CUBE; POLYHEDRON; SOLID (GEOMETRY).

Any bounded plane region has a measure called the area. If a surface is approximated by polygonal regions joined at their edges, an approximation to the area of the surface is obtained by summing the areas of these regions. The area of a surface is the limit of this sum if the number of polygons increases while their areas all approach zero. *See* AREA; CALCULUS; INTEGRATION; PLANE GEOMETRY; POLYGON.

Methods of description. The shape of a surface can be described by several methods. The simplest is to use the commonly accepted name of the surface, such as sphere or cube. In mathematical discussions, surfaces are normally defined by one or more equations, each of which gives information about a relationship that exists between coordinates of points of the surface, using some suitable coordinate system. *See* COORDINATE SYSTEMS.

Some surfaces are conveniently described by explaining how they might be formed. If a curve, called the generator in three-dimensional space, is allowed to move in some manner, then each position the generator occupies during this motion is a collection of points, and the set of all such points constitutes a surface that can be said to be swept out by the generator. In particular, if the generator is a straight line, a ruled surface is formed. If the generator is a straight line and the motion is such that all positions of the generator are parallel, a cylindrical surface (or just cylinder) is formed. If the generator is a straight line and all positions of the generator have a common point of intersection, a conical surface (or just cone) is formed. A ruled surface that could be bent to lie in a plane (the bending to take place without stretching or tearing) is called a developable surface. *See* CONE; CYLINDER.

Dihedron. A dihedron is the surface formed by bending a plane along a line in that plane. More formally, a dihedron is the union of two half-planes that share the same boundary line.

Quadric surfaces. A surface whose implicit equation $F(x,y,z) = 0$ is second degree is a quadric surface, a three-dimensional analog of a conic section. A plane section of a quadric surface is either a conic section or one of its degenerate forms (a point, a line, parallel lines, or intersecting lines). *See* CONIC SECTION; QUADRIC SURFACE.

Surfaces of revolution. When a plane curve (the generator) is revolved about a line in that plane (the axis of revolution, or just axis), a surface of revolution can be said to be swept out. The resulting surface will be symmetric about the axis of revolution.

A circular cylinder (a quadric surface) is formed when the generator and the axis of revolution are distinct parallel lines. If the generator is only a segment of a line (rather than the entire line), a bounded circular cylinder is generated.

A circular cone is a quadric surface formed when a straight-line generator intersects the axis of revolution at an acute angle. The cone consists of two parts, the nappes, joined at the point of intersection, which is the vertex of the cone.

A sphere (a quadric surface) is usually defined as a collection of points in three-dimensional space at a fixed distance (the radius) from a given point (the center). However, a sphere can also be defined as the surface of revolution formed when a semicircle (or the entire circle) is revolved about its diameter.

The intersection of any plane with a sphere will be a circle (except for tangent planes). Such a circle is called, respectively, a great circle or a small circle, depending on whether or not the plane contains the center of the sphere.

If only part of a semicircle is revolved about the diameter, a part of a sphere called a zone is formed. If a semicircle is revolved about its diameter through an angle less than one revolution, the surface swept out is a lune (see illus.). *See* SPHERE.

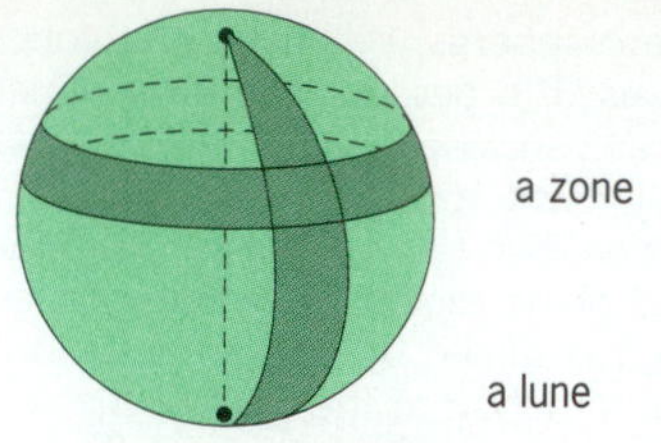

Sphere, with a zone and a lune.

A spheroid (also called an ellipsoid of revolution) is the quadric surface generated when an ellipse is revolved about either its major or minor axis. If the revolving is about the minor axis of the ellipse, the surface can be thought of as a flattened sphere, called an oblate spheroid. If the revolving is about the major axis, the surface can be thought of as a stretched sphere, called a prolate spheroid. A circular paraboloid is the quadric surface formed when a parabola is revolved about its axis. *See* ELLIPSE; PARABOLA.

A circular hyperboloid is the quadric surface formed when a hyperbola is revolved about either its transverse axis or its conjugate axis. The surface will be, respectively, a hyperboloid of one sheet or two sheets, depending on whether the revolving is about the conjugate axis or the transverse axis of the hyperbola. *See* HYPERBOLA.

A torus is generated when a circle is revolved about a line that does not intersect the circle. This doughnut-shaped surface has the property that not all points on the surface have the same sign of curvature. *See* DIFFERENTIAL GEOMETRY; TORUS.

[H.L.Ba.]

Surface acoustic-wave devices Devices which employ surface acoustic waves (SAW) in the analog processing of electronic signals with frequencies in the range 10^7–10^9 Hz.

Surface acoustic waves which contain both compressional and shear components in phase quadrature, propagating nondispersively along and bound to solid surfaces, were discovered by Lord Rayleigh in the 1880s. As an example, earthquakes furnish sources for propagating these waves on the Earth's surface. It is of importance for electronic applications that if the solid is a piezoelectric material, the surface acoustic energy is complemented by a small amount of electric energy. This electric energy provides the physical mechanism for the coupling between conventional electromagnetic signals and propagating SAW. The coupling is attained by means of interdigital transducers (IDT). SAW devices have led to a versatile microminiature technology for analog signal processing in the frequency range 10^7–10^9 Hz. Notable devices include band-pass filters, resonators, oscillators, pulse compression filters, and fast Fourier transform processors. Application areas include the color television consumer market, radar, sonar, communication systems, and nondestructive testing. *See* PIEZOELECTRICITY.

In the basic arrangement, a piezoelectric substrate, often crystalline quartz, has a polished upper surface on which two transducers are deposited. The input transducer is connected, via fine bonded leads, to the electric source through an electrical matching network. The output transducer drives the load, usually 50 ohms, through another electrical matching network. Because these transducers are bidirectional, they lead to devices with at least 6 decibels (dB) loss even in the passband. The unwanted acoustic waves are absorbed by terminations at the ends of the piezoelectric substrate. A metal baffle serves to isolate electromagnetically the two transducers.

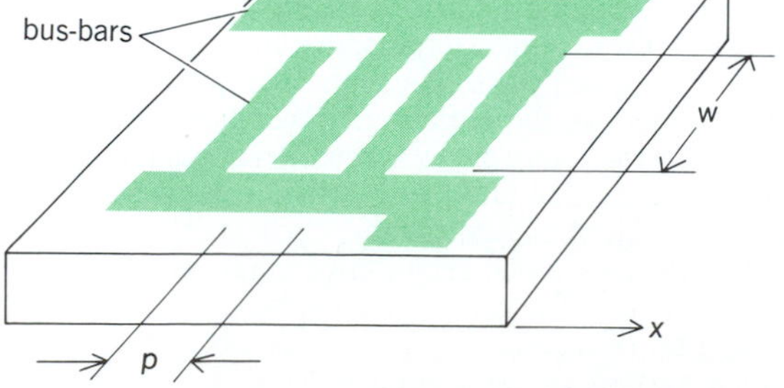

Interdigital construction of surface acoustic-wave transducer. x is a band-pass filter characteristic.

The transducers consist of a set of metal interdigital electrodes, each a few hundred nanometers thick, fed from two bus-bars (see illustration). For this transducer arrangement the period p of the interdigital electrode structure is constant and equals one surface acoustic wavelength λ_0 at the center of frequency f_0 of the response. The width of the metal electrodes is typically $p/4$, being 100 micrometers at 10^7 Hz and 1 μm at 10^9 Hz. The electrode overlap distance w is also constant and defines the acoustic beam width, which is typically 40 wavelengths. *See* TRANSDUCER.

The 100-μm electrodes are readily fabricated by using techniques standard to the semiconductor integrated circuit industry of metallization: photoresist, masking, and chemical etching. The 1-μm electrodes require more sophisticated processing techniques. These include conformable optical masks and x-ray lithography coupled with sputter etching by radio-frequency and ion-beam methods for even finer resolutions. *See* INTEGRATED CIRCUITS.

[J.H.Co.]

Surface and interfacial chemistry Chemical processes that occur at the phase boundary between gas–liquid, liquid–liquid, liquid–solid, or gas–solid interfaces.

The chemistry and physics at surfaces and interfaces govern a wide variety of technologically significant processes. Chemical reactions for the production of low-molecular-weight hydrocarbons for gasoline by the cracking and reforming of the high-molecular-weight hydrocarbons in oil are catalyzed at acidic oxide materials. Surface and interfacial chemistry are also relevant to adhesion, corrosion control, tribology (friction and wear), microelectronics, and biocompatible materials. In the last case, schemes to reduce bacterial adhesion while enhancing tissue integration are critical to the implantation of complex prosthetic devices, such as joint replacements and artificial hearts. *See* CRACKING; HETEROGENEOUS CATALYSIS; MEDICAL CHEMICAL ENGINEERING; PROSTHESIS; SURFACE PHYSICS.

Interactions with the substrate may alter the electronic structure of the adsorbate. Those interactions that lower the activation energy of a chemical reaction result in a catalytic process. Adsorption of reactants on a surface also confines the reaction to two dimensions as opposed to the three dimensions available for a homogeneous process. The two-dimentional confinement of reactants in a bimolecular event seems to drive biochemical processes with higher reaction efficiencies at proteins and lipid membranes. *See* ADSORPTION.

A limitation in the study of surfaces and interfaces rests with the low concentrations of the participants in the chemical process. Concentrations of reactants at surfaces are on the order of 10^{-10} to 10^{-8} mole/cm^2. Such low concentrations pose a sensitivity problem from the perspective of surface analysis. Experimental techniques with high sensitivity are required to examine the low concentrations of a surface species at interfaces.

Electron spectroscopy methods are widely used in the study of surfaces because of the small penetration depth of electrons through solids. This attribute makes electron spectroscopy inherently surface-sensitive, since only a few of the outermost atomic layers are accessible. The methods of electron spectroscopy used in surface studies have several common characteristics (see table). A source provides the incident radiation to the sample, which can be in the form of electrons, x-radiation, or ultraviolet radiation. Electron beams are generated from the

Surface-sensitive experimental techniques

Technique	:Source*	Detectors	Level of information
Auger electron spectroscopy (AES)	Electrons 2–3 keV	Cylindrical mirror of retarding field	Elemental composition
X-ray photoelectron spectroscopy (XPS)	X-rays 1254 eV (Mg) 1487 eV (A1)	Hemispherical or cylindrical mirror	Elemental composition and oxidation state
Ultraviolet photoelectron spectroscopy (UPS)	UV radiation 21 eV He(1) 41 ev He (11)	Hemispherical or cylindrical mirror	Electronic properties of adsorbate and/or bulk material
Energy loss spectroscopy (ELS)	Electrons 50–1000 eV	Electron energy analyzer	Electronic structure of surface
High-resolution electron energy loss spectroscopy (HREELS)	Electrons 1–10 eV	Electron energy analyzer	Vibrational losses
Low-energy electron diffraction (LEED)	Electrons 20–500 eV	Retarding fields and phosphorescent screen	Surface structure or periodicity
Infrared spectroscopy (IRS)	Photons	Mercury-cadmium-telluride or indium antimony	Molecular identity
Optical ellipsometry	Photons	Photomultiplier	Adsorbate layer thickness
Scanning tunneling microscopy (STM)	Tunneling current	Ammeter	Substrate roughness and texture

*Mg = magnesium; Al = aluminum; He = helium.

thermionic emission of metal filaments or metal oxide pellets. The incident radiation induces an excitation at the surface of the sample, which alters the energy distribution of electrons that leave the surface. This distribution provides a diagnostic of the composition or structure of the interface. *See* ELECTRON SPECTROSCOPY.

Optical spectroscopy techniques (visible and infrared) are also useful for probing the chemical composition and molecular arrangement of surface species. Typical application configurations are the transmission and reflection (both external and internal) modes. Transmission spectroscopy relies on the passage of the probe beam through the sample. External and internal reflection spectroscopies involve the reflection of the probe beam from a medium with a lower refractive index to a medium with a higher refractive index, and from a higher to lower refractive index, respectively. The sample support must be optically transparent to the probe beam for the internal reflection mode. In both cases, the substrates are polished to a smooth, mirrorlike finish. *See* SPECTROSCOPY. [M.M.W.; M.D.P.]

Surface and solid of revolution A surface of revolution is generated by revolving a plane curve about a line in its plane. The generating curve may consist of one or more arcs or line segments connected together. A solid of revolution is generated by revolving a connected plane region about a line in its plane not cutting the region. The boundary of a solid of revolution is a surface of revolution. The line about which the generating plane curve or region is revolved is called the axis of revolution, or simply the axis.

In euclidean solid geometry the words cylinder, cone, and sphere commonly refer to solids of revolution, and the corresponding surfaces of revolution are called a cylindrical surface, a conical surface, or a spherical surface. However, in analytic geometry and more generally in modern usage, the words cylinder, cone, and sphere refer to surfaces, not solids. A circular cylinder is a surface generated by revolving an infinite line about a parallel axis, and a circular cone is a surface generated by revolving an infinite line about an intersecting axis. A spheroid is a surface obtained by revolving an ellipse about one of its axes. Paraboloids and hyperboloids of revolution are obtained by revolving parabolas and hyperbolas about their axes. *See* CONE; CYLINDER; ELLIPSOID AND SPHEROID; HYPERBOLOID; PARABOLOID; QUADRIC SURFACE; TORUS. [J.S.F.]

Surface coating A substance applied to other materials to change the surface properties, such as color, gloss, resistance to wear or chemical attack, or permeability, without changing the bulk properties. The term includes such materials as paints, varnishes, enamels, oils, greases, waxes, concrete, lacquers, asbestos, and fire retardant formulations of several varieties. In general, organic coatings are based on a vehicle, usually an oil or resin, which, after being spread out in a relatively thin film, changes to a solid. This change, called drying, may be due entirely to the evaporation of a volatile solvent, or it may be caused by a chemical reaction, such as oxidation or polymerization. Opaque materials called pigments, dispersed in the vehicle, contribute color, opacity, and increased durability and resistance. *See* DRIER (PAINT); DRYING OIL; ELECTROPLATING OF METALS; ENAMEL; JAPANNING; LACQUER; METAL COATINGS; PAINT; PIGMENT; PRIMER (SURFACE COATING); SHELLAC; THINNER; VARNISH. [C.R.Ma.; C.W.Si.]

Surface condenser A heat-transfer device used to condense a vapor, usually steam, by absorbing its latent heat in a cooling fluid, ordinarily water. Most surface condensers consist of a chamber containing a large number of corrosion-resisting alloy tubes through which cooling water flows. The vapor contacts the outside surface of the tubes and is condensed on them. The tubes are arranged so that the cooling water passes through the vapor space one or more times. Air coolers are normally an integral part of the condenser but may be separate and external to it. The condensate is removed by a condensate pump and the noncondensables by a vacuum pump. *See* STEAM CONDENSER; VAPOR CONDENSER. [J.F.Se.]

Surface hardening of steel The selective hardening of the surface layer of a steel product by one of several processes which involve changes in microstructure with or without changes in composition. Surface hardening imparts a combination of properties to the finished product not produced by bulk heat treatment alone. Among these properties are high wear resistance and good toughness or impact properties, increased resistance to failure by fatigue resulting from cyclic loading, and resistance to surface indentation by localized loads. The use of surface hardening frequently is also favored by lower costs and greater flexibility in manufacturing.

The principal surface hardening processes are: (1) carburizing, (2) the modified carburizing processes of carbonitriding, cyaniding, and liquid carburizing, (3) nitriding, (4) flame hardening and induction hardening, and (5) surface working. Carburizing introduces carbon into the surface layer of low-carbon steel parts and converts that layer into high-carbon steel, which can be quench-hardened by appropriate heat treatment. Carbonitriding, cyaniding, and liquid carburizing, in addition to supplying carbon, introduce nitrogen into the surface layer; this element permits lower case-hardening temperatures and has a beneficial effect on the subsequent heat treatment. In nitriding, only nitrogen is supplied, and reacts with special alloy elements present in the steel. Whereas the foregoing processes change the composition of the surface layer, flame hardening and induction hardening depend on a heat treatment applied selectively to the surface layer of a medium-carbon steel. Surface working by shot peening, surface rolling, or prestressing improved fatigue resistance by producing a stronger case, compressive stresses, and a smoother surface. *See* HEAT TREATMENT (METALLURGY); STEEL. [M.B.B.; C.F.F.]

Surface mining A surface method of mining by removing the material overlying the bed and loading the uncovered mineral, usually coal. It is safer than underground mining because neither the workers nor the equipment is subjected to such hazards as roof falls and explosions caused by gas or dust ignitions. Coal near the outcrop or at shallow depth can be stripped not only more cheaply but more completely than by deep mining, and the need for leaving pillars of coal to support the mine roof is eliminated. The roof over coal at shallow depth is weak and difficult to support in underground workings by conventional methods, yet this same weakness, of cover and of coal seam, makes stripping less difficult.

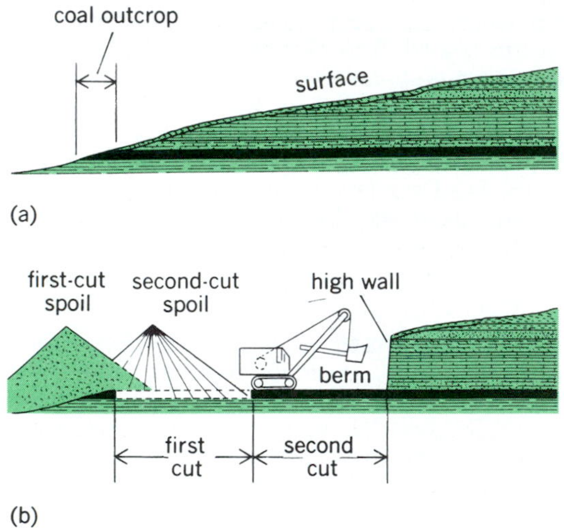

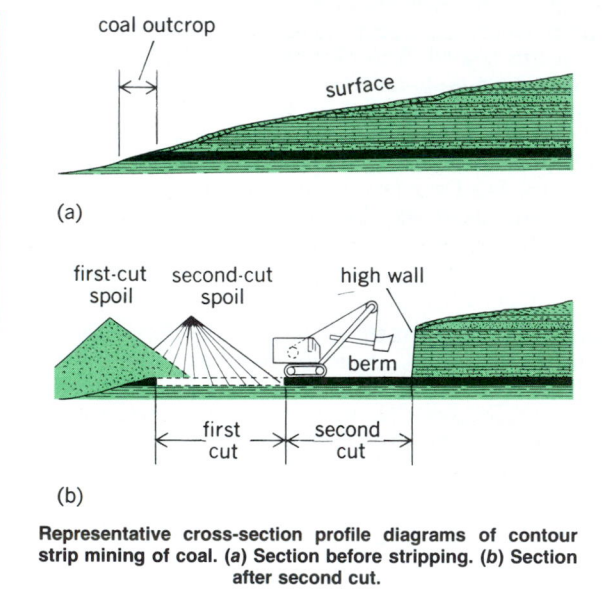

Representative cross-section profile diagrams of contour strip mining of coal. (*a*) Section before stripping. (*b*) Section after second cut.

Power shovels, draglines, bulldozers, and other types of earth-moving equipment slice a cut through the overburden down to the coal. The stripped overburden (spoil) is stacked in a long ridge (spoil bank) parallel with the cut and as far as possible from undisturbed overburden (high wall). The slope of a spoil bank is approximately 1.4:1 and that of a high wall under average conditions is 0.3:1. The uncovered coal (berm) is then fragmented, loaded, and transported from the pit. Spoil from each succeeding cut is stacked overlapping and parallel with the previous ridge and also fills the space left by the coal removed (see illustration). *See* LAND RECLAMATION; MINING. [O.L.B.]

Surface physics The study of the structure and dynamics of atoms and their associated electron clouds in the vicinity of a surface, usually at the boundary between a solid and a low-density gas. Thus, surface physics may be regarded as a branch of solid-state physics which deals with those regions of large and rapid variations of atomic and electron density that occur in the vicinity of an interface between the two "bulk" components of a two-phase system. In conventional usage, surface physics is distinguished from interface physics by the restriction of the scope of the former to interfaces between a solid (or liquid) and a low-density gas, often at ultra-high-vacuum pressures of $p = 10^{-10}$ torr (10^{-8} pascal or 10^{-13} atm). *See* SOLID-STATE PHYSICS.

More specifically, surface physics is concerned with two separate but complementary areas of investigation into the properties of such solid-"vacuum" interfaces. Ultimately, interest centers on the specification and theoretical prediction of surface composition and structure (that is, the masses, charges, and positions of surface species), of the dynamics of surface atoms (such as surface diffusion and vibrational motion), and of the energetics and dynamics of electrons in the vicinity of a surface (such as electron density profiles and localized electronic surface states). As a practical matter, however, the nature and dynamics of surface species must be determined experimentally by scattering and emission measurements involving particles or electromagnetic fields (or both) external to the surface itself. Thus, a second major interest in surface physics is the study of the interaction of external entities (that is, atoms, ions, electrons, and electromagnetic fields) with solids at their vacuum interfaces. It is this aspect of surface physics which most clearly distinguishes it from conventional solid-state physics, because quite different scattering and emission experiments are utilized to examine surface as opposed to bulk properties of a given sample.

Techniques for characterizing the solid-vacuum interface are based on one of two simple physical mechanisms for achieving surface sensitivity. The first, which is the basis for field-emission and field-ionization microscopy, is the achievement of surface sensitivity by utilizing electron tunneling through the potential-energy barrier at a surface. The second mechanism for achieving surface sensitivity is the examination of the elastic scattering or emission of particles which interact strongly with the constituents of matter, for example, "low-energy" ($E \lesssim 10^3$ eV) electrons, thermal atoms and molecules, or "slow" (300 eV $\lesssim E \lesssim 10^3$ eV) ions. Since such entities lose appreciable ($\Delta E \sim 10$ eV) energy in distances of the order of tenths of nanometers, typical electron analyzers with resolutions of tenths of an electronvolt are readily capable of identifying scattering and emission processes which occur in the upper few atomic layers of a solid. *See* ELECTRON DIFFRACTION; ELECTRON SPECTROSCOPY; FIELD-EMISSION MICROSCOPY; PHOTOEMISSION; X-RAY CRYSTALLOGRAPHY.

A major reason for the renaissance in surface physics is the capacity to generate in a vacuum chamber special surfaces which approximate the ideal of being atomically flat. These surfaces are prepared by cycles of fast-ion bombardment, thermal outgassing, and thermal annealing for bulk samples or field evaporation of etched tips for field-ion microscopes. In this fashion, reasonable facsimiles of uncontaminated, atomically flat solid-vacuum interfaces of many simple metals and semiconductors have been prepared and subsequently characterized by various spectroscopic techniques. [C.B.D.]

Surface tension The force acting in the surface of a liquid, tending to minimize the area of the surface. Surface forces, or more generally, interfacial forces, govern such phenomena as the wetting or nonwetting of solids by liquids, the capillary rise of liquids in fine tubes and wicks, and the curvature of free-liquid surfaces. The action of detergents and antifrothing agents and the flotation separation of minerals depend upon the surface tensions of liquids.

In the body of a liquid, the time-averaged force exerted on any given molecule by its neighbors is zero. Even though such a molecule may undergo diffusive displacements because of random collisions with other molecules, there exist no directed forces upon it of long duration. It is equally likely to be momentarily displaced in one direction as in any other. In the surface of a liquid, the situation is quite different; beyond the free surface, there exist no molecules to counteract the forces of attraction exerted by molecules in the interior for molecules in the surface. In consequence, molecules in the surface of a liquid experience a net attraction toward the interior of a drop. These centrally directed forces cause the droplet to assume a

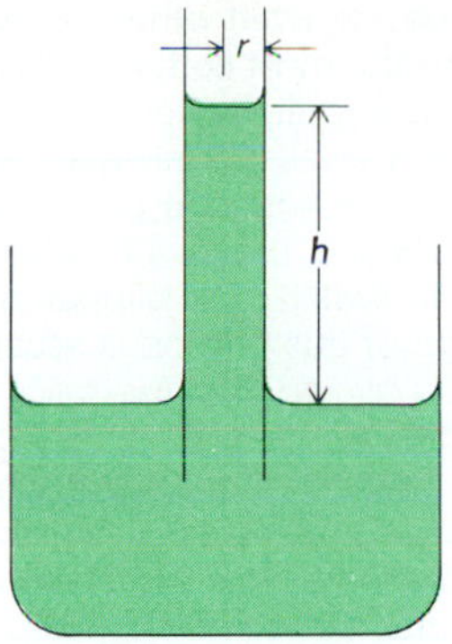

Rise of liquid in capillary tube.

spherical shape, thereby minimizing both the free energy and surface area.

Liquids which wet the walls of fine capillary tubes rise to a height which depends upon the tube radius, the surface tension, the liquid density, and the contact angle between the solid and the liquid (measured through the liquid). In the illustration a liquid of a certain density is shown as having risen to a height h in a capillary whose radius is r. A balance exists between the force exerted by gravity on the mass of liquid raised in the capillary and the opposing force caused by surface tension.

Detergents, soaps, and flotation agents owe their usefulness to their ability to lower the surface tension of water, thereby stabilizing the formation of small bubbles of air. At the same time, the interfacial tension between solid particles and the liquid phase is lowered, so that the particles are more readily wetted and floated after attachment to air bubbles. *See* FLOTATION; INTERFACE OF PHASES; SURFACTANT. [N.H.N.]

Surface water A term commonly used to designate the water flowing in stream channels. The term is sometimes used in a broader sense as opposed to "subsurface water." In this sense, surface water includes water in lakes, marshes, glaciers, and reservoirs as well as that flowing in streams. In the broadest sense, surface water is all the water on the surface of the Earth and thus includes the water of the oceans. Subsurface water includes water in the root zone of the soil and groundwater flowing or stored in the rock mantle of the Earth. Subsurface water differs from surface water in the mechanics of its movement as well as in its location. Surface and subsurface water are two stages of the movement of the Earth's water through the hydrologic cycle. The world's ocean and atmospheric moisture are two other main stages of the grand water cycle of the Earth. *See* HYDROLOGY.

The table gives estimates of the amounts of water in various parts of the hydrologic cycle and their detention periods. It may be noted that surface water on the continents is but a small part of the world's water and that the bulk of that is in fresh-water lakes. However, the detention period is also short. This means that the surface-water part, and especially the water in the streams, is rapidly discharged and replenished. That is why surface water, as well as the shallower groundwater, is called a renewable resource. Water that has a detention period of more than a generation is not renewed within sufficient time to be so considered. *See* GROUNDWATER HYDROLOGY; RIVER.

Precipitation that reaches the Earth is subdivided by processes of evaporation and infiltration into various routes of subsequent travel. Evaporation from wet land surfaces and from vegetation returns some of the water to the atmosphere immediately. Precipitation that falls at rates less than the local rate of infiltration enters the soil. Some of the infiltrated water is retained in the soil, sustaining plant life, and some reaches the groundwater. The precipitation that exceeds the capacity of the soil to absorb water flows overland in the direction of

Distribution of the world's supply of water

Location		Volume of water, 10^9 acre-ft*	Percentage total	Detention period, years
World's oceans		1,060,000	97.39	5,000
Surface water on the continents				
Glaciers and polar ice caps	20,000		1.83	2,000
Fresh-water lakes	100		0.0093	100
Saline lakes and inland seas	68		0.0063	50
Average in stream channels	0.25		0.00002	0.05
Total surface water		20,200		700 av
Subsurface water on the continents				
Root zone of the soil	10		0.00094	0.25
Groundwater above 2500 ft†	3,700		0.339	5
Groundwater below 2500 ft†	4,600		0.425	100
Total subsurface water		8,300		
Atmospheric water		115	0.0011	0.03
Total world water (rounded)		1,088,000	100	3,000

*10^9 acre-ft = 1.233×10^8 ha · m = 1.233×10^{12} m^3.
†2500 ft = 750 m.

the steepest slope and concentrates in rills and minor channels. During storms most of the water in surface streams is derived from that portion of the precipitation which fails to infiltrate the soil. *See* PRECIPITATION (METEOROLOGY).

The distinction between surface and subsurface water, though useful, should not obscure the fact that water on the surface and water underground is physically connected through pores, cracks, and joints in rock and soil material. In many areas, particularly in humid regions, surface water in stream channels is the visible part of a reservoir, which is partly underground; the water surface of a river is the visible extension of the surface of the groundwater. [L.B.L.]

Surfactant A substance that even though present in small amounts, exerts a marked effect on the surface behavior of a system. These agents are essentially responsible for producing great changes in the surface energy of liquid or solid surfaces, and their ability to cause these changes is associated with their tendency to migrate to the interface between two phases. Consequently, surfactants are of potential interest wherever there are solid-solid, solid-liquid, solid-gas, liquid-liquid, or liquid-gas interfaces, and of particular interest at liquid-gas interfaces at which the surfactant is a solute whose presence makes the surface properties of the solution greatly different from those of the solvent. *See* INTERFACE OF PHASES.

Soap, for example, when dissolved in small quantities in water, is responsible for greatly decreasing the surface tension of water, and it is this property of soap that accounts for its ability to act as a detergent. In contrast to soap and other related substances that lower the surface energy of a liquid, other solutes, such as inorganic salts, acids, and bases, may increase the surface tension of a liquid (see illustration), but their effect in increasing the surface tension is not nearly so great as the effect of those agents that decrease the surface tension. Occasionally the term surface-inactive solutes is applied to these substances whose presence causes an increase in surface tension. *See* DETERGENT; MONOMOLECULAR FILM; SOAP.

Surfactants are usually classified in three groups: anionic, cationic, and nonionic types. Anionic types include carboxylate ions such as occur in sodium oleate. Cationic surfactants are usually derived from the amino group where, through either primary, secondary, or tertiary amine salts, the hydrophilic

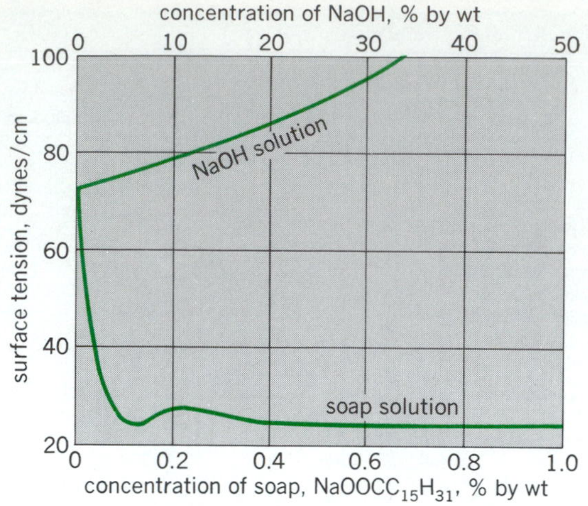

Effect of surfactants on water.

character may be achieved by aliphatic and aromatic groups that may be altered by substituents of varying polarity. The third class of surfactants, the nonionic type, are organic substances which contain groups of varying polarity and which render parts of the molecule lyophilic, whereas other parts of the molecule are lyophobic. In this class are often included certain colloidal substances, such as graphite and clays. *See* LUBRICANT; SURFACE TENSION. [W.H.S.]

Surge arrester A protective device designed primarily for connection between a conductor of an electrical system and ground to limit the magnitude of transient overvoltages on equipment. A lightning arrester is really a voltage-surge arrester.

The valve arrester consists of disks of zinc oxide material that exhibit low resistance at high voltage and high resistance at low voltage. By selecting an appropriate configuration of disk material, the arrester will conduct a low current of a few milliamperes at normal system voltage. During conditions of lightning or switching surge overvoltages, the surge current is limited by the circuit; and for the magnitudes of current that can be delivered to the arrester location, the resulting voltage will be limited to controlled values, and to safe levels as well, when insulation levels of equipment are coordinated with the surge arrester protective characteristics.

A typical surge arrester consists of disks of zinc oxide material sized in cross-sectional area to provide desired energy discharge capability, and in axial length proportional to the voltage capability. The disks are then placed in porcelain enclosures to provide physical support and heat removal, and sealed for isolation from contamination in the electrical environment. *See* LIGHTNING; LIGHTNING AND SURGE PROTECTION. [G.D.B.]

Surge suppressor A device that is designed to offer protection against voltage surges on the power line that supplies electrical energy to the sensitive components in electronic devices and systems. The device offers a limited type of protection to computers, television sets, high-fidelity equipment, and similar types of electronic systems.

A voltage surge is generally considered to be a transient wave of voltage on the power line. The amplitude of the surge may be several thousand volts, and the duration may be as short as 1 or 2 milliseconds or as long as about 100 ms. Typical effects can be damage to the electronics or loss of programs and data in computer memories. Many events can cause the surges, including lightning that strikes the power lines at a considerable distance from the home or office; necessary switching of transmission lines by the utilities; and rapid connections or disconnections of large loads, such as air conditioners and motors, from the power line, or even other appliances in the home. Lightning is perhaps the most common. *See* LIGHTNING.

The suppressor acts to limit the peak voltage applied to the electronic device to a level that normally will not cause either damage to the device or software problems in the computers. The device may include a pilot light, a fuse, a clipping circuit, resistors, and a main switch. The clipper circuit is the principal item, and the design of this portion is usually proprietary information. *See* CLIPPING CIRCUIT; FUSE (ELECTRICITY). [E.C.Jo.]

Surgery That branch of medicine which generally treats diseases by operative intervention. Surgical procedures may involve relieving mechanical obstruction of a tubular organ, such as the intestine; or removing a diseased organ, which cannot be salvaged by medical treatment, such as a gangrenous appendix or inflamed gallbladder; or removing a malignant tumor with a margin of normal tissue; or repairing an injured organ, or removing it if the organ is irreparable and its absence is compatible with survival.

The field of surgery has become increasingly specialized, primarily by organ system, so that surgical diseases of the kidney, bladder, and other components of the urinary tract are treated by surgeons called urologists; surgery of the central nervous system, including the brain and spinal cord, is done by neurosurgeons; reconstructive and cosmetic surgery is done primarily by plastic surgeons; general surgeons continue to do most abdominal surgery, some head and neck surgery, and surgery of the soft tissues of the extremities; and surgical diseases of the bones and joints are treated by orthopedic surgeons. Some specialties within surgery have also developed into specialties that are not limited to one organ system, such as surgical oncology (cancer surgery), so that cancers in most parts of the body may be treated by surgeons with special training in malignant diseases. *See* MEDICINE. [R.F.J.]

Surging A sudden and momentary change of voltage or current in a circuit. It can be due to a sudden change in the applied input signal, a sudden change in the load placed on the circuit, or to the action of a relay, switch, or other device that changes operating conditions within the circuit. The resulting surges or transients are often called pulses or impulses when they have only one polarity. An oscillatory surge includes both positive and negative polarity values. *See* ELECTRIC TRANSIENT. [J.Mar.]

Surveillance radar Ground radar used for traffic control purposes in the approach and landing zone. This radar is used primarily to assist controllers in converting random arrivals to regular landings and in positioning such aircraft so that they may make low approaches by the use of a fixed-beam, low-approach system or a precision radar low-approach system. *See* AIR-TRAFFIC CONTROL. [P.C.S.]

Surveying The measurement of dimensional relationships among points, lines, and physical features on or near the Earth's surface. Basically, surveying determines horizontal distances, elevation differences, directions, and angles. These basic determinations are applied further to the computation of areas and volumes and to the establishment of locations with respect to some coordinate system.

Surveying is typically used to locate and measure property lines; to lay out buildings, bridges, channels, highways, sewers, and pipelines for construction; to locate stations for launching and tracking satellites; and to obtain topographic information for mapping and charting.

Horizontal distances are usually assumed to be parallel to a common plane. Each measurement has both length and direction. Length is expressed in feet or in meters. Direction is expressed as a bearing of the azimuthal angle relationship to a reference meridian, which is the north-south direction. It can be the true meridian, a grid meridian, or some other assumed meridian. The degree-minute-second system of angular expression is standard in the United States. *See* AZIMUTH.

Reference, or control, is a concept that applies to the positions of lines as well as to their directions. In its simplest form, the position control is an identifiable or understood point of origin for the lines of a survey. Conveniently, most coordinate systems have the origin placed west and south of the area to be surveyed so that all coordinates are positive and in the northeast quadrant.

Vertical measurement adds the third dimension to an object's position. This dimension is expressed as the distance above some reference surface, usually mean sea level, called a datum. Mean sea level is determined by averaging high and low tides during a lunar month, but for certainty this must be carried out for as long as 19+ years. *See* SEA-LEVEL DATUM PLANES; SEA-LEVEL FLUCTUATIONS.

Horizontal control. The main framework, or control, of a survey is laid out by traverse, triangulation, or trilateration. Some success has been achieved in locating control points from Doppler measurements of passing satellites, from aerial phototriangulation, from satellites photographed against a star background, and from inertial guidance systems. In traverse, adopted for most ordinary surveying, a line or series of lines is established by directly measuring lengths and angles. In triangulation, used mainly for large areas, angles are again directly measured, but distances are computed trigonometrically. This necessitates triangular patterns of lines connecting intervisible points and starting from a baseline of known length. New baselines are measured at intervals. Trigonometric methods are also used in trilateration, but lengths, rather than angles, are measured. The development of electronic distance measurement (EDM) instruments brought trilateration into significant use.

Distance measurement. Traverse distances are usually measured with a surveyor's tape or by EDM, but also may sometimes be measured by stadia, subtense, or trig-traverse.

Whether on sloping or level ground, it is horizontal distances that must be measured. In taping, horizontal components of hillside distances are measured by raising the downhill end of the tape to the level of the uphill end. On steep ground this technique is used with shorter sections of the tape. The raised end is positioned over the ground point with the aid of a plumb bob. Where slope distances are taped along the ground, the slope angle can be measured with the clinometer. The desired horizontal distance can then be computed. *See* CLINOMETER.

In EDM the time a signal requires to travel from an emitter to a receiver or reflector and back to the sender is converted to a distance readout. The great advantage of electronic distance measuring is its unprecedented precision, speed, and convenience. Further, if mounted directly onto a theodolite, and especially if incorporated into it and electronically coupled to it, the EDM instrument with an internal computer can in seconds measure distance (even slope distance) and direction, then compute the coordinates of the sighted point with all the accuracy required for high-order surveying.

In the stadia technique, a graduated stadia rod is held upright on a point and sighted through a transit telescope set up over another point. The distance between the two points is determined from the length of rod intercepted between two horizontal wires in the telescope. *See* STADIA; TRANSIT (ENGINEERING).

In the subtense technique the transit angle subtended by a horizontal bar of fixed length enables computation of the transit-to-bar distance (Fig. 1). In trig-traverse the subtense bar is replaced by a measured baseline extending at a right angle from the survey line whose distance is desired. The distance calculated in either subtense or trig-traverse is automatically the horizontal distance and needs no correction.

Angular measurement. The most common instrument for measuring angles is the transit or theodolite. It is essentially a telescope that can be rotated a measurable amount about a vertical axis and a horizontal axis. Carefully graduated metal or glass circles concentric with each axis are used to measure the angles. The transit is centered over a point with the aid of

Fig. 1. Subtense bar. (*Lockwood, Kessler, and Bartlett Inc.*)

either a plumb bob suspended by a string from the vertical axis or (on some theodolites) an optical plummet, which enables the operator to sight along the instrument's vertical axis to the ground through a right-angle prism. *See* THEODOLITE.

Elevation differences. Elevations may be measured trigonometrically in conjunction with reduction of slope measurements to horizontal distances, but the resulting elevation differences are of low precision.

Most third-order and all second- and first-order measurements are made by differential leveling, wherein a horizontal line of sight of known elevation is sighted on a graduated rod held vertically on the point being checked (Fig. 2). The transit telescope, leveled, may establish the sight line, but more often a specialized leveling instrument is used. For approximate results a hand level may be used. *See* LEVEL (SURVEYING).

Other methods of measuring elevation include trigonometric leveling which involves calculating height from measurements of horizontal. distance and vertical angle; barometric leveling, a method of determining approximate elevation difference with aid of a barometer; and airborne profiling, in which a radar altimeter on an aircraft is used to obtain ground elevations.

Astronomical observations. To determine meridian direction and geographic latitude, observations are made by a theodolite or transit on Polaris, the Sun, or other stars. Direction of the meridian (geographic north-south line) is needed for direction control purposes; latitude is needed where maps and other sources are insufficient. The simplest meridian determination is made by sighting Polaris at its elongation, as

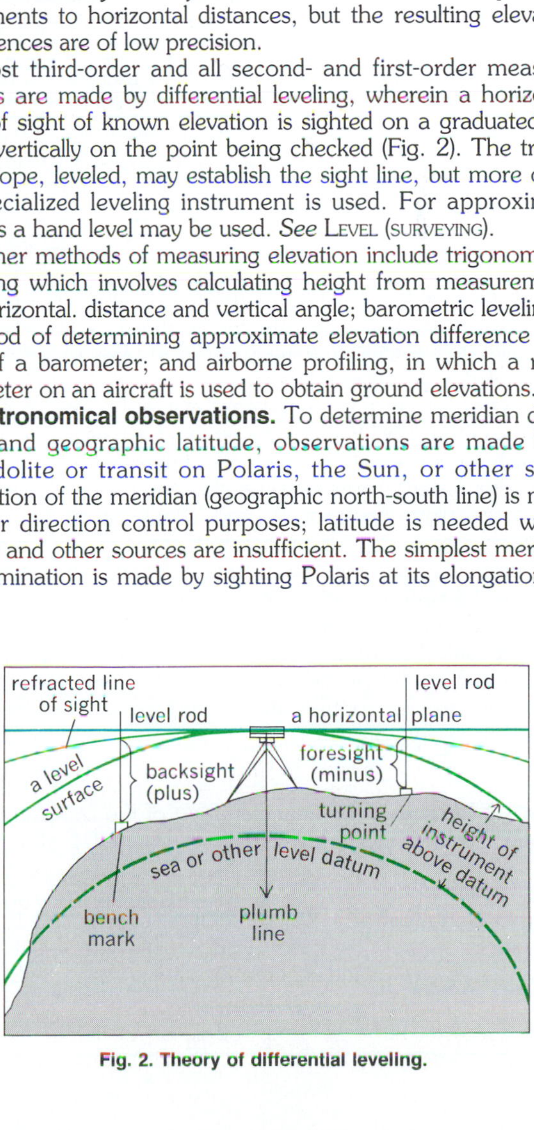

Fig. 2. Theory of differential leveling.

the star is rounding the easterly or westerly extremity of its apparent orbit. An angular correction is applied to the direction of sighting, which is referenced to a line on the ground. The correction value is found in an ephemeris. *See* EPHEMERIS; TOPOGRAPHIC SURVEYING AND MAPPING. [B.A.B.]

Surveying instruments Instruments used in surveying operations to measure vertical angles, horizontal angles, and distance. Such devices were originally mechanical only, but technological advances led to mechanical-optical devices, optical-electronic devices, and finally, electronic-only devices.

Four types of levels are available: optical, automatic, electronic, and laser. An optical level is used to project a line of sight that is at a 90° angle to the direction of gravity. Both dumpy and tilting types use a precision leveling vial to orient to gravity. The dumpy type was used primarily in the United States, while the tilting type was of European origin and used in the remainder of the world. Automatic levels use a pendulum device, in place of the precision vial, for relating to gravity. The pendulum mechanism is called a compensator. The pendulum has a prism or mirror, as part of the telescope, which is precisely positioned by gravity. The electronic level has a compensator similar to that on an automatic level, but the graduated leveling staff is not observed and read by the operator. The operator has only to point the instrument at a bar-code-type staff, which then can be read by the level itself. The laser levels actually employ three different types of light sources: tube laser, infrared diode, and laser diode. The instrument uses a rotating head to project the laser beam in a level 360° plane. *See* LEVEL MEASUREMENT.

The primary purpose of a transit is to measure horizontal and vertical angles. Circles, one vertical and one horizontal, are used for these measurements. The circles are made of metal or glass and have precision graduations engraved or etched on the surface. A vernier is commonly used to improve the accuracy of the circle reading. The theodolite serves the same purpose as the transit, and they have many similar features. The major differences are that the measuring circles are constructed only of glass and are observed through magnifying optics to increase the accuracy of angular readings. The electronic theodolite uses electronic reading circles in place of the optically read ones. *See* VERNIER.

The U.S. Department of Defense installed a satellite system known as the Global Positioning System for navigation and for establishing the position of planes, ships, vehicles, and so forth. This system uses special receivers and sophisticated software to calculate the longitude and latitude of the receiver. It was discovered early in the program that the distance between two nonmoving receivers could be determined very accurately and that the distance between receivers could be many miles apart. This technology has become the standard for highly accurate control surveys, but it is not in general use because of the expense of the precision receivers, the time required for each setup, and the sophistication of the process. *See* SATELLITE NAVIGATION SYSTEMS; SURVEYING. [K.W.K.]

Susceptance The imaginary part of the complex representation for the admittance Y, defined by Eq. (1), of a circuit,

$$Y = G \pm jB \qquad (1)$$

where G is the real part, called the conductance, and B is the susceptance. Since $Y = 1/Z = 1/(R + jX)$ where X is the total reactance, $X_L - X_C$, and R is the resistance, then Eq. (2) holds

$$Y = \frac{R}{R^2 + X^2} - j\frac{X}{R^2 + X^2} \qquad (2)$$

and susceptance $B = X/(R^2 + X^2)$. This is the general expression for susceptance which shows that susceptance is a function involving both resistance and reactance.

These functions find application chiefly in computation of parallel circuits. *See* ADMITTANCE. [B.L.R.]

Swaging A forging method, also called rotary swaging, that consists of two or four dies which are activated radially by blocks in contact with a series of rollers (see illustration). The rotation of the block-die assembly causes the curved ends of the blocks to be in contact with the rollers; relative motion between the blocks and the roller housing then gives a reciprocating motion to the dies. The hammering action on the outer surface of the stock reduces its diameter; the stock is generally prevented from rotating.

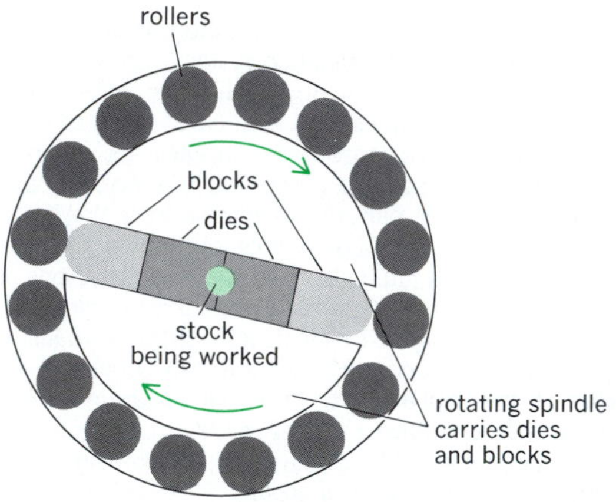

Operation principle of a swaging machine.

A variety of geometries is obtained in this process: (1) reduction of wall thickness of tubes; (2) sinking of tubes, whereby both the outside and inside diameters are reduced, the final wall thickness being controlled by a mandrel; (3) tapered or stepped sections; and (4) a variety of internal geometries, such as fins or hexagonal or spiral grooves, each with the use of a specially designed mandrel. *See* FORGING; METAL FORMING. [S.Ka.]

Swamp, marsh, and bog Wet flatlands, where mesophytic vegetation is really more important than open water, are commonly developed in filled lakes, glacial pits and potholes, or poorly drained coastal plains or floodplains. Swamp is a term usually applied to a wet land where trees and shrubs are an important part of the vegetative association, and bog implies lack of solid foundation. Some bogs consist of a thick zone of vegetation floating on water. Coastal marshes covered with vegetation adapted to saline water are common on all continents. *See* MANGROVE. [L.B.L.]

Sweat gland A coiled, tubular gland found in mammals. There are two kinds, merocrine or eccrine, and apocrine. The latter are generally associated with hair follicles (see illustration). Merocrine glands are distributed extensively over the body in the human, whereas the apoctine variety is restricted to the scalp, nipples, axilla, external auditory meatus, external genitals, and perianal areas. Apocrine sweat glands are more numerous in mammals, with the exception of the chimpanzee and human, in which the merocrine variety predominates. The mammary glands probably represent modified apocrine sweat

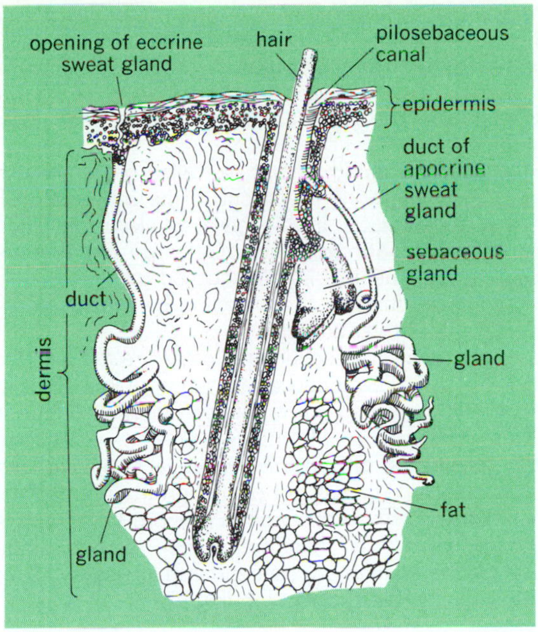

Human skin showing structure of both eccrine and apocrine sweat glands.

glands which grow inward and increase in complexity. *See* GLAND; MAMMARY GLAND. [O.E.N.]

Sweep generator An electronic circuit that generates a voltage or current, usually recurrent, as a prescribed function of time. The resulting waveform is used as a time base to be applied to the deflection system of an electron-beam device, such as a cathode-ray tube. Sweep generators are classified as linear, circular, rotating radial, or hyperbolic.

A linear sweep generator provides a current or voltage that is a linear function of time. The waveform is usually recurrent at uniform periods of time.

A circular sweep is generated by applying a constant-frequency sinusoid to the horizontal component of a deflection system and another of the same frequency but shifted in phase by 90° to the vertical component. A spiral scan is generated if, in addition to the sinusoidal horizontal and vertical components, a linear sawtooth waveform is used to modulate equally both the horizontal and voltage deflection waveform.

A rotating radial sweep may be generated by applying a linear sawtooth of current to the deflection system of a magnetically deflected cathode-ray device and rotating the deflection coil producing the magnetic field at a constant angular velocity. Such a sweep may also be generated by using a fixed position deflection system having separate horizontal and vertical components of deflection. The same combination of linear and sinusoidal modulation as that of the spiral sweep may be used, except that the period of the linear modulation component to the deflection system must be short compared to the period of the angular deflection.

There are applications, particularly in airborne radar systems, where hyperbolic sawtooth sweeps are preferred to linear sawtooth sweeps. Such a hyperbolic sweep used in a rotating radial sweep system provides for true ground range radar mapping in an airborne system. [G.M.G.]

Sweetgum The tree *Liquidambar styraciflua*, also called redgum, a deciduous tree of the southeastern United States. It is found northward as far as southwestern Connecticut, and also grows in Central America. Sweetgum is readily distinguished by its five-lobed, or star-shaped, leaves and by the corky wings or ridges usually developed on the twigs. The erect trunk is a dark gray, but the branches are lighter in color. In winter the persistent, spiny seedballs are an excellent diagnostic feature.

Sweetgum is used for furniture, interior trim, railroad ties, cigar boxes, crates, flooring, barrels, woodenware, and wood pulp, and it is one of the most important materials for plywood manufacture. Sweetgum is one of the most desirable ornamental trees, chiefly because of its brilliant autumn coloration. *See* HAMAMELIDALES. [A.H.G./K.P.D.]

Swim bladder A gas-filled sac found in the body cavities of most bony fishes (Osteichthyes). The swim bladder has various functions in different fishes, acting as a float which gives the fish buoyancy, as a lung, as a hearing aid, and as a sound-producing organ. In many fishes it serves two or three of these functions, and in the African and Asiatic knife fishes (Notopteridae) it may serve all four. The swim bladder contains the same gases that make up air, but often in different proportions. [R.McN.A.]

Swimming birds Birds belonging to the orders Charadriiformes and Pelecaniformes. There are three important families of long-winged swimming birds (Charadriiformes): Stercorariidae, all of which occur in North America, comprising skuas and jaegers; Laridae, the gulls and terns; and Rynchopidae, which contains three species of skimmer. In the six families of pelicaniform birds all four toes are connected by webs. Included are the pelicans, gannets, cormorants, and frigate birds; all are rather large fish-eating birds. The order has a cosmopolitan distribution. All are found near water, some being entirely marine, others inland, fresh-water species. *See* CHARADRIIFORMES; PELECANIFORMES. [C.B.C.]

Swine production An agricultural business whose primary products are pork, lard, hides, and innumerable byproducts of a pharmaceutical nature. Pork and lard supply approximately 15% of the total calories consumed as food in the United States. Pork is more successfully cured and stored than many other meats, and it is estimated that about 60% of the swine carcass is cured by various methods.

Market hogs are produced with emphasis on red meat production while minimizing fat or lard production. Breeders of purebred, hybrid, and crossbred swine provide the seed stock for commercial hog production and are entrusted with the responsibility of changing the carcass to better meet the demands of the consumer. The purebred breeds of major significance include the following: Berkshire, Chester White, Duroc, Hampshire, Landrace, Poland China, Spotted Poland China, and Yorkshire. [A.H.J.; D.E.B.]

Switched capacitor A module consisting of a capacitor with two metal oxide semiconductor (MOS) switches connected as shown in illustration *a*. These elements in the module are easily realized as an integrated circuit on a silicon chip by using MOS technology. The switched capacitor module is approximately equivalent to a resistor, as shown in illustration *b*. The fact that resistors are relatively difficult to implement gives the switched capacitor a great advantage in integrated-circuit applications requiring resistors. Some of the advantages are that the cost is significantly reduced, the chip area needed is reduced, and precision is increased. Although the switched capacitor can be used for any analog circuit realization such as analog-to-digital or digital-to-analog converters, the most notable application has been to voice-frequency filtering.

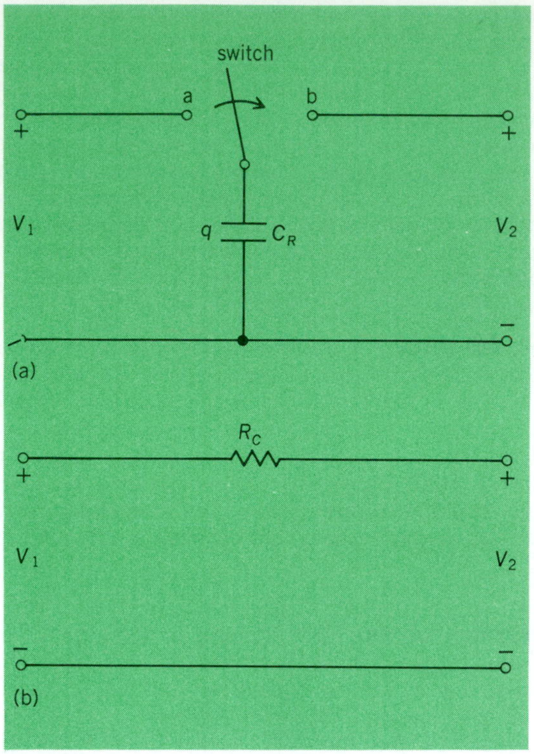

Switched capacitor. (*a*) **Basic circuit.** (*b*) **Equivalent resistive circuit.**

See ANALOG-TO-DIGITAL CONVERTER; DIGITAL-TO-ANALOG CONVERTER; ELECTRIC FILTER; INTEGRATED CIRCUITS. [M.E.V.V.]

Switching circuit A constituent electric circuit of switching or digital data-processing systems. Well-known examples of such systems are digital computers, dial telephone systems, automatic accounting and inventory systems. In these and other switching systems the component circuit units receive, store, and manipulate information in coded (digital) form to accomplish the specified objectives of the system. *See* SWITCHING SYSTEMS (COMMUNICATIONS).

Physically, switching circuits consist of conducting paths interconnecting discrete-valued electrical devices. The most generally used switching circuit devices are two-valued or binary, such as switches and relays in which manual or electromagnetic actuation opens and closes electric contacts; vacuum and gas-filled electronic tubes, semiconductor rectifiers and transistors, which do or do not conduct current; and magnetic structures, which can be saturated in either one of two directions.

The electrical conditions controlling these switching circuit devices are also generally two-valued or binary, such as open versus closed path, full voltage versus no voltage, large current versus small current, and high resistance versus low resistance. Such two-valued electrical conditions, as applied to the input of a switching circuit, represent either (1) a combination of events or situations which exist or do not exist; (2) a sequence of events or situations which occur in a certain order; or (3) both combinations and sequences of events or situations. The switching circuit responds to such inputs by delivering at its output, also in two-valued terms, new information which is functionally related to the input information.

The two fundamental characteristics of switching circuits are logic and memory. A switching circuit embodies such logical relationships as output X is to exist only if inputs A and B occur simultaneously; and output Y is to exist if either input A or input B occurs. The factor of memory, in turn, enables a switching circuit to hold or retain a given state after the condition that produced the state has passed.

Basic combinational circuits. A combinational switching circuit is one in which a particular set of input conditions always establishes the same output, irrespective of the past history of the circuit.

In electronic switching circuits, so-called gates are used to perform logical functions equivalent to the series-parallel networks of switch contacts. In this sense, an electronic gate is an elementary combinational circuit. Gates do not function by physical rearrangement of interconnecting paths, as do switch or relay contacts. Instead, they function by control of voltage or current levels at their output.

The most commonly encountered gates are the AND and the OR gates. The AND gate produces an output only if all its inputs are concurrently present; an OR gate produces an output if any one or any combination of its inputs is present.

Basic sequential circuits. A sequential switching circuit is one whose output depends not only upon the present state of its input, but also on what its input conditions have been in the past. Sequential circuits, therefore, require memory elements. A typical electronic memory element used in sequential circuits is a simple circuit called a flip-flop (Fig. 1). A flip-flop consists of two amplifiers connected so that the output of one amplifier is the input of the other. A voltage pulse will set the flip-flop into one of two states, and that state remains until another voltage pulse resets, or returns the flip-flop to its original state. It can therefore be used to remember that an event has taken place.

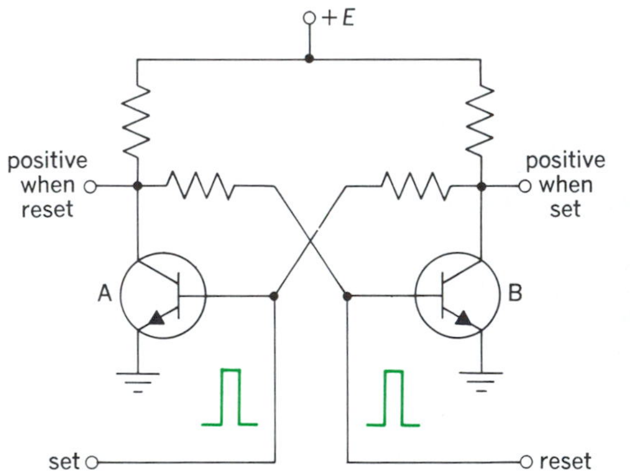

Fig. 1. Transistor switching memory element (flip-flop). When transistor A is conducting transistor B is cut off, and vice versa. E **= voltage source.**

Relays, flip-flop, and similar memory elements provide static, or fixed, memory; they hold the stored information indefinitely, or until they are told to "forget." In contrast, a delay line provides transient memory. A delay line has the property that an electrical signal applied to its input is delayed on its way to the output.

Functional switching circuits. Even in large and complex switching systems the majority of circuit requirements can be met by a relatively small number of types of circuits, each of which performs one or a limited number of somewhat distinct functions. These functional circuits are the basic building blocks of a switching system.

A selecting circuit receives the identity (called the address) of a particular item and selects that item from among a number of similar ones. The selectable items are often represented by terminals or leads. Selection usually involves marking the specified terminal or lead by applying to it some electrical condition,

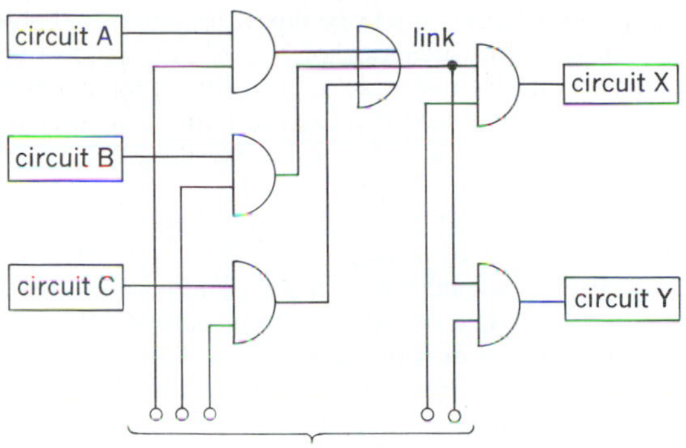

Fig. 2. Connecting circuit using AND and OR gates.

such as a voltage or current pulse, or a steady-state dc signal. By means of this electrical condition, the selected circuit is alerted, seized, or controlled.

A switching system is an aggregate of functional circuit units, some of which must sometimes be directly coupled to each other to interchange information. Such a system needs, therefore, connecting circuits which establish the circuit associations dictated by the momentary needs of the system (Fig. 2).

In switching systems, situations often arise where several similar circuit units are ready at the same instant to request collaboration with another type of functional circuit. Mutual interference among the requesting circuits is prevented by the lockout circuit. In response to concurrent inputs from a number of external circuits, a lockout circuit provides an output indication corresponding to one, and only one, of these circuits at any time, Lockout circuits are sometimes referred to as hunting or finding circuits.

Switching systems process information in coded form; the information they manipulate is generally in the form of numbers. Numerical codes are many and varied, each with its own characteristics and more or less distinct advantages for different switching circuit situations. Therefore, one of the common functional circuits in switching systems is the translating circuit, which translates information received in one code into the same information expressed in another code. These translating circuits are combinational circuits; a given input signal combination representing a code to be translated always produces the same output signals, which represent the desired code.

Information received by a switching system is not always used immediately. It must be stored in register circuits for future use.

In a register circuit the coded information to be stored is applied as input, is retained by memory elements of the circuit and, when needed, the registered information is taken as output in the same code or in a different code. Register circuits are devised with a great variety of memory elements, and have capacities to store from a few to millions of information bits. A frequently encountered form of register circuit is the shift register. This type of register has the ability to shift its stored digital information internally to positions representing higher or lower numerical values in the code employed. For example, in decimal code registration a digit may be shifted from the units to the tens position. An obvious use of such registers is in digital computers when, for example, partial multiplication products have to be lined up for addition.

One of the most frequently encountered circuits in switching systems is the counting circuit whose function, in general, is to detect and count repeated current or voltage pulses which represent incoming information. *See* COUNTING CIRCUIT; LOGIC CIRCUITS. [J.Mes.]

Switching systems (communications) The assemblies of switching and control devices provided so that any station in a communications system may be connected as desired with any other station. A telecommunications network consists of transmission systems, switching systems, and stations. Transmission systems carry messages from an originaling station to one or more distant stations. To enable the transmission facilities to be shared, stations are connected to and reached through switching system nodes that are part of most telecommunications networks. Switching systems act under built-in control to direct messages toward their ultimate destination or address. Most switching systems, known as central or end offices, are used to serve stations. A small number of systems serve as tandem (intermediate) switching offices for large urban areas or toll (long-distance) offices for interurban switching. These end and intermediate office functions are sometimes combined in the same switching system.

Switching system fundamentals. Telecommunications switching systems generally perform three basic functions; they transmit signals over the connection or over separate channels to convey the identity of the called (and sometimes the calling) address (for example, the telephone number), and alert the called station; they establish connections through a switching network for conversational use during the entire call; they process the signal information to control and supervise establishment and disconnection of the switching network connection.

In some data or message switching when realtime communication is not needed, the switching network is replaced by a temporary memory for data storage. This type of switching is known as store-and-forward switching.

Signaling and control. The control of switching systems is accomplished remotely by a specific form of data communications known as signaling. Switching systems are connected with one another by telecommunication channels known as trunks. They are connected with the served stations or terminals by lines. Until 1976 most signals were sent by direct- or alternating-current pulses over the lines and trunks. Then a new form of signaling was introduced for use between toll offices in the United States. It is known as common channel interoffice signaling (CCIS). As its name implies, a network of separate data communication paths is used for transmitting all signaling information between offices.

In some switching systems the signals for a call directly control the switching devices through which the transmission path is established. This direct control was the earliest and is still the most prevalent form of automatic switching around the world.

For most modern switching systems the signals for identifying or addressing the called station are received by a central control that processes calls on a time-shared basis. Central controls receive and interpret signals, select and establish communication paths, and prepare signals for transmission. These signals include addresses for use at succeeding nodes or for alerting (ringing) the called station.

Until the introduction of electronics into switching, the central controls employed complex relay logic circuits known in many systems as markers. Most electronic controls are now designed to process calls not only by complex logic, but also by the use of logic tables or a program of instructions stored in bulk electronic memory. The tabular technique is known as action translator (AT). The latter arrangement is now the most accepted and is known as stored program control (SPC). Either type of control may be distributed among the switching devices rather than residing centrally. *See* COMPUTER STORAGE TECHNOLOGY.

Common channel signaling became practical as a result of processor control. To reduce the number of data channels between all switching nodes, a signaling network with separate switching of signaling data is introduced. These signal switching nodes are fully interconnected and duplicated and are known as signal transfer points (STPs).

Switching connectives. Space and time division are the two basic techniques used in establishing connections. When an individual conductor path is established through a switch for the duration of a call, the system is known as space division. When the transmitted speech signals are sampled and the samples multiplexed in time so that high-speed electronic devices may be used simultaneously by several calls, the switch is known as time division.

Most switching systems now in service are space division and employ some form of electromechanical switch. These switching systems, unless controlled electronically (for example, SPC) are referred to by the name of the electromechanical device (for example, a step-by-step, panel, crossbar, EMD, codebar, XY, or rotary system). These devices are connected together in successive stages to concentrate, distribute, and select idle paths as required to establish the required connections. The simplest arrangement is the one in which, as each digit is received, the call progresses from one stage to the next by making a selection and finding an idle link at each stage until the selection process is completed.

Electronic switching. The invention of the transistor spurred the introduction of electronics and semiconductor technology into switching system design. Stored-program-control switching systems have been placed in service in the United States, serving more than 40,000,000 lines. Stored-program control has become the principal type of control for all types of new switching systems throughout the world, including toll, private branch, data, and Telex systems. About 1200 small systems were earlier placed in service in Canada, Great Britain, and France using action translator (AT) electronic logic and memory controls. *See* TELEPHONE; TELEPHONE SERVICE. [A.E.J.]

Sycamore American sycamore (*Platanus occidentalis*) a member of the plane tree family, known also as American plane tree, buttonball, or buttonwood, and ranging from southern Maine to Nebraska and south into Texas and northern Florida. It has the most massive trunk of any American hardwood. Characteristic are the white patches which are exposed when outer layers of the bark slough off; the simple, large, lobed leaves whose stalks completely cover the conical winter buds; and the spherical fruit heads that are always borne singly in the American species and persist throughout the winter. The tough, coarse-grained wood is difficult to work, but is useful for butchers' blocks, saddle trees, vehicles, tobacco and cigar boxes, crates, and slack cooperage. *See* ROSALES. [A.H.G./K.P.D.]

Sycettida An order of calcareous sponges of the subclass Calcaronea. This order includes the families Sycettidae, Heteropiidae, Grantiidae, Amphoriscidae, and Lelapiidae, Choanocytes occur in flagellated chambers, and the spongocoel is not lined with these cells. *See* CALCAREA. [C.B.C.]

Syenite A phaneritic (visibly crystalline) plutonic rock with granular texture composed largely of alkali feldspar (orthoclase, microcline, usually perthitic) with subordinate plagioclase (oligoclase) and dark-colored (mafic) minerals (biotite, amphibole, and pyroxene). If sodic plagioclase (oligoclase or andesine) exceeds the quantity of alkali feldspar, the rock is called monzonite. Monzonites are generally light to medium gray, but syenites are found in a wide variety of colors (gray, green, pink, red), some of which make the material ideal for use as ornamental stone. Syenite is an uncommon plutonic rock and usually occurs in relatively small bodies (dikes, sills, stocks, and small irregular plutons). *See* IGNEOUS ROCKS. [C.A.C.]

Symbiotic star A stellar object whose optical spectrum displays features indicative of two very different thermal regimes: a stellar spectrum whose flux distribution and absorption lines suggest the presence of a cool star, and emission lines which can be formed only in a much hotter medium. In 1941, P. W. Merrill labeled this kind of composite spectrum a combination spectrum, and the object emitting it a symbiotic star.

This definition is rather vague and, in order to isolate a nearly homogeneous class of objects, "cool star" and "much hotter medium" must be defined more clearly. There is no unanimity in this respect, and as a consequence the number of symbiotic stars listed by various astronomers differs. The most comprehensive lists contain somewhat more than 100 objects. Symbiotic stars are thus quite rare, but they are very important for understanding stellar evolution.

The general consensus is that most (if not all) symbiotic stars are actually binary stars. The "cool" continuum, observed in the optical and infrared spectral regions, is due to a cool star. The "hot" continuum observed in the ultraviolet is due to a much hotter star. The emission lines, observed both in the optical and in the ultraviolet, originate in a plasma which surrounds either component or the whole system. The energy necessary to ionize the plasma and excite the observed emission lines comes from the hot star. But at least one of the components must be postulated to be the source of this radiating plasma. Therefore the system is assumed to be an interacting binary star, in which one of the components is losing mass, which then interacts either with the other component or with its radiation.

A large number of interacting binary stars of very diverse characters are known, but only a very special combination of stars can be observed as a symbiotic object. It is well known that at any given wavelength a hotter star will always radiate more energy per unit area of its surface than a cooler star. Yet in the optical spectrum of the symbiotics, the cool star always dominates, and the continuous radiation of the hot star is practically imperceptible. Thus, the proper combination of objects must pair a large cool star (a red giant or even a supergiant) with a much smaller hot star.

In spite of this restriction, three variants are still possible for the model of an interacting binary star to appear as a symbiotic object. In all three versions, the cool component is a red giant (or even supergiant). According to the nature of the hot object, three types can be distinguished: subdwarf symbiotic, algol symbiotic, and novalike symbiotic.

A subdwarf symbiotic or a PN (planetary nebula) symbiotic would be a combination of a cool red giant with a small hot subdwarf star. These stars are probably the inner cores of for-

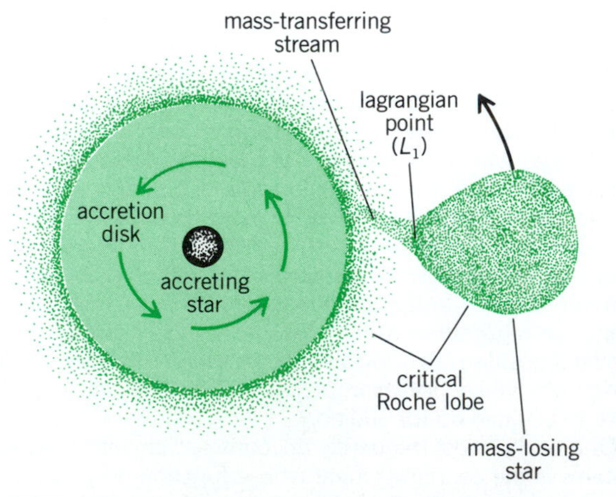

Pole-on view of an interacting binary of the Algol type. (*After M. J. Plavec, IUE looks at the Algol paradox, Sky Telesc., 65:413–416, May 1983*)

mer giants and supergiants, which have shed the cool outer envelope and are now contracting to become genuine white degenerate dwarfs. They are very hot, but their total radiation (luminosity) is small because they have small radii, of the order of 10 earth radii or so. Hot gases may still be streaming out of their surfaces. Thus in this type of a symbiotic the hot star would be both the source of the circumstellar plasma and the source of its luminous energy. The difficulty with this model is that the subdwarf stage is relatively very short, so that these "pure natural" symbiotics should be very rare.

The Algol symbiotic consists of a red giant combined with a main-sequence star. If the red giant loses mass which flows toward the main-sequence star, an accretion disk can form around the latter star. If the rate of mass transfer is sufficiently high (on the order of 10^{-4} solar mass per year or so), the inner parts of the disk become hot enough to provide the ionizing photons, and yet are small. This type of interaction is known to exist, mostly with lower mass-loss rates, in a large class of interacting binaries called the Algol-type semidetached binaries. The illustration shows a pole-on view of an Algol-type interacting binary with an accretion disk. The stars Z Andromedae, CI Cygni, AR Pavonis, and several other symbiotics may be of this type.

The novalike symbiotic consists of a red giant combined with a white dwarf. If again the cool giant loses mass, its accretion onto the white dwarf may ignite thermonuclear reactions of hydrogen and helium in the surface layers of the white dwarf. Low rates of accretion suffice to generate large output of energy in this case, and this energy may be coming in outbursts. This model of a symbiotic is in principle similar to contemporary models of novae. Since many symbiotic stars show a tendency toward irregular or semiregular outbursts in which their brightness increases significantly, this is an important model. *See* NOVA; THERMONUCLEAR REACTION. [M.Pl.]

Symbolic computing The development and use of symbolic systems, that is, computer programs that perform computations with constants and variables according to the rules of algebra, calculus, and other branches of mathematics. Such programs are also called algebraic computation systems, computer algebra systems, or symbolic computation systems. They are used in science and engineering teaching and research to perform nonnumerical calculations that would be done less effectively by hand or numerical means.

Symbolic systems manipulate symbols and equations, unlike conventional scientific systems that manipulate numbers. For example, if a symbolic system is given the expression $x + x$, it will normally return the answer $2x$ and not question the absence of a value for x. Although most provide a language for programming in the conventional sense, they are normally used in the style of a calculator, in which the user types in an expression and the system responds with an answer. Another important characteristic of symbolic systems is that they produce exact results for calculations that, when done by hand, are often tedious, error prone, and difficult. Even the arithmetic can be exact; numbers are manipulated with as many digits as are necessary for a given computation, unlike numerical languages that usually impose a fixed precision on their manipulations. Systems can now solve problems that could not be done in a lifetime by hand, such as advanced calculations in fields such as celestial mechanics, general relativity, and elementary particle physics. Symbolic systems provide a tool whose purpose is to increase scientific and engineering productivity. In most cases, this tool is used in the initial steps of the problem-solving process, with numerical computation as the final step. *See* ALGORITHM; MATHEMATICAL SOFTWARE; NUMERICAL ANALYSIS. [A.C.H.]

Symmetrodonta An order of the extinct mammalian infraclass Pantotheria of Mesozoic times. Long considered to be aberrant triconodonts, symmetrodonts are not recognized as a discrete order of small carnivorous or insectivorous mammals whose earliest representatives were close to the ancestry of all theria and especially close to the origin of the order Pantotheria. *See* MAMMALIA; PANTOTHERIA. [M.C.McK.]

Symmetry breaking A generic term describing the deviation from exact symmetry exhibited by many physical systems. It can occur either explicitly (explicit symmetry breaking) or spontaneously (spontaneous symmetry breaking), with distinct observable consequences characterizing the two cases.

In the explicit case, the system is "not quite" the same for two configurations related by exact symmetry. A simple example is a bicycle wheel with the valve stem sticking out: it is almost symmetric with respect to rotations about the bicycle axis, but the symmetry is explicitly broken by the valve stem. In quantum mechanics the system is characterized by its energy (hamiltonian), and in the case of explicit symmetry breaking the hamiltonian is almost symmetric. Usually explicit symmetry breaking is useful in explaining numerical relations between the energy levels (or masses of nuclear particles) of certain physical systems. *See* ENERGY LEVEL (QUANTUM MECHANICS); QUANTUM MECHANICS.

On the other hand, in the case of spontaneous symmetry breaking, the hamiltonian always displays the exact symmetry, but the state of lowest energy of the system does not share this symmetry. This case can be completely understood only within the framework of quantum mechanics, although a simple classical analog exists: a spinning roulette wheel settling into a state with the ball in one slot, thus breaking the rotational symmetry. The physical consequences of spontaneous symmetry breaking are very different from the explicit case in that they predict domain structure when the broken symmetry is discrete, and new gapless (massless) modes of excitations, called Nambu-Goldstone modes, when the broken symmetry is continuous.

Spontaneous symmetry breaking also serves to describe the different phases of physical systems such as the superconducting phase of conductors, the various phases of ^{3}He, and so forth. In particle physics, the most spectacular effects of spontaneous symmetry breaking occur in the electroweak theory. The electroweak gauge symmetry is broken spontaneously by the vacuum state chosen by nature. As a result, some gauge particles acquire a mass by "eating" the Nambu-Goldstone boson produced by the spontaneous symmetry breaking through a process known as the Higgs mechanism. These massive gauge particles, called W^+, W^-, and Z, were first observed in 1982 and 1983. *See* GAUGE THEORY; INTERMEDIATE VECTOR BOSON; SYMMETRY LAWS (PHYSICS); WEAK NUCLEAR INTERACTIONS; WEINBERG-SALAM MODEL. [P.Ra.]

Symmetry laws (physics) The physical laws which are the expressions of the existing symmetries. A conservation law results from each such symmetry; that is, from each symmetry the existence of a quantity which is conserved (a constant of the motion) can be deduced. Selection rules result from conservation laws. *See* SELECTION RULES (PHYSICS).

Space-time symmetries. A symmetry (or invariance) of the world exists whenever the description of the laws of physics is unaffected by a change in the frame of reference. For instance, the position of the origin of a space coordinate system is quite arbitrary; changing it makes no difference in the description of the motion of bodies because the forces between bodies depend only on their relative positions and not on any absolute position. Equivalently, a system of bodies behaves the same if translated to another place. This symmetry of space to translation implies the conservation of momentum.

Other symmetries of space-time are the irrelevance of (1) the origin of the time coordinate, (2) the orientation of a coordinate system in space, and (3) the velocity of a coordinate system (Lorentz invariance). Each of these implies a conservation law, as shown in the table. All these symmetries are termed continuous because the changes can be arbitrarily small; that is, a finite change can be made bit by bit. The resulting constants of the motion are classical quantities and are additive. *See* FRAME OF REFERENCE; LORENTZ TRANSFORMATIONS; RELATIVITY; SPACE-TIME.

Discrete symmetries (reflections) also exist, for which the irrelevant change is not arbitrarily small. They imply constants of the motion (parities) in quantum mechanics. These parities are multiplicative. For instance, the direction of increasing time is irrelevant; the world is invariant to time reversal (microscopic reversibility). Although, macroscopically, future and past seem distinct, this is merely a result of the disposition of matter (a state of anomalously small entropy at some time in the past) in the same way that a point of space seems distinct by having a particular piece of matter there. Space is also symmetrical to reflection of space or to inversion, the reflection of all three directions of space; it is irrelevant which is the positive direction of a space axis or of all three axes. This amounts to the irrelevance of whether a right-handed or a left-handed coordinate system is used. The resulting conserved quantity (eigenvalue of space inversion) is (space) parity. *See* FUNDAMENTAL INTERACTIONS; PARITY (QUANTUM MECHANICS); STRONG NUCLEAR INTERACTIONS; WEAK NUCLEAR INTERACTIONS.

Internal symmetries. Further symmetries, which are not space-time symmetries, are called internal symmetries. For example, the zero of both scalar and vector electromagnetic potentials is irrelevant; the addition of a constant to an electromagnetic potential is of no consequence (so-called gauge invariance of the first kind). In quantum mechanics this symmetry implies the conservation of charge. There is also a discrete symmetry: It is (nearly) irrelevant which sign of charge is called positive and which is called negative. The qualification "nearly" is necessary here, just as in space inversion, because the weak interactions do not observe the symmetry. Thus the world is (nearly) invariant to charge reversal, the interchange of positive and negative charge. A world with (nearly) the same properties as Earth's would result if all electrons were replaced with positrons and all protons by antiprotons and, in general, if all particles were replaced by their antiparticles. The resulting (nearly) conserved quantity is termed charge conjugation parity or charge parity, *C*.

Although the weak interactions are not invariant under charge reversal or space inversion, they are invariant (with one observed exception, K^0 decay) to the combination of these reflections; they are also invariant under time reversal. Equivalently stated, all interactions appear to very nearly conserve *CP* and also *T* (C = charge conjugation, P = space inversion, T = time reversal).

CP is not exactly conserved in K^0 decay. The smallness of the violation contrasts with the maximal violation in weak interactions of *C* and *P* separately. No other system is as sensitive to *CP* violation as the neutral *K* mesons and no violation of either *CP* or *T* has been observed elsewhere.

The reflection symmetries are correlated by the *CPT* theorem of G. Lüders. This theorem states that a Lorentz invariant field theory is necessarily invariant to the product of the three reflections: charge conjugation *C*, space inversion *P*, and time reversal *T*. Elementary particle theory would be enormously challenged if *CPT* invariance were found not to hold. *See* MESON.

The nuclear force between two protons is found to be identical to the force between two neutrons. This is called the charge symmetry of the nuclear force. More generally, the motion of a system composed of nucleous and pions (the lowest mass quanta of the nuclear force field) is invariant to the operation $p \leftrightarrow n$, $\pi^+ \leftrightarrow \pi^-$, $\pi^0 \leftrightarrow \pi^0$. The combination of charge conjugation and the charge symmetry operation is called isotopic

Invariances (symmetries) and conservation laws

Invariance to	Conserved quantity (arrows indicate range of validity)		Range
Homogeneity of space-time			Exact
Translation of space	Momentum, **p**		Exact
Translation of time	Energy, *E*		Exact
Isotropy of space-time			Exact
Rotation of space	Angular momentum, **J**		Exact
Lorentz transformation	Velocity of the center of energy (center of mass)		Exact
Interchange of identical particles	Symmetry of the wave function (statistics)		Exact
Inversion of time, space, and charge	*CPT*		Exact
Gauge invariance			Exact
Charge U_1	Net charge		Exact
Color SU_3	Color	*a*	Exact
?	Net number of baryons, e-leptons, μ-leptons	*b*	
Reversal of time	Time parity, *T*	*c*	Approx.
Reflection of space and charge	Product of parity and charge parity, *CP*	*c*	Approx.
Reflection (or inversion) of space	Parity, *P*		Approx.
Reflection of charge (charge conjugation)	Charge parity, *C*	*d*	Approx.
?	Net number of up quarks, down quarks, strange quarks, charm quarks	*e*	Approx.
⋮	⋮		Approx.
Flavor SU_2	Isotopic spin, *I*; isotopic parity, *G*	*f*	Approx.
Flavor SU_3	"Unitary spin"	*f*, *g*	Approx.
⋮ ⋮		*f*, *g*	Approx.

[a] A hypothesized hidden symmetry.
[b] Conserved at the present level of detection, but possibly not exactly conserved.
[c] Violation seen only in K_L decay.
[d] Violated by the weak interactions.
[e] Conservation of first three is equivalent to the conservation of baryon number I_3 (the charge axis component of isotopic spin) and hypercharge.
[f] Violated by electromagnetic interactions and by the up-down quark mass difference, $m_u - m_d \approx -5$ MeV.
[g] Violated by the strange-nonstrange quark mass difference, $m_s - m_u \approx m_s - m_d \approx 150$ MeV.

inversion, *G*, and carries each π-meson into itself. The π-meson thus has a *G* parity.

Further, the nuclear force between a neutron and a proton is identical to the force between two protons or two neutrons in the same orbital and spin state (charge independence). *See* NUCLEAR STRUCTURE; SCATTERING EXPERIMENTS (NUCLEI).

The foregoing symmetry can be expressed as the isotropy of a three-dimensional "isotopic space," which implies the conservation of "angular momentum," or isotopic spin **I**, in this space. The component of a particle's isotopic spin along the "third" axis I_3 is related linearly to the charge of the particle. The consequences of charge independence are formally very similar to the consequences of the conservation of angular momentum.

Charge independence can also be described as symmetry with respect to arbitrary unimodular unitary transformations of the proton and neutron. The group of such transformations is called SU_2. The analogous group of transformations on *n* particles is SU_n. *See* UNITARY SYMMETRY.

It appears that hadrons are well described as compounds of so-called quarks, of which there are *n* kinds (flavors): *u*, *d*, *s*, *c*, *b*,.... The net number of quarks of each flavor is changed only in weak interactions. Further, the quarks of different flavors interact in the same way with the fundamental strong interaction between them (the "glue" which binds them together to make hadrons). This means that the strong interactions would have a flavor SU_n internal symmetry, were it not for the fact that the quarks of different flavors have different masses; their masses seem to form roughly a geometrical progression. In addition to this flavor SU_n symmetry of strong interactions, it is believed that there is also a color SU3 symmetry which is exact but hidden. *See* COLOR (QUANTUM MECHANICS); FLAVOR; QUARKS.

[C.J.G.]

Symmorphosis A theory of structural design of biological organisms postulating that structure is quantitatively matched to functional demand as a result of regulated morphogenesis during growth and maintenance. Symmorphosis is a theory of economic design. In biological organisms, all functions depend on structural design, specifically on the morphometric characteristics of the organs. In general terms, the larger the structure the greater the functional capacity. A central postulate of symmorphosis, as a theory of economic design, is that differences in the functional demand on an organ require quantitative adjustments of its structural design parameters in order to match functional capacity to (maximal) functional demand.

The notion that animals, and humans, should be designed economically follows from common sense, but it is also supported by many observations. Blood vessel architecture ensures blood flow distribution with minimal energy loss. Bone structure is patterned according to stress distribution and also quantitatively adapted to total stress. With training, athletes can specifically adjust the structure of their muscles and of their cardiovascular system to higher functional demands, and these modifications are soon reversed when training is stopped.

The theory of symmorphosis postulates ideal adaptation of structural design to functional capacity. This is, however, hardly a reasonable assumption, because good engineering design of complex systems requires some redundancies as safety factors in view of imperfections in functional performance and variable boundary conditions. By using the concept of symmorphosis, such deviations from idealized economic design can be detected.

[E.R.W.]

Sympathetic nervous system The portion of the autonomic nervous system concerned with nonvolitional preparation of the organism for emergency situations. *See* AUTONOMIC NERVOUS SYSTEM.

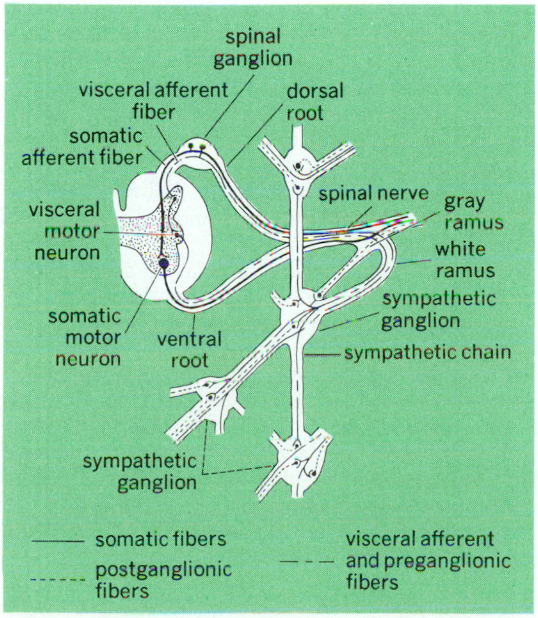

The visceral reflex arc and the sympathetic chain. (*After B. A. Houssay et al., Human Physiology, 2d ed., McGraw-Hill, 1955*)

The sympathetic nervous system is best understood in mammals. It consists of two neuron chains from the thoracic and lumbar regions of the spinal cord to viscera and blood vessels. The first or preganglionic neuron has its cell body in the spinal cord and sends its axon to synapse with a postganglionic sympathetic neuron, which lies either in a chain of sympathetic ganglia paralleling the spinal cord or in a sympathetic ganglion near the base of the large blood vessels vascularizing the alimentary viscera. The postganglionic axons are longer than the preganglionic axons and extend to glands or smooth muscles of viscera and blood vessels. Sensory visceral nerve fibers innervate blood vessels and viscera and carry sensory information to the spinal cord, thus providing a visceral reflex (see illustration). *See* NERVOUS SYSTEM (VERTEBRATE); PARASYMPATHETIC NERVOUS SYSTEM.

[D.B.W.]

Sympathetic vibration The driving of a mechanical or acoustical system at its resonant frequency by energy from an adjacent system vibrating at this same frequency. Examples include the vibration of wall panels by sounds issuing from a loudspeaker, vibration of machinery components at specific frequencies as the speed of a motor increases, and the use of tuned air resonators under the bars of a xylophone to enhance the acoustic output. Increasing the damping of a vibrating system will decrease the amplitude of its sympathetic vibration but at the same time widen the band of frequencies over which it will partake of sympathetic vibration. *See* RESONANCE (ACOUSTICS AND MECHANICS); VIBRATION.

[L.E.K.]

Symphyla A class of the Myriapoda. The symphylans, like the pauropods, are tiny, pale, centipedelike creatures that inhabit humans or soil, or live under debris; in general, they live wherever there is sufficient moisture to preclude excessive water loss. They are similar to the Pauropoda and Diplopoda in being progoneate and anamorphic. Each of their mandibles, like those of millipedes, bears a movable gnathal lobe; at the same time their two pairs of maxillae are more reminiscent of the chilopods and lower insects than of the singly maxillate millipedes and pauropods.

The class consists of three families to which not more than 60 species have been assigned.

[R.E.Cr.]

Synapsida A subclass of extinct reptiles, in general characterized by a temporal fenestra that lies below the junction of the postorbital and squamosal bones, a so-called lower temporal opening. In advanced forms, however, the postorbital-squamosal bridge is absent. Mammals arose from this group of reptiles during the Triassic Period. *See* REPTILIA.

Synapsids first appear in the geological record in the Upper Carboniferous. They flourished during the Late Paleozoic and Early Mesozoic but became extinct at about the end of the Triassic. During this span of time they underwent a broad adaptive radiation on land and made minor invasions of aquatic habitats. There were two major phases of this radiation: one early, by pelycosaurs, and the other later (Late Permian and Triassic), by therapsids. *See* PELYCOSAURIA; REPTILIA; THERAPSIDA.

[E.C.O.]

Synaptic transmission A term applied to the physiological mechanisms by which one nerve cell (neuron) influences the activity of an anatomically adjacent neuron with which it is functionally coupled. The term synapse refers to the region of contact between two neurons which is characterized by unique local morphological specializations. Neurons also make anatomical and functional connection with various kinds of effector cells (muscle, gland, and sensory cells), and here the region of contact is called a neuroeffector junction. Neuroeffector transmission provides the means by which a neuron can exert control over the effector cell to which it is connected. Neuronal coupling by synaptic transmission makes possible the joining of many neurons to form extended pathways or complex networks over which information in the form of nerve impulses can be conducted, processed, and stored. Over such anatomical arrays, information is collected and carried from one region of the nervous systems to another, and it can be passed to various other parts of the organism supplied with neuronal connections.

Typically, as a nerve impulse arrives in presynaptic terminals it initiates a series of events which ultimately produce a change in the postsynaptic membranes of the adjacent neuron. This sequence of steps known as synaptic transmission takes place very rapidly, requiring as little as 0.6 millisecond for its completion. It represents a unidirectional process which flows from pre- to postsynaptic elements, but not in the reverse direction. Thus one important characteristic of a synapse is that it exerts a valvelike action on the transmission of information.

The coupling mechanism which operates at various synapses can be essentially chemical or electrical in nature, but some synapses are known in which both types of transmissions are effective and operate in parallel.

In chemical transmission a nerve impulse which arrives at the presynaptic terminals initiates the rapid release of a chemical intermediary (transmitter) which diffuses through the fluid filling the very narrow space (synaptic cleft) between the pre- and postsynaptic elements. When the transmitter arrives at the postjunctional membrane of the postsynaptic cell it approaches and combines with receptor molecules. When the receptor molecule combines with the transmitter, it undergoes a change in shape, allowing ions to enter the membrane channel. Thus the transmitter can be thought of as a chemical "key" which opens a specific receptor protein gate at the mouth of an ionic channel in the postsynaptic membrane. Examples of relatively well-established transmitters are acetylcholine, morepinephrine, dopamine, serotonin, gamma aminobutyric acid, glutamic acid, and glycine. *See* ACETYLCHOLINE.

In addition to chemical transmission, another type of synaptic coupling exists which is fundamentally electrical in nature. By appropriate anatomical arrangements, the electric currents which accompany the propagated action potentials appearing in presynaptic structures can be channeled to flow across and stimulate postsynaptic membranes, and thereby they may initiate action potentials in the postsynaptic elements. This arrangement, known as electrical transmission, is relatively simple, and since synaptic delay is minimal, it is extremely rapid in its action. For this reason electrical transmission is utilized in animals, such as electric fish, which require virtually synchronous activation of electric organs distributed along the length of the body. The muscle cells which make up the ventricular muscle of the hearts of higher animals are also coupled electrically, but these junctions are not classified as synaptic. *See* ELECTRIC ORGAN (BIOLOGY); NERVOUS SYSTEM (VERTEBRATE).

[W.L.N.]

Synbranchiformes An order of eellike fishes that, unlike true eels, has the premaxillae present as distinct bones. The small gill apertures are often confluent across the breast. The gills are poorly developed, and in some species respiration is accomplished in part by highly vascularized buccopharyngeal pouches. There are no fin spines or pectoral fins, and the median fins, if developed, are continuous. Pelvic fins, if present, are small and located on the throat. The body may be naked or scaled. There is no swim bladder. The group, which has no fossil record, is classified into 2 suborders, 3 families, 7 genera, and 12 species. These serpentine fishes inhabit swamps, caves, and sluggish fresh and brackish waters of tropical America, Australia, eastern and southeastern Asia, the East Indies, and tropical Africa. *See* ACTINOPTERYGII. [R.M.B.]

Syncarida A superorder of the class Crustacea, subclass Malacostraca. There are only a few species of these higher crustaceans. They inhabit special regions such as subterranean wells and springs, as well as mountain lakes. Two orders are recognized, the Anaspidacea and Bathynellacea. A carapace and oostegites are lacking. Eyes may be pedunculate, sessile, or lacking. All thoracic limbs have exopodites, but none are chelate or subchelate. *See* ANASPIDACEA; BATHYNELLACEA; MALACOSTRACA. [H.J.]

Synchronization The process of maintaining one operation in step with another. The commonest example is the electric clock, whose motor rotates at some integral multiple or submultiple of the speed of the alternator in the power station. In television, synchronization is essential in order that the electron beams of receiver picture tubes are at exactly the same spot on the screen at each instant as is the beam in the television camera tube at the transmitter. *See* TELEVISION. [J.Mar.]

Synchronous capacitor A rotating ac machine running with no mechanical load. It is often called a synchronous condenser. These machines generally have salient poles, eight poles being typical. However, two-pole generators are sometimes disconnected from the turbine and operated as synchronous condensers. The purpose of these machines is to provide reactive power to the system or to receive some reactive power. The amount and sign of the reactive power depends on the field current supplied to the rotor. This is an effective method of controlling the voltage at that point in the power system. *See* ELECTRIC POWER MEASUREMENT.

Rotating synchronous condensers are sometimes used at the terminals of dc transmission lines where it has been found necessary to have a connected rotating capacity about 30% of the converter terminal rating. *See* DIRECT-CURRENT TRANSMISSION; SYNCHRONOUS MOTOR. [L.A.K.]

Synchronous converter A synchronous machine used to convert alternating current (ac) to direct current (dc), or vice versa. The ac-to-dc converter has been superseded by the mercury arc rectifier (for reasons of efficiency, lower maintenance costs, and less trouble) or by motor-generator sets. Converters are no longer manufactured, but there are converters still in use. *See* DIRECT-CURRENT GENERATOR; SYNCHRONOUS MOTOR. [L.V.B.]

Synchronous generator An alternating-current generator whose operating speed is proportional to system frequency. These generators usually have the field winding mounted on the rotor and a stationary armature winding mounted on the stator. In small ratings where high reliability is imperative, the magnetic field may be created by permanent magnets. Small alternators are also being built with stationary field windings and rotating armatures with their leads brought out through collector rings or fed directly to rotating rectifiers, which supply direct current to the rotating field of a much larger alternator, as in the brushless excitation system. Still another type, the inductor alternator, has both its field and armature windings in the stator. *See* INDUCTOR ALTERNATOR.

Although synchronous generators may be single-phase, they are usually two- or three-phase; most are three-phase. Single-phase generators are rare because they are larger than polyphase machines of the same kilovolt-ampere ratings, because they have a pulsating torque and are noisy, and because single-phase power is not well suited to self-starting ac motors other than those of fractional horsepower sizes. For theory of synchronous machines *see* SYNCHRONOUS MOTOR.

In the usual type with rotating field windings, a pair of field poles must pass a given point on the armature in 1 cycle. Hence the number of poles required is determined from the frequency f in hertz (Hz) and speed by

$$\text{Number of poles} = 120f/\text{rpm}$$

The speed selected is that best suited to the prime mover. Steam turbines operate most economically at high speed, hence two- or four-pole generators are used for this service, running at 3600 or 1800 rpm, respectively, for an output frequency of 60 Hz. For hydraulic turbine or engine drive, slow-speed machines having many poles are customary. *See* ALTERNATING-CURRENT GENERATOR. [L.T.R.]

Synchronous motor An alternating-current (ac) motor which operates at a fixed synchronous speed proportional to the frequency of the applied ac power. A synchronous machine may operate as a generator, motor, or capacitor depending only on its applied shaft torque (whether positive, negative, or zero) and its excitation. There is no fundamental difference in the theory, design, or construction of a machine intended for any of these roles, although certain design features are stressed for each of them. In use, the machine may change its role from instant to instant. For these reasons it is preferable not to set up separate theories for synchronous generators, motors, and capacitors. It is better to establish a general theory which is applicable to all three and in which the distinction between them is merely a difference in the direction of the currents and the sign of the torque angles. *See* ALTERNATING-CURRENT GENERATOR; ALTERNATING-CURRENT MOTOR; SYNCHRONOUS CAPACITOR. For special types of synchronous motors *see* HYSTERESIS MOTOR; RELUCTANCE MOTOR. [R.T.S.]

Synchroscope An instrument used for indicating whether two alternating-current (ac) generators or other ac voltage sources are synchronized in time phase with each other. In one type, for example, the position of a continuously rotatable pointer indicates the instantaneous phase difference between the two sources at each instant. In more modern synchroscopes, a cathode-ray tube serves as the indicating means.

The term synchroscope is also applied to a special type of cathode-ray oscilloscope designed for observing extremely short pulses, using fast sweeps synchronized with the signal to be observed. *See* ELECTRIC POWER GENERATION; OSCILLOSCOPE. [J.Mar.]

Synchrotron A device for accelerating charged particles (electrons or protons) by directing them along a roughly circular path of fixed radius in a magnetic guide field which varies cyclically and is perpendicular to the plane of the orbit. As the particles pass through rf (radio frequency) accelerating cavities placed along the orbit, their kinetic energy is increased. In this way the accelerating action can be applied repetitively to the same particles, thus multiplying their initial energy by factors of hundreds or thousands.

Electron synchrotron. Electron synchrotrons are used primarily for research in elementary particle physics, but have found increasing use as intense sources of ultraviolet and x-radiation in solid-state and materials science.

For normal injection energies, as a consequence of their low mass, the velocity of the electrons injected into a synchrotron is usually so close to the velocity of light that the change in velocity during acceleration is insignificant. This is in contrast to the situation in proton synchrotrons or cyclotrons where the velocity of the particles being accelerated may vary by a factor of more than 10 during acceleration. As a result, the frequency of revolution of electrons in the synchrotron is a constant, as is the frequency of the power applied to the accelerating cavities. Injection into electron synchrotrons is usually done from a preaccelerator such as a linac or microtron having an output energy of tens or hundreds of MeV.

The orbit radii of machines in use range from a few meters (1 m = 0.3 ft) for a synchrotron which achieves a few hundred MeV peak energy, to 100 m for the largest electron synchrotron, the 12-GeV instrument at Cornell University. Most modern synchrotrons employ the alternating-gradient or strong-focusing principle which minimizes the volume of the magnetic guide field, so that while the circumference of the orbit may be almost a kilometer, the cross-sectional dimensions of the evacuated volume surrounding the beam need be only a few centimeters.

In addition to the energy which must be supplied to the beam by the cavities in order to match the rising strength of the guide field, energy must also be supplied to replenish that which is emitted by the beam as "synchrotron radiation." This electromagnetic radiation is given off by all charged particles which follow a curved path, and is directed in a narrow beam tangent to the path. It is so intense that, for many experiments in atomic and solid-state physics and biology, synchrotron radiation is superior to that produced by ultraviolet lamps and x-ray tubes and is being exploited extensively for these purposes. Serious planning has been undertaken for synchrotrons totally devoted to such studies. *See* SYNCHROTRON RADIATION. [M.Ti.]

Proton synchrotron. The proton synchrotron developed from the cyclotron by way of the synchrocyclotron. This progression has resulted in proton accelerators with higher and higher beam energies. A 400-GeV accelerator is in operation in the United States, while several European countries have constructed a similar facility near Geneva, Switzerland.

The chief limitation on the energy of the synchrotron is the size and cost of the large magnet within which all the beam acceleration takes place. The incorporation of time-varying magnetic fields made it possible to overcome this limitation. The use of numerous but small magnets whose fields could track the proton beam energy led to the major breakthrough that made possible higher proton beam energies at reasonable cost.

The essential feature of a proton synchrotron is that the average radius of the accelerated beam is a constant. This is accomplished by a careful synchronization of the magnetic field and the accelerating radio frequency. The magnets are arranged in an approximate circle and are powered in such a way that the excitation current can be varied from low to high values during acceleration. As the momentum of the proton increases during acceleration, the magnetic field is increased in a manner that keeps the guide path of the protons nearly constant. This permits the bundle of protons to be contained within a relatively small vacuum chamber located within the magnets.

In the proton synchrotron the ion source and initial accelerator system are located away from the accelerator. As seen in the illustration, the proton beam is injected into the ring of magnets at a low energy. At the peak of energy it can be used on internal targets or it can be deflected and extracted from the accelerator to be used externally.

In order to hold the plasma of protons together, it is necessary to have some restoration force that keeps the protons within the vacuum chamber. On older "weak-focusing" acceler-

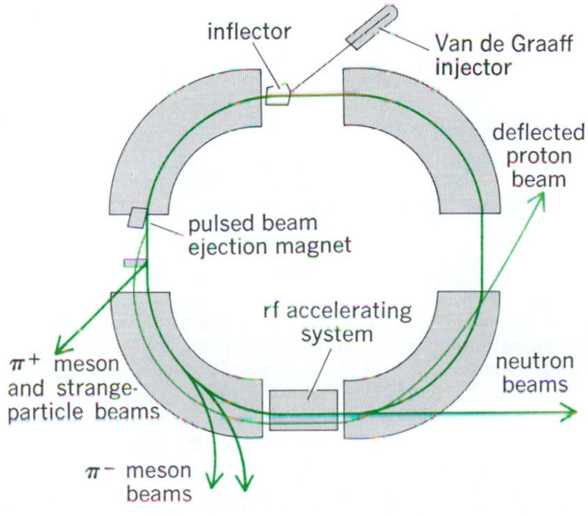

Schematic diagram of the principal components of a proton synchrotron. Note the various possibilities of external neutral beams, charged beams, and extracted primary beams.

ators, focusing of the beam and its maintenance in a circular path are accomplished simultaneously by the gentle shaping of the iron pole faces of the guide field magnets. In order to keep the volume of the magnetic field as small as possible, stronger focusing is needed to reach higher energies. At Fermilab this is accomplished by using quadrupole (four-pole) magnets that act like lenses in keeping particles together. Since their effect is not combined with that of the guide field dipoles which determine the mean orbit shape, the structure is called a separated-function strong-focusing accelerator. *See* PARTICLE ACCELERATOR. [J.R.S.]

Synchrotron radiation Electromagnetic radiation emitted by charged particles in circular motion at relativistic energies. Such emission of light by electrons, guided by celestial magnetic fields, has long been known to astronomers and is responsible, for example, for the beautiful background light in the Crab Nebula. With the construction of high-energy electron synchrotrons and storage rings for high-energy physics research, very powerful sources of synchrotron radiation in the ultraviolet and x-ray parts of the spectrum became available in many countries. *See* CRAB NEBULA; PARTICLE ACCELERATOR; SYNCHROTRON.

As a continuum source, synchrotron radiation produces five or more orders of magnitude higher intensity on an experimental sample than powerful rotating-anode x-ray tubes. The power is radiated in a smooth, featureless continuum without the spikes and structure associated with other sources. For most regions of the ultraviolet and x-ray spectrum, no other source of intense continuum radiation exists, and experiments must work only at the characteristic lines of gas-discharge lamps and x-ray tubes. The radiation is produced in a train of pulses which is different for each source. The pulse duration can be as short as 50 picoseconds, and the interval between pulses can be 1 microsecond or longer. [H.Win.; A.Bi.]

Syncline A fold in which the beds are inclined down and toward the axis. Synclines may be symmetrical, asymmetrical, overturned, or recumbent. Most have elongate trends, with axes that plunge from the extremities toward interior points along the axes. Others, called basins, have no distinct trend. In general, the stratigraphically younger beds are found toward the center of curvature, but in complexly deformed regions such simple concepts may not apply.

Stratigraphic synclines are those folds which, regardless of their observed forms, are inferred from stratigraphic data to

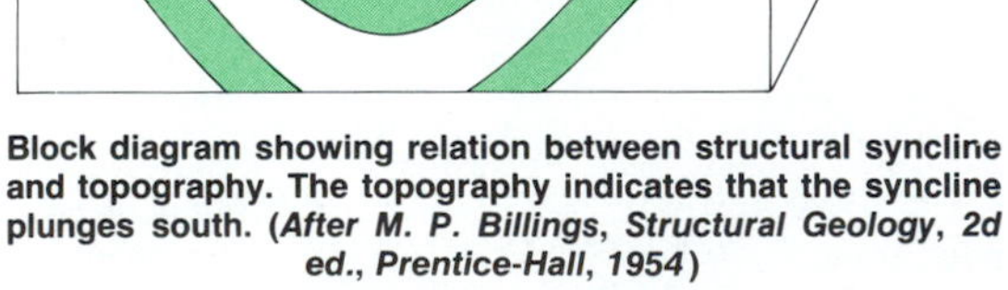

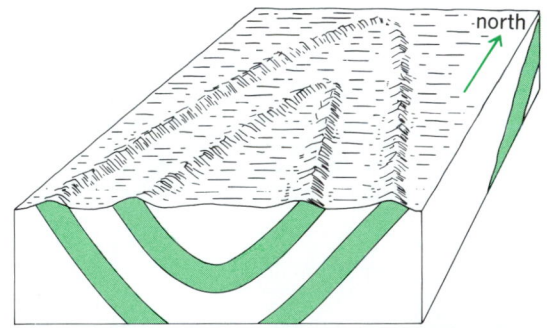

Block diagram showing relation between structural syncline and topography. The topography indicates that the syncline plunges south. (*After M. P. Billings, Structural Geology, 2d ed., Prentice-Hall, 1954*)

have been synclines originally. Structural synclines (see illustration) are those which have synclinal form regardless of their stratigraphic relations. *See* ANTICLINE. [P.H.O.]

Syngamy A process involving the union of sexual cells, or gametes. Customarily these are eggs (female) and spermatozoa (male), and syngamy involves the fusion of a single spermatozoon with a single egg. As such, syngamy is synonymous with fertilization. Accompanying the union of sexual cells are rather profound physiological and structural changes within the egg, but the process is considered terminated when a fusion of the gametic nuclei takes place.

Syngamy is essentially the same in all sexual organisms, although the length of the diploid and haploid phases can vary widely in relation to the entire life cycle. In many lower forms, however, it is not always possible to speak of syngamy as being the union of a spermatozoon and an egg, since the gametes are similar in appearance and undifferentiated as to sex. *See* CELL DIVISION. [C.P.Sw.]

Synthetic fuel A gaseous, liquid, or solid fuel that does not occur naturally; also known as synfuel. Synthetic fuels can be made from coal, oil shale, or tar sands. Included in the category are various fuel gases, such as substitute natural gas and synthesis gas. *See* COAL; OIL SAND; OIL SHALE.

Syncrude is a synthetic crude oil, a complex mixture of hydrocarbons somewhat similar to petroleum. It is obtained from coal (liquefaction), from synthesis gas (a mixture of carbon monoxide and hydrogen), or from oil shale and tar sands. Syncrudes generally differ in composition from petroleum; for example, syncrude from coal usually contains more aromatic hydrocarbons than petroleum. Gaseous fuels can be produced from sources other than petroleum and natural gas. *See* PETROLEUM.

The most important source of synthetic crude oil is the tar sand deposit that occurs in northeastern Alberta, Canada. Tar sand is a common term for oil-impregnated sediments that can be found in almost every continent. The routes by which synthetic fuels can be prepared from coal involve either gasification or liquefaction.

Gasification can yield clean gases for combustion or synthesis gas, which has a controlled ratio of hydrogen to carbon monoxide. Catalytic conversion of synthesis gas to liquids (indirect liquefaction) can be carried out in fixed- and fluidized-bed reactors and in dilute-phase systems. Another method for producing synthetic fuels from coal involves gasification of the coal to a fuel gas that may be used as such or as a source of synthetic liquids. *See* COAL GASIFICATION; FLUIDIZATION.

Coal liquefaction is accomplished by four principal methods: direct catalytic hydrogenation, solvent extraction, pyrolysis, and indirect catalytic hydrogenation (of carbon monoxide). *See* COAL LIQUEFACTION; HYDROGENATION; PYROLYSIS; SOLVENT EXTRACTION.

Shale oil is readily produced by the thermal processing of oil shales. The basic technology is available, and commercial plants are operated in many parts of the world. [J.G.S.]

Syphilis A subacute and chronic infectious disease of humans caused by a spirochete, *Treponema pallidum*, and transmitted principally by sexual intercourse, but occasionally by direct nonsexual contact or from a pregnant woman to her fetus as in congenital syphilis. With the introduction of penicillin, the prevalence of syphilis declined, but in no country is it a negligible health problem.

The disease evolves in a stepwise manner. The primary or initial lesion develops at the site of implantation of treponemes within 1–6 weeks. Characteristically, this lesion, also referred to as the hard chancre, is indurated, often like cartilage, and the lymph node draining the area is enlarged. The primary lesion is most often located on the genitalia; however, in about 10% of cases the primary lesion is extragenital, most commonly on the lips or fingers.

Treponemes are soon widely disseminated by the lymphatics and blood, giving rise after an interval of 3–8 weeks to generalized lesions characteristic of the secondary stage of the disease. Most prominent in this stage are treponeme-containing lesions of the skin and mucous membranes of the lips, mouth, and genitalia. Foci of spirochetal multiplication are present in many organs and may give rise to symptoms of arthritis, iritis, meningitis, nephritis, and hepatitis. Low-grade fever, loss of appetite, and general malaise are common.

Even without treatment, lesions of the secondary stage eventually subside, and the disease enters a latent phase in which infection persists but no lesions are detectable. The latent phase may persist for the rest of the individual's life, but without treatment it is probable that lesions of the tertiary stage will develop in the skin, central nervous system, heart, and great vessels or liver. Formerly, syphilis was an important cause of insanity and cardiovascular disease. *See* SEXUALLY TRANSMITTED DISEASES. [T.B.T.]

Systellommatophora A superorder in the subclass Pulmonata containing three families (Rathouisiidae, Veronicellidae, and Onchidiidae) of sluglike mollusks that lack any trace of a shell, have separate external male and female orifices, lack a mantle cavity, have a posterior anus and excretory pore, and bear eyes on the tops of two contractile, but not retractile, tentacles. *See* PULMONATA. [G.A.S.]

Systems analysis The application of mathematics to the study of systems. The term "operations research" is reserved for the study of part of a system. *See* OPERATIONS RESEARCH.

The basic idea is that a mathematical model of the system under study is constructed, a mathematical analysis is done of the mathematical model, and the results of this analysis are applied to the original system.

A great deal of experience is needed to construct the mathematical model and to interpret the results of the analysis. The mathematical analysis, usually involving a computer, is seldom routine. The procedures are so complex that often an individual systems analysis is required for the systems analysis of a particular system. Often particular parts of the process are carried out by different people. Ideally, there should be an interaction between the mathematical analysis, the construction of the mathematical model, and the interpretation of the results.

Systems analysis is thus a part of applied mathematics. What makes the difference between this and conventional applied mathematics is that the systems studied often involve human beings. The presence of humans and the application of the results to human systems introduce a great deal of complication.

There are three principal difficulties in the application of conventional mathematical techniques: dimensionality, the presence of both "hard" and "soft" variables, and conflicting objectives. Dimensionality refers to the possibility that the description of a system may involve many state variables. This feature presents serious difficulties, but they may be circumvented by means of various techniques. Much more serious is the presence of hard and soft variables and various combinations. Thus, for example, in a study of a business, the number of employees is a hard variable, while the quality of management is a soft variable. Hard and soft variables may be handled by the theory of fuzzy systems. In most situations, there are many objectives. In many cases, these objectives are in partial or complete conflict. Thus, it is impossible to write down a single criterion for performance. Fortunately, these questions can be studied by means of simulation. Simulation allows various possibilities to be studied in electronic time without disturbing the system or the people affected by the system. *See* DECISION THEORY; LINEAR SYSTEM ANALYSIS; OPTIMIZATION; SIMULATION; SYSTEMS ENGINEERING. [R.Be.]

Systems ecology The combined approaches of systems analysis and the ecology of whole ecosystems and subsystems, when viewed as interdependent and functionally interacting components. Among the organisms and organs in any subdivision of vegetation or animal communities or microorganisms, there are many relations from one part to another and from each to the local environment. Physiological functions and other relations can sometimes be expressed in terms of mathematical functions. These need not be quantitative functions (one-to-one correspondence for real numbers) but may express logical alternatives (all-or-nothing) in the direction of flow of material or energy; they may express also alternative pathways of development or direction of change in the organic or inorganic parts of the system. *See* ECOLOGY; ECOSYSTEM; PHYSIOLOGICAL ECOLOGY (PLANT); SYSTEMS ANALYSIS. [J.S.O.]

Systems engineering The design of a complex interconnection of many elements (a system) to maximize an agreed-upon measure of system performance. Systems engineering, also referred to as system engineering, includes two parts: modeling, in which each element of the system and the criterion for measuring performance are described; and optimization in which adjustable elements are set at values that give the best possible performance.

The systems approach can be applied to problems ranging from the very simple to those so complex that the human mind is unable to comprehend the reasons for system behavior. A simple problem is the scheduling of the preparation of a family meal. A much more complex system problem is the control of the timing of hundreds or thousands of traffic lights in a city.

The techniques of systems engineering have been applied to a tremendous range of current problems, from industrial automation to control of weapons and space vehicles.

Modeling refers to the determination of a quantitative picture of the important system characteristics. This model may be in the form of collected data or analytical studies, or it may be a representation of the important system characteristics in a laboratory or computer simulation of the actual system.

A specific and rather narrow systems problem is the control of traffic flow through a tunnel. The designer has to maximize the number of cars permitted per hour through the tunnel. To do this, the designer can vary the speed limit through the tunnel and control the number of cars within the tunnel (or the rate at which cars enter). Before a system design can be decided on, measurements are made of the way in which cars move under various speeds and degrees of congestion. These data are combined with the known behavior characteristics of a driver. On the basis of this quantitative information, a computer simulation of the system can be constructed from which the engineer can study the effects of different operating rules, the results of a breakdown in one lane of the tunnel, or the effects of a driver who stays below the minimum speed limit. *See* SIMULATION.

The engineer attempts to design a system that is optimum according to a quantitative criterion which measures the quality of performance. For example, in the tunnel traffic problem, the entry and speed of cars are controlled in such a way that the tunnel handles the maximum possible number of cars per hour. In many situations, the criterion is probabilistic, and the engineer must optimize on the basis of probable behavior of the system. The routing of emergency vehicles through a network of city streets is a typical example. *See* OPTIMIZATION.

Three technological developments have enormously broadened the scope of problems amenable to the systems approach. The first of these is the understanding of the principles of automatic control (or the principles underlying automation). Systems can be designed in which the desired automatic control is realized by the appropriate choice of input signals and system configuration (including the use of feedback, with the actual system response compared automatically with the desired response and the error used to modify the response toward the desired value). In other words, technology has developed automatic, goal-seeking systems to replace the human role. The automatic control of elevators in a busy office building is a familiar example. The second is the revolution in communications. Enormous quantities of data can be transmitted over great distances with nearly perfect fidelity. The third fundamental technological development has been the high-speed, electronic digital computer. The computer lies at the core of modem systems engineering, since it permits both the modeling and the optimization of systems vastly more complex than the human being alone can handle. *See* AUTOMATION; CONTROL SYSTEMS; DIGITAL COMPUTER. [J.G.Tr.]

Systems integration A discipline that combines processes and procedures from systems engineering, systems management, and product development for the purpose of developing large-scale complex systems. These complex systems involve hardware and software and may be based on existing or legacy systems coupled with new requirements to add significant added functionality. Systems integration generally involves combining products of several contractors to produce the working system. Systems integration applications range from creation of complex inventory tracking systems to designing flight simulation models and reengineering large logistics systems.

Life-cycle activities. Application of systems integration processes and procedures generally follows the life cycle for systems engineering. Minimally, these systems engineering life-cycle phases are requirements definition, design and development, and operations and maintenance. For systems integration, these three phases are usually expanded to include feasibility analysis, program and project plans, logical and physical design, design compatibility and interoperability tests, reviews and evaluations, and graceful system retirement.

Primary uses. Systems integration is essential to the design and development of information systems that automate key operations for business and government. It is required for major procurements for the military services and for private businesses.

Advantages. Systems integration approaches enable early capture of design and implementation needs. The interactions and interfaces across existing system fragments and new requirements are especially critical. It is necessary that interface and intermodule interactions and relationships across components and subsystems that bring together new and existing equipment and software be articulated. The systems integration approach supports this through application of both a top-down and a bottom-up design philosophy; full compliance with audit-trail needs, system-level quality assurance, and risk assessment and evaluation; and definition and documentation of all aspects of the program. It also provides a framework that incorporates appropriate systems management application to all program aspects. A principal advantage of this approach is that it disaggregates large and complex issues and problems into well-defined sequences of simpler problems and issues that are easier to understand, manage, and build. *See* INFORMATION SYSTEMS ENGINEERING; RISK ANALYSIS; SYSTEMS ANALYSIS; SYSTEMS ENGINEERING. [J.D.P.]

Syzygy The alignment of three celestial objects within a solar system (or within any other system of objects in orbit about a star). Syzygy is most often used to refer to the alignment of the Sun, Earth, and Moon at the time of new or full moon. Alignments need not be perfect in order for syzygy to occur: because the orbital planes for any three bodies in the solar system rarely coincide, the geometric centers of three objects that are in syzygy almost never lie along the same line. *See* PHASE; SOLAR SYSTEM.

In general, syzygy occurs whenever an observer on one of the three objects would see the other two objects either in opposition or in conjunction. Opposition occurs when two objects appear 180° apart in the sky as viewed from a third object. Conjunction occurs when two objects appear near one another in the sky as seen from a third object.

Solar and lunar eclipses are dramatic results of syzygy. During a solar eclipse, when the Moon is in its new phase, the alignment of the Sun, Earth, and Moon is so nearly perfect that the Moon's shadow falls on the Earth. During a lunar eclipse, which occurs at the time of the full moon, the Moon passes through the Earth's shadow. *See* ECLIPSE.

An occultation is another type of eclipse that can occur during syzygy. For an Earth-based observer, an occultation occurs when the Moon is seen to pass in front of a planet or other member of the solar system. The occultation of a star by the Moon does not qualify as syzygy, since the star is far beyond the limits of the solar system. *See* OCCULTATION. [H.P.C.]

T

T Tauri star A member of a class of very young, optically visible, solar-mass stars with peculiarities such as variability and evidence for mass loss. T Tauri stars were discovered through their unusually strong emission lines. Many radiate unexpectedly intensely at infrared and ultraviolet wavelengths. Two subclasses have been defined, the classical T Tauri stars, identified from hydrogen-emission-line surveys, and the weak-line, or naked, T Tauri stars, discovered through their x-ray emission. Most of the apparently anomalous properties of T Tauri stars can be attributed to the fact that many are still surrounded by remnants of their parent clouds of gas and dust. *See* VARIABLE STAR.

The youth of the T Tauri stars was originally suspected because of their association with star-forming clouds. Their erratic brightness variations also indicated that they had not yet become stable. Ages of 100,000–10,000,000 years are confirmed by their effective temperatures and luminosities.

The unusually strong emission lines observed in the spectra of classical T Tauri stars often have the extended wings characteristic of significant mass outflow. Outflow velocities are high, typically 100 km/s (60 mi/s). In optical images, high-velocity, oppositely directed jets can often be seen emanating from the poles of the stars themselves. Although they are already visible, T Tauri stars are evidently still in the process of shedding the dust and gas from which they formed. From theories about how stars form, the remnant material is expected to be distributed in disks. Such disks are the intrinsic source of the excess infrared radiation. The excess ultraviolet emission, as well as the very strong emission lines, may derive from the boundary layer between the rapidly rotating disk and the more slowly rotating star.

The T Tauri disks are similar to the primitive solar nebula, before the planets formed. If, as seems likely, the Sun experienced a T Tauri phase in its early history, T Tauri stars may be the birth sites of other planetary system. *See* PROTOSTAR; SOLAR SYSTEM; STELLAR EVOLUTION. [A.I.S.]

Tabulata An extinct Paleozoic order of corals of the subclass Zoantharia in the class Anthozoa. The Tabulata are characterized by their exclusively colonial mode of growth (see illustration) and by secretion of a calcareous exoskeleton of slender tubes crossed by many transverse partitions called tabulae. Septa are of one order and are inconspicuous, represented by a row of spines, or even absent; 12 is a common number. Colony form varies widely, dependent in part on the ecological niche. *See* ZOANTHARIA. [D.H.]

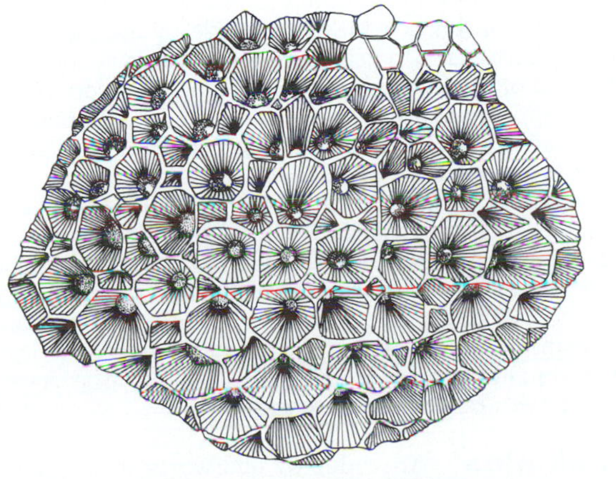

***Michelinia convexa*, seen from above, from the Corniferous Limestone of Ontario. (*After Billings*)**

Tacan A military rho-theta navigation system which allows the distance-measuring equipment (DME) to also provide bearing service without the large antennas or site errors characteristic of the civil very-high-frequency omnidirectional range (VOR). Range and accuracy are the same as DME (300 mi or 480 km, and 0.1 mi or 0.16 km, respectively), with a bearing accuracy of 1°.

To provide the added bearing service, the DME transponder is first arranged to operate at constant duty cycle. This means that the number of output pulses is held constant, whether the beacon is being interrogated by one or a hundred aircraft.

The total output of the transponder is amplitude-modulated by the rotating directional antenna system (see illustration). At the center of this system is the central radiator connected to the DME transponder, just as in the conventional DME. However, rotating around this radiation at 15 revolutions per second are two concentric dielectric cylinders. The inside one contains a single parasitic reflector which imparts a 15-Hz amplitude modulation to the DME replies, and the outside one contains nine parasitic elements which impart a 135-Hz amplitude modulation. On the same rotating shaft are mounted reference pulse generators which additionally modulate the transmitter with

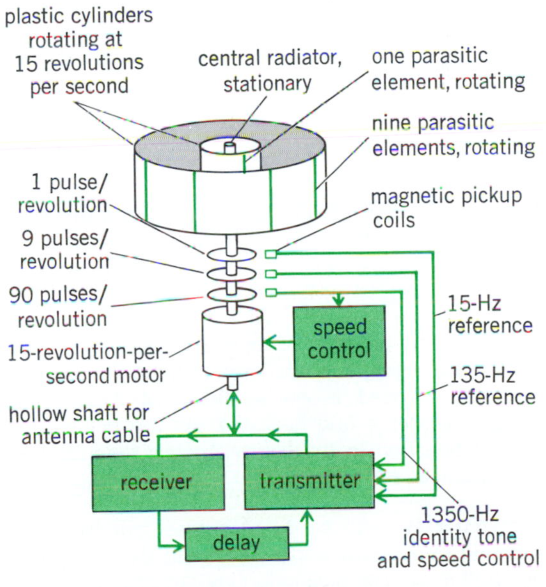

Tacan transponder with rotating directional antenna system.

coded pulses, once per revolution for the 15-Hz signal (called the north reference burst) and nine times per revolution for the 135-Hz signal (called the auxiliary reference bursts).

In the airborne receiver the 15- and 135-Hz sine waves are detected and filtered and compared with the decoded reference bursts to provide a two-speed or fine-coarse bearing display whose accuracy and site freedom are superior to those of conventional VOR, yet with a ground antenna system which is small enough to mount on a ship's mast. *See* DISTANCE-MEASURING EQUIPMENT; ELECTRONIC NAVIGATION SYSTEMS; RADIO RANGE; RHO-THETA SYSTEMS. [S.H.D.]

Tachometer An instrument that measures angular speed, as that of a rotating shaft. The measurement may be in revolutions over an independently measured time interval, as in a revolution counter, or it may be directly in revolutions per minute. The instrument may also indicate the average speed over a time interval or the instantaneous speed. Tachometers are used for direct measurement of angular speed and as elements of control systems to furnish a signal as a function of angular speed. [A.H.W.]

Tachyon A hypothetical faster-than-light particle consistent with the special theory of relativity. According to this theory, a free particle has an energy E and a momentum $\mathbf{p}$ which form a Lorentz four-vector. The length of this vector is a scalar, having the same value in all inertial reference frames. One writes Eq. (1), where c is the speed of light and the para-

$$E^2 - c^2\mathbf{p}^2 = m^2c^4 \qquad (1)$$

meter m^2 is a property of the particle, independent of its momentum and energy. Three cases may be considered: m^2 may be positive, zero, or negative. The case $m^2 > 0$ applies for atoms, nuclei, and the macroscopic objects of everyday experience. The positive root m is called the rest-mass. If $m^2 = 0$, the particle is called massless. A few of these are known: the electron neutrino, the muon neutrino, the photon, and the graviton. The name tachyons (after a Greek word for swift) has been given to particles with $m^2 < 0$.

In general, the particle speed is given by Eq. (2). If $m^2 < 0$,

$$v = \frac{c\mathbf{p}}{E}\,c \qquad (2)$$

Eq. (1) implies $E > c\mathbf{p}$ and Eq. (2) gives $v < c$. If $m^2 = 0$, then $E = c\mathbf{p}$ and $v = c$. In case $m^2 < 0$, one finds $E < c\mathbf{p}$ and $v > c$. Tachyons exist only at faster-than-light speeds. *See* ELEMENTARY PARTICLE; RELATIVITY. [R.H.Go.]

Taconite The name given to the siliceous iron formation from which the high-grade iron ores of the Lake Superior district have been derived. It consists chiefly of fine-grained silica mixed with magnetite and hematite. As the richer iron ores approach exhaustion in the United States, taconite becomes more important as a source of iron. [C.S.Hu.]

Taeniodonta An order of extinct quadrupedal land mammals known from early Cenozoic deposits in North America and Asia. The stylinodontine taeniodonts were adaptively convergent with ground sloths. Stylinodontines developed into cow-sized animals with deep massive lower jaws, a reduced number of peglike upper and lower teeth, and large flattened claws.

Conoryctine taeniodonts were cat-sized to sheep-sized and had less specialization in dentition and skeleton. Conoryctines developed enlarged canines, but the lower jaws were unspecialized and the check teeth, though simplified, were low-crowned, enamel-enclosed, and cuspidate. *See* MAMMALIA. [D.E.S.]

Taiga A zone of forest vegetation encircling the Northern Hemisphere between the arctic-subarctic tundras in the north and the steppes, hardwood forests, and prairies in the south. The chief characteristic of the taiga is the prevalence of forests dominated by conifers. The dominant trees are particular species of spruce, pine, fir, and larch. Other conifers, such as hemlock, white cedar, and juniper, occur locally, and the broad-leaved deciduous trees, birch and poplar, are common associates in the southern taiga regions. Taiga is a Siberian word, equivalent to "boreal forest." *See* TUNDRA.

The northern and southern boundaries of the taiga are determined by climatic factors, of which temperature is most important. However, aridity controls the forest-steppe boundary in central Canada and western Siberia. In the taiga the average temperature in the warmest month, July, is greater than 50°F (10°C), distinguishing it from the forest-tundra and tundra to the north; however, less than four of the summer months have averages above 50°F (10°C), in contrast to the summers of the deciduous forest further south, which are longer and warmer. Taiga winters are long, snowy, and cold—the coldest month has an average temperature below 32°F (0°C). Permafrost occurs in the northern taiga. It is important to note that climate is as significant as vegetation in defining taiga. Thus, many of the world's conifer forests, such as those of the American Pacific Northwest, are excluded from the taiga by their high precipitation and mild winters. [J.C.Ri.]

Tail assembly The assembly of the vertical tail, the horizontal tail, and a small section of the rear of the fuselage (see illustration).

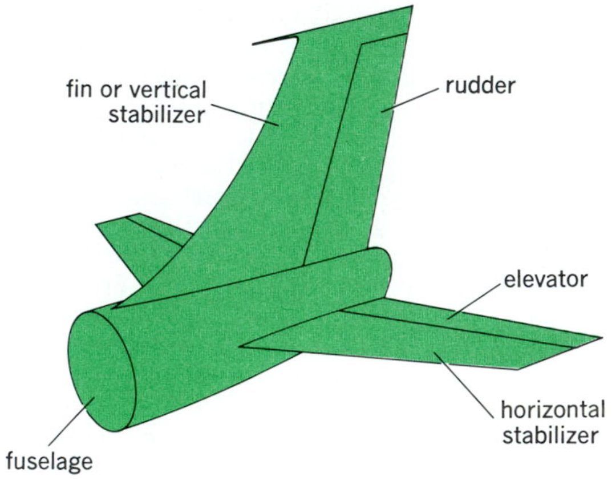

Principal parts of tail assembly.

The vertical tail consists of a fin which is fixed to the fuselage or body and a rudder which is movable by the pilot. The fin (fixed portion) is a symmetrical airfoil which is in line with the center line of the fuselage. In steady flight the rudder is stationary and approximately in line with the fin. The purpose of the vertical tail is to keep the airplane in line with the direction of motion or relative wind velocity.

The span of the horizontal tail lies in a horizontal plane and usually consists of a stabilizer and elevator. The stabilizer and elevator are symmetrical airfoil sections whose usual position is not in line with the fuselage center line. The principal purpose of the horizontal tail is to provide a balancing tail load so that the weight of the airplane, the lift, and the horizontal tail load are in equilibrium for a specific airspeed. *See* AIRPLANE; ELEVATOR (AIRCRAFT); STABILIZER (AIRCRAFT). [R.G.Bo.]

Takakiales An order of liverworts in the subclass Jungermanniidae, consisting of a single genus and two species.

Some authors put the Takakiales in the Calobryales owing to branching from a prostrate branched stem, lack of rhizoids, copious mucilage secretion, and massive frequently scattered archegonia that lack protective envelopes. However, sporophytes have never been seen, and so it is difficult to demonstrate any meaningful relationship.

The members of the Takakiales consist of a very small gametophyte made up of a branched system of prostrate, leafless stolons and erect, radially organized branches with terete appendages. The "leafy" branches are simple or forked and have a weak central strand of narrow cells enclosed by larger cells. The leafy appendages are small and scalelike below, larger and crowded above, and variable in arrangement. The leaf cells lack oil bodies. Mucilage is secreted by simple filaments on leafy branches and by branched filaments clustered on both leafy and stoloniform branches. The archegonia are scattered and not enclosed by protective structures. *See* BRYOPHYTA; CALOBRYALES. [H.Cr.]

Talc A hydrated magnesium silicate with composition close to $Mg_3Si_4O_{10}(OH)_2$. The mineral is commonly pale green, white, or grayish, with a greasy feel and pearly luster. It occurs in foliated masses, fibrous forms, and also in massive form, when it is commonly called soapstone. The mineral is mechanically soft (hardness 1–1½ on Mohs scale).

The major talc-producing states in the United States are Vermont, Texas, Montana, and New York, and the major talc-producing countries are Japan, the United States, the Soviet Union, Korea, China, France, and Brazil.

Talc is widely used in ceramics, paints, plastics, cosmetics, paper, rubber, and in a variety of other applications. Massive talc, cut into slabs, is used for laboratory tables, sinks, sanitary appliances, acid tanks, electrical switchboards, mantels, and hearthstones. It is also used in house-insulating materials. *See* SILICATE MINERALS. [G.W.Br.]

Talus A heap or sloping sheet of loose rock waste at the base of a cliff or steep slope. Talus and its English equivalent, scree, are terms properly applied to the entire form and not to the fragmental material itself, which is rock waste or sliderock. Where the supply of rock waste is funneled downward through a notch or channel, a conelike talus usually develops.

The upper part of a talus is characteristically steep, consists of coarse, angular rock waste, and is easily set in motion by falling blocks or by climbers. In cold climates its interstices may contain snow or ice for much of the year. The lower part is usually less steep and may be partly filled with finer rock debris and soil on which vegetation can gain a foothold. *See* LANDSLIDE. [C.F.S.S.]

Tanaidacea An order of the eumalacostracans of the superorder Peracarida, derived from the genus *Tanais*. These animals have a worldwide distribution and with few exceptions are marine. They occur from the shore down to abyssal depths. They are free-living and benthonic. The order is divided into 2 suborders with 5 families, 44 genera, and about 350 species. The body is linear, more or less cylindrical or dorsoventrally depressed. Thoracic segments 1 and 2 are fused with the head, and the last abdominal segment is fused with the telson. Eight pairs of thoracic legs are present, of which the first pair are maxillipeds, the second pair chelipeds (the first pair of pereiopods), and the following six pairs pereiopods. *See* PERACARIDA. [K.L.]

Tandetron A constant-voltage (dc) particle acceleration system which produces a directed stream of high-energy atoms. The species of these atoms can be from most parts of the periodic table. These accelerators are used to analyze the composition of materials and their structure, and to change the physical and electrical properties of materials.

Most analyses are accomplished by impinging the fast atoms onto a sample and by detecting the energy of those atoms that are scattered in the direction opposite to that of the incident high-velocity ions. Another analytical procedure involves nuclear reactions that can be induced between atoms in the incoming stream and trace nuclei present in the sample. Both methods are used to measure the depth distribution of specific impurities with very high sensitivity. In addition, the physical and electrical properties of materials, including the semiconductors silicon and gallium arsenide, can be dramatically changed by injecting energetic atoms from the tandetron directly into the crystal lattice. *See* ACTIVATION ANALYSIS; ION IMPLANTATION; ION-SOLID INTERACTIONS.

The acceleration principle used to produce these streams of high-energy atoms is based upon a tandem acceleration technique. A unique feature of this acceleration scheme is that the ion source is located at ground potential outside of the high-pressure enclosure, rather than within a high-voltage terminal where electrical sparks can badly damage the equipment. Also, in this scheme, the ion source is always available for adjustment and maintenance without removing pressurized gas from the enclosing tank.

The accelerator system has a very low production rate for nuclear radiations and x-rays. This feature makes it possible to use the tandetron in an ordinary analytical laboratory without the need for concrete shielding blocks or walls. *See* BREMSSTRAHLUNG; X-RAYS.

The 1-MV accelerating voltage is developed by solid-state rectification of a 50-kHz radio-frequency voltage. The solid-state device substitutes for the mechanical charge transport system used in many electrostatic accelerators. The design of this power supply is based on the dynamitron principle, an arrangement which allows over 100 milliamperes of charging current to be reliably produced at voltages as high as several million volts. *See* DYNAMITRON ACCELERATOR; PARTICLE ACCELERATOR; SEMICONDUCTOR RECTIFIER. [K.H.P.]

Tangent A term describing a relationship of two figures (usually of the same dimension) in the neighborhood of a common point. The figures are tangent at a point P if they touch at P but do not intersect in a sufficiently small neighborhood of P.

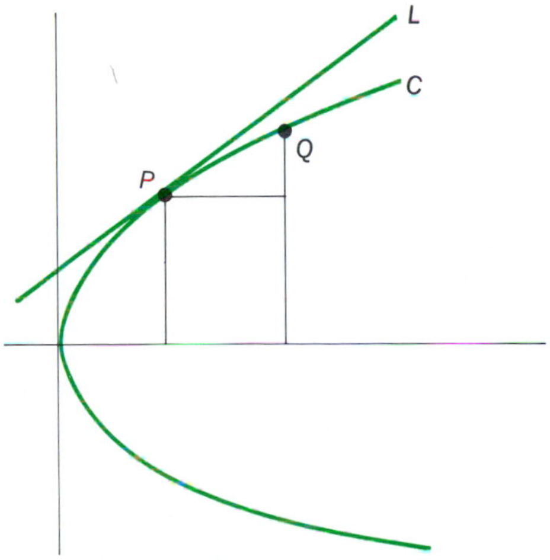

Line L tangent to curve C at P.

To be more precise, if P denotes a point of a curve C (see illustration), a line L is a tangent to C at P provided L is the limit of lines joining P to a variable point Q of C, as Q approaches P along C (that is, for Q sufficiently close to P, the line PQ is arbitrarily close to L). [L.M.Bl.]

Tangerine A name applied to certain varieties of a variable group of loose-skinned citrus fruits belonging to the species *Citrus reticulata*. Although mandarin and tangerine are often used interchangeably to designate the whole group, tangerine is applied more strictly to those varieties (cultivars) having deep-orange or scarlet rinds, whereas the term mandarin is more properly used to include all members of this quite variable group of citrus fruits. *See* MANDARIN.

The fruits are deep orange, loose-skinned, and small to medium-sized, and possess small seeds with green cotyledons. Tangerines are easily peeled and eaten out of hand as fresh fruit. About one-third of the crop is utilized in juice, sherbets, and canned sections. [F.E.G.; C.J.H.]

Tank circuit An inductor and capacitor in parallel. The term is quite often used to denote the parallel resonant circuit in the output stage of a radio transmitter, but it has been applied to any parallel resonant circuit. In many cases the inductance in the tank circuit is one winding of a two-winding, air-core transformer. The secondary is connected to some load, such as an antenna. Power is delivered from the source to the load through the tank circuit, with an effort usually made to adjust the parameters for maximum power transfer. *See* IMPEDANCE MATCHING; RESONANCE (ALTERNATING-CURRENT CIRCUITS); TRANSFORMER. [H.F.K.]

Tannin A generic term for a widely occurring group of substances of vegetable origin, capable of rendering raw hides into leather. Common tannin (tannic acid) occurs in oak gallnuts; tannins are also present in tea, sumac, oak bark, and mangrove bark.

Substances capable of tanning, and hence called tannins, are often of greatly different chemical structure; all tannins, however, have the property of converting the gelatin of hides into insoluble nonputrefying material, thus changing the hide into leather.

Tannins may be divided into three main classes: (1) condensed tannins that cannot be hydrolyzed either by acids or enzymes; (2) hydrolyzable tannins; and (3) tannins of unclassified nature. [E.B.R.]

Tantalite A mineral with composition $(Fe,Mn)Ta_2O_6$, an oxide of iron, manganese, and tantalum. Niobium substitutes for tantalum in all proportions; a complete series extends to columbite $(Fe,Mn)Nb_2O_6$. Pure tantalite is rare. Tantalite is common in short prismatic crystals. The hardness is 6 on Mohs scale, and the specific gravity 7.95 (pure tantalite). The luster is submetallic and the color iron black. Tantalite is the principal ore of tantalum. The chief producing areas are the Congo and Nigeria. *See* NIOBIUM; TANTALUM. [C.S.Hu.]

Tantalum A chemical element, symbol Ta, atomic number 73, and atomic weight 180.948. It is a member of the vanadium group of the periodic table and is in the 5*d* transitional series. Oxidation states of IV, III, and II are also known. *See* TRANSITION ELEMENTS.

Tantalum metal is used in the manufacture of capacitors for electronic equipment, including citizen band radios, smoke detectors, heart pacemakers, and automobiles. It is also used for heat-transfer surfaces in chemical production equipment, especially where extraordinarily corrosive conditions exist. Its chemical inertness has led to dental and surgical applications. Tantalum forms alloys with a large number of metals. Of special importance is ferrotantalum, which is added to austentitic steels to reduce intergranular corrosion.

The metal is quite inert to acid attack except by hydrofluoric acid. It is very slowly oxidized in alkaline solutions. The halogens and oxygen react with it on heating to form the oxidation-state-V halides and oxide. At high temperature it absorbs hydrogen and combines with nitrogen, phosphorus, arsenic, antimony, silicon, carbon, and boron. Tantalum also forms compounds by direct reaction with sulfur, selenium, and tellurium at elevated temperatures. [E.M.L.]

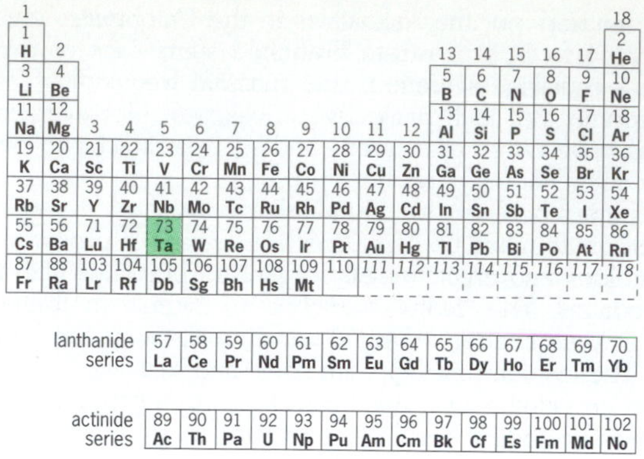

Tantulocarida A subclass of the Maxillopoda. Tantulocarids are minute ectoparasites, less than 0.01 in. (0.3 mm) in length, that infest deep-sea copepods, isopods, tanaids, and ostracods. As a result of their parasitic mode of life, adult females have lost all resemblance to crustaceans; males are free-living but nonfeeding. Infection of the host occurs at the tantalus larval stage, which is believed to occur immediately after the larva is released. The head, or cephalon, of the tantulus larva is covered by a dorsal shield that may protrude anteriorly to form a rostrum. The mechanism of attachment, the mouth tube, is located on the ventral surface with a distal mouth opening, and is surrounded by an oral disk.

Phylogenetic relations of the Tantulocarida are not clearly understood. Although tantulocarids lack the antennules and antennae characteristic of Crustacea, two clusters of sensory hairs in adult males are probably antennulary in origin. The basic tagmosis exhibited and male gonopore position suggest an affinity with some cirriped taxa. Some researchers include the Tantulocarida as a subclass of the Maxillopoda; others accept it as a distinct class. *See* CIRRIPEDIA; COPEPODA; CRUSTACEA; MAXILLOPODA. [P.A.McL.]

Taper pin A tapered self-holding pin used to connect parts together. Standard taper pins have a diametral tape ¼ in. to 12 in. (0.6 cm to 30 cm) and are driven in holes drilled and reamed to fit. The pins are made of soft steel or are cyanide-hardened. They are sometimes used to connect a hub or collar to a shaft. Taper pins are frequently used to maintain the location of one surface with respect to another. A disadvantage of the taper pin is that the holes must be drilled and reamed after assembly of the connected parts; hence they are not interchangeable. *See* COTTER PIN. [P.H.B.]

Taper pipe thread gage A type of gage whereby size is determined by measuring the distance that a plug gage will screw into a coupling or that a ring gage will screw onto a pipe. A manufacturer of pipe or couplings has a master plug and master ring gage. Standards for pitch diameter and taper of a pipe thread were first set to control threads used in petroleum service, where safety depended on accuracy of the thread dimensions. *See* RING GAGES. [R.A.Bo.]

Tapeworm disease The presence in the body of humans of either the larval or adult stages of any of the following cestodes: *Hymenolepis nana*, *Taenia saginata*, *T. solium*, *Dibothriocephalus latus*, or *Echinococcus granulosus*. *See* CESTOIDEA.

Hymenolepis nana is a cosmopolitan parasite that occurs in rodents. The cycle is direct, human to human or rodent to human. Ingestion of the egg results in development of larvae, called cysticercoids, in cysts in the duodenum. After the larvae burst out of the cyst, they attach to the small intestine and become adults. The entire cycle takes a month. Heavy infection may result in nervous disorders.

Taenia saginata, the beef tapeworm, is acquired by ingesting the larval form of the parasite *Cysticercus bovis*, which grows in the intramuscular connective tissues of herbivores. The adult, attached to the small intestine, may grow to a length of many feet. The gravid segments containing the eggs are motile and disengage from the rest of the tapeworm to pass actively through the anus. No symptoms may be exhibited by parasitized individuals, although nervous disorders may arise.

The epidemiology of *T. solium*, the pork tapeworm, is essentially the same as that described for *T. saginata*, except that the parasite uses the hog as intermediate host. The larva of *T. solium* is called *C. cellulosae*.

Dibothriocephalus latus, the fish tapeworm, inhabits the small intestine of humans. While the Baltic and Scandinavian countries are the classical focus, other cold regions are also important ones. The eggs of the worm embryonate and hatch in water. The ciliated embryo is ingested by water fleas which are ingested by many species of fish. Larvae mature in the mammal. Heavy infection may cause intestinal obstruction and possibly toxemia.

Echinococcosis is also known as hydatid disease and is an infection by the larval forms of *E. granulosus*, *Alveococcus multilocularis*, and *E. oligarthus* in humans or other intermediate hosts. Adult *E. granulosus* are minute parasites of the small intestine of carnivores. The carnivore passes eggs in feces so that grazing sheep and other herbivores or rodents become parasitized with the hydatid larva. The cycle is completed by carnivores preying on them. Humans become exposed by petting dogs or ingesting contaminated grass. *See* ECHINOCOCCOSIS. [J.F.M.]

Taphonomy The science of the preservation of fossils; its domain begins with the death of organisms and includes the events that befall their remains until they become part of rocks. Taphonomy covers all events during this transition from the biosphere to the lithosphere, including mode of death, scavenging, surface weathering, transport of organic remains by animals, wind, water, reerosion of buried material, modification of organic remains as tools by hominids, and differential dissolution of mineral tissues within a range of chemical composition. Organic remains bear evidence of their preservational history in their degree of completeness, in damage patterns, in their orientation with respect to sediments and to other organic remains, in mineralization features, and in their abundance in certain sedimentary environments. *See* FOSSIL.

The central aim of taphonomic studies is to elucidate how biological information has been altered from the original living systems. Every fossil assemblage is the result of historically unique living assemblages that have survived certain biological and physical degradative processes. Most taphonomic research has focused on vertebrate fossils, much of it in connection with the study of early human evolution. The three main taxonomic groups of fossils—vertebrates, invertebrates, and plants—pose fundamentally different taphonomic problems.

Current paleontological issues reflect the influence of modern population and community ecology: these issues include the nature of speciation, the origin of new adaptations, changes in species diversity over geologic time, and changes in community organization as major taxonomic replacements occur. As paleontological studies acquire a more ecological perspective, there is increased concern with reconstructing past environments in detail comparable to that of the present. These details include taxonomic diversity, relative abundance of taxa, structural diversity of habitats, and microhabitat preferences of selected taxa. The extent to which these features are discernible in fossil assemblages depends on their taphonomic histories. Taphonomy and paleoecology are closely allied, taphonomy representing the sampling process that have acted upon past ecosystems. Most taphonomic processes are not ecological in character. Taphonomic reconstructions rely on the guidance of well-developed ecological models to recreate information of paleoecological significance. *See* PALEOBOTANY; PALEOECOLOGY; PALEONTOLOGY. [C.Ba.]

Tapir An odd-toed ungulate of the family Tapiridae. These animals, with four species, have a discontinuous distribution in South America and Asia and consequently have a theoretical importance biologically.

These animals have a prehensile, trunklike muzzle, a short tail, and small eyes. The forefoot has four toes, the hindfoot three. The female gives birth to a single young after a gestation period of 13 months. Tapirs are nocturnal, timid animals; they spend the daytime hours in dense thickets. [C.B.C.]

Tardigrada A phylum of microscopic, bilaterally symmetrical invertebrates which are generally less than 0.04 in. (1 mm) in length. About 400 species are known. Commonly called water bears, bear animalcules, or urslets, they are worldwide in distribution and are found in all habitats.

The tardigrade body consists of an anterior prostomium and five segments. The mouth is located in the prostomium in a centroterminal position. A soft, nonchitinous cuticle surrounds the body and lines the fore- and hindgut. The cuticle may be smooth or sculptured and forms innervated cephalic appendages and spines on the trunk and legs. Four pairs of ventrolateral legs arise from the trunk and terminate in claws or other modified structures. The digestive tract is tubular and more or less lobed due to the presence of diverticular dilations. The sexes are separate and the gonads are unpaired dorsal

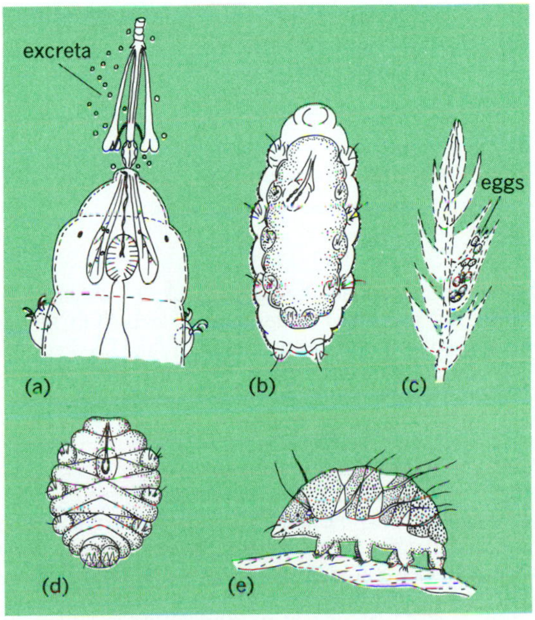

The active and inactive forms of Tardigrada. (*a*) Expulsion of the buccal apparatus and excreta; (*b*) cyst; (*c*) eggs fastened to a moss leaf; (*d*) barrel; (*e*) adult in the walking position.

sacs with paired gonoducts in the male and in theory also in the female.

Food storage cells float in the spacious body cavity, the coelom, which lacks a parietal or visceral peritoneum in the adult. Circulatory and respiratory structures are lacking.

During their active life of 18 months, tardigrades molt about 12 times. They are unable to feed in the 5–10 days of molting; the buccal cavity requires at least 5 days for renewal.

Tardigrada live as active forms, without encystment, only when surrounded by a pellicle of water. When the surrounding medium dries up, most tardigrades continue to live as inactive or anabiotic barrel-shaped structures called cysts without any protective cover (see illustration). Desiccation begins when there is a loss of oxygen from the water. The animal responds by contraction and loss of body water. Dried eggs also survive. Moistened animals usually revive, but anabiosis and revival cannot be repeated indefinitely.

Most species are widely distributed. Dissemination may be by wind, birds, and certain terrestrial animals which transport tardigrade eggs and barrels. The tropics have few species, and members of the family Scutechiniscidae are rare on the Antarctic continent. Most tardigrades are terrestrial. They are found among lichens, liverworts, densely growing soft-leaved mosses, and also in rather hard-leaved Pottiacea and Grimmiacea. *See* Eutardigrada; Heterotardigrada. [Ev.M.]

Target flowmeter A volume flow rate meter that measures the force on a circular obstruction (see illustration). The target-type meter, besides being particularly suitable for dirty fluids, has the important advantage of a built-in secondary device and is without the disadvantage of fluid-filled connection pipes to the secondary device that is characteristic of many volume flow rate meters. *See* Flow measurement; Volume flow rate meter. [M.Br.; L.P.E.]

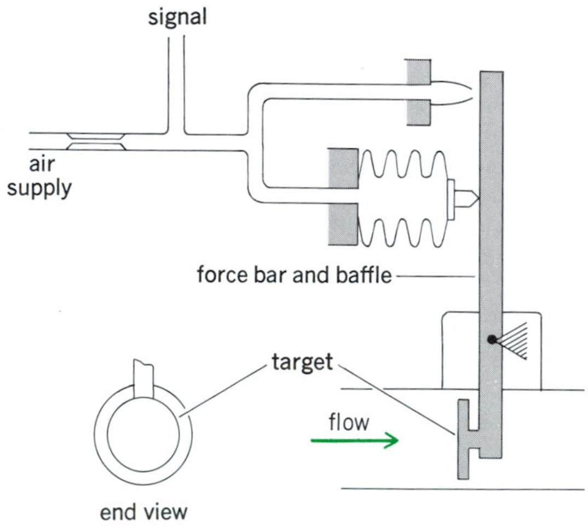

Target flowmeter. (Foxboro Co.)

Tarpon A species of fish, *Megalops atlantica*, which is included together with the ladyfish and machete in the family Elopidae in the order Elopiformes. These fish are widely distributed in tropical and subtropical seas and enter fresh-water streams of the United States.

The tarpon is the largest of the herringlike fishes, weighing up to 300 lb (135 kg) and reaching a length of nearly 8 ft (2.4 m). The body is covered with large scales, and there is a single, soft, rayed dorsal fin, strong jaws, a bony plate under the mouth, and numerous small teeth (see illustration). *See* Elopiformes. [C.B.C.]

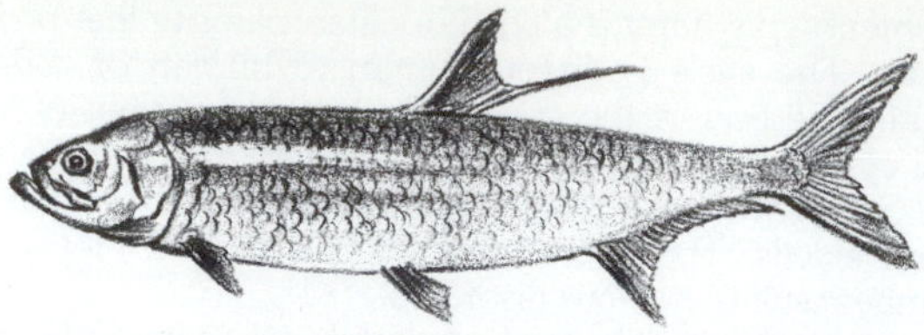
The tarpon has coarse, bony flesh which is eaten in some Central American countries.

Tarragon A herb of the genus *Artemesia* in the aster family (Asteraceae) that is used as a spice. Tarragon is often divided into French and Russian types. French tarragon can be distinguished by its highly aromatic leaves, low seed set or fertility, and compact growth habit. Due to its long history of vegetative propagation to preserve its fine scent characteristics, French tarragon has all but lost its ability to form viable seeds. Russian tarragon does produce seeds, but the plant is much lower in overall oil content and is not as fine-scented.

Tarragon is a perennial that grows in one season, and then dies back to the ground with frost. In spring the plant sprouts from underground rhizomes and resumes its growth. The leaves contain volatile oil with an odor similar to anise and chervil. *See* Asterales; Spice and flavoring. [S.Kir.]

Tarsier A primate of the family Tarsiidae. These mammals are found in the Far East, the Philippines, Malaya, and the Celebes. They exhibit characteristics that relate them to the true monkeys.

The tarsier is insectivorous, thus maintaining what is considered to be a primitive diet. Advanced characteristics are seen in the skull structure in that it is round with a flattened face, and the large eyes are separated from the temporal fossae in the orbital depression. The tail is long and ratlike but not prehensile. The bones of the hindlegs are long, especially the tarsal bones, allowing tarsiers to jump and to leap like frogs. The tarsier is nocturnal and the enormous eyes dilate at night for better vision. Breeding is continuous throughout the year, with a single young born at a time. *See* Mammalia; Primates. [C.B.C.]

Tartaric acid A compound that possesses two similar asymmetric carbon atoms and occurs in four isometric forms: (1) a dextrorotary form, which rotates plane-polarized light to the right; (2) a levorotary form which rotates plane-polarized light to the left; (3) a racemate form, which is an optically inactive equimolecular mixture of separate crystals of the dextro and levo forms; and (4) a meso form, which is optically inactive because of internal compensation. Structural formulas of the dextro, levo, and meso forms are shown. The *dl*-tartaric acid is known as the racemic acid. The principal uses of tartaric acid are in cream of tartar, Rochelle salt, and tartar emetic. *See* Carboxylic acid; Tartrate. [E.B.R.]

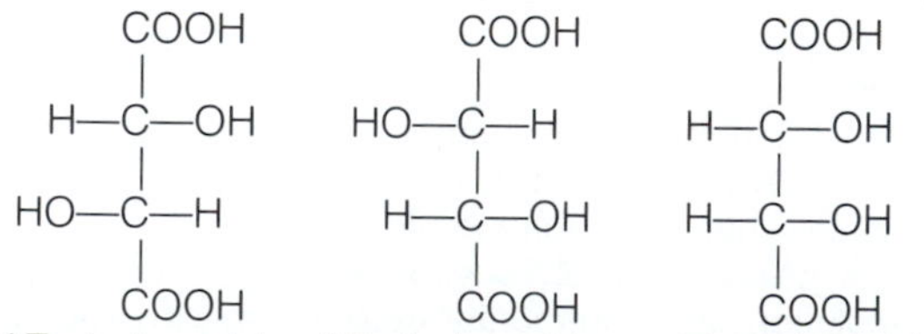
d-Tartaric acid   *l*-Tartaric acid   *m*-Tartaric acid

Tartrate A chemical compound that is a salt or ester of tartaric acid, which is formed by the replacement reaction of the carboxylic hydrogens of tartaric acid by a metal (salt) or organic radical (ester). One or both of the carboxylic hydrogens can be replaced to produce an extensive series of salts, esters, and mixed (double) salts. Many salts have practical applications, as in cream of tartar (monopotassium salt), Rochelle salts (sodium-potassium salt), tartar emetic (potassium-antimony salt), medicines, and textile dyeing (calcium salt). *See* ROCHELLE SALT; TARTARIC ACID. [E.H.H.]

Taste One of the chemical senses. The so-called taste of food is not a simple sensory pattern. It depends not only on the specific taste receptors of the tongue but also upon other sensations supplied by the chemical, tactile, warm, and cold receptors in the mouth, and the smell receptors in the nose. Taste and odor discrimination is reduced if the sense of smell is suppressed, for example, by a cold. *See* OLFACTION.

In humans and mammals, taste buds are located on the tongue in the fungiform, foliate, and circumvallate papillae and in adjacent structures of the throat. Taste buds are goblet-shaped clusters of longer slender cells from the distal end of which a cluster of microvilli protrude into the taste pores that open toward the tongue surface. Individual sensory nerve fibers branch profusely within the taste bud. *See* TONGUE.

No simple relation exists between chemical stimuli and taste quality except, perhaps, for the sourness of acids. The taste qualities of inorganic salts are complex, only sodium chloride giving the purely saline taste. Sweet and bitter tastes occur in many chemical classes. The middle top portion of the tongue surface is insensitive to all tastes. Sensitivity to sweet is greatest at the tip, to sour at the sides, and to bitter at the back; salt sensitivity is relatively more homogeneous around the edges. *See* CHEMICAL SENSES. [C.P.]

Taurus The Bull, in astronomy, a winter constellation. Taurus is the second sign of the zodiac. The group contains two notable star clusters, the Hyades and the Pleiades. The Hyades is a V-shaped cluster, the V forming the head of the charging bull, with the fiery bright star Aldebaran in the right eye (see illustration). This star has long been used in navigation. Farther west lies the compact, beautiful cluster of six stars, the famous Pleiades. *See* CONSTELLATION; HYADES; PLEIADES. [C.-S.Y.]

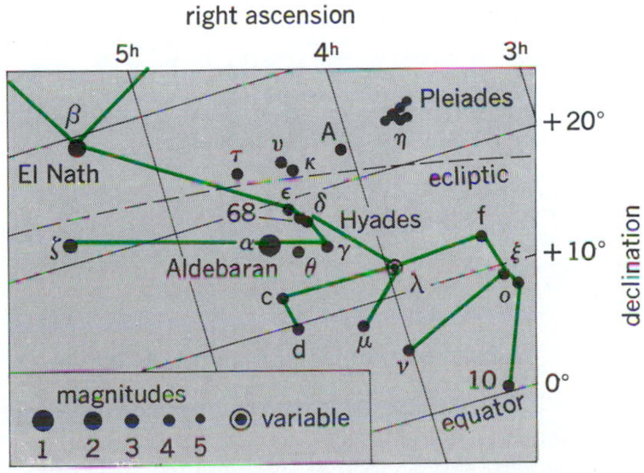

Line pattern of the constellation Taurus. The grid lines represent the coordinates of the sky. The apparent brightness, or magnitude, of the stars is shown by the sizes of the dots, which are graded by appropriate numbers as indicated.

Tautomerism The reversible interconversion of structural isomers of organic chemical compounds. Such interconversions usually involve transfer of a proton, but anionotropic rearrangements may be reversible and so be classed as tautomeric interconversions. A cyclic system containing the grouping —CONH— is called a lactam, and the isomeric form, —COH=N—, a lactim. These terms have been extended to include the same structures in open-chain compounds when considering the shift of the hydrogen from nitrogen to oxygen.

Molecular grouping (**1**) may in certain substances exist partly or wholly as (**2**). The former constitutes the keto form and

—COCH<	—COH=C<	>C=N—	—CN<
(**1**)	(**2**)	(**3**)	(**4**)

the latter the enol form. The existence of an enol in an acyclic system requires that a second carbonyl group (or its equivalent, for example, (**3**) be attached to the same (**4**) as an aldehyde or ketone carbonyl. Thus, ethyl acetoacetate tautomerizes demonstrably, but ethyl malonate does not. Where the enol form includes an aromatic ring such as phenol, the existence of the keto form is often not demonstrable, although in some substances there may be either chemical or spectroscopic evidence for both forms. Closely related to keto-enol tautomerism is the prototropic interconversion of nitro and aci forms of aliphatic nitro compounds such as nitromethane.

In general, tautomeric forms exist in substances possessing functional groups which can interact additively and which are so placed that intramolecular reaction leads to a stable cyclic system. The cyclic form usually predominates (especially if it contains five or six members). *See* MOLECULAR ISOMERISM. [W.R.V.]

Taxales A small order of gymnosperms, division Pinophyta, in the class Pinopsida. The plants are trees or shrubs with evergreen, often needlelike leaves, vegetatively much like the related order Pinales. They differ from the Pinales in having a well-developed fleshy covering surrounding the individual seeds, which are terminal or subterminal on short shoots instead of being in definite cones. The most familiar genus of the Taxales in the Northern Hemisphere is the yew (*Taxus*). *See* PINALES; PINOPHYTA; YEW. [A.Cr.]

Taxis A mechanism of orientation by means of which an animal moves in a direction related to a source of stimulation. There exists a widely accepted terminology in which the nature of the stimulus is indicated by a prefix such as phototaxis (light), chemotaxis (chemical compounds), geotaxis (gravity), thigmotaxis (contact), rheotaxis (water current), and anemotaxis (air current). The directions toward or away from the stimulus are expressed as positive or negative, respectively. Finally, the sensory and locomotory mechanisms by means of which the orientation is achieved are denoted by a second type of prefix forming a compound noun with taxis. Positive phototropotaxis thus describes a mechanism by means of which an animal carries out a directed movement toward a source of light along a path which permits the animal's paired eyes to receive equal intensities of light throughout the movement. [O.E.L.]

Taxodonta An order of bivalves in subclass Lamellibranchia whose hinge dentition is characterized by a series of numerous similar alternating teeth and sockets. Typical genera of so-called ark shells are *Arca*, *Barbatia*, and *Anadara*.

Ark-shell bivalves are probably distantly related to the true marine mussels (Mytilacea), with which they share certain features of gill ciliation and byssal organization. Formerly the name Taxodonta was used to designate a subclass of bivalves consisting of a mixed group of lamellibranch ark shells and primitive protobranchs. *See* BIVALVIA; LAMELLIBRANCHIA; MOLLUSCA; PROTOBRANCHIA. [W.D.R.-H.]

Taxon A monophyletic group of organisms at any level in the taxonomic hierarchy is a taxon (plural, taxa), A clear distinction must be made between the category, which is any one of the levels in the hierarchy, and the taxon, which is the actual group at that categorial level. Thus the genus is a category and *Turdus* is a taxon of passerine birds at the generic level. Categories are words that are defined. Taxa or taxonomic groups are real entities in nature that can be recognized, delimited, and described. Taxa at any level consist of one or more taxa of the next lower level. Members of a taxon are recognized by the joint possession of similar features and are distinguished from other taxa by differences in characteristics. These differences correspond to the presumed gap in evolutionary relationships between the taxa. Members of any taxon are presumed to share a monophyletic origin, and most taxa, but not all, contain more than one taxon of the next-lower rank. *See* TAXONOMIC CATEGORIES.

Species taxa are given special recognition because these groups can be ascertained by criteria other than features shared in common. Members of the same species interbreed with one another and are reproductively isolated from other such species. *See* ANIMAL SYSTEMATICS; PLANT TAXONOMY; SPECIES CONCEPT. [W.J.B.]

Taxonomic categories Any one of the levels in the hierarchy of biological classification of organisms. The categories—for example, species, genus, family—are words that can be defined clearly and definitely. The species category is regarded as the most important because its definition has real biological meaning—that is, there is no breeding between species—and because the categories at higher and lower levels are defined in terms of the species. *See* SPECIES CONCEPT.

A clear distinction must be made between the category and groups of organisms or taxa which are described at that categorial level. *See* TAXON.

Definitions of genera and higher categories are in the following form: A genus is a monophyletic group of one or more species. It is used to indicate the phylogenetic origin of the group and the amount of evolutionary change that has occurred. Groups of similar species are placed together in a genus, similar genera in a family, and so forth, with all similar phyla in a kingdom. All living organisms are divided into a small number of kingdoms, two to five, depending on the authority. *See* ANIMAL SYSTEMATICS.

The respective position of two different animals in this system may be expressed as follows:

CATEGORIES	TAXA (LION)	TAXA (HOUSEFLY)
Kingdom	Animalia	Animalia
Phylum	Chordata	Arthropoda
Class	Mammalia	Insecta
Order	Carnivora	Diptera
Family	Felidae	Muscidae
Genus	*Felis*	*Musca*
Species	*leo*	*domestica*

Note that the binominal (generic and trivial) name is italicized. These seven taxonomic categories are the minimum used to classify animals. In large groups or those with a complex evolutionary history, more categories are utilized. Those which are generally accepted are as follows:

Kingdom	Suborder
Phylum	Superfamily (-oidea)
Subphylum	Family (-idae)
Superclass	Subfamily (-inae)
Class	Tribe (-ini)
Subclass	Genus
Cohort	Subgenus
Superorder	Species
Order	Subspecies

Standardized endings are used for superfamilies, families, subfamilies, and tribes. Names for taxa above the level of the superfamily are not standardized (although -iformes is used for orders in some groups), but usually vary from one group of animals to another without uniform endings. *See* PLANT TAXONOMY; ZOOLOGICAL NOMENCLATURE. [W.J.B.]

Taxonomy A study aimed at producing a hierarchical system of classification of organisms which best reflects the totality of their similarities and differences. Most taxonomists believe that such a classification can be achieved only by heavy reliance on the phylogenetic concept, so that each taxon (a taxonomic group, at whatever rank) represents a unified evolutionary group.

It is customary to recognize two kingdoms of organisms, Plantae (plants) and Animalia (animals), but various other arrangements of two, three, four, or even five kingdoms have been proposed. *See* ANIMAL SYSTEMATICS; PLANT KINGDOM; PLANT PHYLOGENY; PLANT TAXONOMY; TAXON; TAXONOMIC CATEGORIES. [A.Cr.]

Tea A small tree, *Camellia* (*Thea*) *sinensis*. A preparation of its leaves dried and cured by various processes; and a beverage made from these leaves. The plant is an evergreen tree of the Theaceae family, native to southeastern Asia, and does best in a warm climate where the rainfall averages 90–200 in. (2200–5000 mm). China, Japan, Taiwan, India, Sri Lanka, and Indonesia are among the leading tea-producing countries, with China contributing about one-half of the world's supply.

Tea leaves contain caffeine, various tannins, aromatic substances attributed to an essential oil, and other materials of a minor nature, including proteins, gums, and sugars. The tannins provide the astringency, the caffeine the stimulating properties. *See* THEALES. [E.L.C.]

Technetium A chemical element, Tc, atomic number 43. The element was the first made artificially in a cyclotron. It is also produced as a major constituent of nuclear reactor fission products or, alternatively, by action of neutrons on ^{98}Mo.

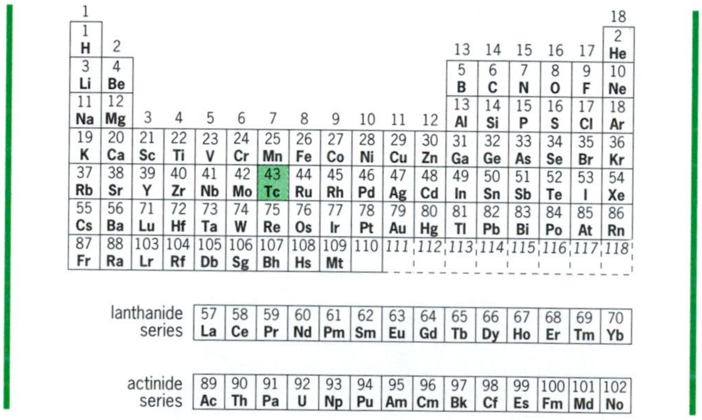

The isotope ^{99}Tc is most suitable for chemical investigation because of its long half-life, 2×10^5 years. The chemistry of technetium is very similar to that of rhenium, and corresponding compounds have been prepared in many cases. *See* NUCLEAR REACTION; RHENIUM; TRANSITION ELEMENTS. [S.F.]

Technology Systematic knowledge and action, usually of industrial processes but applicable to any recurrent activity. Technology is closely related to science and to engineering. Science deals with humans' understanding of the real world about them—the inherent properties of space, matter, energy, and their interactions. Engineering is the application of objective knowledge to the creation of plans, designs, and means for achieving desired objectives. Technology deals with the tools and techniques for carrying out the plans. [R.S.S./H.B.M.]

Tectonophysics A branch of geophysics that deals with the physical processes of inelastic deformations—flow and fracture—of the Earth on all scales, from those of plate tectonics and associated mountain building to crystal-lattice defects. It encompasses some 16 orders of magnitude. High-pressure, high-temperature rheological properties of rocks—elasticity, plasticity, and viscosity—are measured in the laboratory under conditions simulating the natural environment as realistically as possible. The tectonophysicist then uses these measurements to construct analytical and numerical models of the deformations of the crust and upper mantle and test them against field observations by structural geologists, structural petrologists, and geophysicists.

In dealing with such phenomena as folding, faulting, and compaction in the relatively brittle crust, tectonophysics becomes intimately associated with engineering rock mechanics. To treat deep-seated processes, the tectonophysicist teams with geochemists, geodesists, and seismologists to investigate flow in the relatively ductile mantle and coupling of the driving forces in the asthenospheric with the lithospheric plates. *See* ENGINEERING GEOLOGY; GEOPHYSICS; PLATE TECTONICS; ROCK MECHANICS; SEISMOLOGY; STRUCTURAL GEOLOGY. [J.H.]

Tektite A member of one of several groups of objects that are composed almost entirely of natural glass formed from the melting and rapid cooling of terrestrial rocks by the energy accompanying impacts of large extraterrestrial bodies. Tektites are dark brown to green, show laminar to highly contorted flow structure on weathered surfaces and in thin slices, are brittle with excellent conchoidal fracture, and occur in large masses but are mostly small to microscopic in size.

Five major groups of tektites are known: North American, 34,000,000 years old; Czechoslovakian (moldavites), 15,000,000 years; Ivory Coast, 1,300,000 years old; Russian (irgizites) 1,100,000 years old; and (5) Australasian, 700,000 years old. The North American, Ivory Coast, and Australasian tektites also occur as microtektites in oceanic sediment cores near the areas of their land occurrences. In the land occurrences, virtually all of the tektites are found mixed with surface gravels and recent sediments that are younger than their formation ages.

The chemical compositions of tektites differ from those of ordinary terrestrial rocks principally in that they contain less water and have a greater ratio of ferrous to ferric iron, both of which are almost certainly a result of their very-high-temperature history. [E.A.K.]

Telecast A television broadcast, involving the transmission of the picture and sound portions of the program by separate transmitters at assigned carrier frequencies within the channel assigned to a television station. A telecast is intended for reception by the general public, just as is a radio broadcast. *See* TELEVISION; TELEVISION TRANSMITTER. [J.Mar.]

Telecommunications civil defense system A telephone system specially designed to assist in protecting civilian population, limiting casualties and property damage, and preserving maximum civilian support of military effort. Civil defense (CD) requirements for communications fall into two broad categories: preattack warning, and administrative and control communications to alleviate disaster after an attack. Both requirements are usually best met with private-line telephone systems furnished by telephone companies, supplemented by local and interurban telephone service arranged to assure continuity of service.

In the United States the basic facility for attack warning is a private-line telephone system known as the National Warning System (NAWAS), which connects over 1000 warning points in the United States with three Office of Civil Defense (OCD) warning centers.

OCD's primary system of operational communications is the National Communications System (NACOM I). NACOM I network consists of approximately 30,000 mi (48,000 km) of cable and radio facilities for both voice and data communications. Terminal equipment for this network is in place and is exercised daily for administrative purposes. The major function of NACOM I is to furnish the means for coordinating emergency government operation. *See* TELEPHONE SERVICE; TELEPHONE SYSTEMS CONSTRUCTION. [J.A.O.]

Teleconferencing Broadly, the various ways and means by which people communicate with one another over some distance. In a narrow sense, a teleconference is a two-way, interactive meeting, between relatively small groups of people (approximately 1 to 10 at each end), who usually use permanent teleconferencing facilities. A teleconference involves audio communication between the locations, but may also involve video, graphics, or facsimile. *See* COMPUTER GRAPHICS; FACSIMILE; TELEPHONE; TELEVISION.

A teleseminar is utilized for educational purposes; it is primarily one-way communication to many destinations from one source. The facilities are usually built or converted especially for teleseminars.

A telemeeting is often called an ad hoc teleconference, with the ad hoc referring to places, times, participants, and purpose. A telemeeting is similar to a teleseminar in that it is primarily a one-way communication, usually staged or prepared by video program professionals. It is set up to order, using temporary equipment or circuits.

Computer conferencing is a method for people to communicate by using computers. The medium is quite flexible, as it can be used between just two people, between one and many people, or among many people. Basically, it involves typing a message on a computer terminal and transmitting it to one or more destinations electronically. Sophisticated networks are required to accomplish computer conferencing in real time between many users, or a simple data modem and telephone circuit can allow two people at a time to conference. *See* DATA COMMUNICATIONS; WIDE-AREA NETWORK. [J.J.B.]

Telegraphy A method of communication employing electrical signaling impulses produced and received manually or by machines. Telegraph signals are transmitted over open wire or cable land lines, submarine cables, or radio. Telegraphy as a communication technique uses essentially a narrow frequency band and a transmission rate adapted to machine operations. *See* ELECTRICAL COMMUNICATIONS.

Early equipment devised by Samuel F. B. Morse consisted of a mechanical transmitter and receiver or register. Operators soon learned to handle messages faster using simple manual keys and audible sounders. Subsequently, telegraph transmission and reception again became mechanized. *See* TELETYPEWRITER EXCHANGE (TWX) SERVICE; TELEX. For other ways in which telegraphy may be used, *see* FACSIMILE; TELEPHOTOGRAPHY; TELETYPESETTER.

Telegraph facilities for use by the general public to transmit messages both domestically and internationally are provided by communication companies and government administrations. Special telegraph facilities include those for news services, distribution of market prices of securities and commodities, and private lines between such points as the factories and offices of a company for the exchange of messages, orders, payroll data, and inventories. Fire and police alarms are a special form of telegraphy. The armed forces have extensive fixed and mobile telegraph systems.

For manual operation, the telegraph code consists of short dot and long dash signals (see illustration). Most automatic printing telegraph circuits, including American cable operation, use a code of five equally spaced signals or units per letter or other symbol perforated into a paper tape. Machines translate automatic teleprinter code to cable code, cable code to automatic code, or, for special applications, make other translations. For stock quotation systems and teletypesetter operation, a six-unit code is used to provide control of machine action. For data transmission or machine control, a seven- or eight-unit code may be used.

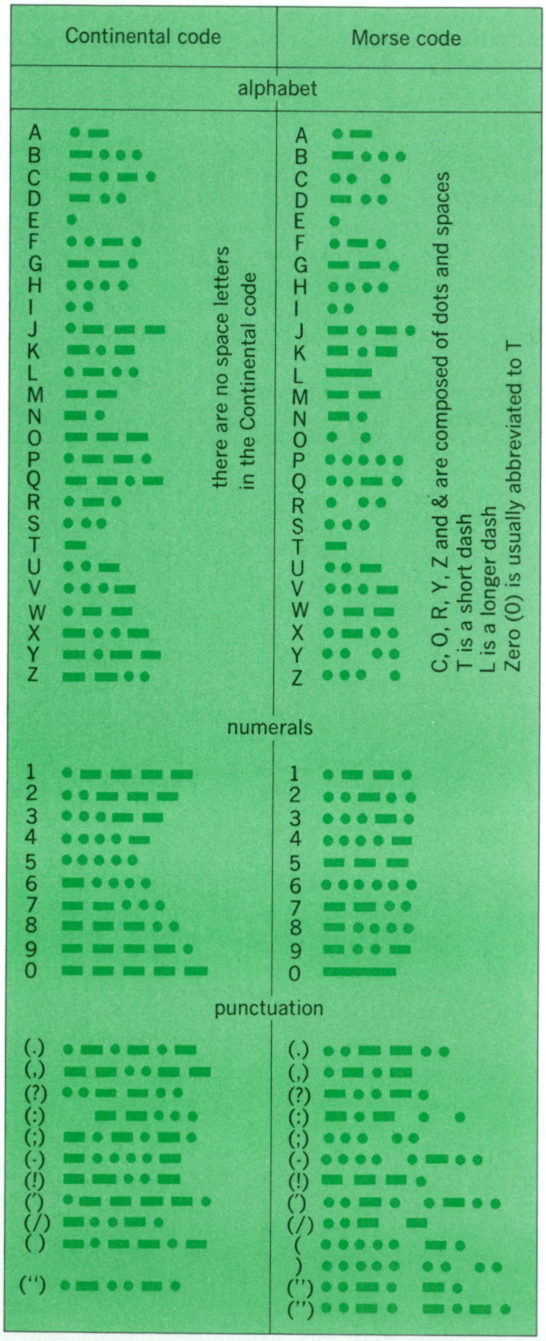

Continental code (International Morse) is commonly used for telegraph communication. Morse code (American Morse) continues in use on a few land lines in the United States and Canada.

A single circuit provides transmission in only one direction at a time. For transmission in both directions simultaneously, a duplex circuit is used. Multiplex (time-division) apparatus provides two, three, or more channels operable in both directions simultaneously over a single circuit. Carrier-current techniques enable several circuits, each comprising one or more communication channels, to operate through the same wide-band wire, cable, or radio facility. *See* MULTIPLEXING.

In automatic transmission, an operator at a manual keyboard, operated like a typewriter, perforates a tape. The tape is fed through an electromechanical or photocell tape reader that drives the tape at a rapid and uniform rate and transmits the electrical code. Interconnecting wire lines or other communication channels carry these code pulses to the receiver. The received impulses may automatically actuate a reperforator to produce a duplicate punched tape for retransmission or later transcription; the impulses may actuate a teletypewriter (also called a teleprinter) to retype the original message; or they may actuate other terminal equipment, such as an accounting machine.

Alternatively, the equipment at both terminals may be teletypewriters, in which keyboard and printing mechanisms are combined in one machine. Such machines use a five-unit code but operate without perforated tape. Business firms that originate and receive numerous messages have such equipment installed at their offices for direct service. Also widely used are facsimile instruments that transmit or reproduce a typed or handwritten message as a picture on electrosensitive paper. [G.Hot.]

To achieve speed and accuracy in handling messages en route, direct circuits may be set up from the originaling keyboard to the distant teleprinter. This is done in TWX and Telex services. This mode of system operation is known as circuit switching. If direct channels are not immediately available, or if the volume of traffic calls for a storage interval, the message is transferred by means of perforated tape at switching centers. This mode is known as message switching or store-and-forward switching. Message switching has the advantage of using the channel capacities of trunk lines more fully than do direct connections. It is used in the United States for public telegram service and on leased-wire switched networks. *See* TELEX. [W.E.G.]

In overseas communications, telegraph messages or telegrams are transmitted over high-frequency radio (3–30 MHz), transoceanic submarine cables, or satellite communication channels. A decreasing number of messages are carried on high-frequency radio. These types of overseas channels are also used for private-line service, which is referred to as leased-channel service, and customer-to-customer teletypewriter service, which is known as Telex in the international service. [E.D.B.]

Telemetering The branch of engineering concerned with the presentation of measured data at a location remote from the source of the data; also called telemetry. An example of a highly complex telemetering system is the equipment utilized to measure the temperature in a space vehicle in flight, radio this temperature to a ground observing station, and provide a meter or visual display to the observers. Telemetry encompasses all remote metering, whether the distances involved be many miles, as in space flights, or only a few feet, as in the measurement of reactivity in the core of a nuclear reactor.

Telemetry involves three separate functions: (1) generation of a signal (electrical or otherwise), which measures the pertinent physical variable and is in suitable form for transmission; (2) transmission of the information to the remote location; and (3)conversion of the data into a form appropriate for display, recording, or application to further data-processing equipment.

Telemetry systems are conventionally classified according to the transmission medium. Although the information can be transmitted in a variety of forms (such as the brightness of a modulated light beam), the great majority of systems fall into three categories: mechanical, electrical, and radio telemetry.

Mechanical telemetry, involving mechanical coupling between the points of measurement and of data utilization, is generally restricted to short distances because of the high attenuation of mechanical media, the low velocity of propagation (of sound, for example), the difficulty of constructing efficient and simple mechanical amplifiers, and the requirement of a continuous mechanical coupling medium throughout the transmission path.

The term electrical telemetry is utilized in contrast to radio telemetry and refers to wired telemetry systems in which the information is transmitted by variations of a voltage or current in the electric circuit. Modern electrical telemetry systems often involve complex electronic equipment, as when the information is transmitted (over wires) in the form of a television picture.

Radio telemetry, originally used to telemeter information from weather balloons to ground observers, employs electromagnetic radiation as the transmission means. [J.G.Tr.]

Teleostei The youngest, largest, and most successful of three organizational levels (infraclasses) of the subclass Actinopterygii or rayfin fishes. The teleosteans are derived, through one or perhaps more evolutionary lines, from the Holostei, which in turn are descendants of the Chondrostei. *See* ACTINOPTERYGII; CHONDROSTEI; HOLOSTEI.

The transition from holostean to teleostean involves attainment of paired bracing bones (uroneurals) in the supporting skeleton of the caudal fin; development of a homocercal caudal fin in which two or more lowermost hypural bones articulate with a single centrum; loss of fulcral scales at the upper and lower base of the caudal fin; development of intermuscular bones; increase from one to two pairs of hypohyal bones; reduction in thickness of scales and gradual loss of the enamel surface (ganoin); development of cycloid scales; development of a supraoccipital bone; fusion of the paired prevomers into a single median bone; reduction in number of bones in the mandible; increased ossification of the vertebral column; and assumption by the swim bladder of a hydrostatic function instead of a respiratory one. [R.M.B.]

Teleostomi This name was originally proposed by R. Owen to include bony fishes with a terminal mouth, excluding the lungfishes or Dipnoi. Later, it was employed in a more inclusive sense to mean a class of fishes, including the bony fishes or subclasses Osteichthyes, and the extinct subclass Acanthodii. Together with class Chondrichthyes and class Placodermi, it forms the group of jawed fishes or Gnathostomata. *See* ACANTHODII; CHONDRICHTHYES; OSTEICHTHYES; PLACODERMI. [R.H.De.]

Telephone The term telephone formerly referred to the telephone receiver, the instrument originally invented by Alexander Graham Bell; it is now commonly applied to the telephone set, which includes a transmitter and an electric network in addition to the receiver. The transmitter and receiver are housed together in a handset. A cord connects the electrical components of the handset to the network in the telephone set.

The transmitter (see illustration) is a transducer which converts acoustical energy into electric energy. In most transmitters an electric current is modulated by the variations in contact resistances of carbon granules. Sound waves impinge on the diaphragm of the transmitter, causing the carbon granules to move closer together, making more contacts and decreasing resistance, or to move further apart, making fewer contacts and higher resistance. The transmitter has a frequency-response range from 250 to 5000 Hz. The frequency response rises uniformly to a broad maximum in the region of 2500 Hz. Because of these characteristics and those of the

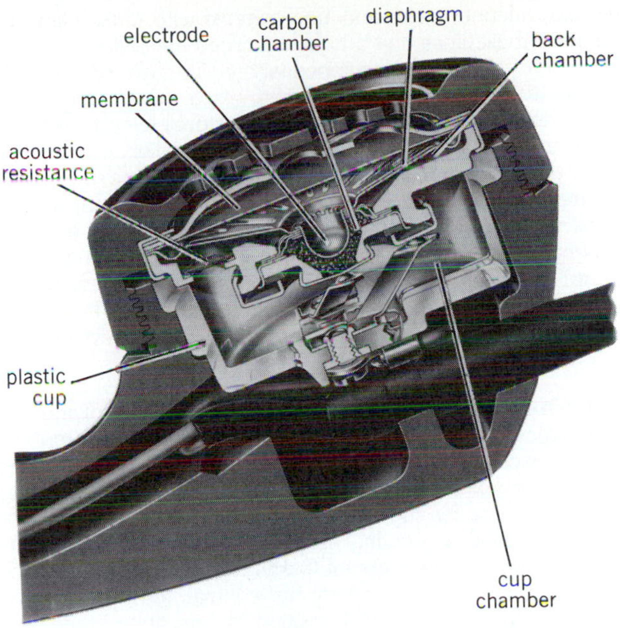

Direct-action granular-carbon transmitter.

telephone receiver, the speech heard by the listener resembles closely that of direct mouth-to-ear speech as heard by a listener a few feet from the person speaking.

The receiver transducer operates on the relatively low power used in the telephone circuit to convert electric energy into acoustical energy. Unlike a loudspeaker, the telephone receiver is designed for close coupling to the ear. There are two types of receiver units, the bipolar receiver and the ring armature receiver. The bipolar receiver is used with operator's head telephone sets. The ring armature receiver is used with the telephone set in which high efficiency is particularly important. The advantages of the ring armature receiver are its low acoustic impedance and high available power response over a wide frequency range (350–3500 Hz). These advantages are achieved by the piston action of a thin, nonmagnetic, lightweight, dome-shaped diaphragm, which is attached to a ring-shaped armature of magnetic material driven at its periphery by a ring-shaped magnetic coil associated with a ring-shaped permanent magnet. *See* TELEPHONE SERVICE. [R.F.D.]

Telephone equipment, customer The apparatus used by the individual telephone subscriber. The equipment used by residential and business customers is the telephone set. The customer may be provided a set as part of telephone service or may purchase a set. Various premium types of service, telephone sets, and accessories can be obtained to supplement or replace the basic ones.

When the serving central office is equipped, the customer can obtain a premium service in which dialing a telephone number is accomplished by pushing buttons on the telephone rather than turning a circular wheel as in rotary-dial service. Each depressed button on the dial generates two audio tones, and each button has its own pair of tones which are recognized by the central-office (or PBX) switching equipment as individual digits of the telephone number being called. Push-button telephones can be arranged to send such pairs of tones over the circuit after a telephone connection has been established. This feature permits users to send information to a remote device such as a computer.

Telephone sets used to provide basic service are the desk type and wall type, with a dial and a handset for talking and listening. Sets used for business are generally connected to more than one telephone line and are equipped with pickup keys to terminate these lines and a hold key. They are called key telephone sets and may be equipped with 6, 10, 12, 18, 20, 24, or 30 buttons.

An automatic dialer is a device which stores telephone numbers in a memory medium and automatically dials the number when the user performs the proper interface function. Automatic dialers are available as an integral part of a telephone set or as adjuncts to existing sets. The memory medium for number storage usually consists of either a perforated plastic card or an integrated circuit memory in which the digits are stored in binary form. *See* PRIVATE BRANCH EXCHANGE; TELEPHONE; TELEPHONE SERVICE; TELEPHONE SIGNALING; TELEPHONE SYSTEMS CONSTRUCTION; VIDEO TELEPHONE. [W.Sch.]

Telephone service The technology of supplying the many types of services offered to the telephone subscriber. These services include local, long-distance, and overseas connections for voice and data, information services, answering services, and the leasing of private lines. Advances in telephony have created a growing number of communications features. These include multipoint (party) telephone conversations (or conference calls) under customer control, temporary transfer of telephone number while visiting friends, abbreviated dialing for commonly used numbers, and an indication to the telephone user when someone else is trying to reach the subscriber's line.

Local service. A local telephone service operates within an exchange, or within a connected system of exchanges in the same locality, and furnishes to subscribers an intercommunicating service. An exchange is a unit established by a telephone company for the administration of telephone service in a specified area, usually a town, a city, or a village and its environs. The exchange consists of one or more central offices together with the associated plant used in furnishing telephone service in that area. An exchange area is the territory served by an exchange.

To provide this intercommunicating service, centrally located switching equipment is used to establish a communication connection between any two telephone subscribers. This switching equipment also provides for connections to distant exchanges.

Local exchange transmission circuits are usually two-wire cable pairs operated at voice frequency, although short-haul analog and digital carrier systems are being more widely used, especially in large metropolitan areas. In the longer transmission circuits, carrier systems are normally employed.

The design of a telephone system involves a consideration of telephone sets, subscriber loops, telephone central offices, interoffice trunks, and numbering plans.

The telephone set converts acoustic energy into the electrical analog signal which is transmitted. In addition, it generates the supervisory (on-hook and off-hook) and the address information used by switching and signaling systems to establish connections. The main telephone set, together with extension telephones, may have exclusive use of a telephone line. This is individual-line service. Many telephone lines terminate on private branch exchanges to service a number of extension telephones. *See* TELEPHONE EQUIPMENT, CUSTOMER; TELEPHONE SYSTEMS CONSTRUCTION.

Subscriber loops are the relatively short circuits, or lines, connecting the customers' telephone sets to a central office. These lines usually employ copper wire.

The telephone central office houses the switching equipment that serves the telephones of its immediate community. The telephone building, housing one or more local central offices, is the "wire center" at which the loops of customers in the vicinity converge and are terminated. The primary function of the central office is to initiate and control the switching and signaling action which, on demand, permits the connection of any local telephone to any other telephone served by the same central office, in the same exchange, in nearby exchanges, and in distant cities. For this purpose, each central office has direct interoffice trunks radiating from it to many of the other local central offices in the same area, to tandem offices, and to long-distance offices for intermediate connection to circuits to more distant places.

A tandem office is an intermediate switching location whose primary function is to interconnect trunk circuits to and from the local central offices in the area. In this manner trunking economies are achieved, particularly in those instances where traffic volumes between two local central offices are not sufficient to justify a large number of direct interoffice trunks between the two local offices.

Interoffice trunks are the speech- and data-carrying telephone circuits between the switching systems in the exchange and its environs. Carrier systems usually provide the transmission medium for the longer trunk circuits. For shorter trunk circuits, under about 5–10 mi (8–16 km), the two-wire circuit operating at voice frequencies continues to be the basic transmission medium. Some carrier systems use the digital transmission technology of time-division multiplex (TDM) and pulse-code modulation (PCM). Time-division multiplex (TDM) techniques have also been introduced into switching systems.

Intercity service. Intercity service provides and operates the equipment for switching and transmitting telephone communications between exchange areas.

Long-distance office. A centralized location in an exchange area is provided to handle long-distance service for that area and often for other nearby exchange areas. Large urban areas may have more than one subsidiary long-distance office when it becomes necessary because of physical limitations at a single office or to protect service by decentralization.

Calls between exchange areas served by two different long-distance offices are often too few in number to justify economically a direct connection between the two long-distance offices, Such calls are routed through a third long-distance office, which can connect directly to the first two offices. At the third long-distance office, these calls are termed through-switched traffic.

Virtually all of the intercity trunks in the United States are equipped for the dialing of calls to completion by either operators or customers. Mechanized offices are equipped with dial-switching systems designed for interurban service. The type of switching system used depends on the number of calls to be handled and the distribution of traffic. Long-distance offices are also provided with the equipment needed for trunks and signaling. *See* TELEPHONE SIGNALING.

Intercity trunks. These are the telephone channels between exchange areas. The distortion and attenuation limits are kept within a narrow range, since these effects are cumulative for the trunks that constitute a connection. At the minimum, there is a connecting circuit between the long-distance office and a local office at each end of an intercity trunk. If an interurban call is between two small towns 3000 mi (5000 km) apart, several intercity trunks in tandem may be used for the connection.

Transmission systems for intercity trunks include voice-frequency circuits on wire cable or open-wire lines, and carrier systems for wire cable, open-wire lines, coaxial cable, fiber-optic cables, and radio relay. Transmission systems for intercity trunks, for the most part, utilize long-haul analog carrier systems and microwave radio relay systems. Development of long-haul digital systems has been completed, and they are being utilized in the network as they become economical.

Electronic switching systems. Many telephone companies are finding it economical to replace existing electromechanical switching systems with electronic ones. Especially in metropolitan areas an electronic switching system has lowered costs, increased productivity, and improved flexibility for tandem switching offices. *See* SWITCHING SYSTEMS (COMMUNICATIONS).

Direct distance dialing (DDD). This service permits customers to dial their own interurban calls. DDD service requires dial-switching equipment for the local office at the originating end and all long-distance offices through which a DDD call may be routed. Automatic equipment is required to record the calling and called telephone numbers and the duration of the conversation. A comprehensive numbering plan is required to give each telephone a unique number to indicate the destination of a call to the dial-switching systems. Finally, a standard roofing pattern and a general switching plan are required. The standard routing pattern provides a basis for the selection of important switching centers, called control switching points (CSPs). The general switching plan provides an orderly method of interconnection over the most direct and economical route available.

Satellite communications. Satellites add circuit capacity and diversity to terrestrial facilities and submarine cable (for overseas circuits). Having diverse transmission systems makes it possible to maintain alternate communications paths to ensure that at least one will be available in the event of facility failures, outages, or natural disaster. In some areas, satellites provide 50–100% of available facilities. For overseas circuits, the cost per channel on satellites is considerably higher than on submarine cable facilities. *See* COMMUNICATIONS SATELLITE.

Light-wave cable. Cable systems that transmit calls on pulses of light, rather than as electrical signals, are coming into use in the telecommunications industry. These systems use hair-thin strands of glass to carry signals translated into pulses of laser light. The progression to commercial feasibility was made possible by the discovery of an efficient means of generating light in microscopic semiconductors and the fact that this light has extremely low attentuation when transmitted through glass fibers, even over many kilometers. *See* OPTICAL COMMUNICATIONS.

Overseas service. Overseas and international telephone service links the telephone systems of about 245 countries and territories throughout the world. Overseas telephone circuits are provided mainly by three transmission systems: tropospheric scatter radio, submarine cable, and space satellites. Each has a definite field of use determined by its own characteristics and limitations.

Tropospheric scatter systems normally operate in the 300–3000-MHz frequency range and provide about 60–100 telephone circuits. Although radio waves at these frequencies are not reflected by the ionosphere, it has been found that, if large amounts of power are radiated, scattered energy will be received over relatively long distances beyond the horizon. By the use of high-powered transmitters (up to 10 kW) and large, high-gain antennas with diameters of 30–60 ft (9–18 m), signals of strength adequate for a multi-channel telephone service can be provided over distances of about 200 mi (320 km). *See* TROPOSPHERIC SCATTER.

Submarine cables generally consist of a single conductor and a grounded coaxial return conductor. Submarine cables over long distances require underwater amplifiers to restore the transmission losses in the cable. These amplifiers, called repeaters, must be capable of operation at 3 or 4 nautical miles (6–7 km) below the surface and must have a long life. *See* SUBMARINE CABLE.

Customer international direct distance dialing (IDDD) from the United States was first introduced in March 1970. It has been expanded to include over 75 countries and has been implemented in certain exchanges in many United States cities. Many foreign countries also permit customer dialing to the United States. Where IDDD service is not available, the international calls are established by the operator. [R.B.Ni.; C.H.Sp.]

Telephone signaling

Telephone signaling A system supplementing or replacing the usual telephone bell, to increase the area covered or otherwise provide more comprehensive means for notifying subscribers that they are wanted at the telephone.

Functionally, these systems are of two types: those which give extended coverage without indicating which individual is wanted, and those which indicate that a particular individual is wanted. The first type of system uses audible or visual signaling systems, generally operated in synchronism with the ringing cycle of the central office or private branch exchange (PBX) ringing source.

Special systems which give an indication that a particular individual is wanted include code calling systems, loudspeaker paging systems, and radio paging systems. Code calling systems sound a prearranged code number throughout the premises, usually by signals on single-stroke musical gongs. Loudspeaker paging systems may use a special microphone or the transmitter of the exchange attendant's regular telephone headset, transferred temporarily to the loudspeaker system by operating a key. Radio paging systems comprise miniature radio receivers which can be carried in a pocket. The receiver gives the wearer an audible signal in response to a radio-frequency signal actuated by a telephone operator at a nearby long-distance switchboard. *See* TELEPHONE SERVICE. [R.F.D.]

Telephone systems construction

Telephone systems construction The process of physically constructing the outside plant portion of a complete telephone system. Principal components consist of (1) subscriber loop facilities from the central office to the customer's premises, (2) interoffice trunks, and (3) interurban cable or wire facilities. This construction includes new construction and reinforcement or replacement of existing facilities. The construction process includes the pole lines and conduit systems, as well as the wires and cables. It also includes joining and testing the circuits to ensure electrical integrity. Circuits are provided over cable or wire. *See* COMMUNICATION CABLES.

Telephone cables consist of conductors that are individually insulated with strip paper, paper pulp, or plastic; the conductors are then twisted into pairs, and the pairs are stranded to produce a cable core. The cores are protected by an outer sheath of plastic or, occasionally for special purposes, of lead. When plastic is used, a metallic shield, generally of aluminum but occasionally of copper, must be placed between the core and the sheath to provide electrical protection.

Wire circuits are open wire, multiple wire, or buried wire. Open wires are individual, bare, spaced wires supported by crossarms or bracket-mounted insulators on poles. Multiple wire consists of insulated conductors that are stranded around an insulated high-strength steel support wire, but does not have the outer sheath of a cable.

There are two predominant types of underground wire. One is composed of two pairs of 20-gage copper steel conductors with 40% conductivity. It is provided with an aluminum shield, protected by a plastic sheath, and is used for service connections from terminal to house in urban situations. The second type is a one-pair facility with 19-gage copper conductors protected by helix-wound steel armor wires covered by a plastic sheath. This facility is used predominantly for long, rural service connections.

Pole line design is based upon the aerial structures to be supported and the average winter weather conditions typical for the location. For economic reasons, communication lines and power lines often share the same poles. This is also done to avoid the objections to two adjacent parallel pole lines.

Underground lines are placed in conduit, which runs between manholes placed at 400- to 1000-ft (100- to 300-m) intervals. Manholes are usually of concrete cast in position; however, they also may be precast. Conduit is available in a number of materials, including concrete, fiber, fiber cement, iron pipe, and vitrified clay. *See* TELEPHONE SERVICE. [G.L.C.]

Telephoto lens

Telephoto lens A photographic lens system specially designed to give a large image of a distant object in a camera of relatively short focal length. A telephoto lens generally consists of a positive lens system and a negative lens system, separated by a considerable distance. *See* CAMERA; LENS (OPTICS). [M.J.H.]

Telephotography The transmission of photogaphs over electrical communication channels. The channels are generally those provided by common carrier communication companies. Individual users furnish telephotograph or facsimile machines to suit their requirements. In the United States such services range from simple arrangements connecting two sets of machines within a city to large networks involving many cities and types of communication facilities. Coordination must thus be effected between the communication companies and the manufacturers of equipment used by customers for their stations to ensure compatibility of equipment and optimum system performance. Unattended reception is made possible by starting and synchronizing receiving machines with pulses or tones generated at the transmitting location.

The transmission of news photographs and weather maps is one example of a use with contrasting requirements involving different types of machines and networks. Normally, news pictures require half-tone reproductions with a number of shades of gray. At the receiving locations photographic reproducing devices are used. In contrast, weather maps and other black-and-white nonshaded material can readily be reproduced by direct recording processes. *See* FACSIMILE. [C.C.Du.]

Teleprocessing A form of information handling in which a data-processing system utilizes communication facilities. The chief example of this mode of computing is computer timesharing, which requires communication support facilities, both hardware and software. Communications support facilities will be considered in terms of the logical communications support required by the user, and the hardware or software which are required to implement communications support functions.

It is important to consider any computer or communications requirement first in terms of the logical process which the user wants to perform independent of the actual physical (that is, hardware and software) means of implementation. Logical process refers to functions, operations, the sequence of operations, and characteristics of the data to be manipulated by the operations.

One of the major user considerations in specifying timesharing communications support is the choice between asynchronous and synchronous communication. The contrast between the two types of communication has both logical and physical aspects. From the logical standpoint, asynchronous communication is characterized by sporadic transmissions of data from terminal to computer—independent bursts in time and content, as in inquiries made for a particular purpose. This mode can be compared to synchronous communication in which data transmissions are neither time- nor content-independent, as during file transfers between a microcomputer and a mainframe machine. *See* DATA COMMUNICATIONS.

The major pieces of communications hardware which are required to support timesharing are a terminal, a modem and communication facility, and a communication controller at the mainframe computer site.

The terminal may have line or full-screen editing capability. Because of the relatively low data transmission rate and the sporadic data entry pattern, line editing terminals are typically asynchronous, whereas the block transmission format of a full-screen editing terminal demands high-speed, clock-controlled transmission as in synchronous communication.

A modem is required to convert the digital signals of the terminal to the analog signals of the telephone network and back again to the digital signals of the computer. Important modem characteristics involve choices between asynchronous or synchronous communication; half-duplex or full-duplex operation; the use of an acoustic coupler or a direct connect; and dial-up, leased line, or limited distance communication.

In the planning of teleprocessing requirements, the characteristics of equipment in the mainframe computer room with which the terminal communicates are frequently overlooked. This may be a controller, front-end processor, or mainframe. The terminal and modem must be compatible with the protocol used by the mainframe or its subordinate unit, such as the controller.

As indicated above, an important addition to teleprocessing communications support is software for providing communications parameter management, menu-selectable commands for invoking a variety of communication protocols, and some constraints on terminal-to-mainframe file transfer. Usually these capabilities are furnished by a microcomputer which takes the place of the terminal with disk storage, an asynchronous communication adapter board installed in the microcomputer, asynchronous communications software operating in the microcomputer and compatible with its operating system, and a modem which is compatible with both the communications software and adapter. *See* DIGITAL COMPUTER; MICROCOMPUTER; MULTIACCESS COMPUTER. [N.F.S.]

Telescope An instrument used to collect, measure, or analyze radiation from a distant object. Most commonly, telescope refers to an assemblage of lenses or mirrors, or both, that enhances the ability of the eye either to see objects more distinctly or to see fainter objects. In its most general meaning, telescope refers to a device that collects radiation, which may be in the form of electromagnetic radiation or particle radiation, from a limited direction in space.

Optical telescopes may be classified as refracting telescopes, which use only lenses; reflecting telescopes, which use mirrors to focus light; and catadioptric telescopes, which use both lenses and mirrors. Special types of reflecting telescopes which depart radically from conventional ground-based systems are the multimirror telescope and the space telescope. *See* OPTICAL TELESCOPE.

As collectors of radiation from a specific direction, telescopes may be classified as focusing and nonfocusing. Nonfocusing telescopes are used for radiation with energies of x-rays and above. Focusing telescopes, intended for nonvisible wavelengths, are similar to optical ones, but they differ in the details of construction. *See* GAMMA-RAY ASTRONOMY; INFRARED ASTRONOMY; RADIO TELESCOPE; X-RAY TELESCOPE. [W.M.S.]

Telestacea An order of the subclass Alcyonaria. Telestacea are typified by *Telesto*, which forms an erect branching colony by lateral budding from the body wall of an elongated primary or axial polyp. The stolon is bandlike or membranous. Sclerites are scattered singly, partly fused, or entirely fused to form a rigid tube. *See* ALCYONARIA. [K.At.]

Teletypesetter A system for automatically operating a linecasting machine (Linotype or Intertype) to produce lines of type at high speed under the control of perforated tape. Basically, this system consists of separating the complex manual operation of a linecasting machine into two simple operations and making one of them fully automatic under the control of perforated tape.

The perforated tape is produced on a Teletypesetter perforator, which has a keyboard layout similar to that of a typewriter, plus additional keys to control the various functions of the linecasting machine. The perforated tape controls the Teletypesetter operating unit, which is attached to the linecasting machine keyboard. Thus the linecasting machine is fully automatic under the control of perforated tape produced on the perforator.

Teletypesetter equipment has been used by most daily and weekly newspapers and by many commercial shops throughout the world. By supplementing Teletypesetting equipment with Teletype sending and receiving equipment operating over special long-distance data telegraph circuits, the perforated tape could be used to produce type automatically in hundreds of printing plants throughout the country. [V.N.V.]

Teletypewriter An electromechanical device, also called a teleprinter, for transmitting and receiving messages over a telegraph circuit. A sending and receiving teletypewriter performs two functions: The keyboard transmitter generates coded electrical signals for transmission over a telegraph circuit; and the typing unit converts such signals into a printed message. A page teletypewriter (see illustration) prints a message in page form, usually on a continuous roll of paper $8\frac{1}{2}$ in. (21.6 cm) wide.

Page teletypewriter.

The Baudot code, once widely used in printing telegraph devices, consists of five code pulses, any one of which may be either marking or spacing. In single-current signals, a marking pulse is an interval of time during which current flows through the circuit, and a spacing pulse is an interval during which no current flows. In polar signals, frequently used over long lines, a marking pulse is an interval during which negative current flows, and a spacing pulse is an interval during which positive current flows. There are 32 possible combinations of the five code pulses. One combination is assigned to each of the 26 letters of the alphabet. Because only 26 code combinations are available for printing in each of the two shift positions, both capital and lowercase letters cannot be used. The uppercase position is reserved for the 10 digits, punctuation marks, and commonly used symbols.

On June 17, 1963, the American Standards Association (now the American National Standards Institute) adopted a new American Standard Code for Information Interchange (ASCII). The code consists of seven code pulses, or bits, instead of five as in the Baudot code. There are thus 2^7, or 128, discrete permutations of the seven bits in the code. Sixty-four of these permutations were assigned to printing characters, including the capital letters of the alphabet, the ten digits, punctuation marks, and special symbols, Thirty-five permutations were assigned to control characters, and the remainder were unassigned. A revised code was approved in 1967. [F.W.S.]

Teletypewriter exchange (TWX) service A teletypewriter communication service provided by Western Union for approximately 53,000 subscribers in the continental United States, except Alaska. Direct interconnection is available to the Canadian TWX network. Canadian Telex subscribers are able to make southbound calls to United States TWX subscribers. Access to Western Union Telex subscribers is on a message-switched basis. *See* TELETYPEWRITER; TELEX.

Any TWX subscriber can communicate directly with any other TWX subscriber. Connections are established in a manner similar to that of the telephone and Telex networks. After a connection is established between two TWX subscribers, communication is carried on in a conversational mode, using the teletypewriter keyboard and printer. TWX subscribers may also have teletypewriters equipped with tape sending and receiving units. This type of teletypewriter allows the subscriber to prepare messages in advance for later transmission. *See* TELEGRAPHY. [E.C.M.]

Television The electrical transmission and reception of transient visual images. Like motion pictures, television consists of a series of successive images, which are registered on the brain as a continuous picture because of the persistence of vision. Each visual image impressed on the eye persists for a fraction of a second. In television in the United States, 30 complete pictures are transmitted each second, which with the use of interlaced scanning is fast enough to avoid evident flicker.

At the television transmitter, minute portions of a scene are sampled individually for brightness (and color for color television), and the information for each portion is transmitted consecutively. At the receiver, each portion is reproduced in its proper position and with correct brightness (and color) to reproduce the original scene. *See* COLOR TELEVISION.

The scene is focused on a photoelectric screen of a camera tube. Each portion of the screen is changed by the photoelectrons to a degree depending upon the brightness of the particular portion. The screen is scanned by an electron beam just as a reader scans a page of printed type, character by character, line by line. When so scanned, an electric current flows with an instantaneous magnitude proportional to the brightness of the portion scanned. *See* TELEVISION CAMERA; TELEVISION CAMERA TUBE.

Variations in the current are transmitted to the receiver, where the process is reversed. An electron beam in the picture tube is varied in intensity (modulated) by the incoming signals as it scans the picture-tube screen in synchronism with the scanning at the transmitter. The photoelectric surface of the picture tube produces light in proportion to the intensity of the electron beam which strikes it. In this way the minute portions of the original scene are re-created in their proper positions, brightness, and (for color transmission) color values. *See* TELEVISION RECEIVER; TELEVISION TRANSMITTER.

In the United States an individual picture (frame) is considered to be made up of 525 lines, each line containing several hundred picture elements. All these lines are scanned and the light values are sent to the receiver so that each second 30 pictures are received.

The bandwidth required for any information transmission system is a function of the number of bits of information, or the detail, to be transmitted per second. The standard of 525 lines sets the vertical detail, and the standard aspect ratio (picture width to height) of 4:3 requires 700 horizontal picture elements for equal horizontal and vertical resolution, or 350 sets of alternate black-and-white squares. The picture is reproduced 30 times a second, for a total of $525 \times 350 \times 30$ or 5,512,500 complete cycles per second. Less detail than this is actually transmitted and received; the highest video frequency actually transmitted is 4.2 MHz. *See* BANDWIDTH REQUIREMENTS (COMMUNICATIONS).

The band of frequencies assigned to a television station for the transmission of synchronized picture and sound signals is called a television channel. In the United States a television channel is 6 MHz wide, with the visual carrier frequency 1.25

MHz above the lower edge of the band and the aural carrier 0.25 MHz below the upper edge of the band.

The sound portion of the program is transmitted by frequency modulation. Maximum frequency deviation (bandwidth) of the sound signals is 25 kHz. Transmission of visual signals uses amplitude modulation and vestigial sideband transmission. *See* AMPLITUDE MODULATION; FREQUENCY MODULATION.

Television programs are recorded for rebroadcast at a later time and for many other reasons. The principal method is video tape recording (VTR) in which television sound and pictures are recorded on magnetic tapes. *See* CLOSED-CIRCUIT TELEVISION; TELEVISION STANDARDS. [S.De.S.]

Television camera An electrooptical system used to pick up and convert a visual image or scene into an electrical signal called video. The video may be transmitted by cable or wireless means to a suitable receiver or monitor some distance from the actual scene.

A television camera may fall within one of several categories: studio, portable, or telecine (Fig. 1). It may also be one of several highly specialized cameras used for remote viewing of inaccessible places, such as the ocean bottom or inside nuclear power reactors. The camera may be capable of producing color or monochrome (black and white) pictures. Modern cameras are entirely solid-state, except for the actual pickup tube. Even the pickup tube may be replaced by solid-state semiconductors called charge-coupled devices. *See* TELEVISION CAMERA TUBE.

Fig. 1. Studio television camera.

Essential elements. Every camera shares certain essential elements: an optical system, one or more picture pickup devices, preamplifiers, scanning circuits, blanking and synchronizing circuits, and video processing circuits and control circuits. Color cameras also include some kind of color-encoding circuit.

A television camera may be one-piece, with all components contained in one assembly, or it may consist of a head and a camera control unit (CCU) connected by a multiple-conductor cable, and be called a camera chain. The head unit comprises the optical system with lens, the picture pickup tubes, and a minimum of electronics necessary to generate and amplify the minute signals from the pickup tubes. The CCU may contain electronics and controls which allow a skilled operator to adjust brightness (luminance), color (balance, saturation, and hue), and certain correction circuits (registration, gamma, and aperture) which improve the picture. A specially calibrated oscilloscope, called a waveform monitor, is generally provided to the CCU operator so that accurate voltage levels may be set. Modern cameras provide for automatic as well as manual setup of some or all of the above adjustments. Some cameras have a built-in microprocessor which enables complete setup of all parameters by simply pushing one button.

Typical configurations. Live studio cameras are equipped with several ancillary systems to enhance their operation. An electronic viewfinder, actually a small television monitor, shows the camera operator what the camera is seeing, making it possible to frame and focus the picture. The tally system consists of one or more red lights which illuminate when the camera's picture is "on the line" so that production and on-camera personnel know which camera is active. Generally an intercom system is built into the camera so that the director can communicate with the camera operator. The camera itself may be mounted upon a tripod, but more often it is on a dolly and pedestal, which allows the camera to be moved around on the studio floor and raised or lowered as desired. A pan head permits the camera to be rotated to the left or right and furnishes the actual mounting plate for the camera. The lens zoom and focus controls are mounted on a panning handle convenient to the operator. A common accessory on studio cameras is the videoprompter. This is a television monitor on which the program script may be displayed and read by the on-camera personnel.

Telecine cameras are used in conjunction with film or slide projectors to televise motion pictures and still images. They

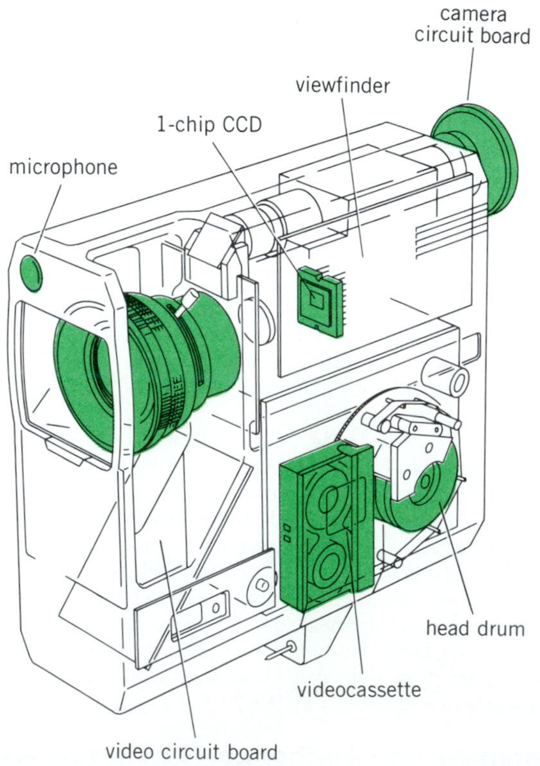

Fig. 2. Prototype portable charge-coupled device (CCD) camera with built-in videotape recorder. (*Sony Corporation*)

generally employ vidicon tubes, one for monochrome and three for color, and many of the usual controls are automatic so as to require less operator attention.

Portable cameras usually combine all of the basic elements into one package and may be used for a multitude of purposes. The units are lightweight, often have built-in microphones for sound pickup, and can be handled by one person. Used in conjunction with modern battery-powered portable videocassette recorders, an entire video and audio pickup system can be carried and operated by one or two people (Fig. 2). *See* COLOR TELEVISION; TELEVISION; TELEVISION RECEIVER; TELEVISION STANDARDS; TELEVISION TRANSMITTER. [E.F.A.]

Television camera tube An electron tube having a light-sensitive receptor that converts an optical image into an electrical television video signal. The tube is used in a television camera to generate a train of electrical pulses representing the light intensities present in an optical image focused on the tube. Each point of this image is interrogated in its proper turn by the tube, and an electrical impulse corresponding to the amount of light at that point of the optical image is generated by the tube. This signal represents the video or picture portion of a television signal. *See* TELEVISION CAMERA.

Image orthicon. The image orthicon made broadcast television practical. It was used for more than 20 years as the primary studio and field camera tube for black and white and color television programming because of its high sensitivity and its ability to handle a wide range of scene contrast and to operate at very low light levels. It is one of the most complicated camera tubes. The image orthicon is divided into an image section, a scanning section, and a multiplier section, within a single vacuum envelope. The image isocon is a further development of the image orthicon.

Photoconductive tubes. These types have a photoconductor as the light-sensitive portion. The name vidicon was applied to the first photoconductive camera tube developed by RCA. It is loosely applied to all photoconductive camera tubes, although some manufacturers adopt their own brand names. The vidicon tube is a small tube that was first developed as a closed-circuit or industrial surveillance television camera tube. The development of new photoconductors has improved its performance to the point where it is now utilized in one form or another in most television cameras. Its small size and simplicity of operation make it well suited for use in systems to be operated by relatively unskilled people.

The vidicon is a simply constructed storage type of camera tube (see illustration). The signal output is developed directly from the target of the tube and is generated by a low-velocity scanning beam from an electron gun. The target generally consists of a transparent signal electrode deposited on the faceplate of the tube and a thin layer of photoconductive material, which is deposited over the electrode. The photoconductive

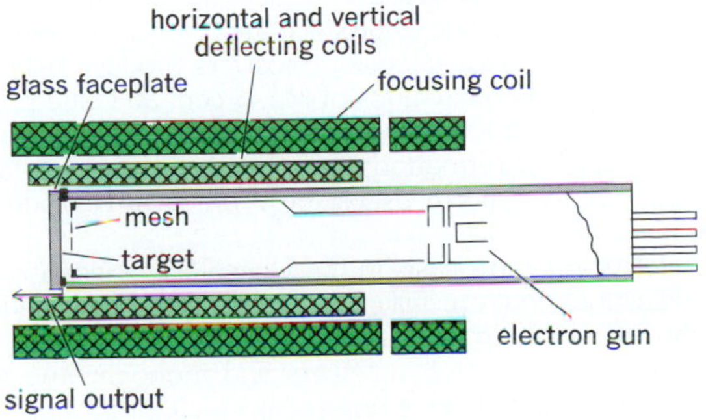

Cross section of a vidicon tube and its associated deflection and focusing coils.

layer serves two purposes. It is the light-sensitive element, and it forms the storage surface for the electrical charge pattern that corresponds to the light image falling on the signal electrode.

Photoconductor properties determine to a large extent the performance of the different types of vidicon tubes. The first and still most widely used photoconductor is porous antimony trisulfide. The latest photoconductors are the lead oxide, selenium-arsenic-tellurium, cadmium selenide, zinc-cadmium telluride, and silicon diode arrays.

Silicon intensifier. The silicon intensifier camera tube utilizes a silicon diode target, but bombards it with a focused image of high-velocity electrons. Each high-energy electron can free thousands of electron carriers in the silicon wafer (compared to one carrier per photon of light on a silicon diode vidicon). This high amplification allows the camera to operate at light levels below that of the dark-adapted eye. The silicon intensifier tube is utilized for nighttime surveillance and other extremely low-light-level television uses in industrial, scientific, and military applications.

Solid-state imagers. These are solid-state devices in which the optical image is projected onto a large-scale integrated-circuit device which detects the light image and develops a television picture signal. Typical of these is the charge-coupled-device imager. The term charge-coupled device (CCD) refers to the action of the device which detects, stores, and then reads out an accumulated electrical charge representing the light on each portion of the image. The device detects light by absorbing it in a photoconductive substrate, such as silicon. The charge carriers generated by the light are accumulated in isolated wells on the surface of the silicon that are formed by voltages applied to an array of electrodes on top of an oxide insulator formed on the surface of the silicon. A practical CCD imager consists of a structure that forms several hundred thousand individual wells or pixels, and transfers the charges accumulated in these pixel wells out to an output amplifier in the proper sequence. *See* CHARGE-COUPLED DEVICES. [R.G.N.]

Television networks Arrangements of communications channels, suitable for transmission of video and accompanying audio signals, which link together groups of television broadcasting stations or closed-circuit television users in different cities so that programs originating at one point can be fed simultaneously to all others.

In the United States, television network service is furnished by the telephone companies. The facilities consist of intercity channels, which interconnect the principal long-distance telephone offices in various cites, and local channels, which connect the telephone offices with the broadcasters' studios or other user locations in each city. In the telephone offices the intercity and local channels are brought together in a television operating center (TOC) where means are provided for testing, monitoring, and connecting the channels in various patterns as required for service.

Use of networks. The principal users of television services are the broadcasting companies. Each of the broadcasting companies contracts on a monthly basis for daily service from its principal originating points to the various broadcasting stations affiliated with its network.

The broadcasting companies frequently order "occasional" channels in addition to their normal networks in order to pick up programs from other than the usual originating points, or to supplement the normal layout with additional channels when it is desired to split the network into several parts, each receiving a different program.

Television networks are also provided, both on a regular daily and occasional basis, for educational television (ETV) users. These are generally state and regional networks interconnecting schools, hospitals, or government agencies as well as ETV broadcasting stations. To keep charges down, these channels frequently run from one user location directly to another, without passing through telephone company operating centers.

Television networks are also used for transmitting programs which are not broadcast but are displayed on screens at the receiving locations. These are called closed-circuit television networks. Their principal use is to distribute sports events to the theaters and hotels and for industrial conferences and displays. These networks are generally temporary. *See* CLOSED-CIRCUIT TELEVISION.

Intercity channels. The facilities for intercity video channels are largely TD-2 radio relay routes which are also used to furnish intercity telephone circuits. TD-2 is a frequency-modulated (FM) microwave radio system capable of providing 12 wide-band channels in each direction in the frequency band from 3700 to 4200 MHz. Each channel is capable of carrying one video signal or up to 1200 voice channels. The relay stations are spaced an average of 25–30 mi (40–48 km) apart.

Local channels. The local channels, which connect the telephone company office with the broadcasters' studios and transmitters and with closed-circuit users' locations, are largely composed of cable facilities. Local channels can also be provided by microwave radio facilities. Economic considerations usually determine whether radio or cable should be used, but the decision may be influenced by the need for a speedy installation, such as for one-time news or sports events.

Satellites. Commercial television service is available over both the Atlantic and Pacific oceans using satellites furnished by the International Satellite Consortium managed by Comsat. Both the American Telephone and Telegraph Co. and the United States international telegraph carriers furnish such services, renting the overseas channels from Comsat. The overseas channels are connected to the television networks in the United States at the TOCs in New York or San Francisco by terrestrial links to the Earth stations furnished by the telephone companies. *See* COMMUNICATIONS SATELLITE; TELEVISION. [J.R.R.]

Television receiver The equipment used to receive the transmitted modulated radio-frequency signals and produce synchronized visual images and sound for entertainment or educational purposes. The radio-frequency portion operates on the superheterodyne principle. *See* AMPLITUDE MODULATION; FREQUENCY MODULATION; RADIO RECEIVER.

The first television receivers to be mass-produced were monochrome; that is, they provided pictures in black and white only. Later, color receivers, which produce pictures in full color as well as black-and-white, became available. *See* TELEVISION.

Monochrome. Since most broadcast television transmissions in the United States are horizontally polarized, the most basic type of television-receiving antenna is the horizontally mounted half-wave dipole. Because the stations serving a given area may operate on widely different frequencies, however, the dipole dimensions must be a compromise that permits reasonable performance on all the desired channels. More complex antennas combine several dipole elements of various lengths, and passive reflectors may be used to achieve some degree of horizontal directivity. *See* ANTENNA (ELECTROMAGNETISM).

The tuner selects the desired channel and converts the frequencies received to lower frequencies within the passband of an intermediate-frequency amplifier. The output from the tuner is applied to this amplifier. Several stages of amplification are required to obtain the desired output signal level and selectivity.

The output of the i-f amplifier consists of two modulated rf signals. One of these, which is amplitude-modulated, provides a varying signal corresponding to the black-and-white portions of the picture, a blanking signal to render the return trace invisible on the picture tube, horizontal sync pulses to initiate the retrace of the beam at the end of each line, and vertical sync pulses to initiate the retrace of the beam at the end of each picture field. The other signal is frequency-modulated and contains the transmitted sound information. These two rf signals are applied to a diode which produces a rectified output that follows the instantaneous peak value of the amplitude-modulated picture carrier.

Coincidentally a 4.5-MHz signal results from the heterodyne beat of the picture and sound carriers. This signal contains the frequency-modulated sound information, which can be further amplified and detected in the sound channel. This is known as the intercarrier sound (ICS) system. For sound reproduction, the intercarrier sound signal formed at the detector is passed through a 4.5-MHz i-f amplifier, through some form of amplitude limiting, and to an FM detector, which converts the sound carrier modulation to an audio-frequency signal. This signal is then passed through an audio amplifier to a loudspeaker and converted to acoustic output.

Since television receivers, like radio receivers, may be subjected to widely varying incoming signal strengths, some form of automatic gain control (AGC) is necessary. *See* AUTOMATIC GAIN CONTROL (AGC).

Picture synchronizing information is obtained from the video signal by means of sync separation circuits, In addition, these circuits must separate this information from noise and interference during the reception of weak signals, particularly if impulse noise is present.

Two independent sweep systems are employed in the vertical and horizontal sweep circuits. Each employs a timing generator, generally of the oscillatory type, controlled by the synchronizing information obtained from the sync separators. The oscillators are followed by drive and waveform-shaping circuits. These are followed by power amplifier stages capable of providing the currents required by the deflection coils of the yoke for picture-tube beam deflection. Substantially different techniques are required for vertical and horizontal scanning.

The display device for a television receiver is a cathode-ray tube, consisting of an evacuated bulb containing an electron gun and a phosphor screen, which emits light when excited by an electron beam. The intensity of the electron beam is controlled by the video signal, which is applied either to the grid or the cathode of the electron gun. The position of the electron beam is controlled by electromagnetic fields produced by the deflection yoke placed around the neck of the tube. *See* CATHODE-RAY TUBE.

Color receivers. Television receivers designed to produce images in full color are necessarily more complex than those designed to produce monochrome images only, because additional information must be handied to produce color. In monochrome systems, the video signal controls only the luminance of the various areas of the image. In color systems, it is necessary to control both the luminance and chrominance of the picture elements.

For color television purposes, chrominance is most frequently expressed quantitatively in terms of the amounts of two hypothetical, zero-luminance primary colors (usually designated *I* and *Q*), which must be added to or subtracted from a neutral color of a given luminance to produce the color in question. As a practical matter, color television receivers produce full-color images as additive combinations of red, green, and blue primary-color images, and it is necessary to process the luminance and chrominance information contained in a color signal in such a way as to make it usable by a practical reproducing device.

Color television broadcasts in the United States employ signal specifications that are fully compatible with those used for monochrome, making it possible for color programs to be received on monochrome receivers and monochrome programs to be received on color receivers. Compatibility is achieved by encoding the color information at the transmitting end of a color television system in such a way that the transmitted signal consists essentially of a normal monochrome signal (conveying luminance information) supplemented by an additional modulated wave conveying chrominance information.

Special decoding circuits are necessary in a color receiver to process the luminance and chrominance information in a color signal so that it can be used for the control of a practical color kinescope utilizing red, green, and blue primary colors.

The great majority of color television receivers employ the shadow-mask color kinescope, in which color images are produced in the form of closely intermingled red, green, and blue dots. The primary-color phosphor dots are excited by three separate electron beams, which are prevented from striking dots of the wrong color by the shadowing effect of an aperture mask located about ½ in. (1.25 cm) behind the special phosphor screen. The three beams in such a kinescope are all deflected simultaneously by the fields produced by a single deflection yoke placed in the conventional position around the neck of the tube. It is necessary, however, to provide an auxiliary deflection system to maintain convergence of the three beams in all areas of the viewing screen, so that the primary-color images are properly registered. A special convergence yoke is commonly placed around the neck of the kinescope, just ahead of the electron guns; this yoke has separate coils for electromagnetic control of the positions of the three beams. *See* COLOR TELEVISION. [C.M.Si./J.W.W.]

Television scanning The process of scrutinizing the brightness of each element of detail contained in the image of a scene to be transmitted by television. In monochrome television, the process is instrumental in converting the brightness of each individual element so scrutinized into a unique voltage-time response suitable for transmission. In color television, the brightness variations of each scene are first separated by red, green, and blue filters, after which the conversion from brightness to voltage on a time basis occurs separately for each of the three colors. Scanning also takes place in the receiver in exact synchronism with the camera-tube scanning, and synchronizing signals are transmitted for that purpose. *See* COLOR TELEVISION; TELEVISION; TELEVISION RECEIVER.

An image is analyzed by scanning it according to a fine structure of parallel, nearly horizontal lines called a scanning raster. The complete raster is rectangular in shape. If all the lines are scanned in sequence, the process is called sequential scanning.

A variation of this kind of scanning, called interlaced scanning, is used to conserve bandwidth in the transmission system without introducing intolerable flicker. Flicker is a function of the frequency of repetition of coverage of the raster. With interlaced scanning, alternate (odd-numbered) lines are scanned first, and the remaining (even-numbered) lines are scanned next. The entire raster area is covered or scanned twice. Therefore, the picture repetition rate for interlaced scanning is twice that of sequential scanning, which is at the same velocity along each line, and a corresponding reduction in the sensation of flicker is obtained. This is called double interlacing and is standard in all broadcast television systems. The entire raster is covered 30 times per second; therefore, the picture area is scanned 60 times a second. [D.G.F.]

Television standards Numerical and graphical criteria which describe essential aspects of a television system, employed in the design and operation of equipment to assure that the various parts of the system (studios, transmitters, networks, and receivers) will operate in cooperative fashion at maximum performance. Television systems have a special need, compared with other communication systems, for definitive standards because television transmitters and receivers must operate in a precise "lock-and-key" relationship. In particular, the scanning of the image in the camera must be matched by the scanning in every associated receiver within a timing precision of approximately one-tenth of a millionth of a second, and with relative positions of picture details correct to a few hundredths of an inch. *See* TELEVISION SCANNING.

To assure that any television receiver can receive programs from any transmitter within range, it is customary to set up a single set of standards within a group of neighboring countries. The systems of Mexico, the United States, Canada, and Cuba operate on one such set of standards. Those of Europe also conform to a single set of standards, which differ in detail from those used in North America. Similar groupings of standards are found in other regions of the world. Television standards are promulgated, and conformity to them by television stations is legally enforced by governmental agencies having responsibility for communications systems in the respective countries. In the United States the responsible agency is the Federal Communications Commission. Conformity is not usually enforced in the case of television receiver design, but is assured by the desire of designers to make best use of available signals.

Television standards are of two kinds. Transmission standards describe the signals radiated by television transmitters and their relation to the intended visual and aural content of the program. Allocation standards define the manner in which the radiated signals occupy the radio spectrum. These affect primarily the distances over which television stations provide satisfactory service in the presence of interference from transmitters, receivers, and other sources. *See* TELEVISION. [D.G.F.]

Television transmitter An electronic device that converts audio and video signals into modulated radio-frequency (rf) energy which can be radiated from an antenna and received on a television receiver. The term can also refer to the entire television transmitting plant, consisting of the transmitter proper, associated visual and aural input and monitoring equipment, transmission line, the antenna with its tower or other support structure, and the building in which the equipment is housed.

A television transmitter is really two separate transmitters integrated into a common cabinet. Video information is transmitted via a visual transmitter, while audio information is transmitted via an aural transmitter. Because video and audio have different characteristics, the two transmitters differ in terms of bandwidth, modulation technique, and output power level. Nevertheless, a common transmitting antenna is generally used, and the two transmitters feed this antenna via an rf diplexer or combiner.

Television stations are licensed to operate on a particular channel, but since it takes a very wide bandwidth to transmit a television picture, these channels are allocated over a broad range of frequencies. Channels 2 through 6 are low-band very-high-frequency (VHF) channels, while channels 7 through 13 are high-band VHF channels. Channels 14 through 83 are ultra-high-frequency (UHF) channels. Because of the wide range of frequencies, television transmitters are designed to work in only one of the foregoing groups, and employ specific circuits which are most efficient for the channels involved.

The horizontal radiation pattern of most television transmitting antennas is circular, providing equal radiated signal strength to all points of the compass. Higher-gain antennas achieve greater power in the direction of the horizon by reducing the power radiated at vertical angles above and below the horizon. Because television signals travel in a "line of sight," transmitting antennas are usually placed as high as possible above ground with respect to the surrounding service area. It is also desirable to locate all of the transmitting antennas serving a given locality in the same place. This allows viewers to orient their receiving antennas in one direction for the best reception from all of the stations. *See* ANTENNA (ELECTROMAGNETISM).

There are two broad classes of VHF visual television transmitter design philosophy. The classical approach modulates the carrier at a moderate power level, amplifies the carrier to rated output power by means of linear amplifiers, and then filters this high-power carrier to obtain the required vestigial-sideband signal. The more contemporary approach employs modulation at a very low power level of an intermediate-frequency (if) signal. The required vestigial-sideband filtering is imposed on this low-level signal, generally by means of a highly stable surface acoustic-wave (SAW) filter, whereupon the signal is upconverted to the carrier frequency and amplified by linear amplifiers to rated output power. *See* AMPLIFIER; SURFACE ACOUSTIC-WAVE DEVICES; TELEVISION. [E.F.A.]

Telex A worldwide teleprinter exchange service providing direct send and receive teleprinter connections between subscribers. The rules for its operation are usually determined within the framework of recommendations by the Consulting Committee for International Telephone and Telegraph (CCITT).

In the continental United States (except Alaska), Western Union provides the domestic Telex service. Direct interconnection is made with the Telex systems of Alaska, Canada, and Mexico. Overseas connections are made through various international companies.

Other services available to domestic Telex subscribers are: TEL(T)EX, which permits subscribers to send messages to selected terminal cities in the United States and Canada; INFOMASTER™, which provides for the delayed delivery of messages, multiple-address messages, collect calls, international and domestic telegrams, MAILGRAM®, and TWX™ messages, on a message-switched basis; FYI News Service, which provides subscribers with news bulletins and financial, market, weather, and other special reports; and Datagram™, which permits telephone-entered messages to be delivered to Telex and TWX subscribers.

Teleprinters or automatic send and receive sets are provided in the domestic Telex system. *See* TELETYPEWRITER; TELETYPEWRITER EXCHANGE (TWX) SERVICE. [E.C.M.]

Tellurium A chemical element, Te, atomic number 52, and chemical atomic weight 127.60. There are eight stable isotopes of natural tellurium. Tellurium makes up approximately 10^{-9}% of the Earth's igneous rock. It is found as the free element, sometimes associated with selenium. It is more often found as the telluride sylvanite (graphic tellurium), nagyagite (black tellurium), hessite, tetradymite, altaite, coloradoite, and other silver-gold tellurides, as well as the oxide, tellurium ocher.

There are two important allotropic modifications of elemental tellurium, the crystalline and the amorphous forms. The crystalline form has a silver-white color and metallic appearance. This form melts at 841.6°F (449.8°C) and boils at 2534°F (1390°C). It has a specific gravity of 6.25, and a hardness of 2.5 on Mohs scale. The amorphous form (brown) has a specific gravity of 6.015. Tellurium burns in air with a blue flame, forming tellurium dioxide, TeO_2. It reacts with halogens, but not sulfur or selenium, and forms, among other products, both the dinegative telluride anion (Te^{2-}), which resembles selenide, and the tetrapositive tellurium cation (Te^{4+}) which resembles platinum(IV).

Tellurium is used primarily as an additive to steel to increase its ductility, as a brightener in electroplating baths, as an additive to catalysts for the cracking of petroleum, as a coloring material for glasses, and as an additive to lead to increase its strength and corrosion resistance. *See* SELENIUM. [S.Ki.]

1	2	3	4	5	6	7	8	9	10	11	12	13	14	15	16	17	18
1 H																	2 He
3 Li	4 Be											5 B	6 C	7 N	8 O	9 F	10 Ne
11 Na	12 Mg											13 Al	14 Si	15 P	16 S	17 Cl	18 Ar
19 K	20 Ca	21 Sc	22 Ti	23 V	24 Cr	25 Mn	26 Fe	27 Co	28 Ni	29 Cu	30 Zn	31 Ga	32 Ge	33 As	34 Se	35 Br	36 Kr
37 Rb	38 Sr	39 Y	40 Zr	41 Nb	42 Mo	43 Tc	44 Ru	45 Rh	46 Pd	47 Ag	48 Cd	49 In	50 Sn	51 Sb	52 Te	53 I	54 Xe
55 Cs	56 Ba	71 Lu	72 Hf	73 Ta	74 W	75 Re	76 Os	77 Ir	78 Pt	79 Au	80 Hg	81 Tl	82 Pb	83 Bi	84 Po	85 At	86 Rn
87 Fr	88 Ra	103 Lr	104 Rf	105 Db	106 Sg	107 Bh	108 Hs	109 Mt	110	111	112	113	114	115	116	117	118

lanthanide series	57 La	58 Ce	59 Pr	60 Nd	61 Pm	62 Sm	63 Eu	64 Gd	65 Tb	66 Dy	67 Ho	68 Er	69 Tm	70 Yb
actinide series	89 Ac	90 Th	91 Pa	92 U	93 Np	94 Pu	95 Am	96 Cm	97 Bk	98 Cf	99 Es	100 Fm	101 Md	102 No

Telosporea A class of the subphylum Sporozoa. These protozoa are divided into two subclasses, the Gregarinia and Coccidia. All members of the group are either intra- or extracellular parasites, and the life cycles have both sexual and asexual phases. The spores lack a polar capsule and develop from an oocyst. The sporozoite is the usual infective stage which initiates the asexual phase in the life cycle. *See* COCCIDIA; GREGARINIA; PROTOZOA; SPOROZOA. [E.R.B./N.D.L.]

Temnopleuroida An order of Eichinacea with a camarodont lantern, smooth or sculptured test, tubercles imperforate or perforate (and usually crenulate), ambulacral plates of diademoid or echinoid type, and branchial slits which are usually shallow. There are three included families, Glyphocyphidae, Temnopleuridae, and Toxopneustidae. *See* ECHINACEA. [H.B.F.]

Temnospondyli The more common of the two major orders of the great extinct subclass Labyrinthodontia. Temnospondyls are differentiated from anthracosaurs by having the cheek suturally attached to the skull table. Their vertebral structure, in which the pleurocentra are seldom prominent and the intercentra are large, resembles that of some members of the rhipidistian crossopterygians, which include the ancestors of all amphibians. Terrestrial temnospondyls achieved their greatest diversity in the upper Carboniferous and lower Permian, during which time they were the most common land vertebrates. The last of the large aquatic temnospondyls occur in the Lower Jurassic of Australia. *See* AMPHIBIA; ANTHRACOSAURIA; CROSSOPTERYGII; LABYRINTHODONTIA. [R.L.C.]

Temperature A concept related to the flow of heat from one object or region of space to another. The term refers not only to the senses of hot and cold but to numerical scales and thermometers as well. Fundamental to the concept of temperature are the absolute scale and absolute zero and the relation of absolute temperatures to atomic and molecular motions.

Thermometers do not measure a special physical quantity. They measure length (as of a mercury column) or pressure or volume (with the gas thermometer at the National Bureau of Standards) or electrical voltage (with a thermocouple). The basic fact is that, if a mercury column has the same length when touching two different, separated objects, when the objects are placed in contact no heat will flow from one to the other. *See* THERMOMETER.

The numbers on the thermometer scales are merely historical choices; they are not scientifically fundamental. The most widely used scales are the Fahrenheit (°F) and the Celsius (°C). The centigrade scale with 0° assigned to ice water (ice point) and 100° assigned to water boiling under one atmosphere pressure (steam point) was formerly used, but it has been succeeded by the Celsius scale, defined in a different way than the centigrade scale. However, on the Celsius scale the temperatures of the ice and steam points differ by only a few hundredths of a degree from 0° and 100°, respectively. The illustration shows how the Celsius and Fahrenheit scales compare and how they fit onto the absolute scales. *See* ICE POINT.

In 1848 William Thomson (Lord Kelvin), following ideas of Sadi Carnot, stated the concept of an absolute scale of temper-

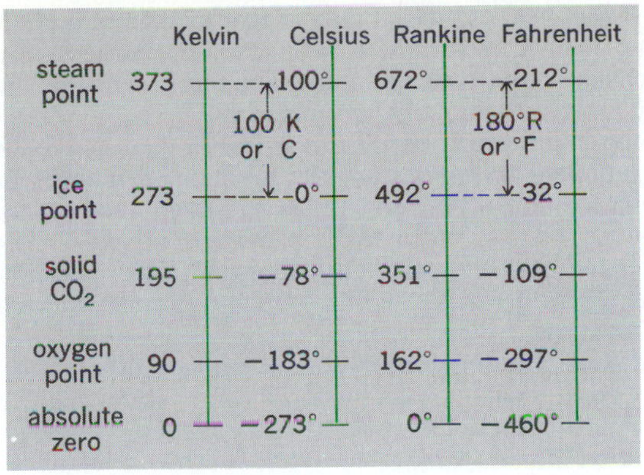

Comparisons of Kelvin, Celsius, Rankine, and Fahrenheit temperature scales. Temperatures are rounded off to nearest degree. (*After M. W. Zemansky, Temperatures Very Low and Very High, Van Nostrand, 1964*)

ature in terms of measuring amounts of heat flowing between objects. Most important, Kelvin conceived of a body which would not give up any heat and which was at an absolute zero of temperature. Experiments have shown that absolute zero corresponds to −273.15°C or −459.7°F. Two absolute scales, shown in the illustration, are the Kelvin (K) and the Rankine (°R).

In practice, absolute temperatures are measured by using low-density helium gas and dilute paramagnetic crystals, the most nearly ideal of real materials. The measurement of a single temperature with a gas or magnetic thermometer is a major scientific event done at a national standards laboratory. Only a few temperatures have been measured, including the freezing point of gold (1337.91 K), and the boiling points of sulfur (717.85 K), oxygen (90.18 K), and helium (4.22 K). Various types of thermometers (platinum, carbon, and doped germanium resistors; thermocouples) are calibrated at these temperatures and used to measure intermediate temperatures. *See* TEMPERATURE MEASUREMENT; THERMODYNAMIC PRINCIPLES. [R.A.Hu.]

Temperature adaptation The ability of animals to survive and function at widely different temperatures as a result of specific physiological adaptations. Temperature is an all-pervasive attribute of the environment that limits the activity, distribution, and survival of animals.

Changes in temperature influence biological systems, both by determining the rate of chemical reactions and by specifying equilibria. Because temperature exerts a greater effect upon the percentage of molecules that possess sufficient energy to react (that is, to exceed the activation energy) than upon the average kinetic energy of the system, modest reductions in temperature (for example, from 77 to 59°F or from 25 to 15°C, corresponding to only a 3% reduction in average kinetic energy) produce a marked depression (two- to threefold) in reaction rate. In addition, temperature specifies the equilibria between the formation and disruption of the noncovalent (electrostatic, hydrophobic, and hydrogen-bonding) interactions that stabilize both the higher levels of protein structure and macromolecular aggregations such as biological membranes. Maintenance of an appropriate structural flexibility is a requirement for both enzyme catalysis and membrane function, yet cold temperatures constrain while warm temperatures relax the conformational flexibility of both proteins and membrane lipids, thereby perturbing biological function. *See* CELL MEMBRANES; ENZYME; OSMOREGULATOR MECHANISMS.

Animals are classified into two broad groups depending on the factors that determine body temperature. For ectotherms, body temperature is determined by sources of heat external to the body; levels of resting metabolism (and heat production) are low, and mechanisms for retaining heat are limited. Such animals are frequently termed poikilothermic or cold-blooded, because the body temperature often conforms to the temperature of the environment. In contrast, endotherms produce more metabolic heat and possess specialized mechanisms for heat retention. Therefore, body temperature is elevated above ambient temperature; some endotherms (termed homeotherms or warm-blooded animals) maintain a relatively constant body temperature. There is no natural taxonomic division between ecto- and endotherms. Most invertebrates, fish, amphibians, and reptiles are ectotherms, while true homeothermy is restricted to birds and mammals. However, flying insects commonly elevate the temperature of their thoracic musculature prior to and during flight (to 96°F or 36°C), and several species of tuna retain metabolic heat in their locomotory musculature via a vascular countercurrent heat exchanger. *See* ADIPOSE TISSUE; HIBERNATION; THERMOREGULATION. [J.R.Ha.]

Temperature-humidity index An index developed by the U.S. Weather Bureau to give a single numerical value, in the general range of 70–80, which would reflect the outdoor atmospheric conditions of temperature and humidity as a measure of comfort (or discomfort) during the warm season of the year. The temperature-humidity index was formerly called the discomfort index. Temperature-humidity index I_{TH} is defined by the equation below.

$$I_{TH} = 0.4\ (\text{dry-bulb temp °F} + \text{wet-bulb temp °F}) + 15$$

When the index is 70 practically all people feel comfortable; when it is 80 no one is comfortable; and when it is 75 about half the population is satisfied. The index has been useful in the prediction and allocation of power-system loads resulting from the impact of air-conditioning equipment operation. *See* DEGREE-DAY; PSYCHROMETRICS; SENSIBLE TEMPERATURE. [T.Ba.]

Temperature inversion The increase of air temperature with height; an atmospheric layer in which the upper portion is warmer than the lower. Such an increase is opposite, or inverse, to the usual decrease of temperature with height, or lapse rate, in the troposphere. However, above the tropopause, temperature increases with height throughout the stratosphere, decreases in the mesosphere, and increases again in the thermosphere. Thus inversion conditions prevail throughout much of the atmosphere much or all of the time, and are not unusual or abnormal. *See* AIR TEMPERATURE; ATMOSPHERE.

Inversions are created by radiative cooling of a lower layer, by subsidence heating of an upper layer, or by advection of warm air over cooler air or of cool air under warmer air. Outgoing radiation, especially at night, cools the Earth's surface, which in turn cools the lowermost air layers, creating a nocturnal surface inversion a few inches to several hundred feet thick.

Inversions effectively suppress vertical air movement, so that smokes and other atmospheric contaminants cannot rise out of the lower layer of air. California smog is trapped under an extensive subsidence inversion; surface radiation inversions, intensified by warm air advection aloft, can create serious pollution problems in valleys throughout the world; radiation and subsidence inversions, when horizontal air motion is sluggish, create widespread pollution potential, especially in autumn over North America and Europe. *See* AIR POLLUTION; SMOG. [A.Cou.]

Temperature measurement Measurement of the hotness of a body relative to a standard scale. The fundamental scale of temperature is the thermodynamic scale, which can be derived from any equation expressing the second law of thermodynamics. Efforts to approximate the thermodynamic scale as closely as possible depend on relating measurements of temperature-dependent physical properties of systems to thermodynamic relations expressed by statistical thermodynamic equations, thus in general linking temperature to the average kinetic energy of the measured system. Temperature-measuring devices, thermometers, are systems with properties that change with temperature in a simple, predictable, reproducible manner. *See* TEMPERATURE; THERMODYNAMIC PRINCIPLES.

In the establishment of a useful standard scale, assigned temperature values of thermodynamic equilibrium fixed points are agreed upon by an international body (General Conference of Weights and Measures), which updates the scale about once every 20 years. Thermometers for interpolating between fixed points and methods for realizing the fixed points are prescribed, providing a scheme for calibrating thermometers used in science and industry.

The scale now in use is the International Temperature Scale of 1990 (ITS-90). Its unit is the kelvin, K, arbitrarily defined as 1/273.16 of the thermodynamic temperature T of the triple point of water (where liquid, solid, and vapor coexist). For temperatures above the water triple point, it is common to use the Celsius scale, with symbol °C, having the same interval as the thermodynamic scale, but with temperatures t defined as $T - T_0$, where $T_0 = 273.15$ K, the ice point, where liquid and solid water coexist at a pressure of 1 atm (101.325 kPa). The ITS-90 is defined by 17 fixed points. *See* TRIPLE POINT.

Primary thermometers are devices which relate the thermodynamic temperature to statistical mechanical formulation. The fixed points of ITS-90 are all based on one or more types of gas thermometry or on spectral radiation pyrometry referenced to gas thermometry. Secondary thermometers are used as reference standards in the laboratory because primary thermometers are often too cumbersome. It is necessary to establish standard secondary thermometers referenced to one or more fixed points for interpolation between fixed points. Lower-order thermometers are used for most practical purposes and, when high accuracy is required, can usually be calibrated against reference standards maintained at laboratories, such as the U.S. National Institute of Standards and Technology, or against portable reference devices (sealed boiling or melting point cells). *See* GAS THERMOMETRY; LOW-TEMPERATURE THERMOMETRY; PYROMETER; THERMISTOR; THERMOCOUPLE; THERMOMETER. [B.W.M.]

Temperature senses The system of sensitivity of the skin and internal organs of the body making possible feelings of warmth and cold. It is not known with certainty whether there are two discrete temperature senses, one taking care of warmth, the other cold, or but a single mechanism yielding warmth under some conditions of simulation and cold under others.

The nervous mechanism responsible for sensations of cold and warmth is unknown. The classical theory is that cold is mediated by Krause's end-bulbs and warmth by Ruffini cylinders. This presumed correlation is based on (1) the shorter reaction time to cold and more superficial location of the Krause ending, and (2) good agreement between sensitivity and end-bulb distribution in the sclera of the eye. A competing, and more modern, theory is the neurovascular theory of temperature sensitivity, according to which both the cold and warmth patterns are handled by nerve endings in the smooth muscle walls of the tiny arterioles of the skin. Cold is a consequence of contraction of blood vessels, warmth the result of their dilation. Thermal pain, occurring both at high (52°C or 126°F) and low (3°C or 37°F) temperatures, comes from spastic contraction of the same muscle elements. *See* CUTANEOUS SENSATION. [F.A.G.]

Tempering The reheating of previously quenched alloy to a predetermined temperature below the critical range, holding the alloy for a specified time at that temperature, and then cooling it at a controlled rate, usually by immediate rapid quenching, to room temperature. The term is broadly applied to any process that toughens a material.

In alloys, if the composition is such that cooling produces a supersaturated solid solution, the resulting material is brittle. Heating the alloy to a temperature only high enough to allow the excess solute to precipitate out and then rapidly cooling the saturated solution fast enough to prevent further precipitation or grain growth result in a microstructure combining hardness and toughness.

With steel, the tempering must be carried out by slow heating to avoid steep temperature gradients, stress relief being one of the objectives. Properties produced by tempering depend on the temperature to which the steel is raised and on its alloy composition. For example, if hardness is to be retained, molybdenum or tungsten is used in the alloy. *See* HEAT TREATMENT (METALLURGY). [F.H.R.]

Temporary structure (engineering) A structure erected to aid in the construction of a permanent project. Temporary structures are used to facilitate the construction of buildings, bridges, tunnels, and other above- and below-ground facilities by providing access, support, and protection for the facility under construction, as well as assuring the safety of the workers and the public. Temporary structures either are dismantled and removed when the permanent works become self-supporting or completed, or are incorporated into the finished work. Temporary structures are also used in inspection, repair, and maintenance work.

The many types of temporary structures include cofferdams; earth-retaining structures; tunneling supports; underpinning; diaphragm/slurry walls; roadway decking; construction ramps, runways, and platforms; scaffolding; shoring; falsework; concrete formwork; bracing and guying; site protection structures such as sidewalk bridges, boards, and nets for protection against falling objects, barricades and fences, and signs; and unique structures that are specially conceived, designed, and erected to aid in a specific construction operation.

These temporary works have a primary influence on the quality, safety, speed, and profitability of all construction projects. More failures occur during construction than during the lifetimes of structures, and most of those construction failures involve temporary structures. However, codes and standards do not provide the same scrutiny as they do for permanent structures. Typical design and construction techniques and some industry practices are well established, but responsibilities and liabilities remain complex and present many contractual and legal pitfalls. [R.T.R.]

Tendon A white, glistening, fibrous cord which joins a muscle to some movable structure such as a bone or cartilage. Tendons permit concentration of muscle force into a small area, allow muscles to act at a distance, and in some cases change the direction of pull, allowing the muscle to act around an articulation. Flattened, sheetlike tendons found in broad muscles are called aponeuroses. Tendons consist almost entirely of white fibrous connective tissue formed by closely packed, parallel bundles of collagen fibrils. Although highly noncompliant tendons are remarkable for their pliability and strength. *See* CONNECTIVE TISSUE.

Surrounding the tendon is the tendon sheath formed by a dense, fibrous connective tissue layer. This cavity contains a small amount of fluid, similar to that in joint cavities, but occasionally this space is filled with a more viscid, mucous material. The sheath and contained fluid facilitate movement of the tendon by reducing friction. Moreover, the sheath is normally the means by which the tendon is held in place. *See* BURSA; MUSCULAR SYSTEM. [W.J.B.]

Tenrec An insectivorous mammal indigenous to Madagascar. There are 30 species in 10 genera. These animals are nocturnally active and feed on insects, worms, and mollusks. All tenrecs are essentially primitive unspecialized mammals, with poor vision. The digits are clawed, and the first digit is not opposable to the others. The body of the tenrec is covered with a mixture of hair, spines, and bristles; the tail is rudimentary; and the toes may be separate or webbed, depending on the species. The female is prolific, and has litters of 12–20 young. *See* INSECTIVORA; MAMMALIA. [C.B.C.]

Tensor analysis The systematic study of tensors which led to an extension and generalization of vectors, begun in 1900 by two Italian mathematicians. G. Ricci and T. Levi-Civita, following G. F. B. Riemann's proposal concerning a generalization of euclidean geometry. The principal aim of the tensor calculus (absolute differential calculus) is to construct relationships which are generally covariant in the sense that these relationships or laws remain valid in all coordinate systems. The differential equations for the geodesics in a Riemannian space are covariant expressions; they yield a description of the geodesics which is valid for all coordinate systems. On the other hand, Newton's equations of motion require a preferred coordinate system for their description, namely, one for which force is proportional to acceleration (an inertial frame of reference). Thus Albert Einstein was led to a study of Riemannian geometry and the tensor calculus in order to construct the general theory of relativity. *See* CALCULUS OF VECTORS; RIEMANNIAN GEOMETRY. [H.La.]

Teratology The study of the deviations from normal growth and development, resulting in abnormal individuals. The abnormality may be either biochemical or anatomic.

Factors responsible for the development of anomalies may be separated into genetic and environmental. Geneticists have studied many conditions in laboratory animals which are caused by the action of mutant genes. The effects may be noted early or late and may be profound (lethal) or trivial. There are also a number of human anomalies, such as polydactyly, syndactyly, achondroplasia, cleft palate, microphthalmia, and spina bifida, which have a known genetic basis. *See* HUMAN GENETICS.

Environmental factors known to produce anomalies in laboratory animals include heavy metals, alcohol, excess or deficiency of hormones, oxygen lack, nutritional conditions including vitamin excess or deficiency, metabolic inhibitors, virtually any chemical, and irradiations of various types. Virus infections may act directly on a fetus or indirectly through alteration of maternal metabolites. The array of agents is so vast that it may be safe to say that virtually any substance or physical agent may produce anomalies when administered in a particular way to a particular embryo. The nature of the anomaly may also vary greatly, and it depends on the agent used, stage of development of the organism when it is first exposed to the agent, the dose and duration of the insult, and the genetic makeup of the exposed individual. The physiological condition of the individual, either germ cell or embryo, at the time of exposure may be important. *See* CONGENITAL ANOMALIES; DEVELOPMENTAL BIOLOGY. [N.K.M.]

Terbium Element number 65, terbium, Tb, is a very rare metallic element of the rare-earth group. Its atomic weight is 158.924, and the stable isotope ^{159}Tb makes up 100% of the naturally occurring element.

1	2	3	4	5	6	7	8	9	10	11	12	13	14	15	16	17	18
1 H																	2 He
3 Li	4 Be											5 B	6 C	7 N	8 O	9 F	10 Ne
11 Na	12 Mg											13 Al	14 Si	15 P	16 S	17 Cl	18 Ar
19 K	20 Ca	21 Sc	22 Ti	23 V	24 Cr	25 Mn	26 Fe	27 Co	28 Ni	29 Cu	30 Zn	31 Ga	32 Ge	33 As	34 Se	35 Br	36 Kr
37 Rb	38 Sr	39 Y	40 Zr	41 Nb	42 Mo	43 Tc	44 Ru	45 Rh	46 Pd	47 Ag	48 Cd	49 In	50 Sn	51 Sb	52 Te	53 I	54 Xe
55 Cs	56 Ba	71 Lu	72 Hf	73 Ta	74 W	75 Re	76 Os	77 Ir	78 Pt	79 Au	80 Hg	81 Tl	82 Pb	83 Bi	84 Po	85 At	86 Rn
87 Fr	88 Ra	103 Lr	104 Rf	105 Db	106 Sg	107 Bh	108 Hs	109 Mt	110	*111*	*112*	*113*	*114*	*115*	*116*	*117*	*118*

lanthanide series	57 La	58 Ce	59 Pr	60 Nd	61 Pm	62 Sm	63 Eu	64 Gd	65 Tb	66 Dy	67 Ho	68 Er	69 Tm	70 Yb
actinide series	89 Ac	90 Th	91 Pa	92 U	93 Np	94 Pu	95 Am	96 Cm	97 Bk	98 Cf	99 Es	100 Fm	101 Md	102 No

The common oxide, Tb_4O_7, is brown and is obtained when its salts are ignited in air. Its salts are all trivalent and white in color and, when dissolved, give colorless solutions. The higher oxides slowly decompose when treated with dilute acid to give the trivalent ions in solution. Although the metal is attacked readily at high temperatures by air, the attack is extremely slow at room temperatures. The metal has a Néel point at about 229 K and a Curie point at about 220 K. For properties of the metal *see* RARE-EARTH ELEMENTS. [F.H.Sp.]

Terebratulida An order of articulate brachiopods whose first representatives occur in beds of Early Devonian age and which is still extant. *See* ARTICULATA (BRACHIOPODA).

The group invariably has a punctate shell structure and is characterized by the possession of a loop extending anteriorly from the crural bases, the loop affording some degree of support for the lophophore. The posterior margin is commonly curved, the ventral interarea reduced, and the dorsal interarea absent. *See* BRACHIOPODA. [A.J.R.]

Termite An insect, also known as the white ant, of the order Isoptera; there are about 1600 species. Termites are soft-bodied and medium-sized. They live in colonies with a caste system comprising three types of functional individuals: sterile workers, sterile soldiers, and the reproductives. Workers are white and wingless and have chewing mouthparts and well-developed legs; but the eyes are very small or lacking. Soldiers resemble the workers except for the head, which is quite large and has massive mandibles. There are two types of reproductive individuals, one wingless or having vestigial wings and the other winged. *See* SOCIAL INSECTS.

Termites feed on cellulose; however, they do not elaborate an enzyme to digest this material. Instead, flagellate protozoan symbionts in the gut of the termites digest the cellulose or at least initiate its breakdown.

Economically, termites are both destructive and beneficial. They serve as reducer organisms in their destruction of dead trees and other vegetation, replenishing the soil with materials necessary for plant growth. On the other hand, termites destroy books, structural materials, fabrics, and other substances. *See* ISOPTERA. [C.B.C.]

Terpene A class of natural products having a structural relationship to isoprene, as shown below. Over 5000 struc-

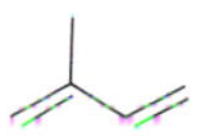

Isoprene unit

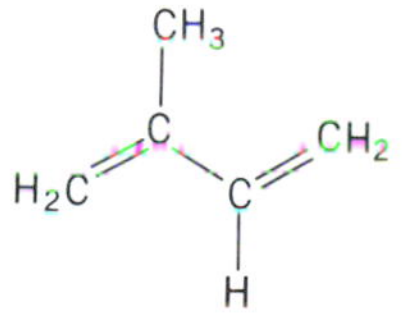

Isoprene

turally determined terpenes are known; many of these have also been synthesized in the laboratory. Historically terpenes have been isolated from green plants, but new compounds structurally related to isoprene continue to be isolated from other sources as well, so the class is also referred to as terpenoids, reflecting the biochemical origin without specification of the natural source. *See* ISOPRENE.

Terpenes are classified according to the number of isoprene units of which they are composed, as follows:

5	hemi	25	ses-
10	mono-	30	tri-
15	sesqui-	40	tetra-
20	di-	$(5)_n$	poly-

Although they may be named according to the systematic nomenclature and numbering systems set by the International Union of Pure and Applied Chemistry for all organic compounds, it is often easier to refer to terpenes by their common names, which usually reflect the botanical or zoological name of their source. [T.Hu.]

Terra-cotta A clay product fired or burned to fuse into a hard unified mass. Being fire- and frost-resistant, terra-cotta is used unglazed for roofing and other exterior finishing. Compacted to a high density after firing, terra-cotta is used structurally. For example, it is used in archways and colonnades. It can also be molded into various shapes having figured surfaces and used decoratively. Handcrafted terra-cotta is used in small sculpture; the finished product has the hardness and durability of stone. [F.H.R.]

Terracing (agriculture) A method of shaping land to control erosion on slopes of rolling land used for cropping and other purposes. In early practice the land was shaped into a series of nearly level benches or steplike formations. Modern practice in terracing, however, consists of the construction of low-graded channels or levees to carry the excess rainfall from the land at nonerosive velocities. The physical principle involved is that, when water is spread in a shallow stream, its flow is retarded by the roughness of the bottom of the channel and its carrying, or erosive, power is reduced. Since direct impact of rainfall on bare land churns up the soil and the stirring effect keeps it in suspension in overland flow and rills, terracing does not prevent sheet erosion. It serves only to prevent destruction of agricultural land by gullying and must be supplemented by other erosion-control practices, such as grass rotation, cover crops, mulching, contour farming, strip cropping, and increased organic matter content. *See* EROSION; SOIL CONSERVATION.

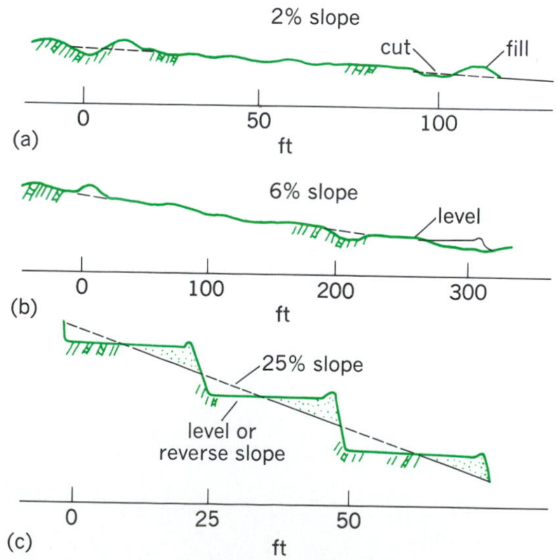

Types of terraces. (*a*) Broadbase. (*b*) Conservation bench. (*c*) Bench. 1 ft = 0.3 m. (*After Soil and Water Conservation Engineering, 2d ed., The Ferguson Foundation Agricultural Engineering Series, John Wiley and Sons, Inc., 1966*)

The two major types of terraces are the bench and the broadbase (see illustration). The bench terrace is essentially a steep-land terrace and consists of an almost vertical retaining wall, called a riser, or a steep vegetative slope to hold the nearly level surface of the soil for cultivation, orchards, vineyards, or landscaping. The broadbase terrace has the distinguishing characteristic of farmability; that is, crops can be grown on this terrace and worked with modern-day machinery. These terraces are constructed either to remove or retain water and, based on their primary function, are classified either as graded or level. *See* LAND DRAINAGE (AGRICULTURE). [C.B.O.]

Terrain-aided navigation The auxiliary-sensor data, obtained by observing specific features of the Earth's terrain, to aid or update an inertial navigation system and thereby improve its navigation accuracy. A position fix is derived by comparing the observed terrain data with an on-board terrain reference map. The position within the reference map where the observation best matches the reference data is taken to be the position of the navigator over the Earth's surface. For proper operation, the terrain feature being observed must be unique so that position can be determined unambiguously from the reference map.

Many different sensor types can be used to observe terrain features: a radar altimeter (in conjunction with an altitude reference) to observe terrain elevation beneath the aircraft; a radiometric sensor to observe the natural electromagnetic radiation of the Earth in a certain frequency band; an optical sensor to observe a visual image; a radar transmitter and receiver to observe terrain reflectivity. *See* ALTIMETER; REMOTE SENSING. [R.D.An.]

Terrain areas Subdivisions of the continental surfaces distinguished from one another on the basis of the form, roughness, and surface composition of the land. The pattern of landform differences is strongly reflected in the arrangement of such other features of the natural environment as climate, soils, and vegetation. These regional associations must be carefully considered in planning of activities as diverse as agriculture, transportation, city development, and military operations.

Eight classes of terrain are distinguished on the basis of steepness of slopes, local relief (the maximum local differences in elevation), cross-sectional form of valleys and divides, and nature of the surface material. Approximate definitions of terms used and percentage figures indicating the fraction of the world's land area occupied by each class are as follows; (1) flat plains: nearly level land, slight relief, 4%; (2) rolling and irregular plains: mostly gently sloping, low relief, 30%; (3) tablelands: upland plains broken at intervals by deep valleys or escarpments, moderate to high relief, 5%; (4) plains with hills or mountains: plains surmounted at intervals by hills or mountains of limited extent, 15%; (5) hills: mostly moderate to steeply sloping land of low to moderate relief, 8%; (6) low mountains: mostly steeply sloping, high relief, 14%; (7) high mountains: mostly steeply sloping, very high relief, 13%; and (8) ice caps: surface material, glacier ice, 11%. [E.H.Ha.]

Terrestrial coordinate systems Methods for establishing relative position on the Earth by utilizing two sets of coordinates perpendicular to each other, called latitude and longitude. The orientation of the coordinates also establishes Earth directions. The fact of Earth rotation establishes the geographic poles, which serve as reference points upon which the coordinate system is based.

Position within the coordinate system, or graticule, is deter-

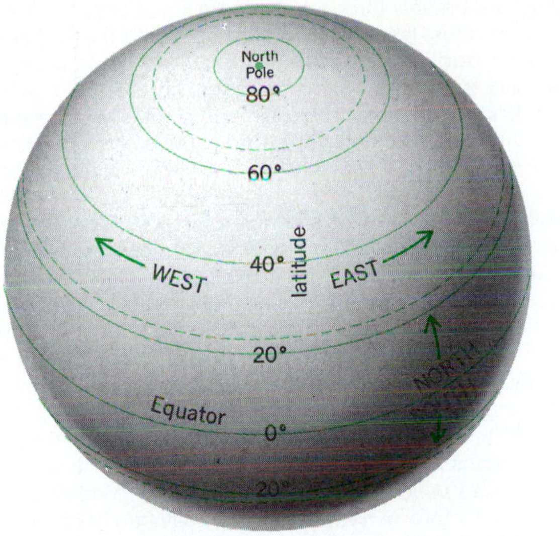

Fig. 1. The parallel system of the Earth's coordinate net establishes directions east and west and provides for designating positions north and south. (*After V. C. Finch et al., Elements of Geography, 4th ed., McGraw-Hill, 1957*)

mined by direct astronomic observation (or by derivatives of it) and is stated in terms of arc distance from reference positions. Position between the poles is designated as latitude and corresponds to a Y value on a cartesian system. Position corresponding to an X value on a cartesian system is designated as longitude.

Latitude. Arc distance of latitude toward the poles is reckoned from a great circle on the Earth which is everywhere midway between the poles. This line, designated as the Equator, is therefore everywhere 90° arc distance from the poles. Any other line a given lesser distance from a pole will be a small circle parallel to the Equator. Direction along such a line is designated as east-west. All such lines on the Earth are termed parallels (Fig. 1).

Because the Earth is not a true sphere but an oblate spheroid of revolution, the quadrant from the Equator to a pole is not a true arc but curves less rapidly near the pole. Consequently, degrees of latitude are slightly shorter near the Equator (68.7 mi or 110 km) and slightly longer near the poles (69.4 mi or 112 km).

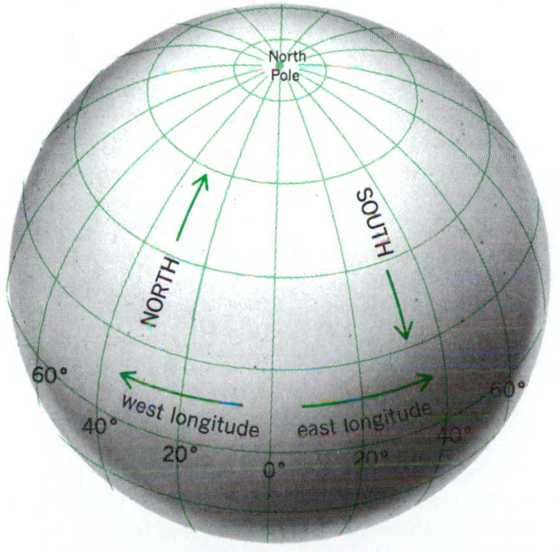

Fig. 2. The meridian system of the Earth's coordinate net plus the parallel system establishes directions north and south and provides a method for designating positions east and west. (*After V. C. Finch et al., Elements of Geography, 4th ed., McGraw-Hill, 1957*)

Latitude is numbered from 0° at the Equator to 90° at the poles. Distance toward the North or South poles is designated as north or south latitude, respectively.

Longitude. Arc distance of longitude is measured along a parallel from some starting point (Fig. 2). Each parallel circle on the Earth contains 360°, and since the Earth rotates on an axis which is perpendicular to the planes of the parallel circles, each parallel therefore rotates 360° in 1 day, or 15° in 1 h. The difference in sun time between two points may be expressed as an angular difference in longitude.

A great circle from pole to pole has everywhere along it the same sun time. Such a line is called a meridian. Direction along such a line is designated as north-south. All meridians are alike and, unlike latitude, there is no natural starting point from which to begin their numbering. The meridian passing through the Royal Astronomical Observatory in Greenwich, near London, is internationally agreed upon as the 0° meridian, or the prime meridian. Longitude is measured and designated as east or west longitude from the prime meridian to 180°, the meridian opposite 0° longitude. The Earth's coordinate system, therefore, "originates" with coordinates numbered zero intersecting in the Gulf of Guinea.

A degree of longitude varies in length with latitude. The length of a degree of longitude varies as the cosine of the latitude. Near the Equator, one degree of longitude is nearly the same length as a degree of latitude. [A.H.Ro.]

Terrestrial ecosystem A term that encompasses ecosystems that exist on the continents and islands of the world and comprehends a series of dynamic open interaction systems that include living forms (animals, plants, and microorganisms) and their nonliving environment (soils, geological formations, and atmospheric constituents) and the activities, interrelations, chemical reactions, physical changes, and all other phenomena of each. Energy that enters these systems, chiefly in the form of sunlight, circulates through the systems and powers the life processes of the organisms, influences the rate and nature of chemical reactions and physical changes, and is partially accumulated in the bodies of organisms and in other chemical and physical states. *See* ECOSYSTEM; FRESH-WATER ECOSYSTEM; MARINE ECOLOGY.

Terrestrial ecosystems differ from aquatic ecosystems in several important respects. The most obvious difference in the abiotic components of the systems is the basic physical contrast between the media. Terrestrial organisms are surrounded by air, a mixture composed of gaseous elements and compounds. Water vapor is present in the atmosphere but forms only a small portion of the total volume of air. Water is often in critically short supply, extremes and rapid changes of temperature are common, mineral elements are limited in occurrence to the substrate (generally in relatively high concentrations), and light is intense, at least at upper levels of the vegetation.

The terrestrial and aquatic ecosystems are connected in a great many direct ways, so that it is most logical to consider the entire Earth and its envelope of gases to be the ultimate ecosystem. However, this world ecosystem is an open system and is dependent upon an external supply of energy, chiefly from the Sun, and is influenced by a variety of forces of external origin. *See* ECOLOGY. [J.S.McC.]

Terrestrial radiation Electromagnetic radiation originating from the Earth and its atmosphere at wavelengths determined by their temperature. It is sometimes called thermal radiation. The units are energy per unit area per unit time, such as calories per square centimeter per second, or watts per square meter. The atmosphere emits, absorbs, and transmits radiation, and the net flux of radiation at any point depends upon the distribution with height of temperature and water vapor. Heating and cooling due to the vertical divergence of terrestrial radiation provide a major part of the potential energy changes necessary to drive the atmospheric wind system. Terrestrial radiation is also responsible for maintaining the air temperature near the ground within limits necessary for comfortable living. *See* Greenhouse effect; Heat balance, terrestrial atmospheric. [L.D.K.]

Terrestrial water The total inventory of water on the Earth. Water is unevenly distributed over the Earth's surface in oceans, rivers, and lakes. In addition, the world's water is distributed throughout the atmosphere and also occurs as soil moisture, groundwater, ice caps, and glaciers. *See* Atmosphere; Glaciology; Groundwater hydrology; Hydrology; Lake; Surface water. [R.L.N.]

Territoriality A pattern of behavior in which one or more animals occupy and defend a definite area or territory. They obtain their food from this area and exclude others of their species from it. Territorial behavior is highly developed among many kinds of birds but appears also, in somewhat different forms, among mammals, reptiles, fish, and social insects.

In many mammals and reptiles, and some other animals, individuals, pairs, or social groups move over a fairly definite area in seeking food or when engaged in other activities. Such an area is termed a home range but not a territory unless it is defended against others of the species. *See* Social animals. [R.H.Wh.]

Tertiary The older major subdivision (period) of the Cenozoic Era, extending from the end of the Cretaceous to the beginning of the Quaternary. The term Tertiary is also applied to all rocks formed during this period and to all the fossils they contain. Tertiary sedimentary rocks include widespread limestones, sandstones, marls, mudstones, and conglomerates; igneous rocks include extrusive and intrusive volcanics and locally some rock of more deep-seated (plutonic) origin. Tertiary life was characterized particularly by (1) a diversification and multiplication of stocks such as pelecypods and gastropods of sea urchins and sand dollars, of microscopic foraminiferans, of some other marine shellfish, and (2) a great development of shrubs, grasses, and other flowering plants; and (3) expansion on both land and sea of birds and mammals, the latter having provided for the Tertiary the more informal designation Age of Mammals.

Configuration of the modern continental land masses developed during Tertiary and Quaternary times. Great mountain-making movements (orogenies) and uplifts of large segments of the Earth's crust (epeirogenies) alternated with fluctuating transgressions and reexpansions of the seas over land areas. Orogenies and uplifts nevertheless grew progressively greater at the expense of the earlier seas as Tertiary time wore on and culminated in the great Late Alpine orogenic uplifts of Pleistocene time. *See* Cenozoic; Cretaceous; Pleistocene. [R.M.Kl.]

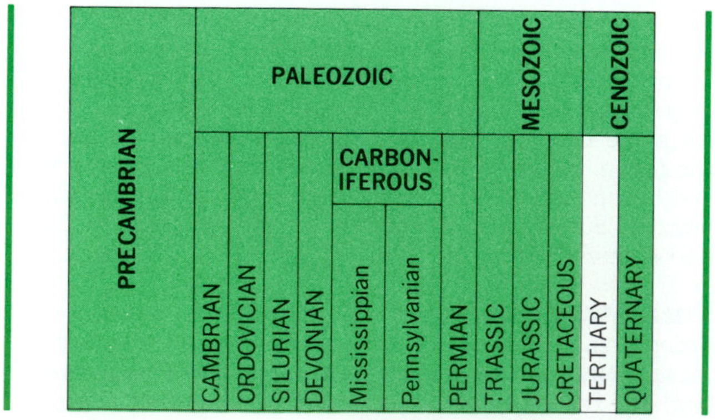

Testis The organ of sperm production. In addition, the testis (testicle) is an organ of endocrine secretion in which male hormone is elaborated. In mammals, the testes are usually ovoid or round. In many species (for example, humans) they are suspended in a pouch (scrotum) outside the main body cavity; in other species they are found in such a pouch only at the reproductive season; in still others the testicles are permanently located in the abdomen (for example, in whales and bats).

Within a firm and thick capsule of connective tissue, the testis contains a varying number of thin but very long seminiferous tubules which are the sites of sperm formation. These tubules open into a network of fine, slitlike canals, the rete testis. From this the sperm drains through a few, narrow ducts, ductuli efferentes, into the epididymis, the storage chamber for sperm. [E.C.R.-R.]

The functions of the testis are dependent on the secretion of gonadotropic hormones, the release of which from the pituitary gland is in turn regulated by the central nervous system. The principal androgenic hormone released by the testis into the bloodstream is testosterone. Testosterone synthesis is normally limited by the rate of pituitary gonadotrophin secretion: Administration of ICSH (also known as luteinizing hormone or LH) or of chorionic gonadotrophin results in increased testosterone synthesis and release within minutes.

At the ambisexual stage of embryonic development, the testis promotes the growth of the paired Wolffian ducts and their differentiation into the epididymis, vasa deferentia, and seminal vesicles; the fetal testis also causes masculinization of the urogenital sinus, fusion of the labioscrotal folds in the midline, and development of the genital tubercle into a phallus.

Testosterone promotes the retention of dietary nitrogen and its incorporation into muscle protein. This effect is responsible for the greater muscular development of the male and contributes to the prepubertal growth spurt of boys. Toward puberty, increased secretion of testosterone stimulates the growth of the penis, scrotum, and male accessory glands responsible for the formation of the seminal plasma, for example, the prostate and seminal vesicles. The hormone brings about the appearance of secondary sex characters, such as the male-type distribution of hair and body fat and lowered pitch of voice in men. Testosterone is probably essential for efficient spermatogenesis, but high doses of testosterone suppress spermatogenesis by inhibiting pituitary gonadotrophin secretion. *See* Reproductive system; Sperm cell; Spermatogenesis. [H.R.L.]

Testudines An order of the Reptilia, subclass Anapsida, including the turtles, terrapins, and tortoises. This order is also known as the Chelonia. The group first appeared in the Triassic and its representatives are among the commonest fossils from that time on. Members of the order are most frequently found in fresh-water streams, lakes, and ponds or in marshy areas. However, a number of strictly terrestrial species are known and several are marine. Turtles occur on all the major continents and continental islands in tropic and temperate regions. The marine forms are basically tropic in distribution but some individuals stray into temperature waters. *See* Anapsida; Reptilia.

The living turtles are usually divided into two major groups, the suborders Pleurodira and Cryptodira, based upon the structures of the head and neck. Both lines are thought to be evolved from primitive forms placed in the suborder Amphichelydia (Triassic to Eocene). This latter group differs most markedly

Families of the Testudines

Group	Common name	Number of species	Distribution
Suborder Pleurodira			
Family Pelomedusidae	Side-necked turtles	14	Southern Africa, Madagascar, South America
Family Chelidae	Side-necked turtles	32	Australia, New Guinea, South America
Suborder Cryptodira			
Family Kinosternidae	Mud and musk turtles	22	North, Middle, and South America
Family Chelydridae	Snapping turtles	3	Middle and northern South America
Family Emydidae	Pond turtles and allies	102	All continents except Australia; only one or two South American species
Family Testudinidae	Tortoises	30	All continents except Australia; two South American species
Family Cheloniidae	Hawksbill, loggerhead, and green sea turtles	6	Tropical and subtropical seas
Family Dermocheylidae	Leatherback turtle	1	Tropical and subtropical seas
Family Trionychidae	Soft-shelled turtles	14	North America, Asia, central and southern Africa
Family Carretochelyidae	Fly River turtle	1	New Guinea

from recent species in having elongate neural spines on the neck vertebrae so that the neck is not retractible into the shell. The table shows the principal groups and distribution of the approximately 225 living species of the order. *See* TURTLE.

The Chelonia differ from most other vertebrates in possessing a hard bony shell which encompasses and protects the body. The shell is made up of a dorsal portion, the carapace, and a ventral segment, the plastron, connected by soft ligamentous tissue or a bony bridge. The rigid bony shell of turtles imposes a basic body plan subject to relatively little variation. Turtles are always recognizable as such and the principal obvious differences between them are in the shell shape, limbs, head, and neck.

The majority of turtles have limbs more or less adapted to aquatic or semiaquatic life with moderate to well-developed palmate webbed feet. However, in strictly terrestrial forms such as the tortoises, the limbs are elephantine with the weight of the body being borne on the flattened soles of the feet.

Associated with the tendency of most turtles toward a life in or near water is the auditory apparatus. Even though an eardrum is present, hearing is adapted to picking up sounds transmitted through the water or substratum. However, there is evidence that airborne sound can be heard. Vision is also important, and turtles have color vision. Sounds are produced by many species through expulsion of air through the glottis. Respiration is by means of the rather rigid lungs, which are inflated and deflated by a series of special muscles. Several species with aquatic habits have special vacularized areas in the mouth or cloaca which act as auxiliary respiratory devices and make possible gaseous exchanges when the turtle is submerged in water. Most forms emit highly pungent oily substances from specialized cloacal glands, and these secretions probably are useful in recognition cues in the water. Most of the species are gregarious and diurnal, and territoriality is unknown in the order.

All turtles lay shelled eggs which are buried in sand or soil in areas where females congregate. The shell is calcareous in most forms, but the marine turtles have leathery shells. Eggs are rather numerous, as many as 200 being laid by a single individual in some species. Incubation takes 60–90 days, and the little turtle cuts its way out of the shell with a small horny egg caruncle at the end of the snout.

Turtles feed on all types of organisms. Aquatic species may eat algae, higher plants, mollusks, crustaceans, insects, or fishes; terrestrial forms are similarly catholic in tastes. Most species are omnivores, but some have very specialized diets. The edges and internal surfaces of the horny beaks of these reptiles are frequently denticulate and modified to form specialized mechanisms adapted to handle particular food items. [J.M.S.]

Tetanus An infectious disease, also known as lockjaw, which is caused by the toxin of *Clostridium tetani*. The bacterium may be isolated from fertile soil and the intestinal tract or fecal material of humans and other animals. Infection commonly follows dirt contamination of deep wounds or other injured tissues.

The incubation period of tetanus is usually 5–10 days, and the disease is characterized by convulsive tonic contraction of voluntary muscles. Prevention of tetanus rests on the proper, prompt surgical care of contaminated wounds and prophylactic use of antitoxin if the individual has not been protected by active immunization with toxoid. *See* IMMUNOLOGY. [L.S.McC.]

Tethered balloon A balloon that is attached to a mooring line and allowed to ascend by releasing a winch mechanism. The tethered balloon is generally configured as an aerodynamic body to provide lift, achieve lower drag, and increase lateral stability. Although 4 mi (6 km) is generally considered the upper altitude for this concept, the National Center for Space Studies in France has reported flying a tethered balloon carrying a 770-lb (350-kg) gondola to an altitude of 10.6 mi (17 km) for 12 h.

The tethered balloon has primarily been used for military purposes to provide acoustic, radar, and optical surveillance, tall antennas, and communication relays. [W.R.N.]

Tetrabranchia A subclass of the Cephalopoda, containing the Nautiloidea and the Ammonoidea. The most numerous group of the cephalopods in the Paleozoic and Mesozoic periods, it is considered to be the most primitive group of cephalopods. The name Tetrabranchia was given to this group because four gills are found in *Nautilus*, the only representative whose anatomy is known. Since no records of the soft parts are preserved in fossil specimens, it is not known if this feature is shared by all of the many genera assigned to the subclass. *See* AMMONOIDEA; NAUTILOIDEA. [G.L.V.]

Tetracycline A broad-spectrum antibiotic. It is one of a group of antibiotics known as tetracyclines; other well-known members of the group are chlortetracycline (aureomycin) and oxytetracycline (terramycin). Tetracycline is produced biosynthetically by a strain of *Streptomyces aureofaciens* (or certain other species) or chemically by hydrogenolysis of chlortetracycline. Tetracycline is particularly useful because of broad antimicrobial action, with low toxicity, in the therapy of infections caused by gram-positive and gram-negative bacteria as well as rickettsiae and the large viruses such as psittacosis-lymphogranuloma viruses. *See* ANTIBIOTIC. [R.E.B.]

Tetractinomorpha A subclass of Demospongiae; the skeletonless genus *Oscarella* has a primitive structure. Cleavage results in a solid mass of cells (morula) which later becomes hollow by cytolysis of the interior cells, rich in food reserves. Upon being freed from the parent, the hollow larva is made up of a single layer of flagellated cells and is known as an amphiblastula. Following metamorphosis, the larva of *Oscarella* assumes the leuconoid grade of construction from the start, with isolated flagellated chambers communicating with inhalant and exhalant canals. The adult *Oscarella* and a related genus, *Plakina*, with two-, three-, and four-rayed spicules, retain the simple leuconoid structure. In form, the species may be thinly encrusting or massive, but often they have spherical or ovoid shapes. Branching species rarely occur. The skeletal system consists of spicules (sclerites), spongin fibers, or both. *See* DEMOSPONGIAE; PORIFERA. [W.D.H.]

Tetraethyllead An organometallic compound that has found wide commercial application as an additive for motor fuels. Small quantities of tetraethyllead markedly reduce the knocking tendencies of gasoline and thus permit use of higher engine compression ratios.

The combustion of gasoline containing tetraethyllead produces deposits of lead and lead oxide along the cylinder walls. Hence, ethylene dibromide and ethylene dichloride are also added as part of the antiknock fluid to remove the lead as lead halide in the exhaust gases. Automobiles equipped with catalytic converters designed to meet antipollution standards must burn "unleaded" gasoline, since lead deposits tend to poison the platinum and palladium oxidation catalysts used in such systems. *See* ANTIKNOCK AGENT; GASOLINE; OCTANE NUMBER; ORGANOMETALLIC COMPOUND. [M.D.R.]

Tetrahedrite A mineral having the composition $(Cu,Fe,Zn,Ag)_{12}Sb_4S_{13}$. It is massive or granular. Its hardness is $3\frac{1}{2}$–4 and the specific gravity varies from 4.6 to 5.1, depending on the composition. The luster is metallic and the color grayish black; thus, in some mining localities, this mineral is called gray copper.

Tetrahedrite is a widely distributed mineral, usually found in silver and copper veins. In some places it has sufficient silver to be a valuable ore, as at Freiberg, Germany, and silver mines of Peru, Bolivia, and Mexico. It is found in silver and copper mines in the western United States. [C.S.Hu.]

Tetrahedron A solid bounded by four planes, or faces (see illustration). It has four vertices (not coplanar) and six edges, the six line segments that join each pair of vertices. As the three-dimensional analog of a triangle, many of its properties are extensions of those of a triangle. Thus, as in the case of a triangle, the medians of a tetrahedron (that is, the lines joining the vertices with the centers of gravity of the opposite faces) are concurrent. On the other hand, the altitudes of a tetrahedron are not always concurrent. *See* POLYHEDRON. [L.M.Bl.]

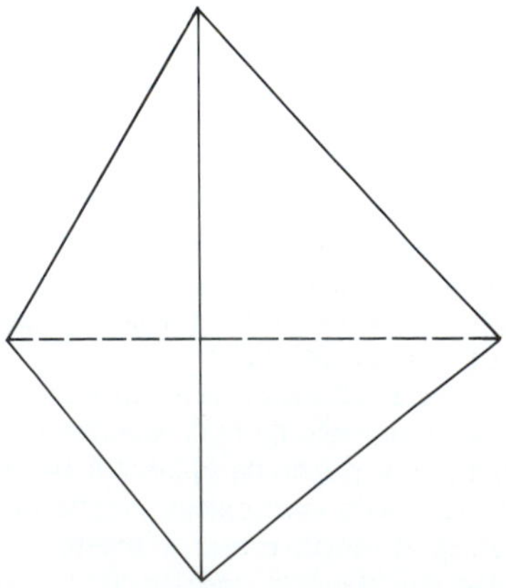

Regular tetrahedron.

Tetraodontiformes An order of specialized teleost fishes, also called Plectognathi, that includes the triggerfishes (see illustration), puffers, trunkfishes, ocean sunfishes, and their allies. It is a group of diverse structure. The body is variably armored with bony plates or spines, encased in bone, prickly, thorny, or naked. Fin spines and some fins are variably well developed or wanting. Some species can inflate the body.

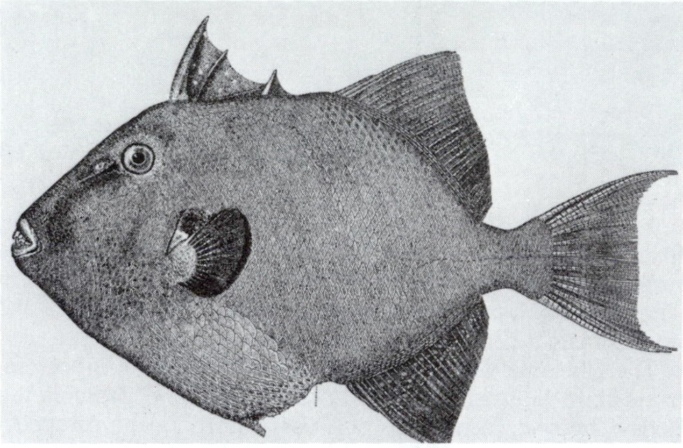

Gray triggerfish (*Balistes capriscus*). (*After G. B. Goode, Fishery Industries of the United States, sect. 1, 1884*)

Tetraodontiforms may be classified conservatively into 7 families, nearly 60 genera, and about 320 Recent species, but some specialists recognize more families and genera. Tetraodontiforms are largely reef and shore fishes of tropical or subtropical seas, but a few are pelagic, enter temperate waters, or ascend tropical rivers. Many are colorful inhabitants of coral reefs. Some are valued as food, but a few have a neurotoxic poison, tetrodotoxin, in the viscera that is sometimes fatal when eaten. *See* ACTINOPTERYGII. [R.M.B.]

Tetraphidales An order of mosses which is composed of the family Tetraphidaceae, the two genera *Tetraphis* and *Tetradontium*, and approximately five species.

Although this order is unique in the scalelike protonema, which produces frondlike growths, the most conspicuous structure distinguishing it from other mosses is the peristome of four rigid, nonsegmented teeth. The plants fruit freely and the capsules and peristome are persistent. The plants occur on humus, moist decaying wood, and sandstone. *See* BRYOPSIDA. [W.H.W.]

Tetraphididae A subclass of the mosses (class Bryopsida) consisting of two families and three genera, especially characterized by growth from protonematal flaps, three-ranked leaves, and peristomes of four teeth made up of whole cells (rather than thickened parts of cells). The Tetraphididae include small acrocarpous mosses with peristome teeth in fours. The plants grow from buds produced on leaflike protonematal flaps. They are erect and simple or merely forked, with oblong-ovate leaves in three rows. The leaves usually have a single costa that ends near the apex. *See* BRYOPHYTA; BRYOPSIDA. [H.Cr.]

Tetrapoda The superclass of the subphylum Vertebrata whose members typically possess limbs in contrast to the superclass, the Pisces (fishes), whose members have fins. *See* PISCES (ZOOLOGY).

The animals making up the Tetrapoda typically live part or all of their lives on land, whereas the members of the Pisces live in water. The classes of the Tetrapoda are Amphibia, Reptilia, Aves, and Mammalia. The term Tetrapoda comes from Greek words meaning "four feet," but there are tetrapods that have only two limbs or none at all, such as some amphibians and reptiles. These forms have, however, evolved from four-footed ancestors. *See* AMPHIBIA; AVES; MAMMALIA; REPTILIA.

[R.G.Z.]

Tetrasporales A heterogeneous and artificial assemblage of fresh-water and marine algae in the division Chlorophyta. These plants are colonial; the cells are embedded but not adjoined in a copious mucilaginous sheath. Many plants have a parietal, cup-shaped chloroplast, a basal pyrenoid, stationary vegetative cells that become flagellated and motile, false swimming organs (pseudocilia), and a red eyespot in vegetative and reproductive cells. Many genera use zoospores in asexual reproduction, and a few are known to have isogamous sexual reproduction. [G.W.P.]

Teuthoidea An order of the class Cephalopoda (subclass Coleoidea) commonly known as squids. They are characterized by 10 appendages (eight arms and two longer tentacles) around the mouth; an elongate, tapered, usually streamlined body; an internal, rod- or blade-like chitinous shell (gladius); and fins on the body. The two tentacles are strongly elastic, contractile, but not reactile into pockets as in cuttlefishes (Sepioidea). Two rows of suckers (infrequently four or six rows) occur on the arms on muscular stalks, with sucker rings that are chitinous, smooth, toothed, or modified as clawlike hooks. The muscular tentacles have terminal clubs with two rows, usually four, ranging up to many rows of suckers (and/or hooks in some families). Adults of the family Octopoteuthidae and genera *Gonatopsis* and *Lepidoteuthis* characteristically lose their tentacles. *See* SEPIOIDEA.

The Teuthoidea are divided into two suborders. The Myopsida, the nearshore, shallow-water squids, have a transparent skin (cornea) covering the two eyes, with a minute pore anteriorly, arms and tentacular clubs with suckers only, never hooks, and a single gonoduct in females, not paired. The Oegopsida, the oceanic squids, have no cornea over the eyes and no anterior pore, arms and tentacular clubs with suckers (and/or hooks in many families), and paired gonoducts in females (some exceptions).

Squids inhabit a wide variety of marine habitats, depending on the species, from very shallow grass flats, mangrove roots, lagoons, bays, and along coasts (myopsids) to the open ocean from the surface of the sea (*Ommastrephes*) to nearly 9600 ft (3000 m) in the deep sea (other oegopsids such as *Bathyteuthis*, *Neotheuthis*, and *Grimalditeuthis*). *See* CEPHALOPODA; COLEOIDA; SQUID. [C.F.E.R.]

Textile A material made mainly of natural or synthetic fibers. Modern textile products may be prepared from a number of combinations of fibers, yards, films, sheets, foams, furs, or leather. They are found in apparel, household and commercial furnishings, vehicles, and industrial products. *See* MANUFACTURED FIBER; NATURAL FIBER.

The term fabric may be defined as a thin, flexible material made of any combination of cloth, fiber, or polymer (film, sheet, or foams); cloth as a thin, flexible material made from yarns; yarn as a continuous strand of fibers; and fiber as a fine, rodlike object in which the length is greater than 100 times the diameter. The bulk of textile products are made from cloth.

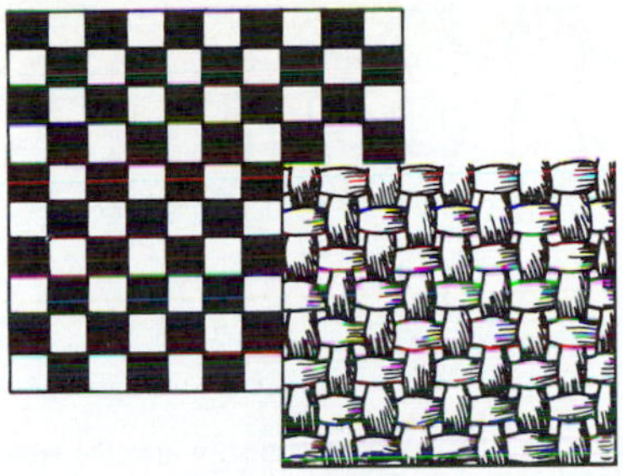

Fig. 1. Construction design for plain weave; filling yarns pass under and over alternate warp yarns, as shown at right. When fabric is closely constructed, there is no distinct pattern. (*After M. D. Potter and B. P. Corbman, Fiber to Fabric, 3d ed., McGraw-Hill, 1959*)

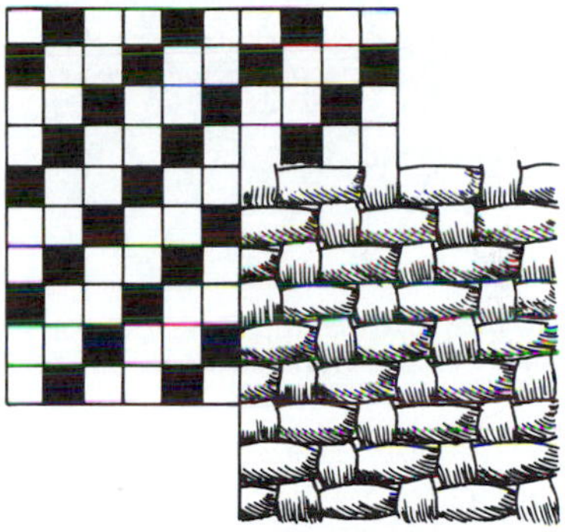

Fig. 2. Three-shaft twill. Two warp yarns are interlaced with one filling yarn. (*After M. D. Potter and B. P. Corbman, Fiber to Fabric, 3d ed., McGraw-Hill, 1959*)

The natural progression from raw material to finished product requires: the cultivation or manufacture of fibers; the twisting of fibers into yarns (spinning); the interlacing (weaving) or interlooping (knitting) of yarns into cloth; and the finishing of cloth prior to sale.

The conversion of staple fiber into yarn (spinning) requires the following steps: picking (sorting, cleaning, and blending), carding and combing (separating and aligning), drawing (reblending), drafting (reblended fibers are drawn out into a long strand), and spinning (drafted fibers are further attenuated and twisted into yarn).

The process of weaving allows a set of yarns running in the machine direction (warp) to be interlaced with another set of yarns running across the machine (filling or weft). The weaving process involves four functions: shedding (raising the warp yarns by means of the appropriate harnesses); picking (inserting the weft yarn); battening (pushing the weft into the cloth with a reed); and taking up and letting off (winding the woven cloth onto the cloth beam and releasing more warp yarn from the warp beam; Figs. 1 and 2).

Knit cloth is produced by interlocking one or more yarns through a series of loops. The lengthwise columns of loops are known as the wales, and the crosswise rows of loops are called courses. Filling (weft) knits (Fig. 3) are those in which the courses are composed of continuous yarns, while in warp knits (Fig. 4) the wale yarns are continuous.

Newly constructed knit or woven fabric must pass through various finishing processes to make it suitable for its intended purpose. Finishing enhances the appearance of fabric and also adds to its serviceability. Finishes can be solely mechanical, solely chemical, or a combination of the two. Those finishes, such as scouring and bleaching, which simply prepare the fabric for further use are known as general finishes. Functional fin-

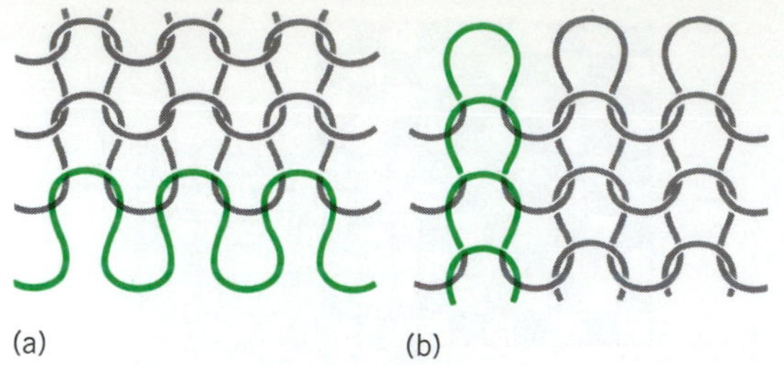

Fig. 3. Interlocking yarns of (*a*) course and (*b*) wale in a Jersey knit cloth. (*After B. P. Corbman, Fiber to Fabric, 5th ed., McGraw-Hill, 1975*)

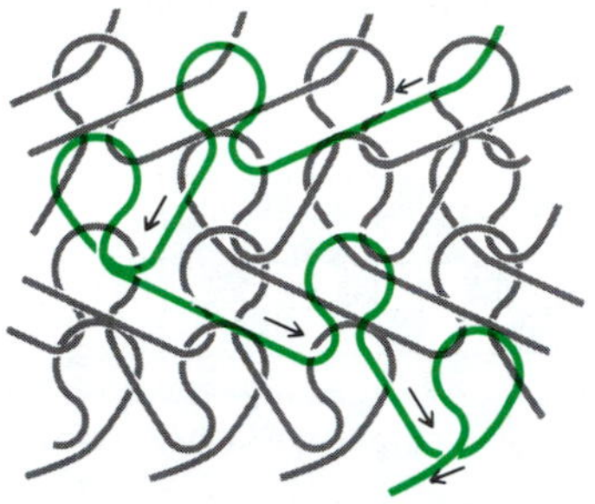

Fig. 4. Single-warp (one-bar) tricot knit. (*After B. P. Corbman, Fiber to Fabric, 5th ed., McGraw-Hill, 1975*)

ishes, such as durable press treatments, impart special characteristics to the cloth. For discussions of important finishing operations *see* BLEACHING; DYEING; TEXTILE CHEMISTRY; TEXTILE PRINTING. [I.Bl.]

Textile chemistry The applied science of textile materials, consisting of the application of the principles of the many basic fields of chemistry to the understanding of textile materials and to their functional and esthetic modification into useful and desirable items. The study of textile chemistry begins with the knowledge of the textile fibers themselves. These are normally divided into three groups: natural, manufactured, and synthetic. *See* MANUFACTURED FIBER; NATURAL FIBER; TEXTILE.

Chemicals. The enormous number of chemicals used in textile processing may be divided broadly into two categories: those intended to remain on the fiber, and those intended to wet or clean the fiber or otherwise function in some related operation. The former includes primarily dyes and finishes. The latter group consists mainly of surface-active agents, commonly known as surfactants. *See* SURFACTANT.

Preparation. Preparation is a term applied to a group of essentially wet chemical processes having as their object the removal of all foreign matter from the fabric. This results in a clean, absorbent substrate, ready for the subsequent coloring and finishing operations.

The operations constituting preparation depend primarily on the fibers being handled. Synthetic fibers contain little or no natural impurities, so that the only materials that normally must be removed are the oils and lubricants or water-soluble sizes needed to facilitate earlier processing. This is generally accomplished by washing with water and a mild detergent capable of emulsifying the oils and waxes. On the other hand, natural fibers contain relatively high amounts of natural impurities, and in addition frequently are sized with materials presenting difficulties in removal. In the case of cotton, prolonged hot treatment with alkali, usually sodium hydroxide, and strong detergent is necessary to break down and remove the naturally occurring impurities. Special scours are necessary for cleaning such materials as wool and silk. The protein fibers are very sensitive to alkali and strong detergents; they are usually washed with mild soap or sulfated alcohols.

After other impurities are removed from the fiber, it is usually desirable to remove any coloring material. This process is known as bleaching. By far the major bleaching agent in use is hydrogen peroxide, which is efficient in color removal, while still being considered relatively controllable and safe for use. *See* BLEACHING; HYDROGEN PEROXIDE.

Mercerization. Mercerization is a special process applied only to cotton. The fabric or yarn is treated with a strong sodium hydroxide solution while being held under tension. This process causes chemical and physical changes within the fiber itself, resulting in a substantial increase in luster and smoothness of the fabric, plus important improvements in dye affinity, stabilization, tensile strength, and chemical reactivity.

Coloring. Although many textiles reach the consumer in their natural color or as a bleached white, most textiles are colored in one way or another. Coloring may be accomplished either by dyeing or printing, and the coloring materials may be either dyes or pigments.

Dyeing essentially consists of immersing the entire fabric in the solution, so that the whole fabric becomes colored. On the other hand, printing may be considered as localized dyeing. In printing, a thickened solution of dyestuff or pigment is used. This thickened solution, or paste, is applied to specific areas of the fabric by means such as engraved rollers or partially porous screens. Application of steam or heat then causes the dyestuff to migrate from the dried paste into the interior of the fiber, but only in those specific areas where it has been originally applied. *See* DYE; DYEING; TEXTILE PRINTING.

Finishing. Finishing includes a group of mechanical and chemical operations which give the fabric its ultimate feel and performance characteristics. Many desirable characteristics may be imparted to the fabric through the application of various chemical agents at this point.

Softeners are used to give a desirable hand or feel to the fabric. These chemicals are generally long fatty chains, with solubilizing groups which may be cationic, nonionic, or occasionally anionic in character. They are essentially surfactants constructed so as to contain a relatively high proportion of fatty material in the molecule. Conversely, certain types of polymeric material such as polyvinyl acetate or polymerized urea formaldehyde resins are used to impart a stiff or crisp hand to a fabric. *See* POLYVINYL RESINS; UREA-FORMALDEHYDE-TYPE RESINS.

It is in finishing that the so-called proof finishes are applied, including fire-retardant and water-repellent finishes. A fire-retardant finish is a chemical or mixture containing a high proportion of phosphorus, nitrogen, chlorine, antimony, or bromine. A truly waterproof fabric may be made by coating with rubber or vinyl, but water-repellent fabrics are produced by treating with hydrophobic materials such as waxes, silicones, or metallic soaps.

Many other types of highly specialized treatments, such as antistatic, antibacterial, or soil-repellent finishes, may be applied to fit the fabric to a particular use. [D.H.A.]

Textile printing The localized application of color on fabrics. In printing textiles, a thick paste of dye or pigment is applied to the fabric by appropriate mechanical means to form a design. The olor is then fixed or transferred from the paste to the fiber itself, maintaining the sharpness and integrity of the design. In a multicolor design, each color must be applied separately and in proper position relative to all other colors. Printing is one of the most complex of all textile operations. *See* DYE.

A design may be applied in three major ways: raising the design in relief on a flat surface (block printing); cutting the design below a flat surface (intaglio or engraved printing); and cutting the design through a flat metal or paper sheet (stencil or screen printing). All three methods have been used for hand printing and reciprocating printing machines. In addition, these methods have been converted into rotary action by replacing blocks or plates with cylinders. Another method of printing utilizes individual computer-controlled nozzles for each color. The nozzles are used to paint a design on the fabric.

Each printing method requires a paste with special characteristics, frequently referred to as flow characteristics. The choice of thickener is dependent not only on the type of dyestuff, but on the type of printing machine on which the printing is to be done, and frequently also on the type of fixation to be used. Most natural thickening agents are based on combinations of starch and gum. The synthetic thickening agents used are generally extremely high-molecular-weight polymers capable of developing a very high viscosity at a relatively low concentration.

The first step is the preparation of print paste, which is made by dissolving the dyestuff and combining it with a solution of the appropriate thickening agent. The fabric is then printed by any of the standard methods and then dried in order to retain a sharp printed mark.

The next operation, steaming, may be likened to a dyeing operation. Before steaming, the bulk of the dyestuff is held in a dried film of thickening agent. During the steaming operation, the printed areas absorb moisture and form a very concentrated dyebath, from which dyeing of the fiber takes place. The thickening agent prevents the dyestuff from spreading outside the area originally printed, because the printed areas act as a concentrated dyebath that exists more in the form of a gel than a solution and restricts any tendency to bleed.

Printed goods are generally washed thoroughly to remove thickening agent, chemicals, and unfixed dyestuff. Drying of the washed goods is the final operation of printing. *See* TEXTILE CHEMISTRY. [D.H.A.]

Thaliacea A small class of pelagic Tunicata especially abundant in warmer seas, This class of animals contains three orders: the Salpida, Doliolida, and Pyrosomida. Oral and atrial apertures occur at opposite ends of the body. Members of the orders Salpida and Doliolida are transparent forms, partly or wholly ringed by muscular bands. The contractions of these bands produce currents used in propulsion, feeding, and respiration. The order Pyrosomida includes species which form tubular swimming colonies and which are often highly luminescent. *See* BIOLUMINESCENCE; TUNICATA. [D.P.A.]

Thallium A chemical element, Tl, atomic number 81, relative atomic weight of 204.37. The valence electron notation corresponding to its ground state term is $6s^2 6p^1$, which accounts for the maximum oxidation state of III in its compounds. Compounds of oxidation state I and apparent oxidation state II are also known.

Thallium occurs in the Earth's crust to the extent of 0.00006%, mainly as a minor constituent in iron, copper, sulfide, and selenide ores. Minerals of thallium are considered rare.

Thallium has significant use in electronic components, such as thallium-activated sodium iodide crystals in photomultiplier tubes. It is also used in low-melting point alloys, optical glass, and in glass seals for enclosing electronic components. Thallium compounds are extremely toxic to humans and other forms of life.

The insolubility of thallium(I) chloride, bromide, and iodide permits their preparation by direct precipitation from aqueous solution; the fluoride, on the other hand, is water-soluble. Thallium(I) chloride resembles silver chloride in its photosensitivity.

Thallium(I) oxide is a black powder which reacts with water to give a solution from which yellow thallium hydroxide can be crystallized. The hydroxide is a strong base and will take up carbon dioxide from the atmosphere.

Thallium also forms organometallic compounds of the following general classes, R_3Tl, R_2TlX, and $RTlX_2$, where R may be an alkyl or aryl group and X a halogen. *See* ORGANOMETALLIC COMPOUND. [E.M.L.]

1	2	3	4	5	6	7	8	9	10	11	12	13	14	15	16	17	18
1 **H**																	2 **He**
3 **Li**	4 **Be**											5 **B**	6 **C**	7 **N**	8 **O**	9 **F**	10 **Ne**
11 **Na**	12 **Mg**											13 **Al**	14 **Si**	15 **P**	16 **S**	17 **Cl**	18 **Ar**
19 **K**	20 **Ca**	21 **Sc**	22 **Ti**	23 **V**	24 **Cr**	25 **Mn**	26 **Fe**	27 **Co**	28 **Ni**	29 **Cu**	30 **Zn**	31 **Ga**	32 **Ge**	33 **As**	34 **Se**	35 **Br**	36 **Kr**
37 **Rb**	38 **Sr**	39 **Y**	40 **Zr**	41 **Nb**	42 **Mo**	43 **Tc**	44 **Ru**	45 **Rh**	46 **Pd**	47 **Ag**	48 **Cd**	49 **In**	50 **Sn**	51 **Sb**	52 **Te**	53 **I**	54 **Xe**
55 **Cs**	56 **Ba**	71 **Lu**	72 **Hf**	73 **Ta**	74 **W**	75 **Re**	76 **Os**	77 **Ir**	78 **Pt**	79 **Au**	80 **Hg**	81 **Tl**	82 **Pb**	83 **Bi**	84 **Po**	85 **At**	86 **Rn**
87 **Fr**	88 **Ra**	103 **Lr**	104 **Rf**	105 **Db**	106 **Sg**	107 **Bh**	108 **Hs**	109 **Mt**	110	*111*	*112*	*113*	*114*	*115*	*116*	*117*	*118*

lanthanide series	57 **La**	58 **Ce**	59 **Pr**	60 **Nd**	61 **Pm**	62 **Sm**	63 **Eu**	64 **Gd**	65 **Tb**	66 **Dy**	67 **Ho**	68 **Er**	69 **Tm**	70 **Yb**
actinide series	89 **Ac**	90 **Th**	91 **Pa**	92 **U**	93 **Np**	94 **Pu**	95 **Am**	96 **Cm**	97 **Bk**	98 **Cf**	99 **Es**	100 **Fm**	101 **Md**	102 **No**

Thallobionta One of the two commonly recognized subkingdoms of plants. In contrast to the more closely knit subkingdom Embryobionta, the Thallobionta (often also called Thallophyta) are diverse in pigmentation, food reserves, cell-wall structure, and flagellar structure. They still form a natural group, however, in the sense that they are all probably derived from ancestors which would be referred to the Thallobionta, without the intervention of any ancestors which would have to be referred to other groups. The Thallobionta are here considered to include eight divisions, the Schizophyta, Rhodophyta, Chlorophyta, Euglenophyta, Pyrrophyta, Chrysophyta, Phaeophyta, and Fungi. All these groups have both modern and fossil representatives. *See* EMBRYOBIONTA; FUNGI; PLANT KINGDOM. [A.Cr.]

Theales An order of flowering plants, division Magnoliophyta (Angiospermae), in the subclass Dilleniidae of the class Magnoliopsida (dicotyledons). The order consists of 13 families and nearly 3000 species. The largest families are the Guttiferae (about 900 species), Theaceae (about 600 species), Dipterocarpaceae (about 400 species), and Ochnaceae (about 400 species). They are mostly woody plants, less often herbaceous, with simple or occasionally compound leaves. The perianth and stamens of the flowers are attached directly to the receptacle; the calyx is arranged in a tight spiral; and the petals are usually separate from each other. The stamens are numerous and initiated in centrifugal sequence or, less often, are few and cyclic. The tea plant (*Thea sinensis*), *Camellia*, and St. John's-wort (*Hypericum*) are familiar members of the Theales. *See* DILLENIIDAE; MAGNOLIOPSIDA; TEA. [A.Cr.]

Thecanephria An order of Pogonophora, a group of elongate, tentaculated, tube-dwelling, sedentary, nonparasitic marine worms lacking a digestive system. In this order the "coelomic" space in the anterior tentacular region is horseshoe-shaped, and the excretory (osmoregulatory) portion of its ducts come close together medially near the median, adneural blood vessel. The species in this order are multitentaculate.

The order includes four families: Polybrachiidae (with seven genera), Sclerolinidae (one genus), Lamellisabellidae (two genera), and Spirobrachiidae (one genus). *See* ATHECANEPHRIA; POGONOPHORA. [E.B.Cu.]

Thecodontia The most primitive order of archosaurian reptiles, confined to the Triassic. They differ from Lepidosauria in the absence of a supratemporal bone, parietal foramen, and palatal teeth, and in the presence of an antorbital fenestra; dorsal intercentra are absent. Primitive features in comparison to other archosaurs include retention of clavicles and interclavicle, and platelike development of pubis and ischium, which, however, show the beginning of elongation associated with bipedalism. The feet were five-toed, and rear limbs were always longer than front limbs. The order is ancestral to crocodiles, pterosaurs, birds, and both saurischian and ornithischian dinosaurs. The Thecodontia includes five highly diverse suborders: Proterosuchia, Pelycosimia, Pseudosuchia, Aetosauria, and Phytosauria. *See* ARCHOSAURIA; LEPIDOSAURIA; REPTILIA. [J.T.G.]

Thelodontida Sometimes called Coelolepida, this is an extinct group of Agnatha or jawless vertebrates known from the Lower Silurian to Middle Devonian of Europe, Asia, Australia, and North America. Since they had no hard skeleton, their structure and relationships are poorly known. Well-preserved specimens of *Logania* suggest a relationship to Heterostraci in their widely spaced, lateral eyes, nearly terminal mouth, pectoral flaps at the posterior end of the head, and hypocercal tail with a downwardly directed main lobe. *Phlebolepis* resembles Anaspida in general form but lacks the dorsal nostril of that group. *See* AGNATHA; HETEROSTRACI; OSTEOSTRACI; PTERASPIDOMORPHA. [R.H.De.]

Theobromine An alkaloid prepared from the dried ripe seed of *Theobroma cacao* or made synthetically. It is often extracted from waste products of the cocoa and chocolate industry. Theobromine is a close chemical relative of caffeine (see illustration).

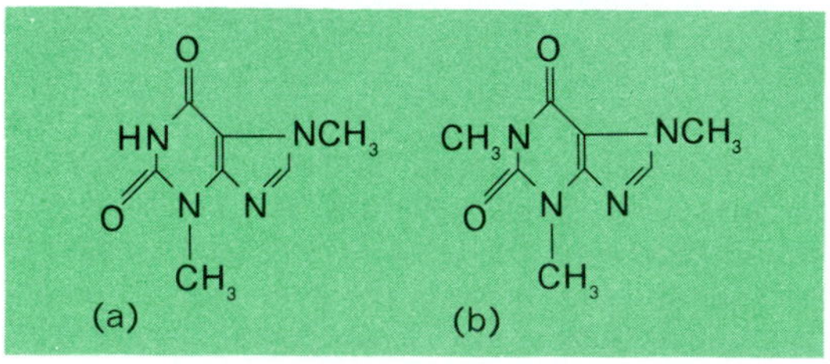

Structural formulas of (*a*) theobromine and (*b*) caffeine.

Theobromine is known to be responsible, in part, for the stimulant action of cocoa and other beverages. In addition, it is used therapeutically as a diuretic in the treatment of various types of edemas. *See* ALKALOID; CACAO; CAFFEINE. [S.M.K.]

Theodolite A surveying instrument for measuring horizontal and vertical angles, similar in principle to the transit (see illustration). In the United States the term implies high precision, the capability of measuring to one second of arc or better. Elsewhere the term is used for an instrument of any precision, and the term transit is not used. Virtually every modern theodolite has glass measuring circles on which fine graduations are scribed. An intensified beam of light passes through the edge of the circle and carries the image of the scribed portion of the circle, which is then greatly magnified for the observer. An optical micrometer assists in reading the finest

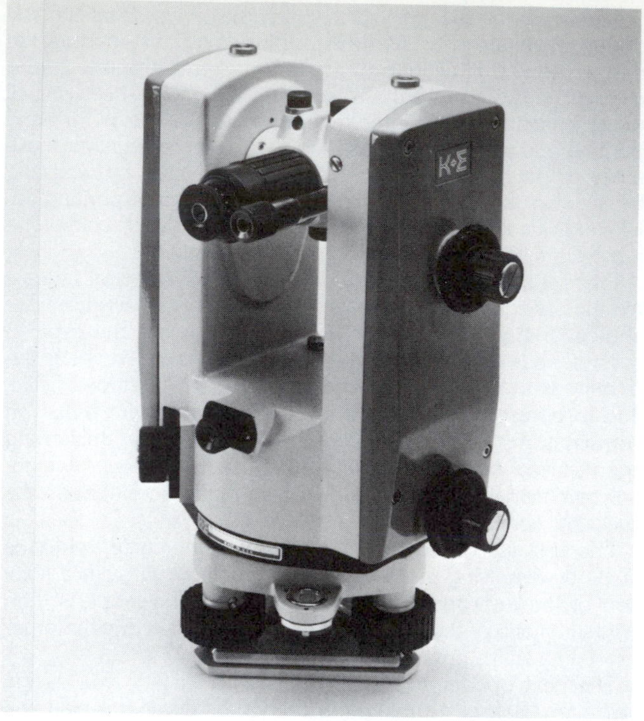

Eagle 60 theodolite. (*Keuffel and Esser Co.*)

divisions. For vertical angles, a sensitive split-image bubble assures precision. *See* SURVEYING; TRANSIT (ENGINEERING). [B.A.B.]

Theorem A proposition arrived at by the methods of logical deduction from a set of basic postulates or axioms accepted as primitive and therefore not subject to deductive proof. So long as a theorem is part of a purely formal system, it is not meaningful to speak about the "truth" of a theorem but only about its "correctness." It becomes true when it, or its consequences, can be shown to be in accord with observable facts. *See* LOGIC. [P.W.Br./H.Ma.]

Theoretical ecology The use of mathematical models to provide a conceptual framework for the analysis of ecological systems. Ecology, the study of the relationship between organisms and their environment, deals with an inherently diverse array of complex systems, rich in idiosyncratic details of natural history. Mathematics is a language that makes it possible to express and think clearly about complex relationships. Thus, mathematical models are considered the essential tools of theoretical ecology. If mathematical ecology is defined as the application of mathematical theory and techniques to ecology, theoretical ecology is the use of such theory and techniques to develop broad conceptual insights into ecological systems. *See* MATHEMATICAL ECOLOGY.

Theoretical ecology cannot as yet rely on universal ecological laws, other than physical laws and the general notion of evolution by natural selection. Rather than aiming at all-inclusive laws, much theoretical work in ecology has more modest goals, for instance clarifying verbal concepts arising from field studies, or sharpening intuition about the possible reasons for qualitative similarities among disparate systems. A wide spectrum of models is used in ecology, ranging from complex tactical models describing specific empirical systems to simpler strategic models that (while admittedly caricaturizing real systems) can be understood in some depth by virtue of their relative simplicity. The term theoretical ecology usually denotes strategic rather than tactical models. In addressing practical problems, where quantitative predictions are essential, large-scale computer models that mimic the behavior or particular systems in a detailed manner are often necessary. *See* ECOLOGY; SYSTEMS ECOLOGY.

The diversity in ecological theories may be categorized in various ways. One obvious way is by the level of organization in the traditional hierarchy of living systems. A given theory may focus on individual organisms, populations, communities, ecosystems, or even the entire biosphere. Particularly rich bodies of theories exist at the organismal, population, and community levels. At the individual level, theoretical models examine the consequences and ultimate evolutionary causes of individual design features, such as leaf size in plants and diet breadth in animals. (A design feature is an organismal trait that can influence rates of death and reproduction, and hence Darwinian fitness.) Such models provide an important link between ecological theory and evolutionary biology. At the population level, theoretical work aims at understanding patterns in distribution and abundance, such as causes for population cycles. Theoretical community ecology is concerned with the factors determining the species composition and functional organization of communities, with a particular emphasis on interspecific interactions such as competition, predation, and mutualism. Theoretical analyses at the ecosystem level attempt to analyze flows of energy and material in entire ecosystems. *See* ECOLOGICAL COMMUNITIES; ECOLOGICAL INTERACTIONS; ORGANIC EVOLUTION.

Ecological theories may be static or dynamic. A static theory does not include time as an explicit variable. Static theories describe compactly the structure of large bodies of information; for that reason they have been particularly common in community ecology, which deals typically with complex, multivariate data sets. In a dynamic theory, variables change through time because of forces embodied in the mathematical structure of the theory. The mathematical formalism may be difference equations, in which variables are computed at discrete time intervals, or differential equations, which are appropriate when variables change continuously. The variables themselves may be continuous (such as leaf temperature) or discrete (such as the number of eagles on an island). *See* DETERMINISM; ECOLOGICAL MODELING; STOCHASTIC PROCESS. [R.Hol.]

Theoretical physics The description of natural phenomena in mathematical form. It is impossible to separate theoretical physics from experimental physics, since a complete understanding of nature can be obtained only by the application of both theory and experiment. There are two main purposes of theoretical physics: the discovery of the fundamental laws of nature and the derivation of conclusions from these fundamental laws.

Physicists aim to reduce the number of laws to a minimum to have as far as possible a unified theory. When the laws are known, it is possible from any given initial conditions of a physical system to derive the subsequent events in the system. Sometimes, especially in quantum theory, only the probability of various events can be predicted. *See* QUANTUM MECHANICS.

The conclusions to be derived from the fundamental laws of nature may be of several different types.

1. Conclusions may be derived in order to test a given theory, particularly a new theory. An example is the derivation of the spectrum of the hydrogen atom from quantum mechanics; the verification of the predictions by accurate measurements is a good test of quantum mechanics. On rather rare occasions experiment has been found to contradict the predictions of an existing theory, and this has then led to the discovery of important new physical laws. An example is the Michelson-Morley experiment on the constancy of the velocity of light, an experiment which led to special relativity theory. *See* ATOMIC STRUCTURE AND SPECTRA; LIGHT.

2. Theory may be required for experiments designed to determine physical constants. Most fundamental physical constants cannot be accurately measured directly. Elaborate theories may be required to deduce the constant from indirect experiments.

3. Predictions of physical phenomena may be made in order to gain understanding of the structure of the physical world. In this category fall theories of the structure of the atom leading to an understanding of the periodic system of elements, or of the structure of the nucleus in which various models are tested (for example, shell model or collective model). In the same category fall applications of theoretical physics to other sciences, for example, to chemistry (theory of the chemical bond and of the rate of chemical reactions), astronomy (theory of planetary motion, internal constitution, and energy production of stars), or biology.

4. Engineering applications may be drawn from fundamental laws. All of engineering may be considered an application of physics, and much of it is an application of mathematical physics, such as elasticity theory, aerodynamics, electricity, and magnetism. *See* ELECTRICITY; MAGNETISM; RADIO-WAVE PROPAGATION.

Apart from the classification of the fields of theoretical physics according to purpose, a classification can also be made according to content. Here one may distinguish three classification principles: type of force, scale of physical phenomena, and type of phenomena. *See* MATHEMATICAL PHYSICS; PHYSICS. [H.A.Be.]

Theralite A dark-colored, phaneritic (visibly crystalline) rock composed chiefly of pyroxene with smaller amounts of calcic plagioclase and nepheline. In a sense it is a nepheline-rich gabbro with abundant dark-colored (mafic) minerals.

Theralite is very uncommon and occurs chiefly in small intrusive bodies (dikes, sills, and laccoliths) where it appears to have formed by differentiation of magma (molten rock) rich in alkalies and potential mafic minerals. It occurs in the Lugar sill in Ayrshire, Scotland. *See* GABBRO; IGNEOUS ROCKS. [C.A.C.]

Therapsida An extinct order of the subclass Synapsida, grouping the majority of advanced mammallike reptiles. Members of this group first appeared in mid-Permian times and persisted until the end of the Triassic. The group is highly ramified and includes several diverse lines of carnivores and herbivores. Most were terrestrial, but one group was aquatic. Therapsids are best known from deposits in South Africa; however, they were worldwide in distribution, with representation in the records of all continents. *See* REPTILIA.

The various evolving lines became increasingly mammalian through time. The most advanced members seem to have invaded uplands of the continental interiors. The ancestors of the mammals lay among the therapsids, with mammals arising from advanced groups known as cynodonts and ictidosaurs. *See* SYNAPSIDA. [E.C.O.]

Theria One of the four subclasses of the class Mammalia, including all living mammals except the monotremes. The Theria were by far the most successful of the several mammalian stocks that arose from the mammallike reptiles in the Triassic. The subclass is divided into three infraclasses: Pantotheria (no living survivors), Metatheria (marsupials), and Eutheria (placentals). Therian mammals are characterized by the distinctive structural history of the molar teeth. The fossil record shows that all the extremely varied therian molar types were derived from a common tribosphenic type in which three main cusps, arranged in a triangle on the upper molar, are opposed to a reversed triangle and basinlike heel on the lower molar. *See* MAMMALIA; THERAPSIDA. [D.D.D./F.S.S.]

Thermal ammeter An ammeter that functions through the medium of the heat generated by the passage of electric

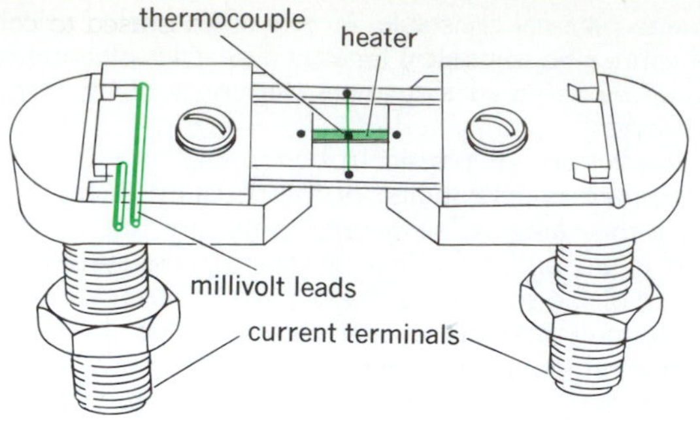

Thermal converter of thermocouple ammeter. (*Weston Instruments, Inc.*)

current through a resistance element. Modern thermal ammeters consist of a thermal converter and a sensitive dc millivoltmeter. The illustration shows such a thermal converter mounted on the instrument terminals. The short and efficient heater is a platinum-alloy wire or tube, to which is welded a thermocouple of two dissimilar metals. When the junction of the thermocouple is heated, a small voltage is generated which is proportional to the temperature rise; this in turn is applied to the millivolt mechanism to deflect its pointer. Typically, the heater is so proportioned as to have a temperature rise of 360°F (200°C) on the desired full-scale current; about 10 millivolts is then available for the dc mechanism. Losses are represented by a potential drop through the heater of 200 mV at full-scale current.

The temperature rise of the heater is a function of the square of the current; thus the instrument scale follows a square law, compressed at the start and expanded at full scale.

Using a 1-mil (25-micrometer) heater wire, full scale represents about 1 ampere. Higher ranges are made with heavier heaters; losses increase correspondingly, with 60 A representing a practical top limit. For measuring currents under 1 A, the thermoelement is placed in vacuum to reduce heat losses; these are now useful as low as 1 millampere. A special bridge form of open thermoelement is also used in the 100-mA range. *See* AMMETER. [J.H.M./J.Be.]

Thermal conduction in solids Thermal conduction in a solid is generally measured by stating the thermal conductivity K, which is the ratio of the steady-state heat flow (heat transfer per unit area per unit time) along a long rod to the temperature gradient along the rod. Thermal conductivity varies widely among different types of solids, and depends markedly on temperature and on the purity and physical state of the solids, particularly at low temperatures.

From the kinetic theory of gases the thermal conductivity can be written as in the equation below, where S is the specific

$$K = (\text{constant})\ Svl$$

heat per unit volume, v is the average particle velocity, and l is the mean free path. In solids, thermal conduction results from conduction by lattice vibrations and from conduction by electrons. In insulating materials, the conduction is by lattice waves; in pure metals, the lattice contribution is negligible and the heat conduction is primarily due to electrons. In many alloys, impure metals, and semiconductors, both conduction mechanisms contribute. *See* CONDUCTION (HEAT); KINETIC THEORY OF MATTER; LATTICE VIBRATIONS; SPECIFIC HEAT.

In superconductors at temperatures below the critical temperature, the electronic conduction is reduced; at sufficiently low temperatures, the thermal conductivity becomes entirely due to lattice waves and is similar to the form of the thermal conductivity of an insulating material. *See* SUPERCONDUCTIVITY. [K.A.McC.]

Thermal converter A device consisting of a conductor heated by an electric current, with one or more hot junctions of a thermocouple attached to it, so that the output emf responds to the temperature rise, and thence the current. Thermal converters are used with external resistors for alternating-current (ac) and voltage measurements over wide ranges.

In the most common form, the conductor is a thin straight wire less than 0.4 in. (1 cm) long, in an evacuated glass bulb, with a single thermocouple junction fastened to the midpoint by a tiny electrically insulating bead. Thermal inertia keeps the temperature of the heater wire constant at frequencies above a few hertz, so that the constant output emf is a true measure of the root-mean-square (rms) heating value of the current. However, the flow of heat energy cannot be controlled precisely, so the temperature, and hence the emf, generally changes with time and other factors. Thus an ordinary thermocouple instrument, consisting of a thermal converter and a millivoltmeter to measure the emf, is accurate only to about 1–3%. *See* THERMOCOUPLE; VOLTMETER.

To overcome this, a thermal converter can be used as an ac-dc transfer instrument (ac-dc comparator) to measure an unknown alternating current or voltage by comparison with a known nearly equal dc quantity. By replacing the millivoltmeter with an adjustable, stable, opposing voltage in series with a microvoltmeter, very small changes in emf can be detected. Accuracies of 0.005% are attainable at audio frequencies. The ultimate accuracy of almost all alternating-current and voltage measurements depends upon thermal converters. *See* ELECTRICAL MEASUREMENTS. [F.L.H.]

Thermal cracking A cracking process in which carbon-to-carbon bonds are severed by the action of heat alone. It consists essentially in the heating of any fraction of petroleum to a temperature at which substantial thermal decomposition takes place through a thermal free-radical mechanism followed by cooling, condensation, and physical separation of the reaction products. There are a number of refinery processes based primarily upon the thermal cracking reaction. They differ primarily in the intensity of the thermal conditions and the feedstock handled.

Visbreaking is a mild thermal cracking operation (850–950°F; 454–510°C) where only 20–25% of the residuum feed is converted to mid-distillate and lighter material. Thermal gas-oil or naphtha cracking is a more severe operation (950–1100°F; 510–593°C) where 45% or more of the feed is converted to lower molecular weight. Steam cracking is an extremely severe operation (1100 to 1400°F; 593–760°C) in which steam is used as a diluent to achieve a very low hydrocarbon partial pressure. In fluid coking the residuum is converted fully to gas-oil products boiling lower than 950°F (510°C) and coke. In delayed coking a residuum is heated and sent to a coke drum, where the liquid has an infinite residence time to convert to lower-molecular-weight hydrocarbons which distill out of the coke drum, and to coke which remains in the drum and must be periodically removed. In fluid coking and delayed coking, there is total conversion of the very heavy high-boiling end of the residuum feed. *See* CRACKING. [E.L.]

Thermal ecology The examination of the independent and interactive biotic and abiotic components of naturally heated environments. Geothermal habitats are present from sea level to the tops of volcanoes and occur as fumaroles, geysers, and hot springs. Hot springs typically possess source pools with overflow, or thermal, streams (rheotherms) or without such streams (limnotherms). Hot spring habitats have existed since life began on Earth, permitting the gradual introduction and evolution of species and communities adapted to each other and to high temperatures. Other geothermal habitats do not have distinct communities.

Hot-spring pools and streams, typified by temperatures higher than the mean annual temperature of the air at the same locality and by benthic mats of various colors, are found on all continents except Antarctica. They are located in regions of geologic activity where meteoric water circulates deep enough to become heated. The greatest densities occur in Yellowstone National Park (Northwest United States), Iceland, and New Zealand. Source waters range from 40°C (104°F) to boiling (around 100°C or 212°F depending on elevation), and may even be superheated at the point of emergence. Few hot springs have pH 5–6; most are either acidic (pH 2–4) or alkaline (pH 7–9). [C.Wi.]

Thermal expansion Solids, liquids, and gases all exhibit dimensional changes for changes in temperature while pressure is held constant. The molecular mechanisms at work and the methods of data presentation are quite different for the three cases.

The temperature coefficient of linear expansion α_l is defined by Eq. (1), where l is the length of the specimen, t is the tem-

$$\alpha_l = \frac{1}{l}\left(\frac{\partial l}{\partial t}\right)_{p\,=\,\mathrm{const}} \qquad (1)$$

perature, and p is the pressure. For each solid there is a Debye characteristic temperature Θ, below which α_l is strongly dependent upon temperature and above which α_l is practically constant. Many common substances are near or above Θ at room temperature and follow approximate equation (2), where

$$l = l_0(1 + \alpha_l t) \qquad (2)$$

l_0 is the length at 0°C and t is the temperature in °C. The total change in length from absolute zero to the melting point has a range of approximately 2% for most substances.

So-called perfect gases follow the relation in Eq. (3), where

$$\frac{pv}{T} = \frac{R}{\text{molecular weight}} \qquad (3)$$

p is absolute pressure, v is specific volume, T is absolute temperature, and R is the so-called gas constant. Real gases often follow this equation closely. *See* GAS CONSTANT.

The coefficient of cubic expansion α_v is defined by Eq. (4), and for a perfect gas this is found to be $1/T$. The behavior of

$$\alpha_v = \frac{1}{v}\left(\frac{\partial v}{\partial t}\right)_{p\,=\,\mathrm{const}} \qquad (4)$$

real gases is largely accounted for by the van der Waals equation. *See* GAS; KINETIC THEORY OF MATTER; VAN DER WAALS EQUATION.

For liquids, α_v is somewhat a function of pressure but is largely determined by temperature. Though αv may often be taken as constant over a sizable range of temperature (as in the liquid expansion thermometer), generally some variation must be accounted for. For example, water contracts with temperature rise from 32 to 39°F (0 to 4°C), above which it expands at an increasing rate. *See* THERMOMETER. [R.A.Bu.]

Thermal flow measurement devices A heated or self-heating sensor used for gas-velocity or gas-mass flow measurement.

In the hot-wire anemometer, a thermopile heated by a constant ac voltage is cooled by the gas flow. The degree of cooling, proportional to local flow velocity, is indicated by the thermopile millivoltage output. *See* THERMOCOUPLE.

In the heat-loss flowmeter, a constant current is supplied to a thermistor bead. The temperature of the bead depends upon the cooling effect of the mass flow of the fluid, and the measure of the voltage drop across the bead is therefore an indication of the flow. A second thermistor bead is used to compensate for the temperature of the fluid. *See* THERMISTOR.

Another heat-loss flowmeter measures the amount of electrical heating that is required to keep the temperature of a platinum resistor constant since the amount of heat carried away changes with the velocity of the fluid. A second sensor is used to compensate for the temperature of the fluid. *See* FLOW MEASUREMENT. [M.Br.; L.P.E.]

Thermal hysteresis A phenomenon in which a physical quantity depends not only on the temperature but also on the preceding thermal history. It is usual to compare the behavior of the physical quantity while heating and the behavior while cooling through the same temperature range. The illustration shows the thermal hysteresis which has been observed

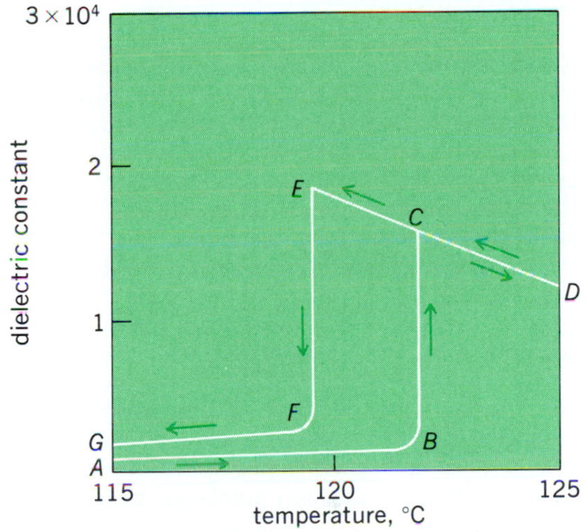

Plot of dielectric constant versus temperature for a single crystal of barium titanate. (*After M. E. Drougard and D. R. Young, Phys. Rev., 95:1152–1153, 1954*)

in the behavior of the dielectric constant of single crystals of barium titanate. On heating, the dielectric constant was observed to follow the path *ABCD*, and on cooling the path *DCEFG*. *See* DIELECTRIC CONSTANT; FERROELECTRICS.

Perhaps the most common example of thermal hysteresis involves a phase change such as solidification from the liquid phase. In many cases these liquids can be dramatically supercooled. Elaborate precautions to eliminate impurities and outside disturbances can be instrumental in supercooling 60 to 80°C. On raising the temperature after freezing, however, the system follows a completely different path, with melting coming at the prescribed temperature for the phase change. *See* CRYSTAL; NUCLEATION; PHASE TRANSITIONS. [H.B.H.; R.K.Mac.C.]

Thermal neutrons Neutrons whose energy distribution is identical with, or similar to, the thermal distribution in the material in which they are found.

Usually, the energy distribution of neutrons in matter is not in thermal equilibrium with the environment, but they are still called thermal distributions of neutrons. Most neutrons are produced at energies above thermal, and the absorption of neutrons interrupts the process of slowing down so that the low-energy neutron spectrum is "hardened" with a greater number of neutrons of higher energy than the maxwellian distribution of energies predicts, and a "tail" of neutrons not yet thermalized. The actual distribution can be fitted to one in which the spectrum is represented as a maxwellian distribution of higher temperature than the environment, with the tail superposed.

See BOLTZMANN STATISTICS; KINETIC THEORY OF MATTER; NUCLEAR FISSION; REACTOR PHYSICS. [B.I.S.]

Thermal stress Mechanical stress induced in a body when some or all of its parts are not free to expand or contract in response to changes in temperature. In most continuous bodies, thermal expansion or contraction cannot occur freely in all directions because of geometry, external constraints, or the existence of temperature gradients, and so stresses are produced. Such stresses caused by a temperature change are known as thermal stresses.

Problems of thermal stress arise in many practical design problems, such as those encountered in the design of steam and gas turbines, diesel engines, jet engines, rocket motors, and nuclear reactors. The high aerodynamic heating rates associated with high-speed flight present even more severe thermal-stress problems for the design of spacecraft and missiles. *See* STRESS AND STRAIN. [S.-Y.C.]

Thermal wind The difference in the geostrophic wind between two heights in the atmosphere over a given position on Earth. It approximates the variation of the actual winds with height for large-scale and slowly changing motions of the atmosphere. Such structure in the wind field is of fundamental importance to the description of the atmosphere and to processes causing its day-to-day changes. The thermal wind embodies a basic relationship between vertical fluctuations of the horizontal wind and horizontal temperature gradients in the atmosphere. This relationship arises from the combination of the geostrophic wind law, the hydrostatic equation, and the gas law.

The geostrophic wind law applies directly to steady, straight, and unaccelerated horizontal motion and is a good approximation for large-scale and slowly changing motions in the atmosphere. The hydrostatic equation combined with the gas law relates the atmospheric pressure and temperature fields. The relationship is accurate for most atmospheric situations but not for small-scale and rapidly changing conditions such as in turbulence and thunderstorms. The equation gives the change of pressure in the vertical direction as a function of pressure and temperature. The key conclusion is that at a given level in the atmosphere the pressure change (decrease) with height is more rapid in cold air than in warm air. *See* ATMOSPHERE; GEOSTROPHIC WIND; HYDROSTATICS; TROPOSPHERE. [D.D.H.]

Thermionic emission The emission of electrons into vacuum by a heated electronic conductor. In its broadest meaning, thermionic emission includes the emission of ions, but this process is quite different from that normally understood by the term. Thermionic emitters are used as cathodes in electron tubes and hence are of great technical and scientific importance. Although in principle all conductors are thermionic emitters, only a few materials satisfy the requirements set by practical applications. Of the metals, tungsten is an important practical thermionic emitter; in most electron tubes, however, the oxide-coated cathode is used to great advantage. For a detailed discussion of practical thermionic emitters *see* VACUUM TUBE.

The thermionic emission of a material may be measured by using the material as the cathode in a vacuum tube and collecting the emitted electrons on a positive anode. If the anode is sufficiently positive relative to the cathode, space charge (a concentration of electrons near the cathode) can be avoided and all electrons emitted can be collected; the saturation thermionic current is then measured. *See* SCHOTTKY EFFECT.

The emission current density J increases rapidly with increasing temperature; this is illustrated by the following approximate values for tungsten:

T (K)	1000	2000	2500	3000
J (amperes/cm^2)	10^{-15}	10^{-3}	0.3	15

The temperature dependence of J is given by the Richardson (or Dushman-Richardson) equation below. Here A is a con-

$$J = AT^2e^{-(\phi - kT)}$$

stant, k is Boltzmann's constant (= 1.38×10^{-23} joule/degree), and ϕ is the work function of the emitter. The work function has the dimensions of energy and is a few electronvolts for thermionic emitters. *See* WORK FUNCTION (ELECTRONICS). [A.J.D.]

Thermionic power generator A device in which heat energy is directly converted to electric energy, and based on the phenomenon of thermionic emission. It is also frequently called a thermionic converter.

Two metallic elements, an emitter and a collector, are the minimum needed for a thermionic converter. The thermionic electron emitter must be capable of yielding electrons to the space that separates the emitter from the electron collector. The collector must be operated at a significantly lower temperature than the emitter so that the collector does not also emit electrons. The general term that describes such a thermionic device is thermionic diode.

If the space between the two elements of the diode is evacuated sufficiently so that the residual gas has no significant influence on the flow of electrons from the emitter to the collector, the device is known as a vacuum thermionic converter. Electrons are negatively charged particles and thus repel each other. The presence of electrons in transit between the emitter and the collector can, therefore, interfere seriously with the free flow of additional charges and thus set up a space-charge limitation on the current density and the efficiency attainable. *See* SPACE CHARGE.

Two methods are used to minimize the space-charge limitation. One depends on a diode construction with extremely close spacing—of the order of 5 micrometers or approximately 0.0002 in. The second method is to introduce an ionized gas. The number density of the ions of the gas must be equal to or locally greater than the electron density in order to neutralize the negative space charge otherwise present. The term plasma is applied to a medium in which the net electric charge is zero. A thermionic converter that depends on the presence of an ionized gas to give good conduction in the space between the emitter and the collector is known as a plasma thermionic converter. *See* THERMIONIC EMISSION. [W.B.N./T.F.S.]

Thermionic tube An electron tube that relies upon thermally emitted electrons from a heated cathode for tube current.

Thermionic emission of electrons means emission by heat. In practical form an electrode, called the cathode because it forms the negative electrode of the tube, is heated until it emits electrons. The cathode may be either a directly heated filament or an indirectly heated surface. With a filamentary cathode, heating current is passed through the wire, which either emits electrons directly or is covered with a material that readily emits electrons. Indirectly heated cathodes have a filament, commonly called the heater, located within the cathode electrode to bring the surface of the cathode to emitting temperature. The majority of all vacuum tubes are thermionic tubes. *See* ELECTRON TUBE; GAS TUBE; THERMIONIC EMISSION; VACUUM TUBE. [L.S.N.]

Thermistor A resistive circuit component, having a high negative temperature coefficient of resistance (as temperature increases, resistance decreases). *See* ELECTRICAL RESISTIVITY.

The typical thermistor is a stable, compact, and rugged two-terminal ceramiclike semiconductor bead, rod, or disk. Thermistors are made of various mixtures of oxides of manganese, nickel, cobalt, copper, uranium, iron, zinc, titanium, and magnesium. The temperature coefficient is determined by the proportions of oxides contained in the mixture.

Thermistors are versatile circuit elements and have many applications in measurement and control. The thermistor's large temperature coefficient of resistance is ideal for temperature measurement. Precise temperature measurement is done with high-resistance thermistors in a resistance bridge. Sensitivity of 0.0009°F (0.0005°C) is readily attained. *See* TEMPERATURE MEASUREMENT; THERMOMETER.

Many electrical and electronic components have positive temperature coefficients which are detrimental to the temperature stability of the circuit. A properly selected thermistor in the circuit containing such a component will provide temperature compensation.

When a thermistor is used as a flowmeter, vacuum gage, or anemometer, a small voltage is applied and the current through the thermistor is measured. The amount of heat dissipated is a function of the degree of vacuum surrounding the device or the velocity of gas passing the device, and the dissipated power reaches a constant value. The measured current is calibrated in terms of vacuum or gas flow. *See* ANEMOMETER; FLOW MEASUREMENT; VACUUM MEASUREMENT.

When a thermistor is self-heated as a result of current passing through it, its resistance decreases. The thermal mass causes the time-rate of decrease to be fixed. The delayed buildup of circuit currents can be used to introduce a fixed time delay between relay operations or to protect equipment during start-up.

The thermistor's resistance versus power characteristic makes it a useful power-measuring device. Microwave power is measured by a bead thermistor mounted in the waveguide and biased so that bead impedance matches the cavity. The thermistor also can be used to measure radiant power, such as infrared or visible light. *See* ELECTRIC POWER MEASUREMENT. [F.H.B.]

Thermoanalysis A group of analytical techniques developed to continuously monitor physical or chemical changes of a sample which occur as the temperature is varied. Thermogravimetry (TG), differential thermal analysis (DTA), and differential scanning calorimetry (DSC) are the three principal thermoanalytical methods. Thermogravimetry involves measuring the changes in mass of a substance, typically a solid, as the substance is being heated. In differential thermal analysis, the temperature difference between a sample and inert reference material is monitored as they are simultaneously heated, or cooled, at a predetermined rate. Differential scanning calorimetry involves heating a sample and a reference material by using separately controlled resistance heaters at a predetermined rate while they are kept isothermal. Modern commercial instruments, many incorporating microprocessors, are available for all three types of thermoanalysis. *See* CALORIMETRY; MICROPROCESSOR. [N.Je.]

Thermochemistry A branch of physical chemistry dealing with the heat effects which accompany chemical reactions, the formation of solutions, and changes in the physical state of substances, such as the fusion of a crystalline solid or the vaporization of a liquid. When these processes evolve heat, they are said to be exothermic. Conversely, those absorbing heat are endothermic. A knowledge of the magnitudes of such heat effects is of practical value to the engineer in solving problems of heating and refrigeration and in controlling chemical reactions at suitable temperatures. Such data are also of great theoretical importance to the scientist who wishes to calculate the chemical affinity or free energy for various reactions, hypothetical or real. *See* CHEMICAL THERMODYNAMICS; FREE ENERGY.

These heat effects are usually measured in calories, partly for historical reasons and partly because the figures thereby obtained are of more convenient magnitude than when the joule, the fundamental unit of energy, is employed. The present-day calorie is arbitrarily defined as being equal to 4.184 absolute joules, rather than in terms of the heat capacity of water, which varies with temperature. *See* CALORIMETRY. [B.J.Z.]

Thermocline A layer of seawater in which the temperature decrease with depth is greater than that of the overlying and underlying water. Such layers are semipermanent features of the oceanic temperature structure, and their depth and thickness show marked variation with season, latitude and longitude, and local environmental conditions. Since the three-dimensional temperature structure has a great effect on many oceanic properties, such as the transmission of sound, the study of the nature and behavior of the thermocline is of extreme importance to many oceanographic interests. In general, two major types of thermocline may be identified: the permanent thermodine (see illustration) and the seasonal thermocline. In addition to these types, shallow thermocline or similar stable layers often occur, owing to diurnal heating of the surface waters. *See* SEAWATER; UNDERWATER SOUND. [J.J.S.]

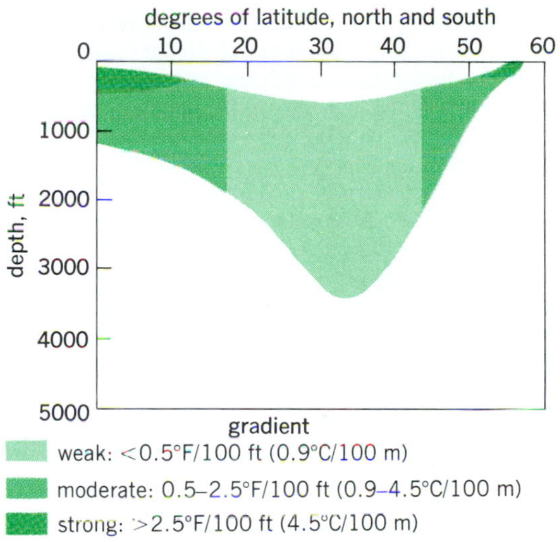

The permanent thermocline, based on averages for depth, thickness, and gradient within thermocline. 1 ft = 0.3 m.

Thermocouple A device that uses the voltage developed by the junction of two dissimilar metals to measure temperature difference. Two wires of dissimilar metals welded together at the ends make up the basic thermocouple (see illustration). One junction, called the sensing or measuring junction, is placed at the point where temperature is to be measured. The other junction, called the reference or cold junction, is maintained at a known reference temperature. The voltage developed between the two junctions is approximately proportional to the difference between the temperatures of the two junctions, and may be measured by including a suitable voltmeter in the circuit. *See* THERMOELECTRICITY.

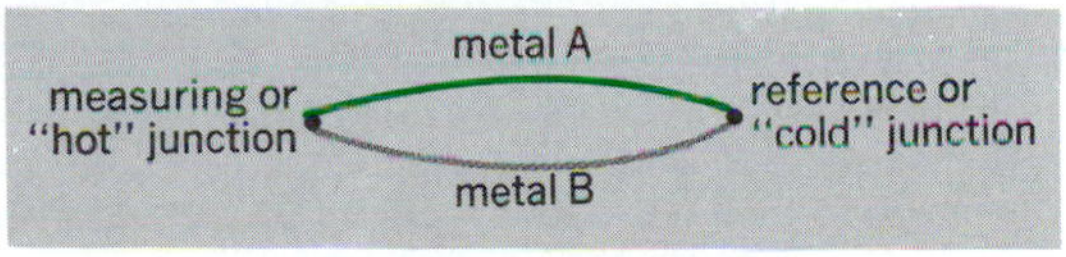

Elements of a thermocouple.

Although all dissimilar metals exhibit the thermoelectric effect, only a few are in wide use. The major characteristics which make certain metals or combinations of metals outstanding for this purpose are stability and reproducibility, constant or controllable compositions, corrosion resistance, sensitivity, ruggedness, and cost.

The thermopile is a number of thermocouples connected in series or parallel. The series circuit provides a higher sensitivity (greater emf per degree) than one thermocouple alone and is often selected for this reason. Either the series or parallel circuit may be used for obtaining an indication which approaches the average of the several temperatures. *See* TEMPERATURE MEASUREMENT. [H.S.B.]

Thermodynamic cycle A procedure or arrangement in which one form of energy, such as heat at an elevated temperature from combustion of a fuel, is in part converted to another form, such as mechanical energy on a shaft, and the remainder is rejected to a lower temperature sink as low-grade heat.

A thermodynamic cycle requires, in addition to the supply of incoming energy, (1) a working substance, usually a gas or vapor; (2) a mechanism in which the processes or phases can be carried through sequentially; and (3) a thermodynamic sink to which the residual heat can be rejected. The cycle itself is a repetitive series of operations.

There is a basic pattern of processes common to power-producing cycles. There is a compression process wherein the working substance undergoes an increase in pressure and therefore density. There is an addition of thermal energy from a source such as a fossil fuel, a fissile fuel, or solar radiation. There is an expansion process during which work is done by the system on the surroundings. There is a rejection process where thermal energy is transferred to the surroundings. The algebraic sum of the energy additions and abstractions is such that some thermal energy is converted into mechanical work. *See* HEAT.

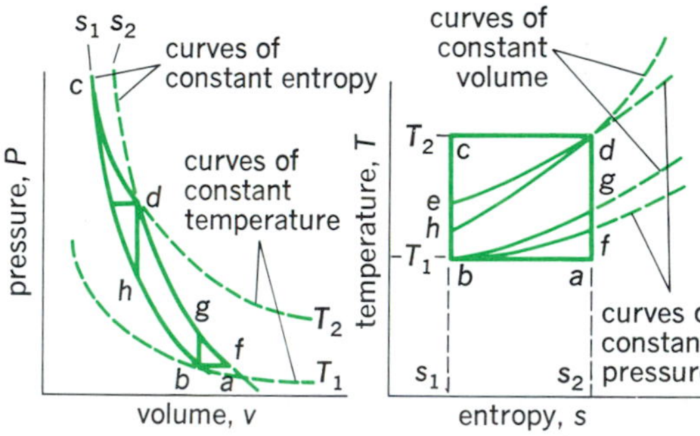

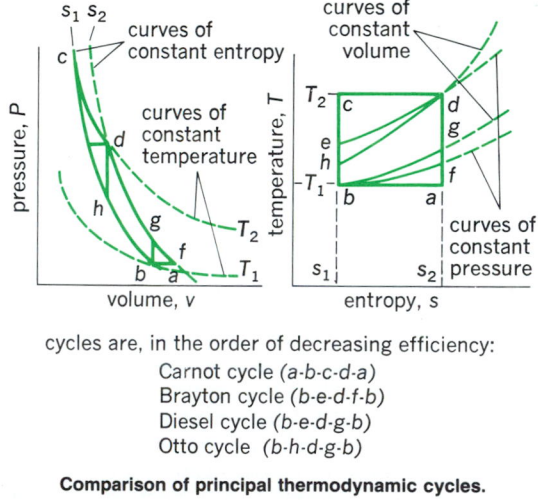

cycles are, in the order of decreasing efficiency:
Carnot cycle (*a-b-c-d-a*)
Brayton cycle (*b-e-d-f-b*)
Diesel cycle (*b-e-d-g-b*)
Otto cycle (*b-h-d-g-b*)

Comparison of principal thermodynamic cycles.

Many cyclic arrangements, using various combinations of phases but all seeking to convert heat into work, have been proposed by many investigators whose names are attached to their proposals, for example, the Diesel, Otto, Rankine, Brayton, Stirling, Ericsson, and Atkinson cycles (see illustration). All proposals are not equally efficient in the conversion of heat into work. However, they may offer other advantages which have led to their practical development for various applications. *See* BRAYTON CYCLE; CARNOT CYCLE; DIESEL CYCLE; OTTO CYCLE; STIRLING ENGINE; THERMODYNAMIC PROCESSES. [T.Ba.]

Thermodynamic principles Laws governing the conversion of energy from one form to another. Among the many consequences of these laws are relationships between the properties of matter and the effects of changes in pressure, temperatures, electric field, magnetic field, and composition. The great practicality of the science arises from the foundations of the subject. Thermodynamics is based upon observations of common experience that have been formulated into the thermodynamic laws. From these few laws all of the remaining laws of the science are deducible by purely logical reasoning.

For exposition, it is necessary to define a few terms. A system is that part of the physical world under consideration. The rest of the world is the surroundings. An open system may exchange mass, heat, and work with the surroundings. A closed system may exchange heat and work but not mass with the surroundings. An isolated system has no exchange with the surroundings. Those parts of a system spatially uniform and homogeneous are called phases. For example, a liquid together with its vapor may be considered a two-phase system. It is a fact of experience that an isolated system approaches a particularly simple terminal state called an equilibrium state.

Temperature. It is within the scope of thermodynamics to refine the primitive notion of hotness and coldness into an operational and precise concept of temperature. The equilibrium states of a single-component, single-phase fluid provide a starting point. For such a fluid the equilibrium state is defined by fixing two of its properties. For example, one could construct a mercury-in-glass thermometer that has its pressure held constant; then only one other property could be varied independently. If the volume (height of mercury) is observed at any equilibrium state, there is a 1:1 correspondence between it and any other property excepting the pressure. The degree of hotness is one of these properties.

Also note that thermal equilibrium between different systems exists. For example, if the mercury thermometer is placed in contact with a body of quiet water, the mercury will either expand or contract. The volume change of the mercury will eventually stop, and the properties of the mercury will be constant, indicating an equilibrium state; moreover, the water will also have the constant properties of an equilibrium state. Bodies in thermal equilibrium are said to have the same temperature. Thus, one arrives at a method of measuring the temperature of a body of water.

Now suppose there is a large body of water in thermal contact through a wall with a large body of another fluid such as alcohol, both at equilibrium. It is an experimental fact that the mercury-in-glass thermometer will register the same volume when placed in either of the fluids. This fact of experience is designated as the zeroth law of thermodynamics: If two bodies A and B are separately in thermal equilibrium with C, then A and B are in equilibrium with each other. Thus, a useful empirical temperature measurement based upon the volume of mercury under constant pressure is established. *See* TEMPERATURE MEASUREMENT.

Internal energy. Thermodynamics does not define the concepts of energy or work but adopts them from the other macroscopic sciences of mechanics and electromagnetism. Also, the conservation of energy is taken as axiomatic. Therefore, if an isolated system is formed from any part of the world, a definite amount of energy will be trapped in the system. The energy resides in the kinetic and potential energy of the trapped molecules. This trapped energy is called the internal energy U.

Because of the conservation of energy, the internal energy of a closed system can be altered only by an exchange of energy with the surroundings. There are only three modes by which the exchange can occur: by mass transfer, heat transfer, or work exchange. So for a closed (no mass transfer), adiabatic (no heat transfer) system the change in internal energy ΔU is equal to the work done by the surroundings on the system, as defined by Eq. (1). Here a convention has been adopted that

$$\Delta U = W_{AD} \quad (1)$$

work done on the system is positive. There is a great mass of experimental information where work has been done on a closed system enclosed within adiabatic walls. Among these are experiments performed by J. Joule more than a century ago. It may be concluded from Joule's experiments that the expenditure of a given quantity of work always causes the same change of state regardless of how the work is carried out. Both W_{AD} and ΔU are independent of the path. It is concluded that U is a state function.

It is known from experience that the same change in state of a system can be effected by either supplying work to the system in an adiabatic enclosure or by contacting the system through a conducting wall with a higher temperature system. The latter method is a different means of transferring energy than work and is termed heat and given the symbol Q.

The first law of thermodynamics, Eq. (2), states that the algebraic sum of heat and work during a process is equal to the

$$\Delta U = Q + W \quad (2)$$

change in the state function U. The term $(Q + W)$ is therefore independent of the path taken between the two states. The differential form of the first law is given by Eq. (3), where q

$$dU = q + w \quad (3)$$

and w represent small quantities.

Reversible and irreversible processes. Any process that occurs in nature is in agreement with the first law, but many processes permissible by the first law never occur. There is an overwhelming preference for processes to proceed in one direction. In one of Joule's experiments, a falling weight caused a paddle to do work on an adiabatically enclosed body of water. The total effect of the experiment was to increase the internal energy of the water and to lower the weight. The surroundings remained unchanged. There is no way one can reverse this process, that is, restore the water to its original state and raise the weight to its original height without also making some additional change in the surroundings. The process is irreversible.

Thermodynamics makes use of an idealization, called a reversible process, that is a limiting case of the natural or irreversible process. The reversible process may be defined as one which can be completely reversed without leaving more than a vanishingly small change in the surroundings. It is a consequence of the definition that a reversible process proceeds through a succession of equilibrium states and may be reversed by an infinitesimal change in the external conditions. Imagine having a cylinder of gas fitted with a frictionless piston. If the piston is moved so slowly that pressure gradients are absent, the gas will be in an equilibrium state at all times. The difference between the gas pressure and the external pressure needs only to be infinitesimal in order to move the frictionless piston. Under the rather restrictive conditions of a reversible process, Eq. (4) may be quite properly written, where p is the gas

$$dW = -pdV \quad (4)$$

pressure and V is the gas volume.

Entropy. The discussion on irreversible processes has led to the second law of thermodynamics, which is just a general statement of the idea that there is a preferred direction for a given process. There are many physical statements of the second law, all being equivalent and leading to the same mathematical statement. The statement of R. Clausius is: "It is *not* possible that, at the end of a cycle of changes, heat has been transferred from a colder to a hotter body without producing some other effect." Lord Kelvin's statement is: "It is *not* possible that, at the end of a cycle of changes, heat has been extracted from a reservoir and an equal amount of work has been produced without producing some other effect."

The most efficient way of developing the mathematical consequences of the second law is to proceed from Caratheodory's principle, which can be either taken as another physical expression of the second law or derived from the Clausius or Kelvin statement. Caratheodory's principle is: "In the neighborhood of any equilibrium state of a system there are states which are not accessible by an adiabatic process."

Caratheodory used this principle together with a mathematical theorem that he developed to infer the existence of a state function S and an integrating factor $1/T$, where T is the thermodynamic temperature such that Eq. (5) holds for a reversi-

$$dQ_{REV} = TdS \quad (5)$$

ble change. The state function S is called the entropy. It can also be shown that the entropy in an adiabatic system increases for an irreversible change and remains constant for a reversible change as in Eq. (6).

$$\Delta S_{AD} \geq 0 \quad (6)$$

The first part of the mathematical statement of the second law allows one to write one of the most important thermodynamic equations, Eq. (7). Although this equation was derived

$$dU = TdS - pdV \quad (7)$$

for reversible changes, it is valid for all changes. All the quantities are functions of state. Therefore, for a change between two states the integral of the equation will be valid even if the path is not reversible.

The second part of the mathematical statement is a concise summary of physical statements on the direction of processes. Also contained in the second part of the entropy statement is the key idea of equilibrium. The equilibrium state of an adiabatic or isolated system is characterized by entropy being at its maximum value consistent with the physical constraints. *See* ADIABATIC PROCESS; ENTROPY. [W.F.J.]

Thermodynamic processes Changes of any property of an aggregation of matter and energy, accompanied by thermal effects.

Systems and processes. To evaluate the results of a process, it is necessary to know the participants that undergo the process, and their mass and energy. A region, or a system, is selected for study, and its contents determined. This region may have both mass and energy entering or leaving during a particular change of conditions, and these mass and energy transfers may result in changes both within the system and within the surroundings which envelop the system.

To establish the exact path of a process, the initial state of the system must be determined, specifying the values of variables such as temperature, pressure, volume, and quantity of material. The number of properties required to specify the state of a system depends upon the complexity of the system. Whenever a system changes from one state to another, a process occurs.

The path of a change of state is the locus of the whole series of states through which the system passes when going from an initial to a final state. For example, suppose a gas expands to twice its volume and that its initial and final temperatures are the same. An extremely large number of paths connect these initial and final states. The detailed path must be specified if the heat or work is to be a known quantity; however, changes in the thermodynamic properties depend only on the initial and final states and not upon the path. A quantity whose change is fixed by the end states and is independent of the path is a point function or a property.

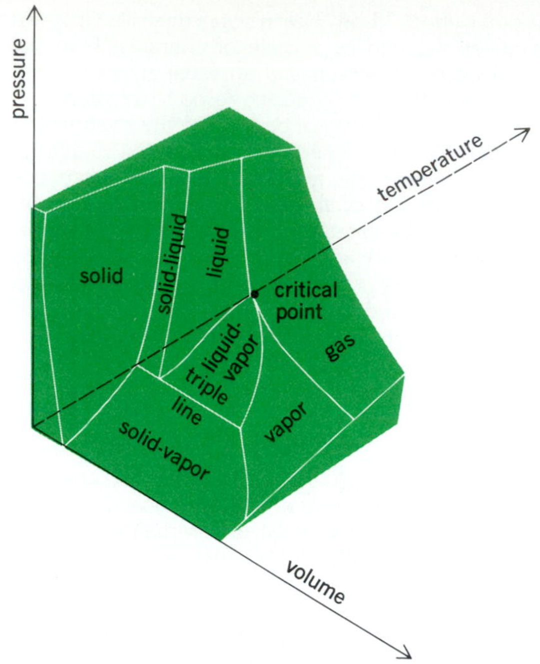

Fig. 1. Portion of pressure-volume-temperature (*P-V-T*) surface for a typical substance.

Pressure-volume-temperature diagram. Whereas the state of a system is a point function, the change of state of a system, or a process, is a path function. Various processes or methods of change of a system from one state to another may be depicted graphically as a path on a plot using thermodynamic properties as coordinates.

The variable properties most frequently and conveniently measured are pressure, volume, and temperature. If any two of these are held fixed (independent variables), the third is determined (dependent variable). To depict the relationship among these physical properties of the particular working substance, these three variables may be used as the coordinates of a three-dimensional space. The resulting surface is a graphic presentation of the equation of state for this working substance, and all possible equilibrium states of the substance lie on this *P-V-T* surface.

Because a *P-V-T* surface represents all equilibrium conditions of the working substance, any line on the surface represents a possible reversible process, or a succession of equilibrium states.

The portion of the *P-V-T* surface shown in Fig. 1 typifies most real substances; it is characterized by contraction of the substance on freezing. Going from the liquid surface to the liquid-solid surface onto the solid surface involves a decrease in both temperature and volume. Water is one of the few exceptions to this condition; it expands upon freezing, and its resultant *P-V-T* surface is somewhat modified where the solid and liquid phases abut.

One can project the three-dimensional surface onto the *P-T* plane as in Fig. 2. The triple point is the point where the three phases are in equilibrium. When the temperature exceeds the critical temperature (at the critical point), only the gaseous phase is possible.

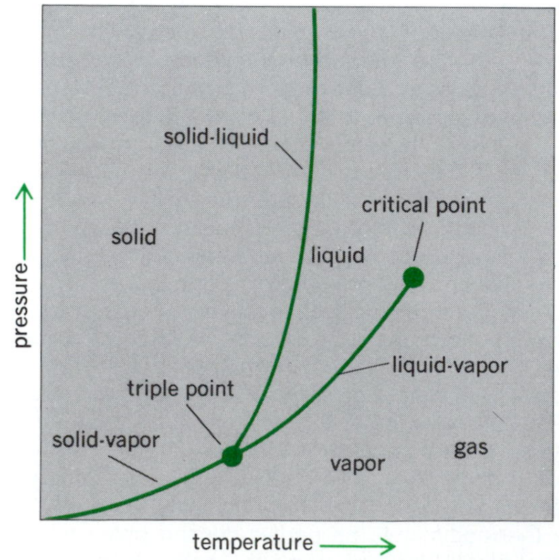

Fig. 2. Portion of equilibrium surface projected on pressure-temperature (*P-T*) plane.

Temperature-entropy diagram. Energy quantities may be depicted as the product of two factors: an intensive property and an extensive one. Examples of intensive properties are pressure, temperature, and magnetic field; extensive ones are volume, magnetization, and mass. Thus, in differential form, work is the product of a pressure exerted against an area which sweeps through an infinitesimal volume, as in Eq. (1).

$$dW = P\,dV \qquad (1)$$

As a gas expands, it is doing work on its environment. However, a number of different kinds of work are known. For example, one could have work of polarization of a dielectric, of magnetization, of stretching a wire, or of making new surface area. In all cases, the infinitesimal work is given by Eq. (2),

$$dW = X\,dx \qquad (2)$$

where X is a generalized applied force which is an intensive quantity, and dx is a generalized displacement of the system and is thus extensive.

By extending this approach, one can depict transferred heat as the product of an intensive property, temperature, and a distributed or extensive property defined as entropy, for which the symbol is S. *See* ENTROPY.

Reversible and irreversible processes. Not all energy contained in or associated with a mass can be converted into useful work. Under ideal conditions only a fraction of the total energy present can be converted into work. The ideal conversions which retain the maximum available useful energy are reversible processes.

Characteristics of a reversible process are that the working substance is always in thermodynamic equilibrium and the process involves no dissipative effects such as viscosity, friction, inelasticity, electrical resistance, or magnetic hysteresis. Thus, reversible processes proceed quasistatically so that the system passes through a series of states of thermodynamic equilibrium, both internally and with its surroundings. This series of states may be traversed just as well in one direction as in the other.

Actual changes of a system deviate from the idealized situation of a quasistatic process devoid of dissipative effects. The extent of the deviation from ideality is correspondingly the extent of the irreversibility of the process. *See* THERMODYNAMIC PRINCIPLES. [P.E.Bl.; W.A.S.]

Thermoelectric power generator A device that converts heat energy directly into electric energy by using the Seebeck effect. A thermoelectric generator is composed of at least two dissimilar materials, one junction of which is in contact with a heat source and the other junction of which is in contact with a heat sink. *See* SEEBECK EFFECT.

The power converted from heat to electricity is dependent

upon the materials used, the temperatures of the heat source and sink, the electrical and thermal design of the thermocouple, and the load of the thermocouple.

The most widely used generator material is lead telluride. It can be doped to produce both *p*- and *n*-type material and has a useful temperature range of about 80–800°F (300–700 K). In segmented couples at the low-temperature end, bismuth telluride and its alloys are sometimes used.

Thermoelectric generators have been built in sizes up to 5 kW, with their primary energy sources being hydrocarbon fuels, radioisotopes, and solar energy.

The maximum theoretical thermal efficiency for materials over a temperature range of 80–1880°F (300–1300 K) is approximately 18%. The best actual thermal efficiency of a physically constructed device is between 6 and 10% and is obtained by operation between 80 and 1250°F (300 and 950 K).

Major problems still exist in the development of materials with higher figures of merit that are capable of operation at higher temperature. Even with present materials, there are severe engineering problems. *See* THERMOELECTRICITY. [D.C.Wh.]

Thermoelectricity The direct conversion of heat into electrical energy, or the reverse, in solid or liquid conductors by means of three interrelated phenomena—the Seebeck effect, the Peltier effect, and the Thomson effect—including the influence of magnetic fields upon each. The Seebeck effect concerns the electromotive force (emf) generated in a circuit composed of two different conductors whose junctions are maintained at different temperatures. The Peltier effect refers to the reversible heat generated at the junction between two different conductors when a current passes through the junction. The Thomson effect involves the reversible generation of heat in a single current-carrying conductor along which a temperature gradient is maintained. Specifically excluded from the definition of thermoelectricity are the phenomena of Joule heating and thermionic emission. *See* JOULE'S LAW; PELTIER EFFECT; SEEBECK EFFECT; THERMIONIC EMISSION; THOMSON EFFECT.

The three thermoelectric effects are described in terms of three coefficients: the absolute thermoelectric power (or thermopower) S, the Peltier coefficient Π, and the Thomson coefficient μ, each of which is defined for a homogeneous conductor at constant temperature. These coefficients are connected by the Kelvin relations, which convert complete information about one into complete information about all three. It is therefore necessary to measure only one of the three coefficients; usually the thermopower S is chosen.

The most important practical application of thermoelectric phenomena is in the accurate measurement of temperature. The phenomenon involved is the Seebeck effect. Of lesser importance are the direct generation of electric power by application of heat (also involving the Seebeck effect) and thermoelectric cooling and heating (involving the Peltier effect).

A basic system suitable for all four applications is illustrated schematically in the illustration. Several thermocouples are connected in series to form a thermopile, a device with increased output (for power generation or cooling and heating) or sensitivity (for temperature measurement) relative to a single thermocouple. The junctions forming one end of the thermopile are all at the same low temperature T_L, and the junctions forming the other end are at the high temperature T_H. The thermopile is connected to a device D which is different for each application. For temperature measurement, the temperature T_L is fixed, for example, by means of a bath; the temperature T_H becomes the running temperature T, which is to be measured; and the device is a potentiometer for measuring the thermoelectric emf generated by the thermopile. For power generation, the temperature T_L is fixed by connection to a heat sink; the temperature T_H is fixed at a value determined by the output of the heat source and the thermal conductivity of the thermopile; and the device is whatever is to be run by the electricity which is generated. For heating or cooling, the device is a current generator which passes current through the thermopile. If the current flows in the proper direction, the junctions at T_H will heat up, and those at T_L will cool down. If T_H is fixed by connection to a heat sink, thermoelectric cooling will be provided by T_L. Alternatively, if T_L is fixed, thermoelectric heating will be provided at T_H. Such a system has the advantage that at any given location it can be converted from a cooler to a heater merely by reversing the direction of the current.

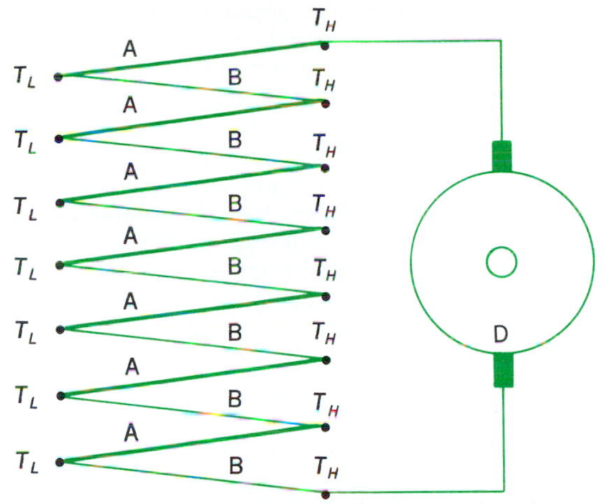

Thermopile, a battery of thermocouples connected in series. *D* is a device appropriate to the particular application; *A* and *B* are the two different conductors.

Thermoelectric systems made from even the best available materials have the disadvantages of relatively low efficiencies and concomitant high cost per unit of output. Their use in power generation, heating, and cooling has therefore been largely restricted to situations in which these disadvantages are outweighed by such advantages as small size, low maintenance due to lack of moving parts, quiet performance, light weight, and long life. *See* THERMOELECTRIC POWER GENERATOR. [J.B.]

Thermography A method that records pictorially the natural emission of infrared radiations emanating from an object or, as in the case of a medical application, from a person. The skin behaves like a blackbody radiator emitting infrared radiations constantly. Two types of thermographic systems are commonly used in medicine. The first one uses a thermographic machine, which contains one or more heat-sensing crystals of either indium antimony or mercury cadmium telluride. Infrared radiations originating from the surface of the skin are focused by an optical system onto the crystal, producing a change in its electrical resistance. This results in the transformation of heat energy into an electronic signal that can be displayed as an image on a cathode-ray tube. The second thermographic system relies on the properties of liquid cholesterol crystals that change color within narrow temperature ranges. These crystals are then applied to a thin foil. Within a few seconds after such a foil is brought into direct contact with the skin, the crystals respond by changing their color, thereby providing a thermal map of the skin.

Thermography has found its greatest application in medicine as a complementary technique for mammography, a method used to evaluate breast carcinoma. *See* MAMMOGRAPHY. [N.-F.B.]

Thermoluminescence A term sometimes used broadly to mean any luminescence appearing in a material due to the application of heat. More frequently the term refers specifically to the luminescence appearing as the temperature of the material is steadily increased.

Many solids that contain luminescent centers often also contain one or more types of centers that can trap electrons or holes. If the solid is exposed to light of sufficiently short wavelength or to x-rays or other high-energy radiation, free electrons and holes are produced in the solid, and some of these charge carriers may be trapped. Very commonly the hole is trapped at the luminescent center itself or at some trapping site nearby while the electron is trapped at a different site, but the reverse situation is also quite possible. If, in the former case, the depth of the electron trap (that is, the amount of energy required to release the electron from the trap) is large and the temperature low, the electron will remain trapped for a long time. If, however, the temperature of the sample is raised slowly, the electron will receive increasing amounts of thermal energy and will eventually escape the trap. An electron thus freed from a trap may go over to a luminescent center and recombine with the hole trapped at or near the center. The energy liberated by the recombination excites the center, causing it to luminesce. For a single type of trap, the glow curve, (plot of thermoluminescence intensity as a function of temperature) first rises, reaches a maximum, and then decreases to zero when all the traps become emptied. *See* HOLE STATES IN SOLIDS; LUMINESCENCE; TRAPS IN SOLIDS. [J.H.S.; C.C.K.]

Thermoluminescent dosimeter A dosimeter that functions on the principle of thermoluminescence—the property of certain substances that release light upon heating after they have been exposed to ionizing radiation, There are four common types of thermoluminescent materials in use: calcium fluoride, lithium fluoride, calcium sulfate, and lithium borate. These materials contain trace quantities of impurity atoms that capture electrons released by the ionizing radiation. These captured electrons can later be released by heating the material and measuring either the peak intensity of the light emitted (the glow curve) or the integrated light; the latter usually gives more accurate results, but the former is a simpler procedure. The thermoluminescent dosimeter has come into wide use since 1971 and is fast replacing the air-capacitor dosimeter and the film badge for personnel monitoring. [K.Z.M.]

Thermomagnetic effects Electrical and thermal phenomena occurring when a conductor or semiconductor which is carrying a thermal current (that is, is in a temperature gradient) is placed in a magnetic field. *See* SEMICONDUCTOR.

Let the temperature gradient be transverse to the magnetic field H_z, for example, along x. Then the following transverse-transverse effects are observed:

1. Ettingshausen-Nernst effect, an electric field along y.
2. Righi-Leduc effect, a temperature gradient along y.
3. An electric potential change along x, amounting to a change of thermoelectric power.
4. A temperature gradient change along x, amounting to a change of thermal resistance.

Let the temperature gradient be along H. Then changes in thermoelectric power and in thermal conductivity are observed in the direction of H.

For related phenomena *see* HALL EFFECT; MAGNETORESISTANCE. [E.A.; F.Ke.]

Thermometer An instrument that measures temperature. Although this broad definition includes all temperature-measuring devices, they are not all called thermometers. Other names have been generally adopted. For a discussion of two such devices *see* PYROMETER; THERMOCOUPLE. For a general discussion of temperature measurement *see* TEMPERATURE MEASUREMENT.

Liquid-in-glass thermometer. This thermometer consists of a liquid-filled glass bulb and a connecting partially filled capillary tube. When the temperature of the thermometer increases, the differential expansion between the glass and the liquid causes the liquid to rise in the capillary. A variety of liquids, such as mercury, alcohol, toluene, and pentane, and a number of different glasses are used in thermometer construction, so that various designs cover diverse ranges between about −300°F and +1200°F (−184° C and +649°C).

Bimetallic thermometer. In this thermometer the differential expansion of thin dissimilar metals, bonded together into a narrow strip and coiled into the shape of a helix or spiral, is used to actuate a pointer. In some designs the pointer is replaced with low-voltage contacts to control, through relays, operations which depend upon temperature, such as furnace controls.

Filled-system thermometer. This type of thermometer has a bourdon tube connected by a capillary tube to a hollow bulb. When the system is designed for and filled with a gas (usually nitrogen or helium) the pressure in the system substantially follows the gas law, and a temperature indication is obtained from the bourdon tube. The temperature-pressure-motion relationship is nearly linear. Atmospheric pressure effects are minimized by filling the system to a high pressure. When the system is designed for and filled with a liquid, the volume change of the liquid actuates the bourdon tube.

Vapor-pressure thermal system. This filled-system thermometer utilizes the vapor pressure of certain stable liquids to measure temperature. The useful portion of any liquid-vapor pressure curve is between approximately 15 psia (100 kilopascals absolute) and the critical pressure, that is, the vapor pressure at the critical temperature, which is the highest temperature for a particular liquid-vapor system. A nonlinear relationship exists between the temperature and the vapor pressure, so the motion of the bourdon tube is greater at the upper end of the vapor-pressure curve. Therefore, these thermal systems are normally used near the upper end of their range, and an accuracy of 1% or better can be expected.

Resistance thermometer. In this type of thermometer the change in resistance of conductors or semiconductors with temperature change is used to measure temperature. Usually, the temperature-sensitive resistance element is incorporated in a bridge network which has a reasonably constant power supply. Although a deflection circuit is occasionally used, almost all instruments of this class use a null-balance system, in which the resistance change is balanced and measured by adjusting at least one other resistance in the bridge. Metals commonly used as the sensitive element in resistance thermometers are platinum, nickel, and copper.

Thermistor. This device is made of a solid semiconductor with a high temperature coefficient of resistance. The thermistor has a high resistance, in comparison with metallic resistors, and is used as one element in a resistance bridge. Since thermistors are more sensitive to temperature changes than metallic resistors, accurate readings of small changes are possible. *See* THERMISTOR. [H.S.B.]

Thermonuclear reaction A nuclear fusion reaction which occurs between various nuclei of the light elements when they are constituents of a gas at very high temperatures. Thermonuclear reactions, the source of energy generation in the Sun and the stable stars, are utilized in the fusion bomb. *See* HYDROGEN BOMB; NUCLEAR FUSION; STELLAR EVOLUTION; SUN.

Thermonuclear reactions occur most readily between isotopes of hydrogen (deuterium and tritium) and less readily among a few other nuclei of higher atomic number. At the temperatures and densities required to produce an appreciable rate of thermonuclear reactions, all matter is completely ionized; that is, it exists only in the plasma state. Thermonuclear fusion reactions may then occur within such an ionized gas when the agitation energy of the stripped nuclei is sufficient to overcome their mutual electrostatic repulsions, allowing the colliding nuclei to approach each other closely enough to react. For this reason, reactions tend to occur much more readily between energy-rich nuclei of low atomic number (small charge) and particularly between those nuclei of the hot gas which have the greatest relative kinetic energy. This latter fact leads to the result that, at the lower fringe of temperatures where thermonuclear reactions may take place, the rate of reactions varies exceedingly rapidly with temperature. *See* CARBON-NITROGEN-OXYGEN CYCLES; KINETIC THEORY OF MATTER; MAGNETOHYDRODYNAMICS; NUCLEAR REACTION; PINCH EFFECT; PLASMA PHYSICS; PROTON-PROTON CHAIN. [R.F.P.]

Thermoregulation A mechanism by which mammals and birds attempt to balance heat gain and heat loss in order to maintain a constant body temperature when exposed to variations in cooling power of the external medium. In the lower animals body temperature varies with changes in the environment.

Among fishes, amphibians, and reptiles, and among invertebrates generally, especially the insects, body heat production while at rest is insufficient to maintain a significant difference between body temperature and ambient. Thus the body temperature will closely approximate the ambient if there is no external source of heat which the animal can utilize, such as radiation. Such forms are termed poikilotherms and, incorrectly in many cases, are said to be cold-blooded. Homeotherms, the birds and mammals, can regulate body temperature independent of ambient conditions due to the high rate of heat production and well-developed insulation. This gives them a great deal of freedom with regard to activity time and place but also imposes a high food demand, especially when temperatures drop.

Hibernation is the solution to this problem. Some mammals and birds have the ability to become poikilothermic during part of the year and thus abandon thermoregulation and its high energy demands. These forms have been termed heterotherms. Abandonment of body temperature regulation at low temperatures, however, also means abandonment of activity. *See* HIBERNATION.

While poikilotherms have little ability to produce body heat metabolically, they are able to use physiological mechanisms to regulate the rates of heat gain and loss. By varying the circulation of the body they are able to modify the rates of heat gain and loss, and some forms may make good use of evaporation by panting to keep the body temperature below ambient levels during periods of heat stress. Also, the skin color of reptiles may change, and this helps the animal to either absorb or reflect more or less of the Sun's rays and thus modify the rate of heat gain.

Automatic physiological temperature regulation is first seen in the primitive mammals, the monotremes (platypus and echidna). In these forms, as well as in certain specialized animals such as the sloth, body temperature may be maintained at a high level only when ambient temperature is considerably above freezing and the body temperature may vary from 80 to 98.5°F (27 to 37°C). In most higher vertebrates, however, body temperature levels are very closely maintained at very specific levels, but even here the limits of ambient temperature fluctuation that can be tolerated vary widely from species to species. Small forms with poor insulation cannot tolerate temperatures of 41°F (5°C), while large animals or those with good coats of fur, feathers, or blubber can survive at the lowest temperatures found on the Earth (about 76°F or − 60°C).

Temperature regulation by animals is accomplished by a careful balancing of the rates of heat loss and gain. The physiological production of heat is accomplished almost entirely by the oxidation of foodstuffs by the metabolic pathways of the body. The heat released in this way is the same as if the foodstuffs were burned. Heat that is produced is lost by evaporation due to sweating or panting, by conduction and convection from the animal's surface, and with the feces and urine as they are voided.

Thermoregulation demands that the temperatures of an animal's body be sensed and, once sensed, be adjusted to some particular level. Sensing is accomplished by the nervous system, and adjustments are made by both the nervous and endocrine systems. There are discrete receptors for heat and cold which sense both absolute levels and changes of temperature. These are scattered in a punctate fashion over the surface of the body. Their output is fed into a thermoregulatory center in the hypothalamus, which is itself affected by temperature. It is possible to elicit sweating or shivering by respectively heating or cooling this area of the brain. *See* HYPOTHERMIA; METABOLISM. [W.Mo./B.T.S.]

Thermosbaenacea An order of small crustaceans in the superorder Pancarida. The order was erected in 1927 by T. Monod and includes two families, the Thermosbaenidae and Monodellidae. Both families are monogeneric. [C.B.C.]

Thermosphere A rarefied portion of the atmosphere, lying in a spherical shell between 50 and 300 mi (80 and 500 km) above the Earth's surface, where the temperature increases dramatically with altitude. The thermosphere responds to the variable outputs of the Sun, the ultraviolet radiation at wavelengths less than 200 nanometers, and the solar wind plasma that flows outward from the Sun and interacts with the Earth's geomagnetic field. This interaction energizes the plasma, accelerates charged particles into the thermosphere, and produces the aurora borealis and aurora australis, which are nearly circular-shaped regions of luminosity that surround the magnetic north and south poles respectively. Embedded within the thermosphere is the ionosphere, a weakly ionized plasma. *See* IONOSPHERE; MAGNETOSPHERE; PLASMA PHYSICS; SOLAR WIND.

In the thermosphere, these molecular species are subjected to intense solar ultraviolet radiation and photodissociation that gradually turns the molecular species into the atomic species oxygen, nitrogen, and hydrogen. Up to above 60 mi (100 km), atmospheric turbulence keeps the atmosphere well mixed, with the molecular concentrations dominating in the lower atmosphere. Above 60 mi, solar ultraviolet radiation most strongly dissociates molecular oxygen, and there is less mixing from atmospheric turbulence. The result is a transition area where molecular diffusion dominates and atmospheric species settle according to their molecular and atomic weights. Above 60 mi, atomic oxygen is the dominant species. *See* ATMOSPHERE.

About 60% of the solar ultraviolet energy absorbed in the thermosphere and ionosphere heats the ambient neutral gas and ionospheric plasma; 20% is radiated out of the thermosphere as airglow from excited atoms and molecules; and 20% is stored as chemical energy of the dissociated oxygen and nitrogen molecules, which is released later when recombination of the atomic species occurs. Most of the neutral gas heating that establishes the basic temperature structure of the thermosphere is derived from excess energy released by the products of ion-neutral and neutral chemical reactions occurring in the thermosphere and ionosphere. *See* AIRGLOW; ULTRAVIOLET RADIATION.

The average vertical temperature profile is determined by a balance of local solar heating by the downward conduction of molecular thermal product to the region of minimum temperature near 50 mi (80 km). For heat to be conducted downward within the thermosphere, the temperature of the thermosphere must increase with altitude. The global mean temperature increases from about 200 K (−100°F) near 50 mi to 700–1400 K (800–2100°F) above 180 mi (300 km), depending upon the intensity of solar ultraviolet radiation reaching the Earth. Above 180 mi, molecular thermal conduction occurs so fast that vertical temperature differences are largely eliminated; the isothermal temperature in the upper thermosphere is called the exosphere temperature.

As the Earth rotates, absorption of solar energy in the thermosphere undergoes a daily variation. Dayside heating causes the atmosphere to expand, and the loss of heat at night causes it to contract. This heating pattern creates pressure differences that drive a global circulation, transporting heat from the warm dayside to the cool nightside. [R.G.R.]

Thermostat An instrument which directly or indirectly controls one or more sources of heating and cooling to maintain a desired temperature. To perform this function a thermostat must have a sensing element and a transducer. The sensing element measures changes in the temperature and produces a desired effect on the transducer. The transducer converts the effect produced by the sensing element into a suitable control of the device or devices which affect the temperature.

The most commonly used principles for sensing changes in temperature are (1) unequal rate of expansion of two dissimilar metals bonded together (bimetals), (2) unequal expansion of two dissimilar metals (rod and tube), (3) liquid expansion (sealed diaphragm and remote bulb or sealed bellows with or without a remote bulb), (4) saturation pressure of a liquid-vapor system (bellows), and (5) temperature-sensitive resistance element.

The most commonly used transducers are (1) switches that make or break an electric circuit, (2) potentiometer with a wiper that is moved by the sensing element, (3) electronic amplifier, and (4) pneumatic actuator.

The most common thermostat application is for room temperature control. The illustration shows a typical on-off heating-cooling room thermostat. In a typical application the thermostat controls a gas valve, oil burner control, electric heat control, cooling compressor control, or damper actuator.

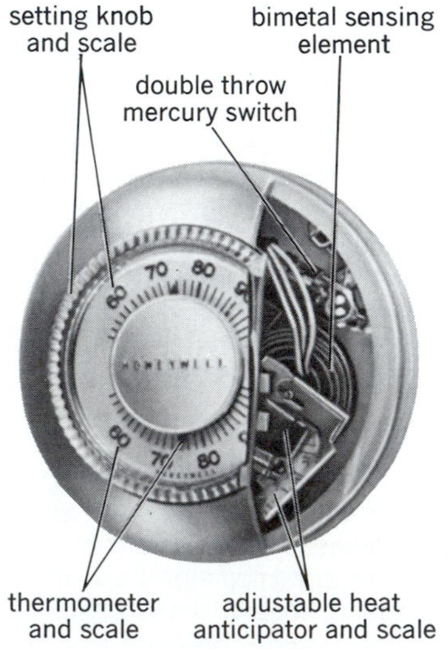

Typical heat-cool thermostat. ***(Honeywell Inc.)***

Thermostats are also used extensively in safety and limit application. Thermostats are generally of the following types: insertion types that are mounted on ducts with the sensing element extending into a duct; immersion types that control a liquid in a pipe or tank with the sensing element extending into the liquid; and surface types in which the sensing element is mounted on a pipe or similar surface. *See* COMFORT HEATING; FURNACE; OIL BURNER. [N.R.]

Thermotherapy The treatment of disease by heat of any kind; it involves the local or general application of heat to the body.

Therapeutic heat may be applied to a very small pinpoint area (cautery), regionally to a part or member of the body, or generally to the entire body. Electrocautery and chemical cautery are commonly used surgical techniques. Regional heating is more commonly used therapeutically and has both local and distant effects. Generalized heating that produces a rise in body temperature is referred to as hyperthermia.

Physiological effects. The application of heat creates soothing, sedative, and analgesic effects, although the exact mechanisms of these effects are unknown. Additionally, heat acts as a vasodilator. Regardless of the method of heat generation, the result within the heated tissues is essentially the same. Heat produces a rise in temperature with an associated increase in metabolism. There is a relative increase in the accumulation of metabolic wastes, such as acid metabolites and carbon dioxide. Increased local circulation due to dilation of arterioles and capillaries leads to improved cellular nutrition, as well as improved exchange of wastes. Also, more phagocytes and antibodies are brought with the blood into the area. The local temperature rises until a point is reached at which maximum vasodilatation occurs; then the local temperature falls slightly.

Regional heating produces remote effects. Thus heating of one leg may cause vasodilatation and an increase in skin temperature in the other leg or in the hands (consensual heating), These are responses to the body mechanisms of heat regulation and are mediated through the central nervous system via the heat-regulatory center of the hypothalamus. *See* THERMOREGULATION.

Methods. Modalities, the clinical methods employed for thermotherapy, vary with regard to techniques of heat production and application. In general, they are classified as follows in accordance with their physical properties: radiant, conductive, and conversive.

Radiant heat is derived from the infrared portion of the electromagnetic spectrum. Specialized luminous and nonluminous generators are readily available for professional use. More commonly used sources such as infrared bulbs or aggregates of bulbs (bakers) provide a soothing if not very penetrating source of heating. However, circulating blood in the area heated may transfer heat to deeper areas. *See* ELECTROMAGNETIC RADIATION.

Any material that has a higher temperature than that of the skin with which it is in contact will transfer heat to the skin by conduction. Hot air, steam, water, paraffin, silicon, and electric heating pads are all familiar clinically used examples. Hot air and steam baths have long been used to induce cutaneous vasodilatation, sweating, relaxation of muscles, and a sense of well-being. However, overexposure to these modalities may precipitate heat stroke and collapse. With its high specific heat, water has been used for centuries for heating. Hot tub baths and natural heat from mineral springs are still being used therapeutically.

Conversive heating involves the conversion of some other form of energy into heat. Most commonly used sources of energy are radio waves in the shortwave and microwave bands of the electromagnetic spectrum and ultrasonic waves (mechanical). Shortwave diathermy utilizes electromagnetic energy at wavelengths of 3–30 m and frequencies of 10–100 MHz. Human tissues are heated with high-frequency currents either through conduction or induction. Heat generated in this manner penetrates to a depth of about 1 in. (2–3 cm) and spreads out over a fairly large area. Microwave diathermy consists of electromagnetic energy therapeutically applied at a wavelength of 12.2 cm and a frequency of 2450 MHz. It effectively penetrates up to a depth of 2 in. (5 cm). This method makes it possible to raise the temperature of fat, muscle, and connective tissue in or near bone. Whereas shortwave diathermy spreads widely in the tissues of the body, microwaves are quasi-optical and can be focused and directed for the heating of small selective areas.

Ultrasonic therapy or ultrasound diathermy involves the application of high-frequency sound waves and the conversion of this mechanical energy to heat. This method is well suited to local thermotherapy. Therapeutic ultrasound generators produce frequencies varying from 0.7 to 1.0 MHz. With sufficient intensity it is possible to penetrate to depths of 2–3 in. (5–8 cm). Ultrasound is selective in its heating properties, is reflected at interfaces, and is mechanically destructive in excessive dosage. [J.R.O'C.]

Thévenin's theorem (electric networks) This theorem, from electric circuit theory, is also known as the Helmholtz or Helmholtz-Thévenin theorem, since H. Helmholtz stated it in an earlier form prior to M. L. Thévenin. Closely related is the Norton theorem, which will also be discussed. Laplace transform notation will be used. *See* LAPLACE TRANSFORM.

Thévenin's theorem states that at a pair of terminals a network composed of lumped, linear circuit elements may, for purposes of analysis of external circuit or terminal behavior, be replaced by a voltage source $V(s)$ in series with a single impedance $Z(s)$. The source $V(s)$ is the Laplace transform of the voltage across the pair of terminals when they are open-circuited; $Z(s)$ is the transform impedance at the two terminals with all independent sources set to zero (Fig. 1).

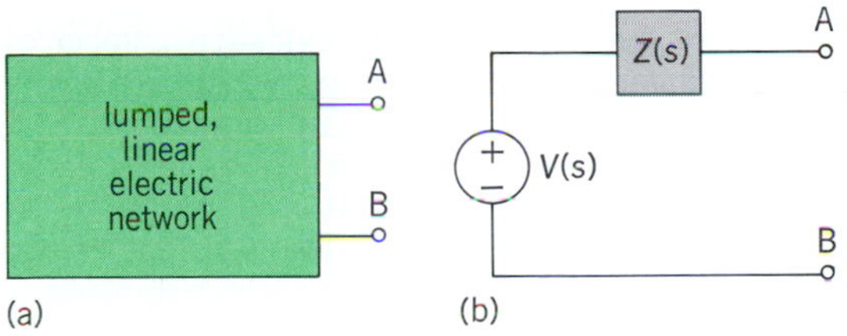

Fig. 1. Network and its Thévenin equivalent. (*a*) Original network. (*b*) Thévenin equivalent circuit.

Norton's theorem states that a second equivalent network consists of a current source $I(s)$ in parallel with an impedance $Z(s)$. The impedance $Z(s)$ is identical with the Thévenin impedance, and $I(s)$ is the Laplace transform of the current between the two terminals when they are short-circuited (Fig. 2).

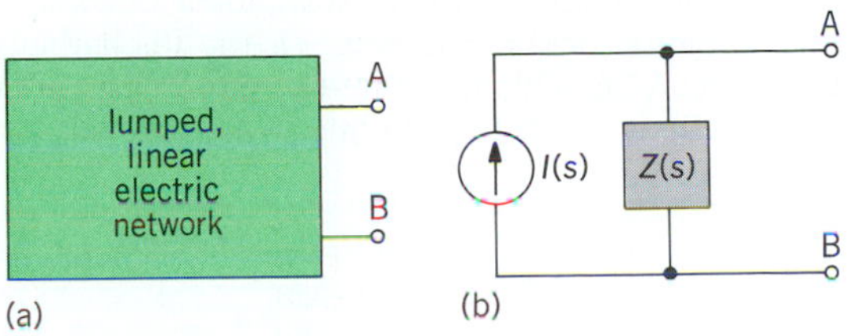

Fig. 2. Network and its Norton equivalent. (*a*) Original network. (*b*) Norton equivalent circuit.

Thévenin's and Norton's equivalent networks are related by the equation $V(s) = Z(s) \cdot I(s)$.

These theorems are useful for the study of the behavior of a load connected to a (possibly complex) system that is supplying electric power to that load. *See* NETWORK (MATHEMATICS); SUPERPOSITION THEOREM (ELECTRIC NETWORKS). [E.C.Jo.]

Thiamine A water-soluble vitamin found in many foods; pork, liver, and whole grains are particularly rich sources. It is also known as vitamin B_1 or aneurin. The structural formula of thiamine is shown below.

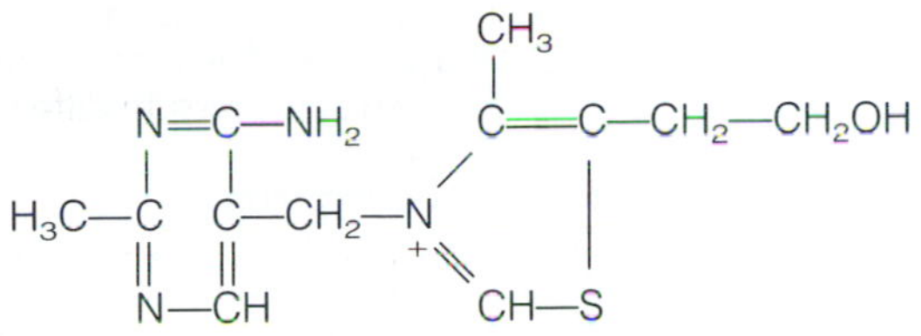

Thiamine deficiency is known as beriberi in humans and polyneuritis in birds. Muscle and nerve tissues are affected by the deficiency, and poor growth is observed. People with beriberi are irritable, depressed, and weak. They often die of cardiac failure. Wernicke's disease observed in alcoholics is associated with a thiamine deficiency. This disease is characterized by brain lesions, liver disease, and partial paralysis, particularly of the motor nerves of the eye. As is the case in all B vitamin diseases, thiamine deficiency is usually accompanied by deficiencies of other vitamins. *See* BERIBERI. [S.N.G.]

Thiazole One of a class of organic heterocyclic compounds (sometimes specified as 1,3-thiazoles) in which a five-membered diunsaturated ring contains one atom of nitrogen and, in a nonadjacent position, one atom of sulfur. Important thiazole derivatives include vitamin B_1 (thiamine) and sulfathiazole. Several valuable dyes and rubber-vulcanizing accelerators contain the benzothiazole nucleus. Penicillin is a thiazolidine derivative. *See* AZOLE; HETEROCYCLIC COMPOUNDS.

The thiazole ring is a resonance-stabilized aromatic system. The ring is relatively resistant to hydrolysis and to disruptive oxidation with nitric acid. The parent compound, thiazole, bp 117°C (243°F), is a colorless, water-soluble liquid with an odor resembling that of pyridine. Although thiazole is a weak base, acids form simple salts, and alkylating agents form quaternary salts. [W.J.Ge.]

Thickening The production of a concentrated slurry from a dilute suspension of solid particles in a liquid. In practice, a thickener is usually expected also to produce a clear liquid, and therefore thickening includes clarification as a concurrent objective. Thickening and clarification are applications of sedimentation, and both are representative of a larger group of industrial processes called mechanical separations. *See* CLARIFICATION; MECHANICAL SEPARATION TECHNIQUES. [W.L.McC.]

Thigmotrichida A restricted group of forms constituting an order of the Holotrichia and generally found in association with mollusks from both fresh and salt water. Some are mouthless. Those species with a cytostome are generally equipped with a buccal ciliature which indicates an advance over the primitive hymenostome arrangement. *See* HOLOTRICHIA; HYMENOSTOMATIDA. [J.O.C.]

Thinner A material used in paints and varnishes to adjust the consistency for application. Thinners are usually solvents for the vehicle used in the coating and are expected to evaporate after application. Water is used as a thinner in emulsion paints and in certain water-soluble paints such as watercolors and calcimines. Petroleum fractions are most commonly used for oil and resin coatings. Stronger solvents contain substantial amounts of aromatic hydrocarbons and may be derived from petroleum or coal tar.

Since numerous coating resins are not sufficiently soluble in hydrocarbons, other materials or mixtures must be used. These include alcohols such as denatured ethyl or isopropyl alcohols for shellac, esters such as amyl acetate for nitrocellulose, and ketones and other compounds for acrylic and vinyl resins. Chlorinated hydrocarbons are used for some materials which are otherwise hard to dissolve, but toxicity limits their usefulness. *See* PAINT; SOLVENT; SURFACE COATING; TURPENTINE; VARNISH. [C.R.Ma; C.W.Si.]

Thio compounds Organosulfur compounds in which one or more sulfur atoms replace oxygen. Because of the frequent use of this nomenclature and the large number of structural variations involved, the thio names may appear confusing but are generally understandable in specific cases. Thiols and thiophenols are the sulfur analogs of alcohols and phenols. Dithio and trithio compounds have two or three sulfur atoms, corresponding to the oxygen analogs. Acyl-SH compounds are the thio acids, for example, thioacetic acid, $CH_3({=}O)SH$, of which many are known. Dithio acids, R—C(=S)—SH, may, for example, be made from RMgX and carbon disulfide, CS_2. Trithiocarbonates are derivatives of $S{=}C(SH)_2$, the sulfur analog of carbonic acid. Dithiocarbamates, related to carbamic acids, are readily obtained in the form of salts.

Many other compounds which carry the thio name are oxygen analogs. For example, thiocyanates, isothiocyanates, thiourea, thioamides, dithioesters, and thio acid chlorides. *See* ORGANOSULFUR COMPOUND. [N.K.]

Thioaldehyde and thioketone Organosulfur compounds of structure

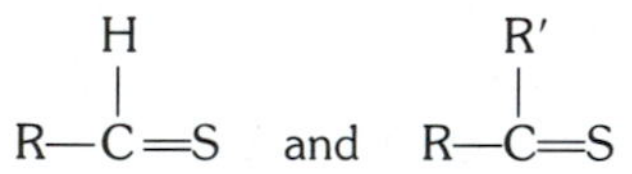

They are isolable only in rare instances because, unlike the oxygen analogs (aldehydes and ketones), thioaldehydes and thioketones tend strongly to polymerize. *See* ORGANOSULFUR COMPOUND. [N.K.]

Thiocyanate One of a group of compounds, both organic and inorganic, which contain the —SCN group and are derived from thiocyanic acid, HSCN. Like cyanic acid, thiocyanic acid may exist in two forms, H—S—C≡N and S=N—H. The latter form is called isothiocyanic acid and gives rise to isothiocyanates.

The inorganic thiocyanates resemble the cyanides and halides because most of the metal salts are water-soluble (except lead, mercury, silver, and copper salts), and many complexes are formed with excess thiocyanate. *See* CYANIDE; SULFUR. [E.E.W.]

Thioether One of a group of organosulfur compounds that are also called sulfides, RSR′. The simplest, dimethyl sulfide, CH_3—S—CH_3, is the precursor of dimethyl sulfoxide, a useful solvent and chemical reactant. Some amino acids, such as methionine and lanthionine, are also sulfides. Mustard gas, $ClCH_2CH_2SCH_2CH_2Cl$, is a well-known vesicant. The thioethers bear a formal resemblance to the oxygen ethers, and may be synthesized by analogous methods. *See* AMINO ACIDS; ETHER; ORGANOSULFUR COMPOUND. [N.K.]

Thiophene An organic heterocyclic compound containing a diunsaturated ring of four carbon atoms and one sulfur atom. *See* HETEROCYCLIC COMPOUNDS.

Thiophene (I), methylthiophenes, and other alkylthiophenes are found in relatively small amounts in coal tar and petroleum. Thiophene accompanies benzene in the fractional distillation of coal tar. 2,5-Dithienylthiophene (II) has been found in

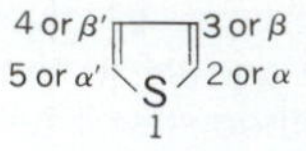

(I)

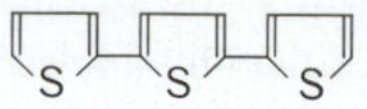

(II)

the marigold plant. Biotin, a water-soluble vitamin, is a tetrahydrothiophene derivative.

The parent compound (I) is nearly insoluble in water, with mp −38.2°C (−36.8°F), bp 84.2°C (183.6°F), and specific gravity (20/4) 1.0644. Thiophene is considered to be an aromatic compound. Thiophenes are stable to alkali and other nucleophilic agents, and are relatively resistant to disruption by acid. *See* AROMATIC HYDROCARBON; ORGANOSULFUR COMPOUND. [W.J.Ge.; M.St.]

Thiosulfate A negative ion having the formula $S_2O_3^{2-}$, which is derived from thiosulfuric acid, $H_2S_2O_3$, an unstable acid. The ion is actually related to the sulfate ion, with the substitution of a sulfur atom for an oxygen, as shown in the formula below. The central sulfur atom has an oxidation number

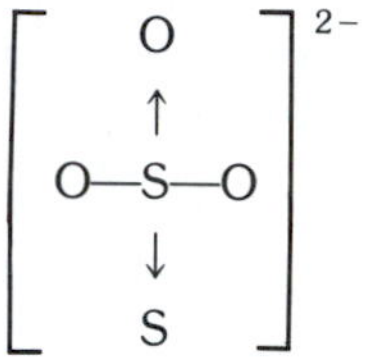

of 6+, whereas the second has an oxidation number of 2−. Sodium thiosulfate, also termed hypo, is used in photography to "fix" films by dissolving the unreacted silver halide. [E.E.W.]

Thiourea A crystalline, colorless solid (prisms or needles) having formula (I) and melting point 180–182°C (356–360°F).

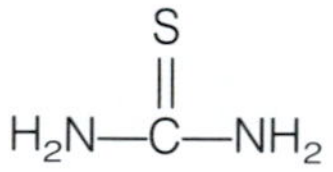

(I)

Thiourea is relatively insoluble in water and only slightly soluble in ether. The three most common methods of preparation are (1) heating ammonium thiocyanate, NH_4CNS, to about 180°C (356°F); (2) the action of hydrogen sulfide at about 180°C (356°F) on calcium cyanamide; and (3) the reaction of dicyandiamide and ammonium sulfide at 60–70°C (140–158°F).

Condensation of thiourea with substituted malonic esters gives thiobarbituric acid derivatives. Of the thiobarbiturates, with 1-methylbutylethylthiobarbiturate as the sodium salt, sodium pentothal (II) is much used in anesthetic premedication to

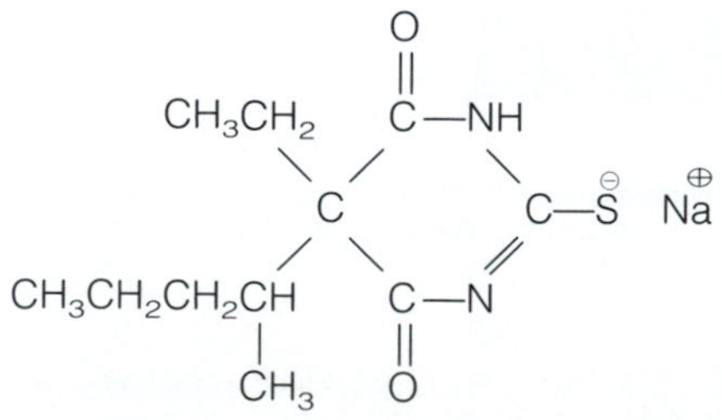

(II)

reduce the amount of general anesthetic needed. [E.B.R.]

Thirst The sensation aroused by a relative or an absolute lack of body water, which normally though not necessarily induces the behavior of drinking water. Theories of thirst in which dryness of the mouth and throat plays an essential role are as old as scientific interest in the subject itself. However, since it was established that there is a thirst center in the hypothalamus, the dry mouth theory has become less influential.

Most evidence is in favor of cellular dehydration as an important cause of thirst. However, loss of extracellular fluid can also give rise to thirst. Three procedures have established that thirst caused by extracellular dehydration is distinct from thirst caused by cellular dehydration: (1) sodium depletion by dieting and other means which results in a secondary loss of extracellular water; (2) the more direct and immediate removal of extracellular fluid by hemorrhage; and (3) mimicking the effects of severe dehydration on the circulation by interfering with the flow of blood in various blood vessels. All cause thirst and a delayed increase in sodium appetite, and all either overhydrate or do not alter the water content of the cellular compartment. Little is known about mechanisms, but extracellular deficits are probably detected by stretch receptors which lie in the walls of the low-pressure capacitance vessels (that is, the large veins near the heart) and in the atria. *See* OSMOREGULATORY MECHANISMS. [E.M.B.]

Thomson effect A phenomenon discovered in 1854 by William Thomson, later Lord Kelvin. He found that there occurs a reversible transverse heat flow into or out of a conductor of a particular metal, the direction depending upon whether a longitudinal electric current flows from colder to warmer metal or from warmer to colder. Any temperature gradient previously existing in the conductor is thus modified if a current is turned on. The Thomson effect does not occur in a current-carrying conductor which is initially at uniform temperature. *See* THERMOELECTRICITY. [J.W.St.]

Thoracica The major order of the crustacean subclass Cirripedia. The adult animals are permanently attached. The mantle is usually reinforced by calcareous plates. Six pairs of biramous cirri are present, and the abdomen is absent or represented by caudal appendages. Antennules are present in the adult, and cement glands are strongly developed. Most species are hermaphroditic. Thoracica are subdivided into three suborders: Lepadomorpha, stalked or goose barnacles; Balanomorpha, the common acorn barnacles; and Verrucomorpha, a rare group of asymmetric barnacles. *See* BALANOMORPHA; BARNACLE; CIRRIPEDIA; LEPADOMORPHA; VERRUCOMORPHA. [H.G.St.]

Thorianite A radioactive mineral with the idealized composition ThO_2 (thorium dioxide) and isostructural with uraninite (pitchblende). Rare earths and uranium are often present in variable amounts, together with small amounts of radiogenic lead. Thorianite usually occurs as worn cubic crystals. The hardness is about 7 on Mohs scale, and the specific gravity is 9.7–9.8. The color is brownish black to reddish brown, and the luster usually is resinous. Thorianite is a primary mineral found chiefly in pegmatites. It is best known as a detrital mineral. It has been obtained commercially from detrital deposits and pegmatites in Madagascar and Ceylon. *See* NUCLEAR FUELS; PEGMATITE; RADIOACTIVE MINERALS; RARE-EARTH ELEMENTS; THORIUM; URANINITE; URANIUM. [C.Fr.]

Thorite A mineral, thorium silicate. The idealized chemical formula of thorite is $ThSiO_4$. All natural material departs widely from this composition owing to the partial substitution of uranium, rare earths, calcium, and iron for thorium. The specific gravity ranges between about 4.3 and 5.4. The hardness on Mohs scale is about 4½. The color commonly is brownish yellow to brownish black and black.

Vein deposits containing thorite occur in Colorado, Idaho, and Montana. A vein deposit of monazite containing thorium is mined at Steenkampskraal near Van Rhynsdorp, Cape Province, South Africa. *See* METAMICT STATE; RADIOACTIVE MINERALS; SILICATE MINERALS; THORIUM. [C.Fr.]

Thorium A chemical element, Th, atomic number 90. Thorium is a member of the actinide series of elements. It is radioactive with a half-life of about 1.4×10^{10} years.

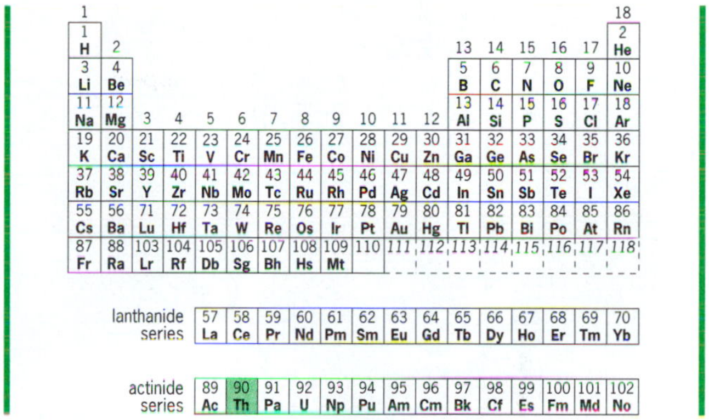

Thorium oxide compounds are used in the production of incandescent gas mantles. Thorium oxide has also been incorporated in tungsten metal, which is used for electric light filaments. It is employed in catalysts for the promotion of certain organic chemical reactions and has special uses as a high-temperature ceramic material. The metal or its oxide is employed in some electronic tubes, photocells, and special welding electrodes. Thorium has important applications as an alloying agent in some structural metals. Perhaps the major use for thorium metal, outside the nuclear field, is in magnesium technology. Thorium can be converted in a nuclear reactor to uranium-233, an atomic fuel. The energy available from the world's supply of thorium has been estimated as greater than the energy available from all of the world's uranium, coal, and oil combined.

Monazite, the most common and commercially most important thorium-bearing mineral, is widely distributed in nature. Monazite is chiefly obtained as a sand, which is separated from other sands by physical or mechanical means. *See* MONAZITE.

Thorium has an atomic weight of 232. The temperature at which pure thorium melts is not known with certainty; it is thought to be about 1750°C (3182°F). Good-quality thorium metal is relatively soft and ductile. It can be shaped readily by any of the ordinary metal-forming operations. The massive metal is silvery in color, but it tarnishes on long exposure to the atmosphere; finely divided thorium has a tendency to be pyrophoric in air.

All of the nonmetallic elements, except the rare gases, form binary compounds with thorium. With minor exceptions, thorium exhibits a valence of 4+ in all of its salts. Chemically, it has some resemblance to zirconium and hafnium. The most common soluble compound of thorium is the nitrate which, as generally prepared, appears to have the formula $Th(NO_3)_4 \cdot 4H_2O$. The common oxide of thorium is ThO_2, thoria. Thorium combines with halogens to form a variety of salts. Thorium sulfate can be obtained in the anhydrous form or as a number of hydrates. Thorium carbonates, phosphates, iodates, chlorates, chromates, molybdates, and other inorganic salts of thorium are well known. Thorium also forms salts with many organic acids, of which the water-insoluble oxalate, $Th(C_2O_4)_2 \cdot 6H_2O$, is important in preparing pure compounds of thorium. *See* ACTINIDE ELEMENTS; RADIOACTIVITY. [H.A.W.]

Threading The forming of a ridge and valley of uniform cross section which spiral about the inner or outer diameter of a cylinder or cone in an even and continuing manner. The work must be produced with sufficient uniformity and accuracy so that the resulting threaded part will accomplish its intended purpose of fastening, transmitting motion or power, or measuring.

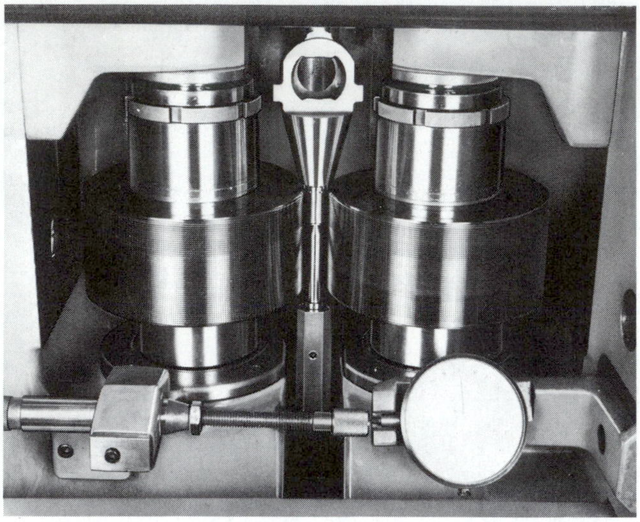

Thread-rolling machine set up to thread a steel aircraft jet engine mount. (*Landis Machine Co.*)

Although machine threading is done on various machine tools, special threading machines may be used for quantity production (see illustration). *See* SCREW THREADS. [A.H.T.]

Threonine An amino acid which is considered essential for normal growth of animals. Threonine is a biosynthetic precursor of isoleucine in microorganisms. Threonine is biosynthesized from aspartic acid.

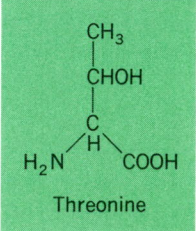

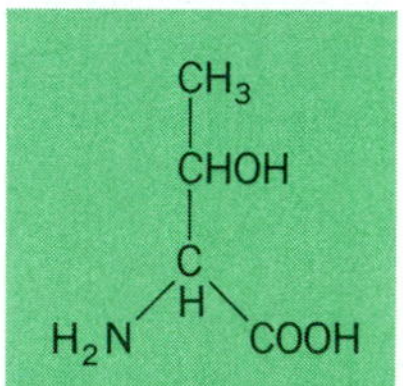

Threonine

The major pathway in metabolic degradation starts with the nonoxidative deamination to α-ketobutyric acid. This keto acid may be oxidatively decarboxylated to propionyl-CoA, or transaminated to form α-aminobutyric acid. Another pathway of threonine degradation involves an initial cleavage to glycine and acetaldehyde. *See* AMINO ACIDS. [E.A.Ad.]

Thrip A small, slender-bodied insect of the order Thysanoptera. There are about 500 species, all in North America. Wings may be present or absent, but when present the usual number is four (see illustration). The mouthparts are suctorial, with a stout proboscis. Males are rare or unknown in some species and parthenogenesis is common. Females lacking an ovipositor lay their eggs under bark or in crevices, while those species possessing an ovipositor deposit the eggs in plant tissues.

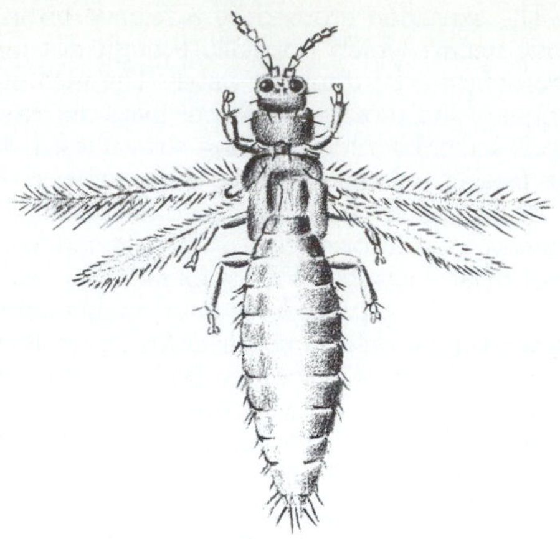

The flower thrip (*Frankliniella tritici*), with two pairs of long featherlike wings.

Many thrips are economically important as serious pests of cultivated plants since they injure or destroy the plants in feeding and breeding activities. Some species harbor plant viruses which they inoculate into plants. A few species are predacious and feed on small arthropods such as mites, aphids, or other species of thrips. Some species may bite humans. *See* THYSANOPTERA. [C.B.C.]

Throat The region that includes the pharynx, the larynx, and related structures. Both the nasal passages and the oral cavity open into the pharynx, which also contains the openings of the Eustachian tubes from the ears. The lower portion of the pharynx leads into the esophagus and the trachea or windpipe. The rather funnel-shaped pharynx is suspended from the base of the skull and the jaws; it is surrounded by three constrictor muscles that function primarily in swallowing. *See* EAR; PHARYNX.

The larynx, or voice box, is marked externally by the shield-shaped thyroid cartilage which forms the Adam's apple. The larynx contains the vocal cords that act as sphincters for air regulation and permit phonation. The lower end of the larynx is continuous with the trachea, a tube composed of cartilaginous rings and supporting tissues. *See* LARYNX.

The term throat is also used in a general sense to denote the front (ventral side) of the neck. [T.S.P.]

Thrombosis The formation of a thrombus. A thrombus is a solid body formed during life, composed of the elements of the blood—platelets, fibrin, red cells, and leukocytes.

Thrombosis is essentially platelet deposition and may occur on any blood vessel wall where the endothelium is damaged. Since platelets release thromboplastinogen, which in turn activates the clotting mechanism, clotting and thrombosis may occur together. Platelets exhibit a tendency to stick together. This tendency is increased if their number is increased, as in thrombocytosis, if the velocity of the stream is decreased below a certain speed, or if the endothelium is roughened. The same factors promote thrombus formation. Heparin, however, causes the platelets to lose their stickiness. The flow of blood is slower in the veins, which may be one reason for the greater frequency of thrombi there. Other determinative factors, as yet unknown, must also come into play.

With occlusion of a vessel, true thrombosis ceases, but clotting continues. The clot may be propagated to the next side branch. The propagated clot may float freely in the vessel, anchored only by the thrombus at the base. Under these conditions a piece may break off and travel in the bloodstream, forming an embolus. *See* BLOOD; EMBOLISM.

Many of the possible sequelae of thrombosis of a vessel are obvious. Thrombi in certain critical vessels can be followed by dire consequences. Arterial thrombosis can result in an infarct of the region supplied by the vessel if the collateral circulation is inadequate. With mesenteric thrombosis, gangrene of the bowel may ensue. Under these circumstances surgical intervention must follow promptly or death will result. Occlusion of any vein causes stasis of blood in the region, with an acute passive congestion of the region being drained. Thrombosis of varices of the rectum or lower extremities may result in local necrosis and ulceration. *See* INFARCTION. [R.A.V.]

Throttled flow Flow which is forced to pass through a restricted area, where the velocity must be increased. This is also known as choked flow. Most of the kinetic energy produced by reduction of pressure in passing through the constriction is generally converted into thermal energy by turbulent eddying. The net result is a loss in mechanical energy in the system. When a gas is throttled, as by a globe valve, the velocity a short distance downstream from the valve is only a little higher than before the throttling section in most cases, the process being one of constant enthalpy. By introducing mechanical energy losses into a flow system by a throttling valve, the amount of flow may be controlled. *See* ISENTROPIC PROCESS. [V.L.S.]

Thrust In aviation context, the force that propels an aircraft or a missile. Thrust is usually expressed in pounds (English gravitational system) or in kilograms (metric gravitational system). It can be expressed in poundals (English absolute system) or newtons or dynes (metric absolute system). The power of turbojets, ramjets, and rockets is usually expressed in thrust terms. The product of thrust times aircraft velocity is thrust power and represents the useful work supplied by the power plant. *See* JET PROPULSION; RAMJET; RECIPROCATING AIRCRAFT ENGINE; ROCKET; TURBOJET. [R.R.H.]

Thulium A chemical element, Tm, atomic number 69, atomic weight 168.934. It is a rare metallic element belonging

1	2	3	4	5	6	7	8	9	10	11	12	13	14	15	16	17	18
1 H																	2 He
3 Li	4 Be											5 B	6 C	7 N	8 O	9 F	10 Ne
11 Na	12 Mg											13 Al	14 Si	15 P	16 S	17 Cl	18 Ar
19 K	20 Ca	21 Sc	22 Ti	23 V	24 Cr	25 Mn	26 Fe	27 Co	28 Ni	29 Cu	30 Zn	31 Ga	32 Ge	33 As	34 Se	35 Br	36 Kr
37 Rb	38 Sr	39 Y	40 Zr	41 Nb	42 Mo	43 Tc	44 Ru	45 Rh	46 Pd	47 Ag	48 Cd	49 In	50 Sn	51 Sb	52 Te	53 I	54 Xe
55 Cs	56 Ba	71 Lu	72 Hf	73 Ta	74 W	75 Re	76 Os	77 Ir	78 Pt	79 Au	80 Hg	81 Tl	82 Pb	83 Bi	84 Po	85 At	86 Rn
87 Fr	88 Ra	103 Lr	104 Rf	105 Db	106 Sg	107 Bh	108 Hs	109 Mt	110	*111*	*112*	*113*	*114*	*115*	*116*	*117*	*118*

lanthanide series	57 La	58 Ce	59 Pr	60 Nd	61 Pm	62 Sm	63 Eu	64 Gd	65 Tb	66 Dy	67 Ho	68 Er	69 Tm	70 Yb
actinide series	89 Ac	90 Th	91 Pa	92 U	93 Np	94 Pu	95 Am	96 Cm	97 Bk	98 Cf	99 Es	100 Fm	101 Md	102 No

to the rare-earth group. The stable isotope ^{169}Tm makes up 100% of the naturally occurring element.

The salts of thulium possess a pale green color and the solutions have a slight greenish tint. The metal has a high vapor pressure at the melting point. When ^{169}Tm is irradiated in a nuclear reactor, ^{170}Tm is formed. The isotope then emits strongly an 84-keV x-ray, and this material is useful in making small portable x-ray units for medical use. *See* RARE-EARTH ELEMENTS. [F.H.Sp.]

Thunder An acoustic wave caused by lightning. When lightning occurs, the surge of current in the return stroke suddenly heats the air in the lightning channel. The resulting expansion of the gaseous medium is similar to an explosion and produces a shock wave. The shock wave can be thought of as an expanding cylinder surrounding the original discharge channel. When the atmospheric pressure along the cylinder axis falls to the levels it had before the lightning flash, the shock wave becomes an acoustic wave called thunder. *See* LIGHTNING; THUNDERSTORM. [L.J.B.]

Thunderstorm A convective storm accompanied by lightning and thunder and a variety of weather such as locally heavy rainshowers, hail, high winds, sudden temperature changes, and occasionally tornadoes. The characteristic cloud is the cumulonimbus or "thunderhead," a towering cloud, often with an anvil-shaped top. A host of accessory clouds, some attached and some detached from the main cloud, are often observed in conjunction with cumulonimbus. *See* CLOUD.

Thunderstorms are manifestations of convective overturning of deep layers in the atmosphere and occur in environments in which the decrease of temperature with height (lapse rate) is sufficiently large to be conditionally unstable and the air at low levels is moist. In such an atmosphere, a rising air parcel, given sufficient lift, becomes saturated and cools less rapidly than it would if it remained unsaturated because the released latent heat of condensation partly counteracts the expansional cooling. The rising parcel reaches levels where it is warmer (by perhaps as much as 50°F or 10°C) and less dense than its surroundings, and buoyancy forces accelerate the parcel upward. The rising parcel is decelerated and its vertical ascent arrested at altitudes where the lapse rate is stable, and the parcel becomes denser than its environment.

Thunderstorms are most frequent in the tropics, and rare poleward of 60° latitude. Thunderstorms occur at all hours of day and night, but are most common during late afternoon because of the influence of diurnal surface heating.

Thunderstorms are considered severe when they produce winds greater than 58 mi/h (26 m/s or 50 knots), hail larger than ¾ in. (19 mm) in diameter, or tornadoes. While thunderstorms are generally beneficial because of their needed rains (except for occasional flash floods), severe storms have the capacity of inflicting utter devastation over narrow swaths of the countryside. *See* HAIL; SQUALL; TORNADO. [R.D.-J.]

A thunderstorm produces lightning and is thus highly electrified. A thunderstorm can be considered as a current generator that separates charge, resulting in strong electric fields and electricity for lightning and various conduction currents within, below, and above the storm. Electric currents from thunderstorms to the earth and ionosphere maintain the fair weather state of the earth and atmosphere. Typical maximum electric fields are 1.8 to 4.6 kV/ft (6 to 15 kV/m) at the earth below the storm and 30 to 120 kV/ft (100 to 400 kV/m) or more inside thunderstorms. Electric fields at the ground are less intense because of corona discharge from pointed objects. Lightning, the most obvious and deadly manifestation of storm electrification, can propagate long distances, with at least one flash 105 mi (170 km) long having been observed with radar. *See* ATMOSPHERIC ELECTRICITY; LIGHTNING; THUNDER. [W.D.R.]

Thyme Any of a large and diverse group of plants in the genus *Thymus* utilized for their essential oil and leaves in both cooking and medicine. Hundreds of different forms, or ecotypes, of thyme are found in the Mediterranean area, where thyme occurs as a wild plant. *Thymus vulgaris*, generally considered to be the true thyme, is the most widely used and cultivated species. Both "French" and "German" thyme are varieties of this species. Most types of thyme are low-growing perennials, typically having small smooth-edged leaves that are closely spaced on stems that become woody with age.

Wild European thyme, the source of much imported material, is usually harvested only once a year, while cultivated plants in the United States are harvested mechanically up to three times a year. As with most herbs, both stems and leaves are harvested and then dehydrated. Dried stems and leaves are separated mechanically. Thyme oil is extracted from fresh material. Thyme is a widely used herb, both alone and in blends such as "fine herbs." Thyme oil is used for flavoring medicines and has strong bactericidal properties. *See* SPICE AND FLAVORING. [S.Kir.]

Thymosin A polypeptide hormone synthesized and secreted by the endodermally derived reticular cells of the thymus gland.

Thymosin exerts its actions in several loci: (1) in the thymus gland, either on precursor stem cells derived from fetal liver or from bone marrow, or on immature thymocytes, and (2) in peripheral sites, on either thymic-derived lymphoid cells or on precursor stem cells. The precursor stem cells, which are immunologically incompetent whether in the thymus or in peripheral sites, have been designated as predetermined T cells or T_0 cells, and mature through stages termed T_1 and T_2, each reflecting varying degrees of immunological competence. Thymosin promotes or accelerates the maturation of T_0 cells to T_1 cells as well as to the final stage of a T_2. In addition to this maturation influence, the hormone also increases the number of total lymphoid cells by accelerating the rate of proliferation of both immature and mature lymphocytes. *See* IMMUNITY; THYMUS GLAND. [A.W.]

Thymus gland An important central lymphoid organ in the neck or upper thorax of all vertebrates from elasmobranchs to mammals. The thymus gland is most prominent during early life. In many laboratory species of mammals and in humans it reaches its greatest relative weight at the time of birth, but its absolute weight continues to increase until the onset of puberty. Thereafter, it begins to undergo an involution and progressively decreases in size throughout adult life.

The thymic stem cells generate a large population of small lymphocytes (thymocytes) through a series of mitotic divisions. Simultaneously these dividing lymphocytes show evidence of cellular differentiation within the special thymic environment. During this division and maturation phase the developing thymocytes undergo an intrathymic migration from the peripheral cortical area to the medullary core of the organ. Some thymocytes degenerate within the organ, but many enter the circulating blood and lymph systems at various stages of maturity. A small percentage of the T lymphocyte population (5–10%) within the thymus is antigenically competent and capable of recognizing antigenic determinents on foreign cells or substances. Some of the T lymphocytes have the capacity to lyse the foreign tissue cells, while others are involved in recognizing the "foreignness" of the antigens and assisting a second subpopulation of bone-marrow-derived lymphocytes (B lymphocytes) to respond to the antigen by producing a specific antibody. These two types of immunocompetent T lymphocytes are called killer cells and helper cells, respectively. They are involved in both tissue transplantation and humoral antibody responses. On the other hand, the vast majority of the thymic lymphocytes are immunologically incompetent (90–95%). Some thymocytes are thought to give rise to the smaller pool of immunocompetent T lymphocytes, but many emigrate into the circulating blood. *See* CELLULAR IMMUNOLOGY; LYMPHATIC SYSTEM; THYMOSIN. [C.E.S.]

Thyrocalcitonin A hormone, the only known secretory product of the parenchymal or C cells of the mammalian thyroid and of the ultimobranchial glands of lower forms.

In conjunction with the parathyroid hormone, thyrocalcitonin is of prime importance in regulating calcium and phosphate metabolism. Its major function is to protect the organism from the dangerous consequences of elevated blood calcium. Its sole known effect is that of inhibiting the resorption of bone. It thus produces a fall in the concentration of calcium and phosphate in the blood plasma because these two minerals are the major constituents of bone mineral and are released into the bloodstream in ionic form when bone is resorbed. *See* BONE; CALCIUM METABOLISM; PARATHYROID GLAND; PARATHYROID HORMONE.

Thyrocalcitonin also causes an increased excretion of phosphate in the urine under certain circumstances, but a question remains as to whether this is a direct effect of the hormone upon the kidney or an indirect consequence of the fall in blood calcium which occurs when the hormone inhibits bone resorption. *See* PHOSPHATE METABOLISM; THYROID GLAND. [H.Ras.]

Thyroid gland An endocrine gland found in all vertebrates that produces, stores, and secretes the thyroid hormones. The primary function of the thyroid, in warm-blooded vertebrates at least, is to regulate the rate of metabolism. In humans, the gland is located in front of, and on either side of, the trachea (see illustration). Thyrocalcitonin, another hormone

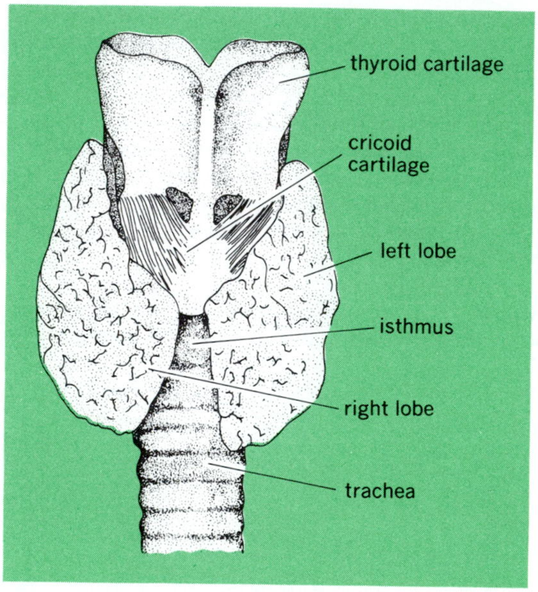

Ventral view of human thyroid gland shown in relation to trachea and larynx. (*After C. K. Weichert, Elements of Chordate Anatomy, 3d ed., McGraw-Hill, 1967*)

of the thyroid gland, assists in regulating serum calcium by reducing its levels. The thyroid gland is capable of accumulating inorganic iodides and uniting them with the amino acid tyrosine to produce iodinated proteins. This activity is regulated by thyrotropic hormone from the anterior lobe of the pituitary gland. *See* THYROID HORMONE. [G.C.Ke.]

Thyroid gland disorders Disorders of the thyroid gland may be classified according to anatomical and functional characteristics. Those thyroid disorders that are primarily anatomical include goiter and neoplasia; those that are primarily functional result in either hyperthyroidism or hypothyroidism.

Thyroid gland enlargement, or goiter, is the most common disorder. Its classification is based upon both the anatomy and function of the gland. An enlarged but normally functioning thyroid gland is termed a nontoxic goiter. This condition affects hundreds of millions of people throughout the world in areas where the diet is deficient in iodine. In other areas it may be caused by subtle disorders in the biosynthesis of thyroid hormone. In both cases, there is compensatory enlargement of the gland that can be diffuse and symmetrical or can produce a multinodular goiter. The multinodular goiter can grow independently from pituitary gland control and produce excess thyroid hormone, causing hyperthyroidism. However, hyperthyroidism is most often the result of Graves' disease. Thyroid enlargement can also be caused by Hashimoto's thyroiditis, in which the individual's immune system develops abnormal antibodies that react with proteins in the thyroid gland. This autoimmunity can make the gland enlarge or become underactive. *See* AUTOIMMUNITY.

Tumors of the thyroid account for a small fraction of human neoplasms and an even smaller fraction of deaths due to cancer. The vast majority of thyroid neoplasms are follicular adenomas, which are benign; however, some thyroid neoplasms are malignant. Rarely, tumors arising elsewhere in the body can metastasize to the thyroid gland.

Hyperthyroidism is the clinical condition that results from excessive levels of the circulating thyroid hormones thyroxine and triiodothyronine, which are secreted by the thyroid gland. Signs and symptoms include weight loss, tachycardia (increased heart rate), heat intolerance, sweating, and tremor. Graves' disease, the most common form of hyperthyroidism, is mediated by an abnormal antibody directed to the thyroid-stimulating hormone (TSH) receptor on the surface of the thyroid cell, which stimulates secretion of thyroid hormone. Unique to Graves' disease is the associated protrusion of the eyes (exophthalmos).

Hypothyroidism is the clinical state that results from subnormal levels of circulating thyroid hormones. Manifestations in infancy and childhood include growth retardation and reduced intelligence; in adults, cold intolerance, dry skin, weight gain, constipation, and fatigue predominate. Individuals with hypothyroidism often have a slow pulse (bradycardia), puffy dry skin, thin hair, and delayed reflexes. In its most extreme form, hypothyroidism can lead to coma and death if untreated. *See* THYROID GLAND; THYROID HORMONE. [L.J.DeG.; D.A.E.]

Thyroid hormone Any of the chemical messengers produced by the thyroid gland, including thyrocalcitonin, a polypeptide, and thyroxine and triiodothyronine which are iodinated thyronines. *See* HORMONE; THYROCALCITONIN; THYROID GLAND; THYROXINE. [H.Ras.]

Thyroxine A hormone secreted by the thyroid gland. Thyroxine is quite similar chemically and in biological activity to triiodothyronine (see illustration). Both are derivatives of the amino acid tyrosine and are unique in being the only iodine-containing compounds of importance in the economy of all higher forms of animal life. The thyroid gland avidly accumulates the small amount of iodine in the diet. This iodine is oxidized to iodide ion in the gland and then reacts with tryosine to form mono- and diiodotyrosine. These latter are then coupled to form either thyroxine or triiodothyronine. *See* THYROID GLAND.

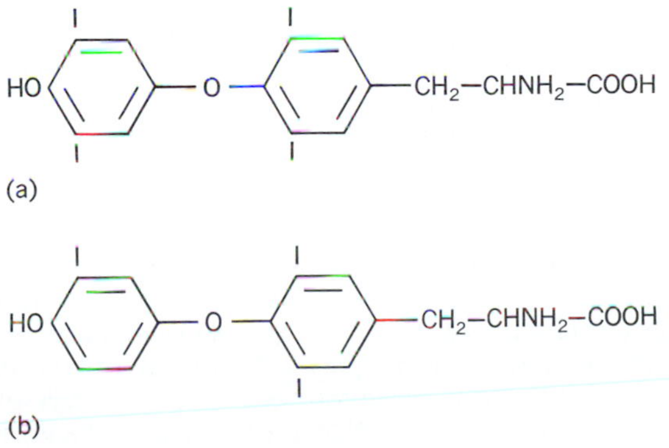

Structural formulas for thyroid hormones (*a*) thyroxine and (*b*) triiodothyronine.

The maintenance of a normal level of thyroxine is critically important for normal growth and development as well as for proper bodily function in the adult. Its absence leads to delayed or arrested development. It is one of the few hormones with general effects upon all tissues. Its lack leads to a decrease in the general metabolism of all cells, most characteristically measured as a decrease in nucleic acid and protein synthesis, and a slowing down of all major metabolic processes. *See* THYROID GLAND DISORDERS. [H.Ras.]

Thysanoptera An order of small, slender insects, commonly called thrips, having exopterygote development, sucking mouthparts, and highly modified wings. The order is a relatively small one, but individuals are often very numerous in favorable environments. *See* THRIP.

The mouthparts are conical and used for scraping, piercing, and sucking. The wings are exceptionally narrow, with few or no veins, and are bordered by long hairs. The tarsi terminate in an inflatable membranous bladder, which has remarkable adhesive properties.

The eggs of thrips are laid on the surface of twigs (suborder Tubulifera) or in small cuts made by the ovipositor (suborder Terebrantia). There are usually four nymphal stages, the last of these being quiescent and pupalike. There are from one to several generations produced in a single year. *See* INSECTA. [F.M.C.]

Thysanura Primarily wingless insects, with soft, fusiform bodies from 0.12 to 0.80 in. (3 to 20 mm) long, often covered with scales forming diverse patterns (see illustration). These insects are considered by some authors to constitute two orders: the true Thysanura (silverfish and allies) and the Microcoryphia (machilids). The mouthparts are free. The mandibles are monocondylous in the machilids, and scraping in function; they are dicondylous—as in all winged insects—in the silverfish, in which they are of the chewing type. Machilids have large compound eyes, but silverfish have simple ommatidia, or lack eyes altogether. Many machilids have coxae with styliform appendages. Tarsi are from two- to five-segmented; there are two or three claws. Abdominal ventral plates are entire, or subdivided into a central sternum and posterolateral coxites. Abdominal styli are present in varied numbers. The females have well-developed ovipositors, and the males have a penis and often one or rarely two pairs of parameres. There are cerci and a median caudal filament.

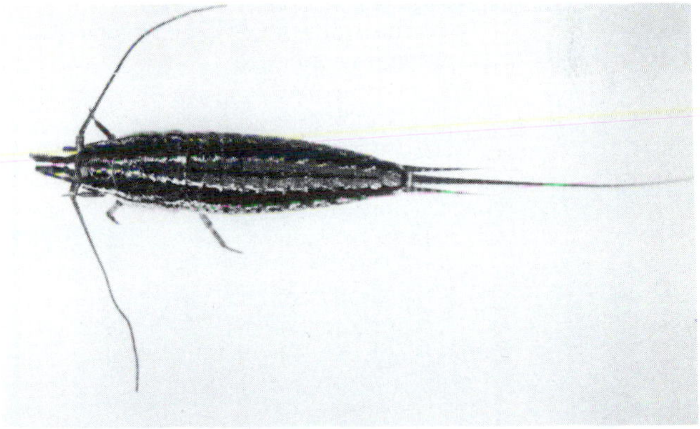

***Machilinus* sp. (Microcoryphia).**

These primitive insects, fossils of which are known as far back as the Devonian (primitive machilids, the Monura), are worldwide in distribution. Machilids are more numerous in temperate climates, the silverfish in the tropics and subtropics. *See* APTERYGOTA; INSECTA. [P.W.W.]

Tick fever A mild, febrile viral disease transmitted by wood ticks. After 4–6 days' incubation, sudden onset comes with chills, fever, severe aches, and sometimes nausea; symptoms and fever are often diphasic. Occasional cases are reported with more severe symptoms involving the central nervous system. [J.L.Me.]

Tick paralysis A loss of muscle function or sensation in humans or certain animals following the prolonged feeding of female ticks. Paralysis, of Landry's type, usually begins in the legs and spreads upward to involve the arms and other parts of the body. Evidence suggests that paralysis is due to a neurotoxin formed by the feeding ticks rather than the result of infection with microorganisms. *See* IXODIDES.

The disease has been reported in North America, Australia, South Africa, and occasionally in some European countries and is caused by appropriate species of indigenous ticks. In Australia, *Ixodes holocyclus* causes frequent cases in dogs, and occasional cases in humans, and paralysis has been known to progress even after removal of ticks. *Ixodes cubicundus* is associated with the disease in South Africa. [C.B.P.]

Tidal bore A part of a tidal rise in a river which is so rapid that water advances as a wall often several feet high. The phenomenon is favored by a substantial tidal range and a channel which shoals and narrows rapidly upstream, but the conditions are so critical that it is not common. Although the bore is a very striking feature, the tide continues to rise after the passage of the bore. Bores may be eliminated by changing channel depth or shape. *See* RIVER TIDES; TIDE.

Tidal bore of the Petitcodiac River, Bay of Fundy, New Brunswick, Canada. Rise of water is about 4 ft (1.2 m). (*New Brunswick Travel Bureau*)

In North America three bores have been observed: at the head of the Bay of Fundy (see illustration), at the head of the Gulf of California, and at the head of Cook Inlet, Alaska. The largest known bore occurs in the Tsientang Kiang, China. At spring tides this bore is a wall of water 15 ft (4.5 m) high moving upstream at 25 ft/s (7.5 m/s). [B.K.]

Tidal power Tidal-electric power is obtained by utilizing the recurring rise and fall of coastal waters. Marginal marine basins are enclosed with dams, making it possible to create differences in the water level between the ocean and the basins. The oscillatory flow of water filling or emptying the basins is used to drive hydraulic turbines which propel electric generators.

Large amounts of electric power could be developed in the world's coastal regions having tides of sufficient range, although even if fully developed this would amount to only a small percentage of the world's potential water (hydroelectric) power. *See* ELECTRIC POWER GENERATION; TIDE. [G.G.A.]

Tide Stresses exerted in a body by the gravitational action of another, and related phenomena resulting from these stresses. Every body in the universe raises tides, to some extent, on every other. This article deals only with tides on the Earth, since these are fundamentally the same as tides on all bodies, and more specifically with variations of sea level, whatever their origin. *See* GRAVITATION; SEA-LEVEL FLUCTUATIONS.

The tide-generating forces arise from the gravitational action of Sun and Moon, the effect of the Moon being about twice as effective as that of the Sun in producing tides. The tidal effects of all other bodies on the Earth are negligible. The tidal forces act to generate stresses in all parts of the Earth and give rise to relative movements of the matter of the solid Earth, ocean, and atmosphere. In the ocean the tidal forces act to generate alternating tidal currents and displacements of the sea surface.

If the Moon attracted every point within the Earth with equal force, there would be no tide. It is the small difference in direction and magnitude of the lunar attractive force, from one point of the Earth's mass to another, which gives rise to the tidal stresses. The tide-generating force is proportional to the mass of the disturbing body (Moon) and to the inverse cube of its distance. This inverse cube law accounts for the fact that the Moon is 2.17 times as important, insofar as tides are concerned, as the Sun, although the latter's direct gravitational pull on the Earth, which is governed by an inverse-square law, is about 180 times the Moon's pull.

At most places in the ocean and along the coasts, sea level rises and falls in a regular manner. The highest level usually occurs twice in any lunar day, the times bearing a constant relationship with the Moon's meridional passage. The time between the Moon's meridional passage and the next high tide is called the lunitidal interval. The difference in level between successive high and low tides, called the range of the tide, is generally greatest near the time of full or new Moon, and smaller near the times of quadrature. The range of the tide usually exhibits a secondary variation, being greater near the time of perigee (when the Moon is closest to the Earth) and smaller at apogee (when the Moon is farthest away).

The above situation is observed at places where the tide is predominantly semidiurnal. At many other places, it is observed that one of the two maxima in any lunar day is higher than the other. This effect is known as the diurnal inequality and represents the presence of an appreciable diurnal variation. At these places, the tide is said to be of the "mixed" type. At a few places, the diurnal tide actually predominates, there generally being only one high and low tide during the lunar day.

The range of the ocean tide varies between wide limits. The highest range is encountered in the Bay of Fundy, where values exceeding 50 ft (15 m) have been observed. In some places in the Mediterranean, South Pacific, and Arctic, the tidal range never exceeds 2 ft (0.6 m).

Owing to the rotation of the Earth, there is a gyroscopic, or Coriolis, force acting perpendicularly to the motion of any water particle in motion. In the Northern Hemisphere this force is to the right of the current vector. The horizontal, or tractive, component of the tidal force generally rotates in the clockwise sense in the Northern Hemisphere. As a result of both these influences the tidal currents in the open ocean generally rotate in the clockwise sense in the Northern Hemisphere, and in the counterclockwise sense in the Southern Hemisphere. *See* CORIOLIS ACCELERATION; EARTH TIDES. [G.W.G.]

Tie rod A tie rod or tie bar, usually circular in cross section, is used in structural parts of machines to tie together or brace connected members, or in moving parts of machines or mechanisms it may connect arms or parts to transmit motion. In the first use the rod ends are usually a threaded fastening, while in the latter they are usually forged into an eye for a pin connection. [P.H.B.]

Tile As a structural material, a burned clay product in which the coring exceeds 25% of the gross volume; as a facing material, any thin, usually flat, square product. Structural tile used for load bearing may or may not be glazed; it may be cored horizontally or vertically. Two principal grades are manufactured: one for exposed masonry construction, and the other for unexposed construction.

As a facing, clay products are formed into thin flat, curved, or embossed pieces, which are then glazed and burned. Commonly used on surfaces subject to water splash or that require frequent cleaning, such vitreous glazed wall tile is fireproof. Unglazed tile is laid as bathroom floor. By extension, any material formed into a size comparable to clay file is called tile. Among the materials formed into tile are asphalt, cork, linoleum, vinyl, and porcelain. [F.H.R.]

Till The generic term for sediment deposited directly from glacier ice. Till is characteristically nonsorted and nonstratified and is deposited by lodgement or melt-out beneath a glacier or by melt-out on the surface of a glacier. The texture of till varies greatly and all tills are characterized by a wide range, of particle sizes. Till contains a variety of rock and mineral fragments which reflect the source material over which the glacier flowed. The particles in the deposit usually show a preferred orientation related to the nature and direction of the ice flow. The overall character of the fill reflects the source material, position and distance of transport, nature and position of deposition, and postdepositional changes. [W.H.J.]

Tillage The mechanical manipulation of the soil to improve its physical condition as a habitat for plants. Tillage includes plowing, inversion, loosening, harrowing, pulverization, packing, and rolling the soil, all to improve aeration and temperature conditions and to produce a firm seedbed. Subsurface tillage is the loosening of soil by sweeps or blades pulled beneath the surface without inversion of the soil. This practice, especially adapted to dry areas, fragments the soil and leaves a mulch of stubble or other plant residues on the soil surface to conserve water and help control erosion. [M.L.N.]

Tillodontia An order of extinct quadrupedal land mammals known from early Cenozoic deposits in the Northern Hemisphere. They progressively developed enlarged, rootless second incisors that became remarkably rodentlike. Tillodont feet have five clawed toes. Tillodonts may have originated in a Paleocene arctocyonid group. If this is true, tillodonts were most closely related to condylarths and to the various hoofed mammals of the later Cenozoic ages. *See* CONDYLARTHRA; MAMMALIA. [D.E.S.]

Time The dimension of the physical universe which, at a given place, orders the sequence of events; also, a designated instant in this sequence, as the time of day, technically known as an epoch.

Measurement. Time measurement consists in counting the repetitions of any recurring phenomenon and, if the interval between successive recurrences is sensible, in subdividing it. The phenomenon most used has been the rotation of Earth, where the counting is by days. Days are measured by observing the meridian passages of stars and are subdivided with the aid of precision clocks. The day is, however, subject to variations in duration; consequently, when the utmost precision is required, years instead of days are measured and subdivided. *See* DAY.

A determination of time is synonymous with the establishment of an epoch; it consists in ascertaining the clock correction, which is the correction that should be applied to the reading of a clock (positive if the clock is slow) at a specified epoch. A time interval may be measured in two ways: as the duration between two known epochs or simply by counting from an arbitrary starting point, as is done with a stopwatch.

Time units are the intervals between successive recurrences of phenomena, such as the period of rotation of Earth, also arbitrary multiples and subdivisions of these intervals, such as the hour, being one twenty-fourth of a day, and the minute, being one-sixtieth of an hour. *See* MONTH; YEAR.

Time bases. Several phenomena are used as the time base to be divided into hours. For astronomical purposes, sidereal time is used; for terrestrial purposes, solar time is used.

The hour angle of the vernal equinox is the measure of sidereal time. It is reckoned from 0 to 24 hours; each hour is subdivided into 60 sidereal minutes and the minutes into 60 sidereal seconds. Sidereal clocks are used for convenience in most astronomical observatories, because a star, or other object outside the solar system, comes to the same place in the sky each night at virtually the same sidereal time.

The hour angle of the Sun is the apparent solar time. The only true indicator of apparent solar time is a sundial. Mean solar time has been devised to eliminate the irregularities in apparent solar time that arise from the obliquity of the ecliptic and the varying speed of Earth in its orbit around the Sun. It is the hour angle of a fictitious mean Sun, an imagined point moving uniformly along the celestial equator at the same rate as the average rate of the actual Sun along the ecliptic.

Because sidereal and solar time are both defined as hour angles, at any instant they vary from place to place on Earth. To avoid the inconvenience of the continuous change of mean solar time with longitude, zone time or civil time is the time generally used. Earth is divided into 24 time zones, each approximately 15° wide and centered on standard longitudes 0, 15, 30°, and so on. Within each of these zones, the time kept is the mean solar time of the standard meridian. Most civilized nations use zone time. Many countries of the world, including the United States, advance their time 1 hour during the summer months, into daylight saving time. *See* ATOMIC TIME; DYNAMICAL TIME. [G.M.C.]

Time, arrow of The uniform and unique direction associated with the apparent inevitable flow of time into the future. There appears to be a fundamental asymmetry in the universe. Herein lies a paradox, for all the laws of physics, whether they are the equations of classical mechanics, classical electromagnetism, general relativity, or quantum mechanics, are time reversible in that they admit solutions in either direction of time. This reversibility raises the question of how these fundamentally time-symmetrical equations can result in the perceived asymmetry of temporally ordered events.

The symmetry breaking of temporal order has not yet been fully explained. There are certain indications that an intrinsic asymmetry exists in temporal evolution. Thus it may be that the fundamental laws of physics are not really time symmetric and that the currently known laws are only symmetrized approximations to the truth. Indeed, the decay of the K^0 meson is not time reversible. However, it is not clear how such a rare and exotic instance of time asymmetry could emerge into the world of essentially macroscopic, electromagnetic phenomena as an everyday observable. *See* TIME REVERSAL INVARIANCE.

Another, more ubiquitous example of a fundamentally time-asymmetric process is the expansion of the universe. It has been speculated that this expansion is the true basis of time asymmetry. *See* BIG BANG THEORY; COSMOLOGY.

Alternatively, even a time-symmetrical universe will have a

statistical behavior in which configurations of molecules and localizations of energy have significant probabilities of recurring only after enormously long time intervals. Indeed, such time intervals are longer than the times required for the ceaseless expansion of the universe and the evolution of its component particles. Time's arrow is destined, either by the nature of space-time or the statistics of large assemblies, to fly into the future. *See* STATISTICAL MECHANICS. [P.W.A.]

Time constant The time required for a physical quantity to change its initial (zero-time) magnitude by the factor $1 - (1/e)$ when the physical quantity is varying as a function of time, $f(t)$, according to the decreasing exponential function as in Eq. (1), or the increasing exponential function as in Eq. (2).

$$f(t) = e^{kt} \quad (1)$$

$$f(t) = 1 - e^{-kt} \quad (2)$$

The numeric e has the value of 2.71828. Therefore, the change in magnitude of $1 - (1/e)$ has the fractional value 0.632121. Thus, after a time lapse of one time constant, starting at zero time, the magnitude of the physical quantity will have changed 63.2%.

When time t is zero, Eq. (1) has the magnitude 1, and when time t is $1/k$, the magnitude is e^{-1}, or $1/e$. The corresponding change in magnitude is $1 - (1/e)$. The specific time required to accomplish this change is shown in Eq. (3), where T is called

$$t = 1/k = T \quad (3)$$

the time constant and is usually expressed in seconds. The same results are obtained for Eq. (2).

The initial rate of change of both the increasing and decreasing functions is equal to the maximum amplitude of the function divided by the time constant.

The concept of time constant is useful when evaluating the presence of transient phenomena. [R.L.R.]

Time-interval measurement Time interval is measured with high precision, most conveniently, with an electronic, digital reading counter. This contains an oscillator, which has a frequency, say, of 10^7 hertz. A starting signal causes the counter to start counting cycles, and a second signal causes the counter to stop. The accumulated number is displayed by digital indicating lights.

Short time intervals may be measured with high precision, to 1 nanosecond (1^{-9} s), with a suitable oscilloscope. Ultrashort phenomena in the picosecond range (1 ps = 10^{-12} s) are studied by use of short pulses generated by lasers (about 0.5–10 ps).

Radioactive decay is used to measure long time intervals concerning human history, the Earth, and the solar system. *See* GEOCHRONOMETRY; RADIOCARBON DATING. [W.M.]

Time-of-flight spectrometer Any of a general class of instruments in which the speed of a particle is determined directly by measuring the time that it takes to travel a measured distance. By knowing the particle's mass, its energy can be calculated. If the particles are uncharged (for example, neutrons), difficulties arise because standard methods of measurement (such as deflection in electric and magnetic fields) are not possible. The time-of-flight method is a powerful alternative, suitable for both uncharged and charged particles.

The time intervals are best measured by counting the number of oscillations of a stable oscillator that occur between the instants that the particle begins and ends its journey. Oscillators operating at 100 MHz are in common use. *See* MASS SPECTROSCOPE; NEUTRON SPECTROMETRY; TIME-INTERVAL MEASUREMENT. [F.W.K.F.]

Time-projection chamber An advanced particle detector for the study of ultra-high-energy collisions of positrons and electrons. The underlying physics of the scattering process can be studied through precise measurements of the momenta, directions, particle species, and correlations of the collision products. The time-projection chamber (TPC) provides a unique combination of capabilities for these studies and other problems in elementary particle physics by offering particle identification over a wide momentum range, and by offering high resolution of intrinsically three-dimensional spatial information for accurate event reconstruction.

The time-projection chamber concept is based on the maximum utilization of ionization information, which is deposited by high-energy charged particles traversing a gas. The ionization trail, a precise image of the particle trajectory, also contains information about the particle velocity. A strong, uniform magnetic field and a uniform electric field are generated within the time-projection chamber active volume in an exactly parallel orientation. The parallel configuration of the fields permits electrons, products of the ionization processes, to drift through the time-projection chamber gas over great distances without distortion; the parallel configuration offers a further advantage in that the diffusion of the electrons during drift can be greatly suppressed by the magnetic field, thus preserving the quality of track information. *See* PARTICLE DETECTOR. [D.R.N.]

Time reversal invariance A symmetry of the fundamental (microscopic) equations of motion of a system; if it holds, the time reversal of any motion of the system is also a motion of the system. With one exception (K_L meson decay), all observations are consistent with time reversal invariance (T invariance).

Time reversal invariance is not evident from casual observation of everyday phenomena. If one takes a movie of a phenomenon, the corresponding time-reversed motion can be exhibited by running the movie backward. The result is usually strange. For instance, water in the ground is not ordinarily observed to collect itself into drops and shoot up into the air. However, if the system is sufficiently well observed, the direction of time is not obvious. For instance, a movie which showed the motion of the planets would look just as right run backward or forward. The apparent irreversibility of everyday phenomena results from the combination of imprecise observation and starting from an improbable situation (a state of low entropy, to use the terminology of statistical mechanics). *See* ENTROPY; STATISTICAL MECHANICS.

If time reversal invariance holds, no particle (a system with a definite mass and spin) can have an electric dipole moment. A polar body, for example, a water (H_2O) molecule, has an electric dipole moment, but its energy and spin eigenstates (which are particles) do not. No particle has been observed to have an electric dipole moment; for instance, the present experimental upper limit on the electric moment of the neutron is approximately 10^{-24} cm times e, where e is the charge of the proton. *See* DIPOLE; NEUTRON; POLAR MOLECULE; SPIN (QUANTUM MECHANICS).

Another test of time reversal invariance is to compare the cross sections for reactions which are inverse to one another. The present experimental upper limit on the relative size of the time reversal invariance–violating amplitude of such reactions is approximately 3×10^{-3}. Another class of tests involves looking at the relative phases of amplitudes. Experiments put an upper limit of approximately 10^{-3} on the relative size of a time reversal invariance-violating amplitude. *See* NUCLEAR REACTION.

If time reversibility holds, then by the CPT theorem CP invariance must hold, that is, invariance of the fundamental equations under the combined operations of charge conjugation C and space inversion P. Consequently if time reversal invariance holds, a self-conjugate particle must be a CP eigenstate. The K_L meson is self-conjugate, but it or its decay state is not a CP eigenstate. *See* MESON.

If this observed CP violation came from the interactions responsible for the K_L decay, they should produce other observable *CP*-violating effects as well, namely an electric dipole moment of the neutron and a change in the relative phase of the amplitudes for K_L decay into pions; neither is seen. Apparently the interactions which violate time reversibility are very weak, and for this reason they are called superweak. *See* FUNDAMENTAL INTERACTIONS; WEAK NUCLEAR INTERACTIONS. [C.J.G.]

Timor Timor, an island in Wallacea 300 mi (480 km) long and varying 10 to 60 mi (16 to 100 km) in width, has an area of about 13,000 mi² (34,000 km²). The island is mountainous, with many peaks above 6000 ft (1800 m). The highest mountain is Tata Mai Lau, which rises to 9780 ft (2981 m). Many mud volcanoes are active. The island has two monsoons and a five-month dry season, which results in lighter vegetation than is usual at these latitudes. *See* EAST INDIES. [M.G.A.-C.]

Timothy A plant, *Phleum pratense*, of the order Cyperales, long the most important hay grass for the cooler temperate humid regions. It is easily established and managed, produces seed abundantly, and grows well in mixtures with alfalfa and clover. It is a short-lived perennial, makes a loose sod, and has moderately leafy stems and a dense cylindrical inflorescence. Timothy-legume mixtures still predominate in hay and pasture seedings for crop rotations in the northern half of the United States. *See* GRASS. [H.B.S.]

Tin A chemical element, symbol Sn, atomic number 50, atomic weight 118.69. Tin forms tin(II) or stannous (Sn^{2+}), and tin(IV) or stannic (Sn^{4+}) compounds, as well as complex salts of the stannite (M_2SnX_4) and stannate (M_2SnX_6) types.

Tin melts at a low temperature, is highly fluid when molten, and has a high boiling point. It is soft and pliable and is corrosion-resistant to many media. An important use of tin has been for tin-coated steel containers (tin cans) used for preserving foods and beverages. Other important uses are solder alloys, bearing metals, bronzes, pewter, and miscellaneous industrial alloys. Tin chemicals, both inorganic and organic, find extensive use in the electroplating, ceramic, plastic, and agricultural industries.

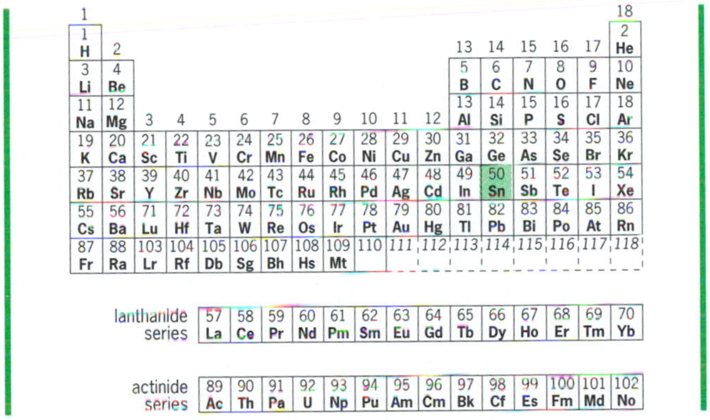

The most important tin-bearing mineral is cassiterite, SnO_2. No high-grade deposits of this mineral are known. The bulk of the world's tin ore is obtained from low-grade alluvial deposits. *See* CASSITERITE.

Two allotropic forms of tin exist: white (β) and gray (α) tin. Tin reacts with both strong acids and strong bases, but it is relatively resistant to solutions that are nearly neutral. In a wide variety of corrosive conditions, hydrogen gas is not evolved from tin and the rate of corrosion becomes controlled by the supply of oxygen or other oxidizing agents. In their absence, corrosion is negligible. A thin film of stannic oxide forms on tin upon exposure to air and provides surface protection. Salts that have an acid reaction in solution, such as aluminum chloride and ferric chloride, attack tin in the presence of oxidizers or air. Most nonaqueous liquids, such as oils, alcohols, or chlorinated hydrocarbons, have slight or no obvious effect on tin. Tin metal and the simple inorganic salts of tin are nontoxic. Some forms of organotin compounds, on the other hand, are toxic. Some important physical constants for tin are shown in the table.

Properties of tin

Property	Value
Melting point, °C	231.9
Boiling point, °C	2270
Specific gravity, α form (gray tin)	5.77
β form (white tin)	7.29
Specific heat, cal/g*, white tin at 25°C	0.053
Gray tin at 10°C	0.049

*1 cal = 4.184 joules.

Stannous oxide, SnO, is a blue-black, crystalline product which is soluble in common acids and strong alkalies. It is used in making stannous salts for plating and glass manufacture. Stannic oxide, SnO_2, is a white powder, insoluble in acids and alkalies. It is an excellent glaze opacifier, a component of pink, yellow, and maroon ceramic stains and of dielectric and refractory bodies. It is an important polishing agent for marble and decorative stones.

Stannous chloride, $SnCl_2$, is the major ingredient in the acid electrotinning electrolyte and is an intermediate for tin chemicals. Stannic chloride, $SnCl_4$, in the pentahydrate form is a white solid. It is used in the preparation of organotin compounds and chemicals to weight silk and to stabilize perfume and colors in soap. Stannous fluoride, SnF_2, a white water-soluble compound, is a toothpaste additive.

Organotin compounds are those compounds in which at least one tin-carbon bond exists, the tin usually being present in the +IV oxidation state. Organotin compounds that find applications in industry are the compounds with the general formula R_4Sn, R_3SnX, R_2SnX_2, and $RSnX_3$. R is an organic group, often methyl, butyl, octyl, or phenyl, while X is an inorganic substituent, commonly chloride, fluoride, oxide, hydroxide, carboxylate, or thiolate. *See* TIN ALLOYS. [J.B.Lo.]

Tin alloys Solid solutions of tin and some other metal or metals. Alloys cover a wide composition range and many applications because tin alloys readily with nearly all metals. *See* TIN; TIN METALLURGY.

Soft solders constitute one of the most widely used and indispensable series of tin-containing alloys. Common solder is an alloy of tin and lead, usually containing 20–70% tin. *See* SOLDERING.

Bronzes form an important group of structural metals. Of the true copper-tin bronzes, up to 10% tin is used in wrought phosphor bronzes, and from 5 to 10% tin in the most common cast bronzes. Many brasses, which are basically copper-zinc alloys, contain 0.75–1.0% tin for additional corrosion resistance. *See* COPPER ALLOYS.

Other useful tin alloys include babbitt metal (tin containing 4–8% each copper and antimony), an alloy with bearing applications; pewter; a tin-base alloy containing small quantifies of antimony and copper; and type metals, which are lead-base alloys containing 3–15% tin. *See* ALLOY. [B.W.G.]

Tin metallurgy The extraction of tin from its ores and its subsequent refining and preparation for use. Tin is comparatively easy to reduce to metal from the oxide by pyrometallurgy. However, this operation differs from the smelting of most common metals because it requires retreatment of the slag to obtain efficient metal recovery.

Smelting of tin concentrates is usually done in reverberatory furnaces, using coke or coal as the reducing agent. Electric arc furnaces and rotary reverberatory furnaces are also used in a few smelters, From alluvial tin ores, very-high-grade cassiterite (natural tin oxide) concentrates are normally obtained which contain 70–77% tin and only minor metal impurities. These are charged directly to the reverberatory furnace, or after roasting if they contain arsenic or sulfur.

Crude tin from smelting is liquated (that is, partially melted) to remove iron, copper, and other impurities which form solid compounds appreciably above the melting point of tin. The refined metal from most smelters is over 99.8% tin.

Secondary tin from metal scrap comes from tin-bearing alloys, and secondary smelters rework them into alloys and chemicals. Some tin metal of highest purity is recovered from the detinning of tinplate scrap by hot caustic solution and electrolytic recovery. *See* PYROMETALLURGY; TIN. [B.W.G.]

Tintinnida An order of the Spirotrichia whose members are conical or trumpet-shaped pelagic forms bearing shells (loricae). These protozoa are planktonic ciliates and are especially abundant in oceans, notably the Pacific. The lorica is composed of a resistant organic compound in which various foreign mineral grains are embedded; its shape may range from trumpet- or bell-form to cylindrical or subspherical.

Fossil tintinnids, representing practically the only fossilized species of ciliate protozoa known to science, are identified on the basis of the shape of the lorica in cross section as seen in randomly oriented thin sections of the rocks in which they are found. Twelve genera of fossil tintinnids have been described from limestones and cherts of the Jurassic and Cretaceous. *See* CILIOPHORA; SPIROTRICHIA. [J.O.C.; D.J.J.]

Tire A continuous pneumatic rubber and fabric cushion encircling and fitting onto the rim of a wheel. In modern tire building, chemicals are compounded into the rubber to help it withstand wear, heat, and aging and to produce desired changes in its characteristics. Fabric (rayon, nylon, or polyester) is used to give the tire body strength and resilience. In belted tires, additional layers of fabric (rayon, fiber glass, finely drawn steel, or aramid) are placed just under the tread rubber to increase mileage and handling. Steel wire is used in the bead that holds the tire to the rim.

A tire is made up of two basic parts: the tread, or road-contacting part, which must provide traction and resist wear and abrasion, and the body or carcass, consisting of rubberized fabric that gives the tire strength and flexibility. In compounding the rubber, large amounts of carbon black are mixed with it to improve abrasion resistance. Other substances, such as sulfur, are added to enable satisfactory processing and vulcanization. *See* RUBBER.

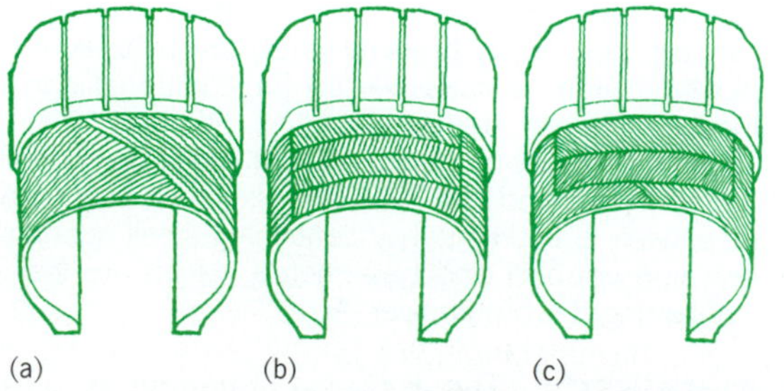

Tire construction. (*a*) Bias-ply. (*b*) Radial-ply. (c) Belted bias-ply. (*Goodyear Tire and Rubber Co.*)

There are three types of tires: bias, belted bias, and radial (see illustration). For bias tires, cords in the piles extend diagonally across the tire from bead to bead. The cords run in opposite directions in each successive ply, resulting in a crisscross pattern. For belted bias tires, plies are placed in a manner similar to that used in the bias-ply tire, with belts of material placed circumferentially around the tire between the plies and the tread rubber. For radial tires, cords in the plies extend transversely from bead to bead, substantially perpendicular to the direction of travel. Belts are placed circumferentially around the tire. [D.B.H.]

Tissue An aggregation of cells more or less similar morphologically and functionally. The animal body is composed of four primary tissues, namely, epithelium, connective tissue (including bone, cartilage, and blood), muscle, and nervous tissue. The process of differentiation and maturation of tissues is called histogenesis. *See* HISTOLOGY. [C.B.C.]

Tissue culture The branch of biology in which tissues or cells of higher animals and plants are grown artificially in a controlled environment. Tissue culture is possible when cells are attached to a solid substrate, such as glass or cellophane, and if the necessary complex nutrient medium is provided. All cultures are now also grown in liquid suspension. Tissue cultures are used in the study of cell growth, multiplication, and differentiation, as well as in cancer research, hereditary mechanisms, radiation biology, all hybridization, and virus studies. *See* MICROBIOLOGICAL METHODS. [T.T.P.]

Titanium A chemical element, Ti, atomic number 22 and atomic weight 47.90. While its chemistry shows many similarities to that of silicon and zirconium, as a first-row transition element, the aqueous solution chemistry, especially of the lower oxidation states, shows some resemblances to that of vanadium and chromium. *See* TRANSITION ELEMENTS.

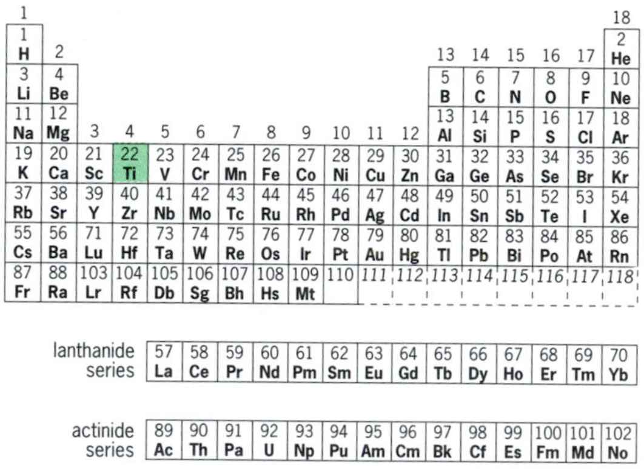
1 H; 2 He
3 Li; 4 Be; 5 B; 6 C; 7 N; 8 O; 9 F; 10 Ne
11 Na; 12 Mg; 13 Al; 14 Si; 15 P; 16 S; 17 Cl; 18 Ar
19 K; 20 Ca; 21 Sc; 22 Ti; 23 V; 24 Cr; 25 Mn; 26 Fe; 27 Co; 28 Ni; 29 Cu; 30 Zn; 31 Ga; 32 Ge; 33 As; 34 Se; 35 Br; 36 Kr
37 Rb; 38 Sr; 39 Y; 40 Zr; 41 Nb; 42 Mo; 43 Tc; 44 Ru; 45 Rh; 46 Pd; 47 Ag; 48 Cd; 49 In; 50 Sn; 51 Sb; 52 Te; 53 I; 54 Xe
55 Cs; 56 Ba; 71 Lu; 72 Hf; 73 Ta; 74 W; 75 Re; 76 Os; 77 Ir; 78 Pt; 79 Au; 80 Hg; 81 Tl; 82 Pb; 83 Bi; 84 Po; 85 At; 86 Rn
87 Fr; 88 Ra; 103 Lr; 104 Rf; 105 Db; 106 Sg; 107 Bh; 108 Hs; 109 Mt; 110; 111; 112; 113; 114; 115; 116; 117; 118
lanthanide series
57 La; 58 Ce; 59 Pr; 60 Nd; 61 Pm; 62 Sm; 63 Eu; 64 Gd; 65 Tb; 66 Dy; 67 Ho; 68 Er; 69 Tm; 70 Yb
actinide series
89 Ac; 90 Th; 91 Pa; 92 U; 93 Np; 94 Pu; 95 Am; 96 Cm; 97 Bk; 98 Cf; 99 Es; 100 Fm; 101 Md; 102 No

The principal valence state is 4+; the 3+ and 2+ states are also known, but are less stable. The element burns in air when heated to give the dioxide, TiO_2, and combines with halogens. It reduces water-vapor to form the dioxide and hydrogen, and reacts similarly with hot concentrated acids, although it forms the trichloride with hydrochloric acid. The metal absorbs hydrogen to give compositions approaching TiH_2 and forms the nitride, TiN, and the carbide, TiC. The sulfide TiS_2 as well as both the lower oxides, Ti_2O_3 and TiO, and sulfides, Ti_2S_3 and TiS, are known. Salts of all three valence states are known.

The dioxide, TiO_2, occurs most commonly in a black or brown form known as rutile. Less prominent naturally occurring forms are anatase and brookite. Both rutile and anatase are white when pure. The black basic oxide, $FeTiO_3$, occurs naturally as the mineral ilmenite; this is a principal commercial source of titanium. See ILMENITE; RUTILE.

Titanium dioxide is widely used as a white pigment for exterior paints because of its chemical inertness, superior covering power, opacity to damaging ultraviolet light, and self-cleaning ability. The dioxide has also been used as a whitening or opacifying agent in numerous situations. Examples would be the use as a filler in paper, a coloring agent for rubber and leather products, a pigment in ink, and a component of ceramics. It has found important use as an opacifying agent in porcelain enamels, giving a finish coat of great brilliance, hardness, and acid resistance.

The alkaline-earth titanates show some remarkable properties. The dielectric constants range from 13 for $MgTiO_3$ to several thousand for solid solutions of $SrTiO_3$ in $BaTiO_3$. Barium titanate itself has a dielectric constant of 10,000 near 120°C (250°F), its Curie point; it has a low dielectric hysteresis. Ceramic transducers containing barium titanate compare favorably with Rochelle salt and quartz, with respect to thermal stability in the first case, and with respect to the strength of the effect and the ability to form the ceramic in various shapes, in the second case. The compound has been used both as a generator for ultrasonic vibrations and as a sound detector. See PIEZOELECTRICITY; TITANIUM METALLURGY. [A.W.A.]

Titanium metallurgy The winning of metallic titanium from its ores followed by alloying and processing into forms and shapes that can be used for structural purposes. Titanium's unique properties—density half that of steel, excellent strength retention to 1000°F (538°C), and atmospheric corrosion immunity superior to that of other metals—make it an ideal construction material for both the engines and airframes.

All commercial production of titanium metal involves chlorination of ore concentrates; reacting TiO_2 with chlorine gas and coke (carbon) in a fluidized-bed reactor forms impure titanium tetrachloride ($TiCl_4$). For the production of acceptable metal, purification of the raw tetrachloride is required to remove other metal chlorides that would contaminate the virgin titanium.

The purified titanium tetrachloride is delivered as a liquid to the reactor vessel. In these vessels, constructed of carbon or stainless steel, the titanium tetrachloride is reacted with either magnesium or sodium to form the pure metal called sponge, so named because of its porous cellular form. To avoid contamination by oxygen or nitrogen, the reaction is carried out in an argon atmosphere.

Titanium sponge is consolidated in the consumable-electrode arc furnace. A mass of sponge, alloy additions, and scrap are mixed, then compressed into compacts and welded together to form a sponge electrode. This is melted by an electric arc into a water-cooled copper crucible in a vacuum or an atmosphere of purified argon or nitrogen. The arc progressively consumes the sponge electrode to form an ingot.

The conversion of the titanium ingot into mill products, such as forging billet, plate, sheet, and tubing, is accomplished for the most part on conventional metalworking equipment. Titanium and its structural alloys are produced in most of the same forms and shapes as stainless steel. See METAL FORMING; STAINLESS STEEL; TITANIUM. [W.W.Mi.]

Titration A quantitative analytical process that is basically volumetric. However, in high-precision titrimetry the titrant solution is sometimes delivered from a weight buret, so that the volumetric aspect is indirect. Generally, a standard solution, that is, one containing a known concentration of substance X (titrant), is progressively added to a measured volume of a solution of a substance Y (titrand) that will react with the titrant. The addition is continued until the end point is reached. Ideally, this is the same as the equivalence point, at which an excess of neither X nor Y remains. If the stoichiometry or exact ratio in which X and Y react is known, it is possible to calculate the amount of Y in the unknown solution. See VOLUMETRIC ANALYSIS.

The normal requirements for the performance of a titration are: a standard titrant solution; calibrated volumetric apparatus, including burets, pipets, and volumetric flasks; and some means of detecting the end point. See BURET; PIPET; VOLUMETRIC FLASK.

Classification by chemical reaction. For the purposes of titrimetry, chemical reactions can be placed in three general categories: acid-base or neutralization, combination, and oxidation-reduction.

Acid-base titrations involve neutralization of an acid by titration with a base, or vice versa. However, the process is often nonspecific; in the titration of a mixture of nitric and hydrochloric acids, only the total acidity can be found without recourse to additional measurements. A salt derived from a strong base and a very weak acid can often be titrated just as if it were a base. See ACID AND BASE.

In titrimetry, attention is usually focused upon the combination of an ion in the titrant with one of the opposite sign in the titrand solution. Sometimes the combination may involve more than two species, some of which may be nonionic. The combinations may result in precipitation or formation of a complex. See COORDINATION COMPLEXES; PRECIPITATION (CHEMISTRY).

In so-called redox titrations the titrant is usually an oxidizing agent, and is used to determine a substance that can be oxidized and hence can act as a reducing agent. See OXIDATION-REDUCTION.

Coulometric titration. The passage of a uniform current for a measured period of time can be used to generate a known amount of a product such as a titrant. This fact is the basis of the technique known as coulometric titration. An obvious requirement is that generation shall proceed with a fixed, preferably 100%, current efficiency. The uniform current is then analogous to the concentration of an ordinary titrant solution, while the total time of passage is analogous to the volume of such a solution that would be needed to reach the end point. See ELECTROLYSIS.

Classification by end-point techniques. The precision and accuracy with which the end point can be detected is a vital factor in all titrations. Because of its simplicity and versatility, chemical indication is quite common, especially in acid-base titrimetry.

Indicators. An acid-base indicator is a weak acid or a weak base that changes color when it is transformed from the molecular to the ionized form, or vice versa. The color change is normally intense, so that only a low concentration of indicator is needed. The working range, or visual color change, of a typical acid-base indicator is spread over about a hundredfold (~2 pH units) change in hydrogen ion concentration. Available indicators have individual working ranges that together cover the entire range of hydrogen ion concentration likely to be encountered in general acid-base titration. See ACID-BASE INDICATOR; HYDROGEN ION; PH.

Sometimes no suitable chemical indicator can be found for a desired titration. Possibly the concentrations involved may be so low that chemical indication functions poorly. Other situations might be the need for high precision or for the automatic arrest of the titration. Recourse is then made to some physical method of end-point detection.

Potentiometric titration. If a pH meter is used, its associated electrodes are first standardized by use of a buffer solution of known pH. By suitable choice of electrodes, potentiometric methods can also be applied to combination titrations and to oxidation-reduction titrations. The advent of modern ion-selective electrodes has greatly extended the scope of potentiometric titration and of other branches of titrimetry. *See* ELECTRODE POTENTIAL; ION-SELECTIVE MEMBRANES AND ELECTRODES.

Conductometric titration. Conductometric titration is sometimes successful when chemical indication fails. The underlying principles of conductometric titration are that the solvent and any molecular species in solution exhibit only negligible conductance; that the conductance of a dilute solution rises as the concentration of ions is increased; and that at a given concentration the hydrogen ion and the hydroxyl ion are much better conductors than any of the other ions. *See* ELECTROLYTIC CONDUCTANCE.

Spectrophotometric titration. The spectrophotometer is an optical device that responds only to radiation within a selected very narrow band of wavelengths in the visual, ultraviolet, or infrared regions of the spectrum. The response can be made both quantitative and linearly related to the concentration of a species that absorbs radiation within this band. Titrations at wavelengths within the visual region are by far the most common. *See* SPECTROPHOTOMETRIC ANALYSIS.

Amperometric titration. By use of a dropping-mercury or other suitable microelectrode, it is possible to find a region of applied electromotive force (emf) in which the current is proportional to the concentration of one or both of the reactants in a titration.

Biamperometric titration is a closely related technique. An emf that is usually small is applied across two identical microelectrodes that dip into the titrand solution. This arrangement, which involves no liquid-liquid junctions, is valuable in nonaqueous titrations, but also finds much use in aqueous titrimetry. *See* POLAROGRAPHIC ANALYSIS.

Thermometric or enthalpimetric titration. Many chemical reactions proceed with the evolution of heat. If one of these is used as the basis of a titration, the temperature first rises progressively and then remains unchanged as the titration is continued past the end point. If the reaction is endothermic, the temperature falls instead of rising. Thermometric titration is applicable to all classes of reactions. *See* THERMOCHEMISTRY.

Nonaqueous titration. This technique is used to perform titrations that give poor or no end points in water. Although applicable in principle to all classes of reactions, acid-base applications have greatly exceeded all others. Nonaqueous titrations in which the solvent is a molten salt or salt mixture are also possible.

Automatic titration. Automation is particularly valuable in routine titrations, which are usually performed repeatedly. One approach is to record the titration curve and to interpret it later. Another method is to stop titrant addition or generation automatically at, or very near to, the end point. Although a constant-delivery device is desirable, an ordinary buret with an electromagnetically controlled valve is often used.

Microcomputer control permits such refinements as the continuous adjustment of the titrant flow rate during the titration. In some cases, it is possible to automate an entire analysis, from the measurement of the sample to the final washout of the titration vessel and the printout of the result of the analysis. *See* ANALYTICAL CHEMISTRY. [J.T.St.]

Toad Any of several species of the amphibian order Anura, which also includes the frogs. The anurans are characterized by their lack of a tail and by hindlimbs that are adapted for jumping. The family Bufonidae contains most of the common species of toads, and 17 genera have been described which occur throughout the world except for Antarctica and Australia. Toads live in drier habitats than frogs do.

Reproductive behavior is varied among the species, but fertilization is usually external. Tubular glands on the thumbs of some toads produce a sticky secretion that assists the male in clinging to the female during amplexus; amplexus may last up to several weeks. *See* AMPHIBIA; ANURA. [C.B.C.]

Tobacco The plant genus *Nicotiana*, certain species in the genus, and dried leaves of these plants are all called tobacco. Most often tobacco means a leaf product containing 1–3% of the alkaloid nicotine, which produces a narcotic effect when smoked, chewed, or snuffed. The plant *N. rustica* provides tobacco in parts of Europe, but the tobacco of world commerce is *N. tabacum*. Tobacco is American in origin. *See* SOLANALES. [G.S.T.]

Todorokite A hydrated manganese oxide mineral containing calcium, barium, potassium, sodium, and sometimes magnesium, general formula $(Na,Ca,K,Ba,Sr)_{0.3-0.7}(Mn,Mg,Al)_6O_{12} \cdot 3.2\text{–}4.5\ H_2O$. Todorokite is a major constituent of manganese nodules, which occur in large quantities ($>10^{12}$ tons) on the ocean floors. It has been shown to host some of the copper, nickel, and cobalt that occur in some manganese nodules in quantities of up to several weight percent. First described in 1934, todorokite is named for its occurrence at the Todoroki mine, Hokkaido, Japan. Terrestrial todorokite has also been found in deposits in Cuba, Portugal, Austria, France, United States, Brazil, and South Africa.

Todorokite is black to brown and is commonly very fine grained. It occurs as massive samples or as fibrous aggregates. Its hardness on the Mohs scale is low (1.5–2.5), and its reported specific gravity ranges from approximately 3.1 to 3.8.

The basic todorokite structure consists of triple chains of manganese-oxygen octahedra that are linked at roughly right angles to form a tunnel structure. This $[3 \times 3]$ structure can accommodate large cations and water in the tunnel sites. *See* CRYSTAL; MANGANESE NODULES. [S.T.]

Toggle Any of a wide variety of mechanisms, many used to open or close electrical contacts abruptly and all characterized by the control of a large force by a small one. The basic action of a toggle mechanism is shown in illustration *a*. When $\alpha = 90°$ the forces P and Q are independent of each other. Again, when $\alpha = 0°$ the forces are isolated, force Q being sustained entirely by the frame, and force P serving only to hold the link in position. At $\alpha = 45°$ from the symmetry $|P| = |Q|$, the mechanism serves to transfer the direction of forces to achieve equilibrium. *See* COUPLING.

Because the simple configuration of illustration *a* requires low-friction sliders, it is impractical. A more useful structure replaces the vertical slider with a second link pinned to the

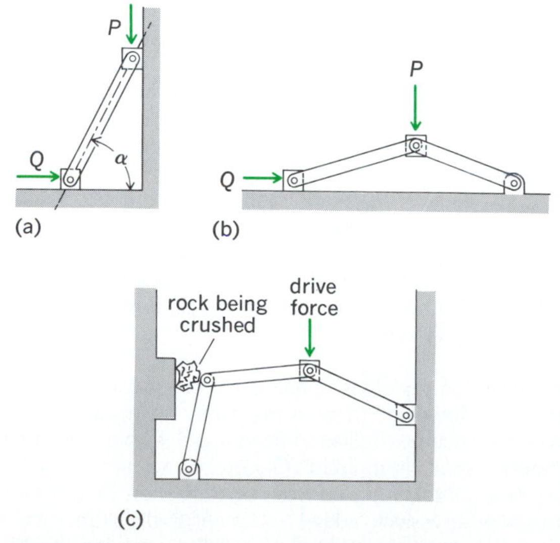

Toggle mechanism. (*a*) Simple structure. (*b*) Traditional configuration. (c) Typical application.

frame (illustration *b*), in which case input *P* sets up forces in both links. A further modification (illustration *c*) replaces the other slider with a link. *See* FOUR-BAR LINKAGE; LINKAGE (MECHANISM). [F.H.R.]

Tolerance Amount of variation permitted or "tolerated" in the size of a machine part. Manufacturing variables make it impossible to produce a part of exact dimensions; hence the designer must be satisfied with manufactured parts that are between a maximum size and a minimum size. Tolerance is the difference between maximum and minimum limits of a basic dimension. For instance, in a shaft and hole fit, when the hole is a minimum size and the shaft is a maximum, the clearance will be the smallest, and when the hole is the maximum size and the shaft the minimum, the clearance will be the largest.

If the initial dimension placed on the drawing represents the size of the part that would be used if it could be made exactly to size, then a consideration of the operating conditions of the pair of mating surfaces shows that a variation in one direction from the ideal would be more dangerous than a variation in the opposite direction. The dimensional tolerance should be in the less dangerous direction. This method of stating tolerance is called unilateral tolerance and has largely displaced bilateral tolerance, in which variations are given from a basic line in plus and minus values. [P.H.B.]

Tomato An important vegetable grown for its edible fruit belonging to the genus *Lycopersicon*, especially *L. esculentum*. *Lycopersicon* species are native to South America, especially Peru, and the Galapagos Islands. The tomato was first domesticated in Mexico, where it received its name from the Aztec word *xitomate*.

It was introduced to Europe in the mid-16th century, and was prominently featured in the early herbals. Known as the "love apple" and "poma d'ora," it was grown for the beauty of its fruit but was not often eaten, except in Italy and Spain, because many considered it to be poisonous like its distant relative the deadly nightshade.

The fruit is a berry with 2 to 12 locules containing many seeds. Most tomato varieties have red fruit, due to the red carotenoid lycopene. There is no basis for the common belief that yellow-fruited tomatoes are low in acidity. Many greenhouse varieties have pink fruit, due to a single gene that prevents formation of the yellow pigment in the epidermis of the fruit.

The tomato is highly esteemed as a source of vitamin C; tomato varieties differ in ascorbic acid content. Most tomato varieties have about 20–25 mg ascorbic acid per 100 g, and one medium-sized tomato provides about half of the required daily allowance of vitamin C for adults. Tomatoes are also a significant source of vitamin A. Red tomatoes have about 1000 international units (IU) of vitamin A per 100 g. Tomatoes are a good source of protein, but most of it is in the seeds. Tomato juice contains 19 amino acids, principally glutamic acid. *See* ASCORBIC ACID; VITAMIN A. [R.W.R.]

Tommotian fauna The oldest fossil fauna including diverse metazoan skeletal fossils. In a strict sense, the Tommotian is a geological stage and age recognized only on the Siberian Platform in Russia, where it is the lowest (oldest) stage of the Cambrian System. The extraordinarily abundant metazoan skeletal fossils occurring in that stage comprise the Tommotian fauna. However, the term Tommotian is frequently used informally in reference to the age of any diverse suites of skeletal fossils that occur before the first trilobites. *See* STRATIGRAPHIC NOMENCLATURE; TRILOBITA.

The base of the stage overlying the Tommotian is defined by the first occurrence of trilobite body fossils. On the Siberian Platform, the Atabanian stage overlies the Tommotian and, like the Tommotian, is often used informally as the second stage of the Phanerozoic and the first containing trilobite fossils (*Profallotaspis* in Siberia, *Fallotaspis* in North America). Chronologically, the base of the Tommotian has an age of approximately 550 million years. Although trilobites are absent in Tommotian-age strata, there is evidence of trilobitelike organisms. The trace fossils *Cruziana* and *Rusophycus* appear in strata equivalent to the upper Tommotian throughout the world. These tracks represent trilobite locomotion and resting traces, respectively.

Tommotian-age faunas are now known from most continents. The precise temporal relationships of these faunas remains unclear, but they appear to be approximately coeval. The faunas from each of these areas share general similarities in composition, although they tend to be endemic (unique) to each region.

Tommotian faunas appear suddenly in the fossil record and tend to be rather long-ranging in duration (most range into much younger strata). Therefore, the presence of a given species or genus conveys little information about the age of the local assemblage. The Tommotian marks an important evolutionary transition from a time dominated by Ediacaran soft-bodied Metazoa to a time dominated by skeletonized organisms more similar to those known today. Many of the modern marine invertebrate clades make their first appearance as fossils in Tommotian-age strata. Also, the advent of the Tommotian marks a dramatic increase in the numbers of animal species.

The rapid appearance of Metazoa in the Tommotian and later stages is nothing less than explosive. The pattern of appearances of new animal families in the fossil record has been analyzed, and it has been proposed that the initial radiation of metazoans was an exponential diversification beginning in the late Proterozoic (Vendian) that slowed as the number of new orders and families approached an equilibrium diversity. This equilibrium was maintained for the remainder of the Paleozoic. This diversification was the most rapid and broad adaptive radiation in the history of life. *See* CAMBRIAN; EDIACARAN FAUNA. [P.W.Si.]

Ton of refrigeration A rate of cooling that is equivalent to the removal of heat at 200 Btu/min, 12,000 Btu/h, or 288,000 Btu/day (3.51 kW). This unit of measure stems from the original use of ice for refrigeration. One pound of ice, in melting at 32°F (0°C), absorbs as latent heat approximately 144 Btu/lb, and 1 ton (2000 lb) of ice, in melting in 24 h, absorbs 288,000 Btu/day. In Europe, where the metric system is used, the equivalent cooling unit is the frigorie, which is a kilogram calorie, or 3.96 Btu. Thus 3000 frigories/h is approximately 1 ton of refrigeration. A standard ton of refrigeration is one developed at standard rating conditions of 5°F (6–15°C) evaporator and 86°F (30°C) condenser temperatures, with 9°F (5°C) liquid subcooling and 9°F (5°C) suction superheat. *See* REFRIGERATION. [C.F.K.]

Tonalite A phaneritic (visibly crystalline) plutonic rock composed chiefly of plagioclase (oligoclase or andesine) and quartz with subordinate dark-colored (mafic) minerals (biotite, amphibole, or pyroxene). The term tonalite is roughly equivalent to quartz diorite. Minor amounts of alkali feldspar may be present, but if this mineral exceeds 5% of the total feldspar, the rock is a granodiorite. As the quartz content decreases, quartz diorite passes into diorite. Tonalite, or quartz diorite, is roughly intermediate between granodiorite and diorite. *See* DIORITE; GRANODIORITE. [C.A.C.]

Tone (music and acoustics) A sound oscillation capable of exciting an auditory sensation having pitch; also, the sensation itself; that is, the word tone is used for both cause and effect. There is not necessarily a complete correspondence between the two; which of the two meanings is intended must be made clear by additional modifiers, context, or units of measurement. For example, if a tone is described as having pitch, it is to be understood that the sound sensation is meant, whereas a tone that has frequency must be a physical oscillation. *See* OSCILLATION; PITCH. [R.W.Y.]

Tongue An organ located in the floor of the mouth. The tongue reaches its highest development in mammals, where it is specialized in form and function. It is rich in papillae and glands and is always extensile, except in whales.

Anatomically the mammalian tongue is a mass of skeletal muscle largely contained within a covering sheet of mucous membrane. The surface epithelium is of the stratified squamous type. It is cornified in some mammals but not in humans. Filiform papillae are best developed in the cat family where they convert the tongue into a rasping organ. Fungiform papillae are more prominent elevations, with a knobbed top somewhat like a button mushroom. Some bear a few taste buds. Conical papillae are considered to be a modified fungiform type. Vallate papillae have a flat top and do not extend much above the lingual surface. Each is encircled by a relatively deep trench. The side surface of a human vallate papilla contains some 200 taste buds, whereas the opposite wall, across the trench, bears about one-fourth as many.

Taste buds are quite similar in all vertebrates. Each is an ovoid specialization paler than the surrounding stratified squamous epithelium; they occupy practically the full thickness of that covering layer. The shape is somewhat like a barrel, but the top usually narrows. Two cell types are recognizable within the taste bud. The taste cells are slender, spindle-shaped elements whose free end terminates in a short, stiff taste hair. These hairs extend into the taste canal and continue to the external taste pore. The supporting cells of the taste bud are mostly at the periphery, something like thick barrel staves. Nerve fibers for the mediation of taste end about both kinds of cells. *See* TASTE.

Lingual glands lie deep in the mucous membrane of the tongue and encroach on the muscle beneath. Mixed mucoserous glands occur under the apex of the tongue. Purely serous glands are restricted to the region of the vallate and foliate papillae, and many discharge on the surface of the tongue. Other glandular ducts open into the trench of each vallate papilla. The root of the tongue contains pure mucous glands, many of which open on the surface but some of which have ducts discharging into the pits of the lingual tonsil.

The mammalian tongue performs a variety of functions. In ruminants it is prehensile and thus is important in browsing. Some mammals use the tongue to lap up liquids. In general, it is of use in directing food to the teeth and pharynx, aiding chewing and swallowing. The dog, lacking sweat glands, cools itself by panting, which draws air over the tongue. The tongue is the most important agent in articulate speech. The sense of taste resides in the taste buds, whose cells respond to substances in solution. [L.B.A.]

Tonsil Localized aggregation of diffuse and nodular lymphoid tissue found in the region where the nasal and oral cavities open into the pharynx. The tonsils are important sources of blood lymphocytes. They often become inflamed and enlarged, necessitating surgical removal. *See* TONSILLITIS.

The two palatine (faucial) tonsils are almond-shaped bodies measuring 1 by 0.5 in. (2.5 by 1.2 cm) and are embedded between folds of tissue connecting the pharynx and posterior part of the tongue with the soft palate. These are the structures commonly known as the tonsils. The lingual tonsil occupies the posterior part of the tongue surface. It is really a collection of 35–100 separate tonsillar units, each having a single crypt surrounded by lymphoid tissue. Each tonsil forms a smooth swelling 0.08–0.16 in. (2–4 mm) in diameter. The pharyngeal tonsil (called adenoids when enlarged) occupies the roof of the nasal part of the pharynx. This tonsil may enlarge to block the nasal passage, forcing mouth breathing. *See* LYMPHATIC SYSTEM. [T.Sn.]

Tonsillitis An inflammation of the tonsil. Tonsillitis is a nonspecific term usually referring to bacterial or viral infection involving all or part of Waldeyer's ring, a collection of lymphatic tissue encircling the pharynx. It consists primarily of the tonsils (palatine tonsils), adenoids (pharyngeal tonsils), and lingual tonsils.

The complication of tonsillitis depend on which tonsil is involved. Recurrent adenoiditis with adenoid hypertrophy is frequently associated with recurrent otitis media, middle-ear fluid, and at times nasal obstruction with mouth breathing and snoring. Acute palatine tonsillitis may be complicated by peritonsillar abscess which may develop lateral to the tonsillar capsule. Removal of the adenoids is considered when there is residual middle-ear fluid. Palatine tonsils must be removed after peritonsillar abscess, but otherwise their removal depends upon the frequency of recurrent attacks of bacterial pharyngotonsillitis in relation to the patient's age. *See* TONSIL. [J.A.D.]

Tooling Auxiliary devices used in manufacturing operations to supplement basic standard and special machine tools. Tooling is used, for example, on drill presses, milling machines, planers, and shapers to facilitate the actions of workers and to adapt machine tools to production of specific parts. All machine tools to a varying degree utilize removable tooling.

The overall category of tooling encompasses a broad spectrum of manufacturing accessories; tooling is further subdivided into classifications such as jigs, fixtures, dies, gages, devices, gadgets, and other less universal designations.

There are no universally accepted definitions for differentiating between these subdivisions of tooling. Nomenclature for similar tools varies from one industry to another; it also varies among similar industries using similar devices. In shop practice, tooling subdivision nomenclature loosely follows the generalities of the table.

Types of tooling

Machine type	Prevailing tool nomenclature
Drill press	Drill jig
Boring machine (horizontal or vertical) cutting tool rotates	Fixture
Turning machine (horizontal or vertical) workpiece rotates	Fixture
Milling	Fixture
Welding	Jig
Assembly operations	Jig
Inspection operations	Gage
Presses (stamping, forming, and drawing)	Dies

Common hand tools and devices and standard perishable tools (consumed as a result of production operation) are excluded from the tooling category.

Operations. A tooling operation may consist of a single function or a series of functions. Most tooling simultaneously performs some combination of a variety of purposes; a few of the many purposes for which tooling is used are locating, clamping, positioning, and cutter guiding.

The trend in machine tools for automation is to design and produce basic machine-tool building blocks, each of which is a separate machine tool, that can be linked with others (with or without special units) to form an automated production line that performs a series of milling, drilling, and similar machining operations. In this way basic machine-tool building blocks perform a variety of jobs. The standard machine-tool building block also offers considerable protection (through retooling or salvage) against obsolescence of an expensive automatic transfer machine. *See* AUTOMATION. [J.I.K.]

Numerical control. Numerical control, in a broad sense, is an extremely versatile method of automatically operating machinery by means of discrete numerical values introduced to the machine through some form of stored-input medium, such as punched cards, punched tape, magnetic tape, or direct control by a digital computer.

By strict definition, numerical control (and certain other processing machines) is the directing of a series of coded instructions for machine operation which is composed largely of numbers, letters of the alphabet, and other symbols.

Numerical control commands for machine tools may range from positioning the spindle, tool, or workpiece to auxiliary functions such as selecting a tool station on a turret, controlling the direction, speed, and feed of a spindle, or turning the coolant off and on. The commands put together and logically organized can often be used to program all the operations required to completely machine a workpiece in one setting. This tremendous versatility has caused numerical control machines to be widely accepted.

The sequence of events in numerical control machining starts with the programmer who studies the part drawing. The programmer must visualize all the machine motions required to make the particular part. Once conceived, they must be documented in logical order on a programming manuscript. The manuscript data are then converted to a medium (often perforated tape) that can be acted upon by the machine control unit.

A schematic closed-loop feedback system is often used in numerical control machine control of a typical machine tool. The machine control unit accepts instructions from the tape reader, converts them into command signals, and transmits the signals to the machine as required. The command signals, which may be in the form of pulses, can produce machine movements of as little as 0.0002 in. (5 micrometers) to a pulse. Alternatively, command signals may be of analog (continuous) nature, and a degree of accuracy as required.

The machine control unit is not generally regarded as a computer, although most numerical control systems incorporate some computer circuits. The more sophisticated contouring systems have machine control units or directors which are essentially process computers.

Many of the advantages of numerical control machining are obvious, such as simpler setups, reduced cutting-tool changing time, less operator skill requirements, less scrap, much greater design flexibility, and better accuracy. However, one of the most important advantages is the increase in production that numerical control affords over conventional machining. Traditionally, it has been observed that parts in the shop spend 95% of the time moving and waiting, with only 5% of the time on the machine. Of the 5% of the time on the machine, less than 30% is spent as actual metalworking time. Through the increased use of numerical control and computer control, the disproportionate nonproductive time can be greatly reduced.

The first numerical control machine generation was classified as hard-wired. Numerical control machines have now progressed to the soft-wired type, where the control can be adapted to several machine tool configurations.

Computer numerical control machines have a minicomputer installed in the control unit. When the control unit is connected to the machine tool, the interface is completed by programming the computer—not by hard-wiring. Thus computer numerical control is referred to as soft-wired control. The large memory of the minicomputer permits addition of many features to the controls.

With direct numerical control, a number of numerical control or computer numerical control machines are connected together to a remote computer. The number of machines connected could be as small as 2 or as great as 256. A big advantage offered by direct numerical control is that the communication between the machine control unit and the computer is two-way. That is, the machine operator can ask for specific part information or communicate with other departments easily and quickly. Thus, quick corrective action can greatly expedite machining operations and reduce machine downtime.

The growth of the direct numerical control hierarchy of machine tools and controls led naturally to complete computer-aided manufacturing (CAM). In direct numerical control the computer can supply information to several machines for individual operation. Computer-aided manufacturing goes a step further in that all pertinent information is contained in the computer, so that it serves as a communication network for the number of pieces machined at any given station, the amount of uptime and downtime, the production rate, and any other pertinent management data. The incoming raw material undergoes succeeding steps of inspection, premachining and so on until it reaches the material inventory and pallet-loading station. As each job order is received from the computer, the pallets are set up with the blanks, cutting tools, and other information. The pallets are coded so that as they go down the line, their paths are controlled by the computer. Depending upon the priority (set by the manager via the computer), the pallets move in queues. The queues may be stacked vertically to take advantage of plant space. The main computer is notified when the individual operations, parts, job lots, or inspections are completed. With input from shop personnel, the main computer can produce an up-to-the-minute status report of everything in the production area. *See* COMPUTER-AIDED DESIGN AND MANUFACTURING; CONTROL SYSTEMS. [R.A.Li.]

Tooth One of the structures found in the mouth of most vertebrates which, in their most primitive form, were conical and were usually used for seizing, cutting up, or chewing food, or for all three of these purposes. The basic tissues that make up the vertebrate tooth are enamel, dentin, cementum, and pulp (see illustration).

Enamel is the hardest tissue in the body because of the very high concentration, about 96%, of mineral salts. The remaining 4% is water and organic matter. The enamel has no nerve supply, although it is nourished to a very slight degree from the dentin it surrounds. The fine, microscopic hexagonal rods (prisms) of apatite which make up the enamel are held together by a cementing substance.

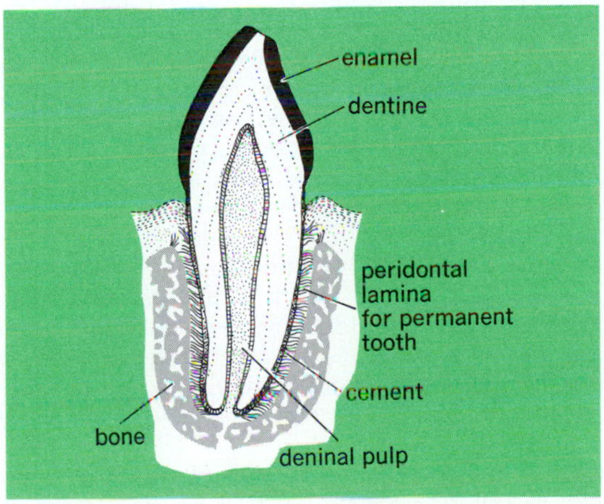

Structure of a tooth.

Dentin, a very bonelike tissue, makes up the bulk of a tooth, consisting of 70% of such inorganic material as calcium and phosphorus, and 30% of water and organic matter, principally collagen. The rich nerve supply makes dentin a highly sensitive tissue; this sensitivity serves no obvious physiological function.

Cement is a calcified tissue, a type of modified bone less hard than dentin, which fastens the roots of teeth to the alveolus, the bony socket into which the tooth is implanted. A miscellaneous tissue, consisting of nerves, fibrous tissue, lymph, and blood vessels, known as the pulp, occupies the cavity of the tooth surrounded by dentin.

The dentition of therian mammals, at least primitively, consists of four different kinds of teeth. The incisors (I) are usually used for nipping and grasping; the canines (C) serve for stabbing or piercing; the premolars (Pm) grasp, slice, or function as additional molars; and the molars (M) do the chewing, cutting, and grinding of the food. Primitively the placentals have 40 teeth and the marsupials 50.

In therian mammals, probably because of the intricacies and vital importance of tooth occlusion, only part of the first (or "milk") dentition is replaced. This second, or permanent, dentition is made up of incisors, canines, and premolars; as a rule only one premolar is replaced in marsupials. Although the molars erupt late in development and are permanent, that is, not replaced, they are part of the first, or deciduous, dentition. *See* TOOTH DISORDERS. [F.S.S.]

Tooth disorders Diseases and disturbances of the teeth and associated structures, including abnormal formation and growth of the teeth and jaws, tooth decay, inflammation of the tissues housing the roots of the teeth, and various diseases of the jaw bones.

Defective formation of dentin and enamel are referred to respectively as dentinogenesis and amelogenesis imperfecta, and may be caused by febrile illness during the period of tooth formation, or by faulty calcium or phosphorus metabolism, such as occurs in rickets. Some or all of the teeth may fail to form completely, or extra or supernumerary teeth may be present.

In most societies, dental decay, or caries, is one of the most important and common tooth disorders. Decay occurring during the adolescent years is caused by a class of microorganisms referred to as the cariogenic streptococci, of which *Streptococcus mutans* is a predominant member. However, a susceptible host and a cariogenic diet containing sucrose are also essential factors. Although the mechanism by which bacteria cause decay is not completely understood, most experts believe that cariogenic organisms, by using sucrose, and to a lesser extent other sugars, produce polymers which bind the organisms to the tooth surface and acids which cause demineralization resulting in cavity formation. Fluoride administration, either in the drinking water or by other routes, effectively reduces dental decay about 50–70%. The mechanism by which fluoride causes decreased decay rates is not understood.

Whereas dental decay is the principal cause of tooth loss in young individuals, inflammatory disease of the tissues surrounding the teeth, referred to as periodontal disease, causes most of the tooth loss in adults. *See* PERIODONTAL DISEASE.

Diseases of the jaws may affect the teeth. The most common jaw disorders fall into four categories: (1) inflammation of the jawbone caused by infections such as osteomyelitis; (2) cysts associated with the teeth as well as those located in bone sutures of the jaws; (3) benign and malignant tumors of the jaws; and (4) systemic diseases such as generalized skeletal abnormalities produced by endocrine disfunction in which the jaws are affected. *See* DENTISTRY; TOOTH. [R.C.P.]

Topaz A mineral best known for its use as a gemstone. Crystals are usually colorless but may be red, yellow, green, blue, or brown. The wine-yellow variety is the one usually cut and most highly prized as a gem. Citrine, a yellow variety of quartz, is the most common substitute and may be sold as quartz topaz.

Topaz is a nesosilicate with chemical composition $Al_2SiO_4(F,OH)$. The mineral is commonly found in well-developed prismatic crystals with pyramidal terminations. Hardness is 8 on Mohs scale; specific gravity is 3.4–3.6. *See* SILICATE MINERALS.

Fine yellow and blue crystals of topaz have come from Siberia and much of the wine-yellow gem material from Minas Gerais, Brazil. In the United States topaz has been found near Florissant, Colorado; Thomas Range, Utah; San Diego County, California; and Topsham, Maine. *See* GEM. [C.S.Hu.]

Tophus A localized swelling occurring specifically in gout. Tophi occur in cartilage and in the connective tissues of the body, usually in or adjacent to the small joints of the hands and feet. Less commonly they are found in the eyelids, ears, muscles, tendons, or heart valves.

The tophus begins as a small deposit of uric acid salts, the irritant properties of which result in a localized region of inflammation characterized by exudation of white blood cells and serum. Once begun the process proceeds slowly but inexorably, gradually eroding the surrounding tissues. The skin overlying a tophus may become ulcerated, discharging semifluid material and chalky fragments of urate deposits, or the lesion may erode into an underlying joint, discharging its contents into the joint cavity. In such cases, the resultant scar formation usually leads to almost complete loss of function in the involved joint. *See* GOUT. [W.R.A.]

Topographic surveying and mapping The measurement of surface features and configuration of an area or region, and the graphic expression of those features. The object of topographic surveying usually is a map of prescribed utility, as for a building site, a highway location, or military operations.

The choice between ground and aerial methods is largely economic. Substantial areas and difficult terrain are surveyed with less cost by aerial surveys. Smaller areas, requiring but one or a few days of field-party time, may be surveyed and mapped for less money than the aerial-survey mobilization costs of flying and photo processing. Visibility of the ground and other features to be mapped also influence the choice. *See* AERIAL PHOTOGRAPH; CARTOGRAPHY; PHOTOGRAMMETRY; SURVEYING. [R.H.Do.]

Topology The study of topological spaces and continuous maps. Precise mathematical definitions of topological space and continuous map require no more than elementary set theory.

Sets. A set is any collection of things or objects. If X is a set and x is a member of the set X, one writes $x \in X$ which reads, in words, x belongs to X. The symbol $\in$ denotes membership. An element or member of a set X is also called a point of X. The rational numbers form a set frequently denoted by Q, and $x \in Q$ means that x is a rational number.

If X and Y are sets, then $X \cup Y$ is the set of elements which belong to either X or Y (including those elements which belong to both X and Y). The notation $X \cup Y$ is read X union Y. Similarly if X and Y are sets, then $X \cap Y$ is the set of elements which belong to both X and Y, and the notation $X \cap Y$ is read X intersect Y.

Suppose that for each element i of a set I there is given a set A_i, the set consisting of all the sets A_i is denoted by $\{A_i\}_{i \in I}$, and is said to be a collection of sets indexed on the set I. If $\{A_i\}_{i \in I}$ is a collection of sets indexed on the set I, then $\cup_{i \in I} Ai$ is the set consisting of those elements which belong to at least one of the sets A_i, and is called the union of $\{A_i\}_{i \in I}$. Similarly $\cap_{i \in I} A_i$ is the set consisting of those elements which belong to every one of the sets A_i, and is called the intersection of $\{A_i\}_{i\in I}$. *See* SET THEORY.

Topological space. This is a set of points X, together with a collection of subsets of X called open subsets of X where the following assumptions are made:

1. ϕ and X are open subsets of X.
2. If A and B are open subsets of X, then $A \cap B$ is an open subset of X.
3. If $\{A_i\}_{i\in I}$ is a collection of open subsets of X, then $\cup_{i\in I} A_i$ is an open subset of X.

Suppose that R is the set of real numbers. If $r \in R$, in other words, if r is a real number, let $|r|$ denote the absolute value of r. This means that, if r is greater than or equal to zero, then $|r| = r$, but if r is less than zero then $|r| = -r$. In order to make the real numbers into a topological space, open subsets of R are defined using the notion of absolute value. Precisely, a subset U of R is open if for every $x \in U$ there is a real number ϵ_x greater than zero having the property that if for some real numbers y, $|y - x| < \epsilon_x$ then $y \in U$. In other words, if $x \in U$, then any real number sufficiently close to x belongs to U also. With this definition of open subset of the real numbers it is not difficult to verify that the axioms for a topological space are satisfied. When one talks of the real numbers in mathematics, one usually means the real numbers as a topological space, that is to say, the set of real numbers together with the collection of open subsets that are defined above.

Continuous map. In order to continue the discussion of topology, it is necessary to introduce some further notions of set theory. First, suppose that X and Y are sets. Define the product of the sets X and Y to be the set consisting of pairs of elements (x, y) such that $x \in X$ and $y \in Y$. The product of X and Y is denoted by $X \times Y$. Now a function f from X to Y is defined to be a subset f of the product $X \times Y$ such that, if $x \in X$, there exists a unique $y \in Y$ such that $(x,y) \in f$. In this case y is said to be the value of f at x, and is denoted by $f(x)$. Intuitively a function from X to Y is thought of as a rule which assigns to each element of the set X an element of the set Y. The standard mathematical notation for a function f from X to Y is $f{:}X \to Y$.

Let $f{:}X \to Y$ be a function and suppose U is a subset of Y, then $f^{-1}(U)$ is the subset of X consisting of those points x such that $f(x) \to U$. If X and Y are topological spaces, then $f{:}X \to Y$ is a continuous map, or continuous function, (1) if f is a function from X to Y, and (2) if, whenever U is an open subset of Y, the set $f^{-1}(U)$ is an open subset of X.

The topological spaces X and Y are said to be homeomorphic if there exist continuous maps $f{:}X \to Y$ and $g{:}Y \to X$ such that $f \circ g{:}Y \to Y$ is the identity map of Y and $g \circ f{:}X \to X$ is the identity map of X. In this case the maps f and g are said to be homeomorphisms, and g is frequently denoted by f^{-1}. The symbol f^{-1} is read f inverse.

Metrics. When the topology on the real numbers was defined, this was a special case of defining a topology by using a metric. Let X be a set. A metric on X is a function $\rho{:}X \times X \to R$ such that the following axioms obtain:

1. If x and x' belong to X, then $\rho(x,x')$ is greater than or equal to zero, and $\rho(x,x') = 0$ if and only if $x = X'$.
2. If x and x' belong to X, then $\rho(x,x') = \rho(x',x)$.
3. If x,x', and x'' belong to X, then $\rho(x,x'')$ is less than or equal to, $\rho(x',x') + \rho(x',x'')$.

If $\rho{:}X \times X \to X$ is a metric, then $\rho(x,x')$ is called the distance from x to x'. Axiom 1 says that the distance between any two points of X is greater than or equal to zero, and is different from zero if the points are different. Axiom 2 says that the distance from x to x' is the same as the distance from x' to x. Axiom 3, the so-called triangle axiom, says intuitively that it is shorter to proceed from x to x'' along a straight line than it is first to proceed from x to x' along a straight line and then proceed from x' to x'' along another straight line.

Construction of topological spaces. Two topological spaces which are homeomorphic cannot be distinguished by the methods of topology. Any topological property of one is also a topological property of the other. Since homeomorphic spaces cannot be distinguished by topological methods, one of the important problems of topology is to determine whether two topological spaces are homeomorphic or not.

The general notion of topological space is not sufficiently restrictive for most purposes. Therefore some additional axioms are almost always assumed—in particular the Hausdorff separation axiom. A topological space X satisfies the Hausdorff separation axiom, or is a Hausdorff space, if for every two distinct points x and x' of X there are open subsets U and V of X such that $x \in U$, $x' \in V$, and $U \cap V$ is empty.

Manifolds. Though Hausdorff spaces form a much more interesting class of spaces than general topological spaces, the most important class of topological spaces, which is still much smaller, consists of the manifolds. An n-manifold is a Hausdorff space X such that the following conditions exist:

1. For every point $X \in X$ there is an open subset U_x of X such that $x \in U_x$ and such that U_x is homeomorphic with an open subset of R^n.
2. There exists a metric $\rho{:}X \times X \to R$ such that the topology on X is induced by the metric.

A topological space X is connected if it cannot be expressed as the union of two disjoint nonempty open subsets. One of the most important problems of topology is the problem of classification of connected n-manifolds. *See* MANIFOLD (MATHEMATICS).

[J.C.Mo.]

Torbanite A variety of coat that resembles a carbonaceous shale in outward appearance. It is fine-grained, brown to black, and tough, and breaks with a conchoidal or subconchoidal fracture. Torbanite is synonymous with boghead coal and is related to cannel coal. It is derived from colonial algae identified with the modern species of *Botryococcus braunii* and antecedent forms. High-assay torbanite yields paraffinic oil, whereas low-assay material yields asphaltic oil. *See* COAL. [I.A.B.]

Torch A gas-mixing and burning tool that produces a hot flame for the welding or cutting of metal. The torch usually delivers acetylene and commercially pure oxygen producing a flame temperature of 5000–6000°F (2750–3300°C), sufficient to melt the metal locally. The torch thoroughly mixes the two gases and permits adjustment and regulation of the flame. Acetylene can produce a higher flame temperature than other fuel gases. *See* ACETYLENE; WELDING AND CUTTING OF METALS.

Torches are of two types: low-pressure and high-pressure. In a low-pressure, or injector, torch, acetylene enters a mixing chamber, where it meets a jet of high-pressure oxygen. The amount of acetylene drawn into the flame is controlled by the velocity of this oxygen jet. In a high-pressure torch both gases are delivered under pressure.

A welding torch mixes the fuel and gas internally and well ahead of the flame. For cutting, the torch delivers an additional jet of pure oxygen to the center of the flame. The oxyacetylene flame produced by the internally mixed gases raises the metal to its ignition temperature. The central oxygen jet oxidizes the metal, the oxide being blown away by the velocity of the gas jet to leave a narrow slit or kerf. [F.H.R.]

Tornado A violently rotating, tall columnar vortex typically 30–330 ft (10–100 m) in diameter. It is characterized by a debris cloud near the ground and a funnel cloud composed of small water droplets pendant from a parent cloud system of cumulus type. The visible funnel cloud need not reach all the way to the ground and is often obscured by debris. The tornado most often develops in association with a severe thunderstorm. Heavy rain and large hailstones often fall just to the north of the tornado's path in the Northern Hemisphere. *See* Thunderstorm.

There is evidence of a characteristic life cycle, especially for large, destructive tornadoes. The tornado life cycle typically consists of five parts: dust-whirl stage (dust swirls upward from the surface, or a short pendant funnel may appear above the ground); organizing stage (the visible funnel intermittently touches the ground with a continuous damage path); mature stage (the tornado is at its largest size and most intense circulation); shrinking stage (the entire funnel decreases to a thin, ropelike column); and decaying stage (the funnel is fragmented and contorted). Even in its final stages, the tornado may retain its destructive power.

Weather conditions that favor tornado formation include abundant moisture in the lowest 0.6 mi (1 km) or so, separated from a dry air mass at intermediate levels by a stable layer or inversion, deep thermodynamic instability, a mechanism to locally remove the stable layer, and winds that veer and increase with height. [J.Go.]

Torque The product of a force and its perpendicular distance to a point of turning, also called the moment of the force. Torque produces torsion and tends to produce rotation. Torque arises from a force or forces acting tangentially to a cylinder or from any force or force system acting about a point. A couple, consisting of two equal, parallel, and oppositely directed forces, produces a torque or moment about the central point. A prime mover such as a turbine exerts a twisting effort on its output shafts, measured as torque. In structures, torque appears as the sum of moments of torsional shear forces acting on a transverse section of a shaft or beam. *See* Couple; Torsion. [N.S.F.]

Torque converter A device for changing the torque-speed ratio or mechanical advantage between an input shaft and an output shaft. A pair of gears is a mechanical torque converter. A hydraulic torque converter is an automatically and continuously variable torque converter, in contrast to a gear shift, whose torque ratio is changed in steps by an external control. *See* Automotive transmission; Gear drive.

A mechanical torque converter transmits power with only incidental losses; thus, the power, which is the product of torque T and rotational speed N, at input I is substantially equal to the power at output O of a mechanical torque converter, or $T_I N_I = kT_O N_O$, where k is the efficiency of the gear train. This equal-power characteristic is in contrast to that of a fluid coupling in which input and output torques are equal during steady-state operations. *See* Fluid coupling.

In a hydraulic torque converter, efficiency depends intimately on the angles at which the fluid enters and leaves the blades of the several parts. Because these angles change appreciably over the operating range, k varies, being by definition zero when the output is stalled, although output torque at stall may be three times engine torque for a single-stage converter and five times engine torque for a three-stage converter. Depending on its input absorption characteristics, the hydraulic torque converter tends to pull down the engine speed toward the speed at which the engine develops maximum torque when the load pulls down the converter output speed toward stall. [H.J.Wir.]

Torricelli's theorem The speed of efflux of a liquid from an opening in a reservoir equals the speed that the liquid would acquire if allowed to fall from rest from the surface of the reservoir to the opening.

Torricelli, a student of Galileo, observed this relationship in 1643. In equation form, $v^2 = 2gh$, in which v is the speed of efflux, h the head (or elevation difference between reservoir surface and center line of opening if in a vertical plane), and g the acceleration due to gravity. (The equation is the same as that for a solid particle dropped a distance h in a vacuum.) The relationship can be derived from the energy equation for flow along a streamline, if energy losses are neglected. *See* Flow measurement. [V.L.S.]

Torsion A straining action produced by couples that act normal to the axis of a member. Torsion is identified by a twisting deformation.

In practice, torsion is often accompanied by bending or axial thrust as in the case of line shafting driving gears or pulleys, or propeller shafts for ship propulsion. Other important examples include springs and machine mechanisms usually having circular sections, either solid or tubular. Members with noncircular sections are of interest in special applications, such as structural members subjected to unsymmetrical bending loads that twist and buckle beams. *See* Spring (machines); Torsion bar.

When subjected only to torque, the member is in pure torsion, which produces pure shear stresses. The shear properties of materials are determined by a torsion test. *See* Shear; Torque. [J.B.S.]

Torsion bar A spring flexed by twisting about its axis. Design of a torsion bar spring is primarily based on the relationships between the torque applied in twisting the spring, the angle through which the torsion bar twists, and the physical dimensions and material (modulus of elasticity in shear) from which the torsion bar is made. The illustration shows the elements of a simple torsion bar and the important dimensions involved in its design. The equation relates these dimensions.

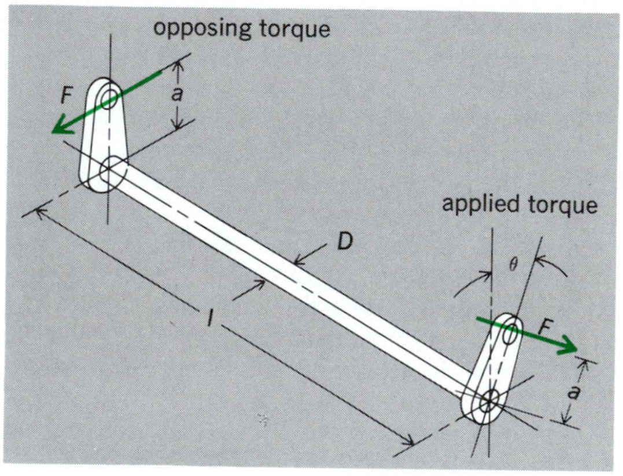

Diagram of torsion bar.

$$\theta = \frac{32Fal}{\pi D^4 G}$$

Here θ is angle of twist in radians, F is force in pounds, a is radius arm of force in inches, l is length of torsion bar in inches, D is diameter of torsion bar in inches, and G is modulus of elasticity in shear in pounds per square inch.

Torsion bar springs are found in the spring suspension of truck and passenger car wheels, in production machines where space limitations are critical, and in high-speed mechanisms where inertia forces must be minimized. *See* Spring (machines). [L.S.L.]

Torus A surface obtained by rotating a circle about a line that lies in its plane, but which has no points in common (see illustration). It is a two-dimensional manifold of genus 1 and

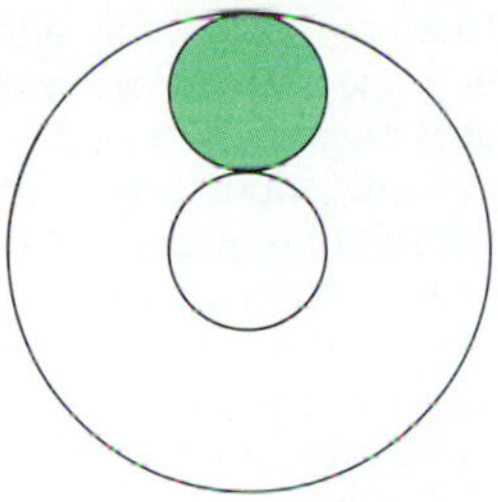

Diagram of a torus.

connectivity 3. *See* MANIFOLD (MATHEMATICS); SURFACE AND SOLID OF REVOLUTION; TOPOLOGY. [L.M.Bl.]

Touch A term for both the generic designation for general bodily feeling and, more narrowly, the array of skin sensations ranging from contact to dull pressure and including light touch, granular pressure, and a host of other cutaneous patterns for which specific names have not entered the English language. The system of pressure sensitivity of the skin yields a wealth of feeling patterns which differ among themselves along the spatial, temporal, and intensitive dimensions. They have also been held to vary within a brightness-dullness continuum, by analogy with visual sensations. For certain of the patterns which occur repeatedly and familiarly, there are terms such as lively contact, tickle, vibration, and deep pressure. Others, though recognizable when felt, are nameless and thus impossible to describe. *See* CUTANEOUS SENSATION; SOMESTHESIS. [F.A.G.]

Tourette's syndrome A neurobehavioral disorder characterized by frequent, recurrent motor and vocal tics. The motor tics include brief, rapid, and darting movements of almost any muscle group, and can include eye blinking, eye rolling or deviations, nose wrinkling, facial grimacing, and head shaking. Some motor tics are more complex, are slow, and appear purposeful such as head turning, shoulder shrugging, touching, hopping, or twirling. Vocal tics are brief guttural sounds such as recurrent sniffing, throat clearing, coughing, and grunting or barking sounds. Complex vocal tics can be more meaningful and include verbal expressions. Tourette's syndrome has been described in nearly every country and ethnic group, with an estimated prevalence of one or two occurrences per 2000 people.

The motor and vocal tics begin in childhood, often worsen during adolescence, and tend to improve during the twenties and thirties. Symptoms increase with stress and excitement and decrease with activities that require focused effort. While the motor and vocal tics are involuntary, they can be suppressed for brief periods of time, giving the false impression that the movements and sounds are voluntary.

The pattern of inheritance is consistent with a single autosomal dominant gene whose expression is variable and dependent on the sex of the person. Tic symptoms can vary from transient tics to Tourette's syndrome and can include obsessive-compulsive symptoms. The complexity of symptoms is likely related to the various brain regions implicated in the development of Tourette's syndrome. Treatment can be targeted toward suppressing tics and the specific associated behavioral problems. Methods for tic suppression include medications that affect the brain by blocking the neurotransmitter dopamine at the site of nerve-to-nerve connections. *See* BRAIN; HUMAN GENETICS; NERVOUS SYSTEM DISORDERS. [J.T.W.; M.A.Ri.]

Tourmaline A cyclosilicate mineral family with (BO_3) triangular groups and a complex chemical composition. The general formula can be written $XY_3Al_6(OH)_4(BO_3)_3(Si_6O_{18})$, in which X = Na, Ca, and Y = Al, Fe^{3+}, Li, Mg, Mn^{2+}. The more common tourmalines are dravite, schorl, uvite, and elbaite. Fluorine commonly substitutes in the hydroxyl position. Tourmaline is a hard (7½ on Mohs scale), varicolored mineral which can be an important semiprecious gemstone. *See* GEM; SILICATE MINERALS. [P.B.M.]

Tower A concrete, metal, or timber structure that is relatively high for its length and width. Towers are constructed for many purposes, including the support of electric power transmission lines, radio and television antennas, and rockets and missiles prior to launching.

Transmission towers are rectangular in plan and are not steadied by guy wires. A transmission tower is subjected to a number of forces; its own weight, the pull of the cables at the top of the tower, the effect of wind and ice on the cable, and the effect of wind on the tower itself.

Radio and television towers are either guyed or freestanding. Freestanding towers are usually rectangular in plan. In addition to their own weight, freestanding towers support the weight of the antenna and accessories and the weight of ice, unless a deicing circuit is installed. Wind forces must also be carefully considered. Guyed towers are usually triangular in plan, with the main structural members, or legs, at the vertexes of the triangle. The legs are usually solid round steel bars. *See* ANTENNA (ELECTROMAGNETISM); TRANSMISSION LINES. [C.M.A.]

Towing tank A tank of water used to determine the hydrodynamic performance of waterborne bodies such as ships and submarines, as well as torpedoes and other underwater forms. In the narrow sense, towing tanks are considered to be experimental facilities used to measure the forces, such as drag, on ship models and in turn to predict the performance of the full-scale prototype. In general, towing tanks are rectangular in planform with a uniform cross section. Different section shapes are used, ranging from rectangular to semicircular.

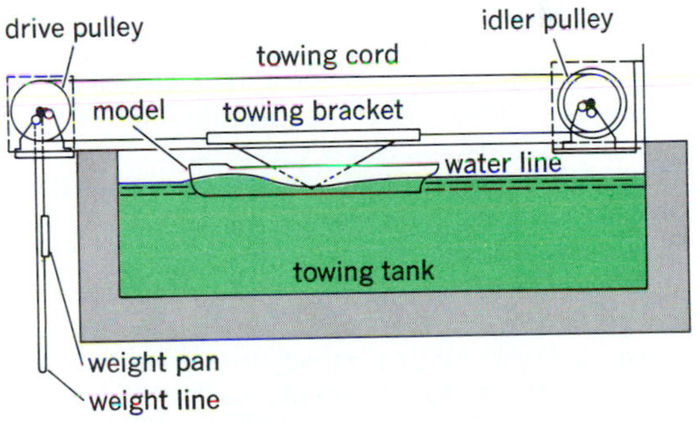

Tank with model towed by falling weight.

The principal measurements made in a towing tank are force measurements, particularly drag or resistance of a towed ship model or other body. One of two principal systems for towing a model is used in most towing tanks. The simpler system consists of a gravity dynamometer and an endless cable attached to the model. A weight provides a constant towing force (see illustration). The time to traverse a fixed distance is measured when the model reaches a constant speed, thus establishing the speed-resistance relationship for the model. This dynamometer is simple and capable of high accuracy, but is limited to the measurement of the drag force of waterborne bodies. It is used in the smaller towing tanks in which the models are generally under 6 ft (1.8 m) in length.

In larger towing tanks the model is towed by a towing carriage mounted on rails at the side of the towing tank or suspended from an overhead track system. Speed can be controlled and measured precisely on these carriages. Most carriages are equipped with a drag dynamometer as a permanent component. [J.B.H.]

Townsend discharge A particular part of the voltage-current characteristic curve for a gaseous discharge device named for J. S. Townsend, who studied it about 1900. It is that part for low current where the discharge cannot be maintained by the field alone. Thus, if the agents producing the initial ionization were removed, conduction would cease. *See* ELECTRICAL CONDUCTION IN GASES. [G.H.M.]

Toxemia A condition in which toxins are found circulating in the blood stream; also, a disease peculiar to pregnancy. Toxins can cause generalized tissue damage, with symptoms such as malaise, weakness, fever, aching, and diaphoresis, or they can specifically injure certain organ systems such as the gastrointestinal tract, nervous tissue, or hematopoietic system. Specific symptoms occur in response to damage to each of these areas, and in some cases this damage is irreversible. *See* BLOOD; ENZYME; FOOD POISONING.

In pregnancy this disease is characterized by acute hypertension after the twenty-fourth week of gestation, accompanied by proteinuria and edema of hands and face (also called preeclampsia-eclampsia). This disease occurs in 6–7% of pregnancies, ranges from mild to severe, and may threaten the life of both fetus and mother. The cause of this condition is unknown, but it involves alterations in the maternal salt-and-water balance, generalized vasoconstriction, compromised kidney function, and increased central nervous system irritability. *See* PREGNANCY. [G.Gu.]

Toxic shock syndrome A potentially life-threatening illness caused by a toxin produced by the bacterium *Staphylococcus aureus*. Toxic shock syndrome is characterized by fever, low blood pressure, nausea and vomiting, diarrhea, muscle tenderness, and a sunburnlike rash that peels during recovery.

Although toxic shock syndrome was first recognized in 1978, it was not until 1980 that it attracted widespread attention because epidemiologic studies showed that more than 95% of cases of toxic shock syndrome were occurring in young women and that the illness was associated with menstruation. Subsequent investigations demonstrated that menstrually associated toxic shock syndrome was related to the use of tampons. However, although approximately 90% of all toxic shock syndrome cases are menstrually associated, several hundred nonmenstrually related cases have been recognized. In each of those cases, a staphylococcal infection at some body site has been found as the source. A major cause of nonmenstrual toxic shock syndrome has been staphylococcal infection of surgical wounds. *See* STAPHYLOCOCCUS. [B.D.]

Toxicology The study of the harmful effects of chemicals and physical agents on living systems. It includes the identification of the stimulating agent and the mechanisms underlying its interactions with biotic systems, the prevention and treatment of negative responses, and in some cases, the recognition of advantageous uses of these interactions. Toxicology is a multidisciplinary field. It includes the use of chemistry, especially biochemistry, and of various branches of the biological sciences, such as physiology, histology, and pathology. Because of the ever-expanding use and manipulation of the environment, there is an increasing demand for the type of knowledge that a toxicologist can provide.

Three major divisions of toxicology are environmental toxicology, clinical toxicology, and forensic toxicology. With increasing populations and industrialization, environmental toxicology has assumed great importance. Among the points of concern to workers in this area are agents which contaminate or are purposefully added to food, air, water, soil, and the working place. Clinical toxicology is a specialized field in medicine and is concerned with the clinical management of diseases that are associated with chemicals of exogenous origin. These diseases include those produced as a result of accidental or purposeful drug overdose, adverse drug interactions, poisoning which is job-related, and toxic responses to naturally occurring agents.

Forensic toxicology involves the medicolegal aspects of a chemical's effects on humans primarily. It has been involved in the determination of the cause of death in order to understand the circumstances leading to the event. It has also played an important role in the identification of specific hazards related to various chemicals. *See* ENVIRONMENTAL PROTECTION; FORENSIC MEDICINE; POISON. [N.K.M.; C.Qu.]

Toxin-antitoxin reaction A term used in serology to denote the combination of a toxic antigen with its corresponding antitoxin. If the antitoxin is derived from any species other than the horse, precipitation occurs over a wide range of reactant ratios, 20 × or more, as in other antigen-antibody reactions. With horse antitoxin, flocculation occurs only if toxin and antitoxin are near equivalence, a twofold excess of either reactant giving soluble complexes. In most instances, the reaction results in partial or complete neutralization of the toxic activity of the antigen. *See* ANTIBODY; ANTIGEN; ANTITOXIN; NEUTRALIZATION REACTION (IMMUNOLOGY); SEROLOGY. [H.P.T.]

Toxoplasmea A class of the subphylum Sporozoa. The organisms are small and crescent-shaped. They move by body flection or gliding and have no flagella or pseudopodia. Characteristic structures are the two-layered pellicle and underlying longitudinal microtubules, micropyle, a conoid, paired organelles, and micronemes.

The most distinguishing characteristic of the Toxoplasmea is the unique means of reproduction. Electron microscope studies indicate that endodyogeny is the sole method. Endodyogeny is an internal budding wherein two daughter cells are produced within a mother cell which is destroyed in the process.

Only two stages are known in the life cycle of most animals. One stage, the proliferative form or trophozoite, occurs singly or in groups within host cells. The other stage, the so-called cysts, consists of a large number of organisms which, with minor differences, are structurally similar to the proliferative forms. *See* SPOROZOA; TOXOPLASMIDA. [H.G.Sh.]

Toxoplasmida An order of the class Toxoplasmea. Four genera, *Toxoplasma*, *Besnoitia*, *Sarcocystis*, and *Encephalitozoon*, make up the order. The organisms are parasites of vertebrates. *Toxoplasma* is often found encysted in nerve tissue, *Besnoitia* in connective tissue, and *Sarcocystis* in muscle. Very little is known about *Encephalitozoon*, but the parasite has been found in the brain of rabbits. *See* TOXOPLASMIDA; TOXOPLASMOSIS. [H.G.Sh.]

Toxoplasmosis An infection of humans and animals by the protozoan *Toxoplasma gondii*. The actively multiplying tachyzoites of *Toxoplasma* are found during acute infection in mammals and birds. Tachyzoites multiply in cells and many cells may be destroyed, resulting in illness. After a number of tachyzoite cycles, cysts develop which persist for months or years. After a cat ingests tissues of an infected animal, at least five multiplicative stages are found in the intestinal epithelium, and then sexual stages appear. Oocysts are shed in the feces of the cat, after which they undergo sporogony. These sporulated oocysts are now infectious to many mammals and birds, which are called intermediary hosts. Cats are the final host since they support the sexual cycle.

Most *Toxoplasma* infections in humans are asymptomatic. The rare clinical symptoms can be characterized according to the stage of infection during which they occur. Acute infection may be associated with pneumonia, hepatitis, a rash, and lymph node inflammation. In these instances, lesions may result from the destruction of many parasitized cells.

Apart from the clear-cut transmission pattern that can be estimated statistically, situations exist that should be avoided since they increase the chance of transmission to the individual. The precautions include the disposal of feces from cats that hunt birds and mice or are fed raw meat. If a litterbox is used, it should be changed and disinfected daily. Hands should be washed after contact with cats and with soil potentially contaminated with cat feces and after handling raw meat. *See* MEDICAL PARASITOLOGY. [J.K.F.]

Trace analysis The determination of the elemental constituents of a sample in which these constituents make up approximately 0.01% of the sample or less. There is no sharp boundary between nontrace and trace constituents. The lower limit is set by the sensitivity of the available analytical methods and, in general, is pushed downward with progress in analytical techniques. A large number of different physical and chemical techniques have been developed for the measurement of the elemental composition at the microgram-per-gram and nanogram-per-gram level, thereby constituting the field of trace analysis.

Included in the goals of trace analysis are the determination of the bulk or total concentrations of the trace elements; the preconcentration of microconstituents into a small, essentially matrix-free sample to improve detectability or remove matrix interferences; the determination of local concentrations using probe techniques in order to establish the topographical distribution of trace elements in solid samples; and the determination of major and minor species in a minute initial sample.

Methods having applicability to trace analysis range from the more classical chemical methods of colorimetric and absorption spectrophotometric analysis to modern instrumental approaches. Of the various criteria used in the selection of an appropriate trace analytical method, sensitivity, accuracy and precision, and selectivity are of prime importance. Other important considerations, such as scope, sampling and standards requirements, cost of equipment, and time of analyses, are of great practical significance.

The application of any analysis technique to trace problems is governed to a large extent by the analytical sensitivity that can be achieved for the species of interest in a given material. Sensitivity reflects the ability to discern a small change in concentration or amount of species of interest. Detection limit is a closely related, but distinct, term used to indicate the lowest concentration or amount that can be determined with a specific degree of confidence. As a result of widely varying pathways of development of many analytical techniques, there is a lack of consistency in the definition of specification of detection limits in the analytical literature. A summary of the experimental values that have been published is given in the table. *See* ACTIVATION ANALYSIS; ANALYTICAL CHEMISTRY; CALIBRATION; CHROMATOGRAPHY; ELECTROCHEMICAL TECHNIQUES; GAS CHROMATOGRAPHY; SPECTROSCOPY. [A.T.Z.]

Trace analysis methods and limits of detection

Method	Limits of detection: Absolute, g	Limits of detection: Concentration, ppm
Chromatography		
Thin-layer	10^{-5} to 10^{-3}	
Gas-liquid		10^{1} to 10^{6}
Liquid-liquid		10^{-3} to 10^{0}
Electrochemical		
Coulometry	10^{-9} to 10^{0}	
Ion-selective electrode		10^{-2} to 10^{2}
Polarography		
Conventional		10^{0} to 10^{3}
Modern		10^{-3} to 10^{3}
Laser probe microanalysis		10^{2} to 10^{4}
Nuclear		
Neutron activation		10^{-3} to 10^{-1}
Electron probe		10^{2} to 10^{3}
Ion probe		10^{-1} to 10^{1}
Mass spectrometry		
Isotope dilution		10^{-5} to 10^{6}
Spark source		10^{-3} to 10^{1}
Organic microanalysis	$>10^{-5}$	
Optical-absorption		
Atomic		
Flame		10^{-3} to 10^{1}
Nonflame	10^{-15} to 10^{-9}	
Molecular		
UV-visible		10^{-3} to 10^{2}
Infrared		10^{3} to 10^{6}
Microwave		10^{0} to 10^{3}
Optical-emission		
AC spark		10^{1} to 10^{3}
DC arc		10^{-2} to 10^{2}
Electrical plasma		
DC jet		10^{-4} to 10^{2}
RF-induced		10^{-4} to 10^{2}
Microwave-induced	10^{-9} to 10^{-6}	10^{-2} to 10^{1}
Optical-fluorescence		
Atomic		
Flame		10^{-3} to 10^{2}
Nonflame	10^{-15} to 10^{-9}	
Molecular		10^{-3} to 10^{-1}
Optical-phosphorescence		10^{-3} to 10^{2}
Optical-Raman		10^{0} to 10^{5}
Spectrometric-resonance		
Nuclear magnetic		10^{1} to 10^{5}
Electron spin	10^{-9} to 10^{-6}	
Thermal analysis	10^{-5} to 10^{-4}	
Wet chemistry		
Gravimetry	10^{-3} to 10^{-2}	
Titrimetry		10^{-2} to 10^{4}
X-ray spectrometry		
Auger		10^{3} to 10^{5}
ESCA		10^{3} to 10^{5}
Fluorescence		10^{-1} to 10^{2}
Mössbauer		10^{0} to 10^{3}
Photoelectron		10^{0} to 10^{3}

Trace fossils Tracks, trails, and burrows made by animals and found in ancient sediments such as sandstones, shales, or limestones. Comparable names are ichnofossils, bioturbate structures, and biogenic structures. An assemblage of trace fossils has been called ichnocoenosis.

Very different biological activities of animals produce these biogenic structures. Study of trace fossils is known as ichnology or palichnology and is a part of the science of paleontology.

Fig. 1. Form of plantlike trace fossils, *Chondrites*, representing feeding burrows of Cretaceous age. (*From R. C. Moore, ed., Treatise on Invertebrate Paleontology, pt. W, University of Kansas Press, 1962*)

The investigations of recent tracks and trails, for instance, on the beach of a sea or on tidal flats, help to interpret the fossil ones.

In the early years of paleontology, most of the ramified burrows, creeping trails, or other trace fossils were considered to be remains of marine algae. This is proven by many names ending with "phycus" given to trace fossils. That misinterpretation was caused by the striking similarity of many of these algae to trace fossils (Fig. 1). Trails resembling worms were sometimes assumed to be the bodies of the worms themselves. Other trace fossils, now recognized to represent the pattern of a grazing activity of animals, were interpreted as the spawn of sea snails (Fig. 2). About 1890 thorough and systematic ichnological observations of the traces of modern marine animals convinced most of the paleontologists that these interpretations were erroneous.

Fig. 2. Feeding (grazing) trail, *Cosmorphaphe*, Tertiary, Pologne. (*From R. C. Moore, ed., Treatise on Invertebrate Paleontology, pt. W, University of Kansas Press, 1962*)

Trace fossils occur in the sedimentary rocks of all ages and of marine, limnic, fluviatile, and even continental environments. Sometimes these fossils alone characterize some stratified rocks in which bodily preserved animals such as mollusks (that is, the shells) are lacking. In addition, the trace fossils are important as direct evidence of many animals, even soft-bodied ones, that once lived within or on these deposits but have not been fossilized.

Generally paleontologists use for trace fossils the same binominal nomenclature, with Latin names, as in naming animals and plants scientifically. The first name then represents the ichnogenus, the second the ichnospecies. That system seems the most suitable one. However, it should never be forgotten that "genus" and "species" in ichnology have different meanings from those of the same terms applied to the actual remains of animals and plants. *See* FOSSIL; PALEONTOLOGY. [W.Ha.]

Tracheophyta A large group of plants, recognized in some systems of classification as a single division, characterized by the presence of specialized conducting tissues called xylem and phloem in the roots, stems, and leaves. Because of their characteristic conducting tissues, the Tracheophyta are often called the vascular plants. *See* PHLOEM; XYLEM.

In the system of classification presented in the encyclopedia, the Tracheophyta are not given formal taxonomic recognition; they are considered instead to consist of seven divisions, the Rhyniophyta, Psilotophyta, Lycopodiophyta, Equisetophyta, Polypodiophyta, Pinophyta, and Magnoliophyta. These seven together with the division Bryophyta constitute the subkingdom Embryobionta. See articles on the individual divisions. *See* EMBRYOBIONTA. [A.Cr.]

Trachoma A chronic conjunctivitis of humans. The disease is endemic in many developing countries; it is the world's leading cause of preventable blindness. The severity of the disease is often related to the socioeconomic status of the community. In severely affected communities the entire population acquires the infection, and more than 10% of adults may be blinded as a result of the disease. Trachoma is caused by an obligate intracellular bacterium, *Chlamydia trachomatis*. This organism, in conjunction with secondary bacterial pathogens, can produce a severe conjunctivitis which results in scarring of the conjunctiva (inner lining of the eyelids). The active disease is seen in children. Blindness occurs in adults and results from contraction of the scars, which turn the eyelids and eyelashes inward so that they abrade the cornea. *See* CHLAMYDIAL DISEASES; CONJUNCTIVITIS. [J.S.]

Trachylina An order of jellyfish of the class Hydrozoa of the phylum Coelenterata. These jellyfish are of moderate size. They differ from other hydrozoan jellyfish in having balancing organs which develop partly from the digestive epithelium and in having only a small polyp stage or none at all. Many authorities recognize three distinct orders of trachylines—Limnomedusae, Trachymedusae, and Narcomedusae—and in this case the older term Trachylina is abandoned. *See* HYDROZOA. [S.Cr.]

Trachyte A light-colored, aphanitic (very finely crystalline) rock of volcanic origin, composed largely of alkali feldspar with minor amounts of dark-colored (mafic) minerals (biotite, hornblende, or pyroxene). If sodic plagioclase (oligoclase or andesine) exceeds the quantity of alkali feldspar, the rock is called latite. Trachyte and latite are chemically equivalent to syenite and monzonite, respectively. *See* LATITE; MONZONITE; SYENITE.

Streaked, banded, and fluidal structures due to flowage of the solidifying lava are commonly visible in many trachytes and may be detected by a parallel arrangement of tabular feldspar phenocrysts. A distinctive microscopic feature is trachytic texture in which the tiny, lath-shaped sanidine crystals of the rock matrix are in parallel arrangement and closely packed.

Trachyte is not an abundant rock, but it is widespread. It occurs as flows, tuffs, or small intrusives (dikes and sills). It may be associated with alkali rhyolite, latite, or phonolite. *See* IGNEOUS ROCKS; MAGMA; SPILITE. [C.A.C.]

Tractor A wheeled, self-propelled vehicle for hauling other vehicles or equipment and for operating the towed implements; also, a crawler which runs on an endless, self-laid track and performs similar functions.

The major components are engine, clutch, and transmission. Unlike passenger-car engines, which are of the high-speed type, tractor engines are relatively low-speed; their maximum horsepower is generated at crankshaft speeds in the neighborhood of 2000 revolutions per minute. These engines have one, two, three, four, six, or eight cylinders and operate on gasoline, kerosine, LP gas, or diesel fuel. They are of the spark-ignition or diesel type, operating on the four-stroke-cycle principle, and are cooled by water or air.

A farm tractor is a multipurpose power unit. It has a drawbar for drawing tillage tools and a power takeoff device for driving implements or operating a belt pulley. All models can be grouped under four general types: four-wheel, row-crop or high-wheel, tricycle, and crawler. Power delivered at the place where it is useful is what counts; therefore, tractors are rated by the horsepower they deliver at the drawbar and at the belt.

The basic design of an industrial tractor for hauling and for operating construction equipment departs little from that of a farm tractor, and differences in design of models fit the vehicle to its intended work. [P.H.Sm.]

Tractrix A plane curve for which the length of any tangent between the curve and a fixed line is constant *c* (see illustration).

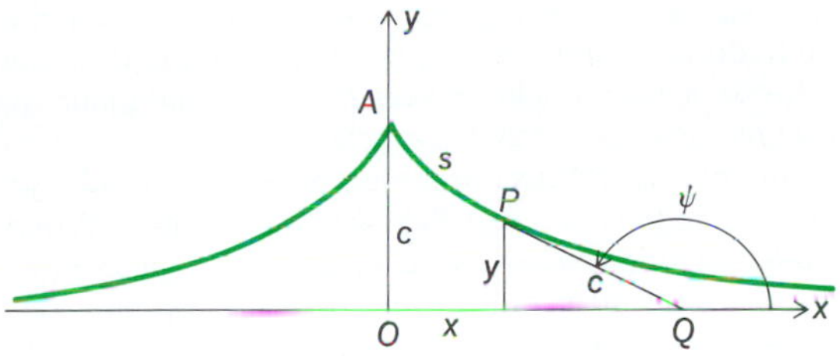

Tractrix (upper half).

If the *x* axis is the fixed line, its differential equation is Eq. (1). With ψ, the inclination of the tangent, as parameter, this yields equations (2). The tractrix has the *x* axis as asymptote and cusps at $(0, \pm c)$. The arc *AP*, measured from a cusp, is as in Eq. (3).

$$(dy/dx)^2 = y^2/(c^2 - y^2) \tag{1}$$

$$x = c \log \tan \tfrac{1}{2}|\psi| + c \cos \psi$$
$$y = c \sin \psi \tag{2}$$

$$s = -c \log \sin \psi = c \log \frac{c}{y} \tag{3}$$

[L.Br.]

Traffic-control systems Systems that act to control the movement of vehicles, such as airplanes, trains, and automobiles. The system of traffic lights used in most urban areas to control automotive traffic is a common example. Traffic-control systems ensure safe, orderly arrival of airplanes at airports during periods of heavy traffic; provide efficient, economical movement of railroad cars from one point to another; and reduce congestion and provide smooth flow of traffic along heavily traveled roads in metropolitan areas. The nature and complexity of a traffic-control system are determined by the responsiveness of the vehicle operator, vehicle type, severity of the traffic problem, and the requirements placed upon the traffic-control system. *See* AIR-TRAFFIC CONTROL; TRAFFIC ENGINEERING; TRANSPORTATION ENGINEERING. [L.J.P.]

Traffic engineering The determination of the required capacity and layout of highway and street facilities that can safely and economically serve vehicular movements between given points.

Major traffic arteries in urban areas are arterial, secondary, and local streets. Arterial routes may be further identified as either major streets or expressways; the latter are also known as freeways. Secondary streets are sometimes described as collectors. Roads in rural areas are usually designated as primary, secondary, or tertiary, depending upon their importance.

In planning improvement in traffic arteries, consideration must be given to many factors, including traffic volumes, desired speeds of travel, vehicular dimensions, driving habits, vehicular performance, type of terrain traversed, and especially expected future changes in all these factors.

Traffic engineering also deals with parking problems. Investigations relating to parking include determinations as to the desired size, type, and location of off-street facilities, plus studies relating to movements into and out of parking areas. *See* HIGHWAY ENGINEERING; TRANSPORTATION ENGINEERING. [A.N.C.]

Trajectory The curve described by a body moving through space, as of a meteor through the atmosphere, a planet around the Sun, a projectile fired from a gun, or a rocket in flight. In general, the trajectory of a body in a gravitational field is a conic section—ellipse, hyperbola, or parabola—depending on the energy of motion. The trajectory of a shell or rocket fired from the ground is a portion of an ellipse with the Earth's center as one focus; however, if the altitude reached is not great, the effect of gravity is essentially constant, and the parabola is a good approximation. *See* BALLISTICS. [J.P.H.]

Tranquilizer A psychopharmacologic drug that tends to have a calming effect, and is unique in inducing drowsiness without impairing ready arousal and in restraining hyperactivity without inducing coma or arrest of respiratory muscles.

According to basic chemical structure, five categories of tranquilizers can be delineated: (1) Phenothiazines (for example, chlorpromazine) and rauwolfia derivatives (for example, reserpine) are major or antipsychotic tranquilizers; (2) propanediols (for example, meprobamate, or Miltown), (3) diphenylmethanes (for example, benactyzine), (4) chlordiazepoxide and derivatives (for example, Librium and Valium), and (5) a miscellaneous group used generally for nighttime sedation (for example, ectylurea, glutethimide, and methylparafynol) are termed minor or sedative antianxiety tranquilizers. The antipsychotic tranquilizers in greatest use are phenothiazines, of which there are many derivatives and trade names. The sedative antianxiety agents are generally used in crisis situations or in neurotic episodes of intense anxiety or panic. Their effectiveness is more difficult to establish and is less specific than that of drugs for the treatment of psychosis. In high dosages such drugs have an addiction liability and can produce convulsions on withdrawals. [D.X.F.]

Transamination The transfer of an amino group from one molecule to another without the intermediate formation of ammonia. Enzymatic reactions of this type play a prominent role in the formation and ultimate breakdown of amino acids by living organisms. Enzymes that catalyze such reactions are widely distributed and are termed transaminases, or aminotransferases. Perhaps the most prominent transamination reactions in higher animals are those in which glutamate is formed from α-ketoglutarate and other amino acids. *See* PROTEIN METABOLISM. [E.E.S.]

Transducer A device which converts energy or information carried by one physical quantity into corresponding values of another quantity. A simple example is a mercury-in-glass thermometer which converts the temperature of the surroundings into the length of a column of mercury. A transducer is to be distinguished from a transformer, which converts one level of a quantity into another level of the same quantity, as in an electrical transformer or a mechanical lever. *See* TRANSFORMER.

Transducers are widely used in measurement and control systems. In this context they are conveniently divided into two classes: sensors and actuators. Sensors are used to measure some property and to convert the measured value into a signal, often electrical, for recording or for processing and use in a control system. A good sensor usually should have high sensitivity and cause little disturbance to the system being measured, or extract a minimum amount of energy. An actuator responds to a command signal and controls the value of some quantity, such as linear or angular position, rate of fluid flow, or rate of generation of heat. Sensitivity is not usually of great importance, but it is often necessary to operate at high power levels. *See* CONTROL SYSTEMS; PHYSICAL MEASUREMENT.

Transducers for the measurement and control of physical, chemical, and biological systems are available in a great variety of types differing in complexity and precision. Many types of transducer are available for the measurement of linear or angular motion. Either one can be readily converted into the other by means of a rack and pinion or lead screw.

Direct measurement of distance can be carried out by means of a linear potentiometer to give an electrical output with moderate accuracy. Interferometric methods of optical, infrared, or microwave wavelengths can give greater accuracy but are more difficult to use. Ultrasonic waves are often employed to measure thickness of biological and other materials, and the absorption of emissions from radioactive sources is used to measure the thickness of steel emerging from a rolling mill. *See* INTERFEROMETRY; RADIOACTIVITY AND RADIATION APPLICATIONS.

A widely used type of transducer for the accurate measurement of small changes in distance is the capacitance gage. Circular potentiometers and rotating capacitors can be used to give a measure of angle, but one of the most widely used types of transducer for angular rotation is the synchro. *See* CAPACITANCE.

Many force and pressure transducers rely on known, or calibrated, deformation characteristics of springs or other mechanical devices. A measurement of the mechanical displacement by any of the means already described then gives a measure of the force or pressure. Capacitance gages are often used, with the variation of capacitance changing the frequency of an oscillator so that a digital output may be derived by counting. A piezoelectric transducer can be used to give a direct electrical output from the application of a force. Strain gages are extensively used in mechanical and aeronautical engineering to determine the distortion of structures in response to applied forces. *See* FLOW MEASUREMENT; FORCE; OSCILLATOR; PRESSURE MEASUREMENT; SPRING (MACHINES); STRAIN GAGE.

Liquid-in-glass thermometers are simple devices that provide a visual indication of temperature. They may have high accuracy but are not suitable for remote measurement or for use in automatic control systems. Almost any property of a material which changes with temperature can in principle be used as the basis of a temperature transducer. In practice the majority of devices rely on thermal expansion, change of electrical resistance, or the generation of electromotive force. *See* TEMPERATURE MEASUREMENT; THERMISTOR; THERMOCOUPLE; THERMOMETER.

Another class of transducers, not used solely for measurement purposes, includes microphones, loudspeakers, and ultrasonic transducers, These convert pressure waves in air, water, or other media into electrical signals or vice versa. *See* LOUDSPEAKER; MICROPHONE; TELEPHONE.

The use of electrical conductors to carry signals from sensors to processors causes problems in some cases, for example, in high-voltage electrical transmission systems or in the human body. The development of optical fibers for information transmission and of sensors capable of feeding information directly into such fibers has made possible new areas of instrumentation. *See* OPTICAL FIBERS. [A.E.Ba.]

Chemical and biological transducers differ from physical transducers in several respects. One distinguishing feature is the enormous diversity of parameters in chemical or biological systems. The parameters in question are essentially the concentrations of possible chemical and biochemical substances, but also include ionic strength, enzyme activity, ligand affinity, and so forth. This diversity in turn implies that there are a considerable number of interfering substances. Obtaining selective sensitivity is therefore a key problem in developing chemical and biological transducers. Most applications of these transducers involve reliable operation in aqueous media such as blood, urine, foods, drinks, and wastewater. An encapsulation of the transducer that is chemically resistant but allows access to the active part of the device is therefore crucial to operational reliability. [C.N.]

Transduction (bacteria) A mechanism for the transfer of genetic material between cells. The material is transferred by particles of bacterial viruses called bacteriophages, or phages. *See* BACTERIOPHAGE; DEOXYRIBONUCLEIC ACID (DNA).

The transduction mechanism has two features to distinguish it from the more usual mechanism of gene recombination, the sexual process. The most striking feature is the transfer of genetic material from cell to cell by phages. The second distinguishing feature is the fact that only a small part of the total genetic material of any one bacterial cell is carried by any particular transducing particle. However, all of the genetic material is distributed among different particles.

Transduction is not accomplished by all bacteriophages. It is done by some that are classified as "temperate." When such temperate bacteriophages infect sensitive bacteria, some of the bacteria respond by producing more bacteriophage particles. These bacteria donate the transducing material. Other bacteria respond to the infection by becoming more or less permanent carriers of the bacteriophage, in a kind of symbiotic relationship; these are called lysogenic bacteria. Bacteria in this latter class survive the infection, and it is among these that transduced cells are found. *See* TRANSFORMATION (BACTERIA); VIRUS. [N.D.Z.]

Transesterification The group of reactions in which an ester (RCOOR′) reacts with another compound to form a different ester (RCOOR″). The other compound may be an acid, an alcohol, or another ester. These reactions are also known as acidolysis, alcoholysis, and ester interchange, respectively. They are all special cases of esterification and have the characteristics of esterification reactions in general. *See* ACIDOLYSIS; ALCOHOLYSIS; ESTER; HYDROLYSIS. [J.M.Wo.]

Transfer cells Plant cells characterized by the elaboration of an unlignifed, secondary cell wall to form fingerlike projections or wall ingrowths which protrude into the cytoplasm of the cell. These ingrowths are enveloped by plasma membrane, forming a wall-membrane apparatus which increases the surface-volume ratio of the cell.

The location of transfer cells within the plant provides circumstantial evidence for their involvement in solute transport. They are situated at many sites in the plant where secretion or absorption takes place. Transfer cells are most frequently associated with the conducting elements of the xylem and phloem, where they may play a role in either loading or unloading solutes from these cells. Four main categories of secretion or absorption involving transfer cells are generally recognized: (1) absorption of solutes from the external environment; (2) secretion of solutes to the external environment; (3) absorption of solutes from internal, extracytoplasmic compartments; and (4) secretion of solutes into internal, extracytoplasmic compartments. *See* PHLOEM; SECRETORY STRUCTURES (PLANT); XYLEM. [R.L.Pe.]

Transform fault A fracture along an offset of a spreading mid-ocean-ridge crest. The recognition that the sea floor

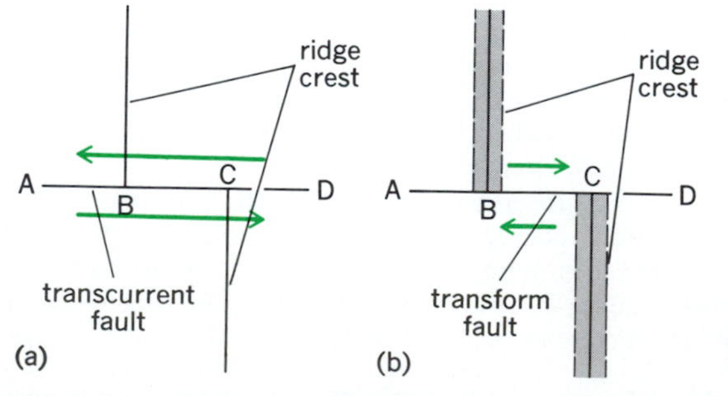

Diagrams of faults. (*a*) Left-lateral transcurrent fault. (*b*) Right-lateral transform fault.

can grow only where the ocean crust was broken—along the rift—means that east-west displacements along the fault occur only between the offset rifts (BC in the illustration) and not along the entire length of the fault (AD). This transform fault differs clearly from a transcurrent fault in crucial respects. First, earthquake activity is restricted to the active portion of the fault (BC) between the rifts and not beyond, where the fault is presently inactive (AB, CD). Second, the sense of displacement is opposite in the transform fault to that of the transcurrent fault. Third, the distance BC between the ridge-crest rifts remains constant for the transform fault and increases with time for the transcurrent fault. *See* FAULT AND FAULT STRUCTURES; MARINE GEOLOGY. [S.K.R.]

Transformation (bacteria) A term reserved almost exclusively for the genetic transformation of bacteria through the action of genetically effective substances extracted from other bacteria. The process is looked upon as the transfer of deoxyribonucleic acid (DNA) from donor bacteria to suitable recipient bacteria. The recipient bacteria are usually closely related to the donor strain. The process may occur in natural conditions, for example, in a host animal infected with two parasitic strains, and it might play a part in the rapid evolution of pathogenic bacteria.

Bacteria competent for transformation, when in certain phases of the growing state, will incorporate into themselves the large DNA molecules present in transforming agents prepared from related strains. Only the purified DNA of all the substances which may be extracted from the donor bacteria need be added in order to produce the changes in the growing recipient bacteria.

That bacterial transformation is true genetic transmission on a small scale, rather than controlled mutation, is demonstrated by the following characteristics: (1) A specific trait is introduced, coming always from donors bearing the trait. (2) The trait is transferred by determinant, genelike material far less complex than whole cells or nuclei, and this material, DNA, is known to be present in gene-carrying chromosomes. (3) The trait is inherited by the progeny of the changed bacteria. (4) The progeny produce, when they grow, increased amounts of DNA carrying the specific property. (5) The traits are transferred as units exactly in the patterns in which they appear or in which they are induced by mutation. (6) The DNA transmits the full potentialities of the donor strain, whether these are in an expressed or in a latent state. (7) The traits are often attributable to the presence of a specific gene-determined enzyme protein. (8) Certain groups of determinants may occur "linked" within DNA molecules, just as genes may be linked. (9) Linked determinants, while transforming a new cell, may become exchanged (recombined) between themselves and their unmarked or unselective alternate forms in such a way that they bring about genetic variation, and in a pattern indicating the existence of larger organized genetic units. *See* GENE; TRANSDUCTION (BACTERIA). [R.D.H.]

Transformer An electrical component used to transfer electric energy from one alternating-current (ac) circuit to another by magnetic coupling. Essentially, it consists of two or more multiturn coils of wire placed in close proximity to cause the magnetic field of one to link the other. In general, the transformer accomplishes one or more of the following between two circuits: (1) a difference in voltage magnitude, (2) a difference in current magnitude, (3) a difference in phase angle, (4) a difference in impedance level, and (5) a difference in voltage insulation level, either between the two circuits or to ground.

Transformers are used to meet a wide range of requirements. Pole-type distribution transformers supply relatively small amounts of power to residences. Power transformers are used at generating stations to step up the generated voltage to high levels for transmission. The transmission voltages are then stepped down by transformers at the substations for local distribution. Instrument transformers are used to measure voltage and currents accurately. Audio- and video-frequency transformers must function over a broad band of frequencies. Radio-frequency (rf) and intermediate-frequency (i-f) transformers transfer energy in narrow frequency bands from one circuit to another. *See* INSTRUMENT TRANSFORMER.

Power transformers. A power transformer consists of two or more multiturn coils wound on a laminated iron core. At least one of these coils serves as the primary winding.

When the primary of a power transformer is connected to an alternating voltage, it produces an alternating flux in the core. The flux generates a primary electromotive force, which is essentially equal and opposite to the voltage supplied to it. It also generates a voltage in the other coil or coils, one of which is called a secondary. This voltage generated in the secondary will supply alternating current to a circuit connected to the terminals of the secondary winding. A current in the secondary winding requires an additional current in the primary. The primary current is essentially self-regulated to meet the power (or volt-ampere) demand of the load connected to the secondary terminals. Thus in normal operation, energy (or volt-amperes) can be transferred from the primary to the secondary electromagnetically.

The illustration shows a transformer with a primary of N_1 turns and a secondary of N_2 turns. A primary voltage V_1 causes a current I_1 to flow through the coil. Since all quantities shown are alternating, the arrows indicate only instantaneous polarities.

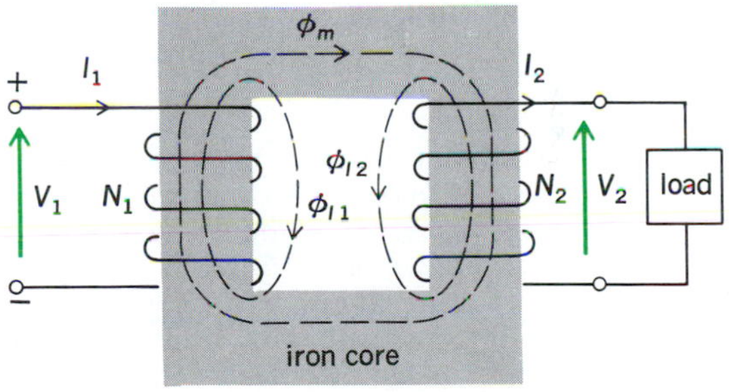

Basic transformer.

The magnetic flux ϕ set up by the primary consists of two components. One part passes completely around the magnetic circuit defined by the iron core, thus linking the secondary coil. This is the mutual flux ϕ_m. The second part is a smaller component of flux that links only the primary coil. This is the primary leakage flux ϕ_{l1}. If the secondary circuit is completed through a load, a secondary current I_2 flows and in turn creates a secondary leakage flux ϕ_{l2}. These leakage fluxes contribute to the impedance of the transformer. If the leakage flux is small, the coupling between primary and secondary is said to be close. The use of an iron core decreases the leakage flux by providing a low-reluctance path for the flux. *See* COUPLED CIRCUITS; MAGNETIC CIRCUITS.

In a power transformer the voltage drops due to winding resistance and leakage are small; therefore V_1 and V_2 are essentially in phase (or 180° out of phase, depending on the choice of polarity). Since the no-load current is small, I_1 and I_2 are essentially in phase (or 180° out of phase). Therefore, Eq. (1) applies, and the voltage ratio is expressed by Eq. (2), in which a is the transformation ratio.

$$V_1 I_1 \simeq V_2 I_2 \quad (1)$$

$$\frac{V_1}{V_2} \simeq a \quad (2)$$

The transmission of power from primary coil to secondary coil is via the magnetic flux. The flux is proportional to the ampere turns in either coil. Since the power in each coil is nearly the same, Eqs. (3) and (4) are obtained.

$$N_1 I_1 \simeq N_2 I_2 \quad (3)$$

$$\frac{N_1}{N_2} \simeq \frac{I_2}{I_1} \simeq a \quad (4)$$

Transformer cores are made of special alloy steels rolled to approximately 0.014 in. (0.35 mm) thick. These thin sheets, or laminations, are stacked to form the transformer core, each sheet being insulated from the others to reduce unwanted eddy-current loss. *See* CORE LOSS; EDDY CURRENT; MAGNETIC HYSTERESIS.

Copper conductors are used almost universally. The conductors are insulated with special paper or cotton covering, with enamel, or with a combination of both. Large outdoor transformers are immersed in oils to obtain good electrical insulation within small spacings and to provide a cooling medium. When lightweight or nonflammable materials are important, transformers may be made with compressed gases as the insulating and cooling medium. [J.R.Su.]

Audio and radio-frequency transformers. Audio or video (broad-band) transformers are used to transfer complex signals containing energy at a large number of frequencies from one circuit to another. They are required to respond uniformly to signal voltages over a frequency range three to five or more decades wide (for example, from 10 to 100,000 Hz), and consequently must be designed so that very nearly all of the magnetic flux threading through one coil also passes through the other.

Audio and video transformers have two resonances (caused by existing stray capacitances) just as many tuned transformers do. One resonance point is near the low-signal-frequency limit; the other is near the high limit. As the coefficient of coupling in a transformer is reduced appreciably below unity by removal of core material and separation of the windings, tuning capacitors are added to provide efficient transfer of energy. The two resonant frequencies combine to one when the coupling is reduced to the value known as critical coupling, then stay relatively fixed as the coupling is further reduced.

The rf and i-f transformers use two or more inductors, loosely coupled together, to limit the band of operating frequencies. Efficient transfer of energy is obtained by resonating one or more of the inductors. By using higher than critical coupling, a wider bandwidth than that from the individual tuned circuits is obtained, while the attenuation of side frequencies is as rapid as with the individual circuits isolated from one another. [K.A.P.]

Transfusion The administration of blood or its components as a part of a medical treatment. There are certain fairly well-delineated indications for the use of some form of transfusion. Hemorrhage, severe burns, and certain forms of shock are perhaps the most important conditions for which blood transfusion is utilized. Other disorders in which hemotherapy may be indicated include hemophilia, leukemia, certain anemias, and rare hereditary or familial disorders in which some portion of the blood is lacking or deficient. *See* HEMATOLOGIC DISORDERS.

In order for a recipient to accept a blood transfusion, the donor blood cells must be immunologically compatible with the recipient. That is, the recipient must recognize certain molecules (antigens) on donor blood cells as "self" and not foreign. The three main antigen systems on blood cells are the ABO, Rh, and HLA. *See* BLOOD GROUPS; RH INCOMPATIBILITY.

Donated blood is tested for blood groups, blood-group antibodies, and laboratory evidence of syphilis, hepatitis B, AIDS, human T-cell lymphotrophic virus type I (HTLV-I, which is associated with adult T-cell leukemia), and hepatitis C. As a result, blood transfusions have become safer. However, persons who may have been exposed to AIDS should not donate blood, because it has been found that blood may test negative for AIDS and yet still be capable of transmitting AIDS to recipients. This situation can arise because there is a period of time during which a recently infected individual has not yet made sufficient antibody to test positive. Because of the concern that AIDS can be transmitted by blood transfusion, patients sometimes request donations from specific family members and friends. In general, such directed donations are statistically no safer than volunteer blood donation. More patients are donating their own blood (autologous blood) before elective surgery, for their own use during and after the surgery. Autologous blood is the safest blood for transfusion. *See* ACQUIRED IMMUNE DEFICIENCY SYNDROME (AIDS).

Shortly before transfusion, the blood of the donor and recipient is tested once again to make sure that the blood groups are compatible. In an emergency, these tests are abbreviated, or type O red blood cells which can be transfused safely to any individual, are used. Transfusion of blood and blood components is essential to support many patients undergoing surgery, treatment for cancer, or organ transplantation, as well as premature infants.

In addition to transmission of infections, adverse effects of blood transfusion are due to immune reactions between donor and recipient. Fever, the most frequent reaction, is caused by reaction of recipient antibody against donor white blood cells. Hives are due to allergic reactions to substances in donor plasma. Destruction of donor red cells (hemolytic reaction) occurs if the wrong type of blood is inadvertently given, or if recipient antibody is not detected prior to transfusion. *See* BLOOD. [P.To.]

Transistor An active component of an electronic circuit which may be used as an amplifier, detector, or switch. A transistor consists of a small block of semiconducting material to which at least three electrical contacts are made. Transistors are of two general types, bipolar and field-effect. The bipolar type involves excess minority current carrier injection. The field-effect type involves only majority current carriers. Historically the bipolar type was developed before the field-effect type. Today both are widely used. The unmodified term transistor usually refers to the bipolar type.

In a bipolar transistor, at least one contact is ohmic (nonrectifying), and at least one contact is rectifying. Usually there are two closely spaced rectifying contacts and one ohmic contact. For a discussion of rectifying contacts *see* JUNCTION DIODE; POINT-CONTACT DIODE; SEMICONDUCTOR RECTIFIER.

The operation of a simple transistor consists of the control of the current flowing in the high-resistance direction through one rectifying contact (called the collector) by the current flowing in the low-resistance direction in the other rectifying contact (called the emitter). The third contact, which is ohmic, is called the base contact.

These contacts usually consist of two or more regions. The regions in which the actual rectification processes take place are called the emitter barrier and collector barrier. The region between these two barriers is called the base region, or simply the base. The regions outside of these barriers are called the emitter and collector regions.

Transistors are used in radio receivers, in electronic computers, in electronic instrumentation and control equipment, and in almost any electronic circuit where the required voltages are not too high.

Classification. Transistors are classified chiefly by four criteria: (1) by the type and number of structural regions of the semiconductor crystal; (2) by the technology used in fabrication; (3) by the semiconductor material used; and (4) by the intended use of the device.

A modern transistor type is the *npn* double-diffused silicon planar passivated transistor. The term double-diffused refers to the fabrication technique in which the base region is formed by diffusion through a mask into the body of the silicon wafer which forms the collector region. *See* DIFFUSION IN SOLIDS.

In turn, the emitter region is formed by diffusion through a second mask into the previously formed base region. The term planar refers to the fact that all three electrical connections are found on a single surface of the device. The term passivated means that the surface to which all junctions return is protected by a layer of naturally grown silicon oxide which, together with an overcoating of glass or other inert material, passivates the surface, electrically minimizing leakage currents. *See* JUNCTION TRANSISTOR; POINT-CONTACT TRANSISTOR.

Action. To explain transistor action in more detail, some of the basic properties of a semiconductor material are first presented. An *n*-type semiconductor contains electrons, and a *p*-type semiconductor contains holes. These are called the majority carriers of the two types. Actually there are always present a small number of holes in an *n*-type semiconductor and a small number of electrons in a *p*-type semiconductor. These are called the minority carriers of the two types. At a given temperature with a given material the product of the densities of the majority and minority carriers is a constant. This means that if there is present a very high density of majority carriers (low-resistivity material), there will be a correspondingly low density of minority carriers.

The emitter current controls the collector current in a simple transistor. To understand this, first consider the magnitude of the collector current in the absence of emitter current. In normal operation the collector barrier is biased in the high-resistance (reverse) direction. Under this condition of bias the majority carriers are stopped by the barrier, and only the minority carriers are free to flow. If the collector barrier is a silicon *pn* junction, the minority-carrier diffusion current is negligible and the reverse-bias leakage current will consist of thermally generated carriers and be in the nanoampere range. If emitter current is present, the portion consisting of carriers entering the base will continue across the collector barrier and thus control the collector current.

Bias. The establishment of an operating point on the transistor volt-ampere characteristics by means of direct voltages and currents is called the bias of a transistor. Since the transistor is a three-terminal device, any one of the three terminals may be used as a common terminal to both input and output. In most transistor circuits the emitter is used as the common terminal and known as a common emitter or grounded emitter. If the transistor is to be used as a linear device, such as an audio amplifier, it must be biased to operate in the active region. In this region the collector is biased in the reverse direction and the emitter in the forward direction. The area in the common-emitter transistor characteristics to the right of the ordinate $V_{CE} = 0$ and above $I_C = 0$ is the active region. Two more biasing regions are of special interest for those cases in which the transistor is intended to operate as a switch. These are the saturation and cutoff regions. The saturation region may be defined as the region where the collector current is independent of base current for given values of V_{CC} and R_L. Thus, the onset of saturation can be considered to take place at the knee of the common-emitter transistor curves. *See* AMPLIFIER.

In saturation the transistor current I_C is nominally V_{CC}/R_L. Since R_L is small, it may be necessary to keep V_{CC} correspondingly small in order to stay within the limitations imposed by the transistor on maximum-current and collector-power dissipation. In the cutoff region it is required that the emitter current I_E be zero, and to accomplish this it is necessary to reverse-bias the emitter junction so that the collector current is approximately equal to the reverse saturation current I_{CO}. A reverse-biasing voltage of the order of 0.1 volt across the emitter junction will ordinarily be adequate to cut off either a germanium or silicon transistor.

The particular method to be used in establishing an operating point on the transistor characteristics depends on whether the transistor is to operate in the active saturation or cutoff regions, on the application under consideration, on the thermal stability of the circuit, and on other factors. [C.C.H.]

Power switching. There are several transistor structures which are used for power switching and make use of current gains greater than unity to achieve a thyraton-like characteristic. These devices are often called four-layer devices since they usually contain four regions of alternating *n*- and *p*-type semiconductor material.

Field-effect transistor (FET). There are two major types of field-effect transistors, the junction-gate FET (JFET) and the insulated-gate FET (IGFET). The IGFET is commonly called MOSFET or MOS transistor. The acronym MOS stands for metal-oxide-semiconductor which describes, in order, the structure of the device from the gate toward the channel. The JFET was developed first, since it involved no technology beyond that of the planar bipolar silicon transistor. The development of the MOSFET was delayed while the technology was extended to stable control of silicon surface potential. The MOSFET is very widely applied in large-scale integration, particularly in implementing large random-access high-speed memories for computers. *See* COMPUTER STORAGE TECHNOLOGY; INTEGRATED CIRCUITS.

Illustration *a* shows a section of a JFET. The channel consists of relatively low-conductivity semiconductor material sandwiched between two regions of high-conductivity material of opposite type. When these junctions are reverse-biased, the junction depletion regions encroach upon the channel and finally, at a high reverse bias, pinch it off entirely. The thickness of the channel, and hence its conductivity, is controlled by the voltage on the two gates.

Illustration *b* shows a section of a MOSFET. Here the source

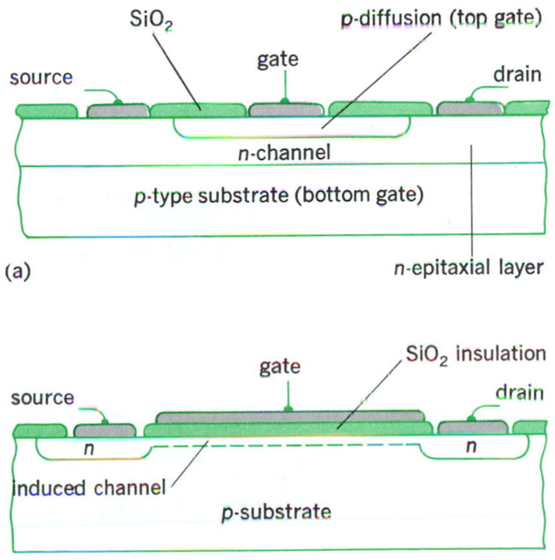

Field effect transistors. (*a*) Junction-gate FET (JFET). (*b*) Insulated-gate FET (IGFET or MOSFET).

and drain regions consist of *n* diffusion in a *p*-type substrate. The gate is a metal film evaporated on a thin SiO insulator spanning the separation between the source and drain. With no voltage on the gate, the source and drain are insulated from each other by their surrounding junctions. When a positive voltage is applied to the gate, electrons are induced to move to the surface of the *p*-type substrate immediately beneath the gate, producing a thin surface of induced *n*-type material which now forms a channel connecting the source and drain. Such a surface layer is called an inversion layer since it is of opposite conductivity type to the substrate. The number of induced electrons is directly proportional to the gate voltage, so that the conductivity of the channel increases with gate voltage. This device is called an *n*-channel enhancement-mode MOSFET. It is normally off at zero gate voltage.

MOSFET devices are fabricated in both *p*-channel and *n*-channel types, as well as for both depletion (normally on) and enhancement (normally off) modes of operation.

Most integrated circuits using MOSFETs are called CMOS integrated circuits, where the C stand for complementary. These circuits use *n*-channel and *p*-channel types together to achieve digital logic. [L.P.H.]

Transistor amplifier An amplifier which depends upon transistors for its operation. The great majority of amplifiers employ transistors because of their low cost, long life, high reliability, better overall characteristics, and small size and because the physical nature of the transistor makes possible circuits that are not possible with vacuum tubes. *See* TRANSISTOR.

A junction transistor consists of a silicon (or germanium) crystal in which a layer of *n*-type silicon is sandwiched between two layers of *p*-type silicon. Alternatively, a transistor may consist of a layer of *p*-type between two layers of *n*-type material. In the former case, the transistor is referred to as a *pnp* transistor, and in the latter case as an *npn* transistor (illustration *a*). The semiconductor sandwich is extremely small and hermetically sealed against moisture inside a metal or plastic case.

The two types of transistor are represented in illustration *a*. The representations employed when transistors are used as circuit elements are shown in illustration *b*. The three portions of a transistor are known as emitter, base, and collector. The arrow on the emitter lead specifies the direction of current flow when the emitter-base junction is biased in the forward direction. In both cases, however, the emitter, base, and collector currents, I_E, I_B, and I_C, respectively, are assumed positive when the currents flow into the transistor. The symbols V_{EB}, V_{CB}, and V_{CE} are the emitter-base, collector-base, and collector-emitter voltages, respectively. (More specifically, V_{EB} represents the voltage drop from emitter to base.) *See* AMPLIFIER. [C.C.H.; C.L.A.]

Transit (astronomy) The apparent passage of a planet across the surface of the Sun, of a satellite across the surface of the parent planet, or of a star, planet, or reference point across an adopted line of reference. Only Mercury and Venus are seen in transit across the surface of the Sun, because they are the only planets orbiting inside the path of Earth.

Passages of stars across the local meridian are observed extensively for solving problems of fundamental positional astronomy, timekeeping, and navigation. At the precise instant of transit of a star across the local meridian, the local sidereal time is exactly equal to the star's right ascension, and the latitude of the observer is equal to the sum of the star's declination and its zenith distance.

Furthermore, the difference between the local sidereal time and the Greenwich sidereal time gives a measure of the longitude of the observer. Any of those quantities may be treated as the unknown to be determined by observations of meridian transits. *See* ASTRONOMICAL COORDINATE SYSTEMS; ECLIPSE. [S.D.G.]

Transit (engineering) A surveying instrument for measuring horizontal and vertical angles. It is also used for prolonging a straight line, establishing a level line of sight, and measuring distances by the stadia method. *See* STADIA.

In the disassembled view (see illustration), three main elements are seen to support the telescope assembly: the alidade or standard with its solid spindle (inner center), which rotates within the hollow spindle (outer center) and graduated circle, which in turn rotates within the leveling head atop its tripod. These are the elements of the azimuth axis. Clamps are used to prevent one rotation while the other is used. This permits an angle to be turned about the azimuth axis and to be measured on the horizontal circle, even several times incrementally for greater precision. Vertical angular measurement is achieved by rotating the telescope about the elevation axis. The leveling screws on the leveling head enable the azimuth axis to be made vertical. Lateral motion on the leveling head permits the instrument to be centered exactly over a ground station by means of a plumb bob on a string or by use of an

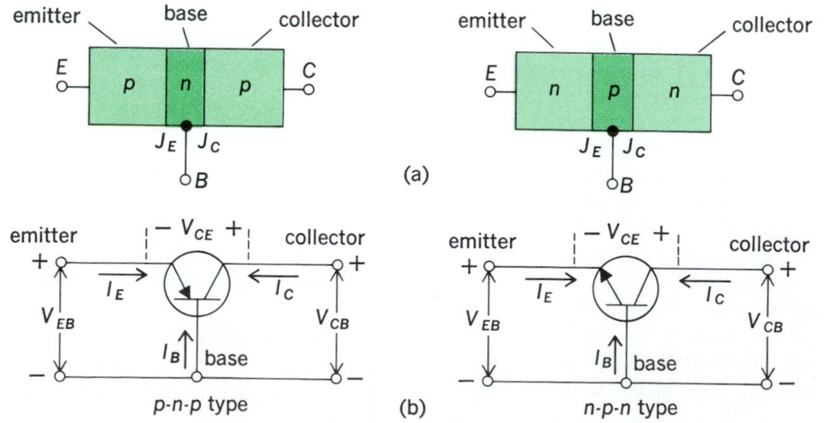

Two types of transistor. (*a*) The *pnp* and *npn* types. The emitter (collector) junction is $J_E(J_C)$. (*b*) Circuit representation of the two types.

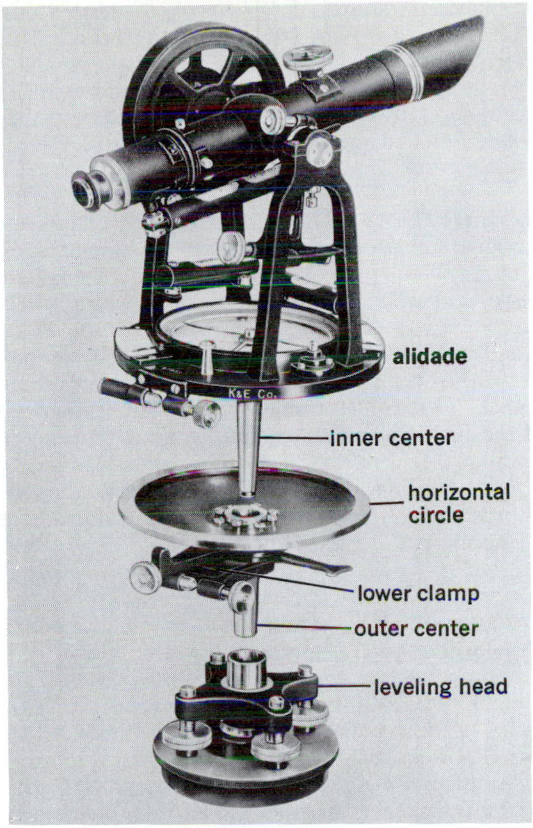

Disassembled transit. (*Keuffel and Esser Co.*)

optical plummet built into the instrument. *See* SURVEYING; THEODOLITE. [B.A.B.]

Transition elements In broad definition, the elements of atomic numbers 21–31, 39–49, and 71–81, inclusive. A more restricted classification of the transition elements, preferred by many chemists, is limited to elements with atomic numbers 22–28, 40–46, and 72–78, inclusive. All of the elements in this classification have one or more electrons present in an unfilled *d* subshell in at least one well-known oxidation state.

All the transition elements are metals and, in general, are characterized by high densities, high melting points, and low vapor pressures. Within a given subgroup, these properties tend to increase with increasing atomic weight. Facility in the formation of metallic bonds is demonstrated also by the existence of a wide variety of alloys between different transition metals.

The transition elements include most metals of major economic importance, such as the relatively abundant iron, nickel, and zinc, on one hand, and the rarer coinage metals, copper, silver, and gold, on the other. Also included are the rare and relatively unfamiliar element, rhenium, and technetium, which is not found naturally in the terrestrial environment, but is available in small amounts as a product of nuclear fission.

In their compounds, the transition elements tend to exhibit multiple valency, the maximum valence increasing from $+3$ at the beginning of a series (Sc, Y, Lu) to $+8$ at the fifth member (Mn, Re). One of the most characteristic features of the transition elements is the ease with which most of them form stable complex ions. Features which contribute to this ability are favorably high charge-to-radius ratios and the availability of unfilled *d* orbitals which may be used in bonding. Most of the ions and compounds of the transition metals are colored, and many of them are paramagnetic. Both color and paramagnetism are related to the presence of unpaired electrons in the *d* subshell. Because of their ability to accept electrons in unoccupied *d* orbitals, transition elements and their compounds frequently exhibit catalytic properties.

Broadly speaking, the properties of the transition elements are intermediate between those of the so-called representative elements, in which the subshells are completely occupied by electrons (alkali metals, halogen elements), and those of the inner or *f* transition elements, in which the subshell orbitals play a much less significant role in influencing chemical properties (rare-earth elements, actinide elements). *See* ACTINIDE ELEMENTS; ATOMIC STRUCTURE AND SPECTRA; RARE-EARTH ELEMENTS. [B.B.Cu.]

Transition point The point at which a substance changes from one state of aggregation to another. This general definition would include the melting point (transition from solid to liquid), boiling point (liquid to gas), or sublimation point (solid to gas); but in practice the term transition point is usually restricted to the transition from one solid phase to another, that is, to the temperature (for a fixed pressure, usually 1 atm or 10^5 pascals) at which a substance changes from one crystal structure to another.

Another kind of transition point is the culmination of a gradual change (for example, the loss of ferromagnetism in iron or nickel) at the lambda point, or Curie point. This behavior is typical of second-order transitions. *See* BOILING POINT; MELTING POINT; PHASE EQUILIBRIUM; SUBLIMATION; TRIPLE POINT. [R.L.S.]

Transition radiation detectors Detectors of energetic charged particles that make use of radiation emitted as the particle crosses boundaries between regions with different indices of refraction. An energetic charged particle moving through matter momentarily polarizes the material nearby. If the particle crosses a boundary where the index of refraction changes, the change in polarization gives rise to the emission of electromagnetic transition radiation. About one photon is emitted for every 100 boundaries crossed, for transitions between air and matter of ordinary density. Transition radiation is emitted even if the velocity of the particle is less than the light velocity of a given wavelength, in contrast to Cerenkov radiation. Consequently, this radiation can take place in the x-ray region of the spectrum where there is no Cerenkov radiation, because the index of refraction is less than one. *See* CERENKOV RADIATION; PARTICLE DETECTOR; REFRACTION OF WAVES. [W.J.W.]

Translucent medium A medium which transmits rays of light so diffused that objects cannot be seen distinctly; that is, the medium is only partially transparent. Familiar examples are various forms of glass which admit considerable light but impede vision. Inasmuch as the term translucent seems to imply seeing, usage of the term is ordinarily limited to the visible region of the spectrum. *See* TRANSPARENT MEDIUM. [M.G.M.]

Transmission lines A system of conductors suitable for conducting electric power of signals between two or more termini. For example, commercial-frequency electric power transmission lines connect electric generating plants, substations, and their loads. Telephone transmission lines interconnect telephone subscribers and telephone exchanges. Radio-frequency transmission lines transmit high-frequency electric signals between antennas and transmitters or receivers.

Although only a short cord is needed to connect an electric lamp to a wall outlet, the cord is, properly speaking, a transmission line. However, in the electrical industry the term transmission line is applied only when both voltage and current at one line terminus may differ appreciably from those at another terminus. Transmission lines are described either as electrically short if the difference between terminal conditions is attributable simply to the effects of conductor series resistance and inductance, or to the effects of a shunt leakage resistance and capacitance, or to both; or as electrically long when the properties of the line result from traveling-wave phenomena.

Classification. Depending on the configuration and number of conductors and the electric and magnetic fields about the conductors, transmission lines are described as open-wire transmission lines, coaxial transmission lines, cables, or waveguide transmission lines.

Open-wire lines may comprise a single wire with an earth (ground) return or two or more conductors. The conductors are supported at more or less evenly spaced points along the line by insulators, with the spacing between conductors maintained as nearly uniform as feasible, except in special-purpose tapered transmission lines. Open-wire construction is used for communication or power transmission whenever practical and permitted, as in open country and where not prohibited by ordinances.

A coaxial transmission line comprises a conducting cylindrical shell, solid tape, or braided conductor surrounding an isolated, concentric, inner conductor which is solid, stranded, or (in certain video cables and delay cables) helically wound on a plastic or ferrite core. The inner conductor is supported by ceramic or plastic beads or washers in air- or gas-dielectric lines, or by a solid polyethylene or polystyrene dielectric. The purpose of this construction is to have the shell prevent radiation losses and interference from external sources. Coaxial lines are widely used in radio, radar, television, and similar applications. *See* COAXIAL CABLE.

Sheathed cables, also termed shielded cables, comprise two or more conductors surrounded by a conducting cylindrical sheath, commonly supported by a continuous solid dielectric. The sheath provides both shielding and mechanical protection.

Theory. When electric power is applied at a terminus of a transmission line, electromagnetic waves are launched and guided along the line. The steady-state and transient electrical properties of transmission lines result from the superposition of such waves, termed direct waves, and the reflected waves which may appear at line discontinuities or at load terminals.

When the electric and magnetic field vectors are perpendicular to one another and transverse to the direction of the transmission line, this condition is called the principal mode or the transverse electromagnetic (TEM) mode. Modes other than the principal mode may exist at any frequency for which conductor spacing exceeds one-half of the wavelength of an electromagnetic wave in the medium separating the conductors. Such high-frequency modes are called waveguide transmission modes. *See* WAVEGUIDE.

In a uniform (nontapered) transmission line, the voltage or current applied at a sending terminal determines the shape of the initial voltage or current wave. In a line with negligible losses the transmitted shape remains unchanged. When losses are present, the shape, unless sinusoidal, is altered, because the phase velocity and attenuation vary with frequency. *See* PHASE VELOCITY. [E.M.W.]

Power lines. In an electric power system the facility used to transfer large amounts of power from one location to a distant location is termed a power transmission line. Power transmission lines are distinguished from subtransmission and distribution lines by their higher voltages, greater power capabilities, and greater lengths. With the exception of a few high-voltage dc lines for satisfying special requirements, power transmission lines employ three-phase alternating currents. Such lines require three conductors. The standard frequency in the United States is 60 Hz. In Europe it is 50 Hz, while in the rest of the world both of these frequencies are used. For transmitting large amounts of power over long distances, high voltages are necessary. Standard transmission voltages in the United States are 69, 115, 138, 161, 230, 345, 500, and 765 kV. These figures refer to the nominal effective voltages between any two of the three conductors. The line conductors are usually placed overhead, supported by poles or towers; however, they may form part of an underground or underwater cable. *See* ELECTRIC DISTRIBUTION SYSTEM; ELECTRIC POWER SYSTEMS. [E.W.K.]

Transmutation The nuclear change of one element into another, either naturally, in radioactive elements, or artificially, by bombarding an element with electrons, deuterons, or α-particles in particle accelerators or with neutrons in atomic piles.

Natural transmutation was first explained by Marie Curie about 1900 as the result of the decay of radioactive elements into others of lower atomic weight. Ernest Rutherford produced the first artificial transmutation (nitrogen into oxygen and hydrogen) in 1919. Artificial transmutation is the method of origin of the heavier, artificial transuranium elements, and also of hundreds of radioactive isotopes of most of the chemical elements in the periodic table. *See* NUCLEAR REACTION; PERIODIC TABLE. [F.H.R.]

Transonic flight In aerodynamics, flight of a vehicle at speeds near the speed of sound. When the velocity of an airplane approaches the speed of sound (roughly 660 mi/h or 1056 km/h at 35,000 ft or 10.7 km altitude), the flight characteristics become radically different from those at subsonic speeds. The drag increases greatly, the lift at a given altitude decreases, the moments acting on the airplane change abruptly, and the vehicle may shake or buffet. Such phenomena usually persist to flight velocities somewhat above the speed of sound. These flight characteristics, as well as the speeds at which they occur, are usually referred to as transonic. The extent of the speed range of these changes depends on the form of the airplane; for configurations designed for subsonic flight they may occur at velocities of 70–110% of the speed of sound (Mach numbers of 0.7–1.1); for airplanes intended for transonic or supersonic flight they may be present only at Mach numbers of 0.95–1.05. *See* MACH NUMBER.

The transonic flight characteristics result from the development of shock waves about the airplane. Because of the accelerations of airflow over the various surfaces, the local velocities become supersonic while the airplane itself is still subsonic. (The flight speed at which such local supersonic flows first occur is called the critical speed.) Shock waves are associated with deceleration of these local supersonic flows to subsonic flight velocities. Such shock waves cause abrupt streamwise increases of pressure on the airplane surfaces. These gradients may cause a reversal and separation of the flow in the boundary layer on the wing surface in roughly the same manner as do similar pressure changes at lower subcritical speeds. *See* AERODYNAMIC FORCE; AERODYNAMIC WAVE DRAG; BOUNDARY-LAYER FLOW; SCHLIEREN PHOTOGRAPHY; SHOCK WAVE.

As for boundary-layer separation at lower speeds, the flow breakdown in this case leads to increases of drag, losses of lift, and changes of aerodynamic moments. The unsteady nature of the separated flow results in an irregular change of the aerodynamic forces acting on the airplane with resultant buffeting and shaking. As the Mach number is increased, the shock waves move aft so that at Mach numbers of about 1.0 or greater, depending on the configurations, they reach the trailing edges of the surfaces. With the shocks in these positions, the associated pressure gradients have relatively little effect on the boundary layer, and the shock-induced separation is greatly reduced. *See* SUBSONIC FLIGHT; SUPERSONIC FLIGHT. [R.T.Wh.]

Transparent medium Ordinarily, a medium which has the property of transmitting rays of light in such a way that the human eye may see through the medium distinctly. It is pervious to light, that is, to the visible region of the electromagnetic spectrum. By extension, a medium may be described as transparent to other regions of the spectrum, such as x-rays and microwaves. Just as a blue filter passes blue rays, an ultraviolet filter might be considered as passing ultraviolet rays, or being transparent to them. *See* TRANSLUCENT MEDIUM. [M.G.M.]

Transplantation biology Transplantation, or grafting, is the artificial removal of part of an organism and its replacement in the body of the same or of a different individual.

The transplantation of cells, tissues, or organs is an important procedure for studying a wide variety of problems. For example, a defining criterion of the endocrine function of a particular organ or tissue is the capacity to restore the effects of its extirpation by returning it to the host at an anatomically abnormal site. In embryology, grafting experiments are used to study the interactions of cells of different types. Transplantation is important in the study of aging since it enables one to produce age chimeras. In cancer research, transplantation is widely employed to propagate malignant tissues in appropriate hosts for biochemical, immunologic, and therapeutic investigations.

Transplantation involves two types of problems: those concerned directly with the act of transplantation itself; and those which stem from the incompatibility of homografts, that is, grafts from a donor transplanted to a genetically dissimilar recipient, and which are of an immunological nature. The success of a therapeutic organ homograft, such as a kidney or a heart, entails overcoming the recipient's reaction against it without seriously impairing his immunological defense mechanism as a whole. The discovery of the principle of immunological tolerance indicated that this problem is soluble, though means of rendering human beings specifically nonreactive in respect to grafts from a donor have yet to be devised. However, by the long-term treatment of patients with so-called immunosuppressive drugs and other agents it is frequently possible to hold their resistance to organ homografts in abeyance for prolonged periods, if not indefinitely, and even to reverse the so-called rejection crises that often develop. Indeed, it is possible that long-term acceptors of homografts do eventually become tolerant of the foreign antigens involved. *See* ACQUIRED IMMUNOLOGICAL TOLERANCE.

As might be inferred, suppression of a recipient's response against a homograft is greatly facilitated if a donor who is relatively compatible can be selected, for there are all degrees of incompatibility. Many different procedures for donor selection have been explored, including histocompatibility testing, or typing for tissue transplantation. The typing system is essentially similar to blood typing, except that lymphocytes are used instead of red cells. *See* HISTOCOMPATIBILITY; IMMUNOLOGY. [R.E.Bi.]

Transport processes The processes whereby mass, energy, or momentum are transported from one region of a material to another under the influence of composition, temperature, or velocity gradients. If a sample of a material in which the chemical composition, the temperature, or the velocity vary from point to point is isolated from its surroundings, the transport processes act so as to eventually render these quantities uniform throughout the material. The nonuniform state required to generate these transport processes causes them to be known also as nonequilibrium processes. Associated with gradients of composition, temperature, and velocity in a material are the transport processes of diffusion, thermal conduction, and viscosity, respectively. *See* DIFFUSION IN GASES AND LIQUIDS; DIFFUSION IN SOLIDS; THERMAL CONDUCTION IN SOLIDS; VISCOSITY. [W.A.W.]

Transportation engineering That branch of engineering relating to the movements of goods and people. Major types of transportation are highway, water, rail, subways, air, and pipeline.

Highway transportation engineering deals with the planning, construction, and operation of roads, streets, bridges, and parking facilities. *See* BRIDGE; HIGHWAY ENGINEERING; PAVEMENT; TRAFFIC ENGINEERING; TUNNEL.

Planning and construction of canals, channels, harbor facilities, navigation aids such as lighthouses, and navigation locks and dams are important concerns of water transportation engineers. Another important field of water transportation relates to the design and production of launches, barges, tugs, ferry boats, and other ships. *See* CANAL; COASTAL ENGINEERING; DAM; RIVER ENGINEERING.

Engineering relating to rail transportation includes the planning and construction of terminals, switchyards, loading and unloading facilities, trackage, bridges, traffic-control and maintenance facilities, and the hauling equipment itself—locomotives and other rolling stock. *See* RAILROAD ENGINEERING.

Engineering work relating to planning, general layout, detailed design, construction, and operation of subways is an important part of transportation engineering. Subways are a major element in mass transit, and as metropolitan areas increase in size and population and vehicular traffic becomes more congested, subways are being given increased consideration as a possible part of the transportation system of a city.

The planning, design, and construction of runways, terminals, aircraft, and navigation aids are the major branches of civilian air transportation engineering. In the larger air terminals, great emphasis is placed on systems for moving passengers and baggage. *See* AIR TRANSPORTATION; AIRPORT ENGINEERING.

Pipelines are used to transport natural gas and petroleum products great distances. Pipeline engineering is concerned with planning and construction of the pipeline, pumping stations to move the material through the line, and any needed storage facilities. Frequently, large bridge structures are required to carry the pipeline across major streams or other barriers. Planning of pumping facilities requires study of power requirements for different types of material moved, standby facilities, spacing of the pumping stations, and related factors. [A.N.C.]

Transposons Units of DNA which can move from one position to another in the same or a different genome by recombination systems other than those involved in general homologous recombination. They are also known as transposable elements. The phenomenon of transposable "controlling elements" was detected first many years ago in plants (*Zea mays*), but most of the detailed molecular knowledge of transposable elements has been obtained in prokaryotic systems. In bacteria several types of transposable elements have been found. The more simple (800–1400 base pairs) "insertion sequences" were recognized by their ability to induce strong polar mutations in the galactose and lactose operons of *Escherichia coli*. Transposons have also been found which carry genes for drug resistance, toxin production, and novel substrate utilization. *See* DEOXYRIBONUCLEIC ACID (DNA); GENE ACTION. [R.Co.; W.Fi.]

Transuranium elements Those synthetic elements with atomic numbers larger than that of uranium (atomic number 92). They are the members of the actinide series, from neptunium (atomic number 93) through lawrencium (atomic number 103), and the transactinide elements (with higher atomic numbers than 103). Of these elements, plutonium, an explosive ingredient for nuclear weapons and a fuel for nuclear power because it is fissionable, has been prepared on the largest (ton) scale, while some of the others have been pro-

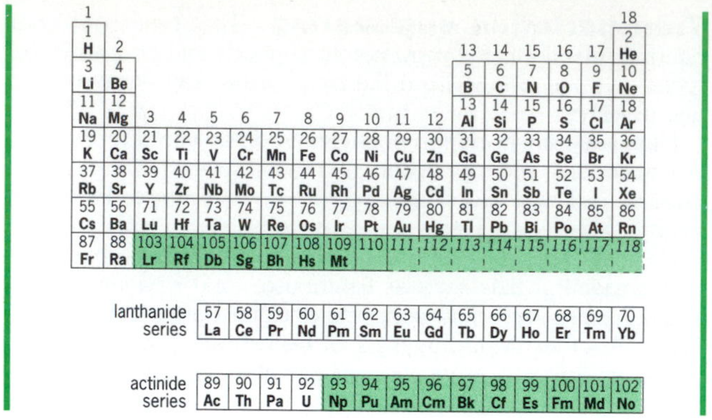

duced in kilograms (neptunium, americium, curium) and in much smaller quantities (berkelium, californium, and einsteinium).

The concept of atomic weight in the sense applied to naturally occurring elements is not applicable to the transuranium elements, since the isotopic composition of any given sample depends on its source. In most cases the use of the mass number of the longest-lived isotope in combination with an evaluation of its availability has been adequate. Good choices at present are neptunium, 237; plutonium, 242; americium, 243; curium, 248; berkelium, 249; californium, 249; einsteinium, 254; fermium, 257; mendelevium, 258; nobelium, 259; lawrencium, 260; rutherfordium, 261; dubnium, 262; and seaborgium, 263. The actinide elements are chemically similar and have a strong chemical resemblance to the lanthanide, or rare-earth, elements (atomic numbers 57–71). The transactinide elements, with atomic numbers 104–118, appear in an expanded periodic table under the row of elements beginning with hafnium, number 72, and ending with radon, number 86. This arrangement allows prediction of the chemical properties of these elements and suggests that they will have a chemical analogy with the elements which appear immediately above them in the periodic table.

The transuranium elements up to and including fermium (atomic number 100) are produced in largest quantity through the successive capture of neutrons in nuclear reactors. The yield decreases with increasing atomic number, and the heaviest to be produced in weighable quantity is einsteinium (number 99). Many additional isotopes are produced by bombardment of heavy target isotopes with charged atomic projectiles in accelerators; beyond fermium, all elements are produced by bombardment with heavy ions.

Although the transactinide elements immediately beyond meitnerium are predicted to have very short half-lives, theoretical considerations suggest increased nuclear stability, compared with preceding and succeeding elements, for a range of elements with larger atomic numbers because of increased stability predicted to result from closed nucleon shells. The element with atomic number 114 (which should have chemical properties similar to those of lead) seems to show special promise of such relative stability, that is, relatively long half-life. It should be possible to synthesize isotopes of such "superheavy" elements through bombardments of heavy-element targets with intense beams of very heavy ions accelerated in specially constructed heavy-ion accelerators. *See* ACTINIDE ELEMENTS; AMERICIUM; BERKELIUM; BOHRIUM; CALIFORNIUM; CURIUM; DUBNIUM; EINSTEINIUM; ELEMENT 110; FERMIUM; HASSIUM; LAWRENCIUM; MEITNERIUM; MENDELEVIUM; NEPTUNIUM; NOBELIUM; NUCLEAR CHEMISTRY; NUCLEAR FISSION; NUCLEAR REACTION; PERIODIC TABLE; PLUTONIUM; RARE-EARTH ELEMENTS; RUTHERFORDIUM; SEABORGIUM. [G.T.S.]

Trapezoid A term used in the United States for a quadrilateral with two sides parallel (see illustration). In Great Britain such a figure is often called a trapezium, a term used in the United States for a general quadrilateral. The parallel sides are called the bases, and the area of a trapezoid is the product of one-half the sum of the bases by the perpendicular distance between them (altitude). If the two nonparallel sides (legs) are equal, the trapezoid is isosceles. [L.M.Bl.]

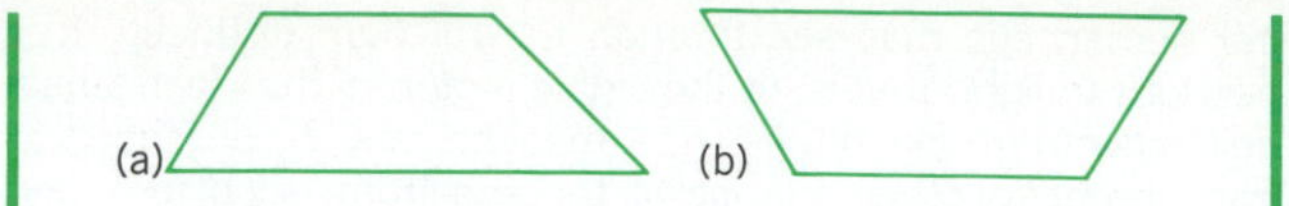

Drawings of (*a*) trapezoid and (*b*) isosceles trapezoid.

Traps in solids Electron traps are defects or chemical impurities in semiconductors and in insulators which capture mobile electrons in a special way. The electrons are immobilized, prevented from recombining with (annihilating) positive holes, and are released some time later as mobile electrons again. Analogous traps act on mobile positive holes. A defect which captures a mobile electron and aids recombination with a hole (or vice versa) is called a recombination center. A defect (nickel in germanium, for example) may be a recombination center at one temperature and a trap at a different one.

Electrons or holes may remain in traps for short times or for as long as months or years. They may be released by heating the solid or by irradiating it with infrared. Traps have an important influence in photoconduction, luminescence, and the photographic process. *See* BAND THEORY OF SOLIDS; HOLE STATES IN SOLIDS; LUMINESCENCE; PHOTOCONDUCTIVITY; PHOTOLYSIS. [L.Ap.]

Trauma Injury to tissue by physical or chemical means. Mechanical injury includes abrasions, contusions, lacerations, and incisions, as well as stab, puncture, and bullet wounds. Trauma to bones and joints results in fractures, dislocations, and sprains. Head injuries are often serious because of the complications of hemorrhage, skull fracture, or concussion.

Thermal, electrical, and chemical burns produce severe damage partly because they coagulate tissue and seal off restorative blood flow. Asphyxiation, including that caused by drowning, produces rapid damage to the brain and respiratory centers, as well as to other organs.

Frequent complications of trauma are shock, the state of collapse precipitated by peripheral circulatory failure, and also hemorrhage, infection, and improper healing. *See* HEMORRHAGE; INFECTION; SHOCK SYNDROME. [E.G.St./N.K.M.]

Traveling-wave tube A microwave electronic tube in which a beam of electrons interacts continuously with a wave that travels along a circuit, the interaction extending over an appreciable distance measured in wavelengths. Traveling-wave tubes are normally used in amplifiers with exceedingly wide bandwidths. Typical bandwidths are 10–100% of the center frequency, with gains of 20–60 dB. Such traveling-wave amplifiers serve at the inputs to sensitive radars or communication receivers. High-power traveling-wave amplifiers operated as the final stages of radars or scatter communication transmitters deliver pulsed powers exceeding a megawatt. For use in satellite transmitters, lightweight traveling-wave amplifiers develop 35 watts continuously at 45% overall efficiency.

In a traveling-wave tube amplifier (see illustration), a thermionic cathode produces the electron beam. An electron gun initially focuses the beam, and additional focusing means retain the electron stream as a beam throughout the length of the tube until the beam strikes the collector electrode. Microwave energy enters the tube near the electron gun and propagates along a slow-wave circuit. The tube delivers amplified microwave energy into a matched load connected near the collector end. The slow-wave circuit serves to propagate the microwave energy along

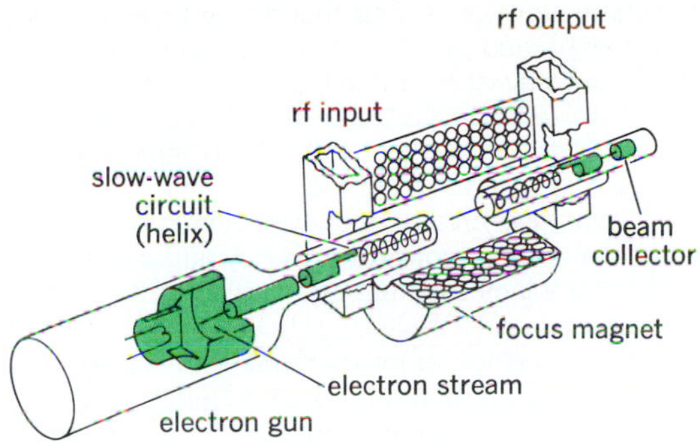

Traveling-wave tube.

the tube at approximately the same velocity as that of the electron beam. Interaction between beam and wave is continuous along the tube with contributions adding in phase.

In principle, operation in a traveling-wave tube is similar to that in a klystron: Velocity modulation of the electrons results in current modulation of the beam because of the nonzero transit time. However, unlike a klystron, the traveling-wave tube provides for the process to take place continuously along the slow-wave circuit. Thus, at one point along the circuit, the microwave field accelerates the electrons axially. As these electrons proceed along the tube, they bunch with those ahead. At the same point along the circuit, but a half-cycle later in time, the microwave field decelerates electrons; these electrons bunch with those behind. Because average velocity of the electrons is made to be slightly faster than the velocity of the energy on the slow-wave circuit, the electron bunches that form drift into a decelerating microwave field, thereby delivering energy to the field. In this manner the original energy in the electron beam is converted into microwave energy and delivered to the slow-wave circuit. Because of the continuous distributed interaction, the circuit wave grows exponentially as it travels along the tube. For the backward-wave class of traveling-wave tubes *see* BACKWARD-WAVE TUBE. For other microwave tubes *see* KLYSTRON; MAGNETRON; MICROWAVE TUBE. [J.W.Ge.]

Travertine A rather dense, banded limestone, sometimes moderately porous, that is formed either by evaporation about springs, as is tufa, or in caves, as stalactites, stalagmites, or dripstone. Where travertine or tufa (calcareous sinter) is deposited by hot springs, it may be the result of the loss of carbon dioxide from the waters as pressure is released upon emerging at the surface; the release of carbon dioxide lowers the solubility of calcium carbonate and it precipitates. High rates of evaporation in hot-spring pools also lead to supersaturation. Travertine formed in caves is simply the result of complete evaporation of waters containing mainly calcium carbonate. *See* LIMESTONE; STALACTITES AND STALAGMITES; TUFA. [R.Si.]

Tree A perennial woody plant at least 20 ft (6 m) in height at maturity, having an erect stem or trunk and a well-defined crown or leaf canopy. However, no sharp lines can be drawn between trees, shrubs, and lianas (woody vines.)

Classification. Almost all existing trees belong to the seed plants. An exception are the giant tree ferns which were more prominent in the forests of the Devonian Period and today exist only in the moist tropical regions. The Spermatophyta are divided into the Pinophyta (gymnosperms) and the flowering plants, Magnoliophyta (angiosperms). The gymnosperms bear their seed naked on modified leaves, called scales, which are usually clustered into structures called cones—for example, pine cones. By contrast the seed of angiosperms is enclosed in a ripened ovary, the fruit. *See* MAGNOLIOPHYTA; PINOPHYTA; POLYPODIALES.

Morphology. The morphology of a tree is similar to that of other higher plants. Its major organs are the stem, or trunk and branches; the leaves; the roots; and the reproductive structures. Almost the entire bulk of a tree is nonliving. Of the trunk, branches, and roots, only the tips and a thin layer of cells just under the bark are alive. Growth occurs only in these meristematic tissues. Meristematic cells are undifferentiated and capable of repeated division. *See* FLOWER; LATERAL MERISTEM; LEAF; PLANT GROWTH; ROOT (BOTANY); STEM. [F.T.L.]

Growth. Trees grow to a larger size at maturity than other woody perennials, have a comparatively long period of development to maturity, and live a long time in the mature state. Trees, like other vascular plants, are made up of cells. Growth is the result of adding more cells through cell division, and of the elongation and maturation of those cells into functional tissues. Cells and the tissues they comprise differ in structure and function. Some tissues, such as the corky bark, mechanically protect the tree and insulate it against rapid temperature changes. Parenchyma tissues are important in photosynthesis and food storage. Xylem tissues conduct water, minerals, and some hormones up the tree, and the phloem tissues conduct photosynthate sugar solutions and other organic molecules in most cases down the tree. The fraction of the xylem and phloem which conducts liquids is termed vascular tissue. There are also other nonconducting cells such as parenchyma within the xylem and phloem tissues.

Tissues may also be compared by looking at their origin. Primary tissues are derived from apical meristems of the stem and root, whereas secondary tissues arise from secondary cambia. Two important secondary cambia in trees are: (1) the vascular cambium which produces cells which become secondary xylem (the wood of commerce) to the inside and secondary phloem tissue to the outside; and (2) the cork cambium or phellogen which produces corky bark tissues to the outside and usually phelloderm tissue to the inside. Collectively, these secondary cambia account for most of the diameter growth in trees.

Shoot and root apices, originating in the seed embryo, are major centers of organization in development of the tree. Cell divisions occur in distinct patterns in these apices. These patterns are the basis for development of tissues from which the mature axial structures of the shoot and root are formed. The shoot apex is the source of cells which develop in very predictable patterns into leaf primordia, axillary buds that can grow into lateral branches, and reproductive structures. The root apex is a less complex structure producing only cells which will develop into the internal tissues of the root. Lateral roots are not formed at the apex, but in pericycle tissues some distance behind the apex. *See* APICAL MERISTEM; PLANT GROWTH.

Diseases. Diseases of both shade and forest trees have the same pathogens, but the trees differ in value, esthetics, and utility. In forests, disease is significant only when large numbers of trees are seriously affected. Diseases with such visible symptoms as leaf spots may be alarming on shade trees but hardly noticed on forest trees. Shade trees with substantial rot may be ornamentals with high value, whereas these trees would be worthless in the forest. Emphasis on disease control for the same tree species thus requires a different approach, depending on location of the tree.

Forest trees. In natural forests, leaf diseases are negligible, but in nurseries and plantations, fungal infections cause severe defoliation, retardation of height growth, or death. Oak wilt is a systemic disease, with the entire tree affected through its water-conducting system. Trees of susceptible species may be killed in a few weeks to a year or more through plugging of the water-conducting vessels. The causal fungus, *Ceratocystis fagacearum*, spreads to nearby healthy trees by root grafts and to trees at longer distances by unrelated insects, including certain bark beetles.

Stem rust diseases occur as cankers or galls on coniferous hosts and as minor lesions on other ones. A few, such as white pine blister rust and southern fusiform rust, are epidemic, lethal, and economically important. Others of less immediate importance (such as western gall rust) are capable of serious, widespread infection, requiring no other hosts to complete their life cycles. Stem infections by numerous fungi, resulting in localized death of cambium and inner bark, range from lesions killing small stems in a year (annual) to gross stem deformities (perennial), where cankers enlarge with stem growth.

All tree species, including decay-resistant ones such as redwood, are subject to ultimate disintegration by fungi. Decay fungi are associated with nondecay fungi and bacteria. These microflora enter the tree through wounds, branch stubs, and roots, and are confined to limited zones of wood by anatomical and wound-stimulated tissue barriers. The extent of decay is limited by compartmentalization of decay in trees. Trees aged beyond maturity are most often invaded by wood-rotting fungi; losses can be minimized by avoiding wounds and by shortening cutting rotations.

Shade trees. Many shade trees are grown under conditions for which they are poorly adapted, and are subject to environmental stresses not common to forest trees. Both native and exotic trees planted out of natural habitats are predisposed to secondary pathogens following environmental stress of noninfectious origins. They are also susceptible to the same infectious diseases as forest trees. Appearance is more important then the wood produced, and individual value is higher per tree than for forest trees. Thus, disease control methods differ from those recommended for forest trees. *See* PLANT PATHOLOGY.

[R.J.C.]

Tree ferns Plants belonging to the families Cyatheaceae and Dicksoniaceae, whose members typically develop tall trunks crowned with leaves (fronds) which often reach some 20 ft (6.1 m) in length and 5–6 ft (1.5–1.8 m) in width. Tree ferns reach their greatest development in the rainforests and cloud forests of the mountainous tropics. *See* RAINFOREST.

Tree fern trunks may reach 65 ft (19.8 m) tall; their diameters vary from 0.4 in. (1.0 cm) to 4 ft (1.2 m). The lower trunk is often densely covered with matted adventitious roots which greatly increase its diameter. Certain specimens branch near the base of the trunk or higher up; perhaps this branching occurs in response to injury.

The degree of division of the leaves varies from simple to four or five pinnate. The leaflets (pinnae) are usually smaller toward the base of the leaf; when these basal leaflets are branched into threadlike divisions, they are called aphlebiae. The Dicksoniaceae have marginal sori terminal on the veins and protected by a bivalved indusium. The Cyatheaceae produce sori well away from the margin, usually seated at the forking of a vein or midway along a simple vein.

The typical vascular system of both families is dictyostelic. The Cyatheaceae have accessory vascular strands in the pith and cortex. Fibrous sheaths around the vascular tissue and just inside the epidermis provide mechanical support. The xylem consists of scalariform tracheids and parenchyma, the phloem of sieve tubes and parenchyma. Numerous mucilage canals are embedded in the pith and cortex. *See* STEM; TREE. [G.J.G.]

Tree physiology The study of growth and development of trees in terms of genetics; biochemistry; cellular, tissue, and organ functions; and interaction with environmental factors. While many physiological processes are similar in trees and other plants, particularly at the molecular and biochemical levels, trees possess unique physiologies that help determine their outward appearance. These physiological processes include carbon relations (photosynthesis, carbohydrate allocation), cold and drought resistance, water relations, and mineral nutrition.

Three characteristics of trees that define their physiology are longevity, height, and simultaneous reproductive and vegetative growth. Trees have physiological processes that are more adaptable than those in the more specialized annual and biennial plants. Height, exceeding 330 ft (100 m) in some species, allows trees to successfully compete for light, but at the same time this advantage creates transport and support problems. These problems were solved by the evolution of the woody stem which combines structure and function into a very strong transport system.

In evergreen species, photosynthesis can occur year-round as long as the air temperature remains above freezing, while some deciduous species can photosynthesize in the bark of twigs and stems during the winter. Carbon dioxide fixed into sugars moves through the tree in the phloem and xylem to tissues of high metabolism, which vary with season and development. At the onset of growth in the spring, sugars are first mobilized from storage sites, primarily in the secondary xylem (wood) and phloem (inner bark) of the woody twigs, branches, stems, and roots. The sugars, stored as starch, are used to build new leaves and twigs, and if present, flowers. *See* PHLOEM; PHOTOSYNTHESIS; XYLEM.

The perennial nature of trees requires them to withstand low temperatures during the winter. Cell damage occurs from ice crystal formation within the cell, which disrupts the plasma membrane. Trees develop resistance to freezing through a process of physiological changes beginning in late summer.

Trees have evolved traits that allow them to avoid desiccation. These traits include using water stored in the stems, stomatal closure, and shedding of leaves to reduce transpirational area. Trees must supply water to the leaves for normal physiological function. If the water potential of the leaves drops too low, the stomata close, reducing photosynthesis. To maintain high water potential, trees use water stored in their stems during the morning, then recharge during the night. *See* PLANT-WATER RELATIONS.

Trees have evolved a means of combining long-distance transport between the roots and foliage with support through the production of secondary xylem (wood) by the vascular cambium. These dead cells function in both transport and support. The interconnecting cells provide an efficient transport system, capable of moving 106 gal (400 liters) of water per day.

Mineral nutrients are required in proper concentrations and proportions for normal growth and development. Tree nutrition is unique because trees require lower concentrations of nutrients than other plant types, and they are able to recycle nutrients within various tissues. Internal cycling keeps nutrients from leaving the tree and accounts for a majority of the nutrients used annually for growth. *See* PLANT MINERAL NUTRITION; PLANT PHYSIOLOGY; TREE. [J.D.Jo.]

Tree-ring hydrology The relationships, or correlations, between ring thicknesses and stream flow via rainfall, and between thicknesses and rainfall via soil moisture. These relationships have validity only if it is possible to identify accurately the annual increments in all the trees employed in a given study. Among the trees where most work has been done, little relation to temperature has been found. *See* DENDROCHRONOLOGY; PALEOCLIMATOLOGY; PLANT-WATER RELATIONS.

Ring thicknesses in the southwestern United States appear to have only a modest correlation with stream flow. Rainfall is in a very broad sense the connecting link, or the common factor, between tree growth and stream flow. In the relationship of ring thicknesses to rainfall, the connecting link is soil moisture. In fact, a strong case can be made for the study of soil moisture in relation to tree growth rather than of the more remote factors of rainfall or of stream flow.

Techniques have been developed for extracting ecologic information and meteorologic or hydrologic information from growth layers. Growth patterns have been offered as an extension, or alternative, to the use of year-by-year comparisons. The use of growth-ring patterns rather than individual rings eliminates the necessity of dating accurately each growth layer. Thus growth patterns reveal information about rainfall through the variability in thickness of successive growth layers, relative and absolute variations in circuit and longitudinal uniformity, areal distribution of partial growth layers, and constitution of the annual increment. [W.S.G.]

Trematoda A loose grouping of acoelomate, parasitic flatworms of the phylum Platyhelminthes formerly accorded class rank and containing the subclasses (or orders) Digenea, Monogenea, and Aspidobothria. These organisms commonly occur as adults in or on all vertebrate groups. They exhibit cephalization, bilateral symmetry, and well-developed anterior and ventral, or anterior and posterior, holdfast structures. The mouth is anterior, and usually a blind, forked gut occurs, as well as three muscle layers. The excretory system consists of flame cells and collecting tubules. These animals are predominantly hermaphroditic and oviparous with operculated egg capsules. The life histories of the Digenea are complex, while those of the Monogenea and Aspidobothria are simple.

Trematodes parasitize a wide variety of invertebrates and vertebrates and occupy almost every available niche within these hosts. The adaptations demanded of the worms for survival are as varied as the characteristics of the microhabitats. Over the millions of years of coevolution of the hosts and their parasites, delicate balances have, for the most part, been attained, and under normal conditions it is probable that trematodes rarely demand more than the host can supply without undue strain. It seems axiomatic that the host must survive until the parasite can again gain access to another host or until the life cycle is completed. Those parasites which cause the least disruption of the host's activities are probably the oldest as well as the most successful. Immunities are sometimes developed by the hosts. Many trematodes seem to possess such rigid requirements and such responses to particular hosts that host specificity is a phenomenon of considerable significance. Monogeneids appear more host-specific than digeneids, and aspidobothreids seem less specific than both. Trematodes are of considerable veterinary and medical importance because under certain conditions they cause debility, even death. *See* ASPIDOGASTREA; DIGENEA; MONOGENEA; PLATYHELMINTHES. [W.J.Ha.]

Tremolite The name given to magnesium-rich monoclinic calcium amphibole $Ca_2Mg_5Si_8O_{22}(OH)_2$ The mineral is white to gray, but colorless in thin section. Unlike other end-member compositions of the calcium amphibole group, very pure tremolite is found in nature. Substitution of Fe for Mg is common, but pure ferrotremolite, $Ca_2Fe_5Si_8O_{22}(OH)_2$, is rare. Intermediate compositions between tremolite and ferrotremolite are referred to as actinolites, and are green in color and encompass a large number of naturally occurring calcium amphiboles. *See* AMPHIBOLE. [B.L.D.]

Trepostomata An extinct order of bryozoans in the class Stenolaemata. Trepostomes possess generally robust colonies, composed of tightly packed, moderately complex, long, slender, tubular or prismatic zooecia, with solid calcareous zooecial walls. Colonies show a moderately gradual transition from endozone to exozone regions. Trepostome colonies range from small and delicate to large and massive; they can be thin encrusting sheets, tabular or nodular or hemispherical or globular masses, or erect twiglike, bushlike, or frondlike growths. *See* BRYOZOA; STENOLAEMATA.

Apparently exclusively marine, the trepostomes first appeared about the start of the Middle Ordovician, and immediately rose to abundance and dominance, which they maintained from mid-Ordovician through Silurian time; in places, trepostomes contributed to building small reefs, as well as being abundant in level-bottom environments. Trepostomes declined during the Devonian, finally dying out in the earliest Triassic. [R.J.Cu.]

Trestle A succession of towers of steel, timber, or reinforced concrete supporting the horizontal beams of a roadway, bridge, or other structure. Little distinction can be made between a trestle and a viaduct, and the terms are used interchangeably by many engineers. A viaduct is defined as a long bridge consisting of a series of short concrete or masonry spans supported on piers or towers, and is used to carry a road or railroad over a valley, a gorge, another roadway, or across an arm of the sea. *See* BRIDGE.

A trestle or a viaduct usually consists of alternate tower spans and spans between towers. For low trestles the spans may be supported on bents, each composed of two columns adequately braced in a transverse direction. A pair of bents braced longitudinally forms a tower. *See* STRUCTURES (ENGINEERING); TOWER. [C.M.A.]

Triangle In the euclidean sense, a triangle is a geometric figure bounded by three (straight) line segments, called sides, which meet pairwise in three points called vertices. It has three angles (each formed by a pair of sides at a vertex), whose three measures have a sum equal to 180° or two right angles. The sum of two sides of a triangle is greater, and the difference less, than the third side. A triangle is called equilateral if three sides are equal, isosceles if at least two sides are equal, and scalene if no two sides are equal (see illustration). The greater of two sides is opposite the greater angle. A triangle is called acute, right, or obtuse, according to whether its largest angle is acute (<90°), right (=90°), or obtuse (>90°).

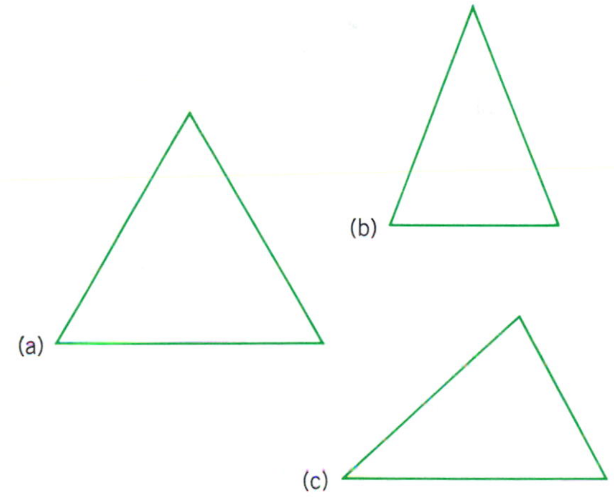

Triangles. (*a*) Equilateral. (*b*) Isosceles. (*c*) Scalene.

The line segment drawn from a vertex of a triangle to the midpoint of the opposite side is called a median of the triangle. The three medians of a triangle intersect in a point called the centroid. *See* EUCLIDEAN GEOMETRY; POLYGON; PYTHAGOREAN THEOREM. [J.S.F.]

Triassic A geological term designating the lowest rock system of the Mesozoic Era. The name is derived from the threefold facies division of a sequence of strata in central Germany into a lower nonmarine redbed facies, a middle marine limestone, sandstone, and shale facies, and an upper nonmarine continental facies similar to the lower division.

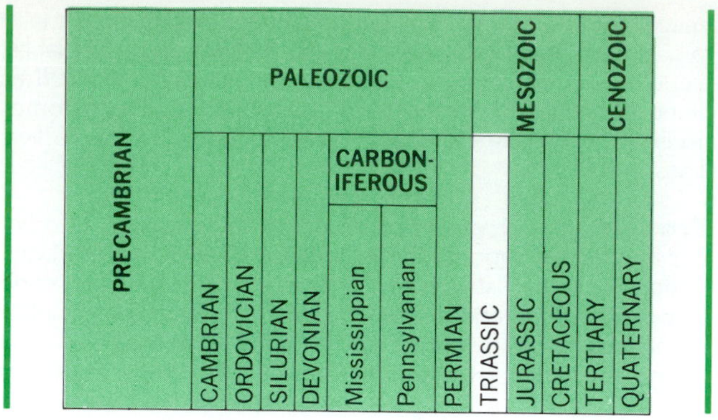

These have been correlated by geologists to facies in other parts of the world.

With the close of the Permian, epicontinental and geosynclinal seas had retreated from most of the continents and the geography of the lands and seas must have been very similar to what it is today; with few exceptions, Triassic seas were marginal to the continents. It was not until the Jurassic that shallow shelf seas began to spread beyond the peripheral geosynclinal belts and inundated parts of Eurasia and North America.

The great restrictions of epicontinental and geosynclinal seas in the late Paleozoic were accompanied by the extinction of a large part of the marine faunas characteristic of that era. This great crisis in the evolutionary history of marine animals is reflected in the composition of Triassic marine faunas. In the Lower Triassic, for instance, corals, foraminiferans, and bryozoans are absent or extremely rare. The dominant marine animals in the Early Triassic seas are molluscans, especially the ammonites. Pelecypods, nautiloids, and gastropods are fairly common but not nearly to the extent of the ammonoids. The crisis of mass extinction did not affect the animals living on land. There are no sharp breaks or interruptions in the evolutionary history of the amphibians or reptiles, and none in the evolutionary development of plants. *See* Extinction (biology). [B.Ku.]

Trichinosis The presence of *Trichinella spiralis*, a nematode, in the body of humans. Wild hogs, rats, and omnivores are natural hosts. Domestic pigs become infected by preying on rats or feeding on raw scraps. Humans are exposed upon eating encysted larvae in raw or partially cooked pork. Ingested larvae develop in a week into adult nematodes in the duodenum. During the 3–16 week period that the adult female parasitizes humans, each one discharges from 300 to 1500 larvae. These enter the lymphatics and are carried to all parts of the body, where they encyst preferably in active muscles. The host reacts by forming a capsule around the larva.

Infection by a few larvae may not occasion a noticeable damage. Heavy exposure results in a syndrome of varied nature. The first stage is related to the intestinal phase of parasitism and simulates acute food poisoning, typhoid, or cholera. The second stage coincides with larval migration, that is, inflammatory processes in the muscle and circulatory systems, swelling or edema, remittent high fever, and hypereosinophilia. The third stage relates to encystation of larvae: cachexia, dehydration, and blood pressure and neuropathic changes.

Treatment is by oral cortisone or thiabendazole. Clinical recovery is dramatic, but the larvae in muscle may live unhampered. Prophylaxis is best accomplished by thoroughly cooking all pork products. A temperature of 66°C (150°F) kills *Trichinella* larva. Curing or pickling pork will not necessarily kill encysted larvae. The larvae will be killed if the pork is kept frozen at −15°C (+5°F) for 21 days or at −30°C (−22°F) for 25 h. [J.F.M.]

Trichomoniasis A localized genitourinary infection of humans caused by *Trichomonas vaginalis*, a flagellate protozoan. Three species of *Trichomonas* inhabit the human body; *T. buccalis* (*T. tenax*) in the mouth and *T. hominis* in the lumen of the intestinal tract, neither of which produces clinical-pathological signs, and the pathogenic *T. vaginalis*. *Trichomonas vaginalis* causes vaginitis, a superficial ulceration and inflammation, and a discharge. It causes a severe itching and burning sensation and produces painful coitus; however, it causes no interference with pregnancy. In the male, *T. vaginalis* may cause a light or severe urethritis and involvement of accessory glands. A chronic carrier state has been found in males. [M.Yo.]

Trichomycetes A polyphyletic class of Eumycota in the subdivision Zygomycotina, containing the orders Amoebidiales, Asellariales, Eccrinales, and Harpellales. Members of these orders exist only as commensals in the mid- and hindgut of arthropods. Classification is based on the type of reproduction; thallus branching pattern, complexity, and septation; and nature of the holdfast.

Asexual reproduction is accomplished by the formation of trichospores (Harpellales), arthrospores (Asellariales), sporgiospores (Eccrinales), or ameboid cells, cystospores, or rigid-walled spores (Amoebidiales). Sexual reproduction (zygospore formation) is known only in Harpellales, although conjugations have been observed in Asellariales. The thallus of Amoebidiales is aseptate, but is regularly septate in the other three orders; septa with plugs in lenticular cavities are produced by Asellariales and Harpellales. Similar septa and plugs are formed by Dimargaritales and Kickxellales (Zygomycetes). *See* Eumycota; Fungi; Zygomycetes. [G.L.Be.]

Trichoptera An aquatic order of the class Insecta commonly known as the caddis flies. The adults have two pairs of well-veined hairy wings, long antennae, and mouthparts capable of lapping only liquids. The larvae are wormlike, with distinct heads, three pairs of legs on the thorax, and a pair of hook-bearing legs at the end of the body. The pupae are delicate, with free appendages held close to the body, and have a pair of sharp mandibles, or jaws, which are used to cut an exit from the cocoon.

The Trichoptera include about 10,000 described species, divided into 34 families, and occur in practically all parts of the world. Except for a brackish-water species in New Zealand and a few moss-inhabiting species in Europe and North America, caddis flies occur only in fresh water. They abound in cold or running water relatively free from pollution. Altogether they compose a large and important segment of the biota of such habitats and of the fish feed economy. *See* Insecta. [H.H.R.]

Trichostomatida An order of the Holotrichia, composed of a small group of ciliates whose species show some advance in structural complexity over the gymnostomes. No true buccal ciliature is present, but a vestibulum, a passageway leading from the outside of the body into the cytostome or true cell-mouth, is characteristically present. These animals exist in a wide range of habitats. A few forms live in cattle, sheep, and horses, but cause no harm to these hosts. *Balantidium* is of some medical significance as it contains the sole ciliated protozoan species parasitic in humans. *Balantidium coli* lives in the large intestine and is the causative agent of the fairly uncommon disease balantidiasis. *See* Balantidiasis; Holotrichia. [J.O.C.]

Trichroism When certain optically anisotropic transparent crystals are subjected to white light, a cube of the material is found to transmit a different color through each of the three pairs of parallel faces. Such crystals are sometimes termed trichroic, and the phenomenon is called trichroism. This expression is used only rarely today since the colors in a particular crystal can appear quite different if the cube is cut with a different orientation with respect to the crystal axes. Accordingly, the term is frequently replaced by the more general term pleochroism. Even this term is being replaced by the phrase linear dichroism or circular dichroism to correspond with linear birefringence or circular birefringence. *See* BIREFRINGENCE; CRYSTAL OPTICS; DICHROISM; PLEOCHROISM. [B.H.Bi.]

Tricladida An order of the Turbellaria known commonly as planaria. They have a diverticulated intestine with a single anterior branch and two posterior branches separated by the plicate pharynx or pharynges. Rhabdites are numerous and, except in cave planarians, two to many eyes are present. *See* PLANARIA; TURBELLARIA. [E.R.J.]

Triconodonta An extinct, nominally mammalian order of small flesh-eating creatures of the Mesozoic Era. They have been found in the Upper Jurassic and Middle Cretaceous of North America. The immediate ancestry of the triconodonts is uncertain, but they may have been related to the primitive Rhaetic genus *Eozostrodon*, often called *Morganucodon*. Apparently the triconodonts represent a sterile offshoot of the cynodont therapsids, and their inclusion with the Mammalia is merely a matter of convenience. *See* DOCODONTA; MAMMALIA; THERAPSIDA. [W.A.C.; M.C.McK.]

Trigger circuit An electronic circuit that generates or modifies an existing waveform to produce a pulse of short time duration with a fast-rising leading edge. This waveform, or trigger, is normally used to initiate a change of state of some relaxation device, such as a multivibrator. The most important characteristic of the waveform generated by a trigger circuit is usually the fast leading edge. The exact shape of the failing portion of the waveform often is of secondary importance, although it is important that the total duration time is not too great. A pulse generator such as a blocking oscillator may also be used and identified as a trigger circuit if it generates sufficiently short pulses. *See* BLOCKING OSCILLATOR; PULSE GENERATOR.

Peaking circuits, which accent the higher-frequency components of a pulse waveform, cause sharp leading and trailing edges and are therefore used as trigger circuits. The simplest form of peaking circuits are the simple *RC* and *RL* networks shown in the illustration. If a steep wavefront of amplitude *V* is

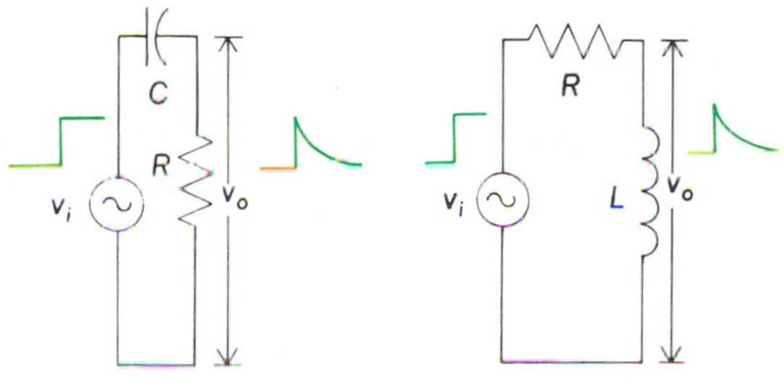

Diagrams of simple peaking circuits. v_i = input voltage; v_o = output voltage.

applied to either of these circuits, the output will be a sudden rise followed by an exponential decay. These circuits are often called differentiating circuits because the outputs are rough approximations of the derivative of the input waveforms, if the *RC* or *R/L* time constant is sufficiently small.

A circuit that is highly underdamped, or oscillatory, and is supplied with a step or pulse input is often referred to as a ringing circuit. When used in the output of a field-effect or bipolar transistor, this circuit can be used as a trigger circuit. *See* WAVE-SHAPING CIRCUITS. [G.M.G.]

Triglyceride A simple lipid. Triglycerides are fatty acid triesters of the trihydroxy alcohol glycerol which are present in plant and animal tissues, particularly in the food storage depots, either as simple esters in which all the fatty acids are the same or as mixed esters in which the fatty acids are different. The triglycerides constitute the main component of natural fats and oils.

The generic formula of a triglyceride is shown below, where

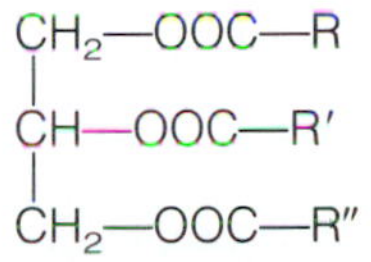

RCO_2H, $R'CO_2H$, and $R''CO_2H$ represent molecules of either the same or different fatty acids, such as butyric or caproic (short chain), palmitic or stearic (long chain), oleic, linoleic, or linolenic (unsaturated). Saponification with alkali releases glycerol and the alkali metal salts of the fatty acids (soaps). The triglycerides in the food storage depots represent a concentrated energy source, since oxidation provides more energy than an equivalent weight of protein or carbohydrate. *See* LIPID METABOLISM; SOAP.

The physical and chemical properties of fats and oils depend on the nature of the fatty acids present. Saturated fatty acids give higher-melting fats and represent the main constituents of solid fats, for example, lard and butter. Unsaturation lowers the melting point of fatty acids and fats. Thus, in the oil of plants, unsaturated fatty acids are present in large amounts, for example, oleic acid in olive oil and linoleic and linolenic acids in linseed soil. *See* FAT AND OIL (FOOD); LIPID. [R.H.G.; H.E.Ca.]

Trigonometric curve The graphical representation of y as a trigonometric function of θ. To obtain the graphs of

$$y = a \sin \theta$$

for the interval $0 \leqq \theta \leqq 2\pi$, one prepares the accompanying table of values of trigonometric functions, plots the corresponding points, and connects them by a smooth curve. This gives the graph shown in the illustration. Since $\sin(\theta + k2\pi) = \sin\theta$, where k is an integer, the complete graph consists of an endless repetition both to the right and to the left of the curve shown. *See* TRIGONOMETRY.

Since $\sin\theta = \cos(\theta - \frac{1}{2}\pi)$, the illustration is also the graph of $y = a\cos(\theta - \frac{1}{2}\pi)$. The graph of $y = a\cos\theta$ has the same

Values of trigonometric functions

θ(radians)	$y = a \sin \theta$	θ(radians)	$y = a \sin \theta$
0	0	$\frac{7}{6}\pi$	$-0.5\ a$
$\frac{1}{6}\pi$	$0.5a$	$\frac{8}{6}\pi$	$-0.87a$
$\frac{2}{6}\pi$	$0.87a$	$\frac{9}{6}\pi$	$-1\ a$
$\frac{3}{6}\pi$	a	$\frac{10}{6}\pi$	$-0.87a$
$\frac{4}{6}\pi$	$0.87a$	$\frac{11}{6}\pi$	$-0.5a$
$\frac{5}{6}\pi$	$0.5a$	$\frac{12}{6}\pi$	0
$\frac{6}{6}\pi$	0		

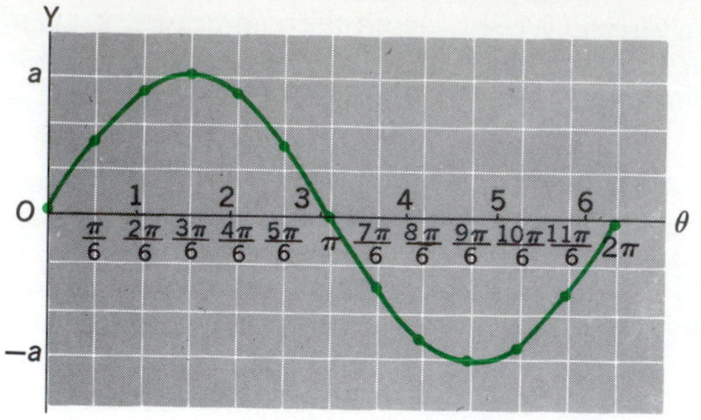

Curve plotted for $y=a \sin \theta$.

shape as that of $y = a \sin \theta$ and could be obtained from the illustration by translating the curve $\frac{1}{2}\pi$ units leftward. [L.M.K.]

Trigonometry The study of triangles and the trigonometric functions. The functions of trigonometry are extremely important because they may be used to represent ranges of values which are repeated again and again. They represent periodic phenomena such as the motions of pendulums or the analysis of alternating-current electricity.

Also, trigonometry is used to find a vast network of lengths that cannot be measured directly. Surveyors use it to find heights of mountains, distances across lakes and countries, and positions of places; engineers use it in the design of large structures and roads; astronomers use it in accurate measurements of time, and in locating the positions of bodies in the sky; and navigators on the sea and in the air use it to find latitudes, longitudes, and directions.

The subject of trigonometry may be divided into plane trigonometry and spherical trigonometry.

Plane trigonometry. Plane trigonometry, for the most part, deals with triangles with straight lines in a plane as sides. Unless otherwise specified, it is understood that all figures considered in this section lie in a plane.

Angles. If a half-line, or ray, having end point *O* (Fig. 1) rotates about *O* from an initial position *OA* to terminal position *OB*, it is said to generate the angle *AOB*. A rotation of $\frac{1}{360}$ part of a complete rotation about a point is called a degree and written 1°. One-sixtieth of a degree is called a minute and written 1′. Generally angles generated by rotation in a clockwise direction are called negative angles, and those generated by counterclockwise turning are called positive angles.

A radian is an angle subtended at the center of a circle by an arc of the circle equal in length to its radius. It is used in purely theoretical discussions because it avoids cumbersome constants. *See* RADIAN MEASURE.

Trigonometric functions. Figure 2 shows an angle θ with vertex *O* at the origin of a system of rectangular coordinates and with initial line *OA* directed in the positive direction of the *x* axis. Let *P*, any point except *O*, on the terminal ray of the angle θ, have coordinates *x* and *y* and be at a distance *r* from *O*. Then the six functions, sine, cosine, tangent, cotangent, secant, and cosecant of θ, abbreviated by sin θ, cos θ, tan θ, cot θ, sec θ, csc θ, are defined by Eqs. (1) and (2), where *r* is

$$\sin \theta = y/r,\ \cos \theta = x/r,\ \tan \theta = y/x \qquad (1)$$

$$\csc \theta = r/y,\ \sec \theta = r/x,\ \cot \theta = x/y \qquad (2)$$

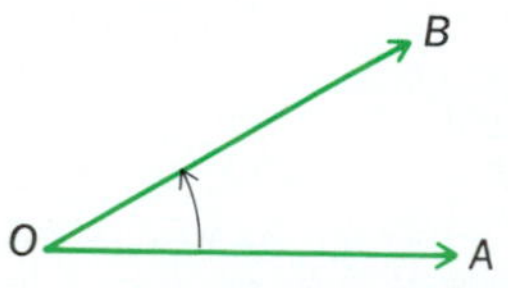

Fig. 1. Generation of angle *AOB*.

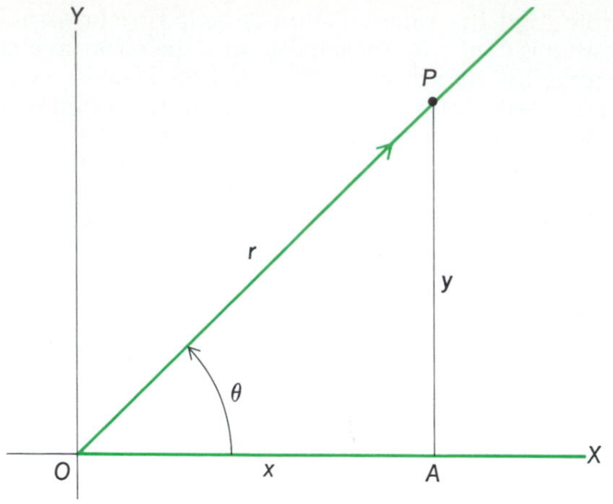

Fig. 2. Angle *AOB* in the rectangular coordinate system.

positive. These ratios are independent of the position of the point (x,y) on the terminal ray. *See* COORDINATE SYSTEMS.

Figure 2 shows that any angle with initial side *OA* obtained by adding an integral multiple of 360° to θ has the same terminal ray as θ, and therefore that Eq. (3) can be written, where

$$\text{fn}\ (\theta + k360°) = \text{fn}\ (\theta) \qquad (3)$$

fn stands for sin, cos, tan, cot, sec, or csc, and *k* is an integer. In other words the trigonometric functions are periodic, and a period is 360° or 2π radians. This is the property that makes them so useful in theoretical considerations involving periodic values.

Figure 3 shows four angles 60°, 120°, 240°, and 300° having terminal lines containing respective points $(1, \sqrt{3})$, $(-1, \sqrt{3})$, $(-1, -\sqrt{3})$, and $(1, -\sqrt{3})$. Since $r = \sqrt{1^2 + (\sqrt{3})^2} = 2$ in all cases the values from Eq. (1) can be tabulated as in notation (4).

	sin	cos	tan	cot	sec	csc
60°	$\frac{1}{2}\sqrt{3}$	$\frac{1}{2}$	$\sqrt{3}$	$\frac{1}{3}\sqrt{3}$	2	$2/\sqrt{3}$
120°	$\frac{1}{2}\sqrt{3}$	$-\frac{1}{2}$	$-\sqrt{3}$	$-\frac{1}{3}\sqrt{3}$	−2	$2/\sqrt{3}$
140°	$-\frac{1}{2}\sqrt{3}$	$-\frac{1}{2}$	$-\sqrt{3}$	$\frac{1}{3}\sqrt{3}$	−2	$-2/\sqrt{3}$
300°	$-\frac{1}{2}\sqrt{3}$	$\frac{1}{2}$	$-\sqrt{3}$	$-\frac{1}{3}\sqrt{3}$	2	$-2/\sqrt{3}$

(4)

In all cases $|x| \leqq r$, $|y| \leqq r$, and therefore the range of the sine and of the cosine is −1 to +1 inclusive. Figure 3 shows the angles 0°, 90°, 180°, and 270° containing on their terminal rays the respective points (1,0), (0,1), (−1,0) and (0, −1). Using Fig. 3 and Eqs, (1) and (3), the values may be tabulated as in notation (5), where *k* is an integer.

	0° + k360°	90° + k360°	180° + k360°	270° + k360°
sin	0	1	0	−1
cos	1	0	−1	0
tan	0	Undefined	0	Undefined

(5)

If the angle θ is near to, but less than, 90°, *x* is small compared to *y* and therefore the tangent is large; in fact tan θ ranges through all positive numbers as angle θ increases from

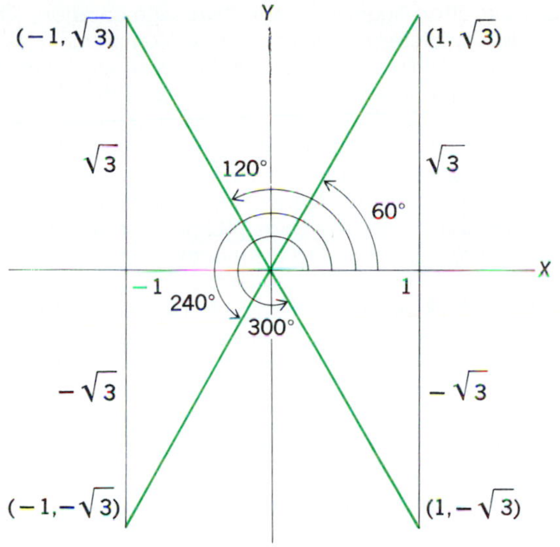

Fig. 3. Four angles in rectangular coordinate.

0° to 90°. The graphs of the trigonometric functions give a clear picture of the ranges and general behavior of these functions. *See* TRIGONOMETRIC CURVE.

From Eqs. (1) and (2), it follows that Eqs. (6) and (7) can

$$\csc\theta = \frac{1}{\sin\theta} \quad \sec\theta = \frac{1}{\cos\theta} \quad \cot\theta = \frac{1}{\tan\theta} \tag{6}$$

$$\tan\theta = \frac{\sin\theta}{\cos\theta} \tag{7}$$

be written. Also from Fig. 2, $y^2 + x^2 = r^2$. Dividing this through by r^2 and using Eq. (1), $(\sin\theta)^2 + (\cos\theta)^2 = 1$, which is written as in Eq. (8). Using Eqs. (6), (7), (8),

$$\sin^2\theta + \cos^2\theta = 1 \tag{8}$$

and others, many complicated expressions can be transformed to simple ones.

Addition formulas. The addition formulas of trigonometry, namely, Eqs. (9) and (10), lead to practically all of the relations between the trigonometric functions.

$$\cos(\phi + \theta) = \cos\phi\cos\theta - \sin\phi\sin\theta \tag{9}$$

$$\sin(\phi + \theta) = \sin\phi\cos\theta + \cos\phi\sin\theta \tag{10}$$

Inverse trigonometric functions. In the equation $x = \sin y$, y is an angle having x as sine. The symbol $\sin^{-1} x$ (or arcsin x) means the angle having x as sine, or the inverse sine of x. Evidently arcsin ½ has the values 30° + k360° and 150° + k360°, where k is an integer. A like notation applies to the other functions. For example, arccos ½ has the values ±60° + k360°. Generally a particular value, called the principal value, is chosen and designated by Arcsin x, Arccos x, Arctan x, etc. Arcsin x and Arctan x are chosen in the range from −90° to +90°, and Arccos x and Arccot x are chosen in the range 0° to 180°.

Solution of rectilinear figures. If a side and another part of a right triangle are known, the triangle can be solved. A rectilinear figure can often be solved by drawing perpendiculars to sides of the figure to divide it into right triangles and then solving the set of right triangles in order. However, various formulas are used for solving rectilinear figures. For convenience, the vertices of a triangle are denoted by capital letters A, B, C, and the respective opposite sides by small letters a, b, c.

Values of trigonometric functions. If θ is an angle in radians and $n! = 1 \cdot 2 \cdot 3 \cdots n$, then sin θ and cos θ are defined for all values of θ by the endless series shown as Eqs. (11) and (12).

$$\sin\theta = \theta - \frac{\theta^3}{3!} + \frac{\theta^5}{5!} - \cdots + (-1)^{n+1}\frac{\theta^{2n+1}}{(2n+1)!} + \cdots \tag{11}$$

$$\cos\theta = 1 - \frac{\theta^2}{2!} + \frac{\theta^4}{4!} - \cdots + (-1)^n\frac{\theta^{2n}}{2n!} + \cdots \tag{12}$$

To find a value of the sine or the cosine of an angle, only as many terms of the series are used as are necessary to ensure required accuracy.

Spherical trigonometry. A great circle on a sphere is the intersection of the sphere with a plane through the center. Spherical trigonometry treats of, for the most part, spherical triangles having as sides arcs of great circles on a sphere. Two cases of these triangles are highly important. In one case the vertices are points on the Earth, and in the other on celestial bodies such as the Sun, planets, and stars. Astronomers, surveyors, and navigators on ships and airplanes apply the formulas of spherical trigonometry to find such values as the time of day, directions of motion, and positions of ships, airplanes, and reference points. Thus spherical trigonometry is basic in astronomy, in certain kinds of surveying, and in navigation. It is also used in mathematics and its applications. [L.M.K.]

Trihedron A geometric figure bounded by three noncoplanar rays called edges that emanate from a common point called the vertex, and by the plane sectors called faces that are formed by each pair of edges (see illustration). A trihedron has

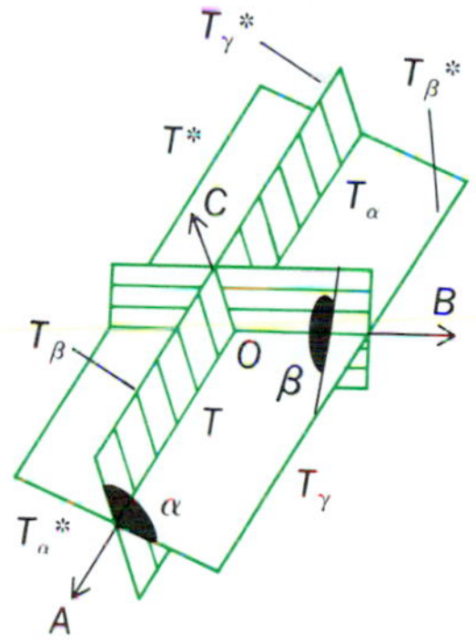

Trihedron and trihedral angles. Three planes having a common point *O* cut space into eight trihedrons, *T*, *T, T_α, T_α*, T_β, T_β*, T_γ, and T_γ*. *A*, *B*, and *C* are directions of common edges of pairs of planes; α, β, and γ are face angles.**

three dihedrons formed by pairs of face planes, and three face angles formed by pairs of edges. [J.S.F.]

Trilobita A class of extinct Paleozoic (Cambrian to Permian) arthropods, occurring with other invertebrates in various types of sedimentary rocks. They are therefore presumed to have inhabited shallow seas but not fresh waters. The exoskeleton (Fig. 1) covered the dorsal surface and extended a short distance over the ventral side as the doublure. It was probably chitinous and strengthened by mineral secretion believed to have been calcite. The mineral matter may be preserved in the rock with little change (Fig. 2), may be replaced by some other mineral, or may be dissolved so that an internal or external mold remains. Trilobites are abundant in many Cambrian sediments, including the earliest, and are used almost exclusively in dating and correlating rocks of that sys-

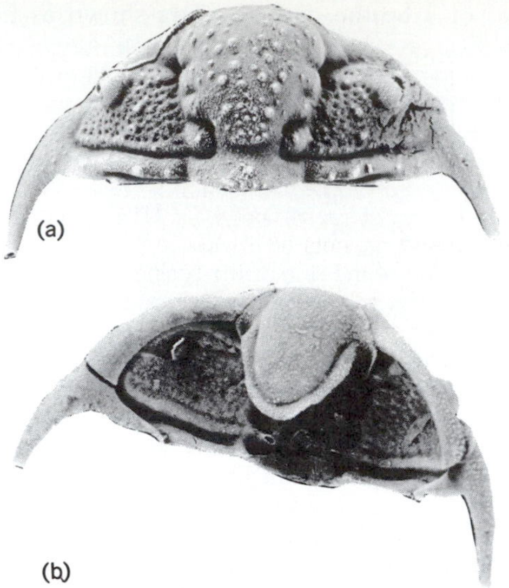

Fig. 1. Trilobite exoskeletal structures, of *Ceraurus whittingtoni*. (*a*) Silicified exoskeleton of cephalon, upper (dorsal) side, Middle Ordovician (Virginia). (*b*) Oblique view of under (ventral) side.

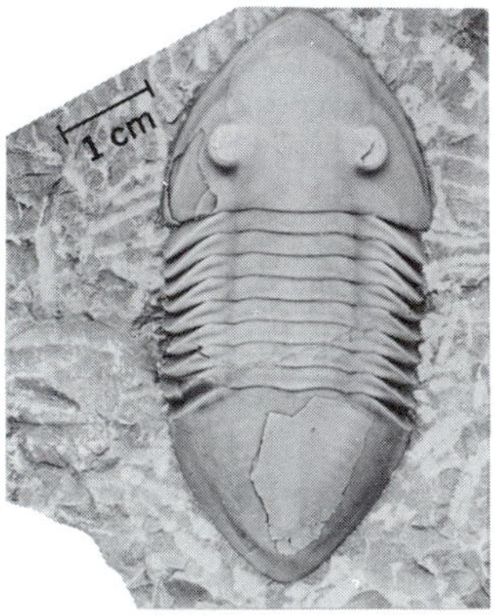

Fig. 2. *Isotelus gigas*, Middle Ordovician (New York). Exoskeleton partly stripped off internal mold of left cheek and of pygidium.

tem. In Ordovician and Silurian rocks, trilobite remains are common, but become fewer in the Devonian and rare in the Carboniferous and Permian. The beauty of trilobites has fascinated collectors, and the evolution, taxonomy, and extinction of this group of more than 1000 genera invite study. An understanding of them is essential in Paleozoic stratigraphy, paleogeography, and animal distribution. *See* ARTHROPODA.

[H.B.W.]

Trilobitomorpha The trilobitomorphs are all extinct and are among the oldest known arthropods. The most dominant group, the trilobites (Trilobita), were in existence at the beginning of the Cambrian Period, and must have lived long before they or any other animals were preserved as fossils; yet they are not the actual ancestors of any modern arthropods. *See* TRILOBITA.

A typical trilobite has a flattened oval shape with an elevation, the rachis, running lengthwise along the middle of the back, giving it a three-part appearance from which the trilobites get their name. More important, however, is the crosswise division of the body into a head, a thorax, and a pygidium. The head is not segmented, but indentations of the median elevation suggest a primitive segmentation. On its upper surface, laterally, it bears a pair of compound eyes. The thorax is completely segmented. On the undersurface the head bears a pair of long antennae and a median lobe, the labrum, that probably covers the mouth. The rest of the undersurface is occupied by a long series of eight-segmented legs, which are all alike and suggest that the diversified appendages of other arthropods were once jointed legs used for locomotion. Mesal spiny lobes of the basal segments probably served for grasping food and passing it forward to the mouth.

[J.G.R.]

Trimerophytopsida Mid-Early-Devonian into Middle-Devonian vascular plants at a higher evolutionary level than Rhyniopsida. Branching was profuse and varied, dichotomous, pseudomonopodial, helical to subopposite and almost whorled, and often trifurcate. Vegetative branches were often in a tight helix, terminated by tiny recurved branchlets simulating leaf precursors. The axes were leafless and glabrous or spiny. Xylem is known only in *Psilophyton*. *See* EMBRYOBIONTA; RHYNIOPSIDA.

[H.P.B.]

Triple point A particular temperature and pressure at which three different phases of one substance can coexist in equilibrium. In common usage these three phases are normally solid, liquid, and gas, although triple points can also occur with two solid phases and one liquid phase, with two solid phases and one gas phase, or with three solid phases.

According to the Gibbs phase rule, a three-phase situation in a one-component system has no degrees of freedom (that is, it is invariant). Consequently, a triple point occurs at a unique temperature and pressure, because any change in either variable will result in the disappearance of at least one of the three phases. Triple points are shown in the illustration of part of the phase diagram for water. *See* PHASE EQUILIBRIUM.

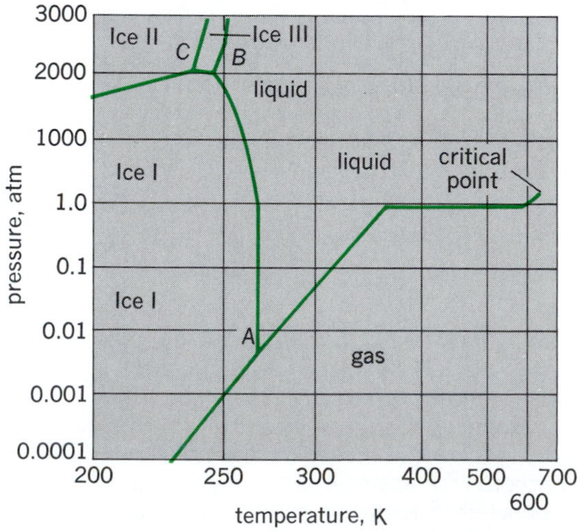

Phase diagram for water, showing gas, liquid, and several solid (ice) phases; triple points at *A*, *B*, and *C*. The pressure scale changes at 1 atm from logarithmic scale at low pressure to linear at high pressure. 1 atm = 10^2 kilopascals.

For most substances the solid-liquid-vapor triple point has a pressure less than 1 atm (10^2 kilopascals); such substances then have a liquid-vapor transition at 1 atm (normal boiling point). However, if this triple point has a pressure above 1 atm, the substance passes directly from solid to vapor at 1 atm. *See* BOILING POINT; ICE POINT; MELTING POINT; SUBLIMATION; TRANSITION POINT; VAPOR PRESSURE; WATER.

[R.L.S.]

Triplet state An electronic state of a molecule that occurs when its total spin angular momentum quantum number S is equal to one. The triplet state is an important intermediate of organic chemistry. In addition to the wide range of triplet molecules available through photochemical excitation techniques, numerous molecules exist in stable triplet ground states, for example, oxygen molecules. *See* ATOMIC STRUCTURE AND SPECTRA; MOLECULAR STRUCTURE AND SPECTRA; SPIN (QUANTUM MECHANICS).

A triplet may result whenever a molecule possesses two electrons which are both orbitally unpaired and spin unpaired. Orbital unpairing of electrons results when a molecule absorbs a photon of visible or ultraviolet light. Direct formation of a triplet as a result of this photon absorption is a very improbable process since both the orbit and spin of the electron would have to change simultaneously. Thus, a singlet state is generally formed by absorption of light. However, quite often the lifetime of this singlet state is sufficiently long to allow the spin of one of the two electrons to invert, thereby producing a triplet. *See* MOLECULAR ORBITAL THEORY. [N.J.T.]

Tripylida An order of nematodes in which the cephalic cuticle is simple and not duplicated; there is no helmet. The body cuticle is smooth or sometimes superficially annulated. Cepahalic sensilla follow the typical pattern in which one whorl is circumoral and the second whorl is often the combination of circlets two and three. The pouchlike amphids have apertures that are inconspicuous or transversally oval. The stoma is variable, being simple, collapsed, funnel shaped or cylindrical, and armed or unarmed. In most taxa the stoma is surrounded by esophageal tissue; that is, it is entirely esophastome. When the stoma is expanded, both the cheilostome and esophastome are evident. Esophagi are cylindrical-conoid. Esophageal glands open anterior to the nerve ring. Males generally have three supplementary organs, more in some taxa. A gubernaculum accompanies the spicules. Caudal glands are generally present.

The two tripylid superfamilies are the Tripyloidea and Ironoidea. The characteristically well-developed cuticular annulation of the Tripyloidea is only rarely seen in other Enoplida. These nematodes are commonly found in fresh water or very moist soils; however, some are found in brackish water and marine habitats. Intestinal contents indicate that their food consists primarily of small microfauna that often include nematodes and rotifers. The Ironoidea contain species (presumably carnivorous) occurring in both fresh-water and soil habitats. *See* NEMATA. [A.R.M.]

Trisomic syndrome Trisomy is an abnormal state characterized by the presence in triplicate of one of the chromosomes of the complement. The normal human fertilized egg has 23 pairs of chromosomes ($2n$), one of each pair provided by each parent (n). In trisomy the total number of chromosomes present is one unit above normal ($2n + 1$), except if the chromosomal material in excess is translocated onto another chromosome. In this case the total number appears normal ($2n$), but one chromosome has an abnormally large size. A trisomic syndrome in humans is a pathological condition resulting from the presence of this extra chromosomal material. Since the error leading to the trisomic state occurs before or shortly after fertilization of the egg by a sperm, the manifestations of the trisomy are present at the birth of the child, in the form of obvious multiple defects. Because these malformations occur in combinations rather specific for each of the possible trisomies, an experienced observer usually can identify any such syndrome on clinical grounds alone. Of course trisomy is responsible for only a relatively small portion of the congenital defects of humans; the majority of such defects are not associated with such gross chromosomal errors. *See* CONGENITAL ANOMALIES.

In 1866 the British physician Langdon Down described a characteristic syndrome of mental retardation and multiple malformations for which he coined the unfortunate denomination of mongolian idiocy. The cause of the disorder remained unknown until 1959, when a team of researchers demonstrated that the syndrome was associated with the presence of an extra chromosome identified as a number 21. Since then mongolism or Down syndrome has been also commonly referred to as trisomy 21. *See* DOWN SYNDROME.

Whereas many patients with trisomy 21 attain an old age, patients with D_1 trisomy, or trisomy 13, usually die early in infancy because of the more severe nature of their malformations. The very few who survive long enough present a very severe mental deficiency. The brain characteristically shows lack of separation of the hemispheres in the frontal region and absence of external olfactory tracts. General reduction in size of the eye, or slits in the eyes, or both are frequent. These infants have seizures and breath-holding spells and often die during such episodes. Their appearance is quite striking, because of microcephaly, abnormal eyes, cleft lip and palate, malformed ears, hands with overlapping fingers and often extra digits, and abnormal penis and scrotum in the males. The internal anomalies, in addition to those of the central nervous system, include various heart defects, renal malformations, malrotation of the colon, accessory spleens, and septate or duplicated Müillerian duct derivatives in the female (uterus, tubes, and upper vagina).

Babies with E trisomy, or trisomy 18, present the following anomalies: severe mental defect and growth failure; muscular hypertonia; cephalic deformities such as small jaw, prominent occiput, abnormal ears, high arched palate; short sternum; heart malformations; diaphragmatic and inguinal hernias; renal malformations; deformities of the extremities such as syndactyly and overriding fingers (tightly clenched fists), hypoplastic nails, short or dorsiflexed big toes, and a posteriorly prominent calcaneus ("rockerbottom foot"); limited hip abduction; and dislocation of the hips (probably related to muscular hypertonia). For unknown reasons this syndrome affects girls much more often than it does boys. The length of survival is, as in trisomy 13, quite limited. *See* HUMAN GENETICS; MUTATION. [P.E.Fe.]

Triterpene A hydrocarbon or its oxygenated analog containing 30 carbon atoms and composed of six isoprene units. Triterpenes form the largest group of terpenoids, but are classified into only a few major categories. Resins and saps contain triterpenes in the free state as well as in the form of esters and glycosides.

Biogenetically triterpenes arise by the cyclization of squalene and subsequent skeletal rearrangements. Apart from the linear squalene itself and some bicyclic, highly substituted skeletons, most triterpenes are either tetracyclic or pentacyclic compounds. The various structural classes are designated by the names of representative members. *See* ISOPRENE; TERPENE. [T.Hu.]

Triticale A cereal grass plant (× *Triticosecale*) obtained from hybridization of wheat (*Triticum*) with rye (*Secale cereale*). It is a crop plant, with a small-seeded cereal grain that is used for human food and livestock feed. Culture of triticale is concentrated in Europe and Australia and to a lesser extent in the United States.

Modern varieties are called secondary triticales because they were selected after interbreeding of various triticales, including primary types. In some triticale varieties, one or more rye chromosomes have been replaced by wheat chromosomes, giving secondary-substituted triticales, as contrasted to complete triticale having all seven rye chromosomes. Triticale is produced by deliberate hybridization of either bread wheat [*Triticum aestivum*; diploid number of chromosomes ($2n$) = 42] or durum wheat (*T. turgidum* var. *durum*; $2n$ = 28) with rye ($2n = 14$), followed by the doubling of the chromosome number of the hybrid plant.

Triticale is grown from seeds sown in soil by using cultivation practices similar to those of wheat or rye. Triticale tends to have a greater ability than wheat to grow in adverse environments, such as saline or acid soils or under droughty conditions.

Being a cereal grain, triticale can be used in food products made from wheat flour. As a livestock feed, triticale is a good source of carbohydrate and protein. Its protein is richer than wheat protein in lysine, one of the amino acids, making it especially desirable for the diets of monogastric animals, such as swine and birds, since these animals cannot synthesize lysine and so must obtain it in their diet. *See* BREEDING (PLANT); RYE; WHEAT. [C.O.Q.]

Tritium The heaviest isotope of the element hydrogen and the only one which is radioactive. Tritium occurs in very small amounts in nature but is generally prepared artificially by processes known as nuclear transmutations. It is widely used as a tracer in chemical and biological research and is a component of the so-called thermonuclear or hydrogen bomb. It is commonly represented by the symbol ${}^3_1H$ indicating that it has an atomic number of 1 and an atomic mass of 3, or by the special symbol T. For information about the other hydrogen isotopes *see* DEUTERIUM; HYDROGEN. *See also* TRANSMUTATION.

Properties of hydrogen and tritium

Property	H_2	T_2
Melting point	−259.20°C (−434.56°F)	−252.54°C (−422.57°F)
Boiling point at 1 atm (10^5pascals)	−252.77°C (−423.00°F)	−248.12°C (−414.62°F)
Heat of vaporization	216 cal/mol (904 J/mol)	333 cal/mol (1390 J/mol)
Heat of sublimation	247 cal/mol (1030 J/mol)	393 cal/mol (1640 J/mol)

Both molecular tritium, T_2, and its counterpart hydrogen, H_2, are gases under ordinary conditions. Because of the great difference in mass, many of the properties of tritium differ substantially from those of ordinary hydrogen, as indicated in the table. Chemically, tritium behaves quite similarly to hydrogen. However, because of its larger mass, many of its reactions take place more slowly than do those of hydrogen.

The nucleus of the tritium atom, often called a triton and symbolized t, consists of a proton and two neutrons. It undergoes radioactive decay by emission of a β-particle to leave a helium nucleus of mass 3. No γ-rays are emitted in this process. The half-life for the decay is 12.26 years. When tritium is bombarded with deuterons of sufficient energy, a nuclear reaction known as fusion occurs and energy considerably greater than that of the bombarding particle is released. This reaction is one of those which supply the energy of the thermonuclear bomb. It is also of major importance in the development of controlled thermonuclear reactors. *See* HEAVY WATER; NUCLEAR FUSION; TRITON. [L.K.]

Triton The nucleus of ${}^2_1H$ (tritium); it is the only known radioactive nuclide belonging to hydrogen. The triton is produced in nuclear reactors by neutron absorption in deuterium (${}^2_1H + {}^1_0n \rightarrow + \gamma$), and decays by β^- emission to ${}^3_2H$ with a half-life of 12.4 years. Much of the interest in producing ${}^3_1H$ arises from the fact that the fusion reaction ${}^3_1H + {}^1_1H \rightarrow {}^4_2H$ releases about 20 MeV of energy. Tritons are also used as projectiles in nuclear bombardment experiments. *See* NUCLEAR REACTION; TRITIUM. [H.E.D.]

Triuridales A small order of flowering plants, division Magnoliophyta (Angiospermae), subclass Alismatidae of the class Liliopsida (monocotyledons). The order consists of 2 families, the Triuridaceae, with about 70 species, and the Petrosaviaceae, with only about 4 species. They are terrestrial, mycotrophic herbs without chlorophyll. They grow in moist, tropical regions, especially of the Old World, and have attracted little botanical attention. *See* ALISMATIDAE; LILIOPSIDA. [A.Cr.]

Trochodendrales An order of flowering plants, division Magnoliophyta (Angiospermae), in the subclass Hamamelidae of the class Magnoliopsida (dicotyledons). The order consists of two families, the Trochodendraceae and Tetracentraceae, each with only a single species. They are trees of eastern and southeastern Asia with primitively vesselless wood and with unique, elongate, often branched idioblasts in the leaves. The flowers have several folded but scarcely sealed, lightly connate carpels and much reduced or no perianth. The group is of botanical interest because it tends to connect the subclass Hamamelidae with the ancestral Magnoliidae. *See* HAMAMELIDAE; MAGNOLIIDAE; MAGNOLIOPSIDA. [A.Cr.]

Trochophore A generalized but distinct type of free-swimming larva found in several invertebrate groups, including nemerteans, marine turbellarians, brachiopods, bryozoans, phoronids, mollusks, sipunculids, and some annelids. The form is somewhat pear-shaped (see illustration), and it is provided externally with a prominent circlet of cilia (a troch) and one or sometimes two accessory circlets. Anterior and posterior apical ciliary tufts and eyespots are often present. The digestive tract is complete and functional; paired nephridia with excretory tubules are present; muscle and nerve fibers, sense organs, and a band of mesoderm complete the internal structure. Presumably, the larva, which develops ontogenetically along many divergent lines, indicates the close evolutionary relationships of the groups it represents. *See* ANNELIDA; BRACHIOPODA; BRYOZOA; MOLLUSCA; PHORONIDA; SIPUNCULIDA; TURBELLARIA. [S.P.P.]

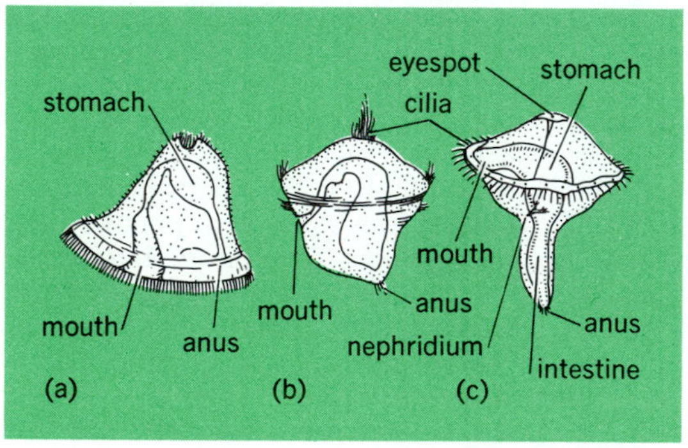

Some trochophore larvae. Pear-shaped form is common in all groups. (*a*) Bryozoan. (*b*) *Patella*, a mollusk. (*c*) *Polygardius*, an annelid. (*After T. I. Storer et al. General Zoology, 4th ed., McGraw-Hill, 1976*)

Trogoniformes A small order of birds that contains only the family Trogonidae, which has 37 species distributed pantropically. The trogons and quetzals are jay-sized birds with large heads, and tails that vary from medium length and squared to elongated and tapered. The dorsal plumage of trogons and quetzals is predominantly metallic green, with blue, violet, red, black, or gray in a few. The ventral feathers are bright red, yellow, or orange. Despite their vivid coloration, the birds are inconspicuous when sitting quietly. The legs of quetzals are short and their feet are weak, with the toes arranged heterodactylously, with the first and second toes reverted. Their flight is rapid, undulating, and brief; trogons rarely walk. Their diet consists of fruit and small invertebrates, and insects caught in flight as the bird darts out from a perch.

Trogons are nonmigratory, arboreal, and sedentary, and they can remain on a perch for hours. The monogamous pairs nest in solitude in a hollow tree or termite nest. After the eggs have been incubated by both parents, the naked hatchlings remain in the nest and are cared for by both. *See* AVES. [W.J.B.]

Trojan asteroids Asteroids located near the equilateral lagrangian stability points of the Sun-Jupiter system (see illustration). As shown by J. L. Lagrange in 1772, these are two of the five stable points in the circular, restricted, three-body system, the other three points being located along a line through the two most massive bodies in the system. In 1906 Max Wolf discovered an asteriod located near the lagrangian point following Jupiter in its orbit. Within a year, two more were found, one of which was located near the preceding lagrangian point. It was quickly decided to name these asteroids after participants in the Trojan War as given in Homer's *Iliad*. *See* ASTEROID. [E.F.T.]

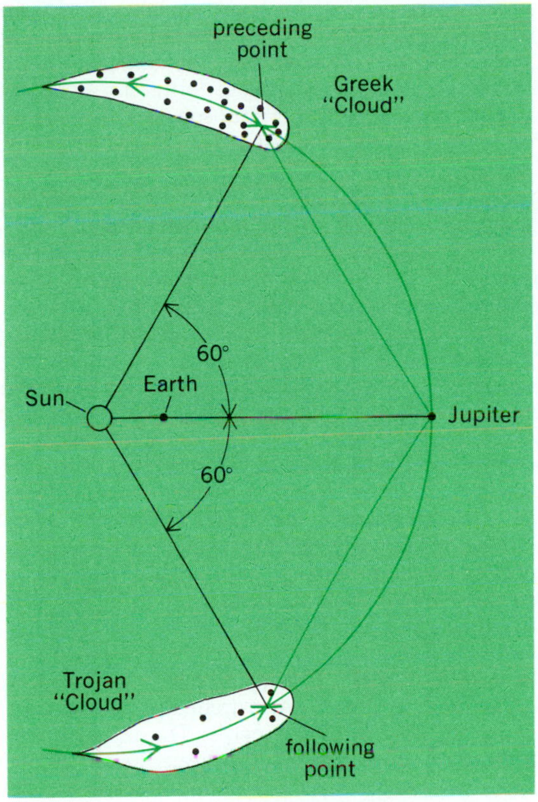

Lagrangian points and Trojan asteroids.

Trombidiformes A suborder of the Acarina (also known as Prostigmata) commonly called the trombidiform mites, more closely related to the Sarcoptiformes than to the other suborders. They are usually distinguished by presence of a respiratory system opening at or near the base of the chelicerae.

The Trombidiformes are probably the most heterogeneous group of mites, both morphologically and ecologically, varying from baglike forms with degenerate legs to the highly evolved, fully developed, parasitic forms. Economically, this group contains two families of plant-feeding mites of great importance to agriculture: the Tetranychidae (spider mites) and the Eriophyidae (bud mites or gall mites). Medically, the Trombiculidae (chiggers or red bugs) are important because the larval forms, which are parasites of vertebrates, can cause intense irritation to their hosts by their feeding. More seriously, some transmit a rickettsial disease, scrub typhus, to humans in the Far East and South Pacific regions. [E.W.B.]

Trophic ecology The study of the feeding relationships of organisms in communities and ecosystems. Trophic links between populations represent flows of organisms, organic energy, and nutrients. Trophic transfers are important in population dynamics, biogeochemistry, and ecosystem energetics. Many influential concepts of community and ecosystem organization have been based on trophic patterns. *See* BIOGEOCHEMISTRY.

The trophic links among populations in a community can be represented in a food web. In a food web diagram, the arrows most commonly represent potential or inferred feeding relationships. In more elaborate representations, the width of each arrow is correlated to the measured amount of energy or mass transferred from the prey to the predator per unit time. Most food webs are drawn as heuristic aids to show the main trophic links in a community. Therefore, minor links are omitted, and populations are often combined into functional groups for convenience (for example, phytoplankton). Trophic links in nature are highly variable through time, and published food webs are intended as averages. *See* FOOD WEB.

Trophic levels are functional groups formed by combining organisms that are the same number of trophic links removed from the base of the food web. Generally, the lowest trophic level, primary producers, is made up of green plants that obtain their energy from sunlight. Herbivores (primary consumers) make up the second trophic level, primary carnivores (secondary consumers) make up the third trophic level, and so forth. Decomposers (saprobes) are often viewed as a trophic level that is distinct from consumers of living prey. Combining the organisms into trophic levels collapses the food web into a food chain. Terrestrial food chains have as many as three or four trophic levels, while aquatic food chains have as many as five or six. *See* BIOLOGICAL PRODUCTIVITY; BIOMASS; ECOLOGY; ECOSYSTEM. [S.R.C.]

Tropic of Cancer The parallel of latitude about $23\frac{1}{2}°$ (23.45°) north of the Equator. The importance of this line lies in the fact that its degree of angle from the Equator is the same as the inclination of the Earth's axis from the vertical to the plane of the ecliptic. Because of this inclination of the axis and the revolution of the Earth in its orbit, the vertical overhead rays of the Sun may progress as far north as $23\frac{1}{2}°$. At no place north of the Tropic of Cancer will the Sun, at noon, be 90° overhead. On June 21, the summer solstice (Northern Hemisphere), the Sun is vertical above the Tropic of Cancer.

The Tropic of Cancer is the northern boundary of the equatorial zone called the tropics, which lies between the Tropic of Cancer and Tropic of Capricorn. *See* MATHEMATICAL GEOGRAPHY; SOLSTICE; TROPIC OF CAPRICORN. [V.H.E.]

Tropic of Capricorn The parallel of latitude approximately $23\frac{1}{2}°$ (23.45°) south of the Equator. It was named for the constellation Capricornus (the goat), for astronomical reasons which no longer prevail.

Because the Earth, in its revolution around the Sun, has its axis inclined $23\frac{1}{2}°$ from the vertical to the plane of the ecliptic, the Tropic of Capricorn marks the southern limit of the zenithal position of the Sun. Thus, on December 22 (Southern Hemisphere summer, but northern winter solstice) the Sun, at noon, is 90° above the horizon.

The Tropic of Capricorn is the southern boundary of the equatorial zone referred to as the tropics, which lies between the Tropic of Capricorn and the Tropic of Cancer. *See* MATHEMATICAL GEOGRAPHY; SOLSTICE; TROPIC OF CANCER. [V.H.E.]

Tropical meteorology The study of atmospheric structure and behavior in those parts of the world that lie astride the Equator—roughly between 25 or 30° north and south latitude, Weather and climate in the tropical belt differ grossly from those experienced farther poleward, partly as a result of the fact that the tropics export heat, moisture, and angular momentum to the higher latitudes. The weather satellite has become an important observing tool for tropical meteorology, covering the atmosphere over wide oceanic and desert regions that were previously virtually unknown. *See* METEOROLOGICAL SATELLITES; METEOROLOGY.

In contrast to the four-season cycle of higher latitudes, the tropics have only small seasonal temperature changes. Distinctions are made between dry and rainy seasons (one or two); monsoon climates with a three-season cycle (cool and dry, hot, rainy humid and cool); and desert climates (cool or hot, always dry). In the highlands the air temperature may be 18°F (10°C) colder than near sea level at which level there is usually only a 1.8°F (1°C) seasonal temperature change. *See* MONSOON METEOROLOGY.

Though rainfall may attain 60–80 in./year (150–200 cm/year) or more in many parts of the tropics, an abundant water supply is relatively rare. The high rate of evaporation caused by the heat greatly reduces the effectiveness of precipitation measured by the precipitation/evaporation ratio, which may be no higher, and even lower, in rainy parts of the tropics than in subarctic latitudes. Most tropical areas derive 70–90% of their annual water supply from a few organized rain systems of a 620-mi (1000-km) extent passing over a location in 1–2 days. It is not uncommon for the largest rainstorm of the year to account for 10% or more of the annual rainfall. [H.Ri.]

Tropolone A member of a class of nonbenzenoid aromatic compounds having the structure below. The acidity of

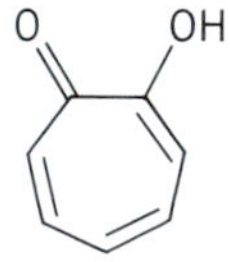

tropolone is intermediate between that of phenol and that of a typical carboxylic acid such as acetic acid.

Natural tropolones include the mold metabolites puberulic and puberulonic acids, sepedonin, and stipitatic acid. The thujaplicins (isopropyltropolones) occur in several species of *Cupressacea* evergreens and possess fungicidal properties that are believed to protect the wood of such trees against rot. The medicinally important alkaloid colchicine, from the meadow saffron (*Colchicum autumnale*), is also a tropolone derivative. *See* ALKALOID. [M.St.]

Tropopause The boundary between the troposphere and the stratosphere in the atmosphere. The tropopause is broadly defined as the lowest level above which the lapse rate (decrease) of temperature with height becomes less than 5.8°F mi^{-1} (2°C km^{-1}). In low latitudes the tropical tropopause is at a height of 9.3–11 mi at about −135°F (15–17 km at about 180 K), and the polar tropopause between tropics and poles is at about 6.2 mi at about −63°F (10 km at about 220 K). There is a well-marked "tropopause gap" or break where the tropical and polar tropopauses overlap at 30–40° latitude. The break is in the region of the subtropical jet stream and is of major importance for the transfer of air and tracers (humidity, ozone, radioactivity) between stratosphere and troposphere. The height of the tropopause varies seasonally and also daily with the weather systems, being higher and colder over anticyclones than over depressions. *See* AIR TEMPERATURE; ATMOSPHERE; STRATOSPHERE; TROPOSPHERE. [R.J.Mu.]

Troposphere The lowest major layer of the atmosphere. The troposphere extends from the Earth's surface to a height of 6–10 mi (10–16 km), the base of the stratosphere. It contains about four-fifths of the mass of the whole atmosphere. *See* ATMOSPHERE.

On the average, the temperature decreases steadily with height throughout this layer, with a lapse rate of about 19°F mi^{-1} (6.5°C km^{-1}), although shallow inversions (temperature increases with height) and greater lapse rates occur, particularly in the boundary layer near the Earth's surface. Appreciable water-vapor contents and clouds are almost entirely confined to the troposphere. Hence it is the seat of all important weather processes and the region where interchange by evaporation and precipitation (rain, snow, and so forth) of water substance between the surface and the atmosphere takes place. *See* CLIMATOLOGY; CLOUD PHYSICS; METEOROLOGY; WEATHER. [R.J.Mu.]

Tropospheric scatter A term applied to propagation of radio waves caused by irregularities in the refractive index of air. The phenomenon is predominant in the lower atmosphere; little or no scattering of importance occurs above the troposphere. Tropospheric scatter propagation provides very useful communication services but also causes harmful interference. For example, it limits the geographic separation required for frequency assignments to services such as television and frequency-modulation broadcasting, very-high-frequency omnidirectional ranges (VOR), and microwave relays. It is used extensively throughout most of the world for long-distance point-to-point services, particularly where high information capacity and high reliability are required. Typical tropospheric scatter relay facilities are commonly 200–300 mi (320–480 km) apart. Some single hops in excess of 500 mi (800 km) are in regular use. High-capacity circuits carry 200–300 voice circuits simultaneously. [R.S.Ki.]

Trunk (anatomy) The main body mass of an animal, exclusive of the head, neck, and extremities. The trunk is divided into thorax, abdomen, and pelvis. The thorax consists of the rib cage area and contains the heart, lungs, great blood vessels, and esophagus. The abdomen lies below the ribs and diaphragm, which separates it from the thorax, and contains the abdominal viscera, including the stomach, intestine, liver, spleen, and kidneys. The pelvis lies within the hipbones and contains the internal genital organs, bladder, and rectum.

The trunk is supported by the vertebral column, back muscles, and muscles of the body wall; the ribs and sternum in the thorax and the pelvic bones also help support the internal structures. The trunk forms appropriate areas of communication and attachment for the neck and limbs. [T.S.P.]

Truss A system of structural members lying in a single plane and joined at their ends to form a stable framework. A truss is used like a beam, particularly for bridge and roof construction. But because a truss can be made deeper than a beam with solid web and yet not weigh more, it is more economical for long spans and heavy loads. *See* BRIDGE; ROOF CONSTRUCTION.

The simplest truss is a triangle composed of three bars with ends pinned together. If small changes in the lengths of the bars are neglected, the relative positions of the joints do not change when loads are applied in the plane of the triangle at the apexes. Such simple trusses as a triangle, perhaps with the

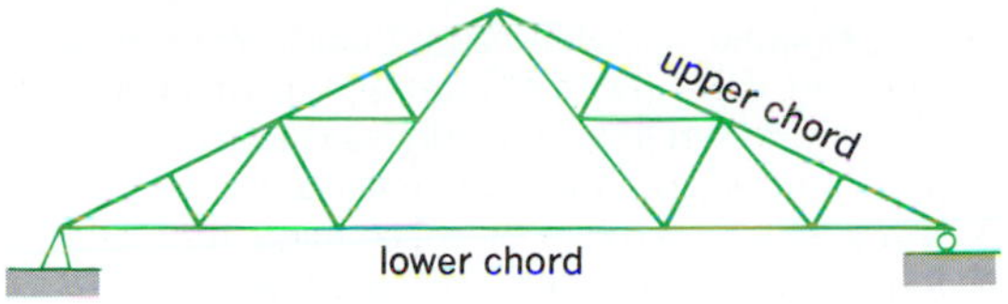

Fig. 1. A Fink truss, used in roof construction.

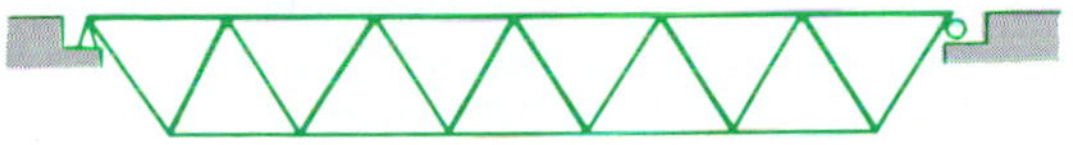

Fig. 2. A deck Warren truss, used in bridge construction. The load is on the upper chord.

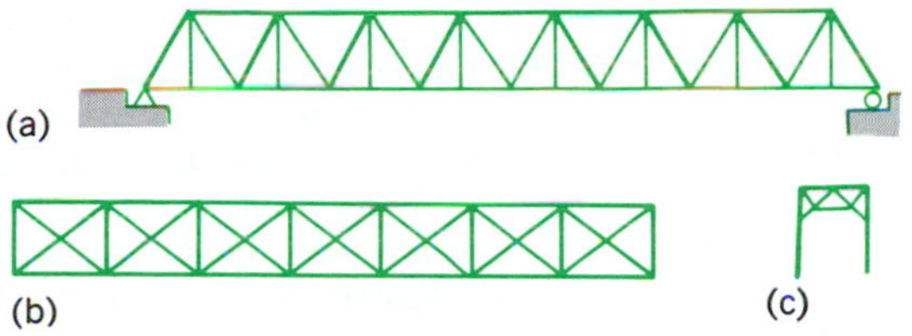

Fig. 3. Loads in through trusses are borne on the lower chord. (*a*) A through Warren truss with verticals. (*b*) Top-chord bracing as seen from above. (*c*) End-on view showing portal bracing.

addition of a vertical bar in the middle, are sometimes used to support peaked roofs of houses and other narrow structures. For longer spans, flat roofs, or bridges many triangles are combined to form a truss, as can be seen in the illustrations.

Roof trusses carry the weight of roof deck and framing and wind loads on the upper chord. They may also support a ceiling or other loads on the lower chord. An example is the Fink truss (Fig. 1). On the other hand, bridge trusses may carry loads on either chord. Deck trusses support loads on the upper chord (Fig. 2); through trusses, on the lower chord (Fig. 3). To maintain stability of truss construction, bracing must be used normal to the planes of the trusses. Usually framing is inserted between the trusses. [H.L.Bo./W.G.B.]

Trypanorhyncha An order of tapeworms of the subclass Cestoda, also known as the Tetrarhynchoidea. All are parasitic in the intestine of elasmobranch fishes. They are distinguished from all other tapeworm groups by having spiny, eversible proboscides on the head. The head also bears two or four shallow, weakly muscular suckers. A complete life history is not known for any trypanorhynchid, although larval forms have been found in the tissues of various marine invertebrates and teleost fishes. *See* CESTODA. [C.P.R.]

Trypanosomatidae A family of Protozoa, order Kinetoplastida, containing flagellated parasites which change their morphology; that is, they exhibit polymorphism during their life cycles. The life cycles of the organisms may involve only an invertebrate host, or an invertebrate and a vertebrate host, or an invertebrate and a plant host. Several distinct morphological forms are recognized: trypanosomal, crithidial, leptomonad, and leishmanial. Differentiation into genera is dependent upon the host infected as well as the morphologic types involved. None of the stages possesses a mouth opening, and nutritive elements are absorbed through the surface of the body; that is, the organisms are saprozoic.

Trypanosoma is the important genus of the family Trypanosomatidae from a number of standpoints. It contains the largest number of species infecting a wide variety of hosts such as mammals, birds, fishes, amphibians, and reptiles. Although most of the species cause no damage to the hosts, there are several which produce serious diseases in humans, domesticated animals, and wild animals. The pathogenic species are prevalent in Africa.

Leishmania is the second most important genus. Three species parasitize humans but have also been found naturally infecting dogs, cats, and perhaps other lower animals. The sand fly, *Phlebotomus*, transmits the parasite from vertebrate to vertebrate. *See* LEISHMANIASIS. [M.M.B./H.W.S.]

Trypanosomiasis A potentially fatal infection caused by parasites of the genus *Trypanosoma*.

The African trypanosomes, the cause of African trypanosomiasis or African sleeping sickness, are flagellated protozoan parasites. *Trypanosoma brucei rhodesiense* and *T. b. gambiense* cause disease in humans. *Trypanosoma brucei* is restricted to domestic and wild animals. The trypanosomes are transmitted by the tsetse fly (*Glossina*), which is restricted to the African continent. The trypanosomes are taken up in a blood meal and grow and multiply within the tsetse gut. After 2–3 weeks, depending upon environmental conditions, they migrate into the salivary glands, where they become mature infective forms, and are then transmitted by the injection of infected saliva into a new host during a blood meal. The survival of the tsetse fly is dependent upon temperature and humidity, and the fly is confined by the Sahara to the north and by the colder drier areas to the south, an area about the size of the United States. Approximately 50 million people live within this endemic area, and 15,000–20,000 new human cases of African trypanosomiasis are reported annually.

In humans and other mammals the trypanosomes are extracellular. During the early stages of infection, the trypanosomes are found in the blood and lymph but not in cerebrospinal fluid. There are fever, malaise, and enlarged lymph nodes. In the absence of treatment the disease becomes chronic and the trypanosomes penetrate into the cerebrospinal fluid and the brain. The symptoms are headaches, behavioral changes, and finally the characteristic sleeping stage. Without treatment the individual sleeps more and more and finally enters a comatose stage which leads to death. Treatment is more difficult if the infection is not diagnosed until the late neurological stage.

Trypanosoma cruzi, the cause of American trypanosomiasis (Chagas' disease), is predominantly an intracellular parasite in the mammalian host. During the intracellular stage, *T. cruzi* loses its flagellum and grows predominantly in cells of the spleen, liver, lymphatic system, and cardiac, smooth, and skeletal muscle. The cells of the autonomic nervous system are also frequently invaded. *See* MEDICAL PARASITOLOGY; PARASITOLOGY. [J.R.See.]

Tryptophan An amino acid considered essential for normal growth of animals. Tryptophan is the precursor of the sev-

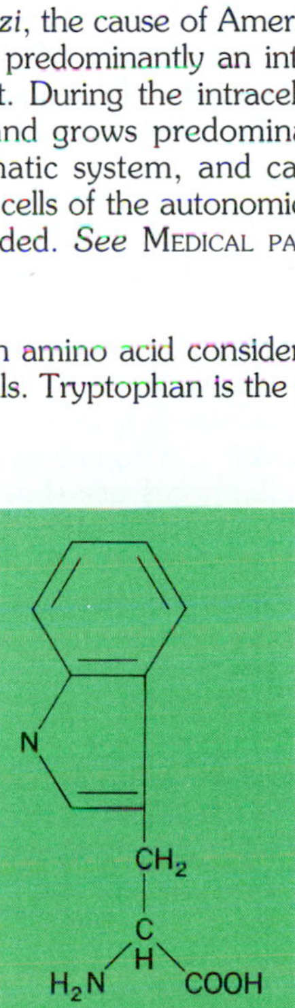

Tryptophan

eral important substances, including the plant growth hormone, indoleacetic acid; the animal hormone, serotonin (5-hydroxytryptamine); the vitamin, nicotinic acid; and certain eye pigments of insects. *See* AMINO ACIDS. [E.A.Ad.]

Tsunami A radially spreading, long-period gravity-wave system caused by any large-scale impulsive sea-surface disturbance. Being only weakly dissipative in deep water, major tsunamis can produce anomalous destructive wave effects at transoceanic distances. Because of their uncertainty of origin, infrequency (10 per century), and sporadicity, tsunami forecasting is impossible, but the progressive implementation, since 1946, of the effective International Tsunami Warning System has greatly reduced human casualties.

While minor tsunamis are occasionally produced by volcanic eruptions or submarine landslides, major events are generated by the sudden quasi-unitized dislocations of large fault blocks associated with the crumpling of slowly moving sea-floor crustal plates, where they abut normally against the continental plates. Such sources are predominately confined to tectonically active ocean margins, the majority of which ring the Pacific from Chile to Japan. As opposed to secular creep, tsunami-producing block dislocations are invariably associated with major shallow-focus (<19 mi or 30 km) earthquakes of intensity greater than 7 on the Gutenberg-Richter scale, followed by swarms of aftershocks of lesser intensity that decay in frequency over a week or two. *See* EARTHQUAKE; FAULT AND FAULT STRUCTURES. [W.G.V.D.]

Tube-still heater Direct-fired equipment used in processing petroleum. The basic feature of these heaters is that the fluid (liquid, vapor, or a mixture of both phases) is heated in carbon- or alloy-steel tubes, which in turn receive their heat from hot products of combustion. The fuel may be liquid, such as heavy fuel oil, or gas, such as natural gas or cracked refinery gas. The modern tube-still heater is the general-purpose tool for supplying heat at the higher temperature levels, that is, over 650–700°F (343–371°C), wherever required in the refinery for distillation, for preheating, for reactions such as pyrolysis, or for reboiling.

Most modern heaters are of one of two types, the box type, usually with horizontal tubes, or the cylindrical type with vertical tubes arranged in a circle. Within the radiant section, the combustion zone in which the burners are located, heat is given up to the tubes from three sources: radiation from the flame, radiation from the hot molecules (carbon dioxide and water vapor), and radiation from the gases within the chamber. The convection section is basically a recuperative zone in which the flue gases can be reduced in temperature while heating the coldest fluid to obtain economic heat recovery. *See* FURNACE; PETROLEUM PROCESSING. [W.E.L.]

Tuberculariaceae A family of fungi of the order Moniliales. The conidiophores are short, forming cushion-shaped aggregates. About 150 genera and 500 species are recognized. The genera are usually arranged into spore groups, depending on characteristics of the spore, such as the number of cells in the spore and shape and pigmentation of the spore. *See* DEUTEROMYCOTINA. [N.F.Bu.]

Tuberculosis A chronic infectious disease in humans and animals caused by tubercle bacilli, *Mycobacterium tuberculosis* or *M. bovis*. For much of the world population and for most of recorded history, tuberculosis has been the most frequent cause of long-lasting sickness and the principal cause of death. Infections may be either of two kinds: without recognized disease manifestations, and with disease production, tuberculosis.

On gaining access to the lung of a new host, tubercle bacilli are quickly engulfed by white blood cells. Within the protoplasm of human cells, the bacilli grow and multiply, although very slowly, invade neighboring cells, and form what is microscopically recognized as a tubercle. The bacillus-containing cells eventually are killed, perhaps by mechanical rupture from the multiplying bacilli, and other cells then phagocytize them. Much enlarged tubercles or groups of them become recognizable macroscopically as nodules. They may become fibrotic and eventually calcified. Tubercle bacilli may remain viable but quiescent in microscopic tubercles or lymph nodes for many decades, often for the life of the host. Consequences of the presence of these bacilli are a hyperactive state made evident by tuberculin sensitivity and an ever ready source of disease production. Breakdown of the body's disease resistance, for example, by stress or poor nutrition, may result in reactivation of the bacilli, and disease.

If the infection spreads into the bloodstream, systemic disease may follow. This is often characterized by a multitude of widely disseminated, very small nodules, the size of millet seeds and therefore designated miliary tuberculosis. Tubercle bacilli may set up infection in almost any part of the body. Tuberculous lymphandenitis involving neck glands of children and infections of genital organs, spleen, and kidneys are well known. Tuberculous meningitis and miliary tuberculosis are rapidly progressive, and fatal unless promptly treated.

A great variety of unusual treatments were used over the years, but all shrunk to relative insignificance following the introduction of effective drugs. Streptomycin (1943), *para*-aminosalicylate (PAS, 1946), and isoniazid (INH, 1952) were the first agents of many hundreds tested that proved to be effective in relatively nontoxic doses. A marked drop in the mortality rate about 1950 reflects the success of treatment with these drugs. Many additional effective drugs are now available. *See* PARA-AMINOSALICYLIC ACID; MYCOBACTERIAL DISEASES; STREPTOMYCIN. [E.H.R.]

Tubeworms The name given to marine polychaete worms (particularly to many species in the family Serpulidae) which construct permanent calcareous tubes on rocks, seaweeds, dock pilings, and ship bottoms. The individual tubes with hard walls of calcite-aragonite are firmly cemented to any hard substrate and to each other. Economically they are among the most important fouling organisms both on ship hulls (where they are second only to barnacles) and inside seawater cooling pipes of power stations.

About 340 valid species of serpulid tubeworms have been described. The majority are truly marine, but several species thrive in brackish waters of low salinity, and one species occurs in fresh waters in Karst limestone caves. The wide geographical distribution of certain abundant species owes much to human transport on the bottoms of relatively fast ships and occurred within the last 120 years. *See* ANNELIDA; BARNACLE; POLYCHAETA. [W.D.R.-H.]

Tubulidentata An order of mammals which contains a single living genus, the aardvark (*Orycteropus*) of Africa. Little is directly known of the origin and history of this peculiar group of mammals. The oldest undoubted record of the Tubulidentata is an early Miocene discovery in East Africa. Pliocene records in India and the Mediterranean border indicate the aardvarks had dispersed from Africa by that time. Tubular dentine reminiscent of the aardvarks occurs in *Tubulodon* from the early Eocene of North America, but the fragmentary material does not permit further comparison. *See* AARDVARK; MAMMALIA. [R.H.T.]

Tufa A spongy, porous limestone formed by precipitation from evaporating spring and river waters; also known as calcareous sinter. Calcium carbonate commonly precipitates from supersaturated waters on the leaves and stems of plants growing around the springs and pools and preserves some of their plant structures. Tufa tends to be fragile and friable. *See* LIMESTONE; TRAVERTINE. [R.Si.]

Tuff Consolidated volcanic ash, composed largely of fragments (less than 0. 16 in. or 4 mm) produced directly by volcanic eruption. Much of the fragmental material represents finely comminuted crystals and rocks. Much, however, was ejected as blobs of liquid lava which rapidly cooled to form volcanic glass. Expansion and solidification to glass may be more or less contemporaneous so that highly vesicular bodies are rapidly shattered to produce vast quantities of tiny particles (ash). Upon consolidation volcanic ash yields tuff. With increase in the size of fragments, tuff passes into lapilli tuff, volcanic breccia, and agglomerate. *See* PYROCLASTIC ROCKS; VOLCANIC GLASS. [C.A.C.]

Tularemia A widespread infection of wild rodents endemic in the Americas, Europe, Asia, and Africa. It is maintained through insect reservoirs and vectors from which the disease may be transmitted to humans. The causative organism, isolated from a plaguelike disease of ground squirrels in Tulare County, California, was named *Pasteurella tularensis* (now *Francisella tularensis*). The infection may be generalized, or it may be localized in the eyes, skin, or lymph nodes or in the respiratory or gastrointestinal tracts. Recovery is followed by reliable immunity. [K.F.M.]

Tulip tree A tree *Liriodendron tulipfera*, also known in forestry as yellow poplar, belonging to the magnolia family, Magnoliaceae. One of the largest and most valuable hardwoods of eastern North America, it is native from southern New England and New York westward to southern Michigan, and south to Louisiana and northern Florida.

Botanically, this tree is distinguished by leaves which are squarish at the tip as if cut off, true terminal buds flattened and covered by two valvate scales, an aromatic odor resembling that of magnolia, and cone-shaped fruit which is persistent in winter. The name tulip refers to the large greenish-yellow and orange-colored flowers.

The wood of the tulip tree is light yellow to brown, hence the common name yellow poplar, which is a misnomer. It is a soft and easily worked wood, used for construction, interior finish, containers (boxes, crates, baskets), woodenware, excelsior, veneer, and sometimes for paper pulp. *See* MAGNOLIALES. [A.H.G./K.P.D.]

Tumbling mill A grinding and pulverizing machine consisting of a shell or drum rotating on a horizontal axis. The material to be reduced in size is fed into one end of the mill. The mill is also charged with grinding material such as iron balls. As the mill rotates, the material and grinding balls tumble against each other, the material being broken chiefly by attrition.

Tumbling mills are variously classified as pebble, ball, or rod depending on the grinding material, and as cylindrical, conical, or tube depending on the shell shape. *See* CRUSHING AND PULVERIZING; GRINDING MILL; PEBBLE MILL. [R.M.H.]

Tumor Literally, a swelling; in the past the term has been used in reference to any swelling of the body no matter what the cause. However, the word is now being used, almost exclusively, to refer to a neoplastic mass, and the more general usage is being discarded. *See* NEOPLASIA.

A neoplastic mass or neoplasm is a pathological lesion characterized by the progressive or uncontrolled proliferation of cells. The cells involved in the neoplastic growth have an intrinsic heritable abnormality such that they are not regulated properly by normal methods. The stimulus which elicits this growth is not usually known.

It is common to divide tumors into benign or malignant. The decision as to which category a tumor should be assigned is usually based on information gained from gross or microscopic examination, or both. Benign neoplasms usually grow slowly, remain so localized that they cause little harm, and generally can be successfully and finally removed. Malignant or cancerous neoplasms tend to grow rapidly, spread throughout the body, and recur if removed. Not all tumors which have been classified as benign are harmless to the host, and some can cause serious problems. Difficulties may occur as a result of mechanical pressure.

The cells of benign tumors are well differentiated. This means that the cells are very like the normal tissue in size, structure, and spatial relationship. The cells forming the tumor usually function normally. Cell proliferation usually is slow enough so that there is not a large number of immature cells. As the cellular mass increases in size, most benign tumors develop a fibrous capsule around them which separates them from the normal tissue. The cells of a benign tumor remain at the site of origin and do not spread throughout the body. Anaplasia is not seen in benign tumors.

The cells of malignant tumors may be well differentiated, but most have some degree of anaplasia. Anaplastic cells tend to be larger than normal and are abnormal, even bizarre, in shape. The nuclei tend to be very large and irregular. Malignant tumors may be partially but never completely encapsulated. The cells of the cancer infiltrate and destroy surrounding tissue. They have the ability to metastasize; that is, cells from the primary tumor are disseminated to other regions of the body where they are able to produce secondary tumors called metastases.

In most cases the formation of a neoplasm is irreversible. It results from a permanent cellular defect which is passed on to daughter cells. Medical appraisal of the tumor is generally recommended. *See* CANCER (MEDICINE); ONCOLOGY; TUMOR VIRUSES. [N.K.M.; C.Qu.]

Tumor viruses Viruses which cause a wide variety of malignant and benign tumors in practically all animal species studied. These viruses are classified in two broad categories: the DNA (deoxyribonucleic acid) and RNA (ribonucleic acid) tumor viruses, depending upon the nucleic acid in the viral genome.

No malignant disease of humans has been unequivocally associated with a virus. However, several human malignancies are suspected to have, at least in part, a viral etiology. These include Burkitt's lymphoma and nasopharyngioma as well as the nonmalignant illness infectious mononucleosis, all of which may be due to a herpes virus-like agent, Epstein-Barr virus; cervical carcinoma may be caused by herpes simplex virus type 2, which infects the genital region; human osteosarcomas have been the source of extracts that cause similar tumors in hamsters. *See* ANIMAL VIRUS; ONCOLOGY; TUMOR. [W.Sch.]

Tuna Any of the large pelagic marine fishes which form the family Thunnidae, usually included in the order Perciformes. These fishes have a worldwide distribution.

The tunas have no scales on the posterior part of the body, and those on the anterior are fused to form an armored covering. The flesh of many is dark red. The body of these fish is streamlined, with a crescent-shaped tail and a narrow caudal peduncle. *See* PERCIFORMES. [C.B.C.]

Tunable laser A laser whose wavelength can be varied continuously over a wide spectral range. The vigorous advances of laser spectroscopy since about 1970 are largely due to the development of laser sources with this property. *See* LASER SPECTROSCOPY.

Tunable coherent sources are available for any wavelength region from the far infrared into the near-vacuum ultraviolet. Tunable infrared lasers include high-pressure molecular gas lasers, semiconductor diode lasers, spin-flip Raman lasers, and color-center lasers. In the visible region, organic dye lasers have proved particularly versatile and powerful spectroscopic sources. Both pulsed and continuous-wave dye lasers cover the entire visible spectrum, including the bordering near-infrared and near-ultraviolet regions. Intense shorter-wave ultraviolet radiation can be generated with excimer lasers over limited regions.

The generation of harmonic frequencies or sum frequencies in nonlinear optical crystals or gases provides a valuable method to produce tunable coherent ultraviolet radiation from laser light of longer wavelengths. Optical parametric oscillators and difference-frequency crystal mixers offer corresponding alternatives for the generation of coherent infrared radiation. Stimulated Raman scattering can be used to shift the frequency of a tunable laser by some integer multiple of a molecular vibration frequency and thus to extend the tuning range. These nonlinear frequency-mixing and shifting techniques work best with the intense radiation from pulsed lasers. *See* LASER; NONLINEAR OPTICS. [T.W.Ha.]

Tundra An area supporting some vegetation beyond the northern limit of trees, between the upper limit of trees and the lower limit of perennial snow on mountains, and on the fringes of the Antarctic continent and its neighboring islands. The term is of Lapp or Russian origin, signifying treeless plains of northern regions. Biologists, and particularly plant ecologists, sometimes use the term tundra in the sense of the vegetation of the tundra landscape. Tundra has distinctive characteristics as a kind of landscape and as a biotic community, but these are expressed with great differences according to the geographic region.

Characteristically tundra has gentle topographic relief, and the cover consists of perennial plants a few inches to a few feet or a little more in height. The general appearance during the growing season is that of a grassy sward in the wetter areas, a matted spongy turf on mesic sites, and a thin or sparsely tufted lawn or lichen heath on dry sites. In winter, snow mantles most of the surface. By far, most tundra occurs where the mean annual temperature is below the freezing point of water, and perennial frost (permafrost) accumulates in the ground below the depth of annual thaw and to depths at least as great as 1600 ft (500 m). *See* PERMAFROST. [W.S.B.]

Tuner A device containing circuits that can be tuned to the carrier frequency of a desired transmitter in the frequency range for which that tuner is designed. The tuner serves to select the desired carrier frequency while rejecting the carrier frequencies of all other stations that may be on the air at that time. A tuner for a television receiver ordinarily contains only the radio-frequency amplifier, local oscillator, and mixer stages. A radio tuner usually contains, in addition, the intermediate-frequency amplifier and second-detector stages so that it can feed an audio signal directly into a separate audio amplifier and loudspeaker system. *See* RADIO RECEIVER; TELEVISION RECEIVER; TUNING. [J.Mar.]

Tung tree The plant *Aleurites fordii*, a species of the spurge family (Euphorbiaceae). The tree, native to central and western China, is the source of tung oil. It has been grown successfully in the southern United States. The globular fruit has three to seven large, hard, rough-coated seeds containing the oil, which is expressed after the seeds have been roasted. Tung oil is used to produce a hard, quick-drying, superior varnish, which is less apt to crack than other kinds. The foliage, sap, fruit, and commercial tung meal contain a toxic saponin, which causes gastroenteritis in animals that eat it. *See* DRYING OIL; EUPHORBIALES; VARNISH. [P.D.St./E.L.C.]

Tungstate One of a group of compounds containing tungsten in the 6+ oxidation state and derived from tungstic acid, H_2WO_4. Both the normal tungstates, such as Na_2WO_4, and the molybdates form condensed ions, such as isopolytungstates. These isopolytungstates are usually classified into metatungstates and paratungstates.

Heteropoly compounds in which another anhydride of an element such as silicon combines with the tungsten to form polymers are also known. Sodium silicotungstate, $Na_4SiW_{12}O_{40}$, is a compound of this type. Most of the polytungstates are soluble, whereas the normal tungstates are sparingly soluble with the exception of the salts of the alkali metals and magnesium.

Tungstates are used to flameproof fabrics and to manufacture phosphorescent screens. *See* TUNGSTEN. [E.E.W.]

Tungsten A chemical element, W, atomic number 74, and atomic weight 183.85. The metal has a body-centered cubic structure and a silver-gray metallic luster. Its melting point of 3410°C (6170°F) is the highest of any metal. The metal exhibits a low vapor pressure, high density, and superior strength at high temperature in the absence of air, and is extremely hard.

1	2	3	4	5	6	7	8	9	10	11	12	13	14	15	16	17	18
1 H																	2 He
3 Li	4 Be											5 B	6 C	7 N	8 O	9 F	10 Ne
11 Na	12 Mg											13 Al	14 Si	15 P	16 S	17 Cl	18 Ar
19 K	20 Ca	21 Sc	22 Ti	23 V	24 Cr	25 Mn	26 Fe	27 Co	28 Ni	29 Cu	30 Zn	31 Ga	32 Ge	33 As	34 Se	35 Br	36 Kr
37 Rb	38 Sr	39 Y	40 Zr	41 Nb	42 Mo	43 Tc	44 Ru	45 Rh	46 Pd	47 Ag	48 Cd	49 In	50 Sn	51 Sb	52 Te	53 I	54 Xe
55 Cs	56 Ba	71 Lu	72 Hf	73 Ta	74 W	75 Re	76 Os	77 Ir	78 Pt	79 Au	80 Hg	81 Tl	82 Pb	83 Bi	84 Po	85 At	86 Rn
87 Fr	88 Ra	103 Lr	104 Rf	105 Db	106 Sg	107 Bh	108 Hs	109 Mt	110	*111*	*112*	*113*	*114*	*115*	*116*	*117*	*118*

lanthanide series	57 La	58 Ce	59 Pr	60 Nd	61 Pm	62 Sm	63 Eu	64 Gd	65 Tb	66 Dy	67 Ho	68 Er	69 Tm	70 Yb
actinide series	89 Ac	90 Th	91 Pa	92 U	93 Np	94 Pu	95 Am	96 Cm	97 Bk	98 Cf	99 Es	100 Fm	101 Md	102 No

Chemically, tungsten is relatively inert. It is not readily attacked by the common acids, alkalies, or aqua regia. It reacts with a mixture of concentrated nitric acids and hydrofluoric acid. Molten oxidizing salts such as sodium nitrite attack tungsten rapidly. Gaseous chlorine, bromine, iodine, carbon dioxide, carbon monoxide, and sulfur react with tungsten only at high temperatures. Carbon, boron, silicon, and nitrogen also form compounds with tungsten at elevated temperatures; hydrogen does not. The table lists several of the important physical properties of tungsten.

Ferrous alloys consume 40% of the tungsten mined. When added to iron or steel, tungsten improves high-temperature

Some physical properties of tungsten

Property	Value
Melting point	3410°C (6170°F)
Boiling point	5930°C (10700°F)
Density, metal powder	2.0–4.8 g/cm³ (1.2–2.8 oz/in.³)
drawn wire	19.3 g/cm³ (11.2 oz/in.³)
Specific heat, 0°C (32°F)	0.03–0.05 cal/g-°C [0.12–0.21 J/(g - °C]
2600°C (4700°F)	0.06 cal/g-°C [0.25 J/(g - °C)]

strength and hardness. Tungsten carbide (representing 38% of all tungsten) has replaced diamond for many die and drilling applications. It is one of the best hard tool materials, retaining its properties at elevated temperatures. Pure tungsten metal in the form of wire, rod, and sheet (15%) is important to the electric lamp, electronics, and electrical industries. Other applications include welding rod, x-ray targets, lead wires, cathodes for power tubes, and automotive and aircraft distributor points.

Tungsten compounds embrace oxidation states of tungsten from 2+ to 6+, the higher oxidation states being most stable. Tungsten chemistry resembles that of chromium and molybdenum, which also occupy the same subgroup in the periodic table. Tungsten aqueous chemistry is complicated by its tendency to form complex ions.

Tungsten forms four well-defined, stable oxides and two carbides. Other important compounds include the carbonyl, nitride, boride, phosphide, silicide, and sulfide. [W.D.Wi.]

Tunicata A group of marine animals forming a subdivision (subphylum Tunicata or Urochordata) of the Chordata. They are characterized by a perforated pharynx or branchial sac used for food collection, a dorsal notochord restricted to the tail of the larva (and the adult in one class), absence of mesodermal segmentation or a recognizable coelom, and secretion of an outer covering (the test or tunic) which contains large amounts of polysaccharides related to cellulose. Three classes are usually recognized: the sessile Ascidiacea (sea squirts or ascidians); planktonic Thaliacea (salps, doliolids, and pyrosomids); and Larvacea (Appendicularia or Copelata), minute planktonic forms with tails living inside a specialized test or house adapted for filtering and food gathering. Approximately 2000 species of Tunicata are recognizable. The group is found in all parts of the ocean. *See* APPENDICULARIA; ASCIDIACEA; CHORDATA; THALIACEA. [I.Go.]

Tuning The process of adjusting the inductance or capacitance, or both, in a tuned circuit—for example in a radio, television, or radar receiver or transmitter—so as to obtain optimum performance at a selected frequency. The tuning procedure that is carried out during manufacture or servicing of equipment generally involves the adjustment of screwdriver-type controls located inside or at the rear of the equipment. These are adjusted in such a way that the main tuning control on the front of the panel will serve to adjust all tuning circuits in the equipment simultaneously when a change of frequency is desired. *See* RADIO RECEIVER; RESONANCE (ALTERNATING-CURRENT CIRCUITS); TELEVISION RECEIVER. [J.Mar.]

Tuning fork A steel instrument consisting of two prongs and a handle which, when struck, emits a tone of fixed pitch. Because of their simple mechanical structure, purity of tone, and constant frequency, tuning forks are widely used as standards of frequency in musical acoustics. In its electrically driven form, a tuning fork serves to control electric circuits by producing frequency standards of high accuracy and stability. A tuning fork is essentially a transverse vibrator (see illustration). *See* VIBRATION. [L.E.K.]

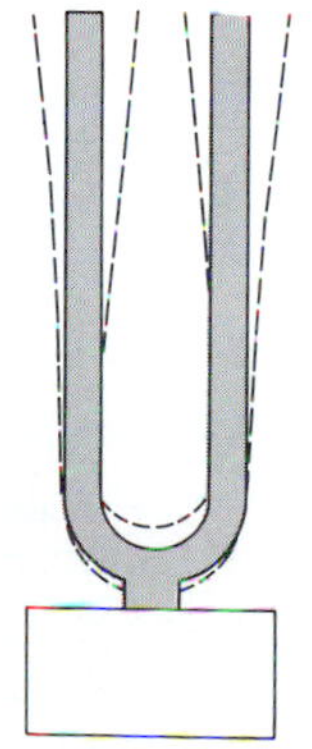

A tuning fork vibrating at its fundamental frequency.

Tunnel A general term for subterranean passages. Tunnels are used for aqueducts and sewers; for carrying railroad and vehicular traffic under rivers, through mountains, and below cities; and for specific kinds of underground installations, such as hydroelectric plants.

The construction of a tunnel begins with excavation of earth or rock. Excavating methods vary with the character of the material to be removed. Rock tunnels require blasting. If the rock is structurally poor, supports must often be placed under the tunnel ceiling, or arch, as the tunnel is being driven. Tunnels driven through earth do not require blasting, but arch supports are almost always necessary. Tunnels under rivers in soft silts are usually driven with the aid of a short cylinder called a shield, which is pushed through the silt ahead of the excavation to provide advance support of the arch. After excavation has been completed, the exposed tunnel wall is usually lined with concrete. Vehicular tunnel construction often requires the added installation of tile walls and of lighting and ventilation equipment. [W.Her.]

Tunnel diode A two-terminal semiconductor junction device (also called the Esaki diode) which does not show rectification in the usual sense, but exhibits a negative resistance region at very low voltage in the forward-bias characteristic and a short circuit in the negative-bias direction.

This device is a version of the semiconductor *pn* junction diode which is made of a *p*-type semiconductor, containing mobile positive charges called holes (which correspond to the vacant electron sites), and an *n*-type semiconductor, containing mobile electrons (the electron has a negative charge). The densities of holes and electrons in the respective regions are made extremely high by doping a large amount of the appropriate impurities with an abrupt transition from one region to the other. In semiconductors, the conduction band for mobile electrons is separated from the valence band for mobile holes by an energy gap, which corresponds to a forbidden region. Therefore, a narrow transition layer from *n*-type to *p*-type, 5 to 15 nanometers thick, consisting of the forbidden region of the energy gap, provides a tunneling barrier. Since the tunnel diode exhibits a negative incremental resistance with a rapid response, it is capable of serving as an active element for amplification, oscillation, and switching in electronic circuits at high frequencies. The discovery of the diode, however, is probably more significant from the scientific aspect because it opened up a new field of research—tunneling in solids. *See* BAND THEORY OF SOLIDS; JUNCTION DIODE; NEGATIVE-RESISTANCE CIRCUITS; SEMICONDUCTOR; SEMICONDUCTOR DIODE; TUNNELING IN SOLIDS. [L.E.]

Tunneling in solids A quantum-mechanical process which permits electrons to penetrate from one side to the other through an extremely thin potential barrier to electron flow. The barrier would be a forbidden region if the electron were treated as a classical particle. A two-terminal electronic device in which such a barrier exists and primarily governs the transport characteristic (current-voltage curve) is called a tunnel junction. *See* QUANTUM MECHANICS.

During the infancy of the quantum theory, L. de Broglie introduced the fundamental hypothesis that matter may be endowed with a dualistic nature—particles such as electrons, alpha particles, and so on, may also have the characteristics of waves. This hypothesis found expression in the definite form now known as the Schrödinger wave equation, whereby an electron or an alpha particle is represented by a solution to this equation. The nature of such solutions implies an ability to penetrate classically forbidden regions of negative kinetic energy and a probability of tunneling from one classically allowed region to another. The concept of tunneling, indeed, arises from this quantum-mechanical result. The subsequent experimental manifestations of this concept, such as high-field electron emission from cold metals, alpha decay, and so on, in the 1920s, can be regarded as one of the early triumphs of the quantum theory. *See* FIELD EMISSION; RADIOACTIVITY; SCHRÖDINGER'S WAVE EQUATION.

The tunnel diode (also called the Esaki diode), discovered in 1957 by L. Esaki, demonstrated the first convincing evidence of electron tunneling in solids. *See* TUNNEL DIODE.

Negative resistance phenomena can be observed in novel tunnel structures in semiconductors. Double tunnel barriers and periodic structures with a combination of semiconductors exhibit resonant tunneling and negative resistance effects. *See* SEMICONDUCTOR HETEROSTRUCTURES.

Tunneling had been considered to be a possible electron transport mechanism between metal electrodes separated by either a narrow vacuum or a thin insulating film usually made of metal oxides. In 1960, I. Giaever demonstrated that, if one or both of the metals were in a superconducting state, the current-voltage curve in such metal tunnel junctions revealed many details of that state.

In 1962, B. Josephson made a penetrating theoretical analysis of tunneling between two superconductors by treating the two superconductors and the coupling process as a single system, which would be valid if the insulating oxide were sufficiently thin, say 2 nanometers. His theory predicted the existence of a supercurrent, arising from tunneling of the bound electron pairs. This led to two startling conclusions: the dc and ac Josephson effects. The dc effect implies that a supercurrent may flow even if no voltage is applied to the junction. The ac effect implies that, at finite voltage V, there is an alternating component of the supercurrent which oscillates at a frequency of 483.6 MHz per microvolt of voltage across the junction, and is typically in the microwave range. *See* JOSEPHSON EFFECT. [L.E.]

Tupelo A tree belonging to the genus *Nyssa* of the sour gum family, Nyssaceae. The most common species is *N. sylvatica*, variously called pepperidge, black gum, or sour gum, the authorized name being black tupelo. Tupelo grows in the easternmost third of the United States; southern Ontario, Canada; and Mexico.

The tree can be identified by the comparatively small, obovate, shiny leaves and by branches that develop at a wide angle from the axis. The fruit is a small blue-black drupe, a popular food for birds. The wood is yellow to light-brown and hard to split because of the twisted grain. Tupelo is used for boxes, baskets, and berry crates, and as backing on which veneers of rarer and more expensive woods are glued. It is also used for flooring, rollers in glass factories, hatters' blocks, and gunstocks. *See* MYRTALES. [A.H.G./K.P.D.]

Turbellaria A class of the phylum Platyhelminthes commonly known as the flatworms. These animals are chiefly free-living and have simple life histories. The bodies are elongate and flat to oval or circular in cross section. Their length ranges from less than 0.04 in. (1 mm) to several inches, but may exceed 20 in. (50 cm) in land planaria. Large forms are often brightly colored. This class, which numbers some 3400 described species, is ordinarily subdivided into the orders Acoela, with 200 species; Rhabdocoela, 1110 species; Alloeocoela, 350 species; Tricladida, 1000 species; and Polycladida, 750 species. Although widely distributed in fresh and salt water and moist soil, they are usually overlooked because of their generally small size, secretive habits, and inconspicuous color. *See* ACOELA; ALLOEOCOELA; POLYCLADIDA; RHABDOCOELA; TRICLADIDA. [E.R.J.]

Turbidite A bed of sediment or sedimentary rock that was deposited from a turbidity current. The term turbidite is fundamentally genetic and interpretive in nature, rather than being a descriptive term (like common rock names). Turbidites are classic sedimentary rocks, but they may be composed of silicic grains (quartz, feldspar, rock fragments) and therefore be a type of sandstone, or they may be composed of carbonate grains and therefore be a type of limestone. A geologist's description of a rock as a turbidite is actually an expression of an opinion that the rock was deposited by a turbidity current, rather than being a description of a particular type of rock. *See* TURBIDITY CURRENT. [G.V.M.]

Turbidity current A submarine flow of sediment or sediment-laden water which occurs when an unstable mass of sediment at the top of a relatively steep slope is jarred loose and slides downslope. As the slide or slump travels downslope, it becomes more fluid because of the loss of internal cohesive strength and because of inmixing of the superadjacent water.

Turbidity currents occur at the edge of the continental slope, in the vicinity of river mouths, and off prominent capes. They are triggered by earthquakes, hurricanes, floods, or simply by the bedload transport of rivers debouching at the edge of continental slopes.

The lowered sea level of the glacial stages must have forced most rivers of the world to empty at or near the top of the continental slope. Under those conditions turbidity currents must have been much more important than at present, and they undoubtedly transported much of the sediment deposited in ancient geosynclines. *See* CONTINENTAL MARGIN; MARINE GEOLOGY; MARINE SEDIMENTS; SUBMARINE CANYON; TURBIDITE. [B.C.He.]

Turbine A machine for generating rotary mechanical power from the energy in a stream of fluid. The energy, originally in the form of head or pressure energy, is converted to velocity energy by passing through a system of stationary and moving blades in the turbine. Changes in the magnitude and direction of the fluid velocity are made to cause tangential forces on the rotating blades, producing mechanical power via the turning rotor. Turbines effect the conversion of fluid to mechanical energy through the principles of impulse, reaction, or a mixture of the two (see illustration).

The fluids most commonly used in turbines are steam, hot air or combustion products, and water. Steam raised in fossil fuel-fired boilers or nuclear reactor systems is widely used in turbines for electrical power generation, ship propulsion, and

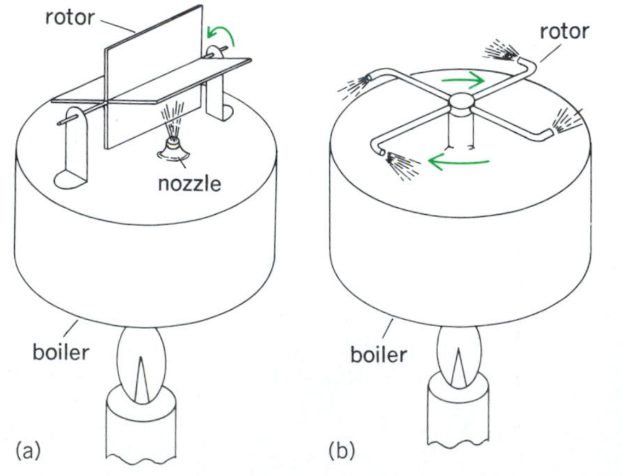

Turbine principles. (*a*) Impulse. (*b*) Reaction.

mechanical drives. The combustion gas turbine has these applications in addition to important uses in aircraft propulsion. Water turbines are used for electrical power generation. *See* GAS TURBINE; HYDRAULIC TURBINE; IMPULSE TURBINE; PELTON WHEEL; REACTION TURBINE; STEAM TURBINE; TURBINE PROPULSION; TURBOJET. [F.G.B.]

Turbine propulsion Propulsion of a vehicle by means of a gas turbine. Gas turbines have come to dominate most areas of common carrier aircraft propulsion, have made significant inroads into the propulsion of surface ships, and are being incorporated into military tanks.

The primary power producer common to all gas turbines used for propulsion is the core or gas generator, operating on a continuous flow of air as working fluid. The air is compressed in a rotating compressor, heated at constant pressure in a combustion chamber burning a liquid hydrocarbon fuel, and expanded through a core turbine which drives the compressor. This manifestation of the Brayton thermodynamic cycle generates a continuous flow of high-pressure, high-temperature gas which is the primary source of power for a large variety of propulsion schemes. The turbine is generally run as an open cycle; that is, the airflow is ultimately exhausted to the atmosphere rather than being recycled to the inlet. *See* BRAYTON CYCLE.

The residual energy available in the high-temperature, high-pressure airstream exiting from the core is used for propulsion in a variety of ways. For traction-propelled vehicles (buses, trucks, automobiles, military tanks, and most railroad locomotives), the core feeds a power turbine which extracts the available energy from the core exhaust and provides torque to a high-speed drive shaft as motive power for the vehicle. With a free-turbine arrangement, this power turbine is a separate shaft, driving at a speed not mechanically linked to the core speed. With a fixed turbine, this power turbine is on the same shaft as the core turbine, and must drive at the same speed as the core spool. In traction vehicles the power turbine generally drives through a transmission system which affords a constant- or a variable-speed reduction to provide the necessary torque-speed characteristics to the traction wheels.

Aircraft, ships, and high-speed land vehicles, which cannot be driven by traction, are propelled by reaction devices. Some of the ambient fluid around the vehicle (that is, the water for most ships, and the air for all other vehicles) is accelerated by some turbomachinery (a ship propeller, aircraft propeller, helicopter rotor, or a fan integrated with the core to constitute a turbofan engine). The reaction forces on this propulsion turbomachinery, induced in the process of accelerating the ambient flow, provide the propulsion thrust to the vehicle. In all these cases, motive power to the propeller or fan is provided by a power turbine extracting power from the gas generator exhaust. In the case of a jet engine, exhaust from the gas generator is accelerated through a jet nozzle, so that the reaction thrust is evolved in the gas generator rather than in an auxiliary propeller or fan. Indeed, in turboprop and turbofan engines, both forms of reaction thrust (from the stream accelerated by the propeller or fan and from the stream accelerated by the core and not fully extracted by the power turbine) are used for propulsion. *See* GAS TURBINE; JET PROPULSION; TURBOFAN; TURBOJET; TURBOPROP. [F.F.E.]

Turbocharger An air compressor or supercharger on an internal combustion piston engine that is driven by the engine exhaust gas to increase or boost the amount of fuel that can be burned in the cylinder, thereby increasing engine power and performance. On an aircraft piston engine, the turbocharger allows the engine to retain its sea-level power rating at higher altitudes despite a decrease in atmospheric pressure. *See* RECIPROCATING AIRCRAFT ENGINE; SUPERCHARGER.

The turbocharger is a turbine-powered centrifugal supercharger. It consists of a radial-flow compressor and turbine mounted on a common shaft. The turbine uses the energy in the exhaust gas to drive the compressor, which draws in outside air, precompresses it, and supplies it to the cylinders at a pressure above atmospheric pressure.

Common turbocharger components include the rotor assembly, bearing housing, and compressor housing. The shaft bearings usually receive oil from the engine lubricating system. Engine coolant may circulate through the housing to aid in cooling. *See* ENGINE COOLING; INTERNAL COMBUSTION ENGINE. [D.L.An.]

Turbodrill A rotary tool used in drilling oil or gas wells in which the bit is rotated by a turbine motor inside the well. The principal difference between rotary and turbodrilling lies in the manner in which power is applied to the rotating bit or cutting tool. In the rotary method, the bit is attached to a drill pipe rotated through power supplied on the surface. In the turbodrill method, power is generated at the bottom of the hole by a mud-operated turbine. *See* ROTARY TOOL DRILL.

The turbodrill (see illustration) consists of four basic components: the upper, or thrust, bearing; the turbine; the lower bearing; and the bit. In operation, mud is pumped through the drill pipe, passing through the thrust bearing and into the turbine. In the turbine, stators attached to the body of the tool divert the mud flow onto rotors attached to the shaft. This causes the

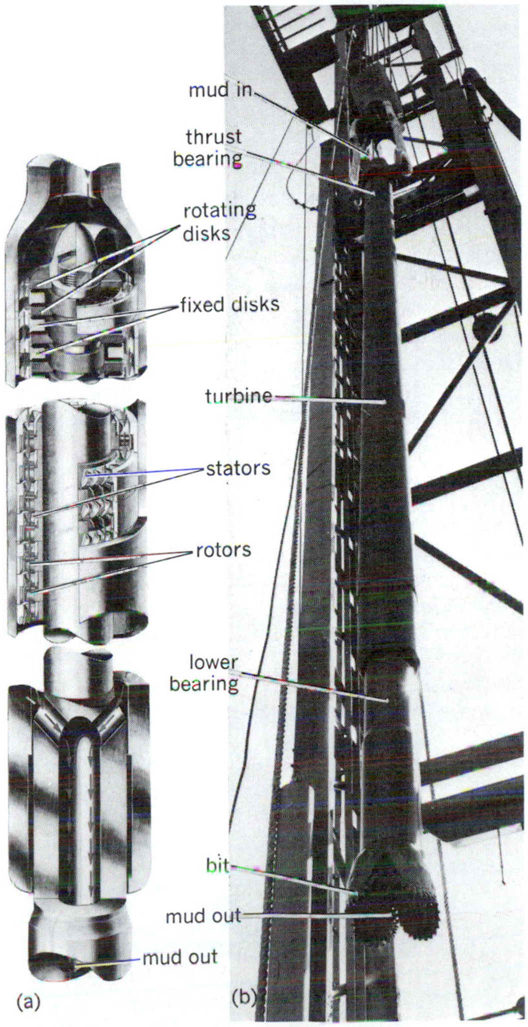

The components of a turbodrill. (*a*) Cutaway view of turbine, bearings, and bit. (*b*) Drill string suspended above hole. (*Dresser Industries*)

shaft, which is connected to the bit, to rotate. The mud passes through a hollow part of the shaft in the lower bearing and through the bit, as in rotary drilling, to remove cuttings, cool the bit, and perform the other functions of drilling fluid. Capacity of the mud pump, which is the power source, determines rotational speed. *See* OIL AND GAS WELL DRILLING; TURBINE. [A.L.P.]

Turbofan An air-breathing aircraft gas turbine engine with operational characteristics between those of the turbojet and the turboprop. Like the turboprop, the turbofan consists of a compressor-combustor-turbine unit, called a core or gas generator, and a power turbine. This power turbine drives a low- or medium-pressure-ratio compressor, called a fan, some or most of whose discharge bypasses the core (see illustration). *See* TURBOJET; TURBOPROP.

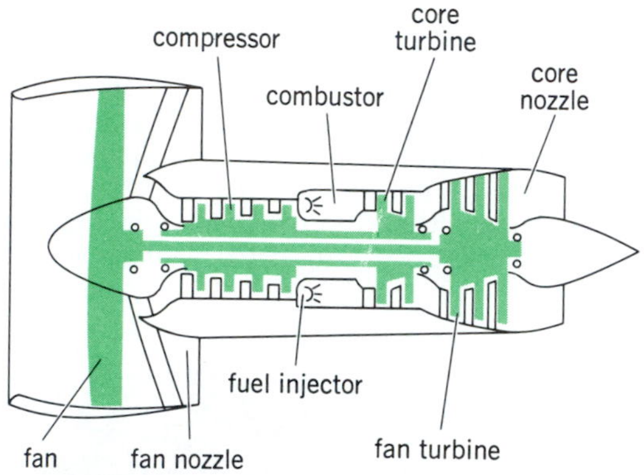

High-bypass, separate-flow turbofan configuration.

The gas generator produces useful energy in the form of hot gas under pressure. Part of this energy is converted by the power turbine and the fan it drives into increased pressure of the fan airflow. This airflow is accelerated to ambient pressure through a fan jet nozzle and is thereby converted into kinetic energy. The residual core energy is converted into kinetic energy by being accelerated to ambient pressure through a separate core jet nozzle. The reaction in the turbomachinery in producing both streams produces useful thrust. *See* GAS TURBINE; TURBINE PROPULSION. [F.F.E.]

Turbojet A propulsion engine used in many high-speed military and commercial aircraft.

In a turbojet, as illustrated, air approaches the inlet diffuser at a relative velocity equal to the flight speed. In passing through the diffuser the velocity of the air is decreased and its pressure is increased. The air pressure is increased further as it passes through the compressor. In the combustion chamber a steady stream of fuel is injected into the air and combustion takes place continuously. The high-pressure hot gas passes through the turbine nozzles, which direct it at high velocity against the buckets on the turbine wheel, thereby causing the wheel to rotate. The turbine wheel drives the compressor to which it is connected through a shaft. This is the sole function of the turbine. *See* DIFFUSER; JET PROPULSION; TURBINE PROPULSION.

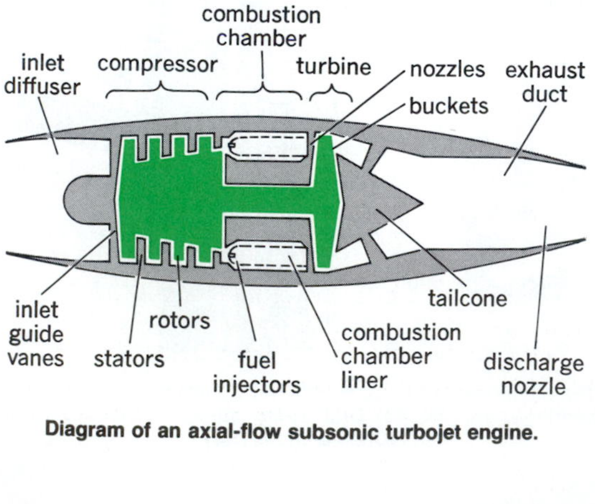

Diagram of an axial-flow subsonic turbojet engine.

After the hot gas leaves the turbine, it is still at a high temperature and at a pressure considerably above atmospheric. The hot gas is discharged rearward through the exhaust nozzle of the engine at a high velocity, thus producing thrust. *See* BRAYTON CYCLE. [B.Pi.]

Turboprop A gas turbine power plant producing shaft power for aircraft using a propeller. The turboprop engine has basic components similar to those of a turbojet: compressor, combustor, and turbine. In addition, it has a power turbine. This power turbine extracts usable shaft power from the engine mass flow and drives a conventional propeller through a reduction gear. The illustration shows the simplest possible turboprop configuration. Here the power turbine is rigidly connected to the main engine rotor; hence this type is commonly called a single-shaft engine. This has certain disadvantages in regard to starting, partial-load fuel consumption, and windmilling drag. *See* TURBINE.

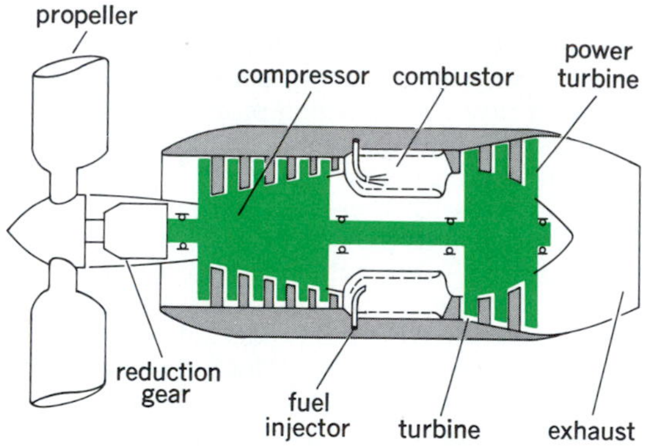

Turboprop engine with connected power turbine.

In the free-turbine engine, a more advanced turboprop type, the power turbine has no mechanical connection with the main engine rotor and consequently operates at a different rotational speed. This arrangement improves the operating characteristics and, especially, the cruise efficiency under partial-load conditions. This form of engine is also used to drive helicopter rotors and is then called a gas turbine. *See* GAS TURBINE; TURBINE PROPULSION. [P.G.K.]

Turboramjet An aircraft engine in which intake air is compressed at high flight speeds by the ram effect of the intake moving into the air ahead of the engine and at low flight speeds by the compressor rotating behind the intake, and in which propulsion is produced by a high-speed exhaust. For operation from standstill to high flight Mach numbers, the turboramjet engine provides useful performance characteristics. The compressor is driven by a turbine for engine operation at low flight speeds. At high supersonic flight speeds, the ram entry of air into the engine makes the compressor and its turbine superfluous. Under this operating condition, the fuel is burned to greater advantage in a combustion chamber downstream of the turbine instead of between compressor and turbine as in the turbojet mode of operation. Also, this second combustion chamber in back of the turbine can provide additional thrust at low speed when the engine is operated as a turbojet. *See* AFTERBURNER; RAMJET; TURBOJET.

The turbojet with afterburning is the simplest of the class of engines called turboramjets. The chamber for afterburning is provided with fuel injection nozzles and flameholders. The temperature of the gas from the primary combustion chamber is limited to a value below stoichiometric to avoid weakening the blades of the turbine. This limits the fuel burned in the primary burner to an amount that consumes about one-third of the available oxygen. The remaining oxygen is used for combustion of the fuel in the secondary combustion chamber (afterburner). *See* TURBINE PROPULSION. [B.Pi.]

Turbulent flow Motion of fluids in which local velocities and pressures fluctuate irregularly. Most flows observed in nature, such as rivers and winds, are turbulent. In turbulent flow, motion of the fluids is steady only insofar as the temporal mean values of velocities and pressures are concerned. The velocity and the pressure distributions in turbulent flows as well as the energy losses are determined mainly by the turbulent fluctuations.

The essential characteristic of turbulent flow is that the fluctuations are random. Hence, the solution of the turbulence problem requires the application of methods of statistical mechanics. In turbulent flow, the most important phenomenon is the transfer of forces by such random motion. The rate of turbulent transfer, such as heat transfer, shearing stress, and diffusion, is much higher than that due to molecular mechanism in laminar flow.

In turbulent motion, even though the fluid is regarded as a continuum with an average overall molecular motion, turbulent velocity fluctuations must be superimposed on the mean motion. The separation of mean motion and turbulent fluctuation depends mainly on the scale of turbulence. Different scales give different descriptions of turbulent flow. [S.-I.P.]

Turing machine An automaton that basically consists of a computing element and a tape. A series of characters, C_1, $C_2, \ldots, C_n$, can be written on the tape so that each character occupies a finite space along the tape, or, say, one square. The computing element can assume a number of states, S_1, $S_2, \ldots, S_m$. The tape passes through the computing element one square at a time so that the machine operation can be described by the quadruple $S_iC_jC_kS_l$, which is read: When the machine is in state S_i and the character C_j is read, then replace the character C_i with character C_k and proceed to state S_l. The character C_k can be considered as being written over the character C_j. In addition, the instruction R or L may be given, calling for a movement of the tape one square to the right or left. Thus the command $S_iC_jRS_l$ is read: When the machine is in state S_i and character C_j is read, then move the tape one square to the right and proceed to state S_l.

Studies with Turing machines have been concerned with the consideration of those functions which are computable with this machine organization. Whether all functions that are computable can be computed on a Turing machine is generally regarded as unprovable. *See* AUTOMATA THEORY; COMPUTER. [N.B.R.]

Turmeric A dye or a spice obtained from the plant *Curcuma longa*, which belongs to the ginger family (Zingiberaceae). It is a stout perennial with short stem, tufted leaves, and short, thick rhizomes which contain the colorful condiment. As a natural dye, turmeric is orange-red or reddish brown, but it changes color in the presence of acids or bases. As a spice, turmeric has a decidedly musky odor and a pungent, bitter taste. It is an important item in curry and is used to flavor and color butter, cheese, pickles, and other food. *See* SPICE AND FLAVORING; ZINGIBERALES. [P.D.St./E.L.C.]

Turn and bank indicator A device used to advise the pilot of the rate at which the aircraft is turning, and that the wings are properly banked to preclude slipping or sliding of the aircraft as it continues in flight. The turn section uses a rate gyroscope as the sensor. This device has a rotor electrically or vacuum driven at high speed with its spin axis horizontal and at 90° to the fore and aft axis of the aircraft. A simple pin-and-lever system causes a vertical pointer to move left or right, proportional to the turn rate.

The bank portion of the device is a simple liquid level that indicates the degree off the aircraft vertical that the combined force due to gravity and flight is acting. *See* GYROSCOPE. [J.W.A.]

Turnbuckle A device for tightening a rod or wire rope. Its parts are a sleeve with a screwed connection at one end and a swivel at the other or, more commonly, a sleeve with screwed connections of opposite hands (left and right) at each end so that by turning the sleeve, the connected parts will be drawn together, taking up slack and producing tension (see illustration). Types of ends are hook, eye, and clevis. The turnbuckle can be connected at any convenient place in the rod or rope, and several may be used in series if required. [P.H.B.]

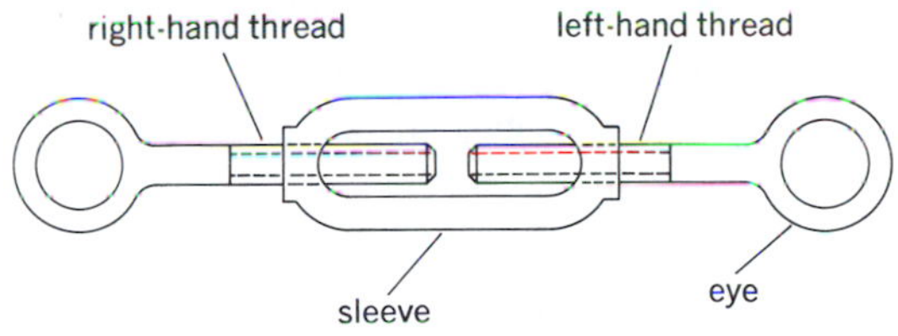

Turnbuckle with eyes.

Turning (woodworking) The shaping of wood by rotating it in a lathe and cutting it with a chisel. The lathe consists essentially of a bed on which are mounted a headstock, a tailstock, and a tool rest, as illustrated. The headstock is rotated by a motor and holds one end of the wood to be turned. The tailstock holds the other end of the wood, allowing it to rotate freely. The tool rest provides a fixed guide along which the operator can handle the chisels if the turning is by hand, or along which the tool is driven if the turning is mechanized. *See* WOODWORKING. [A.H.T.]

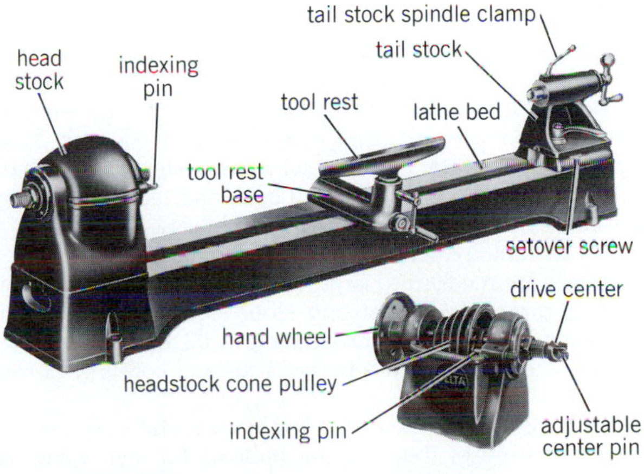

Wood-turning lathe and detail of headstock. (*Delta*)

Turnip The plant *Brassica rapa*, or *B. campestris* var. *rapa*, a cool-season, hardy crucifer of Asiatic origin belonging to the order Capparales which is grown for its enlarged root and its foliage, which are eaten cooked as a vegetable. Popular white-fleshed varieties (cultivars) grown for their roots are Purple Top Globe and White Milan; Yellow Globe and Golden Ball are common yellow-fleshed varieties. Shogoin is a popular variety grown principally in the southern United States for turnip greens. Principal areas of production in the United States are in the South. *See* CAPPARALES. [H.J.C.]

Turpentine An essential oil produced by steam distillation of pine woods and from gum turpentine, an exudate of living trees. It is recovered also in the sulfate process of cooking wood pulp. Slash and longleaf pines grown in the southeastern United States are the chief sources.

Turpentine is used principally as a solvent and a thinner for paints and varnishes. It is a starting material in the production of camphor, borneol, terpineol, and terpin hydrate. *See* CAMPHOR; ESSENTIAL OILS. [W.Mos.]

Turquois A mineral consisting of hydrated phosphate of aluminum and copper, $CuAl_6(PO_4)_4(OH)_8 \cdot 4H_2O$, and prized as a semiprecious stone. Fe^{2+} may substitute for some Cu. Bone turquois or odontolite, similar to turquois, is formed in fossil bones or teeth and consists of microcrystalline apatite colored by a hydrated phosphate of iron (vivianite). *See* GEM.

Generally turquois occurs in veinlets or as crusts of massive, dense, finely granular, concretionary, and stalactic shapes. The color of massive turquois ranges through sky blue, bluish green, apple green, and greenish gray. The robin's-egg blue variety is the most valued. The blue color in turquois is due to the presence of a small amount of copper; the presence of Fe^{2+} results in greenish hues.

Turquois of gem quality is found at one or more localities in Arizona, California, Colorado, New Mexico, and Nevada. Large deposits located in the Los Cerillos Mountains, southwest of Santa Fe, New Mexico, were mined early by Indians and Mexicans and later were extensively exploited by Americans. Fine-quality turquois has been mined for at least 800 years from a deposit located on the south slopes of the Ali-Mirsa-Kuh Mountains near Nishapur, Iran. Siberia, Turkistan, Asia Minor, the Sinai Peninsula, Silesia and Saxony in Germany, and France possess turquois deposits. [W.R.Lo.]

Turtle Any of about 240 species of reptiles which constitute the order Chelonia and include the tortoise. Turtles and tortoises live in marine, terrestrial, and fresh-water environments. The body is enclosed in two bony shells, the dorsal carapace and a ventral plastron, which are laterally connected. The jaw is developed into a horny beak and teeth are lacking. The tongue is not protrusible as in amphibians, and eyelids as well as a nictitating membrane are present.

Tortoises are medium to large terrestrial reptiles with a highly arched carapace. The strong stout legs are covered with scales; the toes are short, terminating in thick claws, and lack webbing. These animals are herbivorous and consume all varieties of plant life.

The sea turtles have excited much interest and study because the loss of some of their natural habitats for egg laying has imperiled the species. The eggs are being protected by wild life services, as are the newly hatched turtles, which are preyed upon by birds as they attempt to reach the sea. *See* REPTILIA; TESTUDINES. [C.B.C.]

Twilight The period between sunset and darkness in the evening, and between darkness and sunrise in the morning. The characteristic light is caused by atmospheric scattering, which transmits sunlight to the observer for some time after sunset and before sunrise. It depends geometrically on latitude, longitude, and elevation of the observer, and on the time of year. Physically it depends also on local conditions, particularly the weather. [G.M.C.]

Twinkling stars Stars exhibiting rapid fluctuations of the brightness and size of the images caused by the Earth's atmosphere; the phenomenon is also called astronomical seeing or scintillation. Irregularities of temperature, density, and water-vapor content, accompanied by changes of refractive index, are carried across the beam of light from the star and produce an irregularly corrugated wave front. Irregularities of the electron density in the ionosphere also change the refractive index at radio frequencies and cause twinkling of extraterrestrial sources of radio noise.

Much seeing is caused by high-altitude turbulence and winds. However, the flow of air over local terrain obstacles and around buildings produces considerable microturbulence, which has been found to affect the performance of telescopes. The best astronomical seeing (at night) gives images about 0.3 second of arc; good seeing gives 1 second; poor seeing gives up to 10 seconds; daytime seeing is generally worse. *See* STAR. [J.L.Gr.]

Twinning (crystallography) A process in which two or more crystals, or parts of crystals, assume orientations such that one may be brought to coincidence with the other by reflection across a plane or by rotation about an axis. Crystal twins represent a particularly symmetric kind of grain boundary; however, the energy of the twin boundary is much lower than that of the general grain boundary because some of the atoms in the twin interface are in the correct positions relative to each other. *See* GRAIN BOUNDARIES. [R.M.T.]

Twins (human) Two babies born to a mother at one birth. Twins may arise from either one (monozygotic) or two (dizygotic) fertilized eggs. Monozygotic (identical) twins have identical sets of nuclear genes, while dizygotic (fraternal) twins are no more similar genetically than siblings, sharing on average only half their genes. Although the exact cause of monozygotic twinning is not known, it appears to result from a developmental abnormality during the first 10–14 days of embryonic life that leads to either separation of the blastomeres or duplication of the inner cell mass. Since monozygotic twinning is a form of asexual or clonal replication, it represents an important exception to the rule that human reproduction occurs exclusively by sexual processes. Typically, dizygotic twins arise when two eggs are released and fertilized.

Monozygotic twins occur at a remarkably constant frequency of about 1 in 240 births; the rate does not appear to be influenced by any of the epidemiologic or demographic variables that have been investigated. In striking contrast, several factors are known to influence the rate of dizygotic twinning. In Asian populations dizygotic twins are less common than monozygotic pairs, while in certain African tribal populations nearly 5% of all confinements are for dizygotic twin gestations. The incidence of dizygotic twins in Caucasian populations is intermediate, and about twice that of monozygotic pairs. Maternal age and parity have independent effects on dizygotic twinning, the risk increasing with parity and with maternal age up to about 35 years, when it drops sharply. Studies have shown that the racial background of the mother is more important than that of the father in determining the frequency of polyovulation, and even within a population group a history of twinning in the mother's family appears to make an important contribution to the risk of recurrence of dizygotic twins. Endocrinologic studies of the mothers of multiple sets of dizygotic twins have demonstrated significant hormonal differences that are consistent with the view that some women are twin-prone; whether this phenotype can best be explained as a sex-limited recessive or a polygenic trait is not clear. In most developed countries there has been a dramatic secular decline in the incidence of twin births during the past 200 years. A major component, but probably not all of this trend, can be explained by demographic changes in maternal age and parity. *See* HUMAN GENETICS. [W.E.N.]

Tylenchida An order of nematodes in which the labial region is variable and may be distinctly set off or smoothly rounded and well developed; the hexaradiate symmetry is most often retained or discernible. The hollow stylet is the product of the cheilostome (conus, guiding apparatus, and framework) and the esophastome (shaft and knobs). Throughout the order the stylet may be present or absent and may be adorned with knobs. The variable esophagus is most often divisible into the corpus, isthmus, and glandular posterior bulb. The corpus is divisible into the procorpus and metacorpus. The metacorpus is generally valved but may not occur in some females and males, and the absence is characteristic of some taxa. The orifice of the dorsal esophageal gland opens either into the anterior procorpus or just anterior to the metacorporal valve. The excretory system is asymmetrical, and there is but one longitudinal collecting tubule. Females have one or two genital branches; when only one branch is present, it is anteriorly directed. Except for sex-reversed males, there is only one genital branch. Males may have one (=phasmid) or more caudal papillae. The spicules are always paired and variable in shape; they may be accompanied by a gubernaculum. The order comprises five superfamilies: Tylenchoidea, Criconematoidea, Saphaerularoidea, Aphelenchoidea, and Aphelenchoidoidea. *See* NEMATA; PLANT PATHOLOGY. [A.R.M.]

Tyndall effect Visible scattering of light along the path of a beam of light as it passes through a system containing discontinuities. The luminous path of the beam of light is called a Tyndall cone. An example is shown in the illustration. In colloidal systems the brilliance of the Tyndall cone is directly dependent on the magnitude of the difference in refractive index between the particle and the medium.

For systems of particles with diameters less than one-twentieth the wavelength of light, the light scattered from a polychromatic beam is predominantly blue in color and is polarized to a degree which depends on the angle between the observer and the incident beam. The blue color of tobacco smoke is an example of Tyndall blue. As particles are increased in size, the blue color of scattered light disappears and the scattered radiation appears white. If this scattered light is received through a nicol prism which is oriented to extinguish the vertically polarized scattered light, the blue color appears again in increased brilliance. This is called residual blue, and its intensity varies as the inverse eighth power of the wavelength. *See* COLLOID; SCATTERING OF ELECTROMAGNETIC RADIATION. [Q.V.W.]

The luminous light path known as the Tyndall cone or Tyndall effect. (*Courtesy of H. Steeves and R. G. Babcock*)

Type (printing) The relief or plane characters used to generate printed characters of various styles and sizes. Type used in printing is divided into three categories: foundry, machine-cast, and photocomposed type. In the first two the face of the letter is raised on one end of a piece of metal. It is from that surface, when inked, that the impression of type is made. In photocomposition, the type is reproduced photographically.

Classification. Foundry type, also known as hand type, is cast as single characters. Machine-cast type is produced by Linotype, Intertype, Ludlow, and Monotype machines. All but the last cast type in lines, or what are known as slugs. The Monotype—in reality two devices, a keyboard and a caster—produces individual types set in lines of desired lengths.

Venetian		ABC
Old Style	French	ABC
	Dutch-English	ABC
Transitional		ABC
Modern		ABC
Contemporary	sans serif	ABC
	square serif	ABC
Scripts		ABC
Black Letter		ABC
Decorative Letters		ABC

Fig. 1. Eight type classifications.

The product of the photocomposing type machine is the image of type on film or on photosensitized paper in negative and in positive form. This method is called cold composition because no hot type metal is needed for its manufacture. The term cold composition or cold type is also applied to text matter produced on a typewriter or similar machine for reproduction on a printing plate, and to words or lines made up of individual printed characters assembled or pasted together for photographic reproduction.

About 3000 type styles are in everyday use throughout the world. The most widely used method for classifying them is the serif-evolution system, based on the different shapes of the terminals or endings of letters. This provides eight classifications (Fig. 1).

Type measurements. Type is cast in sizes from 4 point to 144 point. In the American point system, the unit of type measurement is a point, 0.01384 in. (0.0346 cm) or nearly $\frac{1}{72}$ in., and all type sizes are multiples of this unit. Sizes are measured through the body, lengthwise of the letters. Some type size names used in the 16th century have remained and have been assigned other functions. For example, the word pica is commonly used to denote a unit of space measuring 12 points. It is applied as a dimensional unit to the length of type lines and to the width and depth of pages and blocks of type.

The standard height of type in the United States is 0.918 in. (2.295 cm), a dimension called type-high and one observed by photoengravers and electrotypers who make plates to be combined with cast type in letterpress printing. Printing presses are adjusted to fit this standard height.

Fonts. A complete complement of letters of one size and style from A to Z, together with the arabic numerals, punctuation, and reference marks, is called a font. One style of letters is used in the setting of body, or text, matter; it is available in roman capitals, italic capitals (majuscules), roman and italic lower case (minuscules), and an alphabet of small capitals, which is approximately the height of lowercase letters. A letter from a different font which is included by accident in type composition is known as a wrong font.

Type families. A family of type may be likened to the shades of a color in that it includes all variations made of a given type face or design. Weights are varied, from light to medium, bold, and extra bold; letters are condensed and expanded, as well as outlined, inlined, and shadowed (Fig. 2).

Fig. 2. A family of type.

[E.M.E.]

Typewriter

A self-contained, stand-alone machine that prints characters one after another on paper. Its essential parts are a keyboard, raised type, a means of inking, a platen for holding paper as it is imprinted, and a mechanism for advancing position as imprinting takes place. It operates manually, electromechanically, or electronically.

Basically, the typewriter depends on input from a typist operating a keyboard. The keyboard contains a complete alphabet, along with the numbers and symbols commonly used in various languages and technical disciplines. Strictly speaking, other keyboard machines, such as high-speed printers linked through communications channels to computers, or machines operating automatically in word-processing units, as well as other input or data-capture devices, are not typewriters since they can, in a variety of applications, operate without the direct intervention of a typist and are not, by definition, stand-alone units.

A continuing series of human-factors enhancements and technological advances have both improved the efficiency of the typewriter and simplified its operation. Of particular significance was the use of an electric motor to diminish the amount of effort required by the typist in generating key strokes.

Electronic word processor. (*IBM*)

Through electronics, a working memory (generally not designed for permanent storage) and other capabilities were added to the stand-alone typewriter. A microprocessor-driven electronic typewriter (see illustration) stores and retrieves words, phases, and pages of typing by using a high-density random-access memory chip. *See* MICROPROCESSOR; WORD PROCESSING.

[E.W.G.]

Typhales

An order of flowering plants, division Magnoliophyta (Angiospermae), in the subclass Commelinidae of the class Liliopsida (monocotyledons). The order consists of only two genera: *Typha*, with about 15 species (making up the family Typhaceae), and *Sparganium*, with about 20 species (making up the family Sparganiaceae). These are marsh or aquatic plants with emergent or floating stems and leaves that have well-developed vessels, and with reduced, unisexual, wind-pollinated flowers. Large colonies of cattails (*Typha*) are familiar features of marshlands in many parts of the world. *See* COMMELINIDAE; LILIOPSIDA.

[A.Cr.]

Tyrosine

An amino acid. Tyrosine is a precursor of the hormones epinephrine (adrenalin), norepinephrine (noradrenalin), thyroxin, and triiodothyronine, and of the black pigment melanin. The amino acid is formed from phosphoenolpyruvic acid and D-erythrose-4-phosphate, by way of shikimic acid and prephenic acid. In animals, tyrosine is formed by the oxidation of dietary phenylalanine. The major pathway for metabolic degradation leads to fumaric and acetoacetic acid. *See* AMINO ACIDS.

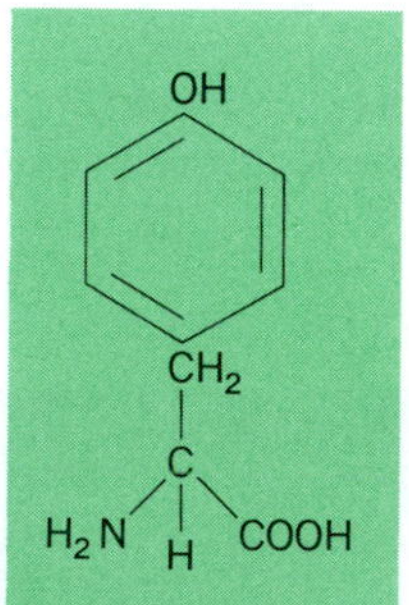

Tyrosine

[E.A.Ad.]

Tyrothricin The first antibiotic to be used for human therapy. It was isolated from a culture of *Bacillus brevis* by René J. Dubos in 1939. This basic discovery laid the groundwork for the many antibiotics which have followed. *See* ANTIBIOTIC.

Tyrothricin is a combination of two separate antibiotics. At least 18% is gramicidin, and 65% is tyrocidine hydrochloride. The remainder consists of other polypeptides. Tyrothricin is principally effective against a wide variety of gram-positive organisms. It is offered commercially in ointments, creams, powders, lotions, sprays, suppositories, tablets, and aerosol preparations. Such products are used for the treatment of dermal infections, infections of the nose, throat and eyes, and for the preparation of solutions used to irrigate the body cavities.

[B.G.M.; E.J.B.]

U

Ulcer Common inflammatory lesion in which there is a loss or destruction of superficial tissue. Ulcers occur in several locations as acute, subacute, chronic, or recurrent types. In each location some initiating factor, such as action of bacterial toxins or lack of oxygen, causes death of the surface tissue. This necrotic tissue then usually sloughs off, leaving the underlying area exposed to further damage. There is some inflammation present.

Peptic ulcers are a common form of ulcer which occur primarily in the stomach and duodenum. The exact causes are obscure, although predisposing factors such as tension, anxiety, and abnormal amounts of acid secretion appear in most cases. *See* PEPTIC ULCER.

Ulcerative colitis of the large intestine is similar, in many respects, to peptic ulcer, because its exact cause is unknown and a relationship exists with certain personality patterns.

In many skin disorders in which a circulatory deficiency is involved, ulcerative lesions may appear. Superficial ulcerations of the mouth, tongue, and other mucosal surfaces may occur as the result of illness, infection, or irritation. Several serious diseases may be accompanied by characteristic ulcers of the gastrointestinal tract, including typhoid fever, tuberculosis, and both amebic and bacillary dysentery. *See* AMEBIASIS; BACILLARY DYSENTERY; TUBERCULOSIS. [E.G.St./N.K.M.]

Ulotrichales A large, artificial order of the Chlorophyta, composed mostly of fresh-water, branched or unbranched filamentous species. Some show basal-distal differentiation and are attached; a few are endophytic or epiphytic whereas others are free-floating. Cells are mostly cylindrical with cellulosic, often mucilaginous walls and, except for *Sphaeroplea*, are uninucleate. The chloroplast is mostly a parietal band or ring. Reproduction is vegetative, or by quadriflagellate zoospores, and sexually by iso-, aniso-, or heterogametes. A conservative but often modified classification recognizes two suborders, Ulotrichineae (ten families) and Sphaeropleineae (one family). [G.W.P.]

Ultracentrifuge A centrifuge of high or low speed which provides convection-free conditions and which is used for quantitative measurement of sedimentation velocity or sedimentation equilibrium or for the separation of solutes in liquid solution. *See* CENTRIFUGATION.

The ultracentrifuge is used (1) to measure molecular weights of solutes and to provide data on molecular weight distributions in polydisperse systems; (2) to determine the frictional coefficients, and thereby the sizes and shapes, of solutes; and (3) to characterize and separate macromolecules on the basis of their buoyant densities in density gradients. *See* MOLECULAR WEIGHT.

The ultracentrifuge is most widely used to study high polymers, particularly proteins, nucleic acids, viruses, and other macromolecules of biological origin. However, it is also used to study solution properties of small solutes. In applications to macromolecules, the analytical ultracentrifuge, which is used for accurate determination of sedimentation velocity or equilibrium, is distinguished from the preparative ultracentrifuge, which is used to separate solutes on the basis of their sedimentation velocities or buoyant densities.

The application of a centrifugal field to a solution causes a net motion of the solute. If the solution is denser than the solvent, the motion will be away from the axis of rotation. The nonuniform concentration distribution produced in this way leads to an opposing diffusion flux tending to reestablish uniformity. In sedimentation-velocity experiments, sedimentation prevails over diffusion, and the solute sediments with finite velocity toward the bottom of the cell, although the concentration profile may be markedly influenced by diffusion. In sedimentation-equilibrium experiments, centrifugal and diffusive forces balance out, and an equilibrium concentration distribution results which may be analyzed by thermodynamic methods. *See* CENTRIFUGAL FORCE. [V.A.B.]

Ultrafast molecular processes Various types of physical and molecular changes occurring on time scales of 10^{-14} to 10^{-9} s that are studied in the field of photophysics, photochemistry, and photobiography. The time scale ranges from near the femtosecond regime (10^{-15} s) on the fast side, embodies the entire picosecond regime (10^{-12} s), and borders the nanosecond regime (10^{-9} s) on the slow side.

A typical experiment is initiated with an ultrashort pulse of energy—radiation (light) or particles (electrons). Rapid changes in the system under study are brought about by the absorption of these ultrashort energy pulses. These changes can be measured by ultrafast detection methods. Such methods are usually based on some type of linear or nonlinear spectroscopic monitoring of the system, or on optical delay lines that employ the speed of light itself to create a yardstick of time (a distance of 3 mm equals a time of 10 picoseconds). These studies are important because elementary motion, such as rotation, vibrational exchange, chemical bond breaking, and charge transfer, takes place on these ultrafast time scales. The relationship between such motions and overall physical, chemical, or biological changes can thus be observed directly. *See* CHEMICAL DYNAMICS; OPTICAL PULSES; SPECTROSCOPY.

Just as a very fast shutter speed is necessary to obtain a sharp photograph of a fast-moving object, energy pulses for the study of ultrafast molecular motions must be extremely narrow in time. The basic technique for producing ultrashort light pulses is mode-locking a laser. One way to do this is to put a nonlinear absorption medium in the laser cavity, which functions somewhat like a shutter. In the time domain, the developing light pulse, bouncing back and forth between the laser cavity reflectors, experiences an intensity-dependent loss each time it passes through the nonlinear absorption medium. High-intensity light penetrates the medium, and its intensity is allowed to build up in the laser cavity; weak-intensity light is blocked. This process shaves off the low-amplitude edges of a pulse, thus shortening it. One of the laser cavity reflectors has less than 100% reflectance at the laser wavelength, so part of the pulse energy is coupled out of the cavity each time the pulse reaches that reflector. In this way, a train of pulses is emitted from the mode-locked laser, each pulse separated from the next by the round-trip time for a pulse traveling at the speed of light in the laser cavity, about 13 nanoseconds for a 2-m (6.6-ft) laser cavity length. *See* LASER; NONLINEAR OPTICS.

In addition to the extremely narrow energy pulses in the

time domain, special ultrafast detection techniques are required in order to preserve the time resolution of an experiment.

A photobiological process, which has received considerable attention by spectroscopists studying the femtosecond regime, is light-driven transmembrane proton pumping in purple bacteria (*Halobacterium halobium*). Each link in the complex chain of molecular processes, such as an ultrafast photoionization event, is "fingerprinted" by a somewhat different absorption spectrum, making it possible to sort out these events by methods of ultrafast laser spectroscopy. The necessity for use of fast primary events in nature concerns overriding unwanted competing chemical and energy loss processes. The faster the wanted process, the less is the likelihood that unwanted, energy-wasting processes will take place. *See* LASER PHOTOBIOLOGY; LASER SPECTROSCOPY. [G.W.R.; N.L.]

Ultrafiltration A filtration process in which particles of colloidal size are retained by a filter medium while solvent and accompanying low-molecular-weight solutes are allowed to pass through. Ultrafilters are used (1) to separate colloid from suspending medium, (2) to separate particles of one size from particles of another size, and (3) to determine the distribution of particle sizes in colloidal systems by the use of filters of graded pore size.

Ultrafilter membranes have been prepared from various types of gel-forming substances: Unglazed porcelain has been impregnated with gels such as gelatin or silicic acid. Filter paper has been impregnated with glacial acetic acid collodions. Another type of ultrafilter membrane is made up of a thin plastic sheet containing millions of tiny pores evenly distributed over its surface. *See* COLLOID; FILTRATION. [Q.V.W.]

Ultralight aircraft A lightweight, single-seat aircraft with low flight speed and power, used for sport or recreation. Ultralights evolved from hang gliders. *See* GLIDER.

There is a tremendous variety of ultralight airframe configurations and control systems. Airframe types include flying wings, canard designs, antique biplane replicas, and traditional, monoplane structures with conventional tail designs. Control systems can be either weight-shift systems, two-axis controls, or conventional three-axis controls. In weight-shift designs pilots must shift their weight, using a movable seat, to change the attitude. Two-axis designs use controls to the elevator and rudder only; there are no ailerons. Three-axis controls resemble those of a standard airplane and include either ailerons or spoilerons (spoiler-type systems used to make turns). Some more sophisticated ultralights are equipped with wing flaps, used to steepen approach profiles and to land at very slow airspeeds. *See* AIRFRAME; FLIGHT CONTROLS.

Typically, ultralight airframes are made of aircraft-grade aluminum tubes covered with Dacron sailcloth. Areas of stress concentration are reinforced with double-sleeving or solid aluminum components. Most ultralights are cable-braced and use aircraft-grade stainless steel cables with reinforced terminals. Wingspans average 30 ft (9 m), and glide ratios range from 7:1 to 10:1, depending primarily on gross weight and wing aspect ratio. *See* AIRFOIL; ASPECT RATIO; WING.

Ultralight engines are lightweight, two-stroke power plants with full-power values in the 28–35-hp (21–26-kW) range. They operate on a mixture of gasoline and oil, and most transmit power to the propeller via a reduction or belt drive, a simple transmission that enables the propeller to rotate at a lower, more efficient rate than the engine shaft. *See* AIRPLANE; PROPELLER (AIRCRAFT); RECIPROCATING AIRCRAFT ENGINE. [T.A.Ho.]

Ultrasonic flow measurement The use of high-frequency sound to measure the rate of fluid flow through a pipe, duct, or open channel. Ultrasonic flowmeters are available in many physical arrangements and are based on a great variety of fundamental principles.

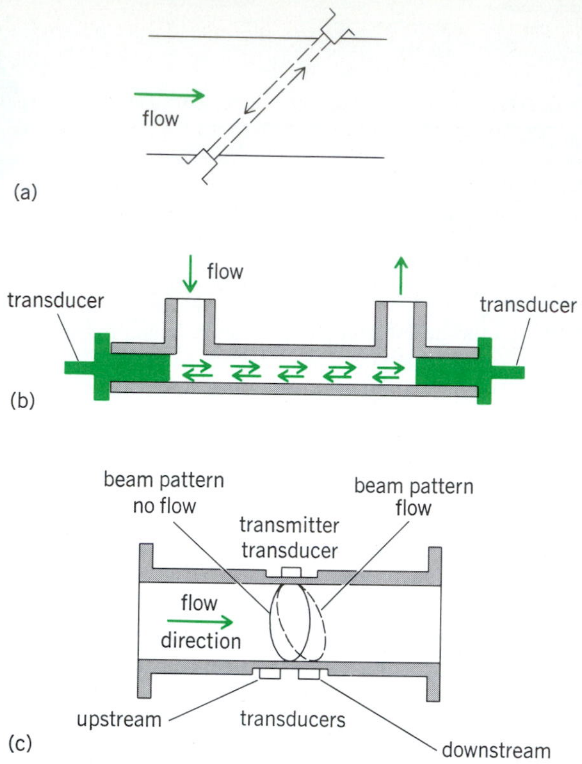

Ultrasonic flowmeters. (a) Contrapropagating type (*from Fischer and Porter Co., Publ. no. 12257, 1957*). (b) Axial-transmission contrapropagating type (*DuPont Co., Industrial Products*). (c) Deflection (drift) type (*after H.E. Dalke and W. Walkowitz, A new ultrasonic flowmeter for industry, ISA J., 7(10):60–63, 1960*).

Contrapropagating diagonal-beam designs (illustration *a*) have transducers mounted in sockets machined in the pipe or welded on wedges, and also as clamp-on elements. The difference between upstream and downstream "flight" times combined with the speed of sound in the fluid will give fluid velocity. Since the flow near the middle of the pipe is faster, a multiple parallel-beam design will give a better average. In a symmetrical profile a beam offset from the center of the pipe may also give a better average.

For small pipes, the short path length across the fluid stream makes measurement difficult. Illustration *b* shows a contrapropagating arrangement for measuring lengthwise over a relatively longer path where timing measurements are more practical.

In the correlation meter, two pairs of elements with beams across the pipe are used with a correlation circuit to detect the time it requires for discontinuities in the fluid stream to pass between the two detectors.

In the deflection or drift meter, an ultrasonic beam directed across the pipe is deflected by an amount depending on the relative velocity of sound in the fluid and the flowing rate of the fluid (illustration *c*). One arrangement detects this deflection with two receivers mounted on the other side of the pipe. The relative intensity of the two receivers is an indication of the flowing-fluid velocity.

In the Doppler meter, flow rate is detected by the Doppler shift of sound waves reflected from particles or discontinuities in the flowing fluid. *See* DOPPLER EFFECT.

In the noise-type meter, the noise generated by the flowing fluid is measured in a selected frequency band, with a useful but relatively inaccurate readout.

A different application of ultrasonics to flow measurement can be found in flumes and weirs where the level of the flowing fluid, measured by ultrasonic reflection, is used to calculate the flow rate. *See* FLOW MEASUREMENT. [M.Br.; L.P.E.]

Ultrasonics The science of sound waves having frequencies above the audible range, that is, above about 20,000 Hz. Original workers in this field adopted the term supersonics. However, this name was also used in the study of airflow for velocities faster than the speed of sound. The present convention is to use the term ultrasonics as defined above. The term silent sound also has been used to denote ultrasonic waves. The term pretersonics is now used to refer to frequencies about 10^{10} Hz. Since there is no marked distinction between the propagation and the uses of sound waves above and below 20,000 Hz, the division is rather artificial. *See* SOUND.

Projectors and detectors. The usual types of generators for air, liquids, and solids are the piezoelectric and magnetostrictive generators. X-cut quartz crystals are used to produce longitudinal waves in gases, liquids, and solids. Y-cut and AC-cut quartz crystals are used to produce shear or transverse waves in solids. Crystals of these types are utilized in instruments such as the acoustic interferometer. *See* ACOUSTIC INTERFEROMETER.

Pulse systems have been used to measure properties of liquids and solids. A short burst of ultrasonic waves is sent into the medium and is reflected back. By timing the received pulse with respect to the transmitted pulse, or by a phasing technique, accurate velocity measurements can be made. For higher powers, ferroelectric ceramics, such as barium titanate, PZT (lead zirconate titanate), and $NaKNbO_3$, or magnetostrictive materials, such as nickel or ferrites, commonly are employed. They generally are used for ultrasonic cutting, wear or fatigue testing, and ultrasonic welding.

A number of shear-wave transducers, most of them employing torsional or shear-wave generators of quartz, have been used to measure the shear viscosity and shear stiffness of liquids. By this means it has been proved that moderately viscous liquids have elastic, as well as viscous, properties. For pretersonic frequencies (above 500 MHz), all liquids are found to have shear elastic properties.

Engineering applications. The engineering applications of ultrasonics can be divided into those dealing with low-amplitude sound waves and those dealing with high-amplitude (now usually called macrosonic) waves.

Low-amplitude. Low-amplitude applications are in sonar (an underwater-detection apparatus), in the measurement of the elastic constants of gases, liquids, and solids by a determination of the velocity of propagation of sound waves, in the measurement of the attenuation of sound waves, and in a number of ultrasonic devices such as delay lines, mechanical filters, inspectoscopes, and thickness gages. *See* DELAY LINE; SONAR.

All these applications depend on the modifications that boundaries and imperfections in the materials cause in wave-propagation properties. The attenuation and scattering of the sound in the mediums are important factors in determining the frequencies used and the sizes of the pieces that can be utilized or investigated.

Mechanical filters are used for separating telephone communications sent simultaneously over one transmission line. The monolithic filter consists of a series of vacuum-deposited electrodes with a ratio of total electrode mass to the mass of the crystal plate between electrodes of $R = 0.02$ to 0.04. By using as many as six sections of platings, this device acts as a bandpass filter. Attenuation characteristics are sufficiently good for the device to be used as a channel filter in a long-distance carrier or submarine cable system. The filter can be made from a single quartz crystal, and hence considerable economies can be obtained.

Ultrasonic inspectoscopes transmit sound waves into a metal casting or other solid piece and determine the presence of flaws by reflections or by an interruption of the sound-wave transmission through the piece. Ultrasonic thickness gages have been used in measuring the thickness of pieces when one side is not accessible, such as in boilers. Surface-acoustic-wave devices involve the use of Rayleigh surface waves generated by equally spaced interdigital electrode transducers. One of the largest uses of these devices is in obtaining dispersive delay lines used in chirp radars. *See* SURFACE-ACOUSTIC-WAVE DEVICES.

An effect related to internal friction, motion of dislocations, and fatigue in materials is the noise in the specimens produced by strain. This is called acoustic emission. Research on this phenomenon has had wide application in such subjects as structural integrity of metals and rocks, flaws in metals, composite materials, concrete, ceramics, ice, soils, wood, integrity of welds, martensite transformations, nuclear power reactors, and leaks in pressure systems. Acoustic emission is used to test the goodness of spot welds on relay contacts. *See* ACOUSTIC EMISSION.

High-amplitude. High-amplitude acoustic waves (macrosonic) have been used in a variety of applications involving gases, liquids, and solids.

Holes (gas-bubble cavities) can be created in a liquid by high-intensity sound waves. When such a cavity collapses, extremely high pressures are produced. The process, called cavitation, is the origin of a number of mechanical, chemical, and biological effects. The cavitation effect can be used to disperse metals and sulfur in solutions, to produce extrafine grain photographic emulsions, and to achieve a finer texture (smaller grain size) and more uniform alloying of a molten metal. In chemistry, cavitation can be used to break long-chain polymers into shorter chains, affording a polymer of more uniform chain length than is possible with other depolymerizing methods. Cavitation forces also can be used to sterilize milk. *See* CAVITATION.

Ultrasound is used widely in the cleaning of metal parts, such as in watches. One of the principal applications of ultrasonics to gases is particle agglomeration. This technique has been used in industry to collect fumes, dust, sulfuric acid mist, carbon black, and other substances.

Analytical uses. In addition to their engineering applications, high-frequency sound waves have been used to determine the specific types of motions that can occur in gaseous, liquid, and solid mediums. Both the velocity and attenuation of a sound wave are functions of the sound frequency. By studying the changes in these properties with changes of frequency, temperature, and pressure, indications of the motions taking place can be obtained. *See* SOUND ABSORPTION.

Ultrasonic methods have also been used to produce light images of ultrasonic beams, outputs from ultrasonic transducers, and visual pictures of transmission through live tissues and have been used for the reproduction of pictures through sound and ultrasound holographic methods. *See* ACOUSTIC MICROSCOPE; ACOUSTICAL HOLOGRAPHY. [W.P.M.]

Ultraviolet astronomy Astronomical observations carried out in the region of the electromagnetic spectrum with wavelengths from approximately 10 to 350 nanometers. Ultraviolet radiation from astronomical sources contains important diagnostic information about the composition and physical conditions of these objects. This information includes atomic absorption and emission lines of all the most abundant elements in many states of ionization. The hydrogen molecule (H_2), the most abundant molecule in the universe, has its absorption and emission lines in the far-ultraviolet. *See* ASTRONOMICAL SPECTROSCOPY; ULTRAVIOLET RADIATION.

Ultraviolet radiation with wavelengths less than 310 nm is strongly absorbed by molecules in the atmosphere of the Earth. Therefore, ultraviolet observations must be carried out by using instrumentation situated above the atmosphere. Ultraviolet astronomy began with instrumentation aboard sounding rockets. The first major ultraviolet satellite observatories to be placed in space were the United States *Orbiting Astronomical Observatories* (*OAOs*). The *International Ultraviolet Explorer* (*IUE*), launched into a geosynchronous orbit in 1978, realized the full potential of ultraviolet astronomy.

Although it had an inauspicious beginning, the Hubble Space Telescope, an observatory designed for visible and ultraviolet measurements, is expected to become the centerpiece of ultraviolet astronomy for the 1990s. *See* ROCKET ASTRONOMY; SATELLITE ASTRONOMY.

The important discoveries of ultraviolet astronomy span all areas of modern astronomy and astrophysics. Some of the notable discoveries in the area of solar system astronomy include new information on the upper atmospheres of the planets, including planetary aurorae and the discovery of the enormous hydrogen halos surrounding comets. In studies of the interstellar medium, ultraviolet astronomy has provided fundamental information about the molecular hydrogen content of cold interstellar clouds along with the discovery of the hot phase of the interstellar medium, which is created by the supernova explosions of stars. In stellar astronomy, ultraviolet measurements led to important insights about the processes of mass loss through stellar winds and have permitted comprehensive studies of the conditions in the outer chromospheric and coronal layers of cool stars. In galactic and extragalactic astronomy, two significant results are the discovery of the hot gaseous halo of the Milky Way Galaxy and new insights about the mysterious energetic sources at the centers of active galaxies and quasars. *See* COMET; GALAXY, EXTERNAL; INTERSTELLAR MATTER; MILKY WAY GALAXY; PLANETARY PHYSICS; QUASAR; SUPERNOVA. [B.D.S.]

Ultraviolet lamp A mercury-vapor lamp designed to produce ultraviolet radiation. Also, some fluorescent lamps and mercury-vapor lamps that produce light are used for ultraviolet effects. *See* FLUORESCENT LAMP; MERCURY-VAPOR LAMP.

Fluorescent and mercury lamps can be filtered so that visible energy is absorbed and emission is primarily in the near-ultraviolet or black-light spectrum (320–400 nanometers). The ultraviolet energy emitted is used to excite fluorescent pigments in paints, dyes, or natural materials.

Mercury-vapor lamps are sometimes designed with pressures that produce maximum radiation in the middle ultraviolet region (280–320 nm), using special glass bulbs that freely transmit this energy. One such lamp type is the sunlamp. Other lamps designed for middle-ultraviolet radiation are known as photochemical lamps. *See* SUNLAMP.

Some radiation in the 220–280-nm wavelength band has the capacity to destroy certain kinds of bacteria. Mercury lamps designed to produce energy in this region (the 253.7-nm mercury line) are electrically identical with fluorescent lamps; they differ from fluorescent lamps in the absence of a phosphor coating and in the use of glass tubes that transmit far ultraviolet. *See* ULTRAVIOLET RADIATION (BIOLOGY). [A.M.]

Ultraviolet radiation Electromagnetic radiation in the wavelength range 4–400 nanometers. The ultraviolet region begins at the short wavelength (violet) limit of visibility and extends to the wavelength of long x-rays. It is loosely divided into the near (400–300 nm), far (300–200 nm), and extreme (below 200 nm) ultraviolet regions (see illustration). In the extreme ultraviolet, strong absorption of the radiation by air requires the use of evacuated apparatus; hence this region is called the vacuum ultraviolet. Important phenomena associated with ultraviolet radiation include biological effects and applications, the generation of fluorescence, and chemical analysis through characteristic absorption or fluorescence. *See* ULTRAVIOLET RADIATION (BIOLOGY).

Sources of ultraviolet radiation include the Sun (although much solar ultraviolet radiation is absorbed in the atmosphere);

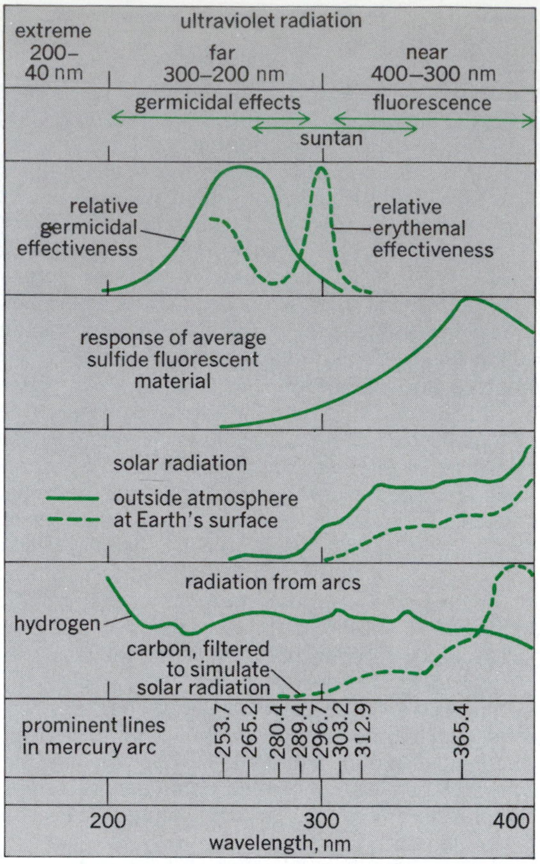

Phenomena associated with ultraviolet radiation. (*After L. R. Koller and General Electric*)

arcs of elements such as carbon, hydrogen, and mercury; and incandescent bodies. *See* ULTRAVIOLET LAMP. [F.W.B.]

Ultraviolet radiation (biology) Radiations between 200 and 300 nanometers are selectively absorbed by organic matter, and produce the best-known effects of ultraviolet radiations on organisms. Ultraviolet radiations, in contrast to x-rays, do not penetrate far into larger organisms; therefore, the effects they produce are surface effects, such as sunburn and development of D vitamins from precursors present in skin or fur. The effects of ultraviolet radiations on life have, therefore, been assayed chiefly with unicellular organisms such as bacteria, yeast, and protozoans, although suspensions of cells of higher organisms, for example, eggs and blood corpuscles, have been useful as well.

Photobiological effects. Only the ultraviolet radiations which are absorbed can produce photobiological action. All life activities are shown to be affected by ultraviolet radiations, the effect depending upon the dosage. Small dosages activate unfertilized eggs of marine animals, reduce the rate of cell division, decrease the synthesis of nucleic acid, especially in the nucleus, reduce the motility of cilia and of contractile vacuoles, and sensitize cells to heat. Large doses increase the permeability of cells to various substances, inhibit most synthetic processes, produce mutations, stop division of cells, decrease the rate of respiration, and may even disrupt cells. The effect of ultraviolet radiations upon cells is invariably deleterious.

Effects on the skin. Erythema is the reddening of the skin following exposure to ultraviolet radiation of wavelength shorter than 320 nm, wavelength 296.7 nm being most effective. These radiations injure cells in the outer layer of the skin, or epidermis, liberating substances which diffuse to the inner layer of the skin, or dermis, causing enlargement of the small blood vessels. A minimal erythemal dose just induces reddening of the skin observed 10 h after exposure. A dose several times the minimal gives a sunburn, killing some cells in the epidermis after which serum and white blood cells accumulate, causing a blister. After the dried blister peels, the epidermis is temporarily thickened and pigment develops in the lower layers of the epidermis, both of these factors serving to protect against subsequent exposure to ultraviolet.

Both thickening of the epidermis and tanning may occur without blistering. Since the pigment in light-skinned races develops chiefly below the sensitive cells in the epidermis, it is not as effective as in dark-skinned races where the pigment is scattered throughout the epidermis. Consequently, the minimal erythemal dose is much higher for the dark- than for the light-skinned races.

Excessive exposure to ultraviolet radiation has been found to lead to cancer in mice, and it is claimed by some to cause cutaneous cancer in humans.

Clinical use. Ultraviolet radiations were once used extensively in the treatment of rickets, many skin diseases, tuberculosis other than pulmonary, especially skin tuberculosis (lupus vulgaris), and of many other diseases. The enthusiasm for sun bathing is, in part, a relic of the former importance of ultraviolet radiation as a clinical tool. Vitamin preparations, synthetic drugs, and antibiotics have either displaced ultraviolet radiations in such therapy or are used in conjunction with the radiations.

Ultraviolet radiations alone are still employed to treat rickets in individuals sensitive to vitamin D preparations. In conjunction with chemicals, they are used in treating skin diseases, for example, psoriasis, pityriasis rosea, and sometimes acne, as well as for the rare cases of sensitivity to visible light. Ultraviolet radiations, however, are more important in research than in clinical practice. [A.C.Gi.]

Ulvales An order of algae in the division Chlorophyta. The thalli are macroscopic, attached tubes or sheets. All genera except *Schizomeris* are marine; however, a few species occur in streams. This order is sometimes included under the Ulotrichales, which it resembles in cell organization and reproductive elements. [G.W.P.]

Umklapp process A concept in the theory of transport properties of solids which has to do with the interaction of three or more waves in the solid, such as lattice waves or electron waves. In a continuum, such interactions occur only among waves described by wave vectors $\mathbf{k}_1$, $\mathbf{k}_2$, and so on, such that the interference condition, given by Eq. (1), is satisfied.

$$\mathbf{k}_1 + \mathbf{k}_2 + \mathbf{k}_3 = 0 \quad (1)$$

The sign of $\mathbf{k}$ depends on whether the wave absorbs or emits energy. Since $\hbar\mathbf{k}$ is the momentum of a quantum (or particle) described by the wave, Eq. (1) corresponds to conservation of momentum. In a crystal lattice further interactions occur, satisfying Eq. (2), where $\mathbf{b}$ is any integral combination

$$\mathbf{k}_1 + \mathbf{k}_2 + \mathbf{k}_3 = \mathbf{b} \quad (2)$$

of the three inverse lattice vectors $\mathbf{b}_i$, defined by $\mathbf{a} \cdot \mathbf{b}_j = 2\pi\delta_{ij}$, the $\mathbf{a}$'s being the periodicity vectors. The group of processes described in Eq. (2) are the Umklapp processes or flip-over processes, so called because the total momentum of the initial particles or quanta is reversed. *See* CRYSTAL. [P.G.Kl.]

Uncertainty principle In quantum mechanics, the precept (W. Heisenberg, 1927) that accurate measurement of an observable quantity necessarily produces uncertainties in one's knowledge of the values of other observables. It is also called the indeterminacy principle.

In particular, for a single particle, Eqs. (1) and (2) hold. In

$$\Delta x \Delta p_x \gtrsim h/2\pi \quad (1)$$

$$\Delta t \Delta E \gtrsim h/2\pi \quad (2)$$

Eq. (1), Δx represents the uncertainty (error) in the location of the x coordinate of the particle at any instant, and Δp_x is the simultaneous uncertainty in its x component of momentum; h = Planck's constant = 6.61×10^{-27} erg-second. Equation (2) is more subtle than Eq. (1), and its interpretation depends somewhat on the circumstances. For instance, Eq. (2) relates the uncertainty ΔE in an energy measurement to the time interval Δt during which the measurement was performed; however, Eq. (2) also relates the uncertainty ΔE in the energy radiated by a system to the uncertainty Δt in the time at which it radiates, that is, the uncertainty in its lifetime. *See* QUANTUM MECHANICS. [E.G.]

Unconformity The relation between adjacent rock strata whose time of deposition was separated by a period of nondeposition or of erosion. The break in the stratigraphic sequence represents a significant change in regimen (in contrast to diastem). For example, marine strata separated by a surface of subaerial erosion are unconformable, but those separated by a surface of mere nondeposition or of erosion by shifting currents which do not represent an overall change of conditions are not. The word unconformity is sometimes also used to denote the surface between the unconformable strata. *See* DIASTEM.

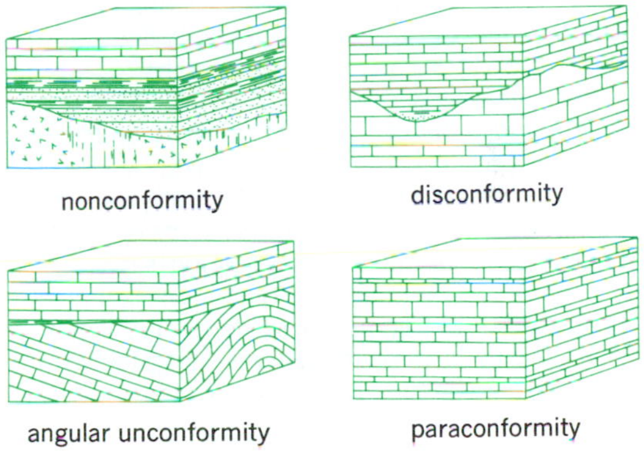

Four types of unconformity. (*After C. O. Dunbar and J. Rodgers, Principles of Stratigraphy, John Wiley and Sons, Inc., 1957*)

In general, there are four main kinds of unconformable relations (see illustration), although the names of each are not universally accepted: (1) Nonconformity—rocks below the break are not stratified, such as massive crystalline rocks. (2) Angular unconformity—rocks below the break are stratified but lie at an angle to those above. (3) Disconformity—strata below are parallel to those above but are separated by an evident surface of erosion. (4) Pseudoconformity or paraconformity—strata are parallel but surface is hardly distinguishable from a simple bedding plane. *See* STRATIGRAPHY. [J.R.]

Underground mining An underground mine is a system of underground workings for the removal of ore from its place of occurrence to the surface, and involves the deployment of miners and services.

There are several basic physical elements in an underground mining system. The passageways (openings) in a mine are called drifts if they are parallel to the geological structure, and cross-cuts if they cut across it. They range in size about 60–200 ft^2 (6–18 m^2) in cross section, depending on their functions. The workings on a level (horizontal plane) are joined with those on another level by passageways of similar cross section, called raises if they are driven upward and winzes if driven downward.

The passageways give access to, and provide transportation routes from, the stopes, which are the excavations where the ore is mined. The stopes are between levels. There may be rooms on the level, such as pump rooms, service shops, and lunchrooms. For other aspects *see* COAL MINING; MINING. [A.V.C.]

Undersea vehicles Small research vessels for underwater transport of people and equipment and for use as underwater platforms for observation, sampling, measurement, and performing various work tasks. They are also referred to as submersibles.

Undersea vehicle systems are designed in accordance with user requirements. A major trend involves design of a completely integrated system, which, in addition to the undersea vehicle, includes support ship, handling gear for launch and retrieval, and logistic and maintenance support. The objective is to obtain an effective, high utilization rate under varying weather conditions. The major vehicle operating problem is handling during launch and retrieval in heavy seas. *See* OCEANOGRAPHIC VESSELS.

Most small undersea vehicles need buoyancy in excess of that produced by their pressure hulls to attain neutral buoyancy while submerged. Low-density solids that may be considered for this purpose include polyethylene, polypropylene, expanded plastics, inorganic foams, and syntactic foam.

As a source of power, the lead-acid storage battery is the most widely employed. It places severe limitations on small vehicles because of its relatively high weight-to-energy ratio. However, its characteristics are well know. *See* STORAGE BATTERY.

Sensors are instruments required on small undersea vehicles to determine position, communicate with other units, and make observations of the environment both for a record and while performing work. Life support systems on the small vehicles take into consideration the following services for atmospheric control and monitoring: breathing mixture supply system; carbon dioxide removal system; hydrogen, carbon monoxide, and toxin removal systems; air purification and filtering system; atmospheric monitoring system; and emergency breathing supply system. [D.C.B.; R.F.Bus.]

Underwater demolition The destruction or fragmentation of underwater obstacles by the use of explosive charges hand-placed by diver personnel.

Diving equipments used by naval personnel in underwater demolition include deep-sea equipment, light-weight equipment, and scuba gear. Standard demolition equipment and techniques are employed. The explosive requirement for a specific task would be judged by the properties of the individual high explosive (rapidity of detonation, sensitivity, brisance, and so on), in light of the problem at hand. *See* DIVING; SHIP SALVAGE. [H.W.Pi.]

Underwater navigation The process of directing the movements of submersible vehicles, and divers, from one point to another. The development of improved submersible vehicles, coupled with advances in saturated diving, has resulted in new requirements for underwater navigation. Various methods which have proved successful include acoustic transponder systems, dead reckoning, surface-referenced navigation from a support ship, homing, and various combinations of these. The choice of the navigation system depends on such factors as precision required, area to be covered, availability of surface vessels, sea state under which they are expected to operate, and for crewed vehicles the redundancy necessary for safety. *See* SUBMARINE; UNDERSEA VEHICLES.

Acoustic transponders. A system of navigation using acoustic transponder beacons represents the most accurate technique for precise location with respect to points on the sea floor. Usually two or more transponders are dropped from the surface vehicle or from a submerged vehicle to establish an array on the bottom. These transponders can be interrogated by surface vessels, submersible vehicles or divers. When navigation is performed from the surface vessel, the submerged vehicle also carries a transponder. The position of the submerged vehicle with respect to the sea-floor transponders is computed. In the case of a crewed submersible the information is forwarded via an acoustic communications link.

A second technique is based on the submerged vehicle's determining its own position by interrogating the array of transponders. This method is widely used both with crewed submersibles and towed vehicles, and provides high accuracy, since the interrogating sonar is located well below the surface layers. The time delay between interrogation and received return (plus a fixed time delay associated with the transponder) is a measure of the range of each transponder from the interrogating vehicle.

Dead reckoning. An undersea vehicle needs sensors to show distance traveled and direction of travel for dead-reckoning estimation of position. The most promising sensor for operation near the sea floor is a Doppler sonar. Ground speed, fore-aft and athwartship, can be determined from the Doppler shift in frequency of signals returned from the sea floor. Pulse-type and continuous-wave Doppler sonars are under development.

In addition to Doppler sonar, some specialized submersibles carry inertial dead reckoning in the form of a small, stabilized platform. *See* DEAD RECKONING; INERTIAL GUIDANCE SYSTEM; SONAR.

Other methods. Most submersible vehicles carry other sensors which can perform a navigation function. The horizontal obstacle sonar (HOS) is normally used to detect objects ahead of the submersible. An altitude-depth sonar provides vertical navigation by furnishing depth and altitude off the bottom. A vertical obstacle sonar (VOS) is used to determine heights of objects in the path of the vehicle.

An acoustic tracking system allows the monitoring and vectoring of the position of a vehicle from a surface vessel, where space and weight are not at a premium. Surface tracking systems are of two types: ultrashort baseline and short baseline. The ultrashort baseline system employs a single transducer, mounting two orthogonal arrays in an assembly less than 10 in. (25 cm) in diameter. The short baseline system consists of an array of four hydrophones mounted on a ship in two orthogonal baselines. The length of each baseline is the longest possible consistent with the ship's geometry, typically 50–100 ft (15–30 m).

The submarine inertial navigation system (SINS) is the primary navigation system in most U.S. Navy nuclear-powered submarines, with multiple systems being employed in ballistic missile-firing submarines for improved reliability and accuracy. The high stability and low drift of electrostatically supported gyroscopes have been employed to monitor SINS performance. [J.A.C.; E.S.G.]

Underwater photography The technique of using photography equipment underwater. Photography is as useful a recording system underwater as it is in air. All forms of photography are in use: single pictures, motion pictures, stereo, shadow, elapsed-time motion pictures, and so forth.

Shallow water. In shallow underwater photography, or hand-held underwater photography as it is more commonly called, the photographer, wearing some type of diving equipment, is submerged with the camera and takes photographs directly (see illustration). This type of underwater photography requires scuba.

Photographer using underwater motion-picture camera to record data on the bottom.

Underwater photography is controlled by many different factors, some of which are available light, clarity of water, differential absorption of light in various wavelengths, film capability, and construction of underwater housing for cameras. The underwater photographer can never expect to take clear pictures at distances greater than 165 ft (50 m). In most instances, the photographer must depend on natural light for film exposure. *See* DIVING. [R.F.Di.]

Deep-sea. Photography in the deep sea involves the design and use of camera and lighting equipment for which several requirements must be fulfilled, including the following.

1. Suitable watertight cases are required, for both the camera and the light source, to withstand the pressure of the sea. For each 30 ft (10 m) of depth, approximately one additional atmosphere of pressure is exerted. The deepest ocean, about 40,000 ft (12,000 m), requires a case to withstand 17,600 lb/in.^{2} (1200 kg/cm^2). The windows for the lens and the electrical seals must likewise be designed for such pressures.
2. Auxiliary lighting is required, especially for deep photography, since daylight is rapidly absorbed with depth, becoming almost useless even at shallow depths of a few hundred meters.
3. Corrected optics for the camera are desirable to compensate for distortions caused by the lens window—water interface when an air lens is used. However, many underwater cameras have lenses designed for use in the air even if some information is lost.
4. The camera must be positioned and triggered to get the desired photograph. Operation from a cable (with sonar sensing equipment) or from deep-diving underwater vehicles is employed. Bottom-sensing switches are often used to operate the deep cameras when bottom photographs are desired.

When a person descends as an observer to great depths in a diving vehicle, a camera can help in remembering and recording what was seen, and provide visual evidence.

A television link from the bottom to the surface is often mentioned as a substitute for underwater photography since the picture is seen immediately and the television screen can be photographed or taped at the surface. However, the quality of the television photograph or tape is not as high as that taken by a camera below because of the limitations of the television link and the lower resolution intrinsic in TV. *See* UNDERWATER TELEVISION. [H.E.E.; S.O.R.]

Underwater sound The production, transmission, and reception of sounds in the ocean are an important part of modern acoustics, Since the ocean is highly absorptive to light and other electromagnetic radiation but not to sound waves, acoustics has found many applications in probing the sea. Acoustical methods of detection, for example, are useful in locating submarines and other submerged objects, and therefore have many applications in naval science. Underwater sound is also important in scientific research: Oceanographers and geophysicists employ sound waves to determine the physical structure of the ocean and its bottom; sound is also used by marine biologists to study life in the ocean. There are also many commercial uses for underwater sound, such as seismic exploration for oil and gas beneath the ocean floor, use of sonar techniques for fish location, navigation and positioning, and communication. *See* SEAWATER; SONAR.

Sound projection and reception. In general, devices which transform one type of energy into another are known as transducers; when the end product is underwater sound, such devices are known as sound projectors. Most projectors use either the magnetostrictive or piezoelectric effect in which electrical energy is transformed into acoustical energy. However, oceanographers also make extensive use of explosives to study oceanographic features. Other projects are based on spark discharge, hydraulic flow, mechanical impact, and pneumatic and other effects.

To obtain the information sought, the acoustic signal must be picked up and transformed into an electrical signal by a receiving transducer called a hydrophone. Current practice favors the use of piezoelectric ceramics for these instruments. Transducer elements are usually arranged in arrays to achieve greater sensitivity, better directionality, and reduction of noise. Once the acoustical signal has been transformed into an electrical signal, the problem of extracting the information from the background noise becomes a problem in signal processing. *See* UNDERWATER TRANSDUCER.

When background noise has no clearly identifiable source, it is referred to as ambient sea noise. It is caused by marine life, seismic disturbances, distant shipping, wave action, cracking ice, rain on the surface, and so on. Ambient noise has been studied over a frequency range from about 0.1 Hz to over 100 kHz.

Speed of sound. The speed of sound in the ocean varies roughly from 4750 to 5100 feet per second (1448 to 1554 meters per second, or mps; about 3% higher than in fresh water), depending on pressure, temperature, and salinity (see table).

Absorption of sound. Sound is absorbed in seawater when an irreversible production of heat occurs during the alternate compression and rarefaction accompanying the sound wave. Absorption is greater in seawater than in fresh water over a certain frequency range because a dissociation-reassociation

Variation of sound speed with temperature, salinity, and depth (pressure)

Parameter	Coefficient*
Temperature, °C (ambient about 20°C)	$\frac{\Delta c}{\Delta T} = 2.7$ mps/°C
Salinity, parts per thousand (ppt)	$\frac{\Delta c}{\Delta S} = 1.2$ mps/ppt
Depth, m	$\frac{\Delta c}{\Delta D} = .017$ mps/m

*c = speed of sound in meters per second (mps).

process occurs with the ions of magnesium sulfate. There is a low-frequency relaxation in the 1-kHz region which is believed to be due to a very small amount of boric acid in seawater.

Transmission loss. In addition to the irreversible losses in the medium, a reduction in signal intensity due to the spreading of the wave also occurs. If there is a simple, omnidirectional source in a homogeneous, unbound medium the intensity varies as the reciprocal of the range squared. When the medium has plane-parallel upper and lower boundaries, the spreading is cylindrical and varies as the reciprocal of the range. The latter type of spreading, with its lower attenuation, occurs when sound is trapped in a "channel."

Refraction of sound. Because of variations of sound speed with depth, sound rays are bent as they propagate from point to point. This effect becomes very important when horizontal propagation distances exceed a few hundred feet. The profile of speed versus depth depends on the geographical location, the season of the year, and even the time of day. An idealized profile is shown in the illustration. Just below the surface is the surface layer in which sound speed is susceptible to daily and local changes due to heating, cooling, and wind action. Below the surface layer lies the seasonal thermocline, where the temperature decreases with depth, giving a negative gradient. This region is close enough to the surface to be affected by seasonal changes. Below this is the main thermocline, which is little affected by the season, where the speed decreases rapidly with depth. Between 3000 and 4000 ft or 0.9 and 1.2 km (at mid-latitudes) a point is reached at which the temperature attains a constant value and the speed a minimum; from there to the bottom the speed increases because of the pressure effect. The region below the axis is called the deep isothermal layer. In the Arctic regions this layer may lie at or near the ice-covered surface. *See* THERMOCLINE.

The depth at which the speed reaches a minimum defines the axis of the deep sound (or sofar) channel. A sound wave generated in this channel remains trapped in it and travels (by cylindrical spreading) for long distances. An explosion in this channel will be heard as a drawn-out signal that ceases abruptly with the passing of the slowest ray, the one which traveled directly along the axis. From this clearly defined time marker the exact location of a deep channel explosion can be determined by triangulating with two monitoring stations. This method was used during World War II by downed aviators, who dropped sofar bombs to give their location. It is also used in determining missile impact locations.

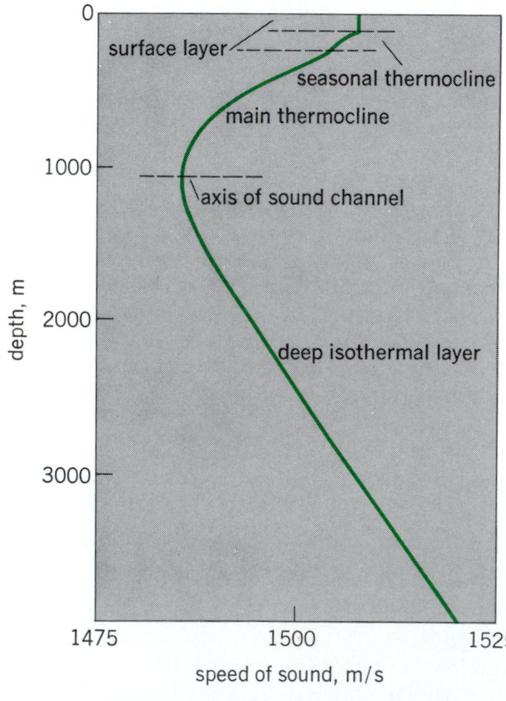

Typical deep-sea sound speed profile divided into layers. (*After R. J. Urick, Principles of Underwater Sound for Engineers, McGraw-Hill, 1967*)

Reflection of sound. Reflection occurs when sound strikes the sea surface, the sea bottom, or large obstacles in the water. If the reflecting surface is flat over distances that are large compared with the wavelength of sound, a fraction of the incident sound energy is reflected at an angle with respect to the normal which is equal to the angle of incidence. The amount of energy reflected depends on the acoustic properties of the materials involved.

The reflection from the surface of the ocean is essentially perfect. However, if the wavelength is small compared to the dimensions of the surface roughness, a well-defined reflected ray does not exist and the energy is scattered.

The reflection of sound from the sea bottom is similar to that from the surface but more complicated. The composition of the bottom ranges from soft mud through sand to hard rock at different locations, causing considerable variation in sound-reflecting and sound-scattering properties.

Scattering of sound. Scattering is the process by which sound energy is redirected from objects (or parts of objects) which have dimensions considerably smaller than a wavelength. The scattering from the surface and bottom has already been mentioned, but scattering from objects suspended in the water is also important. Contributions to such scattering are made by marine organisms, free air bubbles, suspended particles, the thermal microstructure in the sea, and the turbulence of natural or human origin. *See* SCATTERING LAYER. [W.S.Cr.]

Underwater telephone A method of voice communication using underwater sound as the means of transmission. The sound is generated by a projector at the sending point and is received by a hydrophone. Generally, a carrier wave of a fixed frequency is amplitude-modulated by the voice signal and demodulated in the receiving system. The range of an underwater telephone is fairly short (perhaps a few miles) because of transmission losses in the sound propagation. Its most important use is communication with, or between, fully submerged submarines. *See* UNDERWATER SOUND; UNDERWATER TRANSDUCER. [R.W.Mo.]

Underwater television The technique of using television equipment underwater. Underwater television picture quality is limited by clarity of the water, available light, and the sensitivity of the camera tube. Television color cameras require white artificial light at depth to achieve the color image. Monochrome cameras can be used with a light source color-matched to the peak sensitivity of the camera tube. *See* TELEVISION.

Television equipment used in underwater observation is compact and reliable and can be tailored for specific applications. The major advantages of underwater television over conventional photography are that the observer can remain at the surface while viewing a continuously monitored scene in real time. Observations can be made and recorded for long periods on video tape either in real time or by using time-lapse video recorders. Additional information such as numerical data or a time clock can be superimposed on the picture screen, and other information can be recorded on the video tape in the form of sound tracks. Free-swimming divers can operate a television camera in much the same way as a movie camera. *See* UNDERWATER PHOTOGRAPHY.

Underwater television has applications in fishery research for behavioral studies of fish in the wild, and in the oil industry for inspection of oil pipes, well heads, and platform structures and for seabed surveys. Remotely controlled vehicles to carry television cameras exist in great variety and are continuing to be adapted and developed for special purposes. *See* UNDERSEA VEHICLES. [R.Pr.]

Underwater transducer A device used for the generation or reception of underwater sounds. The word projector is applied to a generator of sound, while hydrophonic refers to a receiver. Since the same device may serve both purposes, the word transducer is used as a general designation. Thus a transducer, when a projector, converts electrical energy to motional mechanical energy, whereas a hydrophone converts mechanical energy to electrical energy. This conversion of energy is usually based on one of the following properties of certain materials: piezoelectricity, magnetostriction, and electrostriction. In each case, the application of an alternating electric or magnetic field of a given frequency causes a mechanical vibration of the transducer material at the same frequency. The transducer, being in contact with the water, communicates a similar motion to it, giving rise to a sound wave. With a hydrophone, the vibratory sound pressure causes a mechanical motion of the material, which in turn generates an electrical voltage. This voltage is amplified, then either read on a meter, recorded, or played through a loudspeaker. *See* ELECTROSTRICTION; MAGNETOSTRICTION; PIEZOELECTRICITY; SONAR; TRANSDUCER; UNDERWATER SOUND. [R.W.Mo.]

Uniform flow Fluid flow in which, at a given instant, the velocity is the same in both magnitude and direction at every point in space. This strict definition can have little meaning for the flow of a real fluid when the velocity varies across a section. However, when the size and shape of cross section are the same in any given length, the flow is said to be uniform. Specifically, the flow in a pipe of constant diameter is uniform; the flow in a pipe of varying size is not. Also, the flow in an open canal is uniform if the size and shape of the cross section of the stream are the same at different locations along the canal. *See* OPEN CHANNEL. [R.L.D.]

Unit operations A concept used by chemical engineers to teach and practice industrial processing. The basis of the unit operation approach is that chemical processes, in spite of their great variety, may be resolved into a comparatively small number of units, each of which appears again and again in the various processes. Each unit operation performs a definite function no matter where it is used, and the individual operations cut across industry and process lines (see table). For example, distillation, one of the important operations, is used where it is necessary to separate the constituents of a mixture of volatile liquids and where the separation may be achieved by a system of vaporizations and condensations. This operation is used in petroleum refining, alcohol manufacture, liquid gas production, refining of solvents, and many other industries. Knowledge of the fundamentals of distillation is applicable, then, to all these industries. This is characteristic of all unit operations.

The theoretical foundations of the unit operations are useful only because they lead to sound designs of the equipment in an actual plant. There is greater variety in the equipment than in the theoretical laws. Pumps, piping systems, dryers, heat transfer equipment, distilling columns, crushers, mixers, absorbers, evaporators, conveyors, and centrifuges are typical classes of unit operations equipment. The final objective of the unit operation approach is to provide the plant designer with sound selections and designs of practical apparatus for efficient production. *See* CHEMICAL ENGINEERING. [W.L.McC.]

Unit operations

Name	Function
Operations based on fluid mechanics	
Fluid handling	Storing and transporting fluids; controlling and measuring flow of gas, liquid, and vapor
Mixing and agitation	Combining liquids, solids, and gases into homogeneous fluids or intimate mixtures of phases
Filtration and clarification	Separating solid particles from liquids and gases
Thickening and sedimentation	Concentrating solids from their mixtures with liquids
Classification	Sorting solid particles by size and specific gravity
Centrifugation	Separating solids from liquids and liquids from liquids by centrifugal force.
Operations based on heat transfer	
Heat exchange and condensation	Heating, cooling, and condensing of fluids with or without phase change
Furnaces and kilns	High-temperature heating of materials
Evaporation and boiling	Vaporizing liquids, concentrating solutions of nonvolatile solids, recovering distilled water
Drying	Removing moisture or other liquid from solids by vaporization or other means
Cooling towers	Cooling water for reuse in condensers or for air conditioning
Operations based on mass transfer	
Distillation	Separating miscible liquids by vaporization
Liquid extraction	Separating miscible liquids by differences of solubility
Leaching	Dissolving soluble substances from solids by solvent extraction
Absorption and desorption	Washing soluble gas from its mixture with inert gas by liquid; removing dissolved gas from liquid by inert gas
Adsorption	Selectively removing substances from liquids or gases by reactive solids
Ion exchange	Selectively exchanging ions in solutions of electrolytes by reactive solids
Humidification and dehumidification	Controlling humidity or vapor content of air or gas
Diffusion of gases	Separating mixtures of gases by temperature gradients or by other specialized methods
Operations based on mechanical principles	
Screening	Separating particles according to size by screens
Solids handling	Transporting and sorting solids
Size reduction	Subdividing solids into smaller particles
Flotation	Separating solids by selective aeration
Magnetic and electrostatic separations	Separating, by electrical methods, solids according to chemical composition, or solids from gases

Unit processes Processes that involve making chemical changes to materials, as a result of chemical reaction taking place. For instance, in the combustion of coal, the entering and leaving materials differ from each other chemically: coal and air enter, and flue gases and residues leave the combustion chamber. Combustion is therefore a unit process. Unit processes are also referred to as chemical conversions.

Together with unit operations (physical conversions), unit processes (chemical conversions) form the basic building blocks of a chemical manufacturing process. Most chemical processes consist of a combination of various unit operations and unit processes.

The basic tools of the chemical engineer for the design, study, or improvement of a unit process are the mass balance, the energy balance, kinetic rate of reaction, and position of equilibrium (the last is included only if the reaction does not go to completion). *See* CHEMICAL ENGINEERING; UNIT OPERATIONS.

[W.F.F.]

Unitary symmetry One of the approximate internal symmetry laws obeyed by the strong interactions of elementary particles. A system of particles has an SU_n internal symmetry if all of the particles can be described as compounds of a fundamental multiplet of n particles, and if all physical properties of the system are unchanged by an arbitrary unitary transformation of the fundamental multiplet. *See* SYMMETRY LAWS (PHYSICS).

An analog is the approximate spin independence of electrons under electrostatic forces (as in an atom): There is a fundamental doublet, namely the spin-up electron and the spin-down electron. Denoting these two states by $|u\rangle$ and $|d\rangle$, all physical properties (energy eigenvalues, charge density, and so on) are unchanged by the replacements shown in the equations below, where α and β are complex numbers. The group

$$|u\rangle \rightarrow \alpha|u\rangle + \beta|d\rangle \qquad |d\rangle \rightarrow -\beta^*|u\rangle + \alpha^*|d\rangle$$
$$|\alpha|^2 + |\beta|^2 = 1$$

of all the transformations of two states which preserve their scalar products $[\langle u|d\rangle = 0, \langle u|u\rangle = \langle d|d\rangle = 1]$ is known as the two-dimensional unitary group, U_2; the transformations of the equations above form a subgroup known as SU_2 which merely lacks the uninteresting transformations of the form $|u\rangle \rightarrow e^{i\varphi}|u\rangle$ and $|d\rangle \rightarrow e^{i\varphi}|d\rangle$, that is, an equal change of phase of the two states.

The strong interactions are approximately invariant to such a group; the fundamental doublet can be taken to be the nucleon, with the up and down states proton and neutron. This SU_2 symmetry is known as independence, or, loosely, as *i*-spin conservation, the analog to the electron spin being known as *i*-spin **I**. *See* ISOBARIC SPIN.

When a sufficient number of strange particles had been observed, it was seen that they, together with the old non-strange particles, were grouped into multiplets whose members had the same space-time quantum numbers (except for mass; the masses of the members are only similar, not equal). This suggested the existence of a yet larger symmetry; it has turned out that this symmetry is the group of unitary transformations of a triplet of fundamental particles, SU_3. This symmetry is often loosely called unitary symmetry.

A striking difference in the manifestations of SU_2 and SU_3 is that whereas all possible multiplets of the former appear in nature, only those multiplets of the latter appear which can be regarded as compounds of the fundamental triplet in which the net number of component fundamental particles (number of particles minus number of antiparticles) is an integral multiple of 3. In particular, no particle which could be regarded as the fundamental triplet is found. Despite this nonappearance, it turns out that a great deal about the strongly interacting particles (hadrons) is at least qualitatively explained if they are regarded as physical compounds of a fundamental triplet of particles, to which the name quark has been given. There are interesting theories to explain why single quarks are never observed. *See* HADRON; QUARKS.

Hadrons have the approximate symmetry SU_N, where N is the number of kinds of quarks, or flavors. Five flavors of quark are known. *See* FLAVOR.

In the quark model of hadrons, baryons are correctly described only if quarks carry, in addition to spin and flavor, another quantum number, color, which can take on three values. The resulting symmetry, color SU_3, is thought to be exact; that is, the differently colored quarks of a given flavor are thought to be absolutely equivalent. However, this exact symmetry is rather well hidden, because apparently all free particles belong to the singlet representation of the symmetry. *See* COLOR (QUANTUM MECHANICS); ELEMENTARY PARTICLE.

[C.J.G.]

Units of measurement Values, quantities, or magnitudes in terms of which other such are expressed. Units are grouped into systems, suitable for use in the measurement of physical quantities and in the convenient statement of laws relating physical quantities. A quantity is a measureable attribute of phenomena or matter.

A given physical quantity A, such as length, time, or energy, is the product of a numerical value or measure $\{A\}$ and a unit $[A]$. Thus Eq. (1) holds.

$$A = \{A\}\,[A] \tag{1}$$

The unit $[A]$ can be chosen arbitrarily, but it is desirable to define units in such a way that they are derived from a few base units by equations without numerical factors other than unity, and that the equations between numerical values of quantities have exactly the same form as the equations between the quantities. For example, the kinetic energy E of a body is given in terms of its mass M and speed V by Eq. (2),

$$E = \tfrac{1}{2}MV^2 \tag{2}$$

where $E = \{E\}\,[E]$, $M = \{M\}\,[M]$, $V = \{V\}\,[V]$, and $\tfrac{1}{2}$ is called a definitional factor and is dimensionless. If the units of E, M, and V are defined in such a way that Eq. (3) holds, then

$$[E] = [M]\,[V]^2 \tag{3}$$

the equation between the numerical values is Eq. (4). A sys-

$$\{E\} = \tfrac{1}{2}\{M\}\,\{V\}^2 \tag{4}$$

tem of units defined in this way is called a coherent system. It is constructed by defining the units of a few base quantities independently; these are called base units. The units of all other quantities are defined by equations similar to Eq. (3) with no numerical factors other than unity, and are called derived units.

In 1960 the General Conference on Weights and Measures (CGPM) gave official status to a single practical system, the International System of Units, abbreviated SI in all languages. The system is a modernized version of the metric system. The SI, as subsequently extended, includes seven base units, two supplementary units, and nineteen derived units with special names. These derived units, and others without special names, are derived from the base and supplementary units in a coherent manner. A set of prefixes is used to form decimal multiples and submultiples of the SI units. Certain units which are not part of the SI but which are widely used or are useful in specialized fields have been accepted for use with the SI or for temporary use in those fields. *See* METRIC SYSTEM.

Geometrical units. Units of plane angle and solid angle are purely geometrical. The SI units of plane and solid angle are called supplementary units, and may be regarded either as base units or as derived units.

Plane angle units. The radian (rad), the SI unit of plane angle, is the plane angle between two radii of a circle which cut off on the circumference an arc equal in length to the radius. Since the circumference of a circle is 2π times the radius, the complete angle about a point is 2π rad. *See* RADIAN MEASURE.

The degree and its decimal submultiples can be used with the SI when the radian is not a convenient unit. By definition, 2π rad = 360°. The minute [1′ = (1/60)°] and the second [1″ = (1/60)′] can also be used.

Steradian. The steradian (sr), the SI unit of solid angle, is the solid angle which, having its vertex at the center of a sphere, cuts off an area on the surface of the sphere equal to that of a square with sides of length equal to the radius of the sphere.

Mechanical units. In mechanics, it is convenient to have three base quantities, and two of these are generally chosen to be length and time. Systems of mechanical units may be classified as absolute systems, in which the third base quantity is mass, and gravitational systems, in which the third base quantity is force.

Two absolute systems of metric units are commonly employed, each named for its base units of length, mass, and time: the mks (meter-kilogram-second) absolute system, and the cgs (centimeter-gram-second) absolute system. The mks absolute system is the mechanical portion of the SI. A coherent absolute system of British units is based on the foot, the pound (1 lb ≅ 0.4536 kg), and the second.

Gravitational systems, in which the base quantities are length, force, and time, have been frequently employed by engineers, and are therefore sometimes called technical systems.

Length units. The meter (m) is the SI base unit of length. The use of special names for decimal submultiples of the meter should be avoided, and units formed by attaching appropriate SI prefixes to the meter should be used instead.

The angstrom (Å) is equal to 10^{-10} m. Although it has been accepted for temporary use with the SI, it is preferable to replace this unit with the nanometer, using the relation 1 Å = 0.1 nm.

The nautical mile (nmi), equal to 1852 m, has been accepted for temporary use with the SI in navigations. *See* NAVIGATION.

The foot (ft) is, as discussed above, the unit of length in the British systems of units, and it is also in customary use in the United States. Since 1959 the foot has been defined as exactly 0.3048 m. The yard (yd) is defined as exactly 3 ft or 0.9144 m. *See* LENGTH; MEASURE.

Relative measurements of x-ray wavelengths can be made to a higher accuracy than absolute measurements. Before 1965, most x-ray wavelengths were expressed in terms of the X-unit, which is approximately 10^{-13} m. The X-unit has been superseded by the A* unit, which is based on the tungsten $K\alpha_1$ line as a standard. The peak of this line is defined as exactly 0.2090100 A*. X-ray wavelength tables have been published in terms of this unit. At the time the A* unit was defined, it was thought to equal 10^{-10} m (the angstrom unit, Å) to within 5 parts per million, but the A* unit is now believed to be 20±5 parts per million larger than 10^{-10} m.

Special units whose values are obtained experimentally are used in astronomy. For their definitions *see* ASTRONOMICAL UNIT; LIGHT-YEAR; PARSEC.

Area units. The square meter (m²), the SI unit of area, is the area of a square with sides of length 1 m. Other area units are defined by forming squares of various length units in the same manner. *See* AREA.

Cross sections, which measure the probability of interaction between an atomic nucleus, atom, or molecule and an incident particle, have the dimensions of area, and the appropriate SI unit for expressing them is therefore the square meter. The barn (b), a unit of cross section equal to 10^{-28} m², has been accepted for temporary use with the SI.

Units of volume. The cubic meter (m³), the SI unit of volume, is the volume of a cube with sides of length 1 m. Other units of volume are defined by forming cubes of various length units in the same manner. The liter (symbol L in the United States) is equal to 1 cubic decimeter (1 dm³), or equivalently to 10^{-3} m³. It has been accepted for use with the SI for measuring volumes of liquids and gases.

Time units. The second (s) is the SI base unit of time. However, other units of time in customary use, such as the minute (1 min = 60 s), hour (1 h = 60 min), and day (1 d = 24 h), are acceptable for use with the SI. *See* TIME.

Frequency units. The hertz (Hz), the SI unit of frequency, is equal to 1 cycle per second. A periodic oscillation has a frequency of n hertz if it goes through n cycles in 1 s. *See* FREQUENCY (WAVE MOTION).

Speed and velocity units. The meter per second (m/s), the SI unit of speed or velocity, is the magnitude of the constant velocity at which a body traverses 1 m in 1 s. Other speed and velocity units are defined by dividing a unit of length by a unit of time in the same manner. *See* SPEED; VELOCITY.

The knot (kn) is equal to 1 nautical mile per hour (1 nmi/h); it has been accepted for temporary use with the SI.

Acceleration units. The meter per second squared (m/s²), the SI unit of acceleration, is the magnitude of the constant acceleration of a body whose velocity changes by 1 m/s in 1 s. Other units of acceleration are defined by dividing a unit of velocity by a unit of time in the same manner. *See* ACCELERATION.

The gal or galileo (symbol Gal) is equal to 1 cm/s², or equivalently to 10^{-2} m/s².

Mass units. The kilogram (kg), the SI base unit of mass, is the only SI unit whose name, for historical reasons, contains a prefix. Names of decimal multiples and submultiples of the kilogram are formed by attaching prefixes to the word gram (g). The metric ton (t), which is equal to 10^3 kg or 1 megagram (Mg), is permitted in commercial usage of the SI.

The pound (lb), the unit of mass in the British absolute system, is also in customary use in the United States. In 1959 the pound was defined to be exactly 0.45359237 kg.

The slug is the unit of mass in the British gravitational system. By definition 1 pound force (lbf) acting on a body of mass 1 slug produces an acceleration of 1 foot per second squared (1 ft/s²). The slug is equal to approximately 32.174 lb or 14.594 kg. *See* MASS.

Force units. The newton (N), the SI unit of force, is the force which imparts an acceleration of 1 meter per second squared (1 m/s²) to a body having a mass of 1 kg.

The dyne, the cgs absolute unit of force, is the force which imparts an acceleration of 1 centimeter per second squared (1 cm/s²) to a body having a mass of 1 g.

The unit of force in the British absolute system is the poundal (pdl), the force which imparts an acceleration of 1 foot per second squared (1 ft/s²) when applied to a body of mass 1 lb. One poundal is approximately 0.13825 N.

The units of force in the mks gravitational, cgs gravitational, and British gravitational systems are the forces which impart an acceleration equal to the standard acceleration of gravity, g_n = 9.80665 m/s² ≅ 32.174 ft/s², when applied to bodies having masses of 1 kg, 1 g, and 1 lb, respectively. These units are named the kilogram force (kgf), gram force (gf), and pound force (lbf), respectively. Unfortunately, these units have also been called simply the kilogram, gram, and pound, giving rise to confusion with the mass units of the same name. *See* FORCE.

Pressure and stress units. The pascal (Pa), the SI unit of pressure and stress, is the pressure or stress of 1 newton per square meter (N/m²). Other units of pressure can also be formed by dividing various units of force by various units of area, such as the pound force per square inch (lbf/in.², frequently abbreviated psi).

Pressure has been frequently expressed in terms of the bar and its decimal submultiples, where 1 bar = 10^6 dynes/cm² = 10^5 Pa. Pressures are also frequently expressed in terms of the height of a column of either mercury or water which the pressure will support.

Two other units which have been frequently used for measuring pressure are the standard atmosphere and the torr. The standard atmosphere (atm) is exactly 101,325 Pa, which is approximately the average value of atmospheric pressure at sea level. The torr is exactly 1/760 atmosphere, or approximately 133.322 Pa. To within 1 part per million, the torr equal to the pressure of a column of mercury of height 1 millimeter (1 mmHg) at a temperature of 0°C when the acceleration due to gravity has the standard value $g_n = 9.80665\ m/s^2$. *See* ATMOSPHERE; PRESSURE; PRESSURE MEASUREMENT.

Energy and work units. The joule (J), the SI unit of energy or work, is the work done by a force of magnitude 1 newton when the point at which the force is applied is displaced 1 m in the direction of the force. Thus, joule is a short name for newton-meter (N-m) of energy or work. *See* ENERGY; WORK.

Units of energy or work in other systems are defined by forming the product of a unit of force and a unit of length in precisely the same manner as in the definition of the joule. Thus, the erg, the cgs absolute unit of energy or work, is the product of 1 dyne and 1 cm.

The foot-poundal (ft-pdl), the British absolute unit of energy or work, is the product of 1 poundal and 1 foot. The foot-pound, or, more properly, the foot-pound force (ft-lbf), the British gravitational unit of energy or work, is the product of 1 lbf and 1 ft.

Sometimes energy is measured in units which are products of a unit of power and a unit of time. Since 1 watt (W) of power equals 1 joule per second (1 J/s), as discussed below, the joule is equivalent to 1 watt-second (1 W · s). In electrical power applications, energy is frequently measured in kilowatthours (kWh), where 1 kWh = $(10^3\ W)(3600\ s) = 3.6 \times 10^6\ J$.

The calorie was originally defined as the quantity of heat required to raise the temperature of 1 g of air-free water 1°C under a constant pressure of 1 atm. However, the magnitude of the calorie, so defined, depends on the place on the Celsius temperature scale at which the measurement is made. The International (Steam) Table calorie is defined as exactly 4.1868 J; this is the type of calorie most frequently used in mechanical engineering. The thermochemical calorie, which has been used in thermochemistry in preference to the other types of calorie, is exactly 4.184 J.

The British thermal unit (Btu) was originally defined as the quantity of heat required to raise the temperature of 1 lb of air-free water 1°F under a constant pressure of 1 atm. The International Table Btu is approximately 1055.056 J, and the thermochemical Btu is approximately 1054.350 J.

Power units. The watt (W), the SI unit of power, is the power which gives rise to the production of energy at the rate of 1 joule per second (1 J/s). Other units of power can be defined by forming the ratio of a unit of energy to a unit of time in the same manner. *See* POWER.

The horsepower (hp) is equal to exactly 550 ft · lbf/s, or approximately 745.700 W.

Torque units. The newton-meter (N · m), the SI unit of torque, is the magnitude of the torque produced by a force of 1 newton acting at a perpendicular distance of 1 m from a specified axis of rotation. The joule should never be used as a synonym for this unit.

Units of torque in other systems are defined by forming the product of a unit of force and a unit of length in precisely the same manner as in the definition of the newton-meter.

The foot-poundal (ft · pdl), the British absolute unit of torque, is the product of 1 poundal and 1 foot. The foot-pound (ft · lbf), the British gravitational unit of torque, is the product of 1 lbf and 1 ft. These units are sometimes called the poundal-foot (pdl · ft) and pound-foot (lbf · ft) to distinguish them from the units of energy or work. *See* TORQUE.

Electrical units. For a general discussion of electrical units, including the SI or mks system, three cgs systems [electrostatic system of units (esu), electromagnetic system of units (emu), and gaussian system], and definitions of the SI units ampere (A), volt (V), ohm (Ω), coulomb (C), farad (F), henry (H), weber (Wb), and tesla (T) *see* ELECTRICAL UNITS AND STANDARDS.

This section discusses some additional SI units and some units in the cgs electromagnetic system which are frequently encountered in scientific literature in spite of the fact that their use has been discouraged.

Siemens. The siemens (S), the SI unit of electrical conductance, is the electrical conductance of a conductor in which a current of 1 ampere is produced by an electric potential difference of 1 volt. *See* CONDUCTANCE; ELECTRICAL RESISTANCE.

The siemens was formerly called the mho (℧) to illustrate the fact the unit is the reciprocal of the ohm.

Abampere. The abampere (abA), the cgs electromagnetic unit of current, is that current which, if maintained in two straight, parallel conductors of infinite length, of negligible circular cross section, and placed 1 cm apart in vacuum, would produce between these conductors a force equal to 2 dynes per centimeter of length. The abampere is equal to exactly 10 A.

Abvolt. The abvolt (abV), the cgs electromagnetic unit of electrical potential difference and electromotive force, is the difference of electrical potential between two points of a conductor carrying a constant current of 1 abA, when the power dissipated between these points is equal to 1 erg per second. Then 1 abV = 10^{-8} V.

Maxwell. The maxwell (Mx), the cgs electromagnetic unit of magnetic flux, is the magnetic flux which, linking a circuit of one turn, produces in it an electromotive force of 1 abV as it is reduced to zero in 1 s. Then 1 maxwell = 10^{-8} weber. *See* MAGNETIC FLUX.

Units of magnetic flux density. The gauss (Gs), the cgs electromagnetic unit of magnetic flux density (also called magnetic induction), is a magnetic flux density of 1 maxwell per square centimeter (1 Mx/cm^2). Then 1 gauss = 10^{-4} tesla. *See* MAGNETIC INDUCTION.

Units of magnetic field strength. The SI unit of magnetic field strength is 1 ampere per meter (1 A/m), which is the magnetic field strength at a distance of 1 m from a straight conductor of infinite length and negligible circular cross section which carries a current of 2π A. This definition is based on the definition, in the SI, of the magnetic field strength H_{SI}. At a distance r from a long straight conductor carrying I, H_{SI} is given by Eq. (5) The left-hand side of this equation is the line

$$2\pi r H_{SI} = I \tag{5}$$

integral of H_{SI} around a circular path, all of whose points are at distance r from the conductor. Substituting $r = 1$ m, and $I = 2\pi$ A in this equation gives $H_{SI} = 1$ A/m.

The oersted (Oe), the cgs electromagnetic unit of magnetic field strength, is the magnetic field strength at a distance of 1 cm from a straight conductor of infinite length and negligible circular cross section which carries a current of 0.5 abA. *See* MAGNETIC FIELD.

Units of magnetic potential and mmf. The ampere serves as the SI unit of magnetic potential difference and magnetomotive force (mmf), as well as the unit of current. In the SI, the magnetomotive force around a closed path equals the current passing through a surface enclosed by the path. Thus, 1 A is the magnetomotive force around a closed path when a current of 1 A passes through an enclosed surface.

The gilbert (Gb), the cgs electromagnetic unit of magnetic potential difference and magnetomotive force, is the magnetomotive force around a closed path enclosing a surface through which flows a current of (1/4π) abA. *See* MAGNETOMOTIVE FORCE.

Photometric units. Photometric units involve a new base quantity, luminous intensity. For the definition of the candela (cd), the SI unit of luminous intensity, *see* PHOTOMETRY; PHYSICAL MEASUREMENT.

For a general discussion of photometric units, including units of illuminance (illumination) and luminance, and in particular the SI units lux (lx) and candela per square meter (cd/ m^2), *see* ILLUMINATION. *See also* LUMINANCE.

Lumen. The lumen (lm), the SI unit of luminous flux, is the luminous flux emitted within a unit solid angle (1 steradian) by a point source having a uniform intensity of 1 candela. *See* LUMINOUS FLUX.

Luminous energy units. The lumen-second (lm · s), the SI unit of luminous energy (also called quantity of light), is the luminous energy radiated or received over a period of 1 s by a luminous flux of 1 lumen. This unit is also called the talbot. *See* LUMINOUS ENERGY.

Radiation units. Certain quantities and units are used particularly in the area of ionizing radiation. The special units curie, roentgen, rad, and rem, which were previously adopted for use in this area, are not coherent with the SI, but their temporary use with the SI has been approved while the transition to SI units takes place.

Activity units. The becquerel (Bq), the SI unit of activity (radioactive disintegration rate), is the activity of a radionuclide decaying at the rate of one spontaneous nuclear transition per second. Thus 1 Bq = 1 s^{-1}.

The curie (Ci), the special unit of activity, is equal to 3.7 × 10^{10} Bq. *See* RADIOACTIVITY.

Exposure units. The SI unit of exposure to ionizing radiation, 1 coulomb per kilogram (1 C/kg), is the amount of electromagnetic radiation (x-radiation or gamma radiation) which in 1 kg of pure dry air produces ion pairs carrying 1 coulomb of charge of either sign. (The ionization arising from the absorption of bremsstrahlung emitted by electrons is not to be included in measuring the charge.) *See* BREMSSTRAHLUNG.

The roentgen (R), the special unit of exposure, is equal to 2.58 × 10^{-4} C/kg.

Absorbed dose units. The gray (Gy), the SI unit of absorbed dose, is the absorbed dose when the energy per unit mass imparted to matter by ionizing radiation is 1 joule per kilogram (1 J/kg).

The rad (rd), the special unit of absorbed dose, is equal to 10^{-2} Gy.

Dose equivalent units. Different types of radiation cause slightly different effects in biological tissue. For this reason, a weighted absorbed dose called the dose equivalent is used in comparing the effects of radiation on living systems. The dose equivalent is the product of the absorbed dose and various dimensionless modifying factors.

The sievert (Sv), the SI unit of dose equivalent, is the dose equivalent when the absorbed dose of ionizing radiation multiplied by the stipulated dimensionless factors is 1 joule per kilogram (1 J/kg).

The rem, the special unit of dose equivalent, is equal to 10^{-2} Sv.

Other units. Logarithmic measures may be used with the SI. *See* DECIBEL; NEPER; PH; VOLUME UNIT (VU).

For units of loudness (sone, phon) *see* LOUDNESS. For units of sound absorption by surfaces *see* ARCHITECTURAL ACOUSTICS. For units of pitch *see* PITCH. For quantities and units pertaining to sound transmission *see* SOUND.

For temperature scales *see* TEMPERATURE.

For units measuring quantities in chemistry *see* ATOMIC MASS UNIT; ATOMIC WEIGHT; CONCENTRATION SCALES; ELECTROCHEMICAL EQUIVALENT; GRAM-MOLECULAR WEIGHT; MOLE (CHEMISTRY); MOLECULAR WEIGHT; RELATIVE ATOMIC MASS; RELATIVE MOLECULAR MASS.

For quantities and units pertaining to nuclear reactors *see* REACTOR PHYSICS.

For units of information content *see* BIT; INFORMATION THEORY. [J.F.We.]

Universal joint

A linkage that transmits rotation between two shafts whose axes are coplanar but not coinciding. The universal joint is used in almost every class of machinery: machine tools, instruments, control devices, and, most familiarly, automobiles.

A simple universal joint, known in English-speaking countries as Hooke's joint and in continental Europe as a Cardan joint, is shown in the illustration. It consists of two yokes attached to their respective shafts and connected by means of a spider. The angle between the shafts may have any value up to approximately 35°, if angular velocity is moderate when the angle is large. Although one shaft must make a single revolution for each revolution of the second shaft, the instantaneous angular

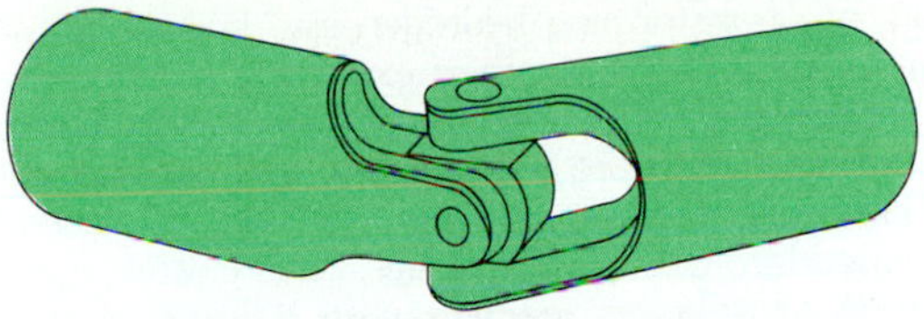

Simple universal joints. ***(After C. W. Ham, E. J. Crane, and W. L. Rogers, Mechanics of Machinery, McGraw-Hill, 1958)***

displacement of the first shaft is the same as that of the second shaft only at the end of each 90° of shaft rotation. Thus, only at four positions during each revolution is angular velocity of both shafts the same.

The variation in angular displacement and angular velocity between driving and driven shafts, which is objectionable in many mechanisms, can be eliminated by using two Hooke's joints, with an intermediate shaft. This arrangement is conventional for an automobile drive shaft. The axes of the driving and driven shafts need not intersect; however, it is necessary that the axes of the two yokes attached to the intermediate shaft lie respectively in planes containing the axes of adjoining shafts. *See* FOUR-BAR LINKAGE. [D.P.Ad.]

Universal motor

A series motor built to operate on either alternating current (ac) or direct current (dc). It is normally designed for capacities less than 1 hp (0.75 kW). It is usually operated at high speed, 3500 revolutions per minutes (rpm) loaded and 8000 to 10,000 revolutions per minute unloaded. For lower speeds, reduction gears are often employed, as in the case of electric hand drills or food mixers. As in all series motors, the rotor speed increases as the load decreases and the no-load speed is limited only by friction and windage. *See* ALTERNATING-CURRENT MOTOR; DIRECT-CURRENT MOTOR. [I.L.K.]

Universe

The sum of all the matter and energy that exist. The true scale of the universe is even now only dimly perceived by astronomers, despite tremendous advances in both observing techniques and equipment. The planet Earth is small compared with the Sun, which is only one of a hundred billion stars in the Milky Way Galaxy. And as large as this "island universe" is, it too is small compared with the clusters of thousands of such galaxies which populate the universe out to and (undoubtedly) beyond the range of astronomical vision. *See* MILKY WAY GALAXY.

The immense size of the universe and the great distances between objects introduce some fundamental limitations upon observations of remote galaxies. The basic restriction is caused by the finite speed of light in space. Light travels at 186,000 mi/s (300,000 km/s). Light from very distant galaxies has been traveling across space for a large fraction of the age of the universe. A galaxy 10^{10} light-years (9.46 × 10^{22} km) away is seen as it was when the universe was half its present age; such a galaxy may have changed significantly during this interval. The universe and its galaxies can therefore be studied at earlier and earlier epochs merely by observing to greater and greater distances.

The observation of spectral lines from distant galaxies shows them to be shifted from their normal laboratory wavelengths toward the red end of the spectrum. This so-called Doppler shift is due to the motion of these galaxies away from the Milky Way. E. P. Hubble showed that this is due to general expansion of the universe, and demonstrated that there is a relation between velocity v of a galaxy and its distance d, given by $v = Hd$, where H, the constant of proportionality, is called the Hubble constant and is a measure of the rate of expansion.

On the scale of atoms, the universe is over 90% hydrogen. Most of these atoms are collected into gaseous stars which, like the Sun, are transforming hydrogen into helium by means of nuclear fusion reactions in their extremely hot interiors. This hydrogen burning is therefore slowly changing the composition of the universe. Stars are collected into vast galaxies; the Milky Way Galaxy alone contains over 10^{11} stars. Galaxies in turn are grouped into clusters of galaxies; each cluster may contain thousands of galaxies, each having billions of stars. *See* GALAXY, EXTERNAL; PLANET; STAR; SUN.

The universe is dynamic. The early observation that all galaxies are receding led to the now well-accepted conclusion that the universe is expanding. This expansion implies that the universe is evolving. There are several ways to determine a rough value for the age of the universe. The universe must be older than the Milky Way Galaxy, whose age is estimated at something like 1.4×10^{10} years, based on studies of the oldest known stars. Extrapolating the expansion backwards (assuming that the observed rate has been constant since the expansion started), all of the observed galaxies would have started from the same point about 2×10^{10} years ago. The expansion rate was probably greater in the early history of the universe; the age of the universe is therefore somewhat less than the 2×10^{10} years given by the Hubble expansion time.

One explanation for the observed expansion of the universe is the "big bang" theory, in which all of the matter and radiant energy of the universe were initially compressed into a single object. At the assumed temperatures of some trillions of degrees, matter could not have existed in its present form, but consisted of subatomic or even subnuclear particles. This "primordial egg" exploded, hurling matter and radiation outward in a slowly decreasing expansion that led to rapidly dropping temperatures. *See* BIG BANG THEORY.

How long the universe will continue to expand will depend on whether or not the "big bang" explosion was stronger than the force of gravity due to all the matter of the universe. If the explosion was more powerful than the gravity of the universe, then the universe has "escape velocity" and will expand forever. Such an "open" universe is infinite in both time and space. The other extreme is given by the reverse situation: the total density of matter in the universe is so great that its gravity is greater than the explosion-induced expansion. In this case the universe will eventually come to rest, then reverse and fall back on itself. Conditions will approach those of the previous "big bang." It seems reasonable that the universe will then "bounce" or explode in another "big bang" and start over. Such an "oscillating" universe is finite in extent but infinite in time. The third possibility is that the expansion exactly balances the force of gravity. The universe will then expand and stop, remaining motionless for all time.

The universe appears to be open, and the best evidence suggests that the expansion will continue. Refutation of this would necessitate the discovery of large amounts of previously unknown mass in the universe. *See* COSMOLOGY. [H.Fre.]

Unsteady flow Fluid flow in which conditions are changing with respect to time and ultimately may become either steady flow or zero flow. This changing rate of flow may take place slowly, as when the action of a valve in a channel produces a gradual change in the rate, or it may take place rapidly as a result of a sudden closure, which produces a phenomenon known as water hammer. It is also found in such a case as the flow from one reservoir to another, in which equilibrium is approached as the two levels approach each other. *See* WATER HAMMER.

Unsteady flow also includes periodic motion such as that of waves on beaches, tidal motion in estuaries, and other oscillations. The difference between such cases and mean steady flow in turbulent flow is that the deviations from the mean are much greater and the time scale is also much longer. *See* FLUID FLOW; TURBULENT FLOW. [R.L.D.]

Upper-atmosphere dynamics The motion of the atmosphere above 30 mi (50 km). The predominant dynamical phenomena of the upper atmosphere are quite different from those encountered in the lower atmosphere. Among those encountered in the lower atmosphere are cyclones, anticyclones, tropical hurricanes, thunderstorms and shower clouds, tornadoes, dust devils, and similar phenomena. Even the largest of these phenomena do not penetrate far into the upper atmosphere. Above an altitude of about 30 mi (50 km), the predominant dynamical phenomena are internal gravity waves, tides, sound waves (including infrasonic), turbulence, and large-scale circulation.

Except under meteorological conditions characterized by convection, the atmosphere is stable against small vertical displacements of small air parcels; this results from buoyancy forces that tend to restore displaced air parcels to their original levels. An air parcel therefore tends to oscillate around its undisturbed position at a frequency known as the Brunt-Vaisala frequency, ω_B. Typical periods for the Brunt-Vaisala oscillation are in the vicinity of 5 min. If pressure waves are generated in the atmosphere with frequencies much greater than ω_B, they propagate as sound waves. For frequencies much less than ω_B, the waves propagate as internal gravity waves; in this case, the restoring forces for the wave motion are provided primarily by buoyancy (that is, gravity) rather than by compression.

Tides are internal gravity waves of particular frequencies. The term tidal usually implies that the exciting force is gravitational attraction by the Moon or Sun. However, it is conventional in the case of atmospheric tides to include also those waves that are excited by solar heating. One is therefore concerned with three separate excitation functions—lunar gravitation, solar gravitation, and solar heating.

Tidal wind patterns in the upper atmosphere generate electrical currents in the ionosphere through a dynamo action. These in turn give rise to diurnal variations in the geomagnetic field that can be observed at the Earth's surface.

Sound waves generated in the lower atmosphere may propagate upward. However, higher temperatures in the upper atmosphere refract most of the energy back toward the Earth's surface, giving rise to the phenomenon known as anomalous propagation. This involves the redirection of upward-moving sound waves back to the surface beyond the point where the source (usually a large explosion) can be heard by waves propagating along the surface. Thus there is normally a zone of silence, beyond which the source can again be heard because of anomalous propagation. Infrasonic waves with periods from 20 to 80 s have been observed occasionally with detectors at the Earth's surface in connection with auroral activity. See SOUND.

There is clear visual evidence of turbulence in the upper atmosphere; this evidence is obtained by examination of vapor trails released from rockets or of long persisting meteor trails. The source of the turbulence is not clear. The atmosphere is thermodynamically stable against vertical displacements throughout the region above the troposphere, and work has to be done against buoyancy forces in order to produce and maintain turbulence. The only apparent source of energy is internal gravity waves, either tidal or of random period.

Meteorological circulations and disturbances extend up to altitudes of about 30 mi (50 km). However, at higher altitudes, internal gravity waves and tides become more prominent. Above 60 mi (100 km), the circulation is not very well understood. [F.S.J.]

Upsilon particles A family of subnuclear particles having about 10 times the mass of the proton and generally described as the atomlike combination of a new quark with its antiquark. The new quark, designated *b* quark (for bottom or beauty), is about five times as massive as a proton and has a charge one-third that of the electron.

The discovery of the *b* quark led to a revision of the ideas associated with quarks and leptons and an ordering into three "generations" of quark-lepton groupings:

Generation I	up quark	electron
	down quark	electron neutrino
Generation II	charmed quark	muon
	strange quark	muon neutrino
Generation III	top quark	tau
	bottom quark	tau neutrino

The tau lepton was discovered in 1977. A vigorous search for the top quark has been unsuccessful so far. It is now frequently assumed that all matter is composed of various combinations of these objects. *See* ELEMENTARY PARTICLE; LEPTON; QUARKS.

[L.M.L.]

Upwelling The phenomenon by which deep, nutrient-rich ocean waters are introduced into the well-mixed surface layer. The net upwelling of deep ocean water must equal the net production of deep ocean water; each year about 0.1% of the ocean water is cooled in the polar regions and sinks to form deep water. This must be offset by an average oceanic upwelling rate of about 13 ft/year (4 m/year). A small but significant proportion of the total oceanic upwelling, perhaps as much as 20%, is concentrated in coastal and equatorial areas where seasonal upwelling rates may exceed 330 ft/month (100 m/month). *See* SEAWATER.

The geologic significance of upwelling areas results from their high productivity of phytoplankton. Fixation of silica by diatoms and its removal from the system by incorporation of the diatom frustules into the sediment constitute a major sink for silica in the ocean. *See* BACILLARIOPHYCEAE; MARINE SEDIMENTS; PHYTOPLANKTON.

[W.W.H.]

Uraninite The chief ore mineral of uranium. Uraninite has the idealized chemical composition UO_2, uranium dioxide. Thorium and rare earths, chiefly cerium, are usually present in variable and sometimes large amounts. Lead always is present by radioactive decay of the thorium and uranium present. Complete solid-solution series extend between UO_2, ThO_2 (thorianite), and CeO_2 (cerianite). *See* RADIOACTIVE MINERALS; RARE-EARTH ELEMENTS; THORIANITE; URANIUM.

The color of uraninite is black, grading to brownish black and dark brown in the more highly oxidized material. The luster of fresh material is steel-gray. The hardness is 5½–6 on Mohs scale. The specific gravity of pure UO_2 is 10.9, but that of most natural material is 9.7–7.5.

[C.Fr.]

Uranium A chemical element, symbol U, atomic number 92, atomic weight 238.03. The melting point is 1132°C (2070°F) and the boiling point is 3818°C (6904°F). Uranium is one of the actinide series. *See* ACTINIDE ELEMENTS.

1	2	3	4	5	6	7	8	9	10	11	12	13	14	15	16	17	18
1 H																	2 He
3 Li	4 Be											5 B	6 C	7 N	8 O	9 F	10 Ne
11 Na	12 Mg											13 Al	14 Si	15 P	16 S	17 Cl	18 Ar
19 K	20 Ca	21 Sc	22 Ti	23 V	24 Cr	25 Mn	26 Fe	27 Co	28 Ni	29 Cu	30 Zn	31 Ga	32 Ge	33 As	34 Se	35 Br	36 Kr
37 Rb	38 Sr	39 Y	40 Zr	41 Nb	42 Mo	43 Tc	44 Ru	45 Rh	46 Pd	47 Ag	48 Cd	49 In	50 Sn	51 Sb	52 Te	53 I	54 Xe
55 Cs	56 Ba	71 Lu	72 Hf	73 Ta	74 W	75 Re	76 Os	77 Ir	78 Pt	79 Au	80 Hg	81 Tl	82 Pb	83 Bi	84 Po	85 At	86 Rn
87 Fr	88 Ra	103 Lr	104 Rf	105 Db	106 Sg	107 Bh	108 Hs	109 Mt	110	*111*	*112*	*113*	*114*	*115*	*116*	*117*	*118*

lanthanide series	57 La	58 Ce	59 Pr	60 Nd	61 Pm	62 Sm	63 Eu	64 Gd	65 Tb	66 Dy	67 Ho	68 Er	69 Tm	70 Yb
actinide series	89 Ac	90 Th	91 Pa	92 U	93 Np	94 Pu	95 Am	96 Cm	97 Bk	98 Cf	99 Es	100 Fm	101 Md	102 No

Uranium in nature is a mixture of three isotopes: ^{234}U, ^{235}U, and ^{238}U. Uranium is believed to be concentrated largely in the Earth's crust, where the average concentration is 4 parts per million (ppm). The total uranium content of the Earth's crust to a depth of 15 mi (25 km) is calculated to be 2.2×10^{17} lb (10^{17} kg); the oceans may contain 2.2×10^{13} lb (10^{13} kg) of uranium. Several hundred uranium-containing minerals have been identified, but only a few are of commercial interest. *See* RADIOACTIVE MINERALS; URANINITE.

Because of the great importance of the fissile isotope ^{235}U, rather sophisticated industrial methods for its separation from the natural isotope mixture have been devised. The gaseous diffusion process, which in the United States is operated in three large plants (at Oak Ridge, Tennessee; Paducah, Kentucky; and Portsmouth, Ohio) has been the established industrial process. Other processes applied to the separation of uranium include the centrifuge process, in which gaseous uranium hexafluoride is separated in centrifuge cascades, the liquid thermal diffusion process, the separation nozzle, and laser excitation. *See* ISOTOPE (STABLE) SEPARATION.

Uranium is a very dense, strongly electropositive, reactive metal; it is ductile and malleable, but a poor conductor of electricity. Many uranium alloys are of great interest in nuclear technology because the pure metal is chemically active and anisotropic and has poor mechanical properties. However, cylindrical rods of pure uranium coated with silicon and canned in aluminum tubes (slugs) are used in production reactors. Uranium alloys can also be useful in diluting enriched uranium for reactors and in providing liquid fuels. Uranium depleted of the fissile isotope ^{235}U has been used in shielded containers for storage and transport of radioactive materials. *See* NUCLEAR FUELS; NUCLEAR REACTOR.

Uranium reacts with nearly all nonmetallic elements and their binary compounds. Uranium dissolves in hydrochloric acid and nitric acid, but nonoxidizing acids, such as sulfuric, phosphoric, or hydrofluoric acid, react very slowly. Uranium metal is inert to alkalies, but addition of peroxide causes formation of water-soluble peruranates. *See* URANIUM METALLURGY.

Uranium reacts reversibly with hydrogen to form UH_3 at 250°C (482°F). Correspondingly, the hydrogen isotopes form uranium deuteride, UD_3, and uranium tritide, UT_3. The uranium-oxygen system is extremely complicated. Uranium monoxide, UO, is a gaseous species which is not stable below 1800°C (3270°F). In the range UO_2 to UO_3, a large number of phases exist. The uranium halides constitute an important group of compounds. Uranium tetrafluoride is an intermediate in the preparation of the metal and the hexafluoride. Uranium hexafluoride, which is the most volatile uranium compound, is used in the isotope separation of ^{235}U and ^{238}U. The halides react with oxygen at elevated temperatures to form uranyl compounds and ultimately U_3O_8.

[F.We.]

Uranium metallurgy The processing treatments for the production of uranium concentrates and the recovery of pure uranium compounds, as well as the conversion chemistry for producing uranium metal and the processes employed for preparing uranium alloys.

The procedures to recover uranium from its ores are numerous, because of the great variety in the nature of uranium minerals and associated materials and the wide range of concentration in the naturally occurring ores. Recovery of uranium requires chemical processing; however, preliminary treatment of the ore may involve a roasting operation, a physical or chemical concentration step, or a combination of these. In general, one of two leaching treatments—acid leaching and carbonate leaching—is used as the initial step in chemical concentration. The choice depends on the nature of the ore, which largely determines the efficiency and the cost of the process employed. *See* LEACHING.

The concentrate, whether obtained by chemical or physical means, is treated chemically to give a uranyl nitrate solution that can be further purified by solvent extraction. The impurities remain in the aqueous phase, while the uranium is extracted into the organic phase. *See* SOLVENT EXTRACTION.

Uranium metal can be obtained from its halides by fused-salt electrolysis or by reduction with more reactive metals. The reaction of UO_2 with calcium yields metal of fair quality. The largest tonnages of good-quality uranium have been produced by metallothermic reduction of finely divided UF_4 with calcium or magnesium in steel bombs lined with fused dolomitic oxide (Ames process). The charge, consisting of an intimate mixture of UF_4 with the reductant metal in granular form together with a suitable booster (usually calcium plus iodine), is placed into the reduction bomb, the lid is bolted down, and the bomb is heated to ignition temperature. The shape of the metal ingot (also referred to as a biscuit) depends on the shape of the reduction bomb.

Uranium alloys are prepared by fusing the components together. All procedures following conventional metallurgical techniques, however, may have to be carried out inside inert-gas glove boxes because many alloys are attacked by oxygen or moisture. *See* NUCLEAR FUELS; NUCLEAR FUELS REPROCESSING; URANIUM. [F.We.]

Uranus The first planet to be discovered with the telescope and the seventh in the order of distance from the Sun. It was found accidentally by W. Herschel in England on March 13, 1781.

The obliquity (inclination of rotational axis to orbit plane) of Uranus is the highest of any known planet: 97.9°. This means that the axis is almost in the plane of the orbit, and thus the seasons on Uranus are very unusual. During summer in one hemisphere, the pole points almost directly toward the Sun while the other hemisphere is in total darkness. Forty-one years later, the situation is reversed.

The linear equatorial diameter is 32,400 mi (51,800 km). The mass is 14.63 times the mass of the Earth. The mean density is greater than that of Saturn even though Uranus is smaller than Saturn. This means that Uranus is richer than Saturn (or Jupiter) in elements heavier than hydrogen and helium but not nearly so rich in these elements as Earth. The same is true for Neptune. *See* JUPITER; NEPTUNE; SATURN.

The temperature Uranus would assume in simple equilibrium with the incident solar radiation is about −360°F (55 K), according to calculations. This is essentially identical with the value that is obtained by direct measurement. There is no evidence for the existence of an internal energy source or for a thermal inversion in the upper atmosphere, as observed for Jupiter, Saturn, Titan, and Neptune.

Through the telescope, Uranus appears as a small, slightly elliptical blue-green disk. Some observers have reported the presence of faint, dusky bands on the disk when the equator is oriented toward the Earth, but these sightings are not fully confirmed. Spectroscopic observations have shown that the atmosphere of the planet contains large amounts of methane and hydrogen.

Uranus has five known satellites that form a remarkably regular system, with low orbital eccentricities and inclinations. In 1977 a system of rings was discovered around Uranus during observations of a stellar occultation. There are at least nine rings, the largest having a width of only 30 mi (50 km). [T.C.O.]

Urban climatology The branch of climatology concerned with urban areas. These locales produce significant changes in the surface of the Earth and the quality of the air. In turn, surface climate in the vicinity of urban sites is altered. The era of urbanization on a worldwide scale has been accompanied by unintentional, measurable changes in city climate. *See* CLIMATOLOGY.

The process of urbanization changes the physical surroundings and induces alterations in the energy, moisture, and motion regime near the surface. Most of these alterations may be traced to causal factors such as air pollution; anthropogenic heat; surface waterproofing; thermal properties of the surface materials; and morphology of the surface and its specific three-dimensional geometry—building spacing, height, orientation, vegetative layering, and the overall dimensions and geography of these elements. Other factors that must be considered are relief, nearness to water bodies, size of the city, population density, and land-use distributions.

In general, cities are warmer than their surroundings, as documented over a century ago. They are islands or spots on the broader, more rural surrounding land. Thus, cities produce a heat island effect on the spatial distribution of temperatures. The timing of a maximum heat island is followed by a lag shortly after sundown, as urban surfaces, which absorbed and stored daytime heat, retain heat and affect the overlying air. Meantime, rural areas cool at a rapid rate.

A number of energy processes are altered to create warming, and various features lead to those alterations. City size, the morphology of the city, land-use configuration, and the geographic setting (such as relief, elevation, and regional climate) dictate the intensity of the heat island, its geographic extent, its orientation, and its persistence through time. Individual causes for heat island formation are related to city geometry, air pollution, surface materials, and anthropogenic heat emission. There are two atmosphere layers in an urban environment, besides the planetary boundary layer outside and extending well above the city: (1) The urban boundary layer is due to the spatially integrated heat and moisture exchanges between the city and its overlying air. (2) The surface of the city corresponds to the level of the urban canopy layer. Fluxes across this plane comprise those from individual units, such as roofs, canyon tops, trees, lawns, and roads, integrated over larger land-use divisions (for example, suburbs). [A.J.B.]

Urea The diamide of carbonic acid. Urea is a white, crystalline, water-soluble compound, melting point 132.7°C, with the formula given below. Important biologically as an end prod-

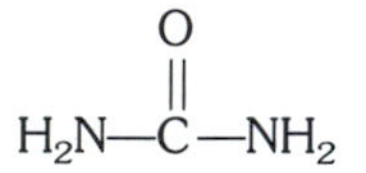

uct of normal animal and human protein metabolism, it is also of synthetic value in urea-formaldehyde plastics and is used as a stabilizer for many explosives.

Urea is used medically as a diuretic, and was used formerly in the treatment of indolent ulcers. Because of its high nitrogen content, it is used as a commercial fertilizer. *See* FERTILIZER. [E.B.R.]

Urea is the excretory end product of amino acid metabolism in mammals, elasmobranchs, amphibia, and chelonia (tortoises and turtles). In the liver, liberated ammonia combines with CO_2 and phosphate to form carbamyl phosphate, which reacts with the amino acid ornithine to become citrulline and then with aspartic acid to form arginosuccinic acid, which spontaneously decomposes to arginine. Finally, the enzyme arginase catalyzes the hydrolysis of arginine to form urea and ornithine, which is recycled. Each molecule of urea is formed from two molecules of ammonia and one molecule of CO_2. This complex reaction involves at least seven enzymes. Blood urea concentration rises in kidney diseases. *See* AMINO ACIDS; LIVER. [M.K.S.]

Urea-formaldehyde-type resins The condensation products obtained by the reaction of urea or melamine with formaldehyde. Resinous condensation products of formaldehyde with other nitrogen-containing compounds, for example, aniline and amides, also belong to this group of resins but have gained only limited utility.

The urea- and melamine-formaldehyde resins (amino resins) possess an excellent combination of physical properties and can be easily fabricated in a variety of colors. They are widely used as adhesives, laminating resins, molding compounds, paper and textile finishes, and surface coatings.

The aniline- and sulfonamide-formaldehyde resins are produced from other compounds containing —NH_2 groups that condense with formaldehyde to form methylol derivatives which are capable of further reaction. These resins have received limited attention except for aniline and *p*-toluenesulfonamide. Because of their resistance to the absorption of water, the aniline-formaldehyde resins have been used in electrical applications, such as insulation or panels, where their natural brown color is not objectionable. The sulfonamide-formaldehyde polymers are less colored than the aniline-type resins and have been employed in surface coatings. *See* CONDENSATION REACTION; FORMALDEHYDE; POLYMERIZATION; UREA. [J.A.M.]

Urediniomycetes A class of fungi in the subdivision Basidiomycotina, commonly known as rust fungi, that cause plant rust diseases. Vegetative hyphae are binucleate. Spores are produced in powdery or waxy, yellow, brown, or black pustules. The hallmark of the rust fungi is their thick-walled teliospores, produced in the terminal state in pustules called telia. Some rust fungi overwinter as telia, others as dormant mycelia. A septate basidium extends from the teliospore as the nuclei fuse and undergo meiosis. Four haploid basidiospores bud off for wind dissemination to infect new plants. Heteroecious rusts require two host species.

Rust fungi are widespread and are feared pathogens because of their pleomorphic life cycles; their specialization on the host species or within the host species; their ability to jump to new, unrelated hosts; and their efficient dispersal of spores. The 3000–5000 rust species parasitize various vascular plants such as ferns, onions, beans, and chrysanthemums. Rust diseases are controlled by fungicides, by eradication of the alternate (noneconomic) host of a heteroecious rust or, most effectively, by breeding resistant host varieties. *See* EUMYCOTA; FUNGI; PLANT PATHOLOGY. [J.W.McC.]

Ureid An acyl derivative of urea, such as acetylurea, CH_3CO—$NNHCONH_2$, and diacetylurea, CH_3CO—NHCONH—$COCH_3$. The most important members are cyclic, containing five- or six-atom hetero rings, and are known as parabanic acid and barbituric acid, respectively. Parabanic acid is formed from urea and oxalyl chloride; barbituric acid or its derivatives (barbiturates) is formed from urea and malonic ester or suitably substituted malonic esters.

The barbiturates are acidic compounds, soluble in base (in which form they are often used). They are widely used in treating insomnia, epilepsy, hysteria, and as preliminary anesthetics. *See* BARBITURATES.

Many naturally occurring substances, for example, the purines (from nucleoproteins of plant and animal cells), and the xanthine alkaloids (from coffee, tea, and cocoa), contain diureid structures, in which two N—C—N units are combined with one C—C—C unit. *See* PURINE; UREA. [E.B.R.]

Urethane Urethane, or ethyl urethan or ethyl carbamate, $H_2NCOOC_2H_5$, is the ethyl ester of carbamic acid. It forms white crystals with melting point 48–50°C (118–122°F), boiling point 182–184°C (360–363°F), and it is freely soluble in water, alcohol, and chloroform. Less toxic than other urethanes, ethyl urethan is a mild hypnotic and weak diuretic; in water, it forms a neutral solution with a cooling, saline taste. With quinine hydrochloride, it is used as a sclerosing agent for treatment of varicose veins.

Urethane has applications as an intermediate in organic synthesis and, when molten, as a solvent for various organic materials. [E.B.R.]

Uric acid The excretory end product in amino acid metabolism by uricotelic species, including birds, terrestrial invertebrates, and snakes; in purine metabolism by most insects, snakes, lizards, birds and primates (including humans). Nonuricotelic animals oxidize uric acid by the enzyme uricase to a more soluble compound, allantoin. The disease gout is associated with elevated uric acid. *See* GOUT; PROTEIN METABOLISM; PURINE. [M.K.S.]

Uridine diphosphoglucose (UDPG) A compound in which α-glucopyranose is esterified, at carbon atom 1, with the terminal phosphate group of uridine-5′-pyrophosphate (that is, uridine diphosphate, UDP). On very mild acid hydrolysis, glucose and UDP are liberated. Uridine diphosphoglucose occurs in animal, plant, and microbial cells and is synthesized enzymatically from uridine triphosphate (UTP) and α-glucose-1-phosphate. This compound functions as a key in the transformation of glucose to other sugars.

In general, UDP-glucose is a prominent member of a family of compounds composed of sugars activated as derivatives of nucleoside diphosphates. In these active forms they can be used directly as glycosyl donors, or they may be transformed to other more complex sugars before their utilization for the biosynthesis of both simple saccharides and complex polysaccharides. *See* CARBOHYDRATE METABOLISM; NUCLEIC ACID. [J.L.Str.]

Urinalysis The laboratory examination of urine. Examination of the urine requires a consideration of normal kidney and lower urinary tract anatomy and physiology, as well as the awareness of the clinical and laboratory conditions under which a urine sample is obtained and examined.

In general, urine is examined for color, specific gravity, pH, and the presence or absence of normal and abnormal materials. The color is often a rough index of the ability of the kidney to concentrate urine and may also afford the first indication of the presence of blood or other abnormal pigmented material. In healthy individuals, the normal range of the specific gravity varies between 1.015 and 1.030 under controlled conditions. Deviations above or below this range may indicate decrease in kidney function or the presence of some specific disease process. The pH reflects the role of the kidney in aiding in the maintenance of acid-base balance. Urine is also examined for the presence of blood, glucose, protein, and bacteria. The presence of any of these substances offers important clues to the diagnosis of a wide variety of disorders. [E.G.St./N.K.M.]

Urinary bladder A distensible, muscular sac in most vertebrates which serves as a reservoir for urine. Snakes, crocodilians, birds (with the exception of the ostrich), most lizards, and a few fish lack a urinary bladder. In these organisms, urine empties directly into the cloaca. Three general types of urinary bladder are recognized among the vertebrates: tubal, cloacal, and allantoic. *See* URINE.

Most fish possess tubal bladders, that is, enlargements of the mesonephric ducts, The cloacal bladder is found in monotremes, amphibians, and some dipnoans. There is no direct connection between the excretory ducts and this type of bladder. The bladder is an outpouching or diverticulum of the cloacal wall. The cloacal opening is closed by a sphincter muscle and the urine which seeps into the cloaca from the excretory ducts is forced into the bilobed bladder. The allantoic bladder is derived from the ventral wall of the cloaca and possibly the allantoic diverticulum. *See* ALLANTOIS.

The bladder drains through the urethra, the opening being controlled by a sphincter. Innervation is by both sympathetic and parasympathetic fibers; stimulation of the parasympathetic causes the bladder muscle to contract and relaxes the internal sphincter. Micturition is a reflex act which is initiated voluntarily except in children. *See* PARASYMPATHETIC NERVOUS SYSTEM; SYMPATHETIC NERVOUS SYSTEM; URINARY SYSTEM. [C.B.C.]

Urinary bladder disorders The urinary bladder is subject to anomalies, obstructions, inflammations, calculi, fistulae, and tumors.

A fairly common, distressing anomaly is a failure of both the lower abdominal wall and anterior bladder wall to close. The urine appears directly through the abdominal wall defect.

Interference with normal complete emptying of the bladder may be the result of various factors. In one form, an elevation of the internal urethral opening above its normal dependent position creates an unemptied pool of residual urine in the bottom of the bladder. This elevation is usually caused in the male by prostatic enlargement and in the female by cystocele, a bulging of the lower bladder wall into the vagina. Another form of obstruction is blockage of the bladder neck or urethra by lesions or tumors. A third major cause of inadequate emptying is neurogenic, usually following spinal cord injury or disease.

Cystitis or inflammation of the urinary bladder is an extremely common affliction; it is annoying because of the accompanying painful, frequent micturition. Calculi usually develop as a result of infection and obstruction. They often reach a considerable size, up to several inches in diameter. They are usually composed of phosphates and urates.

Abnormal openings between the bladder or female urethra and the vagina are not unusual. Such fistulae most often follow injury during childbirth. The chief effects are infection and the constant leakage of urine through the vagina. [R.N.B.]

Urinary system The urinary system consists of the kidneys, urinary ducts, and bladder. Similarities are not particularly evident among the many and varied types of excretory organs found among vertebrates. The variations that are encountered are undoubtedly related to problems with which vertebrates have had to cope in adapting to different environmental conditions.

Kidneys. In reptiles, birds, and mammals three types of kidneys are usually recognized: the pronephros, mesonephros, and metanephros. These appear in succession during embryonic development, but only the metanephros persists in the adult.

The metanephric kidneys of reptiles lie in the posterior part of the abdominal cavity, usually in the pelvic region. They are small, compact, and often markedly lobulated. The posterior portion on each side is somewhat narrower. In some lizards the hind parts may even fuse. The degree of symmetry varies.

The kidneys of birds are situated in the pelvic region of the body cavity; their posterior ends are usually joined. They are lobulated structures with short ureters which open independently into the cloaca.

A rather typical mammalian metanephric kidney (Fig. 1) is a compact, bean-shaped organ attached to the dorsal body wall outside the peritoneum. The ureter leaves the medial side at a depression, the hilum. At this point a renal vein also leaves the kidney and a renal artery and nerves enter it. The kidneys of mammals are markedly lobulated in the embryo, and in many forms this condition is retained throughout life. *See* KIDNEY.

[C.K.W.]

Urinary bladder. At or near the posterior ends of the nephric ducts there frequently is a reservoir for urine. This is the urinary bladder. Actually there are two basic varieties of bladders in vertebrates. One is found in fishes in which the reservoir is no more than an enlargement of the posterior end of each urinary duct. Frequently the urinary ducts are conjoined and a small bladder is formed by expansion of the common duct. The far more common type of bladder is that exhibited by tetrapods. This is a sac which originates embryonically as an outgrowth from the ventral side of the cloaca. Present in all embryonic life, it is exhibited differentially in adults. All amphibians retain the bladder, but it is lacking in snakes, crocodilians, and most lizards; birds, also, with the exception of the ostrich, lack a bladder. It is present in all mammals. *See* URINARY BLADDER. [T.W.T.]

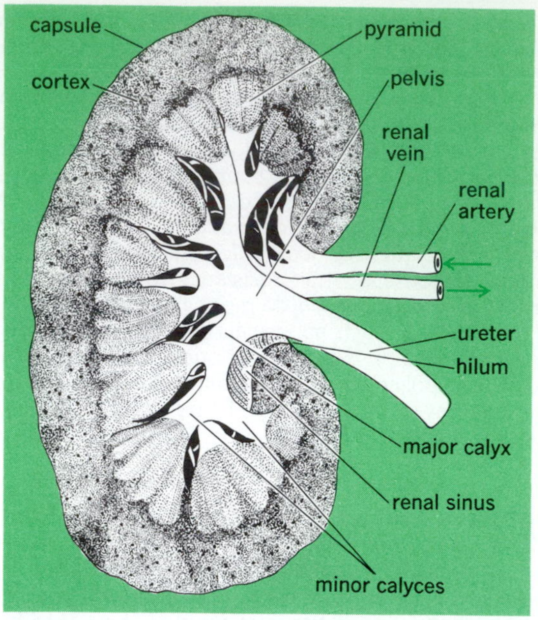

Fig. 1. Sagittal section of a human metanephric kidney (semidiagrammatic). (*After C. K. Weichert and W. Presch, Elements of Chordate Anatomy, 4th ed., McGraw-Hill, 1975*)

Physiology. Urine is produced by individual renal nephron units which are fundamentally similar from fish to mammals (Fig. 2); however, the basic structural and functional pattern of these nephrons varies among representatives of the vertebrate classes in accordance with changing environmental demands. Kidneys serve the general function of maintaining the chemical and physical constancy of blood and other body fluids. The most striking modifications are associated particularly with the relative amounts of water made available to the animal. Alterations in degrees of glomerular development, in the structural complexity of renal tubules, and in the architectural dispo-

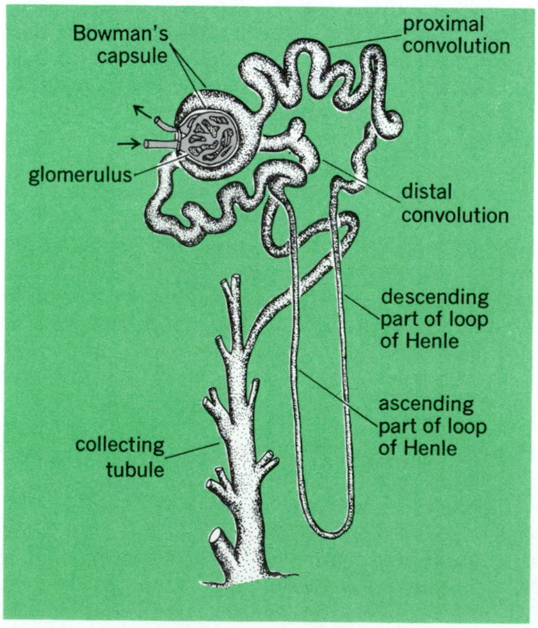

Fig. 2. A mammalian metanephric tubule, showing the renal corpuscle and secretory and collecting portions. (*After C. K. Weichert and W. Presch, Elements of Chordate Anatomy, 4th ed., McGraw-Hill, 1975*)

sition of the various nephrons in relation to one another within the kidneys may all represent adaptations made either to conserve or eliminate water.

Regulation of volume. Except for the primitive marine cyclostome *Myxine*, all modern vertebrates, whether marine, fresh-water, or terrestrial, have concentrations of salt in their blood only one-third or one-half that of seawater. The early development of the glomerulus can be viewed as a device responding to the need for regulating the volume of body fluids. Hence, in a hypotonic fresh-water environment the osmotic influx of water through gills and other permeable body surfaces would be kept in balance by a simple autoregulatory system whereby a rising volume of blood results in increased hydrostatic pressure which in turn elevates the rate of glomerular filtration. Similar devices are found in fresh-water invertebrates where water may be pumped out either as the result of work done by the heart, contractile vacuoles, or cilia found in such specialized "kidneys" as flame bulbs, solenocytes, or nephridia that extract excess water from the body cavity rather than from the circulatory system. Hence, these structures which maintain a constant water content for the invertebrate animal by balancing osmotic influx with hydrostatic output have the same basic parameters as those in vertebrates that regulate the formation of lymph across the endothelial walls of capillaries. *See* OSMOREGULATORY MECHANISMS.

Electrolyte balance. A system that regulates volume by producing an ultrafiltrate of blood plasma must conserve inorganic ions and other essential plasma constituents. The salt-conserving operation appears to be a primary function of the renal tubules which encapsulate the glomerulus. As the filtrate passes along their length toward the exterior, inorganic electrolytes are extracted from them through highly specific active cellular resorptive processes which restore plasma constituents to the circulatory system.

Movement of water. Concentration gradients of water are attained across cells of renal tubules by water following the active movement of salt or other solute. Where water is free to follow the active resorption of sodium and covering anions, as in the proximal tubule, an isosmotic condition prevails. Where water is not free to follow salt, as in the distal segment in the absence of antidiuretic hormone, a hypotonic tubular fluid results.

Nitrogenous end products. Of the major categories of organic foodstuffs, end products of carbohydrate and lipid metabolism are easily eliminated mainly in the form of carbon dioxide and water. Proteins, however, are more difficult to eliminate because the primary derivative of their metabolism, ammonia, is a relatively toxic compound. For animals living in an aquatic environment ammonia can be eliminated rapidly by simple diffusion through the gills. However, when ammonia is not free to diffuse into an effectively limitless aquatic environment, its toxicity presents a problem, particularly to embryos of terrestrial forms that develop wholly within tightly encapsulated eggshells or cases. For these forms the detoxication of ammonia is an indispensable requirement for survival. During evolution of the vertebrates two energy-dependent biosynthetic pathways arose which incorporated potentially toxic ammonia into urea and uric acid molecules, respectively. Both of these compounds are relatively harmless, even in high concentrations, but the former needs a relatively large amount of water to ensure its elimination, and uric acid requires a specific energy-demanding tubular secretory process to ensure its efficient excretion. *See* UREA; URIC ACID.

Urine concentration. The unique functional feature of the mammalian kidney is its ability to concentrate urine. Human urine can have four times the osmotic concentration of plasma, and some desert rats that survive on a diet of seeds without drinking any water have urine/plasma concentration ratios as high as 17. More aquatic forms such as the beaver have correspondingly poor concentrating ability.

The concentration operation depends on the existence of a decreasing gradient of solute concentration that extends from the tips of the papillae in the inner medulla of the kidney outward toward the cortex. The high concentration of medullary solute is achieved by a double hairpin countercurrent multiplier system which is powered by the active removal of salt from urine while it traverses the ascending limb of Henle's loop (Fig. 2). The salt is redelivered to the tip of the medulla after it has diffused back into the descending limb of Henle's loop. In this way a hypertonic condition is established in fluid surrounding the terminations of the collecting ducts. Urine is concentrated by an entirely passive process as water leaves the lumen of collecting ducts to come into equilibrium with the hypertonic fluid surrounding its terminations. *See* URINE; UROGENITAL SYSTEM. [R.P.F.]

Urine An aqueous solution of organic and inorganic substances, mostly waste products of metabolism. The kidneys maintain the internal milieu of the body by excreting these waste products and adjusting the loss of water and electrolytes to keep the body fluids relatively constant in amount and composition. The urine normally is clear and has a specific gravity of 1.017–1.020, depending upon the amount of fluid ingested, perspiration, and diet. The increase in specific gravity above that of water is due to the presence of dissolved solids, about 60% of which are organic substances such as urea, uric acid, creatinine, and ammonia; and 40% of which are inorganic substances such as sodium, chloride, calcium, potassium, phosphates, and sulfates. Its reaction is usually acid (pH 6) but this too varies with the diet. It usually has a faint yellow color due to a urochrome pigment, but the color varies depending upon the degree of concentration, and the ingestion of certain foods (for example, rhubarb) or cathartics. It usually has a characteristic aromatic odor, the cause of which is not known. *See* KIDNEY; UREA; URIC ACID; URINALYSIS; URINARY SYSTEM. [R.Str.]

Urodela One of the orders of the class Amphibia, also known as the Caudata. The members of this order are the tailed amphibians, or salamanders, and are distinguished superficially from the frogs and toads (order Anura) by the possession of a tail, and from the caecilians (order Apoda) by the possession of limbs. Salamanders resemble lizards in that members of both groups normally have a relatively elongate body, four limbs, and a tail. The similarity is however, only superficial. The most obvious external difference, the moist glandular skin of the amphibian and dry scaly skin of the reptile, is underlain by the numerous characters that distinguish reptiles from amphibians. *See* ANURA.

The vast majority of salamanders are well under a foot (0.3 m) in length. The largest is the giant salamander of Japan and China, which may attain a length of 5 ft (1.5 m). A close relative, the hellbender, which lives in streams in the eastern United States, grows to over 2 ft (0.6 m) in length.

Four well-developed limbs are typically present in salamanders, but the legs of the mud eel (*Amphiuma*) of the southeastern United States are very tiny appendages and of no use in locomotion. The sirens (*Siren* and *Pseudobranchus*), also aquatic salamanders of the same region, have undergone even further degeneration of the limbs and retain only tiny forelimbs. Salamanders with normal limbs usually have four toes on the front feet and five on the rear, though a few have only four on all feet. A feature of the Urodela not shared with the Anura is the ability to regenerate limbs.

The moist and highly vascularized skin of a salamander serves as an organ of respiration, and in some forms shares with the buccal region virtually the whole burden. In one entire family of salamanders, the Plethodontidae, lungs are not present. Even in those forms with lungs and gills, dermal respiration plays a very important part. The tympanum is absent in salamanders and the middle ear is degenerate. Hence these animals are deaf to airborne sounds. Undoubtedly correlated with this is the reduction of voice; salamanders are mute or produce only slight squeaking or, rarely, barking sounds when annoyed. However, salamanders can detect sounds carried through ground or water.

Salamanders live in a variety of aquatic or moist habitats. Many forms are wholly aquatic, living in streams, rivers, lakes, and ponds. Others live on land most of the year but must return to water to breed. The most advanced species live out their lives in moist places on land or, in the case of some tropical species, in trees, and never go in the water.

Producing the young alive rather than by laying eggs is very rare among salamanders. The eggs of salamanders may be deposited in water or in moist places on land. Presumably, aquatic breeding is the more primitive. Fertilization is external in most of the Cryptobranchoidea, internal by means of spermatophores in the remaining suborders. *See* NEOTENY.

The greatest concentration of families and genera of salamanders is found in the eastern United States, and these animals are one of the few major groups of vertebrate animals that is distinctly nontropical. No salamander occurs south of the Equator in the Old World, and only 18 species of the evolutionarily advanced family Plethodontidae are found as far south as South America in the New World.

The Urodela are readily arranged in seven families, but authorities differ as to the relationships among these families and their classification into suborders. Three to five suborders may be recognized, and one of these is thought by some to merit ordinal rank, equivalent to the Urodela. About 300 species of salamanders are known to be living today, and these are distributed among about 54 genera. *See* AMPHIBIA. [R.G.Z.]

Urogenital system The combined structures composing the urinary and genital, or reproductive, organs of vertebrates. The terms urinogenital or genitourinary system are equally applicable and just as frequently used. During embryonic development, the urinary and reproductive systems are closely interrelated, as their ducts arise in associated mesodermal regions. Common passages are associated with both these systems in various vertebrates, such as the single orifice for emission of urine and sperm. *See* KIDNEY; REPRODUCTIVE SYSTEM; URINARY SYSTEM. [C.B.C.]

Urokinase An enzyme that is classified as a plasminogen activator; its action is believed to be essential in the dissolution of blood clots. Urokinase is a very specific peptidase whose only function is to cleave a single arginyl-valine bond in the inactive zymogen plasminogen; this is an essential step in the transformation of inactive zymogen into the active proteolytic enzyme plasmin. In blood, this process is believed to be the major pathway for the dissolution of blood clots.

Malignancy in the living organism and the production of plasminogen activator (urokinase) by tumor cells are highly correlated. This is the basis for a theory that fibrin digestion is involved in metastasis. It is thought that fibrin digestion around tumor cells provides increased nutrients and that this digestion modifies circulation in some way to favor the proliferation of tumor cells. These observations have been suggested as a model for the development of selected anticancer drugs and cancer diagnostic tests.

Since the main pathway in living cells for clot dissolution is via the activation of the plasmin system, urokinase has been developed for clinical use. It has been investigated for treatment of myocardial infacts and is also used as an adjunct to cancer chemotherapy. [G.H.Ba.]

Uropygi An order of arachnids, the tailed whip scorpions, comprising about 70 species from tropical and warm temperate Asia and the Americas. Most are dark reddish-brown, of medium to giant size, 0.7–2.6 in. (18–65 mm), the largest one being *Mastigoproctus giganteus* of the southern United States and Mexico. The elongate, flattened body bears in front a pair of greatly thickened, raptorial pedipalps set with many sharp spines and used to hold and crush insect prey. The first pair of legs is elongated and modified into feelers. The abdomen terminates in a slender, many jointed, whiplike flagellum. The uropygids are harmless, nocturnal creatures without poison glands that live in dark places and burrow into the soil. When disturbed, they expel a volatile liquid, with the strong odor of acetic acid, from a gland at the base of the tail. This accounts for the name "vinegaroon" given by many Americans to these much-feared animals. *See* ARACHNIDA. [W.J.Ger.]

Uropygial gland A relatively large, compact, bilobed, secretary organ located at the base of the tail (uropygium) of most birds having a keeled sternum. It is known also as the preen, oil, or scent gland. This is the only true skin gland possessed by this class of vertebrates.

The glandular secretion, predominantly oily and sometimes of offensive odor (musk-duck, hoopoe, petrel) is discharged through an orifice at the tip of a nipplelike protuberance often encircled by short, bristly feathers. The act of preening induces a flow of secretion from the nipple which is transferred by the beak to the body plumage. In water fowl there is some evidence that this oily secretion assists in maintaining the water-repellent quality of the feathers, either directly or by preserving their physical structure. *See* FEATHER; SCENT GLAND. [M.E.R.]

Ursa Major The most widely known and oldest of the astronomical constellations. Ursa Major, or the Great Bear, is a circumpolar group as viewed from the middle latitudes of the Northern Hemisphere. One part of the configuration, a group of seven bright stars, which is pictured as the tail of the Great

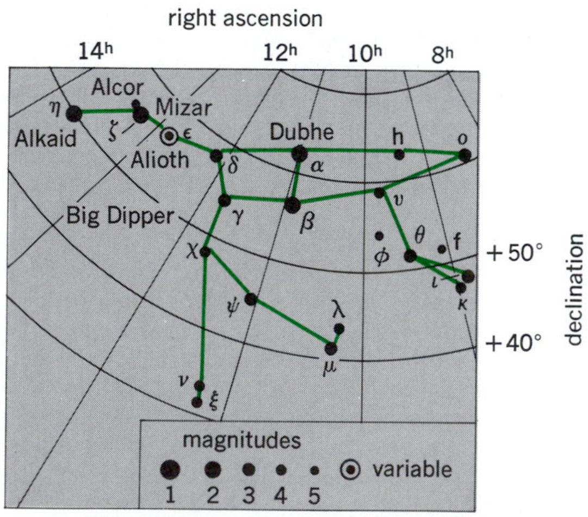

Line pattern of the constellation Ursa Major. The grid lines represent the coordinates of the sky. The apparent brightness, or magnitude, of the stars is shown by the sizes of the dots, which are graded by appropriate numbers as indicated.

Bear, is commonly known in the United States as the Big Dipper which it resembles (see illustration). The two stars α and β at the front of the bowl of the dipper are called pointers, because a line joining them points to Polaris, the North Star. *See* CONSTELLATION. [C.-S.Y.]

Ursa Minor The astronomical constellation Little Bear, Ursa Minor is a circumpolar constellation whose brightest star, Polaris, is almost at the north celestial pole. Seven of the eight stars appear to form a dipper, hence the constellation is alternately known as the Little Dipper (see illustration). The two bright stars β and γ at the front of the bowl are often called the

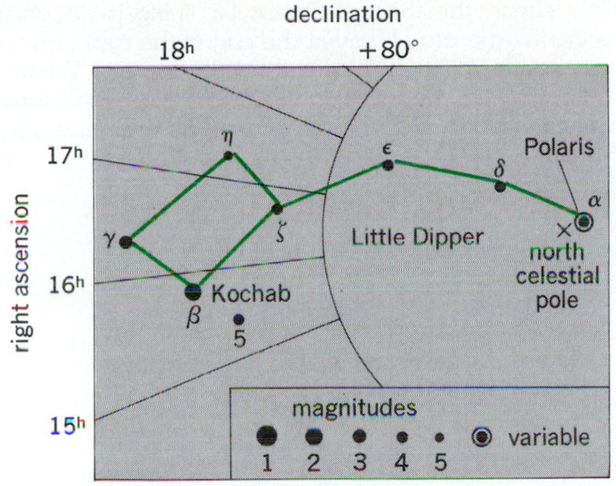

Line pattern of the constellation Ursa Minor. The grid lines represent the coordinates of the sky. The apparent brightness, or magnitude, of the stars is shown by the sizes of the dots, which are graded by appropriate numbers as indicated.

Guardian of the Pole, because they circle about Polaris closer than other conspicuous stars. *See* CONSTELLATION; URSA MAJOR. [C.-S.Y.]

Urticales An order of flowering plants, division Magnoliophyta (Angiospermae), in the subclass Hamamelidae of the class Magnoliopsida (dicotyledons). The order consists of 5 families and nearly 2400 species. The largest families are the Moraceae (about 1500 species), Urticacea (about 700 species), and Ulmaceae (about 150 species). The Urticales are herbs or woody plants with simple, usually stipulate leaves, and reduced, clustered flowers that usually have a vestigial perianth. Figs (*Ficus*, in the Moracea), mulberries (*Morus*), nettles (*Urtica*), elms (*Ulmus*), and Indian hemp (*Cannabis sativa*, family Cannabaceae, the source of a commercial fiber as well as hashish and marijuana) are well-known members of the Urticales. *See* ELM; FIG; HACKBERRY; HAMAMELIDAE; HEMP; HOP; MAGNOLIOPSIDA; MULBERRY; RAMIE. [A.Cr.]

Ustilaginomycetes A class of the subdivision Basidiomycotina. These microscopic plant-parasitic fungi are commonly known as smut fungi. Many species may have a shorter or longer saprophytic life cycle, capable of yeastlike reproduction. Several nonparasitic yeasts have been identified as closely related to the smuts. The Ustilaginales are divided into two families: Ustilaginaceae (with separate basidia producing lateral and apical basidiospores) and Tilletiaceae (with unseptate basidia producing apical basidiospores), together comprising 50 genera. About 1500 true Ustilaginales species are known.

The body (mycelium) of the smut fungi consists of hyphae that are thin, transparent, branched, septate, and binucleate. Usually parasitic, the fungus grows in the host tissues and gives little, if any, evidence of its presence before spore formation sets in. The smut spores (teliospores or ustilospores) are the organs of dispersion and resistance, and are thick walled, pigmented, and variously ornamented. They are formed in great number in the sori, which consist of host tissues, spore masses, and sometimes modified fungal cells or tissues.

The smut fungi develop variable sori in different organs (such as the roots, stems, leaves, inflorescences, flowers, anthers, and seeds) filled by powdery or agglutinated, usually colored (black, brown, violet, or yellow) spore masses. The spores may develop singly, in pairs, or in aggregates of spore balls. These balls may be composed entirely of fertile spores or of a combination of spores, sterile cells, or hyphae. Spore germination results in a basidium (promycelium or ustidium) which gives rise to hyaline basidiospores. *See* BASIDIOMYCOTINA; EUMYCOTA; FUNGI; SMUT (MICROBIOLOGY). [K.V.]

Uterine disorders Uterine and cervical disorders account for a large portion of medical practice. Although the cervix is usually considered separate pathologically, cervical disorders are included in this article. The major categories include congenital, inflammatory, neoplastic, and functional diseases. *See* UTERUS.

Failure of formation or developmental abnormalities are not uncommon, but most aberrations are of minor significance, such as a double uterus.

Inflammations of the uterus are not common but, when present, follow infections resulting from trauma, childbirth, or miscarriage. However, cervicitis, an inflammation of the uterine cervix, is probably the most frequently encountered female disease. Most cervicitis is nonspecific and tends to become chronic. *See* GONORRHEA; SOFT CHANCRE; SYPHILIS; TUBERCULOSIS.

The uterus is frequently the site of both benign and malignant tumor formation, and the three most common growths are as follows. (1) Endometrial polyps are benign simple or multiple masses which are often sources of abnormal bleeding; they are otherwise unimportant. (2) Leiomyomas, or fibroids, are the most common tumors in women, and occur in 10% of women of reproductive age. These are benign smooth-muscle growths, varying in size from minute bodies to huge masses which fill the pelvis and lower abdomen. They may be asymptomatic or may cause abnormal bleeding. The larger fibroids commonly produce signs or symptoms as a result of the mechanical effects on adjacent organs. They may also interfere with pregnancy, labor, or delivery. (3) Carcinoma of the endometrium is an important female malignancy but has a lower incidence than breast or cervical cancer. It is most often found in the 50- to 70-year age groups and in patients with obesity, hypertension, and diabetes mellitus.

Functional disorders of the uterine mucosa produce menstrual irregularities and other symptoms and, for the most part, stem from obscure causes. The complex endocrine regulation of the very reactive endometrium is subject to a great many other individual factors. Hyperplasias, atrophies, and displaced endometrial tissue are seen frequently. The more common hyperplasias and atrophies account for about one-fifth of all gynecologic disorders and are the most frequent cause of abnormal bleeding. *See* MENSTRUATION. [N.K.M.]

Uterus The hollow, muscular womb, being an enlarged portion of the oviduct in the adult female. An adult human uterus, before pregnancy, measures 3 × 2 × 1 in. (7.5 × 5 × 2.5 cm) in size and has the shape of an inverted, flattened pear. The paired Fallopian tubes enter the uterus at its upper

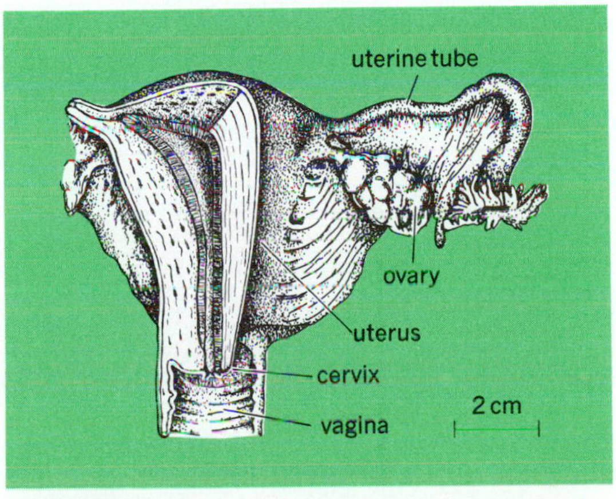

Human uterus and associated structures. (*After L. B. Arey, Developmental Anatomy, 7th ed., Saunders, 1965*)

corners; the lower, narrowed portion, the cervix, projects into the vagina (see illustration). Normally the uterus is tilted slightly forward and lies behind the urinary bladder.

The lining, or mucosa, responds to hormonal stimulation, growing in thickness with a tremendous increase in blood vessels during the first part of the menstrual cycle. If fertilization does not occur, the thickened vascular lining is sloughed off, producing the menstrual flow at the end of the cycle, and a new menstrual cycle begins with growth of the mucosa. When pregnancy occurs, the mucosa continues to thicken and forms an intimate connection with the implanted and enlarging placenta. *See* MENSTRUATION; PREGNANCY; REPRODUCTIVE SYSTEM. [W.J.B.]

V

Vaccination By strict definition, vaccination refers only to the inoculation of vaccinia virus (smallpox vaccination). In common usage, vaccination now implies the active immunization procedures employed against a variety of infectious agents. A more accurate term is immunization.

Active immunization describes the administration of all or part of a biologic agent in an effort to evoke a defensive response in the host. Killed bacteria, components of bacteria, altered toxins, live and killed viruses, and components of viruses have each been utilized in vaccine formulation. The vaccine may be injected, or given by mouth in some instances, with the expectation that antibody against an antigenic component of the vaccine will be elicited. In addition, other immune responses, such as lymphocyte-associated-cell-mediated immunity, may be evoked. The ultimate goal is to protect the host against subsequent challenge by the naturally occurring or "wild" infectious agent. *See* ANTIBODY; ANTIGEN; IMMUNITY. [V.Fu.]

Vacuole A cavity within a cell. In its widest sense the term connotes all the vesicles bound by a single membrane within the cytoplasm, including the endoplasmic reticulum, the nuclear envelope, and the Golgi apparatus or dictyosomes, all structures usually considered as separate entities. The term is more commonly reserved for the membrane-bound cavities which are associated with the central vacuole of plants, food vacuoles, self-digestive vacuoles, gas vacuoles, and the contractile vacuole. Vacuoles are discussed in this article in the more traditional context. The term vacuole is somewhat ambiguous in that it groups structures which are not truly homologous. Some have a single limiting membrane, or unit membrane, while others have a layer of protein granules which separates the contents from the cytoplasm. It is believed that the membrane carries enzymes which are concerned with functions such as transport of substances across the membrane between the cytoplasm and the cavity. *See* CELL MEMBRANES.

The central vacuole of the mature plant cell (see illustration) is formed from the fusion of smaller ones which occur universally in plants. It may contain many diverse substances, including salts, sugars, organic acids, amino acids, proteins, tannins, and pigments. Some of the contents of the vacuole are thought to be waste products from cellular metabolism. The colors of many flowers and some leaves are due to pigments contained in the central vacuole. *See* PLANT CELLS.

Food vacuoles, sometimes called heterophagic vacuoles, are common in most protozoans and some algae. In higher animals they occur in cells of the intestinal epithelium. They form where the surface of the cell contacts a particle of food. *See* ENDOCYTOSIS; PHAGOCYTOSIS.

When there is a need for the cell to digest portions of itself, autophagic vacuoles arise. This often happens in response to starvation. Portions of the cytoplasm, which may contain mitochondria or other cytoplasmic organelles, become isolated within the cytoplasm by a partitioning membrane. These may develop as special vacuoles which contain a number of digestive enzymes. The cell thus uses some of its own substances to obtain the essential materials of metabolism when these are not available from outside. *See* LYSOSOME.

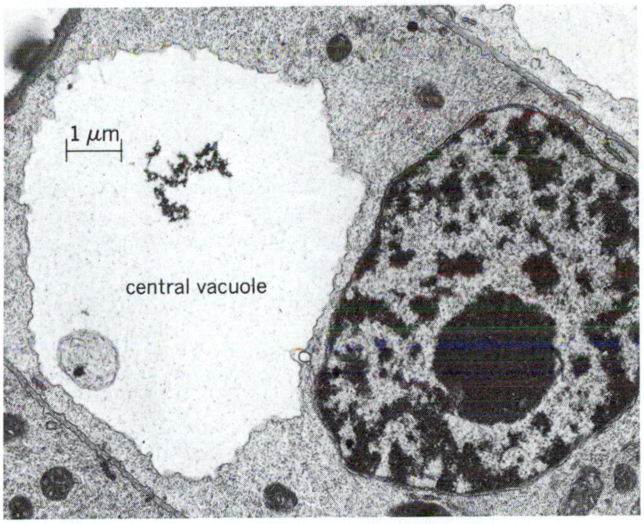

Electron micrograph of cell from a developing root tip showing the central vacuole of a partially mature cell. The osmotic pressure in this vacuole is responsible for the turgor and provides the force to expand the cell to its mature size.

Contractile vacuoles are common in protozoans and are found in some algae. In single-celled organisms they are found only in fresh-water forms, and even in species which grow in both fresh and brackish water they are lacking when the environment is brackish. They expand and collapse, often rhythmically, as fluid is accumulated slowly from the cytoplasm and then rapidly expelled from the cell. *See* ENDOPLASMIC RETICULUM; GOLGI APPARATUS; OSMOREGULATORY MECHANISMS. [M.C.L.]

Vacuum fusion A technique of analytical chemistry for determining the oxygen, hydrogen, and sometimes nitrogen content of metals.

The metal sample is either fused or dissolved in a bath, or flux, of a second metal in a heated graphite crucible supported inside an evacuated glass or quartz vessel. Oxygen is released from the metal as carbon monoxide by reaction of oxides or dissolved oxygen with carbon from the graphite crucible at high temperature. Metal nitrides dissociate to form elemental nitrogen. Hydrogen is evolved as elemental hydrogen. The mixture of carbon monoxide, nitrogen, and hydrogen is analyzed to determine individual component concentrations by various techniques.

Inert gas fusion has been developed for determination of gases in metals. The techniques used are similar to vacuum fusion but substitute inert gases for the vacuum environment. *See* VACUUM METALLURGY. [F.C.B.]

Vacuum measurement The determination of a gas pressure that is less in magnitude than the pressure of the atmosphere. (Atmospheric pressure is 29.92 in. or 760 mm of mercury.) This low pressure can be expressed in terms of the height of a column of mercury which the given pressure (vacuum) will support, referenced to zero pressure. The unit most commonly used is the torr, equal to 1 mm of mercury (mmHg). The unit of pressure in the International System (SI) is the pascal (Pa), equal to 1 newton per square meter (1 torr = 133.322 Pa).

Pressures above 1 torr can be easily measured by familiar pressure gages, such as liquid-column gages, diaphragm-pressure gages, bellows gages, and bourdon-spring gages. At pressures below 1 torr, mechanical effects such as hysteresis, ambient errors, and vibration make these gages impractical. *See* MANOMETER; PRESSURE MEASUREMENT.

Pressures below 1 torr are best measured by gages which infer the pressure from the measurement of some other property of the gas, such as thermal conductivity or ionization. The thermocouple gage, in combination with a hot- or cold-cathode gage (ionization type), is the most widely used method of vacuum measurement. *See* IONIZATION GAGE.

Other gages used to measure vacuum in the range of 1 torr or below are the McLeod gage, the Pirani gage, and the Knudsen gage. The McLeod gage is used as an absolute standard of vacuum measurement in the 10 torr (10^3 Pa) to 10^{-4} torr (10^{-2} Pa) range. *See* MCLEOD GAGE; PIRANI GAGE. [R.C.]

Vacuum metallurgy The making, shaping, and treating of metals, alloys, and intermetallic and refractory metal compounds in a gaseous environment where the composition and partial pressures of the various components of the gas phase are carefully controlled. In many instances, this environment is a vacuum ranging from subatmospheric to ultrahigh vacuum (less than 760 torr or 101 kilopascals to 10^{-12} torr or 10^{-10} pascal). In other cases, reactive gases are deliberately added to the environment to produce the desired reactions, such as in reactive evaporation and sputtering processes and chemical vapor depositon. The processes in vacuum metallurgy involve liquid/solid, vapor/solid, and vapor/liquid/solid transitions. In addition, they include testing of metals in controlled environments.

There are three basic reasons for vacuum processing of metals: elimination of contamination from the processing environment, reduction of the level of impurities in the product, and deposition with a minimum of impurities. Contamination from the processing environment includes the container for the metal and the gas phase surrounding the metal. In the vacuum process, impurities, particularly oxygen, nitrogen, hydrogen, and carbon, are released from the molten metal and pumped away; and metals, alloys, and compounds are deposited with a minimum of entrained impurities. There are numerous and varied application areas for vacuum metallurgy including special areas of extractive metallurgy, melting processes, casting of shaped products, degassing of molten steel, heat treatment, surface treatment, vapor deposition, space processing, and joining processes. *See* ARC WELDING; STEEL MANUFACTURE; WELDING AND CUTTING OF METALS. [R.F.Bu.]

Vacuum pump A device that reduces the pressure of a gas (usually air) in a container. When gas in a closed container is lowered from atmospheric pressure, the operation constitutes an increase in vacuum in this container. *See* PRESSURE.

Vacuum pumps are evaluated for the degree of vacuum they can attain and for how much gas they can pump in a unit of time. In practice, where high vacuum is required, two or more different types of pumps are used in series.

In the rotary oil-seal pump (see illustration), gas is sucked into chamber A through the opening intake port by the rotor. A sliding vane partitions chamber A from chamber B. The compressed gas that has been moved from position A to position B is pushed out of the exit port through the valve, which prevents the gas from flowing back. The valve and the rotor contact point are oil-sealed. Since each revolution sweeps out a fixed volume, it is called a constant-displacement pump. Other mechanical pumps are the rotary blower pump, which operates by the propelling action of one or more rapidly rotating lobelike vanes, and the molecular drag pump, which operates at very high speeds, as much as 16,000 rpm. Pumping is accomplished by imparting a high momentum to the gas molecules by the impingement of the rapidly rotating body.

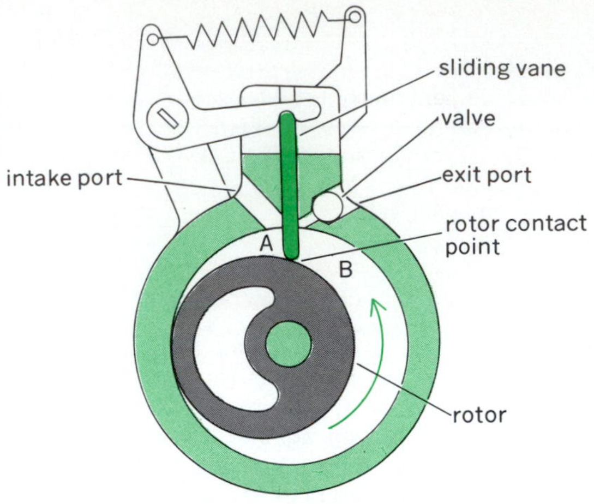

Chief components of a typical mechanical pump, the rotary oil-seal pump.

The water aspirator is an ejector pump. When water is forced under pressure through the jet nozzle, it will force the gas in the inlet chamber to go through the diffuser, thus lowering the pressure in the inlet chamber. When high-pressure steam is used instead of water, it is called a steam ejector.

In addition, a number of pumps have been developed which meet special pumping requirements. The ion pump operates electronically. Electrons that are generated by a high voltage applied to an anode and a cathode are spiraled into a long orbit by a high-intensity magnetic field. These electrons colliding with gas molecules ionize the molecules, imparting a positive charge to them. These are attracted to, and are collected on, the cathode. Thus a pumping action takes place. Sorption pumping is the removal of gases by adsorbing and absorbing them on a granular sorbent material such as a molecular sieve held in a metal container. Cryogenic pumping is accomplished by condensing gases on surfaces that are at extremely low temperatures. [E.S.Ba.]

Vacuum tube An electron tube evacuated to such a degree that its electrical characteristics are essentially unaffected by any residual gas or vapor.

Classification. According to function, vacuum tubes are classified as receiving tubes, transmitting tubes, phototubes, and cathode-ray tubes, microwave tubes, and storage tubes. Structurally, they are classified according to number of electrodes as diodes, triodes, tetrodes, pentodes, and so on.

Prior to the advent of semiconductor devices, hundreds of types of vacuum tubes were developed, many differing only in minor respects. With a few exceptions, most of these have been replaced by semiconductor devices or are used only as replacements in older equipment designed for vacuum tubes. Important exceptions include cathode-ray tubes, microwave tubes, high-power transmitting tubes, and x-ray tubes. *See* SEMICONDUCTOR.

Receiving tubes are low-voltage and low-power tubes designed for use in radio receivers, computers, and sensitive control and measuring equipment. Phototubes are two-electrode electron tubes in which the current is controlled by light flux incident upon one of the electrodes, They are used in sound-film equipment, in light-controlled switches, and in many industrial and scientific applications. Cathode-ray tubes function by virtue of an electron beam that can be focused to a small cross section on a screen and varied in position and intensity by means of electrical signals. They include oscilloscope tubes, television picture tubes, and camera tubes. Oscilloscope tubes make possible the visual examination of electrical signals. Picture tubes transform television signals into pictures upon a luminescent screen. Conversely, camera tubes convert light images into electrical output or, as image converters or intensifiers, produce a visible picture from incident infrared light or from x-rays.

Microwave tubes are designed for operation at frequencies of the order of 3 gigahertz and above. They are used in such diverse applications as radar equipment, space communication and control channels, scientific research, and radio-frequency ovens. Storage tubes, which are electron tubes designed so that information can be introduced and then extracted after a required storage interval, were developed for use in computers. *See* BACKWARD-WAVE TUBE; CATHODE-RAY TUBE; KLYSTRON; MAGNETRON; MICROWAVE TUBE; PHOTOTUBE; PICTURE TUBE; STORAGE TUBE; TELEVISION CAMERA TUBE; TRAVELING-WAVE TUBE; X-RAY TUBE.

Basic operation. Vacuum tubes depend upon two basic physical phenomena for their operation. The first is the emission of electrons by certain elements and compounds when the energy of the surface atoms is raised by the addition of heat (thermionic emission), by incident light photons (photoelectric emission), by kinetic energy of bombarding particles (secondary emission), or by potential energy (field emission). The second phenomenon is the control of the movement of electrons within an evacuated enclosure by forces exerted upon them by electric and magnetic fields. *See* ELECTRON EMISSION.

Vacuum tubes consist of an electrode capable of electron emission and one or more electrodes for collecting the emitted electrons and for establishing variable electric fields in order to control the movement of the electrons between the emitting electrode and the collecting electrode or electrodes. Where magnetic fields are required, they are produced by permanent magnets or by electromagnets, usually external to the evacuated space of the tube. In thermionic vacuum tubes, the primary source of electrons is thermionic emission, and control is by electric fields. The emitting electrode is called the cathode, or filament when it is filamentary in form, and the collecting electrodes are called anodes. The principal anode is usually called the plate. Control electrodes are termed grids. *See* THERMIONIC EMISSION.

Electrode configurations. Because characteristics and functions of vacuum tubes are determined by the number and configurations of the electrodes that the tubes incorporate, one of the most useful classifications of vacuum tubes is by number of electrodes.

The diode. The simplest tube is the diode; it has only two electrodes: the cathode, which emits electrons, and the anode or plate, which collects the electrons. In most diodes, electrodes have the form of concentric cylinders, although plane structures are also used. Utility of the diode rests mainly in the fact that conduction between plate and cathode occurs only when the plate is positive with respect to the cathode; when the plate is negative, electrons emitted by the cathode are prevented from reaching the plate. *See* DIODE.

The triode. A three-electrode tube, called a triode, is formed by the addition of a grid between the cathode and the plate of a diode. Electrons, in moving from the cathode to the plate, can pass through openings in the grid. The principal value of this third electrode is that large plate current and plate-circuit power can be controlled by small variations of grid voltage and with the expenditure of little power in the grid circuit.

The tetrode. In the use of vacuum triodes in very-high-frequency amplifiers, undesirable oscillation may result from the feedback of plate-circuit power to the grid circuit through the capacitance between plate and grid. Although this difficulty can be avoided by making circuit modifications, a much more desirable solution is the reduction of the grid-plate capacitance. This can be accomplished by placing between grid and plate a fourth electrode, which functions as an electrostatic shield. The name of this electrode, screen grid, is descriptive both of its function and of its physical structure. The first grid, which is used to control the plate current, is called the control grid. Such a four-electrode tube is called a tetrode.

The pentode. The negative-slope region of the tetrode plate characteristics, which is undesirable in most applications, can be eliminated by the addition of a third grid located between the screen grid and the plate. The name of this grid, suppressor grid, is indicative of its function in suppressing secondary-electron plate current.

Special tubes containing four, five, or six grids have been developed for special applications. [H.J.R.]

Vacuum-tube amplifier An amplifier which depends upon vacuum tubes for its operation. The basic electron-tube amplifier consists of a vacuum triode tube and associated circuitry, as shown in the illustration. The signal voltage to be amplified is applied to the input, or grid, circuit and the amplified signal appears across the load resistor R_L in the anode (plate), or output, circuit. A gain in voltage, current, and power is thereby provided. By selection of tubes and associated circuitry, these gains may be maximized.

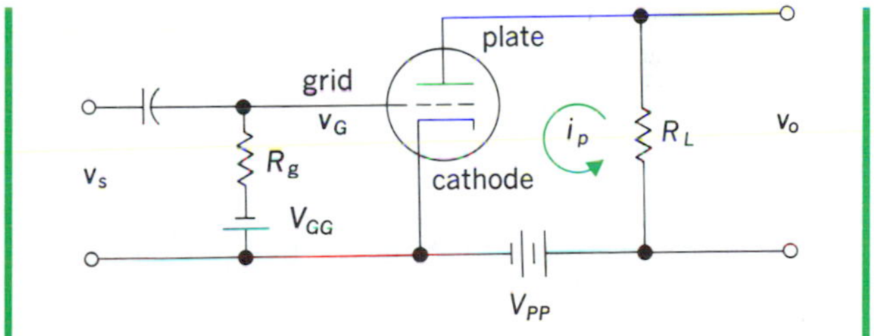

Diagram of a basic vacuum-tube amplifier.

In normal operation a small negative voltage called a bias voltage, V_{GG}, is applied to the grid of tube. This voltage establishes a definite plate-to-cathode current through the tube for a fixed load resistor R_L and high-potential plate-supply voltage V_{PP}. This plate current I_P is called the quiescent current since it is the zero-signal current. A dc voltage drop $I_P R_L$ appears across the output terminals because of the flow of the quiescent current through the load resistor. In the vacuum tube, electrons are emitted by the cathode and pass through the grid to the plate. The direction of "conventional" current is opposite to that of the electron, and therefore the plate current (conventional notation) flows from plate to cathode. *See* VACUUM TUBE.

If a small signal voltage is added to the grid bias voltage, the instantaneous grid voltage v_G is the algebraic sum of the bias voltage V_{GG} and the signal voltage v_G. As the instantaneous grid voltage changes, the plate current i_P changes in proportion to the grid voltage changes. The result is a net plate current i_P that is the algebraic sum of the quiescent plate current I_P and a varying or signal component of plate current e_v. *See* AMPLIFIER. [C.C.H.; C.L.A.]

Vaginal disorders Congenital defects, inflammations, tumors, trauma, and functional disorders of the vagina. *See* REPRODUCTIVE SYSTEM.

Congenital defects are uncommon but may include absence of the vagina, failure of its growth, or abnormal formations such as double vagina. Congenital cysts, although common, are usually unimportant.

Inflammations of the vagina ordinarily are extensions from infections of the external organs, but there are four types of vaginitis which account for a high proportion of gynecologic disease. All forms are marked by pruritus (itching) and inflammation of variable degree, and they often cause a discharge of some type. Gonorrheal vulvovaginitis is seen almost always in children, trichomonal vaginitis results from a protozoan infection, and monilial vaginitis is a fungus infection similar to thrush. Senile vaginitis occurs in elderly women following chronic irritation of the thinned, atrophic mucosa. *See* CANDIDIASIS; GONORRHEA; TRICHOMONIASIS.

The most important primary lesion of the vagina is carcinoma but its incidence is low. However, extension of the much more common cervical carcinoma to the vagina occurs in advanced cases. In both types, the prognosis is dependent largely upon time of diagnosis and extent of spread. *See* ONCOLOGY.

Vaginismus is primarily a functional disorder marked by spasmodic, painful, and involuntary contraction of the lower vaginal muscles. Although usually psychic in origin, it may result from remediable physical causes. [E.G.St./N.K.M.]

Valence A term commonly used by chemists to characterize the combining power of an element for other elements, as measured by the number of bonds to other atoms which one atom of the given element forms upon chemical combination. The term also has come to signify the theory of all the physical and chemical properties of molecules that specially depend on molecular electronic structure.

Thus, in water, H_2O, the valence of each hydrogen atom is 1; the valence of oxygen, 2. In methane, CH_4, the valence of hydrogen again is 1; of carbon, 4. In NaCl and CCl_4 the valence of chlorine is 1, and in CH_2 the valence of carbon is 2. *See* CHEMICAL BONDING.

Most of the simple facts of valence (though certainly not all) follow from the postulate that atoms combine in such a way as to seek closed-shell or inert-gas structures (rule of eight) by the transfer of electrons between them or the sharing of a pair of electrons between them. Many molecular structures may be obtained by inspection using these rules; letting a dot represent an electron:

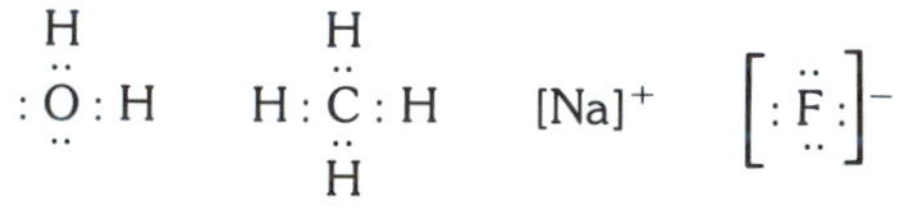

In these electron-dot symbols, the electrons in the *K* shell are not included for atoms after He, nor are the electrons in the *K* and *L* shells for atoms following Ne.

As generally used and here defined, the word valence is ambiguous. Before a value can be assigned to the valence of an atom in a molecule, the electronic structure of the molecule must be exactly known, and this structure must be describable simply in terms of simple bonds. In practice neither of these conditions is ever precisely fulfilled. A term not so ambiguous is oxidation number or valence number. Oxidation numbers are useful for the balancing of oxidation-reduction equations, but they are not related simply to ordinary valences. Thus the valence of carbon in CH_4, $CHCl_3$, and CCl_4 is 4; oxidation numbers of carbon in these three substances are −4, +2, and +4. *See* CHEMICAL STRUCTURES; ELECTRONEGATIVITY; MOLECULAR ORBITAL THEORY; OXIDATION-REDUCTION. [R.G.P.]

Valence band The highest electronic energy band in a semiconductor or insulator which can be filled with electrons. The electrons in the valence band correspond to the valence electrons of the constituent atoms. In a semiconductor or insulator, at sufficiently low temperatures, the valence band is completely filled and the conduction band is empty of electrons. Some of the high energy levels in the valence band may become vacant as a result of thermal excitation of electrons to higher energy bands or as a result of the presence of impurities. The net effect of the valence band is then equivalent to that of a few particles which are equal in number and similar in motion to the missing electrons but each of which carries a positive electronic charge. These "particles" are referred to as holes. *See* BAND THEORY OF SOLIDS; CONDUCTION BAND; ELECTRIC INSULATOR; HOLE STATES IN SOLIDS; SEMICONDUCTOR; VALENCE. [H.Y.F.]

Valine An amino acid considered essential for normal growth of animals. The biosynthetic precursor, as well as the deamination product of valine, is α-ketoisovaleric acid. Valine

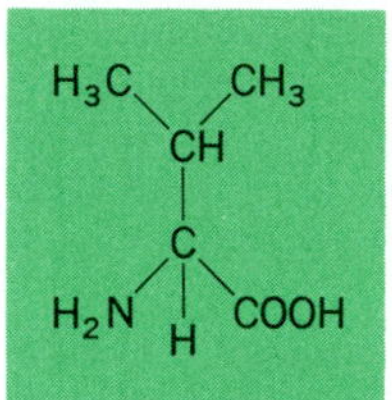

Valine

is biosynthesized from pyruvic acid. Most or all of the enzymes concerned also catalyze the analogous reactions in isoleucine biosynthesis. *See* AMINO ACIDS; ISOLEUCINE. [E.A.Ad.]

Value engineering A thinking system (also called value management or value analysis) used to develop decision criteria when it is important to secure as much as possible of what is wanted from each unit of the resource used. The resource may be money, time, material, labor, space, energy, and so on. The system is unique in that it effectively uses both knowledge and creativity, and provides step-by-step techniques for maximizing the benefits from both. It promotes development of alternatives suitable for the future as well as the present. This is accomplished by identifying and studying each function that is wanted by the customer or user, then applying knowledge and creativity to achieve the desired function. Resources are converted into costs to achieve direct, meaningful comparisons. By using the methods of value engineering, 15 to 40% reduction in the required resources often results.

Value engineering has applications in five broad areas: in design, purchase, and manufacture of products; in administrative groups, private or public, where the task is to achieve accomplishment through people; in all areas of social service work, such as hospitals, insurance services, or colleges; in architectural design and construction; and in development as well as research.

The system is used to improve value in either or both of two situations: (1) The product or service as used or as planned may provide 100% of the functions the user wants, but lower costs may be needed. The system then holds those functions but achieves them at lower cost. (2) The product or service may have deficiencies, that is, it does not perform the desired functions or lacks quality, and so also lacks good value. The system aims at correcting those deficiencies, providing the functions wanted, while at the same time holding the use of resources (costs) at a minimum. *See* INDUSTRIAL COST CONTROL; INDUSTRIAL ENGINEERING; METHODS ENGINEERING; OPERATIONS RESEARCH; OPTIMIZATION; PROCESS ENGINEERING; PRODUCTION ENGINEERING; PRODUCTION PLANNING. [L.D.M.]

Valvatida An order of Asteroidea in which pedicellariae, when present, are valvate and sunk into the skeletal ossicles. The mouth plates are inconspicuous, and the inferomarginal and superomarginal plates correspond and lack intermarginal grooves. There are two rows of tube feet. The order is thought to include seven families. *See* ASTEROIDEA; ECHINODERMATA. [A.C.C.]

Valve A flow-control device. Valves are used to regulate the flow of fluids in piping systems and machinery. In machinery the flow phenomenon is frequently of a pulsating or intermittent character and the valve, with its associated gear, contributes a timing feature.

The valves commonly used in piping systems are gate valves (Fig. 1), usually operated closed or wide open and seldom used for throttling; globe valves, frequently fitted with a renewable disk and adaptable to throttling operations; check valves, for automatically limiting flow in a piping system to a single direction; and plug cocks, for operation in the open or closed position by turning the plug through 90° and with a shearing action to clear foreign matter from the seat. Safety and relief valves are automatic protective devices for the relief of excess pressure. *See* SAFETY VALVE.

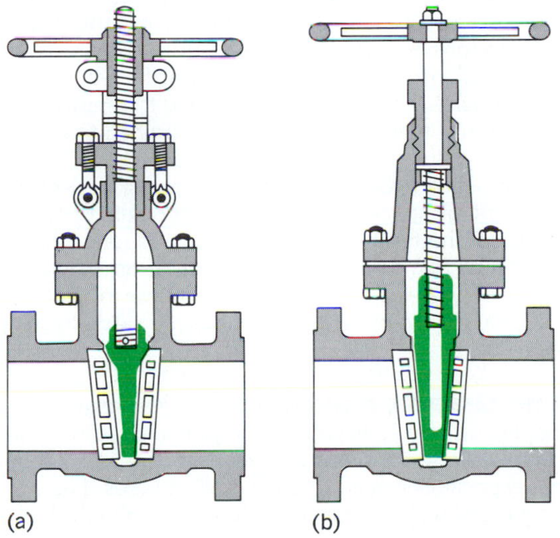

Fig. 1. Gate valves with disk gates shown in color. (*a*) Rising threaded stem shows when valve is open. (*b*) Nonrising stem valve requires less overhead.

For hydraulic turbines and hydroelectric systems, valves and gates control water flow for (1) regulation of power output at sustained efficiency and with minimum wastage of water, and (2) safety under the inertial flow conditions of large masses of water.

To control the kinematics of the cycle, steam-engine valves range from simple D-slide and piston valves to multiported types. Many types of reversing gear have been perfected which use the same slide valve or piston valve for both forward and backward rotation of an engine, as in railroad and marine service. *See* STEAM ENGINE.

Poppet valves are used almost exclusively in internal combustion reciprocating engines because of the demands for tightness with high operating pressures and temperatures (Fig. 2). Two-cycle engines utilize ports, alternately covered and uncovered by the main piston, for inlet or exhaust. *See* CAM MECHANISM; INTERNAL COMBUSTION ENGINE; VALVE TRAIN.

In compressors, valves are usually automatic, operating by pressure difference on the two sides of a movable, spring loaded member and without any mechanical linkage to the moving parts of the compressor mechanism. Like those for compressors, pump valves are usually of the automatic type operating by pressure difference. [T.Ba.]

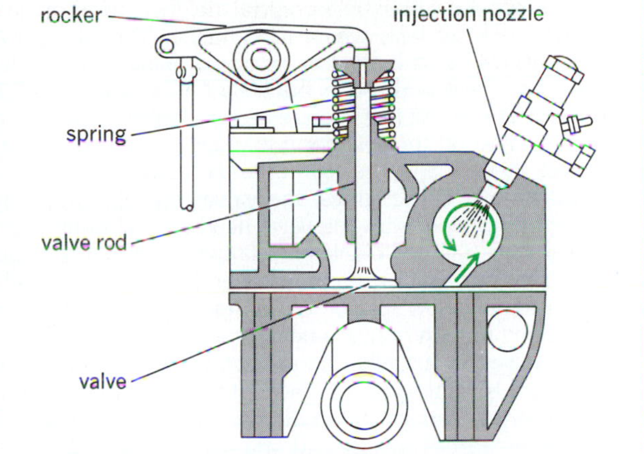

Fig. 2. Poppet valve for internal combustion engine. (*After T. Baumeister, ed., Marks' Standard Handbook for Mechanical Engineers, 8th ed., McGraw-Hill, 1978*)

Valve train The valves and valve-operating mechanism by which an internal combustion engine takes air or a fuel-air mixture into the cylinders and discharges combustion products to the exhaust. *See* VALVE.

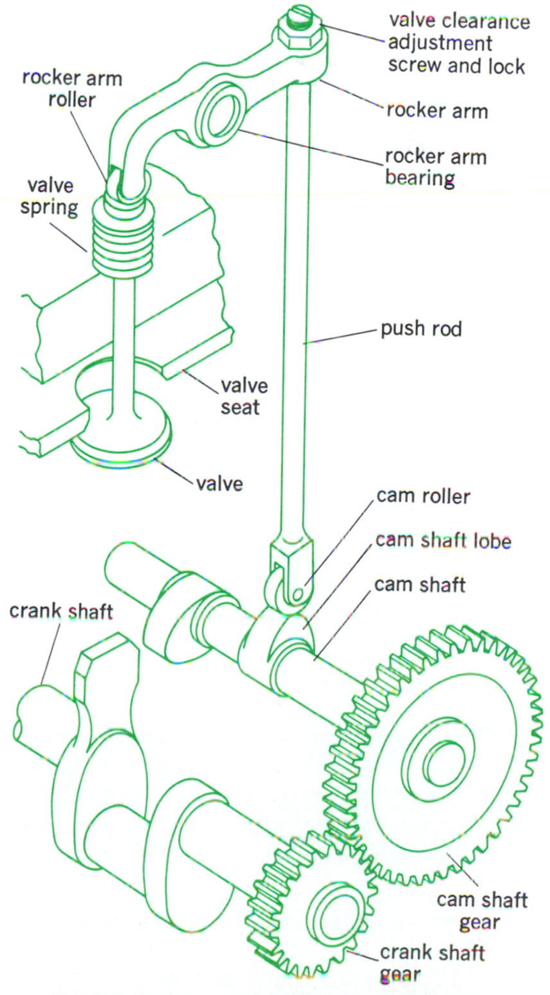

Crankshaft gear meshes with the camshaft gear.

In an internal combustion engine the fuel takes up little space but needs air with which to combine. Thus the power output of an engine is limited by its air-pumping capacity. It is essential that the flow through the engine be restricted as little as possible. This is the first requirement for valves. The second is that they close off the cylinder firmly during the compression and power strokes. *See* INTERNAL COMBUSTION ENGINE.

In most four-stroke engines, the valves are of this inward-opening poppet type, with the valve head ground to fit a conical seat in the cylinder block or cylinder head. Valve action requires that the valve be streamlined and as large as possible to give maximum flow, yet be of low inertia so that it follows the prescribed motion at high engine speed.

Engine valves are usually opened by means of cams. Riding on each cam is a follower, or valve lifter, which may be a flat or slightly convex surface, or a roller (see illustration). The valve is opened by forces applied to the end of the valve stem through a mechanical linkage actuated by the cam follower. In overhead valve engines, the camshaft may be mounted on the cylinder head near the valves. *See* CAM MECHANISM; HYDRAULIC VALVE LIFTER. [A.R.R.]

Valvifera A suborder of isopod crustaceans resembling woodlice and pill bugs. They are distinguished from these and other isopods by having a pair of flat, valvelike uropods which hinge laterally and fold inward beneath the rear part of the body. The uropods open like doors to reveal five pairs of leaflike appendages, or pleopods, which are used for respiration and swimming. Members of the group are almost exclusively marine. Some species are omnivorous and have been found in dense populations on accumulations of broken algae and fish waste. *See* ISOPODA. [E.N.]

Vampyromorpha An order of dibranchiate cephalopod mollusks. *Vampyroteuthis infernalis*, an inhabitant of the deeper waters of tropical and temperate seas, is the only representative of this order. It is considered to be a link form between the decapod and octopod mollusks. *See* CEPHALOPODA. [C.B.C.]

Vanadium A chemical element, V, atomic number 23, atomic weight 50.9414, melting point 1917°C (3483°F), and density 6.1 g/cm^3 (0.22 lb/in.^{3}). Vanadium is a metal which has been used primarily as an alloying element for steels and irons. Several of the compounds of vanadium are used in the chemical industry, notably in making oxidation catalysts, and in the ceramic industry as coloring agents.

Like some other transition elements, vanadium forms numerous and frequently complicated compounds because of its variable valence, It has at least three oxidation states, 2+, 3+, and 5+. It is amphoteric, mostly basic in the lower oxidation states, acidic in the higher ones. It forms derivatives from more or less well-defined radicals such as VO^{2+} and VO^{3+}. *See* METALLOACID ELEMENTS; TRANSITION ELEMENTS.

In its pure form vanadium is soft and ductile. It can be hot- and cold-worked easily, but it must be heated in an inert atmosphere or in a vacuum because it oxidizes readily at temperatures above the melting point of its oxide. The metal retains its strength unusually well at elevated temperatures. The resistance of vanadium to hydrochloric and sulfuric acids is outstanding, and it withstands aerated-saltwater attack better than most stainless steels. However, vanadium cannot withstand nitric acid. *See* VACUUM METALLURGY. [T.W.M.]

1	2	3	4	5	6	7	8	9	10	11	12	13	14	15	16	17	18
1 H																	2 He
3 Li	4 Be											5 B	6 C	7 N	8 O	9 F	10 Ne
11 Na	12 Mg											13 Al	14 Si	15 P	16 S	17 Cl	18 Ar
19 K	20 Ca	21 Sc	22 Ti	23 V	24 Cr	25 Mn	26 Fe	27 Co	28 Ni	29 Cu	30 Zn	31 Ga	32 Ge	33 As	34 Se	35 Br	36 Kr
37 Rb	38 Sr	39 Y	40 Zr	41 Nb	42 Mo	43 Tc	44 Ru	45 Rh	46 Pd	47 Ag	48 Cd	49 In	50 Sn	51 Sb	52 Te	53 I	54 Xe
55 Cs	56 Ba	71 Lu	72 Hf	73 Ta	74 W	75 Re	76 Os	77 Ir	78 Pt	79 Au	80 Hg	81 Tl	82 Pb	83 Bi	84 Po	85 At	86 Rn
87 Fr	88 Ra	103 Lr	104 Rf	105 Db	106 Sg	107 Bh	108 Hs	109 Mt	110	*111*	*112*	*113*	*114*	*115*	*116*	*117*	*118*

lanthanide series	57 La	58 Ce	59 Pr	60 Nd	61 Pm	62 Sm	63 Eu	64 Gd	65 Tb	66 Dy	67 Ho	68 Er	69 Tm	70 Yb
actinide series	89 Ac	90 Th	91 Pa	92 U	93 Np	94 Pu	95 Am	96 Cm	97 Bk	98 Cf	99 Es	100 Fm	101 Md	102 No

Van Allen radiation The high-energy, charged particles which are trapped by the geomagnetic field and form belts of intense radiation in space about the Earth. The belts consist primarily of electrons and protons and extend from a few hundred kilometers above the Earth to a distance of about $8R_e$ (R_e = radius of Earth = 6371 km or 3959 mi). The belts of trapped particles were discovered by James Van Allen and coworkers in 1958 through radiation detectors carried on satellites *Explorer 1* and *Explorer 3*.

A charged particle under the influence of the geomagnetic field follows a trajectory which can be conveniently described as a superposition of three separate motions. The first motion is a rapid spiral about magnetic field lines that is produced by the magnetic force acting at right angles to both the particle velocity and to the magnetic field. As the spiraling particle moves along the field line toward either the North or South Pole, the increase in magnetic field strength causes the particle to be reflected so that it bounces between Northern and Southern hemispheres. Superimposed on the spiral and bounce motions is a slow east-west drift, electrons drifting eastward and protons drifting westward. Thus individual trapped particles move completely around the Earth in a complicated pattern, their motion being constrained to lie on magnetic shells.

The Van Allen belts are one feature of the general plasma environment of space. Plasma from the Sun is continually impinging on the Earth's magnetic field, and ionospheric electrons and ions also form a plasma which at times is accelerated. Some of the electrons and ions of these plasma populations become trapped in the geomagnetic field, forming the Van Allen belts. [M.W.]

Van de Graaff generator A high-voltage electrostatic generator in which electrical charge is carried from ground to a high-voltage terminal by means of an insulating belt. Current capacity is limited to a few milliamperes. Each generator is equipped with an evacuated tube through which charged

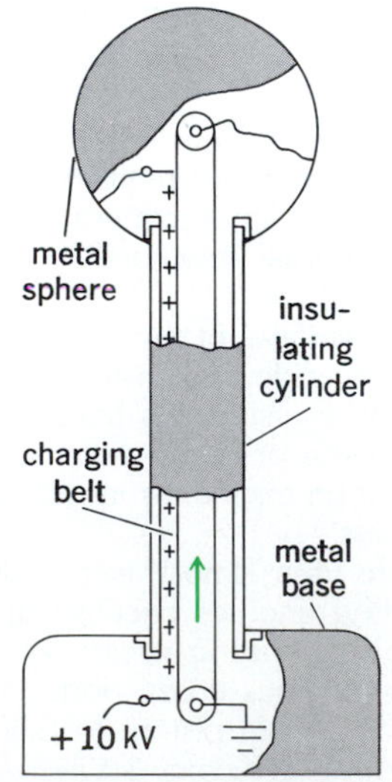

Schematic drawing of Van de Graaff generator for operation in air at atmospheric pressure.

particles of any type may be accelerated. Machines equipped with positive ion sources and arranged for positive ion acceleration are widely used for research applications. Machines equipped with electron sources are used for x-ray therapy, industrial radiography, food and drug sterilization, and research.

Principles of operation of the Van de Graaff generator are most easily understood from study of a simple model as shown in the illustration. The charging belt, which is made of insulating material such as rubberized fabric, is carried by two pulleys. One, at ground potential, is driven by an electric motor, and the second is in the high-voltage terminal. The terminal is a metal enclosure which must be smooth and well rounded over its exterior surface. The terminal is supported by a cylinder of insulating material. The base is enclosed by metal with a smooth, well-rounded exterior surface, and this enclosure is at ground potential. Charge is sprayed onto the belt from a row of needle points or a fine wire held at a potential of about 10 kilovolts. A row of needles near the upper pulley with points close to the belt will remove charge from the belt, and the charge will pass to the outer surface of the terminal. Terminal voltage will rise to some limit determined by sparkover through air, by insulator failure, or by a load such as ion current through an evacuated tube. *See* ELECTROSTATICS; PARTICLE ACCELERATOR. [R.G.H.]

Van der Waals equation An empirical equation of state which takes into account the finite size of the molecules and the attractive forces between them. For a homogeneous gas at pressure p, molar volume v_N, and absolute temperature T, the equation is

$$p = \frac{R_u T}{v_N - b} - \frac{a}{(v_N)^2}$$

or

$$\left(p + \frac{a}{(v_N)^2}\right)(v_N - b) = R_u T$$

The terms a and b are constants and evaluated from experimental data. The molar specific volume v_N is equal to the total volume V divided by the number of moles N. R_u is the universal gas constant. *See* THERMODYNAMIC PRINCIPLES. [G.A.H.]

Vane anemometer A portable instrument for measuring low wind speeds and airspeeds in large ducts. It consists of a number of vanes radiating from a common shaft and set to rotate when facing the wind; a guard ring surrounds the vanes. The vanes operate a counter to indicate the number of rotations, which when timed with a stopwatch serve to determine the speed. *See* ANEMOMETER. [H.Fo.]

Vanilla A choice flavoring obtained from a climbing orchid, *Vanilla fragrans*, a native of tropical American forests. Its fruits are pods called vanilla beans. These are picked at the proper time before they have fully matured.

Vanillin (4-hydroxy-3-methoxybenzaldehyde) is the principal component of vanilla, although other components contribute to the distinctive flavor of the extract compared to synthetic vanilla. When they are harvested, the beans contain no free vanillin; it develops during the curing period from glucosides that break down during the fermentation and sweating of the beans. The sweating process consists of alternately drying the beans in sunlight and bunching them so that they heat and ferment. [P.D.St./E.L.C.]

Vapor condenser A heat-transfer device that reduces a thermodynamic fluid from its vapor phase to its liquid phase. The vapor condenser extracts the latent heat of vaporization from the vapor, as a higher-temperature heat source, by absorption in a heat-receiving fluid of lower temperature. The vapor to be condensed may be wet, saturated, or superheated. The heat receiver is usually water but may be a fluid such as air, a process liquid, or a gas. When the condensing of vapor is primarily used to add heat to the heat-receiving fluid, the condensing device is called a heater and is not within the normal classification of a condenser.

Condensers may be divided into two major classes according to use: those used as part of a processing system (process condensers) and those used for serving engines or turbines in a steam power plant cycle (power cycle condensers). Condensers may be further classified according to mode of operation as surface condensers or as contact condensers. *See* CONTACT CONDENSER; SURFACE CONDENSER.

Condensers are required, almost without exception, to condense impure vapors, that is, vapors containing air or other noncondensable gases. Because most condensers operate at subatmospheric pressures, air leaking into the apparatus or system becomes a common cause for vapor contamination, and a variety of designs have been developed to reduce such problems. Accumulation of noncondensable gases seriously affects heat transfer, and means must be provided to direct them to a suitable outlet. Most surface and contact condensers are arranged with a separate zone of heat-transfer surface within the condenser and located at the outlet end of the vapor flow path for efficient removal of the noncondensable gases through dehumidification. [J.F.Se.]

Vapor cycle A thermodynamic cycle, operating as a heat engine or a heat pump, during which the working substance is in, or passes through, the vapor state. A vapor is a substance at or near its condensation point. It may be wet, dry, or slightly superheated. One hundred percent dryness is an exactly definable condition which is only transiently encountered in practice. *See* HEAT PUMP; THERMODYNAMIC CYCLE.

A steam power plant operates on a vapor cycle where steam is generated by boiling water at high pressure, expanding it in a prime mover, exhausting it to a condenser, where it is reduced to the liquid state at low pressure, and then returning the water by a pump to the boiler (Fig. 1).

In the customary vapor-compression refrigeration plant, the process is essentially reversed with the refrigerant evaporating at low temperature and pressure, being compressed to high pressure, condensed at elevated temperature, and returned as

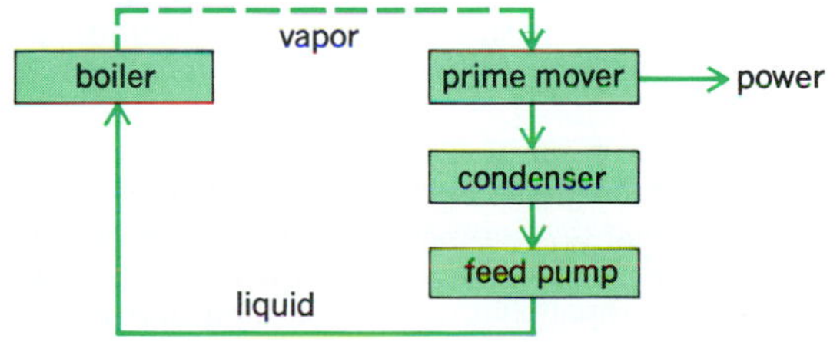

Fig. 1. Rudimentary steam power plant flow diagram.

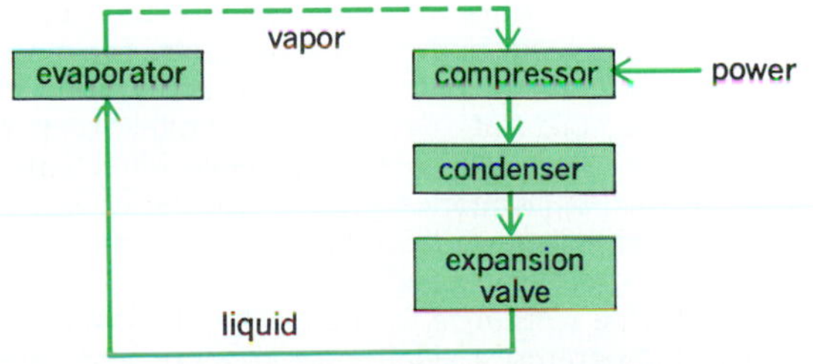

Fig. 2. Rudimentary vapor-compression refrigeration plant flow diagram.

liquid refrigerant through an expansion valve to the evaporating coil (Fig. 2). *See* REFRIGERATION.

The Carnot cycle, between any two temperatures, gives the limit for the efficiency of the conversion of heat into work. This efficiency is independent of the properties of the working fluid. The Rankine cycle is more realistic in describing the ideal performance of steam power plants and vapor-compression refrigeration systems. *See* CARNOT CYCLE; RANKINE CYCLE. [T.Ba.]

Vapor deposition Production of a film of material often on a heated surface and in a vacuum. Vapor deposition technology is used in a large variety of applications. Coatings are produced from a wide range of materials, including metals, alloys, compound, cermets, and composites. The vapor deposition processes can be classified into the two basic groups, physical (evaporation and sputtering) and chemical. In addition, there are hybrid processes such as ion plating, which is a subset of evaporation or sputter deposition. This process involves an electrically biased substrate and uses a working gas such as argon, which results in ion bombardment of the substrate/depositing film that produces changes in microstructure, residual stress, and impurity content. *See* ALLOY, CERMET; COMPOSITE MATERIAL; METAL.

All deposition processes consist of three major steps: generation of the depositing species, transport of the species from source to substrate, and film growth on the substrate. In chemical vapor deposition, these steps occur simultaneously at or near the substrate. Physical vapor deposition processes are more versatile, because the steps occur sequentially and can be controlled independently.

In physical vapor deposition the first step involves generation of the depositing species by evaporation using resistance, induction, electron-beam, or laser-beam heating, or by sputtering using direct-current or radio-frequency plasma generation. The second step involves transport from source to substrate.

The third step is film growth on the substrate. The two basic processes for physical vapor deposition are evaporation deposition and sputter deposition. In evaporation, thermal energy converts a solid or liquid target material to the vapor phase. In sputtering, the target is biased to a negative potential and bombarded by positive ions of the working gas from the plasma, which knock out the target atoms and convert them to vapor by momentum transfer. *See* SPUTTERING.

In the basic thermal chemical vapor deposition process, the reactants flow over a heated substrate and react at or near the substrate surface to deposit a film. The kinetics of the process are dependent on diffusion through the boundary layer between the substrate and the bulk gas-flow region. *See* DIFFUSION IN SOLIDS.

Chemical vapor deposition is one of several process steps for fabrication of integrated circuits. Thus, much effort is being directed at lowering the deposition processing temperatures in order to achieve more economical processing as well as increasing the compatibility with preceding and subsequent steps in device processing. *See* INTEGRATED CIRCUITS.

Coatings provided by vapor deposition have many applications. For optical functions they are used as reflective or transmitting coatings for lenses, mirrors, and headlamps; for energy transmission and control they are used in glass, optical, or selective solar absorbers and in heat blankets; for electrical and magnetic functions they are used for resistors, conductors, capacitors, active solid-state devices, photovoltaic solar cells, and magnetic recording devices; for mechanical functions they are used for solid-state lubricants, tribological coatings for wear and erosion resistance in cutting and forming tools and other engineering surfaces; and for chemical functions they are used to provide resistance to chemical and galvanic corrosion in high-temperature oxidation, corrosion, and catalytic applications as well as to afford protection in the marine environment. Decorative applications include toys, costume jewelry, eyeglass frames, watchcases and bezels, and medallions. Vapor-deposited coatings also act as moisture barriers for paper and polymers. They can be used for bulk or free-standing shapes such as sheet or foil tubing, and they can be formulated into submicrometer powders. *See* VACUUM METALLURGY. [R.F.Bu.]

Vapor lamp A source of radiant energy excited by a supply of electricity which creates a current of ionized gas between electrodes in an enclosure that contains the arc while permitting transmission of the radiant energy. Gaseous-discharge lamps or vapor lamps are given various names relating to the element responsible for the majority of the radiation (mercury, sodium metal-halide, xenon), to the physical attribute of the lamp (short-arc, high-pressure), or, in the case of fluorescent lamps, to the way a phosphor on the bulb wall fluoresces as a result of the lamp's low-pressure mercury-vapor excitation. *See* ARC DISCHARGE; ELECTRICAL CONDUCTION IN GASES.

Gaseous-discharge lamps are broadly used throughout the world because the conversion of electric energy to radiant energy in a gaseous discharge provides radiation in narrow bands within the range of visible light in which the rods and cones of the eye are most sensitive. These light sources have high efficiency in conversion of electricity to light.

For discussions of common types of vapor lamps *see* FLUORESCENT LAMP; MERCURY-VAPOR LAMP; NEON GLOW LAMP; SODIUM-VAPOR LAMP. [T.F.N.]

Vapor lock Interruption of fuel flow to an engine due to blockage of passages in the fuel system by fuel vapor.

To promote easy starting, all gasolines contain volatile constituents which under some conditions, such as high ambient temperature, tend to produce more vapor than the fuel pump and carburetor vents can handle. The very action of the fuel pump, in decreasing the pressure at its inlet, tends to vaporize the fuel. If the vapor forms faster than the pump can draw it from the fuel line, the flow of fuel to the carburetor is effectively stopped and the engine stalls. *See* AIRCRAFT FUEL; CARBURETOR; FUEL SYSTEM; GASOLINE; INTERNAL COMBUSTION ENGINE. [A.R.R.]

Vapor pressure The saturation pressures exerted by vapors which are in equilibrium with their liquid or solid forms. One of the most important physical properties of a liquid, the vapor pressure, enters into many thermodynamic calculations and underlies several methods for the determination of the molecular weights of substances dissolved in liquids. For a discussion of the vapor pressure relationships of solids *see* SUBLIMATION. *See also* MOLECULAR WEIGHT; SOLUTION.

If a liquid is introduced into an evacuated vessel at a given temperature, some of the liquid will vaporize, and the pressure of the vapor will attain a maximum value which is termed the vapor pressure of the liquid at that temperature. Although the quantity of liquid remaining does not diminish thereafter, the process of evaporation does not cease. A dynamic equilibrium is established, in which molecules escape from the liquid phase and return from the vapor phase at equal rates. *See* EVAPORATION.

It is important to make a distinction between the vapor pressure of a liquid, as described above, and the pressure of a vapor. The vapor pressure of a pure liquid is a unique and characteristic property of the liquid and depends only upon the temperature. A gas or vapor may, on the other hand, exert any pressure within reason, depending upon the volume to which it is confined, provided it is not in contact with its liquid phase. *See* PHASE EQUILIBRIUM. [N.H.N.]

Varactor A semiconductor diode designed to maximize the variation of its capacitance with applied reverse bias voltage. Such diodes are also often called variable capacitance diodes. Today they are almost exclusively junction diodes. The free-charge depletion region of any *pn* junction in a semiconductor widens with the application of reverse bias. Since the surfaces of this free-charge depletion region represent the effective plates of the capacitance of the junction, the application of increasing reverse bias voltage will cause a decrease in the capacitance. The design and fabrication of varactors then reduces to the art of doping the two sides of the junction in such a way that the desired voltage variation of depletion region width (and hence capacitance) is obtained. *See* JUNCTION DIODE; SEMICONDUCTOR; SEMICONDUCTOR DIODE. [L.P.H.]

Variable (mathematics) A symbol *x* is usually defined to be a variable if it may denote any member of a set of objects. A variable is discrete or continuous according to whether its range (the set) is discrete (for example, a subset of the natural numbers) or continuous (for example, all real numbers between two real numbers), respectively. *See* ANALYTIC GEOMETRY; NUMBER THEORY; PARAMETRIC EQUATION. [L.M.Bl.]

Variable-area flowmeter A volume flow rate meter based on the principle of a variable restrictor in the flowing stream being forced by the fluid to a position to allow the required flow-through. Those that depend upon gravity provide essentially a constant differential pressure, while those that supplement the gravity with a spring will have a variable differential pressure.

In the tapered-tube rotameter (illustration *a*), the fluid flows upward through a tapered tube, lifting a shaped weight (possibly misnamed a float) to a position where the upward fluid force just balances the float weight, thereby giving a float position that indicates the flow rate.

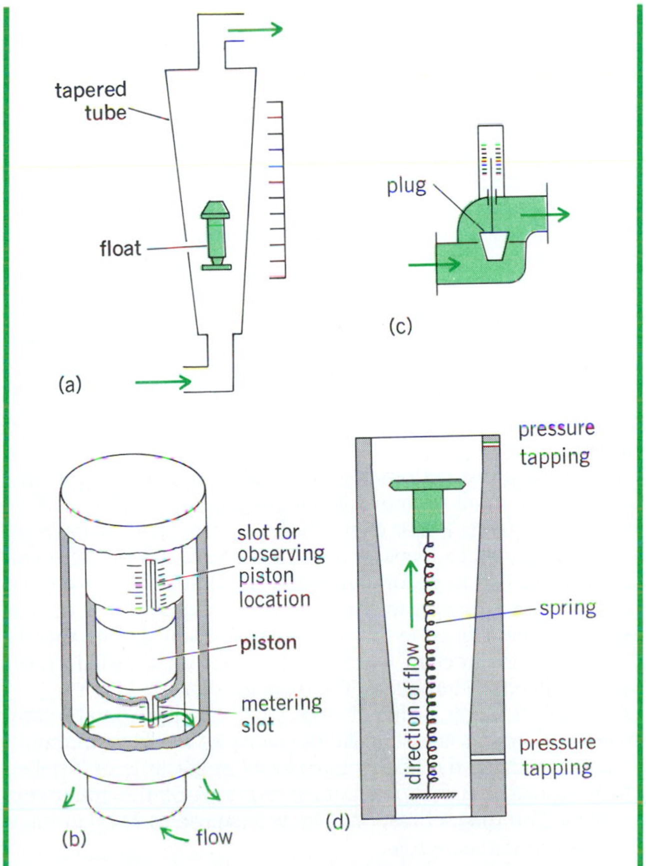

Variable-area flowmeters. (*a*) Rotameter type. (*b*) Piston type. (*c*) Plug type (*from 1969 Guide to process instrument elements, Chem. Eng., 76(12):137–164, 1969*). (*d*) Spring-loaded differential-pressure producer (*from Institute of Mechanical Engineers, How to Choose a Flowmeter, 1975*)

In a piston type (illustration *b*) the buoyant force of the liquid carries the piston upward until a sufficient area has been uncovered in a slot in the side of the vertical tube to allow the liquid to flow through the slot. The position of the piston indicates the flow rate and, by properly shaping the slot, this type of meter may be made with a uniform flow scale.

In the tapered-plug design (illustration *c*), the plug, located in an orifice, is raised until the opening is sufficient to handle the fluid flow. Readouts can be the same as in the tapered-tube design.

A spring-loaded type (illustration *d*) provides a differential pressure for a secondary device. *See* VOLUME FLOW RATE METER. [M.Br.; L.P.E.]

Variable star A star that has a detectable change in its intensity, which is often accompanied by other physical changes. The changes in brightness of a variable star may be a few thousandths of a magnitude up to 20 magnitudes or even more. The cause of the variation may be either an eclipsing effect caused by the motions within a system of two or more bodies or an actual physical change within the star itself or in its surrounding atmosphere.

Most intrinsic variables are either pulsating or eruptive (see table). They may be separated into three main divisions—those which vary with periodic regularity, those with cyclic or semiregular changes, and those which are completely erratic, or irregular. A few variables are unique and cannot be placed in any of these classes, and others seem to be combinations of several classes.

Periodic variables are divided into two main classes: long-period, or Mira-type, variables and cepheids. There are many intermediate examples, so that no definite dividing lines may be drawn. Mira-type variables are by far the most numerous of the known variables in the Milky Way Galaxy. They are red giants with spectra of either class M or one of the carbon types, and usually have hydrogen emission lines. Cepheid variables are characterized by several important relations. They are among the most luminous stars known, and thus can be

Representative types of intrinsic variables

Class	Range of period or cycle*	Color and luminosity	Amplitude (magnitude)	Number†
		Pulsating		
RR Lyrae type	0.05–1.2 d	White	<2	4433
Classical cepheid	1–70 d	High-luminosity yellow	0.1–2	706
Long-period	80–1000 d	Red giants	2.5–6+	4566
Semiregular	30–1000 d	Red giants	<1.5–2	2227
Others				1856
Total				13,788
		Eruptive		
U Gem and Z Cam	10–600 d	Subdwarfs	2–6	235
Flare stars	?	Red dwarfs	1–6	28
Recurrent novae	20 years–?	Dwarfs?	7–9	6
Novae	centuries?	Dwarfs?	8–16	160
Supernovae	?	?	20+	7
Nebular variables	?	Yellow dwarfs		869
Others				311
Total				1616

*d = days.

†From B. V. Kukarkin and P. P. Parenago, *General Catalogue of Variable Stars*, 3d ed., 1969. In addition there are listed 4062 eclipsing stars, 803 unstudied variables, and 183 others unique or constant.

seen at great distances. There is a definite correlation between period and luminosity. As a result of this relationship, they are used as a measuring tape for the universe. Other correlations exist between period and spectrum (temperature), period and radial velocity, and period and form of light curve. Intrinsic variables with the shortest known period are the cluster-type cepheids or RR Lyrae stars. The RR Lyrae stars are fainter than the classical cepheids. They have an absolute photographic magnitude of 0.0 and show no progression of luminosity with period. Therefore, if an RR Lyrae star can be identified in a cluster or galaxy, and its apparent magnitude determined, the distance of that system may be obtained. *See* CEPHEIDS.

Nonperiodic variables may be separated to some extent by their spectra. Semiregular red giants are the most numerous and are closely related to the long-period variables. Amplitudes are small, ranging from less than 2 magnitudes to infinitesimal amounts. Possibly all red stars with molecular bands in their spectra are variable.

Most red variables with periods less than 150 days are subject to great irregularities, and some lose all semblance of periodicity at times; others are completely irregular.

The most violent explosions or eruptions occur in the novae and supernovae. Increases of 10–12 magnitudes have been observed within a few hours. Previous to the explosion, novae are probably dense blue dwarfs. *See* NOVA; SUPERNOVA.

Other stars which have rapid increases of brightness are the U Geminorum and Z Camelopardalis stars. These also are blue stars with dwarf characteristics and are often referred to as dwarf novae. Their eruptions appear at semiperiodic intervals.

A small group of novae, called recurrent novae, have been observed to have two or more explosions. Recurrent novae may be considered to be intermediate between true novae and the U Geminorum stars.

Nebular variables, also known as T Tauri, RW Aurigae, or Orion variables, are subject to rapid changes at irregular and unpredictable intervals. *See* BINARY STAR; ECLIPSING VARIABLE STARS. [M.W.M.]

Variational methods (physics) Methods based on the principle that, among all possible configurations or histories of a physical system, the system realizes the one that minimizes some specified quantity. Variational methods are used in physics both for theory construction and for calculational purposes.

The earliest use of a variational principle for physics is Fermat's principle in optics, which states that when a light ray traverses a medium with nonuniform index of refraction its path is such as to minimize its travel time. An integral expresses the time that the light takes to travel from one point to another along a particular path, and an application of the calculus of variations to this integral makes it possible to determine the particular path for which the travel time is a minimum. This problem is mathematically identical to the variational principle that determines a geodesic, the path of shortest distance, in a given geometry. In that form, the same principle determines the world lines of all objects in the general theory of relativity. *See* RELATIVITY; RIEMANNIAN GEOMETRY.

Similarly, in mechanics, Hamilton's principle for the action is defined for any system of point particles by an integral (called the action) that extends over an arbitrarily prescribed path Γ in configuration space. Hamilton's principle asserts that the trajectories of all the particles are determined by the requirement that Γ be such that, for given initial and final times, the action is a minimum; for this reason it is also called the principle of least action. If the calculus of variations is applied to implement this principle, the corresponding Euler-Lagrange equations are obtained. These are the lagrangian equations of motion, that is, Newton's equations of motion in lagrangian form. *See* ACTION; HAMILTON'S PRINCIPLE; LAGRANGE'S EQUATIONS; LAGRANGIAN FUNCTION; LEAST-ACTION PRINCIPLE; NEWTON'S LAWS OF MOTION.

The principle of least action has been generalized to systems with infinitely many degrees of freedom, that is, fields. A Lagrange density function is then defined, which is a function of the fields and their time derivatives at any given point in space and time. For any field theory, only the Lagrange density needs to be given; the field equations are then derivable as the corresponding Euler-Lagrange equations. A similar technique makes it possible to derive the Schrödinger equation and the Dirac equation in quantum mechanics from specific Lagrange density functions. *See* QUANTUM MECHANICS; QUANTUM THEORY OF MATTER; RELATIVISTIC QUANTUM THEORY.

This method has great procedural advantages. For example, it facilitates a check of whether the theory satisfies certain invariance principles (such as relativistic invariance or rotational invariance) by simply ascertaining whether the Lagrange density satisfies them. The corresponding conservation laws can also be derived directly from the lagrangian. *See* CONSERVATION LAWS (PHYSICS); QUANTUM FIELD THEORY; SYMMETRY LAWS (PHYSICS).

The variational method also plays an important role in quantum-mechanical calculations. For the computation of needed quantities in terms of functions that result from the solution of differential equations, it is always of great advantage to use formulas that have the special form required to make them stationary with respect to small variations of the input functions in the vicinity of the unknown, exact solutions. *See* MINIMAL PRINCIPLES. [R.G.Ne.]

Varicose veins Varicose veins or varicosities are enlarged, tortuous vessels. They occur chiefly in superficial veins and their tributaries in the lower extremities. These veins are vulnerable to dilation because of their location in loose subcutaneous tissues which offer little support to the walls of the veins. Predisposing factors in the development of varices are (1) heredity, (2) venous obstruction, (3) phlebitis and thrombophlebitis, and (4) standing erect for long periods, which increases intravenous pressure. Varicose veins show loss of elastic tissue in their walls. They become dilated and finally may develop incompetent valves.

Hemorrhoids are varicosities of the veins deep to the anal epithelium, and varicoceles are varicosities of the veins surrounding the spermatic cord. Both conditions respond favorably to surgical treatment. *See* HEMORRHOIDS. [F.A.C.]

Variometer A geomagnetic device for detecting and indicating changes in one of the components of the magnetic field vector—usually magnetic declination, the horizontal intensity component, or the vertical intensity component. When used with a suitable recording unit, a set of three variometers forms a magnetograph.

The declination variometer, also called the *D* variometer, consists of a small permanent bar magnet suspended with a plane mirror from a fine quartz fiber. A lens fixed in front of the mirror serves to focus to a point a beam of light reflected from the mirror to recording paper mounted on a rotating drum. With no torsion in the fiber, the magnetic axis of the magnet will be aligned in the magnetic meridian. As the meridian changes its direction, because of normal daily variations or magnetic storm disturbances, the magnet follows.

Although a larger, stiffer fiber is used, the horizontal intensity (*H*) variometer is essentially the same as the *D* instrument. Enough torsion in the fiber is introduced at the time of installation to cause the magnet to turn 90° out of the magnetic meridian. The magnet responds to an increase of *H* by turning about the suspension axis.

Employing a larger permanent magnet because of mechanical construction requirements, the vertical intensity (*Z*) variometer is equipped with very fine steel knife-edges or pivots resting on agate planes or saddles and balanced so that its magnetic axis is horizontal. In this case, the torque due to the reaction of the magnetic moment with the Z component of the field is balanced against a torque due to the action of gravity on the laterally displaced center of mass of the magnet, and the magnet tilts one way or the other when the vertical field changes. *See* MAGNETOMETER. [J.H.N.]

Varistor Any variable resistor whose resistance depends upon voltage, current, or polarity. All semiconducting diodes are varistors. A varistor is generally made to have a symmetrical current-voltage relationship, but it can be nonsymmetrical.

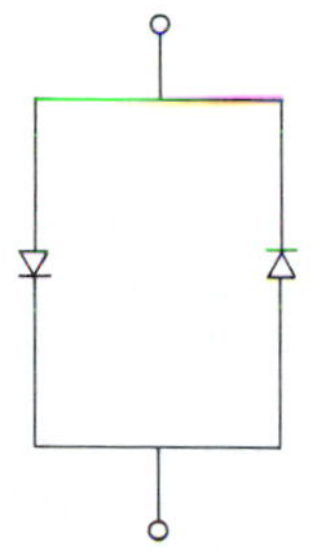

Symmetrical rectifier varistor.

Semiconductor rectifiers, of either the *pn*-junction or Schottky-barrier (hot-carrier) types, are popularly utilized for varsitors. A single rectifier has a nonsymmetrical characteristic. A symmetrical varistor is made, utilizing two rectifiers connected in parallel with opposing polarity (see illustration). *See* JUNCTION DIODE; SEMICONDUCTOR RECTIFIER. [I.A.L.]

Varnish A transparent surface coating which is applied as a liquid and then changes to a hard solid. Varnishes are solutions of resinous materials in a solvent, and dry by the evaporation of the solvent or by a chemical reaction, either with oxygen from the air or by some other means, including absorption of atmospheric moisture.

Spirit varnishes are those in which the evaporation of solvent is the only drying process; the solvent is usually alcohol, although the term is used for similar coatings made with other solvents. Shellac varnish, made by dissolving shellac in alcohol, is the most common of this type. Oleoresinous varnishes are made by treating a drying oil with a resin, usually with heat, and dissolving the reaction product in a solvent, usually a petroleum fraction; drying results from the evaporation of the solvent, followed by polymerization of the drying oil portion, a reaction which is accelerated by metallic driers added to the varnish. For a discussion of the mechanism of this drying action *see* DRIER (PAINT); SOLVENT; SURFACE COATING. [C.R.Ma.; C.N.Si.]

Varnish tree The plant *Rhus vernicifera*, also called lacquer tree, a member of the sumac family (Anacardiaceae). It is a native of China, but has long been cultivated in Japan. When the bark is cut, it exudes a milky juice which darkens and thickens on exposure. This is the lacquer long used in China and Japan. When properly applied, the thin transparent film becomes a varnish of extreme hardness. *See* LACQUER; SAPINDALES. [P.D.St./E.L.C.]

Varve A distinctive, thin annual sedimentary layer, the lower part consisting of coarser, lighter-colored clay and silt that was deposited in summer, and the upper of a finer-grained, darker clay deposited in winter. Numerous successive varves, generally less than 1 in. (2.5 cm) thick, accumulated in temporary lakes near melting glaciers (see illustration). Thicker and thinner varves at different places can be matched like tree rings. Similar layers occur in ancient rocks, but it is difficult to determine whether the layers are annual; many of them represent longer or less regular sedimentary cycles. [J.M.W.]

Varved clay of glacial lake. Dark layers were deposited in winter. (*Photograph by F. T. Thwaites*)

Vector (mathematics) A directed line segment. As such, vectors have magnitude and direction. Many physical quantities, for example, velocity, acceleration, and force, are vectors. Vectors are widely used in mathematical physics. *See* CALCULUS OF VECTORS. [M.Ye.]

Vector methods (physics) Methods that make use of the behavior of physical quantities under coordinate transformations.

From the point of view of physics, the most appropriate definition of a vector in three-dimensional space is a quantity that has three components which transform under rotations of the coordinate system like the coordinates of a point in space. What characterizes rotations is that the distance from the origin,

$$\sqrt{x_1^2 + x_2^2 + x_3^2}$$

of all points $\mathbf{x}$ with cartesian coordinates x_i, $i = 1, 2, 3$, remains unchanged. Specifically, if the rotation takes the x_i to new coordinates x_i', given by Eqs. (1) [in which $\det\{a_{ij}\}$ is the

$$x_i' = \sum_{j=1}^{3} a_{ij} x_j \qquad i = 1, 2, 3$$
$$\sum_{k=1}^{3} a_{ik} a_{jk} = \begin{cases} 1 & \text{if } i = j, \\ 0 & \text{if } i \neq j, \end{cases} \qquad \det\{a_{ij}\} = 1 \tag{1}$$

determinant of the matrix $\{a_{ij}\}$], then the three quantities V_i, $i = 1, 2, 3$, form the components of a vector $\mathbf{V}$, if in the new coordinate system the transformed coordinates are given by Eq. (2).

$$V_i' = \sum_{j=1}^{3} a_{ij} V_j \qquad i = 1, 2, 3 \tag{2}$$

If the coordinate transformation of Eqs. (1) is such that $|a_{ij}| = 1$ for $i = j$ and $a_{ij} = 0$ for $i \neq j$, but such that $\det\{a_{ij}\} = -1$ rather than $+1$ as in Eqs. (1), then it describes a reflection, in which a right-handed coordinate system is replaced by a left-handed one. If, for such a transformation, Eq. (2) also holds, then $\mathbf{V}$ is called a polar vector, whereas if the components of $\mathbf{V}$ do not change sign, it is called an axial vector or pseudovector. The vector $\mathbf{V}$ can also be looked upon as a quantity with a direction, with the magnitude

$$\sqrt{V_1^2 + V_2^2 + V_3^2}$$

pointing from the origin of the coordinate system to the point in space with the cartesian coordinates (V_1, V_2, V_3).

A quantity that remains invariant under a rotation of the coordinate system is called a scalar. The importance of vectors and scalars in physics derives from the assumed isotropy of the universe, which implies that all general physical laws should have the same form in any two coordinate systems that differ only by a rotation. It is therefore useful to classify physical quantities according to their transformation properties under coordinate rotations. Examples of scalars include the mass of an object, its electric charge, its volume, its surface area, the energy of a system, and its temperature. Other quantities have a direction and thus are vectors, such as the force exerted on a body, its velocity, its acceleration, its angular momentum, and the electric and magnetic fields. Since the sum of two scalars is a scalar, and the sum of two vectors is a vector, it is important in the formulation of physical laws not to mix quantities that have different transformation properties under coordinate rotations. The sum of a vector and a scalar has no simple transformation properties; a law that equated a vector to a scalar would have different forms in different coordinate systems and would thus not be acceptable. [R.G.Ne.]

Vectorcardiography An electrocardiographic technique which considers the resultant of all electrical manifestations of the heart beat as equivalent to a single dipole, defined as a positive point source followed by a negative point sink. For any instance of time during cardiac activity, this hypothetical entity has magnitude and direction and can therefore be treated as a vector quantity. By considering the surface electrocardiographic leads as vector components, size and direction of the cardiac vector forces can be obtained either by calculations (vecto electrocardiography) or by recording the resultant of two vector components on an oscilloscope screen. Combining the image of lead components obtained from three planes of the body, a three-dimensional reconstruction of the heart vector time course can be traced (spatial vectorcardiography).

Vectorcardiography has contributed relatively little to the diagnostic advance in electrocardiography, but has clarified the interrelations of lead connections and has allowed a remarkably simplified approach to the understanding of the complicated electrical activity of the heart. [H.H.He.]

Veering wind A wind that changes direction in a clockwise sense; for example, a change from a southerly to a westerly direction. The wind may veer gradually, over a period of hours, or abruptly, in a few minutes or seconds on passage of a wind-shift line. The wind veers when a cyclone passes eastward on a path north of the observer. Veering with height is usually found on the east side of a cyclone, or west of an anticyclone, with warm air to the south. In the Southern Hemisphere the meaning in terms of cardinal directions is reversed. *See* WIND. [C.W.N.]

Vega One of the brightest stars in the sky, apparent magnitude 0.1. Vega, or α Lyrae, is a normal main sequence star of spectral type A0, having an effective temperature near 10,050 K (17,630°F). Its distance is 8 parsecs, its absolute magnitude is +0.5 (40 times as bright as the Sun), and its radius is three times that of the Sun. *See* STAR. [J.L.Gr.]

Vegetable ivory The seed of the tagua palm (*Phytelephas macrocarpa*) of tropical America. Each drupelike fruit contains six to nine bony seeds. The extremely hard endosperm of the seed is used as a substitute for ivory. Vegetable ivory can be carved and tooled to make buttons, chessmen, knobs, inlays, and various ornamental articles. *See* ARECALES. [P.D.St./E.L.C.]

Vein The relatively thin-walled blood vessels that carry blood from the capillaries to the heart. Veins have several characteristics which differentiate them from their counterparts, the arteries. They are much more variable in location and distribution and are usually larger than the corresponding artery. Veins have thinner, less muscular walls and many contain valves which prevent backflow of blood, particularly in areas where gravity exerts a large effort. *See* ARTERY.

As is the case with the arterial system, two major circuits, the pulmonary and the systemic, are recognized. The former carries oxygenated blood from the lungs to the heart; the latter returns blood from all other tissues to the heart, either directly or through the portal system. In the adult human, the veins of the head, neck, upper extremities, and upper thorax empty into the large superior vena cava. Both of the caval vessels empty into the right atrium of the heart. A few veins, such as the coronary veins, drain directly into the heart itself. *See* BLOOD VESSELS; CARDIOVASCULAR SYSTEM; CIRCULATORY SYSTEM. [W.J.B.]

Velocity The time rate of change of position of a body in a particular direction. Linear velocity is velocity along a straight line, and its magnitude is commonly measured in such units as meters per second (m/s), feet per second (ft/s), and miles per hour (mi/h). Since both a magnitude and a direction are implied in a measurement of velocity, velocity is a directed or vector quantity, and to specify a velocity completely, the direction must always be given. The magnitude only is called the speed. *See* SPEED.

A body need not move in a straight line path to possess linear velocity. When a body is constrained to move along a curved path, it possesses at any point an instantaneous linear velocity in the direction of the tangent to the curve at that point. The average value of the linear velocity is defined as the ratio of the displacement to the elapsed time interval during which the displacement took place.

The representation of angular velocity ω as a vector is shown in the illustration. The vector is taken along the axis of spin. Its length is proportional to the angular speed and its

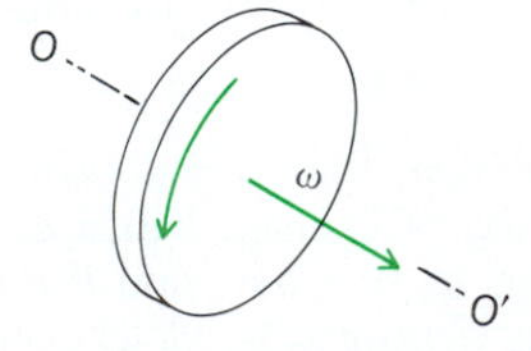

Angular velocity shown as an axial vector. Axis of rotation is OO′.

direction is that in which a right-hand screw would move. If a body rotates simultaneously about two or more rectangular axes, the resultant angular velocity is the vector sum of the individual angular velocities. [R.D.Ru.]

Veneer A thin sheet of wood of uniform thickness produced by peeling, slicing, or sawing. Depending on the manner of production and the portion of wood from which a veneer is made, the grain may be flat, vertical, or biased. Most veneer is rotary-cut from a bolt of wood, called a flitch, centered in chucks of a lathe. A nose bar bears against the flitch parallel to the center line of the lathe and a knife, also extending nearly the length of the lathe, peels off the veneer. Knives near the ends of the flitch cut the edges of the veneer.

Veneers cut from selected hardwoods, burls, crotches, and stumps are used for facings on furniture and the interior decoration of buildings. Other uses of veneer include plate separators in storage batteries, boxes for fruit and vegetables, drums for cheeses, and crates, hampers, and baskets for transportation and storage. Most veneer is used in plywood panels. *See* PLYWOOD; WOOD PRODUCTS. [F.H.R.]

Ventilation The supplying of air motion in a space by circulation or by moving air through the space. Ventilation may be produced by any combination of natural or mechanical supply and exhaust. Such systems may include partial treatment such as heating, humidity control, filtering or purification, and, in some cases, evaporative cooling. More complete treatment of the air is generally called air conditioning. *See* AIR CONDITIONING.

Natural ventilation may be provided by wind force, convection, or a combination of the two. Although largely supplanted by mechanical ventilation and air conditioning, natural ventilation still is widely used in homes, schools, and commercial and industrial buildings.

Mechanical supply ventilation may be of the central type consisting of a central fan system with distributing ducts serving a large space or a number of spaces, or of the unitary type with little or no ductwork, serving a single space or a portion of large space. Outside air connections are generally provided for all ducted systems. Outside air is needed in controlled quantities to remove odors and to replace air exhausted from the various building spaces and equipment.

Exhaust ventilation is required to remove odors, fumes, dust, and heat from an enclosed occupied space. Such exhaust may be of the natural variety or may be mechanical by means of roof or wall exhaust fans or mechanical exhaust systems. The mechanical systems may have minimal ductwork or none at all, or may be provided with extensive ductwork which is used to collect localized hot air, gases, fumes, or dust from process operations. Where it is possible to do so, the process operations are enclosed or hooded to provide maximum collection efficiency with the minimum requirement of exhaust air. [J.H.Cl.]

Venturi tube A device that causes a drop in pressure as a fluid flows through it. Essentially, a venturi tube is a short straight pipe section, or throat, between two tapered sections. Local pressure varies in the vicinity of the constriction; thus, by attaching at the throat a manometer or recording instrument, the drop in pressure can be measured and the flow rate calculated from it, or, by attaching a fuel source, fuel can be drawn into the main flow stream.

The principal advantage of the venturi tube is that not more than 10–20% of the difference in pressure between the inlet and the throat is permanently lost. This is accomplished by the discharge cone gradually decelerating the flow with minimum turbulence. [R.E.Sp.]

Venus The second planet in distance from the Sun. This neighbor of the Earth is very similar to it in such gross characteristics as mass, radius, and density. In other ways Venus is apparently different. Its atmospheric mass is almost a hundred times that of the Earth; its atmosphere is mostly carbon dioxide instead of nitrogen and oxygen; an extensive cloud layer of concentrated sulfuric acid is present; its surface temperature is an unbearable 730 K (850°F); and it rotates with a period of 243 days, and from east to west, in the opposite sense of most other planets. Some of these differences are due more to alternate evolutionary paths of the two planets than to totally different initial conditions.

To the naked eye, Venus is the brightest starlike object in the sky. It is usually visible during the night either soon after sunset or close to sunrise. It can sometimes be seen during the daytime.

The light seen coming from Venus is almost entirely due to sunlight that is reflected from a dense cloud layer. In contrast to the Earth's approximately 50% cloud cover, the clouds of Venus are present over the entire planet. The clouds of Venus consist of a large number of tiny particles that are made of a water solution of concentrated sulfuric acid.

Measurements conducted within Venus's atmosphere from various space probes and remotely from spacecraft and the Earth have provided a good definition of the gases that constitute Venus's atmosphere. By far, the chief gas species is carbon dioxide, which makes up 96% of the atmospheric molecules, while nitrogen accounts for almost all the remainder.

The very similar mean densities of Venus and the Earth imply that Venus is made of rocks similar to those that make up the Earth. Venus's interior may be qualitatively similar to that of the Earth in having a central iron core, a middle mantle made of rocks rich in silicon, oxygen, iron, and magnesium, and a thin outer crust containing rocks enriched in silicon in comparison with the rocks of the mantle. However, in contrast to the situation for the Earth, Venus's core may now be solid, which could account for the absence of a detectable magnetic field.

The ubiquitous cloud layer totally obscures Venus's surface at visible wavelengths. However, since the clouds and atmosphere are transparent to long-wavelength radio waves, observations of Venus's surface can be made with radar telescopes. Crude pictures of its surface show the presence of large, circular features. These may be giant craters produced by impacting meteoroids, analogous to the ones that pockmark the Moon. In addition, there are features that resemble large volcanic constructs, enormous rift valleys, and jagged mountain ranges. There is little topographic relief over most of Venus's surface. However, in a few places, huge continental-size plateaus rise about 6 mi (10 km) above the surrounding plains, while in other places isolated lowlands occur. [J.B.Po.]

Venus' flytrap *Dionaea muscipula*, an insectivorous plant of North and South Carolina (see illustration). The two halves of a leaf blade can move as if they were hinged along the midrib and, swinging upward and inward, the two surfaces come together. Any insect alighting on a leaf triggers this sensitive motor mechanism, and is caught between the closing halves of the leaf blade. In this trap, the insect is slow-

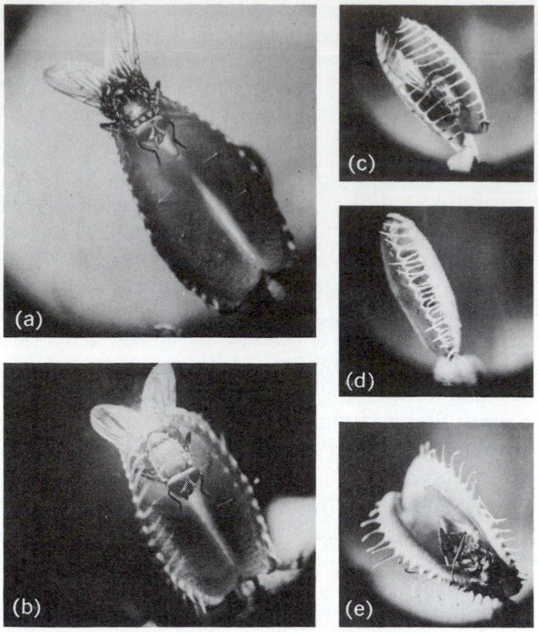

Stages *a-e* in capture and digestion of fly by leaf of Venus' flytrap. (*General Biological Supply House*)

ly digested by enzymes secreted by cells in the leaf. *See* INSECTIVOROUS PLANTS; NEPENTHALES; SECRETORY STRUCTURES (PLANT). [P.D.St./E.L.C.]

Verbal learning A field of experimental psychology which studies the formation of certain types of verbal associations. Verbal learning deals with the acquisition of associations; the study of memory (retention and forgetting) deals with the weakening or decay of associations over time. A sharp demarcation between verbal learning and memory is difficult because laboratory studies of verbal learning test what is remembered from one presentation to the next; eventually what is now termed verbal "learning" will probably be explained and understood in terms of principles of memory. *See* MEMORY.

The laboratory conditions under which verbal learning is studied are characterized by a display of information to the subject under carefully controlled conditions and by essentially continuous measurement of performance at all stages of practice. The stimulus material or information which the subject is required to learn generally consists of digits, consonants, common words, or nonsense syllables (for example, NFP or WIX). Usually there are a number of individual items which are combined into a list of items, and performance on the list as a whole is taken as the basic unit of analysis. The items in the list may be related in some way: similarity in terms of repeated elements (letters), similarity of meaning, and communality of category membership have all been studied experimentally.

One type of association which has been studied is that which would be involved in learning the alphabet, that is, sequential or serial associations. To study the formation of sequential associations, subjects are given serial lists to learn, where each list consists of a number of unrelated items. A second type of association is a direct association between pairs of items such as already exist in semantic memory (for example, *table-chair*, *black-white*). Subjects are given lists of paired associates consisting of A-B items, but A and B are not associatively related as are *table* and *chair*. A third type of task is free recall wherein a list of items is presented to the subject and the task is to recall as many items as possible in any order. The list is repeated again and again (each time with a different ordering of the items) until the subject has attained some criterion level of performance.

The types of associations involved in serial and paired-associate tasks are certainly not completely different. It can be argued that a serial list actually is a succession of linked A-B pairs wherein each item in the list functions first as a response term and then as a stimulus for the next response. In fact, many psychologists use paired-associate tasks because they feel the experimental analysis may be easier when the stimulus and response terms are clearly distinct. However, the empirical phenomenon, conceptual problems, and theoretical models for serial learning, paired-associate learning, and free-recall learning are different. [B.B.M.]

Vermiculite A clay mineral constituent of clay materials. Vermiculites are similar to the montmorillonites in having an expanding structure. They differ, however, in that the expansion can take place only to a limited degree. When rapidly heated, vermiculite produces a lightweight expanded product that is widely used for thermal insulation. *See* CLAY MINERALS.

Vermiculites show a considerable range in chemical composition. Their composition may be like that for some montmorillonites, in which case the only difference would be in the larger particle size of the vermiculites. Like the montmorillonites, the vermiculites have a high cation-exchange capacity. They also absorb certain organic molecules between their mica layers.

Vermiculite is frequently listed as an alteration product of biotite mica, and it is often present along with chloritic mica as a minor constituent in ancient sediments. [R.E.Gr.; F.M.W.]

Vernalization The induction in plants of the competence or ripeness to flower by the influence of cold, that is, temperatures below the optimal temperature for growth. Vernalization thus concerns the first of the three phases of flower formation in plants. In the second stage, for which a certain photoperiod frequently is required, flowers are initiated. In the third stage flowers are unfolded. *See* FLOWER; PHOTOPERIODISM; PLANT GROWTH. [K.N.-Z.]

Vernier A short, auxiliary scale placed along the main instrument scale to permit accurate fractional reading of the

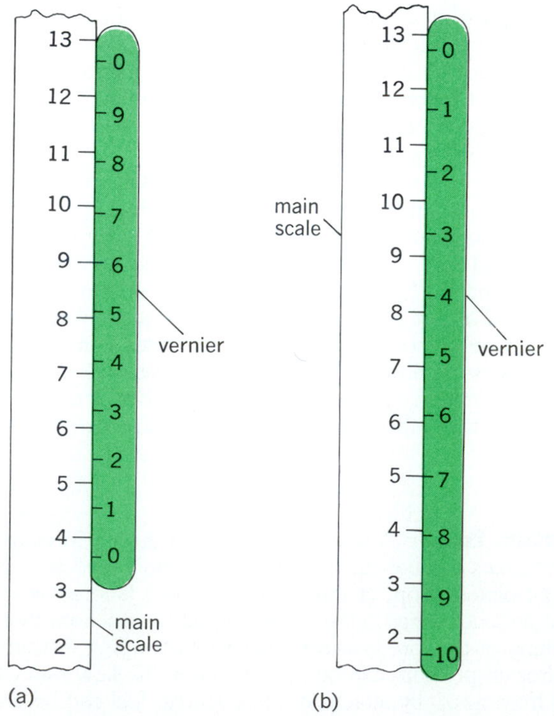

Two types of vernier scales. (*a*) Direct (reading 3.6). (*b*) Retrograde (reading 12.7).

least main division of the main scale. The auxiliary, or vernier, scale is graduated in one or both directions from the fiducial (index) mark in numbered divisions which are fractionally shorter (in a direct vernier) or longer (in a retrograde vernier) than those on the main scale (see illustration). The position of the fiducial mark (the zero mark of the vernier scale) between divisions on the main scale is indicated by the number of the graduation on the vernier scale which lines up exactly with a graduation on the main scale. [W.A.Wi.]

Verrucomorpha A suborder of the Thoracica comprising one Recent genus, *Verruca*, and one fossil, *Proverruca*, from the Lower Devonian. These barnacles are sessile and asymmetrical. In *Verruca* only four plates form the wall, rostrum, carina, and scutum and tergum of one side. In *Proverruca* two additional lateral plates are inserted between the rostrum and the carina. These barnacles are hermaphroditic. Most species live in deep water, though *V. stroemia* is from the lower intertidal and sublittoral. *See* THORACICA. [H.G.St.]

Vertebra The basic unit of the vertebral column. Collectively, the vertebrae surround and protect the spinal cord and provide some type of axial support for the body. The stresses that the vertebral column must meet change somewhat from one end of the animal to the other, and also differ greatly between aquatic and terrestrial vertebrates because the problems of support and locomotion in these media are quite different. Accordingly, vertebral structure varies widely; yet all vertebrae have many features in common. *See* SPINE.

A vertebra from the thoracic region of a mammal illustrates the basic morphology well (see illustration). The ventral portion consists of a disc-shaped mass of bone known as the body or centrum. An arch of bone, the neural arch, extends dorsally from the centrum and encompasses a space, the vertebral canal, in which the spinal cord lies. The bases, or roots, of the arch are narrower than other parts so that clefts, the intervertebral foramina, lie between the arch bases of successive vertebrae. Spinal nerves pass through these foramina.

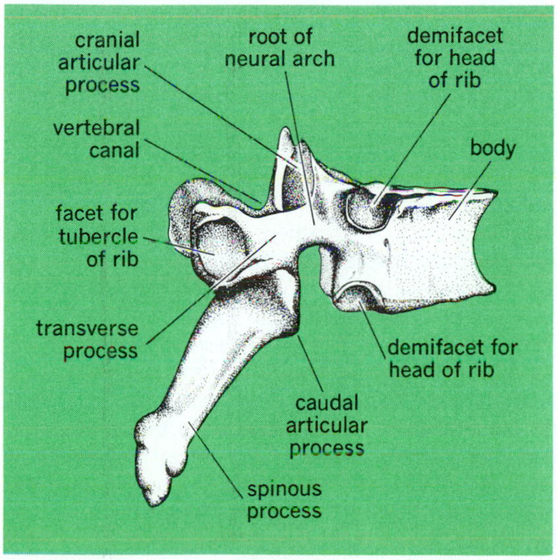

Lateral view of a human thoracic vertebra. Anterior is toward the right.

Certain muscles and ligaments attach onto the spinous process, which extends dorsally from the top of the arch, and onto a pair of transverse processes, which extend laterally from the arch. One pair of articular processes, or zygapophyses, extends forward from the neural arch, and another pair extends posteriorly. Articular processes of successive vertebrae overlap and help to hold the vertebrae together. The centra of adjacent vertebrae are also joined together by invertebral discs of fibrocartilage. Numerous ligaments interlace the vertebrae.

Ribs articulate onto the thoracic vertebrae. Typically, each mammalian rib has two articular surfaces—a terminal head and a tubercle situated a short distance distal to the head. The tubercle articulates with a facet located on the end of the transverse process; the head usually articulates intervertebrally onto the intervertebral disc and adjacent parts of the bordering centra.

In mammals different parts of the column are clearly specialized to subserve certain functions in addition to their general supportive role.

The head moves independently of the trunk, and a distinct neck region, consisting of cervical vertebrae, is present. With few exceptions all mammals from a shrew to a giraffe have seven cervical vertebrae.

Well-developed ribs, which play an important role in respiratory movements, articulate with the anterior trunk or thoracic vertebrae of mammals. The number of thoracic vertebrae varies between species but is on the order of 11 (bat) to 18 or 20 (horse). Humans have 12.

Lumbar vertebrae occupy the posterior part of the mammal trunk region. They are characterized by relatively large transverse processes to which certain of the powerful back muscles attach. Again the number of lumbar vertebrae vary between species but is on the order of 5 (bat) to 8 (whale). Humans have 5.

Correlated with the greater efficiency of terrestrial locomotion and the need for strong support for the powerful hindlegs, the number of sacral vertebrae increases during evolution from the single one of amphibians. Reptiles usually have two, and most mammals have three which are fused together, along with their embryonic rib rudiments, into a solid complex of bone called the sacrum. Humans, which are bipeds, have 5, and certain of the powerful hoofed mammals, for example, the horse, also have 5. Birds, whose hindlegs act as shock absorbers upon landing, have between 10 and 23 vertebrae fused together in their synsacrum.

The tail and caudal musculature no longer play an important role in the locomotion of most mammals (the Cetacea being a conspicuous exception), and the tail is greatly reduced in size. The spinal cord of mammals ends within the lumbar region, and only a few spinal nerves continue through the vertebrae canal into the tail. Caudal vertebrae are small and become progressively incomplete as one moves distally along the tail until only centra are left. Tail length, and hence the number of caudal vertebrae, vary widely. Some opossums have as many as 35, and humans, in which the tail is absent as an external structure, have only 3 to 5 caudal vertebrae. These form an internal coccyx to which certain anal muscles attach.

Vertebrata The major subphylum of the phylum Chordata, including the backboned animals from fish to mammals and humans. This subphylum is often termed the Craniata because of the common possession of a cranium or braincase; some of the primitive agnathan fish lack a vertebral column.

The vertebrates are divided into eight classes which are grouped together into several partly overlapping higher taxa which are generally given informal rank. Often two superclasses are used to group the aquatic and the terrestrial vertebrates. The term Gnathostomata is used to designate the seven classes of jawed vertebrates in contrast to the jawless Agnatha. The classification used herein is as follows (see separate articles for each class):

Superclass Pisces
 Class Agnatha
  Placodermi
  Chondrichthyes
  Osteichthyes
Superclass Tetrapoda
 Class Amphibia
  Reptilia
  Aves
  Mammalia

In the Pisces and Amphibia, the only extraembryonic membrane is the yolk sac, and these groups are restricted to aquatic reproduction. In reptiles, birds, and mammals, three additional membranes are present during development; one surrounds the embryo and is termed the amnion. These groups can reproduce independently of water and are called amniotes, in contrast to anamniotes (fish and amphibians) in which these additional membranes are lacking. *See* ANAMNIA; ANIMAL KINGDOM; CHORDATA; PISCES (ZOOLOGY); TETRAPODA; VERTEBRA. [W.J.B.]

Vertical takeoff and landing (VTOL) A flight technique in which an aircraft rises directly into the air and settles vertically onto the ground. Such aircraft do not need runways, but can operate from a small pad or, in some cases, from an unprepared site. The helicopter was the first aircraft that could hover and take off and land vertically, and is now the most widely used VTOL concept. *See* HELICOPTER.

The tilt-rotor concept is closest to the conventional helicopter and relies heavily on helicopter technology. The rotor disks are horizontal in VTOL operation, and are tilted 90° to act as propellers in cruising flight. The tilt-rotor concept can have cruise efficiencies at least twice those of the helicopter, and therefore the concept is of interest for helicopter missions where greater range, speed, and time on station are desired.

In the vectored-thrust concept a moderate-bypass-ratio engine is fitted with four mechanically interconnected rotating nozzles that deflect the thrust from horizontal for conventional flight to vertical for VTOL operations. Attitude control of the airplane in VTOL operation is provided by high-pressure air bled from the engine and ducted to control nozzles at the extremities of the airplane. These control nozzles are coupled to their corresponding conventional control surfaces so that they are operated by the pilot's normal controls.

A combination of the thrust-deflection and dual-propulsion types is usually referred to as a lift-plus-lift/cruise type. The cruise engine is fitted with a pair of rotating nozzles. In the VTOL mode the lift provided by this engine is augmented and balanced by a pair of special lift engines installed immediately behind the cockpit. [R.E.Ku.]

Vessel traffic service A shore-based, interactive navigation system for regulating vessel transits in selected waters. A vessel traffic service (VTS) system improves order and predictability throughout its service area by providing an organizational structure, a communications network, and standard procedures for enhanced information sharing and decision making. Opportunities for human error are reduced through the acquisition, interpretation, and broadcasting of traffic and navigation safety information and alerts for use by captains, mates, and marine pilots (local-area experts) in directing and controlling the maneuvering of ships, tugs with tows, ferries, and other participating vessels.

Vessel traffic service systems consist of a crewed vessel traffic center or centers, operational policies, a voice radio network with dedicated frequencies, radar surveillance, and sometimes traffic separation scheme. Visual overlooks, closed-circuit television, and multimedia displays are sometimes available. Remote vessel traffic centers and communications and surveillance equipment are linked by microwave or dedicated landline networks, such as telephone lines.

Advanced shipboard and aviation radar technology was developed specifically for vessel traffice service applications. Shipboard collision avoidance radar with automatic target acquisition and tracking features, known as ARPA (automatic radar plotting aid), has been installed in some traffic centers. Advanced multimedia displays have been developed for vessel traffic service use or adapted from technology transferred from defense or air-traffic control applications. The most advanced systems use high-fidelity, multicolor electronic display systems that correlate and fuse radar data from multiple sensors, closed-circuit television video, vessel-specific data, automated data management, electronic charts, and expert-system decision aids. *See* COMPUTER GRAPHICS; ELECTRONIC NAVIAGATION SYSTEMS; EXPERT SYSTEMS; MARINE NAVIGATION; NAVIGATION; SATELLITE NAVIGATION SYSTEMS. [W.Y.]

Vestigial-sideband modulation Modulation whereby in effect the modulated wave to be transmitted is composed of one sideband plus a portion of the other adjoining the carrier. The carrier may or may not be transmitted. Vestigial-sideband modulation (VSB) is like single-sideband modulation (SSB) except in a restricted region around the carrier. The overall frequency response to the wanted sideband and to the vestigial sideband is so proportioned by networks that upon demodulation, preferably but not necessarily by a product demodulator, the original modulating wave will be recovered with adequate accuracy. [H.S.Bl.]

Vestimentifera Giant tube worms; a phylum of benthic marine tube-dwelling worms with voluminous red tentacles. They seem restricted to unique habitats where nutrient-rich water comes up through the ocean floor—hot-water vents and cold-water seeps. Before 1985 the group was considered to be part of the phylum Pogonophora. *See* HYDROTHERMAL VENTS; POGONOPHORA.

The body of the Vestimentifera consists of four parts: (1) at the anterior end, the tentacular plume, with a central axial obturaculum and tentacular lamellae perpendicular to the axis, that looks like a thick, red feather; (2) the winged vestimentum, composed of dorsolateral flaps that close around the plume when it is withdrawn into the tube; (3) the trunk; and (4) a segmented posterior opisthosome with setae.

The phylum presently has two classes based on the condition of the branchial blood vessels and the arrangement of the branchial lamellae: Axonobranchia, with one order, Riftiida, and one family (two species in one genus); and Basibranchia, with the order Lamellibrachiida, containing two families, each with one genus and two species; and the order Tevniida, with two families, one having two genera (one species each), the second with one genus and two species. [E.B.Cu.]

Vetch Any of a group of plants which are mostly annual legumes with weak viny stems terminating in tendrils. Vetches are used mainly for green manure, cover crops, hay, and pasture.

Hairy vetch (*Vicia villosa*) is winter hardy and is most widely grown. Smooth vetch (*V. villosa* var. *glabrescens*) and woollypod vetch (*V. dasycarpa*) are similar to hairy vetch; seeds of the three may be found in mixtures. Bird vetch (*V. cracca*) is a winter-hardy perennial. Identification of vetches is difficult until pods and seeds develop. Common vetch (*V. sativa*) and purple vetch (*V. bengalensis*) are less winter-hardy than hairy vetch. Hungarian vetch (*V. pannonica*) is confined to the Pacific Northwest. Monantha vetch (*V. articulatea*) has fine leaves and stems but, lacking winter hardiness, is confined to warm regions. Bitter vetch (*V. ervilla*) is not grown commercially.

Narrow-leaf vetch (*V. angustifolia*) occurs mostly as a weed in waste places in the United States. *See* ROSALES. [F.V.G.]

Vibration The term used to describe a continuing periodic change in the magnitude of a displacement with respect to a specified central reference. The periodic motion may range from the simple to-and-fro oscillations of a pendulum, through the more complicated vibrations of a steel plate when struck with a hammer, to the extremely complicated vibrations of large structures such as an automobile on a rough road. Vibrations are also experienced by atoms, molecules, and nuclei. *See* PENDULUM.

A mechanical system must possess the properties of mass and stiffness or their equivalents in order to be capable of self-supported free vibration. Stiffness implies that an alteration in the normal configuration of the system will result in a restoring force tending to return it to this configuration. Mass or inertia implies that the velocity imparted to the system in being restored to its normal configuration will cause it to overshoot this configuration. It is in consequence of the interplay of mass and stiffness that periodic vibrations in mechanical systems are possible.

Mechanical vibration is the term used to describe the continuing periodic motion of a solid body at any frequency. When the rate of vibration of the solid body ranges between 20 and 20,000 hertz (Hz), it may also be referred to as an acoustic vibration, for if these vibrations are transmitted to a human ear they will produce the sensation of sound. The vibration of such a solid body in contact with a fluid medium such as air or water induces the molecules of the medium to vibrate in a similar fashion and thereby transmit energy in the form of an acoustic wave. Finally, when such an acoustic wave impinges on a material body, it forces the latter into a similar acoustic vibration. In the case of the human ear it produces the sensation of sound. *See* MECHANICAL VIBRATION; SOUND.

Systems with one degree of freedom are those for which one space coordinate alone is sufficient to specify the system's displacement from its normal configuration. An idealized example known as a simple oscillator consists of a point mass m fastened to one end of a massless spring and constrained to move back and forth in a line about its undisturbed position (Fig. 1). Although no actual acoustic vibrator is identical with this idealized example, the actual behavior of many vibrating systems when vibrating at low frequencies is similar and may be specified by giving values of a single space coordinate.

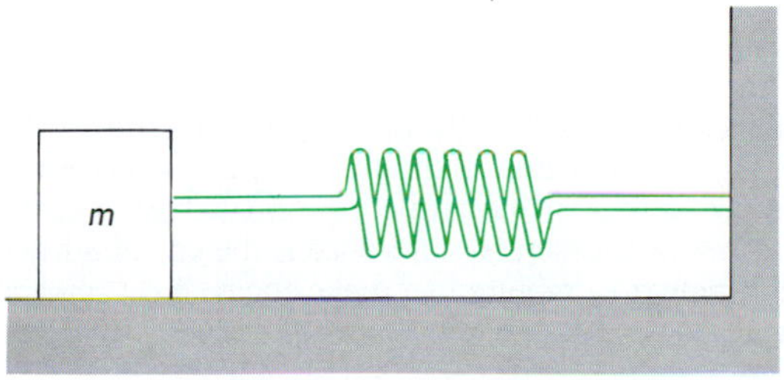

Fig. 1. Simple oscillator.

When the restoring force of the spring of a simple oscillator on its mass is directly proportional to the displacement of the latter from its normal position, the system vibrates in a sinusoidal manner called simple harmonic motion. This motion is identical with the projection of uniform circular motion on a diameter of a circle. *See* HARMONIC MOTION.

When two simple vibrating systems are interconnected by a flexible connection, the combined system has two degrees of freedom (Fig. 2). Such a system has two normal modes of vibration of two frequencies. Both of these frequencies differ from the respective natural frequencies of the individual uncoupled oscillators.

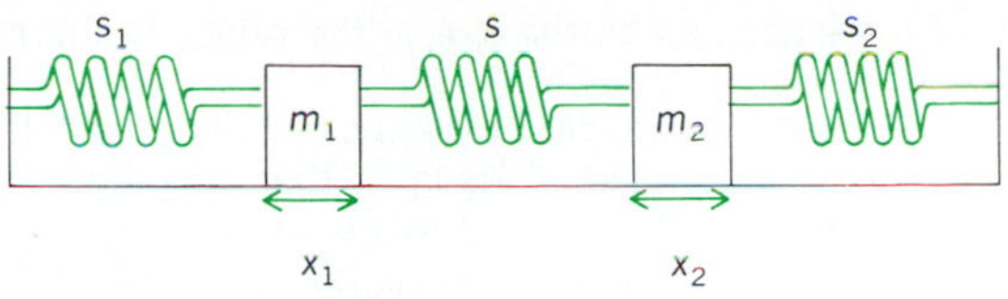

Fig. 2. Simple oscillator with two degrees of freedom. Masses m_1 and m_2, with displacements x_1 and x_2, are connected by springs s_1, s, and s_2.

A vibrating system is said to have several degrees of freedom if many space coordinates are required to describe its motion. One example is n masses $m_1, m_2, \ldots, m_n$ constrained to move in a line and interconnected by $(n-1)$ coupling springs with additional terminal springs leading from m_1 and m_n to rigid supports. This system has n normal modes of vibration, each of a distinct frequency. *See* DAMPING; VIBRATION DAMPING; VIBRATION ISOLATION. [L.E.K.]

Vibration damping The processes and techniques used for converting the mechanical vibrational energy of solids into heat energy. While vibration damping is helpful under conditions of resonance, it may be detrimental in many instances to a system at frequencies above the resonant point. This is due to the fact that the relative motion between the base of the vibration isolator and the mounted body tends to become smaller as the isolator becomes more efficient at the higher frequencies. With damping present, the force transmitted by the elastic element is unable to overcome the damping force; this leads to a resulting increase in transmissibility. *See* DAMPING; VIBRATION; VIBRATION ISOLATION.

All metal springs which include structural members such as brackets and shelves have some damping. However, such damping is insufficient for vibration isolators and must be augmented by special damping devices. *See* SPRING (MACHINES).

Several different types of damping devices have been developed and used successfully. Probably the most familiar is that used on automobiles, which, although known as a shock absorber, is in reality a damper, and functions as a limiter to the spring system of spring constant k. The system is shown in Fig. 1. A piston p is attached to the body m and is arranged to move vertically through the liquid in a cylinder c which is secured to the support s. As the piston moves, the force required to cause the liquid to flow from one side of the piston to the other is approximately proportional to the velocity of the piston in the cylinder. This type of damping is known as viscous damping. The damping force is controlled by the vis-

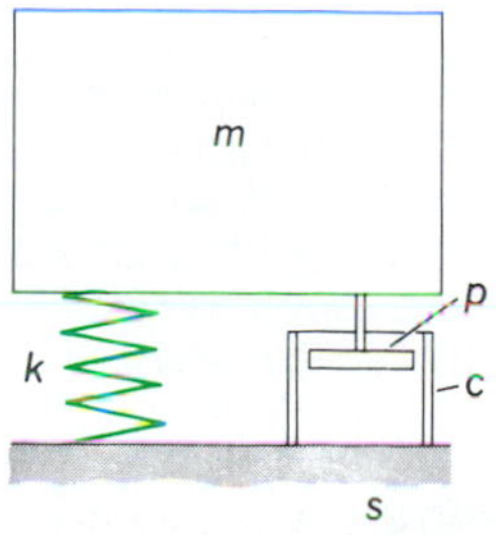

Fig. 1. Automobile shock absorber.

cosity of the liquid and by the size of the orifice in the piston. *See* SHOCK ABSORBER.

Some of the disadvantages of viscous damping may be overcome by using air instead of liquid as the damping medium. Air, being compressible, will add to the effective spring force with large displacements. If the air is housed within a flexible bellows, damping will be attainable horizontally as well as vertically. Such a system is illustrated in Fig. 2. This type of damping has proved very effective in vibration isolators.

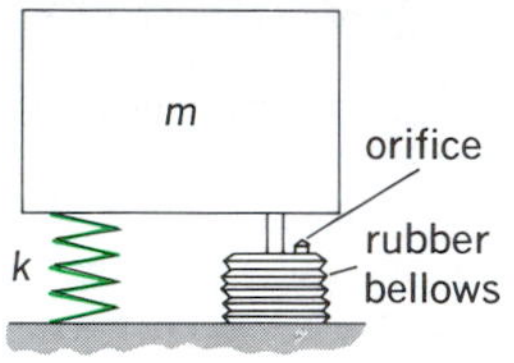

Fig. 2. System employing viscous damping with air.

Damping forces may be generated by causing one dry member to slide on another. This is known as dry friction or coulomb damping. Friction damping is used in several commercially available isolators because it provides a simple means to control the damping forces.

Magnetic damping is attainable as a result of the electric current induced in a conductor moving through a magnetic field. The damping force can be made proportional to the velocity of the conductor moving through the field. *See* ELECTROMAGNETIC INDUCTION. [K.W.J.]

Vibration isolation The isolation, in structures, of those vibrations or motions that are classified as mechanical vibration. Vibration isolation involves the control of the supporting structure, the placement and arrangement of isolators, and control of the internal construction of the equipment to be protected.

The simplest kind of mechanical vibration has the waveform of sinusoidal motion. Vibrations in structures, although generally more complex in waveform, exist wherever movement takes place. Such movement may be caused, for example, by the engine in an automobile, by engines or wind buffeting in aircraft, or by a punch press in a building. Delicate electronic equipment and precision instruments must normally be isolated from these motions if accurate measurements are to be obtained. *See* MECHANICAL VIBRATION; VIBRATION.

Vibration, in most cases, may be effectively isolated by placing a resilient medium, or vibration isolator, between the source of vibration and its surrounding area to reduce the magnitude of the force transmitted from a structure to its support or, alternatively, to reduce the magnitude of motion transmitted from a vibrating support to the structure. Isolating vibration at its source is commonly termed active or source isolation; isolating an instrument from its surroundings is known as passive isolation.

The vibration isolators may be positioned and arranged in many different ways, all variations of three basic types, each of which requires a definite amount of space: (1) isolators attached underneath equipment, known as an underneath mounting system; (2) isolators located in the plane of the center of gravity of the equipment, known as a center-of-gravity system; (3) mountings arranged four on each side in the plane of the radius of gyration, known as a double side-mounted system or radius-of-gyration system. *See* CENTER OF GRAVITY; RADIUS OF GYRATION. [K.W.J.]

Vibration machine A device for subjecting a system to controlled and reproducible mechanical vibration. Vibration machines, commonly called shake tables, are widely used in vibration measurement and analysis. There are three types of vibration machines in general use. These are the mechanical direct-drive type, the mechanical reaction type, and the electrodynamic type. Other types, such as hydraulic excitation devices, resonant systems, piezoelectric vibration generators used for instrument calibration, and machines for testing packages, have only limited or specialized applications. *See* MECHANICAL VIBRATION; VIBRATION. [K.W.J.]

Vibration pickup An electromechanical transducer capable of converting mechanical vibrations into electrical voltages. Depending upon their sensing element and output characteristics, such pickups are referred to as accelerometers, velocity pickups, or displacement pickups.

The accelerometer consists essentially of a mass which is seismically supported with respect to a surrounding case by means of a spring and guided to prevent motions other than those along the seismic direction of support. The mass exerts a force on the spring's support which is directly proportional to the acceleration being measured. This, in turn, is converted into an electrical voltage by means of stresses produced in a piezoelectric crystal. *See* ACCELEROMETER.

The velocity pickup generates a voltage proportional to the relative velocity between two principal elements of the pickup, the two elements usually being a coil of wire and a source of magnetic field (see illustration).

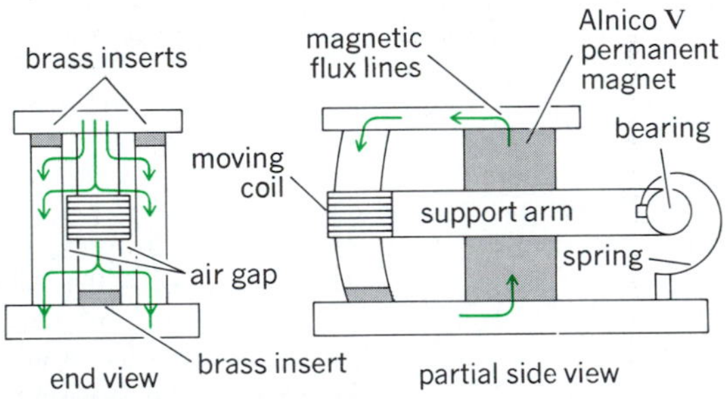

Velocity pickup. The coil swings on one end of an arm which is supported by bearings at the opposite end. The case follows the motion of the structure to which it is attatched. (*MB Manufacturing Co.*)

The displacement pickup is a device that generates an output voltage which is directly proportional to the relative displacement between two elements of the instrument. These pickups are similar in construction and behavior to velocity pickups. The only essential difference is the use of a frequency-weighting network, required to make them direct-reading. *See* VIBRATION. [L.E.K.]

Vibrator An electromechanical device used primarily to convert direct current to alternating current but is also used as a synchronous rectifier. The chopper is a closely related device. *See* CHOPPING.

A vibrator consists of a flat reed, with a soft-iron slug mounted at the tip of the reed adjacent to a driving coil and pole piece (see illustration). The soft-iron slug, or armature, is attracted by the pole piece of the driving coil. The longitudinal axis of the driving coil is usually parallel to the reed axis, but the pole piece is slightly offset from the neutral position of the reed so that the reed is attracted sideways toward the pole piece whenever the coil is energized. The reed also has a set of contacts which hit stationary contacts attached to the frame. The stationary contact points are usually mounted on leaf springs so that they can "give" when the reed contact

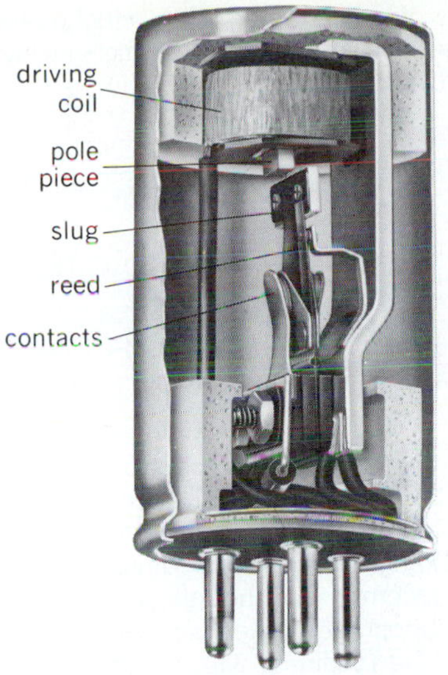

Typical low-power, plug-in vibrator.

hits them. This minimizes contact bounce and lengthens vibrator life.

To energize the drive coil periodically, a power vibrator frequently has a separate coil contact. When the vibrator is at rest, the two main contacts are open and the coil contact is closed. When the dc input is connected, the coil becomes energized and pulls the reed over to one side, opening the coil contact and deenergizing the coil. The reed then moves back, closing the coil contact, and the cycle repeats. The vibrations of the reed build up, and when equilibrium is reached, the reed oscillates vigorously and each of the main contacts is closed for nearly half of the vibrator period.

The main contacts alternately switch current from the dc supply through opposite halves of transformer primary, setting up an alternating flux in the transformer core. Alternating voltage is then available from the transformer secondary. A resistor and capacitor connected across the transformer primary reduce sparking at the contacts. [F.B.T.]

Vibrotaction The response of tactile nerve endings of varying forces on the skin and to oscillatory motion of the skin. Grasping, holding, and tactile exploration of an object are part of everyday experience, and their performance is dependent on vibrotaction.

Knowledge of the neural and psychophysical processes involved in vibrotaction is necessary to develop effective tactile communication systems, such as vibrotactile pagers, and multipin vibrotactile stimulators used to produce images on an extended skin surface. Potential users range from business people to the visually and hearing impaired. Potential applications range from activities involving visual and aural sensory saturation (such as pilots in high stress situations) to those requiring sensory stimulation.

Small electrical impulses, or action potentials, generated by a single nerve ending can be recorded by inserting a microelectrode into the arm of an alert human subject and positioning the electrode tip in a nerve fiber. Studies of the action potentials produced when the skin is locally depressed at a fingertip have identified activity in four networks of distinctive nerve endings lying within 1–2 mm of the skin surface. These nerve endings, which transform skin motion into neural signals, consist of physically and functionally different mechanoreceptors.

Three types of mechanoreceptor in the fingertip have been implicated in vibrotaction. One is a population of slowly adapting mechanoreceptors (SA type I, or SAI); their action potentials persist for some time after a transient skin indentation. The two others are populations of rapidly adapting receptors (FAI and FAII types): FAI receptors respond only to skin indentation within a few millimeters of the nerve ending, whereas FAII receptors also respond to stimuli at a greater distance from the nerve ending. A fourth mechanoreceptor type appears to respond primarily to skin stretch. *See* CUTANEOUS SENSATION; MECHANORECEPTORS. [A.J.Br.]

Vicuna A rare animal whose fiber makes the world's most costly and most exquisite cloth, surpassing all others in fineness and beauty. It is found in an almost inaccessible area of the Andes Mountains.

A single animal yields only approximately $\frac{1}{4}$ lb (0.11 kg) of hair; thus 40 animals are required to provide enough hair for the average coat. The fiber of the vicuna is the softest and most delicate of the known animal fibers; yet it is strong for its weight, is resilient, and has a marked degree of elasticity and surface cohesion. *See* ALPACA; CAMEL'S HAIR; CASHMERE; LLAMA; MOHAIR; NATURAL FIBER; WOOL. [M.D.P.]

Video amplifier A low-pass amplifier having a bandwidth in the range from 2 to 100 MHz. Typical applications are in television receivers, cathode-ray-tube computer terminals, and pulse amplifiers. The function of a video amplifier is to amplify a signal containing high-frequency components without introducing distortion.

Modern video amplifiers use specially designed integrated circuits. With one chip and an external resistor to control the voltage gain, it is possible to make a video amplifier with a bandwidth between 50 and 100 MHz having voltage gains ranging from 20 to 500. *See* AMPLIFIER; INTEGRATED CIRCUITS. [H.F.K.]

Video disk recording A disk system used to reproduce television pictures and sound. The reproduction system consists of three parts: disk, disk player, and television receiver. The disk player rotates the disk, whose information density is several hundred times that of conventional audio long-playing records, at high speed and reads the recorded signal by a sensor. The signal can be seen on the picture tube or heard from the loudspeaker of the television receiver. The recording of television signals is not yet possible in the home; the system operates only for reproduction. Although the video tape recorder (VTR) can record and reproduce the television signal at home, the tape is expensive. A video disk can be mass-produced and is relatively inexpensive. Further, the degradation of recorded signal with elapsed time is far less than that of magnetic tapes. [H.D.]

Video telephone A communications instrument used for the simultaneous exchange of visual images and associated speech. A complete video telephone communication system involves three basic elements: (1) terminal equipment (camera, display, microphone, and speakerphone) to transform both the aural and visual inputs into electrical signals, and vice versa; (2) transmission facilities to carry the electrical signals over large distances; and (3) switching systems to allow a choice of terminals to be interconnected. The equipment that is used as a visual adjunct to telephony may also be arranged to serve many of the applications associated with closed-circuit television and surveillance. *See* CLOSED-CIRCUIT TELEVISION; TELEPHONE EQUIPMENT, CUSTOMER; TELEPHONE SERVICE; TELEVISION. [C.C.C.]

Vincent's angina An inflammation of the tissues of the oral cavity or pharynx caused by fusiform bacilli, *Bacillus fusiformis* (= *Fusobacterium fusiforme*), and a spirochete, *Borrelia vincenti*. The underlying cause, however, is a dietary deficiency in the vitamin B complex which favors the overgrowth of these organisms.

The commonest form of the disease, trench mouth, is characterized by redness, congestion, and edema of the gums with involvement of the entire oral cavity in severe cases. It is observed mainly among institutional or military groups with a deficient diet. [T.B.T.]

Vinegar The product of the incomplete oxidation to acetic acid of ethyl alcohol produced by a primary fermentation of vegetable materials. The oxidation, often called secondary fermentation, proceeds under the influence of various kinds of so-called vinegar bacteria, species of the genera *Acetobacter* and *Acetomonas*. The overall chemical reactions for the production of vinegar are given below:

$$\text{Sugar} \xrightarrow{\text{yeast}} 2CH_3CH_2OH + 2CO_2$$

$$2CH_3CH_2OH + 2O_2 \xrightarrow[\text{bacteria}]{\text{vinegar}} 2CH_3COOH + 2H_2O$$

Vinegar contains not less than 4 g of acetic acid per 100 ml. It is used in the preparation of pickled fruits and vegetables or other large-scale food manufacture such as salad dressings, and as a seasoning agent for table use.

Vinegars are classified as to raw materials used in their manufacture; there are products designated as wine, cider, malt, and sugar vinegar. Wine vinegar is the product obtained from the acetous souring of grape wine. Cider vinegar is derived from the alcoholic fermentation of other fruit juices, usually apple juice, followed by acetous fermentation. Malt vinegar is obtained by the alcoholic fermentation of starchy materials, such as cereal grains or potatoes saccharified, or converted to sugar, by enzymes of malt, followed by acetous fermentation. Distilled vinegars usually belong to the malt vinegar class. Sugar vinegar is obtained by the alcoholic fermentation of sugars such as syrups of glucose, blackstrap molasses, or refiners' molasses. Nondistilled vinegars are characterized by distinctive flavors and odor. Seasoned vinegars are prepared by soaking vegetable seasoning agents, such as tarragon leaves, onions, garlic, peppers, or various spices, in vinegar. [L.B.Lo.]

Vinyl resin An addition polymer made from monomers with the general structure shown, wherein R_1, and R_2, represent

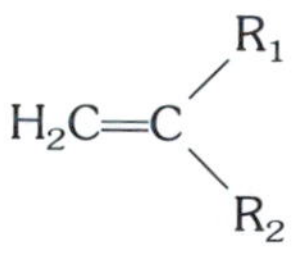

the same or different hydrogen, alkyl, halogen, ether, carboxylate, or other substituents. Vinyl resin often refers to any of a large variety of thermoplastics with the $-CH_2CHR_1-$ repeating group. Popularly, vinyl resin refers to a plasticized polyvinyl chloride used as a film or fabric, as for furniture or automobile seat covers. In common chemical usage, vinyl resin implies a polymer of an olefinically unsaturated monomer, whether the double bond is terminal or not. *See* POLYVINYL RESINS. [F.W.]

Vinylogy A principle used to correlate the properties of organic chemical compounds. Organic compounds that differ from each other by a vinylene linkage ($-CH{=}CH-$) are said to be vinylogs of one another. Thus, ethyl crotonate is a vinylog of ethyl acetate and of the next higher vinylog, ethyl sorbate. Their structures are:

$$CH_3{-}CO_2C_2H_5$$
Ethyl acetate

$$CH_3{-}CH{=}CH{-}CO_2C_2H_5$$
Ethyl crotonate

$$CH_3{-}CH{=}CH{-}CH{=}CH{-}CO_2C_2H_5$$
Ethyl sorbate

These esters are the first members of the vinylogous series $CH_3{-}(CH{=}CH)_n{-}CO_2C_2H_5$. The relationship between the members of such a series is known as vinylogy.

The outstanding characteristic common to the members of a vinylogous system is that the influence of the terminal functional groups upon each other persists throughout the series. For example, the readiness with which the methyl group of ethyl acetate loses a proton is ascribed to the enhanced stability of the resulting anion because of the presence of the ester group. [R.C.F.]

Violales An order of flowering plants, division Magnoliophyta (Angiospermae), in the subclass Dilleniidae of the class Magnoliopsida (dicotyledons). The order consists of 31 families and more than 5200 species. The largest families are the Flacourtiaceae (about 1300 species), Violaceae (about 850 species), Cucurbitaceae (about 850 species), Begoniaceae (about 800 species), and Passifloraceae (about 600 species).

The most characteristic feature of this morphologically heterogeneous order is the unilocular, compound ovary with mostly parietal placentation. When the stamens are numerous they are initiated in centrifugal sequence. The more primitive members of the order, such as some of the Flacourtiaceae, are trees with stipulate, alternate leaves; perfect, polypetalous flowers with numerous centrifugal stamens; a compound pistil with free styles and parietal placentation; and seeds with a well-developed endosperm.

Cucumbers and melons (species of *Cucumis*, in the Cucurbitaceae), pumpkins and squashes (species of *Cucurbita*), begonias, and violets (*Viola*) are familiar members of the Violales. *See* CANTALOUPE; CUCUMBER; DILLENIIDAE; HONEY DEW MELON; MAGNOLIOPSIDA; MUSKMELON; PERSIAN MELON; PUMPKIN; SQUASH; WATERMELON. [A.Cr.]

Viral inclusion bodies Abnormal structures which appear within the cell nucleus, the cytoplasm, or both, during the course of virus multiplication. In general, the inclusion bodies are concerned with the developmental processes of the virus. In some virus infections, such as molluscum contagiosum, inclusion bodies may be simply masses of maturing virus particles. In other infections (herpes simplex), typical inclusion bodies do not appear until after the virus has multiplied. Such inclusions may be remnants of the process of virus multiplication.

The presence of inclusion bodies is often important in diagnosis. A cytoplasmic inclusion in nerve cells, the Negri body, is pathogenic for rabies. *See* RABIES; VIRUS. [J.L.Me.]

Virgo In astronomy, a constellation handed down from antiquity, visible throughout the summer months. This sixth sign of the zodiac represents a maiden (see illustration). The brightest star in the constellation, Spica, is a spectroscopic binary with a massive dark companion. It is a fine first-magni-

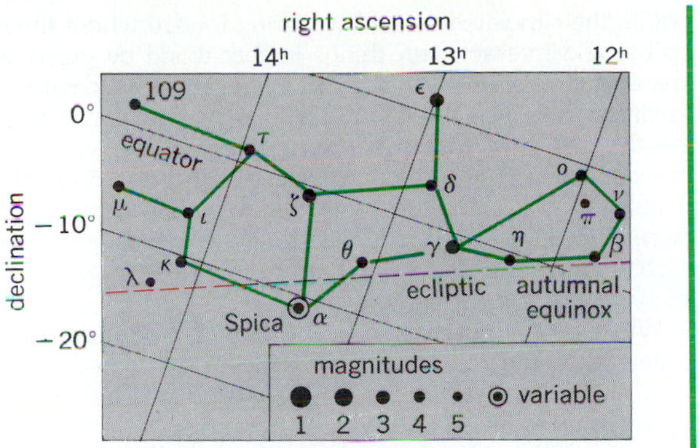

Line pattern of the constellation Virgo. The grid lines represent the coordinates of the sky. The apparent brightness, or magnitude, of the stars is shown by the sizes of the dots, which are graded by appropriate numbers as indicated.

tude star of the purest white tint, and also a navigational star. *See* CONSTELLATION; LIBRA. [C-.S.Y.]

Virgo Cluster The nearest large cluster of galaxies, forming a conspicuous clump in the distribution of easily visible elliptical and spiral galaxies. The clump lies mostly in the constellation of Virgo but also extends into neighboring constellations, especially Coma Berenices. *See* CONSTELLATION; VIRGO.

Besides the 100 or so bright galaxies in Virgo, more than 2000 fainter galaxies show up in deeper searches of the cluster. Furthermore, the cluster includes the Southern Extension, where several dozen additional galaxies lie. G. de Vaucouleurs promoted the idea, now universally recognized, that the Virgo Cluster is so extensive in size that it includes a giant halo of many outlying groups, including the Local Group, to which the Milky Way Galaxy belongs. He called this structure the Local Supercluster. *See* LOCAL GROUP; UNIVERSE.

The Virgo Cluster is an example of a mixed-population cluster, with both early-type (elliptical and S0) and late-type (spiral and irregular) galaxies present. As the nearest large cluster, it provides a good testing ground for theories of how the dense cluster environment affects the properties of galaxies. The most obvious peculiarity of the Virgo galaxies is the limited extent of the disks of the spirals. The spiral galaxies have possibly been stripped of their outer gas by encounters with other galaxies in the cluster. A related phenomenon is the presence of anemic galaxies, spirals that have unusually low surface brightness and inconspicuous spiral arms.

Determining the distance to the Virgo Cluster has been an important task of twentieth-century astronomy. The Virgo Cluster forms an essential stepping stone for the cosmic distance scale, as it provides a connection between local distance criteria and the use of the Hubble law of gauging distances. The best measurements give a mean distance of 16 megaparsecs (5×10^7 light-years). When the infall velocity of the Milky Way Galaxy toward the center of the Virgo Cluster is taken into account, this distance indicates that the local cosmic value of the Hubble constant, which relates expansion velocity to distance, is about 70 kilometers per second per megaparsec. *See* COSMOLOGY; HUBBLE CONSTANT. [P.H.]

Virial equation In thermodynamics, an empirical equation of state with additional terms beyond those for an ideal gas. The additional terms account for some of the differences between real gases and ideal gases. For an ideal gas Eq. (1)

$$pV = NR_uT \text{ or } pv_N = R_uT \quad (1)$$

holds, where p is the pressure, V the total volume, $v_N = V/N$ the molar volume, N the number of moles, R_u the universal gas constant, and T the absolute temperature. To account for departures by real gases from the idealized relation, Eq. (1) may be written as Eq. (2).

$$\frac{pv_N}{R_uT} = B_0 + \frac{B_1(T)}{v_N} + \frac{B_2(T)}{(v_N)^2} + \frac{B_3(T)}{(v_N)^3} + \cdots \quad (2)$$

The *B*s are functions of temperature and are known as the virial coefficients. *See* VAN DER WAALS EQUATION. [G.A.H.]

Virial theorem A theorem in classical mechanics which relates the kinetic energy of a system to the virial of Clausius, as defined below. The theorem can be generalized to quantum mechanics and has widespread application. It connects the average kinetic and potential energies for systems in which the potential is a power of the radius. Since the theorem involves integral quantities such as the total kinetic energy, rather than the kinetic energies of the individual particles that may be involved, it gives valuable information on the behavior of complex systems. For example, in statistical mechanics the virial theorem is intimately connected to the equipartition theorem; in astrophysics it may be used to connect the internal temperature, mass, and radius of a star and to discuss stellar stability. The virial theorem makes possible a very easy derivation of the counterintuitive result that as a star radiates energy and contracts it heats up rather than cooling down. *See* STAR; STATISTICAL MECHANICS; STELLAR EVOLUTION.

The virial theorem states that the time-averaged value of the kinetic energy in a confined system (that is, a system in which the velocities and position vectors of all the particles remain finite) is equal to the virial of Clausius. The virial of Clausius is defined to equal $-1/2$ times the time-averaged value of a sum over all the particles in the system. The term in this sum associated with a particular particle is the dot product of the particle's position vector and the force acting on the particle. Alternatively, this term is the product of the distance, r, of the particle from the origin of coordinates and the radial component of the force acting on the particle.

In the common case that the forces are derivable from a power-law potential, V, proportional to r^k, where k is a constant, the virial is just $-k/2$ times the potential energy. Thus, in this case the virial theorem simply states that the kinetic energy is $k/2$ times the potential energy. For a system connected by Hooke's-law springs, $k = 2$, and the average kinetic and potential energies are equal. For $k = 1$, that is, for gravitational or Coulomb forces, the potential energy is minus twice the kinetic energy. *See* COULOMB'S LAW; GRAVITATION; HARMONIC MOTION. [A.G.P.]

Viroid One of the smallest known agents of infectious disease. The molecular weight of viroids is estimated to be as little as $1.0–1.3 \times 10^4$ daltons, in contrast to the conventional plant virus genomes, which have molecular weights of approximately 2×10^6 daltons. Conventional viruses are made up of nucleic acid encapsulated in protein (capsid), whereas viroids are uniquely characterized by the absence of a capsid. In spite of small size, viroid ribonucleic acids (RNAs) can replicate and produce characteristic disease syndromes when introduced into cells. Although the viroids thus far identified are associated with plants, based on the unusual properties of the infectious agents, they may also be found to affect other forms of life, including humans. Scrapie, kuru, transmissible mink encephalopathy, and Creutzfeld-Jakob disease had been considered candidates for viroid etiology, but current thinking focuses on prions as the causative agents. It is possible that in instances where a viral etiology of disease has been assumed but no causal virus has been identified, viroid infectious agents may be involved. *See* PLANT VIRUSES AND VIROIDS; VIRUS; PRION DISEASE. [R.K.H.]

Virology A specialized part of clinical microbiology. Because of the nature of viruses and their proclivity for growth only in living cell cultures or in eggs, few clinical microbiology laboratories have facilities for the isolation and identification of viruses. Most of these procedures are best done in large hospitals or in state departments of health. There is a growing trend, however, for smaller hospitals to develop screening virology laboratories in which specimens are brought to the laboratory, placed in tissue culture, and preliminary identification steps taken in order to provide a rapid diagnosis of viral disease. Although only a few drugs exist that can be used to treat virus infections, the early recognition of viral disease may allow the clinician to discontinue the use of powerful antibiotics that are only effective for bacterial diseases. Common virus diseases include influenza, herpes, chicken pox, the common cold, and many nonspecific respiratory infections. Much of the viral diagnosis that is done in the laboratory is by the determination of a specific antibody response to viral disease. The routine clinical microbiology laboratory usually does not become involved in such procedures. A special procedure for the detection of viruses and other organisms that affect infants is called the TORCH screen. It is a procedure for the detection of antibodies to the parasitic disease toxoplasmosis and the viral diseases rubella, cytomegalovirus, and herpes. *See* ANIMAL VIRUS; VIRUS. [R.C.T.]

Virtual acoustics The stimulation of the complex acoustic field experienced by a listener within an environment. The technology is also known as three-dimensional sound and auralization. Going beyond the simple left-right volume adjustment of normal stereo techniques, the goal is to process sounds so that they appear to come from particular locations in three-dimensional space. Although loudspeaker systems have been developed, much of the work in the field focuses on using headphones for playback and is the outgrowth of earlier analog techniques. For example, in binaural recording, the sound of an orchestra playing classical music is recorded through small microphones in the two imitation ear canals of an artificial or dummy head placed in the audience of a concert hall. When the recorded piece is played back over headphones, the listener passively experiences the illusion of hearing the violins on the left and the cellos on the right, along with all the associated echoes, resonances, and ambience of the original environment. Techniques use digital signal processing to synthesize the acoustical properties that people use to localize a sound source in space. Thus, they provide the flexibility of a kind of digital dummy head, allowing a more active experience in which a listener can both design and move around or interact with a simulated acoustic environment in real time. *See* BINAURAL SOUND.

The success of virtual acoustics is critically dependent on whether the acoustical cues used by humans to locate sounds have been adequately synthesized. There may be many cumulative effects on the sound as it makes its way to the eardrum, but all of these effects can be expressed as a single filtering operation much like the effects of a graphic equalizer in a stereo system. The exact nature of this filter can be measured by an experiment in which an impulse (a single, very short sound pulse or click) is produced by a loudspeaker at a particular location. The acoustic shaping by the two ears is then measured by recording the outputs of small probe microphones placed inside the ear canals of the individual or an artificial head. If the measurement of the two ears occurs simultaneously, the responses, when taken together as a pair of filters, include an estimate of the interaural differences as well. Thus, this technique makes it possible to measure all of the relevant spatial cues together for a given source location, for a given listener, and in a given room or environment. *See* EAR; EQUALIZER; HEARING; PSYCHOACOUSTICS; SOUND. [E.M.We.]

Virtual reality A form of human-computer interaction in which a real or imaginary environment is simulated and users interact with and manipulate that world. Users travel within the simulated world by moving toward where they want to be, and interact with things in that world by grasping and manipulating simulated objects. In the most successful virtual environments, users feel that they are truly present in the simulated world and that their experience in the virtual world matches what they would experience in the environment being simulated. This sensation is referred to as engagement, immersion, or presence, and it is this quality that distinguishes virtual reality from other forms of human-computer interaction. *See* HUMAN-COMPUTER INTERACTION.

When a user interacts with a virtual environment, the computer-generated graphics display must be updated with each turn of the head or movement of the hand. The virtual environment must be able to generate and display realistic-looking views of the simulated world quickly enough that the interaction feels responsive and natural. *See* COMPUTER GRAPHICS.

Hardware. Virtual reality relies on a variety of specialized input and output devices to achieve this sense of natural interaction.

The most important of the input devices used in a virtual environment, a tracker is capable of reporting its location in space and its orientation. Tracking devices can be optical, magnetic, or acoustic. A tracker is sometimes combined with a traditional computer input device, such as a mouse or a joystick. *See* COMPUTER PERIPHERAL DEVICES.

An attempt to provide a truly natural input device, the data glove is outfitted with sensors that can read the angle of each of the finger joints in the hand. Wearing such a glove, users can interact with the virtual world through hand gestures, such as pointing or making a fist. *See* FIBER-OPTIC SENSOR; STRAIN GAGE.

The real-world visual experience is approximated in virtual environments by using stereoscopic displays. Two views of the simulated world are generated, one for each eye, and a stereoscopic display device is used to show the correct view to each eye.

Applications. Virtual reality can be applied in a variety of ways. In scientific and engineering research, virtual environments are used to visually explore whatever physical world phenomenon is under study. Training personnel for work in dangerous environments or with expensive equipment is best done through simulation. Airplane pilots, for example, train in flight simulators. Virtual reality can enable medical personnel to practice new surgical procedures on simulated individuals. As a form of entertainment, virtual reality is a highly engaging way to experience imaginary worlds and to play games. Virtual reality also provides a way to experiment with prototype designs for new products. *See* AIRCRAFT DESIGN; COMPUTER-AIDED DESIGN AND MANUFACTURING. [M.P.Ba.]

Virtual work principle The principle stating that the total virtual work done by all the forces acting on a system in static equilibrium is zero for a set of infinitesimal virtual displacements from equilibrium. The infinitesimal displacements are called virtual because they need not be obtained by a displacement that actually occurs in the system. The virtual work is the work done by the virtual displacements, which can be arbitrary, provided they are consistent with the constraints of the system. *See* CONSTRAINT.

The principle of virtual work is equivalent to the conditions for static equilibrium of a rigid body expressed in terms of the total forces and torques. That is, the principle of virtual work can be derived from these conditions, and conversely. *See* EQUILIBRIUM OF FORCES; STATICS.

One advantage of the principle of virtual work is that it can serve as a basis for all of statics. In the solution of problems the principle of virtual work is often useful for eliminating the need for consideration of the forces of constraint, since these forces often are perpendicular to the virtual displacements and consequently do no work. [P.W.S.]

Virulence The capacity of microorganisms to inflict injury or death on a susceptible host. The concept of virulent microorganisms actually has no absolute, independent existence as a microbial attribute. The same microorganism may under identical conditions kill one host and be harmless to another; thus the same microorganism is either virulent or avirulent. It follows that virulence acquires reality only when both the microorganism and the host are considered. Certain ancillary considerations contribute to the success or failure of the parasite. These are the numbers of the invading microorganism, the portal of entry, the mode of dissemination in the tissues and fluids of the host, the localization of microorganism in certain tissues of the host, and the natural and induced humoral and cellular defensive reactions of the host. Genotypic variation, or mutation of either microorganism or host, may have a determining effect on the expression of virulence. Thus mutation may result in either a gain or loss of a character responsible for the microorganism's capability to inflict damage on the host. *See* IMMUNITY; VIRUS. [M.T.]

Virus An infectious agent characterized mainly by its small size and chemical simplicity. During a part of its infectious cycle in a cell, and sometimes even outside the cell, a virus consists only of nucleic acid, either deoxyribonucleic acid (DNA) or ribonucleic acid (RNA). However, a mature virus particle normally exists in the form of nucleic acid encased in a protein shell. A virus will not reproduce in nutrient media, but it may multiply within the cells of appropriate tissue cultures. *See* DEOXYRIBONUCLEIC ACID (DNA); RIBONUCLEIC ACID (RNA).

Almost all forms of life are susceptible to virus infection, including humans, animals, plants, and even bacteria (bacteriophage). Viruses cause enormous annual losses of domestic crops and animals, and they have been responsible for countless diseases since antiquity. *See* ANIMAL VIRUS; BACTERIOPHAGE; PLANT VIRUSES AND VIROIDS.

Direct visualization of a virus became common with the development of the electron microscope about 1940. Each virus was found to have a characteristic size and shape. Some are spheroidal bodies, some filamentous, others are loaf-shaped or polyhedral, and some are spermlike. All are exceedingly small, the largest being only a few hundred-thousandths of an inch in diameter. *See* ELECTRON MICROSCOPE.

Chemical composition and structure. Each virus seems to have a definite and characteristic composition. The larger viruses tend to be more complex than the smaller ones both morphologically and chemically. Protein, nucleic acid, lipid, and carbohydrate have all been found in purified virus preparations. Although few viruses contain all these constituents, there are none that lack nucleic acid since this is their genetic material. The simplest viruses consist solely of nucleic acid within a protein coat. *See* CARBOHYDRATE; LIPID; NUCLEIC ACID; PROTEIN.

The protein components of viruses are somewhat acidic, have the usual amino acids found in all proteins, and are largely responsible for the morphology and serological properties of the viruses. A significant feature of viral protein structure is the presence of repeating subunits. Thus, the protein component of tobacco mosaic virus appears to consist of about 2100 identical peptide chains, each having a molecular weight of 17,531. While the proteins of different viruses have distinctive compositions, the proportions of amino acids in virus strains, also called variants or mutants, may or may not be different.

Viruses usually contain either RNA or DNA located toward the interior of the virus particles and more or less closely associated with the peptide chains of the protein. The amounts of nucleic acid range from less than 1% RNA in influenza virus to over 50% DNA in T2 bacteriophage. Different viruses (for example, poliovirus versus tobacco mosaic virus) usually have distinctive proportions of purine and pyrimidine components in their nucleic acids, whereas mutant strains of one virus commonly differ so slightly in proportions of these components that it is very difficult to show by analytical methods that such differences exist. Yet change of a single purine or pyrimidine produces mutants having profound biological consequences which are recognized by the different symptoms they elicit in a given host.

Reproduction and mutation. It appears that a single virus particle can initiate the process of virus multiplication. However, duplication is not by growth and division but by production of virus parts and their subsequent assembly. This process can be initiated by the nucleic acid alone if it can penetrate the cell since nucleic acid is the genetic substance of the virus. In fact, the process of infection begins either by injection of viral nucleic acid into a cell (the case with many bacterial viruses) or by engulfment of whole virus and uncoating of the nucleic acid (animal and plant viruses). Once inside the cell, the viral nucleic acid competes with the genetic material of the host for control of cell processes and interacts with it in various ways, depending on the particular virus. Some bacterial viruses induce synthesis of an enzyme that destroys the host nucleic acid. Thereafter synthesis in the cell is controlled solely by the viral nucleic acid and is directed toward synthesis of new virus particles. In other instances, host material is not destroyed and viral induced functions are supplemented by those of the host nucleic acid with the resultant production of new virus particles. In still other cases, all or part of the viral nucleic acid appears to be physically inserted into the host nucleic acid by a breakage and reunion phenomenon. In some bacterial and animal tumor viruses, the viral nucleic acid may function to some extent to produce certain specific proteins, but whole new virus particles are not produced.

When whole virus is produced in infected cells, it appears to occur by assembly from pools of virus parts that were synthesized under the direction of the viral nucleic acid. Such synthesis occurs in the nucleus, in the cytoplasm, or in both, depending upon the virus. Viral parts are usually assembled into characteristic particles in the cytoplasm, and then are released into the surroundings by a disruptive process (lytic or cytocidal effect). However, some viruses are only completed at the cell membrane, then released by a budding or extrusion process which does not result in the cell's death. [C.A.K.]

Virus, defective A virus that by mutation has lost the ability to be replicated in the host cell without the aid of a helper virus. The virus particles (virions) contain all the viral structural components; they can attach, penetrate, and release their nucleic acid [ribonucleic acid (RNA), deoxyribonucleic acid (DNA)] within the host cell. However, since the mutation has destroyed an essential function, new virions will not be made unless the cell was simultaneously infected with the helper virus, which can provide the missing function. Only then will the cell produce a mixed population of new helper and defective viruses. Occasionally, when their nucleic acids become integrated in the DNA of the host cell, defective viruses persist in nature by propagation from mother cell to daughter cell.

The most important group of defective viruses are deletion mutants. They are derived from their homologous nondefective (wild-type) virus through errors in the nucleic acid replication, which result in the deletion of a fragment in the newly synthesized molecules. The defective nucleic acid must be capable of self-replication, at least in the presence of the wild-type virus, and must combine with other viral components to form a particle in order to exit the cell. The defective RNA tumor viruses are deletion mutants. *See* TUMOR VIRUSES.

Certain deletion mutants, which interfere with the replication of the helper virus, are called defective interfering particles. Cells infected with a mixture of a standard virus and its defective interfering particles produce much less of the standard virus than the same type of cells infected with the standard

Occurrence of defective interfering particles in RNA viruses*

Virus group	Member name
Negative strand	
Rhabdo	Vesicular stomatitis
	Rabies, bovine ephemeral fever, others
Paramyxo	Sendai
	Newcastle disease
	Measles
	Mumps
Orthomyxo	Influenza
	Fowl plague
Arena	Lymphocytic choriomeningitis, Amapair, Parana, Pichinde
	Tacaribe
Bunya	Bunyavera
	LaCrosse
	Turlock
Positive strand	
Picorna	Poliovirus
	Mengovirus
Calici	Feline calcivirus
Toga	Sindbis, Semliki Forest, West Nile
	Japanese encephalitis
Corona	Mouse hepatitis
Double strand	
Reo	Reovirus
Orbi	Bluetongue virus of sheep
Fungal	Yeast-killer, others
Other	Infectious pancreatic necrosis virus

*Defective interfering particles were also observed in DNA viruses of the Papova and Herpes groups. Only one type of adenovirus (adeno-12) seems to generate these particles.
SOURCE: J. Perrault, *Curr. Topics Microbiol. Immunol.*, vol. 93, 1981.

virus in the absence of defective interfering particles. Defective interfering particles have been found in most of the major groups of viruses (see table). The widespread occurrence of defective interfering particles has prompted the suggestion that they must perform an important role in the evolution of viral infection. *See* ANIMAL VIRUS; MUTATION; VIRULENCE; VIRUS INFECTION, LATENT, PERSISTENT, SLOW; VIRUS INTERFERENCE. [M.E.Re.]

Virus classification There is no evidence that viruses possess a common ancestor or are in any way phylogenetically related. Nevertheless, classification along the lines of the Linnean system into families, genera, and species has been partially successful. Based on the organisms they infect, the first broad division is into bacterial, plant, and animal viruses. Within these classes, other criteria for subdivision are used. Among these are: general morphology; envelope or the lack of it; nature of the genome (DNA or RNA); structure of the genome (single- or double-stranded, linear or circular, fragmented or nonfragmented); mechanisms of gene expression and virus replication (positive- or negative-strand RNA); serological relationship; host and tissue susceptibility; pathology (symptoms, type of disease). Information about many of these parameters has been available only since the 1950s, and was acquired after the introduction of tissue culture laboratory techniques.

The families of animal viruses are sometimes divided into viruses that multiply in vertebrates and in invertebrates, but many families have members that multiply in either, and some families contain members that multiply in both types of hosts. Families are sometimes subdivided into subfamilies; the suffix -virinae may then be used. The subgroups of a family or subfamily are equivalent to the genera of the Linnean classification. The animal DNA viruses are divided into five families: Poxviridae, Herpesviridae, Adenoviridae, Papovaviridae, and Parvoviridae. RNA animal viruses may be either single-stranded or double-stranded. The single-stranded are further subdivided into positive-strand and negative-strand RNA viruses, depending on whether the RNA contains the messenger RNA (mRNA) nucleotide sequence or its complement, respectively. Further, the RNA genes may be located on one or several RNA molecules (nonfragmented or fragmented genomes, respectively). The positive-strand RNA animal viruses contain six families: Picornaviridae, Calciviridae, Coronaviridae, Togaviridae, Retroviridae, and Nodamuraviridae. The nucleocapsid of negative-strand RNA animal viruses contains an RNA-dependent RNA polymerase required for the transcription of the negative strand into the positive mRNAs. Virion RNA is neither capped nor polyadenylated. The group is divided into five families: Arenaviridae, Orthomyxoviridae, Paramyxoviridae, Rhabdoviridae, and Bunyaviridae. The double-stranded RNA animal viruses contain only one group, the Reoviridae. *See* ANIMAL VIRUS.

Bacterial viruses are also known as bacteriophages or phages. They may be tailed or nontailed. Nontailed phages are further subdivided into those with envelopes and those without. Tailed phages, which do not have envelopes, are divided into three families: Myoviridae, Styloviridae, and Pedoviridae. The group of nontailed DNA bacteriophages contains seven families, each with a distinctive morphology: Tectiviridae, Corticoviridae, Inoviridae, Microviridae, Leviviridae, Plasmaviridae, and Cystoviridae. Only the latter two families have envelopes. *See* BACTERIOPHAGE.

Plant viruses are divided into groups, rather than families, except those which belong to families of rhabdoviridae and reoviridae. The group, and correspondingly subgroup and type, can be viewed as analogous to family, genus, and species, respectively. Most common among plant viruses are those with a single-stranded, capped but not polyadenylated, positive-strand RNA. There are two groups of DNA plant viruses, one double- and one single-stranded. There are also two groups of enveloped plant viruses. *See* PLANT VIRUSES AND VIROIDS. [M.E.Re.]

Virus infection, latent, persistent, slow
Inapparent infections which are chronic and in which a virus-host equilibrium is established. The adjective latent is best reserved to qualify the noun, infection; the term latent virus should be avoided. Slow and persistent viruses include several members of different recognized taxonomic groups of viruses, as well as a number of agents not yet well enough understood or characterized to be taxonomically placed. They are intermittently or continuously present in the infected animal, and they may cause diseases usually of a chronic and degenerative nature, often associated with the central nervous system.

There are many examples of natural viral infections of animals, plants, bacteria, and insects, as well as humans, in which there are no obvious evidences of injury or illness in the host. At one end of the scale are infections such as the inapparent or subclinical forms of poliomyelitis, influenza, or yellow fever; no illness is manifested, although laboratory tests can easily show the virus to be present and multiplying in the host. The virus presumably does not persist in the host beyond a few days or a few weeks at most. In contrast, after a primary infection—usually in childhood—type 1 herpes simplex virus persists in a latent infection for long periods, sometimes for life, in the presence of antibodies. The virus adversely affects the individual only when nonspecific factors, such as fever, precipitate attacks of cold sores. *See* HERPES; INFLUENZA; POLIOMYELITIS; VIRUS; YELLOW FEVER. [J.L.Me.]

Virus interference A phenomenon which may be defined as protection of host cells against one virus, conferred as a result of prior infection with a different virus. Interference between viruses has been observed in humans, in laboratory animals, and in tissue culture systems. *See* ANIMAL VIRUS; PLANT VIRUSES AND VIROIDS.

Interference is believed to act in one of two ways: (1) The first virus may inactivate surface receptors of the cell and so make them unavailable to the second virus; or (2) the cell materials or enzymes necessary for the growth of the second virus may be taken over by or directed by the first virus.

Interference occurs between related as well as unrelated viruses. It may be unilateral; for example, one virus may prevent infection with a second, but the second virus may not prevent infection with the first. In some cases, inactivated virus may interfere with subsequent infection by live virus. *See* IMMUNOLOGY. [J.L.Me.]

Viscacha Any of four species of viscacha belonging to two genera, *Lagidium* (three species) and *Lagostomus*. The mountain viscacha (*Lagidium viscaccia*) lives in burrows near rocks in the foothills of the Andes. It is larger than the chinchilla and has a poor-quality fur, which is woven with wool into cloth. The plains viscacha (*Lagostomus maximum*) inhabits burrows in the pampas region of Argentina. It is exclusively herbivorous. *See* CHINCHILLA; RODENTIA. [C.B.C.]

Viscosity The resistance that a gaseous or liquid system offers to flow when it is subjected to a shear stress. Viscosity is a measure of the internal friction that arises when there are velocity gradients within the system. For fluids (gases and liquids) its meaning is conceptually and operationally well defined.

Simple gases typically have viscosities in the range of 100 to 200 micropoise at standard temperature and pressure (273 K, 1 atm or 101,325 pascals), whereas simple liquids under the same conditions have coefficients of viscosity about two orders of magnitude larger. The flow characteristics of gases and simple liquids such as water, carbon tetrachloride, and ethyl alcohol are typically those of newtonian fluids. Aqueous suspensions, such as clays, gelatin, and agar, are termed non-newtonian fluids because their viscosities may depend upon the rate of shear and prior treatment. Hydrophilic sols often form extended networks involving water, and their non-newtonian behavior is believed to be due to the breakdown of their structure under shear. *See* FLUID FLOW; LIQUID; NEWTONIAN FLUID; NON-NEWTONIAN FLUID.

The origin of internal friction (viscosity) at the molecular level is the net transfer of momentum between layers of fluid moving with different velocities in parallel flow by the mechanism of molecular collisions. In this process the directed energy of fluid flow is degraded to random thermal energy (heat). It was one of the early triumphs of the kinetic theory of gases that established the relationship between the viscosity of a hard-sphere gas and the mean speed of its molecules. *See* KINETIC THEORY OF MATTER.

Momentum transfer between shearing layers also underlies the viscous behavior of simple liquids, but since the mean free path has little meaning for liquids, no simple relation exists for them. In contrast to the behavior of gases, temperature decreases the viscosities of simple liquids and its effect is much larger. *See* LIQUID; RHEOLOGY. [N.H.N.]

Viscous impingement air filter A type of air filter which is made up of a relatively loosely arranged medium, usually consisting of spun glass fibers, metal screens, or layers of crimped expanded metal. The surfaces of the medium are coated with a tacky oil, generally referred to as an adhesive. The arrangement of the filter medium is such that the airstream is forced to change direction frequently as it passes through the filter. Solid particles, because of their momentum, are thrown against, and adhere to, the viscous coated surfaces. Larger airborne particles, having greater mass, are filtered in this manner, whereas small particles tend to follow the path of the airstream and escape entrapment. [M.A.B.]

Vision The sense of sight, which perceives the form, color, size, movement, and distance of objects. Of all the senses, vision provides the most detailed and extensive information about the environment. *See* VISUAL IMPAIRMENT.

Human vision is most sensitive for light comprising the visible spectrum in the range 380–720 nanometers in wavelength. Sunlight and common sources of artificial light contain substantially all wavelengths in this range but each source has a characteristic spectral energy distribution. In general, light stimuli can be measured by physical means with respect to their energy, dominant wavelength, and spectral purity. These three physical aspects of the light are closely related to the perceived brightness, hue, and saturation, respectively. *See* COLOR; LIGHT.

Anatomical basis for vision. The anatomical structures include the eyes, optic nerves and tracts, optic thalamus, and visual cortex. The eyes are motor organs as well as sensory; that is, each eye can turn directly toward an object to inspect it. The two eyes are coordinated in their inspection of objects, and they are able to converge for near objects and diverge for far ones. Each eye can also regulate the shape of its crystalline lens to focus the rays from the object and to form a sharp image on the retina. Furthermore, the eyes can regulate the amount of light reaching the sensitive cells on the retina by contracting and expanding the pupil of the iris. *See* EYE (VERTEBRATE).

The process of seeing begins when light passes through the eye and is absorbed by the sensitive cells of the retina. These cells are activated by the light in such a way that electrical potentials are generated. Some of these potentials are the signs of electrochemical activity that serves to generate nerve responses in various successive neural cells in the vicinity of excitation. Finally, impulses emerge from the eye in the form of repetitive discharges in the fibers of the optic nerve. The optic nerves from the two eyes traverse the optic chiasma. The fibers from the inner (nasal) half of each retina cross over to the opposite side, while those from the outer (temporal) half do not cross over but remain on the same side.

The visual cortex includes a projection area in the occipital lobe of each hemisphere. Here there appears to be a point-for-point correspondence between the retina of each eye and the cortex. Thus the cortex contains a "map" or projection area, each point of which represents a point on the retina and therefore a point in visual space as seen by each eye. But the map is much too simple a model for cortical function. Vision tells much more than the location at which an object is seen. Visual tests show that other important features of an object such as its color, motion, orientation, and shape are simultaneously perceived.

Scotopic and photopic vision. Night animals, such as the cat or the owl, have eyes that are specialized for seeing with a minimum of light. This type of vision is called scotopic. Day animals such as the horned toad, ground squirrel, or pigeon have predominantly photopic vision. They require much more light for seeing, but their daytime vision is specialized for quick and accurate perception of fine details of color, form, and texture, and location of objects. Color vision, when it is present, is also a property of the photopic system. Human vision is duplex; humans are in the fortunate position of having both photopic and scotopic vision. Some of the chief characteristics of human scotopic and photopic vision are enumerated in the table.

Scotopic vision. This occurs when the rod receptors of the eye are stimulated by light. The outer limbs of the rods contain a photosensitive substance known as visual purple or rhodopsin. This substance is bleached away by the action of

Characteristics of human vision

Characteristic	Scotopic vision	Photopic vision
Photochemical substance	Rhodopsin	Cone pigments
Receptor cells	Rods	Cones
Speed of adaptation	Slow (30 min or more)	Rapid (8 min or less)
Color discrimination	No	Yes
Region of retina	Periphery	Center
Spatial summation	Much	Little
Visual acuity	Low	High
Number of receptors per eye	120,000,000	7,000,000
Cortical representation	Small	Large
Spectral sensitivity peak	505 nm	555 nm

strong light so that the scotopic system is virtually blind in the daytime. Weak light causes little bleaching but generates neural inhibitory signals that lower the overall sensitivity of the eye. In darkness, however, the rhodopsin is regenerated by restorative reactions based on the transport of vitamin A to the retina by the blood. One experiences a temporary blindness upon walking indoors on a bright day, especially into a dark room or dimly lighted theater. As the eyes become accustomed to the dim light the scotopic system gradually begins to function. This process is known as dark adaptation. *See* VITAMIN A.

Photopic vision. Normal photopic vision has the characteristics enumerated in the table. Emphasis is placed on the fovea centralis, a small region at the very center of the retina of each eye.

Foveal vision is achieved by looking directly at objects in the daytime. The image of a small object falls within a region almost exclusively populated by cone receptors. These are closely packed together in the central fovea. Furthermore, each of the cones in the fovea is provided with a series of specialized nerve cells that process the incoming pattern of stimulation and convey it to the cortical projection area.

Peripheral vision takes place outside the fovea centralis. As an example, look directly at a single letter at the center of a printed page. This letter, and a few letters immediately adjacent to it, appear clear and black because they are seen with foveal vision. The rest of the page is a blur in which the lines of print are seen as gray streaks.

There is a simple anatomical explanation for the clarity of foveal vision as compared to peripheral vision. The cone receptors become less and less numerous in the retinal zones that are more and more remote from the fovea. In the extreme periphery there are scarcely any cones, and even the rod receptors are more sparsely distributed. Furthermore, the plentiful neural connections from the foveal cones are replaced in the periphery by network connections in which hundreds of receptors may activate a single optic nerve fiber. This mass action is favorable for the detection of large or dim stimuli in the periphery or at night, but it is unfavorable for visual acuity or color vision, both of which require the brain to differentiate between signals arriving from closely adjacent cone receptors.

Space and time perception. Spatial and temporal effects are clearly apparent in the sense of sight. These two effects enable the individual to become oriented with regard to space and time in the external environment, especially in the perception of motion and distance. *See* PERCEPTION.

Elementary forms of space perception are vernier and stereoscopic discrimination. Here, the eye is required to judge the relative position of one object in relation to another (see illustration). The left eye, for example, sees the lower line as displaced slightly to the right of the upper. This is known as vernier discrimination. If the right eye is presented with similar lines that are oppositely displaced, then the images for the two eyes appear fused into one and the subject sees the lower line as nearer than the upper. This is the principle of the stereoscope. The distance judgment is made not at the level of the retina but at the cortex where the spatial patterns from the separate eyes are fused together. The fineness of vernier and stereoscopic discrimination transcends that of the retinal mosaic and suggests that some averaging mechanism must be operating in space or time or both.

The temporal characteristics of vision are revealed by studying the responses of the eye to various temporal patterns of stimulation. When a light is first turned on, there is a vigorous burst of nerve impulses that travel from the eye to the brain. Continued illumination results in fewer and fewer impulses as the eye adapts itself to the given level of illumination. Turning the light off elicits another strong neural response. Afterimages are often seen at this time. A positive afterimage, resembling the original stimulus, is sometimes seen during the first fraction of a second after the light goes off. This is usually followed by a longer-lasting negative afterimage in which the color of the original object appears to be reversed. A clear afterimage may be produced by staring fixedly at the stimulus for at least a half minute, then turning away and "projecting" the afterimage against a white or gray screen. The afterimage is thought to arise chiefly from the fact that the affected region of the retina has been bleached, or otherwise changed photochemically, in such a way that its responses are different from those of the regions not stimulated. To some extent, however, visual aftereffects are due to central, rather than retinal, processes. *See* COLOR VISION. [L.A.R.]

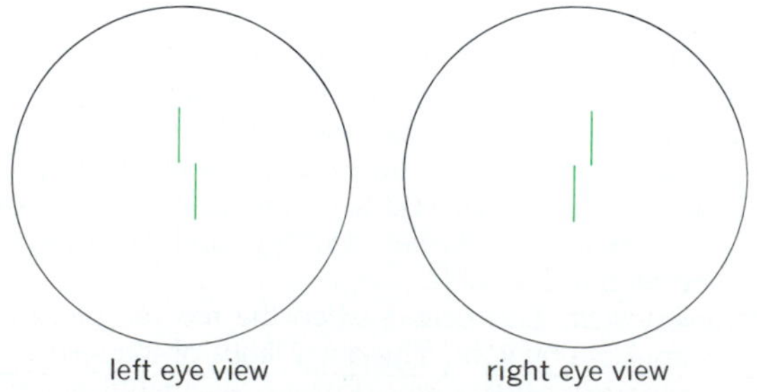

Vernier and stereoscopic discriminations of space.

Visual disorders Vision is a complex process. If the structures within the eye are not properly aligned, or if the eye is too long or too short, or if the crystalline lens is too strong or too weak, certain conditions result. *See* VISION.

Hyperopia (farsightedness) results from too short an eyeball so that parallel rays from infinity, that is, from a distance of 20 ft (6 m) or beyond, come to a focus behind the retina. The condition can also occur because a crystalline lens is underpowered and thus does not bend the rays enough to focus on the retina.

Presbyopia is the progressive loss of ability to accommodate to near vision as a result of aging processes. It usually becomes apparent in the 40–45 age group.

Myopia (nearsightedness) results either from an eye that is too long or from a crystalline lens that is too strong, resulting in a focus of parallel light in front of the retina.

In astigmatism the parallel rays of light do not come to a single point, but rather have a variable focus because the cornea is not spherical and refracts light in different meridians at different distances. It is rare for the human cornea to be completely spherical, so that all people have some degree of astigmatism, and a certain amount is termed physiological. [J.Hart.]

Eyeglass is the general term for optical devices containing corrective lenses for defects in vision or for special purposes.

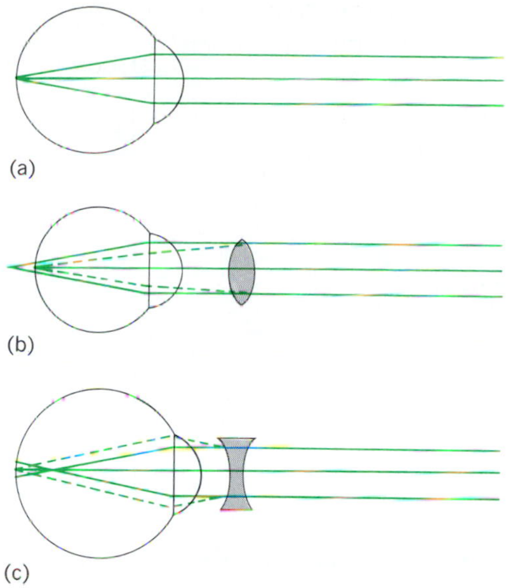

Course of light rays and formation of retinal image in the eye and correction by appropriate lenses. (*a*) Emmetropia (normal). (*b*) Hyperopia. (c) Myopia. (*After B. A. Houssay et al., Human Physiology, 2d ed., McGraw-Hill, 1955*)

Presbyopia may be corrected by convex lenses, usually worn only for close work or reading. Bi-, tri-, or tetrafocal lenses are used under different circumstances to permit rapid shifting from one visual distance to another in failure of accommodation.

Hyperopia may be corrected by accommodation, particularly in young people, but the increased accommodation required for close vision often leads to eyestrain. Convex lenses are used to gain proper focus (illus. *b*).

For myopia, concave lenses are used to permit focusing of far vision (illus. *c*); near vision is often better than that of the normal eye since little accommodation is necessary, less eyestrain is produced with close work, and the retinal image is somewhat larger than in the normal eye.

Cylindrical lenses are ground to correct astigmatism. Most astigmatics also have other defects so that compound lenses are required.

Other lens systems are commonly devised for persons who have unequal vision and to correct for abnormal visual axes and defects in ocular movements caused by muscle imbalance.

Special forms of eyeglasses are made to absorb portions of the light spectrum, such as certain colors and ultraviolet and infrared rays. Aviators, welders, radiologists, furnace workers, and others require lenses of certain physical or optical characteristics. *See* LENS (OPTICS). [E.G.St./N.K.M.]

Visual impairment Abnormal visual acuity. The term is used to describe visual acuity substantially less than normal. The World Health Organization defines visual impairment as acuity less than 20/60 (normal being 20/20). The legal definition of blindness in the United States is visual acuity of 20/200 or worse (or severely restricted peripheral vision). The World Health Organization defines blindness as visual acuity worse than 20/400. Visual impairment and blindness increase substantially with age. The major causes of blindness differ substantially by race. Cataracts, which involve opacification of the normally clear lens, and diabetic retinopathy, which is an accumulation of fluid or the growth of abnormal blood vessels in the retina (most commonly in insulin-dependent diabetics), are important causes of blindness. A large proportion of people having blindness caused by cataracts can be treated by surgery. A growing proportion of blindness caused by diabetic retinopathy can be prevented by laser surgery.

Other important causes of blindness are glaucoma and macular degeneration. Early detection of glaucoma requires routine, careful screening and examination, since many patients remain asymptomatic until much of the optic nerve is destroyed. Macular degeneration involves atrophy of that portion of the retina responsible for fine (reading) vision. People with macular degeneration rarely suffer the total blindness of advanced glaucoma or diabetic retinopathy. *See* DIABETES.

Visual problems are far more common in developing countries, where there are limited resources for dealing with problems that are otherwise readily treated, such as cataract, or prevented, such as trachoma or xerophthalmia ("blinding malnutrition" caused by vitamin A deficiency). *See* CATARACT; GLAUCOMA; VISION. [A.So.]

Vitamin An organic compound essential for the normal functioning of the body, and usually obtained from foods. Vitamins are present in food in minute quantities compared to the other utilizable components of the diet—proteins, fats, carbohydrates, and minerals.

Synthetic and natural vitamins usually have the same biological value. The vitamins are often not related chemically or functionally. They are conventionally divided into a fat-soluble group consisting of vitamins A, D, E and K, and a water-soluble group consisting of the various B vitamins, including thiamine, riboflavin, nicotinic acid, B_6, pantothenic acid, biotin, folic acid, B_{12}, *p*-aminobenzoic acid, inositol, and choline, as well as vitamin C, or ascorbic acid. The vitamins, particularly the water-soluble ones, are found spread almost universally throughout the animal and plant kingdoms (see table).

Vitamin sources and deficiencies

Vitamin*	Best sources	Deficiency and areas primarily affected
	Fat-soluble	
α-, β-carotene	Fish-liver oils, green plants, carrots	Xerophthalmia, nyctalopia; eyes, skin, mouth, respiratory and urogenital tracts
D	Fish oils	Rickets, osteomalacia; bones, teeth
E	Grains and vegetable oils	Reproductive tract, muscles, red blood cells, liver, brain
K	Green vegetables	Blood prothrombin
	Water soluble	
Thiamine	Pork, liver, whole grains	Beriberi; brain, nerves, heart
Riboflavin	Milk, egg white, liver, leafy vegetables	Skin, mouth, eyes, liver, nerves
Niacin	Yeast, wheat germ, meats	Pellagra; gastrointestinal tract, skin, brain
Pantothenic acid	Liver, kidneys, green vegetables, egg yolk	Adrenals, kidneys, skin, brain, spinal cord
B_6	Whole grains, yeast, egg yolk, liver	Skin, red blood cells, brain, kidneys, adrenals
Inositol	Whole grains, liver	
Biotin	Liver, kidney, yeast	Skin, muscles
Choline	Egg yolk, brain, grains	Liver, kidneys, pancreas
p-Aminobenzoic acid	Yeast	
Folic acid	Liver, deep green leafy vegetables	Various macrocytic anemias; red blood cells
B_{12}	Liver, meats	Pernicious anemia; red blood cells
Ascorbic acid	Citrus fruits, fresh vegetables, potatoes	Scurvy; bones, joints, mouth, capillaries

*See individual articles on each vitamin, except α-, β-carotene.

The vitamins function as coenzymes, active parts of enzyme systems, which catalyze many of the various anabolic and catabolic reactions of living organisms necessary for the production of energy, the synthesis of tissue components, hormones, and chemical regulators, and the detoxification and degradation of waste products and toxins. Because of their role in metabolism, the vitamins are generally concentrated in those animal and plant tissues which are most active metabolically. Thus, liver or kidneys are a more potent source of vitamins than muscle or skin, and the germ of a seed contains more vitamins than in other parts. *See* COENZYME; NUTRITION. [S.N.G.]

Vitamin A A pale-yellow alcohol, soluble in fat but not in water. It is readily destroyed by oxidation, which causes significant losses during storage and cooking.

There are three sources of vitamin A activity. Vitamin A itself is found in all animal tissues. It is particularly concentrated in the liver and viscera, and the most important natural sources of the vitamin are the fish-liver oils. Plants contain a number of carotenoid pigments, which can be converted to vitamin A in the tissues of animals such as in the rat intestinal wall. *See* CAROTENOID.

In vitamin A deficiency the epithelial tissues of many organs are affected. Growth failure occurs, particularly in the bones and teeth. Besides severe inflammation of the eyes, vitamin A-deficient animals suffer from night blindness, a condition known as nyctalopia. The role of vitamin A in sight is related to the retinal pigment called rhodopsin. Under the stimulus of light this pigment breaks down. Resynthesis of rhodopsin, necessary for normal vision, is poor in vitamin A deficiency.

Approximately 20 IU (international units) of vitamin A per kilogram of body weight will support growth and prevent symptoms of deficiency. In estimating human requirements, it has been assumed that two-thirds of the vitamin A consumed is in the form of carotenoid pigments. There is evidence that it is of value to ingest considerably more vitamin A than the amount necessary to prevent deficiency signs. The recommended dietary allowances of the National Research Council are approximately twice the minimal requirements. *See* VITAMIN. [S.N.G.]

Vitamin B_6 A vitamin which exists as three chemically related and water-soluble forms found in food; pyridoxine, pyridoxal, and pyridoxamine. All three forms have equal activity for animals and yeast.

Vitamin B_6 deficiency is accompanied by poor growth, dermatitis, microcytic anemia, epileptiform convulsions, and kidney and adrenal lesions. There is evidence that some women in the third trimester of pregnancy may have a special requirement for vitamin B_6 in that its administration often relieves the nausea of pregnancy. Some types of human dermatitis respond to local application of this vitamin.

It is difficult to set requirements for vitamin B_6, since no single set of assay conditions or criteria has received universal acceptance. Adults probably require about 1.5–2 milligrams per day, and a dietary intake of 0.4 milligrams per day would probably be satisfactory for most infants. *See* VITAMIN. [S.N.G.]

Vitamin B_{12} A group of closely related polypyrrole compounds, with the structure shown, containing trivalent cobalt; it is often called cobalamin, The vitamin is a dark-red crystalline compound; in aqueous solution and at room temperature it is most stable at pH 4–7.

Vitamin B_{12} deficiency in animals is characterized primarily by anemia and neuropathy. In humans, this deficiency is called pernicious anemia. People suffering from this disease lack a factor secreted in gastric juice which, by affecting absorption directly and by protecting vitamin B_{12} from intestinal destruction, enables the vitamin to be absorbed. *See* ANEMIA.

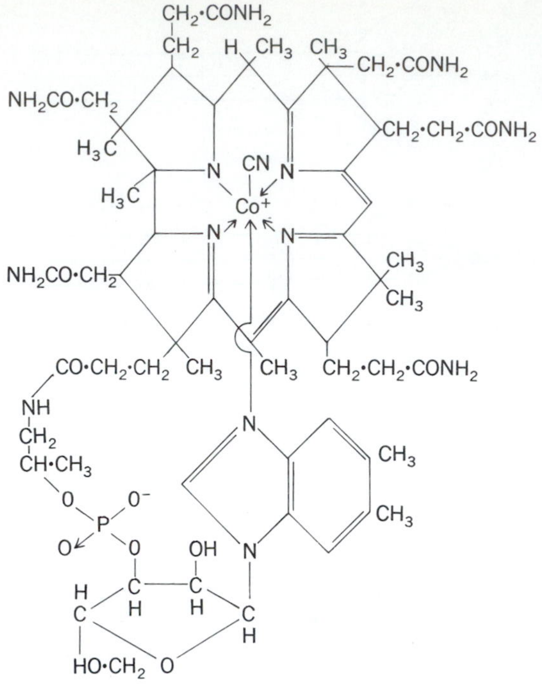

Requirements for vitamin B_{12} are increased by reproduction or hyperthyroidism. Of the known vitamins B_{12} is the most active biologically. A daily injection of 1 microgram of vitamin B_{12} will prevent the recurrence of symptoms in people with pernicious anemia. For normal people a diet containing 3–5 μg per day (providing 1–1.5 μg absorbed) will satisfy vitamin B_{12} requirements. *See* VITAMIN. [S.N.G.]

Vitamin D Either of two chemically similar fat-soluble, sterol-like compounds, calciferol, or vitamin D_2, and activated 7-dehydrocholesterol, or vitamin D_3.

In animals 7-dehydrocholesterol is secreted at the surface of the skin, where it is activated by sunlight, and the vitamin D_3 formed is reabsorbed. Vitamin D_2 is prepared by the ultraviolet irradiation of ergosterol, a vitamin D precursor found in plants. Most naturally occurring foods have little vitamin D activity. Fish oils are an exception, being highly potent sources of the vitamin. Children should be given vitamin D supplements; much of the milk marketed in the United States has been enriched with 400 IU per quart. (An international unit, or IU, of vitamin D is equal to 0.025 microgram of vitamin D_3.)

Vitamin D deficiency in growing animals is called rickets. In adult animals, the disease is called osteomalacia. Vitamin D deficiency is associated with skeletal pathology. Bones and teeth are soft and fracture easily. *See* RICKETS.

The recommended daily dietary allowance of the National Research Council for children and pregnant or lactating women is 400 IU. Normal adults can probably depend on sunshine and the small amount of the vitamin from their food to meet their needs. *See* VITAMIN. [S.N.G.]

Vitamin E A group of compounds called tocopherols. There are eight of these, each differing slightly in its chemical configuration (see illustration). The tocopherols are fat-soluble. They are widespread in nature and are particularly concentrated in vegetable oils. They are stable to heat, acids, and alkalies, but they are easily destroyed by ultraviolet light or oxidizing agents.

Vitamin E deficiency in animals has often been associated with reproductive failure and irreversible testicular damage. In vitamin E deficiency, dystrophy of both striated and cardiac

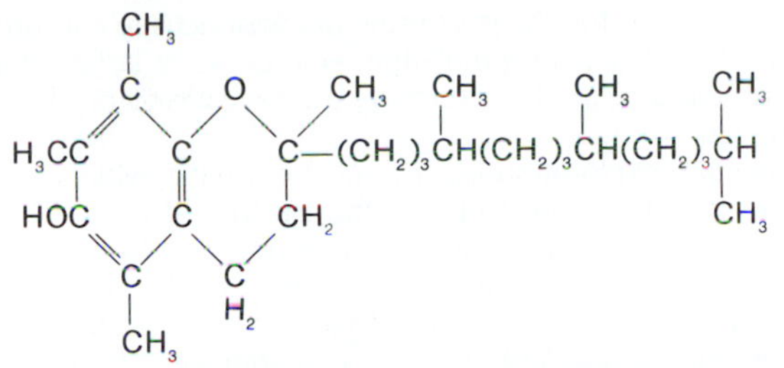
Structural formula for α-tocopherol.

muscles occurs, and death often occurs as a result of cardiac failure. Vitamin E therapy has not been of value in treating muscular dystrophy in humans.

There are few good data concerning the vitamin E requirements of humans. Adult requirements may vary between 10 and 30 mg per day. *See* VITAMIN. [S.N.G.]

Vitamin K A naturally occurring vitamin consisting of two chemically similar yellowish oils known as K_1 and K_2, which are fat-soluble, nonsteroid, nonsaponifiable, and unstable to light.

The only major effect of a vitamin K deficiency in animals is a decrease in blood prothrombin, resulting in prolonged clotting time. Thus, vitamin K deficiency results in subcutaneous and intramuscular hemorrhages.

Vitamin K is so widespread in nature in green leafy vegetables, and its synthesis by intestinal microorganisms is so great that vitamin K deficiency is rare. Liver disease, which may result in decreased prothrombin formation and bile secretion, can result in vitamin K deficiency, because bile secretion is necessary for its absorption. Intestinal disease or the ingestion of large amounts of mineral oil or antibiotics can also affect vitamin K economy. *See* VITAMIN. [S.N.G.]

Vitellogenesis The process by which yolk is formed. In most organisms, the early embryo cannot manufacture its own nutritive material. In almost all instances, nutritive substances called yolk are deposited within the ooplasm of the oocyte during its growth. The kinds and amounts of yolk within eggs vary greatly within the animal kingdom. In some animals, such as echinoderms and mammals, the yolk is produced in small quantities, whereas in others, such as sharks and birds, large quantities are stored in the egg. The formed bodies or platelets composing the yolk of a wide variety of eggs contain protein, lipid, and carbohydrate (including glycogen). [E.An.]

Vivianite A mineral of the vivianite group, other important members of which are annabergite and erythrite.

Vivianite is a hydrated ferrous phosphate, $Fe_3(PO_4)_2 \cdot 8H_2O$; usually ferric iron is present as the result of oxidation. It crystallizes in the monoclinic system, with crystals generally prismatic. Vivianite also occurs in earthy form and as globular and encrusting masses of fibrous structure. Crystals are colorless and transparent when fresh. Oxidation changes the color progressively to pale blue, greenish blue, dark blue, or bluish black. [W.R.Lo.]

Vocal cords The pair of elastic, fibered bands inside the human larynx. The cords are covered with a mucous membrane and pass horizontally backward from the thyroid cartilage (Adam's apple) to insert on the smaller, paired arytenoid cartilages at the back of the larynx. The vocal cords act as sphincters for air regulation and may be vibrated to produce sounds. Separation, approximation, and alteration of tension are produced by action of laryngeal muscles acting on the pivoting arytenoids. Vibration of the cords produces fundamental sounds and overtones. These can be modified by the strength of the air current, the size and shape of the glottis (the opening between the cords), and tension in the cords. *See* LARYNX; SPEECH. [W.J.B.]

Voice response The process of generating an acoustic speech signal that communicates an intended message so that a machine can respond to a request for information by talking to the human user. Communication of speech from the machine to the human requires abilities to convert machine information into a desired word sequence, with the pronunciations of all desired words carefully sequenced and inflected to produce the natural flow of continuous speech waves.

Talking machines permit natural interactions that can be valuable for announcing warnings, reporting machine status, or otherwise informing the computer user, especially when the user cannot view displays, due to concurrent visual tasks, visual handicap, or remote telephone links. The many advantages of voice interactions with machines and the expanding uses of computers in offices, factories, schools, and homes have encouraged the development, sales, and use of voice response systems as voice warning devices, cockpit advisory systems, time-of-day services, bank-by-phone facilities, talking clocks and appliances, toys that speak, and automatic readers of printed texts for the blind.

For simple tasks involving small vocabularies, machine production of speech is not difficult to achieve. Human speech can be stored and reproduced later on demand, using techniques similar to those in familiar devices like audio tape recorders, phonograph records, and other analog or digital magnetic media.

It is only a small conceptual step from such recorders to the simplest form of modern voice response system, called a word concatenation system. Acting like "automated tape recorders," word concatenation systems can retrieve previously spoken versions of words or phrases and carefully concatenate them without pauses to approximate normally spoken word sequences.

The storage of a high-quality digitized waveform may require as much as 64,000 bits per second of speech, so that large computer disk packs can store perhaps at most 1200 seconds of speech, and microcomputer memories might be able to store only hundreds, or perhaps only a few tens, of words. Low-cost concatenation systems are thus restricted to small vocabularies.

There is a general tradeoff between storage requirements and the complexity of coding and decoding of the speech. Storage requirements can be reduced substantially by first extracting informative parameters from the original human speech and later reconstructing speech from such stored parameters. Such parametrized approaches assume that not all the information in the original signal is important to conveying the message. Informative parameters totaling about 16,000 bits/s to as low as 600 bits/s can be stored as simplified instructions for controlling reconstitution of the waveform.

In a phonemic synthesizer, human speech is not the direct basis for stored representation of words or phrases. Rather, each word is abstractly represented as a sequence of expected vowels and consonants (phonemes, or phones if more detailed articulatory contrasts are included). Speech is composed or synthesized by juxtaposing the expected phonemic sequence for each word with the sequences for preceding and following words. Each phonemic unit is used as a set of instructions for the form of acoustic data to be produced at the appropriate times during the utterance.

Problems with the extensive hand-tailoring required for phonemic synthesis have prompted the development of complex rule sets for predicting the needed phonemic states directly from the input message and dictionary pronunciations. Such "synthesis-by-rule" permits virtually unlimited-vocabulary voice response. By providing automatic means to take a specification of any English text at the input and to generate a natural and intelligible acoustic speech signal at the output, a text-to-speech synthesizer represents the most ambitious form of voice response system. Without the need for intervening human sequence selection such as a phonemic synthesizer needs, and with less storage requirement than other synthesizers, the text-to-speech synthesizer is probably the primary choice for large-vocabulary speech generation. *See* PHONETICS; SPEECH RECOGNITION. [W.A.Le.]

Voigt effect An anisotropic substance placed in a magnetic field becomes birefringent (doubly refracting), and its optical properties are similar to those of a uniaxial crystal. The Faraday effect is the result of this birefringence when observations are made parallel to the magnetic field. The analogous observations perpendicular to the magnetic lines of force are more difficult and were not successfully carried out until 1898 because of the smallness of the effect. The transverse magnetooptic birefringence is called the Voigt effect (also called magnetic double refraction).

The Voigt effect can be observed only in the vicinity of sharply defined absorption lines, that is, in gases and in certain crystals having sharp lines, such as rare-earth salts. *See* FARADAY EFFECT; MAGNETOOPTICS. [G.H.Di.; W.W.W.]

Volatilization The process of converting a chemical substance from a liquid or solid state to a gaseous or vapor state. Other terms used to describe the same process are vaporization, distillation, and sublimation. A substance can often be separated from another by volatilization and can then be recovered by condensation of the vapor. The substance can be made to volatilize more rapidly either by heating to increase its vapor pressure or by removal of the vapor using a stream of inert gas or a vacuum pump. Chemical reactions are sometimes utilized to produce volatile products. Volatilization methods are generally characterized by great simplicity and ease of operation, except when high temperatures or highly corrosion-resistant materials are needed. *See* CHEMICAL SEPARATION TECHNIQUES; DISTILLATION; SUBLIMATION; VAPOR PRESSURE. [L.Go./R.W.Mu.]

Volcanic glass A natural glass formed by rapid cooling of lava. The material is opaque except on thin edges, occurs in a variety of colors, and has a conchoidal fracture. Shades of red, brown, black, gray, or green may be displayed in a uniform, banded, or variegated fashion.

Most natural glasses are chemically equivalent to rhyolite, into which they may grade. Types corresponding to trachyte, dacite, andesite, and latite are much less common. Basaltic glass generally is given the name tachylite. Varieties include obsidian, pitchstone, pumice, and perlite. *See* OBSIDIAN; PERLITE; PITCHSTONE; PUMICE; RHYOLITE.

In addition to crystallites, most volcanic glasses carry microlites (microscopically tiny crystals). Somewhat larger and usually well-formed crystals are called phenocrysts. Their presence gives the glass a porphyritic texture. They consist chiefly of quartz, alkali feldspar, and plagioclase and less commonly of mafics (biotite, hornblende, or pyroxene). As their number increases, these porphyritic glasses pass into vitrophyre. *See* LAVA; PHENOCRYST; PORPHYRY. [C.A.C.]

Volcano A mountain or hill, generally steep-sided, formed by accumulation of magma (molten rock with associated gas and crystals) extruded through openings or volcanic vents in the Earth's crust; the term volcano also refers to the vent itself. During the evolution of a long-lived volcano, a permanent shift in the locus of principal vent activity can produce a satellitic volcanic accumulation as large as or larger than the parent volcano, in effect forming a new volcano on the flanks of the old.

Planetary exploration has revealed dramatic evidence of volcanoes and their products on the Earth's Moon, Mars, and Jupiter's moons on a scale much more vast than on Earth. Only the products and landforms of terrestrial volcanic activity are described here. *See* JUPITER; MARS; MOON; VOLCANOLOGY.

Volcanic vents. Volcanic vents, channelways for magma to ascend toward the surface, can be grouped into two general types: fissure and pipelike. Magma consolidating below the surface in fissures or pipes forms a variety of igneous bodies, but that breaking the surface produces fissure or pipe eruptions. Fissures, most of them less than 10 ft (3 m) wide, may form in the summit region of a volcano, on its flanks, or near its base; pipes tend to be restricted to the summit area of a volcano. For some volcanoes or volcanic regions, swarms of fissure vents are clustered in swaths called rift zones.

Volcanic products. Magma erupted onto the Earth's surface is called lava. If the lava is chilled and solidifies quickly, it forms volcanic glass; slower rates of chilling result in greater crystallization before complete solidification. Lava may accrete near the vent to form various minor structures or may pour out in streams called lava flows, which may travel many miles from the vents. During more violent eruptions, lava torn into fragments and hurled into the air is called pyroclastic (fire-broken materials). *See* IGNEOUS ROCKS; LAVA; MAGMA; PYROCLASTIC ROCKS; VOLCANIC GLASS.

Violent volcanic explosions may throw dust high into the stratosphere, where it may drift across the surface of the globe for thousands of miles. Most of the solid fragments in the cloud settle out within a few days, and nearly all settle out within a few weeks, but some finely divided material may remain suspended in the stratosphere for more than a year. Volcanic dust and related sulfate aerosols from major explosive eruptions may screen out solar radiation sufficiently to lower the Earth's surface temperature and thus possibly affect climate on a global scale. On a local scale, the fallout from eruption plumes and associated acid rains can damage crops. *See* AIR POLLUTION.

In general, water vapor is the most abundant constituent in volcanic gases; the water is mostly of meteoric (atmospheric) origin, but in some volcanoes can have a significant magmatic or juvenile component. Excluding water vapor, the most abundant gases are the various species of carbon, sulfur, hydrogen, chlorine, and fluorine.

Mudflows are common on volcanoes where poorly indurated or nonwelded pyroclastic material is abundant. Probably by far the most common cause is heavy rain saturating a thick cover of loose unstable pyroclastic material on the steep slope of the volcano, transforming the material into a mobile, water-saturated "mud," which can rush downslope at a speed as great as 50–55 mi/h (80–90 km/h). Such a fast-moving mass can be highly destructive, sweeping up everything loose in its path.

Volcanic landforms. Much of the Earth's solid surface, on land and below the sea, has been shaped by volcanic activity. Landscape features of volcanic origin may be either positive (constructional) forms, the result of accumulation of volcanic materials, or negative forms, the result of the lack of accumulation or collapse.

Not all volcanoes show a graceful, symmetrical cone shape, such as that exemplified by Mount Fuji, Japan. In reality, most volcanoes, especially those near tectonic plate boundaries, are more irregular, though of grossly conical shape. Such volcanoes, called strato-volcanoes or composite volcanoes, typically erupt explosively. Volcanoes constructed primarily of fluid basaltic lava flows, which may spread great distances from the vents, typically are gentle-sloped, broadly upward convex structures. Such shield volcanoes, a classic example of which is Mauna Kea Volcano, Hawaii, tend to form in oceanic intraplate regions and are associated with hot-spot volcanism. Depending on magma viscosity, eruptive history, and posteruption modifications, a volcano can assume a wide variety of shapes intermediate between the simple forms of composite and shield volcanoes.

Some of the largest volcanic edifices are not shaped like the composite or shield volcanoes. In certain regions of the world, voluminous extrusions of very fluid basaltic lava from dispersed fissure swarms have built broad, nearly flat-topped accumulations. These voluminous outpourings of lava are known as flood basalts or plateau basalts. *See* BASALT.

Submarine volcanism. Volcanic eruptions in shallow water are very similar in character to those on land but, on average, are probably somewhat more explosive, owing to heating of hot water and resultant violent generation of supercritical steam. Much of the ocean basins appears to be floored by basaltic lava. *See* MARINE GEOLOGY; OCEANIC ISLANDS.

Fumaroles and hot springs. Vents at which volcanic gases issue without lava or after the eruption are known as fumaroles. They are found on active volcanoes during and between eruptions and on dormant volcanoes. Fumaroles grade into hot springs and geysers. The water of most, if not all, hot springs is predominantly of meteoric origin, and is not water liberated from magma. *See* GEOTHERMAL POWER; GEYSER.

Distribution of volcanoes. About 500 active volcanoes are known on the Earth, mostly along or near the boundaries of the dozen or so lithospheric plates that compose the Earth's solid surface. These rigid plates, which range in thickness from 30 to 90 mi (50 to 150 km) and consist of both crustal and upper mantle material, form the lithosphere and move relative to one another above a hotter, more plastic zone in the mantle called the asthenosphere. *See* ASTHENOSPHERE; LITHOSPHERE; PLATE TECTONICS. [G.A.M./R.I.T.]

Volcanology

Volcanology The scientific study of volcanic phenomena. Strictly speaking, it refers only to the surface eruption of magmas and related gases, and the structures, deposits, and other effects produced thereby. Commonly, however, it includes much of plutonic igneous geology, since effects at the Earth's surface are the result of events at depth. For a discussion of the surface structures and deposits which result from the activity of volcanos *see* VOLCANO.

Composition of magma. As formed at depth, magma consists of liquid rock containing dissolved gases. Rising toward the surface, it enters zones of lower temperature and pressure. Decreasing temperature tends to bring about crystallization, which produces solid crystals suspended in the liquid. Other solid fragments are incorporated from the walls and roof of the conduit through which the magma is rising. As crystallization progresses, volatiles and the more soluble silicate components are concentrated in the remaining liquid.

At some point in the rise of the magma, decreasing confining pressure and increasing concentration of volatiles in the residual liquid initiate the separation of gas from the liquid. From that point on to its final complete consolidation, the magma consists of all three phases: liquid, solid, and gas.

Part of the gas liberated at volcanoes probably comes from the same deep-seated source as the silicate portion of the magma, but some is of shallower origin. Part of the steam is the result of surficial oxidation of deep-seated hydrogen, and some of the oxidation of the sulfur gases must have taken place close to the surface. *See* MAGMA.

The character of a volcanic eruption is determined largely by the viscosity of the liquid phase of the erupting magma and the abundance and condition of the gas it contains. Viscosity is in turn affected by such factors as the chemical composition and temperature of the liquid, the load of solid crystals and xenoliths it carries, the abundance of gas, and whether the gas is dissolved or separated as bubbles. In very fluid lavas small gas bubbles form gradually, and generally are able to rise through the liquid, coalescing to some extent to form larger bubbles, and escape at the surface with only minor disturbance. In more viscous lavas the escape of gas is less easy, and produces minor explosions as the bubbles burst their way out of the liquid. In still more viscous lavas there appears at times to be a tendency for the formation of large numbers of small bubbles more or less simultaneously throughout a large volume of liquid, and the violent expansion of these may tear the frothy liquid to pieces and throw it into the air as a shower of volcanic ash and dust, accompanied by some larger blocks; or it may produce an outpouring of an emulsion of gas and semisolid bits of glass to form an ash flow. Also, gas may accumulate beneath a solid or highly viscous seal in the vent until it acquires enough pressure to burst its way forth in an explosion that hurls out fragments of the disrupted plug.

Surface deformation. The entire volcanic mountain swells and shrinks, apparently in response to changes in conditions beneath the volcano. The measurements are of two sorts: precise leveling surveys in which the change of altitude of a series of bench marks is related to some point outside the disturbed area, and measurement of the change in inclination of a given point on the slope of the cone by means of instruments known as tiltmeters. Some tiltmeters are capable of detecting changes in inclination, but the total changes of level resulting from tumescence and detumescence of the volcano may amount to several feet.

Volcanic Earth vibrations. Many eruptions are preceded by numerous earthquakes, and these, together with tumescence of the volcano, constitute the principal evidences of coming eruption. For the most part they closely resemble tectonic earthquakes.

Other volcanic Earth vibrations are termed harmonic tremors, which may go on continuously with little or no change for many hours, days, or weeks; and spasmodic tremor, which is simply a closely spaced succession of minute earthquakes. Volcanic explosions produce shock waves in the atmosphere—sometimes seen as flashing arcs rendered visible by refraction of light in the denser wavefront. Explosions or collapses taking place underwater may cause water waves, sometimes of great size. These fall within the class of tsunamis (seismic sea waves), though most tsunamis have other causes. *See* EARTHQUAKE; SEISMOLOGY; TSUNAMI. [G.A.M.]

Vole

Vole A rodent in the tribe Microtini in the family Cricetidae. In addition to the voles, the steppe lemmings and the muskrats are included in this group. There are about 79 species of vole, 42 of which belong to the genus *Microtus*. Voles have a stout body with short legs. The ears are small and are obscured by the coarse long fur.

Commonly, these rodents have four litters of six young each year; there does not appear to be a definite breeding period. The gestation period is about 3 weeks. Nests are built of grass, on the ground surface, in depressions, shallow burrows, or even on grass stems above the ground. However, surface nests are used primarily as shelters, most young being born in underground burrows. These animals do not hibernate, and in the winter they move great distances under the snow in search of food. *See* RODENTIA. [C.B.C.]

Volt-ampere

Volt-ampere The apparent-power index of drive level for sinusoidal alternating-current loads. A circuit branch with E volts across its two terminals, carrying I amperes from one to the other, is said to be receiving EI volt-amperes of apparent power, whatever may be the phase lag θ of current behind

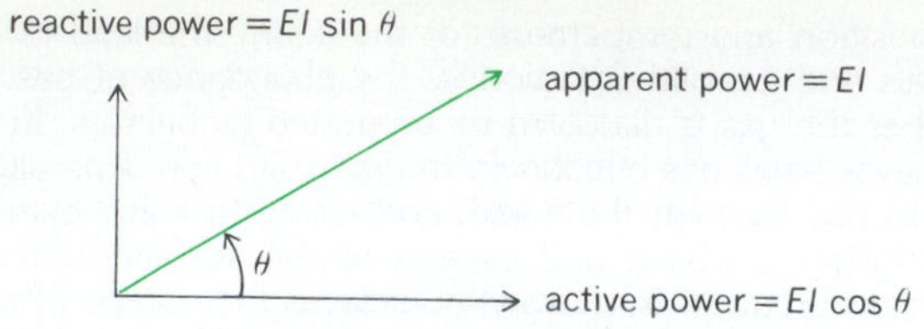

Apparent, active, and reactive power.

voltage. If such a load is driven through a transformer, the volt-amperes into the transformer primary is the same number as the volt-amperes into the load.

Practical computation of real power, reactive power, and apparent power is usually done with complex-number algebra using the geometrical diagram of the illustration. [M.G.F.]

Volt-ohm-milliammeter A self-contained test instrument for measuring a wide range of voltages (both ac and dc), resistances, and currents (usually dc only), particularly in radio, TV, and electronic servicing.

All readings are obtained on a single multiscale indicating instrument. The selection of the correct ranges and function is accomplished either by means of one or two rotary switches, by multiple pin jacks, or by a combination of both.

The terms multimeter and analyzer (or circuit analyzer) are quite commonly used as synonyms for volt-ohm-milliammeter. *See* AMMETER; OHMMETER; VOLTMETER. [I.F.K.]

Voltage amplifier An electronic amplifier that produces, with minimum distortion, an output voltage greater in magnitude than the input voltage. For a general discussion of amplifiers *see* AMPLIFIER.

Voltage amplifiers are built to amplify signals with frequency components in the range from zero hertz to thousands of megahertz. To cover this frequency range, several different types of amplifiers are required. An amplifier which cannot amplify voltage signals with zero frequency components (dc) is called *RC*-coupled. If zero frequency components (dc) are amplified, the amplifier is known as dc or direct-coupled. In the range from a few hertz to 30 kHz, amplifiers are classified as audio. In the frequency range up to 10 MHz for use in television, pulse circuits, and electronic instruments, amplifiers are known as video. For the radio-frequency range from a few hundred kilohertz to hundreds of megahertz, tuned amplifiers are normally used. At ultrahigh frequencies, amplifiers designed around special tubes are employed. *See* KLYSTRON; MAGNETRON; TRAVELING-WAVE TUBE.

RC-coupled amplifier. The *RC*-coupled amplifier is employed in the frequency range from a few hertz to about 10 MHz. However, operation at the higher frequencies in this range is obtained only by employing frequency-compensation networks. The term *RC*-coupled amplifier generally refers to any set of amplifier stages that are coupled by a resistance-capacitance network.

Tuned amplifier. A tuned amplifier amplifies voltage signals in a selected narrow frequency band and suppresses signals outside the desired band. A tuned amplifier is thus a bandpass amplifier with the additional constraint that its bandwidth be a small fraction of the center frequency. Such amplifiers are commonly used as radio-frequency (rf) or intermediate-frequency (i-f) amplifiers for superheterodyne radio receivers. In the case of rf amplifiers it is required that the center frequency be tunable or variable and that, in addition, over the tuning range the overall selectivity should remain constant. Constant selectivity means that the bandwidth should remain constant. In order to obtain the necessary selectivity and high voltage gain, *LC* tank or resonant circuits are used as loads with transistor or field-effect transistor (FET) amplifiers.

Grounded-gate (base) amplifier. If the gate (or base) is grounded and the input signal is applied to the source (or emitter) with the output taken from the drain (or collector), the circuit is called a grounded-gate (or grounded-base) amplifier. This circuit features a low input impedance and a high output impedance. Because of the low input impedance, the amplifier will draw power from the source and must, therefore, be used with a low-impedance source. It has found its chief application in rf amplifiers operating at very high frequencies, such as the tuner in a television receiver; at these frequencies it gives a high signal-to-noise ratio with low input-output capacitance. It does so at the expense of rf selectivity. This excludes it from use in a narrow-band receiver application.

Direct-coupled amplifier. When the coupling between successive stages of an amplifier is such that direct current may flow from the output of one stage to the input of the following stage, the amplifier is said to be direct-coupled. *See* DIRECT-COUPLED AMPLIFIER.

Difference amplifiers. The function of a difference, or differential, amplifier is, in general, to amplify the difference between two signals. The need for differential amplifiers arises in many physical measurements, in medical electronics, in analog computers, and in direct-coupled amplifier applications.

Integrated amplifier. Integrated circuit amplifiers are dc amplifiers commonly referred to as operational amplifiers. The basic building block of an integrated tuned amplifier is an emitter-coupled difference amplifier. *See* INTEGRATED CIRCUITS; OPERATIONAL AMPLIFIER. [C.C.H.]

Voltage measurement Determination of the difference in electrostatic potential between two points. *See* POTENTIALS; VOLTMETER.

In dc circuits the flow of current is unidirectional and the voltmeter must respond only to the magnitude of the potential difference. For voltages greater than 1000 volts, electrostatic voltmeters are often employed. These depend upon electrostatic forces to deflect a pointer against a spring restoring force. In the 1–1000-volt range, voltage is commonly measured by using a large resistance in series with a milliammeter.

To measure voltages below 1 volt, particularly in the microvolt range, these direct deflection instruments often require more power to operate than is available from the circuit under measurement. Extremely sensitive galvanometers have been devised to solve this problem, although use of an electronic amplifier between the circuit under measurement and a deflection-type instrument is more common. The modern amplifier-type voltmeter, called the vacuum-tube voltmeter, is more likely to employ solid-state transistors or integrated circuits than the older vacuum tube. The name vacuum-tube voltmeter, however, has remained.

In the volt range, the voltage to be measured is impressed directly upon a dc amplifier, while in the millivolt or microvolt range, a chopper-type amplifier is used. *See* VIBRATOR.

When greater accuracy than can be obtained with a moving-pointer voltmeter is required, a potentiometric method is often employed. *See* POTENTIOMETER (VOLTAGE METER).

AC voltage waveforms have several measurable attributes. AC voltmeters commonly measure the peak value, the average rectified value, or the root-mean-square (rms) value of the waveform. In the low-frequency range, where sufficient power is available, moving-pointer voltmeters that respond to the rms value are usually employed. If a semiconductor rectifier is utilized directly with a dc voltmeter, an ac voltmeter responding to the average value of the rectified voltage results, which is useful through the audio-frequency range. One of the most widely used average-responding ac voltmeters is the operational rectifier. Peak voltage is measured by using a vacuum-tube diode to charge a capacitor to the peak value, which is then measured by a dc vacuum-tube voltmeter. Voltmeters of this type are employed up to 1 GHz. To extend voltage measurement about 10 GHz, a sampling voltmeter has been used.

For high-accuracy ac-voltage measurement, a thermocouple or dynamometer which responds identically to ac and dc voltage is used as a transfer device. *See* CURRENT MEASUREMENT; ELECTRICAL MEASUREMENTS. [M.C.H.]

Voltage-multiplier circuit A rectifier circuit capable of supplying a dc output voltage that is two or more times the peak value of the ac input voltage. Such circuits are especially useful for high-voltage, low-current supplies. These supplies are usually lighter in weight, smaller in size, and less expensive than the more usual half-wave and full-wave rectifier supplies. They require either no power transformer or a much smaller transformer, but they have the disadvantage of requiring more rectifiers and capacitors. If, for safety or other reasons, input and output must be isolated, a transformer may be required at the input even though the voltage-multiplier circuit does not need it for operation. Common configurations are the half-wave and full-wave voltage-doubling circuits, but tripling and higher orders of multiplication are used. [D.L.W.]

Voltage regulation The change in voltage magnitude that occurs when the load (at a specified power factor) is reduced from the rated or nominal value to zero, with no intentional manual readjustment of any voltage control, expressed in percent of nominal full-load voltage. Voltage regulation is a convenient measure of the sensitivity of a device to changes in loading. *See* GENERATOR; TRANSFORMER; VOLTAGE REGULATOR. [P.M.A.]

Voltage regulator A device or circuit that maintains a load voltage nearly constant over a range of variations of input voltage and load current. Voltage regulators are used wherever the unregulated voltage would vary more than can be tolerated by the electrical equipment using that voltage. Alternating-current distribution feeders use voltage regulators to keep the voltage supplied to the user within a prescribed range. Electronic equipment often has voltage regulators in dc power supplies.

Electronic regulators. In much electronic equipment the dc power supply voltage must remain constant in spite of input ac line voltage variations and output load variations. To perform this function, two types of voltage-regulator circuits are used in electronic power supplies. The first uses a nearly constant voltage device, such as a zener diode or a gas tube, as the voltage regulator; the second uses the variable-resistance properties of a transistor or a vacuum tube. Because of the rapid response of these devices, either type regulator serves not only to compensate for irregular changes in input or output but also to counteract ripple voltage. Thus an electronic regulator makes possible the use of simpler filters. The voltage regulator circuit may also augment the filtering action of the power-supply filters, and it may also raise or lower the internal impedance of the power supply as seen from the load. The trade off for the regulator action is a considerable power loss in the regulator circuit. Most of the power is lost either in a regulator resistor or in the regulator diodes or tubes. *See* ELECTRONIC POWER SUPPLY; RECTIFIER; RIPPLE VOLTAGE. [D.L.W.]

Power-system regulators. Voltage regulators are used on distribution feeders to maintain voltage constant, irrespective of changes in either load current or supply voltage. Voltage variations must be minimized for the efficient operation of industrial equipment and for the satisfactory functioning of domestic appliances, television in particular. Voltage is controlled at the system generators, but this alone is inadequate because each generator supplies many feeders of diverse impedance and load characteristics. Regulators are applied either in substations to control voltage on a bus or individual feeder or on the line to reregulate the outlying portions of the system. These regulators are variable autotransformers with the primary connected across the line. The secondary, in which an adjustable voltage is induced, is connected in series with the line to boost or buck the voltage.

A control and drive provide automatic operation. A voltage-regulating relay senses output voltage. When the voltage is either above or below the band of acceptable voltage maintained by the regulator, this relay causes the motor to operate and change the regulator position to raise or lower the voltage as required to bring it back within the control band. It is desirable to maintain constant voltage at the average load center out on the feeder rather than at the regulator terminals. Hence the control circuit includes a line-drop compensator with resistance and reactance elements that can be adjusted to represent line impedance. These impedances carry current proportional to circuit load, thereby simulating the voltage drop between the regulator and the load center, and modify the voltage sensed by the voltage-regulating relay.

Two principal types of feeder regulators are used: the step regulator, which provides increments or steps of voltage change, and the induction regulator, which provides continuous voltage adjustment. [D.D.MacC.]

Voltmeter An instrument for the measurement in volts of potential difference between two points. Derivatives of the voltmeter are the microvoltmeter, millivoltmeter, and kilovoltmeter, for measurement of voltages with a measurement span of 1,000,000,000:1. *See* VOLTAGE MEASUREMENT.

Voltmeters, in general, are secondary instruments consisting of voltage-to-current transducers and mechanisms which respond to milliamperes or microamperes. In the simpler forms, the transducer consists of a resistor of constant value, which may have taps for several voltage ranges. Since by Ohm's law the current is proportional to the potential difference, calibration of the combination using a primary standard, such as the potentiometer and standard cell, is valid. In the more specialized forms, the transducer consists of a thermal converter (ac to dc), rectifier, or amplifier. The electrostatic instrument is the only voltmeter which measures potential difference directly.

The moving-coil dc voltmeter consists of a permanent magnet, moving coils pivoted in jewel bearings, control springs, and a pointer, as shown in Fig. 1. The current is passed through the moving coil via the two control springs which apply a restraining torque. The deflecting torque is equal to the product of the magnetic moment of the coil and the field intensity.

The moving-magnet dc voltmeter has the magnet attached

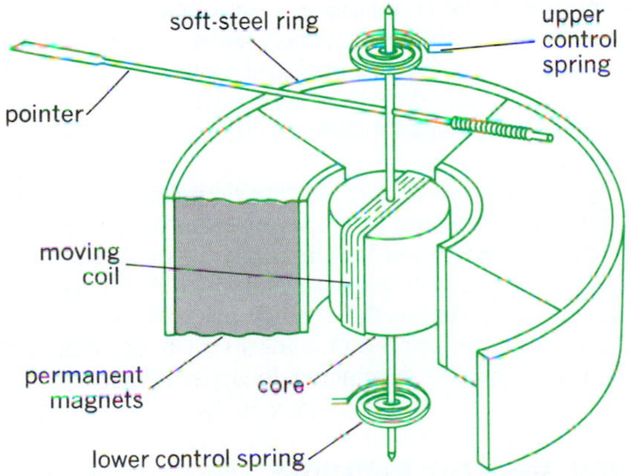

Fig. 1. Direct-current permanent-magnet, moving-coil voltmeter. (*General Electric Co.*)

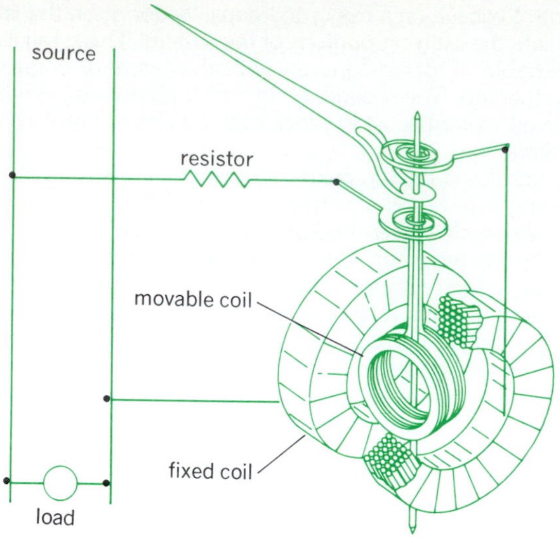

Fig. 2. Electrodynamic mechanism used as voltmeter. (*General Electric Co.*)

to the pivoted shaft which carries the pointer. The magnet is surrounded by the field coil and aligned by a control magnet which supplies the restoring torque. As current is applied to the field coil, the coil and restoring-magnet fields produce a resultant field. The rotor aligns itself with this resultant field and indicates the magnitude of the current and thus the potential difference.

The electrostatic voltmeter action is based on the force of attraction or repulsion between two charged conductors, such as the plates of a variable air capacitor. The moving plate, when charged, tends to move so as to increase the capacitance between it and the fixed plate.

The electrodynamic ac voltmeter has two coils connected in series through a resistor to the source, the potential of which is to be measured, as shown in Fig. 2. The fixed coil provides the field in which the moving coil, supported by a pivoted shaft, operates. The deflection of the moving system is restrained by a control spring. When the circuit is closed, the current which flows through the coils produces a deflecting torque which moves the indicating needle.

In a common form of moving-iron ac voltmeter, the field coil is connected to the line through a series resistor. Inside the field coil are two rectangular vanes, one fixed and the other attached to the shaft which carries the pointer. When a current traverses the coil, the vanes are similarly magnetized, and the moving vane is repelled from the fixed vane, deflecting the pointer in a clockwise direction.

The thermovoltmeter utilizes a thermojunction and dc millivoltmeter. A current from the voltage source is passed through a resistor and a fine vacuum-enclosed platinum heater wire. A thermocouple, attached to the midpoint of the heater, generates millivolts proportional to the temperature. *See* THERMOCOUPLE.

The rectifier voltmeter consists of a dc milliammeter calibrated in volts and connected to the voltage source through a rectifier bridge. The individual rectifiers are connected so that the current through the milliammeter is always in the same direction. This meter measures the average value of the current.

The electronic ac voltmeter consists of a dc milliammeter calibrated in volts and connected to an amplifier-rectifier circuit. [A.J.Co.]

Volume control systems Electronic systems that regulate the signal amplification or limit the output of circuits. Examples are volume compressors, limiters, and expanders. A volume compressor is an electronic system that reduces the amplification of an amplifier when the signal being amplified is large and increases the amplification when the signal is small. Compressors are used to reduce the volume range in sound motion picture and phonograph recording, sound broadcasting, public address and sound reinforcing systems, and so forth.

A volume limiter is an electronic system in which the relationship between the input and output signals is constant up to a certain level, but beyond this point the output remains constant regardless of the input. The limiter is useful for protection against sudden overloads, as, for example, in the audio input to a radio or television transmitter.

A volume expander is an electronic system that increases the amplification of an amplifier when the signal is large and decreases the amplification when the signal is small. In reproduction, a volume expander is used to counteract or complement the effect of the compressor in recording or the transmission of an audio signal.

Volume compressors, limiters, and expanders are electronic amplifiers in which the amplification varies as a function of the average level of the signal. In all the systems, the input signal to the control circuits is rectified and applied to a capacitor-resistance network. The direct-current (dc) voltage across the capacitor is used to vary the amplification of a vacuum tube or transistor amplifier operating in a variable gain mode. *See* AMPLIFIER; AUTOMATIC GAIN CONTROL (AGC). [H.F.O.]

Volume flow rate meter An instrument for measuring the rate of volumetric flow of fluid through a pipe, duct, or open channel. Flow rate meters may modify the flow pattern by insertion of some obstruction or by modification of the shape of the walls of the conduit or channel carrying the fluid, or they may measure the velocity without affecting the flow pattern. *See* FLOW MEASUREMENT.

Differential-producing primary devices (sometimes called head meters) produce a difference in pressure caused by a modification of the flow pattern. The pressure difference is based on the laws of conservation of energy and conservation of mass. The law of conservation of energy states that the total energy at any given point in the stream is equal to the total energy at a second point in the stream, neglecting frictional and turbulent losses between the points. It is possible, however, to convert pressure (potential energy) to velocity (kinetic energy), and vice versa. By use of a restriction in the pipe, such as an orifice plate, venturi tube, flow nozzle, or flow tube, a portion of the potential energy of the stream is temporarily converted to kinetic energy as the fluid speeds up to pass through these primary devices. *See* BERNOULLI'S THEOREM.

The common orifice plate (Fig. 1) is a thin plate inserted between pipe flanges, usually having a round, concentric hole with a sharp, square upstream edge. The extensive empirical data available and its simplicity make it the most common of the primary devices. Segmental and eccentric orifices are useful for measurement of liquids with solids, and of vapors or gases with entrained liquids. *See* METERING ORIFICE.

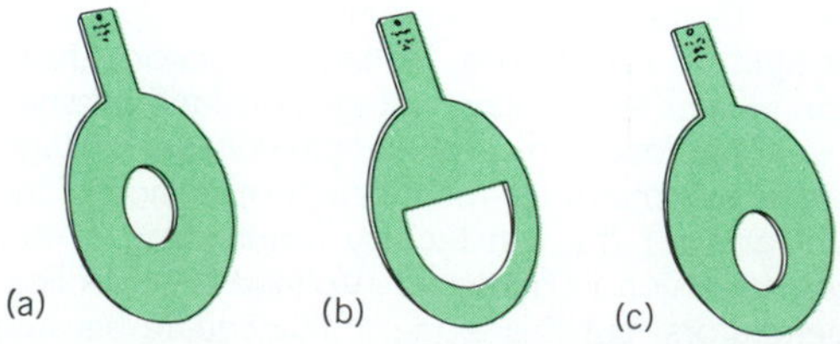

Fig. 1. Orifice plate primary device. (*a*) Concentric plate. (*b*) Segmental plate. (c) Eccentric plate.

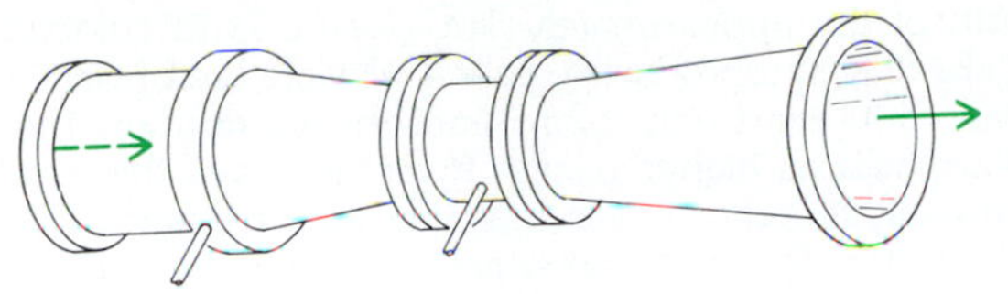

Fig. 2. Venturi tube.

The Venturi tube (Fig. 2) has the advantage of introducing less permanent loss of pressure than the orifice plate. The converging inlet cone permits solids and dirt to be flushed through it, and the outlet cone reduces the turbulent losses. *See* VENTURI TUBE.

The flow nozzle (Fig. 3) also has a converging inlet, but it has no diverging outlet section. The pressure loss is similar to that with an orifice plate for the same differential pressure. It is used where solids are entrained in the liquid and also where fluid velocities are particularly high (it will pass more fluid than an orifice of the same diameter).

In industrial flow measurement, pitot tubes are used in spot checks and for comparative measurements to determine trends. A pitot tube measures the difference between total impact pressure and static pressure. There are a wide variety of tube configurations available for sensing both impact and static pressure or combinations thereof. Several readings should be taken across the pipe along a diameter (preferably two diameters) of the pipe and the average velocity determined. *See* PITOT TUBE.

Fluid flow around the bend in a pipe elbow produces a centrifugal force greater on the outside than on the inside, and thereby produces a difference in pressure between pressure connections on the inside and outside of the bend. This differential pressure is used as an indication of the volume flow rate.

All the primary devices discussed above create a difference in pressure between two points. This differential pressure must be measured and converted by the equations given to obtain flow. Differential pressure may be measured accurately by simple liquid-filled (usually mercury) U-tube manometers or by more refined types of meters, such as the float-operated mechanism or the weight-balanced, ring-type meter, to provide controlling, recording, and totalizing functions. These devices are ordinarily connected directly to the pressure connections on the primary device and are therefore exposed to the process fluids. *See* MANOMETER.

Flow in open channels is determined by measuring the

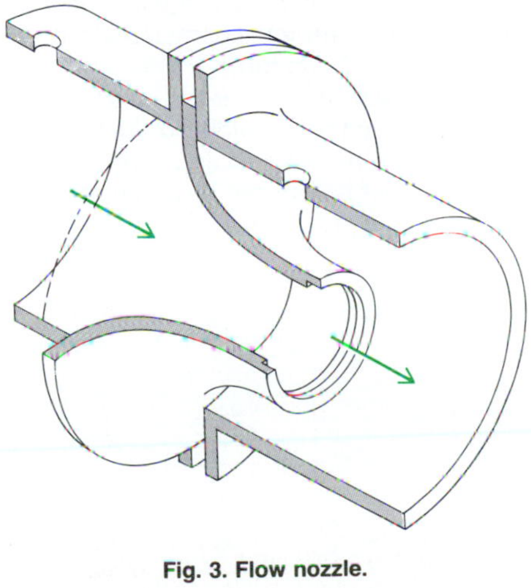

Fig. 3. Flow nozzle.

height of liquid passing through a weir of flume. Consistent results require that the measuring device be protected from direct impact of the flowing stream. Because the height-flow velocity relationship is exponential, the secondary device frequently uses cams to obtain a uniform flow scale. For special types of volume flow rate meters *see* ELECTROMAGNETIC FLOWMETER; TARGET FLOWMETER; VARIABLE-AREA FLOWMETER. [M.Br.; L.P.E.]

Volume unit (VU) A unit for expressing the audio-frequency (af) power level of a complex electric wave, such as that corresponding to speech or music. The volume unit (VU) is defined as 10 times the common logarithm of the power ratio p_2/p_1, where the reference power level p_1 is selected as 1 mW (0.001 watt) and p_2 is the signal power level. Instruments which give readings in volume units, called VU meters, are widely used for monitoring radio broadcasts and for sound recording. [D.E.F.]

Volumetric analysis Volumetric analysis, or possibly more correctly titrimetric analysis, is one of the major divisions of quantitative analytical methods. In this method, as the names imply, a titration is made and the amount of the desired component is determined indirectly from the volume and concentration of solution used in the titration.

In most volumetric procedures, the sample to be analyzed is weighed, or an accurately measured volume of a solution of the sample is taken. The sample is dissolved in an appropriate solvent and subjected to some preliminary treatment or separation if necessary, and then the titration is performed on the resulting solution.

During the titration, a chemical reaction occurs between a constituent of the titrant solution and the substance to be determined. In order to be suitable for volumetric titrations, these reactions must be rapid and quantitative. That is, it is necessary to know exactly how many molecules or ions of the substance to be determined react with each molecule or ion of the reagent in the titrant. In volumetric analysis, it is also necessary to determine when the number of reagent ions or molecules equivalent to the substance being determined has been added. This point in the titration is called the equivalence point. The determination of the amount of titrant corresponding to the equivalence point is usually done either by visual means with colored indicators or reagents, or by some physicochemical measurement. *See* QUANTITATIVE CHEMICAL ANALYSIS; TITRATION; VOLUMETRIC FLASK. [C.E.B.]

Volumetric efficiency In describing an engine or gas compressor, the ratio of volume of working substance actually admitted, measured at specified temperature and pressure, to the full piston displacement volume. For a liquid-fuel engine, such as a diesel engine, volumetric efficiency is the ratio of volume of air drawn into a cylinder to the piston displacement. For a gas-fuel engine, such as a gasoline engine with carburetor, volumetric efficiency is based on the charge of fuel and air drawn into the cylinder. *See* INTERNAL COMBUSTION ENGINE; RECIPROCATING AIRCRAFT ENGINE. [N.MacC.]

Volumetric flask A long narrow-necked glass flask which has been calibrated to contain a given volume of liquid when filled so that the bottom of the meniscus is tangent to the ring which has been etched on the neck of the flask. Some volumetric flasks have a second calibration mark, higher on the neck of the flask, which is used when the flask is needed to deliver a given volume of liquid. The marks etched on volumetric flasks are accurate only at the temperature at which these calibrations were made, because the volume of a flask increases slightly with an increase of temperature and vice versa. Volumetric flasks are used when it is necessary to dilute a weight or a volume of a substance to some accurately known volume. *See* CONCENTRATION SCALES; TITRATION; VOLUMETRIC ANALYSIS. [C.E.B.]

Volvocales A large order of green algae (Chlorophyceae) comprising all forms that normally are flagellate and motile. In zoological classification, it is called Volvocida, in the class Phytomastigophora. The cells are solitary or united into colonies of definite structure (coenobia), often with morphological and functional differentiation among the component cells. Some taxonomists place the unicellular forms in a separate order, Chlamydomonadales. Unicells that have volvocalean cytological features but are nonflagellate and sedentary in their vegetative phase are considered here to constitute the order Tetrasporales. Or these sedentary forms may be retained in the Chlamydomonadales or Volvocales. *See* CHLOROPHYCEAE.

Of the unicellular forms, *Chlamydomona*, which is presumed to be similar to ancestral Volvocales, has a very small cell body (usually 7–40 micrometers in greatest dimension) that is spherical, ellipsoid, or pyriform. The wall (or theca), which consists primarily of glycoproteins, is usually smooth, but may have blunt protuberances. The cell bears two smooth flagella of equal length at its apex, with contractile vacuoles and an eyespot at the anterior end. Chloroplasts contain one or more pyrenoids. As in all Volvocales, the cells are uninucleate. Other genera with chlamydomonad features are distinguished on the basis of flagellar number, presence or absence of photosynthetic pigments, presence or absence of pyrenoids, and cell shape. Together with *Chlamydomonas* they constitute the family Chlamydomonadaceae. Two other families of unicellular Volvocales are recognized: Dunaliellaceae, in which the unicells are naked; and Phacotaceae, in which the unicells are surrounded by a lorica—a rigid, hyaline or dark brown, wall-like structure, which is often impregnated with compounds of calcium, iron, or manganese and which may be sculptured. Unicellular Volvocales are ubiquitous in fresh, brackish, and coastal marine waters and in such terrestrial habitats as soil, snow, and ice. They are especially abundant in organically enriched bodies of fresh water.

Most colonial (coenobial) Volvocales are placed in the family Volvocaceae. In this family, coenobia are composed of *Chlamydomonas*-like biflagellate cells embedded in a firm gelatinous matrix. Coenobia have the form of a slightly curved plate or are ellipsoid to spherical. The component cells are arranged according to a predetermined pattern. Asexual reproduction is by repeated bipartition of the protoplast of an ordinary or specialized cell to form a new coenobium (autocoenobium) that is a miniature of the parent. Sexual reproduction forms a phylogenetic series from isogamy through anisogamy to oogamy. The Volvocaceae are abundant in organically enriched bodies of fresh water, especially temporary pools. [P.C.Si.; R.L.Moe.]

Volvocida An order of the class Phytamastigophorea. The protozoans, also known as the Phytomonadida, are grass-green, but a few are colorless (*Polytoma*). Individual cells may be as small as 8 micrometers. They closely relate the flagellates to the algae. They have one, two, four, or eight flagella; plural flagella are usually equal. The group is large and about one-fourth of the approximately 100 genera form colonies, with flagella only in reproductive cells. Cell walls are of cellulose and often they are thick. Chromatophores contain the same chlorophylls as higher plants. Pyrenoids have the usual form commonly found with starch as the reserve material. *See* CILIA AND FLAGELLA; PHYTAMASTIGOPHOREA. [J.B.L.]

Vortac A ground radio station consisting of a co-located vhf omnidirectional radio range (VOR) and Tacan facility. This station permits obtaining polar coordinates by the use of a VOR receiver and distance-measuring equipment or by Tacan equipment. *See* RADIO RANGE; TACAN. [P.C.S.]

Vortex A line vortex in two-dimensional fluid flow produces a flow or circulation around the line. Consider the effect of rotating a right-circular cylinder of radius r_0 about its axis with a peripheral velocity v_0 in a fluid otherwise at rest. The fluid in contact with the surface of the cylinder rotates with the cylinder. Fluid at greater radius is also set in motion in concentric circles with velocity diminishing as the radius increases. This type of fluid motion, in which the velocity varies inversely as the radius, is referred to as a free vortex. If the cylinder is reduced to zero radius in such a manner that v_0r_0 remains constant in the limit as r_0 approaches zero, a line vortex results. The velocity at the line is infinite, so the line itself must be considered as a singular line, to be excluded from the actual fluid.

Examples of vortices occur frequently in nature. The tornado is an example of a free vortex, with high velocities near its center, and correspondingly low pressure intensities. The waterspout is its counterpart over water. *See* FLUID FLOW; KARMAN VORTEX STREET. [V.L.S.]

Vortex flowmeter A flowmeter based on the phenomenon of vortex shedding or vortex precession. *See* VORTEX.

The vortex-shedding flowmeter is based on the eddies (vortices) forming behind an obstruction in a flowing stream. Close observation has shown that each vortex builds up and breaks loose, followed by a repeat in the opposite rotation on the other side of the obstruction, and so alternately the vortices shed and go downstream in what is called a vortex street. Further observation shows that the distance between successive vortices is directly proportional to the blocking width of the obstruction. The distance is also independent of the fluid velocity over a large range, and thus the frequency of the shedding is directly proportional to the velocity of the fluid. *See* KARMAN VORTEX STREET.

Application of vortex-shedding flowmeters is normally limited to the turbulent-flow regime, and various shedding-element shapes and internal constructions have been developed in order to improve the shedding characteristics and the detection of the shedding. The common characteristic is sharp edges around which the vortices can form.

In the vortex-precession flowmeter, used to measure gas flows, a swirl is imparted to the flowing fluid by a fixed set of radial vanes. As this swirl goes into an expanding tube, it forms a precessing vortex that is detected by a heated thermistor sensor. *See* FLOW MEASUREMENT. [M.Br.; L.P.E.]

W

Wading birds Members of the order Ciconiiformes. These birds have a worldwide distribution, and are characterized by relatively long legs, necks, and bills. There are four toes, the anterior three being webbed to various degrees.

There are about 116 species in 7 families which include: Balaenicipitidae (the shoebill stork, *Balaeniceps rex*), Scopidae (the hammerhead, *Scopus umbretta*), Ardeidae (bitterns, egrets, and herons), Cochleariidae (the boatbill heron, *Cochlearius cochlearius*), Ciconiidae (storks and wood ibises), Threskiornithidae (spoonbills and ibises), and Phoenicopteridae (flamingos). *See* Aves; Ciconiiformes. [C.B.C.]

Wage incentives Wage payment plans designed to provide extra pay for extra effort over and above a "fair day's work." Base hourly, weekly, or monthly wage rates are paid to employees for normal effort, while incentives allow them to earn more through increased effort.

Incentives are used over a broad spectrum of the labor force. Sales commissions, executive bonuses, profit sharing, and suggestion systems are some of the more popular uses. However, the most widely used and the most effective incentives are in the manufacturing areas. Some of the most used incentive devices are: the standard hour plan, the piecework plan, and factor plans.

Some basic principles for an effective incentive plan are that incentives should: pay for extra effort only; be individual wherever possible or for the smallest group that can be arranged; be based on factors which can be controlled by the employee; not be a substitute for management or management effort; and be high enough to provide the motivation to make the extra effort worthwhile. It is generally accepted that 25 to 35% earnings above base is an adequate range of opportunity. *See* Productivity. [W.J.O.]

Wake flow Turbulent eddying flow that occurs downstream from bluff bodies. When fluid flows along the boundary of a solid body, the fluid near the boundary is slowed down by viscous shear stresses exerted at the boundary. This action is progressive along the body and, under certain conditions, fluid near the boundary is brought to rest, which causes the fluid moving near the body to separate from the body (see illustration). A wake develops and produces additional form drag on the body.

A wake is formed downstream from bluff bodies such as bridge piers, smoke stacks, buildings, or trees. For unsteady flow cases, such as the motion of a train or a ship, the fluid behind the moving object has great turbulence remaining in it and, in the case of a ship, is easily discerned for a great distance. *See* Fluid flow. [V.L.S.]

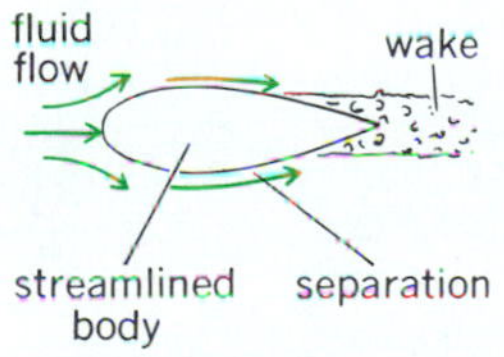

Wake formed downstream from a streamlined body.

Wall construction An acceptable exterior wall for a building must be pleasing in appearance, structurally sound, and durable; it must provide insulation and fire resistance; and it must offer protection against condensation of water vapor. An interior wall must meet the same conditions but to a much lesser extent; it must also provide resistance to sound transmission.

In home construction, both the interior and exterior walls are formed around a basic core of wooden studs. The exterior exposed faces are covered with wood siding, wood or asbestos shingles, brick veneer, or other finish and insulated to retard heat flow. Solid and hollow concrete masonry and brick walls are used extensively for low buildings. Reinforced concrete walls, precast concrete walls, and precast concrete walls erected by the tilt-up method also offer a variety of surface finishes, low maintenance, strength, and durability.

The walls of many tall buildings consist principally of glass, the spandrels and pilasters being of steel or aluminum backed up by brick or tile masonry.

Plain or corrugated sheets of cement-asbestos and corrugated metal sheets are used for the walls of industrial buildings. Prefabricated panels consisting of two sheets of metal with a layer of insulation between provide a complete wall section which can be quickly attached to steel framing. *See* Buildings; Concrete; Masonry. [C.M.A.]

Walnut This name is applied to about a dozen species of large deciduous trees widely distributed over temperate North and South America, southeastern Europe, and central and eastern Asia. The genus (*Juglans*) is characterized by pinnately compound aromatic leaves and chambered or laminate pith. The staminate (male) flowers are borne in unbranched catkins on the previous season's growth, and the pistillate (female) flowers are terminal on the current season's shoots. The shells of the nuts of most species are deeply furrowed or sculptured.

Two species, the black walnut (*J. nigra*) and the Persian or English walnut (*J. regia*), are of primary importance for their timber and nuts. The butternut finds local use in the northeastern United States. The other species are sparingly used as shade trees, as grafting stocks, and as sources of nuts. *See* Juglandales. [L.H.Mac.D.]

Warm-air heating system In a general sense, a heating system which circulates warm air. Under this definition both a parlor stove and a steam blast coil circulate warm air. Strictly speaking, however, a warm-air system is one containing a direct-fired furnace surrounded by a bonnet through which air circulates to be heated.

When air circulation is obtained by natural gravity action, the system is referred to as a gravity warm-air system. If positive air circulation is provided by means of a centrifugal fan (referred to in the industry as a blower), the system is referred to as a forced-air heating system.

Direct-fired furnaces are available for burning of solid, liquid, or gaseous fuels. Furnaces have also been designed which have air circulating over electrical resistance heaters. A completely equipped furnace-blower package consists of furnace, burner, bonnet, blower, filter, and accessories. The furnace shell is usually of welded steel. The burner supplies a positively metered rate of fuel and a proportionate amount of air for combustion. A casing, or jacket, encloses the furnace and provides a passage for the air to be circulated over the heated furnace shell. The casing is insulated and contains openings to which return-air and warm-air ducts can be attached. The blower circulates air against static pressure. The air filter removes dust particles from the circulating air. *See* OIL BURNER.

The complete forced-air heating system consists of the furnace-blower package unit; the return-air intake, or grille, together with return-air ducts leading from the grille to the return-air plenum chamber at the furnace; and the supply trunk duct and branch ducts leading to the registers located in the different spaces to be heated. *See* COMFORT HEATING. [S.Ko.]

Wart A papillomatous growth which occurs singly or in groups on the skin surface and is often classified according to its location, for example, digital, genital, plantar, or laryngeal. It is also called a verruca. The agent responsible is considered to be a virus. [A.E.Mo.]

Washer A flattened, ring-shaped device used to improve the tightness of a screw fastener. Three types of washer are in common use: plain, spring-lock, and antiturn (tooth-lock washers). Standard plain washers are used to protect a part from damage or to provide for a wider distribution of the load. Because a plain washer will not prevent a nut from turning, a locking-type washer should be used to prevent a bolt or nut from loosening under vibration (see illustration). Lock washers create a continuous pressure between the parts and the fastener. The antiturn-type washers may be externally serrated, internally serrated, or both. The bent teeth bite into the bearing surface to prevent the nut from turning and the fastening from loosening under vibration. *See* SCREW FASTENER. [W.J.L.]

Lock washers. ***(After W. J. Luzadder, Fundamentals of Engineering Drawing, 6th ed., Prentice-Hall, 1971)***

Wasp An insect belonging to the Hymenoptera. In general, wasps range in size from the minute (microscopic) parasitic species to those which are 2 in. (5 cm) long. These insects are economically important as parasites or predators of injurious pests and have been useful in biological control of other pest species. Some species of wasps show a complex social organization like that of ants and bees. *See* HYMENOPTERA; SOCIAL INSECTS. [C.B.C.]

Watch A portable timepiece. Its operation may be described as mechanical, electromechanical, or electronic.

In the mechanical watch a mainspring in the barrel stores operating energy; the user retightens the spring daily by means of the winding stem. The wheel train advances at five increments per second under control by the escapement. From there the dial train turns the minute and hour hands across the watch face. *See* CLOCK; ESCAPEMENT; GEAR TRAIN; JEWEL BEARING; SPRING (MACHINES).

Among the variety of features incorporated into modern watches are: self-winding mechanisms, substitution of an electrochemical cell for the mechanical mainspring, and several forms of electromechanical escapement in place of the balance-and-hairspring mechanism.

When solid-state electronic integrated circuits became available in quantity, the all-electronic watch became a commercial reality. In it an electrochemical cell supplies the energy. A chain of binary dividers triggered from a crystal oscillator develops a train of seconds pulses. These pulses drive a digital counter or scaler, which develops minute and hour pulses to activate the digital display. *See* DIGITAL COUNTER; INTEGRATED CIRCUITS; OSCILLATOR; PIEZOELECTRICITY.

One form of readout uses a light-emitting diode (LED). Because illumination of the readout consumes most of the power in an electronic watch, a liquid crystal display (LCD) is used where low power consumption is a first consideration. *See* ELECTRONIC DISPLAY; HOROLOGY; LIGHT-EMITTING DIODE; LIQUID CRYSTALS. [F.H.R.]

Water The chemical compound with two atoms of hydrogen and one atom of oxygen in each of its molecules. It is formed by the direct reaction (1) of hydrogen with oxygen.

$$2H_2+O_2 \rightarrow 2H_2O \quad (1)$$

The other compound of hydrogen and oxygen, hydrogen peroxide, readily decomposes to form water, reaction (2). Water

$$2H_2O_2 \rightarrow 2H_2O + O_2 \quad (2)$$

also is formed in the combustion of hydrogen-containing compounds, in the pyrolysis of hydrates, and in animal metabolism. Some properties of water are given in the table.

Gaseous state. Water vapor consists of water molecules which move nearly independently of each other. The relative positions of the atoms in a water molecule are shown in the illustration. The dotted circles show the effective sizes of the isolated atoms. The atoms are held together in the molecule by chemical bonds which are very polar, the hydrogen end of each bond being electrically positive relative to the oxygen. When two molecules near each other are suitably oriented, the positive hydrogen of one molecule attracts the negative oxygen of the other, and while in this orientation, the repulsion of the like charges is comparatively small. The net attraction is strong enough to hold the molecules together in many circumstances

Properties of water

Property	Value
Freezing point	0°C (32°F)
Density of ice, 0°C (32°F)	0.92 g/cm^3 (0.53 oz/in.^{3})
Density of water, 0°C (32°F)	1.00 g/cm^3 (0.578 oz/in.^{3})
Heat of fusion	80 cal/g (335 J/g)
Boiling point	100°C (212°F)
Heat of vaporization	540 cal/g (2260 J/g)
Critical temperature	347°C (657°F)
Critical pressure	217 atm (22.0 MPa)
Specific electrical conductivity at 25°C	1 × 10^{-7}/ohm-cm
Dielectric constant, 25°C (77°F)	78

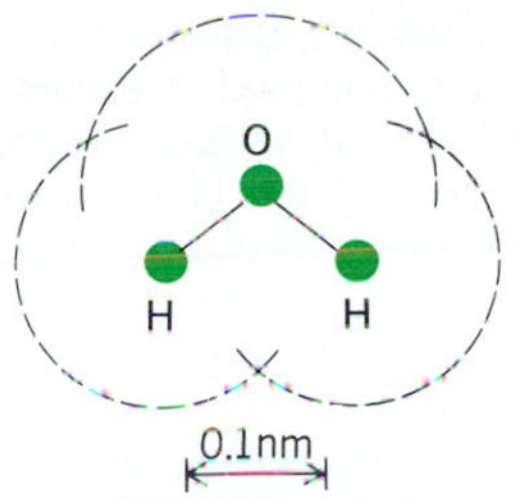

The water molecule.

and is called a hydrogen bond. *See* CHEMICAL BONDING; ELECTRONEGATIVITY; GAS; VALENCE.

Solid state. Ordinary ice consists of water molecules joined together by hydrogen bonds in a regular arrangement. It appears that there is considerable empty space between the molecules. This unusual feature is a result of the strong and directional hydrogen bonds taking precedence over all other intermolecular forces in determining the structure of the crystal. *See* CRYSTAL; HYDROGEN BOND.

Liquid state. The molecules in liquid water also are held together by hydrogen bonds. When ice melts, many of the hydrogen bonds are broken, and those that remain are not numerous enough to keep the molecules in a regular arrangement. As water is heated from 0°C (32°F), it contracts until 4°C (39°F) is reached and then begins the expansion which is normally associated with increasing temperature. This phenomenon and the increase in density when ice melts both result from a breaking down of the open, hydrogen-bonded structure as the temperature is raised. *See* LIQUID.

Properties. The electrical conductivity of water is at least 1,000,000 times larger than that of most other nonmetallic liquids at room temperature. The current in this case is carried by ions produced by the dissociation of water according to reaction (3).

$$H_2O \rightleftharpoons H^+ + OH^- \qquad (3)$$

Water is an excellent solvent for many substances, but particularly for those which dissociate to form ions. Its principal scientific and industrial use as a solvent is to furnish a medium for purifying such substances and for carrying out reactions between them. *See* SOLUTION; SOLVENT.

Water is not a strong oxidizing agent, although it may enhance the oxidizing action of other oxidizing agents, notably oxygen. Water is an even poorer reducing agent than oxidizing agent. One of the few substances that it reduces rapidly is fluorine. *See* OXIDATION-REDUCTION.

Substances with strong acidic or basic character react with water. For example, calcium oxide, a basic oxide, reacts in a process called the slaking of lime, reaction (4). Another type of

$$CaO + H_2O \rightarrow Ca(OH)_2 \qquad (4)$$

substance with strong acidic character is an acid chloride. An example and its reaction with water is boron trichloride, reaction (5). This is termed hydrolysis, as is the reaction of an ester

$$BCl_3 + 3H_2O \rightarrow H_3BO_3 + 3HCl \qquad (5)$$

with water. *See* HYDROLYSIS.

Water also reacts with a variety of substances to form solid compounds in which the water molecule is intact, but in which it becomes a part of the structure of the solid. Such compounds are called hydrates. *See* HYDRATE.

For various aspects of water, its uses, and occurrence *see* HEAVY WATER; HYDROGEN; HYDROLOGY; IRRIGATION (AGRICULTURE); NUTRITION; OXYGEN; PLANT-WATER RELATIONS; PRECIPITATION (METEOROLOGY); SEAWATER; TRIPLE POINT; VAPOR PRESSURE; WATER ANALYSIS; WATER CONSERVATION; WATER MICROBIOLOGY; WATER POLLUTION; WATER PURIFICATION; WATER SOFTENING; WATER SUPPLY ENGINEERING; WATER TABLE; WATER TREATMENT; WATERPOWER. [H.L.F.]

Water analysis A broad field that deals with the specific applications of chemical, physical, and biological methods to the analysis of aqueous samples ranging from the purest distilled water to the most polluted wastewater. One major area of concern is the analysis of drinking water to ensure that the water does not transmit organisms causing human disease, and does not contain chemicals hazardous to human health. Another concern is that the water is esthetically acceptable to consumers in terms of taste, odor, and appearance. Raw or treated wastewaters are analyzed to determine proper treatment and to monitor their discharge to assure public safety and environmental protection. *See* WATER-BORNE DISEASE.

Chemical analysis. The analytical chemistry of water deals with major mineral constituents, both anions and cations, trace organic and inorganic substances, and radionuclides. Insofar as possible, contemporary procedures are instrumental and do not depend on conventional colorimetric or titrimetric methods. Foremost among analytical instruments are the atomic absorption spectrophotometer (for metals) and the gas-liquid chromatograph (for organic constituents). Inorganic nonmetals typically are measured by noninstrumental methods, and radionuclides are determined by a combination of chemical separation followed by instrumental measurement of radioactivity. *See* ANALYTICAL CHEMISTRY.

Physical analysis. Physical analyses are made to assess important characteristics of both water and wastewater. Organoleptic tests are made for taste and odor which are significant in the consumer's perception of drinking-water quality. Color is measured by direct visual comparison with standards or by a spectrophotometric method. Conductivity is measured electrometrically and serves as an indirect indicator of the dissolved residue. Residue also may be measured gravimetrically by determining the weight of a dried sample. The suspended residue (suspended-solids test) is important in evaluating the strength of wastewater and the efficacy of its treatment. Turbidity relates to water clarity and is important in the analysis of drinking water and, for safety reasons, in swimming pools. It is measured visually by determining the depth of water required to obscure the image of a standard candle flame or, better, by nephelometry in which light scatter under standard conditions is measured.

Biological analysis. Biological analyses include a broad spectrum of tests for organisms found in water and for the effects of substances in water on aquatic life. In making biological analyses, the classical methods of visual field observations are coupled with laboratory examinations, often microscopic, for identification and enumeration of plants and animals in water. Analytical results, especially those that are historical or comparative, are important in assessing the effect of pollution on natural waters. Algae may affect the quality of drinking water by modifying taste, odor, and color, and usually require control in water supply reservoirs.

Microbiological tests are used to determine sanitary quality and suitability of water for general use. In the routine analysis of water, including drinking water, enteric pathogens are not enumerated; rather, indicators of fecal pollution are determined. The most commonly used indicator is the coliform group, which includes normal inhabitants of the gastrointestinal tract of warm-blooded animals. Microbiological standards of water quality have been based on coliform counts for more than 50 years and have served well to protect the public health. Organisms other than coliform bacteria may be tested

Maximum contaminant levels (MCL) allowable for drinking water

Contaminant	MCL, mg/liter*
Arsenic	0.05
Barium	1.
Cadmium	0.010
Chromium	0.05
Fluoride	1.4–2.4†
Lead	0.05
Mercury	0.002
Nitrate (as N)	10.
Selenium	0.01
Silver	0.05
Endrin	0.0002
Lindane	0.004
Methoxychlor	0.1
Toxaphene	0.005
2,4-D	0.1
2,4,5-TP Silvex	0.01
Turbidity	1 turbidity unit
Coliform bacteria	1 per 100 ml

*1 mg/liter = 0.0584 grain/gal.
†Level varies inversely with annual average of the maximum daily air temperature.

for under special circumstances, but routine analysis for pathogens in water is not recommended. *See* WATER MICROBIOLOGY.

Water quality standards. For drinking water, regulations prescribe the maximum contaminant level (MCL) for a number of chemical species, turbidity, and coliform bacteria (see table). Experience has demonstrated that water that meets these specifications will, with high probability, not produce human disease. *See* WATER SUPPLY ENGINEERING. [A.E.G.]

Water-borne disease Disease transmitted by drinking water or by contact with potable or bathing water. Certain bacterial, protozoan, and helminthic diseases of humans are commonly or exclusively water-borne (see table). The bacterial and protozoan pathogens are not normal to the water environment and do not multiply in natural waters. They gain entrance to water primarily from pollution by human excrement and less significantly from pollution by domestic and wild animals. In contrast, the helminthic pathogens may have a water animal as an intermediate host and may have a temporary free-living stage in water. *See* EPIDEMIOLOGY. [S.C.R.]

Water-borne disease

Type	Disease	Agent
Bacterial	Typhoid fever	*Salmonella typhosa*
	Asiatic cholera	*Vibrio comma*
	Bacillary dysentery	*Shigella* sp.
Protozoan	Amebic dysentery	*Endamoeba histolytica*
Helminthic	Ascariasis (roundworm infection)	*Ascaris lumbricoides*
	Trichuriasis (whipworm infection)	*Trichuris trichura*
	Schistosomiasis	*Schistosoma* sp.
	Dracontiasis	*Dracunculus medinensis*

Water conservation The protection, development, and efficient management of water resources for beneficial purposes. Water occurs both underground and on the surface. The volume of groundwater greatly exceeds that of all freshwater lakes and reservoirs combined. It occurs in several geologic formations (aquifers) and at various depths. Streams supply most of the water needs of the United States. Lakes, ponds, swamps, and marshes, like reservoirs, represent stored streamflow. The oceans and salty or brackish sounds, bays, bayous, or estuaries represent almost unlimited potential freshwater sources. *See* HYDROLOGY.

Streams have traditionally served for waste disposal. Towns and cities, industries, and mines provide thousands of pollution sources. Pollution dilution requires large amounts of water. Treatment at the source is safer and less wasteful than flushing untreated or poorly treated wastes downstream. However, sufficient flows must be released to permit the streams to dilute, assimilate, and carry away the treated effluents. *See* WATER POLLUTION.

The problems of water management involve economic, social, and intangible values. Efforts to plan and develop river systems for multiple purposes often generate conflicts among different water uses, for example, irrigation versus navigation, or hydropower versus salmon or trout fisheries, wildlife, national park, wilderness, or historic resources.

Water management technology involves the application of biological and engineering principles to attain desired goals. Biological methods include growing upland vegetation having low moisture requirements. Mechanical methods include the practice of water spreading, which is utilized to desilt floodwaters and to promote the percolation of water into the soil for crop use and groundwater recharge. Water that would be a strong pollutant of streams and lakes may be spread on the land. The avoidance of water waste takes several forms. Recycling has permitted huge savings of water, especially in petroleum plants, chemical factories, and steel mills.

Watershed control is an approach to planning, development, and management that rests on the established interdependence of water, land, and people. Conditions on the land are often directly reflected in the behavior of streamflow and in the accumulation of groundwater. Coordination of structures and land-use practices is sought to prevent erosion, promote infiltration, and retard high flows. River basins may be large and complex watersheds. Basin projects may involve systems of multipurpose storage reservoirs, intensive programs of watershed protection, and improvement and management of farm, forest, range, and urban lands. They may call for scientific research, industrial development, health and educational programs, and financial arrangements to stimulate local initiative. [B.F.]

Water hammer A phenomenon in a piping system caused by an abrupt change in flow at some section. The disturbance, which might be caused by changing a valve opening, travels throughout the system at the speed of a sound wave in the fluid.

The magnitude of the pressure wave travels unaltered (except for fluid friction) until it arrives at a change in pipe characteristics, such as a branch, a size change, an orifice, or perhaps a reservoir or a pump. Such a change in conditions is referred to as a boundary condition in the problem solution, and the properties and characteristics of the boundary are solved simultaneously with the pipe disturbance equation to determine how the wave proceeds. *See* HYDRODYNAMICS. [V.L.S.]

Water microbiology An aspect of microbiology that deals with the normal and adventitious microflora of natural and artificial water bodies. Two major lines of investigation have been pursued: the characterization of the normal flora and its activities, and the public health aspects of potable and recreational waters. *See* WATER ANALYSIS; WATER-BORNE DISEASE.

Qualitatively, the normal microbial flora is controlled by such factors as temperature, salinity, and chemical composition of the water, and is largely independent of geographical location. The numbers of bacteria in unpolluted waters are low. Samples collected in the oceans or large lakes remote from land generally give viable counts of less than 100 per milliliter; direct microscopic counts, which include both viable and nonviable bacteria, are higher.

Surface waters may receive large numbers of bacteria, including pathogens, by runoff from the land surfaces, and by pollution from human and animal sources. Bacteria thus introduced are eliminated quite rapidly either by failure to grow in an unfavorable environment, or by such processes as sedimentation and predation. Thus a natural water not subject to sufficient pollution to change its character maintains its indigenous flora; this fact permits, within limits, the deliberate disposal of wastes into a water body. *See* WATER POLLUTION.

The bacterial flora carries out the same types of biochemical processes in water and sediments as it does on land and is presumed responsible, at least in part, for mineralization of organic matter and completion of the cycle of nutrients. In contrast to the situation on land, the microscopic algae are the primary producers of organic matter in the water. *See* MICROBIOLOGY; WATER PURIFICATION. [S.C.R.]

Water pollution Any change in natural waters which may impair their further use, caused by the introduction of organic or inorganic substances, or a change in temperature of the water. The growth of population and the concomitant expansion in industrial and agricultural activities have rapidly increased the importance of the field of water-pollution control. In the attack on environmental pollution, higher standards for water cleanliness have been adopted by state and Federal governments, as well as by interstate organizations.

The field of water-pollution control encompasses a part of the broader field of sanitary or environmental engineering. It includes some aspects of chemistry, hydrology, biology, and bacteriology, in addition to public administration and management.

Relation to water supply. Water-pollution control is closely allied with the water supplies of communities and industries because both generally share the same water resources. There is great similarity in the pipe systems that bring water to each home or business property, and the systems of sewers or drains that subsequently collect the wastewater and conduct it to a treatment facility. Treatment should prepare the flow for return to the environment so that the receiving watercourse will be suitable for beneficial uses such as general recreation, and safe for subsequent use by downstream communities or industries. *See* SEWAGE COLLECTION SYSTEMS; WATER SUPPLY ENGINEERING.

Wastewaters. Domestic wastewaters result from the use of water in dwellings of all types, and include both water after use and the various waste materials added: body wastes, kitchen wastes, household cleaning agents, and laundry soaps and detergents. The character of these waste materials is such that they cause significant degradation of receiving waters, and they may be a major factor in spreading water-borne diseases, notably typhoid and dysentery.

In contrast to the general uniformity of substances found in domestic wastewaters, industrial wastewaters show increasing variation as the complexity of industrial processes rises. The table lists major industrial categories along with the undesirable characteristics of their wastewaters.

General nature of industrial wastewaters

Industry	Processes or waste	Effect
Brewery and distillery	Malt and fermented liquors	Organic load
Chemical	General	Stable organics, phenols, inks
Dairy	Milk processing, bottling, butter and cheese making	Acid
Dyeing	Spent dye, sizings, bleach	Color, acid or alkaline
Food processing	Canning and freezing	Organic load
Laundry	Washing	Alkaline
Leather tanning	Leather cleaning and tanning	Organic load, acid and alkaline
Meat packing	Slaughter, preparation	Organic load
Paper	Pulp and paper manufacturing	Organic load, waste wood fibers
Steel	Pickling, plating, and so on	Acid
Textile manufacture	Wool scouring, dyeing	Organic load, alkaline

Any natural watercourse contains dissolved gases normally found in air in equilibrium with the atmosphere. In this way fish and other aquatic life obtain oxygen for their respiration. The amount of oxygen which the water holds at saturation depends on temperature and follows the law of decreased solubility of gases with a temperature increase. Degradable or oxidizable substances in wastewaters deplete oxygen through the action of bacteria and related organisms which feed on organic waste materials, using available dissolved oxygen for their respiration. If this activity proceeds at a rate fast enough to depress seriously the oxygen level, the natural fauna of a stream is affected; if the oxygen is entirely used up, a condition of oxygen exhaustion occurs which suffocates aerobic organisms in the stream. Under such conditions the stream is said to be septic and is likely to become offensive to the sight and smell.

An increasing amount of attention has been given to thermal pollution, the raising of the temperature of a waterway by heat discharged from the cooling system or effluent wastes of an industrial installation. This rise in temperature may sufficiently upset the ecological balance of the waterway to pose a threat to the native life-forms. Thermal pollution may be combated by allowing the wastewater to cool before it empties into the waterway.

With the passage of time, the waste loads imposed on streams exceeded the ability of the receiving water to assimilate them. The first attempts at wastewater treatment were made by artificially providing means for the purification of wastewaters as observed in nature. These forces included sedimentation and exposure to sunlight and atmospheric oxygen, either by agitated contact or by filling the interstices of large stone beds intermittently as a means of oxidation. In later years testing stations were set up by municipalities and states for experimental work. This led to treatment methods based on modern chemistry and biology.

Current status. Modern water-pollution engineers or chemists have a wealth of published information, both theoretical and practical, to assist them. While research continues, they can draw on established practices for the solution to almost any problem. A challenging problem has been the handling of radioactive wastes. Reduction in volume, containment, and storage constitute the principal attack on this problem. *See* RADIOACTIVE WASTE MANAGEMENT.

Public desire for complete water pollution control continues, but there is an increasing realization that solution to the problem is costly. While cities have had little concern about the initial construction cost because of the large Federal share, the expenditures for operation fall entirely on the local community.

There are strong manifestations of improved quality in the waters throughout the United States. This is apparent not only in chemical and biological measurements, but in more readily observed effects such as better appearance, eliminated smells, and the return of fish life to watercourses which had become "biological deserts" because of the effects of pollution from municipal and industrial wastewaters. *See* WATER PURIFICATION. [R.E.Fu.]

Water purification The treatment of potable water to destroy disease-producing agents. Purification is employed in all but economically underdeveloped areas and, where practiced, has reduced water-borne diseases to the vanishing point. Chemical disinfection, either by itself or in combination with storage, sedimentation, or filtration, is the universal practice. Chlorine is the most economical agent for disinfection and is used in most water systems. Disinfection by physical agents is possible, but cost considerations have limited the application of such procedures. Boiling is the method of choice for treating small quantities of water of dubious quality for individual or household use. Utraviolet irradiation has been employed for the disinfection of swimming-pool waters and for the treatment of military and some municipal water supplies where economic factors are not of primary significance.

Ozone treatment is employed by a limited number of public water supply systems, mainly in France. Ozone is generated by the discharge of high-voltage electricity through dry air between stationary electrodes. The water is sprayed into an atmosphere of ozone, or the ozonized air is discharged into the water in a mixing chamber. *See* CHLORINE; SANITARY ENGINEERING; WATER-BORNE DISEASE; WATER MICROBIOLOGY; WATER TREATMENT. [S.C.R.]

Water softening A water-treatment process by which undesirable cations (of calcium and magnesium) are removed from hard waters. (Hardness is usually expressed as calcium carbonate equivalents in parts per million.) The presence of these cations in water is undesirable for household purposes, boiler feed, food processing, and chemical processing, because of reactions that form soap scum, boiler scale, and unwanted by-products. *See* WATER TREATMENT.

Hard waters may be softened by precipitation processes which involve the use of hydrated lime, sometimes along with soda ash; cation-exchange processes which involve the use of sodium-cation exchangers (zeolites) or hydrogen cation-exchangers (usually resins in bead form); or combinations of these. The choice of a process depends on a number of factors, among which are the composition of the hard water, the end uses of the softened water, the type or types of hardnesses to be removed, the degree of removal required, and the relative processing costs. *See* ION EXCHANGE; MOLECULAR SIEVE; SALINE WATER RECLAMATION; SURFACTANT; WATER TREATMENT; ZEOLITE. [E.No.]

Water supply engineering A branch of civil engineering concerned with the development of sources of supply, transmission, distribution, and treatment of water. The term is used most frequently in regard to municipal waterworks, but applies also to water systems for industry, irrigation, and other purposes.

Sources. Underground waters, rivers, lakes, and reservoirs, the primary sources of fresh water, are replenished by rainfall. Some of this water flows to the sea through surface and underground channels, some is taken up by vegetation, and some is lost by evaporation.

Water obtained from subsurface sources, such as sands and gravels and porous or fractured rocks, is called groundwater. Groundwater flows toward points of discharge in river valleys and, in some areas, along the seacoast. The flow takes place in water-bearing strata known as aquifers. In an unconfined stratum the water table is the top or surface of the groundwater. The water table elevation and artesian pressure may vary substantially with the seasons. If pumpage exceeds recharge for an extended period, the aquifer is depleted and the water supply lost. *See* AQUIFER; ARTESIAN SYSTEMS; GROUNDWATER HYDROLOGY; WATER TABLE.

Springs occur at the base of sloping ground or in depressions where the surface elevation is below the water table, or below the hydraulic gradient in an artesian aquifer from which the water can escape. *See* SPRING (HYDROLOGY).

Wells are vertical openings, excavated or drilled, from the ground surface to a water-bearing stratum or aquifer. Pumping a well lowers the water level in it, which in turn forces water to flow from the aquifer. Dug wells, several feet in diameter, are frequently used to reach shallow aquifers, particularly for small domestic and farm supplies. Large-capacity dug wells or caisson wells, in coarse sand and gravel, are used frequently for municipal supplies. Drilled wells are sometimes several thousand feet deep. *See* WELL.

Natural sources, such as rivers and lakes, and impounding reservoirs are sources of surface water. Water is withdrawn from rivers, lakes, and reservoirs through intakes. The simplest intakes are pipes extending from the shore into deep water. *See* DAM; RESERVOIR; SURFACE WATER.

Transmission and distribution. The water from the source must be transmitted to the community or area to be served and distributed to the individual customers.

The major supply conduits, or feeders, from the source to the distribution system are called mains or aqueducts. The oldest and simplest type of aqueducts, especially for transmitting large quantities of water, are canals. Used to transmit water through ridges or hills, tunnels may follow the hydraulic grade line and flow by gravity or may be built below the grade line to operate under considerable pressure. Pipelines are a common type of transmission main, especially for moderate supplies not requiring large aqueducts or canals. *See* CANAL; PIPELINE; TUNNEL.

Included in the distribution system are the network of smaller mains branching off from the transmission mains, the house services and meters, the fire hydrants, and the distribution storage reservoirs. The network is composed of transmission or feeder mains, usually 12 in. (30 cm) or more in diameter, and lateral mains along each street, or in some cities along alleys between the streets. The mains are installed in grids so that lateral mains can be fed from both ends where possible. Valves at intersections of mains permit a leaking or damaged section of pipe to be shut off with minimum interruption of water service to adjacent areas.

Pumps are required wherever the source of supply is not high enough to provide gravity flow and adequate pressure in the distribution system. Pumping stations usually include two or more pumps, each of sufficient capacity to meet demands when one unit is down for repairs or maintenance. The station must also include piping and valves arranged so that a break can be isolated quickly without cutting the whole station out of service. *See* WATER TREATMENT. [R.H.]

Water table The upper surface of the zone of saturation in permeable rocks not confined by impermeable rocks. It may also be defined as the surface underground at which the water is at atmospheric pressure. Saturated rock may extend a little above this level, but the water in it is held up above the water table by capillarity and is under less than atmospheric pressure; therefore, it is the lower part of the capillary fringe and is not free to flow into a well by gravity. Below the water table, water is free to move under the influence of gravity.

The position of the water table is shown by the level at which water stands in wells penetrating an unconfined water-bearing formation. Where a well penetrates only impermeable material, there is no water table and the well is dry. But if the well passes through impermeable rock into water-bearing material whose hydrostatic head is higher than the level of the bottom of the impermeable rock, water will rise approximately to the level it would have assumed if the whole column of rock penetrated had been permeable. This is called artesian water. *See* ARTESIAN SYSTEMS; GROUNDWATER HYDROLOGY; WELL. [A.N.S./R.K.L.]

Water treatment Physical and chemical processes for making water suitable for human consumption and other purposes. Drinking water must be bacteriologically safe, free from toxic or harmful chemicals or substances, and comparatively free of turbidity, color, and taste-producing substances. Excessive hardness and high concentration of dissolved solids are also undesirable, particularly for boiler feed and industrial purposes. The treatment processes of greatest importance are sedimentation, coagulation, filtration, disinfection, softening, and aeration.

Sedimentation occurs naturally in reservoirs and is accomplished in treatment plants by basins or settling tanks. Plain sedimentation will not remove extremely fine or colloidal material within a reasonable time, and the process is used principally as a preliminary to other treatment methods.

Fine particles and colloidal material are combined into masses by coagulation. These masses, called floc, are large enough to settle in basins and to be caught on the surface of filters.

Suspended solids, colloidal material, bacteria, and other organisms are filtered out by passing the water through a bed of sand or pulverized coal, or through a matrix of fibrous material supported on a perforated core. Soluble materials such as salts and metals in ionic form are not removed by filtration. *See* FILTRATION.

There are several methods of treatment of water to kill living organisms, particularly pathogenic bacteria; the application of chlorine or chlorine compounds is the most common. Less frequently used methods include the use of ultraviolet light, ozone, or silver ions. Boiling is the favorite household emergency measure.

Municipal water softening is common where the natural water has a hardness in excess of 150 parts per million. Two methods are used: (1) The water is treated with lime and soda ash to precipitate the calcium and magnesium as carbonate and hydroxide, after which the water is filtered; (2) the water is passed through a porous cation exchanger which has the ability of substituting sodium ions in the exchange medium for calcium and magnesium in the water. For high-pressure steam boilers or some other industrial processes, almost complete deionization of water is needed, and treatment includes both cation and anion exchangers. *See* ION EXCHANGE.

Aeration is a process of exposing water to air by dividing the water into small drops, by forcing air through the water, or by a combination of both. Aeration is used to add oxygen to water and to remove carbon dioxide, hydrogen sulfide, and taste-producing gases or vapors. *See* WATER POLLUTION; WATER PURIFICATION; WATER SUPPLY ENGINEERING. [R.H.]

Water-tube boiler A steam boiler in which water circulates within tubes and heat is applied from outside the tubes. The outstanding feature of the water-tube boiler is the use of small tubes [usually 1–3 in. (2.5–7.5 cm) outside diameter] exposed to the products of combustion and connected to steam and water drums which are shielded from these high-temperature gases. Thus, possible failure of boiler parts exposed to direct heat transfer is restricted to the small-diameter tubes and, in the event of failure, the energy released is reduced and explosion hazards are minimized.

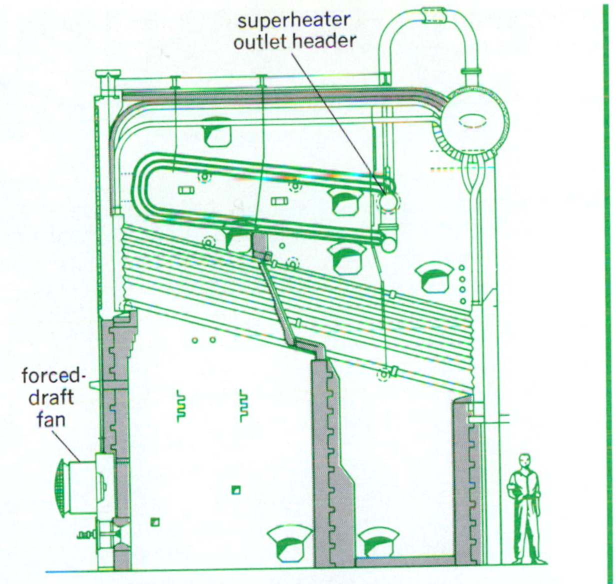

Diagram of a straight-tube-type boiler.

There are many types of water-tube boilers but, in general, they can be grouped into two categories: the straight-tube (see illustration) and the bent-tube types. In essence, both types consist of banks of parallel tubes which are connected to, or by, headers or drums. Most modern water-tube boilers utilize a water-cooled surface in the furnace, and this surface is an integral part of the boiler's circulatory system. Further, modern water-tube boilers generally incorporate the use of superheaters, economizers, or air heaters to utilize more efficiently the heat from the fuel and to provide steam at a high potential for useful work in an engine or turbine. *See* BOILER; STEAM; STEAM-GENERATING FURNACE; STEAM-GENERATING UNIT. [G.W.K.]

Water tunnel A hydrodynamic test and research tool or facility, comprising a well-guided and controlled stream of water. The water tunnel is related to towing tanks but differs in that in the towing tank the test object moves and the water is essentially at rest, whereas in the water tunnel the test object is at rest (or rotating, for example, a propeller) and the water is moving. With respect to arrangement and operation, the water tunnel is similar to the subsonic wind tunnel, except for the difference in the test medium. *See* TOWING TANK; WIND TUNNEL.

A typical water tunnel forms a closed loop of conduit with a carefully designed, strongly contracted test, or working, section, a circulating pump, and corner elbows shaped as they are in wind tunnels to minimize flow distortions and disturbances. The contraction in front of the test section further serves to produce a stream with a uniform velocity distribution. The size of a water tunnel is usually expressed as the diameter of its test section. Existing water tunnels range in size from a few inches to 48 in. (1.2 m). The test section is on the upper leg to facilitate lowering the water level below it for insertion of test objects (see illustration).

An essential characteristic of most water tunnels is the possibility of changing the absolute pressure in the tunnel without necessarily changing the velocity of the test stream. This characteristic is important for investigations involving cavitation. *See* CAVITATION.

Most water tunnels are operated with the entire test circuit filled with water. Consequently they can serve only the investigations of completely submerged bodies or of conduits and machinery such as pumps or turbines that are completely filled with water. A few tunnels have been built to operate with a free water surface in the test section. Such free-surface water

Water tunnel at Pennsylvania State University.

tunnels are used to investigate open-surface structures such as ship models, with the model standing still. [G.F.W.]

Waterfowl Aquatic birds which constitute the order Anseriformes. There are two families: the Anhimidae of the tropical New World, known as the screamers, and the Anatidae, the swans, ducks, and geese. Waterfowl range in size from 1 to 5 ft (0.3 to 1.5 m) and have partially or completely webbed toes and oil glands.

The screamers are unique in having extensive developments of air spaces between the skin and the body. They have two pairs of pointed spurs on the anterior part of each wing. The toes are only slightly webbed at the base. There are three species, all of which frequent marshy areas and savannas in their range.

The best-known and most widely distributed anseriform birds are the ducks, which are grouped into 10 tribes. The bills are long and flat, the necks are elongate, and three of the toes are webbed. The young are covered with a soft down and leave the nest soon after hatching. Since the flight feathers are lost when they molt, the birds are incapable of flying for these short periods. The plumage is waterproof as in most aquatic birds. Some species are good divers as well as excellent fliers and swimmers. Ducks feed on vegetation and fine particulate material in water. *See* ANSERIFORMES. [C.B.C.]

Watermelon The edible fruit of *Citrullus lanatus*, of the family Cucurbitaceae. The plant is an annual prostrate vine with multiple stems.

The numerous cultivars of watermelon are highly diverse in fruit size (5–85 lb; 2.3–38.3 kg), shape (round, oval, oblong-cylindrical), rind color (very light to very dark green and often striped or mottled), flesh color (red, pink, orange, yellow, white), and seed size and color. The flesh contains 6–12% sugar, depending upon variety and condition of growth, with 8% sugar being acceptable on most markets.

Watermelon juice of the sweet cultivars can be reduced to edible sugar and syrup. Watermelon seeds are relished as food in some Near East countries and in China. Fresh watermelon flesh has a unique melting quality that has proved impossible to preserve in palatable form through any processing technique. The flesh consists of water (91%), fiber, and sugar, and little else of obvious nutritional value, but may have some as yet unproved health-promoting value. It is said to have diuretic properties, and both frozen concentrate and canned juice have been available for the treatment of nephritis. The seeds also are said to contain substances effective in the control of hypertension. [C.F.A.]

Waterpower Power developed from movement of masses of water. Such movement is of two kinds: the falling of streams through the force of gravity, and the rising and falling of tides through lunar (and solar) gravitation.

While that part of solar energy expended to lift water vapor against Earth gravity is a minute fraction of the total, the absolute amount of energy that is theoretically recoverable from resulting streams is an enormous but unknown quantity. Of this, but a tiny portion is actually accessible for harnessing.

Hydro plants, with their initial high cost and generally long distances from major load centers, must compete with the fuel-fired stations and large nuclear plants. Large dam sites usually must be justified not alone on the value of the power developed, but also on the benefits from flood control, irrigation, and recreation. Problems of migrating fish, conservation, and preservation of esthetic values are also factors. On the other hand, waterpower developments add greatly to power-system flexibility in meeting peak and emergency loads. *See* ELECTRIC POWER GENERATION; NUCLEAR REACTOR; POWER PLANT.

The capacity of hydro plants cannot be counted on for perpetuity because of gradual filling of reservoirs with sediment. This effect is serious for irrigation, flood control, and navigation. Even when a lake behind a power dam becomes filled completely with silt, electric power can be generated on the run-of-the-river flow, although output would vary with stream flow.

In a pumped-storage hydroelectric system, water is pumped from a stream or lake to a reservoir at a higher elevation. Pumping up to a storage reservoir is most commonly done by reversing the hydraulic turbine and generator. The generator becomes a motor driving the turbine as a pump. Power is drawn from the power system at night or on weekends when demand is low. The cost of pumping power is low, whereas the power generated from pumped storage at peak periods is valuable. *See* HYDRAULIC TURBINE.

A portion of the kinetic energy of the rotation of the Earth appears as ocean tides. The mean tide of all the oceans has been calculated as 2.1 ft (0.64 m), and the mean power as $54{,}000 \times 10^6$ hp (40 terawatts) or, on a yearly basis, the equivalent of 36×10^{12} kWh. Unfortunately, only a minute amount of this is likely to be harnessed for use. *See* ENERGY SOURCES; TIDAL POWER. [C.A.S.]

Waterspout An intensely whirling, funnel-shaped vortex, extending from a cumulus-type cloud down several hundred to several thousand feet to the water surface. The visible funnel consists mostly of atmospheric water vapor condensed because of lower pressure in the vortex: water or salt spray drawn from the underlying surface also contributes to the structure and visibility of the funnel (see illustration). Diameters range from a few feet to several hundred, with winds occasionally strong enough to overturn small vessels.

Most waterspouts occur in moist tropical air under unstable conditions favorable for thunderstorms, from which they frequently hang. Nearly all dissipate on passing inland. Although most common in the tropical oceans, waterspouts are observed in higher latitudes, particularly in summer; they

Waterspout. (*Official U.S. Navy photograph*)

also occur over inland lakes. Many of the larger and more violent spouts are true tornadoes, formed in association with thunderstorms over land and passing out to sea. *See* TORNADO. [C.W.N.]

Watt-hour meter An electricity meter which measures and registers the integral, with respect to time, of the power in the circuit in which it is connected. In effect, it is an electric motor, the torque of which is proportional to the electric power in the circuit. The speed of the rotor is proportional to the torque, making each revolution of the rotor a measurement in watt-hours. Summation of the watt-hours is accomplished by gearing a counter, or register, to the rotor. *See* MOTOR.

The basic elements of a watt-hour meter are the stator, rotor, retarding magnet or magnets, register, and meter housing.

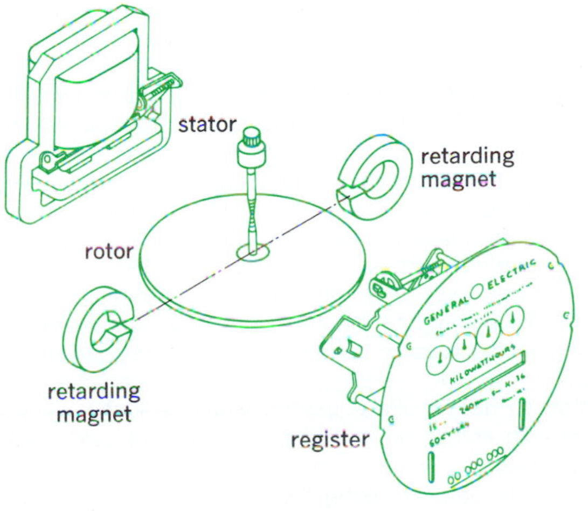

Basic elements of the induction-type watt-hour meter. (*General Electric Co.*)

The mercury and commutator types of waft-hour meter are used for measuring energy on a dc circuit. The commutator type may also be used on ac circuits if all windings are of air-core construction. The induction type is the common meter found in homes. It is used for measuring energy on an ac circuit (see illustration). The potential-circuit winding of the stator is made highly inductive to obtain a quadrature-time relationship between the potential-circuit and the current-circuit working fluxes in the disk air gap. These fluxes, displaced in time and space, produce a rotor-disk torque proportional to the circuit power. Retarding magnets control the rotor speed, making it proportional to the power. The register, which is geared to the rotor, records the watt-hours. *See* ELECTRIC POWER MEASUREMENT. [G.R.St.; J.A.]

Wattmeter An instrument that measures electric power. For a complete discussion of the power in various types of electric circuits *see* ELECTRIC POWER MEASUREMENT.

A variety of wattmeters is available to measure the power in ac circuits. They are generally classified by names descriptive of their operating principles. Determination of power in dc circuits is almost always done by separate measurements of voltage and current. However, some of the instruments described will also function in dc circuits, if desired.

Electrodynamic wattmeter. Probably the most useful instrument in the measurement of ac power at commercial frequencies is the indicating (deflecting) electrodynamic wattmeter. It is similar in principle to the double-coil dc ammeter or voltmeter in that it depends on the interaction of the fields of two sets of coils, one fixed and the other movable. The moving coil is suspended, or pivoted, so that it is free to rotate through a limited angle about an axis perpendicular to that of the fixed coils. As a single-phase wattmeter, the moving (potential) coil carries a current proportional to the voltage applied to the measured circuit, and the fixed (current) coil carries the load current. This arrangement of coils is due to the practical necessity of designing current coils of relatively heavy conductors to carry large values of current. The potential coils can be lighter because the operating current is limited to low values.

Thermal wattmeter. The operation of a thermal wattmeter is based on the heating effect of current, the I^2R relationship. It is applicable to both direct current and alternating current and is usable at frequencies up into the radio range without regard to waveform. Various electronic wattmeters for audio and radio power measurements are largely based on this principle. Thermocouples are commonly used as transducers, and the derived voltage is available for the common types of dc indicators and for potentiometers of either the manual or automatic self-balancing type. A limitation is that it is relatively slow in response and may be affected by temperature.

Electrostatic wattmeter. The elementary quadrant electrometer has been adapted for power measurements. The mechanism consists of two quadrants charged by the voltage drop across a noninductive shunt resistance through which the load current passes. The line voltage is applied between a moving vane and one of the quadrants. The deflection of the indicator is proportional to the average power. [D.B.S.]

Wave (physics) The general term applied to the description of a disturbance which propagates from one point in a medium to other points without giving the medium as a whole any permanent displacement.

Waves are generally described in terms of their amplitude, and how the amplitude varies with both space and time. The actual description of the wave amplitude involves a solution of the wave equation and the particular boundary conditions for the case being studied. *See* WAVE EQUATION; WAVE MOTION.

Acoustic waves, or sound waves, are a particular kind of the general class of elastic waves. Elastic waves are propagated in media having two properties, inertia and elasticity. Electromagnetic waves (for example, light waves and radio waves) are not elastic waves and therefore can travel through a vacuum. The velocity of the wave depends on the medium through which the wave travels. *See* ELECTROMAGNETIC WAVE; STANDING WAVE. [W.J.G.]

Wave equation The name given to certain partial differential equations in classical and quantum physics which relate the spatial and time dependence of physical functions. In this article the classical and quantum wave equations are discussed separately, with the classical equations first for historical reasons.

In classical physics the name wave equation is given to the linear, homogeneous partial differential equations which have the form of Eq. (1). Here v is a parameter with the dimensions

$$\left[\nabla^2 - \frac{1}{v^2}\frac{\partial^2}{\partial t^2}\right] f(\mathbf{r},t) = 0 \quad (1)$$

of velocity; $\mathbf{r}$ represents the space coordinates x, y, z; t is the time; and ∇^2 is Laplace's operator defined by Eq. (2). The

$$\nabla^2 = \frac{\partial^2}{\partial x^2} + \frac{\partial^2}{\partial y^2} + \frac{\partial^2}{\partial z^2} \quad (2)$$

function $f(\mathbf{r},t)$ is a physical observable; that is, it can be measured and consequently must be a real function.

The simplest example of a wave equation in classical physics is that governing the transverse motion of a string under tension and constrained to move in a plane.

A second type of classical physical situation in which the wave equation, Eq. (1), supplies a mathematical description of the physical reality is the propagation of pressure waves in a fluid medium. Such waves are called acoustical waves, the propagation of sound being an example. A third example of a classical physical situation in which Eq. (1) gives a description of the phenomena is afforded by electromagnetic waves. In a region of space in which the charge and current densities are zero, Maxwell's equations for the photon lead to the wave equations, Eqs. (3). Here $\mathbf{E}$ is the electric field strength and $\mathbf{B}$

$$\left[\nabla^2 - \frac{1}{c^2}\frac{\partial^2}{\partial t^2}\right] E(\mathbf{r},t) = 0$$
$$\left[\nabla^2 - \frac{1}{c^2}\frac{\partial^2}{\partial t^2}\right] B(\mathbf{r},t) = 0 \quad (3)$$

is the magnetic flux density; they are both vectors in ordinary space. The parameter c is the speed of light in vacuum. *See* ELECTROMAGNETIC RADIATION; MAXWELL'S EQUATIONS.

The nonrelativistic Schrödinger equation is an example of a quantum wave equation. Relativistic quantum-mechanical wave equations include the Schrödinger-Klein-Gordon equation and the Dirac equation. *See* QUANTUM MECHANICS; RELATIVISTIC QUANTUM THEORY. [D.L.We.]

Wave mechanics The modern theory of matter holding that elementary particles (such as electrons, protons, and neutrons) have wavelike properties. In 1924 Louis de Broglie postulated that the same wave–corpuscle duality which was then known to exist in the case of light might also occur in matter; this hypothesis was subsequently verified experimentally. This theory of matter has become the highly successful quantum mechanics of the present day. *See* QUANTUM MECHANICS. [E.G.]

Wave motion The process by which a disturbance at one point in space is propagated to another point more remote from the source with no net transport of the material of the medium itself. For example, sound is a form of wave motion; wind is not. Wave motion can occur only in a medium in which energy can be stored in both kinetic and potential form. In a mechanical medium, kinetic energy results from inertia and is stored in the velocity of the molecules, while potential energy results from elasticity and is stored in the displacement of the molecules.

Media. In a free traveling wave (as distinguished from a stationary or standing wave) one part of the medium disturbs an adjacent part, thereby imparting energy to it. This portion of the medium, in turn, disturbs another part, thereby causing a flow of energy in a given direction away from the source. More technically, wave propagation is the result of kinetic energy at one point being transferred into potential energy at an adjacent point, and vice versa. The rate of travel of the disturbance, or velocity of propagation, is determined by the constants of the medium. A stationary wave is the combination of two waves of the same frequency and strength traveling in opposite directions so that no net transfer of energy away from the source takes place. A standing wave is the same but with the returning wave (toward the source) being of lesser intensity than the outwardly traveling wave so that a net transfer of energy away from the source does take place. *See* STANDING WAVE.

Wave motion can occur in a vacuum (electromagnetic waves), in gases (sound waves), in liquids (hydrodynamic waves), and in solids (vibration waves). Electromagnetic waves can also travel in gases, liquids, and solids provided that the electrical conductivity of the medium is not perfect or that the imaginary part of the dielectric constant is not infinitely great. By current usage, elastic waves propagated in gases, liquids, and solids, regardless of whether one can hear them or not, are called acoustic waves.

Fundamental relations. A wave is commonly referred to in terms of either its wavelength or its frequency. In any type of wave motion, these two quantities are related to a third quantity, velocity of propagation, by the simple relation in the equation below, where f = frequency, λ = wavelength, and

$$f\lambda = c$$

c = velocity of propagation. The period T is the reciprocal of the frequency, and the amplitude A is the maximum magnitude taken on by the variable of the wave at a given point in space. It is a basic property of wave motion that the frequency of a wave remains constant under all circumstances except for a relative motion between the source of the wave and the observer. The velocity of propagation is dependent on the properties of the medium (and, sometimes, also on the frequency) and the wavelength will vary with the velocity in accordance with the equation above.

Electromagnetic waves. The media in which electromagnetic waves travel possess no elasticity or inertia, but rather the ability to store energy in the electric and magnetic fields. J. C. Maxwell recognized in about 1863 that the basic equations governing these fields could be combined to yield an equation resembling the wave equation for mechanical wave motion. Thus he predicted the existence of electromagnetic waves which had not been suspected theretofore. Later, electromagnetic waves proved to be identical with light waves. *See* ELECTROMAGNETIC RADIATION; LIGHT; MAXWELL'S EQUATIONS; WAVE EQUATION. [L.L.B.]

Motion in fluids. Wave motion within a fluid is generated by successive compression and expansion of adjacent volume elements. Because compression and expansion of an ordinary fluid can only proceed along the direction of propagation of the disturbance, waves within a fluid are mostly longitudinal waves.

Waves in a fluid can be classified as compression waves and expansion waves, according to whether the disturbance is a compression or an expansion. They can further be classified according to the amplitude of the disturbance and the chemical nature of the fluid. For example, waves of small amplitude are called acoustic (or sound) waves; compression waves propagating in chemically inert fluids are called shock waves; waves propagating in the Earth are seismic waves; and waves of large amplitude generated by rapid chemical reactions in explosive fluids are called detonation waves and can propagate much faster than sound waves. Waves in an electrically conducting fluid in the presence of strong magnetic fields are called magnetohydrodynamic waves. *See* MAGNETOHYDRODYNAMICS; SEISMOLOGY; SHOCK WAVE; SOUND. [S.-C.Li.]

Motion in liquids. Disturbances propagated at a gas-liquid interface are primarily dependent upon the gravitational fluid property (surface tension and viscosity being of secondary importance). Wave motions which occur in confined fluids (either liquid or gaseous) are primarily dependent upon the elastic property of the medium.

Oscillatory waves may be generated in a rectangular channel by a simple harmonic translation of a vertical wall forming one end of the flume. A standing wave can be considered to be composed of two equal oscillatory wave trains traveling in opposite directions.

A solitary wave consists of a single crest above the original liquid surface which is neither preceded nor followed by another elevation or depression of the surface. Such a wave is generated by the translation of a vertical wall starting from an initial position at rest and coming to rest again some distance downstream. In practice, solitary waves are generated by a motion of barges in narrow waterways or by a sudden change in the rate of inflow into a river; they are therefore related to a form of flood wave. [D.R.F.H.]

Wave optics The branch of optics which treats of light (or electromagnetic radiation in general) with explicit recognition of its wave nature. The counterpart to wave optics is ray optics or geometrical optics, which does not assume any wave character but treats the propagation of light as a straight-line phenomenon except for changes of direction induced by reflection or refraction. *See* ELECTROMAGNETIC RADIATION; GEOMETRICAL OPTICS; OPTICS.

Any optical phenomenon which is correctly describable in terms of geometrical optics can also be correctly described in terms of wave optics. However, the many phenomena of interference, diffraction, and polarization are incontrovertible evidence of the wave nature of light, and geometrical optics often gives an incomplete or incorrect description of the behavior of light in an optical system. This is especially true if changes of refractive index occur within a space which is of the order of several wavelengths of the light. *See* DIFFRACTION; INTERFERENCE OF WAVES; POLARIZED LIGHT. [R.C.L.]

Wave packet In wave phenomena, a superposition of waves of differing lengths, so phased that the resultant amplitude is negligibly small except in a limited portion of space whose dimensions are the dimensions of the packet. If the reciprocal wavelengths $k = \lambda^{-1}$ forming a one-dimensional packet lie in a band Δk, the minimum dimension of the packet is $\Delta x \cong (2\pi\Delta k)^{-1}$. When all the component waves move in the same direction, the packet speed is the group velocity $v_g = df/dk$ evaluated at the mean k; f is the frequency. When the phase velocity c depends on λ, $v_g \neq c$, and Δx changes with time. *See* GROUP VELOCITY; PHASE VELOCITY; QUANTUM MECHANICS. [E.G.]

Wave-shaping circuits Electronic circuits used to create or modify a specified time-varying electrical quantity, usually voltage or current, using combinations of electronic devices, such as vacuum tubes or transistors, and circuit elements including resistors, capacitors, and inductors.

The square wave can be obtained by amplifying a sinusoidal time-varying voltage and removing all but the section near the zero-voltage axis. Also, square waves of either equal or unequal intervals, sometimes referred to as rectangular waves, can be generated by multivibrators or various other forms of vacuum-tube or transistor switching or gating circuits. *See* CLIPPING CIRCUIT; LIMITER CIRCUIT; MULTIVIBRATOR; RELAXATION OSCILLATOR; WAVEFORM.

Other wave shapes of particular interest in electronics are the linearly increasing function of time or ramp function (which if recurring at equally spaced periods is usually called the linear sawtooth waveform), the hyperbolic waveform, and the rectified sine wave. *See* CLAMPING CIRCUIT; COINCIDENCE AMPLIFIER; ELECTRONIC DISPLAY; ELECTRONIC SWITCH; PULSE GENERATOR; RECTIFIER; SWEEP GENERATOR; TRIGGER CIRCUIT. [G.M.G.]

Waveform The pictorial representation of the form or shape of a wave, obtained by plotting the amplitude of the wave with respect to time. There are an infinite number of possible waveforms; a few of the more common electrical waveforms are shown in the illustration.

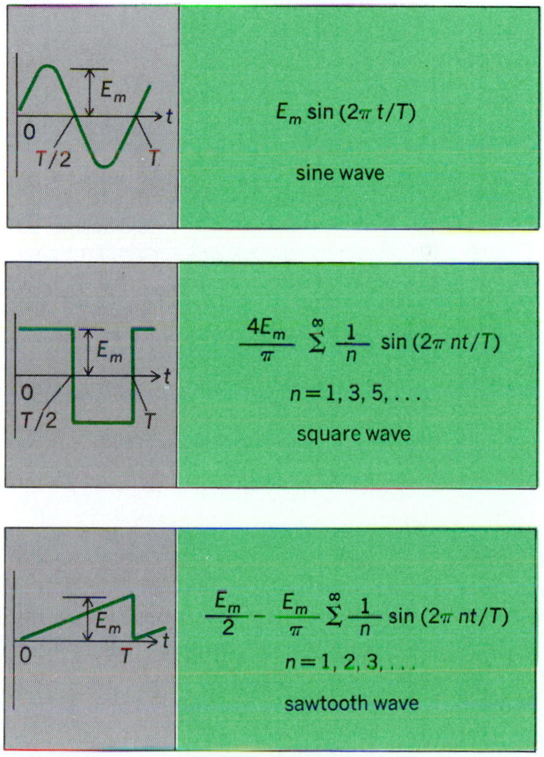

Common waveforms.

It is possible to represent any periodic waveform mathematically as a Fourier series of sine and cosine terms at harmonic frequencies. Any nonsinusoidal wave is composed of a constant or dc term, plus a series of harmonic terms in which the frequencies are integral multiples of the fundamental frequency. The Fourier series for each waveform is given beside each figure as functions of time t, where E_m is the maximum value of the wave and T is the period. *See* NONSINUSOIDAL WAVEFORM; WAVE-SHAPING CIRCUITS. [D.L.W.]

Waveguide A device which constrains or guides the propagation of electromagnetic waves along a path defined by the physical construction of the guide. In a broad sense, devices such as a pair of parallel wires and a coaxial cable can certainly be called waveguides. When used in a more restricted sense, however, the term waveguide usually means a metallic tube which can confine and guide the propagation of electromagnetic waves in the hollow space along the lengthwise direction of the tube. *See* COAXIAL CABLE.

Hollow waveguides of convenient sizes are best adapted to the transmission of microwaves. A sound wave can transmit through a pipe only when its wavelength is comparable to or smaller than the size of the pipe, and one would expect that a similar requirement should hold true for electromagnetic waves. Indeed, if the frequency of an electromagnetic wave is high enough that the wavelength is comparable to or smaller than the waveguide dimension, then wave transmission through the hollow waveguide becomes possible. *See* MICROWAVE TRANSMISSION LINES.

The type of waveguide with a rectangular cross section is not only the commonest in use but also the simplest in theoretical analysis. Next to the rectangular type, metallic waveguides with circular cross-sections are most frequently used. *See* ELECTROMAGNETIC WAVE TRANSMISSION. [C.K.J.]

Wavelength The distance between two points on a wave which have the same value and the same rate of change of the value of a parameter, for example, electric intensity, characterizing the wave. The wavelength is usually designated by the Greek letter λ (see illustration). *See* WAVE (PHYSICS); WAVE MOTION. [W.J.G.]

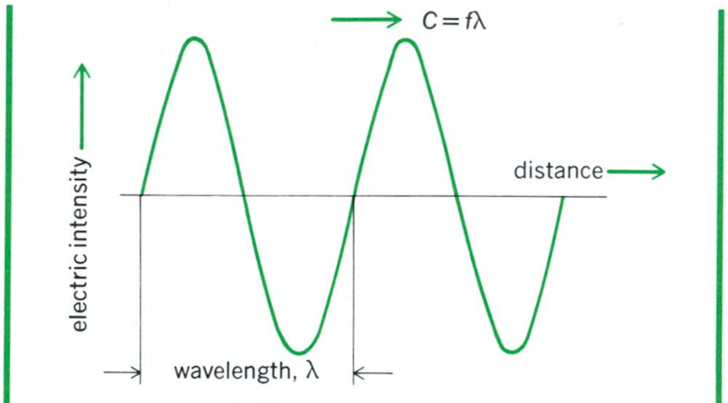

Wavelength λ and related quantities.

Measurement. The wavelength in a transmission system is obtained by measuring the distance between successive waveforms of equal phase. This measurement is most conveniently carried out in a standing-wave field in which the interference occurs between forward-propagating and reverse-propagating, or reflected, waves. A distance equal to one-half wavelength exists between successive minima or maxima in the standing-wave pattern. *See* WAVEMETER.

In order to specify the wavelength of an electromagnetic wave, it is essential to know the medium or device through which the wave is propagating. However, unless otherwise specified, it is general practice to quote the wavelength of an oscillatory electric wave as that in air, or free space. When this convention is used, all that is necessary is a frequency measurement in order to apply the relation $\lambda_0 = c/f$, where c is the velocity of light, λ_0 is the free-space wavelength, and f is frequency.

The standing-wave method of measurement is somewhat similar to the interferometer measurements used in optics. With shorter-wavelength radio waves it is possible to apply optically derived interferometer techniques directly to the measurement of wavelength.

A more convenient, if less precise, measurement method for the determination of wavelength is the Lecher wire wavemeter. With this simple device, wavelength is measured by sliding a short circuit along the line from a first to a second point of equal amplitude of effect, as indicated by absorption of signal being detected by an external detector, or by some similar method. The distance between two successive absorption maxima is one-half wavelength (λ/2). *See* TRANSMISSION LINES.

At lower frequencies, inductance-capacitance resonant circuits may serve similar functions as absorption devices. A variation of the absorption wavemeter, known as a grid-dip meter or grid-dip oscillator, provides in one instrument an absorption wavemeter and a calibrated oscillator, which may be used to determine the resonant frequency (or wavelength) of a passive network, such as an antenna system or an *LC* tuned circuit, by observing the dip in grid current of the oscillator as the oscillator frequency is tuned to the network resonant frequency. *See* RESONANCE (ALTERNATING-CURRENT CIRCUITS).

Microwave wavemeters make use of resonant coaxial-line sections or cavities as tuned elements. The two general types of microwave wavemeters are the absorption, or reaction, type and the transmission type. *See* CAVITY RESONATOR. [F.D.L.]

Standards. Wavelength standards are accurately measured lengths of waves emitted by specified light sources for the purpose of obtaining the wavelengths in other spectra by interpolating between the standards.

Spectra are produced by refraction in prisms or by diffraction from gratings. In either case, the spectra are usually photographed, and in order to identify or describe them reliably it is necessary to measure the positions of the unknown spectral lines relative to known standards. It is customary to photograph the spectrum containing the standards either superposed or juxtaposed on the spectrum under investigation. The wavelengths of hundreds of thousands of spectral lines have already been measured relative to a few hundred standards, and this will continue to be a major activity in spectroscopy. *See* SPECTROSCOPY.

At its meeting in October 1973 the Comité International des Poids et Mesures (CIPM) adopted recommended values for the wavelength of light (from two stabilized lasers) and for the speed of light. A He-Ne laser has been stabilized to an absorption line in methane at 3.39 μm. In addition, a ^{3}He-^{20}Ne laser has been stabilized to hyperfine components in an absorption line of either $^{127}I_2$ or $^{129}I_2$ at 633 nm. The principal gain from a shift to the standards based on the stabilized laser sources is the remarkable agreement in the measurements with independently constructed equipment. *See* ATOMIC STRUCTURE AND SPECTRA; HYPERFINE STRUCTURE; LASER SPECTROSCOPY.

Secondary iron wavelengths recommended in 1962 and later are all emitted by improved low-pressure, low-temperature sources. However, the accuracy of these standard iron wavelengths will not satisfy future precision demands in spectroscopy. In order to measure the wavelengths in very complex spectra with adequate accuracy, better and more uniformly spaced secondary standards are required. These requirements are met by electrodeless thorium-halide lamps excited by microwaves. In 1958 W. F. Meggers and R. W. Stanley published the wavelengths of 222 thorium lines ranging from 328.77885 to 698.96562 nm. This list has been extended to a longer list covering the range 297.14332 to 905.07361 nm.

Tables of other recommended secondary wavelengths have subsequently been published. Accurate compilations of Kr and Xe wavelengths have been made and wavelength standards in the vacuum ultraviolet region have been determined. [W.W.W.]

Wavelets The elementary building blocks in a mathematical tool for analyzing functions. The functions can be very diverse; examples are solutions of a differential equation, and one- and two-dimensional signals. The tool itself, the wavelet transform, is the result of a synthesis of ideas from many different fields, ranging from pure mathematics to quantum physics and electrical engineering.

In many practical applications, it is desirable to extract frequency information from a signal—in particular, which frequencies are present and their respective importance. An example is the decomposition into spectral lines in spectroscopy. The tool that is generally used to achieve this is the Fourier transform. Many applications, however, concern nonstationary signals, in which the makeup of the different frequency components is constantly shifting. An example is music, where this shifting nature has been recognized for centuries by the standard notation, which tells a musician which note (frequency information) to play when and how long (time information). For signals of this nature, a time-frequency representation is needed. *See* FOURIER SERIES.

There exist many different mathematical tools leading to a time-frequency representation of a given signal, each with its own strengths and weaknesses. The wavelet transform is such a time-frequency analysis tool. Its strength lies in its ability to deal well with transient high-frequency phenomena, such as sudden peaks or discontinuities, as well as with the smoother portions of the signal. (An example is a crack in the sound from a damaged record, or the attack at the start of a music note.) The wavelet transform is less well adapted to harmonically oscillating parts in the signal, for which Fourier-type methods are more indicated.

Applications of wavelets include various forms of data compression (such as for images and fingerprints), data analysis (nuclear magnetic resonance, radar, seismograms, and sound), and numerical analysis (fast solvers for partial differential equations). *See* DIFFERENTIAL EQUATION; INTEGRAL TRANSFORM; NUMERICAL ANALYSIS. [I.D.]

Wavellite A hydrated phosphate of aluminum mineral with composition $Al_3(OH)_3(PO_4)_2 \cdot 5H_2O$, in which small amounts of fluorine and iron may substitute for the hydroxyl group (OH) and aluminum (Al), respectively. Wavellite crystallizes in the orthorhombic system. The crystals are stout to long prismatic, but are rare. Wavellite commonly occurs as globular aggregates of fibrous structure and as encrusting and stalactitic masses. Wavellite crystals range in color from colorless and white to different shades of blue, green, yellow, brown, and black. Wavellite is found at many places in Europe and North America; it is also abundant in tin veins at Llallagua, Bolivia. *See* PHOSPHATE MINERALS. [W.R.Lo.]

Wavemeter A device for measuring the geometrical spacing between successive surfaces of equal phase along an electromagnetic wave. At frequencies up to about 100 MHz, wavemeters consist basically of a tuned *LC* circuit and a suitable resonance indicator.

At higher frequencies it is necessary to employ well-defined transmission systems such as open-wire or coaxial lines and waveguides. Any open-circuited or short-circuited section of transmission system can be adjusted in physical size to cause it to resonate at a given wavelength. From a construction standpoint short-circuited sections are preferred and may take the form of Lecher wires, coaxial wavemeters, and cavity resonators. Suitable electronic standing-wave detectors are employed to indicate when the wavemeter is tuned to resonance. *See* CAVITY RESONATOR; STANDING-WAVE DETECTOR; WAVEGUIDE; WAVELENGTH. [R.L.R.]

Wax, animal and vegetable Any of the substances containing esters of higher fatty acids and long-chain monohydric alcohols. From a practical standpoint, this definition is inadequate because it allows liquids such as sperm whale oil and jojoba oil to be called waxes, and it fails to indicate the complexity of waxes. While waxes do contain wax esters, they are seldom if ever pure. They are usually mixtures that may contain high-molecular-weight acids, alcohols, esters, ketones, hydrocarbons, sterols, diesters, hydroxyacids, and so forth, as well as the wax esters. *See* ESTER.

Practical wax formulators use physical, rather than chemical, properties to define waxes. Waxes must: be a solid at 68°F (20°C); be crystalline; melt above 140°F (40°C) without decomposition; have relatively low viscosity above the melting point; have consistency and solubility properties that are strongly dependent upon temperature; and be capable of being polished under slight pressure.

Wax sources are abundant. They have been isolated from the outer layers of bacteria, the roots, stems, leaves, fruit, and flowers of plants, the exudates of insects, the skin and hair of some animals, and the bodies of certain marine and land animals. Carnauba wax is extracted from an exudate on the leaves of the carnauba palm (*Copernicia prunifera*). Candelilla wax is obtained from a coating on the stem of *Euphorbia antisyphilitica*, a leafless desert shrub. Beeswax is an exudate of the honeybee. Wool wax is obtained from sheep wool. Many other waxes have been of commercial importance from time to time, but their availability or cost has forced them from the market. *See* CARNAUBA WAX; FAT AND OIL. [H.M.H.]

Wax, petroleum A substance produced primarily from the dewaxing of lubricating-oil fractions of petroleum. It may be of either the crystalline or microcrystalline type. Petroleum wax has a wide variety of uses. It is used to coat paper products, to blend with other waxes for the manufacture of candles, in the manufacture of electrical equipment and many polishes for home and industry, and as a source material for oxidized products. The softer waxes, such as petroleum jelly, after proper purification, are being used as medicinal products. *See* DEWAXING OF PETROLEUM; PETROLEUM PROCESSING; PETROLEUM PRODUCTS. [W.E.Ku.]

Weakly interacting massive particle (WIMP) A hypothetical elementary particle that could provide simultaneous solutions to at least four long-standing astrophysical puzzles: the solar neutrino and pulsation problems, and the nature of the mysterious galactic dark-matter background and the so-called missing mass of the universe.

The concept dates from 1977, when it was suggested that a suitably massive particle, interacting only via gravity and the weak interaction, would survive the big bang in sufficient numbers to dominate conventional matter and possibly to eventually reverse the expansion of the universe. *See* FUNDAMENTAL INTERACTIONS.

Effects in Sun. WIMPs soon reach a quasithermodynamic equilibrium near the solar center. Energy removed by WIMP collisions from hotter, conventional, central particles is delivered to cooler regions farther out. *See* SUN.

The efficient, gravitationally confined, central energy transport reduces the temperature gradient and thus lowers central temperatures (see illustration) in precisely those regions of the Sun responsible for the highly-temperature-dependent flux of higher-energy neutrinos detectable by the chlorine-37 experiment, and could therefore explain why less than the expected number of solar neutrinos are detected in this experiment. *See* SOLAR NEUTRINOS.

The frequency separations of successive low-degree, p-mode solar oscillations are well observed and sensitive to the sound speed near the solar center. Separations in the standard solar

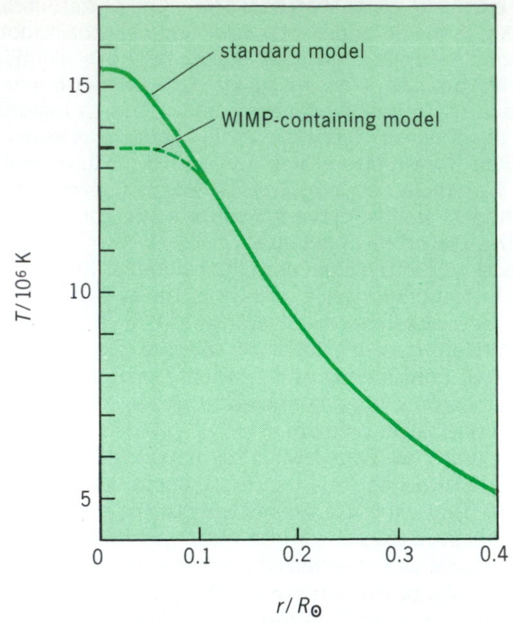

Plot of absolute temperature *T* versus radius *r* (as fraction of solar radius $R_\odot$) in central regions of the current Sun for the standard model and a WIMP-containing model.

model are too large by approximately 10%. Only the WIMP model satisfies both the neutrino and the p-mode separation constraints without any arbitrary adjustment of the initially published model. *See* HELIOSEISMOLOGY.

Galactic dark-matter background. Dynamical arguments indicate the presence, but not the nature, of this background. If it is composed of WIMPs with the very properties needed in the solar neutrino problem, the Sun would indeed accrete the desired number while orbiting in the Milky Way Galaxy for its known lifetime. *See* COSMOLOGY.

Detection. Solar-neutrino-satisfying WIMPs may be detectable with conventional, noncryogenic, silicon diode ionization detectors, if their interaction is nonaxial. If the WIMP interaction is axial, hydrogen or deuterium in a modified low-pressure time-projection chamber could work. *See* ELEMENTARY PARTICLE; JUNCTION DETECTOR; PARTICLE DETECTOR; TIME-PROJECTION CHAMBER; WEAK NUCLEAR INTERACTIONS. [J.F.]

Weak nuclear interactions Fundamental interactions of nature that play a significant role in elementary particle and nuclear physics, and are distinguished from other such interactions by special properties such as participation of all the fundamental fermions and failure to conserve parity. The weak force has very short range (less than 10^{-17} m) and is extremely feeble compared to strong and electromagnetic forces, but can be distinguished from these two by its special character. For example, according to the present view, all of matter consists of certain fundamental spin-½ constituents, the quarks and leptons, collectively called the fundamental fermions. While only the quarks participate in strong interactions, and only the quarks and charged leptons e, μ, and τ participate in electromagnetic interactions, all of the fundamental fermions, including neutrinos, engage in weak interactions. Also, the strong and electromagnetic interactions respect spatial inversion symmetry (they conserve parity) and are also particle-antiparticle (charge conjugation) symmetric, whereas the weak interaction violates these two symmetries. *See* FUNDAMENTAL INTERACTIONS; LEPTON; PARITY (QUANTUM MECHANICS); QUARKS; SYMMETRY LAWS (PHYSICS).

Weak interactions are classified as "charged" or "neutral," depending on whether or not a particle participating in a weak reaction suffers a change of electric charge of one electronic unit. Observed charged weak interactions include nuclear beta decay and electron capture, muon capture on nuclei, and the slow decays of unstable elementary particles such as the μ and τ leptons, π, K, and charmed mesons, and hyperons and charmed baryons. Also, there are the charged neutrino-nucleon and neutrino-lepton scattering reactions. Neutral weak interactions include neutrino-nucleon and neutrino-lepton scattering as well as the electron-nucleon reaction which can also occur by electromagnetic interaction. *See* BARYON; ELEMENTARY PARTICLE; HYPERON; MESON.

The most important development in the study of weak interactions has been the creation of a successful theory based on the principles of local gauge invariance and spontaneous symmetry breaking. This theory proposes a single basis for the weak and electromagnetic interactions, and indeed, despite striking differences in the observed characteristics of strong, electromagnetic, and weak interactions, important theoretical ideas of a similar type suggest that all these interactions possess a common origin. [E.D.C.]

Wear The removal of material from a solid surface as a result of sliding action. It constitutes the main reason why the artifacts of society (automobiles, washing machines, tape recorders, cameras, clothing) become useless and have to be replaced. There are a few uses of the wear phenomenon, but in the great majority of cases wear is a nuisance, and a tremendous expenditure of human and material resources is required to overcome the effects.

Adhesive wear is the only universal form of wear, and in many sliding systems it is also the most important. It arises from the fact that, during sliding, regions of adhesive bonding, called junctions, form between the sliding surfaces. If one of these junctions does not break along its original interface, then a chunk from one of the sliding surfaces will have been transferred to the other surface. In this way, an adhesive wear particle will have been formed. Initially adhering to the other surface, adhesive particles soon come off loose and can disappear from the sliding system. *See* FRICTION.

Abrasive wear is produced by a hard, sharp surface sliding against a softer one and digging out a groove. The abrasive agent may be one of the surfaces (such as a file), or it may be a third component (such as sand particles in a bearing abrading material from each surface).

Corrosive wear arises when a sliding surface is in a corrosive environment, and the sliding action continuously removes the protective corrosion product, thus exposing fresh surface to further corrosive attack. *See* CORROSION.

Surface fatigue wear occurs as a result of the formation and growth of cracks. It is the main form of wear of rolling devices such as ball bearings, wheels on rails, and gears.

While most manifestations of wear are highly objectionable, the phenomenon does have a few uses. Thus, a number of systems for recording information (pencil and paper, chalk, and blackboard) operate via a wear mechanism. Many methods of preparing solid surfaces (filling, sandpapering, sandblasting) also make use of wear. [E.R.]

Weasel The common name for at least 12 species of carnivores which are members of the family Mustelidae. They have a varied distribution in many regions of the world (see table).

The common or European weasel is *Mustela nivalis*. This animal is a slim, voracious carnivore with a long body. The limbs are short and the body is muscular. Although the claws are nonretractile, these animals climb with agility. The gestation period is 5 weeks, and four to six young are born in a nest made in a hollow tree or rabbit burrow. One or two litters are born each year, and the life-span is about 8 years.

The common weasel has a wide distribution throughout

Common names and distribution of 12 species of weasel

Scientific name	Common name	Distribution
Mustela rixosa	Pygmy weasel	North America
M. frenata	Long-tailed weasel	North and South America
M. nivalis	Common weasel	Europe, Asia, and Africa
M. altaica	Alpine weasel	Asia
M. sibirica	Siberian weasel	Asia
M. kathiah	Yellow-bellied weasel	Asia
M. strigidorsa	Back-striped weasel	Southern Asia
M. lutreolina	Java weasel	Java
M. nudipes	Bare-footed weasel	Malaysia
Lyncodon patagonicus	Patagonian weasel	South America
Poecilictis libyca	Libyan striped weasel	North Africa
Poecilogale albinucha	White-naped weasel	Africa

Europe, North Africa, and Asia, and has managed to survive because of its size and ability to escape detection. It preys upon small rodents, such as mice, voles, and rats, and has been incriminated in attacking domestic animals, such as poultry. *See* CARNIVORA; FERRET; FISHER; MARTEN; MINK; OTTER; SKUNK; WOLVERINE. [C.B.C.]

Weather The state of the atmosphere, as determined by the simultaneous occurrence of several meteorological phenomena at a geographical locality or over broad areas of the Earth. When such a collection of weather elements is part of an interrelated physical structure of the atmosphere, it is termed a weather system, and includes phenomena at all elevations above the ground. More popularly, weather refers to a certain state of the atmosphere as it affects humans' activities on the Earth's surface. In this sense, it is often taken to include such related phenomena as waves at sea and floods on land.

A weather element is any individual physical feature of the atmosphere. At a given locality, at least seven such elements may be observed at any one time. These are clouds, precipitation, temperature, humidity, wind, pressure, and visibility. Each of these principal elements is divided into many subtypes. *See* WEATHER MAP.

The various forms of precipitation are included by international agreement among the hydrometeors, which comprise all the visible features in the atmosphere, besides clouds, that are due to water in its various forms. For convenience in processing weather data and information, this definition is made to include some phenomena not due to water, such as dust and smoke. Some of the more common hydrometeors include rain, snow, fog, hail, dew, and frost. *See* PRECIPITATION (METEOROLOGY).

Certain optical and electrical phenomena have long been observed among the weather elements. These include lightning, aurora, solar or lunar corona, and halo. *See* AIR MASS; ATMOSPHERE; AURORA; CLOUD; FRONT; LIGHTNING; METEOROLOGY; STORM; WEATHER OBSERVATIONS; WIND. [P.F.C.]

Weather forecasting and prediction Processes for formulating and disseminating information about future weather conditions based upon the collection and analysis of meteorological observations. Weather forecasts may be classified according to the space and time scale of the predicted phenomena. Atmospheric fluctuations with a length of less than 330 ft (100 m) and a period of less than 100 s are considered to be turbulent. Prediction of turbulence extends only to establishing its statistical properties, insofar as these are determined by the thermal and dynamic stability of the air and by the aerodynamic roughness of the underlying surface. The study of atmospheric turbulence is called micrometeorology; it is of importance for understanding the diffusion of air pollutants and other aspects of the climate near the ground. Standard meteorological observations are made with sampling techniques designed to filter out the influence of turbulence. Common terminology distinguishes among three classes of phenomena with a scale that is larger than the turbulent microscale: the mesoscale, synoptic scale, and planetary scale.

The mesoscale includes all moist convection phenomena, ranging from individual cloud cells up to the convective cloud complexes associated with prefrontal squall lines, tropical storms, and the intertropical convergence zone. Most forecasting methods depend upon empirical rules or the short-range extrapolation of current observations, particularly those provided by radar and geostationary satellites. Since many mesoscale phenomena pose serious threats to life and property, it is the practice to issue advisories of potential occurrence significantly in advance. These "watch" advisories encourage the public to attain a degree of readiness appropriate to the potential hazard. Once the phenomenon is considered to be imminent, the advisory is changed to a "warning," with the expectation that the public will take immediate action to prevent the loss of life.

The next-largest scale of weather events is called the synoptic scale, because the network of meteorological stations making simultaneous, or synoptic, observations serves to define the phenomena. The migratory storm systems of the extratropics are synoptic-scale events, as are the undulating wind currents of the upper-air circulation which accompany the storms. These storms are associated with barometric minima, variously called lows, depressions, or cyclones.

Planetary-scale phenomena are persistent, quasi-stationary perturbations of the global circulation of the air with horizontal dimensions comparable to the radius of the Earth. These dominant features of the general circulation appear to be correlated with the major orographic features of the globe and with the latent and sensible heat sources provided by the oceans. They tend to control the paths followed by the synoptic-scale storms, and to draw upon the synoptic transients for an additional source of heat and momentum. For further information on the observation and processing of weather data *see* METEOROLOGICAL SATELLITES; MICROMETEOROLOGY; WEATHER OBSERVATIONS. [J.P.G.]

Long-range forecasting. Long-range weather forecasts are of two types. Medium- or extended-range forecasts cover periods of from 48 h to a week in advance. Forecasts for longer periods generally extend over periods of a month, a season, or possibly two or more seasons in advance. Although meteorologists have been working on long-range prediction problems for more than 100 years, the degree of accuracy is small for all predictions exceeding a week in advance.

Scientific methods of extended-range prediction take for granted that the further out in time the forecast is projected, the more general must be the nature of the prediction. Short-range forecasts for 48 h or less in advance specify the detail of weather in space and in time. Medium-range forecasts for a week in advance cover average conditions and trends within the week in intervals of a couple of days. Forecasts for a month or more, however, can indicate only the broad-scale features of average or prevailing weather.

Medium-range forecast methods are apt to use one or a combination of dynamic, statistical, and synoptic techniques. Dynamic methods capitalize upon the best physical knowledge of meteorological phenomena. Statistical methods employ empirically derived equations as substitutes for physical knowledge. In the synoptic technique, various hemispheric wind and weather charts are surveyed and interpreted by an experienced meteorologist. [J.N.]

Statistical forecasting. Statistical weather forecasting is the prediction of weather by use of data from past weather developments. Examples of statistical predictions may be the nocturnal minimum temperature tomorrow at a particular site and the probability of precipitation within a given county during the next 36 h. In each case the prediction is based on statistical relations obtained from earlier records and intended for a specific location and forecast period. Because of geographic and atmospheric complexities, extension of these relations to another location or time period tends to produce less accurate forecasts, and a separate treatment usually leads to better success.

The necessity for statistical methods in weather prediction is largely due to a lack of a detailed knowledge of atmospheric processes at any one time. It is presumed that the atmosphere obeys accepted physical laws, under the influence of solar energy, terrain and oceanic effects, and human activities. These laws are the basis for numerical weather predictions of relatively large-scale features of the atmosphere. Because of gaps between observing stations, however, such predictions, while valuable, have limited accuracy. Consequently, statistical weather predictions based on historical data continue to be useful.

By themselves, statistical methods are highly suitable for predicting special local phenomena, such as the occurrence of fog. For preparing prognostic weather maps a day or so in advance, statistical formulas are somewhat useful but are frequently inferior to subjective forecasts. [T.A.Gl.]

Numerical weather prediction. Numerical weather prediction is the prediction of weather phenomena by the numerical solution of the equations governing the motion and changes of condition of the atmosphere. More generally, the term applies to any numerical solution or analysis of the atmospheric equations of motion.

The laws of motion of the atmosphere may be expressed as a set of partial differential equations relating the instantaneous rates of change of the meteorological variables to their instantaneous distribution in space. These are developed in dynamic meteorology. A prediction for a finite time interval is obtained by summing the succession of infinitesimal time changes of the meteorological variables, each of which is determined by their distribution at the preceding instant of time. Although this process of integration may be carried out in principle, the nonlinearity of the equations and the complexity and multiplicity of the data make it impossible in practice. Instead, the prediction is based on finite-difference approximation techniques in which successive changes in the variables are calculated for small, but finite, time intervals at a finite grid of points spanning part or all of the atmosphere. The vast amount of computation requires modern high-speed computers. [R.A.An.]

Centers and offices of forecasting. Weather forecasts for all parts of the United States are prepared by the National Weather Service (NWS), a part of the National Oceanic and Atmospheric Administration (NOAA). Various phases of the forecast work are performed at five working levels: the National Meteorological Center (NMC); National Severe Storms Forecast Center (NSSFC); National Hurricane Center (NHC); River Forecast Centers (RFCs); and Weather Service Forecast Offices (WSFOs).

The National Meteorological Center located in Camp Springs, Maryland, collects weather observations, prepares meteorological analyses, and provides forecast guidance for use by NWS field offices. The aerial coverage of NMC products includes the entire globe, with most products covering the Northern Hemisphere and the tropical regions of the Southern Hemisphere. The World Meteorological Organization (WMO) has designated the NMC as the analysis and forecast arm of the Washington World Meteorological Center, which requires global responsibilities as part of the international efforts and cooperation known as the World Weather Watch.

The National Severe Storms Forecast Center located in Kansas City, Missouri, is responsible for preparing and releasing forecasts of areas of expected severe local storms, including tornadoes. The long-term outlook, prepared shortly after midnight, gives the geographic areas most likely to have severe thunderstorms or tornadoes during the next 24 h.

The National Hurricane Center located in Miami, Florida, prepares hurricane and tropical storm watches and advisories for most of the tropical Atlantic Ocean, the Caribbean, and the Gulf of Mexico. The NHC delegates parts of its public warning responsibilities to three Hurricane Watch Offices (HWO) when the United States coastal areas are threatened.

There are 13 River Forecast Centers which prepare river and flood forecasts and warnings for approximately 2500 communities. This includes forecasts for height of the flood crest as well as times when the river is expected to overflow its banks and when it will recede into its banks.

There are an additional 238 Weather Service Offices, which represent the third echelon of the system. They issue local forecasts which are adaptations of the zone forecasts. They are important in the warning and observation program and are generally located in smaller cities. [A.P.]

Weather map A chart portraying the state of the atmosphere's winds and weather over a wide area at one of several levels above the Earth's surface and at one universal time. It is derived from an analysis of weather observations made at many points in the area near the selected time. Such a map gives the weather forecaster an organized picture of the location, structure, and, when several successive maps are available, the motion and development of the various weather systems. *See* WEATHER OBSERVATIONS.

Many kinds of weather maps are used, depending on the weather elements of immediate interest and their elevation above the ground. At a typical large weather analysis central, as many as 35 different maps are constructed at a given time. Among the more common of these is the surface or sea-level map.

Surface maps are often constructed as a section through the atmosphere along a horizontal surface near sea level, at least over oceans. Over land, observations must be made at the elevation of the observing point, but pressure is extrapolated to sea level. Lines of equal sea-level pressure (isobars) are drawn to portray the sea-level winds. Fronts are drawn, which are relatively narrow transition zones separating cool and warm air masses, along which storms tend to form in extratropical latitudes. Sea-level maps with fronts and air masses are often analyzed subjectively by a meteorologist, and are especially useful in predicting the smaller scales of weather systems over relatively short time intervals. *See* FRONT; ISOBAR (METEOROLOGY).

Upper-level maps may also be constructed by depicting the distribution of weather at fixed elevations above sea level. It is more common, however, to portray the weather at high elevations on constant-pressure surfaces. Lines of constant elevation (isohypses) portray the topography of the selected pressure surface. The contours bear a relation to the winds similar to that of the isobars of a constant-level chart. The true wind is approximated by the geostrophic wind. *See* GEOSTROPHIC WIND; WIND.

Other two-dimensional or one-dimensional atmospheric sections cannot be termed weather maps, but rather meteorological charts and diagrams. These include vertical cross sections through the atmosphere, and thermodynamic diagrams similar to those used in studies of heat engines.

Weather maps may be analyzed subjectively by a trained analyst (usually for limited regions and a few levels) or by a stored program in an electronic computer. In the latter case the analysis program must be supported by a numerical weather prediction model of the atmosphere which must start from a set of digitized data at a geographical network (or grid) of points. *See* WEATHER FORECASTING AND PREDICTION. [P.F.C.]

Weather modification The changing of natural weather phenomena by humans. Weather is the product of the interaction of atmospheric processes on many scales, reaching from the planetary circulation to the microphysical processes in the evolution of cloud droplets and ice crystals. So far, only on the microscale of condensation and freezing nuclei have humans begun to exert modifying influences on weather. These influences may expand to a scale of several hundreds or thousands of square miles, that is, to what is called in meteorology the meso scale of meteorological phenomena.

There are actually four techniques by which humans may affect the natural course of weather: (1) by injection of large amounts of heat into the atmosphere in order to burn off warm fog; (2) by utilizing metastable states in the atmosphere as they happen—for instance, if clouds are being cooled below the freezing point, seeding can be effective; (3) by altering the surface condition, as by deforestation or urbanization; (4) by influencing the weather inadvertently by pollution. The second technique, the seeding of clouds with certain types of nuclei, has been frequently used. *See* Air pollution; Cloud physics; Fog; Hail. [H.K.W.]

Weather observations The measuring, recording and transmitting of data of the variable elements of weather. The National Weather Service (NWS), a component of the National Oceanic and Atmospheric Administration, has as one of its primary responsibilities the acquisition of meteorological information. The data are sent by various communication methods to National Weather Service field stations and to the National Meteorological Center near Washington, D.C.

At the center the raw data are fed into large computers that are programmed to plot, analyze and process the data and also to make prognostic weather charts. The processed data and the forecast guidance are then distributed by facsimile and teletypewriter to field stations, other government agencies, and private meteorologists. They in turn prepare forecasts and warnings based on both processed and raw data.

The Synoptic Weather Program is designed to serve as the backbone for the preparation of forecasts and to provide data for international exchange. Surface observations are taken all over the globe at standard times: 0000, 0600, 1200, and 1800 Greenwich mean time (GMT), and sent in synoptic code.

The Basic Observation Program routinely provides meteorological data every hour. Special observations are taken at any intervening time to report significant weather events or changes. The vast majority of stations involved in either or both programs are operated by National Weather Service and Federal Aviation Administration personnel and are referred to as first-order stations. Observational sites are primarily at airports; a few are in cities. They all report the following weather elements: present weather, air pressure, temperature, humidity, wind speed and direction, visibility, and clouds and ceilings. *See* Air pressure; Cloud; Humidity; Wind.

Upper air observations are made with an instrument called a radiosonde, borne aloft by a helium-inflated balloon. It measures and then telemeters back to ground stations the pressure, temperature, and water vapor existing in the atmosphere up to about 100,000 ft (30 km). Winds aloft are computed by tracking the instrument with precision electronic equipment. Those measurements plus radar and satellite observations provide meteorologists with a three-dimensional picture of the atmosphere. *See* Meteorological satellites; Radar meteorology; Weather forecasting and prediction. [A.E.T.]

Weathering processes The response of geologic materials to the environment (physical, chemical, and biological) at or near the Earth's surface. This response typically results in a reduction in size of the weathering materials; some may become as tiny as ions in solution.

The agents and energies that activate weathering processes and the products resulting therefrom have been classified traditionally as physical and chemical in type. In classic physical weathering, rock materials are broken by action of mechanical forces into smaller fragments without change in chemical composition, whereas in chemical weathering the process is characterized by change in chemical composition. In practice, however, the two processes commonly overlap.

Specific agents of weathering may be recognized and correlated with the types of effects they produce. Important agents of weathering are water in all surface occurrences (rain, soil and ground water, streams, and ocean); the atmosphere (H_2O, O_2, CO_2, wind); temperature (ambient and changing, especially at the freezing point of water); insolation (on large bare surfaces); ice (in soil and glaciers); gravity; plants (bacteria and macroforms); animals (micro and macro, including humans). Human modifications of otherwise geologic weathering that have increased exponentially during recent centuries include construction, tillage, lumbering, use of fire, chemically active industry (fumes, liquid, and solid effluents), and manipulation of geologic water systems.

Products of physical weathering include jointed (horizontal and vertical) rock masses, disintegrated granules, frost-riven soil and surface rock, and rock and soil flows. Products of chemical weathering include the soil, and the clays used in making ceramic structural products, whitewares, refractories, various fillers and coating of paper, portland cement, absorbents, and vanadium. These are the relatively insoluble products of weathering; characteristically they occur in clays, siltstones, and shales. Sand-size particles resulting from both physical and chemical weathering may accumulate as sandstones.

After precipitation, the relatively soluble products of chemical weathering give rise to products and rocks such as limestone, gypsum, rock salt, silica, and phosphate and potassium compounds useful as fertilizers. [W.D.K.]

Wedge A piece of resistant material whose two major surfaces make an acute angle. It is closely related to the inclined plane and is used to multiply the applied force and to change the direction in which it acts (see illustration). *See* Inclined plane.

Force F is the smaller applied force and Q is the larger force to be exerted. In the absence of friction, forces must act normal to their surfaces; thus the actual force on the inclined surface is not Q but a larger force F_n. Summing up forces in the horizontal and vertical directions gives Eqs. (1).

$$F_n \sin\theta - F = 0$$
$$Q - F_n \cos\theta = 0 \qquad (1)$$

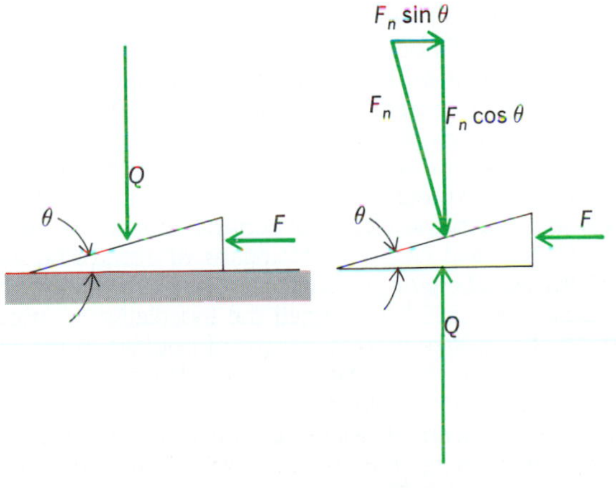

Forces acting on a wedge.

Combining the expressions for F and Q and solving for F gives Eq. (2).

$$F = Q \tan \theta \qquad (2)$$

If angle θ is small, the reaction of Q against F is exceeded by the friction between the face of the wedge and the adjacent body on which it rests. Thus the wedge tends to remain in position even when loaded by a large force Q. *See* SIMPLE MACHINE. [R.M.Ph.]

Weeds Unwanted plants or plants whose negative values outweigh the positive values in a given situation. Weeds impact growers each year in reduced yield and quality of agricultural products. Especially in tropical areas, irrigation systems have become unusable because of clogging with weeds. Weeds can harbor deleterious disease organisms and insects that harm crops and livestock. In addition, weeds can cause allergic reactions and serious skin problems (poison ivy), break up pavement, slow or stop water flow in municipal water supplies, interfere with power lines, cause fire hazards around buildings and along railroad tracks, and produce poisonous plant parts. *See* ALLERGY.

The most serious weeds are those that succeed in invading new areas and surviving at the expense of other plants by monopolizing light, nutrients, and water or by releasing chemicals detrimental to the growth of surrounding vegetation (allelopathy). In the plant kingdom, dozens of species have been shown to release allelopathic chemicals from roots, leaves, and stems. *See* ALLELOPATHY.

Classification. Weeds can be classified as summer annuals, which germinate in the spring, set seed, and die in the fall (crabgrass); winter annuals, which germinate in the fall, set seed, and die in the spring (common chickweek); biennials, which germinate one year, overwinter, set seed, and die the following summer (wild carrot); simple perennials, which live for several years but spread only by seed (dandelion); and creeping perennials, which live for several years and can spread both by seed and by underground roots or rhizomes (field bindweed).

Control methods. Hand pulling, fire, flooding, and tillage are useful for controlling weeds. Insects and pathogens have also been introduced to control certain weed species. Techniques such as herbicides, computerization of spray and tillage technology, remote sensing for weed mapping and identification, and laser treatment have also been explored. Herbicides have resulted in large improvements in the availability and quality of food. They have increased the feasibility of no-till agriculture, leading to significant reductions in soil erosion. They commonly kill weeds by disrupting a physiological process that is not present in animals. Some, however, are moderately high in toxicity and must be used carefully. Rare individual weeds that are genetically resistant to the herbicide have flourished and reproduced, leading to populations of weeds resistant to that herbicide. *See* HERBICIDE. [A.P.A.]

Biological control. Biological control involves the use of natural enemies (parasites, pathogens, and predators) to control pest populations. In the case of weed pests, the primary natural enemy groups utilized are arthropods, fungal pathogens, and vertebrates. The two major approaches are classical and inundative biological control. Biocontrol can be a highly effective and cost-efficient means of controlling weeds without the use of chemical herbicides.

Classical biocontrol (also termed the inoculative or importation method) is based on the principle of population regulation by natural enemies. Most naturalized weeds leave behind their natural enemies when they colonize new areas, and so can increase to significant densities. Classical biocontrol involves the importation of natural enemies, usually from the area of origin of the weed (and preferably from a part of its native range that is a good climatic match with the intended control area), and their field release. Imported biocontrol agents must be host specific to the target weed.

Inundative control is the mass production and periodic release of large numbers of biocontrol agents to achieve controlling densities. It can be used where existing populations of agents are lacking or where existing populations that are not self-sustaining at high, controlling densities can be augmented. A chief advantage of this method is that it can be integrated with conventional farming practices on cultivated croplands.

Arthropods, especially insects, are heavily utilized as imported biocontrol agents in uncultivated environments such as grasslands and aquatic systems.

Rusts (Uredinales) are the fungal group most frequently employed as imported agents in classical biocontrol programs. The rust *Puccinia chondrillina*, imported from Italy, controlled the narrow-leaf form of skeletonweed (*Chondrilla juncea*) in Australia. Formulations of spores of foreign and endemic pathogens can be used as mycoherbicides in inundative applications.

Vertebrate animal agents, such as goats, typically do not possess a high degree of host-plant specificity, and their feeding has to be carefully managed to focus it on the target weeds. *See* ARTHROPODA; FUNGI. [C.E.Tu.]

Weight The gravitational weight of a body is the force with which the Earth attracts the body. By extension, the term is also used for the attraction of the Sun or a planet on a nearby body. This force is proportional to the body's mass and depends on the location. Because the distance from the surface to the center of the Earth decreases at higher latitudes, and because the centrifugal force of the Earth's rotation is greatest at the Equator, the observed weight of a body is smallest at the Equator and largest at the poles. The difference is sizable, about 1 part in 300. At a given location, the weight of a body is highest at the surface of the Earth. Weight is measured by several procedures. *See* BALANCE; MASS; WEIGHT MEASUREMENT. [H.S.B.]

Weight measurement Weight is the resultant force acting on a mass (in a vacuum) due to the Earth's gravitational field corrected for the effect of the Earth's rotation. Units of weight are based upon an acceleration of gravity. When the weight of an unknown is determined by comparison with a known weight, there is no error in the readings due to gravity variations. The varying buoyant effect of the atmosphere is negligible when the density of the unknown is approximately the same as that of the standard. In precision weighing, the buoyant effect of air must be considered. *See* GRAVITY.

The equal-arm balance is probably the most common form of instrument for measuring weight. These balances are made in many designs and sizes; in some the knife-edge fulcrums are replaced with flexure plates; others have arms of unequal length. Conventionally, the unknown weight is placed on one pan, the known weight on the other. The final securing of a balance is done by adjusting the position of a rider, or small weight, on a bar of the balance arm bridge. The condition of balance is indicated when the pointer swings equal distances from its rest point. *See* BALANCE.

The mechanical-type industrial scale incorporates a number of levers with precisely located fulcrums to permit heavy objects to be balanced (weighed) with small, convenient counterweights or counterpoises.

The pendulum-type mechanical scale balances the force of the load by the rotation of a bent lever. With this construction, the deflection of the load on the scale moves the counterweights through the lever system so that their center of gravity is at a greater distance from the final fulcrum. Thus the increased lever arm of the counterweights automatically balances the load.

The spring scale utilizes the deflection of a spring to measure the load. If sensitive enough, such a scale can detect and indicate the effect of differences in the weight of a body due to changes in elevation.

In hydraulic systems, the load applied to the load cell piston is converted to hydraulic pressure. The effective area of the piston must be known. The pressure may be measured at a remote point by a pressure-gage, such as a Bourdon tube.

Pneumatic systems detect the load by a sensitive nozzle and flapper system and balance the load by modulating an air pressure in an opposing capsule.

Electrical weighing systems usually involve the electrical measurement of the elastic deformation of a mechanical element under stress. The strain gage is attached to the weighing element in a manner to produce the maximum resistance change per unit of load. The change in resistance with load is measured and amplified by electronic means, and the load is read on a potentiometer. *See* WEIGHT. [H.S.B.]

Weightlessness A condition induced by the effective lack of resistance to gravitational force on an object or organism, sometimes known as free fall.

Newton proposed the law of universal gravitation, which states that two bodies of matter in the universe attract each other with a force that is directly proportional to the product of their masses and inversely proportional to the square of the distance between their centers. According to this law, even a small increase in the distance between bodies will produce a large decrease in the gravitational force, since the force decreases with the square of the distance. As a body moves from the Earth's surface to a location an infinite distance from the Earth, the gravitational force approaches zero and the body approaches weightlessness. In the true sense, a body can be weightless only when it is an infinite distance from all other objects.

Weightlessness is also defined as a condition in which no acceleration, whether of gravity or any other force, can be detected by an object or organism within the system in question. According to Albert Einstein's principle of equivalence, there is no way to distinguish between the forces of gravitational fields and the forces due to inertial motion. When a gravitational force on a body is opposed by an equal and opposite inertial force, a weightless state is produced. This is based on the fact that the mass that determines the gravitational force of a body is the same as the mass related to the acceleration produced by an inertial force of any kind. These inertial forces have no external physical origin, but are the consequences of an accelerated state of motion. Because of inertia, a moving object always tends to follow a straight line. When a person swings a bucket by the handle in a large circle, he or she feels a pull on his or her hand, because inertial force (also called centrifugal force in this case) tends to keep the bucket moving in a straight line, while the bucket holder exerts a counterforce constraining the bucket to move along the circle. A similar situation exists in a spaceship orbiting the Earth 200 mi (320 km) above the Earth's surface, where the gravitational field is only slightly weaker than at sea level. The ship, in free fall with negligible atmospheric drag is pulled toward the Earth by the Earth's gravitational attraction force, while the inertial or centrifugal force of the moving ship is directed radially outward from the Earth; consequently, the force of gravity on the orbiting ship is opposed and nullified by the centrifugal force, and apparent weightlessness results. *See* GRAVITATION; GRAVITY; INERTIA; NEWTON'S LAWS OF MOTION; SPACE FLIGHT; SPACE PROCESSING. [T.W.H.]

Weinberg-Salam model A theory describing the electromagnetic and the weak forces between the elementary particles known as quarks and leptons that are presently believed to be the fundamental constituents of matter. Before this model there were believed to be four basic forces or interactions between the elementary particles: the strong nuclear forces, the electromagnetic forces, the weak interactions, and gravity. The model has succeeded in unifying the electromagnetic and the weak forces into so-called electroweak interactions, thus reducing the basic number of forces required to describe the interactions of the elementary particles to three. This consolidation may be similar in significance to the work of J. C. Maxwell and others in the nineteenth century that united the electric and the magnetic forces into a single electromagnetic theory. *See* MAXWELL'S EQUATIONS.

One of the most striking predictions of the Weinberg-Salam model was that of neutral current weak interactions. An apparent dilemma, that is, the existence of strangeness-conserving neutral currents, while strangeness-changing neutral currents were absent, was reconciled by a new property of elementary particles called charm (similar to but distinct from the property called strangeness) postulated by S. Glashow, J. Iliopoulos, and L. Maiani. In this scheme, the strangeness-changing neutral currents were exactly canceled out by the charm-changing neutral currents, while the strangeness- (and charm-) conserving neutral currents remain in existence. This postulate also predicted the existence of a new family of elementary particles, called the charmed particles, which were actually discovered in 1974. *See* CHARM; NEUTRAL CURRENT; NEUTRINO.

The Weinberg-Salam model introduced a triplet of intermediate bosons, the W^+, W^0 and W^- and a singlet, the B^0. The neutral bosons form a quantum-mechanical mixture given by the equations below, where θ is a mixing angle called the

$$\gamma = \cos\theta\, B^0 - \sin\theta\, W^0$$

$$Z^0 = \sin\theta\, B^0 + \cos\theta\, W^0$$

Weinberg angle. The γ is the well-known photon that mediates the electromagnetic interactions, and the Z^0 mediates the neutral current weak interactions. The $W^\pm$ mediate the charged current weak interactions. Thus the electromagnetic and the weak interactions are described by a single theory. *See* INTERMEDIATE VECTOR BOSON.

The theory had great predictive power in that it predicted the cross sections (reaction rates) and other properties of a large variety of neutral current processes in terms of a single parameter, the mixing angle θ. This model, together with the Glashow-Iliopoulos-Maiani charm scheme, forms the basis of the current understanding of particle physics, called the standard model. *See* ELEMENTARY PARTICLE; FUNDAMENTAL INTERACTIONS; WEAK NUCLEAR INTERACTIONS. [C.B.]

Welded joint The joining of two or more metallic components by introducing fused metal (welding rod) into a fillet between the components or by raising the temperature of their surfaces or edges to the fusion temperature and applying pressure (flash welding).

Figure 1 shows three types of welded joints. In a lap weld, the edges of a plate are lapped one over the other and the edge of one is welded to the surface of the other. In a butt weld, the edge of one plate is brought in line with the edge of a second plate and the joint is filled with welding metal or the

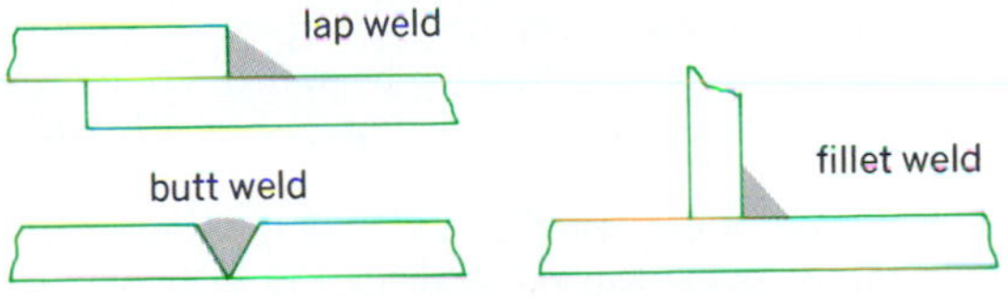

Fig. 1. Three types of welded joints.

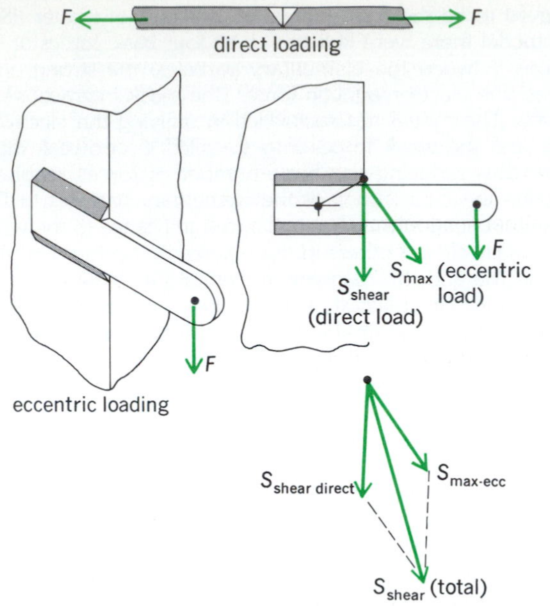

Fig. 2. Loading forces on a welded joint.

two edges are resistance-heated and pressed together to fuse. For a fillet weld, the edge of one plate is brought against the surface of another not in the same plane and welding metal is fused in the corner between the two plates, thus forming a fillet. The joint can be welded on one or both sides.

Because welded joints are usually exposed to a complex stress pattern as a result of the high temperature gradients present when the weld is made, it is customary to design joints by use of arbitrary and simplified equations and generous safety factors. The force F of direct loading, and consequently the stress S, is applied directly along or across a weld. The stress-force equation is then simply $F = SA$, in which A is the area of the plane of failure (Fig. 2). For eccentric loading, the force F causes longitudinal and transverse forces of varying magnitudes along the weld. *See* RIVETED JOINT; STRUCTURAL CONNECTIONS; WELDING AND CUTTING OF METALS. [L.S.L.]

Welding and cutting of metals Processes based on heat to join and sever metals. Welding and cutting are grouped together because, in many manufacturing operations, severing precedes welding and involves the same production personnel. Welding is one of the joining processes, others being riveting, bolting, gluing, and adhesive bonding. *See* BOLTED JOINT; RIVETED JOINT; WELDED JOINT.

The American Welding Society's definition of welding is "a metal-joining process wherein coalescence is produced by heating to suitable temperatures with or without the application of pressure, and with or without the use of filler metal." Brazing is defined as "a group of welding processes wherein coalescence is produced by heating to suitable temperature and by using a filler metal having a liquidus about 800°F (427°C) and below the solidus of the base metals. The filler metal is distributed between the closely fitted surfaces of the joint by capillary attraction." Soldering is similar in principle, except that the melting point of solder is below 800°F (427°C). The adhesion of solder depends not so much on alloying as on its keying into small irregularities in the surfaces to be joined. For comparison of metal joints *see* JOINT (STRUCTURES). *See also* BRAZING; SOLDERING.

Cutting is one of the severing and material-shaping processes, some others being sawing, drilling, and planning. Thermal cutting is defined as a group of cutting processes wherein the severing or removing of metals is effected by melting or by the chemical reaction of oxygen with the metal at elevated temperatures. Welding and cutting are widely used in building ships, machinery, boilers, structures, atomic reactors, aircraft, railroad cars, missiles, buses and trailers, and pressure vessels, as well as in constructing piping and storage tanks of steel, stainless steel, aluminum, nickel, copper, lead, titanium, tantalum, and their alloys. For many products, welding is the only joining process that achieves the desired economy and properties, particularly leak-tightness.

Nearly all industrial welding involves fusion. The edges or surfaces to be welded are brought to the molten state. The liquid metal bridges the gap between the parts. After the source of welding heat has been removed, the liquid solidifies, thus joining or welding the parts together. In the past there were three principal sources of heat for fusion welding: electric arc, electric resistance, and flame. However, other sources must be added to the above: light beams and electrons. Hence there are terms for five fusion welding processes: arc welding, resistance welding, gas welding, laser welding, and electron-beam welding. *See* ARC WELDING; ELECTRON BEAM WELDING; GAS WELDING; INERTIA WELDING; LASER WELDING; RESISTANCE WELDING. [M.M.S.]

Well An artificial excavation made to extract water, oil, gas, brine, or other fluid substance from the earth. Most wells are of the drilled type. Dug wells are almost obsolete, because of the greater speed of drilling and the greater efficiency of drilled wells. *See* ARTESIAN SYSTEMS; OIL AND GAS WELL DRILLING.

Drilled wells, commonly 2–36 in. (5–90 cm) in diameter, usually are fitted with a steel tube or casing inserted in the drilled hole to the desired depth. Where the water-bearing formation is competent to stand without support, the casing is set, or finished, at the top of solid rock. Where there is danger of caving, as in sand or gravel, the casing is carrried below the top of the water-bearing bed, and a perforated pipe or screen extends below the casing to the bottom of the hole. The construction includes a considerable period of pumping, surging, or other treatment (called well development), during which the finer particles of the formation are drawn into the well and removed. This process substantially increases the initial yield of the well.

Most wells of large capacity are equipped with pumps of the deep-well turbine type to lift the water to the surface. When a well is pumped, the pressure head at the well is lowered and a hydraulic gradient toward the well is established which causes water to flow toward the well. This lowering of head is called drawdown. *See* GROUNDWATER HYDROLOGY; PUMPING MACHINERY. [A.N.S./R.K.Li.]

Well logging The technique of making measurements in drill holes with probes designed to measure the physical and chemical properties of rocks and their contained fluids. Much information can be obtained from samples of rock brought to the surface in cores or bit cuttings, or from other clues while drilling, such as penetration rate; however, the greatest amount of information comes from well logs. *See* BOREHOLE LOGGING.

Well logs result from a probe lowered into the borehole at the end of an insulated cable. The resulting measurements are recorded graphically or digitally as a function of depth. These records are known as geophysical well logs, petrophysical logs, or more commonly well logs, or simply logs.

Although the most common uses of logs are for correlation of geological strata and location of hydrocarbon zones, there are many other important subsurface parameters that need to be detected or measured. Also, different borehole and formation conditions can require different tools to measure the same basic property. In petroleum engineering, logs are used to: identify potential reservoir rock; determine bed thickness;

determine porosity; estimate permeability; locate hydrocarbons; estimate water salinity; quantify amount of hydrocarbons; estimate type and rate of fluid production; estimate formation pressure; identify fracture zones; measure borehole inclination and azimuth; measure hole diameter; aid in setting casing; evaluate quality of cement bonding; locate entry, rate, and type of fluid into borehole; and trace material injected into formations (such as artificial fractures). In geology and geophysics, they are used to: correlate between wells; locate faults; determine dip and strike of beds; identify lithology; deduce environmental deposition of sediments; determine thermal and pressure gradients; create synthetic seismograms; calibrate seismic amplitude anomalies to help identify hydrocarbons from surface geophysics; calibrate seismic with velocity surveys; and calibrate gravity surveys with borehole gravity meter. Other applications include: locating fresh-water aquifers; locating solid minerals; and studying soil and rock conditions for foundations of large structures.

Electrical devices employ instruments that generate data based on electrical measurements. The spontaneous potential, usually recorded along with resistivity curves, is a simple but valuable aid to help geologists correlate from one well to another and to assist them in inferring the depositional environment of the sediments. It defines permeable zones from surrounding nonpermeable shales and can be used to estimate the salinity of formation water. Resistivity is one of the most important physical properties to record. A sand filled with salt water will conduct electricity much easier (and therefore has a lower resistivity) than a sand in which most of the salt water is replaced with a nonconducting oil or gas. The primary purpose of resistivity measurements is to determine hydrocarbon saturation.

Quantitative interpretation of hydrocarbon saturation requires a knowledge of porosity. One of the most widely used porosity logs is the acoustic log, also known as a sonic log, acoustilog, or velocity log. The acoustical log measures the shortest time for sound waves to travel through 1 ft (or 1 m) of formation. This log is used for many other purposes, including identification of lithology, prediction of pressures, and assisting with geophysical interpretation. In addition, information from the amplitude of the acoustic wave aids in detection of fractures and gas zones, determination of mechanical properties of rocks, and analysis of cement bond behind pipe.

Nuclear devices are instruments which generate data based on measurements of nuclear particles. In most cases the gamma-ray log may be considered as a shale log; that is, clays or shaly rocks will give a higher radioactive count than clean sands or carbonates. Like the spontaneous potential, this provides a good correlation curve, defines bed thicknesses, and aids interpretation of environmental deposition. Almost all the naturally occurring radioactivity that is detected by the gamma-ray log comes from three elements: potassium, thorium, and uranium.

An important porosity tool is the density log, which emits a beam of gamma rays into the rock and these rays interact with electrons in the formation through Compton scattering. The tool provides a bulk density measurement. Another important device for determination of porosity is the neutron log. These logs respond primarily to hydrogen atoms. Therefore, in clean or shale-free formations the neutron log reflects the amount of liquid-filled porosity. A tool developed to distinguish hydrocarbon from salt water behind casing is the pulsed neutron capture tool which employs a neutron generator that repeatedly emits pulses of high-energy neutrons. After these neutrons are slowed down to the thermal state, they are captured by nuclei of the various atoms surrounding the tool. With each capture a corresponding emission of gamma rays occurs.

Production logging tools are designed to locate fluid movement behind the casing and into the well bore, to detect the type of fluid, and to determine the flow rate. Production logging tools include various types of flow meters, fluid density devices, sensitive thermometers, radioactive tracer devices, noise loggers, and capacitance logs. In addition, these tools usually record some correlation curve such as gamma-ray or collar locator to tie the production curves to the formation and casing. Closely related to production tools are logging devices that either magnetically or mechanically inspect the condition of the casing in the borehole.

Small but rugged and powerful computers designed for field operation record logs on magnetic tapes as they are run and mathematically manipulate them in the logging trucks at the well site. Detailed computations are made quickly on a foot by foot basis to give information on fluid and rock properties. The digital information from practically every curve can be incorporated into some type of analysis or interpretation program.

[R.E.Wy.]

Wellpoint systems A method of keeping an excavated area dry by intercepting the flow of groundwater with pipe wells located around the excavation area. Intercepting the flow before it reaches the excavated area also improves the stability of the edge of the excavation, permitting steeper bank slopes and often eliminating the need for supporting or shoring the banks. *See* CONSTRUCTION METHODS; GROUNDWATER HYDROLOGY; WELL.

Wellpoint systems are most effective in coarse-grained soils, such as gravel or sand. They are not effective in fine soils, such as silts and clays, where the small size of the pores between grains restricts the flow of water.

The basic components of a wellpoint system are the wellpoint, the riser pipe, the header pipe or manifold, and the pump (see illustration). The wellpoint consists of a perforated

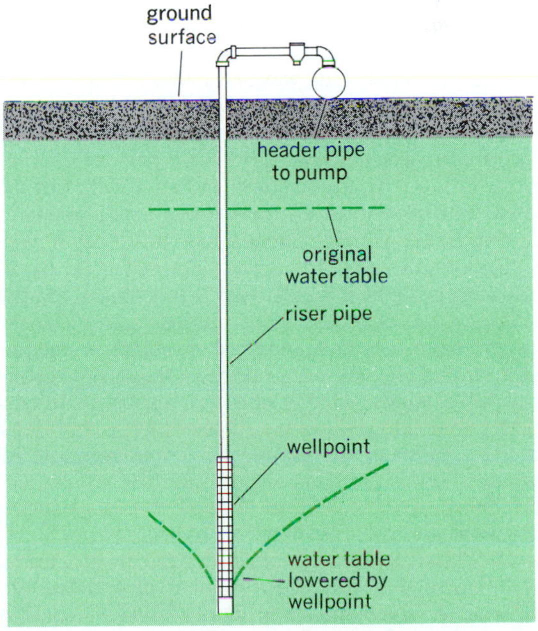

Components of a wellpoint system.

pipe equipped with a ball valve to regulate the flow of water, a screen to prevent the entry of sand during pumping, and a jetting tip. The steel riser pipe brings the groundwater to the surface, where it is collected by the horizontal manifold pipe or header pipe. The pumps are located above the water table and collect the water from the header pipes for discharge away from the excavation area. See PUMP; PUMPING MACHINERY. [W.Her.]

Welwitschiales An order of the class Cycadopsida having one species, *Welwitschia mirabilis*. This plant is native to the very arid deserts of southwestern Africa. Its appearance is bizarre—it has a very short, unbranched, woody stem (sometimes to 3 ft or 1 m in diameter), which is cushion- or saucer-shaped and tapers quickly to a long taproot. There are only two leaves, and these persist throughout the life of the plant. The leaf is broadly strap-shaped, as wide as the stem, firm, and leathery, and gradually splits lengthwise between the veins. The leaf (of indefinite growth) develops from a meristem at this point of connection to the stem. The species is dioecious; the cones are borne on branched axes originating between the crown of the stem and the base of the leaf. The order is known only from the one living species and not from any fossils. *See* CYCADOPSIDA; PINOPHYTA; PLANT KINGDOM. [T.A.Z.]

Wentzel-Kramers-Brillouin method A special technique for obtaining an approximation to the solutions of the one-dimensional time-independent Schrödinger equation, valid when the wavelength of the solution varies slowly with position. It is named after G. Wentzel, H. A. Kramers, and L. Brillouin, who independently in 1926 contributed to its understanding in the quantum-mechanical application. It is also called the WKB method, BWK method, the classical approximation, the quasi-classical approximation, and the phase integral method. *See* QUANTUM MECHANICS; SCHRÖDINGER'S WAVE EQUATION. [R.H.Go.]

West Indies An archipelago, including the Bahamas, the Greater Antilles (including Cuba, Jamaica, Hispaniola—the Dominican Republic and Haiti—and Puerto Rico), the Lesser Antilles, and other islands, curving 2500 mi (4000 km) from Yucatan Peninsula and southeastern Florida to northern Venezuela and enclosing the Caribbean Sea. Situated between latitude 10° and 27°N and longitude 59° and 85°W, in the zone of the northeast trade winds, the West Indies have a subtropical and predominantly oceanic climate, with even warmth and steady breezes. Temperatures vary little from season to season, ranging from means of 80–85°F (27–29°C) in July to 70–78°F (21–26°C) in January at sea level. Freezing is unknown, and the hottest temperatures rarely exceed 90°F (32°C). Precipitation ranges from a low of 25–50 in. (64–127 cm) a year on low-lying islands and drier coasts up to 300 in. (7.6 m) on the highest peaks, which are almost perpetually cloud-capped. At lower elevations, rainfall is erratic from year to year and from season to season, but reaches a maximum in the summer and fall, when the northeast trades are replaced by light, variable winds. This is also the season of hurricanes, destructive tropical cyclones which sweep west and northwest across the Caribbean, sparing only the southernmost islands. The winter months are generally dry, and there is frequently a shorter dry season in July or August. *See* CARIBBEAN SEA; TROPICAL METEOROLOGY.

The West Indian flora is chiefly derived from Central and South America, but there are a number of endemic species, notably palms; many mainland plants failed to colonize the islands, which are floristically poor. The effect of isolation and small size is evident in the meager character of West Indian fauna. Animal species are limited; there are few mammals and no large ones, except for domesticated animals and, especially on Hispaniola, feral cattle, goats, pigs, and horses. [D.Low.]

West Nile fever An acute, usually mild, mosquito-borne virus disease. The infection occurs in the summer, chiefly in Egypt, Israel, Africa, India, and Korea. Subclinical infections are common. In those infections which do produce illness, the usual signs are fever and lymphadenopathy, sometimes with rash; fatality or serious aftereffects are rare.

The infection cycle by which the virus is maintained in nature is thought to be bird → mosquito → bird, with *Culex univittatus* the principal vector. Ticks may play a role as reservoir-vectors of West Nile virus, serving as sources of transmission to birds and rodents. Human infection, as well as that of other vertebrates, occurs seasonally in relation to the presence and density of the infected mosquito population. [J.L.Me.]

Wet cell A primary cell in which there is a substantial amount of free electrolyte in liquid form. Important examples of wet cells are the Lalande or caustic soda cell, the air-depolarized alkaline cell, the Weston standard cell, and the organic electrolyte cell.

The Lalande cell uses a zinc anode, a cupric oxide cathode, and an electrolyte of sodium hydroxide in aqueous solution (caustic soda). The anode reaction in the Lalande cell is the oxidation of zinc to form zinc oxide, which dissolves in the electrolyte to form sodium zincate. The cathodic reaction is the reduction of cupric oxide to metallic copper.

The air-depolarized alkaline cell uses a zinc anode, a porous carbon cathode exposed to air on one face, and an alkaline electrolyte. The carbon cathode utilizes atmospheric oxygen. Its zinc anode and alkaline electrolyte are like those of the Lalande cell, but it has twice the operating voltage and twice the watt-hour output of an equal-size Lalande cell.

The Weston cell of 1893 has become the accepted standard of electromotive force (emf). The Weston normal or saturated cadmium cell has an emf of 1.01864 absolute volts at 68°F (20°C). When purified materials are used, cells having the same emf to within a few microvolts may be made. Reference standards, in daily use for many years, are remarkably constant. *See* VOLTAGE MEASUREMENT.

The cell uses a two-phase amalgam of cadmium as the anode. The cathode is mercurous sulfate, Hg_2SO_4, in contact with mercury. The electrolyte is a saturated solution of cadmium sulfate in equilibrium with the solid phase, $CdSO_4 \cdot {}^8\!/_3H_2O$.

A different class of cells is that based on the use of particularly reactive metals (Li, Ca, Mg) in conjunction with organic electrolytes. The best-known type in this class is the lithium-cupric fluoride cell, theoretically capable of delivering over 700 Wh/lb, more than three times the capacity of the highly ranked Zn-AgO cell. The lithium anode is usually made in sheet form, but variants using lithium powder trapped in appropriate grids are also known. The cathode is a mixture of CuF_2 and various conductive materials, most often graphite and carbon black either together or separately, in order to obtain the electronic conductivity that CuF_2 does not possess. The electrolyte considered most compatible with these electrodes is lithium perchlorate in propylene carbonate or butyrolactone. [L.R.; J.D.]

Wetlands Areas that are inundated or saturated by surface or ground water with sufficient frequency and duration to support vegetation typically adapted for life in saturated soil conditions; wetlands include swamps, marshes, bogs, and similar areas. A wetland system has saturated soils or standing water; hydrophytes or plants adapted to saturated soil conditions; and soils that, if mineral rather than peat, are usually defined as hydric soils, indicating that they have developed the colors, chemistry, and physics of soils deprived of oxygen. Wetlands are often found between drier uplands and deep-water aquatic systems such as lakes, rivers, estuaries, and oceans, and serve as interface systems. They are shallow to intermittently flooded ecosystems, characterized by a unique combination of hydrology, wetland vegetation, and wetland soils. *See* BOG; ESTUARINE OCEANOGRAPHY; HYDROLOGY; SALT MARSH; TUNDRA.

The extent of wetlands in the world is estimated to be 2–3 million square miles (5–8 million square kilometers) or about 4–6% of the land surface. Wetlands occur in humid, cool regions as bogs, fens, and tundra of Canada, Alaska, and northern Europe and Asia; along rivers and streams as riparian wetlands, seasonally flooded forests, and backswamps; in the deltas of the great rivers, including the Amazon, Mekong, Danube, Mississippi, and Rhone; along temperate, subtropical, and tropical coastlines as salt marshes, mud flats, and mangrove swamps; and even in arid regions as inland salt flats, seasonal playas, riparian systems, and vernal pools. They are found on every continent except Antarctica and in every clime from the tropics to the frozen tundra. *See* MANGROVE; PLAYA.

Compared to other ecosystems, wetlands are often an extremely productive part of the landscape and are havens for great biodiversity. Most wetlands support a rich variety of waterfowl and aquatic organisms. From a conservation viewpoint, wetlands are an extremely important habitat for rare and endangered species. For example, over 63% of reptiles, 68% of birds, and 75% of amphibians that are listed as threatened or endangered in the United States are associated at some point in their life cycle with wetlands. *See* ECOSYSTEM. [W.J.M.]

Whale A marine mammal of the order Cetacea. There are about 37 species, with a worldwide distribution. The cetaceans are divided into two distinct groups, the Odontoceti or toothed whales, and the Mysticeti or whalebone whales.

As marine forms, whales are extremely atypical; they are highly adapted to live in the marine environment but this is a secondary adaptation, an example of evolutionary convergence. During the course of evolution, the body became streamlined and the tail transformed into an organ for propulsion and the limbs into balancing structures. There are no external ears, and the small opening which leads to the eardrum has a wax plug. Nostrils, also lacking as external projecting structures, occur as a blowhole which is single in toothed whales and double in whalebone whales and is located on the top of the head. The blowholes lead to the lungs. The thick skin is smooth and hairless, and both sweat and sebaceous glands are absent. There is a thick layer of insulating fat, or blubber, beneath the skin.

The teeth in toothed whales may vary from 260 to a single pair and are simple in structure. The upper left canine of the narwhal may reach a length of 8 ft (2.4 m) or more and has the appearance of a coiled tusk. In place of teeth the whalebone whales have platelike structures of baleen, or whalebone. The toothed whales are referred to as ichthyophagous because they feed on a variety of marine organisms, such as fish and cephalopods. The whalebone whales feed on krill, or microcrustaceans, which they filter from the sea water and swallow by the ton; they are therefore referred to as teuthophagous.

Other important adaptations include huge lungs, a necessary adaptation since these animals may remain submerged for as long as 1 hr and may dive to depths of 300–600 ft (91–182 m). The eyes are small for the size of the animal and are adapted for underwater vision. In the males the testes are internal; the mammary glands of the female are inguinal. A single young, rarely two, is born. The gestation period of the blue whale is about 47 weeks, and that of the sperm whale is 68 weeks. *See* CETACEA. [C.B.C.]

Wharf A structure along a waterfront providing a berth for ships to load and discharge passengers and cargo. A marginal wharf, one parallel to shore, is generally known as a quay or bulkhead. When perpendicular or oblique to shore, a wharf is known as a pier. Construction is essentially that required for retaining earth fill or for embankment protection while at the same time providing a landing platform. Reinforced-concrete or masonry retaining walls are used as well as the familiar pile-supported platforms with sheet piling bulkheads of steel, timber, and concrete. *See* COASTAL ENGINEERING; PIER. [E.J.Q.]

Wheat A food grain crop. Wheat is the most widely grown food crop in the world. It ranks first in world crop production and is the national food staple of 43 countries. The principal food use of wheat is as bread, either leavened or unleavened.

Wheat for milling is classified according to hardness, color, and best use. In the United States, there are seven official market classes of which the following five are the most important: (1) hard red winter, for bread; (2) hard red spring, for bread and rolls; (3) soft red winter, for cake and pastries; (4) white, for bread, breakfast foods, and pastries; and (5) durum, for macaroni products.

Most countries in which wheat is grown have wheat breeding programs in which the objective is to develop more productive and more stable varieties (cultivars). Many methods are combined in these programs, but in nearly all of them specially selected parent types are crossbred followed by pure-line selection among the progeny to develop new combinations of merit. Varieties and genetic types from all over the world become candidate parents to provide the desired recombinations of good quality, winter and drought hardiness, straw strength, yield, and disease resistance. Wheats must be bred for specific milling processes and to provide quality end-use products. Many new varieties have complex pedigrees. *See* GRAIN CROPS. [L.P.R.]

The milling process breaks open the wheat kernel and reduces the particles formed so as to separate the outer and inner portions of the kernel. Bran and germ are almost completely separated from the white interior portions of the kernel in the milling of refined flour. When the entire kernel is ground and no separations are made, the product is whole wheat flour.

Flour milling has become one of the principal industries in terms of value of product processed and of value added to the product during processing. Flours are milled for many purposes such as bread, pastries, cakes, cookies, or macaroni. This necessitates wheat selection and different processing to some extent, although the basic principles are not changed. Flours are manufactured from hard, soft, or durum wheat or combinations of these. [J.A.Sh.]

Wheatstone bridge A device used to measure the electrical resistance of an unknown resistor by comparing it with a known standard resistance. This method was first described by S. H. Christie in 1833. Since 1843 when Sir Charles Wheatstone called attention to Christie's work, Wheatstone's name has been associated with this network.

The Wheatstone bridge network consists of four resistors R_{AB}, R_{BC}, R_{CD}, and R_{AD} interconnected as shown in the illustration to form the bridge. A detector G, having an internal resistance R_G, is connected between the B and D bridge points; and a power supply, having an open-circuit voltage E and internal resistance R_B, is connected between the A and C bridge points. *See* BRIDGE CIRCUIT.

If the network is adjusted so that Eq. (1) is satisfied, the

$$R_{BC}R_{AD} - R_{AB}R_{CD} = 0 \qquad (1)$$

detector current will be zero and this adjustment will be independent of the supply voltage, the supply resistance, and the detector resistance. Thus, when the bridge is balanced, Eq. (2)

$$R_{BC}R_{AD} = R_{AB}R_{CD} \qquad (2)$$

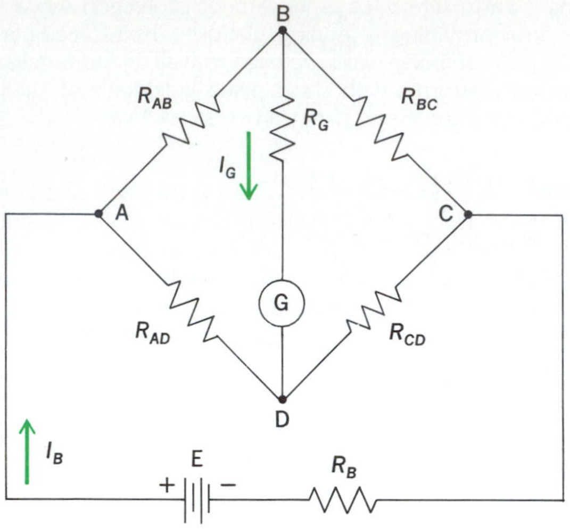

Wheatstone bridge circuit.

holds, and, if it is assumed that the unknown resistance is the one in the CD arm of the bridge, then it is given by Eq. (3).

$$R_{CD} = \left(\frac{R_{BC}}{R_{AB}}\right) \times R_{AD} \quad (3)$$

See Resistance measurement. [C.E.A.]

Wheel and axle A wheel and its axle or, more generally, two wheels with different diameters or a wheel and drum, as in a windlass, rigidly connected together so that they rotate as a unit on a common axis. The principle of operation is the same as that of the lever in that, for static equilibrium, the summation of torques about the axis of rotation equals zero.

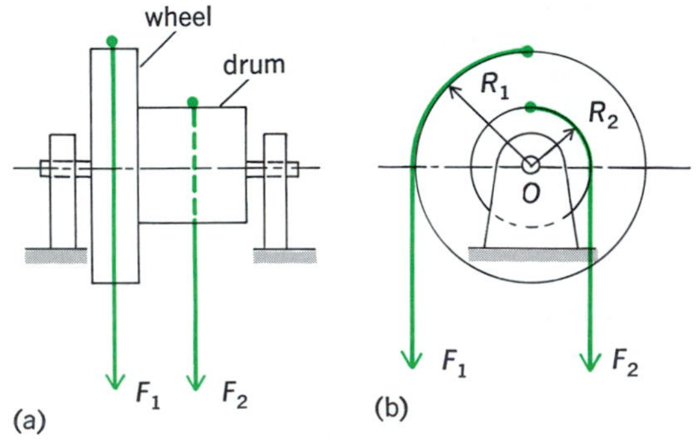

Wheel and axle. (*a*) Side view. (*b*) Front view.

Where flexible members, such as ropes, have been firmly attached to a wheel and drum (see illustration) and the machine is mounted on frictionless bearings, $F_1R_1 - F_2R_2 = 0$.

The main difference between the lever and the wheel and axle is that the wheel and axle permits the forces to operate through a much greater distance. In the illustration the wheel and drum could be allowed to rotate any number of revolutions if the ropes were wrapped the required number of times around each before they were attached. *See* Force; Simple machine. [R.M.Ph.]

Wheel base The distance in the direction of travel from front to rear wheels of a vehicle, measured between centers of ground contact under each wheel. For a vehicle with two rear axles, the rear measuring point is on the ground midway between rear axles. Tread of a vehicle is the distance perpendicular to the direction of travel between front wheels, or between rear wheels, measured from centers of ground contact. [F.H.R.]

Whelk A name given to many species of larger marine snails, mostly belonging to genera such as *Buccinum*, *Busycon*, *Colus*, and *Neptunea* (all in the order Neogastropoda). Their shells may be ovate or fusiform, but always consist of symmetrically enlarging whorls and a large aperture (see illustration). Whelks are found in all oceans, from polar to warm temperate seas. Rarely found above low-tide mark, they are most abundant on harder substrates in the shallow sublittoral, but some species extend downward to depths of 9900 ft (3000 m). Whelks are carnivores and carrion feeders, unlike the even larger-shelled mesogastropod conchs. Their locomotion appears as a smooth gliding over rock and shell surfaces, is based on sequential waves of muscle contraction passing along the sole of the foot, and is considerably faster than that of conchs or limpets. *See* Gastropoda; Snail. [W.D.R.-H.]

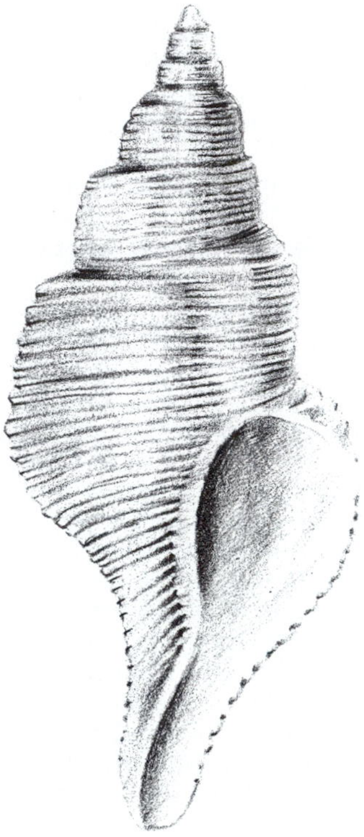

Shell of a typical whelk species, *Neptunea tabulata*, about 3.5 in. (9 cm) long, living in depths of 33–198 ft (10–60 m) off the Pacific coast of North America. The characteristic siphonal canal extends anteriorly from the shell aperture.

Whipworm disease A chronic, wasting diarrhea produced by heavy parasitization of the large intestine by the nematode *Trichuris trichiura*, particularly in undernourished children in the tropics. The worms are 1 in. (2.5 cm) long, with the anterior body attenuated to resemble a whip, and are firmly attached to the intestine, where they irritate the mucosa. Their barrel-shaped eggs pass in the feces. In countries with poor sanitation, the soil becomes contaminated. *See* Nemata. [J.F.M.]

White dwarf star An intrinsically faint star of very small radius and high density. The mass of a typical white dwarf is about 0.7 that of the Sun, and the average radius is 5000 mi (8000 km). Surface temperatures are between 85,000 and 4000 K, and brightness is from 10 to 10^{-4} that of the Sun. Such stars are composed largely of helium, carbon, or heavier elements.

A white dwarf is one final stage of stellar evolution, with thermonuclear energy sources extinct. The heavy nuclei are still mobile, and their thermal energy maintains the star's radiation. The star cools and becomes fainter and redder. *See* STELLAR EVOLUTION.

White dwarfs have been found to be one member of certain close, interacting binary stars of the type that may become classical novae (undergoing a giant explosion), recurrent novae, or dwarf novae. *See* BINARY STAR; BLACK HOLE; NOVA. [J.L.Gr.]

Whooping cough An acute communicable disease of humans, also known as pertussis, caused by *Bordetella pertussis*. It occurs chiefly among infants and children, about one-half of those affected being under 4 years of age.

The course of the disease is roughly 6 weeks, equally divided into catarrhal, paroxysmal and decline periods; each lasts approximately 2 weeks. About 2 weeks after exposure the child begins to cough, usually at night; the cough soon becomes diurnal. It becomes more frequent and severe, occurring in a series of explosive spells with strangling and suffusion of the face. The spell finally ends in the loud inspiratory crow (the whoop) as the child attempts to regain his breath. This is often followed by vomiting.

The rapid decline in the incidence and severity of the disease in the United States is attributed to better nutritional status and to the extensive immunization of infants with immune serum and vaccine. [W.L.Br.]

Wide-area network A system consisting of sets of nodes that are interconnected by sets of links. The nodes and links cover a relatively large geographic area, usually on the order of hundreds of miles. This very general definition provides a framework for describing any one of a large number of network configurations, each of which may differ in application, hardware, software, and other features. For example, the nodes may be used for switching and routing messages in a communications network; the nodes in another network may be host computers which do application processing. The links in one network may be land-line cables; in another network, they may be satellite channels. In contemporary networks, it is common to find a variety of nodes and links; each type is allocated to tasks which it can do effectively. There are many reasons for developing and using wide-area networks (WANs), including message communication, resource sharing, remote computing, file transfer, integrated networking, application performance improvement, and combinations of some or all of these functions. *See* COMMUNICATION CABLES; COMMUNICATIONS SATELLITE; DIGITAL COMPUTER; SWITCHING SYSTEMS (COMMUNICATIONS).

Modern networks are designed according to standards which have been developed by international standards organizations. Standards were developed to facilitate compatibility between networks and to allow for a mix of vendor equipments in a network. The most common standard is the International Standards Organization (ISO) layered model. A layer is a partition of a network which has responsibility for a major function (for example, switching and routing).

There are two major services offered by networks which are related to network architecture: datagram and virtual circuit. Datagram service provides best-effort type of message delivery and is used in noninteractive applications (for example electronic mail and transaction processing where no interaction with the user is necessary). Loosely speaking, this type of service may be considered one-way transmission. In contrast, virtual circuit service is for interactive processing, where sequenced, error-checked message delivery is essential for correct transmission and response of interactive commands (for example, text editing). *See* DATA COMMUNICATIONS. [N.F.S.]

Wiedemann-Franz law An empirical law of physics which states that the ratio of the thermal conductivity of a metal to its electrical conductivity is a constant times the absolute temperature, as given by the equation below. Here K_c

$$K_c = L_0 \sigma T$$

is the thermal conductivity due to the conduction electrons, σ is the electrical conductivity, T is the absolute temperature, and L_0 is known as the Lorentz number. The Wiedemann-Franz law provides an important check on theories of electrical and thermal conductivity. *See* CONDUCTION (HEAT); THERMAL CONDUCTION IN SOLIDS. [F.J.B.]

Wild cherry A tree, *Prunus serotina*, that belongs to the rose family, and is an important Ember species of cherry in the United States. Native in the eastern half of the United States and adjacent Canada, it reaches its largest size in the southern Appalachian Mountains. The deciduous leaves are long-pointed, thickish, and shining above. The edible fruit is a shiny black drupe found in long clusters. The hard, strong wood is a red-brown and is used for furniture, often as an imitation mahogany. *See* ROSALES. [A.H.G./K.P.D.]

Wildlife conservation The science and art of making decisions and taking actions to manipulate the structure, dynamics, and relations of wild animal populations, habitats, and people to achieve specific human benefits from the wildlife resource. Although still called wildlife conservation by some people, wildlife management, the preferred term, has more than its earlier connotations of preservation. Wildlife management is concerned not only with preserving and increasing populations of threatened or endangered species, such as the whooping crane, but also with stabilizing certain populations, such as the American bison, increasing or stabilizing desirable species taken for sport recreation (game), and decreasing some populations of birds and harmful mammals.

The problems of management are complex and involve conflicts between groups with opposing interests. For example, orchard owners want fewer bark-eating rabbits, while hunters want more rabbits of all kinds; fox hunters want increased populations, while other people wish to reduce foxes as vectors of rabies; hunters who want only native game animals differ with hunters who would stock or import foreign species into any area; hunters and farmers are at odds on the proper number of raccoons to have in forests near cornfields; foresters managing their forests to supply deer to hunters do not want deer in their nurseries; city dwellers want birds in parks but not on buildings.

Management task. To achieve a desired abundance of a particular species of wildlife, modern wildlife managers work intensively with the environment in which the animals live. Wildlife populations naturally respond very sensitively to their habitats. When water levels are down, ducks produce fewer young; when dams of beaver are torn out and the rich land within beaver ponds is planted to crops, beaver disappear; where the woody food of deer and elk is not abundant or after the supply of forage has been eaten, populations of these animals are reduced or they have fewer young.

Most wildlife managers now believe that energy is the primary factor limiting the ability of area to support animals. Animals require large amounts of energy to survive. It is provided to them by food, sunlight, and reradiation of sun energy from plants and the ground. It is being removed from them at a rapid rate by natural forces, primarily wind, contact with the cold ground, and evaporation of moisture; of course, movement and producing young require large expenditure of energy. Animals have developed means for conserving this energy so that outgo does not exceed income. Wildlife managers may attempt to supply food energy or reduce energy losses. By providing protection from the wind, preserving areas where animals may sun, protecting them from harassment by dogs, predators, or people, and providing high-energy foods, the modern wildlife manager attempts to improve the energy balance and budgets of wild animal populations.

Where soils are poor, wildlife will be sparse. Richer soils produce more vigorous plants that store sun energy better and make it available to wildlife. The types of plants and animals must be balanced by the wildlife manager, for on very rich soil, plants can grow to heights out of reach of animals. Rich soils can encourage large populations with high food consumption that can temporarily destroy the vegetation on areas and even result in soil erosion. These changes can result in population fluctuations.

Wildlife managers are concerned with the natural ability of populations to reproduce. The reproduction potential is set by genetics but strongly influenced by the quality of food. When food supplies are low or the quality of food is reduced (as a result of soil erosion or leaching of nutrients from soil), reproduction is reduced. Another factor called stress influences populations in several complex ways: as populations become more crowded, reproduction is reduced, more food is required, and diseases have greater effects. Thus there are several interacting natural limits on wildlife population growth imposed both by the environment and by the population itself.

Threats to populations of wildlife usually result from a very complex set of factors. The factor usually at the top of the list is destruction of wildlife homes and habitat. However, establishing areas of complete wildlife protection, stocking, and broad-scale predator control have proved over and over to be insufficient. Habitat management has been a more efficient technique. This approach, a systems approach, is that of selecting a group of techniques specifically designed for benefiting one target species for each area and for achieving the specific objectives of the landowner. The system involves (1) intensive study and research by experts to isolate problems, (2) definition of goals, (3) education of the public (and specialists) by many media, (4) programs to retain and improve the habitat, (5) goal-oriented game harvest regulations or population control programs, (6) manipulation of sex and age ratios, as well as density of animals, to achieve desired population change, (7) actions by agents to deter or apprehend poachers or game-law violators, (8) control of disease and parasites, and of select individual predators and the causes of accidents, and (9) well-researched transplanting of wildlife from areas where it has bred successfully to areas with high potential for population buildups. *See* SYSTEMS ECOLOGY.

Today's needs. The public and the professional wildlife staffs of state and Federal conservation agencies can influence wildlife populations by the following major practices: (1) laws to protect endangered species, (2) regulations to allow a recreational harvest of game animals and manipulation of other populations that put each population in balance with its food supply, (3) protection of wildlife areas, both large land holdings of the government and small areas, such as woodlots and fencerows, needed by certain species of birds and animals, (4) research into the needs of and responses of animals to each other and their environment, (5) investment in select areas for maximum wildlife production, maintaining moderate production cheaply under existing conditions, and (6) education of the public: increasing their appreciation of wildlife (and thus the benefits which they derive from it) and also their awareness of how actions (hunting, voting, destroying food and cover, changing the environment, and contributing to research and law enforcement) influence wildlife and the larger complex of humans' own environment.

Wildland operations research. There has emerged an art called operations research (OR) based on the science of analyzing large and complex systems of all types and designing means to achieve specified goals in some optimal fashion. Wildland operations research is the application of the concepts and methods of operations research to wildland problems. The term wildlands usually refers to all of the resources (space, water, soil, air, minerals, forests, and wildlife) of nonagricultural lands, such as forests, rangelands, marshes, deserts, and undeveloped parks. The differences within and between these categories may be great or indistinct (for example, differences between highly managed tree farms, farm woodlots, and wilderness forests). Every system can be considered a subsystem of a larger system. *See* TERRESTRIAL ECOSYSTEM.

Wildland operations research deals with natural resource systems such as the ecosystem, with populations of fish or other animals, with forests and wood-dependent industries, with rangelands and marshlands, with migratory-bird flyways, and with pest species and their effects. *See* ECOLOGY; OPERATIONS RESEARCH. [R.H.G.]

Willemite A rare nesosilicate mineral, composition Zn_2SiO_4, crystallizing in the hexagonal system. It is usually massive or granular with a vitreous luster; crystals are rare. The mineral may be variously colored, most commonly green, red, or brown. Hardness is $5\frac{1}{2}$ on Mohs scale; specific gravity is 3.9–4.2. Willemite forms a valuable ore of zinc at Franklin, New Jersey. At this famous zinc deposit willemite fluoresces a yellow-green. *See* SILICATE MINERALS. [C.S.Hu.]

Willow A deciduous tree and shrub of the genus *Salix*, order Salicales, common along streams and in wet places in the United States, Europe, and China. The twigs are often yellow-green and bear alternate leaves which are characteristically long, narrow, and pointed, usually with fine teeth along the margins. Flowers occur in catkins. The fruit contains several silky seeds. *See* SALICALES.

Willow lumber is used for fuel and in making charcoal, excelsior, ball bats, boxes, crates, boats, waterwheels, and wicker furniture. The tough, pliable shoots of many species are used to make baskets; the bark of other species is used for tanning. Willows are of great value in checking soil erosion. A few species are ornamental shade trees. [J.F.F.]

Wind The motion of air relative to the Earth's surface. The term usually refers to horizontal air motion, as distinguished from vertical motion, and to air motion averaged over a chosen period of 1–3 min. Micrometeorological circulations (air motion over periods of the order of a few seconds) and others small enough in extent to be obscured by this averaging are thereby eliminated. *See* WIND MEASUREMENT.

The direct effects of wind near the surface of the Earth are manifested by soil erosion, the character of vegetation, damage to structures, and the production of waves on water surfaces. At higher levels wind directly affects aircraft, missile and rocket operations, and dispersion of industrial pollutants, radioactive products of nuclear explosions, dust, volcanic debris, and other material. Directly or indirectly, wind is responsible for the production and transport of clouds and precipitation and for the transport of cold and warm air masses from one region to another. *See* WEATHERING PROCESSES.

Cyclonic and anticyclonic circulation are each a portion of the pattern of airflow within which the streamlines (which indicate the pattern of wind direction at any instant) are curved so as to indicate rotation of air about some central point of the cyclone or anticyclone. The rotation is considered cyclonic if it is in the same sense as the rotation of the surface of the Earth about the local vertical, and is considered anticyclonic if in the opposite sense. Thus, in a cyclonic circulation, the streamlines indicate counterclockwise (clockwise for anticyclonic) rotation of air about a central point on the Northern Hemisphere or clockwise (counterclockwise for anticyclonic) rotation about a point on the Southern Hemisphere. When the streamlines close completely about the central point, the pattern is denoted respectively a cyclone or an anticyclone. Since the gradient wind represents a good approximation to the actual wind, the center of a cyclone tends strongly to be a point of minimum atmospheric pressure on a horizontal surface. Thus the terms cyclone, low-pressure area, or "low" are often used to denote essentially the same phenomenon. *See* GRADIENT WIND; STORM.

Convergent or divergent patterns are said to occur in areas in which the (horizontal) wind flow and distribution of air density is such as to produce a net accumulation or depletion, respectively, of mass of air. The horizontal mass divergence or convergence is intimately related to the vertical component of motion. For example, since local temporal rates of change of air density are relatively small, there must be a net vertical export of mass from a volume in which horizontal mass convergence is taking place. Only thus can the total mass of air within the volume remain approximately constant.

The horizontal mass divergence or convergence is closely related to the circulation. In a convergent wind pattern the circulation of the air tends to become more cyclonic; in a divergent wind pattern the circulation of the air tends to become more anticyclonic. A convergent surface wind field is typical of fronts. As the warm and cold currents impinge at the front, the warm air tends to rise over the cold air, producing the typical frontal band of cloudiness and precipitation. *See* FRONT.

Zonal surface wind patterns result from a longitudinal averaging of the surface circulation. This averaging typically reveals a zone of weak variable winds near the Equator (the doldrums) flanked by northeasterly trade winds in the Northern Hemisphere and southeasterly trade winds in the Southern Hemisphere, extending poleward in each instance to about latitude 30°. The doldrum belt, particularly at places and times at which it is so narrow that the trade winds from the two hemispheres impinge upon it quite sharply, is designated the intertropical convergence zone, or ITC. The resulting convergent wind field is associated with abundant cloudiness and locally heavy rainfall. *See* MONSOON METEOROLOGY.

Longitudinal averaging indicates a predominance of westerly winds. These westerlies typically increase with elevation and culminate in the average jet stream, which is found in lower middle latitudes near the tropopause at elevations between 35,000 and 40,000 ft (10.7 and 12.2 km). *See* JET STREAM.

Local winds commonly represent modifications by local topography of a circulation of large scale. Examples are the mistral which blows down the Rhone Valley in the south of France, the bora which blows down the gorges leading to the coast of the Adriatic Sea, the foehn winds which blow down the Alpine valleys, the williwaws which are characteristic of the fiords of the Alaskan coast and the Aleutian Islands, and the chinook which is observed on the eastern slopes of the Rocky Mountains. *See* CHINOOK. [F.S.]

Wind measurement The determination of three parameters: the size of an air sample, its speed, and its direction of motion.

The size of the air sample is highly dependent on how the measurement is made. When the wind measurement is taken with small, sensitive, and rapid-response instruments, or by the drift of small suspended particles, the air sample can have a scale of a fraction of an inch or less. If the wind measurement is made from the pressure gradient, as measured on a weather map, the scale can be hundreds of miles.

Wind direction is designated as the direction from which the wind is blowing, given in terms of 8, 16, or 32 points of the compass and in degrees or tens of degrees from north, measured clockwise. A wind vane is used to measure the direction of the surface wind. Basically, this is a grossly unsymmetrical body mounted near its center of gravity and free to rotate about a vertical axis.

In terms of its force on common objects such as leaves, smoke, or waves, wind speed can be estimated by the Beaufort scale. The scale of Beaufort numbers ranges from 0 for winds calm with velocity under 1 mi/h or 1.6 km/h (smoke rises vertically) to 12 for hurricane-force winds with velocity over 75 mi/h or 120 km/h. Meteorologists often measure the wind speed in terms of the cause of the wind, that is, the atmospheric pressure gradient, which is inversely proportional to the spacing of the isobars on a weather map. For information concerning instruments which measure surface winds *see* ANEMOMETER. [V.E.S.]

Wind power Kinetic energy in the Earth's atmosphere used to perform useful work. Total atmospheric wind power is of the order of 10^{14} kW. Annual kinetic energy is of the order of 10^{17} kWh (1 kWh = 3.60 × 10^6 joules). Practical land-based wind generators could extract as much as 10^{14} kWh of energy per year worldwide. Energy, and thus productivity of winds, varies markedly with geographic location. Annual energy available to a conversion machine at a site is very reproducible (± 15% variability).

Most large-scale irrigation pumping in the United States consumes natural gas, but wind-powered pump-back could expand the capacity of hundreds of smaller hydroelectric installations. At least 20% of United States energy is consumed in heating buildings. In colder climates, heating demand frequently matches high winds. Wind-generated electricity feeding thermal storage units, in some instances combined with solar thermal collectors feeding adjacent storage units, offers excellent potential for reducing fuel oil consumption. *See* ELECTRIC POWER GENERATION.

A windmill is a rotating machine capable of interchanging (extracting) momentum with particles of air mass that flow through its swept area. Power available in the wind in a swept area varies with the size of that area, the density of the air, and the square of the velocity. Energy extractable from an oncoming wind stream over a period of time varies as the size of swept area, density, and cube of the velocity. Thus, the utility and competitiveness of a wind power system depend upon the wind regime and height at which windmills are placed, size and characteristics of the machine, nature of delivered product, and productivity of that product. *See* ENERGY SOURCES; WIND. [W.He.]

Wind rose A diagram in which statistical information concerning the direction and speed of the wind at a particular

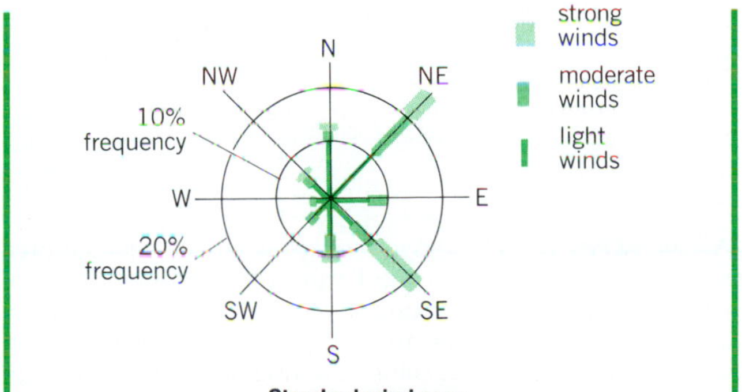

Standard wind rose.

location may be conveniently summarized. In the standard wind rose a line segment is drawn in each of perhaps eight compass directions from a common origin (see illustration). The length of a particular segment is proportional to the frequency with which winds blow from that direction. Parts of a given segment are given various thicknesses, indicating frequencies of occurrence of various classes of wind speed from the given direction. *See* WIND MEASUREMENT. [F.S.]

Wind shear Any change of vertical or horizontal wind divided by the diXstance over which the change is measured. Mathematically, wind shear is a tensor whose nine components correspond to the three orthogonal directions in which variations can occur combined with the three directions from which the wind can be blowing. *See* WIND.

Outside of thunderstorms, wind shear is prominent in temperate latitudes as a result of horizontal contrasts of temperature. This nonthunderstorm shear, manifested as variation of horizontal wind with height and often culminating in jet streams of great altitude, is most intense in frontal zones, where temperature variations are concentrated. Aircraft accidents have resulted from encounters with strong wave motions and turbulence in clear air, often associated with fast airflow over mountain ranges or extraordinary vertical variations of the horizontal wind near jet streams. *See* CLEAR AIR TURBULENCE (CAT); FRONT JET; STREAM.

Most wind shear that is a serious hazard to aircraft occurs in and near thunderstorms. There have been fatal accidents attributed to various phenomena that accompany thunderstorms, including hail, turbulence, lightning, the shear associated with outflowing winds, and excessive precipitation, which restricts visibility and may degrade engine performance when large amounts of water are ingested by the engines. *See* THUNDERSTORM.

Although it has not always been possible to measure precisely the role of each different factor operative in combination, shear manifested as turbulence and vertical drafts has always been important in aircraft accidents involving thunderstorms along air routes. During the 1970s and 1980s, aircraft accidents caused by thunderstorm wind shear during takeoffs and landings became more prominent than before. Detailed analyses of the related weather situations have shown that aircraft traversed regions of thunderstorm outflow with dimensions of just a few miles. Often, the first untoward event was a headwind increase.

A number of actions begun during the 1970s and 1980s are aimed at reducing wind shear accidents near air terminals. A system of anemometers on airport grounds, the Low Level Wind-Shear Alert System (LLWAS), presents a display in the control tower of winds at various points around the airport. More attention is being given to avoiding thunderstorm cells which lie on approach and departure courses; radars on the ground as well as on the aircraft are used, and meteorologists stationed at air-route traffic control centers send critical weather data to terminal controllers. Detection systems are being designed for installation on aircraft to alert flight crews immediately upon a wind shear encounter and to indicate appropriate compensatory responses. In addition, special training courses help pilots learn how to respond more effectively to dangerous shear situations encountered inadvertently or in spite of best efforts at avoidance. [E.K.]

Wind stress The drag or tangential force per unit area exerted on the surface of the Earth by the adjacent layer of moving air. Erosion of ground surfaces and the production of waves on water surfaces are manifestations of wind stress. Surface wind stress determines the exchange of momentum between the Earth and the atmosphere and, together with internal atmospheric viscous stresses, exerts a strong influence on the typical variation of wind through the lowest few thousand feet of the atmosphere. For a discussion of the relation of wind stress to ocean currents and waves *see* OCEAN CIRCULATION; OCEAN WAVES.

Significant stresses arise within the lower atmosphere because of the strong shear of the wind between the slowly moving air near the ground and the more rapidly moving air a few thousand feet above and because of the turbulent nature of the airflow in this region. Such turbulent eddies have characteristic dimensions ranging up to a few hundreds of feet.

Wind pressure is the force exerted by the wind per unit area of solid surface exposed normal to the wind direction. In contrast to shearing stresses, the wind pressure arises from the difference in pressure between the windward and lee sides of the exposed surface. Wind pressure thus represents a substantial force when the wind speed is high. *See* METEOROLOGY; WIND. [F.S.]

Wind tunnel A duct in which the effects of airflow past objects can be determined. The steady-state forces on a body held still in moving air are the same as those when the body moves through still air, given the same body shape, speed, and air properties. Scaling laws permit the use of models rather than full-scale objects, such as aircraft or automobiles. Models are less costly and may be modified more easily, and conditions may be simulated in the wind tunnel that would be impossible or dangerous in full scale.

Most data are secured from wind tunnels through measurement of forces and moments, surface pressures, changes produced in the airstream by the model, local temperatures, and motions of dynamically scaled models, and by visual studies.

A balance system separates and measures the six components of the total force. The three forces taken parallel and perpendicular to a flight path are lift, drag, and side force. The three moments about these axes are yawing moment, rolling moment, and pitching moment, respectively.

Surface pressures are measured by connecting orifices flush with the model surface to pressure-measuring devices. Local air load, total surface load, moment about a control surface hinge line, boundary-layer characteristics, and local Mach number may be obtained from pressure data.

Measurements of stream changes produced by the model may be interpreted in terms of forces and moments on the model. In two-dimensional tunnels, where an aircraft model spans the tunnel, it is possible to determine the lift and center of pressure by measuring the pressure changes on the floor and ceiling of the tunnel. The parasite drag of a wing section may be determined by measuring the total pressure of the air which has passed over the model and calculating its loss of momentum.

Measurements of surface temperatures indicate the rate of heat transfer or define the amount of cooling that may be necessary.

In elastically and dynamically scaled models used for flutter testing, measurements of amplitude and frequency of motion are made by using accelerometers and strain gages in the structure. In free-flight models, such as bomb or missile drop tests, data are frequently obtained photographically.

At low speeds, smoke and tufts are often used to show flow direction. A mixture of lampblack and kerosine painted on the model shows the surface streamlines. A suspension of talcum powder and a detergent in water is also used.

For aircraft at velocities near or above the speed of sound, some flow features may be made visible by optical devices. *See* INTERFEROMETRY; SCHLIEREN PHOTOGRAPHY; SHADOWGRAPH.

The V/STOL wind tunnel is a newer development of low-speed wind tunnels having a large very-low-speed section to permit testing of aircraft designed for vertical or short takeoff and landing (V/STOL) while operating in the region between vertical flight and cruising flight. [R.G.Jo.]

Windings in electric machinery Windings can be classified in two groups: armature windings and field windings. The armature winding is the main current-carrying winding in which the electromotive force (emf) or counter-emf of rotation is induced. The current in the armature winding is known as the armature current. The field winding produces the magnetic field in the machine. The current in the field winding is known as the field or exciting current. *See* ELECTRIC ROTATING MACHINERY; GENERATOR; MOTOR.

The location of the winding depends upon the type of machine. The armature windings of dc motors and generators are located on the rotor, since they must operate in conjunction with the commutator, and the field windings are mounted on stator field poles. *See* DIRECT-CURRENT GENERATOR; DIRECT-CURRENT MOTOR.

Alternating-current synchronous motors and generators are normally constructed with the armature winding on the stator and the field winding on the rotor. There is no clear distinction between the armature and field windings of ac induction motors or generators. One winding may carry the main current of the machine and also establish the magnetic field. It is customary to use the terms stator winding and rotor winding to identify induction motor windings. The word armature, when used with induction motors, applies to the winding connected to the power source (usually the stator). *See* ALTERNATING-CURRENT GENERATOR; ALTERNATING-CURRENT MOTOR; SYNCHRONOUS MOTOR. [A.R.E.]

Wine Alcoholic beverages made by fermentation of the juice of fruits or berries, essentially grape juice. Wines from other materials are always required to show their source on the label, for example, apple wine, berry wine, and cherry wine. California produces the largest percentage of the grape wine made in the United States, and well over a hundred varieties of the cultivated grape (*Vitis vinifera*) are used.

Classification of wine depends on the color, relative sweetness, alcoholic content, presence of carbon dioxide, the variety of grape, and the region where the grapes are grown. Wines may be red or white. The terms dry and sweet refer to the relative sugar content of a wine. Table wines contain less than 14% alcohol by volume, and dessert wines contain over 14%, usually 20%, alcohol. The higher alcohol content of dessert wines is obtained by the addition of brandy, which is called fortification. Sparkling wines such as champagne and sparkling Burgundy contain carbon dioxide.

Grapes are harvested and go through a crusher-stemmer, which crushes the berries, but not the seeds, and removes the stems. The must, still containing seeds and skins, is pumped to fermentation tanks where it receives sulfur dioxide gas. This controls, to a large extent, the growth of bacteria and wild yeasts, whereas the various strains of wine yeast are adapted to this amount of SO_2. A pure yeast starter is then added. In the production of white wines, the juice is separated from the skins and seeds at an early stage of fermentation. *See* FERMENTATION.

After the initial fermentation, the wine is transferred to the storage cellar for completion of fermentation, clarification, aging, stabilization, and bottling. These operations are called cellar practices. The sediment of yeast and other insoluble matter is called lees. The new wine is drawn off, or racked, from the lees to avoid picking up undesirable flavors from the lees. Aging is then continued. Wines are often chilled to precipitate excess cream of tartar, the potassium acid tartrate from the grape, which otherwise might precipitate upon chilling of bottled wine. Champagne is made by allowing a secondary fermentation to occur with a special flocculating type of yeast, either in bottles, by the original French procedure, or in bulk, using pressure tanks followed by bottling. The carbon dioxide, CO_2, of sparkling wines is produced by yeast fermentation. *See* ALCOHOLIC BEVERAGES. [E.M.M.; H.J.P.]

Wing The airfoil which is the principal source of an airplane's sustentation. The lift generated by the motion of the wing through the air supports the airplane. The main geometrical features of a wing are its span, that is, the straight line distance from tip to tip; the area of the wing as seen in a plan view of the airplane; and its sweep angle, the angle made by the wing leading edge and a line perpendicular to the plane of symmetry of the airplane. Another important geometrical characteristic is the wing section, which is the shape of the wing when cut parallel to the plane of symmetry of the airplane. *See* AIRFOIL; AIRPLANE.

The wing has several functions beyond that of providing lift. On larger airplanes the engines are mounted in nacelles either attached to the wing or mounted in the wing. The nacelles also provide a housing for the landing gear when it is retracted. The space within the wing is usually used for fuel storage.

Many of the control functions of the airplane are provided by special devices built into the wing. Most obvious are the ailerons, movable trailing-edge flaps occupying the outer part of the wing. Inboard of the ailerons and also on the trailing edge are the landing flaps, movable portions deflected during landing to augment the lifting capability of the wing. Through their use, the airplane can land more slowly, or the amount of fuel or the number of passengers can be increased at a given landing speed. Some airplanes have brakes to slow the airplane through a drag increase. A similar device used only on one wing at any given time decreases the lift on that wing. Acting like an aileron, it is used for roll control. *See* AILERON; ELEVON.

Loading. Wing loading is a measure of the load carried by an airplane wing per unit of wing area. Commonly used units are pounds per square foot and kilograms per square meter. As a general rule, the lower-speed airplanes are smaller, have lower wing loadings, and are more maneuverable. [J.Bi.]

Structure. In an aircraft, the combination of outside fairing panels that provide the aerodynamic lifting surfaces and the inside supporting members that transmit the lifting force to the fuselage is called the structure of a wing. It is an integration of the environment external to the vehicle wing, the aerodynamic shape of the wing, and the proposed use of the vehicle. The interaction of these three aspects of design leads to the selection of material, to the general structural layout, and, finally, to the detail choice of structural shapes, material thickness, joints, and attachments. The result is a structural framework covered with a metal skin that also contributes to the load-carrying function. *See* SUPERCRITICAL WING.

Wing structure has evolved from the early use of wood, doped canvas, and wire. Today, aluminum alloy outer skins are prime structural elements on all commercial transports and on the great majority of military craft. Magnesium, steel, and titanium are also used in internal primary structure and in local skin areas. The more exotic research craft, such as the X-15, are built at least partly of nickel alloy. Even less conventional materials may be used for reentry structures.

The derivation of the net loads permits a quantitative consideration of the general structural framework. Design of the structure, however, is inseparable from the choice of material. The thickness and shape of the individual pieces also affect strength in lightweight, efficient structures because of compression stability modes.

High static strength and light weight—coupled with reasonable rigidity characteristics, good corrosion resistance, fatigue endurance, cost and formability—are prime considerations in choosing airframe material. Highly loaded, efficient airframes have found these qualities in aluminum. It does appear, however, that the demands for speed interacting with the physical environment will cause aluminum to be replaced by steel, refractories, ceramics, and composites of these. A material's creep characteristics and strength at temperature are now important added parameters.

Prime structural framework. An airframe wing is essentially two cantilever beams joined together. Each wing tip is the free end of the cantilever, and the center line of the vehicle represents the plane where the two fixed ends of the cantilevers are joined. The prime load-carrying portion of these cantilevers is a box beam made up usually of two or more vertical webs, plus a major portion of the upper and lower skins of the wing, which serve as chords of the beam. This box section also provides torsional strength and rigidity (Fig. 1). Normally the prime box is designed to carry all the primary structural loads.

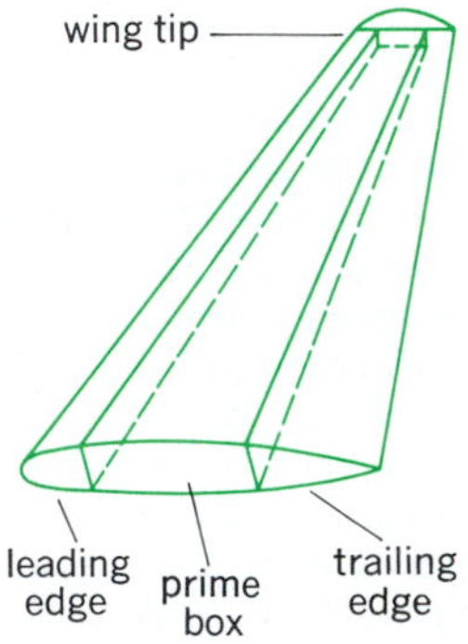

Fig. 1. Prime box of wing half-span.

Leading and trailing edge portions of the wing, forward and aft of the prime box respectively, help to provide the airfoil shape required. These portions are designed to minimize their participation in the major load-carrying function.

At numerous places within the wing box, bulkhead-type structures called ribs are located. These internal structures serve to maintain the rectangular box shape and to cut down the unsupported length of compression cover structures, to separate fuel tanks, and to distribute concentrated loads from guns, bombs, landing gear, or engines into the prime box. They are also located at any wing cross section where major load redistributions occur.

Leading edge structure. The leading edge, or most forward portion of the wing, serves an important aerodynamic function in establishing smooth airflow and efficient lifting power for the wing. This area sustains the highest aerodynamic pressures. On hypersonic vehicles, it is also the hot spot of the design. Structurally, the leading edge is an appendage, a fairing whose local loads must be supported by the prime wing box. This structure takes the form of cantilever beams or arch-type structures from the leading edge back to the front vertical beam. Frequently the leading edge of an aircraft incorporates anti-icing provisions.

Numerous aircraft have leading edge slats. These small airfoil sections are pulled out and forward automatically by air pressures and serve to maintain smooth airflow to a higher angle of attack than the basic large airfoil would sustain. As the angle of attack is decreased, these devices are withdrawn into the basic airfoil by springs.

Trailing edge structures. The trailing edge structure is noteworthy for its various auxiliary devices which assist in aircraft control or in reducing aircraft landing speeds. Various forms of flaps, ailerons, and spoilers, with their hydraulic or electrical control mechanisms, fill the volume of space aft of the rear beam or spar (Fig. 2).

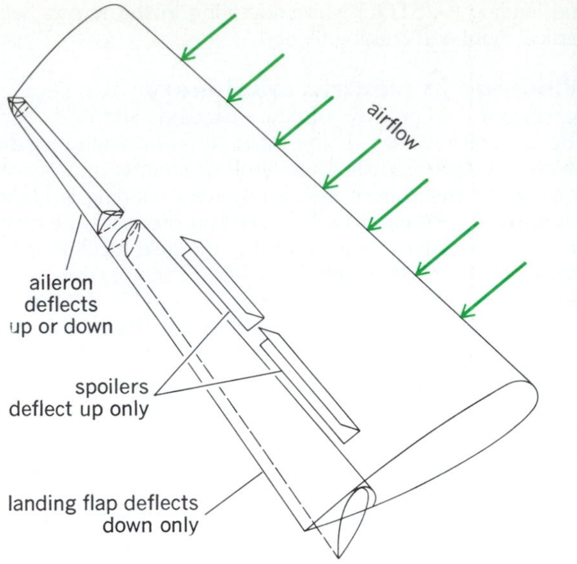

Fig. 2. Trailing edge surfaces.

Landing flaps consist of a movable airfoil-shaped structure located aft of the rear beam or spar. They extend about two-thirds of the span of the wing. Their aerodynamic function is to increase substantially the lift, thereby permitting lower take-off and landing speeds.

Ailerons are located near the tips of the wings in the trailing edge. Their prime aerodynamic function is roll control of the aircraft. Structurally, the aileron is similar to the landing flaps.

Spoilers are a special form of control surface and perform the aerodynamic function of the aileron, that is, roll control. They are located on the upper surface of the trailing edge at about midsemispan and deflect only upward by pilot control.

Wing-to-fuselage structure. The structural heart of the aircraft is the wing-to-fuselage joint. This connection is usually the most complex in form and in analysis. Two major structural elements, the wing and the fuselage, with major loads running at right angles to one another, must be joined and analyzed for consistent deformations. The wing structures which have been described are attached to heavy aluminum ring frames in the fuselage; these rings distribute the wing loads to the fuselage skins. *See* AIRFRAME. [H.J.H.]

Winged bean A plant (*Psophocarpus tetragonolobus*), also known as four-cornered bean, asparagus pea, goa bean, and manila bean, in the family Leguminosae. It is a climbing perennial that is usually grown as an annual. It has been suggested that it originated either in East Africa or Southeast Asia, but there is more evidence to support an African origin. However, Southeast Asia and the highlands of Papua New Guinea represent two foci of its domestication.

Traditionally, winged bean is grown as a backyard vegetable in Southeast Asia and a few islands of the Pacific. It is grown as a field crop in Burma and Papua New Guinea. However, between 1980 and 1990 winged bean was introduced throughout the tropical world.

Almost all parts of this plant are edible and are rich sources of protein. The green pods, tubers, and young leaves can be used as vegetables, and the flowers can be added to salads. The dry seeds are similar to soybeans and can be used for extracting edible oil, feeding animals, and making milk and traditional Southeast Asian foods such as tempeh, tofu, and miso. Flour from the winged bean can also be used as a protein supplement in bread making.

A number of diseases and insect pests may limit winged

bean yield. The most widespread and damaging disease appears to be false rust or orange gall (caused by *Synchytrium psophocarpi*). *See* BREEDING (PLANT); PLANT PATHOLOGY; ROSALES. [T.N.K.]

Wire A thread or slender rod of metal. Wire is usually circular in cross section and is flexible. If it is of such a diameter or composition that it is fairly stiff, it is termed rod. The wire may be of several small twisted or woven strands, but if used for lifting or in a structure, it is classed as cable. Wire may be used structurally in tension, as in a suspension bridge, or as an electrical conductor, as in a power line. The working of metal into wire greatly increases its tensile strength. Thus, a cable of stranded small-diameter wires is stronger as well as more flexible than a corresponding solid rod. *See* MAGNET WIRE. [F.H.R.]

Wiring A system of electric conductors, components, and apparatus for conveying electric power from source to the point of use. In general, electric wiring for light and power must convey energy safely and reliably with low power losses, and must deliver it to the point of use in adequate quantity at rated voltage. Electric wiring systems are designed to provide a practically constant voltage to the load within the capacity limits of the system. There are a few exceptions, notably series street-lighting circuits which operate at constant current. The building wiring system originates at a source of electric power, conventionally the distribution lines or network of an electric utility system. *See* ELECTRICAL CODES.

Systems and service. Wiring systems are generally three-phase to conform to the supply systems. Energy is transformed to the desired voltage levels by a bank of three single-phase transformers. The transformers may be connected in either a delta or Y connection.

Service provided at the primary voltage of the utility distribution system, typically 13,800 or 4160 volts, is termed primary service. Service provided at secondary or utilization voltage, typically 120/208 or 277/480 volts, is called secondary service.

Service at primary voltage levels is often provided for large industrial, commercial, and institutional buildings, where the higher voltage can be used to advantage for power distribution within the buildings. Where primary service is provided, power is distributed at primary voltage from the main switchboard through feeders to load-center substations installed at appropriate locations throughout the building.

Most secondary services in the United States are 120/208 volts, three-phase, four-wire, or 120/240 volts, single-phase, three-wire serving both light and power. For relatively large buildings where the loads are predominantly fluorescent lighting and power (as for air conditioning), the service is often 277/480 volts, three-phase, four-wire, supplying 480 volts for power and 277 volts, phase-to-neutral, for the lighting fixtures.

From the service entrance, power is carried in feeders to the main switchboard, then to distribution panelboards. Smaller feeders extend from the distribution panelboards to light and power panelboards. Branch circuits then carry power to the outlets serving the various lighting fixtures, plug receptacles, motors, or other utilization devices.

Methods. Methods of wiring in common use for light and power circuits are as follows: (1) insulated wires and cables in raceways; (2) nonmetallic sheathed cables; (3) metallic armored cables; (4) busways; (5) copper-jacketed, mineral-insulated cables; (6) aluminum-sheathed cables; (7) nonmetallic sheathed and armored cables in cable support systems; and (8) open insulated wiring on solid insulators (knob and tube).

The selection of the wiring method or methods is governed by a variety of considerations, which usually include code rules limiting the use of certain types of wiring materials; suitability for structural and environmental conditions; installation (exposed or concealed); accessibility for changes and alterations; and costs.

Circuit design. The design of a particular wiring system is developed by considering the various loads, establishing the branch-circuit and feeder requirements, and then determining the service-entrance requirements. Outlets for lighting fixtures, motors, portable appliances, and other utilization devices are indicated on the building plans and the load requirement of each outlet noted in watts or horsepower. Lighting fixtures and plug receptacles are then grouped on branch circuits and connections to the lighting panelboard indicated.

Lighting branch circuits may be loaded to 80% of circuit capacity. However, there is a reasonable probability that the lighting equipment will be replaced at some future time by equipment of higher output and greater load. Therefore, in modern practice, lighting branch circuits are loaded only to about 50% capacity. Lighting branch circuits are usually rated at 20 A. Smaller 15-A branch circuits are used mostly in residences. [W.T.S./J.F.McP.]

Wiring diagram A drawing illustrating electrical and mechanical relationships between parts of a component between which electrical wiring must be connected.

A wiring diagram is distinguished from an electrical schematic in that the arrangement of the schematic bears no necessary relationship to the mechanical arrangement of the electrical elements in the component. The wiring diagram provides an accurate picture of how the wiring on the components and between components should appear in order that the electrical wiring technician can install the wiring in the manner that will best contribute to the optimum performance of the device.

Wiring diagrams also include such information as type of wire, color coding, methods of wire termination, and methods of wire and cable clamping. *See* SCHEMATIC DRAWING. [R.W.M.]

Witherite The mineral form of barium carbonate. Crystals may appear hexagonal in outline. It may be white or gray with yellow, brown, or green tints. Its hardness is $3\frac{1}{4}$ on Mohs scale and its specific gravity 4.3. Witherite is found in many places in Europe, and large crystals occur at Rosiclare, Illinois. *See* BARIUM; CARBONATE MINERALS. [R.I.Ha.]

Wolf-Rayet star A member of a class of very hot stars (100,000–35,000 K or 180,000–63,000°F) which characteristically show broad, bright emission lines in their spectra. Wolf-Rayet stars are classified into two main groups, the WN (nitrogen) and the WC (carbon) stars. The emission lines show very high excitation (lines of He^+ and even up to O^{5+} are strong) and are broadened by an unexplained mechanism. Luminosities must be very high, in the range of 10^4–10^5 times that of the Sun. These stars are probably very young and represent an unstable and short-lived early stage in stellar evolution. Many are close binaries with two stars nearly in contact. *See* STAR; STELLAR EVOLUTION. [J.L.Gr.]

Wolframite A mineral with chemical composition $(Fe,Mn)WO_4$, intermediate between ferberite and huebnerite, which form a complete solid solution series. Wolframite occurs commonly in short, brownish-black, monoclinic, prismatic, bladed crystals. It is probably the most important tungsten mineral. China is the major producer of wolframite. Tungsten minerals of the wolframite series occur in many areas of the western United States; the major producing district is Boulder and northern Gilpin counties in Colorado. *See* HUEBNERITE. [E.C.T.C.]

Wollastonite A mineral inosilicate with composition $CaSiO_3$. Commonly it is massive, or in cleavable to fibrous aggregates. Hardness is 5–$5\frac{1}{2}$ on Mohs scale; specific gravity is 2.85. On the cleavages the luster is pearly or silky; the color is white to gray.

Wollastonite is found in large masses in the Black Forest of Germany; Brittany, France; Chiapas, Mexico; and Willsboro, New York, where it is mined as a ceramic material. *See* SILICATE MINERALS. [C.S.H.]

Wolverine A carnivorous mammal, *Gulo gulo*, also known as the glutton, which is the largest and most vicious member of the family Mustelidae, to which the ferrets, minks, martens, weasels, badgers, otters, and skunks belong. It ranges throughout northern Europe, Asia, and North America, where it is found in coniferous forests during the winter and on the tundras during summer.

This animal is about 40 in. (100 cm) long and weighs about 40 lb (18 kg) (see illustration). It feeds primarily on mice, lemmings, and birds but will attack large animals when they are sick or trapped and exhausted in deep snow. It is a solitary animal, and the male remains with the female only during the courtship period, after which the female is left to rear the cubs. The wolverine lives in a den or burrow and does not hibernate. *See* BADGER; CARNIVORA; FISHER; MARTEN; OTTER; SKUNK; WEASEL. [C.B.C.]

The wolverine (*Gulo gulo*), a stoutly built carnivore with strong teeth and claws.

Wood chemicals Substances derived from wood. Woody plants comprise the greatest part of the organic materials produced by photosynthesis on a renewable basis, and were the precursors of the fossil coal deposits. The derivation of chemicals from wood is carried out wherever technical utility and economic conditions have combined to make it feasible.

Wood is a mixture of three natural polymers—cellulose, hemicelluloses, and lignin—in an approximate abundance of 50:25:25. In addition to these polymeric cell wall components which make up the major portion of the wood, different species contain varying amounts and kinds of extraneous materials called extractives. The nature of the chemicals derived from wood depends on the wood component involved. *See* CELLULOSE; HEMICELLULOSE.

Chemicals derived from wood include: bark products, cellulose, cellulose esters, cellulose ethers, charcoal, dimethyl sulfoxide, ethyl alcohol, fatty acids, furfural, hemicellulose extracts, kraft lignin, lignin sulfonates, pine oil, rayons, rosin, sugars, tall oil, turpentine, and vanillin. Most of these are either direct products or by-products of wood pulping, in which the lignin that cements the wood fibers together and stiffens them is dissolved away from the cellulose. High-purity chemical cellulose or dissolving pulp is the starting material for such polymeric cellulose derivatives as viscose rayon and cellophane, cellulose esters such as the acetate and butyrate for fiber, film, and molding applications, and cellulose ethers such as carboxymethylcellulose, ethylcellulose, and hydroxyethylcellulose for use as gums. *See* CELLOPHANE; DIMETHYL SULFOXIDE; ETHYL ALCOHOL; FURFURAL; ROSIN; TURPENTINE; VANILLA; WAX, ANIMAL AND VEGETABLE; WOOD PRODUCTS. [I.S.G.]

Wood decay A complex series of orderly events that involve the living tree, chemical reactions, bacteria, yeasts, non-decay-causing fungi, and decay-causing fungi. The tree responds to infection in an orderly way by compartmentalizing the infected wood. Many microorganisms are involved in the infection process. Survival of a tree after wounding depends on rapid and effective compartmentalization of the infection. Survival of wood-inhabiting microorganisms after the tree responds to the injury depends on successions.

After a living tree is wounded, it responds in an orderly way since it is a highly compartmented plant, with wood cells defining the compartments. Wood that is injured is never repaired or replaced; in this sense, trees do not heal wounds. Trees thus survive after wounding by walling off, confining, or compartmentalizing the injured and infected wood. The compartmentalization model resolves the apparent paradox that trees can survive only so long as wood-inhabiting microorganisms do not attack, and that wood-inhabiting microorganisms can survive only as long as they attack trees.

There are two basic types of wounds: outer core, where the wound breaks into the wood through the bark, and branch stubs, where the interior of the tree may be exposed. Wounds to the outer core expose the living and dead cells in the wood to the air. The sapwood of trees contains a great number of living cells and most of it, on a number-of-cells basis, is alive. However, on a volume basis, most of the sapwood is dead because the fibers and vessels or tracheids are so very large compared to the small living cells of the ray parenchyma and the marginal parenchyma (the parenchyma that often forms at the end of a growth ring). It is obvious, therefore, that wounds break into or injure the living cells in the wood, killing some immediately, while others are injured and die slowly, and still others, farther from the injury, remain alive.

The wound initiates chemical defense reactions in the living cells near the injury. For one thing, phenol-based substances in the living cells are oxidized to form materials that inhibit many types of microorganisms. Wounds disrupt the normal physiological processes of the cells, and biochemical shunts occur. The tree begins to form a chemical protective zone around the injury. A problem for the tree centers around the many large dead cells—the vessels (in angiosperms) or tracheids (in gymnosperms)—that cannot respond. To counter this, one of the early events in the defense reaction is the plugging of these cells. Plugging serves to stall the vertical spread of the microorganisms. Other changes in the living cells serve to limit the inward and lateral spread of infection. [A.L.Sh.]

Wood engineering design The process of creating products, components, and structural systems with wood and wood-based materials. Wood engineering design applies concepts of engineering in the design of systems and products that must carry loads and perform in a safe and serviceable fashion. Examples include structural systems such as buildings or electric power transmission structures, components such as trusses or prefabricated stressed-skin panels, and products such as furniture or pallets and containers. The design process considers the shape, size, physical and mechanical properties of the materials, type and size of the connections, and the type of system response needed to resist both stationary and moving (dynamic) loads and to function satisfactorily in the end-use environment. *See* ENGINEERING DESIGN; STRUCTURAL DESIGN.

Wood is used in both light frame structures and heavy timber structures. Light frame structures consist of many relatively small wood elements such as lumber covered with a sheathing material such as plywood. The lumber and sheathing are connected to act together as a system in resisting loads; an example is a residential house wood floor system where the plywood is nailed to lumber bending members or joists. In this system, no one joist is heavily loaded because the sheathing spreads the load out over many joists. Service factors such as deflections or vibration often govern the design of floor systems rather than strength. Light frame systems are often designed as diaphragms or shear walls to resist lateral forces resulting from wind or earthquake. *See* FLOOR CONSTRUCTION; WOOD PROCESSING.

In heavy timber construction, such as bridges or industrial buildings, there is less reliance on system action and, in general, large beams or columns carry more load transmitted through decking or panel assemblies. Strength, rather than deflection, often governs the selection of member size and connections. There are many variants of wood construction using poles, wood shells, folded plates, prefabricated panels, logs, and combinations with other materials. *See* WOOD PRODUCTS.

[T.E.McL.]

Wood finishing Any operation performed on wood surfaces aimed at attaining protection against mechanical, physical, and chemical influences or to enhance the natural beauty of the wood. Unfinished wood is dimensionally unstable under climatic influences of varying atmospheric moisture and temperature, and is easily marred and soiled.

Objectives. The main objectives of wood finishing are: (1) stabilization—the protective film is not a perfect barrier against moisture, but the rate of moisture exchange into the wood can be retarded; (2) esthetic quality—the finish should enhance the natural beauty of luster, grain pattern, and color; and (3) sanitation—finishing is intended to close the porous wood surface to reduce contamination by handling, atmospheric grime, and food, and to render it more easily cleaned. The chief deteriorating factors against which wood surfaces need to be protected are weathering and dimensional fluctuations due to environmental conditions such as moisture, irradiation, extreme temperature changes, mechanical erosion, or the combined action of these. About ¼ in. (6 mm) of wood wears away from the exposed surface in a century. *See* WOOD PHYSICS.

Even under indoor conditions, dimensional changes can be considerable as a result of high or low humidity and heating or cooling. Wood swells and shrinks as the moisture content and temperature of the surrounding atmosphere change. The more efficient coatings retard moisture movement sufficiently so that under conditions of frequently changing humidity, swelling and shrinking are reduced to tolerable limits.

Stabilizing treatments usually have two aims: to decrease the rate of swelling and shrinking by minimizing the exchange of moisture into or out of the cell wall structure of wood, and to reduce the final equilibrium swelling and shrinking that can take place. Surface coatings fall under the first category.

Kinds of wood finishes. Wood finishes can be grouped in two main classes: exterior and interior. Exterior finishes are expected to resist chiefly climatic influences, sometimes combined with mechanical abuse, as in outdoor furniture. Interior finishes are expected to fill such exacting standards of appearance and quality as sheen, luster, gloss, or chemical and abrasion resistance. Water resistance is also required in exposed interior locations, such as kitchens, bathrooms, and verandas.

Exterior finishes include: (1) transparent natural finishes showing wood texture, which are further classified as (*a*) surface coatings (transparent finishes consisting of varnishes or synthetics), and (*b*) penetrating finishes including stains, oils, sealers, and water repellents or fungicides; (2) pigmented covering paints, applied on a priming coat containing white lead but no zinc oxide and having a total pigment the same as the finish coat. These covering paints must contain sufficient thinner (solvents) to wet wood quickly and adequate binders (bodied oils and resins) to prevent undue penetration of the vehicle into the wood.

Interior finishes include: (1) transparent penetrating finishes consisting of stains in combination with water-repellent finishes based on waxes and other nonvolatile ingredients; (2) transparent film-forming coating based on shellac, nitrocellulose lacquers, or oleoresinous varnishes in combination with stains, fillers, and sealers; and (3) pigmented covering paints (enamels) used in combination with a priming coat. [J.E.M.]

Wood physics That area of wood science concerned with the physical and mechanical properties of wood and the factors which affect them. Since wood consists of aggregates of long tubular cells, most of which are oriented longitudinally in the tree, whereas the ray cells are oriented radially, wood is an orthotropic material and exhibits different physical behavior in the three main structural directions, longitudinal, radial, and tangential. The orientation of most of the cellulose crystallites is also longitudinal. *See* CELLULOSE; XYLEM.

The cell-wall density for most woods is essentially the same, that is, about 93.6 lb/ft^3 (1.5 g/cm^3) in the oven-dry condition and about 87.4 lb/ft^3 (1.4 g/cm^3) at the fiber saturation point. The main factor governing the density of a given wood sample is the ratio of cell-wall to cell-cavity volume. *See* CELL WALLS (PLANT).

Wood under normal atmospheric conditions contains water which can be removed by drying. The total moisture content is usually defined as the ratio of the weight of water lost, by drying in an oven at 212–221°F (100–105°C), to the oven-dry weight of the wood, expressed as a percentage. Greenwood normally contains water in the cell cavities as well as in the cell walls. The cavity, or free, water can be removed by exposure to relative humidities only slightly lower than 100%, or about 99.9%. The cell-wall, or bound, water is hygroscopic and remains in the wood unless exposed to humidities lower than 100%. The hygroscopic water attains moisture equilibrium with the environment. It also varies with type of wood, extractive content, and exposure history.

The dimensional change associated with loss or gain of bound water is defined respectively as shrinking or swelling.

Owing to the uncertain behavior of the cell cavities, volumetric shrinkage of the gross wood cannot be predicted accurately. On the average, the cavities remain essentially constant in size, as a result of the restraining effects of the inner and outer layers of the secondary cell wall. Longitudinal shrinkage in normal mature wood is almost negligible. Tangential shrinkage is about twice radial shrinkage.

Dry wood is an excellent insulator, but its electrical characteristics change markedly with increasing moisture content and temperature. The dc resistivity decreases almost exponentially with increasing moisture content up to the fiber saturation point, above which there is relatively little change. The dielectric properties of wood in an ac electric field are complex. The dielectric constant and loss factor increase with moisture content and density at almost all frequencies.

Mechanical properties of wood relate to its resistance to deformation by applied forces. Wood is a polymeric material, and therefore its deformation response to external forces is time-dependent or rheological, Grain direction, density, moisture content, temperature, and defects also affect the mechanical behavior of wood. All strength properties increase with wood density at a given moisture content. However, the relative effect of density on strength across the grain is much more pronounced. Temperature affects wood strength in two ways: the immediate or reversible effect, and the permanent or irreversible effect due to thermal degradation. The immediate effect of temperature causes a decrease in strength with increasing temperature. The permanent effect of temperature is caused by thermal degradation of wood exposed to high temperatures, resulting in weight and strength loss when the wood is returned to standard conditions. [C.S.]

Wood preservation The technique of protecting wood, through chemical treatment, from fire and from deterioration by the biological agencies of fungi, insects, and marine borers. Wood preservation increases the useful life of wood and reduces replacement costs. Properly designed wood structures give long service without special protection, but a large economic loss may result when wood in its natural state is used at high temperatures, in structures exposed to salt water, or under climatic conditions that favor the development of harmful fungi and insects. *See* WOOD DECAY.

Wood preservatives in general use are oils, including oil-borne and water-borne chemicals. Oils are mostly for outdoor and wet exposure conditions. They do not swell the wood but may contribute to staining and painting difficulties. Coal tar creosote, solutions of creosote with either coal tar or petroleum oil, and 5% of pentachlorophenol in petroleum oil are used for the treatment of products such as ties, posts, poles, piling, and construction timbers. Solutions of pentachlorophenol and copper naphthenate and other oil-borne chemicals are often sold under proprietary (trade) names for special-purpose treatments of wood. These may contain water repellents when used for such items as millwork. Solvents used with oil-borne preservatives vary from heavy fuel oils to mineral spirits, depending on use requirements.

Water-borne preservatives are usually lower in cost than oils and less likely to involve odor and painting problems. They are used principally for less severe exposure conditions, but several are highly resistant to leaching and perform as well as preservative oils in contact with the soil or fresh water. Water in the treating solutions causes swelling of the wood, however, and redrying to levels meeting use requirements is necessary. [J.O.B.]

Wood processing Peeling, slicing, sawing, and chemically altering hardwoods and softwoods to form finished products such as boards or veneer; particles or chips for making paper, particle, or fiber products; and fuel. *See* PAPER; VENEER.

A high percentage of the weight of freshly cut or green wood is water. Green wood contains free water in the cell cavities and bound water in the cell walls. When all the free water has been extracted and before any of the bound water has been removed, the wood is said to be at the fiber saturation point. As the moisture content falls below the fiber saturation point, the bound water leaves the cell walls and the wood shrinks. During the drying process, differential shrinkage can cause internal stresses in the wood. If not controlled, this can result in defects such as cracks, splits, and warp. Below the fiber saturation point, wood takes on and gives off water molecules depending on the relative humidity of the air around it and swells and shrinks accordingly.

Wood is machined to bring it to a specific size and shape for fastening, gluing, or finishing. With the exception of lasers, which have a limited application at this time, all machining is based on a sharpened wedge that is used to sever wood fibers. Tools for sawing, boring holes, planing, and shaping, as well as the particles in sandpaper, use some version of the sharpened wedge. *See* WOOD ENGINEERING.

Wood is ground to fibers for hardboard, medium-density fiberboard, and paper products. It is sliced and flaked for particle-board products, including wafer boards and oriented strand boards. Whether made from waste products (sawdust, planer shavings, slabs, edgings) or roundwood, the individual particles generally exhibit the anisotropy and hygroscopicity of larger pieces of wood. The negative effects of these properties are minimized to the degree that the three wood directions (longitudinal, tangential, and radial) are distributed more or less randomly. *See* WOOD PRODUCTS. [C.J.Ga.]

Wood products Materials developed from use of the hard fibrous substance (wood) which makes up the greater part of the trunks and limbs of trees.

Solid wood products include lumber, veneer and plywood, furniture, poles, piling, mine timbers, and posts; and composite wood products such as laminated timbers, insulation board, hardboard, and particle board. *See* PLYWOOD.

Fiber wood products can be referred to as those which develop initially from the various processes for pulping wood. All are intended to separate the cellulose fibers one from another in relatively pure form to be recombined into layers of pulp, paper sheets, or paperboards. *See* PAPER.

Chemical wood products result from the chemical modification or conversion of cellulose, lignin, and extractives. Chief among these products are textile fibers such as rayon, and many cellulose plastics products such as cellophane, nitrocellulose, photographic film, telephone parts, and plastic housewares and toys. *See* CELLULOSE; TEXTILE; WOOD CHEMICALS. [W.S.Br.]

Woodpecker A bird of the order Piciformes. Woodpeckers belong to the family Picidae, which is subdivided into three distinct subfamilies: the Picinae or true woodpeckers, the Picumninae or piculets, and the Jynginae or wrynecks. There are about 209 species of woodpeckers, 20 of which occur in the United States.

Most woodpeckers are solitary, and very few species migrate. During the breeding season they pair and remain together during the period of nesting, incubation, and caring for the young. The unlined nest is frequently a hollow in a tree or, in nonarboreal species, in banks of earth. The female lays from two to eight eggs which are incubated by both parents for a period of 11–17 days, depending upon the species. The young are blind at hatching; they are able to leave the nest and climb before they can fly. *See* PICIFORMES. [C.B.C.]

Woodward-Hoffmann rule A concept which can predict or explain the stereochemistry of certain types of reactions in organic chemistry. It is also described as the conservation of orbital symmetry, and is named for its developers, R. B. Woodward and Roald Hoffmann. The rule applies to a limited group of reactions, called pericyclic, which are characterized by being more or less concerted (that is, one-step, without a distinct intermediate between reactants and products) and having a cyclic arrangement of the reacting atoms of the molecule in the transition state. Most pericyclic reactions fall into one of three major classes: electrocyclic, cycloaddition, or sigmatropic. *See* STEREOCHEMISTRY. [D.L.D.]

Woodworking The shaping and assembling of wood and wood products into finished articles such as mold patterns, furniture, window sashes and frames, and boats. The pronounced grain of wood requires modifications in the working techniques when cutting with the grain and when cutting across it. Five principal woodworking operations are sawing, planing, steam bending, gluing, and finishing. To shape round pieces, wood is worked on a lathe. *See* LATHE; SAWING; TURNING (WOODWORKING); WOOD FINISHING. [A.H.T.]

Wool A textile fiber made of the undercoat of various animals, especially sheep. Wool provides warmth and physical comfort that cotton and linen fabrics cannot give. These qualities, combined with its soft resiliency, make wool a necessity for apparel and for rugs and blankets.

Sheep are generally shorn of their fleeces in the spring, but the time of shearing varies in different parts of the world. Each fleece contains different grades, or sorts, of wool, and the raw stock must be carefully graded and segregated according to length, diameter, and quality of fiber. Wool from different parts of the body of the lamb differs greatly. The shoulders and sides generally yield the best quality of wool, because the fibers from those parts are longer, softer, and finer.

Wool's major physical characteristics include fiber diameter (fineness or grade), staple length, and clean wool yield. Also significant are soundness, color, luster, and content of vegetable matter. Grade refers specifically to mean fiber diameter and its variability. Fiber diameter is the most important manufacturing characteristic. Fleeces are commercially graded visually through observation and handling by men of long experience in the industry. Degree of crimp and relative softness of the fleece are important deciding factors employed by the graders.

The inherent advantages of wool have been exploited, and its limitations as a textile fiber have been overcome by the application of technology to manufacturing processes. The use of the insecticide dieldrin as a dye renders wool mothproof for life. Permanent pleats have been imparted to garments which are shrinkproofed and can be home laundered. Each such technological advance enables wool to hold its competitive place in the field of textile manufacture and use. *See* TEXTILE.

[T.D.W.]

Word processing Word processing is a term commonly used to describe a system for accomplishing office work through people, procedures, and technologically advanced equipment. Originally limited to improved production of written communications generated from dictation in an office, word processing has expanded to encompass all business communications generated by an organization, including the entire spectrum of office work, The term fails to reflect this comprehensive expansion into the storage, retrieval, manipulation, and distribution of information and, most recently, a growing interdependence with electronic data processing. As a result, new descriptive terms such as information processing, administrative systems, and office systems are being used.

Word processing represents a further stage in modern society's application of automation, reaching beyond manufacturing and production lines into the office. "Automatic typewriters"—the generic predecessors of multifunction information processors—were the starting point. Now, a wide range of advanced equipment is available to electronically record keystrokes in electronic memories and on a variety of magnetic storage media—tape, card, cassette, or diskette—and electronically assist in correcting, editing, revising, and manipulating text material. Retyping is limited to changes and corrections; perfect copy is automatically produced by pressing a button. Once stored, words, sentences, and paragraphs can be automatically rearranged, reassembled, and retyped repeatedly, and customized versions produced. Distribution of the finished work has also been affected. The effect in the office has become analogous to automation in the factory: higher production, lower costs, higher quality, and more efficient use of people.

[E.W.G.]

Work In physics, the term work refers to the transference of energy that occurs when a force is applied to a body that is moving in such a way that the force has a component in the direction of the body's motion. Thus work is done on a weight that is being lifted, or on a spring that is being stretched or compressed, or on a gas that is undergoing compression in a cylinder.

When the force acting on a moving body is constant in magnitude and direction, the amount of work done is defined as the product of just two factors: the component of the force in the direction of motion, and the distance moved by the point of application of the force. Thus the defining equation for work W is given below, where f and s are the magnitudes of the

$$W = f \cos \phi \cdot s$$

force and displacement, respectively, and ϕ is the angle between these two vector quantities (see illustration). Because

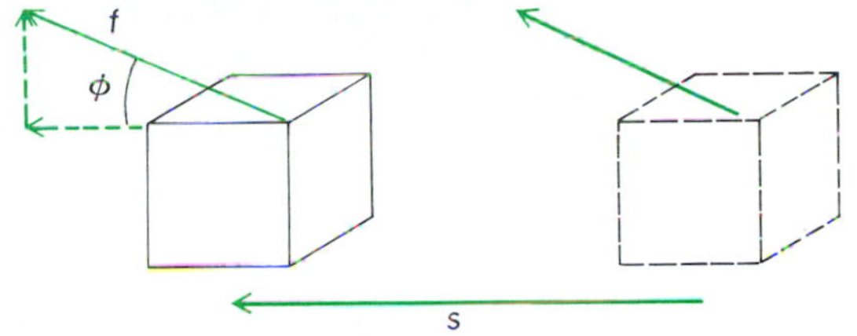

Work of constant force *f* is *fs* cos ϕ.

$f \cos \phi \cdot s = f \cdot s \cos \phi$, work may be defined alternatively as the product of the force and the component of the displacement in the direction of the force.

The work done is positive in sign whenever the force or any component of it is in the same direction as the displacement; one then says that work is being done *by* the agent exerting the force and *on* the moving body. The work is said to be negative whenever the direction of the force or force component is opposite to that of the displacement; then work is said to be done *on* the agent and *by* the moving body. From the point of view of energy, an agent doing positive work is losing energy to the body on which the work is done, and one doing negative work is gaining energy from that body.

The work principle, which is a generalization from experiments on many types of machines, asserts that, during any given time, the work of the forces applied to the machine is equal to the work of the forces resisting the motion of the machine, whether these resisting forces arise from gravity, friction, molecular interactions, or inertia.

The work done by any conservative force, such as a gravitational, elastic, or electrostatic force, during a displacement of a body from one point to another has the important property of being path-independent: Its value depends only on the initial and final positions of the body, not upon the path traversed between these two positions. On the other hand, the work done by any nonconservative force, such as friction due to air, depends on the path followed and not alone on the initial and final positions, for the direction of such a force varies with the path, being at every point of the path tangential to it. *See* ENERGY; FORCE.

[L.N.]

Work function (electronics) A quantity with the dimensions of energy which determines the thermionic emission of a solid at a given temperature. For metals, the work function may also be determined by measuring the photoemission as a function of the frequency of the incident electromagnetic radiation: the work function is then equal to the minimum (threshold) frequency for which electon emission is observed times Planck's constant h ($= 6.63 \times 10^{-34}$ joule second). The work function of a solid is usually expressed in electronvolts.

The work function of metals varies from one crystal plane to another and also varies slightly with temperature. For a metal, the work function has a simple interpretation. At absolute zero, the energy of the most energetic electrons in a metal is referred to as the Fermi energy; the work function of a metal is then equal to the energy required to raise an electron with the Fermi energy to the energy level corresponding to an electron at rest in vacuum. The work function of a semiconductor or an insulator has the same interpretation, but in these materials the Fermi level is in general not occupied by electrons and thus has a more abstract meaning. *See* FIELD EMISSION; PHOTOEMISSION; THERMIONIC EMISSION. [A.J.D.]

Work function (thermodynamics) A thermodynamic function, also called the work content, Helmholtz free energy, or by the European school, simply the free energy. It is defined as the internal energy E of a system minus the temperature-entropy product, TS, and has a characteristic value for each state of a system. In an isothermal process, the maximum work which can be done by a system is equal to the decrease in its work function. When a process such as a chemical reaction occurs spontaneously at constant temperature and volume, it is characterized by a decrease in the work function. Consequently, the criterion for equilibrium under these conditions is that the work function for the system should be at a minimum. *See* CHEMICAL THERMODYNAMICS; FREE ENERGY.

[P.J.B.; W.A.S.]

Work measurement The determination of a set of parameters associated with a task. There are four reasons, common to most organizations whether profit seeking or not, why time, effort, and money are spent to measure the amount of time a job takes. These are cost accounting, evaluation of alternatives, acceptable day's work, and scheduling. The fifth, pay by results, is used only by a minority of organizations.

There are three common ways to determine time per job: stopwatch time study (sequential observations), occurrence sampling (nonsequential observations), and standard data.

Stopwatch time study can be used for almost any existing job. It is reasonable in cost and gives reasonable accuracy. However, it does require the worker to be rated. Once the initial cost of standard data system has been incurred, standard data may be the lowest-cost, most accurate, and most accepted technique.

Occurrence sampling is also called work sampling or ratio-delay sampling. If time study is a "movie," then occurrence sampling is a "series of snapshots." The primary advantage of this approach may be that occurrence sampling standards are obtained from data gathered over a relatively long time period, so the sample is likely to be representative of the universe. That is, the time from the study is likely to be representative of the long-run performance of the worker.

Reuse of previous times (standard data) is an alternative to measuring new times for an operation. There are three levels of detail: micro, elemental, and macro (see table). Micro-level systems have times of the smallest component ranging from about 0.01 to 1 s. Components usually come from a predetermined time system such as methods-time-measurement (MTM) or Work-Factor. Elemental level systems have the time of the smallest component, ranging from about 1 to 1000 s. Components come from time study or micro-level combinations. Macro-level systems have times ranging upward from about 1000 s. Components come from elemental-level combinations, from time studies, and from occurrence sampling. *See* METHODS ENGINEERING; PERFORMANCE RATING; PRODUCTIVITY.

[S.A.K.]

Three levels of detail for standard time systems

Micro system (typical component time range from 0.01 to 1 s; MTM nomenclature)		
Element	*Code*	*Time*
Reach	R10C	12.9 TMU*
Grasp	G4B	9.1
Move	M10B	12.2
Position	P1SE	5.6
Release	RL1	2.0
Elemental system (typical component time range from 1 to 1000 s)		
Element		*Time*
Get equipment		1.5 min
Polish shoes		3.5
Put equipment away		2.0
Macro system (typical component times vary upward from 1000 s)		
Element		*Time*
Load truck		2.5 h
Drive truck 200 km		4.0
Unload truck		3.4

*27.8 TMU = 1 s; 1 s = 0.036 TMU.

SOURCE: S. A. Konz, *Work Design*, published by Grid, 4666 Indianola Avenue, Columbus, OH 43214, 1979.

Work standardization The establishment of uniformity of technical procedures, administrative procedures, working conditions, tools, equipment, workplace arrangements, operation and motion sequences, materials, quality requirements, and similar factors which affect the performance of work. It involves the concepts of design standardization applied to the performance of jobs or operations in industry or business. *See* DESIGN STANDARDS; WORK MEASUREMENT.

Work standardization is part of methods engineering and, where it is practiced, usually precedes the setting of time standards. The objectives of work standardizations are lower costs, greater productivity, improved quality of workmanship, greater safety, and quicker and better development of skills among workers. *See* METHODS ENGINEERING; PRODUCTIVITY.

One of the best known of the more formal techniques of work standardization is group technology. This is the careful description of a heterogeneous lot of machine or other piece parts with a view to discovering as many common features in materials and dimensions as can be identified. It is then possible to start a rather large lot of a basic part through the production process, doing the common operations on all of them. Any changes or additional operations required to produce the final different parts can then be made at a later stage. The economy is realized in being able to do the identical jobs at one time.

There has been considerable progress in computerized systems to facilitate group technology. The techniques are closely linked to computer-aided design and to formalized codes that permit the detailed description of many operations. *See* COMPUTER-AIDED DESIGN AND MANUFACTURING. [J.E.U.]

Worm gears Gears to connect nonparallel, nonintersecting shafts that are at right angles. The worm, ordinarily the driver, is similar to a crossed helical gear except that it has at least one complete tooth (thread) around the pitch surface. The mating gear is the worm wheel or worm gear. Worm gearing is generally used to obtain large velocity reductions with the worm as the driver and the worm wheel as the driven gear, although occasional applications, for example cream separators, have the worm wheel as the driver. [J.R.Z.]

Wrist The flexible link between the forearm and hand, consisting of the wrist joint and adjacent structures. This articulation is composed of a set of synovial joints formed by the distal end of the radius and the four upper (proximal) carpals, or wristbones (see illustration). The four lower (distal) carpals articulate with the hand bones, or metacarpals. All the bony structures are held together by a complex set of deep ligaments, and movement is obtained by forearm muscles acting through

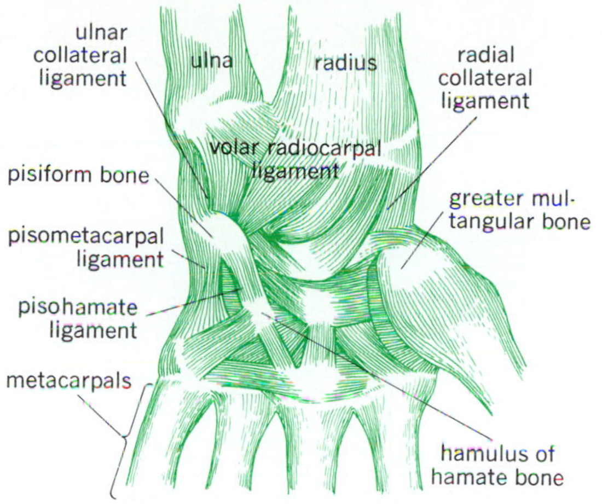

Wrist, volar aspect. (*After W. H. Lewis, ed., Gray's Anatomy of the Human Body, 22d ed., Lea and Febiger, 1930*)

tendons passing to the hand and fingers. The wrist is surrounded by a sheet of very heavy connective tissue and skin. [T.S.P.]

Wrought iron As defined by the American Society for Testing and Materials, "a ferrous material, aggregated from a solidifying mass of pasty particles of highly refined metallic iron, with which, without subsequent fusion, is incorporated a minutely and uniformly distributed quantity of slag." This slag is a ferrous silicate resulting from the oxidizing reactions of refining, and it varies in amount from 1 to 3% in various types of final product. It is in purely mechanical association with the iron base metal, as contrasted with the alloying relationships of the metalloids present in steel. *See* STEEL.

A distinguishing characteristic of wrought iron is a fragmented or irregular fracture, as contrasted with a fibrous or crystalline type in steel. Metallographic analysis shows that this results from the fiberlike slag inclusions. Although wrought iron once held competitive merit in certain uses, notably where corrosion- and shock-resistance were important, very little wrought iron is produced at present. *See* IRON ALLOYS. [J.As./J.F.W.]

Wulfenite A mineral consisting of lead molybdate, $PbMoO_4$. Wulfenite occurs commonly in yellow, orange, red, and grayish-white crystals, with a luster from adamantine to resinous. Wulfenite may also be massive or granular. Its fracture is uneven. Its hardness is 2.7–3 on Mohs scale and its specific gravity 6.5–7. Its streak is white.

Wulfenite is found in numerous localities in the western and southwestern United States. Brilliant orange tabular wulfenite crystals up to 2 in. (5 cm) in size have been found from the Red Cloud and Hamburg mines in Yuma County, Arizona. *See* MOLYBDENUM. [E.C.T.C.]

Wurtzilite A black, infusible carbonaceous substance occurring in Uinta County, Utah. It is insoluble in carbon disulfide, has a density of about 1.05, and consists of 79–80% carbon, 10.5–12.5% hydrogen, 4–6% sulfur, 1.8–2.2% nitrogen, and some oxygen. Wurtzilite is derived from shale beds deposited near the close of Eocene (Green River) time. The material was introduced into the calcareous shale beds as a fluid after which it polymerized to form nodules or veins. *See* ASPHALT AND ASPHALTITE; ELATERITE; IMPSONITE. [I.A.B.]

Wurtzite A mineral with composition ZnS (zinc sulfide). It exists most commonly in fibrous or columnar aggregates and banded crusts with resinous luster and brownish-black color. Hardness is $3\frac{1}{2}$ on Mohs scale and its specific gravity 4.0. Zinc sulfide is dimorphous, crystallizing as both the high-temperature form wurtzite, α-ZnS, and the low-temperature form sphalerite, ß-ZnS. Wurtzite is the rarer and at room temperature the less stable form and, unlike sphalerite, is rarely mined as an ore of zinc. Wurtzite usually contains iron and cadmium, and a complete solid solution series extends to greenockite, CdS. *See* GREENOCKITE; SPHALERITE. [C.S.Hu.]

X

X-ray astronomy The study of x-ray emission from extra-solar sources. It includes the study of virtually all types of astronomical objects, from stars to galaxies and quasars. X-ray astronomy is a recent addition to the ancient science of astronomy.

X-ray astronomy is subdivided into broad bands—soft and hard—depending upon the energy of the radiation being studied. Observations in the soft band (below about 10 keV) must be carried out above the atmosphere with rockets or space satellites, while hard x-ray observations can be made at high altitudes achieved with balloons. Presently, most observations have been in the soft band and have been carried out with a series of scientific satellites. Sky surveys have detected and located hundreds of sources at a sensitivity of about 1/10,000 the strength of the brightest source Sco X-1. There is a concentration of sources along the galactic equator, corresponding to objects which lie in the Milky Way Galaxy, particularly in the disk, which contains most of the galactic stars and the spiral arms. Other x-ray sources are associated mainly with extragalactic objects. *See* MILKY WAY GALAXY; SATELLITE ASTRONOMY; X-RAY TELESCOPE.

The first two identified nonsolar x-ray sources, Sco X-1 and the Crab Nebula, are galactic objects which are unusual and distinct from one another. The Crab is a supernova remnant which is left over from the explosive death of a star. It contains a rapidly rotating neutron star at its center as well as a nebula consisting of hot gas and energetic particles. Other supernova remnants have been observed and found to emit x-rays.

In addition to the supernova remnants there are compact sources which form the majority of galactic emitters. These sources are further classified by their variability, which ranges from time scales as long as years to as short as milliseconds. In some instances the x-rays appear pulsed with a period of a few seconds or less. Another type of time behavior is represented by the "bursters." These sources usually emit at some constant level, with occasional short bursts or flares of increased brightness. It is widely accepted that many of the galactic x-ray sources are different types of collapsed objects—white dwarfs, neutron stars, and even black holes—which are accreting material from their surroundings, and that the in-falling matter heats up and radiates in x-rays as it comes close to the surface of the compact stars. *See* BLACK HOLE; NEUTRON STAR; NOVA; PULSAR; STELLAR EVOLUTION; SUPERNOVA; WHITE DWARF STAR; X-RAY STAR.

Among the more interesting types of extragalactic sources detected have been apparently normal galaxies, galaxies with active nuclei, radio galaxies, clusters of galaxies, and quasars. The nearest spiral galaxy is M31—the Andromeda Galaxy. There appears to be no emission associated with M31 as a whole, but just the summed emission of the stars. For active nuclei galaxies, this is not the case. X-ray emission from these objects is orders of magnitude in excess of that from normal galaxies, and is clearly associated with the galaxy itself. Many astrophysicists believe that galactic nuclei are the sites of massive black holes, at least 10^6 to 10^9 times the mass of the Sun, and that the radiation from these objects is due to gravitational energy released by in-falling material. The quasars may represent the extreme case of this mechanism. *See* ANDROMEDA GALAXY; QUASAR. [S.S.M.]

X-ray crystallography The study of crystal structure by x-ray diffraction techniques. For the experimental aspects of x-ray diffraction *see* X-RAY DIFFRACTION.

Structurally, a crystal is a three-dimensional periodic arrangement in space of atoms, groups of atoms, or molecules. If the periodicity of this pattern extends throughout a given piece of material, one speaks of a single crystal. The exact structure of any given crystal is determined if the locations of all atoms making up the three-dimensional periodic pattern called the unit cell are known. The very close and periodic arrangement of the atoms in a crystal permits it to act as a diffraction grating for x-rays. *See* CRYSTALLOGRAPHY. [L.F.D.]

X-ray diffraction The scattering of x-rays by matter with accompanying variation in intensity in different directions due to interference effects. X-ray diffraction is one of the most important tools of solid-state chemistry, since it constitutes a powerful and readily available method for determining atomic arrangements in matter. X-ray diffraction methods depend upon the fact that x-ray wavelengths of the order of 1 nanometer are readily available and that this is the order of magnitude of atomic dimensions. When an x-ray beam falls on matter, scattered x-radiation is produced by all the atoms. These scattered waves spread out spherically from all the atoms in the sample, and the interference effects of the scattered radiation from the different atoms cause the intensity of the scattered radiation to exhibit maxima and minima in various directions. *See* DIFFRACTION.

Uses. Some of the uses of x-ray diffraction are: (1) differentiation between crystalline and amorphous materials; (2) determination of the structure of crystalline materials (crystal axes, size and shape of the unit cell, positions of the atoms in the unit cell); (3) determination of electron distribution within the atoms, and throughout the unit cell; (4) determination of the orientation of single crystals; (5) determination of the texture of polygrained materials; (6) identification of crystalline phases and measurement of the relative proportions; (7) measurement of limits of solid solubility, and determination of phase diagrams; (8) measurement of strain and small grain size; (9) measurement of various kinds of randomness, disorder, and imperfections in crystals; and (10) determination of radial distribution functions for amorphous solids and liquids.

Techniques. The techniques employed in the study of crystalline substances, gases, and liquids are discussed below.

Laue method. The Laue pattern uses polychromatic x-rays provided by the continuous spectrum from an x-ray tube operated at 35–50 kV. The different diffracted beams have different wavelengths, and their directions are determined solely by the orientations of the set of planes with Miller indices hkl. Transmission Laue patterns were once used for structure determinations, but their many disadvantages have made them practically obsolete. On the other hand, the back-reflection Laue pattern is used a great deal in the study of the orientation of crystals.

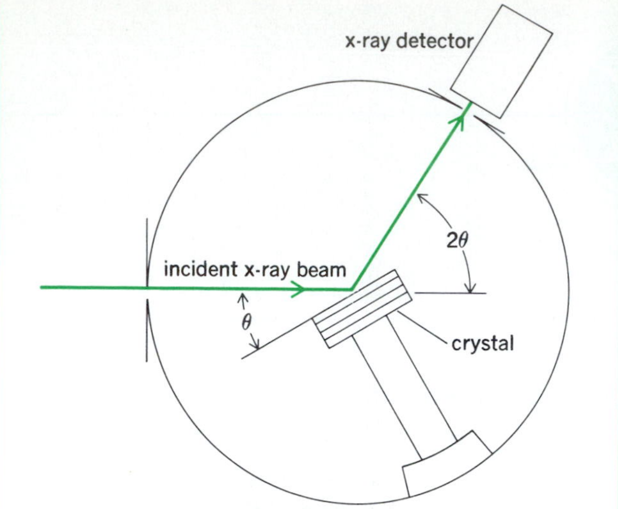

Fig. 1. Schematic of Bragg spectrometer.

Rotating crystal method. The original rotating crystal method was employed in the Bragg spectrometer. A sufficiently monochromatic beam, of wavelength of the order of 1 Å, is collimated by a system of slits and then falls on the large extended face of a single crystal as shown by Fig. 1. The Bragg spectrometer has been used extensively in obtaining quantitative measurements of the integrated intensity from planes parallel to the face of the crystal. The chamber is set at the correct 2θ angle with a slit so wide that all of the radiation reflected from the crystal can enter and be measured. The crystal is turned at constant angular speed through the Bragg law position, and the total diffracted energy received by the ionization chamber during this process is measured. Similar readings with the chamber set on either side of the peak give a background correction.

The rotation camera, which is frequently used for structure determinations, is illustrated in Fig. 2. The monochromatic primary beam $\mathbf{s}_0$ falls on a small single crystal at *O*. The crystal is mounted with one of its axes (say, $\mathbf{a}_3$) vertical, and it rotates with constant velocity about the vertical axis during the exposure. The various diffracted beams are registered on a cylindrical film concentric with the axis of rotation.

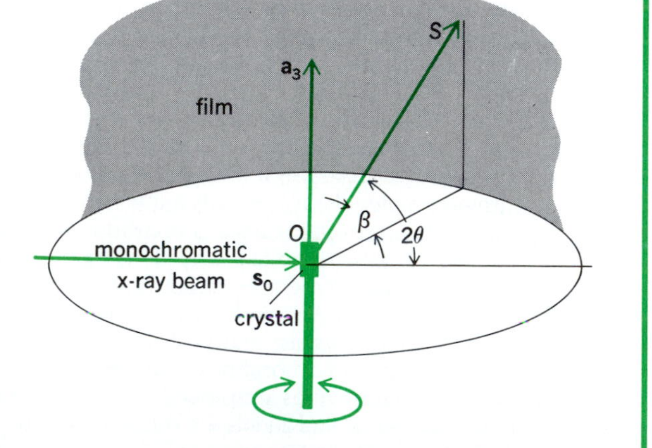

Fig. 2. Schematic of rotation camera.

Powder method. The powder method involves the diffraction of a collimated monochromatic beam from a sample containing an enormous number of tiny crystals having random orientation. Powder pattern studies are made with Geiger counter, or proportional counter, diffractometers. The apparatus is shown schematically in Fig. 3. X-rays diverging from a target at *T* fall on the sample at *O*, the sample being a flat-faced briquet of powder. Diffracted radiation from the sample passes through the receiving slit at *s* and enters the Geiger counter. During the operation the sample turns at angular velocity ω and the counter at 2ω. The distances *TO* and *OS* are made equal to satisfy approximate focusing conditions. A filter *F* before the receiving slit gives the effect of a sufficiently monochromatic beam. A chart recording of the amplified output of the Geiger counter gives directly a plot of intensity versus scattering angle 2θ. [B.E.W.]

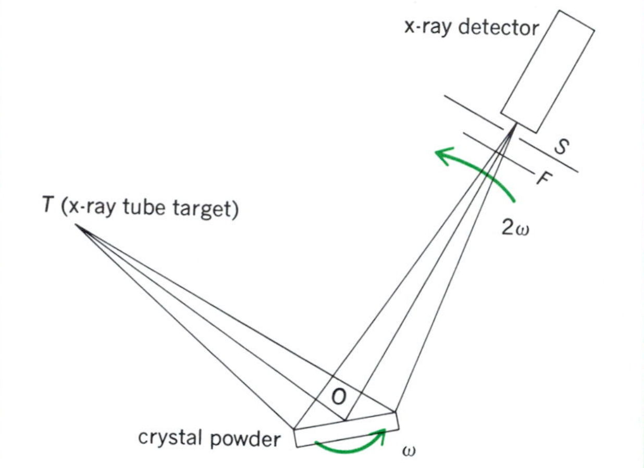

Fig. 3. Schematic representation of the Geiger counter diffractometer for powder samples.

Gases. Gases and liquids are found to give rise to x-ray diffraction patterns characterized by one or more halos or interference rings which are usually somewhat diffuse. These diffraction patterns, which are similar to those for glasses and amorphous solids, are due to interference effects depending both upon the electronic distribution of each of the individual atoms or molecules and upon their relative positions in the system.

For monatomic gases the only appreciable interference effects giving rise to a distribution of scattered intensities are those produced by the electronic distribution about each nucleus. These interference effects giving rise to so-called coherent intensities are the result of the interference of the individual waves scattered by electrons in different parts of the atom. The electronic distribution of an atom is described in terms of a characteristic atomic scattering factor which is defined as the ratio of the resultant amplitude scattered by an atom to the amplitude that a free electron would scatter under the same conditions.

Liquids. One cannot, as in the cases of dilute gases and crystalline solids, derive unambiguous, detailed descriptions of liquid structures from diffraction data. Nevertheless, diffraction studies of liquids do provide most useful information. Instead of comparing the experimental intensity distributions with theoretical distributions computed for various models, the experimental results are usually provided in the form of a radial distribution function which specifies the density of atoms or electrons as a function of the radial distance from any reference atom or electron in the system without any prior assumptions about the structure. From the radial distribution function one can obtain (1) the average interatomic distances most frequently occurring in the structure corresponding to the positions of the first, second, and possibly third nearest neighbors; (2) the distribution of distances; and (3) the average coordination number for each interatomic distance. [L.F.D.]

X-ray fluorescence analysis A nondestructive physical method used for chemical analyses of solids and liquids. The specimen is irradiated by an intense x-ray beam which causes the elements in the specimen to emit (that is, fluoresce) their characteristic x-ray line spectra. The lines of the spectra are diffracted at various angles by a single-crystal plate which is analogous to the diffraction grating of optical spectroscopy. The elements may be identified by the wavelengths of their spectral lines, which vary in a regular manner with atomic number, and their concentrations may be determined from the intensities of the lines. Counter tubes and associated electronic circuits are used to measure the x-ray intensities. *See* FLUORESCENCE.

Unlike x-ray diffraction, the elements (rather than the compounds) are identified, and it is not necessary that the specimen be crystalline. The method supplements optical spectroscopy, being most generally used for concentrations > 0.1%, although with prior wet chemical or physical extraction methods to enhance the peak-to-background ratio, much lower amounts can be detected. *See* X-RAY DIFFRACTION.

The x-ray fluorescence method is less sensitive than the optical method, although with prior specimen treatment, analyses can be made in the range of parts per million. Microgram quantities have been determined in milligram samples with standard deviations of 10–15%, using focusing crystal spectrographs. Optical spectrographic methods become difficult to use for concentrations exceeding about 20%, whereas the x-ray method is applicable up to 100%. The x-ray fluorescence method has proven particularly useful for mixtures of elements of similar chemical properties which are difficult to separate and analyze by conventional chemical methods.

The x-ray spectrograph for fluorescence analysis consists of the primary x-ray tube and its stabilized high-voltage power supply; a ray-proof specimen chamber; the goniometer which carries the crystal analyzer and counter tube; and the electronic circuitry for the counter tube consisting of a power supply, linear amplifier, single-channel pulse-height analyzer, scaler, rate meter, and strip-chart recorder. [W.P.]

X-ray microscope A technique and an instrument or combination of instruments which utilize x-radiation for chemical analysis and for magnification of 100–1000 diameters. The resolution possible is about 0.25 micrometer. The contrast in the x-ray microscopic image is caused by the varying x-ray attenuation in the specimen. The advantage of x-ray microscopy is that it yields quantitative chemical information, besides structural information, about objects, including those which are opaque to light. It is a reliable ultramicrochemical analytical technique by which amounts of elements and weights of samples as small as 10^{-12} to 10^{-14} g can be analyzed with an error of only a few percent. *See* X-RAY OPTICS.

There are four general principles of image formation in x-ray microscopy: (1) contact microradiography (illustration *a*), (2) projection x-ray microscopy (illustration *b*), (3) reflection x-ray microscopy (illustration *c*), and (4) x-ray image spectrography (illustration *d*). *See* MICRORADIOGRAPHY.

In contact microradiography the thin specimen is placed in close contact with an extremely fine-grained photographic emulsion which has a resolution of more than 25,000 lines/in. (1000 lines/mm), and radiographed with x-rays of suitable wavelength. Thus an absorption image in scale 1:1 is obtained, and this image is subsequently viewed in a light microscope. The maximal resolution is that of the optical

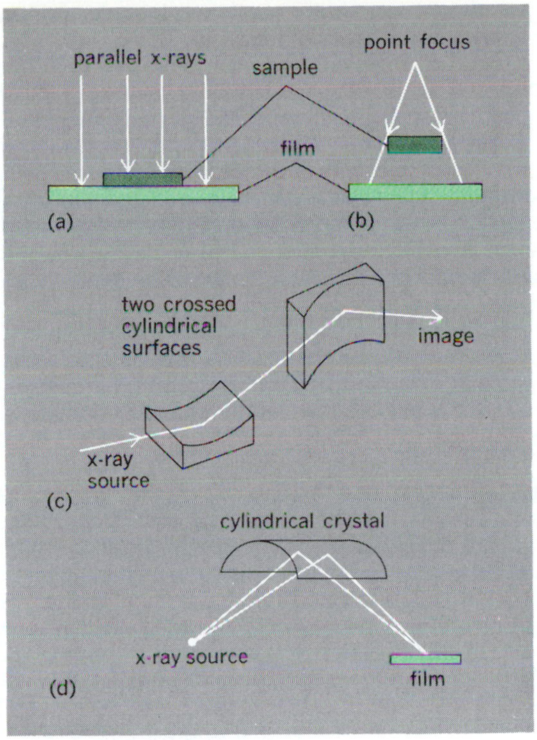

Principles for x-ray microscopy. (*a*) Contact microradiography; (*b*) projection x-ray microscopy; (*c*) reflection x-ray microscopy; and (*d*) x-ray image spectrography.

microscope (0.25 μm), but the image has more information, which can be obtained by examining the microradiogram in the electron microscopes. *See* ELECTRON MICROSCOPE.

Projection x-ray microscopy, or x-ray shadow microscopy, is based on the possibility of producing an extremely fine x-ray focal spot. This is achieved by an electronic lens system similar to that in the electron microscope. The fine focal spot is produced on a very thin metal foil which serves as a transmission target. The x-rays are generated on the target by the impact of the electrons. The sample is placed near the target, and the primary magnification depends on the ratios of the distances from focal spot to sample and sample to film. Resolution is of the same order as the size of the focal spot; the best value is about 0.1 μm in favorable objects.

The method of reflection x-ray microscopy is based on the fact that the refractive index for x-rays in solids is a very small amount less than 1. Thus at grazing incidence (that is, incidence at very small angles), the x-rays are totally reflected, and if the reflecting surface is made cylindrical, there will be a focusing action in one dimension. By crossing two such surfaces a true image formation can be obtained, although with some astigmatism, which can be corrected by giving the surfaces a complicated optical shape. The resolution by this procedure is about 0.5–1 μm.

X-ray image spectrography utilizes Bragg reflections in a cylindrically bent crystal and produces slightly enlarged emission images; this technique is best classified as a micromodification of x-ray fluorescence analysis. The resolution is about 50 μm. *See* X-RAY DIFFRACTION; X-RAY FLUORESCENCE ANALYSIS.

In biology, x-ray microscopy has been utilized for the quantitative determination of the dry weight, water content, and elementary composition of many tissues, especially mineralized tissues. [A.E.]

X-ray optics A title-by-analogy of those phases of x-ray physics in which x-rays demonstrate properties similar to those of light waves. X-ray optics is also called roentgen optics. *See* GEOMETRICAL OPTICS; OPTICS; PHYSICAL OPTICS.

Polarization, diffraction, reflection, and refraction of x-rays were all detected by using crystals having refractive indices that were slightly less than 1. Total reflection or diffraction of x-rays from ruled gratings could occur only with incident beams at very small grazing angles of the order of a few minutes. Until this discovery by A. Compton in 1922 all attempts to concentrate or focus x-rays by lenses and mirrors had failed. *See* COMPTON EFFECT; X-RAYS.

Bent crystals can be used as diffraction gratings to focus diffracted beams and to form enlarged images of a specimen, itself serving as the source of x-rays or bathed by a beam of x-rays, in accordance with the same principles and equations that apply to light optics. *See* X-RAY DIFFRACTION; X-RAY FLUORESCENCE ANALYSIS; X-RAY MICROSCOPE; X-RAY POWDER METHODS. [G.L.Cl.]

X-ray powder methods Physical techniques used for the identification of substances and for other types of analyses, principally for crystalline materials in the solid state. In these techniques, a small collimated beam of x-rays is directed onto a small polycrystalline specimen in the form of powder, producing a diffraction pattern that is recorded on film or with a counter tube. This x-ray pattern is a fundamental and uniquely characteristic property resulting from the atomic arrangement of the diffracting substance. Different substances have different atomic arrangements or crystal structures, and hence no two chemically distinct substances give identical diffraction patterns. Identification may be made by comparing the pattern of the unknown substance with patterns of known substances, in a manner somewhat analogous to the identification of persons by their fingerprints. The analytical information is different from that obtained by chemical or spectrographic analysis. X-ray identification of chemical compounds indicates the constituent elements, and shows how they are combined.

The x-ray powder method is widely used in fundamental and applied research; for instance, it is used in the analysis of raw materials and finished products, in phase-diagram investigations, in following the course of solid-state chemical reactions, and in the study of minerals, ores, rocks, metals, chemicals, and many other types of material. The use of powder methods to determine the actual atomic arrangement, which has been so important in the study of chemical bonds, crystal physics, and crystal chemistry, is described in related articles. *See* X-RAY CRYSTALLOGRAPHY; X-RAY DIFFRACTION; X-RAY FLUORESCENCE ANALYSIS. [W.P.]

X-ray spectrometry A rapid and economical technique for quantitative analysis of the elemental composition of specimens. It differs from x-ray diffraction, whose purpose is the identification of crystalline compounds. It differs from spectrometry in the visible region of the spectrum in that the x-ray photons have energies of thousands of electronvolts and come from tightly bound inner-shell electrons in the atoms, whereas visible photons come from the outer electrons and have energies of only a few electronvolts. *See* X-RAY DIFFRACTION.

In x-ray spectrometry the irradiation of a sample by high-energy electrons, protons, or photons ionizes some of the atoms, which then emit characteristic x-rays whose wavelength depends on the atomic number of the element, and whose intensity is related to the concentration of that element. Generally speaking, the characteristic x-ray lines are independent of the physical state (solid or liquid) and of the type of compound (valence) in which an element is present, because the x-ray emission comes from inner, well-shielded electrons in the atom. The illustration shows the removal of one of the innermost, *K*-shell, electrons by a high-energy photon. The photon energy must be greater than the binding energy of the electron; the difference in energy appears as the kinetic energy of the ejected electron. The *K* ionized atom is unstable, and one of the *L*- or *M*-shell electrons drops into the *K*-shell vacancy. As this transition occurs, a characteristic x-ray photon is emitted with an energy equal to the difference in energy between the *K* and the *L* (or *M*) shell, or an additional electron, called an Auger electron, is ejected from the atom. Either the x-rays or the Auger electrons may be used for analysis. *See* AUGER EFFECT.

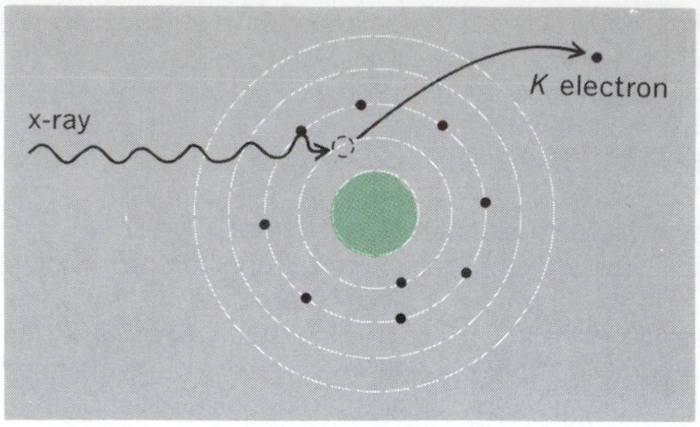

Removal of a *K* election from an atom by a primary x-ray photon. (*After L. S. Birks, X-Ray Spectrochemical Analysis, 2d ed., Wiley-Interscience, 1969*)

X-ray spectrometry generally does not require any separation of elements before measuring, because the x-ray lines are easily resolved. However, preconcentration methods are sometimes useful as a means for improving the limit of detection. One limitation of x-ray spectrometry is the progressive difficulty of measurement below atomic number 11. *See* SPECTROSCOPY. [L.S.B.]

X-ray star A source of x-rays from outside the solar system. Although the Sun emits some x-radiation, it cannot be called an x-ray star because its energy output in the optical region of the spectrum vastly exceeds that in the x-ray region. Observations of the 100 or so known galactic x-ray stars show that these objects have just the reverse characteristic: the overwhelming amount of their energy is given off in the x-ray band, while in visible light they are inconspicuous. *See* STAR; SUN; X-RAY ASTRONOMY.

The majority of galactic x-ray stars are located in a narrow layer coincident with the plane of the Milky Way Galaxy. Because this coincides with the spatial distribution of bright young stars in the Galaxy, there is at least circumstantial evidence that many x-ray stars may be associated with such young stars, or at least may have had similar stages of early evolution. *See* MILKY WAY GALAXY; STELLAR EVOLUTION.

The study of both x-ray and optical data from the region of x-ray stars has led to the conclusion that most, or perhaps even all, such stars are members of binary star systems, consisting of one relatively normal visible star and one subluminous compact star, in a gravitationally bound orbit about each other. In cases where the position of the x-ray source has been very accurately measured, it is sometimes possible to locate a visible star associated with the system. In most cases, studies of the visible component show evidence independent of the x-ray data for the binary nature of these systems.

Most theories assume these objects to be highly evolved and quite compact. The compact star accretes gaseous matter from its nearby visible companion. This in-falling matter reaches a very high temperature before it is stopped at or near the surface of the x-ray star, and the very hot gas then emits x-rays. In effect, a machine is at work to convert gravitational energy into x-radiation. [J.Si./B.M.]

X-ray telescope An instrument designed to collect and detect x-rays emitted from a source outside the Earth's atmosphere and to resolve the x-rays into an image. Absorption by the atmosphere requires that x-ray telescopes be carried to high altitudes. Balloons are used for detection systems designed for higher-energy (harder) x-ray observations, whereas rockets and satellites are required for softer x-ray detectors. *See* X-RAY ASTRONOMY.

An image-forming telescopic lens for x-ray wavelengths can be based on the phenomenon of total external reflection at a surface where the index of refraction changes. In the case of x-rays, the index of refraction in matter is slightly less than unity. By application of Snell's laws, the condition for total external reflection is that the radiation be incident at small grazing angles, less than a critical angle of about 1°, to the reflecting surface. Based on these properties, x-ray mirrors have been constructed which focus an image in two dimensions. Various configurations of surfaces are possible. *See* REFLECTION OF ELECTROMAGNETIC RADIATION; REFRACTION OF WAVES.

The high angular resolution of the grazing-incidence telescope requires a camera which has correspondingly good spatial resolution. One type of detector uses microchannel plates and yields about 20-micrometer resolution for x-rays in the soft energy band (about 200–4000 eV). The microchannel plate is an array of small hollow tubes (about 15 μm in diameter) or channels which are processed to have high secondary electron yield from their inner walls.

Another detector used with x-ray telescopes is a form of gas counter. By operation of the counter in the proportional mode and use of planes of wires to localize the electrical signals, the position and amplitude of each event are recorded. These detectors generally have lower spatial resolution (about 1 mm) than do microchannel plate detectors, but they can be made larger in size and provide some energy measurement which is not possible in the other device. *See* IONIZATION CHAMBER. [S.S.M.]

X-ray tube An electronic device used for the generation of x-rays. X-rays are produced in the x-ray tube by accelerating electrons to a high velocity by an electrostatic field and then suddenly stopping them by collision with a solid body, the so-called target, interposed in their path. The x-rays radiate in all directions from the spot on the target where the collisions take place. The x-rays are due to the mutual interaction of the fast-moving electrons with the electrons and positively charged nuclei which constitute the atoms of the target. Depending upon the method used in generating the electrons, x-ray tubes may all be classified in two general groups, gas tubes and high-vacuum tubes. *See* X-RAYS.

In gas tubes electrons are freed from a cold cathode by positive ion bombardment. For the existence of the positive ions a certain gas pressure is required without which the tube will allow no current to pass. Metals, such as platinum and tungsten, are placed in the path of the electron beam to serve as the target. Concave metal cathodes are used to focus the electrons on a small area of the metal target and increase the sharpness of the resulting shadows on the fluorescent screen or the photographic film. Many designs of gas tubes have been built for useful application, particularly in the medical field.

The operational difficulties and erratic behavior of gas x-ray tubes are inherently associated with the gas itself and the positive ion bombardment that takes place during operation. The high-vacuum x-ray tube eliminates these difficulties by using other means of emitting electrons from the cathode. The original type of high-vacuum x-ray tube had a hot tungsten-filament cathode and a solid tungsten target. This tube permitted stable and reproducible operation with relatively high voltages and large masses of metals. A modern commercial hot-cathode high-vacuum x-ray tube is built with a liquid-cooled, copper-backed tungsten target. [E.E.C.]

X-rays X-rays, or roentgen rays, are electromagnetic waves in which periodically variable electric and magnetic fields are perpendicular to each other and to the direction of propagation. Thus they are identical in nature with visible light and all the other types of radiation that constitute the electromagnetic spectrum. In general, x-rays are generated as the result of energy transitions of atomic electrons caused by the bombardment of a material of high atomic weight by high-energy electrons. *See* ELECTROMAGNETIC RADIATION.

Following W. R. Röntgen's discovery of "a new kind of ray" in 1895, other scientists found the essential experimental conditions to prove that x-rays can be polarized, diffracted by crystals, refracted in prisms and in crystals, reflected by mirrors, and diffracted by ruled gratings. *See* X-RAY OPTICS.

The range of x-rays in the electromagnetic spectrum, as excited in x-ray tubes by the bombardment of anode targets by cathode electrons under a high accelerating potential, overlaps the ultraviolet range on the order of 100 nanometers on the long-wavelength side, and the shortest-wavelength limit moves downward as voltages increase. An accelerating potential of 10^9 volts, now readily generated, produces a wavelength of 10^{-15} m (10^{-6} nm). An average wavelength used in research is 0.1 nm, or about 1/6000 the wavelength of yellow light. *See* X-RAY TUBE.

In diffraction, refraction, polarization, and interference phenomena, x-rays, together with all other related radiations, appear to act as waves. In other phenomena—such as the appearance of sharp spectral lines, a definite short-wavelength limit of the continuous "white" spectrum, the shift in wavelength of x-rays scattered by electrons in atoms (Compton effect), and the photoelectric effect—the energy seems to be propagated and transferred in quanta, called photons. *See* COMPTON EFFECT; ELECTRON DIFFRACTION; NEUTRON DIFFRACTION; PHOTOEMISSION; QUANTUM MECHANICS.

Important uses have been found for x-rays in many fields of scientific endeavor, for example, roentgen spectrometry and roentgen diffractometry. Extensive tables of the wavelengths of x-ray emission lines in series (K, L, M, and so on) and so-called absorption edges, characteristic of the chemical elements, afford the necessary information for chemical analyses, exactly as in the case of optical emission spectra and for derivation of theories of atomic structure to account for the origin of spectra. *See* HISTORADIOGRAPHY; MICRORADIOGRAPHY; RADIATION BIOLOGY; RADIOGRAPHY; RADIOLOGY; X-RAY CRYSTALLOGRAPHY; X-RAY DIFFRACTION; X-RAY FLUORESCENCE ANALYSIS; X-RAY MICROSCOPE; X-RAY POWDER METHODS. [G.L.Cl.]

Xanthophyceae A class of plants, comprising the yellow-green algae, in the chlorophyll *a–c* phylectic line (Chromophycota). Alternate names are Tribophyceae, derived from *Tribonema*, a filamentous member of the class, and Heterokontae, referring to the presence of two kinds of flagella on each motile cell. The absence of starch is an important characteristic in distinguishing yellow-green algae from green algae, which they may resemble superficially in form and color. *See* ALGAE; CHROMOPHYCOTA.

Motile cells have two unequal flagella borne apically or laterally. One (usually the longer) is directed forward and is pleuronematic (hairy); the other is directed backward and is acronematic (smooth). The compound zoospore (synzoospore) of *Vaucheria* is exceptional in having numerous pairs of slightly unequal smooth flagella. There are usually two chloroplasts in each motile cell, the ventral one with a reddish refractile eyespot at its anterior end. One or two contractile vacuoles occur at the anterior end of a motile cell. Cells in most Xanthophyceae are uninucleate, but in some genera they are multinucleate. The mitochondria have tubular cristae. Each vegetative cell contains one to many yellow-green, laminate or discoid, parietal chloroplasts.

Most Xanthophyceae occur in fresh water, especially soft water. Frequently, they are epiphytic on aquatic plants. The chief exception is *Vaucheria*, which has fresh-water, brackish-water, and marine species. Xanthophyceae also occur in and on soil and mud, in snow and ice, and on tree trunks and damp walls. [P.C.Si.; R.L.Moe]

Xenacanthida An order of Paleozoic sharklike fishes abundant in fresh-water deposits of the Carboniferous and Early Permian. Although primitive in cranial structures, the pleuracanths differ notably from other sharks in several regards (see illustration). The teeth are two-pronged; there is a long spine projecting upward and backward from the posterior part of the braincase; the tail extends directly backward, in contrast with the upturned heterocercal caudal fin of other sharks; and the paired fins have the archipterygial skeletal pattern of a central axis and side branches, in contrast with a fan-shaped arrangement in typical sharks. *See* ELASMOBRANCHII. [A.S.R.]

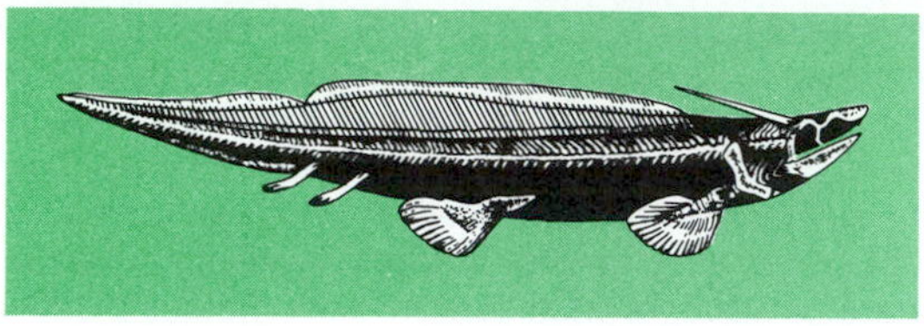

***Xenacanthus* (*Pleuracanthus*), Carboniferous and Permian sharklike form; perhaps 2½ ft (75 cm) long. (*After Fritsch*)**

Xenolith A discrete rock fragment enclosed by magma, lava, or igneous rock. The relation between xenolith and host magma may be close or remote. For example, the consolidated marginal portions of a body of magma may be torn off and incorporated into the unconsolidated central portions of the magma. Xenoliths which are inferred to form in this manner are called cognate xenoliths. Cognate xenoliths in volcanic rocks, like phenocrysts, help scientists find out about the consolidation of bodies of magma underneath volcanoes. On the other hand, many xenoliths are simply pieces of sedimentary and other unrelated igneous rocks. Such xenoliths tend to react with the enclosing magma, so that their constituent minerals become like those in equilibrium with the melt. Reaction is rarely complete, however.

Xenoliths may be angular to round, millimeters to meters in diameter, aligned or haphazard, and sharply or gradationally bounded. Xenoliths are present in most bodies of igneous rock, although they are rare in many igneous rocks. Many rocks are inferred to be igneous because they contain xenoliths. *See* IGNEOUS ROCKS; MAGMA. [A.T.A.]

Xenon A chemical element, Xe, atomic number 54. It is a member of the family of noble gases. Sixteen radioactive isotopes are known. *See* INERT GASES.

Xenon is used to fill a type of flashbulb, called electronic speed light, which produces a white light that has a good balance of all the colors in the visible spectrum, and can be used 10,000 times or more before burning out. A xenon-filled arc lamp gives a light intensity approaching that of the carbon arc; it is particularly valuable in projecting motion pictures.

1	2	3	4	5	6	7	8	9	10	11	12	13	14	15	16	17	18
1 H																	2 He
3 Li	4 Be											5 B	6 C	7 N	8 O	9 F	10 Ne
11 Na	12 Mg											13 Al	14 Si	15 P	16 S	17 Cl	18 Ar
19 K	20 Ca	21 Sc	22 Ti	23 V	24 Cr	25 Mn	26 Fe	27 Co	28 Ni	29 Cu	30 Zn	31 Ga	32 Ge	33 As	34 Se	35 Br	36 Kr
37 Rb	38 Sr	39 Y	40 Zr	41 Nb	42 Mo	43 Tc	44 Ru	45 Rh	46 Pd	47 Ag	48 Cd	49 In	50 Sn	51 Sb	52 Te	53 I	54 Xe
55 Cs	56 Ba	71 Lu	72 Hf	73 Ta	74 W	75 Re	76 Os	77 Ir	78 Pt	79 Au	80 Hg	81 Tl	82 Pb	83 Bi	84 Po	85 At	86 Rn
87 Fr	88 Ra	103 Lr	104 Rf	105 Db	106 Sg	107 Bh	108 Hs	109 Mt	110	111	112	113	114	115	116	117	118

lanthanide series	57 La	58 Ce	59 Pr	60 Nd	61 Pm	62 Sm	63 Eu	64 Gd	65 Tb	66 Dy	67 Ho	68 Er	69 Tm	70 Yb
actinide series	89 Ac	90 Th	91 Pa	92 U	93 Np	94 Pu	95 Am	96 Cm	97 Bk	98 Cf	99 Es	100 Fm	101 Md	102 No

Traces of xenon are found in minerals and meteorites, but the only commercial source of xenon is the air. Xenon constitutes 0.086 part per million by volume of dry air. It is estimated that about 3×10^{-9}% of the weight of the Earth is xenon. Xenon also occurs outside the Earth; the best estimate is that there are about 4 atoms of xenon per 1,000,000 atoms of silicon, silicon being the standard of abundance used for the elements in the universe.

Xenon is colorless, odorless, and tasteless; it is a gas under ordinary conditions. Other properties are shown in the table.

Physical properties of xenon*

Property	Value
Atomic number	54
Atomic weight (atmospheric xenon only)	131.30
Melting point (triple point)	−111.8°C
Boiling point at 1 atm pressure	−108.1°C
Gas density at 0°C and 1 atm pressure	5.8971 g/liter
Liquid density at its boiling point	3.057 g/ml
Solubility in water at 20°C (STP) per 1000 g water at 1 atm partial pressure of xenon	108.1 ml xenon

*1 atm = 10^2 kilopascals. °C = $\frac{5}{9}$ (°F −32).

Xenon is the only one of the nonradioactive noble gases which forms chemical compounds that are stable at room temperature. Xenon also forms weakly bonded clathrates. *See* CLATHRATE COMPOUNDS. [A.W.F.]

Although xenon exhibits all of the even valence states (II, IV, VI, and VIII), and stable compounds of each of these states have been isolated, xenon chemistry is limited to the stable fluorides and oxyfluorides and their complexes, two unstable oxides, and the aqueous species derived from the hydrolysis of the fluorides.

The three fluorides, XeF_2, XeF_4, and XeF_6, are thermodynamically stable compounds at room temperature, and they may be prepared simply by heating mixtures of xenon and fluorine at 300–400°C (570–750°F). All of the fluorides are reduced by hydrogen to form xenon and hydrogen fluoride, and their heats of formation have been measured in this manner.

The reaction of XeF_6 with water gives xenon oxytetrafluoride, $XeOF_4$, a rather stable compound, melting at −46.2°C and boiling at 101°C. If the reaction is allowed to continue, XeO_3 is formed. XeO_3 is a colorless, odorless, and dangerously explosive white solid of low volatility. Gaseous xenon tetroxide, XeO_4, is formed by the reaction of sodium perxenate, Na_4XeO_6, with concentrated H_2SO_4.

The xenon fluorides and XeOF4 form a variety of complex addition compounds. The complexes with metal pentafluorides tend to be powerful fluorinating agents. Xenon difluoride dissolved in antimony pentafluoride can react with elemental xenon itself to form the unexpected dixenon cation Xe_2^+. [J.G.M.; E.H.Ap.]

Xenophyophorida An order of Protozoa in the subclass Granuloreticulosia. The group includes deep-sea forms which are multinucleate at maturity. They develop as discoid to fanshaped or algalike branching forms (illustration *a* and *b*) covered with a hyaline organic layer and sometimes measuring 2 mm or more overall. The aggregate contains many tubes which vary in diameter and form netlike or branching patterns. The lack of detailed information makes the taxonomic status of

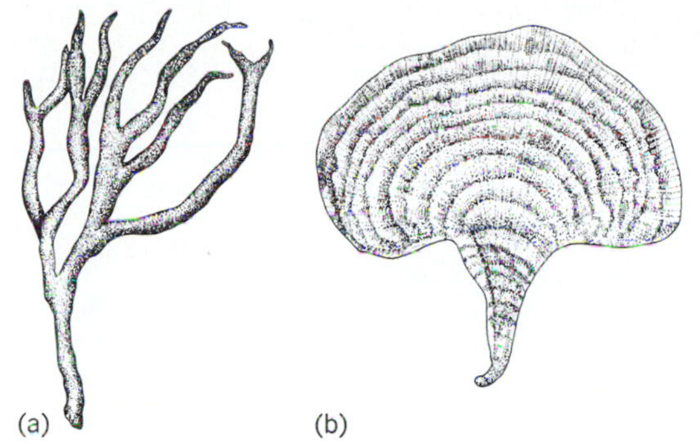

Xenophyophorids. (a) ***Stanomma dendroides.*** **(b)** ***Stannophyllum zonarium.***

these organisms somewhat uncertain. The genera included are *Psammetta*, *Stanomma*, and *Stannophyllum*. *See* GRANULORETICULOSIA. [R.P.H.]

Xenungulata Xenungulates are large, primitive, digitigrade, hoofed, tapirlike mammals, with relatively short, slender limbs and five-toed feet with broad, flat phalanges. They are restricted to the Paleocene deposits of Brazil and Argentina. There is only one family (Carodniidae) in the order and only one genus (*Carodnia*) in the family. *See* EUTHERIA.

Previously assigned to the order Pyrotheria, xenungulates appear to be most closely related to the Pyrotheria and the Dinocerata. *See* DENTITION; DINOCERATA; PYROTHERIA. [G.T.J.]

Xiphosurida An order of the subclass Xiphosura in the class Merostomata. These "sword-tailed" arthropods have three distinct body divisions: a prominent horseshoe-shaped, fused head and thorax, the prosoma; a segmented "abdomen," or opisthosoma; and a spikelike terminal segment, the telson. The Xiphosurida have an extensive fossil record. The two suborders, Synziphosurina and Limulina, span over 500,000,000 years of evolution. The Synziphosurina usually lacked compound eyes. Some species had a distinct axial, or cardiac, region, free segments in the opisthosoma, and a broad, short telson. Superfamilies Belinuracea and Limulacea of the Limulina have the clearly recognizable horseshoe "crab" form.

The carapace (test or shell) is composed of chitin, strengthened internally in the adults by bridgelike struts within the shell margins. Sensory structures on the carapace include chemoreceptors for taste on the bases of the legs and a pair of simple eyes, or ocelli. The multilensed compound eyes form mosaic images and are sensitive to polarized light. The organ systems, except the peripheral blood vessels and nerves, digestive, and reproductive glands, which branch extensively into the lateral portions of the prosoma, are mainly within the prominent axial region. This region contains the musculature of the paired appendages, hinge, and telson, the digestive tube, the coxal (excretory) gland, and the respiratory, central circulatory, and nervous structures. The dorsally situated tubular heart lies within a sinus which receives venous blood from the book gills, the respiratory structures, and the body. The central nervous system and the major nerve branches are sheathed by arteries. A plate of "cartilaginous" chitin, the endocranium, forms a roof over the brain. *See* MEROSTOMATA. [C.N.Sh.]

Xylem The principal water-conducting tissue and the chief supporting system of higher plants. This tissue and the associated phloem constitute the vascular system of vascular plants. Xylem is composed of various kinds of cells, living or nonliving. The structure of these cells differs in their functions, but characteristically all have a rigid and enduring cell wall that is well preserved in fossils.

In terms of their functions, the kinds of cells in xylem are those related principally to conduction and support, tracheids; to conduction, vessel members; to support, fibers; and to food storage, parenchyma. Vessel members and tracheids are often called tracheary elements. The cells in each of the four categories vary widely in structure. *See* PARENCHYMA.

Xylem tissues arise in later stages of embryo development of a given plant and are added to by differentiation of cells derived from the apical meristems of roots and stems. Growth and differentiation of tissues derived from the apical meristem provide the primary body of the plant, and the xylem tissues formed in it are called primary. Secondary xylem, when present, is produced by the vascular cambium. *See* LATERAL MERISTEM.

In the trade, softwood is a name for xylem of gymnosperms (conifers) and hardwood for xylem of angiosperms. The terms do not refer to actual hardness of the wood. Woods of gymnosperms are generally composed only of tracheids, wood parenchyma, and small rays, but differ in detail. Resin ducts are present in many softwoods. Woods of angiosperms show extreme variation in both vertical and horizontal systems, but with few exceptions have vessels. [V.I.C.; K.E.]

Xylene One of a group of three isomeric aromatic hydrocarbons. The three isomeric xylenes and ethylbenzene all have in common a molecular weight of 106.2 and the simplified formula, C_8H_{10}. The names, structural formulas, and boiling and melting points of these compounds are as shown in the table.

Properties of xylenes and ethylbenzene

Name	Structure	Boiling point, °F (°C)	Melting point, °F (°C)
o-Xylene	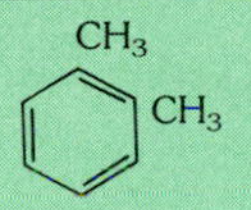	291.6 (144.2)	− 13.4 (25.2)
m-Xylene	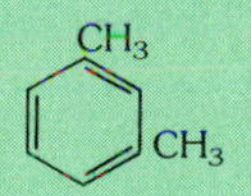	283.4 (139.1)	− 54.22 (−47.9)
p-Xylene	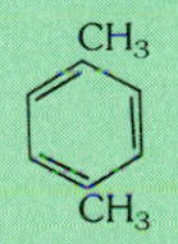	281.1 (138.4)	55.9 (13.3)
Ethylbenzene	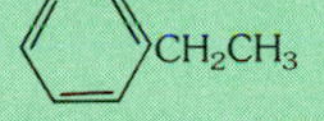	277.2 (136.2)	−139.0 (−95.0)

These compounds are almost exclusively produced from petroleum by hydroforming or catalytic reforming of appropriate naphtha fractions. *See* PETROLEUM PRODUCTS.

o-Xylene is commercially oxidized by air over vanadium oxide catalysts to produce phthalic anhydride, one of the most versatile of organic intermediates. *p*-Xylene, like the ortho isomer, has only one significant chemical use: it is oxidized by one of several processes either to terephthalic acid (TA) or to dimethylterephthalate (DMT), both of which are intermediates in the production of, among other things, the high-polymer materials known commonly as polyesters. *m*-Xylene, nearly twice as abundant as any of the other isomers, is technologically the least important. [R.I.S.; M.St.]

Xylose A pentose sugar, referred to in the early literature as L-xylose. It is present in many woody materials. The polysaccharide xylan, which is closely associated with cellulose, consists practically entirely of D-xylose. Corncobs, cottonseed hulls, pecan shells, and straw contain considerable amounts of this sugar. This pentose sugar is also a component of the hemicelluloses and the rare disaccharide, primeverose. *See* CARBOHYDRATE; POLYSACCHARIDE. [W.Z.H.]

Y

Y-delta transformations Electrically equivalent networks with three terminals, one being connected internally by a Y configuration and the other being connected internally by a Δ configuration.

If a network connects three terminals with one another, it is called a three-terminal network. The simplest configurations of three-terminal networks are the Y or star (Fig. 1*a*) and the Δ or mesh (Fig. 1*b*).

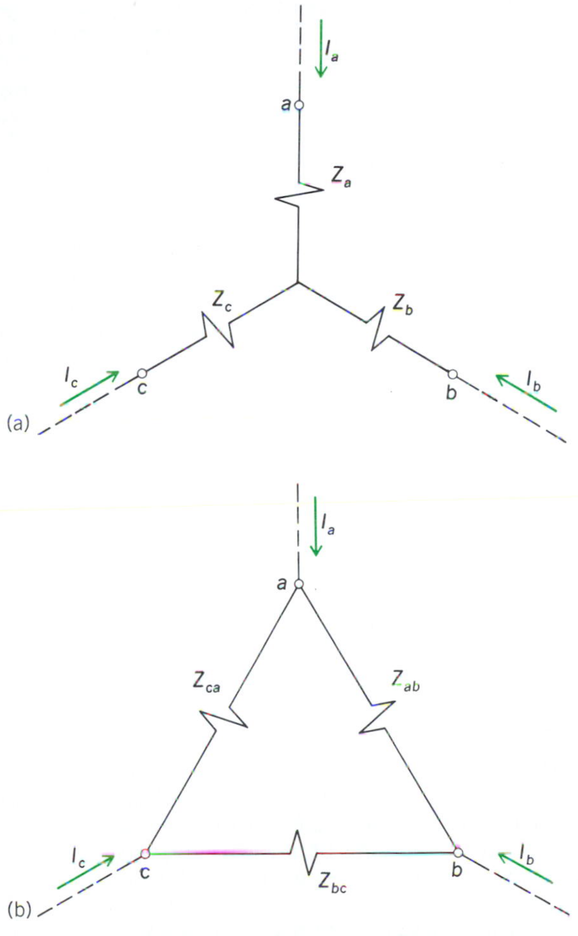

Three-terminal networks. (*a*) Y or star. (*b*) Δ or mesh.

If a given three-terminal network, passive and linear, has a Y configuration, it is possible to determine an equivalent Δ network that could be substituted for the Y without changing the relations of voltage and current at the network terminals, or elsewhere external to the network. Similarly, if a Δ network is given, an equivalent Y network can be found. Impedances of equivalent networks are usually functions of frequency, and realization of these impedances by use of physically possible elements is usually limited to a single frequency. [H.H.Sk.]

Yagi-Uda antenna A combination of a single driven antenna and a closely coupled parasitic element which may function either as a reflector as a result of inductive reactance or as a director as a result of capacitive reactance, depending on both the length and spacing of the parasitic element; also called the Yagi antenna (see illustration). Such structures are

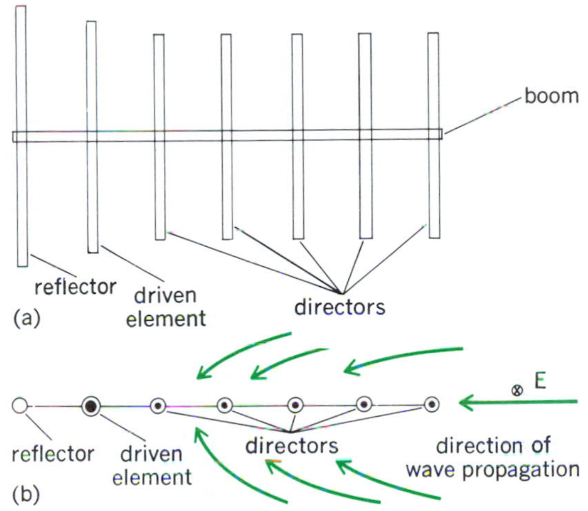

Yagi-Uda antenna. (*a*) View from above. (*b*) Side view; arrows show energy flow close to parasitic elements.

not only feasible but have a rather important place in antenna practice and concept, particularly in very-high-frequency and ultra-high-frequency ranges. *See* ANTENNA (ELECTROMAGNETISM). [K.Ar.]

Yak A heavily built mammal, *Poephagus grunniens*, of the order Artiodactyla that has been domesticated but still exists in the wild state in Tibet and high areas of China, living at elevations of 13,000–20,000 ft (about 4–6 km). It is related to the bison, which it resembles anatomically in having 14 pairs of ribs instead of the 13 pairs more usual of the order. Economically, it is probably the most important animal in Tibet, being used as a beast of burden and for its excellent milk. The dung is used as fuel, the hair is used for making cloth, and the flesh is eaten. Calving occurs in the autumn after a gestation period of about 10 months. *See* ARTIODACTYLA; BISON. [C.B.C.]

Yaw indicator A device that measures the angular direction of the airflow relative to the longitudinal vertical plane of the aircraft. This may be accomplished by a balanced vane or by a differential pressure sensor that aligns the detector to the airflow and, in so doing, transmits the measured angle between the normal axis and the detector as the yaw angle. If the device is used to measure the angle between the horizontal plane and the airflow, it is called an angle-of-attack sensor. *See* AIRCRAFT INSTRUMENTATION; AIRPLANE. [J.W.A.]

Yaws An infectious disease of humans caused by the spirochete *Treponema pertenue*. It is also known as frambesia and is largely confined to the tropics. Usually yaws is contracted in childhood by direct contact or from small flies feeding in succession on infected lesions and open wounds. No race or age possesses natural immunity. [T.B.T.]

Year Any of several units of time based on the revolution of Earth around the Sun. The tropical year, to which the calendar is adjusted, is the period that is required for the mean longitude of the Sun to increase 360°. Its duration is approximately 365.24220 mean solar days. It is also the period after which the seasons repeat themselves.

The sidereal year, 365.25636 mean solar days in duration, is the average period of revolution of Earth with respect to a fixed direction in space.

The anomalistic year, 365.25964 mean solar days in duration, is the average interval between successive closest approaches of Earth to the Sun. *See* TIME. [G.M.C.]

Yeast A collective name for those fungi which possess, under normal conditions of growth, a vegetative body (thallus) consisting, at least in part, of simple, individual, single cells. In addition, the cells making up the thallus may occur in pairs, in groups of three, or in straight or branched chains consisting of as many as 12 or more cells. Yeast plays a large part in industrial fermentation processes such as the production of ethanol, malt beverage, and wine and also in diseases of humans, animals, and plants. *See* FERMENTATION.

The shape and size of the individual cells of some species vary slightly, but in other species the cell morphology is extremely heterogeneous. The shape of yeast cells may be spherical, globose, ellipsoidal, elongate to cylindrical with rounded ends, more or less rectangular, pear-shaped, apiculate or lemon-shaped, ogival or pointed at one end, or tetrahedral. The dimension of a spherical cell may vary from 2 to 10 micrometers in diameter. The length of cylindrical cells is often 20–30 μm and, in some cases, even greater.

The asexual multiplication of yeast cells occurs by a budding process, by the formation of cross walls or fission, and sometimes by a combination of these two processes. Yeast buds are sometimes called blastospores. When yeast reproduces by a fission mechanism, the resulting cells are termed arthrospores.

Yeasts are divided into sporogenous and asporogenous groups, or perfect and imperfect yeasts. The sexual spores of yeasts are ascospores, which are formed in simple structures, often a vegetative cell. Such asci are called naked asci because of the absence of an ascocarp. If the vegetative cells are diploid, a cell may transform directly into an ascus after the $2n$ nucleus undergoes a reduction or meiotic division. Certain haploid yeasts have been shown to be heterothallic; that is, sporulation can occur only if strains of opposite mating type (sometimes indicated by + and − strains) are mixed on sporulation media. If no mating type is known, a yeast is considered to be imperfect. The perfect yeasts are classified in the order Saccharomycetales (Endomycetales), and the imperfect yeasts belong in the class Fungi Imperfecti, order Cryptococcales (Torulopsidales). *See* CRYPTOCOCCALES; DEUTEROMYCOTINA. [E.M.M.; H.J.P.]

Yeast infection An infection mainly caused by fungi of the genus *Candida*. Approximately 200 species of yeast within 25 genera have been isolated from humans, mainly from sputum, skin, and feces. Although most yeasts are transitory, some species, such as *C. albicans* and *C. glabrata*, occur frequently in the indigenous microflora. *Candida albicans* is isolated in a small percentage from the hands of hospital physicians and nurses, and various other yeast species are isolated from food products and beverages. *See* FUNGI; YEAST.

Etiologic agents. Approximately 80 different yeast species within several genera have been isolated from yeast infections, often in single cases. *Candida* species are the most frequent causes of yeast infections; about 20 species have been isolated. *Candida albicans* is the etiological agent in about 70–80% of all cases. The source of infection is always external. The abilities of yeast to adhere to skin and mucosal surfaces, penetrate tissue, produce enzymes, and change physical and genetic characteristics are factors in pathogenesis and virulence.

Predisposing factors. Because yeasts are normally of low virulence, a prerequisite for infection is a decrease of microbial defense mechanisms. Local disruption of tissues (trauma, burns, and maceration) also facilitates infections. Microbial infections, various hormonal disorders, and diseases where the immune system weakens are also predisposing factors. Deficiencies in the humoral immune system, cell-mediated immunity, and the phagocytic cell system (neutrophils, monocytes, macrophages, and eosinophils) confer susceptibility. Physiological states such as pregnancy, infancy, and old age present a high frequency of yeast infections. Surgery (and general anesthesia), especially advanced surgery such as open heart surgery and transplantations, are also risk factors, Recently, yeast infections have gained importance as a complication in individuals with acquired immune deficiency syndrome. Foreign bodies such as intravenous catheters and dentures, to which *Candida albicans* can adhere, may become foci of fungal growth. *See* ACQUIRED IMMUNE DEFICIENCY SYNDROME (AIDS).

Infections and symptoms. Yeast infections are typically characterized by blisters surrounded by reddened, scaly skin, but lesions can have other appearances. Nails with painful reddened swelling along the paronychial edge and discolored and striated nails can be due to yeast; however, other organisms may give similar lesions. Mucosa of the oral cavity covered by a white-gray pseudomembrane indicates a *C. albicans* infection (thrush). Because many symptoms are not particular to yeast, the medical history taken from the individual together with underlying diseases or states will lead the clinician to suspect a yeast infection. In order to make a diagnosis, a sample must be cultured to determine the species and, where such methods have been developed and if hospital-related infections are suspected, the type of fungus. A positive culture from sites that are normally sterile is diagnostic. In individuals with low white blood cell counts, persistent infections have a poor prognosis. Many antifungal agents are used for treatment of yeast infections. *See* MEDICAL MYCOLOGY. [A.Ste.]

Yellow fever An acute, febrile, mosquito-borne viral disease characterized in severe cases by jaundice, albuminuria, and hemorrhage. Inapparent infections also occur.

The agent is an arbovirus of group B. The virus multiplies in many animals, in mosquitoes, in chick embryos, and in tissue cultures from chick or mouse embryos. The virus enters the body through a mosquito bite and multiplies in lymph nodes, circulates in the blood, and localizes in the liver, spleen, kidney, bone marrow, and lymph glands. The severity of the disease and the major signs and symptoms which appear depend upon where the virus localizes and how much cell destruction occurs. *See* ARBOVIRAL ENCEPHALITIDES. [J.L.Me.]

Yew A genus of evergreen trees and shrubs, *Taxus*, with a fruit containing a single seed surrounded by a scarlet, fleshy, cuplike envelope (aril). The leaves are flat and acicular (needle-shaped), green below, with stalks extending downward on the stem. The only native American species of commercial importance is the Pacific yew (*T. brevifolia*), a medium-sized tree of the Pacific Coast and northern Rocky Mountain regions. Although it is not a common tree, its wood is sometimes used for poles, paddles, bows, and small cabinetwork.

The English yew (*T. baccata*), native in Europe, North Africa, and northern Asia, and the Japanese yew (*T. cuspidata*) are much cultivated in the United States as evergreen ornamentals. *See* TAXALES. [A.H.G./K.P.D.]

Ylide A variety of organic compounds that contain two adjacent atoms bearing formal positive and negative charges, and in which both atoms have full octets of electrons. Heteroatoms most commonly utilized as the positive atom are phosphorus, nitrogen, sulfur, selenium and oxygen, and the negative atom usually involves carbon, nitrogen, oxygen, or sulfur. The ylide may be in an alicyclic or cyclic environment, and in the latter the ylide function may be endocyclic or exocyclic. Ylides most useful in organic synthesis are those containing phosphorus or sulfur and an adjacent carbanion.

A phosphorus ylide is known as a phosphorane. Ylides of this type are highly reactive. Sulfur-containing ylides (π-sulfuranes) are of two types, sulfonium ylides and oxosulfonium ylides, which differ in having an oxygen atom attached to the sulfur in the latter. *See* REACTIVE INTERMEDIATES.

The ylide function may also be incorporated into heterocyclic systems. For example, the meso-ionic sydnone molecule contains an azomethine imine ylide, and the mesomeric betaine derived from 3-hydroxypyridine contains an azomethine ylide. These ylides are often referred to as masked ylides. [K.T.P.]

Yolk sac An extraembryonic membrane which extends through the umbilicus in vertebrates. In some elasmobranchs, birds, and reptiles, it is laden with yolk which serves as the nutritive source of embryonic development. The yolk sac of placental mammals is yolkless and highly modified. The yolk sac is involved in the selective transport of serum proteins from mother to fetus. In some mammals, it serves the function of blood-cell formation as well. [G.W.H.]

Young's modulus A constant designated E, the ratio of stress to corresponding strain when the material behaves elastically. Young's modulus is represented by the slope $E = \Delta S/\Delta \epsilon$ of the initial straight segment of the stress-strain diagram. More correctly, E is a measure of stiffness, having the same units as stress: pounds per square inch or pascals. When stress and strain are not directly proportional, E may be represented as the slope of the tangent or the slope of the secant connecting two points on the stress-strain curve. The modulus is then designated as tangent modulus or secant modulus at stated values of stress. The modulus of elasticity applying specifically to tension is called Young's modulus. *See* ELASTICITY; HOOKE'S LAW; STRESS AND STRAIN. [W.J.K./W.G.B.]

Ytterbium A chemical element, Yb, atomic number 70, and atomic weight 173.04. Ytterbium is a metal element of the rare-earth group. There are 7 naturally occurring stable isotopes.

The common oxide, Yb_2O_3, is colorless and dissolves readily in acids to form colorless solutions of trivalent salts which are paramagnetic. Ytterbium also forms a series of divalent com-

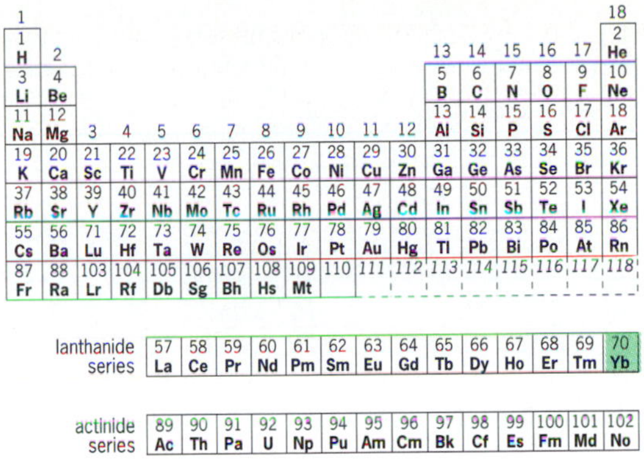

pounds. The divalent salts are soluble in water but react very slowly with water to liberate hydrogen.

The metal is best prepared by distillation. It is a silvery soft metal which corrodes slowly in air and resembles the calcium-strontium-barium series more than the rare-earth series. For a discussion of the properties of the metal and its salts *see* RARE-EARTH ELEMENTS. [F.H.Sp.]

Yttrium A chemical element, Y, atomic number 39, and atomic weight 88.905. Yttrium resembles the rare-earth elements closely. The stable isotope ^{89}Y constitutes 100% of the natural element, which is always found associated with the rare earths and is frequently classified as one.

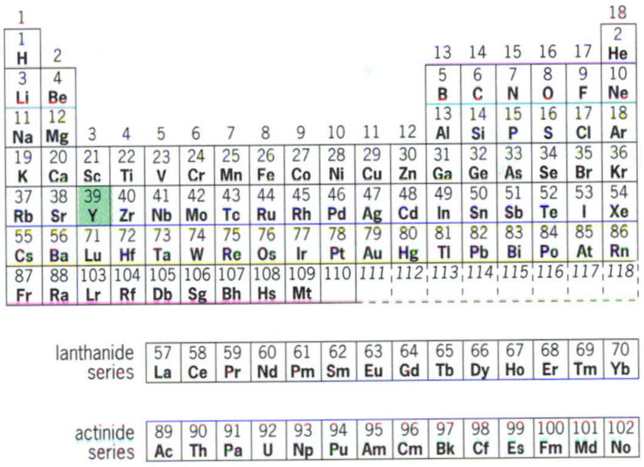

Yttrium metal absorbs hydrogen, and in alloys up to a composition of YH_2 they resemble metals very closely. In fact, in certain composition ranges, the alloy is a better conductor of electricity than the pure metal.

Yttrium forms the matrix for the europium-activated yttrium phosphors which emit a brilliant, clear-red light when excited by electrons. The television industry uses these phosphors in manufacturing television screens.

Yttrium is used commercially in the metal industry for alloy purposes and as a "getter" to remove oxygen and nonmetallic impurities in other metals. For properties of the metal and its salts *see* RARE-EARTH ELEMENTS. [F.H.Sp.]

Z

Z transform The preferred operational-calculus tool for analysis and design of discrete-time systems. (It should not be confused with the *z* transformation.) The role of the *z* transform with regard to discrete-time systems is similar to that of the Laplace transform for continuous systems. In fact, the Laplace transform is a specialized case of the *z* transform. The *z* transform is by far the more insightful tool, and the Laplace transform is just the limiting case of the *z* transform in a practical as well as a conceptual way. *See* CONTROL SYSTEMS; DIGITAL FILTER; LAPLACE TRANSFORM; LINEAR SYSTEM ANALYSIS.

It is useful to consider a band-limited real continuous signal, $x(t)$, with no significant amount of energy above a frequency, f_c. This signal is sampled at uniformly spaced intervals of time, T, $2T$, $3T$, . . . , nT, . . . , where the sampling interval, T, and the sampling frequency, f_s, are reciprocals of one another, and where f_s is greater than $2f_c$, a condition necessary for unambiguous interpretation of the sampled signal. If the common shorthand notation $x(nT) = x_n$ is used, the definition of the *z* transform of $x(t)$ is given by the equation

$$X(z) = x_0 + x_1z^{-1} + x_2z^{-2} + x_3z^{-3} + \cdots$$

The coefficient of z^{-p} is therefore the value of the *p*th sample of the time signal. This gives a great deal of physical insight to the use of the *z* transform, a feature not shared by the Laplace transform. The Laplace transform can be found by evaluating the limit of the *z* transform as T approaches zero. *See* ANALOG-TO-DIGITAL CONVERTER; INFORMATION THEORY. [S.A.Wh.]

Zebra Three species belonging to the family Equidae and indigenous to Africa. These animals are odd-toed ungulates (order Perissodactyla) which are monodactyl; that is, the middle digit is functional while the second and fourth digits are vestigial. The striped coat is considered to be an example of protective coloration since they live on open plains. Zebras are sociable and graze with other animals, such as deer, gnu, and ostriches. The gestation period is 13 months and a single young is born. The maximum life-span is 30 years. *See* PERISSODACTYLA. [C.B.C.]

Zebu A domestic breed of cattle, indigenous to India, belonging to the family Bovidae in the order Artiodactyla. This animal, *Bos indicus*, known as the Brahman in the United States, is protected as the sacred cow in India, where it is used as a draft animal as well as for milk. The ears are long and drooping (see illustration); the color is quite variable, One of the most notable cross breeds is the Santa Gertrudis, in which Brahman bulls are crossed with short-horn cows. This produces one of the heaviest beef breeds. *See* ARTIODACTYLA. [C.B.C.]

The zebu, or Brahman, characterized by a dorsal hump between the shoulders and loose skin which forms a dewlap under the chin.

Zeeman effect A splitting of spectral lines when the light source being studied is placed in a magnetic field. Discovered by P. Zeeman in 1896, the effect furnishes information of prime importance in the analysis of spectra. Each kind of spectral term has its characteristic mode of splitting, and the types of terms are most definitely identified by this property, Furthermore, the effect allows an evaluation of the ratio of charge to mass of the electron and an evaluation of its precise magnetic moment.

The normal Zeeman effect is a splitting into two or three lines, depending on the direction of observation, as shown in the illustration. The light of these components is polarized in ways indicated in the figure. The normal effect is observed for all lines belonging to singlet systems, those for which the spin quantum number $S = 0$. The change of frequency of the shifted components can be evaluated on classical electromagnetic principles.

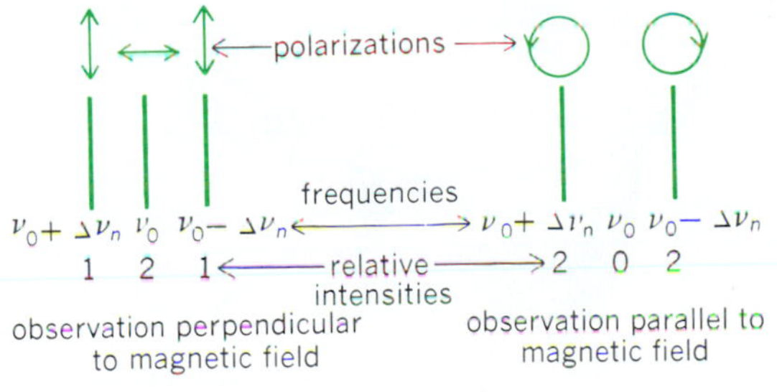

Triplet observed in normal Zeeman effect. ν_0 = unshifted frequency; $\Delta\nu_n$ = frequency shift.

The anomalous Zeeman effect is a more complicated type of line splitting, so named because it did not agree with the predictions of classical theory. It occurs for any spectral line arising from a combination of terms of multiplicity greater than one. Since multiplicity in spectral lines is caused by the presence of a resultant spin vector S of the electrons, the anomalous effect must be attributed to a nonclassical magnetic behavior of the electron spin.

The quadratic Zeeman effect, which depends on the square of the field strength, is of two kinds. The first results from second-order terms, and the second from the diamagnetic reaction of the electron when revolving in large orbits.

The inverse Zeeman effect is the Zeeman effect of absorption lines. It is closely related to the Faraday effect, the rotation of plane-polarized light by matter situated in a magnetic field. *See* FARADAY EFFECT.

The Zeeman effect in molecules is, in general, so small as to be unobservable, even for molecules which have a permanent magnetic moment. An exception occurs for some light molecules where the magnetic moment is coupled so lightly to the frame of the molecule that it can orient itself freely in the magnetic field just as for atoms.

A clear Zeeman effect also can be observed in many crystals with sharp spectrum lines in absorption or fluorescence. Such crystals are found particularly among rare-earth salts.

The magnetic moment of the nucleus causes a Zeeman splitting in atomic spectra which is of an order of magnitude a thousand times smaller than the ordinary Zeeman effect. This Zeeman effect of the hyperfine structure usually is modified by a nuclear Paschen-Back effect. *See* PASCHEN-BACK EFFECT.

[F.A.J.; G.H.Di./W.W.W.]

Zeiformes A small order of teleost fishes, structurally intermediate between the Beryciformes and the Perciformes. This group is also known as the Zeomorphi or Zeoidea; commonly the fish are known as dories. There is no orbitosphenoid bone; the pelvic fin has a spine and from five to nine soft rays; and there is a more or less distinct anterior, spinous anal fin of one to four spines, as well as a spinous dorsal fin (see illustration). The zeiform fishes, which are known from the Paleocene on, are grouped into 6 families, perhaps 12 genera, and fewer than 50 species. All are marine, living in shore waters and chiefly at moderate depths off tropical and temperate coasts. Most are of small size and of minor economic importance. *See* ACTINOPTERYGII; BERYCIFORMES; PERCIFORMES.

[R.M.B.]

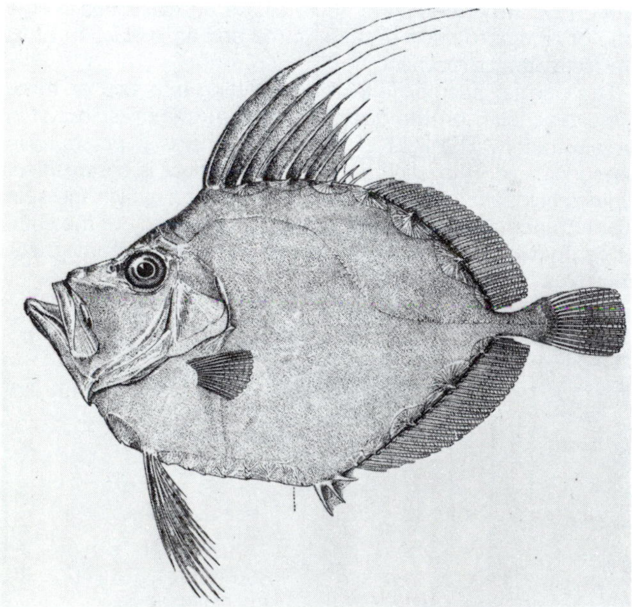

John dory (*Zenopsis ocellata*). (*After G. B. Goode, Fishery Industries of the United States, sect. 1, 1884*)

Zener diode A two-terminal semiconductor junction device with a very sharp voltage breakdown in the reverse-bias region. This device is used principally in voltage regulator circuits to provide a voltage reference. It is named after C. Zener, who first proposed electronic tunneling as the mechanism of electrical breakdown in insulators. *See* SEMICONDUCTOR; VOLTAGE REGULATOR.

For silicon diodes with breakdown voltages of 6 V or less, the Zener mechanism of breakdown is operative and provides excellent voltage reference diodes. Above about 6 V, the breakdown mechanism involves avalanching instead of tunneling, and the breakdown characteristic begins to show more temperature sensitivity and becomes less sharp. Up to about 15 V, the characteristic is still reasonably suitable for voltage reference use, and all such diodes today are called Zener diodes regardless of the actual mechanism of breakdown. *See* JUNCTION DIODE; TUNNELING IN SOLIDS.

[L.P.H.]

Zenith The point directly overhead in the sky. The astronomical zenith, which is that usually meant, is the upper intersection of a plumb line with the celestial sphere. The zenith distance of a celestial object is its angular distance from the astronomical zenith and is identical with the complement of its altitude. The geocentric zenith is the upper intersection with the celestial sphere of an imagined line through the center of Earth and the observer. The point diametrically opposite the zenith is called the nadir. *See* ASTRONOMICAL COORDINATE SYSTEMS.

[G.M.C.]

Zeolite Any mineral belonging to the zeolite family of minerals and synthetic compounds characterized by an alumino-silicate tetrahedral framework, ion-exchangeable large cations, and loosely held water molecules permitting reversible dehydration. Typical large cations are the alkalies and alkaline earths such as Na^+, K^+, Ca^{2+}, Sr^{2+}, and Ba^{2+}. The large cations, coordinated by framework oxygens and water molecules, reside in large cavities in the crystal structure; these cavities and channels may even permit the selective passage of organic molecules. Thus, zeolites are extensively studied from theoretical and technical standpoints because of their potential and actual use as "molecular sieves," catalysts, and water softeners. Dehydrated zeolites can absorb other liquids, such as ammonia, alcohol, and hydrogen sulfide, instead of water. *See* MOLECULAR SIEVE; SILICATE MINERALS; WATER SOFTENING.

Zeolites are low-temperature and low-pressure minerals and commonly occur as late minerals in amygdaloidal basalts, as devitrification products, as authigenic minerals in sandstones and other sediments, and as alteration products of feldspars and nepheline. Zeolites are usually white, but often may be colored pink, brown, red, yellow, or green by inclusions; the hardness is moderate (3–5) and the specific gravity low (2.0–2.5) because of their rather open framework structures.

[P.B.M.]

Zero In mathematics, the concept zero is used in two ways: as a number and as a value of a variable. The positional system of number notation, developed first by the Babylonians (about 500 B.C.) with the base 60, and a millennium later by the Hindus and the Chinese with the base 10, required for greater clarity a special marker of the empty, nonoccupied position.

The zero as a number, however, is a new concept, introduced by the Hindus and Chinese about the same time (6th century). Brahmagupta (born A.D. 598) remarked that the number 0 has special properties: $a \pm 0 = a$, and $a \cdot 0 = 0$, where a may be any number (integer).

In a modern way, zero can be called the identity element of the infinite Abelian additive group of integers. If in an integral domain a product is equal to zero, then at least one factor of the product is zero. In the second concept zero is the value of a variable for which a function is equal to zero. [H.R.; E.Gr.]

Zero-pressure balloon A nonextensible balloon constructed of thin-film polyethylene in a tailored natural shape (inverted teardrop). This is the workhorse of scientific ballooning, with a worldwide launch rate of several hundred per year. At float, it assumes the theoretical shape of a load-bearing balloon with zero circumferential stress. The polyethylene is manufactured by a blown extruding technique which produces a continuous tube of thin film. The thickness of the film can be adjusted by varying the extruder die, temperature, and pressure to produce films as thin as 9 micrometers (0.00035 in.). The tube is slit, unfolded, and rerolled on spools which are approximately 3 m (10 ft) wide. The material is then laid out on an assembly table, which is the full length of the balloon, and is cut to conform to one vertical segment of balloon, which is referred to as a gore. Additional gores are then added, with a polyester load tape heat-sealed to each seam. The bottom of the balloon is left open through an arrangement of ducts, and hence the term zero-pressure balloon.

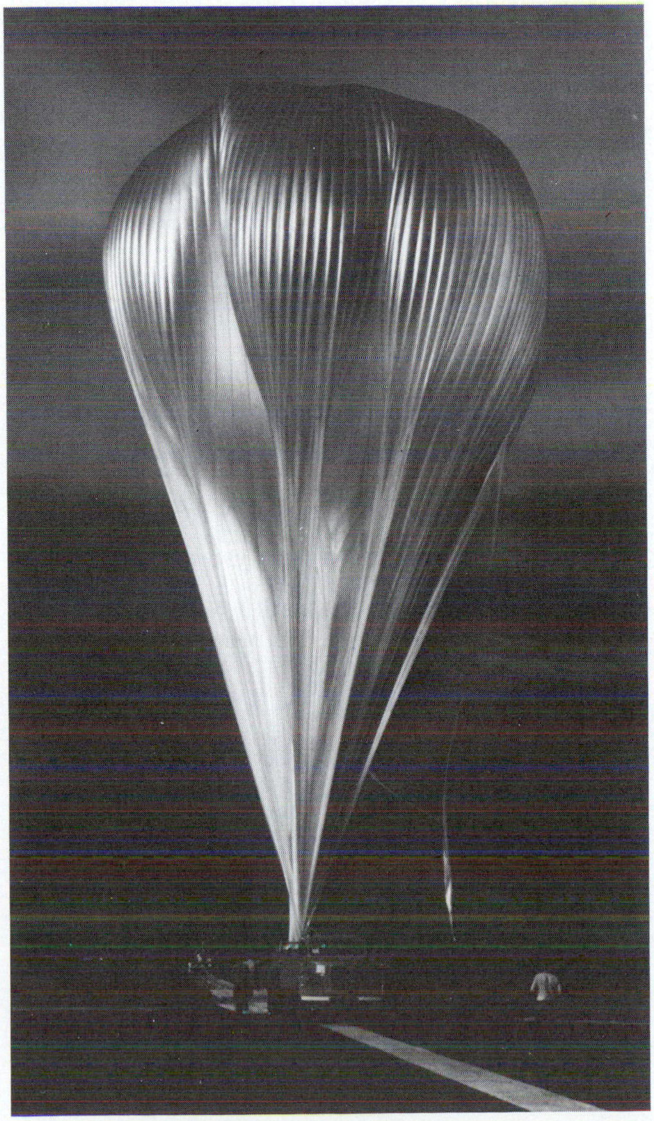

Inflation of zero-pressure balloon. (*NASA*)

During inflation, the lifting gas is contained in a bubble which occupies less than 1% of the total balloon volume (see illustration). Approximately 10% additional lift over that required for neutral buoyancy is added to the balloon to provide the necessary vertical force to achieve an ascent rate of 250 m/min (820 ft/min).

The zero-pressure balloon is capable of carrying scientific instruments weighing thousands of kilograms to an altitude of 42 km (26 mi), which is above 99.8% of the Earth's atmosphere. These scientific instruments have become very sophisticated and resemble a satellite in complexity. They utilize the latest technology in electronics; in fact, new technology is often proved in balloon experiments before being flown on satellites. [W.R.N.]

Zinc A chemical element, Zn, atomic number 30, and atomic weight 65.38. Zinc is a malleable, ductile, gray metal. Fifteen isotopes of zinc are known, of which five are stable, having atomic masses of 64, 66, 67, 68, and 70. About half of ordinary zinc occurs as the isotope of atomic mass 64.

The most important uses of zinc are in its alloys and as a protective coating on other metals. Coating iron or steel with zinc is called galvanizing, and it may be done by immersing the article in melted zinc (hot-dip process), depositing zinc electrolytically onto the article in a plating bath (electrogalvanizing), exposing the article to powdered zinc near its melting point (sherardizing), or spraying the article with melted zinc (metallizing). *See* ELECTROPLATING OF METALS; METAL COATINGS; ZINC ALLOYS.

Zinc is one of the less common elements; it has been estimated to make up 0.0005–0.02% of the Earth's crust. It is twenty-fifth in order of abundance among the elements. The chief ore is zinc blende, marmatite, or sphalerite, ZnS. Zinc is an essential element in the growth of many kinds of organisms, both plant and animal. A deficiency of zinc in the human diet has been found to retard growth and maturity and to produce anemia. The compound insulin is a zinc-containing protein. Zinc is present in most foods, especially those high in protein. The average human body contains about 2 g of zinc.

Pure, freshly polished zinc is bluish white, lustrous, and moderately hard (2.5 on Mohs scale). Moist air brings about a superficial tarnishing to give the metal its usual grayish color. Pure zinc is malleable and ductile enough to be rolled or drawn, but small amounts of other metals present as contaminants may render it brittle. Zinc melts at 420°C (788°F) and boils at 907°C (1665°F). Its density is 7.13 times that of water, so that 1 ft^3 (0.028 m^3) of zinc weighs 445 lb (200 kg).

As a conductor of heat and of electricity, zinc ranks fairly high. As a conductor of heat, zinc is only about one-fourth as

1	2	3	4	5	6	7	8	9	10	11	12	13	14	15	16	17	18
1 H																	2 He
3 Li	4 Be											5 B	6 C	7 N	8 O	9 F	10 Ne
11 Na	12 Mg											13 Al	14 Si	15 P	16 S	17 Cl	18 Ar
19 K	20 Ca	21 Sc	22 Ti	23 V	24 Cr	25 Mn	26 Fe	27 Co	28 Ni	29 Cu	30 Zn	31 Ga	32 Ge	33 As	34 Se	35 Br	36 Kr
37 Rb	38 Sr	39 Y	40 Zr	41 Nb	42 Mo	43 Tc	44 Ru	45 Rh	46 Pd	47 Ag	48 Cd	49 In	50 Sn	51 Sb	52 Te	53 I	54 Xe
55 Cs	56 Ba	71 Lu	72 Hf	73 Ta	74 W	75 Re	76 Os	77 Ir	78 Pt	79 Au	80 Hg	81 Tl	82 Pb	83 Bi	84 Po	85 At	86 Rn
87 Fr	88 Ra	103 Lr	104 Rf	105 Db	106 Sg	107 Bh	108 Hs	109 Mt	110	*111*	*112*	*113*	*114*	*115*	*116*	*117*	*118*

lanthanide series	57 La	58 Ce	59 Pr	60 Nd	61 Pm	62 Sm	63 Eu	64 Gd	65 Tb	66 Dy	67 Ho	68 Er	69 Tm	70 Yb
actinide series	89 Ac	90 Th	91 Pa	92 U	93 Np	94 Pu	95 Am	96 Cm	97 Bk	98 Cf	99 Es	100 Fm	101 Md	102 No

efficient as silver. At 0.91 K zinc is an electrical superconductor. Pure zinc is not ferromagnetic.

Zinc is a fairly active metal chemically. It can be ignited with some difficulty to give a blue-green flame in air and to discharge clouds of zinc oxide smoke. Metallic zinc in an acidic solution will react to liberate hydrogen gas, forming dipositively charged zinc ions, Zn^{2+}. Zinc also dissolves in strongly alkaline solutions, such as sodium hydroxide, to liberate hydrogen and form dinegatively charged tetrahydroxozincate ions, $Zn(OH)_4^{2-}$, sometimes written as ZnO_2^{2-} in the formulas of the zincate compounds.

Zinc is always divalent in its compounds, except for some of those with other metals, which are classed as zinc alloys. Zinc also forms many coordination compounds. In most coordination compounds of zinc, the fundamental structural unit is a central zinc ion surrounded by four coordinated groups arranged spatially at the corners of a regular tetrahedron. *See* ZINC METALLURGY. [W.E.C.]

Zinc alloys Combinations of zinc with one or more other metals. If zinc is the primary constituent of the alloy, it is a zinc-base alloy. Zinc also is commonly used in varying degrees as an alloying component with other base metals, such as copper, aluminum, and magnesium. A familiar example of the latter is the association of varying amounts of zinc (up to 45%) with copper to produce brass. *See* BRASS; COPPER ALLOYS.

Zinc-base alloys have two major uses: for casting and for wrought applications. Casting includes both die casting and gravity casting, which differs from die casting primarily in that no pressure is applied, except the force of gravity, in forcing the molten metal into the mold. *See* METAL CASTING.

The great bulk of zinc die casting is carried out with two alloys, commonly identified as alloys 3 and 5. A third alloy has been added and is commonly designated as alloy 7. The entire group in the United States is often spoken of by its original trade name as Zamak alloys, and in England as the Mazak alloys.

Aluminum, the major alloying constituent in alloys 3 and 5, is added in amounts of about 4%, which has proved to be the optimum composition from the standpoint of strength, ductility, and stability. The aluminum content also sharply reduces the rate of attack of molten zinc on iron containers. Aluminum also avoids "soldering," or sticking of the casting to the die. This permits the casting of these zinc alloys in the more productive hot-chamber (plunger) type of die-casting machine. *See* ALUMINUM.

Alloys 3 and 5 differ primarily in their copper content. Copper additions increase strength and hardness and improve corrosion resistance. The copper content of alloys is held to a specified maximum of 1.25% (in alloy 5) to avoid making the alloy unstable through aging and to avoid reducing the impact strength to a very low value.

Alloy 7 is considered a modification of alloy 3. It has a lower magnesium content, and iron, lead, tin, and cadmium are also held to lower levels. A small amount of nickel is added. When the alloys are compared, alloy 7 exhibits improved casting properties, making it easier to secure high-quality surfaces for finishing (mostly chromium plating) and to obtain higher production rates.

In the wrought zinc area, numerous compositions and alloys are used, depending on ultimate product requirements. Alloying metals can be used to improve various properties, such as stiffness, for special applications. The zinc-copper-titanium alloy has become the dominant wrought-zinc alloy for applications demanding superior performance. *See* ZINC; ZINC METALLURGY. [A.L.P.]

Zinc metallurgy Separating and extracting zinc from ores, refining it, and preparing it into usable forms. Ordinarily, after being mined, ores must first be separated into a concentrated mineral and a waste rock. This concentrate then is reduced to the metal in a metallurgical works. Finally, the metal may be further refined and alloyed to commercially usable form. Commonly, these three extractive metallurgical operations are conducted at separate locations and are broadly categorized as concentrating, smelting, and refining, respectively.

Zinc sulfide ores are usually beneficiated (concentrated or ore-dressed) adjacent to the mine site. First, the ore must be crushed and ground in order to free the mineral lattices from those of the waste rock (gangue). Next, the finely divided ore is mixed into a slurry with water and the mineral and gangue particles are separated utilizing the effect of gravity. The most common means of separation is the froth flotation process. *See* FLOTATION; ORE DRESSING.

The second operation, smelting, involves first the chemical or physical changes required in the source material to prepare it into a crude zinc oxide form and a specified particle size (depending upon the particular smelting method contemplated). Such process steps variously entail the operations of roasting, sintering, or pyroconcentration. The other basic part of smelting is the reduction step, wherein the zinc is reduced from its oxide to its elemental form. Several very successful and varied processes accomplish this function, including horizontal retort, vertical retort, electrothermic furnace, and blast furnace (all of which are pyrometallurgical and use carbon as a reducing agent). *See* PYROMETALLURGY; SINTERING.

Purity of zinc produced by the various smelting processes is largely dependent upon controls and operating procedures practiced during the preparation steps. The quality produced is variously suitable for hot-dip galvanizing, continuous-line galvanizing, and in some cases for brass manufacture and rolled (wrought) zinc; however, for the sizable usage in die-casting alloys, output from this type of smelter must undergo a refining step. Fractional distillation in reflux refining columns is the leading method of upgrading the lower-purity zinc metal. *See* ZINC ALLOYS. [C.H.Co.]

Zincite A mineral with composition ZnO (zinc oxide). It crystallizes in the hexagonal system with a wurtzite-type structure. Thus its principal axis is polar and different forms appear at top and bottom of crystals. Such crystals are rare and the mineral is usually massive. Its hardness is 4 and its specific gravity 5.6. The mineral has a subadamantine luster and a deep-red to orange-yellow color. Zincite is rare except at the zinc deposits at Franklin and Sterling Hill, N.J. There, associated with franklinite and willemite, it is mined as a valuable ore of zinc. *See* FRANKLINITE; WILLEMITE; ZINC. [C.S.Hu.]

Zingiberales An order of flowering plants, division Magnoliophyta (Angiospermae), in the subclass Commelinidae of the class Liliopsida (monocotyledons). The order, which has also been called Scitamineae or Scitaminales, includes 8 families and about 2100 species. The largest families are the Zingiberaceae (about 1300 species), Marantaceae (about 350 species), Costaceae (about 200 species), and Heliconiaceae (about 150 species). The order is morphologically well defined and taxonomically rather isolated. *See* LILIOPSIDA.

The Zingiberales are herbs or scarcely branched trees or shrubs with pinnately veined leaves and irregular flowers that have well-differentiated sepals and petals, an inferior ovary, septal or septal-derived nectaries, and usually either one or five functional stamens. Ginger (*Zingiber officinale*), banana (*Musa*) of the family Musaceae, bird of paradise flower (*Strelitzia*) of the family Strelitziaceae, and *Canna* of the family Cannaceae are familiar members of the Zingiberales. *See*

ABACA; ARROWROOT STARCH; BANANA; CARDAMON; COMMELINIDAE; TURMERIC. [A.Cr.]

Zingiberidae A subclass of Liliopsida (monocotyledons) of the division Magnoliophyta (Angiospermae), the flowering plants, containing two orders (Bromeliales and Zingiberales), with nine families and about 3800 species. The subclass has been associated with the Commelinidae and the Liliidae, sharing some features with each but aberrant in either, and it differs from both in usually having four or more supporting cells around each stomate. The flowers are usually epigynous, usually bisexual, and often irregular, and they are usually showy and adapted to pollination by insects or other animals. The pistil consists of three united carpels, often with septal nectaries opening at the top of the ovary. The seeds usually have well-developed, mealy or starchy endosperm. *See* BROMELIALES; LILIOPSIDA; MAGNOLIOPHYTA; PLANT KINGDOM; SECRETORY STRUCTURES (PLANT); ZINGIBERALES. [T.M.Ba.]

Zircon A mineral with the idealized composition $ZrSiO_4$, one of the chief sources of the element zirconium. Structurally, zircon is a nesosilicate, with isolated SiO_4 groups. *See* SILICATE MINERALS; ZIRCONIUM.

Zircon often occurs as well-formed crystals. The color is variable, usually brown to reddish brown, but also colorless, pale yellowish, green, or blue. The transparent colorless or tinted varieties are popular gemstones. Hardness is $7\frac{1}{2}$ on Mohs scale; specific gravity is 4.7, decreasing in metamict types. *See* METAMICT STATE.

Because of its chemical and physical stability, zircon resists weathering and accumulates in residual deposits and in beach and river sands, from which it has been obtained commercially in Florida and in India, Brazil, and other countries. *See* HEAVY MINERALS. [C.Fr.]

Zirconium A chemical element, Zr, atomic number 40, atomic weight 91.22. Its naturally occurring isotopes are 90, 91, 92, 94, and 96. Zirconium is one of the more abundant

1	2	3	4	5	6	7	8	9	10	11	12	13	14	15	16	17	18
1 H																	2 He
3 Li	4 Be											5 B	6 C	7 N	8 O	9 F	10 Ne
11 Na	12 Mg											13 Al	14 Si	15 P	16 S	17 Cl	18 Ar
19 K	20 Ca	21 Sc	22 Ti	23 V	24 Cr	25 Mn	26 Fe	27 Co	28 Ni	29 Cu	30 Zn	31 Ga	32 Ge	33 As	34 Se	35 Br	36 Kr
37 Rb	38 Sr	39 Y	40 Zr	41 Nb	42 Mo	43 Tc	44 Ru	45 Rh	46 Pd	47 Ag	48 Cd	49 In	50 Sn	51 Sb	52 Te	53 I	54 Xe
55 Cs	56 Ba	71 Lu	72 Hf	73 Ta	74 W	75 Re	76 Os	77 Ir	78 Pt	79 Au	80 Hg	81 Tl	82 Pb	83 Bi	84 Po	85 At	86 Rn
87 Fr	88 Ra	103 Lr	104 Rf	105 Db	106 Sg	107 Bh	108 Hs	109 Mt	110	111	112	113	114	115	116	117	118

lanthanide series	57 La	58 Ce	59 Pr	60 Nd	61 Pm	62 Sm	63 Eu	64 Gd	65 Tb	66 Dy	67 Ho	68 Er	69 Tm	70 Yb
actinide series	89 Ac	90 Th	91 Pa	92 U	93 Np	94 Pu	95 Am	96 Cm	97 Bk	98 Cf	99 Es	100 Fm	101 Md	102 No

elements, and is widely distributed in the Earth's crust. Being very reactive chemically, it is found only in the combined state. Under most conditions, it bonds with oxygen in preference to any other element, and it occurs in the Earth's crust only as the oxide, ZrO_2, baddeleyite, or as part of a complex of oxides as in zircon, elpidite, and eudialyte. Zircon is commercially the most important ore. Zirconium and hafnium are practically indistinguishable in chemical properties, and occur only together. *See* HAFNIUM; ZIRCON.

Most of the zirconium used has been as compounds for the ceramic industry: refractories, glazes, enamels, foundry mold and core washes, abrasive grits, and components of electrical ceramics, The incorporation of zirconium oxide in glass significantly increases its resistance to alkali. The use of zirconium metal is almost entirely for cladding uranium fuel elements for nuclear power plants. Another significant use has been in photo flashbulbs.

Zirconium is a lustrous, silvery metal, with a density of 6.5 g/cm^3 (3.8 $oz/in.^3$) at 20°C (68°F). It melts at about 1850°C (3362°F). Estimates of the boiling point from appropriate data have commonly been of the order of 3600°C (6500°F), but observations suggest about 8600°C (15,500°F). The free energies of formation of its compounds indicate that zirconium should react with any nonmetal, other than the inert gases, at ordinary temperatures. In practice, the metal is found to be nonreactive near room temperature because of an invisible, impervious oxide film on its surface. The film renders the metal passive, and it remains bright and shiny in ordinary air indefinitely. At elevated temperatures it is very reactive to the nonmetallic elements and many of the metallic elements, forming either solid solutions or compounds.

Zirconium generally has normal covalency of 4, and commonly exhibits coordinate covalencies of 5, 6, 7, and 8. Zirconium is at oxidation number 4 in nearly all of its compounds, Halides in which its oxidation numbers are 3 and 2 have been prepared. While zirconium is often part of cationic or anionic complexes, there is no definite evidence for a monatomic zirconium ion in any of its compounds.

Most handling and testing of zirconium compounds have indicated no toxicity. There has generally been no ill consequence of contact of zirconium compounds with the unabraded skin. However, some individuals appear to have allergic sensitivity to zirconium compounds, characteristically manifested by appearance of nonmalignant granulomas. Inhalation of sprays containing some zirconium compounds and of metallic zirconium dusts have had inflammatory effects. [W.B.B.]

Zoantharia A subclass of the Anthozoa. Zoantharians are monomorphic anthozoans, most of which have retractile, simple, tubular tentacles. The tentacles vary from six to several hundred. The mesenteries are six or some multiple of six and may be complete or incomplete in the Actiniaria and Scleractinia, while in other groups they are not constant. When a skeleton is present, it is secreted by ectodermal cells, and free spicules are found. Actiniaria and Scleractinia embrace many species, whereas the other orders, notably the Ceriantheria, include only a few. *See* ANTHOZOA; ZOANTHIDEA. [K.At.]

Zoanthidea An order of the subclass Zoantharia. These animals are mostly colonial, sedentary, skeletonless, anemone-like anthozoans. They live in warm, shallow waters or on coral reefs. The body is in the form of a polyp. The pedal disk is indistinguishable. The column, bearing tubercles, is often encrusted with sand grains, sponge spicules, formaniferous shells, and other detritus. The tentacles are arranged in two cycles. The life cycle is incompletely known. Daughter polyps arise by budding from the stolon or the polyp base, or by longitudinal fission. *See* ZOANTHARIA. [K.At.]

Zodiac An imaginary belt or zone of the sky, 18° wide and centered on the ecliptic (the path of the Sun). Of the constellations in the sky, those along the zodiac are of special significance. Within this zone move the Sun, Moon, and planets. The stars in the zodiac were grouped into constellations most of which have the figures of animals; hence the name zodiac, or animal circle.

The zodiac is divided equally into 12 sections, called signs, in each of which the Sun is situated for 1 month of the year. Each sign, 30° in length, is named from a constellation with which the sign originally coincided. These 12 zodiacal constellations are—beginning from the vernal equinox, the first point of Aries, in order westward—Aries (Ram). Taurus (Bull), Gemini (Twins), Cancer (Crab), Leo (Lion), Virgo (Maiden), Libra (Balance), Scorpius (Scorpion), Sagittarius (Archer), Capricornus (Sea Goat), Aquarius (Water Bearer), and Pisces (Fishes).

When the zodiacal constellations first came into use over 2000 years ago, the vernal equinox was at Aries. Because of the precession of the equinoxes, each of the zodiacal signs has since moved westward 30° along the ecliptic. Thus all the signs are now displaced into the neighboring constellation. For instance, the sign Aries is now in the constellation Pisces. Unsuccessful attempts have been made to change the names of the constellations or to introduce new ones, such as to replace the old pagan nomenclature with Christian names by substituting the names of the 12 Apostles for the 12 zodiacal constellations. *See* CONSTELLATION. [C.-S.Y.]

Zodiacal light A diffuse band of luminosity occasionally visible on the ecliptic. It is sunlight diffracted and reflected by dust particles in the solar system within and beyond the orbit of Earth. Zodiacal light is best seen after evening twilight or before morning twilight when the ecliptic is at right angles to the horizon. In the Northern Hemisphere this is in the evening during springtime, and in the morning during autumn. The light steadily increases from about 170° to 30° away from the Sun. *See* ECLIPTIC.

Sunlight that is scattered within Earth's atmosphere normally prevents observation of zodiacal light at elongations (angular distance from the Sun) of less than 30°. During a solar eclipse, however, the scattered light is reduced and, at positions near the Sun, the zodiacal light may be intense enough to observe.

At elongations of 180°, zodiacal light shows a marked increase in brightness. This patch of light, ordinarily about 10° in diameter, is called the Gegenschein or counterglow. It is attributed to the increased efficiency of reflections at large phase angles by diffusely reflecting particles. [R.E.McC.]

Zone refining One of a number of techniques used in the preparation of high-purity materials. The technique is capable of producing very low impurity levels, namely, parts per million or less in a wide range of materials, including metals, alloys, intermetallic compounds, semiconductors, and inorganic and organic chemical compounds. In principle, zone refining takes advantage of the fact that the solubility level of an impurity is different in the liquid and solid phases of the material being purified; it is therefore possible to segregate or redistribute an impurity within the material of interest. In practice, a narrow molten zone is moved slowly along the complete length of the specimen in order to bring about the impurity segregation. *See* SOLUTION; SOLVENT.

Impurity atoms either raise or lower the melting point of the host material. There is also a difference in the concentration of the impurity in the liquid phase and in the solid phase when the liquid and solid exist together in equilibrium. In zone refining, advantage is taken of this difference, and the impurity atoms are gradually segregated to one end of the starting material. To do this, a molten zone is passed from one end of the impure material to the other, as in illustration *a*, and the process is repeated several times. The end to which the impurities are segregated depends on whether the impurity raises or lowers the melting point of the pure material; a lowering of the melting point is more common, in which case impurities are moved in the direction of travel of the molten zone. The effect of multiple zone passes (in the same direction) on impurity con-

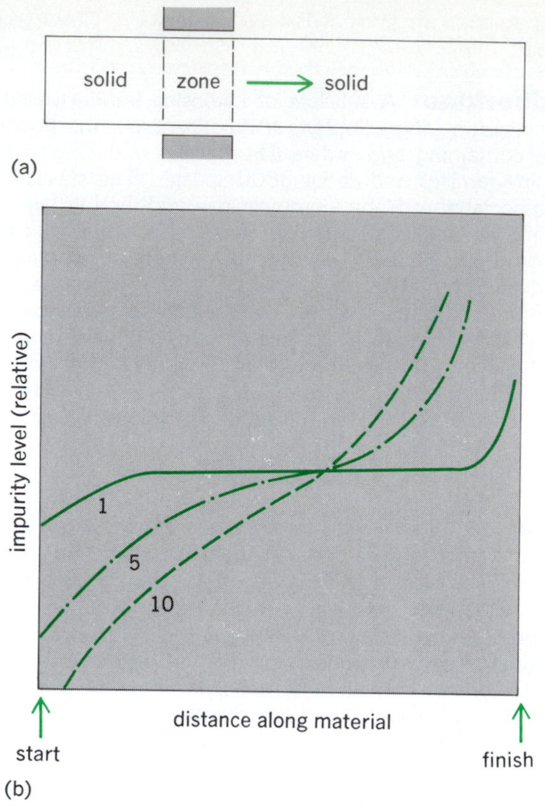

Zone refining. (*a*) Passage of a molten zone along the material to be purified. (*b*) Effect of 1, 5, and 10 zone passes on the impurity distribution along the material.

tent along the material is illustrated in illustration *b*. *See* METALLURGY. [A.La.]

Zoogeography The subdivision of the science of biogeography that is concerned with the detailed description of the distribution of animals and how their past distribution has produced present-day patterns. This field attempts to formulate theories that explain the present distributions as elucidated by geography, physiography, climate, ecological correlates (especially vegetation), geological history, the canons of evolutionary theory, and an understanding of the evolutionary relationships of the particular animals under study.

The field of zoogeography is based upon five observations and two conclusions. The observations are: (1) Each species and higher group of animals has a discrete nonrandom distribution in space and time (for example, the gorilla occurs only in two forest areas in Africa). (2) Different geographical regions have an assemblage of distinctive animals that coexist (for example, the fauna of Africa south of the Sahara with its monkeys, pigs, and antelopes is totally different from the fauna of Australia with its platypuses, kangaroos, and wombats). (3) These differences (and similarities) cannot be explained by the amount of distance between the regions or by the area of the region alone (for example, the faunas of Europe and eastern Asia are strikingly similar although separated by 7140 mi (11,500 km) of land, while the faunas of Borneo and New Guinea are extremely different although separated by a tenth of that distance across land and water. (4) Faunas strikingly different from those found today previously occurred in all geographical regions (for example, dinosaurs existed over much of the world in the Cretaceous). (5) Faunas resembling those found today or their antecedents previously occurred, sometimes at sites far distant from their current range (for example, the subtropical–warm temperate fauna of Eocene Wyoming, including many fresh-water fish, salamander, and turtle groups, is now restricted to the southeastern United States).

The conclusions are: (1) There are recognizible recurrent patterns of animal distribution that are not solely the product of areal extent, distance of separation, or climatic distinctiveness of a geographical region. (2) These patterns represent faunas composed of species and higher groups that have evolved through time in association with one another.

Historical zoogeography recognizes that each major geographical area has a different assemblage of species, that certain systematic groups of organisms tend to cluster geographically, and that the interaction of geography, climate, and evolutionary processes over a long time span is responsible for the patterns or general tracks. Emphasis in this approach is upon the statics and dynamics of major geographical and geological events ranging across vast areas and substantial time intervals of up to millions of years. The approach is based on concordant evolutionary association of diverse groups through time.

Ecological zoogeography is the study of animal distributions in terms of their environments; historical zoogeography is the study of animal distributions in terms of evolutionary history.

[J.M.S.]

Zoological nomenxclature The system of names used to designate the various kinds of organisms which make up the animal kingdom. First formalized by the 18th-century Swedish naturalist Linnaeus, the system has been expanded and modified into a universal procedure for the naming of animal species. *See* ANIMAL SYSTEMATICS.

In the Linnaean system of binominal nomenclature, the species has a dual designation, the binomen, the first element of the name (the generic name) representing the genus (for example, *Canis*, the various dogs), the second (the trivial name), a particular species in that genus (as *lupus*, the wolf). These two elements make up the species name, which must be unique; that is, the particular combination of generic and specific name cannot be used to designate any other species of animal. It is followed by the name of the author. Thus, the correct scientific name for the wolf is *Canis lupus* Linnaeus. When, through inadvertence or through reclassification, two or more such combinations come into existence they become homonyms and the more recent are rejected. Likewise, each species can have only one valid name, the first properly proposed. Subsequent names become synonyms and are rejected (principle of priority). Since Greek and Latin were the common languages of the educated of the 18th century, it was natural to make use of these languages in forming scientific names, the generic names being selected largely from latinized Greek or from combinations of Greek words, the specific names mainly from Latin. To be available for use, a newly proposed name must be accompanied by description of the taxon to which it applies. A name proposed without such a description is a nomen nudum; one with a description which cannot be interpreted satisfactorily is a nomen dubium.

Names of genera and other higher taxa (sing., taxon) in classification are single words (uninominal), and must be unique, not used for other groups of animals. The generic name is singular; those of higher categories are always plural. Names for taxa above the genus and below the order are based upon the root of the name of one of the included genera (the type genus), and their ending indicates the category, as "-inae," subfamily; "-idae," family. Above the family group of names, designations for higher categories, as orders, classes, and phyla, do not have uniform endings, although there is considerable uniformity in certain groups, such as the Insecta, in which the ending "-tera" usually denotes an order. *See* TAXON.

[W.J.B.]

Zoology The science that deals with knowledge of animal life. Together with botany, the science of plants, it forms biology, the science of living things. With the great growth of information about animals, zoology has been much subdivided. Some major fields are anatomy, which deals with gross and microscopic structure; physiology, with living processes in animals; embryology, with development of new individuals; genetics, with heredity and variation; parasitology, with animals living in or on others; natural history, with life and behavior in nature; ecology, with the relation of animals to their environments; evolution, with the origin and differentiation of animal life; and taxonomy, with the classification of animals. *See* ANATOMY, REGIONAL; BIOLOGY; BOTANY; ECOLOGY; EMBRYOLOGY; GENETICS; PARASITOLOGY; PHYLOGENY; PLANT EVOLUTION; TAXONOMY.

[T.I.S.]

Zoom lens A system of lenses in which two or more parts are moved with respect to each other to obtain a continuously variable focal length and hence magnification, while the image is kept in the same image plane.

The system must be constructed so that the errors do not vary too much in the shifting. Thus the designer must strive for a design in which the image errors are small, at least for the beginning and end positions of the "zooming" procedure, and do not become too large for intermediate positions.

Some early variable-focal-length lenses contained 15 or more separated elements, but this number has been reduced to as few as 4. The zoom ratio (the ratio of maximum to minimum power) is in general 3:1, but it has been possible in some designs to increase the range to 4:1 or even more. *See* LENS (OPTICS).

[M.J.H.]

Zoomastigophorea A class of protozoans of the subphylum Sarcomastigophora. Zoomastigophorea, also known as Zoomastigina, are flagellates. Some are simple, some are specialized; some have pseudopodia besides flagella. One group engulfs solid food at any body point; another shows localized ingestive areas. All are colorless. None produce starch or paramylum, and lipids and glycogen are assimilation products. Cells are naked or have delicate membranes. Colony formation is common. Colonies differ greatly in form and may be amorphous, linear, spherical, arboroid, or plane. Flagella vary from none to many.

Nine orders are included in this class: the Choanoflagellida, Bicosoecida, Rhizomastigida, Kinetoplastida, Retortomonadida, Diplomonadida, Oxymonadida, Trichomonadida, and Hypermastigida. See articles on these classes. *See* SARCOMASTIGOPHORA.

[J.B.L.]

Zoonoses Diseases which are biologically adapted to and normally found in lower animals but which under some conditions also infect humans. Diseases primarily human in origin that may temporarily reside in an animal and then reinfect humans would be excluded from this category. Aside from their animal origin, there is little relationship between most of these diseases, since the causative organisms, means of transmission, portals of entry, effects on the host, and outcome of the disease are as varied as those for other infections.

Some common zoonoses

Origin	Disease
Domestic animals	Brucellosis, bovine tuberculosis, Q fever, anthrax, varieties of salmonellosis, glanders, ornithosis-psittacosis
Wild animals	Tularemia, spotted fever, various virus encephalitides, jungle fever
Rodents	Rickettsialpox, bubonic plague, rat-bite fever, salmonellosis

A primary division may be made between zoonoses commonly found in domestic animals with which humans come in frequent contact and those found in wild animals from which transmission to humans is the result of either some chance encounter or some special set of circumstances. In addition to these two clear-cut divisions, there is another group of diseases which primarily involves rodents, animals that cannot be classified as either domestic or wild animals. Examples of disease due to animals of these three types are shown in the table. *See* ANTHRAX; BRUCELLOSIS; DERMATOPHYTOSIS; GLANDERS; PLAGUE; PSITTACOSIS; Q FEVER; RABIES; ROCKY MOUNTAIN SPOTTED FEVER; TICK FEVER; TUBERCULOSIS; YELLOW FEVER. [C.H.D.]

Zooplankton Animals that inhabit the water column of oceans and lakes, and lack the means to counteract transport currents. Zooplankton inhabit all layers of these water bodies to the deepest depths sampled, and constitute a major link between primary production and higher trophic levels in aquatic ecosystems.

Zooplankton can be divided into various operational categories. Size is a common basis of classification, but is primarily related to the collection of plankton, and thus is not always related to biologically meaningful criteria. A commonly accepted size classification scheme includes the groupings: picoplankton (<2 micrometers), nanoplankton (2–20 μm), microplankton (20–200 μm), mesoplankton (0.2–20 mm), macroplankton (20–200 mm), and megaplankton (>200 mm). A classification based on biological criteria divides these animals into meroplankton and holoplankton. Meroplanktonic forms spend part of their life cycles on the bottom, and include larvae of benthic worms, mollusks, crustaceans, echinoderms, coral, and even insects, as well as the eggs and larvae of many fishes. Holoplankton spend essentially their whole existence in the water column. Examples are chaetognaths, pteropods, larvaceans, siphonophores, and many copepods. Nearly every major taxonomic group of animals has either meroplanktonic or holoplanktonic members.

The classic description of the trophic dynamics of plankton is a food chain consisting of algae grazed by crustacean zooplankton which are in turn ingested by fishes. This abstraction may hold true in upwelling areas, but it masks the complexity of most natural food webs. Zooplankton frequently assume different feeding habits as they grow from larval to adult form. They may ingest bacteria or phytoplankton at one stage of their life cycle and become raptorial feeders later. Other zooplankton are primarily herbivorous but can opportunistically become carnivorous. Zooplankton apparently can track changes in the particle-size spectrum of phytoplankton and graze the most abundant size classes. Some phytoplankton are noxious and are avoided by grazers, while others are ingested but not digested.

The importance of understanding the trophic ecology of zooplankton becomes evident when the link to commercial fisheries is considered. Zooplankton are a major food of fishes and also a feasible direct source of protein to humans. Soviet and Japanese fishing crews currently harvest and package Antarctic euphausids (krill) in huge factory ships. Direct exploitation of other planktonic groups is possible if they can be detected in high concentrations. However, exploitation of zooplankton should await a thorough knowledge of their ecology and position in food webs. Repercussions of the uncontrolled harvesting of these animals must be expected—a lesson learned from the collapse of other fisheries in the past. *See* ECOLOGICAL INTERACTIONS; ECOSYSTEM; FRESH-WATER ECOSYSTEM; MARINE ECOLOGY; PHYTOPLANKTON. [R.W.Sa.]

Zoraptera An order of insects which is related to the termites and psocids. Zoraptera are about 0.08–0.1 in. (2–2.5 mm) long and not important economically. They are of great interest because of their rarity, relatively recent discovery in 1913, and intriguing habits and relationships.

Zoraptera live sheltered from light, in decaying wood, usually in logs and stumps, except during the emergence of winged adults. The winged adults are well pigmented, have eyes, and often shed their wings. In the southeastern United States they occur especially on slabs buried in old sawdust piles. They apparently scavenge, mainly on microscopic molds. Most individuals are of the wingless caste, pale in color, and blind. Metamorphosis is gradual.

Zoraptera are distributed nearly worldwide in warm countries. The best known of the two United States species, *Zorothpus hubbardi*, occurs from Pennsylvania to Florida, west to Iowa, Kansas, Missouri, and Texas. Approximately 23 species are known; all are in the genus *Zorotypus* of the family Zorotypidae. *See* INSECTA. [A.B.Gu.]

Zosterophyllopsida The best-defined group of early land vascular plants. It was circumglobal in distribution and extended from Early into Late Devonian time. Like Rhyniopsida, the plants were leafless and rootless, hence simple (see illustration). However, some genera bore spines or teeth or other small emergences. The xylem strand was a solid column but the first cells to mature (protoxylem) were peripheral (exarch maturation). Sporangia were borne along the sides of the stem either in a terminal cluster or more diffusely scattered.

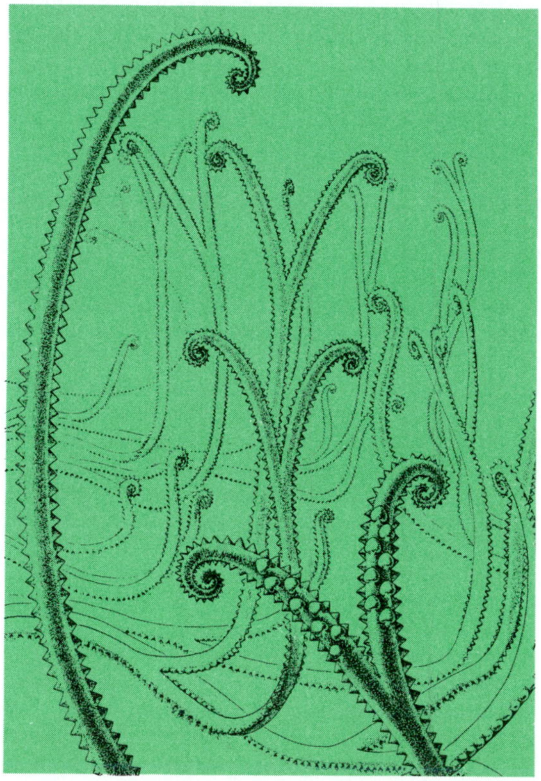

Reconstruction of *Serrulacaulis furcatus*. (After F. M. Hueber and H. P. Banks, *Serrulacaulis furcatus* gen. et sp. nov., a new zosterophyll from the lower Upper Devonian of New York State, *Rev. Palaeobot. Palynol.*, 28:169–189,1979)

Although this group was formerly classified with Rhyniopsida in the class Psilophytales, the characteristics of xylem maturation, sporangial shape and position, and sporangial dehiscence all indicate that it is a distinctly separate group that might have been a precursor to lycopods (club mosses). *See* PSILOPHYTALES; RHYNIOPSIDA. [H.P.B.]

Zygnematales A large order of green algae (Chlorophyceae) that is characterized by the lack of flagellate cells. Sexual reproduction is effected by the fusion of ameboid or passive gametes in conjugation tubes, which give rise to the alternate name Conjugales. Biochemical and ultrastructural features indicate that the Zygnematales are more closely related to charophytes than they are to most other green algae. Members of this order are essentially restricted to fresh-water and subaerial habitats. They are sometimes placed in their own class, with each family elevated to the rank of order. The order Zygnematales is frequently divided into two suborders, Zygnematineae and Desmidiineae. [P.C.Si.; R.L.Moe]

Zygomycetes A class of terrestrial Eumycota (true fungi) in the subdivision Zygomycotina, comprising organisms commonly known as the bread molds. Sexual reproduction is by gametangial copulation and results in the formation of zygospores. Asexual reproduction is by yeast cells, chlamydospores, arthrospores, or endospores (sporangiospores) formed in sporangia or uni- or multispored sporangiola or merosporangia. These fungi occur as haustorial (having food-absorbing cells in the host) or nonhaustorial parasites of fungi, plants, or animals (including humans), or as saprobes, especially in soil and dung; but other substrates containing soluble nutrients may contain Zygomycetes.

The mature spore-bearing structures are dry and readily dispersed by air currents, or wet, and are distributed by direct contact with small animals. Water droplets also may disperse the spores or the intact spore-bearing structures.

Classification is based on mode of nutrition, morphology of the zygospore (if formed), type of asexual reproduction, branching pattern of sporophores, frequency of septa (if formed), and septal morphology. Zygomycetes comprise seven orders: Dimargaritales, Endogonales, Entomophthorales, Glomales, Kickxellales, Mucorales, and Zoopagales. *See* EUMYCOTA; FUNGI; REPRODUCTION (PLANT). [G.L.Be.]

Zygomycotina Subdivision of fungi characterized by distinctive sexual and asexual reproductive stages. The sexual zygospore results from fusion of two gametangia. Asexual reproduction may be sporangial (containing one to many, always nonmotile, sporangiospores) to conidial. The various asexual forms found in different taxa are interpreted as representing evolutionary development from sporangial to conidial. Asexual reproduction is more common than sexual; in fact, sexual reproduction is unknown in many species and within certain orders. Zygomycotina are cosmopolitan. Many are free-living saprobes. Others, even entire orders, are saprophytic, parasitic, or obligate symbionts, especially in arthropods. There are two classes (Zygomycetes and Trichomycetes), but it is uncertain whether they are naturally related or appear similar through convergent evolution. [D.J.S.B.]

Appendix

Bibliographies

ACOUSTICS

Berg, R. E., and D. G. Stork, *The Physics of Sound*, 2d ed., 1994.
Crocker, M., *Handbook of Acoustics*, vols. 1 and 2, 1997.
Frisk, G. V., *Ocean and Seabed Acoustics*, 1994.
Kinsler, L. E., et al., *Fundamentals of Acoustics*, 3d ed., 1982.
Mason, W. P., R. N. Thurston, and A. D. Pierce (eds.), *Physical Acoustics: Principle and Methods*, vols. 1–19, 1964–1990.
Pierce, A. D., *Acoustics: An Introduction to Its Physical Principles and Applications*, 1981, reprint 1989.
Rossing, T. D., *The Science of Sound*, 2d ed., 1990.
Speaks, C. E., *Introduction to Sound Acoustics for the Hearing and Speech Sciences*, 2d ed., 1996.
Urick, R. J., *Principles of Underwater Sound*, 3d ed., 1983, reprint 1996.

Journal of the Acoustical Society of America, American Institute of Physics, monthly.

AERODYNAMICS

Anderson, J. D., Jr., *Fundamentals of Aerodynamics*, 2d ed., 1991.
Bertin, J. J., and M. L. Smith, *Aerodynamics for Engineers*, 3d ed., 1997.
Houghton, E., *Aerodynamics for Engineering Students*, 4th ed., 1993.
Kuethe, A. M., and C. Y. Chow, *Foundations of Aerodynamics: Bases of Aerodynamic Design*, 4th ed., 1986.
McCormick, B. W., *Aerodynamics, Aeronautics and Flight Mechanics*, 2d ed., 1994.
Wegener, P., *What Makes Airplanes Fly?: History, Science and Applications of Aerodynamics*, 2d ed., 1996.

AERONAUTICAL ENGINEERING

Barnard, R. H., *Aircraft Flight*, 2d ed., 1995.
Etkin, B., *Dynamics of Flight Stability and Control*, 3d ed., 1995.
Hancock, G. J., *An Introduction to the Flight Dynamics of Rigid Aeroplanes*, 1995.
Kermode, A. C., *Mechanics of Flight*, 1996.
Kroes, M. J., J. R. Rardon, and R. D. Bent, *Aircraft Basic Science*, 7th ed., 1993.
Peery, D., and J. J. Azar, *Aircraft Structures*, 2d ed., 1982.
Shevell, R. S., *The Fundamentals of Flight*, 2d ed., 1988.
Wagtendonk, W. J., *Principles of Helicopter Flight*, 1996.

AIAA Journal, American Institute of Aeronautics and Astronautics, monthly.
Aviation Week and Space Technology, weekly.

AGRICULTURE

Adams, D., and K. Banford, *Principles of Horticulture*, 1984.
Barrick, R. K., and H. Harmon, *Animal Science*, 1984.
Blakely, J., and D. H. Bade, *The Science of Animal Husbandry*, 5th ed., 1989.
Brady, N. C. (ed.), *Advances in Agronomy*, vols. 28–45, 1976–1991.
Ensminger, M. E. (ed.), *Animal Science*, 9th ed., 1991.
Heath, E., et al., *Forages: The Science of Grassland Agriculture*, 4th ed., 1985.
Lockhart, J. A., and A. J. Wiseman, *Introduction to Crop Husbandry: Including Grassland*, 6th ed., 1988.
Martin, J. H., et al., *Principles of Field Crop Production*, 4th ed., 1986.
Miller, D. A., *Forage Crops*, 1984.

Rechcigl, M., Jr. (ed.), *Handbook of Agricultural Productivity*, vol. 1: *Plant Productivity*, vol. 2: *Animal Productivity*, 1982.
Sparks, D. L. (ed.), *Advances in Agronomy*, vols. 47–49, 1992–1993.

Agronomy Journal, American Society of Agronomy, bimonthly.
Journal of Agricultural Science (England), bimonthly.
Journal of Animal Science, American Society of Animal Science, monthly.

ALGEBRA

Auvil, D., and C. Poluga, *Algebra for College Students*, 1996.
Barnett, R. A., and T. J. Kearns, *Intermediate Algebra: Structure and Use*, 5th ed., 1994.
Birkhoff, G., and S. MacLane, *A Survey of Modern Algebra*, 5th ed., 1996.
Christy, D. T., *College Algebra*, 2d ed., 1993.
Durbin, J. R., *Modern Algebra: An Introduction*, 3d ed., 1991.
Fraleigh, J. B., *A First Course in Abstract Algebra*, 4th ed., 1991.
Hungerford, T. W., *Algebra*, 1992.
Jacobson, N., *Basic Algebra*, vol. 1, 2d ed., 1985, vol. 2, 2d ed., 1989.
Kaufmann, J. E., *Algebra for College Students*, 5th ed., 1996.
Keedy, M. L., and M. L. Bittinger, *Introductory Algebra*, 7th ed., 1995.
Lang, S. A., *Algebra*, 3d ed., 1992.
Rees, P. K., et al., *College Algebra*, 10th ed., 1990.
Siever, N., *Intermediate Algebra*, 1990.
Streeter, J., D. Hutchinson, and L. Hoelzle, *Beginning Algebra*, 4th ed., 1998.

Journal of Algebra, semimonthly.
Journal of Pure and Applied Algebra, approximately 18 issues per year.

ANALYTICAL CHEMISTRY

Christian, G. C., *Analytical Chemistry*, 5th ed., 1993.
Dean, J. A., *Analytical Chemistry Handbook*, 1995.
Kennedy, J. H., *Analytical Chemistry: Principles*, 2d ed., 1990.
Settle, F. A., *Handbook of Instrumental Techniques for Analytical Chemistry*, 1997.
Skoog, D. A., D. M. West, and F. J. Holler, *Fundamentals of Analytical Chemistry*, 6th ed., 1992.
J. D. Winefordner et al., *Treatise on Analytical Chemistry*, 2d ed., vol. 13, p. 1, 1993.

American Laboratory, monthly.
Analytical Chemistry, American Chemical Society, semimonthly.
Trends in Analytical Chemistry, Elsevier, monthly.

ANATOMY

Chee, A. N., *Anatomy and Physiology: A Dynamic Approach*, 4th ed., 1991.
Fong, E., *Body Structures and Functions*, 9th ed., 1998.
Guyton, A., *Physiology of the Human Body*, 6th ed., 1984.
Nicpon-Marieb, E., *Human Anatomy and Physiology*, 3d ed., 1996.
Schlossberg, L., and G. D. Zuidema, *The Johns Hopkins Atlas of Human Functional Anatomy*, 3d ed., 1986.
Tortora, G., *Principles of Anatomy and Physiology*, 8th ed., 1997.
Williams, P. L., et al., *Gray's Anatomy*, 37th rev. ed., 1989.

Anatomical Record, Wiston Institute Press, monthly.
Journal of Anatomy (England), bimonthly.

ANTHROPOLOGY

Campbell, B., *Humankind Emerging*, 6th ed., 1991.
Ember, C. R., and M. Ember, *Anthropology*, 8th ed., 1995.
Fedigan, L. M., *Primate Paradigms, Sex Roles and Social Bonds*, 1992.
Gowlett, J., *Ascent to Civilization: The Archaeology of Early Humans*, 2d ed., 1993.
Leakey, R. E., *Origins Reconsidered: In Search of What Makes Us Human*, 1992.
Whitten, P., and D. E. Hunter, *Anthropology: Contemporary Perspectives*, 7th ed., 1993.

ARCHITECTURAL ENGINEERING

Allen, E., *How Buildings Work: The Natural Order of Architecture*, 2d ed., 1995.
Ambrose, J. E., *Building Structures*, 2d ed., 1993.
Guise, D., *Design and Technology in Architecture*, rev. ed., 1991.
Guthrie, P., *The Architect's Portable Handbook*, 1995.
Harris, C. M. (ed.), *Dictionary of Architecture and Construction*, 2d ed., 1993.
McGraw-Hill, *Architectural Engineer's Solutions Suite* (CD-ROM), 1996.
Roth, L. M., *Understanding Architecture: Its Elements, History, and Meaning*, 1993.
Zampi, G., and C. L. Morgan, *Virtual Architecture*, 1996.

Architectural Record, McGraw-Hill, monthly.

ASTRONOMY

Abell, G. O., D. Morrison, and S. C. Wolff, *Abell's Exploration of the Universe*, 7th ed., 1995.
Audouze, J., and G. Israel (eds.), *The Cambridge Atlas of Astronomy*, 3d ed., 1994.

Beatty, J. K., and A. Chaikin (eds.), *The New Solar System*, 3d ed., 1990.
Foukal, P., *Solar Astrophysics*, 1990.
Hartmann, W. K., *Astronomy: The Cosmic Journey*, 5th ed., 1994.
Henbest, W., and M. Martin, *The New Astronomy*, 2d ed., 1996.
Kaler, J. B., *Astronomy*, 1996.
Kippenhahn, R., *100 Billion Suns: The Birth, Life and Death of the Stars*, 1993.
Pasachoff, J. M., *Astronomy: Earth to Universe*, 5th ed., 1997.
Symposia of the International Astronomical Union, 1963– .
Zeilik, M., *Astronomy: The Evolving Universe*, 8th ed., 1996.

Annual Review of Astronomy and Astrophysics, annually.
Astronomical Journal, American Astronomical Society, American Institute of Physics, monthly.
Astrophysical Journal, American Astronomical Society, three issues per month.
Monthly Notices of the Royal Astronomical Society, three issues per month.
Sky and Telescope, monthly.

ATOMIC PHYSICS

Beiser, A., *Concepts of Modern Physics*, 5th ed., 1995.
Brehm, J. J., and W. J. Mullins, *Introduction to the Structure of Matter: A Course in Modern Physics*, 1989.
Cowan, R. D., *The Theory of Atomic Structure and Spectra*, 1981.
Eisberg, R., and R. Resnick, *Quantum Physics: Of Atoms, Molecules, Solids, Nuclei and Particles*, 2d ed., 1985.
Haken, H., and H. C. Wolf, *The Physics of Atoms and Quanta*, 4th ed., 1994.
Ohanian, H. C., *Modern Physics*, 2d ed., 1995.
Serway, R. A., and J. Faughn, *Modern Physics*, 2d ed., 1996.
Wehr, M. R., J. A. Richards, and T. W. Adair, *Physics of the Atom*, 4th ed., 1984.
Yang, F., and J. H. Hamilton, *Modern Atomic and Nuclear Physics*, 1996.

Physical Review A, American Physical Society, monthly.

AUTOMOTIVE ENGINEERING

Crouse, W. H., and D. L. Anglin, *Automotive Mechanics*, 9th ed., 1985.
Duffy, J. E., *Modern Automotive Technology*, 1993.
Fichter, H., *Brakes, Steering and Suspension*, 1993.

Automotive Engineering, Society of Automotive Engineers, monthly.

BIOCHEMISTRY

Armstrong, F. B., *Biochemistry*, 3d ed., 1989.
Blackstock, J. C., *Guide to Biochemistry*, 1995.
Briggs, T., *Biochemistry*, 3d ed., 1995.
Campbell, M. K., *Biochemistry*, 1991.
Lehninger, A. L., *Principles of Biochemistry*, 2d ed., 1993.
Smith, E. L., et al., *Principles of Biochemistry: General Aspects*, 7th ed., 1983; *Principles of Biochemistry: Mammalian Biochemistry*, 7th ed., 1983.
Stryer, L., *Biochemistry*, 3d ed., 1995.
Van Holde, K. E., *Physical Biochemistry*, 1984.
Voet, D., and J. Voet, *Biochemistry*, 2d ed., 1995.

Annual Review of Biochemistry, annually.
Biochemical Journal, biweekly.
Biochemistry, Centcom Ltd., biweekly.
Journal of Biological Chemistry, American Society of Biological Chemists, semimonthly.
Trends in Biochemical Sciences, monthly.

BIOMEDICAL ENGINEERING

Bronzino, J. D., *The Biomedical Engineering Handbook*, 1995.
Carson, E. R. (ed.), *Advances in Biomedical Measurement*, 1988.
Mikulecky, D. C., and A. M. Clarke (eds.), *Biomedical Engineering: Opening Doors*, 1990.
Profio, A. E., *Biomedical Engineering*, 1993.
Yang, W.-J., and C.-J. Lee (eds.), *Biomedical Engineering: An International Symposium*, 1989.

Advances in Biological and Medical Physics, irregularly.
Annals of Biomedical Engineering, bimonthly.
Physics in Medicine and Biology, American Institute of Physics, bimonthly.

BIOPHYSICS

Austin, R. H., *Biophysics for Physicists*, 1996.
Cerdonio, M., and R. W. Noble, *Introductory Biophysics*, 1986.
Kreighbaum, E., and K. Barthels, *Biomechanics: A Qualitative Approach for Studying Human Movements*, 4th ed., 1995.
Sybesma, C. (ed.), *Biophysics: An Introduction*, 1989.
Valenta, J. (ed.), *Biomechanics*, 1993.
Wainwright, S. A., et al., *Mechanical Design in Organisms*, 1982.

Annual Review of Biophysics and Bioengineering, annually.
Biophysical Journal, monthly.

CALCULUS

Berkey, D. D., and P. Blanchard, *Calculus*, 3d ed., 1992.

Courant, R., and F. John, *Introduction to Calculus and Analysis*, 2 vols., 1993.
Farlow, S. J., and G. Haggard, *Calculus and Its Applications*, 1990.
Grossman, S., and R. B. Lane, *Calculus*, 1993.
Kaplan, W., *Advanced Calculus*, 4th ed., 1991.
Lang, S. A., *A First Course in Calculus*, 5th ed., 1986, reprint 1993.
Leithold, L., *The Calculus*, 7th ed., 1996.
Salas, S. L. and E. Hille, *Salas and Hille's Calculus*, 7th ed., 2 vols., 1995.
Stein, S. K., *Calculus with Analytic Geometry*, 5th ed., 1992.
Taylor, A. E., and R. W. Mann, *Advanced Calculus*, 3d ed., 1982.
Thomas, G. B., *Calculus and Analytic Geometry*, 9th ed., 1996.

CELESTIAL MECHANICS

See Astronomy.

CELL BIOLOGY

Darnell, J. E., *Molecular Cell Biology*, 2d ed., 1995.
Nair, P. K., and K. P. Achar, *A Textbook of Cell Biology, 1990.*
Smith, C. A., and E. J. Wood (eds.), *Cell Biology*, 1992.
Widnell, C., *Essentials of Cell Biology*, 1990.

Cell, monthly.
Cellular and Molecular Biology (including cytoenzymology), bimonthly.
Cytologia/Kitologia, International Association of Cytology, quarterly.
Journal of Cell Biology, monthly.

CHEMICAL ENGINEERING

Epenson, J. H., *Chemical Kinetics and Reaction Mechanisms*, 2d ed., 1995.
Furter, W. F. (ed.), *A Century of Chemical Engineering*, 1980.
Hagerty, D. J., et al., *Chemical Engineering*, 1989.
Himmelblau, D. M., *Basic Principles and Calculations in Chemical Engineering*, 4th ed., 1982.
Humphrey, J. L., *Separation Process Technology*, 1997.
Kirk-Othmer Encyclopedia of Chemical Technology, 3d ed., vols. 1–25, 1991–1997.
McGraw-Hill, *Chemical Engineer's Solutions Suite* (CD-ROM), 1996.
Miller, R. W., *Flow Measurement Engineering Handbook*, 3d ed., 1996.
Perry, R. H., *Perry's Chemical Engineers' Handbook*, 7th ed., 1997.
Schweitzer, P. A., *Handbook of Separation Techniques for Chemical Engineers*, 3d ed., 1997.

Chemical and Engineering News, American Chemical Society, weekly.
Chemical Engineering, monthly.
Chemical Engineering Progress, American Institute of Chemical Engineers, monthly.
Chemtech, American Chemical society, monthly.

CHEMISTRY

See Analytical chemistry; Electrochemistry; General chemistry; Inorganic chemistry; Organic chemistry; Photochemistry; Physical chemistry; Polymer chemistry; Quantum chemistry; Solid-state chemistry; Stereochemistry.

CHROMATOGRAPHY

Fowlis, I., *Gas Chromatography*, 2d. ed., 1995.
Jork, H., *Thin Layer Chromatography*, vol. 1A: *Reagents and Detection Methods*, 1994.
Kalasz, H., and L. Ettre, *Chromatography: The State of the Art*, 1985.
Kitson, F. G., *Gas Chromatography and Mass Spectrometry: A Practical Guide*, 1996.
Lindsay, S., *High Performance Liquid Chromatography*, 2d ed., 1992.
Touchstone, J. C., *Practice of Thin Layer Chromatography*, 3d ed., 1992.

Chromatographia, monthly.
Journal of Chromatography A, Elsevier, biweekly.
Journal of Chromatography B, Elsevier, biweekly.
LC•GC, The Magazine of Separation Science, monthly.

CIVIL ENGINEERING

Ambrose, J., *Simplified Design of Concrete Structures*, 7th ed., 1997.
Ashford, N., H. M. Stanton, and C. A. Moore, *Airport Operations*, 2d ed., 1997.
Derucher, K. N., G. P. Korgiatis, and A. S. Ezeldin, *Materials for Civil and Highway Engineers*, 3d ed., 1993.
Jackson, N., and R. K. Ohir, *Structural Engineering Materials*, 4th ed., 1989.
Mays, L. W., *Water Resources Handbook*, 1996.
McGraw-Hill, *Civil Engineer's Solutions Suite* (CD-ROM), 1996.
Merritt, F. S., M. K. Loftin, and J. T. Ricketts (eds.), *Standard Handbook for Civil Engineers*, 4th ed., 1996.
Scott, J. S., *Dictionary of Civil Engineering*, 4th ed., 1992.

Civil Engineering, American Society of Civil Engineers, monthly.

CLASSICAL MECHANICS

Arya, A. P., *Introduction to Classical Mechanics*, 1990.
Baierlein, R., *Newtonian Dynamics*, 1983.

Barger, V. D., and M. G. Olsson, *Classical Mechanics: A Modern Perspective*, 2d ed., 1995.
Chow, T. L., *Classical Mechanics*, 1995.
Halliday, D., R. Resnick, and K. Krane, *Physics*, 2 vols., 4th ed., 1992.
Reichert, I. F., *A Modern Introduction to Mechanics*, 1990.
Shepley, L., and R. Matzner, *Classical Mechanics*, 1991.
Symon, K. R., *Mechanics*, 3d ed., 1971.

Archive for Rational Mechanics and Analysis, 16 times a year.
Quarterly Journal of Mechanics and Applied Mathematics, quarterly.

CLIMATOLOGY AND METEOROLOGY

Ahrens, C. D., *Meteorology Today*, 5th ed., 1994.
A. L. Berger, S. H. Schneider, and J. C. Duplessy (eds.), *Climate and Geo-Sciences*, 1989.
Cole, F. W., *Introduction to Meteorology*, 3d ed., 1980.
Danielson, E. W., and J. Levin, *Meteorology*, 1997.
Glantz, M. H. (ed.), *The Role of Regional Organizations in Context of Climate Change*, 1993.
Haltiner, G. J., and R. T. Williams, *Numerical Prediction and Dynamic Meteorology*, 2d ed., 1980.
Hidore, J. J., and J. E. Oliver, *Climatology: An Atmospheric Science*, 1993.
Hornak, K. A., *Dictionary of Meteorology and Climatology*, 1996.
Lamb, H. H., *Climatic History and the Future*, 1985.
Lutgens, F., and E. Tarbuck, *The Atmosphere: An Introduction to Meteorology*, 6th ed., 1994.
Rudman, J., *Climatology-Meteorology*, 1994.
Scorer, R. S., *Dynamics of Meteorology and Climate*, 1997.

Journal of Climate, American Meteorological Society, monthly.
Journal of Meteorology, American Meteorological Society, monthly.
Bulletin of the American Meteorological Society, American Meteorological Society, monthly.

COMMUNICATIONS

See Data communications; Radio communications; Telecommunications.

COMPARATIVE PHYSIOLOGY

Dantzler, W. H., *Comparative Physiology*, 1997.
Gorbman, A., et al., *Comparative Endocrinology*, 1983.
Hoar, W. S., *General and Comparative Physiology*, 3d ed., 1983.
Previte, J., *Human Physiology*, 1983.
Prosser, C. L., *Comparative Animal Physiology*, 4th ed., 1991.

American Journal of Physiology, American Physiological Society, monthly.
Journal of Comparative Physiology, A: Sensory, Neural, and Behavioral Physiology, B: Systematic and Environmental Physiology, irregularly.

COMPUTER SCIENCE

Abelson, H., G. Sussman, and J. Sussman, *Structure and Interpretation of Computer Programs*, 2d ed., 1997.
Bartee, T. C., *Computer Architecture and Logic Design*, 1991.
Hamacher, V. C., Z. G. Vranesic, and S. G. Zaky, *Computer Organization*, 4th ed., 1996.
Hayes, J. P., *Computer Architecture and Organization*, 2d ed., 1988.
Hennessy, J. L., and D. A. Patterson, *Computer Organization and Design*, 1994.
Hwang, K., *Advanced Computer Architecture*, 1993.
Long, L., and N. Long, *Computers*, 4th ed., 1996.
Ralston, A., and E. D. Reilly (eds.), *Encyclopedia of Computer Science and Engineering*, 3d rev. ed., 1992.
Tabak, D., *Advanced Microprocessors*, 2d ed., 1995.
Tremblay, J. P., and R. B. Bunt, *Introduction to Computer Science: An Algorithmic Approach*, 2d ed., 1989.

Byte, monthly.
Communications of the Association for Computing Machinery, monthly.
***Computer and Control Abstracts* (*Science Abstracts, Section C*)**, INSPEC, Institution of Electrical Engineers, monthly.
IEEE Spectrum, Institute of Electrical and Electronics Engineers, Inc., monthly.
Journal of the Association for Computing Machinery, bimonthly.

CONSERVATION

Dasmann, R. F., *Environmental Conservation*, 5th ed., 1984.
Foster, A. B., and D. Bosworth, *Approved Practices in Soil Conservation*, 5th ed., 1982.
Kircher, H. B., *Our Natural Resources and Their Conservation*, 7th rev. ed., 1992.
Klee, G., *Conservation of Natural Resources*, 1990.
Owen, O., and D. D. Chiras, *Natural Resource Conservation: An Ecological Approach*, 5th ed., 1990.
Warren, A., B. Goldsmith, and F. B. Goldsmith (eds.), *Conservation in Perspective*, 1984.

American Institute for Conservation Journal, semiannually.
Audubon Magazine, Audubon Society, bimonthly.
The Conservationist, New York Department of Environmental Conservation, bimonthly.

CONTROL SYSTEMS

Clark, R. N., *Control System Dynamics*, 1996.
Close, C. M., and D. K. Frederick, *Modeling and Analysis of Dynamic Systems*, 2d ed., 1994.
D'Azzo, J. J., and C. H. Houpis, *Linear Control System Analysis and Design*, 1995.
Dorf, R. C., and R. H. Bishop, *Modern Control Systems*, 7th ed., 1995.
Grantham, W. J., and T. L. Vincent, *Modern Control Systems Analysis and Design*, 1993.
Kuo, B. C., *Automatic Control Systems*, 7th ed., 1995.
Leondes, C. T. (ed.), *Advances in Control Systems, vols. 1–8, 1964–1971*; *Control and Dynamic Systems: Advances in Theory and Applications*, vols. 9–79, 1973–1996.
Lewis, F. L., and V. L. Symos, *Optimal Control*, 2d ed., 1995.
Phillips, C. L., *Feedback Control Systems*, 3d ed., 1995.
Raven, F., *Automatic Control Engineering*, 5th ed., 1994.

Computer and Control Abstracts (Science Abstracts, Section C), INSPEC, Institution of Electrical Engineers, monthly.
Control Engineering, monthly.

COSMOLOGY

Barrow, J. D., and J. Silk, *The Left Hand of Creation*, 1994.
Kolb, E., and M. Turner, *The Early Universe*, 1990.
Layzer, D., *Cosmogenesis: The Growth of Order in the Universe*, 1991.
Longair, M. S., *Our Evolving Universe*, 1996.
Peebles, P. J., *Principles of Physical Cosmology*, 1993.
Rowan-Robinson, M., *Cosmology*, 3d ed., 1996.
Weinberg, S., *The First Three Minutes*, 2d ed., 1993.

CRYSTALLOGRAPHY

Borchardt-Ott, W., *Crystallography: An Introduction*, 2d ed., 1996.
Engel, P., *An Axiomatic Introduction to Crystallography*, 1986.
Giacovazzo, C. (ed.), *Fundamentals of Crystallography*, 1992.
Glusker, J. P., and K. N. Trueblood, *Crystal Structure Analysis: A Primer*, 2d ed., 1985.
Hammond, C., *The Basics of Crystallography and Diffraction*, 1997.
O'Keefe, M., and B. G. Hyde, *Symmetry and Structures of Crystals*, 1996.
Vainshtein, B. K., *Fundamentals of Crystals*, 2d ed., 1995.

Acta Crystallographica, Sections A, B, and D, bimonthly; ***Section C***, monthly.

CYTOLOGY

See Cell biology.

DATA COMMUNICATIONS

Bertsekas, D. P., and R. Gallager, *Data Networks*, 2d ed., 1991.
Black, U., *Data Communications and Distributed Networks*, 3d ed., 1992.
de Prycker, M., *Asynchronous Transfer Mode: Solution for Broadband ISDN*, 3d ed., 1995.
Ginsburg, D., *Data Communications and Networks*, 1996.
Halsall, F., *Data Communications, Computer Networks, and Open Systems*, 4th ed., 1996.
Held, G., and R. Sarch, *Data Communications: A Comprehensive Approach*, 3d ed., 1995.
Stallings, W., *Data and Computer Communications*, 5th ed., 1996.
Tanenbaum, A. S., *Computer Networks*, 3d ed., 1996.

Data Communications, 18 issues per year.

DATA PROCESSING

Fuori, W. M., and L. Gioia, *Computers and Information Processing*, 4th ed., 1994.
Fuller, F., and W. Manning, *Computers and Information Processing*, 1994.
Harris, M. L., *Introduction to Data Processing*, 3d ed., 1991.
Long, L., *Introduction of Computers and Information Systems*, 4th ed., 1996.
Murdock, E. E., *Computers Today!*, 1995.
O'Brien, J. A., *Introduction to Information Systems*, 8th ed., 1997.
Stern, N. B., and R. A. Stern, *Computing in the Information Age*, 1993.

Datamation, monthly.
IBM Systems Journal, International Business Machines Corp., quarterly.

DENTISTRY

Cawson, R. A., *Essentials of Dental Surgery and Pathology*, 5th ed., 1992.
Cawson, R. A., and R. G. Spector, *Clinical Pharmacology in Dentistry*, 5th ed., 1989.
Liebgott, B., *The Anatomical Basis of Dentistry*, 1986.
Marsh, P., and M. Martin, *Oral Microbiology*, 3d ed., 1992.
Ring, M. E., *Dentistry: An Illustrated History*, 1992.
Seymour, R. A., *Textbook of Dental Pharmacology and Therapeutics*, 1989.

Dental Abstracts, American Dental Association, monthly.
Dental Hygiene, American Dental Hygienists Association, monthly.
Journal of Academy of General Dentistry, bimonthly.
Journal of Dental Research, International Association for Dental Research, 16 issues per year.

DESIGN ENGINEERING

Amirouche, F. M., *Computer-Aided Design and Manufacturing*, 1992.

Dieter, G., *Engineering Design: A Materials and Processing Approach*, 2d ed., 1991.
Earle, J. H., *Engineering Design Graphics*, 9th ed., 1997.
Faupel, J. H., and F. E. Fisher, *Engineering Design: A Synthesis of Stress Analysis and Materials Engineering*, 2d ed., 1981.
Juvinall, R. C., *Fundamentals of Machine Component Design*, 2d ed., 1991.
Pahl, G., and W. Beitz, *Engineering Design*, 2d ed., 1995.
Suh, N. P., *Principles of Design*, 1990.

Journal of Mechanical Design, American Society of Mechanical Engineers, quarterly.

DEVELOPMENTAL BIOLOGY

Beysens, D., G. Forgacs, and F. Gaill, *Interplay of Genetic and Physical Processes in the Development of Biological Form*, 1995.
Browder, L. W., C. A. Erikson, and W. R. Jeffrey, *Developmental Biology*, 3d ed., 1991.
Carlson, B. M., *Human Embryology and Developmental Biology*, 1993.
Gilbert, S. F., *Developmental Biology*, 3d rev. ed., 1991.
Crandell, *Human Development*, 5th ed., 1993.
Hall, B. K., *Evolutionary Developmental Biology*, 1992.
Oppenheimer, S. B., *Introduction to Embryonic Development*, 3d rev. ed., 1988.
Russo, V. E., et al., *Development: The Molecular Genetic Approach*, 1993.

Current Opinion in Genetics and Development, bimonthly.
Developmental Biology, monthly.
Excerpta Medica, Section 21: Developmental Biology and Teratology, 10 issues per year.
Genes and Development, Cold Spring Harbor Laboratory Press, bimonthly.
Roux' Archives of Developmental Biology, semiannually.

ECOLOGY

Bolen, E. G., and W. L. Robinson, *Wildlife Ecology and Management*, 3d ed., 1995.
Carroll, C. R., et al., *Agroecology*, 1990.
Cox, G. W., *General Ecology*, 7th ed., 1995.
Pianka, E. R., *Evolutionary Ecology*, 5th ed., 1993.
Ramade, F., *Ecology of Natural Resources*, 1984.
Ricklefs, R. E., *Ecology*, 3d ed., 1996.
Roughgarden, J., et al. (eds.), *Perspectives in Ecological Theory*, 1989.
Smith, R. L., *Ecology and Field Biology*, 5th ed., 1995.
Westman, W. E., *Ecology, Impact, Assessment and Environmental Planning*, 1985.

Annual Review of Ecology and Systematics, annually.
Ecological Monographs, Ecological Society of America, quarterly.
Ecology, Ecological Society of America, 6 issues per year.
Journal of Environmental Sciences, Institute of Environmental Sciences, bimonthly.

ELECTRICAL POWER ENGINEERING

Bobrow, L. S., *Fundamentals of Electrical Engineering*, 2d ed., 1996.
Chapman, S. J., *Electric Machinery Fundamentals*, 2d ed., 1991.
Fink, D. G., and H. W. Beaty, *Standard Handbook for Electrical Engineers*, 13th ed., 1993.
Fitzgerald, A. E., C. Kingsley, and S. Umans, *Electric Machinery*, 5th ed., 1990.
Grainger, J. J., and W. D. Stevenson, *Power System Analysis*, 1994.
Guru, B. S., and H. R. Hiziroglu, *Electric Machinery and Transformers*, 2d ed., 1994.
Hambley, A. R., *Electrical Engineering: Principles and Applications*, 1996.
McPherson, G., and R. D. Laramore, *An Introduction to Electrical Machines and Transformers*, 2d ed., 1990.
S. A. Nasar, *Electric Machines and Power Systems*, 1995.
Paul, C. R., and S. A. Nasar, *Introduction to Electrical Engineering*, 2d ed., 1992.
Richardson, D. V., and A. J. Caisse, Jr., *Rotating Electric Machinery and Transformer Technology*, 4th ed., 1996.
Ryff, P. F., *Electric Machinery*, 2d ed., 1994.

Electrical and Electronics Abstracts (Science Abstracts, Section B), INSPEC, Institution of Electrical Engineers, monthly.
C&M (Electrical Construction and Maintenance), monthly.
Electrical World, monthly.
IEEE Spectrum, Institute of Electrical and Electronics Engineers, Inc., monthly.

ELECTRICITY AND MAGNETISM

Cottingham, W. N., and D. A. Greenwood, *Electricity and Magnetism*, 1991.
Duffin, W. J., *Electricity and Magnetism*, 4th ed., 1990.
Fowler, E. J., *Electricity: Principles and Applications*, 4th ed., 1993.
Jackson, J. D., *Classical Electrodynamics*, 2d ed., 1975.
Morley, A., and E. Hughes, *Principles of Electricity*, 5th ed., 1994.
Purcell, E. M., *Electricity and Magnetism*, Berkeley Physics Course, vol. 2, 2d ed., 1985.

Journal of Magnetism and Magnetic Materials, 3 issues per month.

ELECTROCHEMISTRY

Bard, A. J., and I. Rubenstein (eds.), *Electroanalytical Chemistry: A Series of Advances*, vol. 19, 1996.
Bockris, J. O., B. E. Conway, and R. E. White (eds.), *Modern Aspects of Electrochemistry*, vol. 29, 1996.
Crow, D. R., *Principles and Applications of Electrochemistry*, 4th ed., 1994.
Koryta, J., *Ions, Electrodes, and Membranes*, 2d ed., 1993.

Newman, J. S., *Electrochemical Systems*, 2d ed., 1991.
Sawyer, D. T., A. Sobkowiak, and J. L. Roberts, *Electrochemistry for Chemists*, 2d ed., 1995.
Stock, J. T., and M. V. Orna (eds.), *Electrochemistry, Past & Present*, 1989.

Journal of the Electrochemical Society, monthly.

ELECTROMAGNETIC FIELDS AND RADIATION

Inan, U., *Fundamentals of Electromagnetics*, 1997.
Lorrain, P., *Electromagnetic Fields and Waves*, 3d ed., 1995.
Neff, H. P., Jr., *Introductory Electromagnetics*, 1991.
Paul, C. R., and S. A. Nasar, *Introduction to Electromagnetic Fields*, 2d ed., 1987.
Rohan, P., *Introduction to Electromagnetic Wave Propagation*, 1991.
Sadiku, M., *Elements of Electromagnetics*, 2d ed., 1994.
Sarwate, V. V., *Electromagnetic Fields and Waves*, 1993.
Smith, G. S., *An Introduction to Classical Electromagnetic Radiation*, 1997.

ELECTRONIC CIRCUITS

Bogart, T. F., *Electronic Devices and Circuits*, 4th ed., 1997.
Boylestadt, R., et al., *Electronic Devices and Circuit Theory*, 6th ed., 1995.
Brophy, J. J., *Basic Electronics for Scientists*, 5th ed., 1990.
Burns, S. G., and P. R. Bond, *Principles of Electronic Circuits*, 2d ed., 1997.
DeMassa, T. A., and Z. Ciccone, *Digital Integrated Circuits*, 1995.
Grey, P. R., and R. G. Meyer, *Analysis and Design of Analog Integrated Circuits*, 3d ed., 1992.
Malvino, A. P., *Electronic Principles*, 5th ed., 1992.
Millman, J., and A. Grabel, *Microelectronics*, 2d ed., 1987.
Paynter, R. T., *Introductory Electronic Devices and Circuits*, 4th ed., 1997.
Sedra, A. S., and K. C. Smith, *Microelectronic Circuits*, 3d ed., 1992.

Electrical and Electronics Abstracts (Science Abstracts, Section B), INSPEC, Institution of Electrical Engineers, monthly.
Electronic Design, biweekly.
Electronics, biweekly.
IEEE Spectrum, Institute of Electrical and Electronics Engineers, Inc., monthly.

ELECTRONICS

See Electronic circuits; Physical Electronics.

ELEMENTARY PARTICLES

Coughlan, G. D., and J. E. Dodd, *The Ideas of Particle Physics: An Introduction for Scientists*, 2d ed., 1991.
Halzen, F., and A. D. Martin, *Quarks and Leptons: An Introductory Course in Modern Particle Physics*, 1984.
Hughes, I. S., *Elementary Particles*, 3d ed., 1991.
Kane, G., *Modern Elementary Particle Physics*, 2d ed., 1993.
Perkins, D. H., *An Introduction to High Energy Physics*, 3d ed., 1987.
Riazuddin, F., *A Modern Introduction to Particle Physics*, 1994.
Rolnick, W. B., *The Fundamental Particles and Their Interactions*, 1994.
Weinberg, S., *The Discovery of Subatomic Particles*, 1996.

Nuclear Physics, Section B, 81 issues per year.
Physical Review D, American Physical Society, semimonthly.

EMBRYOLOGY

See Developmental biology.

ENGINEERING

See Aeronautical engineering; Architectural engineering; Automotive engineering; Biomedical engineering; Chemical engineering; Civil engineering; Design engineering; Electrical power engineering; Environmental engineering; Food engineering; Genetic engineering; Industrial engineering; Marine engineering; Mechanical engineering; Metallurgical engineering; Mining engineering; Nuclear engineering; Petroleum engineering; Production engineering; Systems engineering; Transportation engineering.

ENVIRONMENTAL CHEMISTRY

Armour, M., *Hazardous Laboratory Chemicals Disposal Guide*, 2d ed., 1996.
Bayreuth, O. H. (ed.), *Handbook of Environmental Chemistry*, vol. 2, 1989.
Burke, R., *Hazardous Materials Chemistry for Emergency Responders*, 1997.
Finlayson-Pitts, B. J., and J. M. Pitts, *Atmospheric Chemistry: Fundamentals and Experimental Technique*, 1986.
Furr, K. A., *CRC Handbook of Laboratory Safety*, 4th ed., 1995.
Hathaway, G. J., *Proctor and Hughes' Chemical Hazards of the Workplace*, 4th ed., 1996.
Lund, H. F. (ed.), *The McGraw-Hill Recycling Handbook*, 1993.
Noller, B. N., and M. Chadha (eds.), *Chemistry and the Environment*, 1990.

Pawloski, L., et al., (eds.), *Chemistry for Protection of the Environment*, vol. 51, 1996.
Quigley, D. R., *The Essential Pocket Book of Emergency Chemical Management*, 1996.

Environmental Science and Technology, American Chemical Society, semimonthly.
Journal of Environmental Engineering, American Society of Civil Engineers, monthly.

ENVIRONMENTAL ENGINEERING

Banham, R., *The Architecture of the Well-Tempered Environment*, 2d rev. ed., 1984.
Barrett, G. W., and R. Rosenberg (eds.), *Stress Effects on Natural Ecosystems*, 1982.
Holmes, J. R., *Practical Waste Management*, 1990.
Kiely, G., *Environmental Engineering*, 1996.
Linaweaver, F. P. (ed.), *Environmental Engineering*, 1992.
Revelle, C. S., E. E. Whitlatch, and J. R. Wright, *Civil Engineering Systems*, 1996.
Salvato, J. A., *Environmental Engineering and Sanitation*, 4th ed., 1994.
Wehrli, R., *Environmental Design Research*, 1986.
White, I. D., D. Mottershead, and S. J. Harrison, *Environmental Systems: An Introductory Text*, 2d ed., 1993.

The Diplomate, American Academy of Environmental Engineers, quarterly.
Journal of Environmental Sciences, Institute of Environmental Sciences, bimonthly.

ETHOLOGY

Alcock, J., *Animal Behavior: An Evolutionary Approach*, 5th ed., 1993.
Dawkins, M. S., *Unraveling Animal Behavior*, 2d ed., 1995.
Grier, J. W., and T. Burk, *Biology of Animal Behavior*, 2d ed., 1993.
Lorenz, K., *Evolution and Modification of Behavior*, 1986.
Lorenz, K., *On Aggression*, 1996.
McFarland, D. *Animal Behavior, Psychobiology, Ethology and Evolution*, 2d ed., 1993.
McFarland, D. (ed.), *The Oxford Companion to Animal Behavior*, 1982.
Manning, A., and M. S. Dawkins, *An Introduction to Animal Behavior*, 4th ed., 1992.
Real, L. A., *Behavioral Mechanisms in Evolutionary Biology*, 1994.
Thompson, N. S., *Perspectives in Ethology: Behavior and Evolution*, 1993.

Animal Behavior, Animal Behavior Society, 4 issues per year.
Ethology and Sociobiology, quarterly.

EVOLUTION

Bell, G., *Selection: The Mechanism of Evolution*, 1996.
Cronquist, A., *The Evolution and Classification of Flowering Plants*, 2d ed., 1988.
Gould, S. J., *Evolution and the History of Life: A Science Masters Series Book*, 1998.
Mayr, E., *One Long Argument: Charles Darwin & the Genesis of Modern Evolutionary Thought*, 1991.
Mayr, E., *The Growth of Biological Thought: Diversity, Evolution, and Inheritance*, 1990.
Nitecki, M. H., *Evolutionary Innovations*, 1990.
Strickberger, M., *Evolution*, 2d ed., 1995.
Volpe, P. E., *Understanding Evolution*, 5th ed., 1985.

Evolution, Society for the Study of Evolution, quarterly.

FLUID MECHANICS

Currie, I. G., *Fundamental Mechanics of Fluids*, 2d ed., 1993.
Douglas, J. F., J. M. Gaisorek, and J. A. Swaffield, *Fluid Mechanics*, 3d ed., 1995.
Evett, J., and C. Liu, *Fundamentals of Fluid Mechanics and Hydraulics*, 1987.
Fay, J. A., *Introduction to Fluid Mechanics*, 1994.
Fox, R. W., and A. T. McDonald, *Introduction to Fluid Mechanics*, 4th ed., 1992.
Franzini, J. B., and J. E. Finnemore, *Fluid Mechanics with Engineering Applications*, 9th ed., 1997.
Roberson, J., and C. T. Crowe, *Engineering Fluid Mechanics*, 6th ed., 1996.
Shames, I. H., *Mechanics of Fluids*, 3d ed., 1992.
Streeter, V. L., and E. B. Wylie, *Fluid Mechanics*, 8th ed., 1985.
White, F. M., *Fluid Mechanics*, 3d ed., 1994.

Applied Scientific Research, Central National Organization for Applied Scientific Research in the Netherlands, quarterly.
Physics of Fluids, American Institute of Physics, monthly.

FOOD ENGINEERING

Batty, J. C., and S. L. Folkman, *Food Engineering Fundamentals*, 1983.
Farrall, A. W., *Engineering for Dairy and Food Products*, 2d ed., 1980.
Griffin, Jr., R. C., S. Sacharow, and A. L. Brody, *Principles of Package Development*, 2d ed., 1993.
Heldman, D. R., *Handbook of Food Engineering*, 1992.
Jay, J. M., *Modern Food Microbiology*, 5th ed., 1996.
Roberts, H. R., *Food Safety*, 1981.
Singh, R. P., and D. R. Heldman, *Introduction to Food Engineering*, 2d ed., 1993.

Food Chemistry, Elsevier, monthly.
Food Technology, Institute of Food Technologists, monthly.
Journal of Agricultural and Food Chemistry, American Chemical Society, monthly.

FORENSIC SCIENCE

DeForest, P. R., R. E. Gaenssien, and H. C. Lee, *Forensic Science: An Introduction to Criminalistics*, 2d ed., 1997.
Eckert, W., *Introduction to Forensic Sciences*, 2d ed., 1993.
Kirk, P., and J. I. Thornton, *Crime Investigation*, 2d ed., 1985.
Saferstein, R., *Criminalistics: An Introduction to Forensic Science*, 5th ed., 1994.
Weston, P. B., and K. M. Wells, *Criminal Investigation: Basic Perspectives*, 6th ed., 1993.

Excerpta Medica: Forensic Science, Section 49, bimonthly.
Forensic Science International, monthly.
Journal of Forensic Sciences, American Society for Testing and Materials, quarterly.

FORESTRY

Anderson, D., and I. I. Holland (eds.), *Forests and Forestry*, 4th ed., 1997.
Gholz, H. L. (ed.), *Agroforestry: Realities, Possibilities and Potentials*, 1987.
Landsberg, J. J., and S. T. Gower, *Applications of Physiological Ecology to Forest Management*, 1996.
Leuschner, W. A., *Introduction to Forest Resource Management*, 1992.
Matthews, J. D., *Silvicultural Systems*, 1991.
Sharpe, G. W., and C. Hendes, *Introduction to Forestry*, 6th ed., 1995.
Shugart, H. H., *A Theory of Forest Dynamics: The Ecological Implications of Forest Succession Models*, 1984.
Young, R. A., *Agroforestry for Soil Conservation*, 1989.

Forest Products Journal, Forest Products Research Society, monthly.
Forest Science, Society of American Foresters, quarterly.
Journal of Forestry, Society of American Foresters, monthly.

GENERAL CHEMISTRY

Brady, J. E., and J. R. Holum, *Chemistry: The Study of Matter and Its Changes*, 1992.
Dean, J. A., *Lange's Handbook of Chemistry*, 14th ed., 1992.
Dickson, T. R., *Introduction to Chemistry*, 7th ed., 1995.
Hill, J. W., *Chemistry for Changing times*, 7th ed., 1994.
Holum, J. R., *Fundamentals of General, Organic and Biological Chemistry*, 5th ed., 1993.
Kroschwitz, J., and M. Winokur, *Chemistry: General, Organic, Biological*, 2d ed., 1990.
Pauling, L., *General Chemistry*, 1989.
Rife, W., *Essentials of Chemistry*, 2d ed., 1998.

Journal of the American Chemical Society, weekly.

GENERAL PHYSIOLOGY

Fong, E., *Body Structures and Functions*, 1998.
M. N. Levy (ed.), *Physiology*, 4th ed., 1997.
Jain, M. K., and R. C. Wagner, *Introduction to Biological Membranes*, 2d ed., 1988.
Rhoades, R. A., *Human Physiology*, 3d ed., 1995.
Selkurt, E. E., *Physiology*, 1984.
Shephard, R. J., *Physiology and Biochemistry of Exercise*, 1982.

Annual Review of Physiology, annually.
Journal of General Physiology, monthly.
Journal of Physiology (England), monthly.
Physiological Reviews, American Physiological Society, quarterly.

GENETIC ENGINEERING

Aldridge, S., *The Thread of Life: The Story of Genes and Genetic Engineering*, 1996.
Kammermeyer, K., and V. L. Clark, *Genetic Engineering Fundamentals: An Introduction to Principles and Applications*, 1989.
Maulik, S., and S. Patel, *Molecular Biotechnology: Therapeutic Applications and Strategies*, 1996.
Old, R. W., and S. B. Primrose, *Principles of Gene Manipulation*, 5th ed., 1994.
Setlow, J. K. (ed.), *Genetic Engineering: Principles and Methods*, vols. 11–16, 1986–1994.
Singer, M., *Exploring Genetic Mechanisms*, 1997.
Watson, J. D., *Recombinant DNA*, 2d ed., 1995.

Bio/technology, monthly.
Biotechnology and Bioengineering, monthly.
Genetic Engineering (M. A. Liebert, publisher), monthly.
Genetic Engineering (Plenum), irregularly.
Genetic Technology News, monthly.

GENETICS

Dobzhansky, T., *Genetics and the Origin of Species*, 1982.
Fristrom, J. W., *Principles of Genetics*, 2d ed., 1995.
Hall, J. C., and J. C. Dunlap, *Advances in Genetics*, 1997.
King, R. C., and W. D. Stansfield, *Dictionary of Genetics*, 5th ed., 1996.
Klug, W. S., and M. R. Cummings, *Concepts of Genetics*, 5th ed., 1997.
Kornberg, A., *DNA Replication*, 2d ed., 1995.
Lewin, B., *Genes*, VI, 1997.
Nickerson, R. P., *Genetics: A Guide to Basic Concepts and Problem Solving*, 1990.
Rothwell, N. V., *Understanding Genetics: A Molecular Approach*, 1993.
Russell, P. J., *Fundamentals of Genetics*, 1994.
Scriver, C. R., et al., *The Metabolic Basis of Inherited Disease*, 6th ed., 1989.
Starr, *Cell Biology and Genetics*, 7th ed., 1995.
Strickberger, M. W., *Genetics*, 3d ed., 1985.
Sutton, H. E., *An Introduction to Human Genetics*, 4th ed., 1988.

Watson, J. D., et al., *Molecular Biology of the Gene*, vol. 1, 4th ed., 1987.
Whitehouse, H. L., *Genetic Recombination: Understanding Mechanisms*, 1982.

American Journal of Human Genetics, bimonthly.
Annual Review of Genetics, annually.
Gene, semiweekly.
Genetic Engineering (M. A. Liebert, publisher), monthly.
Genetics, Genetics Society of America, monthly.
Journal of Heredity, American Genetic Association, bimonthly.

GEOCHEMISTRY

Artemyev, V. E., *Geochemistry of Organic Matter in River-Sea System*, 1996.
Brownlow, A. H., *Geochemistry*, 2d ed., 1995.
Chester, R., *Marine Geochemistry*, 1989.
Engel, M., and S. Mako (eds.), *Organic Geochemistry*, 1993.
Hoefs, J., *Stable Isotope Geochemistry*, 4th ed., 1996.

Chemical Geology, Elsevier, monthly.
Geochimica et Cosmochimica Acta (in English, French, and German), Geochemical Society (at Lamont-Doherty and Meteoritical Society, Great Britain), monthly.

GEOGRAPHY

Keates, J. S., *Understanding Maps*, 2d ed., 1996.
Marsh, W. M., and J. Dosier, *Landscape: An Introduction to Physical Geography*, 1986.
Strahler, A. H., and A. N. Strahler, *Introducing Physical Geography*, 1995.
Wieden, F. T., *Physical Geography*, 1995.

Focus, American Geographical Society, bimonthly.
The Professional Geographer, Association of American Geographers, quarterly.

GEOLOGY

Ager, D. V., *Nature of the Stratigraphical Record*, 3d ed., 1993.
Costa, J. E., and V. R. Baker, *Surficial Geology: Building with the Earth*, 1990.
Hallam, A., *Great Geological Controversies*, 2d ed., 1990.
Hay, G. A., and A. L. McAlester, *Physical Geology: Principles and Practices*, 2d ed., 1984.
Lutgens, F. K., and E. J. Tarbuck, *Essentials of Geology*, 4th ed., 1994.
Montgomery, C. W., *Physical Geology*, 3d ed., 1992.
Press, F., and R. Siever, *Earth*, 4th ed., 1995.
Robinson, E. S., *Basic Physical Geology*, 3d ed., 1991.
Skinner, B. J., and S. C. Porter, *Dynamic Earth: An Introduction to Physical Geology*, 3d ed., 1995.
Tarbuck, E. J., and F. K. Lutgens, *The Earth: An Introduction to Physical Geology*, 5th ed., 1995.

Geotimes, American Geological Institute, monthly.
Journal of Geology, bimonthly.

GEOMETRY

Baldwin, H. L., Jr., *Essential Geometry*, 1993.
Coxeter, H. S., *Projective Geometry*, 2d ed., 1994.
Greenberg, M., *Euclidean and Non-Euclidean Geometries: Development and History*, 3d ed., 1995.
Gustafson, R. D., and P. D. Frisk, *Elementary Geometry*, 3d ed., 1991.
Hartshorne, R., *Algebraic Geometry*, rev. ed., 1991.
Jacobs, H. R. (ed.), *Geometry*, 2d ed., 1987.
Klingenberg, W., *Riemannian Geometry*, 2d ed., 1995.
Lang, S., and G. Murrow, *Geometry*, 2d ed., 1991.
O'Daffer, P. G., and S. R. Clemens, *Geometry: An Investigative Approach*, 2d ed., 1992.
O'Neil, B., *Elementary Differential Geometry*, 2d ed., 1997.
Pare, E. G., et al., *Descriptive Geometry*, 9th ed., 1996.
Woodwark, J. (ed.), *Geometric Reasoning*, 1990.

GEOMORPHOLOGY

Bloom, A. L., *Geomorphology: A Systematic Analysis of Late Cenozoic Landforms*, 3d ed., 1991.
Goudie, A., *The Changing Earth: Geomorphological Processes and Time*, 1995.
Leopold, L. B., M. G. Wolmon, and J. P. Miller, *Fluvial Process in Geomorphology*, 1995.
Pitty, A. F. (ed.), *Geomorphology: Themes and Trends*, 1985.
Thomas, D. S. (ed.), *Arid Zone Geomorphology*, 2d ed., 1997.
Tinkler, K. J. (ed.), *The History of Geomorphology*, 1989.
Walker, H. J., and W. E. Grabau (eds.), *The Evolution of Geomorphology: A Nation by Nation Summary of Development*, 1993.

Earth Surface Processes and Landforms, British Geomorphological Research Group, bimonthly.

GEOPHYSICS

Bertotti, B., and P. Farinella, *Physics of the Earth and the Solar System*, 1990.
Elsom, D., *Earth: The Making, Shaping, and Workings of a Planet*, 1992.
Fowler, C. M., *The Solid Earth: An Introduction to Global Geophysics*, 1990.
James, D. E., *Encyclopedia of Geophysics*, 1989.
Sharma, P. V., *Geophysical Methods in Geology*, 2d ed., 1986.
Sheriff, R. E., *Encyclopedia Dictionary of Exploration Geophysics*, 3d ed., 1991.
Telford, W. M., L. P. Geldart, and R. E. Sheriff, *Applied Geophysics*, 2d ed., 1990.
Vogel, S., *Naked Earth: The New Geophysics*, 1995.

Journal of Geophysical Research (issued in three parts), American Geophysical Union, monthly.
Reviews in Geophysics, American Geophysical Union, quarterly.

GRAPHIC ARTS

Adams, J. M., and D. D. Faux, *Printing Technology: A Medium of Visual Communications*, 4th ed., 1995.
Conover, T. E., *Graphic Communications Today*, 3d ed., 1996.
Glassman, A. (ed.), *Printing Fundamentals*, 1985.
Grotta, S. W., and D. Grotta, *The Illustrated Digital Imaging Dictionary*, 1996.
Ihrig, S., and E. Ihrig, *Preparing Digital Images for Print*, 1996.
Levenson, H. R., and T. D. Kinsey, *Graphic Arts and Desktop Publishing Pocket Dictionary*, 1995.
Mulvihill, D. C., *Flexography Primer*, 1985.
Stevenson, G. A., *Graphic Arts Encyclopedia*, 3d ed., revised by W. A. Pakan, 1992.

TAPPI Journal, Technical Association of the Pulp and Paper Industry, monthly.

HYDROLOGY

Barcelona, M., *Contamination of Groundwater: Prevention, Assessment, Restoration*, 1990.
Black, P. E., *Watershed Hydrology*, 2d ed., 1996.
Dunne, T., and L. B. Leopold, *Water in Environmental Planning*, 1995.
Linsley, R. K., et al., *Hydrology for Engineers*, 3d ed., 1982.
Ponce, V. M., *Engineering Hydrology: Principles and Practices*, 1989.
Wilson, E. M., *Engineering Hydrology*, 4th ed., 1990.

Hydrological Sciences Journal, International Association of Hydrological Sciences (Great Britain), quarterly.
Journal of Hydrology, Elsevier, monthly.

ILLUMINATION

Coaton, J., and A. M. Marsden, *Lamps and Lighting*, 4th ed., 1996.
Educational Materials Committee Staff, Illuminating Engineering Society of North America, *Introduction to Light and Lighting*,1991.
Fitt, B., and J. Thornley, *The Control of Light*, 1992.
Gardner, C., and B. Hannaford, *Lighting Design*, 1993.
Grosslight, J., *Light, Light, Light: Effective Use of Daylight and Electric Lighting in Residential and Commercial Spaces*, 2d ed., 1990.
Helms, R., *Lighting for Energy Efficient Luminous Environments*, 2d ed., 1991.
Lindsey, J. L., *Applied Illumination Engineering*, 2d ed., 1996.
Murdoch, J. B., *Illumination Engineering: From Edison's Lamp to the Laser*, 1985, reprint 1995.
Rea, M. (ed.), *IESNA Lighting Handbook*, 8th ed., 1993.
Thumann, A., *Lighting Efficiency Applications*, 2d ed., 1991.

LD&A (Lighting Design & Application), Illuminating Engineering Society, monthly.

IMMUNOLOGY

Baldwin, R. W., and V. S. Byers (eds.), *Monoclonal Antibodies and Immunoconjugates*, 1990.
Benjamin, E., G. Sunshine, and S. Leskowitz, *Immunology: A Short Course*, 3d ed., 1996.
Coleman, R. M., *Fundamental Immunology*, 1992.
Cooper, E. L., and E. Nisbet-Brown, *Developmental Immunology*, 1993.
Davey, B., *Immunology: A Foundation Text*, 1989.
Goding, J. W., *Monoclonal Antibodies: Principles and Practice*, 3d ed., 1996.
Hames, B. D., and D. M. Glover, *Molecular Immunology*, 2d ed., 1996.
Huffer, T. L., et al., *Introduction to Human Immunology*, 1986.
Klein, J., and V. Horseji, *Immunology*, 1997.
Sell, S., *Immunology, Immunopathology and Immunity*, 5th ed., 1996.
Silverstein, A. M., *A History of Immunology*, 1989.
Weir, D. M., and J. Stewart, *Immunology*, 8th ed., 1997.
Widmann, F. K., *Introduction to Clinical Immunology*, 1989.

Advances in Immunology, annually.
Annual Review of Immunology, annually.
Immunobiology, 10 issues per year.
Immunology Abstracts (England), monthly.
Immunology Letters, bimonthly.
Immunology Today, Elsevier, monthly.
Journal of Immunology, American Association of Immunologists, monthly.

INDUSTRIAL CHEMISTRY

Buchner, W., et al., *Industrial Inorganic Chemistry*, 1989.
Chenier, P. J., *Survey of Industrial Chemistry*, 2d ed., 1992.
Heaton, C. A. (ed.), *Introduction to Industrial Chemistry*, 3d ed., 1996.
Kent, J., *Riegel's Handbook of Industrial Chemistry*, 9th ed., 1992.
Kirk-Othmer Encyclopedia of Chemical Technology, Concise Version, 3d ed., 1985.
Merck Index, 12th ed., 1996.
Pletcher, D., and F. C. Walsh, *Industrial Electrochemistry*, 2d ed., 1989.

Industrial and Engineering Chemistry Research, American Chemical Society, monthly.

INDUSTRIAL ENGINEERING

Emerson, H., and D. Naehring, *Origins of Industrial Engineering*, 1988.
Hicks, P. E., *Industrial Engineering and Management: A New Perspective*, 2d ed., 1994.
Philippsborn, H. E., *Elsevier's Dictionary of Industrial Technology*, 1994.
Salvendy, G., *Handbook of Industrial Engineering*, 2d ed., 1991.

Turner, W. C., et al., *Introduction to Industrial and Systems Engineering*, 3d ed., 1992.
Vaugh, R. C., *Introduction to Industrial Engineering*, 3d ed., 1985.

IIE Solutions, Institute of Industrial Engineers, monthly.

INFORMATION THEORY

Blahut, R. E., *Digital Transmission of Information*, 1990.
Cover, T. M., and J. A. Thomas, *Elements of Information Theory*, 1991.
Golomb, S. W., R. E. Peile, and R. A. Scholtz, *Basic Concepts in Information Theory and Coding*, 1994.
Ihara, S., *Information Theory*, 1993.
Kogan, I. M., *Applied Information Theory*, 1988.
Roman, S., *Coding and Information Theory*, 1992.
Sacco, W., et al., *Information Theory: Saving Bits*, 1988.
Usher, M. J., *Information Theory for Information Technologists*, 1984.
Van Der Lubbe, J. C., *Information Theory*, 1997.
Vickery, B. C., and A. Vickery, *Information Sciences in Theory and Practice*, 1990.

IEEE Transactions on Information Theory, Institute of Electrical and Electronics Engineers, Information Theory Group, bimonthly.
Information Sciences, 36 issues per year.

INORGANIC CHEMISTRY

Bowser, J., *Inorganic Chemistry*, 1993.
Cotton, F. A., G. Wilkinson, and P. L. Gans, *Basic Inorganic Chemistry*, 2d ed., 1987.
Jolly, W. J., *Modern Inorganic Chemistry*, 2d ed., 1991.
Katakis, D., and G. Gordon, *Mechanisms of Inorganic Reactions*, 1987.
Porterfield, W. W., *Inorganic Chemistry: A Unified Approach*, 2d ed., 1993.
Sharpe, A. G., *Inorganic Chemistry*, 3d ed., 1992.

Inorganic Chemistry, American Chemical Society, biweekly.

INVERTEBRATE ZOOLOGY

See Zoology.

LASERS

Eberly, J. H., and P. W. Milonni, *Lasers*, 1988.
Hecht, J., *The Laser Guidebook*, 2d ed., 1992.
Hecht, J., *Understanding Lasers: An Entry-Level Guide*, 1988, reprint 1992.
Saleh, B. E. A., and M. C. Teich, *Fundamentals of Photonics*, 1991.
Shimoda, K., *Introduction to Laser Physics*, 2d ed., 1991.
Svelto, O., *Principles of Lasers*, 3d ed., 1989.
Verdeyen, J. T., *Laser Electronics*, 3d ed., 1994.
Yariv, A., *Optical Electronics*, 5th ed., 1995.
Young, M., *Optics and Lasers*, 4th ed., 1993.

Applied Optics, Optical Society of America, American Institute of Physics, 36 issues per year.
Applied Physics B, Deutsche Physikalische Gesellschaft, 12 issues per year.
Journal of the Optical Society of America, A, B, American Institute of Physics, monthly.

LOW-TEMPERATURE PHYSICS

Betts, D. S., *Introduction to Millikelvin Technology*, 1989.
Brewer, D. F., et al. (eds.), *Progress in Low Temperature Physics*, vols. 1–14, 1956–1995.
Clark, A. F., et al. (eds.), *Advances in Cryogenic Engineering*, vols. 1–41, 1954–1996.
Dahl, P. F., *Superconductivity*, 1992.
Flynn, T. M., *Cryogenic Engineering*, 1996.
Khalatnikov, I. M., *Introduction to the Theory of Superfluidity*, 1989.
Pobell, F., *Matter and Methods at Low Temperatures*, 1992.
Poole, C. P., et al. *Superconductivity*, 1995.
Richardson, R. C., and E. N. Smith, *Experimental Techniques in Condensed Matter Physics at Low Temperatures*, 1988.
Tilley, D. R., and J. Tilley, *Superfluidity and Superconductivity*, 3d ed., 1990.
Tinkham, M., *Introduction to Superconductivity*, 2d ed., 1996.
Wilks, J., and D. S. Betts, *An Introduction to Liquid Helium*, 2d ed., 1987, paper 1990.

Cryogenics, monthly.
Journal of Low Temperature Physics, 24 issues per year.

MAGNETISM

See Electricity and magnetism.

MARINE ENGINEERING

Blank, D. A., and A. E. Bock (eds.), *Introduction to Naval Engineering*, 2d ed., 1985.
Harrington, R. L. (ed.), *Marine Engineering*, rev. ed., 1992.
Hunt, E. C. (ed.), *Modern Marine Engineer's Manual*, vol. 2, 2d ed., 1991.
McGeorge, H. D., *Marine Auxiliary Machinery*, 7th ed., 1995.
Neild, A. B. (ed.), *Modern Marine Engineer's Manual*, vol. 1, 2d ed., 1965.
Taylor, D. A., *Introduction to Marine Engineering*, 2d ed., 1990.

Journal of Ship Research, Society of Naval Architects and Marine Engineers, quarterly.

Marine Engineers Review, Institute of Marine Engineers, monthly.
Marine Technology, SNAME NEWS, Society of Naval Architects and Marine Engineers, quarterly.
Transactions of the Society of Naval Architects and Marine Engineers, annually.

MATERIALS SCIENCE

Ashby, M. F., and D. R. Jones, *Engineering Materials*, vol. 1: *An Introduction to Their Properties and Application*, 2d ed., 1996; vol. 2: *An Introduction to Microstructures, Processing and Design*, 1986.
Ashby, M. F., *Materials Selection in Mechanical Design*, 1995.
Bever, M. B. (ed.), *Encyclopedia of Material Science and Engineering.*
Brady, G. S., and H. R. Clauser, *Materials Handbook*, 13th ed., 1991.
Budinski, K. G., *Engineering Materials: Properties and Selection*, 4th ed., 1992.
Carter, G. F., and D. E. Paul, *Materials Science and Engineering*, 1991.
Derby, B., D. A. Hills, and C. Ruiz, *Engineering Materials*, 1992.
Goetzel, C., *Dictionary of Materials and Process Engineering*, 1995.
Higgins, R. A., *Properties of Engineering Materials*, 1990.
Hull, D., *An Introduction to Composite Materials*, 1981.
Singh, S., *Engineering Materials*, 5th rev. ed., 1990.
Van Vlack, L. H., *Elements of Materials Science and Engineering*, 6th ed., 1989.

Materials Science, biweekly.
Chemistry of Materials, American Chemical Society, monthly.
SAMPE, Society for the Advancement of Material and Process Engineering, quarterly.

MATHEMATICS

Bettenger, M., *Essential Mathematics*, 7th ed., 1996.
Doran, E., and R. Boersema, *Fundamental Mathematics*, 2d ed., 1989.
Korn, G. A., and T. M. Korn, *Mathematical Handbook for Scientists and Engineers*, 2d ed., 1968.
Kramer, E. E., *The Nature and Growth of Modern Mathematics*, 1983.
Ledermann, W., et al. (eds.), *Handbook of Applicable Mathematics*, 6 vols., 1980–1985.
Maki, D. P., and M. Thompson, *Finite Mathematics*, 4th ed., 1995.
Mathematical Society of Japan, *Encyclopedic Dictionary of Mathematics*, 2d ed., 4 vols., ed. by K. Ito, 1987.
Research and Education Association Staff, *The Essentials of Numerical Analysis*, nos. 1 and 2, rev. ed., 1994.
Setek, W. M., *Fundamentals of Mathematics*, 7th ed., 1996.

Advances in Mathematics, 16 issues per year.
American Journal of Mathematics, bimonthly.
American Mathematical Monthly, Mathematical Association of America, 10 issues per year.
Annals of Mathematics, bimonthly.
Inventiones Methematicae, monthly.
Journal of the London Mathematical Society, bimonthly.
Mathematical Reviews, American Mathematical Society, 13 issues per year.
Proceedings of the American Mathematical Society, monthly.
Quarterly Journal of Mathematics, quarterly.
Quarterly of Applied Mathematics, American Mathematical society, quarterly.

MECHANICAL ENGINEERING

Avallone, E. A., and T. Baumeister III, *Marks' Standard Handbook for Mechanical Engineers*, 10th ed., 1996.
Konzo, S., and J. W. Bayne, *Mechanical Engineering*, 1991.
Nayler, G. H., *Dictionary of Mechanical Engineering*, 4th ed., 1995.
Rothbart, H. A., *Mechanical Design Handbook*, 1996.
Shigley, J. E., and C. R. Mischke, *Mechanical Engineering Design*, 5th rev. ed., 1989.

Mechanical Engineering, American Society of Mechanical Engineers, monthly.

MECHANICS

See Classical mechanics; Fluid mechanics; Quantum mechanics; Statistical mechanics.

MEDICAL MICROBIOLOGY

Baron, S., et al. (eds.), *Medical Microbiology*, 4d ed., 1996.
Boyd, R. F., and B. G. Hoerl, *Basic Medical Microbiology*, 5th ed., 1995.
Gorbach, *Infectious Diseases*, 2d ed., 1997.
Ingraham, J. L., C. A. Ingraham, and H. Pretiss, *Introduction to Microbiology*, 1995.
Shulman, T. S., et al., *The Biological and Clinical Basis of Infectious Diseases*, 5th ed., 1996.
Thomas, C. G., *Medical Microbiology*, 6th ed., 1988.

Journal of Applied Bacteriology, bimonthly.
Journal of Bacteriology, American Society for Microbiology, monthly.
Journal of Clinical Microbiology, American Society for Microbiology, monthly.

MEDICINE AND PATHOLOGY

Bongiovanni, G., *Essentials of Clinical Gastroenterology*, 2d ed., 1988.
Braunwald, E., et al., *Harrison's Principles of Internal Medicine*, 14th ed., 1997.

Dunnihoo, D. R., *Fundamentals of Gynecology and Obstetrics*, 2d ed., 1993.
Johnson, L. R. (ed.), *Essential Medical Physiology*, 1992.
Kase, N., and A. Weingold, *Principles and Practice of Clinical Gynecology*, 2d ed., 1989.
McMullan, J. T., *Physical Techniques in Medicine*, vol. 1, 1977, vol. 2, 1980.
Schwartz, S. I., et al., *Principles of Surgery*, 6th ed., 1989.
Scriver, C. R., et al., *The Metabolic Basis of Inherited Disease*, 6th ed., 1989.
Stobo, J. D., *Principles and Practice of Medicine*, 23d ed., 1996.
Tamparo, C., and M. A. Lewis, *Diseases of the Human Body*, 2d ed., 1994.
Tannock, I. F., and R. P. Hill (eds.), *The Basic Science of Oncology*, 2d ed., 1992.
Williams, R. H., and J. D. Wilson, *Williams Textbook of Endocrinology*, 9th ed., 1997.

American Journal of Cardiology, monthly.
American Journal of Diseases of Children, American Medical Association, monthly.
American Journal of Medical Technology, American Society for Medical Technology, monthly.
American Journal of Medicine, monthly.
American Journal of Surgery, monthly.
Annals of Internal Medicine, American College of Physicians, monthly.
Cancer, American Cancer Society, semimonthly.
Cardiovascular News, monthly.
Circulation, American Heart Association, monthly.
Contemporary Surgery, monthly.
Endocrinology, monthly.
Human Pathology, monthly.
JAMA: Journal of the American Medical Association, monthly.
Journal of Nuclear Medicine, monthly.
Modern Medicine, monthly.
New England Journal of Medicine, weekly.

METALLURGICAL ENGINEERING

Gilchrist, J. D., *Extraction Metallurgy*, 3d ed., 1989.
Higgins, R. A., *Engineering Metallurgy*, Pt. I: *Applied Physical Metallurgy*, 5th ed., 1990.
Moore, J. C., *Chemical Metallurgy*, 2d ed., 1990.
Rathore, H. S., *Reliability of Metals in Electronics*, 1995.
Reed-Hill, R. E., *Physical Metallurgy*, 3d ed., 1992.
Schlesinger, M. E., and D. E. Mikkola, *Design Education in Metallurgical and Materials Engineering*, 1993.
Van Vlack, L. H., *Elements of Material Science and Engineering*, 7th ed., 1996.

Iron and Steelmaker, Iron and Steel Society (of AIME), monthly.
Metallurgical and Materials Transactions A, American Society for Bimetals, monthly.
Metallurgical and Materials Transactions B, American Society for Bimetals, bimonthly.
Minerals and Metallurgical Processing, Society for Mining, Metallurgy, and Exploration (of AIME), quarterly.

METEOROLOGY

See Climatology and meteorology.

MICROBIOLOGY

Balows, A. (ed.), *Manual of Clinical Microbiology*, 5th ed., 1991.
Brock, T. D., et al., *Basic Microbiology with Application*, 3d ed., 1986.
Gottschalk, G., *Bacterial Metabolism*, 2d ed., 1995.
Gunter-Shlegel, H., *General Microbiology*, 7th ed., 1994.
Laskin, A. I., and S. L. Meidelman, *Advances in Applied Microbiology*, 1996.
Moat, A. G., and J. W. Foster, *Microbial Physiology*, 3d ed., 1995.
Prescott, L., et al., *Microbiology*, 3d ed., 1995.
Tortora, J., *Microbiology: An Introduction*, 5th ed., 1995.
Volk, W. A., and J. C. Brown, *Basic Microbiology*, 8th ed., 1996.

Advances in Applied Microbiology, annually.
Advances in Microbial Physiology, irregularly.
Journal of General Microbiology (England), monthly.
Microbiological Reviews, American Society for Microbiology, quarterly.

MICROSCOPY

Bradbury, S., *An Introduction to the Optical Microscope*, rev. ed., 1989.
Briggs, A., *An Introduction to Scanning Acoustic Microscopy*, 1986.
Duke, P. J., and A. G. Michette (eds.), *Modern Microscopy: Techniques and Applications*, 1989.
Goldstein, J. I., D. E. Newbury, and P. Echlin, *Scanning Electron Microscopy and X-Ray Microanalysis*, 2d ed., 1992.
Johari, O., et al. (eds.), *Scanning Electron Microscopy*, 1983.
Mayer, O., *Clinical Wide-Field Specular Microscopy*, 1984.
Shotton, D. M. (ed.), *Electronic Light Microscopy*, 1992.
Slayter, E. M., and H. S. Slayter, *Light and Electron Microscopy*, 1992.
Wilson, T. C., and J. R. Sheppard, *Theory and Practice of Scanning Optical Microscopy*, 1984.
Wilson, T. (ed.), *Confocal Microscopy*, 1990.
Wischnitzer, S., *Introduction to Electron Microscopy*, 3d ed., 1981.

Micron, quarterly.
Microscope, Cutter Laboratories, 10 issues per year.
The Microscope, McCrone Research Institute, quarterly.

MINERALOGY

Blackburn, W. H., and W. H. Dennen, *Principles of Mineralogy*, 2d ed., 1993.
Hurlbut, C. S., Jr., and R. C. Kammerling, *Gemology*, 2d ed., 1991.
Johnson, W. M., and J. A. Maxwell, *Rock and Mineral Analysis*, 2d ed., 1981, reprint 1989.
Klein, C., and C. S. Hurlbut, Jr., *Manual of Mineralogy (After J. D. Dana)*, 21st ed., 1993.
Manning, D. A., *Introduction to Industrial Minerals*, 1995.
Medenbach, O., and H. Wilk, *The Magic of Minerals*, 1989.
Putnis, A., *An Introduction to Mineral Sciences*, 1992.
Zoltai, T., and J. H. Stout, *Mineralogy: Concepts and Principles*, 2d ed., 1996.

American Mineralogist, Mineralogical Society of America, bimonthly.

MINING ENGINEERING

Hartman, H. L. (ed.), *SME Mining Engineering Handbook*, 2d ed., 2 vols., 1992.
Hartman, H. L., *Introductory Mining Engineering*, 1987.
Peng, S. S., *Surface Subsidence Engineering*, 1992.
Shackleton, W. G., *Economic and Applied Geology: An Introduction*, 1986.

Mining Engineering, Society of Mining Engineers (of AIME), monthly.

MOLECULAR BIOLOGY

Drew, H. R., *Understanding DNA: The Molecular and How It Works*, 2d ed., 1997.
Freifelder, D., *Essentials of Molecular Biology*, 2d ed., 1987.
Lodish, H., and Darnell, *Molecular Cell Biology*, 2d ed., 1995.
Malmberg, R., et al. (eds.), *Molecular Biology of Plants*, 1985.
Ohta, T., and K. Aoki (eds.), *Population of Genetics and Molecular Evolution*, 1986.
Watson, J. D., et al., *Molecular Biology of the Gene*, 4th ed., 1987.

EMBO Journal, European Molecular Biology Organization, bimonthly.
Journal of Molecular Biology, biweekly.
Molecular and Cellular Biology, American Society for Microbiology, monthly.
Molecular Biology, bimonthly.

MOLECULAR PHYSICS

Atkins, P. W., *Molecules*, 1995.
Banwell, C. N., and E. M. McCash, *Fundamentals of Molecular Spectroscopy*, 4th ed., 1994.
Bernath, P. F., *Spectra of Atoms and Molecules*, 1995.
Graybeal, J. D., *Molecular Spectroscopy*, rev. ed., 1993.
R. McWeeny, *Methods of Molecular Quantum Mechanics*, 2d ed., 1992.
Richards, W. G., and P. R. Scott, *Structure and Spectra of Molecules*, 1985.
Struve, W., *Fundamentals of Molecular Spectroscopy*, 1989.
Svanberg, S., *Atomic and Molecular Spectroscopy*, 2d ed., 1992.

Journal of Molecular Structure, 48 issues per year.
Molecular Physics, 18 issues per year.

NAVAL ARCHITECTURE

Benford, H., *Naval Architecture for Non-Naval Architects*, 1991.
Caldwell, J., and G. Ward, *Practical Design of Ships*, 2 vols., 1992.
Gillmer, T., and B. Johnson, *Introduction to Naval Architecture*, 1982.
Lewis, E. V. (ed.), *Principles of Naval Architecture*, 3 vols., 1988.
Taggart, R. (ed.), *Ship Design and Construction*, 3d ed., 1980.
Taylor, D. A., *Merchant Ship Construction*, 3d ed., 1992.
Tupper, E. C., *Introduction Naval Architecture*, 3d ed., 1993.

Naval Architect, The, Royal Institution of Naval Architects, 10 issues per year.
Transactions of the Society of Naval Architects and Marine Engineers, annually.
U.S. Naval Institute Proceedings, monthly.

NAVIGATION

Clausing, D. J., *Aviator's Guide to Navigation*, 3d ed., 1997.
Cunliffe, T., *Coastal and Offshore Navigation*, 2d ed., 1993.
Davies, A., *Celestial Navigation*, 1993.
Hobbs, R. R., *Marine Navigation: Piloting and Celestial Electronic Navigation*, 3d ed., 1990.
Kayton, M., and W. Fried, *Avionics Navigation Systems*, 2d ed., 1996.
Logsdon, T., *Understanding Navstar*, 2d ed., 1995.
Maloney, E. S., *Dutton's Navigation and Piloting*, 14th ed., 1985.
Tetley, L., and C. M. Calcott, *Electronic Aids to Navigation*, 2d ed., 1992.

Navigation, Journal of the Institute of Navigation, quarterly.

NUCLEAR ENGINEERING

Almenas, K., and R. Lee, *Nuclear Engineering: An Introduction*, 1992.

Foster, A. R., and R. L. Wright, Jr., *Basic Nuclear Engineering*, 4th ed., 1983.
Glasstone, S., and W. H. Jordan, *Nuclear Power and Its Environmental Effects*, 1980.
Glasstone, S., and A. Sesonske, *Nuclear Reactor Engineering*, 4th ed., 1995.
Knief, R. A., *Nuclear Engineering*, 2d ed., 1992.
Lamarsh, J. R., *Introduction to Nuclear Engineering*, 3d ed., 1997.
Lewins, J., and M. Becker (eds.), *Advances in Nuclear Science and Technology*, vols. 1–25, 1962–1997.
Ligou, J. P., *Elements of Nuclear Engineering*, 1986.
Murray, R. L., *Nuclear Energy*, 4th ed., 1993.
Weisman, J., *Elements of Nuclear Reactor Design*, 2d ed., 1983.

IEEE Transactions on Nuclear Science, Institute of Electrical and Electronics Engineers, Nuclear and Plasma Sciences Society, monthly.
Nuclear Science and Engineering, American Nuclear Society, 9 issues per year.

NUCLEAR PHYSICS

Burcham, W. E., and M. Jobes, *Nuclear and Particle Physics*, 1994.
Das, A., and T. Ferbel, *Introduction to Nuclear and Particle Physics*, 1993.
Eisenberg, J. M., and W. Greiner, *Nuclear Theory*, vol. 1, 3d ed., 1988; vol. 2, 3d ed., 1988; vol. 3, 2d ed., 1976.
Enge, H. A., and R. P. Redwine, *Introduction to Nuclear Physics*, 2d ed., 1995.
Faessler, A. (ed.), *Progress in Particle and Nuclear Physics*, vols. 3–33, 1980–1994.
Jelley, N. A., *Fundamentals of Nuclear Physics*, 1990.
Krane, K. S., *Introductory Nuclear Physics*, 1987.
Patel, S. B., *Nuclear Physics: An Introduction*, 1991.
Walecka, J. D., *Theoretical Nuclear and Subnuclear Physics*, 1995.

Annual Review of Nuclear and Particle Science, annually.
Nuclear Physics, Section A, 68 issues per year.
Physical Review C, American Physical Society, monthly.

NUTRITION

Adams, C. F., *Nutritive Value of American Food in Common Units*, 1984.
Bender, D. A., and A. E. Bender, *Nutrition: A Reference Handbook*, 1996.
Carroll, K. K. (ed.), *Diet, Nutrition, and Health*, 1990.
Feldman, E. B., *Essentials of Clinical Nutrition*, 1988.
Groff, J., and S. Gropper, *Advanced Nutrition and Human Metabolism*, 2d ed., 1995.
Linder, M. C., *Nutritional Biochemistry and Metabolism, with Clinical Applications*, 2d ed., 1991.
Rechcigl, M., Jr. (ed.), *CRC Handbook of Nutritive Value of Processed Food*, vol. 1, 1982.
Weigley, E. S., D. M. Mueller, and C. H. Robinson, *Basic Nutrition and Diet Therapy*, 8th ed., 1996.
Zeman, F. J., *Clinical Nutrition and Dietetics*, 2d ed., 1990.

Advances in Food Research, irregularly.
Annual Review of Nutrition, annually.
Journal of Nutrition, American Institute of Nutrition, monthly.
Nutrition Reviews, Nutrition Foundation Inc., monthly.
Nutrition Today, bimonthly.

OCEANOGRAPHY

Ansell, A., R. N. Gibson, and H. Barnes (eds.), *Oceanography and Marine Biology: An Annual Review*, vol. 33, 1995.
Davis, R. A., Jr., *Oceanography: An Introduction to the Marine Environment*, 1987.
Duxbury, A. B., and A. C. Duxbury, *Fundamentals of Oceanography*, 2d ed., 1996.
Garrison, T. S., *Oceanography: An Introduction to Marine Science*, 1993.
Gross, M. G., *Oceanography*, 7th ed., 1996.
Ingmanson, D. E., and W. J. Wallace, *Oceanography: An Introduction*, 5th ed., 1989.
Knauss, J. A., *Introduction to Physical Oceanography*, 2d ed., 1996.
Millero, F. J., and M. L. Sohn, *Chemical Oceanography*, 1991.
Pinet, P., *Oceanography: An Introduction to the Planet Oceanus*, 1992.
Ross, D. A., *Introduction to Oceanography*, 4th ed., 1987.
Rowell, B. F., and W. L. Ryan, *Methods in Introductory Oceanography*, 1996.
Stowe, K., *Essentials of Ocean Science*, 1987.
Thurman, H. V., *Essentials of Oceanography*, 5th ed., 1995.

Journal of Physical Oceanography, American Meteorological Society, monthly.
Oceanus, Woods Hole Oceanographic Institute, quarterly.
Sea Frontiers, International Oceanographic Foundation, bimonthly.

OPTICS

Banerjee, P. P., and P. Ting-Chung, *Principles of Applied Optics*, 1991.
Hecht, E., and K. Guardino, *Optics*, 3d ed., 1997.
Jenkins, F. A., and H. E. White, *Fundamentals of Optics*, 4th ed., 1976.
Meyer-Arendt, J., *Introduction to Classical and Modern Optics*, 4th ed., 1994.
Optical Society of America, *Handbook of Optics*, 2d ed., 2 vols., 1995.
Pedrotti, F. L., and S. Pedrotti, *Introduction to Optics*, 2d ed., 1992.
Yu, F. T., and X. Yang, *Introduction to Optical Engineering*, 1997.

Applied Optics, Optical Society of America, American Institute of Physics, 36 issues per year.
Journal of the Optical Society of America, *A*, *B*, American Institute of Physics, monthly.

ORGANIC CHEMISTRY

Baker, D., and R. Engel, *Organic Chemistry*, 1992.
Carey, F. A., *Organic Chemistry*, 3d ed., 1995.
Ege, S. N., *Organic Chemistry: Structure and Reactivity*, 3d ed., 1994.
Holum, J. R., *Fundamentals of General Organic and Biological Chemistry*, 5th ed., 1993.
Mehrotra, R. C., and A. Singh, *Organometallic Chemistry: A Unified Approach*, 1991.
Morrison, R. T., *Organic Chemistry*, 7th ed., 1997.
Smith, M. B., *Organic Synthesis*, 1994.
Solomons, T. W. G., *Fundamentals of Organic Chemistry*, 5th ed., 1996.
Traynham, J. G., *Organic Nomenclature*, 5th ed., 1996.
Weissermel, K., *Industrial Organic Chemistry*, 3d ed., 1997.

Journal of Organic Chemistry, American Chemical Society, biweekly.
Journal of Organometallic Chemistry, Elsevier (Switzerland), 48 issues per year.

PALEONTOLOGY

Colbert, E. H., and M. Morales, *Evolution of the Vertebrates*, 4th ed., 1991.
Gould, S. J., *Evolution and the History of Life*, 1998.
Larsen, C., R. Matter, and D. Gebo, *Human Origins: The Fossil Record*, 1991.
Leakey, R., *Origins Reconsidered: In Search of What Makes Us Human*, 1993.
Raup, D. M., and S. M. Stanley, *Principles of Paleontology*, 2d ed., 1995.
Rudwick, M. J., *The Meaning of Fossils: Episodes in the History of Paleontology*, 2d ed., 1985.
Simpson, G. G., *Fossils and the History of Life*, 1995.
Stearn, C. W., *Paleontology: The Record of Life*, 1989.
Stewart, W. N., and G. W. Rothwell, *Paleobotany and the Evolution of Plants*, 2d ed., 1992.
Taylor, T. N., *The Biology and Evolution of Fossil Plants*, 1993.

Journal of Paleontology, Society of Economic Paleontologists and Mineralogists, bimonthly.
Paleobiology, Paleontological Society, quarterly.
PaleoBios, Museum of Paleontology, University of California, Berkeley, irregularly.

PETROLEUM ENGINEERING

Archer, J. S., and C. G. Wall, *Petroleum Engineering, Principles and Practice*, 1986.
Drew, L. J., *Oil and Gas Forecasting: Reflections of a Petroleum Geologist*, 1990.
Johnson, D. E., and K. E. Pile, *Well Logging for the Nontechnical Person*, 1988.
Mian, M. A., *Petroleum Engineering Handbook for the Practicing Engineer*, 1992.
Nind, T. E., *Hydrocarbon Reservoir and Well Performance*, 1989.
Slider, H. C., *Worldwide Practical Petroleum Reservoir Engineering Methods*, 1983.
Speight, J. G., *The Chemistry and Technology of Petroleum*, 2d rev. ed., 1991.
Terry, R. E., M. Hawkins, and B. C. Craft, *Applied Petroleum Reservoir Engineering*, 2d rev. ed., 1991.
Visher, G., *Exploration Stratigraphy*, 2d ed., 1990.

Hydrocarbon Processing, monthly.
Journal of Petroleum Technology, Society of Petroleum Engineers (of AIME), monthly.
Petroleum Engineer International, monthly.

PETROLOGY

Best, M. G., *Igneous and Metamorphic Petrology*, 1995.
Greensmith, J. T., *Petrology of the Sedimentary Rocks*, 7th ed., 1988.
Hunt, J., *Petroleum Geochemistry and Geology*, 2d ed., 1995.
Hyndman, D. W., *Petrology of Igneous and Metamorphic Rocks*, 2d ed., 1985.
Krumbein, W. C., and F. J. Pettijohn, *Manual of Sedimentary Petrography*, 1988.
McBirney, A. R., *Igneous Petrology*, 2d ed., 1992.
Meyers, R. A., *Handbook of Petroleum Refining Process*, 2d ed., 1996.
Philpotts, A. R., *Principles of Igneous and Metamorphic Petrology*, 1990.
Wilson, M., *Igneous Petrogenesis*, 1988.

Journal of Petrology (England), quarterly.

PHARMACOLOGY

Dipalma, J. R., *Basic Pharmacology in Medicine*, 3d ed., 1990.
Foster, R. W., *Basic Pharmacology*, 4th ed., 1996.
Gilman, A. G., *Goodman and Gilman's The Pharmacological Basis of Therapeutics*, 8th ed., 1990.
Kaminski, A., *Mind-Altering Drugs: A Guide to the History, Uses and Effects of Pyschoactive Drugs*, 1992.
Katzung, B. G. (ed.), *Basic and Clinical Pharmacology*, 6th ed., 1994.
Moore, J. I., *Pharmacology*, 3d ed., 1992.
National Medical Series, *Pharmacology*, 3d ed., 1991.
Neal, M. J., *Medical Pharmacology at a Glance*, 3d ed., 1997.

Advances in Pharmacology and Chemotherapy, irregularly.
American Pharmacy, American Pharmaceutical Association, monthly.
Biochemical Pharmacology, monthly.
Exerpta Medica, Section 80: Pharmacology and Toxicology, 30 issues per year.
Hospital Pharmacy Today, monthly.
Physicians' Desk Reference, 1989.

PHOTOCHEMISTRY

Adamson, A. W., and P. D. Fleischauer (eds.), *Concepts of Inorganic Photochemistry*, 1975, reprint 1984.
Evans, T. R., and A. Weissburger (eds.), *Applications of Lasers to Chemical Problems*, 1982.
Ferraudi, G. J., *Elements of Inorganic Photochemistry*, 1988.
Horspool, W. M., and P.-S. Song (eds.), *Handbook of Organic Photochemistry and Photobiology*, 1995.
Kopecky, J., *Organic Photochemistry: A Visual Approach*, 1992.
Rabek, J. F., *Photochemistry and Photophysics*, vol. 4, 1991.
Suppan, P., *Chemistry and Light*, 1994.
Turro, N. J., *Modern Molecular Photochemistry*, 1981, reprint 1991.

Photochemistry and Photobiology (text in English, French, or German), American Society for Photobiology, monthly.

PHYSICAL CHEMISTRY

Alberty, R. A., and R. J. Silbey, *Physical Chemistry*, 2d ed., 1996.
Atkins, P. W., *Concepts in Physical Chemistry*, 1995.
Barrow, G., *Physical Chemistry*, 6th ed., 1996.
Denbigh, K. G., *The Principles of Chemical Equilibrium*, 4th ed., 1981.
Laidler, K. L., and J. Keith, *Chemical Kinetics*, 3d ed., 1990.
Levine, I. N., *Physical Chemistry*, 4th ed., 1995.
Meites, L., *An Introduction to Chemical Equilibrium and Kinetics*, 1981.
Moore, J. W., and R. G. Pearson, *Kinetics and Mechanism*, 3d ed., 1981.
Reid, C. E., *Chemical Thermodynamics*, 1990.
Servos, J. W., *Physical Chemistry from Ostwald to Pauling: The Making of a Science in America.*
Smith, E. B., *Basic Chemical Thermodynamics*, 4th ed., 1990.

Journal of Physical Chemistry, American Chemical Society, biweekly.

PHYSICAL ELECTRONICS

Chang, C. Y., and S. M. Sze, *ULSI Technology*, 1996.
Einspruch, N. G. (ed.), *VLSI Electronics: Microstructure Science*, vols. 1–22, 1981–1990.
Enderlein, R., and N. J. Horing, *Fundamentals of Semiconductor Physics and Devices*, 1996.
Ferendeci, A. M., *Physical Foundation of Solid State and Electron Devices*, 1991.
Ghandi, S. K., *VLSI Fabrication Principles: Silicon and Gallium Arsenide*, 2d ed., 1994.
Grasserbauer, M., and H. W. Werner (eds.), *Analysis of Microelectronic Materials and Devices*, 1991.
Hawkes, P. W., et al. (eds.), *Advances in Electronics and Electron Physics*, vols. 1–89, 1948–1994.
Neaman, D. A., *Semiconductor Physics and Devices*, 1992.
Ng, K. K., *Guide to Semiconductor Devices*, 1995.
Singh, J., *Semiconductor Devices: An Introduction*, 1994.
Tiyagi, M. S., *Introduction to Semiconductor Materials and Devices*, 1991.

Electrical and Electronics Abstracts, (Science Abstracts, Section B), INSPEC, Institution of Electrical Engineers, monthly.
IEEE Spectrum, Institute of Electrical and Electronics Engineers, Inc., monthly.
IEEE Transactions on Electron Devices, Institute of Electrical and Electronics Engineers, Electron Devices Society, monthly.
Solid State Technology, monthly.

PHYSICS

Beiser, A., *Physics*, 5th ed., 1991.
Brown, L., A. Pais, and B. Pippard, *Twentieth Century Physics*, 3 vols., 1995.
Bueche, F., and D. A. Jerde, *Principles of Physics*, 6th ed., 1994.
Halliday, D., R. Resnick, and J. Walker, *Fundamentals of Physics*, 4th ed., 1993.
Lightman, A. P., *Great Ideas in Physics*, 2d ed., 1997.
Serway, R. A., and J. Faughn, *College Physics*, 4th ed., 1995.
Young, H. D., and R. A. Freedman, *University Physics*, 1996.

American Journal of Physics, American Institute of Physics, monthly.
Annals of Physics, 18 issues per year.
Journal of Applied Physics, American Institute of Physics, monthly.
Journal of Physics, A, B, D, and G, Institute of Physics, Section G, monthly; other sections, semimonthly.
Physical Review, American Physical Society, American Institute of Physics, Section B, 4 issues per month; Section D, semimonthly; Sections A, C, E, monthly.
Physical Review Letters, American Physical Society, American Institute of Physics, weekly.
Physics Abstracts (Science Abstracts, Section A), INSPEC, Institution of Electrical Engineers, biweekly.
Physics Letters, Section A, 78 issues per year; ***Section B***, 108 issues per year.
Physics Reports, 90 issues per year.
Physics Today, American Institute of Physics, monthly.
Proceedings of the Royal Society of London: Series A: Mathematical, Physical and Engineering Sciences, monthly.
Reviews of Modern Physics, American Physical Society, American Institute of Physics, quarterly.

PHYSIOLOGY

See Comparative physiology; General physiology.

PLANT ANATOMY

Fahn, A., *Plant Anatomy*, 4th ed., 1990.
Luttge, U., and H. Behnke, *Progress in Botany: Structural Botany, Physiology, Taxonomy, Geobotany*, 1996.

Mauseth, J. D., *Plant Anatomy*, 1988.
Moore, R. C., *Botany*, 1995.
Northington, D. K., *The Botanical World*, 1994.
Pearson, L. C., *The Diversity of Evolution of Plants*, 1995.
Smith, D., and C. Maier, *Plant Biology*, 1995.
Stern, K. R., *Introductory Plant Biology*, 6th ed., 1995.

PLANT PATHOLOGY

Basu, A. N., *Essentials of Viruses, Vectors and Plant Diseases*, 1993.
Bills, D. D., and S. D. Kung, *Biotechnology and Plant Protection: Viral Pathogenesis and Disease Resistance*, 1995.
Gurr, S. J., and D. J. Bowles, *Molecular Plant Pathology*, 1993.
Lucas, G. B., C. L. Campbell, and L. T. Lucas, *Introduction to Plant Diseases: Identification and Management*, 2d ed., 1992.
Manners, J. G., *Principles of Plant Pathology*, 2d ed., 1993.
Matthews, R. E., *Plant Virology*, 3d ed., 1991.

Annual Review of Phytopathology, annually.
Plant Disease, American Phytopathological Society, monthly.

PLANT PHYSIOLOGY

Baker, N. R., *Photosynthesis and the Environment*, 1996.
Bewley, J. D., and M. Black, *Seeds: Physiology of Development and Germination*, 2d ed., 1994.
Govindjee (ed.), *Molecular Biology of Photosynthesis*, 1989.
Juma, C., *The Gene Hunters: Biotechnology and the Scramble for Seeds*, 1990.
Moore and Salisbury, *Introduction to Plant Physiology*, 1998.
Nobel, P. S., *Physiochemical and Environmental Plant Physiology*, 1991.
Raven, P. H., *Biology of Plants*, 5th ed., 1992.
Salisbury, F. B., and C. W. Ross, *Plant Physiology*, 4th ed., 1992.

Australian Journal of Plant Physiology, 8 issues per year.
Plant Physiology, American Society of Plant Physiologists, monthly.

PLASMA PHYSICS

Chen, F. F., *Introduction to Plasma Physics and Controlled Fusion*, vol. 1: *Plasma Physics*, 2d ed., 1984.
Dendy, R. D. (ed.), *Plasma Physics: An Introduction*, 1994.
Goldston, R., and P. H. Rutherford, *Introduction to Plasma Physics*, 1995.
Leontovich, M. A., and B. B. Kadomtsev (eds.), *Reviews of Plasma Physics*, vols. 1–19, 1965–1996.
Manos, D. M., and D. L. Flam (eds.), *Plasma Etching*, 1989.
Stix, T. H., *Waves in Plasmas*, 1992.
Sturrock, P. A., *Plasma Physics: An Introduction to the Theory of Astrophysical, Geophysical, and Laboratory Plasmas*, 1994.
Wesson, J., *Tokamaks*, 2d ed., 1997.

IEEE Transactions on Plasma Science, Institute of Electrical and Electronics Engineers, Plasma Sciences Society, bimonthly.
Plasma Physics and Controlled Fusion, Institute of Physics, monthly.

POLYMER CHEMISTRY

Elias, H. G., *An Introduction to Polymers*, 1997.
Feldman, D., *Synthetic Polymers; Technology, Properties, and Applications*, 1996.
Fried, J., *Polymer Science and Technology*, 1994.
Kirk-Othmer Encyclopedia of Chemical Technology, 3d ed., vol. 18, 1982.
Mark, H. F., et al., *Encyclopedia of Polymer Science and Technology*, 2d ed., 19 vols., 1984–1989.
Morawetz, H., *Polymers: The Origins and Growth of a Science*, 1995.
Seymour, R. B., and C. E. Carraher, Jr., *Seymour/Carraher's Polymer Chemistry*, 4th ed., 1996.
Stevens, M. P., *Polymer Chemistry*, 2d ed., 1990.
Young, R. J., and P. Lovell, *Introduction to Polymers*, 2d rev. ed., 1991.

Polymer, Butterworth Scientific Ltd. (Great Britain), monthly.

PRODUCTION ENGINEERING

Bolling, G. F., *The Art of Manufacturing Development*, 1995.
Buffa, E. S., and R. K. Sarin, *Modern Production-Operations Management*, 8th ed., 1987.
Groover, M. P., and M. Weiss, *Industrial Robotics: Technology, Programming, and Applications*, 1986.
Kalpakjian, S., *Manufacturing Engineering and Technology*, 3d ed., 1995.
Koenig, D. T., *Manufacturing Engineering: Principles for Optimization*, 2d ed., 1994.
Schmenner, R. W., *Production-Operations Management: Concepts and Situations*, 5th ed., 1992.
Sodhi, R. S., M. Zhou, and S. Das (eds.), *Advances in Manufacturing Systems: Design, Modeling and Analysis*, 1994.
White, J. A., *Production Handbook*, 1987.

PROPULSION

Hill, P., and C. Peterson, *Mechanics and Thermodynamics of Propulsion*, 2d ed., 1992.
Kerrebrock, J. L., *Aircraft Engines and Gas Turbines*, 2d ed., 1992.
Kroes, M. J., J. L. McKinley, and T. W. Wild, *Aircraft Powerplants*, 7th ed., 1994.

Mattingly, J. D., *Elements of Gas Turbine Propulsion*, 1996.
Oates, G. C., *Aircraft Propulsion Systems Technology and Design*, 1989.
Sutton, G. P., *Rocket Propulsion Elements: An Introduction to the Engineering of Rockets*, 6th ed., 1992.
Treager, I. E., *Aircraft Turbine Engine Technology*, 3d ed., 1995.

PSYCHIATRY

American Psychiatric Association, *Diagnostic and Statistical Manual of Mental Disorders*, 4th ed., 1994.
Freud, S., *Outline of Psychoanalysis*, rev. ed., 1989.
Garfield, S. L., *Psychotherapy: An Eclectic Integrative Approach*, 2d ed., 1995.
Garfield, S. L., and A. E. Bergin (eds.), *Handbook of Psychotherapy and Behavior Change: An Empirical Analysis*, 4th ed., 1994.
Hughes, J., *An Outline of Modern Psychiatry*, 3d ed., 1991.
Waldinger, R. J., *Psychiatry for Medical Students*, 3d ed., 1991.

American Journal of Psychiatry, American Psychiatric Association, monthly.
Psychiatric News, American Psychiatric Association, biweekly.

PSYCHOLOGY

Aronson, E., *The Social Animal*, 7th rev. ed., 1991.
Benjamin, L. T., et al., *Psychology*, 3d ed., 1993.
Carlson, N. R., *Physiology of Behavior*, 5th ed., 1993.
Gerow, J. R., *Essentials of Psychology*, 2d ed., 1996.
Gerow, J. R., *Psychology: An Introduction*, 5th ed., 1997.
Kalat, J. W., *Introduction to Psychology*, 3d ed., 1993.

American Psychologist, American Psychological Association Inc., monthly.
Annual Review of Psychology, annually.
Contemporary Psychology, American Psychological Association Inc., monthly.

PUBLIC HEALTH

Bailar, J. C., III, J. Needleman, and B. Berney (eds.), *Assessing Risks to Health: Methodologic Approaches*, 1993.
Donaldson, R. J., and L. J. Donaldson, *Essential Community Medicine: Including Relevant Social Services*, 1983, reprint 1987.
Grant, M., *Handbook of Community Health*, 4th ed., 1987.
Halperin, W., and E. L. Baker (eds.), *Public Health Surveillance*, 1992.
Hurrelman, K., and U. Laoser, *International Handbook of Public Health*, 1996.
Miller, D. F., *Dimensions of Community Health*, 4th ed., 1994.
Raffel, M. (ed.), *Comparative Health Systems: Descriptive Analysis of Fourteen National Health Systems*, 1984.
Unwin, N., J. Leeson, and S. M. Carr, *Public Health and Epidemiology*, 1996.

American Journal of Public Health, American Public Health Association, monthly.
Excerpta Medica, Section 11: Public Health, Social Medicine and Hygiene, 20 issues per year.
Public Health Reports, bimonthly.

QUANTUM CHEMISTRY

Baggott, J., *The Meaning of Quantum Chemistry: A Guide for Students of Chemistry and Physics*, 1992.
J.-L. Calais, *Quantum Chemistry Workbook: Basic Concepts and Procedures in the Theory of the Electronic Structure of Matter*, 1994.
Lowe, J. P., *Quantum Chemistry*, 2d ed., 1993.
McQuarrie, D. A., *Quantum Chemistry*, 1983.
Smith, V. H., Jr., et al., *Applied Quantum Chemistry*, 1986.
Szabo, A., *Modern Quantum Chemistry: Introduction to Advanced Electrical Structure Theory*, 1996.

International Journal of Quantum Chemistry, semimonthly.

QUANTUM MECHANICS

Bohm, A., *Quantum Mechanics: Foundation and Applications*, 3d ed., 1994.
Dirac, P. A. M., *The Principles of Quantum Mechanics*, 4th ed., 1958, reprint 1982.
Eisberg, R., and R. Resnick, *Quantum Physics: Of Atoms, Molecules, Solids, Nuclei and Particles*, 2d ed., 1985.
Gottfried, K., *Quantum Mechanics*, vol. 1: *Fundamentals*, 1966, revised printing 1989.
Greiner, W., *Quantum Mechanics*, 1996.
Liboff, R. L., *Introductory Quantum Mechanics*, 3d ed., 1997.
Mandl, F., *Quantum Mechanics*, 1992, reprint 1994.
Ohanian, H. C., *Principles of Quantum Mechanics*, 1989.
Park, D., *Introduction to the Quantum Theory*, 3d ed., 1991.
Shankar, R., *Principles of Quantum Mechanics*, 2d ed., 1994.

RADAR

Barton, D. K., et al., *Radar Technology Encyclopedia*, 1997.
Cole, H., *Understanding Radar*, 2d ed., 1992.
Cook, C., *Radar Signals: An Introduction to Theory and Application*, 1993.
Edde, B., *Radar: Principles, Technology, Applications*, 1992.

Nathanson, F. E., *Radar Design Principles*, 2d ed., 1991.
Raemer, H. R., *Radar Systems Principles*, 1996.
Rihaczek, A. W., *Principles of High Resolution Radar*, 1996.
Sabatini, S., and M. Tarantino, *Multifunction Array Radar: System Design and Analysis*, 1994.
Sauvageot, H., *Radar Meteorology*, 1992.
Skolnik, M. I., *Radar Applications*, 1988.
Skolnik, M. I. (ed.), *Radar Handbook*, 2d ed., 1990.

***IEE Proceedings: Radar, Sonar and Navigation*,** Institution of Electrical Engineers, semimonthly.

RADIO COMMUNICATIONS

American Radio Relay League, *The ARRL Handbook for the Radio Amateur*, annually.
Carson, R., *Radio Communications Concepts: Analog*, 1990.
Mazda, F., *Radio Technologies*, 1996.
Sabin, W. E., and E. D. Schoenike (eds.), *Single-Sideband Systems and Circuits*, 2d ed., 1995.
Sennitt, A. G. (ed.), *World Radio TV Handbook*, annually.
Shrader, R. I., *Electronic Communications*, 6th ed., 1991.
Valkenburg, M. (ed.), *Reference Data for Engineers: Radio, Electronics, Computer, and Communications*, 8th ed., 1991.
Wilby, P., and A. Conroy, *The Radio Handbook*, 1994.

***IEEE Transactions on Broadcasting*,** Institute of Electrical and Electronics Engineers, Broadcast Technology Society, quarterly.

RADIOCHEMISTRY

Adloff, J.-P., and R. Guillaumont, *Applied Radiochemistry*, 1993.
Arnikar, H. J., *Essentials of Nuclear Chemistry*, 2d ed., 1987.
Choppin, G. R., J.-O. Liljenzin, and J. Rydberg, *Radiochemistry and Nuclear Chemistry: Nuclear Chemistry, Theory and Applications*, 2d ed., 1995.
Friedlander, G., et al., *Nuclear and Radiochemistry*, 3d ed., 1981.
Lambrecht, R. M., and N. Morcos (eds.), *Nuclear and Radiochemistry Applications*, 1982.

***Journal of Labelled Compounds and Radiopharmaceuticals*,** International Isotope Society, monthly.
***Radiation Physics and Chemistry*,** monthly.

RELATIVITY

Bohm, D., *The Special Theory of Relativity: Its Origins, Meanings, and Implications*, 1989.
Misner, C., K. Thorne, and J. Wheeler, *Gravitation*, 1973.
Rindler, W., *Essential Relativity*, rev. ed., 1990.
Stephani, H., *General Relativity*, 2d ed., 1990.
Taylor, E. F., and J. A. Wheeler, *Spacetime Physics: Introduction to Special Relativity*, 2d ed., 1992.
Tourenc, P., *Relativity and Gravitation*, 1997.
Wald, R. M., *General Relativity*, 1984.
Will, C. M., *Was Einstein Right? Putting General Relativity to the Test*, 2d ed., 1993.

***General Relativity and Gravitation*,** International Committee on General Relativity and Gravitation, monthly.

SOIL

Bowles, J. E., *Engineering Properties of Soils and Their Measurement*, 4th ed., 1992.
Brady, N. C., *The Nature and Properties of Soils*, 11th ed., 1995.
Daniels, R. B., and R. D. Hammer, *Soil Geomorphology*, 1992.
Dowdy, R. H., et al. (eds.), *Chemistry in the Soil Environment*, 1981.
Foster, A. B., *Approved Practices in Soil Conservation*, 5th ed., 1982.
Foth, H. D., *Fundamentals of Soil Science*, 8th ed., 1990.
Hanks, R. J., *Applied Soil Physics: Soil Water and Temperature Applications*, 2d ed., 1992.
Lee, J. S., et al., *The Earth and Agriscience*, 1995.
Morgan, R. P. C., *Soil Erosion and Conservation*, 2d ed., 1995.
Plaster, E. J., *Soil Science and Management*, 3d ed., 1997.

***Agronomy Journal*,** American Society of Agronomy, bimonthly.

SOLID-STATE CHEMISTRY

Bruce, P. G. (ed.), *Solid State Electrochemistry*, 1995.
Burdett, J. K., *Chemical Bonding in Solids*, 1995.
Cheetham, A. K., and P. Day, *Solid-State Chemistry*, vol. 1: *Techniques*, 1990, vol. 2: *Compounds*, 1992.
Hoffman, R., *Solids and Surfaces: A Chemist's View of Bonding in Extended Structures*, 1989.
Ladd, M. F. C., *Structure and Bonding in Solid State Chemistry*, 1986.
Schmalzried, H., *Solid State Reactions*, 2d ed., 1981.
West, A. R., *Basic Solid State Chemistry*, 1988.

SOLID-STATE PHYSICS

Chaikin, P. M., and T. C. Lubensky, *Principles of Condensed Matter Physics*, 1995.
Ehrenreich, H., et al. (eds.), *Solid State Physics*, vols. 1-50, 1955–1996.
Hook, J. R., and H. E. Hall, *Solid State Physics*, 2d ed., 1991.
Ibach, H., and H. Luth, *Solid State Physics: An Introduction to Principles of Materials Science*, 1996.
Isihara, A., *Condensed Matter Physics*, 1991.

Kittel, C., *Introduction to Solid State Physics*, 7th ed., 1995.
Myers, H. P., *Introductory Solid State Physics*, 2d ed., 1997.
Omar, M. A., *Elementary Solid State Physics*, 1996.
Rudden, M. N., and J. Wilson, *Elements of Solid State Physics*, 2d ed., 1993.
Tanner, B. K., *Introduction to the Physics of Electrons in Solids*, 1995.

Applied Physics A, Deutsche Physikalische Gesellschaft, 12 issues per year.
Journal of Physics: Condensed Matter, Institute of Physics, 50 issues per year.
Journal of Physics and Chemistry of Solids, monthly.
Physica B, 60 issues per year.

SPACE TECHNOLOGY

Boden, D. S., and W. J. Larson, *Cost-Effective Space Mission Operations*, 1996.
Bouquet, F. L., *Spacecraft Design: Thermal & Radiation*, 1991.
Gordon, C. D., and W. L. Morgan, *Principles of Communications Satellites*, 1993.
Griffin, M. D., and J. B. French, *Space Vehicle Design*, 1991.
Humble, R. W., G. N. Henry, and W. J. Larson, *Space Propulsion Analysis and Design*, 1996.
Pisacane, V. L., and R. C. Moore (eds.), *Space Systems*, 1994.
Rummel, J. D., V. M. Ivanov, and J. Rummel (eds.), *Space and Its Exploration*, 1993.
Sellers, J., and W. J. Larson, *Understanding Space: An Introduction to Astronautics*, 1995.

Aerospace America, American Institute of Aeronautics and Astronautics, monthly.
Journal of Spacecraft and Rockets, American Institute of Aeronautics and Astronautics, bimonthly.

SPECTROSCOPY

Andrews, D. L., and A. A. Demidov (eds.), *An Introduction to Laser Spectroscopy*, 1995.
Bernath, P. F., *Spectra of Atoms and Molecules*, 1995.
Briggs, D., and M. P. Seah (eds.), *Practical Surface Analysis: Auger and X-Ray Photoelectron Spectroscopy*, vol. 1, 2d ed., 1990.
Bruno, T. J., *Spectroscopic Methods*, 1993.
De Hoffman, E., *Mass Spectrometry; Principles and Applications*, 1996.
Dickson, D. P., and J. J. Berry (eds.), *Mössbauer Spectroscopy*, 1987.
Ghosh, P. K., *Introduction to Photoelectron Spectroscopy*, 1983.
Hollas, J. M., *Modern Spectroscopy*, 2d ed., 1991.
Laserna, J. J. (ed.), *Modern Techniques in Raman Spectroscopy*, 1996.
Sanders, J. K., and B. K. Hunter, *Modern NMR Spectroscopy: A Guide for Chemists*, 2d ed., 1993.
Smith, B. C., *Fundamentals of Fourier Transform Infrared Spectroscopy*, 1995.
Stowell, M. C., and T. C. Farrar, *Basic Nuclear Magnetic Resonance Spectroscopy*, 1996.
Thompson, M., et al., *Auger Electron Spectroscopy*, 1985.
Williams, K. L., *Introduction to X-Ray Spectrometry*, 1987.

Applied Spectroscopy, Society for Applied Spectroscopy, monthly.
Journal of Magnetic Resonance, 15 issues per year.
Progress in Nuclear Magnetic Resonance Spectroscopy, 4 issues per year.
Spectrochimica Acta, Parts A and B, monthly.

STATISTICAL MECHANICS

Baracca, A., and R. Livi, *Statistical Mechanics: Foundations, Problems, Perspectives*, 1997.
Betts, D. S., *Introductory Statistical Mechanics*, 1992.
Bowley, R., and M. Sanchez, *Introductory Statistical Mechanics*, 1996.
Garrod, C., *Statistical Mechanics and Thermodynamics*, 1995.
Kubo, R., *Statistical Mechanics*, 1990.
Pathria, R. K., *Statistical Mechanics*, 2d ed., 1996.
Sen, S., *Statistical Mechanics*, 1996.
Ter Harr, D., *Elements of Statistical Mechanics*, 1995.

Journal of Statistical Physics, 12 issues per year.
Physica A, 52 issues per year.

STATISTICS AND PROBABILITY

Dixon, W. J., and F. J. Massey, *Introduction to Statistical Analysis*, 4th ed., 1983.
Hoel, P. G., *Introduction to Mathematical Statistics*, 5th ed., 1984.
Johnson, R. A., and G. K. Bhattacharyya, *Statistics: Principles and Methods*, 3d ed., 1996.
Larsen, R., and M. Marx, *Statistics*, 1990.
Larson, H. J., *Introduction to Probability*, 1995.
Markley, N. G., *Introduction to Probability*, rev. ed., 1991.
Naiman, A., R. Rosenfeld, and G. Zirkel, *Understanding Statistics*, 4th ed., 1996.
Ross, S. M., *A First Course in Probability*, 4th ed., 1994.
Sanders, D. H., *Statistics: A First Course*, 5th ed., 1995.
Spence, J. T., et al., *Elementary Statistics*, 5th ed., 1991.
Triola, M. F., *Elementary Statistics*, 6th ed., 1995.
Weiss, N. A., *Elementary Statistics*, 2d ed., 1993.
Wonnacott, T. H., and R. J. Wonnacott, *Introductory Statistics*, 5th ed., 1990.

Annals of Statistics, Institute of Mathematical Statistics, bimonthly.
Journal of the American Statistical Association (JASA), quarterly.

STEREOCHEMISTRY

Bassindale, A., *The Third Dimension in Organic Chemistry*, 1984.
Burdett, J. K., *Molecular Shapes: Theoretical Models of Inorganic Stereochemistry*, 1980.
Eliel, E., *Stereochemistry of Organic Compounds*, 1994.
Juaristi, E., *Introduction to Stereochemistry and Conformational Analysis*, 1991.
Von Zelewsky, A., *Stereochemistry of Coordination Compounds*, 1996.

SYSTEMS ENGINEERING

Dorny, N. C., *Understanding Dynamic Systems: Approaches to Modeling, Analysis, and Design*, 1993.
Grady, J. O., *System Integration*, 1994.
Hitchins, D., *Putting Systems to Work*, 1993.
Reilly, N. B., *Successful Systems Engineering for Engineers and Managers*, 1993.
Rouse, W. B., *Best Laid Plans*, 1994.
Sage, A. P., *Systems Engineering*, 1992.

TAXONOMY

Cronquist, A., *An Integrated System of Classification of Flowering Plants*, 1992.
Jeffrey, C., *Biological Nomenclature*, 3d ed., 1992.
Jeffrey, C., *An Introduction to Plant Taxonomy*, 2d ed., 1982.
Margulis, L., and K. V. Schwartz, *Five Kingdoms*, 2d ed., 1995.
Scott-Ram, N. R., *Transformed Cladistics, Taxonomy, and Evolution*, 1990.
Stuessy, T. F., *Plant Taxonomy: The Systematic Evaluation of Comparative Data*, 1990.

Systematic Botany, American Society of Plant Taxonomists, quarterly.
Systematic Zoology, Society of Systematic Zoology, quarterly.

TELECOMMUNICATIONS

Freeman, R. L., *Reference Manual for Telecommunications Engineering*, 2d ed., 1993.
Frenzel, L., *Electronic Communication Systems*, 2d ed., 1996.
Haykin, S., *Communication Systems*, 3d ed., 1994.
Johnson, C. D., *Communication Systems*, 1992.
Keiser, G. E., *Optical Fiber Communications*, 2d ed., 1991.
Kazovsky, L. G., S. Benedetto, and A. E. Willner, *Optical Fiber Communication Systems*, 1996.
Lathi, B. P., *Modern Digital and Analog Communication Systems*, 2d ed., 1995.
Pecar, J. A., R. J. O'Connor, and D. A. Garbin, *The McGraw-Hill Telecommunications Factbook*, 1993.
Peterson, R. L., D. E. Borth, and R. E. Ziemer, *An Introduction to Spread Spectrum Communications*, 1995.
Proakis, J. G., *Digital Communications*, 3d ed., 1995.
Roddy, D., *Satellite Communications*, 2d ed., 1996.
Roddy, D., and J. Coolen, *Electronic Communications*, 4th ed., 1995.

Bell Labs Technical Journal, quarterly.
IEEE Spectrum, Institute of Electrical and Electronics Engineers, Inc., monthly.
Telecommunications, monthly.

TELEPHONY

Bellamy, J. C., *Digital Telephony*, 2d ed., 1991.
Chorafras, D. N., *Telephony: Today and Tomorrow*, 1984.
Fike, J. L., G. E. Friend, and S. J. Bigelow, *Understanding Telephone Electronics*, 3d ed., 1991.
Lee, F. E., *Telephone Theory, Principles and Practice*, rev. ed., 1988.
Noll, A. M., *Introduction to Telephones and Telephone Systems*, 2d ed., 1991.
Williams, G. B., *Telephones and Answering Machines*, 1994.

TELEVISION

Benson, K. B., and J. Whitaker, *Television Engineering Handbook: Featuring HDTV Systems*, rev. ed., 1992.
Leak, J. D., *Leak's Television Handbook: Operation and Troubleshooting*, 1995.
National Association of Broadcasters, *NAB Engineering Handbook: Operation and Troubleshooting*, 1995.
Noll, A. M., *Television Technology: Fundamentals and Future Prospects*, 1988.
Prentiss, S., *HDTV: High-Definition Television*, 2d ed., 1994.
Rzeszewski, T., *Digital Video: Concepts and Applications Across Industries*, 1995.
Watkinson, J., *Television Fundamentals*, 1996.
Whitaker, J., *HDTV: The Revolution in Electronic Imaging*, 1997.

TEXTILES AND FIBERS

Joseph, M. L., et al., *Joseph's Introductory Textile Science*, 6th ed., 1993.
Lyle, D. S., *Modern Textiles*, 2d ed., 1982.
Smith, B., and I. Block, *Textiles in Perspective*, 1982.
Tortora, P. G., *Understanding Textiles*, 5th ed., 1991.

Textile Research Journal, Textile Research Institute, monthly.

THEORETICAL PHYSICS

Arfken, G. B., and H. J. Weber (eds.), *Mathematical Methods for Physicists*, 4th ed., 1995.
Courant, R., and D. Hilbert, *Methods of Mathematical Physics*, 2 vols., 1989.
Honerkamp, J., and H. Romer, *Theoretical Physics: A Classical Approach*, 1993.
Lam, L. (ed.), *Nonlinear Physics for Beginners*, 1996.
Landau, L. D., et al., *Course of Theoretical Physics*, 9 vols., 1959–1987.
Moran, M. J., and H. N. Shapiro, *Fundamentals of Engineering Thermodynamics*, 3d ed., 1995.
Reddy, R. G., and N. V. Gokcen, *Thermodynamics*, 2d ed., 1996.
Reed, M., and B. Simon, *Methods of Modern Mathematical Physics*, 4 vols., 1975–1980.
Thirring, W. A., *Course in Mathematical Physics*, 4 vols., 1982–1997.
van Wylen, G. J., and R. E. Sonntag, *Fundamentals of Classical Thermodynamics*, 4th ed., 1993.
Wood, S. E., and R. Battino, *Thermodynamics of Chemical Systems*, 1990.

Communications in Mathematical Physics, 18 issues per year.
Journal of Mathematical Physics, American Institute of Physics, monthly.

THERMODYNAMICS AND HEAT

Black, W. Z., and J. G. Hartley, *Thermodynamics*, 3d ed., 1995.
Holman, J. P., *Thermodynamics*, 4th ed., 1988.
Modell, M., and R. C. Reide, *Thermodynamics and Its Applications*, 3d ed., 1996.
Wark, K., *Thermodynamics*, 5th ed., 1988.
Zemansky, M. W., and R. H. Dittman, *Heat and Thermodynamics*, 7th ed., 1997.

TRANSPORTATION ENGINEERING

Ashford, N., and P. H. Wright, *Airport Engineering*, 3d ed., 1992.
De Old, A. R., et al., *Transportation: Technology of Moving People and Goods*, 1986.
Diebold, J., *Transportation Infostructures: The Development of Intelligent Transportation Systems*, 1995.
Khisty, C. J., *Transportation Engineering: An Introduction*, 1990.
Papacostas, C. S., and P. D. Prevedouros, *Transportation Engineering*, 2d ed., 1992.
Robertson, D., *Manual of Transportation Engineering Studies*, 5th ed., 1994.
Sheets, E., and A. R. De Old, *Activities for Transportation Technology Systems*, 1993.
Wright, P. H., and R. J. Paquette, *Highway Engineering*, 6th ed., 1995.

Journal of Transportation Engineering, monthly.
Transportation Journal, American Society of Traffic and Transportation, quarterly.

TRIGONOMETRY

Drooyan, I., and C. C. Carico, *Trigonometry: An Analytical Approach*, 6th rev. ed., 1990.
Heineman, E. R., and J. D. Tarwater, *Plane Trigonometry*, 7th ed., 1993.
Hungerford, T. W., and R. Mercer, *Trigonometry*, 1992.
Smith, K. J., *Trigonometry for College Students*, 7th ed., 1998.
Sullivan, M., *Trigonometry*, 4th ed., 1996.

VERTEBRATE ZOOLOGY

See Zoology.

VETERINARY MEDICINE

Ettinger, S. J., and E. C. Feldman, *Textbook of Veterinary Internal Medicine*, 4th ed., 1994.
Fenner, W. R., *Quick Reference to Veterinary Medicine*, 2d ed., 1991.
Fraser, C. M. (ed.), *The Merck Veterinary Manual*, 7th ed., 1991.
Parker, W. H., *Health and Disease in Farm Animals: An Introduction to Farm Animal Medicine*, 3d ed., 1980.
Timoney, J. F., et al., *Hagan and Bruner's Microbiology and Infectious Diseases of Domestic Animals*, 8th ed., 1988.

VIROLOGY

Atherton, J. G., and I. R. Holmes, *ICTV Code for the Description of Virus Characters*, 1983.
Belshe, R. B., *Textbook of Human Virology*, 2d ed., 1990.
Diener, T. O. (ed.), *The Viroids*, 1987.
Dimmock, N. J., P. D. Griffiths, and C. R. Madeley (eds.), *Control of Virus Diseases*, 1990.
Fields, B. N., *Fundamental Virology*, 3d ed., 1995.
Fraenkel-Conrat, H., P. C. Kimball, and J. A. Levy, *Virology*, 3d ed., 1994.
Matthews, R. E. (ed.), *Classification and Nomenclature of Viruses*, 1982.
Matthews, R. E., *Fundamentals of Plant Virology*, 1992.
Parker, M. T., et al., *Topley and Wilson's Principles of Bacteriology, Virology and Immunity*, 5 vols., 1990.
Rothschild, H., and J. C. Cohen (eds.), *Virology in Medicine*, 1986.
White, D. O., and F. J. Fenner, *Medical Virology*, 4th ed., 1994.

Advances in Virus Research, irregularly.
Excerpta Medica, Section 47: Virology, 10 issues per year.
Journal of Virology, American Society for Microbiology, monthly.

ZOOLOGY

Barnes, R. D., and E. E. Ruppert, *Invertebrate Zoology*, 6th ed., 1993.
Buchsbaum, M., et al., *Living Invertebrates*, 1987.
Dorit, R., W. F. Walker, and R. D. Barnes, *Zoology*, 1991.
Grzimek, B. (ed.), *Grzimek's Encyclopedia of Mammals*, 5 vols., 2d ed., 1989.
Meglitsch, P. A., and F. R. Schram, *Invertebrate Zoology*, 3d ed., 1991.
Miller, S. A., and J. P. Harley, *Zoology*, 3d ed., 1995.
Pough, F., et al., *Vertebrate Life*, 3d ed., 1989.
Villee, C., et al., *General Zoology*, 6th ed., 1984.

American Zoologist, American Society of Zoologists, quarterly.
Journal of Zoology, monthly.

Scientific Notation

Since the Encyclopedia is a work of science, in the articles it was necessary to follow the scientific style of using symbols, abbreviations, and exact names. This section discusses the more frequently used conventions in the Encyclopedia and includes tables for convenient reference. The relation between the three primary measurement systems—U.S. Customary, metric, and International—is also clarified.

U.S. Customary System and the metric system

Scientists and engineers have been using two major systems of units in measurement. These are commonly called the U.S. Customary System (inherited from the British Imperial System) and the metric system.

In the U.S. Customary System the units yard and pound with their divisions, such as the inch, and multiples, such as the ton, are basic. The metric system was evolved during the eighteenth century and has been adopted for general use by most countries. It is used nearly everywhere for precise measurements in science. The meter and kilogram with their multiples, such as the kilometer, and fractions, such as the gram, are basic to the metric system.

In the U.S. Customary System, units of the same kind are related almost at random. For example, there are the units of length, the inch, yard, and mile. In the metric system the relationships between units of the same kind are strictly decimal (millimeter, meter, and kilometer).

However, to complicate matters, in scientific writing there is no uniformity within each of these two systems as to the choice of units for the same quantities. For example, the hour or the second, the foot or the inch, and the centimeter or the millimeter could be chosen by a scientist as the unit of measurement for the quantities time and length.

Introduction of the International System

To simplify matters and to make communication more understandable, an internationally accepted system of units has come into use. This is termed the International System of Units, which is abbreviated SI in all languages (from the French Système International d'Unités).

Fundamentally the system is metric with the base units derived from scientific formulas or natural constants. For example, the meter in the SI is defined as the length of the path traveled by light in vacuum during a time interval of 1/299 792 458 of a second.

The second in the SI is defined as the duration of 9 192 631 770 periods of the radiation corresponding to the transition between two hyperfine levels of the ground state of the cesium-133 atom.

Interestingly, the kilogram, the SI unit of mass, is still the mass of the kilogram kept at Sèvres, France. However, it is possible that eventually the unit will be redefined in terms of atomic mass.

Although the SI is increasingly used by scientists and engineers, there are some other units in everyday use which will probably remain, for example, minute, hour, day, degree (angle), and liter. The point should be made, however, that these terms will not be employed in a scientific context if the SI is fully adopted.

Because of their extremely common use among scientists, several units are still permitted in conjunction with SI units, for example, the electron-volt, rad, roentgen, barn, and curie. In time their usage might be phased out.

One further point is that in October 1967 the Thirteenth General Conference of Weights and Measures decided to name the SI unit of thermodynamic temperature "kelvin" (symbol K) instead of "degree Kelvin" (symbol °K). For example, the notation is 273 K and not 273°K.

The base units and derived units of the SI are shown in **Tables 1** and **2.**

TABLE 1. Base units of the International System

Quantity	Name of unit	Unit symbol
length	meter	m
mass	kilogram	kg
time	second	s
electric current	ampere	A
temperature	kelvin	K
luminous intensity	candela	cd
amount of substance	mole	mol

TABLE 2. Derived units of the International System

Quantity	Name of unit	Unit symbol, where differing from basic form	Unit expressed in terms of base or supplementary units*
area	square meter		m^2
volume	cubic meter		m^3
frequency	hertz	Hz	s^{-1}
density	kilogram per cubic meter		kg/m^3
velocity	meter per second		m/s
angular velocity	radian per second		rad/s
acceleration	meter per second squared		m/s^2
angular acceleration	radian per second squared		rad/s^2
volumetric flow rate	cubic meter per second		m^3/s
force	newton	N	$kg \cdot m/s^2$
surface tension	newton per meter, joule per square meter	N/m, J/m^2	kg/s^2
pressure	newton per square meter, pascal	N/m^2, Pa	$kg/(m \cdot s^2)$
viscosity, dynamic	newton-second per square meter, pascal-second	$N \cdot s/m^2$, $Pa \cdot s$	$kg/(m \cdot s)$
viscosity, kinematic	meter squared per second		m^2/s
work, torque, energy, quantity of heat	joule, newton-meter, watt-second	J, $N \cdot m$, $W \cdot s$	$kg \cdot m^2/s^2$
power, heat flux	watt, joule per second	W, J/s	$kg \cdot m^2/s^3$
heat flux density	watt per square meter	W/m^2	kg/s^3
volumetric heat release rate	watt per cubic meter	W/m^3	$kg/(m \cdot s^3)$
heat transfer coefficient	watt per square meter kelvin	$W/(m^2 \cdot K)$	$kg/(s^3 \cdot K)$
heat capacity (specific)	joule per kilogram kelvin	$J/(kg \cdot K)$	$m^2/(s^2 \cdot K)$
capacity rate	watt per kelvin	W/K	$kg \cdot m^2/(s^3 \cdot K)$
thermal conductivity	watt per meter kelvin	$W/(m \cdot K)$, $\frac{J \cdot m}{s \cdot m^2 \cdot K}$	$kg \cdot m/(s^3 \cdot K)$
quantity of electricity	coulomb	C	$A \cdot s$
electromotive force	volt	V, W/A	$kg \cdot m^2/(A \cdot s^3)$
electric field strength	volt per meter	V/m	$kg \cdot m/(A \cdot s^3)$
electric resistance	ohm	Ω, V/A	$kg \cdot m^2/(A^2 \cdot s^3)$
electric conductance	siemens	S, A/V	$A^2 \cdot s^3/(kg \cdot m^2)$
electric conductivity	ampere per volt meter	$A/(V \cdot m)$	$A^2 \cdot s^3/(kg \cdot m^3)$
electric capacitance	farad	F, $A \cdot s/V$	$A^2 \cdot s^4/(kg \cdot m^2)$
magnetic flux	weber	Wb, $V \cdot s$	$kg \cdot m^2/(A \cdot s^2)$
inductance	henry	H, $V \cdot s/A$	$kg \cdot m^2/(A^2 \cdot s^2)$
magnetic permeability	henry per meter	H/m	$kg \cdot m/(A^2 \cdot s^2)$
magnetic flux density	tesla, weber per square meter	T, Wb/m^2	$kg/(A \cdot s^2)$
magnetic field strength	ampere per meter		A/m
magnetomotive force	ampere		A
luminous flux	lumen	lm	$cd \cdot sr$
luminance	candela per square meter		cd/m^2
illumination	lux, lumen per square meter	lx, lm/m^2	$cd \cdot sr/m^2$
activity (of radionuclides)	becquerel	Bq	s^{-1}
absorbed dose	gray	Gy, J/kg	m^2/s^2
dose equivalent	sievert	Sv, J/kg	m^2/s^2

*Supplementary units are: plane angle, radian (rad); solid angle, steradian (sr).

TABLE 3. Prefixes for units in the International System

Prefix	Symbol	Power	Example	Prefix	Symbol	Power	Example
yotta	Y	10^{24}		deci	d	10^{-1}	
zetta	Z	10^{21}		centi	c	10^{-2}	centimeter (cm)
exa	E	10^{18}		milli	m	10^{-3}	milligram (mg)
peta	P	10^{15}		micro	μ	10^{-6}	microgram (μg)
tera	T	10^{12}	terawatt (TW)	nano	n	10^{-9}	nanosecond (ns)
giga	G	10^{9}	gigawatt (GW)	pico	p	10^{-12}	picofarad (pF)
mega	M	10^{6}	megahertz (MHz)	femto	f	10^{-15}	femtosecond (fs)
kilo	k	10^{3}	kilometer (km)	atto	a	10^{-18}	
hecto	h	10^{2}		zepto	z	10^{-21}	
deka	da	10^{1}		yocto	y	10^{-24}	

In the SI the prefixes differ from a unit in steps of 10^3. A list of prefix terms, symbols, and their factors is given in **Table 3.** Some examples of the use of these prefixes follow:

1000 m = 1 kilometer = 1 km
1000 V = 1 kilovolt = 1 kV
1 000 000 Ω = 1 megohm = 1 MΩ
0.000 000 001 s = 1 nanosecond = 1 ns

Only one prefix is to be employed for a unit. For example:

1000 kg = 1 Mg not 1 kkg
10^{-9} s = 1 ns not 1 mμs
1 000 000 m = 1 Mm not 1 kkm

Also, when a unit is raised to a power, the power applies to the whole unit including the prefix. For example:

$$\text{km}^2 = (\text{km})^2 = (1000\ \text{m})^2 = 10^6\ \text{m}^2$$
$$\text{not } 1000\ \text{m}^2$$

Some common units defined in terms of SI units are given in **Table 4** (the definitions in the fourth column are exact).

TABLE 4. Some common units defined in terms of SI units

Quantity	Name of unit	Unit symbol	Definition of unit
length	inch	in.	2.54×10^{-2} m
mass	pound (avoirdupois)	lb	0.45359237 kg
force	kilogram-force	kgf	9.80665 N
pressure	atmosphere	atm	101325 Pa
pressure	torr	torr	(101325/760) Pa
pressure	conventional millimeter of mercury*	mmHg	$13.5951 \times 980.665 \times 10^{-2}$ Pa
energy	kilowatt-hour	kWh	3.6×10^{6} J
energy	thermochemical calorie	cal	4.184 J
energy	international steam table calorie	cal_{IT}	4.1868 J
thermodynamic temperature (*T*)	degree Rankine	°R	(5/9) K
customary temperature (*t*)	degree Celsius	°C	$t(°C) = T(K) - 273.15$
customary temperature (*t*)	degree Fahrenheit	°F	$t(°F) = T(°R) - 459.67$
radioactivity	curie	Ci	3.7×10^{10} Bq
energy†	electronvolt	eV	$\text{eV} \approx 1.60218 \times 10^{-19}$ J
mass†	unified atomic mass unit	u	$\text{u} \approx 1.66054 \times 10^{-27}$ kg

*The conventional millimeter of mercury, symbol mmHg (not mm Hg), is the pressure exerted by a column exactly 1 mm high of a fluid of density exactly 13.5951 g · cm^{-3} in a place where the gravitational acceleration is exactly 980.665 cm · s^{-2}. The mmHg differs from the torr by less than 2×10^{-7} torr.

†These units defined in terms of the best available experimental values of certain physical constants may be converted to SI units. The factors for conversion of these units are subject to change in the light of new experimental measurements of the constants involved.

Conversion factors for the measurement systems

The Encyclopedia has retained the U.S. Customary and metric systems, but has incorporated SI units. Conversion factors between the three measurement systems are given in **Table 5** for some prevalent units; in each of the subtables the user proceeds as follows:

To convert a quantity expressed in a unit in the left-hand column to the equivalent in a unit in the top row of a subtable, multiply the quantity by the factor common to both units. For example, to

TABLE 5. Conversion factors for the U.S. Customary System, metric system, and International System

A. Units of length

Units	cm	m	in.	ft	yd	mi
1 cm	=1	0.01*	0.3937008	0.03280840	0.01093613	6.213712×10^{-6}
1 m	= 100.	1	39.37008	3.280840	1.093613	6.213712×10^{-4}
1 in.	= 2.54*	0.0254	1	0.08333333 . . .	0.02777777 . . .	1.578283×10^{-5}
1 ft	= 30.48	0.3048	12.*	1	0.3333333 . . .	$1.893939 \ldots \times 10^{-4}$
1 yd	= 91.44	0.9144	36.	3.*	1	$5.681818 \ldots \times 10^{-4}$
1 mi	$= 1.609344 \times 10^{5}$	1.609344×10^{3}	6.336×10^{4}	5280.*	1760.	1

B. Units of area

Units	cm^2	m^2	$in.^2$	ft^2	yd^2	mi^2
1 cm^2	= 1	10^{-4}*	0.1550003	1.076391×10^{-3}	1.195990×10^{-4}	3.861022×10^{-11}
1 m^2	$= 10^{4}$	1	1550.003	10.76391	1.195990	3.861022×10^{-7}
1 $in.^2$	= 6.4516*	6.4516×10^{-4}	1	$6.944444 \ldots \times 10^{-3}$	7.716049×10^{-4}	2.490977×10^{-10}
1 ft^2	= 929.0304	0.09290304	144.*	1	0.1111111 . . .	3.587007×10^{-8}
1 yd^2	= 8361.273	0.8361273	1296.	9.*	1	3.228306×10^{-7}
1 mi^2	$= 2.589988 \times 10^{10}$	2.589988×10^{6}	4.014490×10^{9}	2.78784×10^{7}*	3.0976×10^{6}	1

C. Units of volume

Units	m^3	cm^3	liter	$in.^3$	ft^3	qt	gal
1 m^3	= 1	10^{6}	10^{3}	6.102374×10^{4}	35.31467×10^{-3}	1.056688	264.1721
1 cm^3	$= 10^{-6}$	1	10^{-3}	0.06102374	3.531467×10^{-5}	1.056688×10^{3}	2.641721×10^{-4}
1 liter	$= 10^{-3}$	1000.*	1	61.02374	0.03531467	1.056688	0.2641721
1 $in.^3$	$= 1.638706 \times 10^{-5}$	16.38706*	0.01638706	1	5.787037×10^{-4}	0.01731602	4.329004×10^{-3}
1 ft^3	$= 2.831685 \times 10^{-2}$	28316.85	28.31685	1728.*	1	2.992208	7.480520
1 qt	$= 9.46353 \times 10^{-4}$	946.353	0.946353	57.75	0.0342014	1	0.25
1 gal (U.S.)	$= 3.785412 \times 10^{-3}$	3785.412	3.785412	231.*	0.1336806	4.*	1

D. Units of mass

Units	g	kg	oz	lb	metric ton	ton
1 g	= 1	10^{-3}	0.03527396	2.204623×10^{-3}	10^{-6}	1.102311×10^{-6}
1 kg	= 1000.	1	35.27396	2.204623	10^{-3}	1.102311×10^{-3}
1 oz (avdp)	= 28.34952	0.02834952	1	0.0625	2.834952×10^{-5}	3.125×10^{-5}
1 lb (avdp)	= 453.5924	0.4535924	16.*	1	4.535924×10^{-4}	$5. \times 10^{-4}$
1 metric ton	$= 10^{6}$	1000.*	35273.96	2204.623	1	1.102311
1 ton	= 907184.7	907.1847	32000.	2000.*	0.9071847	1

TABLE 5. Conversion factors for the U.S. Customary System, metric system, and International System (cont.)

E. Units of density

Units	g · cm^{-3}	g · L^{-1}, kg · m^{-3}	oz · in.$^{-3}$	lb · in.$^{-3}$	lb · ft^{-3}	lb · gal^{-1}
1 g · cm^{-3}	= 1	1000.	0.5780365	0.03612728	62.42795	8.345403
1 g · L^{-1}, kg · m^{-3}	= 10^{-3}	1	5.780365×10^{-4}	3.612728×10^{-5}	0.06242795	8.345403×10^{-3}
1 oz · in.$^{-3}$	= 1.729994	1729.994	1	0.0625	108.	14.4375
1 lb · in.$^{-3}$	= 27.67991	27679.91	16.	1	1728.	231.
1 lb · ft^{-3}	= 0.01601847	16.01847	9.259259×10^{-3}	5.7870370×10^{-4}	1	0.1336806
1 lb · gal^{-1}	= 0.1198264	119.8264	4.749536×10^{-3}	4.3290043×10^{-3}	7.480519	1

F. Units of pressure

Units	Pa, N · m^{-2}	dyn · cm^{-2}	bar	atm	kgf · cm^{-2}	mmHg (torr)	in. Hg	lbf · in.$^{-2}$
1 Pa, 1 N · m^{-2}	= 1	10	10^{-5}	9.869233×10^{-6}	1.019716×10^{-5}	7.500617×10^{-3}	2.952999×10^{-4}	1.450377×10^{-4}
1 dyn · cm^{-2}	= 0.1	1	10^{-6}	9.869233×10^{-7}	1.019716×10^{-6}	7.500617×10^{-4}	2.952999×10^{-5}	1.450377×10^{-5}
1 bar	= 10^{5}*	10^{6}	1	0.9869233	1.019716	750.0617	29.52999	14.50377
1 atm	= 101325.0*	1013250.	1.013250	1	1.033227	760.	29.92126	14.69595
1 kgf · cm^{-2}	= 98066.5	980665.	0.980665	0.9678411	1	735.5592	28.95903	14.22334
1 mmHg (torr)	= 133.3224	1333.224	1.333224×10^{-3}	1.3157895×10^{-3}	$1.35955099 \times 10^{-3}$	1	0.03937008	0.01933678
1 in. Hg	= 3386.388	33863.88	0.03386388	0.03342105	0.03453155	25.4	1	0.4911541
1 lbf · in.$^{-2}$	= 6894.757	68947.57	0.06894757	0.06804596	0.07030696	51.71493	2.036021	1

G. Units of energy

Units	g mass (energy equiv)	J	eV	cal	cal$_{IT}$	Btu$_{IT}$	kWh	hp-h	ft-lbf	ft^3 · lbf in.$^{-2}$	liter-atm
1 g mass (energy equiv)	= 1	8.987552×10^{13}	5.609586×10^{32}	2.148076×10^{-3}	2.146640×10^{13}	8.51855×10^{10}	2.496542×10^{7}	3.347918×10^{7}	6.628878×10^{13}	4.603388×10^{11}	8.870024×10^{11}
1 J	= 1.112650×10^{-14}	1	6.241506×10^{18}	0.2390057	0.2388459	9.478172×10^{-4}	$2.777777\ldots \times 10^{-7}$	3.725062	0.7375622	5.121960×10^{-3}	9.869233×10^{-3}
1 eV	= 1.782663×10^{-33}	1.602177×10^{-19}	1	3.829296×10^{-20}	3.826735×10^{-20}	1.518571×10^{-22}	4.450493×10^{-26}	5.968209×10^{-26}	1.181705×10^{-19}	8.206287×10^{-22}	1.581226×10^{-21}
1 cal	= 4.655328×10^{-14}	4.184*	2.611446×10^{19}	1	0.9993312	3.965667×10^{-3}	$1.1622222\ldots \times 10^{-6}$	1.558562×10^{-6}	3.085960	2.143028×10^{-2}	0.04129287
1 cal$_{IT}$	= 4.658443×10^{-14}	4.1868*	2.613194×10^{19}	1.000669	1	3.968321×10^{-3}	1.163000×10^{-6}	1.559609×10^{-6}	3.088025	2.144462×10^{-2}	0.04132050
1 Btu$_{IT}$	= 1.173908×10^{-11}	1055.056	6.585138×10^{21}	252.1644	251.9958	1	2.930711×10^{-4}	3.930148×10^{-4}	778.1693	5.403953	10.41259
1 kWh	= 4.005540×10^{-8}	3600000.*	2.246942×10^{25}	860420.7	859845.2	3412.142	1	1.341022	2655224.	18349.06	35529.24
1 hp-h	= 2.986931×10^{-8}	2684519.	1.675545×10^{25}	641615.6	641186.5	2544.33	0.7456998	1	1980000.*	13750.	26494.15
1 ft-lbf	= 1.508551×10^{-14}	1.355818	8.462346×10^{18}	0.3240483	0.3238315	1.285067×10^{-3}	3.766161×10^{-7}	$5.050505\ldots \times 10^{-7}$	1	$6.944444\ldots \times 10^{-3}$	0.01338088
1 ft^3 · lbf · in.$^{-2}$	= 2.172313×10^{-12}	195.2378	1.218578×10^{21}	46.66295	46.63174	0.1850497	5.423272×10^{-5}	$7.272727\ldots \times 10^{-5}$	144.*	1	1.926847
1 liter-atm	= 1.127393×10^{-12}	101.3250	6.324206×10^{20}	24.21726	24.20106	0.09603757	2.814583×10^{-5}	3.774419×10^{-5}	74.73349	0.5189825	1

convert 7 ft to the equivalent in meters, go to subtable A, "Units of Length," and find 1 ft in the left-hand column and m in the top row. The conversion factor common to these units is 0.3048. Therefore, 7 ft = 7 × 0.3048 = 2.1336 m.

The conversion factors have been carried out to seven significant figures, as derived from the fundamental constants and the definitions of the units. However, this does not mean that the factors are always known to that accuracy. Numbers followed by ellipses are to be continued indefinitely with repetition of the same pattern of digits. Factors written with fewer than seven significant digits are exact values. Numbers followed by an asterisk are definitions of the relation between the two units.

Units of temperature in measurement systems

Temperature is a basic physical quantity. It is a measure of the thermal energy of random motion of particles in a system. As such it has been chosen as one of the base quantities in the SI. It is to be treated as are the units of length, mass, time, electric current, and luminous intensity. In the SI the unit of length is the meter, the unit of time the second, and so on. The question arises as to the choice of the unit of temperature in the SI.

In the past it was customary to refer to scales of temperature, for example, the Celsius and Fahrenheit scales. On the Celsius scale, 0 designates the freezing point (ice point) and 100 the boiling point (steam point) of water. Corresponding numbers on the Fahrenheit scale are 32 and 212. There are 100 units between the ice point and steam point on the Celsius scale, and 180 units between these points in the Fahrenheit system.

By measuring the volume changes of a gas within the 100-unit interval of the ice point and steam point of water on the Celsius scale, it was found that a numerical value could be assigned for a basic unit of temperature. Careful measurement of this ice-steam interval in a gas thermometer determined that the ice point of water should be assigned the value of 273.15 kelvins. The unit of temperature was thus called the kelvin with the symbol K. Further experiments led to the decision to define the kelvin in the SI along the same lines but in terms of the triple point of water. This is the temperature and pressure at which ice, liquid water, and water vapor coexist at equilibrium. The triple point was chosen because it was a more reproducible value than the ice point. This change led to the SI definition of temperature in terms of the triple point of water, which contains exactly 273.16 kelvins. The scale now in use is the International Temperature Scale of 1990 (ITS-90) [**Table 6**].

The Celsius temperature (°C) is an intermediate scale. It is useful in defining Kelvin temperature in the SI. Celsius temperature (t) is related to Kelvin temperature (K) as follows:

TABLE 6. Defining fixed points of the International Temperature Scale of 1990 (ITS-90)

Equilibrium state*	Temperature	
	T_{90}, K	t_{90}, °C
Vapor pressure equation of helium	3 to 5	−270.15 to −268.15
Triple point of equilibrium hydrogen†	13.8033	−259.3467
Vapor pressure point of equilibrium hydrogen† (or constant volume gas thermometer point of helium)	≈17	≈−256.15
Vapor pressure point of equilibrium hydrogen† (or constant volume gas thermometer point of helium)	≈20.3	≈−252.85
Triple point of neon	24.5561	−248.5939
Triple point of oxygen	54.3584	−218.7916
Triple point of argon	83.8058	−189.3442
Triple point of mercury	234.3156	−38.8344
Triple point of water	273.16	0.01
Melting point of gallium	302.9146	29.7646
Freezing point of indium	429.7485	156.5985
Freezing point of tin	505.078	231.928
Freezing point of zinc	692.677	419.527
Freezing point of aluminum	933.473	660.323
Freezing point of silver	1234.93	961.78
Freezing point of gold	1337.33	1064.18
Freezing point of copper	1357.77	1084.62

*The triple point is the equilibrium temperature at which the solid, liquid, and vapor phases coexist. The freezing point and the melting point are the equilibrium temperatures at which the solid and liquid phases coexist under a pressure of 101,325 Pa, 1 standard atmosphere. The isotopic composition is that naturally occurring.

†Equilibrium hydrogen is hydrogen with the equilibrium distribution of its ortho and para states at the corresponding temperatures. Normal hydrogen at room temperature contains 25% para and 75% ortho hydrogen.

$$t_{\text{ice point}} = 0°C$$

$$t_{\text{steam point}} = 100°C$$

$$0\ K = -273.15°C$$

A summary of the conventions in the SI as proposed in the Thirteenth General Conference of Weights and Measures pertaining to temperature units is given below.

1. The unit of SI temperature is the kelvin, symbol K.
2. The word "scale" is not to be used except in terms of measurement of temperature between certain fixed points on the Celsius scale.
3. The terms "thermodynamic scale" or "absolute scale" are not to be used to describe temperature. The degree sign is to be eliminated with the symbol K.
4. When Celsius temperatures are used (°C), it is understood that the temperature unit is the kelvin.

Not all scientists and engineers have adopted the SI of temperature terminology. For this reason the contributors to the Encyclopedia have retained the term "scale" in relation to thermodynamic temperature. Furthermore, many engineers in the United States still use the Fahrenheit system in discussing practical engineering systems.

In converting Fahrenheit (°F) to Celsius (°C) the following formula applies.

$$°C = \frac{°F - 32°}{1.8}$$

In converting Celsius to Fahrenheit the following formula can be used.

$$°F = (°C \times 1.8) + 32°$$

In changing from Celsius terminoloy (t) to kelvin units (K) the following formula can be used.

$$K = t + 273.15$$

Chemical elements

The mass number, atomic number, number of atoms, and ionic charge of an element are indicated in the Encyclopedia by means of four indices

Table 7. The chemical elements

Name	Symbol	At. no.	Name	Symbol	At. no.	Name	Symbol	At. no.
Actinium	Ac	89	Gold	Au	79	Praseodymium	Pr	59
Aluminum	Al	13	Hafnium	Hf	72	Promethium	Pm	61
Americium	Am	95	Hassium	Hs	108	Protactinium	Pa	91
Antimony	Sb	51	Helium	He	2	Radium	Ra	88
Argon	Ar	18	Holmium	Ho	67	Radon	Rn	86
Arsenic	As	33	Hydrogen	H	1	Rhenium	Re	75
Astatine	At	85	Indium	In	49	Rhodium	Rh	45
Barium	Ba	56	Iodine	I	53	Rubidium	Rb	37
Berkelium	Bk	97	Iridium	Ir	77	Ruthenium	Ru	44
Beryllium	Be	4	Iron	Fe	26	Rutherfordium	Rf	104
Bismuth	Bi	83	Krypton	Kr	36	Samarium	Sm	62
Bohrium	Bh	107	Lanthanum	La	57	Scandium	Sc	21
Boron	B	5	Lawrencium	Lr	103	Seaborgium	Sg	106
Bromine	Br	35	Lead	Pb	82	Selenium	Se	34
Cadmium	Cd	48	Lithium	Li	3	Silicon	Si	14
Calcium	Ca	20	Lutetium	Lu	71	Silver	Ag	47
Californium	Cf	98	Magnesium	Mg	12	Sodium	Na	11
Carbon	C	6	Manganese	Mn	25	Strontium	Sr	38
Cerium	Ce	58	Meitnerium	Mt	109	Sulfur	S	16
Cesium	Cs	55	Mendelevium	Md	101	Tantalum	Ta	73
Chlorine	Cl	17	Mercury	Hg	80	Technetium	Tc	43
Chromium	Cr	24	Molybdenum	Mo	42	Tellurium	Te	52
Cobalt	Co	27	Neodymium	Nd	60	Terbium	Tb	65
Copper	Cu	29	Neon	Ne	10	Thallium	Tl	81
Curium	Cm	96	Neptunium	Np	93	Thorium	Th	90
Dubnium	Db	105	Nickel	Ni	28	Thulium	Tm	69
Dysprosium	Dy	66	Niobium	Nb	41	Tin	Sn	50
Einsteinium	Es	99	Nitrogen	N	7	Titanium	Ti	22
Element 110*		110	Nobelium	No	102	Tungsten	W	74
Erbium	Er	68	Osmium	Os	76	Uranium	U	92
Europium	Eu	63	Oxygen	O	8	Vanadium	V	23
Fermium	Fm	100	Palladium	Pd	46	Xenon	Xe	54
Fluorine	F	889	Phosphorus	P	15	Ytterbium	Yb	70
Francium	Fr	87	Platinum	Pt	78	Yttrium	Y	39
Gadolinium	Gd	64	Plutonium	Pu	94	Zinc	Zn	30
Gallium	Ga	31	Polonium	Po	84	Zirconium	Zr	40
Germanium	Ge	32	Potassium	K	19			

*Element 110 has been reported, but no official name or symbol has yet been assigned.

placed around the symbol. The positions occupied are left upper index, mass number; left lower index, atomic number; right upper index, ionic charge; and right lower index, number of atoms of an element in a molecule or formula unit of a given species: for example, ${}^{12}_{6}C$, Ca^{2+}, O_2, and Al_2O_3. The atomic number, which is redundant, is omitted in most cases; that is ${}^{12}_{6}C$ can be written as ${}^{12}C$.

Ionic charge is indicated by a plus or minus superscript following the symbol of the ion; for multiple charges an arabic superscript number precedes the plus or minus sign, for example, Na^+, No_3^-, Ca^{2+}, PO_4^{3-}.

An alphabetical list of the elements, their symbols, and atomic numbers is shown in **Table 7.**

Chemical nomenclature

The International Union of Pure and Applied Chemistry (IUPAC) has established definitive rules for chemical nomenclature. Chemical species are identified in the Encyclopedia by a systematic name, frequently accompanied by a formula. Occasionally certain well-established so-called trivial names are used.

For inorganic compounds, systematic names of compounds are formed by identifying the constituents and their proportions in a specific order, for example, dinitrogen oxide (N_2O). Also accepted by IUPAC is Stock's system, in which the proportions of the constituents are indicated indirectly, and roman numerals are used to represent the oxidation number or stoichiometric valence of an element, for example, iron(II) chloride ($FeCl_2$). Complex compounds are also named according to rules specified by IUPAC; an example is potassium oxodichloroimidophosphate, $K[POC]_2(NH)]$. Examples of accepted trivial names are diborane (B_2H_6), silane (SiH_4), and ammonia (NH_3).

There also are definitive rules for naming organic compounds. Because of the infinite variety of disciplines and industrial applications involving organic compounds, the rules encompass different types of names. Sometimes a single compound can correctly be identified by a number of names, for example, chloral hydrate is also known as 2,2,2-trichloro-1,1-ethanediol and trichloroacetaldehyde monohydrate.

Symbols in scientific writing

Throughout the Encyclopedia, symbols have been introduced in such a way that their translation into words or phrases will require minimal effort on the part of the reader. In most cases a symbol is defined at its first appearance in an article. For example:

"The energy E in a quantum of radiation of frequency ν (where the frequency is equal to the velocity of the radiation in a given medium divided by its wavelength in the same medium) is directly proportional to the frequency, or inversely proportional to the wavelength, according to the relation given in Eq. (6),

$$E = h\nu \qquad (6)$$

where h is a universal constant known as Planck's constant. The value of h is 6.63×10^{-34} joule-second, and if ν is expressed in s^{-1}, E is given in joules per quantum."

For convenience, symbols commonly encountered in scientific writing are listed in **Table 8.** Symbols following the ellipses and separated by commas are alternatives that are used only when there is some reason for not using the symbol given first.

Some frequently encountered symbols for particles and quanta are as follows:

neutron	n	pion	π
proton	p	muon	μ
deuteron	d	electron	e
triton	t	neutrino	ν
alpha particle	α	photon	γ

Table 8. Commonly used symbols in scientific writing

Space, time, mass, and related quantities

length l
height h
radius r
diameter d
path, length of arc s
plane angle $\alpha,\beta,\gamma,\theta,\phi,\psi$
solid angle ω
area A,S
volume $V \ldots \upsilon$
specific volume υ
wavelength λ
wavenumber σ,ν
time t
period or other characteristic interval T,τ
frequency ν,f
angular frequency ($2\pi\nu$) ω
velocity $\upsilon \ldots u,w$
angular velocity ω
acceleration a
acceleration of free fall g
mass m
moment of inertia I
density ρ
relative density d

Molecular and related quantities

molecular mass m
molar mass M
Avogadro's number N_0,L,N
number of molecules N
number of moles n
mole fraction $x \ldots X,y$
molality m
concentration c
molar concentration of substance B c_B,[B], c(B)
molecular concentration C
partition function Q
statistical weight $g \ldots p$
symmetry number σ
characteristic temperature θ
diameter of molecule $\sigma \ldots D$
mean free path l
diffusion coefficient D
osmotic pressure Π
surface concentration Γ

Mechanical and related quantities

force F
force due to gravity (weight) $G \ldots W$
moment of force M
power P
pressure p,P
traction σ
shear stress τ
modulus of elasticity E
shear modulus G
compressibility κ
compression modulus ($1/\kappa$) K
viscosity η
fluidity ϕ
kinematic viscosity ν
friction coefficient f
surface tension $\gamma \ldots \sigma$
angle of contact θ

Thermodynamic and related quantities

temperature $\theta \ldots t$
temperature, absolute T
gas constant R
Boltzmann constant k
heat q,Q
work w,A
energy (Gibbs ϵ) $E \ldots U$
entropy (Gibbs η) S
*Helmholtz free energy (Gibbs ψ) A
enthalpy (Gibbs χ) H
*Gibbs function (ζ) $G \ldots F$
heat capacity C
specific heat at constant pressure c_p
specific heat at constant volume c_υ
ratio of specific heats c_p/c_υ γ,κ
chemical potential μ
activity, absolute λ
activity, relative a
activity coefficient f,γ
osmotic coefficient g,ϕ
thermal conductivity γ
Joule-Thomson coefficient μ

*The terms for the Helmholtz and Gibbs energies were modified by action of the IUPAC Council, Montreal, August 1961, as follows:
Helmholtz energy (Gibbs $\psi = E - TS$): A Gibbs energy (Gibbs $\zeta = H - TS$): G

(continued)

Table 8. Commonly used symbols in scientific writing (cont.)

Chemical reactions

stoichiometric number of molecules (negative for reactants, positive for products) ν
standard equation of chemical reaction $\Sigma \nu_B B = 0$
affinity ($-\Sigma \nu_B \mu_B$) of a reaction A
equilibrium constant K
equilibrium quotient or equilibrium product (of molalities) Q
extent of reaction ($dn_B = \nu_B d\xi$) ξ
degree of reaction (e.g., degree of dissociation) α
rate constant k
collision number (collisions per unit volume and unit time) Z
rate constant corresponding to the rate Z z
rate of reaction $\nu \ldots r,s,J$

Light

Planck's constant h
Planck's constant divided by 2π $\hbar$
quantity of light Q
radiant power, flux of light (dQ/dt) Φ
luminous intensity ($d\Phi/d\omega$) I
illumination ($d\Phi/dS$) E
luminance L,B
luminous emittance H
absorption factor (fraction of incident radiant power which is absorbed) α
reflection factor (fraction of incident radiant power which is reflected) ρ
transmission factor (fraction of incident radiant power which is transmitted) τ
transmittance ($T = I/I_0$) T
absorption (extinction) coefficient [$\kappa/c = \ln(1/T)$] κ
absorbance (extinction) [$A = \log(1/T)$] $A \ldots E$
absorptivity (specific absorbance) [decadic absorption or extinction coefficient] a
molar absorptivity (molar decadic absorption or extinction coefficient) [$\varepsilon/c = A$] ϵ
refraction index n
refractivity r
angle of optical rotation α

Electricity and magnetism

elementary charge e
quantity of electricity Q
charge density ρ
surface charge density σ
electric current $I \ldots i$
electric current density J
electric potential V
electric field strength E
electric displacement D
electrokinetic potential ζ
capacitance C
permittivity (dielectric constant) ϵ
dielectric polarization P
dipole moment μ
electric polarizability of a molecule α,γ
magnetic field strength H
magnetic induction B
magnetic permeability μ
magnetization M
magnetic susceptibility χ
resistance R
resistivity ρ
self inductance L
mutual inductance M,L_{12}
reactance X
impedance Z
admittance Y

Electrochemistry

Faraday's constant (the faraday) F
charge number of an ion, plus or minus z
degree of electrolytic dissociation α
ionic strength $I \ldots \mu$
electrolytic conductivity (specific conductance) κ
equivalent or molar conductance of electrolyte or ion A
transport number t,T
electromotive force E
overpotential η

The meaning of abbreviated notations for nuclear reactions should be the following:

initial nuclide (incoming particle(s) or quanta, outgoing particle(s) or quanta) final nuclide

Some examples are:

$^{14}N(\alpha, p)^{17}O$ $\qquad$ $^{59}Co(n, \gamma)^{60}Co$
$^{23}Na(\gamma, 3n)^{20}Na$ $\qquad$ $^{31}P(\gamma, pn)^{29}Si$

The Greek alphabet is frequently used to represent terms. A listing of the Greek alphabet is shown in **Table 9.**

Table 9. Greek alphabet

Upper and lower cases	Name	Upper and lower cases	Name
Α α	Alpha	Ν ν	Nu
Β β	Beta	Ξ ξ	Xi
Γ γ	Gamma	Ο ο	Omicron
Δ δ ∂	Delta	Π π	Pi
Ε ϵ ε	Epsilon	Ρ ρ	Rho
Ζ ζ	Zeta	Σ σ	Sigma
Η η	Eta	Τ τ	Tau
Θ θ	Theta	Υ υ	Upsilon
Ι ι	Iota	Φ ϕ φ	Phi
Κ κ	Kappa	Χ χ	Chi
Λ λ	Lambda	Ψ ψ	Psi
Μ μ	Mu	Ω ω	Omega

Mathematical Notation

Common signs and symbols.

$+$	Plus (sign of addition)
$+$	Positive
$-$	Minus (sign of subtraction)
$-$	Negative
$\pm$ ($\mp$)	Plus or minus (minus or plus)
$\times$	Times, by (multiplication sign)
$\cdot$	Multiplied by
$\div$	Sign of division
$/$	Divided by
$:$	Ratio sign, divided by, is to
$::$	Equals, as (proportion)
$<$	Less than
$>$	Greater than
$\ll$	Much less than
$\gg$	Much greater than
$=$	Equals
$\equiv$	Identical with
$\sim$	Similar to
$\approx$	Approximately equals
$\cong$	Approximately equals, congruent
$\leq$	Equal to or less than
$\geq$	Equal to or greater than
$\not\equiv$ $\neq$	Not equal to
$\rightarrow$ $\doteq$	Approaches
$\propto$	Varies as
∞	Infinity
$\sqrt{\ }$	Square root of
$\sqrt[3]{\ }$	Cube root of
$\therefore$	Therefore
$\parallel$	Parallel to
( ) [ ] { }	Parentheses, brackets and braces: quantities enclosed by them to be taken together in multiplying, dividing, etc.
$\overline{AB}$	Length of line from A to B
π	(pi) = 3.14159+
$^\circ$	Degrees
$'$	Minutes
$''$	Seconds
$\angle$	Angle
dx	Differential of x
Δ	(delta) difference
Δx	Increment of x
$\partial u/\partial x$	Partial derivative of u with respect to x
$\int$	Integral of
$\int_b^a$	Integral of, between limits a and b
$\oint$	Line integral around a closed path
Σ	(sigma) summation of
$f(x)$, $F(x)$	Functions of x
∇	Del or nabla, vector differential operator
∇^2	Laplacian operator
£	Laplace operational symbol
$4!$	Factorial $4 = 1 \times 2 \times 3 \times 4$
$\lvert x \rvert$	Absolute value of x
$\dot{x}$	First derivative of x with respect to time
$\ddot{x}$	Second derivative of x with respect to time
$\mathbf{A} \times \mathbf{B}$	Vector product: magnitude of **A** times magnitude of **B** times sine of the angle from **A** to **B;** $AB \sin \overline{AB}$
$\mathbf{A} \cdot \mathbf{B}$	Scalar product of **A** and **B;** magnitude of **A** times magnitude of **B** times cosine of the angle from **A** to **B;** $AB \cos \overline{AB}$

Mathematical logic.

$p, q, P(x)$	Sentences, propositional functions, propositions
$-p$, $\sim p$, non p, Np	Negation, read "not p" ($\neq$: read "not equal")
$p \vee q$, $p + q$, Apq	Disjunction, read "p or q," "p,q," or both
$p \wedge q$, $p \cdot q$, $p \,\&\, q$, Kpq	Conjunction, read "p and q"
$p \rightarrow q$, $p \supset q$, $p \Rightarrow q$, Cpq	Implication, read "p implies q" or "if p then q"
$p \leftrightarrow q$, $p \equiv q$, $p \Leftrightarrow q$, Epq, p if q	Equivalence, read "p is equivalent to q" or "p if and only if q"
n.a.s.c.	Read "necessary and sufficient condition"
(), [], {}, . ., · ·	Parentheses
$\forall$, $\forall$, Σ	Universal quantifier, read "for all" or "for every"
$\exists$, $\exists$, Π	Existential quantifier, read "there is a" or "there exists"
$\vdash$	Assertion sign ($p \vdash q$: read "q follows from p"; $\vdash p$: read "p is or follows from an axiom," or "p is a tautology"
0, 1	Truth, falsity (values)
$=$	Identity
$\stackrel{DF}{=}$, $\stackrel{df}{=}$, $\underset{df}{=}$, $\equiv$	Definitional identity
▮	"End of proof": "QED"

Set theory, relations, functions.

X, Y	Sets
$x \in X$	x is a member of the set X
$x \notin X$	x is not a member of X
$A \subset X$, $A \subseteq X$	Set A is contained in set X
$A \not\subset X$, $A \not\subseteq X$	A is not contained in X
$X \cup Y$, $X + Y$	Union of sets X and Y
$X \cap Y$, $X \cdot Y$	Intersection of sets X and Y
$+$, $\dot{+}$, $\bigcirc$	Symmetric difference of sets
$\cup X_i$, ΣX_i	Union of all the sets X_i
$\cap X_i$, $\cdot \Pi X_i$	Intersection of all the sets X_i
$\varnothing$, 0, $\wedge$	Null set, empty set
X', **C**X, CX	Complement of the set X
$X - Y$, $X \backslash Y$	Difference of sets X and Y

$\hat{x}\,(P(x))$, $\{x \mid P(x)\}$, $\{x{:}P(x)\}$	The set of all x with the property P
(x,y,z), $\langle x,y,z\rangle$	Ordered set of elements x, y, and z: to be distinguished from (x,z,y), for example
$\{x,y,z\}$	Unordered set, the set whose elements are x, y, z, and no others
$\{a_1, a_2, \ldots, a_n\}$, $\{a_i\}_{i=1,2,\ldots,n}$, $\{a_i\}_{i=1}^{n}$	The set whose members are a_i, where i is any whole number from 1 to n
$\{a_1, a_2, \ldots\}$, $\{a_i\}_{i=1,2,\ldots}$, $\{a_i\}_{i=1}^{\infty}$	The set whose members are a_i, where i is any positive whole number
$X \times Y$	Cartesian product, set of all (x,y) such that $x \in X$, $y \in Y$
$\{a_i\}_{i \in I}$	The set whose elements are a_i, where $i \in I$
xRy, $R\{x,y\}$	Relation
$\equiv, \cong, \sim, \simeq$	Equivalence relations, for example, congruence
$\geqq, \geq, >, \succ, \gg, \leqq, \leq, <$	Transitive relations, for example, numerical order
$f\colon X \rightarrow Y$, $X \xrightarrow{f} Y$, $X \rightarrow Y$, $f \in Y^X$	Function, mapping, transformation
f^{-1}, $\overset{-1}{f}$ $X \xleftarrow{f^{-1}} Y$	Inverse mapping
$g \circ f$	Composite functions: $(g \circ f)(x) = g\,(f/x))$
$f(X)$	Image of X by f
$f^{-1}(X)$	Inverse-image set, counter image
1-1, one-one	Read "one-to-one correspondence"
$X \xrightarrow{f} Y$, $\phi \downarrow \; g \; \downarrow \psi$, $W \rightarrow Z$	Diagram: the diagram is commutative in case $\psi \circ f = g \circ \phi$
f/A	Partial mapping, restriction of function f to set A
$\overline{\overline{X}}$, card X, $\lvert X \rvert$	Cardinal of the set A
$\aleph_0$, d	Denumerable infinity
c, $2^{\aleph_0}$	Power of continuum
ω	Order type of the set of positive integers
σ-	Read "countably"

Number, numerical functions.

1.4; 1,4; 1·4	Read "one and four-tenths"
1(1)20(10)100	Read "from 1 to 20 in intervals of 1, and from 20 to 100 in intervals of 10"
const	Constant
$A \geqq 0$	The number A is nonnegative, or, the matrix A is positive definite, or, the matrix A has nonnegative entries
x/y	Read x divides y
$x \equiv y \bmod p$	Read "x congruent to y modulo p"
$a_0 + \dfrac{1}{a_1} + \dfrac{1}{a_2} + \ldots$, $a_0 + \left.\dfrac{1}{\lvert a_1}\right. + \ldots$	Continued fractions

$[a,b]$	Closed interval
$[a,b)$, $[a,b[$	Half-open interval (open at the right)
(a,b), $]a,b[$	Open interval
$[a,\infty)$, $[a,\rightarrow[$	Interval closed at the left, infinite to the right
$(-\infty,\infty)$, $]\leftarrow,\rightarrow[$	Set of all real numbers
$\max_{x \in X} f(x)$, $\max\{f(x) \mid x \in X\}$	Maximum of $f(x)$ when x is in the set X
min	Minimum
sup, l.u.b.	Supremum, least upper bound
inf, g.l.b.	Infimum, greatest lower bound
$\lim_{x \rightarrow a} f(x) = b$, $\lim_{x=a} f(x) = b$, $f(x) \rightarrow b$ as $x \rightarrow a$	b is the limit of $f(x)$ as x approaches a
$\lim_{x \rightarrow a-} f(x)$, $\lim_{x=a-0} f(x)$, $f(a-)$	Limit of f(x) as x approaches a from the left
lim sup, $\overline{\lim}$	Limit superior
lim inf, $\underline{\lim}$	Limit inferior
l.i.m.	Limit in the mean
$z = x + iy = re^{i\theta}$, $\zeta = \xi + i\eta$, $w = u + iv + \rho e^{i\phi}$	Complex variables
z^*	Complex conjugate
Re, $\Re$	Real part
Im, $\Im$	Imaginary part
arg	Argument
$\dfrac{\partial(u,v)}{\partial(x,y)}$, $\dfrac{D(u,v)}{D(x,y)}$	Jacobian, functional determinant
$\int_E f(x), d\mu\,(x)$	Integral (for example, Lebesgue integral) of function f over set E with respect to measure μ
$f(n) \sim \log n$ as $n \rightarrow \infty$	$f(n)/\log n$ approaches 1 as $n \rightarrow \infty$
$f(n) = O(\log n)$ as $n \rightarrow \infty$	$f(n)/\log n$ is bounded as $n \rightarrow \infty$
$f(n) = 0\,(\log n)$	$f(n)/\log n$ approaches zero
$f(x) \nearrow b$, $f(x) \uparrow b$	$f(x)$ increases, approaching the limit b
$f(x) \downarrow b$, $f(x) \searrow b$	$f(x)$ decreases, approaching the limit b
a.e., p.p.	Almost everywhere
ess sup	Essential supremum
C^0, $C^0(X)$, $C(X)$	Space of continuous functions
C^k, $C^k[a,b]$	The class of functions having continuous kth derivative (on $[a,b]$)
Lip_α, Lip α	Lipschitz class of functions
L^p, L_p, $L^p\,[a,b]$	Space of functions having integrable absolute pth power (on $[a,b]$)
(C,α), (C,p)	Cesàro summability

Special functions.

$[x]$	The integral part of x
$\binom{n}{k}$, ${}^nC_k$, ${}_nC_k$	Binomial coefficient $n!/k!\,(n-k)!$

$\left(\frac{n}{p}\right)$	Legendre symbol
e^x, exp x	Exponential function
sinh x, cosh x, tanh x	Hyperbolic functions
sn x, cn x, dn x	Jacobi elliptic functions
$\wp(x)$	Weierstrass elliptic function
$\Gamma(x)$	Gamma function
$J_\nu(x)$	Bessel function
$\chi_X(x)$	Characteristic function of the set X: $\chi_X(x) = 1$ in case $x \in X$, otherwise $\chi_X(x) = 0$
sgn x	Signum: sgn $0 = 0$, while sgn $x = x/\lvert x \rvert$ for $x \neq 0$
$\delta(x)$	Dirac delta function

Algebra, tensors, operators.

$+, \cdot, \times, \circ, \mathrm{T}, \tau$	Laws of composition in algebraic systems
e, 0	Identity, unit, neutral element (of an additive system)
e, I, I	Identity, unit, neutral element (of a general algebraic system)
e, $\mathbf{e}$, E, P	Idempotent
a^{-1}	Inverse of a
Hom(M,N)	Groups of all homomorphisms of M into N
G/H	Factor group, group of cosets
$[K{:}k]$	Dimension of K over k
$\oplus, \dotplus$	Direct sum
$\otimes$	Tensor product, Kronecker product
$\wedge$	Exterior product, Grassmann product
$\vec{x}$, $\mathbf{x}$, $\mathfrak{x}$, x	Vector
$\vec{x} \cdot \vec{y}$, $\mathbf{x} \cdot \mathbf{y}$, $(\mathfrak{x},\mathfrak{y})$	Inner product, scalar product, dot product
$\mathbf{x} \times \mathbf{y}$, $[\mathfrak{x},\mathfrak{y}]$, $\mathbf{x} \wedge \mathbf{y}$	Outer product, vector product, cross product
$\lvert x \rvert$, $\lVert x \rVert$, $\lVert x \rVert_p$	Norm of the vector x
Ax, xA	The image of x under the transformation A
δ_{ij}	Kronecker delta: $\delta_{ij} = 1$, while $\delta_{ij} = 0$ for $i \neq j$
A', tA, A^t, ${}^tA$	Transpose of the matrix A
A^*, $\tilde{A}$	Adjoint, Hermitian conjugate of A
tr A, Sp A	Trace of the matrix A
det A, $\lvert A \rvert$	Determinant of the matrix A
$\Delta^n f(x)$, $\Delta_h{}^n f$, $\underset{h}{\Delta^n} f(x)$	Finite differences
$[x_0, x_1]$, $[x_0, x_1, x_2]$, $[x_0,x_1]_f$	Divided differences
∇f, grad f	Read "gradient of f"
$\nabla \cdot$ v, div v	Read "divergence of v"
$\nabla \times$ v, curl v, rot v	Read "curl of v"
∇^2, ∇, div grad	Laplacian
$[X,Y]$	Poisson bracket, or commutator, or Lie product

GL (n,R)	Full linear group of degree n over field R
O(n,R)	Full orthogonal group
SO (n,R), $\mathrm{O}^+(n,R)$	Special orthogonal group

Geometry.

$\overline{AB}$	Line segment having end points A and B
AB	Length of $\overline{AB}$
$\overleftrightarrow{XY}$	Line containing points X and Y
$\overrightarrow{PQ}$	Ray with end point P and containing Q
$\angle V$	Angle with vertex V
$\angle RST$	Angle formed by $\overrightarrow{SR}$ and $\overrightarrow{ST}$
$\angle x$	Angle named x
$\angle x = 30°$	Angle named x has measure 30°
$\widehat{JK}$	Minor arch with end points J and K
$\widehat{JLK}$	Major arch that contains point L
$\odot C$	Circle with center C
ΔXYZ	Triangle with vertices X, Y, and Z
▭$ABCD$	Rectangle with vertices A, B, C, and D
□$PQRS$	Square with vertices P, Q, R, and S
▱$ABCD$	Parallelogram with vertices A, B, C, and D
$ABCD \ldots X$	Polygon with vertices $A, B, \ldots, X$
$\overline{AB} \perp j$	A segment ($\overline{AB}$) is perpendicular to a line (j)
$\overrightarrow{ZP} \parallel \overline{TV}$	A ray ($\overrightarrow{ZP}$) is parallel to a segment ($\overline{TV}$)
$\overline{AB} \cong \overline{XY}$	Two segments ($\overline{AB}$ and $\overline{XY}$) are congruent
$\triangle ABC \backsim \triangle XYZ$	Two triangles ($\triangle ABC$ and $\triangle XYZ$) are similar

Topology.

E^n	Euclidean n space
S^n	n sphere
$\rho(p,q)$, $d(p,q)$	Metric, distance (between points p and q)
$\overline{X}$, $cl\,X$, X^c	Closure of the set X
FrX, fr$X$, ∂X, bdry X	Frontier, boundary of X
int X, X	Interior of X
T_2 space	Hausdorff space
F_σ	Union of countably many closed sets
G_δ	Intersection of countably many open sets
dim X	Dimensionality, dimension of X
$\pi_1(X)$	Fundamental group of the space X

$\pi_n(X)$, $\pi_n(X,A)$	Homotopy groups
$H_n(X)$, $H_n(X,A;G)$, $H_*(X)$	Homology groups
$H^n(X)$, $H^n(X,A;G)$, $H^*(X)$	Cohomology groups

Probability and statistics.

X, Y	Random variables
$P(X \leqq 2)$, $\Pr\{X \leqq 2\}$	Probability that $X \leqq 2$
$P(X \leqq 2 \mid Y \geqq 1)$	Conditional probability
$E(X)$, $\mathrm{E}(X)$	Expectation of X
$E(X \mid Y \geqq 1)$	Conditional expectation
c.d.f.	Cumulative distribution function
p.d.f.	Probability density function
c.f.	Characteristic function
$\bar{X}$	Mean (especially, sample mean)
σ, s.d.	Standard deviation
σ^2, Var, var	Variance
μ_1, μ_2, μ_3, μ_i, μ_{ij}	Moments of a distribution
ρ	Coefficient of correlation
$\rho_{12\cdot34}$	Partial correlation coefficient

Fundamental constants

1986 recommended values of the fundamental physical constants were compiled by E. R. Cohen and B. N. Taylor under the auspices of the CODATA Task Group on Fundamental Constants. These tables are adapted from a set of values that has been officially adopted by CODATA and is published in CODATA Bulletin no. 63, November 1986, and the *Journal of Research of the National Bureau of Standards,* vol. 92, no. 2, pp. 85–95, 1987.

Quantity	*Symbol*	*Numerical value**	*Relative uncertainty, ppm*	*SI units†*	*cgs units‡*
Speed of light in vacuum	c	299792458	(exact)	$m \cdot s^{-1}$	10^2 $cm \cdot s^{-1}$
Permeability of vacuum	μ_0	4π		10^{-7} $H \cdot m^{-7}$	
		$= 12.566370614. . .$	(exact)	10^{-7} $H \cdot m^{-1}$	
Permittivity of vacuum, $1/\mu_0 c^2$	ϵ_0	8.854187817. . .	(exact)	10^{-12} $F \cdot m^{-1}$	
Fine-structure constant, $[\mu_0 c^2/4\pi](e^2/\hbar c)$	α	7.29735308(33)	0.045	10^{-3}	10^{-3}
	α^{-1}	137.0359895(61)	0.045		
Elementary charge	e	1.6027733(49)	0.30	10^{-19} C	10^{-20} emu
		4.8032068(15)	0.30		10^{-10} esu
Planck constant	h	6.6260755(40)	0.60	10^{-34} $J \cdot s$	10^{-27} $erg \cdot s$
	$\hbar = h/2\pi$	1.0547266(63)	0.60	10^{-34} $J \cdot s$	10^{-27} $erg \cdot s$
Avogadro constant	N_A	6.0221367(36)	0.59	10^{23} mol^{-1}	10^{23} mol^{-1}
Atomic mass unit, 10^{-3} $kg \cdot mol^{-1} N_A^{-1}$	u	1.6605402(10)	0.59	10^{-27} kg	10^{-24} g
Electron rest mass	m_e	9.1093897(54)	0.59	10^{-31} kg	10^{-23} g
		5.48579903(13)	0.023	10^{-4} u	10^{-4} u
Proton rest mass	m_p	1.6726231(10)	0.59	10^{-27} kg	10^{-24} g
		1.007276470(12)	0.012	u	u
Ratio of proton mass to electron mass	m_p/m_e	1836.152701(37)	0.020		
Neutron rest mass	m_n	1.6749286(10)	0.59	10^{-27} kg	10^{-24} g
		1.008664904(14)	0.014	u	u
Electron charge to mass ration	e/m_e	1.75881962(53)	0.30	10^{11} $C \cdot kg^{-1}$	10^7 $emu \cdot g^{-1}$
		5.2728086(16)	0.30		10^{17} $esu \cdot g^{-1}$
Magnetic flux quantum, $[c]^{-1}(hc/2e)$	Φ_0	2.0673461(61)	0.30	10^{-15} Wb	10^{-7} $G \cdot cm^2$
	h/e	4.1356692(12)	0.30	10^{-15} $J \cdot s \cdot C^{-1}$	10^{-7} $erg \cdot s \cdot emu^{-1}$
		1.37951076(41)	0.30		10^{-17} $erg \cdot s \cdot esu^{-1}$
Josephson frequency-voltage ratio	$2e/h$	4.8359767(14)	0.30	10^{14} $Hz \cdot V^{-1}$	
Quantized Hall conductance	e^2/h	3.87404614(17)	0.045	10^{-5} s	
Quantized Hall resistance, $h/e^2 = \mu_0 c/2\alpha$ [SI units]	R_H	2.58128056(12)	0.045	10^4 Ω	
Quantum of circulation	$h/2m_e$	3.63694807(33)	0.089	10^{-4} $J \cdot s \cdot kg^{-1}$	$erg \cdot s \cdot g^{-1}$
	h/m_e	7.27389614(65)	0.089	10^{-4} $J \cdot s \cdot kg^{-1}$	$erg \cdot s \cdot g^{-1}$
Faraday constant, $N_A e$	F	9.6485309(29)	0.30	10^4 $C \cdot mol^{-1}$	10^3 $emu \cdot mol^{-1}$
		2.8925568(9)	0.30		10^{14} $esu \cdot mol^{-1}$
Rydberg constant, $[\mu_0 c^2/4\pi]^2(m_e e^4/4\pi\hbar^3 c)$	R_∞	1.0973731534(13)	0.0012	10^7 m^{-1}	10^5 cm^{-1}
Bohr radius, $[\mu_0 c^2/4\pi]^{-1}(\hbar^2/m_e e^2) = \alpha/4\pi R_\infty$	a_0	5.29177249(24)	0.045	10^{-11} m	10^{-9} cm
Classical electron radius, $[\mu_0 c^2/4\pi](e^2/m_e c^2) = \alpha^3/4\pi R_\infty$	$r_e = \alpha \lambda\!\!\!^{-}_c$	2.81794092(38)	0.13	10^{-15} m	10^{-13} cm
Thomson cross section, $(8/3)\pi r_e^2$	σ_e	0.66524616(18)	0.27	10^{-28} m^2	10^{-24} cm^2
Free electron g factor, or electron magnetic moment in Bohr magnetons	$g_e/2 = \mu_e/\mu_B$	1.001159652193(10)	1×10^{-5}		
Free muon g factor, or muon magnetic moment in units of $[c](e\hbar/2m_\mu c)$	$g_\mu/2$	1.0011659230(84)	0.0084		
Bohr magneton, $[c](e\hbar/2m_e c)$	μ_B	9.2740154(31)	0.34	10^{-24} $J \cdot T^{-1}$	10^{-21} $erg \cdot G^{-1}$
Electron magnetic moment	μ_e	9.2847701(31)	0.34	10^{-24} $J \cdot T^{-1}$	10^{-21} $erg \cdot G^{-1}$
Gyromagnetic ratio of protons in H_2O	γ_p'	2.67515255(81)	0.30	10^8 $s^{-1} \cdot T^{-1}$	10^4 $s^{-1} \cdot G^{-1}$
	$\gamma_p'/2\pi$	4.2576375(13)	0.30	10^7 $Hz \cdot T^{-1}$	10^3 $Hz \cdot G^{-1}$

*See footnotes on page 2213.

(continued)

Fundamental constants (cont.)

Quantity	Symbol	Numerical value*	Relative uncertainty, ppm	SI units†	cgs units‡
γ_p' corrected for diamagnetism of H_2O	γ_p	2.67522128(81)	0.30	$10^8\ s^{-1} \cdot T^{-1}$	$10^4\ s^{-1} \cdot G^{-1}$
	$\gamma_p/2\pi$	4.2577469(13)	0.30	10^7 Hz $\cdot\ T^{-1}$	10^3 Hz $\cdot\ G^{-1}$
Magnetic moment of protons in H_2O in Bohr magnetons	μ_p'/μ_B	1.520993129(17)	0.011	10^{-3}	10^{-3}
Proton magnetic moment in Bohr magnetons	μ_p/μ_B	1.521032202(15)	0.010	10^{-3}	10^{-3}
Ratio of electron and proton magnetic moments	μ_e/μ_p	658.2106881(66)	0.010		
Proton magnetic moment	μ_p	1.41060761(47)	0.34	$10^{-26}\ J \cdot T^{-1}$	10^{-23} erg $\cdot\ G^{-1}$
Magnetic moment of protons in H_2O in nuclear magnetons	μ_p'/μ_N	2.792775642(64)	0.023		
μ_p'/μ_N corrected for diamagnetism of H_2O	μ_p/μ_N	2.792847386(63)	0.023		
Nuclear magneton, $[c](e\hbar/2m_pc)$	μ_N	5.0507866(17)	0.34	$10^{-27}\ J \cdot T^{-1}$	10^{-24} erg $\cdot\ G^{-1}$
Ratio of muon and proton magnetic moments	μ_μ/μ_p	3.18334547(47)	0.15		
Muon magnetic moment	μ_μ	4.4904514(15)	0.33	$10^{-26}\ J \cdot T^{-1}$	10^{-23} erg $\cdot\ G^{-1}$
Ratio of muon mass to electron mass	m_μ/m_p	206.768262(30)	0.15		
Muon rest mass	m_μ	1.8835327(11)	0.61	10^{-23} kg	10^{-25} g
		0.11348913(17)	0.15	u	u
Compton wavelength of the electron, $h/m_ec = \alpha^2/2R_\infty$	λ_C	2.42631058(22)	0.089	10^{-12} m	10^{-10} cm
	$\bar{\lambda}_C = \lambda_C/2\pi = \alpha a_0$	3.8615323(35)	0.089	10^{-13} m	10^{-11} cm
Compton wavelength of the proton, h/m_pc	$\lambda_{C,p}$	1.32141002(12)	0.089	10^{-15} m	10^{-13} cm
	$\bar{\lambda}_{C,p} = \lambda_{C,p}/2\pi$	2.10308937(19)	0.089	10^{-16} m	10^{-14} cm
Compton wavelength of the neutron, h/m_nc	$\lambda_{C,n}$	1.31959110(12)	0.089	10^{-15} m	10^{-13} cm
	$\bar{\lambda}_{C,n} = \lambda_{C,n}/2\pi$	2.10019445(19)	0.089	10^{-16} m	10^{-14} cm
Molar volume of ideal gas at STP	V_m	22.41410(19)	8.4	$10^{-3}\ m^3 \cdot mol^{-1}$	$10^3\ cm^3 \cdot mol^{-1}$
Molar gas constant, V_mp_0/T_0 ($T_0 \equiv$ 273.15 K; $p_0 \equiv$ 101325 Pa $\equiv$ 1atm)	R	8.314510(70)	8.4	$J \cdot mol^{-1} \cdot K^{-1}$	10^7 erg $\cdot\ mol^{-1} \cdot K^{-1}$
		8.205784(69)	8.4	$10^{-5}\ m^3 \cdot atm \cdot mol^{-1} \cdot K^{-1}$	$10\ cm^3 \cdot atm \cdot mol^{-1} \cdot K^{-1}$
Boltzmann constant, R/N_A	k	1.380658(12)	8.5	$10^{-23}\ J \cdot K^{-1}$	10^{-16} erg $\cdot\ K^{-1}$
Stefan-Boltzmann constant, $\pi^2k^1/60\hbar^3c^2$	σ	5.67051(18)	34	$10^{-8}\ W \cdot m^{-2} \cdot K^{-4}$	10^{-5} erg $\cdot\ s^{-1} \cdot cm^{-2} \cdot K^{-4}$
First radiation constant, $2\pi hc^2$	c_1	3.7417749(22)	0.60	$10^{-16}\ W \cdot m^2$	10^{-5} erg $\cdot\ cm^2 \cdot s^{-1}$
Second radiation constant, hc/k	c_2	1.438769(12)	8.4	10^{-2} m $\cdot$ K	cm $\cdot$ K
Gravitational constant	G	6.67259(85)	128	$10^{-11}\ m^3 \cdot s^{-2} \cdot kg^{-1}$	$10^{-3}\ cm^3 \cdot s^{-2} \cdot g^{-1}$
Cu x-unit, $\lambda(CuK\alpha_1) \equiv$ 1537.400 xu	xu($CuK\alpha_1$)	1.00207789(70)	0.70	10^{-13} m	10^{-11} cm
Mo x-unit, $\lambda(MoK\alpha_1) \equiv$ 707.831 xu	xu($MoK\alpha_1$)	1.00209938(45)	0.45	10^{-13} m	10^{-11} cm
Å*, $\lambda(WK\alpha_1) \equiv$ 0.2090100 Å*	Å*	1.0000148(92)	0.92	10^{-10} m	10^{-8} cm
Lattice spacing of Si (in vacuum at 22.5 °C),§	a	5.4310196(11)	0.21	10^{-10} m	10^{-8} cm
$d_{220} = a/\sqrt{8}$	d_{220}	1.92015540(40)	0.21	10^{-10} m	10^{-8} cm
Molar volume of Si, $M(Si)/\rho(Si) = N_Aa^3/8$	V_m(Si)	12.0588179(89)	0.74	$10^{-6}\ m^3 \cdot mol^{-1}$	$cm^3 \cdot mol^{-1}$

*See footnotes on page 2213.

(continued)

Fundamental constants (cont.)

Energy conversion factors and equivalents

Quantity	*Symbol*	*Numerical value**	*Units*	*Relative uncer-tainty, ppm*
1 kilogram (kg · c^2)		8.987551787. . .	10^{16} J	(exact)
		5.6095862(17)	10^{29} MeV	0.30
1 atomic mass unit (u · c^2)		1.49241909(88)	10^{-10} J	0.59
		931.49432(28)	MeV	0.30
1 electron mass ($m_e \cdot c^2$)		8.1871112(48)	10^{-14} J	0.59
		0.51099906(15)	MeV	0.30
1 muon mass ($m_\mu \cdot c^2$)		1.6928348(10)	10^{-11} J	0.61
		105.658389(34)	MeV	0.32
1 proton mass ($m_v \cdot c^2$)		1.5032786(9)	10^{-10} J	0.59
		938.27231(28)	MeV	0.30
1 neutron mass ($m_n \cdot c^2$)		1.5053508(9)	10^{-10} J	0.59
		939.56563(28)	MeV	0.30
1 electronvolt		1.60217733(49)	10^{-19} J	0.30
			10^{-12} erg	0.30
	1 eV/h	2.41798836(72)	10^{14} Hz	0.30
	1 eV/hc	8.0655410(24)	10^{5} m^{-1}	0.30
			10^{3} cm^{-1}	0.30
	1 eV/k	1.160445(10)	10^{4} K	8.4
Voltage-wavelength conversion, hc		1.9864475(12)	10^{-25} J · m	0.60
		1.23984244(37)	10^{-6} eV · m	0.30
			10^{-4} eV · cm	0.30
Rydberg constant	$R_\infty hc$	2.1798741(13)	10^{-18} J	0.60
			10^{-11} erg	0.60
	$R_\infty c$	13.605698(40)	eV	0.30
	$R_\infty hc/k$	3.2898419499(39)	10^{15} Hz	0.0012
		1.578866(13)	10^{5} K	8.4
Bohr magneton	μ_R	9.2740154(31)	10^{-21} J · T^{-1}	0.34
		5.78838263(52)	10^{-5} eV · T^{-1}	0.089
	μ_B/h	1.39962418(42)	10^{10} Hz · T^{-1}	0.30
	μ_B/hc	46.686437(14)	m^{-1} · T^{-1}	0.30
			10^{-2} cm^{-1} · T^{-1}	0.30
	μ_B/k	0.6717099(57)	K · T^{-1}	8.5
Nuclear magneton	μ_N	5.050786(17)	10^{-27} J · T^{-1}	0.34
		3.15245166(28)	10^{-8} eV · T^{-1}	0.089
	μ_N/h	7.6225914(23)	10^{6} Hz · T^{-1}	0.30
	μ_N/hc	2.54262281(77)	10^{-2} m^{-1} · T^{-1}	0.30
			10^{-4} cm^{-1} · T^{-1}	0.30
	μ_N/k	3.658246(31)	10^{-4} K · T^{-1}	8.5

*The numbers in parentheses are the one-standard-deviation uncertainties in the last digits of the quoted value computed on the basis of internal consistency. The unified atomic mass scale ^{12}C ≙ 12 has been used throughout, and u = atomic mass unit, C = coulomb, F = farad, G = gauss, H = henry, Hz = hertz, J = joule, K = kelvin, Pa = pascal = N · m^{-2}, S = siemens = Ω^{-1}, T = tesla (10^4 G), V = volt, Wb = weber = T · m^2, W = watt, and Ω = ohms. In cases where formulas for constants are given (such as R_∞), the relations are written as the product of two factors. The second factor, in parentheses, is the expression to be used when all quantities are expressed in cgs units, with the electron charge in electrostatic units. The first factor, in brackets, is to be included only if all quantities are expressed in SI units. (The formula for the quantized Hall resistance is given only in SI units.) Since the uncertainties of many of these quantities are correlated, a covariance matrix must be used in evaluating the uncertainties of quantities computed from them.

†Quantities given in u and atm are for the convenience of the reader; these units are not part of the International System of Units.

‡In order to avoid separate columns for "electromagnetic" and "electrostatic" units, both are given under the single heading "cgs units." When one is using these units, the elementary charge e in the second column should be understood to be replaced by e_m or e_e, respectively.

§The lattice spacing of single-crystal Si can vary by parts in 10^7 depending on the preparation process. Measurements indicate also the possibility of distortions from exact cubic symmetry of the order of 0.2 ppm.

Fundamental particles[a]

Gauge bosons $J_C^P = 1_-^-$ Self-conjugate except $\overline{W^+} = W^-$.

Name	Symbol	Charge[b]	Mass and width, GeV	Couplings
Photon	γ	0	0	$A \Rightarrow \gamma A$
Gluon[c]	g	0	0	$A \Rightarrow gA'$
Weak bosons				
Charged[d]	$W^\pm$	± 1	80.2, 2.1	$U \Rightarrow W^+D$
Neutral[e]	Z^0	0	91.2, 2.49	$A \Rightarrow Z^0A$

Fermions[f] $J = \frac{1}{2}$ All have distinct antiparticles, except perhaps the neutrinos.

Name	Charge[b]	Symbol	and mass, GeV	Symbol	and mass, GeV	Symbol	and mass, Gev
Leptons							
Neutrinos	0	ν_e	$<10^{-8}$	ν_μ	$<.0003$	ν_τ	$<.035$
Charged leptons[g]	-1	e	.00051	μ	.106[h]	τ	1.78[h]
Quarks[c]							
Up type	$\frac{2}{3}$	u	.005	c	1.4	t	175[i]
Down type	$-\frac{1}{3}$	d	.01	s	.15	b	4.8

[a] The graviton, with $J_C^P = 2_+^+$, has been omitted, since it plays no role in high-energy particle physics.

[b] In units of the proton charge.

[c] The gluon is a color SU_3 octet {8}; each quark is a color triplet {3}. These colored particles are confined constituents of hadrons; they do not appear as free particles.

[d] The branching ratios (%) of the decay modes of W^+ are:

$u\bar{d}$, $c\bar{s}$	34 each
$\nu_e e^+$, $\nu_\nu \mu^+$, $\nu_\tau \tau^+$	11 each

[e] The branching ratios (%) of the decay modes of the Z^0 are:

$d\bar{d}$, $s\bar{s}$, $b\bar{b}$	16.8 each
$u\bar{u}$, $c\bar{c}$	9.7 each
$\nu_e\bar{\nu}_e$, $\nu\mu\bar{\nu}\mu$, $\nu_\tau\bar{\nu}_\tau$	6.7 each
e^+e^-, $\mu^+\mu^-$, $\tau^+\tau^-$	3.4 each

[f] The three known families (generations) of fermions are displayed in three columns.

[g] Any further charged leptons have mass greater than 40 GeV.

[h] The μ and τ leptons are unstable, with the following mean life and principal decay modes (branching ratios in %):

μ	$\tau_\mu = 2.2 \times 10^{-6}$ s	$e\bar{\nu}_e\nu_\mu$ 100
τ	$\tau_\tau = 3 \times 10^{-13}$ s	$\mu\nu_\mu\nu_\tau$ 18, $e\bar{\nu}_e\nu_\tau$ 18, (hadrons)$^-\bar{\nu}_\tau$ 64

[i] The t quark has a width ≈ 2 GeV, with dominant decay to Wb.

Geologic Time Scale*

Eon	Era	Period		Epoch	Dates ($\times 10^6$ years ago)	Age of	Notable events
Phanerozoic	Cenzoic	Tertiary		Holocene	0–0.01	Mammals	Modern humans
				Pleistocene	0.01–2		Early humans
			Neogene	Pliocene	2–5		Large carnivores
				Miocene	5–24		Whales, apes, and grazing animals
			Paleogene	Oligocene	24–37		Browsing animals
				Eocene	37–58		Modern floras
				Paleocene	58–66		Extinction of dinosaurs; first placental mammals
	Mesozoic	Cretaceous			66–144	Reptiles	First flowering plants
		Jurassic			144–208		First birds/ mammals
		Triassic			208–245		First dinosaurs
	Paleozoic	Permian			245–286	Amphibians	Extinction of trilobites
		Carboniferous	Pennsylvanian		286–320		First reptiles
			Mississippian		320–360		Large primitive trees
		Devonian			360–408	Fishes	First amphibians
		Silurian			408–438		First land plant fossil
		Ordovician			438–505	Marine invertebrates	First fish
		Cambrian			505–538		First shells; trilobites dominant
Proterozoic	Sometimes collectively called Precambrian.				538–2500		First multicelled organisms
Archean					2500–3800		First one-celled organisms
Hadean					3800–4600		Approximate age of oldest rocks

*This scale is adapted from G. R. Thompson and J. Turk, *Modern Physical Geology*, Saunders College Publishing, 2d ed., 1997.

Telescopes

Large telescopes

Mirror diameter: Meters	Inches	Observatory	Year completed
		Some of the world's largest reflecting telescopes	
6.0	236	Special Astrophysical Observatory, Zelenchukskaya, Caucasus, Russia	1976
5.1	200	California Institute of Technology/Palomar Observatory, Palomar Mountain, California	1950
4.2	165	William Herschel Telescope, La Palma, Canary Islands	1987
4.0	158	Kitt Peak National Observatory, Arizona	1973
4.0	158	Cerro Tololo Inter-American Observatory, Chile	1976
3.9	153	Anglo-Australian Telescope, Siding Spring Observatory, Australia	1975
3.8	150	United Kingdom Infrared Telescope, Mauna Kea Observatory, Hawaii	1978
3.6	144	Canada-France-Hawaii Telescope, Mauna Kea Observatory, Hawaii	1979
3.6	142	Cerro La Silla European Southern Observatory, Chile	1976
3.6	141	New Technology Telescope, European Southern Observatory, Chile	1989
3.5	138	Calar Alto Observatory, Calar Alto, Spain	1984
3.5	138	Astrophysics Research Consortium, Apache Point, New Mexico	1993
3.5	138	Wisconsin-Indiana-Yale-NOAO (WIYN), Kitt Peak, Arizona	1994
3.2	126	NASA Infrared Telescope, Mauna Kea Observatory, Hawaii	1979
3.0	120	Lick Observatory, Mount Hamilton, California	1959
2.7	107	McDonald Observatory, Fort Davis, Texas	1968
2.6	102	Crimean Astrophysical Observatory, Ukraine	1960
2.6	102	Byurakan Observatory, Yerevan, Armenia	1976
2.56	101	Nordic Optical Telescope, La Palma, Canary Islands	1989
2.5	100	Mount Wilson and Las Campanas, Mount Wilson, California	1917
2.5	100	Cerro Las Campanas, Carnegie Southern Observatory, Chile	1976
2.5	96	Isaac Newton Telescope, La Palma, Canary Islands	1984
2.4	94	University of Michigan–Dartmouth College-Massachusetts Institute of Technology, Kitt Peak, Arizona	1986
2.4	94	Hubble Space Telescope, low Earth orbit	1990
36 × 1.8	6 × 72	Keck Telescope, Mauna Kea, Hawaii	1993
6 × 1.8	6 × 71	Multi-Mirror Telescope, Mount Hopkins, Arizona	1979
		World's largest refracting telescopes	
1.02	40	Yerkes Observatory, Williams Bay, Wisconsin	1897
0.91	36	Lick Observatory, Mount Hamilton, California	1888
0.83	33	Observatoire de Paris, Meudon, France	1893
0.80	32	Astrophysikalisches Observatory, Postdam, Germany	1899
0.76	30	Allegheny Observatory, Pittsburgh, Pennsylvania	1914

Largest Schmidt telescopes

Telescope or institution	Location	Diameter: Corrector plate, in. (m)	Diameter: Spherical mirror, in. (m)	Focal ratio	Date completed
Karl Schwarzschild Observatory	Tautenberg (near Jena), Germany	53 (1.3) (removable)	79 (2.0)	*f*/2	1960
Oschin Telescope, Palomar Observatory	Palomar Mountain, California	48 (1.2)	72 (1.8)	*f*/2.5	1948, 1987
United Kingdom Schmidt, Anglo-Australian Observatory	Warrumbungle National Park, New South Wales, Australia	48 (1.2)	72 (1.8)	*f*/2.5	1973
Tokyo Astronomical Observatory	Kiso Mountains, Japan	41 (1.1)	60 (1.5)	*f*/3.1	1975
European Southern Observatory Schmidt	La Silla, Chile	39 (1.0)	64 (1.6)	*f*/3	1972

Large radio telescopes and synthesis arrays

Institution	Location	Size of reflector, ft (m)
Radio telescopes for meter and centimer wavelengths		
Fully steerable paraboloids		
Max Planck Institut für Radioastronomie	Effelsberg, Germany	330 (100)
Nuffield Radio Astronomy Laboratory	Jodrell Bank, England	250 (76)
CSIRO	Parkes, N.S.W., Australia	211 (64)
Jet Propulsion Laboratory	Goldstone, California	211 (64)
Algonquin Radio Observatory	Lake Traverse, Ontario	152 (46)
National Radio Astronomy Observatory	Green Bank, West Virginia	142 (43)
California Institute of Technology	Big Pine, California	132 (40)
Haystack Observatory	Westford, Massachusetts	122 (37)
Crimean Astrophysical Observatory	Crimea, Russia	73 (22)
Limited-tracking transit telescopes		
Special Astrophysical Observatory	Zelenchukskaya, Russia	33 × 6221 (10 × 1885)
Tata Institute	Ootacamund, India	99 × 1746 (30 × 529)
National Astronomy and Ionosphere Center	Arecibo, Puerto Rico	1007 (305)
Observatory of Paris	Nancy, France	132 × 660 (40 × 200)
Radio telescopes for millimeter wavelengths		
Nobeyama Radio Observatory*	Nobeyama, Japan	148 (45)
Institut de Radio Astronomie Millimetrique	Pico de Veleta, Spain	99 (30)
Onsala Observatory	Gothenburg, Sweden	66 (20)
University of Massachusetts	Amherst, Massachusetts	46 (14)
National Radio Astronomy Observatory	Kitt Peak, Arizona	40 (12)
California Institute of Technology†	Big Pine, California	33 (10)
University of Texas	Fort Davis, Texas	16 (5)
University of California‡	Hat Creek, California	13 (4)
Synthesis arrays		
National Radio Astronomy Observatory (Very Large Array, VLA)	Socorro, New Mexico	Resolution 0.1"
Nuffield Radio Astronomy Laboratory (MERLIN)	Jodrell Bank, England	Resolution 0.1"
Mullard Radio Astronomy Observatory (5-km array)	Cambridge, England	Resolution 0.5"
Westerbork Radio Observatory (WSRT)	Westerbork, Netherlands	Resolution 1"

*Also five-element millimeter-wave interferometer.
†Also three-element millimeter-wave interferometer.
‡Four-element millimeter-wave interferometer.

Some telescopes for submillimeter astronomy

Name	Location	Elevation, ft (m)	Diameter, ft (m)
Existing installations			
Kuiper Airborne Observatory (KAO)	Suborbital	45,000 (13,700)	3.0 (0.9)
Cologne 3-m Telescope	Gomergrat, near Zermatt, Switzerland	10,285 (3,135)	9.8 (3.0)
Caltech Submillimeter Observatory (CSO)	Mauna Kea, Hawaii	13,360 (4,072)	34.1 (10.4)
Swedish-ESO Submillimeter Telescope (SEST)	La Silla, Chile	7,850 (2,400)	49.2 (15.0)
James Clerk Maxwell Telescope (JCMT)	Mauna Kea, Hawaii	13,425 (4,092)	49.2 (15.0)
Max Planck Institute for Radioastronomy/ University of Arizona 10-m Submillimeter Telescope	Mount Graham, Arizona	10,466 (3,190)	32.8 (10.0)
Planned installations			
Stratospheric Observatory for Infrared Astronomy (SOFIA)†	Suborbital	50,000 (15,200)	9.8 (3.0)
Far-Infrared and Submillimeter Telescope (FIRST)‡	Orbital	>200 mil (>321 km)	23.0 (7.0)
Large Deployable Reflector (LDR)§	Orbital	>200 mi (>321 km)	32.8 (10.0)

†Possible completion in 1990s (United States). Optical, infrared, and submillimeter.
‡Possible development in 1990s (European).
§Possible deployment about 2005 (United States). Operating wavelength 30 μm to 1 mm.

Biographical Listing

Abbe, Ernst (1840–1905), German physicist. Developed optical instruments, such as an apochromatic objective and a crystal refractometer.

Abel, Frederick Augustus (1827–1902), English chemist. Expert on the chemistry of explosives; originated the Abel test for determination of the flash point of petroleum.

Abel, John Jacob (1857–1938), American pharmacologist and physiologist. Isolated epinephrine, and insulin in crystal form.

Abel, Niels Henrik (1802–1829), Norwegian mathematician. Contributed to the theory of elliptical functions.

Abell, George Ogden (1927–1983), American astronomer. Research in problems relating to organization, structure, and distribution of galaxies; observational cosmology; and planetary nebulae.

Abney, William de Wiveleslie (1843–1920), English photographic chemist and physicist. Photographed the infrared solar spectrum.

Adams, John Couch (1819–1892), English astronomer. Discovered, independently of U. J. J. Leverrier, Neptune.

Adanson, Michel (1727–1806), French naturalist. Classified plants in his *Les Familles Naturelles des Plantes*.

Addison, Thomas (1793–1860), English physician. Identified pernicious anemia and Addison's disease of the adrenal cortex.

Adler, Alfred (1870–1937), Austrian psychiatrist and psychologist. Founded the school of individual psychology.

Adrian of Cambridge, Edgar Douglas Adrian, Baron (1889–1977), English physiologist. Investigated physiology of nervous system; showed that change in electric potential in electroencephalograph is due to electrical activity of cortex; Nobel Prize, 1932.

Afzelius, Adam (1750–1837), Swedish botanist. Founded the Linnaean Institute.

Agassiz, Jean Louis Rudolphe (1807–1873), Swiss-born American naturalist. Wrote books on ichthyology, especially relating to classification.

Agnesi, Maria Gaetana (1718–1799), Italian mathematician. Author of *Instituzioni Analitiche*, a complete treatment of algebra and analysis; shared in the discovery of a cubic curve ("witch of Agnesi").

Agricola, Georgius, real name Georg Bauer (1494–1555), German physician and mineralogist. Known as the father of systematic mineralogy.

Airy, George Biddell (1801–1892), English astronomer. Discovered inequality in the motions of Venus and the Earth; determined the mass of the Earth.

Aitken, John (1839–1919), Scottish physicist. Studied dust particles in the atmosphere, known as Aitken nuclei.

Aitken, Robert Grant (1864–1951), American astronomer. Discovered more than 3000 binary stars.

Alder, Kurt (1902–1958), German chemist. Codeveloper of the Diels-Alder reaction for diene synthesis; contributed to stereochemistry; Nobel Prize, 1950.

Alembert, Jean le Rond d' (1717–1783), French mathematician. Developed d'Alembert's principle and the calculus of partial differences.

Alfvén, Hannes Olof Gösta (1908–1995), Swedish physicist. Studies in magnetohydrodynamics, planetary physics, antiferromagnetism, and ferrimagnetism; Nobel Prize, 1970.

Alhazen (965–1038), Arab mathematician and astronomer. Provided the first accounts of atmospheric refraction and reflection from concave surfaces; constructed spherical and parabolic mirrors.

al-Khwarizmi (780–?850), Arab mathematician. Wrote treatises on arithmetic and algebra, which were important in the mathematical knowledge of medieval Europe.

Allen, Edgar (1892–1943), American biologist. Discovered estrogen; investigated hormonal mechanisms controlling female reproductive cycle.

Allen, Willard Myron (1904–1993), American physician. With G. W. Corner, discovered progesterone, and proved it necessary for development of embryo in early pregnancy; with O. Wintersteiner, synthesized crystalline progesterone.

Altman, Sidney (1939–   ), American chemist. Discovered an unusual enzyme that contains ribonucleic acid (RNA) in addition to a protein, leading to the discovery that RNA molecules have catalytic properties similar to those of enzymes; Nobel Prize, 1989.

Alvarez, Luis Walter (1911–1988), American physicist. Pioneer in building liquid hydrogen bubble chambers, and in developing measurement devices and computer systems to analyze data from these chambers; discovered large numbers of short-lived elementary particles; Nobel Prize, 1968.

Amagat, Émile (1841–1915), French physicist. Investigated relationship of pressure, density, and temperature in gases and liquids, particularly at high pressure.

Amici, Giovanni Battista (1786–1863), Italian astronomer, optician, and naturalist. Invented the Amici microscope; designed parabolic mirrors for reflecting telescopes.

Ampère, André Marie (1775–1836), French physicist and mathematician. Founder of electrodynamics; formulated Ampère's law; invented the astatic needle.

Anaximander (611–547 B.C.), Greek astronomer and mathematician. Reputed inventor of geographical maps; formulated the concept of the universe as infinite (apeiron).

Anderson, Carl David (1905–1991), American physicist. Discovered the meson in cosmic rays; discovered the positron; Nobel Prize, 1936.

Anderson, Philip Warren (1923–   ), American physicist. Demonstrated existence of electronic localization in disordered solids, and of localized magnetism in metals; Nobel Prize, 1977.

Andrade, Edward Neville da Costa (1887–1971), English physicist. Discovered Andrade's creep law and a law governing variation of viscosity of liquids with temperature.

Andrews, Roy Chapman (1884–1960), American naturalist. Discovered many plant and animal fossils.

Anfinsen, Christian Boehmer (1916–1995), American biochemist. Discovered how three-dimensional structures of ribonuclease and other proteins are formed; Nobel Prize, 1972.

Angström, Anders Jonas (1814–1874), Swedish physicist. Mapped the solar spectrum; discovered hydrogen in the solar atmosphere.

Apollonius of Perga (247–205 B.C.), Greek mathematician. Wrote about conic sections; coined the terms parabola, ellipse, and hyperbola.

Appleton, Edward Victor (1892–1965), English physicist. Demonstrated the existence of the ionosphere and discovered its region known as the Appleton layer; contributed to the development of radar; Nobel Prize, 1947.

Arago, Dominique François (1786–1853), French astronomer and physicist. Discovered the magnetic properties of nonferrous materials, and the production of magnetism by electricity.

Arber, Werner (1929–   ), Swiss molecular biologist. Determined the molecular mechanism of host-controlled restriction modification of bacterial viruses and discovered the restriction enzymes; Nobel Prize, 1978.

Archimedes (287–212 B.C.), Greek physicist and mathematician. Formulated Archimedes' principle; invented the compound pulley and Archimedes' screw.

Argand, Jean Robert (1768–1822), Swiss mathematician. Developed the Argand diagram.

Argelander, Friedrich Wilhelm August (1799–1875), German astronomer. Prepared a star catalog; introduced decimal division of stellar magnitudes.

Aristarchus of Samos (310–250 B.C.), Greek astronomer. Invented the hemispherical sundial; determined the movement of the Earth around the stationary Sun; added a correction factor of 1/1623 of a day to the length of the year.

Aristotle (384–322 B.C.), Greek philosopher. Exponent of the methodology and division of sciences; contributed to physics, astronomy, meteorology, psychology, and biology.

Arkwright, Richard (1732–1792), English inventor. Developed the first practical mechanized spinning frame, utilizing rollers.

Arrhenius, Svante August (1859–1927), Swedish physicist and chemist. Developed theory of electrolytic dissociation; investigated osmosis and viscosity of solutions; Nobel Prize, 1903.

Arsonval, Jacques Arsène d' (1851–1940), French physicist and physiologist. Pioneered in electrotherapy; invented d'Arsonval galvanometer.

Aston, Francis William (1877–1945), English physicist and chemist. Discovered isotopes in nonradioactive elements by using the mass spectrograph he invented; Nobel Prize, 1922.

Atwood, George (1746–1807), English mathematician. Invented the Atwood machine.

Audubon, John James (1785–1851), Haitian-born American ornithologist and artist. Made drawings and paintings of birds and animals.

Auger, Pierre Victor (1899–1993), French physicist. Discovered the Auger effect.

Avicenna (979–1037), Arab physician. Wrote the medical text *Canon Medicinae*.

Avogadro, Amedeo (1776–1856), Italian physicist. Formulated Avogadro's law.

Axelrod, Julius (1912–   ), American biochemist and pharmacologist. Showed that many drugs act by modifying storage of neurotransmitters at nerve terminals; made discoveries concerning metabolism, and mechanisms for formation and inactivation of norepinephrine; Nobel Prize, 1970.

Ayrton, William Edward (1847–1908), English physicist and electrical engineer. Invented the ammeter, voltmeter, and other electrical measuring instruments.

Baade, Walter (1893–1960), German-born American astronomer. Formulated concept of stellar populations; increased distance scale of universe by factor of 2.

Babbage, Charles (1792–1851), English mathematician. Devised a primitive computer to calculate and print mathematical and astronomical tables.

Babcock, Harold Delos (1882–1968) and **Horace Welcome** (1912–   ), American astronomers. Invented the Babcock magnetograph, observed weak solar magnetic fields, and discovered stellar magnetic fields.

Babcock, Stephen Moulton (1843–1931), American agricultural chemist. Pioneer in nutrition; devised the Babcock test to measure fat content in milk.

Babinet, Jacques (1794–1872), French physicist. Invented a polariscope and a goniometer.

Back, Ernst E. A. (1881–1959), German physicist. Developed improved spectrographs; made spectroscopic observations leading to Paschen-Back effect.

Badger, Richard McLean (1896–1974), American physical chemist and spectroscopist. Studied structures of polyatomic molecules; formulated Badger's rule concerning molecular bonds.

Baer, Karl Ernst von (1792–1876), Estonian embryologist. Discovered the mammalian ovum and the notochord; developed the theory of embryonic germ layers.

Baeyer, Johann Friedrich Wilhelm Adolf von (1835–1917), German chemist. Synthesized indigo and hydroaromatic compounds; Nobel Prize, 1905.

Baily, Francis (1774–1844), English astronomer. A founder of the Royal Astronomical Society; first observed phenomenon of Baily's beads.

Baire, René Louis (1874–1932), French mathematician. Contributed to theory of functions of real variables; introduced concept of Baire functions.

Balfour, Francis Maitlant (1851–1882), English biologist. Founder of comparative embryology.

Balmer, Johann Jakob (1825–1898), Swiss physicist. Expressed the mathematical formula for frequencies of hydrogen lines in the visible spectrum.

Baltimore, David (1938–), American virologist. Investigated interaction between ribonucleic acid tumor viruses and genetic material; independently of H. M. Temin, discovered reverse transcriptase; Nobel Prize, 1975.

Banach, Stefan (1892–1945), Polish mathematician. Laid foundations of contemporary functional analysis; introduced concept of Banach space and discovered its fundamental properties.

Bang, Bernhard Laurits Frederik (1848–1932), Danish veterinarian. Discovered method of eradicating bovine tuberculosis; discovered *Brucella abortus*, the agent of contagious abortion (Bang's disease) and brucellosis.

Banting, Frederick Grant (1891–1941), Canadian physician. With J. J. R. Macleod and C. H. Best, discovered insulin and its role in diabetes; Nobel Prize, 1923.

Barany, Robert (1876–1936), Austrian physician. Developed new methods of diagnosing ear diseases; Nobel Prize, 1914.

Bardeen, John (1908–1991), American physicist. With L. N. Cooper and J. R. Schrieffer, formulated a theory of superconductivity; invented the transistor; Nobel Prize, 1956 and 1972.

Barkhausen, Heinrich Georg (1881–1956), German electronic engineer and physicist. Contributed to theory and application of electron tubes; with K. Kurz, developed Barkhausen-Kurz oscillator; discovered Barkhausen effect.

Barkla, Charles Glover (1877–1944), English physicist. Described characteristics of x-rays and other short-wave emissions of elements.

Barnard, Edward Emerson (1857–1923), American astronomer. Discovered 16 comets, the fifth satellite of Jupiter, and dark nebulae; contributed to celestial photography.

Barnett, Samuel Jackson (1873–1956), American physicist. Discovered Barnett effect and used it to measure the gyromagnetic ratio of ferromagnetic materials; gave experimental proof of existence of ionosphere.

Barr, Murray Llewellyn (1908–1995), Canadian anatomist. Discovered the Barr body on the X chromosome of the human female.

Bartholin, Kaspar (1655–1738), Danish physician. Discovered Bartholin's glands of the vagina and a sublingual duct.

Bartholin, Thomas (1616–1680), Danish physician. Discovered lymphatic glands; described the lymphatic system.

Bartlett, James Holly (1904–), American physicist. Introduced concept of Bartlett force; did research on nuclear shell model, electrochemical potentiostat, and restricted three-body problem.

Barton, Derek Harold Richard (1918–), British chemist. Developed and expanded concept of conformation to include large molecules with complex ring systems; Nobel Prize, 1969.

Basov, Nicolai Gennediyevich (1922–), Soviet physicist. Conducted fundamental studies in quantum electronics; with A. M. Prokhorov, developed quantum optical generators; Nobel Prize, 1964.

Bassham, James Alan (1922–), American chemist. Helped to elucidate basic photosynthetic carbon cycle.

Bates, Henry Walter (1825–1892), English naturalist. Discovered Batesian mimicry among butterflies and moths.

Baudot, Émile (1845–1903), French engineer. Invented an improved telegraph transmitter.

Baumé, Antoine (1728–1804), French chemist. Invented a graduated hydrometer which utilizes the Baumé scale.

Bautz, Laura Patricia (1940–), American astronomer. Collaborated with W. W. Morgan in developing Bautz-Morgan classification of galaxy clusters.

Bayer, Johann (1572–1624), German astronomer. Charted 12 constellations; first to use Greek letters to designate the order of brightness of stars in a constellation.

Bayes, Thomas (1702–1761), English mathematician. Formulated a basis for statistical inference.

Bayliss, William Maddock (1860–1924), English physiologist. Did research on electrophysiology of heart action; discovered the hormone secretin.

Beadle, George Wells (1903–1989), American geneticist. With E. L. Tatum, proved that genes affect heredity by controlling cell chemistry; Nobel Prize, 1958.

Beams, Jesse Wakefield (1898–1977), American physicist. Developed vacuum-type ultracentrifuges, used in purification and molecular weight determination of large-molecular-weight substances, isotope separation, and determination of the gravitational constant.

Beattie, James Alexander (1895–1981), American chemist and physicist. Studied ionic theory and thermodynamics; with P. W. Bridgman, proposed Beattie and Bridgman equation for gases.

Beaufort, Francis (1774–1857), English hydrographer. Devised scale of wind velocity.

Beaumont, William (1785–1853), American physician. Did pioneering studies of digestion and gastric juices.

Béchamp, Pierre Jacques Antoine (1816–1908), French chemist. Discovered a method of preparing aniline.

Beckmann, Ernst Otto (1853–1923), German chemist. Discovered Beckmann molecular transformation; invented the Beckmann thermometer.

Becquerel, Antoine César (1788–1878), French physicist. Pioneer in electrochemistry; first to extract metals from ore by electrolysis.

Becquerel, Antoine Henri (1852–1908), French physicist. A discoverer of radioactivity in uranium.

Bednorz, Johannes Georg (1950–), German physicist. With K. A. Müller, discovered high-temperature superconductivity in copper oxide ceramic materials; Nobel Prize, 1987.

Beebe, Charles William (1877–1962), American naturalist. Pioneer in deep-sea exploration; made ornithological collections.

Beer, August (1825–1863), German physicist. Discovered Beer's law of light absorption.

Behring, Emil Adolph von (1854–1917), German bacteriologist. Produced diphtheria and tetanus antitoxins; Nobel Prize, 1901.

Békésy, Georg von (1889–1972), Hungarian-born American physicist. Studied hearing processes, especially inner-ear mechanics; Nobel Prize, 1961.

Bell, Alexander Graham (1847–1922), Scottish-born American inventor. Invented the telephone, photophone, graphophone, and one of the earliest gramophones.

Bell, Charles (1774–1842), Scottish anatomist. Discovered that sensory and motor nerves are anatomically and functionally distinct.

Bellman, Richard Ernest (1920–1984), American mathematician. Research in analytic number theory, differential equations, stochastic processes, dynamic programming, and mathematical biosciences; discovered Bellman's principle of optimality.

Benacerraf, Baruj (1920–), American immunologist. Discovered immune-response (Ir) genes that control specific immune responses to thymus-dependent antigens; Nobel Prize, 1980.

Benioff, Hugo (1899–1968), American geophysicist. Investigated earthquakes, particularly through instrumental seismology.

Bentham, George (1800–1884), English botanist. With J. Hooker, wrote *Genera Plantarum*.

Berg, Paul (1926–), American biochemist. Investigated the biochemistry of deoxyribonucleic acid (DNA) and designed a technique for gene splicing; Nobel Prize, 1980.

Bergey, David Hendricks (1860–1937), American bacteriologist. Authority on classification of bacteria.

Bergius, Friedrich (1884–1949), Polish-born German chemist. Developed Bergius process for hydrogenation of coal to a petroleumlike oil; Nobel Prize, 1931.

Bergström, Sune Carl (1916–), Swedish biochemist and medical scientist. Studied the metabolism of unsaturated fatty acids and determined the chemical structure of prostaglandins; Nobel Prize, 1982.

Bernard, Claude (1813–1878), French physiologist. Studied digestion; discovered that glycogen is produced by the liver.

Bernoulli, Daniel (1700–1782), Swiss mathematician born in the Netherlands. Founder of mathematical physics; worked on hydrodynamics and differential equations; formulated the Bernoulli equation.

Bernoulli, Jacques or Jacob (1654–1705), Swiss mathematician. Contributed to mathematics of curves, calculus, and probability; developed the Bernoulli number.

Bernoulli, Jean or Johann (1667–1748), Swiss mathematician. A founder of calculus of variations; contributed to exponential calculus, complex numbers, geodesics, and trigonometry.

Berthelot, Pierre Eugène Marcellin (1827–1907), French chemist. Founder of thermochemistry; first to synthesize organic compounds; demonstrated nitrogen fixation.

Bertrand, Joseph Louis François (1822–1900), French mathematician. Contributed to analysis, differential geometry, and probability theory.

Berzelius, Jons Jakob (1779–1848), Swedish chemist. Discovered the elements cerium, selenium, thorium, and silicon; developed a system for classification and nomenclature of compounds.

Bessel, Friedrich Wilhelm (1784–1846), German astronomer and mathematician. Developed Bessel functions; determined parallax of the star 61 Cygni; postulated existence of Neptune and dark stars.

Bessemer, Henry (1813–1898), English engineer. Invented the Bessemer process, the first method for manufacturing steel on a large scale.

Best, Charles Herbert (1899–1978), Canadian physiologist and medical researcher. Associated with F. G. Banting and J. J. R. Macleod in the discovery of insulin.

Bethe, Hans Albrecht (1906–), German-born American physicist. Formulated the theory of energy production in stars; research on nuclear physics; Nobel Prize, 1967.

Bhabha, Homi Jehangir (1909–1966), Indian physicist. With W. Heitler, developed theory of cascade showers of cosmic rays; observed slowing of decay rate of high-velocity mesons.

Bianchi, Luigi (1856–1928), Italian mathematician. Contributed to differential geometry and study of noneuclidean geometries; discovered Bianchi identity.

Bichat, Marie François Xavier (1771–1802), French anatomist and physiologist. Founder of animal histology; originated the term "tissues," and distinguished 21 types in his particular scheme.

Bieberbach, Ludwig (1886–1982), German mathematician. Research on complex function theory, differential equations, geometry, and algebra; postulated Bieberbach's conjecture.

Bienaymé, Irénée Jules (1796–1878), French mathematician. Studied calculus of probabilities and its application to financial science; with P. L. Chebyshev, discovered Bienaymé-Chebyshev inequality.

Billet, Felix (1808–1882), French physicist. Invented Billet split lens.

Binet, Alfred (1857–1911), French psychologist. Investigated development and measurement of intelligence.

Binnig, Gerd (1947–), German physicist. With H. Rohrer, developed scanning tunneling microscope; Nobel Prize, 1986.

Biot, Jean Baptiste (1774–1862), French mathematician and physicist. Discovered circular polarization of light; invented a polariscope; with D. Brewster, discovered biaxial crystals; helped formulate Biot-Savart law.

Birkhoff, George David (1884–1944), American mathematician. Investigated differential equations, dynamical systems, ergodic theory, mechanics of fluids, and foundations of relativity and quantum mechanics.

Bishop, John Michael (1936–), American virologist and biochemist. With H. Varmus, researched the genetic basis of human cancers; their work led to the identification of over 50 cellular genes that can become oncogenes; Nobel Prize, 1989.

Bjerknes, Vilhelm Fremann Doren (1862–1951), Norwegian physicist. Research on electric waves; originated the polar-front theory in meteorology.

Black, James (1924–   ), British pharmacologist. Developed the first beta blocker drug, propranolol; also credited with the discovery of another important class of drugs, the H2 antagonists; Nobel Prize, 1988.

Blackett, Patrick Maynard Stuart (1897–1974), English physicist. Built an improved cloud chamber used to photograph tracks of a nuclear disintegration and of a cosmic-ray shower; discovered the positron; Nobel Prize, 1948.

Bloch, Felix (1905–1983), Swiss-born American physicist. Discovered a technique for studying magnetism of atomic nuclei in normal matter; Nobel Prize, 1952.

Bloch, Konrad Emil (1912–   ), German-born American biochemist. Traced the transformations of fat and carbohydrate metabolites to cholesterol; Nobel Prize, 1964.

Bloembergen, Nicholaas (1920–   ), Netherlands-born American physicist. Contributed to development of maser; made extensive contributions to theoretical and experimental development of nonlinear optics; Nobel Prize, 1981.

Blumberg, Baruch Samuel (1925–   ), American physician and biologist. Research leading to a test for hepatitis viruses in blood and to an experimental hepatitis vaccine; Nobel Prize, 1976.

Boas, Franz (1858–1942), American anthropologist. Pioneered in physical anthropology; contributed to stratigraphic archeology in Mexico; emphasized importance of linguistic analysis.

Bobillier, Étienne (1798–1840), French mathematician and physicist. Contributed to geometry and statics; discovered Bobillier's law.

Bode, Johann Elert (1747–1826), German astronomer. Prepared a celestial atlas showing about 17,000 stars; formulated Bode's law.

Boerhaave, Hermann (1668–1738), Dutch physician. A great teacher at the University of Leyden; wrote the physiology textbook *Institutiones Medicae.*

Bohm, David (1917–1992), American-born British physicist. Research in quantum theory and new modes of description in physics.

Bohr, Aage (1922–   ), Danish physicist. With B. R. Mottelson, developed theory which unifies shell and liquid-drop models of the atomic nucleus, and which explains nonspherical nuclei; Nobel Prize, 1975.

Bohr, Niels (1885–1962), Danish physicist. Devised an atomic model; codeveloped the quantum theory, applying it to atomic structure in Bohr's theory; Nobel Prize, 1922.

Boltzmann, Ludwig Eduard (1844–1906), Austrian physicist. An authority on the kinetic theory of gases; demonstrated the Stefan-Boltzmann law of blackbody radiation, Boltzmann's law of energy, and the Boltzmann constant.

Bolyai, János (1802–1860), Hungarian mathematician. Independently of K. F. Gauss and N. I. Lobachevski, originated a system of noneuclidean geometry.

Bolzano, Bernard (1781–1848), Czechoslovakian philosopher, logician, and mathematician. Contributed to theory of real functions; proved Bolzano's theorem and Bolzano-Weierstrass theorem.

Bond, George Phillips (1825–1865), American astronomer. Introduced concept of Bond albedo; pioneered astronomical photography.

Boole, George (1815–1864), English mathematician and logician. Developed new system of mathematical logic, which is known as Boolean algebra.

Borda, Jean Charles (1733–1799), French physicist and mathematician. Introduced Borda mouthpiece; developed instruments for navigation, geodesy, and determination of weights and measures.

Bordet, Jules Jean Baptiste Vincent (1870–1961), Belgian physiologist. Made discoveries in immunology; with O. Gengou, developed the technique of the complement fixation reaction; Nobel Prize, 1919.

Borel, Félix Edouard Émile (1871–1956), French mathematician. Work in infinitesimal calculus and the calculus of probabilities.

Born, Max (1882–1970), German-born British theoretical physicist. Pioneered in the development of quantum mechanics; Nobel Prize, 1954.

Bosch, Carl (1874–1940), German chemist. Developed chemical high-pressure methods, and the Haber-Bosch process for ammonia synthesis; Nobel Prize, 1931.

Bose, Jagadis Chandra (1858–1937), Indian plant physiologist and physicist. Founded the Bose Research Institute in Calcutta; investigated photosynthesis, "nervous mechanism" of plants, and other plant subjects.

Bose, Satyendra Nath (1894–1974), Indian physicist. Originated Bose-Einstein statistics to describe photons.

Bothe, Walter (1891–1957), German physicist. Devised the coincidence method for the investigation of nuclear reactions and cosmic radiation; Nobel Prize, 1954.

Bouguer, Pierre (1698–1758), French geodesist, hydrographer, and physicist. Laid foundations of photometry; discovered Bouguer-Lambert law of light intensity.

Bourdon, Eugène (1808–1884), French inventor. Invented Bourdon pressure gage.

Boussinesq, Joseph Valentin (1842–1929), French mathematical physicist. Research in hydrodynamics; introduced Boussinesq approximation.

Bovet, Daniel (1907–1992), Swiss-born Italian pharmacologist. Research on synthetic compounds that inhibit the action of the vascular system and skeletal muscles; Nobel Prize, 1957.

Bowditch, Nathaniel (1773–1838), American navigator and mathematician. Produced navigation guide; translated and improved P. S. de Laplace's *Mécanique céleste.*

Bowen, Norman Levi (1887–1956), Canadian-born American geologist. Studied physical chemistry of geological processes and phase equilibria of silicates; discovered significance of reaction principle in petrogenesis.

Boyer, Paul D. (1918–   ), American chemist. Made major contributions toward elucidating the enzymatic mechanism underlying the synthesis of adenosine triphosphate (ATP) by proposing the binding change mechanism; Nobel Prize, 1997.

Boyle, Robert (1627–1691), British physicist and chemist. Conducted experiments on properties of the air pump; his law concerning gases is named for him; advanced the atomistic theory of matter.

Bradley, James (1693–1762), English astronomer. Discovered aberration of light due to the Earth's motion, and nutation of the Earth's axis; prepared astronomical tables and star catalogs.

Bragg, William Henry (1862–1942), English physicist. Codeveloper, with W. L. Bragg, of the x-ray spectrometer; used x-ray diffraction to determine crystal structure; Nobel Prize, 1915.

Bragg, William Lawrence (1890–1971), British physicist. With W. H. Bragg, developed x-ray analysis of the atomic arrangement in crystalline structures; Nobel Prize, 1915.

Brahe, Tycho or Tyge (1546–1601), Danish astronomer. Made painstaking observations of the planetary system; wrote *Astronomiae Instauratae Progymnasmata.*

Brattain, Walter Houser (1902–1987), American physicist. Investigated properties of semiconductors; research on surface properties of solids; Nobel Prize, 1956.

Braun, Karl Ferdinand (1850–1918), German physicist. Research on cathode rays and wireless telegraphy; Nobel Prize, 1909.

Bravais, Auguste (1811–1863), French physicist. Studied relationship between crystal form and structure; derived Bravais lattices.

Breit, Gregory (1899–1981), Russian-born American physicist. Research on quantum theory, quantum electrodynamics, hyperfine structure, and ionosphere.

Breuer, Josef (1842–1925), Austrian neurologist. Evolved abreaction method for treatment of neuroses.

Brewster, David (1781–1868), Scottish physicist. Formulated Brewster's law on polarization of light; codiscoverer, with J. B. Biot, of biaxial crystals.

Brianchon, Charles Julien (1783–1864), French mathematician. Proved Brianchon's theorem.

Bridgman, Percy Williams (1882–1961), American physicist. Worked in high-pressure physics and thermodynamics of liquids; Nobel Prize, 1946.

Briggs, Henry (1561–1631), English mathematician. Prepared logarithmic tables (later known as common logarithms); devised sophisticated interpolation techniques.

Bright, Richard (1789–1858), English physician. Made biochemical study of disease; researched Bright's disease of the kidneys.

Brillouin, Leon (1889–1969), French physicist. With G. Wentzel and H. A. Kramers, developed Wentzel-Kramers-Brillouin method; originated concept of Brillouin zones.

Brillouin, Louis Marcel (1854–1948), French physicist. Work on crystal structure, viscosity of liquids and gases, radiotelegraphy, and relativity.

Brinell, Johann August (1849–1925), Swedish engineer. Invented the Brinell machine to measure the hardness of alloys and metals in terms of the Brinell number.

Brockhouse, Bertram N. (1918–   ), Canadian physicist. Developed slow neutron spectroscopy technique for studying dynamics of atoms in solids and liquids; Nobel Prize, 1994.

Broglie, Louis Victor de (1892–1987), French physicist. Worked in nuclear physics; first to link wave and corpuscular theory; Nobel Prize, 1929.

Bromwich, Thomas John I'Anson (1875–1929), English mathematician. Showed how the Heaviside calculus could be developed in a manner acceptable to pure mathematicians through use of contour integrals.

Brönsted, Johannes Nicolaus (1879–1947), Danish chemist. Researched kinetic properties of ions, catalysis, and nitramide; formulated the Brönsted theory of acid-base reactions.

Brown, Herbert Charles (1912–   ), British-born American chemist. Developed methods for chemical synthesis of diborane and organoboranes; Nobel Prize, 1979.

Brown, Michael S. (1941–   ), American biochemist and geneticist. With Joseph L. Goldstein, discovered low-density lipoprotein receptors and their function in cholesterol metabolism; Nobel Prize, 1985.

Browning, John Moses (1855–1926), American inventor. Invented the Browning machine gun.

Brunauer, Stephen (1903–1986), Hungarian-born American chemist. Contributed to surface and colloid chemistry; with P. H. Emmett and E. Teller, developed Brunauer-Emmett-Teller equation for surface area determinations.

Brunel, Isambard Kingdom (1806–1859), English engineer. Constructed great bridges in England; designed important steamships.

Buchner, Eduard (1860–1917), German chemist. Studied alcoholic fermentation of sucrose; Nobel Prize, 1907.

Buckingham, Edgar (1867–1940), American physicist. Worked on thermodynamics and dimensional analysis; derived Buckingham's π theorem.

Buffon, Georges Louis Leclerc, Comte de (1707–1788), French naturalist. Compiled *Histoire Naturelle*, a monumental work on natural history.

Bullen, Keith Edward (1906–1976), Australian applied mathematician. Carried out mathematical studies of earthquake waves; with H. Jeffreys, prepared the Jeffreys-Bullen tables on seismic travel time.

Bunsen, Robert Wilhelm (1811–1899), German chemist. Discovered, with G. R. Kirchhoff, spectrum analysis; invented the Bunsen burner, Bunsen cell, and Bunsen ice calorimeter; formulated law of reciprocity with H. E. Roscoe.

Burbank, Luther (1849–1926), American horticulturist. Experimented on crossing and in-breeding of plant varieties.

Burnet, Frank Macfarlane (1899–1985), Australian immunologist. With P. B. Medawar, studied the body's tolerance of antigenic substances; Nobel Prize, 1960.

Butenandt, Adolph Friedrich Johann (1903–1995), German chemist. Researched sex hormones; Nobel Prize (declined), 1939.

Buys-Ballot, Christoph Hendrik Didericus (1817–1890), Dutch meteorologist. Devised a system of storm signals; formulated Buys-Ballot's law for determination of wind direction.

Cailletet, Louis Paul (1832–1913), French chemist. Researched liquefaction of gases; first to obtain liquid oxygen, hydrogen, nitrogen, and air.

Callendar, Hugh Longbourne (1863–1930), English physicist and engineer. Developed platinum resistance thermometer and continuous-flow calorimeter.

Callow, John Michael (1867–1940), English-born American mining engineer and metallurgist. Invented Callow flotation cell and Callow screen.

Calvin, Melvin (1911–1997), American chemist. With J. A. Bassham, traced the path of carbon in photosynthesis; Nobel Prize, 1961.

Cannizzaro, Stanislao (1826–1910), Italian chemist. Promulgated Avogadro's work as related to atomic weights; discovered Cannizzaro's reaction in organic chemistry.

Cantor, Georg (1845–1918), Russian-born German mathematician. Founded set theory; introduced fundamental concepts in topology; worked on the theory and representations of real numbers.

Carathéodory, Constantin (1873–1950), German mathematician. Developed calculus of variations for curves with corners; introduced Carathéodory outer measure; gave mathematical formulation of second law of thermodynamics (Carathéodory's principle).

Cardano, Geronimo, or Jerome Cardan (1501–1576), Italian physician and mathematician. Wrote on algebra, medicine, and astronomy; invented the Cardan shaft.

Carnot, Nicolas Léonard Sadi (1796–1832), French physicist. Formulated Carnot's theorems in thermodynamics.

Carrel, Alexis (1873–1944), French surgeon and biologist. Worked on transplanting organs, suturing blood vessels, treating deep wounds, and prolonging tissue life; Nobel Prize, 1912.

Carrington, Richard Christopher (1826–1875), English astronomer. Investigated motions of sunspots.

Carver, George Washington (1864–1943), American botanist. Did research on industrial uses of the peanut.

Cassegrain, N. (17th century), French physician. Designed the Cassegrain reflecting telescope.

Cassini, Jean Dominique (1625–1712), Italian-born French astronomer. Director of the Paris Observatory; discovered four new satellites of Saturn and the Cassini division in Saturn's ring; conducted pendulum experiments related to the shape of the Earth.

Castigliano, Carlo Alberto (1847–1884), Italian structural engineer. Proved Castigliano's theorem.

Cauchy, Augustin Louis, Baron (1789–1857), French mathematician. Wrote extensively on wave propagation, calculus, and elasticity.

Cavendish, Henry (1731–1810), French physicist and chemist. Determined the density of the Earth and the composition of the atmosphere; studied properties of carbon dioxide and hydrogen.

Cayley, Arthur (1821–1895), English mathematician. Proposed the theory of matrices; developed the theory of invariants and covariants; worked on quantics and the theory of groups.

Cech, Thomas R. (1947–), American chemist. By studying the single-cell organisms *Tetrahymena* and *Thermophila*, discovered that molecules of ribonucleic acid (RNA) have catalytic properties similar to those of enzymes; Nobel Prize, 1989.

Celsius, Anders (1701–1744), Swedish astronomer. Constructed the thermometer using the Celsius (centigrade) scale.

Cerenkov, Pavel Alexeyevich (1904–1990), Soviet physicist. Discovered the Cerenkov effect of radiation; devised the Cerenkov counter for particle detection; Nobel Prize, 1958.

Cesalpino, Andrea (1519–1603), Italian physician and botanist. First to attempt to classify plants according to characteristics of fruit and seed in his *De Plantis*.

Cesàro, Ernesto (1859–1906), Italian mathematician. Formulated an intrinsic geometry; introduced the Cesàro summation.

Ceva, Giovanni (1647?–1734), Italian mathematician. Formulated a theorem on the concurrency of straight lines passing through the vertices of a triangle.

Chadwick, James (1891–1974), English physicist. Established experimentally the existence of the neutron; Nobel Prize, 1935.

Chain, Ernst Boris (1906–1979), German-born British biochemist. With H. W. Florey, worked on the chemical structure of penicillin and its first clinical trials; Nobel Prize, 1945.

Chamberlain, Owen (1920–), American physicist. With E. G. Segrè, demonstrated the existence of the antiproton; Nobel Prize, 1959.

Chamberlin, Thomas Chrowder (1843–1928), American geologist. Studied the fundamental geology of the solar system.

Chandler, Seth Carlo (1846–1913), American astronomer. Discovered the Chandler wobble.

Chandrasekhar, Subrahmanyan (1910–1995), Indian astrophysicist. Developed a theory of white dwarf stars; Nobel Prize, 1983.

Chaplygin, Sergei Alekseevich (1869–1942), Russian physicist, engineer, and mathematician. Made contributions to fluid mechanics, particularly aerodynamics.

Chapman, Sydney (1888–1970), English mathematician and physicist. Discovered (independently of D. Enskog) gaseous thermal diffusion; studied the daily variations of the geomagnetic field and magnetic storms.

Chaptal, Jean Antoine Claude, Comte de Chanteloup (1756–1832), French chemist. Wrote on technical chemistry; introduced the metric system after the Revolution.

Charcot, Jean Martin (1825–1893), French neurologist. Director of the Salpetrière clinic, where he made systematic clinical studies of chronic nervous disorders, including cerebrospinal disease.

Charles, Jacques Alexandre César (1746–1823), French physicist, chemist, and inventor. Formulated Charles' law, relating gas volume to pressure.

Charpak, Georges (1924–), French physicist. Invented the multiwire proportional chamber, used as a detector in high-energy physics experiments; Nobel Prize, 1992.

Chebyshev, Pafnuti Lvovich (1821–1894), Russian mathematician. Research on convergence of Taylor series, prime numbers, probability theory, quadratic forms, and integral theory.

Chladini, Ernst Florenz Friedrich (1756–1827), German physicist. Discovered Chladini's figures and used them to study vibrations of solid plates.

Christoffel, Elwin Bruno (1829–1900), Swiss mathematician. Worked in higher analysis, geometry, mathematical physics, and geodesy.

Chu, Paul Ching-Wu (1941–), Chinese-born American physicist. Discovered superconductivity at temperatures over 90 K (−298°F) in yttrium-barium-copper-oxygen compounds.

Chu, Steven (1948–), American physicist. Developed a system of opposed laser beams to cool atoms to extremely low temperatures, and a magnetooptical trap to capture them; Nobel Prize, 1997.

Clairaut, Alexis Claude (1713–1765), French mathematician. Studied the shape of the Earth, evolving Clairaut's theorem; made astronomical calculations concerning Halley's comet.

Claisen, Ludwig (1851–1930), German organic chemist. Developed Claisen condensation; contributed to understanding of tautomerism; worked on rearrangement of allyl aryl ethers into phenols.

Clapeyron, Benoit Paul Émile (1799–1864), French engineer. Developed N. L. S. Carnot's concept of a universal function of temperature.

Clarke, Alexander Ross (1828–1914), British geodesist. Worked on the triangulation of the British Isles; proposed Clarke ellipsoids as geodetic standards.

Claude, Albert (1899–1983), American cytologist, born in Luxembourg. Pioneered in applying electron microscopy to cell studies and in using centrifuge to separate cell components; Nobel Prize, 1974.

Clausius, Rudolf Julius Emmanuel (1822–1888), German physicist. A founder of thermodynamics; worked out the Clausius-Clapeyron equation for the universal temperature function.

Clebsch, Rudolf Friedrich Alfred (1833–1872), German mathematician. Contributions to theory of invariants and algebraic geometry.

Cockcroft, John Douglas (1897–1967), English physicist. With E. T. S. Walton, split nuclei by bombarding them with accelerated protons.

Cohen, Stanley (1922–), American biochemist. With R. Levi-Montalcini, made landmark studies of nerve growth factor and its functions; Nobel Prize, 1986.

Cohen-Tannoudji, Claude (1933–), French physicist. Helped develop methods to cool and trap atoms with laser light, and explained how atoms could be cooled to temperatures lower than the previously calculated theoretical limits; Nobel Prize, 1997.

Cohn, Ferdinand Julius (1828–1898), German botanist. A founder of bacteriology; did research in plant pathology; first to classify bacteria according to genus and species.

Cole, Kenneth Stewart (1900–1984), American biophysicist. Research on structure and function of living cell membranes and nerve membranes in particular, concentrating on electrical approach; with brother, R. H. Cole, introduced Cole-Cole plot of dielectric behavior.

Cole, Robert Hugh (1914–1990), American chemist and physicist. Research on dielectric properties of matter and intermolecular forces; with brother, K. S. Cole, introduced Cole-Cole plot of dielectric behavior.

Collins, Samuel Cornette (1898–1984), American engineer. Invented Collins helium liquefier.

Compton, Arthur Holly (1892–1962), American physicist. Discovered the Compton effect of x-rays; studied cosmic rays; helped develop the atomic bomb; Nobel Prize, 1927.

Conant, James Bryant (1893–1978), American chemist. Researched free radicals, hemoglobin, and chlorophyll; contributed to atomic energy development.

Condon, Edward Uhler (1902–1974), American physicist. Contributed to the Franck-Condon principle, by extending and giving quantum-mechanical treatment to J. Franck's concept of nuclear motion to molecules in transition from one energy level to another.

Coolidge, William David (1873–1975), American physicist. Invented Coolidge tube; discovered method for making tungsten strong and ductile.

Cooper, Leon N. (1930–), American physicist. Showed that electrons could form Cooper pairs; with J. R. Schrieffer and J. Bardeen, formulated a theory of superconductivity; Nobel Prize, 1972.

Copernicus, Nicolaus (1473–1543), Polish (or Prussian) astronomer. Proposed the Copernican system, with the Sun as the center of planetary orbits.

Corey, Elias James (1928–), American chemist. Developed theories and methods of organic chemical synthesis that have made possible the production of a wide variety of complex biologically active substances and useful chemicals; Nobel Prize, 1990.

Cori, Carl Ferdinand (1896–1984), and **Cori, Gerty Theresa Radnitz** (1896–1957), Czechoslovakian-born American biochemists. Discovered the enzymatic mechanism of glucose-glycogen interconversion and the effects of hormones on this mechanism; Nobel Prize, 1947.

Coriolis, Gaspard Gustave de (1792–1843), French physicist. Contributed to theoretical and applied mechanics; clarified and supplied concepts of work and kinetic energy; derived Coriolis acceleration.

Cormack, Allan MacLeod (1924–), American physicist. Contributed to the development of computerized axial tomography; Nobel Prize, 1979.

Corner, George Washington (1889–1981), American medical biologist. Contributed to understanding of anatomical details of menstrual cycle and functions of estrogen and progesterone.

Cornforth, John Warcup (1917–), Australian-born British chemist. Investigated stereochemistry of enzyme-catalyzed reactions; Nobel Prize, 1975.

Cornu, Marie Alfred (1841–1902), French physicist. Used Cornu spiral for determination of intensities in interference phenomena.

Coster, Dirk (1889–1950), Dutch physicist. Work in x-ray spectroscopy; with G. von Hevesy, discovered hafnium; with R. Kronig, discovered Coster-Kronig transitions.

Cottreli, Frederick Gardner (1877–1948), American chemist. Invented the Cottrell process for precipitation of particles from gas; researched nitrogen fixation, liquefaction of gases, and recovery of helium.

Coulomb, Charles Augustin de (1736–1806), French physicist. Formulated Coulomb's law of electric charges.

Courant, Richard (1888–1972), German-born American mathematician. Research in geometric function theory, differential equations of mathematical physics, and transition by limiting processes from finite difference equations to differential equations.

Cournand, André Frédéric (1895–1988), French-born American physician. Studied normal and abnormal human cardiovascular and pulmonary functions; Nobel Prize, 1956.

Cowan, Clyde L., Jr. (1919–1974), American physicist. With F. Reines, made first detection of the neutrino, a fundamental particle.

Crafts, James Mason (1839–1917), American chemist. With C. Friedel, discovered the Freidel-Crafts reaction, wherein anhydrous aluminum chloride acts as a catalyst.

Cram, Donald J. (1919–), American chemist. Expanded the field of crown ether chemistry by using crown ethers to synthesize structures that mimic the action of biological molecules; Nobel Prize, 1987.

Cramer, Gabriel (1704–1752), Swiss mathematician. Contributed to Cramer's rule for solving linear equations.

Crick, Francis Harry Compton (1916–), English molecular biologist. With J. D. Watson, proposed a double-helix structure for the deoxyribonucleic acid molecule; Nobel Prize, 1962.

Cronin, James Watson (1931–), American physicist. Collaborated with V. L. Fitch on experiment showing that the principle of time-reversal invariance is violated in the decay of neutral *K* mesons; Nobel Prize, 1980.

Cronstedt, Axel Fredrik, Baron (1722–1765), Swedish mineralogist. Discovered nickel; developed a chemical classification system for minerals.

Crookes, William (1832–1919), English physicist and chemist. Invented Crookes tube to study electrical discharges in high vacuum, and a radiometer; discovered thallium.

Crutzen, Paul J. (1933–), Dutch chemist. Demonstrated that nitrogen oxides react catalytically with ozone, thus accelerating the rate of reduction of the ozone layer; Nobel Prize, 1995.

Curie, Marie, born Marya Sklodowska (1867–1934), Polish physical chemist in France. Explored nature of radioactivity; codiscoverer of radium, and first to separate polonium; Nobel Prize, 1903 and 1911.

Curie, Pierre (1859–1906), French chemist and physicist. Codiscoverer of radium; formulated the Curie point, relating magnetic properties and temperature; discovered the piezoelectric effect; Nobel Prize, 1903.

Curl, Robert F., Jr. (1933–), American chemist. Made major contributions toward the discovery of fullerenes; Nobel Prize, 1996.

Cushing, Harvey (1869–1930), American surgeon. Innovator in neurosurgical techniques; research on function and diseases of the pituitary gland.

Cuvier, Georges Léopold Chrétien Frédéric Dagobert, Baron (1769–1838), French naturalist. Made a detailed classification of the animal kingdom; wrote on comparative anatomy.

Daguerre, Louis Jacques Mandé (1787–1851), French inventor. Invented the daguerreotype photographic process.

Dale, Henry Hallett (1875–1968), British pharmacologist and physiologist. Isolated acetylcholine and recognized its effect to be similar to that brought about by parasympathetic nerves; Nobel Prize, 1936.

Dalén, Nils Gustaf (1869–1937), Swedish physicist. Invented automatic gas lighting for unsupervised lighthouses and railroad signals; Nobel Prize, 1912.

Dalitz, Richard Henry (1925–), Australian-born British theoretical physicist. Research on properties of mesons and baryons and nuclear interactions of the lambda hyperon; proposed models for elementary particles; introduced the Dalitz plot.

Dalton, John (1766–1844), English chemist and physicist. Proposed the atomic theory of chemical reactions; developed the law of partial pressures of gases; studied color-blindness.

Dam, Carl Peter Henrik (1895–1976), Danish biochemist and nutritionist. Discovered vitamin K and studied its role in human hemorrhagic disease; Nobel Prize, 1943.

Danckwerts, Peter Victor (1916–1984), British chemical engineer. Proposed the surface-renewal model of liquids.

Daniell, John Frederic (1790–1845), English physicist and chemist. Invented the Daniell cell.

Darwin, Charles Robert (1809–1882), English naturalist. Proposed far-reaching theory of evolution of species and theory of natural selection in his *Origin of Species*.

Darwin, George Howard (1845–1912), English mathematician and astronomer. Applied detailed dynamical analysis to cosmological and geological problems.

Dausset, Jean (1916–), French biologist and medical scientist. Studied antigen in human leukocytes and their role in transplant acceptance or rejections; Nobel Prize, 1980.

Davisson, Clinton Joseph (1881–1958), American physicist. Studied magnetism, radiant energy, and electricity; independent of G. P. Thomson, discovered electron diffraction by crystals; Nobel Prize, 1937.

Davy, Humphry (1778–1829), English chemist. Discovered potassium and sodium; invented the Davy safety lamp for use in coal mines; proposed theoretical explanations of electrolysis and voltaic action.

Debye, Peter Joseph William (1884–1966), American physical chemist born in the Netherlands. Worked on dipole moments and the diffraction of x-rays in gases; formulated Debye-Hückel theory on the behavior of strong electrolytes; Nobel Prize, 1936.

Dedekind, Julius Wilhelm Richard (1831–1916), German mathematician. Worked in number theory and analysis, particularly with algebraic integers, algebraic functions, and ideals; defined real numbers by Dedekind cuts.

de Duve, Christian René (1917–), Belgian biochemist and cytologist. Refined centrifuge technique for studying cell components; discovered lysosomes; Nobel Prize, 1974.

De Forest, Lee (1873–1961), American inventor. Pioneer in radio technology; invented audio amplifier and the four-electrode valve; incorporated the grid into the thermionic valve.

de Gennes, Pierre-Gilles (1932–), French physicist. Applied physical principles to the study of complex systems, including liquid crystals and polymers; Nobel Prize, 1991.

de Haas, Wander Johannes (1878–1960), Dutch physicist. Demonstrated the Einstein-de Haas effect; worked on production of extremely low temperatures by adiabatic demagnetization; with P. Van Alphen, discovered de Haas-Van Alphen effect.

Dehmelt, Hans Georg (1922–), German-born American physicist. Developed the Penning trap, which uses magnetic and electric fields to hold ions in a small volume; used the traps to isolate a single electron and carry out extremely accurate measurements of atomic properties; Nobel Prize, 1989.

Delbruck, Max (1906–1981), German-born American biologist. Pioneered in molecular biology; research on bacterial viruses; Nobel Prize, 1969.

Demoivre, Abraham (1667–1754), French-born English mathematician. Originated two theorems on expansions of trigonometrical expansions; proposed methods to approximate functions of large numbers, and the concept of the normal distribution curve.

De Morgan, Augustus (1806–1871), English mathematician. Wrote textbooks and treatises on arithmetic, algebra, and trigonometry; formulated the De Morgan theorem.

Desargues, Gérard (1593–1662), French mathematician. A founder of modern geometry; proposed the theory of involution and transversals.

Descartes, René (1596–1650), French mathematician. Originated cartesian, or coordinate, geometry.

de Sitter, Willem (1872–1934), Dutch astronomer. Worked on the application of Einstein's theory to the problems of the universe; computed the size of the universe.

De Vries, Hugo (1848–1935), Dutch botanist. Formulated the mutation theory of evolution.

Dewar, James (1842–1923), British chemist. Made pioneering studies of matter at low temperatures; first to liquefy hydrogen; invented the Dewar vacuum flask.

Dick, George Frederick (1881–1967), American physician and bacteriologist. Isolated scarlet fever streptococci, developed scarlet fever streptococcus antitoxin, and developed Dick test.

Dicke, Robert Henry (1916–1997), American physicist. Developed new relativistic theory of gravitation with C. Brans; investigated cosmic blackbody radiation; worked on development of radar.

Diels, Otto Paul Hermann (1876–1954), German chemist. Codiscoverer of the Diels-Alder reaction (diene synthesis); worked on sterol chemistry; discovered carbon suboxide; Nobel Prize, 1950.

Diesel, Rudolf (1858–1913), German inventor. Designed and built the diesel engine.

Deisenhofer, Johann (1943–), German chemist. With R. Huber and M. Hartmut, elucidated the structure of a bacterial protein that performs photosynthesis; Nobel Prize, 1988.

Diocles (2d century B.C.), Greek mathematician. Contributions to theory of conics and geometry.

Diophantus of Alexandria (3d century), Greek mathematician. Known as the father of algebra; the first to use conventional algebraic notation.

Dirac, Paul Adrien Maurice (1902–1984), English physicist. Worked in quantum mechanics; his theory of negative-energy holes predicted existence of the positron; Nobel Prize, 1933.

Dirichlet, Peter Gustave Lejeuné (1805–1859), German mathematician. Applied higher analysis to the theory of numbers; work on definite integrals.

Dobzhansky, Theodosius (1900–1975), Russian-born American biologist. Elucidated the mechanisms of heredity and variation through studies of *Drosophila*.

Doherty, Peter C. (1940–), Australian-born American immunologist. Collaborated with R. M. Zinkerngagel in the discovery of specificity of cell-mediated immune defense; Nobel Prize, 1996.

Dolsy, Edward Adelbert (1893–1986), American biochemist. Isolated pure crystalline compounds important to human health, such as sex hormones and vitamins; Nobel Prize, 1943.

Domagk, Gerhard (1895–1964), German biochemist. Discovered sulfamidocrysoidin, the first synthetic microbial of broad clinical usefulness. Nobel Prize (declined), 1939.

Donati, Giovanni Battista (1826–1873), Italian astronomer. Studied stellar spectra; discovered six comets.

Donnan, Frederick George (1870–1956), Irish chemist born in Ceylon. Research in chemical kinetics; originated the Donnan theory of membrane equilibrium.

Doppler, Christian Johann (1803–1853), Austrian physicist and mathematician. Formulated Doppler's principle, relating the frequency of wave motion to velocity; described the Doppler effect.

Drake, Frank Donald (1930–), American

astronomer. Research on solar system, 21-centimeter radio line, search for extraterrestrial intelligence (introduced Drake equation), and radio telescope development.

Draper, Henry (1837–1882), American scientist. Research in spectroscopy; the Draper catalog is named in his honor.

Drude, Paul Karl Ludwig (1863–1906), German physicist. Attempted to correlate and account for optical, electrical, thermal, and chemical properties of substances; developed theory of properties of metals based on free electrons treated as a gas.

Duane, William (1872–1935), American physicist and radiologist. Developed treatment of cancer by radioisotopes and x-rays; with F. L. Hunt, discovered Duane-Hunt law of x-rays.

Du Bois-Reymond, Emil (1818–1896), German physiologist. Pioneer work in electrical properties of living tissues, especially nerves.

DuBridge, Lee Alvin (1901–1994), American physicist. Developed Fowler-DuBridge theory of photoelectric emission.

Ducrey, Augusto (1860–1940), Italian dermatologist. Discovered *Hemophilus ducreyi*, the agent of chancroid.

Dufay, Charles François de Cisternay (1698–1739), French chemist. Discovered positive and negative types of electricity.

Dulbecco, Renato (1914–), Italian-born American virologist. Developed techniques for studying animal viruses; investigated interaction between deoxyribonucleic acid tumor viruses and genetic material; Nobel Prize, 1975.

Dulong, Pierre Louis (1785–1838), French chemist and physicist. With A. T. Petit, formulated the law of the constancy of atomic heats; developed the Dulong formula for heat value of fuels.

Dumas, Jean Baptiste André (1800–1884), French chemist. Research on organic compounds; determined many atomic weights.

Dürer, Albrecht (1471–1528), German painter, graphic artist, and mathematician. Work on scientific perspective and mathematical proportion.

Du Vigneaud, Vincent (1901–1978), American biochemist. Synthesized a polypeptide hormone, oxytocin; worked on other biologically important sulfur compounds; Nobel Prize, 1955.

Eccles, John Carew (1903–1997), Australian physiologist. Elucidated the action of nerve impulses across zones of close contact between nerve cells; Nobel Prize, 1963.

Eddington, Arthur Stanley (1882–1944), English astronomer and writer. Theoretical research on stellar movements and internal makeup of stars; wrote on theory of relativity.

Edelman, Gerald Maurice (1929–), American biochemist. Worked to determine chemical structure of immunoglobulins; Nobel Prize, 1972.

Edison, Thomas Alva (1847–1931), American electrician and inventor. Invented the gramophone, carbon transmitter for the telephone, incandescent electric lamp, moving pictures, and the diplex method of telegraphy.

Ehrenfest, Paul (1880–1933), Austrian-born Dutch theoretical physicist. Contributed to statistical mechanics and quantum mechanics; developed Ehrenfest's principle; proved Ehrenfest's theorem.

Ehrlich, Paul (1854–1915), German bacteriologist. Research in chemotherapy, notably the discovery of Salvarsan for treatment of syphilis; pioneered in the study of hematology and immunity; Nobel Prize, 1908.

Eigen, Manfred (1927–), German chemist. Devised relaxation techniques to study high-speed chemical reactions; Nobel Prize, 1967.

Eijkman, Christiaan (1858–1930), Dutch physician. Studied dietary deficiency disease, in particular beriberi; Nobel Prize, 1929.

Einstein, Albert (1879–1955), German-born American physicist. Proposed the theory of relativity; extended the application of quantum theory; Nobel Prize, 1921.

Einthoven, Willem (1860–1927), Dutch physiologist born in Java. Used the string galvanometer to record electrical activity of the heart, thereby inventing the electrocardiograph; Nobel Prize, 1924.

Elion, Gertrude Belle (1918–), American biochemist. With G. H. Hitchings, pioneered research that led them to the development of drugs for the treatment of leukemia, malaria, gout, herpes, bacterial and fungal infections, and autoimmune diseases and organ-transplant rejection; Nobel Prize, 1988.

Elsasser, Walter Maurice (1904–1991), German-born American geophysicist. Formulated the dynamo theory of the Earth's permanent terrestrial magnetic force.

Elster, Johann Philipp Ludwig Julius (1854–1920), German experimental physicist. With H. F. Geitel, studied atmospheric electricity, radioactivity, and photoelectricity, and invented photocell.

Emmett, Paul Hugh (1900–1985), American chemist. Worked on catalysts for ammonia synthesis and the water-gas conversion reaction; with S. Brunauer and E. Teller, formulated Brunauer-Emmett-Teller equation for surface area determinations.

Encke, Johann Franz (1791–1865), German astronomer. Determined period of revolution of Encke's comet, discovered by J. L. Pons; measured the distance of the Sun from the Earth.

Enders, John Franklin (1897–1985), American microbiologist. With F. C. Robbins and T. H. Weller, discovered the capacity of poliomyelitis virus to grow in various tissue cultures; Nobel Prize, 1954.

Enskog, David (1884–1947), Swedish physicist. With S. Chapman, developed the Chapman-Enskog theory for solving the Boltzmann transport equation.

Eötvös, Roland, Baron (1848–1919), Hungarian physicist. Research on gravitation and terrestrial magnetism; formulated a law which relates surface tension to temperature of liquids; designed the Eötvös torsion balance.

Erasistratus (3d century B.C.), Greek physician and anatomist. Founder of physiology; distinguished between the cerebrum and cerebellum, and sensory and motor nerves.

Eratosthenes (3d century B.C.), Greek astronomer. Suggested an extra day in the calendar every fourth year; made a determination of the size of the Earth; measured obliquity of the ecliptic.

Erlanger, Joseph (1874–1965), American physiologist. With H. S. Gasser, created new methods for amplifying and recording electrical impulses in nerves; research on the function of the synapse; Nobel Prize, 1944.

Erlenmeyer, Richard August Carl Emil (1825–1909), German organic chemist. Research on synthesis and constitution of aliphatic compounds; introduced modern structural notation; invented Erlenmeyer flask.

Ernst, Richard R. (1933–), Swiss chemist. Developed methods that transformed nuclear magnetic resonance (NMR) spectroscopy from a tool with a narrow application to a key analytical technique in chemistry as well as many other fields; Nobel Prize, 1991.

Esaki, Leo (1925–), Japanese physicist. Discovered a new negative-resistance characteristic in semiconductor *pn* junctions, leading to the discovery of the tunnel, or Esaki, diode; Nobel Prize, 1973.

Euclid (ca. 330–ca. 275 B.C.), Greek mathematician. Wrote geometry textbooks; euclidean geometry is named after him.

Eudoxus of Cnidus (ca. 408–ca. 355 B.C.). Greek astronomer and mathematician. Expounded theory of motion of planets based on homocentric spheres; developed theory of proportions and methods of measuring areas and volumes of geometrical figures.

Euler, Leonhard (1707–1783), Swiss mathematician. Contributed to algebraic series and differential and integral calculus; realized the significance of coefficients (Euler numbers) of certain trigonometrical expansions.

Euler-Chelpin, Hans Karl August Simon von (1873–1964), German-Swedish chemist. Research on enzyme action and fermentation of sugars; Nobel Prize, 1929.

Ewald, Paul Peter (1888–1985), German-born American physicist. Developed dynamic theory of x-ray interference in crystals.

Ewing, William Maurice (1906–1974), American geophysicist. Made fundamental contributions to seismology, geodesy, oceanography, and submarine geology.

Eyring, Henry (1901–1981), Mexican-born American chemist. Pioneered in the application of quantum and statistical mechanics to chemistry; conceived the theory of absolute reaction rates and the significant structures theory of liquids.

Fabricius, Hieronymus, or Girolamo Fabrizio (ca. 1533–1619), Italian anatomist. Made painstaking descriptions of valves in veins; did comparative research in animal embryology.

Fabry, Charles (1867–1945), French physicist. With A. Pérot, invented Fabry-Pérot interferometer; experimentally verified Doppler broadening and Doppler effect.

Fahrenheit, Gabriel Daniel (1686–1736), German physicist. Constructed thermometers; invented the Fahrenheit temperature scale.

Fallopio, Gabriele (1523–1562), Italian anatomist. Discovered Fallopian tubes; gave first clear description of organs of inner and middle ear.

Fanning, John Thomas (1837–1911), American civil engineer. Designed water works and water supply systems.

Faraday, Michael (1791–1867), English chemist and physicist. Discovered electromagnetic induction; formulated two laws of electrolysis; invented the dynamo.

Fechner, Gustav Theodor (1801–1887), German psychologist. Founded psychophysics; developed the Fechner law concerning intensity of sensation produced by a stimulus.

Fermat, Pierre de (1601–1665), French mathematician. Founder of the modern theory of numbers; originated Fermat's last theorem, and Fermat's principle in optics.

Fermi, Enrico (1901–1954), Italian-born American physicist. Research on producing radioactive isotopes by neutron bombardment; directed construction of the first atomic pile; Nobel Prize, 1938.

Feynman, Richard Phillips (1918–1988), American physicist. Proposed a theory to eliminate difficulties that had arisen in the study of the interaction of electrons, positrons, and radiation; Nobel Prize, 1965.

Fibiger, Johannes Andreas Grib (1867–1928), Danish pathologist. First to produce cancer experimentally; Nobel Prize, 1926.

Finsen, Niels Ryberg (1860–1904), Danish physician. Originated ultraviolet light therapy for certain diseases; Nobel Prize, 1903.

Fischer, Edmond H. (1920–), American biochemist. With E. G. Krebs, discovered phosphorylation processes that play a critical role in cell-protein regulation; they isolated the first protein kinase, a class of enzymes that transfer phosphate from adenosinetriphosphate to proteins; Nobel Prize, 1992.

Fischer, Emil Hermann (1852–1919), German chemist. Synthesized many natural substances, including purines, D-glucose and other sugars, and the first nucleotide; studied polypeptides and proteins; Nobel Prize, 1902.

Fischer, Ernst Otto (1918–), German chemist. Studied how metals and organic molecules combine to form unique molecules with sandwichlike structures; Nobel Prize, 1973.

Fischer, Hans (1881–1945), German organic chemist. Investigated and synthesized pyrrole pigments; studied structure of chlorophylls; Nobel Prize, 1930.

Fisher, Ronald Aylmer (1890–1962), English geneticist and statistician. Developed statistical techniques for analysis of variance, and for use and validation of small samples; developed theory of the evolution of dominance.

Fitch, Val Logsdon (1923–), American physicist. Collaborated with J. W. Cronin on experiment showing that the principle of time-reversal invariance is violated in the decay of neutral *K* mesons; Nobel Prize, 1980.

FitzGerald, George Francis (1851–1901), Irish physicist. Proposed Lorentz-FitzGerald contraction, relating to a material moving through an electromagnetic field.

Fizeau, Armand Hippolyte Louis (1819–1896), French physicist. First to accurately measure the velocity of light; conducted experiments on the velocity of electricity, use of light wavelength to measure length, and measurement of diameter of stars through the method of interference.

Flamsteed, John (1646–1719), English astronomer. Made a trustworthy catalog of stars; invented conical projection in mapmaking.

Fleming, Alexander (1881–1955), British bacteriologist. Discovered lysozyme and penicillin; Nobel Prize, 1945.

Fleming, John Ambrose (1849–1945), English electrical engineer. Invented the thermionic valve; contributed to widespread application of electric lighting and heating.

Florey, Howard Walter (1898–1968), British pathologist born in Australia. Contributed, with E. B. Chain, to development of penicillin as a chemotherapeutic agent; Nobel Prize, 1945.

Flory, Paul John (1910–1985), American physical chemist. Developed analytic techniques to explore properties and molecular structures of long-chain molecules; Nobel Prize, 1974.

Fock (Fok), Vladimir Alexandrovitch (1898–1974), Soviet theoretical physicist. Contributions to quantum electrodynamics, quantum field theory, electromagnetic diffraction and propagation, and general relativity; developed Hartree-Fock approximation of wave functions.

Forbush, Scott Ellsworth (1904–1984), American geophysicist. Discovered the worldwide decrease in cosmic-ray intensity associated with some magnetic storms.

Forssmann, Werner Theodor Otto (1904–1979), German physician. Developed the technique of cardiac catheterization; Nobel Prize, 1956.

Foucault, Jean Bernard Léon (1819–1868), French physicist. Accurately determined the velocity of light; constructed the Foucault pendulum and the Foucault prism; determined experimentally the rotation of the Earth.

Fourier, Jean Baptiste Joseph, Baron (1768–1830), French geometrician and physicist. Proposed the Fourier series on arbitrary functions; formulated the law of heat propagation.

Fowler, Ralph Howard (1889–1944), English physicist. Applied statistical mechanics to matter at high temperatures and high pressures; explained structure of white dwarf stars; with E. A. Guggenheim, R. F. Peierls, and others, developed Ising model.

Fowler, William Alfred (1911–1995), American physicist. Fundamental contributions to understanding of nuclear reactions that generate the energy of stars and synthesize the elements of the universe; Nobel Prize, 1983.

Franck, James (1882–1964), German physicist. With G. Hertz, studied energy transfer in collisions of molecules; formulated Franck-Condon principle of transition from one energy state to another; Nobel Prize, 1925.

Frank, Ilya Milkhallovich (1908–1990), Soviet physicist. With I. Y. Tamm, proposed a theoretical interpretation of Cerenkov radiation; Nobel Prize, 1958.

Franklin, Benjamin (1706–1790), American physicist, oceanographer, meteorologist, and inventor. Formulated a theory of general electrical "action"; introduced principle of conservation of charge; showed that lightning is an electrical phenomenon; invented lighting rod.

Fraunhofer, Joseph von (1787–1826), German optician and physicist. First to study the dark lines in the solar spectrum (Fraunhofer lines); invented a heliometer; improved the spectroscope.

Fredholm, Eric Ivar (1866–1927), Swedish mathematician. Developed the theory of integral equations (Fredholm equations).

Frenet, Jean Frédéric (1816–1900), French mathematician. Helped to develop Frenet-Serret formulas.

Frenkel, Yakov Ilyich (1894–1954), Soviet physicist. Pioneered in modern atomic theory of solids; developed quantum-mechanical explanations for electron mean free path in metals, and for paramagnetism and ferromagnetism; postulated excitons, Frenkel excitons, and Frenkel defects.

Fresnel, Augustin Jean (1788–1827), French physicist. Investigated effects (Fresnel's fringes) due to the interference of light; developed a wave theory of light; originated Fresnel's reflection formula.

Freud, Sigmund (1856–1939), Austrian psychoanalyst. Founder of psychoanalysis, with emphasis on dream interpretation and free association; developed a theory of personality involving id, ego, and superego, and stressing importance of the libido.

Friedel, Charles (1832–1899), French chemist and mineralogist. With J. M. Crafts, described the Friedel-Crafts reaction; work on artificial production of minerals; studied crystals, ketones, and aldehydes.

Friedman, Jerome Isaac (1930–), American physicist. Collaborated in experiments that demonstrated that protons, neutrons, and similar particles are made up of quarks; Nobel Prize, 1990.

Frisch, Karl von (1886–1982), Austrian zoologist. Discovered means by which bees communicate information about the distance and direction of food; Nobel Prize, 1973.

Frobenius, Georg Ferdinand (1849–1917), German mathematician. Developed representation theory of finite groups and method for solving linear homogeneous ordinary differential equations.

Froude, William (1810–1879), English engineer. Discovered the Froude law of comparison, concerning the towing of an object in a liquid.

Fubini, Guido (1879–1943), Italian mathematician. Worked in algebra, analysis, and differential projective geometry; proved Fubini's theorem.

Fukui, Kenichi (1918–1998), Japanese chemist. Developed frontier orbital theory, a quantum-mechanical model useful in prediction of the combinative properties of molecules; Nobel Prize, 1981.

Fuller, R. Buckminster (1895–1983), American engineer and architect. Designed geodesic dome; the carbon molecular form C_{60} was named buckminsterfullerene because of its structured resemblance to the geodesic dome, and the name fullerene was given to any closed-cage molecule containing an even number of carbon atoms.

Gabor, Dennis (1900–1979), Hungarian-born British physicist and engineer. Invented holography; Nobel Prize, 1971.

Gajdusek, Daniel Carleton (1923–), American physician and virologist. Discovered causal virus and transmission mechanism of kuru; Nobel Prize, 1976.

Galen (2d century), Greek physician and medical writer. Wrote treatises long used as textbooks; experimented on animal nervous systems; made anatomical descriptions of structure and functions of body parts.

Galileo Galilei (1564–1642), Italian astronomer. First to use the telescope for observational purposes; made many discoveries related to the planets and the Sun; did theoretical work on classical physics.

Galois, Evariste (1811–1832), French mathematician. Developed Galois theory of polynomials.

Gamow, George (1904–1968), Russian-born American physicist. Made theoretical contributions to nuclear physics, astronomy, and biology; with E. Teller, formulated the selection rule for beta emission; proposed theoretically the genetic code.

Garvey, Gerald Thomas (1935–), American physicist. Research in experimental nuclear physics, particularly nuclear reactions, isobaric spin studies, and weak interactions in nuclear systems.

Gasser, Herbert Spencer (1888–1963), American physiologist. With J. Erlanger, provided a new method for recording electrical impulses of nerves; studied functions of nerve fibers; Nobel Prize, 1944.

Gatterman, Friedrich August Ludwig (1860–1920), German chemist. Originated Gatterman-Koch synthesis of aldehydes; isolated and analyzed nitrogen trichloride; synthesized aromatic carboxylic acids, thionaphthalene, and thioanilide.

Gauss, Karl Friedrich (1777–1855), German mathematician, astronomer, and physicist. Formulated the Gauss theorem in the mathematics of electricity; made many contributions to pure and applied mathematics; determined orbits of planets and comets from observational data.

Gay-Lussac, Joseph Louis (1778–1850), French chemist and physicist. Discovered the law of expansion of gases by heat, and the law of combining volumes of gases; studied chemistry of iodine and cyanogen.

Geiger, Hans Wilhelm (1882–1945), German physicist and inventor. Invented the Geiger counter to detect alpha particles; investigated properties of alpha particles, cosmic rays, and artificial radiation.

Geissler, Johann Heinrich Wilhelm (1815–1879), German instrument maker. Developed Geissler pump and Geissler tube.

Geitel, Hans Friedrich (1855–1923), German experimental physicist. With J. Elster, studied atmospheric electricity, radioactivity, and photoelectricity, and invented photocell.

Gell-Mann, Murray (1929–), American physicist. Proposed law of conservation of strangeness; used unitary symmetry to classify and explain elementary particles; postulated concept of quarks; Nobel Prize, 1969.

Gesner, Konrad von (1516–1565), Swiss naturalist. Wrote *Historia Animalium*, beginning zoology as a science.

Giaever, Ivar (1929–), Norwegian-born American physicist. Discovered that current-voltage characteristics of an electron tunneling across a thin insulating film separating two metals, one or both of which is in a superconducting state, can be used to obtain electron density of states of superconductors; Nobel Prize, 1973.

Giauque, William Francis (1895–1982), Canadian-born American chemist. Developed adiabatic demagnetization technique for production of extremely low temperatures; collaborated in discovery of isotopes of oxygen; Nobel Prize, 1949.

Gibbs, Josiah Willard (1839–1903), American mathematician and physicist. Made a mathematical treatment of chemical subjects, notably thermodynamics; worked on statistical mechanics, leading to the basis for the phase rule of heterogeneous equilibria.

Gilbert, Walter (1932–), American biochemist. Developed methods for determining nucleotide sequence (independently of F. Sanger), advancing the technology of DNA recombination; Nobel Prize, 1980.

Gilman, Alfred G. (1941–), American pharmacologist. With M. Rodbell, discovered G-proteins and their role in cellular signal transduction; Nobel Prize, 1994.

Ginzburg, Vitaly Lazarevich (1916–), Soviet physicist. Developed Ginzburg-Landau and Ginzburg-London theories of superconductivity.

Giorgi, Giovanni (1871–1950), Italian electrical engineer, physicist, and mathematician. Developed the meter-kilogram-second-ampere system of units.

Glaser, Donald Arthur (1926–), American physicist. Invented the bubble chamber for detecting the paths of high-energy atomic particles; Nobel Prize, 1960.

Glashow, Sheldon Lee (1932–), American physicist. Contributed to development of theory uniting electromagnetism and weak nuclear interactions; postulated existence of charmed particles; Nobel Prize, 1979.

Glauber, Johann Rudolf (1604–1670), German chemist. Discovered Glauber's salt (sodium sulfate) and hydrochloric acid; conducted experiments on compounds of mercury, arsenic, and antimony.

Gmelin, Leopold (1788–1853), German chemist. Wrote *Handbuch der Chemie*, first systematic treatment of chemical knowledge; devised Gmelin's test for presence of bile pigments; studied cyanides.

Goldbach, Christian (1690–1764), German-born Russian mathematician. Research on number theory and analysis; proposed the Goldbach conjecture.

Goldhaber, Maurice (1911–), Austrian-born American physicist. With J. Chadwick, discovered photodisintegration and disintegration of light elements by slow neutrons; with L. Grodzins and A. W. Sunyar, discovered that the neutrino has left-handed spin.

Goldstein, Joseph L. (1940–), American biochemist and geneticist. With M. S. Brown, discovered low-density lipoprotein receptors and their function in cholesterol metabolism; Nobel Prize, 1985.

Golgi, Camillo (1843–1926), Italian physician. Pioneered in the study of histology of the nervous system; discovered the Golgi bodies; Nobel Prize, 1906.

Gordan, Paul Albert (1837–1912), German mathematician. Worked on the theory of invariants and on solutions of algebraic equations and their groups of substitutions.

Gordon, Walter (1893–1940), German-born Swedish physicist. Contributed to relativistic quantum theory; with O. B. Klein, originated the Klein-Gordon equation.

Goudsmit, Samuel Abraham (1902–1978), Dutch-born American physicist. With G. E. Uhlenbeck, discovered electron spin.

Gram, Hans Christian Joachim (1853–1938), Danish physician. Developed the Gram method for staining and differentiating bacteria.

Gram, Jorgen Pedersen (1850–1916), Danish mathematician. Worked in number theory and analysis; with E. Schmidt, originated Gram-Schmidt process for obtaining orthogonal set of vectors.

Granit, Rangar Arthur (1900–1991), Finnish-born Swedish physiologist. Research on vision and on motor control by afferent neurons; Nobel Prize, 1967.

Grashof, Franz (1826–1893), German mechanical engineer. Applied mathematics and physics to engineering problems; derived fundamental equations in the theory of elasticity; introduced Grashof number.

Gray, Henry (1825–1861), English anatomist. Wrote *Anatomy of the Human Body.*

Green, George (1793–1841), English mathematician. Worked in analysis; derived Green's theorem and Green's identities; introduced Green's function.

Gregory, James (1638–1675), Scottish geometer. Provided first proof of the theorem of calculus; gave first description of the reflecting telescope; discovered the series from which π can be calculated.

Grignard, François Auguste Victor (1871–1935), French chemist. Discovered organomagnesium compounds, or Grignard reagents, useful in synthesis of organic and organometallic compounds; Nobel Prize, 1912.

Gruneisen, Eduard (1877–1949), German physicist. Formulated laws relating specific heat and other properties of solids.

Guillaume, Charles Édouard (1861–1938), Swiss-born French physicist. Studied nickel-steel alloys and invented Invar; Nobel Prize, 1920.

Guillemin, Ernst Adolph (1898–1970), American electrical engineer. Worked on network analysis and synthesis problems; invented Guillemin line; developed network to produce loran pulses.

Guillemin, Roger Charles Louis (1924–), French-born American physiologist. With A. V. Schally, isolated and analyzed peptide hormones secreted in hypothalamic region of brain which control anterior pituitary hormone secretion; Nobel Prize, 1977.

Gullstrand, Allvar (1862–1930), Swedish ophthalmologist. Discovered intracapsular accommodation of the eye lens; improved techniques for studying eye structure; Nobel Prize, 1911.

Gunn, John Battiscombe (1928–), Egyptian-born American physicist. Discovered Gunn effect and used it to develop Gunn oscillator.

Gunter, Edmund (1581–1626), English mathematician and astronomer. Invented Gunter's chain used in surveying, and the logarithmic scale (Gunter's scale) which is the principle of the slide rule.

Gutenberg, Johann (ca. 1397–1468), German inventor. Invented the movable-type printing press.

Haar, Alfred (1885–1933), Hungarian mathematician. Studied orthogonal systems of functions, complex functions, partial differential equations, and calculus of variations; introduced the Haar measure on groups.

Haber, Fritz (1868–1934), German chemist. Developed the Haber-Boch process for synthesis of ammonia; made electrochemical studies; Nobel Prize, 1918.

Hadamard, Jacques (1865–1963), French mathematician. Proved theorem on the asymptotic behavior of the function giving the number of prime numbers less than a given number; introduced concept of "the problem correctly posed" in the solution of partial differential equations.

Haeckel, Ernst Heinrich (1834–1919), German biologist. Studied various invertebrates; classified animals as uni- and multicellular organisms; proposed the theory of recapitulation: ontogeny repeats phylogeny.

Hagen, Carl Ernst Bessel (1851–1923), German physicist. With H. Rubens, conducted experiments confirming Maxwell's electromagnetic theory of light, permitting determination of electrical conductivity of metals by optical measurements alone.

Hagen, Gotthilf Heinrich Ludwig (1797–1884), German hydraulic engineer. Discovered Hagen-Poiseuille law independently of J. L. M. Poiseuille; directed construction of dikes, harbor installations, and dune fortifications.

Hahn, Hans (1879–1934), Austrian mathematician. With S. Banach, proved the Hahn-Banach theorem of linear functionals.

Hahn, Otto (1879–1968), German chemist. With L. Meitner and F. Strassman, discovered that fission of heavy nuclei was possible by irradiation with neutrons; discovered protactinium with Meitner; Nobel Prize, 1944.

Hahnemann, Christian Friedrich Samuel (1775–1843), German physician. Founder of homeopathy.

Haldane, John Bourdon Sanderson (1892–1964), British geneticist and physiologist. Pioneered in mathematical treatment of population genetics; studied respiration in humans; wrote about enzymes.

Haldane, John Scott (1860–1936), British physiologist. Research on respiration, particularly the effects of high and low atmospheric pressures.

Hale, George Ellery (1868–1938), American astronomer. Established the Mount Wilson and Mount Palomar observatories; proved existence of magnetic fields in sunspots; invented the spectroheliograph.

Hall, Charles Martin (1863–1914), American commercial chemist. Discovered Hall process for extracting aluminum.

Hall, Edwin Herbert (1855–1938), American physicist. Discovered Hall effect and conducted studies of this and other galvanomagnetic and thermomagnetic effects.

Halley, Edmond (1656–1742), English astronomer. Published a southern star catalog; computed orbits of 24 comets; discovered Halley's comet.

Hamel, Georg Karl Wilhelm (1877–1954), German mathematician. Research on analysis and applied mathematics; introduced Hamel basis of vectors.

Hamilton, William Rowan (1805–1865), Irish mathematician and mathematical physicist. Discovered quaternions; developed mathematical theories encompassing wave and particle optics and mechanics; introduced Hamilton's principle and a form of the Hamilton-Jacobi theory.

Hankel, Hermann (1839–1873), German mathematician. Studied complex and hypercomplex numbers, theory of functions, and Hankel functions; proved that no hypercomplex number system can satisfy all the laws of ordinary arithmetic.

Hansen, Armauer (1841–1912), Norwegian bacteriologist. Discovered the bacillus of leprosy, or Hansen's disease.

Harden, Arthur (1865–1940), English chemist. Research on enzymes and alcoholic fermentation; Nobel Prize, 1929.

Hardy, Godfrey Harold (1877–1947), English mathematician. With S. Ramanujan, discovered formula for number of ways of writing a positive integer as the sum of positive integers; proved that the Riemann zeta function has an infinite number of zeros with real part equal to 1/2.

Harker, David (1906–1991), American crystallographer. Completed development of Patterson-Harker method of x-ray diffraction analysis of crystal structure.

Hartley, Ralph Vinton Lyon, (1888–1970), American electrical engineer. Invented Hartley oscillator; developed Hartley principle in information theory.

Hartline, Haldan Keffer (1903–1983), American biophysicist. Elucidated cellular electrical activity in the eye and optic nerve; Nobel Prize, 1967.

Hartmann, Johannes Franz (1865–1936), German astronomer. Derived Hartmann dispersion formula relating index of refraction and wavelengths; devised Hartmann test for telescope mirrors; gave first observational proof of interstellar matter.

Hartree, Douglas Rayner (1897–1958), English mathematician and mathematical physicist. Developed methods of numerical analysis which made it possible to apply Hartree method to calculation of atomic wave functions.

Harvey, William (1578–1657), English anatomist and physician. Described the true circulation of blood and the action of the heart.

Hassel, Odd (1897–1981), Norwegian chemist. Developed concept of conformation by studying three-dimensional structure of cyclohexane molecule, and explaining the orientation of attached atoms or functional groups; Nobel Prize, 1969.

Hauptman, Herbert A. (1917–), American chemist. With J. Karle, developed computer-aided mathematical techniques for use in x-ray crystallography to determine three-dimensional structures of molecules; Nobel Prize, 1985.

Hausdorff, Felix (1868–1942), German mathematician. Founded and advanced general topology and the general theory of metric spaces.

Haüy, René Just, Abbé (1743–1822), French mineralogist. Formulated the geometrical law of crystallization; pioneer in the science of crystallography.

Havers, Clopton (1665–1702), English osteologist. Provided the first full discussion of Haversian lamellae and Haversian canals.

Haworth, Walter Norman (1883–1950), English chemist. Synthesized ascorbic acid; studied carbohydrates, including the structure of sugars; Nobel Prize, 1937.

Heaviside, Oliver (1850–1925), English physicist. Proposed the Heaviside layer in the upper atmosphere.

Hefner-Alteneck, Friedrich Franz von (1845–1904), German engineer. Invented Hefner candle as a standard of luminous intensity.

Heine, Heinrich Eduard (1821–1881), German mathematician. Formulated concept of uniform continuity.

Heisenberg, Werner (1901–1976), German physicist. Founder of quantum mechanics; studied structure of the atom and the Zeeman effect; formulated the principle of indeterminancy in nuclear physics; Nobel Prize, 1932.

Heitler, Watler Heinrich (1904–1981), German-born Swiss theoretical physicist. Developed Heitler-London covalence theory of chemical bonding.

Helmholtz, Hermann Ludwig Ferdinand von (1821–1894), German physicist, anatomist, and physiologist. Physiological research on the nervous system and the human eye and ear, and theoretical work on conservation of force in physics; invented the ophthalmoscope.

Hench, Philip Showalter (1896–1965), American physiologist. Discovered that ACTH and cortisone could be used to treat rheumatoid arthritis; Nobel Prize, 1950.

Henle, Friedrich Gustav Jacob (1809–1885), German pathologist and anatomist. Wrote *Handbuch der Rationellen Pathologie*, integrating the study of physiology and pathology; discovered looped portion of the kidney tubules, and the epithelium.

Henry, Joseph (1797–1878), American physicist. Studied electromagnetic induction, solar phenomena, meteorology, and acoustics.

Henry, William (1775–1836), English chemist and physician. Formulated Henry's law of solubility of gases in liquid.

Henyey, Louis George (1910–1970), American astronomer. Research on reflection nebulae, interstellar matter, stellar atmospheres and evolution, and optical design.

Hermite, Charles (1822–1901), French mathematician. First to solve a fifth-degree equation; investigated *e*, the base of natural logarithms.

Hero of Alexandria (3d century or earlier), Greek mathematician. Wrote on the geometry of plane and solid figures, mechanics, and simple machines; showed that the angle of incidence equals the angle of reflection.

Héroult, Paul Louis Toussaint (1863–1914), French metallurgist. Designed the Héroult furnace for electric steel; developed the Héroult process for aluminum extraction.

Herschbach, Dudley R. (1932–), American chemist. With Y. T. Lee, developed crossed molecular-beam technique for tracing chemical reactions; Nobel Prize, 1986.

Herschel, John Frederick William (1792–1871), English mathematician, physicist, and astronomer. Discovered many nebulae and clusters; pioneered in celestial photography.

Herschel, William or **Friedrich Wilhelm** (1738–1822), German-born English astronomer. Discovered Uranus, two of its satellites, and two satellites of Saturn; discovered the Sun's intrinsic motion; proposed the concept of the form of the Milky Way.

Hershey, Alfred Day (1908–1997), American biologist. With M. Chase, experimented with bacteriophage, confirming earlier indications that the material basis of heredity is contained in nucleic acids; Nobel Prize, 1969.

Hertz, Gustav (1887–1975), German physicist. With J. Franck, studied effects of electron impacts on atoms; Nobel Prize, 1925.

Hertz, Heinrich Rudolph (1857–1894), German physicist. Discovered Hertzian waves in the ether; proved experimentally Maxwell's theories of electricity and magnetism.

Hertzsprung, Ejnar (1873–1967), Danish astronomer. Discovered method of spectroscopic parallax for measuring stellar distances; determined relation between color and luminosity of stars; with H. N. Russell, developed Hertsprung-Russell diagram.

Herzberg, Gerhard (1904–), German-born Canadian physicist. Determined electronic structure and geometry of diatomic and polyatomic molecules, particularly free radicals; Nobel Prize, 1971.

Hess, Victor Franz (1883–1964), Austrian physicist. Studied alpha particles from radium; discovered cosmic rays; Nobel Prize, 1936.

Hess, Walter Rudolf (1881–1973), Swiss physiologist. Discovered the organizer function of the middle brain in coordinating activity of internal organs; developed technique of using electrodes to stimulate localized brain areas; Nobel Prize, 1949.

Hevelius or Hewel, Johannes (1611–1687), Danzig-born astronomer. Discovered four comets; charted surface of the Moon; cataloged numerous stars.

Hevesy, George von (1886–1966), Hungarian chemist. Experimented with radioisotope indication, leading to the technique of isotope tracing of biological and chemical processes; Nobel Prize, 1943.

Hevroský, Jaroslav (1890–1967), Czechoslovakian physical chemist. Developed the technique of polarographic analysis; Nobel Prize, 1959.

Hewish, Antony (1924–), British astronomer. Pioneered in discovery of pulsars, by means of radio telescopes; Nobel Prize, 1974.

Heymans, Corneille (1892–1968), French-Belgian physiologist. Investigated the carotid sinus in connection with the mechanism of breathing; Nobel Prize, 1938.

Hilbert, David (1862–1943), German mathematician. Contributed to theory of numbers and theory of invariants; applied integral equations to physical problems.

Hill, Archibald Vivian (1886–1977), English physiologist. Worked on heat loss and oxygen consumption in muscle contraction; Nobel Prize, 1922.

Hinshelwood, Cyril Norman (1897–1967), British chemist. Elucidated chain reaction and chain branching mechanisms; Nobel Prize, 1956.

Hippias of Elis (5th century B.C.), Greek philosopher and mathematician. Discovered quadratrix.

Hipparchus of Rhodes (fl. 130 B.C.), Greek astronomer. Calculated the inclination of the ecliptic and the precession of the equinoxes; invented trigonometry; made the first star catalog.

Hippocrates (460?–377 B.C.), Greek physician. Known as the father of medicine; writings attributed to him contain clinical observations of diseases, descriptions of surgical practice, and the Hippocratic doctrine of the four humors.

Hitchings, George Herbert (1905–1998), American biochemist. With G. B. Elion, pioneered research that led them to the development of drugs for the treatment of leukemia, malaria, gout, herpes, bacterial and fungal infections, and autoimmune diseases and organ-transplant rejection; Nobel Prize, 1988.

Hittorf, Johann Wilhelm (1824–1914), German physicist. Described effects of Hittorf rays in vacuum tubes; studied electrolysis, and electrical discharge in rarefied gases with the Hittorf tube.

Hodgkin, Alan Lloyd (1914–), British biophysicist. With A. Huxley, devised a system of mathematical equations describing the nerve impulse; presented evidence for the sodium theory of nervous conduction; Nobel Prize, 1963.

Hodgkin, Dorothy Crowfoot (1910–1994), Egyptian-born British chemist. Determined the structure of the vitamin B_{12} molecule through x-ray crystallographic analysis; Nobel Prize, 1964.

Hodgkin, Thomas (1798–1866), English physician. First to describe Hodgkin's disease, a glandular disorder.

Hoffmann, Roald (1937–), Polish-born American chemist. Developed methods for predicting whether a chemical reaction is possible based on molecular orbital models; Nobel Prize, 1981.

Hofmann, August Wilhelm von (1818–1892), German chemist. Studied reactions of derivatives; developed the Hofmann reaction for preparing primary amines.

Hofmeister, Wilhelm Friedrich Benedict (1824–1877), German botanist. Did fundamental work on plant embryology; explained the alternating life cycles of mosses and ferns.

Hofstadter, Robert (1915–1990), American physicist. Investigated the properties and behavior of the proton and neutron; determined the size and shape of many nuclei; discovered the construction scheme of fundamental atomic nuclei; Nobel Prize, 1961.

Hölder, Otto Ludwig (1859–1937), German mathematician. Contributed to analysis and group theory; introduced Hölder condition; proved Hölder inequality and Jordan-Hölder theorem.

Holley, Robert William (1922–1993), American biochemist. With coworkers, made first determination of a nucleotide sequence of a nucleic acid; Nobel Prize, 1968.

Holmberg, Erik Bertil (1908–), Swedish astronomer. Investigations of galaxies, especially photometry of galaxies.

Hooke, Robert (1635–1703), English inventor. Invented the compound microscope, wheel barometer, universal (Hooke's) joint, and the reflecting telescope; formulated theories on light and on the motion of the Earth.

Hooker, Joseph Dalton (1817–1911), English botanist. With G. Bentham, wrote *Genera Plantarum*; prepared works on the flora of New Zealand, Antarctica, and India.

Hopkins, Frederick Gowland (1861–1947), English biochemist. Discovered the amino acid tryptophan and the tripeptide glutathione; did experimental work leading to the discovery of vitamins; Nobel Prize, 1929.

Houdry, Eugene J. (1892–1962), French-born American engineer. Devised catalytic method of producing oil.

Hounsfield, Godfrey Newbold (1919–), British electronics engineer. Invented computerized axial tomography; Nobel Prize, 1979.

Houssay, Bernardo Alberto (1887–1971), Argentine physiologist. Research on the functions and effects of the hypophysis, including its relationship to carbohydrate metabolism; Nobel Prize, 1947.

Howell, William Henry (1860–1945), American physiologist. Discovered heparin; isolated thrombin and thromboplastin; discovered Howell-Jolly bodies; proved that blood platelets are formed in lungs.

Hubble, Edwin Powell (1889–1953), American astronomer. Studied nebulae; formulated Hubble's law of extragalactic nebulae.

Hubel, David Hunter (1926–), American neurobiologist. Contributed to the study of the processing of visual information in the brain; Nobel Prize, 1981.

Huber, Robert (1937–), German chemist. With J. Deisenhofer and M. Hartmut, elucidated the structure of a bacterial protein that performs photosynthesis; Nobel Prize, 1988.

Hückel, Erich (1896–1980), German chemist. With P. J. W. Debye, formulated Debye-Hückel theory of strong electrolytes; devised theoretical explanation of electron properties of aromatic hydrocarbons.

Huggins, Charles Brenton (1901–), Canadian-born American surgeon and cancer researcher. Developed treatment of cancers using endocrinologic methods; Nobel Prize, 1966.

Hughes, David Edward (1831–1900), English-born American inventor. Invented the Hughes electromagnet, a printing telegraph, microphone, and induction balance.

Hugoniot, Pierre Henry (1851–1887), French physicist. Developed theory of shock waves.

Hulse, Russell A. (1950–), American astronomer and physicist. With J. H. Taylor, discovered the binary pulsar and studied it to observe phenomena predicted by general relativity; worked in plasma physics; Nobel Prize, 1993.

Humboldt, Friedrich Heinrich Alexander, Baron von (1769–1859), German naturalist. Founder of physical geography; made scientific explorations of South America and Central Asia.

Hume-Rothery, William (1899–1968), English metallurgist and chemist. Discovered Hume-Rothery rule concerning electron compounds.

Hunt, Franklin Livingston (1883–1973), American physicist. Research on x-ray spectroscopy; with W. Duane, discovered and applied Duane-Hunt law.

Hurwitz, Adolf (1859–1919), German-born Swiss mathematician. Worked on modular functions, number theory, Riemann surfaces, complex function theory, and analytic number theory; formulated condition satisfied by Hurwitz polynomials.

Huxley, Andrew Fielding (1917–), British physiologist. With A. L. Hodgkin, discovered the ionic mechanism involved in excitation in the cell membrane of peripheral nerves; Nobel Prize, 1963.

Huygens, Christiaan (1629–1695), Dutch mathematician, physicist, and astronomer. Discovered Saturn's rings; contributed to dynamics and optics; proposed the wave theory of light.

Hylleraas, Egil Andersen (1898–1965), Norwegian physicist. Applied quantum theory to helium atom, negative hydrogen ion, and other atoms, molecules, and crystals; developed variational method and other methods for mathematical solution of quantum-mechanical problems.

Ingenhousz, Jan (1730–1799), Dutch physician and naturalist. Demonstrated the cycle of photosynthesis in plants.

Ising, Ernest (1900–), German-born American physicist. Introduced Ising model of ferromagnetic material; research in solid-state physics and ferromagnetism.

Itō, Kiyosi (1915–), Japanese mathematician. Studied stochastic processes; introduced Itō's integral and Itō's formula.

Jacob, François (1920–), French biologist. Discovered episomes, a class of genetic elements; with J. Monod, proposed the concepts of messenger ribonucleic acid and of the operon; Nobel Prize, 1965.

Jacobi, Karl Gustav Jacob (1804–1851), German mathematician. Worked on elliptic functions and differential equations; developed the theory of determinants.

Jacquard, Joseph Marie (1752–1834), French inventor. Designed and built the Jacquard loom for figured weaving.

Jaeger, Frans Maurits (1877–1945), Dutch crystallographer and physical chemist. Measured physical properties of molten salts and silicates at extremely high temperatures.

James, William (1842–1910), American psychologist and writer. Coformulator of James-Lange theory that emotions are the perception of physiological changes.

Janet, Pierre Marie Félix (1859–1947), French psychologist. Studied hysteria, obsession, and neurosis; wrote a textbook on the theory of hysteria.

Jeans, James Hopwood (1877–1946), English physicist and astronomer. Worked in stellar dynamics; proposed the tidal theory of the origin of planets.

Jenner, Edward (1749–1823), English physician. Discovered vaccination.

Jensen, J. Hans D. (1906–1973), German physicist. With M. G. Mayer, formulated the nuclear shell model; Nobel Prize, 1963.

Jerne, Niels K. (1911–1994), Swiss immunologist. Formulated three important theories involving the immune system: how the body produces specific antibodies, how the immune system develops and matures, and how the various interrelated aspects of the immune response are coordinated by the body; Nobel Prize, 1984.

Johnson, Harold Lester (1921–1980), American astronomer. Research in astrophysics, astronomical photometry, infrared astronomy, applications of electronics to astronomy, and Fourier transform spectroscopy; collaborated with W. W. Morgan in developing Johnson-Morgan system of stellar magnitudes.

Joliot-Curie, Irène (1897–1956), French physicist. With M. Curie, discovered projection of atomic nuclei by neutrons; with J. F. Joliot-Curie, discovered artificial radiation; Nobel Prize, 1935.

Joliot-Curie, Jean Frédéric (1900–1958), French physicist. With I. Joliot-Curie, produced an artificial radioactive substance by bombarding boron with fast alpha particles; Nobel Prize, 1935.

Jordan, Camille (1838–1921), French mathematician. Discovered many fundamental results in group theory; gave a proof of the Jordan curve theorem (later shown to be incorrect).

Jordan, Pascual (1902–1980), German physicist. Contributed to formulation of quantum mechanics; introduced Jordan algebra in an attempt to generalize quantum mechanics.

Josephson, Brian David (1940–), British physicist. Predicted the Josephson effect concerning electron pairs; Nobel Prize, 1973.

Joukowski, Nikolai Jegorowitch (1847–1921), Russian applied mathematician and aerodynamicist. Helped to introduce concept of Kutta-Joukowski airfoil and to prove Kutta-Joukowski theorem.

Joule, James Prescott (1818–1889), English physicist. Formulated a mechanical theory of heat; demonstrated Joule-Thomson effect relating to the fall in temperature of a gas; first to estimate the velocity of a gas molecule.

Jung, Carl Gustav (1875–1961), Swiss psychologist and psychiatrist. Evolved a theory of complexes; founded the analytical school of psychoanalysis and psychotherapy.

Jussieu, Antoine Laurent de (1748–1836), French botanist. Wrote *Genera Plantarum*, the basis of modern natural botanical classification.

Kalman, Rudolf Emil (1930–), Hungarian-born American mathematician and electrical engineer. Worked on mathematical theory of control systems; developed the Kalman filter; introduced concepts of controllability and observability.

Kaluza, Theodor Franz Eduard (1885–1954), German mathematical physicist. Developed theory which attempted to unify gravitation and electromagnetism.

Kamerlingh Onnes, Heike (1853–1926), Dutch physicist. Research on cryogenics, critical phenomena, and low temperatures; discovered the phenomenon of superconductivity; Nobel Prize, 1913.

Kapitza, Pjotr Leonidovich (1894–1984), Russian physicist. Studied magnetism and low temperature; designed hydrogen and helium liquefaction plants; Nobel Prize, 1978.

Kapteyn, Jacobus Cornelius (1851–1922), Dutch astronomer. Studied the proper motion of stars; with P. J. van Rhijn, evolved a theory of the universe.

Karle, Jerome (1918–), American crystallographer. With H. A. Hauptman, developed computer-aided mathematical techniques for use in x-ray crystallography to determine three-dimensional structures of molecules; Nobel Prize, 1985.

Karrer, Paul (1889–1971), Swiss chemist. Pioneering research on vitamins A and B_2 and on the flavins and carotenoids; Nobel Prize, 1937.

Kastler, Alfred (1902–1984), French physicist. Developed a double-resonance method to study energy levels of atoms in excited states; Nobel Prize, 1966.

Kater, Henry (1777–1835), English geodesist. Developed Kater's reversible pendulum to obtain accurate values of acceleration of gravity.

Katz, Bernard (1911–), German-born British physiologist. Made discoveries concerning mechanism for release of transmitter substances at nerve-muscle junction; Nobel Prize, 1970.

Keesom, Willem Hendrik (1876–1956), Dutch physicist. Worked in low-temperature physics; first to solidify helium; studied molecular structure of liquids and compressed gases.

Kekulé von Stradonitz, Friedrich August (1829–1896), German chemist. A founder of structural organic chemistry; made theoretical proposal of the structure of benzene.

Kelvin, William Thomson, 1st Baron (1824–1907), British mathematician and physicist. Invented the Kelvin balance; formulated Kelvin's laws concerning electric cables; contributed to thermodynamics.

Kendall, Edward Calvin (1886–1972), American biochemist. Chemical investigation of the adrenal cortex, leading to the isolation of crystalline cortical hormones, especially cortisone; Nobel Prize, 1950.

Kendall, Henry Way (1926–), American physicist. Collaborated in experiments that demonstrated that protons, neutrons, and similar particles are made up of quarks; Nobel Prize, 1990.

Kendrew, John Cowdery (1917–1997), British molecular biologist. First to successfully determine the structure of a protein; Nobel Prize, 1962.

Kennelly, Arthur Edwin (1861–1939), American electrical engineer. Discovered the ionized layer in the atmosphere, independently of O. Heaviside; proposed the theory of alternating currents.

Kepler, Johannes (1571–1630), German astronomer. Proposed Kepler's three laws of planetary motion; worked in optics.

Kerr, John (1824–1907), Scottish physicist. Discovered the Kerr magnetooptic effect.

Khorana, Har Gobind (1922–), Indian-born American biochemist. Synthesized complicated nucleic acids; proved that genetic code consists of nonoverlapping triplets of bases without gaps between triplets; Nobel Prize, 1968.

Kirchhoff, Gustav Robert (1824–1887), German physicist. With R. W. Bunsen, discovered method of spectrum analysis; formulated Kirchhoff's law of electric currents and electromotive forces in a network.

Kirkwood, Daniel (1814–1895), American astronomer. Discovered Kirkwood gaps; studied nature, origin, and evolution of solar system, particularly role of asteroids, comets, meteors, and meteorites.

Kitasato, Shibasaburo (1852–1931), Japanese bacteriologist. Independently of A. E. J. Yersin, discovered the bacillus of bubonic plague; isolated bacilli of symptomatic anthrax, dysentery, and tetanus.

Klebs, Edwin (1834–1913), German pathologist. Described diphtheria (Klebs-Löffler) bacillus; studied bacteriology of malaria, anthrax, and tuberculosis.

Klein, Christian Felix (1849–1925), German mathematician. Contributed to function theory, noneuclidean geometry, group therapy, and applied mathematics; introduced Klein bottle.

Klein, Oskar Benjamin (1894–1977), Swedish physicist. Codeveloper of Klein-Gordon equation. Klein-Nishina formula, and Klein-Rydberg method; proposed theory of overall structure of the universe.

Klitzing, Klaus von (1943–), German physicist. Discovered quantum Hall effect; Nobel Prize, 1985.

Klug, Aaron (1926–), South African-born British biochemist. Developed crystallographic electron microscopy and elucidated biologically important nucleic acid-protein complexes; Nobel Prize, 1982.

Knudsen, Martin Hans Christian (1871–1949), Danish physicist and hydrographer. Studied flow and diffusion of gases at low pressure; developed Knudsen cell and Knudsen gage; developed methods to measure the properties of seawater.

Koch, Robert (1843–1910), German physician and bacteriologist. Studied cholera, tuberculosis, and bubonic plague; showed a specific bacillus to be the cause of anthrax; discovered the tubercle bacillus; Nobel Prize, 1905.

Kocher, Emil Theodor (1841–1917), Swiss surgeon. Studied the functions and malfunctions of the thyroid gland; Nobel Prize, 1909.

Köhler, Georges J. F. (1946–1995), Swiss immunologist. With C. Milstein, discovered a laboratory technique for producing monoclonal antibodies, highly uniform immune bodies that are selective in responding to target substances; Nobel Prize, 1984.

Kolbe, Adolf Wilhelm Hermann (1818–1884), German chemist. Contributed to the synthesis concept of compound formation, doing much to eliminate the division of chemistry into two branches: organic and inorganic.

Kolmogorov, Andrei Nikolaevich (1903–1987), Soviet mathematician. Formulated set-theoretic basis of probability theory.

Kornberg, Arthur (1918–), American biochemist. Discovered deoxyribonucleic acid polymerase, providing the first rational enzymatic mechanism for the replication of genetic material of the cell; Nobel Prize, 1959.

Korteweg, Diederik Johannes (1848–1941), Dutch mathematician. Work in applied mathematics, mechanics, and hydrodynamics; with H. de Vries, proposed equation of wave motion with soliton solution.

Kossel, Albrecht (1853–1927), German chemist. Investigated the chemistry of cells and of proteins; Nobel Prize, 1910.

Krafft-Ebing, Richard, Baron von (1840–1902), German neurologist. Authority on psychological disorders and their forensic implications; wrote *Psychopathia Sexualis*, a collection of case histories.

Kramers, Hendrik Anthony (1894–1952), Dutch physicist. Developed quantum theory of dispersion, establishing Kramers-Kronig relation; with G. Wentzel and L. Brillouin, developed Wentzel-Kramers-Brillouin method.

Krebs, Edwin Gerhard (1918–), American biochemist. With E. H. Fischer, discovered phosphorylation processes that play a critical role in cell-protein regulation; they isolated the first protein kinase, a class of enzymes that transfer phosphate from adenosine triphosphate to proteins; Nobel Prize, 1992.

Krebs, Hans Adolf (1900–1981), German-born British biochemist. Elucidated metabolic pathways, including the tricarboxylic acid cycle; Nobel Prize, 1953.

Krogh, Schack August Steenberg (1874–1949), Danish physiologist. Discovered the regulation of the vasomotor mechanism of capillaries; devised the nitrous oxide method for measuring human circulation; Nobel Prize, 1920.

Kronecker, Leopold (1823–1891), German mathematician. Contributed to theory of elliptical functions, algebra, and number theory, and attempted to unify these disciplines; attempted to base all mathematics on integers and finite processes.

Kroto, Harold W. (1939–), British chemist. Studied long-chain molecules found in radioastronomy by using microwave spectroscopy, which ultimately led to the discovery of fullerenes; Nobel Prize, 1996.

Kruskal, Martin David (1925–), American mathematician and physicist. Research in plasma physics, asymptotic phenomena, relativity, and minimal surfaces.

Kuhn, Richard (1900–1967), German chemist. Research on the structures and synthesis of vitamins and carotenoids; Nobel Prize, 1938 (declined).

Kundt, August Adolph (1839–1894), German physicist. Used Kundt tube to determine speed of sound in gases; determined ratio of specific heats of monatomic gases; with W. K. Röntgen, demonstrated Faraday effect in gases.

Kurchatov, Igor Vasilievich (1903–1960), Soviet physicist. Discovered nuclear isomers; studied nuclear reactions; developed nuclear weapons and nuclear power.

Kusch, Polykarp (1911–1993), German-born American physicist. Precisely determined the magnetic moment of the electron; Nobel Prize, 1955.

Kutta, Wilhelm Martin (1867–1944), German applied mathematician. Helped introduce concept of Kutta-Joukowski airfoil, prove Kutta-Joukowski theorem, and develop Runge-Kutta method.

Lacaille, Nicolas Louis de (1713–1762), French astronomer and geodesist. Determined positions of nearly 10,000 stars in southern skies; measured lunar and solar parallax; showed that Earth has equatorial bulge.

Lagrange, Joseph Louis, Count (1736–1813), French geometer and astronomer. Invented the calculus of variations; studied the mathematics of sound; wrote *Mécanique Analytique*, concerning statics and dynamics.

Laguerre, Édmond Nicolas (1834–1886), French mathematician. Discovered Laguerre's differential equations and Laguerre polynomials.

Lamarck, Jean Baptiste Pierre Antoine de Monet, Chevalier de (1744–1829), French naturalist. Proposed theory that changes in animal and plant structure are caused by changes in environment; classified animals into vertebrates and invertebrates.

Lamb, Willis Eugene, Jr. (1913–), American physicist. Made precise atomic measurements leading to a new understanding of the theory of electron interactions and electromagnetic radiation; Nobel Prize, 1955.

Lambert, Johann Heinrich (1728–1777), German physicist. Formulated the Lambert theorem concerning the illumination of a surface.

Lamé, Gabriel (1795–1870), French mathematician, physicist, and engineer. Introduced curvilinear coordinates and applied them to differential equations, elasticity, thermodynamics, and number theory.

Landau, Lev Davydovich (1908–1968), Soviet physicist. Made theoretical explanation of the nature and properties of liquid helium; investigated condensed matter; Nobel Prize, 1962.

Landé, Alfred (1888–1975), German-born American physicist. Introduced Landé g factor; discovered Landé interval rule and Landé Γ-permanence rule.

Landsteiner, Karl (1868–1943), Austrian-born American pathologist. Discovered human blood groups and factors M and N; with A. S. Weiner, discovered the Rh factor; Nobel Prize, 1930.

Lange, Carl Georg (1834–1900), Danish physician and psychologist. With W. James, proposed the James-Lange theory of emotion.

Langerhans, Paul (1847–1888), German pathologist and anatomist. Studied human and animal microscopical anatomy, particularly structures of skin and pancreas; discovered islets of Langerhans.

Langevin, Paul (1872–1946), French physicist. Developed quantitative theories of paramagnetism and diamagnetism; helped to elucidate the theory of relativity; contributed to the development of sonar.

Langley, Samuel Pierpont (1834–1906), American astronomer and airplane pioneer. Studied infrared solar spectrum; constructed in 1896 the first mechanical heavier-than-air machine to fly.

Langmuir, Irving (1881–1957), American chemist. With G. N. Lewis, proposed the Lewis-Langmuir atomic theory; studied surface chemistry and thermionic emission; Nobel Prize, 1932.

Laplace, Pierre Simon, Marquis de (1749–1827), French astronomer and mathematician. Contributed to celestial mechanics, especially to the study of the Moon, Saturn, and Jupiter; formulated the theory of probability; discovered the Laplace differential equation.

Larmor, Joseph (1857–1942), British physicist. Developed electron theory which fused electromagnetic and optical concepts; introduced Larmor precession and derived Larmor formula.

Laue, Max Theodor Felix von (1879–1960), German physicist. Proposed the theory of x-ray diffraction by crystals; developed the Laue method of investigating crystal structure; Nobel Prize, 1914.

Laurent, Pierre Alphonse (1813–1854), French mathematician and physicist. Introduced Laurent series; research on wave theory of light.

Laveran, Charles Louis Alphonse (1845–1922), French physician. Discovered the malaria parasite; researched sleeping sickness; Nobel Prize, 1907.

Lavoisier, Antoine Laurent (1743–1794), French chemist. The founder of modern chemistry; studied combustion and respiration; published a table of the elements.

Lawrence, Ernest Orlando (1901–1958), American physicist. Discovery, development, and use of the cyclotron; Nobel Prize, 1939.

Lebesgue, Henry Léon (1875–1941), French mathematician. Developed theory of measure and integration; studied trigonometric series.

Le Chatelier, Henry Louis (1850–1936), French chemist and metallurgist. Research on cement chemistry, gas combustion, blast furnace reactions, chemical equilibria, alloy properties, and chemistry and metallurgy of iron and steel; formulated Le Chatelier's principle.

Leclanché, Georges (1839–1882), French chemist and electrician. Invented the Leclanché galvanic cell.

Lederberg, Joshua (1925–), American geneticist. With E. L. Tatum, discovered genetic recombination in bacteria and organization of genetic material; Nobel Prize, 1958.

Lederman, Leon Max (1922–), American physicist. Collaborated in experiment that demonstrated the existence of two types of neutrino; led an experiment that discovered the upsilon particle; Nobel Prize, 1988.

Lee, David M. (1931–), American physicist. With D. D. Osheroff and R. C. Richardson, discovered superfluidity in helium-3; Nobel Prize, 1996.

Lee, Tsung-Dao (1926–), Chinese-born American physicist. With C. N. Yang disproved the parity principle; worked on statistical mechanics, astrophysics, nuclear and subnuclear physics, and field theory; Nobel Prize, 1957.

Lee, Yuan T. (1936–), American chemist. With Dudley R. Herschbach, developed crossed molecular-beam technique for tracing chemical reactions; Nobel Prize, 1986.

Legendre, Adrien Marie (1752–?1833), French mathematician. Worked on elliptic functions, the theory of numbers, and the method of least squares.

Lehn, Jean-Marie (1939–), French chemist. Studied crown ethers and developed the synthesis of related structures known as cryptands; Nobel Prize, 1987.

Leibniz, Gottfried Wilhelm, Baron von (1646–1716), German mathematician. Contributed to the development of differential calculus.

Leloir, Luis Federico (1906–1987), French-born Argentine biochemist. Discovered sugar nucleotides and their role in carbohydrate biosynthesis; Nobel Prize, 1970.

Lenard, Phillipp Eduard Anton (1862–1947), Hungarian-born German physicist. Studied cathode rays outside the discharge tube; worked on photoelectricity; Nobel Prize, 1905.

L'Enfant, Pierre Charles (1754–1825), French engineer. Designed Washington, D.C.

Lennard-Jones, John Edward (1894–1954), English physicist and chemist. Proposed Lennard-Jones potential for interatomic forces; contributed to quantum theory of molecular structure and statistical mechanics of liquids, gases, and surfaces.

Lenz, Heinrich Friedrich Emil (1804–1865), German physicist. Formulated Lenz's law governing induced current.

Leverrier, Urbain Jean Joseph (1811–1877), French astronomer. Studied Mercury, Uranus, and Neptune; credited, with J. C. Adams in England, as discoverer of Neptune.

Levi-Civita, Tullio (1873–1941), Italian mathematician and mathematical physicist. With G. Ricci-Curbastro, developed tensor analysis; introduced concept of parallelism in curved spaces.

Levi-Montalcini, Rita (1909–), Italian biologist. With S. Cohen, made landmark studies of nerve growth factor and its functions; Nobel Prize, 1986.

Lewis, Edward B. (1918–), American geneticist. Shared in discoveries showing the genetic involvement of early embryonic development with C. Nusselin-Volhard and E. F. Wieschaus; Nobel Prize, 1995.

Lewis, Gilbert Newton (1875–1946), American chemist. Collaborated in developing the Lewis-Langmuir atomic theory; worked on the electronic theory of valency and chemical thermodynamics.

Leydig, Franz von (1821–1908), German histologist and anatomist. Founder of comparative histology; promoted use of microscope in anatomical study; described cells in testes believed to secrete male hormones.

l'Hospital (l'Hôpital), Guillaume François Antoine de (1661–1704), French mathematician. Wrote first textbook on differential calculus, which gives l'Hospital's rule.

Libby, Willard Frank (1908–1980), American chemist. Developed the method of radiocarbon dating; Nobel Prize, 1960.

Lie, Marius Sophus (1842–1899), Norwegian mathematician. Originated the theory of tangential transformations.

Liebig, Justus, Baron von (1803–1873), German chemist. Discovered chloroform and chloral; founded agricultural chemistry; invented the Liebig condenser.

Linnaeus, Carolus, or Carl von Linné, (1707–1778), Swedish botanist. Developed the Linnaean system of biological classification.

Liouville, Joseph (1809–1882), French mathematician. Proved existence of transcendental functions; developed concept of geodesic curvature; originated theory of doubly periodic functions.

Lipmann, Fritz Albert (1899–1986), German-born American biochemist. Formulated general rules for the biotechnology of energy transmission; discovered coenzyme A; Nobel Prize, 1953.

Lippmann, Gabriel (1845–1921), French physicist. Produced the first colored photograph of the light spectrum; invented the Lippmann capillary electrometer; Nobel Prize, 1908.

Lipscomb, William Nunn, Jr. (1919–), American physical chemist. Studied structure and bonding of boranes, providing insight into nature of chemical bonding; Nobel Prize, 1976.

Lissajous, Jules Antoine (1822–1880), French physicist. Invented the vibration microscope, involving Lissajous figures.

Lister, Joseph, 1st Baron (1827–1912), English surgeon. Introduced antiseptics to surgery; pioneered in bacteriology.

Littlewood, John Endensor (1885–1977), British mathematician. Work on diophantine approximation, Tauberian theorems, Fourier series and associated function theory, the zeta function, additive number theory, and inequalities.

Littrow, Joseph Johann von (1781–1840), Austrian astronomer. Studied light refraction; worked on telescope construction.

Lloyd, Humphrey (1800–1881), Irish physicist. Discovered Lloyd's mirror interference; verified W. R. Hamilton's prediction of conical refraction.

Lobachevski, Nikolai Ivanovich (1793–1856), Russian mathematician. Originated the first comprehensive system of noneuclidean geometry.

Loewi, Otto (1873–1961), German pharmacologist. Investigated nerve impulses; proved the role of acetylcholine in nerve impulse transmission; Nobel Prize, 1936.

Löffler, Friedrich August Johannes (1852–1915). German bacteriologist. Isolated the diphtheria (Klebs-Löffler) bacillus; developed protective serum against foot-and-mouth disease.

London, Fritz (1900–1954), German-born American physicist. Developed, with W. Heitler, theory of covalent bonding; with H. London, theory of superconductivity; and theory of superfluidity.

London, Heinz (1907–1970), German-born English physicist. Research on electrodynamic and thermodynamic behavior of superconductors and properties of superfluid helium.

Lorentz, Hendrik Antoon (1853–1928), Dutch physicist. Proposed the electron theory to explain electromagnetic properties of materials; proposed the Lorentz-FitzGerald contraction and the Lorentz transformation, contributing to the theory of relativity; studied Zeeman effect; Nobel Prize, 1902.

Lorenz, Konrad Zacharias (1903–1989), Austrian zoologist. Pioneered in study of animal behavior patterns; discovered imprinting in birds; Nobel Prize, 1973.

Loschmidt, Johann Joseph (1821–1895), Austrian physicist and chemist, born in Bohemia. Worked on graphical and structural molecular formulas; attempted to estimate size of air molecules and number of air molecules per unit volume.

Lummer, Otto Richard (1860–1925), German physicist. Codeveloper of Lummer-Brodhun sight box and Lummer-Gehrcke plate; constructed an improved bolometer.

Luria, Salvador Edward (1912–1991), Italian-born American biologist. Devised fluctuation test to demonstrate and study spontaneous mutations in bacteria and viruses; Nobel Prize, 1969.

Lwoff, André Michael (1902–1994), French biologist. Explained the phenomenon of lysogeny in bacteria; Nobel Prize, 1965.

Lyapunov, Aleksandr Mikhailovich (1857–1918), Soviet mathematician and physicist. Determined in what cases linear approximations can be used to solve the problem of stability of a mechanical system with a finite number of degrees of freedom; proved existence of various figures of equilibrium for a rotating liquid.

Lyell, Charles (1797–1875), British geologist. Wrote *Principles of Geology*, refuting catastrophic theory of geological changes.

Lyman, Theodore (1874–1954), American physicist. Observed ultraviolet spectra; clarified nature of Lyman ghosts; discovered Lyman series.

Lynen, Feodor (1911–1979), German biochemist. Research on the formation of the cholesterol molecule; discovered chemistry of biotin; Nobel Prize, 1964.

Lyot, Bernard Ferdinand (1879–1952), French astronomer. Invented the coronagraph and developed monochromatic filters that greatly extended knowledge of the solar corona.

McClintock, Barbara (1902–1992), American geneticist. Discovered mobile genetic elements known as jumping genes; Nobel Prize, 1983.

Mach, Ernst (1838–1916), Austrian physicist. Research on supersonic flight, leading to Mach angle and Mach number; studied airflow over objects at high speeds.

Maclaurin, Colin (1698–1746), Scottish mathematician. Systematized and developed Newton's calculus; introduced Maclaurin series and Maclaurin-Cauchy test.

Macleod, John James Rickard (1876–1935), Scottish physiologist. Shared in discovery of insulin with F. G. Banting and C. H. Best; Nobel Prize, 1923.

McMillan, Edwin Mattison (1907–1991), American physicist. Discovered element 93 (neptunium), which led to the creation of element 94 (plutonium); conceived the theory of phase stability; Nobel Prize, 1951.

Magnus, Heinrich Gustav (1802–1872), German physicist and chemist. Made first quantitative analysis of blood gases; showed that arterial blood has higher oxygen content than venous blood; discovered Magnus effect.

Majorana, Ettore (1906–1938), Italian physicist. Studied properties of elementary particles; postulated Majorana force.

Maksutov, Dmitry Dmitrievich (1896–1964), Soviet physicist and astronomer. Developed general theory of aplanatic optical systems; developed Maksutov system.

Malpighi, Marcello (1628–1694), Italian anatomist. Discovered the capillaries; made microscopic studies in embryology; discovered the Malpighian layer of the epidermis and the Malpighian corpuscles in the kidney.

Malus, Étienne Louis (1775–1812), French engineer and physicist. Formulated Malus' cosine-squared law concerning polarized light and Malus' law of rays.

Mandelstam, Stanley (1928–   ), American physicist, born in South Africa. Research in theoretical physics of elementary particles; introduced Mandelstam plane and Mandelstam representation.

Marconi, Guglielmo, Marquis (1874–1937), Italian electrician and inventor. Developed commercial wireless telegraphy; Nobel Prize, 1909.

Mariotte, Edmé (?–1684), French physicist and physiologist. Discovered blind spot; studied circulation of sap in plants, collisions of bodies, properties of air, refraction and color of light, hydrostatics, hydraulics, and meteorology.

Marcus, Rudolph Arthur (1923–   ), Canadian-born American chemist. Developed the mathematical analysis for electron transfer reactions in chemical systems; Nobel Prize, 1992.

Mark, Herman Francis (1895–1992), Austrian-born American chemist. Elucidated molecular structures of natural and synthetic polymers; developed theory of polymerization; studied relation between structure and properties of macromolecular systems.

Markov, Andrei Andreevich (1856–1922), Russian mathematician. Formulated rigorous proofs of law of large numbers and central-limit theorem; introduced Markov chain.

Martin, Archer John Porter (1910–   ), English chemist. With R. L. M. Synge, developed partition chromatography; Nobel Prize, 1952.

Mascheroni, Lorenzo (1750–1800), Italian mathematician. Calculated Euler's constant; proved that all plane construction problems that can be solved with a ruler and compass can also be solved with a compass alone.

Mathieu, Emile Leonard (1835–1890), French mathematician and physicist. Worked on solution of partial differential equations; research in celestial and analytical mechanics; studied Mathieu equation and introduced Mathieu functions.

Matthias, Bernd Teo (1919–1980), German-born American physicist. Tested metals and alloys for superconductivity; developed empirical rules to predict new superconducting materials.

Maunder, Edward Walter (1851–1928), British astronomer. Observations of the sun, sunspots and eclipses.

Maupertuis, Pierre Louis Moreau de (1698–1759), French mathematician and astronomer. Discovered the principle of least action; mathematical writings on the properties of curves.

Maxwell, James Clerk (1831–1879), Scottish physicist. Formulated the electromagnetic theory of light and the Maxwell distribution of molecular velocities of gases; invented the Maxwell disk concerning color vision.

Mayall, Nicholas Ulrich (1906–1993), American astronomer. Research on nebulae, globular star clusters, and external galaxies.

Mayer, Julius Robert von (1814–1878), German physicist. Discovered the principle of conservation of energy.

Mayer, Maria Goeppert (1906–1972), German-born American nuclear physicist. With J. H. D. Jensen, discovered nuclear shell structure; Nobel Prize, 1963.

Meckel, Johann Friedrich (1781–1833), German anatomist and embryologist. Gave first comprehensive description of birth defects; described Meckel's cartilage; discovered Meckel's diverticulum.

Medawar, Peter Brian (1915–1987), Brazilian-born British biologist and medical scientist. Discovered acquired immunological tolerance; Nobel Prize, 1960.

Meissner, Alexander (1883–1958), Austrian-born German radio engineer. Helped develop improved electrical insulators and continuous-wave transmission; invented Meissner oscillator.

Meissner, Walther (1882–1974), German physicist. Research in low-temperature physics; discovered Meissner effect.

Meitner, Lise (1878–1968), German physicist. With O. Hahn, discovered protactinium; found evidence of four other radioactive elements; with Hahn and F. Strassmann, accomplished fission of uranium.

Mendel, Gregor Johann (1822–1884), Austrian botanist. Formulated Mendel's laws of heredity, the foundation of genetics.

Mendeleev, Dmitri Ivanovich (1834–1907), Russian chemist. Formulated Mendeleev's periodic law and table of the elements; research on interatomic and intermolecular forces.

Menelaus of Alexandria (1st century), Greek mathematician and astronomer. Founded spherical trigonometry.

Mercator, Gerhardus, or Gerhard Kremer (1512–1594), Flemish geographer. Created a chart of the world (Mercator projection); made surveying instruments.

Mercer, John (1791–1866), English chemist. Invented the mercerizing process for cotton.

Merrifield, Robert Bruce (1921–   ), American biochemist. Developed methods of protein synthesis, including solid-phase peptide synthesis that produces proteins by assembling amino acids sequentially into peptide chains; Nobel Prize, 1984.

Mersenne, Marin (1588–1648), French physicist. Showed that pitch is proportional to frequency and calculated frequencies of musical notes; discovered Mersenne's law for vibrating strings, and similar relations for wind and percussion instruments.

Messier, Charles (1730–1817), French astronomer. Credited with discovering 21 comets; compiled a catalog of nebulae.

Metchnikoff, Élie (1845–1916), Russian-born French zoologist and bacteriologist. Work on cholera and immunology; Nobel Prize, 1908.

Meusnier de la Place, Jean Baptiste Marie Charles (1754–1793), French mathematician, physicist, and chemist. Derived Meusnier's theorem of curvature of surface curve; with A. L. Lavoisier, did research on analysis and synthesis of water.

Meyerhof, Otto Fritz (1884–1951), German physiologist. Studied the glycogen-lactic acid cycle of muscles; Nobel Prize, 1923.

Michaelis, Leonor (1875–1949), German-born American biochemist. Developed theory of kinetics of enzyme-catalyzed reactions.

Michel, Hartmut (1948–   ), German chemist. With J. Deisenhofer and R. Huber, elucidated the structure of a bacterial protein that performs photosynthesis; Nobel Prize, 1988.

Michelson, Albert Abraham (1852–1931), American physicist. Experimented on the velocity of light with S. Newcomb; invented the Michelson interferometer; performed, with E. W. Morley, an experiment to determine the Earth's motion through the ether; Nobel Prize, 1907.

Mie, Gustav (1868–1957), German physicist. Carried out rigorous electrodynamic calculation of Mie scattering; attempted to formulate theory of matter.

Miller, William Hallowes (1801–1880), British crystallographer and mineralogist. Introduced Miller indices for identifying crystallographic planes.

Millikan, Robert Andrews (1868–1953), American physicist. Determined an accurate value for Planck's constant; originated the "oil drop" experiment to measure electronic charge; work on x-rays and cosmic rays; Nobel Prize, 1923.

Milstein, Cesar (1927–), British immunologist. With Georges J. F. Köhler, discovered a laboratory technique for producing monoclonal antibodies, highly uniform immune bodies that are selective in responding to target substances; Nobel Prize, 1984.

Minkowski, Hermann (1864–1909), Russian-born German mathematician. Studied the mathematical basis of relativity, notably the concept of the space-time continuum.

Minot, George Richards (1885–1950), American physician. With W. P. Murphy, first to recognize the value of liver therapy for pernicious anemia; studied arthritis, cancer, and vitamin B deficiency; Nobel Prize, 1934.

Mitchell, Peter (1920–1992), British chemist. Explained how plant and animal cells store and transfer energy by creating protonic gradients in the oxidative and photosynthetic phosphorylation processes; Nobel Prize, 1978.

Möbius, August Ferdinand (1790–1868), German mathematician and astronomer. Founder of topology; developed the Möbius strip.

Mohl, Hugo von (1805–1872), German botanist. Worked on the anatomy and physiology of higher plant forms; discovered protoplasm.

Mohorovičić, Andrija (1857–1936), Yugoslav meteorologist and seismologist. Discovered Mohorovičić seismic discontinuity.

Mohr, Carl Friedrich (1806–1879), German chemist. Developed titration procedures, including use of Mohr's salt.

Mohr, Christian Otto (1835–1918), German civil engineer. Studied stresses and strains of bodies, and failure of materials; introduced Mohr's stress circle.

Mohs, Friedrich (1773–1839), German mineralogist. Developed Mohs scale of hardness.

Moissan, Ferdinand Frédéric Henri (1852–1907), French chemist. First to isolate fluorine; invented an electric furnace and used it to produce synthetic metal compounds and samples of less common metals; Nobel Prize, 1906.

Molina, Mario J. (1943–), American chemist. Demonstrated that chemically inert chlorofluorocarbon (CFC) could be transported up to the ozone layer and could react with ultraviolet light and deplete the ozone layer; Nobel Prize, 1995.

Mollier, Richard (1863–1935), German physicist and engineer. Presented properties of thermodynamic media in form of charts and diagrams; introduced concept of enthalpy and Mollier diagram.

Moniz, Antonio Egas (1874–1955), Portuguese neurosurgeon. Developed cerebral angiography; introduced the prefrontal lobotomy; Nobel Prize, 1949.

Monod, Jacques (1910–1976), French biologist. With F. Jacob, proposed the concepts of messenger ribonucleic acid and of the operon; Nobel Prize, 1965.

Moody, Lewis Ferry (1880–1953), American hydraulic engineer. Made improvements in hydraulic turbines, pumps, and accessories.

Moore, Stanford (1913–1982), American biochemist. With W. H. Stein, developed technique for determining amino acid sequence in proteins, and applied it to ribonuclease; Nobel Prize, 1972.

Morgagni, Giovanni Battista (1682–1771), Italian anatomist. Founded pathological anatomy; first to describe liver cirrhosis.

Morgan, Thomas Hunt (1866–1945), American geneticist, embryologist, and zoologist. Proposed the chromosome theory of heredity; Nobel Prize, 1933.

Morgan, William Wilson (1906–1994), American astronomer. Developed methods for investigating more precisely the structure of the Milky Way Galaxy and of other galaxies; collaborated in developing the Johnson-Morgan system of stellar magnitudes (with H. L. Johnson) and Bautz-Morgan classification of galaxy clusters (with L. P. Bautz).

Morley, Edward Williams (1838–1923), American chemist and physicist. Associated with A. A. Michelson in an experiment on ether drift; research on variations of atmospheric oxygen content.

Morse, Samuel Finley Breese (1791–1872), American inventor. Invented the receiving and sending instruments for the telegraph, and a code for sending messages.

Moseley, Henry Gwyn Jeffries (1887–1915), English physicist. Discovered Moseley's law for frequency of x-ray spectral lines.

Mössbauer, Rudolf Ludwig (1929–), German physicist. Discovered the property of recoilless resonance absorption, the ability of some nuclei to emit and absorb gamma rays without energy loss; Nobel Prize, 1961.

Mossotti, Ottaviano Fabrizio (1791–1863), Italian physicist. Developed theory of dielectrics, from which he derived the Clausius-Mossotti equation.

Mott, Nevill Francis (1905–1996), British physicist. Applied quantum mechanics to study of charged particle scattering; with R. W. Gurney, developed Gurney-Mott theory of photographic process; introduced fundamental concepts elucidating electronic properties of disordered materials; Nobel Prize, 1977.

Mottelson, Ben Roy (1926–), American-born Danish physicist. With A. Bohr, developed theory which unifies shell and liquid-drop models of atomic nucleus, and which explains nonspherical nuclei; Nobel Prize, 1975.

Muller, Hermann Joseph (1890–1967), American geneticist. Studied genetic mutation rates under natural and artificial conditions; discovered the effect of x-rays on mutation rate; Nobel Prize, 1946.

Müller, Johannes Peter (1801–1858), German physiologist and anatomist. Proposed the principle of specific nerve energies, concerning stimuli to sense organs; discovered the Müllerian duct, an early embryonic structure.

Müller, Karl Alex (1927–), Swiss physicist. With J. G. Bednorz, discovered high-temperature superconductivity in copper oxide ceramic materials; Nobel Prize, 1987.

Müller, Paul Hermann (1899–1965), Swiss chemist. Discovered the insecticidal properties of DDT; Nobel Prize, 1948.

Mulliken, Robert Sanderson (1896–1986), American chemist. Applied principles of quantum mechanics to study of chemical bonding; with F. Hund, systematized electronic states of molecules in terms of molecular orbitals; Nobel Prize, 1966.

Mullis, Kary B. (1944–), American chemist. Invented the polymerase chain reaction (PCR) method used for studying DNA molecules; Nobel Prize, 1993.

Murchison, Roderick Impey (1792–1871), British geologist. Studied the order of rock formations in Great Britain; with A. Sedgwick, differentiated the Silurian and Devonian.

Murphy, William Parry (1892–1987), American physician. With G. R. Minot, first to suggest liver diet as a treatment for pernicious anemia; Nobel Prize, 1934.

Murray, Joseph (1919–), American physician. Performed the first successful transplant of a human organ, a kidney; with E. D. Thomas, helped define and then overcome the immunological mechanisms behind organ rejection; Nobel Prize, 1990.

Napier or Neper, John, Laird of Merchiston (1550–1617), Scottish mathematician. Invented the theory of logarithms and developed methods to compute them.

Nathans, Daniel (1928–), American biologist. Pioneered in the use of restriction enzymes to study the structure and functions of deoxyribonucleic acid (DNA) molecules; Nobel Prize, 1978.

Natta, Giulio (1903–1979), Italian chemist. Discovered stereospecific polymerization, making possible the production of new classes of macromolecules from inexpensive raw materials; Nobel Prize, 1963.

Navier, Claude Louis Marie Henri (1785–1836), French physicist and engineer. Studied analytical mechanics and its application to strength of materials, machines, and motion of solid and liquid bodies; formulated Navier-Stokes equations.

Néel, Louis Eugène Félix (1904–), French physicist. Proposed the theory of behavior of antiferromagnetic and other ferrimagnetic materials in which the cyrstal lattice is divided into one or more sublattices; Nobel Prize, 1970.

Neher, Erwin (1944–), German biophysicist. With B. Sakmann, using the "patch clamp" technique they developed, showed how individual ion channels control the passage of charged ions into and out of cells; Nobel Prize, 1991.

Nernst, Hermann Walther (1864–1941), German chemist. Proposed the heat theorem (third law of thermodynamics); determined the specific heat of solids at low temperatures; proposed the chain reaction theory in photochemistry; Nobel Prize, 1920.

Neumann, Carl Gottfried (1832–1925), German mathematician. Believed to be founder of logarithmic potentials; developed the potential theory.

Newcomb, Simon (1835–1909), American astronomer. With A. A. Michelson, determined the velocity of light; studied the motions of the Moon and planets.

Newton, Isaac (1642–1727), English mathematician. Proposed a dynamical theory of gravitation; discovered three basic laws of motion which are the foundation of practical mechanics; made discoveries in optics and mathematics.

Neyman, Jerzy (1894–1981), Russian-born American statistician. Developed methodology for making decisions based only on results of experiments or observations subject to chance.

Nicholson, Seth Barnes (1891–1963), American astronomer. Discovered four satellites of Jupiter; with E. Petit, invented a thermocouple to measure surface temperature of planets.

Nicol, William (1763–1851), Scottish physicist. Invented the Nicol prism for investigating the polarization of light.

Nicolle, Charles Jules Henri (1866–1936), French physician. Discovered the louse to be the transmission vector of typhus; Nobel Prize, 1928.

Nicomedes (3d century B.C.), Greek mathematician. Discovered the conchoid.

Nirenberg, Marshall Warren (1927–), American biochemist. Pioneered in deciphering genetic code; Nobel Prize, 1968.

Nishina, Yoshio (1890–1951), Japanese physicist. Pioneer in study of cosmic rays; with O. B. Klein, originated the Klein-Nishina formula.

Nobel, Alfred Bernhard (1833–1896), Swedish chemist and engineer. Invented dynamite and a blasting gelatin containing nitroglycerin; established the annual Nobel prizes.

Noguchi, Hideyo (1876–1928), Japanese bacteriologist. First to produce pure cultures of syphilis spirochetes; discovered the parasite of yellow fever.

Norrish, Ronald George Wreyford (1897–1978), British physical chemist. With G. Porter and colleagues, developed methods of flash photolysis and kinetic spectroscopy for the study of very fast reactions; Nobel Prize, 1967.

Northrop, John Howard (1891–1987), American biochemist. Isolated several enzymes and proved them to be proteins; isolated the first bacterial virus; established the chemical nature of enzymes and viruses; Nobel Prize, 1946.

Nusselt, Ernst Kraft Wilhelm (1882–1957), German mechanical engineer and physicist. Used dimensional analysis to derive functional form of solutions to equations for heat flux in a flowing fluid.

Nyquist, Harry (1889–1976), Swedish-born American physicist and engineer. Discovered conditions necessary to keep feedback control circuits stable; determined Nyquist rate for communications channels.

Ochoa, Severo (1905–1993), Spanish-born American biochemist. Discovered a bacterial enzyme that synthesizes ribonucleic acid from nucleoside diphosphates; first to synthesize a ribonucleic acid; Nobel Prize, 1959.

Ockham, William of (ca. 1284–1347), English philosopher and theologian. Developed nominalist school of thought; postulated Ockham's razor.

Oersted, Hans Christian (1777–1851), Danish physicist, chemist, and electromagnetist. Discovered a fundamental principle of electromagnetism: a magnetic needle turns at right angles to an electric current.

Ohm, Georg Simon (1787–1854), German physicist. Discovered Ohm's law relating electrical resistance to voltage and current.

Olah, George A. (1927–), American chemist. Made fundamental contributions in carbocation chemistry; Nobel Prize, 1994.

Olbers, Heinrich Wilhelm Matthias (1758–1840), German astronomer. Devised new method for computing cometary orbits; proposed Olbers' paradox.

Onsager, Lars (1903–1976), Norwegian-born American chemist. Laid the foundation of irreversible thermodynamics; contributed to theories of dielectrics, electrolytes, and cooperative phenomena; Nobel Prize, 1968.

Oort, Jan Hendrik (1900–1992), Dutch astronomer. Research on the structure and dynamics of the galactic system; investigated the origin of comets.

Oppenheimer, J. Robert (1904–1967), American physicist. Research on nuclear disintegration, quantum theory, cosmic rays, and relativity; directed production of the atomic bomb.

Orr, John Boyd, Baron (1880–1971), Scottish physiologist and nutritionist. Work on animal nutrition; pioneer in science of human nutrition; Nobel Peace Prize, 1949.

Osheroff, Douglas D. (1945–), American physicist. With D. M. Lee and R. C. Richardson, discovered superfluidity in helium-3; Nobel Prize, 1996.

Ostwald, Friedrich Wilhelm (1853–1932), German chemist born in Latvia. Researches on affinity and mass action; discovered the Ostwald dilution law; worked on the catalytic oxidation of ammonia; Nobel Prize, 1909.

Otto, Nikolaus August (1832–1891), German inventor. Built the first four-stroke internal combustion engine.

Paget, James (1814–1899), English surgeon and pathologist. Studied pathology of tumors and bone and joint diseases; described osteitis deformans (Paget's disease).

Palade, George Emil (1912–), Rumanian-born American cytologist. Applied electron microscope and centrifuge techniques to study of ultrastructure of cells; discovered ribosomes; Nobel Prize, 1974.

Papanicolaou, George Nicholas (1883–1962), Greek-born American cytologist and anatomist. Developed the Papanicolaou test for diagnosis of uterine cervical and endometrial cancer.

Pappus, Alexandrinus (ca. 3d–4th century), Greek mathematician. Wrote *Mathematical Collection*, an account of Greek geometry; formulated Pappus' theorems.

Paracelsus, Philippus Aureolus, or Theophrastus Bombastus von Hohenheim (1493–1541), Swiss physician. Emphasized use of chemicals in medicine; advocated that diseases are specific and require specific remedies.

Paré, Ambroise (1509–1590), French surgeon. Advocated the treatment of wounds by tying arteries with ligatures rather than by cauterization; proposed improvements in operating methods.

Parkinson, James (1755–1824), English physician and paleontologist. Described parkinsonism.

Parseval des Chenes, Marc Antoine (1755–1836), French mathematician. Introduced an equation from which the theorem now known as Parseval's theorem is derived.

Pascal, Blaise (1623–1662), French mathematician and physicist. Contributed to the geometry of conics; formulated Pascal's law, relating to the pressure of a liquid at rest; applied Pascal's triangle to the calculation of probabilities.

Paschen, Louis Carl Heinrich Friedrich (1865–1947), German physicist. Established Paschen's law; with E. Back, discovered Paschen-Back effect; verified predictions of relativistic fine structure made by Bohr-Sommerfeld theory.

Pasteur, Louis (1822–1895), French biologist. Founder of microbiology; discovered the role of bacteria in fermentation; discovered anaerobic bacteria; developed the pasteurization process; demonstrated the efficacy of vaccination, especially for rabies.

Patterson, Arthur Lindo (1902–1966), New Zealand-born American physicist and crystallographer. Developed Patterson-Harker method of x-ray diffraction analysis of crystal structure.

Paul, Wolfgang (1913–1993), German physicist. Invented the Paul trap, which uses radio-frequency radiation to hold ions in a small volume; Nobel Prize, 1989.

Pauli, Wolfgang (1900–1958), Austrian-born American physicist. Worked on quantum theory; formulated the Pauli exclusion principle; contributed to matrix mechanics; Nobel Prize, 1945.

Pauling, Linus Carl (1901–1994), American chemist. Applied quantum theory to chemistry; research on molecular structure and chemical bonds; contributed to electrochemical theory of valency; Nobel Prize, 1954; Nobel Peace Prize, 1963.

Pavlov, Ivan Petrovich (1849–1936), Russian pathologist. Discovered the nerve fibers affecting heart action and the secretory nerves of the pancreas; research on the physiology of digestive glands; studied conditioned reflexes; Nobel Prize, 1904.

Peano, Giuseppe (1858–1932), Italian mathematician. Pioneer in symbolic logic and foundations of mathematics; promoted axiomatic method in mathematics; formulated postulates for natural numbers.

Pearl, Raymond (1879–1940), American biologist and statistician. Applied statistics to the study of population changes; introduced logistic curve describing population growth.

Pearson, Karl (1857–1936), English applied mathematician, statistician, and biometrician. Pioneered in application of statistics to biology; introduced chi-square test.

Pedersen, Charles J. (1904–1989), American chemist. Developed the synthesis of cyclic polyethers known as crown ethers; Nobel Prize, 1987.

Peierls, Rudolf Ernst (1907–1995). German-born British physicist. Developed theory of heat conduction in nonmetallic crystals; with O. R. Frisch, calculated critical mass of uranium-235.

Peirce, Charles Santiago Sanders (1839–1914), American mathematician, logician, and physicist. Laid foundation for logical analysis of mathematics; contributed to probability theory.

Pelletier, Pierre Joseph (1788–1842), French chemist. Discovered quinine, strychnine, and other alkaloids.

Peltier, Jean Charles Athanase (1785–1845), French physicist. Discovered the Peltier effect in thermoelectricity.

Penrose, Roger (1931–), British mathematician and physicist. Developed twistor theory of space-time geometry; studied singularities in classical general relativity theory.

Penzias, Arno A. (1933–), American astrophysicist. With R. W. Wilson, discovered cosmic background radiation, confirming the big bang theory of the origin of the universe; Nobel Prize, 1978.

Perl, Martin L. (1927–), American physicist. Discovered the tau lepton, a fundamental particle; Nobel Prize, 1995.

Pérot, Jean Baptiste Gaspard Gustav Alfred (1863–1925), French physicist. With C. Fabry, developed Fabry-Pérot interferometer.

Perrin, Jean Baptiste (1870–1942), French physicist. Research on the particle nature of cathode rays; found values for Avogadro's number, thereby proving the existence of molecules; Nobel Prize, 1926.

Perutz, Max Ferdinand (1914–), Austrian-born British crystallographer and molecular biologist. Worked on the structure of hemoglobin; introduced the method of isomorphous replacement with heavy atoms into protein crystallography; Nobel Prize, 1962.

Petit, Alexis Thérèse (1791–1820), French physicist. With P. L. Dulong, formulated the law of constancy of atomic heats; devised methods for determining thermal expansion and specific heats of solids.

Pettit, Edison (1890–1962), American astronomer. Studied the Sun and formulated laws alleged to govern the movement of prominences; constructed the interference polarizing monochromator; with S. B. Nicholson, devised a sensitive thermocouple to measure the surface temperatures of planets.

Pfaff, Johann Friedrich (1765–1825), German mathematician. Developed theory of Pfaffian differential equations, which is basic to general solution of partial differential equations.

Pfeiffer, Richard Friedrich Johann (1858–1945), German bacteriologist. Discovered Pfeiffer's bacillus in influenza; described Pfeiffer's reaction for determination of cholera.

Phillips, William D. (1948), American physicist. Developed a method of slowing and trapping atoms in an atomic beam by using an opposed laser beam and a magnetic trap, and cooled these atoms to temperatures lower than the previously calculated theoretical limits; Nobel Prize, 1997.

Piaget, Jean (1896–1980), Swiss psychologist. Elucidated development of cognitive functions in the child.

Picard, Charles Émile (1856–1941), French mathematician. Formulated Picard's theorem relating to functions.

Piccard, Auguste (1884–1962), Swiss physicist. Conducted a data-collecting exploration of the stratosphere in an airtight gondola of a balloon; constructed and tested a bathysphere for deep-sea exploration.

Pickering, Edward Charles (1846–1919), American astronomer. Invented the meridian photometer; pioneered in stellar spectroscopy.

Pierce, George Washington (1872–1956), American physicist and electronic engineer. Developed theoretical basis of electrical communications; developed Pierce oscillator; with A. E. Kennelly, discovered concept of motional impedance.

Pitzer, Kenneth Sanborn (1914–), American chemist. Pioneered in the development of useful approximations which made possible the calculation of chemical thermodynamic properties of broad classes of chemical substances.

Planck, Max Karl Ernst Ludwig (1858–1947), German physicist. Presented the quantum theory; introduced Planck's constant, or quantum of action.

Planté, Gaston (1834–1889), French physicist. Constructed a storage battery, the first primitive accumulator.

Podolsky, Boris (1896–1966), Russian-born American physicist. Collaborated in formulation of Einstein-Podolsky-Rosen paradox; research on quantum electrodynamics.

Poggendorff, Johann Christian (1796–1877), German physicist. Introduced the small mirror on a suspended system to magnify small deflections of a light beam; invented the galvanometer.

Poincaré, Jules Henri (1854–1912), French mathematician. Worked on the theory of functions, on differential equations, and on the theory of orbits in astronomy.

Poinsot, Louis (1777–1859), French mathematician. Originated theory of couples.

Poiseuille, Jean Léonard Marie (1797–1869), French physiologist and physicist. Studied physiology of arterial circulation; invented improved methods for measuring blood pressure; discovered Hagen-Poiseuille law independently of G. H. L. Hagen.

Poisson, Siméon Denis (1781–1840), French mathematician. Worked on mathematical physics; contributed to the wave theory of light; formulated the Poisson ratio concerning the elasticity of materials.

Polonyi, John C. (1929–), Canadian chemist. Studied chemiluminescence, a phenomenon in which the energy states of excited molecules are revealed by their emission of light; Nobel Prize, 1986.

Pomeranchuk, Isaak Yakolevich (1913–1966), Soviet physicist. Showed that energy of cosmic-ray electrons reaching the atmosphere is limited by their radiation in Earth's magnetic field; proved the Pomeranchuk theorem for scattering cross sections.

Pons, Jean Louis (1761–1831), French astronomer. Discovered 37 comets, including Encke's comet.

Porro, Ignazio (1801–1875), Italian topographer, geodesist, and physicist. Invented optical surveying instruments, Porro prism erecting system, and modern prism binoculars.

Porter, George (1920–), British chemist. With R. G. W. Norrish, developed the technique of flash photolysis to initiate and record very fast chemical reactions; Nobel Prize, 1967.

Porter, Rodney Robert (1917–1985), British biochemist. Research to determine chemical structure of immunoglobulins; Nobel Prize, 1972.

Powell, Cecil Frank (1903–1969), British physicist. Made practical the use of photographic emulsions in nuclear research; with G. P. S. Occhialini and others, discovered and investigated production of pions from cosmic radiation in the Earth's atmosphere; Nobel Prize, 1950.

Poynting, John Henry (1852–1914), English physicist. Determined the constant of gravitation and explained why a comet's tail points away from the Sun.

Prandtl, Ludwig (1875–1953), German physicist. Contributed to fluid mechanics, particularly aerodynamics; introduced concept of boundary layer.

Pregl, Fritz (1869–1930), Austrian chemist. Developed microchemical methods of analysis; Nobel Prize, 1923.

Prelog, Vladimir (1906–1998), Yugoslavian-born Swiss chemist. Investigated stereochemistry of organic molecules and reactions; Nobel Prize, 1975.

Prevost, Pierre (1751–1839), Swiss physicist. Developed theory of exchanges, explaining nature of heat.

Priestley, Joseph (1733–1804), English chemist and physicist. Discovered oxygen, ammonia, oxides of nitrogen, hydrochloric acid gas, nitrogen, carbon monoxide, and sulfur dioxide.

Prigogine, Ilya (1917–), Soviet-born Belgian chemist. Contributed to nonequilibrium thermodynamics, particularly the theory of dissipative structures; Nobel Prize, 1977.

Prokhorov, Aleksandr Mikhailovich (1916–), Soviet physicist. With N. G. Basov, devised a new method for amplifying electromagnetic radiation; Nobel Prize, 1964.

Prout, William (1785–1850), English physician and chemist. Formulated Prout's hypothesis concerning atomic weights.

Prusiner, Stanley B. (1942–), American neurologist. Discovered prions, a new biological agent of infection; Nobel Prize, 1997.

Ptolemy (2d century), Greco-Egyptian astronomer, geographer, and geometer at Alexandria. Proposed the Ptolemaic system, with the Earth as the center of the universe.

Pupin, Michael (1858–1935), Yogoslavian-born American physicist and electrical engineer. Developed inductance coils for telephone lines; contributed to x-ray fluoroscopy, design of radio transmitters, and network theory.

Purcell, Edward Mills (1912–1997), American physicist. Developed the method of nuclear resonance absorption; Nobel Prize, 1952.

Purkinje, Johannes Evangelista (1787–1869), Czech physiologist. Discovered the Purkinje effect in eye physiology and Purkinje cells in the cerebral cortex.

Pythagoras (6th century B.C.), Greek mathematician. Originated a system of geometry, including the Pythagorean theorem.

Quételet, Lambert Adolphe Jacques (1796–1874), Belgian statistician. Did pioneer work on statistics; applied the calculus of probabilities to sociological studies.

Rabi, Isidor Isaac (1898–1988), Austrian-born American physicist. Research on neutrons, magnetism, quantum mechanics, and nuclear physics; Nobel Prize, 1944.

Radon, Johann (1887–1956), Bobemian-born Austrian mathematician. Work in calculus of variations and integration theory.

Rainwater, Leo James (1917–1986), American physicist. Suggested that shell-model potentials of certain atomic nuclei are not spherical but are deformed into spheroids, and proposed mechanism for this distortion; Nobel Prize, 1975.

Raman, Chandrasekhara Venkata (1888–1970), Indian physicist. Research on diffraction and oscillation; discovered the Raman effect; Nobel Prize, 1930.

Ramón y Cajal, Santiago (1852–1934), Spanish histologist. Isolated the neuron and made discoveries concerning nerve cells in gray matter and the spinal cord; Nobel Prize, 1906.

Ramsay, William (1852–1916), British chemist. With J. W. S. Rayleigh, discovered argon; with M. W. Travers, discovered neon, krypton, and xenon; Nobel Prize, 1904.

Ramsden, Jesse (1735–1800), English mathematical-instrument maker. Invented an eyepiece containing cross-wires as a measuring scale; introduced equatorial mounting for telescopes.

Ramsey, Norman Foster (1915–), American physicist. Invented an accurate method of measuring differences between atomic energy levels that formed the basis for the cesium atomic clock; worked on the hydrogen maser; Nobel Prize, 1989.

Rankine, William John Macquorn (1820–1872), Scottish civil engineer. Contributed to thermodynamics and theories of elasticity and waves; wrote textbooks on the steam engine and civil engineering.

Raoult, François Marie (1830–1901), French chemist. Formulated Raoult's law concerning vapor pressure of a solution.

Rathke, Martin Heinrich (1793–1860), German biologist. Discovered gill slits and gill arches in embryo birds and mammals, and Rathke's pocket in developing vertebrates.

Ray or Wray, John (1627?–1706), English naturalist. Identified the difference between mono- and dicotyledons; arranged plants according to their natural form, the foundation of the natural system of classification.

Rayleigh, John William Strutt, 3d Baron (1842–1919), English physicist. Worked on the theory of sound and on physical optics; with W. Ramsay, discovered argon; Nobel Prize, 1904.

Réaumur, René Antoine Ferchault de (1683–1757), French entomologist. Worked in biology and metallurgy; invented the Réaumur thermometer scale.

Regge, Tullio (1931–), Italian physicist. Played a role in introducing the idea of complex angular momenta into elementary particle physics.

Regiomontanus, or Johann Müller (1436–1476), German astronomer. Erected the first European observatory, in 1471 in Nürnberg; produced mathematical tables.

Reichstein, Tadeus (1897–), Polish-born Swiss organic chemist. Isolated about 30 of the 40 substances produced by the adrenal cortex; synthesized and described the structure and properties of many of these substances; Nobel Prize, 1950.

Reines, Frederick (1918–), American physicist. With C. L. Cowan, made first detection of the neutrino, a fundamental particle; Nobel Prize, 1995.

Reynolds, Osborne (1842–1912), British engineer and physicist. Demonstrated streamline and turbulent flow in pipes, and showed that transition between them occurs at a critical velocity determined by Reynolds' number; introduced Reynolds' analogy.

Riccati, Jacopo Francesco (1676–1754), Italian mathematician. Research on analysis, particularly differential equations, and geometry.

Ricci-Curbastro, Gregorio (1853–1924), Italian mathematician and mathematical physicist. Developed theory of tensor analysis, providing mathematical foundation for general relativity.

Richards, Dickinson Woodruff (1895–1973), American physician. With A. F. Cournand, uitlized the technique of cardiac catheterization and proved its value as a diagnostic tool; Nobel Prize, 1956.

Richards, Theodore William (1868–1928), American chemist. Worked on atomic weights; experimentally confirmed the existence of isotopes of lead from uranium and thorium; Nobel Prize, 1914.

Richardson, Owen Willans (1879–1959), English physicist. Studied the emission of electricity from hot bodies and the electron theory of matter; Nobel Prize, 1928.

Richardson, Robert C. (1937–), American physicist. With D. M. Lee and D. D. Osheroff, discovered superfluidity in helium-3; Nobel Prize, 1996.

Richet, Charles Robert (1850–1935), French physiologist. Studied serum therapy and discovered anaphylaxis; Nobel Prize, 1913.

Richter, Burton (1931–), American physicist. Independently of S. C. C. Ting, discovered a new heavy elementary particle, which he named the psi particle; Nobel Prize, 1976.

Richter, Jeremias Benjamin (1762–1807), German chemist. Discovered the law of equivalent proportions.

Riemann, Georg Friedrich Bernhard (1826–1866), German mathematician. Originated Riemannian geometry, a noneuclidean system.

Riesz, Frigyes or Frederic (1880–1956), Hungarian mathematician. Did research on abstract and general theories related to mathematical analysis, particularly functional analysis; independently of E. Fischer, discovered Riesz-Fisher theorem.

Righi, Augusto (1850–1920), Italian physicist. Discovered magnetic hysteresis and Righi-Leduc effect, independently of S. A. Leduc; demonstrated that microwaves have all properties characteristic of light waves.

Ritchey, George Wills (1864–1945), American astronomer. Made important astronomical observations, particularly on the Andromeda nebula; with H. Chrétien, developed Ritchey-Chrétien optics.

Ritz, Walter (1878–1909), Swiss-born German physicist. Introduced Ritz combination principle; developed Ritz method for numerical solution of boundary-value problems.

Robbins, Frederick Chapman (1916–), American microbiologist. Discovered that poliomyelitis virus can be grown in various human tissue cultures; Nobel Prize, 1954.

Roberts, Richard J. (1943–), British geneticist. Independently of P. Sharp, discovered split genes; Nobel Prize, 1993.

Robinson, Robert (1886–1975), English chemist. Worked on plant pigments, alkaloids, and phenanthrene derivatives; Nobel Prize, 1947.

Roche, Edouard Adelbert (1820–1883), French physicist, mathematician, and meteorologist. Studied the internal structure and free-surface form of the celestial bodies; applied results to study of cosmogonic hypotheses.

Rodbell, M. (1925–), American pharmacologist. With A. Gilman, discovered G-proteins and the role of these proteins in cellular signal transduction; Nobel Prize, 1994.

Rohrer, Heinrich (1933–), Swiss physicist. With G. Binnig, developed scanning tunneling microscope; Nobel Prize, 1986.

Rolle, Michel (1652–1719), French mathematician. Worked on Diophantine analysis and algebra of equations.

Röntgen, Wilhelm Konrad (1845–1923), German physicist. Discovered x-rays; Nobel Prize, 1901.

Roscoe, Henry Enfield (1833–1915), English chemist. With R. W. Bunsen, evolved the law of reciprocity and invented the actinometer; first to isolate metallic vanadium.

Rosen, Nathan (1909–1995), American-born Israeli physicist. Collaborated in formulation of Einstein-Podolsky-Rosen paradox; research on general relativity and gravitational waves.

Ross, Ronald (1857–1932), British physician. Proved that malaria is transmitted by the female *Anopheles* mosquito; Nobel Prize, 1902.

Rossby, Carl Gustaf Arvid (1898–1957), Swedish-born American meteorologist. Formulated theories of large-scale air movements; derived the Rossby formula, relating speed of propagation of perturbations to airflow and wavelengths of perturbations; devised the Rossby diagram, used to plot air mass properties.

Rous, Francis Peyton (1879–1970), American physician and virologist. Produced cancer in chickens by inoculating them with filterable virus procured from tissue of chickens with tumors; Nobel Prize, 1966.

Routh, Edward John (1831–1907), British mathematical physicist. Made contributions to classical mechanics, including procedure for eliminating cyclic coordinates from equations of motion.

Roux, Pierre Paul Emile (1853–1933), French physician and bacteriologist. Helped develop modern serum therapeutics, especially concerning diphtheria.

Rowland, F. Sherwood (1927–), American chemist. Demonstrated that chemically inert chlorofluorocarbon (CFC) could be transported up to the ozone layer and could react with ultraviolet light and deplete the ozone layer; Nobel Prize, 1995.

Rowland, Henry Augustus (1848–1901), American physicist. Developed the Rowland grating in spectroscopy; studied electromagnetism and heat.

Rubbia, Carol (1934–), Italian physicist. Principal architect of experiment that first detected intermediate vector bosons, an important step in confirming theory uniting electromagnetic and weak nuclear interactions; Nobel Prize, 1984.

Rubens, Heinrich (1865–1922), German physicist. With E. B. Hagen, conducted electromagnetic experiments; built new types of galvanometer and bolometer.

Rumford, Benjamin Thompson, Count (1753–1814), British physicist. Carried out research on heat.

Runge, Carl David Tolme (1856–1927), German mathematician and physicist. Research on theoretical and experimental spectroscopy, particularly data reduction and development of series formulas; developed methods for numerical and graphical computation, including Runge-Kutta method.

Ruska, Ernst (1906–1988), German electronic engineer. Developed the electron microscope; Nobel Prize, 1986.

Russell Bertrand Arthur William (1872–1970), English mathematician and philosopher. With A. N. Whitehead, pioneered in study of mathematical logic.

Russell, Henry Norris (1877–1957), American astronomer and physicist. Analyzed eclipsing binary stars; with E. Hertzsprung, introduced Hertzsprung-Russell diagram; determined abundance of chemical elements in solar atmosphere; with F. A. Saunders, devised theory of Russell-Saunders coupling.

Rutherford, Ernest, 1st Baron (1871–1937), British physicist. Discovered alpha, beta, and gamma rays; suggested the divisible nuclear atom; effected the transmutation of an atom: Nobel Prize, 1908.

Ružička, Leopold (1887–1976), Swiss chemist born in Croatia. Research on many-membered rings and higher terpenes (including male sex hormones); Nobel Prize, 1939.

Rydberg, Johannes Robert (1854–1919), Swedish physicist. Developed a formula for series of spectral lines, involving Rydberg's constant.

Ryle, Martin (1918–1984), British astronomer. Devised aperture synthesis method in radiotelescopy; designed equipment and made observations in radio astronomy; Nobel Prize, 1974.

Sabatier, Paul (1854–1941), French chemist. Discovered, with J. B. Senderens, the process for catalytic hydrogenation of oils to solid fat; Nobel Prize, 1912.

Sabin, Albert Bruce (1906–1993), Polish born American physician and virologist. Studied nature, mode of transmission, and epidemiology of human poliomyelitis; developed oral polio virus vaccine.

Sabine, Edward (1788–1883), British physicist and astronomer. Headed a magnetic survey of the world which discovered a connection between sunspots and terrestrial magnetic disturbances.

Sabine, Wallace Clement Ware (1868–1919), American physicist. Pioneered in architectural acoustics; discovered law determining reverberation time in acoustics.

Sachs, Julius von (1832–1897), German botanist. Studied the connection between sunlight and chlorophyll; worked on heliotropism and geotropism.

Saha, Meghnad (1894–1956), Indian physicist. Developed theory for degree of ionization of hot gases, a basic component of modern astrophysics.

Sakmann, Bert (1942–), German physiologist. With E. Neher, using the "patch clamp" technique they developed, showed how individual ion channels control the passage of charged ions into and out of cells; Nobel Prize, 1991.

Salam, Abdus (1926–1996), Pakistani physicist. Independently of S. Weinberg, developed theory uniting two of the basic forces of nature, electromagnetism and the weak nuclear interactions; Nobel Prize, 1979.

Salk, Jonas Edward (1914–1995), American physician. Produced killed-virus vaccine effective in preventing poliomyelitis.

Salpeter, Edwin Ernest (1924–), Austrian-born American physicist. Research in quantum theory of atoms, quantum electrodynamics, nuclear theory, energy production of stars, and theoretical astrophysics; with H. A. Bethe, introduced Bethe-Salpeter equation.

Samuelsson, Bengt Ingemar (1934–), Swedish biochemist and medical scientist. Studied prostaglandin metabolism and the formation of prostaglandin from arachidonic acid; Nobel Prize, 1982.

Sanctorius, or Santorio Santorio (1561–1636), Italian physician. Invented the clinical thermometer; experimented with metabolism.

Sanger, Frederick (1918–), English chemist. Determined the exact order of amino acids in insulin; first to establish amino acid sequence for a protein; developed methods for determining nucleotide sequences (independently of W. Gilbert), advancing the technology of DNA recombination; Nobel Prizes, 1958 and 1980.

Savart, Félix (1791–1841), French physicist. Helped formulate the Biot-Savart law in electromagnetism.

Schally, Andrew Victor (1926–), Polish-born American physiologist. With R. Guillemin, isolated and analyzed peptide hormones secreted in hypothalmic region of brain which control anterior pituitary hormone secretion; Nobel Prize, 1977.

Schawlow, Arthur Leonard (1921–), American physicist. Contributed to invention of laser; made numerous contributions to laser spectroscopy, particularly the development of Doppler-free spectroscopy; Nobel Prize, 1981.

Scheele, Karl Wihelm (1742–1786), Swedish chemist. Made many discoveries, including oxygen (independently of J. Priestley), chlorine, and glycerin; synthesized many organic acids.

Schiaparelli, Giovanni Virginio (1835–1910), Italian astronomer. Discovered the connection between comets and meteorites, and the "canals" of Mars.

Schiff, Hugo Josef (1834–1915), German-born Italian organic chemist. Discovered Schiff bases; devised Schiff test; devised an improved nitrometer.

Schmidt, Bernhard Voldemar (1879–1935), Estonian-born German astronomer. Invented Schmidt system for astronomical telescopes.

Schmidt, Erhard (1876–1959), German mathematician. Extended D. Hilbert's work on integral equations; formalized and developed concept of Hilbert space.

Schoenflies, Arthur Moritz (1853–1928), German mathematician and crystallographer. Classified the 230 crystallographic space groups.

Schottky, Walter (1886–1976), Swiss-born German physicist. Discovered Schottky effect; invented screen grid and tetrode; developed Schottky theory of semiconductor-metal junctions.

Schrieffer, John Robert (1931–), American physicist. With J. Bardeen and L. N. Cooper, formulated a theory of superconductivity; Nobel Prize, 1972.

Schrödinger, Erwin (1887–1961), German physicist. Proposed concept of atomic structure based on wave mechanics; contributed to quantum theory and color theory; Nobel Prize, 1933.

Schur, Issai (1875–1941), Russian-born German mathematician. Contributed to representation theory of groups; research on group theory, matrices, algebraic equations, and number theory.

Schwartz, Melvin (1932–), American physicist. Collaborated in an experiment that demonstrated the existence of two types of neutrino; Nobel Prize, 1988.

Schwarz, Hermann Amandus (1843–1921), German mathematician. Introduced Schwarz-reflection principle and Schwarz's lemma while proving the Riemann mapping theorem.

Schwarzschild, Karl (1873–1916), German astronomer. Developed photographic methods for measuring brightness of stars; discovered Schwarzschild solution of equations of general relativity.

Schwarzschild, Martin (1912–1997), German-born American astronomer. Numerical studies of the internal structure and evolution of stars; astronomical observations with balloon-borne telescopes.

Schwinger, Julian Seymour (1918–1994), American physicist. Made fundamental contributions to the quantum theory of radiation; worked out the mathematical formalism of interaction between charged particles and an electromagnetic field; Nobel Prize, 1965.

Seaborg, Glenn Theodore (1912–), American chemist. Synthesized and identified eight transuranium elements and over a hundred isotopes; Nobel Prize, 1951.

Secchi, Pietro Angelo (1818–1878), Italian astronomer. Originated the spectroscopic survey of the heavens; made the first classification of stars according to spectral type.

Sedgwick, Adam (1785–1873), English geologist. With R. I. Murchison, established the Devonian system.

Seebeck, Thomas Johann (1770–1831), German physicist. Investigated thermoelectricity and invented the thermocouple.

Segrè, Emilio Gino (1905–1989). Italian-born American physicist. Codiscovered the elements technetium, astatine, and plutonium, slow neutrons, and the antiproton; Nobel Prize, 1959.

Seidel, Philipp Ludwig von (1821–1896), German astronomer and mathematician. Developed theory of aberrations; made first accurate photometric measurements of stars and planets, and evaluated them with probability theory.

Semenov, Nikolai Nikolaevich (1896–1986), Soviet chemist. Elucidated the mechanisms of chemical reactions, especially the chain mechanism; Nobel Prize, 1956.

Senderens, Jean Baptiste (1856–1937), French chemist. With P. Sabatier, discovered hydrolysis of oils by catalysis.

Serber, Robert (1909–1997), American physicist. Laid foundations of orbit theory of high-energy particle accelerators; introduced Serber potential to describe nuclear forces.

Serret, Joseph Alfred (1819–1885), French mathematician and astronomer. Helped develop Frenet-Serret formulas in the theory of space curves.

Servetus, Michael (1511–1553), Spanish physician. Discovered the pulmonary circulation and the purification of the blood by the lungs.

Shannon, Claude Elwood (1916–), American mathematician. Developed mathematical theory of communication, making use of analogy between concepts of entropy and information.

Shapley, Harlow (1885–1972), American astronomer. Worked on a theory to explain cepheid variables; made an estimate of the size of the universe; provided a description of the universe's form of construction.

Sharp, Phillip A. (1944–), American geneticist. Independently of R. Roberts, discovered split genes; Nobel Prize, 1993.

Sherrington, Charles Scott (1861–1952), English physiologist. Studied the neuron and its function and other aspects of the nervous system; Nobel Prize, 1932.

Shockley, William (1910–1989), English-born American physicist. Discovered the transistor effect for electronic amplification by means of solid-state semiconductors; Nobel Prize, 1956.

Shubnikov, Aleksei Vasilevich (1887–1970). Soviet crystallographer. Classified Shubnikov groups; developed techniques for growing crystals, including synthetic rubies used in lasers.

Shull, Clifford G. (1915–), American physicist. Developed the neutron diffraction technique for studying the atomic structure of solids and liquids; Nobel Prize, 1994.

Siegbahn, Kai Manne Börje (1918–), Swedish physicist. Pioneered the development of high-resolution electron spectroscopy; Nobel Prize, 1981.

Siegbahn, Karl Manne Georg (1886–1978), Swedish physicist. Studied x-ray spectroscopy, in which he discovered the M series; Nobel Prize, 1924.

Siemens, Ernst Werner von (1816–1892), German engineer and electrician. Developed telegraphy and self-acting dynamo.

Siemens, William or Karl Wilhelm (1823–1883), German inventor in London. Made many inventions, including a differential governor, bathometer, dynamometer, and electric furnace.

Simpson, Thomas (1710–1761), English mathematician. Formulated the Simpson rule for finding the area of a figure, given only a limited number of data.

Skou, Jens C. (1918–), Dutch chemist. Discovered an ion-transporting enzyme—sodium, potassium-stimulated adenosine triphosphatase (Na^+, K^+-ATPase)—maintaining the balance of sodium and potassium ions in the living cell; Nobel Prize, 1997.

Slater, John Clarke (1900–1976), American physicist. Introduced Slater determinant describing many-electron systems; developed theory of magnetrons.

Smalley, Richard E. (1943–), American chemist. Designed and built a special laser-supersonic cluster beam apparatus that was used in the discovery of fullerenes; Nobel Prize, 1996.

Smith, Hamilton Othanel (1931–), American geneticist. Isolated a restriction enzyme that cleaves deoxyribonucleic acid (DNA) molecules at a specific site; Nobel Prize, 1978.

Smith, Michael (1927–), Canadian chemist. Made fundamental contribution toward oligonucleotide-based, site-directed mutagenesis and its development for protein studies within DNA-based chemistry; Nobel Prize, 1993.

Smith, Robert (1689–1768), English physicist. Developed a particulate theory of light; developed geometric propositions for computing properties of optical systems; derived a special case of the Smith-Helmholtz law.

Snell, George Davis (1903–1996), American immunogeneticist. Demonstrated the x-ray induction of mutational changes in a mammal; contributed to the study of immunological systems and to the development of transplant immunology; Nobel Prize, 1980.

Snell, Willebrod van Roijen (1591–1626), Dutch mathematician. Formulated Snell laws concerning angles of incidence and refraction; conceived the idea of measuring the Earth by triangulation.

Soddy, Frederick (1877–1956), English chemist. With E. Rutherford, developed theory of atomic disintegration of radioactive substances; research on isotopes; Nobel Prize, 1921.

Solvay, Ernest (1838–1922), Belgian industrial chemist. Developed the Solvay process for production of sodium carbonate.

Sommerfeld, Arnold (1868–1951), German physicist. Developed quantum theory, especially in its application to sprectral lines and the Bohr atomic model.

Sörensen, Sören Peter Lauritz (1868–1939), Danish biochemist. Did pioneer work on hydrogen ion concentration; invented the symbol pH.

Spemann, Hans (1869–1941), German zoologist. Studied embryonic development and discovered the organizer function of certain tissues; Nobel Prize, 1935.

Sperry, Roger Wolcott (1913–1994), American neuroscientist. Discovered the functional split between the left and right hemispheres of the brain; Nobel Prize, 1981.

Spörer, Gustav Friedrich Wilhelm (1822–1895), German astronomer. Observations of the sun and sunspots.

Stanley, Wendell Meredith (1904–1971), American biochemist. Discovered that a virus is a nucleoprotein and can be crystallized; Nobel Prize, 1946.

Stanton, Thomas Ernest (1865–1931), English engineer. Studied surface friction of fluids; built wind tunnel for wind velocity investigations; studied strength of materials, heat transmission, and lubrication.

Stark, Johannes (1874–1957), German physicist. Studied radiation and atomic theory; discovered the Stark effect on spectrum lines and the Doppler effect in canal rays; Nobel Prize, 1919.

Staudinger, Hermann (1881–1965), German chemist. Conceived and elaborated the explanation of phenomenon of polymerization; Nobel Prize, 1953.

Steenrod, Norman Earl (1910–1971), American mathematician. Worked in topology; introduced Steenrod algebra.

Stefan, Josef (1835–1893), Austrian physicist. Originated Stefan's (or Stefan-Boltzmann) law of blackbody radiation; proposed theory of diffusion of gases; studied gas conductivity.

Stein, William Howard (1911–1980), American biochemist. With S. Moore, developed technique for determining amino acid sequence in proteins, and applied it to ribonuclease; Nobel Prize, 1972.

Steinberger, Jack (1931–), German-born American physicist. Collaborated in an experiment that demonstrated the existence of two types of neutrino; Nobel Prize, 1988.

Steinmetz, Charles Proteus (1865–1923), German-born American electrical engineer. Developed complex number technique for analyzing alternating-current circuits; made numerous electrical inventions; applied mathematical methods to solution of electrical engineering problems.

Stern, Otto (1888–1969), German-born American physicist. Developed the molecular beam method and used it to prove directly the existence of the magnetic moment of atoms and nuclei and to measure their magnitudes; Nobel Prize, 1943.

Stieltjes, Thomas Jan (1856–1894), Dutch-born French mathematician. Developed analytic theory of continued fractions, and Stieltjes integral as a tool for their study.

Stirling, James (1692–1770), British mathematician. Discovered Stirling's formula and Stirling's interpolation formula.

Stokes, George Gabriel (1819–1903), British mathematician and physicist. Originated the idea of determining the chemical composition of the Sun and stars from their spectra; studied double refraction and electromagnetic waves.

Stone, Marshall Harvey (1903–1989), American mathematician. Studied structural aspects of mathematical situations having origins in classic problems of analysis, geometry, and logic.

Störmer, Carl Fredrik Mülertz (1874–1957), Norwegian mathematician and geophysicist. Studied atmospheric phenomena; discovered the Störmer cone concerning cosmic rays.

Strassman, Fritz (1902–1980), German chemist. With O. Hahn and L. Meitner, discovered nuclear fission; research on uranium and thorium isotopes.

Strömgren, Bengt Georg Daniel (1908–1987), Swedish-born American astronomer. Developed theory of neubulae consisting of hydrogen ionized by hot stars.

Struve, Friedrich Georg Wilhelm von (1793–1864), German-born Russian astronomer. Authority on double stars and nebulae; one of the first to measure a stellar parallax.

Struve, Otto Wilhelm von (1819–1905), Russian astronomer. Discovered some 500 new double stars; calculated the constant of precession.

Sturgeon, William (1783–1850), English electrician and inventor. Constructed the first useful electromagnet and the first moving-coil galvanometer.

Sturm, Jacques Charles François (1803–1855), French mathematician. Formulated the Sturm theorems, concerning real roots of an equation.

Suhl, Harry (1922–), German-born American physicist. Discovered Suhl effect; invented Suhl amplifier; studied resonance in magnetic materials, superconductivity, and general theory of magnetism.

Sumner, James Batcheller (1887–1955), American biochemist. First to isolate an enzyme in pure, crystalline form and characterize it as a protein; Nobel Prize, 1946.

Sutherland, Earl Wilbur, Jr. (1915–1974), American physiologist. Uncovered intermediary role of cyclic adenylic acid in the mechanism of hormone control over human metabolic activities; Nobel Prize, 1971.

Svedberg, Theodor (1884–1971), Swedish chemist. An authority on colloid chemistry (dispersed phase); developed a centrifuge for colloidal particles and protein molecules; Nobel Prize, 1926.

Sydenham, Thomas (1624–1689), English physician. Gave classic descriptions of gout, venereal disease, fevers, hysteria, and Sydenham's chorea.

Synge, Richard Laurence Millington (1914–1994), English chemist. Developed partition chromatography with A. J. P. Martin; Nobel Prize, 1952.

Szent-Györgyi, Albert von Nagyrapolt (1893–1986), Hungarian biochemist. Isolated vitamin C; research on combustion processes in plant and animal tissues, muscular contraction, and cell division; Nobel Prize, 1937.

Talbot, William Henry Fox (1800–1877), English inventor and mathematician. Invented the calotype photographic process.

Tamm, Igor Yevgenevich (1895–1971), Soviet physicist. With I. M. Frank, formulated the mathematical theory explaining the physical origin and properties of Cerenkov radiation; Nobel Prize, 1958.

Tatum, Edward Lawrie (1909–1975), American biochemist and geneticist. Researched the relation of genes to biochemical reactions in bacterial, yeast, and mold cells; with G. W. Beadle, discovered the phenomenon of genetic recombination in bacteria; Nobel Prize, 1958.

Taube, Henry (1915–), American chemist. Elucidated the mechanisms of electron transfer reactions, especially in metal complexes; Nobel Prize, 1983.

Taylor, Brook (1685–1731), English mathematician. Formulated Taylor's theorem and worked on mathematics of physical problems.

Taylor, Geoffrey Ingram (1886–1975), British mathematician. Work in theoretical hydrodynamics, particularly turbulence and effect of rotation on fluid flow.

Taylor, Joseph H. (1941–), American astronomer. With R. A. Hulse, discovered a binary pulsar and studied it to observe phenomena predicted by general relativity; Nobel Prize, 1993.

Taylor, Richard Edward (1929–), American physicist. Collaborated in experiments that demonstrated that protons, neutrons, and similar particles are made up of quarks; Nobel Prize, 1990.

Teisserenc de Bort, Léon Philippe (1855–1913), French meteorologist. Discovered the stratosphere.

Teller, Edward (1908–), Hungarian-born American physicist. With associates, developed the concept which led to the construction of the first hydrogen bomb; with G. Gamow, proposed the Gamow-Teller interaction and Gamow-Teller selection rules.

Temin, Howard Martin (1934–1994), American virologist. Proposed that genetic information is transferred from ribonucleic acid tumor viruses to deoxyribonucleic acid; independently of D. Baltimore, discovered reverse transcriptase; Nobel Prize, 1975.

Tesla, Nikola (1856–1943), American inventor born in Yugoslavia. Invented a high-frequency electric coil; improved design of dynamos, transformers, and electric bulbs.

Thales (ca. 640–ca. 546 B.C.), Greek mathematician and astronomer. First to scientifically predict an eclipse of the Sun; discovered static electricity; credited with formulating several theorems.

Theiler, Max (1899–1972), South African physician and virologist. Developed a vaccine to prevent human yellow fever; Nobel Prize, 1951.

Theorell, Axel Hugo Teodor (1903–1982), Swedish biochemist. Made discoveries concerning the nature and mode of action of oxidative enzymes; Nobel Prize, 1955.

Thiele, F. K. Johannes (1865–1918), German chemist. Research on nitrogen compounds and the theory of unsaturated organic molecules.

Thomas, Edward Donnall (1920–), American physician. Performed the first successful transfer of bone marrow from one individual to another; with J. Murray, helped define and then overcome the immunological mechanisms behind organ rejection; Nobel Prize, 1990.

Thomas, Llewellyn Hilleth (1903–1992), English-born American physicist. Discovered Thomas precession; with E. Fermi, developed Thomas-Fermi atomic model; developed basic theory for Thomas cyclotron.

Thomson, George Paget (1892–1975), English physicist. Discovered, independently of C. J. Davisson, the diffraction of electrons by crystals; Nobel Prize, 1937.

Thomson, Joseph John (1856–1940), English physicist. Discovered that cathode rays consist of negatively charged particles, or electrons; Nobel Prize, 1906.

Thouless, David James (1934–), British physicist. Studied many-body problem and its applications to nuclear and condensed matter physics, including phase transitions in superfluid helium films and electrons in disordered systems.

Tinbergen, Nikolaas (1907–1988), Dutch-born British zoologist. Pioneered in study of social behavior of animals and their responses to complex stimuli; conducted experimental studies of the effects of selection pressures and evolutionary response to them; Nobel Prize, 1973.

Ting, Samuel C. C. (1936–), American physicist. Independently of B. Richter, discovered a new heavy elementary particle, which he named the J particle; Nobel Prize, 1976.

Tiselius, Arne Wilhelm Kaurin (1902–1971), Swedish biochemist. Research on electrophoresis and absorption analysis; made discoveries concerning the complex nature of serum proteins; Nobel Prize, 1948.

Todd of Trumpington, Alexander Robertus Todd, Baron (1907–1997), British chemist. Worked on the structure and synthesis of nucleotides, and nucleotide coenzymes, and the related problem of phosphorylation; Nobel Prize, 1957.

Tomonaga, Sin-Itiro (1906–1979), Japanese physicist. Showed the modern theory of quantum electrodynamics to be quantitatively consistent with observed physical phenomena; Nobel Prize, 1965.

Tonegawa, Susumu (1939–), Japanese immunologist. Discovered how a limited number of genes are capable of producing a vast number of diverse antibodies, each designed for a specific invading foreign substance; Nobel Prize, 1987.

Torricelli, Evangelista (1608–1647), Italian physicist. Invented the mercury barometer.

Townes, Charles Hard (1915–), American physicist. Invented the maser; Nobel Prize, 1964.

Townsend, John Sealy Edward (1868–1957), British physicist. Developed collision theory of ionization of gases in an electric field.

Travers, Morris William (1872–1961), English chemist. Discovered, with W. Ramsay, krypton, xenon, and neon; investigated low-temperature phenomena.

Ts'ai Lun (fl. 105), Chinese inventor. Invented paper.

Turing, Alan Mathison (1912–1954), English mathematician. Developed concept of Turing machine; research on mathematical logic, group theory, and computer technology.

Tychonoff (Tikhonov), Andrei Nikolaevich (1906–1993), Soviet mathematician and geophysicist. Proved Tychonoff theorem in topology and introduced concept of Tychonoff space.

Tyndall, John (1820–1893), British physicist. Studied temperature waves in metals and diathermancy of gases; discovered the effect of atmospheric density on sound transmission.

Uhlenbeck, George Eugene (1900–1988), Javanese-born American physicist. With S. Goudsmit, developed hypothesis of electron spin.

Urey, Harold Clayton (1893–1981), American chemist. Isolated heavy water and thus discovered the heavy isotope of hydrogen; Nobel Prize, 1934.

Urysohn, Pavel Samuilovich (1898–1924), Soviet mathematician. Proved Urysohn's lemma in topology.

Van Allen, James Alfred (1914–), American physicist. Discovered that the Earth is circled by two high-energy radiation belts, leading to major revisions in concepts of the Earth's atmosphere and magnetic field.

Van de Graaff, Robert Jemison (1901–1967), American physicist. Contributed to the development of the direct particle accelerator and invented the electrostatic belt generator.

van der Meer, Simon (1925–), Dutch physicist. Devised method to ensure frequent and efficient collision of accelerated protons and antiprotons in the superproton synchrotron at CERN, contributing to discovery of intermediate vector bosons; Nobel Prize, 1984.

Vandermonde, Alexandre Théophile (1735–1796), French mathematician. Gave first logical exposition of theory of determinants; developed methods to test solvability of algebraic equations.

van der Waals, Johannes Diderik (1837–1923), Dutch physicist. Formulated van der Waals equation; investigated van der Waals forces, concerning intermolecular attraction; Nobel Prize, 1910.

Vane, John Robert (1927–), English pharmacologist. Discovered prostaglandin X (prostacyclin) and the role of aspirinlike drugs as blocking agents in the prostaglandin synthesis; Nobel Prize, 1982.

van Rhijn, Pieter Johannes (1886–1960), Dutch astrophysicist. With J. C. Kapetyn, evolved a theory of the universe.

van't Hoff, Jacobus Hendricus (1852–1911), Dutch chemist. Pioneered in the study of stereochemistry; studied reaction rates, thermodynamics applied to chemistry, and the theory of dilute solutions; Nobel Prize, 1901.

Van Vleck, Jan Hasbrouck (1899–1980), American mathematical physicist. Pioneer in the development of the modern quantum-mechanical theory of magnetism; Nobel Prize, 1977.

Varmus, Harold (1939–), American researcher in molecular virology and oncogenesis. With J. M. Bishop, researched the genetic basis of human cancers; their work led to the identification of over 50 cellular genes that can become oncogenes; Nobel Prize, 1989.

Vauquelin, Louis Nicola (1763–1829), French chemist. Discovered chromium and its compounds and beryllium compounds.

Vega, George, Baron von (1756–1802), Austrian mathematician. Prepared logarithmic tables.

Verdet, Marcel Émile (1824–1866), French physicist. Determined dependence of Faraday effect on magnetic field strength, wavelength of the light, and index of refraction of the material.

Vernier, Pierre (1580–1637), French technician. Invented the Vernier scale.

Vesalius, Andreas (1514–1564), Belgian anatomist in Italy. Known as the father of modern anatomy; corrected many of Galen's mistaken doctrines.

Viète, François, or Franciscus Vieta (1540–1603), French mathematician. Worked on the solution of equations up to the fourth degree and laid the foundation of modern algebra.

Virtanen, Artturi Ilmari (1895–1973), Finnish biochemist. Research on problems of human nutrition and agriculture; investigated acidity (pH) and biological nitrogen fixation; Nobel Prize, 1945.

Vogel, Hermann Wilhelm (1834–1898), German photochemist. Invented the orthochromatic photographic plate and designed a photometer.

Voigt, Woldemar (1850–1919), German physicist. Introduced transformation equations (later known as Lorentz transformations).

Volta, Alessandro, Count (1745–1827), Italian physicist. Invented the voltaic pile; developed the theory of current electricity.

Volterra, Vito (1860–1940), Italian mathematician. Developed method of solving Volterra equations; pioneered in developing functional analysis.

Von Braun, Wernher (1912–1977). German-born American rocket engineer. Directed development of the German V-2 and Wassefall missiles; instrumental in launch of *Explorer 1*, first American artificial satellite; supervised development of Saturn rockets for the Apollo program.

Von Euler, Ulf Svante (1905–1983), Swedish physiologist. Identified norepinephrine as neurotransmitter of sympathetic nervous system; isolated and characterized norepinephrine storage granules in nerves; Nobel Prize, 1970.

von Kármán, Theodore (1881–1963), American aerodynamicist. Theoretical contributions to aerodynamics; formulated von Kármán's theory of vortex streets, an early step in the mathematical treatment of turbulent motion.

von Neumann, John (1903–1954), Hungarian-born American mathematician. Research in logic, theory of quantum mechanics, theory of high-speed computing machines, and mathematical theory of games and strategy.

Wagner von Jauregg (Wagner-Jauregg), Julius (1857–1940), Austrian psychiatrist. Developed use of malarial infection to treat general paresis; Nobel Prize, 1927.

Waksman, Selman Abraham (1888–1973), Russian-born American bacteriologist. Isolated the antibiotic streptomycin; Nobel Prize, 1952.

Wald, George (1906–1997), American biologist and biochemist. Discovered the role of vitamin A in vision; Nobel Prize, 1967.

Walker, John E. (1941–), British chemist. Made major contributions toward elucidating the enzymatic mechanism underlying the synthesis of adenosine triphosphate (ATP) by clarifying the structural conditions of the enzyme; Nobel Prize, 1997.

Wallace, Alfred Russel (1823–1913), English naturalist. Originated, independently of C. Darwin, theory of natural selection; postulated Wallace's line regarding geographical distribution of animals.

Wallach, Otto (1847–1931), German chemist. Research on essential oils and the terpenes; Nobel Prize, 1910.

Wallis, John (1616–1703), English mathematician. Worked on algebraic curves, interpolation, evaluation of integrals, infinite series, mechanics, and algebra; derived Wallis product and Wallis formulas.

Walton, Ernest Thomas Sinton (1903–1995), British physicist. With J. D. Cockcroft, devised high-voltage apparatus capable of producing fast atomic particles with energies up to 700,000 electronvolts; showed the capability of these particles to disintegrate many light elements; Nobel Prize, 1951.

Wannier, Gregory Hugh (1911–1983), Swiss-born American physicist. Developed harmonization of localized and nonlocalized descriptions of electrons in solids.

Warburg, Otto Heinrich (1883–1970), German physiologist. Worked on chemistry of respiration and on cancer; Nobel Prize, 1931.

Wassermann, August von (1866–1925), German physician. Discovered the Wassermann test for the detection of syphilis.

Watson, James Dewey (1928–), American biochemist. With F. H. C. Crick, determined the double-helix structure of deoxyribonucleic acid; Nobel Prize, 1962.

Watson, John Broadus (1878–1958), American psychologist. Founded the behaviorist school of psychology.

Watt, James (1736–1819), Scottish inventor. Improved the steam engine, making it a commercial success.

Weber, Ernst Heinrich (1795–1878), German anatomist and physiologist. Discovered Weberian apparatus; applied hydrodynamics to study of blood circulation; discovered inhibitory power of vagus nerve; proposed Weber's law of stimuli.

Weber, Heinrich (1842–1913), German mathematician. Introduced Weber differential equation; demonstrated Abel theorem in its most general form; proved that absolute Abelian fields are cyclotomic.

Weber, Wilhelm Eduard (1804–1891), German physicist. Devised instruments for measurement of electrical and magnetic quantities; formulated absolute electrical and magnetic units.

Wegener, Alfred Lothar (1880–1930), German geologist. Presented the idea of continental drift.

Weierstrass, Karl Theodor (1815–1897), German mathematician. Worked on the theory of functions and on the calculus of variations.

Weil, Adolf (1848–1916), German physician. Gave classic description of Weil's disease.

Weinberg, Steven (1933–), American physicist. Independently of A. Salam, developed theory uniting two of the basic forces of nature, electromagnetism and the weak nuclear interactions; Nobel Prize, 1979.

Weismann, August (1834–1914), German biologist. Contributed to the theory of heredity, which he attributed to variations in "germ-plasm."

Weiss, Pierre (1865–1940), French physicist. Developed phenomenological theory of ferromagnetism.

Weizsacker, Carl Friedrich von (1912–), German physicist. Helped develop method for calculating bremsstrahlung in high-energy collisions; developed a theory of origin of solar system.

Weller, Thomas Huckle (1915–), American virologist and parasitologist. Isolated the virus of chickenpox and herpes zoster and proved the common etiology of the two diseases; first to propagate German measles virus; Nobel Prize, 1954.

Wentzel, Gregor (1898–1978), German-born American physicist. Helped develop Wentzel-Kramers-Brillouin method; research on theory of atomic spectra, wave mechanics, quantum electrodynamics, meson field theories, and statistical mechanics of many-body problems, especially superconductivity.

Werner, Alfred (1866–1919), Swiss chemist. Formulated the coordination theory of valency; Nobel Prize, 1913.

Weyl, Hermann (1885–1955), German-born American mathematician and mathematical physicist. Basic research on group representations and Riemann surfaces.

Wheatstone, Charles (1802–1875), English physicist and inventor. Conducted experiments on sound; invented Wheatstone's bridge, an instrument for comparing electrical resistances.

Wheeler, John Archibald (1911–), American physicist. Introduced the concepts of the scattering matrix and resonating group structure into nuclear physics; with N. Bohr, elucidated the mechanism of nuclear fission and predicted the fissibility of plutonium.

Whewell, William (1794–1866), British astronomer. Work on the tides, and on the history and philosophy of science.

Whipple, George Hoyt (1878–1976), American pathologist. Studied anemia and liver treatment; Nobel Prize, 1934.

Whitehead, Alfred North (1861–1947), English mathematician, physicist, and philosopher. With B. Russell, pioneered in mathematical logic and foundations of mathematics.

Whittaker, Edmund Taylor (1873–1956), British mathematician and physicist. Studied special functions of mathematical physics and equations satisfied by them, particularly Whittaker's differential equation; found general integral representation for harmonic functions; made major contributions to analytical dynamics.

Wiedemann, Gustave Heinrich (1826–1899), German physicist and physical chemist. With R. Franz, discovered Wiedemann-Franz law of thermal conductivity of metals; discovered Wiedemann effect.

Wieland, Heinrich (1877–1957), German chemist. Studied bile acids, chlorophyll, and hemoglobin; Nobel Prize, 1927.

Wien, Wilhelm (1864–1928), German physicist. Formulated the two Wien laws pertaining to radiation from blackbodies; Nobel Prize, 1911.

Wiener, Norbert (1894–1964), American mathematician. Formulated a mathematical theory of Brownian motion; founded science of cybernetics.

Wieschaus, Eric F. (1947–), American geneticist. With E. B. Lewis and C. Nusselin-Volhard, discovered the genetic involvement of early embryonic development; Nobel Prize, 1995.

Wiesel, Torsten Nils (1924–), Swedish physiologist. Contributed to the study of the processing of visual information in the brain; Nobel Prize, 1981.

Wigner, Eugene Paul (1902–1995). Hungarian-born American mathematical physicist. With G. Breit, worked out the Breit-Wigner formula for resonant nuclear reactions; proposed the Wigner theorem of conservation of the angular momentum of electron spin; Nobel Prize, 1963.

Wilkins, Maurice Hugh Frederick (1916–), English biophysicist born in New Zealand. Made x-ray diffraction studies that contributed to the structural determination of deoxyribonucleic acid; Nobel Prize, 1962.

Wilkinson, Geoffrey (1921–), British chemist. Research to determine how metals and organic molecules combine to form unique molecules which have sandwichlike structures; Nobel Prize, 1973.

Williamson, William Crawford (1816–1895), English naturalist. Laid the foundation for paleobotany and showed the importance of plant life forms in coal.

Willstätter, Richard (1872–1942), German chemist. Worked on plant pigments; investigated alkaloids and their derivatives; Nobel Prize, 1915.

Wilson, Charles Thomson Rees (1869–1959), British physicist. Worked on ionization; originated the cloud chamber method of studying ionized particles; Nobel Prize, 1927.

Wilson, Kenneth Geddes (1936–), American physicist. Used renormalization group theory to analyze critical phenomena in the behavior of matter at phase transitions; Nobel Prize, 1982.

Wilson, Robert Woodrow (1936–), American astrophysicist. With A. A. Penzias, discovered cosmic background radiation, confirming the big bang theory of the origin of the universe; Nobel Prize, 1978.

Windaus, Adolf (1876–1959), German chemist. Worked on sterols; discovered that ultraviolet light activates ergosterol and gives vitamin D_2; Nobel Prize, 1928.

Wittig, Georg (1897–1987), German chemist. Work on the linking of carbon and phosphorus (Wittig reaction) made it possible to synthesize new types of compounds, including metal-organic complex compounds; Nobel Prize, 1979.

Wöhler, Friedrich (1800–1882), German chemist. First to synthesize an organic compound, urea.

Wolf, Maximilian Franz Joseph Cornelius (1863–1932), German astronomer. Invented the photographic method of discovering asteroids.

Wollaston, William Hyde (1766–1828), English chemist and physicist. Discovered the lines in the solar spectrum; discovered palladium and rhodium; invented the Wollaston lens.

Woodward, Robert Burns (1917–1979), American chemist. Contributed to the development of total synthesis of complex natural products, and structural determination of several complex natural molecules, later confirmed by total synthesis; Nobel Prize, 1965.

Wright, Almroth Edward (1861–1947), British physician and pathologist. Studied parasitic disease; introduced inoculation against typhoid.

Wright, Wilbur (1867–1912) **and Orville** (1871–1948), American pioneers in aviation. Built the first successful airplane, which each flew at Kitty Hawk, North Carolina, on Dec. 17, 1903.

Wundt, Wilhelm Max (1832–1920), German physiologist and psychologist. Founded the first laboratory for experimental psychology.

Wurtz, Charles Adolphe (1817–1884), French chemist. Discovered methyl and ethyl amines; evolved the Wurtz reaction for synthesis of hydrocarbons.

Yalow, Rosalyn Sussman (1921–), American medical physicist. Developed a radioimmunoassay technique to detect and measure minute levels of substances such as hormones in the body; Nobel Prize, 1977.

Yang, Chen Ning (1922–), Chinese-born American physicist. With T. Lee, disproved the law of conservation of parity for weak interactions; Nobel Prize, 1957.

Yersin, Alexandre Émile Jean (1863–1943), Swiss bacteriologist. Discovered the bubonic plague bacillus in Hong Kong, working independently of S. Kitasato, and developed a serum for it.

Young, Thomas (1773–1829), English physicist and physician. Discovered the effect of the ciliary muscle on the shape of the eye lens (the mechanism of accommodation).

Yukawa, Hideki (1907–1981), Japanese physicist. Postulated the existence of a new fundamental particle, the meson; Nobel Prize, 1949.

Zeeman, Pieter (1865–1943), Dutch physicist. Discovered the Zeeman effect in magnetooptics; Nobel Prize, 1902.

Zener, Clarence Melvin (1905–1993), American physicist. Proposed mechanism of Zener breakdown.

Zeno of Elea (ca. 490–425 B.C.), Greek philosopher and mathematician. Formulated a group of paradoxes, important for their stimulation of philosophical and mathematical thought, which appear to deny the possibility of motion.

Zernike, Fritz (1888–1966), Dutch physicist. Developed the phase-contrast microscope, making possible the first microscopic examination of the internal structure of living cells; Nobel Prize, 1953.

Ziegler, Karl (1898–1973), German organic chemist. Developed a low-pressure process for production of polyethylene; Nobel Prize, 1963.

Zinkerngagel, Rolf M. (1944–), Swiss-born American immunologist and physician. Shared in the discovery of the specificity of cell-mediated immune defense with P. C. Doherty; Nobel Prize, 1996.

Zinsser, Hans (1878–1940), American bacteriologist. Developed methods of immunization against typhus.

Zsigmondy, Richard (1865–1929), German chemist. Studied colloidal solutions; introduced the ultramicroscope; Nobel Prize, 1925.

Zworykin, Vladimir Kosma (1889–1982), Russian-born American physicist. Pioneer in the development of television and the electron microscope.

Index

Index

A

C

D

E

F

G

H

I

N

O

P

Q

R

S

T

U

W

X